P9-EDW-200
ARY
PUYALLUP WA 98374
LAKEWOOD WA 98498

REFERENCE ONLY
Do Not Take From
The Library

PIERCE COLLEGE LIBRARY
PUYALLUP WA 98374
LAKEWOOD WA 98498

CRC Handbook of Chemistry and Physics

102nd Edition

REFERENCE ONLY
Do Not Take From
The Library

CRC Handbook of Chemistry and Physics

A Ready-Reference Book of Chemical and Physical Data

HANDBOOK OF CHEMISTRY AND PHYSICS
2021-2022
102nd
EDITION
CRC PRESS

Editor-in-Chief

John R. Rumble, Jr., Ph.D.
R&R Data Services

Associate Editors

Thomas J. Bruno, Ph.D.
(Retired)
National Institute of Standards and Technology

Maria J. Doa, Ph.D.
Washington, District of Columbia

CRC Press
Taylor & Francis Group
Boca Raton London New York

CRC Press is an imprint of the
Taylor & Francis Group, an informa business

CRC Press
Taylor & Francis Group
6000 Broken Sound Parkway NW, Suite 300
Boca Raton, FL 33487-2742

© 2021 by Taylor & Francis Group, LLC
CRC Press is an imprint of Taylor & Francis Group, an Informa business

No claim to original U.S. Government works

Printed on acid-free paper

International Standard Book Number-13: ~~978-0-367-41724-6~~
9780367712600

This book contains information obtained from authentic and highly regarded sources. Reasonable efforts have been made to publish reliable data and information, but the author and publisher cannot assume responsibility for the validity of all materials or the consequences of their use. The authors and publishers have attempted to trace the copyright holders of all material reproduced in this publication and apologize to copyright holders if permission to publish in this form has not been obtained. If any copyright material has not been acknowledged please write and let us know so we may rectify in any future reprint.

Except as permitted under U.S. Copyright Law, no part of this book may be reprinted, reproduced, transmitted, or utilized in any form by any electronic, mechanical, or other means, now known or hereafter invented, including photocopying, microfilming, and recording, or in any information storage or retrieval system, without written permission from the publishers.

For permission to photocopy or use material electronically from this work, please access www.copyright.com (http://www.copyright.com/) or contact the Copyright Clearance Center, Inc. (CCC), 222 Rosewood Drive, Danvers, MA 01923, 978-750-8400. CCC is a not-for-profit organization that provides licenses and registration for a variety of users. For organizations that have been granted a photocopy license by the CCC, a separate system of payment has been arranged.

Trademark Notice: Product or corporate names may be trademarks or registered trademarks, and are used only for identification and explanation without intent to infringe.

Visit the Taylor & Francis Web site at
http://www.taylorandfrancis.com

and the CRC Press Web site at
http://www.crcpress.com

EDITORIAL ADVISORY BOARD 2020–2022

Daniel Alan Anderson
Department of Natural Resources
Bowie State University
Bowie, Maryland

Judith Currano
Chemistry Library
University of Pennsylvania
Philadelphia, Pennsylvania

Steve Freiman
Freiman Consulting
Potomac, Maryland

Jeremy Garritano
Brown Science and Engineering Library
University of Virginia
Charlottesville, Virginia

Daryn Johnson
Science Department
Xavier High School
New York, New York

Ye Li
Chemistry and Chemical Engineering Library
Massachusetts Institute of Technology
Cambridge, Massachusetts

David R. Lide
North Potomac, Maryland

David Martinsen
Martinsen Consulting
Rockville, Maryland

Leah Rae McEwen
Chemistry Library
Cornell University
Ithaca, New York

Donna Wrublewski
Chemistry Library
California Institute of Technology
Pasadena, California

PREFACE

CRC Handbook of Chemistry and Physics
102nd Edition

The Handbook as an Integrated Data Resource for the Future

This 102nd Edition of the *CRC Handbook of Chemistry and Physics* continues our dedication to maintaining the highest standards of quality, completeness of coverage, and diversity. Before looking into the future, let us first look at the diversity of the present *CRC Handbook*.

- 342 different topics are included, with 818 distinct data tables
- 7 new topics added in this 102nd Edition
- 13 topics have major updates in the 102nd Edition
- 193 other topics have also been updated
- Considerable effort has been made to harmonize property and variable names and symbols

The demand for digital access to high-quality data in chemistry and physics keeps growing as knowledge discovery, artificial intelligence, data mining, and advanced modeling have become routine tools. The *CRC Handbook* is as valuable as it has ever been. What are we doing to ensure that the highest quality data is available not only now but also into the future?

Our commitment continues to use the tools of the Information Revolution—advanced database methods and Web technology—to ensure the widest possible distribution and usability of the *CRC Handbook*, which is now available 24 hours a day, every day, regardless of user location. The future is bright. Usage continues to grow, and the *CRC Handbook* is growing to meet today's demands. We continue to explore new developments to meet these growing digital demands.

- Greater discoverability of the *CRC Handbook* contents via independent search engines
- Virtual sections on subjects such as environmental chemistry, laboratory safety, and materials chemistry that will allow users to find the diverse relevant data needed for the complex science in these areas
- Improved linkage of related data sets across topics
- Expanded updates of older data sets reflecting more currency of new measurements
- New emphasis on cross-disciplinary topics for which high-quality chemistry and physics data are of fundamental importance

The success of the *CRC Handbook* is very dependent on feedback from its users. The Editor-in-Chief appreciates any suggestions from readers on proposed new topics for the *CRC Handbook* or comments on how its usefulness may be improved in future editions. Please send your comments to john.rumble@crchbcp.com.

Numerous international experts make key contributions to the *CRC Handbook* every year. Their efforts play a key role in the quality and diversity of the subject matter covered in the *CRC Handbook*. The sound advice and guidance of the Editorial Advisory Board members are very much appreciated.

Fiona Macdonald, Senior Publisher—Chemical & Life Sciences, CRC Press/Taylor & Francis Group has been of great assistance and support in providing oversight to ensure that the *CRC Handbook* continues to satisfy the needs of the user community. Thanks are also due to Linda Manis Leggio and Pam Morrell for their detailed, cooperative work and extreme care in the production of the printed *CRC Handbook*. Special thanks are due to Associate Editors Maria J. Doa and Thomas Bruno for their assistance, especially in the area of laboratory chemistry and chemical safety.

Finally, this 102nd Edition of the *CRC Handbook* is a tribute to David Lide, Editor-in-Chief from 1990 to 2009. His dedication to providing generations of scientists with the highest quality of data is a tradition I hope to continue.

John Rumble
Gaithersburg, MD
john.rumble@crchbcp.com
April 2021

NEW DATA AND UPDATES IN THE 102nd EDITION

The *CRC Handbook* is a dynamic publication, and every year the editors add new topics and data tables and replace existing data tables with new data values when available. In addition, we are constantly improving the usability and accessibility of the *CRC Handbook* with more informative introductions and minor data updates. The new data and updates contained in the 102nd Edition are listed below.

Major Data Updates

Seven new topics have been added in the 102nd Edition, including two on chemical and laboratory safety (marked by *), expanding our commitment to these topics.

- Properties of Controlled Substances
- Summary Tables of Particle Properties
- *Compressed Air Safety
- *Safety in the Use of Cryogens
- Pressure Drop in Open Tubular Gas Chromatographic Columns
- Phase Ratios for Capillary Columns
- Minimum Recommended Injector Split Ratios for Capillary Columns

Many important topics have been updated to reflect best currently available data.

- High-Temperature Superconductors
- Chemical Reaction Rate Constants for Atmospheric Studies
- Thermal Conductivity of Liquids
- Organic Analytical Reagents for the Determination of Inorganic Ions
- Threshold Limits for Airborne Contaminants
- Aqueous Solubility and Henry's Law Constants of Organic Compounds
- Solubility of Selected Gases in Water
- Solubility of Carbon Dioxide in Water at Various Temperatures and Pressures
- Biological Buffers
- Typical pH Values of Biological Materials and Foods
- Bond Dissociation Energies
- Electron Work Function of the Elements
- Interstellar Molecules

Other topics that have been updated with new data values include the following.

- The Elements
- Standard Atomic Weights
- Physical Constants of Organic Compounds
- Physical Constants of Inorganic Compounds
- Structure and Properties of Some Common Drugs
- Atmospheric Concentration of Carbon Dioxide, 1959–2020
- Global Temperature Trend, 1880–2020
- Global Warming Potential of Greenhouse Gases
- Major World Earthquakes

In addition, almost 200 topics have improved introductions, table enhancements, and more harmonized terminology.

CURRENT CONTRIBUTORS

Ian H. Bell
Applied Chemicals and Materials Division
National Institute of Standards and Technology
Boulder, Colorado

Lev I. Berger
California Institute of Electronics and Materials Science
Hemet, California

Peter E. Bradley
Applied Chemicals and Materials Division
National Institute of Standards and Technology
Boulder, Colorado

Thomas J. Bruno
Applied Chemicals and Materials Division
National Institute of Standards and Technology
Boulder, Colorado

Jessica L. Burger
Applied Chemicals and Materials Division
National Institute of Standards and Technology
Boulder, Colorado

James B. Burkholder
Chemical Sciences Division
Earth System Research Laboratory
National Oceanic and Atmospheric Administration (NOAA)
Boulder, Colorado

Charles E. Carraher
Department of Chemistry and Biochemistry
Florida Atlantic University
Boca Raton, Florida

Jin-Pei Cheng
Ministry of Science and Technology
Beijing, China

Robert D. Chirico
Thermodynamics Research Center
Applied Chemicals and Materials Division
National Institute of Standards and Technology
Boulder, Colorado

Ivan Cibulka
Department of Physical Chemistry
Institute of Chemical Technology
Prague, Czech Republic

Christopher J. Cramer
Department of Chemistry
University of Minnesota
Minneapolis, Minnesota

Vladimir Diky
Thermodynamics Research Center
Applied Chemicals and Materials Division
National Institute of Standards and Technology
Boulder, Colorado

Maria J. Doa
Washington, District of Columbia

Michael Frenkel
Naples, Florida

Jeffrey R. Fuhr
Quantum Measurement Division
National Institute of Standards and Technology
Gaithersburg, Maryland

Robert N. Goldberg
Biosystems and Biomaterials Division
National Institute of Standards and Technology
Gaithersburg, Maryland

Thomas W. Grove
Boulder Safety, Health, and Environmental Division
National Institute of Standards and Technology
Boulder, Colorado

Jens Hänisch
Institute for Technical Physics
Karlsruhe Institute of Technology
Karlsruhe, Germany

Allan H. Harvey
Applied Chemicals and Materials Division
National Institute of Standards and Technology
Boulder, Colorado

Steven R. Heller
Chemical and Biochemical Reference Data Division
National Institute of Standards and Technology
Gaithersburg, Maryland

Norman E. Holden
National Nuclear Data Center
Brookhaven National Laboratory
Upton, New York

Marcia L. Huber
Applied Chemicals and Materials Division
National Institute of Standards and Technology
Boulder, Colorado

Andrei Kazakov
Thermodynamics Research Center
Applied Chemicals and Materials Division
National Institute of Standards and Technology
Boulder, Colorado

Daniel E. Kelleher
Quantum Measurement Division
National Institute of Standards and Technology
Gaithersburg, Maryland

M. Naveed Khan
Center for Hydrate Research
Colorado School of Mines
Golden, Colorado
Petroleum Institute, Abu Dhabi, U.A.E.

Carolyn A. Koh
Center for Hydrate Research
Colorado School of Mines
Golden, Colorado

Michael J. Kurylo
Universities Space Research Association /
Goddard Earth Sciences, Technology, and Research
NASA Goddard Space Flight Center
Greenbelt, Maryland

Eric W. Lemmon
Applied Chemicals and Materials Division
National Institute of Standards and Technology
Boulder, Colorado

David R. Lide
North Potomac, Maryland

Frank J. Lovas
Sensor Sciences Division
National Institute of Standards and Technology
Gaithersburg, Maryland

Serguei N. Lvov
Department of Energy and Mineral Engineering
Pennsylvania State University
University Park, Pennsylvania

Manjeera Mantina
Department of Chemistry
University of Minnesota
Minneapolis, Minnesota

Alan D. McNaught
Cambridge, England

Thomas M. Miller
Air Force Research Laboratory/RVBXT
Kirtland AFB, New Mexico

Nasser Moazzen-Ahmadi
Department of Physics and Astronomy
University of Calgary
Calgary, Alberta, Canada

Peter J. Mohr
Quantum Measurement Division
National Institute of Standards and Technology
Gaithersburg, Maryland

Holger S. P. Müller
Astrophysics, First Institute of Physics
University of Cologne
Cologne, Germany

Chris D. Muzny
Thermodynamics Research Center
Applied Chemicals and Materials Division
National Institute of Standards and Technology
Boulder, Colorado

David B. Newell
Quantum Measurement Division
National Institute of Standards and Technology
Gaithersburg, Maryland

Irving Ozier
Department of Physics and Astronomy
University of British Columbia
Vancouver, British Columbia, Canada

Eugene Paulechka
Applied Chemicals and Materials Division
National Institute of Standards and Technology
Boulder, Colorado

Larissa I. Podobedova
Biomolecular Measurement Division
National Institute of Standards and Technology
Gaithersburg, Maryland

Cedric J. Powell
Materials Measurement Science Division
National Institute of Standards and Technology
Gaithersburg, Maryland

Ray Radebaugh
Applied Chemicals and Materials Division
National Institute of Standards and Technology
Boulder, Colorado

Joseph Reader
Quantum Measurement Division
National Institute of Standards and Technology
Gaithersburg, Maryland

E. Dendy Sloan
Center for Hydrate Research
Colorado School of Mines
Golden, Colorado

Beverly L. Smith
Centennial, Colorado

Paris D. N. Svoronos
Queensborough Community College
City University of New York
Bayside, New York

Current Contributors

Barry N. Taylor
Quantum Measurement Division
National Institute of Standards and Technology
Gaithersburg, Maryland

Donald G. Truhlar
Department of Chemistry
University of Minnesota
Minneapolis, Minnesota

Rosendo Valero
Chemistry Department
University of Coimbra
Coimbra, Portugal

Wolfgang L. Wiese
Quantum Measurement Division
National Institute of Standards and Technology
Gaithersburg, Maryland

Mark A. Williams
Institute of Structural and Molecular Biology
Department of Biological Sciences
Birkbeck, University of London
London, England

Stuart C. Wimbush
Robinson Research Institute
Victoria University of Wellington
Wellington, New Zealand

Christian Wohlfarth
Martin Luther University
Institute of Physical Chemistry
Halle (Saale), Germany

TABLE OF CONTENTS

SECTION 1: BASIC CONSTANTS, UNITS, AND CONVERSION FACTORS

CODATA Recommended Values of the Fundamental Physical Constants: 2018 1-1
Standard Atomic Weights 1-10
Atomic Masses and Isotopic Abundances 1-12
Electron Configuration and Ionization Energy of Neutral Atoms in the Ground State 1-16
International Temperature Scale of 1990 (ITS-90) 1-17
International System of Units (SI) 1-18
Units for Magnetic Properties 1-23
Conversion Factors for Energy Units 1-24
Conversion Factors for Pressure Units 1-24
Descriptive Terms for Solubility 1-25
Values of the Gas Constant in Different Unit Systems 1-25
Definition of Ambient 1-26
Conversion Factors for Thermal Conductivity Units 1-26

SECTION 2: SYMBOLS, TERMINOLOGY, AND NOMENCLATURE

Symbols and Terminology for Physical and Chemical Quantities 2-1
Expression of Uncertainty of Measurements 2-13
Nomenclature for Chemical Compounds 2-15
Nomenclature of Inorganic Chemistry 2-16
Representation of Chemical Structures with the IUPAC International Chemical Identifier (InChI) 2-22
Scientific Abbreviations, Acronyms, and Symbols 2-24
Thermodynamic Functions and Relations 2-37
Nobel Laureates in Chemistry and Physics 2-38

SECTION 3: PHYSICAL CONSTANTS OF ORGANIC COMPOUNDS

Physical Constants of Organic Compounds 3-1
Diamagnetic Susceptibility of Selected Organic Compounds 3-56

SECTION 4: PROPERTIES OF THE ELEMENTS AND INORGANIC COMPOUNDS

The Elements 4-1
Physical Constants of Inorganic Compounds 4-40
Physical Properties of the Rare-Earth Metals 4-52
Melting, Boiling, Triple, and Critical Points of the Elements 4-62
Heat Capacity of the Elements at 25 °C 4-65
Vapor Pressure of the Metallic Elements — Equations 4-66
Vapor Pressure of the Metallic Elements — Data 4-69
Density of Molten Elements and Representative Salts 4-71
Magnetic Susceptibility of the Elements and Inorganic Compounds 4-73
Index of Refraction of Inorganic Liquids and Liquid Elements 4-79
Physical and Optical Properties of Minerals 4-80
Crystallographic Data on Minerals 4-87

SECTION 5: THERMOCHEMISTRY, KINETICS, ELECTROCHEMISTRY, AND SOLUTION CHEMISTRY

CODATA Key Values for Thermodynamics 5-1
Standard Thermodynamic Properties of Chemical Substances 5-3
Thermodynamic Properties as a Function of Temperature 5-42
Thermodynamic Properties of Aqueous Ions 5-65
Heat of Combustion 5-67
Energy Content of Fuels 5-68
Chemical Reaction Rate Constants for Atmospheric Studies 5-69
Ionization Constant of Water at Various Temperatures and Pressures 5-92
Ionization Constant of Normal and Heavy Water at Saturated Vapor Pressure 5-93
Electrical Conductivity of Water 5-93
Electrical Conductivity of Aqueous Solutions 5-94
Standard KCl Solutions for Calibrating Electrical Conductivity Cells 5-95
Molar Electrical Conductivity of Aqueous HF, HCl, HBr, and HI 5-96
Molar Electrical Conductivity of Electrolytes in Aqueous Solution 5-97

Ionic Conductivity and Diffusion at Infinite Dilution 5-98
Electrochemical Series 5-101
Dissociation Constants of Inorganic Acids and Bases 5-108
Dissociation Constants of Organic Acids and Bases 5-109
Activity Coefficients of Acids, Bases, and Salts 5-119
Mean Activity Coefficients of Electrolytes as a Function of Concentration 5-121
Enthalpy of Dilution of Acids 5-126
Enthalpy of Solution of Electrolytes 5-127
pH Scale for Aqueous Solutions 5-128
Buffer Solutions Giving Round Values of pH at 25 °C 5-132
Concentrative Properties of Aqueous Solutions 5-133
Solubility of Selected Gases in Water 5-149
Solubility of Carbon Dioxide in Water at Various Temperatures and Pressures 5-151
Aqueous Solubility and Henry's Law Constants of Organic Compounds 5-152
Aqueous Solubility of Inorganic Compounds at Various Temperatures 5-183
Octanol–Water Partition Coefficients 5-189
Solubility Product Constants of Inorganic Salts 5-194
Solubility of Common Salts at Various Temperatures 5-196
Solubility of Hydrocarbons in Seawater 5-197
Solubility of Organic Compounds in Pressurized Hot Water 5-198
Solubility Chart for Inorganic Salts 5-200

SECTION 6: FLUID PROPERTIES

Thermophysical Properties of Water and Steam 6-1
Vapor Pressure and Other Saturation Properties of Water 6-5
Standard Density of Water 6-7
Fixed-Point Properties of H_2O and D_2O 6-9
Properties of Saturated Liquid D_2O 6-10
Properties of Ice and Supercooled Water 6-11
Vapor Pressure of Ice 6-12
Melting Point of Ice as a Function of Pressure 6-13
Permittivity (Dielectric Constant) of Water at Various Frequencies 6-14
Thermophysical Properties of Air 6-15
Thermophysical Properties of Fluids 6-21
Thermophysical Properties of Selected Fluids at Saturation 6-40
Virial Coefficients of Selected Gases 6-51
Mean Free Path and Related Properties of Gases 6-54
Influence of Pressure on Freezing Points 6-55
Critical Constants of Organic Compounds 6-56
Critical Constants of Inorganic Compounds 6-82
Sublimation Pressure of Solids 6-85
Vapor Pressure of Compounds and Elements 6-87
Vapor Pressure of Fluids at Temperatures below 300 K 6-117
Vapor Pressure of Saturated Salt Solutions 6-126
Enthalpy of Vaporization 6-127
Enthalpy of Fusion 6-144
Compressibility and Expansion Coefficients of Liquids 6-154
Temperature and Pressure Dependence of Liquid Density 6-156
Properties of Cryogenic Fluids 6-161
Properties of Liquid Helium 6-163
Properties of Refrigerants 6-164
Properties of Gas Clathrate Hydrates 6-167
Properties of Ionic Liquids 6-178
Surface Tension of Common Liquids 6-182
Surface Tension of Aqueous Mixtures 6-187
Surface Active Chemicals (Surfactants) 6-188
Permittivity (Dielectric Constant) of Liquids 6-193
Permittivity (Dielectric Constant) of Gases 6-216
Azeotropic Data for Binary Mixtures 6-217
Viscosity of Gases 6-233
Viscosity of Liquids 6-235
Thermal Conductivity of Gases and Refrigerants 6-240
Thermal Conductivity of Liquids 6-243

Diffusion in Gases 6-249
Diffusion of Gases in Water 6-251
Diffusion Coefficients in Liquids at Infinite Dilution 6-252

SECTION 7: BIOCHEMISTRY

Properties of Amino Acids 7-1
Structures of Common Amino Acids 7-3
Properties of Purine and Pyrimidine Bases 7-5
The Genetic Code 7-6
Properties of Fatty Acids and Their Methyl Esters 7-7
Composition and Properties of Common Oils and Fats 7-9
Carbohydrate Names and Symbols 7-12
Standard Transformed Gibbs Energies of Formation for Biochemical Reactants 7-14
Apparent Equilibrium Constants for Enzyme-Catalyzed Reactions 7-17
Apparent Equilibrium Thermodynamics of Protein-Ligand Binding Reactions 7-21
Thermodynamic Quantities for the Ionization Reactions of Buffers in Water 7-30
Biological Buffers 7-33
Typical pH Values of Biological Materials and Foods 7-34
Properties and Functions of Common Drugs 7-35
Properties of Controlled Substances 7-57
Chemical Constituents of Human Blood 7-59
Chemical Composition of the Human Body 7-61

SECTION 8: ANALYTICAL CHEMISTRY

Abbreviations and Symbols Used in Analytical Chemistry 8-1
Basic Instrumental Techniques of Analytical Chemistry 8-6
Analytical Standardization and Calibration 8-10
Figures of Merit 8-17
Mass- and Volume-Based Concentration Units 8-18
Detection of Outliers in Measurements 8-19
Properties of Carrier Gases for Gas Chromatography 8-20
Common Symbols Used in Gas and Liquid Chromatographic Schematic Diagrams 8-21
Standard Fittings for Compressed Gas Cylinders 8-22
Stationary Phases for Porous-Layer Open Tubular Columns 8-23
Coolants for Cryotrapping 8-24
Properties of Common Cross-Linked Silicone Stationary Phases 8-25
Detectors for Gas Chromatography 8-27
Varieties of Hyphenated Gas Chromatography with Mass Spectrometry 8-29
Gas Chromatographic Retention Indices 8-31
Pressure Drop in Open Tubular Gas Chromatographic Columns 8-33
Phase Ratios for Capillary Columns 8-34
Minimum Recommended Injector Split Ratios for Capillary Columns 8-35
Eluotropic Values of Solvents on Octadecylsilane and Octylsilane 8-35
Instability of HPLC Solvents 8-36
Detectors for Liquid Chromatography 8-37
Solvents for Ultraviolet Spectrophotometry 8-38
Correlation Table for Ultraviolet Active Functionalities 8-39
Middle-Range Infrared Absorption Correlation Charts 8-42
Common Spurious Infrared Absorption Bands 8-48
Nuclear Spins, Moments, and Other Data Related to NMR Spectroscopy 8-49
Properties of Important NMR Nuclei 8-52
Proton NMR Absorption of Major Chemical Families 8-53
Proton NMR Correlation Chart for Major Organic Functional Groups 8-58
Proton NMR Shifts of Common Organic Liquids 8-59
Proton Chemical Shifts of Contaminants in Deuterated Solvents 8-65
^{13}C-NMR Absorptions of Major Functional Groups 8-66
^{13}C-NMR Chemical Shifts of Common Organic Solvents 8-68
^{15}N-NMR Chemical Shifts of Major Chemical Families 8-69
Natural Abundance of Important Isotopes 8-71
Common Mass Spectral Fragmentation Patterns of Organic Compound Families 8-72
Common Mass Spectral Fragments Lost 8-74
Major Reference Masses in the Spectrum of Heptacosafluorotributylamine (Perfluorotributylamine) 8-75
Mass Spectral Peaks of Common Organic Liquids 8-76

Common Spurious Signals Observed in Mass Spectrometers 8-83
Chlorine–Bromine Combination Isotope Intensities in Mass Spectral Patterns 8-84
Reduction of Weighings in Air to *In Vacuo* 8-85
Standards for Laboratory Weights 8-86
Indicators for Acids and Bases 8-88
Preparation of Special Analytical Reagents 8-90
Organic Analytical Reagents for the Determination of Inorganic Ions 8-95

SECTION 9: MOLECULAR STRUCTURE AND SPECTROSCOPY
Bond Lengths in Crystalline Organic Compounds 9-1
Bond Lengths in Organometallic Compounds 9-17
Structure of Free Molecules in the Gas Phase 9-19
Characteristic Bond Lengths in Free Molecules 9-55
Atomic Radii of the Elements 9-56
Dipole Moments 9-58
Hindered Internal Rotation 9-68
Bond Dissociation Energies 9-73
Electronegativity 9-106
Force Constants for Bond Stretching 9-107
Fundamental Vibrational Frequencies of Small Molecules 9-108
Spectroscopic Constants of Diatomic Molecules 9-111

SECTION 10: ATOMIC, MOLECULAR, AND OPTICAL PHYSICS
Persistent Lines of the Neutral Atomic Elements 10-1
Atomic Transition Probabilities 10-51
Electron Affinities 10-54
Proton Affinities 10-76
Polarizabilities of Atoms and Molecules 10-95
Ionization Energies of Atoms and Atomic Ions 10-112
Ionization Energies of Gas-Phase Molecules 10-116
Attenuation Coefficients for High-Energy Electromagnetic Radiation 10-134
Classification of Electromagnetic Radiation 10-140
Sensitivity of the Human Eye to Light of Different Wavelengths 10-142
Blackbody Radiation 10-143
Characteristics of Infrared Detectors 10-145
Index of Refraction of Inorganic Crystals 10-146
Refractive Index and Transmittance of Representative Glasses 10-150
Index of Refraction of Water 10-151
Index of Refraction of Liquids for Calibration Purposes 10-152
Index of Refraction of Air 10-153
Index of Refraction of Gases 10-154

SECTION 11: NUCLEAR AND PARTICLE PHYSICS
Summary Tables of Particle Properties 11-1
Table of the Isotopes 11-3
Neutron Scattering and Absorption Properties 11-43
Cosmic Radiation 11-56

SECTION 12: PROPERTIES OF SOLIDS
Techniques for Materials Characterization 12-1
Symmetry of Crystals 12-6
Ionic Radii in Crystals 12-11
Polarizabilities of Atoms and Ions in Solids 12-13
Crystal Structures and Lattice Parameters of Allotropes of the Elements 12-15
Phase Transitions in the Solid Elements at Atmospheric Pressure 12-19
The Madelung Constant and Crystal Lattice Energy 12-21
Elastic Constants of Single Crystals 12-22
Electrical Resistivity of Pure Metals 12-27
Electrical Resistivity of Selected Alloys 12-28
Electrical Resistivity of Graphite Materials 12-30
Permittivity (Dielectric Constant) of Inorganic Solids 12-31
Curie Temperature of Selected Ferroelectric Crystals 12-40
Properties of Antiferroelectric Crystals 12-41

Dielectric Constants of Glasses 12-41
Properties of Superconductors 12-42
High-Temperature Superconductors 12-58
Organic Superconductors 12-68
Properties of Semiconductors 12-70
Selected Properties of Semiconductor Solid Solutions 12-83
Properties of Organic Semiconductors 12-85
Diffusion Data for Semiconductors 12-89
Properties of Magnetic Materials 12-97
Organic Magnets 12-105
Electron Inelastic Mean Free Paths 12-106
Electron Stopping Powers 12-108
Electron Work Function of the Elements 12-110
Secondary Electron Emission 12-112
Optical Properties of Selected Elements 12-113
Optical Properties of Selected Inorganic and Organic Solids 12-117
Properties of Selected Materials at Cryogenic Temperatures 12-122
Heat Capacity of Selected Solids 12-128
Thermal and Physical Properties of Pure Metals 12-129
Thermal Conductivity of Metals and Semiconductors as a Function of Temperature 12-131
Thermal Conductivity of Alloys as a Function of Temperature 12-132
Thermal Conductivity of Crystalline Dielectrics 12-134
Thermal Conductivity of Ceramics and Other Insulating Materials 12-136
Thermal Conductivity of Glasses 12-138
Thermoelectric Properties of Metals and Semiconductors 12-141
Fermi Energy and Related Properties of Metals 12-143

SECTION 13: POLYMER PROPERTIES

Abbreviations Used in Polymer Science and Technology 13-1
Physical Properties of Selected Polymers 13-3
Nomenclature for Organic Polymers 13-5
Solvents for Common Polymers 13-9
Glass Transition Temperature for Selected Polymers 13-10
Dielectric Constant of Selected Polymers 13-17
Pressure-Volume-Temperature Relationships for Polymer Melts 13-18
Vapor Pressures (Solvent Activities) for Binary Polymer Solutions 13-22
Solubility Parameters of Selected Polymers 13-27

SECTION 14: GEOPHYSICS, ASTRONOMY, AND ACOUSTICS

Astronomical Constants 14-1
Properties of the Solar System 14-2
Satellites of the Planets 14-6
Interstellar Molecules 14-9
Mass, Dimensions, and Other Parameters of the Earth 14-14
Geological Time Scale 14-16
Acceleration Due to Gravity 14-17
Density, Pressure, and Gravity as a Function of Depth within the Earth 14-17
Ocean Pressure as a Function of Depth and Latitude 14-18
Properties of Seawater 14-19
Abundance of Elements in the Earth's Crust and in the Sea 14-21
Solar Irradiance at the Earth 14-22
U.S. Standard Atmosphere (1976) 14-23
Geographical and Seasonal Variations in Solar Radiation 14-29
Major World Earthquakes 14-30
Infrared Absorption by the Earth's Atmosphere 14-34
Atmospheric Concentration of Carbon Dioxide, 1959–2020 14-35
Global Temperature Trend, 1880–2020 14-36
Global Warming Potential of Greenhouse Gases 14-37
Speed of Sound in Various Media 14-39
Attenuation and Speed of Sound in Air as a Function of Humidity and Frequency 14-41
Speed of Sound in Dry Air 14-42
Allocation of Frequencies in the Radio Spectrum 14-43

SECTION 15: PRACTICAL LABORATORY DATA

Standard ITS-90 Thermocouple Tables 15-1
Reference Points on the ITS-90 Temperature Scale 15-10
Relative Sensitivity of Bayard-Alpert Ionization Gauges to Various Gases 15-12
Laboratory Solvents and Other Liquid Reagents 15-13
Miscibility of Organic Solvents 15-23
Density of Solvents as a Function of Temperature 15-27
Dependence of Boiling Point of Organic Liquids on Pressure 15-28
Ebullioscopic Constants for Calculation of Boiling Point Elevation 15-30
Cryoscopic Constants for Calculation of Freezing Point Depression 15-31
Freezing Point Lowering by Electrolytes in Aqueous Solution 15-32
Correction of Barometer Readings to 0 °C Temperature 15-33
Determination of Relative Humidity from Dew Point 15-34
Determination of Relative Humidity from Wet and Dry Bulb Temperatures 15-35
Constant Humidity Solutions 15-36
Standard Salt Solutions for Humidity Calibration 15-37
Low-Temperature Baths for Maintaining Constant Temperature 15-37
Metals and Alloys with Low-Melting Temperature 15-38
Characteristics of Particles and Particle Dispersoids 15-40
Density of Various Solids 15-41
Density of Sulfuric Acid 15-42
Density of Ethanol–Water Mixtures 15-44
Dielectric Strength of Insulating Materials 15-45
Coefficient of Friction 15-50

SECTION 16: HEALTH AND SAFETY INFORMATION

Chemical Safety 16-1
Abbreviations Used in the Assessment and Presentation of Laboratory Hazards 16-3
Incompatible Chemicals 16-5
Explosion (Shock) Hazards 16-7
Water-Reactive Chemicals 16-8
Testing Requirements for Peroxidizable Compounds 16-8
Tests for the Presence of Peroxides 16-9
Pyrophoric Compounds – Compounds That Are Reactive with Air 16-9
Flammability Hazards of Common Solvents 16-10
Flammability of Chemical Substances 16-12
Materials Compatible with and Resistant to 72% Perchloric Acid 16-30
Selection of Laboratory Gloves 16-31
Selection of Protective Laboratory Garments 16-33
Selection of Respirators and Respirator Cartridges and Filters 16-34
Protective Clothing Levels 16-36
Selection of Hearing Protection Devices 16-37
Chemical Fume Hoods and Biological Safety Cabinets 16-39
Gas Cylinder Safety and Stamped Markings 16-41
Compressed Air Safety 16-42
Safety in the Use of Cryogens 16-43
Nanomaterial Safety Guidelines 16-46
Threshold Limits for Airborne Contaminants 16-48
Chemical Carcinogens 16-76
Laser Hazards in the Laboratory 16-86
General Characteristics of Ionizing Radiation for the Purpose of Practical Application of Radiation Protection 16-88
Radiation Safety Units 16-89
Relative Dose Ranges from Ionizing Radiation 16-91
Annual Limits on Intakes of Radionuclides 16-93

APPENDIX A: SOURCES OF PHYSICAL AND CHEMICAL DATA A-1

INDEX I-1

Section 1
Basic Constants, Units, and Conversion Factors

CODATA Recommended Values of the Fundamental Physical Constants: 2018 1-1
Standard Atomic Weights 1-10
Atomic Masses and Isotopic Abundances 1-12
Electron Configuration and Ionization Energy of Neutral Atoms in the Ground State........ 1-16
International Temperature Scale of 1990 (ITS-90)...................................... 1-17
International System of Units (SI) 1-18
Units for Magnetic Properties.. 1-23
Conversion Factors for Energy Units.. 1-24
Conversion Factors for Pressure Units 1-24
Descriptive Terms for Solubility.. 1-25
Values of the Gas Constant in Different Unit Systems 1-25
Definition of Ambient.. 1-26
Conversion Factors for Thermal Conductivity Units.................................... 1-26

CODATA RECOMMENDED VALUES OF THE FUNDAMENTAL PHYSICAL CONSTANTS: 2018

Eite Tiesinga, Peter J. Mohr, David B. Newell, and Barry N. Taylor

This report gives the 2018 self-consistent set of values of the constants and conversion factors of physics and chemistry recommended by the Committee on Data for Science and Technology (CODATA). These values are based on a least-squares adjustment that take into account all data available up to December 31, 2018. They also reflect the major changes to the SI units made in 2019. See the International System of Units (SI) in this section for further details about these changes.

The tables below are extracted from reports (Refs. 1–4) prepared by the authors under the auspices of the CODATA Task Group on Fundamental Constants and the International Bureau of Weights and Measures (BIPM). The present members of the task group are:

D. B. Newell, National Institute of Standards and Technology, United States of America (Chair)
K. Pachucki, University of Warsaw, Poland (Co-Chair)
F. Cabiati, Istituto Nazionale di Ricerca Metrologica, Italy
K. Fujii, National Metrology Institute of Japan, Japan
S. G. Karshenboim, The Central Astronomical Observatory of the Russian Academy of Sciences at Pulkovo, Russia
H. Margolis, National Physical Laboratory, United Kingdom
E. de Mirandés, Bureau International des Poids et Mesures
P. J. Mohr, National Institute of Standards and Technology, United States of America
F. Nez, Laboratoire Kastler-Brossel, France
R. Pohl, QUANTUM, Johannes Gutenberg-Universitat, Germany
J. Qu, National Institute of Metrology, China
T. J. Quinn, Bureau International des Poids et Mesures (Emeritus)
A. Surzhykov, Physikalisch-Technische Bundesanstalt, Germany
B. N. Taylor, National Institute of Standards and Technology, United States of America (Emeritus)
E. Tiesinga, National Institute of Standards and Technology, United States of America
M. Wang, Institute of Modern Physics, China
B. M. Wood, National Research Council, Canada (Vice-Chair)

References

1. Mohr, P. J., Newell, D. B., and Taylor, B. N., *J. Phys. Chem. Ref. Data* 45, 043102, 2016. <https://doi.org/10.1063/1.4954402>
2. Mohr, P. J., Newell, D. B., and Taylor, B. N., *Rev. Mod. Phys.* 88, 035009, 2016. <https://doi.org/10.1103/RevModPhys.88.035009>
3. Available on the Web at <physics.nist.gov/constants>.
4. Newell, D. B., and Tiesinga, E., *The International System of Units (SI)*, NIST Special Publication 330, 2019 Edition, National Institute of Standards and Technology, Gaithersburg, MD, 2019. <https://doi.org/10.6028/NIST.SP.330-2019>

The 2018 set of CODATA recommended values of the fundamental physical constants is based on the recently adopted definitions of the SI units as derived from a set of seven defining constants, as shown in Table 1.

TABLE 1. The Seven Defining Constants of the SI and the Seven Corresponding Units They Define (Note: All constants are exact)

Defining constant	Symbol	Numerical value	Unit
hyperfine transition of cesium (^{113}Cs)	$\Delta\nu_{Cs}$	9 192 631 770	Hz
speed of light in vacuum	c, c_0	299 792 458	m s^{-1}
Planck constant	h	6.626 070 15 × 10^{-34}	J s
elementary charge	e	1.602 176 634 × 10^{-19}	C
Boltzmann constant	k	1.380 649 × 10^{-23}	J K^{-1}
Avogadro constant	N_A	6.022 140 76 × 10^{23}	mol^{-1}
luminous efficacy	K_{cd}	683	lm W^{-1}

Table 2 is an abbreviated list of the CODATA recomended values of the fundamental constants of physics and chemistry, based on the 2018 adjustment and the recently adopted defining SI constants and units (Table 1). Uncertainties are given when appropriate. See References 3 and 4 for additional information.

TABLE 2. Frequently Used Fundamental Constants of Physics and Chemistry Based on the 2018 CODATA Adjustment

Quantity	Symbol	Numerical value	Unit	Uncertainty
electron volt (e/C) J	eV	1.602 176 634 × 10^{-19}	J	exact
Josephson constant $2e/h$	K_J	483 597.848 4... × 10^{9}	Hz V^{-1}	exact
von Klitzing constant $2\pi h/e^2$	R_K	25 812 807 45...	Ω	exact
molar gas constant	R	8.314 462 618...	J mol^{-1} K^{-1}	exact
Stefan-Boltzmann constant $(\pi^2/60)k^4/\hbar^3c^2$	σ	5.670 374 419... × 10^{-8}	W m^{-2} K^{-4}	exact
(unified) atomic mass unit $(1/12)m(^{12}C)$	u	1.660 539 066 60 × 10^{-27}	kg	0.000 000 000 50 × 10^{-27}
Newtonian constant of gravitation	G	6.674 30 × 10^{-11}	m^3 kg^{-1} s^{-2}	0.000 15 × 10^{-11}
fine-structure constant $e^2/4\pi\epsilon_0\hbar c$	α	7.297 352 5693 × 10^{-3}		0.000 000 0011 × 10^{-3}

Quantity	Symbol	Numerical value	Unit	Uncertainty
inverse fine-structure constant	α^{-1}	137.035 999 084		0.000 000 021
Rydberg constant $\alpha^2 m_e c/2h$	R_∞	10 973 731.568 160	m^{-1}	0.000 021
vac. magnetic permeability $4\pi\alpha\hbar/(e^2c)$	μ_0	$1.256\ 637\ 062\ 12 \times 10^{-6}$	$N\ A^{-2}$	$0.000\ 000\ 000\ 19 \times 10^{-6}$
electron mass	m_e	$9.109\ 383\ 7015 \times 10^{-31}$	kg	$0.000\ 000\ 0028 \times 10^{-31}$
proton mass	m_p	$1.672\ 621\ 923\ 69 \times 10^{-27}$	kg	$0.000\ 000\ 000\ 51 \times 10^{-27}$
proton-electron mass ratio	m_p/m_e	1836.152 673 43		0.000 000 11
reduced Compton wavelength $\hbar/(m_e c)$	C	$3.861\ 592\ 6796 \times 10^{-13}$	m	$0.000\ 000\ 0012 \times 10^{-13}$
Bohr radius $\hbar/(\alpha m_e c)$	α_0	$5.291\ 772\ 109\ 03 \times 10^{-11}$	m	$0.000\ 000\ 000\ 80 \times 10^{-11}$
Bohr magneton $e\hbar/(2m_e)$	μ_B	$9.274\ 010\ 0783 \times 10^{-24}$	$J\ T^{-1}$	$0.000\ 000\ 0028 \times 10^{-24}$
$h/2\pi$	$\hbar$	$1.054\ 571\ 817... \times 10^{-34}$	J s	exact
vac. electric permittivity $1/(\mu_0 c^2)$	ϵ_0	$8.854\ 187\ 8128 \times 10^{-12}$	$F\ m^{-1}$	$0.000\ 000\ 0013 \times 10^{-12}$
magnetic flux quantum $h/2e$	Φ_0	$2.067\ 833\ 848... \times 10^{-15}$	Wb	exact
conductance quantum $2e^2/h$	S	$7.748\ 091\ 729 \times 10^{-5}$	S	exact
Faraday constant $N_A e$	F	96 485.332 12...	$C\ mol^{-1}$	exact

Tables 3, 4, 5, and 8 contain the complete set of fundamental constants of physics and chemistry as recommended by CODATA in their 2018 adjustment. These values also reflect the changes in the International System of Units (SI) as adopted by BIPM in 2019. A number of constants that were previously based on analysis of available experimental measurements have become *exact* given the adoption of defined (exact) values of the seven constants used to define the SI.

TABLE 3. The CODATA Recommended Values of the Fundamental Constants of Physics and Chemistry Based on the 2018 Adjustment

Quantity	Symbol	Numerical value	Unit	Uncertainty
Universal				
vacuum magnetic permeability	μ_0	$4\pi \times 10^{-7}$	$N\ A^{-2}$	
		$1.256\ 637\ 062\ 12 \times 10^{-6}$	$N\ A^{-2}$	$0.000\ 000\ 000\ 19 \times 10^{-6}$
vacuum electric permittivity $1/\mu_0 c^2$	ϵ_0	$8.854\ 187\ 8128 \times 10^{-12}$	$F\ m^{-1}$	$0.000\ 000\ 0013 \times 10^{-12}$
characteristic impedance of vacuum $\mu_0 c$	Z_0	376.730 313 668	Ω	0.000 000 057
Newtonian constant of gravitation	G	$6.674\ 30 \times 10^{-11}$	$m^3\ kg^{-1}\ s^{-2}$	$0.000\ 15 \times 10^{-11}$
	$G/\hbar c$	$6.708\ 83 \times 10^{-39}$	$(GeV/c^2)^{-2}$	$0.000\ 15 \times 10^{-39}$
Planck constant divided by 2π $h/2\pi$	$\hbar$	$1.054\ 571\ 817... \times 10^{-34}$	J s	exact
		$6.582\ 119\ 569... \times 10^{-16}$	eV s	exact
		$4.135\ 667\ 696... \times 10^{-15}$	$eV\ Hz^{-1}$	exact
	$\hbar c$	197.326 980 4...	MeV fm	exact
Planck mass $(\hbar c/G)^{1/2}$	m_P	$2.176\ 434 \times 10^{-8}$	kg	$0.000\ 024 \times 10^{-8}$
energy equivalent	$m_P c^2$	$1.220\ 890 \times 10^{19}$	GeV	$0.000\ 014 \times 10^{19}$
Planck temperature $(\hbar c^5/G)^{1/2}/k$	T_P	$1.416\ 784 \times 10^{32}$	K	$0.000\ 016 \times 10^{32}$
Planck length $\hbar/m_P c = (\hbar G/c^3)^{1/2}$	l_P	$1.616\ 255 \times 10^{-35}$	m	$0.000\ 018 \times 10^{-35}$
Planck time $l_P/c = (\hbar G/c^5)^{1/2}$	t_P	$5.391\ 247 \times 10^{-44}$	s	$0.000\ 060 \times 10^{-44}$
Electromagnetic				
	$e/\hbar$	$1.519\ 267\ 447... \times 10^{15}$	$A\ J^{-1}$	exact
magnetic flux quantum $h/2e$	Φ_0	$2.067\ 833\ 848... \times 10^{-15}$	Wb	exact
conductance quantum $2e2/h$	G_0	$7.748\ 091\ 729... \times 10^{-5}$	S	exact
inverse of conductance quantum	G_0^{-1}	12 906.403 72...	Ω	exact
Josephson constant $2e/h$	K_J	$483\ 597.848\ 4... \times 10^9$	$Hz\ V^{-1}$	exact
von Klitzing constant $h/e^2 = \mu_0 c/2\alpha$	R_K	25 812.807 45...	Ω	exact
Bohr magneton $e\hbar/2m_e$	μ_B	$9.274\ 010\ 0783 \times 10^{-24}$	$J\ T^{-1}$	$0.000\ 000\ 0028 \times 10^{-24}$
		$5.788\ 381\ 8060 \times 10^{-5}$	$eV\ T^{-1}$	$0.000\ 000\ 0017 \times 10^{-5}$
	μ_B/h	$1.399\ 624\ 493\ 61 \times 10^{10}$	$Hz\ T^{-1}$	$0.000\ 000\ 000\ 42 \times 10^{10}$
	μ_B/hc	46.686 447 783	$m^{-1}\ T^{-1}$	0.000 000 014
	μ_B/k	0.671 713 815 63	$K\ T^{-1}$	0.000 000 000 20
nuclear magneton $e\hbar/2m_p$	μ_N	$5.050\ 783\ 7461 \times 10^{-27}$	$J\ T^{-1}$	$0.000\ 000\ 0015 \times 10^{-27}$
		$3.152\ 451\ 258\ 44 \times 10^{-8}$	$eV\ T^{-1}$	$0.000\ 000\ 000\ 96 \times 10^{-8}$
	μ_N/h	7.622 593 2291	$MHz\ T^{-1}$	0.000 000 0023
	μ_N/hc	$2.542\ 623\ 413\ 53 \times 10^{-2}$	$m^{-1}\ T^{-1}$	$0.000\ 000\ 000\ 78 \times 10^{-2}$
	μ_N/k	$3.658\ 267\ 7756 \times 10^{-4}$	$K\ T^{-1}$	$0.000\ 000\ 0011 \times 10^{-4}$

Quantity	Symbol	Numerical value	Unit	Uncertainty
Atomic and Nuclear				
General				
fine-structure constant $e^2/4\pi\epsilon_0\hbar c$	α	7.297 352 5693 × 10^{-3}		0.000 000 0011 × 10^{-3}
inverse fine-structure constant	α^{-1}	137.035 999 084		0.000 000 021
Rydberg constant $\alpha^2 m_e c/2h$	R_∞	10 973 731.568 160	m^{-1}	0.000 021
	$R_\infty c$	3.289 841 960 2508 × 10^{15}	Hz	0.000 000 000 0064 × 10^{15}
	$R_\infty hc$	2.179 872 361 1035 × 10^{-18}	J	0.000 000 000 0042 × 10^{-18}
		13.605 693 122 994	eV	0.000 000 000 026
Bohr radius $\alpha/4\pi R_\infty = 4\pi\epsilon_0\hbar^2/m_e e^2$	a_0	5.291 772 109 03 × 10^{-11}	m	0.000 000 000 80 × 10^{-11}
Hartree energy $e^2/4\pi\epsilon_0 a_0 = 2R_\infty hc = \alpha^2 m_e c^2$	E_h	4.359 744 722 2071 × 10^{-18}	J	0.000 000 000 0085 × 10^{-18}
		27.211 386 245 988	eV	0.000 000 000 053
quantum of circulation	$h/2m_e$	3.636 947 5516 × 10^{-4}	m^2s^{-1}	0.000 000 0011 × 10^{-4}
	h/m_e	7.273 895 1032 × 10^{-4}	m^2s^{-1}	0.000 000 0022 × 10^{-4}
Electroweak				
Fermi coupling constant	$G_F/(\hbar c)^3$	1.166 3787 × 10^{-5}	GeV^{-2}	0.000 006 × 10^{-5}
weak mixing angle θ_W (on-shell scheme) $\sin^2\theta_W = s_W = 1-(m_W/m_Z)^2$	$\sin^2\theta_W$	0.222 90		0.000 30
Electron, e-				
electron mass	m_e	9.109 383 7015 × 10^{-31}	kg	0.000 000 0028 × 10^{-31}
		5.485 799 090 65 × 10^{-4}	u	0.000 000 000 16 × 10^{-4}
energy equivalent	$m_e c^2$	8.187 105 7769 × 10^{-14}	J	0.000 000 0025 × 10^{-14}
		0.510 998 950 00	MeV	0.000 000 000 15
electron relative atomic mass		5.485 799 090 65 × 10^{-4}		0.000 000 000 16 × 10^{-4}
electron-muon mass ratio	m_e/m_μ	4.836 331 69 × 10^{-3}		0.000 000 11 × 10^{-3}
electron-tau mass ratio	m_e/m_τ	2.875 85 × 10^{-4}		0.000 19 × 10^{-4}
electron-proton mass ratio	m_e/m_p	5.446 170 214 87 × 10^{-4}		0.000 000 000 33 × 10^{-4}
electron-neutron mass ratio	m_e/m_n	5.438 673 4424 × 10^{-4}		0.000 000 0026 × 10^{-4}
electron-deuteron mass ratio	m_e/m_d	2.724 437 107 462 × 10^{-4}		0.000 000 000 096 × 10^{-4}
electron-triton mass ratio	m_e/m_t	1.819 200 062 251 × 10^{-4}		0.000 000 000 090 × 10^{-4}
electron-helion mass ratio	m_e/m_h	1.819 543 074 573 × 10^{-4}		0.000 000 000 079 × 10^{-4}
electron to alpha particle mass ratio	m_e/m_α	1.370 933 554 787 × 10^{-4}		0.000 000 000 045 × 10^{-4}
electron charge to mass quotient	$-e/m_e$	-1.758 820 010 76 × 10^{11}	C kg^{-1}	0.000 000 000 53 × 10^{11}
electron molar mass $N_A m_e$	$M(e), M_e$	5.485 799 0888 × 10^{-7}	kg mol^{-1}	0.000 000 0017 × 10^{-7}
Compton wavelength $h/m_e c$	λ_C	2.426 310 238 67 × 10^{-12}	m	0.000 000 000 73 × 10^{-12}
$\lambda_C/2\pi = \alpha a_0 = \alpha^2/4\pi R_\infty$	$\bar{\lambda}_C$	3.861 592 6796 × 10^{-13}	m	0.000 000 0012 × 10^{-13}
classical electron radius $\alpha^2 a_0$	r_e	2.817 940 3262 × 10^{-15}	m	0.000 000 0013 × 10^{-15}
Thomson cross section $(8\pi/3)r_e^2$	σ_e	6.652 458 7321 × 10^{-29}	m^2	0.000 000 0060 × 10^{-29}
electron magnetic moment	μ_e	-9.284 764 7043 × 10^{-24}	J T^{-1}	0.000 000 0028 × 10^{-24}
to Bohr magneton ratio	μ_e/μ_B	-1.001 159 652 181 28		0.000 000 000 000 18
to nuclear magneton ratio	μ_e/μ_N	-1838.281 971 88		0.000 000 11
electron magnetic moment anomaly $\|\mu_e\|/\mu_B-1$	a_e	1.159 652 181 28 × 10^{-3}		0.000 000 000 18 × 10^{-3}
electron g-factor $-2(1+a_e)$	g_e	-2.002 319 304 362 56		0.000 000 000 000 35
electron-muon magnetic moment ratio	μ_e/μ_μ	206.766 9883		0.000 0046
electron-proton magnetic moment ratio	μ_e/μ_p	-658.210 687 89		0.000 000 20
electron to shielded proton magnetic moment ratio (H_2O, sphere, 25 °C)	μ_e/μ_p'	-658.227 5971		0.000 0072
electron-neutron magnetic moment ratio	μ_e/μ_n	960.920 50		0.000 23
electron-deuteron magnetic moment ratio	μ_e/μ_d	-2143.923 4915		0.000 0056
electron to shielded helion magnetic moment ratio (gas, sphere, 25 °C)	μ_e/μ_p'	864.058 257		0.000 010
electron gyromagnetic ratio $2\|\mu_e\|/\hbar$	γ_e	1.760 859 630 23 × 10^{11}	s^{-1} T^{-1}	0.000 000 000 53 × 10^{11}
	$\gamma_e/2\pi$	28 024.951 4242	MHz T^{-1}	0.000 0085
Muon, μ-				
muon mass	m_μ	1.883 531 627 × 10^{-28}	kg	0.000 000 042 × 10^{-28}
		0.113 428 9259	u	0.000 000 0025

Quantity	Symbol	Numerical value	Unit	Uncertainty
energy equivalent	$m_\mu c^2$	$1.692\,833\,804 \times 10^{-11}$	J	$0.000\,000\,038 \times 10^{-11}$
		105.658 3755	MeV	0.000 0023
muon-electron mass ratio	m_μ/m_e	206.768 2830		0.000 0046
muon-tau mass ratio	m_μ/m_τ	$5.946\,35 \times 10^{-2}$		$0.000\,40 \times 10^{-2}$
muon-proton mass ratio	m_μ/m_p	0.112 609 5264		0.000 000 0025
muon-neutron mass ratio	m_μ/m_n	0.112 454 5170		0.000 000 0025
muon molar mass $N_A m_\mu$	$M(\mu), M_\mu$	$1.134\,289\,259 \times 10^{-4}$	kg mol^{-1}	$0.000\,000\,025 \times 10^{-4}$
muon Compton wavelength $h/m_\mu c$	$\lambda_{C,\mu}$	$1.173\,444\,110 \times 10^{-14}$	m	$0.000\,000\,026 \times 10^{-14}$
reduced $\lambda_{C,\mu}/2\pi$	$\bar{\lambda}_{C,\mu}$	$1.867\,594\,306 \times 10^{-15}$	m	$0.000\,000\,042 \times 10^{-15}$
muon magnetic moment	μ_μ	$-4.490\,448\,30 \times 10^{-26}$	J T^{-1}	$0.000\,000\,10 \times 10^{-26}$
to Bohr magneton ratio	μ_μ/μ_B	$-4.841\,970\,47 \times 10^{-3}$		$0.000\,000\,11 \times 10^{-3}$
to nuclear magneton ratio	μ_μ/μ_N	-8.890 597 03		0.000 000 20
muon magnetic moment anomaly				
$\|\mu_\mu\|/(e\hbar/2m_\mu)-1$	a_μ	$1.165\,920\,89 \times 10^{-3}$		$0.000\,000\,63 \times 10^{-3}$
muon g-factor $-2(1+a_\mu)$	g_μ	-2.002 331 8418		0.000 000 0013
muon-proton magnetic moment ratio	μ_μ/μ_p	-3.183 345 142		0.000 000 071
Tau, τ-				
tau mass	m_τ	$3.167\,54 \times 10^{-27}$	kg	$0.000\,21 \times 10^{-27}$
		1.907 54	u	0.000 13
energy equivalent	$m_\tau c^2$	$2.846\,84 \times 10^{-10}$	J	$0.000\,19 \times 10^{-10}$
		1776.86	MeV	0.12
tau-electron mass ratio	m_τ/m_e	3477.23		0.23
tau-muon mass ratio	m_τ/m_μ	16.817		0.0011
tau-proton mass ratio	m_τ/m_p	1.893 76		0.000 13
tau-neutron mass ratio	m_τ/m_n	1.891 15		0.000 13
tau molar mass $N_A m_\tau$	$M(\tau), M_\tau$	$1.907\,54 \times 10^{-3}$	kg mol^{-1}	$0.000\,13 \times 10^{-3}$
tau Compton wavelength $h/m_\tau c$	$\lambda_{C,\tau}$	$6.977\,71 \times 10^{-16}$	m	$0.000\,47 \times 10^{-16}$
reduced $\lambda_{C,\tau}/2\pi$	$\bar{\lambda}_{C,\tau}$	$1.110\,538 \times 10^{-16}$	m	$0.000\,075 \times 10^{-16}$
***Proton,* p**				
proton mass	m_p	$1.672\,621\,923\,69 \times 10^{-27}$	kg	$0.000\,000\,000\,51 \times 10^{-27}$
		1.007 276 466 621	u	0.000 000 000 053
energy equivalent	$m_p c^2$	$1.503\,277\,615\,98 \times 10^{-10}$	J	$0.000\,000\,000\,46 \times 10^{-10}$
		938.272 088 16	MeV	0.000 000 29
proton relative atomic mass		1.007 276 466 621		0.000 000 000 053
proton-electron mass ratio	m_p/m_e	1836.152 673 43		0.000 000 11
proton-muon mass ratio	m_p/m_μ	8.880 243 37		0.000 000 20
proton-tau mass ratio	m_p/m_τ	0.528 051		0.000 036
proton-neutron mass ratio	m_p/m_n	0.998 623 478 12		0.000 000 000 49
proton charge to mass quotient	e/m_p	$9.578\,833\,1560 \times 10^{7}$	C kg^{-1}	$0.000\,000\,0029 \times 10^{7}$
proton molar mass $N_A m_p$	$M(p), M_p$	$1.007\,276\,466\,27 \times 10^{-3}$	kg mol^{-1}	$0.000\,000\,000\,31 \times 10^{-3}$
proton Compton wavelength $h/m_p c$	$\lambda_{C,p}$	$1.321\,409\,855\,39 \times 10^{-15}$	m	$0.000\,000\,000\,40 \times 10^{-15}$
reduced $\lambda_{C,p}/2\pi$	$\bar{\lambda}_{C,p}$	$2.103\,089\,103\,36 \times 10^{-16}$	m	$0.000\,000\,000\,64 \times 10^{-16}$
proton rms charge radius	r_p	8.414×10^{-16}	m	0.019×10^{-16}
proton magnetic moment	μ_p	$1.410\,606\,797\,36 \times 10^{-26}$	J T^{-1}	$0.000\,000\,000\,60 \times 10^{-26}$
to Bohr magneton ratio	μ_p/μ_B	$1.521\,032\,202\,30 \times 10^{-3}$		$0.000\,000\,000\,46 \times 10^{-3}$
to nuclear magneton ratio	μ_p/μ_N	2.792 847 344 63		0.000 000 000 82
proton g-factor $2\mu_p/\mu_N$	g_p	5.585 694 6893		0.000 000 0016
proton-neutron magnetic moment ratio	μ_p/μ_n	-1.459 898 05		0.000 000 34
shielded proton magnetic moment (H_2O, sphere, 25 °C)	μ_p'	$1.410\,570\,560 \times 10^{-26}$	J T^{-1}	$0.000\,000\,015 \times 10^{-26}$
to Bohr magneton ratio	μ_p'/μ_B	$1.520\,993\,128 \times 10^{-3}$		$0.000\,000\,017 \times 10^{-3}$
to nuclear magneton ratio	μ_p'/μ_N	2.792 775 599		0.000 000 030
proton magnetic shielding correction				
$1 - \mu_p' / \mu_p$ (H_2O, sphere, 25 °C)	μ_p'	2.5689×10^{-5}		0.0011×10^{-5}
proton gyromagnetic ratio $2\mu_p/\hbar$	γ_p	$2.675\,221\,8744 \times 10^{8}$	s^{-1} T^{-1}	$0.000\,000\,0011 \times 10^{8}$
	$\gamma_p/2\pi$	42.577 478 518	MHz T^{-1}	0.000 000 018

Quantity	Symbol	Numerical value	Unit	Uncertainty
shielded proton gyromagnetic ratio $2\mu_{p'}/\hbar$ (H_2O, sphere, 25 °C)	$\gamma_{p'}$	2.675 153 151 × 10^{8}	s^{-1} T^{-1}	0.000 000 029 × 10^{8}
	$\gamma_{p'}/2\pi$	42.576 384 74	MHz T^{-1}	0.000 000 46
***Neutron,* n**				
neutron mass	m_n	1.674 927 498 04 × 10^{-27}	kg	0.000 000 000 95 × 10^{-27}
		1.008 664 915 95	u	0.000 000 000 49
energy equivalent	m_nc^2	1.505 349 762 87 × 10^{-10}	J	0.000 000 000 86 × 10^{-10}
		939.565 420 52	MeV	0.000 000 54
neutron relative atomic mass		1.008 664 915 95		0.000 000 000 49
neutron-electron mass ratio	m_n/m_e	1838.683 661 73		0.000 000 89
neutron-muon mass ratio	m_n/m_μ	8.892 484 06		0.000 000 20
neutron-tau mass ratio	m_n/m_τ	0.528 779		0.000 036
neutron-proton mass ratio	m_n/m_p	1.001 378 419 31		0.000 000 000 49
neutron-proton mass difference	m_n-m_p	2.305 574 35 × 10^{-30}	kg	0.000 000 82 × 10^{-30}
		1.388 449 33 × 10^{-3}	u	0.000 000 49 × 10^{-3}
energy equivalent	$(m_n-m_p)c^2$	2.072 146 89 × 10^{-13}	J	0.000 000 74 × 10^{-13}
		1.293 332 36	MeV	0.000 000 46
neutron molar mass N_Am_n	M(n),M_n	1.008 664 915 60 × 10^{-3}	kg mol^{-1}	0.000 000 000 57 × 10^{-3}
neutron Compton wavelength h/m_nc	$\lambda_{C,n}$	1.319 590 905 81 × 10^{-15}	m	0.000 000 000 75 × 10^{-15}
$\lambda_{C,n}/2\pi$	$\bar{\lambda}_{C,n}$	2.100 194 1552 × 10^{-16}	m	0.000 000 0012 × 10^{-16}
neutron magnetic moment	μ_n	-9.662 3651 × 10^{-27}	J T^{-1}	0.000 000 0023 × 10^{-27}
to Bohr magneton ratio	μ_n/μ_B	-1.041 875 63 × 10^{-3}		0.000 000 25 × 10^{-3}
to nuclear magneton ratio	μ_n/μ_N	-1.913 042 73		0.000 000 45
neutron g-factor $2\mu_n/\mu_N$	g_n	-3.826 085 45		0.000 000 90
neutron-electron magnetic moment ratio	μ_n/μ_e	1.040 668 82 × 10^{-3}		0.000 000 25 × 10^{-3}
neutron-proton magnetic moment ratio	μ_n/μ_p	-0.684 979 34		0.000 000 16
neutron to shielded proton magnetic moment ratio (H_2O, sphere, 25 °C)	μ_n/μ_p'	-0.684 996 94		0.000 000 16
neutron gyromagnetic ratio $2\|\mu_n\|/\hbar$	γ_n	1.832 471 71 × 10^{8}	s^{-1} T^{-1}	0.000 000 43 × 10^{8}
	$\gamma_n/2\pi$	29.164 6931	MHz T^{-1}	0.000 0069
***Deuteron,* d**				
deuteron mass	m_d	3.343 583 7724 × 10^{-27}	kg	0.000 000 0010 × 10^{-27}
		2.013 553 212 745	u	0.000 000 000 040
energy equivalent	m_dc^2	3.005 063 231 02 × 10^{-10}	J	0.000 000 000 91 × 10^{-10}
		1875.612 942 57	MeV	0.000 000 57
deuteron relative atomic mass		2.013 553 212 745		0.000 000 000 040
deuteron-electron mass ratio	m_d/m_e	3670.482 967 88		0.000 000 13
deuteron-proton mass ratio	m_d/m_p	1.999 007 501 39		0.000 000 000 11
deuteron molar mass N_Am_d	M(d),M_d	2.013 553 212 05 × 10^{-3}	kg mol^{-1}	0.000 000 000 61 × 10^{-3}
deuteron rms charge radius	r_d	2.127 99 × 10^{-15}	m	0.000 74 × 10^{-15}
deuteron magnetic moment	μ_d	4.330 735 094 × 10^{-27}	J T^{-1}	0.000 000 011 × 10^{-27}
to Bohr magneton ratio	μ_d/μ_B	4.669 754 570 × 10^{-4}		0.000 000 012 × 10^{-4}
to nuclear magneton ratio	μ_d/μ_N	0.857 438 2338		0.000 000 0022
deuteron g-factor μ_d/μ_N	g_d	0.857 438 2338		0.000 000 0022
deuteron-electron magnetic moment ratio	μ_d/μ_e	-4.664 345 551 × 10^{-4}		0.000000012 × 10^{-4}
deuteron-proton magnetic moment ratio	μ_d/μ_p	0.307 012 209 39		0.000 000 000 79
deuteron-neutron magnetic moment ratio	μ_d/μ_n	-0.448 206 53		0.000 000 11
***Triton,* t**				
triton mass	m_t	5.007 356 7446 × 10^{-27}	kg	0.000 000 0015 × 10^{-27}
		3.015 500 716 21	u	0.000 000 000 12
energy equivalent	m_tc^2	4.500 387 8060 × 10^{-10}	J	0.000 000 0014 × 10^{-10}
		2808.921 132 98	MeV	0.000 000 85
triton relative atomic mass		3.015 500 716 21		0.000 000 000 12
triton-electron mass ratio	m_t/m_e	5496.921 535 73		0.000 000 27
triton-proton mass ratio	m_t/m_p	2.993 717 034 14		0.000 000 000 15
triton molar mass N_Am_t	M(t),M_t	3.015 500 715 17 × 10^{-3}	kg mol^{-1}	0.000 000 000 92 × 10^{-3}

Quantity	Symbol	Numerical value	Unit	Uncertainty
triton magnetic moment	μ_t	$1.504\,609\,5202 \times 10^{-26}$	$J\ T^{-1}$	$0.000\,000\,0030 \times 10^{-26}$
to Bohr magneton ratio	μ_t/μ_B	$1.622\,393\,6651 \times 10^{-3}$		$0.000\,000\,0032 \times 10^{-3}$
to nuclear magneton ratio	μ_t/μ_N	2.978 962 4656		0.000 000 0059
triton *g*-factor $2\mu_t/\mu_N$	g_t	5.957 924 931		0.000 000 012
triton to proton magnetic moment ratio		1.066 639 9191		0.000 000 021
***Helion,* h**				
helion mass	m_h	$5.006\,412\,7796 \times 10^{-27}$	kg	$0.000\,000\,0015 \times 10^{-27}$
		3.014 932 247 175	u	0.000 000 000 097
energy equivalent	$m_h c^2$	$4.499\,539\,4125 \times 10^{-10}$	J	$0.000\,000\,0014 \times 10^{-10}$
		2808.391 607 43	MeV	0.000 000 85
helion relative atomic mass		3.014 932 247 175		0.000 000 000 097
helion-electron mass ratio	m_h/m_e	5495.885 280 07		0.000 000 24
helion-proton mass ratio	m_h/m_p	2.993 152 671 67		0.000 000 000 13
helion molar mass $N_A m_h$	$M(h)$, M_h	$3.014\,932\,246\,13 \times 10^{-3}$	$kg\ mol^{-1}$	$0.000\,000\,000\,91 \times 10^{-3}$
helion magnetic moment	μ_h	$-1.074\,617\,532 \times 10^{-26}$	$J\ T^{-1}$	$0.000\,000\,013 \times 10^{-26}$
to Bohr magneton ratio	μ_h/μ_B	$-1.158\,740\,958 \times 10^{-3}$		$0.000\,000\,014 \times 10^{-3}$
to nuclear magneton ratio	μ_h/μ_N	-2.127 625 307		0.000 000 025
helion *g*-factor $2\mu_h/\mu_N$	g_h	-4.255 250 615		0.000 000 050
shielded helion magnetic moment (gas, sphere, 25 °C)	μ_h'	$-1.074\,553\,090 \times 10^{-26}$	$J\ T^{-1}$	$0.000\,000\,013 \times 10^{-26}$
to Bohr magneton ratio	μ_h'/μ_B	$-1.158\,671\,471 \times 10^{-3}$		$0.000\,000\,014 \times 10^{-3}$
to nuclear magneton ratio	$\mu_h'/\mu N$	-2.127 497 719		0.000 000 025
shielded helion to proton magnetic moment ratio (gas, sphere, 25 °C)	μ_h'/μ_p	-0.761 766 5618		0.000 000 0089
shielded helion to shielded proton magnetic moment ratio (gas/H_2O, sphere, 25 °C)	μ_h'/μ_p'	-0.761 786 1313		0.000 000 0033
shielded helion gyromagnetic ratio $2\,\|\mu_h'\|/\hbar$ (gas, sphere, 25 °C)	γ_h'	$2.037\,894\,569 \times 10^{8}$	$s^{-1}\ T^{-1}$	$0.000\,000\,024 \times 10^{8}$
	$\gamma_n'/2\pi$	32.434 099 42	$MHz\ T^{-1}$	0.000 000 38
Alpha particle, α				
helion shielding shift		$5.996\,743 \times 10^{-5}$		$0.000\,010 \times 10^{-5}$
alpha particle mass	m_α	$6.644\,657\,3357 \times 10^{-27}$	kg	$0.000\,000\,0020 \times 10^{-27}$
		4.001 506 179 127	u	0.000 000 000 063
energy equivalent	$m_\alpha c^2$	$5.971\,920\,1914 \times 10^{-10}$	J	$0.000\,000\,0018 \times 10^{-10}$
		3727.379 4066	MeV	0.000 0011
alpha particle relative atomic mass		4.001 506 179 127		0.000 000 000 063
alpha particle to electron mass ratio	m_α/m_e	7294.299 541 42		0.000 000 24
alpha particle to proton mass ratio	m_α/m_p	3.972 599 690 09		0.000 000 000 22
alpha particle molar mass $N_A m_\alpha$	$M(\alpha)$, M_α	$4.001\,506\,1777 \times 10^{-3}$	$kg\ mol^{-1}$	$0.000\,000\,0012 \times 10^{-3}$
Physicochemical				
atomic mass constant				
$m_u = (1/12)\, m(^{12}C) = 1$ u	m_u	$1.660\,539\,066\,60 \times 10^{-27}$	kg	$0.000\,000.000\,50 \times 10^{-27}$
energy equivalent	$m_u c^2$	$1.492\,418\,085\,60 \times 10^{-10}$	J	$0.000\,000\,000\,45 \times 10^{-10}$
		931.494 102 42	MeV	0.000 000 28
Faraday constant $N_A e$	F	96 485.332 12...	$C\ mol^{-1}$	exact
molar Planck constant	$N_A h$	$3.990\,312\,712... \times 10^{-10}$	$J\ s\ mol^{-1}$	exact
molar gas constant	R	8.314 462 618...	$J\ mol^{-1}\ K^{-1}$	exact
molar mass constant		$0.999\,999\,999\,65 \times 10^{-3}$	$kg\ mol^{-1}$	$0.000\,000\,000\,30 \times 10^{-3}$
Boltzmann constant	k	$1.380\,649 \times 10^{-23}$	$J\ K^{-1}$	exact
in $eV\ K^{-1}$		$8.617\,333\,262... \times 10^{-5}$	$eV\ K^{-1}$	exact
in $Hz\ K^{-1}$		$2.083\,661\,912... \times 10^{10}$	$Hz\ K^{-1}$	exact
inverse meter per kelvin		69.503 480 04	$m^{-1}\ K^{-1}$	exact
molar volume of ideal gas RT/p [$T = 273.15$ K, $p = 100$ kPa]	V_m	$22.710\,954\,64... \times 10^{-3}$	$m^3\ mol^{-1}$	exact
Loschmidt constant N_A/V_m (273.15 K, 100 kPa)	n_0	$2.651\,645\,804... \times 10^{25}$	m^{-3}	exact

Quantity	Symbol	Numerical value	Unit	Uncertainty
molar volume of ideal gas RT/p [T = 273.15 K, p = 101.325 kPa]	V_m	22.413 969 54 × 10^{-3}	$m^3\ mol^{-1}$	exact
Loschmidt constant N_A/V_m (273.15 K, 101.325 kPa)	n_0	2.686 780 111 × 10^{25}	m^{-3}	exact
Sackur-Tetrode (absolute entropy) constant[1] $\frac{5}{2}+\ln[(2\pi m_u kT_1/h^2)^{3/2} kT_1/p_0]$				
T_1 = 1 K, p_0 = 100 kPa	S_0/R	-1.151 707 537 06		0.000 000 000 45
T_1 = 1 K, p_0 = 101.325 kPa		-1.164 870 523 58		0.000 000 000 45
Stefan-Boltzmann constant $(\pi^2/60)k^4/\hbar^3c^2$	σ	5.670 374 419... × 10^{-8}	$W\ m^{-2}\ K^{-4}$	exact
first radiation constant $2\pi hc^2$	c_1	3.741 771 852... × 10^{-16}	$W\ m^2$	exact
first radiation constant for spectral radiance $2hc^2$	c_{1L}	1.191 042 972... × 10^{-16}	$W\ m^2\ sr^{-1}$	exact
second radiation constant hc/k	c_2	1.438 776 877... × 10^{-2}	m K	exact
Wien displacement law constants:				
$b = \lambda_{max}T = c_2/4.965\,114\,231...$	b	2.897 771 955 × 10^{-3}	m K	exact
$b' = \nu_{max}/T = 2.821\,439\,372...\ c/c_2$	b'	5.878 925 757... × 10^{10}	$Hz\ K^{-1}$	exact

[1] The entropy of an ideal monoatomic gas of relative atomic mass A_r is given by $S = S_0 + 3/2\,R \ln A_r - R \ln(p/p_0) + 5/2\,R \ln(T/K)$.

Certain quantities related to x-rays are given in Table 4.

TABLE 4. Values of Some X-ray-Related Quantities Based on the 2018 CODATA Adjustment of the Values of the Constants

Quantity	Symbol	Numerical value	Unit	Uncertainty
Cu x unit: $\lambda(\text{CuK}\alpha_1)/1\,537.400$	xu(CuKα_1)	1.002 076 97 × 10^{-13}	m	0.000 000 28 × 10^{-13}
Mo x unit: $\lambda(\text{MoK}\alpha_1)/707.831$	xu(MoKα_1)	1.002 099 52 × 10^{-13}	m	0.000 000 53 × 10^{-13}
ångstrom star: $\lambda(\text{WK}\alpha_1)/0.209\,010\,0$	Å*	1.000 014 95 × 10^{-10}	m	0.000 000 90 × 10^{-10}
lattice parameter[1] of Si (in vacuum, 22.5 °C)	a	5.431 020 511 × 10^{-10}	m	0.000 000 089 × 10^{-10}
{220} lattice spacing of Si $a/\sqrt{8}$ (in vacuum, 22.5 °C)	d_{220}	1.920 155 716 × 10^{-10}	m	0.000 000 032 × 10^{-10}
molar volume of Si $M(\text{Si})/\rho(\text{Si}) = N_A a^3/8$ (in vacuum, 22.5 °C)	V_m(Si)	1.205 883 199 × 10^{-5}	$m^3\ mol^{-1}$	0.000 000 060 × 10^{-5}

[1] This is the lattice parameter (unit cell edge length) of an ideal single crystal of naturally occurring Si free of impurities and imperfections, and is deduced from measurements on extremely pure and nearly perfect single crystals of Si by correcting for the effects of impurities.

Table 5 contains the values in SI units of some non-SI units as based on the 2018 CODATA adjustment of the fundamental constants. Natural units are often used in particle and other subdisciplines of physics (Ref. 1). Atomic units are used in quantum chemical calculations to remove the necessity of carrying constants throughout (Ref. 2).

References

1. Tomilin, K. S., Natural Systems of Units. To the Centenary Anniversary of the Planck System, in *Proceedings: 21st International Workshop on the Fundamental Problems of High-Energy Physics and Field Theory*, Filiminova, I.V., and Petrov, V. A., Eds., Protvino, Russia, June 23-25, 1998.
2. Shull, H., and Hall, G. G., *Nature*, 184.4698, 1559, 1959. <https://doi.org/10.1038/1841559a0>

TABLE 5. The Values in SI Units of Some Non-SI Units Based on the 2018 CODATA Adjustment of the Values of the Constants

Quantity	Symbol	Numerical value	Unit	Uncertainty
Non-SI units accepted for use with the SI				
electron volt: (e/C) J	eV	1.602 176 634 × 10^{-19}	J	exact
(unified) atomic mass unit: $\frac{1}{12}m(^{12}C)$	u	1.660 539 066 60 × 10^{-27}	kg	0.000 000 000 50 × 10^{-27}
Natural units (n.u.)				
n.u. of velocity	c, c_0	299 792 458	$m\ s^{-1}$	exact
n.u. of action: $h/2\pi$	$\hbar$	1.054 571 817... × 10^{-34}	J s	exact
		6.582 119 569... × 10^{-16}	eV s	exact
n.u. of mass	m_e	9.109 383 7015 × 10^{-31}	kg	0.000 000 0028 × 10^{-31}
n.u. of energy	m_ec^2	8.187 105 7769 × 10^{-14}	J	0.000 000 0025 × 10^{-14}
		0.510 998 950 00	MeV	0.000 000 000 15

Pierce College Library

Quantity	Symbol	Numerical value	Unit	Uncertainty
n.u. of momentum	m_ec	$2.730\,924\,530\,75 \times 10^{-22}$	kg m s^{-1}	$0.000\,000\,000\,82 \times 10^{-22}$
		$0.510\,998\,950\,00$	MeV/c	$0.000\,000\,000\,15$
n.u. of length: $\hbar/m_ec$	$\lambda\!\!\!^-_C$	$3.861\,592\,6796 \times 10^{-13}$	m	$0.000\,000\,0012 \times 10^{-13}$
n.u. of time	$\hbar/m_ec^2$	$1.288\,088\,668\,19 \times 10^{-21}$	s	$0.000\,000\,000\,39 \times 10^{-21}$
Atomic units (a.u.)				
a.u. of charge	e	$1.602\,176\,634 \times 10^{-19}$	C	exact
a.u. of mass	m_e	$9.109\,383\,7015 \times 10^{-31}$	kg	$0.000\,000\,0028 \times 10^{-31}$
a.u. of action: $h/2\pi$	$\hbar$	$1.054\,571\,817... \times 10^{-34}$	J s	exact
a.u. of length: Bohr radius (bohr) $\alpha/4\pi R_\infty$	a_0	$5.291\,772\,109\,03 \times 10^{-11}$	m	$0.000\,000\,000\,80 \times 10^{-11}$
a.u. of energy: Hartree energy (hartree) $e^2/4\pi\epsilon_0 a_0 = 2R_\infty hc = \alpha^2 m_ec^2$	E_h	$4.359\,744\,722\,2071 \times 10^{-18}$	J	$0.000\,000\,000\,0085 \times 10^{-18}$
a.u. of time	$\hbar/E_h$	$2.418\,884\,326\,5857 \times 10^{-17}$	s	$0.000\,000\,000\,0047 \times 10^{-17}$
a.u. of force	E_h/a_0	$8.238\,723\,4983 \times 10^{-8}$	N	$0.000\,000\,0012 \times 10^{-8}$
a.u. of velocity: αc	$a_0E_h/\hbar$	$2.187\,691\,263\,64 \times 10^{6}$	m s^{-1}	$0.000\,000\,000\,33 \times 10^{6}$
a.u. of momentum	$\hbar/a_0$	$1.992\,851\,914\,10 \times 10^{-24}$	kg m s^{-1}	$0.000\,000\,000\,30 \times 10^{-24}$
a.u. of current	$eE_h/\hbar$	$6.623\,618\,237\,510 \times 10^{-3}$	A	$0.000\,000\,000\,013 \times 10^{-3}$
a.u. of charge density	e/a_0^3	$1.081\,202\,384\,57 \times 10^{12}$	C m^{-3}	$0.000\,000\,000\,49 \times 10^{12}$
a.u. of electric potential	E_h/e	$27.211\,386\,245\,988$	V	$0.000\,000\,000\,053$
a.u. of electric field	E_h/ea_0	$5.142\,206\,747\,63 \times 10^{11}$	V m^{-1}	$0.000\,000\,000\,78 \times 10^{11}$
a.u. of electric field gradient	E_h/ea_0^2	$9.717\,362\,4292 \times 10^{21}$	V m^{-2}	$0.000\,000\,0029 \times 10^{21}$
a.u. of electric dipole moment	ea_0	$8.478\,353\,6255 \times 10^{-30}$	C m	$0.000\,000\,0013 \times 10^{-30}$
a.u. of electric quadrupole moment	ea_0^2	$4.486\,551\,5246 \times 10^{-40}$	C m^2	$0.000\,000\,0014 \times 10^{-40}$
a.u. of electric polarizability	$e^2a_0^2/E_h$	$1.648\,777\,274\,36 \times 10^{-41}$	C^2 m^2 J^{-1}	$0.000\,000\,000\,50 \times 10^{-41}$
a.u. of 1*st* hyperpolarizability	$e^3a_0^3/E_h^2$	$3.206\,361\,3061 \times 10^{-53}$	C^3 m^3 J^{-2}	$0.000\,000\,0015 \times 10^{-53}$
a.u. of 2*nd* hyperpolarizability	$e^4a_0^4/E_h^3$	$6.235\,379\,9905 \times 10^{-65}$	C^4 m^4 J^{-3}	$0.000\,000\,0038 \times 10^{-65}$
a.u. of magnetic flux density	$\hbar/ea_0^2$	$2.350\,517\,567\,58 \times 10^{5}$	T	$0.000\,000\,000\,71 \times 10^{5}$
a.u. of magnetic dipole moment: $2\mu_B$	$\hbar e/m_e$	$1.854\,802\,015\,66 \times 10^{-23}$	J T^{-1}	$0.000\,000\,000\,56 \times 10^{-23}$
a.u. of magnetizability	$e^2a_0^2/m_e$	$7.891\,036\,6008 \times 10^{-29}$	J T^{-2}	$0.000\,000\,0048 \times 10^{-29}$
a.u. of permittivity: $10^7/c^2$	e^2/a_0E_h	$1.112\,650\,055\,45 \times 10^{-10}$	F m^{-1}	$0.000\,000\,000\,17 \times 10^{-10}$

Tables 6 and 7 provide energy equivalents, including uncertainties, in various units.

TABLE 6. Some Values of Some Energy Equivalents Derived from the Relations $E = mc^2 = hc/\lambda = h\nu = kT$, and Based on the 2018 CODATA Adjustment of the Values of the Constants; 1 eV = (e/C) J, 1 u = $m_u = (1/12)m(^{12}\mathrm{C}) = 10^{-3}$ kg mol$^{-1}/N_A$, and $E_h = 2R_\infty hc = \alpha^2 m_ec^2$ Is the Hartree Energy (hartree); Row below Energy Equivalent Values Contains Uncertainties

	J	kg	m^{-1}	Hz
1 J	(1 J) = 1 J	(1 J)/c^2 = $1.112\,650\,056...\times10^{-17}$ kg exact	(1 J)/hc = $5.034\,116\,567...\times10^{24}$ m^{-1} exact	(1 J)/h = $1.509\,190\,179...\times10^{33}$ Hz exact
1 kg	(1 kg)c^2 = $8.987\,551\,787...\times10^{16}$ J exact	(1 kg) 1 kg	(1 kg)c/h = $4.524\,438\,335...\times10^{41}$ m^{-1} exact	(1 kg)c^2/h = $1.356\,392\,489...\times10^{50}$ Hz exact
1 m^{-1}	(1 m^{-1})hc = $1.986\,445\,857...\times10^{-25}$ J exact	(1 m^{-1})h/c = $2.210\,219\,094...\times10^{-42}$ kg exact	(1 m^{-1}) = 1 m^{-1}	(1 m^{-1})c = 299792458 Hz exact
1 Hz	(1 Hz)h = $6.626\,070\,15\times10^{-34}$ J exact	(1 Hz)h/c^2 = $7.372\,497\,323...\times10^{-51}$ kg exact	(1 Hz)/c = $3.335\,640\,951...\times10^{-9}$ m^{-1} exact	(1 Hz) = 1 Hz
1 K	(1 K)k = $1.380\,649\times10^{-23}$ J exact	(1 K)k/c^2 = $1.536\,179\,187...\times10^{-40}$ kg exact	(1 K)k/hc = 69.503 480 04... m^{-1} exact	(1 K)k/h = $2.083\,661\,912\times10^{10}$ Hz exact
1 eV	(1 eV) = $1.602\,176\,634\times10^{-19}$ J exact	(1 eV)/c^2 = $1.782\,661\,921...\times10^{-36}$ kg exact	(1 eV)/hc = $8.065\,543\,937\times10^{5}$ m^{-1} exact	(1 eV)/h = $2.417\,989\,242...\times10^{14}$ Hz exact
1 u	(1 u)c^2 = $1.492\,418\,085\,60\times10^{-10}$ J	(1 u) = $1.660\,539\,066\,60\times10^{-27}$ kg	(1 u)c/h = $7.513\,006\,6104\times10^{14}$ m^{-1}	(1 u)c^2/h = $2.252\,342\,718\,71\times10^{23}$ Hz

	J	kg	m^{-1}	Hz
	$0.000\,000\,000\,45 \times 10^{-10}$	$0.000\,000\,000\,50 \times 10^{-27}$	$0.000\,000\,0023 \times 10^{14}$	$0.000\,000\,000\,68 \times 10^{23}$
1 E_h	$(1\ E_h) =$ $4.359\,744\,722\,2071 \times 10^{-18}$ J $0.000\,000\,000\,0085 \times 10^{-18}$	$(1\ E_h)/c^2 =$ $4.850\,870\,209\,5432 \times 10^{-35}$ kg $0.000\,000\,000\,0094 \times 10^{-35}$	$(1\ E_h)/hc =$ $2.194\,746\,313\,6320 \times 10^{7}$ m^{-1} $0.000\,000\,000\,0043 \times 10^{7}$	$(1\ E_h)/h =$ $6.579\,683\,920\,502 \times 10^{15}$ Hz $0.000\,000\,000\,013 \times 10^{15}$

TABLE 7. Additional Values of Some Energy Equivalents Derived from the Relations $E = mc^2 = hc/\lambda = h\nu = kT$, and Based on the 2018 CODATA Adjustment of the Values of the Constants; 1 eV = (e/C) J, 1 u = $m_u = (1/12)m(^{12}\mathrm{C}) = 10^{-3}$ kg mol^{-1}/N_A, and $E_h = 2R_\infty hc = \alpha^2 m_e c^2$ Is the Hartree Energy (hartree); Row below Energy Equivalent Values Contains Uncertainties

	K	eV	u	E_h
1 J	$(1\ J)/k =$ $7.242\,970\,516... \times 10^{22}$ K exact	$(1\ J) =$ $6.241\,509\,074... \times 10^{18}$ eV exact	$(1\ J)/c^2 =$ $6.700\,535\,2565 \times 10^{9}$ u $0.000\,000\,0020 \times 10^{9}$	$(1\ J) =$ $2.293\,712\,278\,3963 \times 10^{17} E_h$ $0.000\,000\,000\,0045 \times 10^{17}$
1 kg	$(1\ kg)c^2/k =$ $6.509\,657\,260... \times 10^{39}$ K exact	$(1\ kg)c^2 =$ $5.609\,588\,603... \times 10^{35}$ eV exact	$(1\ kg) =$ $6.022\,140\,7621... \times 10^{26}$ u $0.000\,000\,0018 \times 10^{26}$	$(1\ kg)c^2 =$ $2.061\,485\,788\,7409 \times 10^{34} E_h$ $0.000\,000\,000\,0040 \times 10^{34}$
1 m^{-1}	$(1 m^{-1})hc/k =$ $1.438\,776\,877... \times 10^{-2}$ K exact	$(1 m^{-1})hc =$ $1.239\,841\,984... \times 10^{-6}$ eV exact	$(1 m^{-1})h/c =$ $1.331\,025\,050\,10 \times 10^{-15}$ u $0.000\,000\,000\,40 \times 10^{-15}$	$(1 m^{-1})hc =$ $4.556\,335\,252\,9120 \times 10^{-8} E_h$ $0.000\,000\,000\,0088 \times 10^{-8}$
1 Hz	$(1\ Hz)h/k =$ $4.799\,243\,073... \times 10^{-11}$ K exact	$(1\ Hz)h =$ $4.135\,667\,696... \times 10^{-15}$ eV exact	$(1\ Hz)h/c^2 =$ $4.439\,821\,6652 \times 10^{-24}$ u $0.000\,000\,0013 \times 10^{-24}$	$(1\ Hz)h =$ $1.519\,829\,846\,0570 \times 10^{-16} E_h$ $0.000\,000\,000\,0029 \times 10^{-16}$
1 K	$(1\ K) =$ 1 K	$(1\ K)k =$ $8.617\,333\,262... \times 10^{-5}$ eV exact	$(1\ K)k/c^2 =$ $9.251\,087\,3014 \times 10^{-14}$ u $0.000\,000\,0028 \times 10^{-14}$	$(1\ K)k =$ $3.166\,811\,563\,4556 \times 10^{-6} E_h$ $0.000\,000\,000\,0061 \times 10^{-6}$
1 eV	$(1\ eV)/k =$ $1.160\,451\,812... \times 10^{4}$ K exact	$(1\ eV) =$ 1 eV	$(1\ eV)/c^2 =$ $1.073\,544\,102\,33 \times 10^{-9}$ u $0.000\,000\,000\,32 \times 10^{-9}$	$(1\ eV) =$ $3.674\,932\,217\,5655 \times 10^{-2} E_h$ $0.000\,000\,000\,0071 \times 10^{-2}$
1 u	$(1\ u)c^2/k =$ $1.080\,954\,019\,16 \times 10^{13}$ K $0.000\,000\,000\,33 \times 10^{13}$	$(1\ u)c^2 =$ $9.314\,941\,0242 \times 10^{8}$ eV $0.000\,000\,0028 \times 10^{8}$	$(1\ u) =$ 1 u	$(1\ u)c^2 =$ $3.423\,177\,6874 \times 10^{7} E_h$ $0.000\,000\,0010 \times 10^{7}$
1 E_h	$(1\ E_h)/k =$ $3.157\,750\,248\,0407 \times 10^{5}$ K $0.000\,000\,000\,0061 \times 10^{5}$	$(1\ E_h) =$ $27.211\,386\,245\,988$ eV $0.000\,000\,000\,053$	$(1\ E_h)/c^2 =$ $2.921\,262\,322\,05 \times 10^{-8}$ u $0.000\,000\,000\,88 \times 10^{-8}$	$(1\ E_h) =$ 1 E_h

Table 8 lists several key quantities important in chemistry based on the 2018 CODATA adjustment. It should be noted that the molar mass of carbon-12 (^{12}C) is no longer an integer.

TABLE 8. Values of Various Fundamental and Adopted Quantities Important in Chemistry

Quantity	Symbol	Numerical value	Unit	Relative std. uncert. u_r
molar mass constant	M_u	$0.999\,999\,999\,65 \times 10^{-3}$	kg mol^{-1}	$0.000\,000\,000\,30 \times 10^{-3}$
molar mass of ^{12}C	$M(^{12}\mathrm{C})$	$11.999\,999\,9958 \times 10^{-3}$	kg mol^{-1}	$0.000\,000\,0036 \times 10^{-3}$
standard-state pressure[1]		100	kPa	exact
standard atmosphere[1]		101.325	kPa	exact

[1] Adopted.

STANDARD ATOMIC WEIGHTS

This table of atomic weights includes the changes made through 2019 by the International Union of Pure and Applied Chemistry (IUPAC) Commission on Isotopic Abundances and Atomic Weights (CIAAW) (Refs. 1,5-9).

IUPAC made a significant policy change in 2009 (Refs. 2-4). Each atomic weight had previously been given as a single value with an uncertainty that took into account both the measurement uncertainty and the variation in isotopic abundance in samples of the element from different terrestrial sources. For a variety of reasons (Ref. 3), this fails to give complete information on the natural variability in isotopic abundance of several elements. Therefore, beginning in 2009 the recommendations expressed the atomic weights of several elements as intervals rather than single numbers plus uncertainties. The symbol for these intervals is [*a*, *b*], where *a* is the lower bound of values found in normal materials and *b* the upper bound. For the other elements in the table, a single recommended atomic weight value is given; the number in parentheses following the value gives the uncertainty in the last digit.

Table 1 gives the atomic weights of the elements listed in alphabetical order by name. Table 2 gives reference atomic weights for the 13 elements whose entries in Table 1 are intervals rather than single numbers. These conventional reference values are suggested for use on samples of unspecified origin and for calculation of molecular weights in tables intended to be broadly applicable. They have been selected such that most or all natural terrestrial atomic-weight variation is covered in an interval of plus or minus one in the last digit. It should be emphasized that the conventional values are not simply midpoints of the intervals, but rather represent the best judgment of the data evaluators.

References

1. Wieser, M. E., et al., *Pure Appl. Chem.* 85, 1047, 2013 <https://doi.org/10.1351/PAC-REP-13-03-02>.
2. Wieser, M. E., and Coplen, T. D., *Pure Appl. Chem.* 83, 359, 2011 <https://doi.org/10.1351/PAC-REP-10-09-14>.
3. Coplen, T. B., and Holden, N. E., *Chemistry International*, 33, 2, p. 10, 2011.
4. Berglund, M., and Wieser, M. E., *Pure Appl. Chem.* 83, 397, 2011 <https://doi.org/10.1351/PAC-REP-10-06-02>.
5. *Chemistry International*, 35, 6, p. 17, 2013; <www.ciaaw.org>.
6. *Chemistry International*, 37, 5-6, p. 26, 2015.
7. Meija, J., et al., *Pure Appl. Chem.* 88, 265, 2016 <https://doi.org/10.1515/pac-2015-0305>.
8. Possolo, A., van der Veen, A. M. H., Meija, J., and Hibbert, B., Interpreting and propagating the uncertainty of the standard atomic weights (IUPAC Technical Report), *Pure Appl. Chem.* 90, 395–424, 2018 <https://doi.org/10.1515/pac-2016-0402>.
9. See also *Atomic Weights of the Elements 2019*, <www.ciaaw.org>.

TABLE 1. Standard Atomic Weights 2019

Element	Symbol	Atomic number	Atomic weight	Element	Symbol	Atomic number	Atomic weight
Actinium[u]	Ac	89		Erbium[g]	Er	68	167.259(3)
Aluminum	Al	13	26.9815384(3)	Europium[g]	Eu	63	151.964(1)
Americium[u]	Am	95		Fermium[u]	Fm	100	
Antimony[g]	Sb	51	121.760(1)	Flerovium[u]	Fl	114	
Argon[g, r]	Ar	18	[39.792, 39.963]	Fluorine	F	9	18.998403163(6)
Arsenic	As	33	74.921595(6)	Francium[u]	Fr	87	
Astatine[u]	At	85		Gadolinium[g]	Gd	64	157.25(3)
Barium	Ba	56	137.327(7)	Gallium	Ga	31	69.723(1)
Berkelium[u]	Bk	97		Germanium	Ge	32	72.630(8)
Beryllium	Be	4	9.0121831(5)	Gold	Au	79	196.966570(4)
Bismuth[u]	Bi	83	208.98040(1)	Hafnium[g]	Hf	72	178.486(2)
Bohrium[u]	Bh	107		Hassium[u]	Hs	108	
Boron[m, *]	B	5	[10.806, 10.821]	Helium[g, r]	He	2	4.002602(2)
Bromine[*]	Br	35	[79.901, 79.907]	Holmium	Ho	67	164.930328(7)
Cadmium[g]	Cd	48	112.414(4)	Hydrogen[m, *]	H	1	[1.00784, 1.00811]
Calcium[g]	Ca	20	40.078(4)	Indium	In	49	114.818(1)
Californium[u]	Cf	98		Iodine	I	53	126.90447(3)
Carbon[*]	C	6	[12.0096, 12.0116]	Iridium	Ir	77	192.217(2)
Cerium[g]	Ce	58	140.116(1)	Iron	Fe	26	55.845(2)
Cesium[g]	Cs	55	132.90545196(6)	Krypton[g, m]	Kr	36	83.798(2)
Chlorine[m, *]	Cl	17	[35.446, 35.457]	Lanthanum[g]	La	57	138.90547(7)
Chromium	Cr	24	51.9961(6)	Lawrencium[u]	Lr	103	
Cobalt	Co	27	58.933194(3)	Lead[g, r]	Pb	82	207.2(1)
Copernicium[u]	Cn	112		Lithium[m, *]	Li	3	[6.938, 6.997]
Copper[r]	Cu	29	63.546(3)	Livermorium[u]	Lv	116	
Curium[u]	Cm	96		Lutetium[g]	Lu	71	174.9668(1)
Darmstadtium[u]	Ds	110		Magnesium[*]	Mg	12	[24.304, 24.307]
Dubnium[u]	Db	105		Manganese	Mn	25	54.938043(2)
Dysprosium[g]	Dy	66	162.500(1)	Meitnerium[u]	Mt	109	
Einsteinium[u]	Es	99		Mendelevium[u]	Md	101	

Element	Symbol	Atomic number	Atomic weight
Mercury	Hg	80	200.592(3)
Molybdenum[g]	Mo	42	95.95(1)
Moscovium[u]	Mc	115	
Neodymium[g]	Nd	60	144.242(3)
Neon[g, m]	Ne	10	20.1797(6)
Neptunium[u]	Np	93	
Nickel[r]	Ni	28	58.6934(4)
Nihonium[u]	Nh	113	
Niobium	Nb	41	92.90637(1)
Nitrogen[m, *]	N	7	[14.00643, 14.00728]
Nobelium[u]	No	102	
Oganesson[u]	Og	118	
Osmium[g]	Os	76	190.23(3)
Oxygen[m, *]	O	8	[15.99903, 15.99977]
Palladium[g]	Pd	46	106.42(1)
Phosphorus	P	15	30.973761998(5)
Platinum	Pt	78	195.084(9)
Plutonium[u]	Pu	94	
Polonium[u]	Po	84	
Potassium	K	19	39.0983(1)
Praseodymium	Pr	59	140.90766(1)
Promethium[u]	Pm	61	
Protactinium	Pa	91	231.03588(1)
Radium[u]	Ra	88	
Radon[u]	Rn	86	
Rhenium	Re	75	186.207(1)
Rhodium	Rh	45	102.90549(2)
Roentgenium[u]	Rg	111	
Rubidium[g]	Rb	37	85.4678(3)
Ruthenium[g]	Ru	44	101.07(2)
Rutherfordium[u]	Rf	104	
Samarium[g]	Sm	62	150.36(2)
Scandium	Sc	21	44.955908(5)
Seaborgium[u]	Sg	106	
Selenium[r]	Se	34	78.971(8)
Silicon[*]	Si	14	[28.084, 28.086]
Silver[g]	Ag	47	107.8682(2)
Sodium	Na	11	22.98976928(2)
Strontium[g, r]	Sr	38	87.62(1)
Sulfur[*]	S	16	[32.059, 32.076]
Tantalum	Ta	73	180.94788(2)
Technetium[u]	Tc	43	
Tellurium[g]	Te	52	127.60(3)
Tennessine[u]	Ts	117	
Terbium	Tb	65	158.925354(8)
Thallium[*]	Tl	81	[204.382, 204.385]
Thorium[u]	Th	90	232.0377(4)
Thulium	Tm	69	168.934218(6)
Tin[g]	Sn	50	118.710(7)
Titanium	Ti	22	47.867(1)
Tungsten	W	74	183.84(1)
Uranium[g, m, u]	U	92	238.02891(3)
Vanadium	V	23	50.9415(1)
Xenon[g, m]	Xe	54	131.293(6)
Ytterbium[g]	Yb	70	173.045(10)
Yttrium	Y	39	88.90584(1)
Zinc[r]	Zn	30	65.38(2)
Zirconium[g]	Zr	40	91.224(2)

[g] Geological specimens are known in which the element has an isotopic composition outside the limits for the normal material. The difference between the atomic weight of the element in such specimens and that given in the table may exceed the stated uncertainty.

[m] Modified isotopic compositions may be found in commercially available material because the material has been subjected to an undisclosed or inadvertent isotopic fractionation. Substantial deviations in atomic weight of the element from that given in the table can occur.

[r] Range in isotopic composition of normal terrestrial material prevents a more precise atomic weight being given; the tabulated value and uncertainty should be applicable to any normal material.

[u] Element has no stable isotopes. See "Table of the Isotopes" in Section 11 for individual isotopic masses. However, four such elements (Bi, Th, Pa, and U) do have a characteristic terrestrial isotopic composition, and for these elements standard atomic weights are tabulated.

[*] See Table 2.

TABLE 2. Conventional Atomic Weights 2019

Element	Symbol	Atomic number	Reference atomic weight[a]
Argon	Ar	18	39.948
Boron	B	5	10.81
Bromine	Br	35	79.904
Carbon	C	6	12.011
Chlorine	Cl	17	35.45
Hydrogen	H	1	1.008
Lithium	Li	3	6.94
Magnesium	Mg	12	24.305
Nitrogen	N	7	14.007
Oxygen	O	8	15.999
Silicon	Si	14	28.085
Sulfur	S	16	32.06
Thallium	Tl	81	204.38

[a] For users needing an atomic-weight value for an unspecified sample, such as for trade or commerce, see text.

ATOMIC MASSES AND ISOTOPIC ABUNDANCES

This table lists the mass (in atomic mass relative to $^{12}C = 12$), and the representative abundance (fractional) of the stable nuclides and selected radioactive nuclides. At least one isotope of each of the 118 known elements is included. The atomic masses were taken from the AME 2016 evaluation of the Atomic Mass Data Center, now located at the Institute of Modern Physics in Lanzhou, China (Refs. 1-3). The number in parentheses following the mass value is the uncertainty in the last digit(s) given. The mass values for the artificially produced elements were derived from a combination of experimental data and systematic trends. Mass values that are estimates, without experimental measurements, are indicated by a footnote.

Natural isotopic abundance values were taken from the IUPAC Report "Isotopic Compositions of the Elements 2019" (Ref. 4); these entries are either a range in square brackets or followed by uncertainties in the last digit(s) of the stated values. This uncertainty includes both the estimated measurement uncertainty and the reported range of variation in different terrestrial sources of the element (see Ref. 5 for full details and caveats regarding elements whose abundance is variable). The absence of an entry in the Abundance column indicates a radioactive nuclide is not present in nature or an element whose isotopic composition varies so widely that a meaningful natural abundance cannot be defined.

References 1 and 2 contain mass data on over 3000 nuclides and describe the evaluation procedure in detail. Masses and other properties of nuclides may also be found in Section 11, "Table of the Isotopes" (Ref. 6).

References

1. Huang, W. J., Audi, G., Wang, M., Kondev, F. G., Naimi, S., and Xu, X., *Chinese Physics* C41, 030002, 2017 <https://doi.org/10.1088/1674-1137/41/3/030002>.
2. Wang, M., Audi, G., Kondev, F. G., Huang, W. J., Naimi, S., and Xu, X., *Chinese Physics* C41, 030003, 2017 <https://doi.org/10.1088/1674-1137/41/3/030003>.
3. IAEA Nuclear Data Services, <www-nds.iaea.org/amdc/>.
4. See *Isotopic Compositions of the Elements 2019*, <www.ciaaw.org>.
5. Meija, J., et al, *Pure Appl. Chem.* 88, 293, 2016 <https://doi.org/10.1351/PAC-REP-10-06-02>.
6. Holden, N. E., "Table of the Isotopes," in Rumble, J. R., Ed., *CRC Handbook of Chemistry and Physics, 101st Edition*, CRC Press, Boca Raton, FL, 2020.

Atomic Masses and Isotopic Composition of the Elements 2019

Z	Isotope	Mass (relative to $^{12}C = 12$)	Abundance	Notes
1	^{1}H	1.00782503224(9)	[0.999 72, 0.999 99]	m
	^{2}H	2.01410177811(12)	[0.000 01, 0.000 28]	m
	^{3}H	3.01604928199(23)		
2	^{3}He	3.01602932265(22)	0.000 002(2)	g r
	^{4}He	4.00260325413(6)	0.999 998(2)	g r
3	^{6}Li	6.0151228874(15)	[0.019, 0.078]	m
	^{7}Li	7.016003437(5)	[0.922, 0.981]	m
4	^{9}Be	9.01218307(8)	1	
5	^{10}B	10.012936862(16)	[0.189, 0.204]	m
	^{11}B	11.009305167(13)	[0.796, 0.811]	m
6	^{11}C	11.01143260(6)		
	^{12}C	12	[0.9884, 0.9904]	
	^{13}C	13.00335483521(23)	[0.0096, 0.0116]	
	^{14}C	14.003241988(4)		
7	^{14}N	14.00307400446(21)	[0.995 78, 0.996 63]	
	^{15}N	15.0001088989(6)	[0.003 37, 0.004 22]	
8	^{16}O	15.99491461960(17)	[0.997 38, 0.997 76]	
	^{17}O	16.9991317566(7)	[0.000 367, 0.000 400]	
	^{18}O	17.9991596128(8)	[0.001 87, 0.002 22]	
9	^{18}F	18.0009373(5)		
	^{19}F	18.9984031629(9)	1	
10	^{20}Ne	19.9924401762(17)	0.9048(3)	g m
	^{21}Ne	20.99384669(4)	0.0027(1)	g m
	^{22}Ne	21.991385110(19)	0.0925(3)	g m
11	^{22}Na	21.99443742(18)		
	^{23}Na	22.9897692820(19)	1	
	^{24}Na	23.990963011(18)		
12	^{24}Mg	23.985041697(14)	[0.7888, 0.7905]	
	^{25}Mg	24.98583696(5)	[0.099 88, 0.100 34]	
	^{26}Mg	25.98259297(3)	[0.1096, 0.1109]	
13	^{27}Al	26.98153841(5)	1	
14	^{28}Si	27.9769265350(5)	[0.921 91, 0.923 18]	
	^{29}Si	28.9764946653(6)	[0.046 45, 0.046 99]	
	^{30}Si	29.973770137(23)	[0.030 37, 0.031 10]	
15	^{31}P	30.9737619986(7)	1	
	^{32}P	31.97390764(4)		
16	^{32}S	31.9720711744(14)	[0.9441, 0.9529]	
	^{33}S	32.9714589099(15)	[0.007 29, 0.007 97]	
	^{34}S	33.96786701(5)	[0.0396, 0.0477]	
	^{35}S	34.96903232(4)		
	^{36}S	35.96708070(2)	[0.000 129, 0.000 187]	
17	^{35}Cl	34.96885269(4)	[0.755, 0.761]	m
	^{37}Cl	36.96590258(6)	[0.239, 0.245]	m
18	^{36}Ar	35.967545105(29)	[0.0000, 0.0207]	g r
	^{38}Ar	37.96273210(21)	[0.000, 0.043]	g r
	^{40}Ar	39.9623831238(24)	[0.936, 1.000]	g r
19	^{39}K	38.963706487(5)	0.932 581(44)	
	^{40}K	39.96399817(6)	0.000 117(1)	
	^{41}K	40.961825258(4)	0.067 302(44)	
	^{42}K	41.96240231(11)		
	^{43}K	42.9607347(4)		
20	^{40}Ca	39.962590866(22)	0.096 941(156)	g
	^{42}Ca	41.95861783(16)	0.006 47(23)	g
	^{43}Ca	42.95876643(24)	0.001 35(10)	g
	^{44}Ca	43.9554815(3)	0.020 86(110)	g
	^{45}Ca	44.9561863(4)		
	^{46}Ca	45.9536880(24)	0.000 04(3)	g
	^{47}Ca	46.9545414(24)		
	^{48}Ca	47.95252290(1)	0.001 87(21)	g
21	^{45}Sc	44.9559075(7)	1	
22	^{46}Ti	45.95262686(18)	0.0825(3)	
	^{47}Ti	46.95175775(12)	0.0744(2)	
	^{48}Ti	47.94794093(12)	0.7372(3)	
	^{49}Ti	48.94786463(12)	0.0541(2)	

Z	Isotope	Mass (relative to ^{12}C = 12)	Abundance	Notes
	^{50}Ti	49.94478584(13)	0.0518(2)	
23	^{50}V	49.9471558(4)	0.002 50(10)	
	^{51}V	50.9439569(4)	0.997 50(10)	
24	^{50}Cr	49.9460414(5)	0.043 45(13)	
	^{51}Cr	50.9447647(4)		
	^{52}Cr	51.9405050(4)	0.837 89(18)	
	^{53}Cr	52.9406470(4)	0.095 01(17)	
	^{54}Cr	53.9388780(4)	0.023 65(7)	
25	^{54}Mn	53.9403564(11)		
	^{55}Mn	54.9380432(3)	1	
26	^{52}Fe	51.948115(5)		
	^{54}Fe	53.9396083(4)	0.058 45(105)	
	^{55}Fe	54.9382913(4)		
	^{56}Fe	55.9349356(3)	0.917 54(106)	
	^{57}Fe	56.9353921(3)	0.021 19(29)	
	^{58}Fe	57.9332737(4)	0.002 82(12)	
	^{59}Fe	58.9348736(4)		
27	^{57}Co	56.9362899(6)		
	^{58}Co	57.9357514(12)		
	^{59}Co	58.9331937(4)	1	
	^{60}Co	59.9338157(5)		
28	^{58}Ni	57.9353418(4)	0.680 77(190)	r
	^{59}Ni	58.9343456(4)		
	^{60}Ni	59.9307853(4)	0.262 23(150)	r
	^{61}Ni	60.9310549(4)	0.011 399(13)	r
	^{62}Ni	61.9283449(5)	0.036 345(40)	r
	^{63}Ni	62.9296691(5)		
	^{64}Ni	63.9279663(5)	0.009 256(19)	r
29	^{63}Cu	62.9295972(5)	0.6915(15)	r
	^{64}Cu	63.9297639(5)		
	^{65}Cu	64.9277895(7)	0.3085(15)	r
30	^{64}Zn	63.9291418(7)	0.4917(75)	r
	^{65}Zn	64.9292405(7)		
	^{66}Zn	65.9260337(8)	0.2773(98)	r
	^{67}Zn	66.9271275(8)	0.0404(16)	r
	^{68}Zn	67.9248443(8)	0.1845(63)	r
	^{70}Zn	69.9253192(21)	0.0061(10)	r
	^{67}Ga	66.9282024(13)		
	^{68}Ga	67.9279802(15)		
	^{69}Ga	68.9255735(13)	0.601 08(50)	
	^{71}Ga	70.9247025(9)	0.398 92(50)	
32	^{68}Ge	67.9280953(2)		
	^{70}Ge	69.9242487(9)	0.2052(19)	
	^{72}Ge	71.92207583(8)	0.2745(15)	
	^{73}Ge	72.92345896(6)	0.0776(8)	
	^{74}Ge	73.921177762(13)	0.3652(12)	
	^{76}Ge	75.921402727(19)	0.0775(12)	
33	^{75}As	74.9215946(9)	1	
34	^{74}Se	73.922475935(16)	0.0086(3)	r
	^{75}Se	74.92252287(8)		
	^{76}Se	75.919213704(17)	0.0923(7)	r
	^{77}Se	76.91991415(7)	0.0760(7)	r
	^{78}Se	77.91730924(19)	0.2369(22)	r
	^{79}Se	78.91849925(24)		
	^{80}Se	79.9165218(1)	0.4980(36)	r
	^{82}Se	81.9166995(5)	0.0882(15)	r
35	^{79}Br	78.9183376(11)	[0.505, 0.508]	
	^{81}Br	80.9162882(1)	[0.492, 0.495]	
36	^{78}Kr	77.9203663(3)	0.003 55(3)	g m
	^{80}Kr	79.9163780(7)	0.022 86(10)	g m
	^{82}Kr	81.913481155(6)	0.115 93(31)	g m
	^{83}Kr	82.914126518(1)	0.115 00(19)	g m
	^{84}Kr	83.911497729(4)	0.569 87(15)	g m
	^{86}Kr	85.910610626(4)	0.172 79(41)	g m
37	^{85}Rb	84.911789738(5)	0.7217(2)	g
	^{86}Rb	85.91116744(21)		g
	^{87}Rb	86.909180531(6)	0.2783(2)	g
38	^{84}Sr	83.9134191(13)	0.0056(2)	g r
	^{85}Sr	84.912932(3)		g r
	^{86}Sr	85.909260726(6)	0.0986(20)	g r
	^{87}Sr	86.908877496(5)	0.0700(20)	g r
	^{88}Sr	87.905612256(6)	0.8258(35)	g r
	^{89}Sr	88.90745081(1)		g r
	^{90}Sr	89.9077309(23)		g r
39	^{89}Y	88.9058412(17)	1	
40	^{90}Zr	89.90469876(13)	0.5145(4)	g
	^{91}Zr	90.90564022(11)	0.1122(5)	g
	^{92}Zr	91.90503532(11)	0.1715(3)	g
	^{94}Zr	93.90631252(18)	0.1738(4)	g
	^{96}Zr	95.90827762(12)	0.0280(2)	g
41	^{93}Nb	92.9063732(16)	1	
42	^{92}Mo	91.90680716(17)	0.146 49(106)	g
	^{94}Mo	93.90508359(15)	0.091 87(33)	g
	^{95}Mo	94.90583744(13)	0.158 73(30)	g
	^{96}Mo	95.90467477(13)	0.166 73(8)	g
	^{97}Mo	96.90601690(18)	0.095 82(15)	g
	^{98}Mo	97.90540361(19)	0.242 92(80)	g
	^{99}Mo	98.90770730(25)		g
	^{100}Mo	99.9074680(3)	0.097 44(65)	g
43	^{97}Tc	96.906361(4)		
	^{98}Tc	97.907211(4)		
	^{99}Tc	98.9062497(1)		
44	^{96}Ru	95.90758891(18)	0.0554(14)	g
	^{98}Ru	97.905287(7)	0.0187(3)	g
	^{99}Ru	98.9059303(4)	0.1276(14)	g
	^{100}Ru	99.9042105(4)	0.1260(7)	g
	^{101}Ru	100.9055731(4)	0.1706(2)	g
	^{102}Ru	101.9043403(4)	0.3155(14)	g
	^{104}Ru	103.9054254(27)	0.1862(27)	g
	^{106}Ru	105.907328(6)		g
45	^{103}Rh	102.9054941(25)	1	
46	^{102}Pd	101.9056321(6)	0.0102(1)	g
	^{104}Pd	103.9040304(14)	0.1114(8)	g
	^{105}Pd	104.9050795(12)	0.2233(8)	g
	^{106}Pd	105.9034803(12)	0.2733(3)	g
	^{108}Pd	107.9038918(12)	0.2646(9)	g
	^{110}Pd	109.9051729(7)	0.1172(9)	g
47	^{107}Ag	106.9050915(26)	0.518 39(8)	g
	^{109}Ag	108.9047558(14)	0.481 61(8)	g
48	^{106}Cd	105.9064598(12)	0.012 45(22)	g
	^{108}Cd	107.9041836(12)	0.008 88(11)	g
	^{110}Cd	109.9030075(4)	0.124 70(61)	g
	^{111}Cd	110.9041838(4)	0.127 95(12)	g
	^{112}Cd	111.90276388(27)	0.241 09(7)	g
	^{113}Cd	112.90440810(26)	0.122 27(7)	g
	^{114}Cd	113.90336499(3)	0.287 54(81)	g
	^{116}Cd	115.90476323(17)	0.075 12(54)	g
49	^{111}In	110.905107(4)		
	^{113}In	112.90406045(2)	0.042 81(52)	
	^{115}In	114.903878774(13)	0.957 19(52)	
50	^{112}Sn	111.9048249(3)	000.97(1)	g
	^{113}Sn	112.9051758(17)		g

Z	Isotope	Mass (relative to ^{12}C = 12)	Abundance	Notes
	^{114}Sn	113.90278013(3)	0.0066(1)	g
	^{115}Sn	114.903344697(16)	0.0034(1)	g
	^{116}Sn	115.90174282(1)	0.1454(9)	g
	^{117}Sn	116.9029540(5)	0.0768(7)	g
	^{118}Sn	117.9016066(5)	0.2422(9)	g
	^{119}Sn	118.9033112(8)	0.0859(4)	g
	^{120}Sn	119.9022019(1)	0.3258(9)	g
	^{122}Sn	121.9034440(26)	0.0463(3)	g
	^{124}Sn	123.9052767(11)	0.0579(5)	g
51	^{121}Sb	120.9038101(28)	0.5721(5)	g
	^{123}Sb	122.9042140(16)	0.4279(5)	g
52	^{120}Te	119.904060(3)	0.0009(1)	g
	^{122}Te	121.9030434(16)	0.0255(12)	g
	^{123}Te	122.9042697(16)	0.0089(3)	g
	^{124}Te	123.9028171(16)	0.0474(14)	g
	^{125}Te	124.9044299(16)	0.0707(15)	g
	^{126}Te	125.9033109(16)	0.1884(25)	g
	^{128}Te	127.9044613(9)	0.3174(8)	g
	^{130}Te	129.906222747(12)	0.3408(62)	g
53	^{123}I	122.905589(4)		
	^{125}I	124.9046293(16)		
	^{127}I	126.904472(4)	1	
	^{129}I	128.904984(3)		
	^{131}I	130.9061264(6)		
54	^{124}Xe	123.9058916(19)	0.000 95(5)	g m
	^{126}Xe	125.904297(4)	0.000 89(2)	g m
	^{128}Xe	127.9035310(11)	0.019 10(13)	g m
	^{129}Xe	128.904780859(6)	0.264 01(138)	g m
	^{130}Xe	129.903509349(1)	0.040 71(22)	g m
	^{131}Xe	130.905084136(9)	0.212 32(51)	g m
	^{132}Xe	131.904155087(6)	0.269 09(55)	g m
	^{134}Xe	133.905393034(1)	0.104 36(35)	g m
	^{136}Xe	135.907214476(7)	0.088 57(72)	g m
55	^{129}Cs	128.906066(5)		
	^{133}Cs	132.905451961(9)	1	
	^{134}Cs	133.906718504(18)		
	^{136}Cs	135.9073116(2)		
	^{137}Cs	136.9070895(4)		
56	^{130}Ba	129.9063209(27)	0.0011(1)	
	^{132}Ba	131.9050611(11)	0.0010(1)	
	^{133}Ba	132.9060073(11)		
	^{134}Ba	133.9045084(3)	0.0242(15)	
	^{135}Ba	134.9056886(3)	0.0659(10)	
	^{136}Ba	135.9045760(3)	0.0785(24)	
	^{137}Ba	136.9058274(3)	0.1123(23)	
	^{138}Ba	137.9052472(3)	0.7170(29)	
	^{140}Ba	139.910607(9)		
57	^{138}La	137.907118(3)	0.000 8881(71)	g
	^{139}La	138.9063588(22)	0.999 1119(71)	g
58	^{136}Ce	135.9071294(4)	0.001 85(2)	g
	^{138}Ce	137.905989(5)	0.002 51(2)	g
	^{140}Ce	139.9054464(17)	0.884 50(51)	g
	^{141}Ce	140.9082840(17)		g
	^{142}Ce	141.9092499(27)	0.111 14(51)	g
	^{144}Ce	143.913653(3)		
59	^{141}Pr	140.9076584(18)	1	
60	^{142}Nd	141.9077289(15)	0.271 52(40)	g
	^{143}Nd	142.9098199(15)	0.121 74(26)	g
	^{144}Nd	143.9100929(15)	0.237 98(19)	g
	^{145}Nd	144.9125792(15)	0.082 93(12)	g
	^{146}Nd	145.9131225(15)	0.171 89(32)	g
	^{148}Nd	147.9168991(23)	0.057 56(21)	g
	^{150}Nd	149.9209015(14)	0.056 38(28)	g
61	^{145}Pm	144.912756(3)		
	^{147}Pm	146.9151446(15)		
62	^{144}Sm	143.9120064(17)	0.0308(4)	g
	^{147}Sm	146.9149041(15)	0.1500(14)	g
	^{148}Sm	147.9148290(15)	0.1125(9)	g
	^{149}Sm	148.9171914(14)	0.1382(10)	g
	^{150}Sm	149.9172822(14)	0.0737(9)	g
	^{152}Sm	151.9197390(13)	0.2674(9)	g
	^{154}Sm	153.9222162(16)	0.2274(14)	g
63	^{151}Eu	150.9198569(14)	0.4781(6)	g
	^{153}Eu	152.9212370(14)	0.5219(6)	g
64	^{152}Gd	151.9197988(13)	0.0020(3)	g
	^{154}Gd	153.9208734(13)	0.0218(2)	g
	^{155}Gd	154.9226298(13)	0.1480(9)	g
	^{156}Gd	155.9221306(13)	0.2047(3)	g
	^{157}Gd	156.9239679(13)	0.1565(4)	g
	^{158}Gd	157.9241116(13)	0.2484(8)	g
	^{160}Gd	159.9270615(14)	0.2186(3)	g
65	^{159}Tb	158.9253539(13)	1	
66	^{156}Dy	155.9242840(13)	0.000 56(3)	g
	^{158}Dy	157.9244146(25)	0.000 95(3)	g
	^{160}Dy	159.9252032(8)	0.023 29(18)	g
	^{161}Dy	160.9269391(8)	0.188 89(42)	g
	^{162}Dy	161.9268042(8)	0.254 75(36)	g
	^{163}Dy	162.9287369(8)	0.248 96(42)	g
	^{164}Dy	163.9291805(8)	0.282 60(54)	g
67	^{165}Ho	164.9303280(11)	1	
68	^{162}Er	161.9287870(9)	0.001 39(5)	g
	^{164}Er	163.9292074(8)	0.016 01(3)	g
	^{166}Er	165.9302990(13)	0.33 . 503(36)	g
	^{167}Er	166.9320541(13)	0.228 69(9)	g
	^{168}Er	167.9323762(13)	0.269 78(18)	g
	^{170}Er	169.9354707(17)	0.149 10(36)	g
69	^{169}Tm	168.9342184(9)	1	
70	^{168}Yb	167.9338891(13)	0.001 26(1)	g
	^{169}Yb	168.9351820(13)		g
	^{170}Yb	169.934767246(11)	0.030 23(2)	g
	^{171}Yb	170.936331517(14)	0.142 16(7)	g
	^{172}Yb	171.936386659(15)	0.217 54(10)	g
	^{173}Yb	172.938216215(12)	0.160 98(9)	g
	^{174}Yb	173.938867548(12)	0.318 96(26)	g
	^{176}Yb	175.942574709(16)	0.128 87(30)	g
71	^{175}Lu	174.9407773(13)	0.974 01(13)	g
	^{176}Lu	175.9426918(13)	0.025 99(13)	g
	^{174}Hf	173.9400485(24)	0.001 61(2)	g
	^{176}Hf	175.9414099(16)	0.0524(14)	g
	^{177}Hf	176.9432303(15)	0.1858(9)	g
	^{178}Hf	177.9437085(15)	0.2728(6)	g
	^{179}Hf	178.9458258(15)	0.1363(3)	g
	^{180}Hf	179.9465597(15)	0.3512(16)	g
73	^{180}Ta	179.9474684(21)	0.000 1176(23)	
	^{181}Ta	180.9479993(15)	0.999 8824(23)	
74	^{180}W	179.9467134(15)	0.0012(1)	
	^{182}W	181.9482057(8)	0.2650(16)	
	^{183}W	182.9502245(8)	0.1431(4)	
	^{184}W	183.9509333(8)	0.3064(2)	
	^{186}W	185.9543652(13)	0.2843(19)	
75	^{185}Re	184.9529583(9)	0.3740(5)	
	^{187}Re	186.9557523(8)	0.6260(5)	

Z	Isotope	Mass (relative to ^{12}C = 12)	Abundance	Notes
76	^{184}Os	183.9524929(9)	0.0002(2)	g
	^{186}Os	185.9538377(8)	0.0159(64)	g
	^{187}Os	186.9557496(8)	0.0196(17)	g
	^{188}Os	187.9558374(8)	0.1324(27)	g
	^{189}Os	188.9581460(7)	0.1615(23)	g
	^{190}Os	189.9584455(7)	0.2626(20)	g
	^{192}Os	191.9614789(25)	0.4078(32)	g
77	^{191}Ir	190.9605915(14)	0.3723(9)	
	^{193}Ir	192.9629238(14)	0.6277(9)	
78	^{190}Pt	189.9599499(7)	0.000 12(2)	
	^{192}Pt	191.9610427(28)	0.007 82(24)	
	^{194}Pt	193.9626835(5)	0.328 64(410)	
	^{195}Pt	194.9647944(5)	0.337 75(240)	
	^{196}Pt	195.9649547(5)	0.252 11(340)	
	^{198}Pt	197.9678967(23)	0.073 56(130)	
79	^{197}Au	196.9665701(6)	1	
	^{198}Au	197.9682437(6)		
80	^{196}Hg	195.965833(3)	0.0015(1)	
	^{197}Hg	196.967214(3)		
	^{198}Hg	197.9667692(5)	0.1004(3)	
	^{199}Hg	198.9682810(6)	0.1694(12)	
	^{200}Hg	199.9683269(6)	0.2314(9)	
	^{201}Hg	200.9703030(8)	0.1310(9)	
	^{202}Hg	201.9706436(8)	0.2974(13)	
	^{203}Hg	202.9728723(17)		
	^{204}Hg	203.9734940(5)	0.0682(4)	
81	^{201}Tl	200.970820(15)		
	^{203}Tl	202.9723440(13)	[0.2944, 0.2959]	
	^{205}Tl	204.9744272(13)	[0.7041, 0.7056]	
82	^{204}Pb	203.9730434(12)	0.014(6)	g r
	^{206}Pb	205.9744651(12)	0.241(30)	g r
	^{207}Pb	206.9758967(12)	0.221(50)	g r
	^{208}Pb	207.9766519(12)	0.524(70)	g r
	^{210}Pb	209.9841883(16)		g r
83	^{207}Bi	206.9784705(26)		
	^{209}Bi	208.9803985(15)	1	
84	^{209}Po	208.9824303(19)		
	^{210}Po	209.9828736(12)		
85	^{210}At	209.987147(8)		
	^{211}At	210.9874961(29)		
86	^{211}Rn	210.990601(7)		
	^{220}Rn	220.0113925(19)		
	^{222}Rn	222.0175763(21)		
87	^{223}Fr	223.0197343(21)		
88	^{223}Ra	223.0185007(22)		
	^{224}Ra	224.0202105(19)		
	^{226}Ra	226.0254085(21)		
	^{228}Ra	228.0310687(21)		
89	^{227}Ac	227.0277507(21)		
90	^{228}Th	228.0287398(19)		
	^{230}Th	230.0331324(13)	0.0002(2)	
	^{232}Th	232.0380537(15)	0.9998(2)	
91	^{231}Pa	231.0358826(19)	1	
92	^{233}U	233.0396344(24)		
	^{234}U	234.0409504(12)	0.000 054(5)	g m
	^{235}U	235.0439282(12)	0.007 204(6)	g m
	^{236}U	236.0455662(12)		g m
	^{238}U	238.0507870(16)	0.992 742(10)	g m
93	^{237}Np	237.0481717(12)		
	^{239}Np	239.0529376(14)		
94	^{238}Pu	238.0495583(12)		
	^{239}Pu	239.0521617(12)		
	^{240}Pu	240.0538118(12)		
	^{241}Pu	241.0568497(12)		
	^{242}Pu	242.0587410(13)		
	^{244}Pu	244.0642044(25)		
95	^{241}Am	241.0568274(12)		
	^{243}Am	243.0613799(15)		
96	^{243}Cm	243.0613874(16)		
	^{244}Cm	244.0627507(12)		
	^{245}Cm	245.0654911(12)		
	^{246}Cm	246.0672221(16)		
	^{247}Cm	247.070353(4)		
	^{248}Cm	248.0723491(25)		
97	^{247}Bk	247.070306(6)		
	^{249}Bk	249.0749832(13)		
98	^{249}Cf	249.0748505(13)		
	^{250}Cf	250.0764046(17)		
	^{251}Cf	251.079587(4)		
	^{252}Cf	252.0816265(25)		
99	^{252}Es	252.08298(5)		
100	^{257}Fm	257.095105(5)		
101	^{256}Md	256.09389(13)[a]		
	^{258}Md	258.098430(5)		
102	^{259}No	259.100998(7)		
103	^{262}Lr	262.10961(22)[a]		
104	^{261}Rf	261.10877(5)		
105	^{262}Db	262.11407(15)[a]		
106	^{266}Sg	266.12197(26)[a]		
107	^{272}Bh	272.13826(57)[a]		
108	^{277}Hs	277.15190(58)[a]		
109	^{276}Mt	276.15171(57)[a]		
110	^{281}Ds	281.16472(62)[a]		
111	^{280}Rg	280.16520(57)[a]		
112	^{285}Cn	285.17732(62)[a]		
113	^{285}Nh	285.18007(87)[a]		
114	^{287}Fl	287.18688(66)[a]		
115	^{289}Mc	289.19395(87)[a]		
116	^{291}Lv	291.20117(66)[a]		
117	^{293}Ts	293.20868(87)[a]		
118	^{294}Og	294.21413(71)[a]		

[a] Estimated value, not based on experimental measurement.

[g] Geological materials are known in which the element has an isotopic composition outside the limts for normal material. The difference between the atomic weight of the element in such materials and that given in the table may exceed the stated uncertainty.

[m] Modified isotopic compositions may be found in commercially available material because the material has been subjected to an undisclosed or inadvertent isotopic fractionation. Substantial deviations in atomic weight of the element from that given in the table can occur.

[r] Range in isotopic composition of normal terrestrial material prevents a more precise standard stomic weight being given; the tabulated atomic-weight value an uncertainty should be applicable to normal materials.

ELECTRON CONFIGURATION AND IONIZATION ENERGY OF NEUTRAL ATOMS IN THE GROUND STATE

William C. Martin

The ground-state electron configuration, ground level, and ionization energy of the elements hydrogen through rutherfordium are listed in this table. The electron configurations of elements heavier than neon are shortened by using rare-gas element symbols in brackets to represent the corresponding electrons. See the references for details of the notation for Pa, U, and Np. Ionization energies to higher states (and more precise values of the first ionization energy for certain elements) may be found in the table "Ionization Energies of Atoms and Atomic Ions" in Section 10 of this *CRC Handbook*.

References

1. Martin, W. C., Musgrove, A., Kotochigova, S., and Sansonetti, J. E., NIST Physical Reference Data Web site, <http://www.nist.gov/pml/data/ion_energy.cfm>, June 2013.
2. Martin, W. C., and Wiese, W. L., "Atomic Spectroscopy," in *Atomic, Molecular, & Optical Physics Handbook*, ed. by G.W.F. Drake (AIP, Woodbury, NY, 1996), Chapter 10, pp. 135-153.

Z	Symbol	Element	Ground-state configuration	Ground level	Ionization energy (eV)
1	H	Hydrogen	$1s$	$^2S_{1/2}$	13.5984
2	He	Helium	$1s^2$	$^1S_0$	24.5874
3	Li	Lithium	$1s^2\,2s$	$^2S_{1/2}$	5.3917
4	Be	Beryllium	$1s^2\,2s^2$	$^1S_0$	9.3227
5	B	Boron	$1s^2\,2s^2\,2p$	$^2P^o_{1/2}$	8.2980
6	C	Carbon	$1s^2\,2s^2\,2p^2$	$^3P_0$	11.2603
7	N	Nitrogen	$1s^2\,2s^2\,2p^3$	$^4S^o_{3/2}$	14.5341
8	O	Oxygen	$1s^2\,2s^2\,2p^4$	$^3P_2$	13.6181
9	F	Fluorine	$1s^2\,2s^2\,2p^5$	$^2P^o_{3/2}$	17.4228
10	Ne	Neon	$1s^2\,2s^2\,2p^6$	$^1S_0$	21.5645
11	Na	Sodium	[Ne] $3s$	$^2S_{1/2}$	5.1391
12	Mg	Magnesium	[Ne] $3s^2$	$^1S_0$	7.6462
13	Al	Aluminum	[Ne] $3s^2\,3p$	$^2P^o_{1/2}$	5.9858
14	Si	Silicon	[Ne] $3s^2\,3p^2$	$^3P_0$	8.1517
15	P	Phosphorus	[Ne] $3s^2\,3p^3$	$^4S^o_{3/2}$	10.4867
16	S	Sulfur	[Ne] $3s^2\,3p^4$	$^3P_2$	10.3600
17	Cl	Chlorine	[Ne] $3s^2\,3p^5$	$^2P^o_{3/2}$	12.9676
18	Ar	Argon	[Ne] $3s^2\,3p^6$	$^1S_0$	15.7596
19	K	Potassium	[Ar] $4s$	$^2S_{1/2}$	4.3407
20	Ca	Calcium	[Ar] $4s^2$	$^1S_0$	6.1132
21	Sc	Scandium	[Ar] $3d\,4s^2$	$^2D_{3/2}$	6.5615
22	Ti	Titanium	[Ar] $3d^2\,4s^2$	$^3F_2$	6.8281
23	V	Vanadium	[Ar] $3d^3\,4s^2$	$^4F_{3/2}$	6.7462
24	Cr	Chromium	[Ar] $3d^5\,4s$	$^7S_3$	6.7665
25	Mn	Manganese	[Ar] $3d^5\,4s^2$	$^6S_{5/2}$	7.4340
26	Fe	Iron	[Ar] $3d^6\,4s^2$	$^5D_4$	7.9024
27	Co	Cobalt	[Ar] $3d^7\,4s^2$	$^4F_{9/2}$	7.8810
28	Ni	Nickel	[Ar] $3d^8\,4s^2$	$^3F_4$	7.6399
29	Cu	Copper	[Ar] $3d^{10}\,4s$	$^2S_{1/2}$	7.7264
30	Zn	Zinc	[Ar] $3d^{10}\,4s^2$	$^1S_0$	9.3942
31	Ga	Gallium	[Ar] $3d^{10}\,4s^2\,4p$	$^2P^o_{1/2}$	5.9993
32	Ge	Germanium	[Ar] $3d^{10}\,4s^2\,4p^2$	$^3P_0$	7.8994
33	As	Arsenic	[Ar] $3d^{10}\,4s^2\,4p^3$	$^4S^o_{3/2}$	9.7886
34	Se	Selenium	[Ar] $3d^{10}\,4s^2\,4p^4$	$^3P_2$	9.7524
35	Br	Bromine	[Ar] $3d^{10}\,4s^2\,4p^5$	$^2P^o_{3/2}$	11.8138
36	Kr	Krypton	[Ar] $3d^{10}\,4s^2\,4p^6$	$^1S_0$	13.9996
37	Rb	Rubidium	[Kr] $5s$	$^2S_{1/2}$	4.1771
38	Sr	Strontium	[Kr] $5s^2$	$^1S_0$	5.6949
39	Y	Yttrium	[Kr] $4d\,5s^2$	$^2D_{3/2}$	6.2173
40	Zr	Zirconium	[Kr] $4d^2\,5s^2$	$^3F_2$	6.6339
41	Nb	Niobium	[Kr] $4d^4\,5s$	$^6D_{1/2}$	6.7589
42	Mo	Molybdenum	[Kr] $4d^5\,5s$	$^7S_3$	7.0924
43	Tc	Technetium	[Kr] $4d^5\,5s^2$	$^6S_{5/2}$	7.28
44	Ru	Ruthenium	[Kr] $4d^7\,5s$	$^5F_5$	7.3605
45	Rh	Rhodium	[Kr] $4d^8\,5s$	$^4F_{9/2}$	7.4589
46	Pd	Palladium	[Kr] $4d^{10}$	$^1S_0$	8.3369
47	Ag	Silver	[Kr] $4d^{10}\,5s$	$^2S_{1/2}$	7.5762
48	Cd	Cadmium	[Kr] $4d^{10}\,5s^2$	$^1S_0$	8.9938
49	In	Indium	[Kr] $4d^{10}\,5s^2\,5p$	$^2P^o_{1/2}$	5.7864
50	Sn	Tin	[Kr] $4d^{10}\,5s^2\,5p^2$	$^3P_0$	7.3439
51	Sb	Antimony	[Kr] $4d^{10}\,5s^2\,5p^3$	$^4S^o_{3/2}$	8.6084
52	Te	Tellurium	[Kr] $4d^{10}\,5s^2\,5p^4$	$^3P_2$	9.0096
53	I	Iodine	[Kr] $4d^{10}\,5s^2\,5p^5$	$^2P^o_{3/2}$	10.4513
54	Xe	Xenon	[Kr] $4d^{10}\,5s^2\,5p^6$	$^1S_0$	12.1298
55	Cs	Cesium	[Xe] $6s$	$^2S_{1/2}$	3.8939
56	Ba	Barium	[Xe] $6s^2$	$^1S_0$	5.2117
57	La	Lanthanum	[Xe] $5d\,6s^2$	$^2D_{3/2}$	5.5769
58	Ce	Cerium	[Xe] $4f\,5d\,6s^2$	$^1G^o_4$	5.5387
59	Pr	Praseodymium	[Xe] $4f^3\,6s^2$	$^4I^o_{9/2}$	5.473
60	Nd	Neodymium	[Xe] $4f^4\,6s^2$	$^5I_4$	5.5250
61	Pm	Promethium	[Xe] $4f^5\,6s^2$	$^6H^o_{5/2}$	5.582
62	Sm	Samarium	[Xe] $4f^6\,6s^2$	$^7F_0$	5.6437
63	Eu	Europium	[Xe] $4f^7\,6s^2$	$^8S^o_{7/2}$	5.6704
64	Gd	Gadolinium	[Xe] $4f^7\,5d\,6s^2$	$^9D^o_2$	6.1498
65	Tb	Terbium	[Xe] $4f^9\,6s^2$	$^6H^o_{15/2}$	5.8638
66	Dy	Dysprosium	[Xe] $4f^{10}\,6s^2$	$^5I_8$	5.9389
67	Ho	Holmium	[Xe] $4f^{11}\,6s^2$	$^4I^o_{15/2}$	6.0215
68	Er	Erbium	[Xe] $4f^{12}\,6s^2$	$^3H_6$	6.1077
69	Tm	Thulium	[Xe] $4f^{13}\,6s^2$	$^2F^o_{7/2}$	6.1843
70	Yb	Ytterbium	[Xe] $4f^{14}\,6s^2$	$^1S_0$	6.2542
71	Lu	Lutetium	[Xe] $4f^{14}\,5d\,6s^2$	$^2D_{3/2}$	5.4259
72	Hf	Hafnium	[Xe] $4f^{14}\,5d^2\,6s^2$	$^3F_2$	6.8251
73	Ta	Tantalum	[Xe] $4f^{14}\,5d^3\,6s^2$	$^4F_{3/2}$	7.5496
74	W	Tungsten	[Xe] $4f^{14}\,5d^4\,6s^2$	$^5D_0$	7.8640
75	Re	Rhenium	[Xe] $4f^{14}\,5d^5\,6s^2$	$^6S_{5/2}$	7.8335
76	Os	Osmium	[Xe] $4f^{14}\,5d^6\,6s^2$	$^5D_4$	8.4382
77	Ir	Iridium	[Xe] $4f^{14}\,5d^7\,6s^2$	$^4F_{9/2}$	8.9670
78	Pt	Platinum	[Xe] $4f^{14}\,5d^9\,6s$	$^3D_3$	8.9588
79	Au	Gold	[Xe] $4f^{14}\,5d^{10}\,6s$	$^2S_{1/2}$	9.2255
80	Hg	Mercury	[Xe] $4f^{14}\,5d^{10}\,6s^2$	$^1S_0$	10.4375
81	Tl	Thallium	[Xe] $4f^{14}\,5d^{10}\,6s^2\,6p$	$^2P^o_{1/2}$	6.1082
82	Pb	Lead	[Xe] $4f^{14}\,5d^{10}\,6s^2\,6p^2$	$^3P_0$	7.4167
83	Bi	Bismuth	[Xe] $4f^{14}\,5d^{10}\,6s^2\,6p^3$	$^4S^o_{3/2}$	7.2855
84	Po	Polonium	[Xe] $4f^{14}\,5d^{10}\,6s^2\,6p^4$	$^3P_2$	8.414
85	At	Astatine	[Xe] $4f^{14}\,5d^{10}\,6s^2\,6p^5$	$^2P^o_{3/2}$	
86	Rn	Radon	[Xe] $4f^{14}\,5d^{10}\,6s^2\,6p^6$	$^1S_0$	10.7485
87	Fr	Francium	[Rn] $7s$	$^2S_{1/2}$	4.0727
88	Ra	Radium	[Rn] $7s^2$	$^1S_0$	5.2784
89	Ac	Actinium	[Rn] $6d\,7s^2$	$^2D_{3/2}$	5.3807
90	Th	Thorium	[Rn] $6d^2\,7s^2$	$^3F_2$	6.3067
91	Pa	Protactinium	[Rn] $5f^2(^3H_4)\,6d\,7s^2$	$(4,3/2)_{11/2}$	5.89
92	U	Uranium	[Rn] $5f^3(^4I^o_{9/2})\,6d\,7s^2$	$(9/2,3/2)^o_6$	6.1939
93	Np	Neptunium	[Rn] $5f^4(^5I_4)\,6d\,7s^2$	$(4,3/2)_{11/2}$	6.2657
94	Pu	Plutonium	[Rn] $5f^6\,7s^2$	$^7F_0$	6.0260
95	Am	Americium	[Rn] $5f^7\,7s^2$	$^8S^o_{7/2}$	5.9738
96	Cm	Curium	[Rn] $5f^7\,6d\,7s^2$	$^9D^o_2$	5.9914
97	Bk	Berkelium	[Rn] $5f^9\,7s^2$	$^6H^o_{15/2}$	6.1979
98	Cf	Californium	[Rn] $5f^{10}\,7s^2$	$^5I_8$	6.2817
99	Es	Einsteinium	[Rn] $5f^{11}\,7s^2$	$^4I^o_{15/2}$	6.3676
100	Fm	Fermium	[Rn] $5f^{12}\,7s^2$	$^3H_6$	6.50
101	Md	Mendelevium	[Rn] $5f^{13}\,7s^2$	$^2F^o_{7/2}$	6.58
102	No	Nobelium	[Rn] $5f^{14}\,7s^2$	$^1S_0$	6.65
103	Lr	Lawrencium	[Rn] $5f^{14}\,7s^2\,7p$*	$^2P^o_{1/2}$*	4.96
104	Rf	Rutherfordium	[Rn] $5f^{14}\,6d^2\,7s^2$*	$^3F_2$*	6.0*

* Uncertain.

INTERNATIONAL TEMPERATURE SCALE OF 1990 (ITS-90)

B. W. Mangum

A new temperature scale, the International Temperature Scale of 1990 (ITS-90), was officially adopted by the Comité International des Poids et Mesures (CIPM), meeting 26-28 September 1989 at the Bureau International des Poids et Mesures (BIPM). The ITS-90 was recommended to the CIPM for its adoption following the completion of the final details of the new scale by the Comité Consultatif de Thermométrie (CCT), meeting 12-14 September 1989 at the BIPM in its 17th Session. The ITS-90 became the official international temperature scale on 1 January 1990. The ITS-90 supersedes the previous scales, the International Practical Temperature Scale of 1968 (IPTS-68) and the 1976 Provisional 0.5 to 30 K Temperature Scale (EPT-76).

The ITS-90 (Refs. 1, 2) extends upward from 0.65 K, and temperatures on this scale are in much better agreement with thermodynamic values than are those on the IPTS-68 and the EPT-76. The new scale has subranges and alternative definitions in certain ranges that greatly facilitate its use. Furthermore, its continuity, precision, and reproducibility throughout its ranges are much improved over that of the previous scales. The replacement of the thermocouple with the platinum resistance thermometer at temperatures below 961.78 °C resulted in the biggest improvement in reproducibility.

The ITS-90 is divided into four primary ranges:

1. Between 0.65 and 3.2 K, the ITS-90 is defined by the vapor pressure-temperature relation of ^{3}He, and between 1.25 and 2.1768 K (the λ point) and between 2.1768 and 5.0 K by the vapor pressure–temperature relations of ^{4}He. T_{90} is defined by the vapor pressure equations of the form:

$$T_{90}/K = A_0 + \sum_{i=1}^{9} A_i[(\ln(p/\text{Pa}) - B)/C]^i$$

The values of the coefficients A_i, and of the constants A_0, B, and C of the equations are given below.

2. Between 3.0 and 24.5561 K, the ITS-90 is defined in terms of a ^{3}He or ^{4}He constant volume gas thermometer (CVGT). The thermometer is calibrated at three temperatures — at the triple point of neon (24.5561 K), at the triple point of equilibrium hydrogen (13.8033 K), and at a temperature between 3.0 and 5.0 K, the value of which is determined by using either ^{3}He or ^{4}He vapor pressure thermometry.
3. Between 13.8033 K (−259.3467 °C) and 1234.93 K (961.78 °C), the ITS-90 is defined in terms of the specified fixed points given below, by resistance ratios of platinum resistance thermometers obtained by calibration at specified sets of the fixed points, and by reference functions and deviation functions of resistance ratios which relate to T_{90} between the fixed points.
4. Above 1234.93 K, the ITS-90 is defined in terms of Planck's radiation law, using the freezing-point temperature of either silver, gold, or copper as the reference temperature.

Since the adoption of ITS-90, the isotopic composition of the water and hydrogen whose fixed points appear in the table has been specified (Ref. 3). A Provisional Low Temperature Scale (PLTS-2000) has been developed, covering the region from 0.9 mK to 1 K (Refs. 2, 4). This scale is based on the melting temperature of ^{3}He.

References

1. The International Temperature Scale of 1990, Metrologia 27, 3, 1990; errata in Metrologia 27, 107, 1990.
2. *Mise en pratique* for definition of the kelvin, <www.bipm.org/utils/en/pdf/MeP_K.pdf>, 2011.
3. Technical Annex for the International Temperature Scale of 1990, <www.bipm.org/utils/en/pdf/MeP_K_Technical_Annex.pdf>, 2005.
4. The Provisional Low Temperature Scale from 0.9 mK to 1 K, <www.bipm.org/utils/en/pdf/PLTS-2000.pdf>, 2000.

Defining Fixed Points of the ITS-90

Material[a]	Equilibrium state[b]	Temperature T_{90} (K)	Temperature t_{90} (°C)
He	VP	3 to 5	−270.15 to −268.15
e-H_2	TP	13.8033	−259.3467
e-H_2 (or He)	VP (or CVGT)	≈17	≈ −256.15
e-H_2 (or He)	VP (or CVGT)	≈20.3	≈ −252.85
Ne[c]	TP	24.5561	−248.5939
O_2	TP	54.3584	−218.7916
Ar	TP	83.8058	−189.3442
Hg[c]	TP	234.3156	−38.8344
H_2O	TP	273.16	0.01
Ga[c]	MP	302.9146	29.7646
In[c]	FP	429.7485	156.5985
Sn	FP	505.078	231.928
Zn	FP	692.677	419.527
Al[c]	FP	933.473	660.323
Ag	FP	1234.93	961.78
Au	FP	1337.33	1064.18
Cu[c]	FP	1357.77	1084.62

[a] e-H_2 indicates equilibrium hydrogen, that is, hydrogen with the equilibrium distribution of its ortho and para states. Normal hydrogen at room temperature contains 25% para hydrogen and 75% ortho hydrogen.

[b] VP indicates vapor pressure point; CVGT indicates constant volume gas thermometer point; TP indicates triple point (equilibrium temperature at which the solid, liquid, and vapor phases coexist); FP indicates freezing point; and MP indicates melting point (the equilibrium temperatures at which the solid and liquid phases coexist under a pressure of 101 325 Pa, one standard atmosphere). The isotopic composition is that naturally occurring.

[c] Previously, these were secondary fixed points.

Values of Coefficients in the Vapor Pressure Equations for Helium

Coef. or constant	^{3}He 0.65–3.2 K	^{4}He 1.25–2.1768 K	^{4}He 2.1768–5.0 K
A_0	1.053 447	1.392 408	3.146 631
A_1	0.980 106	0.527 153	1.357 655
A_2	0.676 380	0.166 756	0.413 923
A_3	0.372 692	0.050 988	0.091 159
A_4	0.151 656	0.026 514	0.016 349
A_5	−0.002 263	0.001 975	0.001 826
A_6	0.006 596	−0.017 976	−0.004 325
A_7	0.088 966	0.005 409	−0.004 973
A_8	−0.004 770	0.013 259	0
A_9	−0.054 943	0	0
B	7.3	5.6	10.3
C	4.3	2.9	1.9

INTERNATIONAL SYSTEM OF UNITS (SI)

The International Bureau of Weights and Measures (BIPM) was established by Article 1 of the Metre Convention, which was signed on May 20, 1875. BIPM is charged with providing the basis for a single, coherent system of measurements to be used throughout the world and operates under the authority of the International Committee of Weights and Measures (CIPM). In 1960, the 11th General Conference on Weights and Measures (CGPM) formally defined and established the International System of Units (SI). Since then, the SI has been periodically updated to take into account advances in science and the need for measurements in new domains.

In 2018, the 26th CGPM (2019) decided that the SI would be based on the fixed numerical values of a set of seven defining constants from which the definitions of the seven base units of the SI would be deduced. This major change was made to allow anchoring of the SI to specific experimental realizations of measurable quantities and to remove the base units from dependence on physical artifacts. The change has been taking place over a number of years, and its completion was made possible by realization of the base unit kilogram separate from its previous physical artifact as stored at BIPM in Sevres, France. This discussion of the the newly constituted SI is based on more complete documentation in References 1 and 2. Because of the importance of the SI in science, much of the discussion below is taken verbatim from these references.

The core of the SI is the seven base units for physical quantities as shown in Table 1.

TABLE 1. SI Base Units

Base quantity	Name	Symbol
length	meter	m
mass	kilogram	kg
time	second	s
electric current	ampere	A
thermodynamic temperature	kelvin	K
amount of substance	mole	mol
luminous intensity	candela	cd

As of May 20, 2019, the SI is the system of units in which the base units are now defined by the seven fundamental constants given in Table 2. More complete discussions of these constants and their experimental realizations can be found in References 1 and 2.

TABLE 2. Fundamental Constants Used to Define the SI

Fundamental constant	Symbol	Value
Unperturbed ground-state hyperfine transition frequency of the cesium 133 atom ^{133}Cs	$\Delta\nu_{Cs}$	9 192 631 770 Hz
Speed of light in vacuum	c	299 792 458 m s^{-1}
Planck constant	h	6.626 070 15 × 10^{-34} J s
Elementary charge	e	1.602 176 634 × 10^{-19} C
Boltzmann constant	k	1.380 649 × 10^{-23} J K^{-1}
Avogadro constant	N_A	6.022 140 76 × 10^{23} mol^{-1}
Luminous efficacy of monochromatic radiation of frequency 540 × 1012 Hz	K_{cd}	683 lm W^{-1}

The hertz, joule, coulomb, lumen, and watt are related to the units second, meter, kilogram, ampere, kelvin, mole, and candela as follows in Table 3. (Note sr is steradian.)

TABLE 3. Relationship of Important Units to the Basic SI Units

Unit name	Symbol	Relationship
hertz	Hz	s^{-1}
joule	J	kg m^2 s^{-2}
coulomb	C	A s
lumen	lm	cd m^2 m^{-2} = cd sr
watt	W	kg m^2 s^{-3}

Table 4 provides the definitions of the base quantities. The defintions replace older definitions as discussed in detail in References 1 and 2. These definitions specify the exact numerical value of each constant when its value is expressed in the corresponding SI unit. By fixing the exact numerical value, the unit becomes defined because the product of the numerical value and the unit has to equal the value of the constant, which is invariant. The defining constants have been chosen such that, when taken together, their units cover all of the units of the SI. In general, there is no one-to-one correspondence between the defining constants and the SI base units, except for the cesium frequency ΔvCs and the Avogadro constant N_A. Any SI unit is a product of powers of these seven constants and a dimensionless factor.

TABLE 4. Definitions of the SI Base Units as Adopted by BIPM in 2019

ampere: The ampere, symbol A, is the SI unit of electric current. It is defined by taking the fixed numerical value of the elementary charge e to be 1.602 176 634 × 10^{-19} when expressed in the unit C, which is equal to A s, where the second is defined in terms of $\Delta\nu_{Cs}$.

candela: The candela, symbol cd, is the SI unit of luminous intensity in a given direction. It is defined by taking the fixed numerical value of the luminous efficacy of monochromatic radiation of frequency 540 × 10^{12} Hz, K_{cd}, to be 683, when expressed in the unit lm W^{-1}, which is equal to cd sr W^{-1}, or cd sr kg^{-1} m^{-1} s^3, where the kilogram, meter, and second are defined in terms of h, c, and $\Delta\nu_{Cs}$.

kelvin: The kelvin, symbol K, is the SI unit of thermodynamic temperature. It is defined by taking the fixed numerical value of the Boltzmann constant k to be 1.380 649 × 10^{-23} when expressed in the unit J K^{-1}, which is equal to kg m^2 s^{-2} K^{-1}, where the kilogram, meter, and second are defined in terms of h, c, and $\Delta\nu_{Cs}$.

kilogram: The kilogram, symbol kg, is the SI unit of mass. It is defined by taking the fixed numerical value of the Planck constant h to be 6.626 070 15 × 10^{-34} when expressed in the unit J s, which is equal to kg m^2 s^{-1}, where the meter and second are defined in terms of c and $\Delta\nu_{Cs}$.

meter: The meter, symbol m, is the SI unit of length. It is defined by taking the fixed numerical value of the speed of light in vacuum c to be 299 792 458 when expressed in the unit m s^{-1}, where the second is defined in terms of the cesium frequency $\Delta\nu_{Cs}$.

mole: The mole, symbol mol, is the SI unit of amount of substance. One mole contains exactly 6.022 140 76 × 10^{23} elementary entities. This number is the fixed numerical value of the Avogadro constant, N_A, when expressed in the unit mol^{-1} and is called the Avogadro number. The amount of substance, symbol n, of a system is a measure of the number of specified elementary entities.

An elementary entity may be an atom, a molecule, an ion, an electron, any other particles, or specified group of particles.

second: The second, symbol s, is the SI unit of time. It is defined by taking the fixed numerical value of the cesium frequency $\Delta\nu_{Cs}$, the unperturbed ground-state hyperfine transition frequency of the cesium 133 atom (^{133}Cs), to be 9 192 631 770 when expressed in the unit Hz, which is equal to s^{-1}.

SI derived units

Derived units are units that may be expressed in terms of base units by means of the mathematical symbols of multiplication and division (and, in the case of °C, subtraction). Certain derived units have been given special names and symbols, and these special names and symbols may themselves be used in combination with those for base and other derived units to express the units of other quantities. Table 5 lists some examples of derived units expressed directly in terms of base units.

TABLE 5. Examples of SI Derived Units

Physical quantity	Name	Symbol
area	square meter	m^2
volume	cubic meter	m^3
speed, velocity	meter per second	$m\ s^{-1}$
acceleration	meter per second squared	$m\ s^{-2}$
wave number	reciprocal meter	m^{-1}
density, mass density	kilogram per cubic meter	$kg\ m^{-3}$
specific volume	cubic meter per kilogram	$m^3\ kg^{-1}$
current density	ampere per square meter	$A\ m^{-2}$
magnetic field strength	ampere per meter	$A\ m^{-1}$
concentration (of amount of substance)	mole per cubic meter	$mol\ m^{-3}$
luminance	candela per square meter	$cd\ m^{-2}$
refractive index	(the number) one	1[a]

[a] The symbol "1" (for the number "one") is generally omitted in combination with a numerical value.

For convenience, certain derived units, which are listed in Table 6, have been given special names and symbols. These names and symbols may themselves be used to express other derived units. The special names and symbols are a compact form for the expression of units that are used frequently. The final column shows how the SI units concerned may be expressed in terms of SI base units. In this column, factors such as m^0, kg^0 ..., which are all equal to 1, are not shown explicitly.

TABLE 6. SI Derived Units with Special Names and Symbols

Physical quantity	Name	Symbol	Other SI units	SI base units
plane angle	radian[(a)]	rad	$m \cdot m^{-1} = 1$[(b)]	
solid angle	steradian[(a)]	sr[(c)]	$m^2 \cdot m^{-2} = 1$[(b)]	
frequency	hertz	Hz	s^{-1}	
force	newton	N	$m \cdot kg \cdot s^{-2}$	
pressure, stress	pascal	Pa	$N\ m^{-2}$	$m^{-1} \cdot kg \cdot s^{-2}$
energy, work, quantity of heat	joule	J	$N \cdot m$	$m^2 \cdot kg \cdot s^{-2}$
power, radiant flux	watt	W	$J\ s^{-1}$	$m^2 \cdot kg \cdot s^{-3}$
electric charge, quantity of electricity	coulomb	C	$s \cdot A$	
electric potential difference, electromotive force	volt	V	$W\ A^{-1}$	$m^2 \cdot kg \cdot s^{-3} \cdot A^{-1}$
capacitance	farad	F	$C\ V^{-1}$	$m^{-2} \cdot kg^{-1} \cdot s^4 \cdot A^2$
electric resistance	ohm	Ω	$V\ A^{-1}$	$m^2 \cdot kg \cdot s^{-3} \cdot A^{-2}$
electric conductance	siemens	S	$A\ V^{-1}$	$m^{-2} \cdot kg^{-1} \cdot s^3 \cdot A^2$
magnetic flux	weber	Wb	$V \cdot s$	$m^2 \cdot kg \cdot s^{-2} \cdot A^{-1}$
magnetic flux density	tesla	T	$Wb\ m^{-2}$	$kg \cdot s^{-2} \cdot A^{-1}$
inductance	henry	H	$Wb\ A^{-1}$	$m^2 \cdot kg \cdot s^{-2} \cdot A^{-2}$
Celsius temperature	degree Celsius[(d)]	°C		K
luminous flux	lumen	lm	cd · sr[(c)]	$m^2 \cdot m^{-2} \cdot cd = cd$
illuminance	lux	lx	$lm\ m^{-2}$	$m^2 \cdot m^{-4} \cdot cd = m^{-2} \cdot cd$
activity (of a radionuclide)	becquerel	Bq		s^{-1}
absorbed dose, specific energy (imparted), kerma	gray	Gy	$J\ kg^{-1}$	$m^2 \cdot s^{-2}$
dose equivalent, ambient dose equivalent, directional dose equivalent, personal dose equivalent, organ equivalent dose	sievert	Sv	$J\ kg^{-1}$	$m^2 \cdot s^{-2}$
catalytic activity	katal	kat		$s^{-1} \cdot mol$

(a) The radian and steradian may be used with advantage in expressions for derived units to distinguish between quantities of different nature but the same dimension. Some examples of their use in forming derived units are given in the next table.

(b) In practice, the symbols rad and sr are used where appropriate, but the derived unit "1" is generally omitted in combination with a numerical value.

(c) In photometry, the name steradian and the symbol sr are usually retained in expressions for units.

(d) It is common practice to express a thermodynamic temperature, symbol T, in terms of its difference from the reference temperature $T_0 = 273.15$ K. The numerical value of a Celsius temperature t expressed in degrees Celsius is given by $t/°C = T/K - 273.15$. The unit °C may be used in combination with SI prefixes, e.g., millidegree Celsius, m°C. Note that there should never be a space between the ° sign and the letter C, and that the symbol for kelvin is K, not °K.

The SI derived units with special names may be used in combinations to provide a convenient way to express more complex physical quantities. Examples are given in Table 7.

TABLE 7. Examples of Using SI Derived Units with Special Names to Express Complex Physical Quantities

Physical quantity	Name	Symbol	As SI base units
dynamic viscosity	pascal second	$Pa \cdot s$	$m^{-1} \cdot kg \cdot s^{-1}$
moment of force	newton meter	$N \cdot m$	$m^2 \cdot kg \cdot s^{-2}$
surface tension	newton per meter	$N\ m^{-1}$	$kg \cdot s^{-2}$
angular velocity	radian per second	$rad\ s^{-1}$	$m \cdot m^{-1} \cdot s^{-1} = s^{-1}$
angular acceleration	radian per second squared	$rad\ s^{-2}$	$m \cdot m^{-1} \cdot s^{-2} = s^{-2}$
heat flux density, irradiance	watt per square meter	$W\ m^{-2}$	$kg \cdot s^{-3}$
heat capacity, entropy	joule per kelvin	$J\ K^{-1}$	$m^{-3} \cdot kg \cdot s^{-2} \cdot K^{-1}$
specific heat capacity, specific entropy	joule per kilogram kelvin	$J\ kg^{-1}\ K^{-1}$	$m^2 \cdot s^{-2} \cdot K^{-1}$
specific energy	joule per kilogram	$J\ kg^{-1}$	$m^2 \cdot s^{-2}$
thermal conductivity	watt per meter kelvin	$W\ m^{-1} \cdot K^{-1}$	$m \cdot kg \cdot s^{-3} \cdot K^{-1}$

Physical quantity	Name	Symbol	As SI base units
energy density	joule per cubic meter	$J\ m^{-3}$	$m^{-1} \cdot kg \cdot s^{-2}$
electric field strength	volt per meter	$V\ m^{-1}$	$m \cdot kg \cdot s^{-3} \cdot A^{-1}$
electric charge density	coulomb per cubic meter	$C\ m^{-3}$	$m^{-3} \cdot s \cdot A$
electric flux density	coulomb per square meter	$C\ m^{-2}$	$m^{-2} \cdot s \cdot A$
permittivity	farad per meter	$F\ m^{-1}$	$m^{-3} \cdot kg^{-1} \cdot s^{4} \cdot A^{2}$
permeability	henry per meter	$H\ m^{-1}$	$m \cdot kg \cdot s^{-2} \cdot A^{-2}$
molar energy	joule per mole	$J\ mol^{-1}$	$m^{2} \cdot kg \cdot s^{-2} \cdot mol^{-1}$
molar entropy, molar heat capacity	joule per mole kelvin	$J\ mol^{-1} \cdot K^{-1}$	$m^{2} \cdot kg \cdot s^{-2} \cdot K^{-1} \cdot mol^{-1}$
exposure (x and γ rays)	coulomb per kilogram	$C\ kg^{-1}$	$kg^{-1} \cdot s \cdot A$
absorbed dose rate	gray per second	$Gy\ s^{-1}$	$m^{2} \cdot s^{-3}$
radiant intensity	watt per steradian	$W\ sr^{-1}$	$m^{4} \cdot m^{-2} \cdot kg \cdot s^{-3} = m^{2} \cdot kg \cdot s^{-3}$
radiance	watt per square meter steradian	$W\ m^{-2} \cdot sr^{-1}$	$m^{2} \cdot m^{-2} \cdot kg \cdot s^{-3} = kg \cdot s^{-3}$
catalytic (activity) concentration	katal per cubic meter	$kat\ m^{-3}$	$m^{-3} \cdot s^{-1} \cdot mol$

In practice, with certain quantities, preference is given to the use of certain special unit names, or combinations of unit names, in order to facilitate the distinction between different quantities having the same dimension. For example, the SI unit of frequency is designated the hertz, rather than the reciprocal second, and the SI unit of angular velocity is designated the radian per second rather than the reciprocal second (in this case retaining the word radian emphasizes that angular velocity is equal to 2π times the rotational frequency). Similarly, the SI unit of moment of force is designated the newton meter rather than the joule.

In the field of ionizing radiation, the SI unit of activity is designated the becquerel rather than the reciprocal second, and the SI units of absorbed dose and dose equivalent the gray and sievert, respectively, rather than the joule per kilogram. In the field of catalysis, the SI unit of catalytic activity is designated the katal rather than the mole per second. The special names becquerel, gray, sievert, and katal were specifically introduced because of the dangers to human health which might arise from mistakes involving the units reciprocal second, joule per kilogram, and mole per second.

Units for dimensionless quantities, quantities of dimension one

Certain quantities are defined as the ratios of two quantities of the same kind, and thus have a dimension which may be expressed by the number one. The unit of such quantities is necessarily a derived unit coherent with the other units of the SI and, because it is formed as the ratio of two identical SI units, the unit also may be expressed by the number one. Thus, the SI unit of all quantities having the dimensional product one is the number one. Examples of such quantities are refractive index, relative permeability, and friction factor. Other quantities having the unit 1 include "characteristic numbers" such as the Prandtl number and numbers which represent a count, such as a number of molecules, degeneracy (number of energy levels), and partition function in statistical thermodynamics. All of these quantities are described as being dimensionless, or of dimension one, and have the coherent SI unit 1. Their values are simply expressed as numbers and, in general, the unit 1 is not explicitly shown. In a few cases, however, a special name is given to this unit, mainly to avoid confusion between some compound derived units. This is the case for the radian, steradian, and neper.

SI prefixes

The following prefixes, shown in Table 8, have been approved by the CGPM for use with SI units. Only one prefix may be used before a unit. Thus 10^{-12} farad should be designated pF, not μμF.

TABLE 8. SI Prefixes

Factor	Name	Symbol
10^{24}	yotta	Y
10^{21}	zetta	Z
10^{18}	exa	E
10^{15}	peta	P
10^{12}	tera	T
10^{9}	giga	G
10^{6}	mega	M
10^{3}	kilo	k
10^{2}	hecto	h
10^{1}	deka	da
10^{-1}	deci	d
10^{-2}	centi	c
10^{-3}	milli	m
10^{-6}	micro	μ
10^{-9}	nano	n
10^{-12}	pico	p
10^{-15}	femto	f
10^{-18}	atto	a
10^{-21}	zepto	z
10^{-24}	yocto	y

The kilogram

Among the base units of the International System, the unit of mass is the only one whose name, for historical reasons, contains a prefix. Names and symbols for decimal multiples and submultiples of the unit of mass are formed by attaching prefix names to the unit name "gram" and prefix symbols to the unit symbol "g".

Example: 10^{-6} kg = 1 mg (1 milligram) *but not* 1 μkg (1 microkilogram).

Units used with the SI

Many units that are not part of the SI are important and widely used in everyday life. The CGPM has adopted a classification of non-SI units: (1) units accepted for use with the SI, such as the traditional units of time and of angle (Table 9); (2) units accepted for use with the SI whose values are obtained experimentally (Table 10); and (3) other units currently accepted for use with the SI to satisfy the needs of special interests (Table 11).

TABLE 9. Non-SI Units Accepted for Use with the International System

Name	Symbol	Value in SI units
minute	min	1 min = 60 s

Name	Symbol	Value in SI units
hour	h	1 h = 60 min = 3600 s
day	d	1d = 24 h = 86 400 s
degree	°	1° = (π/180) rad
minute	′	1′ = (1/60)° = (π/10 800) rad
second	″	1″ = (1/60)′ = (π/648 000) rad
liter	l, L	1L = 1 dm^3 = 10^{-3} m^3
metric ton	t	1 t = 10^3 kg
neper[a]	Np	1 Np = 1
bel[b]	B	1 B = (1/2) ln 10 Np

[a] The neper is used to express values of such logarithmic quantities as field level, power level, sound pressure level, and logarithmic decrement. Natural logarithms are used to obtain the numerical values of quantities expressed in nepers. The neper is coherent with the SI, but is not yet adopted by the CGPM as an SI unit. In using the neper, it is important to specify the quantity.

[b] The bel is used to express values of such logarithmic quantities as field level, power level, sound-pressure level, and attenuation. Logarithms to base ten are used to obtain the numerical values of quantities expressed in bels. The submultiple decibel, dB, is commonly used.

TABLE 10. Non-SI Units Accepted for Use with the International System, Whose Values in SI Units Are Obtained Experimentally

Name	Symbol	Value in SI units
electronvolt[b]	eV	1 eV = 1.602 176 565(35) ·10^{-19} J[a]
dalton[c]	Da	1 Da = 1.660 538 921(73) · 10^{-27} kg[a]
unified atomic mass unit[c]	u	1 u = 1 Da
astronomical unit[d]	au	1 au = 149 597 870 700 m (exact)[a]

[a] For the electronvolt and the dalton (unified atomic mass unit), values are quoted from the "CODATA Recommended Values of the Fundamental Physical Constants: 2018" in this section. The value given for the astronomical unit is from Resolution B2 of the XXVIII General Assembly of the International Astronomical Union (Captaine, N., Klioner, S., and McCarthy, D. [2012], "The re-definition of the astronomical unit of length: reasons and consequences", IAU Joint Discussion 7: Space-Time Reference Systems for Future Research, Bibcode: 2012IAUJD 7E 40C). Note the definition of the au in terms of the meter is now exact.

[b] The electronvolt is the kinetic energy acquired by an electron in passing through a potential difference of 1 V in vacuum.

[c] The dalton and unified atomic mass unit are alternative names for the same unit, equal to 1/12 of the mass of an unbound atom of the nuclide ^{12}C, at rest and in its ground state. The dalton may be combined with SI prefixes to express the masses of large molecules in kilodalton, kDa, or megadalton, MDa.

[d] The mean Earth-Sun distance is approximately equal to the astronomical unit.

TABLE 11. Other Non-SI Units Currently Accepted for Use with the International System

Name	Symbol	Value in SI units
nautical mile		1 nautical mile = 1852 m
knot		1 nautical mile per hour = (1852/3600) m s^{-1}
are		1 a = 1 dam^2 = 10^2 m^2
hectare	ha	1 ha = 1 hm^2 = 10^4 m^2
bar	bar	1 bar = 0.1 MPa = 100 kPa = 10^5 Pa
ångström	Å	1 Å = 0.1 nm = 10^{-10} m
barn	b	1 b = 100 fm^2 = 10^{-28} m^2

Other non-SI units

The SI does not encourage the use of cgs units, but these are frequently found in old scientific texts. Table 12 gives the relation of some common cgs units to SI units.

TABLE 12. Relation of cgs Units to SI Units

Name	Symbol	Value in SI units
erg	erg	1 erg = 10^{-7} J
dyne	dyn	1 dyn = 10^{-5} N
poise	P	1P = 1dyn · s cm^{-2} = 0.1 Pa · s
stokes	St	1 St = 1 cm^2 s^{-1} = 10^{-4} m^2 s^{-1}
gauss	G	1G ≙ 10^{-4} T
oersted	Oe	1 Oe ≙ (1000/4π) A m^{-1}
maxwell	Mx	1Mx ≙ 10^{-8} Wb
stilb	sb	1 sb = 1 cd cm^{-2} = 10^4 cd m^{-2}
phot	ph	1 ph = 10^4 lx
gal	Gal	1 Gal = 1 cm s^{-2} = 10^{-2} m s^{-2}

Note: The symbol ≙ should be read as "corresponds to"; these units cannot strictly be equated because of the different dimensions of the electromagnetic cgs and the SI.

Examples of other non-SI units found in the older literature and their relation to the SI are given in Table 13. Use of these units in current texts is discouraged.

TABLE 13. Relation of Other Older Units to SI Units

Name	Symbol	Value in SI units
curie	Ci	1 Ci = 3.7 · 10^{10} Bq
roentgen	R	1 R = 2.58 · 10^{-4} C kg-1
rad	rad	1 rad = 1 cGy = 10^{-2} Gy
rem	rem	1 rem = 1 cSv = 10^{-2} Sv
X unit		1 X unit ≈ 1.002 · 10^{-4} nm
gamma	γ	1 γ = 1 nT = 10^{-9} T
jansky	Jy	1 Jy = 10^{-26} W · m^{-2} · Hz^{-1}
fermi		1 fermi = 1 fm = 10^{-15} m
metric carat		1 metric carat = 200 mg = 2 · 10^{-4} kg
torr	Torr	1 Torr = (101325/760) Pa
standard atmosphere	atm	1 atm = 101325 Pa
calorie[a]	cal	1 cal = 4.184 J
micron	μ	1 μ = 1 μm = 10^{-6} m

[a] Several types of calorie have been used; the value given here is the so-called "thermochemical calorie."

Prefixes for binary multiples

In December 1998 the International Electrotechnical Commission (IEC), the leading international organization for worldwide standardization in electrotechnology, approved as an IEC International Standard names and symbols for prefixes for binary multiples for use in the fields of data processing and data transmission. The prefixes are as given in Table 14. Comparison of these prefixes to those defined by SI is shown in Table 15.

TABLE 14. Prefixes for Binary Multiples

Factor	Name	Symbol	Origin	Derivation
2^{10}	kibi	Ki	kilobinary: (2^{10})^1	kilo: (10^3)^1
2^{20}	mebi	Mi	megabinary: (2^{10})^2	mega: (10^3)^2
2^{30}	gibi	Gi	gigabinary: (2^{10})^3	giga: (10^3)^3
2^{40}	tebi	Ti	terabinary: (2^{10})^4	tera: (10^3)^4
2^{50}	pebi	Pi	petabinary: (2^{10})^5	peta: (10^3)^5
2^{60}	exbi	Ei	exabinary: (2^{10})^6	exa: (10^3)^6

Units

TABLE 15. Examples and Comparisons with SI Prefixes

Multiple	SI equivalent
one kibibit	1 Kibit = 2^{10} bit = 1024 bit
one kilobit	1 kbit = 10^3 bit = 1000 bit
one mebibyte	1 MiB = 2^{20} B = 1 048 576 B
one megabyte	1 MB = 10^6 B = 1 000 000 B
one gibibyte	1 GiB = 2^{30} B = 1 073 741 824 B
one gigabyte	1 GB = 10^9 B = 1 000 000 000 B

It is suggested that in English, the first syllable of the name of the binary-multiple prefix should be pronounced in the same way as the first syllable of the name of the corresponding SI prefix, and that the second syllable should be pronounced as "bee."

It is important to recognize that the new prefixes for binary multiples are not part of the International System of Units (SI), the modern metric system. However, for ease of understanding and recall, they were derived from the SI prefixes for positive powers of ten. As can be seen from the above table, the name of each new prefix is derived from the name of the corresponding SI prefix by retaining the first two letters of the name of the SI prefix and adding the letters "bi," which recalls the word "binary." Similarly, the symbol of each new prefix is derived from the symbol of the corresponding SI prefix by adding the letter "i," which again recalls the word "binary." (For consistency with the other prefixes for binary multiples, the symbol Ki is used for 2^{10} rather than ki.)

References

1. Newell, D. B., and Tiesinga, E., *The International System of Units (SI)*, NIST Special Publication 330, 2019 Edition, National Institute of Standards and Technology, Gaithersburg, MD, 2019 <https://doi.org/10.6028/NIST.SP.330-2019>.
2. Bureau International des Poids et Mesures, *Le Système International d'Unités (SI)*, 9th French and English Edition, BIPM, Sèvres, France, 2019 <www.bipm.org>.
3. Thompson, A., and Taylor, B. N., *Guide for the Use of the International System of Units (SI)*, NIST Special Publication 811, National Institute of Standards and Technology, Gaithersburg, MD, 2008.
4. NIST Physical Reference Data Web site, <physics.nist.gov/cuu/Units/index.html>.
5. Amendment 2 to IEC International Standard IEC 60027-2, 1999-01, Letter symbols to be used in electrical technology – Part 2: Telecommunications and electronics.
6. IEC 60027-2, Second Edition, 2000-11, Letter symbols to be used in electrical technology - Part 2: Telecommunications and electronics.
7. Barrow, B., "A Lesson in Megabytes," *IEEE Stand. Bearer*, January 1997, p. 5.

UNITS FOR MAGNETIC PROPERTIES

Quantity	Symbol	Gaussian & cgs emu [a]	Conversion factor, C [b]	SI & rationalized mks [c]
Magnetic flux density, magnetic induction	B	gauss (G)[d]	10^{-4}	tesla (T), Wb/m^2
Magnetic flux	Φ	maxwell (Mx), G · cm^2	10^{-8}	weber (Wb), volt second (V · s)
Magnetic potential difference, magnetomotive force	U, F	gilbert (Gb)	$10/4\pi$	ampere (A)
Magnetic field strength, magnetizing force	H	oersted (Oe)[e], Gb/cm	$10^3/4\pi$	A/m[f]
(Volume) magnetization[g]	M	emu/cm^3[h]	10^3	A/m
(Volume) magnetization	$4\pi M$	G	$10^3/4\pi$	A/m
Magnetic polarization, intensity of magnetization	J, I	emu/cm^3	$4\pi \times 10^{-4}$	T, Wb/m^2[i]
(Mass) magnetization	σ, M	emu/g	1	A · m^2/kg
			$4\pi \times 10^{-7}$	Wb · m/kg
Magnetic moment	m	emu, erg/G	10^{-3}	A · m^2, joule per tesla (J/T)
Magnetic dipole moment	j	emu, erg/G	$4\pi \times 10^{-10}$	Wb · m[i]
(Volume) susceptibility	χ, κ	dimensionless, emu/cm^3	4π	dimensionless
			$(4\pi)^2 \times 10^{-7}$	henry per meter (H/m), Wb/(A · m)
(Mass) susceptibility	χ_ρ, κ_ρ	cm^3/g, emu/g	$4\pi \times 10^{-3}$	m^3/kg
			$(4\pi)^2 \times 10^{-10}$	H · m^2/kg
(Molar) susceptibility	χ_{mol}, κ_{mol}	cm^3/mol, emu/mol	$4\pi \times 10^{-6}$	m^3/mol
			$(4\pi)^2 \times 10^{-13}$	H · m^2/mol
Permeability	μ	dimensionless	$4\pi \times 10^{-7}$	H/m, Wb/(A · m)
Relative permeability[j]	μ_r	not defined		dimensionless
(Volume) energy density, energy product[k]	W	erg/cm^3	10^{-1}	J/m^3
Demagnetization factor	D, N	dimensionless	$1/4\pi$	dimensionless

[a] Gaussian units and cgs emu are the same for magnetic properties. The defining relation is $B = H + 4\pi M$.
[b] Multiply a number in Gaussian units by C to convert it to SI (e.g., 1 G × 10^{-4} T/G = 10^{-4} T).
[c] SI (*Système International d'Unités*) has been adopted by the National Institute of Standards and Technology. Where two conversion factors are given, the upper one is recognized under, or consistent with, SI and is based on the definition $B = \mu_0(H + M)$, where $\mu_0 = 4\pi \times 10^{-7} H/M$. The lower one is not recognized under SI and is based on the definition $B = \mu_0 H + J$, where the symbol I is often used in place of J.
[d] 1 gauss = 10^5 gamma (γ).
[e] Both oersted and gauss are expressed as cm$^{-1/2}$·g$^{1/2}$·s^{-1} in terms of base units.
[f] A/m was often expressed as "ampere–turn per meter" when used for magnetic field strength.
[g] Magnetic moment per unit volume.
[h] The designation "emu" is not a unit.
[i] Recognized under SI, even though based on the definition $B = \mu_0 H + J$. See footnote c.
[j] $\mu_r = \mu/\mu_0 = 1+\chi$, all in SI. μ_r is equal to Gaussian μ.
[k] $B \cdot H$ and $\mu_0 M \cdot H$ have SI units J/m^3; $M \cdot H$ and $B \cdot H/4\pi$ have Gaussian units erg/cm^3.

Reference

R. B. Goldfarb and F. R. Fickett, U.S. Department of Commerce, National Bureau of Standards, Boulder, Colorado 80303, March 1985, NBS Special Publication 696. Superintendent of Documents, U.S. Government Printing Office, Washington, DC 20402, 1985.

CONVERSION FACTORS FOR ENERGY UNITS

If greater accuracy is required, use the Energy Equivalents section of the "CODATA Recommended Values of the Fundamental Physical Constants: 2018" table in this section. It should be noted that with the change in the SI (see "International System of Units (SI)" in this section), some conversion factors have changed very slightly from previous values in the sixth or seventh significant figure.

Unit	Wavenumber $\bar{\nu}$ cm^{-1}	Frequency ν MHz	Energy E aJ*	Energy E eV	Energy E E_h	Molar energy E_m kJ/mol	Molar energy E_m kcal/mol	Temperature T K
$\bar{\nu}$: 1 cm^{-1}	1	2.99792458×10^{4}	1.986446×10^{-5}	1.239842×10^{-4}	4.556335×10^{-6}	11.96265×10^{-3}	2.85914×10^{-3}	1.438777
ν: 1 MHz	3.33564×10^{-5}	1	6.626070×10^{-10}	4.135668×10^{-9}	1.519830×10^{-10}	3.990311×10^{-7}	9.53707×10^{-8}	4.799243×10^{-5}
E: 1 aJ*	50341.2	1.509190×10^{9}	1	6.241509	0.2293712	602.2139	143.9326	7.242971×10^{4}
E: 1 eV	8065.54	2.417989×10^{8}	0.1602177	1	3.674932×10^{-2}	96.4853	23.06054	1.160452×10^{4}
E: E_h	219474.63	6.579684×10^{9}	4.359745	27.21139	1	2625.499	627.509	3.157750×10^{5}
E_m: 1 kJ/mol	83.59350	2.506070×10^{6}	1.660540×10^{-3}	1.036427×10^{-2}	3.808800×10^{-4}	1	0.2390057	120.2724
E_m: 1 kcal/ mol	349.7552	1.048540×10^{7}	6.947700×10^{-3}	4.336412×10^{-2}	1.593602×10^{-3}	4.184	1	503.2197
T: 1 K	0.6950348	2.083662×10^{4}	1.380649×10^{-5}	8.617333×10^{-5}	3.166812×10^{-6}	8.31446×10^{-3}	1.987204×10^{-3}	1

* attoJoule (10^{-18} J)

Examples of the use of this table:

1 aJ ≙ 50341.1 cm^{-1}
1 ev ≙ 96.4853 kJ mol^{-1}

The symbol ≙ should be read as meaning corresponds to or is equivalent to:

$E = h\nu = hc\bar{\nu} = kT$
$E_m = N_A E$ where N_A is the Avogadro constant
E_h is the Hartree energy

CONVERSION FACTORS FOR PRESSURE UNITS

	Pa	kPa	MPa	bar	atm	Torr	µmHg	psi
Pa	1	0.001	0.000001	0.00001	9.8692×10^{-6}	0.0075006	7.5006	0.0001450377
kPa	1000	1	0.001	0.01	0.0098692	7.5006	7500.6	0.1450377
MPa	1000000	1000	1	10	9.8692	7500.6	7500600	145.0377
bar	100000	100	0.1	1	0.98692	750.06	750060	14.50377
atm	101325	101.325	0.101325	1.01325	1	760	760000	14.69594
Torr	133.322	0.133322	0.000133322	0.00133322	0.00131579	1	1000	0.01933672
µmHg	0.133322	0.000133322	1.33322×10^{-7}	1.33322×10^{-6}	1.31579×10^{-6}	0.001	1	1.933672×10^{-5}
psi	6894.757	6.894757	0.006894757	0.06894757	0.068046	51.7151	51715.1	1

To convert a pressure value from a unit in the left-hand column to a new unit, multiply the value by the factor appearing in the column for the new unit. For example:

1 kPa = 9.8692×10^{-3} atm
1 Torr = 1.33322×10^{-4} MPa

Notes: µmHg is often referred to as "micron"
Torr is essentially identical to mmHg
psi is an abbreviation for the unit pound–force per square inch
psia (as a term for a physical quantity) implies the true (absolute) pressure
psig implies the true pressure minus the local atmospheric pressure

DESCRIPTIVE TERMS FOR SOLUBILITY

This table lists approximate solubility as indicated by one of the following descriptive terms as defined by USP32-NF27 (Ref. 1).

Descriptive term	Parts of solvent required for 1 part of solute
Very soluble	Less than 1
Freely soluble	From 1 to 10
Soluble	From 10 to 30
Sparingly soluble	From 30 to 100
Slightly soluble	From 100 to 1000
Very slightly soluble	From 1000 to 10000
Practically insoluble, or Insoluble	Greater than or equal to 10000

Reference

1. *The Pharmacopeia of the United States of America, 41st Revision,* and *The National Formulary, 36th Edition* (General Notices and Requirements), pp. 1–12, 2009.

VALUES OF THE GAS CONSTANT IN DIFFERENT UNIT SYSTEMS

In the recent (2019) revision of the SI units that anchors the system in a set of defined constants (see this section), the value of the gas constant R is now exact, being the product of the Avogardro constant N_A and the Boltzmann constant k,

$$R = N_A \times k,$$

resulting in the exact values listed below:

$$R = 8.31446261815 \text{ Pa m}^3 \text{ K}^{-1} \text{ mol}^{-1}$$
$$= 8314.46261815 \text{ Pa L K}^{-1} \text{ mol}^{-1}$$
$$= 0.08314461815 \text{ bar L K}^{-1} \text{ mol}^{-1}$$

This table gives the appropriate value of R for use in the ideal gas equation, $PV = nRT$, when the variables are expressed in other units. The change from the previously accepted value of R is in the seventh significant figure and for all practial purposes has resulted in no changes to this table. The following conversion factors for pressure units were used in generating the table.

1 atm = 101325 Pa
1 psi = 6894.757 Pa
1 torr (mmHg) = 133.322 Pa [at 0 °C]
1 in Hg = 3386.38 Pa [at 0 °C]
1 in H_2O = 249.082 Pa [at 4 °C]
1 ft H_2O = 2988.98 Pa [at 4 °C]

Reference

Tiesinga, E., Mohr, P. J., Newell, D. B., and B. N. Taylor, "CODATA Recommended Values of the Fundamental Physical Constants: 2018," Section 1 of this *CRC Handbook.*

Values of the Gas Constant *R* When *V*, *T*, *n*, and *P* Are Expressed in the Indicated Units

V unit	*T* unit	*n* unit	*R* (*P*/psi)	*R* (*P*/kPa)	*R* (*P*/atm)	*R* (*P*/mmHg)	*R* (*P*/inHg)	*R* (P/inH_2O)	*R* (P/ftH_2O)
ft³	K	mol	0.0425863	0.2936224	0.00289784	2.20236	0.0867069	1.17881	0.0982350
ft³	K	lb·mol	19.3168	133.1849	1.31443	998.972	39.3295	534.703	44.5586
ft³	°R	mol	0.0236591	0.1631236	0.00160990	1.22353	0.0481705	0.654899	0.0545750
ft³	°R	lb·mol	10.7316	73.99159	0.730241	554.983	21.8498	297.058	24.7548
cm³	K	mol	1205.91	8314.460	82.0573	62363.7	2455.27	33380.4	2781.71
cm³	K	lb·mol	546992	3771375	37220.5	282878000	1113690	15141100	1261760
cm³	°R	mol	669.950	4619.144	45.5874	34646.5	1364.03	18544.7	1545.39
cm³	°R	lb·mol	303885	2095208	20678.1	15715400	618716	8411729	700978
L	K	mol	1.20591	8.314460	0.0820573	62.3637	2.45527	33.3804	2.78171
L	K	lb·mol	546.992	3771.376	37.2205	28287.8	1113.69	15141.1	1261.76
L	°R	mol	0.669950	4.619144	0.0455874	34.6465	1.36403	18.5447	1.54539
L	°R	lb·mol	303.885	2095.208	20.6781	15715.4	618.716	8411.72	700.978
m³	K	mol	0.00120591	0.008314460	0.0000820573	0.0623637	0.00245527	0.0333804	0.00278171
m³	K	lb·mol	0.546992	3.771376	0.0372205	28.2878	1.11369	15.1411	1.26176
m³	°R	mol	0.000669950	0.004619144	0.0000455874	0.0346465	0.00136403	0.0185447	0.00154539
m³	°R	lb·mol	0.303885	2.095208	0.0206781	15.7154	0.618716	8.41172	0.700978

DEFINITION OF AMBIENT

The term *ambient* is used within the *CRC Handbook*. When it is used, the term *ambient* means normal laboratory conditions, as defined as follows.

ambient pressure is approximately 1 atm or 101 325 Pa
ambient temperature is approximately 20 to 25 °C

In this age of digital experimentation, use of the term ambient to describe laboratory conditions is uncommon.

CONVERSION FACTORS FOR THERMAL CONDUCTIVITY UNITS

MULTIPLY ↓ by appropriate factor to OBTAIN→	**Btu_{IT} $h^{-1}\ ft^{-1}\ °F^{-1}$**	**Btu_{IT} in. $h^{-1}\ ft^{-2}\ °F^{-1}$**	**Btu_{th} $h^{-1}\ ft^{-1}\ °F^{-1}$**	**Btu_{th} in. $h^{-1}\ ft^{-2}\ °F^{-1}$**	**$cal_{IT}\ s^{-1}$ $cm^{-1}\ °C^{-1}$**	**$cal_{th}\ s^{-1}$ $cm^{-1}\ °C^{-1}$**	**$kcal_{th}\ h^{-1}$ $m^{-1}\ °C^{-1}$**	**$J\ s^{-1}$ $cm^{-1}\ K^{-1}$**	**$W\ cm^{-1}\ K^{-1}$**	**$W\ m^{-1}\ K^{-1}$**	**$mW\ cm^{-1}$ K^{-1}**
Btu_{IT} $h^{-1}\ ft^{-1}\ °F^{-1}$	1	12	1.00067	12.0080	4.13379×10^{-3}	4.13656×10^{-3}	1.48916	1.73073×10^{-2}	1.73073×10^{-2}	1.73073	17.3073
Btu_{IT} in $h^{-1}\ ft^{-2}\ °F^{-1}$	8.33333×10^{-2}	1	8.33891×10^{-2}	1.00067	3.44482×10^{-4}	3.44713×10^{-4}	0.124097	1.44228×10^{-3}	1.44228×10^{-3}	0.144228	1.44228
Btu_{th} $h^{-1}\ ft^{-1}\ °F^{-1}$	0.999331	11.9920	1	12	4.13102×10^{-3}	4.13379×10^{-3}	1.48816	1.72958×10^{-2}	1.72958×10^{-2}	1.72958	17.2958
Btu_{th} in. $h^{-1}\ ft^{-2}\ °F^{-1}$	8.32776×10^{-2}	0.999331	8.33333×10^{-2}	1	3.44252×10^{-4}	3.44482×10^{-4}	0.124014	1.44131×10^{-3}	1.44131×10^{-3}	0.144131	1.44131
cal_{IT} $s^{-1}\ cm^{-1}\ °C^{-1}$	2.41909×10^{2}	2.90291×10^{3}	2.42071×10^{2}	2.90485×10^{3}	1	1.00067	3.60241×10^{2}	4.1868	4.1868	4.1868×10^{2}	4.1868×10^{3}
cal_{th} $s^{-1}\ cm^{-1}\ °C^{-1}$	2.41747×10^{2}	2.90096×10^{3}	2.41909×10^{2}	2.90291×10^{3}	0.999331	1	3.6×10^{2}	4.184	4.184	4.184×10^{2}	4.184×10^{3}
$kcal_{th}$ $h^{-1}\ m^{-1}\ °C^{-1}$	0.671520	8.05824	0.671969	8.06363	2.77592×10^{-3}	2.77778×10^{-3}	1	1.16222×10^{-2}	1.16222×10^{-2}	1.16222	11.6222
$J\ s^{-1}\ cm^{-1}\ K^{-1}$	57.7789	6.93347×10^{2}	57.8176	6.93811×10^{2}	0.238846	0.239006	86.0421	1	1	1×10^{2}	1×10^{3}
$W\ cm^{-1}\ K^{-1}$	57.7789	6.93347×10^{2}	57.8176	6.93811×10^{2}	0.238846	0.239006	86.0421	1	1	1×10^{2}	1×10^{3}
$W\ m^{-1}\ K^{-1}$	0.577789	6.93347	0.578176	6.93811	2.38846×10^{-3}	2.39006×10^{-3}	0.860421	1×10^{-2}	1×10^{-2}	1	10
$mW\ cm^{-1}\ K^{-1}$	5.77789×10^{-2}	0.693347	5.78176×10^{-2}	0.693811	2.38846×10^{-4}	2.39006×10^{-4}	8.60421×10^{-2}	1×10^{-3}	1×10^{-3}	0.1	1

Section 2
Symbols, Terminology, and Nomenclature

Symbols and Terminology for Physical and Chemical Quantities 2-1
Expression of Uncertainty of Measurements..2-13
Nomenclature for Chemical Compounds..2-15
Nomenclature of Inorganic Chemistry...2-16
Representation of Chemical Structures with the IUPAC International Chemical Identifier (InChI)..2-22
Scientific Abbreviations, Acronyms, and Symbols.......................................2-24
Thermodynamic Functions and Relations 2-37
Nobel Laureates in Chemistry and Physics 2-38

SYMBOLS AND TERMINOLOGY FOR PHYSICAL AND CHEMICAL QUANTITIES

The International Organization for Standardization (ISO), International Union of Pure and Applied Chemistry (IUPAC), and the International Union of Pure and Applied Physics (IUPAP) have jointly developed a set of recommended symbols for physical and chemical quantities. Consistent use of these recommended symbols helps assure unambiguous scientific communication. The list below is reprinted from Ref. 1 with permission from IUPAC; this list includes updates from the Third Edition of Ref. 1. Full details may be found in the following references:

1. Cohen, E. R. et al, Ed., *Quantities, Units, and Symbols in Physical Chemistry, IUPAC Green Book, Third Edition,* RSC Publishing, Cambridge, UK, 2007.
2. Cohen, E. R., and Giacomo, P., *Symbols, Units, Nomenclature, and Fundamental Constants in Physics,* Document IUPAP–25, 1987; also published in *Physica* 146A, 1–68, 1987.
3. *ISO Standards Handbook 2: Units of Measurement,* International Organization of Standardization, Geneva, 1982.

General Rules

The value of a physical quantity is expressed as the product of a numerical value and a unit, e.g.:

T = 300 K
V = 26.2 cm^3
C_p = 45.3 J mol^{-1} K^{-1}

The symbol for a physical quantity is always given in italic (sloping) type, while symbols for units are given in roman type. Column headings in tables and axis labels on graphs may conveniently be written as the physical quantity symbol divided by the unit symbol, e.g.:

T/K
V/cm^3
C_p/J mol^{-1} K^{-1}

The values in the table or graph axis are then pure numbers. Subscripts to symbols for physical quantities should be italic if the subscript refers to another physical quantity or to a number, e.g.:

C_p – heat capacity at constant pressure
B_n – nth virial coefficient

Subscripts that have other meanings should be in roman type:

m_{p} – mass of the proton
E_{k} – kinetic energy

The following tables give the recommended symbols for the major classes of physical and chemical quantities. The expression in the Definition column is given as an aid in identifying the quantity but is not necessarily the complete or unique definition. The "SI unit" gives one (not necessarily unique) expression for the coherent SI unit for the quantity. Other equivalent unit expressions, including those that involve SI prefixes, may be used.

TABLE 1. Symbols, Units, and Definitions of Basic Physical Quantities

Name	Symbol	Definition	SI unit
Space and Time			
Cartesian space coordinates	x, y, z		m
spherical polar coordinates	r, θ, ϕ		m, 1, 1
generalized coordinate	q, q_i		(varies)
position vector	$\boldsymbol{r}$	$\boldsymbol{r} = x\mathbf{i} + y\mathbf{j} + z\mathbf{k}$	m
length	l		m
special symbols:			
height	h		
breadth	b		
thickness	d, δ		
distance	d		
radius	r		
diameter	d		
path length	s		
length of arc	s		
area	A, A_s, S		m^2
volume	$V, (v)$		m^3
plane angle	$\alpha, \beta, \gamma, \theta, \phi$...	$\alpha = s/r$	rad, 1
solid angle	ω, Ω	$\omega = A/r^2$	sr, 1
time	t		s
period	T	$T = t/N$	s
frequency	ν, f	$\nu = 1/T$	Hz
circular frequency, angular frequency	ω	$\omega = 2\pi\nu$	rad s^{-1}, s^{-1}
characteristic time interval, relaxation time, time constant	τ, T	$\tau = \lvert \mathrm{d}t/\mathrm{dln}x \rvert$	s
angular velocity	ω	$\omega = \mathrm{d}\phi/\mathrm{d}t$	rad s^{-1}, s^{-1}

Name	Symbol	Definition	SI unit
velocity	$\boldsymbol{v}, \boldsymbol{u}, \boldsymbol{w}, \boldsymbol{c}, \dot{\boldsymbol{r}}$	$\boldsymbol{v} = \mathrm{d}\boldsymbol{r}/\mathrm{d}t$	$\mathrm{m\ s^{-1}}$
speed	v, u, w, c	$v = \lvert\boldsymbol{v}\rvert$	$\mathrm{m\ s^{-1}}$
acceleration	$\boldsymbol{a}, (g)$	$\boldsymbol{a} = \mathrm{d}\boldsymbol{v}/\mathrm{d}t$	$\mathrm{m\ s^{-2}}$
Classical Mechanics			
mass	m		kg
reduced mass	μ	$\mu = m_1m_2/(m_1+m_2)$	kg
density, mass density	ρ	$\rho = m/V$	$\mathrm{kg\ m^{-3}}$
relative density	d	$d = \rho/\rho$	1
surface density	ρ_A, ρ_S	$\rho_A = m/A$	$\mathrm{kg\ m^{-2}}$
specific volume	v	$v = V/m = 1/\rho$	$\mathrm{m^3\ kg^{-1}}$
momentum	$\boldsymbol{p}$	$\boldsymbol{p} = m\boldsymbol{v}$	$\mathrm{kg\ m\ s^{-1}}$
angular momentum, action	$\boldsymbol{L}$	$\boldsymbol{L}=\boldsymbol{r}\times\boldsymbol{p}$	J s
moment of inertia	I, J	$I = \Sigma m_i r_i^2$	$\mathrm{kg\ m^2}$
force	$\boldsymbol{F}$	$\boldsymbol{F} = \mathrm{d}\boldsymbol{p}/\mathrm{d}t = m\boldsymbol{a}$	N
torque, moment of a force	$\boldsymbol{T}, (\boldsymbol{M})$	$\boldsymbol{T}=\boldsymbol{r}\times\boldsymbol{F}$	N m
energy	E		J
potential energy	E_p, V, Φ	$E_p = \int \boldsymbol{F}\cdot ds$	J
kinetic energy	E_k, T, K	$E_k = 1/2mv^2$	J
work	W, w	$W = \int \boldsymbol{F}\cdot ds$	J
Hamilton function	H	$H(q, p) = T(q, p) + V(q)$	J
Lagrange function	L	$L(q, \dot{q}) = T(q, \dot{q}) - V(q)$	J
pressure	p, P	$p = F/A$	Pa, $\mathrm{N\ m^{-2}}$
surface tension	γ, σ	$\gamma = \mathrm{d}W/\mathrm{d}A$	$\mathrm{N\ m^{-1}}$, $\mathrm{J\ m^{-2}}$
weight	$G, (W, P)$	$G = mg$	N
gravitational constant	G	$F = Gm_1m_2/r^2$	$\mathrm{N\ m^2\ kg^{-2}}$
normal stress	σ	$\sigma = F/A$	Pa
shear stress	τ	$\tau = F/A$	Pa
linear strain, relative elongation	ε, e	$\varepsilon = \Delta l/l$	1
modulus of elasticity, Young's modulus	E	$E = \sigma/\varepsilon$	Pa
shear strain	γ	$\gamma = \Delta x/d$	1
shear modulus	G	$G = \tau/\gamma$	Pa
volume strain, bulk strain	θ	$\theta = \Delta V/V_0$	1
bulk modulus, compression modulus	K	$K = -V_0(\mathrm{d}p/\mathrm{d}V)$	Pa
viscosity, dynamic viscosity	η, μ	$\tau_{x,z} = \eta(\mathrm{d}v_x/\mathrm{d}z)$	Pa s
fluidity	ϕ	$\phi = 1/\eta$	$\mathrm{m\ kg^{-1}\ s}$
kinematic viscosity	ν	$\nu = \eta/\rho$	$\mathrm{m^2\ s^{-1}}$
friction coefficient	$\mu, (f)$	$F_{frict} = \mu F_{norm}$	1
power	P	$P = \mathrm{d}W/\mathrm{d}t$	W
sound energy flux	P, P_a	$P = \mathrm{d}E/\mathrm{d}t$	W
acoustic factors:			
reflection factor	ρ	$\rho = P_r/P_0$	1
acoustic absorption factor	$\alpha_a, (\alpha)$	$\alpha_a = 1 - \rho$	1
transmission factor	τ	$\tau = P_{tr}/P_0$	1
dissipation factor	δ	$\delta = \alpha_a - \tau$	1
Electricity and Magnetism			
quantity of electricity, electric charge	Q		C
charge density	ρ	$\rho = Q/V$	$\mathrm{C\ m^{-3}}$
surface charge density	σ	$\sigma = Q/A$	$\mathrm{C\ m^{-2}}$
electric potential	V, ϕ	$V = \mathrm{d}W/\mathrm{d}Q$	V, $\mathrm{J\ C^{-1}}$
electric potential difference	$U, \Delta V, \Delta\phi$	$U = V_2 - V_1$	V
electromotive force	E	$E = \int(\boldsymbol{F}/Q)\cdot ds$	V
electric field strength	$\boldsymbol{E}$	$\boldsymbol{E} = \boldsymbol{F}/Q = -\mathrm{grad}\ V$	$\mathrm{V\ m^{-1}}$
electric flux	Ψ	$\Psi = \int \boldsymbol{D}\cdot \mathrm{d}\boldsymbol{A}$	C
electric displacement	$\boldsymbol{D}$	$\boldsymbol{D} = \varepsilon\boldsymbol{E}$	$\mathrm{C\ m^{-2}}$
capacitance	C	$C = Q/U$	F, $\mathrm{C\ V^{-1}}$
permittivity	ε	$\boldsymbol{D} = \varepsilon\boldsymbol{E}$	$\mathrm{F\ m^{-1}}$
permittivity of vacuum	ε_0	$\varepsilon_0 = \mu_0^{-1}c_0^{-2}$	$\mathrm{F\ m^{-1}}$
relative permittivity	ε_r	$\varepsilon_r = \varepsilon/\varepsilon_0$	1
dielectric polarization (dipole moment per volume)	$\boldsymbol{P}$	$\boldsymbol{P} = \boldsymbol{D} - \varepsilon_0\boldsymbol{E}$	$\mathrm{C\ m^{-2}}$

Name	Symbol	Definition	SI unit
electric susceptibility	χ_e	$\chi_e = \varepsilon r - 1$	1
electric dipole moment	$\boldsymbol{p},\boldsymbol{\mu}$	$\boldsymbol{p} = Q\boldsymbol{r}$	C m
electric current	I	$I = \mathrm{d}Q/\mathrm{d}t$	A
electric current density	$\boldsymbol{j},\boldsymbol{J}$	$I = \int \boldsymbol{j}\cdot \mathrm{d}\boldsymbol{A}$	A m^{-2}
magnetic flux density, magnetic induction	$\boldsymbol{B}$	$\boldsymbol{F} = Q\boldsymbol{v}\times\boldsymbol{B}$	T
magnetic flux	Φ	$\Phi = \int \boldsymbol{B}\cdot \mathrm{d}\boldsymbol{A}$	A m^{-2}
magnetic field strength	$\boldsymbol{H}$	$\boldsymbol{B} = \mu\boldsymbol{H}$	A m^{-2}
permeability	μ	$\boldsymbol{B} = \mu\boldsymbol{H}$	N A^{-2}, H m^{-1}
permeability of vacuum	μ_0		H m^{-1}
relative permeability	μ_r	$\mu_r = \mu/\mu_0$	1
magnetization (magnetic dipole moment per volume)	$\boldsymbol{M}$	$\boldsymbol{M} = \boldsymbol{B}/\mu_0 - H$	A m^{-1}
magnetic susceptibility	$\chi, \kappa, (\chi_m)$	$\chi = \mu_r - 1$	1
molar magnetic susceptibility	χ_m	$\chi_m = V_m\chi$	m^3 mol^{-1}
magnetic dipole moment	$\boldsymbol{m},\boldsymbol{\mu}$	$E_p = -\boldsymbol{m}\cdot\boldsymbol{B}$	A m^2, J T^{-1}
electrical resistance	R	$R = U/I$	Ω
conductance	G	$G = 1/R$	S
loss angle	δ	$\delta = (\pi/2) + \phi_I - \phi_U$	1, rad
reactance	X	$X = (U/I)\sin\delta$	Ω
impedance (complex impedance)	Z	$Z = R + iX$	Ω
admittance (complex admittance)	Y	$Y = 1/Z$	S
susceptance	B	$Y = G + iB$	S
resistivity	ρ	$\rho = E/j$	Ω m
conductivity	κ, γ, σ	$\kappa = 1/\rho$	S m^{-1}
self-inductance	L	$E = -L(\mathrm{d}I/\mathrm{d}t)$	H
mutual inductance	M, L_{12}	$E_1 = L_{12}(\mathrm{d}I_2/\mathrm{d}t)$	H
magnetic vector potential	$\boldsymbol{A}$	$\boldsymbol{B} = \nabla\times\boldsymbol{A}$	Wb m^{-1}
Poynting vector	$\boldsymbol{S}$	$\boldsymbol{S} = \boldsymbol{E}\times\boldsymbol{H}$	W m^{-2}

Quantum Mechanics

Name	Symbol	Definition	SI unit
momentum operator	$\dot{\boldsymbol{p}}$	$\dot{\boldsymbol{p}} = -i\hbar\nabla$	m^{-1}J s
kinetic energy operator	$\hat{T}$	$\hat{T} = -(h^2/2m)\nabla^2$	J
Hamiltonian operator	$\hat{H}$	$\hat{H} = \hat{T} + V$	J
wave function*, state function	Ψ, ψ, ϕ	$\hat{H}\psi = E\psi$	($m^{-3/2}$)
probability density	P	$P = \psi^*\psi$	(m^{-3})
charge density of electrons	ρ	$\rho = -eP$	(C m^{-3})
probability current density	$\boldsymbol{S}$	$\boldsymbol{S} = -i\hbar(\psi^*\nabla\psi - \psi\nabla\psi^*)/2m_e$	($m^{-2}s^{-1}$)
electric current density of electrons	$\boldsymbol{j}$	$\boldsymbol{j} = -e\boldsymbol{S}$	(A m^{-2})
matrix element of operator Â	$A_{ij}, \langle i\vert\hat{A}\vert j\rangle$	$A_{ij} = \int\psi_i^*\hat{A}\psi_j\mathrm{d}\tau$	(varies)
expectation value of operator Â	$\langle A\rangle, \bar{A}$	$\langle A\rangle = \int\psi^*\hat{A}\Psi\mathrm{d}\tau$	(varies)
hermitian conjugate of Â	$\hat{A}^\dagger$	$(\hat{A}^\dagger)_{ij} = (A_{ji})^*$	(varies)
commutator of $\hat{A}$ and $\hat{B}$	$[\hat{A},\hat{B}], [\hat{A},\hat{B}]_-$	$[\hat{A},\hat{B}] = \hat{A}\hat{B} - \hat{B}\hat{A}$	(varies)
anticommutator	$[\hat{A}, \hat{B}]_+$	$[\hat{A}, \hat{B}]_+ = \hat{A}\hat{B} + \hat{B}\hat{A}$	(varies)
spin wave function*	$\alpha; \beta$		1
coulomb integral	H_{AA}	$H_{AA} = \int\psi_A^*\hat{H}\psi_A\mathrm{d}\tau$	J
resonance integral	H_{AB}	$H_{AB} = \int\psi_A^*\hat{H}\psi_B\mathrm{d}\tau$	J
overlap integral	S_{AB}	$S_{AB} = \int\psi_A^*\psi_B\mathrm{d}\tau$	1

Atoms and Molecules

Name	Symbol	Definition	SI unit
nucleon number, mass number	A		1
proton number, atomic number	Z		1
neutron number	N	$N = A - Z$	1
electron rest mass	m_e		kg
mass of atom, atomic mass	m_a, m		kg
atomic mass constant	m_u	$m_u = m_a(^{12}C)/12$	kg
mass excess	Δ	$\Delta = m_a - Am_u$	kg
elementary charge, proton charge	e		C
Planck constant	h		J s
Planck constant/2π	$\hbar$	$\hbar = h/2\pi$	J s
Bohr radius	a_0	$a_0 = 4\pi\varepsilon_0\hbar^2/m_e e^2$	m
Hartree energy	E_h	$E_h = \hbar^2/m_e a_0^2$	J
Rydberg constant	R_∞	$R_\infty = E_h/2hc$	m^{-1}
fine-structure constant	α	$\alpha = e^2/4\pi\varepsilon_0\hbar c$	1

Name	Symbol	Definition	SI unit
ionization energy	E_i		J
electron affinity	E_{ea}		J
dissociation energy	E_d, D		J
from the ground state	D_0		J
from the potential minimum	D_e		J
principal quantum number (H atom)	n	$E = -hcR/n^2$	1
angular momentum quantum numbers	see under Spectroscopy		
magnetic dipole moment of a molecule	$\boldsymbol{m}$, $\boldsymbol{\mu}$	$E_p = -\boldsymbol{m} \cdot \boldsymbol{B}$	$J\,T^{-1}$
magnetizability of a molecule	ξ	$\boldsymbol{m} = \xi B$	$J\,T^{-2}$
Bohr magneton	μ_B	$\mu_B = e\hbar/2m_e$	$J\,T^{-1}$
nuclear magneton	μ_N	$\mu_N = (m_e/m_p)\mu_B$	$J\,T^{-1}$
magnetogyric ratio (gyromagnetic ratio)	γ	$\gamma = \mu/L$	$C\,kg^{-1}$
g-factor	g		1
Larmor circular frequency	ω_L	$\omega_L = (e/2m)B$	s^{-1}
Larmor frequency	ν_L	$\nu_L = \omega_L/2\pi$	Hz
longitudinal relaxation time	T_1		s
transverse relaxation time	T_2		s
electric dipole moment of a molecule	$\boldsymbol{p}$, $\boldsymbol{\mu}$	$E_p = -\boldsymbol{p} \cdot \boldsymbol{E}$	C m
quadrupole moment of a molecule	$\boldsymbol{Q}$; $\boldsymbol{\Theta}$	$E_p = 1/2\boldsymbol{Q}{:}V'' = 1/3\boldsymbol{\Theta}{:}V''$	$C\,m^2$
quadrupole moment of a nucleus	eQ	$eQ = 2 \cdot \langle \Theta_{ZZ} \rangle$	$C\,m^2$
electric field gradient tensor	$\boldsymbol{q}$	$q_{\alpha\beta} = -\partial^2 V/\partial\alpha\partial\beta$	$V\,m^{-2}$
quadrupole interaction energy tensor	$\boldsymbol{\chi}$	$\chi_{\alpha\beta} = eQq_{\alpha\beta}$	J
electric polarizability of a molecule	α	p (induced) $= \alpha E$	$C\,m^2\,V^{-1}$
activity (of a radioactive substance)	A	$A = -dN_B/dt$	Bq
decay (rate) constant, disintegration (rate) constant	λ	$A = \gamma N_B$	s^{-1}
half-life**	$t_{1/2}$, $T_{1/2}$		s
mean life	τ		s
level width	Γ	$\Gamma = \hbar/\tau$	J
disintegration energy	Q		J
cross section (of a nuclear reaction)	σ		m^2
Spectroscopy			
total term	T	$T = E_{tot}/hc$	m^{-1}
transition wavenumber	$\tilde{\nu}$, (ν)	$\tilde{\nu} = T' - T''$	m^{-1}
transition frequency	ν	$\nu = (E' - E'')/h$	Hz
electronic term	T_e	$T_e = E_e/hc$	m^{-1}
vibrational term	G	$G = E_{vib}/hc$	m^{-1}
rotational term	F	$F = E_{rot}/hc$	m^{-1}
spin orbit coupling constant	A	$T_{s.o.} = A\langle \hat{\boldsymbol{L}} \cdot \hat{\boldsymbol{S}} \rangle$	m^{-1}
principal moments of inertia	I_A; I_B; I_C	$I_A \le I_B \le I_C$	$kg\,m^2$
rotational constants, in wavenumber	$\tilde{A}$; $\tilde{B}$; $\tilde{C}$	$\tilde{A} = h/8\pi^2 cI_A$	m^{-1}
rotational constants, in frequency	A; B; C	$A = h/8\pi^2 I_A$	Hz
inertial defect	Δ	$\Delta = I_C - I_A - I_B$	$kg\,m^2$
asymmetry parameter	κ	$\kappa = \dfrac{(2B - A - C)}{(A - C)}$	1
centrifugal distortion constants:			
S reduction	D_J; D_{JK}; D_K; d_1; d_2		m^{-1}
A reduction	Δ_J; Δ_{JK}; Δ_K; δ_J; δ_K		m^{-1}
harmonic vibration wavenumber	ω_e; ω_r		m^{-1}
vibrational anharmonicity constant	$\omega_e x_e$; x_{rs}; $g_{u'}$		m^{-1}
vibrational quantum numbers	v_r; l_t		1
Coriolis zeta constant	ζ_{rs}^{α}		1
angular momentum quantum numbers	see additional information below		
degeneracy, statistical weight	g, d, β		1
electric dipole moment of a molecule	$\boldsymbol{p}$, $\boldsymbol{\mu}$	$E_p = -\boldsymbol{p} \cdot \boldsymbol{E}$	C m
transition dipole moment of a molecule	$\boldsymbol{M}$, $\boldsymbol{R}$	$\boldsymbol{M} = \int \psi' \boldsymbol{p} \psi'' d\tau$	C m
molecular geometry, interatomic distances:			
equilibrium distance	r_e		m

Name	Symbol	Definition	SI unit
zero–point average distance	r_z		m
ground-state distance	r_0		m
substitution structure distance	r_s		m
vibrational coordinates:			
internal coordinates	R_i, r_i, θ_j, etc.		(varies)
symmetry coordinates	S_i		(varies)
normal coordinates:			
mass adjusted	Q_r		$kg^{1/2}$ m
dimensionless	q_r		1
vibrational force constants:			
diatomic	f, (k)	$f = \partial^2 V/\partial r^2$	J m^{-2}
polyatomic			
internal coordinates	f_{ij}	$f_{ij} = \partial^2 V/\partial r_i \partial r_j$	(varies)
symmetry coordinates	F_{ij}	$F_{ij} = \partial^2 V/\partial S_i \partial S_j$	(varies)
dimensionless normal coordinates	$\varphi_{rst...}$, $k_{rst...}$		m^{-1}
nuclear magnetic resonance (NMR):			
magnetogyric ratio	γ	$\gamma = \mu/I\hbar$	C kg^{-1}
shielding constant	σ_A	$B_A = (1 - \sigma_A)B$	1
chemical shift, δ scale	δ	$\delta = 10^6(\nu - \nu_0)/\nu_0$	1
(indirect) spin–spin coupling constant	J_{AB}	$\hat{H}/h = J_{AB}\hat{I}_A \cdot \hat{I}_B$	Hz
direct (dipolar) coupling constant	D_{AB}		Hz
longitudinal relaxation time	T_1		s
transverse relaxation time	T_2		s
electron spin resonance, electron paramagnetic resonance (ESR, EPR):			
magnetogyric ratio	γ	$\gamma = \mu/s\hbar$	C kg^{-1}
g-factor	g	$h\nu = g\mu_B B$	1
hyperfine coupling constant			
in liquids	a, A	$\hat{H}_{hfs}/h = a\hat{S} \cdot \hat{I}$	Hz
in solids	T	$\hat{H}_{hfs}/h = \hat{S} \cdot T \cdot \hat{I}$	Hz

* IUPAC recommends "wavefunction."
** IUPAC recommends "half-life."

TABLE 2. Quantum Numbers Describing Angular Momentum

Angular momentum	Operator symbol	Quantum number (total)	Quantum number (Z–axis)	Quantum number (z-axis)
electron orbital	$\hat{L}$	L	M_L	Λ
one electron only	$\hat{l}$	l	m	λ
electron spin	$\hat{S}$	S	M_S	Σ
one electron only	$\hat{s}$	s	m_s	σ
electron orbital + spin	$\hat{L} + \hat{S}$			$\Omega = \Lambda + \Sigma$
nuclear orbital (rotational)	$\hat{R}$	R		K_R, k_R
nuclear spin	$\hat{I}$	I	M_I	
internal vibrational				
spherical top	$\hat{l}$	$l(l\zeta)$		K_l
other	$\hat{j}$, $\hat{\pi}$			$l(l\zeta)$
sum of $R+L(+j)$	$\hat{N}$	N		K, k
sum of $N+S$	$\hat{J}$	J	M_J	K, k
sum of $J+I$	$\hat{F}$	F	M_F	

TABLE 3. Symbols and Terminology in Physical Chemistry

Name	Symbol	Definition	SI unit
Electromagnetic Radiation			
wavelength	λ		m
speed of light			
in vacuum	c_0		m s^{-1}
in a medium	c	$c = c_0/n$	m s^{-1}
wavenumber in vacuum	$\tilde{\nu}$	$\tilde{\nu} = \nu/c_0 = 1/n\lambda$	m^{-1}
wavenumber (in a medium)	σ	$\sigma = 1/\lambda$	m^{-1}
frequency	ν	$\nu = c/\lambda$	Hz
circular frequency, pulsatance	ω	$\omega = 2\pi\nu$	s^{-1}, rad s^{-1}
refractive index	n	$n = c_0/c$	1
Planck constant	h		J s
Planck constant/2π	$\hbar$	$\hbar = h/2\pi$	J s
radiant energy	Q, W		J
radiant energy density	ρ, w	$\rho = Q/V$	J m^{-3}
spectral radiant energy density			
in terms of frequency	ρ_ν, w_ν	$\rho = d\rho/d\nu$	J m^{-3} Hz^{-1}
in terms of wavenumber	$\rho_{\tilde{\nu}}$, $\omega_{\tilde{\nu}}$	$\rho_{\tilde{\nu}} = d\rho \,/\, d\tilde{\nu}$	J m^{-2}
in terms of wavelength	ρ_λ, w_λ	$\rho_\lambda = d\rho/d\lambda$	J m^{-4}
Einstein transition probabilities,			
spontaneous emission	A_{nm}	$dN_n/dt = -A_{nm}N_n$	s^{-1}
stimulated emission	B_{nm}	$dN_n/dt = -\Sigma\, \rho\tilde{\nu}(\tilde{\nu}_{nm}) \times B_{nm}N_n$	s kg^{-1}
stimulated absorption	B_{mn}	$dN_n/dt = -\Sigma\rho\tilde{\nu}(\tilde{\nu}_{nm})B_{mn}N_m$	s kg^{-1}
radiant power, radiant energy per time	Φ, P	$\Phi = dQ/dt$	W
radiant intensity	I	$I = d\Phi/d\Omega$	W sr^{-1}
radiant exitance (emitted radiant flux)	M	$M = d\Phi/dA_{\text{source}}$	W m^{-2}
irradiance (radiant flux received)	E, (I)	$E = d\Phi/dA$	W m^{-2}
emittance	ε	$\varepsilon = M/M_{\text{bb}}$	1
Stefan–Boltzmann constant	σ	$M_{\text{bb}} = \sigma T^4$	W m^{-2} K^{-4}
first radiation constant	c_1	$c_1 = 2\pi hc_0^2$	W m^2
second radiation constant	c_2	$c_2 = hc_0/k$	K m
transmittance, transmission factor	τ, T	$\tau = \Phi_{\text{tr}}/\Phi_0$	1
absorptance, absorption factor	α	$\alpha = \Phi_{\text{abs}}/\Phi_0$	1
reflectance, reflection factor	ρ	$\rho = \Phi_{\text{refl}}/\Phi_0$	1
(decadic) absorbance	A	$A = -\lg(1-\alpha_i)$	1
Napierian absorbance	B	$B = -\ln(1-\alpha_i)$	1
absorption coefficient,			
(linear) decadic	a, K	$a = A/l$	m^{-1}
(linear) napierian	α	$\alpha = B/l$	m^{-1}
molar (decadic)	ε	$\varepsilon = a/c = A/cl$	m^2 mol^{-1}
molar napierian	κ	$\kappa = \alpha/c = B/cl$	m^2 mol^{-1}
absorption index	k	$k = \alpha/4\pi$[MML-7.xml]	1
complex refractive index	$\tilde{n}$	$\tilde{n} = n + ik$	1
molar refraction	R, R_m	$R = \frac{(n^2-1)}{(n^2+2)}V_m$	m^3 mol^{-1}
angle of optical rotation	α		1, rad
Solid State			
lattice vector	$\mathbf{R}$, $\mathbf{R}_0$		m
fundamental translation vectors for the crystal lattice	$\mathbf{a}_1$; $\mathbf{a}_2$; $\mathbf{a}_3$, $\mathbf{a}$; $\mathbf{b}$; $\mathbf{c}$	$\mathbf{R} = n_1\mathbf{a}_1 + n_2\mathbf{a}_2 + n_3\mathbf{a}_3$	m
(circular) reciprocal lattice vector	$\mathbf{G}$	$\mathbf{G}\cdot\mathbf{R} = 2\pi m$	m^{-1}
(circular) fundamental translation vectors for the reciprocal lattice	$\mathbf{b}_1$; $\mathbf{b}_2$; $\mathbf{b}_3$, $\mathbf{a}^*$; $\mathbf{b}^*$; $\mathbf{c}^*$	$\mathbf{ai}\cdot\mathbf{bk} = 2\pi\delta_{ik}$	m^{-1}
lattice plane spacing	d		m
Bragg angle	θ	$n\lambda = 2d \sin\theta$	1, rad
order of reflection	n		1
order parameters			
short range	σ		1
long range	s		1
Burgers vector	$\mathbf{b}$		m

Name	Symbol	Definition	SI unit
particle position vector	$\boldsymbol{r}, \boldsymbol{R}_j$		m
equilibrium position vector of an ion	$\boldsymbol{R}_0$		m
displacement vector of an ion	$\boldsymbol{u}$	$\boldsymbol{u} = \boldsymbol{R} - \boldsymbol{R}_0$	m
Debye–Waller factor	B, D		1
Debye circular wavenumber	q_D		m^{-1}
Debye circular frequency	ω_D		s^{-1}
Grüneisen parameter	γ, Γ	$\gamma = \alpha V/\kappa C_V$	1
Madelung constant	α, M	$E_{coul} = \dfrac{\alpha N_A z_+ z_- e^2}{4\pi\varepsilon_0 R_0}$	1
density of states	N_E	$N_E = dN(E)/dE$	$J^{-1}\ m^{-3}$
(spectral) density of vibrational modes	N_ω, g	$N_\omega = dN(\omega)/d\omega$	$s\ m^{-3}$
resistivity tensor	ρ_{ik}	$\boldsymbol{E} = \boldsymbol{\rho} \cdot \boldsymbol{j}$	$\Omega\ m$
conductivity tensor	σ_{ik}	$\sigma = \rho^{-1}$	$S\ m^{-1}$
thermal conductivity tensor	λ_{ik}	$J_q = -\boldsymbol{\lambda} \cdot \text{grad}\ T$	$W\ m^{-1}\ K^{-1}$
residual resistivity	ρ_R		$\Omega\ m$
relaxation time	τ	$\tau = l/v_F$	s
Lorenz coefficient	L	$L = \lambda/\sigma T$	$V^2\ K^{-2}$
Hall coefficient	$A_H, \boldsymbol{R}_H$	$\boldsymbol{E} = \boldsymbol{\rho} \cdot \boldsymbol{j} + R_H (\boldsymbol{B} \times \boldsymbol{j})$	$m^3\ C^{-1}$
thermoelectric force	E		V
Peltier coefficient	Π		V
Thomson coefficient	$\mu, (\tau)$		$V\ K^{-1}$
work function	Φ	$\Phi = E_\infty - E_F$	J
number density, number concentration	$n, (p)$		m^{-3}
gap energy	E_g		J
donor ionization energy	E_d		J
acceptor ionization energy	E_a		J
Fermi energy	E_F, ε_F		J
circular wave vector, propagation vector	$k, \boldsymbol{q}$	$k = 2\pi/\lambda$	m^{-1}
Bloch function	$u_k(\boldsymbol{r})$	$\psi(\boldsymbol{r}) = u_k(\boldsymbol{r})\exp(i\boldsymbol{k}\cdot\boldsymbol{r})$	$m^{-3/2}$
charge density of electrons	ρ	$\rho(\boldsymbol{r}) = -e\psi^*(\boldsymbol{r})\psi(\boldsymbol{r})$	$C\ m^{-3}$
effective mass	m^*		kg
mobility	μ	$\mu = v_{drift}/E$	$m^2\ V^{-1}\ s^{-1}$
mobility ratio	b	$b = \mu_n/\mu_p$	1
diffusion coefficient	D	$dN/dt = -DA(dn/dx)$	$m^2\ s^{-1}$
diffusion length	L	$L = \sqrt{D\tau}$	m
characteristic (Weiss) temperature	θ, θ_W		K
Curie temperature	T_C		K
Néel temperature	T_N		K
Statistical Thermodynamics			
number of entities	N		1
number density of entities, number concentration	n, C	$n = N/V$	m^{-3}
Avogadro constant	L, N_A		mol^{-1}
Boltzmann constant	k, k_B		$J\ K^{-1}$
gas constant (molar)	R	$R = Lk$	$J\ K^{-1}\ mol^{-1}$
molecular position vector	$\boldsymbol{r}(x, y, z)$		m
molecular velocity vector	$\boldsymbol{c}(c_x, c_y, c_z), \boldsymbol{u}(u_x, u_y, u_z)$	$\boldsymbol{c} = d\boldsymbol{r}/dt$	$m\ s^{-1}$
molecular momentum vector	$\boldsymbol{p}(p_x, p_y, p_z)$	$\boldsymbol{p} = m\boldsymbol{c}$	$kg\ m\ s^{-1}$
velocity distribution function (Maxwell)	$f(c_x)$	$f(c_x) = (m/2\pi kT)^{1/2} \times \exp(-mc_x^2/2kT)$	$m^{-1}\ s$
speed distribution function (Maxwell–Boltzmann)	$F(c)$	$F(c) = (m/2\pi kT)^{3/2} \times 4\pi c^2 \exp(-mc^2/2kT)$	$m^{-1}\ s$
average speed	$\bar{c}, \bar{u}, \langle c \rangle, \langle u \rangle$	$\bar{c} = \int cF(c)dc$	$m\ s^{-1}$
generalized coordinate	q		(m)
generalized momentum	p	$p = \partial L/\partial \dot{q}$	$(kg\ m\ s^{-1})$
volume in phase space	Ω	$\Omega = (1/h)\int p dq$	1
probability	P		1
statistical weight, degeneracy	g, d, W, ω, β		1
density of states	$\rho(E)$	$\rho(E) = dN/dE$	J^{-1}
partition function, sum over states,			
for a single molecule	q, z	$q = \sum_i g_i \exp(-\varepsilon_i / kT)$	1
for a canonical ensemble (system, or assembly)	Q, Z		1

Name	Symbol	Definition	SI unit
microcanonical ensemble	Ω		1
grand (canonical ensemble)	Ξ		1
symmetry number	σ, s		1
reciprocal temperature parameter	β	$\beta = 1/kT$	J^{-1}
characteristic temperature	Θ		K
General Chemistry			
number of entities (e.g. molecules, atoms, ions, formula units)	N		1
amount (of substance)	n	$n_B = N_B/L$	mol
Avogadro constant	L, N_A		mol^{-1}
mass of atom, atomic mass	m_a, m		kg
mass of entity (molecule, or formula unit)	m_f, m		kg
atomic mass constant	m_u	$m_u = m_a(^{12}C)/12$	kg
molar mass	M	$M_B = m/n_B$	kg mol^{-1}
relative molecular mass (relative molar mass, molecular weight)	M_r	$M_{r,B} = m_B/m_u$	1
molar volume	V_m	$V_{m,B} = V/n_B$	$m^3 mol^{-1}$
mass fraction	w	$w_B = m_B/\Sigma m_i$	1
volume fraction	ϕ	$\phi_B = V_B/\Sigma V_i$	1
mole fraction, amount fraction, number fraction	x, y	$x_B = n_B/\Sigma n_i$	1
(total) pressure	p, P		Pa
partial pressure	p_B	$p_B = y_B p$	Pa
mass concentration (mass density)	γ, ρ	$\gamma_B = m_B/V$	kg m^{-3}
number concentration, number density of entities	C, n	$C_B = N_B/V$	m^{-3}
amount concentration, concentration	c	$c_B = n_B/V$	mol m^{-3}
solubility	s	$s_B = c_B$ (saturated solution)	mol m^{-3}
molality (of a solute)	m, (b)	$m_B = n_B/m_A$	mol kg^{-1}
surface concentration	Γ	$\Gamma_B = n_B/A$	mol m^{-2}
stoichiometric number	ν		1
extent of reaction, advancement	ξ	$\Delta\xi = \Delta n_B/\nu_B$	mol
degree of dissociation	α		1
Chemical Thermodynamics			
heat	q, Q		J
work	w, W		J
internal energy	U	$\Delta U = q + w$	J
enthalpy	H	$H = U + pV$	J
thermodynamic temperature	T		K
Celsius temperature	θ, t	$\theta/°C = T/K - 273.15$	°C
entropy	S	$dS \geq dq/T$	J K^{-1}
Helmholtz energy (Helmholtz function)	A	$A = U - TS$	J
Gibbs energy (Gibbs function)	G	$G = H - TS$	J
Massieu function	J	$J = -A/T$	J K^{-1}
Planck function	Y	$Y = -G/T$	J K^{-1}
surface tension	γ, σ	$\gamma = (\partial G/\partial A_s)_{T,p}$	J m^{-2}, N m^{-1}
molar quantity X	X_m	$X_m = X/n$	(varies)
specific quantity X	x	$x = X/m$	(varies)
pressure coefficient	β	$\beta = (\partial p/\partial T)v$	Pa K^{-1}
relative pressure coefficient	α_p	$\alpha_p = (1/p)(\partial p/\partial T)_V$	K^{-1}
compressibility,			
isothermal	κ_T	$\kappa_T = -(1/V)(\partial V/\partial p)_T$	Pa^{-1}
isentropic	κ_S	$\kappa_S = -(1/V)(\partial V/\partial p)_S$	Pa^{-1}
linear expansion coefficient	α_l	$\alpha_l = (1/l)(\partial l/\partial T)$	K^{-1}
cubic expansion coefficient	α, α_V, γ	$\alpha = (1/V)(\partial V/\partial T)_p$	K^{-1}
heat capacity,			
at constant pressure	C_p	$C_p = (\partial H/\partial T)_p$	J K^{-1}
at constant volume	C_V	$C_V = (\partial U/\partial T)_V$	J K^{-1}
ratio of heat capacities	γ, (κ)	$\gamma = C_p/C_V$	1
Joule–Thomson coefficient	μ, μ_{JT}	$\mu = (\partial T/\partial p)_H$	K Pa^{-1}
second virial coefficient	B	$pV_m = RT(1 + B/V_m + \ldots)$	m^3 mol^{-1}
compression factor (compressibility factor)	Z	$Z = pV_m/RT$	1
partial molar quantity X	X_B, (X'_B)	$X_B = (\partial X/\partial n_B)_{T,p,nj \neq B}$	(varies)
chemical potential (partial molar Gibbs energy)	μ	$\mu_B = (\partial G/\partial n_B)_{T,p,nj \neq B}$	J mol^{-1}

Name	Symbol	Definition	SI unit
absolute activity	λ	$\lambda_B = \exp(\mu_B/RT)$	1
standard chemical potential	$\mu^\ominus, \mu^\circ$		J mol⁻¹
standard partial molar enthalpy	$H_B^\ominus$	$H_B^\ominus = \mu_B^\ominus + TS_B^\ominus$	J mol⁻¹
standard partial molar entropy	$S_B^\ominus$	$S_B^\ominus = -(\partial\mu_B^\ominus/\partial T)_p$	J mol⁻¹ K⁻¹
standard reaction Gibbs energy (function)	$\Delta_r G^\ominus$	$\Delta_r G^\ominus = \sum_B \nu_B \mu_B^\ominus$	J mol⁻¹
affinity of reaction	A, $(\mathcal{A})$	$A = -(\partial G/\partial\xi)_{p,T} = -\sum_B \nu_B\mu_B$	J mol⁻¹
standard reaction enthalpy	$\Delta_r H^\ominus$	$\Delta_r H^\ominus = \sum_B \nu_B H_B^\ominus$	J mol⁻¹
standard reaction entropy	$\Delta_r S^\ominus$	$\Delta_r S^\ominus = \sum_B \nu_B S_B^\ominus$	J mol⁻¹ K⁻¹
equilibrium constant	$K^\ominus, K$	$K^\ominus = \exp(-\Delta_r G^\ominus/RT)$	1
equilibrium constant,			
pressure basis	K_p	$K_p = \prod_B p_B^{\nu_B}$	$Pa^{\Sigma\nu}$
concentration basis	K_c	$K_c = \prod_B c_B^{\nu_B}$	$(mol\ m^{-3})^{\Sigma\nu}$
molality basis	K_m	$K_m = \prod_B m_B^{\nu_B}$	$(mol\ kg^{-1})^{\Sigma\nu}$
fugacity	$f, \tilde{p}$	$f_B = \lambda_B \lim_{p\to 0}(p_B/\lambda_B)_T$	Pa
fugacity coefficient	ϕ	$\phi_B = f_B/p_B$	1
activity and activity coefficient referenced to Raoult's law, (relative) activity	a	$a_B = \exp\left[\frac{\mu_B - \mu_B^\varnothing}{RT}\right]$	1
activity coefficient	f	$f_B = a_B/x_B$	1
activities and activity coefficients referenced to Henry's law, (relative) activity,			
molality basis	a_m	$a_{m,B} = \exp\left[\frac{\mu_B - \mu_B^\varnothing}{RT}\right]$	1
concentration basis	a_c	$a_{c,B} = \exp\left[\frac{\mu_B - \mu_B^\varnothing}{RT}\right]$	1
mole fraction basis	a_x	$a_{x,B} = \exp\left[\frac{\mu_B - \mu_B^\varnothing}{RT}\right]$	1
activity coefficient,			
molality basis	γ_m	$a_{m,B} = \gamma_{m,B} m_B/m^\ominus$	1
concentration basis	γ_c	$a_{c,B} = \gamma_{c,B} c_B/c^\ominus$	1
mole fraction basis	γ_x	$a_{x,B} = \gamma_{x,B} x_B$	1
ionic strength,			
molality basis	I_m, I	$I_m = \frac{1}{2}\Sigma m_B z_B^2$	mol kg⁻¹
concentration basis	I_c, I	$I_c = \frac{1}{2}\Sigma c_B z_B^2$	mol m⁻³
osmotic coefficient,			
molality basis	ϕ_m	$\phi_m = (\mu_A^* - \mu_A)/(RTM_A\Sigma m_B)$	1
mole fraction basis	ϕ_x	$\phi_x = (\mu_A - \mu_A^*)/(RT\ln x_A)$	1
osmotic pressure	Π	$\Pi = c_B RT$ (ideal dilute solution)	Pa

TABLE 4. Symbols Used in Kinetics, Electrochemistry, Surface Chemistry, and Transport Properties

Name	Symbol	Definition	SI unit
Chemical Kinetics			
rate of change of quantity X	$\dot{X}$	$\dot{X} = dX/dt$	(varies)
rate of conversion	$\dot{\xi}$	$\dot{\xi} = d\xi/dt$	mol s⁻¹
rate of concentration change (due to chemical reaction)	r_B, v_B	$r_B = dc_B/dt$	mol m⁻³ s⁻¹
rate of reaction (based on amount concentration)	v	$v = \dot{\xi}/V = \nu_B^{-1} dc_B/dt$	mol m⁻³ s⁻¹
partial order of reaction	n_B	$v = k\Pi c_B^{n_B}$	1

Name	Symbol	Definition	SI unit
overall order of reaction	n	$n = \Sigma n_B$	1
rate constant, rate coefficient	k	$v = k\Pi c_B^{nB}$	$(mol^{-1}\ m^3)^{n-1}\ s^{-1}$
Boltzmann constant	k, k_B		$J\ K^{-1}$
half-life	$t_{½}$	$c(t_{½}) = c_0/2$	s
relaxation time	τ	$\tau = 1/(k_1+k_{-1})$	s
energy of activation, activation energy	E_a, E	$E_a = RT^2\ d\ln k/dT$	$J\ mol^{-1}$
pre-exponential factor	A	$k = A\exp(-E_a/RT)$	$(mol^{-1}\ m^3)^{n-1}\ s^{-1}$
volume of activation	$\Delta^{\ddagger}V$	$\Delta^{\ddagger}V = -RT \times (\partial \ln k/\partial p)_T$	$m^3\ mol^{-1}$
collision diameter	d	$d_{AB} = r_A + r_B$	m
collision cross section	σ	$\sigma_{AB} = \pi d_{AB}^2$	m^2
collision frequency	Z_A		s^{-1}
collision number	Z_{AB}, Z_{AA}		$m^{-3}\ s^{-1}$
collision frequency factor	z_{AB}, z_{AA}	$z_{AB} = Z_{AB}/Lc_Ac_B$	$m^3\ mol^{-1}\ s^{-1}$
standard enthalpy of activation	$\Delta^{\ddagger}H^{\ominus}$, $\Delta H^{\ddagger}$		$J\ mol^{-1}$
standard entropy of activation	$\Delta^{\ddagger}S^{\ominus}$, $\Delta S^{\ddagger}$		$J\ mol^{-1}\ K^{-1}$
standard Gibbs energy of activation	$\Delta^{\ddagger}G^{\ominus}$, $\Delta G^{\ddagger}$		$J\ mol^{-1}$
quantum yield, photochemical yield	ϕ		1
Electrochemistry			
elementary charge (proton charge)	e		C
Faraday constant	F	$F = eL$	$C\ mol^{-1}$
charge number of an ion	z	$z_B = Q_B/e$	1
ionic strength	I_c, I	$I_c = ½\Sigma c_i z_i^2$	$mol\ m^{-3}$
mean ionic activity	$a_{\pm}$	$a_{\pm} = m_{\pm}\gamma_{\pm}/m^{\ominus}$	1
mean ionic molality	$m_{\pm}$	$m_{\pm}^{(\nu_+ + \nu_-)} = m_+^{\nu+}m_-^{\nu-}$	$mol\ kg^{-1}$
mean ionic activity coefficient	$\gamma_{\pm}$	$\gamma_{\pm}^{(\nu_+ + \nu_-)} = \gamma_+^{\nu+}\gamma_-^{\nu-}$	1
charge number of electrochemical cell reaction	n, (z)		1
electric potential difference (of a galvanic cell)	ΔV, E, U	$\Delta V = V_R - V_L$	V
emf, electromotive force	E	[MML-23.xml]	V
standard emf, standard potential of the electrochemical cell reaction	$E^{\ominus}$	$E^{\ominus} = -\Delta_r G^{\ominus}/nF = (RT/nF)\ln K^{\ominus}$	V
standard electrode potential	$E^{\ominus}$		V
emf of the cell, potential of the electrochemical cell reaction	E	$E = E^{\ominus} - (RT/nF) \times \Sigma \nu_i \ln a_i$	V
pH	pH	$pH \approx -\lg\left[\frac{c(\mathbf{H}^+)}{mol\ dm^{-3}}\right]$	1
inner electric potential	ϕ	$\nabla\phi = -E$	V
outer electric potential	ψ	$\psi = Q/4\pi\varepsilon_0 r$	V
surface electric potential	χ	$\chi = \phi - \psi$	V
Galvani potential difference	$\Delta\phi$	$\Delta_\alpha^\beta\phi = \phi^\beta - \phi^\alpha$	V
volta potential difference	$\Delta\psi$	$\Delta_\alpha^\beta\psi = \psi^\beta - \psi^\alpha$	V
electrochemical potential	$\tilde{\mu}$	$\tilde{\mu}_B^\alpha = (\partial G/\partial n_B^\alpha)$	$J\ mol^{-1}$
electric current	I	$I = dQ/dt$	A
(electric) current density	j	$j = I/A$	$A\ m^{-2}$
(surface) charge density	σ	$\sigma = Q/A$	$C\ m^{-2}$
electrode reaction rate constant	k	$k_{ox} = I_a/(nFA\prod_i c_i^{n_i})$	(varies)
mass transfer coefficient, diffusion rate constant	k_d	$k_{d,B} = \lvert\nu_B\rvert I_{l,B}/nFcA$	$m\ s^{-1}$
thickness of diffusion layer	δ	$\delta_B = D_B/k_{d,B}$	m
transfer coefficient (electrochemical)	α	$\alpha_c = \frac{-\lvert\nu\rvert RT\ \partial}{nF}\frac{\partial \ln\lvert I_c\rvert}{\partial E}$	1
overpotential	η	$\eta = E_I - E_{I=0} - IR_u$	V
electrokinetic potential (zeta potential)	ζ		V
conductivity	κ, (σ)	$\kappa = j/E$	$S\ m^{-1}$
conductivity cell constant	K_{cell}	$K_{cell} = \kappa R$	m^{-1}
molar conductivity (of an electrolyte)	Λ	$\Lambda_B = \kappa/c_B$	$S\ m^2\ mol^{-1}$
ionic conductivity, molar conductivity of an ion	λ	$\lambda_B = \lvert z_B\rvert F u_B$	$S\ m^2\ mol^{-1}$
electric mobility	u, (μ)	$u_B = v_B/E$	$m^2\ V^{-1}\ s^{-1}$
transport number	t	$t_B = j_B/\Sigma j_i$	1
reciprocal radius of ionic atmosphere	κ	$\kappa = (2F^2I/\varepsilon RT)^{½}$	m^{-1}

Name	Symbol	Definition	SI unit
Colloid and Surface Chemistry			
specific surface area	a, a_s, s	$a = A/m$	$m^2\ kg^{-1}$
surface amount of B, adsorbed amount of B	$n_B{}^s, n_B{}^a$		mol
surface excess of B	$n_B{}^\sigma$		mol
surface excess concentration of B	$\Gamma_B, (\Gamma_B{}^\sigma)$	$\Gamma_B = n_B{}^\sigma/A$	$mol\ m^{-2}$
total surface excess concentration	$\Gamma, (\Gamma^\sigma)$	[MML-27.xml]	$mol\ m^{-2}$
area per molecule	a, σ	$a_B = A/N_B{}^\sigma$	m^2
area per molecule in a filled monolayer	a_m, σ_m	$a_{m,B} = A/N_{m,B}$	m^2
surface coverage	θ	$\theta = N_B{}^\sigma/N_{m,B}$	1
contact angle	θ		1, rad
film thickness	t, h, δ		m
thickness of (surface or interfacial) layer	τ, δ, t		m
surface tension, interfacial tension	γ, σ	$\gamma = (\partial G/\partial A_s)_{T,p}$	$N\ m^{-1}$, $J\ m^{-2}$
film tension	Σ_f	$\Sigma_f = 2\gamma_f$	$N\ m^{-1}$
reciprocal thickness of the double layer	κ	$\kappa = [2F^2I_c/\varepsilon RT]^{1/2}$	m^{-1}
average molar masses			
number–average	M_n	$M_n = \Sigma n_i M_i/\Sigma n_i$	$kg\ mol^{-1}$
mass–average	M_m	$M_m = \Sigma n_i M_i^2/\Sigma n_i M_i$	$kg\ mol^{-1}$
Z–average	M_Z	$M_Z = \Sigma n_i M_i^3/\Sigma n_i M_i^2$	$kg\ mol^{-1}$
sedimentation coefficient	s	$s = v/a$	s
van der Waals constant	λ		J
retarded van der Waals constant	β, B		J
van der Waals–Hamaker constant	A_H		J
surface pressure	π^s, π	$\pi^s = \gamma^0 - \gamma$	$N\ m^{-1}$
Transport Properties			
flux (of a quantity X)	J_X, J	$J_X = A^{-1}\ dX/dt$	(varies)
volume flow rate	$q_V, \dot{V}$	$q_v = dV/dt$	$m^3\ s^{-1}$
mass flow rate	$q_m, \dot{m}$	$q_m = dm/dt$	$kg\ s^{-1}$
mass transfer coefficient	k_d		$m\ s^{-1}$
heat flow rate	ϕ	$\phi = dq/dt$	W
heat flux	J_q	$J_q = \phi/A$	$W\ m^{-2}$
thermal conductance	G	$G = \phi/\Delta T$	$W\ K^{-1}$
thermal resistance	R	$R = 1/G$	$K\ W^{-1}$
thermal conductivity	λ, k	$\lambda = J_q/(dT/dl)$	$W\ m^{-1}\ K^{-1}$
coefficient of heat transfer	$h, (k, K, \alpha)$	$h = J_q/\Delta T$	$W\ m^{-2}\ K^{-1}$
thermal diffusivity	a	$a = \lambda/\rho c_p$	$m^2\ s^{-1}$
diffusion coefficient	D	$D = J_n/(dc/dl)$	$m^2\ s^{-1}$

TABLE 5. Symbols Used to Denote a Chemical Process, Reaction, or Condition

Process	Symbol
Symbols used as subscripts	
vaporization, evaporation (liquid → gas)	vap
sublimation (solid → gas)	sub
melting, fusion (solid → liquid)	fus
transition (between two phases)	trs
mixing of fluids	mix
solution (of solute in solvent)	sol
dilution (of a solution)	dil
adsorption	ads
displacement	dpl
immersion	imm
reaction in general	r
atomization	at
combustion reaction	c
formation reaction	f
Symbols used as superscripts	
standard	$\ominus$, o
pure substance	*
infinite dilution	∞
ideal	id
activated complex, transition state	‡, ≠
excess quantity	E
apparent	app

The following symbols are used in the definitions of the dimensionless quantities: mass (m), time (t), volume (V), area (A), density (ρ), speed (v), length (l), viscosity (η), pressure (p), acceleration of free fall (g), cubic expansion coefficient (α), temperature (T), surface tension (γ), speed of sound (c), mean free path (λ), frequency (f), thermal diffusivity (a), coefficient of heat transfer (h), thermal conductivity (k), specific heat capacity at constant pressure (c_p), diffusion coefficient (D), mole fraction (x), mass transfer coefficient (k_d), permeability (μ), electric conductivity (κ), and magnetic flux density (B).

TABLE 6. Dimensionless Quantities

Name	Symbol	Definition	SI unit
Reynolds number	Re	$Re = pvl/\eta$	1
Euler number	Eu	$Eu = \Delta p/\rho v^2$	1
Froude number	Fr	$Fr = v/(lg)^{1/2}$	1
Grashof number	Gr	$Gr = l^3 g\alpha\Delta T\rho^2/\eta^2$	1
Weber number	We	$We = \rho v^2 l/\gamma$	1
Mach number	Ma	$Ma = v/c$	1
Knudsen number	Kn	$Kn = \lambda/l$	1
Strouhal number	Sr	$Sr = lf/v$	1
Fourier number	Fo	$Fo = at/l^2$	1
Péclet number	Pe	$Pe = vl/a$	1
Rayleigh number	Ra	$Ra = l^3 g\alpha\Delta T\rho/\eta a$	1
Nusselt number	Nu	$Nu = hl/k$	1
Stanton number	St	$St = h/\rho v c_p$	1
Fourier number for mass transfer	Fo^*	$Fo^* = Dt/l^2$	1
Péclet number for mass transfer	Pe^*	$Pe^* = vl/D$	1
Grashof number for mass transfer	Gr^*	$Gr^* = l^3 g\left(\frac{\partial p}{\partial x}\right)_{T,p}\left(\frac{\Delta x p}{\eta}\right)$	1
Nusselt number for mass transfer	Nu^*	$Nu^* = k_d l/D$	1
Stanton number for mass transfer	St^*	$St^* = k_d/v$	1
Prandtl number	Pr	$Pr = \eta/\rho a$	1
Schmidt number	Sc	$Sc = \eta/\rho D$	1
Lewis number	Le	$Le = a/D$	1
magnetic Reynolds number	Rm, Re_m	$Rm = v\mu\kappa l$	1
Alfvén number	Al	$Al = v(\rho\mu)^{1/2}/B$	1
Hartmann number	Ha	$Ha = Bl(\kappa/\eta)^{1/2}$	1
Cowling number	Co	$Co = B^2/\mu\rho v^2$	1

EXPRESSION OF UNCERTAINTY OF MEASUREMENTS

In general, the result of a measurement is only an approximation or estimate of the true value of the quantity subject to measurement, and thus the result is of limited value unless accompanied by a statement of its uncertainty. Much (but not all) of the scientific data appearing in the literature does include some indication of the uncertainty, but this may be stated in many different ways and is often explained poorly. In an effort to encourage consistency in uncertainty statements, the International Committee for Weights and Measures (CIPM) of the Bureau International des Poids et Mesures initiated a project, in collaboration with several other international organizations, to prepare a set of guidelines expressing international consensus on the recommended method of stating uncertainties. This project resulted in the publication of the *Guide to the Expression of Uncertainty in Measurement* (Refs. 1 and 2), which is often referred to as *GUM*. The recommendations of *GUM* have been summarized by the National Institute of Standards and Technology in NIST Technical Note 1297, *Guidelines for Evaluating the Uncertainty of NIST Measurement Results* (Ref. 3).

In the notation of *GUM*, we are concerned with the **measurand**, i.e., the quantity that is being measured. In physics and chemistry this is usually called a **physical quantity** and represents some inherent characteristic of a material, system, or process that can be expressed in numerical terms — specifically as the product of a number and a reference, commonly called a **unit.** Thus, the density of water at room temperature is (approximately) 0.998 g/mL (grams per milliliter) or, alternatively 998 kg m^{-3} (kilograms per meter cubed). This statement gives the most likely value of the measurand, to this level of precision, but gives no information on how much the stated value might differ from the true value. A more detailed discussion of measurement terminology is given in the *International Vocabulary of Metrology* (VIM) (Ref. 4).

It is important to differentiate between the terms **error** and **uncertainty**. The error in a measurement is the difference between the measured value and the true value; the error can be stated if the true value is known (to some level of accuracy). The uncertainty is an estimate of the maximum reasonable extent to which the measured value is believed to deviate from the true value, in a situation where the true value is not known (most often the case). The result of a measurement can unknowably be very close to the true value, and thus have negligible error, even though its uncertainty is large.

The uncertainty of the result of a measurement generally consists of several components, which may be grouped in two types according to the method used to estimate their numerical values:

Type A. Those which are evaluated by statistical methods
Type B. Those which are evaluated by other means

The terms "random uncertainty" and "systematic uncertainty" are often used, but these terms do not always correspond in a simple way to the A and B categories. This is because the nature of an uncertainty component is conditioned by how the quantity appears in the mathematical model that describes the current measurement process. An uncertainty component arising from a systematic effect may in some cases be evaluated by methods of Type A while in other cases by methods of Type B.

In the *GUM* formulation, each component of uncertainty, whether in the A or B category, is represented by an estimated standard deviation, termed **standard uncertainty**, symbol u_i, and equal to the positive square root of the estimated variance u_i^2.

For an uncertainty component of Type A, $u_i = s_i$, where s_i is the statistically estimated standard deviation, as determined from a series of observations by appropriate statistical analysis. Any valid statistical method may be used. Examples are calculating the standard deviation of the mean of a series of independent observations; using the method of least squares to fit a curve to data in order to estimate parameters of the curve and their standard deviations; and carrying out an analysis of variance (ANOVA) in order to identify and quantify random effects in certain types of measurements. Details of statistical analysis are given in Refs. 5–9 and many other places.

In a similar manner, each uncertainty component of Type B is represented by a quantity u_j, which is obtained from an assumed probability distribution based on all the available information about the measurement process. Because u_j is treated like a standard deviation,the standard uncertainty in each Type B component is simply u_j. The evaluation of u_j is usually based on scientific judgment using all the relevant information available, which may include

- Previous measurement data
- Experience with, or general knowledge of, the behavior and properties of relevant materials and instruments
- Manufacturer's specifications
- Data provided in calibrations and other reports
- Uncertainties assigned to reference data taken from handbooks

The specific approach to evaluating the standard uncertainty u_j of a Type B uncertainty will depend on the detailed model of the measurement process. The following are examples of steps that may be used:

1. Convert a quoted uncertainty (for example, in a calibration factor) that is a stated multiple of an estimated standard deviation to a standard uncertainty by dividing the quoted uncertainty by the multiplier.
2. Convert a quoted uncertainty that defines a "confidence interval" having a stated level of confidence, such as 95% or 99%, to a standard uncertainty by treating the quoted uncertainty as if a normal distribution had been used to calculate it (unless otherwise indicated) and dividing it by the appropriate factor for such a distribution. These factors are 1.960 and 2.576 for the two levels of confidence given.
3. Model knowledge of the quantity in question by a normal distribution and estimate lower and upper limits a_- and a_+ such that the best estimated value of the quantity is $(a_+ + a_-)/2$ (i.e., the midpoint of the limits) and there is 1 chance out of 2 (i.e., a 50% probability) that the value of the quantity lies in the interval a_- to a_+. Then $u_j \approx 1.48\ a$, where $a = (a_+ - a_-)/2$ is the half-width of the interval.
4. Model knowledge of the quantity in question by a normal distribution and estimate lower and upper limits a_- and a_+ such that the best estimated value of the quantity is $(a_+ + a_-)/2$ and there is about a 2 out of 3 chance (i.e., a 67% probability) that the value of the quantity lies in the interval a_- to a_+. Then $u_j \approx a$, where $a = (a_+ - a_-)/2$.
5. Estimate lower and upper limits a_- and a_+ for the value of the quantity in question such that the probability that the value lies in the interval a_- to a_+ is, for all practical purposes, 100%. Provided that there is no contradictory in-

formation, treat the quantity as if it is equally probable for its value to lie anywhere within the interval a_- to a_+; that is, model it by a uniform or rectangular probability distribution. The best estimate of the value of the quantity is then $(a_+ + a_-)/2$ with $u_j = a/\sqrt{3}$ where $a = (a_+ - a_-)/2$. If the distribution used to model the quantity is triangular rather than rectangular, then $u_j = a/\sqrt{6}$. The rectangular distribution is a reasonable default model in the absence of any other information. But if it is known that values of the quantity in question near the center of the limits are more likely than values close to the limits, a triangular or a normal distribution may be a better model.

When all the standard uncertainties of Type A and Type B have been determined in this way, they should be combined to produce the **combined standard uncertainty** (suggested symbol u_c), which may be regarded as the estimated standard deviation of the measurement result. This process, often called the *law of propagation of uncertainty* or "root-sum-of-squares," involves taking the square root of the sum of the squares of all the u_i. In many practical measurement situations, the probability distribution characterized by the measurement result y and its combined standard uncertainty $u_c(y)$ is approximately normal (Gaussian). When this is the case, $u_c(y)$ defines an interval $y - u_c(y)$ to $y + u_c(y)$ about the measurement result y within which the value of the measurand Y estimated by y is believed to lie with a level of confidence of approximately 68%. That is, it is believed with an approximate level of confidence of 68% that $y - u_c(y) \le Y \le y + u_c(y)$, which is commonly written as $Y = y \pm u_c(y)$.

In fundamental metrological research (involving physical constants, calibration standards, and the like) the combined standard uncertainty u_c is normally used as the statement of uncertainty in a measurement. In most cases, however, it is desirable to use a measure of uncertainty that defines an interval about the measurement result y within which the value of the measurand Y is confidently believed to lie. The measure of uncertainty intended to meet this requirement is termed **expanded uncertainty**, suggested symbol U, and is obtained by multiplying $u_c(y)$ by a **coverage factor**, suggested symbol k. Thus $U = ku_c(y)$ and it is believed with high confidence that $y - U \le Y \le y + U$, which is commonly written as $Y = y \pm U$. The value of the coverage factor k is chosen on the basis of the desired level of confidence to be associated with the interval defined by $U = ku_c$. Typically, k is in the range 2 to 3. When the normal distribution applies, $U = 2u_c$ (i.e., $k = 2$) defines an interval having a level of confidence of approximately 95%, and $U = 3u_c$ defines an interval having a confidence level greater than 99%. In current international practice it is most common to use $k = 2$, corresponding to about 95% confidence, but the value of k should be stated in each case to avoid confusion. See Refs. 1 and 3 for methods of calculating k when a value other than $k = 2$ is needed for a specific requirement.

It should be noted that the International Union of Pure and Applied Chemistry (IUPAC) is reviewing recommendations on metrological and quality concepts in analytical chemistry (Ref. 10).

Summary of Key Steps

- Group the uncertainty components into Type A (can be evaluated by statistical methods) and Type B (must be evaluated by other means).
- Determine the standard uncertainty for each component of Type A by statistical methods and for each component of Type B by other suitable methods, based on modeling the measurement process.
- Take the square root of the sum of the squares of all the standard uncertainties to get the combined standard uncertainty u_c.
- Specify a coverage factor k which, when multiplied by u_c, gives the expanded uncertainty U. In fundamental metrological research $k = 1$ is usually chosen; in other cases, $k = 2$ (corresponding to a confidence level of about 95%) is the most common choice.

References

1. *Evaluation of Measurement Data — Guide to the Uncertainty in Measurement, JCGM 100:2008*, BIPM, Sevres, 2008, <bipm.org/utils/common/documents/jcgm/JCGM_1002008_E.pdf>
2. ISO, *Guide to the Expression of Uncertainty in Measurement*, International Organization for Standardization, Geneva, Switzerland, 1993. Several supplements have been published; see Bich, W., Cox, M. C., and Harris, P. M., "Evolution of the *Guide to the Expression of Uncertainty in Measurement*," *Metrologia* 43, S161, 2006. <https://doi.org/10.1088/0026-1394/43/4/S01>
3. Taylor, B. N., and Kuyatt, C. E., *Guidelines for Evaluating and Expressing the Uncertainty of NIST Measurement Results*, NIST Technical Note 1297, National Institute of Standards and Technology, Gaithersburg, MD, 1994; available for free download at <physics.nist.gov/cuu/Uncertainty/bibliography.html>.
4. *International Vocabulary of Metrology — Basic and General Concepts and Associated Terms, JCGM 200:2012, Third Edition*, BIPM, Sevres, 2012.
5. Bell, S., *A Beginner's Guide to Uncertainty of Measurement*, National Physical Laboratory, Teddington, Middlesex, UK, 2001; available on the Internet at <www.npl.co.uk/server.php?show=ConWebDoc.1785>.
6. Eisenhart, C., "Realistic Evaluation of the Precision and Accuracy of Instrument Calibration Systems," *J. Res. Natl. Bur. Stand.* (U.S.) 67C, 161, 1963. <https://doi.org/10.6028/jres.067C.015>
7. Mandel, J., *The Statistical Analysis of Experimental Data*, Dover Publishers, New York, 1984.
8. Nantrella, M. G., *Experimental Statistics*, NBS Handbook 91, U.S. Government Printing Office, Washington, DC, 1966.
9. Box, G. E. P., Hunter, J. S., and Hunter, W. G., *Statistics for Experimenters: Design, Innovation, and Discovery, Second Edition*, John Wiley & Sons, Hoboken, NJ, 2005.
10. <https://iupac.org/recommendation/metrological-and-quality-concepts-in-analytical-chemistry/>

NOMENCLATURE FOR CHEMICAL COMPOUNDS

The International Union of Pure and Applied Chemistry (IUPAC) maintains several commissions that deal with the naming of chemical substances. In general, the approach of IUPAC is to present rules for arriving at names in a systematic manner, rather than recommending a unique name for each compound. Thus, there are often several alternative "IUPAC Names," depending on which nomenclature system is used, each of which may have advantages in specific applications. However, each of these names will be unambiguous.

Organizations such as the Chemical Abstracts Service that prepare indexes to the chemical literature must adopt a system for selecting unique names in order to avoid excessive cross referencing. Chemical Abstracts Service uses a system which groups together compounds derived from a single parent compound. Thus, most index names are inverted (e.g., Benzene, bromo rather than bromobenzene; Acetic acid, sodium salt rather than sodium acetate).

Recommended names for the most common substituent groups, ligands, ions, and organic rings are given in the two following tables, "Nomenclature for Inorganic Ions and Ligands" and "Organic Substituent Groups and Ring Systems," which are available in the Online Edition of the *CRC Handbook*. For the basics of macromolecular nomenclature, see "Nomenclature for Organic Polymers" in Section 13.

Some of the most useful recent guides to chemical nomenclature, prepared by IUPAC and other organizations such as the International Union of Biochemistry and Molecular Biology (IUBMB) and the American Chemical Society are listed below. These books contain citations to the more detailed nomenclature documents in each area. Two very useful links to nomenclature documents are:

<iupac.org/what-we-do/nomenclature/>
<www.chem.qmul.ac.uk/iupac/>

Inorganic Chemistry

1. Block, B. P., Powell, W. H., and Fernelius, W. C., *Inorganic Chemical Nomenclature, Principles and Practice*, American Chemical Society, Washington, DC, 1990.
2. Connelly, N. G., Damhus, T., Hartshorn, R. M., and Hutton, A. T., *Nomenclature of Inorganic Chemistry — IUPAC Recommendations 2005,* Royal Society of Chemistry, 2005.
3. Hartshorn, R. M., Hellwich, K. H., Yerin, A., Damhus, T., and Hutton, A. T., A Brief Guide to the Nomenclature of Inorganic Chemistry, *Pure Appl. Chem.* 87, 1039, 2015. <https://doi.org/10.1515/pac-2014-0718>

Organic Chemistry

1. Favre, H. A., and Powell, W. H., *Nomenclature of Organic Chemistry: IUPAC Recommendations and Preferred Names 2013,* Royal Society of Chemistry, 2014.
2. Hellwich, K-. H., Hartshorn, R. M., Yerin, A., Damhus, T., and Hutton, A. T., A Brief Guide to the Nomenclature of Organic Chemistry, IUPAC Technical Report, *Pure Appl. Chem.* 92, 528, 2020.
3. Moss, G. P., Smith, P. A. S., and Tavernier, D., Eds., Glossary of Class Names of Organic Compounds and Reactive Intermediates Based on Structure, *Pure Appl. Chem.* 67, 1307, 1995. <https://doi.org/10.1351/pac199567081307>
4. Moss, G. P., Ed., Basic Terminology of Stereochemistry, *Pure Appl. Chem.* 68, 2193, 1996. <https://doi.org/10.1351/pac199668122193>
5. Panico, R., Powell, W. H., and Richer, J.-C., Eds., *A Guide to IUPAC Nomenclature of Organic Compounds, Recommendations 1993,* Blackwell Scientific Publications, Oxford, 1993.
6. Cooper, C., and Purchase, R., *Organic Chemist's Desk Reference, Third Edition,* CRC Press, 2017. <https://doi.org/10.1201/9781315120768>

Macromolecular Chemistry

1. Hess, M., Jones, R. G., Kahovec, J., Kitayama, T., Metanomski, W. V., Stepto, R., and Wilks, E. S., *Compendium of Polymer Nomenclature,* RSC Publishing, 2009.
2. Hiorns, R. C. et al., A Brief Guide to Polymer Nomenclature, IUPAC Technical Report, *Pure Appl. Chem.* 84, 2167, 2012.
3. Jenkins, A. D., Kratochvil, P., Stepto, R. F. T., and Suter, U. W., Eds., Glossary of Basic Terms in Polymer Science, IUPAC Recommendation 1996, *Pure Appl. Chem.* 68, 2287, 1996. <https://doi.org/10.1351/pac199668122287>

Biochemistry

1. *Biochemical Nomenclature and Related Documents, Second Edition, 1992,* Portland Press, London, 1993; includes recommendations of the IUPAC-IUBMB Joint Commission on Biochemical Nomenclature.
2. *Enzyme Nomenclature, 1992,* Academic Press, Orlando, FL, 1992.
3. McNaught, A. D., Ed., Nomenclature of Carbohydrates, Recommendations 1996: IUPAC-IUBMB Joint Commission on Biochemical Nomenclature, *Pure Appl. Chem.* 68, 1919, 1996. <https://doi.org/10.1351/pac199668101919>
4. Cornish-Bowden, A., Current IUBMB Recommendations on Enzyme Nomenclature and Kinetics, *Perspectives in Science* 1, 74, 2014.

General

1. Leigh, G. J., *Principles of Chemical Nomenclature,* RSC Publishing, 2011.
2. Cahn, R. S. and Dermer, O. C., *Introduction to Chemical Nomenclature,* Butterworth-Heinemann, 2013.

NOMENCLATURE OF INORGANIC CHEMISTRY

R. M. Hartshorn, K.-H. Hellwich, A. Yerin, T. Damhus, and A. T. Hutton

Preamble

The universal adoption of an agreed chemical nomenclature is a key tool for communication in the chemical sciences, for computer-based searching in databases, and for regulatory purposes, such as those associated with health and safety or commercial activity. The International Union of Pure and Applied Chemistry (IUPAC) provides recommendations on the nature and use of chemical nomenclature.[1] The basics of this nomenclature are shown here, and in companion documents on the nomenclature systems for organic chemistry[2] and polymers,[3] with hyperlinks to the original documents. An overall summary of chemical nomenclature can be found in *Principles of Chemical Nomenclature*.[4] Greater detail can be found in the *Nomenclature of Inorganic Chemistry*, colloquially known as the Red Book,[5] and in the related publications for organic compounds (the Blue Book)[6] and polymers (the Purple Book).[7] It should be noted that many compounds may have non-systematic or semi-systematic names (some of which are not accepted by IUPAC for several reasons, for example, because they are ambiguous) and IUPAC rules allow for more than one systematic name in many cases. IUPAC is working towards identification of single names which are to be preferred for regulatory purposes (Preferred IUPAC Names, or PINs). *Note*: In this document, the symbol '=' is used to split names that happen to be too long for the column format, unless there is a convenient hyphen already present in the name.

The boundaries between 'organic' and 'inorganic' compounds are blurred. The nomenclature types described in this document are applicable to compounds, molecules and ions that do not contain carbon, but also to many structures that do contain carbon (Section 2), notably those containing elements of Groups 1–12. Most boron-containing compounds are treated using a special nomenclature.[8]

References

1. Freely available at: (a) http://www.degruyter.com/pac; (b) http://www.chem.qmul.ac.uk/iupac/.
2. Hellwich, K-. H., Hartshorn, R. M., Yerin, A., Damhus, T., and Hutton, A. T., A Brief Guide to the Nomenclature of Organic Chemistry, IUPAC Technical Report, *Pure Appl. Chem.* 92, 528, 2020.
3. R. C. Hiorns, R. J. Boucher, R. Duhlev, K.-H. Hellwich, P. Hodge, A. D. Jenkins, R. G. Jones, J. Kahovec, G. Moad, C. K. Ober, D. W. Smith, R. F. T. Stepto, J.-P. Vairon, J. Vohlídal, *Pure Appl. Chem.* 84(10), 2167–2169 (2012) <https://doi.org/10.1351/PAC-REP-12-03-05>.
4. *Principles of Chemical Nomenclature – A Guide to IUPAC Recommendations, 2011 Edition*, G. J. Leigh (Ed.), Royal Society of Chemistry, Cambridge, U.K., ISBN 978-1-84973-007-5.
5. *Nomenclature of Inorganic Chemistry – IUPAC Recommendations 2005*, N. G. Connelly, T. Damhus, R. M. Hartshorn, A. T. Hutton (Eds.), Royal Society of Chemistry, Cambridge, U.K., ISBN 0-85404-438-8.
6. *Nomenclature of Organic Chemistry – IUPAC Recommendations and Preferred Names 2013*, H. A. Favre, W. H. Powell (Eds.), Royal Society of Chemistry, Cambridge, U.K., ISBN 978-0-85404-182-4.
7. *Compendium of Polymer Terminology and Nomenclature – IUPAC Recommendations 2008*, R. G. Jones, J. Kahovec, R. Stepto, E. S. Wilks, M. Hess, T. Kitayama, W. V. Metanomski (Eds.), Royal Society of Chemistry, Cambridge, U.K., ISBN 978-0-85404-491-7.
8. Reference 4, Chapter 10.
9. Reference 5, Table IX.
10. Reference 4, Table P10.
11. Reference 5, Chapter IR-6.
12. Reference 5, Table X.
13. Reference 5, Section IR-9.2.4.
14. Reference 5, Table IR-10.4.
15. Reference 5, Section IR-10.3.
16. Reference 5, Chapter IR-8.
17. Reference 4, Table P5; Reference 5, Tables IR-9.2 and IR-9.3.
18. Reference 5, Section IR-9.3.3.
19. R. S. Cahn, C. Ingold, V. Prelog, *Angew. Chem., Int. Ed. Engl.*, 5, 385–415 and 511 (1966); V. Prelog, G. Helmchen, *Angew. Chem., Int. Ed. Engl.*, 21, 567–583 (1982) <https://doi.org/10.1002/anie.196603851>.
20. J. Brecher, K. N. Degtyarenko, H. Gottlieb, R. M. Hartshorn, G. P. Moss, P. Murray-Rust, J. Nyitrai, W. Powell, A. Smith, S. Stein, K. Taylor, W. Town, A. Williams, A. Yerin, *Pure Appl. Chem.*, 78(10), 1897–1970 (2006); J. Brecher, K. N. Degtyarenko, H. Gottlieb, R. M. Hartshorn, K.-H. Hellwich, J. Kahovec, G. P. Moss, A. McNaught, J. Nyitrai, W. Powell, A. Smith, K. Taylor, W. Town, A. Williams, A. Yerin, *Pure Appl. Chem.*, 80(2), 277–410 (2008).

To cite, please use: IUPAC, *Pure Appl. Chem.* 87, 1039–1049 (2015). Publication of this document by any means is permitted on condition that it is whole and unchanged. Copyright © IUPAC & De Gruyter 2015.

1 Stoichiometric or Compositional Names

A stoichiometric or compositional name provides information only on the composition of an ion, molecule, or compound, and may be related to either the empirical or molecular formula for that entity. It does not provide any structural information.

For homoatomic entities, where only one element is present, the name is formed (Table 1) by combining the element name with the appropriate multiplicative prefix (Table 2). Ions are named by adding charge numbers in parentheses, e.g., (1+), (3+), (2–), and for (most) homoatomic anion names 'ide' is added in place of the 'en', 'ese', 'ic', 'ine', 'ium', 'ogen', 'on', 'orus', 'um', 'ur', 'y', or 'ygen' endings of element names.[9] Exceptions include Zn and Group 18 elements ending in 'on', where the 'ide' ending is added to the element names. For some elements (e.g., Fe, Ag, Au) a Latin stem is used before the 'ide' ending (cf. Section 2.3).[9] Certain ions may have acceptable traditional names (used without charge numbers).

Table 1: Examples of Homoatomic Entities

Formula	Name	Formula	Name
O_2	dioxygen	Cl^-	chloride(1-) or chloride
S_8	octasulfur	I_3^-	triiodide(1-)
Na^+	sodium(1+)	O_2^{2-}	dioxide(2-) or peroxide
Fe^{3+}	iron(3+)	N_3^-	trinitride(1-) or azide

Table 2: Multiplicative Prefixes for Simple and Complicated Entities

No.	Simple	Complicated	No.	Simple	Complicated
2	di	bis	8	octa	octakis
3	tri	tris	9	nona	nonakis
4	tetra	tetrakis	10	deca	decakis
5	penta	pentakis	11	undeca	undecakis
6	hexa	hexakis	12	dodeca	dodecakis
7	hepta	heptakis	20	icosa	icosakis

Binary compounds (those containing atoms of two elements) are named stoichiometrically by combining the element names and treating, by convention, the element reached first when following the arrow in the element sequence (Figure 1) as if it were an anion. Thus, the name of this formally 'electronegative' element is given an 'ide' ending and is placed after the name of the formally 'electropositive' element followed by a space (Table 3).

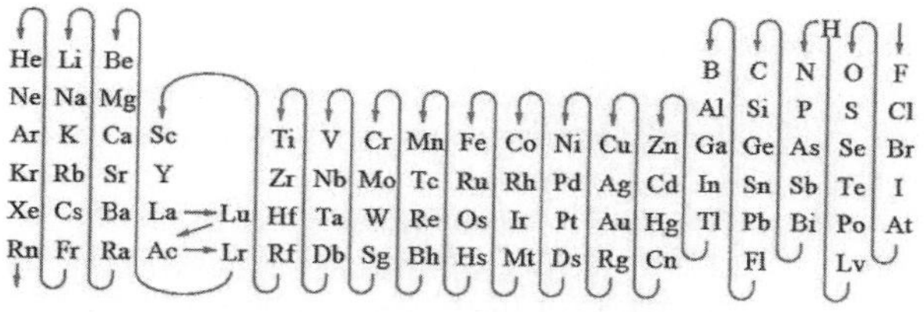

Figure 1: Element Sequence

Table 3: Examples of Binary Compounds

Formula	Name	Formula	Name
GaAs	gallium arsenide	$FeCl_2$	iron dichloride or iron(II) chloride
CO_2	carbon dioxide	$FeCl_3$	iron trichloride or iron(III) chloride
CaF_2	calcium difluoride or calcium fluoride	H_2O_2	dihydrogen dioxide or hydrogen peroxide

Again, multiplicative prefixes (Table 2) are applied as needed, and certain acceptable alternative names[10] may be used. Stoichiometry may be implied in some cases by the use of oxidation numbers, but is often omitted for common cases, such as in calcium fluoride.

Heteropolyatomic entities in general can be named similarly using compositional nomenclature, but often either substitutive[11] or additive nomenclature (Section 2) is used. In the latter case information is also provided about the way atoms are connected. For example, $POCl_3$ (or PCl_3O, compositional name phosphorus trichloride oxide) is given an additive name in Table 10.

Certain ions have traditional short names, which are commonly used and are still acceptable (e.g., ammonium, NH_4^+; hydroxide, OH^-; nitrite, NO_2^-; phosphate, PO_4^{3-}; diphosphate, $P_2O_7^{4-}$).

Inorganic compounds in general can be combinations of cations, anions, and neutral entities. By convention, the name of a compound is made up of the names of its component entities: cations before anions and neutral components last (see examples in Table 4).

The number of each entity present has to be specified in order to reflect the composition of the compound. For this purpose, multiplicative prefixes (Table 2) are added to the name of each entity. The prefixes are 'di', 'tri', 'tetra', etc., for use with names for simple entities, or 'bis( )', 'tris( )', 'tetrakis( )', etc., for names for most entities which themselves contain multiplicative prefixes or locants. Care must also be taken in situations when use of a simple multiplicative prefix may be misinterpreted, e.g., tris(iodide) must be used for $3I^-$ rather than triiodide (which is used for I_3^-), and bis(phosphate) rather than diphosphate (which is used for $P_2O_7^{4-}$). Examples are shown in Table 4. There is no elision of vowels (e.g., tetraaqua, pentaoxide), except in the special case of monoxide.

Names of neutral components are separated by 'em' dashes without spaces. Inorganic compounds may themselves be components in (formal) **addition compounds** (last four examples in Table 4). The ratios of component compounds can be indicated, in general, using a stoichiometric descriptor in parentheses after the name (see the last three examples in Table 4). In the special case of hydrates, multiplicative prefixes can be used with the term 'hydrate'.

Table 4: Use of Multiplicative Prefixes in Compositional Names

Formula	Name
$Ca_3(PO_4)_2$	tricalcium bis(phosphate)
$Ca_2P_2O_7$	dicalcium diphosphate
BaO_2	barium(2+) dioxide(2−) or barium peroxide
$MgSO_4{\cdot}7H_2O$	magnesium sulfate heptahydrate
$CdSO_4{\cdot}6NH_3$	cadmium sulfate—ammonia (1/6)
$AlK(SO_4)_2{\cdot}12H_2O$	aluminum potassium bis(sulfate)—water (1/12) or aluminum potassium bis(sulfate) dodecahydrate
$Al_2(SO_4)_3{\cdot}K_2SO_4{\cdot}24H_2O$	dialuminum tris(sulfate)—dipotassium sulfate—water (1/1/24)

2 Complexes and Additive Nomenclature

2.1 Overall Approach

Additive nomenclature was developed in order to describe the structures of coordination entities, or complexes, but this method is readily extended to other molecular entities as well. Mononuclear complexes are considered to consist of a central atom, often a metal ion, which is bonded to surrounding small molecules or ions, which are referred to as ligands. The names of complexes are constructed (Table 5) by adding the names of the ligands before those of the central atoms, using appropriate multiplicative prefixes. Formulae are constructed by adding the symbols or abbreviations of the ligands *after* the symbols of the central atoms (Section 2.7).

Table 5: Producing Names for Complexes: Simple Ligands

Structure to be named	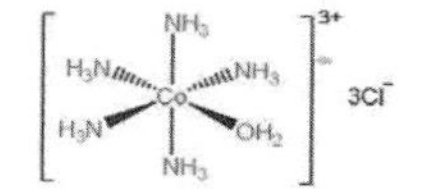	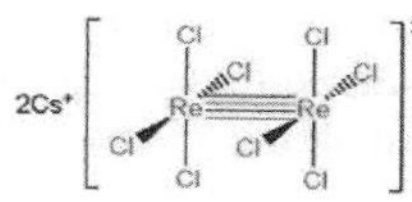
Central atom(s)	cobalt(III)	2 × rhenium
Identify and name ligands	ammonia → ammine water → aqua	chloride → chlorido
Assemble name	pentaammineaqua=cobalt(III) chloride	cesium bis(tetrachlorido=rhenate)(*Re—Re*)(2-)

2.2 Central atom(s) and ligands

The first step is to identify the central atom(s) and thereby also the ligands. By convention, the electrons involved in bonding between the central atom and a ligand are usually treated as belonging to the ligand (and this will determine how it is named).

Each ligand is named as a separate entity, using appropriate nomenclature[4] – usually substitutive nomenclature for organic ligands[2,4,6] and additive nomenclature for inorganic ligands. A small number of common molecules and ions are given special names when present in complexes. For example, a water ligand is represented in the full name by the term 'aqua'. An ammonia ligand is represented by 'ammine', while carbon monoxide bound to the central atom through the carbon atom is represented by the term 'carbonyl' and nitrogen monoxide bound through nitrogen is represented by 'nitrosyl'. Names of **anionic ligands** that end in 'ide', 'ate', or 'ite' are modified within the full additive name for the complex to end in 'ido', 'ato', or 'ito', respectively. Note that the 'ido' ending is now used for halide and oxide ligands as well. By convention, a single coordinated hydrogen atom is always considered anionic and it is represented in the name by the term 'hydrido', whereas coordinated dihydrogen is usually treated as a neutral two-electron donor entity.

2.3 Assembling additive names

Once the ligands have been named, the name can be assembled. This is done by listing the ligand names in alphabetical order before the name of the central atom(s), *without* regard to ligand charge.

If there is more than one ligand of a particular kind bound to a central atom in the same way, the number of such identical ligands is indicated using the appropriate multiplicative prefix for simple or complicated ligands (Table 2), not changing the already established alphabetical order of ligands. The nesting order of enclosing marks, for use in names where more than one set of enclosing marks is required, is: ( ), [( )], {[( )]}, ({[( )]}), etc.

Any **metal-metal bonds** are indicated by placing the central atom symbols in parentheses, in italics and connected by an 'em' dash, after the name of the complex (without spaces). The **charge number** of the complex or the **oxidation number** of the central atom is appended to the name of the complex. For **anions** that are named additively, the name of the central atom is given the 'ate' ending in a similar way to the 'ide' endings of homoatomic anions (Section 1). In some cases, by tradition, the Latin stem is used for the 'ate' names, such as in ferrate (for iron), cuprate (for copper), argentate (for silver), stannate (for tin), aurate (for gold), and plumbate (for lead).[12] Finally, the rules of compositional nomenclature (Section 1) are used to combine the additive names of ionic or neutral coordination entities with the names of any other entities that are part of the compound.

2.4 Specifying connectivity

Some ligands can bind to a central atom through different atoms under different circumstances. Specifying just which ligating (coordinating) atoms are bound in any given complex can be achieved by adding **κ-terms** to the name of the ligand. The κ-term comprises the Greek letter κ followed by the italicized element symbol of the ligating atom. For more complicated ligands the κ-term is often placed within the ligand name following the group to which the κ-term refers. Multiple identical links to a central atom can be indicated by addition of the appropriate numeral as a superscript between the κ and element symbols (see Table 6). These possibilities are discussed in more detail in the Red Book.[13] If the ligating atoms of a ligand are contiguous (i.e., directly bonded to one another), then an **η-term** is used instead, for example, for many organometallic compounds (Section 2.6) and the peroxido complex in Table 6.

A κ-term is required for ligands where more than one coordination mode is possible. Typical cases are thiocyanate, which can be bound through either the sulfur atom (thiocyanato-κ*S*) or the nitrogen atom (thiocyanato-κ*N*), and nitrite, which can be bound through either the nitrogen atom (M–NO_2, nitrito-κ*N*), or an oxygen atom (M–ONO, nitrito-κ*O*). The names pentaammine(nitrito-κ*N*)cobalt(2+) and pentaammine(nitrito-κ*O*)cobalt(2+) are used for each of the isomeric nitrito complex cations. More examples of constructing names using κ-terms to specify the connectivity of ligands are shown in Table 6. A κ-term may also be used to indicate to which central atom a ligand is bound if there is more than one central atom (Section 2.5).

Table 6: Producing Names for Complexes: Complicated Ligands

Structure to be named	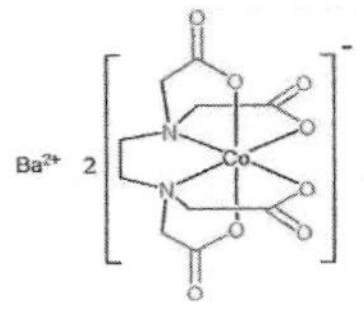	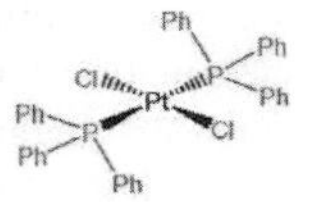
Central atom	cobalt(III) → cobaltate(III)	platinum(II)
Identify and name ligands	2,2′,2″,2‴-(ethane-1,2-diyl=dinitrilo)tetraacetate → 2,2′,2″,2‴-(ethane-1,2-diyl=dinitrilo)tetraacetato	chloride → chlorido triphenylphosphane
Specify ligating atoms	2,2′,2″,2‴-(ethane-1,2-diyl=dinitrilo-κ^{2}*N*)tetraacetato-κ^{4}*O*	*not required for chloride* triphenylphosphane-κ*P*
Assemble name	barium [2,2′,2″,2‴-(ethane-1,2-diyldinitrilo-κ^{2}*N*) tetra=acetato-κ^{4}*O*] cobaltate(III)	dichloridobis(triphenyl=phosphane-κ*P*) platinum(II)
Structure to be named	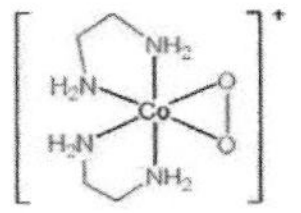	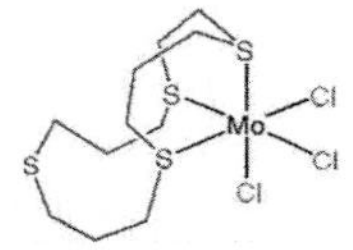
Central atom	cobalt(III)	molybdenum(III)
Identify and name ligands	ethane-1,2-diamine peroxide → peroxido	chloride → chlorido 1,4,8,12-tetrathiacyclo-pentadecane
Specify ligating atoms	ethane-1,2-diamine-κ^{2}*N* η^{2}-peroxido	*not required for chloride* 1,4,8,12-tetrathiacyclo=pentadecane-κ^{3}*S*^{1},*S*^{4},*S*^{8}
Assemble name	bis(ethane-1,2-diamine-κ^{2}*N*)=(η^{2}-peroxido) cobalt(III)	trichlorido(1,4,8,12-tetrathiacyclopentadecane-κ^{3}*S*^{1},*S*^{4},*S*^{8})molybdenum(III)

2.5 Bridging ligands

Bridging ligands are those bound to more than one central atom. They are differentiated in names by the addition of the prefix 'μ' (Greek mu), with the prefix and the name of the bridging ligand being separated from each other, and from the rest of the name, by hyphens. This is sufficient if the ligand is monoatomic, but if the ligand is more complicated it may be necessary to specify which ligating atom of the ligand is attached to which central atom. This is certainly the case if the ligating atoms are of different kinds, and κ-terms can be used for this purpose.

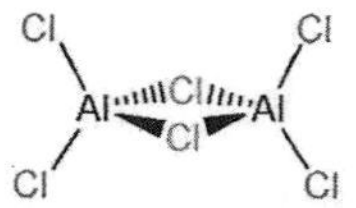

di-μ-chlorido-bis[di=chloridoaluminium(III)]
$[Cl_2Al(\mu\text{-}Cl)_2AlCl_2]$

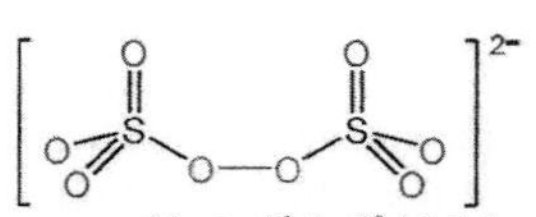

μ-peroxido-1κ*O*^{1},2κ*O*^{2}-bis(tri=oxidosulfate)(2–)
$[O_3S(\mu\text{-}O_2)SO_3]^{2-}$

2.6 Organometallic compounds

Organometallic compounds contain at least one bond between a metal atom and a carbon atom. They are named as coordination compounds, using the additive nomenclature system (see above).

The name for an organic ligand **binding through one carbon atom** may be derived either by treating the ligand as an anion or as a neutral substituent group. The compound $[Ti(CH_2CH_2CH_3)Cl_3]$ is thus named as trichlorido(propan-1-ido)titanium or as trichlorido(propyl)titanium. Similarly, 'methanido' or 'methyl' may be used for the ligand $-CH_3$.

When an organic ligand forms **two or three metal-carbon single bonds** (to one or more metal centers), the ligand may be treated as a di- or tri-anion, with the endings 'diido' or 'triido' being used, with no removal of the terminal 'e' of the name of the parent hydrocarbon. Again, names derived by regarding such ligands as substituent groups and using the suffixes 'diyl' and 'triyl' are still commonly encountered. Thus, the bidentate ligand $-CH_2CH_2CH_2-$ would be named propane-1,3-diido (or propane-1,3-diyl) when chelating a metal centre, and μ-propane-1,3-diido (or μ-propane-1,3-diyl) when bridging two metal atoms.

Organometallic compounds containing a **metal-carbon multiple bond** are given substituent prefix names derived from the parent hydrides which end with the suffix 'ylidene' for a metal-carbon double bond and with 'ylidyne' for a triple bond. These suffixes either replace the ending 'ane' of the parent hydride, or, more generally, are added to the name of the parent hydride with insertion of a locant and elision of the terminal 'e', if present. Thus, the entity $CH_3CH_2CH=$ as a ligand is named propylidene and $(CH_3)_2C=$ is called propan-2-ylidene. The 'diido'/'triido' approach, outlined above, can also be used in this situation. The terms 'carbene' and 'carbyne' are not used in systematic nomenclature.

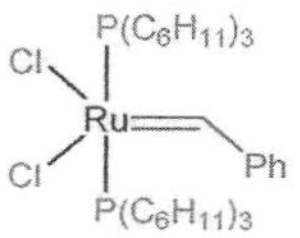

dichlorido(phenylmethylidene)bis(tricyclohexylphosphane-κ*P*)ruthenium, dichlorido(phenylmethanediido)bis(tricyclohexylphosphane-κ*P*)ruthenium, **or** (benzylidene)dichloridobis(tricyclohexylphosphane-κ*P*)ruthenium

The special nature of the bonding to metals of unsaturated hydrocarbons in a 'side-on' fashion via their π-electrons requires the **eta (η) convention**. In this 'hapto' nomenclature, the number of *contiguous* atoms in the ligand coordinated to the metal (the hapticity of the ligand) is indicated by a right superscript on the eta symbol, e.g., η^3 ('eta three' or 'trihapto'). The η-term is added as a prefix to the ligand name, or to that portion of the ligand name most appropriate to indicate the connectivity, with locants if necessary.

A list of many **π-bonding unsaturated ligands**, neutral and anionic, can be found in the Red Book.[14]

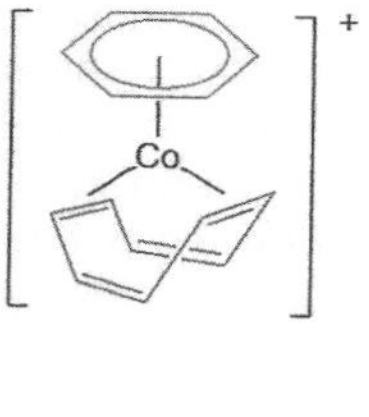

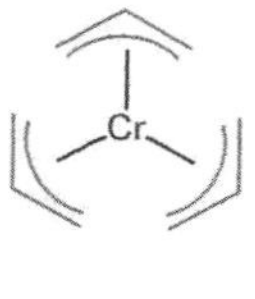

tris(η^3-prop-2-en-1-ido)chromium.

Note that the ubiquitous ligand η5-C_5H_5, strictly η5-cyclopenta-2,4-dien-1-ido, is also acceptably named η5-cyclopentadienido or η5-cyclopentadienyl. When cyclopenta-2,4-dien-1-ido coordinates through one carbon atom via a σ bond, a κ-term is added for explicit indication of that bonding. The symbol η^1 should not be used, as the eta convention applies only to the bonding of contiguous atoms in a ligand.

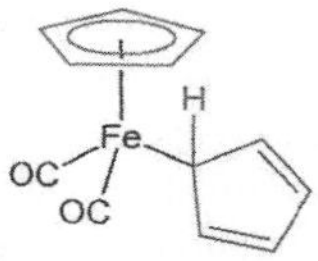

dicarbonyl(η^5-cyclopentadienido)(cyclopenta-2,4-dien-1-ido-κC^1)iron **or** dicarbonyl(η^5-cyclopentadienyl)(cyclopenta-2,4-dien-1-yl-κC^1)iron

Discrete molecules containing two parallel η5-cyclopentadienido ligands in a 'sandwich' structure around a transition metal, as in bis(η5-cyclopentadienido)iron, $[Fe(\eta^5\text{-}C_5H_5)_2]$, are generically called metallocenes and may be given 'ocene' names, in this case ferrocene. These 'ocene' names may be used in the same way as parent hydride names are used in substitutive nomenclature, with substituent group names taking the forms 'ocenyl', 'ocenediyl', 'ocenetriyl' (with insertion of appropriate locants).

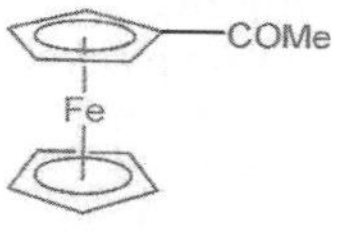

1-ferrocenylethan-1-one

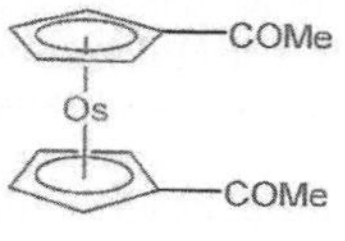

1,1'-(osmocene-1,1'-diyl)di=(ethan-1-one)

By convention, 'organoelement' compounds of the **main group elements** are named by substitutive nomenclature if derived from the elements of Groups 13–16, but by additive nomenclature if derived from the elements of Groups 1 and 2. In some cases compositional nomenclature is used if less structural information is to be conveyed. More detail is provided in the Red Book.[15]

Table 7: Producing Line Formulae for Complexes

Structure	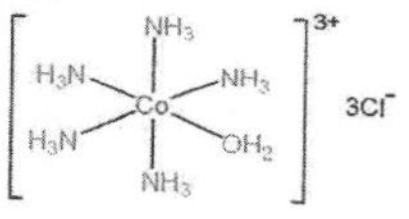	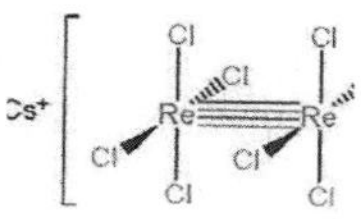
Central atom(s)	Co	2 × Re
Ligands	NH_3, OH_2	Cl
Assemble formula	$[Co(NH_3)_5(OH_2)]Cl_3$	$Cs_2[Cl_4ReReCl_4]$
Structure	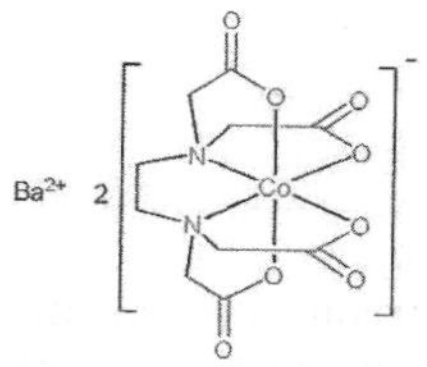	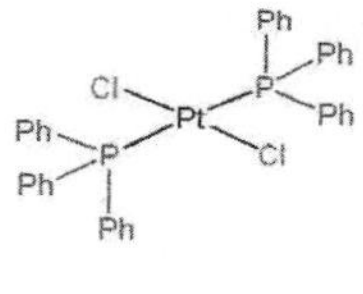
Central atom(s)	Co	Pt
Abbreviate ligands	2,2′,2″,2‴-(ethane-1,2-diyl) dinitrilotetraacetate → edta	Cl triphenylphosphane → PPh_3
Assemble formula	$Ba[Co(edta)]_2$	$[PtCl_2(PPh_3)_2]$

2.7 Formulae of coordination compounds

Line formulae for coordination entities are constructed within square brackets to specify the composition of the entity. The overall process is shown in Table 7. The symbol for the central atom is placed first and is then followed by the symbols or abbreviations for the ligands (in alphabetical order according to the way they are presented in the formula). Where possible the coordinating (ligating) atom should be placed nearer the central atom in order to provide more information about the structure of the complex. If possible, bridging ligands should be placed between central atom symbols for this same reason (see examples in Section 2.5). Generally, ligand formulae and abbreviations are placed within enclosing marks (unless the ligand contains only one atom), remembering that square brackets are reserved to define the coordination sphere. Multiple ligands are indicated by a right subscript following the enclosing marks or ligand symbol.

2.8 Inorganic oxoacids and related compounds

Inorganic oxoacids, and the anions formed by removing the acidic **hydrons** (H^+) from them, have traditional names, many of which are well-known and can be found in many textbooks: sulfuric acid, sulfate; nitric acid, nitrate; nitrous acid, nitrite; phosphoric acid, phosphate; arsenic acid, arsenate; arsinous acid, arsinite; silicic acid, silicate; etc. These names are retained in IUPAC nomenclature, first because they almost invariably are the names used in practice, and second because they play a special role in organic nomenclature when names are needed for organic derivatives. However, all the oxoacids themselves and their derivatives may be viewed as coordination entities and named systematically using additive nomenclature (Table 8).[16]

Table 8: Examples of Inorganic Oxoacids and Derivatives

Formula	Traditional or organic name	Additive name
H_2SO_4 or $[S(O)_2(OH)_2]$	sulfuric acid	dihydroxidodioxidosulfur
$(CH_3)_2SO_4$ or $[S(O)_2(OMe)_2]$	dimethyl sulfate	dimethoxidodioxidosulfur or dimethanolatodioxidosulfur
H_2PHO_3 or $[P(H)(O)(OH)_2]$	phosphonic acid*	hydridodihydroxidooxido-phosphorus
$PhP(O)(OH)_2$	phenylphosphonic acid	dihydroxidooxido(phenyl) phosphorus

* *Note:* The term 'phosphorous acid' has been used in the literature for both the species named phosphonic acid in Table 8 and that with the formula $P(OH)_3$, trihydroxidophosphorus. It is used in organic nomenclature in the latter sense.

The traditional oxoacid names may be modified according to established rules for naming derivatives formed by **functional replacement**[16]: thus 'thio' denotes replacement of =O by =S; prefixes 'fluoro', 'chloro', etc., and infixes 'fluorid', 'chlorid', etc., denote replacement of –OH by –F, –Cl, etc.; 'peroxy'/'peroxo' denote replacement of –O– by –OO–; and so forth (Table 9).

Table 9: Examples of Derivatives of Inorganic Oxoacids and Anions Formed by Functional Replacement

Formula	Name indicating functional replacement	Additive name
H_3PS_4 or $[P(S)(SH)_3]$	tetrathiophosphoric acid or phosphorotetrathioic acid	tris(sulfanido)sulfido=phosphorus
H_2PFO_3 or $[PF(O)(OH)_2]$	fluorophosphoric acid or phosphorofluoridic acid	fluoridodihydroxido=oxidophosphorus
$S_2O_3^{2-}$ or $[S(O)_3(S)]^{2-}$	thiosulfate or sulfurothioate	trioxidosulfido=sulfate(2-)
$[O_3S(\mu\text{-}O_2)SO_3]^{2-}$	peroxydisulfate	see Section 2.5

If all hydroxy groups in an oxoacid are replaced, the compound is no longer an acid and is not named as such, but will have a traditional **functional class name**[16] as, e.g., an acid halide or amide. Such compounds may again be systematically named using additive nomenclature (Table 10).

Table 10: Examples of Functional Class Names and Corresponding Additive Names

Formula	Functional class name	Additive name
PCl_3O	phosphoryl trichloride	trichloridooxido=phosphorus
SCl_2O_2	sulfuryl dichloride	dichloridodioxidosulfur
$S(NH_2)_2O_2$	sulfuric diamide	diamidodioxidosulfur

A special construction is used in **hydrogen names**, which allows the indication of hydrons bound to an anion without specifying exactly where. In such names, the word 'hydrogen' is placed at the front of the name with a multiplicative prefix (if applicable) and with no space between it and the rest of the name, which is placed in parentheses. For example, dihydrogen(diphosphate)(2–) denotes $H_2P_2O_7^{2-}$, a diphosphate ion to which two hydrons have been added, with the positions not known or at least not being specified.

One may view the common names for partially dehydronated oxoacids, such as hydrogenphosphate, HPO_4^{2-}, and dihydrogenphosphate, $H_2PO_4^-$, as special cases of such hydrogen names. In these simplified names, the charge number and the parentheses around the main part of the name are left out. Again, these particular anions may be named systematically by additive nomenclature. The word 'hydrogen' is placed separately in forming analogous names in organic nomenclature, for example, dodecyl hydrogen sulfate, $C_{12}H_{25}OS(O)_2OH$. This difference between the two systems has the consequence that the important carbon-containing ion HCO_3^- can be named equally correctly as 'hydrogen carbonate' and as 'hydrogencarbonate' (but not as bicarbonate).

3 Stereodescriptors

The approximate geometry around the central atom is described using a **polyhedral symbol** placed in front of the name. The symbol is made up of italicized letter codes for the geometry and a number that indicates the coordination number. Frequently used polyhedral symbols are *OC*-6 (octahedral), *SP*-4 (square-planar), *T*-4 (tetrahedral), *SPY*-5 (square-pyramidal), and *TBPY*-5 (trigonal-bipyramidal). More complete lists are available.[17]

The relative positions of ligating groups[19] around a central atom can be described using a **configuration index** that is determined in a particular way for each geometry,[18] based on the Cahn-Ingold-Prelog priorities of the ligating groups,[19] and it may change if the ligands change, even if the geometry remains the same. The absolute configuration can also be described. Generally, configuration indices are used only if there is more than one possibility and a particular stereoisomer is to be identified. The full stereodescriptors for the particular square-planar platinum complexes shown below are (*SP*-4-2) and (*SP*-4-1), for the *cis* and *trans* isomers, respectively. Alternatively, a range of traditional stereodescriptors may be used in particular situations. Thus the isomers that are possible when a square-planar centre is coordinated by two ligating groups of one type and two of another are referred to as *cis*- (when the identical ligands are coordinated next to each other) or *trans*- (when they are coordinated opposite to each other).

cis-diamminedichloridoplatinum(II)

trans-diamminedichloridoplatinum(II)

Octahedral centers with four ligands of one kind and two of another can also be referred to as *cis*- (when the two identical ligands are coordinated next to each other) or *trans*- (when they are coordinated opposite each other). Octahedral centers with three of each of two kinds of ligand can be described as *fac-* (facial), when the three ligands of a particular kind are located at the corners of a face of the octahedron, or *mer-* (meridional), when they are not.

4 Summary

This document provides an outline of the essential nomenclature rules for producing names and formulae for inorganic compounds, coordination compounds, and organometallic compounds. The complementary document for nomenclature systems of organic chemistry[2] will also be useful to the reader.

Names and formulae have only served half their role when they are created and used to describe or identify compounds, for example, in publications. Achieving their full role requires that the reader of a name or formula is able to interpret it successfully, for example, to produce a structural diagram. The present document is also intended to assist in the interpretation of names and formulae.

Finally, we note that IUPAC has produced recommendations on the graphical representation of chemical structures and their stereochemical configurations.[20]

REPRESENTATION OF CHEMICAL STRUCTURES WITH THE IUPAC INTERNATIONAL CHEMICAL IDENTIFIER (InChI)

Stephen R. Heller and Alan D. McNaught

The IUPAC International Chemical Identifier (InChI) is a freely available, non-proprietary identifier for chemical substances that can be used in both printed and electronic data sources. It is generated from a computerized representation of a molecular structure diagram, which can be produced by chemical structure-drawing software. Its use enables linking of diverse data compilations and unambiguous identification of chemical substances. A full description of the Identifier and the software for its generation are available from the IUPAC Web site (Ref. 1), and a helpful compilation of answers to frequently asked questions has been put together (Ref. 2). Commercial structure-drawing software that will generate the Identifier is available from several organizations, listed on the IUPAC Web site.

The conversion of structural information to the Identifier is based on a set of IUPAC structure conventions, and rules for normalization and canonicalization (conversion to a single, predictable sequence) of an input structure representation. The resulting InChI is simply a series of characters that serve to uniquely identify the structure from which it was derived. The InChI uses a layered format to represent all available structural information relevant to compound identity. InChI layers are listed below. Each layer in an InChI representation contains a specific type of structural information. These layers, automatically extracted from the input structure, are designed so that each successive layer adds additional detail to the Identifier. The specific layers generated depend on the level of structural detail available and whether or not allowance is made for tautomerism. Of course, any ambiguities or uncertainties in the original structure will remain in the InChI.

This layered structure design offers a number of advantages. If two structures for the same substance are drawn at different levels of detail, the one with the lower level of detail will, in effect, be contained within the other. Specifically, if one substance is drawn with stereo-bonds and the other without, the layers in the latter will be a subset of the former. The same will hold for compounds treated by one author as tautomers and by another as exact structures with all H-atoms fixed. This can work at a finer level. For example, if one author includes double bond and tetrahedral stereochemistry, but another omits stereochemistry, the latter InChI will be contained in the former.

The InChI layers are

1. Formula
2. Connectivity (no formal bond orders)
   a. disconnected metals
   b. connected metals
3. Isotopes
4. Stereochemistry
   a. double bond (*Z/E*)
   b. tetrahedral (sp^3)
5. Tautomers (on or off)

Charges are not part of the basic InChI, but rather are added at the end of the InChI string.

Two examples of InChI representations are given below. It is important to recognize, however, that InChI strings are intended for use by computers and end users need not understand any of their details. In fact, the open nature of InChI and its flexibility of representation, after implementation into software systems, may allow chemists to be even less concerned with the details of structure representation by computers.

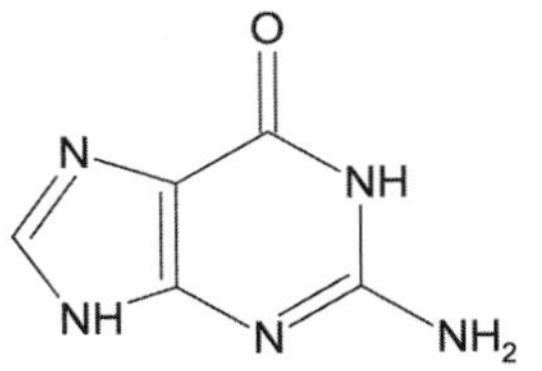

guanine

InChI=1/C5H5N5O/c6-5-9-3-2(4(11)10-5)7-1-8-3/
h1H,(H4,6,7,8,9,10,11)/f/h8,10H,6H2

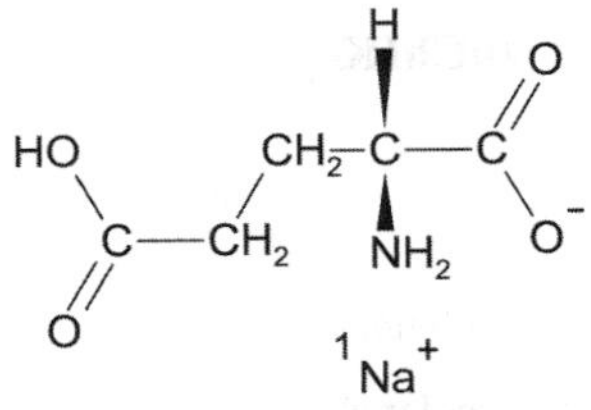

monosodium glutamate

InChI=1/C5H9NO4.Na/c6-3(5(9)10)1-2-4(7)8;/h3H,
1-2,6H2,(H,7,8)(H,9,10);/q;+1/p-1/t3-;/m1./s1/fC5H8NO4.Na/
h7H;/q-1;m

The layers in the InChI string are separated by the '/' character followed by a lowercase letter (except for the first layer, the chemical formula), with the layers arranged in predefined order. In the examples the following segments are included

InChI version number
/- chemical formula
/c connectivity-1.1 (excluding terminal H)
/h connectivity-1.2 (locations of terminal H, including mobile H attachment points)
/q charge
/p proton balance
/t sp^3 (tetrahedral) parity
/m parity inverted to obtain relative stereo (1 = inverted, 0 = not inverted)
/s stereo type (1 = absolute, 2 = relative, 3 = racemic)
/f chemical formula of the fixed-H structure if it is different
/h connectivity-2 (locations of fixed mobile H)
/q charge
/t sp^3 (tetrahedral) parity
/m parity inverted to obtain relative stereo (1 = inverted, 0 = not inverted, **.** = inversion does not affect the parity)
/s stereo type (1 = absolute, 2 = relative, 3 = racemic)

One of the most important applications of InChI is the facility to locate mention of a chemical substance using Internet-based search engines. This is made easier by using a shorter (compressed) form of InChI, known as InChIKey. The InChIKey is a 27-character representation that, because it is compressed, cannot be reconverted into the original structure, but it is not subject to the undesirable and unpredictable breaking of longer character strings by some search engines. The usefulness of the InChIKey as a search tool is enhanced by its derivation from a "standard" InChI, i.e., an InChI produced with standard option settings for features such as tautomerism and stereochemistry. An example is shown below; the "standard" InChI is denoted by the letter "S" after the version number.

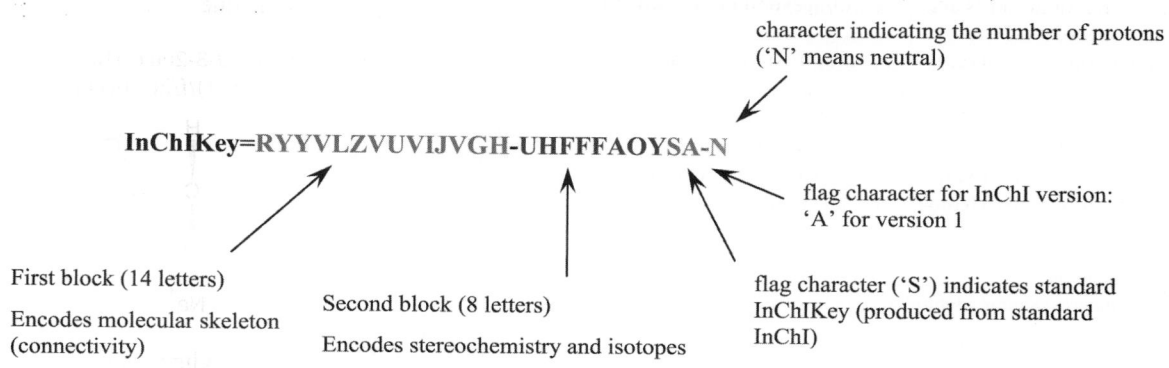

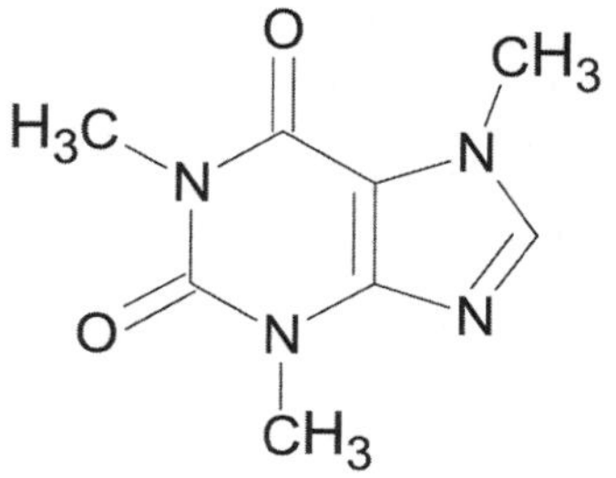

InChI=1S/C8H10N4O2/c1-10-4-9-6-5(10)7(13)12(3)8(14)11(6)2/h4H,1-3H3 (caffeine)

character indicating the number of protons ('N' means neutral)

InChIKey=RYYVLZVUVIJVGH-UHFFFAOYSA-N

flag character for InChI version: 'A' for version 1

First block (14 letters)

Encodes molecular skeleton (connectivity)

Second block (8 letters)

Encodes stereochemistry and isotopes

flag character ('S') indicates standard InChIKey (produced from standard InChI)

Use of InChIKey also allows searches based solely on atomic connectivity (first 14 characters). Software for generating InChIKey is available from the IUPAC Web site (Ref. 1).

More details about the project and algorithm can be found in recent publications (Refs. 3 and 4). The enormous databases compiled by organizations such as PubChem (Ref. 5), which provides InChI-based search access to about 100 million chemical structures from over 600 different public and commercial data sources, the U.S. National Cancer Institute (NCI) (Ref. 6), and ChemSpider (Ref. 7) contain millions of InChIs and InChIKeys, which allow sophisticated searching of these collections. PubChem provides InChI-based structure-search facilities for both identical and similar structures (Ref. 5), and ChemSpider offers both search facilities and Web services enabling a variety of InChI and InChIKey conversions (Ref. 7). The EBI UniChem database (Ref. 8) provides InChI-based search access to over 160 million chemical structures from dozens of different public and commercial data sources.

References

1. https://iupac.org/who-we-are/divisions/division-details/inchi/
2. https://www.inchi-trust.org/technical-faq/
3. Heller, S., McNaught, A., Pletnev, I., Stein, S., Tchekhovskoi, D., and Pletnev, I., InChI – The worldwide chemical structure identifier standard, *Journal of Cheminformatics,* 5, 7, 2013 <https://doi.org/10.1186/1758-2946-5-7>.
4. Heller, S., McNaught, A., Pletnev, I., Stein, S., and Tchekhovskoi, D., InChI – The IUPAC International Chemical Identifier, *Journal of Cheminformatics,* 7, 23, 2015 <https://doi.org/10.1186/s13321-015-0068-4>.
5. https://pubchem.ncbi.nlm.nih.gov/
6. https://cactus.nci.nih.gov/cgi-bin/lookup/search
7. http://www.chemspider.com/
8. https://www.ebi.ac.uk/unichem/info/faq#faq6

SCIENTIFIC ABBREVIATIONS, ACRONYMS, AND SYMBOLS

This table lists some abbreviations, acronyms, and symbols encountered in the physical sciences. Most entries in italic type are symbols for physical quantities; for more details on these, see the table "Symbols and Terminology for Physical and Chemical Quantities" in this section. Additional information on units may be found in the table "International System of Units (SI)" in Section 1. Many of the terms to which these abbreviations refer are included in the tables "Definitions of Scientific Terms" in Section 2 of the Online Edition and "Techniques for Materials Characterization" in Section 12, in both the Print Edition and Online Edition of the *CRC Handbook*. Useful references for further information are given below.

Publication practices vary with regard to the use of capital or lowercase letters for many abbreviations. An effort has been made to follow the most common practices in this table, but much variation is found in the literature. Likewise, policies on the use of periods in an abbreviation vary considerably. Periods are generally omitted in this table unless they are necessary for clarity. Periods should never appear in SI units. The SI prefixes (m, k, M, etc.) are included here, but they should never be used alone. Selected combinations of these prefixes with SI units (e.g., mg, kV, MW) are also included.

Abbreviations are listed in alphabetical order without regard to case. Entries beginning with Greek letters fall at the end of the table.

References

1. *Quantities, Units, and Symbols in Physical Chemistry, Third Edition*, IUPAC 2007, RSC Publishing, 2007.
2. Kotyk, A., *Quantities, Symbols, Units, and Abbreviations in the Life Sciences*, Humana Press, Totawa, NJ, 1999.
3. Cooper, C., and Purchase, R., *Organic Chemist's Desk Reference, Third Edition*, CRC Press/Taylor & Francis, Boca Raton, FL, 2018.
4. Minkin, V., Glossary of Terms used in Theoretical Organic Chemistry, *Pure Appl. Chem.* 71, 1919–1981, 1999.
5. Brown, R. D., Ed., Acronyms Used in Theoretical Chemistry, *Pure Appl. Chem.* 68, 387–456, 1996.
6. *Quantities and Units, ISO Standards Handbook, Third Edition*, International Organization for Standardization, Geneva, 1993.
7. Cohen, E. R., and Giacomo, P., Symbols, Units, Nomenclature, and Fundamental Constants in Physics, *Physica* 146A, 1–68, 1987.
8. *Chemical Acronyms Database*, Indiana University, <www.indiana.edu/~cheminfo/databases.html>.
9. *Acronyms and Symbols*, <www3.interscience.wiley.com/stasa/>.
10. IUPAC Compendium of Chemical Terminology (Gold Book), <goldbook.iupac.org>.
11. IUPAC-IUB Joint Commission on Biochemical Nomenclature, *Pure Appl. Chem.* 56, 595, 1984.

Symbol	Meaning
A	ampere; alanine; adenine (in genetic code)
Å	ångström
A	absorbance; area; Helmholtz energy; mass number
A_H	Hall coefficient
A_r	atomic weight (relative atomic mass)
a	atto (SI prefix for 10^{-18})
a	absorption coefficient; acceleration; activity; van der Waals constant
a_0	Bohr radius
AAA	acetoacetanilide
Aad	2-aminoadipic acid
AAF	2-(acetylamino)fluorene
AAN	aminoacetonitrile
AAO	acetaldehyde oxime
AAS	atomic absorption spectroscopy
ABA	abscisic acid; acrylonitrile-butadiene acrylate
Abe	abequose
ABL	α-acetylbutyrolactone
ABS	acrylonitrile-butadiene-styrene copolymer
abs	absolute
Abu	2-aminobutanoic acid
Ac	acetyl; acetate
ac, AC	alternating current
ACAC	acetylacetone
Aces	2-[(2-amino-2-oxoethyl)amino]ethanesulfonic acid
ACS	acrylonitrile-chlorinated polyethylene-styrene copolymer
ACT	activated complex theory
ACTH	adrenocorticotropic hormone
A/D	analog to digital
Ad	adamantyl
Ada	[(carbamoylmethyl)imino]diacetic acid
Ade	adenine
ADI	acceptable daily intake
Ado	adenosine
ADP	adenosine diphosphate; ammonium dihydrogen phosphate
ads	adsorption
AE	appearance energy
ae	eon (10^9 years)
AEP	1-(2-aminoethyl)piperazine
AEPD	2-amino-2-ethyl-1,3-propanediol
AES	atomic emission spectroscopy; Auger electron spectroscopy
AF	audio frequency
AFM	atomic force microscopy
Ahx	2-aminohexanoic acid
AI	artificial intelligence
AIBN	2,2'-azobis[isobutyronitrile]
AICA	5-amino-1*H*-imidazole-4-carboxamide
AIM	atoms in molecules (method)
AIP	aluminum isopropoxide
Al	Alfén number
Ala	alanine
alc	alcohol
ALE	atomic layer epitaxy
aliph.	aliphatic
alk.	alkaline
All	allose
Alt	altrose
AM	amplitude modulation
Am	amyl
am	amorphous solid
AMA	acrylate maleic anhydride terpolymer
AMMA	acrylate-methyl methacrylate copolymer
AMP	adenosine monophosphate
AMPD	2-amino-2-methyl-1,3-propanediol
AMS	accelerator mass spectrometry
AMTCS	amyltrichlorosilane [trichloropentylsilane]
amu	atomic mass unit (recommended symbol is u)
AN	acetonitrile; acrylonitrile
anh, anhyd	anhydrous

Symbol	Meaning
ANOVA	analysis of variance
antilog	antilogarithm
ANTU	1-naphthalenylthiourea
AO	atomic orbital
AOM	angular overlap model
AP	ethylene-propylene copolymer
APAD	3-acetylpyridine adenine dinucleotide
APAP	acetyl *p*-amino phenol (acetaminophen)
Ape	2-aminopentanoic acid
API	atmospheric pressure ionization
Api	apiose
APM	atomic probe microanalysis
Apm	2-aminopimelic acid
APO	amorphous polyolefin
APPI	atmospheric pressure photoionization
APS	appearance potential spectroscopy; adenosine phosphosulfate
APW	augmented plane wave
aq	aqueous
Ar	aryl
Ara	arabinose
Ara-ol	arabinitol
Arg	arginine
ARPES	angular resolved photoelectron spectroscopy
AS	acrylonitrile styrene copolymer
ASA	acetylsalicylic acid; acrylonitrile-styrene-acrylonitrile block copolymer
ASC	4-(acetylamino)benzenesulfonyl chloride
ASCII	American National Standard Code for Information Interchange
ASE	aromatic stabilization model
Asn	asparagine
Asp	aspartic acid
at	atomization
ATCP	4-amino-3,5,6-trichloro-2-pyridinecarboxlic acid
ATEE	*N*-acetyl-L-tyrosine ethyl ester
ATLC	adsorption thin layer chromatography
atm	standard atmosphere
ATP	adenosine triphosphate
ATR	attenuated total internal reflection
at.wt.	atomic weight
AU	astronomical unit (ua is also used); polyurethane
AUC	area under the time-concentration curve
av	average
avdp	avoirdupois
B	bel; asparagine or aspartic acid (unspecified)
B	magnetic flux density; second virial coefficient; susceptance
b	barn
b	van der Waals constant; molality
BA	benzyladenine
BAL	British anti-Lewisite [2,3-dimercapto-1-propanol]
BAP, BaP	benzo[*a*]pyrene
bar	bar (pressure unit)
bbl	barrel
BBP	benzyl butyl phthalate
BCB	bromocresol blue
bcc	body-centered cubic
BCF	bioconcentration factor
BCG	bromocresol green
BCME	bis(chloromethyl) ether
BCNU	*N,N'*-bis(2-chloroethyl)-*N*-nitrosourea
BCP	bromocresol purple
BCPB	bromochlorophenol blue
BCPE	1,1-bis(4-chlorophenyl)ethanol
BCS	Bardeen-Cooper-Schrieffer (theory)
BDE	bond dissociation energy
BDEA	butyldiethanolamime
BDMA	benzyldimethylamine
Bé	Baumé
BEBO	bond energy bond order (method)
BEI	biological exposure index
BEM	biological effect monitoring
BEP	2-butyl-2-ethyl-1,3-propanediol
BES	2-[bis(2-hydroxyethyl)amino]ethanesulfonic acid
BET	Brunauer-Emmett-Teller (isotherm)
BeV	billion electronvolt
BGE	butyl glycidyl ether
BHA	*tert*-butyl-4-hydroxyanisole
BHC	benzene hexachloride [hexachlorobenzene]
Bhn	Brinell hardness number
BHT	butylated hydroxytoluene [2,6-di-*tert*-butyl-4-methylphenol]
Bi	biot
Bicine	*N,N*-bis(2-hydroxyethyl)glycine
BIRD	blackbody infrared radiative dissociation
Bistris	2-[bis(2-hydroxyethyl)amino]-2-(hydroxymethyl) propane-1,3-diol
Bistris-propane	1,3-bis[tris(hydroxymethyl)methylamino]propane
BLO	γ-butyrolactone
BN	bond number; benzonitrile
BNS	nuclear backscattering spectroscopy
BO	Born-Oppenheimer (approximation); bond order
BOD	biochemical oxygen demand
BON	β-hydroxynaphthoic acid
BP	base peak (in mass spectrometry); benzo[*a*]pyrene
bp	boiling point; base pair
BPB	bromophenol blue
BPG	2,3-bis(phospho)-*D*-glycerate
BPL	β-propiolactone
BPO	benzoyl peroxide
bpy	2,2'-bipyridine
Bq	becquerel
Br	butyryl
BRE	bond resonance energy
BrUrd	5-bromouridine
BS	Birge-Sponer extrapolation
BSE	back scattered electron(s)
BSSE	basis set superposition error
BTMSA	1,2-bis(trimethylsilyl)acetylene
Btu	British thermal unit
BTX	benzene, toluene, and xylene
Bu	butyl
bu	bushel
BVE	butyl vinyl ether
Bz	benzoyl
Bzl	benzyl
C	coulomb; cysteine; cytosine (in genetic code)
°C	degree Celsius
C	capacitance; heat capacity; number concentration
c	centi (SI prefix for 10^{-2}); combustion reaction
c	amount concentration; specific heat; velocity
c_0	speed of light in vacuum
CA	collisional activation; cellulose acetate

Symbol	Meaning
ca.	approximately
CAB	cellulose acetate butyrate
CADD	computer-assisted drug design
cal	calorie
calc	calculated
cAMP	adenosine cyclic 3',5'-(hydrogen phosphate)
CAN	ceric ammonium nitrate
CAR	carbon fiber
CARS	coherent anti-Stokes Raman spectroscopy
CAS	complete active space
CASRN	Chemical Abstracts Service Registry Number
CAT	computerized axial tomography; clear air turbulence
CBE	chemical beam epitaxy
CBS	complete basis set (of orbitals)
CC	coupled cluster; combustion calorimetry
cc	cubic centimeter
CCD	charge-coupled device
CD	circular dichroism
cd	candela; condensed (phase)
CDAA	2-chloro-*N*,*N*-diallylacetamide
CDNO	complete neglect of differential overlap
CDP	cytidine 5'-diphosphate
CDT	1,5,9-cyclododecatriene
CDTA	(1,2-cyclohexylenedinitrilo)tetraacetic acid monohydrate
CDW	charge density waves
CED	cohesive energy density
CEM	channel electron multiplier
CEP	counter electrophoresis
CEPA	coupled electron-pair approximation
cf.	compare
CFC	chlorofluorocarbon compound
cfm	cubic feet per minute
CFRP	carbon reinforced plastics
cgs	centimeter-gram-second system
CHAPS	3-[3-(cholamidopropyl)dimethylammonio]-1-propanesulfonic acid
CHES	2-(*N*-cyclohexylamino)ethanesulfonic acid
CHF	coupled Hartree-Fock (method)
Chl	chlorophyll
Cho	choline
CHT	1,3,5-cycloheptatriene
Ci	curie
CI	configuration interaction; chemical ionization; color index
CID	charge-injection device; collision-induced dissociation
CIDEP	chemically induced dynamic electron polarization
CIDNP	chemically induced dynamic nuclear polarization
CIE	countercurrent immunoelectrophoresis
cir	circular
CKFF	Cotton-Kraihanzel force field
CL	cathode luminescence (spectroscopy)
CLT	central limit theorem
cm	centimeter
c.m.	center of mass
CMC	carboxymethylcellulose
c.m.c.	critical micelle concentration
CMO	canonical molecular orbital
CMP	cytidine 5'-monophosphate; chemical measurement process
CN	coordination number; cellulose nitrate
CNDO	complete neglect of differential overlap
Co	Cowling number
COC	cycloolefin copolymer

Symbol	Meaning
COD	chemical oxygen demand; 1,4-cyclooctadiene
conc	concentrated; concentration
const	constant
COOP	crystal orbital overlap population
cos	cosine
cosh	hyperbolic cosine
COSY	correlation spectroscopy
COT	1,3,5,7-cyclooctatetraene
cot	cotangent
coth	hyperbolic cotangent
CP	chemically pure
Cp	cyclopentadienyl
Cp*	pentamethylcyclopentadienyl
cP	centipoise
cp	candle power
CPA	coherent potential approximation
CPC	centrifugal partition chromatography
cpd	contact potential difference
CPE	chlorinated polyethylene
CPL	circular polarization of luminescence
CPR	chlorophenol red
cps	cycles per second
CPT	charge conjugation/space inversion/time inversion (theorem)
CPU	central processing unit
CPVC	chlorinated poly(vinyl chloride)
CR	chloroprene rubber (neoprene)
cr, cryst	crystalline (phase)
CRF	charge remote fragmentation
CRU	constitutional repeating unit (in polymer nomenclature)
CSA	camphorsulfonic acid
csc	cosecant
CSR	charge stripping reaction
CT	charge transfer
ct	carat
CTA	cellulose triacetate
CTEM	conventional transmission electron microscopy
CTFE	chlorotrifluoroethylene
CTP	cytidine 5'-triphosphate
CTR	controlled thermonuclear reaction
cu	cubic
CV	cyclic voltammetry
CVD	chemical vapor deposition
cw	continuous wave
cwt	hundredweight (112 pounds)
Cy	cyclohexyl
Cya	cysteic acid
Cyd	cytidine
cyl	cylinder
Cys	cysteine
Cyt	cytosine
D	debye unit; aspartic acid
D	diffusion coefficient; dissociation energy; electric displacement
d	day; deuteron; deci (SI prefix for 10^{-1})
d	distance; density; dextrorotatory
2,4-D	2,4-dichlorophenoxyacetic acid
D/A	digital to analog
Da	dalton
DA	donor-acceptor (complex)
da	deka (SI prefix for 10^{1})
DAA	diacetone alcohol

Symbol	Meaning
DAB	4-(dimethylamino)azobenzene
Dab	2,4-diaminobutanoic acid
DACH	*trans*-1,2-diaminocyclohexane
DAIP	diallyl isophthalate plasticizer
DAP	diammonium phosphate
DART	direct analysis in real time mass spectrometry
dB	decibel
DBA	dibenz[*a*,*h*]anthracene; dibenzylamine
DBCP	1,2-dibromo-3-chloropropane
DBED	dibenzyl ethylene diamine
DBM	dibutyl maleate
DBMC	2,4-di-*tert*-butyl-5-methylphenol
DBMS	database management system
DBP	dibutyl phthalate; 2,3-dibromo-1-propanol
DBPC	2,6-di-*tert*-butyl-*p*-cresol
dc, DC	direct current
DCB	dicyanobenzene
DCBP	4,4′-dichlorobenzophenone
DCEE	dichloroethyl ether
DCHA	dicyclohexylamine
DCM	dichloromethane
DCNP	2,6-dichloro-4-nitrophenol
DCP	2,4-dichlorophenol
DCPD	dicyclopentadiene
DDM	4,4′-diaminodiphenylmethane
DDT	dichlorodiphenyltrichloroethane
DE	delocalization energy; delayed extraction
DEA	*N*,*N*-diethylaniline; diethanolamine
Dec	decyl
dec	decomposes
DEET	diethyltoluamide [*N*,*N*-diethyl-3-methylbenzamide]
deg	degree
DEK	diethyl ketone
den	density
DEP	2,2-diethyl-1,3-propanediol
DES	diethyl sulfate
DESI	desorption electrospray ionization (in mass spectrometry)
det	determinant
dev	deviation
DFT	density functional theory
dGlc	2-deoxyglucose
DHBA	2,3-dihydroxybenzoic acid
DHH	dehydroheliotridine
DHR	dehydroretronecine
DHU	dihydrouridine
DI	desorption ionization
diam	diameter
DIBA	diisobutyl adipate
DIBK	diisobutyl ketone
dil	dilute; dilution
DIM	diatomics in molecules (method); digital imaging microscopy
DIPA	diisopropanolamine
dm	decimeter
DMA	*N*,*N*-dimethylaniline
DMAB	4-(dimethylamino)azobenzene
DMAC	*N*,*N*-dimethylacetamide
DMAE	*N*,*N*-dimethylethanolamine
DMBA	7,12-dimethylbenz[*a*]anthracene
DME	1,2-dimethoxyethane
DMF	*N*,*N*-dimethylformamide

Symbol	Meaning
DMP	dimethyl phthalate
DMS	dimethyl sulfide
DMSO	dimethyl sulfoxide
DMT	dimethyl terephthalate; dimethyl tartrate
DN	donor number
DNA	deoxyribonucleic acid
DNase	deoxyribonuclease
DNB	1,3-dinitrobenzene
DNMR	dynamic NMR spectroscopy
DNP	dinitropyrene
Dod	dodecyl
DOP	dioctyl phthalate
DOS	density of states; digital operating system; dioctyl sebacate
doz	dozen
DP, d.p.	degree of polymerization
DPA	diphenylamine
DPG	*N*,*N*′-diphenylguanidine
dpl	displacement
Dpm	2,6-diaminopimelic acid
dpm	disintegrations per minute
dps	disintegrations per second
DPU	*N*,*N*′-diphenylurea
dr	dram
DRE	Dewar resonance energy
dRib	2-deoxyribose
DRIFT	diffuse reflectance infrared Fourier transform
DRP	dynamic reaction path
DRS	diffuse reflectance spectroscopy
DS	degree of substitution
DSC	differential scanning calorimetry
DTA	differential thermal analysis
DTBP	di-*tert*-butyl peroxide
DVB	divinylbenzene
dyn	dyne
DZ	double-zeta (type of basis set)
E	exa (SI prefix for 10^{18}); glutamic acid
E	electric field strength; electromotive force; energy; Young's modulus of elasticity; entgegen (transconfiguration)
E_h	Hartree energy
e	electron; base of natural logarithms
e	elementary charge; linear strain
EA	electron affinity
EAA	ethylene acrylic acid copolymer; ethyl acetoacetate
EAK	ethyl amyl ketone (3-octanone)
EAN	effective atomic number
EC	ethyl cellulose
ECD	electron capture dissociation
ECP	effective core potential
ECR	electron cyclotron resonance
ECTFE	ethylene-chlorotrifluoroethylene copolymer
ED	electron diffraction
EDAX	energy dispersive analysis by x-rays
EDB	ethylene dibromide [1,2-dibromoethane]
EDC	ethylene dichloride [1,2-dichloroethane]
EDI	estimated daily intake
EDS	energy-dispersive x-ray spectroscopy
EDTA	ethylenediaminetetraacetic acid
EEA	ethylene-ethyl acetate copolymer
EEDQ	ethyl 2-ethoxy-1(2*H*)-quinolinecarboxylate
EEL	environmental exposure level
EELS	electron energy loss spectroscopy

Symbol	Meaning
EES	excitation emission spectrum
EFF	empirical force field
EFFF	energy factored force field
EG	equilibrium in the gas phase
EGA	evolved gas analysis
EGG	Einstein-Guth-Gold equation
EHMO, EHT	extended Hückel molecular orbital (theory)
EIMS	electron impact mass spectrometry
EIS	electron impact spectroscopy; electrochemical impedance spectroscopy
ELISA	enzyme-linked immunosorbent assay
ELS	energy loss spectroscopy
EM	extended molarity; electron microscopy
EMAC	ethylene-methyl acrylate copolymer
emf	electromotive force
EMPA, EMA	electron probe microanalysis
emu	electromagnetic unit system
en	ethylenediamine
ENDOR	electron-nuclear double resonance
EOS	equation of state
EP	epoxy resin
EPDS	electron photodetachment spectroscopy
EPM	ethylene-propylene copolymer
EPR	electron paramagnetic resonance; ethylene propylene rubber
EPS	expanded polystyrene
EPT-76	Provisional Low Temperature Scale of 1976
EPTC	dipropylcarbamothioic acid, S-ethyl ester
EPXMA	electron probe x-ray microanalysis
eq, eqn	equation
eqQ	quadrupole coupling constant
erf	error function
erg	erg (energy unit)
ES	equilibrium in solution
ESA	electrostatic energy analyzer
ESCA	electron spectroscopy for chemical analysis
ESD	electron stimulated desorption
e.s.d.	estimated standard deviation
ESI	electrospray ionization
ESR	electron spin resonance
est	estimated
esu	electrostatic unit system
ET	ephemeris time; electron transfer
Et	ethyl
ETA	electrothermal analysis
ETFE	ethylene tetrafluoroethylene polymer
Etn	ethanolamine
ETO	ethylene oxide
ETS	electron tunneling spectroscopy
ETU	ethylene thiourea
EU	polyether polyurethane
Eu	Euler number
e.u.	entropy unit
eV	electronvolt
EVA	ethylene-vinyl acetate copolymer
EVE	ethyl vinyl ether
EXAFS	extended x-ray absorption fine structure (spectroscopy)
EXELFS	extended energy loss fine structure
exp	exponential function
expt	experimental
ext	external
F	farad; phenylalanine
°F	degree Fahrenheit
F	Faraday constant; force; angular momentum
f	formation reaction; femto (SI prefix for 10^{-15})
f	activity coefficient; aperture ratio; focal length; force constant; frequency; fugacity
FAB	fast atom bombardment
FAD	flavine adenine dinucleotide
FAIMS	high-field asymmetric waveform ion mobility spectrometry
FA-SIFT	flowing afterglow – selected ion-flow tube
fcc	face centered cubic
FD	field desorption
FEL	free electron laser
FEM	field emission microscopy
FEMO	free electron molecular orbital
FEP	fluorinated ethylene propylene
FET	field effect transistor
FI	field ionization
fid	free induction decay
FIM	field ion microscopy
FIR	far infrared
fl	fluid (phase)
FM	frequency modulation
Fo	Fourier number
fp	freezing point
fpm	feet per minute
fps	feet per second; foot-pound-second system
Fr	franklin
Fr	Froude number
FRP	fibrous glass reinforced polyester; fiber reinforced plastic
Fru	fructose
FSGO	floating spherical gaussian orbitals
FT	Fourier transform
ft	foot
ft-lb	foot pound
FTIR	Fourier transform infrared spectroscopy
FTMS	Fourier transform mass spectrometry
FTNMR	Fourier transform nuclear magnetic resonance
fus	fusion (melting)
FVP	flash vacuum pyrolysis
FWHM	full width at half maximum
G	gauss; guanine (in genetic code); giga (SI prefix for 10^9); glycine
G	electrical conductance; Gibbs energy; gravitational constant; sheer modulus
g	gram; gas (phase)
g	acceleration due to gravity; degeneracy; Landé g-factor; statistical weight
GABA	γ-aminobutyric acid
Gal	gal; galactose
gal	gallon
GalN	galactosamine
GB	gas-phase basicity
GC	gas chromatography
GC-MS	gas chromatography-mass spectroscopy
GDMS	glow discharge mass spectroscopy
GDP	guanosine 5'-diphosphate
gem	geminal (on the same carbon atom)
GeV	gigaelectronvolt
GF	glass reinforced
GIAO	gauge invariant atomic orbital

Symbol	Meaning
GIBMS	guided ion beam mass spectrometry
gl	glacial
Gla	4-carboxyglutamic acid
GLC	gas-liquid chromatography
Glc	glucose
GlcA	gluconic acid
GlcN	glucosamine
GlcNAc	*N*-acetylglucosamine
GlcU	glucuronic acid
Gln	glutamine
GLP	good laboratory practice
Glu	glutamic acid
Glx	glutamine or glutamic acid (unspecified)
Gly	glycine
GMP	guanosine 5'-monophosphate
GMT	Greenwich mean time
GPC	gel-permeation chromatography
gpm	gallons per minute
gps	gallon per second
Gr	Grashof number
gr	grain
Gra	glyceraldehyde
Gri	glyceric acid
Grn	glycerone [dihydroxyacetone]
Gro	glycerol
GTO	gaussian-type orbital
GTP	guanosine 5'-triphosphate
Gua	guanine
Gul	gulose
Guo	guanosine
GUT	grand unified theory
GVB	generalized valence bond (method)
GWS	Glashow-Weinberg-Salam (theory)
Gy	gray; gigayear
H	henry; histidine
H	enthalpy; Hamiltonian function; magnetic field
H_0	Hubble constant
h	helion; hour; hecto (SI prefix for 10^2)
h	Planck constant
Ha	Hartmann number
ha	hectare
HAM	hydrogenic atoms in molecules
hav	haversine
Hb	hemoglobin
HCA	heterocyclic amine; hexachloroacetone
HCB	hexachlorobenzene
hcp	hexagonal closed packed
Hcy	homocysteine
HCZ, HCTZ	hydrochlorothiazide
HDL	high-density lipoprotein
HDPE	high-density polyethylene
HDS	hydrodesulfurization
HEIS	high-energy ion scattering
HEP	high-energy physics
HEPES	4-(2-hydroxyethyl)-1-piperazineethanesulfonic acid
HEPPS	4-(2-hydroxyethyl)-1-piperazinepropanesulfonic acid
HF	high-frequency; Hartree-Fock (method)
HFA	hexafluoroacetone
HFO	Hartree-Fock orbital
hfs	hyperfine structure

Symbol	Meaning
HHPA	hexahydrophthalic anhydride
HIPS	high-impact polystyrene
His	histidine
HMC	high-strength molding compound
HMDA	hexamethylenediamine
HMO	Hückel molecular orbital
HMT	hexamethylenetetramine
HMX	cyclotetramethylenetetranitramine
HN1	2-chloro-*N*-(2-chloroethyl)-*N*-ethylethanamine
HOAc	acetic acid
HOC	halogenated organic compound(s)
HOMAS	harmonic oscillator model of aromatic stabilization
HOMO	highest occupied molecular orbital
HOSE	harmonic oscillator stabilization energy
Hp	heptyl
hp	horsepower
HPLC	high-performance liquid chromatography
HPMS	high-pressure mass spectrometry
HQ	*p*-hydroquinone
hr	hour
HRE	Hückel resonance energy
HREELS	high-resolution electron energy loss spectroscopy
HREM	high-resolution electron microscopy
HSAB	hard-soft acid-base (theory)
HSE	homodesmotic stabilization energy
Hse	homoserine
HVA	homovanillic acid
Hx	hexyl
Hyl	5-hydroxylysine
Hyp	hypoxanthine; 4-hydroxyproline
Hz	hertz
I	isoleucine; inositol; ionomer
I	electric current; ionic strength; moment of inertia; nuclear spin angular momentum; radiant intensity
i	square root of minus one
i	electric current
I/O	input/output
IAT	international atomic time
IC	integrated circuit
ICD	induced circular dichroism
ICP	inductive-coupled plasma
ICR	ion cyclotron resonance
ICVTST	improved canonical variational transition-state theory
ID	inside diameter
id	ideal (solution)
Ido	iodose
IdoA	iduronic acid
IDP	inosine 5'-diphosphate
IE	ionization energy
i.e.p.	isoelectric point
IEPA	independent electron pair approximation
IF	intermediate frequency
IGLO	individual gauge for localized orbitals
IIR	isobutylene-isoprene rubber (butyl rubber)
IKES	ion kinetic energy spectrometry
Ile	isoleucine
Im	imaginary part
IMFP	inelastic mean free path (of electrons)
imm	immersion
IMP	inosine 5'-monophosphate
IMPATT	impact ionization avalanche transit time

Symbol	Meaning
IMS	ion mobility spectrometry
in.	inch
InChI	IUPAC International Chemical Identifier
INDO	immediate neglect of differential overlap
Ino	inosine
INS	inelastic neutron scattering; ion neutralization spectroscopy
Ins	*myo*-inositol
int	internal
IP	ionization potential
IPA	isopropyl alcohol
IPMA	ion probe microanalysis
IPN	interpenetrating polymer network
IPR	isotope perturbation of resonance
IPTS	International Practical Temperature Scale
IQ	2-amino-3-methyl-3*H*-imidazo(4,5-*f*)quinoline
IR	infrared
IRAS	infrared reflection-absorption spectroscopy
IRC	intrinsic reaction coordinate
IRMPD	infrared multiphoton dissociation
IRMS	isotope ratio mass spectrometry
IRS	infrared spectroscopy
isc	intersystem crossing
ISE	ion-selective electrode; isodesmic stabilization energy
ISS	ion scattering spectroscopy
IT	ion trap; information technology
ITP	inosine 5'-triphosphate
ITS	International Temperature Scale (1990)
IU	international unit
IVE	isobutyl vinyl ether
J	joule; leucine or isoleucine (unspecified)
J	angular momentum; electric current density; flux; Massieu function
j	angular momentum; electric current density
JT	Jahn-Teller (effect)
K	kelvin; lysine
K	absorption coefficient; bulk modulus; equilibrium constant; kinetic energy
k	kilo (SI prefix for 10^3)
k	absorption index; Boltzmann constant; rate constant; thermal conductivity; wave vector
kat	katal (unit of catalytic activity)
kb	kilobar; kilobases (DNA or RNA)
KC-MS	Knudson cell mass spectrometry
kcal	kilocalorie
KDP	potassium dihydrogen phosphate
KE	kinetic energy
KERD	kinetic energy release distributions
keV	kiloelectronvolt
KG	kinetics in the gas phase
kg	kilogram
kgf	kilogram force
KIE	kinetic isotope effect
kJ	kilojoule
km	kilometer
Kn	Knudsen number
kPa	kilopascal
KS	kinetics in solution
kt	karat
KTP	potassium titanium phosphate
kV	kilovolt
kva	kilovolt ampere

Symbol	Meaning
kW	kilowatt
kwh	kilowatt hour
L	liter; lambert; leucine
L	Avogadro constant; inductance; Lagrange function; angular momentum
l	liter; liquid (phase)
l	angular momentum; length; mean free path; levorotatory
Lac	lactose
LAH	lithium aluminum hydride
lat.	latitude
lb	pound
lbf	pound force
LC	liquid chromatography; liquid crystal
LC-MS	liquid chromatography-mass spectrometry
lc	liquid crystal (phase)
LCAO	linear combination of atomic orbitals
LD	lethal dose; laser desorption
LDA	local density approximation; lithium diisopropylamide
LDL	low-density lipoprotein
LDPE	low-density polyethylene
LDV	laser-Doppler velocimetry
Le	Lewis function
LE	localization energy
LEC	liquid exchange chromatography
LED	light-emitting diode
LEED	low-energy electron diffraction
LEIS	low-energy ion scattering
Leu	leucine
LFER	linear free energy relationships
LFL	lower flammable limit
LI	laser ionization
lim	limit
LIMS	laser ionization mass spectroscopy; laboratory information management system
liq	liquid
LIT	linear ion trap
LLCT	ligand to ligand charge transfer
lm	lumen
LMCT	ligand to metal charge transfer
LMMS	laser microprobe mass spectrometry
LMO	localized molecular orbital
LMR	laser magnetic resonance
ln	logarithm (natural)
LNDO	local neglect of differential overlap
log	logarithm (common)
LOMO	lowest occupied molecular orbital
long.	longitude
LPE	linear polyethylene
LPG	liquid petroleum gas
LPHP	laser-powered homogeneous pyrolysis
LPU	law of propagation of uncertainty
LSFE	linear field stabilization energy
LSI	liquid secondary ionization
LST	local sidereal time
LT	local time
LTE	local thermodynamic equilibrium
LUMO	lowest unoccupied molecular orbital
lx	lux
ly	langley
l.y.	light-year
Lys	lysine
Lyx	lyxose

Symbol	Meaning
M	molar (as in 0.1 M solution); mega (SI prefix for 10^6); methionine
M	magnetization; molar mass; mutual inductance; torque; angular momentum component; median
Mr	molecular weight (relative molar mass)
m	meter; molal (as in 0.1 m solution); metastable (isotope); milli (SI prefix for 10^{-3})
m	magnetic dipole moment; mass; molality; angular momentum component; *meta* (locant on aromatic ring)
Ma	Mach number
MA	maleic anhydride
MAAc	methyl amyl acetate
Mal	maltose
Man	mannose
MASNMR	magic angle spinning nuclear magnetic resonance
max	maximum
Mb	myoglobin
MBE	molecular beam epitaxy
MBER	molecular beam electron resonance
MBK	methyl butyl ketone
MBOCA	4,4′-methylenebis[2-chloroaniline]
MBPT	many body perturbation theory
MBS	methyl methacrylate butadiene styrene terpolymer
MC	Monte Carlo (method)
MCAA	monochloroacetic acid
MCD	magnetic circular dichroism
MCP	microchannel plate
MCPA	(4-chloro-2-methylphenoxy)acetic acid
MCPF	modified coupled pair functional
MCS	Monte Carlo simulation
MCSCF	multiconfigurational self-consistent field (approximation)
MD	molecular dynamics (method)
MDI	methylene diphenylisocyanate
MDPE	medium-density polyethylene
Me	methyl
MeCCNU	1-(2-chloroethyl)-3-(4-methylcyclohexyl)-1-nitrosourea
MeIQ	2-amino-3,4-dimethylimidazo[4,5-f]quinoline
MeIQx	2-amino-3,8-dimethylimidazo[4,5-f]quinoxaline
MEK	methyl ethyl ketone
MEP	molecular electrostatic potential
MERP	minimum energy reaction path
Mes	4-morpholineethanesulfonic acid
MESFET	metal-semiconductor field-effect transistor
Met	methionine
MeV	megaelectronvolt
meV	millielectronvolt
MF	molecular formula; melamine-formaldehyde resin
mg	milligram
MHD	magnetohydrodynamics
mi	mile
MIAK	methyl isoamyl ketone
MIBK	methyl isobutyl ketone
MIC	methyl isocyanate
MIK	methyl isobutyl ketone
MIKES	mass-analyzed ion kinetic energy spectrometry
min	minimum; minute
MINDO	modified INDO (method)
MIPK	methyl isopropyl ketone
MIR	mid infrared
misc	miscible
MKS	meter-kilogram-second system
MKSA	meter-kilogram-second-ampere system
mL, ml	milliliter
MM	molecular mechanics
mm	millimeter
MMDR	microwave-microwave double resonance
mmf	magnetomotive force
mmHg	millimeter of mercury
MNA	*m*-nitroaniline
MNDO	modified neglect of diatomic overlap
MNT	*m*-nitrotoluene
MNU	*N*-methyl-*N*-nitrosourea
MO	molecular orbital; methyl orange
MODR	microwave-optical double resonance
mol	mole
mol.wt.	molecular weight
mon	monomeric form
Mops	4-morpholinepropanesulfonic acid
MOS	metal-oxide semiconductor
MOSFET	metal-oxide semiconductor field-effect transistor
mp	melting point
MPa	megapascal
MPA	Mulliken population analysis
Mpc	megaparsec
MPD	2-methyl-2,4-pentanediol
MPI	multiphoton ionization
MPTP	1,2,3,6-tetrahydro-1-methyl-4-phenylpyridine
MR	methyl red
MRD	multireference double substitution (method)
MRI	magnetic resonance imaging
mRNA	messenger RNA
MS	mass spectroscopy
ms	millisecond
MSA	methanesulfonic acid
MSDS	Material Safety Data Sheet
MSF	methanesulfonyl fluoride
MS-K	mass spectroscopy – kinetic method
MSL	mean sea level
MTBE	methyl *tert*-butyl ether
MTD	maximum tolerable dose
Mur	muramic acid
mV	millivolt
MVK	methyl vinyl ketone
MW	megawatt; microwave; molecular weight
mW	milliwatt
MWD	molecular weight distribution
Mx	maxwell
N	newton; asparagine
N	angular momentum; neutron number; number density
N_A	Avogadro constant
n	neutron; nano (SI prefix for 10^{-9})
n	amount of substance; number density; principal quantum number; refractive index; normal (in chemical formulas)
NAA	nuclear activation analysis; 1-naphthaleneacetic acid
NAAD	nicotinic acid adenine dinucleotide
NAD	nicotinamide adenine dinucleotide
NADH	reduced NAD
NADP	NAD phosphate
NANA	*N*-acetylneuraminic acid
NAO	natural atomic orbital
NBO	natural bond orbital
nbp	normal boiling point
NBR	nitrile butadiene rubber [poly(butadiene-*co*-acrylonitrile)]

Symbol	Meaning
NDELA	*N*-nitrosodiethanolamine
NEDOR	nuclear electron double resonance
NEM	*N*-ethylmorpholine
Neu	neuraminic acid
NEXAFS	near-edge x-ray absorption fine structure
ng	nanogram
NHO	natural hybrid orbital
NHOMO	next-to-highest occupied molecular orbital
NICI	negative ion chemical ionization
NICS	nuclear independent chemical shift
NIR	near infrared; ribosylnicotinamide
nm	nanometer
NMN	β-nicotinamide mononucleotide
NMR	nuclear magnetic resonance
Nn	nonyl
NNDO	neglect of nonbonded differential overlap
NO	natural orbital
NOE	nuclear Overhauser effect
NOEL	no-observed-effect level
NOx	nitrogen oxides
NP	nitropyrene
NPA	natural population analysis
NQR	nuclear quadrupole resonance
NR	natural rubber
NRA	nuclear reaction analysis
ns	nanosecond
NSE	neutron spin echo
NTA	nitrilotriacetic acid
NTP	normal temperature and pressure
Nu	nucleophile
Nu	Nusselt number
o	*ortho* (locant on aromatic ring)
OAA	oxaloacetic acid
obs, obsd	observed
Oc	octyl
OD	optical density; outside diameter
ODMR	optically detected magnetic resonance
Oe	oersted
OFGF	outer valence Green's function (method)
ONA	*o*-nitroaniline
ORD	optical rotatory dispersion
Oro	orotate; orotidine
oz	ounce
P	poise; peta (SI prefix for 10^{15}); proline
P	power; pressure; probability; sound energy flux
p	proton; pico (SI prefix for 10^{-12})
p	dielectric polarization; electric dipole moment; momentum; pressure; bond order; *para* (as aromatic ring locant)
Pa	pascal
PA	proton affinity; pyrrolizidine alkaloid; polyamide (nylon)
PAA	poly(acrylic acid)
PABA	*p*-aminobenzoic acid
PABS	*p*-aminobenzenesulfonamide
PAC	photoacoustic calorimetry
PAH	polycyclic aromatic hydrocarbon(s)
PAI	polyamide-imide
PAL	polyaniline
PAM	polyacrylamide
PAN	1-(2-pyridylazo)-2-naphthol; polyacrylonitrile
PAR	4-(2'-pyridylazo)resorcinol
PARA	polyaryl amide

Symbol	Meaning
PAS	photoacoustic spectroscopy; polyarylsulfone
PB	polybutylene
PBA	poly(butyl acrylate)
PBAN	polybutylene-acrylonitrile copolymer
PBB	polybrominated biphenyl
PBD	poly(1,3-butadiene)
PBI	polybenzimidazole
PBMA	poly(butyl methacrylate)
PBS	polybutadiene-styrene copolymer
PBT	poly(butylene terephthalate)
PC	paper chromatography; photocalorimetry; polycarbonate
pc	parsec
PCB	polychlorinated biphenyl
PCHO	paraldehyde (2,4,6-trimethyl-1,3,5-trioxane)
PCL	polycaprolactone
PCM	polarizable continuum model
PCNB	pentachloronitrobenzene
PCP	pentachlorophenol
PCR	polymerase chain reaction
PCT	poly(cyclohexylene terephthalate)
PCTFE	polymonochlorotrifluoroethylene
PD	potential difference
PDB	*p*-dichlorobenzene
pdl	poundal
PDMS	poly(dimethylsiloxane)
PE	polyethylene
Pe	pentyl
Pe	Péclet number
pe	probable error
PEA	poly(ethyl acrylate)
PEEK	poly(ether ether ketone)
PEG	poly(ethylene glycol)
PEI	polyetherimide
PEK	polyetherketone
PEL	permissible exposure limit
PEO	poly(ethylene oxide)
PES	photoelectron spectroscopy; potential energy surface; polyethersulfone
PET	positron emission tomography; poly(ethylene terephthalate); pentaerythritol tetranitrate
peth	petroleum ether
PEX	crosslinked polyethylene
PF	phenol-formaldehyde resin
pf	power factor
PFOA	perfluorooctanoic acid
pg	picogram
Ph	phenyl
pH	negative log of hydrogen ion concentration
Phe	phenylalanine
PhIP	2-amino-1-methyl-6-phenylimidazo[4,5-*b*]pyridine
PHPMS	pulsed high-pressure mass spectrometry
PI	polyimide
pI	isoelectric point
PIB	polyisobutylene
PIMS	photoionization mass spectrometry
PIN	p-intrinsic-n (diode)
Pipes	1,4-piperazinediethanesulfonic acid
PIV	particle-image velocimetry
PIXE	particle induced x-ray emission
p*K*	negative log of ionization constant
PLM	principle of least motion
PLOT	porous-layer open-tabular (column)

Symbol	Meaning
PLS	partial least squares
pm	picometer
PMA	poly(methyl acrylate)
PMAC	phenylmercuric acetate
PMMA	poly(methyl methacrylate)
PMO	perturbation MO (theory)
PMP	polymethylpentene
PMS	polymethylstyrene; *p*-methylstyrene
PNA	*p*-nitroaniline
PNDO	partial neglect of differential overlap
PNO	pair natural orbitals
PNRA	prompt nuclear reaction analysis
PNT	*p*-nitrotoluene
PO	polyolefin
POAV	π-orbital axis vector
pol	polymeric form
POM	polyoxymethylene
POx	phosphorus oxides
PP	polypropylene
ppb	parts per billion
PPC	chlorinated polypropylene
PPE	poly(phenylene ether)
ppm	parts per million
PPO	poly(phenylene oxide)
PPOX	polypropylene oxide
PPP	Pariser-Parr-Pople (method)
PPS	poly(phenylene sulfide)
PPSU	poly(phenylene sulfone)
PPT	poly(propylene terephthalate)
ppt	parts per thousand; precipitate
Pr	propyl
Pr	Prandtl number
PRDDO	partial retention of diatomic differential overlap
Pro	proline
PS	photoelectron spectroscopy; polystyrene
ps	picosecond
PSD	photon-stimulated desorption
psi	pounds per square inch
psia	pounds per square inch absolute
psig	pounds per square inch gage
PT	perturbation theory
pt	pint
PTFE	poly(tetrafluoroethylene)
PTME	poly(tetramethylene terephthalate)
PTMS	propyltrimethoxysilane
PTP	*p*-terphenyl
PTU	phenylthiourea
PU	polyurethane
Pu	purine
PVA	poly(vinyl alcohol)
PVAc	poly(vinyl acetate)
PVC	poly(vinyl chloride)
PVD	physical vapor deposition
PVDC	poly(vinylidene chloride)
PVDF	poly(vinylidene fluoride)
PVF	poly(vinyl fluoride)
PVK	poly(vinyl carbazole)
PVME	poly(methyl vinyl ether)
PVOH	poly(vinyl alcohol)
PVP	poly(vinyl pyrrolidone)
PVT	pressure-volume-temperature

Symbol	Meaning
Py	pyrimidine
PyMS	pyrolysis mass spectrometry
p.z.c.	point of zero charge
Q	electric charge; heat; partition function; quadrupole moment; radiant energy; vibrational normal coordinate; glutamine
q	electric field gradient; flow rate; heat; wave vector (phonons)
QCD	quantum chromodynamics
QCI	quadratic configuration interaction
QCT	quasi-classical trajectory (method)
QED	quantum electrodynamics
Q.E.D.	quod erat demonstrandum (which was to be proved)
QIT	quadrupole ion trap
QMRE	quantum mechanical resonance energy
QMS	quadrupole mass spectrometry
QSAR	quantitative structure-activity relations
QSO	quasi-stellar object
qt	quart
quad	quadrillion BTU ($=1.055 \cdot 10^{18}$ joules)
Qui	quinovose
q.v.	quod vide (which you should see)
R	roentgen; arginine; alkyl radical (in chemical formulas)
°R	degree Rankine
R	electrical resistance; gas constant; molar refraction; Rydberg constant; coefficient of multiple correlation
r	reaction (as in $\Delta_r H$)
r	position vector; radius
RA	right ascension
rad	radian
RAIRS	reflection-absorption infrared spectroscopy
RAM	random access memory
RBS	Rutherford back scattering
Rbu, Rul	ribulose
RCI	ring current index
RDA	rubidium dihydrogen arsenate
RDS	rate determining step
RDX	Royal Demolition Explosive (hexahydro-1,3,5-trinitro-1,3,5-triazine)
Re	real part
RE	resonance energy
RED	radial electron distribution
REELS	reflection electron energy loss spectroscopy
REM	reflection electron microscopy
rem	roentgen equivalent man
REMPI	resonance-enhanced multiphoton ionization
REPE	resonance energy per electron
RF	radio frequency
RGA	residual gas analyzer
Rha	rhamnose
RHEED	reflection high-energy electron diffraction
RHF	restricted Hartree-Fock (theory)
RI	resonance ionization
RIA	radioimmunoassay
Rib	ribose
Ribulo	ribulose
rms	root-mean-square
RNA	ribonucleic acid
RNase	ribonuclease
ROHF	restricted open shell Hartree-Fock
ROM	read only memory
ROMP	ring opening metathesis polymerization
ROP	ring opening polymerization

Symbol	Meaning
RPA	random phase approximation
RPH	reaction path Hamiltonian
RPLC	reversed-phase liquid chromatography
rpm	revolutions per minute
rps	revolutions per second
RRK	Rice-Ramsperger-Kassel (theory)
RRKM	Rice-Ramsperger-Kassel-Marcus (theory)
rRNA	ribosomal RNA
RRS	resonance Raman spectroscopy
RS	Raman spectroscopy
RSC	reaction-solution calorimetry
Ry	Rydberg
S	siemens; serine
S	area; entropy; probability current density; Poynting vector; symmetry coordinate; spin angular momentum
s	second; solid (phase)
s	path length; spin angular momentum; symmetry number; sedimentation coefficient; solubility; symmetrical (as stereochemical descriptor)
SAED	selected area electron diffraction
SALC	symmetry adapted linear combinations
SALI	surface analysis by laser ionization
SAM	scanning Auger microscopy
SAMS	self-assembled monolayers
SANS	small angle neutron scattering
SAR	structure-activity relationship
Sar	sarcosine
sat, satd	saturated
SAXS	small angle x-ray scattering
SB	styrene butadiene copolymer
SBS	styrene butadiene styrene block copolymer
Sc	Schmidt number
SC	spin-coupled (method)
SCD	state correlation diagram
SCE	saturated calomel electrode
SCF	self-consistent field (method); supercritical fluid
SCP	single cell protein
SCR	silicon-controlled rectifier
SCRF	self-consistent reaction field (method)
sd	standard deviation
SDA	sulfadiazine
SDW	spin density wave
SE	strain energy
SEBS	styrene ethylene butylene styrene block copolymer
SEC	size exclusion chromatography
sec	secant; second
sec	secondary (in chemical name)
SECSY	spin-echo correlated spectroscopy
Sed	sedoheptulose
SEELFS	surface extended energy loss fine structure
SEM	scanning electron microscopy; standard error of the mean
sepn	separation
Ser	serine
SERS	surface-enhanced Raman spectroscopy
SET	single electron transfer
SEXAF	surface extended x-ray absorption fine structure
SFC	supercritical fluid chromatography
Sh	Sherwood number
Shy	thiohypoxanthine
SI	International System of Units; surface ionization
SID	surface-induced dissociation
SILAR	successive ionic layer adsorption and reaction

Symbol	Meaning
SIM	selected ion monitoring
SIMS	secondary-ion spectroscopy
sin	sine
sinh	hyperbolic sine
SIPN	semi-interpenetrating polymer network
SIS	styrene isoprene styrene block copolymer
SLAM	scanning laser acoustic microscopy
SLUMO	second-lowest-unoccupied molecular orbital
SMILES	simplified molecular input line entry system
SMMA	styrene methyl methacrylate copolymer
SMO	semiempirical molecular orbital
SMOW	Standard Mean Ocean Water (Vienna)
SNMS	sputtered neutral mass spectroscopy
Sno	thiouridine
SNU	solar neutrino unit
SOJT	second-order Jahn-Teller (effect)
sol	soluble; solution
soln, sln	solution
SOMO	singly occupied molecular orbital
Sor	sorbose
sp gr	specific gravity
SPM	scanned probe microscopy
SPST	single-pulse shock tubes
sq	square
Sr	Strouhal number
sr	steradian
Srd	6-thioinosine
SSMS	source spark mass spectroscopy
St	stoke
St	Stanton number
std, stnd	standard (state)
STEL	short-term exposure limit
STEM	scanning transmission electron microscope
STM	scanning tunneling microscopy
STO	Slater-type orbital
STP	standard temperature and pressure
sub, subl	sublimes; sublimation
Suc, Sac	sucrose
Sur	thiouracil
Sv	sievert
SWIFT	stored waveform inverse Fourier transform
T	tesla; tera (SI prefix for 10^{12}); threonine
T	kinetic energy; period; term value; temperature (thermodynamic); torque; transmittance
t	metric tonne; triton
t	Celsius temperature; thickness; time; transport number
TAC	time-to-amplitude converter
TAI	International Atomic Time
Tal	talose
tan	tangent
tanh	hyperbolic tangent
TAPS	3-{[2-hydroxy-1,1-bis(hydroxymethyl)ethyl]amino}-1-propanesulfonic acid
TBE	1,1,2,2-tetrabromoethane
TBP	tributyl phosphate
TC	titration calorimetry
TCA	trichloroacetic acid
TCB, TCBA	2,3,6-trichlorobenzoic acid
TCE	trichloroethylene
TCG	Geocentric Coordinated Time
TCNE	tetracyanoethylene
TCNQ	tetracyanoquinodimethane

Symbol	Meaning
TCP	tricresyl phosphate
TCSCF	two-configuration self-consistent field
TDA	toluene-2,4-diamine
TDI	toluene diisocyanate
tDNA	transfer DNA
TE	transverse electric
TEA	triethanolamine; triethylamine
TED	transferred electron device; transmission electron diffraction
TEDA	triethylenediamine
TEELS	transmission electron energy loss spectroscopy
TEM	transverse electromagnetic; transmission electron microscope
temp	temperature
TEO	thermoplastic elastic olefin
TEPP	tetraethyl pyrophosphate
tert	tertiary (in chemical name)
TES	2-{[2-hydroxy-1,1-bis(hydroxymethyl)ethyl]amino}-1-propanesulfonic acid
TFD	Thomas-Fermi-Dirac (method)
TFE	tetrafluoroethylene
TGA	thermogravimetric analysis
Thd	ribosylthymine
THEED	transmission high-energy electron diffraction
theor	theoretical
thf, THF	tetrahydrofuran
THQ	1,2,3,4-tetrahydroquinoline
Thr	threonine
Thy	thymine
TI	thermal ionization
TIPA	triisopropanolamine
TL	thermoluminescence
TLC	thin-layer chromatography
TLV	threshold limit value
TM	transverse magnetic
TMAB	tetrabutylammonium bromide
TMAO	trimethylamine oxide
TMCP	tri-*m*-cresyl phosphate
TMEDA	*N,N,N',N'*-tetramethyl-1,2-ethanediamine
TMMV	threshold molecular weight value
TMS	tetramethylsilane
TNA	2,4,6-trinitroaniline
TNB	1,3,5-trinitrobenzene
TNM	tetranitromethane
TNT	2,4,6-trinitrotoluene
TOCP	tri-*o*-cresyl phosphate
TOF	turnover frequency
TOF-MS	time-of-flight mass spectrometer
tol	tolyl
TON	turnover number
TOPO	trioctylphosphine oxide
Torr	torr (pressure unit)
TOTP	tri-*o*-tolyl phosphate
TPE	thermoplastic elastomer
TPTA	triphenyltin acetate
TPTC	triphenyltin chloride
TRE	topological resonance energy
Tre	trehalose
Tricine	*N*-[2-hydroxy-1,1-bis(hydroxymethyl)ethyl]glycine
TRIS	2-amino-2-(hydroxymethyl)-1,3-propanediol
TRMC	time-resolved microwave conductivity
tRNA	transfer RNA
Trp	tryptophan
trs	transition
TS	transition state
TSS	transition state spectroscopy
TST	generalized transition-state theory
TTF	tetrathiofulvalene
Tyr	tyrosine
U	uracil (in genetic code)
U	electric potential difference; internal energy
u	unified atomic mass unit
u	Bloch function; electric mobility; velocity
ua	astronomical unit (AU is also used)
UBFF	Urey-Bradley force field
UDMH	1,1-dimethylhydrazine
UDP	uridine 5'-diphosphate
UHF	ultrahigh frequency; unrestricted Hartree-Fock (method)
UHMWPE	ultrahigh molecular weight polyethylene
ULDPE	ultralow-density polyethylene
ULPE	ultra linear polyethylene
UMP	uridine 5'-monophosphate
uns, unsym	unsymmetrical (as chemical descriptor)
UPS, UPES	ultraviolet photoelectron spectroscopy
Ura	uracil
Urd	uridine
USP	United States Pharmacopeia
UT	Universal Time
UTC	Coordinated Universal Time
UTP	uridine 5'-triphosphate
UV	ultraviolet
V	volt; valine
V	electric potential; potential energy; volume
v	reaction rate; specific volume; velocity; vibrational quantum number; vicinal (as chemical descriptor)
v/v	volume per volume (volume of solute divided by volume of solution, expressed as percent)
VA	vinyl acetate, vanillic acid
Val	valine
vap	vaporization
VAT	vibration assisted tunneling
VB	valence band; valence bond (theory)
VCD	vibrational circular dichroism
VDW	van der Waals interaction
VHF	very high-frequency
vic	vicinal (on adjacent carbon atom)
VIS	visible region of the spectrum
vit	vitreous (phase)
VLDPE	very low-density polyethylene
VLPP	very low-pressure pyrolysis
VMA	vanilmandelic acid
VOC	volatile organic compound(s)
VOFF	valence orbital force field
VPC	vapor phase chromatography
VSEPR	valence shell electron-pair repulsion (method)
VSIP	valence state ionization potential
VSLI	very large-scale integrated (circuit)
VSMOW	Vienna Standard Mean Ocean Water
VTCS	vinyltrichlorosilane
VUV	vacuum ultraviolet
W	watt; tryptophan
W	radiant energy; statistical weight; work
w	energy density; mass fraction; velocity; work

Symbol	Meaning
w/v	weight per volume (mass of solute divided by volume of solution, usually expressed as g/100 mL)
w/w	weight per weight (mass of solute divided by mass of solution, expressed as percent)
WAXS	wide angle x-ray scattering
Wb	weber
We	Weber number
WKB	Wentzel-Kramers-Brillouin (approximation)
WLF	Williams-Landel-Ferry (equation)
WLN	Wiswesser line notation
wt	weight
X	X unit; halogen (in chemical formula)
X	reactance
x	mole fraction
X, Xaa	unspecified amino acid
XAFS	x-ray absorption fine structure
Xan	xanthine
XANES	x-ray absorption near-edge structure
Xao	xanthosine
Xle	leucine or isoleucine (unspecified)
XLPE	crosslinked polyethylene
Xlu, Xul	xylulose
XPS, XPES	x-ray photoelectron spectroscopy
XRD	x-ray diffraction
XRF	x-ray fluorescence
XRS	x-ray spectroscopy
Xyl	xylose
Y	yotta (SI prefix for 10^{24}); tyrosine
Y	admittance; Planck function; Young's modulus
y	yocto (SI prefix for 10^{-24})
y	mole fraction for gas (when *x* refers to liquid phase)
y, yr	year
YAG	yttrium aluminum garnet
yd	yard
YIG	yttrium iron garnet
Z	zetta (SI prefix for 10^{21}); glutamine or glutamic acid (unspecified)
Z	atomic number; compression factor; collision number; impedance; partition function; zusammen (*cis*-configuration)
z	zepto (SI prefix for 10^{-21})
z	charge number (of an ion); collision frequency factor
ZDO	zero differential overlap
ZINDO	Zerner's INDO method
ZPE, ZPVE	zero point vibrational energy
ZULU	Greenwich Mean Time
α	alpha particle

Symbol	Meaning
α	absorption coefficient; degree of dissociation; electric polarizability; expansion coefficient; fine structure constant
β	beta particle
β	reciprocal temperature parameter (= 1/*kT*)
γ	photon; gamma (obsolete mass unit = μg)
γ	activity coefficient; conductivity; magnetogyric ratio; mass concentration; ratio of heat capacities; surface tension
Γ	Grüneisen parameter; level width; surface concentration
Δ	inertial defect; mass excess
δ	chemical shift; Dirac delta function; Kronecker delta; loss angle
ε	emittance; Levi-Civita symbol; linear strain; molar absorption coefficient; permittivity
ζ	Coriolis coupling constant; electrokinetic potential
η	overpotential; viscosity
κ	compressibility; conductivity; magnetic susceptibility; molar absorption coefficient
λ	absolute activity; radioactive decay constant; thermal conductivity; wavelength
Λ	angular momentum; ionic conductivity
μ	muon; micro (SI prefix for 10^{-6})
μ	chemical potential; electric dipole moment; electric mobility; friction coefficient; Joule-Thompson coefficient; magnetic dipole moment; mobility; permeability
μF	microfarad
μg	microgram
μm	micrometer
μs	microsecond
ν	frequency; kinematic velocity; stoichiometric number
ν_e	neutrino
$\tilde{\nu}$	wavenumber
Π	osmotic pressure; Peltier coefficient
π	pion
ρ	density; reflectance; resistivity
σ	electrical conductivity; cross section; normal stress; shielding constant (NMR); Stefan-Boltzmann constant; surface tension; standard deviation
τ	transmittance; chemical shift; shear stress; relaxation time
Φ	magnetic flux; potential energy; radiant power; work function
ϕ	electrical potential; fugacity coefficient; osmotic coefficient; quantum yield; wave function
χ	magnetic susceptibility
χ_e	electric susceptibility
ψ	wave function
Ω	ohm
Ω	axial angular momentum; solid angle
ω	circular frequency; angular velocity; harmonic vibration wavenumber; statistical weight

THERMODYNAMIC FUNCTIONS AND RELATIONS

p = pressure V = volume T = temperature $x_i = n_i/\Sigma_j n_j$ = mole fraction of substance i
n_i = amount of substance i

Energy	U
Entropy	S
Enthalpy	$H = U + pV$
Helmholtz energy	$A = U - TS$
Gibbs energy	$G = U + pV - TS$
Isobaric heat capacity	$C_p = (\partial H/\partial T)_p$
Isochoric heat capacity	$C_V = (\partial U/\partial T)_V$
Isobaric expansivity	$\alpha = V^{-1}(\partial V/\partial T)_p$
Isothermal compressibility	$\kappa_T = -V^{-1}(\partial V/\partial p)_T$
Isentropic compressibility	$\kappa_S = -V^{-1}(\partial V/\partial p)_S$
	$\kappa_T - \kappa_S = T\alpha^2 V/C_p$
	$C_p - C_V = T\alpha^2 V/\kappa_T$
Gibbs-Helmholtz equation	$H = G - T(\partial G/\partial T)_p$
Maxwell relations	$(\partial S/\partial p)_T = -(\partial V/\partial T)_p$
	$(\partial S/\partial V)_T = -(\partial p/\partial T)_V$
Joule-Thomson expansion	$\mu_{JT} = (\partial T/\partial p)_H = -\{V - T(\partial V/\partial T)_p\}/C_p$
	$\phi_{JT} = (\partial H/\partial p)_T = V - T(\partial V/\partial T)_p$
Partial molar quantity	$X_i = (\partial X/\partial n_i)_{T,p,nj\neq i}$
Chemical potential	$\mu_i = (\partial G/\partial n_i)_{T,p,nj\neq i}$
Perfect gas [symbol^{pg}]	$pV = (\Sigma_i n_i)RT$
	$\mu_i^{pg} = \mu_i^\theta + RT\ln(x_i p/p^\theta)$
Fugacity	$f_i = (x_i p)\exp\{(\mu_i - \mu_i^{pg})/RT\}$
Activity coefficient	$\gamma_i = f_i/(x_i f_i^\theta)$
Gibbs-Duhem relation	$0 = SdT - Vdp + \Sigma_i n_i d\mu_i$

Notation for chemical and physical changes ($X = H,S,G$, etc.):

Chemical reaction	$\Delta_r X$
Formation from elements	$\Delta_f X$
Combustion	$\Delta_c X$
Fusion (cry→liq)	$\Delta_{fus} X$
Vaporization (liq→gas)	$\Delta_{vap} X$
Sublimation (cry→gas)	$\Delta_{sub} X$
Phase transition	$\Delta_{trs} X$
Solution	$\Delta_{sol} X$
Mixing	$\Delta_{mix} X$
Dilution	$\Delta_{dil} X$

Note: Superscript θ in above equations indicates standard state.

NOBEL LAUREATES IN CHEMISTRY AND PHYSICS

Full details on nationality and basis of the awards can be found at < http://www.nobelprize.org >.

Chemistry

Year	
2020	Emmanuelle Charpentier, Jennifer A. Doudna
2019	James Peebles, Michel Mayor, Didier Queloz
2018	Frances H. Arnold, George P. Smith, Gregory P. Winter
2017	Jacques Dubochet, Joachim Frank, Richard Henderson
2016	Jean-Pierre Sauvage, J. Fraser Stoddart, Bernard L. Feringa
2015	Tomas Lindahl, Paul Modrich, Aziz Sancar
2014	Eric Betzig, Stefan W. Hell, William E. Moerner
2013	Martin Karplus, Michael Levitt, Arieh Warshel
2012	Robert J. Lefkowitz, Brian K. Kobilka
2011	Dan Shechtman
2010	Richard F. Heck, Ei-ichi Negishi, Akira Suzuki
2009	Venkatraman Ramakrishnan, Thomas A. Steitz, Ada E. Yonath
2008	Martin Chalfie, Osamu Shimomura, Roger Y. Tsien
2007	Gerhard Ertl
2006	Roger D. Kornberg
2005	Yves Chauvin, Robert H. Grubbs, Richard R. Schrock
2004	Aaron Ciechanover, Avram Hershko, Irwin Rose
2003	Peter Agre, Roderick MacKinnon
2002	John B. Fenn, Koichi Tanaka, Kurt Wüthrich
2001	William S. Knowles, Ryoji Noyori, K. Barry Sharpless
2000	Alan Heeger, Alan G. MacDiarmid, Hideki Shirakawa
1999	Ahmed Zewail
1998	Walter Kohn, John Pople
1997	Paul D. Boyer, John E. Walker, Jens C. Skou
1996	Robert F. Curl Jr., Sir Harold Kroto, Richard E. Smalley
1995	Paul J. Crutzen, Mario J. Molina, F. Sherwood Rowland
1994	George A. Olah
1993	Kary B. Mullis, Michael Smith
1992	Rudolph A. Marcus
1991	Richard R. Ernst
1990	Elias James Corey
1989	Sidney Altman, Thomas R. Cech
1988	Johann Deisenhofer, Robert Huber, Hartmut Michel
1987	Donald J. Cram, Jean-Marie Lehn, Charles J. Pedersen
1986	Dudley R. Herschbach, Yuan T. Lee, John C. Polanyi
1985	Herbert A. Hauptman, Jerome Karle
1984	Bruce Merrifield
1983	Henry Taube
1982	Aaron Klug
1981	Kenichi Fukui, Roald Hoffmann
1980	Paul Berg, Walter Gilbert, Frederick Sanger
1979	Herbert C. Brown, Georg Wittig
1978	Peter Mitchell
1977	Ilya Prigogine
1976	William Lipscomb
1975	John Cornforth, Vladimir Prelog
1974	Paul J. Flory
1973	Ernst Otto Fischer, Geoffrey Wilkinson
1972	Christian Anfinsen, Stanford Moore, William H. Stein
1971	Gerhard Herzberg
1970	Luis Leloir
1969	Derek Barton, Odd Hassel
1968	Lars Onsager
1967	Manfred Eigen, Ronald G. W. Norrish, George Porter

Year	
1966	Robert S. Mulliken
1965	Robert B. Woodward
1964	Dorothy Crowfoot Hodgkin
1963	Karl Ziegler, Giulio Natta
1962	Max F. Perutz, John C. Kendrew
1961	Melvin Calvin
1960	Willard F. Libby
1959	Jaroslav Heyrovsky
1958	Frederick Sanger
1957	Lord Todd
1956	Sir Cyril Hinshelwood, Nikolay Semenov
1955	Vincent du Vigneaud
1954	Linus Pauling
1953	Hermann Staudinger
1952	Archer J. P. Martin, Richard L. M. Synge
1951	Edwin M. McMillan, Glenn T. Seaborg
1950	Otto Diels, Kurt Alder
1949	William F. Giauque
1948	Arne Tiselius
1947	Sir Robert Robinson
1946	James B. Sumner, John H. Northrop, Wendell M. Stanley
1945	Artturi Virtanen
1944	Otto Hahn
1943	George de Hevesy
1942	*No prize awarded*
1941	*No prize awarded*
1940	*No prize awarded*
1939	Adolf Butenandt, Leopold Ruzicka
1938	Richard Kuhn
1937	Norman Haworth, Paul Karrer
1936	Peter Debye
1935	Frédéric Joliot, Irène Joliot-Curie
1934	Harold C. Urey
1933	*No prize awarded*
1932	Irving Langmuir
1931	Carl Bosch, Friedrich Bergius
1930	Hans Fischer
1929	Arthur Harden, Hans von Euler-Chelpin
1928	Adolf Windaus
1927	Heinrich Wieland
1926	The Svedberg
1925	Richard Zsigmondy
1924	*No prize awarded*
1923	Fritz Pregl
1922	Francis W. Aston
1921	Frederick Soddy
1920	Walther Nernst
1919	*No prize awarded*
1918	Fritz Haber
1917	*No prize awarded*
1916	*No prize awarded*
1915	Richard Willstätter
1914	Theodore W. Richards
1913	Alfred Werner

Year	
1912	Victor Grignard, Paul Sabatier
1911	Marie Curie
1910	Otto Wallach
1909	Wilhelm Ostwald
1908	Ernest Rutherford
1907	Eduard Buchner
1906	Henri Moissan
1905	Adolf von Baeyer
1904	Sir William Ramsay
1903	Svante Arrhenius
1902	Emil Fischer
1901	Jacobus H. van't Hoff

Physics

Year	
2020	Roger Penrose, Andrea M. Ghenz, Reinhard Genzel
2019	John Goodenough, M. Stanley Whittingham, Akira Yoshino
2018	Arthur Ashkin, Gérard Mourou, Donna Strickland
2017	Rainer Weiss, Barry C. Barish, Kip S. Thorne
2016	David J. Thouless, F. Duncan M. Haldane, J. Michael Kosterlitz
2015	Takaaki Kajita, Arthur B. McDonald
2014	Isamu Akasaki, Hiroshi Amano, Shuji Nakamura
2013	François Englert, Peter Higgs
2012	Serge Haroche, David J. Wineland
2011	Saul Perlmutter, Brian P. Schmidt, Adam G. Riess
2010	Andre Geim, Konstantin Novoselov
2009	Charles K. Kao, Willard S. Boyle, George E. Smith
2008	Makoto Kobayashi, Toshihide Maskawa, Yoichiro Nambu
2007	Albert Fert, Peter Grünberg
2006	John C. Mather, George F. Smoot
2005	Roy J. Glauber, John L. Hall, Theodor W. Hänsch
2004	David J. Gross, H. David Politzer, Frank Wilczek
2003	Alexei A. Abrikosov, Vitaly L. Ginzburg, Anthony J. Leggett
2002	Raymond Davis Jr., Masatoshi Koshiba, Riccardo Giacconi
2001	Eric A. Cornell, Wolfgang Ketterle, Carl E. Wieman
2000	Zhores I. Alferov, Herbert Kroemer, Jack S. Kilby
1999	Gerardus't Hooft, Martinus J. G. Veltman
1998	Robert B. Laughlin, Horst L. Störmer, Daniel C. Tsui
1997	Steven Chu, Claude Cohen-Tannoudji, William D. Phillips
1996	David M. Lee, Douglas D. Osheroff, Robert C. Richardson
1995	Martin L. Perl, Frederick Reines
1994	Bertram N. Brockhouse, Clifford G. Shull
1993	Russell A. Hulse, Joseph H. Taylor Jr.
1992	Georges Charpak
1991	Pierre-Gilles de Gennes
1990	Jerome I. Friedman, Henry W. Kendall, Richard E. Taylor
1989	Norman F. Ramsey, Hans G. Dehmelt, Wolfgang Paul
1988	Leon M. Lederman, Melvin Schwartz, Jack Steinberger
1987	J. Georg Bednorz, K. Alex Müller
1986	Ernst Ruska, Gerd Binnig, Heinrich Rohrer
1985	Klaus von Klitzing
1984	Carlo Rubbia, Simon van der Meer
1983	Subramanyan Chandrasekhar, William A. Fowler
1982	Kenneth G. Wilson
1981	Nicolaas Bloembergen, Arthur L. Schawlow, Kai M. Siegbahn
1980	James Cronin, Val Fitch
1979	Sheldon Glashow, Abdus Salam, Steven Weinberg
1978	Pyotr Kapitsa, Arno Penzias, Robert Woodrow Wilson
1977	Philip W. Anderson, Sir Nevill F. Mott, John H. van Vleck
1976	Burton Richter, Samuel C. C. Ting
1975	Aage N. Bohr, Ben R. Mottelson, James Rainwater
1974	Martin Ryle, Antony Hewish
1973	Leo Esaki, Ivar Giaever, Brian D. Josephson
1972	John Bardeen, Leon N. Cooper, Robert Schrieffer
1971	Dennis Gabor
1970	Hannes Alfvén, Louis Néel
1969	Murray Gell-Mann
1968	Luis Alvarez
1967	Hans Bethe
1966	Alfred Kastler
1965	Sin-Itiro Tomonaga, Julian Schwinger, Richard P. Feynman
1964	Charles H. Townes, Nicolay G. Basov, Aleksandr M. Prokhorov
1963	Eugene Wigner, Maria Goeppert-Mayer, J. Hans D. Jensen
1962	Lev Landau
1961	Robert Hofstadter, Rudolf Mössbauer
1960	Donald A. Glaser
1959	Emilio Segrè, Owen Chamberlain
1958	Pavel A. Cherenkov, Il'ja M. Frank, Igor Y. Tamm
1957	Chen Ning Yang, Tsung-Dao Lee
1956	William B. Shockley, John Bardeen, Walter H. Brattain
1955	Willis E. Lamb, Polykarp Kusch
1954	Max Born, Walther Bothe
1953	Frits Zernike
1952	Felix Bloch, E. M. Purcell
1951	John Cockcroft, Ernest T. S. Walton
1950	Cecil Powell
1949	Hideki Yukawa
1948	Patrick M. S. Blackett
1947	Edward V. Appleton
1946	Percy W. Bridgman
1945	Wolfgang Pauli
1944	Isidor Isaac Rabi
1943	Otto Stern
1942	*No prize awarded*
1941	*No prize awarded*
1940	*No prize awarded*
1939	Ernest Lawrence
1938	Enrico Fermi
1937	Clinton Davisson, George Paget Thomson
1936	Victor F. Hess, Carl D. Anderson
1935	James Chadwick
1934	*No prize awarded*
1933	Erwin Schrödinger, Paul A. M. Dirac
1932	Werner Heisenberg
1931	*No prize awarded*
1930	Sir Venkata Raman
1929	Louis de Broglie
1928	Owen Willans Richardson
1927	Arthur H. Compton, C. T. R. Wilson
1926	Jean Baptiste Perrin
1925	James Franck, Gustav Hertz
1924	Manne Siegbahn
1923	Robert A. Millikan
1922	Niels Bohr
1921	Albert Einstein
1920	Charles Edouard Guillaume
1919	Johannes Stark

Year	
1918	Max Planck
1917	Charles Glover Barkla
1916	*No prize awarded*
1915	William Bragg, Lawrence Bragg
1914	Max von Laue
1913	Heike Kamerlingh Onnes
1912	Gustaf Dalén
1911	Wilhelm Wien
1910	Johannes Diderik van der Waals
1909	Guglielmo Marconi, Ferdinand Braun
1908	Gabriel Lippmann
1907	Albert A. Michelson
1906	J. J. Thomson
1905	Philipp Lenard
1904	Lord Rayleigh
1903	Henri Becquerel, Pierre Curie, Marie Curie
1902	Hendrik A. Lorentz, Pieter Zeeman
1901	Wilhelm Conrad Röntgen

Section 3
Physical Constants of Organic Compounds

Physical Constants of Organic Compounds 3-1
Diamagnetic Susceptibility of Selected Organic Compounds............................ 3-56

PHYSICAL CONSTANTS OF ORGANIC COMPOUNDS

The basic physical constants and structure diagrams for about 1000 important organic compounds are presented in this table. The table is condensed from the full table of 10,867 compounds that appears in the Online Edition of the *CRC Handbook* available at hbcponline.com.

In selecting the compounds to include, an effort has been made to cover the organic compounds most frequently encountered in the laboratory, the workplace, and the environment. Particular emphasis has been given to organic compounds that are considered environmental or human health hazards. Added weight was assigned to the appearance of an organic compound in one or more of the following lists or reference sources:

- Laboratory reagent lists, e.g., the ACS *Reagent Chemicals* volume (Ref. 1)
- The Design Institute for Physical Properties list of industrially important compounds (Ref. 2)
- The Hazardous Substances Data Bank (Ref. 3)
- The Stockholm Convention on Persistent Organic Pollutants (Ref. 4)
- The Toxic Substances Control Act (TSCA) Chemical Substances Inventory (Ref. 5)
- The EPA TSCA Work Plan for Chemical Assessments: 2014 Update (Ref. 6)
- The EPA Integrated Risk Information System (IRIS), a database of human health effects of exposure to chemicals in the environment (Ref. 7)
- Chemicals Restricted under the European Union REACH (Ref. 8)
- REACH Candidate List of Substances of Very High Concern for Authorisation (Ref. 9)
- Compendia of chemicals of biochemical or medical importance, such as *The Merck Index* (Ref. 10)
- Specialized tables in the *CRC Handbook*

It should be noted that the above lists vary widely in their choice of chemical names, and even in the use of Chemical Abstracts Service Registry Numbers. In this table we have used the same names that are used throughout the *CRC Handbook*.

The data in the table have been derived from many sources, including both the primary literature and evaluated compilations. The *Handbook of Data on Organic Compounds, Third Edition* (Ref. 11) and the *Combined Chemical Dictionary* (Ref. 12) were important sources. Other useful sources of physical property data on organic compounds are listed in Refs. 13–22. The values in the table for the normal boiling point and the melting point that are accompanied with uncertainties (in parentheses) have been critically evaluated using the NIST ThermoData Engine (TDE, Ref. 23), designed to implement the dynamic data evaluation concept (Refs. 22–25). This concept requires large electronic databases capable of storing essentially all relevant experimental data known to date with detailed descriptions of metadata and uncertainties. The combination of these electronic databases with expert-system software, designed to automatically generate recommended property values based on available experimental and predicted data, leads to the ability to produce critically evaluated data dynamically or "to order." The uncertainties listed are combined expanded uncertainties (level of confidence, approximately 95%) representing the most comprehensive measure of the overall data reliability (Refs. 28–31).

The table is arranged alphabetically by substance name, which generally is either an IUPAC systematic name or, in the case of pesticides, pharmaceuticals, and other complex compounds, a simple trivial name. Names in ubiquitous use, such as acetic acid and formaldehyde, are adopted rather than their systematic equivalents. Synonyms are given in the column following the primary name, and structure diagrams are given on the page facing the data listing. The explanation of the data columns follows:

No.: An identification number used in the indexes.

Name: Primary name of the substance.

Synonym: A synonym in common use. When the primary name is non-systematic, a systematic name may appear here.

Mol. form.: The molecular formula written in the Hill convention.

CAS Reg. No.: Chemical Abstracts Service Registry Number for the compound. There are cases for which a chemical substance may be associated with more than one CAS Reg. No. (e.g., where there is a CAS No. for a specific compound such as o-xylene and a CAS Reg. No. for a mixture of isomers such as xylenes mixed isomers).

Mol. wt.: Molecular weight (relative molar mass) as calculated with the 2009 IUPAC Standard Atomic Weights.

Physical form: A notation of the physical phase, color, crystal type, or other features of the compound at ambient temperature. Abbreviations are given below.

mp: Normal melting point in °C. A value is sometimes followed by "dec", indicating decomposition is observed at the stated temperature (so that it is probably not a true melting point). The notation "tp" indicates a triple point, where solid, liquid, and gas are in equilibrium. A number in parentheses following the melting point value is the combined expanded uncertainty (see above).

bp: Normal boiling point in °C, if it is available. This is the temperature at which the liquid phase is in equilibrium with the vapor at a pressure of 1 atm (101.325 kPa or 760 mmHg). A number in parentheses following the boiling point value is the combined expanded uncertainty (see above). A notation "sp" following the value indicates a sublimation point, where the vapor pressure of the solid phase reaches 1 atm (760 mmHg). When a notation such as "dec" (decomposes) or "exp" (explodes) follows the value, the temperature may not be a true boiling point. A simple entry "sub" indicates the solid has a significant sublimation pressure at ambient temperatures. When the normal boiling point is not available, a boiling point at reduced pressure may be listed with a superscript indicating the pressure in mmHg.

den: Density (mass per unit volume) in g cm^{-3}. The temperature in °C is indicated by a superscript. Values refer to the liquid or solid phase, and all values are true densities, not specific gravities. The number of decimal places gives a rough estimate of the accuracy of the value.

n_D: Refractive index, at the temperature in °C indicated by the superscript. Unless otherwise indicated, all values refer to a wavelength of 589 nm (sodium D line). Values are given only for liquids and solids.

Solubility: Qualitative indication of solubility in common solvents. Abbreviations are:

i: insoluble
sl: slightly soluble

s: soluble
vs: very soluble
msc: miscible
dec: decomposes

Abbreviations for solvents are given below.

In order to facilitate the location of compounds in the table, an index to synonyms follows the main table. Indexes to Molecular Formulas and CAS Registry Numbers are available in the electronic version of the *CRC Handbook* or as pdf files by request via e-mail (fiona.macdonald@taylorandfrancis.com).

The assistance of members of the Thermodynamics Research Center (TRC) of the National Institute of Standards and Technology (Vladimir Diky, Rob Chirico, Andrei Kazakov) and especially Chris Muzny and Michael Frenkel in the determination of values of the normal boiling-point and melting-point temperatures with uncertainties is greatly appreciated. The editors of the *CRC Handbook* are much indebted to Chris Muzny who spent countless hours in producing these critically evaluated results. The assistance of Fiona Macdonald in checking names and formulas is gratefully acknowledged, as well as the efforts of Janice Shackleton, Trupti Desai, Nazila Kamaly, Matt Griffiths, and Lawrence Braschi in preparing the structure diagrams.

List of Abbreviations

Ac	acetyl
Ac_2O	acetic anhydride
AcOEt	ethyl acetate
ac	acid
ace	acetone
al	alcohol (ethanol)
alk	alkali
amor	amorphous
anh	anhydrous
aq	aqueous
bipym	bipyramidal
bl	blue
blk	black
bp	boiling point
br	brown
bt	bright
Bu	butyl
BuOH	1-butanol
bz	benzene
chl	chloroform
col	colorless
con, conc	concentrated
cry	crystals
ctc	carbon tetrachloride
cy, cyhex	cyclohexane
dec	decomposes
den	density
dil	dilute
diox	dioxane
dk	dark
DMF	dimethylformamide
DMSO	dimethyl sulfoxide
efflor	efflorescent
Et	ethyl
EtOH	ethanol
eth	diethyl ether
exp	explodes
fl	flakes
flr	fluorescent
fum	fumes, fuming
gl	glacial
gr	gray
gran	granular
grn	green
hex	hexagonal
HOAc	acetic acid
hp	heptane
hx	hexane
hyd	hydrate
hyg	hygroscopic
i	insoluble
i-	iso-
iso	isooctane
lf	leaves
lig	ligroin
liq	liquid
lo	long
mcl	monoclinic
Me	methyl
MeCN	acetonitrile
MeOH	methanol
misc	miscible
mp	melting point
n	refractive index
nd	needles
oct	octahedra, octahedral
oran	orange
orth	orthorhombic
os	organic solvents
pa	pale
peth	petroleum ether
Ph	phenyl
PhCl	chlorobenzene
$PhNH_2$	aniline
$PhNO_2$	nitrobenzene
pl	plates
pow	powder
Pr	propyl
PrOH	1-propanol
pr	prisms
purp	purple
py	pyridine
pym	pyramids, pyramidal
reac	reacts
rhom	rhombic
s	soluble
sat	saturated
sc	scales
sl	slightly soluble
soln	solution
sp	sublimation point
stab	stable
sub	sublimes
sulf	sulfuric acid
syr	syrup
tab	tablets
tcl	triclinic
tetr	tetragonal
tfa	trifluoroacetic acid
thf, THF	tetrahydrofuran
tol	toluene
tp	triple point
trg	trigonal
unstab	unstable
vap	vapor
viol	violet
visc	viscous
vol	volatile
vs	very soluble
w	water (number indicates amount of hydration)
wh	white
xyl	xylene
ye	yellow

References

1. American Chemical Society, *Reagent Chemicals, Tenth Edition*, Oxford University Press, New York, 2005.
2. Design Institute for Physical Properties, American Institute of Chemical Engineers, <www.aiche.org/dippr/>.
3. National Library of Medicine, Hazardous Substances Data Bank, now integrated into <pubchem.ncbi.nim.nih.gov>.
4. United Nations Environmental Program, Stockholm Convention on Persistent Organic Pollutants, <www.chm.int//>.
5. Environmental Protection Agency, Toxic Chemical Substances Control Act (TSCA) Toxic Chemical Substance Inventory, <www.epa.gov/tsca-inventory>.
6. Environmental Protection Agency, *TSCA Work Plan for Chemical Assessments: 2014 Update*, <https://www.epa.gov/sites/production/files/2015-01/documents/tsca_work_plan_chemicals_2014_update-final.pdf>.
7. Environmental Protection Agency, Integrated Risk Information System, <www.epa.gov/iris/index.html>.
8. European Chemicals Agency, *Substances Restricted under REACH*, <https://echa.europa.eu/substances-restricted-under-reach>.
9. European Chemicals Agency, *REACH Candidate List of Substances of Very High Concern for Authorisation*, <https://echa.europa.eu/candidate-list-table>.
10. O'Neil, M. J., Editor, *The Merck Index, 14th Edition*, Merck & Co., Whitehouse Station, NJ, 2006.
11. Lide, D. R., and Milne, G. W. A., Editors, *Handbook of Data on Organic Compounds, Third Edition*, CRC Press, Boca Raton, FL, 1993.
12. *Combined Chemical Dictionary*, <ccd.chemnetbase.com/faces/chemical/ChemicalSearch.xhtml>.
13. Linstrom, P. J., and Mallard, W. G., Editors, NIST Chemistry WebBook, NIST Standard Reference Database No. 69, February 2010, National Institute of Standards and Technology, Gaithersburg, MD 20899, <webbook.nist.gov>.
14. Thermodynamic Research Center, National Institute of Standards and Technology, TRC Thermodynamic Tables, <trc.nist.gov>.
15. Stevenson, R. M., and Malanowski, S., *Handbook of the Thermodynamics of Organic Compounds*, Elsevier, New York, 1987. <https://doi.org/10.1007/978-94-009-3173-2>
16. Riddick, J. A., Bunger, W. B., and Sakano, T. K., *Organic Solvents, Fourth Edition*, John Wiley & Sons, New York, 1986.
17. ChemSpider, <www.chemspider.com/>.
18. Crossfire Beilstein, now available as part of Reaxys, <www.reaxys.com>.
19. Springer Materials, The Landolt-Börnstein Database, <www.springermaterials.com>.
20. Vargaftik, N.B., Vinogradov, Y. K., and Yargin, V. S., *Handbook of Physical Properties of Liquids and Gases, Third Edition*, Begell House, New York, 1996.
21. Lide, D. R., and Kehiaian, H. V., *Handbook of Thermophysical and Thermochemical Data*, CRC Press, Boca Raton, FL, 1994.
22. Lide, D. R., Editor, Properties of Organic Compounds, <www.chemnetbase.com/tours/poc/intro.jsf>.
23. Frenkel, M., Chirico, R. D., Diky, V. V., Kazakov, A., and Muzny, C. D., ThermoData Engine, NIST Standard Reference Database 103b, Version 5.0 (Pure Compounds, Binary Mixtures, and Chemical Reactions, TDE-SOURCE Version 5.1), National Institute of Standards and Technology, Gaithersburg, MD — Boulder, CO, 2010, <www.nist.gov/srd/nist103b.cfm>.
24. Frenkel, M., Chirico, R. D., Diky, V., Yan, X., Dong, Q., and Muzny, C., *J. Chem. Inf. Model.* 45, 816, 2005. <https://doi.org/10.1021/ci050067b>
25. Diky, V., Muzny, C. D., Lemmon, E. W., Chirico, R. D., and Frenkel, M., *J. Chem. Inf. Model.* 47, 1713, 2007. <https://doi.org/10.1021/ci700071t>
26. Diky, V., Chirico, R. D., Kazakov, A. F., Muzny, C., and Frenkel, M., *J. Chem. Inf. Model.* 49, 503, 2009. <https://doi.org/10.1021/ci800345e>
27. Diky, V., Chirico, R. D., Kazakov, A. F., Muzny, C., and Frenkel, M., *J. Chem. Inf. Model.* 49, 2883, 2009. <https://doi.org/10.1021/ci900340k>
28. Chirico, R. D., Frenkel, M., Diky, V. V., March, K. N., and Wilhoit, R. C., *J. Chem. Eng. Data* 48, 1344, 2003. <https://doi.org/10.1021/je034088i>
29. *Guide to the Expression of Uncertainty in Measurement*, International Organization for Standardization, Geneva, Switzerland, 1993.
30. *U.S. Guide to the Expression of Uncertainty in Measurement*, ANSI/NCSL, Z540-2-1997, ISBN 1-58464-005-7, NCSL Int., Boulder, CO, 1997.
31. Taylor, B. N., and Kuyatt, C. E., *Guidelines for the Evaluation and Expression of Uncertainty in NIST Measurement Results*, NIST Technical Note 1297, National Institute of Standards and Technology, Gaithersburg, MD, 1994.

No.	Name	Synonym	Mol. form.	CAS Reg. No.	Mol. wt.	Physical form	mp/°C	bp/°C	den/g cm^{-3}	n_D	Solubility
1	Acenaphthene	1,2-Dihydroacenaphthylene	$C_{12}H_{10}$	83-32-9	154.207		93(2)	277.5(8)	1.222^{20}	1.6048^{95}	i H_2O; sl EtOH, chl; vs bz; s HOAc
2	Acenaphthylene	Acenaphthalene	$C_{12}H_8$	208-96-8	152.192		89.4(3)	280	0.8987^{16}		i H_2O; vs EtOH, eth, bz; sl chl
3	Acetaldehyde	Ethanal	C_2H_4O	75-07-0	44.052	vol liq or gas	-123.4(7)	20.8(6)	0.7834^{18}	1.3316^{20}	msc H_2O, EtOH, eth, bz; sl chl
4	Acetamide	Ethanamide	C_2H_5NO	60-35-5	59.067	trg mcl (al-eth)	80.16(4)	222.0	0.9986^{85}	1.4278	vs H_2O, EtOH
5	Acetanilide	*N*-Phenylacetamide	C_8H_9NO	103-84-4	135.163		114.35(4)	292(9)	1.2190^{15}		sl H_2O; vs EtOH, ace; s eth, s bz, tol
6	Acetic acid	Ethanoic acid	$C_2H_4O_2$	64-19-7	60.052	col liq	17(3)	117.9(2)	1.0510^{20}	1.3720^{20}	msc H_2O, EtOH, eth, ace, bz; s chl, CS_2
7	Acetic anhydride	Acetyl acetate	$C_4H_6O_3$	108-24-7	102.089	liq	-73.4(8)	139.5(3)	1.082^{20}	1.3901^{20}	vs H_2O; s EtOH, bz; msc eth; sl ctc
8	Acetone	2-Propanone	C_3H_6O	67-64-1	58.079	liq	-94.9(4)	56.08(7)	0.7902^{20}	1.3588^{20}	msc H_2O, EtOH, eth, ace, bz, chl
9	Acetone cyanohydrin		C_4H_7NO	75-86-5	85.105	liq	-19	180(21)	0.932^{19}	1.3992^{20}	vs H_2O, EtOH, eth; s ace, bz, chl; i peth
10	Acetonitrile	Methyl cyanide	C_2H_3N	75-05-8	41.052	liq	-44(1)	81.6(2)	0.7825^{20}	1.3442^{30}	msc H_2O, EtOH, eth, ace, bz, ctc
11	Acetophenone	Methyl phenyl ketone	C_8H_8O	98-86-2	120.149	mcl pr or pl	19.4(4)	202.1(2)	1.0281^{20}	1.5372^{20}	sl H_2O; s EtOH, eth, ace, bz, con sulf, chl
12	Acetyl chloride	Ethanoyl chloride	C_2H_3ClO	75-36-5	78.497	liq	-112.7(8)	51(2)	1.1051^{20}	1.3886^{20}	msc eth, ace, bz, chl; s ctc
13	Acetylene	Ethyne	C_2H_2	74-86-2	26.037	col gas	-81.5(9)	-84.7 sp	0.377^{25} (p>1 atm)		sl H_2O, EtOH, CS_2; s ace, bz, chl
14	Acridine	Dibenzo[*b,e*]pyridine	$C_{13}H_9N$	260-94-6	179.217	orth nd or pr (al)	110.06(5)	346.9(10)	1.005^{20}		i H_2O; sl ctc; vs EtOH, eth, bz
15	Acrolein	2-Propenal	C_3H_4O	107-02-8	56.063	liq	-87.8(9)	52.3(1)	0.840^{20}	1.4017^{20}	vs H_2O; s EtOH, eth, ace; sl chl
16	Acrylamide	2-Propenamide	C_3H_5NO	79-06-1	71.078	lf (bz)	85(1)	192.6			vs H_2O, chl; s EtOH, eth, ace
17	Acrylic acid	2-Propenoic acid	$C_3H_4O_2$	79-10-7	72.063	acrid liq	13.56(5)	142(2)	1.0511^{20}	1.4224^{20}	msc H_2O, EtOH, eth; s ace, bz, ctc
18	Acrylonitrile	Propenenitrile	C_3H_3N	107-13-1	53.063	liq	-83.51(5)	77.2(2)	0.8007^{25}	1.3911^{20}	s H_2O; vs ace, bz, eth, EtOH
19	Adenosine	β-D-Ribofuranoside, adenine-9	$C_{10}H_{13}N_5O_4$	58-61-7	267.242	n(w+3/2)	235.5				sl H_2O; i EtOH
	Aldrin		$C_{12}H_8Cl_6$	309-00-2	364.910		103.8(3)				i H_2O; s EtOH, eth, ace, bz
20	Allyl acetate	3-Acetoxypropene	$C_5H_8O_2$	591-87-7	100.117			104(2)	0.9275^{20}	1.4049^{20}	sl H_2O; s ace; msc EtOH, eth
21	Allyl alcohol	2-Propen-1-ol	C_3H_6O	107-18-6	58.079	liq	-129	96.9(5)	0.8540^{20}	1.4135^{20}	msc H_2O, EtOH, eth; s chl
22	Allylamine	2-Propen-1-amine	C_3H_7N	107-11-9	57.095	liq	-88.2	54(2)	0.758^{20}	1.4205^{20}	msc H_2O, EtOH, eth; s chl
23	*N*-Allyl-2-propen-1-amine	Diallylamine	$C_6H_{11}N$	124-02-7	97.158			112(3)		1.4387^{20}	s EtOH, eth
24	4-Aminoazobenzene		$C_{12}H_{11}N_3$	60-09-3	197.235	oran mcl nd (al)	125(1)	>360			sl H_2O, lig; s EtOH, eth, bz, chl
25	4-Aminobenzenesulfonic acid	Sulfanilic acid	$C_6H_7NO_3S$	121-57-3	173.190	orth pl or mcl (w+2)	288		1.485^{25}		sl H_2O; i EtOH, eth
26	*N*-(2-Aminoethyl) ethanolamine		$C_4H_{12}N_2O$	111-41-1	104.150			242(5)	1.0286^{20}	1.4863^{20}	msc H_2O, EtOH; s ace; sl bz, lig
27	2-Amino-2-methyl-1-propanol	AMP	$C_4H_{11}NO$	124-68-5	89.136		25.5	163.8(8)	0.934^{20}	1.449^{20}	msc H_2O; s ctc
	Ammonium perfluorooctanoate		$C_8H_4F_{15}NO_2$	3825-26-1	431.100	solid					
28	Aniline	Benzenamine	C_6H_7N	62-53-3	93.127	oily liq	-6.0(1)	184.1(4)	1.0250^{20}	1.5863^{20}	s H_2O, ctc, lig; msc EtOH, eth, ace, bz
29	Aniline hydrochloride	Benzenamine hydrochloride	C_6H_8ClN	142-04-1	129.588	lf or nd	198		1.2215^{4}		vs H_2O, EtOH; i eth, chl; sl DMSO
30	Anisole	Methoxybenzene	C_7H_8O	100-66-3	108.138	liq	-37.3(2)	153.6(2)	0.9940^{20}	1.5174^{20}	i H_2O; s EtOH, eth, chl; vs ace, bz
31	Anthracene		$C_{14}H_{10}$	120-12-7	178.229	tab or mcl pr (al)	216(2)	341.3(4)	1.28^{25}		i H_2O; sl EtOH, eth, ace, bz, chl, ctc
32	9,10-Anthracenedione	Anthraquinone	$C_{14}H_8O_2$	84-65-1	208.213	ye orth nd (al, bz)	284.8(2)	377(2)	1.438^{20}		i H_2O; sl EtOH, eth, bz, chl
33	9(10*H*)-Anthracenone	Anthrone	$C_{14}H_{10}O$	90-44-8	194.228	nd (bz-lig, HOAc)	155(3)				s ace, bz, con sulf, dil alk
34	L-Ascorbic acid	Vitamin C	$C_6H_8O_6$	50-81-7	176.124		191(4)		1.65^{25}		vs H_2O; s EtOH; i eth, bz, chl, peth
35	Benzaldehyde	Benzenecarboxaldehyde	C_7H_6O	100-52-7	106.122	liq	-57.12(5)	178.7(4)	1.0401^{25}	1.5463^{20}	sl H_2O; msc EtOH, eth; vs ace, bz
36	Benzamide	Benzoic acid amide	C_7H_7NO	55-21-0	121.137	mcl pr or pl (w)	128(1)	306(2)	1.0792^{130}		sl H_2O, eth, bz; vs EtOH, ctc, CS_2
37	Benz[*a*]anthracene	1,2-Benzanthracene	$C_{18}H_{12}$	56-55-3	228.288	lf (al)	160(2)	438			i H_2O; vs EtOH
38	Benzene	[6]Annulene	C_6H_6	71-43-2	78.112	orth pr or liq	5.538(2)	80.08(7)	0.8788^{20}	1.5011^{20}	sl H_2O; msc EtOH, eth, ace, chl; s ctc

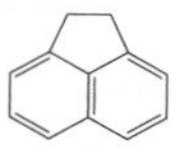
1
Acenaphthene

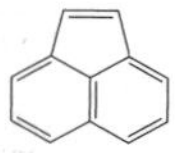
2
Acenaphthylene

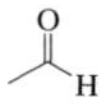
3
Acetaldehyde

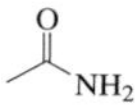
4
Acetamide

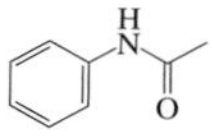
5
Acetanilide

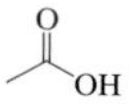
6
Acetic acid

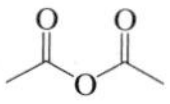
7
Acetic anhydride

8
Acetone

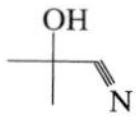
9
Acetone cyanohydrin

10
Acetonitrile

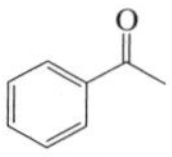
11
Acetophenone

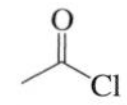
12
Acetyl chloride

13
Acetylene

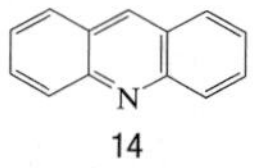
14
Acridine

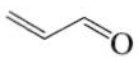
15
Acrolein

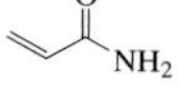
16
Acrylamide

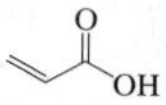
17
Acrylic acid

18
Acrylonitrile

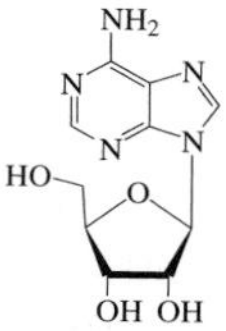
19
Adenosine

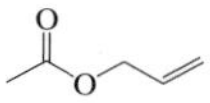
20
Allyl acetate

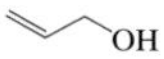
21
Allyl alcohol

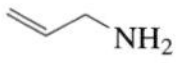
22
Allylamine

23
N-Allyl-2-propen-1-amine

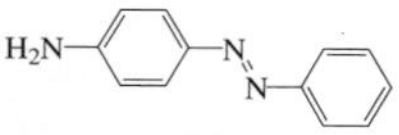
24
4-Aminoazobenzene

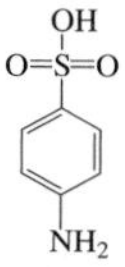
25
4-Aminobenzenesulfonic acid

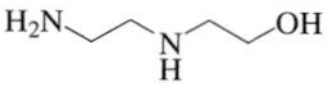
26
N-(2-Aminoethyl)ethanolamine

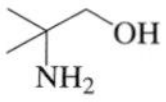
27
2-Amino-2-methyl-1-propanol

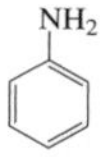
28
Aniline

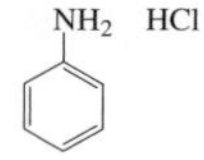
29
Aniline hydrochloride

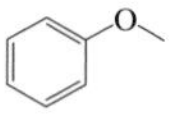
30
Anisole

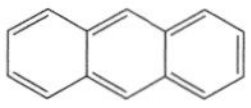
31
Anthracene

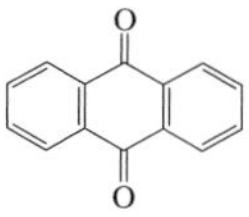
32
9,10-Anthracenedione

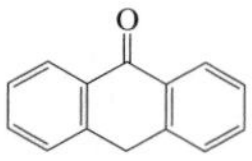
33
9(10H)-Anthracenone

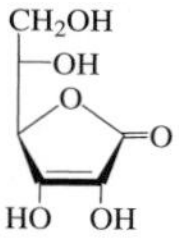
34
L-Ascorbic acid

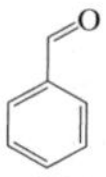
35
Benzaldehyde

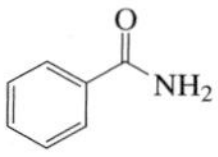
36
Benzamide

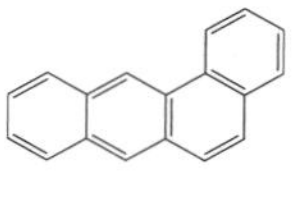
37
Benz[a]anthracene

38
Benzene

No.	Name	Synonym	Mol. form.	CAS Reg. No.	Mol. wt.	Physical form	mp/°C	bp/°C	den/g cm^{-3}	n_D	Solubility
39	Benzeneacetonitrile	Benzyl cyanide	C_8H_7N	140-29-4	117.149	liq	-22.1(5)	232(2)	1.0205^{15}	1.5211^{25}	
40	1,2-Benzenediamine	*o*-Phenylenediamine	$C_6H_8N_2$	95-54-5	108.141	brsh ye lf (w) pl (chl)	103(1)	257(9)			s H_2O, eth, bz, chl; vs EtOH
41	1,3-Benzenediamine	*m*-Phenylenediamine	$C_6H_8N_2$	108-45-2	108.141	orth (al)	65.5(9)	282(18)	1.0096^{58}	1.6339^{58}	vs H_2O; s EtOH, eth, bz
42	1,4-Benzenediamine	*p*-Phenylenediamine	$C_6H_8N_2$	106-50-3	108.141	wh pl (bz, eth)	140.3(6)	267			sl H_2O; s EtOH, eth, bz, chl
43	1,3-Benzenedicarbonyl dichloride		$C_8H_4Cl_2O_2$	99-63-8	203.023	pr(eth)	43.5	276	1.3880^{17}	1.570^{47}	sl H_2O, EtOH; s eth
44	1,4-Benzenedicarbonyl dichloride		$C_8H_4Cl_2O_2$	100-20-9	203.023	nd or pl (lig)	83.5	258			s eth
45	1,4-Benzenedicarboxaldehyde		$C_8H_6O_2$	623-27-8	134.133	nd (w)	117	246			sl H_2O; vs EtOH; s eth, chl, alk
46	Benzeneethanamine	2-Phenylethylamine	$C_8H_{11}N$	64-04-0	121.180	liq	<0	204(4)	0.9640^{25}	1.5290^{25}	s H_2O, ctc; vs EtOH, eth
47	Benzeneethanol	Phenethyl alcohol	$C_8H_{10}O$	60-12-8	122.164	liq	-19(2)	220(3)	1.0202^{20}	1.5325^{20}	sl H_2O; msc EtOH, eth
48	1,2,4,5-Benzenetetracarboxylic acid	Pyromellitic acid	$C_{10}H_6O_8$	89-05-4	254.150	tcl pr (w+2)	271(2)				sl H_2O; s EtOH
49	Benzenethiol	Phenyl mercaptan	C_6H_6S	108-98-5	110.177	liq	-14.87(5)	169.1(2)	1.0775^{20}	1.5893^{20}	i H_2O; s EtOH, eth, bz; sl ctc
50	1,2,4-Benzenetricarboxylic acid	Trimellitic acid	$C_9H_6O_6$	528-44-9	210.140	nd (w) cry (al) cry (HOAc)	219				vs H_2O, eth, EtOH
51	1,2,3-Benzenetriol	Pyrogallol	$C_6H_6O_3$	87-66-1	126.110	lf or nd (bz)	125.5(5)	307(4)	1.453^4	1.561^{134}	vs H_2O, EtOH, eth, NH_3; s ace; i bz
52	*p*-Benzidine	[1,1'-Biphenyl]-4,4'-diamine	$C_{12}H_{12}N_2$	92-87-5	184.236	nd (w)	127.0(5)	401			sl H_2O, eth, DMSO; s EtOH
53	Benzo[*b*]fluoranthene	Benz[*e*]acephenanthrylene	$C_{20}H_{12}$	205-99-2	252.309	nd (bz)	168.4(7)				i H_2O; msc bz
54	Benzoic acid	Benzenecarboxylic acid	$C_7H_6O_2$	65-85-0	122.122	mcl lf or nd	122.352(5)	250.2(6)	1.2659^{15}	1.504^{132}	sl H_2O; vs EtOH, eth; s ace, bz, chl
55	Benzonitrile	Phenyl cyanide	C_7H_5N	100-47-0	103.122	liq	-12.82(2)	191(1)	1.0093^{15}	1.5289^{20}	sl H_2O; msc EtOH; vs ace, bz; s ctc
56	Benzophenone	Diphenyl ketone	$C_{13}H_{10}O$	119-61-9	182.217	(α) orth pr (al); (β) mcl pr	48.0(2)	305.9(2)	1.111^{18}	1.6077^{19}	i H_2O; vs EtOH, eth, chl, ace; s bz
57	Benzo[*a*]pyrene	2,3-Benzopyrene	$C_{20}H_{12}$	50-32-8	252.309		179(2)				i H_2O; vs chl
58	*p*-Benzoquinone	2,5-Cyclohexadiene-1,4-dione	$C_6H_4O_2$	106-51-4	108.095	ye mcl pr (w)	113(2)	subl	1.318^{20}		sl H_2O, peth; s EtOH, eth, chl
59	Benzo[*b*]thiophene	Thianaphthene	C_8H_6S	95-15-8	134.199	lf	31.33(3)	220.9(4)	1.1484^{32}	1.6374^{37}	i H_2O; vs EtOH; s eth, ace, bz; sl chl
60	Benzoyl chloride	Benzoic acid, chloride	C_7H_5ClO	98-88-4	140.567	liq	-0.5(2)	201(8)	1.2120^{20}	1.5537^{20}	msc eth; s bz, ctc, CS_2
61	Benzoyl peroxide		$C_{14}H_{10}O_4$	94-36-0	242.227	orth (eth), pr	104.5(9)	exp		1.543	sl H_2O; s EtOH, eth, ace, bz, CS_2
62	Benzyl acetate	(Acetoxymethyl)benzene	$C_9H_{10}O_2$	140-11-4	150.174	liq	-51.5(4)	215(1)	1.0550^{20}	1.5232^{20}	sl H_2O; msc EtOH; s eth, ace, chl
63	Benzyl alcohol	Benzenemethanol	C_7H_8O	100-51-6	108.138	liq	-15.5(2)	205.3(2)	1.0419^{24}	1.5396^{20}	s H_2O, EtOH, eth, ace, bz, MeOH, chl
64	Benzyl benzoate	Benzyl benzenecarboxylate	$C_{14}H_{12}O_2$	120-51-4	212.244	nd or lf	19(1)	321.3(9)	1.1121^{25}	1.5680^{20}	i H_2O; s EtOH, eth, ace, bz, MeOH, chl
65	Benzyl ethyl ether	(Ethoxymethyl)benzene	$C_9H_{12}O$	539-30-0	136.190			188(4)	0.9478^{20}	1.4955^{20}	i H_2O; msc EtOH, eth
66	Benzyl formate		$C_8H_8O_2$	104-57-4	136.149			203(8)	1.081^{20}	1.5154^{20}	i H_2O; s EtOH, ace; msc eth; sl ctc
67	[1,1'-Bicyclohexyl]-2-one	2-Cyclohexylcyclohexanone	$C_{12}H_{20}O$	90-42-6	180.286	liq	-32	264	0.9696^{25}	1.4877^{25}	
68	Biphenyl	Diphenyl	$C_{12}H_{10}$	92-52-4	154.207	lf (dil al)	68.93(2)	255.2(3)	1.04^{20}	1.588^{77}	i H_2O; s EtOH, eth; vs bz, ctc, MeOH
69	Bis(2-aminoethyl)amine	Diethylenetriamine	$C_4H_{13}N_3$	111-40-0	103.166	ye hyg liq	-39	206.5(3)	0.9569^{20}	1.4810^{25}	msc H_2O, EtOH; i eth; s lig
70	*N,N'*-Bis(2-aminoethyl)-1,2-ethanediamine	Triethylenetetramine	$C_6H_{18}N_4$	112-24-3	146.234	oil	12	266.5		1.4971^{20}	s H_2O, EtOH, acid
71	Bis(2-chloroethyl) ether	Dichloroethyl ether	$C_4H_8Cl_2O$	111-44-4	143.012	liq	-46.9(5)	178(2)	1.22^{20}	1.451^{20}	i H_2O; s EtOH, eth, ace; msc bz
	1,1-Bis(4-chlorophenyl)-2,2,2-trichloroethanol	Dicofol	$C_{14}H_9Cl_5O$	115-32-2	370.485	cry (petr)	75.1(3)				i H_2O, os
72	Bis(2-ethylhexyl) phthalate	Di-*sec*-octyl phthalate	$C_{24}H_{38}O_4$	117-81-7	390.557	liq	-55	384	0.981^{25}	1.4853^{20}	sl ctc
73	Bis(2-ethylhexyl) sodium sulfosuccinate	Docusate sodium	$C_{20}H_{37}NaO_7S$	577-11-7	444.559	waxy solid	155	dec	1.1^{25}		s peth, ctc, eth, ace
74	*N,N*-Bis(2-hydroxyethyl) ethylamine	*N*-Ethyldiethanolamine	$C_6H_{15}NO_2$	139-87-7	133.189	ye liq	-50	248(19)	1.0135^{20}	1.4663^{20}	vs H_2O, EtOH; sl eth
75	*N,N'*-Bis(2-hydroxyethyl) ethylenediamine		$C_6H_{16}N_2O_2$	4439-20-7	148.203		100.1(5)				s H_2O
76	Bis(2-hydroxyethyl) sulfide	2,2'-Thiodiethanol	$C_4H_{10}O_2S$	111-48-8	122.186	liq	-10.2	282	1.1793^{25}	1.5211^{20}	msc H_2O, EtOH, chl, AcOEt; s eth; sl bz
77	2,2-Bis(4-hydroxyphenyl) propane	Bisphenol A	$C_{15}H_{16}O_2$	80-05-7	228.287	cry or fl	160(2)				i H_2O; vs EtOH, eth, bz, alk; s HOAc
78	Bis(1-methyl-1-phenylethyl)peroxide	Dicumyl peroxide	$C_{18}H_{22}O_2$	80-43-3	270.367	cry (EtOH)	40				

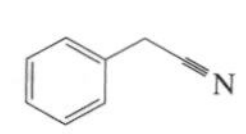

39
Benzeneacetonitrile

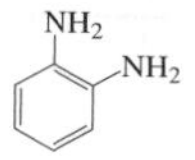

40
1,2-Benzenediamine

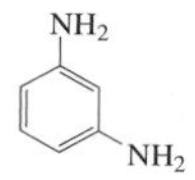

41
1,3-Benzenediamine

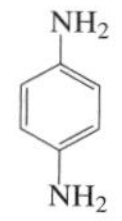

42
1,4-Benzenediamine

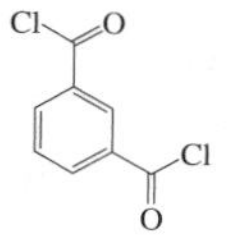

43
1,3-Benzenedicarbonyl dichloride

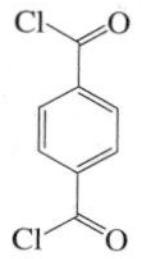

44
1,4-Benzenedicarbonyl dichloride

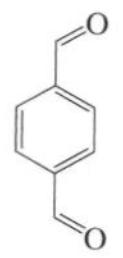

45
1,4-Benzenedicarboxaldehyde

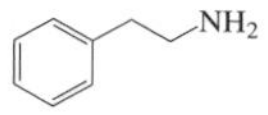

46
Benzeneethanamine

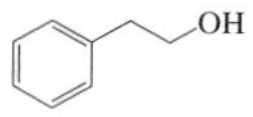

47
Benzeneethanol

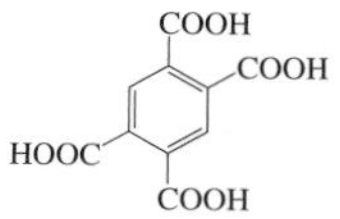

48
1,2,4,5-Benzenetetracarboxylic acid

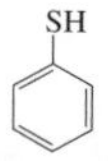

49
Benzenethiol

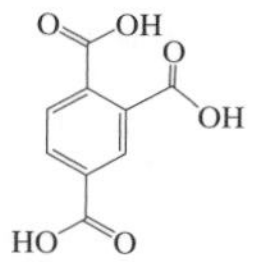

50
1,2,4-Benzenetricarboxylic acid

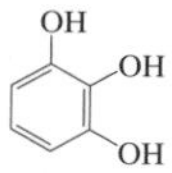

51
1,2,3-Benzenetriol

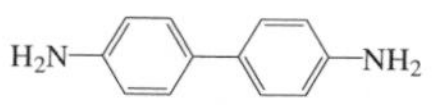

52
p-Benzidine

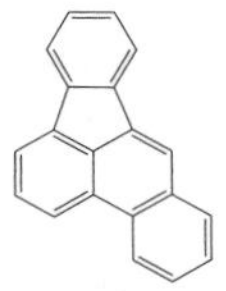

53
Benzo[b]fluoranthene

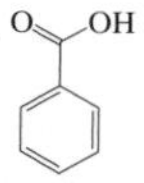

54
Benzoic acid

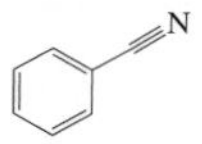

55
Benzonitrile

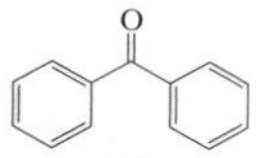

56
Benzophenone

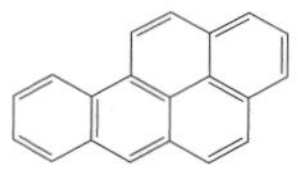

57
Benzo[a]pyrene

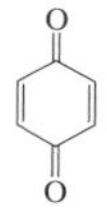

58
p-Benzoquinone

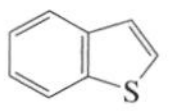

59
Benzo[b]thiophene

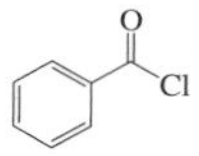

60
Benzoyl chloride

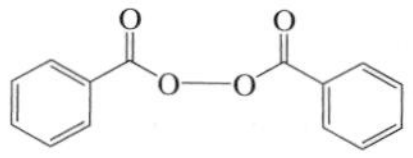

61
Benzoyl peroxide

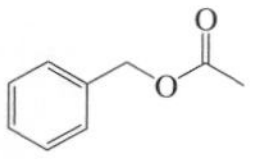

62
Benzyl acetate

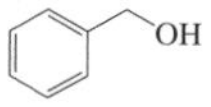

63
Benzyl alcohol

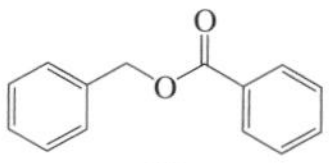

64
Benzyl benzoate

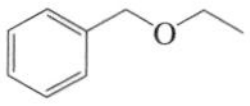

65
Benzyl ethyl ether

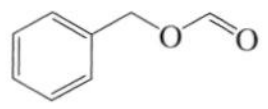

66
Benzyl formate

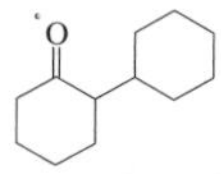

67
[1,1'-Bicyclohexyl]-2-one

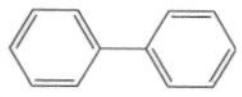

68
Biphenyl

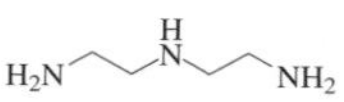

69
Bis(2-aminoethyl)amine

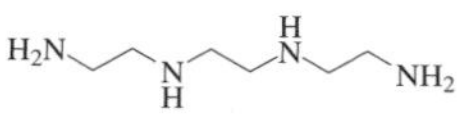

70
N,N'-Bis(2-aminoethyl)-1,2-ethanediamine

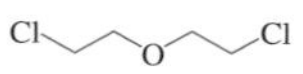

71
Bis(2-chloroethyl) ether

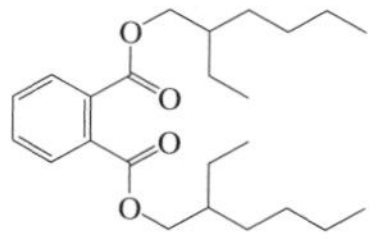

72
Bis(2-ethylhexyl) phthalate

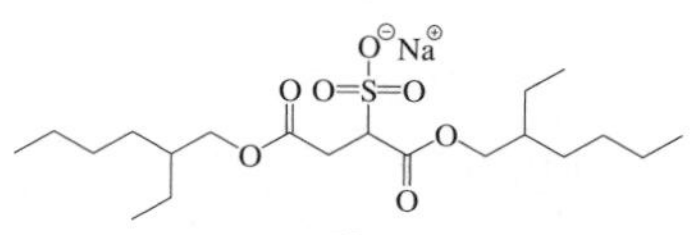

73
Bis(2-ethylhexyl) sodium sulfosuccinate

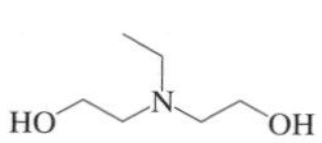

74
N,N-Bis(2-hydroxyethyl)ethylamine

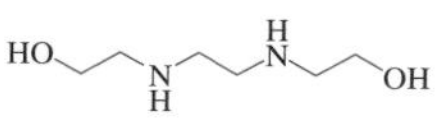

75
N,N'-Bis(2-hydroxyethyl)ethylenediamine

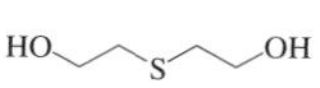

76
Bis(2-hydroxyethyl) sulfide

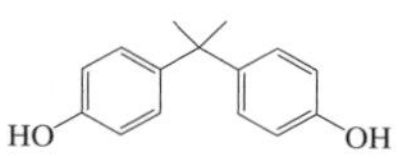

77
2,2-Bis(4-hydroxyphenyl)propane

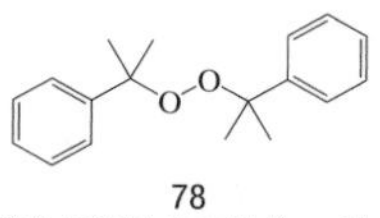

78
Bis(1-methyl-1-phenylethyl)peroxide

No.	Name	Synonym	Mol. form.	CAS Reg. No.	Mol. wt.	Physical form	mp/°C	bp/°C	den/g cm^{-3}	n_D	Solubility
79	Bromobenzene	Phenyl bromide	C_6H_5Br	108-86-1	157.008	liq	-30.74(3)	155.9(2)	1.4950^{20}	1.5597^{20}	i H_2O; vs EtOH, eth, bz; s ctc
80	1-Bromobutane	Butyl bromide	C_4H_9Br	109-65-9	137.018	liq	-112.5(3)	101.4(7)	1.2758^{20}	1.4401^{20}	i H_2O; msc EtOH, eth, ace; sl ctc; s chl
81	1-Bromo-2-chloroethane	2-Chloro-1-bromoethane	C_2H_4BrCl	107-04-0	143.410	liq	-16.7(3)	106(2)	1.7392^{20}	1.4908^{20}	sl H_2O; s EtOH, eth, chl
82	Bromochloromethane	Halon 1011	CH_2BrCl	74-97-5	129.384	liq	-87.9(2)	67.9(4)	1.9344^{20}	1.4838^{20}	i H_2O; s EtOH, eth, ace, bz
83	2-Bromo-2-chloro-1,1,1-trifluoroethane	Halothane	$C_2HBrClF_3$	151-67-7	197.381			50(1)	1.8563^{25}	1.3697^{0}	sl H_2O; s peth
84	Bromoethane	Ethyl bromide	C_2H_5Br	74-96-4	108.965	liq	-118.4(10)	38.2(6)	1.4604^{20}	1.4239^{20}	sl H_2O; msc EtOH, eth, chl
85	Bromoethene	Vinyl bromide	C_2H_3Br	593-60-2	106.949	vol liq or gas	-139.5(2)	16(16)	1.4933^{20}	1.4380^{20}	i H_2O; s EtOH, eth, ace, bz, chl
86	1-Bromoheptane	Heptyl bromide	$C_7H_{15}Br$	629-04-9	179.098	liq	-56.1(3)	179(5)	1.1400^{20}	1.4502^{20}	i H_2O; vs EtOH, eth; sl ctc; s chl
87	Bromomethane	Methyl bromide	CH_3Br	74-83-9	94.939	col gas	-93.7(4)	3.4(1)	1.6755^{20}	1.4218^{20}	sl H_2O; msc EtOH, eth, chl, CS_2
88	1-Bromonaphthalene	1-Naphthyl bromide	$C_{10}H_7Br$	90-11-9	207.067	oily liq	6.1(1)	280(2)	1.4785^{20}	1.658^{20}	s H_2O, ace; msc EtOH, eth, bz; sl ctc
89	1-Bromopentane	Pentyl bromide	$C_5H_{11}Br$	110-53-2	151.045	liq	-88.0(2)	126(3)	1.2182^{20}	1.4447^{20}	i H_2O; s EtOH, bz, chl; sl ctc; msc eth
90	1-Bromopropane	Propyl bromide	C_3H_7Br	106-94-5	122.992	liq	-110.1(3)	70.8(2)	1.3537^{20}	1.4343^{20}	sl H_2O; s EtOH, eth, ace, bz, chl, ctc
91	2-Bromopropane	Isopropyl bromide	C_3H_7Br	75-26-3	122.992	liq	-88.9(5)	59.34(9)	1.3140^{20}	1.4251^{20}	sl H_2O; s ace, bz, chl; msc EtOH, eth
92	4-Bromotoluene	1-Bromo-4-methylbenzene	C_7H_7Br	106-38-7	171.035	cry (al)	26.2(7)	184(4)	1.3959^{35}	1.5477^{20}	i H_2O; s EtOH, eth, ace, bz, chl; sl ctc
93	Bromotrifluoromethane	Halon-1301	$CBrF_3$	75-63-8	148.910	col gas	-174.4(3)	-57.8(4)	1.5800^{20}		i H_2O; vs chl
94	1,2-Butadiene	Methylallene	C_4H_6	590-19-2	54.091	vol liq or gas	-136.20(5)	11.0(2)	0.676^{0}	1.4205^{1}	i H_2O; msc EtOH, eth; vs bz
95	1,3-Butadiene	Divinyl	C_4H_6	106-99-0	54.091	col gas	-108.9(1)	-4.6(2)	0.6149^{25} (p>1 atm)	1.4292^{-25}	i H_2O; s EtOH, eth, bz; vs ace
96	Butanal	Butyraldehyde	C_4H_8O	123-72-8	72.106	liq	-96.86(2)	74.8(2)	0.8016^{20}	1.3843^{20}	s H_2O; msc EtOH; vs ace, bz; sl chl
97	Butane		C_4H_{10}	106-97-8	58.122	col gas	-138.2(2)	-0.5(5)	0.573^{25} (p>1 atm)	1.3326^{20}	i H_2O; vs EtOH, eth, chl
98	1,3-Butanediol	1,3-Butylene glycol	$C_4H_{10}O_2$	107-88-0	90.121	visc liq	-77	208.2(1)	1.0053^{20}	1.4401^{20}	
99	1,4-Butanediol	Tetramethylene glycol	$C_4H_{10}O_2$	110-63-4	90.121		20.43(2)	229.5(4)	1.0171^{20}	1.4460^{20}	msc H_2O; s EtOH, DMSO; sl eth
100	2,3-Butanedione	Diacetyl	$C_4H_6O_2$	431-03-8	86.090	liq	-1.2	87.5(8)	0.9808^{18}	1.3951^{20}	vs H_2O; msc EtOH, eth; s bz, ctc
101	Butanenitrile	Propyl cyanide	C_4H_7N	109-74-0	69.106	liq	-111.76(5)	117.6(4)	0.7936^{20}	1.3842^{20}	sl H_2O, ctc; msc EtOH, eth; s bz
102	1-Butanethiol	Butyl mercaptan	$C_4H_{10}S$	109-79-5	90.187	liq	-115.66(6)	98.4(5)	0.8416^{20}	1.4440^{20}	sl H_2O, chl; vs EtOH, eth
103	1,2,4-Butanetriol		$C_4H_{10}O_3$	3068-00-6	106.120				1.18^{20}	1.4688^{20}	vs H_2O, EtOH
104	Butanoic acid	Butyric acid	$C_4H_8O_2$	107-92-6	88.106	liq	-5.12(9)	163.7(1)	0.9528^{25}	1.3980^{20}	msc H_2O, EtOH, eth; sl ctc
105	Butanoic anhydride	Butyric anhydride	$C_8H_{14}O_3$	106-31-0	158.195	liq	-75.0(6)	195(1)	0.9668^{20}	1.4070^{20}	s eth; sl ctc
106	1-Butanol	Butyl alcohol	$C_4H_{10}O$	71-36-3	74.121	liq	-88.60(2)	117.6(2)	0.8148^{20}	1.3988^{20}	s H_2O, bz; msc EtOH, eth; vs ace
107	2-Butanol	*sec*-Butyl alcohol	$C_4H_{10}O$	78-92-2	74.121	liq	-88.44(7)	99.4(2)	0.8063^{20}	1.3978^{20}	vs H_2O; msc EtOH, eth; s bz, ctc
108	2-Butanone	Methyl ethyl ketone	C_4H_8O	78-93-3	72.106	liq	-86.67(1)	79.6(2)	0.7999^{25}	1.3788^{20}	vs H_2O; msc EtOH, eth, ace, bz; s chl
109	*trans*-2-Butenal	*trans*-Crotonaldehyde	C_4H_6O	123-73-9	70.090	liq	-76.6(3)	102.2(3)	0.8516^{20}	1.4366^{20}	s H_2O, chl; vs EtOH, eth, ace; msc bz
110	1-Butene	1-Butylene	C_4H_8	106-98-9	56.107	col gas	-185.33(2)	-6.3(2)	0.588^{25} (p>1 atm)	1.3962^{20}	i H_2O; vs EtOH, eth; s bz
111	*cis*-2-Butene		C_4H_8	590-18-1	56.107	col gas	-138.89(2)	3.72(8)	0.616^{25} (p>1 atm)	1.3931^{-25}	i H_2O; vs EtOH, eth; s bz
112	*trans*-2-Butene		C_4H_8	624-64-6	56.107	col gas	-105.52(2)	0.88(9)	0.599^{25} (p>1 atm)	1.3848^{-25}	s bz
113	*trans*-2-Butenedinitrile		$C_4H_2N_2$	764-42-1	78.072	nd (bz-peth)	96.0(8)	186	0.9416^{111}	1.4349^{111}	s H_2O, EtOH, eth, ace, bz, chl; sl peth
114	*cis*-2-Butene-1,4-diol		$C_4H_8O_2$	6117-80-2	88.106		11.0(5)	235	1.0698^{20}	1.4782^{20}	s H_2O; vs EtOH
115	*trans*-2-Butene-1,4-diol		$C_4H_8O_2$	821-11-4	88.106		27(1)		1.0700^{20}	1.4755^{20}	vs H_2O, EtOH
116	*cis*-2-Butenenitrile	Isocrotononitrile	C_4H_5N	1190-76-7	67.090	liq		106(6)			
117	*trans*-2-Butenenitrile	Crotononitrile	C_4H_5N	627-26-9	67.090	liq	-51.5	120	0.8239^{20}	1.4225^{20}	s eth, ace
118	*cis*-2-Butenoic acid	Isocrotonic acid	$C_4H_6O_2$	503-64-0	86.090	nd or pr (peth)	15	169	1.0267^{20}	1.4450^{20}	vs H_2O; s EtOH
119	*trans*-2-Butenoic acid	Crotonic acid	$C_4H_6O_2$	107-93-7	86.090	mcl pr or nd (w, lig)	71.3(2)	184.7	0.9604^{77}	1.4249^{77}	vs H_2O, EtOH; s eth, ace, lig

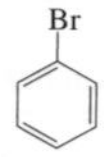

79
Bromobenzene

80
1-Bromobutane

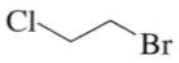

81
1-Bromo-2-chloroethane

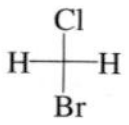

82
Bromochloromethane

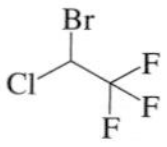

83
2-Bromo-2-chloro-1,1,1-trifluoroethane

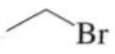

84
Bromoethane

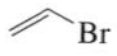

85
Bromoethene

86
1-Bromoheptane

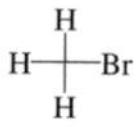

87
Bromomethane

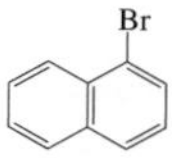

88
1-Bromonaphthalene

89
1-Bromopentane

90
1-Bromopropane

91
2-Bromopropane

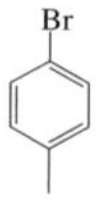

92
4-Bromotoluene

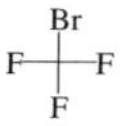

93
Bromotrifluoromethane

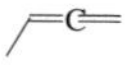

94
1,2-Butadiene

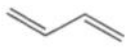

95
1,3-Butadiene

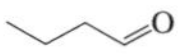

96
Butanal

97
Butane

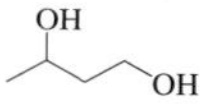

98
1,3-Butanediol

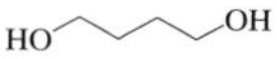

99
1,4-Butanediol

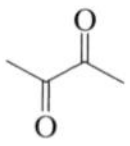

100
2,3-Butanedione

101
Butanenitrile

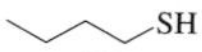

102
1-Butanethiol

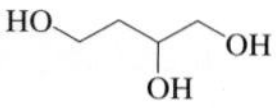

103
1,2,4-Butanetriol

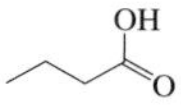

104
Butanoic acid

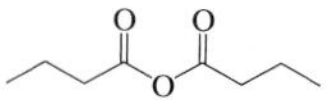

105
Butanoic anhydride

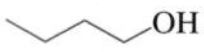

106
1-Butanol

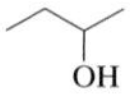

107
2-Butanol

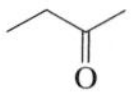

108
2-Butanone

109
trans-2-Butenal

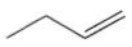

110
1-Butene

111
cis-2-Butene

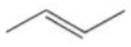

112
trans-2-Butene

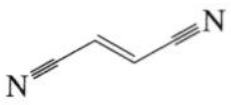

113
trans-2-Butenedinitrile

114
cis-2-Butene-1,4-diol

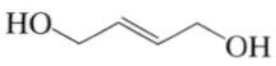

115
trans-2-Butene-1,4-diol

116
cis-2-Butenenitrile

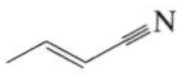

117
trans-2-Butenenitrile

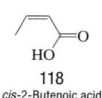

118
cis-2-Butenoic acid

119
trans-2-Butenoic acid

No.	Name	Synonym	Mol. form.	CAS Reg. No.	Mol. wt.	Physical form	mp/°C	bp/°C	den/g cm^{-3}	n_D	Solubility
120	*cis*-2-Buten-1-ol	*cis*-Crotyl alcohol	C_4H_8O	4088-60-2	72.106			123	0.8662^{20}	1.4342^{25}	s H_2O
121	*trans*-2-Buten-1-ol	*trans*-Crotyl alcohol	C_4H_8O	504-61-0	72.106		<-30	121.2	0.8521^{20}	1.4288^{20}	vs H_2O; msc EtOH, eth; s chl
122	3-Buten-2-one	Methyl vinyl ketone	C_4H_6O	78-94-4	70.090			81(4)	0.864^{20}	1.4081^{20}	s H_2O, EtOH, bz; vs eth, ace; sl ctc
123	2-Butoxyethanol	Ethylene glycol monobutyl ether	$C_6H_{14}O_2$	111-76-2	118.174	liq	-74.8	171(2)	0.9015^{20}	1.4198^{20}	msc H_2O, EtOH, eth; sl ctc
124	Butyl acetate		$C_6H_{12}O_2$	123-86-4	116.158	liq	-77.0(1)	126.0(1)	0.8825^{20}	1.3941^{20}	sl H_2O; msc EtOH, eth; s ace, chl
125	*sec*-Butyl acetate	1-Methylpropyl acetate	$C_6H_{12}O_2$	105-46-4	116.158	liq	-98.9	108(4)	0.8748^{20}	1.3888^{20}	sl H_2O, ctc; s EtOH, eth
126	*tert*-Butyl acetate		$C_6H_{12}O_2$	540-88-5	116.158	liq		97.9(10)	0.8665^{20}	1.3855^{20}	s EtOH, eth, chl, HOAc
127	Butylamine	1-Butanamine	$C_4H_{11}N$	109-73-9	73.137	liq	-49(1)	77.0(2)	0.7417^{20}	1.4031^{20}	msc H_2O; s EtOH, eth
128	*sec*-Butylamine	2-Butanamine, (±)-	$C_4H_{11}N$	33966-50-6	73.137	liq	-104.5(6)	62.71(8)	0.7246^{20}	1.3932^{20}	s H_2O, chl; msc EtOH, eth; vs ace
129	*tert*-Butylamine	2-Methyl-2-propanamine	$C_4H_{11}N$	75-64-9	73.137	liq	-66.92(6)	44.02(7)	0.6958^{20}	1.3784^{20}	msc H_2O, EtOH, eth; s chl
130	Butylbenzene	1-Phenylbutane	$C_{10}H_{14}$	104-51-8	134.218	liq	-87.81(5)	183.3(3)	0.8601^{20}	1.4898^{20}	i H_2O; msc EtOH, eth, ace, bz, peth, ctc
131	*sec*-Butylbenzene, (±)-	2-Phenylbutane	$C_{10}H_{14}$	36383-15-0	134.218	liq	-75.5(3)	173.3(4)	0.8621^{20}	1.4902^{20}	i H_2O; msc EtOH, eth, ace, bz, peth, ctc
132	*tert*-Butylbenzene	(1,1-Dimethylethyl)benzene	$C_{10}H_{14}$	98-06-6	134.218	liq	-57.84(4)	169.1(3)	0.8665^{20}	1.4927^{20}	i H_2O; vs EtOH, eth; msc ace, bz
133	*sec*-Butylcyclohexane		$C_{10}H_{20}$	7058-01-7	140.266			179.3(5)	0.8131^{20}	1.4467^{20}	i H_2O; s ace
134	*tert*-Butylcyclohexane		$C_{10}H_{20}$	3178-22-1	140.266	liq	-41.2(3)	171.6(4)	0.8127^{20}	1.4469^{20}	i H_2O
135	1-*tert*-Butyl-4-ethylbenzene		$C_{12}H_{18}$	7364-19-4	162.271	liq	-38.4	211	0.8641^{20}		
136	Butyl ethyl ether	Ethyl butyl ether	$C_6H_{14}O$	628-81-9	102.174	liq	-124	89(2)	0.7495^{20}	1.3818^{20}	i H_2O; msc EtOH, eth; vs ace
137	*tert*-Butyl ethyl ether	Ethyl *tert*-butyl ether	$C_6H_{14}O$	637-92-3	102.174	liq	-94.0(3)	72.7(1)	0.736^{25}	1.3756^{20}	i H_2O; vs EtOH, eth
138	*tert*-Butyl ethyl sulfide	2-Methyl-2-propanethiol	$C_6H_{14}S$	14290-92-7	118.240	liq	-85.9(3)	120.4(6)			
139	*N*-*tert*-Butylformamide		$C_5H_{11}NO$	2425-74-3	101.147	liq	16	202	0.903	1.4330^{20}	
140	Butyl formate		$C_5H_{10}O_2$	592-84-7	102.132	liq	-90.0(4)	106.1(1)	0.8958^{20}	1.3887^{20}	sl H_2O; s ace; msc EtOH, eth
141	*sec*-Butyl formate		$C_5H_{10}O_2$	589-40-2	102.132			93.6(3)	0.8846^{20}	1.3865^{20}	sl H_2O; s ace; msc EtOH, eth
142	*tert*-Butyl formate	1,1-Dimethylethyl formate	$C_5H_{10}O_2$	762-75-4	102.132	liq		83(6)	0.872	1.3790^{20}	
143	*tert*-Butyl hydroperoxide		$C_4H_{10}O_2$	75-91-2	90.121	liq	6	89 dec	0.8960^{20}	1.4015^{20}	s H_2O, EtOH, eth, ctc, chl
144	*tert*-Butyl isobutyl ether		$C_8H_{18}O$	33021-02-2	130.228	liq		112.9(3)			
145	Butyl isocyanate		C_5H_9NO	111-36-4	99.131			125(3)	0.880^{20}	1.4060^{20}	
146	*tert*-Butyl methacrylate		$C_8H_{14}O_2$	585-07-9	142.196			135.2			
147	1-*tert*-Butyl-4-methylbenzene	4-*tert*-Butyltoluene	$C_{11}H_{16}$	98-51-1	148.245	liq	-52.49(8)	193(3)	0.8612^{20}	1.4918^{20}	i H_2O; sl EtOH; vs eth, chl; s ace, bz
148	Butyl methyl ether	1-Methoxybutane	$C_5H_{12}O$	628-28-4	88.148	liq	-115.7(1)	70.1(3)	0.7392^{25}	1.3736^{20}	i H_2O; msc EtOH, eth; s ace
149	*tert*-Butyl methyl sulfide		$C_5H_{12}S$	6163-64-0	104.214	liq		98.9(3)			
150	*tert*-Butyl 3-oxobutanoate		$C_8H_{14}O_3$	1694-31-1	158.195				0.9756^{20}	1.4180^{20}	
151	4-*tert*-Butylphenol		$C_{10}H_{14}O$	98-54-4	150.217	nd (lig)	100(2)	244(5)	0.908^{80}	1.4787^{114}	s H_2O, EtOH, eth, chl, alk
152	Butyl propanoate	Butyl propionate	$C_7H_{14}O_2$	590-01-2	130.185	liq	-89.5(5)	145.1(1)	0.8754^{20}	1.4014^{20}	sl H_2O, ctc; msc EtOH, eth
153	Butyl stearate	Butyl octadecanoate	$C_{22}H_{44}O_2$	123-95-5	340.583		26.56(2)	343	0.854^{25}	1.4328^{50}	i H_2O; s EtOH; vs ace
154	1-*tert*-Butyl-4-vinylbenzene	*p*-*tert*-Butylstyrene	$C_{12}H_{16}$	1746-23-2	160.255	liq	-36.9		0.89^{20}		
155	Butyl vinyl ether	1-(Ethenyloxy)butane	$C_6H_{12}O$	111-34-2	100.158	liq	-92	94(1)	0.7888^{20}	1.4026^{20}	i H_2O; vs EtOH, ace; msc eth; s bz
156	1-Butyne	Ethylacetylene	C_4H_6	107-00-6	54.091	col gas	-125.7(2)	8.1(3)	0.6783^{0}	1.3962^{20}	i H_2O; s EtOH, eth
157	γ-Butyrolactone	Oxolan-2-one	$C_4H_6O_2$	96-48-0	86.090	liq	-43.36(8)	204.6(4)	1.1296^{20}	1.4341^{20}	vs ace, bz, eth, EtOH
158	Camphor, (+)	1,7,7-Trimethylbicyclo[2.2.1]heptan-2-one, (1*R*)-	$C_{10}H_{16}O$	464-49-3	152.233	pl	178.7(5)	209(31)	0.990^{25}	1.5462	i H_2O; vs EtOH, eth; s ace, bz
159	Caprolactam	6-Hexanelactam	$C_6H_{11}NO$	105-60-2	113.157	lf (lig)	69.16(1)	270.8(1)			vs H_2O, bz, EtOH, chl
160	Carbazole	Dibenzopyrrole	$C_{12}H_9N$	86-74-8	167.206	pl or lf	245(2)	354.6(2)	1.297^{25}		i H_2O; sl EtOH, eth, bz, chl; s ace
161	Carbon dioxide	Carbonic anhydride	CO_2	124-38-9	44.010	col gas	-56.561(8) tp	-78.464 sp	1.56^{-79} solid	1.6630^{24} at 6.3 MPa	sl H_2O
162	Carbon disulfide	Carbon bisulfide	CS_2	75-15-0	76.141	col liq	-111.7(3)	46.2(1)	1.2632^{20}	1.6277^{20}	s H_2O, chl; msc EtOH, eth
163	Carbon monoxide	Carbon oxide	CO	630-08-0	28.010	col gas	-205.1(1)	-191.51(9)	$0.8495^{-205.1}$ liq		sl H_2O; s bz, HOAc

120
cis-2-Buten-1-ol

121
trans-2-Buten-1-ol

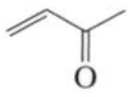

122
3-Buten-2-one

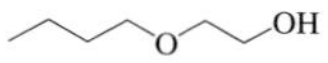

123
2-Butoxyethanol

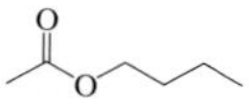

124
Butyl acetate

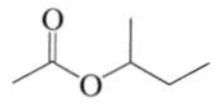

125
sec-Butyl acetate

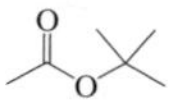

126
tert-Butyl acetate

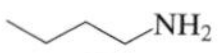

127
Butylamine

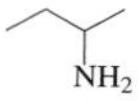

128
sec-Butylamine

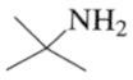

129
tert-Butylamine

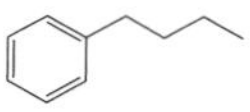

130
Butylbenzene

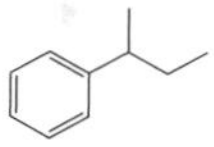

131
sec-Butylbenzene, (±)-

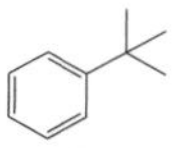

132
tert-Butylbenzene

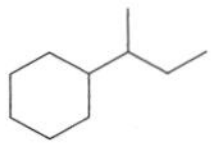

133
sec-Butylcyclohexane

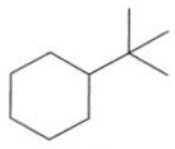

134
tert-Butylcyclohexane

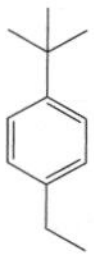

135
1-*tert*-Butyl-4-ethylbenzene

136
Butyl ethyl ether

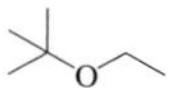

137
tert-Butyl ethyl ether

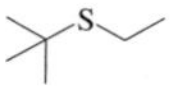

138
tert-Butyl ethyl sulfide

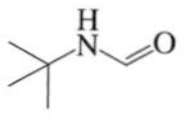

139
N-*tert*-Butylformamide

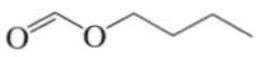

140
Butyl formate

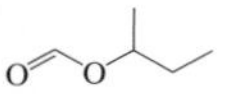

141
sec-Butyl formate

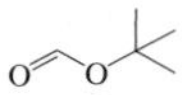

142
tert-Butyl formate

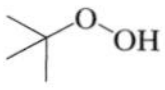

143
tert-Butyl hydroperoxide

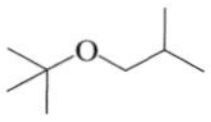

144
tert-Butyl isobutyl ether

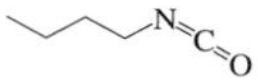

145
Butyl isocyanate

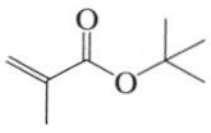

146
tert-Butyl methacrylate

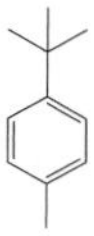

147
1-*tert*-Butyl-4-methylbenzene

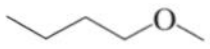

148
Butyl methyl ether

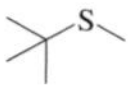

149
tert-Butyl methyl sulfide

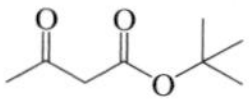

150
tert-Butyl 3-oxobutanoate

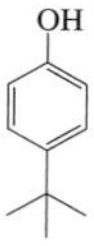

151
4-*tert*-Butylphenol

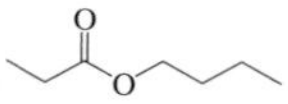

152
Butyl propanoate

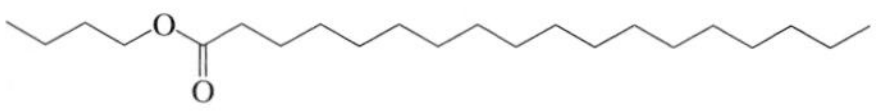

153
Butyl stearate

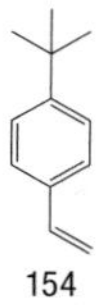

154
1-*tert*-Butyl-4-vinylbenzene

155
Butyl vinyl ether

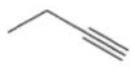

156
1-Butyne

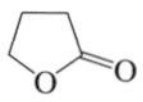

157
γ-Butyrolactone

158
Camphor, (+)

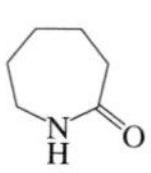

159
Caprolactam

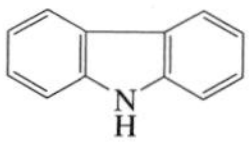

160
Carbazole

O=C=O

161
Carbon dioxide

S=C=S

162
Carbon disulfide

C≡O

163
Carbon monoxide

No.	Name	Synonym	Mol. form.	CAS Reg. No.	Mol. wt.	Physical form	mp/°C	bp/°C	den/g cm^{-3}	n_D	Solubility
164	Carbon oxysulfide	Carbonyl sulfide	COS	463-58-1	60.075	col gas	-138.8(1)	-50.2(3)	0.002456^{25} gas	1.24^{-87}	sl H_2O; s EtOH; vs KOH
165	Carbonyl chloride	Phosgene	CCl_2O	75-44-5	98.916	col gas	-127.77(2)	7.5(4)	0.004043^{25} gas		sl H_2O; s bz, ctc, chl, tol, HOAc
166	β,ψ-Carotene	γ-Carotene	$C_{40}H_{56}$	472-93-5	536.873	red pr (bz-MeOH), viol pr (eth)	153				i H_2O, EtOH; sl eth, peth; s bz, chl
	Chlordene		$C_{10}H_6Cl_6$	3734-48-3	338.873	cry (EtOH)	155				
167	Chloroacetic acid		$C_2H_3ClO_2$	79-11-8	94.497	mcl pl	62.0(7)	189.11(3)	1.4043^{40}	1.4351^{55}	vs H_2O; s EtOH, eth, bz, chl; sl ctc
168	2-Chloroaniline		C_6H_6ClN	95-51-2	127.572	liq	-2.3(9)	209(1)	1.2114^{22}	1.5895^{20}	i H_2O; msc EtOH; s eth, ace
169	3-Chloroaniline		C_6H_6ClN	108-42-9	127.572	liq	-10.3(2)	230(1)	1.2161^{20}	1.5941^{20}	i H_2O; msc EtOH, eth, ace, bz; s chl
170	4-Chloroaniline		C_6H_6ClN	106-47-8	127.572	orth pr	70.4(7)	231(4)	1.429^{19}	1.5546^{87}	s H_2O, EtOH, eth, chl
171	Chlorobenzene	Phenyl chloride	C_6H_5Cl	108-90-7	112.557	liq	-45.2(1)	131.6(2)	1.1058^{20}	1.5241^{20}	i H_2O; msc EtOH, eth; vs bz, ctc
172	2-Chlorobenzoic acid		$C_7H_5ClO_2$	118-91-2	156.567	mcl pr (w)	140.4(7)	274(14)	1.544^{20}		s H_2O, bz; vs EtOH, eth, ace; sl CS_2
173	1-Chlorobutane	Butyl chloride	C_4H_9Cl	109-69-3	92.567	liq	-123.1(2)	78.4(2)	0.8857^{20}	1.4023^{20}	i H_2O; msc EtOH, eth; sl ctc
174	2-Chlorobutane	(±)-*sec*-Butyl chloride	C_4H_9Cl	53178-20-4	92.567	liq	-131.3	71(8)	0.8732^{20}	1.3971^{20}	vs bz, eth, EtOH, chl
175	1-Chloro-1,1-difluoroethane	HCFC-142b	$C_2H_3ClF_2$	75-68-3	100.495	col gas	-130.43(2)	-9.12(7)	1.107^{25}		i H_2O; s bz
176	Chlorodifluoromethane	HCFC-22	$CHClF_2$	75-45-6	86.469	col gas	-157.41(20)	-40.8(5)	1.4909^{-69}		sl H_2O; s eth, ace, chl
177	1-Chloro-2,4-dinitrobenzene		$C_6H_3ClN_2O_4$	97-00-7	202.552	ye orth (eth) nd (al) ye cry	50.2(9)	315	1.4982^{75}	1.5857^{60}	i H_2O; sl EtOH; s eth, bz, CS_2
178	Chloroethane	Ethyl chloride	C_2H_5Cl	75-00-3	64.514	vol liq or gas	-138.4(17)	12.3(2)	0.9239^{0}	1.3676^{20}	sl H_2O, chl; vs EtOH; msc eth
179	2-Chloroethanol	Ethylene chlorohydrin	C_2H_5ClO	107-07-3	80.513	liq	-68(2)	126(2)	1.2019^{20}	1.4419^{20}	msc H_2O, EtOH; sl eth; s chl
180	Chloroethene	Vinyl chloride	C_2H_3Cl	75-01-4	62.498	col gas	-153.84(2)	-13.8(3)	0.9106^{20}	1.3700^{20}	sl H_2O; s EtOH; vs eth
181	Chlorofluoromethane	HCFC-31	CH_2ClF	593-70-4	68.478	col gas	-135.1	-9.1			sl H_2O; vs chl
182	Chloromethane	Methyl chloride	CH_3Cl	74-87-3	50.488	col gas	-97.6(3)	-24.1(3)	0.911^{25} (p>1 atm)	1.3389^{20}	sl H_2O; s EtOH; msc eth, ace, bz, chl
183	(Chloromethyl)benzene	Benzyl chloride	C_7H_7Cl	100-44-7	126.584	liq	-39.4(6)	174(7)	1.1004^{20}	1.5391^{20}	i H_2O; msc EtOH, eth, chl; sl ctc
184	3-(Chloromethyl)heptane	2-Ethylhexyl chloride	$C_8H_{17}Cl$	123-04-6	148.674			171(3)	0.8769^{20}	1.4319^{20}	i H_2O; s EtOH, eth, ace, bz; sl ctc
185	1-Chloro-2-methylpropane	Isobutyl chloride	C_4H_9Cl	513-36-0	92.567	liq	-130.3	69(1)	0.8773^{20}	1.3984^{20}	sl H_2O, ctc; s eth, ace, chl
186	2-Chloro-2-methylpropane	*tert*-Butyl chloride	C_4H_9Cl	507-20-0	92.567	liq	-25.60(2)	50.9(5)	0.8420^{20}	1.3857^{20}	sl H_2O; msc EtOH, eth; s bz, ctc, chl
187	1-Chloronaphthalene	1-Naphthyl chloride	$C_{10}H_7Cl$	90-13-1	162.616	oily liq	-6.0(2)	259(2)	1.1880^{25}	1.6326^{20}	i H_2O; s EtOH, eth, bz, CS_2; sl ctc
188	1-Chloro-2-nitrobenzene	*o*-Chloronitrobenzene	$C_6H_4ClNO_2$	88-73-3	157.555	mcl nd	32.1(3)	246.2(7)	1.368^{242}		i H_2O; s EtOH, eth, bz; vs ace, tol, py
189	1-Chloro-3-nitrobenzene	*m*-Chloronitrobenzene	$C_6H_4ClNO_2$	121-73-3	157.555	pa ye orth pr (al)	43.6(2)	236.5(6)	1.343^{50}	1.5374^{80}	i H_2O; s EtOH, eth, bz, chl, CS_2
190	1-Chloro-4-nitrobenzene	*p*-Chloronitrobenzene	$C_6H_4ClNO_2$	100-00-5	157.555	mcl pr	82.2(7)	238(3)	1.2979^{90}	1.5376^{100}	i H_2O; sl EtOH; s eth, chl, CS_2
191	1-Chlorooctane	Octyl chloride	$C_8H_{17}Cl$	111-85-3	148.674	liq	-57.8	183(3)	0.8734^{20}	1.4309^{20}	i H_2O; vs EtOH, eth; sl ctc
192	Chloropentafluoroethane	CFC-115	C_2ClF_5	76-15-3	154.466	col gas	-99.4(1)	-39.2(2)	1.5678^{-42}	1.2678^{-42}	i H_2O; s EtOH, eth
193	1-Chloropentane	Pentyl chloride	$C_5H_{11}Cl$	543-59-9	106.594	liq	-99.0	107.9(3)	0.8820^{20}	1.4126^{20}	i H_2O; msc EtOH, eth; s bz, ctc; vs chl
194	2-Chlorophenol		C_6H_5ClO	95-57-8	128.556	liq	8(1)	173.4(6)	1.2634^{20}	1.5524^{20}	sl H_2O, chl; s EtOH, eth; vs bz
195	3-Chlorophenol		C_6H_5ClO	108-43-0	128.556		32.5(3)	210(3)	1.245^{45}	1.5565^{40}	sl H_2O, chl; s EtOH, eth; vs bz
196	4-Chlorophenol		C_6H_5ClO	106-48-9	128.556		43.1(7)	219(4)	1.2651^{40}	1.5579^{40}	sl H_2O; vs EtOH, eth, bz; s alk
197	1-Chloropropane	Propyl chloride	C_3H_7Cl	540-54-5	78.541	liq	-122.9(7)	46.2(5)	0.8899^{20}	1.3879^{20}	sl H_2O, ctc; msc EtOH, eth; s bz, chl
198	2-Chloropropane	Isopropyl chloride	C_3H_7Cl	75-29-6	78.541	liq	-117.1(2)	35.0(6)	0.8617^{20}	1.3777^{20}	sl H_2O; msc EtOH, eth; s bz, ctc, chl
199	3-Chloropropene	Allyl chloride	C_3H_5Cl	107-05-1	76.525	liq	-136(2)	44.8(4)	0.9376^{20}	1.4157^{20}	i H_2O; msc EtOH, eth, ace, bz, lig; sl ctc
200	(3-Chloropropyl)trimethoxysilane		$C_6H_{15}ClO_3Si$	2530-87-2	198.720			91	1.077^{25}	1.4183^{25}	
201	3-Chloro-1-propyne	Propargyl chloride	C_3H_3Cl	624-65-7	74.509		-78	56(4)	1.030^{25}	1.4349^{20}	i H_2O; msc EtOH, eth, bz; s ctc

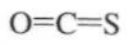

164
Carbon oxysulfide

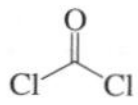
165
Carbonyl chloride

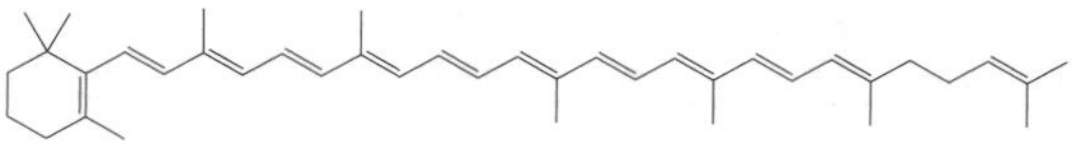
166
β,ψ-Carotene

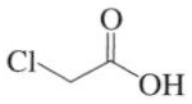
167
Chloroacetic acid

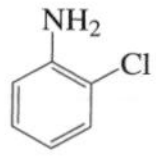
168
2-Chloroaniline

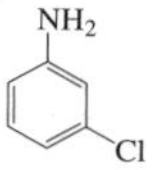
169
3-Chloroaniline

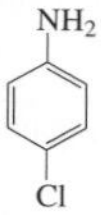
170
4-Chloroaniline

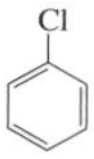
171
Chlorobenzene

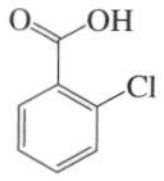
172
2-Chlorobenzoic acid

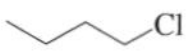
173
1-Chlorobutane

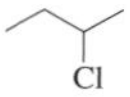
174
2-Chlorobutane

175
1-Chloro-1,1-difluoroethane

176
Chlorodifluoromethane

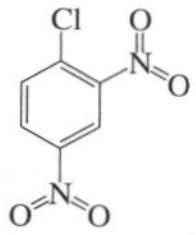
177
1-Chloro-2,4-dinitrobenzene

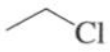
178
Chloroethane

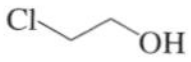
179
2-Chloroethanol

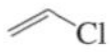
180
Chloroethene

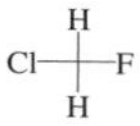
181
Chlorofluoromethane

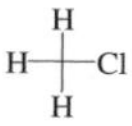
182
Chloromethane

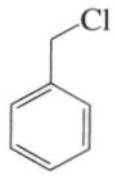
183
(Chloromethyl)benzene

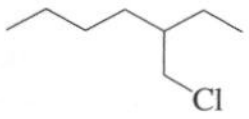
184
3-(Chloromethyl)heptane

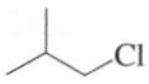
185
1-Chloro-2-methylpropane

186
2-Chloro-2-methylpropane

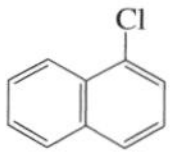
187
1-Chloronaphthalene

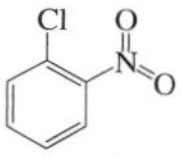
188
1-Chloro-2-nitrobenzene

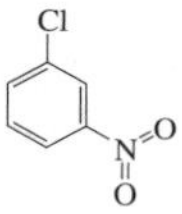
189
1-Chloro-3-nitrobenzene

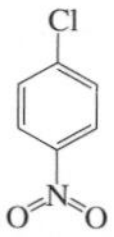
190
1-Chloro-4-nitrobenzene

191
1-Chlorooctane

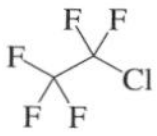
192
Chloropentafluoroethane

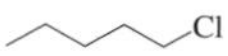
193
1-Chloropentane

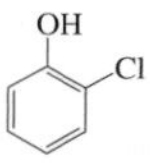
194
2-Chlorophenol

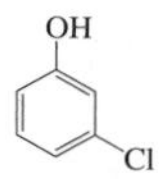
195
3-Chlorophenol

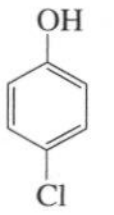
196
4-Chlorophenol

197
1-Chloropropane

198
2-Chloropropane

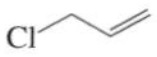
199
3-Chloropropene

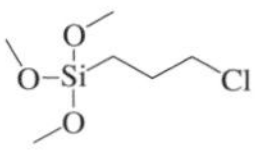
200
(3-Chloropropyl)trimethoxysilane

201
3-Chloro-1-propyne

No.	Name	Synonym	Mol. form.	CAS Reg. No.	Mol. wt.	Physical form	mp/°C	bp/°C	den/g cm^{-3}	n_D	Solubility
202	2-Chlorotoluene	1-Chloro-2-methylbenzene	C_7H_7Cl	95-49-8	126.584	liq	-35.9(7)	158.8(4)	1.0825^{20}	1.5268^{20}	i H_2O; s EtOH, bz; msc eth, ace, chl
203	3-Chlorotoluene	1-Chloro-3-methylbenzene	C_7H_7Cl	108-41-8	126.584	liq	-47.8	162.1(4)	1.075^{20}	1.5214^{19}	i H_2O; s EtOH, bz, ctc, chl; msc eth
204	4-Chlorotoluene	1-Chloro-4-methylbenzene	C_7H_7Cl	106-43-4	126.584	liq	7.4(2)	161.8(2)	1.0697^{20}	1.5150^{20}	i H_2O; s EtOH, ctc, chl; msc eth
205	Chlorotrifluoromethane	CFC-13	$CClF_3$	75-72-9	104.459	col gas	-181.2	-81.48			i H_2O
206	Cholesterol		$C_{27}H_{46}O$	57-88-5	386.653	orth or tcl lf (al), nd (eth)	148.2(8)	459(20)	1.067^{20}		i H_2O; sl EtOH, ace; s bz, HOAc; vs diox
207	Chrysene	Benzo[*a*]phenanthrene	$C_{18}H_{12}$	218-01-9	228.288	red bl fl or orth pl (bz, HOAc)	255.0(1)	448	1.274^{20}		i H_2O; sl EtOH, eth, ace, bz, CS_2; s tol
208	*trans*-Cinnamic acid	3-Phenyl-2-propenoic acid, (*E*)-	$C_9H_8O_2$	140-10-3	148.159	mcl pr (dil al)	134(2)	300	1.2475^{4}		i H_2O, lig; vs EtOH; s eth, ace, bz
209	Citric acid	2-Hydroxy-1,2,3-propanetricarboxylic acid	$C_6H_8O_7$	77-92-9	192.124	orth (w+1)	153	dec	1.665^{20}		vs H_2O, EtOH; s eth, AcOEt; i bz, chl
210	*o*-Cresol	2-Methylphenol	C_7H_8O	95-48-7	108.138		31.0(6)	191.0(1)	1.0327^{35}	1.5386^{35}	s H_2O; vs EtOH, eth; msc ace, bz, ctc
211	*m*-Cresol	3-Methylphenol	C_7H_8O	108-39-4	108.138	liq	12.2(3)	202.2(1)	1.0339^{20}	1.5401^{20}	sl H_2O; msc EtOH, eth, ace, bz, ctc
212	*p*-Cresol	4-Methylphenol	C_7H_8O	106-44-5	108.138	pr	34.77(5)	201.9(1)	1.0185^{40}	1.5312^{20}	sl H_2O; msc EtOH, eth, ace, bz, ctc
213	Cyanogen	Ethanedinitrile	C_2N_2	460-19-5	52.034	col gas	-27.83	-21.1	0.787^{-21} liq		s H_2O, EtOH, eth
214	Cyanogen chloride	Chlorine cyanide	CClN	506-77-4	61.471	col vol liq or gas	-6.55	13	1.186^{20}		s H_2O, EtOH; vs eth
215	Cyanoguanidine	Dicyanodiamide	$C_2H_4N_4$	461-58-5	84.080		207(2)		1.404^{14}		s H_2O, EtOH, ace; i eth, bz, chl
216	Cyclobutane	Tetramethylene	C_4H_8	287-23-0	56.107	vol liq or gas	-90.7(3)	12.5(2)	0.7038^{0}	1.375^{20}	i H_2O; vs EtOH, ace; msc eth; s bz
217	Cycloheptane		C_7H_{14}	291-64-5	98.186	liq	-8.0(2)	118.8(2)	0.8098^{20}	1.4436^{20}	i H_2O; vs EtOH, eth; s bz, chl
218	Cycloheptene		C_7H_{12}	628-92-2	96.170	liq	-55.3(2)	115(3)	0.8228^{20}	1.4552^{20}	i H_2O; s EtOH, eth, bz, chl; sl ctc
219	1,3-Cyclohexadiene		C_6H_8	592-57-4	80.128	liq	-89	80.3(3)	0.8405^{20}	1.4755^{20}	i H_2O; s EtOH, bz, chl, peth; vs eth
220	1,4-Cyclohexadiene	1,4-Dihydrobenzene	C_6H_8	628-41-1	80.128	liq	-49(1)	89.5(2)	0.8471^{20}	1.4725^{20}	i H_2O; msc EtOH, eth; s bz, chl, peth
221	Cyclohexane	Hexahydrobenzene	C_6H_{12}	110-82-7	84.159	liq	6.7(2)	80.7(7)	0.7786^{20}	1.4235^{25}	i H_2O; msc EtOH, eth, ace, bz, lig, ctc
222	*trans*-1,4-Cyclohexanedicarboxylic acid		$C_8H_{12}O_4$	619-82-9	172.179	pr (w)	312.5	300 subl			sl H_2O, eth; vs EtOH; s ace; i chl
223	1,4-Cyclohexanedimethanol		$C_8H_{16}O_2$	105-08-8	144.212		43	283			
224	Cyclohexanol	Cyclohexyl alcohol	$C_6H_{12}O$	108-93-0	100.158	hyg nd	26(1)	160.9(2)	0.9624^{20}	1.4641^{20}	s H_2O, EtOH, eth, ace; msc bz; sl chl
225	Cyclohexanone	Pimelic ketone	$C_6H_{10}O$	108-94-1	98.142	liq	-27.93(5)	155.4(1)	0.9478^{20}	1.4507^{20}	s H_2O, EtOH, eth, ace, bz, chl, ctc
226	Cyclohexanone oxime		$C_6H_{11}NO$	100-64-1	113.157	hex pr (lig)	89.05(9)	208(2)			s H_2O, EtOH, eth, MeOH; sl chl
227	Cyclohexene	Tetrahydrobenzene	C_6H_{10}	110-83-8	82.143	liq	-103.5(4)	82.9(2)	0.8110^{20}	1.4465^{20}	i H_2O; msc EtOH, eth, ace, bz, lig, ctc
228	Cyclohexylamine	Cyclohexanamine	$C_6H_{13}N$	108-91-8	99.174	liq	-17.7(7)	133.6(5)	0.8191^{20}	1.4625^{15}	s H_2O, ctc; vs EtOH; msc eth, ace, bz
229	Cyclohexylbenzene	Phenylcyclohexane	$C_{12}H_{16}$	827-52-1	160.255	pl	7.02(10)	239(2)	0.9427^{20}	1.5329^{20}	i H_2O; vs EtOH; s eth; sl ctc
230	*cis,cis*-1,5-Cyclooctadiene		C_8H_{12}	111-78-4	108.181	liq	-69.2(2)	149(3)	0.883^{20}	1.4905^{25}	vs bz
231	Cyclooctane		C_8H_{16}	292-64-8	112.213		14.82(4)	151.1(1)	0.8349^{20}	1.4586^{20}	i H_2O; s bz, lig
232	1,3,5,7-Cyclooctatetraene	[8]Annulene	C_8H_8	629-20-9	104.150	liq	-4.7(1)	140(3)	0.9206^{20}	1.5381^{20}	s EtOH, eth, ace, bz
233	1,3-Cyclopentadiene	Pyropentylene	C_5H_6	542-92-7	66.102	liq	-96.54(5)	41(1)	0.8021^{20}	1.4440^{20}	i H_2O; msc EtOH, eth, bz; s ace
234	Cyclopentane	Pentamethylene	C_5H_{10}	287-92-3	70.133	liq	-93.4(3)	49.2(1)	0.7457^{20}	1.4065^{20}	i H_2O; msc EtOH, eth, ace, bz, peth, ctc
235	Cyclopentanone	Adipic ketone	C_5H_8O	120-92-3	84.117	liq	-51.70(2)	130.5(2)	0.9487^{20}	1.4366^{20}	i H_2O; s EtOH, ace, ctc, hx; msc eth
236	Cyclopentene		C_5H_8	142-29-0	68.118	liq	-135.02(9)	44.2(2)	0.7720^{20}	1.4225^{20}	i H_2O; s EtOH, eth, bz, ctc, peth
237	Cyclopropane	Trimethylene	C_3H_6	75-19-4	42.080	col gas	-127.6(2)	-31(2)	0.617^{25} (p>1 atm)	1.3799^{-42}	s H_2O, bz, peth; vs EtOH, eth
	Decabromobiphenyl ether	Bis(pentabromophenyl) ether	$C_{12}Br_{10}O$	1163-19-5	959.167	ye pr (tol)	305				i H_2O
238	1,9-Decadiene		$C_{10}H_{18}$	1647-16-1	138.250			164(5)	0.75^{25}	1.4325^{20}	
239	*cis*-Decahydronaphthalene	*cis*-Decalin	$C_{10}H_{18}$	493-01-6	138.250	liq	-42.9(3)	195.8(3)	0.8965^{20}	1.4810^{20}	i H_2O; msc EtOH; vs eth, ace, chl

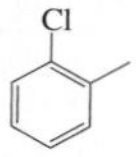
202
2-Chlorotoluene

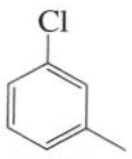
203
3-Chlorotoluene

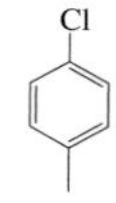
204
4-Chlorotoluene

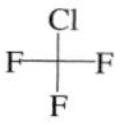
205
Chlorotrifluoromethane

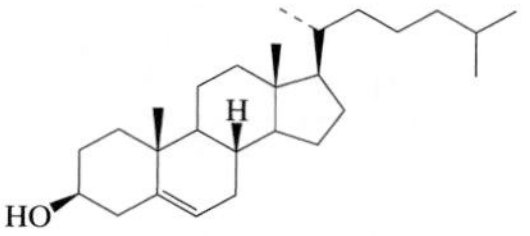
206
Cholesterol

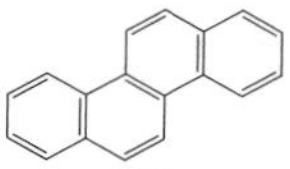
207
Chrysene

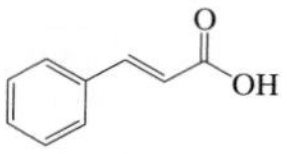
208
trans-Cinnamic acid

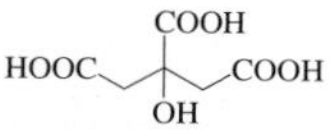
209
Citric acid

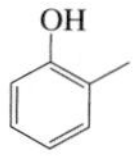
210
o-Cresol

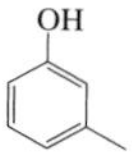
211
m-Cresol

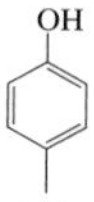
212
p-Cresol

213
Cyanogen

214
Cyanogen chloride

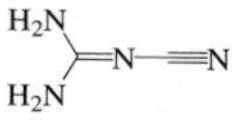
215
Cyanoguanidine

216
Cyclobutane

217
Cycloheptane

218
Cycloheptene

219
1,3-Cyclohexadiene

220
1,4-Cyclohexadiene

221
Cyclohexane

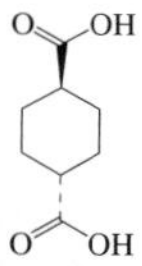
222
trans-1,4-Cyclohexanedicarboxylic acid

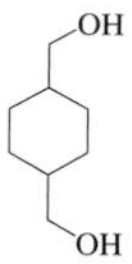
223
1,4-Cyclohexanedimethanol

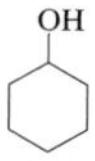
224
Cyclohexanol

225
Cyclohexanone

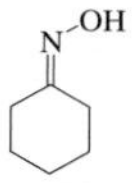
226
Cyclohexanone oxime

227
Cyclohexene

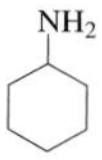
228
Cyclohexylamine

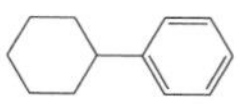
229
Cyclohexylbenzene

230
cis,cis-1,5-Cyclooctadiene

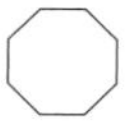
231
Cyclooctane

232
1,3,5,7-Cyclooctatetraene

233
1,3-Cyclopentadiene

234
Cyclopentane

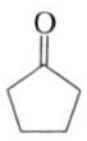
235
Cyclopentanone

236
Cyclopentene

237
Cyclopropane

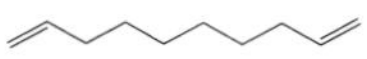
238
1,9-Decadiene

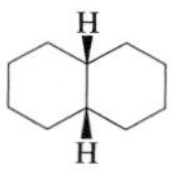
239
cis-Decahydronaphthalene

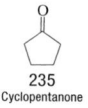

No.	Name	Synonym	Mol. form.	CAS Reg. No.	Mol. wt.	Physical form	mp/°C	bp/°C	den/g cm^{-3}	n_D	Solubility
240	*trans*-Decahydronaphthalene	*trans*-Decalin	$C_{10}H_{18}$	493-02-7	138.250	liq	-30.35(6)	187.3(2)	0.8659^{25}	1.4695^{20}	i H_2O; vs EtOH, eth, ace; msc bz; sl MeOH
241	Decane		$C_{10}H_{22}$	124-18-5	142.282	liq	-29.61(2)	174.1(1)	0.7303^{20}	1.4090^{25}	i H_2O; msc EtOH; s eth; sl ctc
242	Decanedioic acid	Sebacic acid	$C_{10}H_{18}O_4$	111-20-6	202.248	lf	131(1)	374(5)	1.2705^{20}	1.422^{133}	sl H_2O; s EtOH, eth; i bz
243	1,10-Decanediol	Decamethylene glycol	$C_{10}H_{22}O_2$	112-47-0	174.281	nd (w, dil al)	72.4(2)				sl H_2O, eth; vs EtOH; s DMSO; i lig
244	Decanoic acid	Capric acid	$C_{10}H_{20}O_2$	334-48-5	172.265	nd	31.39(2)	270(1)	0.8858^{40}	1.4288^{40}	i H_2O; vs ace, bz, eth, EtOH
245	1-Decanol	Capric alcohol	$C_{10}H_{22}O$	112-30-1	158.281	oily liq	7(1)	229(3)	0.8294^{20}	1.4372^{20}	i H_2O; msc EtOH, eth, ace, bz, chl; s ctc
246	1-Decene		$C_{10}H_{20}$	872-05-9	140.266	liq	-66.21(3)	171(1)	0.7408^{20}	1.4215^{20}	i H_2O; msc EtOH, eth
247	*cis*-2-Decene		$C_{10}H_{20}$	20348-51-0	140.266	col liq		174.2(7)			
248	*trans*-2-Decene		$C_{10}H_{20}$	20063-97-2	140.266	col liq		173.4(5)			
249	Diacetone alcohol	4-Hydroxy-4-methyl-2-pentanone	$C_6H_{12}O_2$	123-42-2	116.158	liq	-47(2)	167.9	0.9387^{20}	1.4213^{20}	msc H_2O, EtOH, eth; s chl
250	*N,N*-Diallyl-2-propen-1-amine	Triallylamine	$C_9H_{15}N$	102-70-5	137.222		94	150(2)	0.809^{20}	1.4502^{20}	s EtOH, eth, ace, bz, acid
251	Diamantane	Congressane	$C_{14}H_{20}$	2292-79-7	188.309	cry	244.73(5)				
252	Dibenzofuran	2,2'-Biphenylene oxide	$C_{12}H_8O$	132-64-9	168.191	lf or nd (al)	82.16(5)	285.2(3)	1.0886^{99}	1.6079^{99}	i H_2O; s EtOH, ace, bz; vs eth, HOAc
253	Dibenzothiophene		$C_{12}H_8S$	132-65-0	184.257	nd (dil al, lig)	98.67(2)	331.6(4)			i H_2O; s chl, MeOH; vs EtOH, bz
254	Dibenzyl ether	Benzyl ether	$C_{14}H_{14}O$	103-50-4	198.260	liq	1.8	298	1.0428^{20}	1.5618^{20}	i H_2O; msc EtOH, eth; s ctc
255	*m*-Dibromobenzene	1,3-Dibromobenzene	$C_6H_4Br_2$	108-36-1	235.904	liq	-6.9(5)	214(14)	1.9523^{20}	1.6083^{17}	i H_2O; s EtOH; msc eth
256	*p*-Dibromobenzene	1,4-Dibromobenzene	$C_6H_4Br_2$	106-37-6	235.904	pl	87.3(1)	222(3)	2.261^{17}	1.5742	i H_2O; s EtOH, bz; vs eth, ace, CS_2
257	1,2-Dibromo-1,1-difluoroethane	Genetron 132b-B2	$C_2H_2Br_2F_2$	75-82-1	223.842	liq	-61.3	92.5	2.2238^{20}	1.4456^{20}	
258	1,1-Dibromoethane	Ethylidene dibromide	$C_2H_4Br_2$	557-91-5	187.861	liq	-63	109(4)	2.0555^{20}	1.5128^{20}	i H_2O; s EtOH, ace, bz; sl chl; vs eth
259	1,2-Dibromoethane	Ethylene dibromide	$C_2H_4Br_2$	106-93-4	187.861	liq	9.8(1)	131.3(3)	2.1683^{25}	1.5356^{25}	vs ace, bz, eth, EtOH
260	Dibromomethane	Methylene bromide	CH_2Br_2	74-95-3	173.835	liq	-52.1(7)	97.0(6)	2.4969^{20}	1.5420^{20}	sl H_2O; msc EtOH, eth, ace; s ctc
261	1,2-Dibromopropane	Propylene dibromide	$C_3H_6Br_2$	78-75-1	201.888	liq	-55.4(3)	140(1)	1.9324^{20}	1.5201^{20}	s EtOH, eth, chl; sl ctc
262	1,3-Dibromopropane		$C_3H_6Br_2$	109-64-8	201.888	liq	-35(1)	164(1)	1.9701^{25}	1.5204^{25}	i H_2O; s EtOH, eth, chl; sl ctc
263	1,2-Dibromotetrafluoroethane	Halon-2402	$C_2Br_2F_4$	124-73-2	259.823	liq	-110.1(12)	47.1(2)	2.149^{25}	1.361^{25}	i H_2O
264	1,2-Dibutoxyethane	Ethylene glycol dibutyl ether	$C_{10}H_{22}O_2$	112-48-1	174.281	liq	-69.1	198(10)	0.8319^{25}	1.4112^{25}	
265	Dibutylamine	*N*-Butylbutanamine	$C_8H_{19}N$	111-92-2	129.244	liq	-61.8(5)	162(2)	0.7670^{20}	1.4177^{20}	s H_2O, ace, bz; vs EtOH, eth
266	1,4-Di-*tert*-butylbenzene		$C_{14}H_{22}$	1012-72-2	190.325	nd (MeOH)	77.63(4)	237.3(5)	0.9850^{20}		i H_2O; s EtOH, eth
267	Dibutyl ether	Butyl ether	$C_8H_{18}O$	142-96-1	130.228	liq	-96(3)	141.6(3)	0.7684^{20}	1.3992^{20}	i H_2O; msc EtOH, eth; vs ace; sl ctc
268	*N,N'*-Di-*tert*-butylethylenediamine	*N,N'*-Di-*tert*-butylethanediamine	$C_{10}H_{24}N_2$	4062-60-6	172.311	cry	53.3	189	0.69		
269	Dibutyl maleate	Butyl *cis*-butenedioate	$C_{12}H_{20}O_4$	105-76-0	228.285		<-80	280			
270	Dibutyl phthalate	Butyl phthalate	$C_{16}H_{22}O_4$	84-74-2	278.344	liq	-35	338(9)	1.0465^{20}	1.4911^{20}	i H_2O; msc EtOH, eth, bz; s ctc
271	Dibutyl sebacate	Butyl sebacate	$C_{18}H_{34}O_4$	109-43-3	314.461	liq	-9.2(5)	356(9)	0.9405^{15}	1.4433^{15}	i H_2O; s eth, ctc
272	Dibutyl sulfide	Butyl sulfide	$C_8H_{18}S$	544-40-1	146.294	liq	-74.97(5)	168(4)	0.8386^{20}	1.4530^{20}	vs eth, EtOH, chl
273	*o*-Dichlorobenzene	1,2-Dichlorobenzene	$C_6H_4Cl_2$	95-50-1	147.002	liq	-17.0(1)	180.2(3)	1.3059^{20}	1.5515^{20}	i H_2O; s EtOH, eth; msc ace, bz, ctc
274	*m*-Dichlorobenzene	1,3-Dichlorobenzene	$C_6H_4Cl_2$	541-73-1	147.002	liq	-24.8(3)	172(2)	1.2884^{20}	1.5459^{20}	i H_2O; s EtOH, eth, bz; msc ace
275	*p*-Dichlorobenzene	1,4-Dichlorobenzene	$C_6H_4Cl_2$	106-46-7	147.002	mcl pr, lf (ace)	53.1(2)	173.9(2)	1.2475^{55}	1.5285^{20}	i H_2O; msc EtOH, ace, bz; s eth, ctc
276	1,2-Dichlorobutane		$C_4H_8Cl_2$	616-21-7	127.013			123.9(8)	1.1116^{25}	1.4450^{20}	i H_2O; s eth, chl; sl ctc
277	1,4-Dichlorobutane		$C_4H_8Cl_2$	110-56-5	127.013	liq	-38.7(4)	155(3)	1.1331^{25}	1.4522^{25}	i H_2O; vs chl
278	*cis*-1,3-Dichloro-2-butene		$C_4H_6Cl_2$	10075-38-4	124.997			127(16)	1.1605^{20}	1.4735^{20}	vs ace, bz, eth, EtOH
279	*cis*-1,4-Dichloro-2-butene		$C_4H_6Cl_2$	1476-11-5	124.997	liq	-42(2)	149(11)	1.188^{25}	1.4887^{25}	vs ace, bz, eth, EtOH
280	*trans*-1,4-Dichloro-2-butene		$C_4H_6Cl_2$	110-57-6	124.997	col liq	3(2)	155.4	1.183^{25}	1.4871^{25}	vs ace, bz, eth, EtOH
281	1,1-Dichloro-2,2-difluoroethene	1,1-Dichloro-2,2-difluoroethylene	$C_2Cl_2F_2$	79-35-6	132.924	vol liq or gas	-116	19	1.555^{-20}	1.383^{-20}	
282	Dichlorodifluoromethane	CFC-12	CCl_2F_2	75-71-8	120.914	col gas	-157.05(1)	-29.8(1)			sl H_2O; s EtOH, eth, HOAc
283	1,1-Dichloroethane	Ethylidene dichloride	$C_2H_4Cl_2$	75-34-3	98.959	liq	-96.93(21)	56.3(7)	1.1757^{20}	1.4164^{20}	sl H_2O; vs EtOH, eth; s ace, bz

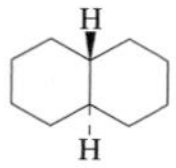
240
trans-Decahydronaphthalene

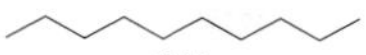
241
Decane

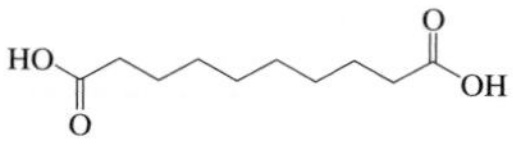
242
Decanedioic acid

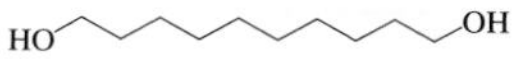
243
1,10-Decanediol

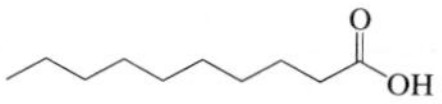
244
Decanoic acid

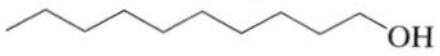
245
1-Decanol

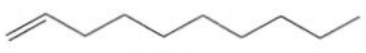
246
1-Decene

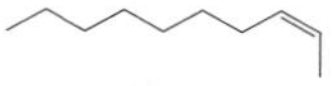
247
cis-2-Decene

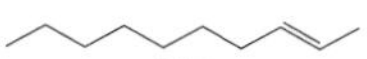
248
trans-2-Decene

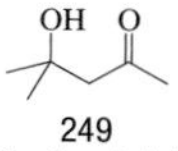
249
Diacetone alcohol

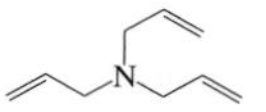
250
N,N-Diallyl-2-propen-1-amine

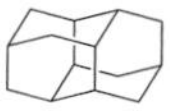
251
Diamantane

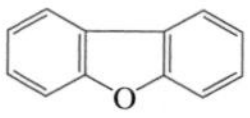
252
Dibenzofuran

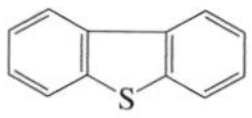
253
Dibenzothiophene

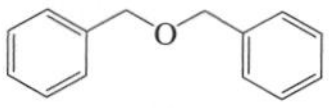
254
Dibenzyl ether

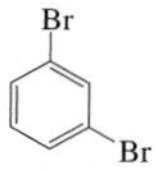
255
m-Dibromobenzene

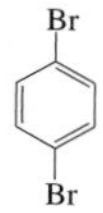
256
p-Dibromobenzene

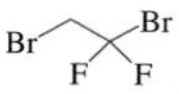
257
1,2-Dibromo-1,1-difluoroethane

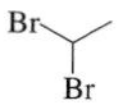
258
1,1-Dibromoethane

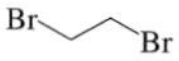
259
1,2-Dibromoethane

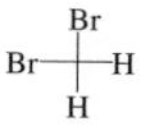
260
Dibromomethane

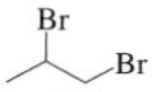
261
1,2-Dibromopropane

262
1,3-Dibromopropane

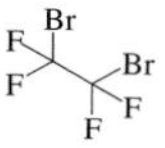
263
1,2-Dibromotetrafluoroethane

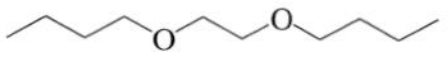
264
1,2-Dibutoxyethane

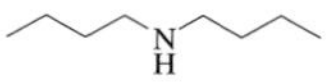
265
Dibutylamine

266
1,4-Di-*tert*-butylbenzene

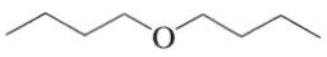
267
Dibutyl ether

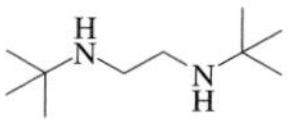
268
N,N'-Di-*tert*-butylethylenediamine

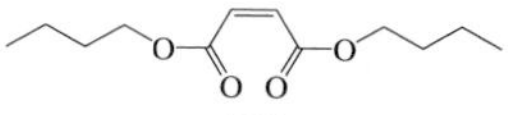
269
Dibutyl maleate

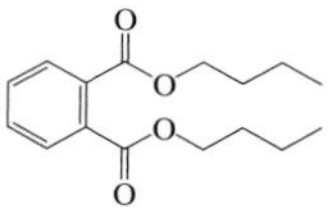
270
Dibutyl phthalate

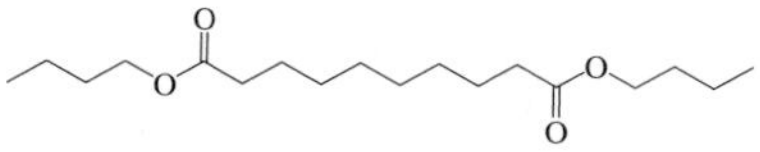
271
Dibutyl sebacate

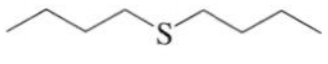
272
Dibutyl sulfide

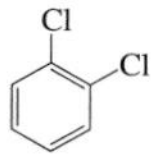
273
o-Dichlorobenzene

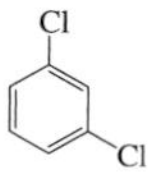
274
m-Dichlorobenzene

275
p-Dichlorobenzene

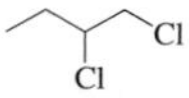
276
1,2-Dichlorobutane

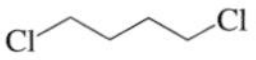
277
1,4-Dichlorobutane

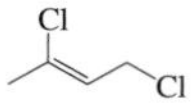
278
cis-1,3-Dichloro-2-butene

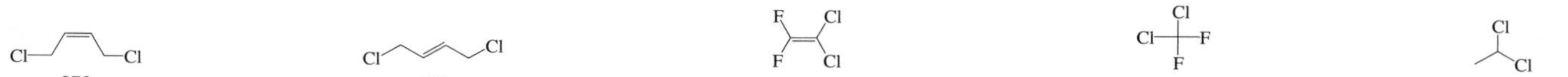
279
cis-1,4-Dichloro-2-butene

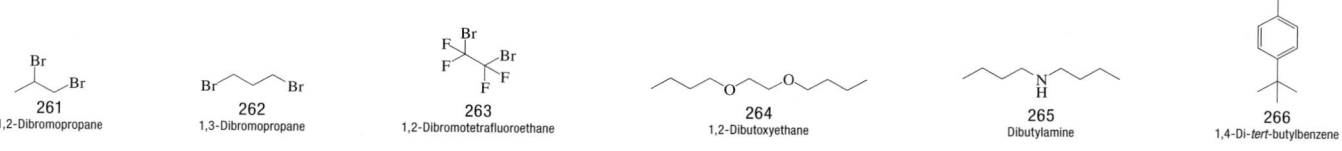

280
trans-1,4-Dichloro-2-butene

281
1,1-Dichloro-2,2-difluoroethene

282
Dichlorodifluoromethane

283
1,1-Dichloroethane

No.	Name	Synonym	Mol. form.	CAS Reg. No.	Mol. wt.	Physical form	mp/°C	bp/°C	den/g cm^{-3}	n_D	Solubility
284	1,2-Dichloroethane	Ethylene dichloride	$C_2H_4Cl_2$	107-06-2	98.959	liq	-35.6(3)	83.4(1)	1.2454^{25}	1.4422^{25}	sl H_2O; vs EtOH; msc eth; s ace, bz, chl
285	1,1-Dichloroethene	Vinylidene chloride	$C_2H_2Cl_2$	75-35-4	96.943	liq	-122.5(1)	31.6(3)	1.213^{20}	1.4249^{20}	i H_2O; s EtOH, ace, bz; vs eth, chl
286	*cis*-1,2-Dichloroethene	*cis*-1,2-Dichloroethylene	$C_2H_2Cl_2$	156-59-2	96.943	liq	-80.0(2)	60(2)	1.2837^{20}	1.4490^{20}	sl H_2O; msc EtOH, eth, ace; vs bz, chl
287	*trans*-1,2-Dichloroethene	*trans*-1,2-Dichloroethylene	$C_2H_2Cl_2$	156-60-5	96.943	liq	-49.8(2)	47.64(8)	1.2565^{20}	1.4454^{20}	sl H_2O; msc EtOH, eth, ace; vs bz, chl
288	1,1-Dichloro-1-fluoroethane	HCFC-141b	$C_2H_3Cl_2F$	1717-00-6	116.949	liq	-103.5(6)	32.05(9)	1.250^{10}	1.3600^{10}	i H_2O
289	Dichlorofluoromethane	HCFC-21	$CHCl_2F$	75-43-4	102.923	col gas	-130.35	8.9	1.405^{9}	1.3724^{9}	i H_2O; s EtOH, eth, ctc, chl, HOAc
290	1,2-Dichloro-1,2,3,3,4,4-hexafluorocyclobutane		$C_4Cl_2F_6$	356-18-3	232.939	liq	-24.2	59.5			
291	1,3-Dichloro-1,1,2,2,3,3-hexafluoropropane	Refrigerant 216	$C_3Cl_2F_6$	662-01-1	220.928	liq	-125.4	35.7	1.573^{20}	1.3030^{20}	
292	Dichloromethane	Methylene chloride	CH_2Cl_2	75-09-2	84.933	liq	-94.9(20)	39.8(3)	1.3232^{20}	1.4242^{20}	sl H_2O; msc EtOH, eth; s ctc
293	(Dichloromethyl)benzene	Benzal chloride	$C_7H_6Cl_2$	98-87-3	161.029	liq	-17.0(5)	205	1.26^{25}	1.5502^{20}	i H_2O; vs eth, EtOH
294	1,2-Dichloro-4-nitrobenzene		$C_6H_3Cl_2NO_2$	99-54-7	192.000	nd (al)	41.0(2)	255.5	1.4558^{75}		i H_2O; s EtOH, eth; sl ctc
295	1,5-Dichloropentane		$C_5H_{10}Cl_2$	628-76-2	141.038	liq	-72.8	182.9(8)	1.0956^{25}	1.4545^{25}	i H_2O; s EtOH, eth, bz, ctc
296	1,1-Dichloropropane	Propylidene chloride	$C_3H_6Cl_2$	78-99-9	112.986			88.4(5)	1.1321^{20}	1.4289^{20}	s EtOH, eth, bz, chl
297	1,2-Dichloropropane, (±)-	Propylene dichloride	$C_3H_6Cl_2$	78-87-5	112.986	liq	-100.53	96.4	1.1560^{20}	1.4394^{20}	sl H_2O; s EtOH, eth, bz, chl
298	1,3-Dichloropropane		$C_3H_6Cl_2$	142-28-9	112.986	liq	-99.5	120.8(3)	1.1785^{25}	1.4455^{25}	sl H_2O; vs EtOH, eth; s bz, chl
299	1,3-Dichloro-2-propanol		$C_3H_6Cl_2O$	96-23-1	128.985			171(4)	1.3506^{17}	1.4837^{20}	vs H_2O, EtOH; msc eth; s ace, chl
300	*cis*-1,3-Dichloropropene	*cis*-1,3-Dichloropropylene	$C_3H_4Cl_2$	10061-01-5	110.970			104(1)	1.224^{20}	1.4682^{20}	i H_2O; s eth, bz, chl
301	*trans*-1,3-Dichloropropene	*trans*-1,3-Dichloropropylene	$C_3H_4Cl_2$	10061-02-6	110.970			111(5)	1.217^{20}	1.4730^{20}	i H_2O; s eth, bz, chl
302	2,3-Dichloropropene		$C_3H_4Cl_2$	78-88-6	110.970	liq	10	93.0(4)	1.211^{20}	1.4603^{20}	i H_2O; msc EtOH; s eth, bz, chl
303	1,1-Dichloro-1,2,2,2-tetrafluoroethane	Refrigerant 114a	$C_2Cl_2F_4$	374-07-2	170.921	col gas	-56.6	3(1)	1.455^{25} (p>1 atm)	1.3092^{0}	vs bz, eth, EtOH
304	1,2-Dichloro-1,1,2,2-tetrafluoroethane	CFC-114	$C_2Cl_2F_4$	76-14-2	170.921	col gas	-92.52(5)	3.6(5)	1.455^{25} (p>1 atm)	1.3092^{0}	i H_2O; vs eth, EtOH
305	2,4-Dichlorotoluene	2,4-Dichloro-1-methylbenzene	$C_7H_6Cl_2$	95-73-8	161.029	liq	-13.5	200(8)	1.2476^{20}	1.5511^{20}	i H_2O; s ctc
306	1,2-Dichloro-1,1,2-trifluoroethane	HCFC-123a	$C_2HCl_2F_3$	354-23-4	152.930	vol liq or gas	-78	30.0(1)	1.50^{25}		
307	2,2-Dichloro-1,1,1-trifluoroethane	HCFC-123	$C_2HCl_2F_3$	306-83-2	152.930	vol liq or gas	-107.15	27.8(6)	1.4638^{25}		sl H_2O
308	1,1-Dichloro-1,2,2-trifluoroethane	Refrigerant 123b	$C_2HCl_2F_3$	812-04-4	152.930			30.2			
	Dieldrin		$C_{12}H_8Cl_6O$	60-57-1	380.909		178.8(3)		1.75^{25}		i H_2O; sl EtOH; s ace, bz
309	Diethanolamine	Bis(2-hydroxyethyl)amine	$C_4H_{11}NO_2$	111-42-2	105.136		27.9(2)	271.2(7)	1.0966^{20}	1.4776^{20}	vs H_2O, EtOH; sl eth, bz
310	1,1-Diethoxyethane	Acetal	$C_6H_{14}O_2$	105-57-7	118.174	liq	-106.1(6)	102(2)	0.8254^{20}	1.3834^{20}	s H_2O, chl; msc EtOH, eth; vs ace
311	1,2-Diethoxyethane	Ethylene glycol diethyl ether	$C_6H_{14}O_2$	629-14-1	118.174	liq	-74.0(2)	120.6(7)	0.8351^{25}	1.3898^{25}	vs ace, bz, eth, EtOH
312	Diethylamine	*N*-Ethylethanamine	$C_4H_{11}N$	109-89-7	73.137	liq	-50(2)	55.4(1)	0.7056^{20}	1.3864^{20}	vs H_2O; msc EtOH; s eth, ctc
313	*N*,*N*-Diethylaniline		$C_{10}H_{15}N$	91-66-7	149.233	ye oil	-21.3(2)	216(1)	0.9307^{20}	1.5409^{20}	sl H_2O; s EtOH, ace, ctc; vs eth, chl
314	*o*-Diethylbenzene	1,2-Diethylbenzene	$C_{10}H_{14}$	135-01-3	134.218	liq	-31.4(3)	183.4(4)	0.8800^{20}	1.5035^{20}	i H_2O; msc EtOH, eth, ace, bz, lig, ctc
315	*m*-Diethylbenzene	1,3-Diethylbenzene	$C_{10}H_{14}$	141-93-5	134.218	liq	-83.9(2)	181.1(5)	0.8602^{20}	1.4955^{20}	i H_2O; msc EtOH, eth, ace, bz, lig, ctc
316	*p*-Diethylbenzene	1,4-Diethylbenzene	$C_{10}H_{14}$	105-05-5	134.218	liq	-43.3(4)	184(1)	0.8620^{20}	1.4967^{20}	i H_2O; msc EtOH, eth, ace, bz, lig, ctc
317	Diethyl carbonate	Ethyl carbonate	$C_5H_{10}O_3$	105-58-8	118.131	liq	-43	125.9(9)	0.9692^{25}	1.3845^{20}	i H_2O; s EtOH, eth, chl
318	1,1-Diethylcyclohexane		$C_{10}H_{20}$	78-01-3	140.266			178(6)			
319	Diethylene glycol	Diglycol	$C_4H_{10}O_3$	111-46-6	106.120	liq	-10.3(3)	245.5(2)	1.1197^{15}	1.4472^{20}	s H_2O, EtOH, eth, chl
320	Diethylene glycol dibutyl ether	Bis(2-butoxyethyl) ether	$C_{12}H_{26}O_3$	112-73-2	218.332	liq	-60.2(2)	255(4)	0.885^{25}	1.4235^{20}	
321	Diethylene glycol diethyl ether	Bis(2-ethoxyethyl) ether	$C_8H_{18}O_3$	112-36-7	162.227	liq	-44.3(2)	185(4)	0.9063^{20}	1.4115^{20}	vs H_2O, EtOH; s eth
322	Diethylene glycol dimethyl ether	Diglyme	$C_6H_{14}O_3$	111-96-6	134.173	liq	-64.0(1)	162(2)	0.9434^{20}	1.4097^{20}	msc H_2O, EtOH, eth
323	Diethylene glycol monoethyl ether	Carbitol	$C_6H_{14}O_3$	111-90-0	134.173	hyg liq		202(3)	0.9885^{20}	1.4300^{20}	msc H_2O, EtOH, ace, bz; vs eth

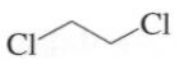
284
1,2-Dichloroethane

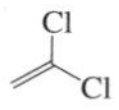
285
1,1-Dichloroethene

286
cis-1,2-Dichloroethene

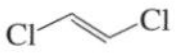
287
trans-1,2-Dichloroethene

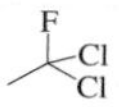
288
1,1-Dichloro-1-fluoroethane

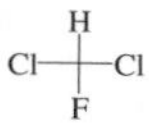
289
Dichlorofluoromethane

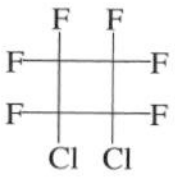
290
1,2-Dichloro-1,2,3,3,4,4-hexafluorocyclobutane

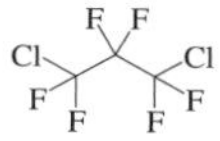
291
1,3-Dichloro-1,1,2,2,3,3-hexafluoropropane

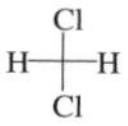
292
Dichloromethane

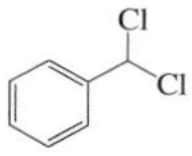
293
(Dichloromethyl)benzene

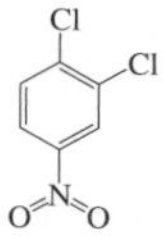
294
1,2-Dichloro-4-nitrobenzene

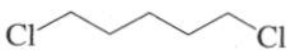
295
1,5-Dichloropentane

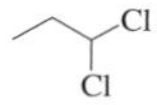
296
1,1-Dichloropropane

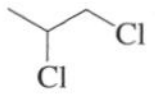
297
1,2-Dichloropropane, (±)-

298
1,3-Dichloropropane

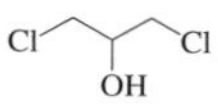
299
1,3-Dichloro-2-propanol

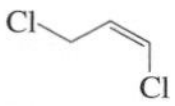
300
cis-1,3-Dichloropropene

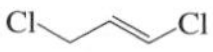
301
trans-1,3-Dichloropropene

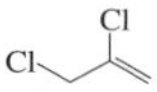
302
2,3-Dichloropropene

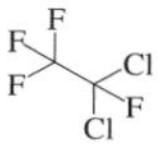
303
1,1-Dichloro-1,2,2,2-tetrafluoroethane

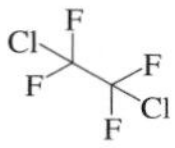
304
1,2-Dichloro-1,1,2,2-tetrafluoroethane

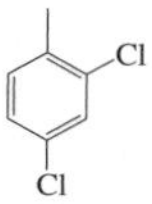
305
2,4-Dichlorotoluene

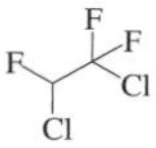
306
1,2-Dichloro-1,1,2-trifluoroethane

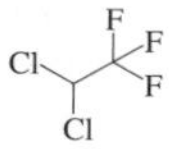
307
2,2-Dichloro-1,1,1-trifluoroethane

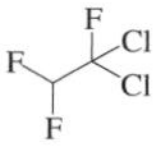
308
1,1-Dichloro-1,2,2-trifluoroethane

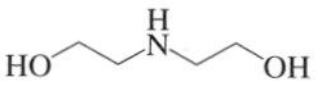
309
Diethanolamine

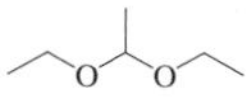
310
1,1-Diethoxyethane

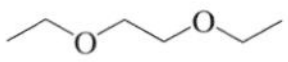
311
1,2-Diethoxyethane

312
Diethylamine

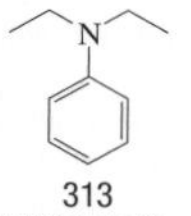
313
N,N-Diethylaniline

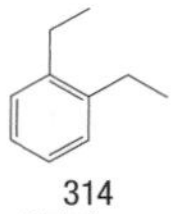
314
o-Diethylbenzene

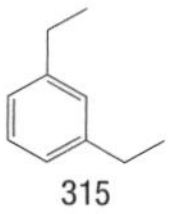
315
m-Diethylbenzene

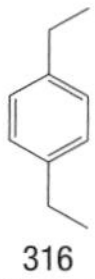
316
p-Diethylbenzene

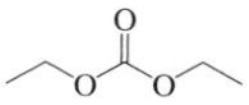
317
Diethyl carbonate

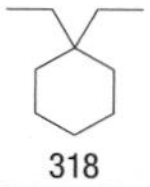
318
1,1-Diethylcyclohexane

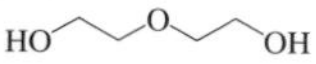
319
Diethylene glycol

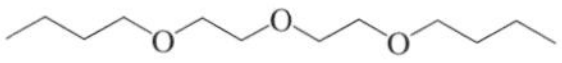
320
Diethylene glycol dibutyl ether

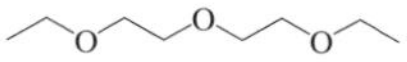
321
Diethylene glycol diethyl ether

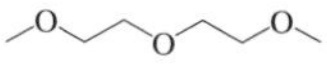
322
Diethylene glycol dimethyl ether

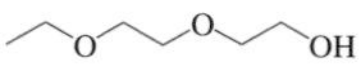
323
Diethylene glycol monoethyl ether

No.	Name	Synonym	Mol. form.	CAS Reg. No.	Mol. wt.	Physical form	mp/°C	bp/°C	den/g cm⁻³	n_D	Solubility
324	Diethylene glycol monoethyl ether acetate	Carbitol acetate	$C_8H_{16}O_4$	112-15-2	176.211	liq	-25	218(1)	1.0096^{20}	1.4213^{20}	vs H_2O, ace, eth, EtOH
325	Diethylene glycol monohexyl ether	2-[2-(Hexyloxy)ethoxy]ethanol	$C_{10}H_{22}O_3$	112-59-4	190.280	col liq	-28	259(2)			
326	Diethylene glycol monomethyl ether	2-(2-Methoxyethoxy)ethanol	$C_5H_{12}O_3$	111-77-3	120.147			194(2)	1.035^{20}	1.4264^{20}	msc H_2O, ace; vs EtOH, eth
327	Diethyl ether	Ethyl ether	$C_4H_{10}O$	60-29-7	74.121	liq	-116.22(21)	34.4(5)	0.7135^{20}	1.3526^{20}	sl H_2O; msc EtOH, bz, eth; vs ace
328	Diethyl maleate		$C_8H_{12}O_4$	141-05-9	172.179	liq	-8.8	222(8)	1.0662^{20}	1.4416^{20}	i H_2O; s EtOH, eth; sl chl
329	Diethyl malonate	Ethyl malonate	$C_7H_{12}O_4$	105-53-3	160.168	liq	-50(2)	200(3)	1.0551^{20}	1.4139^{20}	sl H_2O; msc EtOH, eth; vs ace, bz
330	Diethyl oxalate	Ethyl oxalate	$C_6H_{10}O_4$	95-92-1	146.141	liq	-40.6(3)	186(1)	1.0785^{20}	1.4101^{20}	sl H_2O; msc EtOH, eth, ace; s ctc
331	3,3-Diethylpentane	Tetraethylmethane	C_9H_{20}	1067-20-5	128.255	liq	-33.04(6)	146.2(3)	0.7536^{20}	1.4206^{20}	i H_2O; s eth, bz
332	Diethyl phthalate		$C_{12}H_{14}O_4$	84-66-2	222.237	liq	-40.5	298(2)	1.232^{14}	1.5000^{21}	i H_2O; msc EtOH, eth; s ace, bz, ctc
333	Diethyl succinate	Ethyl succinate	$C_8H_{14}O_4$	123-25-1	174.195	liq	-21.6(8)	217(1)	1.0402^{20}	1.4201^{20}	i H_2O; msc EtOH, eth; s ace, chl
334	Diethyl sulfide	Ethyl sulfide	$C_4H_{10}S$	352-93-2	90.187	liq	-103.9(1)	92.1(2)	0.8362^{20}	1.4430^{20}	sl H_2O, ctc; s EtOH, eth
335	*o*-Difluorobenzene	1,2-Difluorobenzene	$C_6H_4F_2$	367-11-3	114.093	liq	-47.1(1)	93.9(5)	1.1599^{18}	1.4451^{18}	i H_2O; s ace, bz, chl
336	*m*-Difluorobenzene	1,3-Difluorobenzene	$C_6H_4F_2$	372-18-9	114.093	liq	-69.11(1)	83.0(5)	1.1572^{20}	1.4374^{20}	i H_2O; s ace, bz
337	*p*-Difluorobenzene	1,4-Difluorobenzene	$C_6H_4F_2$	540-36-3	114.093	liq	-23.5(2)	88.9(3)	1.1701^{20}	1.4422^{20}	i H_2O; s ace, bz; sl ctc
338	1,1-Difluoroethane	HFC-152a	$C_2H_4F_2$	75-37-6	66.050	col gas	-118.59	-24.02	0.896^{25} (p>1 atm)	1.3011^{-72}	
339	1,2-Difluoroethane	Ethylene difluoride	$C_2H_4F_2$	624-72-6	66.050	vol liq		26			vs bz, eth, chl
340	1,1-Difluoroethene	Vinylidene fluoride	$C_2H_2F_2$	75-38-7	64.034	col gas	-144	-85.5(8)			vs eth, EtOH
341	Diglycolic acid	2,2'-Oxydiacetic acid	$C_4H_6O_5$	110-99-6	134.088	mcl pr (w + 1)	148	269(18)			vs H_2O, eth, EtOH
342	Dihexyl ether	Hexyl ether	$C_{12}H_{26}O$	112-58-3	186.333			220(4)	0.7936^{20}	1.4204^{20}	i H_2O; s eth; sl ctc
343	Diiodomethane	Methylene iodide	CH_2I_2	75-11-6	267.836	ye nd or lf	6.0(2)	182	3.3211^{20}	1.7411^{20}	sl H_2O, ctc; s EtOH, eth, bz, chl
344	Diisobutyl ether	1,1'-Oxybis[2-methylpropane]	$C_8H_{18}O$	628-55-7	130.228			122.7(7)	0.761^{15}		i H_2O; msc EtOH, eth
345	Diisobutyl phthalate		$C_{16}H_{22}O_4$	84-69-5	278.344			296.5	1.0490^{15}		s ctc
346	Diisopentyl ether	Diisoamyl ether	$C_{10}H_{22}O$	544-01-4	158.281	col liq		172(2)	0.7777^{20}	1.4085^{20}	i H_2O; vs ace, EtOH, chl
347	Diisopropylamine	*N*-Isopropyl-2-propanamine	$C_6H_{15}N$	108-18-9	101.190	liq	-61	84(3)	0.7153^{20}	1.3924^{20}	vs ace, bz, eth, EtOH
348	1,3-Diisopropylbenzene		$C_{12}H_{18}$	99-62-7	162.271	liq	-63(2)	203(3)	0.8559^{20}	1.4883^{20}	i H_2O; msc EtOH, eth, ace, bz, ctc
349	1,4-Diisopropylbenzene		$C_{12}H_{18}$	100-18-5	162.271	liq	-17.0(1)	210.3(2)	0.8568^{20}	1.4898^{20}	i H_2O; msc EtOH, eth, ace, bz, ctc
350	*p*-Diisopropylbenzene hydroperoxide		$C_{12}H_{18}O_2$	98-49-7	194.270	waxy cry	30.1		0.9932^{20}		i H_2O
351	Diisopropyl ether	Isopropyl ether	$C_6H_{14}O$	108-20-3	102.174	liq	-85.37(5)	68.4(2)	0.7192^{25}	1.3658^{25}	sl H_2O; msc EtOH, eth; s ace, ctc
352	Dimethoxane	2,6-Dimethyl-1,3-dioxan-4-ol acetate	$C_8H_{14}O_4$	828-00-2	174.195	liq			1.0655^{20}	1.4310^{20}	msc H_2O; s os
353	1,2-Dimethoxybenzene	Veratrole	$C_8H_{10}O_2$	91-16-7	138.164	liq	22.5(2)	206(1)	1.0810^{25}	1.5827^{21}	sl H_2O; s EtOH, eth, ctc
354	1,2-Dimethoxyethane	Ethylene glycol dimethyl ether	$C_4H_{10}O_2$	110-71-4	90.121	liq	-69.0(2)	85.0(1)	0.8637^{25}	1.3770^{25}	s H_2O, EtOH, eth, ace, bz, chl, ctc
355	Dimethoxymethane	Methylal	$C_3H_8O_2$	109-87-5	76.095	liq	-105.11(3)	42.3(2)	0.8593^{20}	1.3513^{20}	s H_2O; vs ace, bz, eth, EtOH
356	*N,N*-Dimethylacetamide	*N,N*-Dimethylethanamide	C_4H_9NO	127-19-5	87.120	liq	-19(1)	165.9(2)	0.9372^{25}	1.4341^{25}	msc H_2O, EtOH, eth, ace, bz, chl
357	Dimethylamine	*N*-Methylmethanamine	C_2H_7N	124-40-3	45.084	col gas	-93(2)	7.3(4)	0.6804^{0}	1.350^{17}	vs H_2O; s EtOH, eth
358	4-(Dimethylamino) benzaldehyde	Ehrlich's reagent	$C_9H_{11}NO$	100-10-7	149.189	lf (w)	73.1(8)		1.0254^{100}		sl H_2O, chl; s EtOH, eth, ace, bz
359	2,4-Dimethylaniline	2,4-Xylidine	$C_8H_{11}N$	95-68-1	121.180	liq	-13(2)	215(2)	0.9723^{20}	1.5569^{20}	sl H_2O, ctc; s EtOH, eth, bz
360	2,5-Dimethylaniline	2,5-Xylidine	$C_8H_{11}N$	95-78-3	121.180	ye lf (lig)	6(1)	214	0.9790^{21}	1.5591^{21}	sl H_2O; s eth, ctc
361	*N,N*-Dimethylaniline	*N,N*-Dimethylbenzenamine	$C_8H_{11}N$	121-69-7	121.180	pa ye	2.1(5)	193(1)	0.9562^{20}	1.5582^{20}	sl H_2O; s EtOH, eth, ace, bz; vs chl
362	2,2-Dimethylbutane	Neohexane	C_6H_{14}	75-83-2	86.175	liq	-99.0(4)	49.7(2)	0.6444^{25}	1.3688^{20}	i H_2O; s EtOH, eth; vs ace, bz, peth, ctc
363	2,3-Dimethylbutane	Diisopropyl	C_6H_{14}	79-29-8	86.175	liq	-128.1(2)	58.0(3)	0.6616^{20}	1.3750^{20}	i H_2O; s EtOH, eth; vs ace, bz, peth, ctc
364	2,3-Dimethyl-2-butanol	Isopropyldimethylcarbinol	$C_6H_{14}O$	594-60-5	102.174	liq	-10.5(3)	118.7(5)	0.8236^{20}	1.4176^{20}	s H_2O; msc EtOH, eth
365	3,3-Dimethyl-2-butanol, (±)-		$C_6H_{14}O$	20281-91-8	102.174		5.6	120.4	0.8122^{25}	1.4148^{20}	sl H_2O; vs EtOH, eth
366	3,3-Dimethyl-2-butanone	Pinacolone	$C_6H_{12}O$	75-97-8	100.158	liq	-51.40(5)	106.1(2)	0.7229^{25}	1.3952^{20}	sl H_2O; s EtOH, eth, ace, ctc

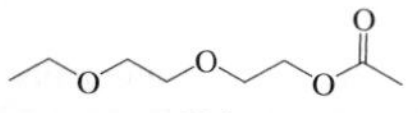
324
Diethylene glycol monoethyl ether acetate

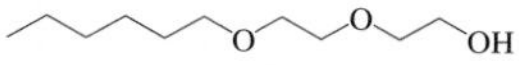
325
Diethylene glycol monohexyl ether

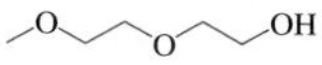
326
Diethylene glycol monomethyl ether

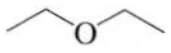
327
Diethyl ether

328
Diethyl maleate

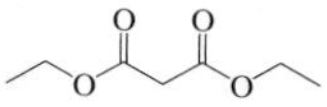
329
Diethyl malonate

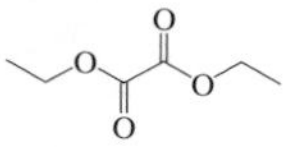
330
Diethyl oxalate

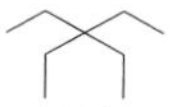
331
3,3-Diethylpentane

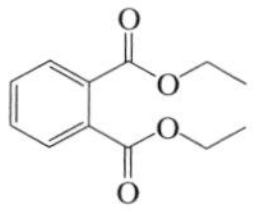
332
Diethyl phthalate

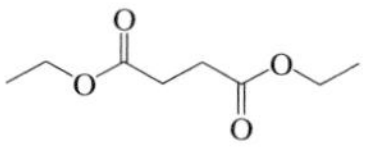
333
Diethyl succinate

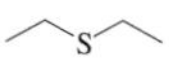
334
Diethyl sulfide

335
o-Difluorobenzene

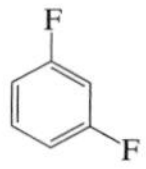
336
m-Difluorobenzene

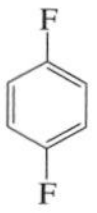
337
p-Difluorobenzene

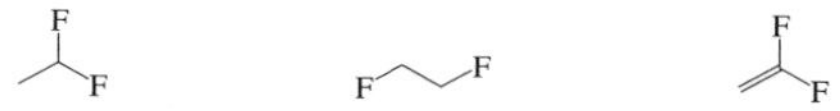
338
1,1-Difluoroethane

339
1,2-Difluoroethane

340
1,1-Difluoroethene

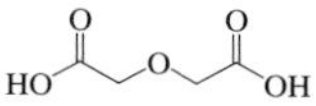
341
Diglycolic acid

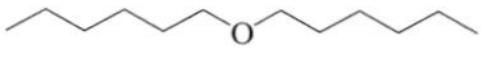
342
Dihexyl ether

343
Diiodomethane

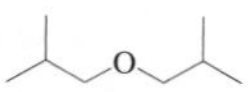
344
Diisobutyl ether

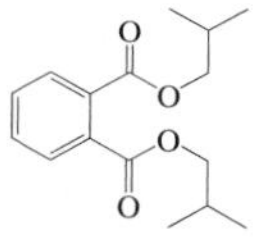
345
Diisobutyl phthalate

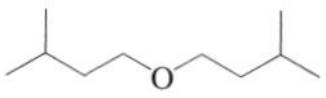
346
Diisopentyl ether

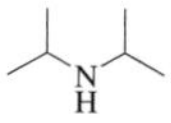
347
Diisopropylamine

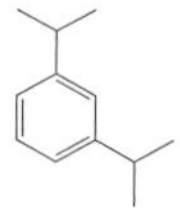
348
1,3-Diisopropylbenzene

349
1,4-Diisopropylbenzene

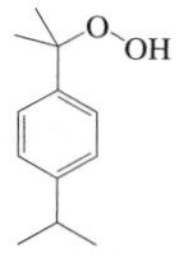
350
p-Diisopropylbenzene hydroperoxide

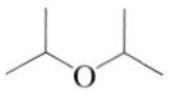
351
Diisopropyl ether

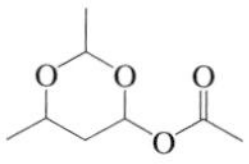
352
Dimethoxane

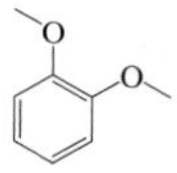
353
1,2-Dimethoxybenzene

354
1,2-Dimethoxyethane

355
Dimethoxymethane

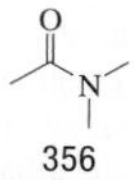
356
N,N-Dimethylacetamide

357
Dimethylamine

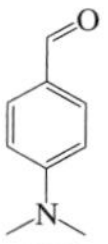
358
4-(Dimethylamino)benzaldehyde

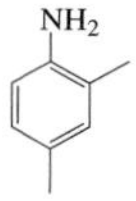
359
2,4-Dimethylaniline

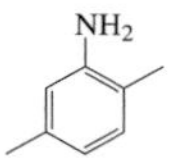
360
2,5-Dimethylaniline

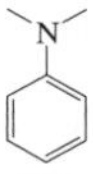
361
N,N-Dimethylaniline

362
2,2-Dimethylbutane

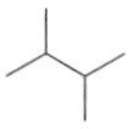
363
2,3-Dimethylbutane

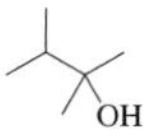
364
2,3-Dimethyl-2-butanol

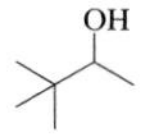
365
3,3-Dimethyl-2-butanol, (±)-

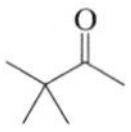
366
3,3-Dimethyl-2-butanone

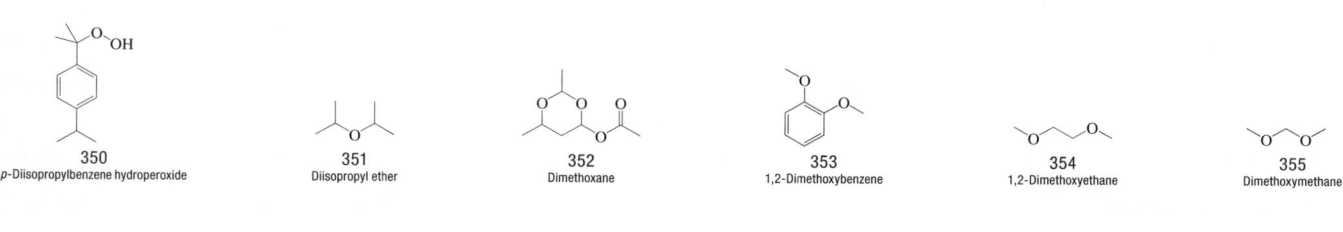

No.	Name	Synonym	Mol. form.	CAS Reg. No.	Mol. wt.	Physical form	mp/°C	bp/°C	den/g cm^{-3}	n_D	Solubility
367	2,3-Dimethyl-1-butene		C_6H_{12}	563-78-0	84.159	liq	-157.27(9)	55.59(4)	0.6803^{20}	1.3995^{20}	i H_2O; s EtOH, eth, ace, ctc, CS_2
368	1,1-Dimethylcyclohexane		C_8H_{16}	590-66-9	112.213	liq	-33.31(4)	119.5(3)	0.7809^{20}	1.4290^{20}	i H_2O; s EtOH, eth, ace, bz; msc ctc
369	*cis*-1,2-Dimethylcyclohexane		C_8H_{16}	2207-01-4	112.213	liq	-49.83(4)	129.7(6)	0.7963^{20}	1.4360^{20}	i H_2O; s EtOH, bz, ctc; msc eth, ace
370	*trans*-1,2-Dimethylcyclohexane		C_8H_{16}	6876-23-9	112.213	liq	-88.12(2)	123.4(3)	0.7760^{20}	1.4270^{20}	i H_2O; s EtOH, eth; msc ace, bz; vs lig
371	*cis*-1,3-Dimethylcyclohexane		C_8H_{16}	638-04-0	112.213	liq	-75.51(3)	124.4(6)	0.7660^{20}	1.4229^{20}	i H_2O; msc EtOH, eth, ace, bz, lig, ctc
372	*trans*-1,3-Dimethylcyclohexane		C_8H_{16}	2207-03-6	112.213	liq	-90.05(3)	120.1(7)	0.79^{15}	1.4284^{25}	
373	*cis*-1,4-Dimethylcyclohexane		C_8H_{16}	624-29-3	112.213	liq	-87.4(3)	124.3(7)	0.7829^{20}	1.4230^{20}	i H_2O; msc EtOH, eth, ace, bz, lig, ctc
374	*trans*-1,4-Dimethylcyclohexane		C_8H_{16}	2207-04-7	112.213	liq	-36.9(2)	119.3(5)	0.77^{15}	1.4185^{25}	i H_2O
375	1,1-Dimethylcyclopentane		C_7H_{14}	1638-26-2	98.186	liq	-69.43(1)	87.8(3)	0.7499^{25}	1.4136^{20}	
376	*cis*-1,2-Dimethylcyclopentane		C_7H_{14}	1192-18-3	98.186	liq	-53.67(2)	99.5(3)	0.7680^{25}	1.4222^{20}	
377	*trans*-1,2-Dimethylcyclopentane		C_7H_{14}	822-50-4	98.186	liq	-118(1)	91.9(4)	0.7468^{25}	1.4120^{20}	
378	*cis*-1,3-Dimethylcyclopentane		C_7H_{14}	2532-58-3	98.186	liq	-133.67(3)	91.7(5)	0.7402^{25}	1.4089^{20}	
379	*trans*-1,3-Dimethylcyclopentane		C_7H_{14}	1759-58-6	98.186	liq	-133.9(1)	90.7(6)	0.7443^{25}	1.4107^{20}	
380	Dimethyl disulfide	Methyl disulfide	$C_2H_6S_2$	624-92-0	94.199	liq	-84.67(10)	109.72(8)	1.0625^{20}	1.5289^{20}	i H_2O; msc EtOH, eth
381	*N,N*-Dimethylethanolamine	Deanol	$C_4H_{11}NO$	108-01-0	89.136	liq	-65(1)	130.7(5)	0.8866^{20}	1.4300^{20}	msc H_2O, EtOH, eth; s chl
382	Dimethyl ether	Methyl ether	C_2H_6O	115-10-6	46.068	col gas	-141.49(21)	-24.8(2)			s H_2O, EtOH, eth, ace, chl; sl bz
383	*N,N*-Dimethylformamide	DMF	C_3H_7NO	68-12-2	73.094	liq	-60.3(2)	152.8(5)	0.9445^{25}	1.4305^{20}	msc H_2O, EtOH, eth, ace, bz; sl lig
384	Dimethyl glutarate	Methyl glutarate	$C_7H_{12}O_4$	1119-40-0	160.168	liq	-42.5	216(4)	1.0876^{20}	1.4242^{20}	vs EtOH, eth; s chl
385	Dimethylglyoxime		$C_4H_8N_2O_2$	95-45-4	116.119	nd (to or dil al)	245.5	234 subl			i H_2O; vs EtOH, eth; sl bz, tol
386	2,2-Dimethylheptane		C_9H_{20}	1071-26-7	128.255	liq	-113.05(10)	133(1)	0.7105^{20}	1.4016^{20}	i H_2O; s eth, ctc; vs ace, chl; msc bz
387	2,6-Dimethyl-4-heptanone	Diisobutyl ketone	$C_9H_{18}O$	108-83-8	142.238	liq	-46.0(2)	157(3)	0.8062^{20}	1.412^{21}	i H_2O; msc EtOH, eth; s ctc
388	2,2-Dimethylhexane		C_8H_{18}	590-73-8	114.229	liq	-121.19(7)	106.8(4)	0.6953^{20}	1.3935^{20}	vs ace, bz, eth, EtOH
389	2,3-Dimethylhexane		C_8H_{18}	584-94-1	114.229			115.6(5)	0.6912^{25}	1.4011^{20}	vs ace, bz, EtOH, lig
390	2,4-Dimethylhexane		C_8H_{18}	589-43-5	114.229			109.4(4)	0.6962^{25}	1.3929^{25}	
391	2,5-Dimethylhexane	Biisobutyl	C_8H_{18}	592-13-2	114.229	liq	-91.14(2)	109.1(7)	0.6901^{25}	1.3925^{20}	i H_2O; msc EtOH, ace, bz; s eth
392	3,3-Dimethylhexane		C_8H_{18}	563-16-6	114.229	liq	-126.2(1)	111.9(6)	0.7100^{20}	1.4001^{20}	i H_2O; msc EtOH; vs eth, ace, bz
393	3,4-Dimethylhexane		C_8H_{18}	583-48-2	114.229			117.7(4)	0.7151^{25}	1.4041^{20}	i H_2O; s eth; msc EtOH, ace, bz
394	Dimethyl maleate	Methyl *cis*-butenedioate	$C_6H_8O_4$	624-48-6	144.126	liq	-19(1)	202(2)	1.1606^{20}	1.4416^{20}	sl H_2O, lig; s eth, ctc
395	Dimethyl malonate	Methyl malonate	$C_5H_8O_4$	108-59-8	132.116	liq	-62(1)	181.1(6)	1.1595^{20}	1.4135^{20}	sl H_2O; msc EtOH; vs ace, bz; s chl
396	1,2-Dimethylnaphthalene		$C_{12}H_{12}$	573-98-8	156.223		-3.0(7)	267(5)	1.0179^{20}	1.6166^{20}	i H_2O; s eth, bz
397	1,6-Dimethylnaphthalene		$C_{12}H_{12}$	575-43-9	156.223	liq	-16.2(5)	263(4)	1.0021^{20}	1.6166^{20}	i H_2O; s eth, bz
398	2,6-Dimethylnaphthalene		$C_{12}H_{12}$	581-42-0	156.223		110.1(2)	253(3)	1.003^{20}		i H_2O
399	1,3-Dimethyl-5-nitrobenzene		$C_8H_9NO_2$	99-12-7	151.163	nd (al)	75	274			i H_2O; vs EtOH, eth
400	2,2-Dimethylpentane		C_7H_{16}	590-35-2	100.202	liq	-123.71(4)	79.2(3)	0.6739^{20}	1.3822^{20}	i H_2O; s EtOH, eth; msc ace, bz, hp, chl
401	2,3-Dimethylpentane		C_7H_{16}	565-59-3	100.202			89.8(6)	0.6908^{25}	1.3894^{25}	i H_2O; s EtOH, eth; msc ace, bz, chl
402	2,4-Dimethylpentane		C_7H_{16}	108-08-7	100.202	liq	-119.16(2)	80.4(5)	0.6727^{20}	1.3815^{20}	i H_2O; s EtOH, eth; msc ace, bz, chl, hp
403	3,3-Dimethylpentane		C_7H_{16}	562-49-2	100.202	liq	-134.4(4)	86.0(6)	0.6936^{20}	1.3909^{20}	i H_2O; s EtOH, eth; msc ace, bz, hp, chl
404	2,4-Dimethyl-3-pentanone	Diisopropyl ketone	$C_7H_{14}O$	565-80-0	114.185	liq	-68.4(6)	125.2(3)	0.8108^{20}	1.3999^{20}	sl H_2O; msc EtOH, eth; s bz; sl ctc
405	Dimethyl phthalate	Methyl phthalate	$C_{10}H_{10}O_4$	131-11-3	194.184	pa ye	1.03(2)	282.7(2)	1.1905^{20}	1.5138^{20}	i H_2O; msc EtOH, eth; s bz; sl ctc
406	*N,N*-Dimethyl-1,3-propanediamine		$C_5H_{14}N_2$	109-55-7	102.178			129(13)	0.8272^{20}		
407	2,2-Dimethyl-1-propanol	Neopentyl alcohol	$C_5H_{12}O$	75-84-3	88.148		55(3)	112(1)	0.812^{20}		sl H_2O; vs EtOH, eth; s ctc

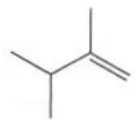
367
2,3-Dimethyl-1-butene

368
1,1-Dimethylcyclohexane

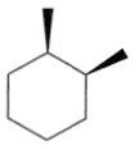
369
cis-1,2-Dimethylcyclohexane

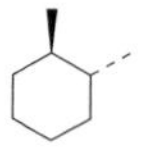
370
trans-1,2-Dimethylcyclohexane

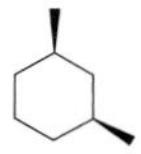
371
cis-1,3-Dimethylcyclohexane

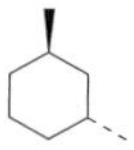
372
trans-1,3-Dimethylcyclohexane

373
cis-1,4-Dimethylcyclohexane

374
trans-1,4-Dimethylcyclohexane

375
1,1-Dimethylcyclopentane

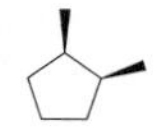
376
cis-1,2-Dimethylcyclopentane

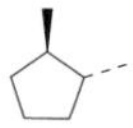
377
trans-1,2-Dimethylcyclopentane

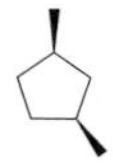
378
cis-1,3-Dimethylcyclopentane

379
trans-1,3-Dimethylcyclopentane

380
Dimethyl disulfide

381
N,N-Dimethylethanolamine

382
Dimethyl ether

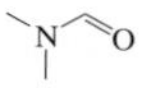
383
N,N-Dimethylformamide

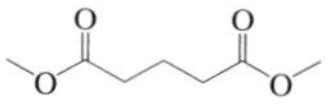
384
Dimethyl glutarate

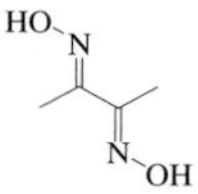
385
Dimethylglyoxime

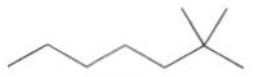
386
2,2-Dimethylheptane

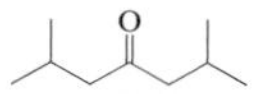
387
2,6-Dimethyl-4-heptanone

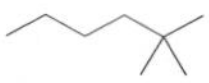
388
2,2-Dimethylhexane

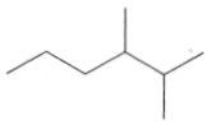
389
2,3-Dimethylhexane

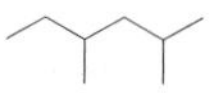
390
2,4-Dimethylhexane

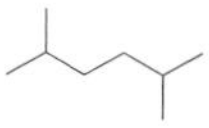
391
2,5-Dimethylhexane

392
3,3-Dimethylhexane

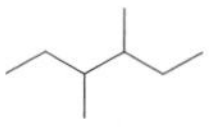
393
3,4-Dimethylhexane

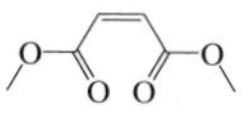
394
Dimethyl maleate

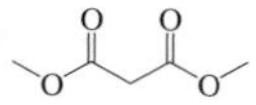
395
Dimethyl malonate

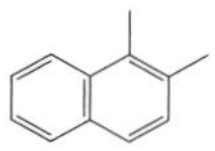
396
1,2-Dimethylnaphthalene

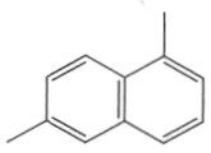
397
1,6-Dimethylnaphthalene

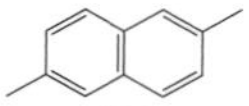
398
2,6-Dimethylnaphthalene

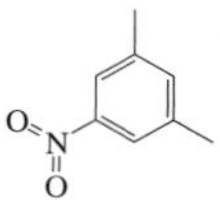
399
1,3-Dimethyl-5-nitrobenzene

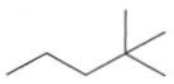
400
2,2-Dimethylpentane

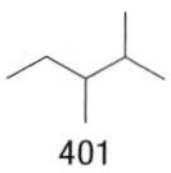
401
2,3-Dimethylpentane

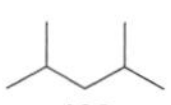
402
2,4-Dimethylpentane

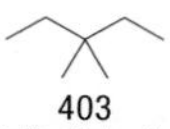
403
3,3-Dimethylpentane

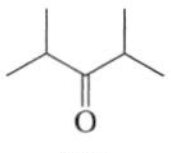
404
2,4-Dimethyl-3-pentanone

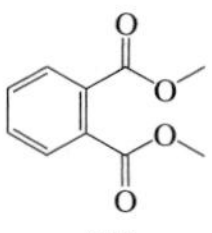
405
Dimethyl phthalate

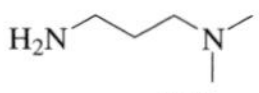
406
N,N-Dimethyl-1,3-propanediamine

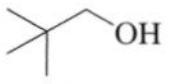
407
2,2-Dimethyl-1-propanol

No.	Name	Synonym	Mol. form.	CAS Reg. No.	Mol. wt.	Physical form	mp/°C	bp/°C	den/g cm^{-3}	n_D	Solubility
408	4-(1,1-Dimethylpropyl) phenol	*p-tert*-Pentylphenol	$C_{11}H_{16}O$	80-46-6	164.244		92.7(10)	262(9)			
409	2,6-Dimethylpyridine	2,6-Lutidine	C_7H_9N	108-48-5	107.153	liq	-6.12(3)	144.0(1)	0.9226^{20}	1.4953^{20}	msc H_2O; sl EtOH; s eth, ace, chl
410	Dimethyl succinate	Methyl succinate	$C_6H_{10}O_4$	106-65-0	146.141		18.6(6)	197(1)	1.1198^{20}	1.4197^{20}	sl H_2O, ctc; s EtOH, ace; vs eth
411	Dimethyl sulfate		$C_2H_6O_4S$	77-78-1	126.132	liq	-31.8(4)	186(3)	1.3322^{20}	1.3874^{20}	s H_2O, eth, bz, ctc; msc EtOH; i CS_2
412	Dimethyl sulfide	2-Thiapropane	C_2H_6S	75-18-3	62.134	liq	-98.26(4)	37.32(5)	0.8483^{20}	1.4438^{20}	sl H_2O; s EtOH, eth
413	Dimethyl sulfoxide	DMSO	C_2H_6OS	67-68-5	78.133		18.52(5)	191.9(9)	1.1010^{25}	1.4793^{20}	s H_2O, EtOH, eth, ace, ctc, AcOEt
414	Dimethyl terephthalate		$C_{10}H_{10}O_4$	120-61-6	194.184		140.602(4)	288	1.075^{141}		sl H_2O, EtOH, MeOH; s eth, chl
415	*N,N'*-Dimethylurea		$C_3H_8N_2O$	96-31-1	88.108	orth bipym (chl-eth)	106(2)	269	1.142^{25}		vs H_2O, EtOH; i eth; sl chl
416	1,2-Dinitrobenzene	*o*-Dinitrobenzene	$C_6H_4N_2O_4$	528-29-0	168.107	nd (bz), pl (al)	115.8(6)	319(3)	1.3119^{120}	1.565^{17}	i H_2O; s EtOH, bz, chl, AcOEt; sl DMSO
417	1,3-Dinitrobenzene	*m*-Dinitrobenzene	$C_6H_4N_2O_4$	99-65-0	168.107	orth pl (al)	89.2(5)	296(2)	1.5751^{18}		sl H_2O; vs EtOH, ace, py; s eth, tol
418	1,4-Dinitrobenzene	*p*-Dinitrobenzene	$C_6H_4N_2O_4$	100-25-4	168.107	nd (al)	171.1(9)	297	1.625^{18}		i H_2O; sl EtOH, chl; s ace, bz, tol
419	2,4-Dinitrophenol		$C_6H_4N_2O_5$	51-28-5	184.106	pa ye pl or lf (w)	114(2)	subl	1.683^{24}		sl H_2O; s EtOH, eth, ace, bz, tol, chl, py
420	1,3-Dioxane	1,3-Dioxacyclohexane	$C_4H_8O_2$	505-22-6	88.106	liq	-45	105(2)	1.0286^{25}	1.4165^{20}	msc H_2O, EtOH, eth, ace, bz
421	1,4-Dioxane	1,4-Dioxacyclohexane	$C_4H_8O_2$	123-91-1	88.106	col liq	11.75(6)	101.2(3)	1.0337^{20}	1.4224^{20}	msc H_2O, EtOH, eth, ace, bz; s ctc
422	1,3-Dioxolane	1,3-Dioxacyclopentane	$C_3H_6O_2$	646-06-0	74.079	liq	-97.21(2)	75.3(1)	1.060^{20}	1.3974^{20}	msc H_2O; s EtOH, eth, ace
423	Dipentyl ether	Amyl ether	$C_{10}H_{22}O$	693-65-2	158.281	liq	-69.2(5)	187(2)	0.7833^{20}	1.4119^{20}	i H_2O; msc EtOH, eth; s chl
424	Diphenylamine	*N*-Phenylbenzenamine	$C_{12}H_{11}N$	122-39-4	169.222	mcl lf(dil al)	53.2(3)	305.1(10)	1.158^{22}		i H_2O; vs EtOH, ace; s eth; sl chl
425	*N,N'*-Diphenyl-1,4-benzenediamine	*N,N'*-Diphenyl-*p*-phenylenediamine	$C_{18}H_{16}N_2$	74-31-7	260.333		150				sl EtOH, eth, bz, chl; i acid
426	1,1-Diphenylethane		$C_{14}H_{14}$	612-00-0	182.261	liq	-18.0(1)	285.8(2)	0.9997^{20}	1.5756^{20}	i H_2O; msc EtOH, eth; s bz
427	1,2-Diphenylethane	Dibenzyl	$C_{14}H_{14}$	103-29-7	182.261	mcl pr (MeOH)	51.18(6)	280(3)	0.9780^{25}	1.5476^{60}	i H_2O; s EtOH, eth, CS_2
428	Diphenyl ether	Oxybisbenzene	$C_{12}H_{10}O$	101-84-8	170.206		26.865(3)	258.0(1)	1.0661^{30}	1.5787^{25}	i H_2O; s EtOH, eth, bz, HOAc; sl chl
429	1,2-Diphenylhydrazine	Hydrazobenzene	$C_{12}H_{12}N_2$	122-66-7	184.236	tab (al-eth)	128.7(5)		1.158^{16}		vs EtOH; sl bz, DMSO; i HOAc
430	Diphenylmethane	Benzylbenzene	$C_{13}H_{12}$	101-81-5	168.234	pr nd	25.22(2)	264.2(3)	1.001^{26}	1.5753^{20}	i H_2O; s EtOH, eth, chl
431	1,3-Diphenyl-1-triazene	Diazoaminobenzene	$C_{12}H_{11}N_3$	136-35-6	197.235	ye lf or pr (al)	98				i H_2O; vs EtOH, eth, bz, py
432	Dipropylamine	*N*-Propyl-1-propanamine	$C_6H_{15}N$	142-84-7	101.190	liq	-63	107.5(9)	0.7400^{20}	1.4050^{20}	s H_2O, EtOH; msc eth; vs ace, bz
433	Dipropyl ether	Propyl ether	$C_6H_{14}O$	111-43-3	102.174	liq	-114.8(4)	90.1(3)	0.7466^{20}	1.3809^{20}	sl H_2O; vs eth, EtOH
434	*m*-Divinylbenzene	1,3-Divinylbenzene	$C_{10}H_{10}$	108-57-6	130.186		-52.2(2)		0.9294^{20}	1.5760^{20}	s ace, bz
435	*p*-Divinylbenzene	1,4-Divinylbenzene	$C_{10}H_{10}$	105-06-6	130.186		30(2)		0.913^{40}	1.5835^{25}	s ace, bz
436	Divinyl ether	Vinyl ether	C_4H_6O	109-93-3	70.090	vol liq or gas	-101(1)	28(3)	0.773^{20}	1.3989^{20}	i H_2O; msc EtOH, eth, ace, chl
437	Docosane		$C_{22}H_{46}$	629-97-0	310.600	pl(to), cry (eth)	43.8(3)	369(5)	0.7944^{20}	1.4455^{20}	i H_2O; s EtOH, chl; vs eth
438	*cis*-13-Docosenoic acid	Erucic acid	$C_{22}H_{42}O_2$	112-86-7	338.567	nd (al)	33.0(5)		0.8532^{70}	1.4758^{20}	i H_2O; s EtOH, ctc; vs eth, MeOH
439	Dodecane		$C_{12}H_{26}$	112-40-3	170.334	liq	-9.55(2)	216.3(2)	0.7495^{20}	1.4210^{20}	i H_2O; vs EtOH, eth, ace, ctc, chl
440	Dodecanedioic acid		$C_{12}H_{22}O_4$	693-23-2	230.301		126.6(8)	348(10)	1.15^{25}		s tfa
441	Dodecanoic acid	Lauric acid	$C_{12}H_{24}O_2$	143-07-7	200.318	nd (al)	43.82(2)	299(1)	0.8679^{50}	1.4183^{82}	i H_2O; vs EtOH, eth; s ace; msc bz
442	1-Dodecanol	Lauryl alcohol	$C_{12}H_{26}O$	112-53-8	186.333	lf (dil al)	24.2(3)	264.1(3)	0.8309^{24}		i H_2O; s EtOH, eth; sl bz
443	1-Dodecene		$C_{12}H_{24}$	112-41-4	168.319	liq	-35.19(5)	213.4(9)	0.7584^{20}	1.4300^{20}	i H_2O; s EtOH, eth, ace, ctc, peth
444	Eicosane	Icosane	$C_{20}H_{42}$	112-95-8	282.547	lf (al)	36.48(1)	344.1(9)	0.7886^{20}	1.4425^{20}	i H_2O; s eth, peth, bz; sl chl; vs ace
	Endosulfan		$C_9H_6Cl_6O_3S$	115-29-7	406.925		106		1.745^{20}		
	Endrin		$C_{12}H_8Cl_6O$	72-20-8	380.909	cry	245 dec				vs ace, bz, xyl; s ctc, hx
445	Epichlorohydrin	(Chloromethyl)oxirane	C_3H_5ClO	13403-37-7	92.524	liq	-26	111.99(5)	1.1812^{20}	1.4358^{25}	sl H_2O; msc EtOH, eth; s bz, ctc
446	Epinephrine	D-Adrenaline	$C_9H_{13}NO_3$	51-43-4	183.204	br (in air)	211.5				sl H_2O; i EtOH; s HOAc, acid
447	1,2-Epoxybutane	Ethyloxirane	C_4H_8O	106-88-7	72.106	liq	-150	63.4(2)	0.8297^{20}	1.3851^{20}	vs EtOH, ace; msc eth

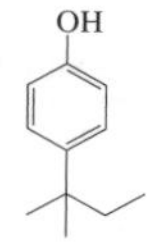
408
4-(1,1-Dimethylpropyl)phenol

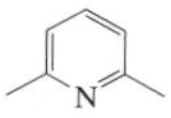
409
2,6-Dimethylpyridine

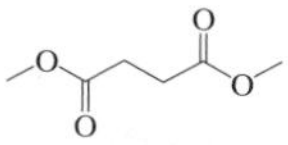
410
Dimethyl succinate

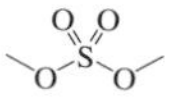
411
Dimethyl sulfate

412
Dimethyl sulfide

413
Dimethyl sulfoxide

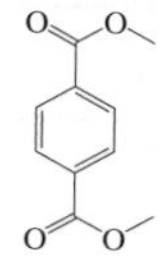
414
Dimethyl terephthalate

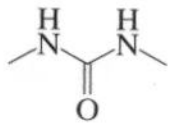
415
N,N'-Dimethylurea

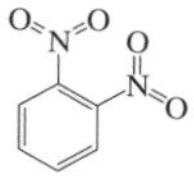
416
1,2-Dinitrobenzene

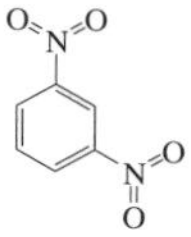
417
1,3-Dinitrobenzene

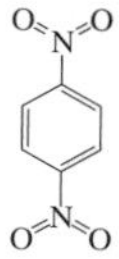
418
1,4-Dinitrobenzene

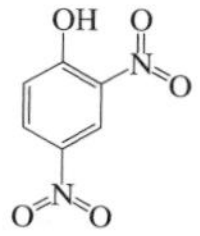
419
2,4-Dinitrophenol

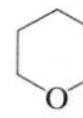
420
1,3-Dioxane

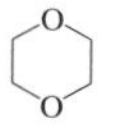
421
1,4-Dioxane

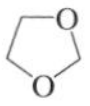
422
1,3-Dioxolane

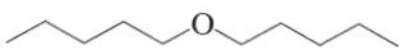
423
Dipentyl ether

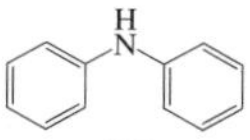
424
Diphenylamine

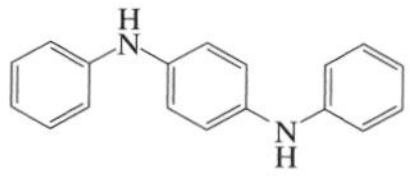
425
N,N'-Diphenyl-1,4-benzenediamine

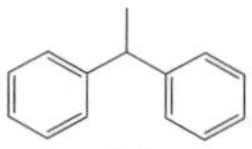
426
1,1-Diphenylethane

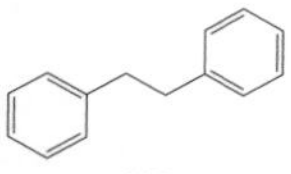
427
1,2-Diphenylethane

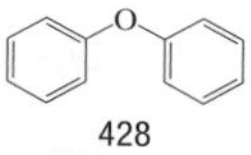
428
Diphenyl ether

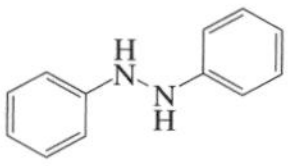
429
1,2-Diphenylhydrazine

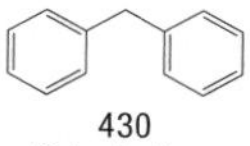
430
Diphenylmethane

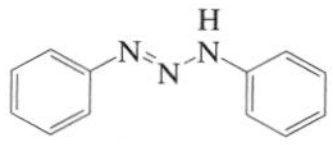
431
1,3-Diphenyl-1-triazene

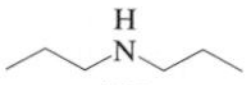
432
Dipropylamine

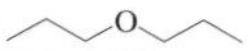
433
Dipropyl ether

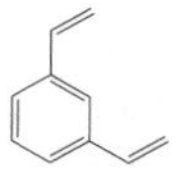
434
m-Divinylbenzene

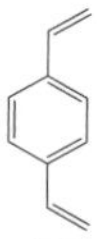
435
p-Divinylbenzene

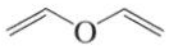
436
Divinyl ether

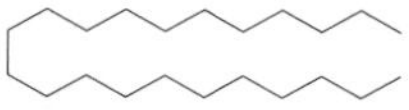
437
Docosane

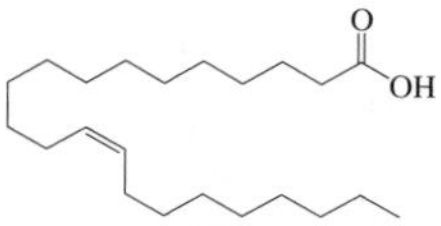
438
cis-13-Docosenoic acid

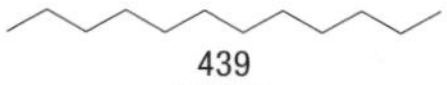
439
Dodecane

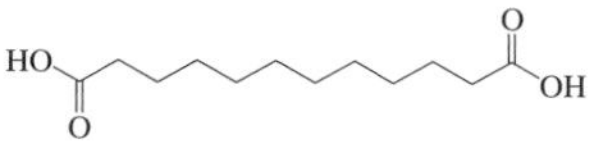
440
Dodecanedioic acid

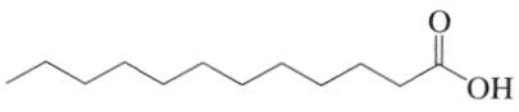
441
Dodecanoic acid

442
1-Dodecanol

443
1-Dodecene

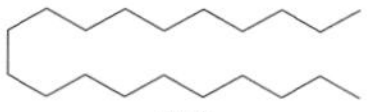
444
Eicosane

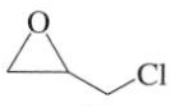
445
Epichlorohydrin

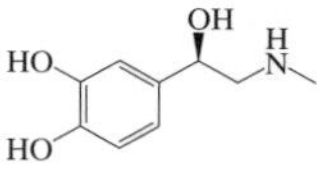
446
Epinephrine

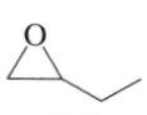
447
1,2-Epoxybutane

No.	Name	Synonym	Mol. form.	CAS Reg. No.	Mol. wt.	Physical form	mp/°C	bp/°C	den/g cm^{-3}	n_D	Solubility
448	Ethane		C_2H_6	74-84-0	30.069	col gas	-182.77(1)	-88.6(4)	0.5446^{-89} liq		i H_2O; vs bz
449	1,2-Ethanediamine	Ethylenediamine	$C_2H_8N_2$	107-15-3	60.098	liq	11.14(5)	116.9(5)	0.8979^{20}	1.4565^{20}	vs H_2O; msc EtOH; i eth, bz; s ctc
450	1,2-Ethanediol	Ethylene glycol	$C_2H_6O_2$	107-21-1	62.068	liq	-13(1)	197.5(1)	1.1135^{20}	1.4318^{20}	msc H_2O, EtOH, ace; s eth, chl; sl bz
451	1,1-Ethanediol, diacetate	Ethylidene diacetate	$C_6H_{10}O_4$	542-10-9	146.141		18.9	168(3)	1.070^{25}	1.3985^{25}	vs eth, EtOH
452	1,2-Ethanediol, diacetate	Ethylene glycol diacetate	$C_6H_{10}O_4$	111-55-7	146.141	liq	-31	184(4)	1.1043^{20}	1.4159^{20}	vs H_2O; msc EtOH, eth, ace, bz, CS_2
453	1,2-Ethanediol, diacrylate	Ethylene glycol diacrylate	$C_8H_{10}O_4$	2274-11-5	170.163	liq			1.0935^{26}		
454	1,2-Ethanediol, dinitrate	Ethylene glycol dinitrate	$C_2H_4N_2O_6$	628-96-6	152.062	ye liq	-22.5(6)	199(3)	1.4918^{20}		vs eth, EtOH
455	1,2-Ethanediphosphonic acid	1,2-Diphosphonoethane	$C_2H_8O_6P_2$	6145-31-9	190.029	nd (EtOH/eth)	223				
456	1,2-Ethanedithiol	Ethylene dimercaptan	$C_2H_6S_2$	540-63-6	94.199	liq	-41.2	144(3)	1.234^{20}	1.5590^{20}	i H_2O; s EtOH, eth, ace, bz; vs alk
457	Ethanol	Ethyl alcohol	C_2H_6O	64-17-5	46.068	liq	-114.14(3)	78.24(9)	0.7893^{20}	1.3611^{20}	msc H_2O, EtOH, eth, ace, chl; s bz
458	Ethanolamine	Glycinol	C_2H_7NO	141-43-5	61.083	liq	10.4(2)	170.3(4)	1.0180^{20}	1.4541^{20}	msc H_2O, EtOH; sl eth, lig, bz; s chl
459	Ethoxybenzene	Phenetole	$C_8H_{10}O$	103-73-1	122.164	liq	-29.6(4)	169.8(2)	0.9651^{20}	1.5076^{20}	i H_2O; s EtOH, eth, ctc
460	2-Ethoxyethanol	Ethylene glycol monoethyl ether	$C_4H_{10}O_2$	110-80-5	90.121	liq	-70	134.7(2)	0.9253^{25}	1.4054^{25}	vs H_2O, ace, eth, EtOH
461	2-Ethoxyethyl acetate	Ethylene glycol monoethyl ether acetate	$C_6H_{12}O_3$	111-15-9	132.157	liq	-61.7	156.6(4)	0.9740^{20}	1.4054^{20}	vs H_2O, ace, eth, EtOH
462	Ethyl acetate		$C_4H_8O_2$	141-78-6	88.106	liq	-83.8(3)	77.1(2)	0.9006^{20}	1.3723^{20}	s H_2O; msc EtOH, eth; vs ace, bz
463	Ethyl acetoacetate	Ethyl 3-oxobutanoate	$C_6H_{10}O_3$	141-97-9	130.141	liq	-45	180(2)	1.0368^{10}	1.4171^{20}	s H_2O; msc EtOH, eth; s bz, chl
464	Ethyl acrylate	Ethyl propenoate	$C_5H_8O_2$	140-88-5	100.117	liq	-71(2)	98.9(6)	0.9234^{20}	1.4068^{20}	sl H_2O, DMSO; msc EtOH, eth; s chl
465	Ethylamine	Ethanamine	C_2H_7N	75-04-7	45.084	vol liq or gas	-81(2)	16.6(2)	0.689^{15}	1.3663^{20}	msc H_2O, EtOH, eth
466	*N*-Ethylaniline		$C_8H_{11}N$	103-69-5	121.180	liq	-63.4(4)	204(1)	0.9625^{20}	1.5559^{20}	i H_2O; msc EtOH, eth; vs ace, bz; s ctc
467	Ethylbenzene	Phenylethane	C_8H_{10}	100-41-4	106.165	liq	-94.95(2)	136.2(4)	0.8668^{20}	1.4930^{25}	i H_2O; msc EtOH, eth; sl chl
468	Ethyl benzoate	Ethyl benzenecarboxylate	$C_9H_{10}O_2$	93-89-0	150.174	liq	-34.5(3)	212.5(2)	1.0415^{25}	1.5007^{20}	i H_2O; s EtOH, ace, bz; msc eth; sl ctc
469	Ethyl butanoate	Ethyl butyrate	$C_6H_{12}O_2$	105-54-4	116.158	liq	-97(6)	121.1(4)	0.8735^{25}	1.3898^{25}	sl H_2O, ctc; s EtOH, eth
470	2-Ethyl-1-butanol	2-Ethylbutyl alcohol	$C_6H_{14}O$	97-95-0	102.174	liq	<-15	155(2)	0.8326^{20}	1.4220^{20}	sl H_2O; s EtOH, eth, chl
471	Ethylcyanoacetate		$C_5H_7NO_2$	105-56-6	113.116	liq	-26.1(1)	216(4)	1.0654^{20}	1.4175^{20}	s H_2O; vs eth, EtOH
472	Ethylcyclohexane		C_8H_{16}	1678-91-7	112.213	liq	-111.28(10)	131.8(4)	0.7880^{20}	1.4330^{20}	i H_2O; s EtOH, ace, bz; vs lig; msc ctc
473	Ethylcyclopentane		C_7H_{14}	1640-89-7	98.186	liq	-138.42(2)	103.5(6)	0.7665^{20}	1.4198^{20}	i H_2O; msc EtOH, eth, ace; s bz, tol
474	Ethyl decanoate	Ethyl caprate	$C_{12}H_{24}O_2$	110-38-3	200.318	liq	-20(2)	242(1)	0.8650^{20}	1.4256^{20}	i H_2O; vs eth, EtOH, chl
475	Ethylene	Ethene	C_2H_4	74-85-1	28.053	col gas	-169.15(2)	-103.8(3)	0.5678^{-104}	1.363^{-100}	i H_2O; sl EtOH, bz, ace; s eth
476	Ethylene carbonate	1,3-Dioxolan-2-one	$C_3H_4O_3$	96-49-1	88.062	mcl pl (al)	36.331(4)	246(1)	1.3214^{39}	1.4148^{50}	msc H_2O, EtOH, eth, bz, chl, AcOEt
477	Ethylenediaminetetraacetic acid	EDTA	$C_{10}H_{16}N_2O_8$	60-00-4	292.242	cry (w)	245 dec				
478	Ethyleneimine	Aziridine	C_2H_5N	151-56-4	43.068	liq	-78.0(5)	54(1)	0.832^{25}		msc H_2O; s EtOH; vs eth; sl chl
479	Ethyl formate		$C_3H_6O_2$	109-94-4	74.079	liq	-79.6(5)	54.09(10)	0.9218^{20}	1.3609^{20}	s H_2O; msc EtOH, eth; vs ace; sl ctc
480	3-Ethylhexane		C_8H_{18}	619-99-8	114.229			118.5(5)	0.7136^{20}	1.4018^{20}	i H_2O; msc EtOH, eth, ace, bz, chl; s ctc
481	2-Ethyl-1-hexanol		$C_8H_{18}O$	104-76-7	130.228	liq	-70	186.2(2)	0.8319^{25}	1.4300^{20}	i H_2O; s EtOH, eth, ace, bz, chl
482	2-Ethylhexyl acetate		$C_{10}H_{20}O_2$	103-09-3	172.265	liq	-80	200(1)	0.8718^{20}	1.4204^{20}	i H_2O; s EtOH, eth
483	2-Ethylhexylamine	2-Ethyl-1-hexanamine	$C_8H_{19}N$	104-75-6	129.244			172(12)			sl H_2O
484	2-[(2-Ethylhexyl)oxy]ethanol	Ethylene glycol mono(2-ethylhexyl) ether	$C_{10}H_{22}O_2$	1559-35-9	174.281			231(4)			
485	*N*-Ethyl-*N*-hydroxyethanamine	*N,N*-Diethylhydroxylamine	$C_4H_{11}NO$	3710-84-7	89.136		10	131.0(3)	0.8669^{20}	1.4195^{20}	
486	Ethyl isopropyl ether		$C_5H_{12}O$	625-54-7	88.148			54(3)	0.720^{25}	1.3698^{25}	s H_2O, ace, chl; msc EtOH, eth
487	*N*-Ethyl-2-methylallylamine	*N*-Ethyl-2-methyl-2-propen-1-amine	$C_6H_{13}N$	18328-90-0	99.174	liq		104.7	0.753	1.4221^{20}	msc H_2O
488	Ethyl 3-methylbutanoate	Ethyl isovalerate	$C_7H_{14}O_2$	108-64-5	130.185	liq	-99.3	135(3)	0.8656^{20}	1.3962^{20}	sl H_2O; vs EtOH, eth
489	Ethyl methyl ether	Methyl ethyl ether	C_3H_8O	540-67-0	60.095	col gas	-113	6(2)	0.7251^{0}	1.3420^{4}	s H_2O, ace, chl; msc EtOH, eth

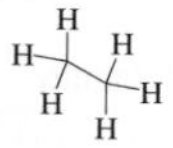
448
Ethane

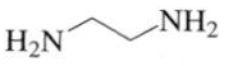
449
1,2-Ethanediamine

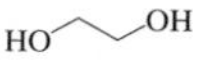
450
1,2-Ethanediol

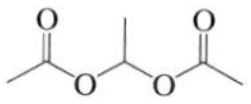
451
1,1-Ethanediol, diacetate

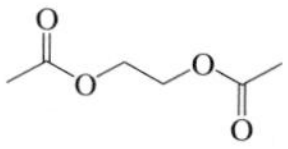
452
1,2-Ethanediol, diacetate

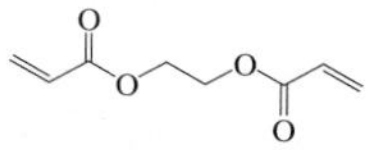
453
1,2-Ethanediol, diacrylate

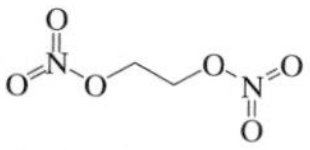
454
1,2-Ethanediol, dinitrate

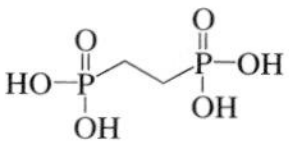
455
1,2-Ethanediphosphonic acid

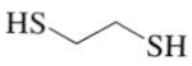
456
1,2-Ethanedithiol

457
Ethanol

458
Ethanolamine

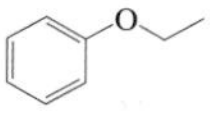
459
Ethoxybenzene

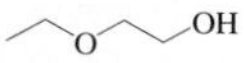
460
2-Ethoxyethanol

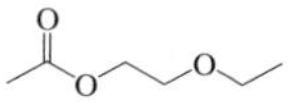
461
2-Ethoxyethyl acetate

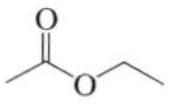
462
Ethyl acetate

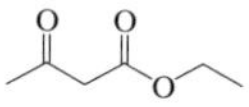
463
Ethyl acetoacetate

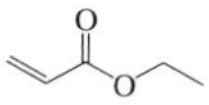
464
Ethyl acrylate

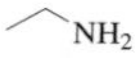
465
Ethylamine

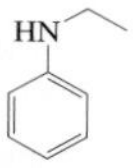
466
N-Ethylaniline

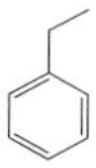
467
Ethylbenzene

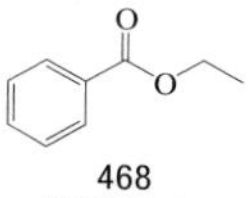
468
Ethyl benzoate

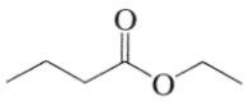
469
Ethyl butanoate

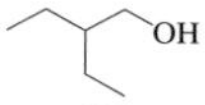
470
2-Ethyl-1-butanol

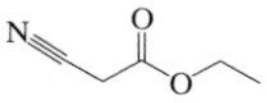
471
Ethyl cyanoacetate

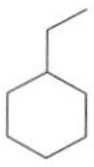
472
Ethylcyclohexane

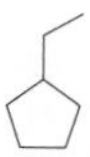
473
Ethylcyclopentane

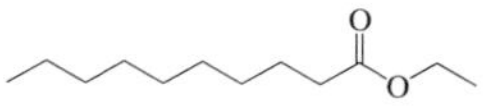
474
Ethyl decanoate

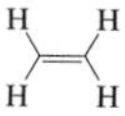
475
Ethylene

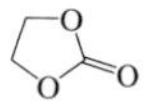
476
Ethylene carbonate

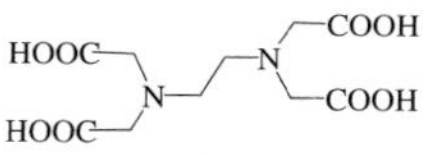
477
Ethylenediaminetetraacetic acid

478
Ethyleneimine

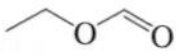
479
Ethyl formate

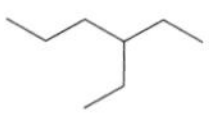
480
3-Ethylhexane

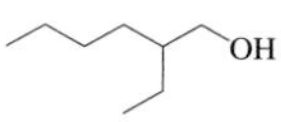
481
2-Ethyl-1-hexanol

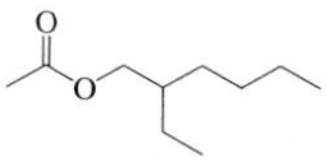
482
2-Ethylhexyl acetate

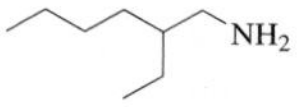
483
2-Ethylhexylamine

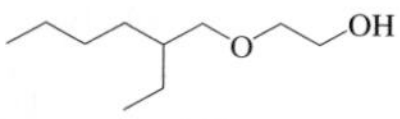
484
2-[(2-Ethylhexyl)oxy]ethanol

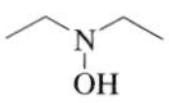
485
N-Ethyl-*N*-hydroxyethanamine

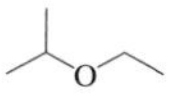
486
Ethyl isopropyl ether

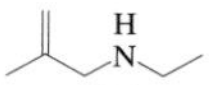
487
N-Ethyl-2-methylallylamine

488
Ethyl 3-methylbutanoate

489
Ethyl methyl ether

No.	Name	Synonym	Mol. form.	CAS Reg. No.	Mol. wt.	Physical form	mp/°C	bp/°C	den/g cm^{-3}	n_D	Solubility
490	3-Ethyl-2-methylpentane	2-Methyl-3-ethylpentane	C_8H_{18}	609-26-7	114.229	liq	-115.0(1)	115.6(6)	0.7193^{20}	1.4040^{20}	i H_2O; s eth; msc EtOH, ace, bz
491	3-Ethyl-3-methylpentane	3-Methyl-3-ethylpentane	C_8H_{18}	1067-08-9	114.229	liq	-90.8(1)	118.2(9)	0.7274^{20}	1.4078^{20}	i H_2O; s eth; msc EtOH, ace, bz
492	*N*-Ethylmorpholine		$C_6H_{13}NO$	100-74-3	115.173			145(5)	0.8996^{20}	1.4400^{20}	msc H_2O, EtOH, eth; s ace, bz
493	1-Ethylnaphthalene		$C_{12}H_{12}$	1127-76-0	156.223	liq	-13.9(2)	258(3)	1.0082^{20}	1.6062^{20}	i H_2O; msc EtOH, eth
494	2-Ethylnaphthalene		$C_{12}H_{12}$	939-27-5	156.223	liq	-7.4(9)	259(2)	0.9922^{20}	1.5999^{20}	i H_2O; msc EtOH, eth; sl chl
495	Ethyl octanoate		$C_{10}H_{20}O_2$	106-32-1	172.265	liq	-44.7(3)	206(1)	0.866^{18}	1.4178^{20}	i H_2O; vs EtOH, eth; sl ctc
496	3-Ethylpentane		C_7H_{16}	617-78-7	100.202	liq	-118.55(1)	93.4(4)	0.6982^{20}	1.3934^{20}	i H_2O; s EtOH, eth; msc ace, bz, hp, chl
497	4-Ethylphenol		$C_8H_{10}O$	123-07-9	122.164	nd	45.0(3)	217.97(6)		1.5239^{25}	sl H_2O, chl; vs EtOH, eth, bz; s ace
498	1-(4-Ethylphenyl)-2-phenylethane		$C_{16}H_{18}$	7439-15-8	210.314	cry		294	1.028^{50}		
499	Ethyl propanoate	Ethyl propionate	$C_5H_{10}O_2$	105-37-3	102.132	liq	-73.6(5)	98.9(2)	0.8895^{20}	1.3839^{20}	sl H_2O, ctc; msc EtOH, eth; s ace
500	Ethyl propyl ether	1-Ethoxypropane	$C_5H_{12}O$	628-32-0	88.148	liq	-127.5(1)	63(3)	0.7386^{20}	1.3695^{20}	vs eth, EtOH, HOAc
501	*S*-Ethyl thioacetate		C_4H_8OS	625-60-5	104.171			114(3)	0.9792^{20}	1.4583^{21}	i H_2O; vs EtOH, eth
502	2-Ethyltoluene		C_9H_{12}	611-14-3	120.191	liq	-80.7(4)	165.1(4)	0.8807^{20}	1.5046^{20}	i H_2O; msc EtOH, eth, ace, bz, peth, ctc
503	4-Ethyltoluene		C_9H_{12}	622-96-8	120.191	liq	-62.7(5)	162.0(6)	0.8614^{20}	1.4959^{20}	i H_2O; vs EtOH, eth; msc ace, bz
504	Ethyl vinyl ether	Ethoxyethylene	C_4H_8O	109-92-2	72.106	liq	-115.8(4)	36(2)	0.7589^{20}	1.3767^{20}	sl H_2O, ctc; s EtOH; msc eth
505	Eucalyptol	Cineole	$C_{10}H_{18}O$	470-82-6	154.249	oil	1.4(3)	176(4)	0.9267^{20}	1.4586^{20}	i H_2O; s EtOH, eth, chl; sl ctc
506	Fluoranthene	1,2-(1,8-Naphthylene) benzene	$C_{16}H_{10}$	206-44-0	202.250	pa ye nd or pl (al)	110.2(2)	380(5)	1.252^{0}		i H_2O; s EtOH, eth, bz, chl, CS_2
507	9*H*-Fluorene	2,2'-Methylenebiphenyl	$C_{13}H_{10}$	86-73-7	166.218	lf (al)	114.76(3)	294(2)	1.203^{0}		i H_2O; sl EtOH; s eth, ace, bz, CS_2
508	Fluorescein	3',6'-Dihydroxyspiro[isobenzo-furan-1(3*H*),9'-[9*H*]xanthen]-3-one	$C_{20}H_{12}O_5$	2321-07-5	332.306	red orth pr	315 dec				sl H_2O, EtOH, eth; vs ace; s py, MeOH
509	Fluorobenzene	Phenyl fluoride	C_6H_5F	462-06-6	96.102	liq	-42.18(5)	84.7(3)	1.0225^{20}	1.4684^{30}	sl H_2O; vs bz, eth, EtOH, lig
510	Fluoroethane	HFC-161	C_2H_5F	353-36-6	48.059	col gas	-143.2	-37.7(3)	0.7182^{20} (p>1 atm)	1.2656^{20}	sl H_2O; vs EtOH, eth
511	Fluoroethene	Vinyl fluoride	C_2H_3F	75-02-5	46.043	col gas	-160.5	-72			i H_2O; s EtOH, ace
512	Fluoromethane	Methyl fluoride	CH_3F	593-53-3	34.033	col gas	-143.33	-78.31	0.557^{25} (p>1 atm)	1.1674^{25}	sl H_2O, bz, chl; vs EtOH, eth
513	Formaldehyde	Methanal	CH_2O	50-00-0	30.026	col gas	-92	-19.1(5)	0.815^{-20}		s H_2O, EtOH, chl; msc eth, ace, bz
514	Formamide	Methanamide	CH_3NO	75-12-7	45.041	col liq	2.57(2)	217(3)	1.1334^{20}	1.4472^{20}	msc H_2O, EtOH; sl eth; s ace; i bz, chl
515	Formic acid	Methanoic acid	CH_2O_2	64-18-6	46.026	col liq	8.3(2)	101	1.220^{20}	1.3714^{20}	msc H_2O, EtOH, eth; vs ace; s bz, tol
516	Fumaric acid	*trans*-2-Butenedioic acid	$C_4H_4O_4$	110-17-8	116.073	nd, mcl pr or lf (w)	289.4(5)	165 subl	1.635^{20}		sl H_2O, eth, ace; s EtOH, con sulf
517	Furan	Oxacyclopentadiene	C_4H_4O	110-00-9	68.074	liq	-85.58(5)	31.3(2)	0.9514^{20}	1.4214^{20}	sl H_2O, chl; vs EtOH, eth; s ace, bz
518	Furfural	2-Furaldehyde	$C_5H_4O_2$	98-01-1	96.085	liq	-38.3(8)	161.5(3)	1.1594^{20}	1.5261^{20}	s H_2O, bz, chl; vs EtOH, ace; msc eth
519	Furfuryl alcohol	2-Furanmethanol	$C_5H_6O_2$	98-00-0	98.101	col-ye liq	-14.5(2)	168(2)	1.1296^{20}	1.4869^{20}	msc H_2O; vs EtOH, eth; s chl
520	D-Galactose		$C_6H_{12}O_6$	59-23-4	180.155	pl or pr (al)pr or nd (w+1)	170				vs H_2O; sl EtOH; i eth, bz; s py
521	D-Glucitol	Sorbitol	$C_6H_{14}O_6$	50-70-4	182.171	nd (w)	97(3)		1.489^{20}	1.3330^{20}	vs H_2O, ace
522	L-Glutamic acid	(*S*)-2-Aminopentanedioic acid	$C_5H_9NO_4$	56-86-0	147.130	orth (dil al)	160 dec	175 subl	1.538^{20}		sl H_2O
523	Glycerol	1,2,3-Propanetriol	$C_3H_8O_3$	56-81-5	92.094	syr, orth pl	18.2(4)	289(3)	1.2613^{20}	1.4746^{20}	msc H_2O, EtOH; sl eth; i bz, ctc, chl
524	Glycerol triacetate	Triacetin	$C_9H_{14}O_6$	102-76-1	218.203	col oily liq	-78	259(2)	1.1583^{20}	1.4301^{20}	sl H_2O; msc EtOH, eth, bz; vs ace
525	Glycerol trioleate	Triolein	$C_{57}H_{104}O_6$	122-32-7	885.432	col-ye oil	5.3(6)		0.915^{15}	1.4676^{15}	i H_2O; sl EtOH; vs eth; s chl, peth
526	Glycerol tristearate	Tristearin	$C_{57}H_{110}O_6$	555-43-1	891.479				0.8559^{90}	1.4395^{80}	i H_2O, EtOH; sl bz, ctc; s ace, chl
527	Glycine	Aminoacetic acid	$C_2H_5NO_2$	56-40-6	75.067	mcl or trg pr (dil al)	290 dec		1.607 (α form)		vs H_2O; i EtOH, eth; sl ace, py
528	Glycolic acid		$C_2H_4O_3$	79-14-1	76.051	orth nd (w) lf (eth)	79.5	100			s H_2O, EtOH, eth
529	Glyoxal	Ethanedial	$C_2H_2O_2$	107-22-2	58.036	ye pr	15	50.4	1.14^{20}	1.3826^{20}	vs H_2O; s EtOH, eth
	Heptachlor		$C_{10}H_5Cl_7$	76-44-8	373.318	wh cry	95.8(3)		1.57^{9}		vs bz, eth, EtOH, lig

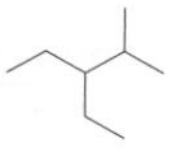
490
3-Ethyl-2-methylpentane

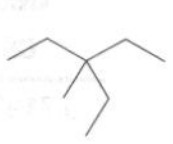
491
3-Ethyl-3-methylpentane

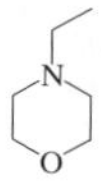
492
N-Ethylmorpholine

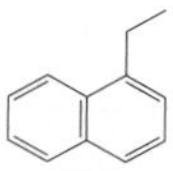
493
1-Ethylnaphthalene

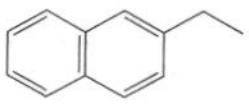
494
2-Ethylnaphthalene

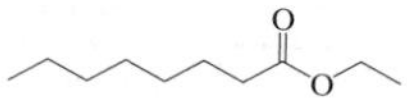
495
Ethyl octanoate

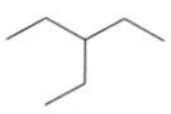
496
3-Ethylpentane

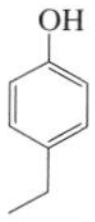
497
4-Ethylphenol

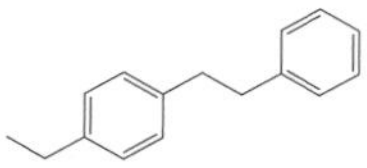
498
1-(4-Ethylphenyl)-2-phenylethane

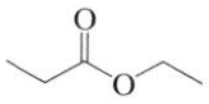
499
Ethyl propanoate

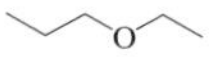
500
Ethyl propyl ether

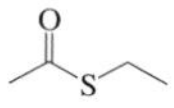
501
S-Ethyl thioacetate

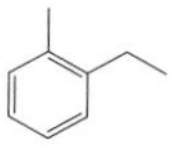
502
2-Ethyltoluene

503
4-Ethyltoluene

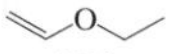
504
Ethyl vinyl ether

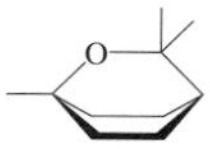
505
Eucalyptol

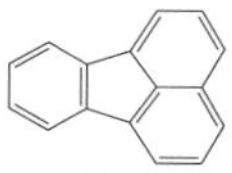
506
Fluoranthene

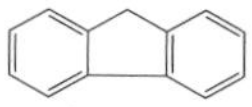
507
9*H*-Fluorene

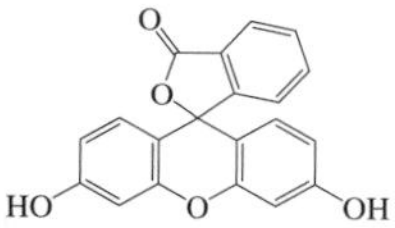
508
Fluorescein

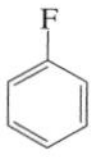
509
Fluorobenzene

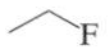
510
Fluoroethane

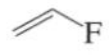
511
Fluoroethene

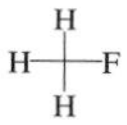
512
Fluoromethane

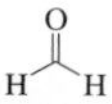
513
Formaldehyde

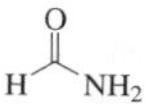
514
Formamide

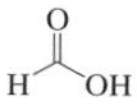
515
Formic acid

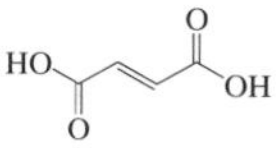
516
Fumaric acid

517
Furan

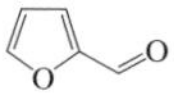
518
Furfural

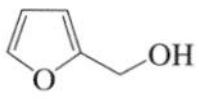
519
Furfuryl alcohol

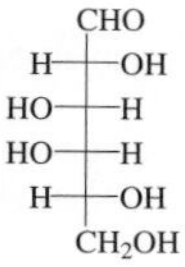
520
D-Galactose

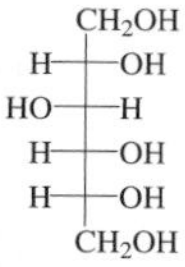
521
D-Glucitol

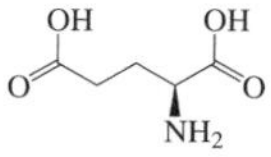
522
L-Glutamic acid

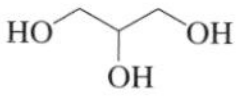
523
Glycerol

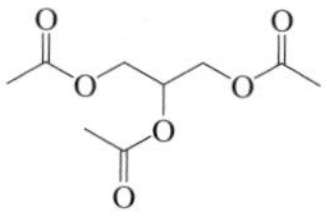
524
Glycerol triacetate

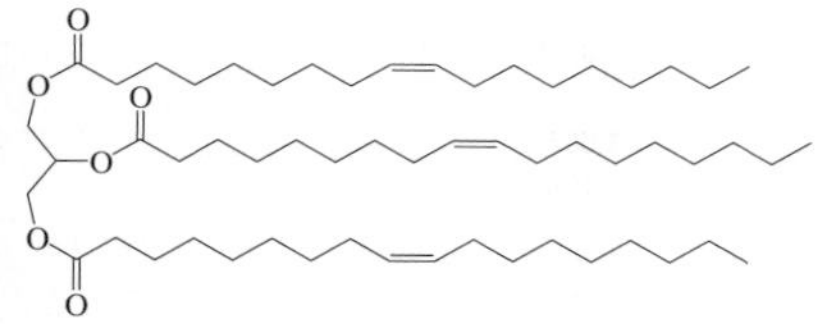
525
Glycerol trioleate

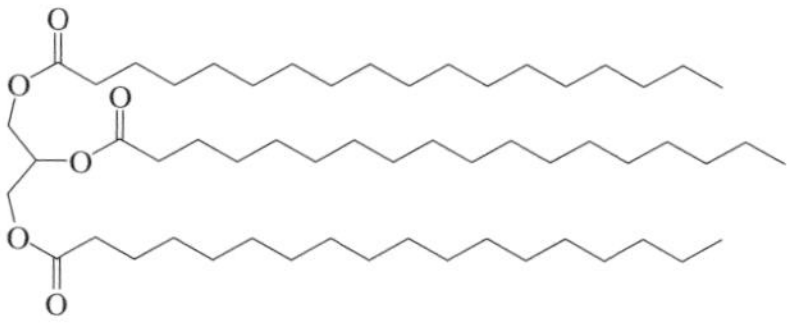
526
Glycerol tristearate

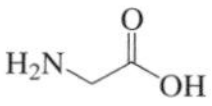
527
Glycine

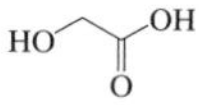
528
Glycolic acid

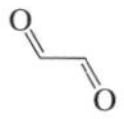
529
Glyoxal

Organic

No.	Name	Synonym	Mol. form.	CAS Reg. No.	Mol. wt.	Physical form	mp/°C	bp/°C	den/g cm-3	n_D	Solubility
530	Heptadecanoic acid	Margaric acid	$C_{17}H_{34}O_2$	506-12-7	270.451	pl (peth)	61.08(4)	362(4)	0.8532^{60}	1.4342^{60}	i H_2O; sl EtOH; s eth, ace, bz, chl
531	1,1,1,2,3,3,3-Heptafluoropropane	HFC-227ea	C_3HF_7	431-89-0	170.029	col gas	-126.80	-16.34			
532	Heptanal	Heptaldehyde	$C_7H_{14}O$	111-71-7	114.185	liq	-43.94(2)	153(3)	0.8132^{25}	1.4113^{20}	sl H_2O, ctc; msc EtOH, eth
533	Heptane		C_7H_{16}	142-82-5	100.202	liq	-90.549(2)	98.38(7)	0.6837^{20}	1.3855^{25}	i H_2O; vs EtOH; msc eth, bz, chl; s ctc
534	Heptanedioic acid	Pimelic acid	$C_7H_{12}O_4$	111-16-0	160.168	pr (w)	104.4(3)	342.0	1.329^{15}		s H_2O, EtOH, eth; i bz
535	Heptanoic acid	Enanthic acid	$C_7H_{14}O_2$	111-14-8	130.185	liq	-7.17(5)	222(2)	0.9181^{20}	1.4170^{20}	sl H_2O, ctc; s EtOH, eth, ace
536	1-Heptanol	Heptyl alcohol	$C_7H_{16}O$	111-70-6	116.201	liq	-33.2(1)	178(1)	0.8219^{20}	1.4249^{20}	sl H_2O, ctc; msc EtOH, eth
537	2-Heptanol, (±)-		$C_7H_{16}O$	52390-72-4	116.201			159	0.8167^{20}	1.4210^{20}	sl H_2O, ctc; s EtOH, eth
538	3-Heptanol, (*S*)-	Ethylbutylcarbinol	$C_7H_{16}O$	26549-25-7	116.201	liq	-70	163(2)	0.8227^{20}	1.4201^{20}	sl H_2O, ctc; s EtOH, eth
539	2-Heptanone	Methyl pentyl ketone	$C_7H_{14}O$	110-43-0	114.185	liq	-34.7(4)	151.0(3)	0.8111^{20}	1.4088^{20}	vs H_2O; s EtOH, eth
540	3-Heptanone	Ethyl butyl ketone	$C_7H_{14}O$	106-35-4	114.185	liq	-37.2(4)	146(2)	0.8183^{20}	1.4057^{20}	sl H_2O, ctc; msc EtOH, eth
541	4-Heptanone	Dipropyl ketone	$C_7H_{14}O$	123-19-3	114.185	liq	-32.1(8)	144(1)	0.8174^{20}	1.4069^{20}	i H_2O; msc EtOH, eth; s ctc
542	1-Heptene		C_7H_{14}	592-76-7	98.186	liq	-118.83(6)	94(1)	0.6970^{20}	1.3998^{20}	i H_2O; s EtOH, eth; sl ctc
543	*cis*-2-Heptene		C_7H_{14}	6443-92-1	98.186			97(2)	0.708^{20}	1.406^{20}	i H_2O; s EtOH, eth, ace, bz, chl; sl ctc
544	*trans*-2-Heptene		C_7H_{14}	14686-13-6	98.186	liq	-109.47(9)	98(2)	0.7012^{20}	1.4045^{20}	i H_2O; s EtOH, eth, ace, bz, peth, chl
545	*cis*-3-Heptene		C_7H_{14}	7642-10-6	98.186	liq	-136.6	96(2)	0.7030^{20}	1.4059^{20}	i H_2O; s EtOH, eth, ace, bz, peth, chl
546	*trans*-3-Heptene		C_7H_{14}	14686-14-7	98.186	liq	-136.64(9)	96(2)	0.6981^{20}	1.4043^{20}	i H_2O; s EtOH, eth, ace, bz, chl; sl ctc
547	1-Heptyne		C_7H_{12}	628-71-7	96.170	liq	-80.9(2)	99.8(3)	0.7328^{20}	1.4087^{20}	sl H_2O; msc EtOH, eth; s bz, chl, peth
	1,2,5,6,9,10-Hexabromocyclododecane		$C_{12}H_{18}Br_6$	3194-55-6	641.695	cry	167				
548	Hexachlorobenzene	Perchlorobenzene	C_6Cl_6	118-74-1	284.782	nd (sub)	230(3)	325	2.044^{23}	1.5691^{23}	i H_2O; sl EtOH; s eth, chl; vs bz
549	Hexachloro-1,3-butadiene	Perchlorobutadiene	C_4Cl_6	87-68-3	260.761	liq	-21	216(2)	1.556^{25}	1.5542^{20}	i H_2O; s EtOH, eth
	1,2,3,4,5,6-Hexachlorocyclohexane, (1α,2α,3β,4α,5β,6β)	α-Hexachlorocyclohexane	$C_6H_6Cl_6$	319-84-6	290.830	cry	157.4(7)				
	1,2,3,4,5,6-Hexachlorocyclohexane, (1α,2β,3α,4β,5α,6β)	β-Hexachlorocyclohexane	$C_6H_6Cl_6$	319-85-7	290.830	cry (bz, al, xyl)			1.89^{19}		i H_2O; sl EtOH, bz, chl, HOAc
	1,2,3,4,5,6-Hexachlorocyclohexane, (1α,2α,3α,4β,5α,6β)	δ-Lindane	$C_6H_6Cl_6$	319-86-8	290.830	pl	137.0(2)				
550	Hexachloroethane	Perchloroethane	C_2Cl_6	67-72-1	236.739	orth (al-eth)	186.8(2)	184.7 sp	2.091^{20}		i H_2O; vs EtOH, eth; s bz; sl liq HF
551	Hexacosane		$C_{26}H_{54}$	630-01-3	366.707	mcl, tcl or orth (bz) cry (eth)	56.09(4)	415(11)	0.7783^{60}	1.4357^{60}	vs bz, lig, chl
552	Hexadecane	Cetane	$C_{16}H_{34}$	544-76-3	226.441	lf (HOAc)	18.18(2)	286.9(7)	0.7701^{25}	1.4329^{25}	i H_2O; sl EtOH; msc eth; s ctc
553	Hexadecanoic acid	Palmitic acid	$C_{16}H_{32}O_2$	57-10-3	256.424	nd (al)	62.49(2)	351(6)	0.8487^{70}	1.43345^{60}	i H_2O; s EtOH, ace, bz; msc eth; vs chl
554	1-Hexadecanol	Cetyl alcohol	$C_{16}H_{34}O$	36653-82-4	242.440	fl (AcOEt)	49.30(1)	325(2)	0.8187^{50}	1.4283^{79}	i H_2O; sl EtOH; vs eth, bz, chl; s ace
555	1,2-Hexadiene	Propylallene	C_6H_{10}	592-44-9	82.143			76(3)	0.7149^{20}	1.4282^{20}	vs eth, chl
556	*trans*-1,3-Hexadiene		C_6H_{10}	20237-34-7	82.143	liq	-102.4(6)	71.5(8)	0.6995^{25}	1.4406^{20}	
557	1,5-Hexadiene	Biallyl	C_6H_{10}	592-42-7	82.143	liq	-140.7(1)	59.2(4)	0.6878^{25}	1.4042^{20}	i H_2O; s EtOH, eth, bz, chl; sl ctc
558	*trans,cis*-2,4-Hexadiene		C_6H_{10}	5194-50-3	82.143	liq	-96.1(4)	83(3)	0.7185^{25}	1.4560^{20}	i H_2O; s EtOH, eth, chl
559	*trans,trans*-2,4-Hexadiene		C_6H_{10}	5194-51-4	82.143	liq	-44.9(4)	82.4(7)	0.7101^{25}	1.4510^{20}	i H_2O; s EtOH, eth, chl
560	Hexafluorobenzene	Perfluorobenzene	C_6F_6	392-56-3	186.054	liq	5.10(1)	80.2(2)	1.6175^{20}	1.3777^{20}	
561	1,1,2,3,4,4-Hexafluoro-1,3-butadiene		C_4F_6	685-63-2	162.033	col gas	-132	5.4(2)	1.473^{5}	1.378^{-20}	
562	1,1,1,2,3,3-Hexafluoropropane	HFC-236ea	$C_3H_2F_6$	431-63-0	152.038	col gas		6.20	1.5026^{0}		
563	1,1,1,3,3,3-Hexafluoropropane	HFC-236fa	$C_3H_2F_6$	690-39-1	152.038	col gas	-93.63	-1.4(2)	1.4343^{0}		
564	Hexahydro-1,3,5-trinitro-1,3,5-triazine	Cyclonite	$C_3H_6N_6O_6$	121-82-4	222.116	orth cry (ace)	203.4(9)		1.82^{20}		i H_2O, EtOH, bz; sl eth, MeOH; s ace, HOAc
565	Hexamethylenetetramine	Methenamine	$C_6H_{12}N_4$	100-97-0	140.186	orth (al)	>250	subl	1.331^{-5}		vs H_2O; s EtOH, ace, chl; sl eth, bz
566	Hexamethylphosphoric triamide	Tris(dimethylamino)phosphine oxide	$C_6H_{18}N_3OP$	680-31-9	179.200	col liq	7.2	235(13)	1.03^{20}	1.4579^{20}	s EtOH, eth
567	Hexane		C_6H_{14}	110-54-3	86.175	liq	-95.27(2)	68.72(6)	0.6593^{20}	1.3727^{25}	i H_2O; vs EtOH; s eth, chl
568	1,6-Hexanediamine	Hexamethylenediamine	$C_6H_{16}N_2$	124-09-4	116.204	orth bipym pl	38.8(6)	197(2)			vs H_2O; s EtOH, bz

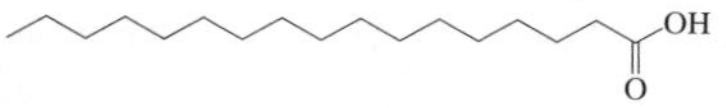
530
Heptadecanoic acid

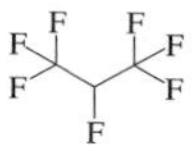
531
1,1,1,2,3,3,3-Heptafluoropropane

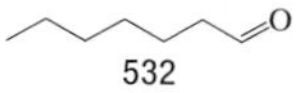
532
Heptanal

533
Heptane

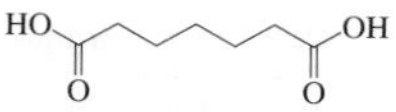
534
Heptanedioic acid

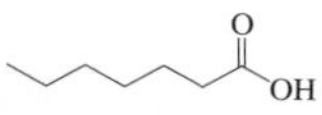
535
Heptanoic acid

536
1-Heptanol

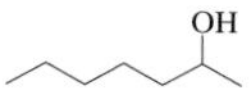
537
2-Heptanol, (±)-

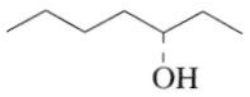
538
3-Heptanol, (S)-

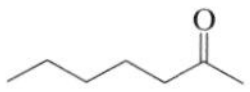
539
2-Heptanone

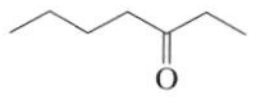
540
3-Heptanone

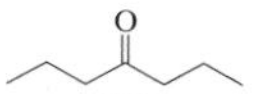
541
4-Heptanone

542
1-Heptene

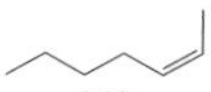
543
cis-2-Heptene

544
trans-2-Heptene

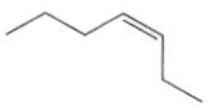
545
cis-3-Heptene

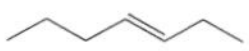
546
trans-3-Heptene

547
1-Heptyne

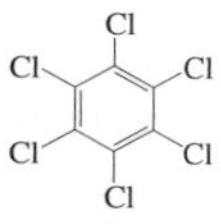
548
Hexachlorobenzene

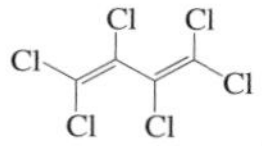
549
Hexachloro-1,3-butadiene

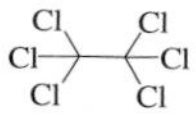
550
Hexachloroethane

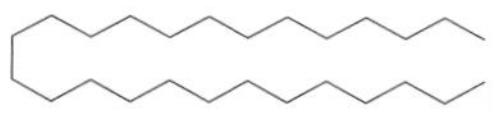
551
Hexacosane

552
Hexadecane

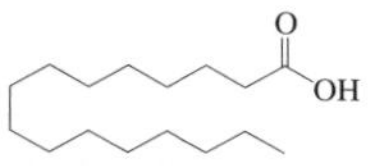
553
Hexadecanoic acid

554
1-Hexadecanol

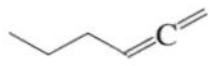
555
1,2-Hexadiene

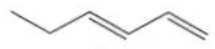
556
trans-1,3-Hexadiene

557
1,5-Hexadiene

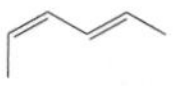
558
trans,cis-2,4-Hexadiene

559
trans,trans-2,4-Hexadiene

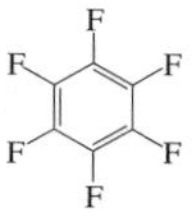
560
Hexafluorobenzene

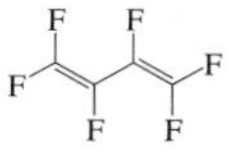
561
1,1,2,3,4,4-Hexafluoro-1,3-butadiene

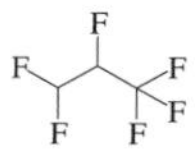
562
1,1,1,2,3,3-Hexafluoropropane

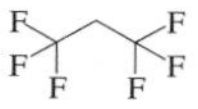
563
1,1,1,3,3,3-Hexafluoropropane

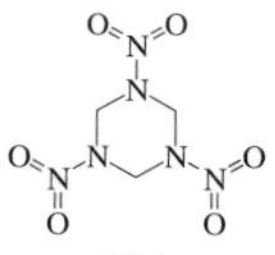
564
Hexahydro-1,3,5-trinitro-1,3,5-triazine

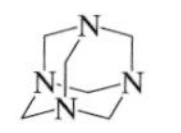
565
Hexamethylenetetramine

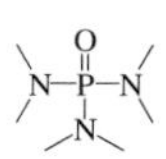
566
Hexamethylphosphoric triamide

567
Hexane

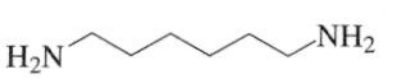
568
1,6-Hexanediamine

No.	Name	Synonym	Mol. form.	CAS Reg. No.	Mol. wt.	Physical form	mp/°C	bp/°C	den/g cm-3	n_D	Solubility
569	Hexanedinitrile	Adiponitrile	$C_6H_8N_2$	111-69-3	108.141	nd (eth)	2.2(4)	295	0.9676[20]	1.4380[20]	sl H_2O, eth; s chl, EtOH
570	1,6-Hexanedioic acid	Adipic acid	$C_6H_{10}O_4$	124-04-9	146.141	mcl pr (w, ace, lig)	151.5(6)	337.5	1.360[25]		sl H_2O; vs EtOH; s eth; i HOAc, lig
571	1,6-Hexanediol	Hexamethylene glycol	$C_6H_{14}O_2$	629-11-8	118.174		41.5(5)	208	1.079[25]	1.4579[25]	s H_2O, EtOH, ace; sl eth; i bz
572	Hexanenitrile	Capronitrile	$C_6H_{11}N$	628-73-9	97.158	liq	-80.3	163.5(3)	0.8051[20]	1.4068[20]	i H_2O; s EtOH, eth; sl chl
573	1,2,6-Hexanetriol	1,2,6-Trihydroxyhexane	$C_6H_{14}O_3$	106-69-4	134.173				1.1049[20]	1.58[20]	
574	Hexanoic acid	Caproic acid	$C_6H_{12}O_2$	142-62-1	116.158	liq	-4.1(7)	204.9(6)	0.9274[20]	1.4163[20]	sl H_2O; s EtOH, eth, chl
575	1-Hexanol	Caproyl alcohol	$C_6H_{14}O$	111-27-3	102.174	liq	-46.4(9)	156.9(7)	0.8195[20]	1.4178[20]	sl H_2O; s EtOH, ace, chl; msc eth, bz
576	2-Hexanol		$C_6H_{14}O$	20281-86-1	102.174			138(6)	0.8159[20]	1.4144[20]	sl H_2O, ctc; s EtOH, eth
577	3-Hexanol		$C_6H_{14}O$	17015-11-1	102.174			143(2)	0.8182[20]	1.4167[20]	sl H_2O; s EtOH, ace; msc eth
578	2-Hexanone	Butyl methyl ketone	$C_6H_{12}O$	591-78-6	100.158	liq	-55.45(5)	127.6(1)	0.8113[20]	1.4007[20]	sl H_2O; s ace; msc EtOH, eth
579	3-Hexanone	Ethyl propyl ketone	$C_6H_{12}O$	589-38-8	100.158	liq	-55.4(2)	123.5(3)	0.8118[20]	1.4004[20]	sl H_2O; s ace; msc EtOH, eth
580	Hexatriacontane		$C_{36}H_{74}$	630-06-8	506.973		75.81(4)	504(7)	0.7803[80]	1.4397[80]	
581	1-Hexene		C_6H_{12}	592-41-6	84.159	liq	-139.76(5)	63.4(1)	0.6685[25]	1.3852[25]	i H_2O; vs bz, eth, EtOH, peth
582	*cis*-2-Hexene		C_6H_{12}	7688-21-3	84.159	liq	-141.12(4)	68.9(5)	0.6824[25]	1.3979[20]	i H_2O; s EtOH, eth, bz, chl, lig
583	*trans*-2-Hexene		C_6H_{12}	4050-45-7	84.159	liq	-133.1(3)	67.85(9)	0.6733[25]	1.3936[20]	i H_2O; s EtOH, eth, bz, chl, lig
584	*cis*-3-Hexene		C_6H_{12}	7642-09-3	84.159	liq	-138.7(7)	66.4(5)	0.6778[20]	1.3947[20]	i H_2O; s EtOH, eth, bz, chl, lig
585	*trans*-3-Hexene		C_6H_{12}	13269-52-8	84.159	liq	-113.7(5)	67.06(9)	0.6772[20]	1.3943[20]	i H_2O; s EtOH, eth, bz, chl, lig
586	*trans*-3-Hexenedinitrile	*trans*-1,4-Dicyano-2-butene	$C_6H_6N_2$	1119-85-3	106.125	cry	76				
587	Hexyl acetate		$C_8H_{16}O_2$	142-92-7	144.212	liq	-61.0(2)	171.1(7)	0.8779[15]	1.4092[20]	i H_2O; vs eth, EtOH
588	*sec*-Hexyl acetate	4-Methyl-2-pentyl acetate	$C_8H_{16}O_2$	108-84-9	144.212			147.5	0.8805[25]	1.3980[20]	sl H_2O; vs eth, EtOH
589	Hexylbenzene		$C_{12}H_{18}$	1077-16-3	162.271	liq	-63.4(2)	226(2)	0.8575[20]	1.4864[20]	i H_2O; msc eth; s bz, peth
590	1-Hexyne	Butylacetylene	C_6H_{10}	693-02-7	82.143	liq	-132.1(4)	71.2(3)	0.7155[25]	1.3989[20]	i H_2O; s EtOH, eth, bz, chl; sl ctc
591	Hydrogen cyanide	Hydrocyanic acid	CHN	74-90-8	27.026	vol liq or gas	-13.28(9)	25.63(4)	0.6876[20]	1.2614[20]	msc H_2O, EtOH, eth
592	*p*-Hydroquinone	1,4-Benzenediol	$C_6H_6O_2$	123-31-9	110.111	mcl pr (sub) nd(w) pr (MeOH)	173(2)	288(5)	1.330[20]	1.632[25]	s H_2O, eth; vs EtOH, ace; i bz
593	4-Hydroxybenzaldehyde	4-Formylphenol	$C_7H_6O_2$	123-08-0	122.122	nd (w)	116.0(2)		1.129[130]	1.5705[130]	sl H_2O, ace; vs EtOH, eth; s bz
594	2-Hydroxybenzoic acid	Salicylic acid	$C_7H_6O_3$	69-72-7	138.121	nd (w), mcl pr (al)	158.6(5)		1.443[20]	1.565	sl H_2O, bz, chl, ctc; vs EtOH, eth, ace
595	1-Hydroxy-1,1-diphosphonoethane	Etidronic acid	$C_2H_8O_7P_2$	2809-21-4	206.028	cry (w)	105				s H_2O, EtOH, MeOH
596	4-Hydroxy-3-methoxybenzaldehyde	Vanillin	$C_8H_8O_3$	121-33-5	152.148	tetr (w, lig)	81(1)	285	1.056[25]		sl H_2O; vs EtOH, eth, ace; s bz, lig
597	1-(4-Hydroxy-3-methoxyphenyl)ethanone	Apocynin	$C_9H_{10}O_3$	498-02-2	166.173	pr (w)	115	297			sl H_2O; s EtOH, ace, bz; vs eth, chl
598	*N*-(4-Hydroxyphenyl) acetamide	Acetaminophen	$C_8H_9NO_2$	103-90-2	151.163	mcl pr (w)	168.0(5)		1.293[21]		i H_2O; vs EtOH
599	1-(2-Hydroxyphenyl) ethanone	2-Hydroxyacetophenone	$C_8H_8O_2$	118-93-4	136.149		2.5	218	1.1307[20]	1.5584[20]	vs eth, EtOH, HOAc
600	1-(4-Hydroxyphenyl) ethanone	4-Hydroxyacetophenone	$C_8H_8O_2$	99-93-4	136.149	nd (eth, dil al)	108.2(5)		1.1090[109]	1.5577[109]	sl H_2O, DMSO; vs EtOH, eth
601	3-Hydroxypropanenitrile	Hydracrylonitrile	C_3H_5NO	109-78-4	71.078	liq	-46	218(5)	1.0404[25]	1.4248[20]	msc H_2O, EtOH; sl eth; s chl; i CS_2
602	Imidazole	1,3-Diazole	$C_3H_4N_2$	288-32-4	68.077	mcl pr (bz)	89.52(4)	257	1.0303[101]	1.4801[101]	vs H_2O, EtOH; s eth, ace, py; sl bz
603	Indan		C_9H_{10}	496-11-7	118.175	liq	-51.34(2)	177.8(4)	0.9639[20]	1.5378[20]	i H_2O; msc EtOH, eth; sl chl
604	1*H*-Indole	2,3-Benzopyrrole	C_8H_7N	120-72-9	117.149	lf (w, peth) cry (eth)	52.3(6)	254(3)	1.22[25]		s H_2O, bz; vs EtOH, eth, tol; sl ctc
605	*myo*-Inositol	(1α,2α,3α,4β,5α,6β)-Cyclohexanehexol	$C_6H_{12}O_6$	87-89-8	180.155	cry (w)	223.8(4)		1.752		s H_2O
606	Iodobenzene	Phenyl iodide	C_6H_5I	591-50-4	204.008	liq	-30.7(5)	188.5(6)	1.8308[20]	1.6200[20]	i H_2O; s EtOH; msc eth, ace, bz, ctc
607	1-Iodobutane	Butyl iodide	C_4H_9I	542-69-8	184.018	liq	-103.5(4)	130(2)	1.6154[20]	1.5001[20]	i H_2O; msc EtOH, eth; vs chl
608	Iodoethane	Ethyl iodide	C_2H_5I	75-03-6	155.965	liq	-111.0(4)	72(1)	1.9357[20]	1.5133[20]	sl H_2O; msc EtOH; s eth, chl

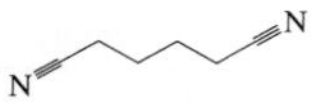
569
Hexanedinitrile

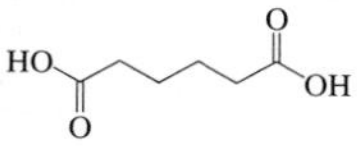
570
1,6-Hexanedioic acid

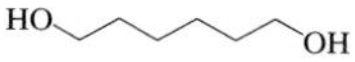
571
1,6-Hexanediol

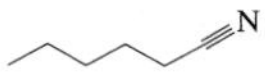
572
Hexanenitrile

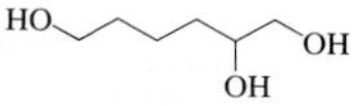
573
1,2,6-Hexanetriol

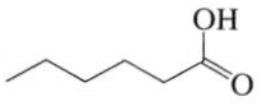
574
Hexanoic acid

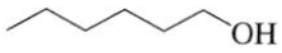
575
1-Hexanol

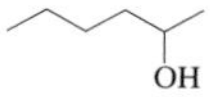
576
2-Hexanol

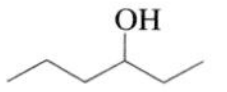
577
3-Hexanol

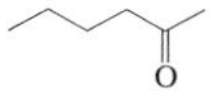
578
2-Hexanone

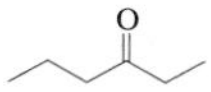
579
3-Hexanone

580
Hexatriacontane

581
1-Hexene

582
cis-2-Hexene

583
trans-2-Hexene

584
cis-3-Hexene

585
trans-3-Hexene

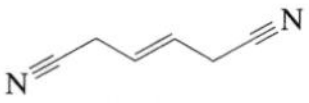
586
trans-3-Hexenedinitrile

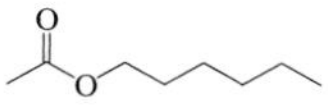
587
Hexyl acetate

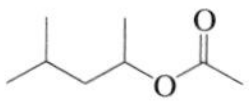
588
sec-Hexyl acetate

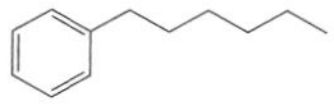
589
Hexylbenzene

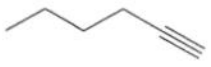
590
1-Hexyne

$HC\equiv N$
591
Hydrogen cyanide

592
p-Hydroquinone

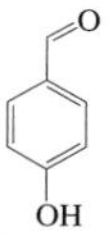
593
4-Hydroxybenzaldehyde

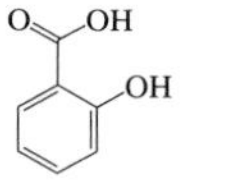
594
2-Hydroxybenzoic acid

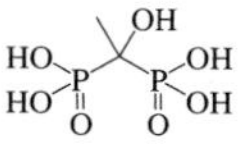
595
1-Hydroxy-1,1-diphosphonoethane

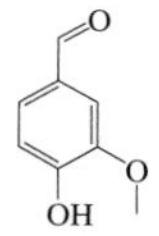
596
4-Hydroxy-3-methoxybenzaldehyde

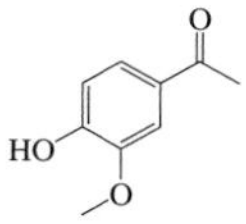
597
1-(4-Hydroxy-3-methoxyphenyl)ethanone

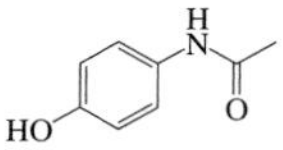
598
N-(4-Hydroxyphenyl)acetamide

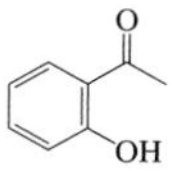
599
1-(2-Hydroxyphenyl)ethanone

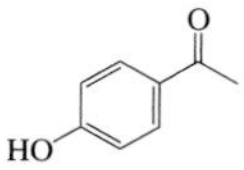
600
1-(4-Hydroxyphenyl)ethanone

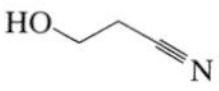
601
3-Hydroxypropanenitrile

602
Imidazole

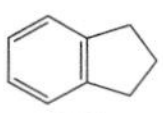
603
Indan

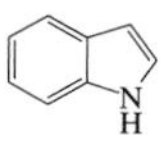
604
1*H*-Indole

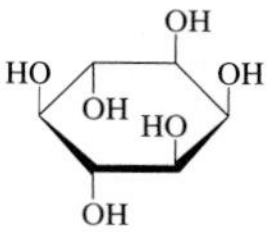
605
myo-Inositol

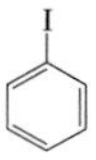
606
Iodobenzene

607
1-Iodobutane

608
Iodoethane

No.	Name	Synonym	Mol. form.	CAS Reg. No.	Mol. wt.	Physical form	mp/°C	bp/°C	den/g cm^{-3}	n_D	Solubility
609	Iodomethane	Methyl iodide	CH_3I	74-88-4	141.939	liq	-66(2)	42.4(2)	2.2789^{20}	1.5308^{20}	sl H_2O; s ace, bz, chl; msc EtOH, eth
610	1-Iodopropane	Propyl iodide	C_3H_7I	107-08-4	169.992	liq	-101.4(4)	102(2)	1.7489^{20}	1.5058^{20}	sl H_2O, ctc; msc EtOH, eth
611	2-Iodopropane	Isopropyl iodide	C_3H_7I	75-30-9	169.992	liq	-90.4(9)	89(3)	1.7042^{20}	1.5028^{20}	sl H_2O; msc EtOH, eth, bz, chl
612	1(3*H*)-Isobenzofuranone		$C_8H_6O_2$	87-41-2	134.133	nd or pl (w)	75	292(2)	1.1636^{99}	1.536^{99}	s H_2O; vs EtOH, eth; sl chl
613	Isobutanal	2-Methyl-1-propanal	C_4H_8O	78-84-2	72.106	liq	-72.1(2)	64.1(2)	0.7891^{20}	1.3730^{20}	s H_2O, eth, ace, chl; sl ctc
614	Isobutane	2-Methylpropane	C_4H_{10}	75-28-5	58.122	col gas	-159.38(20)	-11.7(5)	0.5510^{25} (p>1 atm)	1.3518^{-25}	sl H_2O; s EtOH, eth, chl
615	Isobutene	2-Methyl-1-propene	C_4H_8	115-11-7	56.107	col gas	-140.7(2)	-7.0(2)	0.589^{25} (p>1 atm)	1.3926^{-25}	i H_2O; vs EtOH, eth; s bz, sulf
616	Isobutyl acetate	2-Methylpropyl acetate	$C_6H_{12}O_2$	110-19-0	116.158	liq	-97.1(5)	116.9(6)	0.8712^{20}	1.3902^{20}	sl H_2O, ctc; msc EtOH, eth; s ace
617	Isobutylamine	2-Methyl-1-propanamine	$C_4H_{11}N$	78-81-9	73.137	liq	-86(1)	68.8(1)	0.724^{25}	1.3988^{19}	
618	Isobutylbenzene	(2-Methylpropyl)benzene	$C_{10}H_{14}$	538-93-2	134.218	liq	-51.6(2)	172.7(4)	0.8532^{20}	1.4866^{20}	i H_2O; msc EtOH, eth, ace, bz, peth, ctc
619	Isobutyl formate	2-Methylpropyl formate	$C_5H_{10}O_2$	542-55-2	102.132	liq	-95.5(7)	98.4(3)	0.8776^{20}	1.3857^{20}	sl H_2O, chl; msc EtOH, eth; vs ace
620	Isobutyl isobutanoate		$C_8H_{16}O_2$	97-85-8	144.212	liq	-80.6(2)	148(3)	0.8542^{20}	1.3999^{20}	sl H_2O, ctc; s EtOH, ace; msc eth
621	Isopentane	2-Methylbutane	C_5H_{12}	78-78-4	72.149	vol liq or gas	-159.8(2)	27.83(6)	0.6201^{20}	1.3537^{20}	i H_2O; msc EtOH, eth
622	Isopentyl acetate	Isoamyl acetate	$C_7H_{14}O_2$	123-92-2	130.185	liq	-78.5	141.6(7)	0.876^{15}	1.4000^{20}	sl H_2O; msc EtOH, eth; s ace, chl
623	Isopentyl isopentanoate	Isopentyl isovalerate	$C_{10}H_{20}O_2$	659-70-1	172.265			190(3)	0.8583^{19}	1.4130^{19}	
624	Isophorone	3,5,5-Trimethyl-2-cyclo-hexen-1-one	$C_9H_{14}O$	78-59-1	138.206	liq	-8.1	214.8(7)	0.9255^{20}	1.4766^{18}	
625	Isophthalic acid	1,3-Benzenedicarboxylic acid	$C_8H_6O_4$	121-91-5	166.132	nd (w, al)	348.0(2)	subl	1.538^{25}		sl H_2O; s EtOH, HOAc; i eth, bz, lig
626	Isopropenylbenzene	α-Methyl styrene	C_9H_{10}	98-83-9	118.175	liq	-22.36(5)	165.4(3)	0.9106^{20}	1.5386^{20}	i H_2O; s EtOH, eth; msc ace, bz, ctc
627	*p*-Isopropenylstyrene		$C_{11}H_{12}$	16262-48-9	144.213	liq		242	0.93	1.5684^{20}	
628	Isopropyl acetate	1-Methylethyl acetate	$C_5H_{10}O_2$	108-21-4	102.132	liq	-73.4	88.6(2)	0.8662^{25}	1.3746^{25}	s H_2O, EtOH, ace, chl; msc eth
629	Isopropylamine	2-Propanamine	C_3H_9N	75-31-0	59.110	liq	-95.119(1)	31.8(2)	0.6891^{20}	1.3742^{20}	msc H_2O, EtOH, eth; vs ace; s bz, chl
630	Isopropylbenzene	Cumene	C_9H_{12}	98-82-8	120.191	liq	-96.01(5)	152.4(2)	0.8615^{20}	1.4915^{20}	i H_2O; msc EtOH, eth, ace, bz, peth, ctc
631	Isopropylbenzene hydroperoxide	Cumene hydroperoxide	$C_9H_{12}O_2$	80-15-9	152.190	liq		153 expl	1.03^{20}		
632	1-Isopropyl-2-methylben-zene	*o*-Cymene	$C_{10}H_{14}$	527-84-4	134.218	liq	-71.5(1)	178(2)	0.8766^{20}	1.5006^{20}	i H_2O; msc EtOH, eth, ace, bz, peth, ctc
633	1-Isopropyl-3-methylben-zene	*m*-Cymene	$C_{10}H_{14}$	535-77-3	134.218	liq	-63.8(1)	175(4)	0.8610^{20}	1.4930^{20}	i H_2O; msc EtOH, eth, ace, bz, peth, ctc
634	1-Isopropyl-4-methylben-zene	*p*-Cymene	$C_{10}H_{14}$	99-87-6	134.218	liq	-68.1(3)	177(2)	0.8573^{20}	1.4909^{20}	i H_2O; msc EtOH, eth, ace, bz, peth, ctc
635	Isoquinoline	Benzo[*c*]pyridine	C_9H_7N	119-65-3	129.159	hyg pl	26.46(1)	243.2(6)	1.0910^{30}	1.6148^{20}	i H_2O; vs EtOH, chl; msc eth, bz
636	Isosorbide		$C_6H_{10}O_4$	652-67-5	146.141		63				
	Kepone	Chlordecone	$C_{10}Cl_{10}O$	143-50-0	490.636		350 dec		1.61^{25}		
637	DL-Lactic acid	2-Hydroxypropanoic acid, (±)-	$C_3H_6O_3$	598-82-3	90.078	ye cry	16.9(2)		1.2060^{21}	1.4392^{20}	vs H_2O, EtOH; sl eth
638	α-Lactose		$C_{12}H_{22}O_{11}$	14641-93-1	342.296	wh pow	222.8				vs H_2O; sl EtOH; i eth, chl
639	*d*-Limonene	*p*-Mentha-1,8-diene, (*R*)-	$C_{10}H_{16}$	5989-27-5	136.234	oil	-74.0(6)	177.6(5)	0.8411^{20}	1.4730^{20}	i H_2O; msc EtOH, eth; s ctc
640	*l*-Limonene	*p*-Mentha-1,8-diene, (*S*)-	$C_{10}H_{16}$	5989-54-8	136.234	oil		178(1)	0.843^{20}	1.4746^{20}	i H_2O; vs eth, EtOH
641	Maleic acid	*cis*-2-Butenedioic acid	$C_4H_4O_4$	110-16-7	116.073	mcl pr (w)	143.5(5)		1.590^{20}		vs H_2O, EtOH, ace; s eth; i bz, chl
642	Malonic acid	Propanedioic acid	$C_3H_4O_4$	141-82-2	104.062	tcl (al)	135(1)	subl	1.619^{10}		vs H_2O, py; s EtOH, eth; i bz
643	Malononitrile		$C_3H_2N_2$	109-77-3	66.061		31.83(2)	219(3)	1.1910^{20}	1.4146^{34}	s H_2O, ace, bz, chl; vs EtOH, eth
644	α-Maltose		$C_{12}H_{22}O_{11}$	4482-75-1	342.296	nd (al)	162.5		1.546^{20}		vs H_2O
645	D-Mannitol	Cordycepic acid	$C_6H_{14}O_6$	69-65-8	182.171	orth nd or pr (w)	164.1(1)		1.489^{20}	1.3330	vs H_2O; sl EtOH, py; i eth
646	D-Mannose	Seminose	$C_6H_{12}O_6$	3458-28-4	180.155	nd or orth pr (al)	118(2)		1.539^{20}		vs H_2O; sl EtOH, MeOH; i eth, bz
647	*p*-Menthane hydroperoxide	1-Methyl-1-(4-methylcyclo-hexyl)ethyl hydroperoxide	$C_{10}H_{20}O_2$	80-47-7	172.265			259	0.92		

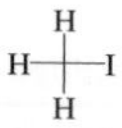
609
Iodomethane

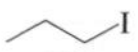
610
1-Iodopropane

611
2-Iodopropane

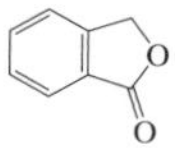
612
1(3H)-Isobenzofuranone

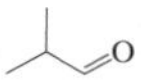
613
Isobutanal

614
Isobutane

615
Isobutene

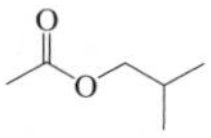
616
Isobutyl acetate

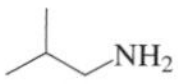
617
Isobutylamine

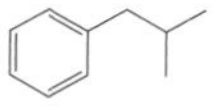
618
Isobutylbenzene

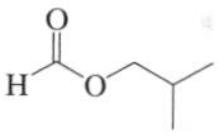
619
Isobutyl formate

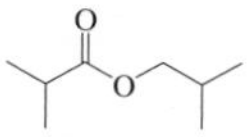
620
Isobutyl isobutanoate

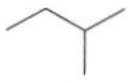
621
Isopentane

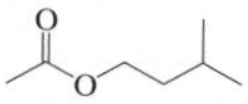
622
Isopentyl acetate

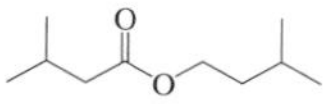
623
Isopentyl isopentanoate

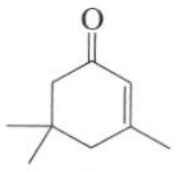
624
Isophorone

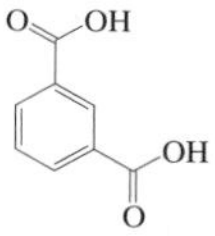
625
Isophthalic acid

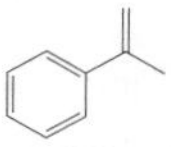
626
Isopropenylbenzene

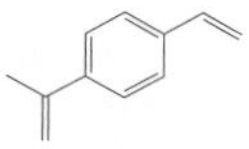
627
p-Isopropenylstyrene

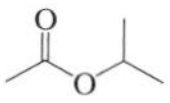
628
Isopropyl acetate

629
Isopropylamine

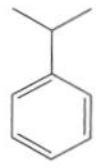
630
Isopropylbenzene

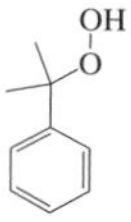
631
Isopropylbenzene hydroperoxide

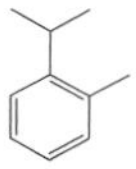
632
1-Isopropyl-2-methylbenzene

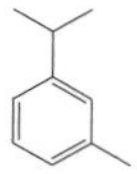
633
1-Isopropyl-3-methylbenzene

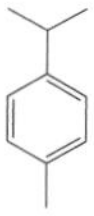
634
1-Isopropyl-4-methylbenzene

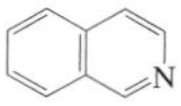
635
Isoquinoline

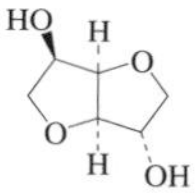
636
Isosorbide

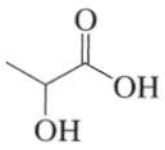
637
dl-Lactic acid

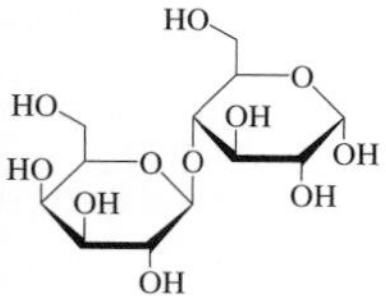
638
α-Lactose

639
d-Limonene

640
l-Limonene

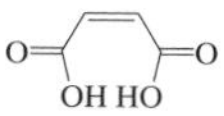
641
Maleic acid

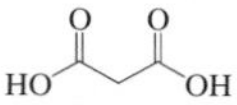
642
Malonic acid

643
Malononitrile

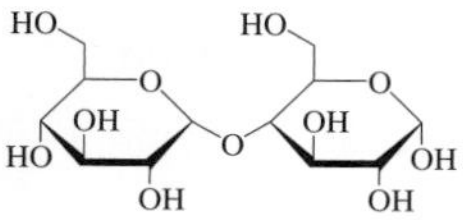
644
α-Maltose

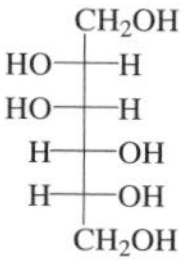
645
D-Mannitol

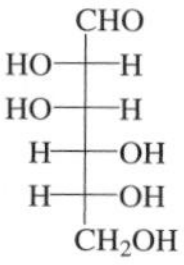
646
D-Mannose

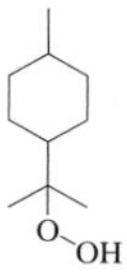
647
p-Menthane hydroperoxide

No.	Name	Synonym	Mol. form.	CAS Reg. No.	Mol. wt.	Physical form	mp/°C	bp/°C	den/g cm^{-3}	n_D	Solubility
648	(-)-Menthol	2-Isopropyl-5-methylcyclohexanol, [1*R*-(1α,2β,5α)]-	$C_{10}H_{20}O$	2216-51-5	156.265	nd (MeOH)	42.1(8)	214(12)	0.903^{15}	1.460^{22}	sl H_2O; vs EtOH, eth, ace, bz; s peth
649	Mesityl oxide	Isobutenyl methyl ketone	$C_6H_{10}O$	141-79-7	98.142	liq	-52.8(4)	129.7(4)	0.8653^{20}	1.4440^{20}	s H_2O, ace; msc EtOH, eth
650	Methacrylic acid	2-Methylpropenoic acid	$C_4H_6O_2$	79-41-4	86.090	pr	14.6(8)	160(1)	1.0153^{20}	1.4314^{20}	s H_2O, chl; msc EtOH, eth
651	Methane		CH_4	74-82-8	16.043	col gas	-182.475(5)	-161.5(2)	$0.4224^{-161.5}$ liq		sl H_2O, ace; s EtOH, eth, bz, tol, MeOH
652	Methanol	Methyl alcohol	CH_4O	67-56-1	32.042	liq	-97.5(1)	64.5(7)	0.7909^{20}	1.3288^{20}	msc H_2O, EtOH, eth, ace; vs bz; s chl
653	2-Methoxyaniline	*o*-Anisidine	C_7H_9NO	90-04-0	123.152	ye liq	6.2	221(7)	1.0923^{20}	1.5715^{10}	sl H_2O; s EtOH, eth, ace, bz
654	2-Methoxyethanol	Ethylene glycol monomethyl ether	$C_3H_8O_2$	109-86-4	76.095	liq	-85.1	124.3(1)	0.9647^{20}	1.4024^{20}	msc H_2O, eth, bz; vs EtOH; s ace; sl chl
655	2-Methoxy-2-methylbutane	Methyl *tert*-pentyl ether	$C_6H_{14}O$	994-05-8	102.174			86.4(1)	0.7660^{25}	1.3862^{25}	sl H_2O; vs eth, EtOH
656	4-Methoxyphenol	Mequinol	$C_7H_8O_2$	150-76-5	124.138	pl	54(3)	253(2)			s H_2O, bz, ctc; vs EtOH, eth
657	*trans*-2-Methoxy-4-(1-propenyl)phenol		$C_{10}H_{12}O_2$	5932-68-3	164.201		33.5		1.0852^{20}	1.5784^{20}	sl H_2O; s EtOH, eth, chl
658	*N*-Methylacetamide		C_3H_7NO	79-16-3	73.094		30.6(1)	208(2)	0.9371^{25}	1.4301^{20}	vs ace, bz, eth, EtOH
659	Methyl acetate		$C_3H_6O_2$	79-20-9	74.079	liq	-98.2(2)	56.7(2)	0.9346^{20}	1.3614^{20}	vs H_2O, eth, EtOH
660	Methyl acetoacetate	Methyl 3-oxobutanoate	$C_5H_8O_3$	105-45-3	116.116		27.5	168(3)	1.0762^{20}	1.4184^{20}	vs H_2O; msc EtOH, eth; s ctc
661	Methyl acrylate	Methyl propenoate	$C_4H_6O_2$	96-33-3	86.090	liq	-75.6(3)	80.1(6)	0.9535^{20}	1.4040^{20}	sl H_2O; s EtOH, eth, ace, bz, chl
662	2-Methylacrylonitrile	2-Methylpropenenitrile	C_4H_5N	126-98-7	67.090	liq	-35.8	90(3)	0.8001^{20}	1.4003^{20}	sl H_2O, chl; msc EtOH, eth, ace, tol
663	Methylamine	Methanamine	CH_5N	74-89-5	31.058	col gas	-93.42(9)	-6.4(3)	0.656^{25} (p>1 atm)		vs H_2O; s EtOH, ace, bz; msc eth
664	2-Methylaniline	*o*-Toluidine	C_7H_9N	95-53-4	107.153	liq	-14.41(2)	200.0(4)	0.9984^{20}	1.5725^{20}	sl H_2O; msc EtOH, eth, ctc
665	3-Methylaniline	*m*-Toluidine	C_7H_9N	108-44-1	107.153	liq	-30.8(5)	203.3(5)	0.9889^{20}	1.5681^{20}	vs ace, bz, eth, EtOH
666	4-Methylaniline	*p*-Toluidine	C_7H_9N	106-49-0	107.153	lf (w+1)	43.3(8)	201(1)	0.9619^{20}	1.5534^{45}	sl H_2O; vs EtOH, py; s eth, ace, ctc
667	*N*-Methylaniline	*N*-Methylbenzenamine	C_7H_9N	100-61-8	107.153	liq	-57(1)	197(1)	0.9859^{20}	1.5684^{20}	i H_2O; s EtOH, eth, ctc, chl
668	Methyl benzoate	Methyl benzenecarboxylate	$C_8H_8O_2$	93-58-3	136.149	liq	-12.35(6)	199(2)	1.0837^{25}	1.5164^{20}	i H_2O; s EtOH, ctc, MeOH; msc eth
669	2-Methyl-1,3-butadiene	Isoprene	C_5H_8	78-79-5	68.118	liq	-146.1(5)	34.0(3)	0.679^{20}	1.4219^{20}	i H_2O; msc EtOH, eth, ace, bz
670	Methyl butanoate		$C_5H_{10}O_2$	623-42-7	102.132	liq	-85.8	101.9(1)	0.8984^{20}	1.3878^{20}	sl H_2O, ctc; msc EtOH, eth
671	3-Methylbutanoic acid	Isovaleric acid	$C_5H_{10}O_2$	503-74-2	102.132	liq	-29.6(7)	176.5(2)	0.931^{20}	1.4033^{20}	s H_2O; msc EtOH, eth, chl
672	3-Methyl-1-butanol	Isopentyl alcohol	$C_5H_{12}O$	123-51-3	88.148	liq	-117.2	130.8(3)	0.8104^{20}	1.40711^{20}	sl H_2O; vs ace, eth, EtOH
673	2-Methyl-2-butanol	*tert*-Pentyl alcohol	$C_5H_{12}O$	75-85-4	88.148	liq	-8.7(6)	102.4	0.8096^{20}	1.4052^{20}	s H_2O, bz, chl; msc EtOH, eth; vs ace
674	3-Methyl-2-butanol, (±)-	Isopropylethanol	$C_5H_{12}O$	70116-68-6	88.148	liq		113.7(4)	0.8180^{20}	1.4089^{20}	sl H_2O; msc EtOH, eth; vs ace; s bz, ctc
675	3-Methyl-2-butanone	Methyl isopropyl ketone	$C_5H_{10}O$	563-80-4	86.132	liq	-93.13(5)	94.2(2)	0.8051^{20}	1.3880^{20}	sl H_2O; msc EtOH, eth; vs ace; s ctc
676	3-Methyl-1-butene		C_5H_{10}	563-45-1	70.133	vol liq or gas	-168.41(2)	20.1(2)	0.6213^{25}	1.3643^{20}	i H_2O; msc EtOH, eth; s bz
677	2-Methyl-2-butene		C_5H_{10}	513-35-9	70.133	liq	-133.72(2)	38.5(4)	0.6623^{20}	1.3874^{20}	i H_2O; s EtOH, eth, bz, ctc; vs lig
678	*cis*-2-Methyl-2-butenedioic acid	Citraconic acid	$C_5H_6O_4$	498-23-7	130.100	nd (eth-lig) tcl pr (eth-bz)	83.2(6)		1.617^{25}		vs H_2O; sl eth, chl; i bz, CS_2
679	Methyl *tert*-butyl ether	*tert*-Butyl methyl ether	$C_5H_{12}O$	1634-04-4	88.148	liq	-108.6(1)	55.1(1)	0.7353^{25}	1.3664^{25}	s H_2O; vs EtOH, eth
680	Methyl cyanoacetate		$C_4H_5NO_2$	105-34-0	99.089	liq	-13.1(2)	203(4)	1.1225^{25}	1.4176^{20}	vs eth, EtOH
681	Methylcyclohexane		C_7H_{14}	108-87-2	98.186	liq	-126.6(4)	100.9(1)	0.7694^{20}	1.4231^{20}	i H_2O; s EtOH, eth; msc ace, bz, lig
682	1-Methylcyclohexanol		$C_7H_{14}O$	590-67-0	114.185		26.2(1)	155	0.9194^{20}	1.4595^{20}	i H_2O; s EtOH, bz, chl
683	*cis*-2-Methylcyclohexanol		$C_7H_{14}O$	615-38-3	114.185		7	165	0.9360^{20}	1.4640^{20}	vs EtOH
684	*trans*-2-Methylcyclohexanol, (±)-		$C_7H_{14}O$	615-39-4	114.185	liq	-2.0	168.4(2)	0.9247^{20}	1.4616^{20}	vs eth, EtOH
685	*cis*-3-Methylcyclohexanol, (±)-		$C_7H_{14}O$	5454-79-5	114.185	liq	-5.5	168	0.9155^{20}	1.4752^{20}	vs eth, EtOH
686	*trans*-3-Methylcyclohexanol, (±)-		$C_7H_{14}O$	7443-55-2	114.185	liq	-0.5	167	0.9214^{30}	1.4580^{20}	vs eth, EtOH
687	*cis*-4-Methylcyclohexanol		$C_7H_{14}O$	7731-28-4	114.185	liq	-9.2	174(10)	0.9170^{20}	1.4614^{20}	vs eth, EtOH

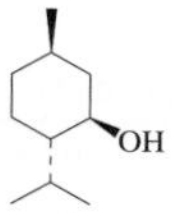
648
(-)-Menthol

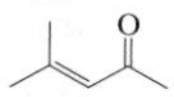
649
Mesityl oxide

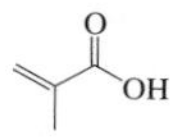
650
Methacrylic acid

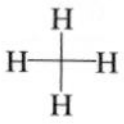
651
Methane

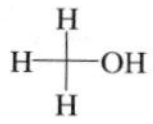
652
Methanol

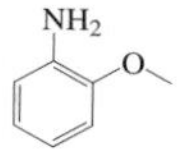
653
2-Methoxyaniline

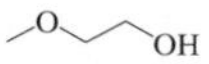
654
2-Methoxyethanol

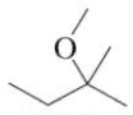
655
2-Methoxy-2-methylbutane

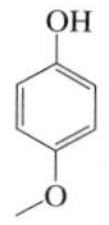
656
4-Methoxyphenol

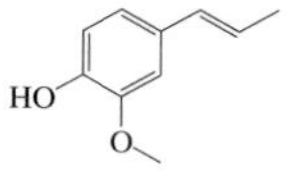
657
trans-2-Methoxy-4-(1-propenyl)phenol

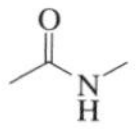
658
N-Methylacetamide

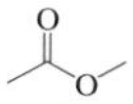
659
Methyl acetate

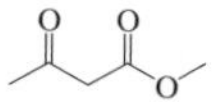
660
Methyl acetoacetate

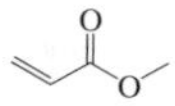
661
Methyl acrylate

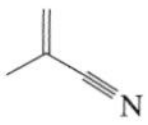
662
2-Methylacrylonitrile

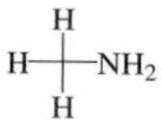
663
Methylamine

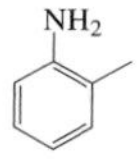
664
2-Methylaniline

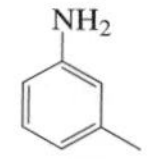
665
3-Methylaniline

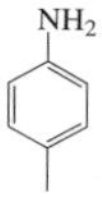
666
4-Methylaniline

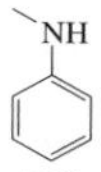
667
N-Methylaniline

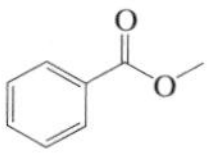
668
Methyl benzoate

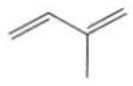
669
2-Methyl-1,3-butadiene

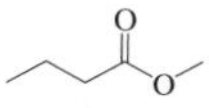
670
Methyl butanoate

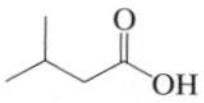
671
3-Methylbutanoic acid

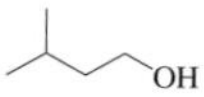
672
3-Methyl-1-butanol

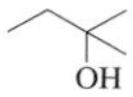
673
2-Methyl-2-butanol

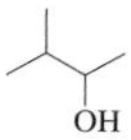
674
3-Methyl-2-butanol, (±)-

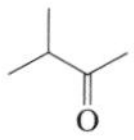
675
3-Methyl-2-butanone

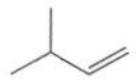
676
3-Methyl-1-butene

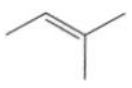
677
2-Methyl-2-butene

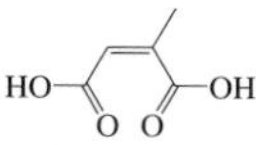
678
cis-2-Methyl-2-butenedioic acid

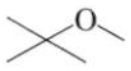
679
Methyl *tert*-butyl ether

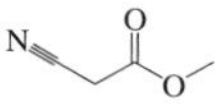
680
Methyl cyanoacetate

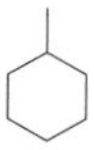
681
Methylcyclohexane

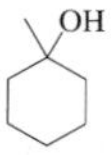
682
1-Methylcyclohexanol

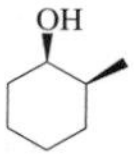
683
cis-2-Methylcyclohexanol

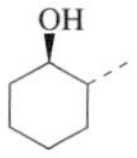
684
trans-2-Methylcyclohexanol, (±)-

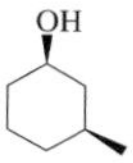
685
cis-3-Methylcyclohexanol, (±)-

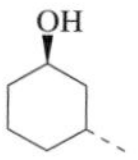
686
trans-3-Methylcyclohexanol, (±)-

687
cis-4-Methylcyclohexanol

No.	Name	Synonym	Mol. form.	CAS Reg. No.	Mol. wt.	Physical form	mp/°C	bp/°C	den/g cm^{-3}	n_D	Solubility
688	*trans*-4-Methylcyclohexanol		$C_7H_{14}O$	7731-29-5	114.185			175(3)	0.9118^{21}	1.4561^{20}	sl H_2O; msc EtOH; s eth
689	Methylcyclopentane		C_6H_{12}	96-37-7	84.159	liq	-142.419(2)	71.8(2)	0.7486^{20}	1.4097^{20}	i H_2O; msc EtOH, eth, ace, bz, lig, ctc
690	1-Methyl-2,4-dinitrobenzene	2,4-Dinitrotoluene	$C_7H_6N_2O_4$	121-14-2	182.134	ye nd or mcl pr (CS_2)	69.6(9)	300 dec	1.3208^{71}	1.442	i H_2O; s EtOH, eth, chl, bz; vs ace, py
691	*N*-Methyl-2-ethanolamine		C_3H_9NO	109-83-1	75.109			159.24(4)	0.937^{20}	1.4385^{20}	msc H_2O, EtOH, eth
692	1-(1-Methylethoxy)butane	Butyl isopropyl ether	$C_7H_{16}O$	1860-27-1	116.201			107(9)	0.7594^{15}	1.3870^{15}	i H_2O; s EtOH, eth, ace, con sulf
693	1-(1-Methylethoxy)-2-propanol	1-Isopropoxy-2-propanol	$C_6H_{14}O_2$	3944-36-3	118.174			137.5	0.879^{20}	1.4070^{20}	
694	*N*-Methylformamide		C_2H_5NO	123-39-7	59.067	liq	-2.5(7)	186(5)	1.011^{19}	1.4319^{20}	vs H_2O, ace, EtOH
695	Methyl formate		$C_2H_4O_2$	107-31-3	60.052	liq	-99.7(4)	31.6(3)	0.9739^{20}	1.3419^{20}	vs H_2O; msc EtOH; s eth, chl, MeOH
696	2-Methylheptane		C_8H_{18}	592-27-8	114.229	liq	-109(4)	117.6(9)	0.6980^{20}	1.3949^{20}	i H_2O; msc EtOH, ace, bz; s eth, ctc
697	3-Methylheptane		C_8H_{18}	589-81-1	114.229	col liq	-120.48(9)	118.9(6)	0.7017^{25}	1.3961^{25}	i H_2O; s EtOH, eth; msc ace, bz, chl
698	4-Methylheptane		C_8H_{18}	589-53-7	114.229	liq	-121.0(1)	117.7(5)	0.7046^{20}	1.3979^{20}	i H_2O; s eth; msc EtOH, ace, bz
699	2-Methylhexane		C_7H_{16}	591-76-4	100.202	liq	-118.23(4)	90.0(8)	0.6787^{20}	1.3848^{20}	i H_2O; s EtOH; msc eth, ace, bz, lig, chl
700	3-Methylhexane		C_7H_{16}	78918-91-9	100.202	liq	-119.4	92	0.687^{21}	1.3854^{25}	i H_2O; s EtOH; msc eth, ace, bz, lig, chl
701	5-Methyl-2-hexanone	Methyl isopentyl ketone	$C_7H_{14}O$	110-12-3	114.185			139(2)	0.888^{20}	1.4062^{20}	sl H_2O; msc EtOH; vs ace, bz; s ctc
702	Methyl methacrylate	Methyl 2-methyl-2-propenoate	$C_5H_8O_2$	80-62-6	100.117	liq	-47.55(2)	100.6(2)	0.9377^{25}	1.4142^{20}	sl H_2O; msc EtOH, eth, ace; s chl
703	1-Methylnaphthalene		$C_{11}H_{10}$	90-12-0	142.197	liq	-30.43(7)	244.4(9)	1.0202^{20}	1.6170^{20}	i H_2O; vs EtOH, eth; s bz
704	2-Methylnaphthalene		$C_{11}H_{10}$	91-57-6	142.197	mcl (al)	34.6(4)	241.1(3)	1.0058^{20}	1.6015^{40}	i H_2O; vs EtOH, eth; s bz, chl
705	Methyl *cis*-9-octadecenoate	Methyl oleate	$C_{19}H_{36}O_2$	112-62-9	296.488		-19.7(5)	347(5)	0.8711^{25}	1.4522^{20}	i H_2O; msc EtOH, eth; s chl
706	2-Methyloctane		C_9H_{20}	3221-61-2	128.255	liq	-80.3(3)	143(1)	0.7095^{25}	1.4031^{20}	i H_2O; s EtOH, eth; sl ctc; vs peth
707	3-Methyloctane		C_9H_{20}	2216-33-3	128.255	liq	-108.0(2)	144(2)	0.717^{25}	1.4040^{25}	
708	4-Methyloctane		C_9H_{20}	2216-34-4	128.255	liq	-116(5)	142(1)	0.716^{25}	1.4039^{25}	i H_2O
709	Methyloxirane	1,2-Propylene oxide	C_3H_6O	75-56-9	58.079	liq	-111.9	35	0.859^{0}	1.3660^{20}	vs H_2O, EtOH, eth; s chl
710	2-Methylpentane	Isohexane	C_6H_{14}	107-83-5	86.175	liq	-153.60(9)	60.21(9)	0.650^{25}	1.3715^{20}	i H_2O; s EtOH, eth; msc ace, bz, chl
711	3-Methylpentane		C_6H_{14}	96-14-0	86.175	liq	-162.89(5)	63.3(5)	0.6598^{25}	1.3765^{20}	i H_2O; s EtOH, ctc; msc eth, ace, bz, hp
712	2-Methyl-2,4-pentanediol	Hexylene glycol	$C_6H_{14}O_2$	107-41-5	118.174	liq	-50	197.9(9)	0.923^{15}	1.4276^{20}	s H_2O, EtOH, eth; sl ctc
713	2-Methyl-1-pentanol		$C_6H_{14}O$	105-30-6	102.174	liq		157(6)	0.8263^{20}	1.4182^{20}	sl H_2O; s EtOH, eth, ace, ctc
714	2-Methyl-2-pentanol		$C_6H_{14}O$	590-36-3	102.174	liq	-103	121(1)	0.8350^{16}	1.4100^{20}	sl H_2O; s EtOH, eth
715	3-Methyl-2-pentanol		$C_6H_{14}O$	565-60-6	102.174			142(2)	0.8307^{20}	1.4182^{20}	sl H_2O; s EtOH, eth
716	4-Methyl-2-pentanol	Methyl isobutyl carbinol	$C_6H_{14}O$	108-11-2	102.174	liq	-90	132.0(5)	0.8075^{20}	1.4100^{20}	sl H_2O, ctc; s EtOH, eth
717	2-Methyl-3-pentanol		$C_6H_{14}O$	565-67-3	102.174			127.9(2)	0.8243^{20}	1.4175^{20}	sl H_2O; msc EtOH, eth
718	3-Methyl-3-pentanol		$C_6H_{14}O$	77-74-7	102.174	liq	-23.6(3)	129(4)	0.8286^{20}	1.4186^{20}	sl H_2O, ctc; msc EtOH, eth
719	4-Methyl-2-pentanone	Isobutyl methyl ketone	$C_6H_{12}O$	108-10-1	100.158	liq	-85(2)	115.7(2)	0.7965^{25}	1.3962^{20}	sl H_2O; msc EtOH, eth, ace, bz; s chl
720	2-Methyl-3-pentanone	Ethyl isopropyl ketone	$C_6H_{12}O$	565-69-5	100.158			118(1)	0.814^{18}	1.3975^{20}	sl H_2O; vs EtOH, bz; msc eth, ace; s chl
721	2-Methyl-1-pentene		C_6H_{12}	763-29-1	84.159	liq	-135.7(4)	62.1(6)	0.6799^{20}	1.3920^{20}	i H_2O; s EtOH, bz, chl; sl ctc
722	4-Methyl-1-pentene		C_6H_{12}	691-37-2	84.159	liq	-153.9(2)	54(2)	0.6642^{20}	1.3828^{20}	i H_2O; s EtOH, bz, chl, peth
723	*N*-Methylpropanamide	*N*-Methylpropionamide	C_4H_9NO	1187-58-2	87.120	liq	-30.9	207(3)	0.9305^{25}	1.4345^{25}	
724	2-Methylpropanenitrile	Isopropyl cyanide	C_4H_7N	78-82-0	69.106	liq	-71.5(4)	102(2)	0.7704^{20}	1.3720^{20}	sl H_2O; vs EtOH, eth, ace, chl
725	Methyl propanoate	Methyl propionate	$C_4H_8O_2$	554-12-1	88.106	liq	-87.5(5)	78.6(2)	0.9150^{20}	1.3775^{20}	sl H_2O; msc EtOH, eth; s ace, ctc
726	2-Methylpropanoic acid	Isobutyric acid	$C_4H_8O_2$	79-31-2	88.106	liq	-46(1)	154.4(2)	0.9681^{20}	1.3930^{20}	vs H_2O; msc EtOH, eth; sl ctc
727	2-Methyl-1-propanol	Isobutyl alcohol	$C_4H_{10}O$	78-83-1	74.121	liq	-101.96(1)	107.84(7)	0.8018^{20}	1.3955^{20}	s H_2O, EtOH, eth, ace, ctc
728	2-Methyl-2-propanol	*tert*-Butyl alcohol	$C_4H_{10}O$	75-65-0	74.121		25.81(4)	82.3(1)	0.7887^{20}	1.3878^{20}	msc H_2O, EtOH, eth; s chl
729	*cis*-(1-Methyl-1-propenyl)benzene		$C_{10}H_{12}$	767-99-7	132.202			175(5)	0.9191^{25}	1.5402^{25}	i H_2O; s bz, chl

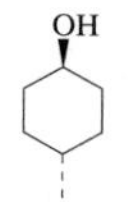
688
trans-4-Methylcyclohexanol

689
Methylcyclopentane

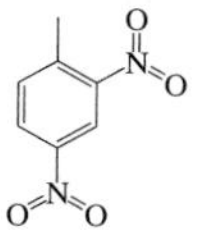
690
1-Methyl-2,4-dinitrobenzene

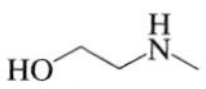
691
N-Methyl-2-ethanolamine

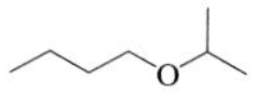
692
1-(1-Methylethoxy)butane

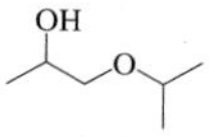
693
1-(1-Methylethoxy)-2-propanol

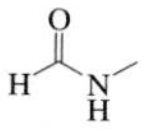
694
N-Methylformamide

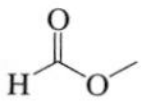
695
Methyl formate

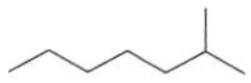
696
2-Methylheptane

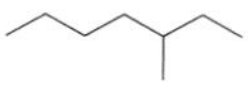
697
3-Methylheptane

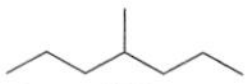
698
4-Methylheptane

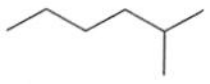
699
2-Methylhexane

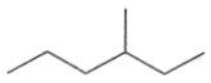
700
3-Methylhexane

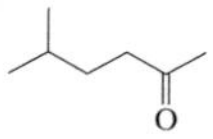
701
5-Methyl-2-hexanone

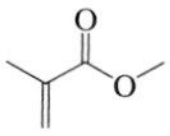
702
Methyl methacrylate

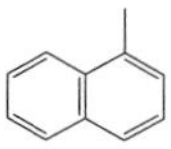
703
1-Methylnaphthalene

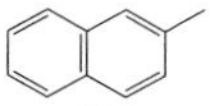
704
2-Methylnaphthalene

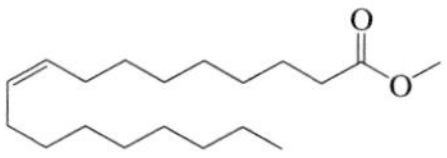
705
Methyl *cis*-9-octadecenoate

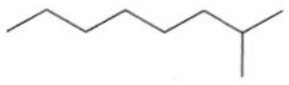
706
2-Methyloctane

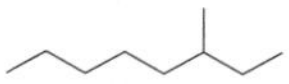
707
3-Methyloctane

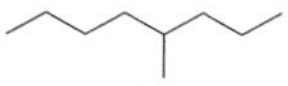
708
4-Methyloctane

709
Methyloxirane

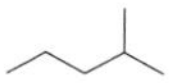
710
2-Methylpentane

711
3-Methylpentane

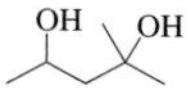
712
2-Methyl-2,4-pentanediol

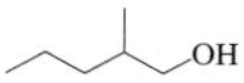
713
2-Methyl-1-pentanol

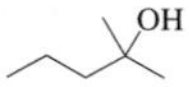
714
2-Methyl-2-pentanol

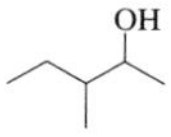
715
3-Methyl-2-pentanol

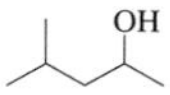
716
4-Methyl-2-pentanol

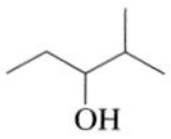
717
2-Methyl-3-pentanol

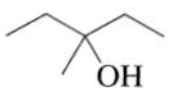
718
3-Methyl-3-pentanol

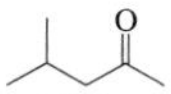
719
4-Methyl-2-pentanone

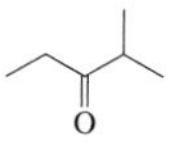
720
2-Methyl-3-pentanone

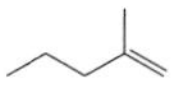
721
2-Methyl-1-pentene

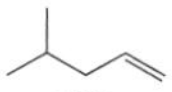
722
4-Methyl-1-pentene

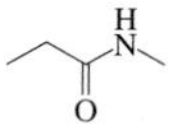
723
N-Methylpropanamide

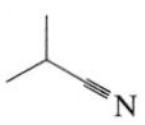
724
2-Methylpropanenitrile

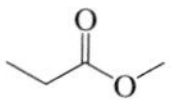
725
Methyl propanoate

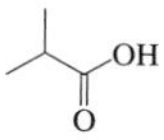
726
2-Methylpropanoic acid

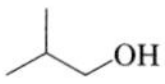
727
2-Methyl-1-propanol

728
2-Methyl-2-propanol

729
cis-(1-Methyl-1-propenyl)benzene

No.	Name	Synonym	Mol. form.	CAS Reg. No.	Mol. wt.	Physical form	mp/°C	bp/°C	den/g cm^{-3}	n_D	Solubility
730	*trans*-(1-Methyl-1-propenyl)benzene		$C_{10}H_{12}$	768-00-3	132.202	liq	-23.5(3)	196(3)	0.9138^{25}	1.5425^{20}	i H_2O; s bz, chl
731	Methyl propyl ether	1-Methoxypropane	$C_4H_{10}O$	557-17-5	74.121			38.5(10)	0.7356^{13}	1.3579^{25}	s H_2O, ace; msc EtOH, eth
732	2-Methylpyridine	2-Picoline	C_6H_7N	109-06-8	93.127	liq	-66.65(3)	129.4(2)	0.9443^{20}	1.4957^{20}	vs H_2O, ace; msc EtOH, eth; s ctc
733	3-Methylpyridine	3-Picoline	C_6H_7N	108-99-6	93.127	liq	-18.1(3)	144.1(1)	0.9566^{20}	1.5040^{20}	msc H_2O, EtOH, eth; vs ace; s ctc
734	4-Methylpyridine	4-Picoline	C_6H_7N	108-89-4	93.127	liq	3.68(2)	145.3(1)	0.9548^{20}	1.5037^{20}	msc H_2O, EtOH, eth; s ace, ctc
735	*N*-Methylpyrrolidine		$C_5H_{11}N$	120-94-5	85.148			206.1(3)	0.8188^{20}	1.4247^{20}	vs H_2O, eth
736	*N*-Methyl-2-pyrrolidinethione		C_5H_9NS	10441-57-3	115.197	oil					
737	*N*-Methyl-2-pyrrolidinone	1-Methyl-2-pyrrolidinone	C_5H_9NO	872-50-4	99.131	liq	-24.0(6)	204.2(3)	1.0230^{25}	1.4684^{20}	vs H_2O; s eth, ace, chl
738	Methyl salicylate	Methyl 2-hydroxybenzoate	$C_8H_8O_3$	119-36-8	152.148	liq	-8.5(5)	222.6(5)	1.181^{25}	1.535^{20}	sl H_2O; vs eth, EtOH, chl
739	2-Methyltetrahydrofuran	2-Methyloxolane	$C_5H_{10}O$	96-47-9	86.132			80(1)	0.8552^{20}	1.4059^{21}	s H_2O; vs EtOH, eth, ace, bz; sl ctc
740	*N*-Methyl-*N*,2,4,6-tetranitroaniline	Nitramine	$C_7H_5N_5O_8$	479-45-8	287.144	ye pr (al)	128(3)	exp	1.57^{10}		i H_2O; sl EtOH, eth, chl; s ace, bz, py
741	(Methylthio)benzene	Methyl phenyl sulfide	C_7H_8S	100-68-5	124.204			194.3(2)	1.0579^{20}	1.5868^{20}	i H_2O; s EtOH; vs ace
742	2-Methylthiophene		C_5H_6S	554-14-3	98.167	liq	-63.3(2)	112.5(4)	1.0193^{20}	1.5203^{20}	i H_2O; msc EtOH, eth, ace, bz, hp, ctc
743	3-Methylthiophene		C_5H_6S	616-44-4	98.167	liq	-68.91(6)	115.4(4)	1.0218^{20}	1.5204^{20}	i H_2O; msc EtOH, eth, ace, bz; vs chl
744	*N*-Methylurea		$C_2H_6N_2O$	598-50-5	74.081	orth pr (w, al)	101.3(8)	dec	1.2040^{0}		vs H_2O, EtOH; i eth, bz; s CS_2, lig
	Mirex	Hexachloropentadiene dimer	$C_{10}Cl_{12}$	2385-85-5	545.543	cry (bz)	485 dec				vs bz, diox
745	Morpholine	Tetrahydro-1,4-oxazine	C_4H_9NO	110-91-8	87.120	hyg liq	-4.8(2)	128.2(2)	1.0005^{20}	1.4548^{20}	msc H_2O; s EtOH, eth, ace, bz; sl chl
746	β-Myrcene	7-Methyl-3-methylene-1,6-octadiene	$C_{10}H_{16}$	123-35-3	136.234	oil		171(3)	0.8013^{15}	1.4722^{20}	i H_2O; s EtOH, eth, bz, chl, HOAc
747	Naphthacene	2,3-Benzanthracene	$C_{18}H_{12}$	92-24-0	228.288	oran-ye lf (bz, xyl)	354(4)	subl			i H_2O; sl bz; s con sulf
748	Naphthalene		$C_{10}H_8$	91-20-3	128.171	mcl pl (al)	80.22(9)	218.0(1)	1.0253^{20}	1.5898^{25}	i H_2O; s EtOH; vs eth, ace, bz, CS_2
749	1,5-Naphthalene diisocyanate	1,5-Diisocyanatonaphthalene	$C_{12}H_6N_2O_2$	3173-72-6	210.188	cry	127				
750	Neopentane	2,2-Dimethylpropane	C_5H_{12}	463-82-1	72.149	col gas	-16.37(4)	9.50(6)	0.5852^{25} (p>1 atm)	1.3476^{6}	i H_2O; s EtOH, eth, ctc
751	2-Nitroaniline		$C_6H_6N_2O_2$	88-74-4	138.124		71(1)	285(2)	0.9015^{25}		sl H_2O; s EtOH; vs eth, ace, bz, chl
752	3-Nitroaniline	*m*-Nitroaniline	$C_6H_6N_2O_2$	99-09-2	138.124		112(2)	312(3)	0.9011^{25}		sl H_2O, bz; s EtOH, eth, ace; vs MeOH
753	4-Nitroaniline		$C_6H_6N_2O_2$	100-01-6	138.124	pa ye mcl nd (w)	147.7(5)	328(8)	1.424^{20}		i H_2O; s EtOH, eth, ace; sl bz, DMSO
754	2-Nitroanisole	1-Methoxy-2-nitrobenzene	$C_7H_7NO_3$	91-23-6	153.136	oily liq	9.4(9)	272(7)	1.2540^{20}	1.5161^{20}	i H_2O; msc EtOH, eth; s ctc
755	Nitrobenzene		$C_6H_5NO_2$	98-95-3	123.110	oily liq	5.65(7)	210.7(3)	1.2037^{20}	1.5562^{20}	sl H_2O, ctc; vs EtOH, eth, ace, bz
756	Nitroethane		$C_2H_5NO_2$	79-24-3	75.067	liq	-89.42(7)	114.1(2)	1.0448^{25}	1.3917^{20}	sl H_2O; msc EtOH, eth; s ace, chl
757	Nitromethane	Nitrocarbol	CH_3NO_2	75-52-5	61.041	liq	-28.7(8)	101.19(10)	1.1371^{20}	1.3817^{20}	s H_2O, EtOH, eth, ace, ctc, alk
758	1-Nitropropane		$C_3H_7NO_2$	108-03-2	89.094	liq	-104.3(6)	131.2(5)	0.9961^{25}	1.4018^{20}	sl H_2O; msc EtOH, eth; s chl
759	2-Nitropropane	Isonitropropane	$C_3H_7NO_2$	79-46-9	89.094	liq	-91.3(2)	120.2(2)	0.9821^{25}	1.3944^{20}	sl H_2O; s chl
760	*N*-Nitrosodiethanolamine	2,2'-(Nitrosoimino)ethanol	$C_4H_{10}N_2O_3$	1116-54-7	134.133	wh-ye oil				1.4849^{20}	
761	*N*-Nitrosodimethylamine	Dimethylnitrosamine	$C_2H_6N_2O$	62-75-9	74.081	ye liq		146(2)	1.0048^{20}	1.4368^{20}	vs H_2O, EtOH, eth; s chl
762	2-Nitrotoluene	1-Methyl-2-nitrobenzene	$C_7H_7NO_2$	88-72-2	137.137	liq	-3.6(8)	220.9(7)	1.1611^{19}	1.5450^{20}	i H_2O; msc EtOH, eth; s ctc
763	3-Nitrotoluene	1-Methyl-3-nitrobenzene	$C_7H_7NO_2$	99-08-1	137.137	pa ye	15.9(6)	232.1(4)	1.1581^{20}	1.5466^{20}	i H_2O; s EtOH, bz, ctc; msc eth
764	4-Nitrotoluene	1-Methyl-4-nitrobenzene	$C_7H_7NO_2$	99-99-0	137.137	orth cry (al, eth)	51.7(3)	238.66(9)	1.1038^{75}		i H_2O; s EtOH; vs eth, ace, bz, chl
765	Nonane		C_9H_{20}	111-84-2	128.255	liq	-53.47(3)	150.8(2)	0.7179^{20}	1.4058^{20}	i H_2O; vs EtOH, eth; msc ace, bz, hp
766	Nonanedioic acid	Azelaic acid	$C_9H_{16}O_4$	123-99-9	188.221	lf or nd	106.5	357.1	1.225^{25}	1.4303^{111}	sl H_2O, eth, bz, DMSO; s EtOH
767	Nonanoic acid	Pelargonic acid	$C_9H_{18}O_2$	112-05-0	158.238	liq	12.38(4)	256(1)	0.9052^{20}	1.4343^{19}	i H_2O; s EtOH, eth, chl
768	1-Nonanol	Nonyl alcohol	$C_9H_{20}O$	143-08-8	144.254	liq	-5.0(2)	213.7(4)	0.8280^{20}	1.4333^{20}	i H_2O; s EtOH, eth; sl ctc
769	2-Nonanone	Heptyl methyl ketone	$C_9H_{18}O$	821-55-6	142.238	liq	-7.4(2)	194(1)	0.8208^{20}	1.4210^{20}	i H_2O; s EtOH, eth, bz; vs ace, chl

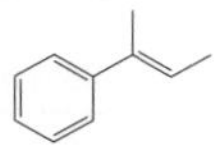
730
trans-(1-Methyl-1-propenyl)benzene

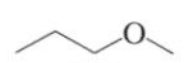
731
Methyl propyl ether

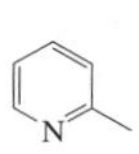
732
2-Methylpyridine

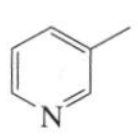
733
3-Methylpyridine

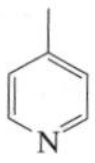
734
4-Methylpyridine

735
N-Methylpyrrolidine

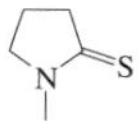
736
N-Methyl-2-pyrrolidinethione

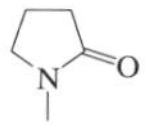
737
N-Methyl-2-pyrrolidinone

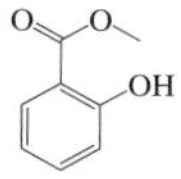
738
Methyl salicylate

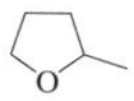
739
2-Methyltetrahydrofuran

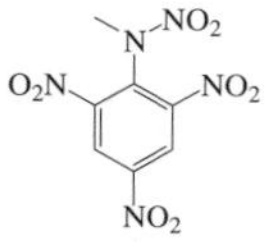
740
N-Methyl-*N*,2,4,6-tetranitroaniline

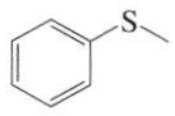
741
(Methylthio)benzene

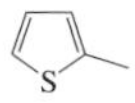
742
2-Methylthiophene

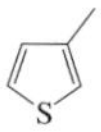
743
3-Methylthiophene

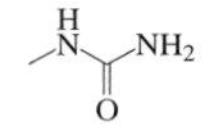
744
N-Methylurea

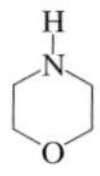
745
Morpholine

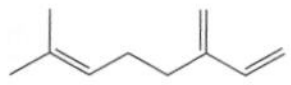
746
β-Myrcene

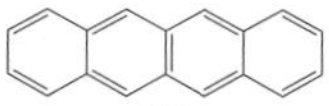
747
Naphthacene

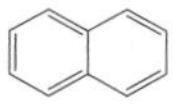
748
Naphthalene

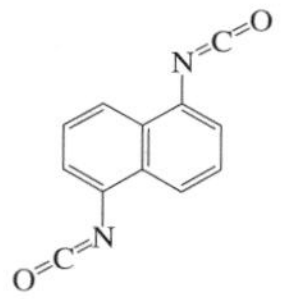
749
1,5-Naphthalene diisocyanate

750
Neopentane

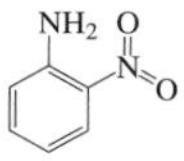
751
2-Nitroaniline

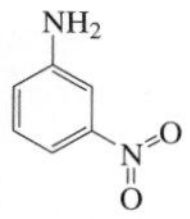
752
3-Nitroaniline

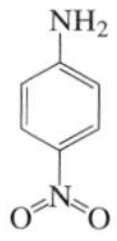
753
4-Nitroaniline

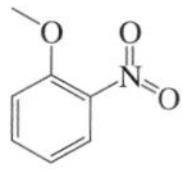
754
2-Nitroanisole

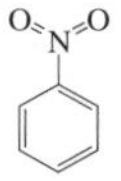
755
Nitrobenzene

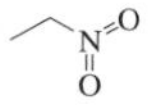
756
Nitroethane

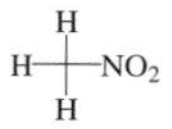
757
Nitromethane

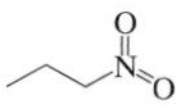
758
1-Nitropropane

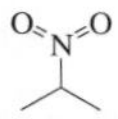
759
2-Nitropropane

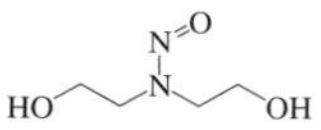
760
N-Nitrosodiethanolamine

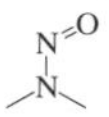
761
N-Nitrosodimethylamine

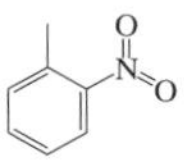
762
2-Nitrotoluene

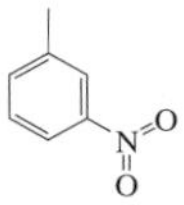
763
3-Nitrotoluene

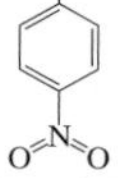
764
4-Nitrotoluene

765
Nonane

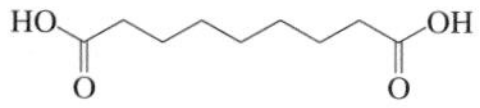
766
Nonanedioic acid

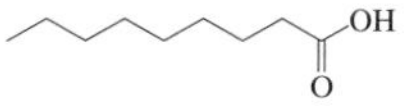
767
Nonanoic acid

768
1-Nonanol

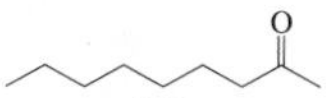
769
2-Nonanone

Organic

No.	Name	Synonym	Mol. form.	CAS Reg. No.	Mol. wt.	Physical form	mp/°C	bp/°C	den/g cm^{-3}	n_D	Solubility
770	3-Nonanone	Ethyl hexyl ketone	$C_9H_{18}O$	925-78-0	142.238	liq	-8(4)	187(4)	0.8241^{20}	1.4208^{20}	i H_2O; s EtOH, eth, bz, chl; vs ace
771	5-Nonanone	Dibutyl ketone	$C_9H_{18}O$	502-56-7	142.238	liq	-3.84(5)	188.4(3)	0.8217^{20}	1.4195^{20}	i H_2O; s EtOH; vs eth, chl
772	1-Nonene	1-Nonylene	C_9H_{18}	124-11-8	126.239	liq	-81.24(4)	146.9(6)	0.7253^{25}	1.4257^{20}	
773	Nonyl formate		$C_{10}H_{20}O_2$	5451-92-3	172.265	liq	-33	216(13)	0.86	1.4216^{20}	
774	1-Nonyne	Heptylacetylene	C_9H_{16}	3452-09-3	124.223	liq	-50	150.8(4)	0.7658^{20}	1.4217^{20}	i H_2O; s eth, bz, ctc
775	Octacosane		$C_{28}H_{58}$	630-02-4	394.761	mcl or orth (bz-al)	61.3(1)	432(6)	0.8067^{20}	1.4330^{70}	i H_2O; msc ace; s bz, chl
776	*cis,cis*-9,12-Octadecadienoic acid	Linoleic acid	$C_{18}H_{32}O_2$	60-33-3	280.446	col liq	-6.9(7)		0.9022^{20}	1.4699^{20}	vs ace, bz, eth, EtOH
777	Octadecane		$C_{18}H_{38}$	593-45-3	254.495	nd (al, eth-MeOH)	28.17(5)	316(2)	0.7768^{28}	1.4390^{20}	i H_2O; sl EtOH; s eth, ace, chl, lig
778	Octadecanoic acid	Stearic acid	$C_{18}H_{36}O_2$	57-11-4	284.478	mcl lf (al)	69.3(2)	371(3)	0.9408^{20}	1.4299^{80}	i H_2O; sl EtOH, bz; s ace, chl, CS_2
779	1-Octadecanol	Stearyl alcohol	$C_{18}H_{38}O$	112-92-5	270.494	lf (al)	58.0(2)	351(2)	0.8124^{59}		i H_2O; s EtOH, eth; sl ace, bz
780	*cis,cis,cis*-9,12,15-Octadecatrienoic acid	α-Linolenic acid	$C_{18}H_{30}O_2$	463-40-1	278.430		-10(2)		0.9164^{20}	1.4800^{20}	i H_2O; s EtOH, eth; sl bz
781	*cis*-9-Octadecenoic acid	Oleic acid	$C_{18}H_{34}O_2$	112-80-1	282.462	liq	14(1)	360	0.8935^{20}	1.4582^{20}	i H_2O; msc EtOH, eth, ace, bz, chl, ctc
782	*cis*-9-Octadecen-1-ol	Oleyl alcohol	$C_{18}H_{36}O$	143-28-2	268.478		0(3)		0.8489^{20}	1.4606^{20}	i H_2O; s EtOH, eth; sl ctc
783	Octane		C_8H_{18}	111-65-9	114.229	liq	-56.73(2)	125.62(10)	0.7022^{20}	1.3944^{25}	i H_2O; s eth; msc EtOH, ace, bz
784	Octanedioic acid	Suberic acid	$C_8H_{14}O_4$	505-48-6	174.195	lo nd or pl (w)	142.3(3)	345.5			i H_2O; msc eth, bz; sl DMSO
785	Octanoic acid	Caprylic acid	$C_8H_{16}O_2$	124-07-2	144.212		16.51(2)	240(1)	0.9106^{20}	1.4285^{20}	sl H_2O; msc EtOH, chl, CH_3CN
786	1-Octanol	Capryl alcohol	$C_8H_{18}O$	111-87-5	130.228	liq	-14.7(4)	194.7(8)	0.8262^{25}	1.4295^{20}	i H_2O; msc EtOH, eth; s ctc
787	2-Octanol	(±)-*sec*-Caprylic alcohol	$C_8H_{18}O$	4128-31-8	130.228	liq	-31.6	179(7)	0.8193^{20}	1.4203^{20}	sl H_2O; s EtOH, eth, ace
788	2-Octanone	Hexyl methyl ketone	$C_8H_{16}O$	111-13-7	128.212	liq	-20.31(8)	173(3)	0.820^{20}	1.4151^{20}	sl H_2O; msc EtOH, eth
789	3-Octanone	Ethyl pentyl ketone	$C_8H_{16}O$	106-68-3	128.212			166(4)	0.822^{25}	1.4150^{20}	i H_2O; msc EtOH, eth
790	1-Octene	Caprylene	C_8H_{16}	111-66-0	112.213	liq	-101.66(5)	121.3(2)	0.7149^{20}	1.4087^{20}	i H_2O; msc EtOH; s eth, ace; sl ctc
791	*cis*-2-Octene		C_8H_{16}	7642-04-8	112.213	liq	-101.3(8)	125.6(5)	0.7243^{20}	1.4150^{20}	i H_2O; s EtOH, eth, ace, bz, chl
792	*trans*-2-Octene		C_8H_{16}	13389-42-9	112.213	liq	-88(2)	124.9(5)	0.7199^{20}	1.4132^{20}	i H_2O; s EtOH, eth, ace, bz; vs chl
793	*cis*-3-Octene		C_8H_{16}	14850-22-7	112.213	liq	-126(1)	123(2)	0.7159^{20}	1.4135^{20}	vs ace, bz, eth, EtOH
794	*trans*-3-Octene		C_8H_{16}	14919-01-8	112.213	liq	-110.0(3)	123.2(6)	0.7152^{20}	1.4126^{20}	i H_2O; s EtOH, eth, ace, bz, lig, ctc
795	*cis*-4-Octene		C_8H_{16}	7642-15-1	112.213	liq	-119.1(4)	122.6(4)	0.7212^{20}	1.4148^{20}	vs ace, bz, eth, EtOH
796	*trans*-4-Octene		C_8H_{16}	14850-23-8	112.213	liq	-93.8(1)	122.4(5)	0.7141^{20}	1.4114^{20}	i H_2O; s EtOH, eth, ace, bz, lig; sl ctc
797	Octyl acetate		$C_{10}H_{20}O_2$	112-14-1	172.265	liq	-38(2)	210(3)	0.8705^{20}	1.4150^{20}	i H_2O; s EtOH, eth; sl ctc
798	1-Octyne	Hexylacetylene	C_8H_{14}	629-05-0	110.197	liq	-79.4(2)	126.2(3)	0.7461^{20}	1.4159^{20}	i H_2O; s EtOH, eth
799	Oxalic acid		$C_2H_2O_4$	144-62-7	90.035	orth pym or oct	189.5 dec	157 subl	1.900^{17}		s H_2O; vs EtOH; sl eth; i bz, chl, peth
800	4-Oxopentanoic acid	Levulinic acid	$C_5H_8O_3$	123-76-2	116.116	lf or pl	33.0(7)	245 dec	1.1335^{20}	1.4396^{20}	vs H_2O, EtOH, eth; s chl
801	Paraldehyde	2,4,6-Trimethyl-1,3,5-trioxane	$C_6H_{12}O_3$	123-63-7	132.157	liq	12(1)	124(2)	0.9943^{20}	1.4049^{20}	sl H_2O; msc EtOH, eth, chl
	2,3,4,5,6-Pentachloroanisole	Methyl pentachlorophenyl ether	$C_7H_3Cl_5O$	1825-21-4	280.363	nd MeOH	108.5				
	Pentachlorobenzene		C_6HCl_5	608-93-5	250.337	nd (al)	84.2(6)	279(4)	1.8342^{16}		i H_2O, EtOH; sl eth, bz, chl, CS_2
802	Pentachloroethane	Refrigerant 120	C_2HCl_5	76-01-7	202.294	liq	-29.0(3)	161(4)	1.6796^{20}	1.5025^{20}	i H_2O; msc EtOH, eth
	Pentachlorophenol		C_6HCl_5O	87-86-5	266.336	mcl pr (al + 1w) nd (bz)	189.5(4)	310 dec	1.978^{22}		i H_2O; sl lig; vs EtOH, eth; s bz
	Pentadecafluorooctanoic acid	Perfluorooctanoic acid	$C_8HF_{15}O_2$	335-67-1	414.069		54.3	192(1)			
803	Pentadecanoic acid	Pentadecyclic acid	$C_{15}H_{30}O_2$	1002-84-2	242.398	pl (dil al, HOAc) cry (peth)	52.52(4)	339(4)	0.8423^{80}	1.4254^{80}	i H_2O; vs EtOH, ace; s eth; sl tfa
804	1-Pentadecanol		$C_{15}H_{32}O$	629-76-5	228.414		43.8(2)	318(2)	0.8347^{25}		i H_2O
805	1,2-Pentadiene	Ethylallene	C_5H_8	591-95-7	68.118	liq	-137.27(8)	44(2)	0.6926^{20}	1.4209^{20}	msc EtOH, eth, ace, bz, ctc, hp
806	*cis*-1,3-Pentadiene	*cis*-Piperylene	C_5H_8	1574-41-0	68.118	liq	-140.81(9)	44.0(7)	0.6910^{20}	1.4363^{20}	msc EtOH, eth, ace, bz, ctc, hp
807	*trans*-1,3-Pentadiene	*trans*-Piperylene	C_5H_8	2004-70-8	68.118	liq	-87.5(2)	42.0(3)	0.6710^{25}	1.4301^{20}	
808	1,4-Pentadiene		C_5H_8	591-93-5	68.118	vol liq or gas	-148.3(3)	25.9(3)	0.6608^{20}	1.3888^{20}	i H_2O; vs EtOH, eth, ace, bz
809	Pentaerythritol		$C_5H_{12}O_4$	115-77-5	136.147	cry (dil HCl)	258	subl		1.548	s H_2O; i eth, bz
810	Pentaerythritol tetranitrate		$C_5H_8N_4O_{12}$	78-11-5	316.138	tetr (ace) pr (ace-al)	140.9(8)		1.773^{20}		sl H_2O, EtOH, eth; vs ace; s bz, py
811	1,1,1,2,2-Pentafluoropropane	HFC-245cb	$C_3H_3F_5$	1814-88-6	134.048	col gas		-18.0(3)			
812	Pentanal	Valeraldehyde	$C_5H_{10}O$	110-62-3	86.132	liq	-81.5(2)	103(2)	0.8095^{20}	1.3944^{20}	sl H_2O; s EtOH, eth

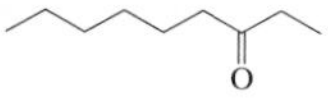
770
3-Nonanone

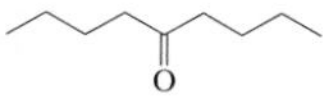
771
5-Nonanone

772
1-Nonene

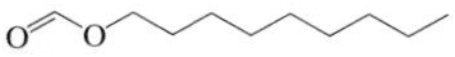
773
Nonyl formate

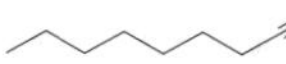
774
1-Nonyne

775
Octacosane

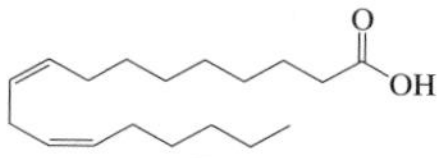
776
cis,cis-9,12-Octadecadienoic acid

777
Octadecane

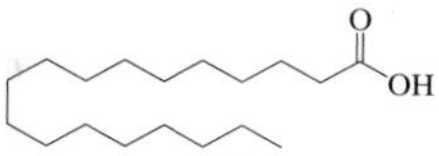
778
Octadecanoic acid

779
1-Octadecanol

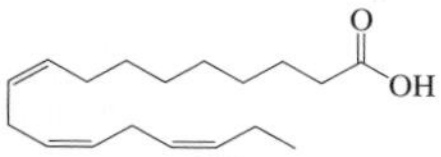
780
cis,cis,cis-9,12,15-Octadecatrienoic acid

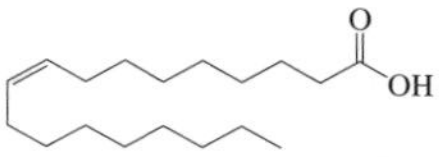
781
cis-9-Octadecenoic acid

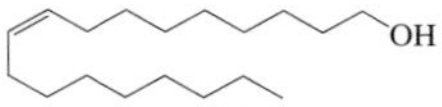
782
cis-9-Octadecen-1-ol

783
Octane

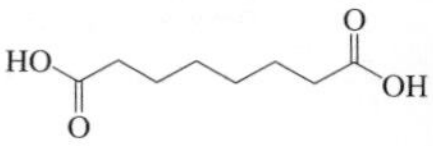
784
Octanedioic acid

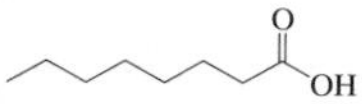
785
Octanoic acid

786
1-Octanol

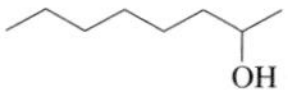
787
2-Octanol

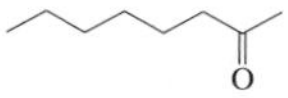
788
2-Octanone

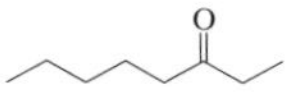
789
3-Octanone

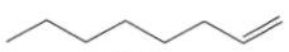
790
1-Octene

791
cis-2-Octene

792
trans-2-Octene

793
cis-3-Octene

794
trans-3-Octene

795
cis-4-Octene

796
trans-4-Octene

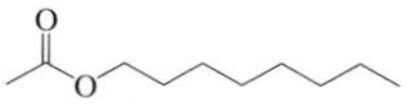
797
Octyl acetate

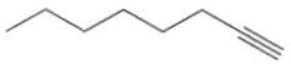
798
1-Octyne

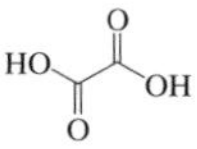
799
Oxalic acid

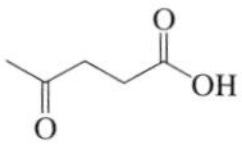
800
4-Oxopentanoic acid

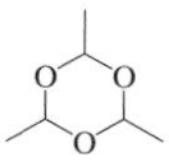
801
Paraldehyde

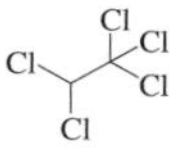
802
Pentachloroethane

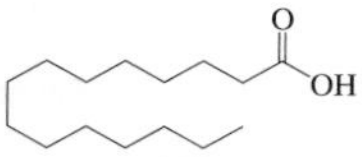
803
Pentadecanoic acid

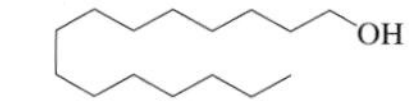
804
1-Pentadecanol

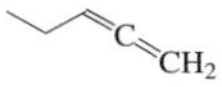
805
1,2-Pentadiene

806
cis-1,3-Pentadiene

807
trans-1,3-Pentadiene

808
1,4-Pentadiene

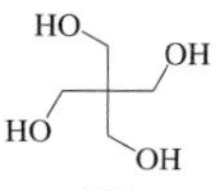
809
Pentaerythritol

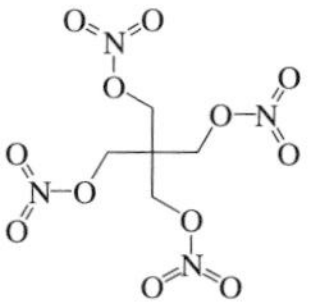
810
Pentaerythritol tetranitrate

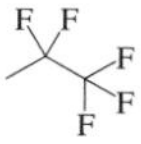
811
1,1,1,2,2-Pentafluoropropane

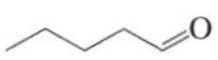
812
Pentanal

No.	Name	Synonym	Mol. form.	CAS Reg. No.	Mol. wt.	Physical form	mp/°C	bp/°C	den/g cm^{-3}	n_D	Solubility
813	Pentane		C_5H_{12}	109-66-0	72.149	liq	-129.67(4)	36.06(7)	0.6262^{20}	1.3575^{20}	sl H_2O; msc EtOH, eth, ace, bz, chl; s ctc
814	Pentanedial	Glutaraldehyde	$C_5H_8O_2$	111-30-8	100.117	oil	-14	176(8)		1.4330^{25}	msc H_2O, EtOH; s bz
815	1,5-Pentanediamine	Cadaverine	$C_5H_{14}N_2$	462-94-2	102.178		11.8(4)	178(6)	0.873^{25}	1.463^{20}	s H_2O, EtOH; sl eth
816	Pentanedioic acid	Glutaric acid	$C_5H_8O_4$	110-94-1	132.116	nd (bz)	97.9(3)	273(10)	1.429^{15}	1.4188^{106}	vs H_2O, EtOH, eth; i bz; s chl, lig
817	1,4-Pentanediol		$C_5H_{12}O_2$	626-95-9	104.148			202	0.9883^{20}	1.4452^{23}	vs H_2O, EtOH, chl
818	1,5-Pentanediol	Pentamethylene glycol	$C_5H_{12}O_2$	111-29-5	104.148	liq	-20(7)	241(2)	0.9914^{20}	1.4494^{20}	s H_2O, EtOH; sl eth, bz
819	2,4-Pentanedione	Acetylacetone	$C_5H_8O_2$	123-54-6	100.117	liq	-18.3(2)	140.7(5)	0.9721^{25}	1.4494^{20}	vs H_2O; msc EtOH, eth, ace, chl
820	Pentanenitrile	Valeronitrile	C_5H_9N	110-59-8	83.132	liq	-96.2(5)	140(1)	0.8008^{20}	1.3971^{20}	s eth, ace, bz; sl ctc
821	Pentanoic acid	Valeric acid	$C_5H_{10}O_2$	109-52-4	102.132	liq	-33.63(5)	186.1(3)	0.9389^{20}	1.4085^{20}	s H_2O, EtOH, eth; sl ctc
822	1-Pentanol	Amyl alcohol	$C_5H_{12}O$	71-41-0	88.148	liq	-77.58(4)	137.6(4)	0.8144^{20}	1.4101^{20}	sl H_2O; msc EtOH, eth; s ace, chl
823	2-Pentanol	*sec*-Amyl alcohol	$C_5H_{12}O$	6032-29-7	88.148	liq	-73	119.1(5)	0.8094^{20}	1.4053^{20}	sl H_2O; s EtOH, eth, ctc, chl
824	3-Pentanol	Diethyl carbinol	$C_5H_{12}O$	584-02-1	88.148	liq	-69.9(4)	123(2)	0.8203^{20}	1.4104^{20}	sl H_2O; s EtOH, eth, ace, ctc
825	2-Pentanone	Methyl propyl ketone	$C_5H_{10}O$	107-87-9	86.132	liq	-76.83(5)	102.2(1)	0.809^{20}	1.3895^{20}	sl H_2O, ctc; msc EtOH, eth
826	3-Pentanone	Diethyl ketone	$C_5H_{10}O$	96-22-0	86.132	liq	-38.98(5)	101.9(1)	0.8098^{25}	1.3905^{25}	s H_2O, ctc; msc EtOH, eth
827	1-Pentene	α-Amylene	C_5H_{10}	109-67-1	70.133	vol liq or gas	-165.13(1)	30.0(3)	0.6405^{20}	1.3715^{20}	i H_2O; msc EtOH, eth; s bz; sl ctc
828	*cis*-2-Pentene	*cis*-β-Amylene	C_5H_{10}	627-20-3	70.133	liq	-151.35(2)	36.9(2)	0.6556^{20}	1.3830^{20}	i H_2O; msc EtOH, eth; s bz, dil sulf
829	*trans*-2-Pentene	*trans*-β-Amylene	C_5H_{10}	646-04-8	70.133	liq	-140.20(2)	36.3(6)	0.6431^{25}	1.3793^{20}	i H_2O; msc EtOH, eth; s bz; vs dil sulf
830	Pentyl acetate	Amyl acetate	$C_7H_{14}O_2$	628-63-7	130.185	liq	-70.9(7)	149.4(3)	0.8756^{20}	1.4023^{20}	sl H_2O; msc EtOH, eth; s ctc
831	Pentylamine	Amylamine	$C_5H_{13}N$	110-58-7	87.164	liq	-51(15)	104.7(2)	0.7544^{20}	1.448^{20}	msc H_2O, EtOH, eth; vs ace, bz; sl chl
832	Pentylbenzene	Amylbenzene	$C_{11}H_{16}$	538-68-1	148.245	liq	-78.2(4)	203(3)	0.8585^{20}	1.4878^{20}	i H_2O; msc EtOH, eth, ace, bz, peth, ctc
833	1-Pentyne	Propylacetylene	C_5H_8	627-19-0	68.118	liq	-106.2(5)	39.9(10)	0.6901^{20}	1.3852^{20}	i H_2O; vs EtOH; msc eth; s bz, chl; sl ctc
834	Perfluorobutane	Decafluorobutane	C_4F_{10}	355-25-9	238.027	col gas	-129(1)	-2.1(8)	1.6484^{25}		s bz, chl
835	Perfluorocyclobutane	Octafluorocyclobutane	C_4F_8	115-25-3	200.030	col gas	-40.16(22)	-5.91	1.500^{25} (p>1 atm)		i H_2O; s eth
836	Perfluorodecane		$C_{10}F_{22}$	307-45-9	538.072			135(3)			i H_2O
837	Perfluoroheptane		C_7F_{16}	335-57-9	388.049	liq	-51.3(1)	82.5(2)	1.7333^{20}	1.2618^{20}	i H_2O; vs ace, eth, EtOH, chl
838	Perfluorohexane	PFC-5-1-14	C_6F_{14}	355-42-0	338.042	liq	-86.1(3)	57.2(2)	1.6910^{20}	1.2515^{20}	i H_2O; s eth, bz, chl
839	Perfluorooctane		C_8F_{18}	307-34-6	438.057			105(2)	1.73^{20}	1.282^{20}	i H_2O
840	Perfluoropentane	PFC-4-1-12	C_5F_{12}	678-26-2	288.035	vol liq or gas	-10	29.2(2)			i H_2O
841	Perfluoropropane	Octafluoropropane	C_3F_8	76-19-7	188.019	col gas	-147.7(1)	-36.8(3)			i H_2O
842	Perfluoropropene	Hexafluoropropene	C_3F_6	116-15-4	150.022	col gas	-156.5	-30.2(5)		1.583^{-40}	i H_2O
843	α-Phellandrene	2-Methyl-5-(1-methylethyl)-1,3-cyclohexadiene	$C_{10}H_{16}$	99-83-2	136.234		238	174.9	0.8410^{20}	1.471^{25}	i H_2O; s eth
844	β-Phellandrene	*p*-Mentha-1(7),2-diene	$C_{10}H_{16}$	555-10-2	136.234			177(3)	0.8520^{20}	1.4788^{20}	i H_2O, EtOH; s eth
845	Phenanthrene		$C_{14}H_{10}$	85-01-8	178.229	mcl pl (al), lf (sub)	99(2)	338.4(9)	0.9800^{4}	1.5943	i H_2O; s EtOH, eth, ace, bz, CS_2
846	1,10-Phenanthroline monohydrate	*o*-Phenanthroline monohydrate	$C_{12}H_{10}N_2O$	5144-89-8	198.219	wh cry pow	93				s EtOH, ace; sl bz
847	Phenol	Hydroxybenzene	C_6H_6O	108-95-2	94.111		40.89(1)	181.8(1)	1.0545^{45}	1.5408^{41}	s H_2O, EtOH; vs eth; msc ace, bz
848	Phenolphthalein	3,3-Bis(4-hydroxyphenyl)-1(3*H*)-isobenzofuranone	$C_{20}H_{14}O_4$	77-09-8	318.323	wh orth nd	262(2)		1.277^{32}		i H_2O, bz; vs EtOH, ace; s eth, chl
849	10*H*-Phenothiazine	Thiodiphenylamine	$C_{12}H_9NS$	92-84-2	199.271	ye pr (al) ye lf or pl (tol)	184.9(6)	371			vs ace, bz, eth, EtOH
850	Phenyl acetate		$C_8H_8O_2$	122-79-2	136.149			195(1)	1.0780^{20}	1.5035^{20}	sl H_2O; msc EtOH, eth, chl; s ctc
851	L-Phenylalanine	α-Aminobenzenepropanoic acid, (*S*)-	$C_9H_{11}NO_2$	63-91-2	165.189	pr (w)	283 dec				sl H_2O; i EtOH, eth, bz, acid
852	*N*-Phenyl-1,4-benzenediamine	*p*-Aminodiphenylamine	$C_{12}H_{12}N_2$	101-54-2	184.236	nd(al)	75(1)	354			sl H_2O, chl; vs EtOH; s eth, lig
853	*N*-Phenylformamide	Formanilide	C_7H_7NO	103-70-8	121.137	mcl pr (lig-xyl)	46.5(6)	271(5)	1.1186^{50}		s H_2O, eth, bz; vs EtOH
854	Phenyl salicylate		$C_{13}H_{10}O_3$	118-55-8	214.216		41.82(4)		1.2614^{30}		i H_2O; vs EtOH, ace, bz; s eth, HOAc

813
Pentane

814
Pentanedial

815
1,5-Pentanediamine

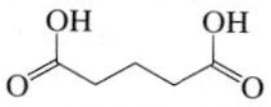

816
Pentanedioic acid

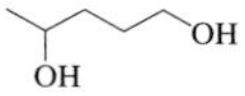

817
1,4-Pentanediol

818
1,5-Pentanediol

819
2,4-Pentanedione

820
Pentanenitrile

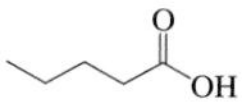

821
Pentanoic acid

822
1-Pentanol

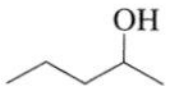

823
2-Pentanol

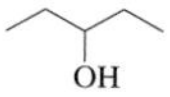

824
3-Pentanol

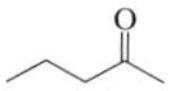

825
2-Pentanone

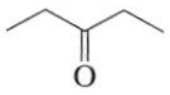

826
3-Pentanone

827
1-Pentene

828
cis-2-Pentene

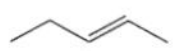

829
trans-2-Pentene

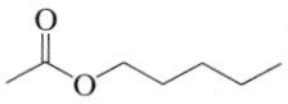

830
Pentyl acetate

831
Pentylamine

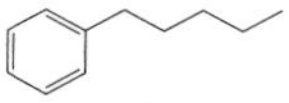

832
Pentylbenzene

833
1-Pentyne

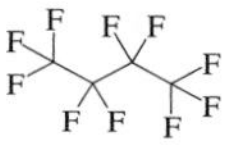

834
Perfluorobutane

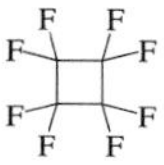

835
Perfluorocyclobutane

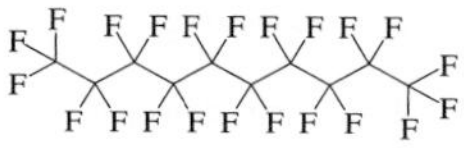

836
Perfluorodecane

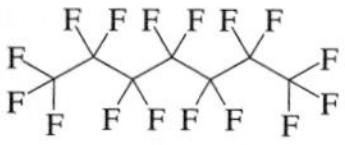

837
Perfluoroheptane

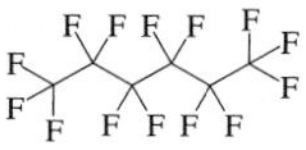

838
Perfluorohexane

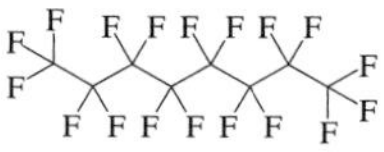

839
Perfluorooctane

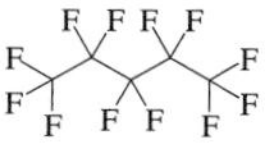

840
Perfluoropentane

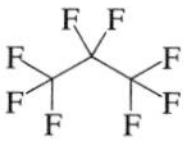

841
Perfluoropropane

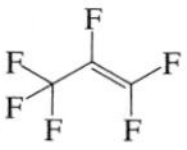

842
Perfluoropropene

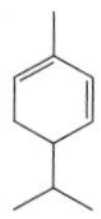

843
α-Phellandrene

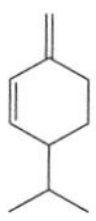

844
β-Phellandrene

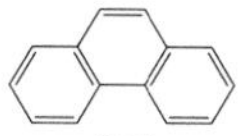

845
Phenanthrene

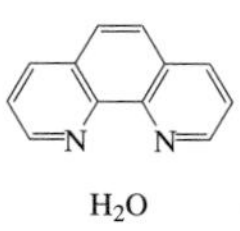

846
1,10-Phenanthroline monohydrate

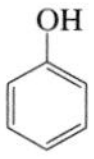

847
Phenol

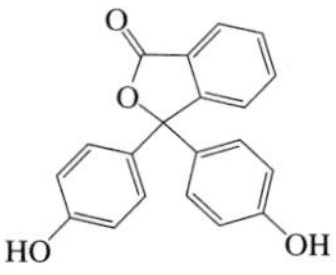

848
Phenolphthalein

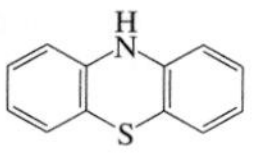

849
10*H*-Phenothiazine

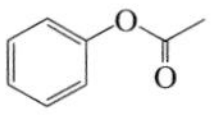

850
Phenyl acetate

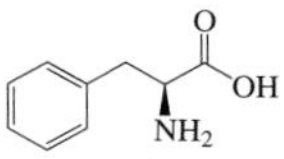

851
L-Phenylalanine

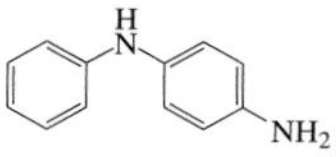

852
N-Phenyl-1,4-benzenediamine

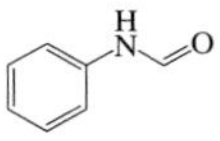

853
N-Phenylformamide

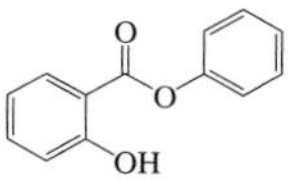

854
Phenyl salicylate

No.	Name	Synonym	Mol. form.	CAS Reg. No.	Mol. wt.	Physical form	mp/°C	bp/°C	den/g cm^{-3}	n_D	Solubility
855	Phthalic acid	1,2-Benzenedicarboxylic acid	$C_8H_6O_4$	88-99-3	166.132	pl (w)	207(3)	dec	2.18^{191}		sl H_2O, eth; i chl; s EtOH
856	Phthalic anhydride		$C_8H_4O_3$	85-44-9	148.116	wh nd (al, bz)	131.4(3)	285.3(8)	1.527^{4}		sl H_2O, eth; s EtOH, ace, bz
857	α-Pinene	2-Pinene	$C_{10}H_{16}$	80-56-8	136.234	liq	-74(4)	156.3(5)	0.8539^{25}	1.4632^{25}	i H_2O; msc EtOH, eth, chl
858	β-Pinene	Nopinene	$C_{10}H_{16}$	127-91-3	136.234	liq	-50.0(2)	165.8(6)	0.860^{25}	1.4768^{25}	i H_2O; s bz, EtOH, eth, chl
859	1,4-Piperazinediethanol		$C_8H_{18}N_2O_2$	122-96-3	174.241		132(2)				
860	1,4-Piperazinedipropan-amine	1,4-Bis(3-aminopropyl) piperazine	$C_{10}H_{24}N_4$	7209-38-3	200.325		15		0.973^{25}	1.5015^{20}	
861	Piperidine	Azacyclohexane	$C_5H_{11}N$	110-89-4	85.148	liq	-11.05(3)	106.19(9)	0.8606^{20}	1.4530^{20}	msc H_2O, EtOH; s eth, ace, bz, chl
862	Propanal	Propionaldehyde	C_3H_6O	123-38-6	58.079	liq	-80	48.0(2)	0.8657^{25}	1.3636^{20}	s H_2O; msc EtOH, eth
863	Propane	LPG	C_3H_8	74-98-6	44.096	col gas	-187.62(5)	-42.11(9)	0.493^{25} (p>1 atm)		s H_2O, EtOH; vs eth, bz; sl ace
864	1,3-Propanediamine	1,3-Diaminopropane	$C_3H_{10}N_2$	109-76-2	74.124	liq	-10.9(3)	139.2(7)	0.884^{25}	1.4600^{20}	s H_2O; msc EtOH, eth
865	1,2-Propanediol	1,2-Propylene glycol	$C_3H_8O_2$	57-55-6	76.095	liq	-60	187.3(2)	1.0361^{20}	1.4324^{20}	msc H_2O, EtOH; s eth, bz, chl
866	1,3-Propanediol	1,3-Propylene glycol	$C_3H_8O_2$	504-63-2	76.095	liq	-27.6(1)	214.7(3)	1.0538^{20}	1.4398^{20}	msc H_2O, EtOH; vs eth; sl bz
867	1,2-Propanediol 1-methacrylate	2-Hydroxypropyl methacrylate	$C_7H_{12}O_3$	923-26-2	144.168				1.066^{25}	1.4458^{20}	
868	1,2-Propanedione	Pyruvaldehyde	$C_3H_4O_2$	78-98-8	72.063	ye hyg liq		72(15)	1.0455^{20}	1.4002^{18}	s EtOH, eth, bz
869	Propanenitrile	Ethyl cyanide	C_3H_5N	107-12-0	55.079	liq	-93(1)	97.3(4)	0.7818^{20}	1.3655^{20}	vs H_2O; s EtOH, eth, ace, bz, ctc
870	Propanoic acid	Propionic acid	$C_3H_6O_2$	79-09-4	74.079	liq	-20.5(5)	141.5(2)	0.9882^{25}	1.3809^{20}	msc H_2O, EtOH; s eth; sl chl
871	Propanoic anhydride	Propionic anhydride	$C_6H_{10}O_3$	123-62-6	130.141	liq	-45.0(5)	168(1)	1.0110^{20}	1.4038^{20}	msc eth; sl ctc
872	1-Propanol	Propyl alcohol	C_3H_8O	71-23-8	60.095	liq	-124.39(2)	97.04(9)	0.8048^{20}	1.3850^{20}	msc H_2O, EtOH, eth; s ace, chl; vs bz
873	2-Propanol	Isopropyl alcohol	C_3H_8O	67-63-0	60.095	liq	-87.91(4)	82.21(9)	0.7855^{20}	1.3776^{20}	msc H_2O, EtOH, eth; s ace, chl; vs bz
874	Propargyl alcohol	3-Hydroxy-1-propyne	C_3H_4O	107-19-7	56.063	liq	-51.8(4)	113(3)	0.9478^{20}	1.4322^{20}	s H_2O, chl; msc EtOH, eth
875	Propene	Propylene	C_3H_6	115-07-1	42.080	col gas	-185.19(2)	-47.6(1)	0.505^{25} (p>1 atm)	1.3567^{-70}	sl H_2O; vs EtOH, HOAc
876	1-Propene-2,3-dicarbox-ylic acid	Itaconic acid	$C_5H_6O_4$	97-65-4	130.100	rhom (bz)	165.7(9)	dec	1.632^{25}		s H_2O, EtOH, ace; sl eth, bz, peth
877	*cis*-1-Propenylbenzene		C_9H_{10}	766-90-5	118.175	liq	-61.7(3)	169(4)	0.9088^{20}	1.5420^{20}	i H_2O; msc EtOH, eth, ace, bz, peth, ctc
878	*trans*-1-Propenylbenzene		C_9H_{10}	873-66-5	118.175	liq	-29.6(7)	179(2)	0.9023^{25}	1.5506^{20}	i H_2O; msc EtOH, eth, ace, bz
879	Propyl acetate		$C_5H_{10}O_2$	109-60-4	102.132	liq	-93(2)	101.0(2)	0.8885^{20}	1.3828^{25}	sl H_2O; msc EtOH, eth; s ctc
880	Propylamine	1-Propanamine	C_3H_9N	107-10-8	59.110	liq	-84.78(4)	47.21(8)	0.7173^{20}	1.3870^{20}	msc H_2O; vs EtOH, ace; s bz, chl; sl ctc
881	Propylbenzene	Isocumene	C_9H_{12}	103-65-1	120.191	liq	-99.52(5)	159.2(5)	0.8619^{20}	1.4895^{25}	i H_2O; msc EtOH, eth, ace, bz, peth, ctc
882	Propyl benzoate	Propyl benzenecarboxylate	$C_{10}H_{12}O_2$	2315-68-6	164.201	liq	-51.6	231(2)	1.0230^{20}	1.5000^{20}	i H_2O; msc EtOH, eth
883	Propyl butanoate		$C_7H_{14}O_2$	105-66-8	130.185	liq	-95.2(5)	144(2)	0.8730^{20}	1.4001^{20}	sl H_2O; msc EtOH, eth
884	Propylcyclopentane		C_8H_{16}	2040-96-2	112.213	liq	-117.34(6)	130.9(8)	0.7763^{20}	1.4266^{20}	i H_2O; msc EtOH, eth, ace; s bz; vs ctc
885	Propylene carbonate	4-Methyl-1,3-dioxolan-2-one	$C_4H_6O_3$	108-32-7	102.089	liq	-48.8	241.6(7)	1.2047^{20}	1.4189^{20}	vs H_2O, EtOH, eth, ace, bz
886	1,2-Propylene glycol 2-*tert*-butyl ether	2-(1,1-Dimethylethoxy)-1-propanol	$C_7H_{16}O_2$	94023-15-1	132.201	liq		152	0.87		
887	1,2-Propylene glycol monomethyl ether acetate	2-Acetoxy-1-methoxypropane	$C_6H_{12}O_3$	108-65-6	132.157	liq		146.0(4)			
888	Propyl formate		$C_4H_8O_2$	110-74-7	88.106	liq	-92.9(4)	80.6(2)	0.9053^{20}	1.377^{20}	sl H_2O, ctc; msc EtOH, eth
889	Propyl propanoate	Propyl propionate	$C_6H_{12}O_2$	106-36-5	116.158	liq	-75.9(4)	122.2(1)	0.8755^{25}	1.3909^{25}	sl H_2O, ctc; msc EtOH, eth; s ace
890	Propyne	Methylacetylene	C_3H_4	74-99-7	40.064	col gas	-103.0(5)	-23.2	0.607^{25} (p>1 atm)	1.3863^{-40}	sl H_2O; vs EtOH; s bz, chl
891	Pyrazine	1,4-Diazine	$C_4H_4N_2$	290-37-9	80.088	pr (w)	52.30(4)	116.3(1)	1.0311^{61}	1.4953^{61}	s H_2O, EtOH, eth, ace; sl ctc
892	1*H*-Pyrazole	1,2-Diazole	$C_3H_4N_2$	288-13-1	68.077	nd or pr (lig)	59.9(3)	187(5)		1.4203	s H_2O, EtOH, eth, bz; sl chl
893	Pyrene	Benzo[*def*]phenanthrene	$C_{16}H_{10}$	129-00-0	202.250	pa ye pl (to, sub)	150.62(4)	394(6)	1.271^{23}		i H_2O; s EtOH, eth, bz, tol; sl ctc

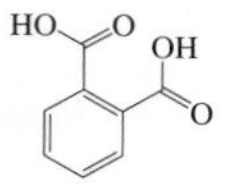
855
Phthalic acid

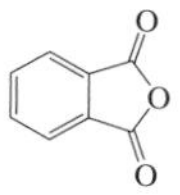
856
Phthalic anhydride

857
α-Pinene

858
β-Pinene

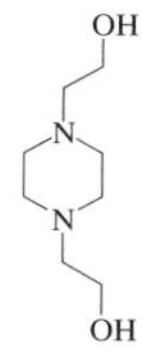
859
1,4-Piperazinediethanol

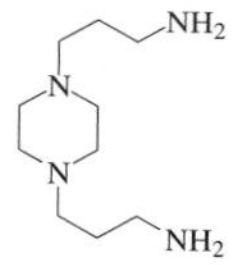
860
1,4-Piperazinedipropanamine

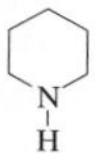
861
Piperidine

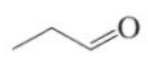
862
Propanal

863
Propane

864
1,3-Propanediamine

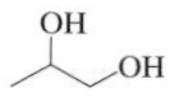
865
1,2-Propanediol

866
1,3-Propanediol

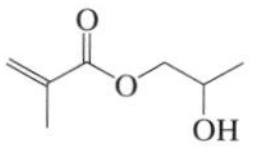
867
1,2-Propanediol 1-methacrylate

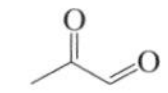
868
1,2-Propanedione

869
Propanenitrile

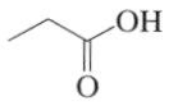
870
Propanoic acid

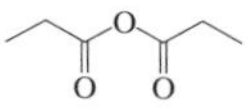
871
Propanoic anhydride

872
1-Propanol

873
2-Propanol

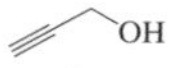
874
Propargyl alcohol

875
Propene

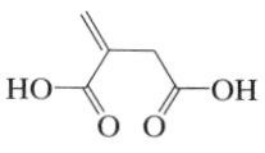
876
1-Propene-2,3-dicarboxylic acid

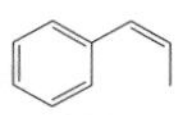
877
cis-1-Propenylbenzene

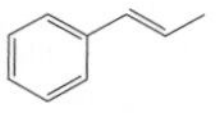
878
trans-1-Propenylbenzene

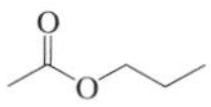
879
Propyl acetate

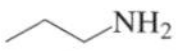
880
Propylamine

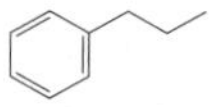
881
Propylbenzene

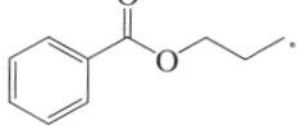
882
Propyl benzoate

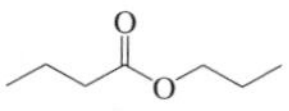
883
Propyl butanoate

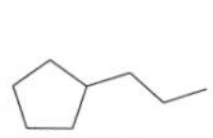
884
Propylcyclopentane

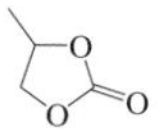
885
Propylene carbonate

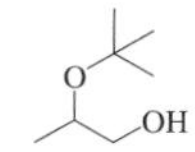
886
1,2-Propylene glycol 2-*tert*-butyl ether

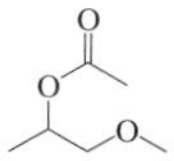
887
1,2-Propylene glycol monomethyl ether acetate

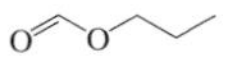
888
Propyl formate

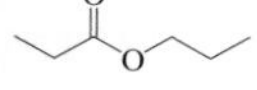
889
Propyl propanoate

890
Propyne

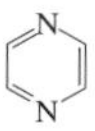
891
Pyrazine

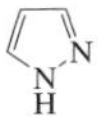
892
1*H*-Pyrazole

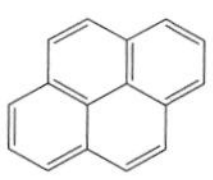
893
Pyrene

No.	Name	Synonym	Mol. form.	CAS Reg. No.	Mol. wt.	Physical form	mp/°C	bp/°C	den/g cm^{-3}	n_D	Solubility
894	Pyridine	Azine	C_5H_5N	110-86-1	79.101	liq	-41.63(3)	115.2(1)	0.9819^{20}	1.5095^{20}	msc H_2O, EtOH, eth, ace, bz, chl
895	3-Pyridinecarboxylic acid	Nicotinic acid	$C_6H_5NO_2$	59-67-6	123.110	nd (al, w)	237(1)	subl	1.473^{25}		sl H_2O, EtOH, eth
896	1-(2-Pyridylazo)-2-naphthol	PAN	$C_{15}H_{11}N_3O$	85-85-8	249.267	red-br cry	130				i H_2O; s EtOH, eth, chl
897	Pyrocatechol	1,2-Benzenediol	$C_6H_6O_2$	120-80-9	110.111	cry	104.6(3)	246(1)	1.344^{20}	1.604^{25}	vs H_2O, bz, eth, EtOH
898	Pyrrole	Imidole	C_4H_5N	109-97-7	67.090	liq	-23.39(2)	129.74(4)	0.9698^{20}	1.5085^{20}	sl H_2O; s EtOH, eth, ace, bz, chl
899	Pyrrolidine	Azacyclopentane	C_4H_9N	123-75-1	71.121	col liq	-57.79(3)	86.6(1)	0.8586^{20}	1.4431^{20}	msc H_2O; s EtOH, eth; sl bz, chl
900	2-Pyrrolidone	γ-Butyrolactam	C_4H_7NO	616-45-5	85.105	cry (peth)	25.92(1)	251.2(1)	1.120^{20}	1.4806^{30}	vs H_2O, EtOH, eth, bz, chl, CS_2
901	Quinoline	1-Azanaphthalene	C_9H_7N	91-22-5	129.159	liq	-14.78(5)	237.1(5)	1.0977^{15}	1.6268^{20}	sl H_2O; msc EtOH, eth, ace, bz, CS_2; s ctc
902	8-Quinolinol	8-Hydroxyquinoline	C_9H_7NO	148-24-3	145.158	nd (dil al)	74(2)	267	1.034^{20}		i H_2O, eth; vs EtOH, bz, chl; s ace
903	Resorcinol	1,3-Benzenediol	$C_6H_6O_2$	108-46-3	110.111	nd (bz), pl (w)	109.8(4)	280(2)	1.278^{20}	1.578^{25}	vs H_2O, ctc; s EtOH, eth; sl bz, chl
904	Saccharin		$C_7H_5NO_3S$	81-07-2	183.185	nd (ace) pr (al), lf (w)	227(2)	subl	0.828^{25}		sl H_2O, bz, eth, chl; s ace, EtOH
905	Salicylaldehyde	2-Hydroxybenzaldehyde	$C_7H_6O_2$	90-02-8	122.122	liq	-7(2)	208(5)	1.1674^{20}	1.5740^{20}	sl H_2O, chl; msc EtOH; vs ace, bz
906	Sodium diethyldithiocarbamate	Dithiocarb sodium	$C_5H_{10}NNaS_2$	148-18-5	171.260	cry (EtOH)	95				s H_2O, EtOH, MeOH, ace; i eth, bz
907	Sodium methanolate	Sodium methoxide	CH_3NaO	124-41-4	54.024	wh hyg tetr cry	300				reac H_2O; s MeOH, EtOH
908	Sodium phenolate	Sodium phenoxide	C_6H_5NaO	139-02-6	116.093	hyg cry	384				vs H_2O; s EtOH, thf
909	*cis*-Stilbene	*cis*-1,2-Diphenylethene	$C_{14}H_{12}$	645-49-8	180.245		-5		1.0143^{20}	1.6130^{20}	i H_2O; s EtOH, eth, ace, bz, peth, chl
910	*trans*-Stilbene	*trans*-1,2-Diphenylethene	$C_{14}H_{12}$	103-30-0	180.245	cry (al)	124.82(2)	307	0.9707^{20}	1.6264^{17}	i H_2O; sl EtOH, chl; vs eth, bz
911	Styrene	Vinylbenzene	C_8H_8	100-42-5	104.150	liq	-30.65(6)	145.3(6)	0.9016^{25}	1.5440^{25}	i H_2O; s EtOH, eth, ace; msc bz; sl ctc
912	Succinic acid	Butanedioic acid	$C_4H_6O_4$	110-15-6	118.089	tcl or mcl pr	185(3)	234(3)	1.572^{25}	1.450	sl H_2O, DMSO; s EtOH, eth, ace; i bz
913	Succinonitrile	Butanedinitrile	$C_4H_4N_2$	110-61-2	80.088		57.985(1)	266	0.9867^{60}	1.4173^{60}	vs H_2O; s EtOH, ace, bz, chl; sl eth
914	Sucrose		$C_{12}H_{22}O_{11}$	57-50-1	342.296	mcl	181(8)		1.5805^{17}	1.5376	s H_2O, py; sl EtOH; i eth
915	Sulfolane	Tetrahydrothiophene, 1-1-dioxide	$C_4H_8O_2S$	126-33-0	120.171		28.45(8)	286(2)	1.2723^{18}	1.4833^{18}	s chl
916	DL-Tartaric acid	2,3-Dihydroxybutanedioic acid, (*R**,*R**)-(±)-	$C_4H_6O_6$	133-37-9	150.087	mcl pr (w, al +1w)	206		1.788^{25}		s H_2O, EtOH; sl eth; i bz
917	L-Tartaric acid	2,3-Dihydroxybutanedioic acid, [*R*-(*R**,*R**)]-	$C_4H_6O_6$	87-69-4	150.087		169				
918	Terephthalic acid	1,4-Benzenedicarboxylic acid	$C_8H_6O_4$	100-21-0	166.132	nd (sub)		300 sp	1.519^{25}		i H_2O, EtOH, eth, chl, HOAc; sl ctc
919	*o*-Terphenyl		$C_{18}H_{14}$	84-15-1	230.304	mcl pr (MeOH)	56.19(5)	337(5)	1.1^{25}		i H_2O; s ace, bz, chl, MeOH
920	*m*-Terphenyl		$C_{18}H_{14}$	92-06-8	230.304	ye nd (al)	86.9(8)	375(1)	1.199^{20}		i H_2O; s EtOH, eth, bz, HOAc; sl chl
921	*p*-Terphenyl		$C_{18}H_{14}$	92-94-4	230.304		213.8(7)	376	1.2^{25}		i H_2O; sl EtOH; s eth, bz, CS_2
922	α-Terpinene	4-Isopropyl-1-methyl-1,3-cyclohexadiene	$C_{10}H_{16}$	99-86-5	136.234	oil		174(4)	0.8375^{19}	1.477^{19}	i H_2O; msc EtOH, eth
923	α-Terpineol		$C_{10}H_{18}O$	2438-12-2	154.249	cry (peth)	35(4)	218(11)	0.9337^{20}	1.4831^{20}	sl H_2O; vs ace, bz, eth, EtOH
924	Terpinolene	*p*-Mentha-1,4(8)-diene	$C_{10}H_{16}$	586-62-9	136.234			187(3)	0.8632^{15}	1.4883^{20}	i H_2O; msc EtOH, eth; s bz, ctc
925	1,1,2,2-Tetrabromoethane	Acetylene tetrabromide	$C_2H_2Br_4$	79-27-6	345.653	ye visc liq	0	248(11)	2.9655^{20}	1.6353^{20}	i H_2O; msc EtOH, eth; s ace, bz; sl ctc
926	1,1,1,2-Tetrachloro-2,2-difluoroethane	Tetrachloro-1,1-difluoroethane	$C_2Cl_4F_2$	76-11-9	203.830		41.0(3)	96(3)	1.649^{25}		i H_2O; s EtOH, eth, chl
927	1,1,2,2-Tetrachloro-1,2-difluoroethane	Tetrachloro-1,2-difluoroethane	$C_2Cl_4F_2$	76-12-0	203.830		26.54(20)	92.83(7)	1.5951^{50}	1.4130^{25}	i H_2O; s EtOH, eth, chl
928	1,1,1,2-Tetrachloroethane		$C_2H_2Cl_4$	630-20-6	167.849	liq	-70.2	130.2(2)	1.5406^{20}	1.4821^{20}	sl H_2O; s ace, bz, chl; msc EtOH, eth
929	1,1,2,2-Tetrachloroethane	Acetylene tetrachloride	$C_2H_2Cl_4$	79-34-5	167.849	liq	-42.4(3)	146.0(3)	1.5953^{20}	1.4940^{20}	sl H_2O; s ace, bz, chl; msc EtOH, eth
930	Tetrachloroethene	Perchloroethylene	C_2Cl_4	127-18-4	165.833	liq	-22.2(1)	121.2(3)	1.6230^{20}	1.5059^{20}	i H_2O; msc EtOH, eth, bz
931	Tetrachloromethane	Carbon tetrachloride	CCl_4	56-23-5	153.823	liq	-22.8(3)	76.7(2)	1.5940^{20}	1.4601^{20}	i H_2O; s EtOH, ace; msc eth, bz, chl
932	Tetracosane		$C_{24}H_{50}$	646-31-1	338.654	cry (eth)	50.3(7)	391(5)	0.7991^{20}	1.4283^{70}	i H_2O; sl EtOH; vs eth

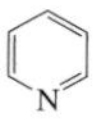
894
Pyridine

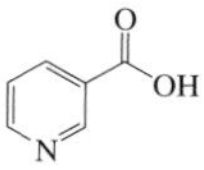
895
3-Pyridinecarboxylic acid

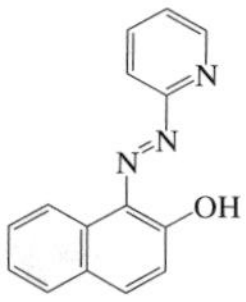
896
1-(2-Pyridylazo)-2-naphthol

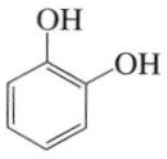
897
Pyrocatechol

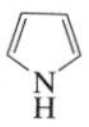
898
Pyrrole

899
Pyrrolidine

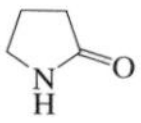
900
2-Pyrrolidone

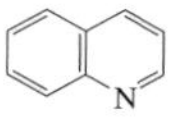
901
Quinoline

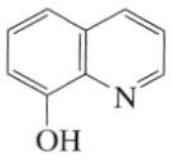
902
8-Quinolinol

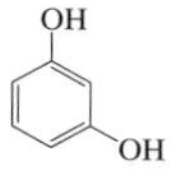
903
Resorcinol

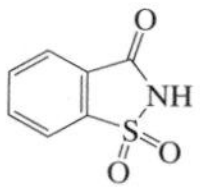
904
Saccharin

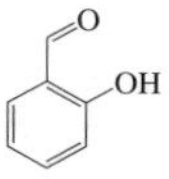
905
Salicylaldehyde

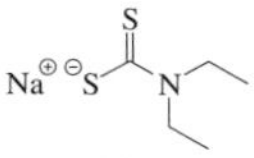
906
Sodium diethyldithiocarbamate

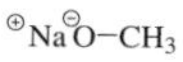
907
Sodium methanolate

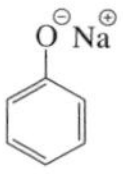
908
Sodium phenolate

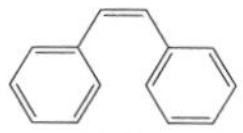
909
cis-Stilbene

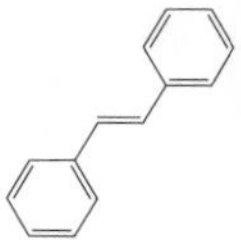
910
trans-Stilbene

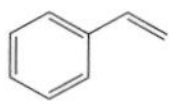
911
Styrene

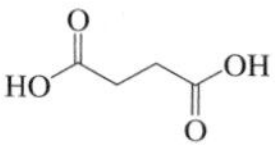
912
Succinic acid

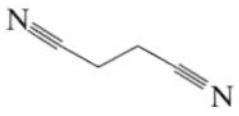
913
Succinonitrile

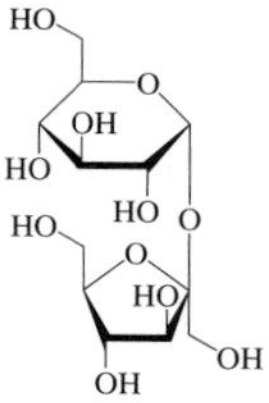
914
Sucrose

915
Sulfolane

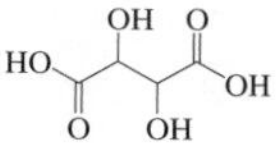
916
dl-Tartaric acid

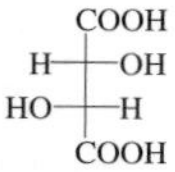
917
L-Tartaric acid

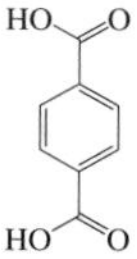
918
Terephthalic acid

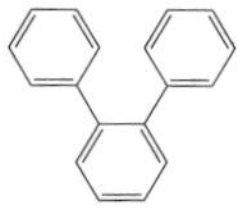
919
o-Terphenyl

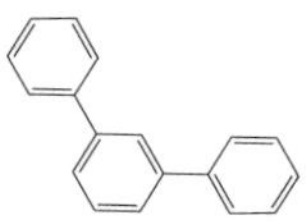
920
m-Terphenyl

921
p-Terphenyl

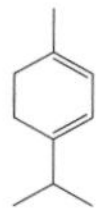
922
α-Terpinene

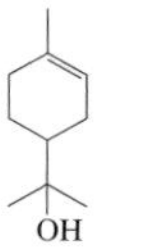
923
α-Terpineol

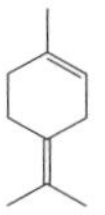
924
Terpinolene

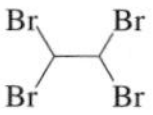
925
1,1,2,2-Tetrabromoethane

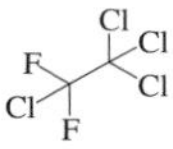
926
1,1,1,2-Tetrachloro-2,2-difluoroethane

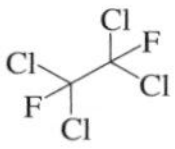
927
1,1,2,2-Tetrachloro-1,2-difluoroethane

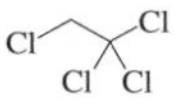
928
1,1,1,2-Tetrachloroethane

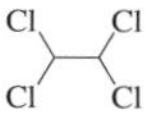
929
1,1,2,2-Tetrachloroethane

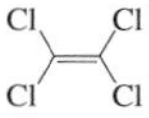
930
Tetrachloroethene

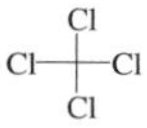
931
Tetrachloromethane

932
Tetracosane

No.	Name	Synonym	Mol. form.	CAS Reg. No.	Mol. wt.	Physical form	mp/°C	bp/°C	den/g cm^{-3}	n_D	Solubility
933	Tetradecane		$C_{14}H_{30}$	629-59-4	198.388	col liq	5.87(2)	253.5(4)	0.7596^{20}	1.4290^{20}	i H_2O; vs EtOH, eth; s ctc
934	Tetradecanoic acid	Myristic acid	$C_{14}H_{28}O_2$	544-63-8	228.371	lf (eth)	54.16(2)	326(1)	0.8622^{54}	1.4723^{70}	i H_2O; s EtOH, ace, chl; sl eth; vs bz
935	1-Tetradecanol	Tetradecyl alcohol	$C_{14}H_{30}O$	112-72-1	214.387	lf	37.7(7)	295.8(4)	0.8236^{38}		i H_2O; vs EtOH, eth, ace, bz, chl
936	1,2,3,5-Tetraethylbenzene		$C_{14}H_{22}$	38842-05-6	190.325			250(3)			
937	Tetraethylene glycol	3,6,9-Trioxaundecane-1,11-diol	$C_8H_{18}O_5$	112-60-7	194.226	liq	-9.4(7)	315(7)	1.1285^{15}	1.4577^{20}	vs H_2O; s EtOH, eth, ctc, diox
938	Tetraethylsilane		$C_8H_{20}Si$	631-36-7	144.331			153.4(7)	0.7658^{20}	1.4268^{20}	i H_2O
939	1,1,1,2-Tetrafluoroethane	HFC-134a	$C_2H_2F_4$	811-97-2	102.031	col gas	-103.30(1)	-26.1(1)	1.2072^{25} (p>1 atm)		i H_2O; s eth
940	1,1,2,2-Tetrafluoroethane	HFC-134	$C_2H_2F_4$	359-35-3	102.031	col gas	-89	-20(1)			
941	Tetrafluoroethene	Tetrafluoroethylene	C_2F_4	116-14-3	100.015	col gas	-131.14(2)	-76(1)	1.519^{-76}		i H_2O
942	Tetrafluoromethane	Carbon tetrafluoride	CF_4	75-73-0	88.005	col gas	-183.58(20)	-127.9(1)	3.034^{25}		i H_2O; s bz, chl
943	Tetrahydrofuran	Tetramethylene oxide	C_4H_8O	109-99-9	72.106	liq	-108.38(2)	66.0(1)	0.8833^{25}	1.4050^{25}	s H_2O, chl; vs EtOH, eth, ace, bz
944	Tetrahydrofurfuryl alcohol	Tetrahydro-2-furanmethanol	$C_5H_{10}O_2$	97-99-4	102.132	liq	<-80	176.3(7)	1.0524^{20}	1.4520^{20}	vs ace, eth
945	1,2,3,4-Tetrahydro-1-methylnaphthalene		$C_{11}H_{14}$	1559-81-5	146.229			221(6)	0.9583^{20}	1.5353^{20}	
946	1,2,3,4-Tetrahydronaphthalene	Tetralin	$C_{10}H_{12}$	119-64-2	132.202	liq	-35.76(6)	207.2(3)	0.9645^{25}	1.5413^{20}	i H_2O; vs EtOH, eth; s chl, $PhNH_2$
947	Tetrahydropyran	Oxane	$C_5H_{10}O$	142-68-7	86.132	liq	-49.1(2)	88.0(4)	0.8814^{20}	1.4200^{20}	s EtOH, eth, bz, ctc
948	Tetrahydrothiophene	Thiacyclopentane	C_4H_8S	110-01-0	88.172	liq	-96.13(5)	121.1(2)	0.9987^{20}	1.4871^{18}	i H_2O; msc EtOH, eth, ace, bz; s chl
949	1,2,3,4-Tetramethylbenzene		$C_{10}H_{14}$	488-23-3	134.218	liq	-6.7(10)	205(1)	0.9052^{20}	1.5203^{20}	i H_2O; msc EtOH, eth, ace, bz, peth, ctc
950	1,2,3,5-Tetramethylbenzene	Isodurene	$C_{10}H_{14}$	527-53-7	134.218	liq	-23.8(2)	198(1)	0.8903^{20}	1.5130^{20}	i H_2O; msc EtOH, eth, ace, bz, peth, ctc
951	1,2,4,5-Tetramethylbenzene	Durene	$C_{10}H_{14}$	95-93-2	134.218		79.2(2)	197(1)	0.8380^{81}	1.4790^{81}	i H_2O; msc EtOH, eth, ace, bz, peth, ctc
952	2,2,3,3-Tetramethylbutane		C_8H_{18}	594-82-1	114.229	lf (eth)	100.79(5)	106.32(4)	0.8242^{20}	1.4695^{20}	i H_2O; s eth, chl
953	1,2,3,4-Tetramethylcyclohexane		$C_{10}H_{20}$	3726-45-2	140.266				0.8219^{20}	1.4531^{20}	
954	*N,N,N',N'*-Tetramethyl-1,2-ethanediamine	1,2-Dimethylaminoethane	$C_6H_{16}N_2$	110-18-9	116.204	liq	-58.0(3)	121(1)	0.77^{25}	1.4179^{20}	
955	2,2,3,3-Tetramethylpentane		C_9H_{20}	7154-79-2	128.255	liq	-9.75(5)	140.2(4)	0.7530^{25}	1.4236^{20}	
956	2,2,3,4-Tetramethylpentane		C_9H_{20}	1186-53-4	128.255	liq	-121.3(7)	133.0(8)	0.7389^{20}	1.4147^{20}	
957	2,2,4,4-Tetramethylpentane	Di-*tert*-butylmethane	C_9H_{20}	1070-87-7	128.255	liq	-66.53(5)	122.2(10)	0.7195^{20}	1.4069^{20}	i H_2O; vs EtOH, bz
958	2,3,3,4-Tetramethylpentane		C_9H_{20}	16747-38-9	128.255	liq	-102.1(1)	141.5(7)	0.7547^{20}	1.4222^{20}	
959	Tetramethylsilane	TMS	$C_4H_{12}Si$	75-76-3	88.224	vol liq or gas	-99.063(5)	26.7(5)	0.648^{19}	1.3587^{20}	i H_2O; vs EtOH, eth; i sulf
960	1,1,2,2-Tetraphenylethane		$C_{26}H_{22}$	632-50-8	334.453	cry (bz), orth nd (chl)	211(2)	360			sl EtOH; s bz, HOAc
961	1,1,2,2-Tetraphenylethene		$C_{26}H_{20}$	632-51-9	332.437	mcl or orth (bz-eth or chl-al)	224.9(3)	420	1.155^{0}		i H_2O; sl EtOH, chl, eth; vs bz
962	Thiophene	Thiofuran	C_4H_4S	110-02-1	84.140	liq	-38.12(5)	84.1(1)	1.0649^{20}	1.5289^{20}	msc EtOH, eth, ace, bz, ctc, diox, py; sl chl
963	Thiourea	Thiocarbamide	CH_4N_2S	62-56-6	76.121	orth (al)	176(3)		1.405^{25}		s H_2O, EtOH; i eth
964	Thymol	2-Isopropyl-5-methylphenol	$C_{10}H_{14}O$	89-83-8	150.217		49.6(3)	233(3)	0.970^{25}	1.5227^{20}	i H_2O; vs EtOH, eth, chl, AcOEt
965	Toluene	Methylbenzene	C_7H_8	108-88-3	92.139	liq	-95.0(2)	110.60(7)	0.8668^{20}	1.4941^{25}	i H_2O; msc EtOH, eth; s ace, CS_2
966	Toluene-2,4-diisocyanate		$C_9H_6N_2O_2$	584-84-9	174.156		20.5	251	1.2244^{20}		vs ace, bz, eth
967	Toluene-2,6-diisocyanate		$C_9H_6N_2O_2$	91-08-7	174.156		18.3				dec H_2O; s ace, bz
968	*p*-Toluenesulfonic acid		$C_7H_8O_3S$	104-15-4	172.202	hyg pl (w+1) mcl lf or pl	104.5				vs H_2O; s EtOH, eth
969	*o*-Toluic acid	2-Methylbenzoic acid	$C_8H_8O_2$	118-90-1	136.149	pr or nd (w)	103.4(2)	259.5(6)	1.062^{115}	1.512^{115}	i H_2O; vs EtOH, eth; s chl
970	*m*-Toluic acid	3-Methylbenzoic acid	$C_8H_8O_2$	99-04-7	136.149		109.3(7)		1.054^{112}	1.509	sl H_2O, chl; vs EtOH, eth
971	*p*-Toluic acid	4-Methylbenzoic acid	$C_8H_8O_2$	99-94-5	136.149		180(1)				i H_2O; vs EtOH, eth, MeOH; sl tfa
972	1,3,5-Triazine-2,4,6-triamine	Melamine	$C_3H_6N_6$	108-78-1	126.120	mcl pr (w)	343(4)	subl	1.573^{16}	1.872^{20}	sl H_2O, EtOH; i eth
973	Tribromomethane	Bromoform	$CHBr_3$	75-25-2	252.731	liq	8.69(2)	149.2(5)	2.8788^{25}	1.5948^{25}	sl H_2O; msc EtOH, eth; s bz, lig, chl

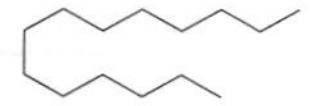
933
Tetradecane

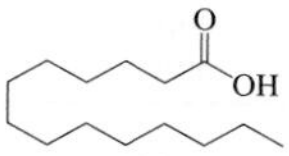
934
Tetradecanoic acid

935
1-Tetradecanol

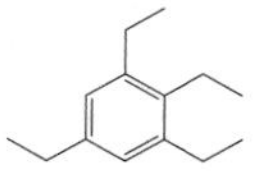
936
1,2,3,5-Tetraethylbenzene

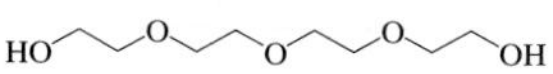
937
Tetraethylene glycol

938
Tetraethylsilane

939
1,1,1,2-Tetrafluoroethane

940
1,1,2,2-Tetrafluoroethane

941
Tetrafluoroethene

942
Tetrafluoromethane

943
Tetrahydrofuran

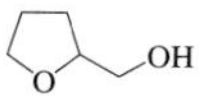
944
Tetrahydrofurfuryl alcohol

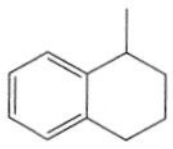
945
1,2,3,4-Tetrahydro-1-methylnaphthalene

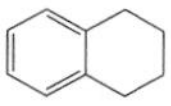
946
1,2,3,4-Tetrahydronaphthalene

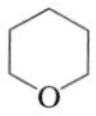
947
Tetrahydropyran

948
Tetrahydrothiophene

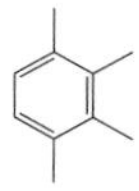
949
1,2,3,4-Tetramethylbenzene

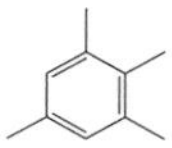
950
1,2,3,5-Tetramethylbenzene

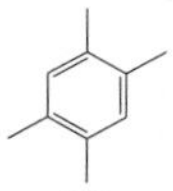
951
1,2,4,5-Tetramethylbenzene

952
2,2,3,3-Tetramethylbutane

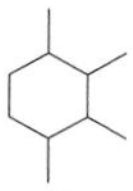
953
1,2,3,4-Tetramethylcyclohexane

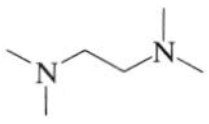
954
N,N,N',N'-Tetramethyl-1,2-ethanediamine

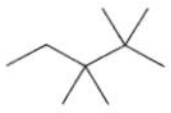
955
2,2,3,3-Tetramethylpentane

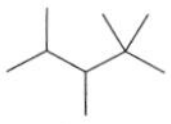
956
2,2,3,4-Tetramethylpentane

957
2,2,4,4-Tetramethylpentane

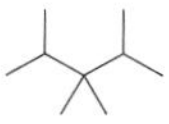
958
2,3,3,4-Tetramethylpentane

959
Tetramethylsilane

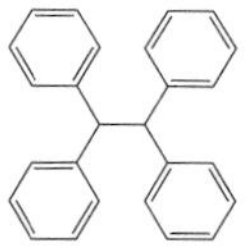
960
1,1,2,2-Tetraphenylethane

961
1,1,2,2-Tetraphenylethene

962
Thiophene

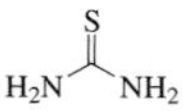
963
Thiourea

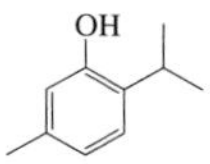
964
Thymol

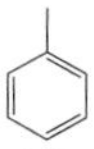
965
Toluene

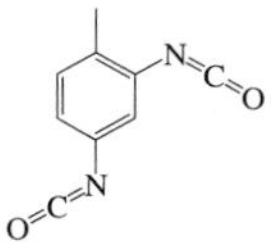
966
Toluene-2,4-diisocyanate

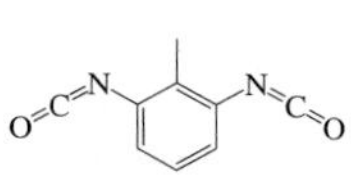
967
Toluene-2,6-diisocyanate

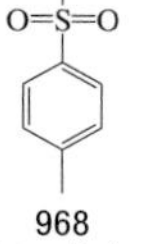
968
p-Toluenesulfonic acid

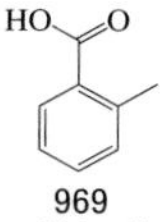
969
o-Toluic acid

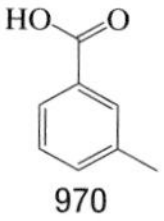
970
m-Toluic acid

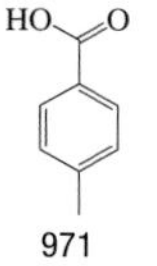
971
p-Toluic acid

972
1,3,5-Triazine-2,4,6-triamine

973
Tribromomethane

Organic

No.	Name	Synonym	Mol. form.	CAS Reg. No.	Mol. wt.	Physical form	mp/°C	bp/°C	den/g cm^{-3}	n_D	Solubility
974	Tributylamine	*N,N*-Dibutyl-1-butanamine	$C_{12}H_{27}N$	102-82-9	185.349	liq	-70	207(2)	0.7770^{20}	1.4299^{20}	sl H_2O, ctc; vs EtOH, eth; s ace, bz
975	Tributyl borate	Butyl borate	$C_{12}H_{27}BO_3$	688-74-4	230.151	oil	<-70	233.8(10)	0.8567^{20}	1.4106^{18}	s EtOH, bz; vs eth, MeOH
976	Tributyl phosphate	Butyl phosphate	$C_{12}H_{27}O_4P$	126-73-8	266.313			289	0.9727^{25}	1.4224^{25}	s H_2O, eth, bz, CS_2; msc EtOH
977	Trichloroacetaldehyde	Chloral	C_2HCl_3O	75-87-6	147.387	liq	-57.5	98(2)	1.512^{20}	1.4580^{20}	vs H_2O; s EtOH, eth
978	Trichloroacetic acid		$C_2HCl_3O_2$	76-03-9	163.387	hyg cry	59.1(1)	198.2(1)	1.6126^{64}	1.4603^{61}	vs H_2O; s EtOH, eth; sl ctc
979	Trichloroacetyl chloride		C_2Cl_4O	76-02-8	181.832			118.2(3)	1.6202^{20}	1.4695^{20}	msc eth
980	1,2,3-Trichlorobenzene		$C_6H_3Cl_3$	87-61-6	181.447	pl (al)	53(1)	219(3)	1.4533^{25}		i H_2O; sl EtOH, chl; vs eth, bz
981	1,2,4-Trichlorobenzene		$C_6H_3Cl_3$	120-82-1	181.447	orth	17.0(4)	213.5(3)	1.459^{25}	1.5717^{20}	i H_2O; sl EtOH, chl; vs eth
982	1,3,5-Trichlorobenzene		$C_6H_3Cl_3$	108-70-3	181.447	nd	62.8(7)	209(1)			i H_2O; sl EtOH; vs eth, bz; s chl
	1,1,1-Trichloro-2,2-bis(4-chlorophenyl)ethane	Dichlorodiphenyltrichloroethane (DDT)	$C_{14}H_9Cl_5$	50-29-3	354.486	nd (al)	109(2)	260			i H_2O; sl EtOH; vs eth, ace, bz, py
983	1,1,1-Trichloroethane	Methyl chloroform	$C_2H_3Cl_3$	71-55-6	133.404	liq	-30(2)	74.02(8)	1.3390^{20}	1.4379^{20}	sl H_2O; s EtOH, chl; msc eth
984	1,1,2-Trichloroethane	Vinyl trichloride	$C_2H_3Cl_3$	79-00-5	133.404	liq	-36.3(5)	113(1)	1.4397^{20}	1.4714^{20}	i H_2O; s EtOH, eth, chl
985	Trichloroethene	Trichloroethylene	C_2HCl_3	79-01-6	131.388	liq	-84.7(3)	86.8(1)	1.4642^{20}	1.4773^{20}	sl H_2O, ctc; msc EtOH, eth; s ace
986	Trichloroethylsilane	Ethyltrichlorosilane	$C_2H_5Cl_3Si$	115-21-9	163.506	liq	-105.6	98.7(7)	1.2373^{20}	1.4256^{20}	s ctc
987	1,1,1-Trichloro-2-fluoroethane	Refrigerant 131b	$C_2H_2Cl_3F$	2366-36-1	151.394	liq		86.5			
988	Trichlorofluoromethane	CFC-11	CCl_3F	75-69-4	137.368	vol liq or gas	-110.44(2)	23.7(6)	1.4879^{20}		i H_2O
989	Trichloromethane	Chloroform	$CHCl_3$	67-66-3	119.378	liq	-63.3(6)	61.2(1)	1.4890^{20}	1.4459^{20}	sl H_2O; msc EtOH, eth, bz; s ace, ctc
990	(Trichloromethyl)benzene	Benzotrichloride	$C_7H_5Cl_3$	98-07-7	195.474	liq	-17.0(6)	221	1.3723^{20}	1.5580^{20}	i H_2O; s EtOH, eth, bz
991	Trichloromethylsilane	Methyltrichlorosilane	CH_3Cl_3Si	75-79-6	149.480	liq	-75.77(5)	66(2)	1.273^{20}	1.4106^{20}	dec H_2O, EtOH
992	1,2,3-Trichloropropane	Allyl trichloride	$C_3H_5Cl_3$	96-18-4	147.431	liq	-13.8(6)	158(2)	1.3889^{20}	1.4852^{20}	sl H_2O, ctc; s EtOH, eth; vs chl
993	1,1,1-Trichloro-2,2,2-trifluoroethane		$C_2Cl_3F_3$	354-58-5	187.375		14.37(20)	46(1)	1.5790^{20}	1.3610^{35}	i H_2O; s EtOH, eth, chl
994	1,1,2-Trichloro-1,2,2-trifluoroethane	CFC-113	$C_2Cl_3F_3$	76-13-1	187.375	liq		47.6(2)	1.5635^{25}	1.3557^{25}	i H_2O; s EtOH; msc eth, bz
995	Tri-*o*-cresyl phosphate	Tri-*o*-tolyl phosphate	$C_{21}H_{21}O_4P$	78-30-8	368.363	col or pa ye	11	410	1.1955^{20}	1.5575^{20}	i H_2O; vs EtOH, eth, ctc, tol; s HOAc
996	Tri-*p*-cresyl phosphate	Tri-*p*-tolyl phosphate	$C_{21}H_{21}O_4P$	78-32-0	368.363	nd (al), tab (eth)	77.5		1.247^{25}		s EtOH, eth, bz, chl, HOAc
997	Tridecane		$C_{13}H_{28}$	629-50-5	184.361	liq	-5.35(2)	235.4(4)	0.7564^{20}	1.4256^{20}	i H_2O; vs EtOH, eth; s ctc
998	Tridecanoic acid	Tridecylic acid	$C_{13}H_{26}O_2$	638-53-9	214.344	cry (peth ace)	41.85(4)		0.8458^{80}	1.4286^{60}	i H_2O; vs EtOH, eth, HOAc; s ace
999	1-Tridecene		$C_{13}H_{26}$	2437-56-1	182.345	liq	-23.07(9)	232.8(7)	0.7658^{20}	1.4340^{20}	i H_2O; vs EtOH, eth; s bz
1000	Tris(2-hydroxyethyl)amine	TEA	$C_6H_{15}NO_3$	102-71-6	149.188	hyg cry	21.5(4)	350(5)	1.1242^{20}	1.4852^{20}	msc H_2O, EtOH; sl eth, bz; s chl
1001	Triethylaluminum	Hexaethyldialuminum	$C_6H_{15}Al$	97-93-8	114.165	col hyg liq	-48.14(2)	193(1)	0.832^{25}		
1002	Triethylamine	*N,N*-Diethylethanamine	$C_6H_{15}N$	121-44-8	101.190	liq	-114.7(2)	88.8(2)	0.7275^{20}	1.4010^{20}	s H_2O, EtOH, eth, ctc; vs ace, bz, chl
1003	1,2,3-Triethylbenzene		$C_{12}H_{18}$	42205-08-3	162.271	col liq	-26	172			
1004	1,2,4-Triethylbenzene		$C_{12}H_{18}$	877-44-1	162.271			217(3)	0.8738^{20}	1.5024^{20}	i H_2O; s EtOH, eth
1005	1,3,5-Triethylbenzene		$C_{12}H_{18}$	102-25-0	162.271	liq	-66.4(1)	215.8(9)	0.8631^{20}	1.4969^{20}	i H_2O; vs EtOH, eth
1006	Triethylene glycol	Triglycol	$C_6H_{14}O_4$	112-27-6	150.173	hyg liq	-9.4(5)	288.6(2)	1.1274^{15}	1.4531^{20}	msc H_2O, EtOH, bz; sl eth, chl; i peth
1007	Triethyl phosphate	Ethyl phosphate	$C_6H_{15}O_4P$	78-40-0	182.154	liq	-56.4	216(11)	1.0695^{20}	1.4053^{20}	s H_2O, eth, bz; vs EtOH; sl chl
1008	Trifluoroacetic acid		$C_2HF_3O_2$	76-05-1	114.023	liq	-15.2	72(2)	1.5351^{25}		s H_2O, EtOH, eth, ace
1009	1,1,1-Trifluoroethane	HFC-143a	$C_2H_3F_3$	420-46-2	84.040	col gas	-111.6(5)	-47.2(1)			s eth, chl
1010	1,1,2-Trifluoroethane	HFC-143	$C_2H_3F_3$	430-66-0	84.040	col gas	-84(5)	3.5(7)			
1011	Trifluoromethane	HFC-23	CHF_3	75-46-7	70.014	col gas	-155.18(20)	-82.0(1)	0.673^{25} (p>1 atm)		s H_2O, ace, bz; vs EtOH; sl chl
1012	(Trifluoromethyl)benzene	Benzotrifluoride	$C_7H_5F_3$	98-08-8	146.110	liq	-28.99(6)	102.0(2)	1.1884^{20}	1.4146^{20}	msc EtOH, eth, ace, bz, ctc
1013	3,4,5-Trihydroxybenzoic acid	Gallic acid	$C_7H_6O_5$	149-91-7	170.120	pr (w+1)	262(1)		1.694^{6}		sl H_2O, eth; vs EtOH; s ace; i bz, chl
1014	1,3,5-Triisopropylbenzene		$C_{15}H_{24}$	717-74-8	204.352	liq	-9(2)	249(7)	0.8545^{20}	1.4882^{20}	s ace, bz, chl
1015	Trimethylamine	*N,N*-Dimethylmethanamine	C_3H_9N	75-50-3	59.110	col gas	-117.1(2)	2.8(2)	0.627^{25} (p>1 atm)	1.3631^{0}	vs H_2O, chl, tol; s EtOH, eth, bz
1016	1,2,3-Trimethylbenzene	Hemimellitene	C_9H_{12}	526-73-8	120.191	liq	-25.32(4)	176.0(4)	0.8944^{20}	1.5139^{20}	i H_2O; msc EtOH, eth, ace, bz, peth, ctc

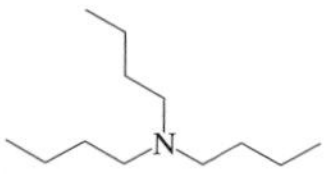
974
Tributylamine

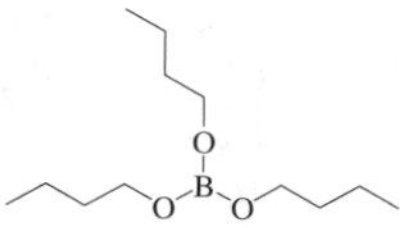
975
Tributyl borate

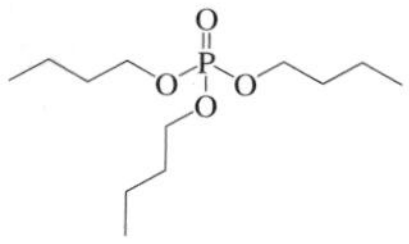
976
Tributyl phosphate

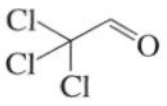
977
Trichloroacetaldehyde

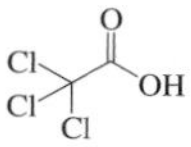
978
Trichloroacetic acid

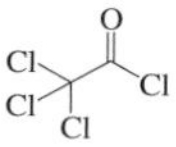
979
Trichloroacetyl chloride

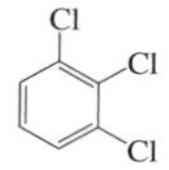
980
1,2,3-Trichlorobenzene

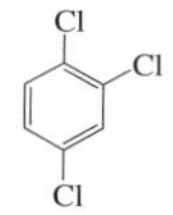
981
1,2,4-Trichlorobenzene

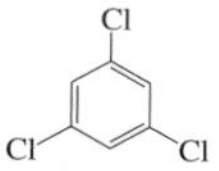
982
1,3,5-Trichlorobenzene

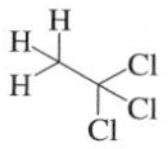
983
1,1,1-Trichloroethane

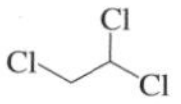
984
1,1,2-Trichloroethane

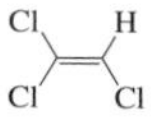
985
Trichloroethene

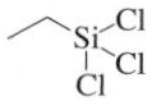
986
Trichloroethylsilane

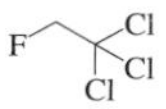
987
1,1,1-Trichloro-2-fluoroethane

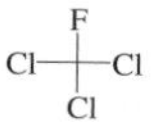
988
Trichlorofluoromethane

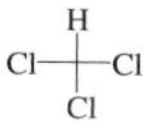
989
Trichloromethane

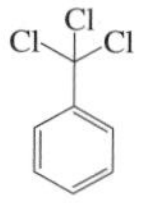
990
(Trichloromethyl)benzene

991
Trichloromethylsilane

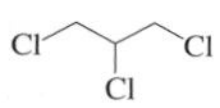
992
1,2,3-Trichloropropane

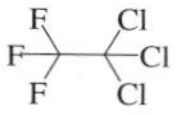
993
1,1,1-Trichloro-2,2,2-trifluoroethane

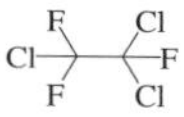
994
1,1,2-Trichloro-1,2,2-trifluoroethane

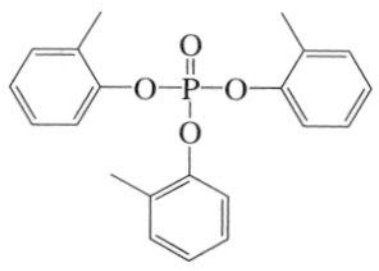
995
Tri-*o*-cresyl phosphate

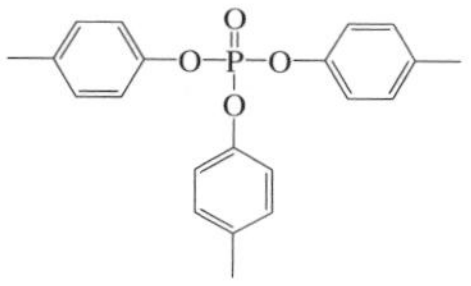
996
Tri-*p*-cresyl phosphate

997
Tridecane

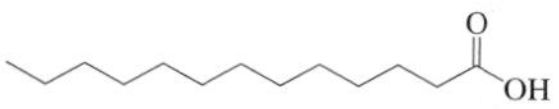
998
Tridecanoic acid

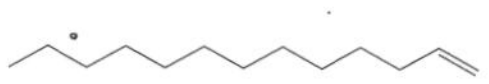
999
1-Tridecene

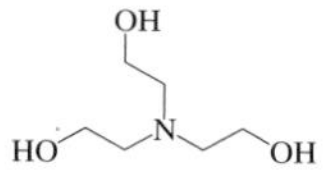
1000
Triethanolamine

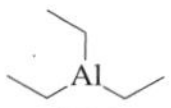
1001
Triethylaluminum

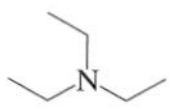
1002
Triethylamine

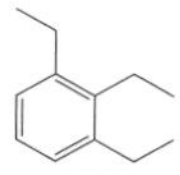
1003
1,2,3-Triethylbenzene

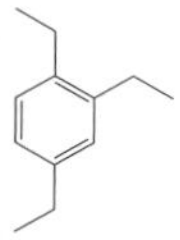
1004
1,2,4-Triethylbenzene

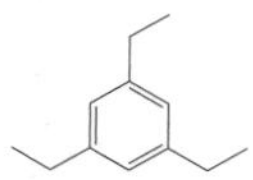
1005
1,3,5-Triethylbenzene

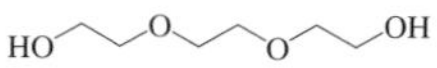
1006
Triethylene glycol

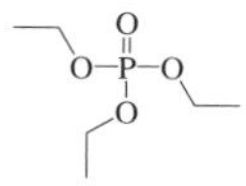
1007
Triethyl phosphate

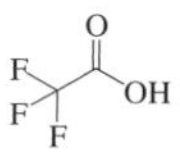
1008
Trifluoroacetic acid

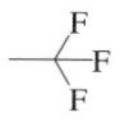
1009
1,1,1-Trifluoroethane

1010
1,1,2-Trifluoroethane

1011
Trifluoromethane

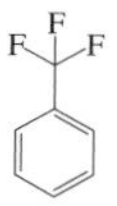
1012
(Trifluoromethyl)benzene

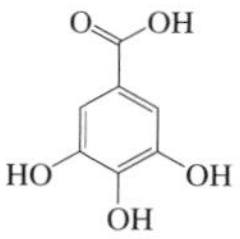
1013
3,4,5-Trihydroxybenzoic acid

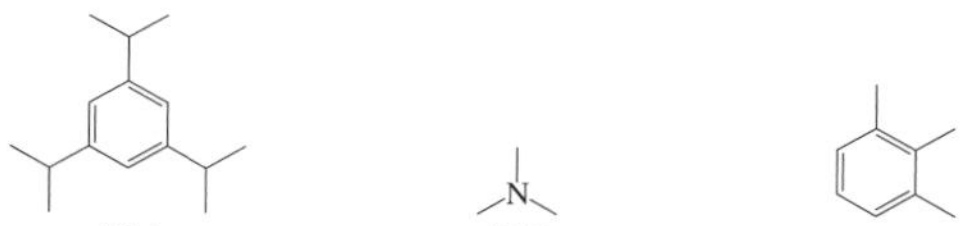
1014
1,3,5-Triisopropylbenzene

1015
Trimethylamine

1016
1,2,3-Trimethylbenzene

No.	Name	Synonym	Mol. form.	CAS Reg. No.	Mol. wt.	Physical form	mp/°C	bp/°C	den/g cm^{-3}	n_D	Solubility
1017	1,2,4-Trimethylbenzene	Pseudocumene	C_9H_{12}	95-63-6	120.191	liq	-43.8(1)	169.4(3)	0.8758^{20}	1.5048^{20}	i H_2O; msc EtOH, eth, ace, bz, peth, ctc
1018	1,3,5-Trimethylbenzene	Mesitylene	C_9H_{12}	108-67-8	120.191	liq	-44.69(5)	164.7(3)	0.8615^{25}	1.4994^{20}	i H_2O; msc EtOH, eth, ace, bz, peth, ctc
1019	2,2,3-Trimethylbutane	Triptane	C_7H_{16}	464-06-2	100.202	liq	-24.56(5)	80.8(1)	0.6901^{20}	1.3864^{20}	i H_2O; s EtOH, eth; vs ace, bz, peth, ctc
1020	1,1,3-Trimethylcyclohexane		C_9H_{18}	3073-66-3	126.239	liq	-68.5(5)	136.6(4)	0.7749^{25}	1.4295^{20}	i H_2O
1021	1,1,2-Trimethylcyclopentane		C_8H_{16}	4259-00-1	112.213	liq	-21.7(2)	113.7(2)	0.7660^{20}	1.4199^{20}	
1022	1,1,3-Trimethylcyclopentane		C_8H_{16}	4516-69-2	112.213	liq	-142.5(2)	104.9(6)	0.7439^{25}	1.4112^{20}	i H_2O
1023	2,2,5-Trimethylhexane		C_9H_{20}	3522-94-9	128.255	liq	-105.9(1)	124(2)	0.7072^{20}	1.3997^{20}	i H_2O; vs EtOH, eth, ace, bz; s ctc
1024	1,2,3-Trimethylindene		$C_{12}H_{14}$	4773-83-5	158.239	liq			0.9714^{20}	1.5521^{20}	
1025	2,2,3-Trimethylpentane	2-*tert*-Butylbutane	C_8H_{18}	564-02-3	114.229	liq	-112.4(3)	109.8(4)	0.7161^{20}	1.4030^{20}	i H_2O; msc EtOH, eth, ace, hp; s bz
1026	2,2,4-Trimethylpentane	Isooctane	C_8H_{18}	540-84-1	114.229	liq	-107.36(4)	99.2(2)	0.6919^{20}	1.3884^{25}	i H_2O; msc EtOH, ace, hp; s eth, ctc
1027	2,3,3-Trimethylpentane		C_8H_{18}	560-21-4	114.229	liq	-101.2(3)	114.7(3)	0.7262^{20}	1.4075^{20}	i H_2O; vs EtOH; msc eth, ace, bz, hp
1028	2,3,4-Trimethylpentane		C_8H_{18}	565-75-3	114.229	liq	-109.3(2)	113.4(3)	0.7191^{20}	1.4042^{20}	i H_2O; vs EtOH; msc eth, ace, bz; sl ctc
1029	Trimethyl phosphate	Methyl phosphate	$C_3H_9O_4P$	512-56-1	140.074	liq	-46	197.2	1.2144^{20}	1.3967^{20}	vs H_2O; sl EtOH; s eth
1030	2,4,6-Trimethylpyridine	2,4,6-Collidine	$C_8H_{11}N$	108-75-8	121.180	liq	-44.3(3)	170(1)	0.9166^{22}	1.4959^{25}	s H_2O, EtOH, eth, ace, ctc
1031	1,3,5-Trinitrobenzene	*sym*-Trinitrobenzene	$C_6H_3N_3O_6$	99-35-4	213.104	orth pl (bz) lf (w)	121.3(4)	315	1.4775^{152}		sl H_2O, EtOH, eth; vs ace; s bz, py
1032	Trinitroglycerol	Nitroglycerin	$C_3H_5N_3O_9$	55-63-0	227.087	pa ye tcl or orth	12.8(2)	218 exp	1.5931^{20}	1.4786^{12}	sl H_2O; s EtOH, bz; msc eth; vs ace, chl
1033	2,4,6-Trinitrophenol	Picric acid	$C_6H_3N_3O_7$	88-89-1	229.104	ye lf (w), pr (eth) pl (al)	121(2)	exp		1.763	sl H_2O; s EtOH, eth, bz, chl; vs ace
1034	2,4,6-Trinitrotoluene	2-Methyl-1,3,5-trinitrobenzene	$C_7H_5N_3O_6$	118-96-7	227.131	orth (al)	80.9(8)	350(8)	1.654^{25}		i H_2O; sl EtOH; s eth; vs ace, bz
1035	1,3,5-Trioxane	Formaldehyde, trimer	$C_3H_6O_3$	110-88-3	90.078	orth nd (eth)	60(2)	116(3)	1.17^{65}		vs H_2O; s EtOH, eth, bz, CS_2; i peth
1036	Triphenylene	Benzo[*l*]phenanthrene	$C_{18}H_{12}$	217-59-4	228.288	nd (al, chl, bz)	197.82(4)	425			i H_2O; s EtOH, HOAc; vs bz, chl
1037	1,1,2-Triphenylethane		$C_{20}H_{18}$	1520-42-9	258.357	mcl lf (dil al), nd (al)	54.5(8)				i H_2O; vs EtOH, eth, bz; sl MeOH
1038	1,1,2-Triphenylethene		$C_{20}H_{16}$	58-72-0	256.341	lf (al)	67.80(4)		1.0373^{78}	1.6292^{78}	i H_2O; s EtOH, chl, MeOH; vs eth
1039	Triphenylmethane		$C_{19}H_{16}$	519-73-3	244.330	orth (al)	92.0(7)	359	1.014^{99}	1.5839^{99}	i H_2O; sl EtOH; vs eth, py, chl; s bz
1040	Triphenyl phosphate		$C_{18}H_{15}O_4P$	115-86-6	326.283	cry (lig), pr (al), nd (eth)	49.39(4)		1.2055^{50}		i H_2O; s EtOH; vs eth, bz, ctc, chl
1041	Tris(hydroxymethyl)methylamine	TRIZMA	$C_4H_{11}NO_3$	77-86-1	121.135		170.5(2)				vs H_2O; s MeOH
1042	Undecane	Hendecane	$C_{11}H_{24}$	1120-21-4	156.309	liq	-25.54(5)	195.9(3)	0.7402^{20}	1.4164^{20}	i H_2O; msc EtOH, eth
1043	1-Undecene		$C_{11}H_{22}$	821-95-4	154.293	liq	-49.12(4)	192.7(5)	0.7503^{20}	1.4261^{20}	i H_2O; s eth, chl, lig
1044	Urea	Carbamide	CH_4N_2O	57-13-6	60.055	tetr pr (al)	132.4(5)	dec	1.3230^{20}	1.484	vs H_2O, EtOH; i eth, bz; s HOAc, py
1045	Vinyl acetate		$C_4H_6O_2$	108-05-4	86.090	liq	-100.2(4)	72.6(3)	0.9256^{25}	1.3926^{25}	sl H_2O; msc EtOH; s eth, ace, bz, chl
1046	4-Vinylcyclohexene		C_8H_{12}	100-40-3	108.181	liq	-108.9	130(4)	0.8299^{20}	1.4639^{20}	i H_2O; s eth, bz, peth
1047	*o*-Xylene	1,2-Dimethylbenzene	C_8H_{10}	95-47-6	106.165	liq	-25.16(2)	144.4(4)	0.8802^{20}	1.5018^{25}	i H_2O; msc EtOH, eth, ace, bz, peth, ctc
1048	*m*-Xylene	1,3-Dimethylbenzene	C_8H_{10}	108-38-3	106.165	liq	-47.85(3)	139.1(4)	0.8641^{20}	1.4944^{25}	i H_2O; msc EtOH, eth, ace, bz; s chl
1049	*p*-Xylene	1,4-Dimethylbenzene	C_8H_{10}	106-42-3	106.165	mcl pr or liq	13.3(1)	138.3(5)	0.8610^{20}	1.4929^{25}	i H_2O; msc EtOH, eth, ace, bz; s chl
1050	2,3-Xylenol	2,3-Dimethylphenol	$C_8H_{10}O$	526-75-0	122.164	nd (w, dil al)	72.7(3)	216.88(5)		1.5420^{20}	sl H_2O; s EtOH, eth
1051	2,4-Xylenol	2,4-Dimethylphenol	$C_8H_{10}O$	105-67-9	122.164	nd (w)	25(1)	210.94(3)	0.9650^{20}	1.5420^{14}	sl H_2O; msc EtOH, eth; s ctc
1052	2,5-Xylenol	2,5-Dimethylphenol	$C_8H_{10}O$	95-87-4	122.164	nd (w), pr (al-eth)	74.9(2)	211.14(8)			s H_2O, EtOH; vs eth; sl chl
1053	2,6-Xylenol	2,6-Dimethylphenol	$C_8H_{10}O$	576-26-1	122.164	lf or nd (al)	45.4(4)	201.03(5)			s H_2O, EtOH, eth, ctc
1054	3,4-Xylenol	3,4-Dimethylphenol	$C_8H_{10}O$	95-65-8	122.164		65.1(2)	227.31(5)	0.9830^{20}		sl H_2O; s EtOH, ctc; msc eth
1055	3,5-Xylenol	3,5-Dimethylphenol	$C_8H_{10}O$	108-68-9	122.164	nd (w, peth)	63.4(3)	221.71(5)	0.9680^{20}		s H_2O, EtOH, ctc

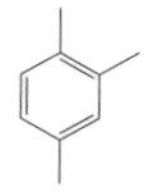
1017
1,2,4-Trimethylbenzene

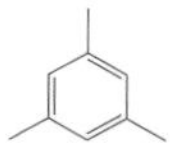
1018
1,3,5-Trimethylbenzene

1019
2,2,3-Trimethylbutane

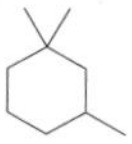
1020
1,1,3-Trimethylcyclohexane

1021
1,1,2-Trimethylcyclopentane

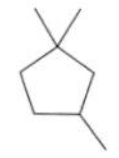
1022
1,1,3-Trimethylcyclopentane

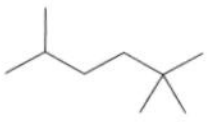
1023
2,2,5-Trimethylhexane

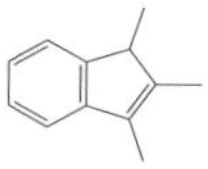
1024
1,2,3-Trimethylindene

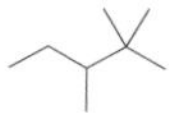
1025
2,2,3-Trimethylpentane

1026
2,2,4-Trimethylpentane

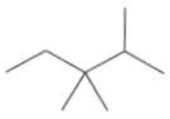
1027
2,3,3-Trimethylpentane

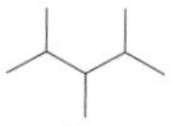
1028
2,3,4-Trimethylpentane

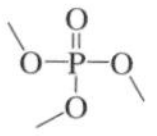
1029
Trimethyl phosphate

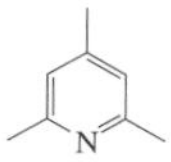
1030
2,4,6-Trimethylpyridine

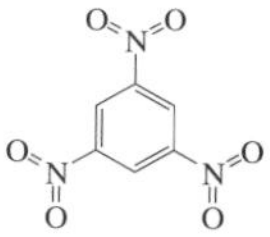
1031
1,3,5-Trinitrobenzene

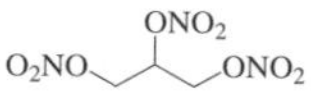
1032
Trinitroglycerol

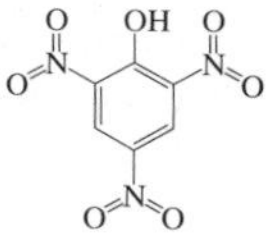
1033
2,4,6-Trinitrophenol

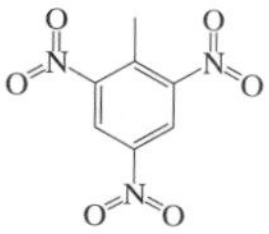
1034
2,4,6-Trinitrotoluene

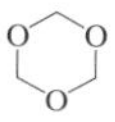
1035
1,3,5-Trioxane

1036
Triphenylene

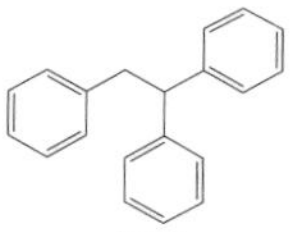
1037
1,1,2-Triphenylethane

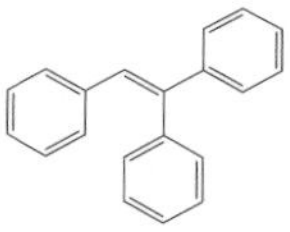
1038
1,1,2-Triphenylethene

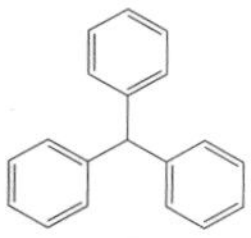
1039
Triphenylmethane

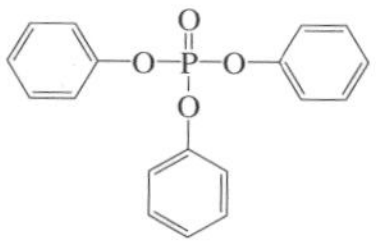
1040
Triphenyl phosphate

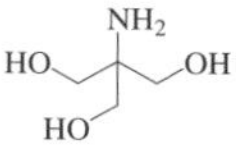
1041
Tris(hydroxymethyl)methylamine

1042
Undecane

1043
1-Undecene

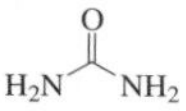
1044
Urea

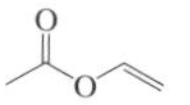
1045
Vinyl acetate

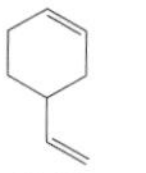
1046
4-Vinylcyclohexene

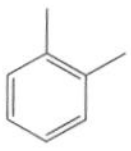
1047
o-Xylene

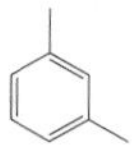
1048
m-Xylene

1049
p-Xylene

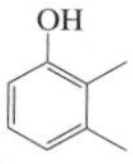
1050
2,3-Xylenol

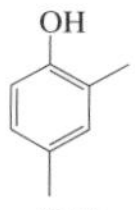
1051
2,4-Xylenol

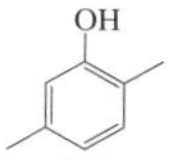
1052
2,5-Xylenol

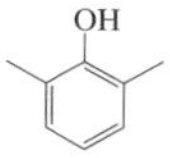
1053
2,6-Xylenol

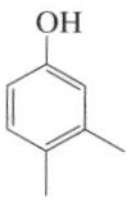
1054
3,4-Xylenol

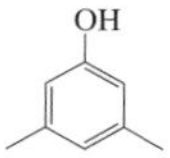
1055
3,5-Xylenol

DIAMAGNETIC SUSCEPTIBILITY OF SELECTED ORGANIC COMPOUNDS

When a material is placed in a magnetic field ***H***, a magnetization ***M*** is induced in the material which is related to ***H*** by ***M***= κ***H***, where κ is called the volume susceptibility. Since ***H*** and ***M*** have the same dimensions, κ is dimensionless. A more useful parameter is the molar susceptibility χ_m, defined by

$$\chi_m = \kappa V_m = \kappa M/\rho$$

where V_m is the molar volume of the substance, M the molar mass, and ρ the mass density. When the cgs system is used, the customary unit for χ_m is cm^3 mol^{-1}; the corresponding SI base unit is m^3 mol^{-1}. Substances with no unpaired electrons are called diamagnetic; they have negative values of χ_m.

This table gives values of the diamagnetic susceptibility for about 400 common organic compounds. Values are given for both the molar susceptibility χ_m and the volume susceptibility κ. Note that the tabulated numbers are the negatives of the susceptibility values. Most values of the volume susceptibility κ refer to solids or liquids at room temperature; an asterisk * following the value indicates a liquefied gas. A more extensive table may be found in Ref. 1.

In keeping with customary practice, the molar susceptibility is given here in units appropriate to the cgs system. These values should be multiplied by 4π to obtain values for use in SI equations (where the magnetic field strength ***H*** has units of A m^{-1}). Column definitions are as follows.

Column heading	Definition
Name	Name of the compound; compounds are ordered alphabetically by name
Mol. form.	Molecular formula of the compound
$-\chi_m$	The negative of the molar susceptibility, χ_m, in units of 10^{-6} cm^3 mol^{-1}
$-\kappa$	The negative of the volume susceptibility, κ, in units of 10^{-6}

References

1. *Landolt-Börnstein, Numerical Data and Functional Relationships in Science and Technology, New Series,* II/16, *Diamagnetic Susceptibility,* Gupta, R. R., Ed., Springer-Verlag, Heidelberg, 1986.
2. Barter, C., Meisenheimer, R. G., and Stevenson, D. P., *J. Phys. Chem.* 64, 1312, 1960. [https://doi.org/10.1021/j100838a045]
3. Broersma, S., *J. Chem. Phys.* 17, 873, 1949. [https://doi.org/10.1063/1.1747080]

Molar and Volume Susceptibility of Organic Compounds

Name	Mol. form.	$-\chi_m/10^{-6}$ cm^3 mol^{-1}	$-\kappa/10^{-6}$
Acenaphthene	$C_{12}H_{10}$	109.9	0.871
Acenaphthylene	$C_{12}H_8$	111.6	0.659
Acetaldehyde	C_2H_4O	22.2	0.395
Acetamide	C_2H_5NO	33.9	0.573
Acetic acid	$C_2H_4O_2$	31.8	0.553
Acetic anhydride	$C_4H_6O_3$	52.8	0.560
Acetone	C_3H_6O	33.8	0.457
Acetonitrile	C_2H_3N	27.8	0.532
Acetophenone	C_8H_8O	72.5	0.620
Acetyl chloride	C_2H_3ClO	39.3	0.553
Acetylene	C_2H_2	20.8	0.301*
Acridine	$C_{13}H_9N$	118.8	0.666
Allene	C_3H_4	25.3	0.369*
Allyl alcohol	C_3H_6O	36.7	0.540
Allylamine	C_3H_7N	40.1	0.532
Aniline	C_6H_7N	62.4	0.685
Anisole	C_7H_8O	72.2	0.664
Anthracene	$C_{14}H_{10}$	129.8	0.932
9,10-Anthracenedione	$C_{14}H_8O_2$	113.0	0.780
trans-Azobenzene	$C_{12}H_{10}N_2$	106.8	0.705
Azulene	$C_{10}H_8$	123.7	
Benzaldehyde	C_7H_6O	60.7	0.595
Benzamide	C_7H_7NO	72.0	0.641
Benzene	C_6H_6	54.8	0.615
Benzeneacetic acid	$C_8H_8O_2$	82.4	0.743
Benzeneacetonitrile	C_8H_7N	76.9	0.670
1,2-Benzenediamine	$C_6H_8N_2$	72.5	
1,3-Benzenediamine	$C_6H_8N_2$	70.4	0.657
1,4-Benzenediamine	$C_6H_8N_2$	70.7	
Benzil	$C_{14}H_{10}O_2$	106.8	0.551
Benzonitrile	C_7H_5N	65.2	0.638
Benzophenone	$C_{13}H_{10}O$	109.6	0.668
p-Benzoquinone	$C_6H_4O_2$	36	0.439
Benzyl acetate	$C_9H_{10}O_2$	93.2	0.655
Benzyl alcohol	C_7H_8O	71.8	0.692
Benzyl benzoate	$C_{14}H_{12}O_2$	132.2	0.693
Biphenyl	$C_{12}H_{10}$	103.3	0.697
Bromobenzene	C_6H_5Br	78.1	0.744
1-Bromobutane	C_4H_9Br	77.1	0.718
Bromochloromethane	CH_2BrCl	55.1	0.824
Bromodichloromethane	$CHBrCl_2$	66.3	0.801
Bromoethane	C_2H_5Br	54.7	0.733
Bromomethane	CH_3Br	42.8	0.755*
1-Bromo-2-methylpropane	C_4H_9Br	79.9	0.742
1-Bromonaphthalene	$C_{10}H_7Br$	115.9	0.828
1-Bromopropane	C_3H_7Br	65.6	0.722
2-Bromopropane	C_3H_7Br	65.1	0.696
3-Bromopropene	C_3H_5Br	58.6	0.677
4-Bromotoluene	C_7H_7Br	88.7	0.724
Bromotrichloromethane	$CBrCl_3$	73.2	0.743
1,2-Butadiene	C_4H_6	35.6	0.445
1,3-Butadiene	C_4H_6	32.1	0.365*
Butanal	C_4H_8O	45.9	0.510
Butane	C_4H_{10}	50.3	0.496*
1,3-Butanediol	$C_4H_{10}O_2$	61.8	0.689
1,4-Butanediol	$C_4H_{10}O_2$	61.8	0.697
Butanenitrile	C_4H_7N	50.4	0.579
1-Butanethiol	$C_4H_{10}S$	70.2	0.655
Butanoic acid	$C_4H_8O_2$	55.1	0.596
1-Butanol	$C_4H_{10}O$	55.9	0.611
2-Butanol	$C_4H_{10}O$	57.6	0.627
2-Butanone	C_4H_8O	45.6	0.506
1-Butene	C_4H_8	41.0	0.430*
cis-2-Butene	C_4H_8	42.6	0.468*

Name	Mol. form.	$-\chi_m/10^{-6}$ $cm^3\ mol^{-1}$	$-\kappa/10^{-6}$
trans-2-Butene	C_4H_8	43.3	0.462*
Butylamine	$C_4H_{11}N$	58.9	0.597
Butylbenzene	$C_{10}H_{14}$	100.7	0.645
tert-Butylbenzene	$C_{10}H_{14}$	101.8	0.657
Butyl formate	$C_5H_{10}O_2$	65.8	0.577
4-*tert*-Butylphenol	$C_{10}H_{14}O$	108.0	0.653
Butyl propanoate	$C_7H_{14}O_2$	89.1	0.599
Camphor, (+)	$C_{10}H_{16}O$	103.0	0.670
Carbazole	$C_{12}H_9N$	119.9	0.930
Carbonyl chloride	CCl_2O	47.9	0.664*
2-Chloroaniline	C_6H_6ClN	79.5	0.755
3-Chloroaniline	C_6H_6ClN	76.6	0.730
4-Chloroaniline	C_6H_6ClN	76.7	0.859
Chlorobenzene	C_6H_5Cl	69.6	0.684
1-Chlorobutane	C_4H_9Cl	67.1	0.642
2-Chlorobutane	C_4H_9Cl	67.4	0.636
Chloroethene	C_2H_3Cl	35.9	0.523*
Chloromethane	CH_3Cl	32.0	0.577*
(Chloromethyl)benzene	C_7H_7Cl	81.2	0.706
1-Chloronaphthalene	$C_{10}H_7Cl$	107.6	0.786
1-Chloro-2-nitrobenzene	$C_6H_4ClNO_2$	75.5	0.656
1-Chloro-3-nitrobenzene	$C_6H_4ClNO_2$	77.2	0.658
1-Chloro-4-nitrobenzene	$C_6H_4ClNO_2$	74.7	0.615
1-Chlorooctane	$C_8H_{17}Cl$	114.9	0.675
2-Chlorophenol	C_6H_5ClO	77.3	0.760
3-Chlorophenol	C_6H_5ClO	77.6	0.752
4-Chlorophenol	C_6H_5ClO	77.7	0.765
1-Chloropropane	C_3H_7Cl	56.0	0.635
2-Chloropropene	C_3H_5Cl	47.8	0.563
3-Chloropropene	C_3H_5Cl	47.8	0.586
2-Chlorotoluene	C_7H_7Cl	82.4	0.705
3-Chlorotoluene	C_7H_7Cl	79.7	0.677
4-Chlorotoluene	C_7H_7Cl	80.3	0.679
Chlorotrifluoroethene	C_2ClF_3	49.1	0.649*
Chlorotrifluoromethane	$CClF_3$	45.3	
Chrysene	$C_{18}H_{12}$	148.0	0.826
o-Cresol	C_7H_8O	70.8	0.676
m-Cresol	C_7H_8O	71.9	0.687
p-Cresol	C_7H_8O	71.9	0.677
Cyanamide	CH_2N_2	24.8	0.756
Cyanogen	C_2N_2	21.6	0.396*
Cyanogen chloride	$CClN$	32.4	0.625
Cyclobutane	C_4H_8	40.0	0.502
Cycloheptane	C_7H_{14}	73.9	0.609
1,4-Cyclohexadiene	C_6H_8	48.7	0.515
Cyclohexane	C_6H_{12}	66.1	0.608
Cyclohexanol	$C_6H_{12}O$	73.4	0.705
Cyclohexanone	$C_6H_{10}O$	62.0	0.599
Cyclohexene	C_6H_{10}	58.0	0.573
Cyclooctane	C_8H_{16}	85.3	0.635
Cyclopentane	C_5H_{10}	59.2	0.629
Cyclopentanol	$C_5H_{10}O$	64.0	0.705
Cyclopentanone	C_5H_8O	51.6	0.582
Cyclopropane	C_3H_6	39.2	0.575*
cis-Decahydronaphthalene	$C_{10}H_{18}$	107.0	0.694
trans-Decahydronaphthalene	$C_{10}H_{18}$	107.6	0.674
Decane	$C_{10}H_{22}$	119.5	0.610
1,2-Dibromoethane	$C_2H_4Br_2$	78.8	0.910
Dibromomethane	CH_2Br_2	65.1	0.935
Dibutylamine	$C_8H_{19}N$	103.7	0.615

Name	Mol. form.	$-\chi_m/10^{-6}$ $cm^3\ mol^{-1}$	$-\kappa/10^{-6}$
Dichloroacetyl chloride	C_2HCl_3O	69.0	0.717
o-Dichlorobenzene	$C_6H_4Cl_2$	84.4	0.750
m-Dichlorobenzene	$C_6H_4Cl_2$	84.1	0.737
p-Dichlorobenzene	$C_6H_4Cl_2$	81.7	0.693
Dichlorodifluoromethane	CCl_2F_2	52.2	
1,1-Dichloroethane	$C_2H_4Cl_2$	57.4	0.682
1,2-Dichloroethane	$C_2H_4Cl_2$	59.6	0.750
1,1-Dichloroethene	$C_2H_2Cl_2$	49.2	0.616
cis-1,2-Dichloroethene	$C_2H_2Cl_2$	51.0	0.675
trans-1,2-Dichloroethene	$C_2H_2Cl_2$	48.9	0.634
Dichloromethane	CH_2Cl_2	46.6	0.728
1,1-Diethoxyethane	$C_6H_{14}O_2$	81.4	0.569
Diethylamine	$C_4H_{11}N$	56.8	0.548
N,N-Diethylaniline	$C_{10}H_{15}N$	107.9	0.673
Diethyl carbonate	$C_5H_{10}O_3$	75.4	0.619
Diethyl ether	$C_4H_{10}O$	55.1	0.531
Diethyl malonate	$C_7H_{12}O_4$	91.8	0.605
Diethyl oxalate	$C_6H_{10}O_4$	81.7	0.603
Diethyl phthalate	$C_{12}H_{14}O_4$	127.5	0.707
Diethyl succinate	$C_8H_{14}O_4$	105.0	0.627
Diiodomethane	CH_2I_2	93.1	1.154
Dimethoxymethane	$C_3H_8O_2$	47.3	0.534
N,N-Dimethylaniline	$C_8H_{11}N$	85.1	0.671
2,2-Dimethylbutane	C_6H_{14}	76.2	0.570
2,3-Dimethylbutane	C_6H_{14}	76.2	0.585
2,3-Dimethyl-2-butene	C_6H_{12}	65.9	0.554
Dimethyl ether	C_2H_6O	26.3	
2,6-Dimethyl-4-heptanone	$C_9H_{18}O$	104.3	0.591
3,4-Dimethylhexane	C_8H_{18}	99.1	0.620
Dimethyl oxalate	$C_4H_6O_4$	55.7	0.553
2,2-Dimethylpentane	C_7H_{16}	87.0	0.585
2,3-Dimethylpentane	C_7H_{16}	87.5	0.603
2,4-Dimethylpentane	C_7H_{16}	87.5	0.587
3,3-Dimethylpentane	C_7H_{16}	89.5	0.620
2,4-Dimethyl-3-pentanone	$C_7H_{14}O$	81.1	0.576
2,4-Dimethylpyridine	C_7H_9N	71.3	0.619
2,6-Dimethylpyridine	C_7H_9N	72.5	0.624
Dimethyl sulfide	C_2H_6S	44.9	0.613
Dimethyl terephthalate	$C_{10}H_{10}O_4$	101.6	0.562
1,4-Dioxane	$C_4H_8O_2$	52.2	0.612
Diphenylacetylene	$C_{14}H_{10}$	116	0.629
Diphenylamine	$C_{12}H_{11}N$	108.4	0.742
1,2-Diphenylethane	$C_{14}H_{14}$	127.8	0.686
Diphenylmethane	$C_{13}H_{12}$	116.0	0.690
Dipropyl ether	$C_6H_{14}O$	79.4	0.580
Dodecanoic acid	$C_{12}H_{24}O_2$	113.0	0.490
1,2-Epoxybutane	C_4H_8O	54.8	0.631
Ethane	C_2H_6	26.8	0.485*
1,2-Ethanediamine	$C_2H_8N_2$	46.5	0.695
1,2-Ethanediol	$C_2H_6O_2$	38.8	0.696
Ethanethiol	C_2H_6S	47.0	0.629
Ethanol	C_2H_6O	33.6	0.576
Ethoxybenzene	$C_8H_{10}O$	84.5	0.668
Ethyl acetate	$C_4H_8O_2$	54.1	0.553
Ethyl acetoacetate	$C_6H_{10}O_3$	71.7	0.571
N-Ethylaniline	$C_8H_{11}N$	85.6	0.680
Ethylbenzene	C_8H_{10}	77.2	0.627
Ethyl benzoate	$C_9H_{10}O_2$	93.8	0.651
Ethyl carbamate	$C_3H_7NO_2$	57.0	0.631
Ethylcyanoacetate	$C_5H_7NO_2$	67.3	0.634

Name	Mol. form.	$-\chi_m/10^{-6}$ cm³ mol⁻¹	$-\kappa/10^{-6}$
Ethylene	C_2H_4	18.8	0.381*
Ethyl formate	$C_3H_6O_2$	43.0	0.534
3-Ethylhexane	C_8H_{18}	97.8	0.611
Ethyl 3-methylbutanoate	$C_7H_{14}O_2$	91.1	0.606
3-Ethylpentane	C_7H_{16}	86.2	0.601
Ethyl propanoate	$C_5H_{10}O_2$	66.3	0.574
Ethyl vinyl ether	C_4H_8O	47.9	0.504
Fluorobenzene	C_6H_5F	58.4	0.621
Fluoromethane	CH_3F	17.8	0.291*
Formaldehyde	CH_2O	18.6	0.505*
Formamide	CH_3NO	23.0	0.579
Formic acid	CH_2O_2	19.9	0.527
Fumaric acid	$C_4H_4O_4$	49.1	0.692
Furan	C_4H_4O	43.1	0.602
Furfural	$C_5H_4O_2$	47.2	0.570
Furfuryl alcohol	$C_5H_6O_2$	61.0	0.702
D-Glucitol	$C_6H_{14}O_6$	107.8	0.881
Glycerol	$C_3H_8O_3$	57.1	0.782
Glycine	$C_2H_5NO_2$	39.6	0.612
Heptanal	$C_7H_{14}O$	81.0	0.577
Heptane	C_7H_{16}	85.2	0.578
Heptanoic acid	$C_7H_{14}O_2$	86.6	0.607
1-Heptanol	$C_7H_{16}O$	91.7	0.649
4-Heptanol	$C_7H_{16}O$	92.1	0.649
2-Heptanone	$C_7H_{14}O$	80.5	0.572
3-Heptanone	$C_7H_{14}O$	80.7	0.578
4-Heptanone	$C_7H_{14}O$	80.5	0.576
1-Heptene	C_7H_{14}	77.8	0.552
Hexachlorobenzene	C_6Cl_6	147.0	1.055
Hexachloroethane	C_2Cl_6	112.8	0.996
Hexadecane	$C_{16}H_{34}$	187.6	0.638
Hexadecanoic acid	$C_{16}H_{32}O_2$	198.6	0.660
1-Hexadecanol	$C_{16}H_{34}O$	183.5	0.620
1,5-Hexadiene	C_6H_{10}	55.1	0.461
Hexamethylbenzene	$C_{12}H_{18}$	122.5	0.802
Hexanal	$C_6H_{12}O$	69.4	0.578
Hexane	C_6H_{14}	74.1	0.568
1,6-Hexanediol	$C_6H_{14}O_2$	84.3	0.770
Hexanoic acid	$C_6H_{12}O_2$	78.1	0.619
1-Hexanol	$C_6H_{14}O$	79.2	0.631
2-Hexanone	$C_6H_{12}O$	69.2	0.561
3-Hexanone	$C_6H_{12}O$	69.0	0.559
1-Hexene	C_6H_{12}	66.4	0.527
Hexyl acetate	$C_8H_{16}O_2$	100.9	0.614
1-Hexyne	C_6H_{10}	64.5	0.562
p-Hydroquinone	$C_6H_6O_2$	64.7	0.781
2-Hydroxybenzoic acid	$C_7H_6O_3$	75	0.784
Indene	C_9H_8	83	0.712
1*H*-Indole	C_8H_7N	85.0	0.885
Iodobenzene	C_6H_5I	92.0	0.826
1-Iodobutane	C_4H_9I	93.6	0.822
Iodoethane	C_2H_5I	69.1	0.858
Iodomethane	CH_3I	57.2	0.918
1-Iodopropane	C_3H_7I	84.3	0.867
Isobutane	C_4H_{10}	50.5	0.479*
Isobutene	C_4H_8	40.8	0.428*
Isobutyl acetate	$C_6H_{12}O_2$	78.7	0.590
Isobutylamine	$C_4H_{11}N$	59.8	0.592
Isobutylbenzene	$C_{10}H_{14}$	101.7	0.646
Isobutyl formate	$C_5H_{10}O_2$	66.8	0.574
Isopentane	C_5H_{12}	64.4	0.553
Isopentyl acetate	$C_7H_{14}O_2$	89.4	0.602
Isopentyl formate	$C_6H_{12}O_2$	78.4	0.592
Isophthalic acid	$C_8H_6O_4$	84.6	0.783
Isopropenylbenzene	C_9H_{10}	80.0	0.616
Isopropyl acetate	$C_5H_{10}O_2$	67.0	0.568
Isopropylbenzene	C_9H_{12}	89.5	0.643
1-Isopropyl-4-methylbenzene	$C_{10}H_{14}$	102.8	0.657
Isoquinoline	C_9H_7N	83.9	0.709
d-Limonene	$C_{10}H_{16}$	98.0	0.605
Maleic acid	$C_4H_4O_4$	49.6	0.679
Maleic anhydride	$C_4H_2O_3$	35.8	0.480
Methane	CH_4	17.4	0.459*
Methanol	CH_4O	21.4	0.529
2-Methoxyaniline	C_7H_9NO	79.1	0.702
Methylamine	CH_5N	27.0	0.570*
2-Methylaniline	C_7H_9N	74.9	0.698
3-Methylaniline	C_7H_9N	74.6	0.688
4-Methylaniline	C_7H_9N	72.5	0.651
N-Methylaniline	C_7H_9N	74.1	0.684
Methyl benzoate	$C_8H_8O_2$	81.6	0.650
2-Methyl-1,3-butadiene	C_5H_8	46.0	0.459
3-Methylbutanoic acid	$C_5H_{10}O_2$	67.7	0.617
2-Methyl-2-butene	C_5H_{10}	54.7	0.517
Methylcyclohexane	C_7H_{14}	78.9	0.618
Methylcyclopentane	C_6H_{12}	70.2	0.624
Methyl formate	$C_2H_4O_2$	31.1	0.503
4-Methylheptane	C_8H_{18}	97.3	0.600
Methyl methacrylate	$C_5H_8O_2$	57.3	0.537
1-Methylnaphthalene	$C_{11}H_{10}$	102.9	0.738
2-Methylnaphthalene	$C_{11}H_{10}$	102.7	0.726
Methyloxirane	C_3H_6O	42.5	0.629
2-Methylpentane	C_6H_{14}	75.3	0.568
3-Methylpentane	C_6H_{14}	75.5	0.578
4-Methyl-2-pentanol	$C_6H_{14}O$	80.4	0.635
4-Methyl-2-pentanone	$C_6H_{12}O$	69.7	0.554
Methyl propanoate	$C_4H_8O_2$	54.5	0.566
2-Methylpropanoic acid	$C_4H_8O_2$	56.1	0.616
2-Methyl-1-propanol	$C_4H_{10}O$	57.6	0.623
2-Methyl-2-propanol	$C_4H_{10}O$	56.6	0.602
4-Methylpyridine	C_6H_7N	59.8	0.613
Methyl salicylate	$C_8H_8O_3$	86.3	0.670
Morpholine	C_4H_9NO	55.0	0.632
Naphthalene	$C_{10}H_8$	91.6	0.733
1-Naphthol	$C_{10}H_8O$	96.2	0.733
2-Naphthol	$C_{10}H_8O$	96.8	0.859
1-Naphthylamine	$C_{10}H_9N$	92.5	0.661
2-Naphthylamine	$C_{10}H_9N$	98.0	1.123
Neopentane	C_5H_{12}	63.0	0.511*
2-Nitroaniline	$C_6H_6N_2O_2$	67.4	0.440
3-Nitroaniline	$C_6H_6N_2O_2$	69.7	0.455
4-Nitroaniline	$C_6H_6N_2O_2$	68.0	0.701
Nitrobenzene	$C_6H_5NO_2$	61.8	0.604
Nitroethane	$C_2H_5NO_2$	35.4	0.493
Nitromethane	CH_3NO_2	21.0	0.391
2-Nitrophenol	$C_6H_5NO_3$	68.9	0.641
3-Nitrophenol	$C_6H_5NO_3$	65.9	0.606
4-Nitrophenol	$C_6H_5NO_3$	66.9	0.711
1-Nitropropane	$C_3H_7NO_2$	45.0	0.503
2-Nitropropane	$C_3H_7NO_2$	45.4	0.500

Name	Mol. form.	$-\chi_m/10^{-6}$ cm³ mol⁻¹	$-\kappa/10^{-6}$
2-Nitrotoluene	$C_7H_7NO_2$	72.2	0.611
3-Nitrotoluene	$C_7H_7NO_2$	72.7	0.614
4-Nitrotoluene	$C_7H_7NO_2$	73.3	0.590
Nonane	C_9H_{20}	108.1	0.606
1-Nonene	C_9H_{18}	100.1	0.575
Octadecanoic acid	$C_{18}H_{36}O_2$	220.8	0.730
cis-9-Octadecenoic acid	$C_{18}H_{34}O_2$	208.5	0.660
Octane	C_8H_{18}	96.6	0.591
Octanoic acid	$C_8H_{16}O_2$	99.5	0.626
1-Octanol	$C_8H_{18}O$	101.6	0.645
1-Octene	C_8H_{16}	88.8	0.566
Oxirane	C_2H_4O	30.5	0.611
Paraldehyde	$C_6H_{12}O_3$	86.1	0.648
Pentachloroethane	C_2HCl_5	99.1	0.823
Pentanal	$C_5H_{10}O$	57.5	0.540
Pentane	C_5H_{12}	63.1	0.548
1,5-Pentanediol	$C_5H_{12}O_2$	73.5	0.700
2,4-Pentanedione	$C_5H_8O_2$	54.9	0.533
Pentanoic acid	$C_5H_{10}O_2$	66.5	0.608
1-Pentanol	$C_5H_{12}O$	67.0	0.619
2-Pentanol	$C_5H_{12}O$	69.1	0.634
2-Pentanone	$C_5H_{10}O$	57.5	0.540
3-Pentanone	$C_5H_{10}O$	57.7	0.542
1-Pentene	C_5H_{10}	54.6	0.499
Pentyl acetate	$C_7H_{14}O_2$	88.9	0.598
Pentylamine	$C_5H_{13}N$	69.3	0.600
Perylene	$C_{20}H_{12}$	167.5	0.896
Phenanthrene	$C_{14}H_{10}$	127.6	0.702
Phenol	C_6H_6O	60.6	0.679
Phthalic acid	$C_8H_6O_4$	83.6	1.097
Phthalic anhydride	$C_8H_4O_3$	66.7	0.688
α-Pinene	$C_{10}H_{16}$	100.7	0.631
β-Pinene	$C_{10}H_{16}$	101.9	0.643
Piperidine	$C_5H_{11}N$	64.2	0.649
Propanal	C_3H_6O	34.2	0.510
Propane	C_3H_8	38.6	0.432*
1,3-Propanediol	$C_3H_8O_2$	50.2	0.695
Propanenitrile	C_3H_5N	38.6	0.548
Propanoic acid	$C_3H_6O_2$	43.5	0.580
1-Propanol	C_3H_8O	45.2	0.601
2-Propanol	C_3H_8O	45.7	0.594
Propene	C_3H_6	30.7	0.368*
Propyl acetate	$C_5H_{10}O_2$	65.9	0.569
Propylbenzene	C_9H_{12}	89.1	0.637
Propyl formate	$C_4H_8O_2$	55.0	0.566
Propyl propanoate	$C_6H_{12}O_2$	77.7	0.586
Pyrazine	$C_4H_4N_2$	37.8	0.487
Pyrene	$C_{16}H_{10}$	147	0.924
Pyridine	C_5H_5N	48.7	0.605
Pyrimidine	$C_4H_4N_2$	43.1	0.581
Pyrocatechol	$C_6H_6O_2$	68.2	0.832
Pyrrole	C_4H_5N	48.6	0.703
Pyrrolidine	C_4H_9N	54.8	0.662
Quinoline	C_9H_7N	86.1	0.732
Resorcinol	$C_6H_6O_2$	67.2	0.780
Safrole	$C_{10}H_{10}O_2$	97.5	0.661
Salicylaldehyde	$C_7H_6O_2$	66.8	0.639
Styrene	C_8H_8	68.2	0.590
Succinic acid	$C_4H_6O_4$	58.0	0.772
Succinic anhydride	$C_4H_4O_3$	47.5	0.570
Terephthalic acid	$C_8H_6O_4$	83.5	0.763
o-Terphenyl	$C_{18}H_{14}$	150.4	0.731
m-Terphenyl	$C_{18}H_{14}$	155.5	0.810
p-Terphenyl	$C_{18}H_{14}$	156.0	0.813
1,1,2,2-Tetrabromoethane	$C_2H_2Br_4$	123.4	1.059
Tetrabromomethane	CBr_4	93.7	0.837
1,1,2,2-Tetrachloroethane	$C_2H_2Cl_4$	89.8	0.853
Tetrachloroethene	C_2Cl_4	81.6	0.799
Tetrachloromethane	CCl_4	66.6	0.690
Tetradecane	$C_{14}H_{30}$	166.2	0.636
Tetradecanoic acid	$C_{14}H_{28}O_2$	176.0	0.664
Tetrahydrofurfuryl alcohol	$C_5H_{10}O_2$	69.4	0.715
Tetraiodomethane	CI_4	136	1.107
1,2,4,5-Tetramethylbenzene	$C_{10}H_{14}$	101.2	0.632
Tetranitromethane	CN_4O_8	43.0	0.359
Thiophene	C_4H_4S	57.3	0.725
Toluene	C_7H_8	66.1	0.619
o-Toluic acid	$C_8H_8O_2$	84.3	0.658
m-Toluic acid	$C_8H_8O_2$	83.0	0.643
p-Toluic acid	$C_8H_8O_2$	82.4	
Tribromomethane	$CHBr_3$	82.6	0.941
Trichloroacetaldehyde	C_2HCl_3O	73.0	0.749
Trichloroacetic acid	$C_2HCl_3O_2$	73.0	0.720
Trichloroethene	C_2HCl_3	65.8	0.733
Trichlorofluoromethane	CCl_3F	58.7	0.636
Trichloromethane	$CHCl_3$	59.3	0.735
Trichloronitromethane	CCl_3NO_2	75.3	0.759
Tridecane	$C_{13}H_{28}$	153.7	0.631
Triethylamine	$C_6H_{15}N$	83.3	0.599
Trifluoroacetic acid	$C_2HF_3O_2$	43.3	0.583
Triiodomethane	CHI_3	117.1	1.192
1,3,5-Trimethylbenzene	C_9H_{12}	92.3	0.662
2,2,4-Trimethylpentane	C_8H_{18}	98.3	0.592
2,3,4-Trimethylpentane	C_8H_{18}	99.8	0.628
2,4,6-Trimethylpyridine	$C_8H_{11}N$	83.1	0.629
Undecane	$C_{11}H_{24}$	131.8	0.624
Urea	CH_4N_2O	33.5	0.738
Vinyl acetate	$C_4H_6O_2$	46.4	0.499
Vinyl formate	$C_3H_4O_2$	34.7	0.465
o-Xylene	C_8H_{10}	77.8	0.642
m-Xylene	C_8H_{10}	76.6	0.620
p-Xylene	C_8H_{10}	76.8	0.620

* Value refers to a liquefied gas.

Section 4
Properties of the Elements and Inorganic Compounds

The Elements 4-1
Physical Constants of Inorganic Compounds 4-40
Physical Properties of the Rare-Earth Metals 4-52
Melting, Boiling, Triple, and Critical Points of the Elements 4-62
Heat Capacity of the Elements at 25 °C 4-65
Vapor Pressure of the Metallic Elements — Equations 4-66
Vapor Pressure of the Metallic Elements — Data 4-69
Density of Molten Elements and Representative Salts 4-71
Magnetic Susceptibility of the Elements and Inorganic Compounds 4-73
Index of Refraction of Inorganic Liquids and Liquid Elements 4-79
Physical and Optical Properties of Minerals 4-80
Crystallographic Data on Minerals 4-87

THE ELEMENTS

C. R. Hammond and David R. Lide

Each chemical element is characterized by the number of protons, Z, contained in the atomic nucleus. As of 2021, a total of 118 elements have been positively identified and assigned names and symbols. Elements can occur as two or more isotopes, differing in the number of neutrons, N, in the nucleus. An isotope is characterized by the sum of the protons and neutrons in its nucleus, which is called the mass number $A = Z + N$. Including those produced artificially, over 3000 isotopes have been identified, most of which are unstable to radioactive decay. A total of 288 stable isotopes of 84 elements exist on earth.[a]

The table below summarizes key properties of the 118 elements. More detailed information can be found in the tables of atomic weights and physical properties. Individual descriptions of the elements follow the table. The definitions of the data columns in the table are as follows.

Column heading	Definition
Element	Name of the element; elements are listed alphabetically
Symbol	IUPAC symbol for the element
Z	Atomic number (number of protons)
Atomic wt.	Standard atomic weight; see also "Standard Atomic Weights" table in Section 1
CAS Reg. No.	Chemical Abstracts Service Registry Number for the element
State at NTP	Physical state of the element at normal room temperature (25 °C) and one atmosphere pressure (101.325 kPa)
Primary valences	The primary valence (oxidation) states the elements in forming stoichiometric chemical compounds
Mass no. of stable isotopes	Mass number (protons plus neutrons) of stable isotopes
Melting point	Normal melting point in °C; the notation "tp" indicates a triple point, where solid, liquid, and gas are in equilibrium at the indicated pressure; other notes indicate the phase involved
Boiling point	Normal boiling point in °C; this is the temperature at which the liquid phase is in equilibrium with the vapor at a pressure of 760 mmHg (101.325 kPa). A notation "sp" following the value indicates a sublimation point, where the vapor pressure of the solid phase reaches 760 mmHg
Density	Density (mass per unit volume) in g cm^{-3}; the temperature in °C is indicated by a superscript; values refer to the liquid or solid phase, and all values are true densities, not specific gravities; the number of decimal places gives a rough estimate of the accuracy of the value

[a] These numbers include radioactive isotopes whose half-lives are comparable to or greater than the age of the earth, making it possible to define a terrestrial isotopic abundance. See Meija, J., et al., Isotopic Compositions of the Elements 2013, *Pure Appl. Chem.* 2016; 88(3): 293–306.

Properties of the Elements

Element	Symbol	Z	Atomic wt.	CAS Reg. No.	State at NTP	Primary valences	Mass no. of stable isotopes[a]	Melting point in °C	Boiling point in °C	Density in g cm^{-3}
Actinium	Ac	89		7440-34-8	sol	+3		1050	≈3200	10
Aluminum	Al	13	26.9815384(3)	7429-90-5	sol	+3	27	660.323	2519	2.70
Americium	Am	95		7440-35-9	sol	+3, +4, +5, +6		1176	≈2011	12
Antimony	Sb	51	121.760(1)	7440-36-0	sol (gray Sb)	+3, +5	121, 123	630.628	1587	6.68
Argon	Ar	18	[39.792, 39.963]	7440-37-1	gas	0	36, 38, 40	-189.34	-185.848	$1.395^{-185.8}$ liq
Arsenic	As	33	74.921595(6)	7440-38-2	sol (gray As)	+3, +5, -3	75	817	616 sp	5.75
Astatine	At	85		7440-68-8	sol			302		
Barium	Ba	56	137.327(7)	7440-39-3	sol	+2	130, 132, 134, 135, 136, 137, 138	727	≈1845	3.62
Berkelium	Bk	97		7440-40-6	sol	+3, +4		986		13.25
Beryllium	Be	4	9.0121831(5)	7440-41-7	sol	+2	9	1287	2468	1.85
Bismuth	Bi	83	208.98040(1)	7440-69-9	sol	+3, +5	209	271.402	1564	9.79
Bohrium	Bh	107		54037-14-8	sol[b]	+7[b]				
Boron	B	5	[10.806, 10.821]	7440-42-8	sol	+3	10, 11	2077	4000	2.34
Bromine	Br	35	[79.901, 79.907]	7726-95-6	liq	+1, +5, -1	79, 81	-7.2	58.8	3.1028
Cadmium	Cd	48	112.414(4)	7440-43-9	sol	+2	106, 108, 110, 111, 112, 113, 114, 116	321.069	767	8.69
Calcium	Ca	20	40.078(4)	7440-70-2	sol	+2	40, 42, 43, 44, 46, 48	842	1484	1.54
Californium	Cf	98		7440-71-3	sol	+3		900		15.1
Carbon	C	6	[12.0096, 12.0116]	7440-44-0	sol (graphite)	+2, +4, -4	12, 13	4489 tp (10.3 MPa)	3825 sp	2.2
Carbon	C	6	[12.0096, 12.0116]	7440-44-0	sol (diamond)	+2, +4, -4	12, 13	4440 (12.4 GPa)		3.513
Cerium	Ce	58	140.116(1)	7440-45-1	sol	+3, +4	136, 138, 140, 142	799	3443	6.770

Element	Symbol	Z	Atomic wt.	CAS Reg. No.	State at NTP	Primary valences	Mass no. of stable isotopes[a]	Melting point in °C	Boiling point in °C	Density in g cm^{-3}
Cesium	Cs	55	132.90545196(6)	7440-46-2	sol	+1	133	28.5	671	1.873
Chlorine	Cl	17	[35.446, 35.457]	7782-50-5	gas	+1, +5, +7, -1	35,37	-101.5	-34.04	$1.565^{-34.0}$ liq
Chromium	Cr	24	51.9961(6)	7440-47-3	sol	+2, +3, +6	50, 52, 53, 54	1907	2671	7.15
Cobalt	Co	27	58.933194(3)	7440-48-4	sol	+2, +3	59	1495	2927	8.86
Copernicium	Cn	112		54084-26-3	liq/gas[b]	+2, +4[b]				
Copper	Cu	29	63.546(3)	7440-50-8	sol	+1, +2	63, 65	1084.62	2560	8.96
Curium	Cm	96		7440-51-9	sol	+3		1345		13.51
Darmstadtium	Ds	110		54083-77-1	sol[b]	+2,+4,+6[b]				
Dubnium	Db	105		53850-35-4	sol[b]	+5[b]				
Dysprosium	Dy	66	162.500(1)	7429-91-6	sol	+3	156, 158, 160, 161, 162, 163, 164	1412	2567	8.55
Einsteinium	Es	99		7429-92-7	sol	+3		860		
Erbium	Er	68	167.259(3)	7440-52-0	sol	+3	162, 164, 166, 167, 168, 170	1529	2868	9.07
Europium	Eu	63	151.964(1)	7440-53-1	sol	+2, +3	151, 153	822	1529	5.24
Fermium	Fm	100		7440-72-4	sol	+3		1527		
Flerovium	Fl	114		54085-16-4	liq/gas[b]	+2, +4[b]				
Fluorine	F	9	18.998403163(6)	7782-41-4	gas	-1	19	-219.67	-188.11	$1.5127^{-188.1}$ liq
Francium	Fr	87		7440-73-5	sol/liq[b]	+1		≈21		
Gadolinium	Gd	64	157.25(3)	7440-54-2	sol	+3	152, 154, 155, 156, 157, 158, 160	1313	3273	7.90
Gallium	Ga	31	69.723(1)	7440-55-3	sol	+3	69, 71	29.7646	2229	5.91
Germanium	Ge	32	72.630(8)	7440-56-4	sol	+2, +4	70, 72, 73, 74, 76	938.25	2833	5.3234
Gold	Au	79	196.966570(4)	7440-57-5	sol	+1, +3	197	1064.18	2836	19.3
Hafnium	Hf	72	178.486(2)	7440-58-6	sol	+4	174, 176, 177, 178, 179, 180	2233	4600	13.3
Hassium	Hs	108		54037-57-9	sol[b]	+8[b]				
Helium	He	2	4.002602(2)	7440-59-7	gas	0	3, 4		-268.928	$0.1250^{-268.9}$ liq
Holmium	Ho	67	164.930328(7)	7440-60-0	sol	+3	165	1472	2700	8.80
Hydrogen	H	1	[1.00784, 1.00811]	1333-74-0	gas	+1	1, 2	-259.16	-252.879	$0.07083^{-252.9}$ liq
Indium	In	49	114.818(1)	7440-74-6	sol	+3	113, 115	156.5985	2027	7.31
Iodine	I	53	126.90447(3)	7553-56-2	sol	+1, +5, +7, -1	127	113.7	184.4	4.933
Iridium	Ir	77	192.217(2)	7439-88-5	sol	+3, +4	191, 193	2446	4428	22.562^{20}
Iron	Fe	26	55.845(2)	7439-89-6	sol	+2, +3	54, 56, 57, 58	1538	2861	7.87
Krypton	Kr	36	83.798(2)	7439-90-9	gas	0	78, 80, 82, 83, 84, 86	-157.37	-153.415	$2.417^{-153.4}$ liq
Lanthanum	La	57	138.90547(7)	7439-91-0	sol	+3	138, 139	920	3464	6.15
Lawrencium	Lr	103		22537-19-5	sol	+3		1627		
Lead	Pb	82	207.2(1)	7439-92-1	sol	+2, +4	204, 206, 207, 208	327.462	1749	11.3
Lithium	Li	3	[6.938, 6.997]	7439-93-2	sol	+1	6, 7	180.50	1342	0.534
Livermorium	Lv	116		54100-71-9	sol[b]	+2, +4[b]				
Lutetium	Lu	71	174.9668(1)	7439-94-3	sol	+3	175, 176	1663	3402	9.84
Magnesium	Mg	12	[24.304, 24.307]	7439-95-4	sol	+2	24, 25, 26	650	1090	1.74
Manganese	Mn	25	54.938043(2)	7439-96-5	sol	+2, +3, +4, +7	55	1246	2061	7.3
Meitnerium	Mt	109		54038-01-6	sol[b]	+1, +3, +6[b]				
Mendelevium	Md	101		7440-11-1	sol	+2, +3		827		
Mercury	Hg	80	200.592(3)	7439-97-6	liq	+1, +2	196, 198, 199, 200, 201, 202, 204	-38.8290	356.619	13.53359^{25} liq
Molybdenum	Mo	42	95.95(1)	7439-98-7	sol	+6	92, 94, 95, 96, 97, 98	2622	4639	10.2
Moscovium	Mc	115		54085-64-2	sol[b]	+1, +3[b]				
Neodymium	Nd	60	144.242(3)	7440-00-8	sol	+3	142, 143, 144, 145, 146, 148, 150	1016	3074	7.01
Neon	Ne	10	20.1797(6)	7440-01-9	gas	0	20, 21, 22	-248.59	-246.046	$1.207^{-246.0}$ liq

Element	Symbol	Z	Atomic wt.	CAS Reg. No.	State at NTP	Primary valences	Mass no. of stable isotopes[a]	Melting point in °C	Boiling point in °C	Density in g cm^{-3}
Neptunium	Np	93		7439-99-8	sol	+3, +4, +5, +6		644	≈3902	20.2
Nickel	Ni	28	58.6934(4)	7440-02-0	sol	+2, +3	58, 60, 61, 62, 64	1455	2913	8.90
Nihonium	Nh	113		54084-70-7	sol[b]	+1, +3[b]				
Niobium	Nb	41	92.90637(1)	7440-03-1	sol	+3, +5	93	2477	4741	8.57
Nitrogen	N	7	[14.00643, 14.00728]	7727-37-9	gas	+1, +2, +3, +4, +5, -1, -2, -3	14, 15	-210.0	-195.795	$0.8061^{-195.8}$ liq
Nobelium	No	102		10028-14-5	sol	+2, +3		827		
Oganesson	Og	118		54144-19-3	sol[b]	0, +2, +4[b]				
Osmium	Os	76	190.23(3)	7440-04-2	sol	+3, +4	184, 186, 187, 188, 189, 190, 192	3033	5008	22.587^{20}
Oxygen	O	8	[15.99903, 15.99977]	7782-44-7	gas	-2	16, 17, 18	-218.79	-182.962	$1.141^{-183.0}$ liq
Palladium	Pd	46	106.42(1)	7440-05-3	sol	+2, +3	102, 104, 105, 106, 108, 110	1554.8	2963	12.0
Phosphorus	P	15	30.973761998(5)	7723-14-0	sol (white P)	+3, +5, -3	31	44.15	280.5	1.823
Phosphorus	P	15	30.973761998(5)	7723-14-0	sol (red P)	+3, +5, -3	31	579.2	431 sp	2.16
Platinum	Pt	78	195.084(9)	7440-06-4	sol	+2, +4	190, 192, 194, 195, 196, 198	1768.2	3825	21.5
Plutonium	Pu	94		7440-07-5	sol	+3, +4, +5, +6		640	3228	19.7
Polonium	Po	84		7440-08-6	sol	+2, +4		254	962	9.20
Potassium	K	19	39.0983(1)	7440-09-7	sol	+1	39, 40, 41	63.5	759	0.89
Praseodymium	Pr	59	140.90766(1)	7440-10-0	sol	+3	141	931	3520	6.77
Promethium	Pm	61		7440-12-2	sol	+3		1042	≈3000	7.26
Protactinium	Pa	91	231.03588(1)	7440-13-3	sol	+4, +5	231	1572		15.4
Radium	Ra	88		7440-14-4	sol	+2		696		5
Radon	Rn	86		10043-92-2	gas	0		-71	-61.7	0.009074^{25} gas
Rhenium	Re	75	186.207(1)	7440-15-5	sol	+4, +6, +7	185, 187	3185	5590	20.8
Rhodium	Rh	45	102.90549(2)	7440-16-6	sol	+3	103	1963	3695	12.4
Roentgenium	Rg	111		54386-24-2	sol[b]	+3, +5, -1[b]				
Rubidium	Rb	37	85.4678(3)	7440-17-7	sol	+1	85, 87	39.30	688	1.53
Ruthenium	Ru	44	101.07(2)	7440-18-8	sol	+3	100, 101, 102, 104	2333	4147	12.1
Rutherfordium	Rf	104		53850-36-5	sol[b]	+4				
Samarium	Sm	62	150.36(2)	7440-19-9	sol	+2, +3	144, 147, 148, 149, 150, 152, 154	1072	1794	7.52
Scandium	Sc	21	44.955908(5)	7440-20-2	sol	+3	45	1541	2836	2.99
Seaborgium	Sg	106		54038-81-2	sol[b]	+6[b]				
Selenium	Se	34	78.971(8)	7782-49-2	sol (gray Se)	+4, +6, -2	74, 76, 77, 78, 80, 82	220.8	685	4.809
Silicon	Si	14	[28.084, 28.086]	7440-21-3	sol	+2, +4, -4	28, 29	1414	3265	2.3296
Silver	Ag	47	107.8682(2)	7440-22-4	sol	+1	107, 109	961.78	2162	10.5
Sodium	Na	11	22.98976928(2)	7440-23-5	sol	+1	23	97.794	882.940	0.97
Strontium	Sr	38	87.62(1)	7440-24-6	sol	+2	84, 86, 87, 88	777	1377	2.64
Sulfur	S	16	[32.059, 32.076]	7704-34-9	sol (monoclinic S)	+4, +6, -2	32, 33, 34, 36	115.21	444.61	2.00
Sulfur	S	16	[32.059, 32.076]	7704-34-10	sol (rhombic S)	+4, +6, -2	32, 33, 34, 36	95.2 trans monocl	444.61	2.07
Tantalum	Ta	73	180.94788(2)	7440-25-7	sol	+5	180, 181	3017	5455	16.4
Technetium	Tc	43		7440-26-8	sol	+4, +6, +7		2157	4262	11
Tellurium	Te	52	127.60(3)	13494-80-9	sol	+4, +6, -2	120, 122, 123, 124, 125, 126, 128, 130	449.51	988	6.232
Tennessine	Ts	117		54101-14-3	sol[b]	+1, +3, +5[b]				
Terbium	Tb	65	158.925354(8)	7440-27-9	sol	+3	159	1359	3230	8.23
Thallium	Tl	81	[204.382, 204.385]	7440-28-0	sol	+1, +3	203, 205	304	1473	11.8
Thorium	Th	90	232.0377(4)	7440-29-1	sol	+4	232	1750	4785	11.7
Thulium	Tm	69	168.934218(6)	7440-30-4	sol	+3	169	1545	1950	9.32

Element	Symbol	Z	Atomic wt.	CAS Reg. No.	State at NTP	Primary valences	Mass no. of stable isotopes[a]	Melting point in °C	Boiling point in °C	Density in g cm^{-3}
Tin	Sn	50	118.710(7)	7440-31-6	sol (white Sn)	+2, +4	112, 114, 115, 116, 117, 118, 119, 120, 122, 124	231.928	2586	7.287
Tin	Sn	50	118.710(7)	7440-31-5	sol (gray Sn)	+2, +4	112, 114, 115, 116, 117, 118, 119, 120, 122, 124	13.2 trans white Sn	2586	5.769
Titanium	Ti	22	47.867(1)	7440-32-6	sol	+2, +3, +4	46, 47, 48, 49, 50	1670	3287	4.506
Tungsten	W	74	183.84(1)	7440-33-7	sol	+6	180, 182, 183, 184, 186	3414	5555	19.3
Uranium	U	92	238.02891(3)	7440-61-1	sol	+3, +4, +5, +6	234. 235, 238	1135	4131	19.1
Vanadium	V	23	50.9415(1)	7440-62-2	sol	+2, +3, +4, +5	50, 51	1910	3407	6.0
Xenon	Xe	54	131.293(6)	7440-63-3	gas	0	124, 126, 128, 129, 130, 131, 132, 134, 136	-111.75	-108.099	$2.942^{-108.1}$ liq
Ytterbium	Yb	70	173.045(10)	7440-64-4	sol	+2, +3	168, 170, 171, 172, 173, 174, 176	824	1196	6.90
Yttrium	Y	39	88.90584(1)	7440-65-5	sol	+3	89	1522	3345	4.47
Zinc	Zn	30	65.38(2)	7440-66-6	sol	+2	64, 66, 67, 68, 70	419.527	907	7.134
Zirconium	Zr	40	91.224(2)	7440-67-7	sol	+4	90, 91, 92, 94, 96	1854	4406	6.52

[a] Includes radioactive isotopes of very long lifetimes, whose terrestrial abundance is effectively stable.
[b] Predicted.

Actinium — (Gr. *aktis, aktinos,* beam or ray), Ac. Discovered by Andre Debierne in 1899 and independently by F. Giesel in 1902. Occurs naturally in association with uranium minerals. Thirty-seven isotopes and isomers are now recognized. All are radioactive. Actinium-227, a decay product of uranium-235, is an alpha and beta emitter with a 21.77-year half-life. Its principal decay products are thorium-227 (18.72-day half-life), radium-223 (11.4-day half-life), and a number of short-lived products including radon, bismuth, polonium, and lead isotopes. In equilibrium with its decay products, it is a powerful source of alpha rays. Actinium metal has been prepared by the reduction of actinium fluoride with lithium vapor at about 1100 to 1300 °C. The chemical behavior of actinium is similar to that of the rare earths, particularly lanthanum. Purified actinium comes into equilibrium with its decay products at the end of 185 days, and then decays according to its 21.77-year half-life. It is about 150 times as active as radium, making it of value in the production of neutrons.

Aluminum — (L. *alumen, alum*), Al. The ancient Greeks and Romans used *alum* in medicine as an astringent, and as a mordant in dyeing. In 1761, de Morveau proposed the name *alumine* for the base in alum, and Lavoisier, in 1787, thought this to be the oxide of a still undiscovered metal. Wöhler is generally credited with having isolated the metal in 1827, although an impure form was prepared by Oersted two years earlier. In 1807, Davy proposed the name *alumium* for the metal, undiscovered at that time, and later agreed to change it to *aluminum*. Shortly thereafter, the name *aluminium* was adopted to conform with the "ium" ending of most elements, and this spelling is now in use elsewhere in the world. *Aluminium* was also the accepted spelling in the U.S. until 1925, at which time the American Chemical Society officially decided to use the name *aluminum* thereafter in their publications. The method of obtaining aluminum metal by the electrolysis of alumina dissolved in cryolite was discovered in 1886 by Hall in the U.S. and at about the same time by Heroult in France. Cryolite, a natural ore found in Greenland, is no longer widely used in commercial production, but has been replaced by an artificial mixture of sodium, aluminum, and calcium fluorides. Bauxite, an impure hydrated oxide ore, is found in large deposits in Jamaica, Australia, Suriname, Guyana, Russia, Arkansas, and elsewhere. The Bayer process is most commonly used today to refine bauxite so it can be accommodated in the Hall–Heroult refining process used to make most aluminum. Aluminum can now be produced from clay, but the process is not economically feasible at present. Aluminum is the most abundant metal to be found in the Earth's crust (8.1%) but is never found free in nature. In addition to the minerals mentioned above, it is found in feldspars, granite, and in many other common minerals. Natural aluminum is made of one isotope, ^{27}Al. Pure aluminum, a silvery-white metal, possesses many desirable characteristics. It is light, nontoxic, has a pleasing appearance, can easily be formed, machined, or cast, has a high thermal conductivity, and has excellent corrosion resistance. It is nonmagnetic and non-sparking, stands second among metals in the scale of malleability, and sixth in ductility. It is extensively used for kitchen utensils, outside building decoration, and in thousands of industrial applications where a strong, light, easily constructed material is needed. Although its electrical conductivity is only about 60% that of copper, it is used in electrical transmission lines because of its light weight. Even with its low critical temperature of just 1.175 K, aluminum is important to the field of superconducting electronics owing to its compatibility with standard microelectronic fabrication techniques and its ability to form a naturally insulating oxide surface layer that aids in Josephson junction formation. Pure aluminum is soft and lacks strength, but it can be alloyed with small amounts of copper, magnesium, silicon, manganese, and other elements to impart a variety of useful properties. These alloys are of vital importance in the construction of modern aircraft and rockets. Aluminum, evaporated in a vacuum, forms a highly reflective coating for both visible light and radiant heat. These coatings soon form a thin layer of the protective oxide and do not deteriorate as do silver coatings. They have found application in coatings for telescope mirrors, in making decorative paper, packages, toys, and in many other uses. The compounds of greatest importance are aluminum oxide, the sulfate, and the soluble sulfate with potassium (alum). The oxide, alumina, occurs naturally as ruby, sapphire, corundum, and emery, and is used in glassmaking and refractories. Synthetic ruby and sapphire have found application in the construction of lasers.

Americium — (the Americas), Am. Americium was the fourth transuranium element to be discovered; the isotope ^{241}Am was identified by Seaborg, James, Morgan, and Ghiorso late in 1944 at the wartime Metallurgical Laboratory of the University of Chicago as the result of successive neutron capture reactions by plutonium isotopes in a nuclear reactor. Since the isotope ^{241}Am can be prepared in relatively pure form by extraction as a decay product over a period of years from strongly neutron-bombarded plutonium, ^{241}Pu, this isotope is used for much of the chemical investigation of this element. Better suited is the isotope ^{243}Am due to its longer half-life (7370 years as compared to 432.2 years for ^{241}Am). A mixture of the isotopes ^{241}Am, ^{242}Am, and ^{243}Am can be prepared by intense neutron irradiation of ^{241}Am according to the reactions $^{241}Am\ (n, \gamma) \rightarrow {}^{242}Am\ (n, \gamma) \rightarrow {}^{243}Am$. Nearly isotopically pure ^{243}Am can be prepared by a sequence of neutron bombardments and chemical separations starting with ^{242}Pu. Twenty-three radioactive isotopes and isomers are now recognized. Americium metal has been prepared by reducing the trifluoride with barium vapor at 1000 to 1200 °C or reducing the dioxide by lanthanum metal. The luster of freshly prepared americium metal is white and more silvery than plutonium or neptunium prepared in the same manner. It appears to be more malleable than uranium or neptunium and tarnishes slowly in dry air at room temperature. Americium is thought to exist in two forms: an alpha form which has a double hexagonal close-packed structure and a loose-packed cubic beta form. Americium must be handled with great care to avoid personal contamination. As little as 0.03 μCi of ^{241}Am is the maximum permissible total body burden. The alpha activity from ^{241}Am is about three times that of radium. When gram quantities of ^{241}Am are handled, the intense gamma activity makes exposure a serious problem. Americium dioxide, AmO_2, is the most important oxide. AmF_3, AmF_4, $AmCl_3$, $AmBr_3$, AmI_3, and other compounds have been prepared. The isotope ^{241}Am has been used as a portable source for gamma radiography. It has also been used as a radioactive glass thickness gage for the flat glass industry and as a source of ionization for smoke detectors.

Antimony — (Gr. *anti* plus *monos* - a metal not found alone), Sb. Antimony was recognized in compounds by the ancients and was known as a metal at the beginning of the 17th century and possibly much earlier. It is not abundant but is found in over 100 mineral species. It is sometimes found native but more frequently as the sulfide, *stibnite* (Sb_2S_3); it is also found as antimonides of the heavy metals, and as oxides. It is extracted from the sulfide by roasting to the oxide, which is reduced by salt and scrap iron; from its oxides it is also prepared by reduction with carbon. Two allotropic forms of antimony exist: the normal stable, metallic form (gray antimony), and the amorphous black form. The so-called explosive antimony is an ill-defined material always containing an appreciable amount of halogen; therefore, it no longer warrants consideration as a separate allotrope. The yellow form, obtained by oxidation of *stibine*, SbH_3, is probably impure, and is not a distinct form. Natural antimony is made of two stable isotopes, ^{121}Sb and ^{123}Sb. Metallic antimony is an extremely brittle metal of a flaky, crystalline texture. It is bluish white and has a metallic luster. It is not acted on by air at room temperature but burns brilliantly when heated with the formation of white fumes of Sb_2O_3. It is a poor conductor of heat and electricity and has a hardness of 3 to 3.5. Antimony, available commercially with a purity of 99.999 + %, is finding use in semiconductor technology for making infrared detectors, diodes, and Hall-effect devices. Commercial-grade antimony is widely used in alloys with percentages ranging from 1% to 20%. It greatly increases the hardness and mechanical strength of lead. Batteries, antifriction alloys, type metal, small arms and tracer bullets, cable sheathing, and minor products use about half the metal produced. Compounds taking up the other half are oxides, sulfides, sodium antimonate, and antimony trichloride. These are used in manufacturing flame-proofing compounds, paints, ceramic enamels, glass, and pottery. Tartar emetic (hydrated potassium antimonyl tartrate) has been used in medicine. Antimony and many of its compounds are toxic.

Argon — (Gr. *argos*, inactive), Ar. Its presence in air was suspected by Cavendish in 1785, discovered by Lord Rayleigh and Sir William Ramsay in 1894. The gas is prepared by fractionation of liquid air, the atmosphere containing 0.934% argon. The atmosphere of Mars contains 1.6% of ^{40}Ar and 5 ppm of ^{36}Ar. Argon is two and one-half times as soluble in water as nitrogen, having about the same solubility as oxygen. It is recognized by the characteristic lines in the red end of the spectrum. It is used in electric light bulbs and in fluorescent tubes at a pressure of about 400 Pa, and in filling photo tubes, glow tubes, etc. Argon is also used as an inert gas shield for arc welding and cutting; as a blanket for the production of titanium and other reactive elements; and as a protective atmosphere for growing silicon and germanium crystals. Argon is colorless and odorless, both as a gas and liquid. It is available in high-purity form. Argon is considered to be more inert than the other rare gases, although a compound HArF has been prepared at extremely low temperatures. However, it does form a hydrate having a dissociation pressure of 105 atm at 0 °C. Ion molecules such as $(ArKr)^+$, $(ArXe)^+$, $(NeAr)^+$ have been observed spectroscopically. Argon also forms a clathrate with β-hydroquinone. This clathrate is stable and can be stored for a considerable time, but a true chemical bond does not exist.

Arsenic — (L. *arsenicum,* Gr. *arsenikon,* yellow orpiment, identified with *arsenikos,* male, from the belief that metals were different sexes; Arabic, *Az-zernikh,* the orpiment from Persian *zerni-zar,* gold), As. It is believed that Albertus Magnus obtained the element in 1250 A.D. In 1649, Schroeder published two methods of preparing the element. It is found native, in the sulfides *realgar* and *orpiment,* as arsenides and sulfarsenides of heavy metals, as the oxide, and as arsenates. *Mispickel,* arsenopyrite, (FeSAs) is the most common mineral, from which on heating the arsenic sublimes leaving ferrous sulfide. The most common form of the element, gray arsenic, is a steel gray, very brittle, crystalline, semimetallic solid; it tarnishes in air, and when heated is rapidly oxidized to arsenous oxide (As_2O_3) with the odor of garlic. Two other allotropes, black arsenic and yellow arsenic, can be made by condensing arsenic vapor. Arsenic and its compounds are extremely poisonous. Arsenic is also used in bronzing, pyrotechny (fireworks manufacture), and for hardening and improving the sphericity of shot. The most important compounds are white arsenic (As_2O_3); the sulfide; Paris green, $3Cu(AsO_2)_2 \cdot Cu(C_2H_3O_2)_2$; calcium arsenate; and lead arsenate. The last three have been used as agricultural insecticides and poisons. Marsh's test for arsenic compounds makes use of the formation and ready decomposition of arsine (AsH_3); it has long been used in forensics. Arsenic is finding increasing uses as a doping agent in solid-state devices such as transistors. Gallium arsenide is used as a laser material and in various electronic devices.

Astatine — (Gr. *astatos,* unstable), At. Synthesized in 1940 by D.R. Corson, K.R. MacKenzie, and E. Segre at the University of California by bombarding bismuth with alpha particles. The longest-lived isotope, ^{210}At, has a half-life of only 8.1 hours. Over 50 other isotopes and isomers are now known. Minute quantities of ^{215}At, ^{218}At, and ^{219}At exist in equilibrium in nature with naturally occurring uranium and thorium isotopes, and traces of ^{217}At are in equilibrium with ^{233}U and ^{239}Np resulting from interaction of thorium and uranium with naturally produced neutrons. The total amount of astatine present in the Earth's crust, however, is probably less than 30 g. Only about 0.05 μg of astatine has been prepared to date. Mass spectrometers have been used to confirm that this highly radioactive halogen behaves chemically very much like other halogens, particularly iodine. The interhalogen compounds AtI, AtBr, and AtCl are known to form, but it is not yet known if astatine forms diatomic astatine molecules. HAt and CH_3At (methyl astatide) have been detected. Astatine is said to be more metallic that iodine, and, like iodine, it probably accumulates in the thyroid gland.

Barium — (Gr. *barys,* heavy), Ba. Baryta was distinguished from lime by Scheele in 1774; the element was discovered by Sir Humphrey Davy in 1808. It is found only in combination with other elements, chiefly in *barite* or *heavy spar* (sulfate) and *witherite* (carbonate) and is prepared by electrolysis of the chloride. Large deposits of barite are found in China, Germany, India, Morocco, and in the U.S. Barium is a metallic element, soft, and when pure is silvery white like lead; it belongs to the alkaline earth group, resembling calcium chemically. The metal oxidizes very easily and should be kept under petroleum or other suitable oxygen-free liquids to exclude air. It is decomposed by water or alcohol. The metal is used as a "getter" in vacuum tubes. The most important compounds are the peroxide (BaO_2), chloride, sulfate, carbonate, nitrate, and chlorate. Lithopone, a pigment containing barium sulfate and zinc sulfide, has good covering power, and does not darken in the presence of sulfides. The sulfate, as permanent white or *blanc fixe,* is also used in paint, in x-ray diagnostic work, and in glassmaking. *Barite* is extensively used as a weighting agent in oil well drilling fluids, and also in making rubber. The carbonate has been used as a rat poison, while the nitrate and chlorate give green colors in pyrotechny (fireworks manufacture). The impure sulfide phosphoresces after exposure to the light. All barium compounds that are water or acid soluble are poisonous.

Berkelium — (*Berkeley,* home of the University of California), Bk. Berkelium, the eighth member of the actinide transition series, was discovered in December 1949 by Thompson, Ghiorso, and Seaborg, and was the fifth transuranium element synthesized. It was produced by cyclotron bombardment of milligram amounts of ^{241}Am with helium ions at Berkeley, California. The first isotope produced had a mass number of 243 and decayed with a half-life of 4.5 hours. Thirteen isotopes are now known and have been synthesized. The existence of ^{249}Bk, with a half-life of 320 days, makes it feasible to isolate berkelium in weighable amounts so that its properties can be investigated with macroscopic quantities. One of the first visible amounts of a pure berkelium compound, berkelium chloride, was produced in 1962. It weighed 3 ng. Elemental berkelium is expected to be a silvery metal, easily soluble in dilute mineral acids, and readily oxidized by air or oxygen at elevated temperatures to form the oxide. X-ray diffraction methods have been used to identify the following compounds: BkO_2, BkO_3, BkF_3, BkCl, and BkOCl. Berkelium most likely resembles terbium with respect to chemical properties. As with other actinide elements, berkelium tends to accumulate in the skeletal system. Because of its rarity, berkelium presently has no commercial or technological use.

Beryllium — (Gr. *beryllos, beryl;* also called Glucinium or Glucinum, Gr. *glykys,* sweet), Be. Discovered as the oxide by Vauquelin in beryl and in emeralds in 1798. The metal was isolated in 1828 by Wöhler and by Bussy independently by the action of potassium on beryllium chloride. Beryllium is found in some 30 mineral species, the most important of which are *bertrandite, beryl, chrysoberyl,* and *phenacite. Aquamarine* and *emerald* are precious forms of *beryl.* Beryllium minerals are found in the U.S., Brazil, Russia, Kazakhstan, and elsewhere. Colombia is known for its emeralds. *Beryl* ($3BeO \cdot Al_2O_3 \cdot 6SiO_2$) and *bertrandite* ($4BeO \cdot 2SiO_2 \cdot H_2O$) are the most important commercial sources of the element and its compounds. Most of the metal is now prepared by reducing beryllium fluoride with magnesium metal. Beryllium metal did not become readily available to industry until 1957. The metal, steel gray in color, has many desirable properties. It is one of the lightest of all metals and has one of the highest melting points of the light metals. Its modulus of elasticity is about one third greater than that of steel. It resists attack by concentrated nitric acid, has excellent thermal conductivity, and is nonmagnetic. It has a high permeability to x-rays, and when bombarded by alpha particles, as from radium or polonium, neutrons are produced in the ratio of about 30 neutrons/million alpha particles. At ordinary temperatures beryllium resists oxidation in air, although its ability to scratch glass is probably due to the formation of a thin layer of the oxide. Beryllium is used as an alloying agent in producing beryllium copper, which is extensively used for springs, electrical contacts, spot-welding electrodes, and non-sparking tools. It has found application as a structural material for high-speed aircraft, missiles, spacecraft, and communication satellites. It was used in the windshield frame, brake discs, support beams, and other structural components of the space shuttle. Because beryllium is relatively transparent to x-rays, ultrathin Be-foil is finding use in x-ray lithography for reproduction of microminiature integrated circuits. Natural beryllium is 100% ^{9}Be. Beryllium is used in nuclear reactors as a reflector or moderator for it has a low thermal neutron absorption cross section. It is used in gyroscopes, computer parts, and instruments where lightness, stiffness, and dimensional stability are required. The oxide has a very high-melting point and is also used in nuclear technology and ceramic applications. Beryllium and its salts are toxic and should be handled with the greatest of care. They should not be tasted to verify the sweetish nature of beryllium (as did early experimenters). The metal, its alloys, and its salts can be handled safely if certain work codes are observed, but no attempt should be made to work with beryllium before becoming familiar with proper safeguards.

Bismuth — (Ger. *Weisse Masse,* white mass; later *Wisuth* and *Bisemutum*), Bi. In early times, bismuth was confused with tin and lead. Claude Geoffroy the Younger showed it to be distinct from lead in 1753. It is a white crystalline, brittle metal with a pinkish tinge. It occurs native. The most important ores are *bismuthinite* or bismuth glance (Bi_2S_3) and *bismite* (Bi_2O_3). Peru, Japan, Mexico, Bolivia, and Canada are major bismuth producers. Much of the bismuth produced in the U.S. is obtained as a by-product in refining lead, copper, tin, silver, and gold ores. Bismuth is the most diamagnetic of all metals, and the thermal conductivity is lower than any metal, except mercury. It has a high electrical resistance and has the highest Hall effect of any metal (i.e., greatest increase in electrical resistance when placed in a magnetic field). "Bismanol" is a permanent magnet of high coercive force, made of MnBi. Bismuth expands 3.32% on solidification. This property makes bismuth alloys particularly suited to the making of sharp castings of objects subject to damage by high temperatures. With other metals such as tin, cadmium, etc., bismuth forms low-melting alloys that are extensively used for safety devices in fire detection and extinguishing systems. Bismuth is used in producing malleable irons and is finding use as a catalyst for making acrylic fibers. When bismuth is heated in air it burns with a blue flame, forming yellow fumes of the oxide. The metal is also used as a thermocouple material and has found application as a carrier for U^{235} or U^{233} fuel in atomic reactors. Its soluble salts are characterized by forming insoluble basic salts on the addition of water, a property sometimes used in detection work. Bismuth oxychloride is used extensively in cosmetics. Bismuth subnitrate and subcarbonate are used in medicine. Natural bismuth contains only one isotope ^{209}Bi.

Bohrium — (Named after Niels Bohr [1885–1962], Danish atomic and nuclear physicist), Bh. Bohrium is expected to have chemical properties similar to rhenium. This element was synthesized and unambiguously identified in 1981 using the Universal Linear Accelerator (UNILAC) at the Gesellschaft für Schwerionenforschung (G.S.I.) in Darmstadt, Germany. The discovery team was led by Armbruster and Münzenberg. The reaction producing the element was proposed and applied earlier by a Dubna Group led by Oganessian in 1976. A target of ^{209}Bi was bombarded by a beam of ^{54}Cr ions. In 1983 experiments at Dubna using the 157-inch cyclotron, produced ^{262}Bh by the reaction ^{209}Bi + ^{54}Cr. IUPAC adopted the name Bohrium with the symbol Bh for Element 107 in August 1997. Ten isotopes of bohrium are now recognized, two of which have lifetimes of around 1 minute.

Boron — (Ar. *Buraq,* Pers. *Burah*), B. Boron compounds have been known for thousands of years, but the element was not discovered until 1808 by Sir Humphry Davy and by Gay-Lussac and Thenard. The element is not found free in nature, but occurs as orthoboric acid, usually in certain volcanic spring waters, and as borates in *borax* and *colemanite. Ulexite,* another boron mineral, is interesting as it is nature's own version of fiber optics. Important sources of boron are the ores *rasorite (kernite)* and *tincal (borax ore).* Both of these ores are found in the Mojave Desert. *Tincal* is the most important source of boron from the Mojave. Extensive *borax* deposits are also found in Turkey. Boron exists naturally as 19.9% ^{10}B isotope and 80.1% ^{11}B isotope. High-purity crystalline boron may be prepared by the vapor phase reduction of boron trichloride or tribromide with hydrogen on electrically heated filaments. The impure, or amorphous, boron, a brownish-black powder, can be obtained by heating the trioxide with magnesium powder. Boron of 99.9999% purity has been produced and is available commercially. Elemental boron has an energy band gap of 1.50 to 1.56 eV, which is higher than that of either silicon or germanium. It has interesting optical characteristics, transmitting portions of the infrared, and is a poor conductor of electricity at room temperature, but a good conductor at high temperature. Amorphous boron is used in pyrotechnic flares to provide a distinctive green color and in rockets as an igniter. By far the most commercially important boron compound in terms of dollar sales is $Na_2B_4O_7 \cdot 5H_2O$. This pentahydrate is used in very large quantities in the manufacture of insulation fiberglass and sodium perborate bleach. Boric acid is also an important boron compound with major markets in textile fiberglass and in cellulose insulation as a flame retardant. Next in order of importance is borax ($Na_2B_4O_7 \cdot 10H_2O$) which is used principally in laundry products. Use of borax as a mild antiseptic is minor in terms of dollars and tons. Boron compounds are also extensively used in the manufacture of borosilicate glasses. The isotope ^{10}B is used as a control for nuclear reactors, as a shield for nuclear radiation, and in instruments used for detecting neutrons. Boron nitride has remarkable properties and can be used to make a material as hard as diamond. The nitride also behaves like an electrical insulator but conducts heat like a metal. It also has lubricating properties similar to graphite. The hydrides are easily oxidized with considerable energy liberation and have been studied for use as rocket fuels. Demand is increasing for boron filaments, a high-strength, lightweight material chiefly employed for advanced aerospace structures. Boron is similar to carbon in that it has a capacity to form stable covalently bonded molecular networks. Carboranes, metalloboranes, phosphacarboranes, and other families comprise thousands of compounds. Elemental boron and the borates are not considered to be toxic, and they do not require special care in handling. However, some of the more exotic boron hydrogen compounds are toxic and do require care.

Bromine — (Gr. *bromos,* stench), Br. Discovered by Balard in 1826, but not prepared in quantity until 1860. A member of the halogen group of elements, it is obtained from natural brines from wells in Michigan and Arkansas. Little bromine is extracted today from seawater, which contains only about 85 ppm. Bromine is the only liquid nonmetallic element. It is a heavy, mobile, reddish-brown liquid, volatilizing readily at room temperature to a red vapor with a strong disagreeable odor, resembling chlorine, and having a very irritating effect on the eyes and throat. It is readily soluble in water or carbon disulfide, forming a red solution; is less active than chlorine but more so than iodine. It unites readily with many elements and has a bleaching action; when spilled on the skin it produces painful sores. It presents a serious health hazard, and maximum safety precautions should be taken when handling it. Much of the bromine output in the U.S. was used in the production of ethylene dibromide, a lead scavenger used in making gasoline antiknock compounds. Since lead in gasoline was eliminated because of environmental considerations, production of bromine has decreased. Bromine is also used in making fumigants, flameproofing agents, water purification compounds, dyes, medicinals, sanitizers, etc. Organic bromides are also important.

Cadmium — (L. *cadmia;* Gr. *kadmeia* - ancient name for calamine, zinc carbonate), Cd. Discovered by Stromeyer in 1817 from an impurity in zinc carbonate. Cadmium most often occurs in small quantities associated with zinc ores, such as *sphalerite* (ZnS). *Greenockite* (CdS) is the only mineral of any consequence bearing cadmium. Almost all cadmium is obtained as a by-product in the treatment of zinc, copper, and lead ores. It is a soft, bluish-white metal which is easily cut with a knife. It is similar in many respects to zinc. It is a component of some of the lowest melting alloys; it is used in bearing alloys with low coefficients of friction and great resistance to fatigue; it is used extensively in electroplating, which accounts for about 60% of its use. It is also used in many types of solder, for standard E.M.F. cells, for Ni-Cd batteries, and as a barrier to control atomic fission. The market for Ni-Cd batteries has grown significantly. Cadmium compounds are used in blue and green phosphors for color TV tubes. It forms a number of salts, of which the sulfate is most common; the sulfide is used as a yellow pigment. Cadmium and solutions of its compounds are toxic. Failure to appreciate the toxic properties of cadmium may cause workers to be unwittingly exposed to dangerous fumes. Some silver solders, for example, contain cadmium and should be handled with care. Serious toxicity problems have been found from long-term exposure and work with cadmium plating baths. Cadmium is present in certain phosphate rocks. This has raised concerns that the long-term use of certain phosphate fertilizers might pose a health hazard from levels of cadmium that might enter the food chain. At one time the International Conference on Weights and Measures defined the meter in terms of the wavelength of the red cadmium spectral line.

Calcium — (L. *calx,* lime), Ca. Though lime was prepared by the Romans in the first century under the name calx, the metal was not discovered until 1808. After learning that Berzelius and Pontin prepared calcium amalgam by electrolyzing lime in mercury, Davy was able to isolate the impure metal. Calcium is a metallic element, fifth in abundance in the Earth's crust, of which it forms more than 3%. It is an essential constituent of leaves, bones, teeth, and shells. Never found in nature uncombined, it occurs abundantly as *limestone* ($CaCO_3$), *gypsum* ($CaSO_4 \cdot 2H_2O$), and *fluorite* (CaF_2); *apatite* is the fluorophosphate or chlorophosphate of calcium. The metal has a silvery color, is rather hard, and is prepared by electrolysis of the fused chloride to which calcium fluoride is added to lower the melting point. Chemically it is one of the alkaline earth elements; it readily forms a white coating of oxide in air, reacts with water, burns with a yellow-red flame, largely forming the oxide. The element under pressure has the highest superconducting critical temperature (29 K) exhibited to date by any elemental material. The metal is used as a reducing agent in preparing other metals such as thorium, uranium, zirconium, etc., and is used as a deoxidizer, desulfurizer, and inclusion modifier for various ferrous and nonferrous alloys. It is also used as an alloying agent for aluminum, beryllium, copper, lead, and magnesium alloys, and serves as a "getter" for residual gases in vacuum tubes. Its natural and prepared compounds are widely used. Quicklime (CaO), made by heating limestone and changed into slaked lime by the careful addition of water, is the great cheap base of the chemical industry with countless uses. Mixed with sand it hardens as mortar and plaster by taking up carbon dioxide from the air. Calcium from limestone is an important element in Portland cement. The solubility of the carbonate in water containing carbon dioxide causes the formation of caves with stalactites and stalagmites and is responsible for hardness in water. Other important compounds are the carbide (CaC_2), chloride ($CaCl_2$), cyanamide ($CaCN_2$), hypochlorite ($Ca(OCl)_2$), nitrate ($Ca(NO_3)_2$), and sulfide (CaS). Calcium sulfide is phosphorescent after being exposed to light.

Californium — (State and University of California), Cf. Californium, the sixth transuranium element to be discovered, was produced by Thompson, Street, Ghioirso, and Seaborg in 1950 by bombarding microgram quantities of ^{242}Cm with 35 MeV helium ions in the Berkeley 60-inch cyclotron. The existence of the isotopes ^{249}Cf, ^{250}Cf, ^{251}Cf, and ^{252}Cf makes it feasible to isolate californium in weighable amounts so that its properties can be investigated. Californium-252 is a very strong neutron emitter. One microgram releases 170 million neutrons per minute, which presents biological hazards. Proper safeguards should be used in handling californium. Twenty isotopes of californium are now recognized. ^{249}Cf and ^{252}Cf have half-lives of 351 years and 900 years, respectively. Californium (III) is the only ion stable in aqueous solutions, all attempts to reduce or oxidize californium (III) having failed. In 1960, a few tenths of a microgram of californium trichloride, $CfCl_3$, californium oxychloride, CfOCl, and californium oxide, Cf_2O_3, were first prepared. Because californium is a very efficient source of neutrons, many new uses are expected for it. It has already found use in neutron moisture gages and in well-logging (the determination of water and oil-bearing layers). It is also being used as a portable neutron source for discovery of metals such as gold or silver by on-the-spot activation analysis.

Carbon — (L. *carbo,* charcoal), C. Carbon, an element of prehistoric discovery, is very widely distributed in nature. It is found in abundance in the sun, stars, comets, and atmospheres of most planets. Carbon in the form of microscopic diamonds is found in some meteorites. Natural diamonds are found in *kimberlite* or *lamporite* of ancient formations called "pipes," such as found in South Africa, Arkansas, and elsewhere. Diamonds are being recovered from the ocean floor off the Cape of Good Hope. About 30% of all industrial diamonds used in the U.S. are now made synthetically. The energy of the sun and stars can be attributed at least in part to the carbon-nitrogen cycle of nuclear reactions. Carbon is found free in nature in three allotropic forms: amorphous, graphite, and diamond. Graphite is one of the softest known materials while diamond is one of the hardest. Graphite exists in two forms: alpha and beta. These have identical physical properties, except for their crystal structure. Naturally occurring graphites are reported to contain as much as 30% of the rhombohedral (beta) form, whereas synthetic materials contain only the alpha form. The hexagonal alpha type can be converted to the beta by mechanical treatment, and the beta form reverts to the alpha on heating it above 1000 °C. Fullerenes or "buckyballs" have a number of unusual properties. These molecules, consisting of 60 or 70 carbon atoms linked together, are capable of withstanding great pressure and trapping foreign atoms inside their network of carbon. Some fullerides have shown to be superconductive at temperatures as high as 33 K. In combination, carbon is found as carbon dioxide in the atmosphere of the Earth and dissolved in all natural waters. On average, the earth's atmosphere contains about 0.04% CO_2, while the atmosphere of Mars contains 95.32% CO_2. It is a component of great rock masses in the form of carbonates of calcium (limestone), magnesium, and iron. Coal, petroleum, and natural gas are chiefly hydrocarbons. Carbon is unique among the elements in the vast number and variety of compounds it can form. With hydrogen, oxygen, nitrogen, and other elements, it forms a very large number of compounds, carbon atom often being linked to carbon atom. Many thousands of carbon compounds are vital to organic and life processes. Without carbon, the basis for life as we know it would be impossible. While it has been thought that silicon might take the place of carbon in forming a host of similar compounds, it is now not possible to form stable compounds with very long chains of silicon atoms. Some of the compounds of carbon of greatest industrial importance are carbon dioxide (CO_2), carbon monoxide (CO), carbon disulfide (CS_2), chloroform ($CHCl_3$), carbon tetrachloride (CCl_4), methane (CH_4), ethylene (C_2H_4), acetylene (C_2H_2), benzene (C_6H_6), ethyl alcohol (C_2H_5OH), acetic acid (CH_3COOH), and their derivatives. Natural carbon consists of 98.93% ^{12}C and 1.07% ^{13}C. In 1961, the International Union of Pure and Applied Chemistry adopted the isotope carbon-12 as the basis for atomic weights. Carbon-14, an isotope with a half-life of 5715 years, has been widely used to date such materials as wood, archeological specimens, etc. A brittle form of carbon, known as "glassy" or vitreous carbon, has been developed. It has a high resistance to corrosion, has good thermal stability, and is structurally impermeable to both gases and liquids. It has a randomized structure, making it useful in applications such as crystal growing, crucibles for high-temperature use, prosthetic devices, etc.

Cerium — (named for the asteroid *Ceres,* which was discovered in 1801 only 2 years before the element), Ce. Discovered in 1803 by Klaproth and by Berzelius and Hisinger; metal prepared by Hillebrand and Norton in 1875. Cerium is the most abundant of the metals of the so-called rare earths. It is found in a number of minerals including *allanite* (also known as *orthite*), *monazite, bastnasite, cerite,* and *samarskite.* Monazite and bastnasite are presently the two most important sources of cerium. Large deposits of monazite found on the beaches of Travancore, India, in river sands in Brazil, and deposits of *allanite* in the western United States, and *bastnasite* in Southern California will supply cerium, thorium, and the other rare-earth metals for many years to come. Metallic cerium is prepared by metallothermic reduction techniques, such as by reducing cerous fluoride with calcium, or by electrolysis of molten cerous chloride or other cerous halides. The metallothermic technique is used to produce high-purity cerium. Cerium is especially interesting because of its variable electronic structure. The energy of the inner 4f level is nearly the same as that of the outer or valence electrons, and only small amounts of energy are required to change the relative occupancy of these electronic levels. This gives rise to dual valency states. For example, a volume change of about 10% occurs when cerium is subjected to high pressures or low temperatures. It appears that the valence changes from about 3 to 4 when it is cooled or compressed. The low-temperature behavior of cerium is complex. Four allotropic modifications are thought to exist: cerium at room temperature and at atmospheric pressure is known as γ cerium. Upon cooling to –16 °C, γ cerium changes to β cerium. The remaining γ cerium starts to change to α cerium when cooled to –172 °C, and the transformation is complete at –269 °C. α Cerium has a density of 8.16; δ cerium exists above 726 °C. At atmospheric pressure, liquid cerium is more dense than its solid form at the melting point. Cerium is an iron-gray lustrous metal. It is malleable, and oxidizes very readily at room temperature, especially in moist air.

Except for europium, cerium is the most reactive of the "rare-earth" metals. It slowly decomposes in cold water, and rapidly in hot water. Alkali solutions and dilute and concentrated acids attack the metal rapidly. The pure metal is likely to ignite if scratched with a knife. Ceric salts are orange red or yellowish; cerous salts are usually white. Cerium is a component of misch metal, which is extensively used in the manufacture of pyrophoric alloys for cigarette lighters, etc. Natural cerium is stable and contains four isotopes. Thirty-two other radioactive isotopes and isomers are known. While cerium is not radioactive, the impure commercial grade may contain traces of thorium, which is radioactive. The oxide is an important constituent of incandescent gas mantles and it is emerging as a hydrocarbon catalyst in "self-cleaning" ovens. In this application it can be incorporated into oven walls to prevent the collection of cooking residues. As ceric sulfate it finds extensive use as a volumetric oxidizing agent in quantitative analysis. Cerium compounds are used in the manufacture of glass, both as a component and as a decolorizer. The oxide is finding increased use as a glass polishing agent instead of rouge, for it is much faster than rouge in polishing glass surfaces. Cerium compounds are finding use in automobile exhaust catalysts. Cerium is also finding use in making permanent magnets. Cerium, with other rare earths, is used in carbon-arc lighting, especially in the motion picture industry. It is also finding use as an important catalyst in petroleum refining and in metallurgical and nuclear applications. In small lots, cerium costs about $5/g (99.9%).

Cesium — (L. *caesius,* sky blue), Cs. Cesium was discovered spectroscopically by Bunsen and Kirchhoff in 1860 in mineral water from Durkheim. Cesium, an alkali metal, occurs in *lepidolite, pollucite* (a hydrated silicate of aluminum and cesium), and in other sources. One of the world's richest sources of cesium is located at Bernic Lake, Manitoba. The deposits are estimated to contain 300,000 tons of pollucite, averaging 20% cesium. It can be isolated by electrolysis of the fused cyanide and by a number of other methods. Very pure, gas-free cesium can be prepared by thermal decomposition of cesium azide. The metal is characterized by a spectrum containing two bright lines in the blue along with several others in the red, yellow, and green. It is silvery white, soft, and ductile. It is the most electropositive and most alkaline element. Cesium, gallium, and mercury are the only three metals that are liquid near room temperature. Cesium reacts explosively with cold water and reacts with ice at temperatures above –116 °C. Cesium hydroxide, the strongest base known, attacks glass. Because of its great affinity for oxygen the metal is used as a "getter" in electron tubes. It is also used in photoelectric cells, as well as a catalyst in the hydrogenation of certain organic compounds. The metal has recently found application in ion propulsion systems. Cesium is used in atomic clocks; a second of time is now defined as being the duration of 9,192,631,770 periods of the radiation corresponding to the transition between the two hyperfine levels of the ground state of the cesium-133 atom. Its chief compounds are the chloride and the nitrate salts.

Chlorine — (Gr. *chloros,* greenish yellow), Cl. Discovered in 1774 by Scheele, who thought it contained oxygen; named in 1810 by Davy, who insisted it was an element. In nature it is found in the combined state only, chiefly with sodium and potassium as common salt (NaCl), *carnallite* ($KMgCl_3 \cdot 6H_2O$), and *sylvite* (KCl). It is a member of the halogen (salt-forming) group of elements and is obtained from chlorides by the action of oxidizing agents and more often by electrolysis; it is a greenish-yellow gas, combining directly with nearly all elements. At 10 °C one volume of water dissolves 3.10 volumes of chlorine, at 30 °C only 1.77 volumes. Chlorine is widely used in making many everyday products. It is used for producing safe drinking water the world over. Even the smallest water supplies are now usually chlorinated. It is also extensively used in the production of paper products, dyestuffs, textiles, petroleum products, medicines, antiseptics, insecticides, foodstuffs, solvents, paints, plastics, and many other consumer products. Most of the chlorine produced is used in the manufacture of chlorinated compounds for sanitation, pulp bleaching, disinfectants, and textile processing. Further use is in the manufacture of chlorates, chloroform, carbon tetrachloride, and in the extraction of bromine. Organic chemistry demands much from chlorine, both as an oxidizing agent and in substitution, since it often brings desired properties in an organic compound when substituted for hydrogen, as in one form of synthetic rubber. Chlorine is a respiratory irritant. The gas irritates the mucous membranes and the liquid burns the skin. As little as 3.5 ppm can be detected as an odor, and 1000 ppm is likely to be fatal after a few deep breaths. It was used as a war gas in 1915. Natural chlorine contains two isotopes, ^{35}Cl and ^{37}Cl.

Chromium — (Gr. *chroma,* color), Cr. Discovered in 1797 by Vauquelin, who prepared the metal the next year, chromium is a steel-gray, lustrous, hard metal that takes a high polish. The principal ore is *chromite* ($FeCr_2O_4$), which is found in Zimbabwe, Russia, South Africa, Turkey, Iran, Albania, Finland, Madagascar, the Philippines, and elsewhere. The U.S. has no appreciable chromite ore reserves. The metal is usually produced by reducing the oxide with aluminum. Chromium is used to harden steel, to manufacture stainless steel, and to form many useful alloys. Much is used in plating to produce a hard, beautiful surface and to prevent corrosion. Chromium is used to give glass an emerald green color. It finds wide use as a catalyst. All compounds of chromium are colored; the most important are the chromates of sodium and potassium (K_2CrO_4) and the dichromates ($K_2Cr_2O_7$) and the potassium and ammonium chrome alums, as $KCr(SO_4)_2 \cdot 12H_2O$. The dichromates are used as oxidizing agents in quantitative analysis, also in tanning leather. Other compounds are of industrial value; lead chromate is chrome yellow, a valued pigment. Chromium compounds are used in the textile industry as mordants, and by the aircraft and other industries for anodizing aluminum. The refractory industry has found chromite useful for forming bricks and shapes, as it has a high-melting point, moderate thermal expansion, and stability of crystalline structure. Chromium is an essential trace element for human health. Many chromium compounds, however, are acutely or chronically toxic, and some are carcinogenic. They should be handled with proper safeguards.

Cobalt — (*Kobald,* from the German, goblin or evil spirit, *cobalos,* Greek, mine), Co. Discovered by Brandt about 1735. Cobalt occurs in the mineral *cobaltite, smaltite,* and *erythrite,* and is often associated with nickel, silver, lead, copper, and iron ores, from which it is most frequently obtained as a by-product. It is also present in meteorites. Important ore deposits are found in Congo-Kinshasa, Australia, Zambia, Russia, Canada, and elsewhere. The U.S. Geological Survey has announced that the bottom of the north central Pacific Ocean may have cobalt-rich deposits at relatively shallow depths in waters close to the Hawaiian Islands and other U.S. Pacific territories. Cobalt is a brittle, hard metal, closely resembling iron and nickel in appearance. It has a magnetic permeability of about two thirds that of iron. Cobalt tends to exist as a mixture of two allotropes over a wide temperature range; the β-form predominates below 400 °C, and the α above that temperature. The transformation is sluggish and accounts in part for the wide variation in reported data on physical properties of cobalt. It is alloyed with iron, nickel and other metals to make Alnico, an alloy of unusual magnetic strength with many important uses. Stellite alloys, containing cobalt, chromium, and tungsten, are used for high-speed, heavy-duty, high-temperature cutting tools, and for dies. Cobalt is also used in other magnet steels and stainless steels, and in alloys used in jet turbines and gas turbine generators. The metal is used in electroplating because of its appearance, hardness, and resistance to oxidation. The salts have been used for centuries for the production of brilliant and permanent blue colors in porcelain, glass, pottery, tiles, and enamels. It is the principal ingredient in Sevre's and Thenard's blue. A solution of the chloride ($CoCl_2 \cdot 6H_2O$) is used as sympathetic ink. The cobalt ammines are of interest; the oxide and the nitrate are important. Cobalt carefully used in the form of the chloride, sulfate, acetate, or nitrate has been found effective in correcting a certain mineral deficiency disease in animals. Soils should contain 0.13 to 0.30 ppm of cobalt for proper animal nutrition. Cobalt is found in Vitamin B-12, which is essential for human nutrition. Cobalt-60, an artificial isotope, is an important gamma ray source, and is extensively used as a tracer and a radiotherapeutic agent.

Copernicium — (After the astronomer Nicolaus Copernicus), Cn. In late February 1996, Siguard Hofmann and his collaborators at GSI Darmstadt announced their discovery of Element-112, having 112 protons and 165 neutrons, with an atomic mass of 277. This element was made by bombarding a lead target with high-energy zinc ions. A single nucleus of Element-112 was detected, which was identified by its decay pattern. Evidence indicates that nuclei with 162 neutrons are held together more strongly than nuclei with a smaller or larger number of neutrons. This suggests a narrow "peninsula" of relatively stable isotopes around Element-114. In 2010 IUPAC approved the name and symbol Copernicium, Cn, for Element-112. Theoretical calculations and limited experimental measurements suggest that copernecium will have properties similar to mercury, possibly existing as a gas at ambient temperature.

Copper — (L. *cuprum,* from the island of Cyprus), Cu. The discovery of copper dates from prehistoric times. It is said to have been mined for more than 5000 years. It is one of man's most important metals. Copper is reddish colored, takes on a bright metallic luster, and is malleable, ductile, and a good conductor of heat and electricity (second only to silver in electrical conductivity). The electrical industry is one of the greatest users of copper. Copper occasionally occurs native, and is found in many minerals such as *cuprite, malachite, azurite, chalcopyrite,* and *bornite.* Large copper ore deposits are found in the U.S., Chile, Zambia, Zaire, Peru, and Canada. The most important copper ores are the sulfides, oxides, and carbonates. From these, copper is obtained by smelting, leaching, and by electrolysis. Its alloys, brass and bronze, long used, are still very important; all American coins are now copper alloys; monel and gun metals also contain copper. The most important compounds are the oxide and the sulfate, blue vitriol; the latter has wide use as an agricultural poison and as an algicide in water purification. Copper compounds such as Fehling's solution are widely used in analytical chemistry in tests for sugar. High-purity copper (99.999 + %) is readily available commercially.

Curium — (Pierre and Marie Curie), Cm. Although curium follows americium in the periodic system, it was actually known before americium and was the third transuranium element to be discovered. It was identified by Seaborg, James, Morgan, and Ghiorso in 1944 at the wartime Metallurgical Laboratory in Chicago as a result of helium-ion bombardment of ^{239}Pu in the Berkeley, California, 60-inch cyclotron. Visible amounts (30 μg) of ^{242}Cm, in the form of the hydroxide, were first isolated by Werner and Perlman of the University of California in 1947. In 1950, Crane, Wallmann, and Cunningham found that the magnetic susceptibility of microgram samples of CmF_3 was of the same magnitude as that of GdF_3. This provided direct experimental evidence for assigning an electronic configuration to Cm^{+3}. In 1951, the same workers prepared curium in its elemental form for the first time. Sixteen isotopes of curium are now known. The most stable, ^{247}Cm, with a half-life of 16 million years, is so short compared to the Earth's age that any primordial curium must have disappeared long ago from the natural scene. Minute amounts of curium probably exist in natural deposits of uranium, as a result of a sequence of neutron captures and β decays sustained by the very low flux of neutrons naturally present in uranium ores. ^{242}Cm and ^{244}Cm are available in multigram quantities. ^{248}Cm has been produced only in milligram amounts. Curium is similar in some regards to gadolinium, its rare-earth homolog, but it has a more complex crystal structure. Curium is silver in color, is chemically reactive, and is more electropositive than aluminum. CmO_2, Cm_2O_3, CmF_3, CmF_4, $CmCl_3$, $CmBr_3$, and CmI_3 have been prepared. Most compounds of trivalent curium are faintly yellow in color. ^{242}Cm generates about three watts of thermal energy per gram. This compares to one-half watt per gram of ^{238}Pu. This suggests use for curium as a power source. Curium absorbed into the body accumulates in the bones and is therefore very toxic as its radiation destroys the red-cell forming mechanism. The maximum permissible total body burden of ^{244}Cm (soluble) in a human being is 0.3 μCi.

Darmstadtium — (Darmstadt, city in Germany), Ds. In 1987, Oganessian et al., at Dubna, claimed discovery of this element. Their experiments indicated the spontaneous fissioning nuclide $^{272}110$ with a half-life of 10 ms. More recently a group led by Armbruster at G.S.I. in Darmstadt, Germany, reported evidence of $^{269}110$, which was produced by bombarding lead for many days with more than 10^{18} nickel atoms. A detector searched each collision for Element 110's distinct decay sequence. On November 9, 1994, evidence of 110 was detected. In 2003 IUPAC approved the name darmstadtium, symbol Ds, for Element 110.

Dubnium — (named after the Joint Institute of Nuclear Research in Dubna, Russia), Db. In 1967, G. N. Flerov reported that a Soviet team working at Dubna may have produced a few atoms of $^{260}105$ and $^{261}105$ by bombarding ^{243}Am with ^{22}Ne. Later, it was reported that Dubna scientists synthesized Element 105 and determined its decay pattern and chemical properties. In April 1970, it was announced that Ghiorso, Nurmia, Harris, K. A. Y. Eskola, and P. L. Eskola, working at the University of California at Berkeley, had positively identified Element 105. When a ^{15}N nucleus is absorbed by a ^{249}Cf nucleus, four neutrons are emitted and a new atom of $^{260}105$ with a half-life of 1.6 s is formed. Soon after the discovery the names Hahnium and Joliotium, named after Otto Hahn and Jean-Frederic Joliot and Mme. Joliot-Curie, were suggested as names for Element 105. The name Hahnium was used in the literature for several years, leading to considerable confusion. The IUPAC in August 1997 finally resolved the issue, naming Element 105 Dubnium with the symbol Db. Dubnium is thought to have properties similar to tantalum.

Dysprosium — (Gr. *dysprositos,* hard to get at), Dy. Dysprosium was discovered in 1886 by Lecoq de Boisbaudran, but not isolated. Neither the oxide nor the metal was available in relatively pure form until the development of ion-exchange separation and metallographic reduction techniques by Spedding and associates about 1950. Dysprosium occurs along with other so-called rare-earth or lanthanide elements in a variety of minerals such as *xenotime, fergusonite, gadolinite, euxenite, polycrase,* and *blomstrandine.* The most important sources, however, are from *monazite* and *bastnasite.* Dysprosium can be prepared by reduction of the trifluoride with calcium. The element has a metallic, bright silver luster. It is relatively stable in air at room temperature, and is readily attacked and dissolved, with the evolution of hydrogen, by dilute and concentrated mineral acids. The metal is soft enough to be cut with a knife and can be machined without sparking if overheating is avoided. Small amounts of impurities can greatly affect its physical properties. While dysprosium has not yet found many applications, its thermal neutron absorption cross section and high-melting point suggest metallurgical uses in nuclear control applications and for alloying with special stainless steels. A dysprosium oxide-nickel cermet has found use in cooling nuclear reactor rods. This cermet absorbs neutrons readily without swelling or contracting under prolonged neutron bombardment. In combination with vanadium, dysprosium has been used in making laser materials. Dysprosium-cadmium chalcogenides, as sources of infrared radiation, have been used for studying chemical reactions.

Einsteinium — (Albert Einstein [1879–1955]), Es. Einsteinium, the seventh transuranic element of the actinide series to be discovered, was identified by Ghiorso and co-workers at Berkeley in December 1952 in debris from the first large thermonuclear explosion, which took place in the Pacific in November 1952. The isotope produced was the 20-day ^{253}Es isotope. In 1961, a sufficient amount of einsteinium was produced to permit separation of a macroscopic amount of ^{253}Es. This sample weighed about 0.01 μg. A special magnetic-type balance was used in making this determination. ^{253}Es so produced was used to produce mendelevium. About 3 μg of einsteinium has been produced at Oak Ridge National Laboratories by irradiating kilogram quantities of ^{239}Pu for several years in a reactor to produce ^{242}Pu. This was then fabricated into pellets of plutonium oxide and aluminum powder and loaded into target rods for an initial 1-year irradiation at the Savannah River Plant, followed by irradiation in a HFIR (High Flux Isotopic Reactor). After 4 months in the HFIR the targets were removed for chemical separation of the einsteinium from californium. Twenty-one isotopes and isomers of einsteinium are now recognized. ^{254}Es has the longest half-life (276 days). Tracer studies using ^{253}Es show that einsteinium has chemical properties typical of a heavy trivalent, actinide element. Einsteinium is extremely radioactive. Great care must be taken when handling it.

Erbium — (*Ytterby,* a town in Sweden), Er. Erbium, one of the rare-earth elements of the lanthanide series, is found in the minerals mentioned under dysprosium above. In 1842, Mosander separated "yttria," found in the mineral *gadolinite,* into three fractions which he called *yttria, erbia,* and *terbia.* The names *erbia* and *terbia* became confused in this early period. After 1860, Mosander's *terbia* was known as *erbia,* and after 1877, the earlier known *erbia* became *terbia.* The *erbia* of this period was later shown to consist of five oxides, now known as *erbia, scandia, holmia, thulia,* and *ytterbia.* By 1905 Urbain and James independently succeeded in isolating fairly pure Er_2O_3. Klemm and Bommer first produced reasonably pure erbium metal in 1934 by reducing the anhydrous chloride with potassium vapor. The pure metal is soft and malleable and has a bright, silvery, metallic luster. As with other rare-earth metals, its properties depend to a certain extent on the impurities present. The metal is fairly stable in air and does not oxidize as rapidly as some of the other rare-earth metals. Recent production techniques, using ion-exchange reactions, have resulted in much lower prices of the rare-earth metals and their compounds. Erbium is finding nuclear and metallurgical uses. Added to vanadium, for example, erbium lowers the hardness and improves workability. Most of the rare-earth oxides have sharp absorption bands in the visible, ultraviolet, and near infrared. This property, associated with the electronic structure, gives beautiful pastel colors to many of the rare-earth salts. Erbium oxide gives a pink color and has been used as a colorant in glasses and porcelain enamel glazes.

Europium — (Europe), Eu. In 1890, Boisbaudran obtained basic fractions from samarium-gadolinium concentrates that had spark spectral lines not accounted for by samarium or gadolinium. These lines subsequently were shown to belong to europium. The discovery of europium is generally credited to Demarcay, who separated the rare earth in reasonably pure form in 1901. The pure metal was not isolated until much later. Europium is now prepared by mixing Eu_2O_3 with a 10% excess of lanthanum metal and heating the mixture in a tantalum crucible under high vacuum. The element is collected as a silvery-white metallic deposit on the walls of the crucible. As with other rare-earth metals, except for lanthanum, europium ignites in air at about 150 to 180 °C. Europium is about as hard as lead and is quite ductile. It is the most reactive of the rare-earth metals, quickly oxidizing in air. It resembles calcium in its reaction with water. *Bastnasite* and *monazite* are the principal ores containing europium. Europium has been identified spectroscopically in the sun and certain stars. Europium isotopes are good neutron absorbers and are being studied for use in nuclear control applications. Europium oxide is now widely used as a phosphor activator and europium-activated yttrium vanadate is in commercial use as the red phosphor in color TV tubes. Europium-doped plastic has been used as a laser material. Europium is one of the rarest and most costly of the rare-earth metals, but with the development of ion-exchange techniques and special processes, the cost of the metal has been greatly reduced in recent years

Fermium — (Enrico Fermi [1901–1954], nuclear physicist), Fm. Fermium, the eighth transuranium element of the actinide series to be discovered, was identified by Ghiorso and co-workers in 1952 in the debris from a thermonuclear explosion in the Pacific in work involving the University of California Radiation Laboratory, the Argonne National Laboratory, and the Los Alamos Scientific Laboratory. The isotope produced was the 20-hour ^{255}Fm. During 1953 and early 1954, while discovery of elements 99 and 100 was withheld from publication for security reasons, a group from the Nobel Institute of Physics in Stockholm bombarded ^{238}U with ^{16}O ions, and isolated a 30-min α-emitter, which they ascribed to $^{250}100$, without claiming discovery of the element. This isotope has since been identified positively, and the 30-min half-life confirmed. The chemical properties of fermium have been studied solely with trace amounts, and in normal aqueous media only the (III) oxidation state appears to exist. The isotope ^{254}Fm and heavier isotopes can be produced by intense neutron irradiation of lower elements such as plutonium by a process of successive neutron capture interspersed with beta decays until these mass numbers and atomic numbers are reached. Twenty-two isotopes and isomers of fermium are known to exist. ^{257}Fm, with a half-life of 100.5 days, is the longest lived. ^{250}Fm, with a half-life of 28 min, has been shown to be a product of decay of ^{254}No. It was by chemical identification of ^{250}Fm that production of nobelium was confirmed. Fermium would probably have chemical properties resembling erbium.

Flerovium — (After the *Flerov* Laboratory of Nuclear Reactions, Russia), Fl. Although the synthesis of Element-114 was first reported in 1998 by a team at the Joint Institute for Nuclear Research (JINR) in Dubna, Russia, further experiments led to confusion, until the discovery was finally confirmed by groups in the U.S. and Germany in 2009-2011. IUPAC approved the name Flerovium in 2011. The most stable isotope so far discovered in ^{285}Fl, with a half-life of 0.13 s. Based on its electronic structure, flerovium is predicted to be a dense, volatile metal, possibly a gas at room temperature. It is calculated to be similar to lead in its chemistry, but probably less reactive.

Fluorine — (L. and F. *fluere*, flow, or flux), F. In 1529, Georgius Agricola described the use of fluorspar as a flux, and as early as 1670 Schwandhard found that glass was etched when exposed to fluorspar treated with acid. Scheele and many later investigators, including Davy, Gay-Lussac, Lavoisier, and Thenard, experimented with hydrofluoric acid, some experiments ending in tragedy. The element was finally isolated in 1886 by Moisson after nearly 74 years of continuous effort. Fluorine occurs chiefly in *fluorspar* (CaF_2) and *cryolite* (Na_2AlF_6) and is in *topaz* and other minerals. It is a member of the halogen family of elements and is obtained by electrolyzing a solution of potassium hydrogen fluoride in anhydrous hydrogen fluoride in a vessel of metal or transparent fluorspar. Modern commercial production methods are essentially variations on the procedures first used by Moisson. Fluorine is the most electronegative and reactive of all elements. It is a pale yellow, corrosive gas, which reacts with practically all organic and inorganic substances. Finely divided metals, glass, ceramics, carbon, and even water burn in fluorine with a bright flame. Until World War II, there was no commercial production of elemental fluorine. The atomic bomb project and nuclear energy applications, however, made it necessary to produce large quantities. Safe handling techniques have now been developed and it is possible at present to transport liquid fluorine by the ton. Fluorine and its compounds are used in producing enriched uranium (from the hexafluoride) and more than 100 commercial fluorochemicals, including many well-known high-temperature plastics. Hydrofluoric acid is extensively used for etching the glass of light bulbs, etc. Fluorochlorohydrocarbons were extensively used in air conditioning and refrigeration systems, but these compounds have been replaced fluorohydrocarbons to reduce the depletion of stratospheric ozone. The presence of fluorine as a soluble fluoride in drinking water to the extent of 2 ppm may cause mottled enamel in teeth, when used by children acquiring permanent teeth; in smaller amounts, however, fluorides are beneficial and are used in water supplies to prevent dental cavities. Elemental fluorine is used as an oxidizer in rocket engines. A number of fluorides of the rare gases xenon, radon, and krypton have been prepared. Elemental fluorine and the fluoride ion are highly toxic. The free element has a characteristic pungent odor, detectable in concentrations as low as 20 ppb, which is below the safe working level.

Francium — (France), Fr. Discovered in 1939 by Mlle. Marguerite Perey of the Curie Institute, Paris. Francium, the heaviest known member of the alkali metal series, occurs from alpha disintegration of actinium. It can also be made artificially by bombarding thorium with protons. While it occurs naturally in uranium minerals, there is probably less than 30 g of francium at any time in the total crust of the earth. It has the highest equivalent weight of any element and is the most unstable of the first 101 elements of the periodic system. Forty-six isotopes and isomers of francium are recognized. The longest lived, ^{223}Fr, a daughter of ^{227}Ac, has a half-life of 22 min. This is the only isotope of francium occurring in nature. Because all known isotopes are highly unstable, knowledge of the chemical properties of this element comes from radiochemical techniques. No weighable quantity of the element has been prepared or isolated. The chemical properties of francium should most closely resemble cesium, and francium salts are found to coprecipitate with cesium salts from aqueous solution. The predicted melting point is around 27 °C, so elemental francium might be either solid or liquid.

Gadolinium — (*gadolinite,* a mineral named for Gadolin, a Finnish chemist), Gd. Gadolinia, the oxide of gadolinium, was separated by Marignac in 1880, and Lecoq de Boisbaudran independently isolated the element from Mosander's "yttria" in 1886. The element was named for the mineral *gadolinite* from which this rare earth was originally obtained. Gadolinium is found in several other minerals, including *monazite* and *bastnasite,* which are of commercial importance. The element has been purified only in recent years. With the development of ion-exchange and solvent extraction techniques, the availability and price of gadolinium and the other rare-earth metals have greatly improved. The metal can be prepared by the reduction of the anhydrous fluoride with metallic calcium. As with other related rare-earth metals, it is silvery white, has a metallic luster, and is malleable and ductile. At room temperature, gadolinium crystallizes in the hexagonal, close-packed α form. Upon heating to 1235 °C, α-gadolinium transforms into the β form, which has a body-centered cubic structure. The metal is relatively stable in dry air, but in moist air it tarnishes with the formation of a loosely adhering oxide film which splits off and exposes more surface to oxidation. The metal reacts slowly with water and is soluble in dilute acid. Gadolinium has the highest thermal neutron capture cross section of any known element (49,000 barns). Natural gadolinium is a mixture of seven isotopes. Two of these, ^{155}Gd and ^{157}Gd, have excellent capture characteristics, but they are present naturally in low concentrations. As a result, gadolinium has a very fast burnout rate and has limited use as a nuclear control rod material. It has been used in making gadolinium yttrium garnets, which have important electronic applications. Compounds of gadolinium are used in making phosphors for color TV tubes. As little as 1% gadolinium has been found to improve the workability and resistance of iron, chromium, and related alloys to high temperatures and oxidation. Gadolinium ethyl sulfate has extremely low noise characteristics and may find use in duplicating the performance of amplifiers, such as the maser. The metal is ferromagnetic. Gadolinium is unique for its high magnetic moment and for its special Curie temperature (above which ferromagnetism vanishes) lying just at room temperature. This suggests uses as a magnetic component that senses hot and cold.

Gallium — (L. *Gallia,* France), Ga. Predicted and described by Mendeleev as eka-aluminum, and discovered spectroscopically by Lecoq de Boisbaudran in 1875, who in the same year obtained the free metal by electrolysis of a solution of the hydroxide in KOH, Gallium is often found as a trace element in *diaspore, sphalerite, germanite, bauxite,* and *coal.* Some flue dusts from burning coal have been shown to contain as much as 1.5% gallium. It is the only metal, except for mercury, cesium, and rubidium, which can be liquid near room temperature; this makes possible its use in high-temperature thermometers. It has one of the longest liquid ranges of any metal and has a low vapor pressure even at high temperatures. There is a strong tendency for gallium to supercool below its freezing point. Therefore, seeding may be necessary to initiate solidification. Ultrapure gallium has a beautiful, silvery appearance, and the solid metal exhibits a conchoidal fracture similar to glass. The metal expands 3.1% on solidifying; therefore, it should not be stored in glass or metal containers, as they may break as the metal solidifies. Gallium wets glass or porcelain and forms a brilliant mirror when it is painted on glass. It is widely used in doping semiconductors and producing solid-state devices such as transistors. Gallium arsenide is capable of converting electricity directly into coherent light. High-purity gallium is attacked slowly only by mineral acids. Magnesium gallate containing divalent impurities such as Mn^{+2} is finding use in commercial ultraviolet activated powder phosphors. Gallium nitride has been used to produce blue light-emitting diodes such as those used in CD and DVD readers. Gallium was the key ingredient in the detector for the Gallex Experiment in the 1990s. In this experiment, 30.3 tons of gallium in the form of a $GaCl_3$-HCl solution was used to detect solar neutrinos. Gallium readily alloys with most metals and has been used as a component in low-melting alloys. Its toxicity appears to be of a low order, but it should be handled with care until more data are forthcoming.

Germanium — (L. *Germania,* Germany), Ge. Predicted by Mendeleev in 1871 as eka-silicon and discovered by Winkler in 1886. The metal is found in *argyrodite,* a sulfide of germanium and silver; in *germanite,* which contains 8% of the element; in zinc ores; in coal; and in other minerals. The element is frequently obtained commercially from flue dusts of smelters processing zinc ores and has been recovered from the by-products of combustion of certain coals. Its presence in coal ensures a large reserve of the element in the years to come. Germanium can be separated from other metals by fractional distillation of its volatile tetrachloride. The tetrachloride may then be hydrolyzed to give GeO_2; the dioxide can be reduced with hydrogen to give the metal. The element is a gray-white metalloid, and in its pure state is crystalline and brittle, retaining its luster in air at room temperature. It is a very important semiconductor material. Zone-refining techniques have led to production of crystalline germanium for semiconductor use with an impurity of only one part in 10^{10}. Doped with arsenic, gallium, or other elements, it is used in thousands of electronic applications. Its application in fiber optics and infrared optical systems now provides the largest use for germanium. Germanium is also finding many other applications including use as an alloying agent, as a phosphor in fluorescent lamps, and as a catalyst. Germanium and germanium oxide are transparent to the infrared and are used in infrared spectrometers and other optical equipment, including extremely sensitive infrared detectors. Germanium oxide's high index of refraction and dispersion make it useful as a component of glasses used in wide-angle camera lenses and microscope objectives. The field of organogermanium chemistry is becoming increasingly important. Certain germanium compounds have a low mammalian toxicity, but a marked activity against certain bacteria, which makes them of interest as chemotherapeutic agents.

Gold — (Sanskrit *Jval;* Anglo-Saxon *gold*), Au (L. *aurum,* gold). Known and highly valued from earliest times, gold is found in nature as the free metal and in tellurides; it is very widely distributed and is almost always associated with quartz or pyrite. It occurs in veins and alluvial deposits and is often separated from rocks and other minerals by sluicing and panning operations. About 25% of the world's gold output comes from South Africa, and about two-thirds of the total U.S. production now comes from South Dakota and Nevada. The metal is recovered from its ores by cyaniding, amalgamating, and smelting processes. Refining is also frequently done by electrolysis. Gold occurs in seawater to the extent of 0.1 to 2 mg/ton, depending on the location where the sample is taken. As of yet, no method has been found for recovering gold from seawater profitably. It is estimated that all the gold in the world, so far refined, could be placed in a single cube 60 ft on a side. Of all the elements, gold in its pure state is undoubtedly the most beautiful. It is metallic, having a yellow color when in a mass, but when finely divided it may be black, ruby, or purple. The Purple of Cassius is a delicate test for auric gold. It is the most malleable and ductile metal; 1 oz. of gold can be beaten out to 300 ft^2. It is a soft metal and is usually alloyed to give it more strength. The specific gravity of gold has been found to vary considerably depending on temperature, how the metal is precipitated, and how it is cold-worked. Gold is a good conductor of heat and electricity and is unaffected by air and most reagents. It is used in coinage and is a standard for monetary systems in many countries. It is also extensively used for jewelry, decoration, dental work, and for plating. It is used for coating certain space satellites, as it is a good reflector of infrared radiation and is inert. Gold, like other precious metals, is measured in troy weight; when alloyed with other metals, the term *carat* is used to express the amount of gold present, 24 carats being pure gold. For many years the value of gold was set by the U.S. at $20.67/troy ounce; in 1934 this value was fixed by law at $35.00/troy ounce, 9/10th fine. On March 17, 1968, because of a financial crisis, a two-tiered pricing system was established whereby gold was still used to settle international accounts at the old $35.00/troy ounce price while the price of gold on the private market would be allowed to fluctuate. Since that time, the price of gold on the free market has fluctuated widely. The most common gold compounds are auric chloride ($AuCl_3$) and chlorauric acid ($HAuCl_4$), the latter being used in photography for toning the silver image. ^{198}Au, with a half-life of 2.7 days, is used for treating cancer and other diseases. Disodium aurothiomalate is administered intramuscularly as a treatment for arthritis. A mixture of one part nitric acid with three parts hydrochloric acid is called *aqua regia* (because it dissolved gold, the King of Metals). For many years the temperature assigned to the freezing point of gold was 1063.0 °C; this served as a calibration fixed point for the International Temperature Scales (ITS-27 and ITS-48) and the International Practical Temperature Scale (IPTS-48). In 1968, a new International Practical Temperature Scale (IPTS-68) was adopted, which required that the freezing point of gold be changed to 1064.43 °C. In 1990, a new International Temperature Scale (ITS-90) was adopted bringing the triple point of H_2O to 0.01 °C and the freezing point of gold to 1064.18 °C.

Hafnium — (*Hafnia,* Latin name for Copenhagen), Hf. Hafnium was thought to be present in various minerals and concentrations many years prior to its discovery in 1923, credited to D. Coster and G. von Hevesey. On the basis of the Bohr theory, the new element was expected to be associated with zirconium. It was finally identified in *zircon* from Norway, by means of x-ray spectroscopic analysis. It was named in honor of the city in which the discovery was made. Most zirconium minerals contain 1% to 5% hafnium. It was originally separated from zirconium by repeated recrystallization of the double ammonium or potassium fluorides by von Hevesey and Jantzen. Metallic hafnium was first prepared by van Arkel and deBoer by passing the vapor of the tetraiodide over a heated tungsten filament. Almost all hafnium metal now produced is made by reducing the tetrachloride with magnesium or with sodium (Kroll Process). Hafnium is a ductile metal with a brilliant silver luster. Its properties are considerably influenced by the impurities of zirconium present. Of all the elements, zirconium and hafnium are two of the most difficult to separate. Their chemistry is almost identical; however, the density of zirconium is about half that of hafnium. Very pure hafnium has been produced, with zirconium being the major impurity. Because hafnium has a good absorption cross section for thermal neutrons (almost 600 times that of zirconium), has excellent mechanical properties, and is extremely corrosion resistant, it is used for reactor control rods. Hafnium has been successfully alloyed with iron, titanium, niobium, tantalum, and other metals. Hafnium carbide is the most refractory binary composition known, and the nitride is the most refractory of all known metal nitrides (mp 3310 °C). Hafnium is used in gas-filled and incandescent lamps and is an efficient "getter" for scavenging oxygen and nitrogen. Finely divided hafnium is pyrophoric and can ignite spontaneously in air. Care should be taken when machining the metal or when handling hot sponge hafnium. At 700 °C hafnium rapidly absorbs hydrogen to form the composition $HfH_{1.86}$. Hafnium is resistant to concentrated alkalis, but at elevated temperatures reacts with oxygen, nitrogen, carbon, boron, sulfur, and silicon. Halogens react directly to form tetrahalides.

Hassium — (named for the German state, Hesse) Hs. This element was first synthesized and identified in 1964 by the same G.S.I. Darmstadt Group who first identified *Bohrium* and *Meitnerium*. Presumably this element has chemical properties similar to osmium. Isotope $^{265}108$ was produced using a beam of ^{58}Fe projectiles, produced by the Universal Linear Accelerator (UNILAC) to bombard a ^{208}Pb target. Discovery of *Bohrium* and *Meitnerium* was made using detection of isotopes with odd proton and neutron numbers. Elements having even atomic numbers have been thought to be less stable against spontaneous fusion than odd elements. The production of $^{265}108$ in the same reaction as was used at G.S.I. was confirmed at Dubna with detection of the seventh member of the decay chain ^{253}Es. Isotopes of *Hassium* are believed to decay by spontaneous fission, explaining why 109 was produced before 108. IUPAC adopted the name *Hassium* after the German state of Hesse in September 1997. In June 2001 it was announced that hassium is the heaviest element to have its chemical properties analyzed. A research team at the UNILAC heavy-ion accelerator in Darmstadt, Germany built an instrument to detect and analyze hassium. Atoms of curium-248 were collided with atoms of magnesium-26, producing about 6 atoms of hassium with a half-life of 9 s. This was sufficiently long to obtain data showing that hassium atoms react with oxygen to form hassium oxide molecules. These condensed at a temperature consistent with the behavior of Group 8 elements. This experiment appears to confirm hassium's location under osmium in the periodic table.

Helium — (Gr. *helios,* the sun), He. Evidence of the existence of helium was first obtained by Janssen during the solar eclipse of 1868 when he detected a new line in the solar spectrum; Lockyer and Frankland suggested the name *helium* for the new element; in 1895, Ramsay discovered helium in the uranium mineral *cleveite,* and it was independently discovered in cleveite by the Swedish chemists Cleve and Langlet about the same time. Rutherford and Royds in 1907 demonstrated that α particles are helium nuclei. Except for hydrogen, helium is the most abundant element found throughout the universe. Helium is extracted from natural gas; all natural gas contains at least trace quantities of helium. It has been detected spectroscopically in great abundance, especially in the hotter stars, and it is an important component in both the proton–proton reaction and the carbon cycle, which account for the energy of the sun and stars. The fusion of hydrogen into helium provides the energy of the hydrogen bomb. The helium content of the atmosphere is about 1 part in 200,000. It is present in various radioactive minerals as a decay product. Much of the world's supply of helium is obtained from wells in Texas, Colorado, and Kansas. Helium has the lowest triple-point temperature of any element and has found wide use in cryogenic research, as its boiling point is near 4 K. Its use in the study of cryogenics is vital. Liquid helium (^4He) exists in two forms: He^4I and He^4II, with a sharp transition point at 2.174 K (3.83 cm Hg). ^4HeI (above this temperature) is a normal liquid, but ^4HeII (below it) is unlike any other known substance. It expands on cooling; its conductivity for heat is enormous; and neither its heat conduction nor viscosity obeys normal rules. It has other peculiar properties. Helium is the only liquid that cannot be solidified by lowering the temperature at atmospheric pressure. It remains liquid down to absolute zero at ordinary pressures, but it can readily be solidified by increasing the pressure. Solid ^3He and ^4He are unusual in that both can readily be changed in volume by more than 30% by application of pressure. The specific heat of helium gas is unusually high. The density of helium vapor at the normal boiling point is also very high, with the vapor expanding greatly when heated to room temperature. Containers filled with helium gas at 5 to 10 K should be treated as though they contained liquid helium due to the large increase in pressure resulting from warming the gas to room temperature. While helium normally has a 0 valence, it seems to have a weak tendency to combine with certain other elements. Means of preparing helium difluoride have been studied, and species such as HeNe and the molecular ions He^+ and He^{++} have been investigated. Helium is widely used as an inert gas shield for arc welding; as a protective gas in growing silicon and germanium crystals, and in titanium and zirconium production; as a cooling medium for nuclear reactors, and as a gas for supersonic wind tunnels. A mixture of helium and oxygen is used as an artificial atmosphere for divers and others working under pressure. Different ratios of He/O_2 are used for different depths at which the diver is operating. Helium is extensively used for filling balloons as it is a much safer gas than hydrogen. One of the recent largest uses for helium has been for pressurizing liquid fuel rockets. A Saturn booster such as used on the Apollo lunar missions required about 13 million ft^3 of helium for a firing, plus more for checkouts. Liquid helium's use in magnetic resonance imaging (MRI) continues to increase as the use of this diagnostic technique expands. Lifting gas applications are increasing. Various companies in addition to Goodyear, are now using "blimps" for advertising. The Navy and the Air Force are investigating the use of airships to provide early warning systems to detect low-flying cruise missiles. NASA is currently using helium-filled balloons to sample the atmosphere in Antarctica to study holes in the ozone layer.

Holmium — (L. *Holmia* , for Stockholm), Ho. The spectral absorption bands of holmium were noticed in 1878 by the Swiss chemists Delafontaine and Soret, who announced the existence of an "Element X." Cleve, of Sweden, later independently discovered the element while working on erbia earth. The element is named after Cleve's native city. Pure holmia, the yellow oxide, was prepared by Homberg in 1911. Holmium occurs in *gadolinite, monazite,* and in other rare-earth minerals. It is commercially obtained from monazite, occurring in that mineral to the extent of about 0.05%. It has been isolated by the reduction of its anhydrous chloride or fluoride with calcium metal. Pure holmium has a metallic to bright silver luster. It is relatively soft and malleable, and is stable in dry air at room temperature, but rapidly oxidizes in moist air and at elevated temperatures. Its high magnetic permeability leads to applications in high-field magnets. It is also used in near-infrared lasers.

Hydrogen — (Gr. *hydro,* water, and *genes,* forming), H. Hydrogen was prepared many years before it was recognized as a distinct substance by Cavendish in 1766. It was named by Lavoisier. Hydrogen is the most abundant of all elements in the universe, and it is thought that the heavier elements were, and still are, being built from hydrogen and helium. It has been estimated that hydrogen makes up more than 90% of all the atoms or three-quarters of the mass of the universe. It is found in the sun and most stars and plays an important part in the proton–proton reaction and carbon–nitrogen cycle, which accounts for the energy of the sun and stars. It is thought that hydrogen is a major component of the planet Jupiter and that at some depth in the planet's interior the pressure is so great that solid molecular hydrogen is converted into solid metallic hydrogen. On Earth, hydrogen occurs chiefly in combination with oxygen in water, but it is also present in organic matter such as living plants, petroleum, coal, etc. It is present as the free element in the atmosphere, but only to the extent of less than 1 ppm by volume. It is the lightest of all gases, and combines with other elements, sometimes explosively, to form compounds. Great quantities of hydrogen are required commercially for the fixation of nitrogen from the air in the Haber ammonia process and for the hydrogenation of fats and oils. It is also used in large quantities in methanol production, in hydrodealkylation, hydrocracking, and hydrodesulfurization. It is also used as a rocket fuel, for cutting and welding of metals, for production of hydrochloric acid, for the reduction of metallic ores, and for filling balloons. The lifting power of 1 ft^3 of hydrogen gas is about 0.076 lb at 0 °C, 760 mm pressure. Production of hydrogen in the U.S. alone amounts to about 3 billion cubic feet per year. It is prepared by the action of steam on heated carbon, by decomposition of certain hydrocarbons with heat, by the electrolysis of water, or by the displacement from acids by certain metals. It is also produced by the action of sodium or potassium hydroxide on aluminum. Liquid hydrogen is important in cryogenics as its melting point is only about 20 K. Hydrogen consists of three isotopes, most of which is ^{1}H. In 1932, Urey announced the discovery of a stable isotope, deuterium (^{2}H or D). Deuterium is present in natural hydrogen to the extent of about 0.01%. Two years later an unstable isotope, tritium (^{3}H), was discovered. Tritium has a half-life of about 12.31 years. Tritium atoms are also present in natural hydrogen but in a much smaller proportion. Tritium is readily produced in nuclear reactors and is used in the production of the hydrogen bomb. It is also used as a radioactive agent in making luminous paints, and as a tracer. Heavy water, deuterium oxide (D_2O), is used as a moderator to slow down neutrons. Hydrogen gas under ordinary conditions is a mixture of two kinds of molecules, known as *ortho-* and *para*-hydrogen, which differ from one another by the spins of their electrons and nuclei. Normal hydrogen at room temperature contains 25% of the *para* form and 75% of the *ortho* form. The *ortho* form cannot be prepared in the pure state. Since the two forms differ in energy, the physical properties also differ. The melting and boiling points of *para*-hydrogen are about 0.1 °C lower than those of normal hydrogen. Both stationary and mobile power sources using hydrogen combustion offer a major advantage for reducing atmospheric carbon dioxide and retarding global temperature rise. The "hydrogen economy" is expected to grow rapidly as steps are taken to reduce the combustion of fossil fuels.

Indium — (from the brilliant indigo line in its spectrum), In. Discovered spectroscopically in 1863 by Reich and Richter, who later isolated the metal. Indium is most frequently associated with zinc materials, and it is from these that most commercial indium is now obtained; however, it is also found in iron, lead, and copper ores. Until 1924, a gram or so constituted the world's supply of this element in isolated form. It is probably about as abundant as silver. About 4 million troy ounces of indium are now produced annually, much of it from Canada. It is available in ultrapure form. Indium is a very soft, silvery-white metal with a brilliant luster. The pure metal gives a high-pitched "cry" when bent. In liquid form it wets glass, as does gallium. It has found application in making low-melting alloys; an alloy of 24% indium–76% gallium is liquid at room temperature. Indium is used in making bearing alloys, and in the form of various semiconductors (InSb, InAs, etc.) it has many applications in transistors, rectifiers, thermistors, liquid crystal displays, batteries, solar cells, and photoconductors. It can be plated onto metal and evaporated onto glass, forming a mirror as good as that made with silver but with more resistance to atmospheric corrosion. Indium has a low order of toxicity.

Iodine — (Gr. *iodes,* violet), I. Discovered by Courtois in 1811. Iodine, a halogen, occurs sparingly in the form of iodides in seawater from which it is assimilated by seaweeds, in Chilean saltpeter and nitrate-bearing earth, known as *caliche* in brines from old sea deposits, and in brackish waters from oil and salt wells. Ultrapure iodine can be obtained from the reaction of potassium iodide with copper sulfate. Several other methods of isolating the element are known. Iodine is a bluish-black, lustrous solid, volatilizing at ordinary temperatures into a blue-violet gas with an irritating odor; it forms compounds with many elements, but is less active than the other halogens, which displace it from iodides. Iodine exhibits some metallic-like properties. It dissolves readily in chloroform, carbon tetrachloride, or carbon disulfide to form beautiful purple solutions. It is only slightly soluble in water. Iodine compounds are important in organic chemistry and very useful in medicine. The artificial radioisotope ^{131}I, with a half-life of 8 days, has been used in treating the thyroid gland. The most common compounds are the iodides of sodium and potassium (Na I and KI) and the iodates (KIO_3). Lack of iodine is the cause of goiter. Iodides and thyroxin, which contains iodine, are used internally in medicine, and a solution of KI and iodine in alcohol is used for external wounds. Potassium iodide finds use in photography. The deep blue color with starch solution is characteristic of the free element. Care should be taken in handling and using iodine, as contact with the skin can cause lesions; iodine vapor is intensely irritating to the eyes and mucous membranes.

Iridium — (L. *iris,* rainbow), Ir. Discovered in 1803 by Tennant in the residue left when crude platinum is dissolved by aqua regia. The name iridium is appropriate, for its salts are highly colored. Iridium, a metal of the platinum family, is white, similar to platinum, but with a slight yellowish cast. It is very hard and brittle, making it hard to machine, form, or work. It is the most corrosion-resistant metal known and was used in making the original standard meter bar, which is a 90% platinum–10% iridium alloy. Iridium is not attacked by any of the acids nor by aqua regia, but is attacked by molten salts, such as NaCl and NaCN. Iridium occurs uncombined in nature with platinum and other metals of this family in alluvial deposits. It is recovered as a by-product from the nickel mining industry. The largest reserves and production of the platinum group of metals, which includes iridium, is in South Africa, followed by Russia and Canada. Meteorites contain small amounts of iridium. The presence of iridium has recently been used in examining the Cretaceous-Tertiary (K-T) boundary. Because iridium is found widely distributed at the K-T boundary, it has been suggested that a large meteorite or asteroid collided with the Earth, killing the dinosaurs, and creating a large dust cloud and crater. Searches for such a crater point to one in the Yucatan, known as Chicxulub. Iridium has found use in making crucibles and apparatus for use at high temperatures. It is also used for electrical contacts. Its principal use is as a hardening agent for platinum. With osmium, it forms an alloy that is used for tipping pens and compass bearings. The density of iridium is only slightly lower than that of osmium, which has been generally credited as being the heaviest known element. Calculations of the densities of iridium and osmium from the space lattices give values of 22.65 and 22.61 g/cm^3, respectively. These values may be more reliable than actual physical measurements. At present, therefore, we know that either iridium or osmium is the densest known element, but the data do not yet allow selection between the two.

Iron — (Anglo-Saxon, *iron*), Fe (L. *ferrum*). The use of iron is prehistoric. Genesis mentions that Tubal-Cain, seven generations from Adam, was "an instructor of every artificer in brass and iron." A remarkable iron pillar, dating to about A.D. 400, remains standing today in Delhi, India. This solid shaft of wrought iron is about 7¼ m high by 40 cm in diameter. Corrosion to the pillar has been minimal although it has been exposed to the weather since its erection. Iron is a relatively abundant element in the universe. It is found in the sun and many types of stars in considerable quantity. It has been suggested that the iron we have here on Earth may have originated in a supernova. Iron is a very difficult element to produce in ordinary nuclear reactions, such as would take place in the sun. Iron is found native as a principal component of a class of iron–nickel meteorites known as *siderites* and is a minor constituent of the other two classes of meteorites. The core of the Earth, 2150 miles in radius, is thought to be largely composed of iron with about 10% occluded hydrogen. The metal is the fourth most abundant element, by weight, making up the crust of the Earth. The most common ore is *hematite* (Fe_2O_3). Magnetite (Fe_3O_4) is frequently seen as *black sands* along beaches and banks of streams. *Lodestone* is another form of magnetite. *Taconite* is becoming increasingly important as a commercial ore. Iron is a vital constituent of plant and animal life, and appears in hemoglobin. It has a rich chemistry, both inorganic and organometallic. The pure metal is not often encountered in commerce, but is usually alloyed with carbon or other metals. The pure metal is very reactive chemically and rapidly corrodes, especially in moist air or at elevated temperatures. It has four principal allotropic forms, known as α, γ, δ, and ε. The α-γ transition point is 912 °C; γ-δ is 1394 °C; and the ε phase appears at high pressures. The α form is magnetic, but the magnetism disappears above 771 °C, although the lattice remains unchanged. The relations of these forms are complex. Pig iron is an alloy containing about 3% carbon with varying amounts of S, Si, Mn, and P. It is hard, brittle, fairly fusible, and is used to produce other alloys, including steel. Wrought iron contains only a few tenths of a percent of carbon, is tough, malleable, less fusible, and usually has a "fibrous" structure. Carbon steel is an alloy of iron with carbon, with small amounts of Mn, S, P, and Si. Alloy steels are carbon steels with other additives such as nickel, chromium, vanadium, etc. Iron is the cheapest and most abundant, useful, and important of all metals.

Krypton — (Gr. *kryptos*, the hidden one), Kr. Discovered in 1898 by Ramsay and Travers in the residue left after liquid air had nearly boiled away, krypton is present in the air to the extent of about 1 ppm. The atmosphere of Mars has been found to contain 0.3 ppm of krypton. It is one of the noble gases. It is characterized by its brilliant green and orange spectral lines. The spectral lines of krypton are easily produced, and some are very sharp; at one time the meter was defined in terms of the wavelength of a krypton line. Solid krypton is a white crystalline substance with a face-centered cubic structure that is common to all the rare gases. While krypton was first thought of as a noble gas that does not combine with other elements, the existence of several krypton compounds has been established. Krypton difluoride has been prepared in gram quantities and can be made by several methods. ^{85}Kr has found application in chemical analysis. By imbedding the isotope in various solids, *kryptonates* are formed. The activity of these kryptonates is sensitive to chemical reactions at the surface. Estimates of the concentration of reactants are therefore made possible. Krypton is used in certain photographic flash lamps for high-speed photography.

Lanthanum — (Gr. *lanthanein*, to lie hidden), La. Mosander in 1839 extracted a new earth *lanthana*, from impure cerium nitrate, and recognized the new element. Lanthanum is found in rare-earth minerals such as *cerite, monazite, allanite,* and *bastnasite*. Monazite and bastnasite are principal ores in which lanthanum occurs in percentages up to 25% and 38%, respectively. Misch metal, used in making lighter flints, contains about 25% lanthanum. Lanthanum was first isolated in relatively pure form in 1923. Ion-exchange and solvent extraction techniques have led to much easier isolation of the rare-earth elements, resulting in improved availability. The metal can be produced by reducing the anhydrous fluoride with calcium. Lanthanum is silvery white, malleable, ductile, and soft enough to be cut with a knife. It is one of the most reactive of the rare-earth metals. It oxidizes rapidly when exposed to air. Cold water attacks lanthanum slowly, and hot water attacks it much more rapidly. The metal reacts directly with elemental carbon, nitrogen, boron, selenium, silicon, phosphorus, sulfur, and with halogens. At 310 °C, lanthanum changes from a hexagonal to a face-centered cubic structure, and at 865 °C it again transforms into a body-centered cubic structure. Rare-earth compounds containing lanthanum are extensively used in carbon lighting applications, especially by the motion picture industry for studio lighting and projection. This application consumes about 25% of the rare-earth compounds produced. La_2O_3 improves the alkali resistance of glass and is used in making special optical glasses. Small amounts of lanthanum, as an additive, can be used to produce nodular cast iron. Hydrogen sponge alloys containing lanthanum can take up to 400 times their own volume of hydrogen gas, and thus may be useful in energy storage systems. Lanthanum and its compounds have a low to moderate acute toxicity rating; therefore, care should be taken in handling them.

Lawrencium — (Ernest O. Lawrence [1901–1958], inventor of the cyclotron), Lr. This member of the 5f transition elements (actinide series) was discovered in March 1961 by A. Ghiorso, T. Sikkeland, A. E. Larsh, and R. M. Latimer. A 3-μg californium target, consisting of a mixture of isotopes of mass number 249, 250, 251, and 252, was bombarded with either ^{10}B or ^{11}B. The electrically charged transmutation nuclei recoiled with an atmosphere of helium and were collected on a thin copper conveyor tape which was then moved to place the collected atoms in front of a series of solid-state detectors. The isotope of element 103 produced in this way decayed by emitting an 8.6-MeV alpha particle with a half-life of 8 s. In 1967, Flerov and associates of the Dubna Laboratory reported their inability to detect an alpha emitter with a half-life of 8 s which was assigned by the Berkeley group to $^{257}103$. This assignment has been changed to ^{258}Lr or ^{259}Lr. In 1965, the Dubna workers found a longer-lived lawrencium isotope, ^{256}Lr, with a half-life of 35 s. In 1968, Ghiorso and associates at Berkeley were able to use a few atoms of this isotope to study the oxidation behavior of lawrencium. Using solvent extraction techniques and working very rapidly, they extracted lawrencium ions from a buffered aqueous solution into an organic solvent, completing each extraction in about 30 s. It was found that lawrencium behaves differently from dipositive nobelium and more like the tripositive elements earlier in the actinide series.

Lead — (Anglo-Saxon *lead*), Pb (L. *plumbum*). Long known, mentioned in Exodus. The alchemists believed lead to be the oldest metal and associated it with the planet Saturn. Native lead occurs in nature, but it is rare. Lead is obtained chiefly from *galena* (PbS) by a roasting process. *Anglesite* ($PbSO_4$), *cerussite* ($PbCO_3$), and *minim* (Pb_3O_4) are other common lead minerals. Lead is a bluish-white metal of bright luster, is very soft, highly malleable, ductile, and a poor conductor of electricity. It is very resistant to corrosion; lead pipes bearing the insignia of Roman emperors, used as drains from the baths, are still in service. It is used in containers for corrosive liquids (such as sulfuric acid) and may be toughened by the addition of a small percentage of antimony or other metals. Its alloys include solder, type metal, and various antifriction metals. Great quantities of lead, both as the metal and as the dioxide, are used in storage batteries. Lead is also used for cable covering, plumbing, and ammunition. The metal is very effective as a sound absorber, is used as a radiation shield around x-ray equipment and nuclear reactors, and is used to absorb vibration. Lead, alloyed with tin, is used in making organ pipes. White lead, the basic carbonate, sublimed white lead ($PbSO_4$), chrome yellow ($PbCrO_4$), red lead (Pb_3O_4), and other lead compounds are used extensively in paints, although in recent years the use of lead in paints has been drastically curtailed to eliminate or reduce health hazards. Lead oxide is used in producing fine "crystal glass" and "flint glass" of a high index of refraction for achromatic lenses. The nitrate and the acetate are soluble salts. Lead salts such as lead arsenate have been used as insecticides, but their use in recent years has been practically eliminated in favor of less harmful organic compounds. Care must be used in handling lead as it is a cumulative poison. Environmental concern with lead poisoning led to elimination of lead tetraethyl in gasoline. The U.S. Occupational Safety and Health Administration (OSHA) has recommended that industries limit airborne lead in the workplace to 50 μg/m^3.

Lithium — (Gr. *lithos*, stone), Li. Discovered by Arfvedson in 1817. Lithium is the lightest of all metals, with a density only about half that of water. It does not occur free in nature; combined it is found in small amounts in nearly all igneous rocks and in the waters of many mineral springs. *Lepidolite, spodumene, petalite,* and *amblygonite* are the more important minerals containing it. Lithium is presently being recovered from brines of Searles Lake, in California, and from Nevada, Chile, and Argentina. Large deposits of spodumene are found in North Carolina. The metal is produced electrolytically from the fused chloride. Lithium is silvery in appearance, much like Na, K, and other members of the alkali metal series. It reacts with water, but not as vigorously as sodium. Lithium imparts a beautiful crimson color to a flame, but when the metal burns strongly the flame is a dazzling white. In recent years, the production of lithium metal and its compounds has increased greatly. Because the metal has the highest specific heat of any solid element, it has found use in heat transfer applications; however, it is corrosive and requires special handling. The metal has been used as an alloying agent, is of interest in synthesis of organic compounds, has important pharmaceutical uses, and has nuclear applications. It ranks as a leading contender as a battery anode material because it has a high electrochemical potential. Lithium is also used in special glasses and ceramics. Lithium chloride is one of the most hygroscopic materials known, and it, as well as lithium bromide, is used in air conditioning and industrial drying systems. Lithium stearate is used as an all-purpose and high-temperature lubricant. The use of lithium-ion storage batteries in applications ranging from cell phones to automobiles is growing rapidly.

Livermorium — (Lawrence *Livermore* National Laboratory), Lv. A single atom of Element-116 was first synthesized in 2000 at Lawrence Livermore Laboratory in collaboration with workers at JINR, Dubna. The name Livermorium was adopted by IUPAC in 2012. Livermorium falls at the top of the chalcogen group: oxygen, sulfur, selenium, tellurium, and polonium. No information is available on its physical and chemical properties. The most stable isotope, ^{293}Lv, has a half-life of 61 ms.

Lutetium — (Lutetia, ancient name for Paris, sometimes called *cassiopeium* in Germany), Lu. In 1907, Urbain described a process by which Marignac's ytterbium (1879) could be separated into the two elements, ytterbium (neoytterbium) and lutetium. These elements were identical with "aldebaranium" and "cassiopeium," independently discovered by von Welsbach about the same time. Charles James of the University of New Hampshire also independently prepared the very pure oxide, *lutecia*, at this time. The spelling of the element was changed from *lutecium* to *lutetium* in 1949. Lutetium occurs in very small amounts in nearly all minerals containing yttrium, and is present in *monazite,* which is a commercial source, to the extent of about 0.003%. The pure metal was only isolated later and is one of the most difficult to prepare. It can be prepared by the reduction of anhydrous $LuCl_3$ or LuF_3 by an alkali or alkaline earth metal. The metal is silvery white and relatively stable in air. While new techniques, including ion-exchange reactions, have been developed to separate the various rare-earth elements, lutetium is still the most costly of all rare earths. Stable lutetium isotopes, which emit pure beta radiation after thermal neutron activation, can be used as catalysts in cracking, alkylation, hydrogenation, and polymerization. Virtually no other commercial uses have been found. Lutetium, like other rare-earth metals, is thought to have a low toxicity rating.

Magnesium — (*Magnesia,* district in Thessaly), Mg. Compounds of magnesium have long been known. Black recognized magnesium as an element in 1755. It was isolated by Davy in 1808 and prepared in coherent form by Bussy in 1831. Magnesium is the eighth most abundant element in the Earth's crust. It does not occur uncombined but is found in large deposits in the form of *magnesite, dolomite,* and other minerals. The metal is now principally obtained in the U.S. by electrolysis of fused magnesium chloride derived from brines, wells, and seawater. Magnesium is a light, silvery-white, and fairly tough metal. It tarnishes slightly in air, and finely divided magnesium readily ignites upon heating in air and burns with a dazzling white flame. It has been used in flashlight photography, flares, and pyrotechnics, including incendiary bombs. It is one third lighter than aluminum, and in alloys is essential for airplane and missile construction. The metal improves the mechanical, fabrication, and welding characteristics of aluminum when used as an alloying agent. Magnesium is used in producing nodular graphite in cast iron and is used as an additive to conventional propellants. It is also used as a reducing agent in the production of pure uranium and other metals from their salts. The hydroxide (*milk of magnesia*), chloride, sulfate (*Epsom salts*), and citrate are used in medicine. Dead-burned magnesite is employed for refractory purposes such as brick and liners in furnaces and converters. Calcined magnesia is also used for water treatment and in the manufacture of rubber, paper, etc. Organic magnesium compounds (Grignard's reagents) are important. Magnesium is an important element in both plant and animal life. Chlorophylls are magnesium-centered porphyrins. The adult daily requirement of magnesium is about 300 mg/day, but this is affected by various factors. Great care should be taken in handling magnesium metal, especially in the finely divided state, as serious fires can occur. Water should not be used on burning magnesium or on magnesium fires.

Manganese — (L. *magnes,* magnet, from magnetic properties of pyrolusite; It. *manganese,* corrupt form of *magnesia*), Mn. Recognized by Scheele, Bergman, and others as an element and isolated by Gahn in 1774 by reduction of the dioxide with carbon. Manganese minerals are widely distributed; oxides, silicates, and carbonates are the most common. The discovery of large quantities of manganese nodules on the floor of the oceans holds promise as a source of manganese. These nodules contain about 24% manganese together with many other elements in lesser abundance. Most manganese today is obtained from ores found in Ukraine, Brazil, Australia, Republic of So. Africa, Gabon, China, and India. *Pyrolusite* (MnO_2) and *rhodochrosite* ($MnCO_3$) are among the most common manganese minerals. The metal is obtained by reduction of the oxide with sodium, magnesium, aluminum, or by electrolysis. It is gray-white, resembling iron but is harder and very brittle. The metal is reactive chemically and decomposes in cold water slowly. Manganese is used to form many important alloys. In steel, manganese improves the rolling and forging qualities, strength, toughness, stiffness, wear resistance, hardness, and hardenability. With aluminum and antimony, especially with small amounts of copper, it forms highly ferromagnetic alloys. Manganese metal is ferromagnetic only after special treatment. The pure metal exists in four allotropic forms. The alpha form is stable up to 727 °C; gamma manganese, which is metastable at ordinary temperatures, is soft, easily cut, and capable of being bent. The dioxide (pyrolusite) is used as a depolarizer in dry cells, and is used to "decolorize" glass that is colored green by impurities of iron. Manganese by itself colors glass an amethyst color and is responsible for the color of true amethyst. The dioxide is also used in the preparation of oxygen and chlorine, and in drying black paints. The permanganate is a powerful oxidizing agent and is used in quantitative analysis and in medicine. Manganese is widely distributed throughout the animal kingdom. It is an important trace element and may be essential for utilization of vitamin B_1.

Meitnerium — (Lise Meitner [1878–1968], Austrian–Swedish physicist and mathematician), Mt. In 1982, Element 109 was made and identified by physicists at the Heavy Ion Research Laboratory (G.S.I.), Darmstadt, Germany, by bombarding a target of ^{209}Bi with accelerated nuclei of ^{58}Fe. In August 1997 IUPAC adopted the name meitnerium for this element, honoring Lise Meitner. The production of Mt has been extremely small. It took a week of target bombardment (10^{11} nuclear encounters) to produce a single atom. Oganessian and his team at Dubna in 1984 repeated the Darmstadt experiment using a tenfold irradiation dose. One fission event from seven alpha decays of Mt was observed, thus indirectly confirming the existence of isotope ^{266}Mt.

Mendelevium — (Dmitri Mendeleev [1834–1907]), Md. Mendelevium, the ninth transuranium element of the actinide series to be discovered, was first identified by Ghiorso, Harvey, Choppin, Thompson, and Seaborg early in 1955 as a result of the bombardment of the isotope ^{253}Es with helium ions in the Berkeley 60-inch cyclotron. The isotope produced was ^{256}Md, which has a half-life of 78 min. This first identification was notable in that ^{256}Md was synthesized on a one-atom-at-a-time basis. ^{258}Md has a half-life of 51.5 days. It now appears possible that eventually enough ^{258}Md can be made so that some of its physical properties can be determined. ^{256}Md has been used to elucidate some of the chemical properties of mendelevium in aqueous solution.

Experiments seem to show that the element possesses a moderately stable dipositive (II) oxidation state in addition to the tripositive (III) oxidation state, which is characteristic of actinide elements.

Mercury — (Planet *Mercury*), Hg (*hydrargyrum*, liquid silver). Known to ancient Chinese and Hindus; found in Egyptian tombs of 1500 B.C. Mercury is the only common metal liquid at ordinary temperatures. It only rarely occurs free in nature. The chief ore is *cinnabar* (HgS). Spain and China produce about 75% of the world's supply of the metal. The commercial unit for handling mercury is the "flask," which weighs 76 lb (34.46 kg). The metal is obtained by heating cinnabar in a current of air and by condensing the vapor. It is a heavy, silvery-white metal; a rather poor conductor of heat, as compared with other metals, and a fair conductor of electricity. It easily forms alloys with many metals, such as gold, silver, and tin, which are called *amalgams*. Its ease in amalgamating with gold is made use of in the recovery of gold from its ores. The metal is widely used in laboratory work for making thermometers, barometers, diffusion pumps, and many other instruments. It is used in making mercury-vapor lamps and advertising signs, etc., and is used in mercury switches and other electrical apparatus. Other uses are in making pesticides, mercury cells for caustic soda and chlorine production, antifouling paint, batteries, and catalysts. The previously heavy use in dental preparations is being phased out because of toxicity concerns. The most important salts are mercuric chloride $HgCl_2$ (corrosive sublimate — a violent poison), mercurous chloride Hg_2Cl_2 (calomel, occasionally still used in medicine), mercury fulminate ($Hg(ONC)_2$), a detonator widely used in explosives, and mercuric sulfide (HgS, vermillion, a high-grade paint pigment). Organic mercury compounds are important. It has been found that an electrical discharge causes mercury vapor to combine with neon, argon, krypton, and xenon. These products, held together with van der Waals' forces, correspond to HgNe, HgAr, HgKr, and HgXe. Mercury is a virulent poison and is readily absorbed through the respiratory tract, the gastrointestinal tract, or through unbroken skin. It acts as a cumulative poison and dangerous levels are readily attained in air. Air saturated with mercury vapor at 20 °C contains a concentration that exceeds the toxic limit many times. The danger increases at higher temperatures. *It is therefore important that mercury be handled with great care.* Containers of mercury should be securely covered, and spillage should be avoided. If it is necessary to heat mercury or mercury compounds, it should be done in a well-ventilated hood. Methyl mercury is a dangerous pollutant and is now widely found in water and streams. The triple point of mercury, −38.8344 °C, is a fixed point on the International Temperature Scale (ITS-90).

Molybdenum — (Gr. *molybdos*, lead), Mo. Before Scheele recognized molybdenite as a distinct ore of a new element in 1778, it was confused with graphite and lead ore. The metal was prepared in an impure form in 1782 by Hjelm. Molybdenum does not occur native but is obtained principally from *molybdenite* (MoS_2). *Wulfenite* ($PbMoO_4$) and *powellite* ($Ca(Mo,W)O_4$) are also minor commercial ores. Molybdenum is also recovered as a by-product of copper and tungsten mining operations. The U.S., Canada, Chile, and China produce most of the world's molybdenum ores. The metal is prepared from the powder made by the hydrogen reduction of purified molybdic trioxide or ammonium molybdate. The metal is silvery white, very hard, but is softer and more ductile than tungsten. It has a high elastic modulus, and only tungsten and tantalum, of the more readily available metals, have higher melting points. It is a valuable alloying agent, as it contributes to the hardenability and toughness of quenched and tempered steels. It also improves the strength of steel at high temperatures. It is used in certain nickel-based alloys, such as the Hastelloys® which are heat-resistant and corrosion-resistant to chemical solutions. Molybdenum oxidizes at elevated temperatures. The metal has found application as electrodes for electrically heated glass furnaces and forehearths. It is also used in nuclear energy applications and for missile and aircraft parts. Molybdenum is valuable as a catalyst in the refining of petroleum. It has found application as a filament material in electronic and electrical applications. Molybdenum is an essential trace element in plant nutrition. Some lands are barren for lack of this element in the soil. Molybdenum sulfide is useful as a lubricant, especially at high temperatures where oils would decompose. Almost all ultrahigh strength steels with minimum yield points up to 300,000 lb/in^2 contain molybdenum in amounts from 0.25% to 8%.

Moscovium — (*Moscow* Oblast, location of Dubna), Mc. First detected in 2003 at JNR, Dubna, Russia. The initial claim could not be verified, but subsequent work in 2013-2015 in Germany, the U.S., and Russia provided the necessary confirmation. IUPAC approved the name Moscovium in 2016. Its most stable isotope is ^{289}Mc, with a half-life of 220 ms. Its position in the periodic table places it at the top of the group nitrogen, phosphorus, arsenic, antimony, and bismuth. However, no information is available yet of its actual chemical properties.

Neodymium — (Gr. *neos*, new, and *didymos*, twin), Nd. In 1841, Mosander extracted from *cerite* a new rose-colored oxide, which he believed contained a new element. He named the element *didymium*, as it was *an inseparable twin brother of lanthanum*. In 1885, von Welsbach separated didymium into two new elemental components, *neodymia* and *praseodymia*, by repeated fractionation of ammonium didymium nitrate. While the free metal is in *misch metal*, long known and used as a pyrophoric alloy for light flints, the element was not isolated in relatively pure form until 1925. Neodymium is present in misch metal to the extent of about 18%. It is present in the minerals *monazite* and *bastnasite*, which are principal sources of rare-earth metals. The element may be obtained by separating neodymium salts from other rare earths by ion-exchange or solvent extraction techniques, and by reducing anhydrous halides such as NdF_3 with calcium metal. Other separation techniques are possible. The metal has a bright silvery metallic luster. Neodymium is one of the more reactive rare-earth metals and quickly tarnishes in air, forming an oxide that splits off and exposes metal to oxidation. The metal, therefore, should be kept under light mineral oil or sealed in a plastic material. Neodymium exists in two allotropic forms, with a transformation from a double hexagonal to a body-centered cubic structure taking place at 855 °C. Didymium, of which neodymium is a component, is used for coloring glass to make welder's goggles. By itself, neodymium colors glass delicate shades ranging from pure violet through wine-red and warm gray. Light transmitted through such glass shows unusually sharp absorption bands. The glass has been used in astronomical work to produce sharp bands by which spectral lines may be calibrated. Glass containing neodymium can be used as a laser material. Neodymium salts are used as a colorant for enamels. The element is also being used with iron and boron to produce extremely strong magnets.

Neon — (Gr. *neos*, new), Ne. Discovered by Ramsay and Travers in 1898. Neon is a rare gaseous element present in the atmosphere to the extent of 1 part in 65,000 of air. It is obtained by liquefaction of air and separated from the other gases by fractional distillation. It is very inert element; no stable compounds of neon have been prepared. In a vacuum discharge tube, neon glows reddish orange. Of all the rare gases, the discharge of neon is the most intense at ordinary voltages and currents. The largest use of neon is in advertising signs. It is also used to make high-voltage indicators and lightning arrestors. Neon and helium were used in the first gas lasers. Liquid neon finds some use as an economical cryogenic refrigerant. It has over 40 times more refrigerating capacity per unit volume than liquid helium and more than three times that of liquid hydrogen.

Neptunium — (Planet *Neptune*), Np. Neptunium was the first synthetic transuranium element of the actinide series discovered; the isotope ^{239}Np was produced by McMillan and Abelson in 1940 at Berkeley, California, as the result of bombarding uranium with cyclotron-produced neutrons. The isotope ^{237}Np (half-life of 2.14×10^6 years) is currently obtained in gram quantities as a by-product from the production of plutonium in nuclear reactors. Trace quantities of the element are actually found in nature due to transmutation reactions in uranium ores produced by the neutrons which are present. Neptunium metal is prepared by the reduction of NpF_3 with barium or lithium vapor at about 1200 °C. It has a silvery appearance, is chemically reactive, and exists in at least three structural modifications: α-neptunium, orthorhombic; β-neptunium (above 280 °C), tetragonal; and γ-neptunium (above 576 °C), cubic. Neptunium has four ionic oxidation states in solution: Np^{+3} (pale purple), analogous to the rare earth ion Pm^{+3}; Np^{+4} (yellow green); NpO^+ (green blue); and NpO^{++} (pale pink). These latter oxygenated species are in contrast to the rare earths that exhibit only simple ions of the (II), (III), and (IV) oxidation states in aqueous solution. The element forms tri- and tetrahalides such as NpF_3, NpF_4, $NpCl_4$, $NpBr_3$, NpI_3, and oxides of various compositions such as are found in the uranium-oxygen system, including Np_3O_8 and NpO_2.

Nickel — (Ger. *Nickel*, Satan or Old Nick's and from *kupfernickel*, Old Nick's copper), Ni. Discovered by Cronstedt in 1751 in kupfernickel (*niccolite*). Nickel is found as a constituent in most meteorites and often serves as one of the criteria for distinguishing a meteorite from other minerals. Iron meteorites, or *siderites*, may contain iron alloyed with from 5% to nearly 20% nickel. Nickel is obtained commercially from *pentlandite* and *pyrrhotite* of the Sudbury region of Ontario, a district that produces much of the world's nickel. It is now thought that the Sudbury deposit is the result of an ancient meteorite impact. Large deposits of nickel, cobalt, and copper have been developed at Voisey's Bay, Labrador. Other deposits of nickel are found in Russia, New Caledonia, Australia, Cuba, Indonesia, and elsewhere. Nickel is silvery white and takes on a high polish. It is hard, malleable, ductile, somewhat ferromagnetic, and a fair conductor of heat and electricity. It belongs to the iron-cobalt group of metals and is chiefly valuable for the alloys it forms. It is extensively used for making stainless steel and other corrosion-resistant alloys such as Invar®, Monel®, Inconel®, and the Hastelloys®. Tubing made of a copper-nickel alloy is used in desalination plants for converting seawater into fresh water. Nickel is also now used extensively in coinage and in making nickel steel for armor plate and burglarproof vaults, and is a component in Nichrome®, Permalloy®, and constantan. Nickel added to glass gives a green color. Nickel plating is often used to provide a protective coating for other metals, and finely divided nickel is a catalyst for hydrogenating vegetable oils. It is also used in ceramics, in the manufacture of Alnico magnets, and in batteries. The sulfate and the oxides are important compounds. Nickel sulfide fume and dust, as well as other nickel compounds, are carcinogens.

Nihonium — (*Nihon*, a word for Japan), Nh. Priority for the discovery of Element 113 has been given to the RIKEN in Japan, which published evidence in 2004. Similar evidence was obtained at Dubna in Russia at about the same time. IUPAC confirmed the name nihonium in 2016. The most stable isotope, ^{286}Nh, has a half-life of 20 s. Nihonium falls in the boron, aluminum, etc., group., but here is no chemical evidence yet on the extent to which its properties differ from the lower members of the group.

Niobium — (*Niobe*, daughter of Tantalus); sometimes called Columbium (*Columbia*, name for America), Nb. Discovered in 1801 by Hatchett in an ore sent to England more than a century before by John Winthrop the Younger, first governor of Connecticut. The metal was first prepared in 1864 by Blomstrand, who reduced the chloride by heating it in a hydrogen atmosphere. The name *niobium* was adopted by the International Union of Pure and Applied Chemistry in 1950 after 100 years of controversy. Most leading chemical societies and government organizations refer to it by this name. Some metallurgists and commercial producers, however, still refer to the metal as "columbium." The element is found in *niobite* (or *columbite*), *niobite-tantalite*, *pyrochlore*, and *euxenite*. Large deposits of niobium have been found associated with carbonatites (carbon-silicate rocks), as a constituent of pyrochlore. Extensive ore reserves are found in Canada, Brazil, Congo-Kinshasa, Rwanda, and Australia. The metal can be isolated from tantalum and prepared in several ways. It is a shiny, white, soft, and ductile metal, and takes on a bluish cast when exposed to air for a long time at room temperature. The metal starts to oxidize in air at 200 °C, and when processed at even moderate temperatures must be placed in a protective atmosphere. It is used in arc-welding rods for stabilized grades of stainless steel. Thousands of pounds of niobium have been used in advanced air frame systems such as were used in the Gemini space program. It has also found use in super-alloys for applications such as jet engine components, rocket subassemblies, and heat-resisting equipment. Niobium has the highest superconducting critical temperature (9.25 K) of any element at ambient pressure and is an essential component of the two primary low temperature superconducting wires developed for applications, Nb-Ti and Nb-Sn.

Nitrogen — (L. *nitrum*, Fr. *nitre*, native soda; *-genes*, producing, N. Discovered by Daniel Rutherford in 1772, but Scheele, Cavendish, Priestley, and others about the same time studied "burnt or dephlogisticated air," as air without oxygen was then called. Nitrogen makes up 78% of the air, by volume. The atmosphere of Mars, by comparison, is 2.7% nitrogen. The estimated amount of this element in our atmosphere is more than 4000 trillion tons. From this inexhaustible source it can be obtained by liquefaction and fractional distillation. Nitrogen molecules give the orange-red, blue-green, blue-violet, and deep violet shades to the aurora. The element is so inert that Lavoisier named it *azote*, meaning without life, yet its compounds are so active as to be most important in foods, poisons, fertilizers, and explosives. Nitrogen can be also easily prepared by heating a water solution of ammonium nitrite. Nitrogen, as a gas, is colorless, odorless, and a generally inert element. As a liquid it is also colorless and odorless and is similar in appearance to water. Two allotropic forms of solid nitrogen exist, with the transition from the α to the β form taking place at –237 °C. When nitrogen is heated, it combines directly with magnesium, lithium, or calcium; when mixed with oxygen and subjected to electric sparks, it forms first nitric oxide (NO) and then the dioxide (NO_2); when heated under pressure with a catalyst with hydrogen, ammonia is formed (Haber process). The ammonia thus formed is of the utmost importance as it is used in fertilizers, and it can be oxidized to nitric acid (Ostwald process). The ammonia industry is the largest consumer of nitrogen. Large amounts are also used by the electronics industry, which uses the gas as a blanketing medium during production of such components as transistors, diodes, etc. Large quantities of nitrogen are used in annealing stainless steel and other steel mill products. The drug industry also uses large quantities. Nitrogen is used as a refrigerant both for the immersion freezing of food products and for transportation of foods. Liquid nitrogen is also used in missile work as a purge for components, insulators for space chambers, etc., and by the oil industry to build up pressure in wells to force crude oil upward. Sodium and potassium nitrates are formed by the decomposition of organic matter with compounds of the metals present. In certain dry areas of the world, these saltpeters are found in quantity. Ammonia, nitric acid, the nitrates, the five oxides (N_2O, NO, N_2O_3, NO_2, and N_2O_5), TNT, the cyanides, etc., are but a few of the important compounds.

Nobelium — (Alfred Nobel [1833–1896], inventor of dynamite), No. Several claims to the discovery of Element-102 were made in the 1957 to 1966 period, leading to a considerable amount of controversy and disputes over naming the element. The disputes were settled in 1966 with definitive evidence produced at the Joint Institute for Nuclear Research at Dubna. Although it had been used much earlier, the name Nobelium was finally ratified by IUPAC in 1994. ^{259}No has a half-life of 58 minutes. The chemical behavior is expected to be similar to ytterbium.

Oganesson — (Yuri *Oganessian*, Russian leader in heavy element research), Og. Element-118, the heaviest element so far found in the periodic table, was first detected in 2002 by Russian and American workers at JINR, Dubna. The name oganesson was approved by IUPAC in 2016. Only three or four atoms on ^{295}Og (half-life about 1 ms) have been produced. Considering its position in the periodic table, where it falls in the noble gases group, one might expect oganesson to be an unreactive gas at normal conditions. However, theoretical calculations suggest that it is a solid and possibly very reactive. Thus, it might exhibit valences of +2 and +4, as well a zero.

Osmium — (Gr. *osme*, a smell), Os. Discovered in 1803 by Tennant in the residue left when crude platinum is dissolved by *aqua regia*. Osmium occurs in *iridosmine* and in platinum-bearing river sands of the Urals, North America, and South America. It is also found in the nickel-bearing ores of the Sudbury, Ontario, region along with other platinum metals. While the quantity of platinum metals in these ores is very small, the large tonnages of nickel ores processed make commercial recovery possible. The metal is lustrous, bluish white, extremely hard, and brittle even at high temperatures. It has the highest melting point and the lowest vapor pressure of the platinum group. The metal is very difficult to fabricate, but the powder can be sintered in a hydrogen atmosphere at a temperature of 2000 °C. The solid metal is not affected by air at room temperature, but the powdered or spongy metal slowly gives off osmium tetroxide, which is a powerful oxidizing agent and has a strong smell. The tetroxide is highly toxic, and boils at 130 °C. Concentrations in air as low as 10^{-7} g/m^3 can cause lung congestion, skin damage, or eye damage. The tetroxide has been used to detect fingerprints and to stain fatty tissue for microscope slides. The metal is almost entirely used with other metals of the platinum group to produce very hard alloys for fountain pen tips, instrument pivots, and electrical contacts. The measured densities of iridium and osmium seem to indicate that osmium is slightly more dense than iridium, so osmium is generally credited with being the heaviest known element.

Oxygen — (Gr. *oxys*, sharp, acid, and *-genes*, forming; acid former), O. For many centuries, workers occasionally realized air was composed of more than one component. The behavior of oxygen and nitrogen as components of air led to the advancement of the phlogiston theory of combustion, which captured the minds of chemists for a century. Oxygen was prepared by several workers, including Bayen and Borch, but they did not know how to collect it, did not study its properties, and did not recognize it as an elementary substance. Priestley is generally credited with its discovery, although Scheele also discovered it independently. Oxygen is the third most abundant element found in the sun, and it plays a part in the carbon–nitrogen cycle, one process thought to give the sun and stars their energy. Oxygen under excited conditions is responsible for the bright red and yellow-green colors of the aurora. Oxygen, as a gaseous element, forms 21% of the atmosphere by volume from which it can be obtained by liquefaction and fractional distillation. The atmosphere of Mars contains about 0.13% oxygen. The element and its compounds make up 49.2%, by weight, of the Earth's crust. About two-thirds of the mass of the human body and nine-tenths of water is oxygen. In the laboratory it can be prepared by the electrolysis of water or by heating potassium chlorate with manganese dioxide as a catalyst. The gas is colorless, odorless, and tasteless. The liquid and solid forms are a pale blue color and are strongly paramagnetic. Ozone (O_3), a highly active compound, is formed by the action of an electrical discharge or ultraviolet light on oxygen. Ozone's presence in the atmosphere (amounting to the equivalent of a layer 3 mm thick at ordinary pressures and temperatures) is of vital importance in preventing the harmful ultraviolet rays of the sun from reaching the Earth's surface. The use of certain halogenated compounds that have a detrimental effect on this ozone layer has been phased out by international agreement. Ozone is toxic and exposure should not exceed 0.1 ppm. Undiluted ozone has a bluish color. Liquid ozone is bluish black, and solid ozone is violet-black. Oxygen is very reactive and capable of combining with most elements. It is a component of hundreds of thousands of organic compounds. It is essential for respiration of all plants and animals and for practically all combustion. In hospitals it is frequently used to aid respiration of patients. Its atomic weight was used as a standard until 1961 when the International Union of Pure and Applied Chemistry adopted carbon 12 as the new basis. Commercial oxygen consumption in the U.S. is estimated to be 20 million short tons per year and the demand is expected to increase. Oxygen enrichment of steel blast furnaces accounts for the greatest use of the gas. Large quantities are also used in making synthesis gas for ammonia and methanol, ethylene oxide, and for oxy-acetylene welding. Air separation plants produce about 99% of the gas, electrolysis plants about 1%.

Palladium — (named after the asteroid *Pallas*, discovered about the same time; Gr. *Pallas*, goddess of wisdom), Pd. Discovered in 1803 by Wollaston. Palladium is found along with platinum and other metals of the platinum group in deposits of Russia, South Africa, Canada (Ontario), and elsewhere. It is frequently found associated with the nickel-copper deposits such as those found in Ontario. Its separation from the platinum metals depends upon the type of ore in which it is found. It is a steel-white metal, does not tarnish in air, and is the least dense and lowest melting of the platinum group of metals. When annealed, it is soft and ductile; cold working greatly increases its strength and hardness. Palladium is attacked by nitric and sulfuric acid. At room temperatures the metal has the unusual property of absorbing up to 900 times its own volume of hydrogen. Hydrogen readily diffuses through heated palladium and this provides a means of purifying the gas. Finely divided palladium is a good catalyst for hydrogenation and dehydrogenation reactions. It is alloyed and used in jewelry trades. White gold is an alloy of gold decolorized by the addition of palladium. Like gold, palladium can be beaten into leaf as thin as 1/250,000 in. The metal is used in dentistry, watchmaking, and in making surgical instruments and electrical contacts. Palladium has been substituted for higher priced platinum in some catalytic converters. Palladium, however, is less resistant to poisoning by sulfur and lead than platinum.

Phosphorus — (Gr. *phosphoros*, light bearing; ancient name for the planet Venus when appearing before sunrise), P. Discovered in 1669 by Brand, who prepared it from urine. Phosphorus exists in several allotropic forms, the most common of which are white (or yellow), red, and black (or violet). White phosphorus has two modifications: α and β with a transition temperature at –76.9 °C. Red and black phosphorus have amorphous structures. Never found free in nature, it is widely distributed in combination with minerals. *Phosphate* rock, which contains the mineral *apatite*, an impure tricalcium phosphate, is an important source of the element. Large deposits are found in Russia, China, Morocco, and in the United States in Florida, Tennessee, Utah, and Idaho. Phosphorus is an essential ingredient of all cell protoplasm, nervous tissue, and bones, and phosphate groups provide a key linkage in the DNA chain. Ordinary phosphorus is a waxy white solid; when pure it is colorless and transparent. It is insoluble in water, but soluble in carbon disulfide. It takes fire spontaneously in air, burning to the pentoxide. It is very poisonous, 50 mg constituting an approximate fatal dose. White phosphorus should be kept under water, as it is dangerously reactive in air, and it should be handled with forceps, as contact with the skin may cause severe burns. When exposed to sunlight or when heated in its own vapor to 250 °C, it is converted to the red variety, which does not phosphoresce in air as does the white form. Red phosphorus does not ignite spontaneously, and it is not as dangerous as white phosphorus. It should, however, be handled with care as it does convert to the white form at some temperatures and it emits highly toxic fumes of the oxides of phosphorus when heated. The red modification is fairly stable, sublimes with a vapor pressure of 1 atm at 431 °C, and is used in the manufacture of safety matches, pyrotechnics, pesticides, incendiary shells, smoke bombs, tracer bullets, etc. White phosphorus may be made by several methods. By one process, tricalcium phosphate, the essential ingredient of phosphate rock, is heated in the presence of carbon and silica in an electric furnace or fuel-fired furnace. Elementary phosphorus is liberated as vapor and may be collected under water. If desired, the phosphorus vapor and carbon monoxide produced by the reaction can be oxidized at once in the presence of moisture to produce phosphoric acid, an important compound in making super-phosphate fertilizers. In recent years, concentrated phosphoric acids, which may contain as much as 70% to 75% P_2O_5, have become of great importance to agriculture and farm production. Worldwide demand for fertilizers has caused record phosphate production. Phosphates are used in the production of special glasses, such as those used for sodium lamps. Bone-ash, calcium phosphate, is also used to produce fine chinaware and monocalcium phosphate is used in baking powder. Phosphorus is also important in the production of steels, phosphor bronze, and many other products. Trisodium phosphate is important as a cleaning agent, as a water softener, and for preventing boiler scale and corrosion of pipes and boiler tubes. Organic compounds of phosphorus are important.

Platinum — (It. *platina*, silver), P. Discovered in South America by Ulloa in 1735 and by Wood in 1741. The metal was used by pre-Columbian Indians. Platinum occurs native, accompanied by small quantities of iridium, osmium, palladium, ruthenium, and rhodium, all belonging to the same group of metals. These are found in the alluvial deposits of the Ural Mountains and in Colombia. *Sperrylite* ($PtAs_2$), occurring with the nickel-bearing deposits of Sudbury, Ontario, is a source of a considerable amount of the metal. The large production of nickel offsets there being only one part of the platinum metals in 2 million parts of ore. The largest supplier of the platinum group of metals is South Africa, followed by Russia and Canada. Platinum is a beautiful silvery-white metal when pure and is malleable and ductile. It has a coefficient of expansion almost equal to that of soda–lime–silica glass and is therefore used to make sealed electrodes in glass systems. The metal does not oxidize in air at any temperature, but is corroded by halogens, cyanides, sulfur, and caustic alkalis. It is insoluble in hydrochloric and nitric acid, but dissolves when they are mixed as *aqua regia*, forming chloroplatinic acid (H_2PtCl_6), an important compound. The metal is used extensively in jewelry, wire, and vessels for laboratory use, and in many valuable instruments including thermocouple elements. It is also used for electrical contacts, corrosion-resistant apparatus, and in dentistry. Platinum–cobalt alloys have magnetic properties. One such alloy made of 76.7% Pt and 23.3% Co, by weight, is a more powerful magnet than Alnico V. Platinum resistance wires are used for constructing high-temperature electric furnaces. The metal is used for coating missile nose cones, jet engine fuel nozzles, etc., which must perform reliably for long periods of time at high temperatures. The metal, like palladium, absorbs large volumes of hydrogen, retaining it at ordinary temperatures but giving it up at red heat. In the finely divided state platinum is an excellent catalyst, having long been used in the contact process for producing sulfuric acid. It is also used as a catalyst in cracking petroleum products. There is also much current interest in the use of platinum as a catalyst in fuel cells and in its use in catalytic convertors for automobiles. Platinum anodes are extensively used in cathodic protection systems for large ships and ocean-going vessels, pipelines, steel piers, etc. Hydrogen and oxygen explode in the presence of platinum. Pure platinum wire will glow red hot when placed in the vapor of methyl alcohol. It acts here as a catalyst, converting the alcohol to formaldehyde.

Plutonium — (planet *Pluto*), Pu. Plutonium was the second transuranium element of the actinide series to be discovered. The isotope ^{238}Pu was produced in 1940 by Seaborg, McMillan, Kennedy, and Wahl by deuteron bombardment of uranium in the 60-inch cyclotron at Berkeley, California. Plutonium also exists in trace quantities in naturally occurring uranium ores. It is formed in much the same manner as neptunium, by irradiation of natural uranium with the neutrons that are present. By far of greatest importance is the isotope Pu^{239}, with a half-life of 24,100 years, produced in extensive quantities in nuclear reactors from natural uranium. Plutonium has assumed the position of dominant importance among the transuranium elements because of its use as an explosive ingredient in nuclear weapons and the place it holds as a key material in the development of industrial sources of nuclear power. One kilogram is equivalent to about 22 million kilowatt hours of heat energy. The complete detonation of a kilogram of plutonium produces an explosion equal to about 20,000 tons of chemical explosive. Its importance depends on the nuclear property of being readily fissionable with neutrons and its availability in quantity. The various nuclear applications of plutonium are well known. ^{238}Pu was used in the Apollo lunar missions to power seismic and other equipment on the lunar surface. As with neptunium and uranium, plutonium metal can be prepared by reduction of the trifluoride with alkaline-earth metals. The metal has a silvery appearance and takes on a yellow tarnish when slightly oxidized. It is chemically reactive. A relatively large piece of plutonium is warm to the touch because of the energy given off in alpha decay. Larger pieces will produce enough heat to boil water. The metal readily dissolves in concentrated hydrochloric acid, hydroiodic acid, or perchloric acid with formation of the Pu^{+3} ion. The metal exhibits six allotropic modifications having various crystalline structures. The densities of these vary from 16.0 to 19.7 g/cm^3. Plutonium also exhibits four ionic valence states in aqueous solutions: Pu^{+3} (blue lavender), Pu^{+4} (yellow brown), PuO^+ (pink?), and PuO^{+2} (pink orange). The ion PuO^+ is unstable in aqueous solutions, disproportionating into Pu^{+4} and PuO^{+2}. The Pu^{+4} thus formed, however, oxidizes the PuO^+ into PuO^{+2}, itself being reduced to Pu^{+3}, giving finally Pu^{+3} and PuO^{+2}. Plutonium forms binary compounds with oxygen (PuO, PuO_2, and intermediate oxides of variable composition); with the halides (PuF_3, PuF_4, $PuCl_3$, $PuBr_3$, PuI_3); and with carbon, nitrogen, and silicon (PuC, PuN, $PuSi_2$). Oxyhalides are also well known: PuOCl, PuOBr, PuOI. Because of the high rate of emission of alpha particles and the element being specifically absorbed by bone marrow, plutonium, as well as all of the other transuranium elements except neptunium, are radiological poisons and must be handled with very special equipment and precautions. Plutonium is a very dangerous radiological hazard. Precautions must also be taken to prevent the unintentional formation of a critical mass. Plutonium in liquid solution is more likely to become critical than solid plutonium.

Polonium — (Poland, native country of Mme. Curie [1867–1934]), Po. Polonium was the first element discovered by Mme. Curie in 1898, while seeking the cause of radioactivity of pitchblende from Joachimsthal, Bohemia. Polonium is a very rare natural element. Uranium ores contain only about 100 μg of the element per ton. Its abundance is only about 0.2% of that of radium. In 1934, it was found that when natural bismuth (^{209}Bi) was bombarded by neutrons, ^{210}Bi, the parent of polonium, was obtained. Milligram amounts of polonium may now be prepared this way, by using the high neutron fluxes of nuclear reactors. Polonium is a low-melting, fairly volatile metal, 50% of which is vaporized in air in 45 hours at 55 °C. It is an alpha emitter with a half-life of 138.39 days. A milligram emits as many alpha particles as 5 g of radium. The energy released by its decay is so large

(140 W/g) that a capsule containing about half a gram reaches a temperature above 500 °C. The capsule also presents a contact gamma-ray dose rate of 0.012 Gy/h. A few curies (1 curie = 3.7×10^{10} Bq) of polonium exhibit a blue glow, caused by excitation of the surrounding gas. Because almost all alpha radiation is stopped within the solid source and its container, giving up its energy, polonium has attracted attention for uses as a lightweight heat source for thermoelectric power in space satellites. Polonium-210 is the most readily available isotope. Metallic polonium has been prepared from polonium hydroxide and some other polonium compounds in the presence of concentrated aqueous or anhydrous liquid ammonia. Two allotropic modifications are known to exist. Polonium is readily dissolved in dilute acids but is only slightly soluble in alkalis. Polonium salts of organic acids char rapidly; halide amines are reduced to the metal. Polonium can be mixed or alloyed with beryllium to provide a source of neutrons. It has been used in devices for eliminating static charges in textile mills, etc.; however, beta sources are more commonly used and are less dangerous. It is also used on brushes for removing dust from photographic films. The polonium for these is carefully sealed and controlled, minimizing hazards to the user. Polonium-210 is very dangerous to handle in even milligram or microgram amounts, and special equipment and strict control are necessary. Damage arises from the complete absorption of the energy of the alpha particles into tissue. The maximum permissible body burden for ingested polonium is only 0.03 μCi, which represents a particle weighing only 6.8×10^{-12} g. Weight for weight it is about 2.5×10^{11} times as toxic as hydrocyanic acid.

Potassium — (English, *potash* — pot ashes; L. *kalium*, Arab. *qali*, alkali), K. Discovered in 1807 by Davy, who obtained it from caustic potash (KOH); this was the first metal isolated by electrolysis. The metal is the seventh most abundant and makes up about 2.4% by weight of the Earth's crust. Most potassium minerals are insoluble, and the metal is obtained from them only with great difficulty. Certain minerals, however, such as *sylvite, carnallite, langbeinite,* and *polyhalite* are found in ancient lake and seabeds and form rather extensive deposits from which potassium and its salts can readily be obtained. Potash is mined in Germany, New Mexico, California, Utah, and elsewhere. Large deposits of potash, found at a depth of some 1000 m in Saskatchewan, promise to be important in coming years. Potassium is also found in the ocean but is present only in relatively small amounts compared to sodium. The greatest demand for potash has been in its use for fertilizers. Potassium is an essential constituent for plant growth, and it is found in most soils. Potassium is never found free in nature, but is obtained by electrolysis of the hydroxide, much in the same manner as prepared by Davy. Thermal methods also are commonly used to produce potassium (such as by reduction of potassium compounds with CaC_2, C, Si, or Na). It is one of the most reactive and electropositive of metals. Except for lithium, it is the lightest known metal. It is soft, easily cut with a knife, and is silvery in appearance immediately after a fresh surface is exposed. It rapidly oxidizes in air and should be preserved in a mineral oil. As with other metals of the alkali group, it decomposes in water with the evolution of hydrogen. It catches fire spontaneously on water. Potassium and its salts impart a violet color to flames. An alloy of sodium and potassium (NaK) is used as a heat-transfer medium. Many potassium salts are of utmost importance, including the hydroxide, nitrate, carbonate, chloride, chlorate, bromide, iodide, cyanide, sulfate, chromate, and dichromate.

Praseodymium — (Gr. *prasios*, green, and *didymos*, twin), Pr. In 1841, Mosander extracted the rare-earth *didymia* from *lanthana*; in 1879, Lecoq de Boisbaudran isolated a new earth, *samaria*, from didymia obtained from the mineral *samarskite*. Six years later, in 1885, von Welsbach separated didymia into two others, *praseodymia* and *neodymia*, which gave salts of different colors. As with other rare-earth elements, compounds of these elements in solution have distinctive sharp spectral absorption bands or lines, some of which are only a few Ångströms wide. The element occurs along with other rare-earth elements in a variety of minerals. *Monazite* and *bastnasite* are the two principal commercial sources of the rare-earth metals. Ion-exchange and solvent extraction techniques have led to much easier isolation of the rare earths and the cost has dropped greatly. Praseodymium can be prepared by several methods, such as by calcium reduction of the anhydrous chloride or fluoride. Misch metal, used in making cigarette lighters, contains about 5% praseodymium metal. Praseodymium is soft, silvery, malleable, and ductile. It was first prepared in relatively pure form in 1931. It is somewhat more resistant to corrosion in air than europium, lanthanum, cerium, or neodymium, but it does develop a green oxide coating that splits off when exposed to air. As with other rare-earth metals it should be kept under a light mineral oil or sealed in plastic. The rare-earth oxides, including Pr_2O_3, are among the most refractory substances known. Salts of praseodymium are used to color glasses and enamels; when mixed with certain other materials, praseodymium produces an intense and unusually clean yellow color in glass. Didymium glass, of which praseodymium is a component, is a colorant for welder's goggles.

Promethium — (*Prometheus*, who, according to mythology, stole fire from heaven), Pm. In 1902, Branner predicted the existence of an element between neodymium and samarium, and this prediction was confirmed by Moseley in 1914. Unsuccessful searches were made for this predicted element over two decades, and various investigators proposed the names "illinium," "florentium," and "cyclonium" for this element. In 1941, workers at Ohio State University irradiated neodymium and praseodymium with neutrons, deuterons, and alpha particles, and produced several new radioactivities, which most likely were those of Element 61. Wu and Segre, and Bethe, in 1942, confirmed the formation; however, chemical proof of the production of Element 61 was lacking because of the difficulty in separating the rare earths from each other at that time. In 1945, Marinsky, Glendenin, and Coryell made the first chemical identification by using ion-exchange chromatography. Their work was done by fission of uranium and by neutron bombardment of neodymium. These investigators named the newly discovered element. Searches

Inorganic

for the element on Earth have been fruitless, and it now appears that promethium is completely missing from the Earth's crust. Promethium, however, has been reported to be in the spectrum of the star HR^{465} in Andromeda. It must be formed near the star's surface, for no known isotope of promethium has a half-life longer than 17.7 years. That isotope, Promethium-145, is the most useful. Promethium-145 has a specific activity of 940 Ci/g. It is a soft beta emitter; although no gamma rays are emitted, X-radiation can be generated when beta particles impinge on elements of a high atomic number, and great care must be taken in handling it. Promethium salts luminesce in the dark with a pale blue or greenish glow, due to their high radioactivity. Ion-exchange methods led to the preparation of about 10 g of promethium from atomic reactor fuel processing wastes in early 1963. Little is yet known about the properties of metallic promethium. Two allotropic modifications exist. The element has applications as a beta source for thickness gages, and its radiation can be absorbed by a phosphor to produce light. Light produced in this manner can be used for signs or signals that require dependable operation; it can be used as a nuclear-powered battery by capturing light in photocells that convert it into electric current. Such a battery, using ^{147}Pm, would have a useful life of about 5 years. It is being used for fluorescent lighting starters and coatings for self-luminous watch dials. Promethium shows promise as a portable x-ray source, and it may become useful as a heat source to provide auxiliary power for space probes and satellites. More than 30 promethium compounds have been prepared. Most are colored.

Protactinium — (Gr. *protos*, first), Pa. The first isotope of Element 91 to be discovered was ^{234}Pa, also known as UX_2, a short-lived member of the naturally occurring ^{238}U decay series. It was identified by K. Fajans and O. H. Gohring in 1913 and they named the new element *brevium*. When the longer-lived isotope ^{231}Pa was identified by Hahn and Meitner in 1918, the name protoactinium was adopted as being more consistent with the characteristics of the most abundant isotope. Soddy, Cranson, and Fleck were also active in this work. The name *protoactinium* was shortened to *protactinium* in 1949. In 1927, Grosse prepared 2 mg of a white powder, which was shown to be Pa_2O_5. Later, in 1934, from 0.1 g of pure Pa_2O_5 he isolated the element by two methods, one of which was by converting the oxide to an iodide and "cracking" it in a high vacuum by an electrically heated filament by the reaction $2PaI_5 \rightarrow 2Pa + 5I_2$. Protactinium has a bright metallic luster that it retains for some time in air. The element occurs in *pitchblende* to the extent of about 1 part ^{231}Pa to 10 million parts of ore. Ores from Congo-Kinshasa have about 3 ppm. The most common isotope is ^{231}Pr with a half-life of 32,500 years. A number of protactinium compounds are known, some of which are colored. The element is a dangerous toxic material and requires precautions similar to those used when handling plutonium. Protactinium is one of the rarest and most expensive naturally occurring elements.

Radium — (L. *radius*, ray), Ra. Radium was discovered in 1898 by Pierre and Marie Curie in the *pitchblende* or *uraninite* of North Bohemia (Czech Republic), where it occurs. There is about 1 g of radium in 7 tons of pitchblende. The element was isolated in 1911 by Mme. Curie and Debierne by the electrolysis of a solution of pure radium chloride, employing a mercury cathode; on distillation in an atmosphere of hydrogen this amalgam yielded the pure metal. Originally, radium was obtained from the rich pitchblende ore found at Joachimsthal, Bohemia. The *carnotite* sands of Colorado furnish some radium, but richer ores are found in the Democratic Republic of Congo and the Great Bear Lake region of Canada. Radium is present in all uranium minerals, and could be extracted, if desired, from the extensive wastes of uranium processing. Large uranium deposits are located in Ontario, New Mexico, Utah, Australia, and elsewhere. Radium is obtained commercially as the bromide or chloride; it is doubtful if any appreciable stock of the pure element now exists. The pure metal is brilliant white when freshly prepared, but blackens on exposure to air, probably due to formation of the nitride. It exhibits luminescence, as do its salts; it decomposes in water and is somewhat more volatile than barium. It is a member of the alkaline-earth group of metals. Radium imparts a carmine red color to a flame. Radium emits alpha, beta, and gamma rays and when mixed with beryllium produce neutrons. One gram of ^{226}Ra undergoes 3.7×10^{10} disintegrations per second. The *curie* (*Ci*) is defined as that amount of radioactivity which has the same disintegration rate as 1 g of ^{226}Ra. Radium-226, the most common isotope, has a half-life of 1599 years. One gram of radium produces about 0.0001 mL (stp) of emanation, or radon gas, per day. This is pumped from the radium and sealed in minute tubes, which are used in the treatment of cancer. One gram of radium yields about 4186 kJ of energy per year. Radium is used in producing neutron sources and in medicine for the treatment of cancer. Some of the more recently discovered radioisotopes, such as ^{60}Co, are now being used in place of radium. Some of these sources are much more powerful, and others are safer to use. Radium loses about 1% of its activity in 25 years, being transformed into elements of lower atomic weight. Lead is a final product of disintegration. Stored radium should be ventilated to prevent build-up of radon. Inhalation, injection, or body exposure to radium can cause cancer and other body disorders.

Radon — (from *radium*; called *niton* at first, L. *nitens*, shining), Rn. The element was discovered in 1900 by Dorn, who called it *radium emanation*. In 1908, Ramsay and Gray, who named it *niton*, isolated the element and determined its density, finding it to be the heaviest known gas. It is essentially inert and occupies the last place in the zero group of gases in the Periodic Table. Since 1923, it has been called radon. Radon-222, coming from radium, has a half-life of 3.823 days and is an alpha emitter; Radon-220, emanating naturally from thorium and called *thoron*, has a half-life of 55.6 s and is also an alpha emitter. Radon-219 emanates from actinium and is called *actinon*. It has a half-life of 3.96 s and is also an alpha emitter. It is estimated that every square mile of soil to a depth of 6 inches contains about 1 g of radium, which releases radon in tiny amounts to the atmosphere. Radon is present in some spring waters, such as those at Hot Springs, Arkansas. On the average, one part of radon is present to 1×10^{21} part of air. At ordinary temperatures radon is a colorless gas; when cooled below the freezing point, radon exhibits a brilliant phosphorescence which becomes yellow as the temperature is lowered and orange-red at the temperature of liquid air. It has been reported that fluorine reacts with radon, forming radon fluoride. Radon clathrates have also been reported. Radon is still produced for therapeutic use by a few hospitals by pumping it from a radium source and sealing it in minute tubes, called seeds or needles, for application to patients. This practice has now been largely discontinued as hospitals can order the seeds directly from suppliers, who make up the seeds with the desired activity for the day of use. Care must be taken in handling radon, as with other radioactive materials. The main hazard is from inhalation of the element and its solid daughters, which are collected on dust in the air. Good ventilation should be provided where radium, thorium, or actinium is stored to prevent build-up of this element. Radon build-up is a health consideration in uranium mines, and build-up in homes is a major concern in some areas. Many deaths from lung cancer are caused by radon exposure.

Rhenium — (L. *Rhenus*, Rhine), Re. Discovery of rhenium is generally attributed to Noddack, Tacke, and Berg, who announced in 1925 they had detected the element in platinum ores and *columbite*. They also found the element in *gadolinite* and *molybdenite*. By working up 660 kg of molybdenite they were able in 1928 to extract 1 g of rhenium. Rhenium does not occur free in nature or as a compound in a distinct mineral species. It is, however, widely spread throughout the Earth's crust to the extent of about 0.001 ppm. Commercial rhenium in the U.S. today is obtained from molybdenite roaster-flue dusts obtained from copper-sulfide ores mined in the vicinity of Miami, Arizona, and other places in Arizona and Utah. Some molybdenites contain from 0.002% to 0.2% rhenium. It is estimated that in 1999 about 16,000 kg of rhenium was being produced. The total estimated world reserve of rhenium is 11,000,000 kg. Natural rhenium is a mixture of two isotopes, one of which is radioactive with a very long half-life. Rhenium metal is prepared by reducing ammonium perrhenate with hydrogen at elevated temperatures. The element is silvery white with a metallic luster; its density is exceeded by that of only platinum, iridium, and osmium, and its melting point is exceeded only by that of tungsten and carbon. It has other useful properties. The usual commercial form of the element is a powder, but it can be consolidated by pressing and resistance-sintering in a vacuum or hydrogen atmosphere. This produces a compact shape in excess of 90% of the density of the metal. Annealed rhenium is very ductile, and can be bent, coiled, or rolled. Rhenium is used as an additive to tungsten and molybdenum-based alloys to impart useful properties. It is widely used for filaments for mass spectrographs and ion gages. Rhenium is also used as an electrical contact material as it has good wear resistance and withstands arc corrosion. Thermocouples made of Re-W are used for measuring temperatures up to 2200 °C, and rhenium wire has been used in photoflash lamps for photography. Rhenium catalysts are exceptionally resistant to poisoning from nitrogen, sulfur, and phosphorus, and are used for hydrogenation of fine chemicals, hydrocracking, reforming, and disproportionation of olefins. Rhenium has become especially important as a catalyst for petroleum refining and in making superalloys for jet engines.

Rhodium — (Gr. *rhodon*, rose), Rh. Wollaston discovered rhodium in 1803–1804 in crude platinum ore he presumably obtained from South America. Rhodium occurs native with other platinum metals in river sands of the Urals and in North and South America. It is also found with other platinum metals in the copper-nickel sulfide ores of the Sudbury, Ontario region. Although the quantity occurring here is very small, the large tonnages of nickel processed make the recovery commercially feasible. The annual world production of rhodium in 2000 was only about 9000 kg. The metal is silvery white and at red heat slowly changes in air to the sesquioxide. At higher temperatures it reverts to the element. Rhodium has a higher melting point and lower density than platinum. Its major use is as an alloying agent to harden platinum and palladium. Such alloys are used for furnace windings, thermocouple elements, bushings for glass fiber production, and laboratory crucibles. It is useful as an electrical contact material as it has a low electrical resistance, a low and stable contact resistance, and is highly resistant to corrosion. Plated rhodium, produced by electroplating or evaporation, is exceptionally hard and is used for optical instruments. It has a high reflectance and is hard and durable. Rhodium is also used for jewelry, for decoration, and as a catalyst.

Roentgenium — (Wilhelm Roentgen, discoverer of x-rays), Rg. On December 20, 1994, scientists at GSI Darmstadt, Germany announced they had detected three atoms of a new element with 111 protons and 161 neutrons. This element was made by bombarding ^{83}Bi with ^{28}Ni. Signals of Element 111 appeared for less than 0.002 s, then decayed into lighter elements including Element ^{268}109 and Element ^{264}107. These isotopes had not previously been observed. In 2004 IUPAC approved the name roentgenium for Element 111. Roentgenium is expected to have properties similar to gold.

Rubidium — (L. *rubidus*, deepest red), Rb. Discovered in 1861 by Bunsen and Kirchhoff in the mineral *lepidolite* by use of spectroscopy. The element is much more abundant than was thought at one time. It is now considered to be the 16th most abundant element in the Earth's crust. Rubidium occurs in *pollucite, carnallite, leucite,* and *zinnwaldite*, which contains traces up to 1%, in the form of the oxide. It is found in lepidolite to the extent of about 1.5% and is recovered commercially from this source. Potassium minerals, such as those found at Searles Lake, California, and potassium chloride recovered from brines in Michigan also contain the element and are commercial sources. It is also found along with cesium in the extensive deposits of *pollucite* at Bernic Lake, Manitoba. Rubidium can be liquid at room temperature. It is a soft, silvery-white metallic element of the alkali group and is the second most electropositive and alkaline element. It ignites spontaneously in air and reacts violently in water, setting fire to the liberated hydrogen. As with other alkali metals, it forms amalgams with mercury and alloys with gold, cesium, sodium, and potassium. It colors a flame yellowish violet. Rubidium metal can be prepared by reducing rubidium chloride with calcium, and by a number of other methods. It must be kept under a dry mineral oil or in a vacuum or inert atmosphere. Naturally occurring rubidium is made of two isotopes, ^{85}Rb and ^{87}Rb. Rubidium-87 is present to the extent of 27.83% in natural rubidium and is a beta emitter with a half-life of 4.9×10^{10} years. Ordinary rubidium is sufficiently radioactive to expose a photographic film in about 30 to 60 days. Rubidium forms four oxides: Rb_2O, Rb_2O_2, Rb_2O_3, Rb_2O_4. Because rubidium can be easily ionized, it has been considered for use in ion engines for space vehicles; however, cesium is somewhat more efficient for this purpose. It has also been proposed for use as a working fluid for vapor turbines and for use in a thermoelectric generator using magnetohydrodynamic principles. Rubidium has been used as a getter in vacuum tubes and as a photocell component. It has been used in making special glasses. $RbAg_4I_5$ has the highest room-temperature conductivity of any known ionic crystal. This suggests use in thin film batteries and other applications.

Ruthenium — (L. *Ruthenia*, Russia), Ru. Berzelius and Osann in 1827 examined the residues left after dissolving crude platinum from the Ural Mountains in *aqua regia*. While Berzelius found no unusual metals, Osann thought he found three new metals, one of which he named ruthenium. In 1844 Klaus, generally recognized as the discoverer, showed that Osann's ruthenium oxide was very impure and that it contained a new metal. Klaus obtained 6 g of ruthenium from the portion of crude platinum that is insoluble in *aqua regia*. A member of the platinum group, ruthenium occurs native with other members of the group of ores found in the Ural Mountains and in North and South America. It is also found along with other platinum metals in small but commercial quantities in *pentlandite* of the Sudbury, Ontario, nickel-mining region, and in *pyroxinite* deposits of South Africa. The metal is isolated commercially by a complex chemical process, the final stage of which is the hydrogen reduction of ammonium ruthenium chloride, which yields a powder. The powder is consolidated by powder metallurgy techniques or by argon-arc welding. Ruthenium is a hard, white metal and has four crystal modifications. It does not tarnish at room temperatures, but oxidizes in air at about 800 °C. The metal is not attacked by hot or cold acids or *aqua regia*, but when potassium chlorate is added to the solution, it oxidizes explosively. It is attacked by halogens and by hydroxides. Ruthenium can be plated by electrodeposition or by thermal decomposition methods. The metal is one of the most effective hardeners for platinum and palladium and is alloyed with these metals to make electrical contacts for severe wear resistance. The corrosion resistance of titanium is improved a hundredfold by addition of 0.1% ruthenium. It is a versatile catalyst. Hydrogen sulfide can be split catalytically by light using an aqueous suspension of CdS particles loaded with ruthenium dioxide. This may have application to removal of H_2S in oil refining and other industrial processes. Compounds in at least eight oxidation states have been found, but of these, the +2, +3, and +4 states are the most common. Ruthenium compounds show a marked resemblance to those of osmium. Ruthenium tetroxide, like osmium tetroxide, is highly toxic and may explode.

Rutherfordium — (Ernest Rutherford [1871–1937], New Zealand, Canadian, and British physicist); Rf. An isotope of Element-104 was produced in 1964 at the Joint Nuclear Research Institute at Dubna by bombardment of plutonium with accelerated neon ions through the following reaction:

$$^{242}Pu + ^{22}Ne \rightarrow ^{260}104 + 4n$$

Other isotopes were later produced at Berkeley. A long-standing naming controversy was finally settled by IUPAC in 1997 through adoption of the name Rutherfordium.

Element 104, the first *transactinide* element, is expected to have chemical properties similar to those of hafnium. Experiments have shown that it forms volatile tetrahalides, $RfCl_4$ and $RfBr_4$, as well as the oxychloride, $RfOCl_2$. More complex compounds in which it shows a valence of 6 have also been produced. The pure element is expected to be a dense metal. The most stable isotope is probably ^{267}Rf, with a half-life of 1.3 minutes.

Samarium — (*Samarskite*, a mineral), Sm. Discovered spectroscopically by its sharp absorption lines in 1879 by Lecoq de Boisbaudran in the mineral *samarskite*, named in honor of a Russian mine official, Col. Samarski. Samarium is found along with other members of the rare-earth elements in many minerals, including *monazite* and *bastnasite*, which are commercial sources. The largest producer of rare-earth minerals is now China, followed by the U.S., India, and Russia. It occurs in monazite to the extent of 2.8%. While *misch metal* containing about 1% of samarium metal has long been used, samarium was not isolated in relatively pure form until 1901. Samarium metal can be produced by reducing the oxide with barium or lanthanum. Samarium has a bright silver luster and is reasonably stable in air. Three crystal modifications of the metal exist, with transformations at 734 and 922 °C. The metal ignites in air at about 150 °C. Natural samarium is a mixture of seven isotopes, three of which are unstable but have long half-lives. The sulfide has excellent high-temperature stability and good thermoelectric efficiencies up to 1100 °C. $SmCo_5$ has been used in making a permanent magnet material with a very high resistance to demagnetization. Samarium oxide has been used in optical glass to absorb infrared radiation. Samarium is used to dope calcium fluoride crystals for use in lasers. Compounds of the metal act as sensitizers for phosphors excited in the infrared; the oxide exhibits catalytic properties in the dehydration and dehydrogenation of ethyl alcohol. It is used as a neutron absorber in nuclear reactors. Little is known of the toxicity of samarium; therefore, it should be handled carefully.

Scandium — (L. *Scandia*, Scandinavia), Sc. On the basis of the Periodic System, Mendeleev predicted the existence of *ekaboron*, which would have an atomic weight between 40 of calcium and 48 of titanium. The element was discovered by Nilson in 1878 in the minerals *euxenite* and *gadolinite*, which had not yet been found anywhere except in Scandinavia. By processing 10 kg of euxenite and other residues of rare-earth minerals, Nilson was able to prepare about 2 g of scandium oxide of high purity. Cleve later pointed out that Nilson's scandium was identical with Mendeleev's ekaboron. Scandium is apparently a much more abundant element in the sun and certain stars than here on Earth. It is about the 23rd most abundant element in the sun, compared to the 50th most abundant on Earth. It is widely distributed on Earth, occurring in very minute quantities in over 800 mineral species. The blue color of beryl (aquamarine variety) is said to be due to scandium. It occurs as a principal component in the rare mineral *thortveitite*, found in Scandinavia and Malagasy. It is also found in the residues remaining after the extraction of tungsten from Zinnwald *wolframite*, and in *wiikite* and *bazzite*. Most scandium is presently being recovered from *thortveitite* or is extracted as a by-product from uranium mill tailings. Metallic scandium was first prepared in 1937 by Fischer, Brunger, and Grieneisen, who electrolyzed a eutectic melt of potassium, lithium, and scandium chlorides at 700 to 800 °C. Tungsten wire and a pool of molten zinc served as the electrodes in a graphite crucible. Pure scandium is now produced by reducing scandium fluoride with calcium metal. The production of the first pound of 99% pure scandium metal was announced in 1960. Scandium is a silver-white metal that develops a slightly yellowish or pinkish cast upon exposure to air. It is relatively soft and resembles yttrium and the rare-earth metals more than it resembles aluminum or titanium. It is a very light metal and has a much higher melting point than aluminum, making it of interest to designers of spacecraft. Scandium is not attacked by a 1:1 mixture of conc. HNO_3 and 48% HF. Scandium reacts rapidly with many acids. About 20 kg of scandium (as Sc_2O_3) are now being used yearly in the U.S. to produce high-intensity lights, and the radioactive isotope ^{46}Sc is used as a tracing agent in crude-oil refining. Scandium iodide added to mercury vapor lamps produces a highly efficient light source resembling sunlight. Little is yet known about the toxicity of scandium; therefore, it should be handled with care.

Seaborgium — (Glenn T. Seaborg [1912–1999], American chemist and nuclear physicist), Sg. The discovery of *Seaborgium*, Element 106, took place in 1974 almost simultaneously at the Lawrence-Berkeley Laboratory and at the Joint Institute for Nuclear Research at Dubna, Russia. The Berkeley Group, under direction of Ghiorso, used the Super-Heavy Ion Linear Accelerator (Super HILAC) as a source of heavy ^{18}O ions to bombard a 259 μg target of ^{249}Cf. This resulted in the production and positive identification of $^{263}106$, which decayed with a half-life of 0.8 s by the emission of alpha particles. The Dubna Team, directed by Flerov and Organessian, produced heavy ions of ^{54}Cr with their 310-cm heavy-ion cyclotron to bombard ^{207}Pb and ^{208}Pb and found a product that decayed with a half-life of 7 ms. They assigned $^{259}106$ to this isotope. Other isotopes of *Seaborgium* were later identified. Two of these, ^{269}Sg and ^{271}Sg, have half-lives of about 2 min. *Seaborgium* most likely would have properties resembling tungsten. IUPAC adopted the name *Seaborgium* in August 1997. Normally the naming of an element is not given until after the death of the person for which the element is named; however, in this case, it was named while Dr. Seaborg was still alive.

Selenium — (Gr. *Selene*, moon), Se. Discovered by Berzelius in 1817, who found it associated with tellurium, named for the Earth. Selenium is found in a few rare minerals, such as *crooksite* and *clausthalite*. In years past it has been obtained from flue dusts remaining from processing copper sulfide ores, but the anode muds from electrolytic copper refineries now provide the source of most of the world's selenium. Selenium is recovered by roasting the muds with soda or sulfuric acid, or by smelting them with soda and niter. Elemental selenium exists in several allotropic forms. Three are generally recognized, but as many as six have been claimed. Selenium can be prepared with either an amorphous or crystalline structure. The color of amorphous selenium is either red, in powder form, or black, in vitreous form. Crystalline monoclinic selenium is a deep red; crystalline hexagonal selenium, the most stable variety, is a metallic gray. The element is a member of the sulfur family and resembles sulfur both in its various forms and in its compounds. Selenium exhibits both photovoltaic action, where light is converted directly into electricity, and photoconductive action, where the electrical resistance decreases with increased illumination. These properties have made selenium useful in the production of photocells and exposure meters for photographic use, as well as solar cells. Selenium is also able to convert a.c. electricity to d.c. and is used in rectifiers. Below its melting point, selenium is a p-type semiconductor and has found uses in electronic and solid-state applications. It was formerly used in xerography for reproducing and copying documents, but it has been replaced in this application organic photoconductors. It is used by the glass industry to decolorize glass and to make ruby-colored glasses and enamels. It is also used as a photographic toner, and as an additive to stainless steel. Elemental selenium has low toxicity, and selenium is considered to be an essential trace element for humans; however, hydrogen selenide and other selenium compounds are extremely toxic, and resemble arsenic in their physiological reactions. Hydrogen selenide in a concentration of 1.5 ppm is intolerable to man. Selenium occurs in some soils in amounts sufficient to produce serious effects on animals feeding on plants, such as locoweed, grown in such soils.

Silicon — (L. *silex, silicis*, flint), Si. Davy in 1800 thought silica to be a compound and not an element; later in 1811, Gay Lussac and Thenard probably prepared impure amorphous silicon by heating potassium with silicon tetrafluoride. Berzelius, generally credited with the discovery, in 1824 succeeded in preparing amorphous silicon by the same general method as used earlier, but he purified the product by removing the fluosilicates by repeated washings. Deville, in 1854, first prepared crystalline silicon, the second allotropic form of the element. Silicon is present in the sun and stars and is a principal component of a class of meteorites known as "aerolites." It is also a component of *tektites*, a natural glass of uncertain origin. Silicon makes up 25.7% of the Earth's crust, by weight, and is the second most abundant element, being exceeded only by oxygen. Silicon is not found free in nature but occurs chiefly as the oxide and as silicates. *Sand, quartz, rock crystal, amethyst, agate, flint, jasper,* and *opal* are some of the forms in which the oxide appears. *Granite, hornblende, asbestos, feldspar, clay mica,* etc., are but a few of the numerous silicate minerals. Silicon is prepared commercially by heating silica and carbon in an electric furnace, using carbon electrodes. Several other methods can be used for preparing the element. Amorphous silicon can be prepared as a brown powder, which can be easily melted or vaporized. Crystalline silicon has a metallic luster and grayish color. The Czochralski process is commonly used to produce single crystals of silicon used for solid-state devices. Hyperpure silicon can be prepared by the thermal decomposition of ultrapure trichlorosilane in a hydrogen atmosphere, and by a vacuum float zone process. This product can be doped with boron, gallium, phosphorus, or arsenic to produce semiconductors for use in transistors, solar cells, rectifiers, and other solid-state devices that are used extensively in the electronics industries. Silicon is a relatively inert element, but it is attacked by halogens and dilute alkali. Most acids, except hydrofluoric, do not affect it. Hydrolysis and condensation of various substituted chlorosilanes can be used to produce a very great number of polymeric products, or silicones, ranging from liquids to hard, glasslike solids with many useful properties. Elemental silicon transmits more than 95% of all wavelengths of infrared light, from 1.3 to 6.7 μm. Silicon is one of man's most useful elements. In the form of sand and clay it is used to make concrete and brick; it is a useful refractory material for high-temperature work, and in the form of silicates it is used in making enamels, pottery, etc. Silica, as sand, is a principal ingredient of glass, one of the most inexpensive of materials with excellent mechanical, optical, thermal, and electrical properties. Glass can be made in a very great variety of shapes, and is used as containers, window glass, insulators, and thousands of other uses. Silicon is important in plant and animal life. Diatoms in both fresh and salt water extract silica from the water to build up their cell walls. Silica is present in ashes of plants and in the human skeleton. Silicon is an important ingredient in steel; silicon carbide is one of the most important abrasives. Miners, stonecutters, and other engaged in work where siliceous dust is breathed in large quantities often develop a serious lung disease known as *silicosis.*

Silver — (Anglo-Saxon, *Seolfor siolfur*), Ag (L. *argentum*). Silver has been known since ancient times. It is mentioned in Genesis. Slag dumps in Asia Minor and on islands in the Aegean Sea indicate that man learned to separate silver from lead as early as 3000 B.C. Silver occurs native and in ores such as *argentite* (Ag_2S) and *horn silver* (AgCl); lead, lead-zinc, copper, gold, and copper-nickel ores are principal sources. Mexico, Canada, Peru, and the U.S. are the principal silver producers in the western hemisphere. Silver is also recovered during electrolytic refining of copper. Commercial fine silver contains at least 99.9% silver. Purities of 99.999+% are available commercially. Pure silver has a brilliant white metallic luster. It is a little harder than gold and is very ductile and malleable, being exceeded only by gold and perhaps palladium. Pure silver has the highest electrical and thermal conductivity of all metals and possesses the lowest contact resistance. It is stable in pure air and water, but tarnishes when exposed to ozone, hydrogen sulfide, or air containing sulfur. The alloys of silver are important. Sterling silver is used for jewelry, silverware, etc., where appearance is paramount. This alloy contains 92.5% silver, the remainder being copper or some other metal. Silver was once of great importance in photography, including production of x-ray images, but that application has diminished as digital photography became popular. It is used for dental alloys and in making solder and brazing alloys, electrical contacts, and high capacity silver–zinc and silver–cadmium batteries. Silver paints are used for making printed circuits. It is used in mirror production and may be deposited on glass or metals by chemical deposition, electrodeposition, or by evaporation. When freshly deposited, it is the best reflector of visible light known, but is rapidly tarnishes and loses much of its reflectance. It is a poor reflector of ultraviolet. Silver fulminate ($Ag_2C_2N_2O_2$), a powerful explosive, is sometimes formed during the silvering process. Silver nitrate, or *lunar caustic,* is an important silver compound. While silver itself is not considered to be toxic, most of its salts are poisonous. Silver compounds can be absorbed in the circulatory system and reduced silver deposited in the various tissues of the body. A condition, known as *argyria,* results with a grayish pigmentation of the skin and mucous membranes. Silver has germicidal effects and kills many lower organisms effectively without harm to higher animals. Silver has been used for coinage since ancient times. However, in the 20th century the U.S. and most other countries abandoned silver as a monetary standard and replaced silver coins by other metals, often alloys of copper, nickel, and zinc.

Sodium — (English, *soda*; Medieval Latin, *sodanum,* headache remedy), Na (L. *natrium*). Long recognized in compounds, sodium was first isolated by Davy in 1807 by electrolysis of caustic soda. Sodium is present in fair abundance in the sun and stars. The D lines of sodium are among the most prominent in the solar spectrum. Sodium is the sixth most abundant element on earth, comprising about 2.6% of the Earth's crust; it is the most abundant of the alkali group of metals of which it is a member. The most common compound is sodium chloride, but sodium occurs in many other minerals, such as *soda niter, cryolite, amphibole, zeolite, sodalite,* etc. It is a very reactive element and is never found free in nature. It is now obtained commercially by the electrolysis of absolutely dry fused sodium chloride. This method is much cheaper than that of electrolyzing sodium hydroxide, as was used in the past. Sodium is a soft, bright, silvery metal that floats on water, decomposing it with the evolution of hydrogen and the formation of the hydroxide. It may or may not ignite spontaneously on water, depending on the amount of oxide and metal exposed to the water. It normally does not ignite in air at temperatures below 115 °C. Sodium should be handled with respect, as it can be dangerous when improperly handled. Metallic sodium is vital in the manufacture of sodamide and esters, and in the preparation of organic compounds. The metal may be used to improve the structure of certain alloys, to descale metal, to purify molten metals, and as a heat transfer agent. An alloy of sodium with potassium, NaK, is also an important heat transfer agent. Sodium compounds are important to the paper, glass, soap, textile, petroleum, chemical, and metal industries. Soap is generally a sodium salt of certain fatty acids. The importance of common salt to animal nutrition has been recognized since prehistoric times. Among the many compounds that are of the greatest industrial importance are common salt (NaCl), soda ash (Na_2CO_3), baking soda ($NaHCO_3$), caustic soda (NaOH), Chile saltpeter ($NaNO_3$), di- and tri-sodium phosphates, sodium thiosulfate (hypo, $Na_2S_2O_3 \cdot 5H_2O$), and borax ($Na_2B_4O_7 \cdot 10H_2O$). Sodium metal should be handled with great care. It should be kept in an inert atmosphere and contact with water and other substances with which sodium reacts should be avoided.

Strontium — (*Strontian*, town in Scotland), Sr. Isolated by Davey by electrolysis in 1808; however, Adair Crawford in 1790 recognized a new mineral (*strontianite*) as differing from other barium minerals (*baryta*). Strontium is found chiefly as *celestite* ($SrSO_4$) and *strontianite* ($SrCO_3$). *Celestite* is found in Mexico, Turkey, Iran, Spain, Algeria, and in the U.K. The U.S. has no active *celestite* mines. The metal can be prepared by electrolysis of the fused chloride mixed with potassium chloride or is made by reducing strontium oxide with aluminum in a vacuum at a temperature at which strontium distills off. Three allotropic forms of the metal exist, with transition points at 235 and 540 °C. Strontium is softer than calcium and decomposes water more vigorously. It does not absorb nitrogen below 380 °C. It should be kept under mineral oil to prevent oxidation. Freshly cut strontium has a silvery appearance, but rapidly turns a yellowish color with the formation of the oxide. The finely divided metal ignites spontaneously in air. Volatile strontium salts impart a beautiful crimson color to flames, and these salts are used in pyrotechnics and in the production of flares. The major use for strontium at present is in producing glass for color television picture tubes. All color TV and cathode ray tubes sold in the U.S. are required by law to contain strontium in the face plate glass to block x-ray emission. Strontium also improves the brilliance of the glass and the quality of the picture. It has also found use in producing ferrite magnets and in refining zinc. Strontium titanate is an interesting optical material as it has an extremely high refractive index and an optical dispersion greater than that of diamond. It has been used as a gemstone, but it is very soft. It does not occur naturally.

Sulfur — (Sanskrit, *sulvere*; L. *sulphurium*), S. Known to the ancients; referred to in Genesis as *brimstone*. Sulfur is found in meteorites. A dark area near the crater Aristarchus on the moon has been studied by R. W. Wood with ultraviolet light. This study suggests strongly that it is a sulfur deposit. Sulfur occurs native in the vicinity of volcanoes and hot springs. It is widely distributed in nature as *iron pyrites, galena, sphalerite, cinnabar, stibnite, gypsum, Epsom salts, celestite, barite,* etc. Sulfur is commercially recovered from wells sunk into the salt domes along the Gulf Coast of the U.S. It is obtained from these wells by the Frasch process, which forces heated water into the wells to melt the sulfur, which is then brought to the surface. Sulfur compounds also occur in natural gas and petroleum crudes and must be removed from these products. Formerly this was done chemically, which wasted the sulfur. New processes now permit recovery, and these sources promise to be very important. Large amounts of sulfur are being recovered from Alberta gas fields. Sulfur is a pale yellow, odorless, brittle solid that is insoluble in water but soluble in carbon disulfide. In every state, whether gas, liquid or solid, elemental sulfur occurs in more than one allotropic form or modification; these present a confusing multitude of forms whose relations are not yet fully understood. Amorphous or "plastic" sulfur is obtained by fast cooling of the crystalline form. X-ray studies indicate that amorphous sulfur may have a helical structure with eight atoms per spiral. Crystalline sulfur seems to be made of rings, each containing eight sulfur atoms that fit together to give a normal x-ray pattern. A finely divided form of sulfur, known as *flowers of sulfur*, is obtained by sublimation. Sulfur readily forms sulfides with many elements. Sulfur is a component of black gunpowder,and is used in the vulcanization of natural rubber and as a fungicide. It is also used extensively is making phosphate fertilizers. A large tonnage is used to produce sulfuric acid, the most important manufactured sulfur compound. It is used in making sulfite paper and other papers, as a fumigant, and in the bleaching of dried fruits. The element is a good electrical insulator. Organic compounds containing sulfur are very important. Calcium sulfate, ammonium sulfate, carbon disulfide, sulfur dioxide, and hydrogen sulfide are but a few of the many other important compounds of sulfur. Sulfur is essential to life. It is a minor constituent of fats, body fluids, and skeletal minerals. Carbon disulfide, hydrogen sulfide, and sulfur dioxide should be handled carefully. Hydrogen sulfide in small concentrations can be metabolized, but in higher concentrations it can quickly cause death by respiratory paralysis. It is insidious in that it quickly deadens the sense of smell. Sulfur dioxide is a dangerous component in atmospheric pollution.

Tantalum — (Gr. *Tantalos*, mythological character, father of *Niobe*), Ta. Discovered in 1802 by Ekeberg, but many chemists thought niobium and tantalum were identical elements until Rose, in 1844, and Marignac, in 1866, showed that niobic and tantalic acids were two different acids. The early investigators only isolated the impure metal. The first relatively pure ductile tantalum was produced by von Bolton in 1903. Tantalum occurs principally in the mineral *columbite-tantalite* (Fe,Mn)$(Nb,Ta)_2O_6$. Tantalum ores are found in Australia, Brazil, Rwanda, Zimbabwe, Congo-Kinshasa, Nigeria, and Canada. Separation of tantalum from niobium requires several complicated steps. Several methods are used to commercially produce the element, including electrolysis of molten potassium fluorotantalate, reduction of potassium fluorotantalate with sodium, or reacting tantalum carbide with tantalum oxide. Natural tantalum contains two isotopes, one of which is radioactive with a very long half-life. Tantalum is a gray, heavy, and very hard metal. When pure, it is ductile and can be drawn into fine wire, which is used as a filament for evaporating metals such as aluminum. Tantalum is almost completely immune to chemical attack at temperatures below 150 °C, and is attacked only by hydrofluoric acid, acidic solutions containing the fluoride ion, and free sulfur trioxide. Alkaline solutions attack it only slowly. At high temperatures, tantalum becomes much more reactive. The melting point of tantalum is exceeded only by tungsten and rhenium. Tantalum is used to make a variety of alloys with desirable properties such as high-melting point, high strength, good ductility, etc. A tantalum carbide graphite composite material is one of the hardest materials ever made. Tantalum has good "gettering" ability at high temperatures, and tantalum oxide films are stable and have good rectifying and dielectric properties. Tantalum is used to make electrolytic capacitors and vacuum furnace parts, which account for about 60% of its production. The metal is also widely used to fabricate chemical process equipment, nuclear reactors, and aircraft and missile parts. Tantalum is completely immune to body liquids and is a nonirritating metal. It has, therefore, found wide use in making surgical appliances. Tantalum oxide is used to make special glass with a high index of refraction for camera lenses. The metal has many other uses.

Technetium — (Gr. *technetos*, artificial), Tc. Element 43 was predicted on the basis of the periodic table, and was erroneously reported as having been discovered in 1925, at which time it was named *masurium*. The element was actually discovered by Perrier and Segre in Italy in 1937. It was found in a sample of molybdenum that was bombarded by deuterons in the Berkeley cyclotron and which E. Lawrence sent to these investigators. Technetium was the first element to be produced artificially. Since its discovery, searches for the element in terrestrial materials have been made without success. If it does exist, the concentration must be very small. Technetium has been found in the spectrum of S-, M-, and N-type stars, and its presence in stellar matter is leading to new theories of the production of heavy elements in the stars. ^{97}Tc has a half-life of 2.6 × 10^6 years. ^{98}Tc has a half-life of 4.2 × 10^6 years. The isomeric isotope ^{95m}Tc, with a half-life of 61 days, is useful for tracer work, as it produces energetic gamma rays. Technetium metal has been produced in kilogram quantities. The metal was first prepared by passing hydrogen gas at 1100 °C over Tc_2S_7. It is now conveniently prepared by the reduction of ammonium pertechnetate with hydrogen. Technetium is a silvery-gray metal that tarnishes slowly in moist air. The chemistry of technetium is similar to that of rhenium. Technetium dissolves in nitric acid, aqua regia, and concentrated sulfuric acid, but is not soluble in hydrochloric acid of any strength. The element is a remarkable corrosion inhibitor for steel. It is reported that mild carbon steels may be effectively protected by as little as 55 ppm of $KTcO_4$ in aerated distilled water at temperatures up to 250 °C. This corrosion protection is limited to closed systems, since technetium is radioactive and must be confined. ^{99}Tc is a contamination hazard and should be handled in a glove box.

Tellurium — (L. *tellus*, earth), Te. Discovered by Muller von Reichenstein in 1782; named by Klaproth, who isolated it in 1798. Tellurium is occasionally found native, but is more often found as the telluride of gold (*calaverite*) and combined with other metals. It is recovered commercially from the anode muds produced during the electrolytic refining of blister copper. The U.S., Canada, Peru, and Japan are the largest producers of the element. Crystalline tellurium has a silvery-white appearance, and when pure exhibits a metallic luster. It is brittle and easily pulverized. Amorphous tellurium is formed by precipitating tellurium from a solution of telluric or tellurous acid. Whether this form is truly amorphous, or made of minute crystals, is open to question. Tellurium is a p-type semiconductor, and shows greater conductivity in certain directions, depending on alignment of the atoms. Its conductivity increases slightly with exposure to light. It can be doped with silver, copper, gold, tin, or other elements. In air, tellurium burns with a greenish-blue flame, forming the dioxide. Molten tellurium corrodes iron, copper, and stainless steel. Tellurium and its compounds are probably toxic and should be handled with care. Workmen exposed to as little as 0.01 mg/m^3 of air, or less, develop "tellurium breath," which has a garlic-like odor. Natural tellurium consists of eight isotopes, two of which are radioactive with very long half-lives. Tellurium improves the machinability of copper and stainless steel, and its addition to lead decreases the corrosive action of sulfuric acid on lead and improves its strength and hardness. Tellurium catalysts are used in the oxidation of organic compounds and are used in hydrogenation and halogenation reactions. Tellurium is also used in electronic and semiconductor devices and is a basic ingredient in blasting caps, and it is added to cast iron for chill control. Tellurium is used in ceramics. Bismuth telluride has been used in thermoelectric devices.

Tennessine — (*Tennessee*, location of Oak Ridge National Laboratory), Ts. A U.S.-Russian collaboration in 2010 detected Element-117 as a product of the fusion reaction of ^{48}Ca and ^{249}Bk. Confirming experiments were done in 2013. IUPAC approved the name Tennessine for Element-117 in 2016. In the periodic table the element falls in the halogen group, but relativistic effects are predicted to alter its chemical behavior beyond what would be expected from simple extrapolation from the lower halogens. The most stable isotope produced so far, ^{293}Ts, has a half-life of about 22 ms.

Terbium — (*Ytterby*, village in Sweden), Tb. Discovered by Mosander in 1843. Terbium is a member of the lanthanide or "rare earth" group of elements. It is found in *cerite, gadolinite*, and other minerals along with other rare earths. It is recovered commercially from *monazite* in which it is present to the extent of 0.03%, from *xenotime*, and from *euxenite*, a complex oxide containing 1% or more of terbia. As with other rare earths, elemental terbium can be produced by reducing the anhydrous chloride or fluoride with calcium metal in a tantalum crucible. Calcium and tantalum impurities can be removed by vacuum remelting. Other methods of isolation are possible. Terbium is reasonably stable in air. It is a silver-gray metal, and is malleable, ductile, and soft enough to be cut with a knife. Two crystal modifications exist, with a transformation temperature of 1289 °C. The oxide is a chocolate or dark maroon color. Sodium terbium borate is used as a laser material and emits coherent light at 0.546 μm. Terbium is used to dope calcium fluoride, calcium tungstate, and strontium molybdate, used in solid-state devices. The oxide has potential application as an activator for green phosphors. It can be used with ZrO_2 as a crystal stabilizer of fuel cells that operate at elevated temperature.

Few other uses have been found. Little is known of the toxicity of terbium. It should be handled with care as with other lanthanide elements.

Thallium — (Gr. *thallos*, a green shoot or twig), Tl. Thallium was discovered spectroscopically in 1861 by Crookes. The element was named after the beautiful green spectral line, which identified the element. The metal was isolated both by Crookes and Lamy in 1862 about the same time. Thallium occurs in *crooksite, lorandite,* and *hutchinsonite.* It is also present in *pyrites* and is recovered from the roasting of this ore in connection with the production of sulfuric acid. It is also obtained from the smelting of lead and zinc ores. Extraction is somewhat complex and depends on the source of the thallium. Manganese nodules, found on the ocean floor, contain thallium. When freshly exposed to air, thallium exhibits a metallic luster, but soon develops a bluish-gray tinge, resembling lead in appearance. A heavy oxide builds up on thallium if left in air, and in the presence of water the hydroxide is formed. The metal is very soft and malleable. It can be cut with a knife. The element and its compounds are toxic and should be handled carefully. Contact of the metal with skin is dangerous, and when melting the metal adequate ventilation should be provided. Thallium is suspected of carcinogenic potential for man. Thallium sulfate has been widely employed as a rodenticide and ant killer. It is odorless and tasteless, giving no warning of its presence. Its use, however, has been prohibited in the U.S. since 1975 as a household insecticide and rodenticide. The electrical conductivity of thallium sulfide changes with exposure to infrared light, and this compound is used in photocells. Thallium bromide-iodide crystals have been used as infrared optical materials. Thallium has been used, with sulfur or selenium and arsenic, to produce low-melting glasses which become fluid between 125 and 150 °C. These glasses have properties at room temperatures similar to ordinary glasses and are said to be durable and insoluble in water. Thallium oxide has been used to produce glasses with a high index of refraction. Thallium has been used in treating ringworm and other skin infections; however, its use has been limited because of the narrow margin between toxicity and therapeutic benefits. A mercury–thallium alloy, which forms a eutectic at 8.5% thallium, freezes at −60 °C, some 20° below the freezing point of mercury.

Thorium — (*Thor*, Scandinavian god of war), Th. Discovered by Berzelius in 1828. Thorium occurs in *thorite* ($ThSiO_4$) and in *thorianite* ($ThO_2 + UO_2$). Large deposits of thorium minerals have been reported in New England and elsewhere, but these have not yet been exploited. Thorium is now thought to be about three times as abundant as uranium and about as abundant as lead or molybdenum. The metal can a source of nuclear power. There is probably more energy available for use from thorium in the minerals of the Earth's crust than from both uranium and fossil fuels. However, any sizable demand for thorium as a nuclear fuel is still in the future. Work has been done in developing thorium cycle converter-reactor systems. Several prototypes, including the HTGR (high-temperature gas-cooled reactor) and MSRE (molten salt converter reactor experiment), have operated. While the HTGR reactors are efficient, they are not expected to become important commercially for many years because of certain operating difficulties. Thorium is recovered commercially from the mineral *monazite*, which contains from 3% to 9% ThO_2 along with rare-earth minerals. Much of the internal heat the Earth produces has been attributed to thorium and uranium. Several methods are available for producing thorium metal: it can be obtained by reducing thorium oxide with calcium, by electrolysis of anhydrous thorium chloride in a fused mixture of sodium and potassium chlorides, by calcium reduction of thorium tetrachloride mixed with anhydrous zinc chloride, and by reduction of thorium tetrachloride with an alkali metal. Thorium was originally assigned a position in Group IV of the periodic table. Because of its atomic weight, valence, etc., it is now considered to be the second member of the *actinide* series of elements. When pure, thorium is a silvery-white metal which is air stable and retains its luster for several months. When contaminated with the oxide, thorium slowly tarnishes in air, becoming gray and finally black. The physical properties of thorium are greatly influenced by the degree of contamination with the oxide. The purest specimens often contain several tenths of a percent of the oxide. High-purity thorium has been made. Pure thorium is soft, very ductile, and can be cold-rolled, swaged, and drawn. Thorium is dimorphic, changing at 1360 °C from a cubic to a body-centered cubic structure. Thorium oxide has a melting point of 3300 °C, which is the highest of all oxides. Only a few elements, such as tungsten, and a few compounds, such as tantalum carbide, have higher melting points. Thorium is slowly attacked by water, but does not dissolve readily in most common acids, except hydrochloric. Powdered thorium metal is often pyrophoric and should be carefully handled. When heated in air, thorium turnings ignite and burn brilliantly with a white light. A former use of thorium was in the preparation of the Welsbach mantle, used for portable gaslights. These mantles, consisting of thorium oxide with about 1% cerium oxide and other ingredients, glow with a dazzling light when heated in a gas flame. Thorium is an important alloying element in magnesium, imparting high strength and creep resistance at elevated temperatures. Because thorium has a low work function and high electron emission, it is used to coat tungsten wire used in electronic equipment. The oxide is also used to control the grain size of tungsten used for electric lamps; it is also used for high-temperature laboratory crucibles. Glasses containing thorium oxide have a high refractive index and low dispersion. Consequently, they find application in high-quality lenses for cameras and scientific instruments. Thorium oxide has also found use as a catalyst in the conversion of ammonia to nitric acid, in petroleum cracking, and in producing sulfuric acid. Its radioactive nature and its handling and disposal problems limit other applications. ^{232}Th is sufficiently radioactive to expose a photographic plate in a few hours. Thorium disintegrates with the production of ^{220}Rn, which is an alpha emitter and presents a radiation hazard. Good ventilation of areas where thorium is stored or handled is therefore essential.

Thulium — (*Thule*, the earliest name for Scandinavia), Tm. Discovered in 1879 by Cleve. Thulium occurs in small quantities along with other rare earths in a number of minerals. It is obtained commercially from *monazite*, which contains about 0.007% of the element. Thulium is the least abundant of the rare-earth elements, but with new sources recently discovered, it is now considered to be about as rare as silver, gold, or cadmium. Ion-exchange and solvent extraction techniques have recently permitted much easier separation of the rare earths, with much lower costs. Thulium can be isolated by reduction of the oxide with lanthanum metal or by calcium reduction of the anhydrous fluoride. The pure metal has a bright, silvery luster. It is reasonably stable in air, but the metal should be protected from moisture in a closed container. The element is silver-gray, soft, malleable, and ductile, and can be cut with a knife. Because of the relatively high price of the metal, thulium has not yet found many practical applications. ^{169}Tm bombarded in a nuclear reactor can be used as a radiation source in portable x-ray equipment. ^{171}Tm is potentially useful as an energy source. Natural thulium also has possible use in *ferrites* (ceramic magnetic materials) used in microwave equipment. As with other lanthanides, thulium has a low-to-moderate acute toxicity rating. It should be handled with care.

Tin — (Anglo-Saxon, *tin*), Sn (L. *stannum*). Known to the ancients. Tin is found chiefly in *cassiterite* (SnO_2). Most of the world's supply comes from China, Indonesia, Peru, Brazil, and Bolivia. Cornwall was a major source in older periods. The U.S. produces almost none, although occurrences have been found in Alaska and Colorado. Tin is obtained by reducing the ore with coal in a reverberatory furnace. Ordinary tin is a silver-white metal, is malleable, somewhat ductile, and has a highly crystalline structure. Due to the breaking of these crystals, a "tin cry" is heard when a bar is bent. The element has two allotropic forms at normal pressure. On warming, gray, or α-tin, with a cubic structure, changes at 13.2 °C into white, or β-tin, the ordinary form of the metal. White tin has a tetragonal structure. When tin is cooled below 13.2 °C, it changes slowly from white to gray. This change is affected by impurities such as aluminum and zinc and can be prevented by small additions of antimony or bismuth. This change from the α to β form is called the tin pest. Tin–lead alloys are used to make organ pipes. There are few if any uses for gray tin. Tin takes a high polish and is used to coat other metals to prevent corrosion or other chemical action. Such tin plate over steel is used in the so-called tin can for preserving food. Alloys of tin are very important. Soft solder, type metal, fusible metal, pewter, bronze, bell metal, Babbitt metal, white metal, die casting alloy, and phosphor bronze are some of the important alloys using tin. Tin resists distilled sea and soft tap water, but is attacked by strong acids, alkalis, and acid salts. Oxygen in solution accelerates the attack. When heated in air, tin forms SnO_2, which is feebly acid, forming stannate salts with basic oxides. The most important salt is the chloride ($SnCl_2 \cdot H_2O$), which is used as a reducing agent and as a mordant in calico printing. Tin salts sprayed onto glass are used to produce electrically conductive coatings. These have been used for panel lighting and for frost-free windshields. Most window glass is now made by floating molten glass on molten tin (float glass) to produce a flat surface (Pilkington process). The small amount of tin found in canned foods is quite harmless. The agreed limit of tin content in U.S. foods is 300 mg/kg. The trialkyl and triaryl tin compounds are used as biocides and must be handled carefully.

Titanium — (L. *Titans*, the first sons of the Earth, myth), Ti. Discovered by Gregor in 1791; named by Klaproth in 1795. Impure titanium was prepared by Nilson and Pettersson in 1887; however, the pure metal (99.9%) was not made until 1910 by Hunter by heating $TiCl_4$ with sodium in a steel bomb. Titanium is present in meteorites and in the sun. Rocks obtained during the Apollo 17 lunar mission showed the presence of 12.1% TiO_2. Analyses of rocks obtained during earlier Apollo missions show lower percentages. Titanium oxide bands are prominent in the spectra of M-type stars. The element is the ninth most abundant in the crust of the Earth. Titanium is almost always present in igneous rocks and in the sediments derived from them. It occurs in the minerals *rutile, ilmenite,* and *sphene*, and is present in titanates and in many iron ores. Deposits of ilmenite and rutile are found in Florida, California, Tennessee, and New York. Australia, Norway, Malaysia, India, and China are also large suppliers of titanium minerals. Titanium is present in the ash of coal, in plants, and in the human body. The metal was a laboratory curiosity until Kroll, in 1946, showed that titanium could be produced commercially by reducing titanium tetrachloride with magnesium. This method is largely used for producing the metal today. The metal can be purified by decomposing the iodide. Titanium, when pure, is a lustrous, white metal. It has a low density, good strength, is easily fabricated, and has excellent corrosion resistance. It is ductile only when it is free of oxygen. The metal burns in air and is the only element that burns in nitrogen. Titanium is resistant to dilute sulfuric and hydrochloric acids, most organic acids, moist chlorine gas, and chloride solutions. The metal is dimorphic. The hexagonal α form changes to the cubic β form very slowly at 882 °C. The metal combines with oxygen at red heat, and with chlorine at 550 °C. Titanium is important as an alloying agent with aluminum, molybdenum, manganese, iron, and other metals. Alloys of titanium are principally used for aircraft and missiles where lightweight strength and ability to withstand extremes of temperature are important. Titanium is as strong as steel, but 45% lighter. It is 60% heavier than aluminum, but twice as strong. Titanium has potential use in desalination plants for converting seawater into fresh water. The metal has excellent resistance to seawater and is used for propeller shafts, rigging, and other parts of ships exposed to salt water. A titanium anode coated with platinum has been used to provide cathodic protection from corrosion by salt water. Titanium metal is considered to be physiologically inert; however, titanium powder may be a carcinogenic hazard. When pure, titanium dioxide is relatively clear and has an extremely high index of refraction with an optical dispersion higher than diamond. It is produced artificially for use as a gemstone, but it is relatively soft. Star sapphires and rubies exhibit their asterism as a result of the presence of TiO_2. Titanium dioxide is extensively used for both house paint and artist's paint, as it is permanent and has good covering power. Titanium oxide pigment accounts for the largest use of the element. Titanium paint is an excellent reflector of infrared and is extensively used in solar observatories where heat causes poor seeing conditions. Titanium tetrachloride is used to iridize glass. This compound fumes strongly in air and has been used to produce smoke screens.

Tungsten — (Swedish, *tung sten*, heavy stone); also known as *wolfram* (from *wolframite*, said to be named from *wolf rahm* or *spumi lupi*, because the ore interfered with the smelting of tin and was supposed to devour the tin), W. In 1779, Peter Woulfe examined the mineral now known as *wolframite* and concluded it must contain a new substance. Scheele, in 1781, found that a new acid could be made from *tung sten* (a name first applied about 1758 to a mineral now known as *scheelite*). Scheele and Berman suggested the possibility of obtaining a new metal by reducing this acid. The de Elhuyar brothers found an acid in *wolframite* in 1783 that was identical to the acid of *tungsten* (tungstic acid) of Scheele, and in that year, they succeeded in obtaining the element by reduction of this acid with charcoal. Tungsten occurs in *wolframite*, $(Fe,Mn)WO_4$; *scheelite*, $CaWO_4$; *huebnerite*, $MnWO_4$; and *ferberite*, $FeWO_4$. Important deposits of tungsten occur in California, Colorado, Bolivia, Russia, and Portugal. China is reported to have about 75% of the world's tungsten resources. The metal is obtained commercially by reducing tungsten oxide with hydrogen or carbon. Pure tungsten is a steel-gray to tin-white metal. Very pure tungsten can be cut with a hacksaw, and can be forged, spun, drawn, and extruded. The impure metal is brittle and can be worked only with difficulty. Tungsten has the highest melting point of all metals, and at temperatures over 1650 °C has the highest tensile strength. The metal oxidizes in air and must be protected at elevated temperatures. It has excellent corrosion resistance and is attacked only slightly by most mineral acids. The thermal expansion is about the same as borosilicate glass, which makes the metal useful for glass-to-metal seals. Tungsten and its alloys are used extensively for filaments for electric light bulbs and for metal evaporation work; for electrical contact points for automobile distributors; x-ray targets; windings and heating elements for electrical furnaces; and for numerous spacecraft and high-temperature applications. High-speed tool steels, Hastelloy®, Stellite®, and many other alloys contain tungsten. Tungsten carbide is of great importance to the metal-working, mining, and petroleum industries. Calcium and magnesium tungstates are widely used in fluorescent lighting; other salts of tungsten are used in the chemical and tanning industries. Tungsten disulfide is a dry, high-temperature lubricant, stable to 500 °C. Tungsten bronzes and other tungsten compounds are used in paints. Zirconium tungstate has found recent applications (see under Zirconium).

Uranium — (Planet *Uranus*), U. Yellow-colored glass, containing more than 1% uranium oxide and dating back to 79 A.D., has been found near Naples, Italy. Klaproth recognized an unknown element in *pitchblende* and attempted to isolate the metal in 1789. The metal apparently was first isolated in 1841 by Peligot, who reduced the anhydrous chloride with potassium. Uranium is not as rare as it was once thought. It is now considered to be more plentiful than mercury, antimony, silver, or cadmium, and is about as abundant as molybdenum or arsenic. It occurs in numerous minerals such as *pitchblende, uraninite, carnotite, autunite, uranophane, davidite,* and *tobernite.* It is also found in *phosphate rock, lignite, monazite sands,* and can be recovered commercially from these sources. Large deposits of uranium ore occur in Utah, Colorado, New Mexico, Canada, and elsewhere. Uranium can be made by reducing uranium halides with alkali or alkaline earth metals or by reducing uranium oxides by calcium, aluminum, or carbon at high temperatures. The metal can also be produced by electrolysis of KUF_5 or UF_4, dissolved in a molten mixture of $CaCl_2$ and NaCl. High-purity uranium can be prepared by the thermal decomposition of uranium halides on a hot filament. Uranium metal exhibits three crystallographic modifications α, β, and γ, with transitions at 669 °C and 776 °C. Uranium is a heavy, silvery-white metal that is pyrophoric when finely divided. It is a little softer than steel and is attacked by cold water in a finely divided state. It is malleable, ductile, and slightly paramagnetic. In air, the metal becomes coated with a layer of oxide. Acids dissolve the metal, but it is unaffected by alkalis. Natural uranium is sufficiently radioactive to expose a photographic plate in an hour or so. Much of the internal heat of the Earth is thought to be attributable to the presence of uranium and thorium. ^{238}U, with a half-life of 4.46×10^9 years, has been used to estimate the age of igneous rocks. The origin of uranium, the highest member of the naturally occurring elements — except perhaps for traces of neptunium or plutonium — is not clearly understood, although it has been suggested that uranium might be a decay product of elements of higher atomic weight, which may have once been present on Earth or elsewhere in the universe. These original elements may have been formed as a result of the primordial big bang, in a supernova, or in some other stellar processes. The fact that recent studies show that most transuranic elements are extremely rare with very short half-lives indicates that it may be necessary to find some alternative explanation for the very large quantities of radioactive uranium we find on Earth. Studies of meteorites from other parts of the solar system show a relatively low radioactive content, compared to terrestrial rocks. Uranium is of great importance as a nuclear fuel. ^{238}U can be converted into fissionable plutonium in breeder reactors, where it is possible to produce more new fissionable material than the fissionable material used in maintaining the chain reaction. ^{235}U is of even greater importance, for it is the key to the utilization of uranium. ^{235}U, while occurring in natural uranium to the extent of only 0.72%, is so fissionable with slow neutrons that a self-sustaining fission chain reaction can be made to occur in a reactor constructed from natural uranium and a suitable moderator, such as heavy water or graphite, alone. ^{235}U can be concentrated by gaseous diffusion and other physical processes and used directly as a nuclear fuel, instead of natural uranium, or used as an explosive. Natural uranium, slightly enriched with ^{235}U by a small percentage, is used to fuel nuclear power reactors for the generation of electricity. One pound of completely fissioned uranium has the fuel value of over 1500 tons of coal. However, there are serious problems with nuclear waste disposal that have not been resolved. Depleted uranium has been used for inertial guidance devices, gyrocompasses, counterweights for aircraft control surfaces, ballast for missile reentry vehicles, and as a shielding material for tanks, etc. Concerns, however, have been raised over its low radioactive properties. Uranium metal is used for x-ray targets for production of high-energy x-rays. The nitrate has been used as photographic toner, and the acetate is used in analytical chemistry. Crystals of uranium nitrate are triboluminescent. Uranium salts have also been used for producing yellow "vaseline" glass and glazes. Uranium and its compounds are highly toxic, both from a chemical and radiological standpoint. Finely divided uranium metal, being pyrophoric, presents a fire hazard.

Vanadium — (Scandinavian goddess, *Vanadis*), V. Vanadium was first discovered by del Rio in 1801. Unfortunately, a French chemist incorrectly declared that del Rio's new element was only impure chromium; del Rio thought himself to be mistaken and accepted the French chemist's statement. The element was rediscovered in 1830 by Sefstrom, who named the element in honor of the Scandinavian goddess *Vanadis* because of its beautiful multicolored compounds. It was isolated in nearly pure form by Roscoe, in 1867, who reduced the chloride with hydrogen. Vanadium of 99.3% to 99.8% purity was not produced until 1927. Vanadium is found in about 65 different minerals among which *carnotite, roscoelite, vanadinite,* and *patronite* are important sources of the metal. Vanadium is also found in phosphate rock and certain iron ores and is present in some crude oils in the form of organic complexes. It is also found in small percentages in meteorites. Commercial production from petroleum ash holds promise as an important source of the element. China, South Africa, and Russia supply much of the world's vanadium ores. High-purity ductile vanadium can be obtained by reduction of vanadium trichloride with magnesium or with magnesium–sodium mixtures. Much of the vanadium metal being produced is now made by calcium reduction of V_2O_5 in a pressure vessel, an adaptation of a process developed by McKechnie and Seybolt. Pure vanadium is a bright white metal and is soft and ductile. It has good corrosion resistance to alkalis, sulfuric and hydrochloric acids, and salt water, but the metal oxidizes readily above 660 °C. The metal has good structural strength and a low-fission neutron cross section, making it useful in nuclear applications. Vanadium is used in producing rust-resistant spring and high-speed tool steels. It is an important carbide stabilizer in making steels. About 80% of the vanadium now produced is used as ferrovanadium or as a steel additive. Vanadium foil is used as a bonding agent in cladding titanium to steel. Vanadium pentoxide is used in ceramics and as a catalyst. Vanadium and its compounds are toxic and should be handled with care. Ductile vanadium is commercially available.

Xenon — (Gr. *xenon*, stranger), Xe. Discovered by Ramsay and Travers in 1898 in the residue left after evaporating liquid air components. Xenon is a member of the so-called noble or "inert" gases. It is present in the atmosphere at a concentration of 0.0114%. Xenon is present in the Martian atmosphere to the extent of 0.08 ppm. The element is found in the gases evolved from certain mineral springs and is commercially obtained by extraction from liquid air. Before 1962, it had generally been assumed that xenon and other noble gases were unable to form compounds. However, it is now known that xenon, as well as other members of the zero valence elements, do form compounds. Among the compounds of xenon now reported are the difluoride, tetrafluoride, and hexafluoride; xenon hydrate, sodium perxenate, xenon deuterate, and $XePtF_6$ and $XeRhF_6$. Xenon trioxide, which is highly explosive, has been prepared. More than 80 xenon compounds have been made with xenon chemically bonded to fluorine and oxygen. Some xenon compounds are colored. Metallic xenon has also been produced, using several hundred kilobars of pressure. Xenon in a vacuum tube produces a beautiful blue glow when excited by an electrical discharge. The gas is used in making stroboscopic lamps, bactericidal lamps, and lamps used to excite ruby lasers. Xenon is used in the nuclear energy field in bubble chambers, probes, and other applications where its high molecular weight is of value. The perxenates are used in analytical chemistry as oxidizing agents. ^{133}Xe and ^{135}Xe are produced by neutron irradiation in air-cooled nuclear reactors. ^{133}Xe has useful applications as a radioisotope. Xenon is not toxic, but its compounds are highly toxic because of their strong oxidizing characteristics.

Ytterbium — (*Ytterby*, village in Sweden), Yb. Marignac in 1878 discovered a new component, which he called *ytterbia*, in the Earth then known as *erbia*. In 1907, Urbain separated ytterbia into two components, which he called *neoytterbia* and *lutecia*. The elements in these earths are now known as *ytterbium* and *lutetium*, respectively. These elements are identical with *aldebaranium* and *cassiopeium*, discovered independently and at about the same time by von Welsbach. Ytterbium occurs along with other rare earths in a number of rare minerals. It is commercially recovered principally from *monazite sand*, which contains about 0.03%. Ion-exchange and solvent extraction techniques developed in recent years have greatly simplified the separation of the rare earths from one another. The element was first prepared by Klemm and Bonner in 1937 by reducing ytterbium trichloride with potassium. Their metal was mixed, however, with KCl. Daane, Dennison, and Spedding prepared a much purer form in 1953 from which the chemical and physical properties of the element could be determined. Ytterbium has a bright silvery luster, is soft, malleable, and quite ductile. While the element is fairly stable, it should be kept in closed containers to protect it from air and moisture. Ytterbium is readily attacked and dissolved by dilute and concentrated mineral acids and reacts slowly with water. Ytterbium has three allotropic forms with transformation points at 3° and 795 °C. The beta form is a room-temperature, face-centered, cubic modification, while the high-temperature gamma form is a body-centered cubic form. Another body-centered cubic phase has recently been found to be stable at high pressures at room temperatures. The beta form ordinarily has metallic-type conductivity, but becomes a semiconductor when the pressure is increased above 16,000 atm. The electrical resistance increases tenfold as the pressure is increased to 39,000 atm and drops to about

80% of its standard temperature-pressure resistivity at a pressure of 40,000 atm. Ytterbium metal has possible use in improving the grain refinement, strength, and other mechanical properties of stainless steel. Few other uses have been found. Ytterbium has a low acute toxicity rating.

Yttrium — (*Ytterby*, village in Sweden near Vauxholm), Y. *Yttria*, which is an earth containing yttrium, was discovered by Gadolin in 1794. *Ytterby* is the site of a quarry which yielded many unusual minerals containing rare earths and other elements. This small town, near Stockholm, bears the honor of giving names to *erbium, terbium*, and *ytterbium* as well as *yttrium*. In 1843, Mosander showed that yttria could be resolved into the oxides (or earths) of three elements. The name yttria was reserved for the most basic one; the others were named *erbia* and *terbia*. Yttrium occurs in nearly all of the rare-earth minerals. Analysis of lunar rock samples obtained during the Apollo missions show a relatively high yttrium content. It is recovered commercially from *monazite sand*, which contains about 3%, and from *bastnasite*, which contains about 0.2%. Wohler obtained the impure element in 1828 by reduction of the anhydrous chloride with potassium. The metal is now produced commercially by reduction of the fluoride with calcium metal. It can also be prepared by other techniques. Yttrium has a silver-metallic luster and is relatively stable in air. Turnings of the metal, however, ignite in air if their temperature exceeds 400 °C, and finely divided yttrium is very unstable in air. Yttrium oxide is one of the most important compounds of yttrium and accounts for the largest use. At one time it was used in making YVO_4 europium, and Y_2O_3 europium phosphors to give the red color in color television tubes. Yttrium oxide also is used to produce yttrium iron garnets, which are very effective microwave filters. Yttrium iron, aluminum, and gadolinium garnets, with formulas such as $Y_3Fe_5O_{12}$ and $Y_3Al_5O_{12}$, have interesting magnetic properties. Yttrium iron garnet is also exceptionally efficient as both a transmitter and transducer of acoustic energy. These yttrium garnets have many uses in the electronics industry. Yttrium aluminum garnet, with a hardness of 8.5, is also finding use as a gemstone (simulated diamond). Small amounts of yttrium (0.1% to 0.2%) can be used to reduce the grain size in chromium, molybdenum, zirconium, and titanium, and to increase strength of aluminum and magnesium alloys. Alloys with other useful properties can be obtained by using yttrium as an additive. The metal can be used as a deoxidizer for vanadium and other nonferrous metals. Yttrium has been considered for use as a nodulizer for producing nodular cast iron, in which the graphite forms compact nodules instead of the usual flakes. Such iron has increased ductility. Yttrium is also finding application in laser systems and as a catalyst for ethylene polymerization. It also has potential use in ceramic and glass formulas, as the oxide has a high-melting point and imparts shock resistance and low expansion characteristics to glass.

Zinc — (Ger. *Zink*, of obscure origin), Zn. Centuries before zinc was recognized as a distinct element, zinc ores were used for making brass. Tubal-Cain, seven generations from Adam, is mentioned as being an "instructor in every artificer in brass and iron." An alloy containing 87% zinc has been found in prehistoric ruins in Transylvania. Metallic zinc was produced in the 13th century A.D. in India by reducing calamine with organic substances such as wool. The metal was rediscovered in Europe by Marggraf in 1746, who showed that it could be obtained by reducing *calamine* with charcoal. The principal ores of zinc are *sphalerite* or *blende* (sulfide), *smithsonite* (carbonate), *calamine* (silicate), and *franklinite* (zinc, manganese, iron oxide). Canada, Japan, Belgium, Germany, and the Netherlands are suppliers of zinc ores. Zinc is also mined in Alaska, Tennessee, Missouri, and elsewhere in the U.S. Zinc can be obtained by roasting its ores to form the oxide and by reduction of the oxide with coal or carbon, with subsequent distillation of the metal. Other methods of extraction are possible. Zinc is a bluish-white, lustrous metal. It is brittle at ordinary temperatures but malleable at 100 to 150 °C. It is a fair conductor of electricity and burns in air at high red heat with evolution of white clouds of the oxide. The metal is employed to form numerous alloys with other metals. Brass, nickel silver, typewriter metal, commercial bronze, spring brass, German silver, soft solder, and aluminum solder are some of the more important alloys. Large quantities of zinc are used to produce die castings, used extensively by the automotive, electrical, and hardware industries. An alloy called *Prestal*®, consisting of 78% zinc and 22% aluminum, is reported to be almost as strong as steel but as easy to mold as plastic. Zinc is also extensively used to galvanize other metals such as iron to prevent corrosion. Neither zinc nor zirconium is ferromagnetic; but $ZrZn_2$ exhibits ferromagnetism at temperatures below 35 K. Zinc oxide is a unique and very useful material to modern civilization. It is widely used in the manufacture of paints, rubber products, cosmetics, pharmaceuticals, floor coverings, plastics, printing inks, soap, storage batteries, textiles, electrical equipment, and other products. It has unusual electrical, thermal, optical, and solid-state properties that have not yet been fully investigated. Lithopone, a mixture of zinc sulfide and barium sulfate, is an important pigment. Zinc sulfide is used in making luminous dials and fluorescent lights. The chloride and chromate are also important compounds. Zinc is an essential element in the growth of human beings and animals. Tests show that zinc-deficient animals require 50% more food to gain the same weight as an animal supplied with sufficient zinc. Zinc is not considered to be toxic, but when freshly formed ZnO is inhaled a disorder known as the *oxide shakes* or *zinc chills* sometimes occurs. It is recommended that where zinc oxide is encountered good ventilation be provided.

Zirconium — (Syriac, *zargun*, color of gold), Zr. The name *zircon* may have originated from the Syriac word *zargono*, which describes the color of certain gemstones now known as *zircon, jargon, hyacinth, jacinth*, or *ligure*. This mineral, or its variations, is mentioned in biblical writings. These minerals were not known to contain this element until Klaproth, in 1789, analyzed a *jargon* from Sri Lanka and found a new earth, which Werner named zircon (*silex circonius*), and Klaproth called *Zirkonerde (zirconia)*. The impure metal was first isolated by Berzelius in 1824 by heating a mixture of potassium and potassium zirconium fluoride in a small iron tube. Pure zirconium was first prepared in 1914. Very pure zirconium was first produced in 1925 by van Arkel and de Boer by an iodide decomposition process they developed. Zirconium is found in abundance in S-type stars and has been identified in the sun and meteorites. Analyses of lunar rock samples obtained during the various Apollo missions to the moon show a surprisingly high zirconium oxide content, compared with terrestrial rocks. *Zircon*, $ZrSiO_4$, the principal ore, is found in deposits in Florida, South Carolina, Australia, South Africa, and elsewhere. *Baddeleyite*, found in Brazil, is an important zirconium mineral. It is principally pure ZrO_2 in crystalline form having a hafnium content of about 1%. Zirconium also occurs in some 30 other recognized mineral species. Zirconium is produced commercially by reduction of the chloride with magnesium (the Kroll Process), and by other methods. It is a grayish-white lustrous metal. When finely divided, the metal may ignite spontaneously in air, especially at elevated temperatures. The solid metal is much more difficult to ignite. The inherent toxicity of zirconium compounds is low. Hafnium is invariably found in zirconium ores, and the separation is difficult. Commercial-grade zirconium contains from 1% to 3% hafnium. Zirconium has a low absorption cross section for neutrons, and is therefore used for nuclear energy applications, such as for cladding fuel elements. Commercial nuclear power generation now takes more than 90% of zirconium metal production. Reactor-grade zirconium is essentially free of hafnium. *Zircaloy*® is an important alloy developed specifically for nuclear applications. Zirconium is exceptionally resistant to corrosion by many common acids and alkalis, by seawater, and by other agents. It is used extensively by the chemical industry where corrosive agents are employed. Zirconium is used as a getter in vacuum tubes, as an alloying agent in steel, in surgical appliances, photoflash bulbs, explosive primers, rayon spinnerets, lamp filaments, etc. It is used in poison ivy lotions in the form of the carbonate as it combines with *urushiol*. Alloyed with zinc, zirconium becomes magnetic at temperatures below 35 K. Zirconium oxide has a high index of refraction and is used as a gem material. The impure oxide, zirconia, is used for laboratory crucibles that will withstand heat shock, for linings of metallurgical furnaces, and by the glass and ceramic industries as a refractory material. Its use as a refractory material accounts for a large share of all zirconium consumed. Zirconium tungstate is an unusual material that shrinks, rather than expands, when heated. A few other compounds are known to possess this property, but they tend to shrink in one direction, while they stretch out in others in order to maintain an overall volume. Zirconium tungstate shrinks in all directions over a wide temperature range of from near absolute zero to +777 °C.

PHYSICAL CONSTANTS OF INORGANIC COMPOUNDS

This table contains physical constants for about 450 important inorganic compounds. The table is condensed from the longer version of 3220 compounds that appears in the Online Edition of the *CRC Handbook*.

The compounds were selected on the basis of their laboratory and industrial importance. Many, if not most, of the compounds that are solids at ambient temperature can exist in more than one crystalline modification. In the absence of other information, the data given here can be assumed to apply to the most stable or common crystalline form. In many cases, however, two or more forms are of practical importance, and separate entries will be found in the table.

Compounds are arranged primarily in alphabetical order by the most commonly used name; however, closely related compounds are often grouped together for convenience (e.g., silane, disilane, trisilane, etc.). An index to CAS Registry Numbers is available by request via e-mail (fiona.macdonald@taylorandfrancis.com).

Column definitions are as follows.

Column heading	Definition
Name	Systematic name for the substance; the valence state of a metallic element is indicated by a Roman numeral, e.g., copper in the +1 state is written as copper(I) rather than cuprous, iron in the +3 state is iron(III) rather than ferric
Mol. form.	The simplest descriptive formula is given, but this does not necessarily specify the actual structure of the compound; for example, aluminum chloride is designated as $AlCl_3$, even though a more accurate representation of the structure in the solid phase (and, under some conditions, in the gas phase) is Al_2Cl_6; a few exceptions are made, such as the use of Hg_2^{+2} for the mercury(I) ion
CAS Reg. No.	Chemical Abstracts Service Registry Number; an asterisk* following the CAS Reg. No. for a hydrate indicates that the number refers to the anhydrous compound; in most cases the generic CAS Reg. No. for the compound is given rather than the number for a specific crystalline form or mineral
Mol. wt.	Molecular weight (relative molar mass) as calculated with the 2009 IUPAC Recommended Atomic Weights; the number of decimal places corresponds to the number of places in the atomic weight of the least accurately known element (e.g., one place for lead compounds, two places for compounds of selenium, germanium, etc.); a maximum of three places is given; for compounds of radioactive elements for which IUPAC makes no recommendation, the mass number of the isotope with longest half-life is used
Phys. form	The crystal system is given, when available, for compounds that are solid at room temperature, together with color and other descriptive features; abbreviations are listed below
mp	Normal melting point in °C; the notation "tp" indicates the temperature where solid, liquid, and gas are in equilibrium at a pressure greater than one atmosphere (i.e., the normal melting point does not exist); when available, the triple-point pressure is listed
bp	Normal boiling point in °C (referred to 101.325 kPa or 760 mmHg pressure); the notation "sp" following the number indicates the temperature where the pressure of the vapor in equilibrium with the solid reaches 101.325 kPa (see Ref. 8, p. 23 for further discussion of sublimation points and triple points); a notation "sublimes" without a temperature being given indicates that there is a perceptible sublimation pressure above the solid at ambient temperatures
ρ	Density values for solids and liquids are always in units of g cm^{-3} and can be assumed to refer to temperatures near room (ambient) temperature unless otherwise stated; values for gases are the calculated ideal gas densities in g l^{-1} at 25 °C and 101.325 kPa; the unit is always specified for a gas value
Sol.	Aqueous solubility is expressed as the number of grams of the compound (excluding any water of hydration) that will dissolve in 100 grams of water (g/100 g H_2O); the temperature in °C is given as a superscript; solubility at other temperatures can be found for many compounds in the table "Aqueous Solubility of Inorganic Compounds at Various Temperatures" in Section 5
Qualitative solubility	Qualitative information on the solubility in other solvents (and in water, if quantitative data are unavailable) is given here; the abbreviations are: **i** = insoluble in (solvent); **sl** = slightly soluble in (solvent); **s** = soluble in (solvent); **vs** = very soluble in (solvent); **reac** = reacts with (solvent as specified)

Data were taken from a wide variety of reliable sources, including monographs, treatises, review articles, evaluated compilations and databases, and in many cases the primary literature. Some of the most useful references for the properties covered here are listed below.

List of Abbreviations

Ac: acetyl
ace: acetone
acid: acid solutions
alk: alkaline solutions
amorp: amorphous
anh: anhydrous
aq: aqueous
blk: black
brn: brown
bz: benzene
chl: chloroform
col: colorless
conc: concentrated
cry: crystals, crystalline
cub: cubic
cyhex: cyclohexane
dec: decomposes
dil: dilute
diox: dioxane
eth: ethyl ether
EtOH: ethanol
exp: explodes, explosive
extrap: extrapolated
flam: flammable
gl: glass, glassy
grn: green
hc: hydrocarbon solvents
hex: hexagonal, hexane
hp: heptane
HT: high temperature
hyd: hydrate
hyg: hygroscopic
i: insoluble in
liq: liquid
LT: low temperature
MeOH: methanol
monocl: monoclinic
octahed: octahedral
oran: orange
orth: orthorhombic

os: organic solvents
peth: petroleum ether
pow: powder
prec: precipitate
pur: purple
py: pyridine
reac: reacts with
refrac: refractory
rhom: rhombohedral
r.t.: room temperature
s: soluble in
silv: silvery
sl: slightly soluble in
soln: solution
sp: sublimation point
stab: stable
subl: sublimes
temp: temperature
tetr: tetragonal
thf: tetrahydrofuran
tol: toluene
tp: triple point
trans: transition, transformation
tricl: triclinic
trig: trigonal
unstab: unstable
viol: violet
visc: viscous
vs: very soluble in
wh: white
xyl: xylene
yel: yellow

References

1. Phillips, S. L., and Perry, D. L., *Handbook of Inorganic Compounds*, CRC Press, Boca Raton, FL, 1995; Second Edition by Perry, D. L., 2011.
2. Trotman-Dickenson, A. F., Executive Editor, *Comprehensive Inorganic Chemistry*, Vol. 1-5, Pergamon Press, Oxford, 1973.
3. Greenwood, N. N., and Earnshaw, A., *Chemistry of the Elements, Second Edition*, Butterworth-Heinemann, Oxford, 1997.
4. Wiberg, N., Wiberg, E., and Holleman, H. F., *Inorganic Chemistry, 34th Edition*, Academic Press, San Diego, 2001.
5. *GMELIN Handbook of Inorganic and Organometallic Chemistry*, Springer-Verlag, Heidelberg.
6. Chase, M.W., Davies, C.A., Downey, J.R., Frurip, D. J., McDonald, R.A., and Syverud, A.N.; *JANAF Thermochemical Tables, Third Edition, J. Phys. Chem. Ref. Data*, Vol. 14, Suppl. 1, 1985; Chase, M. W., *NISTJANAF Thermochemical Tables, Fourth Edition, J. Phys. Chem. Ref. Data*, Monograph No. 9, 1998.
7. *Landolt-Börnstein, Numerical Data and Functional Relationships in Science and Technology, New Series*, IV/19A, "Thermodynamic Properties of Inorganic Materials compiled by SGTE", Springer-Verlag, Heidelberg; Part 1, 1999; Part 2; 1999; Part 3, 2000; Part 4, 2001.
8. Lide, D. R., and Kehiaian, H.V., *CRC Handbook of Thermophysical and Thermochemical Data*, CRC Press, Boca Raton, FL, 1994.
9. *Kirk-Othmer Concise Encyclopedia of Chemical Technology*, Wiley-Interscience, New York, 1985.
10. *Dictionary of Inorganic Compounds*, Chapman & Hall, New York, 1992.
11. Massalski, T. B., Ed., *Binary Alloy Phase Diagrams, 2nd Edition*, ASM International, Metals Park, Ohio, 1990.
12. Dinsdale, A.T., "SGTE Data for Pure Elements", *CALPHAD*, 15, 317– 425, 1991. <https://doi.org/10.1016/0364-5916(91)90030-N>
13. Madelung, O., *Semiconductors: Group IV Elements and III-IV Compounds*, Springer-Verlag, Heidelberg, 1991.
14. Lidin, R. A., Andreeva, L. L., and Molochko, V. A., *Constants of Inorganic Substances*, Begell House, New York, 1995.
15. Gurvich, L. V., Veyts, I. V., and Alcock, C. B., *Thermodynamic Properties of Individual Substances, Fourth Edition*, Hemisphere Publishing Corp., New York, 1989.
16. *The Combined Chemical Dictionary on CDROM, Version 9:1*, Chapman & Hall/CRC, Boca Raton, FL, 2005.
17. Macdonald, F., Editor, *Chapman & Hall/CRC Combined Chemical Dictionary*, <ccd.chemnetbase.com/>.
18. Sangeeta, G., and LaGraff, J. R., *Inorganic Materials Chemistry, Second Edition*, CRC Press, Boca Raton, FL, 2005. <https://doi.org/10.1201/9781420041422>
19. Stern, K. H., *High Temperature Properties and Thermal Decomposition of Inorganic Salts with Oxyanions*, CRC Press, Boca Raton, FL, 2001. <https://doi.org/10.1201/9781420042344>
20. Donnay, J.D.H., and Ondik, H.M., *Crystal Data Determinative Tables, Third Edition, Volumes 2 and 4, Inorganic Compounds*, Joint Committee on Powder Diffraction Standards, Swarthmore, PA, 1973.
21. Robie, R., Bethke, P. M., and Beardsley, K. M., *Selected X-ray Crystallographic Data, Molar Volumes, and Densities of Minerals and Related Substances*, U.S. Geological Survey Bulletin 1248, 1967. <https://doi.org/10.3133/ofr66113>
22. Carmichael, R. S., *Practical Handbook of Physical Properties of Rocks and Minerals*, CRC Press, Boca Raton, FL, 1989.
23. Deer, W. A., Howie, R.A., and Zussman, J., An *Introduction to the Rock-Forming Minerals, Second Edition*, Longman Scientific & Technical, Harlow, Essex, 1992.
24. Linstrom, P. J., and Mallard, W. G., Editors, NIST Chemistry WebBook, NIST Standard Reference Database No. 69, June 2005, National Institute of Standards and Technology, Gaithersburg, MD 20899, <webbook.nist.gov>.
25. *Phase Diagrams for Ceramists, Volumes 1–8; ACerS-NIST Phase Equilibrium Diagrams, Volumes 9–13*, American Ceramic Society, Westerville, Ohio, 1964–2001.

Physical Constants of Inorganic Compounds

No.	Name	Mol. form.	CAS Reg. No.	Mol. wt.	Phys. form	mp/°C	bp/°C	ρ/g cm^{-3}	Sol./g/100 g H_2O	Qualitative solubility
1	Actinium	Ac	7440-34-8	227	silv metal; cub	1050	≈3200	10		
2	Aluminum	Al	7429-90-5	26.982	silv-wh metal; cub cry	660.323	2519	2.70		i H_2O; s acid, alk
3	Aluminum ammonium sulfate dodecahydrate	$NH_4Al(SO_4)_2 \cdot 12H_2O$	7784-26-1	453.329	col cry or powder	94.5	>280 dec	1.65		s H_2O; i EtOH
4	Aluminum chloride	$AlCl_3$	7446-70-0	133.341	wh hex cry or powder; hyg	192.6	180 sp	2.48	45.1^{25}	s bz, ctc, chl
5	Chlorodiethylaluminum	$AlCl(C_2H_5)_2$	96-10-6	120.557	col liq	-74		0.96		reac H_2O
6	Chlorodiisobutylaluminum	$AlCl(C_4H_9)_2$	1779-25-5	176.664	hyg col liq	-40		0.95^{20}		s eth, hx
7	Aluminum hydroxide	$Al(OH)_3$	21645-51-2	78.004	wh amorp powder			2.42		i H_2O; s alk, acid
8	Aluminum nitrate nonahydrate	$Al(NO_3)_3 \cdot 9H_2O$	7784-27-2	375.134	wh hyg monocl cry	73	135 dec	1.72	68.9^{25}	vs EtOH; i pyr
9	Aluminum oxide (α)	Al_2O_3	1344-28-1	101.961	wh powder; hex	2053	2977	3.99		i H_2O, os; sl alk
10	Aluminum oxide (γ)	Al_2O_3	1344-28-1	101.961	soft wh pow	1200 trans corundum		3.97		i H_2O; s acid; sl alk
11	Aluminum phosphate	$AlPO_4$	7784-30-7	121.953	wh rhomb plates	>1460		2.56		i H_2O; sl acid
12	Aluminum sulfate	$Al_2(SO_4)_3$	10043-01-3	342.151	wh cry	1040 dec			38.5^{25}	i EtOH
13	Americium	Am	7440-35-9	243	silv metal; hex or cub	1176	≈2011	12		s acid

No.	Name	Mol. form.	CAS Reg. No.	Mol. wt.	Phys. form	mp/°C	bp/°C	ρ/g cm^{-3}	Sol./g/100 g H_2O	Qualitative solubility
14	Ammonia	NH_3	7664-41-7	17.031	col gas	-77.65	-33.33	$0.7329^{-77.7}$ liq		vs H_2O; s EtOH, eth
15	Ammonium acetate	$NH_4C_2H_3O_2$	631-61-8	77.083	wh hyg cry	114		1.073	148^{4}	s EtOH; sl ace
16	Ammonium benzoate	$NH_4C_7H_5O_2$	1863-63-4	139.152	wh cry or powder	198		1.26		s H_2O; sl EtOH
17	Ammonium carbonate	$(NH_4)_2CO_3$	506-87-6	96.086	col cry powder	58 dec			100^{15}	
18	Ammonium chloride	NH_4Cl	12125-02-9	53.492	col cub cry	520.1 tp (dec)	338 sp	1.519	39.5^{25}	
19	Ammonium dihydrogen phosphate	$NH_4H_2PO_4$	7722-76-1	115.026	wh tetr cry	190		1.80	40.4^{25}	sl EtOH; i ace
20	Ammonium fluoride	NH_4F	12125-01-8	37.037	wh hex cry; hyg	238		1.015	83.5^{25}	sl EtOH
21	Ammonium hydrogen carbonate	NH_4HCO_3	1066-33-7	79.056	col or wh prisms	107 dec		1.586	24.8^{25}	i EtOH, bz
22	Ammonium hydrogen phosphate	$(NH_4)_2HPO_4$	7783-28-0	132.055	wh cry	155 dec		1.619	69.5^{25}	i EtOH, ace
23	Ammonium hydrogen sulfate	NH_4HSO_4	7803-63-6	115.110	wh hyg cry	147		1.78	100^{20}	i EtOH, ace, py
24	Ammonium hydrogen sulfide	NH_4HS	12124-99-1	51.112	wh tetr or orth cry	dec		1.17	128^{0}	sl ace; i bz, eth
25	Ammonium hydrogen sulfite	NH_4HSO_3	10192-30-0	99.110	col cry	dec		2.03	71.8^{0}	
26	Ammonium hydroxide	NH_4OH	1336-21-6	35.046	exists only in soln					
27	Ammonium iodide	NH_4I	12027-06-4	144.943	wh tetr cry; hyg	551 dec	405 sp	2.514	178^{25}	sl EtOH, MeOH
28	Ammonium iron(III) sulfate dodecahydrate	$NH_4Fe(SO_4)_2 \cdot 12H_2O$	7783-83-7	482.192	col to viol cry	≈37		1.71		vs H_2O; i EtOH
29	Ammonium metavanadate	NH_4VO_3	7803-55-6	116.979	wh-yel cry	200 dec		2.326	4.8^{20}	
30	Ammonium nitrate	NH_4NO_3	6484-52-2	80.043	wh hyg cry; orth	169.7	200 -260 dec	1.72	213^{25}	sl MeOH
31	Ammonium oxalate	$(NH_4)_2C_2O_4$	1113-38-8	124.096	col sol			1.5	5.20^{25}	
32	Ammonium perchlorate	NH_4ClO_4	7790-98-9	117.490	wh orth cry	dec, exp		1.95	24.5^{25}	s MeOH; sl EtOH, ace; i eth
33	Ammonium phosphate trihydrate	$(NH_4)_3PO_4 \cdot 3H_2O$	10361-65-6	203.133	wh prisms				25.0^{25}	i ace
34	Ammonium sulfamate	$NH_4NH_2SO_3$	7773-06-0	114.124	wh hyg cry	131	160 dec			vs H_2O; sl EtOH
35	Ammonium sulfate	$(NH_4)_2SO_4$	7783-20-2	132.140	wh or brn orth cry	280 dec		1.77	76.4^{25}	i EtOH, ace
36	Ammonium sulfide	$(NH_4)_2S$	12135-76-1	68.142	yel-oran cry	≈0 dec				s H_2O, EtOH, alk
37	Ammonium thiocyanate	NH_4SCN	1762-95-4	76.121	col hyg cry	≈149	dec	1.30	181^{25}	vs EtOH; s ace; i chl
38	Antimony (gray)	Sb	7440-36-0	121.760	silv metal; hex	630.628	1587	6.68		i dil acid
39	Antimony (black)	Sb	7440-36-0	121.760	blk amorp solid	0 trans gray				
40	Stibine	SbH_3	7803-52-3	124.784	col gas; flam	-88	-17	0.005100^{25} gas		sl H_2O; s EtOH
41	Antimony(III) chloride	$SbCl_3$	10025-91-9	228.119	col orth cry; hyg	73.4	220.3	3.14	987^{25}	s acid, EtOH, bz, ace
42	Argon	Ar	7440-37-1	39.948	col gas	-189.34	-185.848	$1.395^{-185.8}$ liq		sl H_2O
43	Arsenic (gray)	As	7440-38-2	74.922	gray metal; rhomb	817	616 sp	5.75		i H_2O
44	Arsenic (black)	As	7440-38-2	74.922	blk amorp solid	270 trans gray As		4.9		
45	Arsenic (yellow)	As	7440-38-2	74.922	soft yel cub cry	358 trans gray As		1.97		s CS_2
46	Arsine	AsH_3	7784-42-1	77.946	col gas	-116	-62.5	0.003186^{25} gas		sl H_2O
47	Arsenious acid	H_3AsO_3	13464-58-9	125.944	exists only in soln					
48	Arsenic(III) chloride	$AsCl_3$	7784-34-1	181.281	col liq	-16	130	2.150		reac H_2O; vs chl, ctc, eth
49	Arsenic(III) fluoride	AsF_3	7784-35-2	131.917	col liq	-5.9	57.13	2.7		reac H_2O; s EtOH, eth, bz
50	Arsenic(III) oxide (arsenolite)	As_2O_3	1327-53-3	197.841	wh cub cry	274	460	3.86	2.05^{25}	
51	Arsenic(III) oxide (claudetite)	As_2O_3	1327-53-3	197.841	wh monocl cry	314	460	3.74	2.05^{25}	s dil acid, alk; i EtOH
52	Arsenic(III) sulfide	As_2S_3	1303-33-9	246.038	yel-oran monocl cry	312	707	3.46		i H_2O; s alk
53	Arsenic(V) chloride	$AsCl_5$	22441-45-8	252.187	stab at low temp	≈-50 dec				
54	Arsenic(V) oxide	As_2O_5	1303-28-2	229.840	wh amorp powder	730		4.32	65.8^{20}	vs EtOH
55	Astatine	At	7440-68-8	210	cry	302				s HNO_3, os
56	Barium	Ba	7440-39-3	137.327	silv-yel metal; cub	727	≈1845	3.62		reac H_2O; sl EtOH
57	Barium acetate	$Ba(C_2H_3O_2)_2$	543-80-6	255.416	wh powder			2.47	79.2^{25}	
58	Barium carbonate	$BaCO_3$	513-77-9	197.336	wh orth cry	1555		4.308	0.0014^{20}	s acid
59	Barium chloride	$BaCl_2$	10361-37-2	208.233	wh orth cry; hyg	961	1560	3.9	37.0^{25}	

No.	Name	Mol. form.	CAS Reg. No.	Mol. wt.	Phys. form	mp/°C	bp/°C	ρ/g cm^{-3}	Sol./g/100 g H_2O	Qualitative solubility
60	Barium hydroxide octahydrate	$Ba(OH)_2·8H_2O$	12230-71-6	315.464	wh monocl cry	78 dec		2.18	4.91^{25}	
61	Barium nitrate	$Ba(NO_3)_2$	10022-31-8	261.336	wh cub cry	590		3.24	10.3^{25}	sl EtOH, ace
62	Berkelium (α form)	Bk	7440-40-6	247	hex cry	930 trans β		14.78		
63	Berkelium (β form)	Bk	7440-40-6	247	cub cry	986		13.25		
64	Beryllium	Be	7440-41-7	9.012	hex cry	1287	2468	1.85		s acid, alk
65	Beryllium oxide	BeO	1304-56-9	25.011	wh hex cry	2578		3.01		i H_2O; sl acid, alk
66	Bismuth	Bi	7440-69-9	208.980	gray-wh soft metal	271.402	1564	9.79		s acid
67	Bismuth nitrate pentahydrate	$Bi(NO_3)_3·5H_2O$	10035-06-0	485.071	col tricl cry; hyg	≈75 dec		2.83		reac H_2O; s ace; i EtOH
68	Bismuth oxide	Bi_2O_3	1304-76-3	465.959	yel monocl cry or powder	825	1890	8.9		i H_2O; s acid
69	Boron	B	7440-42-8	10.81	blk rhomb cry	2077	4000	2.34		i H_2O
70	Diborane	B_2H_6	19287-45-7	27.670	col gas; flam	-164.85	-92.49	0.001131^{25} gas		reac H_2O
71	Tetraborane(10)	B_4H_{10}	18283-93-7	53.323	unstab col gas	-120	18	0.002180^{25} gas		reac H_2O
72	Boric acid	H_3BO_3	10043-35-3	61.833	col tricl cry	170.9		1.5	5.80^{25}	sl EtOH
73	Boron carbide	B_4C	12069-32-8	55.255	hard blk cry	2350	>3500	2.50		i H_2O, acid
74	Boron tribromide	BBr_3	10294-33-4	250.523	col liq; hyg	-46	91.3	2.6		reac H_2O, EtOH
75	Boron trichloride	BCl_3	10294-34-5	117.170	col liq or gas	-107.3	12.5	0.004789^{25} gas		reac H_2O, EtOH
76	Bromine	Br_2	7726-95-6	159.808	red liq	-7.2	58.8	3.1028		sl H_2O
77	Bromine monoxide	Br_2O	21308-80-5	175.807	unstab brn solid	-17.5 dec				
78	Bromine trifluoride	BrF_3	7787-71-5	136.899	col hyg liq	8.77	125.8	2.803		reac H_2O
79	Cadmium	Cd	7440-43-9	112.411	silv-wh metal	321.069	767	8.69		i H_2O; reac acid
80	Cadmium acetate	$Cd(C_2H_3O_2)_2$	543-90-8	230.500	col cry	255		2.34		s H_2O, EtOH
81	Cadmium chloride	$CdCl_2$	10108-64-2	183.317	rhom cry; hyg	568	964	4.08	120^{25}	s ace; sl EtOH; i eth
82	Cadmium hydroxide	$Cd(OH)_2$	21041-95-2	146.426	wh trig or hex cry	130 dec		4.79	0.00015^{20}	s dil acid
83	Cadmium nitrate tetrahydrate	$Cd(NO_3)_2·4H_2O$	10022-68-1	308.482	col orth cry; hyg	59.5		2.45	156^{25}	s EtOH, ace
84	Cadmium sulfate	$CdSO_4$	10124-36-4	208.474	col orth cry	1000		4.69	76.7^{25}	i EtOH
85	Calcium	Ca	7440-70-2	40.078	silv-wh metal	842	1484	1.54		reac H_2O; i bz
86	Calcium carbide	CaC_2	75-20-7	64.099	gray-blk orth cry	2300		2.22		reac H_2O
87	Calcium carbonate (aragonite)	$CaCO_3$	471-34-1	100.087	wh orth cry or powder	450 trans calcite		2.930	0.00066^{20}	s dil acid
88	Calcium carbonate (calcite)	$CaCO_3$	471-34-1	100.087	wh hex cry or powder	800		2.710	0.00066^{20}	s dil acid
89	Calcium carbonate (vaterite)	$CaCO_3$	471-34-1	100.087	col hex cry			2.653	0.0011^{25}	s dil acid
90	Calcium chloride	$CaCl_2$	10043-52-4	110.984	wh cub cry or powder; hyg	775	1935	2.15	81.3^{25}	vs EtOH
91	Calcium fluoride	CaF_2	7789-75-5	78.075	wh cub cry or powder	1418	2500	3.18	0.0016^{25}	sl acid
92	Calcium hydroxide	$Ca(OH)_2$	1305-62-0	74.093	soft hex cry			≈2.2	0.160^{20}	s acid
93	Calcium hypochlorite	$Ca(OCl)_2$	7778-54-3	142.983	powder	100		2.350		
94	Calcium nitrate tetrahydrate	$Ca(NO_3)_2·4H_2O$	13477-34-4	236.149	wh cry	≈40 dec		1.82	144^{25}	s EtOH, ace
95	Calcium oxide	CaO	1305-78-8	56.077	gray-wh cub cry	2613		3.34		reac H_2O; s acid
96	Calcium sulfate	$CaSO_4$	7778-18-9	136.141	orth cry	1460		2.96	0.205^{25}	
97	Calcium sulfate dihydrate	$CaSO_4·2H_2O$	10101-41-4	172.171	monocl cry or powder	150 dec		2.32	0.205^{20}	i os
98	Californium	Cf	7440-71-3	251	hex or cub metal	900		15.1		
99	Carbon (diamond)	C	7782-40-3	12.011	col cub cry	4440 (12.4 GPa)		3.513		i H_2O
100	Carbon (graphite)	C	7782-42-5	12.011	soft blk hex cry	4489 tp (10.3 MPa)	3825 sp	2.2		i H_2O
101	Carbon black	C	1333-86-4	12.011	fine blk pow					i H_2O
102	Carbon [fullerene-C_{60}]	C_{60}	99685-96-8	720.642	yel needles or plates	>280				s os
103	Carbon [fullerene-C_{70}]	C_{70}	115383-22-7	840.749	red-brn solid	>280				s bz, tol
104	Fullerene fluoride	$C_{60}F_{60}$	134929-59-2	1860.550	col plates	287				vs ace; s thf; i chl
105	Carbon monoxide	CO	630-08-0	28.010	col gas	-205.1	-191.51	$0.8495^{-205.1}$ liq		sl H_2O; s chl, EtOH
106	Carbon dioxide	CO_2	124-38-9	44.010	col gas	-56.561 tp	-78.464 sp	1.56^{-79} solid		s H_2O
107	Carbon disulfide	CS_2	75-15-0	76.141	col or yel liq	-111.7	46.2	1.2632^{20}		i H_2O; vs EtOH, bz, os
108	Carbon oxysulfide	OCS	463-58-1	60.075	col gas	-138.8	-50.2	0.002456^{25} gas		s H_2O, EtOH

No.	Name	Mol. form.	CAS Reg. No.	Mol. wt.	Phys. form	mp/°C	bp/°C	ρ/g cm^{-3}	Sol./g/100 g H_2O	Qualitative solubility
109	Carbonyl chloride	$COCl_2$	75-44-5	98.916	col gas	-127.77	7.5	0.004043^{25} gas		sl H_2O; s bz, tol
110	Carbonyl fluoride	COF_2	353-50-4	66.007	col gas	-114	-84.5	0.002698^{25} gas		reac H_2O
111	Cyanogen	C_2N_2	460-19-5	52.034	col gas	-27.83	-21.1	0.787^{-21} liq		sl H_2O, eth; s EtOH
112	Cyanogen chloride	ClCN	506-77-4	61.471	col vol liq or gas	-6.55	13	1.186^{20}		s H_2O, EtOH, eth
113	Cerium	Ce	7440-45-1	140.116	silv metal; cub or hex	799	3443	6.770		s dil acid
114	Cerium(III) ammonium nitrate tetrahydrate	$(NH_4)_2Ce(NO_3)_5{\cdot}4H_2O$	13083-04-0	558.279	col monocl cry	74				vs H_2O
115	Cesium	Cs	7440-46-2	132.905	silv-wh metal	28.5	671	1.873		reac H_2O
116	Cesium hydroxide	CsOH	21351-79-1	149.912	wh-yel hyg cry	342.3		3.68	300^{30}	s EtOH
117	Chlorine	Cl_2	7782-50-5	70.90	grn-yel gas	-101.5	-34.04	$1.565^{-34.0}$ liq		sl H_2O
118	Hypochlorous acid	HOCl	7790-92-3	52.460	grn-yel; stable only in aq soln					s H_2O
119	Perchloric acid	$HClO_4$	7601-90-3	100.459	col hyg liq	-112	≈90 dec	1.77		s H_2O
120	Chlorine dioxide	ClO_2	10049-04-4	67.452	oran-grn gas	-59	11	0.002757^{25} gas		sl H_2O
121	Chlorine trifluoride	ClF_3	7790-91-2	92.448	gas	-76.34	11.75	0.003779^{25} gas		reac H_2O
122	Perchloryl fluoride	ClO_3F	7616-94-6	102.449	col gas	-147	-46.75	0.004188^{25} gas		
123	Chromium	Cr	7440-47-3	51.996	blue-wh metal; cub	1907	2671	7.15		reac dil acid
124	Chromium(III) potassium sulfate dodecahydrate	$CrK(SO_4)_2{\cdot}12H_2O$	7788-99-0	499.403	viol-blk cub cry	89 dec		1.83		s H_2O; i EtOH
125	Chromium(VI) oxide	CrO_3	1333-82-0	99.994	red orth cry	197	≈250 dec	2.7	169^{25}	
126	Cobalt	Co	7440-48-4	58.933	gray metal; hex or cub	1495	2927	8.86		s dil acid
127	Cobalt(II) acetate tetrahydrate	$Co(C_2H_3O_2)_2{\cdot}4H_2O$	6147-53-1	249.082	red monocl cry			1.705		s H_2O, EtOH, dil acid
128	Cobalt(II) chloride hexahydrate	$CoCl_2{\cdot}6H_2O$	7791-13-1	237.930	pink-red monocl cry	87 dec		1.924	56.2^{25}	s EtOH, ace, eth
129	Cobalt(II) nitrate hexahydrate	$Co(NO_3)_2{\cdot}6H_2O$	10026-22-9	291.034	red monocl cry; hyg	≈55		1.88	103^{25}	s EtOH
130	Copper	Cu	7440-50-8	63.546	red metal; cub	1084.62	2560	8.96		sl dil acid
131	Copper(I) chloride	CuCl	7758-89-6	98.999	wh cub cry	423	1490	4.14	0.0047^{20}	i EtOH, ace
132	Copper(II) acetate	$Cu(C_2H_3O_2)_2$	142-71-2	181.635	blue-grn hyg powder					
133	Copper(II) chloride	$CuCl_2$	7447-39-4	134.452	yel-brn monocl cry; hyg	598	993	3.4	75.7^{25}	s EtOH, ace
134	Copper(II) chloride dihydrate	$CuCl_2{\cdot}2H_2O$	10125-13-0	170.483	grn-blue orth cry; hyg	100 dec		2.51	75.7^{20}	vs EtOH, MeOH; s ace; i eth
135	Copper(II) nitrate	$Cu(NO_3)_2$	3251-23-8	187.555	blue-grn orth cry; hyg	255	subl		145^{25}	s diox; reac eth
136	Copper(II) oxide	CuO	1317-38-0	79.545	blk powder or monocl cry	1227		6.31		i H_2O, EtOH; s dil acid
137	Copper(II) sulfate	$CuSO_4$	7758-98-7	159.609	wh-grn amorp powder or rhomb cry	560 dec		3.60	22.0^{25}	i EtOH
138	Copper(II) sulfate pentahydrate	$CuSO_4{\cdot}5H_2O$	7758-99-8	249.685	blue tricl cry	110 dec		2.286	22.0^{25}	s MeOH; sl EtOH
139	Curium	Cm	7440-51-9	247	silv metal; hex or cub	1345		13.51		
140	Dysprosium	Dy	7429-91-6	162.500	silv metal; hex	1412	2567	8.55		s dil acid
141	Einsteinium	Es	7429-92-7	252	metal; cub	860				
142	Erbium	Er	7440-52-0	167.259	silv metal; hex	1529	2868	9.07		i H_2O; s acid
143	Europium	Eu	7440-53-1	151.964	soft silv metal; cub	822	1529	5.24		reac H_2O
144	Fermium	Fm	7440-72-4	257	metal	1527				
145	Fluorine	F_2	7782-41-4	37.997	pale yel gas	-219.67	-188.11	$1.5127^{-188.1}$ liq		reac H_2O
146	Francium	Fr	7440-73-5	223	short-lived alkali metal	≈21				
147	Gadolinium	Gd	7440-54-2	157.25	silv metal; hex	1313	3273	7.90		s dil acid
148	Gallium	Ga	7440-55-3	69.723	silv liq or gray orth cry	29.7646	2229	5.91		reac alk
149	Gallium arsenide	GaAs	1303-00-0	144.645	gray cub cry	1238		5.3176		
150	Gallium(III) chloride	$GaCl_3$	13450-90-3	176.082	col needles or gl solid	77.9	201	2.47		

No.	Name	Mol. form.	CAS Reg. No.	Mol. wt.	Phys. form	mp/°C	bp/°C	ρ/g cm^{-3}	Sol./g/100 g H_2O	Qualitative solubility
151	Germanium	Ge	7440-56-4	72.63	gray-wh cub cry	938.25	2833	5.3234		i H_2O, dil acid, alk
152	Germane	GeH_4	7782-65-2	76.67	col gas; flam	-165	-88.1	0.003134^{25} gas		i H_2O
153	Germanium(IV) chloride	$GeCl_4$	10038-98-9	214.45	col liq	-51.50	86.55	1.88		reac H_2O; s bz, eth, EtOH, ctc
154	Germanium(IV) fluoride	GeF_4	7783-58-6	148.63	col gas	-15 tp	-36.5 sp	0.006075^{25} gas		reac H_2O
155	Germanium(IV) oxide	GeO_2	1310-53-8	104.64	wh hex cry	1116		4.25		i H_2O
156	Gold	Au	7440-57-5	196.967	soft yel metal	1064.18	2836	19.3		s aqua regia
157	Hafnium	Hf	7440-58-6	178.49	gray metal; hex	2233	4600	13.3		s HF
158	Helium	He	7440-59-7	4.003	col gas		-268.928	$0.1250^{-268.9}$ liq		sl H_2O; i EtOH
159	Holmium	Ho	7440-60-0	164.930	silv metal; hex	1472	2700	8.80		s dil acid
160	Hydrazine	N_2H_4	302-01-2	32.045	col oily liq	1.54	113.55	1.0036		vs H_2O, EtOH, MeOH
161	Hydrazine sulfate	$N_2H_4 \cdot H_2SO_4$	10034-93-2	130.124	col orth cry	254		1.378		sl H_2O; i EtOH
162	Hydrazoic acid	HN_3	7782-79-8	43.028	col liq; exp	-80	35.7			s H_2O
163	Hydroxylamine	NH_2OH	7803-49-8	33.030	wh orth flakes or needles	33.1	58	1.21		vs H_2O, MeOH
164	Hydrogen	H_2	1333-74-0	2.016	col gas; flam	-259.16	-252.879	$0.07083^{-252.9}$ liq		sl H_2O
165	Hydrogen-d_2	D_2	7782-39-0	4.028	col gas	-254.42	-249.48	0.000165^{25} gas		
166	Hydrogen bromide	HBr	10035-10-6	80.912	col gas	-86.80	-66.38	2.603^{-84} liq		vs H_2O; s EtOH
167	Hydrogen chloride	HCl	7647-01-0	36.461	col gas	-114.17	-85	$1.187^{-114.1}$ liq		vs H_2O
168	Hydrogen cyanide	HCN	74-90-8	27.026	col liq or gas	-13.28	25.63	0.6876^{20}		vs H_2O, EtOH; sl eth
169	Hydrogen fluoride	HF	7664-39-3	20.006	col gas	-83.36	20	1.002^{0} liq		vs H_2O, EtOH; sl eth
170	Hydrogen iodide	HI	10034-85-2	127.912	col or yel gas	-50.76	-35.55	2.85^{-47} liq		vs H_2O; s os
171	Hydrogen peroxide	H_2O_2	7722-84-1	34.015	col liq	-0.43	150.2	1.44		vs H_2O
172	Hydrogen selenide	H_2Se	7783-07-5	80.98	col gas; flam	-65.73	-41.25	0.003310^{25} gas		s H_2O
173	Hydrogen sulfide	H_2S	7783-06-4	34.081	col gas; flam	-85.5	-59.55	$0.9923^{-85.5}$ liq		s H_2O
174	Indium	In	7440-74-6	114.818	soft wh metal	156.5985	2027	7.31		s acid
175	Indium antimonide	InSb	1312-41-0	236.578	blk cub cry	524		5.7747		
176	Indium arsenide	InAs	1303-11-3	189.740	gray cub cry	942		5.67		i acid
177	Iodine	I_2	7553-56-2	253.809	blue-blk plates	113.7	184.4	4.933	0.03^{20}	s bz, EtOH, eth, ctc, chl
178	Iodic acid	HIO_3	7782-68-5	175.910	col orth cry	110 dec		4.63	308^{25}	i EtOH, eth
179	Iodine chloride	ICl	7790-99-0	162.357	red cry or oily liq	27.38	97.0 dec	3.24		reac H_2O; s EtOH
180	Iridium	Ir	7439-88-5	192.217	silv-wh metal; cub	2446	4428	22.562^{20}		s aqua regia
181	Iron	Fe	7439-89-6	55.845	silv-wh or gray met	1538	2861	7.87		s dil acid
182	Ferrocene	$Fe(C_5H_5)_2$	102-54-5	186.031	oran needles	175	249			i H_2O; s EtOH, eth, bz, dil HNO_3
183	Iron pentacarbonyl	$Fe(CO)_5$	13463-40-6	195.896	yel oily liq; flam	-20	103	1.5^{20}		i H_2O; s eth, bz, ace
184	Iron(II) chloride	$FeCl_2$	7758-94-3	126.751	wh hex cry; hyg	677	1023	3.16	65.0^{25}	vs EtOH, ace; sl bz
185	Iron(II) oxide	FeO	1345-25-1	71.844	blk cub cry	1377		6.0		i H_2O, alk; s acid
186	Iron(II) sulfate	$FeSO_4$	7720-78-7	151.908	wh orth cry; hyg			3.65	29.5^{25}	
187	Iron(II) sulfate heptahydrate	$FeSO_4 \cdot 7H_2O$	7782-63-0	278.014	blue-grn monocl cry	≈60 dec		1.895	29.5^{25}	i EtOH
188	Iron(III) chloride	$FeCl_3$	7705-08-0	162.204	grn hex cry; hyg	307.6	≈316	2.90	91.2^{25}	s EtOH, eth, ace
189	Iron(III) chloride hexahydrate	$FeCl_3 \cdot 6H_2O$	10025-77-1	270.295	yel-oran monocl cry; hyg	37 dec		1.82	91.2^{25}	s EtOH, eth, ace
190	Iron(III) nitrate nonahydrate	$Fe(NO_3)_3 \cdot 9H_2O$	7782-61-8	403.997	viol-gray hyg cry	47 dec		1.68	82.5^{20}	vs EtOH, ace
191	Iron(III) oxide	Fe_2O_3	1309-37-1	159.688	red-brn hex cry	1539		5.25		i H_2O; s acid
192	Iron(III) sulfate	$Fe_2(SO_4)_3$	10028-22-5	399.878	gray-wh rhomb cry; hyg			3.10	440^{20}	sl EtOH; i ace
193	Krypton	Kr	7439-90-9	83.798	col gas	-157.37	-153.415	$2.417^{-153.4}$ liq		sl H_2O
194	Lanthanum	La	7439-91-0	138.905	silv metal; hex	920	3464	6.15		s dil acid
195	Lawrencium	Lr	22537-19-5	262	metal	1627				
196	Lead	Pb	7439-92-1	207.2	soft silv-gray metal; cub	327.462	1749	11.3		s conc acid

No.	Name	Mol. form.	CAS Reg. No.	Mol. wt.	Phys. form	mp/°C	bp/°C	ρ/g cm⁻³	Sol./g/100 g H_2O	Qualitative solubility
197	Lead(II) acetate	$Pb(C_2H_3O_2)_2$	301-04-2	325.3	wh cry	280	dec	3.25	44.3^{20}	
198	Lead(II) acetate trihydrate	$Pb(C_2H_3O_2)_2 \cdot 3H_2O$	6080-56-4	379.3	col cry	75 dec		2.55		vs H_2O; sl EtOH
199	Lead(II) carbonate	$PbCO_3$	598-63-0	267.2	col orth cry	≈315 dec		6.582		i H_2O
200	Lead(II) chromate	$PbCrO_4$	7758-97-6	323.2	yel-oran monocl cry	844		6.12	0.000017^{20}	s alk, dil acid
201	Lead(II) nitrate	$Pb(NO_3)_2$	10099-74-8	331.2	col cub cry	470		4.53	59.7^{25}	sl EtOH
202	Lead(II) oxide (litharge)	PbO	1317-36-8	223.2	red tetr cry	489 trans massicot		9.35		i H_2O, EtOH; s dil HNO_3
203	Lead(II) perchlorate trihydrate	$Pb(ClO_4)_2 \cdot 3H_2O$	13637-76-8	460.1	wh cry	100 dec		2.6	441^{25}	s EtOH
204	Lead(IV) oxide	PbO_2	1309-60-0	239.2	red tetr cry or brn powder	290 dec		9.64		
205	Lithium	Li	7439-93-2	6.94	soft silv-wh metal	180.50	1342	0.534		reac H_2O
206	Lithium acetate	$LiC_2H_3O_2$	546-89-4	65.985	cry	286			45.0^{25}	vs EtOH
207	Lithium carbonate	Li_2CO_3	554-13-2	73.891	wh monocl cry	732	1300 dec	2.11	1.30^{25}	s acid; i EtOH
208	Lithium chloride	LiCl	7447-41-8	42.394	wh cub cry or powder; hyg	610	1383	2.07	84.5^{25}	s EtOH, ace, py
209	Lithium dihydrogen phosphate	LiH_2PO_4	13453-80-0	103.928	col hyg cry	>100		2.461	126^{0}	
210	Lithium hydroxide	LiOH	1310-65-2	23.948	col tetr cry	473	1626	1.45	12.5^{25}	sl EtOH
211	Lithium iodide	LiI	10377-51-2	133.845	wh cub cry; hyg	469	1171	4.06	165^{25}	
212	Lithium metaborate	$LiBO_2$	13453-69-5	49.751	wh monocl cry; hyg	844		2.18	2.6^{20}	sl H_2O; s EtOH
213	Lithium perchlorate	$LiClO_4$	7791-03-9	106.392	wh orth cry or powder	236	430 dec	2.428	58.7^{25}	s EtOH, ace, eth
214	Lithium sulfate	Li_2SO_4	10377-48-7	109.945	wh monocl cry; hyg	860		2.21	34.2^{25}	
215	Lithium tetraborate	$Li_2B_4O_7$	12007-60-2	169.122	wh tetr cry	917			2.9^{20}	sl H_2O
216	Lutetium	Lu	7439-94-3	174.967	silv metal; hex	1663	3402	9.84		s dil acid
217	Magnesium	Mg	7439-95-4	24.305	silv-wh metal	650	1090	1.74		s dil acid
218	Magnesium acetate	$Mg(C_2H_3O_2)_2$	142-72-3	142.394	wh orth/monocl cry	323 dec		1.50	65.6^{25}	
219	Magnesium chloride	$MgCl_2$	7786-30-3	95.211	wh hex leaflets; hyg	714	1412	2.325	56.0^{25}	
220	Magnesium nitrate	$Mg(NO_3)_2$	10377-60-3	148.314	wh cub cry			≈2.3	71.2^{25}	
221	Magnesium oxide	MgO	1309-48-4	40.304	wh cub cry	2825	3600	3.6		sl H_2O; i EtOH
222	Magnesium perchlorate	$Mg(ClO_4)_2$	10034-81-8	223.206	wh hyg powder	250 dec		2.2	100^{25}	
223	Magnesium sulfate	$MgSO_4$	7487-88-9	120.368	col orth cry	1137		2.66	35.7^{25}	
224	Manganese	Mn	7439-96-5	54.938	hard gray metal	1246	2061	7.3		s dil acids
225	Manganese(II) chloride	$MnCl_2$	7773-01-5	125.844	pink trig cry; hyg	650	1190	2.977	77.3^{25}	s py, EtOH; i eth
226	Manganese(II) sulfate	$MnSO_4$	7785-87-7	151.001	wh orth cry	700	850 dec	3.25	63.7^{25}	
227	Mendelevium	Md	7440-11-1	258	metal	827				
228	Mercury	Hg	7439-97-6	200.59	heavy silv liq	-38.8290	356.619	13.53359^{25} liq		i H_2O
229	Dimethyl mercury	$Hg(CH_3)_2$	593-74-8	230.66	liq		93	3.17^{25}		i H_2O; vs EtOH, eth
230	Mercury(I) chloride	Hg_2Cl_2	10112-91-1	472.09	wh tetr cry	525 tp	383 sp	7.16	0.0004^{25}	i EtOH, eth
231	Mercury(I) nitrate	$Hg_2(NO_3)_2$	10415-75-5	525.19	cry					sl H_2O
232	Mercury(II) acetate	$Hg(C_2H_3O_2)_2$	1600-27-7	318.68	wh-yel cry or powder	179 dec		3.28	25^{10}	s EtOH
233	Mercury(II) bromide	$HgBr_2$	7789-47-1	360.40	wh rhomb cry or powder	241	318	6.05	0.61^{25}	sl chl; s EtOH, MeOH
234	Mercury(II) chloride	$HgCl_2$	7487-94-7	271.50	wh orth cry	277	304	5.6	7.31^{25}	sl bz; s EtOH, MeOH, ace, eth
235	Mercury(II) iodide (red)	HgI_2	7774-29-0	454.40	red pow	127 trans yel			0.006^{25}	sl EtOH, ace, eth, chl
236	Mercury(II) nitrate	$Hg(NO_3)_2$	10045-94-0	324.60	col hyg cry	79		4.3		s H_2O; i EtOH
237	Mercury(II) oxide	HgO	21908-53-2	216.59	red or yel orth cry	500 dec		11.14		i H_2O, EtOH; s dil acid
238	Mercury(II) sulfate	$HgSO_4$	7783-35-9	296.65	wh monocl cry			6.47		reac H_2O
239	Molybdenum	Mo	7439-98-7	95.96	gray-blk metal; cub	2622	4639	10.2		i H_2O, dil acid, alk
240	Molybdenum(III) oxide	Mo_2O_3	1313-29-7	239.88	gray-blk powder					i H_2O; sl acid
241	Molybdenum(VI) oxide	MoO_3	1313-27-5	143.94	wh-yel rhomb cry	802	1155	4.70	0.14^{20}	sl H_2O; s alk, acid
242	Neodymium	Nd	7440-00-8	144.242	silv metal; hex	1016	3074	7.01		
243	Neon	Ne	7440-01-9	20.180	col gas	-248.59	-246.046	$1.207^{-246.0}$ liq		sl H_2O
244	Neptunium	Np	7439-99-8	237	silv metal	644	≈3902	20.2		s HCl
245	Nickel	Ni	7440-02-0	58.693	wh metal; cub	1455	2913	8.90		i H_2O; sl dil acid
246	Nickel carbonyl [$Ni(CO)_4$]	$Ni(CO)_4$	13463-39-3	170.734	col liq	-19.3	43 (exp 60)	1.31^{25}		i H_2O; s EtOH, bz, ace, ctc

No.	Name	Mol. form.	CAS Reg. No.	Mol. wt.	Phys. form	mp/°C	bp/°C	ρ/g cm^{-3}	Sol./g/100 g H_2O	Qualitative solubility
247	Nickel(II) sulfate	$NiSO_4$	7786-81-4	154.756	grn-yel orth cry	840 dec		4.01	40.4^{25}	
248	Niobium	Nb	7440-03-1	92.906	gray metal; cub	2477	4741	8.57		i acid
249	Nitrogen	N_2	7727-37-9	28.014	col gas	-210.0	-195.795	$0.8061^{-195.8}$ liq		sl H_2O; i EtOH
250	Nitric acid	HNO_3	7697-37-2	63.013	col liq; hyg	-41.6	83	1.5129^{20}		vs H_2O
251	Nitrous acid	HNO_2	7782-77-6	47.014	stab only in soln					
252	Nitrous oxide	N_2O	10024-97-2	44.012	col gas	-90.8	-88.48	0.001799^{25} gas		sl H_2O; s EtOH, eth
253	Nitric oxide	NO	10102-43-9	30.006	col gas	-163.6	-151.74	0.001226^{25} gas		sl H_2O
254	Nitrogen dioxide	NO_2	10102-44-0	46.006	brn gas; equil with N_2O_4		see N_2O_4	0.001880^{25} gas		reac H_2O
255	Nitrogen trioxide	N_2O_3	10544-73-7	76.011	blue solid or liq (low temp)	-101.1	≈3 dec	1.4^{2} liq		reac H_2O
256	Nitrogen tetroxide	N_2O_4	10544-72-6	92.011	col liq; equil with NO_2	-9.3	21.15	1.45^{20}		reac H_2O
257	Nitrogen pentoxide	N_2O_5	10102-03-1	108.010	col hex cry		33 sp	2.0		s chl; sl ctc
258	Nitrogen trichloride	NCl_3	10025-85-1	120.366	yel oily liq; exp	-40	71	1.653		i H_2O; s CS_2, bz, ctc
259	Nitrogen trifluoride	NF_3	7783-54-2	71.002	col gas	-206.79	-128.75	0.002902^{25} gas		i H_2O
260	Tetrafluorohydrazine	N_2F_4	10036-47-2	104.007	col gas	-164.5	-74	0.004251^{25} gas		
261	Nitrosyl chloride	NOCl	2696-92-6	65.459	yel gas	-59.6	-5.5	0.002676^{25} gas		reac H_2O
262	Nitrosyl fluoride	NOF	7789-25-5	49.004	col gas	-132.5	-59.9	0.002003^{25} gas		
263	Nobelium	No	10028-14-5	259	metal	827				
264	Osmium	Os	7440-04-2	190.23	blue-wh metal; hex	3033	5008	22.587^{20}		s aqua regia
265	Oxygen	O_2	7782-44-7	31.998	col gas	-218.79	-182.962	$1.141^{-183.0}$ liq		sl H_2O, EtOH, os
266	Ozone	O_3	10028-15-6	47.998	blue gas	-193	-111.35	0.001962^{25} gas		sl H_2O
267	Palladium	Pd	7440-05-3	106.42	silv-wh metal; cub	1554.8	2963	12.0		s aqua regia
268	Phosphorus (white)	P	7723-14-0	30.974	col waxlike cub cry	44.15	280.5	1.823		i H_2O; sl bz, EtOH, chl; s CS_2
269	Phosphorus (red)	P	7723-14-0	30.974	red-viol amorp powder	579.2	431 sp	2.16		i H_2O, os
270	Phosphorus (black)	P	7723-14-0	30.974	blk orth cry or amorp solid	610		2.69		i os
271	Phosphine	PH_3	7803-51-2	33.998	col gas; flam	-133.8	-87.75	0.001390^{25} gas		i H_2O; sl EtOH, eth
272	Phosphoric acid	H_3PO_4	7664-38-2	97.995	col visc liq	42.4	407		548^{20}	s EtOH
273	Phosphonic acid	H_3PO_3	13598-36-2	81.996	wh hyg cry	74.4	200	1.65	309^{0}	vs EtOH
274	Phosphinic acid	HPH_2O_2	6303-21-5	65.997	hyg cry or col oily liq	26.5	130	1.49		vs H_2O, EtOH, eth
275	Phosphorus(III) fluoride	PF_3	7783-55-3	87.969	col gas	-151.5	-101.8	0.003596^{25} gas		reac H_2O
276	Phosphorus(III) oxide	P_2O_3	1314-24-5	109.946	col monocl cry or liq	23.8	173	2.13		reac H_2O
277	Phosphorus(V) chloride	PCl_5	10026-13-8	208.239	wh-yel tetr cry; hyg	167 tp	160 sp	2.1		reac H_2O; s CS_2, ctc
278	Phosphorus(V) fluoride	PF_5	7647-19-0	125.966	col gas	-93.8	-84.6	0.005149^{25} gas		reac H_2O
279	Phosphorus(V) oxide	P_2O_5	1314-56-3	141.945	wh orth cry; hyg	562	605	2.30		reac H_2O, EtOH
280	Phosphorus(V) sulfide	P_2S_5	1314-80-3	222.273	grn-yel hyg cry	285	515	2.03		reac H_2O; s CS_2
281	Phosphoryl chloride	$POCl_3$	10025-87-3	153.332	col liq	1.18	105.5	1.645		reac H_2O, EtOH
282	Phosphorothioc trichloride	$PSCl_3$	3982-91-0	169.398	fuming liq	-36.2	125	1.635		reac H_2O; s bz, ctc, chl, CS_2
283	Platinum	Pt	7440-06-4	195.084	silv-gray metal; cub	1768.2	3825	21.5		i acid; s aqua regia
284	Hexachloroplatinic acid	H_2PtCl_6	16941-12-1	409.818	hyg yel-brn cry	60				s H_2O, EtOH
285	Plutonium	Pu	7440-07-5	244	silv-wh metal; monocl	640	3228	19.7		
286	Polonium	Po	7440-08-6	209	silv metal; cub	254	962	9.20		
287	Potassium	K	7440-09-7	39.098	soft silv-wh metal; cub	63.5	759	0.89		reac H_2O
288	Potassium acetate	$KC_2H_3O_2$	127-08-2	98.142	wh hyg cry	309		1.57	269^{25}	s EtOH; i eth
289	Potassium bromate	$KBrO_3$	7758-01-2	167.000	wh hex cry	434 dec		3.27	8.17^{25}	i EtOH
290	Potassium bromide	KBr	7758-02-3	119.002	col cub cry; hyg	734	1435	2.74	67.8^{25}	sl EtOH

No.	Name	Mol. form.	CAS Reg. No.	Mol. wt.	Phys. form	mp/°C	bp/°C	ρ/g cm^{-3}	Sol./g/100 g H_2O	Qualitative solubility
291	Potassium carbonate	K_2CO_3	584-08-7	138.206	wh monocl cry; hyg	899	dec	2.29	111^{25}	i EtOH
292	Potassium chlorate	$KClO_3$	3811-04-9	122.549	wh monocl cry	357	dec	2.34	8.61^{25}	
293	Potassium chloride	KCl	7447-40-7	74.551	wh cub cry	771		1.988	35.5^{25}	i eth, ace
294	Potassium chromate	K_2CrO_4	7789-00-6	194.191	yel orth cry	974		2.73	65.0^{25}	
295	Potassium cyanide	KCN	151-50-8	65.116	wh cub cry; hyg	622		1.55	69.9^{20}	sl EtOH
296	Potassium dichromate	$K_2Cr_2O_7$	7778-50-9	294.185	oran-red tricl cry	398	≈500 dec	2.68	15.1^{25}	
297	Potassium dihydrogen phosphate	KH_2PO_4	7778-77-0	136.085	wh tetr cry	253 dec		2.34	25.0^{25}	sl EtOH
298	Potassium ferricyanide	$K_3Fe(CN)_6$	13746-66-2	329.244	red cry	dec		1.89	48.8^{25}	
299	Potassium ferrocyanide trihydrate	$K_4Fe(CN)_6 \cdot 3H_2O$	14459-95-1	422.388	yel monocl cry	60 dec		1.85	36.0^{25}	i EtOH, eth
300	Potassium fluoride	KF	7789-23-3	58.096	wh cub cry	858	1502	2.48	102^{25}	
301	Potassium hydrogen carbonate	$KHCO_3$	298-14-6	100.115	col monocl cry	≈100 dec		2.17	36.2^{25}	i EtOH
302	Potassium hydrogen iodate	$KH(IO_3)_2$	13455-24-8	389.911	col cry	dec			1.3^{15}	sl H_2O; i EtOH
303	Potassium hydrogen sulfate	$KHSO_4$	7646-93-7	136.169	wh monocl cry; hyg	≈200		2.32	50.6^{25}	
304	Potassium hydroxide	KOH	1310-58-3	56.105	wh rhomb cry; hyg	406	1327	2.044	121^{25}	s EtOH; s MeOH
305	Potassium iodate	KIO_3	7758-05-6	214.001	wh monocl cry	560 dec		3.89	9.22^{25}	
306	Potassium iodide	KI	7681-11-0	166.003	col cub cry	681	1323	3.12	148^{25}	sl EtOH
307	Potassium nitrate	KNO_3	7757-79-1	101.103	col orth cry or powder	334	400 dec	2.105	38.3^{25}	i EtOH
308	Potassium nitrite	KNO_2	7758-09-0	85.104	wh hyg cry	438	537 exp	1.915	312^{25}	sl EtOH
309	Potassium oxalate	$K_2C_2O_4$	583-52-8	166.216	wh pwd					sl H_2O
310	Potassium perchlorate	$KClO_4$	7778-74-7	138.549	col orth cry; hyg	525		2.52	2.08^{25}	
311	Potassium periodate	KIO_4	7790-21-8	230.001	col tetr cry	582	exp	3.618	0.51^{25}	
312	Potassium permanganate	$KMnO_4$	7722-64-7	158.034	purp orth cry	dec		2.7	7.60^{25}	reac EtOH
313	Potassium peroxide	K_2O_2	17014-71-0	110.196	yel amorp solid	545				reac H_2O
314	Potassium persulfate	$K_2S_2O_8$	7727-21-1	270.322	col cry	≈100 dec		2.48	4.7^{20}	
315	Potassium phosphate	K_3PO_4	7778-53-2	212.266	wh orth cry; hyg	1340		2.564	106^{25}	i EtOH
316	Potassium pyrosulfate	$K_2S_2O_7$	7790-62-7	254.323	col needles	≈325		2.28		s H_2O
317	Potassium sulfate	K_2SO_4	7778-80-5	174.260	wh orth cry	1069		2.66	12.0^{25}	i EtOH
318	Potassium sulfite	K_2SO_3	10117-38-1	158.260	col hex cry				106^{25}	sl EtOH
319	Potassium thiocyanate	$KSCN$	333-20-0	97.181	col tetr cry; hyg	173	500 dec	1.88	238^{25}	s EtOH
320	Praseodymium	Pr	7440-10-0	140.908	silv metal; hex	931	3520	6.77		
321	Promethium	Pm	7440-12-2	145	silv metal; hex	1042	≈3000	7.26		
322	Protactinium	Pa	7440-13-3	231.036	shiny metal; tetr or cub	1572		15.4		
323	Radium	Ra	7440-14-4	226	wh metal; cub	696		5		
324	Radon	Rn	10043-92-2	222	col gas	-71	-61.7	0.009074^{25} gas		sl H_2O
325	Rhenium	Re	7440-15-5	186.207	silv-gray metal	3185	5590	20.8		i HCl
326	Rhodium	Rh	7440-16-6	102.906	silv-wh metal; cub	1963	3695	12.4		i acid, sl aqua regia
327	Rubidium	Rb	7440-17-7	85.468	soft silv metal; cub	39.30	688	1.53		reac H_2O
328	Ruthenium	Ru	7440-18-8	101.07	silv-wh metal; hex	2333	4147	12.1		i acid, aqua regia
329	Samarium	Sm	7440-19-9	150.36	silv metal; rhomb	1072	1794	7.52		
330	Scandium	Sc	7440-20-2	44.956	silv metal; hex	1541	2836	2.99		
331	Selenium (gray)	Se	7782-49-2	78.96	gray metallic cry; hex	220.8	685	4.809		i H_2O, CS_2
332	Selenium (α form)	Se	7782-49-2	78.96	red monocl cry	>120 trans gray Se	685	4.39		i H_2O, EtOH; sl eth
333	Selenium (vitreous)	Se	7782-49-2	78.96	blk amorp solid	180 trans gray Se	685	4.28		i H_2O; sl CS_2
334	Silicon	Si	7440-21-3	28.085	gray cry or brn amorp solid	1414	3265	2.3296		i H_2O, acid; s alk
335	Silane	SiH_4	7803-62-5	32.118	col gas; flam	-185	-111.9	0.001313^{25} gas		reac H_2O; i EtOH, bz
336	Disilane	Si_2H_6	1590-87-0	62.219	col gas; flam	-129.4	-14.8	0.002543^{25} gas		reac H_2O, ctc, chl; s EtOH, bz
337	Trisilane	Si_3H_8	7783-26-8	92.321	flam col liq	-117.4	52.9	0.739		reac H_2O
338	Tetrasilane	Si_4H_{10}	7783-29-1	122.421	col liq; flam	-89.9	108.1	0.792		reac H_2O
339	Pentasilane	Si_5H_{12}	14868-53-2	152.523	col liq	-72.8	153.2	0.827		reac H_2O
340	Chlorosilane	SiH_3Cl	13465-78-6	66.563	col gas	-118	-30.4	0.002721^{25} gas		
341	Dichlorosilane	SiH_2Cl_2	4109-96-0	101.008	col gas; flam	-122	8.3	0.004129^{25} gas		reac H_2O
342	Trichlorosilane	$SiHCl_3$	10025-78-2	135.453	fuming liq	-128.2	33	1.331		reac H_2O
343	Tetrachlorosilane	$SiCl_4$	10026-04-7	169.898	col fuming liq	-68.74	57.65	1.5		reac H_2O

No.	Name	Mol. form.	CAS Reg. No.	Mol. wt.	Phys. form	mp/°C	bp/°C	ρ/g cm^{-3}	Sol./g/100 g H_2O	Qualitative solubility
344	Hexachlorodisilane	Si_2Cl_6	13465-77-5	268.889	col liq	2.5	146			reac H_2O
345	Tetrafluorosilane	SiF_4	7783-61-1	104.080	col gas	-90.2	-86	0.004254^{25} gas		reac H_2O
346	Hexachlorodisiloxane	$(SiCl_3)_2O$	14986-21-1	284.888	liq	-28	137			
347	Methylsilane	SiH_3CH_3	992-94-9	46.145	col gas	-156.5	-57.5			
348	Silicon carbide (hexagonal)	SiC	409-21-2	40.097	hard grn-black hex cry	2830		3.16		i H_2O, EtOH
349	Silicon dioxide (α-quartz)	SiO_2	14808-60-7	60.085	col hex cry	573 trans beta quartz	2950	2.648		i H_2O, acid; s HF
350	Silicon dioxide (β-quartz)	SiO_2	14808-60-7	60.085	col hex cry	867 trans tridymite	2950	2.533^{600}		i H_2O, acid; s HF
351	Silver	Ag	7440-22-4	107.868	silv metal; cub	961.78	2162	10.5		
352	Silver(I) chloride	AgCl	7783-90-6	143.321	wh cub cry	455	1547	5.56	0.00019^{25}	
353	Silver(I) nitrate	$AgNO_3$	7761-88-8	169.873	col rhomb cry	210	440 dec	4.35	234^{25}	sl EtOH, ace
354	Silver(I) sulfate	Ag_2SO_4	10294-26-5	311.799	col cry or powder	660		5.45	0.84^{25}	
355	Sodium	Na	7440-23-5	22.990	soft silv met; cub	97.794	882.940	0.97		reac H_2O
356	Sodium acetate	$NaC_2H_3O_2$	127-09-3	82.034	col cry	328.2		1.528	50.4^{25}	
357	Sodium amide	$NaNH_2$	7782-92-5	39.013	wh-grn orth cry	210	500 dec	1.39		reac H_2O
358	Sodium azide	NaN_3	26628-22-8	65.010	col hex cry	300 dec		1.846	40.8^{20}	sl EtOH; i eth
359	Sodium borohydride	$NaBH_4$	16940-66-2	37.833	wh cub cry; hyg	≈400 dec		1.07	55^{20}	reac EtOH
360	Sodium bromide	NaBr	7647-15-6	102.894	wh cub cry	747	1390	3.200	94.6^{25}	s EtOH
361	Sodium carbonate	Na_2CO_3	497-19-8	105.989	wh hyg powder	856		2.54	30.7^{25}	i EtOH
362	Sodium chlorate	$NaClO_3$	7775-09-9	106.441	col cub cry	248	630 dec	2.5	100^{25}	sl EtOH
363	Sodium chloride	NaCl	7647-14-5	58.443	col cub cry	802.018	1465	2.17	36.0^{25}	sl EtOH
364	Sodium chromate	Na_2CrO_4	7775-11-3	161.974	yel orth cry	794		2.72	87.6^{25}	sl EtOH
365	Sodium cyanide	NaCN	143-33-9	49.008	wh cub cry; hyg	562		1.6	58.2^{20}	sl EtOH
366	Sodium dichromate	$Na_2Cr_2O_7$	10588-01-9	261.968	red hyg cry	357	400 dec		187^{25}	
367	Sodium dihydrogen phosphate	NaH_2PO_4	7558-80-7	119.977	col monocl cry	200 dec			94.9^{25}	
368	Sodium dithionate	$Na_2S_2O_4$	7775-14-6	174.108	gray-wh powder	52 dec			24.1^{20}	sl EtOH
369	Sodium fluoride	NaF	7681-49-4	41.988	col cub or tetr cry	996	1704	2.78	4.13^{25}	i EtOH
370	Sodium formate	$NaCHO_2$	141-53-7	68.008	wh hyg cry	257.3	dec	1.92	94.9^{25}	sl EtOH
371	Sodium hydrogen carbonate	$NaHCO_3$	144-55-8	84.007	wh monocl cry	527		2.20	10.3^{25}	i EtOH
372	Sodium hydrogen phosphate	Na_2HPO_4	7558-79-4	141.959	wh hyg powder			1.7	11.8^{25}	
373	Sodium hydrogen sulfate	$NaHSO_4$	7681-38-1	120.061	wh hyg cry	≈315		2.43	28.5^{25}	
374	Sodium hydrogen sulfite	$NaHSO_3$	7631-90-5	104.061	wh cry			1.48		s H_2O; sl EtOH
375	Sodium hydroxide	NaOH	1310-73-2	39.997	wh orth cry; hyg	323	1388	2.13	100^{25}	s EtOH, MeOH
376	Sodium hypochlorite	NaClO	7681-52-9	74.442	stab in aq soln	anh form exp			79.9^{25}	
377	Sodium iodide	NaI	7681-82-5	149.894	wh cub cry; hyg	661	1304	3.67	184^{25}	s EtOH, ace
378	Sodium metasilicate	Na_2SiO_3	6834-92-0	122.064	wh amorp solid; hyg	1089		2.61		s cold H_2O; reac hot H_2O; i EtOH
379	Sodium nitrate	$NaNO_3$	7631-99-4	84.995	col hex cry; hyg	306.5		2.261	91.2^{25}	sl EtOH, MeOH
380	Sodium nitrite	$NaNO_2$	7632-00-0	68.996	wh orth cry; hyg	284	>320 dec	2.17	84.8^{25}	sl EtOH; reac acid
381	Sodium nitroferricyanide dihydrate	$Na_2[Fe(CN)_5NO]\cdot 2H_2O$	13755-38-9	297.949	red cry			1.72	40^{16}	sl EtOH
382	Sodium perchlorate	$NaClO_4$	7601-89-0	122.441	wh orth cry; hyg	482 dec		2.52	205^{25}	
383	Sodium periodate	$NaIO_4$	7790-28-5	213.892	wh tetr cry	≈300 dec		3.86	14.4^{25}	s acid
384	Sodium peroxide	Na_2O_2	1313-60-6	77.979	yel hyg powder	675		2.805		reac H_2O
385	Sodium phosphate	Na_3PO_4	7601-54-9	163.940	col cry	1583		2.54	14.5^{25}	s H_2O
386	Sodium pyrophosphate	$Na_4P_2O_7$	7722-88-5	265.902	col cry	988		2.53	7.09^{25}	
387	Sodium sulfate	Na_2SO_4	7757-82-6	142.043	wh orth cry or powder	884		2.7	28.1^{25}	i EtOH
388	Sodium sulfide	Na_2S	1313-82-2	78.045	wh cub cry; hyg	1172		1.856	20.6^{25}	sl EtOH; i eth
389	Sodium sulfite	Na_2SO_3	7757-83-7	126.043	wh hex cry	911		2.63	30.7^{25}	i EtOH
390	Sodium tetraborate	$Na_2B_4O_7$	1330-43-4	201.220	col gl solid; hyg	743	1575	2.4	3.17^{25}	sl MeOH
391	Sodium tetraborate decahydrate	$Na_2B_4O_7\cdot 10H_2O$	1303-96-4	381.373	wh monocl cry	75 dec		1.73	3.17^{25}	i EtOH
392	Sodium thiocyanate	NaSCN	540-72-7	81.073	col hyg cry	287			151^{25}	
393	Sodium thiosulfate	$Na_2S_2O_3$	7772-98-7	158.108	col monocl cry	100 dec		1.69	76.4^{25}	i EtOH
394	Sodium tripolyphosphate	$Na_5P_3O_{10}$	7758-29-4	367.864	wh hyg powder	622			20^{25}	
395	Sodium tungstate	Na_2WO_4	13472-45-2	293.82	wh rhom cry	695		4.18	74.2^{25}	
396	Strontium	Sr	7440-24-6	87.62	silv-wh metal; cub	777	1377	2.64		reac H_2O; s EtOH
397	Strontium chloride	$SrCl_2$	10476-85-4	158.53	wh cub cry; hyg	874	1250	3.052	54.7^{25}	
398	Strontium nitrate	$Sr(NO_3)_2$	10042-76-9	211.63	wh cub cry	570		2.99	80.2^{25}	sl EtOH, ace
399	Sulfur (rhombic)	S	7704-34-9	32.06	yel orth cry	95.2 trans monocl	444.61	2.07		i H_2O; sl EtOH, bz, eth; s CS_2

No.	Name	Mol. form.	CAS Reg. No.	Mol. wt.	Phys. form	mp/°C	bp/°C	ρ/g cm^{-3}	Sol./g/100 g H_2O	Qualitative solubility
400	Sulfur (monoclinic)	S	7704-34-9	32.06	yel monocl needles, stable 95.3-120	115.21	444.61	2.00		i H_2O; sl EtOH, bz, eth; s CS_2
401	Sulfuric acid	H_2SO_4	7664-93-9	98.079	col oily liq	10.31	337	1.8305^{20}		vs H_2O
402	Nitrosylsulfuric acid	$HNOSO_4$	7782-78-7	127.077	prisms	73 dec				reac H_2O; s H_2SO_4
403	Chlorosulfonic acid	$SO_2(OH)Cl$	7790-94-5	116.524	col-yel liq	-80	152	1.75		reac H_2O; s py
404	Fluorosulfonic acid	$SO_2(OH)F$	7789-21-1	100.069	col liq	-89	163	1.726		reac H_2O
405	Sulfurous acid	H_2SO_3	7782-99-2	82.079	exists only in aq soln					
406	Sulfamic acid	H_2NSO_3H	5329-14-6	97.094	orth cry	≈205 dec		2.15	14.7^{0}	sl ace; i eth
407	Sulfur dioxide	SO_2	7446-09-5	64.064	col gas	-75.45	-10.02	$1.620^{-75.5}$ liq		s H_2O, EtOh, eth, chl
408	Sulfur trioxide (α-form)	SO_3	7446-11-9	80.063	wh needles	62.2	subl			reac H_2O
409	Sulfur dichloride	SCl_2	10545-99-0	102.971	red visc liq	-122	59.6	1.62		reac H_2O
410	Sulfur hexafluoride	SF_6	2551-62-4	146.055	col gas	-49.596 tp	-63.8 sp	$1.845^{-49.6}$ liq		sl H_2O; s EtOH
411	Sulfuryl chloride	SO_2Cl_2	7791-25-5	134.970	col liq	-51	69.4	1.680		reac H_2O; s bz, tol, eth
412	Thionyl chloride	$SOCl_2$	7719-09-7	118.970	yel fuming liq	-101	75.6	1.631		reac H_2O; s bz, ctc, chl
413	Tantalum	Ta	7440-25-7	180.948	gray metal; cub	3017	5455	16.4		reac HF
414	Technetium	Tc	7440-26-8	98	hex cry	2157	4262	11		
415	Tellurium dioxide	TeO_2	7446-07-3	159.60	wh orth cry	733	1245	5.9		i H_2O; s alk, acid
416	Terbium	Tb	7440-27-9	158.925	silv metal; hex	1359	3230	8.23		
417	Thallium	Tl	7440-28-0	204.38	soft blue-wh metal	304	1473	11.8		i H_2O; reac acid
418	Thorium	Th	7440-29-1	232.038	soft gray-wh metal; cub	1750	4785	11.7		s acid
419	Thulium	Tm	7440-30-4	168.934	silv metal; hex	1545	1950	9.32		s dil acid
420	Tin (gray)	Sn	7440-31-5	118.710	cub cry	13.2 trans white Sn	2586	5.769		
421	Tin (white)	Sn	7440-31-5	118.710	silv tetr cry	231.928	2586	7.287		
422	Tin(II) chloride	$SnCl_2$	7772-99-8	189.616	wh orth cry	247.0	623	3.90	178^{10}	s EtOH, ace, eth; i xyl
423	Tin(II) oxide	SnO	21651-19-4	134.709	blue-blk tetr cry	977		6.45		i H_2O, EtOH; s acid
424	Tin(IV) chloride	$SnCl_4$	7646-78-8	260.522	col fuming liq	-34.07	114.15	2.234		reac H_2O; s EtOH, ctc, bz, ace
425	Titanium	Ti	7440-32-6	47.867	gray metal; hex	1670	3287	4.506		
426	Titanium(II) oxide	TiO	12137-20-1	63.866	yel cub cry	1770	3227	4.95		
427	Titanium(III) oxide	Ti_2O_3	1344-54-3	143.732	blk hex cry	1842		4.486		s hot HF
428	Titanium(III,IV) oxide	Ti_3O_5	12065-65-5	223.598	blk monocl cry	1777		4.24		
429	Titanium(IV) chloride	$TiCl_4$	7550-45-0	189.679	col or yel liq	-24.12	136.45	1.73		reac H_2O; s EtOH
430	Titanium(IV) oxide (anatase)	TiO_2	1317-70-0	79.866	brn tetr cry	1560		3.9		
431	Titanium(IV) oxide (brookite)	TiO_2	12188-41-9	79.866	wh orth cry			4.17		
432	Titanium(IV) oxide (rutile)	TiO_2	1317-80-2	79.866	wh tetr cry	1912	≈3000	4.25		i H_2O, dil acid; s conc acid
433	Tungsten	W	7440-33-7	183.84	gray-wh metal; cub	3414	5555	19.3		
434	Uranium	U	7440-61-1	238.029	silv-wh orth cry	1135	4131	19.1		
435	Uranium(VI) fluoride	UF_6	7783-81-5	352.019	wh monocl solid	64.06 tp	56.5 sp	5.09		reac H_2O; s ctc, chl
436	Uranyl acetate dihydrate	$UO_2(C_2H_3O_2)_2 \cdot 2H_2O$	6159-44-0	424.146	ye cry (HOAc)	80 dec		2.89		sl EtOH
437	Uranyl nitrate hexahydrate	$UO_2(NO_3)_2 \cdot 6H_2O$	13520-83-7	502.129	yel orth cry; hyg	60	118 dec	2.81	127^{25}	s EtOH, eth
438	Vanadium	V	7440-62-2	50.942	gray-wh metal; cub	1910	3407	6.0		i H_2O; s acid
439	Vanadium(IV) chloride	VCl_4	7632-51-1	192.754	red-brn liq	-28	151	1.816		reac H_2O; s EtOH, eth
440	Vanadyl trichloride	$VOCl_3$	7727-18-6	173.300	fuming red-yel liq	-79	127	1.829		reac H_2O; s MeOH, eth, ace
441	Water	H_2O	7732-18-5	18.015	col liq	0.00	99.974	0.9970^{25}		vs EtOH, MeOH, ace
442	Water-d_2	D_2O	7789-20-0	20.027	col liq	3.82	101.42	1.1044^{25}		
443	Water-t_2	T_2O	14940-65-9	22.032	col liq	4.48	101.51	1.2138^{25}		
444	Xenon	Xe	7440-63-3	131.293	col gas	-111.75	-108.099	$2.942^{-108.1}$ liq		sl H_2O
445	Ytterbium	Yb	7440-64-4	173.054	silv metal; cub	824	1196	6.90		s dil acid
446	Yttrium	Y	7440-65-5	88.906	silv metal; hex	1522	3345	4.47		reac H_2O; s dil acid

No.	Name	Mol. form.	CAS Reg. No.	Mol. wt.	Phys. form	mp/°C	bp/°C	ρ/g cm⁻³	Sol./g/100 g H_2O	Qualitative solubility
447	Zinc	Zn	7440-66-6	65.38	blue-wh metal; hex	419.527	907	7.134		s acid, alk
448	Zinc acetate dihydrate	$Zn(C_2H_3O_2)_2 \cdot 2H_2O$	5970-45-6	219.527	wh powder	237 dec		1.735	30.0^{20}	s EtOH
449	Zinc chloride	$ZnCl_2$	7646-85-7	136.315	wh hyg cry	325	732	2.907	408^{25}	vs H_2O; s EtOH, ace
450	Zinc oxide	ZnO	1314-13-2	81.408	wh powder; hex	1974		5.6		i H_2O; s dil acid
451	Zinc sulfate	$ZnSO_4$	7733-02-0	161.472	col orth cry	680 dec		3.8	57.7^{25}	
452	Zinc sulfate heptahydrate	$ZnSO_4 \cdot 7H_2O$	7446-20-0	287.578	col orth cry	100 dec		1.97	57.7^{25}	i EtOH
453	Zinc sulfide (sphalerite)	ZnS	1314-98-3	97.474	gray-wh cub cry	1020 trans wurtzite		4.04		i H_2O, EtOH; s dil acid
454	Zinc sulfide (wurtzite)	ZnS	1314-98-3	97.474	wh hex cry	1827	subl	4.09		i H_2O; s dil acid
455	Zirconium	Zr	7440-67-7	91.224	gray-wh metal; hex	1854	4406	6.52		s hot conc acid

[a] This is the estimated diamond-graphite-liquid carbon triple point (Ref. 12).

PHYSICAL PROPERTIES OF THE RARE-EARTH METALS

K. A. Gschneidner

These tables contain physical property data for the rare-earth metals, which are important in many applications, including magnetic materials. They are often in short supply due to international trade considerations.

Atomic Properties of Rare-Earth Metals

Table 1 contains data for the trivalent ions of the rare-earth elements; its data columns are defined as follows.

Table 1 Column heading	Definition
Rare-earth metal	Name of the rare-earth metal
Symbol	Symbol of the rare-earth metal
Atomic no.	Atomic number
Atomic wt.	Standard atomic weight; see also "Standard Atomic Weights" table in Section 1
No. 4f elec. R^{3+} ion	Number of 4f electrons in the trivalent ion of the element
S	Spin quantum number of the R^{3+} (trivalent) ion
L	Azimuthal quantum number of the R^{3+} (trivalent) ion
J	Total angular momentum quantum number of the R^{3+} (trivalent) ion
Spect. ground st.	The spectroscopic symbol for the ground state of the R^{3+} (trivalent) ion

TABLE 1. Data for the Trivalent Ions of the Rare-Earth Elements

Rare-earth metal	Symbol	Atomic no.	Atomic wt.[a]	No. 4f elec. R^{3+} ion	S^b	L^b	J^b	Spect. ground st.
Scandium	Sc	21	44.955908	0				
Yttrium	Y	39	88.90584	0				
Lanthanum	La	57	138.90547	0				
Cerium	Ce	58	140.116	1	1/2	3	5/2	$^2F_{5/2}$
Praseodymium	Pr	59	140.90766	2	1	5	4	$^3H_4$
Neodymium	Nd	60	144.242	3	3/2	6	9/2	$^4I_{9/2}$
Promethium	Pm	61	145[c]	4	2	6	4	$^5I_4$
Samarium	Sm	62	150.36	5	5/2	5	5/2	$^6H_{5/2}$
Europium	Eu	63	151.964	6	3	3	0	$^7F_0$
Gadolinium	Gd	64	157.25	7	7/2	0	7/2	$^8S_{7/2}$
Terbium	Tb	65	158.92535	8	3	3	6	$^7F_6$
Dysprosium	Dy	66	162.500	9	5/2	5	15/2	$^6H_{15/2}$
Holmium	Ho	67	164.93033	10	2	6	8	$^5I_8$
Erbium	Er	68	167.259	11	3/2	6	15/2	$^4I_{15/2}$
Thulium	Tm	69	168.93422	12	1	5	6	$^3H_6$
Ytterbium	Yb	70	173.045	13	1/2	3	7/2	$^2F_{7/2}$
Lutetium	Lu	71	174.9668	14				

[a] 2013 Standard Atomic Weights.
[b] In R^{3+} ion.
[c] Mass number of longest-lived isotope.
Note: For additional information, see Goldschmidt, Z.B., in *Handbook on the Physics and Chemistry of Rare Earths*, Vol. 1, Gschneidner, K.A., Jr. and Eyring, L., Eds., North-Holland Physics, Amsterdam, 1978; DeLaeter, J.R., and Heumann, K.G., *J. Phys. Chem. Ref. Data* 20, 1313, 1991; *Pure Appl. Chem.* 66, 2423, 1994.

Crystallographic Data

Tables 2 and 3 contain crystallographic data for the rare-earth metals at 297 K or below and high temperatures, respectively. The data columns are defined as follows.

Table 2 Column heading	Definition
Rare-earth metal	Symbol of the rare-earth metal
Crystal structure	Crystal structure symbol, as defined in the table below
a_0	Primary lattice vector, in Å
b_0	Secondary lattice vector, in Å
c_0	Tertiary lattice vector, in Å
r_{metal}(CN12)	Ionic radius of the metallic atom with 12 nearest neighbors (coordination number = 12), in Å
V_A	Volume of one mole of the metal at standard conditions; in units of cm^3 mol^{-1}
ρ	Density (mass per unit volume); in units of g cm^{-3}

Table 3 Column heading	Definition
Rare-earth metal	Symbol of the rare-earth metal
Crystal structure	Crystal structure symbol, as defined in the table below
Lattice parameter	Lattice vector, in Å
Temp.	Measurement temperature, in °C
r_{metal}(CN8)	Ionic radius of the metallic atom with 8 nearest neighbors (coordination number = 8), in Å
r_{metal}(CN12)	Ionic radius of the metallic atom with 12 nearest neighbors (coordination number = 12), in Å
V_A	Volume of one mole of the metal at standard conditions; in units of cm^3 mol^{-1}
ρ	Density (mass per unit volume); in units of g cm^{-3}

Crystal structure symbol	Definition
hcp	Hexagonal close-packed; $P6_3/mmc$, hP2, A3, Mg-type
dhcp	Double-c hexagonal close-packed; $P6_3/mmc$, hP4, A3′, αLa-type
fcc	Face-centered cubic; Fm3¯m, cF4, A1, Cu-type
rhomb	Rhombohedral; R3¯m, hR3, αSm-type
bcc	Body-centered cubic; Im3¯m, cI2, A2, W-type
ortho	Orthorhombic; Cmcm, oC4, α′ Dy-type

TABLE 2. Crystallographic Data for the Rare-Earth Metals at 24 °C (297 K) or Below

Rare-earth metal	Crystal structure	a_0/Å	b_0/Å	c_0/Å	r_{metal}(CN12)/Å	V_A/cm^3 mol^{-1}	ρ/g cm^{-3}
αSc	hcp	3.3088		5.2680	1.6406	15.039	2.989
αY	hcp	3.6482		5.7318	1.8012	19.893	4.469
αLa	dhcp	3.7740		12.171	1.8791	22.602	6.146
αCe[a]	fcc	4.85[a]			1.72[a]	17.2[a]	8.16[a]
βCe	dhcp	3.6810		11.857	1.8321	20.947	6.689
γCe[b]	fcc	5.1610			1.8247	20.696	6.770
αPr	dhcp	3.6721		11.8326	1.8279	20.803	6.773
αNd	dhcp	3.6582		11.7966	1.8214	20.583	7.008
αPm	dhcp	3.65		11.65	1.811	20.24	7.264
αSm	rhomb[c]	3.6290[c]		26.207	1.8041	20.000	7.520
Eu	bcc	4.5827			2.0418	28.979	5.244
αGd	hcp	3.6336		5.7810	1.8013	19.903	7.901
α′Tb[d]	ortho	3.605[d]	6.244[d]	5.706[d]	1.784[d]	19.34[d]	8.219[d]
αTb	hcp	3.6055		5.6966	1.7833	19.310	8.230
α′Dy[e]	ortho	3.595[e]	6.184[e]	5.678[e]	1.774[e]	19.00[e]	8.551[e]
αDy	hcp	3.5915		5.6501	1.7740	19.004	8.551
Ho	hcp	3.5778		5.6178	1.7661	18.752	8.795
Er	hcp	3.5592		5.5850	1.7566	18.449	9.066
Tm	hcp	3.5375		5.5540	1.7462	18.124	9.321
αYb[f]	hcp	3.8799[f]		6.3859[f]	1.9451[f]	25.067[f]	6.903[f]
βYb	fcc	5.4848			1.9392	24.841	6.966
Lu	hcp	3.5052		5.5494	1.7349	17.779	9.841

[a] At 77 K (−196 °C).
[b] Equilibrium room temperature (standard state) phase.
[c] Rhombohedral is the primitive cell. Lattice parameters given are for the nonprimitive hexagonal cell.
[d] At 220 K (−53 °C).
[e] At 86 K (−187 °C).
[f] At 23 °C.

Note: For additional information, see Gschneidner, K.A., Jr. and Calderwood, F.W., in *Handbook on the Physics and Chemistry of Rare Earths*, Vol. 8, Gschneidner, K.A., Jr. and Eyring, L., Eds., North-Holland Physics, Amsterdam, 1986; Gschneidner, K.A., Jr., Pecharsky, V.K., Cho, Jaephil and Martin, S.W., *Scripta Mater.*, 34, 1717, 1996. <https://doi.org/10.1016/1359-6462(96)00035-8>

TABLE 3. Crystallographic Data for Rare-Earth Metals at High Temperature

Rare-earth metal	Crystal structure	Lattice parameter/Å	Temp./°C	r_{metal}(CN8)/Å	r_{metal}(CN12)/Å	V_A/cm³ mol⁻¹	ρ/g cm⁻³
βSc	bcc	3.73[d]	1337	1.62	1.66	15.6	2.88
βY	bcc	4.10[a]	1478	1.78	1.83	20.8	4.28
βLa	fcc	5.303	325		1.875	22.45	6.187
γLa	bcc	4.26	887	1.84	1.90	23.3	5.97
δCe	bcc	4.12	757	1.78	1.84	21.1	6.65
βPr	bcc	4.13	821	1.79	1.84	21.2	6.64
βNd	bcc	4.13	883	1.79	1.84	21.2	6.80
βPm	bcc	4.10[d]	890	1.78	1.83	20.8	6.99
βSm	hcp	[e]	450[b]		1.8176	20.450	7.353
γSm	bcc	4.10[d]	922	1.77	1.82	20.8	7.25
βGd	bcc	4.06	1265	1.76	1.81	20.2	7.80
βTb	bcc	4.07[a]	1289	1.76	1.81	20.3	7.82
βDy	bcc	4.03[a]	1381	1.75	1.80	19.7	8.23
γYb	bcc	4.44	763[c]	1.92	1.98	26.4	6.57

[a] Determined by extrapolation to 0% solute of a vs. composition data for R-Mg alloys at 24 °C and corrected for thermal expansion to temperature given.
[b] The hcp phase was stabilized by impurities and the temperature of measurement was below the equilibrium transition temperature (see Table 4).
[c] The bcc phase was stabilized by impurities and the temperature of measurement was below the equilibrium transition temperature (see Table 4).
[d] Estimated.
[e] $a = 3.6630$, $c = 5.8448$.
Note: The rare earths Eu, Ho, Er, Tm, and Lu are monomorphic. For additional information, see Gschneidner, K.A., Jr. and Calderwood, F.W., in *Handbook on the Physics and Chemistry of Rare Earths*, Vol. 8, Gschneidner, K.A., Jr. and Eyring, L., Eds., North-Holland Physics, Amsterdam, 1986, 1.

Phase Transition Data

Tables 4 and 5 have data on the high-temperature and low-temperature transitions (respectively) for rare-earth metals. The data columns are defined as follows.

Table 4 Column heading	Definition
Rare-earth metal	Symbol of the rare-earth metal
CASRN	Chemical Abstracts Service Registry Number for the rare-earth metal
$T_{\alpha\beta}$	Temperature for transition (α – β), in °C
Phases (αβ)	Crystal symbols for phases involved in transition (α – β)
$T_{\beta\gamma}$	Temperature for transition (β – γ), in °C
Phases (βγ)	Crystal symbols for phases involved in transition (β – γ)
Melt. pt.	Normal melting point, in °C

Table 5 Column heading	Definition
Rare-earth metal	Symbol of the rare-earth metal
Cooling transformation	Crystal symbols for phases involved in cooling transformation
t_C	Cooling transformation temperature, in °C
T_C	Cooling transformation temperature, in K
Heating transformation	Crystal symbols for phases involved in heating transformation
t_H	Heating transformation temperature, in °C
T_H	Heating transformation temperature, in K

TABLE 4. High-Temperature Transition Temperatures and Melting Point of Rare-Earth Metals

Rare-earth metal	CASRN	$T_{\alpha\beta}$/°C	Phases (αβ)[a]	$T_{\beta\gamma}$/°C	Phases (βγ)[a]	Melt. pt./°C
Sc	7440-20-2	1337	hcp ⇌ bcc			1541
Y	7440-65-5	1478	hcp ⇌ bcc			1522
La[b]	7439-91-0	310	dhcp ⇌ fcc	865	fcc ⇌ bcc	920
Ce[c, d]	7440-45-1	139	dhcp ⇌ fcc (β - γ)	726	fcc ⇌ bcc (γ - δ)	799
Pr	7440-10-0	795	dhcp ⇌ bcc			931
Nd	7440-00-8	863	dhcp ⇌ bcc			1016
Pm	7440-12-2	890	dhcp ⇌ bcc			1042
Sm[e]	7440-19-9	734	rhom ⇌ hcp	922	hcp ⇌ bcc	1072
Eu	7440-53-1					822
Gd	7440-54-2	1235	hcp ⇌ bcc			1313
Tb	7440-27-9	1289	hcp ⇌ bcc			1359
Dy	7429-91-6	1381	hcp ⇌ bcc			1412
Ho	7440-60-0					1472
Er	7440-52-0					1529

Rare-earth metal	CASRN	$T_{\alpha\beta}$/°C	Phases (αβ)[a]	$T_{\beta\gamma}$/°C	Phases (βγ)[a]	Melt. pt./°C
Tm	7440-30-4					1545
Yb	7440-64-4	795	fcc ⇌ bcc (β - γ)			824
Lu	7439-94-3					1663

[a] For all the transformations listed, unless otherwise noted.
[b] On cooling, fcc → dhcp (β → α), 260 °C.
[c] The β ⇌ γ equilibrium transition temperature is 10 ± 5 °C.
[d] On cooling, fcc → dhcp (γ → β), −16 °C.
[e] On cooling, hcp → rhomb (β → α), 727 °C.

Note: For additional information, see Gschneidner, K.A., Jr. and Calderwood, F.W., in *Handbook on the Physics and Chemistry of Rare Earths*, Vol. 8, Gschneidner, K.A., Jr. and Eyring, L., Eds., North-Holland Physics, Amsterdam, 1986; Gschneidner, K.A., Jr., Pecharsky, V.K., Cho, Jaephil and Martin, S.W., *Scripta Mater.*, 34, 1717, 1996. <https://doi.org/10.1016/1359-6462(96)00035-8>

TABLE 5. Low-Temperature Transition Temperatures of the Rare-Earth Metals

Rare-earth metal	Cooling transformation	t_C/°C	T_C/K	Heating transformation	t_H/°C	T_H/K
Ce	γ → β[a]	-16	257	α → β	-148	125
Ce	γ → α	-172	101	α → β + γ	-104	169
Ce	β → α	-228	45	β → γ[a]	139	412
Tb	α → α′	-53	220			
Dy	α → α′	-187	86			
Yb	β → α	-13	260	α → β	7	280

[a] The β ⇌ γ equilibrium transition temperature is 10 ± 5 °C (283 ± 5K).

Note: For additional information, see Beaudry, B.J. and Gschneidner, K.A., Jr., in *Handbook on the Physics and Chemistry of Rare Earths*, Vol. 1, Gschneidner, K.A., Jr. and Eyring, L., Eds., North-Holland Physics, Amsterdam, 1978, 173; Koskenmaki, D.C. and Gschneidner, K.A., Jr., 1978, in *Handbook on the Physics and Chemistry of Rare Earths*, Vol. 1, Gschneidner, K.A., Jr. and Eyring, L., Eds., North-Holland Physics, Amsterdam, 1978, 337; Gschneidner, K.A., Jr., Pecharsky, V.K., Cho, Jaephil and Martin, S.W., *Scripta Mater.*, 34, 1717, 1996.

Thermodynamic, Thermal, Electrical, and Magnetic Property Data

The following tables have data on various thermodynamic and thermal properties for the rare-earth metals. Table 6 contains heat capacity, standard entropy, heat of formation, and heat of fusion data. Table 7 has vapor pressure, boiling point, and heat of sublimation data for these elements. Table 8 has data on various magnetic properties and Table 9 has thermal and electrical property data. Table 10 contains data characterizing the various properties of these elements near their superconducting temperature. Column definitions for the tables are given below. For definitions of the a-axis and c-axis, refer to Table 2.

Table 6 Column heading	Definition
Rare-earth metal	Symbol of the rare-earth metal
C_v	Heat capacity at 298 K; in units of J mol^{-1} K^{-1}
S°_{298}	Entropy content of one mole of the element under standard conditions at 298 K, in standard state; in units of J mol^{-1} K^{-1}
Trans. 1	Phase transition 1
$\Delta_{tr}H^1$	Heat of transformation for phase transition 1; in units of kJ mol^{-1}
Trans. 2	Phase transition 2
$\Delta_{tr}H^2$	Heat of transformation for phase transition 2; in units of kJ mol^{-1}
$\Delta_{fus}H$	Heat of fusion; in units of kJ mol^{-1}

Table 7 Column heading	Definition
Rare-earth metal	Symbol of the rare-earth metal
$T(10^{-8}$ atm)	Temperature at which vapor pressure reaches 10^{-8} atm (0.001 Pa), in °C
$T(10^{-6}$ atm)	Temperature at which vapor pressure reaches 10^{-6} atm (0.101 Pa), in °C
$T(10^{-4}$ atm)	Temperature in at which vapor pressure reaches 10^{-4} atm (10.1 Pa), °C

Table 7 Column heading (*continued*)	Definition
$T(10^{-2}$ atm)	Temperature at which vapor pressure reaches 10^{-2} atm (1013 Pa), in °C
t_{bp}	Normal boiling point, in °C. This is the temperature at which the liquid phase is in equilibrium with the vapor at a pressure of 760 mmHg (101.325 kPa)
$\Delta_{sub}H$	Heat of sublimation at 25 °C; in units of kJ mol^{-1}

Table 8 Column heading	Definition
Rare-earth metal	Symbol of the rare-earth metal
χ_A at 298 K	Magnetic susceptibility at 298 K; in units of 10^6 emu mol^{-1}
μ_{the} (paramag.)	Theoretical effective magnetic moment for paramagnetic regime, dimensionless
μ_{obs} (paramag.)	Observed effective magnetic moment for paramagnetic regime, dimensionless
μ_{the} (ferromag.)	Theoretical effective magnetic moment for ferromagnetic regime, dimensionless
μ_{obs} (ferromag.)	Observed effective magnetic moment for ferromagnetic regime, dimensionless

Table 8 Column heading (*continued*)	Definition
Easy axis	Direction inside a crystal, along which a small applied magnetic field is sufficient to reach saturation magnetization
T_N hex sites	Néel temperature for hexagonal sites, in K
T_N cubic sites	Néel temperature for cubic sites, in K
T_C	Curie temperature, in K
$\Theta_D(\|\|c)$	Debye temperature in direction parallel to c-axis, in K
$\Theta_D(\perp c)$	Debye temperature in direction perpendicular to c-axis, in K
Θ_D(polycryst. or avg.)	Debye temperature for polycrystalline forms or averaged over different directions, in K

Table 9 Column heading	Definition
Rare-earth metal	Symbol of the rare-earth metal
α_a	Coefficient of thermal expansion in direction of a-axis; in units of $10^6 \times °C^{-1}$
α_c	Coefficient of thermal expansion in direction of c-axis; in units of $10^6 \times °C^{-1}$
α_{poly}	Coefficient of thermal expansion averaged over polycrystals; in units of $10^6 \times °C^{-1}$
k	Thermal conductivity; in units of W cm^{-1} K^{-1}
ρ_a	Electrical resistivity in direction of a-axis; in units of μΩ cm
ρ_c	Electrical resistivity in direction of c-axis; in units of μΩ cm
ρ_{poly}	Electrical resistivity averaged over polycrystals; in units of μΩ cm
R_a	Hall coefficient in direction of a-axis; in units of 10^{12} V cm A^{-1} Oe^{-1}
R_c	Hall coefficient in direction of c-axis; in units of 10^{12} V cm A^{-1} Oe^{-1}
R_{poly}	Hall coefficient averaged over polycrystals; in units of 10^{12} V cm A^{-1} Oe^{-1}

Table 10 Column heading	Definition
Rare-earth metal	Symbol of the rare-earth metal
γ	Electron specific heat constant; in units of mJ mol^{-1} K^{-2}
μ	Electron–electron (Coulomb) coupling constant, dimensionless
λ	Electron–phonon coupling constant, dimensionless
$\Theta_D(C_v)$	Debye temperature at 0 K from heat capacity, in K
Θ_D(ec)	Debye temperature at 0 K from elastic constants, in K
T_{sup}	Superconducting transition temperature, in K

Inorganic

TABLE 6. Heat Capacity, Standard Entropy, and Heats of Transformation and Fusion of Rare-Earth Metals

Rare-earth metal	C_v/J mol^{-1} K^{-1}	$S°_{298}$/J mol^{-1} K^{-1}	Trans. 1	$\Delta_{tr}H^1$/kJ mol^{-1}	Trans. 2	$\Delta_{tr}H^2$/kJ mol^{-1}	$\Delta_{fus}H$/kJ mol^{-1}
Sc	25.5	34.6	$\alpha \rightleftharpoons \beta$	4.00			14.1
Y	26.5	44.4	$\alpha \rightleftharpoons \beta$	4.99			11.4
La	27.1	56.9	$\alpha \rightleftharpoons \beta$	0.36	$\beta \rightleftharpoons \gamma$	3.12	6.20
Ce	26.9	72.0	$\beta \rightleftharpoons \gamma$	0.05	$\gamma \rightleftharpoons \delta$	2.99	5.46
Pr	27.2	73.2	$\alpha \rightleftharpoons \beta$	3.17			6.89
Nd	27.5	71.5	$\alpha \rightleftharpoons \beta$	3.03			7.14
Pm	27.3[a]	71.6[a]	$\alpha \rightleftharpoons \beta$	3.0[a]			7.7[a]
Sm	29.5	69.6	$\alpha \rightleftharpoons \beta$	0.2[a]	$\beta \rightleftharpoons \gamma$	3.11	8.62
Eu	27.7	77.8					9.21
Gd	37.0	68.1	$\alpha \rightleftharpoons \beta$	3.91			10.0
Tb	28.9	73.2	$\alpha \rightleftharpoons \beta$	5.02			10.79
Dy	27.7	75.6	$\alpha \rightleftharpoons \beta$	4.16			11.06
Ho	27.2	75.3					17.0[a]
Er	28.1	73.2					19.9
Tm	27.0	74.0					16.8
Yb	26.7	59.9	$\beta \rightleftharpoons \gamma$	1.75			7.66
Lu	26.9	51.0					22[a]

[a] Estimated.

Note: For additional information, see Hultgren, R., Desai, P.D., Hawkins, D.T., Gleiser, M., Kelley, K.K., and Wagman, D.D., *Selected Values of the Thermodynamic Properties of the Elements*, ASM International, Metals Park, Ohio, 1973; Wagman, D.D., Evans, W.H., Parker, V.B., Schumm, R.H., Halow, I., Bailey, S.M., Churney, K.L., and Nuttall, R.L., *The NBS Tables of Chemical Thermodynamic Properties*, *J. Phys. Chem. Ref. Data*, Vol. 11, Suppl. 2, 1982; Amitin, E.B., Bessergenev, W.G., Kovalevskaya, Yu. A., and Paukov, I.E., *J. Chem. Thermodyn.*, 15, 181, 1983. <https://doi.org/10.1016/0021-9614(83)90158-1>

TABLE 7. Vapor Pressures (Temperature[a] in °C at Which Vapor Pressure of Rare-Earth Metal Reaches Indicated Value), Boiling Points, and Heats of Sublimation of Rare-Earth Metals

Rare-earth metal	$T(10^{-8}$ atm)/°C	$T(10^{-6}$ atm)/°C	$T(10^{-4}$ atm)/°C	$T(10^{-2}$ atm)/°C	$t_{bp}{}^{a}$/°C	$\Delta_{sub}H$/kJ mol^{-1}
Sc	1036	1243	1533	1999	2836	377.8
Y	1222	1460	1812	2360	3345	424.7
La	1301	1566	1938	2506	3464	431.0
Ce	1290	1554	1926	2487	3443	422.6
Pr	1083	1333	1701	2305	3520	355.6
Nd	955	1175	1500	2029	3074	327.6
Pm					≈3000	348[b]
Sm	508	642	835	1150	1794	206.7
Eu	399	515	685	964	1529	175.3
Gd	1167	1408	1760	2306	3273	397.5
Tb	1124	1354	1698	2237	3230	388.7
Dy	804	988	1252	1685	2567	290.4
Ho	845	1036	1313	1771	2700	300.8
Er	908	1113	1405	1896	2868	317.1
Tm	599	748	964	1300	1950	232.2
Yb	301	400	541	776	1196	152.1
Lu	1241	1483	1832	2387	3402	427.6

[a] International Temperature Scale of 1990 (ITS-90) values.
[b] Estimated.

Note: For additional information, see Hultgren, R., Desai, P.D., Hawkins, D.T., Gleiser, M., Kelley, K.K., and Wagman, D.D., *Selected Values of the Thermodynamic Properties of the Elements*, ASM International, Metals Park, Ohio, 1973; Beaudry, B.J. and Gschneidner, K.A., Jr., in *Handbook on the Physics and Chemistry of Rare Earths*, Vol. 1, Gschneidner, K.A., Jr. and Eyring, L., Eds., North-Holland Physics, Amsterdam, 1978, 173.

TABLE 8. Magnetic Properties of the Rare-Earth Metals: Effective Magnetic Moment (μ), Néel Temperature (T_N), Curie Temperature (T_C), and Debye Temperature (Θ_D)

Rare-earth metal	χ_A at 298 K/10^6 × emu/mol	μ_{the} (paramag.)	μ_{obs} (paramag.)	μ_{the} (ferromag.)	μ_{obs} (ferromag.)	Easy axis	T_N/K hex sites	T_N/K cubic sites	T_C/K	Θ_D(‖c)/K	Θ_D(⊥c)/K	Θ_D(polycryst. or avg.)/K
αSc	295.2											
αY	187.7											
αLa	95.9											
βLa	105											
γCe	2,270	2.54	2.52		2.14			14.4				-50
βCe	2,500	2.54	2.61		2.14		13.7	12.5				-41
αPr	5,530	3.58	3.56	2.7[c]	3.20	**a**	0.03					0
αNd	5,930	3.62	3.45	2.2[c]	3.27	**b**	19.9	7.5		0	5	3.3
αPm		2.68			2.40							
αSm	1,278[d]	0.85	1.74	0.5[c]	0.71	**a**	109	14.0				
Eu	30,900	7.94	8.48	5.9	7.0	<110>		90.4				100
αGd	185,000[e]	7.94	7.98	7.63	7.0	30° to **c**			293.4	317	317	317
αTb	170,000	9.72	9.77				230.0			195	239	224
α′Tb				9.34	9.0	**b**			219.5			
αDy	98,000	10.64	10.83				180.2			121	169	153
α′Dy				10.33	10.0	**a**			90.5[f]			
Ho	72,900	10.60	11.2	10.34	10.0	**b**	132		19.5	73.0	88.0	83.0
Er	48,000	9.58	9.9	9.1	9.0	30° to **c**	85		18.7	61.7	32.5	42.2
Tm	24,700	7.56	7.61	7.14	7.0	**c**	58		32.0	41.0	-17.0	2.3
βYb	67[d]											
Lu	182.9											

[a] $g[J(J+1)]^{1/2}$.
[b] gJ.
[c] At 38 T and 4.2 K.
[d] At 290 K.
[e] At 350 K.
[f] On cooling T_C = 89.6 K and on warming T_C = 91.5 K.

Note: For additional information, see McEwen, K.A., in *Handbook on the Physics and Chemistry of Rare Earths*, Vol. 1, Gschneidner, K.A., Jr. and Eyring, L., Eds., North-Holland Physics, Amsterdam, 1978, 411; Legvold, S., in *Ferromagnetic Materials*, Vol. 1, Wohlfarth, E.P., Ed., North-Holland Physics, Amsterdam, 1980, 183; Pecharsky, V.K., Gschneidner, K.A., Jr. and Fort, D., *Phys. Rev. B*, 47, 5063, 1993; Pecharsky, V.K., Gschneidner, K.A., Jr. and Fort, D., 1996, to be published; Steward, A.M. and Collocott, S.J., *J. Phys.: Condens. Matter*, 1, 677, 1988.

TABLE 9. Room Temperature Coefficient of Thermal Expansion (α), Thermal Conductivity (k), Electrical Resistivity (ρ), and Hall Coefficient (R) of Rare-Earth Metals

Rare-earth metal	$\alpha_a \times 10^6/°C^{-1}$	$\alpha_c \times 10^6/°C^{-1}$	$\alpha_{poly} \times 10^6/°C^{-1}$	k/W cm^{-1} K^{-1}	ρ_a/μΩ cm	ρ_c/μΩ cm	ρ_{poly}/μΩ cm	$R_a \times 10^{12}$/V cm A^{-1} Oe^{-1}	$R_c \times 10^{12}$/V cm A^{-1} Oe^{-1}	$R_{poly} \times 10^{12}$/V cm A^{-1} Oe^{-1}
αSc	7.6	15.3	10.2	0.158	70.9	26.9	56.2[a]			-0.13
αY	6.0	19.7	10.6	0.172	72.5	35.5	59.6	-0.27	-1.6	
αLa	4.5	27.2	12.1	0.134			61.5			-0.35
βCe							82.8			
γCe	6.3		6.3	0.113			74.4			1.81
αPr	4.5	11.2	6.7	0.125			70.0			0.709
αNd	7.6	13.5	9.6	0.165			64.3			0.971
αPm	9[b]	16[b]	11[b]	0.15[b]			75[b]			
αSm	9.6	19.0	12.7	0.133			94.0			-0.21
Eu	35.0		35.0	0.139[b]			90.0			24.4
αGd	9.1[c]	10.0[c]	9.4[c]	0.105	135.1	121.7	131.0	-10	-54	4.48[d]
αTb	9.3	12.4	10.3	0.111	123.5	101.5	115.0	-1.0	-3.7	
αDy	7.1	15.6	9.9	0.107	111.0	76.6	92.6	-0.3	-3.7	
Ho	7.0	19.5	11.2	0.162	101.5	60.5	81.4	0.2	-3.2	
Er	7.9	20.9	12.2	0.145	94.5	60.3	86.0	0.3	-3.6	
Tm	8.8	22.2	13.3	0.169	88.0	47.2	67.6			-1.8
βYb	26.3		26.3	0.385			25.0			3.77
Lu	4.8	20.0	9.9	0.164	76.6	34.7	58.2	0.45	-2.6	-0.535

[a] Calculated from single crystal values.
[b] Estimated.
[c] At 100 °C.
[d] At 77 °C.

Note: For additional information, see Beaudry, B. J. and Gschneidner, K.A., Jr., in *Handbook on the Physics and Chemistry of Rare Earths*, Vol. 1, Gschneidner, K.A., Jr. and Eyring, L., Eds., North-Holland Physics, Amsterdam, 1978, 173; McEwen, K.A., in *Handbook on the Physics and Chemistry of Rare Earths*, Vol. 1, Gschneidner, K.A., Jr. and Eyring, L., Eds., North-Holland Physics, Amsterdam, 1978, 411.

TABLE 10. Electronic Specific Heat Constant (γ), Electron–Electron (Coulomb) Coupling Constant (μ^*), Electron–Phonon Coupling Constant (λ), Debye Temperature at 0 K (Θ_D), and Superconducting Transition Temperature of Rare-Earth Metals

Rare-earth metal	γ/mJ mol^{-1} K^{-2}	μ^*	λ	$\Theta_D(C_V)$/K	Θ_D(ec)/K	T_{sup}/K
αSc	10.334	0.16	0.30	345.3		0.050[a]
αY	7.878	0.15	0.30	244.4	258	1.3[b]
αLa	9.45	0.08	0.76	150	154	5.10
βLa	11.5			140		6.00
αCe	12.8			179		0.022[c]
αPr	20		1.07[d]	155[e]	153	
αNd	[f]		0.86[d]	157[e]	163	
αPm				159[e]		
αSm	8.1[g]		0.81[d]	162[e, f]	169	
Eu	[f]			[f]	118	
αGd	4.48		0.30	169	182	
α′Tb	3.71		0.34[d]	169.6	177	
α′Dy	4.9		0.32[d]	192	183	
Ho	2.1		0.30[d]	175[e]	190	
Er	8.7		0.33[d]	176.9	188	
Tm	[f]		0.36[d]	179[e]	200	
αYb	3.30			117.6	118	
βYb	8.36			109		
Lu	8.194	0.14	0.31	183.2	185	0.022[h]

[a] At 18.6 GPa.
[b] At 11 GPa.
[c] At 2.2 GPa.
[d] Calculated value.
[e] Estimated.
[f] Heat capacity results have been reported, but the resultant γ and Θ_D values are unreliable because of the presence of impurities and/or there was no reliable procedure or model to correct for the magnetic contribution to the heat capacity.
[g] ±1.5; based on the values reported for the purer Sm sample (IV).
[h] At 4.5 GPa.

Note: For additional information, see Sundström, L.J., in *Handbook on the Physics and Chemistry of Rare Earths*, Vol. 1, Gschneidner, K.A., Jr., and Eyring, L., Eds., North-Holland Physics, Amsterdam, 1978, 379; Scott, T., in *Handbook on the Physics and Chemistry of Rare Earths*, Vol. 1, Gschneidner, K.A., Jr. and Eyring, L., Eds., North-Holland Physics, Amsterdam, 1978, 591; Probst, C. and Wittig, J., in *Handbook on the Physics and Chemistry of Rare Earths*, Vol. 1, Gschneidner, K.A., Jr. and Eyring, L., Eds., North-Holland Physics, Amsterdam, 1978, 749; Tsang, T.-W.E., Gschneidner, K.A., Jr., Schmidt, F.A., and Thome, D.K., *Phys. Rev., B*, 31, 235, 1985; Collocott, S.J., Hill, R.W. and Stewart, A.M., *J. Phys. F*, 18, L223, 1988; Hill, R.W. and Gschneidner, K.A., Jr., *J. Phys. F*, 18, 2545, 1988; Skriver, H.L. and Mertig, I., *Phys. Rev. B*, 41, 6553, 1990; Collocott, S.J. and Stewart, A.M., *J. Phys.: Condens. Matter*, 4, 6743, 1992; Pecharsky, V.K., Gschneidner, K.A., Jr. and Fort, D., *Phys. Rev. B*, 47, 5063, 1993.

Mechanical, Thermal, Electrical, Magnetic, and Other Properties of Solid and Liquid Rare-Earth Metals

Table 11 has elastic and mechanical properties of rare-earth metals, and Table 12 has properties of liquid rare-earth metals near the melting point. Data columns are defined below.

Table 11 Column heading	Definition
Rare-earth metal	Symbol of the rare-earth metal
Young's (elastic) modulus	Young's (tensile elastic) modulus at room temperature, dimensionless
Shear modulus	Shear (transverse) modulus at room temperature, dimensionless
Bulk modulus	Bulk (compressibility) modulus at room temperature, dimensionless
Poisson's ratio	Poisson's ratio at room temperature; absolute value of the ratio of the transverse strain to the corresponding axial strain resulting from uniformly distributed axial stress below the proportional limit, dimensionless
Yield strength, 0.2% offset	Yield strength at room temperature, stress at which a material exhibits a specified deviation (here chosen as 0.2% for metals) from proportionality of stress and strain, in units of N m^{-2}
Ultimate tensile strength	Ultimate tensile strength at room temperature, the maximum stress that a material can withstand without being elongated; in units of N m^{-2}
Uniform elongation	Uniform elongation at room temperature, maximum elongation before necking, in percent (%)
Reduction in area	Reduction in area at room temperature, the proportional reduction of the cross-sectional area at the plane of fracture measured after fracture, in percent (%)
Recryst. temp.	Recrystallization temperature, temperature at which the liquid rare-earth metal crystalizes upon cooling, in °C

Table 12 Column heading	Definition
Rare-earth metal	Symbol of the rare-earth metal
ρ	Density (mass per unit volume); in units of g cm^{-3}
γ	Surface tension; in units of N m^{-1}
η	Viscosity, in units of centipoise
C_v	Molar heat capacity; in units of J mol^{-1} K^{-1}
λ	Thermal conductivity; in units of W cm^{-1} K^{-1}
χ	Magnetic susceptibility χ; in units 10^4 emu mol^{-1}
$\rho_{elec\ res}$	Electrical resistivity; in units of μΩ cm
$\Delta V(l \rightarrow s)$	Volume change from liquid phase to solid phase (upon freezing), in percent (%)
Spect. emit.	Spectral emittance at 645 nm wavelength, in percent (%)
Temp. range	Temperature range, in °C

TABLE 11. Room Temperature Elastic Moduli and Mechanical Properties of Rare-Earth Metals

Rare-earth metal	Young's (elastic) modulus	Shear modulus	Bulk modulus	Poisson's ratio	Yield strength 0.2% offset	Ultimate tensile strength	Uniform elongation/%	Reduction in area/%	Recryst. temp./°C
Sc	74.4	29.1	56.6	0.279	173[a]	255[a]	5.0[a]	8.0[a]	550
Y	63.5	25.6	41.2	0.243	42	129	34.0		550
αLa	36.6	14.3	27.9	0.280	126[a]	130	7.9[a]		300
βCe					86	138		24.0	
γCe	33.6	13.5	21.5	0.24	28	117	22.0	30.0	325
αPr	37.3	14.8	28.8	0.281	73	147	15.4	67.0	400
αNd	41.4	16.3	31.8	0.281	71	164	25.0	72.0	400
αPm	46[b]	18[b]	33[b]	0.28[b]					400[b]
αSm	49.7	19.5	37.8	0.274	68	156	17.0	29.5	440
Eu	18.2	7.9	8.3	0.152					300
αGd	54.8	21.8	37.9	0.259	15	118	37.0	56.0	500
αTb	55.7	22.1	38.7	0.261					500
αDy	61.4	24.7	40.5	0.247	43	139	30.0	30.0	550
Ho	64.8	26.3	40.2	0.231					520
Er	69.9	28.3	44.4	0.237	60	136	11.5	11.9	520
Tm	74.0	30.5	44.5	0.213					600
βYb	23.9	9.9	30.5	0.207	7	58	43.0	92.0	300
Lu	68.6	27.2	47.6	0.261					600

[a] Value is questionable.

[b] Estimated.

Note: For additional information, see Scott, T., in *Handbook on the Physics and Chemistry of Rare Earths*, Vol. 1, Gschneidner, K.A., Jr. and Eyring, L., Eds., North-Holland Physics, Amsterdam, 1978, 591.

TABLE 12. Properties of Liquid Rare-Earth Metals near the Melting Point

Rare-earth metal	ρ/g cm^{-3}	γ/N m^{-1}	η/centipoise	C_V/J mol^{-1} K^{-1}	λ/W cm^{-1} K^{-1}	$\chi\times10^4$/ emu mol^{-1}	$\rho_{elec\ res}$/μΩ cm	ΔV (l→s)[a]/%	Spect. emit.[c]/%	Temp. range/°C
Sc	2.80	0.954		44.2[b]						
Y	4.24	0.871		43.1					36.8	1522–1647
La	5.96	0.718	2.65	34.3	0.238	1.20	133	-0.6	25.4	920–1287
Ce	6.68	0.706	3.20	37.7	0.210	9.37	130	1.1	32.2	877–1547
Pr	6.59	0.707	2.85	43.0	0.251	17.3	139	-0.02	28.4	931–1537
Nd	6.72	0.687		48.8	0.195	18.7	151	-0.9	39.4	1021–1567
Pm	6.9[b]	0.680[b]		50[b]			160[b]			
Sm	7.16	0.431		50.2[b]		18.3	182	-3.6	43.7	1075
Eu	4.87	0.264		38.1		97	242	-4.8		
Gd	7.4	0.664		37.2	0.149	67	195	-2.0	34.2	1313–1600
Tb	7.65	0.669		46.5		82	193	-3.1		
Dy	8.2	0.648		49.9	0.187	95	210	-4.5	29.7	1412–1437
Ho	8.34	0.650		43.9		88	221	-7.4		
Er	8.6	0.637		38.7		69	226	-9.0	37.2	1529–1587
Tm	9.0[b]			41.4		41	235[b]	-6.9		
Yb	6.21	0.320	2.67	36.8			113	-5.1		
Lu	9.3	0.940		47.9[b]			224	-3.6		

[a] Volume change on freezing.
[b] Estimated.
[c] At 645 nm wavelength.

Note: For additional information, see Van Zytveld, J., in *Handbook on the Physics and Chemistry of Rare Earths,* Vol. 12, Gschneidner, K.A., Jr. and Eyring, L., Eds., North-Holland Physics, Amsterdam, 1989, 357; Stretz, L.A. and Bautista, R.G., in *Temperature, Its Measurement and Control in Science and Industry,* Vol. 4, part I, H.H. Plumb, Ed., Instrument Society of America, Pittsburgh, 1972, 489; King, T.S., Baria, D.N., and Bautista, R.G., *Met. Trans. B,* 7, 411, 1976; Baria, D.N., King, T.S., and Bautista, R.G., *Met. Trans. B,* 7, 577, 1976. <https://doi.org/10.1007/BF02698590>

Ionization Potentials and Ionic Radii

Tables 13 and 14 have important atomic properties for rare-earth metals. Table 13 contains data for the first five ionization potentials for these elements. Table 14 has effective ionic radii data for the elements in oxidation states +2, +3, and +4 for different coordination numbers. The data columns for these tables are defined below.

Table 13 Column heading	Definition
Rare-earth metal	Symbol of the rare-earth metal (R), in eV
IP (R^0)	Neutral atom ionization potential for the neutral atom (R^0), in eV
IP (R^{+1})	Singly ionized ionization potential for the singly ionized atom (R^{+1}), in eV
IP (R^{+2})	Doubly ionized ionization potential for the doubly ionized atom (R^{+2}), in eV
IP (R^{+3})	Triply ionized ionization potential for the triply ionized atom (R^{+3}), in eV
IP (R^{+4})	Quadruply ionized ionization potential for the quadruply ionized atom (R^{+4}), in eV

Table 14 Column heading	Definition
Rare-earth metal	Symbol of the rare-earth metal
***r* (R^{2+} CN6)**	Effective radii of doubly ionized rare-earth element with coordination number = 6, in Å
***r* (R^{2+} CN8)**	Effective radii of doubly ionized rare-earth element with coordination number = 8, in Å
***r* (R^{3+} CN6)**	Effective radii of triply ionized rare-earth element with coordination number = 6, in Å
***r* (R^{3+} CN8)**	Effective radii of triply ionized rare-earth element with coordination number = 8, in Å
***r* (R^{3+} CN12)**	Effective radii of triply ionized rare-earth element with coordination number = 12, in Å
***r* (R^{4+} CN6)**	Effective radii of quadruply ionized rare-earth element with coordination number = 6, in Å
***r* (R^{4+} CN8)**	Effective radii of quadruply ionized rare-earth element with coordination number = 8, in Å

TABLE 13. Ionization Potentials (eV) of Rare-Earth Elements

Rare-earth metal	IP (R^0)/eV	IP (R^{+1})/eV	IP (R^{+2})/eV	IP (R^{+3})/eV	IP (R^{+4})/eV
Sc	6.56144	12.79967	24.75666	73.4894	91.65
Y	6.217	12.24	20.52	60.597	77.0
La	5.5770	11.060	19.1773	49.95	61.6
Ce	5.5387	10.85	20.198	36.758	65.55
Pr	5.464	10.55	21.624	38.98	57.53
Nd	5.5250	10.73	22.1	40.41	
Pm	5.554	10.90	22.3	41.1	
Sm	5.6437	11.07	23.4	41.4	
Eu	5.6704	11.241	24.92	42.7	
Gd	6.1500	12.09	20.63	44.0	
Tb	5.8639	11.52	21.91	39.79	
Dy	5.9389	11.67	22.8	41.47	
Ho	6.0216	11.80	22.84	42.5	
Er	6.1078	11.93	22.74	42.7	
Tm	6.18431	12.05	23.68	42.7	
Yb	6.25416	12.1761	25.05	43.56	
Lu	5.42585	13.9	20.9594	45.25	66.8

Note: For references, see the table "Ionization Energies of Atoms and Atomic Ions" in Section 10 of this *CRC Handbook*.

TABLE 14. Effective Ionic Radii (Å) of Rare-Earth Ions

Rare-earth ion	r (R^{2+} CN6)/Å	r (R^{2+} CN8)/Å	r (R^{3+} CN6)/Å	r (R^{3+} CN8)/Å	r (R^{3+} CN12)/Å	r (R^{4+} CN6)/Å	r (R^{4+} CN8)/Å
Sc			0.745	0.87	1.116		
Y			0.900	1.015	1.220		
La			1.045	1.18	1.320		
Ce			1.010	1.14	1.290	0.80	0.97
Pr			0.997	1.14	1.286	0.78	0.96
Nd			0.983	1.12	1.276		
Pm			0.97	1.10	1.267		
Sm	1.19	1.27	0.958	1.09	1.260		
Eu	1.17	1.25	0.947	1.07	1.252		
Gd			0.938	1.06	1.246		
Tb			0.923	1.04	1.236	0.76	0.88
Dy			0.912	1.03	1.228		
Ho			0.901	1.02	1.221		
Er			0.890	1.00	1.214		
Tm			0.880	0.99	1.207		
Yb	1.00	1.07	0.868	0.98	1.199		
Lu			0.861	0.97	1.194		

Note: Radius of O^{2-} is 1.40 Å; for a coordination number (CN) of 6.
Note: For additional information, see Shannon, R.D. and Prewitt, C.T., *Acta Cryst.*, 25, 925, 1969; Shannon, R.D. and Prewitt, C.T., *Acta Cryst.*, 26, 1046, 1970. <https://doi.org/10.1107/S0567740870003576>

Inorganic

MELTING, BOILING, TRIPLE, AND CRITICAL POINTS OF THE ELEMENTS

This table summarizes the significant points on the phase diagrams for the elements. When reliable data are available, values are given for the solid-liquid-gas triple-point temperature and pressure; normal melting point at 101.325 kPa pressure; normal boiling point; and critical temperature and pressure. All temperatures are on the ITS-90 scale. The column definitions are as follows.

Column heading	Definition
Name	Element name; common phase names are in parentheses if more than one allotrope is included
Formula	Formula; for elements normally occuring as diatomic gases, the molecular formula is given
t_{tp}	Solid-liquid-gas triple-point temperature, in °C
p_{tp}	Solid-liquid-gas triple-point pressure, in kPa
t_m	Normal melting point, in °C at 101.325 kPa pressure; the notation "trans" in the melting point column indicates the temperature of the transition to the crystalline form immediately below that entry; transition temperatures between allotropic forms are included for several elements
t_b	Normal boiling point, in °C; an "sp" notation in the boiling point column indicates a sublimation point, where the vapor pressure of the solid phase reaches 101.325 kPa (1 atm)
t_c	Critical temperature, in °C
p_c	Critical pressure, in MPa

An asterisk* indicates an extrapolated or estimated value. The major data sources are listed below.

References

1. The International Temperature Scale of 1990, *Metrologia* 27, 3, 1990; errata in *Metrologia* 27, 107, 1990. <https://doi.org/10.1088/0026-1394/27/1/002>
2. Bedford, R. E., Bonnier, G., Maas, H., and Pavese, F., *Metrologia* 33, 133, 1996. <https://doi.org/10.1088/0026-1394/33/2/3>
3. Lemmon, E. W., Huber, M. L., and McLinden, M. O., NIST Standard Reference Database 23: Reference Fluid Thermodynamic and Transport Properties-REFPROP, Version 9.0, National Institute of Standards and Technology, Standard Reference Data Program, Gaithersburg, MD, 2007 <http://www.nist.gov/nist23.cfm>. [Ar, F, H, He, Kr, N, Ne, O, Xe]
4. Dinsdale, A.T., SGTE Data for Pure Elements, *CALPHAD* 15, 317-425, 1991. <https://doi.org/10.1016/0364-5916(91)90030-N>
5. *Landolt-Börnstein, Numerical Data and Functional Relationships in Science and Technology, New Series,* IV/19A, "Thermodynamic Properties of Inorganic Materials compiled by SGTE," Springer-Verlag, Heidelberg; Part 1, 1999.
6. Chase, M. W., *NIST-JANAF Thermochemical Tables, Fourth Edition, J. Phys. Chem. Ref. Data*, Monograph No. 9, 1998.
7. Gurvich, L. V., Veyts, I. V., and Alcock, C. B., *Thermodynamic Properties of Individual Substances, Fourth Edition,* Hemisphere Publishing Corp., New York, 1989.
8. Greenwood, N. N., and Earnshaw, A., *Chemistry of the Elements, Second Edition,* Butterworth-Heinemann, Oxford, 1997.
9. Hultgren, R. R., *Selected Values of Thermodynamic Properties of the Elements,* American Society of Metals, 1973.
10. Geiger, F., Busse, C. A., and Loehrke, R. I., *Int. J. Thermophys.* 8, 425, 1987. [Boiling points of In, Ag, Ga, Cu, Sn, and Au] <https://doi.org/10.1007/BF00567103>
11. Velasco, S., Roman, F. L., White, J. A., and Mulero, A., *Fluid Phase Equilib.* 244, 11, 2006. [Critical temperatures of Ag, Be, Bi, Ge, Fe, Mn, Te] <https://doi.org/10.1016/j.fluid.2006.03.017>
12. Gustafson, P., *Carbon* 24, 169, 1986. [Graphite and diamond] <https://doi.org/10.1016/0008-6223(86)90113-2>
13. Michels, A., and Prins, C., *Physica* 28, 101, 1962. [Triple point of Ar, Kr, Xe] <https://doi.org/10.1016/0031-8914(62)90096-4>

Melting, Boiling, Triple, and Critical Point Temperatures of the Elements

Name	Formula	t_{tp}/°C	p_{tp}/kPa	t_m/°C	t_b/°C	t_c/°C	p_c/MPa
Actinium	Ac			1050	≈3200		
Aluminum	Al			660.323	2519	6427 *	
Americium	Am			1176	≈2011		
Antimony (gray)	Sb			630.628	1587		
Argon	Ar	-189.3442	68.89	-189.34	-185.848	-122.463	4.863
Arsenic (gray)	As	817	3700	817	616 sp	1400	22.3
Astatine	At			302			
Barium	Ba			727	≈1845		
Berkelium (β form)	Bk			986			
Beryllium	Be			1287	2468	4932 *	
Bismuth	Bi			271.402	1564	4347 *	
Boron	B			2077	4000		
Bromine	Br_2	-7.25	5.879	-7.2	58.8	315	10.34
Cadmium	Cd			321.069	767		
Calcium	Ca			842	1484		
Californium	Cf			900			
Carbon (graphite)	C	4489	10300	4489 tp (10.3 MPa)	3825 sp		
Carbon (diamond)	C			4440 (12.4 GPa)**			
Cerium	Ce			799	3443		
Cesium	Cs			28.5	671	1665	9.4
Chlorine	Cl_2	-100.98	1.392	-101.5	-34.04	143.9	7.991

Name	Formula	t_{tp}/°C	p_{tp}/kPa	t_m/°C	t_b/°C	t_c/°C	p_c/MPa
Chromium	Cr			1907	2671		
Cobalt	Co			1495	2927		
Copper	Cu			1084.62	2560		
Curium	Cm			1345			
Dysprosium	Dy			1412	2567		
Einsteinium	Es			860			
Erbium	Er			1529	2868		
Europium	Eu			822	1529		
Fermium	Fm			1527			
Fluorine	F_2	-219.67	90	-219.67	-188.11	-128.74	5.1724
Francium	Fr			≈21			
Gadolinium	Gd			1313	3273		
Gallium	Ga	29.7666		29.7646	2229		
Germanium	Ge			938.25	2833	9529 *	
Gold	Au			1064.18	2836		
Hafnium	Hf			2233	4600		
Helium	He	-270.973	5.043		-268.928	-267.9550	0.22746
Holmium	Ho			1472	2700		
Hydrogen	H_2	-259.3467	7.041	-259.16	-252.879	-240.212	1.2858
Indium	In	156.5936		156.5985	2027		
Iodine	I_2	113.6	12.11	113.7	184.4	546	
Iridium	Ir			2446	4428		
Iron	Fe			1538	2861	9067 *	
Krypton	Kr	-157.375	73.53	-157.37	-153.415	-63.67	5.525
Lanthanum	La			920	3464		
Lawrencium	Lr			1627			
Lead	Pb			327.462	1749		
Lithium	Li			180.50	1342	2950 *	67 *
Lutetium	Lu			1663	3402		
Magnesium	Mg			650	1090		
Manganese	Mn			1246	2061	4052 *	
Mendelevium	Md			827			
Mercury	Hg	-38.8344		-38.8290	356.619	1491	167
Molybdenum	Mo			2622	4639		
Neodymium	Nd			1016	3074		
Neon	Ne	-248.5939	43.37	-248.59	-246.046	-228.6580	2.6786
Neptunium	Np			644	≈3902		
Nickel	Ni			1455	2913		
Niobium	Nb			2477	4741		
Nitrogen	N_2	-209.999	12.52	-210.0	-195.795	-146.958	3.3958
Nobelium	No			827			
Osmium	Os			3033	5008		
Oxygen	O_2	-218.7916	0.1463	-218.79	-182.962	-118.569	5.0430
Palladium	Pd			1554.8	2963		
Phosphorus (white)	P			44.15	280.5	721	
Phosphorus (red)	P			579.2	431 sp	721	
Platinum	Pt			1768.2	3825		
Plutonium	Pu			640	3228		
Polonium	Po			254	962		
Potassium	K			63.5	759	1950 *	16 *
Praseodymium	Pr			931	3520		
Promethium	Pm			1042	≈3000		
Protactinium	Pa			1572			
Radium	Ra			696			
Radon	Rn			-71	-61.7	104	6.28
Rhenium	Re			3185	5590		
Rhodium	Rh			1963	3695		
Rubidium	Rb	39.26		39.30	688	1820 *	16 *
Ruthenium	Ru			2333	4147		
Samarium	Sm			1072	1794		
Scandium	Sc			1541	2836		

Name	Formula	t_{tp}/°C	p_{tp}/kPa	t_m/°C	t_b/°C	t_c/°C	p_c/MPa
Selenium (vitreous)	Se			180 trans gray Se	685	1493	
Selenium (gray)	Se			220.8	685	1493	27.2
Silicon	Si			1414	3265		
Silver	Ag			961.78	2162	6137 *	
Sodium	Na			97.794	882.940	2300 *	35 *
Strontium	Sr			777	1377		
Sulfur (rhombic)	S			95.2 trans monocl	444.61	1041	20.7
Sulfur (monoclinic)	S			115.21	444.61	1041	
Tantalum	Ta			3017	5455		
Technetium	Tc			2157	4262		
Tellurium	Te			449.51	988	2056 *	
Terbium	Tb			1359	3230		
Thallium	Tl			304	1473		
Thorium	Th			1750	4785		
Thulium	Tm			1545	1950		
Tin (gray)	Sn			13.2 trans white Sn	2586		
Tin (white)	Sn			231.928	2586		
Titanium	Ti			1670	3287		
Tungsten	W			3414	5555		
Uranium	U			1135	4131		
Vanadium	V			1910	3407		
Xenon	Xe	-111.745	81.77	-111.75	-108.099	16.583	5.8420
Ytterbium	Yb			824	1196		
Yttrium	Y			1522	3345		
Zinc	Zn			419.527	907		
Zirconium	Zr			1854	4406		

* Extrapolated or estimated value.
** This is the estimated diamond-graphite-liquid carbon triple point (Ref. 12).

Inorganic

HEAT CAPACITY OF THE ELEMENTS AT 25 °C

This table gives the specific heat capacity at constant pressure (c_p) and the molar heat capacity (C_p) at a temperature of 25 °C and a pressure of 100 kPa (1 bar or 0.987 standard atmospheres) for all the elements for which reliable data are available. The column definitions are as follows.

Column heading	Definition
Name	Name of the element
Mol. form.	Formula; for elements normally occuring as diatomic gases at 25 °C, the molecular formula is given
State	Physical state of the element at 25 °C and a pressure of 100 kPa (1 bar or 0.987 standard atmosphere)
c_p	Specific heat capacity at constant pressure at 25 °C and a pressure of 100 kPa (1 bar or 0.987 standard atmosphere), in units of J g^{-1} K^{-1}
C_p	Molar heat capacity at constant pressure at 25 °C and a pressure of 100 kPa (1 bar or 0.987 standard atmosphere), in units of J mol^{-1} K^{-1}

Heat Capacity of the Elements at 25 °C

Name	Mol. form.	State	c_p/J g^{-1} K^{-1}	C_p/J mol^{-1} K^{-1}
Actinium	Ac	solid	0.120	27.2
Aluminum	Al	solid	0.897	24.20
Antimony (gray)	Sb	solid	0.207	25.23
Argon	Ar	gas	0.520	20.786
Arsenic (gray)	As	solid	0.329	24.64
Barium	Ba	solid	0.204	28.07
Beryllium	Be	solid	1.825	16.443
Bismuth	Bi	solid	0.122	25.52
Boron	B	solid	1.026	11.087
Bromine	Br_2	liquid	0.474	75.69
Cadmium	Cd	solid	0.231	26.020
Calcium	Ca	solid	0.647	25.929
Carbon (graphite)	C	solid	0.709	8.517
Cerium	Ce	solid	0.192	26.94
Cesium	Cs	solid	0.242	32.210
Chlorine	Cl_2	gas	0.479	33.949
Chromium	Cr	solid	0.449	23.35
Cobalt	Co	solid	0.421	24.81
Copper	Cu	solid	0.385	24.440
Dysprosium	Dy	solid	0.173	28.16
Erbium	Er	solid	0.168	28.12
Europium	Eu	solid	0.182	27.66
Fluorine	F_2	gas	0.824	31.304
Gadolinium	Gd	solid	0.235	37.03
Gallium	Ga	solid	0.373	26.03
Germanium	Ge	solid	0.320	23.222
Gold	Au	solid	0.129	25.418
Hafnium	Hf	solid	0.144	25.73
Helium	He	gas	5.193	20.786
Holmium	Ho	solid	0.165	27.15
Hydrogen	H_2	gas	14.304	28.836
Indium	In	solid	0.233	26.74
Iodine	I_2	solid	0.214	54.43
Iridium	Ir	solid	0.131	25.10
Iron	Fe	solid	0.449	25.10
Krypton	Kr	gas	0.248	20.786
Lanthanum	La	solid	0.195	27.11
Lead	Pb	solid	0.130	26.84
Lithium	Li	solid	3.582	24.860
Lutetium	Lu	solid	0.154	26.86
Magnesium	Mg	solid	1.023	24.869
Manganese	Mn	solid	0.479	26.32
Mercury	Hg	liquid	0.140	27.983
Molybdenum	Mo	solid	0.251	24.06
Neodymium	Nd	solid	0.190	27.45
Neon	Ne	gas	1.030	20.786
Nickel	Ni	solid	0.444	26.07
Niobium	Nb	solid	0.265	24.60
Nitrogen	N_2	gas	1.040	29.124
Osmium	Os	solid	0.130	24.7
Oxygen	O_2	gas	0.918	29.378
Palladium	Pd	solid	0.244	25.98
Phosphorus (white)	P	solid	0.769	23.824
Platinum	Pt	solid	0.133	25.86
Potassium	K	solid	0.757	29.600
Praseodymium	Pr	solid	0.193	27.20
Radon	Rn	gas	0.094	20.786
Rhenium	Re	solid	0.137	25.48
Rhodium	Rh	solid	0.243	24.98
Rubidium	Rb	solid	0.363	31.060
Ruthenium	Ru	solid	0.238	24.06
Samarium	Sm	solid	0.196	29.54
Scandium	Sc	solid	0.568	25.52
Selenium (gray)	Se	solid	0.321	25.363
Silicon	Si	solid	0.712	19.99
Silver	Ag	solid	0.235	25.350
Sodium	Na	solid	1.228	28.230
Strontium	Sr	solid	0.306	26.79
Sulfur (rhombic)	S	solid	0.708	22.70
Tantalum	Ta	solid	0.140	25.36
Tellurium	Te	solid	0.202	25.73
Terbium	Tb	solid	0.182	28.91
Thallium	Tl	solid	0.129	26.32
Thorium	Th	solid	0.118	27.32
Thulium	Tm	solid	0.160	27.03
Tin (white)	Sn	solid	0.227	26.99
Titanium	Ti	solid	0.524	25.060
Tungsten	W	solid	0.132	24.27
Uranium	U	solid	0.116	27.665
Vanadium	V	solid	0.489	24.89
Xenon	Xe	gas	0.158	20.786
Ytterbium	Yb	solid	0.155	26.74
Yttrium	Y	solid	0.298	26.53
Zinc	Zn	solid	0.388	25.390
Zirconium	Zr	solid	0.278	25.36

VAPOR PRESSURE OF THE METALLIC ELEMENTS — EQUATIONS

C. B. Alcock

This table gives coefficients in an equation for the vapor pressure of 65 metallic elements in both the solid and liquid states. Vapor pressures in the range 10^{-10} to 10^{2} Pa (10^{-15} to 10^{-3} atm) are covered. The equation is:

for p in atmospheres: $\log(p/\text{atm}) = A + BT^{-1} + C\log T + DT^{-3}$
for p in pascals: $\log(p/\text{Pa}) = 5.006 + A + BT^{-1} + C\log T + DT^{-3}$
for p in torr (mmHg): $\log(p/\text{torr}) = 2.881 + A + BT^{-1} + C\log T + DT^{-3}$

where T is the temperature in K.

The column definitions are as follows.

Column heading	Definition
Name	Name of element; the metallic elements are listed alphabetically by name
Symbol	Atomic symbol of element
Phase	Phase for which the vapor pressure equation applies
A	Numerical constant in vapor pressure equations above
B	Coefficient of T^{-1} term in vapor pressure equations above
C	Coefficient of $\log T$ term in vapor pressure equations above
D	Coefficient of T^{-3} term in vapor pressure equations above
Range	Temperature range of applicability of vapor pressure equations, in K
mp	Melting point of solid phase, in K

This equation reproduces the observed vapor pressures to an accuracy of 5% or better.

The table following this one gives values of the vapor pressure at several temperatures in the 400 K to 2400 K range, as calculated from these equations.

Reprinted with permission of the publisher, Pergamon Press.

Reference

Alcock, C. B., Itkin, V. P., and Horrigan, M. K., *Canadian Metallurgical Quarterly,* 23, 309, 1984. <https://doi.org/10.1179/cmq.1984.23.3.309>

Vapor Pressure of Metallic Elements: Equation Coefficients

Name	Symbol	Phase	*A*	*B*	*C*	*D*	Range/K	mp/K
Aluminum	Al	solid	9.459	-17342	-0.7927		298-mp	933
Aluminum	Al	liquid	5.911	-16211			mp-1800	933
Americium	Am	solid	11.311	-15059	-1.3449		298-mp	1449
Barium	Ba	solid	12.405	-9690	-2.2890		298-mp	1000
Barium	Ba	liquid	4.007	-8163			mp-1200	1000
Beryllium	Be	solid	8.042	-17020	-0.4440		298-mp	1560
Beryllium	Be	liquid	5.786	-15731			mp-1800	1560
Cadmium	Cd	solid	5.939	-5799			298-mp	594
Cadmium	Cd	liquid	5.242	-5392			mp-650	594
Calcium	Ca	solid	10.127	-9517	-1.4030		298-mp	1115
Cerium	Ce	solid	6.139	-21752			298-mp	1072
Cerium	Ce	liquid	5.611	-21200			mp-2450	1072
Cesium	Cs	solid	4.711	-3999			298-mp	302
Cesium	Cs	liquid	4.165	-3830			mp-550	302
Chromium	Cr	solid	6.800	-20733	0.4391	-0.4094	298-2000	2180
Cobalt	Co	solid	10.976	-22576	-1.0280		298-mp	1768
Cobalt	Co	liquid	6.488	-20578			mp-2150	1768
Copper	Cu	solid	9.123	-17748	-0.7317		298-mp	1358
Copper	Cu	liquid	5.849	-16415			mp-1850	1358
Curium	Cm	solid	8.369	-20364	-0.5770		298-mp	1618
Curium	Cm	liquid	5.223	-18292			mp-2200	1618
Dysprosium	Dy	solid	9.579	-15336	-1.1114		298-mp	1685
Erbium	Er	solid	9.916	-16642	-1.2154		298-mp	1802
Erbium	Er	liquid	4.668	-14380			mp-1900	1802
Europium	Eu	solid	9.240	-9459	-1.1661		298-mp	1095
Gadolinium	Gd	solid	8.344	-20861	-0.5775		298-mp	1586
Gadolinium	Gd	liquid	5.557	-19389			mp-2250	1586
Gallium	Ga	solid	6.657	-14208			298-mp	303
Gallium	Ga	liquid	6.754	-13984	-0.3413		mp-1600	303
Gold	Au	solid	9.152	-19343	-0.7479		298-mp	1337
Gold	Au	liquid	5.832	-18024			mp-2050	1337

Name	Symbol	Phase	*A*	*B*	*C*	*D*	Range/K	mp/K
Hafnium	Hf	solid	9.445	-32482	-0.6735		298-mp	2506
Holmium	Ho	solid	9.785	-15899	-1.1753		298-mp	1745
Indium	In	solid	5.991	-12548			298-mp	430
Indium	In	liquid	5.374	-12276			mp-1500	430
Iridium	Ir	solid	10.506	-35099	-0.7500		298-2500	2719
Iron	Fe	solid	7.100	-21723	0.4536	-0.5846	298-mp	1811
Iron	Fe	liquid	6.347	-19574			mp-2100	1811
Lanthanum	La	solid	7.463	-22551	-0.3142		298-mp	1193
Lanthanum	La	liquid	5.911	-21855			mp-2450	1193
Lead	Pb	solid	5.643	-10143			298-mp	600
Lead	Pb	liquid	4.911	-9701			mp-1200	600
Lithium	Li	solid	5.667	-8310			298-mp	454
Lithium	Li	liquid	5.055	-8023			mp-1000	454
Lutetium	Lu	solid	8.793	-22423	-0.6200		298-mp	1936
Lutetium	Lu	liquid	5.648	-20302			mp-2350	1936
Magnesium	Mg	solid	8.489	-7813	-0.8253		298-mp	923
Manganese	Mn	solid	12.805	-15097	-1.7896		298-mp	1519
Mercury	Hg	liquid	5.116	-3190			298-400	234
Molybdenum	Mo	solid	11.529	-34626	-1.1331		298-2500	2895
Neodymium	Nd	solid	8.996	-17264	-0.9519		298-mp	1289
Neodymium	Nd	liquid	4.912	-15824			mp-2000	1289
Neptunium	Np	solid	19.643	-24886	-3.9991		298-mp	917
Neptunium	Np	liquid	10.076	-23378	-1.3250		mp-2500	917
Nickel	Ni	solid	10.557	-22606	-0.8717		298-mp	1728
Nickel	Ni	liquid	6.666	-20765			mp-2150	1728
Niobium	Nb	solid	8.882	-37818	-0.2575		298-2500	2750
Osmium	Os	solid	9.419	-41198	-0.3896		298-2500	3306
Palladium	Pd	solid	9.502	-19813	-0.9258		298-mp	1828
Palladium	Pd	liquid	5.426	-17899			mp-2100	1828
Platinum	Pt	solid	4.882	-29387	1.1039	-0.4527	298-mp	2041
Platinum	Pt	liquid	6.386	-26856			mp-2500	2041
Plutonium	Pu	solid	26.160	-19162	-6.6675		298-600	913
Plutonium	Pu	solid	18.858	-18460	-4.4720		500-913	913
Plutonium	Pu	liquid	3.666	-16658			mp-2450	913
Potassium	K	solid	4.961	-4646			298-mp	337
Potassium	K	liquid	4.402	-4453			mp-600	337
Praseodymium	Pr	solid	8.859	-18720	-0.9512		298-mp	1204
Praseodymium	Pr	liquid	4.772	-17315			mp-2200	1204
Protactinium	Pa	solid	10.552	-34869	-1.0075		298-mp	1845
Protactinium	Pa	liquid	6.177	-32874			mp-2500	1845
Rhenium	Re	solid	11.543	-40726	-1.1629		298-2500	3458
Rhodium	Rh	solid	10.168	-29010	-0.7068		298-mp	2236
Rhodium	Rh	liquid	6.802	-26792			mp-2500	2236
Rubidium	Rb	solid	4.857	-4215			298-mp	312
Rubidium	Rb	liquid	4.312	-4040			mp-550	312
Ruthenium	Ru	solid	9.755	-34154	-0.4723		298-mp	2606
Samarium	Sm	solid	9.988	-11034	-1.3287		298-mp	1345
Scandium	Sc	solid	6.650	-19721	0.2885	-0.3663	298-mp	1814
Scandium	Sc	liquid	5.795	-17681			mp-2000	1814
Silver	Ag	solid	9.127	-14999	-0.7845		298-mp	1235
Silver	Ag	liquid	5.752	-13827			mp-1600	1235
Sodium	Na	solid	5.298	-5603			298-mp	371
Sodium	Na	liquid	4.704	-5377			mp-700	371
Strontium	Sr	solid	9.226	-8572	-1.1926		298-mp	1050
Tantalum	Ta	solid	16.807	-41346	-3.2152	0.7437	298-2500	3290
Terbium	Tb	solid	9.510	-20457	-0.9247		298-mp	1632
Terbium	Tb	liquid	5.411	-18639			mp-2200	1632
Thallium	Tl	solid	5.971	-9447			298-mp	577
Thallium	Tl	liquid	5.259	-9037			mp-1100	577
Thorium	Th	solid	8.668	-31483	-0.5288		298-mp	2023
Thorium	Th	liquid	-18.453	-24569	6.6473		mp-2500	2023

Name	Symbol	Phase	*A*	*B*	*C*	*D*	Range/K	mp/K
Thulium	Tm	solid	8.882	-12270	-0.9564		298-1400	1818
Tin (white)	Sn	solid	6.036	-15710			298-mp	505
Tin (white)	Sn	liquid	5.262	-15332			mp-1850	505
Titanium	Ti	solid	11.925	-24991	-1.3376		298-mp	1943
Titanium	Ti	liquid	6.358	-22747			mp-2400	1943
Tungsten	W	solid	2.945	-44094	1.3677		298-2350	3687
Tungsten	W	solid	54.527	-57687	-12.2231		2200-2500	3687
Uranium	U	solid	0.770	-27729	2.6982	-1.5471	298-mp	1408
Uranium	U	liquid	20.735	-28776	-4.0962		mp-2500	1408
Vanadium	V	solid	9.744	-27132	-0.5501		298-mp	2183
Vanadium	V	liquid	6.929	-25011			mp-2500	2183
Ytterbium	Yb	solid	9.111	-8111	-1.0849		298-900	1097
Yttrium	Y	solid	9.735	-22306	-0.8705		298-mp	1795
Yttrium	Y	liquid	5.795	-20341			mp-2300	1795
Zinc	Zn	solid	6.102	-6776			298-mp	693
Zinc	Zn	liquid	5.378	-6286			mp-750	693
Zirconium	Zr	solid	10.008	-31512	-0.7890		298-mp	2127
Zirconium	Zr	liquid	6.806	-30295			mp-2500	2127

VAPOR PRESSURE OF THE METALLIC ELEMENTS — DATA

The following values of the vapor pressure of metallic elements are calculated from the equations in the preceding table. All values are given in pascals. For conversion, note that 1 Pa = 7.50 μm Hg = $9.87 \cdot 10^{-6}$ atm. The column definitions are as follows.

Column heading	Definition
Name	Name of element
T_{mp}	Melting point, in K
***p* (400 K)**	Vapor pressure of element at 400 K, in Pa
***p* (600 K)**	Vapor pressure of element at 600 K, in Pa
***p* (800 K)**	Vapor pressure of element at 800 K, in Pa
***p* (1000 K)**	Vapor pressure of element at 1000 K, in Pa
***p* (1200 K)**	Vapor pressure of element at 1200 K, in Pa
***p* (1400 K)**	Vapor pressure of element at 1400 K, in Pa
***p* (1600 K)**	Vapor pressure of element at 1600 K, in Pa
***p* (1800 K)**	Vapor pressure of element at 1800 K, in Pa
***p* (2000 K)**	Vapor pressure of element at 2000 K, in Pa
***p* (2200 K)**	Vapor pressure of element at 2200 K, in Pa
***p* (2400 K)**	Vapor pressure of element at 2400 K, in Pa

Vapor Pressure of Metallic Elements at Selected Temperatures

Name	T_{mp}/K	*p* (400 K)/Pa	*p* (600 K)/Pa	*p* (800 K)/Pa	*p* (1000 K)/Pa	*p* (1200 K)/Pa	*p* (1400 K)/Pa	*p* (1600 K)/Pa	*p* (1800 K)/Pa	*p* (2000 K)/Pa	*p* (2200 K)/Pa	*p* (2400 K)/Pa
Aluminum	933			$3.06 \cdot 10^{-10}$	$5.08 \cdot 10^{-6}$	0.00256	0.218	6.10	81.4			
Americium	1449			$3.88 \cdot 10^{-7}$	0.00167	0.423	21.35					
Barium	1000		$7.97 \cdot 10^{-6}$	0.0450	7.11	162						
Beryllium	1560			$3.04 \cdot 10^{-10}$	$4.96 \cdot 10^{-6}$	0.00314	0.312	9.12	113			
Cadmium	594	0.000280	18.2									
Calcium	1115		$2.36 \cdot 10^{-5}$	0.146	25.5							
Cerium	1072				$2.47 \cdot 10^{-11}$	$8.91 \cdot 10^{-8}$	$2.97 \cdot 10^{-5}$	0.00233	0.0691	1.04	9.56	60.8
Cesium	302	0.394										
Chromium	2180				$2.45 \cdot 10^{-8}$	$7.59 \cdot 10^{-5}$	0.0239	1.80	52.1	774		
Cobalt	1768				$2.09 \cdot 10^{-10}$	$1.00 \cdot 10^{-6}$	0.000419	0.0379	1.15	16.0		
Copper	1358			$6.60 \cdot 10^{-11}$	$1.53 \cdot 10^{-6}$	0.00122	0.135	3.94	54.4			
Curium	1618				$1.90 \cdot 10^{-9}$	$4.24 \cdot 10^{-6}$	0.00103	0.0629	1.17	12.1	82.1	
Dysprosium	1685			$1.54 \cdot 10^{-8}$	$8.21 \cdot 10^{-5}$	0.0241	1.362	27.5				
Erbium	1802			$3.90 \cdot 10^{-10}$	$4.30 \cdot 10^{-6}$	0.00205	0.163	4.23	52.5			
Europium	1095		$1.74 \cdot 10^{-5}$	0.109	19.4							
Gadolinium	1586				$5.70 \cdot 10^{-10}$	$1.54 \cdot 10^{-6}$	0.000429	0.0279	0.618	7.39	56.2	
Gallium	303			$1.94 \cdot 10^{-7}$	0.000565	0.114	4.98	84.4				
Gold	1337				$3.72 \cdot 10^{-8}$	$5.44 \cdot 10^{-5}$	0.00920	0.374	6.68	67.0		
Hafnium	2506						$1.35 \cdot 10^{-11}$	$9.81 \cdot 10^{-9}$	$1.63 \cdot 10^{-6}$	$9.69 \cdot 10^{-5}$	0.00272	0.0437
Holmium	1745			$3.20 \cdot 10^{-9}$	$2.32 \cdot 10^{-5}$	0.00837	0.546	12.3				
Indium	430		$8.31 \cdot 10^{-11}$	$1.08 \cdot 10^{-5}$	0.0127	1.413	40.9					
Iridium	2719							$1.48 \cdot 10^{-9}$	$3.72 \cdot 10^{-7}$	$3.06 \cdot 10^{-5}$	0.00112	0.0225
Iron	1811				$5.54 \cdot 10^{-9}$	$2.51 \cdot 10^{-5}$	0.0104	0.961	32.7	36.8		
Lanthanum	1193					$5.09 \cdot 10^{-8}$	$2.02 \cdot 10^{-5}$	0.00181	0.0596	0.976	9.61	64.7
Lead	600		$5.54 \cdot 10^{-7}$	0.00618	1.64	68.1						
Lithium	454	$7.90 \cdot 10^{-11}$	0.000489	1.08	109							
Lutetium	1936				$3.28 \cdot 10^{-11}$	$1.59 \cdot 10^{-7}$	$6.79 \cdot 10^{-5}$	0.00628	0.211	3.18	26.7	
Magnesium	923	$6.53 \cdot 10^{-9}$	0.0152	21.5								
Manganese	1519			$5.55 \cdot 10^{-7}$	0.00221	0.524	24.9					
Mercury	234	140										
Molybdenum	2895							$1.83 \cdot 10^{-9}$	$4.07 \cdot 10^{-7}$	$3.03 \cdot 10^{-5}$	0.00102	0.0189
Neodymium	1289			$4.55 \cdot 10^{-11}$	$7.62 \cdot 10^{-7}$	0.000483	0.0412	1.07	13.4	101		
Neptunium	917					$3.31 \cdot 10^{-9}$	$1.63 \cdot 10^{-6}$	0.000168	0.00604	0.105	1.06	7.28
Nickel	1728				$2.19 \cdot 10^{-10}$	$1.09 \cdot 10^{-6}$	0.000471	0.0438	1.37	19.5		
Niobium	2750							$2.32 \cdot 10^{-11}$	$9.54 \cdot 10^{-9}$	$1.17 \cdot 10^{-6}$	$5.98 \cdot 10^{-5}$	0.00158
Osmium	3306								$1.85 \cdot 10^{-10}$	$3.46 \cdot 10^{-8}$	$2.49 \cdot 10^{-6}$	$8.75 \cdot 10^{-5}$
Palladium	1828				$8.27 \cdot 10^{-9}$	$1.40 \cdot 10^{-5}$	0.00277	0.144	3.07	30.4		
Platinum	2041						$2.34 \cdot 10^{-8}$	$1.14 \cdot 10^{-5}$	0.00143	0.0689	0.153	1.59
Plutonium	913				$1.03 \cdot 10^{-8}$	$6.17 \cdot 10^{-6}$	0.000594	0.0182	0.262	2.20	12.6	53.8
Potassium	337	0.0188	96.9									
Praseodymium	1204				$1.95 \cdot 10^{-8}$	$2.16 \cdot 10^{-5}$	0.00257	0.0904	1.44	13.2	80.8	
Protactinium	1845							$3.44 \cdot 10^{-10}$	$8.06 \cdot 10^{-8}$	$5.57 \cdot 10^{-6}$	0.000174	0.00306
Rhenium	3458								$1.37 \cdot 10^{-10}$	$2.22 \cdot 10^{-8}$	$1.41 \cdot 10^{-6}$	$4.45 \cdot 10^{-5}$
Rhodium	2236						$1.69 \cdot 10^{-8}$	$5.99 \cdot 10^{-6}$	0.000571	0.0217	0.422	4.41

Name	T_{mp}/ K	p (400 K)/ Pa	p (600 K)/ Pa	p (800 K)/ Pa	p (1000 K)/ Pa	p (1200 K)/ Pa	p (1400 K)/ Pa	p (1600 K)/ Pa	p (1800 K)/ Pa	p (2000 K)/ Pa	p (2200 K)/ Pa	p (2400 K)/ Pa
Rubidium	312	0.165										
Ruthenium	2606							$7.96 \cdot 10^{-9}$	$1.77 \cdot 10^{-6}$	0.000133	0.00455	0.0858
Samarium	1345		$8.17 \cdot 10^{-8}$	0.00221	0.942	51.0						
Scandium	1814				$6.31 \cdot 10^{-8}$	0.000129	0.0300	1.80	43.6	91.3		
Silver	1235			$1.27 \cdot 10^{-7}$	0.000603	0.165	7.61	131				
Sodium	371	0.000185	5.60									
Strontium	1050	$4.99 \cdot 10^{-11}$	0.000429	1.134	121							
Tantalum	3290									$3.36 \cdot 10^{-10}$	$1.87 \cdot 10^{-8}$	$5.21 \cdot 10^{-7}$
Terbium	1632				$1.92 \cdot 10^{-9}$	$4.18 \cdot 10^{-6}$	0.000988	0.0585	1.15	12.5	88.0	
Thallium	577		$1.59 \cdot 10^{-5}$	0.0931	16.9							
Thorium	2023						$3.33 \cdot 10^{-11}$	$2.00 \cdot 10^{-8}$	$2.89 \cdot 10^{-6}$	0.000154	0.00401	0.0610
Thulium	1818		$6.03 \cdot 10^{-10}$	$5.94 \cdot 10^{-5}$	0.0561	5.22	130					
Tin	505			$1.26 \cdot 10^{-9}$	$8.62 \cdot 10^{-6}$	0.00310	0.207	4.85	56.3			
Titanium	1943					$9.69 \cdot 10^{-9}$	$7.44 \cdot 10^{-6}$	0.00106	0.0493	0.978	10.6	76.9
Tungsten	3687									$2.62 \cdot 10^{-10}$	$3.01 \cdot 10^{-8}$	$1.59 \cdot 10^{-6}$
Uranium	1408					$9.47 \cdot 10^{-10}$	$2.87 \cdot 10^{-6}$	$4.27 \cdot 10^{-6}$	0.000263	0.00678	0.0933	0.803
Vanadium	2183					$2.79 \cdot 10^{-10}$	$4.35 \cdot 10^{-7}$	0.000107	0.00769	0.233	3.68	32.6
Ytterbium	1097	$1.03 \cdot 10^{-9}$	0.00384	6.74								
Yttrium	1795				$6.66 \cdot 10^{-11}$	$2.96 \cdot 10^{-7}$	0.000117	0.0102	0.316	4.27	35.9	
Zinc	693	$1.47 \cdot 10^{-6}$	0.653									
Zirconium	2127						$1.05 \cdot 10^{-10}$	$6.17 \cdot 10^{-8}$	$8.68 \cdot 10^{-6}$	0.000450	0.0110	0.155

Inorganic

DENSITY OF MOLTEN ELEMENTS AND REPRESENTATIVE SALTS

This table lists the liquid density at the melting point, ρ_m, for elements that are solid at room temperature, as well as for some representative salts of these elements. Densities at higher temperatures (up to the t_{max} given in the last column) may be estimated from the equation

$$\rho(t) = \rho_m - k(t - t_m)$$

where t_m is the melting point and k is the derivative of density with respect to temperature. If a value of t_{max} is not given, the equation should not be used to extrapolate more than about 20 °C beyond the melting point. Column definitions are as follows.

Column heading	Definition
Name	Name of molten element or molten salt
Mol. form	Molecular formula
t_m	Melting point, in °C
ρ_m	Liquid density at the melting point, in units of g cm^{-3}
k	Extrapolation factor (derivative of density with respect to temperature), in units of g cm^{-3} °C^{-1}
t_{max}	Maximum temperature for validity of extrapolation equation, in °C

Data for the elements were selected from the primary literature. The molten salt data were derived from Ref. 1.

References

1. Janz, G. J., Thermodynamic and Transport Properties of Molten Salts: Correlation Equations for Critically Evaluated Density, Surface Tension, Electrical Conductance, and Viscosity Data, *J. Phys. Chem. Ref. Data* 17, Suppl. 2, 1988.
2. Nasch, P. M., and Steinemann, S. G., *Phys. Chem. Liq.* 29, 43, 1995. <https://doi.org/10.1080/00319109508030263>
3. Assael, M. J., Kakosimos, K., Banish, R. M., Brillo, J., Egry, I., Brooks, R., Quested, P. N., Mills, K. C., Nagashima, A., Sato, Y., and Wakeham, W. A., *J. Phys. Chem. Ref. Data* 35, 285, 2006. [Al, Fe] <https://doi.org/10.1063/1.2149380>
4. Assael, M. J., Kalyva, A. E., Antoniadis, K. D., Banish, R. M., Egry, I., Wu, J., Kaschnitz, E., and Wakeham, W. A., *J. Phys. Chem. Ref. Data* 39, 033105-1, 2010. [Cu, Sn] <https://doi.org/10.1063/1.3467496>

Density of Molten Elements and Salts

Name	Mol. form	t_m/°C	ρ_m/g cm^{-3}	k/g cm^{-3} °C^{-1}	t_{max}/°C
Aluminum	Al	660.323	2.377	0.000311	917
Aluminum bromide	$AlBr_3$	97.5	2.647	0.002435	267
Aluminum chloride	$AlCl_3$	192.6	1.302	0.002711	296
Aluminum iodide	AlI_3	188.28	3.223	0.0025	240
Antimony (gray)	Sb	630.628	6.53	0.00067	745
Antimony(III) chloride	$SbCl_3$	73.4	2.681	0.002293	77
Antimony(III) iodide	SbI_3	171	4.171	0.002483	322
Antimony(V) chloride	$SbCl_5$	4	2.37	0.001869	77
Arsenic (gray)	As	817	5.22	0.000544	
Barium	Ba	727	3.338	0.000299	1550
Barium bromide	$BaBr_2$	857	3.991	0.000924	900
Barium chloride	$BaCl_2$	961	3.174	0.000681	1081
Barium fluoride	BaF_2	1368	4.14	0.000999	1727
Barium iodide	BaI_2	711	4.26	0.000977	975
Beryllium	Be	1287	1.690	0.00011	
Beryllium chloride	$BeCl_2$	415	1.54	0.0011	473
Beryllium fluoride	BeF_2	552	1.96	0.000015	850
Bismuth	Bi	271.402	10.05	0.00135	800
Bismuth tribromide	$BiBr_3$	219	4.76	0.002637	927
Bismuth trichloride	$BiCl_3$	234	3.916	0.0023	350
Boron	B	2077	2.08		
Cadmium	Cd	321.069	7.996	0.001218	500
Cadmium bromide	$CdBr_2$	568	4.075	0.00108	720
Cadmium chloride	$CdCl_2$	568	3.392	0.00082	807
Cadmium iodide	CdI_2	388	4.396	0.001117	700
Calcium	Ca	842	1.378	0.000230	1484
Calcium bromide	$CaBr_2$	742	3.111	0.0005	791
Calcium chloride	$CaCl_2$	775	2.085	0.000422	950
Calcium fluoride	CaF_2	1418	2.52	0.000391	2027
Calcium iodide	CaI_2	783	3.443	0.000751	1028
Cerium	Ce	799	6.55	0.000710	1460
Cerium(III) chloride	$CeCl_3$	807	3.25	0.00092	950
Cerium(III) fluoride	CeF_3	1430	4.659	0.000936	1927
Cesium	Cs	28.5	1.843	0.000556	510
Cesium bromide	CsBr	636	3.133	0.001223	860
Cesium chloride	CsCl	646	2.79	0.001065	906
Cesium fluoride	CsF	703	3.649	0.001282	912
Cesium iodide	CsI	632	3.197	0.001183	907
Cesium nitrate	$CsNO_3$	409	2.820	0.001166	491
Cesium sulfate	Cs_2SO_4	1005	3.1	0.00095	1530
Chromium	Cr	1907	6.3	0.0011	2100
Cobalt	Co	1495	7.75	0.00165	1580
Copper	Cu	1084.62	7.997	0.000819	2227
Copper(I) chloride	CuCl	423	3.692	0.00076	585
Dysprosium	Dy	1412	8.37	0.00143	1540
Dysprosium(III) chloride	$DyCl_3$	718	3.62	0.00068	987
Erbium	Er	1529	8.86	0.00157	1700
Europium	Eu	822	5.13	0.0028	980
Gadolinium	Gd	1313	7.4		
Gadolinium(III) chloride	$GdCl_3$	602	3.56	0.000671	1007
Gadolinium(III) iodide	GdI_3	930	4.12	0.000908	1032
Gallium	Ga	29.7646	6.08	0.00062	400
Gallium(III) bromide	$GaBr_3$	123	3.116	0.00246	135
Gallium(III) chloride	$GaCl_3$	77.9	2.053	0.002083	141
Gallium(III) iodide	GaI_3	212	3.630	0.002377	252
Germanium	Ge	938.25	5.60	0.00055	1600
Gold	Au	1064.18	17.31	0.001343	1200
Hafnium	Hf	2233	12		
Holmium	Ho	1472	8.34		
Indium	In	156.5985	7.02	0.000836	500
Indium(III) bromide	$InBr_3$	420	3.121	0.0015	528
Indium(III) chloride	$InCl_3$	583	2.140	0.0021	666
Indium(III) iodide	InI_3	207	3.820	0.0015	360
Iridium	Ir	2446	19		

Name	Mol. form	t_m/°C	ρ_m/g cm^{-3}	k/g cm^{-3} °C^{-1}	t_{max}/°C
Iron	Fe	1538	7.035	0.000926	2207
Iron(II) chloride	$FeCl_2$	677	2.348	0.000555	877
Lanthanum	La	920	5.94	0.00061	1600
Lanthanum bromide	$LaBr_3$	788	4.933	0.000096	912
Lanthanum chloride	$LaCl_3$	858	3.209	0.000777	973
Lanthanum fluoride	LaF_3	1493	4.589	0.000682	2177
Lanthanum iodide	LaI_3	778	4.29	0.001110	907
Lead	Pb	327.462	10.66	0.00122	700
Lead(II) bromide	$PbBr_2$	371	5.73	0.00165	600
Lead(II) chloride	$PbCl_2$	501	4.951	0.0015	710
Lead(II) iodide	PbI_2	410	5.691	0.001594	697
Lithium	Li	180.50	0.512	0.00052	285
Lithium bromide	LiBr	550	2.528	0.000652	739
Lithium chloride	LiCl	610	1.502	0.000432	781
Lithium fluoride	LiF	848.2	1.81	0.000490	1047
Lithium iodide	LiI	469	3.109	0.000917	667
Lithium nitrate	$LiNO_3$	253	1.781	0.000546	441
Lithium sulfate	Li_2SO_4	860	2.003	0.000407	1214
Lutetium	Lu	1663	9.3		
Magnesium	Mg	650	1.584	0.000234	900
Magnesium bromide	$MgBr_2$	711	2.62	0.000478	935
Magnesium chloride	$MgCl_2$	714	1.68	0.000271	826
Magnesium iodide	MgI_2	634	3.05	0.000651	888
Manganese	Mn	1246	5.95	0.00105	1590
Manganese(II) chloride	$MnCl_2$	650	2.353	0.000437	850
Mercury(II) bromide	$HgBr_2$	241	5.126	0.003233	319
Mercury(II) chloride	$HgCl_2$	277	4.368	0.002862	304
Mercury(II) iodide (yellow)	HgI_2	256	5.222	0.003235	354
Molybdenum	Mo	2622	9.33		
Neodymium	Nd	1016	6.89	0.00076	1350
Nickel	Ni	1455	7.81	0.000726	1700
Nickel(II) chloride	$NiCl_2$	1031	2.653	0.00066	1057
Osmium	Os	3033	20		
Palladium	Pd	1554.8	10.38	0.001169	1700
Platinum	Pt	1768.2	19.77	0.0024	2200
Plutonium	Pu	640	16.63	0.001419	950
Potassium	K	63.5	0.828	0.000232	500
Potassium bromide	KBr	734	2.127	0.000825	930
Potassium chloride	KCl	771	1.527	0.000583	939
Potassium fluoride	KF	858	1.910	0.000651	1037
Potassium iodide	KI	681	2.448	0.000956	904
Potassium nitrate	KNO_3	334	1.865	0.000723	457
Praseodymium	Pr	931	6.50	0.00093	1460
Praseodymium(III) chloride	$PrCl_3$	786	3.23	0.00074	977
Rhenium	Re	3185	18.9		
Rhodium	Rh	1963	10.7	0.000895	2200
Rubidium	Rb	39.30	1.46	0.000451	800
Rubidium bromide	RbBr	692	2.715	0.001072	907
Rubidium carbonate	Rb_2CO_3	873	2.84	0.000640	1007
Rubidium chloride	RbCl	724	2.248	0.000883	923
Rubidium fluoride	RbF	795	2.87	0.00102	1067
Rubidium iodide	RbI	656	2.904	0.001143	902
Rubidium nitrate	$RbNO_3$	310	2.519	0.001068	417
Rubidium sulfate	Rb_2SO_4	1066	2.56	0.000665	1545
Ruthenium	Ru	2333	10.65		
Samarium	Sm	1072	7.16		
Scandium	Sc	1541	2.68	0.00054	1680
Selenium (gray)	Se	220.8	3.99		
Silicon	Si	1414	2.57	0.00036	1500
Silver	Ag	961.78	9.320	0.0009	1500
Silver(I) bromide	AgBr	430	5.577	0.001035	667
Silver(I) chloride	AgCl	455	4.83	0.00094	627
Silver(I) iodide	AgI	558	5.58	0.00101	802
Silver(I) nitrate	$AgNO_3$	210	3.970	0.001098	360
Silver(I) sulfate	Ag_2SO_4	660	4.84	0.001089	770
Sodium	Na	97.794	0.927	0.00023	600
Sodium bromide	NaBr	747	2.342	0.000816	945
Sodium carbonate	Na_2CO_3	856	1.972	0.000448	1004
Sodium chloride	NaCl	802.018	1.556	0.000543	1027
Sodium fluoride	NaF	996	1.948	0.000636	1097
Sodium iodide	NaI	661	2.742	0.000949	912
Sodium nitrate	$NaNO_3$	306.5	1.90	0.000715	370
Sodium sulfate	Na_2SO_4	884	2.069	0.000483	1077
Strontium	Sr	777	6.980		
Strontium bromide	$SrBr_2$	657	3.70	0.000745	1004
Strontium chloride	$SrCl_2$	874	2.727	0.000578	1037
Strontium fluoride	SrF_2	1477	3.470	0.000751	1927
Strontium iodide	SrI_2	538	4.085	0.000885	1026
Sulfur (monoclinic)	S	115.21	1.819	0.00080	160
Tantalum	Ta	3017	15		
Tantalum(V) chloride	$TaCl_5$	216.6	2.700	0.004316	457
Tellurium	Te	449.51	5.70	0.00035	600
Terbium	Tb	1359	7.65		
Thallium	Tl	304	11.22	0.00144	600
Thallium(I) bromide	TlBr	460	5.98	0.001755	647
Thallium(I) chloride	TlCl	431	5.628	0.0018	642
Thallium(I) iodide	TlI	441.7	6.15	0.001761	737
Thallium(I) nitrate	$TlNO_3$	206	4.91	0.001873	279
Thallium(I) sulfate	Tl_2SO_4	632	5.62	0.00130	927
Thorium(IV) chloride	$ThCl_4$	770	3.363	0.0014	847
Thorium(IV) fluoride	ThF_4	1110	6.058	0.000759	1378
Thulium	Tm	1545	8.56	0.00050	1675
Tin (white)	Sn	231.928	6.979	0.000652	1650
Tin(II) chloride	$SnCl_2$	247.0	3.36	0.001253	480
Tin(IV) chloride	$SnCl_4$	-34.07	2.37	0.002687	138
Titanium	Ti	1670	4.11		
Titanium(IV) chloride	$TiCl_4$	-24.12	1.807	0.001735	137
Tungsten	W	3414	17.6		
Uranium	U	1135	17.3		
Uranium(III) chloride	UCl_3	837	4.84	0.007943	1057
Uranium(IV) chloride	UCl_4	590	3.572	0.001945	667
Uranium(IV) fluoride	UF_4	1036	6.485	0.000992	1341
Vanadium	V	1910	5.5		
Ytterbium	Yb	824	6.21		
Yttrium	Y	1522	4.18	0.00029	1685
Yttrium chloride	YCl_3	721	2.510	0.0005	845
Zinc	Zn	419.527	6.57	0.0011	700
Zinc bromide	$ZnBr_2$	402	3.47	0.000959	602
Zinc chloride	$ZnCl_2$	325	2.54	0.00053	557
Zinc iodide	ZnI_2	450	3.878	0.00136	588
Zinc sulfate	$ZnSO_4$	680 dec	3.14	0.00047	987
Zirconium	Zr	1854	5.8		
Zirconium(IV) chloride	$ZrCl_4$	437 tp	1.643	0.007464	492

MAGNETIC SUSCEPTIBILITY OF THE ELEMENTS AND INORGANIC COMPOUNDS

When a material is placed in a magnetic field ***H***, a magnetization (magnetic moment per unit volume) ***M*** is induced in the material which is related to ***H*** by ***M*** = κ***H***, where κ is called the volume susceptibility. Because ***H*** and ***M*** have the same dimensions, κ is dimensionless. A more useful parameter is the molar susceptibility χ_m, defined by

$$\chi_m = \kappa V_m = \kappa M/\rho$$

where V_m is the molar volume of the substance, M the molar mass, and ρ the mass density. When the cgs system is used, the customary units for χ_m are $cm^3 mol^{-1}$; the corresponding SI units are $m^3 mol^{-1}$.

Substances that have no unpaired electron orbital or spin angular momentum generally have negative values of χ_m and are called diamagnetic. Their molar susceptibility varies only slightly with temperature. Substances with unpaired electrons, which are termed paramagnetic, have positive χ_m and show a much stronger temperature dependence, varying roughly as $1/T$. The net susceptibility of a paramagnetic substance is the sum of the paramagnetic and diamagnetic contributions, but the former almost always dominates. Ferromagnetic substances, indicated here by a footnote, have more complex magnetic behavior.

This table gives values of χ_m for the elements and selected inorganic compounds. All values refer to nominal room temperature (285 to 300 K) unless otherwise indicated.

In keeping with customary practice, the molar susceptibility is given here in units appropriate to the cgs system. These values should be multiplied by 4π to obtain values for use in SI equations (where the magnetic field strength ***H*** has units of A m^{-1}). Columns are defined as follows.

Column heading	Definition
Name	Name of element or inorganic compound; substances are arranged in alphabetical order by the most common name
Mol. form.	Molecular formula
Phase	Phase; when the phase is indicated as "solid," the most common crystalline form is understood
χ_m	Molar magnetic suspectibility, in units of 10^{-6} cm^3 mol^{-1}

References

1. *Landolt-Börnstein, Numerical Data and Functional Relationships in Science and Technology, New Series,* II/16, *Diamagnetic Susceptibility,* Springer-Verlag, Heidelberg, 1986.
2. *Landolt-Börnstein, Numerical Data and Functional Relationships in Science and Technology, New Series,* III/19, Subvolumes a to i2, *Magnetic Properties of Metals,* Springer-Verlag, Heidelberg, 1986-1992.
3. *Landolt-Börnstein, Numerical Data and Functional Relationships in Science and Technology, New Series,* II/2, II/8, II/10, II/11, and II/12a, *Coordination and Organometallic Transition Metal Compounds,* Springer-Verlag, Heidelberg, 1966-1984.
4. *Tables de Constantes et Données Numérique, Volume 7, Relaxation Paramagnetique,* Masson, Paris, 1957.

Magnetic Susceptibility of the Elements and Inorganic Materials

Name	Mol. form.	Phase	$\chi_m/10^{-6}$ cm^3 mol^{-1}
Aluminum	Al	solid	16.5
Aluminum fluoride	AlF_3	solid	-13.9
Aluminum oxide (α)	Al_2O_3	solid	-37
Aluminum sulfate	$Al_2(SO_4)_3$	solid	-93
Ammonia	NH_3	gas	-16.3
Ammonia	NH_3	aq. soln.	-18.3
Ammonium acetate	$NH_4C_2H_3O_2$	solid	-41.1
Ammonium bromide	NH_4Br	solid	-47
Ammonium carbonate	$(NH_4)_2CO_3$	solid	-42.5
Ammonium chlorate	NH_4ClO_3	solid	-42.1
Ammonium chloride	NH_4Cl	solid	-36.7
Ammonium fluoride	NH_4F	solid	-23
Ammonium iodate	NH_4IO_3	solid	-62.3
Ammonium iodide	NH_4I	solid	-66
Ammonium nitrate	NH_4NO_3	solid	-33
Ammonium sulfate	$(NH_4)_2SO_4$	solid	-67
Ammonium thiocyanate	NH_4SCN	solid	-48.1
Antimony	Sb	solid	-99
Antimony(III) bromide	$SbBr_3$	solid	-111.4
Antimony(III) chloride	$SbCl_3$	solid	-86.7
Antimony(V) chloride	$SbCl_5$	solid	-120.5
Antimony(III) fluoride	SbF_3	solid	-46
Antimony(III) iodide	SbI_3	solid	-147.2
Antimony(III) oxide (senarmontite)	Sb_2O_3	solid	-69.4
Antimony(III) sulfide	Sb_2S_3	solid	-86
Argon	Ar	gas	-19.32
Arsenic (gray)	As	solid	-5.6
Arsenic(III) bromide	$AsBr_3$	solid	-106
Arsenic(III) chloride	$AsCl_3$	solid	-72.5
Arsenic(III) iodide	AsI_3	solid	-142.2
Arsenic(III) oxide (arsenolite)	As_2O_3	solid	-30.34
Arsenic(III) sulfide	As_2S_3	solid	-70
Arsenic (yellow)	As	solid	-23.2
Arsine	AsH_3	gas	-35.2
Barium	Ba	solid	20.6
Barium bromide	$BaBr_2$	solid	-92
Barium bromide dihydrate	$BaBr_2·2H_2O$	solid	-119.3
Barium carbonate	$BaCO_3$	solid	-58.9
Barium chloride	$BaCl_2$	solid	-72.6
Barium chloride dihydrate	$BaCl_2·2H_2O$	solid	-100
Barium fluoride	BaF_2	solid	-51
Barium hydroxide	$Ba(OH)_2$	solid	-53.2
Barium iodate	$Ba(IO_3)_2$	solid	-122.5
Barium iodide	BaI_2	solid	-124.4
Barium iodide dihydrate	$BaI_2·2H_2O$	solid	-163
Barium nitrate	$Ba(NO_3)_2$	solid	-66.5
Barium oxide	BaO	solid	-29.1
Barium peroxide	BaO_2	solid	-40.6
Barium sulfate	$BaSO_4$	solid	-65.8
Beryllium	Be	solid	-9.0
Beryllium chloride	$BeCl_2$	solid	-26.5
Beryllium hydroxide (α)	$Be(OH)_2$	solid	-23.1
Beryllium oxide	BeO	solid	-11.9

Name	Mol. form.	Phase	$\chi_m/10^{-6}$ cm³ mol⁻¹
Beryllium sulfate	$BeSO_4$	solid	-37
Bismuth	Bi	solid	-280.1
Bismuth hydroxide	$Bi(OH)_3$	solid	-65.8
Bismuth nitrate pentahydrate	$Bi(NO_3)_3{\cdot}5H_2O$	solid	-159
Bismuth oxide	Bi_2O_3	solid	-83
Bismuth phosphate	$BiPO_4$	solid	-77
Bismuth sulfate	$Bi_2(SO_4)_3$	solid	-199
Bismuth sulfide	Bi_2S_3	solid	-123
Bismuth tribromide	$BiBr_3$	solid	-147
Bismuth trichloride	$BiCl_3$	solid	-26.5
Bismuth trifluoride	BiF_3	solid	-61.2
Bismuth triiodide	BiI_3	solid	-200.5
Boric acid	H_3BO_3	solid	-34.1
Boron	B	solid	-6.7
Boron oxide	B_2O_3	solid	-38.7
Boron trichloride	BCl_3	gas	-59.9
Bromine	Br_2	liquid	-56.4
Bromine	Br_2	gas	-73.5
Bromine pentafluoride	BrF_5	liquid	-45.1
Bromine trifluoride	BrF_3	gas	-33.9
Cadmium	Cd	solid	-19.7
Cadmium bromide	$CdBr_2$	solid	-87.3
Cadmium bromide tetrahydrate	$CdBr_2{\cdot}4H_2O$	solid	-131.5
Cadmium carbonate	$CdCO_3$	solid	-46.7
Cadmium chloride	$CdCl_2$	solid	-68.7
Cadmium chromate	$CdCrO_4$	solid	-16.8
Cadmium cyanide	$Cd(CN)_2$	solid	-54
Cadmium fluoride	CdF_2	solid	-40.6
Cadmium hydroxide	$Cd(OH)_2$	solid	-41
Cadmium iodate	$Cd(IO_3)_2$	solid	-108.4
Cadmium iodide	CdI_2	solid	-117.2
Cadmium nitrate	$Cd(NO_3)_2$	solid	-55.1
Cadmium nitrate tetrahydrate	$Cd(NO_3)_2{\cdot}4H_2O$	solid	-140
Cadmium oxide	CdO	solid	-30
Cadmium sulfate	$CdSO_4$	solid	-59.2
Cadmium sulfide	CdS	solid	-50
Calcium	Ca	solid	40
Calcium bromide	$CaBr_2$	solid	-73.8
Calcium carbonate (calcite)	$CaCO_3$	solid	-38.2
Calcium chloride	$CaCl_2$	solid	-54.7
Calcium fluoride	CaF_2	solid	-28
Calcium hydroxide	$Ca(OH)_2$	solid	-22
Calcium iodate	$Ca(IO_3)_2$	solid	-101.4
Calcium iodide	CaI_2	solid	-109
Calcium oxide	CaO	solid	-15.0
Calcium sulfate	$CaSO_4$	solid	-49.7
Calcium sulfate dihydrate	$CaSO_4{\cdot}2H_2O$	solid	-74
Carbon (diamond)	C	solid	-5.9
Carbon dioxide	CO_2	gas	-21.0
Carbon (graphite)	C	solid	-6.0
Carbon monoxide	CO	gas	-9.8
Cerium	Ce	β form	2500
Cerium(III) chloride	$CeCl_3$	solid	2490
Cerium(III) fluoride	CeF_3	solid	2190
Cerium(IV) oxide	CeO_2	solid	26
Cerium(IV) sulfate tetrahydrate	$Ce(SO_4)_2{\cdot}4H_2O$	solid	-97
Cerium(II) sulfide	CeS	solid	2110
Cerium(III) sulfide	Ce_2S_3	solid	5080
Cesium	Cs	solid	29
Cesium bromate	$CsBrO_3$	solid	-75.1
Cesium bromide	CsBr	solid	-67.2
Cesium carbonate	Cs_2CO_3	solid	-103.6
Cesium chlorate	$CsClO_3$	solid	-65
Cesium chloride	CsCl	solid	-56.7
Cesium fluoride	CsF	solid	-44.5
Cesium iodide	CsI	solid	-82.6
Cesium sulfate	Cs_2SO_4	solid	-116
Cesium superoxide	CsO_2	solid	1534
Chlorine	Cl_2	liquid	-40.4
Chlorine trifluoride	ClF_3	gas	-26.5
Chromium	Cr	solid	167
Chromium(II) chloride	$CrCl_2$	solid	7230
Chromium(III) chloride	$CrCl_3$	solid	6350
Chromium(III) fluoride	CrF_3	solid	4370
Chromium(III) oxide	Cr_2O_3	solid	1960
Chromium(VI) oxide	CrO_3	solid	40
Chromium(III) sulfate	$Cr_2(SO_4)_3$	solid	11800
Cobalt	Co	solid	[a]
Cobalt(II) bromide	$CoBr_2$	solid	13000
Cobalt(II) chloride	$CoCl_2$	solid	12660
Cobalt(II) chloride hexahydrate	$CoCl_2{\cdot}6H_2O$	solid	9710
Cobalt(II) cyanide	$Co(CN)_2$	solid	3825
Cobalt(II) fluoride	CoF_2	solid	9490
Cobalt(III) fluoride	CoF_3	solid	1900
Cobalt(II) iodide	CoI_2	solid	10760
Cobalt(II,III) oxide	Co_3O_4	solid	7380
Cobalt(III) oxide	Co_2O_3	solid	4560
Cobalt(II) sulfate	$CoSO_4$	solid	10000
Cobalt(II) sulfide	CoS	solid	225
Copper	Cu	solid	-5.46
Copper(I) bromide	CuBr	solid	-49
Copper(II) bromide	$CuBr_2$	solid	685
Copper(I) chloride	CuCl	solid	-40
Copper(II) chloride	$CuCl_2$	solid	1080
Copper(II) chloride dihydrate	$CuCl_2{\cdot}2H_2O$	solid	1420
Copper(I) cyanide	CuCN	solid	-24
Copper(II) fluoride	CuF_2	solid	1050
Copper(II) fluoride dihydrate	$CuF_2{\cdot}2H_2O$	solid	1600
Copper(II) hydroxide	$Cu(OH)_2$	solid	1170
Copper(I) iodide	CuI	solid	-63
Copper(II) nitrate hexahydrate	$Cu(NO_3)_2{\cdot}6H_2O$	solid	1625
Copper(II) nitrate trihydrate	$Cu(NO_3)_2{\cdot}3H_2O$	solid	1570
Copper(I) oxide	Cu_2O	solid	-20
Copper(II) oxide	CuO	solid	238
Copper(II) sulfate	$CuSO_4$	solid	1330
Copper(II) sulfate pentahydrate	$CuSO_4{\cdot}5H_2O$	solid	1460
Copper(II) sulfide	CuS	solid	-2.0
Diborane	B_2H_6	gas	-21.0
Disilane	Si_2H_6	gas	-37.3
Dysprosium	Dy	α form	98000
Dysprosium(III) oxide	Dy_2O_3	solid	89600
Dysprosium(III) sulfide	Dy_2S_3	solid	95200
Erbium	Er	solid	48000
Erbium oxide	Er_2O_3	solid	73920
Erbium sulfate octahydrate	$Er_2(SO_4)_3{\cdot}8H_2O$	solid	74600
Erbium sulfide	Er_2S_3	solid	77200
Europium	Eu	solid	30900
Europium(II) bromide	$EuBr_2$	solid	26800
Europium(II) chloride	$EuCl_2$	solid	26500
Europium(II) fluoride	EuF_2	solid	23750

Name	Mol. form.	Phase	$\chi_m/10^{-6}$ $cm^3\ mol^{-1}$
Europium(II) iodide	EuI_2	solid	26000
Europium(III) oxide	Eu_2O_3	solid	10100
Europium(III) sulfate	$Eu_2(SO_4)_3$	solid	10400
Europium(II) sulfide	EuS	solid	23800
Gadolinium	Gd	solid, 350 K	185000
Gadolinium(III) chloride	$GdCl_3$	solid	27930
Gadolinium(III) oxide	Gd_2O_3	solid	53200
Gadolinium(III) sulfate octahydrate	$Gd_2(SO_4)_3 \cdot 8H_2O$	solid	53280
Gadolinium(III) sulfide	Gd_2S_3	solid	55500
Gallium	Ga	solid	-21.6
Gallium(III) chloride	$GaCl_3$	solid	-63
Gallium suboxide	Ga_2O	solid	-34
Gallium(II) sulfide	GaS	solid	-23
Gallium(III) sulfide	Ga_2S_3	solid	-80
Germane	GeH_4	gas	-29.7
Germanium	Ge	solid	-11.6
Germanium(IV) chloride	$GeCl_4$	solid	-72
Germanium(IV) fluoride	GeF_4	solid	-50
Germanium(IV) iodide	GeI_4	solid	-171
Germanium(II) oxide	GeO	solid	-28.8
Germanium(IV) oxide	GeO_2	solid	-34.3
Germanium(II) sulfide	GeS	solid	-40.9
Germanium(IV) sulfide	GeS_2	solid	-53.9
Gold	Au	solid	-28
Gold(I) bromide	AuBr	solid	-61
Gold(I) chloride	AuCl	solid	-67
Gold(III) chloride	$AuCl_3$	solid	-112
Gold(I) iodide	AuI	solid	-91
Hafnium	Hf	solid	71
Hafnium(IV) oxide	HfO_2	solid	-23
Helium	He	gas	-2.02
Holmium	Ho	solid	72900
Holmium oxide	Ho_2O_3	solid	88100
Hydrazine	N_2H_4	liquid	-201
Hydrogen	H_2	liquid, 20.3 K	-5.44
Hydrogen	H_2	gas	-3.99
Hydrogen chloride	HCl	liquid	-22.6
Hydrogen chloride	HCl	aq. soln.	-22
Hydrogen fluoride	HF	liquid	-8.6
Hydrogen fluoride	HF	aq. soln.	-9.3
Hydrogen iodide	HI	solid, 195 K	-47.3
Hydrogen iodide	HI	liquid, 233 K	-48.3
Hydrogen iodide	HI	aq. soln.	-50.2
Hydrogen peroxide	H_2O_2	liquid	-17.3
Hydrogen sulfide	H_2S	gas	-25.5
Indium	In	solid	-10.2
Indium(III) bromide	$InBr_3$	solid	-107
Indium(I) chloride	InCl	solid	-30
Indium(II) chloride	$InCl_2$	solid	-56
Indium(III) chloride	$InCl_3$	solid	-86
Indium(III) oxide	In_2O_3	solid	-56
Indium(II) sulfide	InS	solid	-28
Indium(III) sulfide	In_2S_3	solid	-98
Iodic acid	HIO_3	solid	-48
Iodine	I_2	solid	-90
Iodine chloride	ICl	liquid	-54.6
Iodine pentafluoride	IF_5	liquid	-58.1
Iodine pentoxide	I_2O_5	solid	-79.4
Iodine trichloride	ICl_3	solid	-90.2
Iridium	Ir	solid	25
Iridium(III) chloride	$IrCl_3$	solid	-14.4
Iridium(IV) oxide	IrO_2	solid	224
Iron	Fe	solid	[a]
Iron(II) bromide	$FeBr_2$	solid	13600
Iron(II) carbonate	$FeCO_3$	solid	11300
Iron(II) chloride	$FeCl_2$	solid	14750
Iron(III) chloride	$FeCl_3$	solid	13450
Iron(III) chloride hexahydrate	$FeCl_3 \cdot 6H_2O$	solid	15250
Iron(II) chloride tetrahydrate	$FeCl_2 \cdot 4H_2O$	solid	12900
Iron(II) fluoride	FeF_2	solid	9500
Iron(III) fluoride	FeF_3	solid	13760
Iron(III) fluoride trihydrate	$FeF_3 \cdot 3H_2O$	solid	7870
Iron(II) iodide	FeI_2	solid	13600
Iron(III) nitrate nonahydrate	$Fe(NO_3)_3 \cdot 9H_2O$	solid	15200
Iron(II) oxide	FeO	solid	7200
Iron(II) sulfate	$FeSO_4$	solid	12400
Iron(II) sulfate heptahydrate	$FeSO_4 \cdot 7H_2O$	solid	11200
Iron(II) sulfate monohydrate	$FeSO_4 \cdot H_2O$	solid	10500
Iron(II) sulfide	FeS	solid	1074
Krypton	Kr	gas	-29.0
Lanthanum	La	α form	95.9
Lanthanum oxide	La_2O_3	solid	-78
Lanthanum sulfate nonahydrate	$La_2(SO_4)_3 \cdot 9H_2O$	solid	-262
Lanthanum sulfide	La_2S_3	solid	-37
Lead	Pb	solid	-23
Lead(II) acetate	$Pb(C_2H_3O_2)_2$	solid	-89.1
Lead(II) bromide	$PbBr_2$	solid	-90.6
Lead(II) carbonate	$PbCO_3$	solid	-61.2
Lead(II) chloride	$PbCl_2$	solid	-73.8
Lead(II) chromate	$PbCrO_4$	solid	-18
Lead(II) fluoride	PbF_2	solid	-58.1
Lead(II) iodate	$Pb(IO_3)_2$	solid	-131
Lead(II) iodide	PbI_2	solid	-126.5
Lead(II) nitrate	$Pb(NO_3)_2$	solid	-74
Lead(II) oxide (massicot)	PbO	solid	-42
Lead(II) phosphate	$Pb_3(PO_4)_2$	solid	-182
Lead(II) sulfate	$PbSO_4$	solid	-69.7
Lead(II) sulfide	PbS	solid	-83.6
Lithium	Li	solid	14.2
Lithium bromide	LiBr	solid	-34.3
Lithium carbonate	Li_2CO_3	solid	-27
Lithium chloride	LiCl	solid	-24.3
Lithium fluoride	LiF	solid	-10.1
Lithium hydride	LiH	solid	-4.6
Lithium hydroxide	LiOH	aq. soln.	-12.3
Lithium iodide	LiI	solid	-50
Lithium sulfate	Li_2SO_4	solid	-41.6
Lutetium	Lu	solid	182.9
Magnesium	Mg	solid	13.1
Magnesium bromide	$MgBr_2$	solid	-72
Magnesium carbonate	$MgCO_3$	solid	-32.4
Magnesium chloride	$MgCl_2$	solid	-47.4
Magnesium fluoride	MgF_2	solid	-22.7
Magnesium hydroxide	$Mg(OH)_2$	solid	-22.1
Magnesium iodide	MgI_2	solid	-111
Magnesium oxide	MgO	solid	-10.2

Name	Mol. form.	Phase	$\chi_m/10^{-6}$ cm³ mol⁻¹
Magnesium sulfate	$MgSO_4$	solid	-42
Magnesium sulfate heptahydrate	$MgSO_4·7H_2O$	solid	-135.7
Magnesium sulfate monohydrate	$MgSO_4·H_2O$	solid	-61
Manganese	Mn	solid	511
Manganese(II) bromide	$MnBr_2$	solid	13900
Manganese(II) carbonate	$MnCO_3$	solid	11400
Manganese(II) chloride	$MnCl_2$	solid	14350
Manganese(II) chloride tetrahydrate	$MnCl_2·4H_2O$	solid	14600
Manganese(II) fluoride	MnF_2	solid	10700
Manganese(III) fluoride	MnF_3	solid	10500
Manganese(II) hydroxide	$Mn(OH)_2$	solid	13500
Manganese(II) iodide	MnI_2	solid	14400
Manganese(II) oxide	MnO	solid	4850
Manganese(II,III) oxide	Mn_3O_4	solid	12400
Manganese(III) oxide	Mn_2O_3	solid	14100
Manganese(IV) oxide	MnO_2	solid	2280
Manganese(II) sulfate	$MnSO_4$	solid	13660
Manganese(II) sulfate monohydrate	$MnSO_4·H_2O$	solid	14200
Manganese(II) sulfate tetrahydrate	$MnSO_4·4H_2O$	solid	14600
Manganese(II) sulfide (α form)	MnS	α form	5630
Manganese(II) sulfide (β form)	MnS	β form	3850
Mercury	Hg	solid, 234 K	-24.1
Mercury	Hg	liquid	-33.5
Mercury(I) bromide	Hg_2Br_2	solid	-105
Mercury(II) bromide	$HgBr_2$	solid	-94.2
Mercury(I) chloride	Hg_2Cl_2	solid	-120
Mercury(II) chloride	$HgCl_2$	solid	-82
Mercury(II) cyanide	$Hg(CN)_2$	solid	-67
Mercury(I) fluoride	Hg_2F_2	solid	-106
Mercury(II) fluoride	HgF_2	solid	-57.3
Mercury(I) iodide	Hg_2I_2	solid	-166
Mercury(II) iodide (red)	HgI_2	solid	-165
Mercury(I) nitrate	$Hg_2(NO_3)_2$	solid	-121
Mercury(II) nitrate	$Hg(NO_3)_2$	solid	-74
Mercury(I) oxide	Hg_2O	solid	-76.3
Mercury(II) oxide	HgO	solid	-46
Mercury(I) sulfate	Hg_2SO_4	solid	-123
Mercury(II) sulfate	$HgSO_4$	solid	-78.1
Mercury(II) sulfide (red)	HgS	solid	-55.4
Mercury(II) thiocyanate	$Hg(SCN)_2$	solid	-96.5
Molybdenum	Mo	solid	72
Molybdenum(III) bromide	$MoBr_3$	solid	525
Molybdenum(IV) bromide	$MoBr_4$	solid	520
Molybdenum(III) chloride	$MoCl_3$	solid	43
Molybdenum(IV) chloride	$MoCl_4$	solid	1750
Molybdenum(V) chloride	$MoCl_5$	solid	990
Molybdenum(VI) fluoride	MoF_6	solid	-26.0
Molybdenum(III) oxide	Mo_2O_3	solid	-42.0
Molybdenum(IV) oxide	MoO_2	solid	41
Molybdenum(VI) oxide	MoO_3	solid	3
Neodymium	Nd	α form	5930
Neodymium(III) fluoride	NdF_3	solid	4980
Neodymium(III) oxide	Nd_2O_3	solid	10200
Neodymium(III) sulfate	$Nd_2(SO_4)_3$	solid	9990
Neodymium(III) sulfide	Nd_2S_3	solid	5550

Name	Mol. form.	Phase	$\chi_m/10^{-6}$ cm³ mol⁻¹
Neon	Ne	gas	-6.96
Neptunium	Np	solid	575
Nickel	Ni	solid	[a]
Nickel(II) bromide	$NiBr_2$	solid	5600
Nickel(II) chloride	$NiCl_2$	solid	6145
Nickel(II) chloride hexahydrate	$NiCl_2·6H_2O$	solid	4240
Nickel(II) fluoride	NiF_2	solid	2410
Nickel(II) hydroxide	$Ni(OH)_2$	solid	4500
Nickel(II) iodide	NiI_2	solid	3875
Nickel(II) nitrate hexahydrate	$Ni(NO_3)_2·6H_2O$	solid	4300
Nickel(II) oxide	NiO	solid	660
Nickel subsulfide	Ni_3S_2	solid	1030
Nickel(II) sulfate	$NiSO_4$	solid	4005
Nickel(II) sulfide	NiS	solid	190
Niobium	Nb	solid	208
Niobium(V) oxide	Nb_2O_5	solid	-10
Nitric acid	HNO_3	liquid	-19.9
Nitric oxide	NO	solid, 90 K	19.8
Nitric oxide	NO	liquid, 118 K	114.2
Nitric oxide	NO	gas	1461
Nitrogen	N_2	gas	-12.0
Nitrogen dioxide	NO_2	gas, 408 K	150
Nitrogen tetroxide	N_2O_4	gas	-23.0
Nitrogen trioxide	N_2O_3	gas	-16
Nitrous oxide	N_2O	gas	-18.9
Osmium	Os	solid	11
Oxygen	O_2	solid, 54 K	10200
Oxygen	O_2	liquid, 90 K	7699
Oxygen	O_2	gas	3415
Ozone	O_3	liquid	6.7
Palladium	Pd	solid	540
Palladium(II) chloride	$PdCl_2$	solid	-38
Phosphine	PH_3	gas	-26.2
Phosphonic acid	H_3PO_3	aq. soln.	-42.5
Phosphoric acid	H_3PO_4	aq. soln.	-43.8
Phosphorus (white)	P	solid	-26.66
Phosphorus (red)	P	solid	-20.77
Phosphorus(III) chloride	PCl_3	liquid	-63.4
Platinum	Pt	solid	193
Platinum(II) chloride	$PtCl_2$	solid	-54
Platinum(III) chloride	$PtCl_3$	solid	-66.7
Platinum(IV) chloride	$PtCl_4$	solid	-93
Platinum(IV) fluoride	PtF_4	solid	445
Plutonium	Pu	solid	525
Plutonium(IV) fluoride	PuF_4	solid	1760
Plutonium(VI) fluoride	PuF_6	solid	173
Plutonium(IV) oxide	PuO_2	solid	730
Potassium	K	solid	20.8
Potassium bromate	$KBrO_3$	solid	-52.6
Potassium bromide	KBr	solid	-49.1
Potassium carbonate	K_2CO_3	solid	-59
Potassium chlorate	$KClO_3$	solid	-42.8
Potassium chloride	KCl	solid	-38.8
Potassium chromate	K_2CrO_4	solid	-3.9
Potassium cyanide	KCN	solid	-37
Potassium ferricyanide	$K_3Fe(CN)_6$	solid	2290

Name	Mol. form.	Phase	$\chi_m/10^{-6}$ $cm^3\ mol^{-1}$
Potassium ferrocyanide trihydrate	$K_4Fe(CN)_6 \cdot 3H_2O$	solid	-172.3
Potassium fluoride	KF	solid	-23.6
Potassium hydrogen sulfate	$KHSO_4$	solid	-49.8
Potassium hydroxide	KOH	aq. soln.	-22
Potassium iodate	KIO_3	solid	-63.1
Potassium iodide	KI	solid	-63.8
Potassium nitrate	KNO_3	solid	-33.7
Potassium nitrite	KNO_2	solid	-23.3
Potassium permanganate	$KMnO_4$	solid	20
Potassium sulfate	K_2SO_4	solid	-67
Potassium sulfide	K_2S	solid	-60
Potassium superoxide	KO_2	solid	3230
Potassium thiocyanate	KSCN	solid	-48
Praseodymium	Pr	α form	5530
Praseodymium(III) chloride	$PrCl_3$	solid	44.5
Praseodymium(III) oxide	Pr_2O_3	solid	8994
Praseodymium(III) sulfide	Pr_2S_3	solid	10770
Protactinium	Pa	solid	277
Rhenium	Re	solid	67
Rhenium(V) chloride	$ReCl_5$	solid	1225
Rhenium(IV) oxide	ReO_2	solid	44
Rhenium(VI) oxide	ReO_3	solid	16
Rhenium(VII) oxide	Re_2O_7	solid	-16
Rhenium(IV) sulfide	ReS_2	solid	38
Rhodium	Rh	solid	102
Rhodium(III) chloride	$RhCl_3$	solid	-7.5
Rhodium(III) oxide	Rh_2O_3	solid	104
Rubidium	Rb	solid	17
Rubidium bromide	RbBr	solid	-56.4
Rubidium carbonate	Rb_2CO_3	solid	-75.4
Rubidium chloride	RbCl	solid	-46
Rubidium fluoride	RbF	solid	-31.9
Rubidium iodide	RbI	solid	-72.2
Rubidium nitrate	$RbNO_3$	solid	-41
Rubidium sulfate	Rb_2SO_4	solid	-88.4
Rubidium superoxide	RbO_2	solid	1527
Ruthenium	Ru	solid	39
Ruthenium(III) chloride	$RuCl_3$	solid	1998
Ruthenium(IV) oxide	RuO_2	solid	162
Samarium	Sm	α form	1278
Samarium(II) bromide	$SmBr_2$	solid	5337
Samarium(III) bromide	$SmBr_3$	solid	972
Samarium(III) oxide	Sm_2O_3	solid	1988
Samarium(III) sulfate octahydrate	$Sm_2(SO_4)_3 \cdot 8H_2O$	solid	1710
Samarium(III) sulfide	Sm_2S_3	solid	3300
Scandium	Sc	α form	295.2
Selenium (gray)	Se	solid	-25
Selenium bromide	Se_2Br_2	liquid	-113
Selenium chloride	Se_2Cl_2	liquid	-94.8
Selenium dioxide	SeO_2	solid	-27.2
Selenium hexafluoride	SeF_6	gas	-51
Silane	SiH_4	gas	-20.4
Silicon	Si	solid	-3.12
Silicon carbide (hexagonal)	SiC	solid	-12.8
Silicon dioxide (α-quartz)	SiO_2	solid	-29.6
Silver	Ag	solid	-19.5
Silver(I) bromide	AgBr	solid	-61
Silver(I) carbonate	Ag_2CO_3	solid	-80.90
Silver(I) chloride	AgCl	solid	-49
Silver(I) chromate	Ag_2CrO_4	solid	-40
Silver(I) cyanide	AgCN	solid	-43.2
Silver(I) fluoride	AgF	solid	-36.5
Silver(I) iodide	AgI	solid	-80
Silver(I) nitrate	$AgNO_3$	solid	-45.7
Silver(I) nitrite	$AgNO_2$	solid	-42
Silver(I) oxide	Ag_2O	solid	-134
Silver(II) oxide	AgO	solid	-19.6
Silver(I) phosphate	Ag_3PO_4	solid	-120
Silver(I) sulfate	Ag_2SO_4	solid	-92.90
Silver(I) thiocyanate	AgSCN	solid	-61.8
Sodium	Na	solid	16
Sodium acetate	$NaC_2H_3O_2$	solid	-37.6
Sodium bromate	$NaBrO_3$	solid	-44.2
Sodium bromide	NaBr	solid	-41
Sodium carbonate	Na_2CO_3	solid	-41
Sodium chlorate	$NaClO_3$	solid	-34.7
Sodium chloride	NaCl	solid	-30.2
Sodium dichromate	$Na_2Cr_2O_7$	solid	55
Sodium fluoride	NaF	solid	-15.6
Sodium hydrogen phosphate	Na_2HPO_4	solid	-56.6
Sodium hydroxide	NaOH	aq. soln.	-15.8
Sodium iodate	$NaIO_3$	solid	-53
Sodium iodide	NaI	solid	-57
Sodium nitrate	$NaNO_3$	solid	-25.6
Sodium nitrite	$NaNO_2$	solid	-14.5
Sodium oxide	Na_2O	solid	-19.8
Sodium peroxide	Na_2O_2	solid	-28.10
Sodium sulfate	Na_2SO_4	solid	-52
Sodium sulfate decahydrate	$Na_2SO_4 \cdot 10H_2O$	solid	-184
Sodium sulfide	Na_2S	solid	-39
Sodium tetraborate	$Na_2B_4O_7$	solid	-85
Stibine	SbH_3	gas	-34.6
Strontium	Sr	solid	92
Strontium bromide	$SrBr_2$	solid	-86.6
Strontium bromide hexahydrate	$SrBr_2 \cdot 6H_2O$	solid	-160
Strontium carbonate	$SrCO_3$	solid	-47
Strontium chlorate	$Sr(ClO_3)_2$	solid	-73
Strontium chloride	$SrCl_2$	solid	-61.5
Strontium chloride hexahydrate	$SrCl_2 \cdot 6H_2O$	solid	-145
Strontium chromate	$SrCrO_4$	solid	-5.1
Strontium fluoride	SrF_2	solid	-37.2
Strontium hydroxide	$Sr(OH)_2$	solid	-40
Strontium iodate	$Sr(IO_3)_2$	solid	-108
Strontium iodide	SrI_2	solid	-112
Strontium nitrate	$Sr(NO_3)_2$	solid	-57.2
Strontium oxide	SrO	solid	-35
Strontium peroxide	SrO_2	solid	-32.3
Strontium sulfate	$SrSO_4$	solid	-57.9
Sulfur (rhombic)	S	solid	-15.5
Sulfur (monoclinic)	S	solid	-14.9
Sulfur chloride [ClSSCl]	S_2Cl_2	liquid	-62.2
Sulfur dichloride	SCl_2	liquid	-49.4
Sulfur dioxide	SO_2	gas	-18.2
Sulfur hexafluoride	SF_6	gas	-44
Sulfuric acid	H_2SO_4	liquid	-39
Sulfur trioxide (α-form)	SO_3	liquid	-28.54
Tantalum	Ta	solid	154
Tantalum(V) chloride	$TaCl_5$	solid	140

Name	Mol. form.	Phase	$\chi_m/10^{-6}$ cm^3 mol^{-1}
Tantalum(V) oxide	Ta_2O_5	solid	-32
Technetium	Tc	solid	115
Tellurium	Te	solid	-38
Tellurium dibromide	$TeBr_2$	solid	-106
Tellurium dichloride	$TeCl_2$	solid	-94
Tellurium hexafluoride	TeF_6	gas	-66
Terbium	Tb	α form	170000
Terbium(III) oxide	Tb_2O_3	solid	78340
Tetrabromosilane	$SiBr_4$	liquid	-126
Tetrachlorosilane	$SiCl_4$	liquid	-87.5
Tetraethylsilane	$(C_2H_5)_4Si$	liquid	-120.2
Tetramethylsilane	$(CH_3)_4Si$	liquid	-74.80
Thallium	Tl	solid	-50
Thallium(I) bromate	$TlBrO_3$	solid	-75.9
Thallium(I) bromide	TlBr	solid	-63.9
Thallium(I) carbonate	Tl_2CO_3	solid	-101.6
Thallium(I) chlorate	$TlClO_3$	solid	-65.5
Thallium(I) chloride	TlCl	solid	-57.8
Thallium(I) chromate	Tl_2CrO_4	solid	-39.3
Thallium(I) cyanide	TlCN	solid	-49
Thallium(I) fluoride	TlF	solid	-44.4
Thallium(I) iodate	$TlIO_3$	solid	-86.8
Thallium(I) iodide	TlI	solid	-82.2
Thallium(I) nitrate	$TlNO_3$	solid	-56.5
Thallium(I) nitrite	$TlNO_2$	solid	-50.8
Thallium(I) sulfate	Tl_2SO_4	solid	-112.6
Thallium(I) sulfide	Tl_2S	solid	-88.8
Thionyl chloride	$SOCl_2$	liquid	-44.3
Thorium	Th	solid	97
Thorium(IV) oxide	ThO_2	solid	-16
Thulium	Tm	solid	24700
Thulium(III) oxide	Tm_2O_3	solid	51444
Tin(IV) bromide	$SnBr_4$	solid	-149
Tin(II) chloride	$SnCl_2$	solid	-69
Tin(IV) chloride	$SnCl_4$	liquid	-115
Tin(II) chloride dihydrate	$SnCl_2 \cdot 2H_2O$	solid	-91.4
Tin (gray)	Sn	solid	-37.4
Tin(II) oxide	SnO	solid	-19
Tin(IV) oxide	SnO_2	solid	-41
Titanium	Ti	solid	151
Titanium(II) bromide	$TiBr_2$	solid	720
Titanium(III) bromide	$TiBr_3$	solid	660
Titanium(II) chloride	$TiCl_2$	solid	484
Titanium(III) chloride	$TiCl_3$	solid	1110
Titanium(IV) chloride	$TiCl_4$	solid	-54
Titanium(III) fluoride	TiF_3	solid	1300
Titanium(II) iodide	TiI_2	solid	1790
Titanium(III) oxide	Ti_2O_3	solid	132
Titanium(IV) oxide (rutile)	TiO_2	solid	5.9
Titanium(II) sulfide	TiS	solid	432
Tungsten	W	solid	53
Tungsten(V) bromide	WBr_5	solid	270
Tungsten carbide	WC	solid	10
Tungsten(II) chloride	WCl_2	solid	-25
Tungsten(V) chloride	WCl_5	solid	387
Tungsten(VI) chloride	WCl_6	solid	-71
Tungsten(VI) fluoride	WF_6	gas	-53
Tungsten(IV) oxide	WO_2	solid	57
Tungsten(VI) oxide	WO_3	solid	-15.8
Tungsten(IV) sulfide	WS_2	solid	5850
Uranium	U	solid	409
Uranium(III) bromide	UBr_3	solid	4740
Uranium(IV) bromide	UBr_4	solid	3530
Uranium(III) chloride	UCl_3	solid	3460
Uranium(IV) chloride	UCl_4	solid	3680
Uranium(IV) fluoride	UF_4	solid	3530
Uranium(VI) fluoride	UF_6	solid	43
Uranium(III) hydride	UH_3	solid	6244
Uranium(III) iodide	UI_3	solid	4460
Uranium(IV) oxide	UO_2	solid	2360
Uranium(VI) oxide	UO_3	solid	128
Vanadium	V	solid	285
Vanadium(II) bromide	VBr_2	solid	3230
Vanadium(III) bromide	VBr_3	solid	2910
Vanadium(II) chloride	VCl_2	solid	2410
Vanadium(III) chloride	VCl_3	solid	3030
Vanadium(IV) chloride	VCl_4	liquid	1215
Vanadium(III) fluoride	VF_3	solid	2757
Vanadium(III) oxide	V_2O_3	solid	1976
Vanadium(IV) oxide	VO_2	solid	99
Vanadium(V) oxide	V_2O_5	solid	128
Vanadium(III) sulfide	V_2S_3	solid	1560
Water	H_2O	solid, 273 K	-12.63
Water	H_2O	liquid, 293 K	-12.96
Water	H_2O	liquid, 373 K	-13.09
Water	H_2O	gas, 373 K	-13.1
Xenon	Xe	gas	-45.5
Ytterbium	Yb	β form	67
Yttrium	Y	α form	187.7
Yttrium oxide	Y_2O_3	solid	44.4
Yttrium sulfide	Y_2S_3	solid	100
Zinc	Zn	solid	-9.15
Zinc carbonate	$ZnCO_3$	solid	-34
Zinc chloride	$ZnCl_2$	solid	-55.33
Zinc cyanide	$Zn(CN)_2$	solid	-46
Zinc fluoride	ZnF_2	solid	-34.3
Zinc hydroxide	$Zn(OH)_2$	solid	-67
Zinc iodide	ZnI_2	solid	-108
Zinc oxide	ZnO	solid	-27.2
Zinc phosphate	$Zn_3(PO_4)_2$	solid	-141
Zinc sulfate	$ZnSO_4$	solid	-47.8
Zinc sulfate heptahydrate	$ZnSO_4 \cdot 7H_2O$	solid	-138
Zinc sulfate monohydrate	$ZnSO_4 \cdot H_2O$	solid	-63
Zinc sulfide (sphalerite)	ZnS	solid	-25
Zirconium	Zr	solid	120
Zirconium carbide	ZrC	solid	-26
Zirconium(IV) nitrate pentahydrate	$Zr(NO_3)_4 \cdot 5H_2O$	solid	-77
Zirconium(IV) oxide	ZrO_2	solid	-13.8

[a] Ferromagnetic.

INDEX OF REFRACTION OF INORGANIC LIQUIDS AND LIQUID ELEMENTS

This table gives the index of refraction n of several inorganic substances, including a few elements, in the liquid state. The measurements refer to ambient atmospheric pressure except for substances whose normal boiling points are greater than the indicated temperature; in this case the pressure is the saturated vapor pressure of the substance. All values refer to a wavelength of 589 nm unless otherwise indicated in a footnote. Column definitions are as follows.

Column heading	Definition
Name	Name of inorganic liquid or liquid element; entries are arranged in alphabetical order by chemical name
Mol. form.	Molecular formula of inorganic liquid or liquid element
n	Index of refraction; the temperature in °C is indicated by a superscript on n

Data on the index of refraction at other temperatures and wavelengths may be found in Ref. 1.

References

1. Wohlfarth, C., and Wohlfarth, B., *Landolt-Börnstein, Numerical Data and Functional Relationships in Science and Technology, New Series*, III/38A, Martienssen, W., Editor, Springer-Verlag, Heidelberg, 1996.
2. Francis, A.W., *J. Chem. Eng. Data*, 5, 534, 1960. <https://doi.org/10.1021/je60008a034>

Index of Refraction of Inorganic Liquids and Liquid Elements

Name	Mol. form.	n
Ammonia[b,d]	NH_3	1.3944^{-77}
Antimony(V) chloride	$SbCl_5$	1.5925^{22}
Argon	Ar	1.2312^{-188}
Arsenic(III) chloride	$AsCl_3$	1.604^{16}
Boron tribromide	BBr_3	1.312^{16}
Bromine	Br_2	1.659^{15}
Bromine pentafluoride	BrF_5	1.3529^{25}
Bromine trifluoride	BrF_3	1.4536^{25}
Carbon dioxide	CO_2	1.6630^{24} at 6.3 MPa
Carbon disulfide	CS_2	1.6277^{20}
Carbon oxysulfide	OCS	1.24^{-87}
Carbon suboxide	C_3O_2	1.4538^{0}
Chlorine	Cl_2	1.3834^{20}
Chromium(VI) dichloride dioxide	CrO_2Cl_2	1.524^{23}
Germanium(IV) bromide	$GeBr_4$	1.6269^{26}
Germanium(IV) chloride	$GeCl_4$	1.4614^{25}
Helium[c]	He	1.02451^{-269}
Hydrazine	N_2H_4	1.470^{22}
Hydrogen	H_2	1.1096^{-253}
Hydrogen bromide	HBr	1.325^{10}
Hydrogen chloride[a]	HCl	1.3287^{18}
Hydrogen cyanide	HCN	1.2614^{20}
Hydrogen disulfide	H_2S_2	1.630^{20}
Hydrogen fluoride	HF	1.1574^{25}
Hydrogen iodide	HI	1.466^{16}
Hydrogen peroxide	H_2O_2	1.4061^{28}
Hydrogen sulfide[e]	H_2S	1.460^{-80}
Iron pentacarbonyl	$Fe(CO)_5$	1.5196^{20}
Krypton[c]	Kr	1.3032^{-157}
Nitric acid	HNO_3	1.393^{25}
Nitric oxide	NO	1.330^{-90}
Nitrogen	N_2	1.19876^{-196}
Nitrous oxide	N_2O	1.238^{25} at high pressure
Oxygen	O_2	1.2243^{-183}
Perchloric acid	$HClO_4$	1.3819^{50}
Phosphine	PH_3	1.317^{17}
Phosphorus(III) bromide	PBr_3	1.687^{25}
Phosphorus(III) chloride	PCl_3	1.5122^{21}
Phosphorus(III) oxide	P_2O_3	1.540^{27}
Sulfur (rhombic)	S	1.9170^{125}
Sulfur chloride [ClSSCl]	S_2Cl_2	1.671^{20}
Sulfur dichloride	SCl_2	1.557^{14}
Sulfur dioxide	SO_2	1.3396^{25}
Sulfur hexafluoride	SF_6	1.167^{25}
Sulfuric acid	H_2SO_4	1.4183^{20}
Sulfur trioxide (α-form)	SO_3	1.40965^{20}
Sulfuryl chloride	SO_2Cl_2	1.444^{12}
Tetrabromosilane	$SiBr_4$	1.5685^{31}
Tetrachlorosilane	$SiCl_4$	1.41156^{25}
Thionyl chloride	$SOCl_2$	1.527^{10}
Tin(IV) bromide	$SnBr_4$	1.6628^{31}
Tin(IV) chloride	$SnCl_4$	1.5086^{25}
Titanium(IV) chloride	$TiCl_4$	1.6076^{18}
Water	H_2O	1.33336^{20}
Xenon	Xe	1.3918^{-112}

[a] At 581 nm.
[b] At 578 nm.
[c] At 546 nm.
[d] The value for compressed liquid NH_3 at 20 °C and 589 nm wavelength is 1.3327.
[e] The value for compressed liquid H_2S at 20 °C is 1.3682.

PHYSICAL AND OPTICAL PROPERTIES OF MINERALS

The chemical formula, crystal system, density, hardness, and index of refraction of some common minerals are given in this table. The column definitions are as follows.

Column heading	Definition
Name	Common name of the mineral, arranged alphabetically
Mol. form.	Chemical formula for a typical sample of the mineral; composition often varies considerably with the origin of the sample
Crystal system	Crystal system of the mineral
ρ	Typical density, in g cm^{-3}; individual samples may vary by a few percent
Hardness	Hardness on the Mohs scale (range of 1 to 10, with talc = 1 and diamond = 10)
n_α	Index of refraction, least value; light of wavelength in the neighborhood of 589 nm; for cubic crystal, there is only a single value
n_β	Index of refraction, intermediate value; light of wavelength in the neighborhood of 589 nm
n_γ	Index of refraction, greatest value; light of wavelength in the neighborhood of 589 nm

See Ref. 1 for details on the axis systems. Variations of several percent for the index of refraction values are common, depending on the origin and exact composition of the sample.

References

1. Deer, W. A., Howie, R. A., and Zussman, J., *An Introduction to the Rock-Forming Minerals, Second Edition*, Longman Scientific & Technical, Harlow, Essex, 1992.
2. Carmichael, R. S., *Practical Handbook of Physical Properties of Rocks and Minerals*, CRC Press, Boca Raton, FL, 1989.
3. Donnay, J. D. H., and Ondik, H. M., *Crystal Data Determinative Tables, Third Edition, Volume 2, Inorganic Compounds*, Joint Committee on Powder Diffraction Standards, Swarthmore, PA, 1973.

Density, Hardness, and Refractive Index of Common Minerals

Name	Mol. form.	Crystal system	ρ/g cm^{-3}	Hardness	n_α	n_β	n_γ
Acanthite	Ag_2S	orthorhombic	7.2	2.3			
Actinolite	$Ca_2(Mg,Fe)_5Si_8O_{22}(OH,F)_2$	monoclinic	3.23	5.5	1.624	1.655	1.664
Aegirine	$NaFe(SiO_3)_2$	monoclinic	3.58	6	1.763	1.800	1.815
Akermanite	$Ca_2MgSi_2O_7$	tetragonal	2.94	5.5	1.632	1.640	
Alabandite	MnS	cubic	4.0	3.8			
Albite	$NaAlSi_3O_8$	triclinic	2.63	6.3	1.527	1.531	1.538
Allanite	$(Ca,Mn,Ce,La,Y,Th)_2(Fe,Ti)(Al,Fe)O{\cdot}OH(Si_2O_7)(SiO_4)$	monoclinic	3.8	5.8	1.75	1.78	1.80
Allemontite	SbAs	hexagonal	6.0	3.5			
Almandine	$Fe_3Al_2Si_3O_{12}$	cubic	4.32	6.8	1.830		
Altaite	PbTe	cubic	8.16	3			
Aluminite	$Al_2(SO_4)(OH)_4{\cdot}7H_2O$	monoclinic	1.74	1.5	1.459	1.464	1.470
Alunite	$KAl_3(SO_4)_2(OH)_6$	rhombohedral	2.8	3.8	1.572	1.592	
Alunogen	$Al_2(SO_4)_3{\cdot}18H_2O$	monoclinic	1.69	1.8	1.467	1.47	1.478
Amblygonite	$(Li,Na)Al(PO_4)(F,OH)$	triclinic	3.1	5.8	1.591	1.604	1.613
Analcite	$NaAl(SiO_3)_2{\cdot}H_2O$	cubic	2.27	5.5	1.486		
Anatase	TiO_2	tetragonal	4.23	5.8	2.488	2.561	
Andalusite	Al_2OSiO_4	orthorhombic	3.15	7.5	1.635	1.639	1.644
Andesine	$NaAlSi_3O_8{\cdot}CaAl_2Si_2O_8$	triclinic	2.67	6.3	1.550	1.553	1.557
Andorite	$PbAgSb_3S_6$	rhombohedral	2.67	3.3			
Andradite	$Ca_3Fe_2Si_3O_{12}$	cubic	3.86	6.8	1.887		
Anglesite	$PbSO_4$	orthorhombic	6.29	2.8	1.877	1.883	1.894
Anhydrite	$CaSO_4$	orthorhombic	2.96	3.5	1.570	1.575	1.614
Ankerite	$Ca(Fe,Mg,Mn)(CO_3)_2$	rhombohedral	3.0	3.8	1.529	1.720	
Anorthite	$CaAl_2Si_2O_8$	triclinic	2.76	6.3	1.577	1.585	1.590
Anorthoclase	$(Na,K)AlSi_3O_8$	triclinic	2.58	6	1.523	1.528	1.529
Anthophyllite	$Mg_7Si_8O_{22}(OH)_2$	rhombohedral	3.21	5.8	1.645	1.658	1.668
Apatite	$Ca_5(PO_4)_3(OH,F,Cl)$	hexagonal	3.2	5	1.645	1.648	
Apophyllite	$KFCa_4Si_8O_{20}{\cdot}8H_2O$	tetragonal	2.35	4.8	1.535	1.536	
Aragonite	$CaCO_3$	orthorhombic	2.83	3.5	1.531	1.680	1.686
Arcanite	K_2SO_4	orthorhombic	2.66		1.494	1.494	1.497
Argentite	Ag_2S	orthorhombic	7.2	2.3			
Arsenolite	As_2O_3	cubic	3.86	1.5	1.755		
Arsenopyrite	FeAsS	monoclinic	6.1	5.8			
Atacamite	$Cu_2(OH)_3Cl$	rhombohedral	3.76	3.3	1.831	1.861	1.880

Name	Mol. form.	Crystal system	ρ/g cm^{-3}	Hardness	n_α	n_β	n_γ
Augelite	$Al_2(PO_4)(OH)_3$	monoclinic	2.70	4.8	1.574	1.576	1.588
Augite	$(Ca,Mg,Fe,Ti,Al)_2(Si,Al)_2O_6$	monoclinic	3.38	6	1.703	1.707	1.738
Autunite	$Ca(UO_2)_2(PO_4)_2{\cdot}10H_2O$	tetragonal	3.2	2.3	1.553	1.577	
Axinite	$(Ca,Mn,Fe)_3Al_2BO_3Si_4O_{12}(OH)$	triclinic	3.31	6.8	1.684	1.691	1.694
Azurite	$Cu_3(OH)_2(CO_3)_2$	monoclinic	3.77	3.8	1.730	1.758	1.838
Baddeleyite	ZrO_2	monoclinic	5.7	6.5	2.13	2.19	2.20
Barite	$BaSO_4$	orthorhombic	4.49	3.3	1.636	1.637	1.648
Benitoite	$BaTi(SiO_3)_3$	rhombohedral	3.65	6.3	1.757	1.804	
Bertrandite	$Be_4Si_2O_7(OH)_2$	rhombohedral	2.6	6	1.589	1.602	1.613
Beryl	$Be_3Al_2(SiO_3)_6$	hexagonal	2.64	7.8	1.582	1.589	
Beryllonite	$NaBe(PO)_4$	monoclinic	2.81	5.8	1.552	1.558	1.561
Biotite	$K(Mg,Fe)_3AlSi_3O_{10}(OH,F)_2$	monoclinic	3.0	2.8	1.595	1.651	1.651
Bismuthinite	Bi_2S_3	orthorhombic	6.78	2			
Bixbyite	$(Mn,Fe)_2O_3$	cubic	4.95	6.3			
Bloedite	$Na_2Mg(SO_4)_2{\cdot}4H_2O$	monoclinic	2.25	2.8	1.483	1.486	1.487
Boehmite	$AlO(OH)$	orthorhombic	3.44	3.8	1.64	1.65	1.66
Boracite	$Mg_3B_7O_{13}Cl$	rhombohedral	2.94	7.3	1.66	1.66	1.67
Borax	$Na_2B_4O_7{\cdot}10H_2O$	monoclinic	1.73	2.3	1.447	1.469	1.472
Bornite	Cu_5FeS_4	cubic	5.07	3			
Boulangerite	$Pb_5Sb_4S_{11}$	monoclinic	6.1	2.8			
Bournonite	$PbCuSbS_3$	rhombohedral	5.83	2.8			
Braggite	PtS	tetragonal	10.2				
Braunite	$(Mn,Si)_2O_3$	tetragonal	4.78	6.3			
Bravoite	$(Ni,Fe)S_2$	cubic	4.62	5.8			
Breithauptite	$NiSb$	hexagonal	≈8.7	5.5			
Brochantite	$Cu_4SO_4(OH)_6$	monoclinic	3.79	3.8	1.728	1.771	1.800
Bromyrite	$AgBr$	cubic	6.47	2.5	2.253		
Brookite	TiO_2	orthorhombic	4.23	5.8	2.583	2.584	2.700
Brucite	$Mg(OH)_2$	hexagonal	2.37	2.5	1.575	1.59	
Bunsenite	NiO	cubic	6.72	5.5			
Cacoxenite	$Fe_4(PO_4)_3(OH)_3{\cdot}12H_2O$	hexagonal	2.3	3.5	1.580	1.646	
Calcite	$CaCO_3$	hexagonal	2.71	3	1.486	1.658	
Caledonite	$Cu_2Pb_5(SO_4)_3(CO_3)(OH)_6$	rhombohedral	5.76	2.8	1.818	1.866	1.909
Calomel	Hg_2Cl_2	tetragonal	7.16	1.5	1.973	2.656	
Cancrinite	$(Na,Ca,K)_7[Al_6Si_6O_{24}](CO_3,SO_4,Cl,OH)_2{\cdot}H_2O$	hexagonal	2.42	5.5	1.495	1.509	
Carbon (diamond)	C	cubic	3.513	10	2.418		
Carnalite	$KMgCl_3{\cdot}6H_2O$	rhombohedral	1.60	2.5	1.466	1.475	1.494
Carnotite	$K_2(UO_2)_2(VO_4)_2{\cdot}3H_2O$	rhombohedral		1.5	1.75	1.92	1.95
Cassiterite	SnO_2	tetragonal	6.85	6.5	2.006	2.097	
Celestite	$SrSO_4$	orthorhombic	3.96	3.3	1.622	1.624	1.631
Celsian	$BaAl_2Si_2O_8$	monoclinic	3.25	6.3	1.583	1.588	1.594
Cerargyrite	$AgCl$	cubic	5.56	2.5	2.071		
Cerussite	$PbCO_3$	orthorhombic	6.6	3.3	1.804	2.076	2.079
Cervantite	Sb_2O_4	orthorhombic	6.64	4.5			
Chabazite	$CaAl_2Si_4O_{12}{\cdot}6H_2O$	trig	2.08	4.5	1.482		
Chalcanthite	$CuSO_4{\cdot}5H_2O$	triclinic	2.29	2.5	1.514	1.537	1.543
Chalcocite	Cu_2S	orthorhombic	5.6	2.8			
Chalcopyrite	$CuFeS_2$	tetragonal	4.2	3.8			
Chiolite	$Na_5Al_3F_{14}$	tetragonal	3.00	3.8	1.342	1.349	
Chlorite	$(Mg,Al,Fe)_{12}(Si,Al)_8O_{20}(OH)_{16}$	monoclinic	3.0	2.5	1.61	1.62	1.62
Chloritoid	$FeAl_4O_2(SiO_4)_2(OH)_4$	monoclinic	3.66	6.5	1.717	1.721	1.726
Chondrodite	$2Mg_2SiO_4{\cdot}MgF_2$	monoclinic	3.21	6.5	1.604	1.615	1.634
Chromite	$FeCr_2O_4$	cubic	5.0	5.5	2.16		
Chrysoberyl	$BeAl_2O_4$	orthorhombic	3.65	8.5	1.746	1.748	1.756
Chrysocolla	$CuSiO_3{\cdot}2H_2O$	rhombohedral	2.4	2	1.575	1.597	1.598
Cinnabar	HgS	hexagonal	8.17	2.3	2.814	3.143	
Claudetite	As_2O_3	monoclinic	3.74	2.5	1.87	1.92	2.01
Clinohumite	$4Mg_2SiO_4{\cdot}MgF_2$	monoclinic	3.21	6	1.633	1.647	1.668
Clinozoisite	$Ca_2Al_3(SiO_4)_3OH$	monoclinic	3.30	6.5	1.693	1.700	1.712
Cobaltite	$CoAsS$	cubic	≈6.1	5.5			
Colemanite	$Ca_2B_6O_{11}{\cdot}5H_2O$	monoclinic	2.42	4.5	1.586	1.592	1.614

Name	Mol. form.	Crystal system	ρ/g cm^{-3}	Hardness	n_α	n_β	n_γ
Columbite	$(Fe,Mn)(Nb,Ta)_2O_6$	rhombohedral	5.20	6			
Connellite	$Cu_{19}(SO_4)Cl_4(OH)_{32}\cdot 3H_2O$	hexagonal	3.36	3	1.731	1.752	
Copiapite	$(Fe,Mg)Fe_4(SO_4)_6(OH)_2\cdot 20H_2O$	triclinic	2.13	2.8	1.52	1.54	1.59
Coquimbite	$Fe_2(SO_4)_3\cdot 9H_2O$	hexagonal	2.1	2.5	1.54	1.56	
Cordierite	$Al_3(Mg,Fe)_2Si_5AlO_{18}$	rhombohedral	2.66	7	1.540	1.549	1.553
Corundum	Al_2O_3	hexagonal	3.97	9	1.761	1.769	
Cotunnite	$PbCl_2$	orthorhombic	5.98	2.5	2.199	2.217	2.260
Covellite	CuS	hexagonal	4.8	1.8			
Cristobalite	SiO_2	hexagonal	2.33	6.5	1.484	1.487	
Crocoite	$PbCrO_4$	monoclinic	6.12	2.8	2.29	2.36	2.66
Cryolite	Na_3AlF_6	monoclinic	2.97	2.5	1.338	1.338	1.339
Cryolithionite	$Na_3Li_3Al_2F_{12}$	cubic	2.77	2.8	1.340		
Cubanite	$CuFe_2S_3$	rhombohedral	4.11	3.5			
Cummingtonite	$(Mg,Fe,Mn)_7(Si_4O_{11})_2(OH)_2$	monoclinic	3.4	5.5	1.650	1.660	1.676
Cuprite	Cu_2O	cubic	6.0	3.8			
Danburite	$CaB_2Si_2O_8$	rhombohedral	3.0	7	1.63	1.63	1.63
Datolite	$CaBSiO_4(OH)$	monoclinic	2.98	5.3	1.624	1.652	1.668
Daubreelite	Cr_2FeS_4	cubic	3.81				
Derbylite	$Fe_6Ti_6Sb_2O_{23}$	rhombohedral	4.53	5	2.45	2.45	2.51
Diaspore	$AlO(OH)$	orthorhombic	3.4	6.8	1.694	1.715	1.741
Digenite	$Cu_{1.79}S$	cubic	5.55	2.8			
Diopside	$CaMg(SiO_3)_2$	monoclinic	3.30	6	1.680	1.687	1.708
Dioptase	$CuSiO_2(OH)_2$	rhombohedral	3.5	5	1.65	1.70	
Dolomite	$CaMg(CO_3)_2$	rhombohedral	2.86	3.5	1.500	1.679	
Douglasite	$K_2FeCl_4\cdot 2H_2O$	orthorhombic	2.16		1.488	1.500	
Dyscrasite	Ag_3Sb	rhombohedral	9.74	3.8			
Eddingtonite	$BaAl_2Si_3O_{10}\cdot 4H_2O$	rhombohedral	2.8		1.541	1.553	1.557
Eglestonite	Hg_4OCl_2	cubic	8.4	2.5	2.49		
Emplectite	$CuBiS_2$	rhombohedral	6.38	2			
Enargite	Cu_3AsS_4	rhombohedral	4.5	3			
Enstatite	$MgSiO_3$	monoclinic	3.19	5.5	1.656	1.662	1.669
Epidote	$Ca_2Al_2FeOH(SiO_4)_3$	monoclinic	3.44	6	1.733	1.755	1.765
Epsomite	$MgSO_4\cdot 7H_2O$	orthorhombic	1.67	2.3	1.433	1.455	1.461
Erythrite	$(Co,Ni)_3(AsO_4)_2\cdot 8H_2O$	monoclinic	3.06	2	1.626	1.661	1.699
Eucairite	$AgCuSe$	orthorhombic	7.7	2.5			
Euclasite	$BeAlSiO_4(OH)$	monoclinic	3.1	7.5	1.651	1.655	1.671
Eudialite	$(Na,Ca,Ce)_5(Fe,Mn)(Zr,Ti)(Si_3O_9)_2(OH,Cl)$	hexagonal	3.0	5.5	1.623	1.600	1.615
Eulytite	$Bi_4(SiO_4)_3$	cubic	6.6	4.5	2.05		
Euxenite	$(Y,Ca,Ce,U,Th)(Nb,Ta,Ti)_2O_6$	rhombohedral	5.5	6	2.2		
Fayalite	Fe_2SiO_4	orthorhombic	4.30	6.5	1.827	1.869	1.879
Ferberite	$FeWO_4$	monoclinic	7.51	4.3			
Fergussonite	$(Y,Er,Ce,Fe)(Nb,Ta,Ti)O_4$	tetragonal	5.7	6	2.1		
Fluorite	CaF_2	cubic	3.18	4	1.434		
Forsterite	Mg_2SiO_4	orthorhombic	3.21	7	1.635	1.651	1.670
Franklinite	$ZnFe_2O_4$	cubic	5.21	6	2.36		
Gahnite	$ZnAl_2O_4$	cubic	4.62	7.8	1.805		
Galaxite	$MnAl_2O_4$	cubic	4.04	7.8	1.92		
Galena	PbS	cubic	7.60	2.5	3.91		
Galenabismuthite	$PbBi_2S_4$	rhombohedral	7.04	3			
Ganomalite	$(Ca,Pb)_{10}(OH,Cl)_2(Si_2O_7)_3$	hexagonal	5.6	3.5	1.910	1.945	
Gaylussite	$Na_2Ca(CO_3)_2\cdot 5H_2O$	monoclinic	1.99	2.8	1.444	1.516	1.523
Gehlenite	$Ca_2Al_2SiO_7$	tetragonal	3.04	5.5	1.658	1.669	
Geikielite	$MgTiO_3$	hexagonal	3.85	5.5	1.95	2.31	
Gibbsite	$Al(OH)_3$	monoclinic	2.42	3	1.57	1.57	1.59
Glauberite	$Na_2Ca(SO_4)_2$	monoclinic	2.80	2.8	1.515	1.535	1.536
Glauconite	$(K,Na,Ca)_{1.6}(Fe,Al,Mg)_{4.0}Si_{7.3}Al_{0.7}O_{20}(OH)_4$	monoclinic	2.7	2	1.60	1.63	1.63
Glaucophane	$Na_2Mg_3Al_2[Si_8O_{22}](OH)_2$	monoclinic	3.19	6	1.634	1.645	1.648
Gmelinite	$(Ca,Na_2)[Al_2Si_4O_{12}]\cdot 6H_2O$	hexagonal	2.10	4.5	1.477	1.485	
Goethite	$FeO(OH)$	orthorhombic	4.3	5.3	2.268	2.401	2.457
Goslarite	$ZnSO_4\cdot 7H_2O$	orthorhombic	1.97	2.3	1.457	1.480	1.484
Greenockite	CdS	hexagonal	4.8	3.3	2.506	2.529	

Name	Mol. form.	Crystal system	ρ/g cm^{-3}	Hardness	n_α	n_β	n_γ
Grossularite	$Ca_3Al_2(SiO_4)_3$	cubic	3.59	6.8	1.734		
Gummite	$UO_3{\cdot}H_2O$	orthorhombic	7.05	3.8			
Gypsum	$CaSO_4{\cdot}2H_2O$	monoclinic	2.32	2	1.520	1.525	1.530
Halite	$NaCl$	cubic	2.17	2	1.544		
Hambergite	$Be_2(OH,F)BO_3$	rhombohedral	2.36	7.5	1.56	1.59	1.63
Hanksite	$Na_{22}K(SO_4)_9(CO_3)_2Cl$	hexagonal	2.56	3.3	1.461	1.481	
Harmotome	$Ba[Al_2Si_6O_{16}]{\cdot}6H_2O$	monoclinic	2.44	4.5	1.506	1.507	1.511
Hausmannite	Mn_3O_4	tetragonal	4.84	5.5	2.15	2.46	
Haüyne	$(Na,Ca)_{4\text{-}8}Al_6Si_6O_{24}(SO_4,S)_{1\text{-}2}$	cubic	2.47	5.8	1.502		
Hedenbergite	$CaFe(SiO_3)_2$	monoclinic	3.53	6	1.721	1.727	1.746
Helvite	$Mn_4Be_3Si_3O_{12}S$	cubic	3.32	6	1.739		
Hematite	Fe_2O_3	hexagonal	5.25	6	2.91	3.19	
Hemimorphite	$Zn_4(OH)_2Si_2O_7{\cdot}H_2O$	rhombohedral	3.45	5	1.614	1.617	1.636
Hercynite	$Fe(AlO_2)_2$	cubic	4.3	7.8	1.835		
Herderite	$CaBe(PO_4)(Fe,OH)$	monoclinic	2.98	5.3	1.592	1.612	1.621
Hessite	Ag_2Te	orthorhombic	8.4	2.5			
Heulandite	$(Ca,Na_2,K_2)[Al_2Si_7O_{18}]{\cdot}6H_2O$	monoclinic	2.2	3.8	1.498	1.498	1.506
Hopeite	$Zn_3(PO_4)_2{\cdot}4H_2O$	orthorhombic	3.0	3.2	1.58	1.59	1.59
Hornblende	$Ca_2(Mg,Fe)_4Al(Si_7AlO_{22})(OH)_2$	monoclinic	3.24	5.5	1.67	1.67	1.69
Huebnerite	$MnWO_4$	monoclinic	7.2	4.3	2.17	2.22	2.32
Humite	$Mg(OH,F)_2{\cdot}3Mg_2SiO_4$	orthorhombic	3.3	6	1.625	1.636	1.657
Huntite	$Mg_3Ca(CO_3)_4$	trigonal-trapezohedral	2.70				
Hydrogrossularite	$Ca_3Al_2Si_2O_8(SiO_4)_{1\text{-}m}(OH)_{4m}$	cubic	3.4	6.8	1.70		
Hydromagnesite	$3MgCO_3{\cdot}Mg(OH)_2{\cdot}3H_2O$	monoclinic	2.24	3.5	1.523	1.527	1.545
Illite	$KAl_4[Si_7AlO_{20}](OH)_4$	monoclinic	2.8	1.5	1.56	1.59	1.59
Ilmenite	$FeTiO_3$	rhombohedral	4.72	5.5			
Iodyrite	AgI	hexagonal	5.68	1.5	2.21	2.22	
Jacobsite	$MnFe_2O_4$	cubic	4.87	7.8	2.3		
Jadeite	$NaAl(SiO_3)_2$	monoclinic	3.34	6	1.649	1.654	1.663
Jamesonite	$Pb_4FeSb_6S_{14}$	monoclinic	5.63	2.5			
Jarosite	$KFe_3(SO_4)_2(OH)_6$	rhombohedral	3.09	3	1.715	1.820	
Kainite	$KMg(SO_4)Cl{\cdot}3H_2O$	monoclinic	2.15	2.8	1.494	1.505	1.516
Kaliophilite	$KAlSiO_4$	hexagonal	2.61	6	1.532	1.537	
Kaolinite	$Al_2Si_2O_5(OH)_4$	triclinic	2.65	2.3	1.549	1.564	1.565
Kernite	$Na_2B_4O_7{\cdot}4H_2O$	monoclinic	1.95	2.5	1.454	1.472	1.488
Kieserite	$MgSO_4{\cdot}H_2O$	monoclinic	2.57	3.5	1.520	1.533	1.584
Kyanite	Al_2OSiO_4	triclinic	3.59	6.3	1.715	1.722	1.731
Lanarkite	$Pb_2(SO_4)O$	monoclinic	6.92	2.3	1.928	2.007	2.036
Lanthanite	$(La,Ce)_2(CO_3)_3{\cdot}8H_2O$	rhombohedral	2.72	2.8	1.52	1.587	1.613
Laumontite	$Ca_4[Al_8Si_{16}O_{48}]{\cdot}16H_2O$	monoclinic	2.3	3.3	1.508	1.517	1.519
Laurionite	$Pb(OH)Cl$	rhombohedral	6.24	3.3	2.08	2.12	2.16
Lawsonite	$CaAl_2Si_2O_7(OH)_2{\cdot}H_2O$	rhombohedral	3.08	6	1.655	1.675	1.685
Lazulite	$(Mg,Fe)Al_2(PO_4)_2(OH)_2$	monoclinic	3.23	5.8	1.615	1.64	1.650
Lazurite	$Na_4SSi_3Al_3O_{12}$	cubic	2.42	5.3	1.500		
Leadhillite	$Pb_4(SO_4)(CO_3)_2(OH)_2$	monoclinic	6.55	2.8	1.87	2.00	2.01
Lepidocrocite	$FeO(OH)$	orthorhombic	4.26	5	1.94	2.20	2.51
Lepidolite	$K_2Al_3Li_2AlSi_7O_{20}(OH)_4$	monoclinic	2.85	3.3	1.536	1.565	1.566
Leucite	$KAlSi_2O_6$	tetragonal	2.49	5.8	1.510		
Levyne	$(Ca,Na_2)Al_2Si_4O_{12}{\cdot}6H_2O$	rhombohedral	2.10	4.5	1.496	1.501	
Litharge	PbO	tetragonal	9.35	2	2.535	2.665	
Loellingite	$FeAs_2$	rhombohedral	7.40	5.3			
Maghemite	Fe_2O_3	cubic	4.88	7.8	2.63		
Magnesite	$MgCO_3$	hexagonal	3.05	4	1.536	1.741	
Magnetite	Fe_3O_4	cubic	5.17	6	2.42		
Malachite	$Cu_2(OH)_2CO_3$	monoclinic	4.05	3.8	1.655	1.875	1.909
Manganite	$MnO(OH)$	monoclinic	≈4.3	4	2.25	2.25	2.53
Manganosite	MnO	cubic	5.37	5.5			
Marcasite	FeS_2	cubic	5.02	6.3			
Marialite	$Na_4Al_3Si_9O_{24}Cl$	tetragonal	2.56	5.5	1.541	1.548	
Marshite	CuI	cubic	5.67	2.5	2.346		
Mascagnite	$(NH_4)_2SO_4$	orthorhombic	1.77	2.3	1.520	1.523	1.533

Name	Mol. form.	Crystal system	ρ/g cm^{-3}	Hardness	n_α	n_β	n_γ
Matlockite	$PbClF$	tetragonal	7.05	2.8	2.006	2.145	
Meionite	$Ca_4Al_6Si_6O_{24}CO_3$	tetragonal	2.78	5.5	1.559	1.595	
Melanterite	$FeSO_4{\cdot}7H_2O$	monoclinic	1.89	2	1.47	1.48	1.49
Melilite	$(Ca,Na)_2(Mg,Fe,Al,Si)_3O_7$	tetragonal	3.00	5.5	1.639	1.645	
Mellite	$Al_2C_{12}O_{12}{\cdot}18H_2O$	tetragonal	1.64	2.3	1.511	1.539	
Mendipite	$Pb_3O_2Cl_2$	rhombohedral	7.24	2.5	2.24	2.27	2.31
Mesolite	$Na_2Ca_2(Al_2Si_3O_{10})_3{\cdot}8H_2O$	orthorhombic	2.26	5	1.506		
Metacinnabar	HgS	cubic	7.70	3			
Microcline	$KAlSi_3O_8$	monoclinic	2.56	6.3	1.522	1.526	1.530
Miersite	AgI	hexagonal	5.68	2.5	2.20		
Millerite	NiS	hexagonal	5.5	3.3			
Mimetite	$Pb_5(AsO_4,PO_4)_3Cl$	hexagonal	7.24	3.8	2.128	2.147	
Minium	Pb_3O_4	tetragonal	8.9	2.5			
Mirabilite	$Na_2SO_4{\cdot}10H_2O$	monoclinic	1.46	1.8	1.394	1.396	1.398
Moissanite	SiC	hexagonal	3.16	9.5	2.648	2.691	
Molybdenite	MoS_2	hexagonal	5.06	1.3			
Monazite	$(Ce,La,Th)PO_4$	monoclinic	5.2	5	1.787	1.789	1.840
Monetite	$CaHPO_4$	triclinic	2.92	3.5	1.587	1.61	1.640
Monticellite	$CaMgSiO_4$	orthorhombic	3.18	5.5	1.647	1.655	1.664
Montmorillonite	$(0.5Ca,Na)_{0.7}(Al,Mg,Fe)_4[(Si,Al)_8O_{20}](OH)_4{\cdot}nH_2O$	monoclinic	2.5	1.5	1.55	1.57	1.57
Montroydite	HgO	orthorhombic	11.14	2.5	2.37	2.50	2.65
Mordenite	$(Na,K,Ca)[Al_2Si_{10}O_{24}]{\cdot}7H_2O$	orthorhombic	2.13	3.5	1.478	1.480	1.482
Muscovite	$KAl_2AlSi_3O_{10}(OH)_2$	monoclinic	2.83	2.8	1.563	1.596	1.602
Nantokite	$CuCl$	cubic	4.14	2.5	1.930		
Natrolite	$Na_2Al_2Si_3O_{10}{\cdot}2H_2O$	orthorhombic	2.23	5	1.478	1.481	1.491
Nepheline	$Na_3KAl_4Si_4O_{16}$	hexagonal	2.61	5.8	1.534	1.538	
Newberyite	$MgHPO_4{\cdot}3H_2O$	orthorhombic	2.13	3.3	1.514	1.517	1.533
Niccolite	$NiAs$	hexagonal	7.77	5.3			
Norbergite	$Mg(OH,F)_2{\cdot}Mg_2SiO_4$	orthorhombic	3.21	6.5	1.565	1.573	1.592
Nosean	$Na_8Al_6Si_6O_{24}SO_4$	cubic	2.35	5.5	1.495		
Oldhamite	CaS	cubic	2.59	4	2.137		
Oligoclase	$([NaSi]_{0.9\text{-}0.7}[CaAl]_{0.1\text{-}0.3})AlSi_2O_8$	triclinic	2.64	6.3	1.539	1.543	1.547
Olivenite	$Cu_2(AsO_4)(OH)$	rhombohedral	4.2	3	1.77	1.80	1.85
Olivine	$(Mg,Fe)SiO_4$	rhombohedral	3.81	6.8	1.73	1.76	1.78
Opal	$SiO_2{\cdot}nH_2O$	amorp	1.9	5	1.44		
Orpiment	As_2S_3	monoclinic	3.46	1.8	2.40	2.81	3.02
Orthoclase	$KAlSi_3O_8$	monoclinic	2.56	6	1.523	1.527	1.531
Orthopyroxene	$(Mg,Fe)SiO_3$	rhombohedral	3.6	5.5	1.709	1.712	1.723
Paragonite	$NaAl_2AlSi_3O_{10}(OH)_2$	monoclinic	2.85	2.5	1.572	1.602	1.605
Parisite	$(Ce,La,Na)FCO_3{\cdot}CaCO_3$	hexagonal	4.42	4.5	1.672	1.771	
Pectolite	$Ca_2NaH(SiO_3)_3$	triclinic	2.88	4.8	1.603	1.610	1.639
Penfieldite	$Pb_4Cl_6(OH)_2$	hexagonal	6.6		2.13	2.21	
Pentlandite	$Fe_{4.75}Ni_{5.25}S_8$	cubic	4.8	3.8			
Percylite	$PbCuCl_2(OH)_2$	cubic		2.5	2.05		
Periclase	MgO	cubic	3.6	5.5	1.735		
Perovskite	$CaTiO_3$	cubic	3.98	5.5	2.34		
Petalite	$LiAlSi_4O_{10}$	monoclinic	2.42	6.5	1.506	1.511	1.519
Pharmacosiderite	$Fe_3(AsO_4)_2(OH)_3{\cdot}5H_2O$	cubic	2.80	2.5	1.690		
Phenakite	Be_2SiO_4	rhombohedral	2.98	7.5	1.654	1.670	
Phillipsite	$K(Ca_{0.5},Na)_2[Al_3Si_5O_{16}]{\cdot}6H_2O$	monoclinic	2.2	4.3	1.494	1.497	1.505
Phlogopite	$KMg_3AlSi_3O_{10}(OH)_2$	monoclinic	2.83	2.3	1.560	1.597	1.598
Phosgenite	$Pb_2(CO_3)Cl_2$	tetragonal	6.13	2.5	2.118	2.145	
Piemontite	$Ca_2Al_{1.5}Mn_{1.5}(SiO_4)_3OH$	monoclinic	3.49	6	1.762	1.773	1.796
Pigeonite	$(Mg,Fe,Ca)(Mg,Fe)Si_2O_6$	monoclinic	3.38	6	1.702	1.703	1.728
Pollucite	$CsAlSi_2O_6$	tetragonal	2.9	6.5	1.517		
Polybasite	$(Ag,Cu)_{16}Sb_2S_{11}$	monoclinic	6.1	2.5			
Powellite	$Ca(Mo,W)O_4$	tetragonal	4.35	3.8	1.971	1.980	
Prehnite	$Ca_2Al_2Si_3O_{10}(OH)_2$	rhombohedral	2.93	6.3	1.622	1.628	1.648
Proustite	Ag_3AsS_3	rhombohedral	5.57	2.3	2.792	3.088	
Pseudobrookite	Fe_2TiO_5	rhombohedral	4.36	6	2.38	2.39	2.42
Psilomelane	$BaMn_9O_{16}(OH)_4$	rhombohedral	4.71	5.5			

Name	Mol. form.	Crystal system	ρ/g cm^{-3}	Hardness	n_α	n_β	n_γ
Pumpellyite	$Ca_2Al_2(Al,Fe,Mg)[Si_2(O,OH)_7](SiO_4)(OH,O)_3$	monoclinic	3.21	5.5	1.688	1.695	1.705
Pyrargyrite	Ag_3SbS_3	rhombohedral	5.85	2.5	2.88	3.08	
Pyrite	FeS_2	cubic	5.02	6.3			
Pyrochlore	$NaCaNb_2O_6F$	cubic	5.3	5.3			
Pyrochroite	$Mn(OH)_2$	hexagonal	3.26	2.5	1.68	1.72	
Pyrolusite	MnO_2	tetragonal	5.08	6.3			
Pyromorphite	$Pb_5(PO_4,AsO_4)_3Cl$	hexagonal	7.04	3.8	2.048	2.058	
Pyrope	$Mg_3Al_2Si_3O_{12}$	cubic	3.58	6.8	1.714		
Pyrophyllite	$Al_2Si_4O_{10}(OH)_2$	monoclinic	2.78	1.5	1.545	1.579	1.599
Pyrrhotite	$Fe_{0.885}S$	hexagonal	4.62	4			
Quartz (α)	SiO_2	hexagonal	2.65	7	1.544	1.553	
Rammelsbergite	$NiAs_2$	orthorhombic	7.1	5.8			
Raspite	$PbWO_4$	monoclinic	8.46	2.8	1.27	1.27	1.30
Realgar	As_4S_4	monoclinic	3.5	1.8	2.538	2.684	2.704
Rhodochrosite	$MnCO_3$	hexagonal	3.70	3.8	1.597	1.816	
Rhodonite	$MnSiO_3$	orthorhombic	3.48	6	1.725	1.729	1.737
Riebeckite	$Na_2Fe_5FSi_8O_{22}(OH)_2$	monoclinic	3.3	5	1.675	1.683	1.694
Rutile	TiO_2	tetragonal	4.25	6.2	2.609	2.900	
Safflorite	$(Co,Fe)As_2$	rhombohedral	7.3	4.8			
Samarskite	$(Y,Er,Ce,U,Ca,Fe,Pb,Th)(Nb,Ta,Ti,Sn)_2O_6$	rhombohedral	5.69	5.5	2.200		
Sapphirine	$Mg_2Al_4O_6SiO_4$	monoclinic	3.49	7.5	1.709	1.712	1.715
Scapolite	$(Na,Ca)_4Al_3(Al,Si)_3Si_6O_{24}(Cl,F,OH,CO_3,SO_4)$	tetragonal	2.64	5.5	1.551	1.573	
Scheelite	$CaWO_4$	tetragonal	6.06	4.8	1.920	1.936	
Scolecite	$CaAl_2Si_3O_{10}{\cdot}3H_2O$	monoclinic	2.27	5	1.510	1.518	1.519
Scorodite	$Fe(AsO_4){\cdot}2H_2O$	rhombohedral	3.28	3.8	1.784	1.795	1.814
Sellaite	MgF_2	tetragonal	3.15	5	1.378	1.390	
Senarmontite	Sb_2O_3	cubic	5.58	2.3	2.087		
Serpentine	$Mg_3Si_2O_5(OH)_4$	monoclinic	2.55	3	1.55	1.56	1.56
Siderite	$FeCO_3$	hexagonal	3.9	4.3	1.635	1.875	
Sillimanite	Al_2OSiO_4	rhombohedral	3.25	7	1.658	1.660	1.660
Co-Skutterudite	$(Co,Ni)As_3$	cubic	6.8	5.8			
Smithsonite	$ZnCO_3$	rhombohedral	4.4	4.3	1.621	1.848	
Sodalite	$Na_8Al_6Si_6O_{24}Cl_2$	cubic	2.30	5.8	1.485		
Sperrylite	$PtAs_2$	cubic	10.58	6.5			
Spessartite	$Mn_3Al_2Si_3O_{12}$	cubic	4.19	6.8	1.800		
Sphalerite	ZnS	cubic	4.0	3.8	2.369		
Sphene	$CaTiSiO_5$	monoclinic	3.50	5	1.90	1.95	2.03
Spinel	$MgAl_2O_4$	cubic	3.55	7.8	1.719		
Spodumene	$LiAl(SiO_3)_2$	monoclinic	3.13	6.8	1.656	1.662	1.671
Stannite	Cu_2FeSn_4	tetragonal	4.4	4			
Staurolite	$Fe_2Al_9Si_4O_{22}(OH)_2$	monoclinic	3.79	7.5	1.743	1.747	1.755
Stercorite	$Na(NH_4)H(PO_4){\cdot}4H_2O$	triclinic	1.62	2	1.439	1.442	1.469
Stibiotantalite	$Sb(Ta,Nb)O_4$	rhombohedral	6.6	5.5	2.38	2.41	2.46
Stibnite	Sb_2S_3	orthorhombic	4.56	2			
Stilbite	$NaCa_2[Al_5Si_{13}O_{36}]{\cdot}14H_2O$	monoclinic	2.2	3.8	1.492	1.499	1.503
Stilpnomelane	$(K,Na,Ca)_{0.6}(Fe,Mg)_6Si_8Al(O,OH)_{27}{\cdot}2H_2O$	monoclinic	2.8	3.5	1.585	1.665	1.665
Stolzite	$PbWO_4$	tetragonal	8.2	2.8	2.19	2.27	
Strengite	$FePO_4{\cdot}2H_2O$	orthorhombic	2.87	4	1.707	1.719	1.741
Strontianite	$SrCO_3$	orthorhombic	3.5	3.5	1.518	1.666	1.668
Struvite	$Mg(NH_4)(PO_4){\cdot}6H_2O$	rhombohedral	1.71	2	1.495	1.496	1.504
Sulfur (orthorhombic)	S_8	orthorhombic	2.07	2	1.958	2.038	2.245
Sylvanite	$(Ag,Au)Te_2$	monoclinic	8.16	1.8			
Sylvite	KCl	cubic	1.99	2	1.490		
Talc	$3MgO{\cdot}4SiO_2{\cdot}H_2O$	monoclinic	2.71	1	1.545	1.592	1.595
Tantalite	$(Fe,Mn)(Ta,Nb)_2O_6$	rhombohedral	7.95	6.5	2.26	2.32	2.43
Tapiolite	$FeTa_2O_6$	tetragonal	7.9	6.3	2.27	2.42	
Tellurobismuthite	Bi_2Te_3	hexagonal	7.74	1.8			
Terlinguaite	Hg_2OCl	monoclinic	8.73	2.5	2.35	2.64	2.66
Tetrahedrite	$(Cu,Fe)_{12}Sb_4S_{13}$	cubic	4.9	3.8			
Thenardite	Na_2SO_4	orthorhombic	2.7	2.8	1.468	1.475	1.483
Thermonatrite	$Na_2CO_3{\cdot}H_2O$	orthorhombic	2.25	1.3	1.420	1.506	1.524

Name	Mol. form.	Crystal system	ρ/g cm^{-3}	Hardness	n_α	n_β	n_γ
Thomsenolite	$NaCaAlF_6 \cdot H_2O$	monoclinic	2.98	2	1.407	1.414	1.415
Thorianite	ThO_2	cubic	10.0	6.5	2.200		
Thorite	$ThSiO_4$	tetragonal	6.7	4.8	1.8		
Topaz	$Al_2(SiO_4)(F,OH)_2$	rhombohedral	3.53	8	1.618	1.620	1.627
Torbernite	$Cu(UO_2)_2(PO_4)_2 \cdot 8H_2O$	tetragonal	3.22	2.3	1.582	1.592	
Tourmaline	$Na(Mg,Fe,Mn,Li,Al)_3Al_6Si_6O_{18}(BO_3)_3$	rhombohedral	3.14	7	1.62	1.65	
Tremolite	$Ca_2Mg_5Si_8O_{22}(OH)_2$	monoclinic	3.0	5.5	1.599	1.612	1.622
Trevorite	$NiFe_2O_4$	cubic	5.33	7.8	2.3		
Tridymite	SiO_2	hexagonal	2.27	7	1.475	1.476	1.479
Triphyllite-Lithiophyllite	$Li(Fe,Mn)PO_4$	rhombohedral	3.46	4.5	1.68	1.68	1.69
Troegerite	$(UO_2)_3(AsO_4)_2 \cdot 12H_2O$	tetragonal		2.5	1.59	1.630	
Troilite	FeS	hexagonal	4.7	4			
Trona	$Na_3H(CO_3)_2 \cdot 2H_2O$	monoclinic	2.14	2.8	1.412	1.492	1.540
Turquois	$CuAl_6(PO_4)_4(OH)_8 \cdot 4H_2O$	triclinic	2.9	5.3	1.70	1.73	1.75
Ullmannite	NiSbS	cubic	6.65	5.3			
Uraninite	UO_2	cubic	11.0	5.5			
Uvarovite	$Ca_3Cr_2Si_3O_{12}$	cubic	3.83	6.8	1.865		
Valentinite	Sb_2O_3	orthorhombic	5.7	2.8	2.18	2.35	2.35
Vanadinite	$Pb_5(VO_4)_3Cl$	hexagonal	6.8	2.9	2.350	2.416	
Variseite-Strengite	$(Al,Fe)(PO_4) \cdot 2H_2O$	rhombohedral	2.72	4	1.635	1.654	1.668
Vaterite	$CaCO_3$	hexagonal	2.71		1.550	1.645	
Vermiculite	$(Mg,Ca)_{0.7}(Mg,Fe,Al)_6[(Al,Si)_8O_{20}](OH)_4 \cdot 8H_2O$	monoclinic	2.3	1.5	1.542	1.556	1.556
Vesuvianite	$Ca_{10}(Mg,Fe)_2Al_4(Si_2O_7)_2(SiO_4)_5(OH,F)_4$	tetragonal	3.33	6.5	1.72	1.73	
Villiaumite	NaF	cubic	2.78	2.3	1.327		
Vivianite	$Fe_3(PO_4)_2 \cdot 8H_2O$	monoclinic	2.58	1.8	1.598	1.629	1.652
Wagnerite	$Mg_2(PO_4)F$	monoclinic	3.15	5.3	1.568	1.572	1.582
Wavellite	$Al_3(OH)_3(PO_4)_2 \cdot 5H_2O$	rhombohedral	2.36	3.6	1.527	1.535	1.553
Whewellite	$CaC_2O_4 \cdot H_2O$	cubic	2.2	2.8	1.491	1.554	1.650
Willemite	Zn_2SiO_4	hexagonal	4.1	5.5	1.691	1.719	
Witherite	$BaCO_3$	orthorhombic	4.29	3.5	1.529	1.676	1.677
Wolframite	$Fe_{0.5}Mn_{0.5}WO_4$	monoclinic	7.3	4.3	2.26	2.32	2.42
Wollastonite	$CaSiO_3$	monoclinic	2.92	4.8	1.628	1.639	1.642
Wulfenite	$PbMoO_4$	tetragonal	6.7	2.9	2.283	2.403	
Wurtzite	ZnS	hexagonal	4.09	3.8	2.356	2.378	
Xenotime	YPO_4	tetragonal	4.8	4.5	1.721	1.816	
Zeunerite	$Cu(UO_2)_2(AsO_4)_2 \cdot 10H_2O$	tetragonal			1.606		
Zincite	ZnO	hexagonal	5.6	4	2.013	2.029	
Zircon	$ZrSiO_4$	tetragonal	4.6	7.5	1.94	1.99	
Zoisite	$Ca_2Al_3(SiO_4)_3OH$	rhombohedral	3.26	6	1.695	1.699	1.711

CRYSTALLOGRAPHIC DATA ON MINERALS

This table contains x-ray crystallographic data on about 400 common minerals, as well as selected crystalline elements. The column definitions are as follows.

Column heading	Definition
Name	Common name of the mineral, arranged alphabetically
Mol. form.	Chemical formula for a typical sample of the mineral; composition often varies considerably with the origin of the sample
Crystal system	Crystal system of the mineral
Structure type	Prototype for the structural arrangement of the crystallographic cell
Z	Number of formula units per unit cell
a	Length of the *a* cell edge, in Å (1 Å = 10^{-8} cm)
b	Length of the *b* cell edge, in Å (1 Å = 10^{-8} cm)
c	Length of the *c* cell edge, in Å (1 Å = 10^{-8} cm)
α	*α* angle between cell axes, in degrees
β	*β* angle between cell axes, in degrees
γ	*γ* angle between cell axes, in degrees

References

1. Robie, R. A., Bethke, P. M., and Beardsley, K. M., *U.S. Geological Survey Bulletin 1248*, U.S. Government Printing Office, Washington, DC.
2. Donnay, J. D. H., and Ondik, H. M., *Crystal Data Determinative Tables, Third Edition, Volume 2, Inorganic Compounds*, Joint Committee on Powder Diffraction Standards, Swarthmore, PA, 1973.
3. Deer, W. A., Howie, R. A., and Zussman, J., *An Introduction to the Rock-Forming Minerals, Second Edition*, Longman Scientific & Technical, Harlow, Essex, 1992.

Crystallographic Data for Minerals

Name	Mol. form.	Crystal system	Structure type	*Z*	*a*/Å	*b*/Å	*c*/Å	*α*/°	*β*/°	*γ*/°
Acanthite	Ag_2S	monoclinic		4	4.228	6.928	7.862		99.58	
Acmite	$NaFe(SiO_3)_2$	monoclinic	diopside	4	9.658	8.795	5.294		107.42	
Akermanite	$Ca_2MgSi_2O_7$	tetragonal	melilite	2	7.8435		5.010			
Alabandite	MnS	cubic	rock salt	4	5.223					
Almandine	$Fe_3Al_2Si_3O_{12}$	cubic	garnet	8	11.526					
Altaite	PbTe	cubic	rock salt	4	6.4606					
Aluminum	Al	cubic	copper	4	4.049					
Alunite	$KAl_3(SO_4)_2(OH)_6$	rhombohedral		3	6.982		17.32			
Analcite	$NaAl(SiO_3)_2{\cdot}H_2O$	cubic		16	13.733					
Anatase	TiO_2	tetragonal		4	3.785		9.514			
Andalusite	Al_2OSiO_4	orthorhombic		4	7.7959	7.8983	5.5583			
Andradite	$Ca_3Fe_2Si_3O_{12}$	cubic	garnet	8	12.048					
Anglesite	$PbSO_4$	orthorhombic	barite	4	8.480	5.398	6.958			
Anhydrite	$CaSO_4$	orthorhombic	anhydrite	4	6.991	6.996	6.238			
Annite	$KFe_3[AlSi_3O_{10}](OH)_2$	monoclinic	1M mica	2	10.29	9.33	5.39		105.1	
Anorthite	$CaAl_2Si_2O_8$	triclinic	primitive cell	8	8.177	12.877	14.169	93.17	115.85	91.22
Anthophyllite	$Mg_7Si_8O_{22}(OH)_2$	orthorhombic		4	18.61	18.01	5.24			
Antimony	Sb	rhombohedral	arsenic	6	4.2996		11.2516			
Aragonite	$CaCO_3$	orthorhombic	aragonite	4	5.741	7.968	4.959			
Arcanite	K_2SO_4	orthorhombic	arcanite	4	5.772	10.072	7.483			
Argentite	Ag_2S	cubic		2	4.870					
Argentopyrite	$AgFe_2S_3$	orthorhombic		4	6.64	11.47	6.45			
Arsenic (gray)	As	rhombohedral	arsenic	6	3.760		10.555			
Arsenolite	As_2O_3	cubic	diamond	16	11.074					
Arsenopyrite	FeAsS	triclinic		4	5.760	5.690	5.785	90.00	112.23	90.00
Azurite	$Cu_3(OH)_2(CO_3)_2$	monoclinic		2	5.008	5.844	10.336		92.45	
Baddeleyite	ZrO_2	monoclinic	baddeleyite	4	5.1454	5.2075	5.3107		99.23	
Banalsite	$BaNa_2Al_4Si_4O_{16}$	orthorhombic		4	8.50	9.97	16.72			
Barite	$BaSO_4$	orthorhombic	barite	4	8.878	5.450	7.152			
Berlinite	$AlPO_4$	hexagonal	α-quartz	3	4.942		10.97			
Beryl	$Be_3Al_2(SiO_3)_6$	hexagonal	beryl	2	9.215		9.192			
Berzelianite	Cu_2Se	cubic		4	5.85					
Bismite	Bi_2O_3	monoclinic	pseudo-orth	4	7.48	8.14	5.83		112.9	
Bismuth	Bi	rhombohedral	arsenic	6	4.5367		11.8383			
Bismuthinite	Bi_2S_3	orthorhombic	stibnite	4	11.150	11.300	3.981			

Name	Mol. form.	Crystal system	Structure type	Z	a/Å	b/Å	c/Å	α/°	β/°	γ/°
Bixbyite	$(Mn,Fe)_2O_3$	cubic	thallium trioxide	16	9.411					
Boehmite	$AlO(OH)$	orthorhombic	lepidocrocite	4	2.868	12.227	3.700			
Borax	$Na_2B_4O_7{\cdot}10H_2O$	monoclinic		4	11.858	10.674	12.197		106.68	
Bornite (metastable)	Cu_5FeS_4	cubic		8	10.94					
Braggite	PtS	tetragonal		2	3.4699		6.1098			
Breithauptite	$NiSb$	hexagonal	niccolite	2	3.942		5.155			
Brochantite	$Cu_4SO_4(OH)_6$	monoclinic		4	13.066	9.85	6.022		103.27	
Bromellite	BeO	hexagonal	zincite	2	2.6979		4.3772			
Bromyrite	$AgBr$	cubic	rock salt	4	5.7745					
Brookite	TiO_2	orthorhombic		8	5.456	9.182	5.143			
Brucite	$Mg(OH)_2$	hexagonal	cadmium iodide	1	3.147		4.769			
Bunsenite	NiO	cubic	rock salt	4	4.177					
Bustamite	$CaMn(SiO_3)_2$	triclinic		6	7.736	7.157	13.824	90.52	94.58	103.87
Cadmium telluride	$CdTe$	cubic	sphalerite	4	6.4805					
Cadmoselite	$CdSe$	hexagonal	zincite	2	4.2977		7.0021			
Calcite	$CaCO_3$	rhombohedral	calcite	6	4.9899		17.064			
Calomel	Hg_2Cl_2	tetragonal		4	4.478		10.910			
Carbonate-apatite	$Ca_{10}(PO_4)_6CO_3{\cdot}H_2O$	hexagonal	apatite	1	9.436		6.883			
Carbon (diamond)	C	cubic	diamond	8	3.5670					
Carbon (graphite)	C	hexagonal	graphite	4	2.4612		6.7079			
Cassiterite	SnO_2	tetragonal	rutile	2	4.738		3.188			
Cattierite	CoS_2	cubic	pyrite	4	5.5345					
Celestite	$SrSO_4$	orthorhombic	barite	4	8.359	5.352	6.866			
Celsian	$BaAl_2Si_2O_8$	monoclinic		8	8.627	13.045	14.408		115.20	
Cerargyrite	$AgCl$	cubic	rock salt	4	5.5491					
Cerianite	CeO_2	cubic	fluorite	4	5.4110					
Cerussite	$PbCO_3$	orthorhombic	aragonite	4	6.152	8.436	5.195			
Cervantite	Sb_2O_4	cubic		8	10.305					
Chalcanthite	$CuSO_4{\cdot}5H_2O$	triclinic		2	6.1045	10.72	5.949	97.57	107.28	77.43
Chalcocite	Cu_2S	orthorhombic		96	11.881	27.323	13.491			
Chalcopyrite	$CuFeS_2$	tetragonal		4	5.2988		10.434			
Chlorapatite	$Ca_5(PO_4)_3Cl$	hexagonal	apatite	2	9.629		6.777			
Chloritoid	$FeAl_4O_2(SiO_4)_2(OH)_4$	monoclinic		8	9.48	5.48	18.18		101.77	
Chloromagnesite	$MgCl_2$	rhombohedral		3	3.632		17.795			
Chondrodite	$2Mg_2SiO_4{\cdot}MgF_2$	monoclinic		2	7.89	4.743	10.29		109.03	
Chrysoberyl	$BeAl_2O_4$	orthorhombic	olivine	4	5.4756	9.4041	4.4267			
Cinnabar	HgS	hexagonal	cinnabar	3	4.149		9.495			
Claudetite	As_2O_3	monoclinic		4	5.339	12.984	4.5405		94.27	
Clausthalite	$PbSe$	cubic	rock salt	4	6.1255					
Clinoenstatite	$MgSiO_3$	monoclinic		8	9.620	8.825	5.188		108.33	
Clinoferrosilite	$FeSiO_3$	monoclinic		8	9.7085	9.0872	5.2284		108.43	
Clinohumite	$4Mg_2SiO_4{\cdot}MgF_2$	monoclinic		2	13.68	4.75	10.27		100.83	
Clinozoisite	$Ca_2Al_3(SiO_4)_3OH$	monoclinic		2	8.887	5.581	10.14		115.93	
Cobalticalcite	$CoCO_3$	rhombohedral	calcite	6	4.6581		14.958			
Cobaltite	$CoAsS$	cubic	NiSbS	4	5.60					
Cobalt olivine	Co_2SiO_4	orthorhombic	olivine	4	4.782	10.301	6.003			
Cobalt(II) oxide	CoO	cubic	rock salt	4	4.260					
Cobalt(II) sulfide	CoS	cubic	sphalerite	4	5.339					
Cobalt(II) titanate	$CoTiO_3$	rhombohedral	ilmenite	6	5.066		13.918			
Coesite	SiO_2	monoclinic		16	7.152	12.379	7.152		120.00	
Coffinite	$USiO_4$	tetragonal	zircon	4	6.995		6.263			
Colemanite	$Ca_2B_6O_{11}{\cdot}5H_2O$	monoclinic		4	8.743	11.264	6.102		110.12	
Coloradoite	$HgTe$	cubic	sphalerite	4	6.4600					
Copper	Cu	cubic	face-centered cubic	4	3.6150					
Fe-Cordierite	$Fe_2Al_3(AlSi_5O_{18})$	orthorhombic	cordierite	4	9.726	17.065	9.287			
Corundum	Al_2O_3	rhombohedral	corundum	6	4.7591		12.9894			
Cotunnite	$PbCl_2$	orthorhombic		4	4.535	7.62	9.05			
Covellite	CuS	hexagonal		6	3.792		16.34			
Cristobalite (α)	SiO_2	tetragonal		4	4.971		6.918			
Cristobalite (β)	SiO_2	cubic		8	7.1382					
Cryolite	Na_3AlF_6	monoclinic		2	5.40	5.60	7.776		90.18	

Name	Mol. form.	Crystal system	Structure type	Z	a/Å	b/Å	c/Å	α/°	β/°	γ/°
Cubanite	$CuFe_2S_3$	orthorhombic		4	6.46	11.12	6.23			
Cummingtonite	$(Mg,Fe,Mn)_7(Si_4O_{11})_2(OH)_2$	monoclinic	tremolite	2	9.522	18.223	5.332		101.92	
Cuprite	Cu_2O	cubic		2	4.2696					
Danburite	$CaB_2Si_2O_8$	orthorhombic		4	8.04	8.77	7.74			
Datolite	$CaBSiO_4(OH)$	monoclinic		4	9.62	7.60	4.84		90.15	
Daubreeite	$FeCr_2S_4$	cubic	spinel	8	9.966					
Diaspore	$AlO(OH)$	orthorhombic		4	4.401	9.421	2.845			
Dickite	$Al_2Si_2O_5(OH)_4$	monoclinic		4	5.150	8.940	14.736		103.58	
Digenite	$Cu_{1.79}S$	cubic	deformed fluorite	4	5.5695					
Diopside	$CaMg(SiO_3)_2$	monoclinic	diopside	4	9.743	8.923	5.251		105.93	
Dioptase	$CuSiO_2(OH)_2$	rhombohedral	phenacite	18	14.61		7.80			
Dolerophanite	$Cu_2O(SO_4)$	monoclinic		4	8.334	6.312	7.628		108.4	
Dolomite	$CaMg(CO_3)_2$	rhombohedral	calcite	3	4.8079		16.010			
Dravite	$NaMg_3Al_6B_3Si_6O_{27}(OH)_4$	rhombohedral	tourmaline	3	15.942		7.224			
Elbaite	$NaLiAl_{7.67}B_3Si_6O_{27}(OH)_4$	rhombohedral	tourmaline	3	15.842		7.009			
Enargite	Cu_3AsS_4	orthorhombic		2	6.426	7.422	6.144			
Enstatite	$MgSiO_3$	orthorhombic		16	8.829	18.22	5.192			
Epidote	$Ca_2Al_2FeOH(SiO_4)_3$	monoclinic		2	8.89	5.63	10.19		115.40	
Epsomite	$MgSO_4{\cdot}7H_2O$	orthorhombic		4	11.86	11.99	6.858			
Eskolaite	Cr_2O_3	rhombohedral	corundum	6	4.9607		13.599			
Eucairite	AgCuSe	orthorhombic		10	4.105	20.35	6.31			
Euclase	$AlBeSiO_4(OH)$	monoclinic		4	4.763	14.29	4.618		100.25	
Famatimite	Cu_3SbS_4	tetragonal		2	5.384		10.770			
Fayalite	Fe_2SiO_4	orthorhombic	olivine	4	4.817	10.477	6.105			
Ferberite	$FeWO_4$	monoclinic	wolframite	2	4.732	5.708	4.965		90.00	
Ferriannite	$KFe_3[FeSi_3O_{10}](OH)_2$	monoclinic		2	5.430	9.404	10.341		100.07	
Ferroselite	$FeSe_2$	orthorhombic	marcasite	2	4.801	5.778	3.587			
Ferrotremolite	$Ca_2Fe_5[Si_8O_{22}](OH)_2$	monoclinic	tremolite	2	9.97	18.34	5.30		104.50	
Fluorapatite	$Ca_5(PO_4)_3F$	hexagonal	apatite	2	9.3684		6.8841			
Fluor-edenite	$NaCa_2Mg_5[AlSi_7O_{22}]F_2$	monoclinic	tremolite	2	9.847	18.00	5.282		104.83	
Fluor-humite	$3Mg_2SiO_4{\cdot}MgF_2$	orthorhombic		4	10.243	20.72	4.735			
Fluorite	CaF_2	cubic	fluorite	4	5.4638					
Fluor-norbergite	$Mg_2SiO_4{\cdot}MgF_2$	orthorhombic		4	8.727	10.271	4.709			
Fluor-phlogopite	$KMg_3[AlSi_3O_{10}]F_2$	monoclinic	1M mica	2	5.299	9.188	10.135		99.92	
Fluor-richterite	$Na_2CaMg_5[Si_8O_{22}]F_2$	monoclinic	tremolite	2	9.823	17.96	5.268		104.33	
Fluor-tremolite	$Ca_2Mg_5[Si_8O_{22}]F_2$	monoclinic	tremolite	2	9.781	18.01	5.267		104.52	
Forsterite	Mg_2SiO_4	orthorhombic	olivine	4	4.758	10.214	5.984			
Frohbergite	$FeTe_2$	orthorhombic	marcasite	2	5.265	6.265	3.869			
Gahnite	$ZnAl_2O_4$	cubic	spinel	8	8.0848					
Galaxite	$MnAl_2O_4$	cubic	spinel	8	8.258					
Galena	PbS	cubic	rock salt	4	5.9360					
Gallium(III) oxide	Ga_2O_3	rhombohedral	corundum	6	4.9793		13.429			
Gehlenite	$Ca_2Al_2SiO_7$	tetragonal	melilite	2	7.690		5.0675			
Fe-Gehlenite	$Ca_2Fe_2SiO_7$	tetragonal	melilite	2	7.54		4.855			
Geikielite	$MgTiO_3$	rhombohedral	ilmenite	6	5.054		13.898			
Gerhardite	$Cu_2(NO_3)(OH)_3$	orthorhombic		4	6.075	13.812	5.592			
Gersdorfite	NiAsS	cubic		4	5.693					
Gibbsite	$Al(OH)_3$	monoclinic		8	9.719	5.0705	8.6412		94.57	
Glauchroite	$CaMnSiO_4$	orthorhombic	olivine	4	4.944	11.19	6.529			
Glaucodot	(Co,Fe)AsS	orthorhombic		24	6.64	28.39	5.64			
Glaucophane I	$Na_2Mg_3Al_2[Si_8O_{22}](OH)_2$	monoclinic	tremolite	2	9.748	17.915	5.273		102.78	
Glaucophane II	$Na_2Mg_3Al_2[Si_8O_{22}](OH)_2$	monoclinic	tremolite	2	9.663	17.696	5.277		103.67	
Goethite	$FeO(OH)$	orthorhombic		4	4.596	9.957	3.021			
Gold	Au	cubic	face-centered cubic	4	4.0786					
Goldmanite	$Ca_3V_2Si_3O_{12}$	cubic	garnet	8	12.070					
Goslarite	$ZnSO_4{\cdot}7H_2O$	orthorhombic	epsomite	4	11.779	12.050	6.822			
Greenockite	CdS	hexagonal	zincite	2	4.1354		6.7120			
Greigite	Fe_3S_4	cubic	spinel	8	9.876					
Grossularite	$Ca_3Al_2(SiO_4)_3$	cubic	garnet	8	11.851					
Grunerite	$Fe_7[Si_8O_{22}](OH)_2$	monoclinic	tremolite	2	9.572	18.44	5.342		101.77	

Name	Mol. form.	Crystal system	Structure type	*Z*	*a*/Å	*b*/Å	*c*/Å	*α*/°	*β*/°	*γ*/°
Gudmundite	FeSbS	monoclinic		8	10.00	5.93	6.73		90.00	
Gypsum	$CaSO_4{\cdot}2H_2O$	monoclinic		4	5.68	15.18	6.29		113.83	
Hafnia	HfO_2	monoclinic	baddeleyite	4	5.1156	5.1722	5.2948		99.18	
Halite	NaCl	cubic	rock salt	4	5.6402					
Hambergite	$Be_2(OH,F)BO_3$	orthorhombic		8	9.755	12.201	4.426			
Hardystonite	$Ca_2ZnSi_2O_7$	tetragonal	melilite	2	7.87		5.01			
Hauerite	MnS_2	cubic	pyrite	4	6.1014					
Hausmannite	Mn_3O_4	tetragonal		8	8.136		9.422			
Hawleyite	CdS	cubic	sphalerite	4	5.833					
Heazelwoodite	Ni_3S_2	rhombohedral		3	5.746		7.134			
Hedenbergite	$CaFe(SiO_3)_2$	monoclinic	diopside	4	9.854	9.024	5.263		104.23	
Hematite	Fe_2O_3	rhombohedral	corundum	6	5.025		13.735			
Hemimorphite	$Zn_4(OH)_2Si_2O_7{\cdot}H_2O$	orthorhombic		2	8.370	10.719	5.120			
Hercynite	$Fe(AlO_2)_2$	cubic	spinel	8	8.150					
Herzenbergite	SnS	orthorhombic	germanium sulfide	4	4.328	11.190	3.978			
Hessite	Ag_2Te	monoclinic		4	8.13	4.48	8.09		111.9	
Hexahydrite	$MgSO_4{\cdot}6H_2O$	monoclinic		8	10.110	7.212	24.41		98.30	
High albite	$NaAlSi_3O_8$	triclinic		4	8.160	12.870	7.106	93.54	116.36	90.19
High argentite	Ag_2S	cubic		4	6.269					
High bornite	Cu_5FeS_4	cubic		1	5.50					
High carnegeite	$NaAlSiO_4$	cubic		4	7.325					
High chalcocite	Cu_2S	hexagonal		2	3.961		6.722			
High clinoenstatite	$MgSiO_3$	triclinic		8	10.000	8.934	5.170	88.27	70.03	91.01
High digenite	Cu_2S	cubic		4	5.725					
High germania	GeO_2	hexagonal	α-quartz	3	4.987		5.652			
High leucite	$KAlSi_2O_6$	cubic		16	13.43					
High naumanite	Ag_2Se	cubic		2	4.993					
High sanidine	$KAlSi_3O_8$	monoclinic		4	8.615	13.031	7.177		115.98	
Huebnerite	$MnWO_4$	monoclinic	wolframite	2	4.834	5.758	4.999		91.18	
Huntite	$Mg_3Ca(CO_3)_4$	rhombohedral	calcite	3	9.498		7.816			
Hydroxylapatite	$Ca_5(PO_4)_3OH$	hexagonal	apatite	2	9.418		6.883			
Ice	H_2O	hexagonal		4	4.5212		7.3666			
Ilmenite	$FeTiO_3$	rhombohedral	ilmenite	6	5.093		14.055			
Indialite	$Mg_2Al_3(AlSi_5O_{18})$	hexagonal	beryl	2	9.7698		9.3517			
Fe-Indialite	$Fe_2Al_3(AlSi_5O_{18})$	hexagonal	beryl	2	9.860		9.285			
Mn-Indialite	$Mn_2Al_3(AlSi_5O_{18})$	hexagonal	beryl	2	9.925		9.297			
Iodargyrite	AgI	hexagonal	zincite	2	4.5955		7.5005			
Iron	Fe	cubic	body-centered cubic	2	2.8664					
Jacobsite	$MnFe_2O_4$	cubic	spinel	8	8.499					
Jadeite	$NaAl(SiO_3)_2$	monoclinic	diopside	4	9.409	8.564	5.220		107.50	
Jalpaite	$Ag_{1.55}Cu_{0.45}S$	tetragonal		16	8.673		11.756			
Johannsenite	$CaMn(SiO_3)_2$	monoclinic	diopside	4	9.83	9.04	5.27		105.00	
Kaliophilite	$KAlSiO_4$	hexagonal		54	26.930		8.522			
Kalsilite	$KAlSiO_4$	hexagonal		2	5.1597		8.7032			
Kaolinite	$Al_2Si_2O_5(OH)_4$	triclinic		2	5.155	8.959	7.407	91.68	104.87	89.93
Karelianite	V_2O_3	rhombohedral	corundum	6	4.952		14.002			
Keatite	SiO_2	tetragonal		12	7.456		8.604			
Kernite	$Na_2B_4O_7{\cdot}4H_2O$	monoclinic		4	7.022	9.151	15.676		108.83	
Kerschsteinite	$CaFeSiO_4$	orthorhombic	olivine	4	4.886	11.146	6.434			
Klockmannite	CuSe	hexagonal	deformed covellite	78	14.206		17.25			
Knebelite	$MnFeSiO_4$	orthorhombic	olivine	4	4.854	10.602	6.162			
Kyanite	Al_2OSiO_4	triclinic		4	7.123	7.848	5.564	89.92	101.25	105.97
Larnite	Ca_2SiO_4	monoclinic		4	5.48	6.76	9.28		94.55	
Laurite	RuS_2	cubic	pyrite	4	5.60					
Lawrencite	$FeCl_2$	rhombohedral		3	3.593		17.58			
Lawsonite	$CaAl_2Si_2O_7(OH)_2{\cdot}H_2O$	orthorhombic		4	8.787	5.836	13.123			
Lead	Pb	cubic	face-centered cubic	4	4.9505					
Leonhardtite	$MgSO_4{\cdot}4H_2O$	monoclinic		4	5.922	13.604	7.905		90.85	
Lepidocrocite	FeO(OH)	orthorhombic		4	3.868	12.525	3.066			
Lepidolite	$K_2Al_3Li_2AlSi_7O_{20}(OH)_4$	monoclinic	2M2 mica	2	9.2	5.3	20.0		98.00	
Leucite	$KAlSi_2O_6$	tetragonal		16	13.074		13.738			

Name	Mol. form.	Crystal system	Structure type	Z	a/Å	b/Å	c/Å	α/°	β/°	γ/°
Fe-Leucite	$KFeSi_2O_6$	tetragonal		16	13.205		13.970			
Lime	CaO	cubic	rock salt	4	4.8108					
Lime olivine	Ca_2SiO_4	orthorhombic	olivine	4	5.091	11.371	6.782			
Linnaeite	Co_3S_4	cubic	spinel	8	9.401					
Litharge	PbO	tetragonal		2	3.9759		5.023			
Loellingite	$FeAs_2$	orthorhombic	marcasite	2	5.300	5.981	2.882			
Low albite	$NaAlSi_3O_8$	triclinic		4	8.139	12.788	7.160	94.27	116.57	87.68
Low bornite	Cu_5FeS_4	tetragonal		16	10.94		21.88			
Low cordierite	$Mg_2Al_3(AlSi_5O_{18})$	orthorhombic		4	9.721	17.062	9.339			
Low germania	GeO_2	tetragonal	rutile	2	4.3963		2.8626			
Low nepheline	$NaAlSiO_4$	hexagonal		8	9.986		8.330			
Luzonite	Cu_3AsS_4	tetragonal		2	5.289		10.440			
Mackinawite	FeS	tetragonal		2	3.675		5.030			
Magnesioriebeckite	$Na_2Mg_3Fe_2[Si_8O_{22}](OH)_2$	monoclinic	tremolite	2	9.733	17.946	5.299		103.30	
Magnesite	$MgCO_3$	rhombohedral	calcite	6	4.6330		15.016			
Magnetite	Fe_3O_4	cubic	spinel	8	8.3940					
Malachite	$Cu_2(OH)_2CO_3$	monoclinic		4	9.502	11.974	3.240		98.75	
Maldonite	Au_2Bi	cubic		8	7.958					
Manganese(II) sulfide (β form)	MnS	cubic	sphalerite	4	5.611					
Manganese(II) sulfide (γ form)	MnS	hexagonal	zincite	2	3.976		6.432			
Manganosite	MnO	cubic	rock salt	4	4.4448					
Marcasite	FeS_2	orthorhombic	marcasite	2	4.443	5.423	3.3876			
Margarite	$CaAl_2[AlSi_2O_{10}](OH)_2$	monoclinic	2M mica	4	5.13	8.92	19.50		95.00	
Marialite	$Na_4Al_3Si_9O_{24}Cl$	tetragonal		2	12.064		7.514			
Marshite	CuI	cubic	sphalerite	4	6.0507					
Mascagnite	$(NH_4)_2SO_4$	orthorhombic	arcanite	4	7.782	5.993	10.636			
Massicot	PbO	orthorhombic		4	5.489	4.755	5.891			
Matlockite	PbClF	tetragonal		2	4.106		7.23			
Maucherite	$Ni_{11}As_8$	tetragonal		4	6.870		21.81			
Meionite	$Ca_4Al_6Si_6O_{24}CO_3$	tetragonal		2	12.174		7.652			
Melanophlogite	SiO_2	cubic	clathrate type	46	13.402					
Melanterite	$FeSO_4 \cdot 7H_2O$	monoclinic		4	14.072	6.503	11.041		105.57	
Melonite	$NiTe_2$	hexagonal	cadmium iodide	1	3.869		5.308			
Metacinnabar	HgS	cubic	sphalerite	4	5.8517					
Miargyrite	$AgSbS_2$	monoclinic		8	12.862	4.111	13.220		98.63	
Microcline	$KAlSi_3O_8$	triclinic		4	8.582	12.964	7.222	90.62	115.92	87.68
Fe-Microcline	$KFeSi_3O_8$	triclinic		4	8.68	13.10	7.340	90.75	116.05	86.23
Miersite	AgI	cubic	sphalerite	4	6.4963					
Millerite	NiS	rhombohedral		9	9.616		3.152			
Minium	Pb_3O_4	tetragonal		4	8.815		6.565			
Minnesotaite	$Fe_3Si_4O_{10}(OH)_2$	monoclinic		4	5.4	9.42	19.4		100.00	
Mirabilite	$Na_2SO_4 \cdot 10H_2O$	monoclinic		4	11.51	10.38	12.83		107.75	
Molybdenite	MoS_2	hexagonal	molybdenite	2	3.1604		12.295			
Molybdenum	Mo	cubic		2	3.1653					
Molybdite	MoO_3	orthorhombic		4	3.962	13.858	3.697			
Monteponite	CdO	cubic	rock salt	4	4.6953					
Monticellite	$CaMgSiO_4$	orthorhombic	olivine	4	4.827	11.084	6.376			
Montroydite	HgO	orthorhombic		4	6.608	5.518	3.519			
Mullite (2:1)	$2Al_2O_3 \cdot SiO_2$	orthorhombic		6	7.5788	7.6909	2.8883			
Mullite (3:2)	$3Al_2O_3 \cdot 2SiO_2$	orthorhombic		3	7.557	7.6876	2.8842			
Muscovite	$KAl_2AlSi_3O_{10}(OH)_2$	monoclinic	2M2 mica	4	5.203	8.995	20.030		94.47	
Nacrite	$Al_2Si_2O_5(OH)_4$	monoclinic		4	8.909	5.146	15.697		113.70	
Nantokite	CuCl	cubic	sphalerite	4	5.416					
Natroalunite	$NaAl_3(SO_4)_2(OH)_6$	rhombohedral		3	6.974		16.69			
Natrolite	$Na_2Al_2Si_3O_{10} \cdot 2H_2O$	orthorhombic		8	18.30	18.63	6.60			
Neighborite	$NaMgF_3$	orthorhombic	perovskite	4	5.363	7.676	5.503			
Niccolite	NiAs	hexagonal	niccolite	2	3.618		5.034			
Nickel	Ni	cubic	face-centered cubic	4	3.5238					
Nickel(II) carbonate	$NiCO_3$	rhombohedral	calcite	6	4.5975		14.723			

Name	Mol. form.	Crystal system	Structure type	Z	a/Å	b/Å	c/Å	α/°	β/°	γ/°
Nickel olivine	Ni_2SiO_4	orthorhombic	olivine	4	4.727	10.121	5.915			
Nickel selenide [$NiSe_2$]	$NiSe_2$	cubic	pyrite	4	5.9604					
Niter	KNO_3	orthorhombic	aragonite	4	6.431	9.164	5.414			
Norsethite	$BaMg(CO_3)_2$	rhombohedral	calcite	3	5.020		16.75			
Oldhamite	CaS	cubic	rock salt	4	5.689					
Orpiment	As_2S_3	monoclinic		4	11.49	9.59	4.25		90.45	
Orthoclase	$KAlSi_3O_8$	monoclinic		4	8.562	12.996	7.193		116.02	
Orthoferrosilite	$FeSiO_3$	orthorhombic	enstatite	16	9.080	18.431	5.238			
Otavite	$CdCO_3$	rhombohedral	calcite	6	4.9204		16.298			
Paracelsian	$BaAl_2Si_2O_8$	monoclinic		4	8.58	9.583	9.08		90.00	
Paragonite	$NaAl_2AlSi_3O_{10}(OH)_2$	monoclinic	2M1 mica	4	5.13	8.89	19.32		95.17	
Pararammelsbergite	$NiAs_2$	orthorhombic		8	5.75	5.82	11.428			
Paratellurite	TeO_2	tetragonal		4	4.810		7.613			
Parawollastonite	$CaSiO_3$	monoclinic		12	15.417	7.321	7.066		95.40	
Pectolite	$Ca_2NaH(SiO_3)_3$	triclinic		2	7.99	7.04	7.02	90.05	95.27	102.47
Pentlandite	$Fe_{4.75}Ni_{5.25}S_8$	cubic		4	10.095					
Periclase	MgO	cubic	rock salt	4	4.2117					
Perovskite	$CaTiO_3$	orthorhombic	perovskite	4	5.3670	7.6438	5.4439			
Petalite	$LiAlSi_4O_{10}$	monoclinic		2	11.32	5.14	7.62		105.90	
Petzite	Ag_3AuTe_2	cubic		8	10.38					
Phenacite	Be_2SiO_4	rhombohedral	phenacite	18	12.472		8.252			
Phlogopite	$KMg_3AlSi_3O_{10}(OH)_2$	monoclinic	1M mica	2	5.326	9.210	10.311		100.17	
Picrochromite	$MgCr_2O_4$	cubic	spinel	8	8.333					
Piemontite	$Ca_2Al_{1.5}Mn_{1.5}(SiO_4)_3OH$	monoclinic		2	8.95	5.70	9.41		115.70	
Platinum	Pt	cubic	face-centered cubic	4	3.9231					
Polymidite	Ni_3S_4	cubic	spinel	8	9.480					
Portlandite	$Ca(OH)_2$	hexagonal	cadmium iodide	1	3.5933		4.9086			
Powellite	$Ca(Mo,W)O_4$	tetragonal	scheelite	4	5.226		11.43			
Protoenstatite	$MgSiO_3$	orthorhombic		8	9.25	8.74	5.32			
Proustite	Ag_3AsS_3	rhombohedral		6	10.816		8.6948			
Pseudowollastonite	$CaSiO_3$	triclinic		24	6.90	11.78	19.65	90.00	90.80	90.00
Pyrargyrite	Ag_3SbS_3	rhombohedral		6	11.052		8.7177			
Pyrite	FeS_2	cubic	pyrite	4	5.4175					
Pyrolusite	MnO_2	tetragonal	rutile	2	4.388		2.865			
Pyrope	$Mg_3Al_2Si_3O_{12}$	cubic	garnet	8	11.459					
Pyrophanite	$MnTiO_3$	rhombohedral	ilmenite	6	5.155		14.18			
Pyrophyllite	$Al_2Si_4O_{10}(OH)_2$	monoclinic	2M1 mica	4	5.14	8.90	18.55		99.92	
Pyroxmangite	$MnFe(SiO_3)_2$	triclinic		7	7.56	17.45	6.67	84.00	94.30	113.70
Pyrrhotite	$Fe_{0.885}S$	hexagonal	defect niccolite	2	3.440		5.709			
Quartz (α)	SiO_2	hexagonal		3	4.9136		5.4051			
Quartz (β)	SiO_2	hexagonal		3	4.999		5.4592			
Rammelsbergite	$NiAs_2$	orthorhombic	marcasite	2	4.757	5.797	3.542			
Realgar	As_4S_4	monoclinic		16	9.29	13.53	6.57		106.55	
Retgersite	$NiSO_4{\cdot}4H_2O$	tetragonal		4	6.782		18.28			
Rhodochrosite	$MnCO_3$	rhombohedral	calcite	6	4.7771		15.664			
Rhodonite	$MnSiO_3$	triclinic		10	7.682	11.818	6.707	92.36	93.95	105.66
Riebeckite	$Na_2Fe_5FSi_8O_{22}(OH)_2$	monoclinic	tremolite	2	9.729	18.065	5.334		103.31	
Rutile	TiO_2	tetragonal		2	4.5937		2.9618			
Safflorite	$(Co,Fe)As_2$	orthorhombic	marcasite	2	5.231	5.953	2.962			
Fe-Sanidine	$KFeSi_3O_8$	monoclinic		4	8.689	13.12	7.319		116.10	
Sanmartinite	$ZnWO_4$	monoclinic	wolframite	2	4.691	5.720	4.925		89.36	
Sapphirine	$Mg_2Al_4O_6SiO_4$	monoclinic		8	9.96	28.60	9.85		110.5	
Scacchite	$MnCl_2$	rhombohedral		3	3.711		17.59			
Scheelite	$CaWO_4$	tetragonal	scheelite	4	5.242		11.372			
Schorl	$NaFe_3Al_6B_3Si_6O_{27}(OH)_4$	rhombohedral	tourmaline	3	16.032		7.149			
Selenium (gray)	Se	hexagonal		3	4.3642		4.9588			
Selenolite	SeO_2	tetragonal		8	8.35		5.05			
Sellaite	MgF_2	tetragonal	rutile	2	4.621		3.050			
Senarmontite	Sb_2O_3	cubic	arsenic trioxide	16	11.152					
Shandite	$Ni_3Pb_2S_2$	rhombohedral		3	5.576		13.658			
Shortite	$Na_2Ca_2(CO_3)_3$	orthorhombic		2	4.961	11.03	7.12			

Name	Mol. form.	Crystal system	Structure type	Z	a/Å	b/Å	c/Å	α/°	β/°	γ/°
Siderite	$FeCO_3$	rhombohedral	calcite	6	4.6887		15.373			
Silicon	Si	cubic	diamond	8	5.4305					
Sillimanite	Al_2OSiO_4	orthorhombic		4	7.4843	7.6730	5.7711			
Silver	Ag	cubic	face-centered cubic	4	4.0862					
Silver telluride I	Ag_2Te	cubic		2	5.29					
Silver telluride II	Ag_2Te	cubic		4	6.585					
Fe-Skutterudite	$FeAs_{2.95}$	cubic		8	8.1814					
Ni-Skutterudite	$NiAs_{2.95}$	cubic		8	8.3300					
Smithsonite	$ZnCO_3$	rhombohedral	calcite	6	4.6528		15.025			
Soda niter	$NaNO_3$	rhombohedral	calcite	6	5.0696		16.829			
Sodium melilite	$NaCaAlSi_2O_7$	tetragonal	melilite	2	8.511		4.809			
Sperrylite	$PtAs_2$	cubic	pyrite	4	5.968					
Spessartite	$Mn_3Al_2Si_3O_{12}$	cubic	garnet	8	11.621					
Sphalerite	ZnS	cubic	sphalerite	4	5.4093					
Sphene	$CaTiSiO_5$	monoclinic		4	7.07	8.72	6.56		113.95	
Spinel	$MgAl_2O_4$	cubic	spinel	8	8.080					
Spodumene	$LiAl(SiO_3)_2$	monoclinic	diopside	4	9.451	8.387	5.208		110.07	
Spodumene (β)	$LiAl(SiO_3)_2$	tetragonal		4	7.5332		9.1540			
Staurolite	$Fe_2Al_9Si_4O_{22}(OH)_2$	monoclinic		2	7.90	16.65	5.63		90.00	
Sternbergite	$AgFe_2S_3$	orthorhombic		8	11.60	12.675	6.63			
Stibnite	Sb_2S_3	orthorhombic	stibnite	4	11.229	11.310	3.8389			
Stilleite	ZnSe	cubic	sphalerite	4	5.6685					
Stishovite	SiO_2	tetragonal	rutile	2	4.1790		2.6649			
Stolzite	$PbWO_4$	tetragonal	scheelite	4	5.4616		12.046			
Stromeyerite	$Ag_{0.93}Cu_{1.07}S$	orthorhombic		4	4.066	6.628	7.972			
Strontianite	$SrCO_3$	orthorhombic	aragonite	4	6.029	8.414	5.107			
Sulfur (monoclinic)	S	monoclinic	S8 ring molecules	48	11.04	10.98	10.92		96.73	
Sulfur (rhombohedral)	S_6	rhombohedral	S6 ring molecules	18	10.818		4.280			
Sulfur (orthorhombic)	S_8	orthorhombic	S8 ring molecules	128	10.4646	12.8660	24.4860			
Sylvite	KCl	cubic	rock salt	4	6.2931					
Syngenite	$K_2Ca(SO_4)_2{\cdot}H_2O$	monoclinic		2	9.775	7.156	6.251		104.00	
Synthetic anorthite (hexagonal)	$CaAl_2Si_2O_8$	hexagonal		2	5.10		14.72			
Synthetic anorthite (orthorhombic)	$CaAl_2Si_2O_8$	orthorhombic		2	8.22	8.60	4.83			
Talc	$3MgO{\cdot}4SiO_2{\cdot}H_2O$	monoclinic	2M1 mica	4	5.287	9.158	18.95		99.50	
Tantalum	Ta	cubic	tungsten	2	3.3058					
Teallite	$PbSnS_2$	orthorhombic	germanium sulfide	2	4.266	11.419	4.090			
Tellurite	TeO_2	orthorhombic	tellurite	8	5.607	12.034	5.463			
Tellurium	Te	hexagonal	selenium	3	4.4570		5.9290			
Tellurobismuthite	Bi_2Te_3	rhombohedral		3	4.3835		30.487			
Tennantite	$Cu_{12}As_4S_{13}$	cubic	tetrahedrite	2	10.190					
Tenorite	CuO	monoclinic		4	4.684	3.425	5.129		99.47	
Tephroite	Mn_2SiO_4	orthorhombic	olivine	4	4.871	10.636	6.232			
Tetrahedrite	$(Cu,Fe)_{12}Sb_4S_{13}$	cubic	tetrahedrite	2	10.327					
Thenardite	Na_2SO_4	orthorhombic	thenardite	8	5.863	12.304	9.821			
Thorianite	ThO_2	cubic	fluorite	4	5.5952					
Thorite	$ThSiO_4$	tetragonal	zircon	4	7.143		6.327			
Tiemannite	HgSe	cubic	sphalerite	4	6.0853					
Tin (white)	Sn	tetragonal		4	5.8315		3.1813			
Titanium	Ti	hexagonal		2	2.953		4.729			
Titanium(III) oxide	Ti_2O_3	rhombohedral	corundum	6	5.149		13.642			
Topaz	$Al_2(SiO_4)(F,OH)_2$	orthorhombic		4	8.394	8.792	4.649			
Tremolite	$Ca_2Mg_5Si_8O_{22}(OH)_2$	monoclinic	tremolite	2	9.840	18.052	5.275		104.70	
Trevorite	$NiFe_2O_4$	cubic	spinel	8	8.339					
Tridymite	SiO_2	hexagonal		4	5.0463		8.2563			
Trogtalite	$CoSe_2$	cubic	pyrite	4	5.8588					
Troilite	FeS	hexagonal	niccolite	2	3.446		5.877			
Tschermakite	$CaAl_2SiO_6$	monoclinic	diopside	4	9.615	8.661	5.272		106.12	
Tungsten	W	cubic		2	3.1653					
Tungstenite	WS_2	hexagonal	molybdenite	2	3.154		12.362			

Name	Mol. form.	Crystal system	Structure type	*Z*	*a*/Å	*b*/Å	*c*/Å	*α*/°	*β*/°	*γ*/°
Turquois	$CuAl_6(PO_4)_4(OH)_8{\cdot}4H_2O$	triclinic		1	7.424	7.629	9.910	68.61	69.71	65.08
Umangite	Cu_3Se_2	tetragonal		2	6.402		4.276			
Uraninite	UO_2	cubic	fluorite	4	5.4682					
Ureyite	$NaCr(SiO_3)_2$	monoclinic	diopside	4	9.550	8.712	5.273		107.44	
Uvarovite	$Ca_3Cr_2Si_3O_{12}$	cubic	garnet	8	11.999					
Uvite	$CaMg_4Al_5B_3Si_6O_{27}(OH)_4$	rhombohedral	tourmaline	3	15.86		7.19			
Vaesite	NiS_2	cubic	pyrite	4	5.6873					
Valentinite	Sb_2O_3	orthorhombic	antimony trioxide	4	4.914	12.468	5.421			
Vanthoffite	$MgSO_4{\cdot}3Na_2SO_4$	monoclinic		2	9.797	9.217	8.199		113.50	
Vaterite	$CaCO_3$	hexagonal		6	7.135		8.524			
Villiaumite	NaF	cubic	rock salt	4	4.6342					
Violarite	$FeNi_2S_4$	cubic	spinel	8	9.464					
Willemite	Zn_2SiO_4	rhombohedral	phenacite	18	13.94		9.309			
Witherite	$BaCO_3$	orthorhombic	aragonite	4	6.430	8.904	5.314			
Wolframite	$Fe_{0.5}Mn_{0.5}WO_4$	monoclinic	wolframite	2	4.782	5.731	4.982		90.57	
Wollastonite	$CaSiO_3$	triclinic		6	7.94	7.32	7.07	90.03	95.37	103.43
Wulfenite	$PbMoO_4$	tetragonal	scheelite	4	5.435		12.110			
Wurtzite	ZnS	hexagonal	zincite	2	3.8230		6.2565			
Wustite	$Fe_{0.953}O$	cubic	defect rock salt	4	4.3088					
Xenotime	YPO_4	tetragonal	zircon	4	6.885		5.982			
Zinc	Zn	hexagonal	hexagonal close pack	2	2.665		4.947			
Zincite	ZnO	hexagonal	zincite	2	3.2495		5.2069			
Zinc telluride	ZnTe	cubic	sphalerite	4	6.1020					
Zinkosite	$ZnSO_4$	orthorhombic	barite	4	8.588	6.740	4.770			
Zircon	$ZrSiO_4$	tetragonal	zircon	4	6.604		5.979			
Zoisite	$Ca_2Al_3(SiO_4)_3OH$	orthorhombic		4	16.15	5.581	10.06			

Section 5
Thermochemistry, Kinetics, Electrochemistry, and Solution Chemistry

CODATA Key Values for Thermodynamics 5-1
Standard Thermodynamic Properties of Chemical Substances 5-3
Thermodynamic Properties as a Function of Temperature 5-42
Thermodynamic Properties of Aqueous Ions 5-65
Heat of Combustion 5-67
Energy Content of Fuels 5-68
Chemical Reaction Rate Constants for Atmospheric Studies 5-69
Ionization Constant of Water at Various Temperatures and Pressures..................... 5-92
Ionization Constant of Normal and Heavy Water at Saturated Vapor Pressure 5-93
Electrical Conductivity of Water 5-93
Electrical Conductivity of Aqueous Solutions... 5-94
Standard KCl Solutions for Calibrating Electrical Conductivity Cells 5-95
Molar Electrical Conductivity of Aqueous HF, HCl, HBr, and HI 5-96
Molar Electrical Conductivity of Electrolytes in Aqueous Solution........................ 5-97
Ionic Conductivity and Diffusion at Infinite Dilution.................................... 5-98
Electrochemical Series 5-101
Dissociation Constants of Inorganic Acids and Bases...................................5-108
Dissociation Constants of Organic Acids and Bases 5-109
Activity Coefficients of Acids, Bases, and Salts..5-119
Mean Activity Coefficients of Electrolytes as a Function of Concentration.................5-121
Enthalpy of Dilution of Acids 5-126
Enthalpy of Solution of Electrolytes.. 5-127
pH Scale for Aqueous Solutions 5-128
Buffer Solutions Giving Round Values of pH at 25 °C 5-132
Concentrative Properties of Aqueous Solutions 5-133
Solubility of Selected Gases in Water 5-149
Solubility of Carbon Dioxide in Water at Various Temperatures and Pressures.............5-151
Aqueous Solubility and Henry's Law Constants of Organic Compounds...................5-152
Aqueous Solubility of Inorganic Compounds at Various Temperatures....................5-183
Octanol–Water Partition Coefficients...5-189
Solubility Product Constants of Inorganic Salts.. 5-194
Solubility of Common Salts at Various Temperatures................................... 5-196
Solubility of Hydrocarbons in Seawater 5-197
Solubility of Organic Compounds in Pressurized Hot Water 5-198
Solubility Chart for Inorganic Salts.. 5-200

CODATA KEY VALUES FOR THERMODYNAMICS

The Committee on Data for Science and Technology (CODATA) conducted a project to establish internationally agreed values for the thermodynamic properties of key chemical substances. This table presents the final results of the project. Use of these recommended, internally consistent values is encouraged in the analysis of thermodynamic measurements, data reduction, and preparation of other thermodynamic tables.

The table includes the standard enthalpy of formation at 298.15 K, the entropy at 298.15 K, and the quantity $H°(298.15\ K)–H°(0)$. A value of 0 in the $\Delta_f H°$ column for an element indicates the reference state for that element. The standard state pressure is 100,000 Pa (1 bar). See the reference for information on the dependence of gas-phase entropy on the choice of standard state pressure.

Column definitions are as follows.

Column heading	Definition
Name	Chemical substance name (Online Edition only)
Mol. form.	Molecular formula; substances are listed in alphabetical order of their chemical formulas when written in the most common form
Phys. state	Physical state; g = gas; cr = crystalline; aq = aqueous; l = liquid; for some physical states, the crystal system or other information is also given
$\Delta_f H°(298)$	Standard enthalpy of formation at 298.15 K
$S°(298)$	Entropy at 298.15 K
$H°(298)$-$H°(0)$	Enthalpy difference between 298.15 K and 0 K

The table is reprinted with permission of CODATA.

Reference

Cox, J. D., Wagman, D. D., and Medvedev, V. A., *CODATA Key Values for Thermodynamics,* Hemisphere Publishing Corp., New York, 1989.

CODATA Key Values for Thermodynamics

Mol. form.	Phys. state	$\Delta_f H°(298)$/ kJ mol^{-1}	$S°(298)$/ J mol^{-1} K^{-1}	$H°(298)$-$H°(0)$/ kJ mol^{-1}
Ag	cr	0	42.55 ± 0.20	5.745 ± 0.020
Ag	g	284.9 ± 0.8	172.997 ± 0.004	6.197 ± 0.001
Ag^+	aq	105.79 ± 0.08	73.45 ± 0.40	
AgCl	cr	-127.01 ± 0.05	96.25 ± 0.20	12.033 ± 0.020
Al	cr	0	28.30 ± 0.10	4.540 ± 0.020
Al	g	330.0 ± 4.0	164.554 ± 0.004	6.919 ± 0.001
Al^{+3}	aq	-538.4 ± 1.5	-325 ± 10	
AlF_3	cr	-1510.4 ± 1.3	66.5 ± 0.5	11.62 ± 0.04
Al_2O_3	cr, corundum	-1675.7 ± 1.3	50.92 ± 0.10	10.016 ± 0.020
Ar	g	0	154.846 ± 0.003	6.197 ± 0.001
B	cr, rhombic	0	5.90 ± 0.08	1.222 ± 0.008
B	g	565 ± 5	153.436 ± 0.015	6.316 ± 0.002
BF_3	g	-1136.0 ± 0.8	254.42 ± 0.20	11.650 ± 0.020
B_2O_3	cr	-1273.5 ± 1.4	53.97 ± 0.30	9.301 ± 0.040
Be	cr	0	9.50 ± 0.08	1.950 ± 0.020
Be	g	324 ± 5	136.275 ± 0.003	6.197 ± 0.001
BeO	cr	-609.4 ± 2.5	13.77 ± 0.04	2.837 ± 0.008
Br^-	aq	-121.41 ± 0.15	82.55 ± 0.20	
Br	g	111.87 ± 0.12	175.018 ± 0.004	6.197 ± 0.001
Br_2	l	0	152.21 ± 0.30	24.52 ± 0.01
Br_2	g	30.91 ± 0.11	245.468 ± 0.005	9.725 ± 0.001
C	cr, graphite	0	5.74 ± 0.10	1.050 ± 0.020
C	g	716.68 ± 0.45	158.100 ± 0.003	6.536 ± 0.001
CO	g	-110.53 ± 0.17	197.660 ± 0.004	8.671 ± 0.001
CO_2	g	-393.51 ± 0.13	213.785 ± 0.010	9.365 ± 0.003
CO_2	aq, undissoc.	-413.26 ± 0.20	119.36 ± 0.60	
CO_3^{-2}	aq	-675.23 ± 0.25	-50.0 ± 1.0	
Ca	cr	0	41.59 ± 0.40	5.736 ± 0.040
Ca	g	177.8 ± 0.8	154.887 ± 0.004	6.197 ± 0.001
Ca^{+2}	aq	-543.0 ± 1.0	-56.2 ± 1.0	
CaO	cr	-634.92 ± 0.90	38.1 ± 0.4	6.75 ± 0.06
Cd	cr	0	51.80 ± 0.15	6.247 ± 0.015
Cd	g	111.80 ± 0.20	167.749 ± 0.004	6.197 ± 0.001
Cd^{+2}	aq	-75.92 ± 0.60	-72.8 ± 1.5	
CdO	cr	-258.35 ± 0.40	54.8 ± 1.5	8.41 ± 0.08
$CdSO_4$•(8/3)H_2O	cr	-1729.30 ± 0.80	229.65 ± 0.40	35.56 ± 0.04
Cl^-	aq	-167.080 ± 0.10	56.60 ± 0.20	
Cl	g	121.301 ± 0.008	165.190 ± 0.004	6.272 ± 0.001
ClO_4^-	aq	-128.10 ± 0.40	184.0 ± 1.5	
Cl_2	g	0	223.081 ± 0.010	9.181 ± 0.001
Cs	cr	0	85.23 ± 0.40	7.711 ± 0.020
Cs	g	76.5 ± 1.0	175.601 ± 0.003	6.197 ± 0.001
Cs^+	aq	-258.00 ± 0.50	132.1 ± 0.5	
Cu	cr	0	33.15 ± 0.08	5.004 ± 0.008
Cu	g	337.4 ± 1.2	166.398 ± 0.004	6.197 ± 0.001
Cu^{+2}	aq	64.9 ± 1.0	-98 ± 4	
$CuSO_4$	cr	-771.4 ± 1.2	109.2 ± 0.4	16.86 ± 0.08
F^-	aq	-335.35 ± 0.65	-13.8 ± 0.8	
F	g	79.38 ± 0.30	158.751 ± 0.004	6.518 ± 0.001
F_2	g	0	202.791 ± 0.005	8.825 ± 0.001
Ge	cr	0	31.09 ± 0.15	4.636 ± 0.020
Ge	g	372 ± 3	167.904 ± 0.005	7.398 ± 0.001
GeF_4	g	-1190.20 ± 0.50	301.9 ± 1.0	17.29 ± 0.10
GeO_2	cr, tetragonal	-580.0 ± 1.0	39.71 ± 0.15	7.230 ± 0.020
H	g	217.998 ± 0.006	114.717 ± 0.002	6.197 ± 0.001
H^+	aq	0	0	
HBr	g	-36.29 ± 0.16	198.700 ± 0.004	8.648 ± 0.001
HCO_3^-	aq	-689.93 ± 0.20	98.4 ± 0.5	
HCl	g	-92.31 ± 0.10	186.902 ± 0.005	8.640 ± 0.001
HF	g	-273.30 ± 0.70	173.779 ± 0.003	8.599 ± 0.001
HI	g	26.50 ± 0.10	206.590 ± 0.004	8.657 ± 0.001
HPO_4^{-2}	aq	-1299.0 ± 1.5	-33.5 ± 1.5	
HSO_4^-	aq	-886.9 ± 1.0	131.7 ± 3.0	
H_2	g	0	130.680 ± 0.003	8.468 ± 0.001
H_2O	l	-285.830 ± 0.040	69.95 ± 0.03	13.273 ± 0.020
H_2O	g	-241.826 ± 0.040	188.835 ± 0.010	9.905 ± 0.005
$H_2PO_4^-$	aq	-1302.6 ± 1.5	92.5 ± 1.5	
H_2S	g	-20.6 ± 0.5	205.81 ± 0.05	9.957 ± 0.010
H_2S	aq, undissoc.	-38.6 ± 1.5	126 ± 5	
H_3BO_3	cr	-1094.8 ± 0.8	89.95 ± 0.60	13.52 ± 0.04
H_3BO_3	aq, undissoc.	-1072.8 ± 0.8	162.4 ± 0.6	
He	g	0	126.153 ± 0.002	6.197 ± 0.001
Hg	l	0	75.90 ± 0.12	9.342 ± 0.008
Hg	g	61.38 ± 0.04	174.971 ± 0.005	6.197 ± 0.001
Hg^{+2}	aq	170.21 ± 0.20	-36.19 ± 0.80	
HgO	cr, red	-90.79 ± 0.12	70.25 ± 0.30	9.117 ± 0.025
Hg_2^{+2}	aq	166.87 ± 0.50	65.74 ± 0.80	
Hg_2Cl_2	cr	-265.37 ± 0.40	191.6 ± 0.8	23.35 ± 0.20

Mol. form.	Phys. state	$\Delta_f H°(298)$/ kJ mol^{-1}	$S°(298)$/ J mol^{-1} K^{-1}	$H°(298)-H°(0)$/ kJ mol^{-1}
Hg_2SO_4	cr	-743.09 ± 0.40	200.70 ± 0.20	26.070 ± 0.030
I^-	aq	-56.78 ± 0.05	106.45 ± 0.30	
I	g	106.76 ± 0.04	180.787 ± 0.004	6.197 ± 0.001
I_2	cr	0	116.14 ± 0.30	13.196 ± 0.040
I_2	g	62.42 ± 0.08	260.687 ± 0.005	10.116 ± 0.001
K	cr	0	64.68 ± 0.20	7.088 ± 0.020
K	g	89.0 ± 0.8	160.341 ± 0.003	6.197 ± 0.001
K^+	aq	-252.14 ± 0.08	101.20 ± 0.20	
Kr	g	0	164.085 ± 0.003	6.197 ± 0.001
Li	cr	0	29.12 ± 0.20	4.632 ± 0.040
Li	g	159.3 ± 1.0	138.782 ± 0.010	6.197 ± 0.001
Li^+	aq	-278.47 ± 0.08	12.24 ± 0.15	
Mg	cr	0	32.67 ± 0.10	4.998 ± 0.030
Mg	g	147.1 ± 0.8	148.648 ± 0.003	6.197 ± 0.001
Mg^{+2}	aq	-467.0 ± 0.6	-137 ± 4	
MgF_2	cr	-1124.2 ± 1.2	57.2 ± 0.5	9.91 ± 0.06
MgO	cr	-601.60 ± 0.30	26.95 ± 0.15	5.160 ± 0.020
N	g	472.68 ± 0.40	153.301 ± 0.003	6.197 ± 0.001
NH_3	g	-45.94 ± 0.35	192.77 ± 0.05	10.043 ± 0.010
NH_4^+	aq	-133.26 ± 0.25	111.17 ± 0.40	
NO_3^-	aq	-206.85 ± 0.40	146.70 ± 0.40	
N_2	g	0	191.609 ± 0.004	8.670 ± 0.001
Na	cr	0	51.30 ± 0.20	6.460 ± 0.020
Na	g	107.5 ± 0.7	153.718 ± 0.003	6.197 ± 0.001
Na^+	aq	-240.34 ± 0.06	58.45 ± 0.15	
Ne	g	0	146.328 ± 0.003	6.197 ± 0.001
O	g	249.18 ± 0.10	161.059 ± 0.003	6.725 ± 0.001
OH^-	aq	-230.015 ± 0.040	-10.90 ± 0.20	
O_2	g	0	205.152 ± 0.005	8.680 ± 0.002
P	cr, white	0	41.09 ± 0.25	5.360 ± 0.015
P	g	316.5 ± 1.0	163.199 ± 0.003	6.197 ± 0.001
P_2	g	144.0 ± 2.0	218.123 ± 0.004	8.904 ± 0.001
P_4	g	58.9 ± 0.3	280.01 ± 0.50	14.10 ± 0.20
Pb	cr	0	64.80 ± 0.30	6.870 ± 0.030
Pb	g	195.2 ± 0.8	175.375 ± 0.005	6.197 ± 0.001
Pb^{+2}	aq	0.92 ± 0.25	18.5 ± 1.0	
$PbSO_4$	cr	-919.97 ± 0.40	148.50 ± 0.60	20.050 ± 0.040

Mol. form.	Phys. state	$\Delta_f H°(298)$/ kJ mol^{-1}	$S°(298)$/ J mol^{-1} K^{-1}	$H°(298)-H°(0)$/ kJ mol^{-1}
Rb	cr	0	76.78 ± 0.30	7.489 ± 0.020
Rb	g	80.9 ± 0.8	170.094 ± 0.003	6.197 ± 0.001
Rb^+	aq	-251.12 ± 0.10	121.75 ± 0.25	
S	cr, rhombic	0	32.054 ± 0.050	4.412 ± 0.006
S	g	277.17 ± 0.15	167.829 ± 0.006	6.657 ± 0.001
SH^-	aq	-16.3 ± 1.5	67 ± 5	
SO_2	g	-296.81 ± 0.20	248.223 ± 0.050	10.549 ± 0.010
SO_4^{-2}	aq	-909.34 ± 0.40	18.50 ± 0.40	
S_2	g	128.60 ± 0.30	228.167 ± 0.010	9.132 ± 0.002
Si	cr	0	18.81 ± 0.08	3.217 ± 0.008
Si	g	450 ± 8	167.981 ± 0.004	7.550 ± 0.001
SiF_4	g	-1615.0 ± 0.8	282.76 ± 0.50	15.36 ± 0.05
SiO_2	cr, alpha quartz	-910.7 ± 1.0	41.46 ± 0.20	6.916 ± 0.020
Sn	cr, white	0	51.18 ± 0.08	6.323 ± 0.008
Sn	g	301.2 ± 1.5	168.492 ± 0.004	6.215 ± 0.001
Sn^{+2}	aq	-8.9 ± 1.0	-16.7 ± 4.0	
SnO	cr, tetragonal	-280.71 ± 0.20	57.17 ± 0.30	8.736 ± 0.020
SnO_2	cr, tetragonal	-577.63 ± 0.20	49.04 ± 0.10	8.384 ± 0.020
Th	cr	0	51.8 ± 0.5	6.35 ± 0.05
Th	g	602 ± 6	190.17 ± 0.05	6.197 ± 0.003
ThO_2	cr	-1226.4 ± 3.5	65.23 ± 0.20	10.560 ± 0.020
Ti	cr	0	30.72 ± 0.10	4.824 ± 0.015
Ti	g	473 ± 3	180.298 ± 0.010	7.539 ± 0.002
$TiCl_4$	g	-763.2 ± 3.0	353.2 ± 4.0	21.5 ± 0.5
TiO_2	cr, rutile	-944.0 ± 0.8	50.62 ± 0.30	8.68 ± 0.05
U	cr	0	50.20 ± 0.20	6.364 ± 0.020
U	g	533 ± 8	199.79 ± 0.10	6.499 ± 0.020
UO_2	cr	-1085.0 ± 1.0	77.03 ± 0.20	11.280 ± 0.020
UO_2^{+2}	aq	-1019.0 ± 1.5	-98.2 ± 3.0	
UO_3	cr, gamma	-1223.8 ± 1.2	96.11 ± 0.40	14.585 ± 0.050
U_3O_8	cr	-3574.8 ± 2.5	282.55 ± 0.50	42.74 ± 0.10
Xe	g	0	169.685 ± 0.003	6.197 ± 0.001
Zn	cr	0	41.63 ± 0.15	5.657 ± 0.020
Zn	g	130.40 ± 0.40	160.990 ± 0.004	6.197 ± 0.001
Zn^{+2}	aq	-153.39 ± 0.20	-109.8 ± 0.5	
ZnO	cr	-350.46 ± 0.27	43.65 ± 0.40	6.933 ± 0.040

STANDARD THERMODYNAMIC PROPERTIES OF CHEMICAL SUBSTANCES

This table gives the standard state chemical thermodynamic properties of about 2500 individual substances in the crystalline, liquid, and gaseous states. For each state, the standard molar enthalpy of formation, standard molar Gibbs energy of formation, standard molar entropy, and molar heat capacity at constant pressure, all at 298.15 K, are givern in consecutive columns. The column definitions are as follows.

Column heading	Definition
Name	Name of chemical substance; substances are listed alphabetically by chemical name
Mol. form.	Molecular formula
$\Delta_f H°$	Standard molar enthalpy (heat) of formation at 298.15 K, in units of kJ/mol; in the parentheses, (c) = crystalline state, (l) = liquid state, and (g) = gaseous state
$\Delta_f G°$	Standard molar Gibbs energy of formation at 298.15 K, in units of kJ/mol; in the parentheses, (c) = crystalline state, (l) = liquid state, and (g) = gaseous state
$S°$	Standard molar entropy at 298.15 K in units of J/mol K; in the parentheses, (c) = crystalline state, (l) = liquid state, and (g) = gaseous state
C_p	Molar heat capacity at constant pressure at 298.15 K in units of J/mol K, in the parentheses, (c) = crystalline state, (l) = liquid state, and (g) = gaseous state

The standard state pressure is 100 kPa (1 bar). The standard states are defined for different phases by:

- The standard state of a pure gaseous substance is that of the substance as a (hypothetical) ideal gas at the standard state pressure.
- The standard state of a pure liquid substance is that of the liquid under the standard state pressure.
- The standard state of a pure crystalline substance is that of the crystalline substance under the standard state pressure.

An entry of 0.0 for $\Delta_f H°$ for an element indicates the reference state of that element. See Refs. 1 and 2 for further information on reference states. A blank means no value is available.

The data are derived from the sources listed in the references, other papers appearing in the *Journal of Physical and Chemical Reference Data*, and the primary research literature. We are indebted to M. V. Korobov for providing data on fullerene compounds.

References

1. Cox, J. D., Wagman, D. D., and Medvedev, V. A., *CODATA Key Values for Thermodynamics*, Hemisphere Publishing Corp., New York, 1989.
2. Wagman, D. D., Evans, W. H., Parker, V. B., Schumm, R. H., Halow, I., Bailey, S. M., Churney, K. L., and Nuttall, R. L., *The NBS Tables of Chemical Thermodynamic Properties, J. Phys. Chem. Ref. Data*, Vol. 11, Suppl. 2, 1982.
3. Chase, M. W., Davies, C. A., Downey, J. R., Frurip, D. J., McDonald, R. A., and Syverud, A. N., *JANAF Thermochemical Tables, Third Edition, J. Phys. Chem. Ref. Data*, Vol. 14, Suppl. 1, 1985.
4. Chase, M. W., *NIST-JANAF Thermochemical Tables, Fourth Edition, J. Phys. Chem. Ref. Data*, Monograph 9, 1998.
5. Daubert, T. E., Danner, R. P., Sibul, H. M., and Stebbins, C. C., *Physical and Thermodynamic Properties of Pure Compounds: Data Compilation*, extant 1994 (core with 4 supplements), Taylor & Francis, Bristol, PA.
6. Pedley, J. B., Naylor, R. D., and Kirby, S. P., *Thermochemical Data of Organic Compounds, Second Edition*, Chapman & Hall, London, 1986. <https://doi.org/10.1007/978-94-009-4099-4_1>
7. Pedley, J. B., *Thermochemical Data and Structures of Organic Compounds*, Thermodynamic Research Center, Texas A & M University, College Station, TX, 1994.
8. Domalski, E. S., and Hearing, E. D., Heat Capacities and Entropies of Organic Compounds in the Condensed Phase, Volume III, *J. Phys. Chem. Ref. Data*, 25, 1–525, 1996. <https://doi.org/10.1063/1.555985>
9. Zabransky, M., Ruzicka, V., Majer, V., and Domalski, E. S., *Heat Capacity of Liquids, J. Phys. Chem. Ref. Data*, Monograph No. 6, 1996.
10. Gurvich, L. V., Veyts, I.V., and Alcock, C. B., *Thermodynamic Properties of Individual Substances, Fourth Edition,* Vol. 1, Hemisphere Publishing Corp., New York, 1989.
11. Gurvich, L. V., Veyts, I.V., and Alcock, C. B., *Thermodynamic Properties of Individual Substances, Fourth Edition,* Vol. 3, CRC Press, Boca Raton, FL, 1994.
12. *NIST Chemistry WebBook*, <webbook.nist.gov>.

Standard Thermodynamic Properties of Chemical Substances

Name	Mol. form.	$\Delta_f H°$(c)/ kJ mol^{-1}	$\Delta_f G°$(c)/ kJ mol^{-1}	$S°$(c)/J mol^{-1} K^{-1}	C_p(c)/J mol^{-1} K^{-1}	$\Delta_f H°$(l)/ kJ mol^{-1}	$\Delta_f G°$(l)/ kJ mol^{-1}	$S°$(l)/J mol^{-1} K^{-1}	C_p(l)/J mol^{-1} K^{-1}	$\Delta_f H°$(g)/ kJ mol^{-1}	$\Delta_f G°$(g)/ kJ mol^{-1}	$S°$(g)/J mol^{-1} K^{-1}	C_p(g)/J mol^{-1} K^{-1}
Acenaphthene	$C_{12}H_{10}$	70.3		188.9	190.4					156.0			
Acenaphthylene	$C_{12}H_8$	186.7			166.4					259.7			
Acetaldehyde	C_2H_4O					-192.2	-127.6	160.2	89.0	-166.2	-133.0	263.8	55.3
Acetamide	C_2H_5NO	-317.0		115.0	91.3					-238.3			
Acetanilide	C_8H_9NO	-209.4			179.3								
Acetic acid	$C_2H_4O_2$					-484.3	-389.9	159.8	123.3	-432.2	-374.2	283.5	63.4
Acetic anhydride	$C_4H_6O_3$					-624.4				-572.5			
Acetone	C_3H_6O					-248.4		199.8	126.3	-217.1	-152.7	295.3	74.5
Acetone cyanohydrin	C_4H_7NO					-120.9							
Acetonitrile	C_2H_3N					40.6	86.5	149.6	91.5	74.0	91.9	243.4	52.2
Acetophenone	C_8H_8O					-142.5				-86.7			
Acetyl bromide	C_2H_3BrO					-223.5				-190.4			
Acetyl chloride	C_2H_3ClO					-272.9	-208.0	200.8	117.0	-242.8	-205.8	295.1	67.8
Acetylene	C_2H_2									227.4	209.9	200.9	44.0

Name	Mol. form.	$\Delta_f H°$(c)/ kJ mol^{-1}	$\Delta_f G°$(c)/ kJ mol^{-1}	$S°$(c)/J mol^{-1} K^{-1}	C_p(c)/J mol^{-1} K^{-1}	$\Delta_f H°$(l)/ kJ mol^{-1}	$\Delta_f G°$(l)/ kJ mol^{-1}	$S°$(l)/J mol^{-1} K^{-1}	C_p(l)/J mol^{-1} K^{-1}	$\Delta_f H°$(g)/ kJ mol^{-1}	$\Delta_f G°$(g)/ kJ mol^{-1}	$S°$(g)/J mol^{-1} K^{-1}	C_p(g)/J mol^{-1} K^{-1}
Acetyl fluoride	C_2H_3FO					-467.2				-442.1			
Acetyl iodide	C_2H_3IO					-163.5				-126.4			
2-(Acetyloxy)benzoic acid	$C_9H_8O_4$	-815.6											
9-Acridinamine	$C_{13}H_{10}N_2$	159.2											
Acridine	$C_{13}H_9N$	179.4								273.9			
Acrolein	C_3H_4O												71.3
Acrylamide	C_3H_5NO	-212.1			110.6	-224.0				-130.2			
Acrylic acid	$C_3H_4O_2$					-383.8			145.7				
Acrylonitrile	C_3H_3N					147.1				180.6			
Actinium	Ac	0.0		56.5	27.2					406.0	366.0	188.1	20.8
Adenine	$C_5H_5N_5$	96.9			147.0					205.7			
DL-Alanine	$C_3H_7NO_2$	-563.6											
D-Alanine	$C_3H_7NO_2$	-561.2											
L-Alanine	$C_3H_7NO_2$	-604.0								-465.9			
β-Alanine	$C_3H_7NO_2$	-558.0								-424.0			
Allene	C_3H_4									190.5			
Allyl acetate	$C_5H_8O_2$								184.1				
Allyl alcohol	C_3H_6O					-171.8			138.9	-124.5			
Allylamine	C_3H_7N					-10.0							
Allylcyclopentane	C_8H_{14}					-64.5				-24.1			
Aluminum	Al	0.0		28.3	24.2					330.0	289.4	164.6	21.4
Aluminum borohydride	AlB_3H_{12}					-16.3	145.0	289.1	194.6	13.0	147.0	379.2	
Aluminum bromide	$AlBr_3$	-527.2		180.2	100.6					-425.1			
Aluminum chloride	$AlCl_3$	-704.2	-628.8	109.3	91.1					-583.2			
Aluminum dichloride	$AlCl_2$									-331.0			
Aluminum fluoride	AlF_3	-1510.4	-1431.1	66.5	75.1					-1204.6	-1188.2	277.1	62.6
Aluminum hexabromide	Al_2Br_6									-970.7			
Aluminum hexachloride	Al_2Cl_6									-1290.8	-1220.4	490.0	
Aluminum hexafluoride	Al_2F_6									-2628.0			
Aluminum hexaiodide	Al_2I_6									-516.7			
Aluminum hydride	AlH_3	-46.0		30.0	40.2								
Aluminum iodide	AlI_3	-302.9		195.9	98.7					-289.4		223.6	
Aluminum monobromide	AlBr									-4.0	-42.0	239.5	35.6
Aluminum monochloride	AlCl									-47.7	-74.1	228.1	35.0
Aluminum monofluoride	AlF									-258.2	-283.7	215.0	31.9
Aluminum monohydride	AlH									259.2	231.2	187.9	29.4
Aluminum monoiodide	AlI									65.5			36.0
Aluminum monosulfide	AlS									200.9	150.1	230.6	33.4
Aluminum monoxide	AlO									91.2	65.3	218.4	30.9
Aluminum nitride	AlN	-318.0	-287.0	20.2	30.1								
Aluminum oxide (α)	Al_2O_3	-1675.7	-1582.3	50.9	79.0								
Aluminum oxide (Al_2O)	Al_2O									-130.0	-159.0	259.4	45.7
Aluminum phosphate	AlO_4P	-1733.8	-1617.9	90.8	93.2								
Aluminum phosphide	AlP	-166.5											
Aluminum sulfide	Al_2S_3	-724.0		116.9	105.1								
Americium	Am	0.0											
Amidogen	H_2N									184.9	194.6	195.0	33.9
Aminetrifluoroboron	BF_3H_3N	-1353.9											
2-Aminobiphenyl	$C_{12}H_{11}N$	93.8								184.4			
4-Aminobiphenyl	$C_{12}H_{11}N$	81.0											
4-Aminobutanoic acid	$C_4H_9NO_2$	-581.0								-441.0			
N-(Aminocarbonyl)acetamide	$C_3H_6N_2O_2$	-544.2								-441.2			
6-Aminohexanoic acid	$C_6H_{13}NO_2$	-637.3											
3-Amino-1-nitroguanidine	$CH_5N_5O_2$	22.1											
5-Aminopentanoic acid	$C_5H_{11}NO_2$	-604.1								-460.0			
Amminetrimethylboron	$C_3H_{12}BN$	-284.1	-79.3	218.0									
Ammonia	H_3N									-45.9	-16.4	192.8	35.1
Ammonium azide	H_4N_4	115.5	274.2	112.5									
Ammonium bromide	BrH_4N	-270.8	-175.2	113.0	96.0								
Ammonium chloride	ClH_4N	-314.4	-202.9	94.6	84.1								
Ammonium cyanide	CH_4N_2	0.4			134.0								
Ammonium fluoride	FH_4N	-464.0	-348.7	72.0	65.3								
Ammonium hexafluorosilicate	$F_6H_8N_2Si$	-2681.7	-2365.3	280.2	228.1								
Ammonium hydrogen carbonate	CH_5NO_3	-849.4	-665.9	120.9									
Ammonium hydrogen phosphate	$H_9N_2O_4P$	-1566.9			188.0								
Ammonium hydrogen sulfate	H_5NO_4S	-1027.0											

Name	Mol. form.	$\Delta_f H°$(c)/ kJ mol^{-1}	$\Delta_f G°$(c)/ kJ mol^{-1}	$S°$(c)/J mol^{-1} K^{-1}	C_p(c)/J mol^{-1} K^{-1}	$\Delta_f H°$(l)/ kJ mol^{-1}	$\Delta_f G°$(l)/ kJ mol^{-1}	$S°$(l)/J mol^{-1} K^{-1}	C_p(l)/J mol^{-1} K^{-1}	$\Delta_f H°$(g)/ kJ mol^{-1}	$\Delta_f G°$(g)/ kJ mol^{-1}	$S°$(g)/J mol^{-1} K^{-1}	C_p(g)/J mol^{-1} K^{-1}
Ammonium hydrogen sulfite	H_5NO_3S	-768.6											
Ammonium hydroxide	H_5NO					-361.2	-254.0	165.6	154.9				
Ammonium iodide	H_4IN	-201.4	-112.5	117.0									
Ammonium nitrate	$H_4N_2O_3$	-365.6	-183.9	151.1	139.3								
Ammonium nitrite	$H_4N_2O_2$	-256.5											
Ammonium oxalate	$C_2H_8N_2O_4$	-1123.0			226.0								
Ammonium perchlorate	ClH_4NO_4	-295.3	-88.8	186.2									
Ammonium phosphate	$H_{12}N_3O_4P$	-1671.9											
Ammonium sulfate	$H_8N_2O_4S$	-1180.9	-901.7	220.1	187.5								
Aniline	C_6H_7N					31.6			191.9	87.5	-7.0	317.9	107.9
Aniline-2-carboxylic acid	$C_7H_7NO_2$	-401.1								-296.0			
Aniline-3-carboxylic acid	$C_7H_7NO_2$	-417.3								-283.6			
Aniline-4-carboxylic acid	$C_7H_7NO_2$	-410.0			177.8					-296.7			
Anisole	C_7H_8O					-114.8				-67.9			
Anthracene	$C_{14}H_{10}$	129.2		207.5	210.5					230.9			
9,10-Anthracenedione	$C_{14}H_8O_2$	-188.5								-75.7			
Antimony	Sb	0.0		45.7	25.2					262.3	222.1	180.3	20.8
Antimony(III) bromide	Br_3Sb	-259.4	-239.3	207.1						-194.6	-223.9	372.9	80.2
Antimony(III) chloride	Cl_3Sb	-382.2	-323.7	184.1	107.9								
Antimony(III) fluoride	F_3Sb	-915.5											
Antimony(III) iodide	I_3Sb	-100.4											
Antimony(V) oxide	O_5Sb_2	-971.9	-829.2	125.1									
α-D-Arabinopyranose	$C_5H_{10}O_5$	-1057.9											
D-Arginine	$C_6H_{14}N_4O_2$	-623.5		250.6	232.0								
Argon	Ar									0.0		154.8	20.8
Arsenic (gray)	As	0.0		35.1	24.6					302.5	261.0	174.2	20.8
Arsenic acid	AsH_3O_4	-906.3											
Arsenic(III) bromide	$AsBr_3$	-197.5								-130.0	-159.0	363.9	79.2
Arsenic(III) chloride	$AsCl_3$					-305.0	-259.4	216.3		-261.5	-248.9	327.2	75.7
Arsenic(III) fluoride	AsF_3					-821.3	-774.2	181.2	126.6	-785.8	-770.8	289.1	65.6
Arsenic(III) iodide	AsI_3	-58.2	-59.4	213.1	105.8							388.3	80.6
Arsenic monoxide	AsO									70.0			
Arsenic(V) oxide	As_2O_5	-924.9	-782.3	105.4	116.5								
Arsenic(III) sulfide	As_2S_3	-169.0	-168.6	163.6	116.3								
Arsenic (yellow)	As	14.6											
Arsine	AsH_3									66.4	68.9	222.8	38.1
L-Ascorbic acid	$C_6H_8O_6$	-1164.6											
L-Asparagine	$C_4H_8N_2O_3$	-789.4											
L-Asparagine, monohydrate	$C_4H_{10}N_2O_4$	-1086.6											
L-Aspartic acid	$C_4H_7NO_4$	-973.3											
Astatine	At	0.0											
Atrazine	$C_8H_{14}ClN_5$	-125.4											
2,2'-Azobis[isobutyronitrile]	$C_8H_{12}N_4$	246.0			237.6								0.0
Azobutane	$C_8H_{18}N_2$					-40.1				9.2			
Azopropane	$C_6H_{14}N_2$					11.5				51.3			
trans-Azoxybenzene	$C_{12}H_{10}N_2O$	243.4								342.0			
Azulene	$C_{10}H_8$	212.3								289.1			
Barbituric acid	$C_4H_4N_2O_3$	-634.7											
Barium	Ba	0.0		62.5	28.1					180.0	146.0	170.2	20.8
Barium bromide	$BaBr_2$	-757.3	-736.8	146.0									
Barium carbonate	$CBaO_3$	-1213.0	-1134.4	112.1	86.0								
Barium chloride	$BaCl_2$	-855.0	-806.7	123.7	75.1								
Barium chloride dihydrate	$BaCl_2H_4O_2$	-1456.9	-1293.2	203.0									
Barium fluoride	BaF_2	-1207.1	-1156.8	96.4	71.2								
Barium hydride	BaH_2	-177.0	-138.2	63.0	46.0								
Barium hydroxide	BaH_2O_2	-944.7											
Barium iodide	BaI_2	-602.1											
Barium nitrate	BaN_2O_6	-988.0	-792.6	214.0	151.4								
Barium nitrite	BaN_2O_4	-768.2											
Barium oxide	BaO	-548.0	-520.3	72.1	47.3					-112.0			
Barium sulfate	BaO_4S	-1473.2	-1362.2	132.2	101.8								
Barium sulfide	BaS	-460.0	-456.0	78.2	49.4								
Benzaldehyde	C_7H_6O					-87.0		221.2	172.0	-36.7			
Benzamide	C_7H_7NO	-202.6								-100.9			
Benz[*a*]anthracene	$C_{18}H_{12}$	170.8								293.0			
Benzene	C_6H_6					49.1	124.5	173.4	136.0	82.9	129.7	269.2	82.4

Name	Mol. form.	$\Delta_f H°(c)$/ kJ mol^{-1}	$\Delta_f G°(c)$/ kJ mol^{-1}	$S°(c)$/J mol^{-1} K^{-1}	$C_p(c)$/J mol^{-1} K^{-1}	$\Delta_f H°(l)$/ kJ mol^{-1}	$\Delta_f G°(l)$/ kJ mol^{-1}	$S°(l)$/J mol^{-1} K^{-1}	$C_p(l)$/J mol^{-1} K^{-1}	$\Delta_f H°(g)$/ kJ mol^{-1}	$\Delta_f G°(g)$/ kJ mol^{-1}	$S°(g)$/J mol^{-1} K^{-1}	$C_p(g)$/J mol^{-1} K^{-1}
1,2-Benzenediamine	$C_6H_8N_2$	-0.3											
1,3-Benzenediamine	$C_6H_8N_2$	-7.8		154.5	159.6								
1,4-Benzenediamine	$C_6H_8N_2$	3.0											
Benzeneethanol	$C_8H_{10}O$								252.6				
Benzenethiol	C_6H_6S					63.7		222.8	173.2	111.3			
1,2,3-Benzenetriol	$C_6H_6O_3$	-551.1								-434.2			
1,2,4-Benzenetriol	$C_6H_6O_3$	-563.8								-444.0			
1,3,5-Benzenetriol	$C_6H_6O_3$	-584.6								-452.9			
p-Benzidine	$C_{12}H_{12}N_2$	70.7											
Benzil	$C_{14}H_{10}O_2$	-153.9								-55.5			
1*H*-Benzimidazole	$C_7H_6N_2$	79.5								181.7			
Benzoic acid	$C_7H_6O_2$	-385.2		167.6	146.8					-294.0			
Benzonitrile	C_7H_5N					163.2		209.1	165.2	215.7			
Benzophenone	$C_{13}H_{10}O$	-34.5			224.8					54.9			
Benzo[*a*]pyrene	$C_{20}H_{12}$												254.8
Benzo[*f*]quinoline	$C_{13}H_9N$	150.6								233.7			
p-Benzoquinone	$C_6H_4O_2$	-185.7			129.0					-122.9			
Benzo[*b*]thiophene	C_8H_6S	100.6								166.3			
1*H*-Benzotriazole	$C_6H_5N_3$	236.5								335.5			
Benzoxazole	C_7H_5NO	-24.2								44.8			
Benzoyl chloride	C_7H_5ClO					-158.0				-103.2			
Benzoyl peroxide	$C_{14}H_{10}O_4$	-369.4								-281.7			
Benzyl acetate	$C_9H_{10}O_2$								148.5				
Benzyl alcohol	C_7H_8O					-160.7		216.7	217.9	-100.4			
Benzylamine	C_7H_9N					34.2			207.2	94.4			0.0
N-Benzylaniline	$C_{13}H_{13}N$	101.4											
Berkelium (β form)	Bk	0.0											
Beryllium	Be	0.0		9.5	16.4					324.0	286.6	136.3	20.8
Beryllium bromide	$BeBr_2$	-353.5		108.0	69.4								
Beryllium carbonate	$CBeO_3$	-1025.0		52.0	65.0								
Beryllium chloride	$BeCl_2$	-490.4	-445.6	75.8	62.4								
Beryllium fluoride	BeF_2	-1026.8	-979.4	53.4	51.8								
Beryllium hydroxide (α)	BeH_2O_2	-902.5	-815.0	45.5	62.1								
Beryllium iodide	BeI_2	-192.5		121.0	71.1								
Beryllium oxide	BeO	-609.4	-580.1	13.8	25.6								
Beryllium sulfate	BeO_4S	-1205.2	-1093.8	77.9	85.7								
Beryllium sulfide	BeS	-234.3		34.0	34.0								
9,9'-Bianthracene	$C_{28}H_{18}$	326.2								454.3			
Bicyclo[2.2.1]heptane	C_7H_{12}	-95.1			151.0					-54.8			0.0
1,1'-Bicyclopentyl	$C_{10}H_{18}$					-178.9							
Biphenyl	$C_{12}H_{10}$	99.4		209.4	198.4					181.4			
Bis(2-aminoethyl)amine	$C_4H_{13}N_3$								254.0				
Bis(2-chloroethyl) ether	$C_4H_8Cl_2O$								220.9				
Bis(2-cyanoethyl) sulfide	$C_6H_8N_2S$					96.3							
Bis(2-ethylhexyl) phthalate	$C_{24}H_{38}O_4$								704.7				
2,2-Bis(4-hydroxyphenyl)propane	$C_{15}H_{16}O_2$	-368.6											
Bismuth	Bi	0.0		56.7	25.5					207.1	168.2	187.0	20.8
Bismuth hydroxide	BiH_3O_3	-711.3											
Bismuth oxide	Bi_2O_3	-573.9	-493.7	151.5	113.5								
Bismuth oxychloride	BiClO	-366.9	-322.1	120.5									
Bismuth sulfate	$Bi_2O_{12}S_3$	-2544.3											
Bismuth sulfide	Bi_2S_3	-143.1	-140.6	200.4	122.2								
Bismuth trichloride	$BiCl_3$	-379.1	-315.0	177.0	105.0					-265.7	-256.0	358.9	79.7
Bismuth triiodide	BiI_3		-175.3										
Borane(1)	BH									442.7	412.7	171.8	29.2
Borane(3)	BH_3									89.2	93.3	188.2	36.0
Borane carbonyl	CH_3BO									-111.2	-92.9	249.4	59.5
Borazine	$B_3H_6N_3$					-541.0	-392.7	199.6					
Boric acid	BH_3O_3	-1094.3	-968.9	90.0	86.1					-994.1			
Boron	B	0.0		5.9	11.1					565.0	521.0	153.4	20.8
Boron dioxide	BO_2									-300.4	-305.9	229.6	43.0
Boron monosulfide	BS									342.0	288.8	216.2	30.0
Boron monoxide	BO									25.0	-4.0	203.5	29.2
Boron nitride	BN	-254.4	-228.4	14.8	19.7					647.5	614.5	212.3	29.5
Boron oxide	B_2O_3	-1273.5	-1194.3	54.0	62.8					-843.8	-832.0	279.8	66.9
Boron sulfide	B_2S_3	-240.6		100.0	111.7					67.0			

Name	Mol. form.	$\Delta_f H°$(c)/ kJ mol^{-1}	$\Delta_f G°$(c)/ kJ mol^{-1}	$S°$(c)/J mol^{-1} K^{-1}	C_p(c)/J mol^{-1} K^{-1}	$\Delta_f H°$(l)/ kJ mol^{-1}	$\Delta_f G°$(l)/ kJ mol^{-1}	$S°$(l)/J mol^{-1} K^{-1}	C_p(l)/J mol^{-1} K^{-1}	$\Delta_f H°$(g)/ kJ mol^{-1}	$\Delta_f G°$(g)/ kJ mol^{-1}	$S°$(g)/J mol^{-1} K^{-1}	C_p(g)/J mol^{-1} K^{-1}
Boron tribromide	BBr_3					-239.7	-238.5	229.7		-205.6	-232.5	324.2	67.8
Boron trichloride	BCl_3					-427.2	-387.4	206.3	106.7	-403.8	-388.7	290.1	62.7
Boron trifluoride	BF_3									-1136.0	-1119.4	254.4	
Boron triiodide	BI_3									71.1	20.7	349.2	70.8
Bromine	Br_2					0.0		152.2	75.7	30.9	3.1	245.5	36.0
Bromine (atomic)	Br									111.9	82.4	175.0	20.8
Bromine chloride	BrCl									14.6	-1.0	240.1	35.0
Bromine dioxide	BrO_2									152.0	155.0	271.1	45.4
Bromine fluoride	BrF									-93.8	-109.2	229.0	33.0
Bromine monoxide	BrO									125.8	109.6	233.0	34.2
Bromine pentafluoride	BrF_5					-458.6	-351.8	225.1		-428.9	-350.6	320.2	99.6
Bromine trifluoride	BrF_3					-300.8	-240.5	178.2	124.6	-255.6	-229.4	292.5	66.6
Bromoacetic acid	$C_2H_3BrO_2$									-383.5	-338.3	337.0	80.5
Bromoacetone	C_3H_5BrO									-181.0			
Bromoacetylene	C_2HBr											253.7	55.7
Bromobenzene	C_6H_5Br					60.9		219.2	154.3				
Bromoborane(1)	BBr									238.1	195.4	225.0	32.9
1-Bromobutane	C_4H_9Br					-143.8				-107.1			
2-Bromobutane, (±)-	C_4H_9Br					-154.9				-120.3			
Bromochlorodifluoromethane	$CBrClF_2$											318.5	74.6
1-Bromo-2-chloroethane	C_2H_4BrCl								130.1				
Bromochlorofluoromethane	CHBrClF											304.3	63.2
Bromochloromethane	CH_2BrCl											287.6	52.7
1-Bromo-2-chloro-1,1,2-trifluoroethane	$C_2HBrClF_3$					-675.3				-644.8			
2-Bromo-2-chloro-1,1,1-trifluoroethane	$C_2HBrClF_3$					-720.0				-690.4			
Bromodichlorofluoromethane	$CBrCl_2F$											330.6	80.0
Bromodichloromethane	$CHBrCl_2$											316.4	67.4
Bromodifluoromethane	$CHBrF_2$									-424.9		295.1	58.7
1-Bromododecane	$C_{12}H_{25}Br$					-344.7				-269.9			
Bromoethane	C_2H_5Br					-90.5	-25.8	198.7	100.8	-61.9	-23.9	286.7	64.5
Bromoethene	C_2H_3Br									79.2	81.8	275.8	55.5
Bromofluoromethane	CH_2BrF											276.3	49.2
Bromogermane	$BrGeH_3$											274.8	56.4
1-Bromoheptane	$C_7H_{15}Br$					-218.4				-167.8			
1-Bromohexadecane	$C_{16}H_{33}Br$					-444.5				-350.2			
1-Bromohexane	$C_6H_{13}Br$					-194.2		453.0	204.0	-148.3			
Bromomethane	CH_3Br					-59.8				-35.4	-26.3	246.4	42.4
2-Bromo-2-methylpropane	C_4H_9Br					-164.4				-132.4			
1-Bromooctane	$C_8H_{17}Br$					-245.1				-189.3			
Bromopentafluoroethane	C_2BrF_5									-1064.4			
1-Bromopentane	$C_5H_{11}Br$					-170.2				-128.9			
1-Bromopropane	C_3H_7Br					-121.9				-87.0			
2-Bromopropane	C_3H_7Br					-130.5				-99.4			
cis-1-Bromopropene	C_3H_5Br					7.9				40.8			
3-Bromopropene	C_3H_5Br					12.2				45.2			
Bromosilane	BrH_3Si											262.4	52.8
Bromosilyldyne	BrSi									209.0			38.6
Bromosilylene	BrHSi									-464.4			
N-Bromosuccinimide	$C_4H_4BrNO_2$	-335.9											
4-Bromotoluene	C_7H_7Br					12.0							
Bromotrichloromethane	$CBrCl_3$									-41.1			85.3
Bromotrichlorosilane	$BrCl_3Si$											350.1	90.9
2-Bromo-1,1,1-trifluoroethane	$C_2H_2BrF_3$									-694.5			
Bromotrifluoromethane	$CBrF_3$									-648.3			69.3
Bromotrinitromethane	$CBrN_3O_6$					32.5				80.3			
1,2-Butadiene	C_4H_6					138.6				162.3			
1,3-Butadiene	C_4H_6					88.5		199.0	123.6	110.0			
Butanal	C_4H_8O					-239.2		246.6	163.7	-204.8		343.7	103.4
Butanamide	C_4H_9NO	-364.8								-282.0			
Butane	C_4H_{10}					-147.3			140.9	-125.7			
1,2-Butanediol, (±)-	$C_4H_{10}O_2$					-523.6							
1,3-Butanediol	$C_4H_{10}O_2$					-501.0				-433.2			
1,4-Butanediol	$C_4H_{10}O_2$					-505.3		223.4	200.1	-428.7			
2,3-Butanediol	$C_4H_{10}O_2$					-541.5			213.0	-482.3			

Name	Mol. form.	$\Delta_f H°$(c)/ kJ mol^{-1}	$\Delta_f G°$(c)/ kJ mol^{-1}	$S°$(c)/J mol^{-1} K^{-1}	C_p(c)/J mol^{-1} K^{-1}	$\Delta_f H°$(l)/ kJ mol^{-1}	$\Delta_f G°$(l)/ kJ mol^{-1}	$S°$(l)/J mol^{-1} K^{-1}	C_p(l)/J mol^{-1} K^{-1}	$\Delta_f H°$(g)/ kJ mol^{-1}	$\Delta_f G°$(g)/ kJ mol^{-1}	$S°$(g)/J mol^{-1} K^{-1}	C_p(g)/J mol^{-1} K^{-1}
1,4-Butanedithiol	$C_4H_{10}S_2$					-105.7				-50.6			
Butanenitrile	C_4H_7N					-5.8				33.6			
1-Butanethiol	$C_4H_{10}S$					-124.7			171.2	-88.0			
2-Butanethiol	$C_4H_{10}S$					-131.0				-96.9			
Butanoic acid	$C_4H_8O_2$					-533.8		222.2	178.6	-475.9			
Butanoic anhydride	$C_8H_{14}O_3$								283.7				
1-Butanol	$C_4H_{10}O$					-327.3		225.8	177.2	-274.9			
2-Butanol	$C_4H_{10}O$					-342.6		214.9	196.9	-292.8		359.5	112.7
2-Butanone	C_4H_8O					-273.3		239.1	158.7	-238.5		339.9	101.7
trans-2-Butenal	C_4H_6O					-138.7				-100.6			
1-Butene	C_4H_8					-20.8		227.0	118.0	0.1			
cis-2-Butene	C_4H_8					-29.8		219.9	127.0	-7.1			
trans-2-Butene	C_4H_8					-33.3				-11.4			
trans-2-Butenedinitrile	$C_4H_2N_2$	268.2								340.2			
trans-2-Butenenitrile	C_4H_5N					95.1				134.3			
3-Butenenitrile	C_4H_5N					117.8				159.7			
trans-2-Butenoic acid	$C_4H_6O_2$												
2-Butoxyethanol	$C_6H_{14}O_2$								281.0				
N-Butylacetamide	$C_6H_{13}NO$					-380.9				-305.9			
Butyl acetate	$C_6H_{12}O_2$					-529.2			227.8	-485.3			
tert-Butyl acetate	$C_6H_{12}O_2$					-554.5			231.0	-516.5			
Butyl acrylate	$C_7H_{12}O_2$					-422.6			251.0	-375.3			
Butylamine	$C_4H_{11}N$					-127.6			179.2	-91.9			
sec-Butylamine	$C_4H_{11}N$					-137.5				-104.6			
tert-Butylamine	$C_4H_{11}N$					-150.6			192.1	-121.0			
Butylbenzene	$C_{10}H_{14}$					-63.2		321.2	243.4	-11.8			
sec-Butylbenzene, (±)-	$C_{10}H_{14}$					-66.4				-18.4			
tert-Butylbenzene	$C_{10}H_{14}$					-71.9				-23.0			
Butyl chloroacetate	$C_6H_{11}ClO_2$					-538.4				-487.4			
Butyl 2-chloropropanoate	$C_7H_{13}ClO_2$					-571.7				-517.3			
Butyl 3-chloropropanoate	$C_7H_{13}ClO_2$					-557.9				-502.3			
Butylcyclohexane	$C_{10}H_{20}$					-263.1		345.0	271.0	-213.7			
Butyl dichloroacetate	$C_6H_{10}Cl_2O_2$					-550.1				-497.8			
1-*tert*-Butyl-3,5-dimethylbenzene	$C_{12}H_{18}$					-146.5							
5-Butyldocosane	$C_{26}H_{54}$					-713.5				-587.6			
11-Butyldocosane	$C_{26}H_{54}$					-716.0				-593.4			
Butyl ethyl ether	$C_6H_{14}O$								159.0				
tert-Butyl ethyl ether	$C_6H_{14}O$									-313.9			
Butyl ethyl sulfide	$C_6H_{14}S$					-172.3				-127.8			
Butyl formate	$C_5H_{10}O_2$								200.2				
tert-Butyl hydroperoxide	$C_4H_{10}O_2$					-293.6				-245.9			
tert-Butyl isobutyl ether	$C_8H_{18}O$					-409.1				-369.0			
tert-Butyl isopropyl ether	$C_7H_{16}O$					-392.8				-358.1			
1-*tert*-Butyl-3-methylbenzene	$C_{11}H_{16}$					-109.7							
1-*tert*-Butyl-4-methylbenzene	$C_{11}H_{16}$					-109.7				-57.0			
Butyl methyl ether	$C_5H_{12}O$					-290.6		295.3	192.7	-258.1			
Butyl methyl sulfide	$C_5H_{12}S$					-142.9		307.5	200.9	-102.4			
tert-Butyl methyl sulfide	$C_5H_{12}S$					-157.1		276.1	199.9	-121.3			
Butyl oleate	$C_{22}H_{42}O_2$					-816.9							
Butyl pentanoate	$C_9H_{18}O_2$					-613.3				-560.2			
sec-Butyl pentanoate	$C_9H_{18}O_2$					-624.2				-573.2			
N-Butylpiperidine	$C_9H_{19}N$					-171.8							
Butyl stearate	$C_{22}H_{44}O_2$												
Butyl trichloroacetate	$C_6H_9Cl_3O_2$					-545.8				-492.3			
Butylurea	$C_5H_{12}N_2O$	-419.5											
tert-Butylurea	$C_5H_{12}N_2O$	-417.4											
Butyl vinyl ether	$C_6H_{12}O$					-218.8			232.0	-182.6			
1-Butyne	C_4H_6					141.4				165.2			
2-Butyne	C_4H_6					119.1				145.7			
2-Butynedinitrile	C_4N_2					500.4				529.2			
2-Butynedioic acid	$C_4H_2O_4$	-577.3											
γ-Butyrolactone	$C_4H_6O_2$					-420.9			141.4	-366.5			
Cadmium	Cd	0.0		51.8	26.0					111.8		167.7	20.8
Cadmium bromide	Br_2Cd	-316.2	-296.3	137.2	76.7								
Cadmium carbonate	$CCdO_3$	-750.6	-669.4	92.5									
Cadmium chloride	$CdCl_2$	-391.5	-343.9	115.3	74.7								

Name	Mol. form.	$\Delta_f H°(c)$/ kJ mol^{-1}	$\Delta_f G°(c)$/ kJ mol^{-1}	$S°(c)$/J mol^{-1} K^{-1}	$C_p(c)$/J mol^{-1} K^{-1}	$\Delta_f H°(l)$/ kJ mol^{-1}	$\Delta_f G°(l)$/ kJ mol^{-1}	$S°(l)$/J mol^{-1} K^{-1}	$C_p(l)$/J mol^{-1} K^{-1}	$\Delta_f H°(g)$/ kJ mol^{-1}	$\Delta_f G°(g)$/ kJ mol^{-1}	$S°(g)$/J mol^{-1} K^{-1}	$C_p(g)$/J mol^{-1} K^{-1}
Cadmium fluoride	CdF_2	-700.4	-647.7	77.4									
Cadmium hydroxide	CdH_2O_2	-560.7	-473.6	96.0									
Cadmium iodide	CdI_2	-203.3	-201.4	161.1	80.0								
Cadmium oxide	CdO	-258.4	-228.7	54.8	43.4								
Cadmium sulfate	CdO_4S	-933.3	-822.7	123.0	99.6								
Cadmium sulfide	CdS	-161.9	-156.5	64.9									
Cadmium telluride	$CdTe$	-92.5	-92.0	100.0									
Calcium	Ca	0.0		41.6	25.9					177.8	144.0	154.9	20.8
Calcium bromide	Br_2Ca	-682.8	-663.6	130.0									
Calcium carbide	C_2Ca	-59.8	-64.9	70.0	62.7								
Calcium carbonate (calcite)	$CCaO_3$	-1207.6	-1129.1	91.7	83.5								
Calcium carbonate (aragonite)	$CCaO_3$	-1207.8	-1128.2	88.0	82.3								
Calcium chloride	$CaCl_2$	-795.4	-748.8	108.4	72.9								
Calcium cyanide	C_2CaN_2	-184.5											
Calcium fluoride	CaF_2	-1228.0	-1175.6	68.5	67.0								
Calcium hydride	CaH_2	-181.5	-142.5	41.4	41.0								
Calcium hydroxide	CaH_2O_2	-985.2	-897.5	83.4	87.5								
Calcium iodide	CaI_2	-533.5	-528.9	142.0									
Calcium nitrate	CaN_2O_6	-938.2	-742.8	193.2	149.4								
Calcium oxalate	C_2CaO_4	-1360.6											
Calcium oxide	CaO	-634.9	-603.3	38.1	42.0								
Calcium phosphate	$Ca_3O_8P_2$	-4120.8	-3884.7	236.0	227.8								
Calcium sulfate	CaO_4S	-1434.5	-1322.0	106.5	99.7								
Calcium sulfide	CaS	-482.4	-477.4	56.5	47.4								
Californium	Cf	0.0											
Camphor, (±)-	$C_{10}H_{16}O$	-319.4			271.2					-267.5			
Caprolactam	$C_6H_{11}NO$	-329.4			156.8					-239.6			
Carbazole	$C_{12}H_9N$	101.7								200.7			
Carbon (diamond)	C	1.9	2.9	2.4	6.1								
Carbon dioxide	CO_2									-393.5	-394.4	213.8	37.1
Carbon diselenide	CSe_2					164.8							
Carbon disulfide	CS_2					89.0	64.6	151.3	76.4	116.7	67.1	237.8	45.4
Carbon [fullerene-C_{60}]	C_{60}	2327.0	2302.0	426.0	520.0					2502.0	2442.0	544.0	512.0
Carbon [fullerene-C_{70}]	C_{70}	2555.0	2537.0	464.0	650.0					2755.0	2692.0	614.0	585.0
Carbon (graphite)	C	0.0		5.7	8.5					716.7	671.3	158.1	20.8
Carbon monosulfide	CS									280.3	228.8	210.6	29.8
Carbon monoxide	CO									-110.5	-137.2	197.7	29.1
Carbon oxysulfide	COS									-142.0	-169.2	231.6	41.5
Carbonyl bromide	CBr_2O					-127.2				-96.2	-110.9	309.1	61.8
Carbonyl chloride	CCl_2O									-219.1	-204.9	283.5	57.7
Carbonyl chloride fluoride	$CClFO$											276.7	52.4
Carbonyl fluoride	CF_2O									-639.8			46.8
Cerium	Ce	0.0		72.0	26.9					423.0	385.0	191.8	23.1
Cerium(III) bromide	Br_3Ce	-891.4											
Cerium(III) chloride	$CeCl_3$	-1060.5	-984.8	151.0	87.4								
Cerium(III) iodide	CeI_3	-669.3											
Cerium(III) oxide	Ce_2O_3	-1796.2	-1706.2	150.6	114.6								
Cerium(IV) oxide	CeO_2	-1088.7	-1024.6	62.3	61.6								
Cerium(II) sulfide	CeS	-459.4	-451.5	78.2	50.0								
Cesium	Cs	0.0		85.2	32.2					76.5	49.6	175.6	20.8
Cesium amide	CsH_2N	-118.4											
Cesium bromide	$BrCs$	-405.8	-391.4	113.1	52.9								
Cesium carbonate	CCs_2O_3	-1139.7	-1054.3	204.5	123.9								
Cesium chloride	$ClCs$	-443.0	-414.5	101.2	52.5								
Cesium fluoride	CsF	-553.5	-525.5	92.8	51.1								
Cesium hydride	CsH	-54.2											
Cesium hydrogen carbonate	$CHCsO_3$	-966.1											
Cesium hydrogen fluoride	CsF_2H	-923.8	-858.9	135.2	87.3								
Cesium hydrogen sulfate	$CsHO_4S$	-1158.1											
Cesium hydroxide	$CsHO$	-416.2	-371.8	104.2	69.9					-256.0	-256.5	254.8	49.7
Cesium iodide	CsI	-346.6	-340.6	123.1	52.8								
Cesium metaborate	$BCsO_2$	-972.0	-915.0	104.4	80.6								
Cesium nitrate	$CsNO_3$	-506.0	-406.5	155.2									
Cesium oxide	Cs_2O	-345.8	-308.1	146.9	76.0								
Cesium perchlorate	$ClCsO_4$	-443.1	-314.3	175.1	108.3								
Cesium sulfate	Cs_2O_4S	-1443.0	-1323.6	211.9	134.9								

Name	Mol. form.	$\Delta_f H°$(c)/ kJ mol⁻¹	$\Delta_f G°$(c)/ kJ mol⁻¹	$S°$(c)/J mol⁻¹ K⁻¹	C_p(c)/J mol⁻¹ K⁻¹	$\Delta_f H°$(l)/ kJ mol⁻¹	$\Delta_f G°$(l)/ kJ mol⁻¹	$S°$(l)/J mol⁻¹ K⁻¹	C_p(l)/J mol⁻¹ K⁻¹	$\Delta_f H°$(g)/ kJ mol⁻¹	$\Delta_f G°$(g)/ kJ mol⁻¹	$S°$(g)/J mol⁻¹ K⁻¹	C_p(g)/J mol⁻¹ K⁻¹
Cesium sulfide	Cs_2S	-359.8											
Cesium sulfite	Cs_2O_3S	-1134.7											
Cesium superoxide	CsO_2	-286.2											
Chlorine	Cl_2									0.0		223.1	33.9
Chlorine (atomic)	Cl									121.3	105.3	165.2	21.8
Chlorine dioxide	ClO_2									102.5	120.5	256.8	42.0
Chlorine fluoride	ClF									-50.3	-51.8	217.9	32.1
Chlorine monoxide	Cl_2O									80.3	97.9	266.2	45.4
Chlorine oxide (ClO)	ClO									101.8	98.1	226.6	31.5
Chlorine superoxide [ClOO]	ClO_2									89.1	105.0	263.7	46.0
Chlorine trifluoride	ClF_3					-189.5				-163.2	-123.0	281.6	63.9
Chloroacetic acid	$C_2H_3ClO_2$	-509.7								-427.6	-368.5	325.9	78.8
Chloroacetyl chloride	$C_2H_2Cl_2O$					-283.7				-244.8			
Chloroacetylene	C_2HCl											242.0	54.3
2-Chloroaniline	C_6H_6ClN					-4.6							
3-Chloroaniline	C_6H_6ClN					-20.3			198.7				
4-Chloroaniline	C_6H_6ClN	-33.3			147.3								
Chlorobenzene	C_6H_5Cl					11.1			150.1	52.0			
2-Chlorobenzoic acid	$C_7H_5ClO_2$	-404.5								-325.0			
3-Chlorobenzoic acid	$C_7H_5ClO_2$	-424.3								-342.3			
4-Chlorobenzoic acid	$C_7H_5ClO_2$	-428.9			163.2					-341.0			
3-Chlorobenzoyl chloride	$C_7H_4Cl_2O$					-189.7							
Chloroborane(1)	BCl									149.5	120.9	213.2	31.7
1-Chlorobutane	C_4H_9Cl					-188.1				-154.4			
2-Chlorobutane	C_4H_9Cl					-192.8				-161.1			
2-Chlorobutanoic acid	$C_4H_7ClO_2$					-575.5							
3-Chlorobutanoic acid	$C_4H_7ClO_2$					-556.3							
4-Chlorobutanoic acid	$C_4H_7ClO_2$					-566.3							
Chlorocyclohexane	$C_6H_{11}Cl$					-207.2				-163.7			
Chlorodibromomethane	$CHBr_2Cl$											327.7	69.2
1-Chloro-1,1-difluoroethane	$C_2H_3ClF_2$											307.2	82.5
1-Chloro-2,2-difluoroethene	C_2HClF_2									-315.5	-289.1	303.0	72.1
Chlorodifluoromethane	$CHClF_2$									-482.6		280.9	55.9
1-Chlorododecane	$C_{12}H_{25}Cl$					-392.3				-321.1			
Chloroethane	C_2H_5Cl					-136.8	-59.3	190.8	104.3	-112.1	-60.4	276.0	62.8
2-Chloroethanol	C_2H_5ClO					-295.4							
Chloroethene	C_2H_3Cl	-94.1			59.4	14.6				37.2	53.6	264.0	53.7
2-Chloroethyl ethyl ether	C_4H_9ClO					-335.6				-301.3			
2-Chloroethyl vinyl ether	C_4H_7ClO					-208.1				-170.1			
1-Chloro-1-fluoroethane	C_2H_4ClF									-313.4			
Chlorofluoromethane	CH_2ClF											264.4	47.0
Chlorogermane	$ClGeH_3$											263.7	54.7
2-Chlorohexane	$C_6H_{13}Cl$					-246.1				-204.3			
Chloromethane	CH_3Cl									-81.9		234.6	40.8
(Chloromethyl)benzene	C_7H_7Cl					-32.5				18.9			
1-Chloro-3-methylbutane	$C_5H_{11}Cl$					-216.0				-179.7			
2-Chloro-2-methylbutane	$C_5H_{11}Cl$					-235.7				-202.2			
2-Chloro-3-methylbutane	$C_5H_{11}Cl$					-226.6				-185.1			
1-Chloro-2-methylpropane	C_4H_9Cl					-191.1				-159.3			
2-Chloro-2-methylpropane	C_4H_9Cl					-211.3				-182.2			
1-Chloronaphthalene	$C_{10}H_7Cl$					54.6			212.6	119.8			
2-Chloronaphthalene	$C_{10}H_7Cl$	55.4								137.4			
1-Chloro-4-nitrobenzene	$C_6H_4ClNO_2$	-48.7			250.2								
1-Chlorooctadecane	$C_{18}H_{37}Cl$					-544.1				-446.0			
1-Chlorooctane	$C_8H_{17}Cl$					-291.3				-238.9			
Chloropentafluorobenzene	C_6ClF_5	-858.4								-809.3			
Chloropentafluoroethane	C_2ClF_5									-1118.8			184.2
1-Chloropentane	$C_5H_{11}Cl$					-213.2				-174.9			
2-Chlorophenol	C_6H_5ClO								188.7				
3-Chlorophenol	C_6H_5ClO	-206.4				-189.3							
4-Chlorophenol	C_6H_5ClO	-197.7				-181.3							
1-Chloropropane	C_3H_7Cl					-160.5				-131.9			
2-Chloropropane	C_3H_7Cl					-172.3				-144.9			
3-Chloro-1,2-propanediol	$C_3H_7ClO_2$					-525.3							
2-Chloro-1,3-propanediol	$C_3H_7ClO_2$					-517.5							
2-Chloropropanoic acid	$C_3H_5ClO_2$					-522.5				-475.8			

Name	Mol. form.	$\Delta_f H°$(c)/ kJ mol^{-1}	$\Delta_f G°$(c)/ kJ mol^{-1}	$S°$(c)/J mol^{-1} K^{-1}	C_p(c)/J mol^{-1} K^{-1}	$\Delta_f H°$(l)/ kJ mol^{-1}	$\Delta_f G°$(l)/ kJ mol^{-1}	$S°$(l)/J mol^{-1} K^{-1}	C_p(l)/J mol^{-1} K^{-1}	$\Delta_f H°$(g)/ kJ mol^{-1}	$\Delta_f G°$(g)/ kJ mol^{-1}	$S°$(g)/J mol^{-1} K^{-1}	C_p(g)/J mol^{-1} K^{-1}
3-Chloropropanoic acid	$C_3H_5ClO_2$	-549.3											
2-Chloropropene	C_3H_5Cl									-21.0			
3-Chloropropene	C_3H_5Cl								125.1				
Chlorosilane	ClH_3Si											250.7	51.0
Chlorosilylidyne	ClSi									189.9			36.9
N-Chlorosuccinimide	$C_4H_4ClNO_2$	-357.9											
2-Chlorotoluene	C_7H_7Cl								166.8				
1-Chloro-2,2,2-trifluoroethane	$C_2H_2ClF_3$											326.5	89.1
Chlorotrifluoroethene	C_2ClF_3					-522.7				-505.5	-523.8	322.1	83.9
Chlorotrifluoromethane	$CClF_3$									-706.3			66.9
Chlorotrinitromethane	$CClN_3O_6$					-27.1				18.4			
Chlorooxoborane	BClO									-314.0			
Chromium	Cr	0.0		23.8	23.4					396.6	351.8	174.5	20.8
Chromium(II) bromide	Br_2Cr	-302.1											
Chromium(II) chloride	Cl_2Cr	-395.4	-356.0	115.3	71.2								
Chromium(III) chloride	Cl_3Cr	-556.5	-486.1	123.0	91.8								
Chromium(VI) dichloride dioxide	Cl_2CrO_2					-579.5	-510.8	221.8		-538.1	-501.6	329.8	84.5
Chromium(II) fluoride	CrF_2	-778.0											
Chromium(III) fluoride	CrF_3	-1159.0	-1088.0	93.9	78.7								
Chromium(II) iodide	CrI_2	-156.9											
Chromium(III) iodide	CrI_3	-205.0											
Chromium iron oxide	Cr_2FeO_4	-1444.7	-1343.8	146.0	133.6								
Chromium(II,III) oxide	Cr_3O_4	-1531.0											
Chromium(III) oxide	Cr_2O_3	-1139.7	-1058.1	81.2	118.7								
Chromium(IV) oxide	CrO_2	-598.0											
Chromium(VI) oxide	CrO_3									-292.9		266.2	56.0
Chrysene	$C_{18}H_{12}$	145.3								269.8			
Citric acid	$C_6H_8O_7$	-1543.8											
Cobalt	Co	0.0		30.0	24.8					424.7	380.3	179.5	23.0
Cobalt(II) bromide	Br_2Co	-220.9			79.5								
Cobalt(II) carbonate	$CCoO_3$	-713.0											
Cobalt(II) chloride	Cl_2Co	-312.5	-269.8	109.2	78.5								
Cobalt(II) fluoride	CoF_2	-692.0	-647.2	82.0	68.8								
Cobalt(II) hydroxide	CoH_2O_2	-539.7	-454.3	79.0									
Cobalt(II) iodide	CoI_2	-88.7											
Cobalt(II) nitrate	CoN_2O_6	-420.5											
Cobalt(II) oxide	CoO	-237.9	-214.2	53.0	55.2								
Cobalt(II,III) oxide	Co_3O_4	-891.0	-774.0	102.5	123.4								
Cobalt(II) sulfate	CoO_4S	-888.3	-782.3	118.0									
Cobalt(II) sulfide	CoS	-82.8											
Cobalt(III) sulfide	Co_2S_3	-147.3											
Copper	Cu	0.0		33.2	24.4					337.4	297.7	166.4	20.8
Copper(I) bromide	BrCu	-104.6	-100.8	96.1	54.7								
Copper(II) bromide	Br_2Cu	-141.8											
Copper(I) chloride	ClCu	-137.2	-119.9	86.2	48.5								
Copper(II) chloride	Cl_2Cu	-220.1	-175.7	108.1	71.9								
Copper(I) cyanide	CCuN	96.2	111.3	84.5									
Copper(II) fluoride	CuF_2	-542.7											
Copper(II) hydroxide	CuH_2O_2	-449.8											
Copper(I) iodide	CuI	-67.8	-69.5	96.7	54.1								
Copper(II) nitrate	CuN_2O_6	-302.9											
Copper(I) oxide	Cu_2O	-168.6	-146.0	93.1	63.6								
Copper(II) oxide	CuO	-157.3	-129.7	42.6	42.3								
Copper(II) selenide	CuSe	-39.5											
Copper(II) sulfate	CuO_4S	-771.4	-662.2	109.2									
Copper(I) sulfide	Cu_2S	-79.5	-86.2	120.9	76.3								
Copper(II) sulfide	CuS	-53.1	-53.6	66.5	47.8								
Copper(II) tungstate	CuO_4W	-1105.0											
Creatine	$C_4H_9N_3O_2$	-537.2											
Creatinine	$C_4H_7N_3O$	-238.5											
o-Cresol	C_7H_8O	-204.6		165.4	154.6					-128.6			
m-Cresol	C_7H_8O					-194.0		212.6	224.9	-132.3			
p-Cresol	C_7H_8O	-199.3		167.3	150.2					-125.4			
Curium	Cm	0.0											
Cyanamide	CH_2N_2	58.8											
Cyanide	CN									437.6	407.5	202.6	29.2

Name	Mol. form.	$\Delta_f H°$(c)/ kJ mol⁻¹	$\Delta_f G°$(c)/ kJ mol⁻¹	$S°$(c)/J mol⁻¹ K⁻¹	C_p(c)/J mol⁻¹ K⁻¹	$\Delta_f H°$(l)/ kJ mol⁻¹	$\Delta_f G°$(l)/ kJ mol⁻¹	$S°$(l)/J mol⁻¹ K⁻¹	C_p(l)/J mol⁻¹ K⁻¹	$\Delta_f H°$(g)/ kJ mol⁻¹	$\Delta_f G°$(g)/ kJ mol⁻¹	$S°$(g)/J mol⁻¹ K⁻¹	C_p(g)/J mol⁻¹ K⁻¹
Cyanogen	C_2N_2					285.9				306.7		241.9	56.8
Cyanogen bromide	CBrN	140.5								186.2	165.3	248.3	46.9
Cyanogen chloride	CClN					112.1				138.0	131.0	236.2	45.0
Cyanogen fluoride	CFN											224.7	41.8
Cyanogen iodide	CIN	166.2	185.0	96.2						225.5	196.6	256.8	48.3
Cyclobutanamine	C_4H_9N					5.6				41.2			
Cyclobutane	C_4H_8					3.7				27.7			
Cyclobutanecarbonitrile	C_5H_7N					103.0				147.4			
Cyclobutene	C_4H_6									156.7			
Cycloheptane	C_7H_{14}					-156.6				-118.1			
Cyclohexane	C_6H_{12}					-156.4			154.9	-123.4			
Cyclohexanecarbonitrile	$C_7H_{11}N$					-47.2				4.8			
Cyclohexanethiol	$C_6H_{12}S$					-140.7		255.6	192.6	-96.2			
Cyclohexanol	$C_6H_{12}O$					-348.2			208.2	-286.2			
Cyclohexanone	$C_6H_{10}O$					-271.2			182.2	-226.1			
Cyclohexene	C_6H_{10}					-38.5		214.6	148.3	-5.0			
1-Cyclohexenecarbonitrile	C_7H_9N					48.1				101.6			
Cyclohexylamine	$C_6H_{13}N$					-147.6				-104.0			
Cyclohexylbenzene	$C_{12}H_{16}$					-76.6				-16.7			
Cyclohexylcyclohexane	$C_{12}H_{22}$					-273.7				-215.7			
Cyclooctane	C_8H_{16}					-167.7				-124.4			
1,3-Cyclopentadiene	C_5H_6					105.9				134.3			
Cyclopentane	C_5H_{10}					-105.1		204.5	128.8	-76.4			
Cyclopentanecarbonitrile	C_6H_9N					0.7				44.1			
cis-1,2-Cyclopentanediol	$C_5H_{10}O_2$	-485.0											
trans-1,2-Cyclopentanediol	$C_5H_{10}O_2$	-490.1											
Cyclopentanethiol	$C_5H_{10}S$					-89.5		256.9	165.2	-48.1			
Cyclopentanol	$C_5H_{10}O$					-300.1		204.1	182.5	-242.5		362.9	
Cyclopentanone	C_5H_8O					-235.9				-192.1			
Cyclopentene	C_5H_8					4.3		201.2	122.4	34.0			
1-Cyclopentenecarbonitrile	C_6H_7N					111.5				156.5			
Cyclopentylamine	$C_5H_{11}N$					-95.1		241.0	181.2	-54.9			
Cyclopentyl methyl sulfide	$C_6H_{12}S$					-109.8				-64.7			
Cyclopropane	C_3H_6					35.2				53.3	104.5	237.5	55.6
Cyclopropanecarbonitrile	C_4H_5N					140.8				182.8			
Cyclopropene	C_3H_4									277.1			
Cyclopropylamine	C_3H_7N					45.8		187.7	147.1	77.0			
Cyclopropylbenzene	C_9H_{10}					100.3				150.5			
Cyclotetramethylenetetranitramine	$C_4H_8N_8O_8$									187.9		568.8	275.5
L-Cysteine	$C_3H_7NO_2S$	-534.1											
L-Cystine	$C_6H_{12}N_2O_4S_2$	-1032.7											
Cytosine	$C_4H_5N_3O$	-221.3			132.6								
Decaborane(14)	$B_{10}H_{14}$									47.3	232.8	350.7	186.1
cis-Decahydronaphthalene	$C_{10}H_{18}$					-219.4		265.0	232.0	-169.2			
trans-Decahydronaphthalene	$C_{10}H_{18}$					-230.6		264.9	228.5	-182.1			
Decane	$C_{10}H_{22}$					-300.9			314.4	-249.5			
Decanedioic acid	$C_{10}H_{18}O_4$	-1082.6								-921.9			
1,10-Decanediol	$C_{10}H_{22}O_2$	-678.9											
Decanenitrile	$C_{10}H_{19}N$					-158.4				-91.5			
1-Decanethiol	$C_{10}H_{22}S$	-309.9				-276.5		476.1	350.4	-211.5			
Decanoic acid	$C_{10}H_{20}O_2$	-713.7				-684.3				-594.9			
1-Decanol	$C_{10}H_{22}O$					-478.1			370.6	-396.6			
1-Decene	$C_{10}H_{20}$					-173.8		425.0	300.8	-123.3			
Decylbenzene	$C_{16}H_{26}$					-218.3				-138.6			
Decylcyclopentane	$C_{15}H_{30}$					-367.3							
11-Decylheneicosane	$C_{31}H_{64}$					-848.0				-705.8			
Diacetone alcohol	$C_6H_{12}O_2$								221.3				
Dialuminum	Al_2									485.9	433.3	233.2	36.4
4,4'-Diaminodiphenylmethane	$C_{13}H_{14}N_2$				270.9								
Diantimony	Sb_2									235.6	187.0	254.9	36.4
Diarsenic	As_2									222.2	171.9	239.4	35.0
Diazomethane	CH_2N_2											242.9	52.5
Dibenz[*a,h*]anthracene	$C_{22}H_{14}$												283.9
Dibenzofuran	$C_{12}H_8O$	-5.3								83.4			
Dibenzothiophene	$C_{12}H_8S$	120.0								205.1			
Dibismuth	Bi_2									219.7			36.9

Name	Mol. form.	$\Delta_f H°$(c)/ kJ mol⁻¹	$\Delta_f G°$(c)/ kJ mol⁻¹	$S°$(c)/J mol⁻¹ K⁻¹	C_p(c)/J mol⁻¹ K⁻¹	$\Delta_f H°$(l)/ kJ mol⁻¹	$\Delta_f G°$(l)/ kJ mol⁻¹	$S°$(l)/J mol⁻¹ K⁻¹	C_p(l)/J mol⁻¹ K⁻¹	$\Delta_f H°$(g)/ kJ mol⁻¹	$\Delta_f G°$(g)/ kJ mol⁻¹	$S°$(g)/J mol⁻¹ K⁻¹	C_p(g)/J mol⁻¹ K⁻¹
Diborane	B_2H_6									36.4	87.6	232.1	56.7
Diboron	B_2									830.5	774.0	201.9	30.5
Diboron dioxide	B_2O_2									-454.8	-462.3	242.5	57.3
1,2-Dibromobutane	$C_4H_8Br_2$					-142.1				-91.6			
1,3-Dibromobutane	$C_4H_8Br_2$					-148.0							
1,4-Dibromobutane	$C_4H_8Br_2$					-140.3				-87.8			
2,3-Dibromobutane	$C_4H_8Br_2$					-139.6				-102.0			
Dibromochlorofluoromethane	CBr_2ClF											342.8	82.4
1,2-Dibromo-1-chloro-1,2,2-trifluoroethane	$C_2Br_2ClF_3$					-691.7				-656.6			
1,2-Dibromo-1,2-dichloroethane	$C_2H_2Br_2Cl_2$									-36.9			
Dibromodichloromethane	CBr_2Cl_2											347.8	87.1
Dibromodifluoromethane	CBr_2F_2											325.3	77.0
1,1-Dibromoethane	$C_2H_4Br_2$					-66.2						327.7	80.8
1,2-Dibromoethane	$C_2H_4Br_2$					-79.2		223.3	136.0	-37.5			
cis-1,2-Dibromoethene	$C_2H_2Br_2$											311.3	68.8
trans-1,2-Dibromoethene	$C_2H_2Br_2$											313.5	70.3
Dibromofluoromethane	$CHBr_2F$											316.8	65.1
1,2-Dibromoheptane	$C_7H_{14}Br_2$					-212.3				-157.9			
Dibromomethane	CH_2Br_2											293.2	54.7
2,3-Dibromo-2-methylbutane	$C_5H_{10}Br_2$									-137.6			
1,2-Dibromo-2-methylpropane	$C_4H_8Br_2$					-156.6				-113.3			
1,2-Dibromopropane	$C_3H_6Br_2$					-113.6				-71.6			
Dibromosilane	Br_2H_2Si											309.7	65.5
1,2-Dibromotetrafluoroethane	$C_2Br_2F_4$					-817.7				-789.1			
1,2-Dibutoxyethane	$C_{10}H_{22}O_2$								350.0				
Dibutylamine	$C_8H_{19}N$					-206.0			292.9	-156.6			
1,3-Di-*tert*-butylbenzene	$C_{14}H_{22}$					-188.8							
1,4-Di-*tert*-butylbenzene	$C_{14}H_{22}$	-212.0											
Dibutyl disulfide	$C_8H_{18}S_2$					-222.9				-160.6			
Di-*tert*-butyl disulfide	$C_8H_{18}S_2$					-255.2				-201.0			
cis-1,2-Di-*tert*-butylethene	$C_{10}H_{20}$					-163.6							
Dibutyl ether	$C_8H_{18}O$					-377.9			278.2	-332.8			
Di-*sec*-butyl ether	$C_8H_{18}O$					-401.5				-360.6			
Di-*tert*-butyl ether	$C_8H_{18}O$					-399.6			276.1	-362.0			
1,3-Di-*tert*-butyl-5-methylbenzene	$C_{15}H_{24}$	-245.8											
2,6-Di-*tert*-butyl-4-methylphenol	$C_{15}H_{24}O$	-410.0								-296.9			
Dibutyl phthalate	$C_{16}H_{22}O_4$					-842.6				-750.9			
Dibutyl sebacate	$C_{18}H_{34}O_4$								619.0				
Dibutyl sulfide	$C_8H_{18}S$					-220.7		405.1	284.3	-167.7			
Di-*sec*-butyl sulfide	$C_8H_{18}S$					-220.7				-167.7			
Di-*tert*-butyl sulfide	$C_8H_{18}S$					-232.6				-188.8			
Dibutyl sulfite	$C_8H_{18}O_3S$					-693.1				-625.3			
Dicarbon	C_2									831.9	775.9	199.4	43.2
Dichloroacetic acid	$C_2H_2Cl_2O_2$					-496.3							
Dichloroacetyl chloride	C_2HCl_3O					-280.4				-241.0			
3,4-Dichloroaniline	$C_6H_5Cl_2N$	-89.1											
o-Dichlorobenzene	$C_6H_4Cl_2$					-17.5			162.4	30.2			
m-Dichlorobenzene	$C_6H_4Cl_2$					-20.7				25.7			
p-Dichlorobenzene	$C_6H_4Cl_2$	-42.3		175.4	147.8					22.5			
1,3-Dichlorobutane	$C_4H_8Cl_2$					-237.3				-195.0			
1,4-Dichlorobutane	$C_4H_8Cl_2$					-229.8				-183.4			
Dichlorodifluoromethane	CCl_2F_2									-477.4	-439.4	300.8	72.3
1,1-Dichloroethane	$C_2H_4Cl_2$					-158.4	-73.8	211.8	126.3	-127.7	-70.8	305.1	76.2
1,2-Dichloroethane	$C_2H_4Cl_2$					-166.8			128.4	-126.4		308.4	78.7
1,1-Dichloroethene	$C_2H_2Cl_2$					-23.9	24.1	201.5	111.3	2.8	25.4	289.0	67.1
cis-1,2-Dichloroethene	$C_2H_2Cl_2$					-26.4		198.4	116.4	4.6		289.6	65.1
trans-1,2-Dichloroethene	$C_2H_2Cl_2$					-24.3	27.3	195.9	116.8	5.0	28.6	290.0	66.7
1,1-Dichloro-1-fluoroethane	$C_2H_3Cl_2F$											320.2	88.7
1,1-Dichloro-2-fluoroethene	C_2HCl_2F											313.9	76.5
Dichlorofluoromethane	$CHCl_2F$											293.1	60.9
Dichloromethane	CH_2Cl_2					-124.2		177.8	101.2	-95.4		270.2	51.0
2,4-Dichlorophenol	$C_6H_4Cl_2O$	-226.4								-156.3			
1,2-Dichloropropane, (±)-	$C_3H_6Cl_2$					-198.8			149.1	-162.8			
1,3-Dichloropropane	$C_3H_6Cl_2$					-199.9				-159.2			
2,2-Dichloropropane	$C_3H_6Cl_2$					-205.8				-173.2			

Name	Mol. form.	$\Delta_f H°(c)$/ kJ mol^{-1}	$\Delta_f G°(c)$/ kJ mol^{-1}	$S°(c)$/J mol^{-1} K^{-1}	$C_p(c)$/J mol^{-1} K^{-1}	$\Delta_f H°(l)$/ kJ mol^{-1}	$\Delta_f G°(l)$/ kJ mol^{-1}	$S°(l)$/J mol^{-1} K^{-1}	$C_p(l)$/J mol^{-1} K^{-1}	$\Delta_f H°(g)$/ kJ mol^{-1}	$\Delta_f G°(g)$/ kJ mol^{-1}	$S°(g)$/J mol^{-1} K^{-1}	$C_p(g)$/J mol^{-1} K^{-1}
2,3-Dichloro-1-propanol	$C_3H_6Cl_2O$					-381.5				-316.3			
1,3-Dichloro-2-propanol	$C_3H_6Cl_2O$					-385.3				-318.4			
2,3-Dichloropropene	$C_3H_4Cl_2$					-73.3							
Dichlorosilane	Cl_2H_2Si											285.7	60.5
1,2-Dichloro-1,1,2,2-tetrafluoroethane	$C_2Cl_2F_4$					-960.2			111.7	-937.0			
2,2-Dichloro-1,1,1-trifluoroethane	$C_2HCl_2F_3$											352.8	102.5
Dicopper	Cu_2									484.2	431.9	241.6	36.6
Diethanolamine	$C_4H_{11}NO_2$	-493.8			233.5					-397.1			
1,1-Diethoxyethane	$C_6H_{14}O_2$					-491.4				-453.5			
1,2-Diethoxyethane	$C_6H_{14}O_2$					-451.4			259.4	-408.1			
Diethoxymethane	$C_5H_{12}O_2$					-450.5				-414.7			
2,2-Diethoxypropane	$C_7H_{16}O_2$					-538.9				-506.9			
Diethylamine	$C_4H_{11}N$					-103.7			169.2	-72.2			
Diethylamine hydrochloride	$C_4H_{12}ClN$	-358.6											
2-Diethylaminoethanol	$C_6H_{15}NO$					-305.9							
o-Diethylbenzene	$C_{10}H_{14}$					-68.5							
m-Diethylbenzene	$C_{10}H_{14}$					-73.5							
p-Diethylbenzene	$C_{10}H_{14}$					-72.8							
Diethyl carbonate	$C_5H_{10}O_3$					-681.5				-637.9			
Diethyl 3,5-dimethylpyrrole-2,4-dicarboxylate	$C_{12}H_{17}NO_4$	-916.7											
Diethyl disulfide	$C_4H_{10}S_2$					-120.1		305.0	204.0	-79.4			
Diethylene glycol	$C_4H_{10}O_3$					-628.5			244.8	-571.2			
Diethylene glycol dibutyl ether	$C_{12}H_{26}O_3$								452.0				
Diethylene glycol diethyl ether	$C_8H_{18}O_3$								341.4				
Diethylene glycol dimethyl ether	$C_6H_{14}O_3$								274.1				
Diethylene glycol monobutyl ether	$C_8H_{18}O_3$								354.9				
Diethylene glycol monoethyl ether	$C_6H_{14}O_3$								301.0				
Diethylene glycol monomethyl ether	$C_5H_{12}O_3$								271.1				
Diethyl ether	$C_4H_{10}O$					-279.5		253.5	172.5	-252.1		342.7	119.5
Diethyl malonate	$C_7H_{12}O_4$								285.0				
Diethyl mercury	$C_4H_{10}Hg$					30.1			182.8	75.3			
Diethyl oxalate	$C_6H_{10}O_4$					-805.5				-742.0			
3,3-Diethylpentane	C_9H_{20}					-275.4			278.2	-233.3			
Diethyl phthalate	$C_{12}H_{14}O_4$					-776.6		425.1	366.1	-688.4			
Diethyl sulfate	$C_4H_{10}O_4S$					-813.2				-756.3			
Diethyl sulfide	$C_4H_{10}S$					-119.4		269.3	171.4	-83.5		368.1	117.0
Diethyl sulfite	$C_4H_{10}O_3S$					-600.7				-552.2			
Diethyl sulfoxide	$C_4H_{10}OS$					-268.0				-205.6			
N,N-Diethylurea	$C_5H_{12}N_2O$	-372.2											
Difluoramine	F_2HN											252.8	43.4
Difluorine dioxide	F_2O_2									19.2	58.2	277.2	62.1
Difluoroamidogen	F_2N									43.1	57.8	249.9	41.0
o-Difluorobenzene	$C_6H_4F_2$					-330.0		222.6	159.0	-293.8			
m-Difluorobenzene	$C_6H_4F_2$					-343.9		223.8	159.1	-309.2			
p-Difluorobenzene	$C_6H_4F_2$					-342.3			157.5	-306.7			
cis-Difluorodiazine	F_2N_2									69.5			
trans-Difluorodiazine	F_2N_2									82.0			
1,1-Difluoroethane	$C_2H_4F_2$									-497.0		282.5	67.8
1,1-Difluoroethene	$C_2H_2F_2$									-335.0		266.2	60.1
cis-1,2-Difluoroethene	$C_2H_2F_2$											268.3	58.2
Difluoromethane	CH_2F_2									-452.3		246.7	42.9
Difluorosilylene	F_2Si									-619.0	-628.0	252.7	43.9
Digallium	Ga_2									438.5			
Digermane	Ge_2H_6					137.3				162.3			
Digermanium	Ge_2									473.1	416.3	252.8	35.6
Digold	Au_2									515.1			36.9
Dihydro-5-methyl-2(3*H*)-furanone	$C_5H_8O_2$					-461.3				-406.5			
1,2-Dihydronaphthalene	$C_{10}H_{10}$					71.6							
1,4-Dihydronaphthalene	$C_{10}H_{10}$					84.2							
2,3-Dihydrothiophene	C_4H_6S					52.9				90.7	133.5	303.5	79.8
2,5-Dihydrothiophene	C_4H_6S					47.0				86.9	131.6	297.1	83.3
1,4-Dihydroxy-9,10-anthracenedione	$C_{14}H_8O_4$	-595.8								-471.7			
Diindium	In_2									380.9			

Name	Mol. form.	$\Delta_f H°$(c)/ kJ mol^{-1}	$\Delta_f G°$(c)/ kJ mol^{-1}	$S°$(c)/J mol^{-1} K^{-1}	C_p(c)/J mol^{-1} K^{-1}	$\Delta_f H°$(l)/ kJ mol^{-1}	$\Delta_f G°$(l)/ kJ mol^{-1}	$S°$(l)/J mol^{-1} K^{-1}	C_p(l)/J mol^{-1} K^{-1}	$\Delta_f H°$(g)/ kJ mol^{-1}	$\Delta_f G°$(g)/ kJ mol^{-1}	$S°$(g)/J mol^{-1} K^{-1}	C_p(g)/J mol^{-1} K^{-1}
Diiodoacetylene	C_2I_2											313.1	70.3
1,4-Diiodobutane	$C_4H_8I_2$					-30.0							
1,2-Diiodoethane	$C_2H_4I_2$	9.3								75.0			
cis-1,2-Diiodoethene	$C_2H_2I_2$									-207.4			
Diiodomethane	CH_2I_2					68.5	90.4	174.1	134.0	119.5	95.8	309.7	57.7
1,2-Diiodopropane	$C_3H_6I_2$									35.6			
1,3-Diiodopropane	$C_3H_6I_2$					-9.0							
Diisobutylamine	$C_8H_{19}N$					-218.5				-179.2			
Diisobutyl sulfide	$C_8H_{18}S$					-229.2				-180.5			
Diisopentyl ether	$C_{10}H_{22}O$								379.0				
Diisopentyl sulfide	$C_{10}H_{22}S$					-281.8				-221.5			
Diisopropylamine	$C_6H_{15}N$					-178.5				-143.8			
Diisopropyl ether	$C_6H_{14}O$					-351.5			216.8	-319.2			
Diisopropyl sulfide	$C_6H_{14}S$					-181.6		313.0	232.0	-142.0			
Diketene	$C_4H_4O_2$					-233.1				-190.3			
Dilithium	Li_2									215.9	174.4	197.0	36.1
Dimagnesium	Mg_2									287.7			
Dimercury	Hg_2									108.8	68.2	288.1	37.4
1,2-Dimethoxybenzene	$C_8H_{10}O_2$					-290.3				-223.3			
1,2-Dimethoxyethane	$C_4H_{10}O_2$					-376.6			193.3				
Dimethoxymethane	$C_3H_8O_2$					-377.8		244.0	162.0	-348.5			
1,1-Dimethoxypropane	$C_5H_{12}O_2$					-443.6							
2,2-Dimethoxypropane	$C_5H_{12}O_2$					-459.4				-429.9			
Dimethylacetal	$C_4H_{10}O_2$					-420.6				-389.7			
N,N-Dimethylacetamide	C_4H_9NO					-278.3			175.6	-228.0			
Dimethylamine	C_2H_7N					-43.9	70.0	182.3	137.7	-18.8	68.5	273.1	70.7
Dimethylamine hydrochloride	C_2H_8ClN	-289.3											
2,4-Dimethylaniline	$C_8H_{11}N$					-39.2							
2,5-Dimethylaniline	$C_8H_{11}N$					-38.9							
2,6-Dimethylaniline	$C_8H_{11}N$								238.9				
N,N-Dimethylaniline	$C_8H_{11}N$					46.0				100.5			
2,2-Dimethylbutane	C_6H_{14}					-213.8		272.5	191.9	-185.9			
2,3-Dimethylbutane	C_6H_{14}					-207.4		287.8	189.7	-178.1			
2,3-Dimethyl-2-butanethiol	$C_6H_{14}S$					-187.1				-147.9			
3,3-Dimethyl-2-butanone	$C_6H_{12}O$					-328.6				-290.6			
2,3-Dimethyl-1-butene	C_6H_{12}					-93.2				-62.4			
3,3-Dimethyl-1-butene	C_6H_{12}					-87.5				-60.3			
2,3-Dimethyl-2-butene	C_6H_{12}					-101.4		270.2	174.7	-68.1			
3,3-Dimethyl-1-butyne	C_6H_{10}					78.4							
Dimethyl cadmium	C_2H_6Cd					63.6	139.0	201.9	132.0	101.6	146.9	303.0	
1,1-Dimethylcyclohexane	C_8H_{16}					-218.7		267.2	209.2	-180.9			
cis-1,2-Dimethylcyclohexane	C_8H_{16}					-211.8		274.1	210.2	-172.1			
trans-1,2-Dimethylcyclohexane	C_8H_{16}					-218.2		273.2	209.4	-179.9			
cis-1,3-Dimethylcyclohexane	C_8H_{16}					-222.9		272.6	209.4	-184.6			
trans-1,3-Dimethylcyclohexane	C_8H_{16}					-215.7		276.3	212.8	-176.5			
cis-1,4-Dimethylcyclohexane	C_8H_{16}					-215.6		271.1	212.1	-176.6			
trans-1,4-Dimethylcyclohexane	C_8H_{16}					-222.4		268.0	210.2	-184.5			
1,1-Dimethylcyclopentane	C_7H_{14}					-172.1				-138.2			
cis-1,2-Dimethylcyclopentane	C_7H_{14}					-165.3		269.2		-129.5			
trans-1,2-Dimethylcyclopentane	C_7H_{14}					-171.2				-136.6			
cis-1,3-Dimethylcyclopentane	C_7H_{14}					-170.1				-135.8			
trans-1,3-Dimethylcyclopentane	C_7H_{14}					-168.1				-133.6			
1,1-Dimethylcyclopropane	C_5H_{10}					-33.3				-8.2			
cis-1,2-Dimethylcyclopropane	C_5H_{10}					-26.3							
trans-1,2-Dimethylcyclopropane	C_5H_{10}					-30.7							
Dimethyl disulfide	$C_2H_6S_2$					-62.6		235.4	146.1	-24.7			
N,N-Dimethylethanolamine	$C_4H_{11}NO$					-253.7				-203.6			
Dimethyl ether	C_2H_6O					-203.3				-184.1	-112.6	266.4	64.4
N,N-Dimethylformamide	C_3H_7NO					-239.3			150.6	-192.4			
3,4-Dimethyl-2,5-furandione	$C_6H_6O_3$	-581.4											
Dimethylglyoxime	$C_4H_8N_2O_2$	-199.7											
2,2-Dimethylheptane	C_9H_{20}					-288.1							
2,6-Dimethyl-4-heptanone	$C_9H_{18}O$					-408.5			297.3	-357.6			
2,2-Dimethylhexane	C_8H_{18}					-261.9				-224.5			
2,3-Dimethylhexane	C_8H_{18}					-252.6				-213.8			
2,4-Dimethylhexane	C_8H_{18}					-257.0				-219.2			

Name	Mol. form.	$\Delta_f H°$(c)/ kJ mol^{-1}	$\Delta_f G°$(c)/ kJ mol^{-1}	$S°$(c)/J mol^{-1} K^{-1}	C_p(c)/J mol^{-1} K^{-1}	$\Delta_f H°$(l)/ kJ mol^{-1}	$\Delta_f G°$(l)/ kJ mol^{-1}	$S°$(l)/J mol^{-1} K^{-1}	C_p(l)/J mol^{-1} K^{-1}	$\Delta_f H°$(g)/ kJ mol^{-1}	$\Delta_f G°$(g)/ kJ mol^{-1}	$S°$(g)/J mol^{-1} K^{-1}	C_p(g)/J mol^{-1} K^{-1}
2,5-Dimethylhexane	C_8H_{18}					-260.4			249.2	-222.5			
3,3-Dimethylhexane	C_8H_{18}					-257.5			246.6	-219.9			
3,4-Dimethylhexane	C_8H_{18}					-251.8				-212.8			
2,5-Dimethyl-2,5-hexanediol	$C_8H_{18}O_2$	-681.7											
cis-2,2-Dimethyl-3-hexene	C_8H_{16}					-126.4				-89.3			
trans-2,2-Dimethyl-3-hexene	C_8H_{16}					-144.9				-107.7			
1,1-Dimethylhydrazine	$C_2H_8N_2$					48.9	206.4	198.0	164.1	84.1			
1,2-Dimethylhydrazine	$C_2H_8N_2$					52.7				92.2			
1,1-Dimethylindan	$C_{11}H_{14}$					-53.6				-1.6			
Dimethyl isophthalate	$C_{10}H_{10}O_4$	-730.9											
Dimethyl maleate	$C_6H_8O_4$								263.2				
Dimethyl mercury	C_2H_6Hg					59.8	140.3	209.0		94.4	146.1	306.0	83.3
cis, cis-2,6-Dimethyl-2,4,6-octatriene	$C_{10}H_{16}$					-24.0							
Dimethyl oxalate	$C_4H_6O_4$	-756.3								-708.9			
3,3-Dimethyloxetane	$C_5H_{10}O$					-182.2				-148.2			
2,2-Dimethylpentane	C_7H_{16}					-238.3		300.3	221.1	-205.7			
2,3-Dimethylpentane	C_7H_{16}					-233.1				-198.7			
2,4-Dimethylpentane	C_7H_{16}					-234.6		303.2	224.2	-201.6			
3,3-Dimethylpentane	C_7H_{16}					-234.2				-201.0			
2,2-Dimethyl-3-pentanone	$C_7H_{14}O$					-356.1				-313.6			
2,4-Dimethyl-3-pentanone	$C_7H_{14}O$					-352.9		318.0	233.7	-311.3			
2,4-Dimethyl-1-pentene	C_7H_{14}					-117.0				-83.8			
4,4-Dimethyl-1-pentene	C_7H_{14}					-110.6				-81.6			
2,4-Dimethyl-2-pentene	C_7H_{14}					-123.1				-88.7			
cis-4,4-Dimethyl-2-pentene	C_7H_{14}					-105.3				-72.6			
trans-4,4-Dimethyl-2-pentene	C_7H_{14}					-121.7				-88.8			
Dimethyl phthalate	$C_{10}H_{10}O_4$								303.1				
2,2-Dimethylpropanamide	$C_5H_{11}NO$	-399.7								-313.1			
2,2-Dimethyl-1,3-propanediol	$C_5H_{12}O_2$	-551.2											
2,2-Dimethylpropanenitrile	C_5H_9N					-39.8		232.0	179.4	-2.3			
2,2-Dimethyl-1-propanethiol	$C_5H_{12}S$					-165.4				-129.0			
2,2-Dimethylpropanoic acid	$C_5H_{10}O_2$	-564.5								-491.3			
2,2-Dimethyl-1-propanol	$C_5H_{12}O$					-399.4							
2,3-Dimethylpyridine	C_7H_9N					19.4		243.7	189.5	67.1			
2,4-Dimethylpyridine	C_7H_9N					16.1		248.5	184.8	63.6			
2,5-Dimethylpyridine	C_7H_9N					18.7		248.8	184.7	66.5			
2,6-Dimethylpyridine	C_7H_9N					12.7		244.2	185.2	58.1			
3,4-Dimethylpyridine	C_7H_9N					18.3		240.7	191.8	68.8			
3,5-Dimethylpyridine	C_7H_9N					22.5		241.7	184.5	72.0			
2,4-Dimethylpyrrole	C_6H_9N	-422.3											
2,5-Dimethylpyrrole	C_6H_9N					-16.7				39.8			
Dimethyl sulfate	$C_2H_6O_4S$					-735.5				-687.0			
Dimethyl sulfide	C_2H_6S					-65.3		196.4	118.1	-37.4		286.0	74.1
Dimethyl sulfite	$C_2H_6O_3S$					-523.6				-483.4			
Dimethyl sulfone	$C_2H_6O_2S$	-450.1	-302.4	142.0						-373.1	-272.7	310.6	100.0
Dimethyl sulfoxide	C_2H_6OS					-204.2	-99.9	188.3	153.0	-151.3			
Dimethyl terephthalate	$C_{10}H_{10}O_4$	-732.6			261.1								
N,N-Dimethylurea	$C_3H_8N_2O$	-319.1											
N,N'-Dimethylurea	$C_3H_8N_2O$	-312.1											
Dimethyl zinc	C_2H_6Zn					23.4		201.6	129.2	53.0			
2,3-Dinitroaniline	$C_6H_5N_3O_4$	-11.7											
2,4-Dinitroaniline	$C_6H_5N_3O_4$	-67.8											
2,5-Dinitroaniline	$C_6H_5N_3O_4$	-44.3											
2,6-Dinitroaniline	$C_6H_5N_3O_4$	-50.6											
3,5-Dinitroaniline	$C_6H_5N_3O_4$	-38.9											
1,2-Dinitrobenzene	$C_6H_4N_2O_4$	-2.0			200.4								
1,3-Dinitrobenzene	$C_6H_4N_2O_4$	-27.0			197.5	-36.0							
1,4-Dinitrobenzene	$C_6H_4N_2O_4$	-38.0			200.0								
3,5-Dinitrobenzoic acid	$C_7H_4N_2O_6$	-409.8											
1,4-Dinitrobutane	$C_4H_8N_2O_4$					-237.5							
1,1-Dinitroethane	$C_2H_4N_2O_4$					-148.2							
1,2-Dinitroethane	$C_2H_4N_2O_4$					-165.2							
Dinitromethane	$CH_2N_2O_4$					-104.9				-61.5		358.1	86.4
1,5-Dinitronaphthalene	$C_{10}H_6N_2O_4$	29.8											
1,8-Dinitronaphthalene	$C_{10}H_6N_2O_4$	39.7											

Name	Mol. form.	$\Delta_f H°(c)$/ kJ mol⁻¹	$\Delta_f G°(c)$/ kJ mol⁻¹	$S°(c)$/J mol⁻¹ K⁻¹	$C_p(c)$/J mol⁻¹ K⁻¹	$\Delta_f H°(l)$/ kJ mol⁻¹	$\Delta_f G°(l)$/ kJ mol⁻¹	$S°(l)$/J mol⁻¹ K⁻¹	$C_p(l)$/J mol⁻¹ K⁻¹	$\Delta_f H°(g)$/ kJ mol⁻¹	$\Delta_f G°(g)$/ kJ mol⁻¹	$S°(g)$/J mol⁻¹ K⁻¹	$C_p(g)$/J mol⁻¹ K⁻¹
2,4-Dinitrophenol	$C_6H_4N_2O_5$	-232.7								-128.1			
1,1-Dinitropropane	$C_3H_6N_2O_4$					-163.2				-100.7			
1,3-Dinitropropane	$C_3H_6N_2O_4$					-207.1							
2,2-Dinitropropane	$C_3H_6N_2O_4$					-181.2							
1,3-Dioxane	$C_4H_8O_2$					-379.7			143.9	-340.6			
1,4-Dioxane	$C_4H_8O_2$					-353.9		270.2	152.1	-315.3			
1,3-Dioxolane	$C_3H_6O_2$					-333.5			118.0	-298.0			
1,3-Dioxol-2-one	$C_3H_2O_3$					-459.9				-418.6			
Dipentene	$C_{10}H_{16}$					-50.8			249.4	-2.6			
Dipentyl ether	$C_{10}H_{22}O$								250.0				
Dipentyl sulfide	$C_{10}H_{22}S$					-266.4				-204.9			
Diphenylacetylene	$C_{14}H_{10}$	312.4			225.9								
Diphenylamine	$C_{12}H_{11}N$	130.2								219.3			
9,10-Diphenylanthracene	$C_{26}H_{18}$	308.7								465.6			
1,1-Diphenylethane	$C_{14}H_{14}$					48.7							
1,2-Diphenylethane	$C_{14}H_{14}$	51.5								142.9			
Diphenyl ether	$C_{12}H_{10}O$	-32.1		233.9	216.6	-14.9				52.0			
Diphenylmethane	$C_{13}H_{12}$	71.5		239.3		89.7				139.0			
Diphenyl phthalate	$C_{20}H_{14}O_4$	-489.2											
Diphosphine	H_4P_2					-5.0				20.9			
Diphosphoric acid	$H_4O_7P_2$	-2241.0				-2231.7							
Diphosphorus	P_2									144.0	103.5	218.1	32.1
Dipotassium	K_2									123.7	87.5	249.7	37.9
Dipropylamine	$C_6H_{15}N$					-156.1				-116.0			
Dipropyl disulfide	$C_6H_{14}S_2$					-171.5				-118.3			
Dipropyl ether	$C_6H_{14}O$					-328.8		323.9	221.6	-293.0			
Dipropyl sulfate	$C_6H_{14}O_4S$					-859.0				-792.0			
Dipropyl sulfoxide	$C_6H_{14}OS$					-329.4				-254.9			
2,2'-Dipyrrolylmethane	$C_9H_{10}N_2$	126.2											
Diselenium	Se_2									146.0	96.2	252.0	35.4
Disilane	H_6Si_2									80.3	127.3	272.7	80.8
Disilicon	Si_2									594.0	536.0	229.9	34.4
Disilver	Ag_2									410.0	358.8	257.1	37.0
Disodium	Na_2									142.1	103.9	230.2	37.6
Disulfur	S_2									128.6	79.7	228.2	32.5
Ditellurium	Te_2									168.2	118.0	268.1	36.7
1,3-Dithiane	$C_4H_8S_2$									-10.0	72.4	333.5	110.4
1,4-Dithiane	$C_4H_8S_2$									0.0	84.5	326.2	109.7
1,2-Dithiolane	$C_3H_6S_2$									0.0	47.7	313.5	86.5
1,3-Dithiolane	$C_3H_6S_2$									10.0	54.7	323.3	84.7
Diuron	$C_9H_{10}Cl_2N_2O$	-329.0											
cis-1,2-Divinylcyclobutane	C_8H_{12}					124.3				166.5			
trans-1,2-Divinylcyclobutane	C_8H_{12}					101.3				143.5			
Divinyl ether	C_4H_6O					-39.8				-13.6			
trans-13-Docosenoic acid	$C_{22}H_{42}O_2$	-960.7											
3,9-Dodecadiyne	$C_{12}H_{18}$					197.8							
5,7-Dodecadiyne	$C_{12}H_{18}$					181.5							
Dodecane	$C_{12}H_{26}$					-350.9			375.8	-289.4			
Dodecanedioic acid	$C_{12}H_{22}O_4$	-1130.0								-976.9			
Dodecanoic acid	$C_{12}H_{24}O_2$	-774.6			404.3	-737.9				-642.0			
1-Dodecanol	$C_{12}H_{26}O$					-528.5			438.1	-436.6			
1-Dodecene	$C_{12}H_{24}$					-226.2		484.8	360.7	-165.4			
Dotriacontane	$C_{32}H_{66}$	-968.3								-697.2			
Dysprosium	Dy	0.0		75.6	28.2					290.4	254.4	196.6	20.8
Dysprosium(III) bromide	Br_3Dy	-836.2											
Dysprosium(III) chloride	Cl_3Dy	-1000.0											
Dysprosium(III) iodide	DyI_3	-620.5											
Dysprosium(III) oxide	Dy_2O_3	-1863.1	-1771.5	149.8	116.3								
Eicosanoic acid	$C_{20}H_{40}O_2$	-1011.9			545.1	-940.0				-812.4			
Einsteinium	Es	0.0											
Epichlorohydrin	C_3H_5ClO					-148.4			131.6	-107.8			
1,2-Epoxybutane	C_4H_8O					-168.9		230.9	147.0				
Erbium	Er	0.0		73.2	28.1					317.1	280.7	195.6	20.8
Erbium chloride	Cl_3Er	-998.7			100.0								
Erbium fluoride	ErF_3	-1711.0											
Erbium oxide	Er_2O_3	-1897.9	-1808.7	155.6	108.5								

Thermochem

Name	Mol. form.	$\Delta_f H°$(c)/ kJ mol^{-1}	$\Delta_f G°$(c)/ kJ mol^{-1}	$S°$(c)/J mol^{-1} K^{-1}	C_p(c)/J mol^{-1} K^{-1}	$\Delta_f H°$(l)/ kJ mol^{-1}	$\Delta_f G°$(l)/ kJ mol^{-1}	$S°$(l)/J mol^{-1} K^{-1}	C_p(l)/J mol^{-1} K^{-1}	$\Delta_f H°$(g)/ kJ mol^{-1}	$\Delta_f G°$(g)/ kJ mol^{-1}	$S°$(g)/J mol^{-1} K^{-1}	C_p(g)/J mol^{-1} K^{-1}
Ethane	C_2H_6									-84.0	-32.0	229.2	52.5
Ethanedial dioxime	$C_2H_4N_2O_2$	-90.5											
1,2-Ethanediamine	$C_2H_8N_2$					-63.0			172.6	-18.0			
1,2-Ethanediol	$C_2H_6O_2$					-460.0		163.2	148.6	-392.2		303.8	82.7
1,2-Ethanediol, diacetate	$C_6H_{10}O_4$								310.0				
Ethanedithioamide	$C_2H_4N_2S_2$	-20.8								83.0			
1,2-Ethanedithiol	$C_2H_6S_2$					-54.3				-9.7			
Ethanethiol	C_2H_6S					-73.6	-5.5	207.0	117.9	-46.1	-4.8	296.2	72.7
Ethanol	C_2H_6O					-277.6	-174.8	160.7	112.3	-234.8	-167.9	281.6	65.6
Ethanolamine	C_2H_7NO								195.5				
Ethoxybenzene	$C_8H_{10}O$					-152.6			228.5	-101.6			
2-Ethoxyethanol	$C_4H_{10}O_2$								210.8				
2-Ethoxyethyl acetate	$C_6H_{12}O_3$								376.0				
(Ethoxymethyl)oxirane	$C_5H_{10}O_2$					-296.5							
Ethyl acetate	$C_4H_8O_2$					-479.3		257.7	170.7	-443.6			
Ethyl acetoacetate	$C_6H_{10}O_3$								248.0				
Ethyl acrylate	$C_5H_8O_2$					-370.6				-354.2			
Ethylamine	C_2H_7N					-74.1			130.0	-47.5	36.3	283.8	71.5
N-Ethylaniline	$C_8H_{11}N$					8.2				56.3			
Ethylbenzene	C_8H_{10}					-12.3			183.2	29.9			
Ethyl benzoate	$C_9H_{10}O_2$								246.0				
Ethyl butanoate	$C_6H_{12}O_2$								228.0				
2-Ethyl-1-butene	C_6H_{12}					-87.1				-56.0			
Ethyl *trans*-2-butenoate	$C_6H_{10}O_2$					-420.0				-375.6			
Ethyl carbamate	$C_3H_7NO_2$	-517.1			156.4	-497.3				-446.3			
Ethyl 4-chlorobutanoate	$C_6H_{11}ClO_2$					-566.5				-513.8			
Ethyl chloroformate	$C_3H_5ClO_2$					-505.3				-462.9			
Ethylcyanoacetate	$C_5H_7NO_2$								220.2				
Ethylcyclobutane	C_6H_{12}					-59.0				-27.5			
Ethylcyclohexane	C_8H_{16}					-212.1		280.9	211.8	-171.5			
Ethylcyclopentane	C_7H_{14}					-163.4		279.9		-126.9			
1-Ethylcyclopentene	C_7H_{12}					-53.3				-19.8			
Ethylcyclopropane	C_5H_{10}					-24.8							
1-Ethyl-2,4-dimethylbenzene	$C_{10}H_{14}$					-84.1							
1-Ethyl-3,5-dimethylbenzene	$C_{10}H_{14}$					-87.8							
2-Ethyl-1,3-dimethylbenzene	$C_{10}H_{14}$					-80.1							
2-Ethyl-1,4-dimethylbenzene	$C_{10}H_{14}$					-84.8							
3-Ethyl-1,2-dimethylbenzene	$C_{10}H_{14}$					-80.5							
4-Ethyl-1,2-dimethylbenzene	$C_{10}H_{14}$					-86.0							
3-Ethyl-2,2-dimethylpentane	C_9H_{20}					-272.7							
3-Ethyl-2,4-dimethylpentane	C_9H_{20}					-269.7							
Ethyl 2,2-dimethylpropanoate	$C_7H_{14}O_2$					-577.2				-536.0			
Ethyl 3,5-dimethylpyrrole-2-carboxylate	$C_9H_{13}NO_2$	-474.5											
Ethyl 2,4-dimethylpyrrole-3-carboxylate	$C_9H_{13}NO_2$	-463.2											
Ethyl 2,5-dimethylpyrrole-3-carboxylate	$C_9H_{13}NO_2$	-478.7											
Ethyl 4,5-dimethylpyrrole-3-carboxylate	$C_9H_{13}NO_2$	-470.3											
Ethylene	C_2H_4									52.4	68.4	219.3	42.9
Ethylene carbonate	$C_3H_4O_3$					-682.8			133.9	-508.4			
Ethylenediaminetetraacetic acid	$C_{10}H_{16}N_2O_8$	-1759.5											
Ethyleneimine	C_2H_5N					91.9				126.5			
Ethyl formate	$C_3H_6O_2$								149.3				
2-Ethylhexanal	$C_8H_{16}O$					-348.5				-299.6			
3-Ethylhexane	C_8H_{18}					-250.4				-210.7			
2-Ethylhexanoic acid	$C_8H_{16}O_2$					-635.1				-559.5			
2-Ethyl-1-hexanol	$C_8H_{18}O$					-432.8		347.0	317.5	-365.3			
Ethylidenecyclohexane	C_8H_{14}					-103.5				-59.5			
Ethyl isocyanide	C_3H_5N					108.6				141.7			
Ethyl isopropyl sulfide	$C_5H_{12}S$					-156.1				-118.3			
Ethyl lactate	$C_5H_{10}O_3$								254.0				
Ethyl 3-methylbutanoate	$C_7H_{14}O_2$					-571.0				-527.0			
2-Ethyl-3-methyl-1-butene	C_7H_{14}					-114.1				-79.5			
1-Ethyl-1-methylcyclopentane	C_8H_{16}					-193.8							

Name	Mol. form.	$\Delta_f H°$(c)/ kJ mol^{-1}	$\Delta_f G°$(c)/ kJ mol^{-1}	$S°$(c)/J mol^{-1} K^{-1}	C_p(c)/J mol^{-1} K^{-1}	$\Delta_f H°$(l)/ kJ mol^{-1}	$\Delta_f G°$(l)/ kJ mol^{-1}	$S°$(l)/J mol^{-1} K^{-1}	C_p(l)/J mol^{-1} K^{-1}	$\Delta_f H°$(g)/ kJ mol^{-1}	$\Delta_f G°$(g)/ kJ mol^{-1}	$S°$(g)/J mol^{-1} K^{-1}	C_p(g)/J mol^{-1} K^{-1}
cis-1-Ethyl-2-methylcyclopentane	C_8H_{16}					-190.8							
trans-1-Ethyl-2-methylcyclopentane	C_8H_{16}					-195.1				-156.2			
cis-1-Ethyl-3-methylcyclopentane	C_8H_{16}					-194.4							
trans-1-Ethyl-3-methylcyclopentane	C_8H_{16}					-196.0							
Ethyl methyl ether	C_3H_8O									-216.4		309.2	93.3
3-Ethyl-2-methylpentane	C_8H_{18}					-249.6				-211.0			
3-Ethyl-3-methylpentane	C_8H_{18}					-252.8				-214.8			
3-Ethyl-2-methyl-1-pentene	C_8H_{16}					-137.9				-100.3			
Ethyl methyl sulfide	C_3H_8S					-91.6		239.1	144.6	-59.6			
Ethyl nitrate	$C_2H_5NO_3$					-190.4				-154.1			
Ethyl nitroacetate	$C_4H_7NO_4$					-487.1							
2-Ethyl-2-nitro-1,3-propanediol	$C_5H_{11}NO_4$	-606.4											
Ethyl *cis*-9-octadecenoate	$C_{20}H_{38}O_2$					-775.8							
Ethyl *trans*-9-octadecenoate	$C_{20}H_{38}O_2$					-773.3							
3-Ethylpentane	C_7H_{16}					-224.9		314.5	219.6	-189.5			
Ethyl pentanoate	$C_7H_{14}O_2$					-553.0				-505.9			
Ethyl 2-pentynoate	$C_7H_{10}O_2$					-301.8				-250.3			
2-Ethylphenol	$C_8H_{10}O$					-208.8				-145.2			
3-Ethylphenol	$C_8H_{10}O$					-214.3				-146.1			
4-Ethylphenol	$C_8H_{10}O$	-224.4			206.9					-144.1			
Ethyl propanoate	$C_5H_{10}O_2$					-502.7				-463.4			
Ethyl propyl ether	$C_5H_{12}O$					-303.6		295.0	197.2	-272.0			
Ethyl propyl sulfide	$C_5H_{12}S$					-144.8		309.5	198.4	-104.8			
Ethyl silicate	$C_8H_{20}O_4Si$							533.1	364.4				
S-Ethyl thioacetate	C_4H_8OS					-268.2				-228.1			
2-Ethyltoluene	C_9H_{12}					-46.4				1.3			
3-Ethyltoluene	C_9H_{12}					-48.7				-1.8			
4-Ethyltoluene	C_9H_{12}					-49.8				-3.2			
3-Ethyl-2,4,5-trimethylpyrrole	$C_9H_{15}N$	-89.2											
N-Ethylurea	$C_3H_8N_2O$	-357.8											
Ethyl vinyl ether	C_4H_8O					-167.4				-140.8			
Ethynylsilane	C_2H_4Si											269.4	72.6
Europium	Eu	0.0		77.8	27.7					175.3	142.2	188.8	20.8
Europium(III) chloride	Cl_3Eu	-936.0											
Europium(II,III) oxide	Eu_3O_4	-2272.0	-2142.0	205.0									
Europium(III) oxide	Eu_2O_3	-1651.4	-1556.8	146.0	122.2								
Fermium	Fm	0.0											
Fluoranthene	$C_{16}H_{10}$	189.9		230.6	230.2					289.0			
9*H*-Fluorene	$C_{13}H_{10}$	89.9		207.3	203.1					175.0			173.1
Fluorine	F_2									0.0		202.8	31.3
Fluorine (atomic)	F									79.4	62.3	158.8	22.7
Fluorine monoxide	F_2O									24.5	41.8	247.5	43.3
Fluorine oxide	FO									109.0	105.3	216.4	32.0
Fluorine superoxide [FOO]	FO_2									25.4	39.4	259.5	44.5
Fluoroacetylene	C_2HF											231.7	52.4
Fluorobenzene	C_6H_5F					-150.6		205.9	146.4	-115.9			
Fluoroborane(1)	BF									-122.2	-149.8	200.5	29.6
Fluoroethane	C_2H_5F											264.5	58.6
Fluoroethene	C_2H_3F									-138.8			
Fluorogermane	$FGeH_3$											252.8	51.6
Fluoromethane	CH_3F											222.9	37.5
Fluorooxoborane	BFO									-607.0			
1-Fluoropropane	C_3H_7F									-285.9			
2-Fluoropropane	C_3H_7F									-293.5			
Fluorosilane	FH_3Si											238.4	47.4
Fluorosilylidyne	FSi									7.1	-24.3	225.8	32.6
4-Fluorotoluene	C_7H_7F					-186.9			171.2	-147.4			
Formaldehyde	CH_2O									-108.6	-102.5	218.8	35.4
Formamide	CH_3NO					-254.0				-193.9			
Formic acid	CH_2O_2					-425.0	-361.4	129.0	99.0	-378.7			
Formyl	CHO									43.1	28.0	224.7	34.6
Formyl fluoride	CHFO											246.6	39.9
Francium	Fr	0.0		95.4									
β-D-Fructose	$C_6H_{12}O_6$	-1265.6											
Fumaric acid	$C_4H_4O_4$	-811.7		168.0	142.0					-675.8			
Furan	C_4H_4O					-62.3		177.0	114.8	-34.8		267.2	65.4

Name	Mol. form.	$\Delta_f H°$(c)/ kJ mol^{-1}	$\Delta_f G°$(c)/ kJ mol^{-1}	$S°$(c)/J mol^{-1} K^{-1}	C_p(c)/J mol^{-1} K^{-1}	$\Delta_f H°$(l)/ kJ mol^{-1}	$\Delta_f G°$(l)/ kJ mol^{-1}	$S°$(l)/J mol^{-1} K^{-1}	C_p(l)/J mol^{-1} K^{-1}	$\Delta_f H°$(g)/ kJ mol^{-1}	$\Delta_f G°$(g)/ kJ mol^{-1}	$S°$(g)/J mol^{-1} K^{-1}	C_p(g)/J mol^{-1} K^{-1}
2-Furancarboxylic acid	$C_5H_4O_3$	-498.4								-390.0			
3-(2-Furanyl)-2-propenal	$C_7H_6O_2$	-182.0								-105.9			
Furfural	$C_5H_4O_2$					-201.6			163.2	-151.0			
Furfuryl alcohol	$C_5H_6O_2$					-276.2			204.0	-211.8			
Gadolinium	Gd	0.0		68.1	37.0					397.5	359.8	194.3	27.5
Gadolinium(III) chloride	Cl_3Gd	-1008.0			88.0								
Gadolinium(III) fluoride	F_3Gd									-1297.0			
Gadolinium(III) oxide	Gd_2O_3	-1819.6			106.7								
Galactitol	$C_6H_{14}O_6$					-1317.0							
D-Galactose	$C_6H_{12}O_6$	-1286.3											
Gallium	Ga	0.0	0.0	40.8	26.0	5.6				272.0	233.7	169.0	25.3
Gallium antimonide	GaSb	-41.8	-38.9	76.1	48.5								
Gallium arsenide	AsGa	-71.0	-67.8	64.2	46.2								
Gallium(III) bromide	Br_3Ga	-386.6	-359.8	180.0									
Gallium(III) chloride	Cl_3Ga	-524.7	-454.8	142.0									
Gallium(III) fluoride	F_3Ga	-1163.0	-1085.3	84.0									
Gallium(III) hydroxide	GaH_3O_3	-964.4	-831.3	100.0									
Gallium(III) iodide	GaI_3	-238.9		205.0	100.0								
Gallium monofluoride	FGa									-251.9			33.3
Gallium monoxide	GaO									279.5	253.5	231.1	32.1
Gallium nitride	GaN	-110.5											
Gallium(III) oxide	Ga_2O_3	-1089.1	-998.3	85.0	92.1								
Gallium phosphide	GaP	-88.0											
Gallium suboxide	Ga_2O	-356.0											
Germane	GeH_4									90.8	113.4	217.1	45.0
Germanium	Ge	0.0		31.1	23.3					372.0	331.2	167.9	30.7
Germanium(IV) bromide	Br_4Ge					-347.7	-331.4	280.7		-300.0	-318.0	396.2	101.8
Germanium(IV) chloride	Cl_4Ge					-531.8	-462.7	245.6		-495.8	-457.3	347.7	96.1
Germanium(IV) fluoride	F_4Ge									-1190.2	-1150.0	301.9	
Germanium(IV) iodide	GeI_4	-141.8	-144.3	271.1						-56.9	-106.3	428.9	104.1
Germanium monobromide	BrGe									235.6			37.1
Germanium monochloride	ClGe									155.2	124.2	247.0	36.9
Germanium monofluoride	FGe									-33.4			34.7
Germanium(II) oxide	GeO	-261.9	-237.2	50.0						-46.2	-73.2	224.3	30.9
Germanium(IV) oxide	GeO_2	-580.0	-521.4	39.7	52.1								
Germanium phosphide	GeP	-21.0	-17.0	63.0									
Germanium(II) sulfide	GeS	-69.0	-71.5	71.0						92.0	42.0	234.0	33.7
Germanium(II) telluride	GeTe	20.0											
α-D-Glucose	$C_6H_{12}O_6$	-1273.3											
α-D-Glucose pentaacetate	$C_{16}H_{22}O_{11}$	-2249.4											
β-D-Glucose pentaacetate	$C_{16}H_{22}O_{11}$	-2232.6											
D-Glutamic acid	$C_5H_9NO_4$	-1005.3											
L-Glutamic acid	$C_5H_9NO_4$	-1009.7											
L-Glutamine	$C_5H_{10}N_2O_3$	-826.4											
Glycerol	$C_3H_8O_3$					-669.6		206.3	218.9	-577.9			
Glycerol 1-acetate	$C_5H_{10}O_4$					-909.2							
Glycerol triacetate	$C_9H_{14}O_6$					-1330.8		458.3	384.7	-1245.0			
Glycine	$C_2H_5NO_2$	-528.5								-392.1			
Glycolic acid	$C_2H_4O_3$									-583.0	-504.9	318.6	87.1
N-Glycylglycine	$C_4H_8N_2O_3$	-747.7											
Glyoxal	$C_2H_2O_2$									-212.0	-189.7	272.5	60.6
Gold	Au	0.0		47.4	25.4					366.1	326.3	180.5	20.8
Gold(I) bromide	AuBr	-14.0											
Gold(III) bromide	$AuBr_3$	-53.3											
Gold(I) chloride	AuCl	-34.7											
Gold(III) chloride	$AuCl_3$	-117.6											
Gold(III) fluoride	AuF_3	-363.6											
Gold hydride	AuH									295.0	265.7	211.2	29.2
Gold(I) iodide	AuI	0.0											
Guanidine	CH_5N_3	-56.0											
Guanine	$C_5H_5N_5O$	-183.9											
Hafnium	Hf	0.0		43.6	25.7					619.2	576.5	186.9	20.8
Hafnium(IV) chloride	Cl_4Hf	-990.4	-901.3	190.8	120.5					-884.5			
Hafnium(IV) fluoride	F_4Hf	-1930.5	-1830.4	113.0						-1669.8			
Hafnium(IV) oxide	HfO_2	-1144.7	-1088.2	59.3	60.3								
Helium	He									0.0		126.2	20.8

Name	Mol. form.	$\Delta_f H°$(c)/ kJ mol^{-1}	$\Delta_f G°$(c)/ kJ mol^{-1}	$S°$(c)/J mol^{-1} K^{-1}	C_p(c)/J mol^{-1} K^{-1}	$\Delta_f H°$(l)/ kJ mol^{-1}	$\Delta_f G°$(l)/ kJ mol^{-1}	$S°$(l)/J mol^{-1} K^{-1}	C_p(l)/J mol^{-1} K^{-1}	$\Delta_f H°$(g)/ kJ mol^{-1}	$\Delta_f G°$(g)/ kJ mol^{-1}	$S°$(g)/J mol^{-1} K^{-1}	C_p(g)/J mol^{-1} K^{-1}
Heptadecanoic acid	$C_{17}H_{34}O_2$	-924.4			475.7	-865.6							
Heptanal	$C_7H_{14}O$					-311.5		335.4	230.1	-263.8			
Heptane	C_7H_{16}					-224.2			224.7	-187.6			
1,7-Heptanediol	$C_7H_{16}O_2$					-574.2							
Heptanenitrile	$C_7H_{13}N$					-82.8				-31.0			
1-Heptanethiol	$C_7H_{16}S$					-200.5				-149.9			
Heptanoic acid	$C_7H_{14}O_2$					-610.2			265.4	-536.2			
1-Heptanol	$C_7H_{16}O$					-403.3			272.1	-336.5			
2-Heptanone	$C_7H_{14}O$								232.6				
3-Heptanone	$C_7H_{14}O$									-297.1			
4-Heptanone	$C_7H_{14}O$									-298.3			
1-Heptene	C_7H_{14}					-97.9		327.6	211.8	-62.3			
cis-2-Heptene	C_7H_{14}					-105.1							
trans-2-Heptene	C_7H_{14}					-109.5							
cis-3-Heptene	C_7H_{14}					-104.3							
trans-3-Heptene	C_7H_{14}					-109.3							
Hexaborane(10)	B_6H_{10}					56.3				94.6	211.3	296.8	125.7
Hexabromoethane	C_2Br_6											441.9	139.3
Hexachlorobenzene	C_6Cl_6	-127.6		260.2	201.2					-35.5			
Hexachloro-1,3-butadiene	C_4Cl_6					-24.5							
Hexachloroethane	C_2Cl_6	-202.8		237.3	198.2					-143.6			
Hexadecane	$C_{16}H_{34}$					-456.1			501.6	-374.8			
Hexadecanoic acid	$C_{16}H_{32}O_2$	-891.5		452.4	460.7	-838.1				-737.1			
1-Hexadecanol	$C_{16}H_{34}O$	-686.5			422.0					-517.0			
1-Hexadecene	$C_{16}H_{32}$					-328.7		587.9	488.9	-248.4			
1,5-Hexadiene	C_6H_{10}					54.1				84.2			
1,5-Hexadiyne	C_6H_6					384.2							
Hexafluoroacetylacetone	$C_5H_2F_6O_2$	-2286.7											
Hexafluorobenzene	C_6F_6					-991.3		280.8	221.6	-955.4			
Hexafluorodisilane	F_6Si_2	-2427.0	-2299.7	219.1	129.5					-2383.3	-2307.3	391.0	129.9
Hexafluoroethane	C_2F_6									-1344.2		332.3	106.7
Hexahydro-1,3,5-trinitro-1,3,5-triazine	$C_3H_6N_6O_6$									192.0		482.4	230.2
Hexamethylbenzene	$C_{12}H_{18}$	-162.4		306.3	245.6					-77.4			
Hexamethyldisiloxane	$C_6H_{18}OSi_2$					-815.0	-541.5	433.8	311.4	-777.7	-534.5	535.0	238.5
Hexamethylphosphoric triamide	$C_6H_{18}N_3OP$								321.0				
Hexanal	$C_6H_{12}O$							280.3	210.4				
Hexanamide	$C_6H_{13}NO$	-423.0								-324.2			
Hexane	C_6H_{14}					-198.7			195.6	-166.9			
1,6-Hexanediamine	$C_6H_{16}N_2$	-205.0											
Hexanedinitrile	$C_6H_8N_2$					85.1			128.7	149.5			
1,6-Hexanedioic acid	$C_6H_{10}O_4$	-994.3											
1,2-Hexanediol	$C_6H_{14}O_2$					-577.1				-490.1			
1,6-Hexanediol	$C_6H_{14}O_2$	-569.9				-548.6				-461.2			
1-Hexanethiol	$C_6H_{14}S$					-175.7				-129.9			
Hexanoic acid	$C_6H_{12}O_2$					-583.8				-511.9			
1-Hexanol	$C_6H_{14}O$					-377.5		287.4	240.4	-315.9			
2-Hexanol	$C_6H_{14}O$					-392.0				-333.5			
3-Hexanol	$C_6H_{14}O$					-392.4			286.2				
2-Hexanone	$C_6H_{12}O$					-322.0			213.3	-278.9			
3-Hexanone	$C_6H_{12}O$					-320.2		305.3	216.9	-277.6			
1-Hexene	C_6H_{12}					-74.2		295.2	183.3	-43.5			
cis-2-Hexene	C_6H_{12}					-83.9				-52.3			
trans-2-Hexene	C_6H_{12}					-85.5				-53.9			
cis-3-Hexene	C_6H_{12}					-78.9				-47.6			
trans-3-Hexene	C_6H_{12}					-86.1				-54.4			
Hexyl acetate	$C_8H_{16}O_2$								282.8				
L-Histidine	$C_6H_9N_3O_2$	-466.7											
Holmium	Ho	0.0		75.3	27.2					300.8	264.8	195.6	20.8
Holmium chloride	Cl_3Ho	-1005.4			88.0								
Holmium fluoride	F_3Ho	-1707.0											
Holmium oxide	Ho_2O_3	-1880.7	-1791.1	158.2	115.0								
Hydrazine	H_4N_2					50.6	149.3	121.2	98.9	95.4	159.4	238.5	48.4
Hydrazinecarbothioamide	CH_5N_3S	24.7											
1,2-Hydrazinedicarboxamide	$C_2H_6N_4O_2$	-498.7											
Hydrazoic acid	HN_3					264.0	327.3	140.6		294.1	328.1	239.0	43.7

Name	Mol. form.	$\Delta_fH°$(c)/ kJ mol⁻¹	$\Delta_fG°$(c)/ kJ mol⁻¹	$S°$(c)/J mol⁻¹ K⁻¹	C_p(c)/J mol⁻¹ K⁻¹	$\Delta_fH°$(l)/ kJ mol⁻¹	$\Delta_fG°$(l)/ kJ mol⁻¹	$S°$(l)/J mol⁻¹ K⁻¹	C_p(l)/J mol⁻¹ K⁻¹	$\Delta_fH°$(g)/ kJ mol⁻¹	$\Delta_fG°$(g)/ kJ mol⁻¹	$S°$(g)/J mol⁻¹ K⁻¹	C_p(g)/J mol⁻¹ K⁻¹
Hydrogen (atomic)	H									218.0	203.3	114.7	20.8
Hydrogen	H_2									0.0		130.7	28.8
Hydrogen bromide	BrH									-36.3	-53.4	198.7	29.1
Hydrogen chloride	ClH									-92.3	-95.3	186.9	29.1
Hydrogen cyanide	CHN					108.9	125.0	112.8	70.6	135.1	124.7	201.8	35.9
Hydrogen disulfide	H_2S_2					-18.1			84.1	15.5			51.5
Hydrogen fluoride	FH					-299.8				-273.3	-275.4	173.8	
Hydrogen iodide	HI									26.5	1.7	206.6	29.2
Hydrogen peroxide	H_2O_2					-187.8	-120.4	109.6	89.1	-136.3	-105.6	232.7	43.1
Hydrogen selenide	H_2Se									29.7	15.9	219.0	34.7
Hydrogen sulfide	H_2S									-20.6	-33.4	205.8	34.2
Hydrogen telluride	H_2Te									99.6			
Hydroperoxy	HO_2									10.5	22.6	229.0	34.9
p-Hydroquinone	$C_6H_6O_2$	-364.5			136.0					-265.3			
2-Hydroxybenzoic acid	$C_7H_6O_3$	-589.9								-494.8			
Hydroxyl	HO									39.0	34.2	183.7	29.9
Hydroxylamine	H_3NO	-114.2											
2-(Hydroxymethyl)-2-methyl-1,3-propanediol	$C_5H_{12}O_3$	-744.6											
Hypochlorous acid	$ClHO$									-78.7	-66.1	236.7	37.2
Hypoxanthine	$C_5H_4N_4O$	-110.8		145.6	134.5								
Imidazole	$C_3H_4N_2$	49.8								132.9			
Imidogen	HN									351.5	345.6	181.2	29.2
Iminodiacetic acid	$C_4H_7NO_4$	-932.6											
Indan	C_9H_{10}					11.5		234.4	190.2	60.3			
1*H*-Indazole	$C_7H_6N_2$	151.9								243.0			
Indene	C_9H_8					110.6		215.3	186.9	163.4			
Indium	In	0.0		57.8	26.7					243.3	208.7	173.8	20.8
Indium antimonide	$InSb$	-30.5	-25.5	86.2	49.5					344.3			
Indium arsenide	$AsIn$	-58.6	-53.6	75.7	47.8								
Indium(I) bromide	$BrIn$	-175.3	-169.0	113.0						-56.9	-94.3	259.5	36.7
Indium(III) bromide	Br_3In	-428.9								-282.0			
Indium(I) chloride	$ClIn$	-186.2								-75.0			
Indium(III) chloride	Cl_3In	-537.2								-374.0			
Indium(I) fluoride	FIn									-203.4			
Indium(I) iodide	IIn	-116.3	-120.5	130.0						7.5	-37.7	267.3	36.8
Indium(III) iodide	I_3In	-238.0								-120.5			
Indium monoxide	InO									387.0	364.4	236.5	32.6
Indium(III) oxide	In_2O_3	-925.8	-830.7	104.2	92.0								
Indium phosphide	InP	-88.7	-77.0	59.8	45.4								
Indium(II) sulfide	InS	-138.1	-131.8	67.0						238.0			
Indium(III) sulfide	In_2S_3	-427.0	-412.5	163.6	118.0								
Indium(IV) telluride	In_2Te_5	-175.3											
1*H*-Indole	C_8H_7N	86.6								156.5			
1*H*-Indole-2,3-dione	$C_8H_5NO_2$	-268.2											
Iodic acid	HIO_3	-230.1											
Iodine	I_2	0.0		116.1	54.4					62.4	19.3	260.7	36.9
Iodine (atomic)	I									106.8	70.2	180.8	20.8
Iodine bromide	BrI									40.8	3.7	258.8	36.4
Iodine chloride	ClI					-23.9	-13.6	135.1		17.8	-5.5	247.6	35.6
Iodine fluoride	FI									-95.7	-118.5	236.2	33.4
Iodine monoxide	IO									126.0	102.5	239.6	32.9
Iodine pentafluoride	F_5I					-864.8				-822.5	-751.7	327.7	99.2
Iodoacetone	C_3H_5IO									-130.5			
Iodobenzene	C_6H_5I					117.2		205.4	158.7	164.9			
Iodoethane	C_2H_5I					-40.0	14.7	211.7	115.1	-8.1	19.2	306.0	66.9
Iodoethene	C_2H_3I											285.0	57.9
Iodogermane	GeH_3I											283.2	57.5
Iodomethane	CH_3I					-13.6		163.2	126.0	14.4		254.1	44.1
1-Iodo-2-methylpropane	C_4H_9I								162.3				
2-Iodo-2-methylpropane	C_4H_9I					-107.5				-72.1			
1-Iodonaphthalene	$C_{10}H_7I$					161.5				233.8			
2-Iodonaphthalene	$C_{10}H_7I$	144.3								235.1			
1-Iodopropane	C_3H_7I					-66.0				-30.0			
2-Iodopropane	C_3H_7I					-74.8				-40.3			
3-Iodopropanoic acid	$C_3H_5IO_2$	-460.0											

Name	Mol. form.	$\Delta_f H°$(c)/ kJ mol^{-1}	$\Delta_f G°$(c)/ kJ mol^{-1}	$S°$(c)/J mol^{-1} K^{-1}	C_p(c)/J mol^{-1} K^{-1}	$\Delta_f H°$(l)/ kJ mol^{-1}	$\Delta_f G°$(l)/ kJ mol^{-1}	$S°$(l)/J mol^{-1} K^{-1}	C_p(l)/J mol^{-1} K^{-1}	$\Delta_f H°$(g)/ kJ mol^{-1}	$\Delta_f G°$(g)/ kJ mol^{-1}	$S°$(g)/J mol^{-1} K^{-1}	C_p(g)/J mol^{-1} K^{-1}
3-Iodopropene	C_3H_5I					53.7				91.5			
Iodosilane	H_3ISi											270.9	54.4
Iridium	Ir	0.0		35.5	25.1					665.3	617.9	193.6	20.8
Iridium(III) chloride	Cl_3Ir	-245.6											
Iridium(VI) fluoride	F_6Ir	-579.7	-461.6	247.7						-544.0	-460.0	357.8	121.1
Iridium(IV) oxide	IrO_2	-274.1			57.3								
Iridium(III) sulfide	Ir_2S_3	-234.0											
Iridium(IV) sulfide	IrS_2	-138.0											
Iron	Fe	0.0		27.3	25.1					416.3	370.7	180.5	25.7
Iron(II) bromide	Br_2Fe	-249.8	-238.1	140.6									
Iron(III) bromide	Br_3Fe	-268.2											
Iron carbide	CFe_3	25.1	20.1	104.6	105.9								
Iron(II) carbonate	$CFeO_3$	-740.6	-666.7	92.9	82.1								
Iron(II) chloride	Cl_2Fe	-341.8	-302.3	118.0	76.7								
Iron(III) chloride	Cl_3Fe	-399.5	-334.0	142.3	96.7								
Iron disulfide	FeS_2	-178.2	-166.9	52.9	62.2								
Iron(II) fluoride	F_2Fe	-711.3	-668.6	87.0	68.1								
Iron(II) iodide	FeI_2	-113.0											
Iron(III) iodide	FeI_3									71.0			
Iron(II) molybdate	$FeMoO_4$	-1075.0	-975.0	129.3	118.5								
Iron(II) orthosilicate	Fe_2O_4Si	-1479.9	-1379.0	145.2	132.9								
Iron(II) oxide	FeO	-272.0											
Iron(II,III) oxide	Fe_3O_4	-1118.4	-1015.4	146.4	143.4								
Iron(III) oxide	Fe_2O_3	-824.2	-742.2	87.4	103.9								
Iron pentacarbonyl	C_5FeO_5					-774.0	-705.3	338.1	240.6				
Iron(II) sulfate	FeO_4S	-928.4	-820.8	107.5	100.6								
Iron(II) sulfide	FeS	-100.0	-100.4	60.3	50.5								
Iron(II) tungstate	FeO_4W	-1155.0	-1054.0	131.8	114.6								
Isobutanal	C_4H_8O					-247.3				-215.7			
Isobutane	C_4H_{10}					-154.2				-134.2			
Isobutene	C_4H_8					-37.5				-16.9			
Isobutyl acetate	$C_6H_{12}O_2$								233.8				
Isobutylamine	$C_4H_{11}N$					-132.6			183.2	-98.7			
Isobutylbenzene	$C_{10}H_{14}$					-69.8				-21.9			
Isobutyl 2-chloropropanoate	$C_7H_{13}ClO_2$					-603.1				-549.6			
Isobutyl 3-chloropropanoate	$C_7H_{13}ClO_2$					-572.6				-517.3			
Isobutyl isobutanoate	$C_8H_{16}O_2$					-587.4				-542.9			
Isobutyl pentanoate	$C_9H_{18}O_2$					-620.0				-568.6			
Isobutyl trichloroacetate	$C_6H_9Cl_3O_2$					-553.4				-500.2			
Isocyanic acid (HNCO)	CHNO											238.0	44.9
Isocyanomethane	C_2H_3N					130.8	159.5	159.0		163.5	165.7	246.9	52.9
DL-Isoleucine	$C_6H_{13}NO_2$	-635.3											
L-Isoleucine	$C_6H_{13}NO_2$	-637.8											
Isopentane	C_5H_{12}					-178.4		260.4	164.8	-153.6			
Isopentyl acetate	$C_7H_{14}O_2$								248.5				
Isopentyl trichloroacetate	$C_7H_{11}Cl_3O_2$					-580.9				-523.1			
Isophorone	$C_9H_{14}O$								253.5				
Isophthalic acid	$C_8H_6O_4$	-803.0								-696.3			
Isopropyl acetate	$C_5H_{10}O_2$					-518.9			199.4	-481.6			
Isopropylamine	C_3H_9N					-112.3		218.3	163.8	-83.7	32.2	312.2	97.5
Isopropylbenzene	C_9H_{12}					-41.1			210.7	4.0			
Isopropylbenzene hydroperoxide	$C_9H_{12}O_2$					-148.3				-78.4			
1-Isopropyl-2-methylbenzene	$C_{10}H_{14}$					-73.3							
1-Isopropyl-3-methylbenzene	$C_{10}H_{14}$					-78.6							
1-Isopropyl-4-methylbenzene	$C_{10}H_{14}$					-78.0			236.4				
Isopropyl methyl ether	$C_4H_{10}O$					-278.8		253.8	161.9	-252.0			
Isopropyl methyl sulfide	$C_4H_{10}S$					-124.7		263.1	172.4	-90.5			
Isopropyl nitrate	$C_3H_7NO_3$					-229.7				-191.0			
Isopropyl pentanoate	$C_8H_{16}O_2$					-592.2				-544.9			
2-Isopropylphenol	$C_9H_{12}O$					-233.7				-182.2			
3-Isopropylphenol	$C_9H_{12}O$					-252.5				-196.0			
4-Isopropylphenol	$C_9H_{12}O$	-270.0								-175.3			
Isoquinoline	C_9H_7N					144.3		216.0	196.2	204.6			
Isothiocyanic acid	CHNS									127.6	113.0	247.8	46.9
Isoxazole	C_3H_3NO					42.1				78.6			
Ketene	C_2H_2O					-67.9				-47.5	-48.3	247.6	51.8

Name	Mol. form.	$\Delta_f H°$(c)/ kJ mol^{-1}	$\Delta_f G°$(c)/ kJ mol^{-1}	$S°$(c)/J mol^{-1} K^{-1}	C_p(c)/J mol^{-1} K^{-1}	$\Delta_f H°$(l)/ kJ mol^{-1}	$\Delta_f G°$(l)/ kJ mol^{-1}	$S°$(l)/J mol^{-1} K^{-1}	C_p(l)/J mol^{-1} K^{-1}	$\Delta_f H°$(g)/ kJ mol^{-1}	$\Delta_f G°$(g)/ kJ mol^{-1}	$S°$(g)/J mol^{-1} K^{-1}	C_p(g)/J mol^{-1} K^{-1}
Krypton	Kr									0.0		164.1	20.8
β-D-Lactose	$C_{12}H_{22}O_{11}$	-2236.7											
α-Lactose monohydrate	$C_{12}H_{24}O_{12}$	-2484.1											
Lanthanum	La	0.0		56.9	27.1					431.0	393.6	182.4	22.8
Lanthanum chloride	Cl_3La	-1072.2			108.8								
Lanthanum iodide	I_3La	-668.9											
Lanthanum monosulfide	LaS	-456.0	-451.5	73.2	59.0								
Lanthanum oxide	La_2O_3	-1793.7	-1705.8	127.3	108.8								
Lawrencium	Lr	0.0											
Lead	Pb	0.0		64.8	26.8					195.2	162.2	175.4	20.8
Lead(II) bromide	Br_2Pb	-278.7	-261.9	161.5	80.1								
Lead(II) carbonate	CO_3Pb	-699.1	-625.5	131.0	87.4								
Lead(II) chloride	Cl_2Pb	-359.4	-314.1	136.0									
Lead(IV) chloride	Cl_4Pb					-329.3							
Lead(II) chromate	CrO_4Pb	-930.9											
Lead(II) fluoride	F_2Pb	-664.0	-617.1	110.5									
Lead(IV) fluoride	F_4Pb	-941.8											
Lead(II) iodide	I_2Pb	-175.5	-173.6	174.9	77.4								
Lead(II) metasilicate	O_3PbSi	-1145.7	-1062.1	109.6	90.0								
Lead(II) molybdate	MoO_4Pb	-1051.9	-951.4	166.1	119.7								
Lead(II) nitrate	N_2O_6Pb	-451.9											
Lead(II) orthosilicate	O_4Pb_2Si	-1363.1	-1252.6	186.6	137.2								
Lead(II) oxalate	C_2O_4Pb	-851.4	-750.1	146.0	105.4								
Lead(II,II,IV) oxide	O_4Pb_3	-718.4	-601.2	211.3	146.9								
Lead(IV) oxide	O_2Pb	-277.4	-217.3	68.6	64.6								
Lead(II) oxide (litharge)	OPb	-219.0	-188.9	66.5	45.8								
Lead(II) oxide (massicot)	OPb	-217.3	-187.9	68.7	45.8								
Lead(II) selenate	O_4PbSe	-609.2	-504.9	167.8									
Lead(II) selenide	PbSe	-102.9	-101.7	102.5	50.2								
Lead(II) sulfate	O_4PbS	-920.0	-813.0	148.5	103.2								
Lead(II) sulfide	PbS	-100.4	-98.7	91.2	49.5								
Lead(II) sulfite	O_3PbS	-669.9											
Lead(II) telluride	PbTe	-70.7	-69.5	110.0	50.5								
DL-Leucine	$C_6H_{13}NO_2$	-640.6											
D-Leucine	$C_6H_{13}NO_2$	-637.3											
L-Leucine	$C_6H_{13}NO_2$	-637.4			200.1					-486.8			
d-Limonene	$C_{10}H_{16}$					-54.5			249.0				
Lithium	Li	0.0		29.1	24.9					159.3	126.6	138.8	20.8
Lithium aluminum hydride	AlH_4Li	-116.3	-44.7	78.7	83.2								
Lithium amide	H_2LiN	-179.5											
Lithium borohydride	BH_4Li	-190.8	-125.0	75.9	82.6								
Lithium bromide	BrLi	-351.2	-342.0	74.3									
Lithium carbonate	CLi_2O_3	-1215.9	-1132.1	90.4	99.1								
Lithium chloride	ClLi	-408.6	-384.4	59.3	48.0								
Lithium fluoride	FLi	-616.0	-587.7	35.7	41.6								
Lithium hydride	HLi	-90.5	-68.3	20.0	27.9								
Lithium hydroxide	HLiO	-487.5	-441.5	42.8	49.6					-229.0	-234.2	214.4	46.0
Lithium iodide	ILi	-270.4	-270.3	86.8	51.0								
Lithium metaborate	$BLiO_2$	-1032.2	-976.1	51.5	59.8								
Lithium metasilicate	Li_2O_3Si	-1648.1	-1557.2	79.8	99.1								
Lithium nitrate	$LiNO_3$	-483.1	-381.1	90.0									
Lithium nitrite	$LiNO_2$	-372.4	-302.0	96.0									
Lithium oxide	Li_2O	-597.9	-561.2	37.6	54.1								
Lithium perchlorate	$ClLiO_4$	-381.0											
Lithium peroxide	Li_2O_2	-634.3											
Lithium phosphate	Li_3O_4P	-2095.8											
Lithium sulfate	Li_2O_4S	-1436.5	-1321.7	115.1	117.6								
Lithium sulfide	Li_2S	-441.4											
Lutetium	Lu	0.0		51.0	26.9					427.6	387.8	184.8	20.9
Lutetium chloride	Cl_3Lu	-945.6								-649.0			
Lutetium iodide	I_3Lu	-548.0											
Lutetium oxide	Lu_2O_3	-1878.2	-1789.0	110.0	101.8								
DL-Lysine	$C_6H_{14}N_2O_2$	-678.7											
Magnesium	Mg	0.0		32.7	24.9					147.1	112.5	148.6	20.8
Magnesium bromide	Br_2Mg	-524.3	-503.8	117.2									
Magnesium carbonate	$CMgO_3$	-1095.8	-1012.1	65.7	75.5								

Name	Mol. form.	$\Delta_f H°$(c)/ kJ mol^{-1}	$\Delta_f G°$(c)/ kJ mol^{-1}	$S°$(c)/J mol^{-1} K^{-1}	C_p(c)/J mol^{-1} K^{-1}	$\Delta_f H°$(l)/ kJ mol^{-1}	$\Delta_f G°$(l)/ kJ mol^{-1}	$S°$(l)/J mol^{-1} K^{-1}	C_p(l)/J mol^{-1} K^{-1}	$\Delta_f H°$(g)/ kJ mol^{-1}	$\Delta_f G°$(g)/ kJ mol^{-1}	$S°$(g)/J mol^{-1} K^{-1}	C_p(g)/J mol^{-1} K^{-1}
Magnesium chloride	Cl_2Mg	-641.3	-591.8	89.6	71.4								
Magnesium fluoride	F_2Mg	-1124.2	-1071.1	57.2	61.6								
Magnesium hydride	H_2Mg	-75.3	-35.9	31.1	35.4								
Magnesium hydroxide	H_2MgO_2	-924.5	-833.5	63.2	77.0								
Magnesium iodide	I_2Mg	-364.0	-358.2	129.7									
Magnesium nitrate	MgN_2O_6	-790.7	-589.4	164.0	141.9								
Magnesium orthosilicate	Mg_2O_4Si	-2174.0	-2055.1	95.1	118.5								
Magnesium oxalate	C_2MgO_4	-1269.0											
Magnesium oxide	MgO	-601.6	-569.3	27.0	37.2								
Magnesium selenate	MgO_4Se	-968.5											
Magnesium sulfate	MgO_4S	-1284.9	-1170.6	91.6	96.5								
Magnesium sulfide	MgS	-346.0	-341.8	50.3	45.6								
Maleic acid	$C_4H_4O_4$	-789.4		160.8	137.0					-679.4			
Maleic anhydride	$C_4H_2O_3$	-469.8								-398.3			
Malononitrile	$C_3H_2N_2$	186.4								265.5			
Manganese	Mn	0.0		32.0	26.3					280.7	238.5	173.7	20.8
Manganese(II) bromide	Br_2Mn	-384.9											
Manganese(II) carbonate	$CMnO_3$	-894.1	-816.7	85.8	81.5								
Manganese(II) chloride	Cl_2Mn	-481.3	-440.5	118.2	72.9								
Manganese(II) metasilicate	MnO_3Si	-1320.9	-1240.5	89.1	86.4								
Manganese(II) nitrate	MnN_2O_6	-576.3											
Manganese(II) orthosilicate	Mn_2O_4Si	-1730.5	-1632.1	163.2	129.9								
Manganese(II) oxide	MnO	-385.2	-362.9	59.7	45.4								
Manganese(II,III) oxide	Mn_3O_4	-1387.8	-1283.2	155.6	139.7								
Manganese(III) oxide	Mn_2O_3	-959.0	-881.1	110.5	107.7								
Manganese(IV) oxide	MnO_2	-520.0	-465.1	53.1	54.1								
Manganese(II) selenide	MnSe	-106.7	-111.7	90.8	51.0								
Manganese(II) sulfide (α form)	MnS	-214.2	-218.4	78.2	50.0								
D-Mannitol	$C_6H_{14}O_6$					-1314.5							
D-Mannose	$C_6H_{12}O_6$	-1263.0											
Mendelevium	Md	0.0											
Mercapto	HS									142.7	113.3	195.7	32.3
Mercury	Hg					0.0		75.9	28.0	61.4	31.8	175.0	20.8
Mercury(I) bromide	Br_2Hg_2	-206.9	-181.1	218.0									
Mercury(II) bromide	Br_2Hg	-170.7	-153.1	172.0									
Mercury(I) carbonate	CHg_2O_3	-553.5	-468.1	180.0									
Mercury(I) chloride	Cl_2Hg_2	-265.4	-210.7	191.6									
Mercury(II) chloride	Cl_2Hg	-224.3	-178.6	146.0									
Mercury(I) iodide	Hg_2I_2	-121.3	-111.0	233.5									
Mercury(II) iodide (red)	HgI_2	-105.4	-101.7	180.0									
Mercury(II) oxalate	C_2HgO_4	-678.2											
Mercury(II) oxide	HgO	-90.8	-58.5	70.3	44.1								
Mercury(I) sulfate	Hg_2O_4S	-743.1	-625.8	200.7	132.0								
Mercury(II) sulfate	HgO_4S	-707.5											
Mercury(II) sulfide (red)	HgS	-58.2	-50.6	82.4	48.4								
Mercury(II) telluride	HgTe	-42.0											
Mesityl oxide	$C_6H_{10}O$								212.5				
Metaboric acid (β form)	BHO_2	-794.3	-723.4	38.0						-561.9	-551.0	240.1	42.2
Metaphosphoric acid	HO_3P	-948.5											
Metasilicic acid	H_2O_3Si	-1188.7	-1092.4	134.0									
Methacrylic acid	$C_4H_6O_2$								161.1				
Methane	CH_4									-74.6	-50.5	186.3	35.7
Methanethiol	CH_4S					-46.7	-7.7	169.2	90.5	-22.9	-9.3	255.2	50.3
Methanol	CH_4O					-239.2	-166.6	126.8	81.1	-201.0	-162.3	239.9	44.1
L-Methionine	$C_5H_{11}NO_2S$	-577.5								-413.5			
2-Methoxyethanol	$C_3H_8O_2$								171.1				
2-Methoxyethyl acetate	$C_5H_{10}O_3$								310.0				
Methyl	CH_3									145.7	147.9	194.2	38.7
Methyl acetate	$C_3H_6O_2$					-445.9			141.9	-413.3		324.4	86.0
Methyl acetoacetate	$C_5H_8O_3$					-623.2							
Methyl acrylate	$C_4H_6O_2$					-362.2		239.5	158.8	-333.0			
2-Methylacrylonitrile	C_4H_5N								126.3				
Methylamine	CH_5N					-47.3	35.7	150.2	102.1	-22.5	32.7	242.9	50.1
Methylamine hydrochloride	CH_6ClN	-298.1											
2-Methylaniline	C_7H_9N					-6.3				56.4	167.6	351.0	130.2
3-Methylaniline	C_7H_9N					-8.1				54.6	165.4	352.5	125.5

Name	Mol. form.	$\Delta_f H°(c)$/ kJ mol^{-1}	$\Delta_f G°(c)$/ kJ mol^{-1}	$S°(c)$/J mol^{-1} K^{-1}	$C_p(c)$/J mol^{-1} K^{-1}	$\Delta_f H°(l)$/ kJ mol^{-1}	$\Delta_f G°(l)$/ kJ mol^{-1}	$S°(l)$/J mol^{-1} K^{-1}	$C_p(l)$/J mol^{-1} K^{-1}	$\Delta_f H°(g)$/ kJ mol^{-1}	$\Delta_f G°(g)$/ kJ mol^{-1}	$S°(g)$/J mol^{-1} K^{-1}	$C_p(g)$/J mol^{-1} K^{-1}
4-Methylaniline	C_7H_9N	-23.5								55.3	167.7	347.0	126.2
N-Methylaniline	C_7H_9N								207.1				
Methyl benzoate	$C_8H_8O_2$					-343.5			221.3	-287.9			
1-Methylbicyclo[3.1.0]hexane	C_7H_{12}					-33.2				1.7			
3-Methyl-1,2-butadiene	C_5H_8					101.2							
2-Methyl-1,3-butadiene	C_5H_8					48.2		229.3	152.6	75.5			
2-Methyl-1-butanethiol, (+)	$C_5H_{12}S$					-154.4				-114.9			
3-Methyl-1-butanethiol	$C_5H_{12}S$					-154.4				-114.9			
2-Methyl-2-butanethiol	$C_5H_{12}S$					-162.8		290.1	198.1	-127.1			
3-Methyl-2-butanethiol	$C_5H_{12}S$					-158.8				-121.3			
Methyl butanoate	$C_5H_{10}O_2$								198.2				
2-Methylbutanoic acid	$C_5H_{10}O_2$					-554.5							
3-Methylbutanoic acid	$C_5H_{10}O_2$					-561.6				-510.0			
2-Methyl-1-butanol, (±)-	$C_5H_{12}O$					-356.6				-301.4			
3-Methyl-1-butanol	$C_5H_{12}O$					-356.4				-300.7			
2-Methyl-2-butanol	$C_5H_{12}O$					-379.5			247.1	-329.3			
3-Methyl-2-butanol, (±)-	$C_5H_{12}O$					-366.6				-313.5			
3-Methyl-2-butanone	$C_5H_{10}O$					-299.5		268.5	179.9	-262.6			
2-Methyl-1-butene	C_5H_{10}					-61.1		254.0	157.2	-35.2			
3-Methyl-1-butene	C_5H_{10}					-51.5		253.3	156.1	-27.5			
2-Methyl-2-butene	C_5H_{10}					-68.6		251.0	152.8	-41.7			
Methyl *trans*-2-butenoate	$C_5H_8O_2$					-382.9				-341.9			
3-Methylbutyl 2-chloropropanoate	$C_8H_{15}ClO_2$					-627.3				-575.0			
3-Methylbutyl 3-chloropropanoate	$C_8H_{15}ClO_2$					-593.4				-539.4			
Methyl *tert*-butyl ether	$C_5H_{12}O$					-313.6		265.3	187.5	-283.7			
9-Methyl-9*H*-carbazole	$C_{13}H_{11}N$	105.5								201.0			
Methyl chloroacetate	$C_3H_5ClO_2$					-487.0				-444.0			
Methylcyclobutane	C_5H_{10}					-44.5							
Methyl cyclobutanecarboxylate	$C_6H_{10}O_2$					-395.0				-350.2			
Methylcyclohexane	C_7H_{14}					-190.1			184.8	-154.7			
cis-2-Methylcyclohexanol	$C_7H_{14}O$					-390.2				-327.0			
trans-2-Methylcyclohexanol, (±)-	$C_7H_{14}O$					-415.7				-352.5			
cis-3-Methylcyclohexanol, (±)-	$C_7H_{14}O$					-416.1				-350.9			
trans-3-Methylcyclohexanol, (±)-	$C_7H_{14}O$					-394.4				-329.1			
cis-4-Methylcyclohexanol	$C_7H_{14}O$					-413.2				-347.5			
trans-4-Methylcyclohexanol	$C_7H_{14}O$					-433.3				-367.2			
Methylcyclopentane	C_6H_{12}					-137.9				-106.2			
cis-2-Methylcyclopentanol	$C_6H_{12}O$					-345.5							
2-Methylcyclopentanone	$C_6H_{10}O$					-265.2							
1-Methylcyclopentene	C_6H_{10}					-36.4				-3.8			
3-Methylcyclopentene	C_6H_{10}					-23.7				7.4			
4-Methylcyclopentene	C_6H_{10}					-17.6				14.6			
Methylcyclopropane	C_4H_8					1.7							
Methyl decanoate	$C_{11}H_{22}O_2$					-640.5				-573.8			
Methyl 2,2-dimethylpropanoate	$C_6H_{12}O_2$					-530.0			257.9	-491.2			
1-Methyl-2,4-dinitrobenzene	$C_7H_6N_2O_4$	-66.4								33.2			
4-Methyl-1,3-dioxane	$C_5H_{10}O_2$					-416.1				-376.9			
2-Methyl-1,3-dioxolane	$C_4H_8O_2$					-386.9				-352.0			
Methyl dodecanoate	$C_{13}H_{26}O_2$					-693.0				-614.9			
Methylene	CH_2									390.4	372.9	194.9	33.8
Methylenecyclobutane	C_5H_8					93.8				121.6			
Methylenecyclohexane	C_7H_{12}					-61.3				-25.2			
2-Methylenecyclohexanol	$C_7H_{12}O$					-277.6							
N-Methylformamide	C_2H_5NO								123.8				
Methyl formate	$C_2H_4O_2$					-386.1			119.1	-357.4		285.3	64.4
3-Methyl-2,5-furandione	$C_5H_4O_3$					-504.5				-447.2			
Methyl α-D-glucopyranoside	$C_7H_{14}O_6$	-1233.3											
2-Methylheptane	C_8H_{18}					-255.0		356.4	252.0	-215.3			
3-Methylheptane	C_8H_{18}					-252.3		362.6	250.2	-212.5			
4-Methylheptane	C_8H_{18}					-251.6			251.1	-211.9			
Methyl heptanoate	$C_8H_{16}O_2$					-567.1			285.1	-515.5			
2-Methylhexane	C_7H_{16}					-229.5		323.3	222.9	-194.5			
3-Methylhexane	C_7H_{16}					-226.4				-191.3			
Methyl hexanoate	$C_7H_{14}O_2$					-540.2				-492.2			
5-Methyl-1-hexene	C_7H_{14}					-100.0				-65.7			
cis-3-Methyl-3-hexene	C_7H_{14}					-115.9				-79.4			

Name	Mol. form.	$\Delta_f H°$(c)/ kJ mol^{-1}	$\Delta_f G°$(c)/ kJ mol^{-1}	$S°$(c)/J mol^{-1} K^{-1}	C_p(c)/J mol^{-1} K^{-1}	$\Delta_f H°$(l)/ kJ mol^{-1}	$\Delta_f G°$(l)/ kJ mol^{-1}	$S°$(l)/J mol^{-1} K^{-1}	C_p(l)/J mol^{-1} K^{-1}	$\Delta_f H°$(g)/ kJ mol^{-1}	$\Delta_f G°$(g)/ kJ mol^{-1}	$S°$(g)/J mol^{-1} K^{-1}	C_p(g)/J mol^{-1} K^{-1}
trans-3-Methyl-3-hexene	C_7H_{14}					-112.7				-76.8			
Methyl 2-hexynoate	$C_7H_{10}O_2$					-242.7							
Methylhydrazine	CH_6N_2					54.2	180.0	165.9	134.9	94.7	187.0	278.8	71.1
Methylidyne	CH									595.8			
Methyl isocyanate	C_2H_3NO					-92.0							
Methyl isothiocyanate	C_2H_3NS	79.4											
Methyl methacrylate	$C_5H_8O_2$								191.2				
1-Methylnaphthalene	$C_{11}H_{10}$					56.3		254.8	224.4				
2-Methylnaphthalene	$C_{11}H_{10}$	44.9		220.0	196.0					106.7			
Methyl nitrate	CH_3NO_3					-156.3	-43.4	217.1	157.3	-122.0		305.8	76.6
Methyl nitrite	CH_3NO_2									-66.1			
Methyl nitroacetate	$C_3H_5NO_4$					-464.0							
2-Methyl-2-nitropropane	$C_4H_9NO_2$					-217.2				-177.1			
2-Methyl-2-nitro-1-propanol	$C_4H_9NO_3$	-410.1											
2-Methylnonane	$C_{10}H_{22}$					-309.8		420.1	313.3	-260.2			
5-Methylnonane	$C_{10}H_{22}$					-307.9		423.8	314.4	-258.6			
Methyl nonanoate	$C_{10}H_{20}O_2$					-616.2				-554.2			
Methyl *cis*-9-octadecenoate	$C_{19}H_{36}O_2$					-734.5				-649.9			
Methyl *trans*-9-octadecenoate	$C_{19}H_{36}O_2$					-737.0							
Methyl octanoate	$C_9H_{18}O_2$					-590.3				-533.9			
2-Methyl-2-oxazoline	C_4H_7NO					-169.5				-130.5			
Methyloxirane	C_3H_6O					-123.0		196.5	120.4	-94.7		286.9	72.6
Methyl pentadecanoate	$C_{16}H_{32}O_2$					-771.0				-680.0			
2-Methylpentane	C_6H_{14}					-204.6		290.6	193.7	-174.6			
3-Methylpentane	C_6H_{14}					-202.4		292.5	190.7	-171.9			
2-Methyl-2,4-pentanediol	$C_6H_{14}O_2$								336.0				
2-Methyl-2-pentanethiol	$C_6H_{14}S$					-188.3				-148.3			
Methyl pentanoate	$C_6H_{12}O_2$					-514.2			229.3	-471.1			
2-Methyl-1-pentanol	$C_6H_{14}O$								248.0				
3-Methyl-2-pentanol	$C_6H_{14}O$								275.9				
4-Methyl-2-pentanol	$C_6H_{14}O$					-394.7			273.0				
2-Methyl-3-pentanol	$C_6H_{14}O$					-396.4							
3-Methyl-3-pentanol	$C_6H_{14}O$								293.4				
4-Methyl-2-pentanone	$C_6H_{12}O$								213.3				
2-Methyl-3-pentanone	$C_6H_{12}O$					-325.9				-286.0			
2-Methyl-1-pentene	C_6H_{12}					-90.0				-59.4			
3-Methyl-1-pentene	C_6H_{12}					-78.2				-49.5			
4-Methyl-1-pentene	C_6H_{12}					-80.0				-51.3			
2-Methyl-2-pentene	C_6H_{12}					-98.5				-66.9			
3-Methyl-*cis*-2-pentene	C_6H_{12}					-94.5				-62.3			
3-Methyl-*trans*-2-pentene	C_6H_{12}					-94.6				-63.1			
4-Methyl-*cis*-2-pentene	C_6H_{12}					-87.0				-57.5			
4-Methyl-*trans*-2-pentene	C_6H_{12}					-91.6				-61.5			
Methyl pentyl sulfide	$C_6H_{14}S$					-167.1				-121.8			
2-Methylpiperidine, (±)-	$C_6H_{13}N$					-124.9				-84.4			
1-Methyl-2-piperidinone	$C_6H_{11}NO$					-293.0							
2-Methylpropanamide	C_4H_9NO	-368.6								-282.6			
N-Methylpropanamide	C_4H_9NO								179.0				
2-Methyl-1,2-propanediamine	$C_4H_{12}N_2$					-133.9				-90.3			
2-Methyl-1,2-propanediol	$C_4H_{10}O_2$					-539.7							
2-Methylpropanenitrile	C_4H_7N					-13.8				23.4			
2-Methyl-1-propanethiol	$C_4H_{10}S$					-132.0				-97.3			
2-Methyl-2-propanethiol	$C_4H_{10}S$					-140.5				-109.6			
Methyl propanoate	$C_4H_8O_2$								171.2				
2-Methylpropanoic acid	$C_4H_8O_2$								173.0				
2-Methyl-1-propanol	$C_4H_{10}O$					-334.7		214.7	181.5	-283.8			
2-Methyl-2-propanol	$C_4H_{10}O$					-359.2		193.3	218.6	-312.5		326.7	113.6
Methyl propyl ether	$C_4H_{10}O$					-266.0		262.9	165.4	-238.1			
Methyl propyl sulfide	$C_4H_{10}S$					-118.5		272.5	171.6	-82.2			
2-Methylpyridine	C_6H_7N					56.7			158.6	99.2			
3-Methylpyridine	C_6H_7N					61.9		216.3	158.7	106.4			
4-Methylpyridine	C_6H_7N					59.2		209.1	159.0	103.8			
1-Methylpyrrole	C_5H_7N					62.4				103.1			
2-Methylpyrrole	C_5H_7N					23.3				74.0			
3-Methylpyrrole	C_5H_7N					20.5				70.2			
N-Methyl-2-pyrrolidinone	C_5H_9NO					-262.2			307.8				

Name	Mol. form.	$\Delta_f H°$(c)/ kJ mol^{-1}	$\Delta_f G°$(c)/ kJ mol^{-1}	$S°$(c)/J mol^{-1} K^{-1}	C_p(c)/J mol^{-1} K^{-1}	$\Delta_f H°$(l)/ kJ mol^{-1}	$\Delta_f G°$(l)/ kJ mol^{-1}	$S°$(l)/J mol^{-1} K^{-1}	C_p(l)/J mol^{-1} K^{-1}	$\Delta_f H°$(g)/ kJ mol^{-1}	$\Delta_f G°$(g)/ kJ mol^{-1}	$S°$(g)/J mol^{-1} K^{-1}	C_p(g)/J mol^{-1} K^{-1}
Methyl salicylate	$C_8H_8O_3$								249.0				
Methylsilane	CH_6Si											256.5	65.9
Methyl tetradecanoate	$C_{15}H_{30}O_2$					-743.9				-656.9			
4-Methylthiazole	C_4H_5NS					67.9				111.8			
Methylthiirane	C_3H_6S					11.3				45.8			
2-Methylthiophene	C_5H_6S					44.6		218.5	149.8	83.5			
3-Methylthiophene	C_5H_6S					43.1				82.5			
Methyl tridecanoate	$C_{14}H_{28}O_2$					-717.9				-635.3			
Methyl undecanoate	$C_{12}H_{24}O_2$					-665.2				-593.8			
N-Methylurea	$C_2H_6N_2O$	-332.8											
Molybdenum	Mo	0.0		28.7	24.1					658.1	612.5	182.0	20.8
Molybdenum carbonyl	C_6MoO_6	-982.8	-877.7	325.9	242.3					-912.1	-856.0	490.0	205.0
Molybdenum(VI) fluoride	F_6Mo					-1585.5	-1473.0	259.7	169.8	-1557.7	-1472.2	350.5	120.6
Molybdenum(IV) oxide	MoO_2	-588.9	-533.0	46.3	56.0								
Molybdenum(VI) oxide	MoO_3	-745.1	-668.0	77.7	75.0								
Molybdenum silicide (Mo_3Si)	Mo_3Si	-125.2	-125.7	106.3	93.1								
Molybdenum(IV) sulfide	MoS_2	-235.1	-225.9	62.6	63.6								
Morpholine	C_4H_9NO								164.8				
β-Myrcene	$C_{10}H_{16}$					14.5							
Naphthalene	$C_{10}H_8$	78.5	201.6	167.4	165.7					150.6	224.1	333.1	131.9
1-Naphthaleneacetic acid	$C_{12}H_{10}O_2$	-359.2											
2-Naphthaleneacetic acid	$C_{12}H_{10}O_2$	-371.9											
1-Naphthalenecarboxylic acid	$C_{11}H_8O_2$	-333.5								-223.1			
2-Naphthalenecarboxylic acid	$C_{11}H_8O_2$	-346.1								-232.5			
1-Naphthol	$C_{10}H_8O$	-121.5			166.9					-30.4			149.4
2-Naphthol	$C_{10}H_8O$	-124.1		179.0	172.8					-29.9		366.6	147.8
1-Naphthylamine	$C_{10}H_9N$	67.8								132.8			
2-Naphthylamine	$C_{10}H_9N$	60.2								134.3			
Neodymium	Nd	0.0		71.5	27.5					327.6	292.4	189.4	22.1
Neodymium(III) chloride	Cl_3Nd	-1041.0			113.0								
Neodymium(III) fluoride	F_3Nd	-1657.0											
Neodymium(III) oxide	Nd_2O_3	-1807.9	-1720.8	158.6	111.3								
Neon	Ne									0.0		146.3	20.8
Neopentane	C_5H_{12}					-190.2				-168.0			
Nickel	Ni	0.0		29.9	26.1					429.7	384.5	182.2	23.4
Nickel(II) bromide	Br_2Ni	-212.1											
Nickel carbonyl [$Ni(CO)_4$]	C_4NiO_4					-633.0	-588.2	313.4	204.6	-602.9	-587.2	410.6	145.2
Nickel(II) chloride	Cl_2Ni	-305.3	-259.0	97.7	71.7								
Nickel(II) fluoride	F_2Ni	-651.4	-604.1	73.6	64.1								
Nickel(II) hydroxide	H_2NiO_2	-529.7	-447.2	88.0									
Nickel(II) iodide	I_2Ni	-78.2											
Nickel(III) oxide	Ni_2O_3	-489.5											
Nickel(II) sulfate	NiO_4S	-872.9	-759.7	92.0	138.0								
Nickel(II) sulfide	NiS	-82.0	-79.5	53.0	47.1								
Niobium	Nb	0.0		36.4	24.6					725.9	681.1	186.3	30.2
Niobium(V) chloride	Cl_5Nb	-797.5	-683.2	210.5	148.1					-703.7	-646.0	400.6	120.8
Niobium(V) fluoride	F_5Nb	-1813.8	-1699.0	160.2	134.7					-1739.7	-1673.6	321.9	97.1
Niobium(II) oxide	NbO	-405.8	-378.6	48.1	41.3								
Niobium(IV) oxide	NbO_2	-796.2	-740.5	54.5	57.5								
Niobium(V) oxide	Nb_2O_5	-1899.5	-1766.0	137.2	132.1								
Nitramide	$H_2N_2O_2$	-89.5											
Nitric acid	HNO_3					-174.1	-80.7	155.6	109.9	-133.9	-73.5	266.9	54.1
Nitric oxide	NO									91.3	87.6	210.8	29.9
Nitrilotriacetic acid	$C_6H_9NO_6$	-1311.9											
Nitroacetone	$C_3H_5NO_3$					-278.6							
2-Nitroaniline	$C_6H_6N_2O_2$	-26.1			166.0	-9.4				63.8			
3-Nitroaniline	$C_6H_6N_2O_2$	-38.3			158.8	-14.4				58.4			
4-Nitroaniline	$C_6H_6N_2O_2$	-42.0			167.0	-20.7				58.8			
Nitrobenzene	$C_6H_5NO_2$					12.5			185.8	68.5		348.8	120.4
2-Nitrobenzoic acid	$C_7H_5NO_4$	-378.8											
3-Nitrobenzoic acid	$C_7H_5NO_4$	-394.7											
4-Nitrobenzoic acid	$C_7H_5NO_4$	-392.2											
1-Nitrobutane	$C_4H_9NO_2$					-192.5				-143.9		369.9	115.1
3-Nitro-2-butanol	$C_4H_9NO_3$					-390.0							
1-Nitrodecane	$C_{10}H_{21}NO_2$					-351.5							
N-Nitrodiethylamine	$C_4H_{10}N_2O_2$					-106.2				-53.0			

Name	Mol. form.	$\Delta_f H°$(c)/ kJ mol^{-1}	$\Delta_f G°$(c)/ kJ mol^{-1}	$S°$(c)/J mol^{-1} K^{-1}	C_p(c)/J mol^{-1} K^{-1}	$\Delta_f H°$(l)/ kJ mol^{-1}	$\Delta_f G°$(l)/ kJ mol^{-1}	$S°$(l)/J mol^{-1} K^{-1}	C_p(l)/J mol^{-1} K^{-1}	$\Delta_f H°$(g)/ kJ mol^{-1}	$\Delta_f G°$(g)/ kJ mol^{-1}	$S°$(g)/J mol^{-1} K^{-1}	C_p(g)/J mol^{-1} K^{-1}
Nitroethane	$C_2H_5NO_2$					-143.9			134.4	-103.8		320.5	79.0
2-Nitroethanol	$C_2H_5NO_3$					-350.7							
Nitroethene	$C_2H_3NO_2$									33.3		300.5	73.7
2-Nitrofuran	$C_4H_3NO_3$	-104.1								-28.8			
5-Nitro-2-furancarboxylic acid	$C_5H_3NO_5$	-516.8											
Nitrogen	N_2									0.0		191.6	29.1
Nitrogen (atomic)	N									472.7	455.5	153.3	20.8
Nitrogen dioxide	NO_2									33.2	51.3	240.1	37.2
Nitrogen pentoxide	N_2O_5	-43.1	113.9	178.2	143.1					13.3	117.1	355.7	95.3
Nitrogen tetroxide	N_2O_4					-19.5	97.5	209.2	142.7	11.1	99.8	304.4	79.2
Nitrogen trichloride	Cl_3N					230.0							
Nitrogen trifluoride	F_3N									-132.1	-90.6	260.8	53.4
Nitrogen trioxide	N_2O_3					50.3				86.6	142.4	314.7	72.7
Nitroguanidine	$CH_4N_4O_2$	-92.4											
Nitromethane	CH_3NO_2					-112.6	-14.4	171.8	106.6	-80.8		282.9	55.5
(Nitromethyl)benzene	$C_7H_7NO_2$					-22.8				30.7			
1-Nitronaphthalene	$C_{10}H_7NO_2$	42.6								111.2			
1-Nitropentane	$C_5H_{11}NO_2$					-215.4				-164.4		390.9	137.1
2-Nitrophenol	$C_6H_5NO_3$	-202.4											
N-Nitropiperidine	$C_5H_{10}N_2O_2$					-93.0				-44.5			
1-Nitropropane	$C_3H_7NO_2$					-167.2				-124.3		350.0	104.1
2-Nitropropane	$C_3H_7NO_2$					-180.3			170.3	-138.9			
N-Nitrosodiphenylamine	$C_{12}H_{10}N_2O$	227.2											
N-Nitrosopiperidine	$C_5H_{10}N_2O$					-31.1				16.6			
Nitrosyl bromide	BrNO									82.2	82.4	273.7	45.5
Nitrosyl chloride	ClNO									51.7	66.1	261.7	44.7
Nitrosyl fluoride	FNO									-66.5	-51.0	248.1	41.3
2-Nitrotoluene	$C_7H_7NO_2$					-9.7							
3-Nitrotoluene	$C_7H_7NO_2$					-31.5							
4-Nitrotoluene	$C_7H_7NO_2$	-48.1			172.3					31.0			
Nitrous acid	HNO_2									-79.5	-46.0	254.1	45.6
Nitrous oxide	N_2O									81.6	103.7	220.0	38.6
Nitryl chloride	$ClNO_2$									12.6	54.4	272.2	53.2
Nitryl fluoride	FNO_2											260.4	49.8
Nobelium	No	0.0											
Nonaborane(15)	B_9H_{15}									158.4	357.5	364.9	187.0
Nonane	C_9H_{20}					-274.7			284.4	-228.2			
Nonanedioic acid	$C_9H_{16}O_4$	-1054.3											
1,9-Nonanediol	$C_9H_{20}O_2$	-657.6											
Nonanoic acid	$C_9H_{18}O_2$					-659.7			362.4	-577.3			
1-Nonanol	$C_9H_{20}O$					-453.4				-376.5			
2-Nonanone	$C_9H_{18}O$					-397.2				-340.7			
5-Nonanone	$C_9H_{18}O$					-398.2		401.4	303.6	-344.9			
1-Nonyne	C_9H_{16}					16.3				62.3			
L-Norleucine	$C_6H_{13}NO_2$	-639.1											
Octadecane	$C_{18}H_{38}$	-567.4		480.2	485.6					-414.6			
Octadecanoic acid	$C_{18}H_{36}O_2$	-947.7			501.5	-884.7				-781.2			
cis-9-Octadecenoic acid	$C_{18}H_{34}O_2$								577.0				
1,7-Octadiyne	C_8H_{10}					334.4							
Octanal	$C_8H_{16}O$									-291.9		365.4	
Octanamide	$C_8H_{17}NO$	-473.2								-362.7			
Octane	C_8H_{18}					-250.1			254.6	-208.5			
1,8-Octanediol	$C_8H_{18}O_2$	-626.6											
Octanenitrile	$C_8H_{15}N$					-107.3				-50.5			
Octanoic acid	$C_8H_{16}O_2$					-636.0			297.9	-554.3			
1-Octanol	$C_8H_{18}O$					-426.5			305.2	-355.6			
2-Octanol	$C_8H_{18}O$								330.1				
2-Octanone	$C_8H_{16}O$								273.3				
1-Octene	C_8H_{16}					-124.5			241.0	-81.3			
cis-2-Octene	C_8H_{16}					-135.7			239.0				
trans-2-Octene	C_8H_{16}					-135.7			239.0				
1-Octen-3-yne	C_8H_{12}					140.7							
Octyldimethylamine	$C_{10}H_{23}N$					-232.8							
Orthosilicic acid	H_4O_4Si	-1481.1	-1332.9	192.0									
Osmium	Os	0.0		32.6	24.7					791.0	745.0	192.6	20.8
Osmium(III) chloride	Cl_3Os	-190.4											

Name	Mol. form.	$\Delta_f H°$(c)/ kJ mol^{-1}	$\Delta_f G°$(c)/ kJ mol^{-1}	$S°$(c)/J mol^{-1} K^{-1}	C_p(c)/J mol^{-1} K^{-1}	$\Delta_f H°$(l)/ kJ mol^{-1}	$\Delta_f G°$(l)/ kJ mol^{-1}	$S°$(l)/J mol^{-1} K^{-1}	C_p(l)/J mol^{-1} K^{-1}	$\Delta_f H°$(g)/ kJ mol^{-1}	$\Delta_f G°$(g)/ kJ mol^{-1}	$S°$(g)/J mol^{-1} K^{-1}	C_p(g)/J mol^{-1} K^{-1}
Osmium(VI) fluoride	F_6Os			246.0								358.1	120.8
Osmium(VIII) oxide	O_4Os	-394.1	-304.9	143.9						-337.2	-292.8	293.8	74.1
Oxalic acid	$C_2H_2O_4$	-829.9		109.8	91.0					-731.8	-662.7	320.6	86.2
Oxalyl chloride	$C_2Cl_2O_2$					-367.6				-335.8			
Oxalyl dihydrazide	$C_2H_6N_4O_2$	-295.2											
Oxamic acid	$C_2H_3NO_3$	-661.2								-552.3			
Oxamide	$C_2H_4N_2O_2$	-504.4								-387.1			
Oxazole	C_3H_3NO					-48.0				-15.5			
Oxetane	C_3H_6O					-110.8				-80.5			
2-Oxetanone	$C_3H_4O_2$					-329.9		175.3	122.1	-282.9			
Oxirane	C_2H_4O					-78.0	-11.8	153.9	88.0	-52.6	-13.0	242.5	47.9
Oxygen	O_2									0.0		205.2	29.4
Oxygen (atomic)	O									249.2	231.7	161.1	21.9
Oxymethurea	$C_3H_8N_2O_3$	-717.0											
Ozone	O_3									142.7	163.2	238.9	39.2
Palladium	Pd	0.0		37.6	26.0					378.2	339.7	167.1	20.8
Palladium(II) oxide	OPd	-85.4			31.4					348.9	325.9	218.0	
Palladium(II) sulfide	PdS	-75.0	-67.0	46.0									
Paraformaldehyde	$(CH_2O)x$	-177.6											
Paraldehyde	$C_6H_{12}O_3$					-673.1				-631.7			
Pentaborane(9)	B_5H_9					42.7	171.8	184.2	151.1	73.2	173.6	280.6	99.6
Pentaborane(11)	B_5H_{11}					73.2				103.3	230.6	321.0	130.3
Pentachloroethane	C_2HCl_5					-187.6			173.8	-142.0			
Pentachlorophenol	C_6HCl_5O	-292.5		253.2	202.0								
Pentadecanoic acid	$C_{15}H_{30}O_2$	-861.7			443.3	-811.7				-699.0			
1-Pentadecanol	$C_{15}H_{32}O$	-658.2											
1,2-Pentadiene	C_5H_8									140.7			
cis-1,3-Pentadiene	C_5H_8									81.4			
trans-1,3-Pentadiene	C_5H_8									76.1			
1,4-Pentadiene	C_5H_8									105.7			
2,3-Pentadiene	C_5H_8									133.1			
Pentaerythritol	$C_5H_{12}O_4$	-920.6								-776.7			
Pentaerythritol tetranitrate	$C_5H_8N_4O_{12}$	-538.6								-387.0		614.7	294.8
Pentafluorobenzene	C_6HF_5	-852.7				-841.8				-806.5			
Pentafluoroethane	C_2HF_5									-1100.4			
Pentafluorophenol	C_6HF_5O	-1024.1				-1007.7							
2,3,4,5,6-Pentafluorotoluene	$C_7H_3F_5$					-883.8		306.4	225.8	-842.7			
Pentamethylbenzene	$C_{11}H_{16}$	-144.6								-67.2			
Pentanal	$C_5H_{10}O$					-267.2				-228.4			
Pentanamide	$C_5H_{11}NO$	-379.5								-290.2			
Pentane	C_5H_{12}					-173.5			167.2	-146.9			
Pentanedioic acid	$C_5H_8O_4$	-960.0											
1,5-Pentanediol	$C_5H_{12}O_2$					-528.8				-450.8			
2,4-Pentanedione	$C_5H_8O_2$					-423.8				-382.0			
Pentanenitrile	C_5H_9N					-33.1				10.5			
1-Pentanethiol	$C_5H_{12}S$					-151.3				-110.0			
Pentanoic acid	$C_5H_{10}O_2$					-559.4		259.8	210.3	-491.9			
1-Pentanol	$C_5H_{12}O$					-351.6			208.1	-294.6			
2-Pentanol	$C_5H_{12}O$					-365.2				-311.0			
3-Pentanol	$C_5H_{12}O$					-368.9			239.7	-314.9			
2-Pentanone	$C_5H_{10}O$					-297.3			184.1	-258.8			
3-Pentanone	$C_5H_{10}O$					-296.5		266.0	190.9	-257.9			
1-Pentene	C_5H_{10}					-46.9		262.6	154.0	-21.1			
cis-2-Pentene	C_5H_{10}					-53.7		258.6	151.7	-27.6			
trans-2-Pentene	C_5H_{10}					-58.2		256.5	157.0	-31.9			
trans-3-Pentenenitrile	C_5H_7N					80.9				125.7			
4-Pentenoic acid	$C_5H_8O_2$					-430.6							
cis-3-Penten-1-yne	C_5H_6					226.5							
trans-3-Penten-1-yne	C_5H_6					228.2							
Pentetic acid	$C_{14}H_{23}N_3O_{10}$	-2225.2											
Pentyl acetate	$C_7H_{14}O_2$								261.0				
Pentylamine	$C_5H_{13}N$								218.0				
Perchloric acid	$ClHO_4$					-40.6							
Perchloryl fluoride	$ClFO_3$									-23.8	48.2	279.0	64.9
Perfluorobutane	C_4F_{10}								127.2				
Perfluorocyclobutane	C_4F_8									-1542.6			

Name	Mol. form.	$\Delta_f H°$(c)/ kJ mol^{-1}	$\Delta_f G°$(c)/ kJ mol^{-1}	$S°$(c)/J mol^{-1} K^{-1}	C_p(c)/J mol^{-1} K^{-1}	$\Delta_f H°$(l)/ kJ mol^{-1}	$\Delta_f G°$(l)/ kJ mol^{-1}	$S°$(l)/J mol^{-1} K^{-1}	C_p(l)/J mol^{-1} K^{-1}	$\Delta_f H°$(g)/ kJ mol^{-1}	$\Delta_f G°$(g)/ kJ mol^{-1}	$S°$(g)/J mol^{-1} K^{-1}	C_p(g)/J mol^{-1} K^{-1}
Perfluorocyclohexane	C_6F_{12}					-2406.3				-2370.4			
Perfluorocyclohexene	C_6F_{10}					-1963.5				-1932.7			
Perfluoroheptane	C_7F_{16}					-3420.0		561.8	419.0	-3383.6			
Perfluoromethylcyclohexane	C_7F_{14}					-2931.1			353.1	-2897.2			
Perfluoropropane	C_3F_8									-1783.2			
Perfluorotoluene	C_7F_8					-1311.1		355.5	262.3				
Peroxyacetic acid	$C_2H_4O_3$												82.4
Perrhenic acid	HO_4Re	-762.3	-656.4	158.2									
Perylene	$C_{20}H_{12}$	182.8		264.6	274.9								
Phenanthrene	$C_{14}H_{10}$	116.2		215.1	220.6					207.5			
9,10-Phenanthrenedione	$C_{14}H_8O_2$	-154.7								-46.6			
Phenanthridine	$C_{13}H_9N$	141.9								240.5			
Phenazine	$C_{12}H_8N_2$	237.0								328.8			
Phenol	C_6H_6O	-165.1		144.0	127.4					-96.4			
L-Phenylalanine	$C_9H_{11}NO_2$	-466.9		213.6	203.0					-312.9			
Phenylhydrazine	$C_6H_8N_2$					141.0			217.0	202.9			
Phenylurea	$C_7H_8N_2O$	-218.6											
Phenyl vinyl ether	C_8H_8O					-26.2				22.7			
Phosphine	H_3P									5.4	13.5	210.2	37.1
Phosphinic acid	H_3O_2P	-604.6				-595.4							
Phosphonic acid	H_3O_3P	-964.4											
Phosphonium chloride	ClH_4P	-145.2											
Phosphoric acid	H_3O_4P	-1284.4	-1124.3	110.5	106.1	-1271.7	-1123.6	150.8	145.0				
Phosphorus (white)	P	0.0		41.1	23.8					316.5	280.1	163.2	20.8
Phosphorus (red)	P	-17.6		22.8	21.2								
Phosphorus (black)	P	-39.3											
Phosphorus(III) bromide	Br_3P					-184.5	-175.7	240.2		-139.3	-162.8	348.1	76.0
Phosphorus(V) bromide	Br_5P	-269.9											
Phosphorus(III) chloride	Cl_3P					-319.7	-272.3	217.1		-287.0	-267.8	311.8	71.8
Phosphorus(V) chloride	Cl_5P	-443.5								-374.9	-305.0	364.6	112.8
Phosphorus dioxide	O_2P									-279.9	-281.6	252.1	39.5
Phosphorus(III) fluoride	F_3P									-958.4	-936.9	273.1	58.7
Phosphorus(V) fluoride	F_5P									-1594.4	-1520.7	300.8	84.8
Phosphorus(III) iodide	I_3P	-45.6										374.4	78.4
Phosphorus monoxide	OP									-28.5	-51.9	222.8	31.8
Phosphorus nitride	NP	-63.0								171.5	149.4	211.1	29.7
Phosphoryl bromide	Br_3OP	-458.6										359.8	89.9
Phosphoryl chloride	Cl_3OP					-597.1	-520.8	222.5	138.8	-558.5	-512.9	325.5	84.9
Phosphoryl fluoride	F_3OP									-1254.3	-1205.8	285.4	68.8
Phthalic acid	$C_8H_6O_4$	-782.0		207.9	188.1								
Phthalic anhydride	$C_8H_4O_3$	-460.1		180.0	160.0					-371.4			
α-Pinene	$C_{10}H_{16}$					-16.4				28.3			
β-Pinene	$C_{10}H_{16}$					-7.7				38.7			
Piperazine	$C_4H_{10}N_2$	-45.6											
2,5-Piperazinedione	$C_4H_6N_2O_2$	-446.5											
Piperidine	$C_5H_{11}N$					-86.4		210.0	179.9	-47.1			
2-Piperidinone	C_5H_9NO	-306.6											
Platinum	Pt	0.0		41.6	25.9					565.3	520.5	192.4	25.5
Platinum(II) bromide	Br_2Pt	-82.0											
Platinum(III) bromide	Br_3Pt	-120.9											
Platinum(IV) bromide	Br_4Pt	-156.5											
Platinum(II) chloride	Cl_2Pt	-123.4											
Platinum(III) chloride	Cl_3Pt	-182.0											
Platinum(IV) chloride	Cl_4Pt	-231.8											
Platinum(VI) fluoride	F_6Pt			235.6								348.3	122.8
Platinum(IV) iodide	I_4Pt	-72.8											
Platinum(II) sulfide	PtS	-81.6	-76.1	55.1	43.4								
Platinum(IV) sulfide	PtS_2	-108.8	-99.6	74.7	65.9								
Plutonium	Pu	0.0											
Polonium	Po	0.0											
Potassium	K	0.0		64.7	29.6					89.0	60.5	160.3	20.8
Potassium acetate	$C_2H_3KO_2$	-723.0											
Potassium aluminum hydride	AlH_4K	-183.7											
Potassium amide	H_2KN	-128.9											
Potassium borohydride	BH_4K	-227.4	-160.3	106.3	96.1								
Potassium bromate	$BrKO_3$	-360.2	-271.2	149.2	105.2								

Name	Mol. form.	$\Delta_f H°$(c)/ kJ mol^{-1}	$\Delta_f G°$(c)/ kJ mol^{-1}	$S°$(c)/J mol^{-1} K^{-1}	C_p(c)/J mol^{-1} K^{-1}	$\Delta_f H°$(l)/ kJ mol^{-1}	$\Delta_f G°$(l)/ kJ mol^{-1}	$S°$(l)/J mol^{-1} K^{-1}	C_p(l)/J mol^{-1} K^{-1}	$\Delta_f H°$(g)/ kJ mol^{-1}	$\Delta_f G°$(g)/ kJ mol^{-1}	$S°$(g)/J mol^{-1} K^{-1}	C_p(g)/J mol^{-1} K^{-1}
Potassium bromide	BrK	-393.8	-380.7	95.9	52.3								
Potassium carbonate	CK_2O_3	-1151.0	-1063.5	155.5	114.4								
Potassium chlorate	$ClKO_3$	-397.7	-296.3	143.1	100.3								
Potassium chloride	ClK	-436.5	-408.5	82.6	51.3					-214.6	-233.3	239.1	36.5
Potassium cyanide	CKN	-113.0	-101.9	128.5	66.3								
Potassium dihydrogen phosphate	H_2KO_4P	-1568.3	-1415.9	134.9	116.6								
Potassium fluoride	FK	-567.3	-537.8	66.6	49.0								
Potassium formate	$CHKO_2$	-679.7											
Potassium hexafluorosilicate	F_6K_2Si	-2956.0	-2798.6	226.0									
Potassium hydride	HK	-57.7											
Potassium hydrogen carbonate	$CHKO_3$	-963.2	-863.5	115.5									
Potassium hydrogen fluoride	F_2HK	-927.7	-859.7	104.3	76.9								
Potassium hydrogen sulfate	HKO_4S	-1160.6	-1031.3	138.1									
Potassium hydroxide	HKO	-424.6	-379.4	81.2	68.9					-232.0	-229.7	238.3	49.2
Potassium iodate	IKO_3	-501.4	-418.4	151.5	106.5								
Potassium iodide	IK	-327.9	-324.9	106.3	52.9								
Potassium metaborate	BKO_2	-981.6	-923.4	80.0	66.7								
Potassium nitrate	KNO_3	-494.6	-394.9	133.1	96.4								
Potassium nitrite	KNO_2	-369.8	-306.6	152.1	107.4								
Potassium oxalate	$C_2K_2O_4$	-1346.0											
Potassium oxide	K_2O	-361.5											
Potassium perbromate	$BrKO_4$	-287.9	-174.4	170.1	120.2								
Potassium perchlorate	$ClKO_4$	-432.8	-303.1	151.0	112.4								
Potassium periodate	IKO_4	-467.2	-361.4	175.7									
Potassium permanganate	$KMnO_4$	-837.2	-737.6	171.7	117.6								
Potassium peroxide	K_2O_2	-494.1	-425.1	102.1									
Potassium phosphate	K_3O_4P	-1950.2											
Potassium sodium	KNa					6.3							
Potassium sulfate	K_2O_4S	-1437.8	-1321.4	175.6	131.5								
Potassium sulfide	K_2S	-380.7	-364.0	105.0									
Potassium superoxide	KO_2	-284.9	-239.4	116.7	77.5								
Potassium thiocyanate	CKNS	-200.2	-178.3	124.3	88.5								
Praseodymium	Pr	0.0		73.2	27.2					355.6	320.9	189.8	21.4
Praseodymium(III) chloride	Cl_3Pr	-1056.9			100.0								
Praseodymium(III) oxide	O_3Pr_2	-1809.6			117.4								
L-Proline	$C_5H_9NO_2$	-515.2								-366.2			
Promethium	Pm	0.0										187.1	24.3
Propanal	C_3H_6O					-215.6				-185.6		304.5	80.7
Propanamide	C_3H_7NO	-338.2								-259.0			
Propane	C_3H_8					-120.9				-103.8	-23.4	270.3	73.6
Propanediamide	$C_3H_6N_2O_2$	-546.1											
1,2-Propanediamine, (±)-	$C_3H_{10}N_2$					-97.8				-53.6			
1,2-Propanediol	$C_3H_8O_2$					-501.0			190.8	-429.8			
1,3-Propanediol	$C_3H_8O_2$					-480.8				-408.0			
1,2-Propanedione	$C_3H_4O_2$					-309.1				-271.0			
1,3-Propanedithiol	$C_3H_8S_2$					-79.4				-29.8			
Propanenitrile	C_3H_5N					15.5			119.3	51.7			
1-Propanethiol	C_3H_8S					-99.9		242.5	144.6	-67.8			
2-Propanethiol	C_3H_8S					-105.9		233.5	145.3	-76.2			
Propanoic acid	$C_3H_6O_2$					-510.7		191.0	152.8	-455.7			
Propanoic anhydride	$C_6H_{10}O_3$					-679.1				-626.5			
1-Propanol	C_3H_8O					-302.6		193.6	143.9	-255.1		322.6	85.6
2-Propanol	C_3H_8O					-318.1		181.1	156.5	-272.6		309.2	89.3
Propene	C_3H_6					4.0				20.0			
trans-1-Propene-1,2-dicarboxylic acid	$C_5H_6O_4$	-824.4											
cis-1-Propene-1,2,3-tricarboxylic acid	$C_6H_6O_6$	-1224.4											
trans-1-Propene-1,2,3-tricarboxylic acid	$C_6H_6O_6$	-1232.7											
Propyl acetate	$C_5H_{10}O_2$								196.2				
Propylamine	C_3H_9N					-101.5			164.1	-70.1	39.9	325.4	91.2
Propylamine hydrochloride	$C_3H_{10}ClN$	-354.7											
Propylbenzene	C_9H_{12}					-38.3		287.8	214.7	7.9			
Propyl carbamate	$C_4H_9NO_2$	-552.6								-471.4			
Propyl chloroacetate	$C_5H_9ClO_2$					-515.5				-467.0			

Name	Mol. form.	$\Delta_f H°$(c)/ kJ mol^{-1}	$\Delta_f G°$(c)/ kJ mol^{-1}	$S°$(c)/J mol^{-1} K^{-1}	C_p(c)/J mol^{-1} K^{-1}	$\Delta_f H°$(l)/ kJ mol^{-1}	$\Delta_f G°$(l)/ kJ mol^{-1}	$S°$(l)/J mol^{-1} K^{-1}	C_p(l)/J mol^{-1} K^{-1}	$\Delta_f H°$(g)/ kJ mol^{-1}	$\Delta_f G°$(g)/ kJ mol^{-1}	$S°$(g)/J mol^{-1} K^{-1}	C_p(g)/J mol^{-1} K^{-1}
Propyl 2-chlorobutanoate	$C_7H_{13}ClO_2$					-630.7				-578.4			
Propyl chlorocarbonate	$C_4H_7ClO_2$					-533.4				-492.7			
Propyl 3-chloropropanoate	$C_6H_{11}ClO_2$					-537.6				-485.7			
Propylcyclohexane	C_9H_{18}					-237.4		311.9	242.0	-192.3			
Propylcyclopentane	C_8H_{16}					-188.8		310.8	216.3	-147.7			
Propylene carbonate	$C_4H_6O_3$					-613.2			218.6	-582.5			
Propyl formate	$C_4H_8O_2$					-500.3				-462.7			
Propyl nitrate	$C_3H_7NO_3$					-214.5				-174.1		362.6	123.2
Propyl pentanoate	$C_8H_{16}O_2$					-583.0				-533.6			
S-Propyl thioacetate	$C_5H_{10}OS$					-294.5				-250.4			
2-Propyn-1-amine	C_3H_5N					205.7							
Propyne	C_3H_4									184.9			
2-Propynoic acid	$C_3H_2O_2$					-193.2							
Protactinium	Pa	0.0		51.9						607.0	563.0	198.1	22.9
Protactinium(IV) bromide	Br_4Pa	-824.0	-787.8	234.0									
Protactinium(IV) chloride	Cl_4Pa	-1043.0	-953.0	192.0									
Protactinium(V) chloride	Cl_5Pa	-1145.0	-1034.0	238.0									
1*H*-Purine	$C_5H_4N_4$	169.4											
Pyrazine	$C_4H_4N_2$	139.8								196.1			
1*H*-Pyrazole	$C_3H_4N_2$	105.4			81.0					179.4			
Pyrene	$C_{16}H_{10}$	125.5		224.9	229.7					225.7			
Pyridazine	$C_4H_4N_2$					224.9				278.3			
Pyridine	C_5H_5N					100.2			132.7	140.4			
3-Pyridinecarboxylic acid	$C_6H_5NO_2$	-344.9								-221.5			
Pyrimidine	$C_4H_4N_2$					145.9				195.7			
Pyrocatechol	$C_6H_6O_2$	-354.1								-267.5			
Pyrrole	C_4H_5N					63.1		156.4	127.7	108.2			
1*H*-Pyrrole-2-carboxaldehyde	C_5H_5NO	-106.4											
Pyrrolidine	C_4H_9N					-41.1		204.1	156.6	-3.6			
2-Pyrrolidone	C_4H_7NO					-286.2							
Quinoline	C_9H_7N					141.2				200.5			
2-Quinolinecarbonitrile	$C_{10}H_6N_2$	246.5											
3-Quinolinecarbonitrile	$C_{10}H_6N_2$	242.3											
2-Quinolinol	C_9H_7NO	-144.9								-25.5			
8-Quinolinol	C_9H_7NO	82.1											
Radium	Ra	0.0		71.0						159.0	130.0	176.5	20.8
Radium nitrate	N_2O_6Ra	-992.0	-796.1	222.0									
Radium oxide	ORa	-523.0											
Radium sulfate	O_4RaS	-1471.1	-1365.6	138.0									
Radon	Rn									0.0		176.2	20.8
Resorcinol	$C_6H_6O_2$	-368.0								-274.7			
Rhenium	Re	0.0		36.9	25.5					769.9	724.6	188.9	20.8
Rhenium(III) bromide	Br_3Re	-167.0											
Rhenium(III) chloride	Cl_3Re	-264.0	-188.0	123.8	92.4								
Rhenium(VII) oxide	O_7Re_2	-1240.1	-1066.0	207.1	166.1					-1100.0	-994.0	452.0	
Rhodium	Rh	0.0		31.5	25.0					556.9	510.8	185.8	21.0
Rhodium(III) chloride	Cl_3Rh	-299.2											
Rhodium monoxide	ORh									385.0			
Rhodium(III) oxide	O_3Rh_2	-343.0			103.8								
D-Ribose	$C_5H_{10}O_5$	-1047.2											
Rubidium	Rb	0.0		76.8	31.1					80.9	53.1	170.1	20.8
Rubidium amide	H_2NRb	-113.0											
Rubidium bromide	BrRb	-394.6	-381.8	110.0	52.8								
Rubidium carbonate	CO_3Rb_2	-1136.0	-1051.0	181.3	117.6								
Rubidium chloride	ClRb	-435.4	-407.8	95.9	52.4								
Rubidium fluoride	FRb	-557.7											
Rubidium hydride	HRb	-52.3											
Rubidium hydrogen fluoride	F_2HRb	-922.6	-855.6	120.1	79.4								
Rubidium hydrogen sulfate	HO_4RbS	-1159.0											
Rubidium hydroxide	HORb	-418.8	-373.9	94.0	69.0					-238.0	-239.1	248.5	49.5
Rubidium iodide	IRb	-333.8	-328.9	118.4	53.2								
Rubidium metaborate	BO_2Rb	-971.0	-913.0	94.3	74.1								
Rubidium nitrate	NO_3Rb	-495.1	-395.8	147.3	102.1								
Rubidium nitrite	NO_2Rb	-367.4	-306.2	172.0									
Rubidium oxide	ORb_2	-339.0											
Rubidium perchlorate	ClO_4Rb	-437.2	-306.9	161.1									

Name	Mol. form.	$\Delta_f H°$(c)/ kJ mol⁻¹	$\Delta_f G°$(c)/ kJ mol⁻¹	$S°$(c)/J mol⁻¹ K⁻¹	C_p(c)/J mol⁻¹ K⁻¹	$\Delta_f H°$(l)/ kJ mol⁻¹	$\Delta_f G°$(l)/ kJ mol⁻¹	$S°$(l)/J mol⁻¹ K⁻¹	C_p(l)/J mol⁻¹ K⁻¹	$\Delta_f H°$(g)/ kJ mol⁻¹	$\Delta_f G°$(g)/ kJ mol⁻¹	$S°$(g)/J mol⁻¹ K⁻¹	C_p(g)/J mol⁻¹ K⁻¹
Rubidium peroxide	O_2Rb_2	-472.0											
Rubidium sulfate	O_4Rb_2S	-1435.6	-1316.9	197.4	134.1								
Rubidium superoxide	O_2Rb	-278.7											
Ruthenium	Ru	0.0		28.5	24.1					642.7	595.8	186.5	21.5
Ruthenium(III) bromide	Br_3Ru	-138.0											
Ruthenium(III) chloride	Cl_3Ru	-205.0											
Ruthenium(III) iodide	I_3Ru	-65.7											
Ruthenium(IV) oxide	O_2Ru	-305.0											
Ruthenium(VIII) oxide	O_4Ru	-239.3	-152.2	146.4									
Salicylaldehyde	$C_7H_6O_2$								222.0				
Salicylaldoxime	$C_7H_7NO_2$	-183.7											
Samarium	Sm	0.0		69.6	29.5					206.7	172.8	183.0	30.4
Samarium(III) chloride	Cl_3Sm	-1025.9											
Samarium(III) fluoride	F_3Sm	-1778.0											
Samarium(III) oxide	O_3Sm_2	-1823.0	-1734.6	151.0	114.5								
Sarcosine	$C_3H_7NO_2$	-513.3								-367.3			
Scandium	Sc	0.0		34.6	25.5					377.8	336.0	174.8	22.1
Scandium bromide	Br_3Sc	-743.1											
Scandium chloride	Cl_3Sc	-925.1											
Scandium fluoride	F_3Sc	-1629.2	-1555.6	92.0						-1247.0	-1234.0	300.5	67.8
Scandium oxide	O_3Sc_2	-1908.8	-1819.4	77.0	94.2								
Selenic acid	H_2O_4Se	-530.1											
Selenium (gray)	Se	0.0		42.4	25.4					227.1	187.0	176.7	20.8
Selenium (α form)	Se	6.7								227.1			
Selenium (vitreous)	Se	5.0								227.1			
Selenium dibromide	Br_2Se									-21.0			
Selenium dioxide	O_2Se	-225.4											
Selenium hexafluoride	F_6Se									-1117.0	-1017.0	313.9	110.5
Selenium monoxide	OSe									53.4	26.8	234.0	31.3
DL-Serine	$C_3H_7NO_3$	-739.0											
L-Serine	$C_3H_7NO_3$	-732.7											
Silane	H_4Si									34.3	56.9	204.6	42.8
Silicon	Si	0.0		18.8	20.0					450.0	405.5	168.0	22.3
Silicon carbide (hexagonal)	CSi	-62.8	-60.2	16.5	26.7								
Silicon carbide (cubic)	CSi	-65.3	-62.8	16.6	26.9								
Silicon dioxide (α-quartz)	O_2Si	-910.7	-856.3	41.5	44.4					-322.0			
Silicon monosulfide	SSi									112.5	60.9	223.7	32.3
Silicon monoxide	OSi									-99.6	-126.4	211.6	29.9
Silicon nitride [Si_3N_4]	N_4Si_3	-743.5	-642.6	101.3									
Silver	Ag	0.0		42.6	25.4					284.9	246.0	173.0	20.8
Silver(I) bromate	$AgBrO_3$	-10.5	71.3	151.9									
Silver(I) bromide	AgBr	-100.4	-96.9	107.1	52.4								
Silver(I) carbonate	CAg_2O_3	-505.8	-436.8	167.4	112.3								
Silver(I) chlorate	$AgClO_3$	-30.3	64.5	142.0									
Silver(I) chloride	AgCl	-127.0	-109.8	96.3	50.8								
Silver(I) chromate	Ag_2CrO_4	-731.7	-641.8	217.6	142.3								
Silver(I) cyanide	CAgN	146.0	156.9	107.2	66.7								
Silver(I) fluoride	AgF	-204.6											
Silver(II) fluoride	AgF_2	-360.0											
Silver(I) iodate	$AgIO_3$	-171.1	-93.7	149.4	102.9								
Silver(I) iodide	AgI	-61.8	-66.2	115.5	56.8								
Silver(I) nitrate	$AgNO_3$	-124.4	-33.4	140.9	93.1								
Silver(I) oxide	Ag_2O	-31.1	-11.2	121.3	65.9								
Silver[II] oxide [Ag_2O_2]	Ag_2O_2	-24.3	27.6	117.0	88.0								
Silver(III) oxide	Ag_2O_3	33.9	121.4	100.0									
Silver(I) perchlorate	$AgClO_4$	-31.1											
Silver(I) sulfate	Ag_2O_4S	-715.9	-618.4	200.4	131.4								
Silver(I) sulfide	Ag_2S	-32.6	-40.7	144.0	76.5								
Silylidyne	HSi									361.0			
Sodium	Na	0.0		51.3	28.2					107.5	77.0	153.7	20.8
Sodium acetate	$C_2H_3NaO_2$	-708.8	-607.2	123.0	79.9								
Sodium aluminum hydride	AlH_4Na	-115.5											
Sodium amide	H_2NNa	-123.8	-64.0	76.9	66.2								
Sodium azide	N_3Na	21.7	93.8	96.9	76.6								
Sodium borohydride	BH_4Na	-188.6	-123.9	101.3	86.8								
Sodium bromate	$BrNaO_3$	-334.1	-242.6	128.9									

Name	Mol. form.	$\Delta_fH°$(c)/ kJ mol^{-1}	$\Delta_fG°$(c)/ kJ mol^{-1}	$S°$(c)/J mol^{-1} K^{-1}	C_p(c)/J mol^{-1} K^{-1}	$\Delta_fH°$(l)/ kJ mol^{-1}	$\Delta_fG°$(l)/ kJ mol^{-1}	$S°$(l)/J mol^{-1} K^{-1}	C_p(l)/J mol^{-1} K^{-1}	$\Delta_fH°$(g)/ kJ mol^{-1}	$\Delta_fG°$(g)/ kJ mol^{-1}	$S°$(g)/J mol^{-1} K^{-1}	C_p(g)/J mol^{-1} K^{-1}
Sodium bromide	$BrNa$	-361.1	-349.0	86.8	51.4					-143.1	-177.1	241.2	36.3
Sodium carbonate	CNa_2O_3	-1130.7	-1044.4	135.0	112.3								
Sodium chlorate	$ClNaO_3$	-365.8	-262.3	123.4									
Sodium chloride	$ClNa$	-411.2	-384.1	72.1	50.5								
Sodium chlorite	$ClNaO_2$	-307.0											
Sodium cyanate	$CNNaO$	-405.4	-358.1	96.7	86.6								
Sodium cyanide	$CNNa$	-87.5	-76.4	115.6	70.4								
Sodium fluoride	FNa	-576.6	-546.3	51.1	46.9								
Sodium formate	$CHNaO_2$	-666.5	-599.9	103.8	82.7								
Sodium hexafluorosilicate	F_6Na_2Si	-2909.6	-2754.2	207.1	187.1								
Sodium hydride	HNa	-56.3	-33.5	40.0	36.4								
Sodium hydrogen carbonate	$CHNaO_3$	-950.8	-851.0	101.7	87.6								
Sodium hydrogen fluoride	F_2HNa	-920.3	-852.2	90.9	75.0								
Sodium hydrogen phosphate	HNa_2O_4P	-1748.1	-1608.2	150.5	135.3								
Sodium hydrogen sulfate	$HNaO_4S$	-1125.5	-992.8	113.0									
Sodium hydroxide	$HNaO$	-425.8	-379.7	64.4	59.5					-191.0	-193.9	229.0	48.0
Sodium iodate	$INaO_3$	-481.8			92.0								
Sodium iodide	INa	-287.8	-286.1	98.5	52.1								
Sodium metaborate	$BNaO_2$	-977.0	-920.7	73.5	65.9								
Sodium metasilicate	Na_2O_3Si	-1554.9	-1462.8	113.9									
Sodium molybdate	$MoNa_2O_4$	-1468.1	-1354.3	159.7	141.7								
Sodium nitrate	$NNaO_3$	-467.9	-367.0	116.5	92.9								
Sodium nitrite	$NNaO_2$	-358.7	-284.6	103.8									
Sodium oxalate	$C_2Na_2O_4$									-1318.0			
Sodium oxide	Na_2O	-414.2	-375.5	75.1	69.1								
Sodium perchlorate	$ClNaO_4$	-383.3	-254.9	142.3									
Sodium periodate	$INaO_4$	-429.3	-323.0	163.0									
Sodium permanganate	$MnNaO_4$	-1156.0											
Sodium peroxide	Na_2O_2	-510.9	-447.7	95.0	89.2								
Sodium sulfate	Na_2O_4S	-1387.1	-1270.2	149.6	128.2								
Sodium sulfide	Na_2S	-364.8	-349.8	83.7									
Sodium sulfite	Na_2O_3S	-1100.8	-1012.5	145.9	120.3								
Sodium superoxide	NaO_2	-260.2	-218.4	115.9	72.1								
Sodium tetraborate	$B_4Na_2O_7$	-3291.1	-3096.0	189.5	186.8								
Sodium tetrafluoroaluminate	AlF_4Na									-1869.0	-1827.5	345.7	105.9
Sodium tetrafluoroborate	BF_4Na	-1844.7	-1750.1	145.3	120.3								
L-Sorbose	$C_6H_{12}O_6$	-1271.5											
Spiro[2.2]pentane	C_5H_8					157.5		193.7	134.5	185.2			
Spiro[5.5]undecane	$C_{11}H_{20}$					-244.5				-188.3			
Stannane	H_4Sn									162.8	188.3	227.7	49.0
Stibine	H_3Sb									145.1	147.8	232.8	41.1
cis-Stilbene	$C_{14}H_{12}$					183.3				252.3			
trans-Stilbene	$C_{14}H_{12}$	136.9								236.1			
Strontium	Sr	0.0		55.0	26.8					164.4	130.9	164.6	20.8
Strontium bromide	Br_2Sr	-717.6	-697.1	135.1	75.3								
Strontium carbonate	CO_3Sr	-1220.1	-1140.1	97.1	81.4								
Strontium chloride	Cl_2Sr	-828.9	-781.1	114.9	75.6								
Strontium fluoride	F_2Sr	-1216.3	-1164.8	82.1	70.0								
Strontium formate	$C_2H_2O_4Sr$	-1393.3											
Strontium hydride	H_2Sr	-180.3											
Strontium hydroxide	H_2O_2Sr	-959.0											
Strontium iodide	I_2Sr	-558.1			81.6								
Strontium metasilicate	O_3SiSr	-1633.9	-1549.7	96.7	88.5								
Strontium nitrate	N_2O_6Sr	-978.2	-780.0	194.6	149.9								
Strontium nitrite	N_2O_4Sr	-762.3											
Strontium orthosilicate	O_4SiSr_2	-2304.5	-2191.1	153.1	134.3								
Strontium oxide	OSr	-592.0	-561.9	54.4	45.0					1.5			
Strontium selenide	$SeSr$	-385.8											
Strontium sulfate	O_4SSr	-1453.1	-1340.9	117.0									
Strontium sulfide	SSr	-472.4	-467.8	68.2	48.7								
Styrene	C_8H_8					103.8			182.0	147.9			
Succinamide	$C_4H_8N_2O_2$	-581.2											
Succinic acid	$C_4H_6O_4$	-940.5		167.3	153.1					-823.0			
Succinic anhydride	$C_4H_4O_3$	-608.6								-527.9			
Succinimide	$C_4H_5NO_2$	-459.0								-375.4			
Succinonitrile	$C_4H_4N_2$	139.7		191.6	145.6					209.7			

Thermochem

Name	Mol. form.	$\Delta_f H°$(c)/ kJ mol^{-1}	$\Delta_f G°$(c)/ kJ mol^{-1}	$S°$(c)/J mol^{-1} K^{-1}	C_p(c)/J mol^{-1} K^{-1}	$\Delta_f H°$(l)/ kJ mol^{-1}	$\Delta_f G°$(l)/ kJ mol^{-1}	$S°$(l)/J mol^{-1} K^{-1}	C_p(l)/J mol^{-1} K^{-1}	$\Delta_f H°$(g)/ kJ mol^{-1}	$\Delta_f G°$(g)/ kJ mol^{-1}	$S°$(g)/J mol^{-1} K^{-1}	C_p(g)/J mol^{-1} K^{-1}
Sucrose	$C_{12}H_{22}O_{11}$	-2226.1											
Sulfolane	$C_4H_8O_2S$								180.0				
Sulfur (rhombic)	S	0.0		32.1	22.6					277.2	236.7	167.8	23.7
Sulfur (monoclinic)	S	0.3											
Sulfur bromide [BrSSBr]	Br_2S_2					-13.0							
Sulfur chloride [ClSSCl]	Cl_2S_2					-59.4							
Sulfur chloride pentafluoride	ClF_5S					-1065.7							
Sulfur dichloride	Cl_2S					-50.0							
Sulfur dioxide	O_2S					-320.5				-296.8	-300.1	248.2	39.9
Sulfur hexafluoride	F_6S									-1220.5	-1116.5	291.5	97.0
Sulfuric acid	H_2O_4S					-814.0	-690.0	156.9	138.9				
Sulfur monoxide	OS									6.3	-19.9	222.0	30.2
Sulfur tetrafluoride	F_4S									-763.2	-722.0	299.6	77.6
Sulfur trioxide (α-form)	O_3S	-454.5	-374.2	70.7		-441.0	-373.8	113.8		-395.7	-371.1	256.8	50.7
Sulfuryl chloride	Cl_2O_2S					-394.1			134.0	-364.0	-320.0	311.9	77.0
Sulfuryl fluoride	F_2O_2S											284.0	66.0
Tantalum	Ta	0.0		41.5	25.4					782.0	739.3	185.2	20.9
Tantalum(V) bromide	Br_5Ta	-598.3											
Tantalum(V) chloride	Cl_5Ta	-859.0											
Tantalum(V) fluoride	F_5Ta	-1903.6											
Tantalum hydride (Ta_2H)	HTa_2	-32.6	-69.0	79.1	90.8								
Tantalum(V) oxide	O_5Ta_2	-2046.0	-1911.2	143.1	135.1								
Technetium	Tc	0.0								678.0		181.1	20.8
Tellurium	Te	0.0		49.7	25.7					196.7	157.1	182.7	20.8
Tellurium dioxide	O_2Te	-322.6	-270.3	79.5									
Tellurium hexafluoride	F_6Te									-1318.0			
Tellurium tetrabromide	Br_4Te	-190.4											
Tellurium tetrachloride	Cl_4Te	-326.4			138.5								
Terbium	Tb	0.0		73.2	28.9					388.7	349.7	203.6	24.6
Terbium(III) chloride	Cl_3Tb	-997.0											
Terbium(III) oxide	O_3Tb_2	-1865.2			115.9								
Terephthalic acid	$C_8H_6O_4$	-816.1								-717.9			
o-Terphenyl	$C_{18}H_{14}$			298.8	274.8			337.1	369.1				
p-Terphenyl	$C_{18}H_{14}$	163.0		285.6	278.7					279.0			
α-Terpinene	$C_{10}H_{16}$									-20.6			
Tetraborane(10)	B_4H_{10}									66.1	184.3	280.3	93.2
1,1,2,2-Tetrabromoethane	$C_2H_2Br_4$								165.7				
Tetrabromoethene	C_2Br_4											387.1	102.7
Tetrabromomethane	CBr_4	29.4	47.7	212.5	144.3					83.9	67.0	358.1	91.2
Tetrabromosilane	Br_4Si					-457.3	-443.9	277.8		-415.5	-431.8	377.9	97.1
Tetrabutylammonium iodide	$C_{16}H_{36}IN$	-498.6											
Tetrachlorodiborane	B_2Cl_4					-523.0	-464.8	262.3	137.7	-490.4	-460.6	357.4	95.4
1,1,1,2-Tetrachloro-2,2-difluoroethane	$C_2Cl_4F_2$									-489.9	-407.0	382.9	123.4
1,1,2,2-Tetrachloro-1,2-difluoroethane	$C_2Cl_4F_2$								173.6				
1,1,1,2-Tetrachloroethane	$C_2H_2Cl_4$											356.0	102.7
1,1,2,2-Tetrachloroethane	$C_2H_2Cl_4$					-195.0		246.9	162.3	-149.2		362.8	100.8
Tetrachloroethene	C_2Cl_4					-50.6	3.0	266.9	143.4	-10.9			
Tetrachloromethane	CCl_4					-128.2			130.7	-95.7			83.3
1,1,1,3-Tetrachloropropane	$C_3H_4Cl_4$					-208.7							
1,2,2,3-Tetrachloropropane	$C_3H_4Cl_4$					-251.8							
Tetrachlorosilane	Cl_4Si					-687.0	-619.8	239.7	145.3	-657.0	-617.0	330.7	90.3
Tetracyanoethene	C_6N_4	623.8								705.0			
Tetradecanenitrile	$C_{14}H_{27}N$					-260.2				-174.9			
Tetradecanoic acid	$C_{14}H_{28}O_2$	-833.5			432.0	-788.8				-693.7			
1-Tetradecanol	$C_{14}H_{30}O$	-629.6			388.0	-580.6							
Tetraethylammonium bromide	$C_8H_{20}BrN$	-342.7											
Tetraethylene glycol	$C_8H_{18}O_5$					-981.7			428.8	-883.0			
Tetraethyl lead	$C_8H_{20}Pb$					52.7		464.6	307.4	109.6			
Tetraethylsilane	$C_8H_{20}Si$								298.1				
Tetraethylurea	$C_9H_{20}N_2O$					-380.0				-316.4			
1,2,4,5-Tetrafluorobenzene	$C_6H_2F_4$					-683.8							
Tetrafluorodiborane	B_2F_4									-1440.1	-1410.4	317.3	79.1
Tetrafluoroethene	C_2F_4	-820.5								-658.9		300.1	80.5
Tetrafluorohydrazine	F_4N_2									-8.4	79.9	301.2	79.2

Name	Mol. form.	$\Delta_f H°(c)$/ kJ mol⁻¹	$\Delta_f G°(c)$/ kJ mol⁻¹	$S°(c)$/J mol⁻¹ K⁻¹	$C_p(c)$/J mol⁻¹ K⁻¹	$\Delta_f H°(l)$/ kJ mol⁻¹	$\Delta_f G°(l)$/ kJ mol⁻¹	$S°(l)$/J mol⁻¹ K⁻¹	$C_p(l)$/J mol⁻¹ K⁻¹	$\Delta_f H°(g)$/ kJ mol⁻¹	$\Delta_f G°(g)$/ kJ mol⁻¹	$S°(g)$/J mol⁻¹ K⁻¹	$C_p(g)$/J mol⁻¹ K⁻¹
Tetrafluoromethane	CF_4									-933.6		261.6	61.1
2,2,3,3-Tetrafluoro-1-propanol	$C_3H_4F_4O$					-1114.9				-1061.3			
Tetrafluorosilane	F_4Si									-1615.0	-1572.8	282.8	73.6
Tetrahydrofuran	C_4H_8O					-216.2		204.3	124.0	-184.1		302.4	76.3
Tetrahydrofurfuryl alcohol	$C_5H_{10}O_2$					-435.7				-369.1			
1,2,3,4-Tetrahydronaphthalene	$C_{10}H_{12}$					-29.2			217.5	26.0			
Tetrahydropyran	$C_5H_{10}O$					-258.3				-223.4			
Tetrahydro-2*H*-pyran-2-one	$C_5H_8O_2$					-436.7				-379.6			
1,2,5,6-Tetrahydropyridine	C_5H_9N					33.5							
Tetrahydrothiophene	C_4H_8S					-72.9				-34.1	45.8	309.6	92.5
Tetraiodoethene	C_2I_4	305.0											
Tetraiodomethane	CI_4	-392.9								474.0		391.9	95.9
Tetraiodosilane	I_4Si	-189.5											
1-Tetralone	$C_{10}H_{10}O$	-209.6											
Tetramethylammonium bromide	$C_4H_{12}BrN$	-251.0											
Tetramethylammonium chloride	$C_4H_{12}ClN$	-276.4											
Tetramethylammonium iodide	$C_4H_{12}IN$	-203.9											
1,2,4,5-Tetramethylbenzene	$C_{10}H_{14}$	-119.9		245.6	215.1								
2,2,3,3-Tetramethylbutane	C_8H_{18}	-269.0		273.7	239.2					-226.0			
2,2,4,4-Tetramethyl-1,3-cyclobutanedione	$C_8H_{12}O_2$	-379.9								-307.6			
1,1,2,2-Tetramethylcyclopropane	C_7H_{14}					-119.8							
Tetramethyl lead	$C_4H_{12}Pb$					97.9				135.9			
N,N,N',N'-Tetramethylmethanediamine	$C_5H_{14}N_2$					-51.1				-18.2			
2,2,3,3-Tetramethylpentane	C_9H_{20}					-278.3			271.5	-237.1			
2,2,3,4-Tetramethylpentane	C_9H_{20}					-277.7				-236.9			
2,2,4,4-Tetramethylpentane	C_9H_{20}					-280.0			266.3	-241.6			
2,3,3,4-Tetramethylpentane	C_9H_{20}					-277.9				-236.1			
2,2,6,6-Tetramethylpiperidine	$C_9H_{19}N$					-206.9				-159.9			
2,2,6,6-Tetramethyl-4-piperidinone	$C_9H_{17}NO$	-334.2								-273.4			
Tetramethylsilane	$C_4H_{12}Si$					-264.0	-100.0	277.3	204.1	-239.1	-99.9	359.0	143.9
Tetramethylstannane	$C_4H_{12}Sn$					-52.3				-18.8			
Tetramethylthiourea	$C_5H_{12}N_2S$	-38.1								44.9			
Tetramethylurea	$C_5H_{12}N_2O$					-262.2							
Tetranitromethane	CN_4O_8					38.4				82.4		503.7	176.1
Tetraphosphorus	P_4									58.9	24.4	280.0	67.2
Thallium	Tl	0.0		64.2	26.3					182.2	147.4	181.0	20.8
Thallium(I) bromide	BrTl	-173.2	-167.4	120.5						-37.7			
Thallium(I) carbonate	CO_3Tl_2	-700.0	-614.6	155.2									
Thallium(I) chloride	ClTl	-204.1	-184.9	111.3	50.9					-67.8			
Thallium(III) chloride	Cl_3Tl	-315.1											
Thallium(I) fluoride	FTl	-324.7								-182.4			
Thallium(I) hydroxide	HOTl	-238.9	-195.8	88.0									
Thallium(I) iodide	ITl	-123.8	-125.4	127.6						7.1			
Thallium(I) nitrate	NO_3Tl	-243.9	-152.4	160.7	99.5								
Thallium(I) oxide	OTl_2	-178.7	-147.3	126.0									
Thallium(I) selenide	$SeTl_2$	-59.0	-59.0	172.0									
Thallium(I) sulfate	O_4STl_2	-931.8	-830.4	230.5									
Thallium(I) sulfide	STl_2	-97.1	-93.7	151.0									
Thiacyclohexane	$C_5H_{10}S$					-106.3		218.2	163.3	-63.5	53.1	323.0	109.7
Thianthrene	$C_{12}H_8S_2$	182.0								286.0			
Thiepane	$C_6H_{12}S$									-65.8	79.4	363.5	131.3
Thietane	C_3H_6S					24.7		184.9		60.6	107.1	285.0	68.3
Thiirane	C_2H_4S					51.6				82.0	96.8	255.2	53.3
Thiirene	C_2H_2S									300.0	275.8	255.3	54.7
Thioacetamide	C_2H_5NS	-71.7								11.4			
Thioacetic acid	C_2H_4OS					-216.9				-175.1			
Thiolactic acid	$C_3H_6O_2S$					-468.4							
Thionitrosyl fluoride (NSF)	FNS											259.8	44.1
Thionyl chloride	Cl_2OS					-245.6			121.0	-212.5	-198.3	309.8	66.5
Thionyl fluoride	F_2OS											278.7	56.8
Thiophene	C_4H_4S					80.2		181.2	123.8	114.9	126.1	278.8	72.8
Thiourea	CH_4N_2S	-89.1								22.9			
Thiram	$C_6H_{12}N_2S_4$	40.2			301.7								
Thorium	Th	0.0		51.8	27.3					602.0	560.7	190.2	20.8

Name	Mol. form.	$\Delta_f H°$(c)/ kJ mol^{-1}	$\Delta_f G°$(c)/ kJ mol^{-1}	$S°$(c)/J mol^{-1} K^{-1}	C_p(c)/J mol^{-1} K^{-1}	$\Delta_f H°$(l)/ kJ mol^{-1}	$\Delta_f G°$(l)/ kJ mol^{-1}	$S°$(l)/J mol^{-1} K^{-1}	C_p(l)/J mol^{-1} K^{-1}	$\Delta_f H°$(g)/ kJ mol^{-1}	$\Delta_f G°$(g)/ kJ mol^{-1}	$S°$(g)/J mol^{-1} K^{-1}	C_p(g)/J mol^{-1} K^{-1}
Thorium(IV) chloride	Cl_4Th	-1186.2	-1094.1	190.4	120.3					-964.4	-932.0	390.7	107.5
Thorium(III) fluoride	F_3Th									-1166.1	-1160.6	339.2	73.3
Thorium(IV) fluoride	F_4Th	-2097.8	-2003.4	142.0	110.7					-1759.0	-1724.0	341.7	93.0
Thorium hydride	H_2Th	-139.7	-100.0	50.7	36.7								
Thorium(IV) oxide	O_2Th	-1226.4	-1169.2	65.2	61.8								
DL-Threonine	$C_4H_9NO_3$	-758.8											
L-Threonine	$C_4H_9NO_3$	-807.2											
Thulium	Tm	0.0		74.0	27.0					232.2	197.5	190.1	20.8
Thulium(III) chloride	Cl_3Tm	-986.6											
Thulium(III) oxide	O_3Tm_2	-1888.7	-1794.5	139.7	116.7								
Thymine	$C_5H_6N_2O_2$	-462.8			150.8					-328.7			
Thymol	$C_{10}H_{14}O$	-309.7								-218.5			
Tin (gray)	Sn	-2.1	0.1	44.1	25.8								
Tin (white)	Sn	0.0		51.2	27.0					301.2	266.2	168.5	21.3
Tin(II) bromide	Br_2Sn	-243.5											
Tin(IV) bromide	Br_4Sn	-377.4	-350.2	264.4						-314.6	-331.4	411.9	103.4
Tin(II) chloride	Cl_2Sn	-325.1											
Tin(IV) chloride	Cl_4Sn					-511.3	-440.1	258.6	165.3	-471.5	-432.2	365.8	98.3
Tin(II) hydroxide	H_2O_2Sn	-561.1	-491.6	155.0									
Tin(II) iodide	I_2Sn	-143.5											
Tin(IV) iodide	I_4Sn				84.9							446.1	105.4
Tin(II) oxide	OSn	-280.7	-251.9	57.2	44.3					15.1	-8.4	232.1	31.6
Tin(IV) oxide	O_2Sn	-577.6	-515.8	49.0	52.6								
Tin(II) sulfide	SSn	-100.0	-98.3	77.0	49.3								
Titanium	Ti	0.0		30.7	25.1					473.0	428.4	180.3	24.4
Titanium(II) bromide	Br_2Ti	-402.0											
Titanium(III) bromide	Br_3Ti	-548.5	-523.8	176.6	101.7								
Titanium(IV) bromide	Br_4Ti	-616.7	-589.5	243.5	131.5					-549.4	-568.2	398.4	100.8
Titanium(II) chloride	Cl_2Ti	-513.8	-464.4	87.4	69.8								
Titanium(III) chloride	Cl_3Ti	-720.9	-653.5	139.7	97.2								
Titanium(IV) chloride	Cl_4Ti					-804.2	-737.2	252.3	145.2	-763.2	-726.3	353.2	95.4
Titanium(IV) iodide	I_4Ti	-375.7	-371.5	249.4	125.7					-277.8			
Titanium(II) oxide	OTi	-519.7	-495.0	50.0	40.0								
Titanium(III) oxide	O_3Ti_2	-1520.9	-1434.2	78.8	97.4								
Titanium(III,IV) oxide	O_5Ti_3	-2459.4	-2317.4	129.3	154.8								
Titanium(IV) oxide (rutile)	O_2Ti	-944.0	-888.8	50.6	55.0								
Toluene	C_7H_8					12.4			157.3	50.5			
Toluene-2,4-diisocyanate	$C_9H_6N_2O_2$								287.8				
o-Toluic acid	$C_8H_8O_2$	-416.5			174.9								
m-Toluic acid	$C_8H_8O_2$	-426.1			163.6								
p-Toluic acid	$C_8H_8O_2$	-429.2			169.0								
Triacetamide	$C_6H_9NO_3$					-610.5				-550.1			
1*H*-1,2,4-Triazol-3-amine	$C_2H_4N_4$	76.8											
Tribromochloromethane	CBr_3Cl											357.8	89.4
Tribromochlorosilane	Br_3ClSi											377.1	95.3
Tribromofluoromethane	CBr_3F											345.9	84.4
Tribromomethane	$CHBr_3$					-22.3	-5.0	220.9	130.7	23.8	8.0	330.9	71.2
Tribromosilane	Br_3HSi					-355.6	-336.4	248.1		-317.6	-328.5	348.6	80.8
Tributylamine	$C_{12}H_{27}N$					-281.6							
1,3,5-Tri-*tert*-butylbenzene	$C_{18}H_{30}$	-320.0											
Tributyl phosphate	$C_{12}H_{27}O_4P$								379.4				
Trichloroacetaldehyde	C_2HCl_3O					-234.5			151.0	-196.6			
2,2,2-Trichloroacetamide	$C_2H_2Cl_3NO$	-358.0											
Trichloroacetic acid	$C_2HCl_3O_2$	-503.3											
Trichloroacetonitrile	C_2Cl_3N											336.6	96.1
Trichloroacetyl chloride	C_2Cl_4O					-280.8				-239.8			
1,2,3-Trichlorobenzene	$C_6H_3Cl_3$	-70.8								3.8			
1,2,4-Trichlorobenzene	$C_6H_3Cl_3$					-63.1				-8.1			
1,3,5-Trichlorobenzene	$C_6H_3Cl_3$	-78.4								-13.4			
1,1,1-Trichloroethane	$C_2H_3Cl_3$					-177.4		227.4	144.3	-144.4		323.1	93.3
1,1,2-Trichloroethane	$C_2H_3Cl_3$					-190.8		232.6	150.9	-151.3		337.2	89.0
Trichloroethene	C_2HCl_3					-43.6		228.4	124.4	-9.0		324.8	80.3
Trichlorofluoromethane	CCl_3F					-301.3	-236.8	225.4	121.6	-268.3			78.1
Trichloromethane	$CHCl_3$					-134.1	-73.7	201.7	114.2	-102.7	6.0	295.7	65.7
Trichloromethyl	CCl_3									59.0			
Trichloromethylsilane	CH_3Cl_3Si							262.8	163.1	-528.9		351.1	102.4

Name	Mol. form.	$\Delta_f H°$(c)/ kJ mol^{-1}	$\Delta_f G°$(c)/ kJ mol^{-1}	$S°$(c)/J mol^{-1} K^{-1}	C_p(c)/J mol^{-1} K^{-1}	$\Delta_f H°$(l)/ kJ mol^{-1}	$\Delta_f G°$(l)/ kJ mol^{-1}	$S°$(l)/J mol^{-1} K^{-1}	C_p(l)/J mol^{-1} K^{-1}	$\Delta_f H°$(g)/ kJ mol^{-1}	$\Delta_f G°$(g)/ kJ mol^{-1}	$S°$(g)/J mol^{-1} K^{-1}	C_p(g)/J mol^{-1} K^{-1}
1,2,3-Trichloropropane	$C_3H_5Cl_3$					-230.6			183.6	-182.9			
1,2,3-Trichloro-1-propene	$C_3H_3Cl_3$					-101.8							
Trichlorosilane	Cl_3HSi					-539.3	-482.5	227.6		-513.0	-482.0	313.9	75.8
1,1,2-Trichloro-1,2,2-trifluoroethane	$C_2Cl_3F_3$					-745.0			170.1	-716.8			
Tri-*o*-cresyl phosphate	$C_{21}H_{21}O_4P$			570.0	578.0								
Tridecane	$C_{13}H_{28}$								406.7				
Tridecanedioic acid	$C_{13}H_{24}O_4$	-1148.3											
1-Tridecanol	$C_{13}H_{28}O$	-599.4											
1-Tridecene	$C_{13}H_{26}$								391.8				
Tris(2-hydroxyethyl)amine	$C_6H_{15}NO_3$	-664.2			389.0					-558.3			
Triethylamine	$C_6H_{15}N$					-127.7			219.9	-92.7			
Triethylborane	$C_6H_{15}B$					-194.6	9.4	336.7	241.2	-157.7	16.1	437.8	
Triethylene glycol	$C_6H_{14}O_4$					-804.3				-725.0			
Trifluoroacetic acid	$C_2HF_3O_2$					-1069.9				-1031.4			
Trifluoroacetonitrile	C_2F_3N									-497.9		298.1	77.9
1,1,1-Trifluoroethane	$C_2H_3F_3$									-744.6		279.9	78.2
1,1,2-Trifluoroethane	$C_2H_3F_3$									-730.7			
2,2,2-Trifluoroethanol	$C_2H_3F_3O$					-932.4				-888.4			
Trifluoroethene	C_2HF_3									-490.5			
1,1,1-Trifluoro-2-iodoethane	$C_2H_2F_3I$									-644.5			
Trifluoroiodomethane	CF_3I									-587.8		307.4	70.9
Trifluoromethane	CHF_3									-695.4		259.7	51.0
Trifluoromethyl	CF_3									-477.0	-464.0	264.5	49.6
(Trifluoromethyl)benzene	$C_7H_5F_3$								188.4				
1,1,1-Trifluoro-2,4-pentanedione	$C_5H_5F_3O_2$					-1040.2				-993.3			
3,3,3-Trifluoro-1-propene	$C_3H_3F_3$									-614.2			
Trifluorosilane	F_3HSi											271.9	60.5
Trigermane	Ge_3H_8					193.7				226.8			
Trihexylamine	$C_{18}H_{39}N$					-433.0							
Trihydro(phosphorus trifluoride) boron	BF_3H_3P									-854.0			
Triiodomethane	CHI_3	-181.1								251.0		356.2	75.0
Trimethyl aluminum	C_3H_9Al					-136.4	-9.9	209.4	155.6	-74.1			
Trimethylamine	C_3H_9N					-45.7		208.5	137.9	-23.6		287.1	91.8
Trimethylamine borane	$C_3H_{12}BN$	-142.5	70.7	187.0									
Trimethylamine hydrochloride	$C_3H_{10}ClN$	-282.9											
1,2,3-Trimethylbenzene	C_9H_{12}					-58.5		267.9	216.4	-9.5			
1,2,4-Trimethylbenzene	C_9H_{12}					-61.8			215.0	-13.8			
1,3,5-Trimethylbenzene	C_9H_{12}					-63.4			209.3	-15.9			
Trimethylborane	C_3H_9B					-143.1	-32.1	238.9		-124.3	-35.9	314.7	88.5
Trimethyl borate	$C_3H_9BO_3$								189.9				
2,2,3-Trimethylbutane	C_7H_{16}					-236.5		292.2	213.5	-204.4			
2,3,3-Trimethyl-1-butene	C_7H_{14}					-117.7				-85.5			
Trimethylchlorosilane	C_3H_9ClSi					-382.8	-246.4	278.2		-352.8	-243.5	369.1	
1α,3α,5β-1,3,5-Trimethylcyclohexane	C_9H_{18}									-212.1			
1,1,2-Trimethylcyclopropane	C_6H_{12}					-96.2							
2,2,3-Trimethylhexane	C_9H_{20}					-282.7							
2,2,4-Trimethylhexane	C_9H_{20}					-282.8							
2,2,5-Trimethylhexane	C_9H_{20}					-293.3							
2,3,3-Trimethylhexane	C_9H_{20}					-281.1							
2,3,5-Trimethylhexane	C_9H_{20}					-284.0				-242.6			
2,4,4-Trimethylhexane	C_9H_{20}					-280.2							
3,3,4-Trimethylhexane	C_9H_{20}					-277.5							
Trimethylolpropane	$C_6H_{14}O_3$	-750.9											
2,2,3-Trimethylpentane	C_8H_{18}					-256.9				-220.0			
2,2,4-Trimethylpentane	C_8H_{18}					-259.2			239.1	-224.0			
2,3,3-Trimethylpentane	C_8H_{18}					-253.5			245.6	-216.3			
2,3,4-Trimethylpentane	C_8H_{18}					-255.0		329.3	247.3	-217.3			
2,2,4-Trimethyl-3-pentanone	$C_8H_{16}O$					-381.6				-338.3			
2,4,4-Trimethyl-1-pentene	C_8H_{16}					-145.9				-110.5			
2,4,4-Trimethyl-2-pentene	C_8H_{16}					-142.4				-104.9			
Trimethylsilane	$C_3H_{10}Si$											331.0	117.9
Trimethylurea	$C_4H_{10}N_2O$	-330.5											
Trinitroacetonitrile	$C_2N_4O_6$					183.7							
1,3,5-Trinitrobenzene	$C_6H_3N_3O_6$	-37.0			214.6								
2,4,6-Trinitro-1,3-benzenediol	$C_6H_3N_3O_8$	-467.5											

Name	Mol. form.	$\Delta_f H°$(c)/ kJ mol^{-1}	$\Delta_f G°$(c)/ kJ mol^{-1}	$S°$(c)/J mol^{-1} K^{-1}	C_p(c)/J mol^{-1} K^{-1}	$\Delta_f H°$(l)/ kJ mol^{-1}	$\Delta_f G°$(l)/ kJ mol^{-1}	$S°$(l)/J mol^{-1} K^{-1}	C_p(l)/J mol^{-1} K^{-1}	$\Delta_f H°$(g)/ kJ mol^{-1}	$\Delta_f G°$(g)/ kJ mol^{-1}	$S°$(g)/J mol^{-1} K^{-1}	C_p(g)/J mol^{-1} K^{-1}
Trinitroglycerol	$C_3H_5N_3O_9$					-370.9				-279.1		545.9	234.2
Trinitromethane	CHN_3O_6					-32.8				-13.4		435.6	134.1
2,4,6-Trinitrophenol	$C_6H_3N_3O_7$	-217.9			239.7								
2,4,6-Trinitrotoluene	$C_7H_5N_3O_6$	-63.2			243.3								
Trioctylamine	$C_{24}H_{51}N$					-585.0							
1,3,5-Trioxane	$C_3H_6O_3$	-522.5		133.0	111.4					-465.9			
Triphenylamine	$C_{18}H_{15}N$	234.7								326.8			
Triphenylmethanol	$C_{19}H_{16}O$	-2.5											
Triphenyl phosphate	$C_{18}H_{15}O_4P$			397.5	356.2								
Triphenylphosphine	$C_{18}H_{15}P$				312.5								
Tripropylamine	$C_9H_{21}N$					-207.1				-161.0			
Tris(hydroxymethyl)methylamine	$C_4H_{11}NO_3$	-717.8											
Trisilane	H_8Si_3					92.5				120.9			
Tris(perfluorobutyl)amine	$C_{12}F_{27}N$								418.4				
1,3,5-Trithiane	$C_3H_6S_3$									80.0	130.4	336.4	111.3
Trithiocarbonic acid	CH_2S_3					24.0							
L-Tryptophan	$C_{11}H_{12}N_2O_2$	-415.3		251.0	238.1								
Tungsten	W	0.0		32.6	24.3					849.4	807.1	174.0	21.3
Tungsten(VI) bromide	Br_6W	-348.5											
Tungsten(VI) chloride	Cl_6W	-602.5								-513.8			
Tungsten(VI) fluoride	F_6W					-1747.7	-1631.4	251.5		-1721.7	-1632.1	341.1	119.0
Tungsten(IV) oxide	O_2W	-589.7	-533.9	50.5	56.1								
Tungsten(VI) oxide	O_3W	-842.9	-764.0	75.9	73.8								
L-Tyrosine	$C_9H_{11}NO_3$	-685.1		214.0	216.4								
Undecane	$C_{11}H_{24}$					-327.2			344.9	-270.8			
1-Undecanol	$C_{11}H_{24}O$					-504.8							
1-Undecene	$C_{11}H_{22}$								344.9				
Uracil	$C_4H_4N_2O_2$	-429.4			120.5					-302.9			
Uranium	U	0.0		50.2	27.7					533.0	488.4	199.8	23.7
Uranium(III) chloride	Cl_3U	-866.5	-799.1	159.0	102.5								
Uranium(IV) chloride	Cl_4U	-1019.2	-930.0	197.1	122.0					-809.6	-786.6	419.0	
Uranium(VI) chloride	Cl_6U	-1092.0	-962.0	285.8	175.7					-1013.0	-928.0	431.0	
Uranium(III) fluoride	F_3U	-1502.1	-1433.4	123.4	95.1					-1058.5	-1051.9	331.9	74.3
Uranium(IV) fluoride	F_4U	-1914.2	-1823.3	151.7	116.0					-1598.7	-1572.7	368.0	91.2
Uranium(VI) fluoride	F_6U	-2197.0	-2068.5	227.6	166.8					-2147.4	-2063.7	377.9	129.6
Uranium(III) hydride	H_3U	-127.2	-72.8	63.7	49.3								
Uranium(II) oxide	OU									21.0			
Uranium(IV) oxide	O_2U	-1085.0	-1031.8	77.0	63.6					-465.7	-471.5	274.6	51.4
Uranium(IV,V) oxide	O_9U_4	-4510.4	-4275.1	334.1	293.3								
Uranium(IV,VI) oxide	O_7U_3	-3427.1	-3242.9	250.5	215.5								
Uranium(V,VI) oxide	O_8U_3	-3574.8	-3369.5	282.6	238.4								
Uranium(VI) oxide	O_3U	-1223.8	-1145.7	96.1	81.7								
Uranyl chloride	Cl_2O_2U	-1243.9	-1146.4	150.5	107.9								
Uranyl fluoride	F_2O_2U	-1653.5	-1557.4	135.6	103.2								
Urea	CH_4N_2O	-333.1								-245.8			
Uric acid	$C_5H_4N_4O_3$	-618.8		173.2	166.1								
DL-Valine	$C_5H_{11}NO_2$	-628.9											
L-Valine	$C_5H_{11}NO_2$	-617.9								-455.1			
Vanadium	V	0.0		28.9	24.9					514.2	754.4	182.3	26.0
Vanadium(IV) bromide	Br_4V									-336.8			
Vanadium(III) chloride	Cl_3V	-580.7	-511.2	131.0	93.2								
Vanadium(IV) chloride	Cl_4V					-569.4	-503.7	255.0		-525.5	-492.0	362.4	96.2
Vanadium(IV) fluoride	F_4V	-1403.3											
Vanadium(V) fluoride	F_5V					-1480.3	-1373.1	175.7		-1433.9	-1369.8	320.9	98.6
Vanadium(IV) iodide	I_4V									-122.6			
Vanadium(II) oxide	OV	-431.8	-404.2	38.9	45.4								
Vanadium(III) oxide	O_3V_2	-1218.8	-1139.3	98.3	103.2								
Vanadium(III,IV) oxide	O_5V_3	-1933.0	-1803.0	163.0									
Vanadium(V) oxide	O_5V_2	-1550.6	-1419.5	131.0	127.7								
Vanadyl chloride	ClOV	-607.0	-556.0	75.0									
Vanadyl trichloride	Cl_3OV					-734.7	-668.5	244.3		-695.6	-659.3	344.3	89.9
Vinyl acetate	$C_4H_6O_2$					-349.2				-314.4			
Vinylcyclopentane	C_7H_{12}					-34.8							
2-Vinylfuran	C_6H_6O					-10.3				27.8			
Water	H_2O					-285.8	-237.1	70.0	75.3	-241.8	-228.6	188.8	33.6
Xanthine	$C_5H_4N_4O_2$	-379.6		161.1	151.3								

Name	Mol. form.	$\Delta_fH°$(c)/ kJ mol^{-1}	$\Delta_fG°$(c)/ kJ mol^{-1}	$S°$(c)/J mol^{-1} K^{-1}	C_p(c)/J mol^{-1} K^{-1}	$\Delta_fH°$(l)/ kJ mol^{-1}	$\Delta_fG°$(l)/ kJ mol^{-1}	$S°$(l)/J mol^{-1} K^{-1}	C_p(l)/J mol^{-1} K^{-1}	$\Delta_fH°$(g)/ kJ mol^{-1}	$\Delta_fG°$(g)/ kJ mol^{-1}	$S°$(g)/J mol^{-1} K^{-1}	C_p(g)/J mol^{-1} K^{-1}
Xanthone	$C_{13}H_8O_2$	-191.5											
Xenon	Xe									0.0		169.7	20.8
Xenon tetrafluoride	F_4Xe	-261.5											
o-Xylene	C_8H_{10}					-24.4			186.1	19.1			
m-Xylene	C_8H_{10}					-25.4			183.0	17.3			
p-Xylene	C_8H_{10}					-24.4			181.5	18.0			
2,3-Xylenol	$C_8H_{10}O$	-241.1								-157.2			
2,4-Xylenol	$C_8H_{10}O$					-228.7				-163.8			
2,5-Xylenol	$C_8H_{10}O$	-246.6								-161.6			
2,6-Xylenol	$C_8H_{10}O$	-237.4								-162.1			
3,4-Xylenol	$C_8H_{10}O$	-242.3								-157.3			
3,5-Xylenol	$C_8H_{10}O$	-244.4								-162.4			
Xylitol	$C_5H_{12}O_5$	-1118.5											
D-Xylose	$C_5H_{10}O_5$	-1057.8											
Ytterbium	Yb	0.0		59.9	26.7					152.3	118.4	173.1	20.8
Ytterbium(III) chloride	Cl_3Yb	-959.8											
Ytterbium(III) oxide	O_3Yb_2	-1814.6	-1726.7	133.1	115.4								
Yttrium	Y	0.0		44.4	26.5					421.3	381.1	179.5	25.9
Yttrium chloride	Cl_3Y	-1000.0								-750.2			75.0
Yttrium fluoride	F_3Y	-1718.8	-1644.7	100.0						-1288.7	-1277.8	311.8	70.3
Yttrium oxide	O_3Y_2	-1905.3	-1816.6	99.1	102.5								
Zinc	Zn	0.0		41.6	25.4					130.4	94.8	161.0	20.8
Zinc bromide	Br_2Zn	-328.7	-312.1	138.5									
Zinc carbonate	CO_3Zn	-812.8	-731.5	82.4	79.7								
Zinc chloride	Cl_2Zn	-415.1	-369.4	111.5	71.3					-266.1			
Zinc fluoride	F_2Zn	-764.4	-713.3	73.7	65.7								
Zinc hydroxide	H_2O_2Zn	-641.9	-553.5	81.2									
Zinc iodide	I_2Zn	-208.0	-209.0	161.1									
Zinc nitrate	N_2O_6Zn	-483.7											
Zinc orthosilicate	O_4SiZn_2	-1636.7	-1523.2	131.4	123.3								
Zinc oxide	OZn	-350.5	-320.5	43.7	40.3								
Zinc selenide	SeZn	-163.0	-163.0	84.0									
Zinc sulfate	O_4SZn	-982.8	-871.5	110.5	99.2								
Zinc sulfide (sphalerite)	SZn	-206.0	-201.3	57.7	46.0								
Zinc sulfide (wurtzite)	SZn	-192.6											
Zirconium	Zr	0.0		39.0	25.4					608.8	566.5	181.4	26.7
Zirconium(IV) bromide	Br_4Zr	-760.7											
Zirconium(II) chloride	Cl_2Zr	-502.0											
Zirconium(IV) chloride	Cl_4Zr	-980.5	-889.9	181.6	119.8								
Zirconium(IV) fluoride	F_4Zr	-1911.3	-1809.9	104.6	103.7								
Zirconium(II) hydride	H_2Zr	-169.0	-128.8	35.0	31.0								
Zirconium(IV) iodide	I_4Zr	-481.6											
Zirconium(IV) orthosilicate	O_4SiZr	-2033.4	-1919.1	84.1	98.7								
Zirconium(IV) oxide	O_2Zr	-1100.6	-1042.8	50.4	56.2								
Zirconium(IV) sulfate	O_8S_2Zr	-2217.1			172.0								
Zirconium titanate	O_4TiZr	-2024.1	-1915.8	116.7	114.0								

THERMODYNAMIC PROPERTIES AS A FUNCTION OF TEMPERATURE

L. V. Gurvich, V. S. Iorish, V. S. Yungman, and O. V. Dorofeeva

The thermodynamic properties (standard heat capacity at constant pressure, standard entropy, Gibbs energy function, change of enthalpy above the reference temperature) and the formation properties (enthalpy of formation, Gibbs energy of formation, and equilibirum constant) are tabulated as functions of temperature in the range 298.15 to 1500 K for 80 important substances in the standard state. The reference temperature, T_r, is equal to 298.15 K. The standard state pressure is taken as 1 bar (100,000 Pa). The numerical data are extracted from IVTANTHERMO databases (Ref. 1) except for C_2H_4O, C_3H_6O, C_6H_6, C_6H_6O, $C_{10}H_8$, and CH_5N, which are based upon TRC Tables (Ref. 2).

The column definitions are as follows.

Column heading	Definition
T	Temperature, in K
Cp°	Standard heat capacity at constant pressure; in units J K^{-1} mol^{-1}
S°	Standard entropy, in units J K^{-1} mol^{-1}
-(*G°*-*H°*(*T*))	Negative of the Gibbs energy funcion, in units J K^{-1} mol^{-1}
(*H*-*H°*(*T*))	Change of enthalpy above the reference temperature (also called high-temperature molar heat content), in units kJ mol^{-1}
$\Delta_f H°$	Enthalpy of formation, in units kJ mol^{-1}
$\Delta_f G°$	Gibbs energy of formation, in kJ mol^{-1}
Log K_f	Log base 10 of equilibrium constant

Table 1 contains a cross-reference of chemical substance and subtables of Table 2. Table 2 contains the thermodynamic data for each chemical substance at a wide variety of temperatures. Phase transitions when they occur are clearly indicated. The tables are presented in the JANAF Thermochemical Tables format (Ref. 3). See the references for information on standard states and other details.

References

1. Gurvich, L. V., Veyts, I. V., and Alcock, C. B., Eds., *Thermodynamic Properties of Individual Substances, Fourth Edition*, Hemisphere Publishing Corp., New York, 1989.
2. *TRC Thermodynamic Tables - Hydrocarbons*, Ed., Frenkel, M., Standard Reference Data Program Publication Series NSRDS-NIST-75, Gaithersburg, MD, 2010, and references therein.
3. Chase, M. W., *NIST-JANAF Thermochemical Tables, Fourth Edition*, *J. Phys. Chem. Ref. Data*, Monograph No. 9, 1998.

Thermochem

TABLE 1. Thermodynamic Properties as a Function of Temperature: Order of Listing

No.	Mol. form.	Name	State
1	Ar	Argon	g
2	Br	Bromine (atomic)	g
3	Br_2	Bromine	g
4	BrH	Hydrogen bromide	g
5	C	Carbon (graphite)	cr
6	C	Carbon (diamond)	cr
7	C_2	Dicarbon	g
8	C_3	Propadienediylidene	g
9	CO	Carbon monoxide	g
10	CO_2	Carbon dioxide	g
11	CH_4	Methane	g
12	C_2H_2	Acetylene	g
13	C_2H_4	Ethylene	g
14	C_2H_6	Ethane	g
15	C_3H_6	Cyclopropane	g
16	C_3H_8	Propane	g
17	C_6H_6	Benzene	l
18	C_6H_6	Benzene	g
19	$C_{10}H_8$	Naphthalene	cr, l
20	$C_{10}H_8$	Naphthalene	g
21	CH_2O	Formaldehyde	g
22	CH_4O	Methanol	g
23	C_2H_4O	Acetaldehyde	g
24	C_2H_6O	Ethanol	g
25	$C_2H_4O_2$	Acetic acid	g
26	C_3H_6O	Acetone	g
27	C_6H_6O	Phenol	g
28	CF_4	Tetrafluoromethane	g
29	CHF_3	Trifluoromethane	g
30	$CClF_3$	Chlorotrifluoromethane	g
31	CCl_2F_2	Dichlorodifluoromethane	g
32	$CHClF_2$	Chlorodifluoromethane	g
33	CH_5N	Methylamine	g
34	Cl	Chlorine (atomic)	g
35	Cl_2	Chlorine	g
36	ClH	Hydrogen chloride	g
37	Cu	Copper	cr, l
38	Cu	Copper	g
39	CuO	Copper(II) oxide	cr
40	Cu_2O	Copper(I) oxide	cr
41	Cl_2Cu	Copper(II) chloride	cr, l
42	Cl_2Cu	Copper(II) chloride	g
43	F	Fluorine (atomic)	g
44	F_2	Fluorine	g
45	FH	Hydrogen fluoride	g
46	Ge	Germanium	cr, l
47	Ge	Germanium	g
48	GeO_2	Germanium(IV) oxide	cr, l
49	Cl_4Ge	Germanium(IV) chloride	g
50	H	Hydrogen (atomic)	g
51	H_2	Hydrogen	g
52	HO	Hydroxyl	g
53	H_2O	Water	l
54	H_2O	Water	g

No.	Mol. form.	Name	State
55	I	Iodine (atomic)	g
56	I_2	Iodine	cr, l
57	I_2	Iodine	g
58	HI	Hydrogen iodide	g
59	K	Potassium	cr, l
60	K	Potassium	g
61	K_2O	Potassium oxide	cr, l
62	HKO	Potassium hydroxide	cr, l
63	HKO	Potassium hydroxide	g
64	ClK	Potassium chloride	cr, l
65	ClK	Potassium chloride	g
66	N_2	Nitrogen	g
67	NO	Nitric oxide	g
68	NO_2	Nitrogen dioxide	g
69	H_3N	Ammonia	g
70	O	Oxygen	g
71	O_2	Oxygen	g
72	S	Sulfur (rhombic)	cr, l
73	S	Sulfur (rhombic)	g
74	S_2	Disulfur	g
75	S_8	Sulfur (orthorhombic)	g
76	O_2S	Sulfur dioxide	g
77	Si	Silicon	cr
78	Si	Silicon	g
79	O_2Si	Silicon dioxide (α-quartz)	cr
80	Cl_4Si	Tetrachlorosilane	g

TABLE 2. Thermodynamic Properties as a Function of Temperature

T/K	$C_p°$/ J K⁻¹mol⁻¹	$S°$/ J K⁻¹ mol⁻¹	-($G°$-$H°(T)$)/ J K⁻¹ mol⁻¹	(H-$H°(T)$)/ kJ mol⁻¹	$\Delta_f H°$/ kJ mol⁻¹	$\Delta_f G°$/ kJ mol⁻¹	Log K_f
1. ARGON Ar (g)							
298.15	20.786	154.845	154.845	0.000	0.000	0.000	0.000
300	20.786	154.973	154.845	0.038	0.000	0.000	0.000
400	20.786	160.953	155.660	2.117	0.000	0.000	0.000
500	20.786	165.591	157.200	4.196	0.000	0.000	0.000
600	20.786	169.381	158.924	6.274	0.000	0.000	0.000
700	20.786	172.585	160.653	8.353	0.000	0.000	0.000
800	20.786	175.361	162.322	10.431	0.000	0.000	0.000
900	20.786	177.809	163.909	12.510	0.000	0.000	0.000
1000	20.786	179.999	165.410	14.589	0.000	0.000	0.000
1100	20.786	181.980	166.828	16.667	0.000	0.000	0.000
1200	20.786	183.789	168.167	18.746	0.000	0.000	0.000
1300	20.786	185.453	169.434	20.824	0.000	0.000	0.000
1400	20.786	186.993	170.634	22.903	0.000	0.000	0.000
1500	20.786	188.427	171.773	24.982	0.000	0.000	0.000
2. BROMINE Br (g)							
298.15	20.786	175.017	175.017	0.000	111.870	82.379	-14.432
300	20.786	175.146	175.018	0.038	111.838	82.196	-14.311
400	20.787	181.126	175.833	2.117	96.677	75.460	-9.854
500	20.798	185.765	177.373	4.196	96.910	70.129	-7.326
600	20.833	189.559	179.097	6.277	97.131	64.752	-5.637
700	20.908	192.776	180.827	8.364	97.348	59.338	-4.428
800	21.027	195.575	182.499	10.461	97.568	53.893	-3.519
900	21.184	198.061	184.093	12.571	97.796	48.420	-2.810
1000	21.365	200.302	185.604	14.698	98.036	42.921	-2.242
1100	21.559	202.347	187.034	16.844	98.291	37.397	-1.776
1200	21.752	204.231	188.390	19.010	98.560	31.850	-1.386
1300	21.937	205.980	189.676	21.195	98.844	26.279	-1.056
1400	22.107	207.612	190.900	23.397	99.141	20.686	-0.772
1500	22.258	209.142	192.065	25.615	99.449	15.072	-0.525
3. DIBROMINE Br_2 (g)							
298.15	36.057	245.467	245.467	0.000	30.910	3.105	-0.544
300	36.074	245.690	245.468	0.067	30.836	2.933	-0.511
332.25[a]	36.340	249.387	245.671	1.235			
400	36.729	256.169	246.892	3.711	0.000	0.000	0.000
500	37.082	264.406	249.600	7.403	0.000	0.000	0.000
600	37.305	271.188	252.650	11.123	0.000	0.000	0.000
700	37.464	276.951	255.720	14.862	0.000	0.000	0.000
800	37.590	281.962	258.694	18.615	0.000	0.000	0.000
900	37.697	286.396	261.530	22.379	0.000	0.000	0.000
1000	37.793	290.373	264.219	26.154	0.000	0.000	0.000

T/K	$C_p°$/ J K^{-1}mol^{-1}	$S°$/ J K^{-1} mol^{-1}	-($G°$-$H°(T)$)/ J K^{-1} mol^{-1}	(H-$H°(T)$)/ kJ mol^{-1}	$\Delta_f H°$/ kJ mol^{-1}	$\Delta_f G°$/ kJ mol^{-1}	Log K_f
1100	37.883	293.979	266.763	29.938	0.000	0.000	0.000
1200	37.970	297.279	269.170	33.730	0.000	0.000	0.000
1300	38.060	300.322	271.451	37.532	0.000	0.000	0.000
1400	38.158	303.146	273.615	41.343	0.000	0.000	0.000
1500	38.264	305.782	275.673	45.164	0.000	0.000	0.000
4. HYDROGEN BROMIDE HBr (g)							
298.15	29.141	198.697	198.697	0.000	-36.290	-53.360	9.348
300	29.141	198.878	198.698	0.054	-36.333	-53.466	9.309
400	29.220	207.269	199.842	2.971	-52.109	-55.940	7.305
500	29.454	213.811	202.005	5.903	-52.484	-56.854	5.939
600	29.872	219.216	204.436	8.868	-52.844	-57.694	5.023
700	30.431	223.861	206.886	11.882	-53.168	-58.476	4.363
800	31.063	227.965	209.269	14.957	-53.446	-59.214	3.866
900	31.709	231.661	211.555	18.095	-53.677	-59.921	3.478
1000	32.335	235.035	213.737	21.298	-53.864	-60.604	3.166
1100	32.919	238.145	215.816	24.561	-54.012	-61.271	2.909
1200	33.454	241.032	217.799	27.880	-54.129	-61.925	2.696
1300	33.938	243.729	219.691	31.250	-54.220	-62.571	2.514
1400	34.374	246.261	221.499	34.666	-54.291	-63.211	2.358
1500	34.766	248.646	223.230	38.123	-54.348	-63.846	2.223
5. CARBON (GRAPHITE) C (cr; graphite)							
298.15	8.536	5.740	5.740	0.000	0.000	0.000	0.000
300	8.610	5.793	5.740	0.016	0.000	0.000	0.000
400	11.974	8.757	6.122	1.054	0.000	0.000	0.000
500	14.537	11.715	6.946	2.385	0.000	0.000	0.000
600	16.607	14.555	7.979	3.945	0.000	0.000	0.000
700	18.306	17.247	9.113	5.694	0.000	0.000	0.000
800	19.699	19.785	10.290	7.596	0.000	0.000	0.000
900	20.832	22.173	11.479	9.625	0.000	0.000	0.000
1000	21.739	24.417	12.662	11.755	0.000	0.000	0.000
1100	22.452	26.524	13.827	13.966	0.000	0.000	0.000
1200	23.000	28.502	14.968	16.240	0.000	0.000	0.000
1300	23.409	30.360	16.082	18.562	0.000	0.000	0.000
1400	23.707	32.106	17.164	20.918	0.000	0.000	0.000
1500	23.919	33.749	18.216	23.300	0.000	0.000	0.000
6. CARBON (DIAMOND) C (cr; diamond)							
298.15	6.109	2.362	2.362	0.000	1.850	2.857	-0.501
300	6.201	2.400	2.362	0.011	1.846	2.863	-0.499
400	10.321	4.783	2.659	0.850	1.645	3.235	-0.422
500	13.404	7.431	3.347	2.042	1.507	3.649	-0.381
600	15.885	10.102	4.251	3.511	1.415	4.087	-0.356
700	17.930	12.709	5.274	5.205	1.361	4.537	-0.339
800	19.619	15.217	6.361	7.085	1.338	4.993	-0.326
900	21.006	17.611	7.479	9.118	1.343	5.450	-0.316
1000	22.129	19.884	8.607	11.277	1.372	5.905	-0.308
1100	23.020	22.037	9.731	13.536	1.420	6.356	-0.302
1200	23.709	24.071	10.842	15.874	1.484	6.802	-0.296
1300	24.222	25.990	11.934	18.272	1.561	7.242	-0.291
1400	24.585	27.799	13.003	20.714	1.646	7.675	-0.286
1500	24.824	29.504	14.047	23.185	1.735	8.103	-0.282
7. DICARBON C_2 (g)							
298.15	43.548	197.095	197.095	0.000	830.457	775.116	-135.795
300	43.575	197.365	197.096	0.081	830.506	774.772	-134.898
400	42.169	209.809	198.802	4.403	832.751	755.833	-98.700
500	39.529	218.924	201.959	8.483	834.170	736.423	-76.933
600	37.837	225.966	205.395	12.342	834.909	716.795	-62.402
700	36.984	231.726	208.758	16.078	835.148	697.085	-52.016
800	36.621	236.637	211.943	19.755	835.020	677.366	-44.227

T/K	C_p°/ J K^{-1}mol^{-1}	S°/ J K^{-1} mol^{-1}	-(G°-H°(T))/ J K^{-1} mol^{-1}	(H-H°(T))/ kJ mol^{-1}	$\Delta_f H$°/ kJ mol^{-1}	$\Delta_f G$°/ kJ mol^{-1}	Log K_f
900	36.524	240.943	214.931	23.411	834.618	657.681	-38.170
1000	36.569	244.793	217.728	27.065	834.012	638.052	-33.328
1100	36.696	248.284	220.349	30.728	833.252	618.492	-29.369
1200	36.874	251.484	222.812	34.406	832.383	599.006	-26.074
1300	37.089	254.444	225.133	38.104	831.437	579.596	-23.288
1400	37.329	257.201	227.326	41.824	830.445	560.261	-20.903
1500	37.589	259.785	229.405	45.570	829.427	540.997	-18.839
8. TRICARBON C_3 (g)							
298.15	42.202	237.611	237.611	0.000	839.958	774.249	-135.643
300	42.218	237.872	237.611	0.078	839.989	773.841	-134.736
400	43.383	250.164	239.280	4.354	841.149	751.592	-98.147
500	44.883	260.003	242.471	8.766	841.570	729.141	-76.172
600	46.406	268.322	246.104	13.331	841.453	706.659	-61.519
700	47.796	275.582	249.807	18.042	840.919	684.230	-51.057
800	48.997	282.045	253.440	22.884	840.053	661.901	-43.217
900	50.006	287.876	256.948	27.835	838.919	639.698	-37.127
1000	50.844	293.189	260.310	32.879	837.572	617.633	-32.261
1100	51.535	298.069	263.524	37.999	836.059	595.711	-28.288
1200	52.106	302.578	266.593	43.182	834.420	573.933	-24.982
1300	52.579	306.768	269.524	48.417	832.690	552.295	-22.191
1400	52.974	310.679	272.326	53.695	830.899	530.793	-19.804
1500	53.307	314.346	275.006	59.010	829.068	509.421	-17.739
9. CARBON OXIDE CO (g)							
298.15	29.141	197.658	197.658	0.000	-110.530	-137.168	24.031
300	29.142	197.838	197.659	0.054	-110.519	-137.333	23.912
400	29.340	206.243	198.803	2.976	-110.121	-146.341	19.110
500	29.792	212.834	200.973	5.930	-110.027	-155.412	16.236
600	30.440	218.321	203.419	8.941	-110.157	-164.480	14.319
700	31.170	223.067	205.895	12.021	-110.453	-173.513	12.948
800	31.898	227.277	208.309	15.175	-110.870	-182.494	11.915
900	32.573	231.074	210.631	18.399	-111.378	-191.417	11.109
1000	33.178	234.538	212.851	21.687	-111.952	-200.281	10.461
1100	33.709	237.726	214.969	25.032	-112.573	-209.084	9.928
1200	34.169	240.679	216.990	28.426	-113.228	-217.829	9.482
1300	34.568	243.430	218.920	31.864	-113.904	-226.518	9.101
1400	34.914	246.005	220.763	35.338	-114.594	-235.155	8.774
1500	35.213	248.424	222.527	38.845	-115.291	-243.742	8.488
10. CARBON DIOXIDE CO_2 (g)							
298.15	37.135	213.783	213.783	0.000	-393.510	-394.373	69.092
300	37.220	214.013	213.784	0.069	-393.511	-394.379	68.667
400	41.328	225.305	215.296	4.004	-393.586	-394.656	51.536
500	44.627	234.895	218.280	8.307	-393.672	-394.914	41.256
600	47.327	243.278	221.762	12.909	-393.791	-395.152	34.401
700	49.569	250.747	225.379	17.758	-393.946	-395.367	29.502
800	51.442	257.492	228.978	22.811	-394.133	-395.558	25.827
900	53.008	263.644	232.493	28.036	-394.343	-395.724	22.967
1000	54.320	269.299	235.895	33.404	-394.568	-395.865	20.678
1100	55.423	274.529	239.172	38.893	-394.801	-395.984	18.803
1200	56.354	279.393	242.324	44.483	-395.035	-396.081	17.241
1300	57.144	283.936	245.352	50.159	-395.265	-396.159	15.918
1400	57.818	288.196	248.261	55.908	-395.488	-396.219	14.783
1500	58.397	292.205	251.059	61.719	-395.702	-396.264	13.799
11. METHANE CH_4 (g)							
298.15	35.695	186.369	186.369	0.000	-74.600	-50.530	8.853
300	35.765	186.590	186.370	0.066	-74.656	-50.381	8.772
400	40.631	197.501	187.825	3.871	-77.703	-41.827	5.462
500	46.627	207.202	190.744	8.229	-80.520	-32.525	3.398
600	52.742	216.246	194.248	13.199	-82.969	-22.690	1.975

T/K	C_p°/ J K^{-1}mol^{-1}	S°/ J K^{-1} mol^{-1}	-(G°-H°(T))/ J K^{-1} mol^{-1}	(H-H°(T))/ kJ mol^{-1}	$\Delta_f H$°/ kJ mol^{-1}	$\Delta_f G$°/ kJ mol^{-1}	Log K_f
700	58.603	224.821	198.008	18.769	-85.023	-12.476	0.931
800	64.084	233.008	201.875	24.907	-86.693	-1.993	0.130
900	69.137	240.852	205.773	31.571	-88.006	8.677	-0.504
1000	73.746	248.379	209.660	38.719	-88.996	19.475	-1.017
1100	77.919	255.607	213.511	46.306	-89.698	30.358	-1.442
1200	81.682	262.551	217.310	54.289	-90.145	41.294	-1.797
1300	85.067	269.225	221.048	62.630	-90.367	52.258	-2.100
1400	88.112	275.643	224.720	71.291	-90.390	63.231	-2.359
1500	90.856	281.817	228.322	80.242	-90.237	74.200	-2.584
12. ACETYLENE C_2H_2 (g)							
298.15	44.036	200.927	200.927	0.000	227.400	209.879	-36.769
300	44.174	201.199	200.927	0.082	227.397	209.770	-36.524
400	50.388	214.814	202.741	4.829	227.161	203.928	-26.630
500	54.751	226.552	206.357	10.097	226.846	198.154	-20.701
600	58.121	236.842	210.598	15.747	226.445	192.452	-16.754
700	60.970	246.021	215.014	21.704	225.968	186.823	-13.941
800	63.511	254.331	219.418	27.931	225.436	181.267	-11.835
900	65.831	261.947	223.726	34.399	224.873	175.779	-10.202
1000	67.960	268.995	227.905	41.090	224.300	170.355	-8.898
1100	69.909	275.565	231.942	47.985	223.734	164.988	-7.835
1200	71.686	281.725	235.837	55.067	223.189	159.672	-6.950
1300	73.299	287.528	239.592	62.317	222.676	154.400	-6.204
1400	74.758	293.014	243.214	69.721	222.203	149.166	-5.565
1500	76.077	298.218	246.709	77.264	221.774	143.964	-5.013
13. ETHYLENE C_2H_4 (g)							
298.15	42.883	219.316	219.316	0.000	52.400	68.358	-11.976
300	43.059	219.582	219.317	0.079	52.341	68.457	-11.919
400	53.045	233.327	221.124	4.881	49.254	74.302	-9.703
500	62.479	246.198	224.864	10.667	46.533	80.887	-8.450
600	70.673	258.332	229.441	17.335	44.221	87.982	-7.659
700	77.733	269.770	234.393	24.764	42.278	95.434	-7.121
800	83.868	280.559	239.496	32.851	40.655	103.142	-6.734
900	89.234	290.754	244.630	41.512	39.310	111.036	-6.444
1000	93.939	300.405	249.730	50.675	38.205	119.067	-6.219
1100	98.061	309.556	254.756	60.280	37.310	127.198	-6.040
1200	101.670	318.247	259.688	70.271	36.596	135.402	-5.894
1300	104.829	326.512	264.513	80.599	36.041	143.660	-5.772
1400	107.594	334.384	269.225	91.223	35.623	151.955	-5.669
1500	110.018	341.892	273.821	102.107	35.327	160.275	-5.581
14. ETHANE C_2H_6 (g)							
298.15	52.487	229.161	229.161	0.000	-84.000	-32.015	5.609
300	52.711	229.487	229.162	0.097	-84.094	-31.692	5.518
400	65.459	246.378	231.379	5.999	-88.988	-13.473	1.759
500	77.941	262.344	235.989	13.177	-93.238	5.912	-0.618
600	89.188	277.568	241.660	21.545	-96.779	26.086	-2.271
700	99.136	292.080	247.835	30.972	-99.663	46.800	-3.492
800	107.936	305.904	254.236	41.334	-101.963	67.887	-4.433
900	115.709	319.075	260.715	52.525	-103.754	89.231	-5.179
1000	122.552	331.628	267.183	64.445	-105.105	110.750	-5.785
1100	128.553	343.597	273.590	77.007	-106.082	132.385	-6.286
1200	133.804	355.012	279.904	90.131	-106.741	154.096	-6.708
1300	138.391	365.908	286.103	103.746	-107.131	175.850	-7.066
1400	142.399	376.314	292.178	117.790	-107.292	197.625	-7.373
1500	145.905	386.260	298.121	132.209	-107.260	219.404	-7.640
15. CYCLOPROPANE C_3H_6 (g)							
298.15	55.571	237.488	237.488	0.000	53.300	104.514	-18.310
300	55.941	237.832	237.489	0.103	53.195	104.832	-18.253
400	76.052	256.695	239.924	6.708	47.967	122.857	-16.043

T/K	$C_p°$/ J K^{-1}mol^{-1}	$S°$/ J K^{-1} mol^{-1}	$-(G°-H°(T))$/ J K^{-1} mol^{-1}	$(H-H°(T))$/ kJ mol^{-1}	$\Delta_f H°$/ kJ mol^{-1}	$\Delta_f G°$/ kJ mol^{-1}	Log K_f
500	93.859	275.637	245.177	15.230	43.730	142.091	-14.844
600	108.542	294.092	251.801	25.374	40.405	162.089	-14.111
700	120.682	311.763	259.115	36.854	37.825	182.583	-13.624
800	130.910	328.564	266.755	49.447	35.854	203.404	-13.281
900	139.658	344.501	274.516	62.987	34.384	224.441	-13.026
1000	147.207	359.616	282.277	77.339	33.334	245.618	-12.830
1100	153.749	373.961	289.965	92.395	32.640	266.883	-12.673
1200	159.432	387.588	297.538	108.060	32.249	288.197	-12.545
1300	164.378	400.549	304.967	124.257	32.119	309.533	-12.437
1400	168.689	412.892	312.239	140.915	32.215	330.870	-12.345
1500	172.453	424.662	319.344	157.976	32.507	352.193	-12.264
16. PROPANE C_3H_8 (g)							
298.15	73.597	270.313	270.313	0.000	-103.847	-23.458	4.110
300	73.931	270.769	270.314	0.136	-103.972	-22.959	3.997
400	94.014	294.739	273.447	8.517	-110.33	15.029	-0.657
500	112.591	317.768	280.025	18.872	-115.658	34.507	-3.605
600	128.700	339.753	288.162	30.955	-119.973	64.961	-5.655
700	142.674	360.668	297.039	44.540	-123.384	96.065	-7.168
800	154.766	380.528	306.245	59.427	-126.016	127.603	-8.331
900	165.352	399.381	315.555	75.444	-127.982	159.430	-9.253
1000	174.598	417.293	324.841	92.452	-129.380	191.444	-10.000
1100	182.673	434.321	334.026	110.325	-130.296	223.574	-10.617
1200	189.745	450.526	343.064	128.954	-130.802	255.770	-11.133
1300	195.853	465.961	351.929	148.241	-130.961	287.993	-11.572
1400	201.209	480.675	360.604	168.100	-130.829	320.217	-11.947
1500	205.895	494.721	369.080	188.460	-130.445	352.422	-12.272
17. BENZENE C_6H_6 (l)							
298.15	135.950	173.450	173.450	0.000	49.080	124.521	-21.815
300	136.312	174.292	173.453	0.252	49.077	124.989	-21.762
400	161.793	216.837	179.082	15.102	48.978	150.320	-19.630
500	207.599	257.048	190.639	33.204	50.330	175.559	-18.340
18. BENZENE C_6H_6 (g)							
298.15	82.430	269.190	269.190	0.000	82.880	129.750	-22.731
300	83.020	269.700	269.190	0.153	82.780	130.040	-22.641
400	113.510	297.840	272.823	10.007	77.780	146.570	-19.140
500	139.340	326.050	280.658	22.696	73.740	164.260	-17.160
600	160.090	353.360	290.517	37.706	70.490	182.680	-15.903
700	176.790	379.330	301.360	54.579	67.910	201.590	-15.042
800	190.460	403.860	312.658	72.962	65.910	220.820	-14.418
900	201.840	426.970	324.084	92.597	64.410	240.280	-13.945
1000	211.430	448.740	335.473	113.267	63.340	259.890	-13.575
1100	219.580	469.280	346.710	134.827	62.620	277.640	-13.184
1200	226.540	488.690	357.743	157.137	62.200	299.320	-13.029
1300	232.520	507.070	368.534	180.097	62.000	319.090	-12.821
1400	237.680	524.490	379.056	203.607	61.990	338.870	-12.643
1500	242.140	541.040	389.302	227.607	62.110	358.640	-12.489
19. NAPHTHALENE $C_{10}H_8$ (cr, l)							
298.15	165.720	167.390	167.390	0.000	78.530	201.585	-35.316
300	167.001	168.419	167.393	0.308	78.466	202.349	-35.232
353.43	208.722	198.948	169.833	10.290	96.099	224.543	-33.186
Phase transition: $\Delta_{trs}H$ = 18.980 kJ/mol, $\Delta_{trs}S$ = 53.702 J/K·mol, cr–l							
353.43	217.200	252.650	169.833	29.270	96.099	224.543	-33.186
400	241.577	280.916	181.124	39.917	96.067	241.475	-31.533
470	276.409	322.712	199.114	58.091	97.012	266.859	-29.658
20. NAPHTHALENE $C_{10}H_8$ (g)							
298.15	131.920	333.150	333.150	0.000	150.580	224.100	-39.260

T/K	$C_p°$/ J K^{-1}mol^{-1}	$S°$/ J K^{-1} mol^{-1}	$-(G°-H°(T))$/ J K^{-1} mol^{-1}	$(H-H°(T))$/ kJ mol^{-1}	$\Delta_f H°$/ kJ mol^{-1}	$\Delta_f G°$/ kJ mol^{-1}	Log K_f
300	132.840	333.970	333.157	0.244	150.450	224.560	-39.098
400	180.070	378.800	338.950	15.940	144.190	250.270	-32.681
500	219.740	423.400	351.400	36.000	139.220	277.340	-28.973
600	251.530	466.380	367.007	59.624	135.350	305.330	-26.581
700	277.010	507.140	384.146	86.096	132.330	333.950	-24.919
800	297.730	545.520	401.935	114.868	130.050	362.920	-23.696
900	314.850	581.610	419.918	145.523	128.430	392.150	-22.759
1000	329.170	615.550	437.806	177.744	127.510	421.700	-22.027
1100	341.240	647.500	455.426	211.281	127.100	450.630	-21.398
1200	351.500	677.650	472.707	245.932	126.960	480.450	-20.913
1300	360.260	706.130	489.568	281.531	127.060	509.770	-20.482
1400	367.780	733.110	506.009	317.941	127.390	539.740	-20.137
1500	374.270	758.720	522.019	355.051	127.920	568.940	-19.812
21. FORMALDEHYDE H_2CO (g)							
298.15	35.387	218.760	218.760	0.000	-108.700	-102.667	17.987
300	35.443	218.979	218.761	0.066	-108.731	-102.630	17.869
400	39.240	229.665	220.192	3.789	-110.438	-100.340	13.103
500	43.736	238.900	223.028	7.936	-112.073	-97.623	10.198
600	48.181	247.270	226.381	12.534	-113.545	-94.592	8.235
700	52.280	255.011	229.924	17.560	-114.833	-91.328	6.815
800	55.941	262.236	233.517	22.975	-115.942	-87.893	5.739
900	59.156	269.014	237.088	28.734	-116.889	-84.328	4.894
1000	61.951	275.395	240.603	34.792	-117.696	-80.666	4.213
1100	64.368	281.416	244.042	41.111	-118.382	-76.929	3.653
1200	66.453	287.108	247.396	47.655	-118.966	-73.134	3.183
1300	68.251	292.500	250.660	54.392	-119.463	-69.294	2.784
1400	69.803	297.616	253.833	61.297	-119.887	-65.418	2.441
1500	71.146	302.479	256.915	68.346	-120.249	-61.514	2.142
22. METHANOL CH_3OH (g)							
298.15	44.101	239.865	239.865	0.000	-201.000	-162.298	28.434
300	44.219	240.139	239.866	0.082	-201.068	-162.057	28.216
400	51.713	253.845	241.685	4.864	-204.622	-148.509	19.393
500	59.800	266.257	245.374	10.442	-207.750	-134.109	14.010
600	67.294	277.835	249.830	16.803	-210.387	-119.125	10.371
700	73.958	288.719	254.616	23.873	-212.570	-103.737	7.741
800	79.838	298.987	259.526	31.569	-214.350	-88.063	5.750
900	85.025	308.696	264.455	39.817	-215.782	-72.188	4.190
1000	89.597	317.896	269.343	48.553	-216.916	-56.170	2.934
1100	93.624	326.629	274.158	57.718	-217.794	-40.050	1.902
1200	97.165	334.930	278.879	67.262	-218.457	-23.861	1.039
1300	100.277	342.833	283.497	77.137	-218.936	-7.624	0.306
1400	103.014	350.367	288.007	87.304	-219.261	8.644	-0.322
1500	105.422	357.558	292.405	97.729	-219.456	24.930	-0.868
23. ACETALDEHYDE C_2H_4O (g)							
298.15	55.318	263.840	263.840	0.000	-166.190	-133.010	23.302
300	55.510	264.180	263.837	0.103	-166.250	-132.800	23.122
400	66.282	281.620	266.147	6.189	-169.530	-121.130	15.818
500	76.675	297.540	270.850	13.345	-172.420	-108.700	11.356
600	85.942	312.360	276.550	21.486	-174.870	-95.720	8.334
700	94.035	326.230	282.667	30.494	-176.910	-82.350	6.145
800	101.070	339.260	288.938	40.258	-178.570	-68.730	4.487
900	107.190	351.520	295.189	50.698	-179.880	-54.920	3.187
1000	112.490	363.100	301.431	61.669	-180.850	-40.930	2.138
1100	117.080	374.040	307.537	73.153	-181.560	-27.010	1.283
1200	121.060	384.400	313.512	85.065	-182.070	-12.860	0.560
1300	124.500	394.230	319.350	97.344	-182.420	1.240	-0.050
1400	127.490	403.570	325.031	109.954	-182.640	15.470	-0.577
1500	130.090	412.460	330.571	122.834	-182.750	29.580	-1.030

T/K	$C_p°$/ J K^{-1}mol^{-1}	$S°$/ J K^{-1} mol^{-1}	-($G°$-$H°(T)$)/ J K^{-1} mol^{-1}	(H-$H°(T)$)/ kJ mol^{-1}	$\Delta_f H°$/ kJ mol^{-1}	$\Delta_f G°$/ kJ mol^{-1}	Log K_f
24. ETHANOL C_2H_5OH (g)							
298.15	65.652	281.622	281.622	0.000	-234.800	-167.874	29.410
300	65.926	282.029	281.623	0.122	-234.897	-167.458	29.157
400	81.169	303.076	284.390	7.474	-239.826	-144.216	18.832
500	95.400	322.750	290.115	16.318	-243.940	-119.820	12.517
600	107.656	341.257	297.112	26.487	-247.260	-94.672	8.242
700	118.129	358.659	304.674	37.790	-249.895	-69.023	5.151
800	127.171	375.038	312.456	50.065	-251.951	-43.038	2.810
900	135.049	390.482	320.276	63.185	-253.515	-16.825	0.976
1000	141.934	405.075	328.033	77.042	-254.662	9.539	-0.498
1100	147.958	418.892	335.670	91.543	-255.454	36.000	-1.709
1200	153.232	431.997	343.156	106.609	-255.947	62.520	-2.721
1300	157.849	444.448	350.473	122.168	-256.184	89.070	-3.579
1400	161.896	456.298	357.612	138.160	-256.206	115.630	-4.314
1500	165.447	467.591	364.571	154.531	-256.044	142.185	-4.951
25. ACETIC ACID $C_2H_4O_2$ (g)							
298.15	63.438	283.470	283.470	0.000	-432.249	-374.254	65.567
300	63.739	283.863	283.471	0.118	-432.324	-373.893	65.100
400	79.665	304.404	286.164	7.296	-436.006	-353.840	46.206
500	93.926	323.751	291.765	15.993	-438.875	-332.950	34.783
600	106.181	341.988	298.631	26.014	-440.993	-311.554	27.123
700	116.627	359.162	306.064	37.169	-442.466	-289.856	21.629
800	125.501	375.331	313.722	49.287	-443.395	-267.985	17.497
900	132.989	390.558	321.422	62.223	-443.873	-246.026	14.279
1000	139.257	404.904	329.060	75.844	-443.982	-224.034	11.702
1100	144.462	418.429	336.576	90.039	-443.798	-202.046	9.594
1200	148.760	431.189	343.933	104.707	-443.385	-180.086	7.839
1300	152.302	443.240	351.113	119.765	-442.795	-158.167	6.355
1400	155.220	454.637	358.105	135.146	-442.071	-136.299	5.085
1500	157.631	465.432	364.903	150.793	-441.247	-114.486	3.987
26. ACETONE C_3H_6O (g)							
298.15	74.517	295.349	295.349	0.000	-217.150	-152.716	26.757
300	74.810	295.809	295.349	0.138	-217.233	-152.339	26.521
400	91.755	319.658	298.498	8.464	-222.212	-129.913	16.962
500	107.864	341.916	304.988	18.464	-226.522	-106.315	11.107
600	122.047	362.836	312.873	29.978	-230.120	-81.923	7.133
700	134.306	382.627	321.470	42.810	-233.049	-56.986	4.252
800	144.934	401.246	330.265	56.785	-235.350	-31.673	2.068
900	154.097	418.860	339.141	71.747	-237.149	-6.109	0.353
1000	162.046	435.513	347.950	87.563	-238.404	19.707	-1.030
1100	168.908	451.286	356.617	104.136	-239.283	45.396	-2.157
1200	174.891	466.265	365.155	121.332	-239.827	71.463	-3.110
1300	180.079	480.491	373.513	139.072	-240.120	97.362	-3.912
1400	184.556	493.963	381.596	157.314	-240.203	123.470	-4.607
1500	188.447	506.850	389.533	175.975	-240.120	149.369	-5.202
27. PHENOL C_6H_6O (g)							
298.15	103.220	314.810	314.810	0.000	-96.400	-32.630	5.720
300	103.860	315.450	314.810	0.192	-96.490	-32.230	5.610
400	135.790	349.820	319.278	12.217	-100.870	-10.180	1.330
500	161.910	383.040	328.736	27.152	-104.240	12.970	-1.360
600	182.480	414.450	340.430	44.412	-106.810	36.650	-3.190
700	198.840	443.860	353.134	63.508	-108.800	60.750	-4.530
800	212.140	471.310	366.211	84.079	-110.300	85.020	-5.550
900	223.190	496.950	379.327	105.861	-111.370	109.590	-6.360
1000	232.490	520.960	392.302	128.658	-111.990	134.280	-7.010
1100	240.410	543.500	405.033	152.314	-112.280	158.620	-7.530
1200	247.200	564.720	417.468	176.703	-112.390	183.350	-7.980
1300	253.060	584.740	429.568	201.723	-112.330	208.070	-8.360

T/K	$C_p°$/ J K^{-1}mol^{-1}	S°/ J K^{-1} mol^{-1}	-(G°-H°(T))/ J K^{-1} mol^{-1}	(H-H°(T))/ kJ mol^{-1}	$\Delta_f H°$/ kJ mol^{-1}	$\Delta_f G°$/ kJ mol^{-1}	Log K_f
1400	258.120	603.680	441.331	227.288	-112.120	233.050	-8.700
1500	262.520	621.650	452.767	253.325	-111.780	257.540	-8.970
28. CARBON TETRAFLUORIDE CF_4 (g)							
298.15	61.050	261.455	261.455	0.000	-933.200	-888.518	155.663
300	61.284	261.833	261.456	0.113	-933.219	-888.240	154.654
400	72.399	281.057	264.001	6.822	-933.986	-873.120	114.016
500	80.713	298.153	269.155	14.499	-934.372	-857.852	89.618
600	86.783	313.434	275.284	22.890	-934.490	-842.533	73.348
700	91.212	327.162	281.732	31.801	-934.431	-827.210	61.726
800	94.479	339.566	288.199	41.094	-934.261	-811.903	53.011
900	96.929	350.842	294.542	50.670	-934.024	-796.622	46.234
1000	98.798	361.156	300.695	60.460	-933.745	-781.369	40.814
1100	100.250	370.643	306.629	70.416	-933.442	-766.146	36.381
1200	101.396	379.417	312.334	80.500	-933.125	-750.952	32.688
1300	102.314	387.571	317.811	90.687	-932.800	-735.784	29.564
1400	103.059	395.181	323.069	100.957	-932.470	-720.641	26.887
1500	103.671	402.313	328.116	111.295	-932.137	-705.522	24.568
29. TRIFLUOROMETHANE CHF_3 (g)							
298.15	51.069	259.675	259.675	0.000	-696.700	-662.237	116.020
300	51.258	259.991	259.676	0.095	-696.735	-662.023	115.267
400	61.148	276.113	261.807	5.722	-698.427	-650.186	84.905
500	69.631	290.700	266.149	12.275	-699.715	-637.969	66.647
600	76.453	304.022	271.368	19.593	-700.634	-625.528	54.456
700	81.868	316.230	276.917	27.519	-701.253	-612.957	45.739
800	86.201	327.455	282.542	35.930	-701.636	-600.315	39.196
900	89.719	337.818	288.116	44.732	-701.832	-587.636	34.105
1000	92.617	347.426	293.572	53.854	-701.879	-574.944	30.032
1100	95.038	356.370	298.879	63.240	-701.805	-562.253	26.699
1200	97.084	364.730	304.022	72.849	-701.629	-549.574	23.922
1300	98.833	372.571	308.997	82.647	-701.368	-536.913	21.573
1400	100.344	379.952	313.804	92.607	-701.033	-524.274	19.561
1500	101.660	386.921	318.449	102.709	-700.635	-511.662	17.817
30. CHLOROTRIFLUOROMETHANE $CClF_3$ (g)							
298.15	66.886	285.419	285.419	0.000	-707.800	-667.238	116.896
300	67.111	285.834	285.421	0.124	-707.810	-666.986	116.131
400	77.528	306.646	288.187	7.383	-708.153	-653.316	85.313
500	85.013	324.797	293.734	15.532	-708.170	-639.599	66.818
600	90.329	340.794	300.271	24.314	-707.975	-625.901	54.489
700	94.132	355.020	307.096	33.547	-707.654	-612.246	45.686
800	96.899	367.780	313.897	43.106	-707.264	-598.642	39.087
900	98.951	379.317	320.536	52.903	-706.837	-585.090	33.957
1000	100.507	389.827	326.947	62.880	-706.396	-571.586	29.856
1100	101.708	399.465	333.108	72.993	-705.950	-558.126	26.503
1200	102.651	408.357	339.013	83.213	-705.505	-544.707	23.710
1300	103.404	416.604	344.668	93.517	-705.064	-531.326	21.349
1400	104.012	424.290	350.084	103.889	-704.628	-517.977	19.326
1500	104.512	431.484	355.273	114.316	-704.196	-504.660	17.574
31. DICHLORODIFLUOROMETHANE CCl_2F_2 (g)							
298.15	72.476	300.903	300.903	0.000	-486.000	-447.030	78.317
300	72.691	301.352	300.905	0.134	-486.002	-446.788	77.792
400	82.408	323.682	303.883	7.919	-485.945	-433.716	56.637
500	89.063	342.833	309.804	16.514	-485.618	-420.692	43.949
600	93.635	359.500	316.729	25.663	-485.136	-407.751	35.497
700	96.832	374.189	323.909	35.196	-484.576	-394.897	29.467
800	99.121	387.276	331.027	44.999	-483.984	-382.126	24.950
900	100.801	399.053	337.942	55.000	-483.388	-369.429	21.441
1000	102.062	409.742	344.596	65.146	-482.800	-356.799	18.637
1100	103.030	419.517	350.969	75.402	-482.226	-344.227	16.346

T/K	$C_p°$/ J K⁻¹mol⁻¹	$S°$/ J K⁻¹ mol⁻¹	-($G°$-$H°(T)$)/ J K⁻¹ mol⁻¹	(H-$H°(T)$)/ kJ mol⁻¹	$\Delta_f H°$/ kJ mol⁻¹	$\Delta_f G°$/ kJ mol⁻¹	Log K_f
1200	103.786	428.515	357.061	85.745	-481.667	-331.706	14.439
1300	104.388	436.847	362.882	96.154	-481.121	-319.232	12.827
1400	104.874	444.602	368.445	106.618	-480.588	-306.799	11.447
1500	105.270	451.851	373.767	117.126	-480.065	-294.404	10.252
32. CHLORODIFLUOROMETHANE $CHClF_2$ (g)							
298.15	55.853	280.915	280.915	0.000	-475.000	-443.845	77.759
300	56.039	281.261	280.916	0.104	-475.028	-443.652	77.246
400	65.395	298.701	283.231	6.188	-476.390	-432.978	56.540
500	73.008	314.145	287.898	13.123	-477.398	-422.001	44.086
600	78.940	328.003	293.448	20.733	-478.103	-410.851	35.767
700	83.551	340.533	299.294	28.867	-478.574	-399.603	29.818
800	87.185	351.936	305.172	37.411	-478.870	-388.299	25.353
900	90.100	362.379	310.956	46.280	-479.031	-376.967	21.878
1000	92.475	371.999	316.586	55.413	-479.090	-365.622	19.098
1100	94.433	380.908	322.033	64.761	-479.068	-354.276	16.823
1200	96.066	389.196	327.289	74.289	-478.982	-342.935	14.927
1300	97.438	396.941	332.352	83.966	-478.843	-331.603	13.324
1400	98.601	404.206	337.228	93.769	-478.661	-320.283	11.950
1500	99.593	411.044	341.923	103.681	-478.443	-308.978	10.759
33. METHYLAMINE CH_5N (g)							
298.15	50.053	242.881	242.881	0.000	-22.529	32.734	-5.735
300	50.227	243.196	242.893	0.091	-22.614	33.077	-5.759
400	60.171	258.986	244.975	5.604	-26.846	52.294	-6.829
500	70.057	273.486	249.244	12.121	-30.431	72.510	-7.575
600	78.929	287.063	254.431	19.579	-33.364	93.382	-8.129
700	86.711	299.826	260.008	27.873	-35.712	114.702	-8.559
800	93.545	311.865	265.749	36.893	-37.548	136.316	-8.900
900	99.573	323.239	271.511	46.555	-38.949	158.138	-9.178
1000	104.886	334.006	277.220	56.786	-39.967	180.098	-9.407
1100	109.576	344.233	282.861	67.509	-40.681	201.822	-9.584
1200	113.708	353.944	288.374	78.685	-41.136	224.240	-9.761
1300	117.341	363.190	293.775	90.239	-41.376	246.364	-9.899
1400	120.542	372.012	299.061	102.131	-41.451	268.504	-10.018
1500	123.353	380.426	304.209	114.326	-41.381	290.639	-10.121
34. CHLORINE Cl (g)							
298.15	21.838	165.190	165.190	0.000	121.302	105.306	-18.449
300	21.852	165.325	165.190	0.040	121.311	105.207	-18.318
400	22.467	171.703	166.055	2.259	121.795	99.766	-13.028
500	22.744	176.752	167.708	4.522	122.272	94.203	-9.841
600	22.781	180.905	169.571	6.800	122.734	88.546	-7.709
700	22.692	184.411	171.448	9.074	123.172	82.813	-6.179
800	22.549	187.432	173.261	11.337	123.585	77.019	-5.029
900	22.389	190.079	174.986	13.584	123.971	71.175	-4.131
1000	22.233	192.430	176.615	15.815	124.334	65.289	-3.410
1100	22.089	194.542	178.150	18.031	124.675	59.368	-2.819
1200	21.959	196.458	179.597	20.233	124.996	53.416	-2.325
1300	21.843	198.211	180.963	22.423	125.299	47.439	-1.906
1400	21.742	199.826	182.253	24.602	125.587	41.439	-1.546
1500	21.652	201.323	183.475	26.772	125.861	35.418	-1.233
35. DICHLORINE Cl_2 (g)							
298.15	33.949	223.079	223.079	0.000	0.000	0.000	0.000
300	33.981	223.290	223.080	0.063	0.000	0.000	0.000
400	35.296	233.263	224.431	3.533	0.000	0.000	0.000
500	36.064	241.229	227.021	7.104	0.000	0.000	0.000
600	36.547	247.850	229.956	10.736	0.000	0.000	0.000
700	36.874	253.510	232.926	14.408	0.000	0.000	0.000
800	37.111	258.450	235.815	18.108	0.000	0.000	0.000
900	37.294	262.832	238.578	21.829	0.000	0.000	0.000

T/K	$C_p°$/ J K^{-1}mol^{-1}	$S°$/ J K^{-1} mol^{-1}	-($G°$-$H°$(T))/ J K^{-1} mol^{-1}	(H-$H°$(T))/ kJ mol^{-1}	$\Delta_f H°$/ kJ mol^{-1}	$\Delta_f G°$/ kJ mol^{-1}	Log K_f
1000	37.442	266.769	241.203	25.566	0.000	0.000	0.000
1100	37.567	270.343	243.692	29.316	0.000	0.000	0.000
1200	37.678	273.617	246.052	33.079	0.000	0.000	0.000
1300	37.778	276.637	248.290	36.851	0.000	0.000	0.000
1400	37.872	279.440	250.416	40.634	0.000	0.000	0.000
1500	37.961	282.056	252.439	44.426	0.000	0.000	0.000
36. HYDROGEN CHLORIDE HCl (g)							
298.15	29.136	186.902	186.902	0.000	-92.310	-95.298	16.696
300	29.137	187.082	186.902	0.054	-92.314	-95.317	16.596
400	29.175	195.468	188.045	2.969	-92.587	-96.278	12.573
500	29.304	201.990	190.206	5.892	-92.911	-97.164	10.151
600	29.576	207.354	192.630	8.835	-93.249	-97.983	8.530
700	29.988	211.943	195.069	11.812	-93.577	-98.746	7.368
800	30.500	215.980	197.435	14.836	-93.879	-99.464	6.494
900	31.063	219.604	199.700	17.913	-94.149	-100.145	5.812
1000	31.639	222.907	201.858	21.049	-94.384	-100.798	5.265
1100	32.201	225.949	203.912	24.241	-94.587	-101.430	4.816
1200	32.734	228.774	205.867	27.488	-94.760	-102.044	4.442
1300	33.229	231.414	207.732	30.786	-94.908	-102.645	4.124
1400	33.684	233.893	209.513	34.132	-95.035	-103.235	3.852
1500	34.100	236.232	211.217	37.522	-95.146	-103.817	3.615
37. COPPER Cu (cr, l)							
298.15	24.440	33.150	33.150	0.000	0.000	0.000	0.000
300	24.460	33.301	33.150	0.045	0.000	0.000	0.000
400	25.339	40.467	34.122	2.538	0.000	0.000	0.000
500	25.966	46.192	35.982	5.105	0.000	0.000	0.000
600	26.479	50.973	38.093	7.728	0.000	0.000	0.000
700	26.953	55.090	40.234	10.399	0.000	0.000	0.000
800	27.448	58.721	42.322	13.119	0.000	0.000	0.000
900	28.014	61.986	44.328	15.891	0.000	0.000	0.000
1000	28.700	64.971	46.245	18.726	0.000	0.000	0.000
1100	29.553	67.745	48.075	21.637	0.000	0.000	0.000
1200	30.617	70.361	49.824	24.644	0.000	0.000	0.000
1300	31.940	72.862	51.501	27.769	0.000	0.000	0.000
1358	32.844	74.275	52.443	29.647	0.000	0.000	0.000
Phase transition: $\Delta_{trs}H$ = 13.141 kJ/mol, $\Delta_{trs}S$ = 9.676 J/K·mol, cr–l							
1358	32.800	83.951	52.443	42.788	0.000	0.000	0.000
1400	32.800	84.950	53.403	44.166	0.000	0.000	0.000
1500	32.800	87.213	55.583	47.446	0.000	0.000	0.000
38. COPPER Cu (g)							
298.15	20.786	166.397	166.397	0.000	337.600	297.873	-52.185
300	20.786	166.525	166.397	0.038	337.594	297.626	-51.821
400	20.786	172.505	167.213	2.117	337.179	284.364	-37.134
500	20.786	177.143	168.752	4.196	336.691	271.215	-28.333
600	20.786	180.933	170.476	6.274	336.147	258.170	-22.475
700	20.786	184.137	172.205	8.353	335.554	245.221	-18.298
800	20.786	186.913	173.874	10.431	334.913	232.359	-15.171
900	20.786	189.361	175.461	12.510	334.219	219.581	-12.744
1000	20.786	191.551	176.963	14.589	333.463	206.883	-10.806
1100	20.788	193.532	178.380	16.667	332.631	194.265	-9.225
1200	20.793	195.341	179.719	18.746	331.703	181.726	-7.910
1300	20.803	197.006	180.986	20.826	330.657	169.270	-6.801
1400	20.823	198.548	182.186	22.907	316.342	157.305	-5.869
1500	20.856	199.986	183.325	24.991	315.146	145.987	-5.084
39. COPPER OXIDE CuO (cr)							
298.15	42.300	42.740	42.740	0.000	-162.000	-134.277	23.524
300	42.417	43.002	42.741	0.078	-161.994	-134.105	23.349

T/K	$C_p°$/ J K⁻¹mol⁻¹	S°/ J K⁻¹ mol⁻¹	-(G°-H°(T))/ J K⁻¹ mol⁻¹	(H-H°(T))/ kJ mol⁻¹	$\Delta_f H°$/ kJ mol⁻¹	$\Delta_f G°$/ kJ mol⁻¹	Log K_f
400	46.783	55.878	44.467	4.564	-161.487	-124.876	16.307
500	49.190	66.596	47.852	9.372	-160.775	-115.803	12.098
600	50.827	75.717	51.755	14.377	-159.973	-106.883	9.305
700	52.099	83.651	55.757	19.526	-159.124	-98.102	7.320
800	53.178	90.680	59.691	24.791	-158.247	-89.444	5.840
900	54.144	97.000	63.491	30.158	-157.356	-80.897	4.695
1000	55.040	102.751	67.134	35.617	-156.462	-72.450	3.784
1100	55.890	108.037	70.615	41.164	-155.582	-64.091	3.043
1200	56.709	112.936	73.941	46.794	-154.733	-55.812	2.429
1300	57.507	117.507	77.118	52.505	-153.940	-47.601	1.913
1400	58.288	121.797	80.158	58.295	-166.354	-39.043	1.457
1500	59.057	125.845	83.070	64.163	-165.589	-29.975	1.044
40. DICOPPER OXIDE Cu_2O (cr)							
298.15	62.600	92.550	92.550	0.000	-173.100	-150.344	26.339
300	62.721	92.938	92.551	0.116	-173.102	-150.203	26.152
400	67.587	111.712	95.078	6.654	-173.036	-142.572	18.618
500	70.784	127.155	99.995	13.580	-172.772	-134.984	14.101
600	73.323	140.291	105.643	20.789	-172.389	-127.460	11.096
700	75.552	151.764	111.429	28.235	-171.914	-120.009	8.955
800	77.616	161.989	117.121	35.894	-171.363	-112.631	7.354
900	79.584	171.245	122.629	43.755	-170.750	-105.325	6.113
1000	81.492	179.729	127.920	51.809	-170.097	-98.091	5.124
1100	83.360	187.584	132.992	60.052	-169.431	-90.922	4.317
1200	85.202	194.917	137.850	68.480	-168.791	-83.814	3.648
1300	87.026	201.808	142.507	77.092	-168.223	-76.756	3.084
1400	88.836	208.324	146.978	85.885	-194.030	-68.926	2.572
1500	90.636	214.515	151.276	94.858	-193.438	-60.010	2.090
41. COPPER DICHLORIDE $CuCl_2$ (cr, l)							
298.15	71.880	108.070	108.070	0.000	-218.000	-173.826	30.453
300	71.998	108.515	108.071	0.133	-217.975	-173.552	30.218
400	76.338	129.899	110.957	7.577	-216.494	-158.962	20.758
500	78.654	147.204	116.532	15.336	-214.873	-144.765	15.123
600	80.175	161.687	122.884	23.282	-213.182	-130.901	11.396
675	81.056	171.183	127.732	29.329	-211.185	-120.693	9.340
Phase transition: $\Delta_{trs}H$ = 0.700 kJ/mol, $\Delta_{trs}S$ = 1.037 J/K·mol, crII–crI							
675	82.400	172.220	127.732	30.029	-211.185	-120.693	9.340
700	82.400	175.216	129.375	32.089	-210.719	-117.350	8.757
800	82.400	186.219	135.808	40.329	-208.898	-104.137	6.799
871	82.400	193.226	140.207	46.179	-192.649	-94.893	5.691
Phase transition: $\Delta_{trs}H$ = 15.001 kJ/mol, $\Delta_{trs}S$ = 17.221 J/K·mol, crI–l							
871	100.000	210.447	140.207	61.180	-192.649	-94.893	5.691
900	100.000	213.723	142.523	64.080	-191.640	-91.655	5.319
1000	100.000	224.259	150.179	74.080	-188.212	-80.730	4.217
1100	100.000	233.790	157.353	84.080	-184.873	-70.144	3.331
1130.75	100.000	236.547	159.470	87.155	-183.867	-66.951	3.093
42. COPPER DICHLORIDE $CuCl_2$ (g)							
298.15	56.814	278.418	278.418	0.000	-43.268	-49.883	8.739
300	56.869	278.769	278.419	0.105	-43.271	-49.924	8.692
400	58.992	295.456	280.679	5.911	-43.428	-52.119	6.806
500	60.111	308.752	285.010	11.871	-43.606	-54.271	5.670
600	60.761	319.774	289.911	17.918	-43.814	-56.385	4.909
700	61.168	329.173	294.865	24.015	-44.060	-58.462	4.362
800	61.439	337.360	299.677	30.147	-44.349	-60.500	3.950
900	61.630	344.608	304.274	36.301	-44.688	-62.499	3.627
1000	61.776	351.109	308.638	42.471	-45.088	-64.457	3.367
1100	61.900	357.003	312.771	48.655	-45.566	-66.372	3.152
1200	62.022	362.394	316.685	54.851	-46.139	-68.239	2.970

T/K	$C_p°$/ J K^{-1}mol^{-1}	$S°$/ J K^{-1} mol^{-1}	$-(G°-H°(T))$/ J K^{-1} mol^{-1}	$(H-H°(T))$/ kJ mol^{-1}	$\Delta_f H°$/ kJ mol^{-1}	$\Delta_f G°$/ kJ mol^{-1}	Log K_f
1300	62.159	367.364	320.395	61.060	-46.829	-70.053	2.815
1400	62.325	371.976	323.916	67.284	-60.784	-71.404	2.664
1500	62.531	376.283	327.265	73.526	-61.613	-72.133	2.512

43. FLUORINE F (g)

T/K	$C_p°$/ J K^{-1}mol^{-1}	$S°$/ J K^{-1} mol^{-1}	$-(G°-H°(T))$/ J K^{-1} mol^{-1}	$(H-H°(T))$/ kJ mol^{-1}	$\Delta_f H°$/ kJ mol^{-1}	$\Delta_f G°$/ kJ mol^{-1}	Log K_f
298.15	22.746	158.750	158.750	0.000	79.380	62.280	-10.911
300	22.742	158.891	158.750	0.042	79.393	62.173	-10.825
400	22.432	165.394	159.639	2.302	80.043	56.332	-7.356
500	22.100	170.363	161.307	4.528	80.587	50.340	-5.259
600	21.832	174.368	163.161	6.724	81.046	44.246	-3.852
700	21.629	177.717	165.008	8.897	81.442	38.081	-2.842
800	21.475	180.595	166.780	11.052	81.792	31.862	-2.080
900	21.357	183.117	168.458	13.193	82.106	25.601	-1.486
1000	21.266	185.362	170.039	15.324	82.391	19.308	-1.009
1100	21.194	187.386	171.525	17.447	82.654	12.986	-0.617
1200	21.137	189.227	172.925	19.563	82.897	6.642	-0.289
1300	21.091	190.917	174.245	21.675	83.123	0.278	-0.011
1400	21.054	192.479	175.492	23.782	83.335	-6.103	0.228
1500	21.022	193.930	176.673	25.886	83.533	-12.498	0.435

44. DIFLUORINE F_2 (g)

T/K	$C_p°$/ J K^{-1}mol^{-1}	$S°$/ J K^{-1} mol^{-1}	$-(G°-H°(T))$/ J K^{-1} mol^{-1}	$(H-H°(T))$/ kJ mol^{-1}	$\Delta_f H°$/ kJ mol^{-1}	$\Delta_f G°$/ kJ mol^{-1}	Log K_f
298.15	31.304	202.790	202.790	0.000	0.000	0.000	0.000
300	31.337	202.984	202.790	0.058	0.000	0.000	0.000
400	32.995	212.233	204.040	3.277	0.000	0.000	0.000
500	34.258	219.739	206.453	6.643	0.000	0.000	0.000
600	35.171	226.070	209.208	10.117	0.000	0.000	0.000
700	35.839	231.545	212.017	13.669	0.000	0.000	0.000
800	36.343	236.365	214.765	17.279	0.000	0.000	0.000
900	36.740	240.669	217.409	20.934	0.000	0.000	0.000
1000	37.065	244.557	219.932	24.625	0.000	0.000	0.000
1100	37.342	248.103	222.334	28.346	0.000	0.000	0.000
1200	37.588	251.363	224.619	32.093	0.000	0.000	0.000
1300	37.811	254.381	226.794	35.863	0.000	0.000	0.000
1400	38.019	257.191	228.866	39.654	0.000	0.000	0.000
1500	38.214	259.820	230.843	43.466	0.000	0.000	0.000

45. HYDROGEN FLUORIDE HF (g)

T/K	$C_p°$/ J K^{-1}mol^{-1}	$S°$/ J K^{-1} mol^{-1}	$-(G°-H°(T))$/ J K^{-1} mol^{-1}	$(H-H°(T))$/ kJ mol^{-1}	$\Delta_f H°$/ kJ mol^{-1}	$\Delta_f G°$/ kJ mol^{-1}	Log K_f
298.15	29.137	173.776	173.776	0.000	-273.300	-275.399	48.248
300	29.137	173.956	173.776	0.054	-273.302	-275.412	47.953
400	29.149	182.340	174.919	2.968	-273.450	-276.096	36.054
500	29.172	188.846	177.078	5.884	-273.679	-276.733	28.910
600	29.230	194.169	179.496	8.804	-273.961	-277.318	24.142
700	29.350	198.683	181.923	11.732	-274.277	-277.852	20.733
800	29.549	202.614	184.269	14.676	-274.614	-278.340	18.174
900	29.827	206.110	186.505	17.645	-274.961	-278.785	16.180
1000	30.169	209.270	188.626	20.644	-275.309	-279.191	14.583
1100	30.558	212.163	190.636	23.680	-275.652	-279.563	13.275
1200	30.974	214.840	192.543	26.756	-275.988	-279.904	12.184
1300	31.403	217.336	194.355	29.875	-276.315	-280.217	11.259
1400	31.831	219.679	196.081	33.037	-276.631	-280.505	10.466
1500	32.250	221.889	197.729	36.241	-276.937	-280.771	9.777

46. GERMANIUM Ge (cr, l)

T/K	$C_p°$/ J K^{-1}mol^{-1}	$S°$/ J K^{-1} mol^{-1}	$-(G°-H°(T))$/ J K^{-1} mol^{-1}	$(H-H°(T))$/ kJ mol^{-1}	$\Delta_f H°$/ kJ mol^{-1}	$\Delta_f G°$/ kJ mol^{-1}	Log K_f
298.15	23.222	31.090	31.090	0.000	0.000	0.000	0.000
300	23.249	31.234	31.090	0.043	0.000	0.000	0.000
400	24.310	38.083	32.017	2.426	0.000	0.000	0.000
500	24.962	43.582	33.798	4.892	0.000	0.000	0.000
600	25.452	48.178	35.822	7.414	0.000	0.000	0.000
700	25.867	52.133	37.876	9.980	0.000	0.000	0.000
800	26.240	55.612	39.880	12.586	0.000	0.000	0.000
900	26.591	58.723	41.804	15.227	0.000	0.000	0.000
1000	26.926	61.542	43.639	17.903	0.000	0.000	0.000

T/K	$C_p°$/ J K^{-1}mol^{-1}	S°/ J K^{-1} mol^{-1}	-(G°-H°(T))/ J K^{-1} mol^{-1}	(H-H°(T))/ kJ mol^{-1}	$\Delta_f H°$/ kJ mol^{-1}	$\Delta_f G°$/ kJ mol^{-1}	Log K_f
1100	27.252	64.124	45.386	20.612	0.000	0.000	0.000
1200	27.571	66.509	47.048	23.353	0.000	0.000	0.000
1211.4	27.608	66.770	47.232	23.668	0.000	0.000	0.000
Phase transition: $\Delta_{trs}H$ = 37.030 kJ/mol, $\Delta_{trs}S$ = 30.568 J/K·mol, cr–l							
1211.4	27.600	97.338	47.232	60.698	0.000	0.000	0.000
1300	27.600	99.286	50.714	63.143	0.000	0.000	0.000
1400	27.600	101.331	54.258	65.903	0.000	0.000	0.000
1500	27.600	103.236	57.460	68.663	0.000	0.000	0.000
47. GERMANIUM Ge (g)							
298.15	30.733	167.903	167.903	0.000	367.800	327.009	-57.290
300	30.757	168.094	167.904	0.057	367.814	326.756	-56.893
400	31.071	177.025	169.119	3.162	368.536	312.959	-40.868
500	30.360	183.893	171.415	6.239	369.147	298.991	-31.235
600	29.265	189.334	173.965	9.222	369.608	284.914	-24.804
700	28.102	193.758	176.487	12.090	369.910	270.773	-20.205
800	27.029	197.439	178.882	14.845	370.060	256.598	-16.754
900	26.108	200.567	181.122	17.501	370.073	242.414	-14.069
1000	25.349	203.277	183.205	20.072	369.969	228.234	-11.922
1100	24.741	205.664	185.141	22.575	369.763	214.069	-10.165
1200	24.264	207.795	186.941	25.025	369.471	199.928	-8.703
1300	23.898	209.722	188.621	27.432	332.088	188.521	-7.575
1400	23.624	211.483	190.192	29.807	331.704	177.492	-6.622
1500	23.426	213.105	191.666	32.159	331.296	166.491	-5.798
48. GERMANIUM DIOXIDE GeO_2 (cr, l)							
298.15	50.166	39.710	39.710	0.000	-580.200	-521.605	91.382
300	50.475	40.021	39.711	0.093	-580.204	-521.242	90.755
400	61.281	56.248	41.850	5.759	-579.893	-501.610	65.503
500	66.273	70.519	46.191	12.164	-579.013	-482.134	50.368
600	69.089	82.872	51.299	18.943	-577.915	-462.859	40.295
700	70.974	93.671	56.597	25.952	-576.729	-443.776	33.115
800	72.449	103.247	61.841	33.125	-575.498	-424.866	27.741
900	73.764	111.857	66.928	40.436	-574.235	-406.113	23.570
1000	75.049	119.696	71.819	47.877	-572.934	-387.502	20.241
1100	76.378	126.910	76.504	55.447	-571.582	-369.024	17.523
1200	77.796	133.616	80.987	63.155	-570.166	-350.671	15.264
1300	79.332	139.903	85.279	71.010	-605.685	-329.732	13.249
1308	79.460	140.390	85.615	71.646	-584.059	-328.034	13.100
Phase transition: $\Delta_{trs}H$ = 21.500 kJ/mol, $\Delta_{trs}S$ = 16.437 J/K·mol, crII–crI							
1308	80.075	156.827	85.615	93.146	-584.059	-328.034	13.100
1388	81.297	161.617	89.858	99.601	-565.504	-312.415	11.757
Phase transition: $\Delta_{trs}H$ = 17.200 kJ/mol, $\Delta_{trs}S$ = 12.392 J/K·mol, crI–l							
1388	78.500	174.009	89.858	116.801	-565.504	-312.415	11.757
1400	78.500	174.685	90.582	117.743	-565.328	-310.228	11.575
1500	78.500	180.100	96.372	125.593	-563.882	-292.057	10.170
49. GERMANIUM TETRACHLORIDE $GeCl_4$ (g)							
298.15	95.918	348.393	348.393	0.000	-500.000	-461.582	80.866
300	96.041	348.987	348.395	0.178	-499.991	-461.343	80.326
400	100.750	377.342	352.229	10.045	-499.447	-448.540	58.573
500	103.206	400.114	359.604	20.255	-498.845	-435.882	45.536
600	104.624	419.067	367.980	30.652	-498.234	-423.347	36.855
700	105.509	435.266	376.463	41.162	-497.634	-410.914	30.662
800	106.096	449.396	384.715	51.744	-497.057	-398.565	26.023
900	106.504	461.917	392.611	62.375	-496.509	-386.287	22.419
1000	106.799	473.155	400.113	73.041	-495.993	-374.068	19.539
1100	107.020	483.344	407.224	83.733	-495.512	-361.899	17.185
1200	107.189	492.664	413.961	94.444	-495.067	-349.772	15.225

T/K	$C_p°$/ J K^{-1}mol^{-1}	$S°$/ J K^{-1} mol^{-1}	-($G°$-$H°(T)$)/ J K^{-1} mol^{-1}	(H-$H°(T)$)/ kJ mol^{-1}	$\Delta_f H°$/ kJ mol^{-1}	$\Delta_f G°$/ kJ mol^{-1}	Log K_f
1300	107.320	501.249	420.349	105.169	-531.677	-334.973	13.459
1400	107.425	509.206	426.416	115.907	-531.265	-319.857	11.934
1500	107.509	516.621	432.185	126.654	-530.861	-304.771	10.613
50. HYDROGEN H (g)							
298.15	20.786	114.716	114.716	0.000	217.998	203.276	-35.613
300	20.786	114.845	114.716	0.038	218.010	203.185	-35.377
400	20.786	120.824	115.532	2.117	218.635	198.149	-25.875
500	20.786	125.463	117.071	4.196	219.253	192.956	-20.158
600	20.786	129.252	118.795	6.274	219.867	187.639	-16.335
700	20.786	132.457	120.524	8.353	220.476	182.219	-13.597
800	20.786	135.232	122.193	10.431	221.079	176.712	-11.538
900	20.786	137.680	123.780	12.510	221.670	171.131	-9.932
1000	20.786	139.870	125.282	14.589	222.247	165.485	-8.644
1100	20.786	141.852	126.700	16.667	222.806	159.781	-7.587
1200	20.786	143.660	128.039	18.746	223.345	154.028	-6.705
1300	20.786	145.324	129.305	20.824	223.864	148.230	-5.956
1400	20.786	146.864	130.505	22.903	224.360	142.393	-5.313
1500	20.786	148.298	131.644	24.982	224.835	136.522	-4.754
51. DIHYDROGEN H_2 (g)							
298.15	28.836	130.680	130.680	0.000	0.000	0.000	0.000
300	28.849	130.858	130.680	0.053	0.000	0.000	0.000
400	29.181	139.217	131.818	2.960	0.000	0.000	0.000
500	29.260	145.738	133.974	5.882	0.000	0.000	0.000
600	29.327	151.078	136.393	8.811	0.000	0.000	0.000
700	29.440	155.607	138.822	11.749	0.000	0.000	0.000
800	29.623	159.549	141.172	14.702	0.000	0.000	0.000
900	29.880	163.052	143.412	17.676	0.000	0.000	0.000
1000	30.204	166.217	145.537	20.680	0.000	0.000	0.000
1100	30.580	169.113	147.550	23.719	0.000	0.000	0.000
1200	30.991	171.791	149.460	26.797	0.000	0.000	0.000
1300	31.422	174.288	151.275	29.918	0.000	0.000	0.000
1400	31.860	176.633	153.003	33.082	0.000	0.000	0.000
1500	32.296	178.846	154.653	36.290	0.000	0.000	0.000
52. HYDROXYL OH (g)							
298.15	29.886	183.737	183.737	0.000	39.349	34.631	-6.067
300	29.879	183.922	183.738	0.055	39.350	34.602	-6.025
400	29.604	192.476	184.906	3.028	39.384	33.012	-4.311
500	29.495	199.067	187.104	5.982	39.347	31.422	-3.283
600	29.513	204.445	189.560	8.931	39.252	29.845	-2.598
700	29.655	209.003	192.020	11.888	39.113	28.287	-2.111
800	29.914	212.979	194.396	14.866	38.945	26.752	-1.747
900	30.265	216.522	196.661	17.874	38.763	25.239	-1.465
1000	30.682	219.731	198.810	20.921	38.577	23.746	-1.240
1100	31.135	222.677	200.848	24.012	38.393	22.272	-1.058
1200	31.603	225.406	202.782	27.149	38.215	20.814	-0.906
1300	32.069	227.954	204.621	30.332	38.046	19.371	-0.778
1400	32.522	230.347	206.374	33.562	37.886	17.941	-0.669
1500	32.956	232.606	208.048	36.836	37.735	16.521	-0.575
53. WATER H_2O (l)							
298.15	75.300	69.950	69.950	0.000	-285.830	-237.141	41.546
300	75.281	70.416	69.951	0.139	-285.771	-236.839	41.237
373.21	76.079	86.896	71.715	5.666	-283.454	-225.160	31.513
54. WATER H_2O (g)							
298.15	33.598	188.832	188.832	0.000	-241.826	-228.582	40.046
300	33.606	189.040	188.833	0.062	-241.844	-228.500	39.785
400	34.283	198.791	190.158	3.453	-242.845	-223.900	29.238
500	35.259	206.542	192.685	6.929	-243.822	-219.050	22.884

T/K	$C_p°$/ J K⁻¹mol⁻¹	$S°$/ J K⁻¹ mol⁻¹	$-(G°-H°(T))$/ J K⁻¹ mol⁻¹	$(H-H°(T))$/ kJ mol⁻¹	$\Delta_f H°$/ kJ mol⁻¹	$\Delta_f G°$/ kJ mol⁻¹	Log K_f
600	36.371	213.067	195.552	10.509	-244.751	-214.008	18.631
700	37.557	218.762	198.469	14.205	-245.620	-208.814	15.582
800	38.800	223.858	201.329	18.023	-246.424	-203.501	13.287
900	40.084	228.501	204.094	21.966	-247.158	-198.091	11.497
1000	41.385	232.792	206.752	26.040	-247.820	-192.603	10.060
1100	42.675	236.797	209.303	30.243	-248.410	-187.052	8.882
1200	43.932	240.565	211.753	34.574	-248.933	-181.450	7.898
1300	45.138	244.129	214.108	39.028	-249.392	-175.807	7.064
1400	46.281	247.516	216.374	43.599	-249.792	-170.132	6.348
1500	47.356	250.746	218.559	48.282	-250.139	-164.429	5.726
55. IODINE I (g)							
298.15	20.786	180.787	180.787	0.000	106.760	70.172	-12.294
300	20.786	180.915	180.787	0.038	106.748	69.945	-12.178
400	20.786	186.895	181.602	2.117	97.974	58.060	-7.582
500	20.786	191.533	183.142	4.196	75.988	50.202	-5.244
600	20.786	195.323	184.866	6.274	76.190	45.025	-3.920
700	20.786	198.527	186.594	8.353	76.385	39.816	-2.971
800	20.787	201.303	188.263	10.432	76.574	34.579	-2.258
900	20.789	203.751	189.851	12.510	76.757	29.319	-1.702
1000	20.795	205.942	191.352	14.589	76.936	24.038	-1.256
1100	20.806	207.924	192.770	16.669	77.109	18.740	-0.890
1200	20.824	209.735	194.110	18.751	77.277	13.426	-0.584
1300	20.851	211.403	195.377	20.835	77.440	8.098	-0.325
1400	20.889	212.950	196.577	22.921	77.596	2.758	-0.103
1500	20.936	214.392	197.717	25.013	77.745	-2.592	0.090
56. DIIODINE I_2 (cr, l)							
298.15	54.440	116.139	116.139	0.000	0.000	0.000	0.000
300	54.518	116.476	116.140	0.101	0.000	0.000	0.000
386.75	61.531	131.039	117.884	5.088	0.000	0.000	0.000
Phase transition: $\Delta_{trs}H$ = 15.665 kJ/mol, $\Delta_{trs}S$ = 40.504 J/K·mol, cr–l							
386.75	79.555	171.543	117.884	20.753	0.000	0.000	0.000
400	79.555	174.223	119.706	21.807	0.000	0.000	0.000
457.67	79.555	184.938	127.266	26.395	0.000	0.000	0.000
57. DIIODINE I_2 (g)							
298.15	36.887	260.685	260.685	0.000	62.420	19.324	-3.385
300	36.897	260.913	260.685	0.068	62.387	19.056	-3.318
400	37.256	271.584	262.138	3.778	44.391	5.447	-0.711
457.67[a]	37.385	276.610	263.652	5.931			
500	37.464	279.921	264.891	7.515	0.000	0.000	0.000
600	37.613	286.765	267.983	11.269	0.000	0.000	0.000
700	37.735	292.573	271.092	15.037	0.000	0.000	0.000
800	37.847	297.619	274.099	18.816	0.000	0.000	0.000
900	37.956	302.083	276.965	22.606	0.000	0.000	0.000
1000	38.070	306.088	279.681	26.407	0.000	0.000	0.000
1100	38.196	309.722	282.249	30.220	0.000	0.000	0.000
1200	38.341	313.052	284.679	34.047	0.000	0.000	0.000
1300	38.514	316.127	286.981	37.890	0.000	0.000	0.000
1400	38.719	318.989	289.166	41.751	0.000	0.000	0.000
1500	38.959	321.668	291.245	45.635	0.000	0.000	0.000
58. HYDROGEN IODIDE HI (g)							
298.15	29.157	206.589	206.589	0.000	26.500	1.700	-0.298
300	29.158	206.769	206.589	0.054	26.477	1.546	-0.269
400	29.329	215.176	207.734	2.977	17.093	-6.289	0.821
500	29.738	221.760	209.904	5.928	-5.481	-9.946	1.039
600	30.351	227.233	212.348	8.931	-5.819	-10.806	0.941
700	31.070	231.965	214.820	12.002	-6.101	-11.614	0.867
800	31.807	236.162	217.230	15.145	-6.323	-12.386	0.809

T/K	$C_p°$/ J K^{-1}mol^{-1}	$S°$/ J K^{-1} mol^{-1}	-($G°$-$H°$(T))/ J K^{-1} mol^{-1}	(H-$H°$(T))/ kJ mol^{-1}	$\Delta_f H°$/ kJ mol^{-1}	$\Delta_f G°$/ kJ mol^{-1}	Log K_f
900	32.511	239.950	219.548	18.362	-6.489	-13.133	0.762
1000	33.156	243.409	221.763	21.646	-6.608	-13.865	0.724
1100	33.735	246.597	223.878	24.991	-6.689	-14.586	0.693
1200	34.249	249.555	225.896	28.391	-6.741	-15.302	0.666
1300	34.703	252.314	227.823	31.839	-6.775	-16.014	0.643
1400	35.106	254.901	229.666	35.330	-6.797	-16.723	0.624
1500	35.463	257.336	231.430	38.858	-6.814	-17.432	0.607

59. POTASSIUM K (cr, l)

298.15	29.600	64.680	64.680	0.000	0.000	0.000	0.000
300	29.671	64.863	64.681	0.055	0.000	0.000	0.000
336.86	32.130	68.422	64.896	1.188	0.000	0.000	0.000

Phase transition: $\Delta_{trs}H$ = 2.321 kJ/mol, $\Delta_{trs}S$ = 6.891 J/K·mol, cr–l

336.86	32.129	75.313	64.896	3.509	0.000	0.000	0.000
400	31.552	80.784	66.986	5.519	0.000	0.000	0.000
500	30.741	87.734	70.469	8.632	0.000	0.000	0.000
600	30.158	93.283	73.824	11.675	0.000	0.000	0.000
700	29.851	97.905	76.943	14.673	0.000	0.000	0.000
800	29.838	101.887	79.818	17.655	0.000	0.000	0.000
900	30.130	105.415	82.470	20.651	0.000	0.000	0.000
1000	30.730	108.618	84.927	23.691	0.000	0.000	0.000
1039.4	31.053	109.812	85.847	24.908	0.000	0.000	0.000

60. POTASSIUM K (g)

298.15	20.786	160.340	160.340	0.000	89.000	60.479	-10.596
300	20.786	160.468	160.340	0.038	88.984	60.302	-10.499
400	20.786	166.448	161.155	2.117	85.598	51.332	-6.703
500	20.786	171.086	162.695	4.196	84.563	42.887	-4.480
600	20.786	174.876	164.419	6.274	83.599	34.643	-3.016
700	20.786	178.080	166.148	8.353	82.680	26.557	-1.982
800	20.786	180.856	167.817	10.431	81.776	18.601	-1.215
900	20.786	183.304	169.404	12.510	80.859	10.759	-0.624
1000	20.786	185.494	170.905	14.589	79.897	3.021	-0.158
1039.4[a]	20.786	186.297	171.474	15.408			
1100	20.786	187.475	172.323	16.667	0.000	0.000	0.000
1200	20.786	189.284	173.662	18.746	0.000	0.000	0.000
1300	20.789	190.948	174.929	20.825	0.000	0.000	0.000
1400	20.793	192.489	176.129	22.904	0.000	0.000	0.000
1500	20.801	193.923	177.268	24.983	0.000	0.000	0.000

61. DIPOTASSIUM OXIDE K_2O (cr, l)

298.15	72.000	96.000	96.000	0.000	-361.700	-321.171	56.267
300	72.130	96.446	96.001	0.133	-361.704	-320.920	55.876
400	79.154	118.158	98.914	7.698	-366.554	-306.416	40.013
500	86.178	136.575	104.647	15.964	-366.043	-291.423	30.444
590	92.500	151.348	110.662	24.005	-364.204	-278.079	24.619

Phase transition: $\Delta_{trs}H$ = 0.700 kJ/mol, $\Delta_{trs}S$ = 1.186 J/K·mol, crIII–crII

590	100.000	152.534	110.662	24.705	-364.204	-278.079	24.619
600	100.000	154.215	111.374	25.705	-363.968	-276.621	24.082
645	100.000	161.447	114.618	30.205	-358.901	-270.109	21.874

Phase transition: $\Delta_{trs}H$ = 4.000 kJ/mol, $\Delta_{trs}S$ = 6.202 J/K·mol, crII–crI

645	100.000	167.649	114.618	34.205	-358.901	-270.109	21.874
700	100.000	175.832	119.111	39.705	-357.592	-262.592	19.595
800	100.000	189.185	127.054	49.705	-355.224	-249.183	16.270
900	100.000	200.963	134.625	59.705	-352.919	-236.067	13.701
1000	100.000	211.499	141.794	69.705	-350.732	-223.202	11.659
1013	100.000	212.791	142.697	71.005	-323.459	-221.546	11.424

Phase transition: $\Delta_{trs}H$ = 27.000 kJ/mol, $\Delta_{trs}S$ = 26.654 J/K·mol, crI–l

1013	100.000	239.444	142.697	98.005	-323.459	-221.546	11.424

T/K	$C_p°$/ J K^{-1}mol^{-1}	$S°$/ J K^{-1} mol^{-1}	-($G°$-$H°(T)$)/ J K^{-1} mol^{-1}	(H-$H°(T)$)/ kJ mol^{-1}	$\Delta_f H°$/ kJ mol^{-1}	$\Delta_f G°$/ kJ mol^{-1}	Log K_f
1100	100.000	247.684	150.679	106.705	-479.439	-203.633	9.670
1200	100.000	256.385	159.131	116.705	-475.371	-178.740	7.780
1300	100.000	264.389	166.924	126.705	-471.321	-154.185	6.195
1400	100.000	271.800	174.154	136.705	-467.287	-129.941	4.848
1500	100.000	278.699	180.896	146.705	-463.268	-105.986	3.691
62. POTASSIUM HYDROXIDE KOH (cr, l)							
298.15	64.900	78.870	78.870	0.000	-424.580	-378.747	66.354
300	65.038	79.272	78.871	0.120	-424.569	-378.463	65.895
400	72.519	99.007	81.512	6.998	-426.094	-362.765	47.372
500	80.000	115.993	86.745	14.624	-424.572	-347.093	36.260
520	81.496	119.159	87.931	16.239	-417.725	-344.002	34.555
Phase transition: $\Delta_{trs}H$ = 6.450 kJ/mol, $\Delta_{trs}S$ = 12.404 J/K·mol, crII–crI							
520	79.000	131.563	87.931	22.689	-417.725	-344.002	34.555
600	79.000	142.868	94.520	29.009	-416.274	-332.766	28.969
678	79.000	152.523	100.649	35.171	-405.464	-321.998	24.807
Phase transition: $\Delta_{trs}H$ = 9.400 kJ/mol, $\Delta_{trs}S$ = 13.865 J/K·mol, crI–l							
678	83.000	166.388	100.649	44.571	-405.464	-321.998	24.807
700	83.000	169.038	102.757	46.397	-404.981	-319.297	23.826
800	83.000	180.121	111.750	54.697	-402.808	-307.206	20.058
900	83.000	189.897	119.901	62.997	-400.694	-295.383	17.143
1000	83.000	198.642	127.345	71.297	-398.668	-283.791	14.824
1100	83.000	206.553	134.192	79.597	-475.618	-267.780	12.716
1200	83.000	213.775	140.527	87.897	-472.711	-249.014	10.839
1300	83.000	220.418	146.421	96.197	-469.843	-230.490	9.261
1400	83.000	226.569	151.929	104.497	-467.011	-212.184	7.917
1500	83.000	232.296	157.098	112.797	-464.217	-194.080	6.758
63. POTASSIUM HYDROXIDE KOH (g)							
298.15	49.184	238.283	238.283	0.000	-227.989	-229.685	40.239
300	49.236	238.588	238.284	0.091	-228.007	-229.696	39.993
400	51.178	253.053	240.243	5.124	-231.377	-229.667	29.991
500	52.178	264.591	243.998	10.296	-232.309	-229.129	23.937
600	52.804	274.163	248.251	15.547	-233.145	-228.413	19.885
700	53.296	282.340	252.551	20.853	-233.934	-227.562	16.981
800	53.758	289.487	256.730	26.206	-234.708	-226.599	14.795
900	54.229	295.846	260.730	31.605	-235.495	-225.538	13.090
1000	54.713	301.585	264.533	37.052	-236.322	-224.388	11.721
1100	55.203	306.823	268.143	42.548	-316.077	-218.535	10.377
1200	55.686	311.647	271.570	48.092	-315.925	-209.674	9.127
1300	56.153	316.122	274.827	53.684	-315.764	-200.826	8.069
1400	56.598	320.300	277.927	59.322	-315.595	-191.991	7.163
1500	57.016	324.220	280.884	65.003	-315.420	-183.169	6.378
64. POTASSIUM CHLORIDE KCl (cr, l)							
298.15	51.300	82.570	82.570	0.000	-436.490	-408.568	71.579
300	51.333	82.887	82.571	0.095	-436.481	-408.395	71.107
400	52.977	97.886	84.605	5.312	-438.463	-398.651	52.058
500	54.448	109.867	88.498	10.685	-437.990	-388.749	40.612
600	55.885	119.921	92.919	16.201	-437.332	-378.960	32.991
700	57.425	128.649	97.413	21.865	-436.502	-369.295	27.557
800	59.205	136.430	101.812	27.694	-435.505	-359.760	23.490
900	61.361	143.523	106.058	33.719	-434.337	-350.360	20.334
1000	64.032	150.121	110.138	39.983	-432.981	-341.100	17.817
1044	65.405	152.908	111.882	42.830	-485.450	-336.720	16.847
Phase transition: $\Delta_{trs}H$ = 26.320 kJ/mol, $\Delta_{trs}S$ = 25.210 J/K·mol, cr–l							
1044	72.000	178.118	111.882	69.150	-485.450	-336.720	16.847
1100	72.000	181.880	115.351	73.182	-483.633	-328.790	15.613
1200	72.000	188.145	121.160	80.382	-480.393	-314.856	13.705

T/K	$C_p°$/ J K^{-1}mol^{-1}	$S°$/ J K^{-1} mol^{-1}	-($G°$-$H°(T)$)/ J K^{-1} mol^{-1}	(H-$H°(T)$)/ kJ mol^{-1}	$\Delta_f H°$/ kJ mol^{-1}	$\Delta_f G°$/ kJ mol^{-1}	Log K_f
1300	72.000	193.908	126.537	87.582	-477.158	-301.192	12.102
1400	72.000	199.244	131.542	94.782	-473.928	-287.778	10.737
1500	72.000	204.211	136.223	101.982	-470.704	-274.594	9.562
65. POTASSIUM CHLORIDE KCl (g)							
298.15	36.505	239.091	239.091	0.000	-214.575	-233.320	40.876
300	36.518	239.317	239.092	0.068	-214.594	-233.436	40.644
400	37.066	249.904	240.532	3.749	-218.112	-239.107	31.224
500	37.384	258.212	243.267	7.473	-219.287	-244.219	25.513
600	37.597	265.048	246.344	11.222	-220.396	-249.100	21.686
700	37.769	270.857	249.441	14.991	-221.461	-253.799	18.938
800	37.907	275.910	252.441	18.775	-222.509	-258.347	16.868
900	38.041	280.382	255.302	22.572	-223.568	-262.764	15.250
1000	38.162	284.397	258.014	26.383	-224.667	-267.061	13.950
1100	38.279	288.039	260.581	30.205	-304.696	-266.627	12.661
1200	38.401	291.375	263.010	34.039	-304.821	-263.161	11.455
1300	38.518	294.454	265.312	37.885	-304.941	-259.684	10.434
1400	38.639	297.313	267.496	41.743	-305.053	-256.199	9.559
1500	38.761	299.983	269.574	45.613	-305.159	-252.706	8.800
66. DINITROGEN N_2 (g)							
298.15	29.124	191.608	191.608	0.000	0.000	0.000	0.000
300	29.125	191.788	191.608	0.054	0.000	0.000	0.000
400	29.249	200.180	192.752	2.971	0.000	0.000	0.000
500	29.580	206.738	194.916	5.911	0.000	0.000	0.000
600	30.109	212.175	197.352	8.894	0.000	0.000	0.000
700	30.754	216.864	199.812	11.936	0.000	0.000	0.000
800	31.433	221.015	202.208	15.046	0.000	0.000	0.000
900	32.090	224.756	204.509	18.222	0.000	0.000	0.000
1000	32.696	228.169	206.706	21.462	0.000	0.000	0.000
1100	33.241	231.311	208.802	24.759	0.000	0.000	0.000
1200	33.723	234.224	210.801	28.108	0.000	0.000	0.000
1300	34.147	236.941	212.708	31.502	0.000	0.000	0.000
1400	34.517	239.485	214.531	34.936	0.000	0.000	0.000
1500	34.842	241.878	216.275	38.404	0.000	0.000	0.000
67. NITRIC OXIDE NO (g)							
298.15	29.862	210.745	210.745	0.000	91.277	87.590	-15.345
300	29.858	210.930	210.746	0.055	91.278	87.567	-15.247
400	29.954	219.519	211.916	3.041	91.320	86.323	-11.272
500	30.493	226.255	214.133	6.061	91.340	85.071	-8.887
600	31.243	231.879	216.635	9.147	91.354	83.816	-7.297
700	32.031	236.754	219.168	12.310	91.369	82.558	-6.160
800	32.770	241.081	221.642	15.551	91.386	81.298	-5.308
900	33.425	244.979	224.022	18.862	91.405	80.036	-4.645
1000	33.990	248.531	226.298	22.233	91.426	78.772	-4.115
1100	34.473	251.794	228.469	25.657	91.445	77.505	-3.680
1200	34.883	254.811	230.540	29.125	91.464	76.237	-3.318
1300	35.234	257.618	232.516	32.632	91.481	74.967	-3.012
1400	35.533	260.240	234.404	36.170	91.495	73.697	-2.750
1500	35.792	262.700	236.209	39.737	91.506	72.425	-2.522
68. NITROGEN DIOXIDE NO_2 (g)							
298.15	37.178	240.166	240.166	0.000	34.193	52.316	-9.165
300	37.236	240.397	240.167	0.069	34.181	52.429	-9.129
400	40.513	251.554	241.666	3.955	33.637	58.600	-7.652
500	43.664	260.939	244.605	8.167	33.319	64.882	-6.778
600	46.383	269.147	248.026	12.673	33.174	71.211	-6.199
700	48.612	276.471	251.575	17.427	33.151	77.553	-5.787
800	50.405	283.083	255.107	22.381	33.213	83.893	-5.478
900	51.844	289.106	258.555	27.496	33.3340	90.221	-5.236
1000	53.007	294.631	261.891	32.741	33.495	96.534	-5.042

T/K	$C_p°$/ J K⁻¹mol⁻¹	$S°$/ J K⁻¹ mol⁻¹	-($G°$-$H°(T)$)/ J K⁻¹ mol⁻¹	(H-$H°(T)$)/ kJ mol⁻¹	$\Delta_f H°$/ kJ mol⁻¹	$\Delta_f G°$/ kJ mol⁻¹	Log K_f
1100	53.956	299.729	265.102	38.090	33.686	102.828	-4.883
1200	54.741	304.459	268.187	43.526	33.898	109.105	-4.749
1300	55.399	308.867	271.148	49.034	34.124	115.363	-4.635
1400	55.960	312.994	273.992	54.603	34.360	121.603	-4.537
1500	56.446	316.871	276.722	60.224	34.604	127.827	-4.451
69. AMMONIA NH_3 (g)							
298.15	35.630	192.768	192.768	0.000	-45.940	-16.407	2.874
300	35.678	192.989	192.769	0.066	-45.981	-16.223	2.825
400	38.674	203.647	194.202	3.778	-48.087	-5.980	0.781
500	41.994	212.633	197.011	7.811	-49.908	4.764	-0.498
600	45.229	220.578	200.289	12.174	-51.430	15.846	-1.379
700	48.269	227.781	203.709	16.850	-52.682	27.161	-2.027
800	51.112	234.414	207.138	21.821	-53.695	38.639	-2.523
900	53.769	240.589	210.516	27.066	-54.499	50.231	-2.915
1000	56.244	246.384	213.816	32.569	-55.122	61.903	-3.233
1100	58.535	251.854	217.027	38.309	-55.589	73.629	-3.496
1200	60.644	257.039	220.147	44.270	-55.920	85.392	-3.717
1300	62.576	261.970	223.176	50.432	-56.136	97.177	-3.905
1400	64.339	266.673	226.117	56.779	-56.251	108.975	-4.066
1500	65.945	271.168	228.971	63.295	-56.282	120.779	-4.206
70. OXYGEN O (g)							
298.15	21.911	161.058	161.058	0.000	249.180	231.743	-40.600
300	21.901	161.194	161.059	0.041	249.193	231.635	-40.331
400	21.482	167.430	161.912	2.207	249.874	225.677	-29.470
500	21.257	172.197	163.511	4.343	250.481	219.556	-22.937
600	21.124	176.060	165.290	6.462	251.019	213.319	-18.571
700	21.040	179.310	167.067	8.570	251.500	206.997	-15.446
800	20.984	182.115	168.777	10.671	251.932	200.610	-13.098
900	20.944	184.584	170.399	12.767	252.325	194.171	-11.269
1000	20.915	186.789	171.930	14.860	252.686	187.689	-9.804
1100	20.893	188.782	173.372	16.950	253.022	181.173	-8.603
1200	20.877	190.599	174.733	19.039	253.335	174.628	-7.601
1300	20.864	192.270	176.019	21.126	253.630	168.057	-6.753
1400	20.853	193.815	177.236	23.212	253.908	161.463	-6.024
1500	20.845	195.254	178.389	25.296	254.171	154.851	-5.392
71. DIOXYGEN O_2 (g)							
298.15	29.378	205.148	205.148	0.000	0.000	0.000	0.000
300	29.387	205.330	205.148	0.054	0.000	0.000	0.000
400	30.109	213.873	206.308	3.026	0.000	0.000	0.000
500	31.094	220.695	208.525	6.085	0.000	0.000	0.000
600	32.095	226.454	211.045	9.245	0.000	0.000	0.000
700	32.987	231.470	213.612	12.500	0.000	0.000	0.000
800	33.741	235.925	216.128	15.838	0.000	0.000	0.000
900	34.365	239.937	218.554	19.244	0.000	0.000	0.000
1000	34.881	243.585	220.878	22.707	0.000	0.000	0.000
1100	35.314	246.930	223.096	26.217	0.000	0.000	0.000
1200	35.683	250.019	225.213	29.768	0.000	0.000	0.000
1300	36.006	252.888	227.233	33.352	0.000	0.000	0.000
1400	36.297	255.568	229.162	36.968	0.000	0.000	0.000
1500	36.567	258.081	231.007	40.611	0.000	0.000	0.000
72. SULFUR S (cr, l)							
298.15	22.690	32.070	32.070	0.000	0.000	0.000	0.000
300	22.737	32.210	32.070	0.042	0.000	0.000	0.000
368.3	24.237	37.030	32.554	1.649	0.000	0.000	0.000
Phase transition: $\Delta_{trs}H$ = 0.401 kJ/mol, $\Delta_{trs}S$ = 1.089 J/K·mol, crII–crI							
368.3	24.773	38.119	32.553	2.050	0.000	0.000	0.000
388.36	25.180	39.444	32.875	2.551	0.000	0.000	0.000

T/K	$C_p°$/ J K^{-1}mol^{-1}	$S°$/ J K^{-1} mol^{-1}	$-(G°-H°(T))/$ J K^{-1} mol^{-1}	$(H-H°(T))/$ kJ mol^{-1}	$\Delta_f H°$/ kJ mol^{-1}	$\Delta_f G°$/ kJ mol^{-1}	Log K_f
Phase transition: $\Delta_{trs}H$ = 1.722 kJ/mol, $\Delta_{trs}S$ = 4.431 J/K·mol, crI–l							
388.36	31.710	43.875	32.872	4.273	0.000	0.000	0.000
400	32.369	44.824	33.206	4.647	0.000	0.000	0.000
500	38.026	53.578	36.411	8.584	0.000	0.000	0.000
600	34.371	60.116	39.842	12.164	0.000	0.000	0.000
700	32.451	65.278	43.120	15.511	0.000	0.000	0.000
800	32.000	69.557	46.163	18.715	0.000	0.000	0.000
882.38	32.000	72.693	48.496	21.351	0.000	0.000	0.000
73. SULFUR S (g)							
298.15	23.673	167.828	167.828	0.000	277.180	236.704	-41.469
300	23.669	167.974	167.828	0.044	277.182	236.453	-41.170
400	23.233	174.730	168.752	2.391	274.924	222.962	-29.115
500	22.741	179.860	170.482	4.689	273.286	210.145	-21.953
600	22.338	183.969	172.398	6.942	271.958	197.646	-17.206
700	22.031	187.388	174.302	9.160	270.829	185.352	-13.831
800	21.800	190.314	176.125	11.351	269.816	173.210	-11.309
900	21.624	192.871	177.847	13.522	215.723	162.258	-9.417
1000	21.489	195.142	179.465	15.677	216.018	156.301	-8.164
1100	21.386	197.185	180.985	17.821	216.284	150.317	-7.138
1200	21.307	199.043	182.413	19.955	216.525	144.309	-6.282
1300	21.249	200.746	183.759	22.083	216.743	138.282	-5.556
1400	21.209	202.319	185.029	24.206	216.940	132.239	-4.934
1500	21.186	203.781	186.231	26.325	217.119	126.182	-4.394
74. DISULFUR S2 (g)							
298.15	32.505	228.165	228.165	0.000	128.600	79.696	-13.962
300	32.540	228.366	228.165	0.060	128.576	79.393	-13.823
400	34.108	237.956	229.462	3.398	122.703	63.380	-8.276
500	35.133	245.686	231.959	6.863	118.296	49.031	-5.122
600	35.815	252.156	234.800	10.413	114.685	35.530	-3.093
700	36.305	257.715	237.686	14.020	111.599	22.588	-1.685
800	36.697	262.589	240.501	17.671	108.841	10.060	-0.657
882.38[a]	36.985	266.200	242.734	20.706			
900	37.045	266.932	243.201	21.358	0.000	0.000	0.000
1000	37.377	270.852	245.773	25.079	0.000	0.000	0.000
1100	37.704	274.430	248.218	28.833	0.000	0.000	0.000
1200	38.030	277.725	250.541	32.620	0.000	0.000	0.000
1300	38.353	280.781	252.751	36.439	0.000	0.000	0.000
1400	38.669	283.635	254.856	40.290	0.000	0.000	0.000
1500	38.976	286.314	256.865	44.173	0.000	0.000	0.000
75. OCTASULFUR S8 (g)							
298.15	156.500	432.536	432.536	0.000	101.277	48.810	-8.551
300	156.768	433.505	432.539	0.290	101.231	48.484	-8.442
400	167.125	480.190	438.834	16.542	80.642	32.003	-4.179
500	173.181	518.176	451.022	33.577	66.185	21.409	-2.237
600	177.936	550.180	464.951	51.137	55.101	13.549	-1.180
700	182.441	577.948	479.152	69.157	46.349	7.343	-0.548
800	186.764	602.596	493.071	87.620	39.177	2.263	-0.148
900	190.595	624.821	506.495	106.494	-392.062	6.554	-0.380
1000	193.618	645.067	519.355	125.712	-387.728	50.614	-2.644
1100	195.684	663.625	531.639	145.185	-383.272	94.233	-4.475
1200	196.825	680.707	543.359	164.817	-378.786	137.444	-5.983
1300	197.195	696.480	554.539	184.524	-374.356	180.283	-7.244
1400	196.988	711.089	565.206	204.237	-370.048	222.785	-8.312
1500	196.396	724.662	575.389	223.909	-365.905	264.984	-9.227
76. SULFUR DIOXIDE SO2 (g)							
298.15	39.842	248.219	248.219	0.000	-296.810	-300.090	52.574
300	39.909	248.466	248.220	0.074	-296.833	-300.110	52.253

T/K	$C_p°$/ J K^{-1}mol^{-1}	$S°$/ J K^{-1} mol^{-1}	-($G°$-$H°(T)$)/ J K^{-1} mol^{-1}	(H-$H°(T)$)/ kJ mol^{-1}	$\Delta_f H°$/ kJ mol^{-1}	$\Delta_f G°$/ kJ mol^{-1}	Log K_f
400	43.427	260.435	249.828	4.243	-300.240	-300.935	39.298
500	46.490	270.465	252.978	8.744	-302.735	-300.831	31.427
600	48.938	279.167	256.634	13.520	-304.699	-300.258	26.139
700	50.829	286.859	260.413	18.513	-306.308	-299.386	22.340
800	52.282	293.746	264.157	23.671	-307.691	-298.302	19.477
900	53.407	299.971	267.796	28.958	-362.075	-295.987	17.178
1000	54.290	305.646	271.301	34.345	-362.012	-288.647	15.077
1100	54.993	310.855	274.664	39.810	-361.934	-281.314	13.358
1200	55.564	315.665	277.882	45.339	-361.849	-273.989	11.926
1300	56.033	320.131	280.963	50.920	-361.763	-266.671	10.715
1400	56.426	324.299	283.911	56.543	-361.680	-259.359	9.677
1500	56.759	328.203	286.735	62.203	-361.605	-252.053	8.777

77. SILICON Si (cr)

T/K	$C_p°$/ J K^{-1}mol^{-1}	$S°$/ J K^{-1} mol^{-1}	-($G°$-$H°(T)$)/ J K^{-1} mol^{-1}	(H-$H°(T)$)/ kJ mol^{-1}	$\Delta_f H°$/ kJ mol^{-1}	$\Delta_f G°$/ kJ mol^{-1}	Log K_f
298.15	19.789	18.810	18.810	0.000	0.000	0.000	0.000
300	19.855	18.933	18.810	0.037	0.000	0.000	0.000
400	22.301	25.023	19.624	2.160	0.000	0.000	0.000
500	23.610	30.152	21.231	4.461	0.000	0.000	0.000
600	24.472	34.537	23.092	6.867	0.000	0.000	0.000
700	25.124	38.361	25.006	9.348	0.000	0.000	0.000
800	25.662	41.752	26.891	11.888	0.000	0.000	0.000
900	26.135	44.802	28.715	14.478	0.000	0.000	0.000
1000	26.568	47.578	30.464	17.114	0.000	0.000	0.000
1100	26.974	50.130	32.138	19.791	0.000	0.000	0.000
1200	27.362	52.493	33.737	22.508	0.000	0.000	0.000
1300	27.737	54.698	35.265	25.263	0.000	0.000	0.000
1400	28.103	56.767	36.728	28.055	0.000	0.000	0.000
1500	28.462	58.719	38.130	30.883	0.000	0.000	0.000

78. SILICON Si (g)

T/K	$C_p°$/ J K^{-1}mol^{-1}	$S°$/ J K^{-1} mol^{-1}	-($G°$-$H°(T)$)/ J K^{-1} mol^{-1}	(H-$H°(T)$)/ kJ mol^{-1}	$\Delta_f H°$/ kJ mol^{-1}	$\Delta_f G°$/ kJ mol^{-1}	Log K_f
298.15	22.251	167.980	167.980	0.000	450.000	405.525	-71.045
300	22.234	168.117	167.980	0.041	450.004	405.249	-70.559
400	21.613	174.416	168.843	2.229	450.070	390.312	-50.969
500	21.316	179.204	170.456	4.374	449.913	375.388	-39.216
600	21.153	183.074	172.246	6.497	449.630	360.508	-31.385
700	21.057	186.327	174.032	8.607	449.259	345.682	-25.795
800	21.000	189.135	175.748	10.709	448.821	330.915	-21.606
900	20.971	191.606	177.375	12.808	448.329	316.205	-18.352
1000	20.968	193.815	178.911	14.904	447.791	301.553	-15.751
1100	20.989	195.815	180.358	17.002	447.211	286.957	-13.626
1200	21.033	197.643	181.723	19.103	446.595	272.416	-11.858
1300	21.099	199.329	183.014	21.209	445.946	257.927	-10.364
1400	21.183	200.895	184.236	23.323	445.268	243.489	-9.085
1500	21.282	202.360	185.396	25.446	444.563	229.101	-7.978

79. SILICON DIOXIDE SiO2 (cr)

T/K	$C_p°$/ J K^{-1}mol^{-1}	$S°$/ J K^{-1} mol^{-1}	-($G°$-$H°(T)$)/ J K^{-1} mol^{-1}	(H-$H°(T)$)/ kJ mol^{-1}	$\Delta_f H°$/ kJ mol^{-1}	$\Delta_f G°$/ kJ mol^{-1}	Log K_f
298.15	44.602	41.460	41.460	0.000	-910.700	-856.288	150.016
300	44.712	41.736	41.461	0.083	-910.708	-855.951	149.032
400	53.477	55.744	43.311	4.973	-910.912	-837.651	109.385
500	60.533	68.505	47.094	10.705	-910.540	-819.369	85.598
600	64.452	79.919	51.633	16.971	-909.841	-801.197	69.749
700	68.234	90.114	56.414	23.590	-908.958	-783.157	58.439
800	76.224	99.674	61.226	30.758	-907.668	-765.265	49.966
848	82.967	104.298	63.533	34.569	-906.310	-756.747	46.613

Phase transition: $\Delta_{trs}H$ = 0.411 kJ/mol, $\Delta_{trs}S$ = 0.484 J/K·mol, crII–crII′

T/K	$C_p°$/ J K^{-1}mol^{-1}	$S°$/ J K^{-1} mol^{-1}	-($G°$-$H°(T)$)/ J K^{-1} mol^{-1}	(H-$H°(T)$)/ kJ mol^{-1}	$\Delta_f H°$/ kJ mol^{-1}	$\Delta_f G°$/ kJ mol^{-1}	Log K_f
848	67.446	104.782	63.532	34.980	-906.310	-756.747	46.613
900	67.953	108.811	66.033	38.500	-905.922	-747.587	43.388
1000	68.941	116.021	70.676	45.345	-905.176	-730.034	38.133
1100	69.940	122.639	75.104	52.289	-904.420	-712.557	33.836
1200	70.947	128.768	79.323	59.333	-901.382	-695.148	30.259

T/K	$C_p°$/ J K^{-1}mol^{-1}	$S°$/ J K^{-1} mol^{-1}	-($G°$-$H°$(T))/ J K^{-1} mol^{-1}	(H-$H°$(T))/ kJ mol^{-1}	$\Delta_f H°$/ kJ mol^{-1}	$\Delta_f G°$/ kJ mol^{-1}	Log K_f
Phase transition: $\Delta_{trs}H$ = 2.261 kJ/mol, $\Delta_{trs}S$ = 1.883 J/K·mol, crII′–crI							
1200	71.199	130.651	79.323	61.594	-901.382	-695.148	30.259
1300	71.743	136.372	83.494	68.742	-900.574	-677.994	27.242
1400	72.249	141.707	87.463	75.941	-899.782	-660.903	24.658
1500	72.739	146.709	91.248	83.191	-899.004	-643.867	22.421
80. SILICON TETRACHLORIDE $SiCl_4$ (g)							
298.15	90.404	331.446	331.446	0.000	-662.200	-622.390	109.039
300	90.562	332.006	331.448	0.167	-662.195	-622.143	108.323
400	96.893	359.019	335.088	9.572	-661.853	-608.841	79.505
500	100.449	381.058	342.147	19.456	-661.413	-595.637	62.225
600	102.587	399.576	350.216	29.616	-660.924	-582.527	50.713
700	103.954	415.500	358.432	39.948	-660.417	-569.501	42.496
800	104.875	429.445	366.455	50.392	-659.912	-556.548	36.338
900	105.523	441.837	374.155	60.914	-659.422	-543.657	31.553
1000	105.995	452.981	381.490	71.491	-658.954	-530.819	27.727
1100	106.349	463.101	388.456	82.109	-658.515	-518.027	24.599
1200	106.620	472.366	395.068	92.758	-658.107	-505.274	21.994
1300	106.834	480.909	401.347	103.431	-657.735	-492.553	19.791
1400	107.003	488.833	407.316	114.123	-657.400	-479.860	17.904
1500	107.141	496.220	413.000	124.830	-657.104	-467.189	16.269

[a] Boiling point at 1 bar pressure.

THERMODYNAMIC PROPERTIES OF AQUEOUS IONS

This table contains standard state thermodynamic properties of positive and negative ions in aqueous solution. It includes enthalpy and Gibbs energy of formation, entropy, and heat capacity, and thus serves as a companion to the preceding table, "Standard Thermodynamic Properties of Chemical Substances." The standard state is the hypothetical ideal solution with molality $m = 1$ mol/kg (mean ionic molality $m_\pm$ in the case of a species that is assumed to dissociate at infinite dilution). Further details on conventions may be found in Ref. 1.

All values refer to standard conditions of 25 °C and 100 kPa pressure. Table 1 has data for cations listed in alphabetical order by formula. Table 2 is for anions, also listed in alphabetical order by formula. The column definitions are as follows.

References

1. Wagman, D. D., Evans, W. H., Parker, V. B., Schumm, R. H., Halow, I., Bailey, S. M., Churney, K. L., and Nuttall, R. L., *The NBS Tables of Chemical Thermodynamic Properties*, *J. Phys. Chem. Ref. Data*, Vol. 11, Suppl. 2, 1982.
2. Zemaitis, J. F., Clark, D. M., Rafal, M., and Scrivner, N. C., *Handbook of Aqueous Electrolyte Thermodynamics*, American Institute of Chemical Engineers, New York, 1986. [https://doi.org/10.1002/9780470938416]

Column heading	Definition
Mol. form.	Molecular formula of aqueous ion
$\Delta_f H°$(aq)	Enthalpy of formation, in units kJ mol^{-1}
$\Delta_f G°$(aq)	Gibbs energy of formation, in units kJ mol^{-1}
$S°$(aq)	Entropy, in units J K^{-1} mol^{-1}
C_p(aq)	Heat capacity at constant pressure, in units J K^{-1} mol^{-1}

TABLE 1. Thermodynamic Properties of Aqueous Cations

Mol. form.	$\Delta_f H°$(aq)/ kJ mol^{-1}	$\Delta_f G°$(aq)/ kJ mol^{-1}	$S°$(aq)/ J K^{-1} mol^{-1}	C_p(aq)/ J K^{-1}mol^{-1}
Ag^{+}	105.6	77.1	72.7	21.8
Al^{+3}	-531.0	-485.0	-321.7	
$AlOH^{+2}$		-694.1		
$BaOH^{+}$		-730.5		
Ba^{+2}	-537.6	-560.8	9.6	
Be^{+2}	-382.8	-379.7	-129.7	
Bi^{+3}		82.8		
$BiOH^{+2}$		-146.4		
Ca^{+2}	-542.8	-553.6	-53.1	
$CaOH^{+}$		-718.4		
Cd^{+2}	-75.9	-77.6	-73.2	
$CdOH^{+}$		-261.1		
Ce^{+3}	-696.2	-672.0	-205.0	
Ce^{+4}	-537.2	-503.8	-301.0	
Co^{+2}	-58.2	-54.4	-113.0	
Co^{+3}	92.0	134.0	-305.0	
Cr^{+2}	-143.5			
Cs^{+}	-258.3	-292.0	133.1	-10.5
Cu^{+}	71.7	50.0	40.6	
Cu^{+2}	64.8	65.5	-99.6	
Dy^{+3}	-699.0	-665.0	-231.0	21.0
Er^{+3}	-705.4	-669.1	-244.3	21.0
Eu^{+2}	-527.0	-540.2	-8.0	
Eu^{+3}	-605.0	-574.1	-222.0	8.0
Fe^{+2}	-89.1	-78.9	-137.7	
Fe^{+3}	-48.5	-4.7	-315.9	
$FeOH^{+}$	-324.7	-277.4	-29.0	
$FeOH^{+2}$	-290.8	-229.4	-142.0	
$Fe(OH)_2^{+}$		-438.0		
Ga^{+2}		-88.0		
Ga^{+3}	-211.7	-159.0	-331.0	
$GaOH^{+2}$		-380.3		
$Ga(OH)_2^{+}$		-597.4		
Gd^{+3}	-686.0	-661.0	-205.9	
H^{+}	0	0	0	0
Hg^{+2}	171.1	164.4	-32.2	
$HgOH^{+}$	-84.5	-52.3	71.0	
Hg_2^{+2}	172.4	153.5	84.5	
Ho^{+3}	-705.0	-673.7	-226.8	17.0
In^{+}		-12.1		
$InOH^{+2}$	-370.3	-313.0	-88.0	
$In(OH)_2^{+}$	-619.0	-525.0	25.0	
In^{+2}		-50.7		
In^{+3}	-105.0	-98.0	-151.0	
K^{+}	-252.4	-283.3	102.5	21.8
La^{+3}	-707.1	-683.7	-217.6	-13.0
Li^{+}	-278.5	-293.3	13.4	68.6
Lu^{+3}	-665.0	-628.0	-264.0	25.0
LuF^{+2}		-931.4		
Mg^{+2}	-466.9	-454.8	-138.1	
$MgOH^{+}$		-626.7		
Mn^{+2}	-220.8	-228.1	-73.6	50.0
$MnOH^{+}$	-450.6	-405.0	-17.0	
NH_4^{+}	-132.5	-79.3	113.4	79.9
$N_2H_5^{+}$	-7.5	82.5	151.0	70.3
Na^{+}	-240.1	-261.9	59.0	46.4
Nd^{+3}	-696.2	-671.6	-206.7	-21.0
Ni^{+2}	-54.0	-45.6	-128.9	
$NiOH^{+}$	-287.9	-227.6	-71.0	
PH_4^{+}		92.1		
Pa^{+4}	-619.0			
Pb^{+2}	-1.7	-24.4	10.5	
$PbOH^{+}$		-226.3		
Pd^{+2}	149.0	176.5	-184.0	
Po^{+2}		71.0		
Po^{+4}		293.0		
Pr^{+3}	-704.6	-679.1	-209.0	-29.0
Pt^{+2}		254.8		
Ra^{+2}	-527.6	-561.5	54.0	
Rb^{+}	-251.2	-284.0	121.5	

Mol. form.	$\Delta_f H°$(aq)/ kJ mol^{-1}	$\Delta_f G°$(aq)/ kJ mol^{-1}	$S°$(aq)/ J K^{-1} mol^{-1}	C_p(aq)/ J K^{-1}mol^{-1}
Re^{+}		-33.0		
Sc^{+3}	-614.2	-586.6	-255.0	
$ScOH^{+2}$	-861.5	-801.2	-134.0	
Sm^{+2}		-497.5		
Sm^{+3}	-691.6	-666.6	-211.7	-21.0
Sn^{+2}	-8.8	-27.2	-17.0	
$SnOH^{+}$	-286.2	-254.8	50.0	
Sr^{+2}	-545.8	-559.5	-32.6	
$SrOH^{+}$		-721.3		
Tb^{+3}	-682.8	-651.9	-226.0	17.0
$Te(OH)_3^{+}$	-608.4	-496.1	111.7	
Th^{+4}	-769.0	-705.1	-422.6	
$Th(OH)^{+3}$	-1030.1	-920.5	-343.0	
$Th(OH)_2^{+2}$	-1282.4	-1140.9	-218.0	
Tl^{+}	5.4	-32.4	125.5	
Tl^{+3}	196.6	214.6	-192.0	
$TlOH^{+2}$		-15.9		
$Tl(OH)_2^{+}$		-244.7		
Tm^{+3}	-697.9	-662.0	-243.0	25.0
U^{+3}	-489.1	-476.2	-188.0	
U^{+4}	-591.2	-531.9	-410.0	
Y^{+3}	-723.4	-693.8	-251.0	
$Y(OH)^{+2}$		-879.1		
$Y_2(OH)_2^{+4}$		-1780.3		
Yb^{+2}		-527.0		
Yb^{+3}	-674.5	-644.0	-238.0	25.0
Zn^{+2}	-153.9	-147.1	-112.1	46.0
$ZnOH^{+}$		-330.1		

TABLE 2. Thermodynamic Properties of Aqueous Anions

Mol. form.	$\Delta_f H°$(aq)/ kJ mol^{-1}	$\Delta_f G°$(aq)/ kJ mol^{-1}	$S°$(aq)/ J K^{-1} mol^{-1}	C_p(aq)/ J K^{-1}mol^{-1}
AlO_2^{-}	-930.9	-830.9	-36.8	
$Al(OH)_4^{-}$	-1502.5	-1305.3	102.9	
AsO_2^{-}	-429.0	-350.0	40.6	
AsO_4^{-3}	-888.1	-648.4	-162.8	
BF_4^{-}	-1574.9	-1486.9	180.0	
BH_4^{-}	48.2	114.4	110.5	
BO_2^{-}	-772.4	-678.9	-37.2	
$B_4O_7^{-2}$		-2604.8		
BeO_2^{-2}	-790.8	-640.1	-159.0	
Br^{-}	-121.6	-104.0	82.4	-141.8
BrO^{-}	-94.1	-33.4	42.0	
BrO_3^{-}	-67.1	18.6	161.7	
BrO_4^{-}	13.0	118.1	199.6	
$CHOO^{-}$	-425.6	-351.0	92.0	-87.9
CH_3COO^{-}	-486.0	-369.3	86.6	-6.3
CN^{-}	150.6	172.4	94.1	
CO_3^{-2}	-677.1	-527.8	-56.9	
$C_2O_4^{-2}$	-825.1	-673.9	45.6	
$C_2O_4H^{-}$	-818.4	-698.3	149.4	
Cl^{-}	-167.2	-131.2	56.5	-136.4
ClO^{-}	-107.1	-36.8	42.0	
ClO_2^{-}	-66.5	17.2	101.3	
ClO_3^{-}	-104.0	-8.0	162.3	
ClO_4^{-}	-129.3	-8.5	182.0	
CrO_4^{-2}	-881.2	-727.8	50.2	
$Cr_2O_7^{-2}$	-1490.3	-1301.1	261.9	
F^{-}	-332.6	-278.8	-13.8	-106.7
$Fe(CN)_6^{-3}$	561.9	729.4	270.3	
$Fe(CN)_6^{-4}$	455.6	695.1	95.0	
$HB_4O_7^{-}$		-2685.1		
HCO_3^{-}	-692.0	-586.8	91.2	
HF_2^{-}	-649.9	-578.1	92.5	
HPO_3F^{-}		-1198.2		
HPO_4^{-2}	-1292.1	-1089.2	-33.5	
$HP_2O_7^{-3}$	-2274.8	-1972.2	46.0	
HSO_3^{-}	-626.2	-527.7	139.7	
HSO_4^{-}	-887.3	-755.9	131.8	-84.0
$HS_2O_4^{-}$		-614.5		
HSe^{-}	15.9	44.0	79.0	
$HSeO_3^{-}$	-514.6	-411.5	135.1	
$HSeO_4^{-}$	-581.6	-452.2	149.4	
$H_2AsO_3^{-}$	-714.8	-587.1	110.5	
$H_2AsO_4^{-}$	-909.6	-753.2	117.0	
$H_2PO_4^{-}$	-1296.3	-1130.2	90.4	
$H_2P_2O_7^{-2}$	-2278.6	-2010.2	163.0	
I^{-}	-55.2	-51.6	111.3	-142.3
IO^{-}	-107.5	-38.5	-5.4	
IO_3^{-}	-221.3	-128.0	118.4	
IO_4^{-}	-151.5	-58.5	222.0	
MnO_4^{-}	-541.4	-447.2	191.2	-82.0
MnO_4^{-2}	-653.0	-500.7	59.0	
MoO_4^{-2}	-997.9	-836.3	27.2	
NO_2^{-}	-104.6	-32.2	123.0	-97.5
NO_3^{-}	-207.4	-111.3	146.4	-86.6
N_3^{-}	275.1	348.2	107.9	
OCN^{-}	-146.0	-97.4	106.7	
OH^{-}	-230.0	-157.2	-10.8	-148.5
PO_4^{-3}	-1277.4	-1018.7	-220.5	
$P_2O_7^{-4}$	-2271.1	-1919.0	-117.0	
Re^{-}	46.0	10.1	230.0	
S^{-2}	33.1	85.8	-14.6	
SCN^{-}	76.4	92.7	144.3	-40.2
SH^{-}	-17.6	12.1	62.8	
SO_3^{-2}	-635.5	-486.5	-29.0	
SO_4^{-2}	-909.3	-744.5	20.1	-293.0
S_2^{-2}	30.1	79.5	28.5	
$S_2O_3^{-2}$	-652.3	-522.5	67.0	
$S_2O_4^{-2}$	-753.5	-600.3	92.0	
$S_2O_8^{-2}$	-1344.7	-1114.9	244.3	
Se^{-2}		129.3		
SeO_3^{-2}	-509.2	-369.8	13.0	
SeO_4^{-2}	-599.1	-441.3	54.0	
VO_3^{-}	-888.3	-783.6	50.0	
VO_4^{-3}		-899.0		
WO_4^{-2}	-1075.7			

HEAT OF COMBUSTION

The heat of combustion of a substance at 25 °C can be calculated from the enthalpy of formation ($\Delta_f H^\circ$) data in the table "Standard Thermodynamic Properties of Chemical Substances" in this section. We can write the general combustion reaction as

$$X + O_2 \rightarrow CO_2(g) + H_2O\ (l) + \text{other products}$$

For a compound containing only carbon, hydrogen, and oxygen, the reaction is simply

$$C_aH_bO_c + \quad a + \frac{1}{4}b - \frac{1}{2}c \quad O_2 \rightarrow a\ CO_2\ (g) + \frac{1}{2}b\ H_2O\ (l)$$

and the standard heat of combustion $\Delta_c H^\circ$, which is defined as the negative of the enthalpy change for the reaction (i.e., the heat released in the combustion process), is given by

$$\Delta_c H^o = -a\Delta_f H^o(CO_2, g) - \frac{1}{2}b\Delta_f H^o(H_2O, l) + \Delta_f H^o(C_aH_bO_c)$$

$$= 393.51a + 142.915b + \Delta_f H^o(C_aH_bO_c)$$

This equation applies if the reactants start in their standard states (25 °C and one atmosphere pressure) and the products return to the same conditions. The same equation applies to a compound containing another element if that element ends in its standard reference state (e.g., nitrogen, if the product is N_2); in general, however, the exact products containing the other elements must be known in order to calculate the heat of combustion.

The following table gives the standard heat of combustion calculated in this manner for a few representative substances. Data for a much longer list of substances is given in the Online Edition of this *CRC Handbook*. The definitions of the columns are as follows.

Column heading	Definition
Name	Name of chemical; given in alphabetical order
Mol. form.	Chemical formula; written in Hill format
State	Physical state of substance being combusted; liq = liquid; cry = crystalline; gas = gas
$\Delta_c H^\circ$	Standard heat of combustion at 25 °C and 1 atm pressure, in units kJ mol^{-1}

Heat of Combustion of Selected Compounds

Name	Mol. form.	State	$\Delta_c H^\circ$/ kJ mol^{-1}
Acetaldehyde	C_2H_4O	liq	1167
Acetamide	C_2H_5NO	cry	1185
Acetic acid	$C_2H_4O_2$	liq	874
Acetone	C_3H_6O	liq	1790
Acetonitrile	C_2H_3N	liq	1256
Acetylene	C_2H_2	gas	1300
L-Alanine	$C_3H_7NO_2$	cry	1577
Ammonia	H_3N	gas	383
Aniline	C_6H_7N	liq	3393
Anthracene	$C_{14}H_{10}$	cry	7068
Benzene	C_6H_6	liq	3268
Benzoic acid	$C_7H_6O_2$	cry	3228.2
1,3-Butadiene	C_4H_6	gas	2542
Butane	C_4H_{10}	gas	2878
2-Butanone	C_4H_8O	liq	2444
1-Butene	C_4H_8	gas	2718
cis-2-Butene	C_4H_8	gas	2710
trans-2-Butene	C_4H_8	gas	2706
Carbon (graphite)	C	cry	394
Carbon monoxide	CO	gas	283
Cyclobutene	C_4H_6	gas	2588
Cyclohexane	C_6H_{12}	liq	3920
Cyclopropane	C_3H_6	gas	2091
Decane	$C_{10}H_{22}$	liq	6778
Diethyl ether	$C_4H_{10}O$	liq	2724
Dimethyl ether	C_2H_6O	gas	1460
Ethane	C_2H_6	gas	1561
1,2-Ethanediol	$C_2H_6O_2$	liq	1185
Ethanol	C_2H_6O	liq	1367
Ethyl acetate	$C_4H_8O_2$	liq	2238
Ethylene	C_2H_4	gas	1411
Formaldehyde	CH_2O	gas	571
Formic acid	CH_2O_2	liq	254
Glycerol	$C_3H_8O_3$	liq	1654

Name	Mol. form.	State	$\Delta_c H^\circ$/ kJ mol^{-1}
Heptane	C_7H_{16}	liq	4817
Hexane	C_6H_{14}	liq	4163
Hydrazine	H_4N_2	liq	622
Hydrogen	H_2	gas	286
Hydrogen cyanide	CHN	gas	672
Ketene	C_2H_2O	gas	1025
Methane	CH_4	gas	891
Methanol	CH_4O	liq	726
Methyl acetate	$C_3H_6O_2$	liq	1592
Methylamine	CH_5N	gas	1086
Methyl formate	$C_2H_4O_2$	liq	973
Naphthalene	$C_{10}H_8$	cry	5157
Nitric oxide	NO	gas	91
Nitrobenzene	$C_6H_5NO_2$	liq	3088
Nitromethane	CH_3NO_2	liq	710
Nitrous oxide	N_2O	gas	82
Nonane	C_9H_{20}	liq	6125
Octane	C_8H_{18}	liq	5470
Pentane	C_5H_{12}	liq	3509
1-Pentanol	$C_5H_{12}O$	liq	3331
Phenanthrene	$C_{14}H_{10}$	cry	7055
Phenol	C_6H_6O	cry	3054
Propanal	C_3H_6O	liq	1822
Propane	C_3H_8	gas	2220
Propanenitrile	C_3H_5N	liq	1911
1-Propanol	C_3H_8O	liq	2021
2-Propanol	C_3H_8O	liq	2006
Propene	C_3H_6	gas	2058
Pyridine	C_5H_5N	liq	2782
Toluene	C_7H_8	liq	3910
Trimethylamine	C_3H_9N	gas	2443
2,4,6-Trinitrotoluene	$C_7H_5N_3O_6$	cry	3406
Urea	CH_4N_2O	cry	632.7

ENERGY CONTENT OF FUELS

Several fuels are compared in this table with respect to their energy content per unit mass and the amount of CO_2 released per unit of available energy. The energy content is taken to be the negative of the standard enthalpy of combustion (see the table "Heat of Combustion" in this section for more details). The energy is assumed to be released by combustion with oxygen at normal atmospheric pressure, with products of gaseous CO_2 and liquid H_2O at room temperature. This quantity is often called the "gross heat of combustion" (sometimes called "higher heating value or HHV") to distinguish it from the "net heat of combustion," for which the water remains in the gas state. The latter quantity is typically 5% to 10% less than the values given here.

The energy content is given both in SI units of MJ/kg and conventional units of BTU/lb. Values for the fossil fuels and other materials are typical; individual samples show wide variations. The data refer to substances in their normal physical state at ambient temperature and pressure.

Column heading	Definition
Name	Name of fuel; fossil fuels and other materials are typical composition; individual samples show wide variations
***E*(SI)**	Energy content per unit mass, in units of MJ kg^{-1} (SI units)
***E*(conv.)**	Energy content per unit mass, in units of 10^3·BTU lb^{-1} (conventional units)
CO_2 rel.	Mass of carbon released as carbon dioxide (CO_2) per megajoule of energy, in units g MJ^{-1}

Examination of the table shows that the minimum CO_2 release occurs for fuels that have a high ratio of hydrogen to carbon. Furthermore, fuels containing oxygen have a lower energy content and higher CO_2 release than hydrocarbons with the same number of carbon atoms.

References

1. Domalski, E. S., Jobe, T. L., and Milne, T. A., *Thermodynamic Data for Biomass Conversion and Waste Incineration,* SERI/SP-271-2839, Solar Technical Information Program, U.S. Department of Energy, September 1986; see also NBSIR 78-1479, National Bureau of Standards, August 1978. <https://doi.org/10.2172/7038865>
2. Green, D. W., and Ackers, D. E, *Perry's Chemical Engineers' Handbook, Eighth Edition,* McGraw-Hill, New York, 2007.
3. *Transportation Energy Data Book,* Edition 35, U S. Department of Energy, October 2016. <http://cta.ornl.gov/data/index.shtml>
4. *Chemical Composition of Natural Gas,* Union Gas Limited, www.uniongas.com/aboutus/aboutng/composition.asp.
5. Alternative Fuels Data Center-Fuel Properties Comparison, www.afdc.energy.gov/fuels/fuel_comparison_chart.pdf, October 2014.

Energy Produced and Carbon Released from Fuels

Name	*E*(SI)/ MJ kg^{-1}	*E*(conv.)/ 10^3·BTU lb^{-1}	CO_2 rel./ g MJ^{-1}
Pure compounds			
Hydrogen	141.8	61.0	0.0
Methane	55.5	23.9	13.5
Ethane	51.9	22.3	15.4
Propane	50.3	21.7	16.2
Butane	49.5	21.3	16.7
Pentane	48.6	20.9	17.1
Hexane	48.3	20.8	17.3
Heptane	48.1	20.7	17.5
Octane	47.9	20.6	17.6
Methanol	22.7	9.7	16.5
Ethanol	29.7	12.8	17.6
1-Propanol	33.6	14.5	17.8
1-Butanol	36.1	15.5	18.0
1-Octanol	40.7	17.5	18.1
Methyl *tert*-butyl ether	38.2	16.4	17.8
Fossil fuels			
Natural gas[a]	54.0	23.2	13.9
Gasoline	46.5	20.0	17.6
E85 fuel[b]	32.1	13.8	17.6
Diesel fuel[c]	46.3	19.9	17.9
Kerosene	46.4	20.0	18.5
Fuel oil	40.9	17.6	21.3
Coal, high bituminous	36.3	15.6	23.5
Coal, low bituminous	28.9	12.4	26.3
Coal, anthracite	34.6	14.9	27.3
Other materials			
Wood, oak	18.9	8.1	25.3
Wood, locust	19.7	8.5	25.7
Wood, ponderosa pine	20.0	8.6	24.6
Wood, redwood	20.7	8.9	24.4
Charcoal, wood	34.7	14.9	26.8
Newsprint	18.6	8.0	26.5
Cellulose	17.3	7.5	25.6
Grass[d]	19.3	8.3	24.9

[a] Natural gas is assumed to be 95% methane, 2.5% ethane, and 2.5% inert compounds; however, the actual composition varies widely (see Ref. 4).

[b] E85 fuel (also called "ethanol flex fuel" or FFV fuel) has a nominal composition of 85% by volume ethanol and 15% by volume gasoline. In the United States the composition is allowed to vary from 51% ethanol to 83% ethanol.

[c] Diesel fuel from petroleum typically consists of C_8 to C_{25} aliphatic and aromatic hydrocarbons, with a density of around 0.85 kg/L (compared to about 0.75 kg/L for gasoline).

[d] Lawn clippings.

CHEMICAL REACTION RATE CONSTANTS FOR ATMOSPHERIC STUDIES

James B. Burkholder and Michael J. Kurylo

These tables present evaluated rate constants and equilibrium constants for bimolecular and termolecular reactions of a wide variety of species important in understanding and modeling the chemistry of the different components of the Earth's atmosphere. The data in these tables are taken from the latest data tables produced by the NASA Panel for Data Evaluation (Ref. 20). For a complete history of the Panel and its predecessors, please see the Introduction to this document in the Online Edition of the *CRC Handbook* available at <http://www.hbcponline.com/> or Refs. 1-20 below.

Bimolecular reaction rate constants

In Tables 1a through 1l, rate constants for bimolecular reactions are grouped into the following classes, with reaction constants for each class given in a separate table as follows.

Table No.	Contents
1a	O_x
1b	$O(^1D)$
1c	Singlet O_2
1d	HO_x
1e	NO_x
1f	Organic compounds
1g	FO_x
1h	ClO_x
1i	BrO_x
1j	IO_x
1k	SO_x
1l	Metals

The tables contain recommended bimolecular reaction rate constants $k(T)$ at temperature T in Arrhenius form,

$$k(T) = A \times \exp^{(-E/RT)} \qquad (1)$$

The parameters for equation 1 given in Tables 1a to 1l are defined as follows.

Column heading	Definition
Reaction	Molecular formulas of chemical reactants and products; tables are divided into reactant families; for $O(^1D)$, branching ratios are given for each reaction pathway
A	Arrhenius pre-exponential factor, in units of cm^3 $molecule^{-1}$ s^{-1}
E/R	Recommended temperature dependence ("activation temperature"), in K
k_{298}	Recommended rate constant at 298 K, in of units cm^3 $molecule^{-1}$ s^{-1}
f_{298}	Rate constant uncertainty factor at 298 K
g	Parameter used to calculate the rate constant uncertainty at temperatures other than 298 K

The temperature ranges shown for each reaction indicate the range for the available experimental data. This is not necessarily the range of temperature over which the recommended Arrhenius parameters are applicable. See the corresponding note in Ref. 20 for each reaction for such information.

The parameters f_{298} and g can be used to calculate the estimated rate constant uncertainty over the recommended temperature range from the following expression:

$$f(T) = f_{298} \times \exp \left| g \times \{1/T - 1/298\} \right| \qquad (2)$$

where the exponent is an absolute value.

Note that f_{298} and g have been defined to correspond to approximately one standard deviation. Hence, $f(T)$ yields a similar uncertainty interval. The more commonly used 95% confidence limits at a given temperature can be obtained by multiplying and dividing the recommended value of the rate constant at that temperature by the factor $f^2(T)$. It should be emphasized that the parameter g has been defined for use with f_{298} in the above expression and should not be interpreted as the uncertainty in the Arrhenius activation temperature (E/R).

The uncertainty represented by $f(T)$ is normally symmetric; i.e., the rate constant may be greater than or less than the recommended value, $k(T)$ by the factor $f(T)$. In a few cases, asymmetric uncertainties are given in the temperature coefficient. For these cases, the factors by which a rate constant is to be multiplied or divided to obtain, respectively, the upper and lower limits are not equal, except at 298 K where the factor is simply f_{298}.

Termolecular reaction rate constants

Table 2.1 gives rate constants for the termolecular reactions, which have a more complicated dependence on temperature and pressure. Hence, recommendations are made for the low-pressure limiting rate constant $k_0(T)$ at temperature T and for the high-pressure limiting rate constant $k_\infty(T)$, also at temperature T, as well as the parameters required to obtain the effective second-order rate constant for a given temperature and pressure.

For the low-pressure limit, the following equation holds:

$$k_0(T) = k_0^{298}\,(T/298)^{-n} \qquad (3)$$

Similarly, for the high-pressure limit, the equation is as follows:

$$k_\infty(T) = k_\infty^{298}\,(T/298)^{-m} \qquad (4)$$

The parameters for equations 3 and 4 are given in Table 2.1 and are defined as follows.

Column heading	Definition
Reaction	Molecular formulas of chemical reactants and products; the table is divided into reactant families
k_0^{298}	Low-pressure rate constant for the termolecular reaction at 298 K, in units of cm^6 $molecule^{-2}$ s^{-2}
n	Exponent for low-pressure temperature dependence
k_∞^{298}	High-pressure rate constant for the termolecular reaction at 298 K, in units of cm^6 $molecule^{-2}$ s^{-2}
m	Exponent for high-pressure temperature dependence
$k(T,[M])$ (298 K, 1 Atm)	Bimolecular rate coefficient calculated at 298 K and 1 Atm using the coefficients given in the other columns, in units cm^3 $molecule^{-1}$ s^{-1}
f_{298}	Rate constant uncertainty factor at 298 K
g	Parameter used to calculate the rate constant uncertainty at temperatures other than 298 K

More complete details on the equations in which these parameters are used are available in Ref. 20.

Rate constants for chemical activation reactions

Table 2.2 contains rate constants for chemical activation reactions.

For the low-pressure limit, the following equation holds:

$$k_0(T) = k_0^{298}\,(298/T)^{n} \qquad (5)$$

Similarly, for the high-pressure limit, the equation is as follows.

$$k_\infty(T) = k_\infty^{298}\,(298/T)^{m} \qquad (6)$$

Some association reactions produce not only recombination product(s) but also additional products that appear to originate from a simple bimolecular reaction. In these cases, the total rate constant k_{Total} is equal to the sum of the rate constant for the association reaction $k_f(T,[M])$ and the rate constant for the formation of products resulting from chemical activation $k_f^{CA}(T,[M])$. More details on this model are given in Ref. 20. $k_f(T,[M])$ is a function of $k_0(T)$ and $k_\infty(T)$ (Ref. 20). $k_f^{CA}(T,[M])$ is dependent on the second order rate constant at [M] = 0 (i.e., the zero-pressure intercept), given by the Arrhenius expression as follows.

$$k_{int}(T) = A e^{-BT} \qquad (7)$$

The parameters for equations 5, 6, and 7 are given in Table 2.2 and are defined as follows.

Column heading	Definition
Reactants, reaction type, and products	Molecular formulas of chemical reactants; second line has products for association pathway; third line has products for dissociation pathway
k_0^{298}	Low-pressure rate constant for the termolecular reaction at 298 K, in units of $cm^6\ molecule^{-2}\ s^{-2}$
n	Exponent for low-pressure temperature dependence
k_∞^{298}	High-pressure rate constant for the termolecular reaction at 298 K, in units of $cm^6\ molecule^{-2}\ s^{-2}$
m	Exponent for high-pressure temperature dependence
A	Arrhenius pre-exponential parameter for the rate coefficient in the zero-pressure limit, in units $cm^3\ molecule^{-1}\ s^{-1}$
B	Arrhenius temperature dependence parameter for the rate coefficient in the zero-pressure limit, in units K
k_{Total} **(298 K, 1 Atm)**	Total rate constant at 298 K and 1 Atm, calculated with data in the other columns, in units $cm^3\ molecule^{-1}\ s^{-1}$
f_{298}	Rate constant uncertainty factor at 298 K
g	A parameter used to calculate the rate constant uncertainty at temperatures other than 298 K

Equilibrium constants

Some three-body reactions form products that are thermally unstable at atmospheric temperatures. In such cases, the thermal decomposition reaction may compete with other loss processes, such as photodissociation or radical attack. Table 3 contains recommended equilibrium constants $K_{eq}(T)$ for several reactions that may fall into this category. Each recommended equilibrium constant is given in Arrhenius form and contains the information shown below.

$$K_{eq}(T) = A\,\exp^{(B/T)} \text{ for temperatures } 200\text{ K} < T < 300\text{ K} \qquad (8)$$

The parameters for equation 8 are given in Table 3 and are defined as follows.

Column heading	Definition
Reaction	Molecular formulas of chemical reactants and products
A	Arrhenius pre-exponential factor, in units of $cm^3\ molecule^{-1}$
B	Recommended temperature dependence ("activation temperature"), in K
K_{eq}**(298)**	Recommended equilibrium constant at 298 K, in units of $cm^3\ molecule^{-1}$
f_{298}	Equilibrium constant uncertainty factor at 298 K
g	Parameter used to calculate the rate constant uncertainty (corresponding to approximately one standard deviation) at temperatures other than 298 K

As for bimolecular reactions, f_{298} is the approximate one standard deviation uncertainty factor in the equilibrium constant at 298 K, and g is the parameter to be used to calculate the equilibrium constant uncertainty at temperatures other than 298 K.

The process of evaluating chemical kinetic data does not conform to a simple set of mathematical rules. There is no "one-size-fits-all" algorithm that can be applied and each reaction must be examined on a case-by-case basis. Consideration of uncertainties in the kinetic and photochemical parameters used in atmospheric models plays a key role in determining the reliability of and uncertainty in the model results. Quite often the cause(s) of differences in experimental results from various laboratories cannot be determined with confidence and making recommendations for the uncertainties of the rate constant is often more difficult than making recommendations for the Arrhenius parameters themselves. In many cases, investigators suggest possible qualitative reasons for disagreements among data sets. Thus, data evaluators necessarily must consider a variety of factors in assigning a recommendation, including such aspects as the chemical complexity of the system, sensitivities and shortcomings of the experimental techniques employed, similarities or trends in reactivity, and the level of agreement among studies using different techniques.

These data are the recommendations that appear in the most current NASA JPL document (JPL 19-5) (Ref. 20). Readers should consult the Online Edition of the *CRC Handbook* (see footnotes) for the descriptive notes and bibliography associated with each line entry. In particular, the notes contain important details about the latest revision date, changes from previous evaluations, data formats, units, and the actual use of the recommendations and their indicated uncertainties.

References

1. Hudson, R. D., Ed., *Chlorofluoromethanes and the Stratosphere*, NASA Reference Publication 1010, NASA, Washington, DC, 1977.
2. Hudson, R. D., Reed, E. I., Eds., *The Stratosphere: Present and Future*, NASA Reference Publication 1049, NASA, Washington, DC, 1979.
3. DeMore, W. B., Golden, D. M., Hampson, R. F., Howard, C. J., Kurylo, M. J., Margitan, J. J., Molina, M. J., Ravishankara, A. R., Watson, R. T., *Chemical Kinetics and Photochemical Data for Use in Stratospheric Modeling, Evaluation, Number 7*, Jet Propulsion Laboratory, California Institute of Technology, Pasadena, CA, JPL Publication 85-37, 1985. <https://doi.org/10.1002/kin.550171010>
4. DeMore, W. B., Golden, D. M., Hampson, R. F., Howard, C. J., Kurylo, M. J., Molina, M. J., Ravishankara, A. R., Sander, S. P., *Chemical Kinetics and Photochemical Data for Use in Stratospheric Modeling, Evaluation Number 8*, Jet Propulsion Laboratory, California Institute of Technology, Pasadena, CA, JPL Publication 87-41, 1987.

5. DeMore, W. B., Golden, D. M., Hampson, R. F., Howard, C. J., Kurylo, M. J., Molina, M. J., Ravishankara, A. R., Sander, S. P., *Chemical Kinetics and Photochemical Data for Use in Stratospheric Modeling, Evaluation Number 9*, Jet Propulsion Laboratory, California Institute of Technology, Pasadena, CA, JPL Publication 90-1, 1990.
6. DeMore, W. B., Golden, D. M., Hampson, R. F., Howard, C. J., Kurylo, M. J., Molina, M. J., Ravishankara, A. R., Sander, S. P., *Chemical Kinetics and Photochemical Data for Use in Stratospheric Modeling, Evaluation Number 10*, Jet Propulsion Laboratory, California Institute of Technology, Pasadena, CA, JPL Publication 92-20, 1992.
7. DeMore, W. B., Golden, D. M., Hampson, R. F., Howard, C. J., Kurylo, M. J., Molina, M. J., Ravishankara, A. R., Sander, S. P., *Chemical Kinetics and Photochemical Data for Use in Stratospheric Modeling, Evaluation Number 11*, Jet Propulsion Laboratory, California Institute of Technology, Pasadena, CA, JPL Publication 94-26, 1994.
8. DeMore, W. B., Golden, D. M., Hampson, R. F., Howard, C. J., Kurylo, M. J., Molina, M. J., Ravishankara, A. R., Watson, R. T., *Chemical Kinetics and Photochemical Data for Use in Stratospheric Modeling, Evaluation Number 5*, Jet Propulsion Laboratory, California Institute of Technology, Pasadena CA, JPL Publication 82-57, 1982.
9. DeMore, W. B., Golden, D. M., Hampson, R. F., Howard, C. J., Kurylo, M. J., Molina, M. J., Ravishankara, A. R., Watson, R. T., *Chemical Kinetics and Photochemical Data for Use in Stratospheric Modeling, Evaluation Number 6*, Jet Propulsion Laboratory, California Institute of Technology, Pasadena CA, JPL Publication 83-62, 1983.
10. DeMore, W. B., Golden, D. M., Hampson, R. F., Kurylo, M. J., Margitan, J. J., Molina, M. J., Stief, L. J., Watson, R. T., *Chemical Kinetics and Photochemical Data for Use in Stratospheric Modeling, Evaluation Number 4*, Jet Propulsion Laboratory, California Institute of Technology, Pasadena, CA, JPL Publication 81-3, 1981.
11. DeMore, W. B., Sander, S. P., Golden, D. M., Hampson, R. F., Kurylo, M. J., Howard, C. J., Ravishankara, A. R., Kolb, C. E., Molina, M. J., *Chemical Kinetics and Photochemical Data for Use in Stratospheric Modeling, Evaluation Number 12*, Jet Propulsion Laboratory, California Institute of Technology, Pasadena, CA, JPL Publication 97-4, 1997.
12. DeMore, W. B., Stief, L. J., Kaufman, F., Golden, D. M., Hampson, R. F., Kurylo, M. J., Margitan, J. J., Molina, M. J., Watson, R. T., *Chemical Kinetics and Photochemical Data for Use in Stratospheric Modeling, Evaluation Number 2*, Jet Propulsion Laboratory, California Institute of Technology, Pasadena, CA, JPL Publication 79-27, 1979.
13. Ko, M. K. W., Newman, P. A., Reimann, S., Strahan, S. E., Plumb, R. A., Stolarski, R. S., Burkholder, J. B., Mellouki, W., Engel, A., Atlas, E. L., Chipperfield, M., Liang, Q., *Lifetimes of Stratospheric Ozone-Depleting Substances, Their Replacements, and Related Species*, SPARC Report No. 6, WCRP-15/2013, 2013.
14. Sander, S. P., Abbatt, J., Barker, J. R., Burkholder, J. B., Friedl, R. R., Golden, D. M., Huie, R. E., Kolb, C. E., Kurylo, M. J., Moortgat, G. K., Orkin, V. L., Wine, P. H., *Chemical Kinetics and Photochemical Data for Use in Atmospheric Studies, Evaluation Number 16*, Jet Propulsion Laboratory, California Institute of Technology, Pasadena, CA, JPL Publication 09-24, 2009.
15. Sander, S. P., Abbatt, J., Barker, J. R., Burkholder, J. B., Friedl, R. R., Golden, D. M., Huie, R. E., Kolb, C. E., Kurylo, M. J., Moortgat, G. K., Orkin, V. L., Wine, P. H., *Chemical Kinetics and Photochemical Data for Use in Atmospheric Studies, Evaluation Number 17*, Jet Propulsion Laboratory, California Institute of Technology, Pasadena, CA, JPL Publication 10-6, 2011.
16. Sander, S. P., Finlayson-Pitts, B. J., Friedl, R. R., Golden, D. M., Huie, R. E., Keller-Rudek, H., Kolb, C. E., Kurylo, M. J., Molina, M. J., Moortgat, G. K., Orkin, V. L., Ravishankara, A. R., Wine, P. H., *Chemical Kinetics and Photochemical Data for Use in Atmospheric Studies, Evaluation Number 15*, Jet Propulsion Laboratory, California Institute of Technology, Pasadena, CA, JPL Publication 06-2, 2006.
17. Sander, S. P., Finlayson-Pitts, B. J., Friedl, R. R., Golden, D. M., Huie, R. E., Kolb, C. E., Kurylo, M. J., Molina, M. J., Moortgat, G. K., Orkin, V. L., Ravishankara, A. R., *Chemical Kinetics and Photochemical Data for Use in Atmospheric Studies, Evaluation Number 14*, Jet Propulsion Laboratory, California Institute of Technology, Pasadena, CA, JPL Publication 02-25, 2002.
18. Sander, S. P., Friedl, R. R., DeMore, W. B., Golden, D. M., Kurylo, M. J., Hampson, R. F., Huie, R. E., Moortgat, G. K., Ravishankara, A. R., Kolb, C. E., Molina, M. J., *Chemical Kinetics and Photochemical Data for Use in Stratospheric Modeling, Evaluation Number 13*, Jet Propulsion Laboratory, California Institute of Technology, Pasadena, CA, JPL Publication 00-3, 2000.
19. Burkholder, J. B., Sander, S. P., Barker, J. R., Huie, R. E., Kolb, C. E., Kurylo, M. J., Orkin, V. L., Wilmouth, D. M., and Wine, P. H., *Chemical Kinetics and Photochemical Data for Use in Atmospheric Studies, Evaluation Number 18*, JPL Publication 15-10, Jet Propulsion Laboratory, California Institute of Technology, Pasadena, CA, 2015.
20. Burkholder, J. B., Sander, S. P., Abbatt, J., Cappa, C., Crounse, J. D., Dibble, T. S., Huie, R. E., Kolb, C. E., Kurylo, M. J., Orkin, V. L., Percival, C. J., Wilmouth, D. M., and Wine, P. H., *Chemical Kinetics and Photochemical Data for Use in Atmospheric Studies, Evaluation Number 19*, JPL Publication 19-5, Jet Propulsion Laboratory, California Institute of Technology, Pasadena, CA, 2019. <http://jpldataeval.jpl.nasa.gov/>

Refs. 3-12 and 14-20 are available at http://jpldataeval.jpl.nasa.gov.

Ref. 13 is available at http://www.sparc-climate.org/publications/sparc-reports/sparc-report-no6/.

TABLE 1a. O_x Bimolecular Reaction Rates of Atmospheric Importance

Reaction	T/K	A/cm³ molecule⁻¹ s⁻¹	E/R/K	k_{298}/cm³ molecule⁻¹ s⁻¹	f_{298}	g
$O + O_3 \rightarrow O_2 + O_2$	220–409	8.0×10^{-12}	2060	8.0×10^{-15}	1.1	200

TABLE 1b. $O(^1D)$ Bimolecular Reaction Rates of Atmospheric Importance

Reaction	Branching ratio*	T/K	A*/cm³ molecule⁻¹ s⁻¹	E/R/K*	k_{298}*/cm³ molecule⁻¹ s⁻¹	f_{298}	g
$O(^1D) + O_2$		104–424	3.3×10^{-11}	-55	3.95×10^{-11}	1.1	10
$\rightarrow O(^3P) + O_2$	*0*						
$\rightarrow O(^3P) + O_2(^1\Sigma)$	0.80 ± 0.20						
$\rightarrow O(^3P) + O_2(^1\Delta)$	*0.20 (0.40–0)*						
$O(^1D) + O_3$		103–393	2.4×10^{-10}	0	2.4×10^{-10}	1.2	50
$\rightarrow O(^3P) + O_3$	*0*						
$\rightarrow O_2 + O_2$	0.50 ± 0.03						
$\rightarrow O_2 + O(^3P) + O(^3P)$	0.50 ± 0.03						
$O(^1D) + H_2$		204–420	1.2×10^{-10}	0	1.2×10^{-10}	1.15	50

Thermochem

Reaction	Branching ratio*	T/K	A*/cm³ molecule⁻¹ s⁻¹	E/R/K*	k_{298}*/cm³ molecule⁻¹ s⁻¹	f_{298}	g
→ $O(^3P) + H_2$	<0.01						
→ OH + H	1.0 (+0/-0.01)						
$O(^1D) + H_2O$		217–453	1.63×10^{-10}	-60	2.0×10^{-10}	1.08	20
→ $O(^3P) + H_2O$	<0.003						
→ $O_2 + H_2$	0.006 ± 0.006						
→ OH + OH	1.0 (+0/-0.015)						
$O(^1D) + N_2$		104–673	2.15×10^{-11}	-110	3.1×10^{-11}	1.1	20
→ $O(^3P) + N_2$	1						
$O(^1D) + N_2O$		195–719	1.19×10^{-10}	-20	1.27×10^{-10}	1.1	25
→ $O(^3P) + N_2O$	*<0.01*						
→ $N_2 + O_2$	0.39 (0.36–0.42)						
→ NO + NO	0.61 ± 0.03						
$O(^1D) + NH_3$		204–354	2.5×10^{-10}	0	2.5×10^{-10}	1.2	25
→ $O(^3P) + NH_3$	*0*						
→ products	*1*						
$O(^1D)$ + HCN		193–430	1.08×10^{-10}	-105	1.54×10^{-10}	1.2	0
→ $O(^3P)$ + HCN	0.15 × exp(200/T)						
→ products	0.93 × exp(-82/T)						
$O(^1D) + CH_3CN$		193–430	2.54×10^{-10}	24	2.34×10^{-10}	1.2	0
→ $O(^3P) + CH_3CN$	0.035 (+0.05/-0.035)						
→ products	0.965 (+0.035/-0.05)						
$O(^1D) + CO_2$		195–370	7.5×10^{-11}	-115	1.1×10^{-10}	1.15	20
→ $O(^3P) + CO_2$	1.0 (+0/-0.01)						
$O(^1D) + CH_4$		198–413	1.75×10^{-10}	0	1.75×10^{-10}	1.15	25
→ $O(^3P) + CH_4$	<0.005						
→ CH_3 + OH	0.75 ± 0.15						
→ CH_3O or CH_2OH + H	0.20 ± 0.10						
→ $CH_2O + H_2$	0.05 ± 0.05						
$O(^1D)$ + HCl		199–379	1.5×10^{-10}	0	1.5×10^{-10}	1.1	25
→ $O(^3P)$ + HCl	0.12 ± 0.04						
→ H + ClO	0.22 ± 0.05						
→ Cl + OH	0.66 (0.61–0.71)						
$O(^1D)$ + HF		298	*5.0×10^{-11}*	*0*	5.0×10^{-11}	1.5	25
→ $O(^3P)$ + HF	0.70 ± 0.05						
→ F + OH	0.30 ± 0.05						
$O(^1D) + NF_3$		199–356	2.0×10^{-11}	-44	2.3×10^{-11}	1.1	0
→ $O(^3P) + NF_3$	0.07 (+0.21/-0.07)						
→ products	0.93 ((+0.07/-0.21)						
$O(^1D)$ + HBr		297	*1.5×10^{-10}*	*0*	1.5×10^{-10}	1.5	25
→ $O(^3P)$ + HBr	0.20 ± 0.07						
→ H + BrO	0.20 ± 0.04						
→ Br + OH	0.60 (0.49–0.71)						
$O(^1D) + Cl_2$		298	2.7×10^{-10}	0	2.7×10^{-10}	1.1	25
→ $O(^3P) + Cl_2$	0.25 ± 0.10						
→ Cl + ClO	0.75 ± 0.07						
$O(^1D) + CCl_2O$		194–429	2.2×10^{-10}	-30	2.4×10^{-10}	1.1	25
→ $O(^3P) + CCl_2O$	0.20 ± 0.04						
→ products	0.80 ± 0.04						
$O(^1D)$ + CClFO		298	*1.9×10^{-10}*	*0*	*1.9×10^{-10}*	1.5	25
→ $O(^3P)$ + CClFO	*0.2*						
→ products	*0.8*						
$O(^1D) + CF_2O$		298	*7.4×10^{-11}*	*0*	7.4×10^{-11}	1.5	25
→ $O(^3P) + CF_2O$	0.35						
→ products	0.65 ± 0.10						
$O(^1D) + CH_3Cl$		298	*2.6×10^{-10}*	*0*	2.6×10^{-10}	1.3	50
→ $O(^3P) + CH_3Cl$	0.1						
→ ClO + products	0.46 ± 0.06						
→ Cl + products	*0.35*						
→ H + products	*0.09*						
$O(^1D) + CCl_4$ (CFC-10)		203–343	3.30×10^{-10}	0	3.30×10^{-10}	1.15	0

Reaction	Branching ratio*	T/K	A*/cm³ molecule⁻¹ s⁻¹	E/R/K*	k₂₉₈*/cm³ molecule⁻¹ s⁻¹	f_{298}	g
→ $O(^3P) + CCl_4$	0.21 ± 0.04						
→ ClO + products	0.79 ± 0.04						
$O(^1D) + CH_3CCl_3$		298	*3.25 × 10⁻¹⁰*	*0*	*3.25 × 10⁻¹⁰*	1.4	0
→ $O(^3P) + CH_3CCl_3$	*0.1*						
→ products	*0.9*						
$O(^1D) + CH_3Br$		297	*1.8 × 10⁻¹⁰*	*0*	1.8 × 10⁻¹⁰	1.15	50
→ $O(^3P) + CH_3Br$	*0 (+0.07/-0)*						
→ BrO + products	0.44 ± 0.05						
→ OH + products	*0.56 (0.44–0.61)*						
$O(^1D) + CH_2Br_2$		297	*2.7 × 10⁻¹⁰*	*0*	2.7 × 10⁻¹⁰	1.2	25
→ $O(^3P) + CH_2Br_2$	0.05 ± 0.07						
→ products	0.95 (+0.05/-0.10)						
$O(^1D) + CHBr_3$		297	*6.6 × 10⁻¹⁰*	*0*	6.6 × 10⁻¹⁰	1.3	25
→ $O(^3P) + CHBr_3$	0.30 ± 0.10						
→ products	0.70 ± 0.10						
$O(^1D) + CH_3F$ (HFC-41)		298	*1.5 × 10⁻¹⁰*	*0*	1.5 × 10⁻¹⁰	1.15	50
→ $O(^3P) + CH_3F$	0.18 ± 0.07						
→ products	0.82 ± 0.07						
$O(^1D) + CH_2F_2$ (HFC-32)		298	*5.1 × 10⁻¹¹*	*0*	5.1 × 10⁻¹¹	1.2	50
→ $O(^3P) + CH_2F_2$	0.70 ± 0.11						
→ products	0.30 ± 0.11						
$O(^1D) + CHF_3$ (HFC-23)		217–372	8.7 × 10⁻¹²	-30	9.6 × 10⁻¹²	1.05	0
→ $O(^3P) + CHF_3$	0.75 ± 0.05						
→ products	0.25 ± 0.05						
$O(^1D) + CHCl_2F$ (HCFC-21)		188–343	1.9 × 10⁻¹⁰	0	1.9 × 10⁻¹⁰	1.15	50
→ $O(^3P) + CHCl_2F$	*0.20 ± 0.05*						
→ ClO + products	0.74 ± 0.06						
→ OH + products	*0.06 (0–0.17)*						
$O(^1D) + CHClF_2$ (HCFC-22)		173–373	1.02 × 10⁻¹⁰	0	1.02 × 10⁻¹⁰	1.07	0
→ $O(^3P) + CHClF_2$	0.25 ± 0.05						
→ ClO + products	0.56 ± 0.03						
→ OH + products	0.05 ± 0.02						
→ Other products	*0.14 (0.04–0.24)*						
$O(^1D) + CHF_2Br$		211–425	1.75 × 10⁻¹⁰	-70	2.2 × 10⁻¹⁰	1.15	25
→ $O(^3P) + CHF_2Br$	0.40 ± 0.06						
→ BrO + products	0.39 ± 0.07						
→ Other products	*0.21 (0.08–0.34)*						
$O(^1D) + CCl_3F$ (CFC-11)		173–372	2.30 × 10⁻¹⁰	0	2.30 × 10⁻¹⁰	1.1	0
→ $O(^3P) + CCl_3F$	0.10 ± 0.07						
→ ClO + products	0.79 ± 0.04						
→ Other products	*0.11 (0.0–0.22)*						
$O(^1D) + CCl_2F_2$ (CFC-12)		173–373	1.40 × 10⁻¹⁰	-25	1.52 × 10⁻¹⁰	1.15	0
→ $O(^3P) + CCl_2F_2$	0.14 ± 0.07						
→ ClO + products	0.76 ± 0.06						
→ Other products	*0.10 (0–0.23)*						
$O(^1D) + CClF_3$ (CFC-13)		298	*8.7 × 10⁻¹¹*	*0*	8.7 × 10⁻¹¹	1.2	50
→ $O(^3P) + CClF_3$	0.18 ± 0.06						
→ ClO + products	0.82 ± 0.06						
$O(^1D)$ + 1,2-c-C4Cl_2F_6 (E,Z)		296	*1.56 × 10⁻¹⁰*	*0*	1.56 × 10⁻¹⁰	1.1	0
→ $O(^3P)$ + 1,2-c-C4Cl_2F_6 (E,Z)	*0.12 ± 0.12*						
→ products	0.88 (+0.12/-0.15)						
$O(^1D) + CClBrF_2$ (Halon-1211)		297	*1.50 × 10⁻¹⁰*	*0*	1.50 × 10⁻¹⁰	1.2	50
→ $O(^3P) + CClBrF_2$	0.35 ± 0.04						
→ BrO + products	0.31 ± 0.06						
→ Other products	*0.34 (0.24–0.44)*						
$O(^1D) + CBr_2F_2$ (Halon-1202)		297	*2.20 × 10⁻¹⁰*	*0*	2.20 × 10⁻¹⁰	1.2	50
→ $O(^3P) + CBr_2F_2$	0.55 ± 0.06						
→ products	0.45 ± 0.06						
$O(^1D) + CBrF_3$ (Halon-1301)		297	*1.00 × 10⁻¹⁰*	*0*	1.00 × 10⁻¹⁰	1.2	50
→ $O(^3P) + CBrF_3$	0.55 ± 0.08						

Reaction	Branching ratio*	T/K	A*/cm³ molecule⁻¹ s⁻¹	E/R/K*	k_{298}*/cm³ molecule⁻¹ s⁻¹	f_{298}	g
→ BrO + products	0.45 ± 0.08						
$O(^1D) + CF_4$ (PFC-14)		297			$<2 \times 10^{-14}$		
→ $O(^3P) + CF_4$							
$O(^1D) + CH_3CH_2F$ (HFC-161)		297	2.6×10^{-10}	0	2.6×10^{-10}	1.2	25
→ $O(^3P) + CH_3CH_2F$	0.18 ± 0.05						
→ products	0.82 ± 0.05						
$O(^1D) + CH_3CHF_2$ (HFC-152a)		297	1.75×10^{-10}	0	1.75×10^{-10}	1.2	50
→ $O(^3P) + CH_3CHF_2$	0.45 ± 0.15						
→ OH + products	0.15 ± 0.02						
→ Other products	*0.4 (0.18–0.62)*						
$O(^1D) + CH_3CCl_2F$ (HCFC-141b)		297	2.60×10^{-10}	0	2.60×10^{-10}	1.2	50
→ $O(^3P) + CH_3CCl_2F$	0.31 ± 0.05						
→ products	0.69 ± 0.05						
$O(^1D) + CH_3CClF_2$ (HCFC-142b)		217–373	2.0×10^{-10}	0	2.0×10^{-10}	1.1	0
→ $O(^3P) + CH_3CClF_2$	0.35 ± 0.10						
→ products	0.65 ± 0.10						
$O(^1D) + CH_3CF_3$ (HFC-143a)		217–373	5.6×10^{-11}	-20	6.0×10^{-11}	1.2	0
→ $O(^3P) + CH_3CF_3$	0.35 ± 0.05						
→ OH + products	0.38 ± 0.06						
→ Other products	*0.27 (0.16–0.38)*						
$O(^1D) + CH_2ClCClF_2$ (HCFC-132b)		297	1.6×10^{-10}	0	1.6×10^{-10}	1.5	50
→ $O(^3P) + CH_2ClCClF_2$	*0.1*						
→ products	*0.9*						
$O(^1D) + CH_2ClCF_3$ (HCFC-133a)		297	1.2×10^{-10}	0	1.2×10^{-10}	1.25	50
→ $O(^3P) + CH_2ClCF_3$	0.20 ± 0.05						
→ products	0.80 ± 0.05						
$O(^1D) + CH_2FCF_3$ (HFC-134a)		297	4.9×10^{-11}	0	4.9×10^{-11}	1.15	50
→ $O(^3P) + CH_2FCF_3$	0.65 ± 0.06						
→ OH + products	0.24 ± 0.04						
→ Other products	*0.11 (0.01–0.21)*						
$O(^1D) + CHCl_2CF_3$ (HCFC-123)		297	2.0×10^{-10}	0	2.0×10^{-10}	1.2	50
→ $O(^3P) + CHCl_2CF_3$	0.21 ± 0.08						
→ products	0.79 ± 0.08						
$O(^1D) + CHClFCF_3$ (HCFC-124)		297	8.6×10^{-11}	0	8.6×10^{-11}	1.2	50
→ $O(^3P) + CHClFCF_3$	0.31 ± 0.10						
→ products	0.69 ± 0.10						
$O(^1D) + CHF_2CF_3$ (HFC-125)		217–373	9.5×10^{-12}	-25	1.03×10^{-11}	1.07	0
→ $O(^3P) + CHF_2CF_3$	0.25 ± 0.05						
→ OH + products	0.60 ± 0.10						
→ Other products	*0.15 (0–0.30)*						
$O(^1D) + CCl_3CF_3$ (CFC-113a)		296	2.6×10^{-10}	0	2.6×10^{-10}	1.25	0
→ $O(^3P) + CCl_3CF_3$	*0.1*						
→ ClO + products	0.79 ± 0.05						
→ Other products	*0.11 (0–0.16)*						
$O(^1D) + CCl_2FCClF_2$ (CFC-113)		217–373	2.32×10^{-10}	0	2.32×10^{-10}	1.1	0
→ $O(^3P) + CCl_2FCClF_2$	*0.1*						
→ ClO + products	0.80 ± 0.05						
→ Other products	*0.10 (0–0.15)*						
$O(^1D) + CCl_2FCF_3$ (CFC-114a)		296	1.6×10^{-10}	0	1.6×10^{-10}	1.2	0
→ $O(^3P) + CCl_2FCF_3$	*0.1*						
→ ClO + products	0.80 ± 0.05						
→ Other products	*0.10 (0–0.15)*						
$O(^1D) + CClF_2CClF_2$ (CFC-114)		217–373	1.30×10^{-10}	-25	1.41×10^{-10}	1.1	0
→ $O(^3P) + CClF_2CClF_2$	*0.1*						
→ ClO + products	0.85 ± 0.06						
→ Other products	*0.05 (0–0.1)*						
$O(^1D) + CClF_2CF_3$ (CFC-115)		217–373	5.4×10^{-11}	-30	6.0×10^{-11}	1.15	0
→ $O(^3P) + CClF_2CF_3$	0.14 ± 0.06						
→ products	0.86 ± 0.06						
$O(^1D) + CBrF_2CBrF_2$ (Halon-2402)		297	1.60×10^{-10}	0	1.60×10^{-10}	1.2	50

Reaction	Branching ratio*	T/K	A*/cm³ molecule⁻¹ s⁻¹	E/R/K*	k₂₉₈*/cm³ molecule⁻¹ s⁻¹	f_{298}	g
→ $O(^3P) + CBrF_2CBrF_2$	0.25 ± 0.07						
→ products	0.75 ± 0.07						
$O(^1D) + CF_3CF_3$ (CFC-116)		297			1.5×10^{-13}		
→ $O(^3P) + CF_3CF_3$							
→ products	<0.2						
$O(^1D) + CF_3CHFCF_3$ (HFC-227ea)		217–373	7.9×10^{-12}	-70	1.0×10^{-11}	1.1	0
→ $O(^3P) + CF_3CHFCF_3$	0.28 ± 0.07						
→ products	0.72 ± 0.07						
$O(^1D) + CHF_2CH_2CF_3$ (HFC-245fa)			*1.5×10^{-10}*	*0*	1.5×10^{-10}	1.3	0
→ $O(^3P) + CHF_2CH_2CF_3$	*0.5*						
→ products	*0.5*						
$O(^1D) + CHF_2CF_2CF_2CHF_2$ (HFC-338pcc)		297	*1.8×10^{-11}*	*0*	1.8×10^{-11}	1.3	50
→ $O(^3P) + CHF_2CF_2CF_2CHF_2$	0.95 (+0.05/-0.09)						
→ products	0.05 (+0.09/-0.05)						
$O(^1D) + c\text{-}C_4F_8$		297			8×10^{-13}		
→ $O(^3P) + c\text{-}C_4F_8$							
→ products	<0.04						
$O(^1D) + CF_3CHFCHFCF_2CF_3$ (HFC-43-10mee)		297	*2.1×10^{-10}*	*0*	2.1×10^{-10}	2	50
→ $O(^3P) + CF_3CHFCHFCF_2CF_3$	0.90 ± 0.10						
→ products	0.10 ± 0.10						
$O(^1D) + C_5F_{12}$ (PFC-41-12)		297			4×10^{-13}		
→ $O(^3P) + C_5F_{12}$							
→ products	<0.12						
$O(^1D) + C_6F_{14}$ (PFC-51-14)		297			1×10^{-12}		
→ $O(^3P) + C_6F_{14}$							
→ products	<0.16						
$O(^1D) + 1,2\text{-}(CF_3)2c\text{-}C_4F_6$		297			$<3 \times 10^{-13}$		
→ $O(^3P) + 1,2\text{-}(CF_3)2c\text{-}C_4F_6$							
→ products							
$O(^1D) + C_4F_{10}$		297			$<5 \times 10^{-13}$		
→ $O(^3P) + C_4F_{10}$							
→ products							
$O(^1D) + SF_6$		297			1.8×10^{-14}		
→ $O(^3P) + SF_6$							
→ products	<0.7						
$O(^1D) + SO_2$		298	*2.2×10^{-10}*	*0*	2.2×10^{-10}	1.3	30
→ $O(^3P) + SO_2$	0.24 ± 0.07						
→ products	0.76 ± 0.07						
$O(^1D) + SO_2F_2$		199–351	9×10^{-11}	-100	1.26×10^{-10}	1.3	30
→ $O(^3P) + SO_2F_2$	0.45 ± 0.04						
→ products	0.55 ± 0.04						
$O(^1D) + SF_5CF_3$		296–300			2×10^{-13}		
→ $O(^3P) + SF_5CF_3$							
→ products	<0.3						

* Italicized values are estimated.

TABLE 1c. Singlet O_2 Bimolecular Reaction Rates of Atmospheric Importance

Reaction	T/K	A/cm³ molecule⁻¹ s⁻¹	E/R/K	k₂₉₈/cm³ molecule⁻¹ s⁻¹	f_{298}	g
$O_2(^1\Delta) + O \rightarrow$ products	298			$<2 \times 10^{-16}$		
$O_2(^1\Delta) + O_2 \rightarrow$ products	100–450	3.6×10^{-18}	220	1.7×10^{-18}	1.2	100
$O_2(^1\Delta) + O_3 \rightarrow O + 2O_2$	283–360	5.2×10^{-11}	2840	3.8×10^{-15}	1.2	500
$O_2(^1\Delta) + H_2O \rightarrow$ products	298			4.8×10^{-18}	1.5	
$O_2(^1\Delta) + N \rightarrow NO + O$	195–300			$<9 \times 10^{-17}$		
$O_2(^1\Delta) + N_2 \rightarrow$ products	298			$<10^{-20}$		
$O_2(^1\Delta) + CO_2 \rightarrow$ products	298			$<2 \times 10^{-20}$		
$O_2(^1\Sigma) + O \rightarrow$ products	300			8×10^{-14}	5	
$O_2(^1\Sigma) + O_2 \rightarrow$ products	298			3.9×10^{-17}	1.5	

Reaction	*T*/K	*A*/cm³ molecule⁻¹ s⁻¹	*E*/*R*/K	k_{298}/cm³ molecule⁻¹ s⁻¹	f_{298}	*g*
$O_2(^1\Sigma) + O_3 \rightarrow$ products	210–370	3.5×10^{-11}	135	2.2×10^{-11}	1.15	50
$O_2(^1\Sigma) + H_2 \rightarrow$ products	173–393	6.4×10^{-12}	600	8.5×10^{-13}	1.15	100
$O_2(^1\Sigma) + H_2 \rightarrow 2OH$				$<4 \times 10^{-17}$		
$O_2(^1\Sigma) + H_2O \rightarrow O_2 + H_2O$	250–370	3.9×10^{-12}	-125	5.9×10^{-12}	1.3	100
$O_2(^1\Sigma) + N \rightarrow$ products	300			$<10^{-13}$		
$O_2(^1\Sigma) + N_2 \rightarrow$ products	203–370	1.8×10^{-15}	-45	2.1×10^{-15}	1.1	100
$O_2(^1\Sigma) + N_2O \rightarrow$ products	210–370	7.0×10^{-14}	-75	9.0×10^{-14}	1.3	50
$O_2(^1\Sigma) + N_2O \rightarrow NO + NO_2$				$<2 \times 10^{-17}$		
$O_2(^1\Sigma) + CO_2 \rightarrow$ products	245–362	4.2×10^{-13}	0	4.2×10^{-13}	1.2	200

TABLE 1d. HO_x Bimolecular Reaction Rates of Atmospheric Importance

Reaction	*T*/K	*A*/cm³ molecule⁻¹ s⁻¹	*E*/*R*/K	k_{298}/cm³ molecule⁻¹ s⁻¹	f_{298}	*g*
$OH + O \rightarrow O_2 + H$	136–515	1.8×10^{-11}	-180	3.3×10^{-11}	1.15	50
$HO_2 + O \rightarrow OH + O_2$	229–391	3.0×10^{-11}	-200	5.9×10^{-11}	1.05	50
$H_2O_2 + O \rightarrow OH + HO_2$	283–386	1.4×10^{-12}	2000	1.7×10^{-15}	1.2	100
$H + O_3 \rightarrow OH + O_2$	196–424	1.4×10^{-10}	470	2.9×10^{-11}	1.1	40
$HO_2 + H \rightarrow 2OH$	245–300	7.2×10^{-11}	0	7.2×10^{-11}	1.2	100
$HO_2 + H \rightarrow O + H_2O$	245–300	1.6×10^{-12}	0	1.6×10^{-12}	1.5	100
$HO_2 + H \rightarrow H_2 + O_2$	245–300	6.9×10^{-12}	0	6.9×10^{-12}	1.4	100
$OH + O_3 \rightarrow HO_2 + O_2$	190–357	1.7×10^{-12}	940	7.3×10^{-14}	1.15	50
$OH + H_2 \rightarrow H_2O + H$	200–1050	2.8×10^{-12}	1800	6.7×10^{-15}	1.05	100
$OH + HD \rightarrow$ products	248–418	5.0×10^{-12}	2130	4.0×10^{-15}	1.15	50
$OH + OH \rightarrow H_2O + O$	233–580	1.8×10^{-12}	0	1.8×10^{-12}	1.25	50
$OH + HO_2 \rightarrow H_2O + O_2$	252–420	4.8×10^{-11}	-250	1.1×10^{-10}	1.15	50
$OH + H_2O_2 \rightarrow H_2O + HO_2$						
$HO_2 + O_3 \rightarrow OH + 2O_2$	197–413	1.0×10^{-14}	490	1.9×10^{-15}	1.15	80
$HO_2 + HO_2 \rightarrow H_2O_2 + O_2$	222–1120	3.0×10^{-13}	-460	1.4×10^{-12}	1.15	100
$HO_2 + HO_2 (+M) \rightarrow H_2O_2 + O_2$	222–1120	2.1×10^{-33} [M]	-920	4.6×10^{-32} [M]	1.2	200
$HO_2 + HO_2 \cdot H_2O \rightarrow$ products	298–350	5.4×10^{-11}	410	1.4×10^{-11}	2	100

TABLE 1e. NO_x Bimolecular Reaction Rates of Atmospheric Importance

Reaction	*T*/K	*A*/cm³ molecule⁻¹ s⁻¹	*E*/*R*/K	k_{298}/cm³ molecule⁻¹ s⁻¹	f_{298}	*g*
$NO_3 + O \rightarrow O_2 + NO_2$	298–329	1.3×10^{-11}	0	1.3×10^{-11}	1.5	150
$N_2O_5 + O \rightarrow$ products	223–300			$<3.0 \times 10^{-16}$		
$HNO_3 + O \rightarrow OH + NO_3$	298			$<3.0 \times 10^{-17}$		
$HO_2NO_2 + O \rightarrow$ products	228–297	7.8×10^{-11}	3400	8.6×10^{-16}	3	750
$NO_2 + H \rightarrow OH + NO$	195–2000	1.35×10^{-10}	0	1.35×10^{-10}	1.2	150
$NO_3 + OH \rightarrow$ products	298			2.0×10^{-11}	1.5	
$HONO + OH \rightarrow H_2O + NO_2$	278–1400	3.0×10^{-12}	-250	6.9×10^{-12}	1.3	100
$HO_2NO_2 + OH \rightarrow$ products	218–335	4.5×10^{-13}	-610	3.5×10^{-12}	1.3	50
$NH_3 + OH \rightarrow H_2O + NH_2$	228–2360	1.7×10^{-12}	710	1.6×10^{-13}	1.15	100
$NO + HO_2 \rightarrow NO_2 + OH$	183–1270	3.44×10^{-12}	-260	8.2×10^{-12}	1.1	0
$NO_2^* + H_2O \rightarrow OH + HONO$						
$NO_2 + HO_2 \rightarrow HONO + O_2$	296			$< 5 \times 10^{-16}$		
$NO_3 + HO_2 \rightarrow$ products	263–338			3.5×10^{-12}	1.5	
$NH_2 + HO_2 \rightarrow$ products	298			4.8×10^{-11}	2	
$N + O_2 \rightarrow NO + O$	280–1220	3.3×10^{-12}	3150	8.5×10^{-17}	1.25	400
$N + O_3 \rightarrow NO + O_2$	298			$<2.0 \times 10^{-16}$		
$N + NO \rightarrow N_2 + O$	196–3660	2.1×10^{-11}	-100	3.0×10^{-11}	1.3	100
$N + NO_2 \rightarrow N_2O + O$	223–700	5.8×10^{-12}	-220	1.2×10^{-11}	1.5	100
$NO + O_3 \rightarrow NO_2 + O_2$	195–443	3.0×10^{-12}	1500	1.9×10^{-14}	1.07	130
$NO + NO_3 \rightarrow 2NO_2$	209–703	1.7×10^{-11}	-125	2.6×10^{-11}	1.15	50
$NO_2 + O_3 \rightarrow NO_3 + O_2$	231–362	1.2×10^{-13}	2450	3.2×10^{-17}	1.15	150
$NO_2 + NO_3 \rightarrow NO + NO_2 + O_2$	236–538					
$NO_3 + NO_3 \rightarrow 2NO_2 + O_2$	298–1100	8.5×10^{-13}	2450	2.3×10^{-16}	1.5	500

Reaction	T/K	A/cm^3 molecule^{-1} s^{-1}	E/R/K	k_{298}/cm^3 molecule^{-1} s^{-1}	f_{298}	g
$NH_2 + O_2 \rightarrow$ products	295–2300			$<6.0 \times 10^{-21}$		
$NH_2 + O_3 \rightarrow$ products	248–380	4.3×10^{-12}	930	1.9×10^{-13}	2	200
$NH_2 + NO \rightarrow$ products	200–2500	4.0×10^{-12}	-450	1.8×10^{-11}	1.3	150
$NH_2 + NO_2 \rightarrow$ products	295–910	6.0×10^{-12}	-385	2.2×10^{-11}	1.3	300
$NH + NO \rightarrow$ products	269–3350	4.9×10^{-11}	0	4.9×10^{-11}	1.5	300
$NH + NO_2 \rightarrow$ products	269–377	3.5×10^{-13}	-1140	1.6×10^{-11}	2	100
$HNO_2 + O_3 \rightarrow O_2 + HNO_3$	226–300			$<5.0 \times 10^{-19}$		
$N_2O_5 + H_2O \rightarrow 2HNO_3$	290–298			$<2.0 \times 10^{-21}$		
$N_2(A,v) + O_2 \rightarrow$ products	80–560			2.5×10^{-12} [b]	1.5	
$N_2(A,v) + O_3 \rightarrow$ products	298			4.1×10^{-11} [b]	2	

[a] Asymmetric uncertainties, see Introduction.
[b] $v = 0$ state only.

TABLE 1f. Organic Bimolecular Reaction Rates of Atmospheric Importance

Reaction	T/K	A*/cm^3 molecule^{-1} s^{-1}	E/R/K*	k_{298}/cm^3 molecule^{-1} s^{-1}	f_{298}	g
$CH_3 + O \rightarrow$ products	259–341	1.1×10^{-10}	0	1.1×10^{-10}	1.3	250
$HCN + O \rightarrow$ products	469–900	1.0×10^{-11}	4000	1.5×10^{-17}	10	0
$C_2H_2 + O \rightarrow$ products	(See Note)	3.0×10^{-11}	1600	1.4×10^{-13}	1.3	250
$H_2CO + O \rightarrow$ products	250–748	3.4×10^{-11}	1600	1.6×10^{-13}	1.25	250
$CH_3CHO + O \rightarrow CH_3CO + OH$	298–475	1.8×10^{-11}	1100	4.5×10^{-13}	1.25	200
$HOCO + O_2 \rightarrow HO_2 + CO_2$	298			2×10^{-12}	2	
$C_2H_2 + O_3 \rightarrow$ products	243–323	*1.0×10^{-14}*	4100	1.0×10^{-20}	3	500
$C_2H_4 + O_3 \rightarrow$ products	178–373	1.2×10^{-14}	2630	1.7×10^{-18}	1.25	100
$C_3H_6 + O_3 \rightarrow$ products	183–362	6.5×10^{-15}	1900	1.1×10^{-17}	1.15	200
$CH_2{=}C(CH_3)CHO + O_3 \rightarrow$ products	240–324	1.5×10^{-15}	2110	1.3×10^{-18}	1.1	200
$CH_3C(O)CH{=}CH_2 + O_3 \rightarrow$ products	240–324	8.5×10^{-16}	1520	5.2×10^{-18}	1.15	100
$CH_2{=}C(CH_3)CH{=}CH_2 + O_3 \rightarrow$ products	242–353	1.1×10^{-14}	2000	1.3×10^{-17}	1.1	200
$CH_4 + OH \rightarrow CH_3 + H_2O$	178–2025	2.45×10^{-12}	1775	6.3×10^{-15}	1.1	100
$^{13}CH_4 + OH \rightarrow {}^{13}CH_3 + H_2O$	273–353					
$CH_3D + OH \rightarrow$ products	249–420	3.5×10^{-12}	1950	5.0×10^{-15}	1.15	200
$H_2CO + OH \rightarrow H_2O + HCO$	228–2500	5.5×10^{-12}	-125	8.5×10^{-12}	1.15	50
$CH_3OH + OH \rightarrow$ products	210–1350	2.9×10^{-12}	345	9.1×10^{-13}	1.1	60
$CH_3OOH + OH \rightarrow$ products	203–423	3.8×10^{-12}	-200	7.4×10^{-12}	1.4	150
$HC(O)OH + OH \rightarrow$ products	296–445	4.0×10^{-13}	0	4.0×10^{-13}	1.2	100
$HC(O)C(O)H + OH \rightarrow$ products	210–390	1.15×10^{-11}	*0*	1.15×10^{-11}	1.5	200
$HOCH_2CHO + OH \rightarrow$ products	233–362	1.1×10^{-11}	*0*	1.1×10^{-11}	1.2	200
$HCN + OH \rightarrow$ products	296–563	1.2×10^{-13}	400	3.1×10^{-14}	3	150
$C_2H_6 + OH \rightarrow H_2O + C_2H_5$	226–2000	7.66×10^{-12}	1020	2.5×10^{-13}	1.05	20
$CH_3CHO + OH \rightarrow$ products	202–900	4.63×10^{-12}	-350	1.5×10^{-11}	1.05	20
$CH_3OC(O)H + OH \rightarrow$ products	233–372	8.43×10^{-12}	460	1.8×10^{-13}	1.15	50
$CH_3CH_2OH + OH \rightarrow$ products	216–498	3.35×10^{-12}	0	3.35×10^{-12}	1.05	20
$CH_3C(O)OH + OH \rightarrow$ products	229–802	3.15×10^{-14}	-920	6.9×10^{-13}	1.15	100
$C_3H_8 + OH \rightarrow$ products	190–908	9.19×10^{-12}	630	1.11×10^{-12}	1.03	20
$C_2H_5CHO + OH \rightarrow C_2H_5CO + H_2O$	240–372	4.9×10^{-12}	-405	1.9×10^{-11}	1.05	80
$1\text{–}C_3H_7OH + OH \rightarrow$ products	263–372	4.4×10^{-12}	-70	5.6×10^{-12}	1.05	80
$2\text{–}C_3H_7OH + OH \rightarrow$ products	220–745	2.10×10^{-12}	-270	5.2×10^{-12}	1.03	20
$C_2H_5C(O)OH + OH \rightarrow$ products	240–445	1.2×10^{-12}	0	1.2×10^{-12}	1.1	200
$CH_3C(O)CH_3 + OH \rightarrow H_2O + CH_3C(O)CH_2$						
$CH_3C(O)CH_3 + OH \rightarrow CH_3 + CH_3C(O)OH$						
$CH_3OCH_2OCH_3 + OH \rightarrow$ products	252–372	2.12×10^{-12}	-250	4.9×10^{-12}	1.1	50
n-$C_4H_{10} + OH \rightarrow$ products	185–509	1.02×10^{-11}	430	2.4×10^{-12}	1.03	20
iso-$C_4H_{10} + OH \rightarrow$ products	213–864	4.86×10^{-12}	250	2.1×10^{-12}	1.05	20
n-$C_5H_{12} + OH \rightarrow$ products	224–900	1.49×10^{-11}	400	3.9×10^{-12}	1.05	50
iso-$C_5H_{12} + OH \rightarrow$ products	213–407	6.55×10^{-12}	170	3.7×10^{-12}	1.07	20
cyclo-$C_5H_{10} + OH \rightarrow$ products	209–491	1.68×10^{-11}	370	4.85×10^{-12}	1.05	50
$CH_2{=}C(CH_3)CHO + OH \rightarrow$ products	234–423	9.6×10^{-12}	-360	3.2×10^{-11}	1.15	100
$CH_3C(O)CH{=}CH_2 + OH \rightarrow$ products	232–424	2.7×10^{-12}	-580	1.9×10^{-11}	1.07	100

Reaction	T/K	A*/cm³ molecule⁻¹ s⁻¹	E/R/K*	k_{298}/cm³ molecule⁻¹ s⁻¹	f_{298}	g
$CH_2=C(CH_3)CH=CH_2 + OH \rightarrow$ products	240–422	3.0×10^{-11}	-360	9.9×10^{-11}	1.1	50
$CH_3CN + OH \rightarrow$ products	256–424	7.8×10^{-13}	1050	2.3×10^{-14}	1.5	200
$CH_3ONO_2 + OH \rightarrow$ products	221–423	8.0×10^{-13}	1000	2.8×10^{-14}	1.7	200
$CH_3C(O)O_2NO_2$ (PAN) + OH → products	273–299			$<4 \times 10^{-14}$		
$C_2H_5ONO_2 + OH \rightarrow$ products	298–373	1.0×10^{-12}	490	2.0×10^{-13}	1.4	150
$1-C_3H_7ONO_2 + OH \rightarrow$ products	298–368	7.1×10^{-13}	0	7.1×10^{-13}	1.5	200
$2-C_3H_7ONO_2 + OH \rightarrow$ products	295–299	1.2×10^{-12}	320	4.1×10^{-13}	1.5	200
$CH_2O + HO_2 \rightarrow$ adduct	273–373	6.7×10^{-15}	-600	5.0×10^{-14}	5	600
$CH_3O_2 + HO_2 \rightarrow CH_3OOH + O_2$	228–700	4.1×10^{-13}	-750	5.2×10^{-12}	1.3	150
$C_2H_5O_2 + HO_2 \rightarrow C_2H_5OOH + O_2$	210–480	7.5×10^{-13}	-700	8.0×10^{-12}	1.5	250
$CH_3C(O)O_2 + HO_2 \rightarrow$ products	253–403	4.3×10^{-13}	-1040	1.4×10^{-11}	2	500
$CH_3C(O)CH_2O_2 + HO_2 \rightarrow$ products	298	8.6×10^{-13}	-700	9.0×10^{-12}	2	300
$CO + NO_3 \rightarrow$ products	295–298			$<4.0 \times 10^{-19}$		
$CH_2O + NO_3 \rightarrow$ products	295–298			5.8×10^{-16}	1.3	
$CH_3CHO + NO_3 \rightarrow$ products	263–433	1.4×10^{-12}	1900	2.4×10^{-15}	1.3	300
$CH_2=C(CH_3)CHO + NO_3 \rightarrow$ products	296–323			3.4×10^{-15}	1.25	
$CH_3C(O)CH=CH_2 + NO_3 \rightarrow$ products	296–323			$<4 \times 10^{-16}$		
$CH_2=C(CH_3)CH=CH_2 + NO_3 \rightarrow$ products	251–381	3.5×10^{-12}	450	7.8×10^{-13}	1.25	100
$CH_3 + O_2 \rightarrow$ products	298–3000			$<3.0 \times 10^{-16}$		
$CH_3 + O_3 \rightarrow$ products	221–384	5.4×10^{-12}	220	2.6×10^{-12}	2	150
$HCO + O_2 \rightarrow CO + HO_2$	200–2000	5.2×10^{-12}	0	5.2×10^{-12}	1.4	100
$CH_2OH + O_2 \rightarrow CH_2O + HO_2$	215–2000	9.1×10^{-12}	0	9.1×10^{-12}	1.3	200
$CH_3O + O_2 \rightarrow CH_2O + HO_2$	296–900	3.9×10^{-14}	900	1.9×10^{-15}	1.5	300
$CH_3O + NO \rightarrow CH_2O + HNO$						
$CH_3O + NO_2 \rightarrow CH_2O + HONO$	298–1100	1.1×10^{-11}	1200	2.0×10^{-13}	5	600
$CH_3O_2 + O_3 \rightarrow$ products	298	*2.9×10^{-16}*	1000	1.0×10^{-17}	3	500
$CH_3O_2 + CH_3O_2 \rightarrow$ products	248–700	9.5×10^{-14}	-390	3.5×10^{-13}	1.2	100
$CH_3O_2 + NO \rightarrow CH_3O + NO_2$	193–429	2.8×10^{-12}	-300	7.7×10^{-12}	1.15	100
$CH_3O_2 + CH_3C(O)O_2 \rightarrow$ products	253–368	2.0×10^{-12}	-500	1.1×10^{-11}	1.5	250
$CH_3O_2 + CH_3C(O)CH_2O_2 \rightarrow$ products	298	*7.5×10^{-13}*	-500	4.0×10^{-12}	2	300
$C_2H_5 + O_2 \rightarrow C_2H_4 + HO_2$				$<2.0 \times 10^{-14}$		
$C_2H_5O + O_2 \rightarrow CH_3CHO + HO_2$	225–411	6.3×10^{-14}	550	1.0×10^{-14}	1.5	200
$C_2H_5O_2 + C_2H_5O_2 \rightarrow$ products	228–460	6.8×10^{-14}	0	6.8×10^{-14}	2	300
$C_2H_5O_2 + NO \rightarrow$ products	220–355	2.6×10^{-12}	-365	8.7×10^{-12}	1.2	150
$CH_3C(O)O_2 + CH_3C(O)O_2 \rightarrow$ products	253–368	2.9×10^{-12}	-500	1.5×10^{-11}	1.5	150
$CH_3C(O)O_2 + NO \rightarrow$ products	218–402	8.1×10^{-12}	-270	2.0×10^{-11}	1.5	100
$CH_3C(O)CH_2O_2 + NO \rightarrow$ products	298	*2.9×10^{-12}*	-300	8.0×10^{-12}	1.5	300
$HOC_5H_8OO + NO \rightarrow$ products	295–300	2.7×10^{-12}	-350	8.8×10^{-12}	1.15	300
Criegee Intermediate Reactions						
$CH_2OO + H_2O \rightarrow$ products	293–298			2.8×10^{-16}	1.7	
$CH_2OO + (H_2O)_2 \rightarrow$ products	283–324	2.8×10^{-16}	-3010	6.8×10^{-12}	1.1	400
syn-$CH_3CHOO + H_2O \rightarrow$ products	298			$<2 \times 10^{-16}$		
anti-$CH_3CHOO + H_2O \rightarrow$ products	298			2.4×10^{-14}	1.3	
syn-$CH_3CHOO + (H_2O)_2 \rightarrow$ products						
anti-$CH_3CHOO + (H_2O)_2 \rightarrow$ products						
$CH_2OO + NO \rightarrow$ products	298			$<6.4 \times 10^{-14}$		
$CH_2OO + NO_2 \rightarrow$ products	298			4.2×10^{-12}	1.7	
$CH_2OO + HNO_3 \rightarrow$ products	298			5.4×10^{-10}	1.2	
$CH_2OO + CH_2OO \rightarrow$ products	298			7.12×10^{-11}	1.4	
$CH_2OO + HC(O)OH \rightarrow$ products	298			1.1×10^{-10}	1.1	
$CH_2OO + CH_3CHO \rightarrow$ products	298–494	2.5×10^{-14}	-1180	1.3×10^{-11}	1.4	100
$CH_2OO + CH_3C(O)OH \rightarrow$ products	298			1.25×10^{-10}	1.1	
$CH_2OO + CH_2=CH_2 \rightarrow$ products	298–494	1.1×10^{-14}	850	6.6×10^{-16}	1.2	100
$CH_2OO + CH_3C(O)CH_3 \rightarrow$ products	298–494	6×10^{-15}	-1132	2.9×10^{-13}	1.1	85
$CH_2OO + CH_2=CHCH_3 \rightarrow$ products	298–494	8.0×10^{-15}	450	1.75×10^{-15}	1.2	100
$CH_2OO + CH_2=CHCH_2CH_3 \rightarrow$ products	298–494	5.3×10^{-15}	380	1.5×10^{-15}	1.2	100
$CH_2OO + (CH_3)_2C=CH_2 \rightarrow$ products	298–494	5.4×10^{-15}	400	1.4×10^{-15}	1.2	100
$CH_2OO + CH_3CH=CHCH_3 \rightarrow$ products	298–494	2.5×10^{-15}	400	6.5×10^{-16}	1.2	100
$CH_2OO + CF_3C(O)OH \rightarrow$ products	240–340	1.2×10^{-10}	-300	3.3×10^{-10}	1.2	200
$CH_2OO + CF_3C(O)CF_3$(HFA) → products	298			3.17×10^{-11}	1.1	

Reaction	T/K	A*/cm³ molecule⁻¹ s⁻¹	E/R/K*	k_{298}/cm³ molecule⁻¹ s⁻¹	f_{298}	g
$CH_2OO + HCl \rightarrow$ products	298			4.6×10^{-11}	1.2.	
$CH_2OO + H_2S \rightarrow$ products	278–318	1×10^{-14}	-580	4.6×10^{-11}	1.2	200
$CH_2OO + SO_2 \rightarrow$ products	293–298			3.8×10^{-11}	1.1	
anti-$CH_3CHOO + SO_2 \rightarrow$ products	298			2.2×10^{-10}	1.2	
syn-$CH_3CHOO + SO_2 \rightarrow$ products	298			2.65×10^{-11}	1.1	

* Italicized values are estimated.

TABLE 1g. FO_x Bimolecular Reaction Rates of Atmospheric Importance

Reaction	T/K	A*/cm³ molecule⁻¹ s⁻¹	E/R/K*	k_{298}*/cm³ molecule⁻¹ s⁻¹	f_{298}	g
$FO + O \rightarrow F + O_2$	298	*2.7×10^{-11}*	*0*	*2.7×10^{-11}*	3	250
$FO_2 + O \rightarrow FO + O_2$		*5.0×10^{-11}*	*0*	*5.0×10^{-11}*	5	
CH_3F (HFC–41) + OH $\rightarrow CH_2F + H_2O$	243–480	2.2×10^{-12}	1400	2.0×10^{-14}	1.1	150
CH_2F_2 (HFC-32) + OH $\rightarrow CHF_2 + H_2O$	220–492	1.7×10^{-12}	1500	1.1×10^{-14}	1.07	100
CHF_3 (HFC-23) + OH $\rightarrow CF_3 + H_2O$	253–1663	6.1×10^{-13}	2260	3.1×10^{-16}	1.15	100
CH_3CH_2F (HFC-161) + OH $\rightarrow$ products	210–480	2.5×10^{-12}	730	2.2×10^{-13}	1.07	50
CH_3CHF_2 (HFC-152a) + OH $\rightarrow$ products	210–480	8.7×10^{-13}	975	3.3×10^{-14}	1.07	50
CH_2FCH_2F (HFC-152) + OH $\rightarrow CHFCH_2F + H_2O$	210–480	1.05×10^{-12}	710	9.7×10^{-14}	1.07	100
CH_3CF_3 (HFC-143a) + OH $\rightarrow CH_2CF_3 + H_2O$	261–425	1.07×10^{-12}	2000	1.3×10^{-15}	1.1	100
CH_2FCHF_2 (HFC-143) + OH $\rightarrow$ products	278–441	3.9×10^{-12}	1620	1.7×10^{-14}	1.2	200
CH_2FCF_3 (HFC-134a) + OH $\rightarrow CHFCF_3 + H_2O$	220–473	1.03×10^{-12}	1620	4.5×10^{-15}	1.1	100
CHF_2CHF_2 (HFC-134) + OH $\rightarrow CF_2CHF_2 + H_2O$	294–434	1.6×10^{-12}	1660	6.1×10^{-15}	1.2	200
CHF_2CF_3 (HFC-125) + OH $\rightarrow CF_2CF_3 + H_2O$	220–441	5.16×10^{-13}	1670	1.9×10^{-15}	1.1	100
CH_3CHFCH_3 (HFC-281ea) + OH $\rightarrow$ products	288–394	3.0×10^{-12}	490	5.8×10^{-13}	1.2	100
$CH_3CH_2CF_3$ (HFC-263fb) + OH $\rightarrow$ products	238–375	3.7×10^{-12}	1290	4.9×10^{-14}	1.15	100
$CH_2FCF_2CHF_2$ (HFC-245ca) + OH $\rightarrow$ products	260–365	2.1×10^{-12}	1620	9.2×10^{-15}	1.2	150
$CH_3CF_2CF_3$ (HFC-245cb) + OH $\rightarrow$ products	298–370	4.2×10^{-13}	1680	1.5×10^{-15}	1.1	200
$CHF_2CHFCHF_2$ (HFC-245ea) + OH $\rightarrow$ products	238–375	1.53×10^{-12}	1340	1.7×10^{-14}	1.1	150
$CH_2FCHFCF_3$ (HFC-245eb) + OH $\rightarrow$ products	238–375	1.16×10^{-12}	1260	1.7×10^{-14}	1.15	100
$CHF_2CH_2CF_3$ (HFC-245fa) + OH $\rightarrow$ products	273–370	6.1×10^{-13}	1330	7.0×10^{-15}	1.15	100
$CH_2FCF_2CF_3$ (HFC-236cb) + OH $\rightarrow CHFCF_2CF_3 + H_2O$	251–314	*1.03×10^{-12}*	*1620*	*4.5×10^{-15}*	2	200
CHF_2CHFCF_3 (HFC-236ea) + OH $\rightarrow$ products	251–380	9.4×10^{-13}	1550	5.2×10^{-15}	1.2	200
$CF_3CH_2CF_3$ (HFC–236fa) + OH $\rightarrow CF_3CHCF_3 + H_2O$	251–413	1.45×10^{-12}	2500	3.3×10^{-16}	1.15	150
CF_3CHFCF_3 (HFC-227ea) + OH $\rightarrow CF_3CFCF_3 + H_2O$	250–463	4.8×10^{-13}	1680	1.7×10^{-15}	1.15	75
$CH_3CF_2CH_2CF_3$ (HFC-365mfc) + OH $\rightarrow$ products	269–373	1.8×10^{-12}	1660	6.9×10^{-15}	1.3	100
$CF_3CH_2CH_2CF_3$ (HFC-356mff) + OH $\rightarrow$ products	260–365	3.4×10^{-12}	1820	7.6×10^{-15}	1.2	300
$CH_2FCH_2CF_2CF_3$ (HFC-356mcf) + OH $\rightarrow$ products	252–346	1.7×10^{-12}	1100	4.2×10^{-14}	1.3	150
$CHF_2CF_2CF_2CF_2H$ (HFC-338pcc) + OH $\rightarrow$ products	232–419	7.7×10^{-13}	1540	4.4×10^{-15}	1.2	150
$CHF_2CF_2CF_2CF_3$ (HFC-329p) + OH $\rightarrow$ products	296	4.62×10^{-13}	1670	1.7×10^{-15}	1.15	0
$CHF_2(CF_2)_4CF_3$ (HFC-52-13p) + OH $\rightarrow$ products	250–430	5.16×10^{-13}	1670	1.9×10^{-15}	1.15	0
$CHF_2(CF_2)_6CF_3$ (HFC-72-17p) + OH $\rightarrow$ products	253–328	6.92×10^{-13}	1670	2.55×10^{-15}	1.2	0
$CF_3CH_2CF_2CH_2CF_3$ (HFC-458mfcf) + OH $\rightarrow$ products	278–354	1.1×10^{-12}	1800	2.6×10^{-15}	1.5	200
$CF_3CHFCHFCF_2CF_3$ (HFC-43-10mee) + OH $\rightarrow$ products	250–400	5.2×10^{-13}	1500	3.4×10^{-15}	1.2	150
$CF_3CF_2CH_2CH_2CF_2CF_3$ (HFC–55-10-mcff) + OH $\rightarrow$ products	298	*3.5×10^{-12}*	*1800*	8.3×10^{-15}	1.5	300
$CH_2{=}CHF + OH \rightarrow$ products	220–426	1.77×10^{-12}	-310	5.0×10^{-12}	1.07	20
$CH_2{=}CF_2 + OH \rightarrow$ products	219–373	1.7×10^{-12}	-150	2.8×10^{-12}	1.07	20
$CHF{=}CF_2 + OH \rightarrow$ products	215–375	2.96×10^{-12}	-300	8.1×10^{-12}	1.07	20
$CF_2{=}CF_2 + OH \rightarrow$ products	250–370	3.4×10^{-12}	-320	1.0×10^{-11}	1.15	100
$CH_2{=}CHCH_2F + OH \rightarrow$ products	228–388	6.0×10^{-12}	-290	1.6×10^{-11}	1.3	100
$CH_2{=}CHCF_3 + OH \rightarrow$ products	252–370	8.2×10^{-13}	-170	1.45×10^{-12}	1.07	50
$CH_2{=}CFCF_3 + OH \rightarrow$ products	206–380	1.1×10^{-12}	0	1.1×10^{-12}	1.05	0
(E)-$CHF{=}CHCF_3 + OH \rightarrow$ products	220–370	6.1×10^{-13}	-40	7.0×10^{-13}	1.05	20
(Z)-$CHF{=}CHCF_3 + OH \rightarrow$ products	253–328	8.6×10^{-13}	-135	1.35×10^{-13}	1.1	50
(E)-$CHF{=}CFCF_3 + OH \rightarrow$ products	296	*1.65×10^{-12}*	*-100*	2.3×10^{-12}	1.3	50
(Z)-$CHF{=}CFCF_3 + OH \rightarrow$ products	200–380	7.5×10^{-13}	-165	1.3×10^{-12}	1.07	50
$CF_2{=}CFCF_3 + OH \rightarrow$ products	250–489	5.34×10^{-13}	-415	2.15×10^{-12}	1.05	50
$CF_2{=}CFOCF_3 + OH \rightarrow$ products	250–430	1.03×10^{-12}	-320	3.0×10^{-12}	1.1	50
$CF_2{=}CFCF{=}CF_2 + OH \rightarrow$ products	298	*4.0×10^{-12}*	*-300*	1.1×10^{-11}	1.15	100

Reaction	T/K	A*/cm³ molecule⁻¹ s⁻¹	E/R/K*	k_{298}*/cm³ molecule⁻¹ s⁻¹	f_{298}	g
CH_2=$CHCF_2CF_3$ + OH → products	296	*7.9×10^{-13}*	*-170*	1.4×10^{-12}	1.2	50
(*E*)-CF_3CH=$CHCF_3$ + OH → products	211–373	7.0×10^{-13}	500	1.3×10^{-13}	1.1	0
(Z)-CF_3CH=$CHCF_3$ + OH → products	212–374	2.46×10^{-13}	-200	4.8×10^{-13}	1.07	20
(E)-CF_3CF=$CFCF_3$ + OH → products	230–370	3.17×10^{-13}	-180	5.8×10^{-13}	1.1	20
(Z)-CF_3CF=$CFCF_3$ + OH → products	230–370	2.9×10^{-13}	-80	3.8×10^{-13}	1.1	20
CF_2=$CFCF_2CF_3$ + OH → products	296	4.47×10^{-13}	-415	1.8×10^{-12}	1.15	50
cyclo-CH=$CHCF_2CF_2$- + OH → products	253–328	3.27×10^{-13}	190	1.73×10^{-13}	1.1	0
cyclo-CF=$CHCF_2CF_2$- + OH → products	253–328	4.8×10^{-13}	610	6.2×10^{-14}	1.1	0
cyclo-CF=$CFCF_2F_2$- + OH → products	253–328	3.0×10^{-13}	615	3.8×10^{-14}	1.1	0
cyclo-CH=$CHCF_2CF_2CF_2$- + OH → products	253–328	3.7×10^{-13}	370	1.07×10^{-13}	1.15	0
cyclo-CF=$CFCF_2CF_2CF_2$- + OH → products	253–328	3.46×10^{-13}	625	4.25×10^{-14}	1.1	0
(*E*)-CHF=$CHCF(CF_3)_2$+ OH → products	214–380	3.2×10^{-13}	0	3.2×10^{-13}	1.05	0
$(CF_3)_2C$=$CFCF_2CF_3$	250–370	7.1×10^{-14}	0	7.1×10^{-14}	1.05	50
CF_3OH + OH → CF_3O + H_2O				*$<2 \times 10^{-17}$*		
CH_2FCH_2OH + OH → products	220–370	1.82×10^{-12}	210	9.0×10^{-13}	1.05	50
CHF_2CH_2OH + OH → products	220–370	1.4×10^{-12}	500	2.6×10^{-13}	1.05	100
CF_3CH_2OH + OH → products	220–430	8.6×10^{-13}	640	1.0×10^{-13}	1.05	100
$CF_3CH_2CH_2OH$ + OH → products	263–358	1.82×10^{-12}	210	9.0×10^{-13}	1.1	50
$CF_3CF_2CH_2OH$ + OH → products	250–430	*1.16×10^{-12}*	*730*	1.0×10^{-13}	1.05	100
$(CF_3)_2CHOH$ + OH → products	220–430	4.05×10^{-13}	830	2.5×10^{-14}	1.05	100
$CF_3CHFCF_2CH_2OH$ + OH → products	230–430	1.07×10^{-12}	640	1.25×10^{-13}	1.1	100
$CF_3CF_2CF_2CF_2CH_2OH$ + OH → products	296	1.16×10^{-12}	730	1.0×10^{-13}	1.1	200
CH_3OCHF_2 (HFE-152a) + OH → products	298–460	1.05×10^{-11}	1700	3.5×10^{-14}	1.3	200
CH_3OCF_3 (HFE-143a) + OH → CH_2OCF_3 + H_2O	268–460	1.84×10^{-12}	1500	1.2×10^{-14}	1.1	150
CHF_2OCHF_2 (HFE-134) + OH → CF_2OCHF_2 +H_2O	251–464	1.1×10^{-12}	1830	2.4×10^{-15}	1.1	150
CHF_2OCF_3 (HFE-125) + OH → CF_2OCF_3 + H_2O	298–393	4.6×10^{-13}	2040	4.9×10^{-16}	1.2	200
$CH_3OCHFCF_3$ + OH → products	253–328	2.05×10^{-12}	760	1.6×10^{-13}	1.15	50
$CH_2FOCH(CF_3)_2$ (Sevoflurane) + OH → products	241–422	8.77×10^{-13}	960	3.5×10^{-14}	1.15	100
$CH_3OCF_2CHF_2$ + OH → products	250–430	1.7×10^{-12}	1300	2.2×10^{-14}	1.3	200
$CH_3OCF_2CF_3$ + OH → products	250–430	1.1×10^{-12}	1370	1.1×10^{-14}	1.2	150
$CHF_2OCH_2CF_3$ (HFE-245fa2) + OH → products	292–460	2.9×10^{-12}	1660	1.1×10^{-14}	1.15	200
$CHF_2OCHFCF_3$ (Desflurane) + OH → products	241–298	8.15×10^{-13}	1570	4.2×10^{-15}	1.15	100
$CHF_2OCF_2CHF_2$ + OH → products	253–407	5.8×10^{-13}	1600	2.7×10^{-15}	1.2	50
$CF_3OCHFCF_3$ + OH → products	250–430	3.1×10^{-13}	1680	1.1×10^{-15}	1.15	100
$CH_3OCF_2CF_2CF_3$ + OH → products	250–430	1.4×10^{-12}	1440	1.1×10^{-14}	1.15	150
$CH_3OCH(CF_3)_2$ (HFE-356mm1) + OH → products	230–370	1.14×10^{-12}	470	2.35×10^{-13}	1.05	50
$CH_3OCF(CF_3)_2$ + OH → products	250–430	1.3×10^{-12}	1330	1.5×10^{-14}	1.1	100
$CH_3OC_4F_9$ + OH → products	253–328	1.17×10^{-12}	1390	1.1×10^{-14}	1.1	100
$CHF_2OCH_2CF_2CHF_2$ + OH → products	250–430	1.82×10^{-12}	1410	1.6×10^{-14}	1.2	200
$CHF_2OCH_2CF_2CF_3$ + OH → products	250–430	1.6×10^{-12}	1510	1.0×10^{-14}	1.3	200
$CHF_2OCH(CF_3)_2$ + OH → products	284–398	1.03×10^{-12}	1760	2.8×10^{-15}	1.2	150
$CH_3CH_2OCF_2CHF_2$ + OH → products	250–430	2.1×10^{-12}	670	2.2×10^{-13}	1.1	100
$CF_3CH_2OCH_2CF_3$ + OH → products	268–409	2.8×10^{-12}	890	1.4×10^{-13}	1.1	100
$CF_3CH_2OCF_2CHF_2$ (HFE-347pcf2) + OH → products	250–430	1.32×10^{-12}	1470	9.5×10^{-15}	1.07	50
$CF_3CHFCF_2OCH_2CF_2CF_3$ + OH → products	253–328	5.36×10^{-12}	1345	5.9×10^{-15}	1.15	0
$CHF_2OCF_2OCHF_2$ + OH → products	295	*1.0×10^{-12}*	*1800*	2.4×10^{-15}	1.4	200
$CHF_2OCF_2CF_2OCHF_2$ + OH → products	295	*2.0×10^{-12}*	*1800*	4.7×10^{-15}	1.5	200
$CHF_2OCF_2CF_2OCF_2OCHF_2$ + OH → products	295	*1.9×10^{-12}*	*1800*	4.6×10^{-15}	1.5	200
F + O_3 → FO + O_2	253–365	2.2×10^{-11}	230	1.0×10^{-11}	1.5	200
F + H_2 → HF + H	77–765	1.4×10^{-10}	500	2.6×10^{-11}	1.2	200
F + H_2O → HF + OH	240–373	1.4×10^{-11}	0	1.4×10^{-11}	1.3	200
F + HNO_3 → HF + NO_3	260–373	6.0×10^{-12}	-400	2.3×10^{-11}	1.3	200
F + CH_4 → HF + CH_3	139–423	1.6×10^{-10}	260	6.7×10^{-11}	1.4	200
FO + O_3 → products	298			$<1 \times 10^{-14}$		
FO + NO → NO_2 + F	298–845	8.2×10^{-12}	-300	2.2×10^{-11}	1.5	200
FO + FO → 2F + O_2	298–435	1.0×10^{-11}	0	1.0×10^{-11}	1.5	250
FO_2 + O_3 → products	298			$<3.4 \times 10^{-16}$		
FO_2 + NO → FNO + O_2	190–298	7.5×10^{-12}	690	7.5×10^{-13}	2	400
FO_2 + NO_2 → products	260–315	3.8×10^{-11}	2040	4.0×10^{-14}	2	500
FO_2 + CO → products	298			$<5.1 \times 10^{-16}$		

Reaction	T/K	A*/cm³ molecule⁻¹ s⁻¹	E/R/K*	k_{298}*/cm³ molecule⁻¹ s⁻¹	f_{298}	g
$FO_2 + CH_4 \rightarrow$ products	298			$<2 \times 10^{-16}$		
$CF_3O + O_2 \rightarrow FO_2 + CF_2O$	373	*$<3 \times 10^{-11}$*	*5000*	*$<1.5 \times 10^{-18}$*		
$CF_3O + O_3 \rightarrow CF_3O_2 + O_2$	210–353	*2×10^{-12}*	*1400*	1.8×10^{-14}	1.3	600
$CF_3O + H_2O \rightarrow OH + CF_3OH$	296–381	*3×10^{-12}*	*>3600*	*$<2 \times 10^{-17}$*		
$CF_3O + NO \rightarrow CF_2O + FNO$	213–393	3.7×10^{-11}	-110	5.4×10^{-11}	1.2	70
$CF_3O + NO_2 \rightarrow$ products	222–302			–		
$CF_3O + CO \rightarrow$ products	233–332			$<2 \times 10^{-15}$		
$CF_3O + CH_4 \rightarrow CH_3 + CF_3OH$	231–573	2.6×10^{-12}	1420	2.2×10^{-14}	1.1	200
$CF_3O + C_2H_6 \rightarrow C_2H_5 + CF_3OH$	233–573	4.9×10^{-12}	400	1.3×10^{-12}	1.2	100
$CF_3O_2 + O_3 \rightarrow CF_3O + 2O_2$	210–353			$<3 \times 10^{-15}$		
$CF_3O_2 + CO \rightarrow CF_3O + CO_2$	296			$<5 \times 10^{-16}$		
$CF_3O_2 + NO \rightarrow CF_3O + NO_2$	230–430	5.4×10^{-12}	-320	1.6×10^{-11}	1.1	150

* Italicized values are estimated.

TABLE 1h. ClO_x Bimolecular Reaction Rates of Atmospheric Importance

Reaction	T/K	A*/cm³ molecule⁻¹ s⁻¹	E/R/K*	k_{298}*/cm³ molecule⁻¹ s⁻¹	f_{298}	g
$ClO + O \rightarrow Cl + O_2$	220–1250	2.8×10^{-11}	-85	3.7×10^{-11}	1.05	50
$OClO + O \rightarrow ClO + O_2$	243–400	2.4×10^{-12}	960	1.0×10^{-13}	2	300
$Cl_2O + O \rightarrow ClO + ClO$	230–297	2.7×10^{-11}	530	4.5×10^{-12}	1.2	100
$HCl + O \rightarrow OH + Cl$	293–3197	1.0×10^{-11}	3300	1.5×10^{-16}	1.5	350
$HOCl + O \rightarrow OH + ClO$	213–298	1.7×10^{-13}	0	1.7×10^{-13}	3	300
$ClONO_2 + O \rightarrow$ products	202–325	3.6×10^{-12}	840	2.1×10^{-13}	1.2	100
$OClO + O_3 \rightarrow$ products	262–298	2.1×10^{-12}	4700	3.0×10^{-19}	2.5	1000
$Cl_2O_2 + O_3 \rightarrow$ products	195–217			$<1 \times 10^{-19}$		
$Cl_2 + OH \rightarrow HOCl + Cl$	231–836	2.6×10^{-12}	1100	6.5×10^{-14}	1.1	200
$ClO + OH \rightarrow Cl + HO_2$	208–373	7.4×10^{-12}	-270	1.8×10^{-11}	1.2	50
$ClO + OH \rightarrow HCl + O_2$	208–373	6.0×10^{-13}	-230	1.3×10^{-12}	1.7	100
$OClO + OH \rightarrow HOCl + O_2$	242–473	1.4×10^{-12}	-600	1.0×10^{-11}	1.5	150
$Cl_2O + OH \rightarrow HOCl + ClO$	223–383	4.7×10^{-12}	-140	7.5×10^{-12}	1.2	100
$Cl_2O_2 + OH \rightarrow HOCl + ClOO$	223–318	6.0×10^{-13}	-670	5.7×10^{-12}	1.3	100
$HCl + OH \rightarrow H_2O + Cl$	138–1060	1.8×10^{-12}	250	7.8×10^{-13}	1.1	50
$HOCl + OH \rightarrow H_2O + ClO$	298	3.0×10^{-12}	500	5.0×10^{-13}	3	500
$ClNO_2 + OH \rightarrow HOCl + NO_2$	259–348	2.4×10^{-12}	1250	3.6×10^{-14}	2	300
$ClONO_2 + OH \rightarrow$ products	245–387	1.2×10^{-12}	330	3.9×10^{-13}	1.5	200
$CH_3Cl + OH \rightarrow CH_2Cl + H_2O$	224–955	1.96×10^{-12}	1200	3.5×10^{-14}	1.1	50
$CH_2Cl_2 + OH \rightarrow CHCl_2 + H_2O$	219–955	1.92×10^{-12}	880	1.0×10^{-13}	1.15	100
$CHCl_3 + OH \rightarrow CCl_3 + H_2O$	249–775	2.2×10^{-12}	920	1.0×10^{-13}	1.15	150
$CCl_4 + OH \rightarrow$ products	298	*1×10^{-11}*	*>6200*	*$<1 \times 10^{-20}$*		
CH_2FCl (HCFC-31) + OH $\rightarrow CHFCl + H_2O$	245–486	2.4×10^{-12}	1210	4.1×10^{-14}	1.15	200
$CHFCl_2$ (HCFC-21) + OH $\rightarrow CFCl_2 + H_2O$	241–810	1.52×10^{-12}	1170	3.0×10^{-14}	1.1	150
CHF_2Cl (HCFC-22) + OH $\rightarrow CF_2Cl + H_2O$	241–807	9.2×10^{-13}	1560	4.9×10^{-15}	1.07	100
$CFCl_3$ (CFC-11) + OH $\rightarrow$ products	297–434	*1×10^{-11}*	*>9700*	*$<1 \times 10^{-25}$*		
CF_2Cl_2 (CFC-12) + OH $\rightarrow$ products	293–480	*1×10^{-11}*	*>11900*	*$<1 \times 10^{-28}$*		
CCl_2FCClF_2 (CFC-113) + OH $\rightarrow$ products	298	*1×10^{-11}*	*>6200*	*$<1 \times 10^{-20}$*		
$CClF_2CClF_2$ (CFC-114) + OH $\rightarrow$ products	296	*1×10^{-11}*	*>6200*	*$<1 \times 10^{-20}$*		
CF_3CClF_2 (CFC-115) + OH $\rightarrow$ products	–	*1×10^{-11}*	*>6200*	*$<1 \times 10^{-20}$*		
$CH_3CH_2Cl + OH \rightarrow$ products	223–789	5.4×10^{-12}	800	3.7×10^{-13}	1.2	100
$CH_2ClCH_2Cl + OH \rightarrow$ products	292–775	1.14×10^{-11}	1150	2.4×10^{-13}	1.1	200
$CH_3CCl_3 + OH \rightarrow CH_2CCl_3 + H_2O$	243–761	1.64×10^{-12}	1520	1.0×10^{-14}	1.1	50
CH_3CFCl_2 (HCFC-141b) + OH $\rightarrow CH_2CFCl_2 + H_2O$	220–479	1.25×10^{-12}	1600	5.8×10^{-15}	1.07	100
CH_3CF_2Cl (HCFC-142b) + OH $\rightarrow CH_2CF_2Cl + H_2O$	220–808	1.3×10^{-12}	1770	3.4×10^{-15}	1.15	50
CH_2ClCF_2Cl (HCFC-132b) + OH $\rightarrow CHClCF_2Cl + H_2O$	249–788	3.6×10^{-12}	1600	1.7×10^{-14}	1.5	200
CH_2FCFCl_2 (HCFC-132c) + OH $\rightarrow CHFCFCl_2 + H_2O$	298–370	8.2×10^{-13}	1250	1.23×10^{-14}	1.15	100
CH_2ClCF_3 (HCFC-133a) + OH $\rightarrow CHClCF_3 + H_2O$	263–866	9.4×10^{-13}	1300	1.2×10^{-14}	1.1	100
$CHCl_2CF_2Cl$ (HCFC-122) + OH $\rightarrow CCl_2CF_2Cl + H_2O$	298–460	7.7×10^{-13}	810	5.1×10^{-14}	1.2	150
$CHFClCFCl_2$ (HCFC-122a) + OH $\rightarrow CFClCFCl_2 + H_2O$	298–460	9.0×10^{-13}	1200	1.6×10^{-14}	1.1	100
$CHCl_2CF_3$ (HCFC-123) + OH $\rightarrow CCl_2CF_3 + H_2O$	213–866	7.4×10^{-13}	900	3.6×10^{-14}	1.1	100

Reaction	T/K	A*/cm³ molecule⁻¹ s⁻¹	E/R/K*	k_{298}*/cm³ molecule⁻¹ s⁻¹	f_{298}	g
$CHFClCF_2Cl$ (HCFC-123a) + OH → $CFClCF_2Cl$ + H_2O	298–460	8.6×10^{-13}	1250	1.3×10^{-14}	1.3	200
$CHFClCF_3$ (HCFC-124) + OH → $CFClCF_3$ + H_2O	210–867	7.1×10^{-13}	1300	9.0×10^{-15}	1.15	100
$CH_3CH_2CH_2Cl$ + OH → products	253–372	5.8×10^{-12}	530	9.8×10^{-13}	1.15	50
$CH_3CHClCH_3$ + OH → products	233–372	2.35×10^{-12}	365	6.9×10^{-13}	1.15	50
$CH_3CF_2CFCl_2$ (HCFC-243cc) + OH → products	295–367	7.7×10^{-13}	1720	2.4×10^{-15}	1.3	200
$CHCl_2CF_2CF_3$ (HCFC-225ca) + OH → products	251–400	6.3×10^{-13}	960	2.5×10^{-14}	1.2	200
CF_2ClCF_2CHFCl (HCFC-225cb) + OH → products	270–400	5.5×10^{-13}	1230	8.9×10^{-15}	1.1	150
$CF_3CH_2CFCl_2$ (HCFC-234fb) + OH → $CF_3CHCFCl_2$ + H_2O	298	*1.8×10^{-12}*	*2300*	8.0×10^{-16}	1.5	200
CH_2=CHCl + OH → products	293–1173	1.3×10^{-12}	-500	6.9×10^{-12}	1.2	100
(E)-CHCl=CHCl + OH → products	240–720	1.06×10^{-12}	-230	2.3×10^{-12}	1.1	100
(Z)-CHCl=CHCl + OH → products	240–400	2.04×10^{-12}	-70	2.6×10^{-12}	1.1	100
CH_2=CCl_2 + OH → products	240–750	2.17×10^{-12}	-470	1.05×10^{-11}	1.1	100
CHCl=CCl_2 + OH → products	234–1500	8.0×10^{-13}	-300	2.2×10^{-12}	1.2	100
CCl_2=CCl_2 + OH → products	296–720	4.7×10^{-12}	990	1.7×10^{-13}	1.2	200
CF_2=CFCl + OH → products	296–364	1.06×10^{-12}	-580	7.4×10^{-12}	1.2	100
(E)-CHCl=$CHCF_3$ + OH → products	213–376	9.0×10^{-13}	280	3.5×10^{-13}	1.07	20
(Z)-CHCl=$CHCF_3$ + OH → products	213–376	3.67×10^{-13}	-280	9.4×10^{-13}	1.1	20
CH_3OCF_2CHClF + OH → products	250–430	1.64×10^{-12}	1130	3.7×10^{-14}	1.2	0
$CHF_2OCHClCF_3$ (Isoflurane) + OH → products	250–430	1.1×10^{-12}	1275	1.5×10^{-14}	1.07	50
CHF_2OCF_2CHFCl (Enflurane) + OH → products	250–430	6.73×10^{-13}	1200	1.2×10^{-14}	1.1	100
CH_3OCl + OH → products	250–431	2.5×10^{-12}	370	7.1×10^{-13}	2	150
CCl_3CHO + OH → H_2O + CCl_3CO	298–520	9.1×10^{-12}	580	1.3×10^{-12}	1.3	200
Cl + HO_2 → HCl + O_2	226–420	1.4×10^{-11}	-270	3.5×10^{-11}	1.2	100
Cl + HO_2 → OH + ClO	226–420	3.6×10^{-11}	375	1.0×10^{-11}	1.4	150
ClO + HO_2 → HOCl + O_2	203–364	2.6×10^{-12}	-290	6.9×10^{-12}	1.2	150
$ClONO_2$ + H_2O → products	298			$<2 \times 10^{-21}$		
OClO + NO → NO_2 + ClO	220–367	*2.5×10^{-12}*	*600*	3.4×10^{-13}	2	300
Cl_2O_2 + NO → products	220–298			$<1 \times 10^{-15}$		
HCl + NO_3 → HNO_3 + Cl	298–473			$<5 \times 10^{-17}$		
HCl + HO_2NO_2 → products	296			$<1 \times 10^{-21}$		
Cl + O_3 → ClO + O_2	184–1350	2.3×10^{-11}	200	1.2×10^{-11}	1.15	50
Cl + H_2 → HCl + H	199–3000	3.05×10^{-11}	2270	1.5×10^{-14}	1.1	100
Cl + H_2O_2 → HCl + HO_2	265–424	1.1×10^{-11}	980	4.1×10^{-13}	1.3	300
Cl + NO_3 → ClO + NO_2	278–338	2.4×10^{-11}	0	2.4×10^{-11}	1.5	400
Cl + N_2O → ClO + N_2	773–1030					
Cl + HNO_3 → products	243–633			$<2 \times 10^{-16}$		
Cl + HO_2NO_2 → products	296–399			$<1 \times 10^{-13}$		
Cl + CH_4 → HCl + CH_3	181–1550	7.1×10^{-12}	1270	1.0×10^{-13}	1.05	50
Cl + CH_3D → products	223–343	7.46×10^{-12}	1400	6.8×10^{-14}	1.07	50
Cl + H_2CO → HCl + HCO	200–500	8.1×10^{-11}	30	7.3×10^{-11}	1.15	100
Cl + HC(O)OH → products	298			2.0×10^{-13}	1.5	
Cl + CH_3O_2 → products	298			1.6×10^{-10}	1.5	
Cl + CH_3OH → CH_2OH + HCl	200–573	5.5×10^{-11}	0	5.5×10^{-11}	1.2	100
Cl + CH_3OOH → products	295			5.7×10^{-11}	2	
Cl + CH_3ONO_2 → products	298	*1.3×10^{-11}*	*1200*	2.3×10^{-13}	1.5	300
Cl + C_2H_6 → HCl + C_2H_5	48–1400	7.2×10^{-11}	70	5.7×10^{-11}	1.07	20
Cl + $C_2H_5O_2$ → ClO + C_2H_5O	298			7.4×10^{-11}	2	
Cl + $C_2H_5O_2$ → HCl + $C_2H_4O_2$	298			7.7×10^{-11}	2	
Cl + CH_3CH_2OH → products	266–600	9.6×10^{-11}	0	9.6×10^{-11}	1.2	100
Cl + $CH_3C(O)OH$ → products	298			2.8×10^{-14}	2	
Cl + CH_3CN → products	274–728	1.6×10^{-11}	2140	1.2×10^{-14}	2	300
Cl + $C_2H_5ONO_2$ → products	298	*1.5×10^{-11}*	*400*	3.9×10^{-12}	1.5	200
Cl + $CH_3CO_3NO_2$ → products	298			$<1 \times 10^{-14}$		
Cl + C_3H_8 → HCl + CH_3CHCH_3	48–1400	6.54×10^{-11}	-60	8.0×10^{-11}	1.1	20
Cl + C_3H_8 → HCl + $CH_2CH_2CH_3$	48–1400	8.12×10^{-11}	90	6.0×10^{-11}	1.05	20
Cl + $CH_3C(O)CH_3$ → $CH_3C(O)CH_2$ + HCl	210–440	1.63×10^{-11}	610	2.1×10^{-12}	1.07	50
Cl + CH_2=$C(CH_3)CHO$ → products	296–298			2.2×10^{-10}	1.3	
Cl + $CH_3C(O)CH$=CH_2 → products	296–298			2.1×10^{-10}	1.1	
Cl + CH_2=$C(CH_3)CH$=CH_2 → products	233–320	7.6×10^{-11}	-500	4.1×10^{-10}	1.15	100

Reaction	T/K	A*/cm³ molecule⁻¹ s⁻¹	E/R/K*	k_{298}*/cm³ molecule⁻¹ s⁻¹	f_{298}	g
$Cl + C_2H_5CO_3NO_2 \rightarrow$ products	295			1.1×10^{-12}	2	
$Cl + 1\text{-}C_3H_7ONO_2 \rightarrow$ products	295			2.3×10^{-11}	1.5	
$Cl + 2\text{-}C_3H_7ONO_2 \rightarrow$ products	295			4.0×10^{-12}	2	
$Cl + OClO \rightarrow ClO + ClO$	229–588	3.4×10^{-11}	-160	5.8×10^{-11}	1.25	200
$Cl + ClOO \rightarrow Cl_2 + O_2$	160–306	2.3×10^{-10}	0	2.3×10^{-10}	2	200
$Cl + ClOO \rightarrow ClO + ClO$	160–306	1.2×10^{-11}	0	1.2×10^{-11}	2	200
$Cl + Cl_2O \rightarrow Cl_2 + ClO$	233–373	6.2×10^{-11}	-130	9.6×10^{-11}	1.2	130
$Cl + Cl_2O_2 \rightarrow$ products	217–298	7.6×10^{-11}	-65	1.0×10^{-10}	1.2	100
$Cl + HOCl \rightarrow$ products	243–365	3.4×10^{-12}	130	2.2×10^{-12}	1.3	200
$Cl + ClNO \rightarrow NO + Cl_2$	220–450	5.8×10^{-11}	-100	8.1×10^{-11}	1.5	200
$Cl + ClONO_2 \rightarrow$ products	195–298	6.5×10^{-12}	-135	1.0×10^{-11}	1.1	50
$Cl + CH_3Cl \rightarrow HCl + CH_2Cl$	233–843	2.03×10^{-11}	1110	4.9×10^{-13}	1.07	50
$Cl + CH_2Cl_2 \rightarrow HCl + CHCl_2$	273–790	7.4×10^{-12}	910	3.5×10^{-13}	1.07	100
$Cl + CHCl_3 \rightarrow HCl + CCl_3$	220–1010	3.3×10^{-12}	990	1.2×10^{-13}	1.15	100
$Cl + CH_3F$ (HFC-41) $\rightarrow HCl + CH_2F$	216–368	1.96×10^{-11}	1200	3.5×10^{-13}	1.15	150
$Cl + CH_2F_2$ (HFC-32) $\rightarrow HCl + CHF_2$	253–553	7.6×10^{-12}	1630	3.2×10^{-14}	1.08	100
$Cl + CHF_3$ (HFC-23) $\rightarrow HCl + CF_3$	298			$<5 \times 10^{-16}$		
$Cl + CH_2FCl$ (HCFC-31) $\rightarrow HCl + CHFCl$	273–298	*5.9×10^{-12}*	*1200*	1.05×10^{-13}	1.1	200
$Cl + CHFCl_2$ (HCFC-21) $\rightarrow HCl + CFCl_2$	294–433	6.0×10^{-12}	1700	2.0×10^{-14}	1.2	200
$Cl + CHF_2Cl$ (HCFC-22) $\rightarrow HCl + CF_2Cl$	296–411	5.9×10^{-12}	2430	1.7×10^{-15}	1.1	150
$Cl + CH_3CCl_3 \rightarrow HCl + CH_2CCl_3$	253–423	3.0×10^{-12}	1730	9.0×10^{-15}	1.15	100
$Cl + CH_3CH_2F$ (HFC-161) $\rightarrow HCl + CH_3CHF$	264–368	1.82×10^{-11}	330	6.0×10^{-12}	1.1	100
$Cl + CH_3CH_2F$ (HFC-161) $\rightarrow HCl + CH_2CH_2F$	264–368	1.4×10^{-11}	940	6.0×10^{-13}	1.15	100
$Cl + CH_3CHF_2$ (HFC-152a) $\rightarrow HCl + CH_3CF_2$	264–368	6.0×10^{-12}	950	2.5×10^{-13}	1.1	100
$Cl + CH_3CHF_2$ (HFC-152a) $\rightarrow HCl + CH_2CHF_2$	264–368	6.5×10^{-12}	2320	2.7×10^{-15}	1.15	200
$Cl + CH_2FCH_2F$ (HFC-152) $\rightarrow HCl + CHFCH_2F$	280–368	2.27×10^{-11}	1050	6.7×10^{-13}	1.15	200
$Cl + CH_3CFCl_2$ (HCFC-141b) $\rightarrow HCl + CH_2CFCl_2$	276–429	3.5×10^{-12}	2200	2.2×10^{-15}	1.15	200
$Cl + CH_3CF_2Cl$ (HCFC-142b) $\rightarrow HCl + CH_2CF_2Cl$	295–673	1.35×10^{-12}	2400	4.3×10^{-16}	1.15	200
$Cl + CH_3CF_3$ (HFC-143a) $\rightarrow HCl + CH_2CF_3$	281–368	1.64×10^{-11}	3900	3.4×10^{-17}	1.5	300
$Cl + CH_2FCHF_2$ (HFC-143) $\rightarrow HCl + CH_2FCF_2$	281–368	6.8×10^{-12}	1670	2.5×10^{-14}	1.3	200
$Cl + CH_2FCHF_2$ (HFC-143) $\rightarrow HCl + CHFCHF_2$	281–368	9.1×10^{-12}	1770	2.4×10^{-14}	1.3	200
$Cl + CH_2ClCF_3$ (HCFC-133a) $\rightarrow HCl + CHClCF_3$	296–500	1.83×10^{-12}	1680	6.5×10^{-15}	1.2	200
$Cl + CH_2FCF_3$ (HFC-134a) $\rightarrow HCl + CHFCF_3$	253–423	2.1×10^{-12}	2160	1.5×10^{-15}	1.1	200
$Cl + CHF_2CHF_2$ (HFC-134) $\rightarrow HCl + CF_2CHF_2$	280–368	7.0×10^{-12}	2430	2.0×10^{-15}	1.2	200
$Cl + CHCl_2CF_3$ (HCFC-123) $\rightarrow HCl + CCl_2CF_3$	276–382	5.0×10^{-12}	1800	1.2×10^{-14}	1.15	200
$Cl + CHFClCF_3$ (HCFC-124) $\rightarrow HCl + CFClCF_3$	276–376	1.13×10^{-12}	1800	2.7×10^{-15}	1.2	200
$Cl + CHF_2CF_3$ (HFC-125) $\rightarrow HCl + CF_2CF_3$	295–399	*1.85×10^{-12}*	*2600*	3.0×10^{-16}	1.5	300
$ClO + O_3 \rightarrow ClOO + O_2$	223–413	2×10^{-12}	>3600	$<1.4 \times 10^{-17}$		
$ClO + O_3 \rightarrow OClO + O_2$	223–413	*1×10^{-12}*	*>4000*	$<1.0 \times 10^{-18}$		
$ClO + H_2 \rightarrow$ products	294–670	*1×10^{-12}*	*>4800*	*$<1 \times 10^{-19}$*		
$ClO + NO \rightarrow NO_2 + Cl$	202–415	6.4×10^{-12}	-290	1.7×10^{-11}	1.15	100
$ClO + NO_3 \rightarrow ClOO + NO_2$	210–353	4.7×10^{-13}	0	4.7×10^{-13}	1.5	400
$ClO + N_2O \rightarrow$ products	587	*1×10^{-12}*	*>4300*	*$<6 \times 10^{-19}$*		
$ClO + CO \rightarrow$ products	587	*1×10^{-12}*	*>3700*	*$<4 \times 10^{-18}$*		
$ClO + CH_4 \rightarrow$ products	670	*1×10^{-12}*	*>3700*	*$<4 \times 10^{-18}$*		
$ClO + H_2CO \rightarrow$ products	298	*1×10^{-12}*	*>2100*	*$<1 \times 10^{-15}$*		
$ClO + CH_3O_2 \rightarrow$ products	197–355	1.8×10^{-12}	600	2.4×10^{-12}	1.1	-15
$ClO + ClO \rightarrow Cl_2 + O_2$	260–1250	1.0×10^{-12}	1590	4.8×10^{-15}	1.5	+300 (-20)
$ClO + ClO \rightarrow ClOO + Cl$	260–710	3.0×10^{-11}	2450	8.0×10^{-15}	1.5	500
$ClO + ClO \rightarrow OClO + Cl$	254–390	3.5×10^{-13}	1370	3.5×10^{-15}	1.5	300
$HCl + ClONO_2 \rightarrow$ products	298			$<1 \times 10^{-20}$		
$CH_2ClO + O_2 \rightarrow CHClO + HO_2$	264–336			6×10^{-14}	5	
$CH_2ClO_2 + HO_2 \rightarrow CH_2ClO_2H + O_2$	251–600	3.3×10^{-13}	-820	5.2×10^{-12}	1.5	200
$CH_2ClO_2 + NO \rightarrow CH_2ClO + NO_2$	295	*7×10^{-12}*	*-300*	1.9×10^{-11}	1.5	200
$CCl_3O_2 + NO \rightarrow CCl_2O + NO_2 + Cl$	228–413	7.3×10^{-12}	-270	1.8×10^{-11}	1.3	200
$CCl_2FO_2 + NO \rightarrow CClFO + NO_2 + Cl$	228–413	4.5×10^{-12}	-350	1.5×10^{-11}	1.3	200
$CClF_2O_2 + NO \rightarrow CF_2O + NO_2 + Cl$	228–413	3.8×10^{-12}	-400	1.5×10^{-11}	1.2	200

* Italicized values are estimated.

TABLE 1i. BrO_x Bimolecular Reaction Rates of Atmospheric Importance

Reaction	*T*/K	*A**/cm³ molecule⁻¹ s⁻¹	*E*/*R*/K*	k_{298}*/cm³ molecule⁻¹ s⁻¹	f_{298}	*g*
$BrO + O \rightarrow Br + O_2$	231–328	1.9×10^{-11}	-230	4.1×10^{-11}	1.5	150
$HBr + O \rightarrow OH + Br$	221–554	5.8×10^{-12}	1500	3.8×10^{-14}	1.3	200
$HOBr + O \rightarrow OH + BrO$	233–423	1.2×10^{-10}	430	2.8×10^{-11}	3	300
$BrONO_2 + O \rightarrow NO_3 + BrO$	227–339	1.9×10^{-11}	-215	3.9×10^{-11}	1.25	40
$Br_2 + OH \rightarrow HOBr + Br$	230–360	2.1×10^{-11}	-240	4.6×10^{-11}	1.1	50
$BrO + OH \rightarrow$ products	230–355	1.7×10^{-11}	-250	3.9×10^{-11}	1.4	100
$HBr + OH \rightarrow H_2O + Br$	230–360	5.5×10^{-12}	-200	1.1×10^{-11}	1.1	100
$CH_3Br + OH \rightarrow CH_2Br + H_2O$	233–400	1.42×10^{-12}	1150	3.0×10^{-14}	1.07	100
$CH_2Br_2 + OH \rightarrow CHBr_2 + H_2O$	244–375	2.0×10^{-12}	840	1.2×10^{-13}	1.1	100
$CHBr_3 + OH \rightarrow CBr_3 + H_2O$	230–370	9.0×10^{-13}	360	2.7×10^{-13}	1.05	20
$CH_2ClBr + OH \rightarrow CHClBr + H_2O$	230–376	2.1×10^{-12}	880	1.1×10^{-13}	1.07	100
$CHClBr_2 + OH \rightarrow CClBr_2 + H_2O$	230–330	9.0×10^{-13}	420	2.2×10^{-13}	1.07	20
$CHCl_2Br + OH \rightarrow CCl_2Br + H_2O$	230–330	9.4×10^{-13}	510	1.7×10^{-13}	1.07	20
CHF_2Br (Halon-1201) + $OH \rightarrow CF_2Br + H_2O$	233–460	7.85×10^{-13}	1300	1.0×10^{-14}	1.07	100
CF_2Br_2 (Halon-1202) + $OH \rightarrow$ products	298	*1×10^{-12}*	*>2200*	*$<5 \times 10^{-16}$*		
CF_3Br (Halon-1301) + $OH \rightarrow$ products	460	*1×10^{-12}*	*>3600*	*$<6 \times 10^{-18}$*		
CF_2ClBr (Halon-1211) + $OH \rightarrow$ products	373	*1×10^{-12}*	*>3500*	*$<8 \times 10^{-18}$*		
$CH_3CH_2Br + OH \rightarrow$ products	233–422	2.9×10^{-12}	640	3.4×10^{-13}	1.2	150
$CH_2BrCH_2Br + OH \rightarrow$ products	292–366	1.75×10^{-11}	1290	2.3×10^{-13}	1.15	200
CH_2BrCF_3 (Halon-2301) + $OH \rightarrow CHBrCF_3 + H_2O$	280–460	9.5×10^{-13}	1200	1.7×10^{-14}	1.2	150
$CHFBrCF_3$ (Halon-2401) + $OH \rightarrow CFBrCF_3 + H_2O$	279–460	7.3×10^{-13}	1120	1.7×10^{-14}	1.2	100
$CHClBrCF_3$ (Halothane, Halon-2311) + $OH \rightarrow CClBrCF_3 + H_2O$	298–460	1.1×10^{-12}	940	4.7×10^{-14}	1.2	150
$CHFClCF_2Br + OH \rightarrow CFClCF_2Br + H_2O$	315–372	8.4×10^{-13}	1220	1.4×10^{-14}	1.3	200
CF_2BrCF_2Br (Halon-2402) + $OH \rightarrow$ products	460	*1×10^{-12}*	*>3600*	*$<6 \times 10^{-18}$*		
$CH_2BrCH_2CH_3 + OH \rightarrow$ products	210–480	3.0×10^{-12}	330	1.0×10^{-12}	1.05	50
$CH_3CHBrCH_3 + OH \rightarrow$ products	210–480	1.85×10^{-12}	270	7.5×10^{-13}	1.05	50
$CHBr{=}CF_2 + OH \rightarrow$ products	250–370	1.3×10^{-12}	-370	4.5×10^{-12}	1.1	20
$CFBr{=}CF_2 + OH \rightarrow$ products	250–370	2.0×10^{-12}	-400	7.6×10^{-12}	1.1	20
$CH_2{=}CBrCF_3 + OH \rightarrow$ products	220–370	1.06×10^{-12}	-380	3.8×10^{-12}	1.05	20
$CH_2{=}CBrCF_2CF_3 + OH \rightarrow$ products	250–370	9.5×10^{-12}	-370	3.3×10^{-12}	1.05	20
$CH_2{=}CHCF_2CF_2Br + OH \rightarrow$ products	250–370	8.7×10^{-12}	-200	1.7×10^{-12}	1.1	20
$CH_2{=}CHCFClCF_2Br + OH \rightarrow$ products	230–370	6.6×10^{-13}	-420	2.7×10^{-12}	1.1	0
$Br + HO_2 \rightarrow HBr + O_2$	230–355	4.8×10^{-12}	310	1.7×10^{-12}	1.3	150
$BrO + HO_2 \rightarrow$ products	210–360	4.5×10^{-12}	-460	2.1×10^{-11}	1.15	100
$HBr + NO_3 \rightarrow HNO_3 + Br$	298			$<1.0 \times 10^{-16}$		
$CH_3Br + Cl \rightarrow HCl + CH_2Br$	210–700	1.46×10^{-11}	1040	4.45×10^{-13}	1.03	50
$CH_2Br_2 + Cl \rightarrow HCl + CHBr_2$	222–395	6.8×10^{-12}	830	4.2×10^{-13}	1.1	50
$CHBr_3 + Cl \rightarrow CBr_3 + HCl$	273–363	4.85×10^{-12}	850	2.8×10^{-13}	1.3	200
$CH_2ClBr + Cl \rightarrow HCl + CHClBr$	298	*6.8×10^{-12}*	*870*	3.7×10^{-13}	1.2	100
$Br + O_3 \rightarrow BrO + O_2$	195–422	1.6×10^{-11}	780	1.2×10^{-12}	1.15	100
$Br + H_2O_2 \rightarrow HBr + HO_2$	298–378	*1×10^{-11}*	*>3000*	$<5 \times 10^{-16}$		
$Br + NO_3 \rightarrow BrO + NO_2$	298			1.6×10^{-11}	2	
$Br + H_2CO \rightarrow HBr + HCO$	223–480	1.7×10^{-11}	800	1.1×10^{-12}	1.2	125
$Br + CH_2{=}C(CH_3)CHO \rightarrow$ products	301			2.3×10^{-11}[a]	1.5	
$Br + CH_3C(O)CH{=}CH_2 \rightarrow$ products	301			1.9×10^{-11}[a]	1.5	
$Br + CH_2{=}C(CH_3)CH{=}CH_2 (+M) \leftrightarrow X (+O_2) \rightarrow$ products						
$Br + CH_2{=}C(CH_3)CH{=}CH_2 \rightarrow CH_2{=}C(\cdot CH_2)CH{=}CH_2 + HBr$	526–673	1.2×10^{-11}	2100	1.0×10^{-14}	2	200
$Br{-}CH_2{=}C(CH_3)CH{=}CH_2 + O_2 \rightarrow$ products	297			3.2×10^{-13}	1.5	
$Br + OClO \rightarrow BrO + ClO$	267–423	2.6×10^{-11}	1300	3.4×10^{-13}	2	300
$Br + Cl_2O \rightarrow BrCl + ClO$	220–402	2.1×10^{-11}	470	4.3×10^{-12}	1.3	150
$Br + Cl_2O_2 \rightarrow$ products	223–298	5.9×10^{-12}	170	3.3×10^{-12}	1.3	200
$BrO + O_3 \rightarrow$ products	298	*1×10^{-12}*	*>3200*	$<2 \times 10^{-18}$		
$BrO + NO \rightarrow NO_2 + Br$	224–425	8.8×10^{-12}	-260	2.1×10^{-11}	1.15	130
$BrO + NO_3 \rightarrow$ products	298			1.0×10^{-12}	3	
$BrO + ClO \rightarrow Br + OClO$	200–400	9.5×10^{-13}	-550	6.0×10^{-12}	1.2	100
$BrO + ClO \rightarrow Br + ClOO$	200–400	2.3×10^{-12}	-260	5.5×10^{-12}	1.2	100
$BrO + ClO \rightarrow BrCl + O_2$	200–400	4.1×10^{-13}	-290	1.1×10^{-12}	1.2	100
$BrO + BrO \rightarrow$ products	220–348	1.5×10^{-12}	-230	3.2×10^{-12}	1.15	150
$OBrO + O_3 \rightarrow$ products	298			$<1.5 \times 10^{-15}$		
$OBrO + NO \rightarrow$ products	240–350	2.4×10^{-13}	-610	1.8×10^{-12}	3	200
$CH_2BrO_2 + NO \rightarrow CH_2O + NO_2 + Br$	298	*4×10^{-12}*	*-300*	1.1×10^{-11}	1.5	200

[a] 1 atm air.

* Italicized values are estimated.

TABLE 1j. IO_x Bimolecular Reaction Rates of Atmospheric Importance

Reaction	*T*/K	*A*/cm³ molecule⁻¹ s⁻¹	*E/R*/K	k_{298}/cm³ molecule⁻¹ s⁻¹	f_{298}	*g*
$I_2 + O \rightarrow IO + I$	298	1.3×10^{-10}	0	1.3×10^{-10}	1.2	250
$IO + O \rightarrow O_2 + I$	298	1.4×10^{-10}	0	1.4×10^{-10}	1.2	0
$I_2 + OH \rightarrow HOI + I$	294–298			1.8×10^{-10}	2	
$HI + OH \rightarrow H_2O + I$	298			3.0×10^{-11}	2	
$CH_3I + OH \rightarrow H_2O + CH_2I$	271–423	2.9×10^{-12}	1100	7.2×10^{-14}	1.5	300
$CF_3I + OH \rightarrow HOI + CF_3$	271–450	2.5×10^{-11}	2070	2.4×10^{-14}	1.3	200
$I + HO_2 \rightarrow HI + O_2$	283–353	1.5×10^{-11}	1090	3.8×10^{-13}	2	500
$IO + HO_2 \rightarrow$ products	273–373	1.3×10^{-11}	-570	8.8×10^{-11}	1.15	100
$HI + NO_3 \rightarrow HNO_3 + I$						
$CH_3I + Cl \rightarrow CH_2I + HCl$	273–363	2.9×10^{-11}	1000	1.0×10^{-12}	1.5	250
$I + O_3 \rightarrow IO + O_2$	231–337	2.0×10^{-11}	830	1.2×10^{-12}	1.2	100
$I + BrO \rightarrow IO + Br$	298			1.2×10^{-11}	2	
$IO + NO \rightarrow I + NO_2$	240–370	8.6×10^{-12}	-230	1.9×10^{-11}	1.1	50
$IO + ClO \rightarrow I + OClO$	200–362	2.7×10^{-12}	-280	6.9×10^{-12}	1.3	150
$IO + ClO \rightarrow I + Cl + O_2$	200–362	1.2×10^{-12}	-280	3.1×10^{-12}	1.3	150
$IO + ClO \rightarrow ICl + O_2$	200–362	0.92×10^{-12}	-280	2.3×10^{-12}	1.3	150
$IO + BrO \rightarrow Br + OIO$	204–388	4.4×10^{-12}	-760	5.6×10^{-11}	1.25	300
$IO + BrO \rightarrow Br + I + O_2$	204–388	5.5×10^{-13}	-760	7.0×10^{-12}	1.5	300
$IO + BrO \rightarrow I + OBrO$	204–388	4.4×10^{-13}	-760	5.6×10^{-12}	1.5	300
$IO + BrO \rightarrow IBr + O_2$	204–388	1.1×10^{-13}	-760	1.4×10^{-12}	2	300
$IO + IO \rightarrow$ products	250–373	1.5×10^{-11}	-500	8.0×10^{-11}	1.5	500
$INO + INO \rightarrow I_2 + 2NO$	320–450	8.4×10^{-11}	2620	1.3×10^{-14}	2.5	600
$INO_2 + INO_2 \rightarrow I_2 + 2NO_2$	320–450	2.9×10^{-11}	2600	4.7×10^{-15}	3	1000

TABLE 1k. SO_x Bimolecular Reaction Rates of Atmospheric Importance

Reaction	*T*/K	*A*/cm³ molecule⁻¹ s⁻¹	*E/R*/K	k_{298}/cm³ molecule⁻¹ s⁻¹	f_{298}	*g*
$SH + O \rightarrow SO + H$	295			1.6×10^{-10}	5	
$CS + O \rightarrow CO + S$	150–305	2.7×10^{-10}	760	2.1×10^{-11}	1.1	250
$H_2S + O \rightarrow OH + SH$	205–502	9.2×10^{-12}	1800	2.2×10^{-14}	1.7	550
$OCS + O \rightarrow CO + SO$	239–808	2.1×10^{-11}	2200	1.3×10^{-14}	1.15	150
$CS_2 + O \rightarrow CS + SO$	218–543	3.2×10^{-11}	650	3.6×10^{-12}	1.2	150
$CH_3SCH_3 + O \rightarrow CH_3SO + CH_3$	252–557	1.3×10^{-11}	-410	5.0×10^{-11}	1.1	100
$CH_3SSCH_3 + O \rightarrow CH_3SO + CH_3S$	270–571	3.9×10^{-11}	-290	1.03×10^{-10}	1.1	100
$CH_3S(O)CH_3 + O \rightarrow$ products	266–383	2.0×10^{-12}	-440	8.8×10^{-12}	1.2	200
$H_2S + O_3 \rightarrow$ products	~298			$<2.0 \times 10^{-20}$		
$CH_3SCH_3 + O_3 \rightarrow$ products	301			$<1.5 \times 10^{-19}$		
$SO_2 + O_3 \rightarrow SO_3 + O_2$	300	3.0×10^{-12}	>7000	$<2.0 \times 10^{-22}$		
$SO_2F_2 + O_3 \rightarrow$ products	294-296			$<1.0 \times 10^{-23}$		
$H_2S + OH \rightarrow SH + H_2O$	228–885	3.3×10^{-12}	-100	4.6×10^{-12}	1.1	50
$OCS + OH \rightarrow$ products	255–517	7.2×10^{-14}	1070	2.0×10^{-15}	2	200
$CS_2 + OH \rightarrow SH + OCS$	251–520			$<2.0 \times 10^{-15}$		
$CS_2 + OH \rightarrow CS_2OH\ (+O_2) \rightarrow$ products				1.2×10^{-12}	1.25	
$CS_2OH + O_2 \rightarrow$ products	249–348	2.8×10^{-14}	0	2.8×10^{-14}	1.2	100
$CH_3SH + OH \rightarrow CH_3S + H_2O$	244–430	9.9×10^{-12}	-360	3.3×10^{-11}	1.07	75
$CH_3SCH_3 + OH \rightarrow H_2O + CH_2SCH_3$	248–573	1.2×10^{-11}	280	4.7×10^{-12}	1.1	100
$CH_3SCH_3 + OH\ (+M) \leftrightarrow (CH_3)_2SOH\ (+O_2) \rightarrow$ products				2.0×10^{-12}[a]	1.2	
$(CH_3)_2SOH + O_2 \rightarrow$ products	222–297	8.5×10^{-13}	0	8.5×10^{-13}	1.25	0
$CH_3SCH_2Cl + OH \rightarrow$ products	298			2.5×10^{-12}	2	
$CH_3SSCH_3 + OH \rightarrow$ products	249–368	6.0×10^{-11}	-400	2.3×10^{-10}	1.2	200
$CH_3S(O)CH_3 + OH \rightarrow$ products	298–401	6.1×10^{-12}	-800	8.9×10^{-11}	1.2	500
$CH_3S(O)OH + OH \rightarrow$ products	298			9.0×10^{-11}	1.4	
$CH_3SO_3H + OH \rightarrow CH_3SO_3 + H_2O$						
$CH_3SC_2H_5 + OH$ H-abstraction rxns.	242–299	8.0×10^{-12}	0	8.0×10^{-12}	1.1	50
$CH_3SC_2H_5 + OH\ (+M) \leftrightarrow CH_3S(OH)C_2H5\ (+O_2) \rightarrow$ products	242–346			2.3×10^{-12}[a]	1.3	
$CH_3S(OH)C_2H_5 + O_2 \rightarrow$ products	242–296	1.0×10^{-12}	0	1.0×10^{-12}	1.5	0
$S + OH \rightarrow H + SO$	298			6.6×10^{-11}	3	

Reaction	T/K	A/cm^3 molecule^{-1} s^{-1}	E/R/K	k_{298}/cm^3 molecule^{-1} s^{-1}	f_{298}	g
$SO + OH \rightarrow H + SO_2$	295–703	2.6×10^{-11}	-330	7.9×10^{-11}	1.15	150
$SO_2F_2 + OH \rightarrow$ products	294–333			$<1.0 \times 10^{-16}$		
$H_2S + HO_2 \rightarrow$ products	298			$<3.0 \times 10^{-15}$		
$CH_3SH + HO_2 \rightarrow$ products	298			$<4.0 \times 10^{-15}$		
$CH_3SCH_3 + HO_2 \rightarrow$ products	298			$<5.0 \times 10^{-15}$		
$SO_2 + HO_2 \rightarrow$ products	295–300			$<1.0 \times 10^{-18}$		
$SO_2 + NO_2 \rightarrow$ products	298			$<2.0 \times 10^{-26}$		
$H_2S + NO_3 \rightarrow$ products	298			$<8.0 \times 10^{-16}$		
$OCS + NO_3 \rightarrow$ products	297			$<1.0 \times 10^{-16}$		
$CS_2 + NO_3 \rightarrow$ products	297–298			$<4.0 \times 10^{-16}$		
$CH_3SH + NO_3 \rightarrow$ products	254–367	4.4×10^{-13}	-210	8.9×10^{-13}	1.25	200
$CH_3SCH_3 + NO_3 \rightarrow CH_3SCH_2 + HNO_3$	256–376	1.9×10^{-13}	-530	1.1×10^{-12}	1.1	150
$CH_3SSCH_3 + NO_3 \rightarrow$ products	280–382	5.0×10^{-13}	-60	6.1×10^{-13}	1.25	150
$CH_3S(O)CH_3 + NO_3 \rightarrow$ products	295–298			2.9×10^{-13}	1.6	
$CH_3SC_2H_5 + NO_3 \rightarrow$ products	298			2.5×10^{-12}	1.5	
$SO_2 + NO_3 \rightarrow$ products	298–303			$<7.0 \times 10^{-21}$		
$CH_3SCH_3 + N_2O_5 \rightarrow$ products	296			$<1.0 \times 10^{-17}$		
$SO_2 + CH_3O_2 \rightarrow$ products	298–423			$<5.0 \times 10^{-17}$		
$CH_3SCH_3 + F \rightarrow$ products	298			2.4×10^{-10}	2	
$H_2S + Cl \rightarrow HCl + SH$	202–914	3.4×10^{-11}	-225	7.3×10^{-11}	1.1	50
$SO_2F_2 + Cl \rightarrow$ products	296–333			$<1.5 \times 10^{-18}$		
$OCS + Cl \rightarrow$ products	298			$<1.0 \times 10^{-16}$		
$CS_2Cl + O_2 \rightarrow$ products	230–298			$<2.5 \times 10^{-16}$		
$CH_3SH + Cl \rightarrow CH_3S + HCl$	193–430	1.2×10^{-10}	-150	2.0×10^{-10}	1.1	100
$CH_3SCH_3 + Cl \rightarrow CH_3SCH_2 + HCl$	240–421	9.4×10^{-11}	-190	1.8×10^{-10}	+1.2/ -2.5[b]	0
$CH_3SCH_3 + Cl \rightarrow$ products (1 atm)	240–356	3.5×10^{-10}	0	3.5×10^{-10}	1.2	0
$(CH_3)_2SCl + O_2 \rightarrow$ products	298			$<4.0 \times 10^{-18}$		
$(CH_3)_2SCl + NO \rightarrow$ products	298			1.2×10^{-11}	1.25	
$(CH_3)_2SCl + NO_2 \rightarrow$ products	298			2.7×10^{-11}	1.25	
$CH_3S(O)CH_3 + Cl \rightarrow CH_3S(O)CH_2 + HCl$	270–571	1.4×10^{-11}	0	1.4×10^{-11}	1.2	150
$CH_3S(O)CH_3 + Cl\ (+M) \rightarrow CH_3(Cl)S(O)CH_3$						
$CH_3(Cl)S(O)CH_3 + O_2 \rightarrow$ products	296			$<3.0 \times 10^{-18}$		
$CH_3(Cl)S(O)CH_3 + NO \rightarrow$ products	296			1.2×10^{-11}	1.5	
$CH_3(Cl)S(O)CH_3 + NO_2 \rightarrow$ products	296			2.1×10^{-11}	1.5	
$CH_3SC_2H_5 + Cl \rightarrow$ products (1 atm)	298			3.8×10^{-10}	1.2	
$CH_3SCH_3 + Cl_2 \rightarrow$ products	294–298			$<5.0 \times 10^{-14}$		
$OCS + ClO \rightarrow$ products	298			$<2.0 \times 10^{-16}$		
$CH_3SCH_3 + ClO \rightarrow$ products	259–335	2.1×10^{-15}	-340	6.6×10^{-15}	1.5	300
$CH_3S(O)CH_3 + ClO \rightarrow$ products	298			$<2.0 \times 10^{-14}$		
$SO + ClO \rightarrow Cl + SO_2$	248–363	2.8×10^{-11}	0	2.8×10^{-11}	1.3	50
$SO_2 + ClO \rightarrow Cl + SO_3$	298			$<4.0 \times 10^{-18}$		
$H_2S + Br \rightarrow HBr + SH$	319–431	1.4×10^{-11}	2750	1.4×10^{-15}	2	300
$CH_3SH + Br \rightarrow CH_3S + HBr$	273–431	9.2×10^{-12}	390	2.5×10^{-12}	2	100
$CH_3SCH_3 + Br \rightarrow CH_3SCH_2 + HBr$	386–604	9.0×10^{-11}	2390	3.0×10^{-14}	1.4	150
$CH_3S(O)CH_3 + Br \rightarrow$ products	298			1.2×10^{-14}	1.5	
$CH_3SH + BrO \rightarrow$ products						
$CH_3SCH_3 + BrO \rightarrow$ products	233–333	1.4×10^{-14}	-950	3.4×10^{-13}	1.25	200
$CH_3SSCH_3 + BrO \rightarrow$ products	296			1.5×10^{-14}	2	
$CH_3S(O)CH_3 + BrO \rightarrow$ products	296–298			1.0×10^{-14}	2	
$SO + BrO \rightarrow Br + SO_2$	298			5.7×10^{-11}	1.4	
$CH_3SH + IO \rightarrow$ products	298			6.6×10^{-16}	2	
$CH_3SCH_3 + IO \rightarrow$ products	256–468	2.5×10^{-12}	1500	1.6×10^{-14}	1.5	400
$S + O_2 \rightarrow SO + O$	252–878	1.6×10^{-12}	-100	2.2×10^{-12}	1.2	100
$S + O_3 \rightarrow SO + O_2$	298			1.2×10^{-11}	2	
$SO + O_2 \rightarrow SO_2 + O$	230–585	1.6×10^{-13}	2280	7.6×10^{-17}	1.2	200
$SO + O_3 \rightarrow SO_2 + O_2$	230–420	3.4×10^{-12}	1100	8.4×10^{-14}	1.1	150
$SO + NO_2 \rightarrow SO_2 + NO$	210–363	1.4×10^{-11}	0	1.4×10^{-11}	1.2	50
$SO + OClO \rightarrow SO_2 + ClO$	298			1.9×10^{-12}	3	
$SO_3 + 2H_2O \rightarrow$ products					1.2	200
$SO_3 + NO_2 \rightarrow$ products	298			1.0×10^{-19}	10	

Reaction	T/K	A/cm^3 molecule^{-1} s^{-1}	E/R/K	k_{298}/cm^3 molecule^{-1} s^{-1}	f_{298}	g
$SH + O_2 \rightarrow OH + SO$	295–298			$<4.0 \times 10^{-19}$		
$SH + O_3 \rightarrow HSO + O_2$	296–431	9.0×10^{-12}	280	3.5×10^{-12}	1.2	200
$SH + H_2O_2 \rightarrow$ products	298			$<5.0 \times 10^{-15}$		
$SH + NO_2 \rightarrow HSO + NO$	221–415	2.9×10^{-11}	-250	6.7×10^{-11}	1.1	50
$SH + N_2O \rightarrow HSO + N_2$	298			$<5.0 \times 10^{-16}$		
$SH + Cl_2 \rightarrow ClSH + Cl$	273–373	2.1×10^{-11}	800	1.4×10^{-12}	1.15	150
$SH + BrCl \rightarrow$ products	298–373	1.9×10^{-11}	-290	5.0×10^{-11}	2	150
$SH + Br_2 \rightarrow BrSH + Br$	273–373	5.7×10^{-11}	-160	9.7×10^{-11}	1.5	150
$SH + F_2 \rightarrow FSH + F$	298–373	5.3×10^{-11}	1440	4.2×10^{-13}	1.5	200
$HSO + O_2 \rightarrow$ products	296			$<2.0 \times 10^{-17}$		
$HSO + O_3 \rightarrow$ products	273–423			1.0×10^{-13}	1.3	
$HSO + NO \rightarrow$ products	293–298			$<1.0 \times 10^{-15}$		
$HSO + NO_2 \rightarrow HSO_2 + NO$	293–298			9.6×10^{-12}	2	
$HSO_2 + O_2 \rightarrow HO_2 + SO_2$	296			3.0×10^{-13}	3	
$HOSO_2 + O_2 \rightarrow HO_2 + SO_3$	297–423	1.3×10^{-12}	330	4.3×10^{-13}	1.15	200
$CS + O_2 \rightarrow OCS + O$	293–495			2.9×10^{-19}	2	
$CS + O_3 \rightarrow OCS + O_2$	298			3.0×10^{-16}	3	
$CS + NO_2 \rightarrow OCS + NO$	298			7.6×10^{-17}	3	
$CH_3S + O_2 \rightarrow$ products	298			$<3.0 \times 10^{-18}$		
$CH_3S + O_3 \rightarrow$ products	259–381	1.5×10^{-12}	-360	5.0×10^{-12}	1.15	100
$CH_3S + NO \rightarrow$ products	295–503			$<1.0 \times 10^{-13}$		
$CH_3S + NO_2 \rightarrow CH_3SO + NO$	222–511	3.0×10^{-11}	-240	6.7×10^{-11}	1.2	150
$CH_3S + CO \rightarrow$ products	208–295			$<1.4 \times 10^{-16}$		
$CH_3S + Br_2 \rightarrow CH_3SBr + Br$	298			1.7×10^{-10}	1.5	
$CH_2SH + O_2 \rightarrow$ products	298			6.5×10^{-12}	2	
$CH_2SH + O_3 \rightarrow$ products	298			3.5×10^{-11}	2	
$CH_2SH + NO \rightarrow$ products	298			1.9×10^{-11}	2	
$CH_2SH + NO_2 \rightarrow$ products	298			5.2×10^{-11}	2	
$CH_3SO + O_3 \rightarrow$ products	300			4.0×10^{-13}	1.5	
$CH_3SO + NO_2 \rightarrow CH_3SO_2 + NO$	297–300			1.2×10^{-11}	1.2	
$CH_3SOO + O_3 \rightarrow$ products	227			$<8.0 \times 10^{-13}$		
$CH_3SOO + NO \rightarrow$ products	227–256	1.1×10^{-11}	0	1.1×10^{-11}	2	100
$CH_3SOO + NO_2 \rightarrow$ products	227–246	2.2×10^{-11}	0	2.2×10^{-11}	2	100
$CH_3SO_2 + NO_2 \rightarrow$ products	298			2.2×10^{-12}	2	
$CH_3SCH_2 + NO_3 \rightarrow$ products	298			3.0×10^{-10}	2	
$CH_3SCH_2O_2 + NO \rightarrow CH_3S + CH_2O + NO_2$	261–400	4.9×10^{-12}	-260	1.2×10^{-11}	1.3	200
$CH_3SCH_2O_2 + CH_3SCH_2O_2 \rightarrow$ products	298			1.0×10^{-11}	1.25	
$CH_3SS + O_3 \rightarrow$ products	300			4.6×10^{-13}	2	
$CH_3SS + NO_2 \rightarrow$ products	297			1.8×10^{-11}	2	
$CH_3SSO + NO_2 \rightarrow$ products	297			4.5×10^{-12}	2	
$SO_2 + CH_2OO \rightarrow$ products	293–298			3.8×10^{-11}	1.1	
SO_2 + anti-$CH_3CHOO \rightarrow$ products	298			2.2×10^{-10}	1.2	
SO_2 + syn-$CH_3CHOO \rightarrow$ products	298			2.65×10^{-11}	1.1	

[a] 1 atm air

[b] Asymmetric uncertainties, see Introduction.

TABLE 1l. Metal Bimolecular Reaction Rates of Atmospheric Importance

Reaction	T/K	A/cm^3 molecule^{-1} s^{-1}	E/R/K	k_{298}/cm^3 molecule^{-1} s^{-1}	f_{298}	g
$Na + O_3 \rightarrow NaO + O_2$	208–377	1.0×10^{-9}	95	7.3×10^{-10}	1.2	50
$Na + O_3 \rightarrow NaO_2 + O$	208–377			$<4.0 \times 10^{-11}$		
$Na + N_2O \rightarrow NaO + N_2$	240–850	2.8×10^{-10}	1600	1.3×10^{-12}	1.2	400
$Na + Cl_2 \rightarrow NaCl + Cl$	294	7.3×10^{-10}	0	7.3×10^{-10}	1.3	200
$NaO + O \rightarrow Na + O_2$	290–573	4.4×10^{-10}	0	4.4×10^{-10}	1.5	200
$NaO + O_3 \rightarrow NaO_2 + O_2$	207–377	1.1×10^{-9}	570	1.6×10^{-10}	1.5	300
$NaO + O_3 \rightarrow Na + 2O_2$	296	6.0×10^{-11}	0	6.0×10^{-11}	3	800
$NaO + H_2 \rightarrow NaOH + H$	296	2.6×10^{-11}	0	2.6×10^{-11}	2	600
$NaO + H_2O \rightarrow NaOH + OH$	260–716	5.06×10^{-10}	240	2.3×10^{-10}	1.5	200

Reaction	T/K	A/cm^3 molecule^{-1} s^{-1}	E/R/K	k_{298}/cm^3 molecule^{-1} s^{-1}	f_{298}	g
$NaO + NO \rightarrow Na + NO_2$	296	1.5×10^{-10}	0	1.5×10^{-10}	4	400
$NaO + CO \rightarrow Na + CO_2$	294–300	1.6×10^{-10}	0	1.6×10^{-10}	1.5	100
$NaO + HCl \rightarrow$ products	308	2.8×10^{-10}	0	2.8×10^{-10}	3	400
$NaO_2 + O \rightarrow NaO + O_2$	300	2.2×10^{-11}	0	2.2×10^{-11}	5	600
$NaO_2 + NO \rightarrow NaO + NO_2$	296			$<10^{-14}$		
$NaO_2 + HCl \rightarrow$ products	295	2.3×10^{-10}	0	2.3×10^{-10}	3	400
$NaOH + H \rightarrow Na + H_2O$	230–298	3.8×10^{-11}	0	3.8×10^{-11}	1.5	200
$NaOH + HCl \rightarrow NaCl + H_2O$	308	2.8×10^{-10}	0	2.8×10^{-10}	3	400
$NaHCO_3 + H \rightarrow Na + H_2O + CO_2$	227–307	1.4×10^{-11}	1000	5×10^{-13}	2	100

TABLE 2.1 Termolecular Reaction Rates of Atmospheric Importance

Reaction	Low-pressure limit k_0^{298}/cm^6 molecule^{-2} s^{-2}	n	High-pressure limit k_∞^{298}/cm^6 molecule^{-2} s^{-2}	m	$k(T,[M])$ (298 K, 1 Atm)/ cm^3 molecule^{-1} s^{-1}	f_{298}	g
O_x Reactions							
$O + O_2$ (+M)$\rightarrow O_3$	6.0×10^{-34}	2.4			1.5×10^{-14}	1.1	50
$O(^1D)$ Reaction							
$O(^1D) + N_2$ (+M)$\rightarrow N_2O$	2.8×10^{-36}	0.9			6.9×10^{-17}	1.3	75
HO_x Reactions							
$H + O_2$ (+M)$\rightarrow HO_2$	5.3×10^{-32}	1.8	9.5×10^{-11}	-0.4	1.15×10^{-12}	1.2	50
$OH + OH$ (+M)$\rightarrow H_2O_2$	6.9×10^{-31}	1	2.6×10^{-11}	0	6.3×10^{-12}	1.5	100
NO_x Reactions							
$NO + O$ (+M)$\rightarrow NO_2$	9.0×10^{-32}	1.5	3.0×10^{-11}	0	1.7×10^{-12}	1.2	100
$NO + OH$ (+M)$\rightarrow HONO$	7.0×10^{-31}	2.6	3.6×10^{-11}	0.1	7.4×10^{-12}	1.2	50
$NO_2 + OH$ (+M)$\rightarrow HONO_2$	1.8×10^{-30}	3	2.8×10^{-11}	0	1.1×10^{-11}	1.3	100
$NO_2 + OH$ (+M)$\rightarrow HOONO$	9.1×10^{-32}	3.9	4.2×10^{-11}	0.5	1.8×10^{-12}	1.5	200
$NO + HO_2$ (+M)$\rightarrow HONO_2$							
$NO_2 + HO_2$ (+M)$\rightarrow HO_2NO_2$	1.9×10^{-31}	3.4	4.0×10^{-12}	0.3	1.3×10^{-12}	1.06	400
$NO_2 + NO_3$ (+M)$\rightarrow N_2O_5$	2.4×10^{-30}	3	1.6×10^{-12}	-0.1	1.3×10^{-12}	1.1	100
NO_3 (+M)$\rightarrow NO + O_2$							
Hydrocarbon Reactions							
$CH_3 + O_2$ (+M)$\rightarrow CH_3O_2$	4.0×10^{-31}	3.6	1.2×10^{-12}	-1.1	8.1×10^{-13}	1.1	50
$C_2H_2 + OH$ (+M)$\rightarrow HOCHCH$	5.5×10^{-30}	0	8.3×10^{-13}	-2	1.35×10^{-10}	1.1	50
$C_2H_4 + OH$ (+M)$\rightarrow HOCH_2CH_2$	1.1×10^{-28}	3.5	8.4×10^{-12}	1.75	7.9×10^{-12}	1.15	90
$CH_2{=}CHCH_3 + OH$ (+M)$\rightarrow HOC_3H_6$	4.6×10^{-27}	4	2.6×10^{-11}	1.3	2.5×10^{-11}	1.2	50
iso-Butene + OH (+M)$\rightarrow$ products			5.4×10^{-11}	1.4	5.5×10^{-11}	1.2	200
1-Butene + OH (+M) $\rightarrow$ products			3.2×10^{-11}	1.4	3.2×10^{-11}	1.1	50
cis-2-Butene + OH (+M)$\rightarrow$ products			5.5×10^{-11}	1.2	5.5×10^{-11}	1.1	50
trans-2-Butene + OH (+M)$\rightarrow$ products			6.9×10^{-11}	1.2	7.0×10^{-11}	1.1	50
$CH_3O + NO$ (+M)$\rightarrow CH_3ONO$	2.3×10^{-29}	2.8	3.8×10^{-11}	0.6	2.9×10^{-11}	1.3	100
$CH_3O + NO_2$ (+M)$\rightarrow CH_3ONO_2$	5.3×10^{-29}	4.4	1.9×10^{-11}	1.8	1.7×10^{-11}	1.1	0
$C_2H_5O + NO$ (+M)$\rightarrow C_2H_5ONO$	2.8×10^{-27}	4	5.0×10^{-11}	0.2	4.8×10^{-11}	1.2	50
$C_2H_5O + NO_2$ (+M)$\rightarrow C_2H_5ONO_2$	2.0×10^{-27}	4	2.8×10^{-11}	1	2.7×10^{-11}	1.1	100
$CH_3O_2 + NO_2$ (+M)$\rightarrow CH_3O_2NO_2$	1.0×10^{-30}	4.8	7.2×10^{-12}	2.1	3.8×10^{-12}	1.5	100
$C_2H_5O_2 + NO_2$ (+M)$\rightarrow C_2H_5O_2NO_2$	1.2×10^{-29}	4	9.0×10^{-12}	0	7.5×10^{-12}	1.3	50
$CH_3C(O)O_2 + NO_2$ (+M)$\rightarrow CH_3C(O)O_2NO_2$	7.3×10^{-29}	4.1	9.5×10^{-12}	1.6	8.7×10^{-12}	1.1	0
$CH_3CH_2C(O)O_2 + NO_2$ (+M)$\rightarrow CH_3CH_2C(O)O_2NO_2$	9.0×10^{-28}	8.9	7.7×10^{-12}	0.2	7.4×10^{-12}	2	100
$CH_3C(O)CH_2 + O_2$ (+M)$\rightarrow CH_3C(O)CH_2O_2$	3×10^{-29}		1.0×10^{-12}		9.4×10^{-12}	1.3	
$CH_2OO + CH_2OO \rightarrow$ products	0	0	7×10^{-11}	0	7.0×10^{-11}	1.5	100
$HCN + OH$ (+M) $\rightarrow HC(OH)N$	6.1×10^{-33}	1.5	9.8×10^{-15}	-4.6	7.4×10^{-15}	1.25	0
FO_x Reactions							
$F + O_2$ (+M)$\rightarrow FO_2$	5.8×10^{-33}	1.7	1×10^{-10}	0	1.4×10^{-13}	1.3	100
$F + NO$ (+M)$\rightarrow FNO$	1.2×10^{-31}	0.5	2.8×10^{-10}	0	2.6×10^{-12}	1.4	200
$F + NO_2$ (+M)$\rightarrow FNO_2$	1.5×10^{-30}	2	1.0×10^{-11}	0	5.4×10^{-12}	1.3	100

	Low-pressure limit		High-pressure limit				
Reaction	k_0^{298}/cm^6 molecule^{-2} s^{-2}	n	k_∞^{298}/cm^6 molecule^{-2} s^{-2}	m	$k(T,[M])$ (298 K, 1 Atm)/ cm^3 molecule^{-1} s^{-1}	f_{298}	g
$FO + NO_2$ (+M)→ $FONO_2$	2.6 × 10^{-31}	1.3	2.0 × 10^{-11}	1.5	3.2 × 10^{-12}	3	200
$CF_3 + O_2$ (+M)→ CF_3O_2	3.0 × 10^{-29}	4	3.0 × 10^{-12}	1	2.8 × 10^{-12}	1.2	100
$CF_3O + NO_2$ (+M)→ CF_3ONO_2	1.7 × 10^{-28}	6.9	1.1 × 10^{-11}	1	1.0 × 10^{-11}	1.1	50
$CF_3O_2 + NO_2$ (+M)→ $CF_3O_2NO_2$	1.5 × 10^{-29}	2.2	9.6 × 10^{-12}	1	8.1 × 10^{-12}	1.1	50
$CF_3O + CO$ (+M)→ CF_3OCO	2.5 × 10^{-31}	2	6.8 × 10^{-14}	-1.2	6.0 × 10^{-14}	1.2	500
CF_3O (+M)→ $CF_2O + F$							
ClO_x Reactions							
$Cl + O_2$ (+M)→ $ClOO$	2.2 × 10^{-33}	3.1	1.8 × 10^{-10}	0	5.3 × 10^{-14}	1.1	50
$Cl + NO$ (+M)→ $ClNO$	7.6 × 10^{-32}	1.8			1.9 × 10^{-12}	1.2	50
$Cl + NO_2$ (+M)→ $ClONO$	1.3 × 10^{-30}	2	1 × 10^{-10}	1	1.6 × 10^{-11}	1.2	100
$Cl + NO_2$ (+M)→ $ClNO_2$	1.8 × 10^{-31}	2	1 × 10^{-10}	1	3.6 × 10^{-12}	1.3	100
$Cl + CO$ (+M)→ $ClCO$	1.3 × 10^{-33}	3.8			3.3 × 10^{-14}	1.1	50
$Cl + C_2H_2$ (+M)→ ClC_2H_2	5.2 × 10^{-30}	2.4	2.2 × 10^{-10}	0.7	5.0 × 10^{-11}	1.1	50
$Cl + C_2H_4$ (+M)→ ClC_2H_4	1.6 × 10^{-29}	3.3	3.1 × 10^{-10}	1	1.1 × 10^{-10}	1.5	50
$Cl + C_2Cl_4$ (+M)→ C_2Cl_5	1.4 × 10^{-28}	8.5	4.0 × 10^{-11}	1.2	3.6 × 10^{-11}	1.12	50
$ClO + NO_2$ (+M)→ $ClONO_2$	1.8 × 10^{-31}	3.4	1.5 × 10^{-11}	1.9	2.3 × 10^{-12}	1.3	50
$OClO + NO_3$ (+M)→ O_2ClONO_2							
$ClO + ClO$ (+M)→ Cl_2O_2	1.9 × 10^{-32}	3.6	3.7 × 10^{-12}	1.6	3.2 × 10^{-13}	1.15	0
$ClO + OClO$ (+M)→ Cl_2O_3	6.2 × 10^{-32}	4.7	2.4 × 10^{-11}	0	1.2 × 10^{-12}	1.1	25
$OClO + O$ (+M)→ ClO_3	2.9 × 10^{-31}	3.1	8.3 × 10^{-12}	0	2.3 × 10^{-12}	1.1	100
$CH_2Cl + O_2$ (+M)→ CH_2ClO_2	1.9 × 10^{-30}	3.2	2.9 × 10^{-12}	1.2	2.2 × 10^{-12}	1.1	125
$CHCl_2 + O_2$ (+M)→ $CHCl_2O_2$	1.3 × 10^{-30}	4	2.8 × 10^{-12}	1.4	2.0 × 10^{-12}	1.1	125
$CCl_3 + O_2$ (+M)→ CCl_3O_2	8 × 10^{-31}	6	3.5 × 10^{-12}	1	2.2 × 10^{-12}	1.2	50
$CFCl_2 + O_2$ (+M)→ $CFCl_2O_2$	5.0 × 10^{-30}	4	6.0 × 10^{-12}	1	4.8 × 10^{-12}	1.3	200
$CF_2Cl + O_2$ (+M)→ CF_2ClO_2	5.2 × 10^{-29}	5.6	1.0 × 10^{-11}	0.8	9.1 × 10^{-12}	2	300
$CCl_3O_2 + NO_2$ (+M)→ $CCl_3O_2NO_2$	2.9 × 10^{-29}	6.8	1.3 × 10^{-11}	1	1.1 × 10^{-11}	1.1	50
$CFCl_2O_2 + NO_2$ (+M)→ $CFCl_2O_2NO_2$	2.2 × 10^{-29}	5.8	1.0 × 10^{-11}	1	8.7 × 10^{-12}	1.1	50
$CF_2ClO_2 + NO_2$ (+M)→ $CF_2ClO_2NO_2$	1.1 × 10^{-29}	4.6	1.7 × 10^{-11}	1.2	1.3 × 10^{-11}	2	300
BrO_x Reactions							
$Br + NO_2$ (+M)→ products	4.2 × 10^{-31}	2.4	2.7 × 10^{-11}	0	4.9 × 10^{-12}	1.1	50
$BrO + NO_2$ (+M)→ $BrONO_2$	5.4 × 10^{-31}	3.1	6.5 × 10^{-12}	2.9	2.8 × 10^{-12}	1.2	400
$Br + CH_2{=}CHCH{=}CH_2$ (+M)→ products	1.1 × 10^{-28}	4.8	2.1 × 10^{-10}	0.6	1.6 × 10^{-10}	1.5	100
$Br + CH_2{=}C(CH_3)CH{=}CH_2$ (+M) → products	1.1 × 10^{-27}	2.5	2.0 × 10^{-10}	1.1	3.5 × 10^{-10}	1.5	100
IO_x Reactions							
$I + NO$ (+M)→ INO	1.8 × 10^{-32}	1	1.7 × 10^{-11}	0	3.8 × 10^{-13}	1.3	150
$I + NO_2$ (+M)→ INO_2	3.0 × 10^{-31}	1	6.6 × 10^{-11}	0	5.1 × 10^{-12}	1.2	300
$IO + NO_2$ (+M)→ $IONO_2$	7.5 × 10^{-31}	3.5	7.6 × 10^{-12}	1.5	3.5 × 10^{-12}	1.3	50
SO_x Reactions							
$HS + NO$ (+M)→ $HSNO$	2.4 × 10^{-31}	2.5	2.7 × 10^{-11}	0	3.4 × 10^{-12}	1.2	100
$CH_3S + NO$ (+M)→ CH_3SNO	3.2 × 10^{-29}	4	3.5 × 10^{-11}	1.8	2.8 × 10^{-11}	1.2	100
$SO_2 + O$ (+M)→ SO_3	1.8 × 10^{-33}	-2	4.2 × 10^{-14}	-1.8	1.3 × 10^{-14}	2	100
$SO_2 + OH$ (+M)→ $HOSO_2$	2.9 × 10^{-31}	4.1	1.7 × 10^{-12}	-0.2	9.5 × 10^{-13}	1.08	100
$CH_3S + O_2$ (+M)→ CH_3SOO							
$CH_3SCH_2 + O_2$ (+M)→ $CH_3SCH_2O_2$							
$SO_3 + CH_3$ (+M) → CH_3SO_3			2.2 × 10^{-13}	0.2	2.2 × 10^{-13}	10	50
SO_x Reactions							
$SO_3 + NH_3$ (+M)→ H_3NSO_3	3.6 × 10^{-30}	6.1	4.3 × 10^{-11}	0	1.9 × 10^{-11}	1.2	200
$CS_2 + HO$ (+M)→ HO---CS_2	4.9 × 10^{-31}	3.5	1.4 × 10^{-11}	1	4.0 × 10^{-12}	1.5	100
$CS_2 + Cl$ (+M)→ Cl---CS_2	5.9 × 10^{-31}	3.6	4.6 × 10^{-10}	0	1.2 × 10^{-11}	1.1	50
$(CH_3)_2S + HO$ (+M)→ HO---$(CH_3)_2S$	2.9 × 10^{-31}	6.24			7.4 × 10^{-12}	1.2	
$(CH_3)_2S + Cl$ (+M)→ Cl---$(CH_3)_2S$	4 × 10^{-28}	7	2 × 10^{-10}	1	1.7 × 10^{-10}	1.1	50
$(CH_3)_2S + Br$ (+M)→ Br---$(CH_3)_2S$	3.7 × 10^{-29}	5.3	1.5 × 10^{-10}	2	9.6 × 10^{-11}	1.1	100
Metal Reactions							
$Na + O_2$ (+M)→ NaO_2	3.2 × 10^{-30}	1.4	6.0 × 10^{-10}	0	5.3 × 10^{-11}	1.3	200
$NaO + O_2$ (+M)→ NaO_3	3.5 × 10^{-30}	2	5.7 × 10^{-10}	0	5.6 × 10^{-11}	1.3	200

Reaction	Low-pressure limit k_0^{298}/cm^6 molecule^{-2} s^{-2}	n	High-pressure limit k_∞^{298}/cm^6 molecule^{-2} s^{-2}	m	$k(T,[M])$ (298 K, 1 Atm)/ cm^3 molecule^{-1} s^{-1}	f_{298}	g
$NaO + CO_2$ (+M)→ $NaCO_3$	8.7×10^{-28}	3	6.5×10^{-10}	0.2	5.4×10^{-10}	1.3	200
$NaOH + CO_2$ (+M)→ $NaHCO_3$	1.0×10^{-28}	4	6.8×10^{-10}	0.2	3.6×10^{-10}	1.5	200
Hg + Br (+M)→ HgBr	1.5×10^{-32}	1.9			3.7×10^{-13}	1.5	100

Table 2.2 Rate Constants for Chemical Activation Reactions of Atmospheric Importance

Reactant	Products (Upper → Association)/ (Lower → Dissociation)	Low-pressure limit k_0^{298}/cm^6 molecule^{-2} s^{-2}	n	High-pressure limit k_∞^{298}/cm^6 molecule^{-2} s^{-2}	m	A/cm^3 molecule^{-1} s^{-1}	B/K	k_{Total} (298 K,1 Atm)/ cm^3 molecule^{-1} s^{-1}	f_{298}	g
NO_2	NO_2 + O (+M) → → NO_3 → NO + O_2	3.4×10^{-31}	1.6	2.3×10^{-11}	0.2	5.3×10^{-12}	-200	1.26×10^{-11}	1.06	80
$HONO_2$	$HONO_2$ + OH (+M) → → OH-$HONO_2$ → H_2O + NO_2	3.9×10^{-31}	7.2	1.5×10^{-13}	4.8	3.7×10^{-14}	-240	1.4×10^{-13}	1.12	50
CO	CO + OH (+M) → → HOCO → H + CO_2	6.9×10^{-33}	2.1	1.1×10^{-12}	-1.3	1.85×10^{-13}	65	2.4×10^{-13}	1.05	25
CH_3CH_2	CH_3CH_2 + O_2 (+M) → → $CH_3CH_2O_2$ → C_2H_4 + HO_2	1.3×10^{-28}	4.2	7.6×10^{-12}	1.4	1.8×10^{-13}	0	7.1×10^{-12}	1.15	50
$CH_3C(O)$	$CH_3C(O)$ + O_2 (+M) → → $CH_3C(O)OO$ → $CH_2C(O)O$ + OH	2.0×10^{-29}	2.6	5.3×10^{-12}	0.7	6.6×10^{-14}	-800	4.8×10^{-12}	1.05	50
CH_2I	CH_2I + O_2 (+M) → → CH_2IO_2 → CH_2OO + I	4.9×10^{-31}	2.5	1.6×10^{-12}	1.6	2.9×10^{-13}	-460	1.5×10^{-12}	1.08	50

TABLE 3. Equilibrium Constants for Reactions of Atmospheric Importance

Reaction	A/cm^3 molecule^{-1}	B/°K	K_{eq}(298)/cm^3 molecule^{-1}	f_{298}	g
$H_2O + H_2O \leftrightarrow (H_2O)_2$	8.9×10^{-24}	1622	2.0×10^{-21}	1.05	0
$HO + NO_2 \leftrightarrow HOONO$	3.5×10^{-27}	10135	2.2×10^{-12}	1.4	200
$HO_2 + NO_2 \leftrightarrow HO_2NO_2$	2.1×10^{-27}	10900	1.6×10^{-11}	1.3	100
$HO_2 + H_2O \leftrightarrow HO_2{\bullet}H_2O$	2.4×10^{-25}	4350	5.2×10^{-19}	2	200
$HO_2 + CH_3OH \leftrightarrow HO_2{\bullet}CH_3OH$	1.1×10^{-24}	4093	1.0×10^{-18}	1.2[a]	100
$HO_2 + CH_2O \leftrightarrow HOCH_2OO$	6.3×10^{-27}	7488	5.2×10^{-16}	1.2	300
$HO_2 + CH_3CHO \leftrightarrow CH_3CH(OH)OO$	5.0×10^{-28}	7130	1.7×10^{-17}	2	0
$HO_2 + CH_3C(O)CH_3 \leftrightarrow HO_2{\bullet}CH_3C(O)CH_3$	1.8×10^{-24}	4040	1.4×10^{-18}	1.6	0
$NO + NO_2 \leftrightarrow N_2O_3$	3.3×10^{-27}	4667	2.1×10^{-20}	2	100
$NO_2 + NO_2 \leftrightarrow N_2O_4$	5.9×10^{-29}	6643	2.8×10^{-19}	1.4	100
$NO_2 + NO_3 \leftrightarrow N_2O_5$	5.8×10^{-27}	10840	2.8×10^{-11}	1.2	75
$CH_3O_2 + NO_2 \leftrightarrow CH_3O_2NO_2$	9.5×10^{-29}	11234	2.2×10^{-12}	1.3	500
$CH_3C(O)O_2 + NO_2 \leftrightarrow CH_3C(O)O_2NO_2$	9.0×10^{-29}	14000	2.3×10^{-8}	1.2	200
$CH_3CH_2C(O)O_2 + NO_2 \leftrightarrow CH_3CH_2C(O)O_2NO_2$	9.0×10^{-29}	14000	2.3×10^{-8}	10	800
$CH_3C(O)CH_2 + O_2 \leftrightarrow CH_3C(O)CH_2O_2$	6.3×10^{-27}	12200	3.8×10^{-9}	1.5	0
$F + O_2 \leftrightarrow FOO$	4.5×10^{-25}	6118	3.7×10^{-16}	1.5	300
$Cl + O_2 \leftrightarrow ClOO$	1.24×10^{-24}	2370	3.2×10^{-21}	1.7	100
$Cl + CO \leftrightarrow ClCO$	3.5×10^{-25}	3730	9.6×10^{-20}	1.2	200
$ClO + O_2 \leftrightarrow ClO.O_2$	2.9×10^{-26}	<3700	$<7.2 \times 10^{-21}$		
$ClO + ClO \leftrightarrow Cl_2O_2$	2.16×10^{-27}	8537	6.0×10^{-15}	1.2	20
$ClO + ClO \leftrightarrow ClOClO$	6.5×10^{-27}	4460	2.0×10^{-20}	2	600
$Cl + OClO \leftrightarrow ClClO_2$	2.2×10^{-26}	13100	3.1×10^{-7}	2	600
$ClO + OClO \leftrightarrow Cl_2O_3$	1.5×10^{-27}	7170	4.2×10^{-17}	1.2	400
$OClO + NO_3 \leftrightarrow O_2ClONO_2$	6.6×10^{-29}	3971	4.0×10^{-23}	2.3	250
$Br + CH_2{=}CHC(CH_3){=}CH_2$ (Isoprene)	4.2×10^{-24}	7480	3.3×10^{-13}	1.3	100
$Br + CH_2{=}CHCH{=}CH_2$ (1,3-Butadiene)	1.1×10^{-24}	7520	8.6×10^{-14}	1.3	100

Reaction	A/cm³ molecule⁻¹	B/°K	K_{eq}(298)/cm³ molecule⁻¹	f_{298}	g
$OH + CS_2 \leftrightarrow CS_2OH$	4.5×10^{-25}	5140	1.4×10^{-17}	1.4	300
$CH_3S + O_2 \leftrightarrow CH_3SO_2$	1.8×10^{-27}	5545	2.2×10^{-19}	1.4	300
$Cl + CS_2 \leftrightarrow Cl{-}{-}{-}CS_2$	1.8×10^{-25}	4982	3.3×10^{-18}	1.3	150
$CH_3S + O_2 \leftrightarrow CH_3SOO$	1.4×10^{-26}	4850	1.7×10^{-19}	1.2	0
$OH + (CH_3)_2S \leftrightarrow HO{-}{-}{-}(CH_3)_2S$	9.6×10^{-27}	5376	6.6×10^{-19}	1.4	0
$Br + (CH_3)_2S \leftrightarrow Br{-}{-}{-}(CH_3)_2S$	3.4×10^{-25}	3021	4.6×10^{-15}	1.2	100
$IO + NO_2 \leftrightarrow IONO_2$	5.0×10^{-28}	14120	1.9×10^{-7}	2.5	300
$Hg + Br \leftrightarrow HgBr$	9.1×10^{-24}	7800	2.1×10^{-12}	2	200

[a] At 230 K.

IONIZATION CONSTANT OF WATER AT VARIOUS TEMPERATURES AND PRESSURES

Serguei N. Lvov and Allan H. Harvey

This table gives values of $pK_w = -\log_{10}(K_w)$, where K_w is the ionization constant (also known as the equilibrium constant) of the reaction $2H_2O \leftrightarrow H_3O^+(aq) + OH^-(aq)$. K_w is defined as

$$K_w = a_{H_3O^+} a_{OH^-} / a_{H_2O}^2$$

where a_i is the dimensionless activity of species i. The activities are on the molality basis for ions and mole fraction basis for water molecules. It is assumed that the activity of $H_3O^+(aq)$ is the same as the activity of $H^+(aq)$, so that K_w is numerically equal to

$$a_{H^+} a_{OH^-} / a_{H_2O}^2$$

the equilibrium constant for the ionization reaction of water, $H_2O = H^+(aq) + OH^-(aq)$, that is most commonly used in the literature.

Values in the table are calculated using an analytical equation given in Refs. 1-3 where K_w is presented as a function of temperature and density from 0 °C to 800 °C and 0 g cm^{-3} to 1.25 g cm^{-3}.

References

1. International Association for the Properties of Water and Steam, *Release on the Ionization Constant of H_2O* (2007), available from <www.iapws.org>.
2. Bandura, A.V., and Lvov, S. N., *J. Phys. Chem. Ref. Data* 35, 15, 2006. <https://doi.org/10.1063/1.1928231>
3. International Association for the Properties of Steam and Water, IAPWS R11-07, *Revised Release on the Ionization Constant of H_2O* (2019), available from <www.iapws.org/relguide/ionization.pdf>.

Ionization Constant of Water as a Function of Temperature and Pressure — Values of $pK_w = -\log_{10}K_w$

p/MPa	0 °C	25 °C	50 °C	75 °C	100 °C	150 °C	200 °C	250 °C	300 °C	400 °C	600 °C	700 °C	800 °C
0.1[a]	14.946	13.995	13.264	12.696	12.252[a]	11.641[a]	11.310[a]	11.205[a]	11.339[a]	47.961	46.384	43.925	40.785
25	14.848	13.908	13.181	12.613	12.165	11.543	11.189	11.050	11.125	16.566	19.425	19.829	20.113
50	14.754	13.824	13.102	12.533	12.084	11.450	11.076	10.898	10.893	11.557	15.621	16.279	16.693
75	14.665	13.745	13.026	12.458	12.006	11.364	10.974	10.769	10.715	11.045	13.507	14.301	14.791
100	14.580	13.668	12.953	12.385	11.933	11.283	10.880	10.655	10.568	10.744	12.296	13.040	13.544
150	14.422	13.524	12.815	12.249	11.795	11.135	10.713	10.458	10.327	10.345	11.117	11.613	12.032
200	14.278	13.390	12.687	12.123	11.668	11.000	10.564	10.289	10.131	10.063	10.513	10.853	11.171
250	14.145	13.265	12.567	12.004	11.549	10.876	10.430	10.140	9.963	9.839	10.112	10.360	10.609
300	14.021	13.148	12.453	11.892	11.437	10.760	10.306	10.005	9.814	9.651	9.810	9.998	10.199
350	13.906	13.037	12.346	11.786	11.331	10.651	10.191	9.881	9.679	9.487	9.567	9.712	9.877
400	13.797	12.932	12.243	11.685	11.230	10.548	10.083	9.766	9.555	9.341	9.361	9.475	9.613
500	13.595	12.736	12.052	11.496	11.042	10.356	9.884	9.557	9.332	9.086	9.024	9.094	9.191
600	13.411	12.556	11.875	11.322	10.868	10.181	9.703	9.369	9.135	8.866	8.749	8.790	8.861
700	13.240	12.389	11.710	11.159	10.705	10.018	9.537	9.197	8.956	8.670	8.514	8.536	8.587
800	13.080	12.233	11.556	11.006	10.553	9.865	9.381	9.037	8.791	8.493	8.308	8.314	8.352
900	12.930	12.085	11.410	10.861	10.410	9.721	9.236	8.888	8.638	8.330	8.122	8.117	8.144
1000	12.788	11.946	11.272	10.725	10.273	9.585	9.098	8.748	8.495	8.178	7.952	7.939	7.957

[a] *Note*: Pressure for first row is 0.1 MPa at $t < 100$ °C and $t \geq 400$ °C, or p_s (saturated liquid) for 100 °C $\leq t \leq 350$ °C.

IONIZATION CONSTANT OF NORMAL AND HEAVY WATER AT SATURATED VAPOR PRESSURE

Serguei N. Lvov and Allan H. Harvey

This table gives the ionization constant for liquid H_2O and D_2O at temperatures from 0 °C to 100 °C at the saturated vapor pressure. The quantity tabulated is $pK_w = -\log_{10}(K_w)$, where

$$K_w = a_{H_3O^+} a_{OH^-} / a_{H_2O}^2$$

for H_2O and

$$K_w = a_{D_3O^+} a_{OD^-} / a_{D_2O}^2$$

or D_2O. Values in the table are calculated using analytical equations given in Refs. 1-3 for H_2O and Ref. 4 for D_2O. Column definitions are as follows.

Column heading	**Definition**
Temperature, in °C	
$pK_w(H_2O)$	$-\log_{10}(K_w)$ for normal water (H_2O)
$pK_w(D_2O)$	$-\log_{10}(K_w)$ for heavy water (D_2O)

References

1. International Association for the Properties of Water and Steam, *Release on the Ionization Constant of H_2O* (2007), available from <www.iapws.org>.
2. Bandura, A. V., and Lvov, S. N., *J. Phys. Chem. Ref. Data* 35, 15, 2006. <https://doi.org/10.1063/1.1928231>
3. International Association for the Properties of Steam and Water, IAPWS R11-07, *Revised Release on the Ionization Constant of H_2O* (2019), available from <www.iapws.org/relguide/ionization.pdf>.
4. Mesmer, R. E., and Herting, D. L., *J. Solution Chem.* 7, 901, 1978. <https://doi.org/10.1007/BF00645300>

Ionization Constant of Normal and Heavy Water as a Function of Temperature — Values of $pK_w = -\log_{10}K_w$

t/°C	$pK_w(H_2O)$	$pK_w(D_2O)$
0	14.947	15.972
5	14.734	15.743
10	14.534	15.527
15	14.344	15.324
20	14.165	15.132
25	13.995	14.951
30	13.833	14.779
35	13.680	14.616
40	13.535	14.462
45	13.396	14.316
50	13.265	14.176
55	13.140	14.044
60	13.020	13.918
65	12.907	13.798
70	12.799	13.683
75	12.696	13.574
80	12.598	13.470
85	12.505	13.371
90	12.417	13.276
95	12.332	13.186
100	12.252	13.099

ELECTRICAL CONDUCTIVITY OF WATER

This table gives the electrical conductivity (κ) of highly purified water over a range of temperature and pressure. The column definitions are as follows.

Column heading	**Definition**
t	Temperature in °C
κ	Electrical conductivity, in units μS cm^{-1}; value in first column is at its own vapor pressure; in other columns, value is for pressure indicated within parentheses

Equations for calculating the conductivity at any temperature and pressure may be found in the references.

Reference

1. Marshall, W. L., *J. Chem. Eng. Data* 32, 221, 1987. <https://doi.org/10.1021/je00048a027>
2. International Association for the Properties of Water and Steam (IAPWS) Guideline Statement G1-90, *Electrolytic Conductivity (Specific Conductance) of Liquid and Dense Supercritical Water from 0 °C to 800 °C and Pressures up to 1000 MPa* (1990), available from <www.iapws.org/relguide/conduct.pdf>.

Electrical Conductivity of Water at Various Temperatures and Pressures

t/°C	κ(sat. vap.)/μS cm^{-1}	κ(50 MPa)/μS cm^{-1}	κ(100 MPa)/μS cm^{-1}	κ(200 MPa)/μS cm^{-1}	κ(400 MPa)/μS cm^{-1}	κ(600 MPa)/μS cm^{-1}
0	0.0115	0.0150	0.0189	0.0275	0.0458	0.0667
25	0.0550	0.0686	0.0836	0.117	0.194	0.291
100	0.765	0.942	1.13	1.53	2.45	3.51
200	2.99	4.08	5.22	7.65	13.1	19.5
300	2.41	4.87	7.80	14.1	28.9	46.5
400		1.17	4.91	14.3	39.2	71.3
600			0.134	4.65	33.8	85.7

ELECTRICAL CONDUCTIVITY OF AQUEOUS SOLUTIONS

The following table gives the electrical conductivity (κ) of aqueous solutions of some acids, bases, and salts as a function of concentration. All values refer to 20 °C. The conductivity (often called specific conductance in older literature) is the reciprocal of the resistivity. The molar conductivity Λ is related to this by $\Lambda = \kappa/c$, where c is the amount-of-substance concentration of the electrolyte. Thus, if κ has units of millisiemens per centimeter (mS/cm), as in this table, and c is expressed in mol/L, then Λ has units of S cm^2 mol^{-1}. For these electrolytes, the concentration c corresponding to the mass percent values given here can be found in the table "Concentrative Properties of Aqueous Solutions" in this section.

Column definitions are as follows.

Column heading	Definition
Name	Name of solute
Mol. form.	Solute formula
κ(mass %)	Electrical conductivity at 20 °C, in units mS cm^{-1}; concentration in mass percent indicated by value in parentheses

References

1. *CRC Handbook of Chemistry and Physics, 70th Edition*, Weast, R. C., Ed., CRC Press, Boca Raton, FL, 1989, p. D-221.
2. Wolf, A. V., *Aqueous Solutions and Body Fluids*, Harper and Row, New York, 1966.

Electrical Conductivity κ of Aqueous Solutions in mS/cm at 20 °C for Indicated Concentration in Mass Percent

Name	Mol. form.	κ(0.5%)	κ(1%)	κ(2%)	κ(5%)	κ(10%)	κ(15%)	κ(20%)	κ(25%)	κ(30%)	κ(40%)	κ(50%)
Acetic acid	CH_3COOH	0.3	0.6	0.8	1.2	1.5	1.7	1.7	1.6	1.4	1.1	0.8
Ammonia	NH_3	0.5	0.7	1.0	1.1	1.0	0.7	0.5	0.4			
Ammonium chloride	NH_4Cl	10.5	20.4	40.3	95.3	180						
Ammonium sulfate	$(NH_4)_2SO_4$	7.4	14.2	25.7	57.4	105	147	185	215			
Barium chloride	$BaCl_2$	4.7	9.1	17.4	40.4	76.7	109.0	137.0				
Calcium chloride	$CaCl_2$	8.1	15.7	29.4	67.0	117	157	177	183	172	106	
Cesium chloride	$CsCl$	3.8	7.4	13.8	32.9	65.8	102	142				
Citric acid	$C(OH)(COOH)_3$	1.2	2.1	3.0	4.7	6.2	7.0	7.2	7.1			
Copper(II) sulfate	$CuSO_4$	2.9	5.4	9.3	19.0	32.2	42.3					
Formic acid	$HCOOH$	1.4	2.4	3.5	5.6	7.8	9.0	9.9	10.4	10.5	9.9	8.6
Hydrogen chloride	HCl	45.1	92.9	183								
Lithium chloride	$LiCl$	10.1	19.0	34.9	76.4	127	155	170	165	146		
Magnesium chloride	$MgCl_2$	8.6	16.6	31.2	66.9	108	129	134	122	98		
Magnesium sulfate	$MgSO_4$	4.1	7.6	13.3	27.4	42.7	54.2	51.1	44.1			
Manganese(II) sulfate	$MnSO_4$		6.2	10.6	21.6	34.5	43.7	47.6				
Nitric acid	HNO_3	28.4	56.1	108								
Oxalic acid	$H_2C_2O_4$	14.0	21.8	35.3	65.6							
Phosphoric acid	H_3PO_4	5.5	10.1	16.2	31.5	59.4	88.4	118	146	173	209	
Potassium bromide	KBr	5.2	10.2	19.5	47.7	95.6	144	194				
Potassium carbonate	K_2CO_3	7.0	13.6	25.4	58.0	109	152	188	223			
Potassium chloride	KCl	8.2	15.7	29.5	71.9	143	208					
Potassium dihydrogen phosphate	KH_2PO_4	3.0	5.9	11.0	25.0	44.6						
Potassium hydrogen carbonate	$KHCO_3$	4.6	8.9	17.0	38.8	72.4	101	128				
Potassium hydrogen phosphate	K_2HPO_4	5.2	9.9	18.3	40.3							
Potassium hydroxide	KOH	20.0	38.5	75.0	178							
Potassium iodide	KI	3.8	7.5	14.2	35.2	71.8	110	188	224			
Potassium nitrate	KNO_3	5.5	10.7	20.1	47.0	87.3	124	157	182			
Potassium permanganate	$KMnO_4$	3.5	6.9	13.0	30.5							
Potassium sulfate	K_2SO_4	5.8	11.2	21.0	48.0	88.6						
Silver(I) nitrate	$AgNO_3$	3.1	6.1	12.0	26.7	49.8	72.0	92.8	112	129	162	
Sodium acetate	$NaC_2H_3O_2$	3.9	7.6	14.4	30.9	53.4	64.1	69.3	69.2	64.3		
Sodium bromide	$NaBr$	5.0	9.7	18.4	44.0	84.6	122	157	191	216		
Sodium carbonate	Na_2CO_3	7.0	13.1	23.3	47.0	74.4	88.6					
Sodium chloride	$NaCl$	8.2	16.0	30.2	70.1	126	171	204	222			
Sodium dihydrogen phosphate	NaH_2PO_4	2.2	4.4	9.1	21.0	33.2	43.3	49.6	53.1	54.0	46.1	
Sodium hydrogen carbonate	$NaHCO_3$	4.2	8.2	15.0	31.4							
Sodium hydrogen phosphate	Na_2HPO_4	4.6	8.7	15.6	31.4							
Sodium hydroxide	$NaOH$	24.8	48.6	93.1	206							
Sodium nitrate	$NaNO_3$	5.4	10.6	20.4	46.2	82.6	111	134	152	165	178	
Sodium phosphate	Na_3PO_4	7.3	14.1	22.7	43.5							
Sodium sulfate	Na_2SO_4	5.9	11.2	19.8	42.7	71.3	91.1	109				
Sodium thiosulfate	$Na_2S_2O_3$	5.7	10.7	19.5	43.3	76.7	104	123	134	136	118	
Strontium chloride	$SrCl_2$	5.9	11.4	22.0	49.1	91.5	127	153	168	178		
Sulfuric acid	H_2SO_4	24.3	47.8	92	211							
Trichloroacetic acid	CCl_3COOH	10.3	19.6	37.2	84.7	148	193	221				
Trisodium citrate			7.4	12.8	26.2	42.1	52.0	57.1	57.3	53.5		
Zinc sulfate	$ZnSO_4$	2.8	5.4	10.0	20.5	33.7	43.3					

Thermochem

STANDARD KCl SOLUTIONS FOR CALIBRATING ELECTRICAL CONDUCTIVITY CELLS

This table presents recommended electrIcal conductivity (κ) values for aqueous potassium chloride solutions with molalities of 0.01 mol kg^{-1}, 0.1 mol kg^{-1}, and 1.0 mol kg^{-1} at temperatures from 0 °C to 50 °C. The values, which are based on measurements at the National Institute of Standards and Technology, provide primary standards for the calibration of conductivity cells. The measurements at 0.01 and 0.1 molal are described in Ref. 1, while those at 1.0 molal are in Ref. 2. Temperatures are given on the ITS-90 scale. The uncertainty in the conductivity is about 0.03% for the 0.01 molal values and about 0.04% for the 0.1 and 1.0 molal values.

Column heading	Definition
t	Temperature, in °C
κ(concentration)	Electrical conductivity, in units μS m^{-1}; in columns 2-4, concentration values are indicated in parentheses; in the last column, the value is for water saturated with atmospheric CO_2

Conductivity values in the last column were subtracted from the original measurements to give the values in the preceding columns for KCl (aq).

The assistance of Kenneth W. Pratt is appreciated.

References

1. Wu, Y. C., Koch, W. F., and Pratt, K. W., *J. Res. Natl. Inst. Stand. Technol.* 96, 191, 1991. <https://doi.org/10.6028/jres.096.008>
2. Wu, Y. C., Koch, W. F., Feng, D., Holland, L. A., Juhasz, E., Arvay, E., and Tomek, A., *J. Res. Natl. Inst. Stand. Technol.* 99, 241, 1994. <https://doi.org/10.6028/jres.099.019>
3. Pratt, K. W., Koch, W. F., Wu, Y. C., and Berezansky, P. A., *Pure Appl. Chem.* 73, 1783, 2001. <https://doi.org/10.1351/pac200173111783>

Electrical Conductivity of Aqueous Potassium Chloride at Various Concentrations and Temperatures

t/°C	κ(0.01 m)/ μS m^{-1}	κ(0.1 m)/ μS m^{-1}	κ(1.0 m)/ μS m^{-1}	$\kappa(H_2O)$/ μS m^{-1}
0	772.92	7116.85	63488	0.58
5	890.96	8183.70	72030	0.68
10	1013.95	9291.72	80844	0.79
15	1141.45	10437.1	89900	0.89
18	1219.93	11140.6		0.95
20	1273.03	11615.9	99170	0.99
25	1408.23	12824.6	108620	1.10
30	1546.63	14059.2	118240	1.20
35	1687.79	15316.0	127970	1.30
40	1831.27	16591.0	137810	1.40
45	1976.62	17880.6	147720	1.51
50	2123.43	19180.9	157670	1.61

MOLAR ELECTRICAL CONDUCTIVITY OF AQUEOUS HF, HCl, HBr, AND HI

These tables contain values for the molar electrical conductivity Λ of aqueous hydrohalogen acids. The molar electrical conductivity Λ of an electrolyte solution is defined as the electrical conductivity divided by amount-of-substance concentration. The customary unit is S cm^2 mol^{-1} (i.e., Ω^{-1} cm^2 mol^{-1}). Table 1 gives the molar electrical conductivity of the hydrohalogen acids at 25 °C as a function of the concentration. Tables 2 and 3 give both the temperature and concentration dependence of Λ for HCl and HBr, respectively. More extensive tables and mathematical representations may be found in the reference.

Reference

Hamer, W. J., and DeWane, H. J., *Electrolytic Conductance and the Conductances of the Hydrohalogen Acids in Water,* Natl. Stand. Ref. Data Sys., Natl. Bur. Standards (U.S.), No. 33, 1970. <https://doi.org/10.6028/NBS.NSRDS.33>

TABLE 1. Molar Electrical Conductivity Λ of Aqueous Hydrohalogen Acids at 25 °C in Units of S cm^2 mol^{-1} for the Specified Concentrations Expressed as Molarity (mol/L)

	Inf. dil.	0.0001 M	0.001 M	0.005 M	0.01 M	0.05 M	0.1 M	0.5 M	1.0 M	2.0 M	3.0 M	4.0 M	5.0 M	6.0 M	7.0 M	8.0 M	9.0 M	10.0 M
HF	405.1			128.1	96.1	50.1	39.1	26.3	24.3									
HCl	426.1	424.5	421.2	415.7	411.9	398.9	391.1	360.7	332.2	281.4	237.6	200.0	167.4	139.7	116.9	98.2	83.1	70.7
HBr	427.7	425.9	422.9	417.6	413.7	400.4	391.9	361.9	334.5	281.7	236.8	199.4	166.5	138.2	114.2	94.4		
HI	426.4	424.6	421.7	416.4	412.8	400.8	394.0	369.8	343.9	288.9	237.9	195.1	160.4	131.7	105.7			

TABLE 2. Temperature and Concentration Dependence of Molar Electrical Conductivity Λ for HCl in Units of S cm^2 mol^{-1} for HCl

c/mol L^{-1}	−20 °C	−10 °C	0 °C	10 °C	20 °C	30 °C	40 °C	50 °C
0.5			228.7	283.0	336.4	386.8	436.9	482.4
1.0			211.7	261.6	312.2	359.0	402.9	445.3
1.5			196.2	241.5	287.5	331.1	371.6	410.8
2.0			182.0	222.7	262.9	303.3	342.4	378.2
2.5		131.7	168.5	205.1	239.8	277.0	315.2	347.6
3.0		120.8	154.6	188.5	219.3	253.3	289.3	319.0
3.5	85.5	111.3	139.6	172.2	201.6	232.9	263.9	292.1
4.0	79.3	102.7	129.2	158.1	185.6	214.2	242.2	268.2
4.5	73.7	94.9	119.5	145.4	170.6	196.6	222.5	246.7
5.0	68.5	87.8	110.3	133.5	156.6	180.2	204.1	226.5
5.5	63.6	81.1	101.7	122.5	143.6	165.0	187.1	207.7
6.0	58.9	74.9	93.7	112.3	131.5	151.0	171.3	190.3
6.5	54.4	69.1	86.2	103.0	120.4	138.2	156.9	174.3

c/mol L^{-1}	−20 °C	−10 °C	0 °C	10 °C	20 °C	30 °C	40 °C	50 °C
7.0	50.2	63.7	79.3	94.4	110.2	126.4	143.3	159.7
7.5	46.3	58.6	73.0	86.5	100.9	115.7	131.6	146.2
8.0	42.7	54.0	67.1	79.4	92.4	106.1	120.6	134.0
8.5	39.4	49.8	61.7	72.9	84.7	97.3	110.7	123.0
9.0	36.4	45.9	56.8	67.1	77.8	89.4	101.7	112.9
9.5	33.6	42.3	52.3	61.8	71.5	82.3	93.6	103.9
10.0	31.2	39.1	48.2	57.0	65.8	75.9	86.3	95.7
10.5	28.9	36.1	44.5	52.7	60.7	70.1	79.6	88.4
11.0	26.8	33.4	41.1	48.8	56.1	64.9	73.6	81.7
11.5	24.9	31.0	38.0	45.3	51.9	60.1	68.0	75.6
12.0	23.1	28.7	35.3	42.0	48.0	55.6	62.8	70.0
12.5	21.4	26.7	32.7	39.0	44.4	51.4	57.9	64.8

TABLE 3. Temperature and Concentration Dependence of Molar Electrical Conductivity Λ of HBr in Units of S cm^2 mol^{-1} for HBr

c/mol L^{-1}	−20 °C	−10 °C	0 °C	10 °C	20 °C	30 °C	40 °C	50 °C
0.5			240.9	295.9	347.0	398.9	453.6	496.8
1.0			229.6	276.0	329.0	380.4	418.6	465.2
1.5			209.5	254.9	298.9	340.6	381.8	421.4
2.0		150.8	188.6	231.3	271.8	314.1	350.5	387.4
2.5		136.8	171.7	208.3	244.8	281.7	316.0	349.1
3.0		125.7	157.2	189.5	222.2	255.0	287.8	318.6
3.5		116.1	144.1	174.6	203.2	234.4	263.7	291.9

c/mol L^{-1}	−20 °C	−10 °C	0 °C	10 °C	20 °C	30 °C	40 °C	50 °C
4.0	84.0	107.5	132.3	160.2	186.8	214.2	239.7	266.9
4.5	78.0	99.0	123.0	146.4	171.2	195.1	218.8	242.6
5.0	72.3	91.4	112.6	134.0	155.7	178.2	199.6	221.3
5.5	67.0	84.2	103.1	122.7	142.1	162.8	181.4	201.8
6.0	61.8	77.2	94.3	112.0	129.6	148.0	165.4	183.4
6.5	56.8	70.7	86.0	102.0	118.0	134.1	150.5	166.3
7.0	51.9	64.6	78.4	92.6	107.1	121.4	136.3	150.8

MOLAR ELECTRICAL CONDUCTIVITY OF ELECTROLYTES IN AQUEOUS SOLUTION

Petr Vanýsek

This table gives the molar (equivalent) electrical conductivity Λ at 25 °C for some common electrolytes in aqueous solution at concentrations up to 0.1 M (0.1 mol/L).

For very dilute solutions, the molar electrical conductivity for any electrolyte of concentration c can be approximately calculated using the Debye–Hückel–Onsager equation, which can be written for a symmetrical (equal charge on cation and anion) electrolyte as

$$\Lambda = \Lambda^\circ - (A + B\Lambda^\circ)c^{1/2}$$

For a solution at 25 °C and both cation and anion with charge |1|, the constants are A = 60.20 and B = 0.229. Λ° can be found from the "inf. dil." column in the table below or calculated from the ion values in the more extensive table "Ionic Conductivity and Diffusion at Infinite Dilution" in this section. The equation is reliable for c <0.001 mol/L; with higher concentration, the error increases.

Column definitions for the table are as follows.

Column heading	Definition
Mol. form.	Electrolyte molecular formula; electrolytes are listed alphabetically by molecular formula
Λ°(inf. dil.)	Molar electrical conductivity at infinite dilution; in units S cm^2 mol^{-1}
Λ(concentration)	Molar electrical conductivity at concentration as indicated within parentheses; in units cm^2 S mol^{-1}

Molar Electrical Conductivity of Electrolytes in Aqueous Solution at 25 °C as a Function of Concentration

Mol. form.	Λ°(inf. dil.)/ S cm^2 mol^{-1}	Λ(0.0005 M)/ S cm^2 mol^{-1}	Λ(0.001 M)/ S cm^2 mol^{-1}	Λ(0.005 M)/ S cm^2 mol^{-1}	Λ(0.01 M)/ S cm^2 mol^{-1}	Λ(0.02 M)/ S cm^2 mol^{-1}	Λ(0.05 M)/ S cm^2 mol^{-1}	Λ(0.1 M)/ S cm^2 mol^{-1}
$AgNO_3$	133.29	131.29	130.45	127.14	124.70	121.35	115.18	109.09
$1/2BaCl_2$	139.91	135.89	134.27	127.96	123.88	119.03	111.42	105.14
$1/2CaCl_2$	135.77	131.86	130.30	124.19	120.30	115.59	108.42	102.41
$1/2Ca(OH)_2$	258			233	226	214		
$CuSO_4$	133.6	121.6	115.20	94.02	83.08	72.16	59.02	50.55
HCl	425.95	422.53	421.15	415.59	411.80	407.04	398.89	391.13
KBr	151.9	149.8	148.9	146.02	143.36	140.41	135.61	131.32
KCl	149.79	147.74	146.88	143.48	141.20	138.27	133.30	128.90
$KClO_4$	139.97	138.69	137.80	134.09	131.39	127.86	121.56	115.14
$1/3K_3Fe(CN)_6$	174.5	166.4	163.1	150.7				
$1/4K_4Fe(CN)_6$	184		167.16	146.02	134.76	122.76	107.65	97.82
$KHCO_3$	117.94	116.04	115.28	112.18	110.03	107.17		
KI	150.31	148.2	143.32	144.30	142.11	139.38	134.90	131.05
KIO_4	127.86	125.74	124.88	121.18	118.45	114.08	106.67	98.2
$KMnO_4$	134.8	132.7	131.9		126.5			113
KNO_3	144.89	142.70	141.77	138.41	132.75	132.34	126.25	120.34
KOH	271.5		234	230	228		219	213
$KReO_4$	128.20	126.03	125.12	121.31	118.49	114.49	106.40	97.40
$1/3LaCl_3$	145.9	139.6	137.0	127.5	121.8	115.3	106.2	99.1
LiCl	114.97	113.09	112.34	109.35	107.27	104.60	100.06	95.81
$LiClO_4$	105.93	104.13	103.39	100.52	98.56	96.13	92.15	88.52
$1/2MgCl_2$	129.34	125.55	124.15	118.25	114.49	109.99	103.03	97.05
NH_4Cl	149.6	147.5	146.7	143.9	141.21	138.25	133.22	128.69
$NaC_2H_3O_2$	91.0	89.2	88.5	85.68	83.72	81.20	76.88	72.76
Na picrate	80.45	78.7	78.6	75.7	73.7		66.3	61.8
NaCl	126.39	124.44	123.68	120.59	118.45	115.70	111.01	106.69
$NaClO_4$	117.42	115.58	114.82	111.70	109.54	106.91	102.35	98.38
NaI	126.88	125.30	124.19	121.19	119.18	116.64	112.73	108.73
NaOH	247.7	245.5	244.6	240.7	237.9			
$1/2Na_2SO_4$	129.8	125.68	124.09	117.09	112.38	106.73	97.70	89.94
$1/2SrCl_2$	135.73	131.84	130.27	124.18	120.23	115.48	108.20	102.14
$ZnSO_4$	132.7	121.3	114.47	95.44	84.87	74.20	61.17	52.61

IONIC CONDUCTIVITY AND DIFFUSION AT INFINITE DILUTION

Petr Vanýsek

This table gives the molar (equivalent) conductivity λ for common ions at infinite dilution. All values refer to aqueous solutions at 25 °C. It also lists the diffusion coefficient D of the ion in dilute aqueous solution, which is related to λ through the equation

$$D = (RT/F^2)(\lambda/|z|)$$

where R is the molar gas constant, T the temperature, F the Faraday constant, and z the charge on the ion. The variation with temperature is fairly sharp; for typical ions, λ and D increase by 2 to 3% per degree as the temperature increases from 25 °C. Column definitions for the table are as follows.

Column heading	Definition
Mol. form.	Molecular formula of ion; ions are listed alphabetically by molecular formula; in the table, values are categorized in the following order: Inorganic Cations, Inorganic Anions, Organic Cations, Organic Anions
$\Lambda_\pm$	Molar (equivalent) ionic conductivity per unit charge, in units 10^{-4} m^2 S mol^{-1}
D	Diffusion coefficient, in units 10^{-5} cm^2 s^{-1}

The diffusion coefficient for a salt, D_{salt}, may be calculated from the D_+ and D_- values of the constituent ions by the relation

$$D_{salt} = \frac{(z_+ + |z_-|)D_+D_-}{z_+D_+ + |z_-|D_-}$$

For solutions of simple, pure electrolytes (one positive and one negative ionic species), such as NaCl, equivalent ionic conductivity $\Lambda°$, which is the molar conductivity per unit concentration of charge, is defined as

$$\Lambda° = \Lambda_+ + \Lambda_-$$

where Λ_+ and Λ_- are equivalent ionic conductivities of the cation and anion. The more general formula is

$$\Lambda° = \nu_+\Lambda_+ + \nu_-\Lambda_-$$

where ν_+ and ν_- refer to the number of moles of cations and anions, respectively, to which one mole of the electrolyte gives rise in solution.

References

1. Gray, D. E., Ed., *American Institute of Physics Handbook*, McGraw-Hill, New York, 1972, 2–226.
2. Robinson, R. A., and Stokes, R. H., *Electrolyte Solutions*, Butterworths, London, 1959.
3. Lobo, V. M. M., and Quaresma, J. L., *Handbook of Electrolyte Solutions*, Physical Science Data Series 41, Elsevier, Amsterdam, 1989. <https://doi.org/10.1016/S0003-2670(00)83965-3>
4. Conway, B. E., *Electrochemical Data*, Elsevier, Amsterdam, 1952.
5. Milazzo, G., *Electrochemistry: Theoretical Principles and Practical Applications*, Elsevier, Amsterdam, 1963.

Ionic Conductivity and Diffusion Coefficients of Inorganic and Organic Anions and Cations at Infinite Dilution

Mol. form.	$\Lambda_\pm$/ 10^{-4} m^2 S mol^{-1}	D/ 10^{-5} cm^2 s^{-1}
Inorganic Cations		
Ag^+	61.9	1.648
$1/3Al^{3+}$	61	0.541
$1/2Ba^{2+}$	63.6	0.847
$1/2Be^{2+}$	45	0.599
$1/2Ca^{2+}$	59.47	0.792
$1/2Cd^{2+}$	54	0.719
$1/3Ce^{3+}$	69.8	0.620
$1/2Co^{2+}$	55	0.732
$1/3[Co(NH_3)_6]^{3+}$	101.9	0.904
$1/3[Co(en)_3]^{3+}$*	74.7	0.663
$1/6[Co_2(trien)_3]^{6+}$**	69	0.306
$1/3Cr^{3+}$	67	0.595
Cs^+	77.2	2.056
$1/2Cu^{2+}$	53.6	0.714
D^+	249.9	6.655
$1/3Dy^{3+}$	65.6	0.582
$1/3Er^{3+}$	65.9	0.585
$1/3Eu^{3+}$	67.8	0.602
$1/2Fe^{2+}$	54	0.719
$1/3Fe^{3+}$	68	0.604
$1/3Gd^{3+}$	67.3	0.597
H^+	349.65	9.311
$1/2Hg^{2+}$	68.6	0.913
$1/2Hg^{2+}$	63.6	0.847
$1/3Ho^{3+}$	66.3	0.589
K^+	73.48	1.957
$1/3La^{3+}$	69.7	0.619
Li^+	38.66	1.029
$1/2Mg^{2+}$	53.0	0.706
$1/2Mn^{2+}$	53.5	0.712
NH_4^+	73.5	1.957
$N_2H_5^+$	59	1.571
Na^+	50.08	1.334
$1/3Nd^{3+}$	69.4	0.616
$1/2Ni^{2+}$	49.6	0.661
$1/4[Ni_2(trien)_3]^{4+}$**	52	0.346
$1/2Pb^{2+}$	71	0.945
$1/3Pr^{3+}$	69.5	0.617
$1/2Ra^{2+}$	66.8	0.889
Rb^+	77.8	2.072
$1/3Sc^{3+}$	64.7	0.574
$1/3Sm^{3+}$	68.5	0.608
$1/2Sr^{2+}$	59.4	0.791
Tl^+	74.7	1.989
$1/3Tm^{3+}$	65.4	0.581
$1/2UO_2^{2+}$	32	0.426

Mol. form.	$\Lambda_\pm$/ 10^{-4} m^2 S mol^{-1}	D/ 10^{-5} cm^2 s^{-1}
$1/3Y^{3+}$	62	0.550
$1/3Yb^{3+}$	65.6	0.582
$1/2Zn^{2+}$	52.8	0.703
Inorganic Anions		
$Au(CN)_2^-$	50	1.331
$Au(CN)_4^-$	36	0.959
$B(C_6H_5)_4^-$	21	0.559
Br^-	78.1	2.080
Br_3^-	43	1.145
BrO_3^-	55.7	1.483
CN^-	78	2.077
CNO^-	64.6	1.720
$1/2CO_3^{2-}$	69.3	0.923
Cl^-	76.31	2.032
ClO_2^-	52	1.385
ClO_3^-	64.6	1.720
ClO_4^-	67.3	1.792
$1/3[Co(CN)_6]^{3-}$	98.9	0.878
$1/2CrO_4^{2-}$	85	1.132
F^-	55.4	1.475
$1/4[Fe(CN)_6]^{4-}$	110.4	0.735
$1/3[Fe(CN)_6]^{3-}$	100.9	0.896
$H_2AsO_4^-$	34	0.905
HCO_3^-	44.5	1.185
HF_2^-	75	1.997
$1/2HPO_4^{2-}$	57	0.759
$H_2PO_4^-$	36	0.959
$H_2PO_2^-$	46	1.225
HS^-	65	1.731
HSO_3^-	58	1.545
HSO_4^-	52	1.385
$H_2SbO_4^-$	31	0.825
I^-	76.8	2.045
IO_3^-	40.5	1.078
IO_4^-	54.5	1.451
MnO_4^-	61.3	1.632
$1/2MoO_4^{2-}$	74.5	1.984
$N(CN)_2^-$	54.5	1.451
NO_2^-	71.8	1.912
NO_3^-	71.42	1.902
$NH_2SO_3^-$	48.3	1.286
N_3^-	69	1.837
OCN^-	64.6	1.720
OD^-	119	3.169
OH^-	198	5.273
PF_6^-	56.9	1.515
$1/2PO_3F^{2-}$	63.3	0.843
$1/3PO_4^{3-}$	92.8	0.824
$1/4P_2O_7^{4-}$	96	0.639
$1/3P_3O_9^{3-}$	83.6	0.742
$1/5P_3O_{10}^{5-}$	109	0.581
ReO_4^-	54.9	1.462
SCN^-	66	1.758
$1/2SO_3^{2-}$	72	0.959
$1/2SO_4^{2-}$	80.0	1.065
$1/2S_2O_3^{2-}$	85.0	1.132
$1/2S_2O_4^{2-}$	66.5	0.885
$1/2S_2O_6^{2-}$	93	1.238
$1/2S_2O_8^{2-}$	86	1.145
$Sb(OH)_6^-$	31.9	0.849
$SeCN^-$	64.7	1.723
$1/2SeO_4^{2-}$	75.7	1.008
$1/2WO_4^{2-}$	69	0.919
Organic Cations		
Benzyltrimethylammonium$^+$	34.6	0.921
Butyltrimethylammonium$^+$	33.6	0.895
Decylpyridinium$^+$	29.5	0.786
Decyltrimethylammonium$^+$	24.4	0.650
Diethylammonium$^+$	42.0	1.118
Dimethylammonium$^+$	51.8	1.379
Dipropylammonium$^+$	30.1	0.802
Dodecylammonium$^+$	23.8	0.634
Dodecyltrimethylammonium$^+$	22.6	0.602
Ethanolammonium$^+$	42.2	1.124
Ethylammonium$^+$	47.2	1.257
Ethyltrimethylammonium$^+$	40.5	1.078
Hexadecyltrimethylammonium$^+$	20.9	0.557
Hexyltrimethylammonium$^+$	29.6	0.788
Histidyl$^+$	23.0	0.612
Hydroxyethyltrimethylarsonium$^+$	39.4	1.049
Isobutylammonium	38	1.012
Methylammonium$^+$	58.7	1.563
Octadecylpyridinium$^+$	20	0.533
Octadecyltributylammonium$^+$	16.6	0.442
Octadecyltriethylammonium$^+$	17.9	0.477
Octadecyltrimethylammonium$^+$	19.9	0.530
Octadecyltripropylammonium$^+$	17.2	0.458
Octyltrimethylammonium$^+$	26.5	0.706
Pentylammonium$^+$	37	0.985
Piperidinium$^+$	37.2	0.991
Propylammonium$^+$	40.8	1.086
Pyrilammonium$^+$	24.3	0.647
Tetrabutylammonium$^+$	19.5	0.519
Tetradecyltrimethylammonium$^+$	21.5	0.573
Tetraethylammonium$^+$	32.6	0.868
Tetramethylammonium$^+$	44.9	1.196
Tetraisopentylammonium$^+$	17.9	0.477
Tetrapentylammmonium$^+$	17.5	0.466
Tetrapropylammonium$^+$	23.4	0.623
Triethylammonium$^+$	34.3	0.913
Triethylsulfonium$^+$	36.1	0.961
Trimethylammonium$^+$	47.23	1.258
Trimethylhexylammonium$^+$	34.6	0.921
Trimethylsulfonium$^+$	51.4	1.369
Tripropylammonium$^+$	26.1	0.695
Organic Anions		
Acetate$^-$	40.9	1.089
p-Anisate$^-$	29.0	0.772
1/2Azelate^{2-}	40.6	0.541
Benzoate$^-$	32.4	0.863
Bromoacetate$^-$	39.2	1.044
Bromobenzoate$^-$	30	0.799
Butyrate$^-$	32.6	0.868
Chloroacetate$^-$	39.8	1.060
m-Chlorobenzoate$^-$	31	0.825
o-Chlorobenzoate$^-$	30.2	0.804
1/3Citrate^{3-}	70.2	0.623
Crotonate$^-$	33.2	0.884
Cyanoacetate$^-$	43.4	1.156

Thermochem

Mol. form.	$\Lambda_{\pm}$/ 10^{-4} m^2 S mol^{-1}	D/ 10^{-5} cm^2 s^{-1}
Cyclohexane carboxylate$^-$	28.7	0.764
1/2 1,1-Cyclopropanedicarboxylate^{2-}	53.4	0.711
Decylsulfate$^-$	26	0.692
Dichloroacetate$^-$	38.3	1.020
1/2Diethylbarbiturate^{2-}	26.3	0.350
Dihydrogencitrate$^-$	30	0.799
1/2Dimethylmalonate^{2-}	49.4	0.658
3,5-Dinitrobenzoate$^-$	28.3	0.754
Dodecylsulfate$^-$	24	0.639
Ethylmalonate$^-$	49.3	1.313
Ethylsulfate$^-$	39.6	1.055
Fluoroacetate$^-$	44.4	1.182
Fluorobenzoate$^-$	33	0.879
Formate$^-$	54.6	1.454
1/2Fumarate^{2-}	61.8	0.823
1/2Glutarate^{2-}	52.6	0.700
Hydrogenoxalate$^-$	40.2	1.070
Isovalerate$^-$	32.7	0.871
Iodoacetate$^-$	40.6	1.081
Lactate$^-$	38.8	1.033
1/2Malate^{2-}	58.8	0.783
1/2Maleate^{2-}	61.9	0.824
1/2Malonate^{2-}	63.5	0.845
Methylsulfate$^-$	48.8	1.299
Naphthylacetate$^-$	28.4	0.756
1/2Oxalate^{2-}	74.11	0.987
Octylsulfate$^-$	29	0.772
Phenylacetate$^-$	30.6	0.815
1/2*o*-Phthalate^{2-}	52.3	0.696
1/2*m*-Phthalate^{2-}	54.7	0.728
Picrate$^-$	30.37	0.809
Pivalate$^-$	31.9	0.849
Propionate$^-$	35.8	0.953
Propylsulfate$^-$	37.1	0.988
Salicylate$^-$	36	0.959
1/2Suberate^{2-}	36	0.479
1/2Succinate^{2-}	58.8	0.783
p-Sulfonate	29.3	0.780
1/2Tartarate^{2-}	59.6	0.794
Trichloroacetate$^-$	35	0.932

* Ethylenediamine.
** Tris(ethylenediamine).

ELECTROCHEMICAL SERIES

Petr Vanýsek

There are three tables for the electrochemical series. Each table lists standard reduction potentials, $E°$ values, at 298.15 K (25 °C), and at a pressure of 101.325 kPa (1 atm) (not the standard pressure of 1 bar). The activity of all soluble species is assumed to be 1.000 mol/L. This is particularly important when pH (H^+ or OH^-) takes part in the equilibrium. Table 1 is an alphabetical listing of the elements and compounds, according to the symbol of the elements. Thus, data for silver (Ag) precede those for aluminum (Al). Table 2 lists only those reduction reactions that have $E°$ values positive with respect to the standard hydrogen electrode. In Table 2, the reactions are listed in the order of increasing positive potential, and they range from 0.0000 V to +3.4 V. Table 3 lists only those reduction potentials that have $E°$ negative with respect to the standard hydrogen electrode. In Table 3, the reactions are listed in the order of decreasing potential and range from 0.0000 V to −4.10 V. The reliability of the potentials is not the same for all the data. Typically, the values with fewer significant figures have lower reliability. The values of reduction potentials, in particular those of less common reactions, are not definite; they are subject to occasional revisions.

Abbreviations: ac = acetate; bipy = 2,2'-dipyridine, or bipyridine; en = ethylenediamine; phen = 1,10-phenanthroline.

References

1. Milazzo, G., Caroli, S., and Sharma, V. K., *Tables of Standard Electrode Potentials, Fourth Edition*, Wiley, Chichester, 1978.
2. Bard, A. J., Parsons, R., and Jordan, J., *Standard Potentials in Aqueous Solutions*, Marcel Dekker, New York, 1985.
3. Bratsch, S. G., *J. Phys. Chem. Ref. Data*, 18, 1–21, 1989. <https://doi.org/10.1063/1.555839>

TABLE 1. Electrochemical Series – Alphabetical Listing

Reaction	$E°$/V
$Ac^{3+} + 3\,e \rightleftharpoons Ac$	-2.20
$Ag^{+} + e \rightleftharpoons Ag$	0.7996
$Ag^{2+} + e \rightleftharpoons Ag^{+}$	1.980
$Ag(ac) + e \rightleftharpoons Ag + (ac)^{-}$	0.643
$AgBr + e \rightleftharpoons Ag + Br^{-}$	0.07133
$AgBrO_3 + e \rightleftharpoons Ag + BrO_3^{-}$	0.546
$Ag_2C_2O_4 + 2\,e \rightleftharpoons 2\,Ag + C_2O_4^{2-}$	0.4647
$AgCl + e \rightleftharpoons Ag + Cl^{-}$	0.22233
$AgCN + e \rightleftharpoons Ag + CN^{-}$	-0.017
$Ag_2CO_3 + 2\,e \rightleftharpoons 2\,Ag + CO_3^{2-}$	0.47
$Ag_2CrO_4 + 2\,e \rightleftharpoons 2\,Ag + CrO_4^{2-}$	0.4470
$AgF + e \rightleftharpoons Ag + F^{-}$	0.779
$Ag_4[Fe(CN)_6] + 4e \rightleftharpoons 4Ag + [Fe(CN)_6]^{4-}$	0.1478
$AgI + e \rightleftharpoons Ag + I^{-}$	-0.15224
$AgIO_3 + e \rightleftharpoons Ag + IO_3^{-}$	0.354
$Ag_2MoO_4 + 2\,e \rightleftharpoons 2\,Ag + MoO_4^{2-}$	0.4573
$AgNO_2 + e \rightleftharpoons Ag + 2\,NO_2^{-}$	0.564
$Ag_2O + H_2O + 2\,e \rightleftharpoons 2\,Ag + 2\,OH^{-}$	0.342
$Ag_2O_3 + H_2O + 2\,e \rightleftharpoons 2\,AgO + 2\,OH^{-}$	0.739
$Ag^{3+} + 2\,e \rightleftharpoons Ag^{+}$	1.9
$Ag^{3+} + e \rightleftharpoons Ag^{2+}$	1.8
$Ag_2O_2 + 4\,H^{+} + e \rightleftharpoons 2\,Ag + 2\,H_2O$	1.802
$2\,AgO + H_2O + 2\,e \rightleftharpoons Ag_2O + 2\,OH^{-}$	0.607
$AgOCN + e \rightleftharpoons Ag + OCN^{-}$	0.41
$Ag_2S + 2\,e \rightleftharpoons 2\,Ag + S^{2-}$	-0.691
$Ag_2S + 2\,H^{+} + 2\,e \rightleftharpoons 2\,Ag + H_2S$	-0.0366
$AgSCN + e \rightleftharpoons Ag + SCN^{-}$	0.08951
$Ag_2SeO_3 + 2\,e \rightleftharpoons 2\,Ag + SeO_4^{2-}$	0.3629
$Ag_2SO_4 + 2\,e \rightleftharpoons 2\,Ag + SO_4^{2-}$	0.654
$Ag_2WO_4 + 2\,e \rightleftharpoons 2\,Ag + WO_4^{2-}$	0.4660
$Al^{3+} + 3\,e \rightleftharpoons Al$	-1.676
$Al(OH)_3 + 3\,e \rightleftharpoons Al + 3\,OH^{-}$	-2.30
$Al(OH)_4^{-} + 3\,e \rightleftharpoons Al + 4\,OH^{-}$	-2.310
$H_2AlO_3^{-} + H_2O + 3\,e \rightleftharpoons Al + 4\,OH^{-}$	-2.33
$AlF_6^{3-} + 3\,e \rightleftharpoons Al + 6\,F^{-}$	-2.069
$Am^{4+} + e \rightleftharpoons Am^{3+}$	2.60
$Am^{2+} + 2\,e \rightleftharpoons Am$	-1.9
$Am^{3+} + 3\,e \rightleftharpoons Am$	-2.048
$Am^{3+} + e \rightleftharpoons Am^{2+}$	-2.3
$As + 3\,H^{+} + 3\,e \rightleftharpoons AsH_3$	-0.608
$As_2O_3 + 6\,H^{+} + 6\,e \rightleftharpoons 2\,As + 3\,H_2O$	0.234
$HAsO_2 + 3\,H^{+} + 3\,e \rightleftharpoons As + 2\,H_2O$	0.248
$AsO_2^{-} + 2\,H_2O + 3\,e \rightleftharpoons As + 4\,OH^{-}$	-0.68
$H_3AsO_4 + 2\,H^{+} + 2\,e \rightleftharpoons HAsO_2 + 2\,H_2O$	0.560
$AsO_4^{3-} + 2\,H_2O + 2\,e \rightleftharpoons AsO_2^{-} + 4\,OH^{-}$	-0.71
$At_2 + 2\,e \rightleftharpoons 2\,At^{-}$	0.2
$Au^{+} + e \rightleftharpoons Au$	1.692
$Au^{3+} + 2\,e \rightleftharpoons Au^{+}$	1.401
$Au^{3+} + 3\,e \rightleftharpoons Au$	1.498
$Au^{2+} + e^{-} \rightleftharpoons Au^{+}$	1.8
$AuOH^{2+} + H^{+} + 2\,e \rightleftharpoons Au^{+} + H_2O$	1.32
$AuBr_2^{-} + e \rightleftharpoons Au + 2\,Br^{-}$	0.959
$AuBr_4^{-} + 3\,e \rightleftharpoons Au + 4\,Br^{-}$	0.854
$AuCl_4^{-} + 3\,e \rightleftharpoons Au + 4\,Cl^{-}$	1.002
$Au(OH)_3 + 3\,H^{+} + 3\,e \rightleftharpoons Au + 3\,H_2O$	1.45
$H_2BO_3^{-} + 5\,H_2O + 8\,e \rightleftharpoons BH_4^{-} + 8\,OH^{-}$	-1.24
$H_2BO_3^{-} + H_2O + 3\,e \rightleftharpoons B + 4\,OH^{-}$	-1.79
$H_3BO_3 + 3\,H^{+} + 3\,e \rightleftharpoons B + 3\,H_2O$	-0.8698
$B(OH)_3 + 7\,H^{+} + 8\,e \rightleftharpoons BH_4^{-} + 3\,H_2O$	-0.481
$Ba^{2+} + 2\,e \rightleftharpoons Ba$	-2.912
$Ba^{2+} + 2\,e \rightleftharpoons Ba(Hg)$	-1.570
$Ba(OH)_2 + 2\,e \rightleftharpoons Ba + 2\,OH^{-}$	-2.99
$Be^{2+} + 2\,e \rightleftharpoons Be$	-1.847
$Be_2O_3^{2-} + 3\,H_2O + 4\,e \rightleftharpoons 2\,Be + 6\,OH^{-}$	-2.63
p–benzoquinone + $2\,H^{+} + 2\,e \rightleftharpoons$ hydroquinone	0.6992
$Bi^{+} + e \rightleftharpoons Bi$	0.5
$Bi^{3+} + 3\,e \rightleftharpoons Bi$	0.308
$Bi^{3+} + 2\,e \rightleftharpoons Bi^{+}$	0.2
$Bi + 3\,H^{+} + 3\,e \rightleftharpoons BiH_3$	-0.8
$BiCl_4^{-} + 3\,e \rightleftharpoons Bi + 4\,Cl^{-}$	0.16
$Bi_2O_3 + 3\,H_2O + 6\,e \rightleftharpoons 2\,Bi + 6\,OH^{-}$	-0.46
$Bi_2O_4 + 4\,H^{+} + 2\,e \rightleftharpoons 2\,BiO^{+} + 2\,H_2O$	1.593
$BiO^{+} + 2\,H^{+} + 3\,e \rightleftharpoons Bi + H_2O$	0.320
$BiOCl + 2\,H^{+} + 3\,e \rightleftharpoons Bi + Cl^{-} + H_2O$	0.1583
$Bk^{4+} + e \rightleftharpoons Bk^{3+}$	1.67
$Bk^{2+} + 2\,e \rightleftharpoons Bk$	-1.6
$Bk^{3+} + e \rightleftharpoons Bk^{2+}$	-2.8
$Br_2(aq) + 2\,e \rightleftharpoons 2\,Br^{-}$	1.0873
$Br_2(l) + 2\,e \rightleftharpoons 2\,Br^{-}$	1.066
$HBrO + H^{+} + 2\,e \rightleftharpoons Br^{-} + H_2O$	1.331
$HBrO + H^{+} + e \rightleftharpoons 1/2\,Br_2(aq) + H_2O$	1.574
$HBrO + H^{+} + e \rightleftharpoons 1/2\,Br_2(l) + H_2O$	1.596
$BrO^{-} + H_2O + 2\,e \rightleftharpoons Br^{-} + 2\,OH^{-}$	0.761
$BrO_3^{-} + 6\,H^{+} + 5\,e \rightleftharpoons 1/2\,Br_2 + 3\,H_2O$	1.482
$BrO_3^{-} + 6\,H^{+} + 6\,e \rightleftharpoons Br^{-} + 3\,H_2O$	1.423
$BrO_3^{-} + 3\,H_2O + 6\,e \rightleftharpoons Br^{-} + 6\,OH^{-}$	0.61
$(CN)_2 + 2\,H^{+} + 2\,e \rightleftharpoons 2\,HCN$	0.373
$2\,HCNO + 2\,H^{+} + 2\,e \rightleftharpoons (CN)_2 + 2\,H_2O$	0.330
$(CNS)_2 + 2\,e \rightleftharpoons 2\,CNS^{-}$	0.77
$CO_2 + 2\,H^{+} + 2\,e \rightleftharpoons HCOOH$	-0.199
$Ca^{+} + e \rightleftharpoons Ca$	-3.80
$Ca^{2+} + 2\,e \rightleftharpoons Ca$	-2.868
$Ca(OH)_2 + 2\,e \rightleftharpoons Ca + 2\,OH^{-}$	-3.02
Calomel electrode, 1 molal KCl	0.2800
Calomel electrode, 1 molar KCl (NCE)	0.2801
Calomel electrode, 0.1 molar KCl	0.3337
Calomel electrode, saturated KCl (SCE)	0.2412
Calomel electrode, saturated NaCl (SSCE)	0.2360
$Cd^{2+} + 2\,e \rightleftharpoons Cd$	-0.4030
$Cd^{2+} + 2\,e \rightleftharpoons Cd(Hg)$	-0.3521
$Cd(OH)_2 + 2\,e \rightleftharpoons Cd(Hg) + 2\,OH^{-}$	-0.809
$CdSO_4 + 2\,e \rightleftharpoons Cd + SO_4^{2-}$	-0.246
$Cd(OH)_4^{2-} + 2\,e \rightleftharpoons Cd + 4\,OH^{-}$	-0.658
$CdO + H_2O + 2\,e \rightleftharpoons Cd + 2\,OH^{-}$	-0.783

Reaction	$E°/V$
$Ce^{3+} + 3 e \rightleftharpoons Ce$	-2.336
$Ce^{3+} + 3 e \rightleftharpoons Ce(Hg)$	-1.4373
$Ce^{4+} + e \rightleftharpoons Ce^{3+}$	1.72
$CeOH^{3+} + H^+ + e \rightleftharpoons Ce^{3+} + H_2O$	1.715
$Cf^{4+} + e \rightleftharpoons Cf^{3+}$	3.3
$Cf^{3+} + e \rightleftharpoons Cf^{2+}$	-1.6
$Cf^{3+} + 3 e \rightleftharpoons Cf$	-1.94
$Cf^{2+} + 2 e \rightleftharpoons Cf$	-2.12
$Cl_2(g) + 2 e \rightleftharpoons 2 Cl^-$	1.35827
$HClO + H^+ + e \rightleftharpoons 1/2 Cl_2 + H_2O$	1.611
$HClO + H^+ + 2 e \rightleftharpoons Cl^- + H_2O$	1.482
$ClO^- + H_2O + 2 e \rightleftharpoons Cl^- + 2 OH^-$	0.81
$ClO_2 + H^+ + e \rightleftharpoons HClO_2$	1.277
$HClO_2 + 2 H^+ + 2 e \rightleftharpoons HClO + H_2O$	1.645
$HClO_2 + 3 H^+ + 3 e \rightleftharpoons 1/2 Cl_2 + 2 H_2O$	1.628
$HClO_2 + 3 H^+ + 4 e \rightleftharpoons Cl^- + 2 H_2O$	1.570
$ClO_2^- + H_2O + 2 e \rightleftharpoons ClO^- + 2 OH^-$	0.66
$ClO_2^- + 2 H_2O + 4 e \rightleftharpoons Cl^- + 4 OH^-$	0.76
$ClO_2(aq) + e \rightleftharpoons ClO_2^-$	0.954
$ClO_3^- + 2 H^+ + e \rightleftharpoons ClO_2 + H_2O$	1.152
$ClO_3^- + 3 H^+ + 2 e \rightleftharpoons HClO_2 + H_2O$	1.214
$ClO_3^- + 6 H^+ + 5 e \rightleftharpoons 1/2 Cl_2 + 3 H_2O$	1.47
$ClO_3^- + 6 H^+ + 6 e \rightleftharpoons Cl^- + 3 H_2O$	1.451
$ClO_3^- + H_2O + 2 e \rightleftharpoons ClO_2^- + 2 OH^-$	0.33
$ClO_3^- + 3 H_2O + 6 e \rightleftharpoons Cl^- + 6 OH^-$	0.62
$ClO_4^- + 2 H^+ + 2 e \rightleftharpoons ClO_3^-H_2O$	1.189
$ClO_4^- + 8 H^+ + 7 e \rightleftharpoons 1/2 Cl_2 + 4 H_2O$	1.39
$ClO_4^- + 8 H^+ + 8 e \rightleftharpoons Cl^- + 4 H_2O$	1.389
$ClO_4^- + H_2O + 2 e \rightleftharpoons ClO_3^- + 2 OH^-$	0.36
$Cm^{4+} + e \rightleftharpoons Cm^{3+}$	3.0
$Cm^{3+} + 3 e \rightleftharpoons Cm$	-2.04
$Co^{2+} + 2 e \rightleftharpoons Co$	-0.28
$Co^{3+} + e \rightleftharpoons Co^{2+}$	1.92
$[Co(NH_3)_6]^{3+} + e \rightleftharpoons [Co(NH_3)_6]^{2+}$	0.108
$Co(OH)_2 + 2 e \rightleftharpoons Co + 2 OH^-$	-0.73
$Co(OH)_3 + e \rightleftharpoons Co(OH)_2 + OH^-$	0.17
$Cr^{2+} + 2 e \rightleftharpoons Cr$	-0.913
$Cr^{3+} + e \rightleftharpoons Cr^{2+}$	-0.407
$Cr^{3+} + 3 e \rightleftharpoons Cr$	-0.744
$Cr_2O_7^{2-} + 14 H^+ + 6 e \rightleftharpoons 2 Cr^{3+} + 7 H_2O$	1.36
$CrO_2^- + 2 H_2O + 3 e \rightleftharpoons Cr + 4 OH^-$	-1.2
$HCrO_4^- + 7 H^+ + 3 e \rightleftharpoons Cr^{3+} + 4 H_2O$	1.350
$CrO_2 + 4 H^+ + e \rightleftharpoons Cr^{3+} + 2H_2O$	1.48
$Cr(V) + e \rightleftharpoons Cr(IV)$	1.34
$CrO_4^{2-} + 4H_2O + 3e \rightleftharpoons Cr(OH)_3 + 5OH^-$	-0.13
$Cr(OH)_3 + 3 e \rightleftharpoons Cr + 3 OH^-$	-1.48
$Cs^+ + e \rightleftharpoons Cs$	-3.026
$Cu^+ + e \rightleftharpoons Cu$	0.521
$Cu^{2+} + e \rightleftharpoons Cu^+$	0.153
$Cu^{2+} + 2 e \rightleftharpoons Cu$	0.3419
$Cu^{2+} + 2 e \rightleftharpoons Cu(Hg)$	0.345
$Cu^{3+} + e \rightleftharpoons Cu^{2+}$	2.4
$Cu_2O_3 + 6 H^+ + 2e \rightleftharpoons 2Cu^{2+} + 3 H_2O$	2.0
$Cu^{2+} + 2 CN^- + e \rightleftharpoons [Cu(CN)_2]^-$	1.103
$CuI_2^- + e \rightleftharpoons Cu + 2 I^-$	0.00
$Cu_2O + H_2O + 2 e \rightleftharpoons 2 Cu + 2 OH^-$	-0.360
$Cu(OH)_2 + 2 e \rightleftharpoons Cu + 2 OH^-$	-0.222
$2 Cu(OH)_2 + 2 e \rightleftharpoons Cu_2O + 2 OH^- + H_2O$	-0.080
$2 D^+ + 2 e \rightleftharpoons D_2$	-0.013
$Dy^{2+} + 2 e \rightleftharpoons Dy$	-2.2
$Dy^{3+} + 3 e \rightleftharpoons Dy$	-2.295
$Dy^{3+} + e \rightleftharpoons Dy^{2+}$	-2.6
$Er^{2+} + 2 e \rightleftharpoons Er$	-2.0
$Er^{3+} + 3 e \rightleftharpoons Er$	-2.331
$Er^{3+} + e \rightleftharpoons Er^{2+}$	-3.0
$Es^{3+} + e \rightleftharpoons Es^{2+}$	-1.3
$Es^{3+} + 3 e \rightleftharpoons Es$	-1.91
$Es^{2+} + 2 e \rightleftharpoons Es$	-2.23
$Eu^{2+} + 2 e \rightleftharpoons Eu$	-2.812
$Eu^{3+} + 3 e \rightleftharpoons Eu$	-1.991
$Eu^{3+} + e \rightleftharpoons Eu^{2+}$	-0.36
$F_2 + 2 H^+ + 2 e \rightleftharpoons 2 HF$	3.053
$F_2 + 2 e \rightleftharpoons 2 F^-$	2.866
$F_2O + 2 H^+ + 4 e \rightleftharpoons H_2O + 2 F^-$	2.153
$Fe^{2+} + 2 e \rightleftharpoons Fe$	-0.447
$Fe^{3+} + 3 e \rightleftharpoons Fe$	-0.037
$Fe^{3+} + e \rightleftharpoons Fe^{2+}$	0.771
$2 HFeO_4^- + 8 H^+ + 6 e \rightleftharpoons Fe_2O_3 + 5 H_2O$	2.09
$HFeO_4^- + 4 H^+ + 3 e \rightleftharpoons FeOOH + 2 H_2O$	2.08
$HFeO_4^- + 7 H^+ + 3 e \rightleftharpoons Fe^{3+} + 4 H_2O$	2.07
$Fe_2O_3 + 4 H^+ + 2 e \rightleftharpoons 2 FeOH^+ + H_2O$	0.16
$[Fe(CN)_6]^{3-} + e \rightleftharpoons [Fe(CN)_6]^{4-}$	0.358
$FeO_4^{2-} + 8 H^+ + 3 e\ Fe^{3+} + 4 H_2O$	2.20
$[Fe(bipy)_2]^{3+} + e \rightleftharpoons Fe(bipy)_2]^{2+}$	0.78
$[Fe(bipy)_3]^{3+} + e \rightleftharpoons Fe(bipy)_3]^{2+}$	1.03
$Fe(OH)_3 + e \rightleftharpoons Fe(OH)_2 + OH^-$	-0.56
$[Fe(phen)_3]^{3+} + e \rightleftharpoons [Fe(phen)_3]^{2+}$	1.147
$[Fe(phen)_3]^{3+} + e \rightleftharpoons [Fe(phen)_3]^{2+}$(1 molar H_2SO_4)	1.06
$[Ferricinium]^+ + e \rightleftharpoons$ ferrocene	0.400
$Fm^{3+} + e \rightleftharpoons Fm^{2+}$	-1.1
$Fm^{3+} + 3 e \rightleftharpoons Fm$	-1.89
$Fm^{2+} + 2 e \rightleftharpoons Fm$	-2.30
$Fr^+ + e \rightleftharpoons Fr$	-2.9
$Ga^{3+} + 3 e \rightleftharpoons Ga$	-0.549
$Ga^+ + e \rightleftharpoons Ga$	-0.2
$GaOH^{2+} + H^+ + 3 e \rightleftharpoons Ga + H_2O$	-0.498
$H_2GaO^-_3 + H_2O + 3 e \rightleftharpoons Ga + 4 OH^-$	-1.219
$Gd^{3+} + 3 e \rightleftharpoons Gd$	-2.279
$Ge^{2+} + 2 e \rightleftharpoons Ge$	0.24
$Ge^{4+} + 4 e \rightleftharpoons Ge$	0.124
$Ge^{4+} + 2 e \rightleftharpoons Ge^{2+}$	0.00
$GeO_2 + 2 H^+ + 2 e \rightleftharpoons GeO + H_2O$	-0.118
$H_2GeO_3 + 4 H^+ + 4 e \rightleftharpoons Ge + 3 H_2O$	-0.182
$2 H^+ + 2 e \rightleftharpoons H_2$	0.00000
$H_2 + 2 e \rightleftharpoons 2 H^-$	-2.23
$HO_2 + H^+ + e \rightleftharpoons H_2O_2$	1.495
$2 H_2O + 2 e \rightleftharpoons H_2 + 2 OH^-$	-0.8277
$H_2O_2 + 2 H^+ + 2 e \rightleftharpoons 2 H_2O$	1.776
$Hf^{4+} + 4 e \rightleftharpoons Hf$	-1.55
$HfO^{2+} + 2 H^+ + 4 e \rightleftharpoons Hf + H_2O$	-1.724
$HfO_2 + 4 H^+ + 4 e \rightleftharpoons Hf + 2 H_2O$	-1.505
$HfO(OH)_2 + H_2O + 4 e \rightleftharpoons Hf + 4 OH^-$	-2.50
$Hg^{2+} + 2 e \rightleftharpoons Hg$	0.851
$2 Hg^{2+} + 2 e \rightleftharpoons Hg_2^{2+}$	0.920
$Hg_2^{2+} + 2 e \rightleftharpoons 2 Hg$	0.7973
$Hg_2(ac)_2 + 2 e \rightleftharpoons 2 Hg + 2(ac)^-$	0.51163
$Hg_2Br_2 + 2 e \rightleftharpoons 2 Hg + 2 Br^-$	0.13923
$Hg_2Cl_2 + 2 e \rightleftharpoons 2 Hg + 2 Cl^-$	0.26808
$Hg_2HPO_4 + 2 e \rightleftharpoons 2 Hg + HPO_4^{2-}$	0.6359
$Hg_2I_2 + 2 e \rightleftharpoons 2 Hg + 2 I^-$	-0.0405
$Hg_2O + H_2O + 2 e \rightleftharpoons 2 Hg + 2 OH^-$	0.123
$HgO + H_2O + 2 e \rightleftharpoons Hg + 2 OH^-$	0.0977
$Hg(OH)_2 + 2 H^+ + 2 e \rightleftharpoons Hg + 2 H_2O$	1.034
$Hg_2SO_4 + 2 e \rightleftharpoons 2 Hg + SO_4^{2-}$	0.6125
$Ho^{2+} + 2 e \rightleftharpoons Ho$	-2.1
$Ho^{3+} + 3 e \rightleftharpoons Ho$	-2.33
$Ho^{3+} + e \rightleftharpoons Ho^{2+}$	-2.8
$I_2 + 2 e \rightleftharpoons 2 I^-$	0.5355
$I_3^- + 2 e \rightleftharpoons 3 I^-$	0.536
$H_3IO_6^{2-} + 2 e \rightleftharpoons IO^-_3 + 3 OH^-$	0.7
$H_5IO_6 + H^+ + 2 e \rightleftharpoons IO_3^- + 3 H_2O$	1.601
$2 HIO + 2 H^+ + 2 e \rightleftharpoons I_2 + 2 H_2O$	1.439
$HIO + H^+ + 2 e \rightleftharpoons I^- + H_2O$	0.987
$IO^- + H_2O + 2 e \rightleftharpoons I^- + 2 OH^-$	0.485
$2 IO_3^- + 12 H^+ + 10 e \rightleftharpoons I_2 + 6 H_2O$	1.195
$IO_3^- + 6 H^+ + 6 e \rightleftharpoons I^- + 3 H_2O$	1.085
$IO_3^- + 2 H_2O + 4 e \rightleftharpoons IO^- + 4 OH^-$	0.15
$IO_3^- + 3 H_2O + 6 e \rightleftharpoons IO^- + 6 OH^-$	0.26
$In^+ + e \rightleftharpoons In$	-0.14
$In^{2+} + e \rightleftharpoons In^+$	-0.40
$In^{3+} + e \rightleftharpoons In^{2+}$	-0.49
$In^{3+} + 2 e \rightleftharpoons In^+$	-0.443
$In^{3+} + 3 e \rightleftharpoons In$	-0.3382
$In(OH)_3 + 3 e \rightleftharpoons In + 3 OH^-$	-0.99
$In(OH)_4^- + 3 e \rightleftharpoons In + 4 OH^-$	-1.007
$In_2O_3 + 3 H_2O + 6 e \rightleftharpoons 2 In + 6 OH^-$	-1.034
$Ir^{3+} + 3 e \rightleftharpoons Ir$	1.156
$[IrCl_6]^{2-} + e \rightleftharpoons [IrCl_6]^{3-}$	0.8665
$[IrCl_6]^{3-} + 3 e \rightleftharpoons Ir + 6 Cl^-$	0.77
$Ir_2O_3 + 3 H_2O + 6 e \rightleftharpoons 2 Ir + 6 OH^-$	0.098
$K^+ + e \rightleftharpoons K$	-2.931
$La^{3+} + 3 e \rightleftharpoons La$	-2.379
$La(OH)_3 + 3 e \rightleftharpoons La + 3 OH^-$	-2.90
$Li^+ + e \rightleftharpoons Li$	-3.0401
$Lr^{3+} + 3 e \rightleftharpoons Lr$	-1.96
$Lu^{3+} + 3 e \rightleftharpoons Lu$	-2.28
$Md^{3+} + e \rightleftharpoons Md^{2+}$	-0.1
$Md^{3+} + 3 e \rightleftharpoons Md$	-1.65
$Md^{2+} + 2 e \rightleftharpoons Md$	-2.40
$Mg^+ + e \rightleftharpoons Mg$	-2.70
$Mg^{2+} + 2 e \rightleftharpoons Mg$	-2.372
$Mg(OH)_2 + 2 e \rightleftharpoons Mg + 2 OH^-$	-2.690
$Mn^{2+} + 2 e \rightleftharpoons Mn$	-1.185
$Mn^{3+} + e \rightleftharpoons Mn^{2+}$	1.5415
$MnO_2 + 4 H^+ + 2 e \rightleftharpoons Mn^{2+} + 2 H_2O$	1.224
$MnO_4^- + e \rightleftharpoons MnO_4^{2-}$	0.558
$MnO_4^- + 4 H^+ + 3 e \rightleftharpoons MnO_2 + 2 H_2O$	1.679
$MnO_4^- + 8 H^+ + 5 e \rightleftharpoons Mn^{2+} + 4 H_2O$	1.507
$MnO_4^- + 2H_2O + 3e \rightleftharpoons MnO_2 + 4OH^-$	0.595
$MnO_4^- + 4 H_2O + 5 e \rightleftharpoons Mn(OH)_2 + 6 OH^-$	0.34
$MnO_4^{2-} + 2H_2O + 2e \rightleftharpoons MnO_2 + 4OH^-$	0.60
$Mn(OH)_2 + 2 e \rightleftharpoons Mn + 2 OH^-$	-1.56
$Mn(OH)_3 + e \rightleftharpoons Mn(OH)_2 + OH^-$	0.15

Reaction	$E°$/V
$Mn_2O_3 + 6 H^+ + e \rightleftharpoons 2 Mn^{2+} + 3 H_2O$	1.485
$Mo^{3+} + 3 e \rightleftharpoons Mo$	-0.200
$MoO_2 + 4 H^+ + 4 e \rightleftharpoons Mo + 4 H_2O$	-0.152
$H_3Mo_7O_{24}^{3-} + 45 H^+ + 42 e \rightleftharpoons 7 Mo + 24 H_2O$	0.082
$MoO_3 + 6 H^+ + 6 e \rightleftharpoons Mo + 3 H_2O$	0.075
$N_2 + 2 H_2O + 6 H^+ + 6 e \rightleftharpoons 2 NH_4OH$	0.092
$3 N_2 + 2 H^+ + 2 e \rightleftharpoons 2 HN_3$	-3.09
$N_5^+ + 3 H^+ + 2 e \rightleftharpoons 2 NH_4^+$	1.275
$N_2O + 2 H^+ + 2 e \rightleftharpoons N_2 + H_2O$	1.766
$H_2N_2O_2 + 2 H^+ + 2 e \rightleftharpoons N_2 + 2 H_2O$	2.65
$N_2O_4 + 2 e \rightleftharpoons 2 NO_2^-$	0.867
$N_2O_4 + 2 H^+ + 2 e \rightleftharpoons 2 NHO_2$	1.065
$N_2O_4 + 4 H^+ + 4 e \rightleftharpoons 2 NO + 2 H_2O$	1.035
$2 NH_3OH^+ + H^+ + 2 e \rightleftharpoons N_2H_5^+ + 2 H_2O$	1.42
$2 NO + 2 H^+ + 2 e \rightleftharpoons N_2O + H_2O$	1.591
$2 NO + H_2O + 2 e \rightleftharpoons N_2O + 2 OH^-$	0.76
$HNO_2 + H^+ + e \rightleftharpoons NO + H_2O$	0.983
$2 HNO_2 + 4 H^+ + 4 e \rightleftharpoons H_2N_2O_2 + 2 H_2O$	0.86
$2 HNO_2 + 4 H^+ + 4 e \rightleftharpoons N_2O + 3 H_2O$	1.297
$NO_2^- + H_2O + e \rightleftharpoons NO + 2 OH^-$	-0.46
$2 NO_2^- + 2 H_2O + 4 e \rightleftharpoons N_2O_2^{2-} + 4 OH^-$	-0.18
$2 NO_2^- + 3 H_2O + 4 e \rightleftharpoons N_2O + 6 OH^-$	0.15
$NO_3^- + 3 H^+ + 2 e \rightleftharpoons HNO_2 + H_2O$	0.934
$NO_3^- + 4 H^+ + 3 e \rightleftharpoons NO + 2 H_2O$	0.957
$2 NO_3^- + 4 H^+ + 2 e \rightleftharpoons N_2O_4 + 2 H_2O$	0.803
$NO_3^- + H_2O + 2 e \rightleftharpoons NO_2^- + 2 OH^-$	0.01
$2 NO_3^- + 2 H_2O + 2 e \rightleftharpoons N_2O_4 + 4 OH^-$	-0.85
$Na^+ + e \rightleftharpoons Na$	-2.71
$Nb^{3+} + 3 e \rightleftharpoons Nb$	-1.099
$NbO_2 + 2 H^+ + 2 e \rightleftharpoons NbO + H_2O$	-0.646
$NbO_2 + 4 H^+ + 4 e \rightleftharpoons Nb + 2 H_2O$	-0.690
$NbO + 2 H^+ + 2 e \rightleftharpoons Nb + H_2O$	-0.733
$Nb_2O_5 + 10 H^+ + 10 e \rightleftharpoons 2 Nb + 5 H_2O$	-0.644
$Nd^{3+} + 3 e \rightleftharpoons Nd$	-2.323
$Nd^{2+} + 2 e \rightleftharpoons Nd$	-2.1
$Nd^{3+} + e \rightleftharpoons Nd^{2+}$	-2.7
$Ni^{2+} + 2 e \rightleftharpoons Ni$	-0.257
$Ni(OH)_2 + 2 e \rightleftharpoons Ni + 2 OH^-$	-0.72
$NiO_2 + 4 H^+ + 2 e \rightleftharpoons Ni^{2+} + 2 H_2O$	1.678
$NiO_2 + 2 H_2O + 2 e \rightleftharpoons Ni(OH)_2 + 2 OH^-$	-0.490
$No^{3+} + e \rightleftharpoons No^{2+}$	1.4
$No^{3+} + 3 e \rightleftharpoons No$	-1.20
$No^{2+} + 2 e \rightleftharpoons No$	-2.50
$Np^{3+} + 3 e \rightleftharpoons Np$	-1.856
$Np^{4+} + e \rightleftharpoons Np^{3+}$	0.147
$NpO_2 + H_2O + H^+ + e \rightleftharpoons Np(OH)_3$	-0.962
$O_2 + 2 H^+ + 2 e \rightleftharpoons H_2O_2$	0.695
$O_2 + 4 H^+ + 4 e \rightleftharpoons 2 H_2O$	1.229
$O_2 + H_2O + 2 e \rightleftharpoons HO_2^- + OH^-$	-0.076
$O_2 + 2 H_2O + 2 e \rightleftharpoons H_2O_2 + 2 OH^-$	-0.146
$O_2 + 2 H_2O + 4 e \rightleftharpoons 4 OH^-$	0.401
$O_3 + 2 H^+ + 2 e \rightleftharpoons O_2 + H_2O$	2.076
$O_3 + H_2O + 2 e \rightleftharpoons O_2 + 2 OH^-$	1.24
$O(g) + 2 H^+ + 2 e \rightleftharpoons H_2O$	2.421
$OH + e \rightleftharpoons OH^-$	2.02
$HO_2^- + H_2O + 2 e \rightleftharpoons 3 OH^-$	0.878
$OsO_4 + 8 H^+ + 8 e \rightleftharpoons Os + 4 H_2O$	0.838
$OsO_4 + 4 H^+ + 4 e \rightleftharpoons OsO_2 + 2 H_2O$	1.02
$[Os(bipy)_2]^{3+} + e \rightleftharpoons [Os(bipy)_2]^{2+}$	0.81
$[Os(bipy)_3]^{3+} + e \rightleftharpoons [Os(bipy)_3]^{2+}$	0.80
$P(red) + 3 H^+ + 3 e \rightleftharpoons PH_3(g)$	-0.111
$P(white) + 3 H^+ + 3 e \rightleftharpoons PH_3(g)$	-0.063
$P + 3 H_2O + 3 e \rightleftharpoons PH_3(g) + 3 OH^-$	-0.87
$H_2P_2^- + e \rightleftharpoons P + 2 OH^-$	-1.82
$H_3PO_2 + H^+ + e \rightleftharpoons P + 2 H_2O$	-0.508
$H_3PO_3 + 2 H^+ + 2 e \rightleftharpoons H_3PO_2 + H_2O$	-0.499
$H_3PO_3 + 3 H^+ + 3 e \rightleftharpoons P + 3 H_2O$	-0.454
$HPO_3^{2-} + 2 H_2O + 2 e \rightleftharpoons H_2PO_2^- + 3 OH^-$	-1.65
$HPO_3^{2-} + 2 H_2O + 3 e \rightleftharpoons P + 5 OH^-$	-1.71
$H_3PO_4 + 2 H^+ + 2 e \rightleftharpoons H_3PO_3 + H_2O$	-0.276
$PO_4^{3-} + 2 H_2O + 2 e \rightleftharpoons HPO_3^{2-} + 3 OH^-$	-1.05
$Pa^{3+} + 3 e \rightleftharpoons Pa$	-1.34
$Pa^{4+} + 4 e \rightleftharpoons Pa$	-1.49
$Pa^{4+} + e \rightleftharpoons Pa^{3+}$	-1.9
$Pb^{2+} + 2 e \rightleftharpoons Pb$	-0.1262
$Pb^{2+} + 2 e \rightleftharpoons Pb(Hg)$	-0.1205
$PbBr_2 + 2 e \rightleftharpoons Pb + 2 Br^-$	-0.284
$PbCl_2 + 2 e \rightleftharpoons Pb + 2 Cl^-$	-0.2675
$PbF_2 + 2 e \rightleftharpoons Pb + 2 F^-$	-0.3444
$PbHPO_4 + 2 e \rightleftharpoons Pb + HPO_4^{2-}$	-0.465
$PbI_2 + 2 e \rightleftharpoons Pb + 2 I^-$	-0.365
$PbO + H_2O + 2 e \rightleftharpoons Pb + 2 OH^-$	-0.580
$PbO_2 + 4 H^+ + 2 e \rightleftharpoons Pb^{2+} + 2 H_2O$	1.455
$HPbO_2^- + H_2O + 2 e \rightleftharpoons Pb + 3 OH^-$	-0.537
$PbO_2 + H_2O + 2 e \rightleftharpoons PbO + 2 OH^-$	0.247
$PbO_2 + SO_4^{2-} + 4 H^+ + 2 e \rightleftharpoons PbSO_4 + 2 H_2O$	1.6913
$PbSO_4 + 2 e \rightleftharpoons Pb + SO_4^{2-}$	-0.3588
$PbSO_4 + 2 e \rightleftharpoons Pb(Hg) + SO_4^{2-}$	-0.3505
$Pd^{2+} + 2 e \rightleftharpoons Pd$	0.951
$[PdCl_4]^{2-} + 2 e \rightleftharpoons Pd + 4 Cl^-$	0.591
$[PdCl_6]^{2-} + 2 e \rightleftharpoons [PdCl_4]^{2-} + 2 Cl^-$	1.288
$Pd(OH)_2 + 2 e \rightleftharpoons Pd + 2 OH^-$	0.07
$Pm^{2+} + 2 e \rightleftharpoons Pm$	-2.2
$Pm^{3+} + 3 e \rightleftharpoons Pm$	-2.30
$Pm^{3+} + e \rightleftharpoons Pm^{2+}$	-2.6
$Po^{2+} + 2 e \rightleftharpoons Po$	0.368
$PoO_2 + 4 H^+ + 2 e \rightleftharpoons Po^{2+} + 2 H_2O$	1.095
$Pr^{4+} + e \rightleftharpoons Pr^{3+}$	3.2
$Pr^{2+} + 2 e \rightleftharpoons Pr$	-2.0
$Pr^{3+} + 3 e \rightleftharpoons Pr$	-2.353
$Pr^{3+} + e \rightleftharpoons Pr^{2+}$	-3.1
$Pt^{2+} + 2 e \rightleftharpoons Pt$	1.18
$[PtCl_4]^{2-} + 2 e \rightleftharpoons Pt + 4 Cl^-$	0.755
$[PtCl_6]^{2-} + 2 e \rightleftharpoons [PtCl_4]^{2-} + 2 Cl^-$	0.68
$Pt(OH)_2 + 2 e \rightleftharpoons Pt + 2 OH^-$	0.14
$PtO_3 + 2 H^+ + 2 e \rightleftharpoons PtO_2 + H_2O$	1.7
$PtO_3 + 4 H^+ + 2 e \rightleftharpoons Pt(OH)_2^{2+} + H_2O$	1.5
$PtOH^+ + H^+ + 2 e \rightleftharpoons Pt + H_2O$	1.2
$PtO_2 + 2 H^+ + 2 e \rightleftharpoons PtO + H_2O$	1.01
$PtO_2 + 4 H^+ + 4 e \rightleftharpoons Pt + 2 H_2O$	1.00
$Pu^{3+} + 3 e \rightleftharpoons Pu$	-2.031
$Pu^{4+} + e \rightleftharpoons Pu^{3+}$	1.006
$Pu^{5+} + e \rightleftharpoons Pu^{4+}$	1.099
$PuO_2(OH)_2 + 2 H^+ + 2 e \rightleftharpoons Pu(OH)_4$	1.325
$PuO_2(OH)_2 + H^+ + e \rightleftharpoons PuO_2OH + H_2O$	1.062
$Ra^{2+} + 2 e \rightleftharpoons Ra$	-2.8
$Rb^+ + e \rightleftharpoons Rb$	-2.98
$Re^{3+} + 3 e \rightleftharpoons Re$	0.300
$ReO_4^- + 4 H^+ + 3 e \rightleftharpoons ReO_2 + 2 H_2O$	0.510
$ReO_2 + 4 H^+ + 4 e \rightleftharpoons Re + 2 H_2O$	0.2513
$ReO_4^- + 2 H^+ + e \rightleftharpoons ReO_3 + H_2O$	0.768
$ReO_4^- + 4 H_2O + 7 e \rightleftharpoons Re + 8 OH^-$	-0.604
$ReO_4^- + 8 H^+ + 7 e \rightleftharpoons Re + 4 H_2O$	0.34
$Rh^+ + e \rightleftharpoons Rh$	0.600
$Rh^{3+} + 3 e \rightleftharpoons Rh$	0.758
$[RhCl_6]^{3-} + 3 e \rightleftharpoons Rh + 6 Cl^-$	0.431
$RhOH^{2+} + H^+ + 3 e \rightleftharpoons Rh + H_2O$	0.83
$Ru^{2+} + 2 e \rightleftharpoons Ru$	0.455
$Ru^{3+} + e \rightleftharpoons Ru^{2+}$	0.2487
$RuO_2 + 4 H^+ + 2 e \rightleftharpoons Ru^{2+} + 2 H_2O$	1.120
$RuO_4^- + e \rightleftharpoons RuO_4^{2-}$	0.59
$RuO_4 + e \rightleftharpoons RuO_4^-$	1.00
$RuO_4 + 6 H^+ + 4 e \rightleftharpoons Ru(OH)_2^{2+} + 2 H_2O$	1.40
$RuO_4 + 8 H^+ + 8 e \rightleftharpoons Ru + 4 H_2O$	1.038
$[Ru(bipy)_3]^{3+} + e \rightleftharpoons [Ru(bipy)_3]^{2+}$	1.24
$[Ru(H_2O)_6]^{3+} + e \rightleftharpoons [Ru(H_2O)_6]^{2+}$	0.23
$[Ru(NH_3)_6]^{3+} + e \rightleftharpoons [Ru(NH_3)_6]^{2+}$	0.10
$[Ru(en)_3]^{3+} + e \rightleftharpoons [Ru(en)_3]^{2+}$	0.210
$[Ru(CN)_6]^{3-} + e \rightleftharpoons [Ru(CN)_6]^{4-}$	0.86
$S + 2 e \rightleftharpoons S^{2-}$	-0.47627
$S + 2H^+ + 2 e \rightleftharpoons H_2S(aq)$	0.142
$S + H_2O + 2 e \rightleftharpoons SH^- + OH^-$	-0.478
$2 S + 2 e \rightleftharpoons S_2^{2-}$	-0.42836
$S_2O_6^{2-} + 4 H^+ + 2 e \rightleftharpoons 2 H_2SO_3$	0.564
$S_2O_8^{2-} + 2 e \rightleftharpoons 2 SO_4^{2-}$	2.010
$S_2O_8^{2-} + 2 H^+ + 2 e \rightleftharpoons 2 HSO_4^-$	2.123
$S_4O_6^{2-} + 2 e \rightleftharpoons 2 S_2O_3^{2-}$	0.08
$2 H_2SO_3 + H^+ + 2 e \rightleftharpoons HS_2O_4^- + 2 H_2O$	-0.056
$H_2SO_3 + 4 H^+ + 4 e \rightleftharpoons S + 3 H_2O$	0.449
$2 SO_3^{2-} + 2 H_2O + 2 e \rightleftharpoons S_2O_4^{2-} + 4 OH^-$	-1.12
$2 SO_3^{2-} + 3 H_2O + 4 e \rightleftharpoons S_2O_3^{2-} + 6 OH^-$	-0.571
$SO_4^{2-} + 4 H^+ + 2 e \rightleftharpoons H_2SO_3 + H_2O$	0.172
$2 SO_4^{2-} + 4 H^+ + 2 e \rightleftharpoons S_2O_6^{2-} + H_2O$	-0.22
$SO_4^{2-} + H_2O + 2 e \rightleftharpoons SO_3^{2-} + 2 OH^-$	-0.93
$Sb + 3 H^+ + 3 e \rightleftharpoons SbH_3$	-0.510
$Sb_2O_3 + 6 H^+ + 6 e \rightleftharpoons 2 Sb + 3 H_2O$	0.152
$Sb_2O_5(senarmontite) + 4 H^+ + 4 e \rightleftharpoons Sb_2O_3 + 2 H_2O$	0.671
$Sb_2O_5(valentinite) + 4 H^+ + 4 e \rightleftharpoons Sb_2O_3 + 2 H_2O$	0.649
$Sb_2O_5 + 6 H^+ + 4 e \rightleftharpoons 2 SbO^+ + 3 H_2O$	0.581
$SbO^+ + 2 H^+ + 3 e \rightleftharpoons Sb + 2 H_2O$	0.212
$SbO_2^- + 2 H_2O + 3 e \rightleftharpoons Sb + 4 OH^-$	-0.66
$SbO_3^- + H_2O + 2 e \rightleftharpoons SbO_2^- + 2 OH^-$	-0.59
$Sc^{3+} + 3 e \rightleftharpoons Sc$	-2.077
$Se + 2 e \rightleftharpoons Se^{2-}$	-0.670
$Se + 2 H^+ + 2 e \rightleftharpoons H_2Se(aq)$	-0.399
$H_2SeO_3 + 4 H^+ + 4 e \rightleftharpoons Se + 3 H_2O$	0.74
$Se + 2 H^+ + 2 e \rightleftharpoons H_2Se$	-0.082
$SeO_3^{2-} + 3 H_2O + 4 e \rightleftharpoons Se + 6 OH^-$	-0.366
$SeO_4^{2-} + 4 H^+ + 2 e \rightleftharpoons H_2SeO_3 + H_2O$	1.151
$SeO_4^{2-} + H_2O + 2 e \rightleftharpoons SeO_3^{2-} + 2 OH^-$	0.05
$SiF_6^{2-} + 4 e \rightleftharpoons Si + 6 F^-$	-1.24

Reaction	$E°$/V
$SiO + 2 H^+ + 2 e \rightleftharpoons Si + H_2O$	-0.8
$SiO_2(quartz) + 4 H^+ + 4 e \rightleftharpoons Si + 2 H_2O$	0.857
$SiO_3^{2-} + 3 H_2O + 4 e \rightleftharpoons Si + 6 OH^-$	-1.697
$Sm^{3+} + e \rightleftharpoons Sm^{2+}$	-1.55
$Sm^{3+} + 3 e \rightleftharpoons Sm$	-2.304
$Sm^{2+} + 2 e \rightleftharpoons Sm$	-2.68
$Sn^{2+} + 2 e \rightleftharpoons Sn$	-0.1375
$Sn^{4+} + 2 e \rightharpoonup Sn^{2+}$	0.151
$Sn(OH)_3^+ + 3 H^+ + 2 e \rightleftharpoons Sn^{2+} + 3 H_2O$	0.142
$SnO_2 + 4 H^+ + 2 e \rightleftharpoons Sn^{2+} + 2 H_2O$	-0.094
$SnO_2 + 4 H^+ + 4 e \rightleftharpoons Sn + 2 H_2O$	-0.117
$SnO_2 + 3 H^+ + 2 e \rightleftharpoons SnOH^+ + H_2O$	-0.194
$SnO_2 + 2 H_2O + 4 e \rightleftharpoons Sn + 4 OH^-$	-0.945
$HSnO_2^- + H_2O + 2 e \rightleftharpoons Sn + 3 OH^-$	-0.909
$Sn(OH)_6^{2-} + 2 e \rightleftharpoons HSnO_2^- + 3 OH^- + H_2O$	-0.93
$Sr^+ + e \rightleftharpoons Sr$	-4.10
$Sr^{2+} + 2 e \rightleftharpoons Sr$	-2.899
$Sr^{2+} + 2 e \rightleftharpoons Sr(Hg)$	-1.793
$Sr(OH)_2 + 2 e \rightleftharpoons Sr + 2 OH^-$	-2.88
$Ta_2O_5 + 10 H^+ + 10 e \rightleftharpoons 2 Ta + 5 H_2O$	-0.750
$Ta^{3+} + 3 e \rightleftharpoons Ta$	-0.6
$Tc^{2+} + 2 e \rightleftharpoons Tc$	0.400
$TcO_4^- + 4 H^+ + 3 e \rightleftharpoons TcO_2 + 2 H_2O$	0.738
$Tc^{3+} + e \rightleftharpoons Tc^{2+}$	0.3
$TcO_4^- + 8 H^+ + 7 e \rightleftharpoons Tc + 4 H_2O$	0.472
$Tb^{4+} + e \rightleftharpoons Tb^{3+}$	3.1
$Tb^{3+} + 3 e \rightleftharpoons Tb$	-2.28
$Te + 2 e \rightleftharpoons Te^{2-}$	-1.143
$Te + 2 H^+ + 2 e \rightleftharpoons H_2Te$	-0.793
$Te^{4+} + 4 e \rightleftharpoons Te$	0.568
$TeO_2 + 4 H^+ + 4 e \rightleftharpoons Te + 2 H_2O$	0.593
$TeO_3^{2-} + 3 H_2O + 4 e \rightleftharpoons Te + 6 OH^-$	-0.57
$TeO_4^- + 8 H^+ + 7 e \rightleftharpoons Te + 4 H_2O$	0.472
$H_6TeO_6 + 2 H^+ + 2 e \rightleftharpoons TeO_2 + 4 H_2O$	1.02
$Th^{4+} + 4 e \rightleftharpoons Th$	-1.899
$ThO_2 + 4 H^+ + 4 e \rightleftharpoons Th + 2 H_2O$	-1.789
$Th(OH)_4 + 4 e \rightleftharpoons Th + 4 OH^-$	-2.48
$Ti^{2+} + 2 e \rightleftharpoons Ti$	-1.628
$Ti^{3+} + e \rightleftharpoons Ti^{2+}$	-0.369
$TiO_2 + 4 H^+ + 2 e \rightleftharpoons Ti^{2+} + 2 H_2O$	-0.502
$Ti^{3+} + 3 e \rightleftharpoons Ti$	-1.209
$TiOH^{3+} + H^+ + e \rightleftharpoons Ti^{3+} + H_2O$	-0.055
$Tl^+ + e \rightleftharpoons Tl$	-0.336
$Tl^+ + e \rightleftharpoons Tl(Hg)$	-0.3338
$Tl^{3+} + 2 e \rightleftharpoons Tl^+$	1.252
$Tl^{3+} + 3 e \rightleftharpoons Tl$	0.741
$TlBr + e \rightleftharpoons Tl + Br^-$	-0.658
$TlCl + e \rightleftharpoons Tl + Cl^-$	-0.5568
$TlI + e \rightleftharpoons Tl + I^-$	-0.752
$Tl_2O_3 + 3 H_2O + 4 e \rightleftharpoons 2 Tl^+ + 6 OH^-$	0.02
$TlOH + e \rightleftharpoons Tl + OH^-$	-0.34
$Tl(OH)_3 + 2 e \rightleftharpoons TlOH + 2 OH^-$	-0.05
$Tl_2SO_4 + 2 e \rightleftharpoons Tl + SO_4^{2-}$	-0.4360
$Tm^{3+} + e \rightleftharpoons Tm^{2+}$	-2.2
$Tm^{3+} + 3 e \rightleftharpoons Tm$	-2.319
$Tm^{2+} + 2 e \rightleftharpoons Tm$	-2.4
$U^{3+} + 3 e \rightleftharpoons U$	-1.66
$U^{4+} + e \rightleftharpoons U^{3+}$	-0.52
$UO_2^+ + 4 H^+ + e \rightleftharpoons U^{4+} + 2 H_2O$	0.612
$UO_2^{2+} + e \rightleftharpoons UO^{+}_2$	0.16
$UO_2^{2+} + 4 H^+ + 2 e \rightleftharpoons U^{4+} + 2 H_2O$	0.327
$UO_2^{2+} + 4 H^+ + 6 e \rightleftharpoons U + 2 H_2O$	-1.444
$V^{2+} + 2 e \rightleftharpoons V$	-1.175
$V^{3+} + e \rightleftharpoons V^{2+}$	-0.255
$VO^{2+} + 2 H^+ + e \rightleftharpoons V^{3+} + H_2O$	0.337
$VO_2^+ + 2 H^+ + e \rightleftharpoons VO^{2+} + H_2O$	0.991
$V_2O_5 + 6 H^+ + 2 e \rightleftharpoons 2 VO^{2+} + 3 H_2O$	0.957
$V_2O_5 + 10 H^+ + 10 e \rightleftharpoons 2 V + 5 H_2O$	-0.242
$V(OH)_4^+ + 2 H^+ + e \rightleftharpoons VO^{2+} + 3 H_2O$	1.00
$V(OH)_4^+ + 4 H^+ + 5 e \rightleftharpoons V + 4 H_2O$	-0.254
$[V(phen)_3]^{3+} + e \rightleftharpoons [V(phen)_3]^{2+}$	0.14
$W^{3+} + 3 e \rightleftharpoons W$	0.1
$W_2O_5 + 2 H^+ + 2 e \rightleftharpoons 2 WO_2 + H_2O$	-0.031
$WO_2 + 4 H^+ + 4 e \rightleftharpoons W + 2 H_2O$	-0.119
$WO_3 + 6 H^+ + 6 e \rightleftharpoons W + 3 H_2O$	-0.090
$WO_3 + 2 H^+ + 2 e \rightleftharpoons WO_2 + H_2O$	0.036
$2 WO_3 + 2 H^+ + 2 e \rightleftharpoons W_2O_5 + H_2O$	-0.029
$H_4XeO_6 + 2 H^+ + 2 e \rightleftharpoons XeO_3 + 3 H_2O$	2.42
$XeO_3 + 6 H^+ + 6 e \rightleftharpoons Xe + 3 H_2O$	2.10
$XeF + e \rightleftharpoons Xe + F^-$	3.4
$Y^{3+} + 3 e \rightleftharpoons Y$	-2.372
$Yb^{3+} + e \rightleftharpoons Yb^{2+}$	-1.05
$Yb^{3+} + 3 e \rightleftharpoons Yb$	-2.19
$Yb^{2+} + 2 e \rightleftharpoons Yb$	-2.76
$Zn^{2+} + 2 e \rightleftharpoons Zn$	-0.7618
$Zn^{2+} + 2 e \rightleftharpoons Zn(Hg)$	-0.7628
$ZnO_2^{2-} + 2 H_2O + 2 e \rightleftharpoons Zn + 4 OH^-$	-1.215
$ZnSO_4 \cdot 7 H_2O + 2 e = Zn(Hg) + SO_4^{2-} + 7 H_2O$ (Saturated $ZnSO_4$)	-0.7993
$ZnOH^+ + H^+ + 2 e \rightleftharpoons Zn + H_2O$	-0.497
$Zn(OH)_4^{2-} + 2 e \rightleftharpoons Zn + 4 OH^-$	-1.199
$Zn(OH)_2 + 2 e \rightleftharpoons Zn + 2 OH^-$	-1.249
$ZnO + H_2O + 2 e \rightleftharpoons Zn + 2 OH^-$	-1.260
$ZrO_2 + 4 H^+ + 4 e \rightleftharpoons Zr + 2 H_2O$	-1.553
$ZrO(OH)_2 + H_2O + 4 e \rightleftharpoons Zr + 4 OH^-$	-2.36
$Zr^{4+} + 4 e \rightleftharpoons Zr$	-1.45

TABLE 2. Reduction Reactions Having $E°$ Values More Positive Than That of the Standard Hydrogen Electrode

Reaction	$E°$/V
$2 H^+ + 2 e \rightleftharpoons H_2$	0.00000
$CuI_2^- + e \rightleftharpoons Cu + 2 I^-$	0.00
$Ge^{4+} + 2 e \rightleftharpoons Ge^{2+}$	0.00
$NO_3^- + H_2O + 2 e \rightleftharpoons NO_2^- + 2 OH^-$	0.01
$Tl_2O_3 + 3 H_2O + 4 e \rightleftharpoons 2 Tl^+ + 6 OH^-$	0.02
$SeO_4^{2-} + H_2O + 2 e \rightleftharpoons SeO_3^{2-} + 2 OH^-$	0.05
$WO_3 + 2 H^+ + 2 e \rightleftharpoons WO_2 + H_2O$	0.036
$Pd(OH)_2 + 2 e \rightleftharpoons Pd + 2 OH^-$	0.07
$AgBr + e \rightleftharpoons Ag + Br^-$	0.07133
$MoO_3 + 6 H^+ + 6 e \rightleftharpoons Mo + 3 H_2O$	0.075
$S_4O_6^{2-} + 2 e \rightleftharpoons 2 S_2O_3^{2-}$	0.08
$H_3Mo_7O_{24}^{3-} + 45 H^+ + 42 e \rightleftharpoons 7 Mo + 24 H_2O$	0.082
$AgSCN + e \rightleftharpoons Ag + SCN^-$	0.8951
$N_2 + 2 H_2O + 6 H^+ + 6 e \rightleftharpoons 2 NH_4OH$	0.092
$HgO + H_2O + 2 e \rightleftharpoons Hg + 2 OH^-$	0.0977
$Ir_2O_3 + 3 H_2O + 6 e \rightleftharpoons 2 Ir + 6 OH^-$	0.098
$2 NO + 2 e \rightleftharpoons N_2O_2^{2-}$	0.10
$[Ru(NH_3)_6]^{3+} + e \rightleftharpoons [Ru(NH_3)_6]^{2+}$	0.10
$W^{3+} + 3 e \rightleftharpoons W$	0.1
$[Co(NH_3)_6]^{3+} + e \rightleftharpoons [Co(NH_3)_6]^{2+}$	0.108
$Hg_2O + H_2O + 2 e \rightleftharpoons 2 Hg + 2 OH^-$	0.123
$Ge^{4+} + 4 e \rightleftharpoons Ge$	0.124
$Hg_2Br_2 + 2 e \rightleftharpoons 2 Hg + 2 Br^-$	0.13923
$Pt(OH)_2 + 2 e \rightleftharpoons Pt + 2 OH^-$	0.14
$[V(phen)_3]^{3+} + e \rightleftharpoons [V(phen)_3]^{2+}$	0.14
$S + 2H^+ + 2 e \rightleftharpoons H_2S(aq)$	0.142
$Sn(OH)_3^+ + 3 H^+ + 2 e \rightleftharpoons Sn^{2+} + 3 H_2O$	0.142
$Np^{4+} + e \rightleftharpoons Np^{3+}$	0.147
$Ag_4[Fe(CN)_6] + 4e \rightleftharpoons 4Ag + [Fe(CN)_6]^{4-}$	0.1478
$IO_3^- + 2 H_2O + 4 e \rightleftharpoons IO^- + 4 OH^-$	0.15
$Mn(OH)_3 + e \rightleftharpoons Mn(OH)_2 + OH^-$	0.15
$2 NO_2^- + 3 H_2O + 4 e \rightleftharpoons N_2O + 6 OH^-$	0.15
$Sn^{4+} + 2 e \rightleftharpoons Sn^{2+}$	0.151
$Sb_2O_3 + 6 H^+ + 6 e \rightleftharpoons 2 Sb + 3 H_2O$	0.152
$Cu^{2+} + e \rightleftharpoons Cu^+$	0.153
$BiOCl + 2 H^+ + 3 e \rightleftharpoons Bi + Cl^- + H_2O$	0.1583
$BiCl_4^- + 3 e \rightleftharpoons Bi + 4 Cl^-$	0.16
$Fe_2O_3 + 4 H^+ + 2 e \rightleftharpoons 2 FeOH^+ + H_2O$	0.16
$UO_2^{2+} + e = UO_2^+$	0.16
$Co(OH)_3 + e \rightleftharpoons Co(OH)_2 + OH^-$	0.17
$SO_4^{2-} + 4 H^+ + 2 e \rightleftharpoons H_2SO_3 + H_2O$	0.172
$Bi^{3+} + 2 e \rightleftharpoons Bi^+$	0.2
$At_2 + 2 e \rightleftharpoons 2 At^-$	0.2
$[Ru(en)_3]^{3+} + e \rightleftharpoons [Ru(en)_3]^{2+}$	0.210
$SbO^+ + 2 H^+ + 3 e \rightleftharpoons Sb + 2 H_2O$	0.212
$AgCl + e \rightleftharpoons Ag + Cl^-$	0.22233
$[Ru(H_2O)_6]^{3+} + e \rightleftharpoons [Ru(H_2O)_6]^{2+}$	0.23
$As_2O_3 + 6 H^+ + 6 e \rightleftharpoons 2 As + 3 H_2O$	0.234
Calomel electrode, saturated NaCl (SSCE)	0.2360
$Ge^{2+} + 2 e \rightleftharpoons Ge$	0.24
$Ru^{3+} + e \rightleftharpoons Ru^{2+}$	0.24
Calomel electrode, saturated KCl	0.2412
$PbO_2 + H_2O + 2 e \rightleftharpoons PbO + 2 OH^-$	0.247
$HAsO_2 + 3 H^+ + 3 e \rightleftharpoons As + 2 H_2O$	0.248
$Ru^{3+} + e \rightharpoonup Ru^{2+}$	0.2487
$ReO_2 + 4 H^+ + 4 e \rightleftharpoons Re + 2 H_2O$	0.2513
$IO_3^- + 3 H_2O + 6 e \rightleftharpoons I^- + OH^-$	0.26
$Hg_2Cl_2 + 2 e \rightleftharpoons 2 Hg + 2 Cl^-$	0.26808
Calomel electrode, 1 molal KCl	0.2800
Calomel electrode, 1 molar KCl (NCE)	0.2801
$Re^{3+} + 3 e \rightleftharpoons Re$	0.300
$Tc^{3+} + e \rightleftharpoons Tc^{2+}$	0.3
$Bi^{3+} + 3 e \rightleftharpoons Bi$	0.308

Reaction	E°/V
$BiO^+ + 2 H^+ + 3 e \rightleftharpoons Bi + H_2O$	0.320
$UO_2^{2+} + 4 H^+ + 2 e \rightleftharpoons U^{4+} + 2 H_2O$	0.327
$ClO_3^- + H_2O + 2 e \rightleftharpoons ClO_2^- + 2 OH^-$	0.33
$2 HCNO + 2 H^+ + 2 e \rightleftharpoons (CN)_2 + 2 H_2O$	0.330
Calomel electrode, 0.1 molar KCl	0.3337
$VO^{2+} + 2 H^+ + e \rightleftharpoons V^{3+} + H_2O$	0.337
$MnO_4^- + 4 H_2O + 5 e \rightleftharpoons Mn(OH)_2 + 6 OH^-$	0.34
$ReO_4^- + 8 H^+ + 7 e \rightleftharpoons Re + 4 H_2O$	0.34
$Cu^{2+} + 2 e \rightleftharpoons Cu$	0.3419
$Ag_2O + H_2O + 2 e \rightleftharpoons 2 Ag + 2 OH^-$	0.342
$Cu^{2+} + 2 e \rightleftharpoons Cu(Hg)$	0.345
$AgIO_3 + e \rightleftharpoons Ag + IO_3^-$	0.354
$[Fe(CN)_6]^{3-} + e \rightleftharpoons [Fe(CN)_6]^{4-}$	0.358
$ClO_4^- + H_2O + 2 e \rightleftharpoons ClO^-{}_3 + 2 OH^-$	0.36
$Ag_2SeO_3 + 2 e \rightleftharpoons 2 Ag + SeO_3^{2-}$	0.3629
$Po^{2+} + 2 e \rightleftharpoons Po$	0.368
$(CN)_2 + 2 H^+ + 2 e \rightleftharpoons 2 HCN$	0.373
$[Ferrocenium]^+ + e \rightleftharpoons$ ferrocene	0.400
$Tc^{2+} + 2 e \rightleftharpoons Tc$	0.400
$O_2 + 2 H_2O + 4 e \rightleftharpoons 4 OH^-$	0.401
$AgOCN + e \rightleftharpoons Ag + OCN^-$	0.41
$[RhCl_6]^{3-} + 3 e \rightleftharpoons Rh + 6 Cl^-$	0.431
$Ag_2CrO_4 + 2 e \rightleftharpoons 2 Ag + CrO_4^{2-}$	0.4470
$H_2SO_3 + 4 H^+ + 4 e \rightleftharpoons S + 3 H_2O$	0.449
$Ru^{2+} + 2 e \rightleftharpoons Ru$	0.455
$Ag_2MoO_4 + 2 e \rightleftharpoons 2 Ag + MoO_4^{2-}$	0.4573
$Ag_2C_2O_4 + 2 e \rightleftharpoons 2 Ag + C_2O_4^{2-}$	0.4647
$Ag_2WO_4 + 2 e \rightleftharpoons 2 Ag + WO_4^{2-}$	0.4660
$Ag_2CO_3 + 2 e \rightleftharpoons 2 Ag + CO_3^{2-}$	0.47
$TcO_4^- + 8 H^+ + 7 e \rightleftharpoons Tc + 4 H_2O$	0.472
$TeO_4^- + 8 H^+ + 7 e \rightleftharpoons Te + 4 H_2O$	0.472
$IO^- + H_2O + 2 e \rightleftharpoons I^- + 2 OH^-$	0.485
$NiO_2 + 2 H_2O + 2 e \rightleftharpoons Ni(OH)_2 + 2 OH^-$	0.490
$Bi^+ + e \rightleftharpoons Bi$	0.5
$ReO_4^- + 4 H^+ + 3 e \rightleftharpoons ReO_2 + 2 H_2O$	0.510
$Hg_2(ac)_2 + 2 e \rightleftharpoons 2 Hg + 2(ac)^-$	0.51163
$Cu^+ + e \rightleftharpoons Cu$	0.521
$I_2 + 2 e \rightleftharpoons 2 I^-$	0.5355
$I_3^- + 2 e \rightleftharpoons 3 I^-$	0.536
$AgBrO_3 + e \rightleftharpoons Ag + BrO_3^-$	0.546
$MnO_4^- + e \rightleftharpoons MnO_4^-$	0.558
$H_3AsO_4 + 2 H^+ + 2 e \rightleftharpoons HAsO_2 + 2 H_2O$	0.560
$S_2O_6^{2-} + 4 H^+ + 2 e \rightleftharpoons 2 H_2SO_3$	0.564
$AgNO_2 + e \rightleftharpoons Ag + NO_2^-$	0.564
$Te^{4+} + 4 e \rightleftharpoons Te$	0.568
$Sb_2O_5 + 6 H^+ + 4 e \rightleftharpoons 2 SbO^+ + 3 H_2O$	0.581
$RuO_4^- + e \rightleftharpoons RuO_4^{2-}$	0.59
$[PdCl_4]^{2-} + 2 e \rightleftharpoons Pd + 4 Cl^-$	0.591
$TeO_2 + 4 H^+ + 4 e \rightleftharpoons Te + 2 H_2O$	0.593
$MnO_4^- + 2 H_2O + 3 e \rightleftharpoons MnO_2 + 4 OH^-$	0.595
$Rh^{2+} + 2 e \rightleftharpoons Rh$	0.600
$Rh^+ + e \rightleftharpoons Rh$	0.600
$MnO_4^{2-} + 2 H_2O + 2 e \rightleftharpoons MnO_2 + 4 OH^-$	0.60
$2 AgO + H_2O + 2 e \rightleftharpoons Ag_2O + 2 OH^-$	0.607
$BrO_3^- + 3 H_2O + 6 e \rightleftharpoons Br^- + 6 OH^-$	0.61
$UO_2^+ + 4 H^+ + e \rightleftharpoons U^{4+} + 2 H_2O$	0.612
$Hg_2SO_4 + 2 e \rightleftharpoons 2 Hg + SO_4^{2-}$	0.6125
$ClO_3^- + 3 H_2O + 6 e \rightleftharpoons Cl^- + 6 OH^-$	0.62
$Hg_2HPO_4 + 2 e \rightleftharpoons 2 Hg + HPO_4^{2-}$	0.6359
$Ag(ac) + e \rightleftharpoons Ag + (ac)^-$	0.643
Sb_2O_5(valentinite) $+ 4 H^+ + 4 e \rightleftharpoons Sb_2O_3 + 2 H_2O$	0.649
$Ag_2SO_4 + 2 e \rightleftharpoons 2 Ag + SO_4^{2-}$	0.654
$ClO_2^- + H_2O + 2 e \rightleftharpoons ClO^- + 2 OH^-$	0.66
Sb_2O_5(senarmontite) $+ 4 H^+ + 4 e \rightleftharpoons Sb_2O_5 + 2 H_2O$	0.671
$[PtCl_6]^{2-} + 2 e \rightleftharpoons [PtCl_4]^{2-} + 2 Cl^-$	0.68
$O_2 + 2 H^+ + 2 e \rightleftharpoons H_2O_2$	0.695
p–benzoquinone $+ 2 H^+ + 2 e \rightleftharpoons$ hydroquinone	0.6992
$H_3IO_6^{2-} + 2 e \rightleftharpoons IO_3^- + 3 OH^-$	0.7
$TcO_4^- + 4 H^+ + 3 e \rightleftharpoons TcO_2 + 2 H_2O$	0.738
$Ag_2O_3 + H_2O + 2 e \rightleftharpoons 2 AgO + 2 OH^-$	0.739
$Tl^{3+} + 3 e \rightleftharpoons Tl$	0.741
$[PtCl_4]^{2-} + 2 e \rightleftharpoons Pt + 4 Cl^-$	0.755
$Rh^{3+} + 3 e \rightleftharpoons Rh$	0.758
$ClO_2^- + 2 H_2O + 4 e \rightleftharpoons Cl^- + 4 OH^-$	0.76
$2 NO + H_2O + 2 e \rightleftharpoons N_2O + 2 OH^-$	0.76
$BrO^- + H_2O + 2 e \rightleftharpoons Br^- + 2 OH^-$	0.761
$ReO_4^- + 2 H^+ + e \rightleftharpoons ReO_3 + H_2O$	0.768
$(CNS)_2 + 2 e \rightleftharpoons 2 CNS^-$	0.77
$[IrCl_6]^{3-} + 3 e \rightleftharpoons Ir + 6 Cl^-$	0.77
$Fe^{3+} + e \rightleftharpoons Fe^{2+}$	0.771
$AgF + e \rightleftharpoons Ag + F^-$	0.779
$[Fe(bipy)_2]^{3+} + e \rightleftharpoons [Fe(bipy)_2]^{2+}$	0.78
$Hg_2^{2+} + 2 e \rightleftharpoons 2 Hg$	0.7973
$Ag^+ + e \rightleftharpoons Ag$	0.7996
$[Os(bipy)_3]^{3+} + e \rightleftharpoons [Os(bipy)_3]^{2+}$	0.80
$2 NO_3^- + 4 H^+ + 2 e \rightleftharpoons N_2O_4 + 2 H_2O$	0.803
$[Os(bipy)_2]^{3+} + e \rightleftharpoons [Os(bipy)_2]^{2+}$	0.81
$RhOH^{2+} + H + 3 e \rightleftharpoons Rh + H_2O$	0.83
$OsO_4 + 8 H^+ + 8 e \rightleftharpoons Os + 4 H_2O$	0.838
$ClO^- + H_2O + 2 e \rightleftharpoons Cl^- + 2 OH^-$	0.841
$Hg^{2+} + 2 e \rightleftharpoons Hg$	0.851
$AuBr_4^- + 3 e \rightleftharpoons Au + 4 Br^-$	0.854
SiO_2(quartz) $+ 4 H^+ + 4 e \rightleftharpoons Si + 2 H_2O$	0.857
$2 HNO_2 + 4 H^+ + 4 e \rightleftharpoons H_2N_2O_2 + H_2O$	0.86
$[Ru(CN)_6]^{3-} + e \rightleftharpoons [Ru(CN)_6]^{4-}$	0.86
$[IrCl_6]^{2-} + e \rightleftharpoons [IrCl_6]^{3-}$	0.8665
$N_2O_4 + 2 e \rightleftharpoons 2 NO_2^-$	0.867
$HO_2^- + H_2O + 2 e \rightleftharpoons 3 OH^-$	0.878
$2 Hg^{2+} + 2 e \rightleftharpoons Hg_2^{2+}$	0.920
$NO_3^- + 3 H^+ + 2 e \rightleftharpoons HNO_2 + H_2O$	0.934
$Pd^{2+} + 2 e \rightleftharpoons Pd$	0.951
$ClO_2(aq) + e \rightleftharpoons ClO_2^-$	0.954
$NO_3^- + 4 H^+ + 3 e \rightleftharpoons NO + 2 H_2O$	0.957
$V_2O_5 + 6 H^+ + 2 e \rightleftharpoons 2 VO^{2+} + 3 H_2O$	0.957
$AuBr_2^- + e \rightleftharpoons Au + 2 Br^-$	0.959
$HNO_2 + H^+ + e \rightleftharpoons NO + H_2O$	0.983
$HIO + H^+ + 2 e \rightleftharpoons I^- + H_2O$	0.987
$VO_2^+ + 2 H^+ + e \rightleftharpoons VO^{2+} + H_2O$	0.991
$PtO_2 + 4 H^+ + 4 e \rightleftharpoons Pt + 2 H_2O$	1.00
$RuO_4 + e \rightleftharpoons RuO_4^-$	1.00
$V(OH)_4^+ + 2 H^+ + e \rightleftharpoons VO^{2+} + 3 H_2O$	1.00
$AuCl_4^- + 3 e \rightleftharpoons Au + 4 Cl^-$	1.002
$Pu^{4+} + e \rightleftharpoons Pu^{3+}$	1.006
$PtO_2 + 2 H^+ + 2 e \rightleftharpoons PtO + H_2O$	1.01
$OsO_4 + 4 H + 4 e \rightleftharpoons OsO_2 + 2 H_2O$	1.02
$H_6TeO_6 + 2 H^+ + 2 e \rightleftharpoons TeO_2 + 4 H_2O$	1.02
$[Fe(bipy)_3]^{3+} + e \rightleftharpoons [Fe(bipy)_3]^{2+}$	1.03
$Hg(OH)_2 + 2 H^+ + 2 e \rightleftharpoons Hg + 2 H_2O$	1.034
$N_2O_4 + 4 H^+ + 4 e \rightleftharpoons 2 NO + 2 H_2O$	1.035
$RuO_4 + 8 H^+ + 8 e \rightleftharpoons Ru + 4H_2O$	1.038
$[Fe(phen)_3]^{3+} + e \rightleftharpoons [Fe(phen)_3]^{2+}$(1 molar H_2SO_4)	1.06
$PuO_2(OH)_2 + H^+ + e \rightleftharpoons PuO_2OH + H_2O$	1.062
$N_2O_4 + 2 H^+ + 2 e \rightleftharpoons 2 HNO_2$	1.065
$Br_2(l) + 2 e \rightleftharpoons 2Br^-$	1.066
$IO_3^- + 6 H^+ + 6 e \rightleftharpoons I^- + 3 H_2O$	1.085
$Br_2(aq) + 2 e \rightleftharpoons 2Br^-$	1.0873
$PoO_3 + 4 H^+ + 3 e \rightleftharpoons Po^{2+} + 2 H_2O$	1.095
$Pu^{5+} + e \rightleftharpoons Pu^{4+}$	1.099
$Cu^{2+} + 2 CN^- + e \rightleftharpoons [Cu(CN)_2]^-$	1.103
$RuO_2 + 4 H^+ + 2 e \rightleftharpoons Ru^{2+} + 2 H_2O$	1.120
$[Fe(phen)_3]^{3+} + e \rightleftharpoons [Fe(phen)_3]^{2+}$	1.147
$SeO_4^{2-} + 4 H^+ + 2 e \rightleftharpoons H_2SeO_3 + H_2O$	1.151
$ClO_3^- + 2 H^+ + e \rightleftharpoons ClO_2 + H_2O$	1.152
$Ir^{3+} + 3 e \rightharpoonup Ir$	1.156
$Pt^{2+} + 2 e \rightleftharpoons Pt$	1.18
$ClO_4^- + 2 H^+ + 2 e \rightleftharpoons ClO_3^- + H_2O$	1.189
$2 IO_3^- + 12 H^+ + 10 e \rightleftharpoons I_2 + 6 H_2O$	1.195
$PtOH^+ + H^+ + 2 e \rightleftharpoons Pt + H_2O$	1.2
$ClO_3^- + 3 H^+ + 2 e \rightleftharpoons HClO_2 + H_2O$	1.214
$MnO_2 + 4 H^+ + 2 e \rightleftharpoons Mn^{2+} + 2 H_2O$	1.224
$O_2 + 4 H^+ + 4 e \rightleftharpoons 2 H_2O$	1.229
$O_3 + H_2O + 2 e \rightleftharpoons O_2 + 2 OH^-$	1.24
$[Ru(bipy)_3]^{3+} + e \rightleftharpoons [Ru(bipy)_3]^{2+}$	1.24
$Tl^{3+} + 2 e \rightleftharpoons Tl^+$	1.252
$N_2H_5^+ + 3 H^+ + 2 e \rightleftharpoons 2 NH_4^+$	1.275
$ClO_2 + H^+ + e \rightleftharpoons HClO_2$	1.277
$[PdCl_6]^{2-} + 2 e \rightleftharpoons [PdCl_4]^{2-} + 2 Cl^-$	1.288
$2 HNO_2 + 4 H^+ + 4 e \rightleftharpoons N_2O + 3 H_2O$	1.297
$AuOH^{2+} + H^+ + 2 e \rightleftharpoons Au^+ + H_2O$	1.32
$PuO_2(OH)_2 + 2 H^- + 2 e \rightleftharpoons Pu(OH)_4$	1.325
$HBrO + H^+ + 2 e \rightleftharpoons Br^- + H_2O$	1.331
$Cr(V) + e \rightleftharpoons Cr(IV)$	1.34
$HCrO_4^- + 7 H^+ + 3 e \rightleftharpoons Cr^{3+} + 4 H_2O$	1.350
$Cl_2(g) + 2 e \rightleftharpoons 2Cl^-$	1.35827
$Cr_2O_7^{2-} + 14 H^+ + 6 e \rightleftharpoons 2 Cr^{3+} + 7 H_2O$	1.36
$ClO_4^- + 8 H^+ + 8 e \rightleftharpoons Cl^- + 4 H_2O$	1.389
$ClO_4^- + 8 H^+ + 7 e \rightleftharpoons 1/2 Cl_2 + 4 H_2O$	1.39
$No^{3+} + e \rightleftharpoons No^{2+}$	1.4
$RuO_4 + 6 H^+ + 4 e \rightleftharpoons Ru(OH)_2^{2+} + 2 H_2O$	1.40
$Au^{3+} + 2 e \rightleftharpoons Au^+$	1.401
$2 NH_3OH^+ + H^+ + 2 e \rightleftharpoons N_2H_5^+ + 2 H_2O$	1.42
$BrO_3^- + 6 H^+ + 6 e \rightleftharpoons Br^- + 3 H_2O$	1.423
$2 HIO + 2 H^+ + 2 e \rightleftharpoons I_2 + 2 H_2O$	1.439
$Au(OH)_3 + 3 H^+ + 3 e \rightleftharpoons Au^- + 3 H_2O$	1.45
$3IO_3^- + 6 H^+ + 6 e \rightleftharpoons Cl^- + 3 H_2O$	1.451
$PbO_2 + 4 H^+ + 2 e \rightleftharpoons Pb^{2+} + 2 H_2O$	1.455

Reaction	$E°$/V
$ClO_3^- + 6 H^+ + 5 e \rightleftharpoons 1/2 Cl_2 + 3 H_2O$	1.47
$CrO_2 + 4 H^+ + e \rightleftharpoons Cr^{3+} + 2 H_2O$	1.48
$BrO_3^- + 6 H^+ + 5 e \rightleftharpoons 1/2 Br_2 + 3 H_2O$	1.482
$HClO + H^+ + 2 e \rightleftharpoons Cl^- + H_2O$	1.482
$Mn_2O_3 + 6 H^+ + e \rightleftharpoons 2 Mn^{2+} + 3 H_2O$	1.485
$HO_2 + H^+ + e \rightleftharpoons H_2O_2$	1.495
$Au^{3+} + 3 e \rightleftharpoons Au$	1.498
$PtO_3 + 4 H^+ + 2 e \rightleftharpoons Pt(OH)_2^{2+} + H_2O$	1.5
$MnO_4^- + 8 H^+ + 5 e \rightleftharpoons Mn^{2+} + 4 H_2O$	1.507
$Mn^{3+} + e \rightleftharpoons Mn^{2-}$	1.5415
$HClO_2 + 3 H^+ + 4 e \rightleftharpoons Cl^- + 2 H_2O$	1.570
$HBrO + H^+ + e \rightleftharpoons 1/2 Br_2(aq) + H_2O$	1.574
$2 NO + 2 H^+ + 2 e \rightleftharpoons N_2O + H_2O$	1.591
$Bi_2O_4 + 4 H^+ + 2 e \rightleftharpoons 2 BiO^+ + 2 H_2O$	1.593
$HBrO + H^+ + e \rightleftharpoons 1/2 Br_2(l) + H_2O$	1.596
$H_5IO_6 + H^+ + 2 e \rightleftharpoons IO_3^- + 3 H_2O$	1.601
$HClO + H^+ + e \rightleftharpoons 1/2 Cl_2 + H_2O$	1.611
$HClO_2 + 3 H^+ + 3 e \rightleftharpoons 1/2 Cl_2 + 2 H_2O$	1.628
$HClO_2 + 2 H^+ + 2 e \rightleftharpoons HClO + H_2O$	1.645

Reaction	$E°$/V
$Bk^{4+} + e \rightleftharpoons Bk^{3+}$	1.67
$NiO_2 + 4 H^+ + 2 e \rightleftharpoons Ni^{2+} + 2 H_2O$	1.678
$MnO_4^- + 4 H^+ + 3 e \rightharpoonup MnO_2 + 2 H_2O$	1.679
$PbO_2 + SO_4^{2-} + 4 H^+ + 2 e \rightleftharpoons PbSO_4 + 2 H_2O$	1.6913
$Au^+ + e \rightleftharpoons Au$	1.692
$PtO_3 + 2 H^+ + 2 e \rightleftharpoons PtO_2 + H_2O$	1.7
$CeOH^{3+} + H^+ + e \rightleftharpoons Ce^{3+} + H_2O$	1.715
$Ce^{4+} + e \rightleftharpoons Ce^{3+}$	1.72
$N_2O + 2 H^+ + 2 e \rightleftharpoons N_2 + H_2O$	1.766
$H_2O_2 + 2 H^+ + 2 e \rightleftharpoons 2 H_2O$	1.776
$Ag^{3+} + e \rightleftharpoons Ag^{2+}$	1.8
$Au^{2+} + e \rightleftharpoons Au^+$	1.8
$Ag_2O_2 + 4 H^+ + e \rightleftharpoons 2 Ag + 2 H_2O$	1.802
$Co^{3+} + e \rightleftharpoons Co^{2-}$(2 molar H_2SO_4)	1.83
$Ag^{3+} + 2 e \rightleftharpoons Ag^+$	1.9
$Co^{3+} + e \rightleftharpoons Co^{2+}$	1.92
$Ag^{2+} + e \rightleftharpoons Ag^+$	1.980
$Cu_2O_3 + 6 H^+ + 2 e \rightleftharpoons 2 Cu^{2+} + 3 H_2O$	2.0
$S_2O_8^{2-} + 2 e \rightleftharpoons 2 SO_4^{2-}$	2.010
$OH + e \rightleftharpoons OH^-$	2.02

Reaction	$E°$/V
$HFeO_4^- + 7 H^+ + 3 e \rightleftharpoons Fe^{3+} + 4 H_2O$	2.07
$O_3 + 2 H^+ + 2 e \rightleftharpoons O_2 + H_2O$	2.076
$HFeO_4^- + 4 H^+ + 3 e \rightleftharpoons FeOOH + 2 H_2O$	2.08
$2 HFeO_4^- + 8 H^+ + 6 e \rightleftharpoons Fe_2O_3 + 5 H_2O$	2.09
$XeO_3 + 6 H^+ + 6 e \rightleftharpoons Xe + 3 H_2O$	2.10
$S_2O_8^{2-} + 2 H^+ + 2 e \rightleftharpoons 2 HSO_4^-$	2.123
$F_2O + 2 H^+ + 4 e \rightleftharpoons H_2O + 2 F^-$	2.153
$FeO_4^{2-} + 8 H^+ + 3 e \rightleftharpoons Fe^{3+} + 4 H_2O$	2.20
$Cu^{3+} + e \rightleftharpoons Cu^{2+}$	2.4
$H_4XeO_6 + 2 H^+ + 2 e \rightleftharpoons XeO_3 + 3 H_2O$	2.42
$O(g) + 2 H^+ + 2 e \rightleftharpoons H_2O$	2.421
$Am^{4+} + e \rightleftharpoons Am^{3+}$	2.60
$H_2N_2O_2 + 2 H^+ + 2 e \rightleftharpoons N_2 + 2 H_2O$	2.65
$F_2 + 2 e \rightleftharpoons 2 F^-$	2.866
$Cm^{4+} + e \rightleftharpoons Cm^{3+}$	3.0
$F_2 + 2 H^+ + 2 e \rightleftharpoons 2 HF$	3.053
$Tb^{4+} + e \rightleftharpoons Tb^{3+}$	3.1
$Pr^{4+} + e \rightleftharpoons Pr^{3+}$	3.2
$Cf^{4+} + e \rightleftharpoons Cf^{3+}$	3.3
$XeF + e \rightleftharpoons Xe + F^-$	3.4

TABLE 3. Reduction Reactions Having $E°$ Values More Negative Than That of the Standard Hydrogen Electrode

Reaction	$E°$/V
$2 H^+ + 2 e \rightleftharpoons H_2$	0.00000
$2 D^+ + 2 e \rightleftharpoons D_2$	-0.013
$AgCN + e \rightleftharpoons Ag + CN^-$	-0.017
$2 WO_3 + 2 H^+ + 2 e \rightleftharpoons W_2O_5 + H_2O$	-0.029
$W_2O_5 + 2 H^+ + 2 e \rightleftharpoons 2 WO_2 + H_2O$	-0.031
$Ag_2S + 2 H^+ + 2 e \rightleftharpoons 2 Ag + H_2S$	-0.0366
$Fe^{3+} + 3 e \rightleftharpoons Fe$	-0.037
$Hg_2I_2 + 2 e \rightleftharpoons 2 Hg + 2 I^-$	-0.0405
$Tl(OH)_3 + 2 e \rightleftharpoons TlOH + 2 OH^-$	-0.05
$TiOH^{3+} + H^+ + e \rightleftharpoons Ti^{3+} + H_2O$	-0.055
$2 H_2SO_3 + H^+ + 2 e \rightleftharpoons HS_2O_4^- + 2 H_2O$	-0.056
$P(white) + 3 H^+ + 3 e \rightleftharpoons PH_3(g)$	-0.063
$O_2 + H_2O + 2 e \rightleftharpoons HO_2^- + OH^-$	-0.076
$2 Cu(OH)_2 + 2 e \rightleftharpoons Cu_2O + 2 OH^- + H_2O$	-0.080
$Se + 2 H^+ + 2 e \rightleftharpoons H_2Se$	-0.082
$WO_3 + 6 H^+ + 6 e \rightleftharpoons W + 3 H_2O$	-0.090
$SnO_2 + 4 H^+ + 2 e \rightleftharpoons Sn^{2+} + 2 H_2O$	-0.094
$Md^{3+} + e \rightleftharpoons Md^{2+}$	-0.1
$P(red) + 3 H^+ + 3 e \rightleftharpoons PH_3(g)$	-0.111
$SnO_2 + 4 H^+ + 4 e \rightleftharpoons Sn + 2 H_2O$	-0.117
$GeO_2 + 2 H^+ + 2 e \rightleftharpoons GeO + H_2O$	-0.118
$WO_2 + 4 H^+ + 4 e \rightleftharpoons W + 2 H_2O$	-0.119
$Pb^{2+} + 2 e \rightleftharpoons Pb(Hg)$	-0.1205
$Pb^{2+} + 2 e \rightleftharpoons Pb$	-0.1262
$CrO_4^{2-} + 4 H_2O + 3 e \rightleftharpoons Cr(OH)_3 + 5 OH^-$	-0.13
$Sn^{2-} + 2 e \rightleftharpoons Sn$	-0.1375
$In^+ + e \rightleftharpoons In$	-0.14
$O_2 + 2 H_2O + 2 e \rightleftharpoons H_2O_2 + 2 OH^-$	-0.146
$MoO_2 + 4 H^+ + 4 e \rightleftharpoons Mo + 4 H_2O$	-0.152
$AgI + e \rightleftharpoons Ag + I^-$	-0.15224
$2 NO_2^- + 2 H_2O + 4 e \rightleftharpoons N_2O_2^{2-} + 4 OH^-$	-0.18

Reaction	$E°$/V
$H_2GeO_3 + 4 H^+ + 4 e \rightleftharpoons Ge + 3 H_2O$	-0.182
$SnO_2 + 3 H^+ + 2 e \rightleftharpoons SnOH^+ + H_2O$	-0.194
$CO_2 + 2 H^+ + 2 e \rightleftharpoons HCOOH$	-0.199
$Mo^{3+} + 3 e \rightleftharpoons Mo$	-0.200
$Ga^+ + e \rightleftharpoons Ga$	-0.2
$2 SO_3^{2-} + 4 H^+ + 2 e \rightleftharpoons S_2O_6^{2-} + H_2O$	-0.22
$Cu(OH)_2 + 2 e \rightleftharpoons Cu + 2 OH^-$	-0.222
$V_2O_5 + 10 H^+ + 10 e \rightleftharpoons 2 V + 5 H_2O$	-0.242
$CdSO_4 + 2 e \rightleftharpoons Cd + SO_4^{2-}$	-0.246
$V(OH)_4^+ + 4 H^+ + 5 e \rightleftharpoons V + 4 H_2O$	-0.254
$V^{3+} + e \rightleftharpoons V^{2+}$	-0.255
$Ni^{2+} + 2 e \rightleftharpoons Ni$	-0.257
$PbCl_2 + 2 e \rightleftharpoons Pb + 2 Cl^-$	-0.2675
$H_3PO_4 + 2 H^+ + 2 e \rightleftharpoons H_3PO_3 + H_2O$	-0.276
$Co^{2+} + 2 e \rightleftharpoons Co$	-0.28
$PbBr_2 + 2 e \rightleftharpoons Pb + 2 Br^-$	-0.284
$Tl^+ + e \rightleftharpoons Tl(Hg)$	-0.3338
$Tl^+ + e \rightleftharpoons Tl$	-0.336
$In^{3+} + 3 e \rightleftharpoons In$	-0.3382
$TlOH + e \rightleftharpoons Tl + OH^-$	-0.34
$PbF_2 + 2 e \rightleftharpoons Pb + 2 F^-$	-0.3444
$PbSO_4 + 2 e \rightleftharpoons Pb(Hg) + SO_4^{2-}$	-0.3505
$Cd^{2+} + 2 e \rightleftharpoons Cd(Hg)$	-0.3521
$PbSO_4 + 2 e \rightleftharpoons Pb + SO_4^{2-}$	-0.3588
$Cu_2O + H_2O + 2 e \rightleftharpoons 2 Cu + 2 OH^-$	-0.360
$Eu^{3+} + e \rightleftharpoons Eu^{2+}$	-0.36
$PbI_2 + 2 e \rightleftharpoons Pb + 2 I^-$	-0.365
$SeO_3^{2-} + 3 H_2O + 4 e \rightleftharpoons Se + 6 OH^-$	-0.366
$Ti^{3+} + e \rightleftharpoons Ti^{2+}$	-0.369
$Se + 2 H^+ + 2 e \rightleftharpoons H_2Se(aq)$	-0.399
$In^{2+} + e \rightleftharpoons In^+$	-0.40
$Cd^{2+} + 2 e \rightleftharpoons Cd$	-0.4030
$Cr^{3+} + e \rightleftharpoons Cr^{2+}$	-0.407

Reaction	$E°$/V
$2 S + 2 e \rightleftharpoons S_2^{2-}$	-0.42836
$Tl_2SO_4 + 2 e \rightleftharpoons Tl + SO_4^{2-}$	-0.4360
$In^{3+} + 2 e \rightleftharpoons In^+$	-0.443
$Fe^{2+} + 2 e \rightleftharpoons Fe$	-0.447
$H_3PO_3 + 3 H^+ + 3 e \rightleftharpoons P + 3 H_2O$	-0.454
$Bi_2O_3 + 3 H_2O + 6 e \rightleftharpoons 2 Bi + 6 OH^-$	-0.46
$NO_2^- + H_2O + e \rightleftharpoons NO + 2 OH$	-0.46
$PbHPO_4 + 2 e \rightleftharpoons Pb + HPO_4^{2-}$	-0.465
$S + 2 e \rightleftharpoons S^{2-}$	-0.47627
$S + H_2O + 2 e \rightleftharpoons HS^- + OH^-$	-0.478
$B(OH)_3 + 7 H^+ + 8 e \rightleftharpoons BH_4^- + 3 H_2O$	-0.481
$In^{3+} + e \rightleftharpoons In^{2+}$	-0.49
$ZnOH^+ + H^+ + 2 e \rightleftharpoons Zn + H_2O$	-0.497
$GaOH^{2+} + H^+ + 3 e \rightleftharpoons Ga + H_2O$	-0.498
$H_3PO_3 + 2 H^+ + 2 e \rightleftharpoons H_3PO_2 + H_2O$	-0.499
$TiO_2 + 4 H^+ + 2 e \rightleftharpoons Ti^{2+} + 2 H_2O$	-0.502
$H_3PO_2 + H^+ + e \rightleftharpoons P + 2 H_2O$	-0.508
$Sb + 3 H^+ + 3 e \rightleftharpoons SbH_3$	-0.510
$U^{4+} + e \rightleftharpoons U^{3+}$	-0.52
$HPbO_2^- + H_2O + 2 e \rightleftharpoons Pb + 3 OH^-$	-0.537
$Ga^{3+} + 3 e \rightleftharpoons Ga$	-0.549
$TlCl + e \rightleftharpoons Tl + Cl^-$	-0.5568
$Fe(OH)_3 + e \rightleftharpoons Fe(OH)_2 + OH^-$	-0.56
$TeO_3^{2-} + 3 H_2O + 4 e \rightleftharpoons Te + 6 OH^-$	-0.57
$2 SO_3^{2-} + 3 H_2O + 4 e \rightleftharpoons S_2O_3^{2-} + 6 OH^-$	-0.571
$PbO + H_2O + 2 e \rightleftharpoons Pb + 2 OH^-$	-0.580
$SbO_3^- + H_2O + 2 e \rightleftharpoons SbO_2^- + 2 OH^-$	-0.59
$Ta^{3+} + 3 e \rightleftharpoons Ta$	-0.6
$ReO_4^- + 4 H_2O + 7 e \rightleftharpoons Re + 8 OH^-$	-0.604
$As + 3 H^+ + 3 e \rightleftharpoons AsH_3$	-0.608
$Nb_2O_5 + 10 H^+ + 10 e \rightleftharpoons 2 Nb + 5 H_2O$	-0.644
$NbO_2 + 2 H^+ + 2 e \rightleftharpoons NbO + H_2O$	-0.646

Reaction	$E°/V$
$Cd(OH)_4^{2-} + 2 e \rightleftharpoons Cd + 4 OH^-$	-0.658
$TlBr + e \rightleftharpoons Tl + Br^-$	-0.658
$SbO_2^- + 2 H_2O + 3 e \rightleftharpoons Sb + 4 OH^-$	-0.66
$Se + 2 e \rightleftharpoons Se^{2-}$	-0.670
$AsO_2^- + 2 H_2O + 3 e \rightleftharpoons As + 4 OH^-$	-0.68
$NbO_2 + 4 H^+ + 4 e \rightleftharpoons Nb + 2 H_2O$	-0.690
$Ag_2S + 2 e \rightleftharpoons 2 Ag + S^{2-}$	-0.691
$AsO_4^{3-} + 2 H_2O + 2 e \rightleftharpoons AsO_2^- + 4 OH^-$	-0.71
$Ni(OH)_2 + 2 e \rightleftharpoons Ni + 2 OH^-$	-0.72
$Co(OH)_2 + 2 e \rightleftharpoons Co + 2 OH^-$	-0.73
$NbO + 2 H^+ + 2 e \rightleftharpoons Nb + H_2O$	-0.733
$H_2SeO_3 + 4 H^+ + 4 e \rightleftharpoons Se + 3 H_2O$	-0.74
$Cr^{3+} + 3 e \rightleftharpoons Cr$	-0.744
$Ta_2O_5 + 10 H^+ + 10 e \rightleftharpoons 2 Ta + 5 H_2O$	-0.750
$TlI + e \rightleftharpoons Tl + I^-$	-0.752
$Zn^{2+} + 2 e \rightleftharpoons Zn$	-0.7618
$Zn^{2+} + 2 e \rightleftharpoons Zn(Hg)$	-0.7628
$CdO + H_2O + 2 e \rightleftharpoons Cd + 2 OH^-$	-0.783
$Te + 2 H^+ + 2 e \rightleftharpoons H_2Te$	-0.793
$ZnSO_4 \cdot 7H_2O + 2 e \rightleftharpoons Zn(Hg) + SO_4^{2-} + 7 H_2O$ (Saturated $ZnSO_4$)	-0.7993
$Bi + 3 H^+ + 3 e \rightleftharpoons BiH_3$	-0.8
$SiO + 2 H^+ + 2 e \rightleftharpoons Si + H_2O$	-0.8
$Cd(OH)_2 + 2 e \rightleftharpoons Cd(Hg) + 2 OH^-$	-0.809
$2 H_2O + 2 e \rightleftharpoons H_2 + 2 OH^-$	-0.8277
$2 NO_3^- + 2 H_2O + 2 e \rightleftharpoons N_2O_4 + 4 OH^-$	-0.85
$H_3BO_3 + 3 H^+ + 3 e \rightleftharpoons B + 3 H_2O$	-0.8698
$P + 3 H_2O + 3 e \rightleftharpoons PH_3(g) + 3 OH^-$	-0.87
$HSnO_2^- + H_2O + 2 e \rightleftharpoons Sn + 3 OH^-$	-0.909
$Cr^{2+} + 2 e \rightleftharpoons Cr$	-0.913
$SO_4^{2-} + H_2O + 2 e \rightleftharpoons SO_3^{2-} + 2 OH^-$	-0.93
$Sn(OH)_6^{2-} + 2 e \rightleftharpoons HSnO_2^- + 3 OH^- + H_2O$	-0.93
$SnO_2 + 2 H_2O + 4 e \rightleftharpoons Sn + 4 OH^-$	-0.945
$In(OH)_3 + 3 e \rightleftharpoons In + 3 OH^-$	-0.99
$NpO_2 + H_2O + H^+ + e \rightleftharpoons Np(OH)_3$	-0.962
$In(OH)_4^- + 3 e \rightleftharpoons In + 4 OH^-$	-1.007
$In_2O_3 + 3 H_2O + 6 e \rightleftharpoons 2 In + 6 OH^-$	-1.034
$PO_4^{3-} + 2 H_2O + 2 e \rightleftharpoons HPO_3^{2-} + 3 OH^-$	-1.05
$Yb^{3+} + e \rightleftharpoons Yb^{2+}$	-1.05
$Nb^{3+} + 3 e \rightleftharpoons Nb$	-1.099
$Fm^{3+} + e \rightleftharpoons Fm^{2+}$	-1.1
$2 SO_3^{2-} + 2 H_2O + 2 e \rightleftharpoons S_2O_4^{2-} + 4 OH^-$	-1.12
$Te + 2 e \rightleftharpoons Te^{2-}$	-1.143
$V^{2+} + 2 e \rightleftharpoons V$	-1.175
$Mn^{2+} + 2 e \rightleftharpoons Mn$	-1.185
$Zn(OH)_4^{2-} + 2 e \rightleftharpoons Zn + 4 OH^-$	-1.199
$CrO_2 + 2 H_2O + 3 e \rightleftharpoons Cr + 4 OH^-$	-1.2
$No^{3+} + 3 e \rightleftharpoons No$	-1.20
$Ti^{3+} + 3 e \rightleftharpoons Ti$	-1.209
$ZnO_2^- + 2 H_2O + 2 e \rightleftharpoons Zn + 4 OH^-$	-1.215
$H_2GaO_3^- + H_2O + 3 e \rightleftharpoons Ga + 4 OH^-$	-1.219
$H_2BO_3^- + 5 H_2O + 8 e \rightleftharpoons BH_4^- + 8 OH^-$	-1.24
$SiF_6^{2-} + 4 e \rightleftharpoons Si + 6 F^-$	-1.24
$Zn(OH)_2 + 2 e \rightleftharpoons Zn + 2 OH^-$	-1.249
$ZnO + H_2O + 2 e \rightleftharpoons Zn + 2 OH^-$	-1.260
$Es^{3+} + e \rightleftharpoons Es^{2+}$	-1.3
$Pa^{3+} + 3 e \rightleftharpoons Pa$	-1.34
$Ce^{3+} + 3 e \rightleftharpoons Ce(Hg)$	-1.4373
$UO_2^{2+} + 4 H^+ + 6 e \rightleftharpoons U + 2 H_2O$	-1.444
$Zr^{4+} + 4 e \rightleftharpoons Zr$	-1.45
$Cr(OH)_3 + 3 e \rightleftharpoons Cr + 3 OH^-$	-1.48
$Pa^{4+} + 4 e \rightleftharpoons Pa$	-1.49
$HfO_2 + 4 H^+ + 4 e \rightleftharpoons Hf + 2 H_2O$	-1.505
$Hf^{4+} + 4 e \rightleftharpoons Hf$	-1.55
$Sm^{3+} + e \rightleftharpoons Sm^{2+}$	-1.55
$ZrO_2 + 4 H^+ + 4 e \rightleftharpoons Zr + 2 H_2O$	-1.553
$Mn(OH)_2 + 2 e \rightleftharpoons Mn + 2 OH^-$	-1.56
$Ba^{2+} + 2 e \rightleftharpoons Ba(Hg)$	-1.570
$Bk^{2+} + 2 e \rightleftharpoons Bk$	-1.6
$Cf^{3+} + e \rightleftharpoons Cf^{2+}$	-1.6
$Ti^{2+} + 2 e \rightleftharpoons Ti$	-1.628
$Md^{3+} + 3 e \rightleftharpoons Md$	-1.65
$HPO_3^{2-} + 2 H_2O + 2 e \rightleftharpoons H_2PO_2^- + 3 OH^-$	-1.65
$U^{3+} + 3 e \rightleftharpoons U$	-1.66
$Al^{3+} + 3 e \rightleftharpoons Al$	-1.676
$SiO_3^{2-} + H_2O + 4 e \rightleftharpoons Si + 6 OH^-$	-1.697
$HPO_3^{2-} + 2 H_2O + 3 e \rightleftharpoons P + 5 OH^-$	-1.71
$HfO^{2+} + 2 H^+ + 4 e \rightleftharpoons Hf + H_2O$	-1.724
$ThO_2 + 4 H^+ + 4 e \rightleftharpoons Th + 2 H_2O$	-1.789
$H_2BO_3^- + H_2O + 3 e \rightleftharpoons B + 4 OH^-$	-1.79
$Sr^{2+} + 2 e \rightleftharpoons Sr(Hg)$	-1.793
$H_2PO_2^- + e \rightleftharpoons P + 2 OH^-$	-1.82
$Be^{2+} + 2 e \rightleftharpoons Be$	-1.847
$Np^{3+} + 3 e \rightleftharpoons Np$	-1.856
$Fm^{3+} + 3 e \rightleftharpoons Fm$	-1.89
$Th^{4+} + 4 e \rightleftharpoons Th$	-1.899
$Am^{2+} + 2 e \rightleftharpoons Am$	-1.9
$Pa^{4+} + e \rightleftharpoons Pa^{3+}$	-1.9
$Es^{3+} + 3 e \rightleftharpoons Es$	-1.91
$Cf^{3+} + 3 e \rightleftharpoons Cf$	-1.94
$Lr^{3+} + 3 e \rightleftharpoons Lr$	-1.96
$Eu^{3+} + 3 e \rightleftharpoons Eu$	-1.991
$Er^{2+} + 2 e \rightleftharpoons Er$	-2.0
$Pr^{2+} + 2 e \rightleftharpoons Pr$	-2.0
$Pu^{3+} + 3 e \rightleftharpoons Pu$	-2.031
$Cm^{3+} + 3 e \rightleftharpoons Cm$	-2.04
$Am^{3+} + 3 e \rightleftharpoons Am$	-2.048
$AlF_6^{3-} + 3 e \rightleftharpoons Al + 6 F^-$	-2.069
$Sc^{3+} + 3 e \rightleftharpoons Sc$	-2.077
$Ho^{2+} + 2 e \rightleftharpoons Ho$	-2.1
$Nd^{2+} + 2 e \rightleftharpoons Nd$	-2.1
$Cf^{2+} + 2 e \rightleftharpoons Cf$	-2.12
$Yb^{3+} + 3 e \rightleftharpoons Yb$	-2.19
$Ac^{3+} + 3 e \rightleftharpoons Ac$	-2.20
$Dy^{2+} + 2 e \rightleftharpoons Dy$	-2.2
$Tm^{3+} + e \rightleftharpoons Tm^{2+}$	-2.2
$Pm^{2+} + 2 e \rightleftharpoons Pm$	-2.2
$Es^{2+} + 2 e \rightleftharpoons Es$	-2.23
$H_2 + 2 e \rightleftharpoons 2 H^-$	-2.23
$Gd^{3+} + 3 e \rightleftharpoons Gd$	-2.279
$Tb^{3+} + 3 e \rightleftharpoons Tb$	-2.28
$Lu^{3+} + 3 e \rightleftharpoons Lu$	-2.28
$Dy^{3+} + 3 e \rightleftharpoons Dy$	-2.295
$Am^{3+} + e \rightleftharpoons Am^{2+}$	-2.3
$Fm^{2+} + 2 e \rightleftharpoons Fm$	-2.30
$Pm^{3+} + 3 e \rightleftharpoons Pm$	-2.30
$Al(OH)_3 + 3 e \rightleftharpoons Al + 3 OH^-$	-2.30
$Sm^{3+} + 3 e \rightleftharpoons Sm$	-2.304
$Al(OH)^- + 3 e \rightleftharpoons Al + 4 OH^-$	-2.310
$Tm^{3+} + 3 e \rightleftharpoons Tm$	-2.319
$Nd^{3+} + 3 e \rightleftharpoons Nd$	-2.323
$H_2AlO_3^- + H_2O + 3 e \rightleftharpoons Al + 4 OH^-$	-2.33
$Ho^{3+} + 3 e \rightleftharpoons Ho$	-2.33
$Er^{3+} + 3 e \rightleftharpoons Er$	-2.331
$Ce^{3+} + 3 e \rightleftharpoons Ce$	-2.336
$Pr^{3+} + 3 e \rightleftharpoons Pr$	-2.353
$ZrO(OH)_2 + H_2O + 4 e \rightleftharpoons Zr + 4 OH^-$	-2.36
$Mg^{2+} + 2 e \rightleftharpoons Mg$	-2.372
$Y^{3+} + 3 e \rightleftharpoons Y$	-2.372
$La^{3+} + 3 e \rightleftharpoons La$	-2.379
$Tm^{2+} + 2 e \rightleftharpoons Tm$	-2.4
$Md^{2+} + 2 e \rightleftharpoons Md$	-2.40
$Th(OH)_4 + 4 e \rightleftharpoons Th + 4 OH^-$	-2.48
$HfO(OH)_2 + H_2O + 4 e \rightleftharpoons Hf + 4 OH^-$	-2.50
$No^{2+} + 2 e \rightleftharpoons No$	-2.50
$Dy^{3+} + e \rightleftharpoons Dy^{2+}$	-2.6
$Pm^{3+} + e \rightleftharpoons Pm^{2+}$	-2.6
$Be_2O_3^{2-} + 3 H_2O + 4 e \rightleftharpoons 2 Be + 6 OH^-$	-2.63
$Sm^{2+} + 2 e \rightleftharpoons Sm$	-2.68
$Mg(OH)_2 + 2 e \rightleftharpoons Mg + 2 OH^-$	-2.690
$Nd^{3+} + e \rightleftharpoons Nd^{2+}$	-2.7
$Mg^+ + e \rightleftharpoons Mg$	-2.70
$Na^+ + e \rightleftharpoons Na$	-2.71
$Yb^{2+} + 2 e \rightleftharpoons Yb$	-2.76
$Bk^{3+} + e \rightleftharpoons Bk^{2+}$	-2.8
$Ho^{3+} + e \rightleftharpoons Ho^{2+}$	-2.8
$Ra^{2+} + 2 e \rightleftharpoons Ra$	-2.8
$Eu^{2+} + 2 e \rightleftharpoons Eu$	-2.812
$Ca^{2+} + 2 e \rightleftharpoons Ca$	-2.868
$Sr(OH)_2 + 2 e \rightleftharpoons Sr + 2 OH^-$	-2.88
$Sr^{2+} + 2 e \rightleftharpoons Sr$	-2.899
$Fr^+ + e \rightleftharpoons Fr$	-2.9
$La(OH)_3 + 3 e \rightleftharpoons La + 3 OH^-$	-2.90
$Ba^{2+} + 2 e \rightleftharpoons Ba$	-2.912
$K^+ + e \rightleftharpoons K$	-2.931
$Rb^+ + e \rightleftharpoons Rb$	-2.98
$Ba(OH)_2 + 2 e \rightleftharpoons Ba + 2 OH^-$	-2.99
$Er^{3+} + e \rightleftharpoons Er^{2+}$	-3.0
$Ca(OH)_2 + 2 e \rightleftharpoons Ca + 2 OH^-$	-3.02
$Cs^+ + e \rightleftharpoons Cs$	-3.026
$Li^+ + e \rightleftharpoons Li$	-3.0401
$3 N_2 + 2 H^+ + 2 e \rightleftharpoons 2 HN_3$	-3.09
$Pr^{3+} + e \rightleftharpoons Pr^{2+}$	-3.1
$Ca^+ + e \rightleftharpoons Ca$	-3.80
$Sr^+ + e \rightleftharpoons Sr$	-4.10

DISSOCIATION CONSTANTS OF INORGANIC ACIDS AND BASES

The data in this table are presented as values of pK_a, defined as the negative logarithm of the acid dissociation constant K_a for the reaction

$$BH \rightleftharpoons B^- + H^+$$

Thus pK_a = –log K_a , and the hydrogen ion concentration [H⁺] can be calculated from

$$K_a = \frac{[H^+][B^-]}{[BH]}$$

In the case of bases, the entry in the table is for the conjugate acid; e.g., ammonium ion for ammonia. The OH^- concentration in the system

$$NH_3 + H_2O \rightleftharpoons NH_4^+ + OH^-$$

can be calculated from the equation

$$K_b = K_{water} / K_a = \frac{[OH^-][NH_4^+]}{[NH_3]}$$

where $K_{water} = 1.01 \times 10^{-14}$ at 25 °C. Note that pK_a + pK_b = pK_{water}.
Similarly, for the elemental ions, the relevant reaction is

$$NaOH + H_2O_{(l)} \rightleftharpoons Na^+_{(aq)} + OH^-_{(aq)}$$

where (aq) indicates both ions are surrounded by water molecules.
All values refer to dilute aqueous solutions at zero ionic strength at the temperature indicated. The table is arranged alphabetically by compound name.

Reference

1. Perrin, D. D., *Ionization Constants of Inorganic Acids and Bases in Aqueous Solution, Second Edition,* Pergamon, Oxford, 1982. <https://doi.org/10.1021/ed060pA151.2>

Acid Dissociation Constant (as Negative Logarithm) for Inorganic Acids and Bases at Specified Temperatures

Name	Formula	Step	t/°C	pK_a
Aluminum ion [Al^{+3}]	Al^{+3}		25	5.0
Ammonia	NH_3		25	9.25
Arsenic acid	H_3AsO_4	1	25	2.26
		2	25	6.76
		3	25	11.29
Arsenious acid	H_3AsO_3		25	9.29
Barium ion [Ba^{+2}]	Ba^{+2}		25	13.4
Boric acid	H_3BO_3	1	20	9.27
		2	20	14[a]
Calcium ion [Ca^{+2}]	Ca^{+2}		25	12.6
Carbonic acid	H_2CO_3	1	25	6.35
		2	25	10.33
Chlorous acid	$HClO_2$		25	1.94
Chromic acid	H_2CrO_4	1	25	0.74
		2	25	6.49
Cyanic acid	HOCN		25	3.46
Diphosphoric acid	$H_4P_2O_7$	1	25	0.91
		2	25	2.10
		3	25	6.70
		4	25	9.32
Germanic acid	H_2GeO_3	1	25	9.01
		2	25	12.3
Hydrazine	N_2H_4		25	8.1
Hydrazoic acid	HN_3		25	4.6
Hydrogen cyanide	HCN		25	9.21
Hydrogen fluoride	HF		25	3.20
Hydrogen peroxide	H_2O_2		25	11.62
Hydrogen selenide	H_2Se	1	25	3.89
		2	25	11.0
Hydrogen sulfide	H_2S	1	25	7.05
		2	25	19
Hydrogen telluride	H_2Te	1	18	2.6
		2	25	11
Hydroxylamine	H_2NOH		25	5.94
Hypobromous acid	HOBr		25	8.55
Hypochlorous acid	HOCl		25	7.40
Hypoiodous acid	HIO		25	10.5
Iodic acid	HIO_3		25	0.78
Lithium ion [Li^+]	Li^+		25	13.8
Magnesium ion [Mg^{+2}]	Mg^{+2}		25	11.4
Nitrous acid	HNO_2		25	3.25
Orthosilicic acid	H_4SiO_4	1	30	9.9
		2	30	11.8
		3	30	12
		4	30	12
Perchloric acid	$HClO_4$		20	-1.6
Periodic acid	HIO_4		25	1.64
Phosphonic acid	H_3PO_3	1	20	1.3
		2	20	6.70
Phosphoric acid	H_3PO_4	1	25	2.16
		2	25	7.21
		3	25	12.32
Selenic acid	H_2SeO_4	2	25	1.7
Selenous acid	H_2SeO_3	1	25	2.62
		2	25	8.32
Sodium ion [Na^+]	Na^+		25	14.8
Strontium ion [Sr^{+2}]	Sr^{+2}		25	13.2
Sulfamic acid	H_2NSO_3H		25	1.05
Sulfuric acid	H_2SO_4	2	25	1.99
Sulfurous acid	H_2SO_3	1	25	1.85
		2	25	7.2
Telluric(VI) acid	H_6TeO_6	1	18	7.68
		2	18	11.0
Tellurous acid	H_2TeO_3	1	25	6.27
		2	25	8.43
Tetrafluoroboric acid	HBF_4		25	0.5
Thiocyanic acid	HCNS		25	-1.8
Water	H_2O		25	13.995

[a] Lower limit.

DISSOCIATION CONSTANTS OF ORGANIC ACIDS AND BASES

This table lists the dissociation (ionization) constants of over 1070 organic acids, bases, and amphoteric compounds. All data apply to dilute aqueous solutions and are presented as values of pK_a, which is defined as the negative of the logarithm of the equilibrium constant K_a for the reaction

$$HA \rightleftharpoons H^+ + A^-$$

i.e.,

$$K_a = [H^+][A^-]/[HA]$$

where [H$^+$], etc., represent the concentrations of the respective species in mol/L. It follows that pK_a = pH + log[HA] – log[A$^-$], so that a solution with 50% dissociation has pH equal to the pK_a of the acid.

Data for bases are presented as pK_a values for the conjugate acid, i.e., for the reaction

$$BH^+ \rightleftharpoons H^+ + B$$

In older literature, an ionization constant K_b was used for the reaction B + $H_2O \rightleftharpoons BH^+ + OH^-$. This is related to K_a by

$$pK_a + pK_b = pK_{water} = 14.00 \text{ (at 25 °C)}$$

Compounds are listed alphabetically by name.

References

1. Perrin, D. D., *Dissociation Constants of Organic Bases in Aqueous Solution*, Butterworths, London, 1965; Supplement, 1972.
2. Serjeant, E. P., and Dempsey, B., *Ionization Constants of Organic Acids in Aqueous Solution*, Pergamon, Oxford, 1979.
3. Albert, A., "Ionization Constants of Heterocyclic Substances", in Katritzky, A. R., Ed., *Physical Methods in Heterocyclic Chemistry*, Academic Press, New York, 1963.
4. Sober, H. A., Ed., *CRC Handbook of Biochemistry*, CRC Press, Boca Raton, FL, 1968.
5. Perrin, D. D., Dempsey, B., and Serjeant, E. P., pK_a *Prediction for Organic Acids and Bases*, Chapman and Hall, London, 1981. <https://doi.org/10.1007/978-94-009-5883-8>
6. Albert, A., and Serjeant, E. P., *The Determination of Ionization Constants, Third Edition*, Chapman and Hall, London, 1984. <https://doi.org/10.1007/978-94-009-5548-6>
7. O'Neil, M. J., Ed., *The Merck Index, 14th Edition*, Merck & Co., Whitehouse Station, NJ, 2006.

Acid Dissociation Constants (as Negative Logarithms) of Organic Acids and Bases at Specified Temperatures

Name	Mol. form.	Step	t/°C	pK_a
Acetaldehyde	C_2H_4O		25	13.57
Acetamide	C_2H_5NO		25	15.1
Acetanilide	C_8H_9NO		25	0.5
Acetazolamide	$C_4H_6N_4O_3S_2$			7.2
Acetic acid	$C_2H_4O_2$		25	4.756
Acetoacetic acid	$C_4H_6O_3$		25	3.6
Acetohydroxamic acid	$C_2H_5NO_2$			8.70
N-Acetylglycine	$C_4H_7NO_3$		25	3.67
2-(Acetyloxy)benzoic acid	$C_9H_8O_4$		25	3.48
Aconine	$C_{25}H_{41}NO_9$			9.52
Aconitine	$C_{34}H_{47}NO_{11}$			5.88
9-Acridinamine	$C_{13}H_{10}N_2$		20	9.99
Acridine	$C_{13}H_9N$		20	5.58
3,6-Acridinediamine	$C_{13}H_{11}N_3$		20	9.65
Acrylic acid	$C_3H_4O_2$		25	4.25
Adenine	$C_5H_5N_5$	1		4.3
		2		9.83
Adenosine	$C_{10}H_{13}N_5O_4$	1	25	3.6
		2	25	12.4
Adenosine 5'-monophosphate	$C_{10}H_{14}N_5O_7P$	1		3.8
		2		6.2
Adipamic acid	$C_6H_{11}NO_3$		25	4.63
Agaritine	$C_{12}H_{17}N_3O_4$	1		3.4
		2		8.86
L-Alanine	$C_3H_7NO_2$	1	25	2.34
		2	25	9.87
β-Alanine	$C_3H_7NO_2$	1	25	3.55
		2	25	10.24
Allantoin	$C_4H_6N_4O_3$		25	8.96
Allopurinol	$C_5H_4N_4O$			10.2
Alloxanic acid	$C_4H_4N_2O_5$		25	6.64
Allyl alcohol	C_3H_6O		25	15.5
Allylamine	C_3H_7N		25	9.49
N-Allylaniline	$C_9H_{11}N$		25	4.17
Aminoacetonitrile	$C_2H_4N_2$		25	5.34
2-Aminoadipic acid	$C_6H_{11}NO_4$	1	25	2.14
		2	25	4.21
		3	25	9.77
4-Aminoazobenzene	$C_{12}H_{11}N_3$		25	2.82
2-Aminobenzenesulfonic acid	$C_6H_7NO_3S$		25	2.46
3-Aminobenzenesulfonic acid	$C_6H_7NO_3S$		25	3.74
4-Aminobenzenesulfonic acid	$C_6H_7NO_3S$		25	3.23
2-Aminobenzonitrile	$C_7H_6N_2$		25	0.77
3-Aminobenzonitrile	$C_7H_6N_2$		25	2.75
4-Aminobenzonitrile	$C_7H_6N_2$		25	1.74
2-Aminobiphenyl	$C_{12}H_{11}N$		25	3.83
3-Aminobiphenyl	$C_{12}H_{11}N$		18	4.25
4-Aminobiphenyl	$C_{12}H_{11}N$		18	4.35
DL-2-Aminobutanoic acid	$C_4H_9NO_2$	1	25	2.29
		2	25	9.83
4-Aminobutanoic acid	$C_4H_9NO_2$	1	25	4.031
		2	25	10.556
4-(2-Aminoethyl)phenol	$C_8H_{11}NO$	1	25	9.74
		2	25	10.52
6-Aminohexanoic acid	$C_6H_{13}NO_2$	1	25	4.37
		2	25	10.80
4-Amino-2-hydroxybenzoic acid	$C_7H_7NO_3$			3.25
2-Amino-4-hydroxypteridine	$C_6H_5N_5O$	1	20	2.27
		2	20	7.96
2-Amino-2-methylpropanoic acid	$C_4H_9NO_2$	1	25	2.36
		2	25	10.21
5-Amino-1-naphthol	$C_{10}H_9NO$		25	3.97

Name	Mol. form.	Step	t/°C	pK_a
5-Amino-4-oxopentanoic acid	$C_5H_9NO_3$	1	25	4.05
		2	25	8.90
5-Aminopentanoic acid	$C_5H_{11}NO_2$	1	25	4.27
		2	25	10.77
2-Aminophenol	C_6H_7NO	1	20	4.78
		2	20	9.97
3-Aminophenol	C_6H_7NO	1	20	4.37
		2	20	9.82
4-Aminophenol	C_6H_7NO	1	25	5.48
		2	25	10.30
3-Aminopropanenitrile	$C_3H_6N_2$		20	7.80
4-Amino-3,5,6-trichloro-2-pyridinecarbox	$C_6H_3Cl_3N_2O_2$			3.6
Amiodarone	$C_{25}H_{29}I_2NO_3$		25	6.56
Amitriptyline	$C_{20}H_{23}N$			9.4
Amobarbital	$C_{11}H_{18}N_2O_3$		25	8.0
Aniline	C_6H_7N		25	4.87
Aniline-2-carboxylic acid	$C_7H_7NO_2$	1	25	2.17
		2	25	4.85
Aniline-3-carboxylic acid	$C_7H_7NO_2$	1	25	3.07
		2	25	4.79
Aniline-4-carboxylic acid	$C_7H_7NO_2$	1	25	2.50
		2	25	4.87
Apomorphine	$C_{17}H_{17}NO_2$	1		7.0
		2		8.92
Arecoline	$C_8H_{13}NO_2$			6.84
L-Arginine	$C_6H_{14}N_4O_2$	1	25	1.82
		2	25	8.99
		3	25	12.5
L-Argininosuccinic acid	$C_{10}H_{18}N_4O_5$	1	25	1.62
		2	25	2.70
		3	25	4.26
		4	25	9.58
L-Ascorbic acid	$C_6H_8O_6$	1	25	4.04
		2	16	11.7
L-Asparagine	$C_4H_8N_2O_3$	1	20	2.1
		2	20	8.80
L-Aspartic acid	$C_4H_7NO_4$	1	25	1.99
		2	25	3.90
		3	25	9.90
Aspergillic acid	$C_{12}H_{20}N_2O_2$			5.5
Atenolol	$C_{14}H_{22}N_2O_3$			9.6
Atisine	$C_{22}H_{33}NO_2$			12.2
Azaserine	$C_5H_7N_3O_4$			8.55
Azathioprine	$C_9H_7N_7O_2S$			8.2
6-Azauridine	$C_8H_{11}N_3O_6$			6.70
Azetidine	C_3H_7N		25	11.29
Barbital	$C_8H_{12}N_2O_3$		25	7.43
Barbituric acid	$C_4H_4N_2O_3$		25	4.01
Benzaldehyde	C_7H_6O		25	14.90
Benzamide	C_7H_7NO		25	13[a]
Benzeneacetic acid	$C_8H_8O_2$		25	4.31
Benzeneboronic acid	$C_6H_7BO_2$			8.83
Benzenebutanoic acid	$C_{10}H_{12}O_2$		25	4.76
1,2-Benzenediamine	$C_6H_8N_2$	1	20	4.57
		2	20	0.80
1,3-Benzenediamine	$C_6H_8N_2$	1	20	5.11
		2	20	2.50
1,4-Benzenediamine	$C_6H_8N_2$	1	20	6.31
		2	20	2.97
Benzeneethanamine	$C_8H_{11}N$		25	9.83
Benzenemethanethiol	C_7H_8S		25	9.43
Benzenepropanoic acid	$C_9H_{10}O_2$		25	4.66
Benzenesulfinic acid	$C_6H_6O_2S$		20	1.3
Benzenesulfonic acid	$C_6H_6O_3S$		25	0.70
Benzenethiol	C_6H_6S		25	6.62
p-Benzidine	$C_{12}H_{12}N_2$	1	20	4.65
		2	20	3.43
1*H*-Benzimidazole	$C_7H_6N_2$		25	5.53
Benzoic acid	$C_7H_6O_2$		25	4.204
1*H*-Benzotriazole	$C_6H_5N_3$		20	1.6
N-Benzoylglycine	$C_9H_9NO_3$		25	3.62
Benzpiperylon	$C_{22}H_{25}N_3O$	1		6.73
		2		9.13
Benzylamine	C_7H_9N		25	9.34
4-Benzylaniline	$C_{13}H_{13}N$		25	2.17
2-Benzylpyridine	$C_{12}H_{11}N$		25	5.13
Betaine	$C_5H_{11}NO_2$		0	1.83
Biguanide	$C_2H_7N_5$	1		11.52
		2		2.93
N,N-Bis(2-hydroxyethyl)glycine	$C_6H_{13}NO_4$	2	20	8.35
Bithionol	$C_{12}H_6Cl_4O_2S$	1		4.82
		2		10.50
Bornylamine	$C_{10}H_{19}N$		25	10.17
Bromadiolone	$C_{30}H_{23}BrO_4$		21	4.04
Bromoacetic acid	$C_2H_3BrO_2$		25	2.90
2-Bromoaniline	C_6H_6BrN		25	2.53
3-Bromoaniline	C_6H_6BrN		25	3.53
4-Bromoaniline	C_6H_6BrN		25	3.89
2-Bromobenzoic acid	$C_7H_5BrO_2$		25	2.85
3-Bromobenzoic acid	$C_7H_5BrO_2$		25	3.81
4-Bromobenzoic acid	$C_7H_5BrO_2$		25	3.96
Bromocresol Green	$C_{21}H_{14}Br_4O_5S$			4.7
Bromocresol Purple	$C_{21}H_{16}Br_2O_5S$			6.3
4-Bromo-*N,N*-dimethylaniline	$C_8H_{10}BrN$		25	4.23
2-Bromophenol	C_6H_5BrO		25	8.45
3-Bromophenol	C_6H_5BrO		25	9.03
4-Bromophenol	C_6H_5BrO		25	9.37
Bromophenol Blue	$C_{19}H_{10}Br_4O_5S$			4.0
3-Bromopropanoic acid	$C_3H_5BrO_2$		25	4.00
3-Bromopyridine	C_5H_4BrN		25	2.84
3-Bromoquinoline	C_9H_6BrN		25	2.69
Bromothymol Blue	$C_{27}H_{28}Br_2O_5S$			7.0
Brucine	$C_{23}H_{26}N_2O_4$	1		6.04
		2		11.07
1,4-Butanediamine	$C_4H_{12}N_2$	1	25	10.80
		2	25	9.63
1,2,3,4-Butanetetrol	$C_4H_{10}O_4$			13.9
Butanoic acid	$C_4H_8O_2$		25	4.83
trans-2-Butenoic acid	$C_4H_6O_2$		25	4.69
3-Butenoic acid	$C_4H_6O_2$		25	4.34
Butylamine	$C_4H_{11}N$		25	10.60
sec-Butylamine	$C_4H_{11}N$		25	10.56
tert-Butylamine	$C_4H_{11}N$		25	10.68
N-tert-Butylaniline	$C_{10}H_{15}N$		25	7.00
2-*tert*-Butylbenzoic acid	$C_{11}H_{14}O_2$		25	3.54
3-*tert*-Butylbenzoic acid	$C_{11}H_{14}O_2$		25	4.20
4-*tert*-Butylbenzoic acid	$C_{11}H_{14}O_2$		25	4.38
Butylcyclohexylamine	$C_{10}H_{21}N$		25	11.23
2-*tert*-Butylphenol	$C_{10}H_{14}O$		25	10.62
3-*tert*-Butylphenol	$C_{10}H_{14}O$		25	10.12
4-*tert*-Butylphenol	$C_{10}H_{14}O$		25	10.23

Name	Mol. form.	Step	t/°C	pK_a
N-Butylpiperidine	$C_9H_{19}N$		23	10.47
Butylpropanedioic acid	$C_7H_{12}O_4$	1	5	2.96
2-Butynoic acid	$C_4H_4O_2$		25	2.62
Captopril	$C_9H_{15}NO_3S$	1		3.7
		2		9.8
Carbamodithioic acid	CH_3NS_2		25	2.95
Carbendazim	$C_9H_9N_3O_2$			4.48
L-γ-Carboxyglutamic acid	$C_6H_9NO_6$	1	25	1.7
		2	25	3.2
		3	25	4.75
		4	25	9.9
Carnitine	$C_7H_{15}NO_3$		25	3.80
Carnosine	$C_9H_{14}N_4O_3$	1	20	2.73
		2	20	6.87
		3	20	9.73
Cephalexin	$C_{16}H_{17}N_3O_4S$	1		5.2
		2		7.3
Cephaloridine	$C_{19}H_{17}N_3O_4S_2$			3.2
Cephradine	$C_{16}H_{19}N_3O_4S$	1		2.63
		2		7.27
Chloroacetic acid	$C_2H_3ClO_2$		25	2.87
2-Chloroaniline	C_6H_6ClN		25	2.66
3-Chloroaniline	C_6H_6ClN		25	3.52
4-Chloroaniline	C_6H_6ClN		25	3.98
2-Chlorobenzeneacetic acid	$C_8H_7ClO_2$		25	4.07
3-Chlorobenzeneacetic acid	$C_8H_7ClO_2$		25	4.14
4-Chlorobenzeneacetic acid	$C_8H_7ClO_2$		25	4.19
2-Chlorobenzoic acid	$C_7H_5ClO_2$		25	2.90
3-Chlorobenzoic acid	$C_7H_5ClO_2$		25	3.84
4-Chlorobenzoic acid	$C_7H_5ClO_2$		25	4.00
2-Chlorobutanoic acid	$C_4H_7ClO_2$			2.86
3-Chlorobutanoic acid	$C_4H_7ClO_2$			4.05
4-Chlorobutanoic acid	$C_4H_7ClO_2$			4.52
trans-o-Chlorocinnamic acid	$C_9H_7ClO_2$		25	4.23
trans-m-Chlorocinnamic acid	$C_9H_7ClO_2$		25	4.29
trans-p-Chlorocinnamic acid	$C_9H_7ClO_2$		25	4.41
Chlorodiazepoxide	$C_{16}H_{14}ClN_3O$			4.8
3-Chloro-*N,N*-dimethylaniline	$C_8H_{10}ClN$		20	3.83
4-Chloro-*N,N*-dimethylaniline	$C_8H_{10}ClN$		20	4.39
2-Chlorophenol	C_6H_5ClO		25	8.56
3-Chlorophenol	C_6H_5ClO		25	9.12
4-Chlorophenol	C_6H_5ClO		25	9.41
2-Chlorophenoxyacetic acid	$C_8H_7ClO_3$		25	3.05
3-Chlorophenoxyacetic acid	$C_8H_7ClO_3$		25	3.10
3-(2-Chlorophenyl)propanoic acid	$C_9H_9ClO_2$		25	4.58
3-(3-Chlorophenyl)propanoic acid	$C_9H_9ClO_2$		25	4.59
3-(4-Chlorophenyl)propanoic acid	$C_9H_9ClO_2$		25	4.61
2-Chloropropanoic acid	$C_3H_5ClO_2$		25	2.83
3-Chloropropanoic acid	$C_3H_5ClO_2$		25	3.98
2-Chloropyridine	C_5H_4ClN		25	0.49
3-Chloropyridine	C_5H_4ClN		25	2.81
4-Chloropyridine	C_5H_4ClN		25	3.83
Chlorothiazide	$C_7H_6ClN_3O_4S_2$	1		6.85
		2		9.45
Cholic acid	$C_{24}H_{40}O_5$		20	4.98
Choline	$C_5H_{14}NO$		25	13.9
Cinchonidine	$C_{19}H_{22}N_2O$	1		5.80
		2		10.03
Cinchonine	$C_{19}H_{22}N_2O$	1		5.85
		2		9.92
cis-Cinnamic acid	$C_9H_8O_2$		25	3.88
trans-Cinnamic acid	$C_9H_8O_2$		25	4.44
Cinnoline	$C_8H_6N_2$		20	2.37
Citric acid	$C_6H_8O_7$	1	25	3.13
		2	25	4.76
		3	25	6.40
Citrulline	$C_6H_{13}N_3O_3$	1	25	2.43
		2	25	9.69
Clindamycin	$C_{18}H_{33}ClN_2O_5S$			7.6
Clonazepam	$C_{15}H_{10}ClN_3O_3$	1		1.5
		2		10.5
Clozapine	$C_{18}H_{19}ClN_4$	1		3.70
		2		7.60
Codeine	$C_{18}H_{21}NO_3$			8.21
Colchicine	$C_{22}H_{25}NO_6$		20	12.36
Creatine	$C_4H_9N_3O_2$	1	25	2.63
		2	25	14.3
Creatinine	$C_4H_7N_3O$	1	25	4.8
		2		9.2
o-Cresol	C_7H_8O		25	10.29
m-Cresol	C_7H_8O		25	10.09
p-Cresol	C_7H_8O		25	10.26
o-Cresolphthalein	$C_{22}H_{18}O_4$			9.4
Cresol Red	$C_{21}H_{18}O_5S$			8.3
Cupreine	$C_{19}H_{22}N_2O_2$			6.57
Cyanamide	CH_2N_2		29	1.1
Cyanic acid	CHNO		25	3.7
Cyanoacetic acid	$C_3H_3NO_2$		25	2.47
3-Cyanobenzoic acid	$C_8H_5NO_2$		25	3.60
4-Cyanobenzoic acid	$C_8H_5NO_2$		25	3.55
4-Cyanobutanoic acid	$C_5H_7NO_2$		25	2.42
(2-Cyanophenoxy)acetic acid	$C_9H_7NO_3$		25	2.98
(3-Cyanophenoxy)acetic acid	$C_9H_7NO_3$		25	3.03
(4-Cyanophenoxy)acetic acid	$C_9H_7NO_3$		25	2.93
Cyanuric acid	$C_3H_3N_3O_3$	1		6.88
		2		11.40
		3		13.5
Cyclohexanecarboxylic acid	$C_7H_{12}O_2$		25	4.91
cis-1,2-Cyclohexanediamine	$C_6H_{14}N_2$	1	20	9.93
		2	20	6.13
trans-1,2-Cyclohexanediamine	$C_6H_{14}N_2$	1	20	9.94
		2	20	6.47
1,3-Cyclohexanedione	$C_6H_8O_2$		25	5.26
Cyclohexylamine	$C_6H_{13}N$		25	10.64
Cyclopentanecarboxylic acid	$C_6H_{10}O_2$		25	4.99
Cyclopropanecarboxylic acid	$C_4H_6O_2$		25	4.83
1,1-Cyclopropanedicarboxylic acid	$C_5H_6O_4$	1	25	1.82
		2	25	7.43
Cysteamine	C_2H_7NS	1	25	8.27
		2	25	10.53
DL-Cysteic acid	$C_3H_7NO_5S$	1	25	1.3
		2	25	1.9
		3	25	8.70
L-Cysteine	$C_3H_7NO_2S$	1	25	1.5
		2	25	8.7
		3	25	10.2
L-Cystine	$C_6H_{12}N_2O_4S_2$	1		1
		2		2.1

Name	Mol. form.	Step	t/°C	pK_a
		3		8.02
		4		8.71
Cytidine	$C_9H_{13}N_3O_5$	1		4.22
		2		12.5
3'-Cytidylic acid	$C_9H_{14}N_3O_8P$	1		0.8
		2		4.28
		3		6.0
Cytisine	$C_{11}H_{14}N_2O$	1		6.11
		2		13.08
Cytosine	$C_4H_5N_3O$	1		4.60
		2		12.16
Decanedioic acid	$C_{10}H_{18}O_4$	1		4.59
		2		5.59
Decylamine	$C_{10}H_{23}N$		25	10.64
D-2-Deoxyribose	$C_5H_{10}O_4$		25	12.61
2,4-Diaminobutanoic acid	$C_4H_{10}N_2O_2$	1	25	1.85
		2	25	8.24
		3	25	10.44
1,3-Diamino-2-propanol	$C_3H_{10}N_2O$	1	20	9.69
		2	20	7.93
Dibenzepin	$C_{18}H_{21}N_3O$			8.25
3,5-Dibromoaniline	$C_6H_5Br_2N$		25	2.34
3,5-Dibromo-4-hydroxybenzonitrile	$C_7H_3Br_2NO$			4.06
3,5-Dibromo-L-tyrosine	$C_9H_9Br_2NO_3$	1		2.17
		2		6.45
		3		7.60
Dibutylamine	$C_8H_{19}N$		21	11.25
2,6-Di-*tert*-butylpyridine	$C_{13}H_{21}N$			3.58
Dichloroacetic acid	$C_2H_2Cl_2O_2$		25	1.35
2,4-Dichloroaniline	$C_6H_5Cl_2N$		22	2.05
2,3-Dichlorophenol	$C_6H_4Cl_2O$		25	7.44
Dicyclohexylamine	$C_{12}H_{23}N$			10.4
2',3'-Dideoxyinosine	$C_{10}H_{12}N_4O_3$			9.12
Diethylamine	$C_4H_{11}N$		25	10.84
N,N-Diethylaniline	$C_{10}H_{15}N$		25	6.57
Diethylmethylamine	$C_5H_{13}N$		25	10.35
N,N-Diethyl-2-methylaniline	$C_{11}H_{17}N$		25	7.24
1,7-Dihydro-6*H*-purine-6-thione	$C_5H_4N_4S$	1		7.77
		2		11.17
2,5-Dihydroxybenzeneacetic acid	$C_8H_8O_4$		25	4.40
2,4-Dihydroxybenzoic acid	$C_7H_6O_4$	1	25	3.11
		2	25	8.55
		3	25	14.0
2,5-Dihydroxybenzoic acid	$C_7H_6O_4$	1	25	2.97
3,4-Dihydroxybenzoic acid	$C_7H_6O_4$	1	25	4.48
		2	25	8.83
		3	25	12.6
3,5-Dihydroxybenzoic acid	$C_7H_6O_4$	1	25	4.04
3,12-Dihydroxycholan-24-oic acid, (3α,5β,12α)	$C_{24}H_{40}O_4$		20	5.15
Dihydroxytartaric acid	$C_4H_6O_8$		25	1.92
3,5-Diiodo-L-tyrosine	$C_9H_9I_2NO_3$	1	25	2.12
		2	25	5.32
		3	25	9.48
Diisopropylamine	$C_6H_{15}N$		25	11.05
Dimethylamine	C_2H_7N		25	10.73
4-(Dimethylamino)benzoic acid	$C_9H_{11}NO_2$	1		6.03
		2		11.49
2-Dimethylaminopurine	$C_7H_9N_5$	1	20	4.00
		2	20	10.24
2,6-Dimethylaniline	$C_8H_{11}N$		25	3.89
N,N-Dimethylaniline	$C_8H_{11}N$		25	5.07
Dimethylarsinic acid	$C_2H_7AsO_2$	1	25	1.57
		2	25	6.27
3,5-Dimethylbenzoic acid	$C_9H_{10}O_2$		25	4.32
5,5-Dimethyl-1,3-cyclohexanedione	$C_8H_{12}O_2$		25	5.15
2,2-Dimethyl-1,3-dioxane-4,6-dione	$C_6H_8O_4$			5.1
N,N-Dimethylglycine	$C_4H_9NO_2$		25	9.89
2,4-Dimethyl-1*H*-imidazole	$C_5H_8N_2$		25	8.36
Dimethylmalonic acid	$C_5H_8O_4$		25	3.15
N,N-Dimethyl-1-naphthylamine	$C_{12}H_{13}N$		25	4.83
N,N-Dimethyl-2-naphthylamine	$C_{12}H_{13}N$		25	4.566
N,N-Dimethyl-3-nitroaniline	$C_8H_{10}N_2O_2$		25	2.62
5,5-Dimethyl-2,4-oxazolidinedione	$C_5H_7NO_3$		37	6.13
cis-2,5-Dimethylpiperazine	$C_6H_{14}N_2$	1	25	9.66
		2	25	5.20
1,2-Dimethylpiperidine, (±)-	$C_7H_{15}N$		25	10.22
2,2-Dimethylpropanoic acid	$C_5H_{10}O_2$		20	5.03
2,2-Dimethylpropylamine	$C_5H_{13}N$		25	10.15
2,3-Dimethylpyridine	C_7H_9N		25	6.57
2,4-Dimethylpyridine	C_7H_9N		25	6.99
2,5-Dimethylpyridine	C_7H_9N		25	6.40
2,6-Dimethylpyridine	C_7H_9N		25	6.65
3,4-Dimethylpyridine	C_7H_9N		25	6.46
3,5-Dimethylpyridine	C_7H_9N		25	6.15
4,6-Dimethyl-2-pyrimidinamine	$C_6H_9N_3$		20	4.82
1,2-Dimethylpyrrolidine	$C_6H_{13}N$		26	10.20
2,4-Dinitrobenzoic acid	$C_7H_4N_2O_6$		25	1.43
2,4-Dinitrophenol	$C_6H_4N_2O_5$		25	4.07
2,5-Dinitrophenol	$C_6H_4N_2O_5$		15	5.15
Dinoseb	$C_{10}H_{12}N_2O_5$			4.62
Diphenylamine	$C_{12}H_{11}N$		25	0.79
4-[(Dipropylamino)sulfonyl] benzoic acid	$C_{13}H_{19}NO_4S$			5.8
Dodecylamine	$C_{12}H_{27}N$		25	10.63
Dopamine	$C_8H_{11}NO_2$	1	25	8.9
		2	25	10.6
Droperidol	$C_{22}H_{22}FN_3O_2$			7.64
Emetine	$C_{29}H_{40}N_2O_4$	1		5.77
		2		6.64
Ephedrine, (1*S*,2*R*)	$C_{10}H_{15}NO$		10	10.139
Ephedrine, (1*R*,2*S*)	$C_{10}H_{15}NO$		10	9.958
Epinephrine	$C_9H_{13}NO_3$	1	25	8.66
		2	25	9.95
Ergometrinine	$C_{19}H_{23}N_3O_2$			7.3
Ergonovine	$C_{19}H_{23}N_3O_2$			6.8
Erythromycin	$C_{37}H_{67}NO_{13}$			8.8
Ethacrynic acid	$C_{13}H_{12}Cl_2O_4$			3.50
1,2-Ethanediamine	$C_2H_8N_2$	1	25	9.92
		2	25	6.86
1,2-Ethanediol	$C_2H_6O_2$		25	15.1
Ethanimidamide	$C_2H_6N_2$		25	12.1
Ethanol	C_2H_6O		25	15.5
Ethanolamine	C_2H_7NO		25	9.50
Ethoxyacetic acid	$C_4H_8O_3$		18	3.65
2-Ethoxyaniline	$C_8H_{11}NO$		28	4.43
3-Ethoxyaniline	$C_8H_{11}NO$		25	4.18
4-Ethoxyaniline	$C_8H_{11}NO$		28	5.20

Name	Mol. form.	Step	t/°C	pK_a
Ethyl acetoacetate	$C_6H_{10}O_3$		25	10.68
Ethylamine	C_2H_7N		25	10.65
Ethyl 4-aminobenzoate	$C_9H_{11}NO_2$			2.5
N-Ethylaniline	$C_8H_{11}N$		25	5.12
2-Ethyl-1*H*-benzimidazole	$C_9H_{10}N_2$		25	6.18
O-Ethyl *S*-[2-(diisopropylamino) ethyl] methylphosphonothioate	$C_{11}H_{26}NO_2PS$			7.9
Ethyleneimine	C_2H_5N		25	8.04
α-Ethylglutamic acid	$C_7H_{13}NO_4$	1	25	3.846
		2	25	7.838
N-Ethylmorpholine	$C_6H_{13}NO$		25	7.67
1-Ethylpiperidine	$C_7H_{15}N$		23	10.45
2-Ethylpyridine	C_7H_9N		25	5.89
Etoposide	$C_{29}H_{32}O_{13}$			9.8
Fluoroacetic acid	$C_2H_3FO_2$		25	2.59
2-Fluoroaniline	C_6H_6FN		25	3.20
3-Fluoroaniline	C_6H_6FN		25	3.59
4-Fluoroaniline	C_6H_6FN		25	4.65
2-Fluorobenzoic acid	$C_7H_5FO_2$		25	3.27
3-Fluorobenzoic acid	$C_7H_5FO_2$		25	3.86
4-Fluorobenzoic acid	$C_7H_5FO_2$		25	4.15
5-Fluorocytosine	$C_4H_4FN_3O$			3.26
2-Fluorophenol	C_6H_5FO		25	8.73
3-Fluorophenol	C_6H_5FO		25	9.29
4-Fluorophenol	C_6H_5FO		25	9.89
2-Fluoropyridine	C_5H_4FN		25	-0.44
Folinic acid	$C_{20}H_{23}N_7O_7$	1		3.1
		2		4.8
		3		10.4
Formaldehyde	CH_2O		25	13.27
Formic acid	CH_2O_2		25	3.75
3-Formylbenzoic acid	$C_8H_6O_3$		25	3.84
4-Formylbenzoic acid	$C_8H_6O_3$		25	3.77
β-D-Fructose	$C_6H_{12}O_6$		25	12.27
Fumaric acid	$C_4H_4O_4$	1	25	3.02
		2	25	4.38
2-Furancarboxylic acid	$C_5H_4O_3$		25	3.16
3-Furancarboxylic acid	$C_5H_4O_3$		25	3.9
Furethidine	$C_{21}H_{31}NO_4$			7.48
Gibberellic acid	$C_{19}H_{22}O_6$			4.0
α-D-Glucose	$C_6H_{12}O_6$		25	12.46
L-Glutamic acid	$C_5H_9NO_4$	1	25	2.13
		2	25	4.31
		3		9.67
L-Glutamine	$C_5H_{10}N_2O_3$	1	25	2.17
		2	25	9.13
Glutathione	$C_{10}H_{17}N_3O_6S$	1	25	2.12
		2	25	3.59
		3	25	8.75
		4	25	9.65
Glyceric acid	$C_3H_6O_4$		25	3.52
Glycerol	$C_3H_8O_3$		25	14.15
Glycine	$C_2H_5NO_2$	1	25	2.35
		2	25	9.78
Glycocholic acid	$C_{26}H_{43}NO_6$			4.4
Glycolic acid	$C_2H_4O_3$		25	3.83
Glycylalanine	$C_5H_{10}N_2O_3$		25	3.15
L-Glycylasparagine	$C_6H_{11}N_3O_4$	1	25	2.942
		2	18	8.44
N-Glycylglycine	$C_4H_8N_2O_3$	1	25	3.14
		2		8.17
N-(*N*-Glycylglycyl)glycine	$C_6H_{11}N_3O_4$	1	25	3.225
		2	25	8.09
N-Glycyl-L-leucine	$C_8H_{16}N_2O_3$		25	3.18
N-Glycylserine, (*DL*)-	$C_5H_{10}N_2O_4$	1	25	2.98
		2	25	8.38
Glyoxylic acid	$C_2H_2O_3$		25	3.18
Guanidine	CH_5N_3		25	13.6
Guanidinoacetic acid	$C_3H_7N_3O_2$		25	2.82
Guanine	$C_5H_5N_5O$		40	9.92
Haloperidol	$C_{21}H_{23}ClFNO_2$			8.3
Harmaline	$C_{13}H_{14}N_2O$			4.2
Harmine	$C_{13}H_{12}N_2O$			7.70
2-Heptanamine	$C_7H_{17}N$		19	10.7
Heptanedioic acid	$C_7H_{12}O_4$	1	25	4.71
		2	25	5.58
Heptanoic acid	$C_7H_{14}O_2$		25	4.89
Heptylamine	$C_7H_{17}N$		25	10.67
Hexadecylamine	$C_{16}H_{35}N$		25	10.61
2,4-Hexadienoic acid	$C_6H_8O_2$		25	4.76
Hexamethyldisilazane	$C_6H_{19}NSi_2$			7.55
1,6-Hexanediamine	$C_6H_{16}N_2$	1	0	11.86
		2	0	10.76
1,6-Hexanedioic acid	$C_6H_{10}O_4$	1	18	4.41
		2	18	5.41
Hexanoic acid	$C_6H_{12}O_2$		25	4.85
Hexylamine	$C_6H_{15}N$		25	10.56
Histamine	$C_5H_9N_3$	1	25	6.04
		2	25	9.75
L-Histidine	$C_6H_9N_3O_2$	1	25	1.80
		2	25	6.04
		3	25	9.33
DL-Homocysteine	$C_4H_9NO_2S$	1	25	2.22
		2	25	8.87
		3	25	10.86
Homocystine	$C_8H_{16}N_2O_4S_2$	1	25	1.59
		2	25	2.54
		3	25	8.52
		4	25	9.44
L-Homoserine	$C_4H_9NO_3$	1	25	2.71
		2	25	9.62
Hydrastine	$C_{21}H_{21}NO_6$			7.8
Hydrastinine	$C_{11}H_{13}NO_3$			11.38
Hydrochlorothiazide	$C_7H_8ClN_3O_4S_2$	1		7.9
		2		9.2
Hydroflumethiazide	$C_8H_8F_3N_3O_4S_2$	1		8.9
		2		9.7
Hydroquinine	$C_{20}H_{26}N_2O_2$			5.33
p-Hydroquinone	$C_6H_6O_2$	1	25	9.85
		2	25	11.4
3-Hydroxybenzaldehyde	$C_7H_6O_2$		25	8.98
4-Hydroxybenzaldehyde	$C_7H_6O_2$		25	7.61
α-Hydroxybenzeneaceticacid,(±)-	$C_8H_8O_3$		25	3.37
3-Hydroxybenzenesulfonic acid	$C_6H_6O_4S$		25	9.07
4-Hydroxybenzenesulfonic acid	$C_6H_6O_4S$		25	9.11
2-Hydroxybenzoic acid	$C_7H_6O_3$	1	20	2.98
		2	20	13.6
3-Hydroxybenzoic acid	$C_7H_6O_3$	1	25	4.08
		2	19	9.92
4-Hydroxybenzoic acid	$C_7H_6O_3$	1	25	4.57
		2	25	9.46
2-Hydroxybenzonitrile	C_7H_5NO		25	6.86

Name	Mol. form.	Step	t/°C	pK_a
3-Hydroxybenzonitrile	C_7H_5NO		25	8.61
4-Hydroxybenzonitrile	C_7H_5NO		25	7.97
2-Hydroxybiphenyl	$C_{12}H_{10}O$		25	10.01
3-Hydroxybiphenyl	$C_{12}H_{10}O$		25	9.64
4-Hydroxybiphenyl	$C_{12}H_{10}O$		25	9.55
3-Hydroxybutanoic acid, (±)-	$C_4H_8O_3$		25	4.70
4-Hydroxybutanoic acid	$C_4H_8O_3$		25	4.72
1-Hydroxy-1,1-diphosphonoethane	$C_2H_8O_7P_2$	1		1.35
		2		2.87
		3		7.03
		4		11.3
5-Hydroxy-2-(hydroxymethyl)-4*H*-pyran-4-one	$C_6H_6O_4$			7.9
α-Hydroxy-α-methylbenzeneaceti	$C_9H_{10}O_3$		25	3.47
α-Hydroxy-α-phenylbenzeneacetic acid	$C_{14}H_{12}O_3$		25	3.04
1-(2-Hydroxyphenyl)ethanone	$C_8H_8O_2$		25	10.06
1-(3-Hydroxyphenyl)ethanone	$C_8H_8O_2$		25	9.19
1-(4-Hydroxyphenyl)ethanone	$C_8H_8O_2$		25	8.05
trans-4-Hydroxy-L-proline	$C_5H_9NO_3$	1	25	1.82
		2	25	9.66
Hydroxypropanedioic acid	$C_3H_4O_5$	1		2.42
		2		4.54
3-Hydroxypropanoic acid	$C_3H_6O_3$		25	4.51
5-Hydroxytryptamine	$C_{10}H_{12}N_2O$	1	25	9.8
		2	25	11.1
Hyoscyamine	$C_{17}H_{23}NO_3$		21	9.7
Hypoxanthine	$C_5H_4N_4O$		25	8.7
Imazapyr	$C_{13}H_{15}N_3O_3$	1		1.9
		2		3.6
Imazethapyr	$C_{15}H_{19}N_3O_3$	1		2.1
		2		3.9
Imidazole	$C_3H_4N_2$		25	6.99
Iminodiacetic acid	$C_4H_7NO_4$	1		2.98
		2		9.89
1-Indanamine	$C_9H_{11}N$		22	9.21
1*H*-Indole-3-acetic acid	$C_{10}H_9NO_2$			4.75
Indomethacin	$C_{19}H_{16}ClNO_4$			4.5
Iocetamic acid	$C_{12}H_{13}I_3N_2O_3$			4
Iodoacetic acid	$C_2H_3IO_2$		25	3.18
2-Iodoaniline	C_6H_6IN		25	2.54
3-Iodoaniline	C_6H_6IN		25	3.58
4-Iodoaniline	C_6H_6IN		25	3.81
2-Iodobenzoic acid	$C_7H_5IO_2$		25	2.86
3-Iodobenzoic acid	$C_7H_5IO_2$		25	3.87
4-Iodobenzoic acid	$C_7H_5IO_2$		25	4.00
2-Iodophenol	C_6H_5IO		25	8.51
3-Iodophenol	C_6H_5IO		25	9.03
4-Iodophenol	C_6H_5IO		25	9.33
L-3-Iodotyrosine	$C_9H_{10}INO_3$	1	25	2.2
		2	25	8.7
		3	25	9.1
Iopanoic acid	$C_{11}H_{12}I_3NO_2$			4.8
Isocitric acid	$C_6H_8O_7$	1	25	3.29
		2	25	4.71
		3	25	6.40
L-Isoleucine	$C_6H_{13}NO_2$	1	25	2.32
		2	25	9.76
Isophthalic acid	$C_8H_6O_4$	1	25	3.70
		2	25	4.60
Isopropylamine	C_3H_9N		25	10.63
N-Isopropylaniline	$C_9H_{13}N$		25	5.77
Isoproterenol	$C_{11}H_{17}NO_3$			8.64
1-Isoquinolinamine	$C_9H_8N_2$		20	7.62
3-Isoquinolinamine	$C_9H_8N_2$		20	5.05
Isoquinoline	C_9H_7N		20	5.40
7-Isoquinolinol	C_9H_7NO	1	20	5.68
		2	20	8.90
Isoxazole	C_3H_3NO		25	-2.0
Ketamine	$C_{13}H_{16}ClNO$			7.5
D-Lactic acid	$C_3H_6O_3$		25	3.86
L-Leucine	$C_6H_{13}NO_2$	1	25	2.33
		2	25	9.74
N-Leucylglycine	$C_8H_{16}N_2O_3$	1	25	3.25
		2	25	8.2
Levodopa	$C_9H_{11}NO_4$	1	25	2.32
		2	25	8.72
		3	25	9.96
		4	25	11.79
Lisinopril	$C_{21}H_{35}N_3O_7$	1		2.5
		2		4.0
		3		6.7
		4		10.1
Lycodine	$C_{16}H_{22}N_2$	1		3.97
		2		8.08
Lysergic acid	$C_{16}H_{16}N_2O_2$	1		3.44
		2		7.68
L-Lysine	$C_6H_{14}N_2O_2$	1	25	2.16
		2	25	9.06
		3	25	10.54
Maleic acid	$C_4H_4O_4$	1	25	1.92
		2	25	6.23
Malic acid	$C_4H_6O_5$	1	25	3.40
		2	25	5.11
Malonic acid	$C_3H_4O_4$	1	25	2.85
		2	25	5.70
α-Maltose	$C_{12}H_{22}O_{11}$		21	12.05
D-Mannitol	$C_6H_{14}O_6$		18	13.5
D-Mannose	$C_6H_{12}O_6$		25	12.08
Mefenamic acid	$C_{15}H_{15}NO_2$			4.2
Mefluidide	$C_{11}H_{13}F_3N_2O_3S$			4.6
2-Mercaptoethanol	C_2H_6OS		25	9.72
Methanethiol	CH_4S		25	10.33
Methanol	CH_4O		25	15.5
Metharbital	$C_9H_{14}N_2O_3$			8.45
Methazolamide	$C_5H_8N_4O_3S_2$			7.30
L-Methionine	$C_5H_{11}NO_2S$	1	25	2.13
		2	25	9.27
Methoxamine hydrochloride	$C_{11}H_{18}ClNO_3$		25	9.2
2-Methoxyaniline	C_7H_9NO		25	4.53
3-Methoxyaniline	C_7H_9NO		25	4.20
4-Methoxyaniline	C_7H_9NO		25	5.36
2-Methoxybenzoic acid	$C_8H_8O_3$		25	4.08
3-Methoxybenzoic acid	$C_8H_8O_3$		25	4.10
4-Methoxybenzoic acid	$C_8H_8O_3$		25	4.50
2-Methoxyethanol	$C_3H_8O_2$		25	14.8
2-Methoxyethylamine	C_3H_9NO		25	9.40
2-(2-Methoxyethyl)pyridine	$C_8H_{11}NO$			5.5
2-Methoxyphenol	$C_7H_8O_2$		25	9.98
3-Methoxyphenol	$C_7H_8O_2$		25	9.65
4-Methoxyphenol	$C_7H_8O_2$		25	10.21

Name	Mol. form.	Step	t/°C	pK_a
2-Methoxypyridine	C_6H_7NO		20	3.28
3-Methoxypyridine	C_6H_7NO		25	4.78
4-Methoxypyridine	C_6H_7NO		25	6.58
6-Methoxyquinoline	$C_{10}H_9NO$		20	5.03
Methyclothiazide	$C_9H_{11}Cl_2N_3O_4S_2$			9.4
Methylamine	CH_5N		25	10.66
2-(Methylamino)benzoic acid	$C_8H_9NO_2$		25	5.34
3-(Methylamino)benzoic acid	$C_8H_9NO_2$		25	5.10
4-(Methylamino)benzoic acid	$C_8H_9NO_2$		25	5.04
2-Methylaniline	C_7H_9N		25	4.45
3-Methylaniline	C_7H_9N		25	4.71
4-Methylaniline	C_7H_9N		25	5.08
N-Methylaniline	C_7H_9N		25	4.85
α-Methylbenzeneacetic acid, (±)-	$C_9H_{10}O_2$		25	4.64
2-Methyl-1*H*-benzimidazole	$C_8H_8N_2$		25	6.19
3-Methyl-1-butanamine	$C_5H_{13}N$		25	10.60
2-Methyl-2-butanamine	$C_5H_{13}N$		19	10.85
2-Methylbutanoic acid	$C_5H_{10}O_2$		25	4.80
3-Methylbutanoic acid	$C_5H_{10}O_2$		25	4.77
trans-o-Methylcinnamic acid	$C_{10}H_{10}O_2$		25	4.50
trans-m-Methylcinnamic acid	$C_{10}H_{10}O_2$		25	4.44
trans-p-Methylcinnamic acid	$C_{10}H_{10}O_2$		25	4.56
α-Methylenebenzeneacetic acid	$C_9H_8O_2$			4.35
Methyl α-*D*-glucopyranoside	$C_7H_{14}O_6$		25	13.71
N-Methyl-2-heptanamine	$C_8H_{19}N$		17	10.99
L-1-Methylhistidine	$C_7H_{11}N_3O_2$	1	25	1.69
		2	25	6.48
		3	25	8.85
L-3-Methylhistidine	$C_7H_{11}N_3O_2$	1	25	1.92
		2	25	6.56
		3	25	8.73
O-Methylhydroxylamine	CH_5NO			12.5
1-Methylimidazol	$C_4H_6N_2$		25	6.95
Methylmalonic acid	$C_4H_6O_4$	1	25	3.07
		2	25	5.76
4-Methylmorpholine	$C_5H_{11}NO$		25	7.38
Methyl-1-naphthylamine	$C_{11}H_{11}N$		27	3.67
3-Methylpentanedioic acid	$C_6H_{10}O_4$		25	4.24
4-Methylpentanoic acid	$C_6H_{12}O_2$		18	4.84
Methylphenidate	$C_{14}H_{19}NO_2$			8.9
1-Methylpiperidine	$C_6H_{13}N$		25	10.38
2-Methylpropanoic acid	$C_4H_8O_2$		20	4.84
3-Methylpyrazinamine	$C_5H_7N_3$		25	3.39
2-Methylpyrazine	$C_5H_6N_2$		27	1.45
N-Methyl-4-pyridinamine	$C_6H_8N_2$		20	9.65
2-Methylpyridine	C_6H_7N		25	6.00
3-Methylpyridine	C_6H_7N		25	5.70
4-Methylpyridine	C_6H_7N		25	5.99
N-Methylpyrrolidine	$C_5H_{11}N$		25	10.46
2-Methylquinoline	$C_{10}H_9N$		20	5.83
4-Methylquinoline	$C_{10}H_9N$		20	5.67
5-Methylquinoline	$C_{10}H_9N$		20	5.20
Methyl Red	$C_{15}H_{15}N_3O_2$	1		2.5
		2		9.5
Methylsuccinic acid	$C_5H_8O_4$	1	25	4.13
		2	25	5.64
(Methylthio)acetic acid	$C_3H_6O_2S$		25	3.66
2-(Methylthio)aniline	C_7H_9NS		25	3.45
4-(Methylthio)aniline	C_7H_9NS		25	4.35
4-(Methylthio)phenol	C_7H_8OS		25	9.53
6-Methyl-1,2,4-triazine-3,5(2*H*,4*H*)-dione	$C_4H_5N_3O_2$			7.6
Minoxidil	$C_9H_{15}N_5O$			4.61
Morphine	$C_{17}H_{19}NO_3$	1	25	8.21
		2	20	9.85
Morpholine	C_4H_9NO		25	8.50
Mycophenolic acid	$C_{17}H_{20}O_6$			4.5
Nadolol	$C_{17}H_{27}NO_4$			9.67
1-Naphthalenecarboxylic acid	$C_{11}H_8O_2$		25	3.69
2-Naphthalenecarboxylic acid	$C_{11}H_8O_2$		25	4.16
1-Naphthol	$C_{10}H_8O$		25	9.39
2-Naphthol	$C_{10}H_8O$		25	9.63
1-Naphthylamine	$C_{10}H_9N$		25	3.92
2-Naphthylamine	$C_{10}H_9N$		25	4.16
Neobornylamine	$C_{10}H_{19}N$		25	10.01
Neutral Red	$C_{15}H_{17}ClN_4$			6.7
L-Nicotine	$C_{10}H_{14}N_2$	1		8.02
		2		3.12
Nitrilotriacetic acid	$C_6H_9NO_6$	1	20	3.03
		2	20	3.07
		3	20	10.70
Nitroacetic acid	$C_2H_3NO_4$		24	1.48
2-Nitroaniline	$C_6H_6N_2O_2$		25	-0.25
3-Nitroaniline	$C_6H_6N_2O_2$		25	2.46
4-Nitroaniline	$C_6H_6N_2O_2$		25	1.02
Nitrobenzene	$C_6H_5NO_2$		0	3.98
2-Nitrobenzeneacetic acid	$C_8H_7NO_4$		25	4.00
3-Nitrobenzeneacetic acid	$C_8H_7NO_4$		25	3.97
4-Nitrobenzeneacetic acid	$C_8H_7NO_4$		25	3.85
2-Nitrobenzoic acid	$C_7H_5NO_4$		25	2.17
3-Nitrobenzoic acid	$C_7H_5NO_4$		25	3.46
4-Nitrobenzoic acid	$C_7H_5NO_4$		25	3.43
Nitroethane	$C_2H_5NO_2$		25	8.46
Nitrofurantoin	$C_8H_6N_4O_5$			7.2
Nitromethane	CH_3NO_2		25	10.21
2-Nitrophenol	$C_6H_5NO_3$		25	7.23
3-Nitrophenol	$C_6H_5NO_3$		25	8.36
4-Nitrophenol	$C_6H_5NO_3$		25	7.15
3-(2-Nitrophenyl)propanoic acid	$C_9H_9NO_4$		25	4.50
3-(4-Nitrophenyl)propanoic acid	$C_9H_9NO_4$		25	4.47
4-Nitropyridine	$C_5H_4N_2O_2$		25	1.61
5-Nitropyrimidinamine	$C_4H_4N_4O_2$		20	0.35
Nizatidine	$C_{12}H_{21}N_5O_2S_2$	1		2.1
		2		6.8
Nonanedioic acid	$C_9H_{16}O_4$	1	25	4.53
		2	25	5.33
Nonanoic acid	$C_9H_{18}O_2$		25	4.96
Nonylamine	$C_9H_{21}N$		25	10.64
Norepinephrine	$C_8H_{11}NO_3$	1	25	8.64
		2	25	9.70
L-Norleucine	$C_6H_{13}NO_2$	1	25	2.34
		2	25	9.83
DL-Norvaline	$C_5H_{11}NO_2$	1		2.36
		2		9.72
L-Norvaline	$C_5H_{11}NO_2$	1	25	2.32
		2	25	9.81
Noscapine	$C_{22}H_{23}NO_7$			7.8
Novobiocin	$C_{31}H_{36}N_2O_{11}$	1		4.3
		2		9.1

Name	Mol. form.	Step	t/°C	pKa
cis,cis-9,12-Octadecadienoic acid	$C_{18}H_{32}O_2$		25	4.77
Octadecylamine	$C_{18}H_{39}N$		25	10.60
1,8-Octanediamine	$C_8H_{20}N_2$	1	20	11.00
		2	20	10.1
Octanedioic acid	$C_8H_{14}O_4$	1	25	4.52
Octanoic acid	$C_8H_{16}O_2$		25	4.89
Octylamine	$C_8H_{19}N$		25	10.65
Oleanolic acid	$C_{30}H_{48}O_3$			2.52
L-Ornithine	$C_5H_{12}N_2O_2$	1	25	1.71
		2	25	8.69
		3	25	10.76
Oxalic acid	$C_2H_2O_4$	1	25	1.25
		2	25	3.81
Oxaloacetic acid	$C_4H_4O_5$	1	25	2.55
		2	25	4.37
		3	25	13.03
Oxazole	C_3H_3NO		33	0.8
2-Oxobutanoic acid	$C_4H_6O_3$		25	2.50
2-Oxoglutaric acid	$C_5H_6O_5$	1	25	2.47
		2	25	4.68
Papaverine	$C_{20}H_{21}NO_4$			6.4
Pentadecylamine	$C_{15}H_{33}N$		25	10.61
Pentafluorobenzoic acid	$C_7HF_5O_2$		25	1.75
1,2,2,6,6-Pentamethylpiperidine	$C_{10}H_{21}N$		30	11.25
3-Pentanamine	$C_5H_{13}N$		17	10.59
1,5-Pentanediamine	$C_5H_{14}N_2$	1	25	10.05
		2	25	10.93
Pentanedioic acid	$C_5H_8O_4$	1	18	4.32
		2	25	5.42
Pentanoic acid	$C_5H_{10}O_2$		20	4.83
trans-3-Pentenoic acid	$C_5H_8O_2$		25	4.51
Pentostatin	$C_{11}H_{16}N_4O_4$			5.2
Pentylamine	$C_5H_{13}N$		25	10.63
Perfluidone	$C_{14}H_{12}F_3NO_4S_2$			2.5
1*H*-Perimidine	$C_{11}H_8N_2$		20	6.35
Phenanthridine	$C_{13}H_9N$		20	5.58
1,10-Phenanthroline	$C_{12}H_8N_2$		25	4.84
Phenazine	$C_{12}H_8N_2$		20	1.20
Phenobarbital	$C_{12}H_{12}N_2O_3$	1		7.3
		2		11.8
Phenol	C_6H_6O		25	9.99
Phenolphthalein	$C_{20}H_{14}O_4$		25	9.7
Phenol Red	$C_{19}H_{14}O_5S$			7.9
Phenoxyacetic acid	$C_8H_8O_3$		25	3.17
2-Phenoxybenzoic acid	$C_{13}H_{10}O_3$		25	3.53
3-Phenoxybenzoic acid	$C_{13}H_{10}O_3$		25	3.95
4-Phenoxybenzoic acid	$C_{13}H_{10}O_3$		25	4.57
2-(3-Phenoxyphenyl)propanoic acid, (±)-	$C_{15}H_{14}O_3$			4.5
L-Phenylalanine	$C_9H_{11}NO_2$	1	25	2.20
		2	25	9.31
α-Phenylbenzeneacetic acid	$C_{14}H_{12}O_2$		25	3.94
2-Phenylbenzimidazole	$C_{13}H_{10}N_2$	1	25	5.23
		2	25	11.91
2-Phenylbenzoic acid	$C_{13}H_{10}O_2$		25	3.46
Phenyl biguanide	$C_8H_{11}N_5$	1		10.76
		2		2.13
Phenylbutazone	$C_{19}H_{20}N_2O_2$			4.5
N-Phenylglycine	$C_8H_9NO_2$	1	25	1.83
		2		4.39
Phenylhydrazine	$C_6H_8N_2$		15	8.79
Phenylpropanolamine hydrochloride	$C_9H_{14}ClNO$			9.44
Phthalazine	$C_8H_6N_2$		20	3.47
Phthalic acid	$C_8H_6O_4$	1	25	2.943
		2	25	5.432
Physostigmine	$C_{15}H_{21}N_3O_2$	1		6.12
		2		12.24
Pilocarpine	$C_{11}H_{16}N_2O_2$	1	25	1.6
		2	25	6.9
Piperazine	$C_4H_{10}N_2$	1	25	9.73
		2	25	5.33
Piperidine	$C_5H_{11}N$		25	11.123
2-Piperidinecarboxylic acid	$C_6H_{11}NO_2$	1	25	2.28
		2	25	10.72
Piperine	$C_{17}H_{19}NO_3$		18	12.22
L-Proline	$C_5H_9NO_2$	1	25	1.95
		2	25	10.64
1,2-Propanediamine, (±)-	$C_3H_{10}N_2$	1	25	9.82
		2	25	6.61
1,3-Propanediamine	$C_3H_{10}N_2$	1	25	10.55
		2	25	8.88
1,2,3-Propanetriamine	$C_3H_{11}N_3$	1	20	9.59
		2	20	7.95
Propanoic acid	$C_3H_6O_2$		25	4.87
2-Propanone oxime	C_3H_7NO		25	12.42
Propargyl alcohol	C_3H_4O		25	13.6
trans-1-Propene-1,2-dicarboxylic	$C_5H_6O_4$	1	25	3.09
		2	25	4.75
1-Propene-2,3-dicarboxylic acid	$C_5H_6O_4$	1	25	3.85
		2	25	5.45
cis-1-Propene-1,2,3-tricarboxylic	$C_6H_6O_6$		25	1.95
trans-1-Propene-1,2,3-tricarboxylic acid	$C_6H_6O_6$	1	25	2.80
		2	25	4.46
Propylamine	C_3H_9N		25	10.54
N-Propylglycine	$C_5H_{11}NO_2$	1	25	2.35
		2	25	10.19
2-Propylpentanoic acid	$C_8H_{16}O_2$			4.6
2-Propylpiperidine, (*S*)-	$C_8H_{17}N$			10.9
trans-6-Propyl-3-piperidinol,(3*S*)-	$C_8H_{17}NO$			10.3
Propyl 3,4,5-trihydroxybenzoate	$C_{10}H_{12}O_5$			8.11
2-Propynoic acid	$C_3H_2O_2$		25	1.84
Protriptyline	$C_{19}H_{21}N$			8.2
Pseudotropine	$C_8H_{15}NO$		15	3.80
Pteridine	$C_6H_4N_4$		20	4.05
1*H*-Purine	$C_5H_4N_4$	1	20	2.30
		2	20	8.96
Pyrazine	$C_4H_4N_2$		20	0.65
Pyrazinecarboxamide	$C_5H_5N_3O$			0.5
1*H*-Pyrazole	$C_3H_4N_2$		25	2.49
Pyridazine	$C_4H_4N_2$		20	2.24
2-Pyridinamine	$C_5H_6N_2$		20	6.82
3-Pyridinamine	$C_5H_6N_2$		25	6.04
4-Pyridinamine	$C_5H_6N_2$		25	9.11
Pyridine	C_5H_5N		25	5.23
2-Pyridinecarboxaldehyde	C_6H_5NO		25	12.68
4-Pyridinecarboxaldehyde	C_6H_5NO		30	12.05
2-Pyridinecarboxaldehyde oxime	$C_6H_6N_2O$	1	20	3.59

Name	Mol. form.	Step	t/°C	pK_a
		2	20	10.18
3-Pyridinecarboxamide	$C_6H_6N_2O$		20	3.3
2-Pyridinecarboxylic acid	$C_6H_5NO_2$	1	20	0.99
		2	20	5.39
3-Pyridinecarboxylic acid	$C_6H_5NO_2$	1	25	2.00
		2	25	4.82
4-Pyridinecarboxylic acid	$C_6H_5NO_2$	1	25	1.77
		2	25	4.84
2,5-Pyridinediamine	$C_5H_7N_3$		20	6.48
2,3-Pyridinedicarboxylic acid	$C_7H_5NO_4$	1	25	2.43
		2	25	4.78
2,4-Pyridinedicarboxylic acid	$C_7H_5NO_4$	1	25	2.15
2,6-Pyridinedicarboxylic acid	$C_7H_5NO_4$	1	25	2.16
		2	25	4.76
3,5-Pyridinedicarboxylic acid	$C_7H_5NO_4$	1	25	2.80
Pyridine-1-oxide	C_5H_5NO		24	0.79
2-Pyridinol	C_5H_5NO	1	20	0.75
		2	20	11.65
3-Pyridinol	C_5H_5NO	1	20	4.79
		2	20	8.75
4-Pyridinol	C_5H_5NO	1	20	3.20
		2	20	11.12
2(1*H*)-Pyridinone	C_5H_5NO	1	20	0.75
		2	20	11.65
2-Pyrimidinamine	$C_4H_5N_3$		20	3.45
4-Pyrimidinamine	$C_4H_5N_3$		20	5.71
Pyrimidine	$C_4H_4N_2$		20	1.23
2,4,6-Pyrimidinetriamine	$C_4H_7N_5$		20	6.84
Pyrocatechol	$C_6H_6O_2$	1	25	9.34
		2	25	12.6
L-Pyroglutamic acid	$C_5H_7NO_3$		25	3.32
Pyrrole	C_4H_5N		25	-3.8
1*H*-Pyrrole-2-carboxylic acid	$C_5H_5NO_2$		20	4.45
1*H*-Pyrrole-3-carboxylic acid	$C_5H_5NO_2$		20	5.00
Pyrrolidine	C_4H_9N		25	11.31
Pyruvic acid	$C_3H_4O_3$		25	2.39
Quinazoline	$C_8H_6N_2$		29	3.43
Quinidine	$C_{20}H_{24}N_2O_2$	1	20	5.4
		2	20	10.0
Quinine	$C_{20}H_{24}N_2O_2$	1	25	8.52
		2	25	4.13
2-Quinolinamine	$C_9H_8N_2$		20	7.34
3-Quinolinamine	$C_9H_8N_2$		20	4.91
4-Quinolinamine	$C_9H_8N_2$		20	9.17
Quinoline	C_9H_7N		20	4.90
8-Quinolinecarboxylic acid	$C_{10}H_7NO_2$		25	1.82
2-Quinolinol	C_9H_7NO	1	20	-0.31
		2	20	11.76
3-Quinolinol	C_9H_7NO	1	20	4.28
		2	20	8.08
4-Quinolinol	C_9H_7NO	1	20	2.23
		2	20	11.28
6-Quinolinol	C_9H_7NO	1	20	5.15
		2	20	8.90
8-Quinolinol	C_9H_7NO	1	25	4.91
		2	25	9.81
Quinoxaline	$C_8H_6N_2$		20	0.56
Reserpine	$C_{33}H_{40}N_2O_9$			6.6
Resorcinol	$C_6H_6O_2$	1	25	9.32
		2	25	11.1
Riboflavin	$C_{17}H_{20}N_4O_6$	1		1.7

Name	Mol. form.	Step	t/°C	pK_a
		2	25	9.69
L-Ribose	$C_5H_{10}O_5$		25	12.22
Rifampin	$C_{43}H_{58}N_4O_{12}$	1		1.7
		2		7.9
Saccharin	$C_7H_5NO_3S$		18	11.68
Salicylaldehyde	$C_7H_6O_2$		25	8.37
Sarcosine	$C_3H_7NO_2$	1	25	2.21
		2	25	10.1
L-Serine	$C_3H_7NO_3$	1	25	2.19
		2	25	9.21
Solanine	$C_{45}H_{73}NO_{15}$		15	6.66
Sparteine	$C_{15}H_{26}N_2$	1	20	2.24
		2	20	9.46
Strychnine	$C_{21}H_{22}N_2O_2$		25	8.26
Succinic acid	$C_4H_6O_4$	1	25	4.21
		2	25	5.64
Succinimide	$C_4H_5NO_2$		25	9.62
Sucrose	$C_{12}H_{22}O_{11}$		25	12.7
Sulfabenzamide	$C_{13}H_{12}N_2O_3S$		25	4.57
Sulfacytine	$C_{12}H_{14}N_4O_3S$			6.9
Sulfamethazine	$C_{12}H_{14}N_4O_2S$	1		7.4
		2		2.65
Sulfamethoxypyridazine	$C_{11}H_{12}N_4O_3S$			6.7
Sulfathiazole	$C_9H_9N_3O_2S_2$			7.2
Sulfisoxazole	$C_{11}H_{13}N_3O_3S$			5
DL-Tartaric acid	$C_4H_6O_6$	1	25	3.03
		2	25	4.37
meso-Tartaric acid	$C_4H_6O_6$	1	25	3.17
		2	25	4.91
L-Tartaric acid	$C_4H_6O_6$	1	25	2.98
		2	25	4.34
Taurine	$C_2H_7NO_3S$	1	25	1.5
		2	25	9.06
Taurocholic acid	$C_{26}H_{45}NO_7S$			1.4
Teniposide	$C_{32}H_{32}O_{13}S$			10.13
Terephthalic acid	$C_8H_6O_4$	1	25	3.54
		2	25	4.34
Tetradecylamine	$C_{14}H_{31}N$		25	10.62
5,6,7,8-Tetrahydro-2-naphthol	$C_{10}H_{12}O$		25	10.48
N,N,N',N'-Tetramethyl-1,2-ethanediamine	$C_6H_{16}N_2$	1	25	10.40
		2	25	8.26
2,2,6,6-Tetramethylpiperidine	$C_9H_{19}N$		25	11.07
Tetramethylurea	$C_5H_{12}N_2O$			2
Tetrodotoxin	$C_{11}H_{17}N_3O_8$			8.76
Thebaine	$C_{19}H_{21}NO_3$		15	6.05
Theobromine	$C_7H_8N_4O_2$		18	7.89
Theophylline	$C_7H_8N_4O_2$	1	25	8.77
2-Thiazolamine	$C_3H_4N_2S$		20	5.36
Thiazole	C_3H_3NS		25	2.52
Thioacetic acid	C_2H_4OS		25	3.33
Thioctic acid	$C_8H_{14}O_2S_2$			5.4
Thioglycolic acid	$C_2H_4O_2S$		25	3.68
2-Thiophenecarboxylic acid	$C_5H_4O_2S$		25	3.49
3-Thiophenecarboxylic acid	$C_5H_4O_2S$		25	4.1
Thiourea	CH_4N_2S		25	-1
L-Threonine	$C_4H_9NO_3$	1	25	2.09
		2	25	9.10
Thymine	$C_5H_6N_2O_2$		25	9.94
L-Thyroxine	$C_{15}H_{11}I_4NO_4$	1	25	2.2
		2	25	6.45

Name	Mol. form.	Step	t/°C	pK_a
		3	25	10.1
Tolazamide	$C_{14}H_{21}N_3O_3S$		25	3.6
o-Toluic acid	$C_8H_8O_2$		25	3.91
m-Toluic acid	$C_8H_8O_2$		25	4.25
p-Toluic acid	$C_8H_8O_2$		25	4.37
1,3,5-Triazine-2,4,6-triamine	$C_3H_6N_6$		25	5.00
1*H*-1,2,3-Triazole	$C_2H_3N_3$		20	1.17
1*H*-1,2,4-Triazole	$C_2H_3N_3$		20	2.27
Trichloroacetaldehyde	C_2HCl_3O		25	10.04
Trichloroacetic acid	$C_2HCl_3O_2$		20	0.66
2,2,2-Trichloroethanol	$C_2H_3Cl_3O$		25	12.24
Triclopyr	$C_7H_4Cl_3NO_3$			2.68
(Tridecyl)amine	$C_{13}H_{29}N$		25	10.63
Triethanolamine	$C_6H_{15}NO_3$		25	7.76
Triethylamine	$C_6H_{15}N$		25	10.75
Triethylenediamine	$C_6H_{12}N_2$	1		3.0
		2		8.7
Trifluoroacetic acid	$C_2HF_3O_2$		25	0.52
2,2,2-Trifluoroethanol	$C_2H_3F_3O$		25	12.37
3-(Trifluoromethyl)aniline	$C_7H_6F_3N$		25	3.49
4-(Trifluoromethyl)aniline	$C_7H_6F_3N$		25	2.45
2-(Trifluoromethyl)phenol	$C_7H_5F_3O$		25	8.95
3-(Trifluoromethyl)phenol	$C_7H_5F_3O$		25	8.68
2,4,6-Trihydroxybenzoic acid	$C_7H_6O_5$		25	1.68
3,4,5-Trihydroxybenzoic acid	$C_7H_6O_5$		25	4.41
Trimethoprim	$C_{14}H_{18}N_4O_3$			6.6
Trimethylamine	C_3H_9N		25	9.80
Trimethylamine oxide	C_3H_9NO		20	4.65
2,2,4-Trimethylpiperidine	$C_8H_{17}N$		30	11.04
2,4,6-Trimethylpyridine	$C_8H_{11}N$		25	7.43
2,4,6-Trinitrobenzoic acid	$C_7H_3N_3O_8$		25	0.65
2,4,6-Trinitrophenol	$C_6H_3N_3O_7$		24	0.42
Tris(hydroxymethyl) methylamine	$C_4H_{11}NO_3$		20	8.3
Tropacocaine	$C_{15}H_{19}NO_2$		15	4.32
Tropine	$C_8H_{15}NO$		15	3.80
Tryptamine	$C_{10}H_{12}N_2$		25	10.2
L-Tryptophan	$C_{11}H_{12}N_2O_2$	1	25	2.46
		2	25	9.41
L-Tyrosine	$C_9H_{11}NO_3$	1	25	2.20
		2	25	9.11
		3	25	10.1
Tyrosineamide	$C_9H_{12}N_2O_2$		25	7.33
Undecylamine	$C_{11}H_{25}N$		25	10.63
Uracil	$C_4H_4N_2O_2$		25	9.45
Urea	CH_4N_2O		25	0.10
Uric acid	$C_5H_4N_4O_3$		12	3.89
5'-Uridylic acid	$C_9H_{13}N_2O_9P$	1		6.4
		2		9.5
L-Valine	$C_5H_{11}NO_2$	1	25	2.29
		2	25	9.74
Valium	$C_{16}H_{13}ClN_2O$			3.4
Verapamil	$C_{27}H_{38}N_2O_4$			8.6
Veratridine	$C_{36}H_{51}NO_{11}$			9.54
Vinblastine	$C_{46}H_{58}N_4O_9$	1		5.4
		2		7.4
Vincristine	$C_{46}H_{56}N_4O_{10}$			5.4
Xanthopterin	$C_6H_5N_5O_2$	2	20	6.59
		3	20	9.31
D-Xylose	$C_5H_{10}O_5$		18	12.14

[a] Approximate value.

Thermochem

ACTIVITY COEFFICIENTS OF ACIDS, BASES, AND SALTS

Petr Vanýsek

This table gives mean activity coefficients γ at 25 °C for molalities in the range 0.1 to 1.0 mol/kg. See the following table "Mean Activity Coefficients of Electrolytes as a Function of Concentration" in this section, for definitions, references, and data over a wider concentration range. Compounds are listed alphabetically by formula.

Activity Coefficients of Acids, Bases, and Salts at 25 °C and the Indicated Molalities

Formula	γ(0.1 m)	γ(0.2 m)	γ(0.3 m)	γ(0.4 m)	γ(0.5 m)	γ(0.6 m)	γ(0.7 m)	γ(0.8 m)	γ(0.9 m)	γ(1.0 m)
$AgNO_3$	0.734	0.657	0.606	0.567	0.536	0.509	0.485	0.464	0.446	0.429
$AlCl_3$	0.337	0.305	0.302	0.313	0.331	0.356	0.388	0.429	0.479	0.539
$Al_2(SO_4)_3$	0.035	0.0225	0.0176	0.0153	0.0143	0.014	0.0142	0.0149	0.0159	0.0175
$BaCl_2$	0.500	0.444	0.419	0.405	0.397	0.391	0.391	0.391	0.392	0.395
$BeSO_4$	0.150	0.109	0.0885	0.0769	0.0692	0.0639	0.0600	0.0570	0.0546	0.0530
$CaCl_2$	0.518	0.472	0.455	0.448	0.448	0.453	0.460	0.470	0.484	0.500
$CdCl_2$	0.2280	0.1638	0.1329	0.1139	0.1006	0.0905	0.0827	0.0765	0.0713	0.0669
$Cd(NO_3)_2$	0.513	0.464	0.442	0.430	0.425	0.423	0.423	0.425	0.428	0.433
$CdSO_4$	0.150	0.103	0.0822	0.0699	0.0615	0.0553	0.0505	0.0468	0.0438	0.0415
$CoCl_2$	0.522	0.479	0.463	0.459	0.462	0.470	0.479	0.492	0.511	0.531
$CrCl_3$	0.331	0.298	0.294	0.300	0.314	0.335	0.362	0.397	0.436	0.481
$Cr(NO_3)_3$	0.319	0.285	0.279	0.281	0.291	0.304	0.322	0.344	0.371	0.401
$Cr_2(SO_4)_3$	0.0458	0.0300	0.0238	0.0207	0.0190	0.0182	0.0181	0.0185	0.0194	0.0208
CsBr	0.754	0.694	0.654	0.626	0.603	0.586	0.571	0.558	0.547	0.538
$CsC_2H_3O_2$	0.799	0.771	0.761	0.759	0.762	0.768	0.776	0.783	0.792	0.802
CsCl	0.756	0.694	0.656	0.628	0.606	0.589	0.575	0.563	0.553	0.544
CsI	0.754	0.692	0.651	0.621	0.599	0.581	0.567	0.554	0.543	0.533
$CsNO_3$	0.733	0.655	0.602	0.561	0.528	0.501	0.478	0.458	0.439	0.422
CsOH	0.795	0.761	0.744	0.739	0.739	0.742	0.748	0.754	0.762	0.771
Cs_2SO_4	0.456	0.382	0.338	0.311	0.291	0.274	0.262	0.251	0.242	0.235
$CuCl_2$	0.508	0.455	0.429	0.417	0.411	0.409	0.409	0.410	0.413	0.417
$Cu(NO_3)_2$	0.511	0.460	0.439	0.429	0.426	0.427	0.431	0.437	0.445	0.455
$CuSO_4$	0.150	0.104	0.0829	0.0704	0.0620	0.0559	0.0512	0.0475	0.0446	0.0423
$FeCl_2$	0.5185	0.473	0.454	0.448	0.450	0.454	0.463	0.473	0.488	0.506
HBr	0.805	0.782	0.777	0.781	0.789	0.801	0.815	0.832	0.850	0.871
HCl	0.796	0.767	0.756	0.755	0.757	0.763	0.772	0.783	0.795	0.809
$HClO_4$	0.803	0.778	0.768	0.766	0.769	0.776	0.785	0.795	0.808	0.823
HI	0.818	0.807	0.811	0.823	0.839	0.860	0.883	0.908	0.935	0.963
HNO_3	0.791	0.754	0.735	0.725	0.720	0.717	0.717	0.718	0.721	0.724
H_2SO_4	0.2655	0.2090	0.1826		0.1557		0.1417			0.1316
KBr	0.772	0.722	0.693	0.673	0.657	0.646	0.636	0.629	0.622	0.617
$KC_2H_3O_2$	0.796	0.766	0.754	0.750	0.751	0.754	0.759	0.766	0.774	0.783
KCl	0.770	0.718	0.688	0.666	0.649	0.637	0.626	0.618	0.610	0.604
$KClO_3$	0.749	0.681	0.635	0.599	0.568	0.541	0.518			
KF	0.775	0.727	0.700	0.682	0.670	0.661	0.654	0.650	0.646	0.645
KH_2PO_4	0.731	0.653	0.602	0.561	0.529	0.501	0.477	0.456	0.438	0.421
KI	0.778	0.733	0.707	0.689	0.676	0.667	0.660	0.654	0.649	0.645
KNO_3	0.739	0.663	0.614	0.576	0.545	0.519	0.496	0.476	0.459	0.443
KOH	0.798	0.760	0.742	0.734	0.732	0.733	0.736	0.742	0.749	0.756
KSCN	0.769	0.716	0.685	0.663	0.646	0.633	0.623	0.614	0.606	0.599
K_2CrO_4	0.456	0.382	0.340	0.313	0.292	0.276	0.263	0.253	0.243	0.235
K_2SO_4	0.441	0.360	0.316	0.286	0.264	0.246	0.232			
$K_3Fe(CN)_6$	0.268	0.212	0.184	0.167	0.155	0.146	0.140	0.135	0.131	0.128
$K_4Fe(CN)_6$	0.139	0.0993	0.0808	0.0693	0.0614	0.0556	0.0512	0.0479	0.0454	
LiBr	0.796	0.766	0.756	0.752	0.753	0.758	0.767	0.777	0.789	0.803
$LiC_2H_3O_2$	0.784	0.742	0.721	0.709	0.700	0.691	0.689	0.688	0.688	0.689
LiCl	0.790	0.757	0.744	0.740	0.739	0.743	0.748	0.755	0.764	0.774
$LiClO_4$	0.812	0.794	0.792	0.798	0.808	0.820	0.834	0.852	0.869	0.887
LiI	0.815	0.802	0.804	0.813	0.824	0.838	0.852	0.870	0.888	0.910
$LiNO_3$	0.788	0.752	0.736	0.728	0.726	0.727	0.729	0.733	0.737	0.743

Formula	γ(0.1 m)	γ(0.2 m)	γ(0.3 m)	γ(0.4 m)	γ(0.5 m)	γ(0.6 m)	γ(0.7 m)	γ(0.8 m)	γ(0.9 m)	γ(1.0 m)
LiOH	0.760	0.702	0.665	0.638	0.617	0.599	0.585	0.573	0.563	0.554
Li_2SO_4	0.468	0.398	0.361	0.337	0.319	0.307	0.297	0.289	0.282	0.277
$MgCl_2$	0.529	0.489	0.477	0.475	0.481	0.491	0.506	0.522	0.544	0.570
$MgSO_4$	0.150	0.107	0.0874	0.0756	0.0675	0.0616	0.0571	0.0536	0.0508	0.0485
$MnCl_2$	0.516	0.469	0.450	0.442	0.440	0.443	0.448	0.455	0.466	0.479
$MnSO_4$	0.150	0.105	0.0848	0.0725	0.0640	0.0578	0.0530	0.0493	0.0463	0.0439
NH_4Cl	0.770	0.718	0.687	0.665	0.649	0.636	0.625	0.617	0.609	0.603
NH_4NO_3	0.740	0.677	0.636	0.606	0.582	0.562	0.545	0.530	0.516	0.504
$(NH_4)_2SO_4$	0.439	0.356	0.311	0.280	0.257	0.240	0.226	0.214	0.205	0.196
NaBr	0.782	0.741	0.719	0.704	0.697	0.692	0.689	0.687	0.687	0.687
$NaC_2H_3O_2$	0.791	0.757	0.744	0.737	0.735	0.736	0.740	0.745	0.752	0.757
NaCl	0.778	0.735	0.710	0.693	0.681	0.673	0.667	0.662	0.659	0.657
$NaClO_3$	0.772	0.720	0.688	0.664	0.645	0.630	0.617	0.606	0.597	0.589
$NaClO_4$	0.775	0.729	0.701	0.683	0.668	0.656	0.648	0.641	0.635	0.629
NaF	0.765	0.710	0.676	0.651	0.632	0.616	0.603	0.592	0.582	0.573
NaH_2PO_4	0.744	0.675	0.629	0.593	0.563	0.539	0.517	0.499	0.483	0.468
NaI	0.787	0.751	0.735	0.727	0.723	0.723	0.724	0.727	0.731	0.736
$NaNO_3$	0.762	0.703	0.666	0.638	0.617	0.599	0.583	0.570	0.558	0.548
NaOH	0.766	0.727	0.708	0.697	0.690	0.685	0.681	0.679	0.678	0.678
NaSCN	0.787	0.750		0.720	0.715	0.712	0.710	0.710	0.711	0.712
Na_2CrO_4	0.464	0.394	0.353	0.327	0.307	0.292	0.280	0.269	0.261	0.253
Na_2SO_4	0.445	0.365	0.320	0.289	0.266	0.248	0.233	0.221	0.210	0.201
$NiCl_2$	0.522	0.479	0.463	0.460	0.464	0.471	0.482	0.496	0.515	0.563
$NiSO_4$	0.150	0.105	0.0841	0.0713	0.0627	0.0562	0.0515	0.0478	0.0448	0.0425
$Pb(NO_3)_2$	0.395	0.308	0.260	0.228	0.205	0.187	0.172	0.160	0.150	0.141
RbBr	0.763	0.706	0.673	0.650	0.632	0.617	0.605	0.595	0.586	0.578
$RbC_2H_3O_2$	0.796	0.767	0.756	0.753	0.755	0.759	0.766	0.773	0.782	0.792
RbCl	0.764	0.709	0.675	0.652	0.634	0.620	0.608	0.599	0.590	0.583
RbI	0.762	0.705	0.671	0.647	0.629	0.614	0.602	0.591	0.583	0.575
$RbNO_3$	0.734	0.658	0.606	0.565	0.534	0.508	0.485	0.465	0.446	0.430
Rb_2SO_4	0.451	0.374	0.331	0.301	0.279	0.263	0.249	0.238	0.228	0.219
$SrCl_2$	0.511	0.462	0.442	0.433	0.430	0.431	0.434	0.441	0.449	0.461
$TlClO_4$	0.730	0.652	0.599	0.527						
$TlNO_3$	0.702	0.606	0.545	0.500						
UO_2Cl_2	0.544	0.510	0.520	0.505	0.517	0.532	0.549	0.571	0.595	0.620
UO_2SO_4	0.150	0.102	0.0807	0.0689	0.0611	0.0566	0.0515	0.0483	0.0458	0.0439
$ZnCl_2$	0.515	0.462	0.432	0.411	0.394	0.380	0.369	0.357	0.348	0.339
$Zn(NO_3)_2$	0.531	0.489	0.474	0.469	0.473	0.480	0.489	0.501	0.518	0.535
$ZnSO_4$	0.150	0.10	0.0835	0.0714	0.0630	0.0569	0.0523	0.0487	0.0458	0.0435

MEAN ACTIVITY COEFFICIENTS OF ELECTROLYTES AS A FUNCTION OF CONCENTRATION

The mean activity coefficient γ_{mean} of an electrolyte X_aY_b is defined as

$$\gamma = (\gamma_+^a \gamma_-^b)^{1/(a+b)}$$

where γ_+ and γ_- are activity coefficients of the individual ions (which cannot be directly measured). This table gives the mean activity coefficients of about 100 electrolytes in aqueous solution as a function of concentration, expressed in molality terms. All values refer to a temperature of 25 °C. Electrolytes are arranged alphabetically by molecular formula.

References

1. Hamer, W. J., and Wu, Y. C., *J. Phys.Chem. Ref. Data,* 1, 1047, 1972. <https://doi.org/10.1063/1.3253108>
2. Staples, B. R., *J. Phys. Chem. Ref. Data,* 6, 385, 1977; 10, 767, 1981; 10, 779, 1981.
3. Goldberg, R. N. et al., *J. Phys. Chem. Ref. Data,* 7, 263, 1978; 8, 923, 1979; 8, 1005, 1979; 10, 1, 1981; 10, 671, 1981.

Mean Activity Coefficients of Electrolytes as a Function of Concentration at 25 °C

Mol. form.	m/mol kg^{-1}	γ_{mean}	Mol. form.	m/mol kg^{-1}	γ_{mean}	Mol. form.	m/mol kg^{-1}	γ_{mean}	Mol. form.	m/mol kg^{-1}	γ_{mean}
$AgNO_3$	0.001	0.964		0.500	0.536		0.050	0.451		0.005	0.783
	0.002	0.950		1.000	0.664		0.100	0.344		0.010	0.724
	0.005	0.924		2.000	1.242		0.200	0.247		0.020	0.661
	0.010	0.896	$CaBr_2$	0.001	0.890		0.500	0.148		0.050	0.582
	0.020	0.859		0.002	0.853		1.000	0.098		0.100	0.540
	0.050	0.794		0.005	0.791		2.000	0.069		0.200	0.527
	0.100	0.732		0.010	0.735		5.000	0.054		0.500	0.596
	0.200	0.656		0.020	0.674	$Cd(NO_3)_2$	0.001	0.888		1.000	0.845
	0.500	0.536		0.050	0.594		0.002	0.851		2.000	2.287
	1.000	0.430		0.100	0.540		0.005	0.787		5.000	55.3
	2.000	0.316		0.200	0.502		0.010	0.728		10.000	196
	5.000	0.181		0.500	0.500		0.020	0.664	$Co(NO_3)_2$	0.001	0.888
	10.000	0.108		1.000	0.604		0.050	0.576		0.002	0.850
	15.000	0.085		2.000	1.125		0.100	0.515		0.005	0.786
$BaBr_2$	0.001	0.881		5.000	18.7		0.200	0.465		0.010	0.728
	0.002	0.850	$CaCl_2$	0.001	0.888		0.500	0.428		0.020	0.663
	0.005	0.785		0.002	0.851		1.000	0.437		0.050	0.576
	0.010	0.727		0.005	0.787		2.000	0.517		0.100	0.516
	0.020	0.661		0.010	0.727	$CoBr_2$	0.001	0.890		0.200	0.469
	0.050	0.573		0.020	0.664		0.002	0.854		0.500	0.446
	0.100	0.517		0.050	0.577		0.005	0.794		1.000	0.492
	0.200	0.463		0.100	0.517		0.010	0.740		2.000	0.722
	0.500	0.435		0.200	0.469		0.020	0.681		5.000	3.338
	1.000	0.470		0.500	0.444		0.050	0.605	CsBr	0.001	0.965
	2.000	0.654		1.000	0.495		0.100	0.556		0.002	0.951
$BaCl_2$	0.001	0.887		2.000	0.784		0.200	0.523		0.005	0.925
	0.002	0.849		5.000	5.907		0.500	0.538		0.010	0.898
	0.005	0.782		10.000	43.1		1.000	0.685		0.020	0.864
	0.010	0.721	CaI_2	0.001	0.890		2.000	1.421		0.050	0.806
	0.020	0.653		0.002	0.853		5.000	13.9		0.100	0.752
	0.050	0.559		0.005	0.791	$CoCl_2$	0.001	0.889		0.200	0.691
	0.100	0.492		0.010	0.736		0.002	0.852		0.500	0.605
	0.200	0.436		0.020	0.677		0.005	0.789		1.000	0.540
	0.500	0.391		0.050	0.600		0.010	0.732		2.000	0.485
	1.000	0.393		0.100	0.552		0.020	0.670		5.000	0.454
BaI_2	0.001	0.890		0.200	0.524		0.050	0.586	CsCl	0.001	0.965
	0.002	0.853		0.500	0.554		0.100	0.528		0.002	0.951
	0.005	0.792		1.000	0.729		0.200	0.483		0.005	0.925
	0.010	0.737	$Cd(NO_2)_2$	0.001	0.881		0.500	0.465		0.010	0.898
	0.020	0.678		0.002	0.837		1.000	0.532		0.020	0.864
	0.050	0.600		0.005	0.759		2.000	0.864		0.050	0.805
	0.100	0.551		0.010	0.681	CoI_2	0.001	0.887		0.100	0.751
	0.200	0.520		0.020	0.589		0.002	0.849		0.200	0.691

Mol. form.	m/mol kg^{-1}	γ_{mean}
	0.500	0.607
	1.000	0.546
	2.000	0.496
	5.000	0.474
	10.000	0.508
CsF	0.001	0.965
	0.002	0.952
	0.005	0.929
	0.010	0.905
	0.020	0.876
	0.050	0.830
	0.100	0.792
	0.200	0.755
	0.500	0.721
	1.000	0.726
	2.000	0.803
CsI	0.001	0.965
	0.002	0.951
	0.005	0.925
	0.010	0.898
	0.020	0.863
	0.050	0.804
	0.100	0.749
	0.200	0.688
	0.500	0.601
	1.000	0.534
	2.000	0.470
$CsNO_3$	0.001	0.964
	0.002	0.951
	0.005	0.924
	0.010	0.897
	0.020	0.860
	0.050	0.796
	0.100	0.733
	0.200	0.655
	0.500	0.529
	1.000	0.421
CsOH	0.001	0.966
	0.002	0.953
	0.005	0.930
	0.010	0.906
	0.020	0.878
	0.050	0.836
	0.100	0.802
	0.200	0.772
	0.500	0.755
	1.000	0.782
Cs_2SO_4	0.001	0.885
	0.002	0.845
	0.005	0.775
	0.010	0.709
	0.020	0.634
	0.050	0.526
	0.100	0.444
	0.200	0.369
	0.500	0.285
	1.000	0.233
$CuBr_2$	0.001	0.889
	0.002	0.853
	0.005	0.791
	0.010	0.735
	0.020	0.674
	0.050	0.594
	0.100	0.541
	0.200	0.504
	0.500	0.503
	1.000	0.591
	2.000	0.859
$CuCl_2$	0.001	0.887
	0.002	0.849
	0.005	0.783
	0.010	0.722
	0.020	0.654
	0.050	0.561
	0.100	0.495
	0.200	0.441
	0.500	0.401
	1.000	0.405
	2.000	0.453
	5.000	0.601
$Cu(ClO_4)_2$	0.001	0.890
	0.002	0.854
	0.005	0.795
	0.010	0.741
	0.020	0.685
	0.050	0.613
	0.100	0.572
	0.200	0.553
	0.500	0.617
	1.000	0.892
	2.000	2.445
$Cu(NO_3)_2$	0.001	0.888
	0.002	0.851
	0.005	0.787
	0.010	0.729
	0.020	0.664
	0.050	0.577
	0.100	0.516
	0.200	0.466
	0.500	0.431
	1.000	0.456
	2.000	0.615
	5.000	2.083
$FeCl_2$	0.001	0.888
	0.002	0.850
	0.005	0.785
	0.010	0.725
	0.020	0.659
	0.050	0.570
	0.100	0.509
	0.200	0.462
	0.500	0.443
	1.000	0.500
	2.000	0.782
HBr	0.001	0.966
	0.002	0.953
	0.005	0.930
	0.010	0.907
	0.020	0.879
	0.050	0.837
	0.100	0.806
	0.200	0.783
	0.500	0.790
	1.000	0.872
	2.000	1.167
	5.000	3.800
	10.000	33.4
HCl	0.001	0.965
	0.002	0.952
	0.005	0.929
	0.010	0.905
	0.020	0.876
	0.050	0.832
	0.100	0.797
	0.200	0.768
	0.500	0.759
	1.000	0.811
	2.000	1.009
	5.000	2.380
	10.000	10.4
$HClO_4$	0.001	0.966
	0.002	0.953
	0.005	0.929
	0.010	0.906
	0.020	0.878
	0.050	0.836
	0.100	0.803
	0.200	0.776
	0.500	0.769
	1.000	0.826
	2.000	1.055
	5.000	3.100
	10.000	30.8
	15.000	323
HF	0.001	0.551
	0.002	0.429
	0.005	0.302
	0.010	0.225
	0.020	0.163
	0.050	0.106
	0.100	0.0766
	0.200	0.0550
	0.500	0.0352
	1.000	0.0249
	2.000	0.0175
	5.000	0.0110
	10.000	0.0085
	15.000	0.0077
	20.000	0.0075
HI	0.001	0.966
	0.002	0.953
	0.005	0.931
	0.010	0.909
	0.020	0.884
	0.050	0.847
	0.100	0.823
	0.200	0.811
	0.500	0.845
	1.000	0.969
	2.000	1.363
	5.000	4.760
	10.000	49.100
HNO_3	0.001	0.965
	0.002	0.952
	0.005	0.929
	0.010	0.905
	0.020	0.875
	0.050	0.829
	0.100	0.792
	0.200	0.756
	0.500	0.725
	1.000	0.730
	2.000	0.788
	5.000	1.063
	10.000	1.644
	15.000	2.212
	20.000	2.607
H_2SO_4	0.001	0.804
	0.002	0.740
	0.005	0.634
	0.010	0.542
	0.020	0.445
	0.050	0.325
	0.100	0.251
	0.200	0.195
	0.500	0.146
	1.000	0.125
	2.000	0.119
	5.000	0.197
	10.000	0.527
	15.000	1.077
	20.000	1.701
KBr	0.001	0.965
	0.002	0.952
	0.005	0.927
	0.010	0.902
	0.020	0.870
	0.050	0.817
	0.100	0.771
	0.200	0.722
	0.500	0.658
	1.000	0.617
	2.000	0.593
	5.000	0.626
KCl	0.001	0.965
	0.002	0.951
	0.005	0.927
	0.010	0.901
	0.020	0.869
	0.050	0.816
	0.100	0.768
	0.200	0.717
	0.500	0.649
	1.000	0.604
	2.000	0.573
	5.000	0.593
$KClO_3$	0.001	0.965
	0.002	0.951
	0.005	0.926
	0.010	0.899
	0.020	0.865
	0.050	0.805
	0.100	0.749
	0.200	0.681

Mol. form.	m/mol kg^{-1}	γ_{mean}
	0.500	0.569
KF	0.001	0.965
	0.002	0.952
	0.005	0.927
	0.010	0.902
	0.020	0.870
	0.050	0.818
	0.100	0.773
	0.200	0.726
	0.500	0.670
	1.000	0.645
	2.000	0.658
	5.000	0.871
	10.000	1.715
	15.000	3.120
KH_2PO_4	0.001	0.964
	0.002	0.950
	0.005	0.924
	0.010	0.896
	0.020	0.859
	0.050	0.793
	0.100	0.730
	0.200	0.652
	0.500	0.529
	1.000	0.422
KI	0.001	0.965
	0.002	0.952
	0.005	0.927
	0.010	0.902
	0.020	0.871
	0.050	0.820
	0.100	0.776
	0.200	0.731
	0.500	0.676
	1.000	0.646
	2.000	0.638
KNO_3	0.001	0.964
	0.002	0.950
	0.005	0.924
	0.010	0.896
	0.020	0.860
	0.050	0.797
	0.100	0.735
	0.200	0.662
	0.500	0.546
	1.000	0.444
	2.000	0.332
KOH	0.001	0.965
	0.002	0.952
	0.005	0.927
	0.010	0.902
	0.020	0.871
	0.050	0.821
	0.100	0.779
	0.200	0.740
	0.500	0.710
	1.000	0.733
	2.000	0.860
	5.000	1.697
	10.000	6.110
	15.000	19.9
	20.000	46.4
KSCN	0.001	0.965
	0.002	0.951
	0.005	0.927
	0.010	0.901
	0.020	0.869
	0.050	0.815
	0.100	0.768
	0.200	0.716
	0.500	0.647
	1.000	0.598
	2.000	0.556
	5.000	0.525
K_2CrO_4	0.001	0.886
	0.002	0.847
	0.005	0.779
	0.010	0.715
	0.020	0.643
	0.050	0.539
	0.100	0.460
	0.200	0.385
	0.500	0.296
	1.000	0.239
	2.000	0.199
K_2HPO_4	0.001	0.886
	0.002	0.847
	0.005	0.779
	0.010	0.715
	0.020	0.643
	0.050	0.538
	0.100	0.457
	0.200	0.379
	0.500	0.283
K_2SO_4	0.001	0.885
	0.002	0.844
	0.005	0.772
	0.010	0.704
	0.020	0.625
	0.050	0.511
	0.100	0.424
	0.200	0.343
	0.500	0.251
LiBr	0.001	0.965
	0.002	0.952
	0.005	0.929
	0.010	0.905
	0.020	0.877
	0.050	0.832
	0.100	0.797
	0.200	0.767
	0.500	0.754
	1.000	0.803
	2.000	1.012
	5.000	2.696
	10.000	20.0
	15.000	147
	20.000	486
LiCl	0.001	0.965
	0.002	0.952
	0.005	0.928
	0.010	0.904
	0.020	0.874
	0.050	0.827
	0.100	0.789
	0.200	0.756
	0.500	0.739
	1.000	0.775
	2.000	0.924
	5.000	2.000
	10.000	9.600
	15.000	30.9
$LiClO_4$	0.001	0.966
	0.002	0.953
	0.005	0.931
	0.010	0.908
	0.020	0.882
	0.050	0.843
	0.100	0.815
	0.200	0.795
	0.500	0.806
	1.000	0.887
	2.000	1.161
LiI	0.001	0.966
	0.002	0.953
	0.005	0.930
	0.010	0.908
	0.020	0.882
	0.050	0.843
	0.100	0.817
	0.200	0.802
	0.500	0.824
	1.000	0.912
	2.000	1.197
$LiNO_3$	0.001	0.965
	0.002	0.952
	0.005	0.928
	0.010	0.904
	0.020	0.874
	0.050	0.827
	0.100	0.788
	0.200	0.753
	0.500	0.726
	1.000	0.743
	2.000	0.837
	5.000	1.298
	10.000	2.500
	15.000	3.960
	20.000	4.970
LiOH	0.001	0.964
	0.002	0.950
	0.005	0.923
	0.010	0.895
	0.020	0.858
	0.050	0.794
	0.100	0.735
	0.200	0.668
	0.500	0.579
	1.000	0.522
	2.000	0.484
	5.000	0.493
Li_2SO_4	0.001	0.887
	0.002	0.847
	0.005	0.780
	0.010	0.716
	0.020	0.645
	0.050	0.544
	0.100	0.469
	0.200	0.400
	0.500	0.325
	1.000	0.284
	2.000	0.270
$MgBr_2$	0.001	0.889
	0.002	0.852
	0.005	0.790
	0.010	0.733
	0.020	0.672
	0.050	0.593
	0.100	0.543
	0.200	0.512
	0.500	0.540
	1.000	0.715
	2.000	1.590
	5.000	36.1
$MgCl_2$	0.001	0.889
	0.002	0.852
	0.005	0.790
	0.010	0.734
	0.020	0.672
	0.050	0.590
	0.100	0.535
	0.200	0.493
	0.500	0.485
	1.000	0.577
	2.000	1.065
	5.000	14.40
MgI_2	0.001	0.889
	0.002	0.853
	0.005	0.791
	0.010	0.736
	0.020	0.677
	0.050	0.602
	0.100	0.556
	0.200	0.535
	0.500	0.594
	1.000	0.858
	2.000	2.326
	5.000	109.8
$MnBr_2$	0.001	0.889
	0.002	0.853
	0.005	0.791
	0.010	0.735
	0.020	0.674
	0.050	0.595
	0.100	0.543
	0.200	0.508
	0.500	0.519
	1.000	0.650
	2.000	1.224
	5.000	6.697
$MnCl_2$	0.001	0.888
	0.002	0.850
	0.005	0.786
	0.010	0.727

Mol. form.	m/mol kg^{-1}	γ_{mean}
	0.020	0.662
	0.050	0.574
	0.100	0.513
	0.200	0.464
	0.500	0.437
	1.000	0.477
	2.000	0.661
	5.000	1.539
$Mn(ClO_4)_2$	0.001	0.892
	0.002	0.858
	0.005	0.801
	0.010	0.752
	0.020	0.700
	0.050	0.637
	0.100	0.604
	0.200	0.596
	0.500	0.686
	1.000	1.030
	2.000	3.072
NH_4Cl	0.001	0.965
	0.002	0.952
	0.005	0.927
	0.010	0.901
	0.020	0.869
	0.050	0.816
	0.100	0.769
	0.200	0.718
	0.500	0.649
	1.000	0.603
	2.000	0.569
	5.000	0.563
NH_4ClO_4	0.001	0.964
	0.002	0.950
	0.005	0.924
	0.010	0.895
	0.020	0.859
	0.050	0.794
	0.100	0.734
	0.200	0.663
	0.500	0.560
	1.000	0.479
	2.000	0.399
NH_4NO_3	0.001	0.964
	0.002	0.951
	0.005	0.925
	0.010	0.897
	0.020	0.862
	0.050	0.801
	0.100	0.744
	0.200	0.678
	0.500	0.582
	1.000	0.502
	2.000	0.419
	5.000	0.303
	10.000	0.220
	15.000	0.179
	20.000	0.154
$(NH_4)_2HPO_4$	0.001	0.882
	0.002	0.839
	0.005	0.763
	0.010	0.688

Mol. form.	m/mol kg^{-1}	γ_{mean}
	0.020	0.600
	0.050	0.469
	0.100	0.367
	0.200	0.273
	0.500	0.171
	1.000	0.114
	2.000	0.074
NaBr	0.001	0.965
	0.002	0.952
	0.005	0.928
	0.010	0.903
	0.020	0.873
	0.050	0.824
	0.100	0.783
	0.200	0.742
	0.500	0.697
	1.000	0.687
	2.000	0.730
	5.000	1.083
$NaBrO_3$	0.001	0.965
	0.002	0.951
	0.005	0.926
	0.010	0.900
	0.020	0.867
	0.050	0.811
	0.100	0.759
	0.200	0.698
	0.500	0.605
	1.000	0.528
	2.000	0.449
NaCl	0.001	0.965
	0.002	0.952
	0.005	0.928
	0.010	0.903
	0.020	0.872
	0.050	0.822
	0.100	0.779
	0.200	0.734
	0.500	0.681
	1.000	0.657
	2.000	0.668
	5.000	0.874
$NaClO_3$	0.001	0.965
	0.002	0.952
	0.005	0.927
	0.010	0.902
	0.020	0.870
	0.050	0.818
	0.100	0.771
	0.200	0.719
	0.500	0.646
	1.000	0.590
	2.000	0.537
$NaClO_4$	0.001	0.965
	0.002	0.952
	0.005	0.928
	0.010	0.903
	0.020	0.872
	0.050	0.821
	0.100	0.777
	0.200	0.729

Mol. form.	m/mol kg^{-1}	γ_{mean}
	0.500	0.668
	1.000	0.630
	2.000	0.608
	5.000	0.648
NaF	0.001	0.965
	0.002	0.951
	0.005	0.926
	0.010	0.901
	0.020	0.868
	0.050	0.813
	0.100	0.764
	0.200	0.710
	0.500	0.633
	1.000	0.573
NaI	0.001	0.965
	0.002	0.952
	0.005	0.928
	0.010	0.904
	0.020	0.874
	0.050	0.827
	0.100	0.789
	0.200	0.753
	0.500	0.722
	1.000	0.734
	2.000	0.823
	5.000	1.402
	10.000	4.011
$NaNO_3$	0.001	0.965
	0.002	0.951
	0.005	0.926
	0.010	0.900
	0.020	0.866
	0.050	0.810
	0.100	0.759
	0.200	0.701
	0.500	0.617
	1.000	0.550
	2.000	0.480
	5.000	0.388
	10.000	0.329
NaOH	0.001	0.965
	0.002	0.952
	0.005	0.927
	0.010	0.902
	0.020	0.870
	0.050	0.819
	0.100	0.775
	0.200	0.731
	0.500	0.685
	1.000	0.674
	2.000	0.714
	5.000	1.076
	10.000	3.258
	15.000	9.796
	20.000	19.410
Na_2CO_3	0.001	0.887
	0.002	0.847
	0.005	0.780
	0.010	0.716
	0.020	0.644
	0.050	0.541

Mol. form.	m/mol kg^{-1}	γ_{mean}
	0.100	0.462
	0.200	0.385
	0.500	0.292
	1.000	0.229
	2.000	0.182
Na_2CrO_4	0.001	0.887
	0.002	0.849
	0.005	0.783
	0.010	0.722
	0.020	0.653
	0.050	0.554
	0.100	0.479
	0.200	0.406
	0.500	0.318
	1.000	0.261
	2.000	0.231
Na_2HPO_4	0.001	0.887
	0.002	0.848
	0.005	0.780
	0.010	0.717
	0.020	0.644
	0.050	0.539
	0.100	0.456
	0.200	0.373
	0.500	0.266
	1.000	0.191
	2.000	0.133
Na_2SO_3	0.001	0.887
	0.002	0.847
	0.005	0.779
	0.010	0.716
	0.020	0.644
	0.050	0.540
	0.100	0.462
	0.200	0.386
	0.500	0.296
	1.000	0.237
	2.000	0.196
Na_2SO_4	0.001	0.886
	0.002	0.846
	0.005	0.777
	0.010	0.712
	0.020	0.637
	0.050	0.529
	0.100	0.446
	0.200	0.366
	0.500	0.268
	1.000	0.204
	2.000	0.155
Na_2WO_4	0.001	0.886
	0.002	0.846
	0.005	0.777
	0.010	0.712
	0.020	0.638
	0.050	0.534
	0.100	0.457
	0.200	0.388
	0.500	0.320
	1.000	0.291
	2.000	0.291
$NiBr_2$	0.001	0.889

Mol. form.	m/mol kg^{-1}	γ_{mean}
	0.002	0.853
	0.005	0.791
	0.010	0.735
	0.020	0.675
	0.050	0.596
	0.100	0.546
	0.200	0.514
	0.500	0.535
	1.000	0.692
	2.000	1.476
$NiCl_2$	0.001	0.889
	0.002	0.852
	0.005	0.789
	0.010	0.732
	0.020	0.669
	0.050	0.584
	0.100	0.527
	0.200	0.482
	0.500	0.465
	1.000	0.538
	2.000	0.915
	5.000	4.785
$Ni(ClO_4)_2$	0.001	0.891
	0.002	0.855
	0.005	0.797
	0.010	0.745
	0.020	0.690
	0.050	0.621
	0.100	0.582
	0.200	0.567
	0.500	0.639
	1.000	0.946
	2.000	2.812
$Ni(NO_3)_2$	0.001	0.889
	0.002	0.851
	0.005	0.787
	0.010	0.730
	0.020	0.666
	0.050	0.581
	0.100	0.524
	0.200	0.481
	0.500	0.467
	1.000	0.528
	2.000	0.797
$Pb(ClO_4)_2$	0.001	0.889
	0.002	0.851
	0.005	0.787
	0.010	0.729
	0.020	0.666
	0.050	0.580
	0.100	0.522
	0.200	0.476
	0.500	0.458
	1.000	0.516
	2.000	0.799
	5.000	4.043
	10.000	33.8
$Pb(NO_3)_2$	0.001	0.882
	0.002	0.840
	0.005	0.764
	0.010	0.690
	0.020	0.604
	0.050	0.476
	0.100	0.379
	0.200	0.291
	0.500	0.195
	1.000	0.136
RbBr	0.001	0.965
	0.002	0.951
	0.005	0.926
	0.010	0.900
	0.020	0.866
	0.050	0.811
	0.100	0.760
	0.200	0.705
	0.500	0.630
	1.000	0.578
	2.000	0.535
	5.000	0.514
RbCl	0.001	0.965
	0.002	0.951
	0.005	0.926
	0.010	0.900
	0.020	0.867
	0.050	0.811
	0.100	0.761
	0.200	0.707
	0.500	0.633
	1.000	0.583
	2.000	0.546
	5.000	0.544
RbF	0.001	0.965
	0.002	0.952
	0.005	0.927
	0.010	0.902
	0.020	0.871
	0.050	0.821
	0.100	0.780
	0.200	0.739
	0.500	0.701
	1.000	0.697
	2.000	0.724
RbI	0.001	0.965
	0.002	0.951
	0.005	0.926
	0.010	0.900
	0.020	0.866
	0.050	0.810
	0.100	0.759
	0.200	0.703
	0.500	0.627
	1.000	0.574
	2.000	0.532
	5.000	0.517
$RbNO_3$	0.001	0.964
	0.002	0.950
	0.005	0.924
	0.010	0.896
	0.020	0.859
	0.050	0.795
	0.100	0.733
	0.200	0.657
	0.500	0.536
	1.000	0.430
	2.000	0.320
Rb_2SO_4	0.001	0.886
	0.002	0.845
	0.005	0.776
	0.010	0.710
	0.020	0.635
	0.050	0.526
	0.100	0.443
	0.200	0.365
	0.500	0.274
	1.000	0.217
$SrBr_2$	0.001	0.889
	0.002	0.852
	0.005	0.790
	0.010	0.734
	0.020	0.673
	0.050	0.591
	0.100	0.535
	0.200	0.492
	0.500	0.476
	1.000	0.545
	2.000	0.921
$SrCl_2$	0.001	0.888
	0.002	0.850
	0.005	0.785
	0.010	0.725
	0.020	0.659
	0.050	0.569
	0.100	0.506
	0.200	0.455
	0.500	0.421
	1.000	0.451
	2.000	0.650
SrI_2	0.001	0.890
	0.002	0.854
	0.005	0.793
	0.010	0.740
	0.020	0.681
	0.050	0.606
	0.100	0.557
	0.200	0.526
	0.500	0.542
	1.000	0.686
UO_2Cl_2	0.001	0.888
	0.002	0.851
	0.005	0.787
	0.010	0.729
	0.020	0.666
	0.050	0.583
	0.100	0.529
	0.200	0.493
	0.500	0.501
	1.000	0.601
	2.000	0.948
$UO_2(NO_3)_2$	0.001	0.888
	0.002	0.849
	0.005	0.784
	0.010	0.726
	0.020	0.663
	0.050	0.583
	0.100	0.535
	0.200	0.509
	0.500	0.532
	1.000	0.673
	2.000	1.223
	5.000	3.020
$ZnBr_2$	0.001	0.890
	0.002	0.854
	0.005	0.794
	0.010	0.741
	0.020	0.683
	0.050	0.606
	0.100	0.553
	0.200	0.515
	0.500	0.516
	1.000	0.558
	2.000	0.578
	5.000	0.788
	10.000	2.317
	15.000	5.381
	20.000	7.965
$ZnCl_2$	0.001	0.887
	0.002	0.847
	0.005	0.781
	0.010	0.719
	0.020	0.652
	0.050	0.561
	0.100	0.499
	0.200	0.447
	0.500	0.384
	1.000	0.330
	2.000	0.283
	5.000	0.342
	10.000	0.876
	15.000	1.914
	20.000	2.968
ZnI_2	0.001	0.893
	0.002	0.859
	0.005	0.804
	0.010	0.757
	0.020	0.708
	0.050	0.644
	0.100	0.601
	0.200	0.574
	0.500	0.635
	1.000	0.836
	2.000	1.062
	5.000	1.546
	10.000	4.698

[a] The anion is $H_2PO_4^-$.
[b] The anion is HPO_4^{-2}.

ENTHALPY OF DILUTION OF ACIDS

The quantity given in this table is $-\Delta_{dil}H$, the negative of the enthalpy (heat) of dilution to infinite dilution for aqueous solutions of several common acids; i.e., the negative of the enthalpy change when a solution of molality m at a temperature of 25 °C is diluted with an infinite amount of water. The tabulated numbers thus represent the heat produced (or, if the value is negative, the heat absorbed) when the acid is diluted from the initial molality m to infinite dilution.

It is sometimes useful to have the dilution ratio, which is the number of moles of water that must be added to one mole of acid to produce a solution of desired molality. This may be calculated from the relation: dilution ratio = 55.506/m, where m is in mol/kg.

Reference

Parker, V. B., *Thermal Properties of Aqueous Uni-univalent Electrolytes*, Natl. Stand. Ref. Data Ser. — Natl. Bur. Stand. (U.S.) 2, U.S. Government Printing Office, 1965. <https://doi.org/10.6028/NBS.NSRDS.2>

Enthalpy of Dilution of Acids ($-\Delta_{dil}H$ in kJ/mol) at 25 °C for the Specified Molality m in mol/kg

Mol. form.	0.000111 m	0.00111 m	0.01110 m	0.05 m	0.1 m	0.2 m	0.5 m	1 m	2 m	3 m	4 m	5 m	9 m	20 m	55.506 m
HF	1.255	5.439	9.874	12.24	12.80	13.09	13.20	13.30	13.40	13.45	13.53	13.62	13.81	14.88	
HCl	0.021	0.067	0.197	0.406	0.556	0.761	1.172	1.695	2.623	3.506	4.402	5.318	9.213	19.87	45.61
HBr	0.021	0.054	0.184	0.372	0.498	0.649	0.941	1.314	1.996	2.611	3.330	4.113	7.719	19.92	48.83
HI	0.021	0.050	0.172	0.339	0.439	0.536	0.711	0.933	1.318	1.787	2.460	3.197	6.569	21.71	
HNO_3	0.021	0.063	0.176	0.305	0.372	0.439	0.498	0.506	0.527	0.665	0.958	1.310	3.368	9.498	19.73
$HClO_4$	0.021	0.059	0.167	0.259	0.272	0.247	0.075	-0.201	-0.623	-0.782	-0.787	-0.628	1.280	13.81	
HCOOH	0.038	0.084	0.109	0.121	0.134	0.146	0.176	0.226	0.276	0.289	0.289	0.289	0.230	0.038	0.046
CH_3COOH	0.167	0.222	0.255	0.272	0.289	0.331	0.406	0.544	0.803	1.025	1.218	1.393	1.782	2.075	2.167

ENTHALPY OF SOLUTION OF ELECTROLYTES

This table gives the molar enthalpy (heat) of solution at infinite dilution for some common uni-univalent electrolytes. A uni-univalent electrolyte is one that dissociates into two univalent ions. Molar enthalpy of solution is the enthalpy change when 1 mol of solute in its standard state is dissolved in an infinite amount of water. Values are given in kilojoules per mole at 25 °C. Column definitions are as follows.

Column heading	Definition
Mol. form.	Molecular formula of the electrolyte
State	Physical state of the electrolyte
$\Delta_{sol}H$	Enthalpy of solution, in units kJ mol^{-1}

Reference

Parker, V. B., *Thermal Properties of Aqueous Uni-univalent Electrolytes,* Natl. Stand. Ref. Data Ser. — Natl. Bur. Stand. (U.S.) 2, U.S. Government Printing Office, 1965. <https://doi.org/10.6028/NBS.NSRDS.2>

Molar Enthalpy of Solution of Uni-Univalent Electrolytes at 25 °C

Mol. form.	State	$\Delta_{sol}H$/ kJ mol^{-1}
Acids		
HF	gas	-61.50
HCl	gas	-74.84
HBr	gas	-85.14
HI	gas	-81.67
$HClO_4$	liquid	-88.76
$HClO_4 \cdot H_2O$	crystal	-32.95
HIO_3	crystal	8.79
HNO_3	liquid	-33.28
HCOOH	liquid	-0.86
CH_3COOH	liquid	-1.51
Silver Compounds		
$AgClO_4$	crystal	7.36
$AgNO_2$	crystal	36.94
$AgNO_3$	crystal	22.59
Cesium Compounds		
CsBr	crystal	25.98
$CsBrO_3$	crystal	50.46
CsCl	crystal	17.78
$CsClO_4$	crystal	55.44
CsF	crystal	-36.86
$CsF \cdot H_2O$	crystal	-10.46
$CsF \cdot 1.5H_2O$	crystal	-5.44
CsI	crystal	33.35
$CsNO_3$	crystal	40.00
CsOH	crystal	-71.55
$CsOH \cdot H_2O$	crystal	-20.50
Potassium Compounds		
KBr	crystal	19.87
$KBrO_3$	crystal	41.13
KCN	crystal	11.72
KCNO	crystal	20.25
$KC_2H_3O_2$	crystal	-15.33
KCl	crystal	17.22
$KClO_3$	crystal	41.38
$KClO_4$	crystal	51.04
KF	crystal	-17.73
$KF \cdot 2H_2O$	crystal	6.97
KI	crystal	20.33
KIO_3	crystal	27.74
$KMnO_4$	crystal	43.56
KNO_2	crystal	13.35
KNO_3	crystal	34.89
KOH	crystal	-57.61
$KOH \cdot H_2O$	crystal	-14.64
$KOH \cdot 1.5H_2O$	crystal	-10.46
KSCN	crystal	24.23
Lithium Compounds		
LiBr	crystal	-48.83
$LiBr \cdot H_2O$	crystal	-23.26
$LiBr \cdot 2H_2O$	crystal	-9.41
$LiBrO_3$	crystal	1.42
LiCl	crystal	-37.03
$LiCl \cdot H_2O$	crystal	-19.08
$LiClO_4$	crystal	-26.55
$LiClO_4 \cdot 3H_2O$	crystal	32.61
LiF	crystal	4.73
LiI	crystal	-63.30
$LiI \cdot H_2O$	crystal	-29.66
$LiI \cdot 2H_2O$	crystal	-14.77
$LiI \cdot 3H_2O$	crystal	0.59
$LiNO_2$	crystal	-11.00
$LiNO_2 \cdot H_2O$	crystal	7.03
$LiNO_3$	crystal	-2.51
LiOH	crystal	-23.56
$LiOH \cdot H_2O$	crystal	-6.69
Ammonium Compounds		
NH_3	gas	-30.50
NH_4Br	crystal	16.78
NH_4Cl	crystal	14.78
NH_4I	crystal	13.72
NH_4CN	crystal	17.57
$NH_4C_2H_3O_2$	crystal	-2.38
NH_4ClO_4	crystal	33.47
NH_4IO_3	crystal	31.80
NH_4NO_2	crystal	19.25
NH_4NO_3	crystal	25.69
NH_4SCN	crystal	22.59
$N(CH_3)_4Br$	crystal	24.27
$N(CH_3)_4Cl$	crystal	4.08
$N(CH_3)_4I$	crystal	42.07
$(CH_3)_3N \cdot HCl$	crystal	1.46
$CH_3NH_2 \cdot HCl$	crystal	5.77
Sodium Compounds		
NaBr	crystal	-0.60
$NaBr \cdot 2H_2O$	crystal	18.64
$NaBrO_3$	crystal	26.90
NaCN	crystal	1.21
$NaCN \cdot 0.5H_2O$	crystal	3.31
$NaCN \cdot 2H_2O$	crystal	18.58
NaCNO	crystal	19.20
$NaC_2H_3O_2$	crystal	-17.32
$NaC_2H_3O_2 \cdot 3H_2O$	crystal	19.66
NaCl	crystal	3.88
$NaClO_2$	crystal	0.33
$NaClO_2 \cdot 3H_2O$	crystal	28.58
$NaClO_3$	crystal	21.72
$NaClO_4$	crystal	13.88
$NaClO_4 \cdot H_2O$	crystal	22.51
NaF	crystal	0.91
NaI	crystal	-7.53
$NaI \cdot 2H_2O$	crystal	16.13
$NaIO_3$	crystal	20.29
$NaNO_2$	crystal	13.89
$NaNO_3$	crystal	20.50
NaOH	crystal	-44.51
$NaOH \cdot H_2O$	crystal	-21.41
NaSCN	crystal	6.83
Rubidium Compounds		
RbBr	crystal	21.88
$RbBrO_3$	crystal	48.95
RbCl	crystal	17.28
$RbClO_3$	crystal	47.74
$RbClO_4$	crystal	56.74
RbF	crystal	-26.11
$RbF \cdot H_2O$	crystal	-0.42
$RbF \cdot 1.5H_2O$	crystal	1.34
RbI	crystal	25.10
$RbNO_3$	crystal	36.48
RbOH	crystal	-62.34
$RbOH \cdot H_2O$	crystal	-17.99
$RbOH \cdot 2H_2O$	crystal	0.88

pH SCALE FOR AQUEOUS SOLUTIONS

A. K. Covington

A Working Party of IUPAC, after extensive considerations over five years, has produced a report (1) which sets pH firmly within the International System of Units (SI). A summary of these important developments is given below.

The concept of pH is unique among the commonly encountered physicochemical quantities in that, in terms of its definition,

$$\mathrm{pH} = -\lg a_{\mathrm{H}} \quad (1)$$

it involves a single ion quantity, the activity of the hydrogen ion, which is immeasurable by any thermodynamically valid method and requires a convention for its evaluation.

pH was originally defined by Sørensen (Ref. 2) in terms of the concentration of hydrogen ions (in modern nomenclature) as pH $= -\lg (c_{\mathrm{H}}/c^{\mathrm{o}})$ where c_{H} is the hydrogen ion concentration in mol dm^{-3}, and $c^{\mathrm{o}} = 1\ \mathrm{mol\ dm}^{-3}$ is the standard amount concentration. Subsequently (Ref. 3), it was accepted as more satisfactory to define pH in terms of the relative activity of hydrogen ions in solution

$$\mathrm{pH} = -\lg a_{\mathrm{H}} = -\lg (m_{\mathrm{H}}\gamma_{\mathrm{H}}/m^{\mathrm{o}}) \quad (2)$$

where a_{H} is the relative (molality basis) activity and γ_{H} is the molal activity coefficient of the hydrogen ion H^+ at the molality m_{H}, and m^{o} the standard molality. The quantity pH is intended to be a measure of the activity of hydrogen ions in solution. However, since it is defined in terms of a quantity that cannot be measured by a thermodynamically valid method, eqn. (2) can only be considered a *notional definition* of pH.

pH being a single ion quantity, it is not determinable in terms of a fundamental (or base) unit of any measurement system, and there is difficulty providing a proper basis for the traceability of pH measurements. A satisfactory approach is now available in that pH determinations can be incorporated into the International System (SI) if they can be traced to measurements made using a method that fulfils the definition of a 'primary method of measurement' (Ref. 4).

The essential feature of a primary method is that it must operate according to a well-defined measurement equation in which all of the variables can be determined experimentally in terms of SI units. Any limitation in the determination of the experimental variables, or in the theory, must be included within the estimated uncertainty of the method if traceability to the SI is to be established. If a convention were used without an estimate of its uncertainty, true traceability to SI would not be established. The electrochemical cell without liquid junction, known as the Harned cell (Ref. 5), fulfills the definition of a primary method for the measurement of the acidity function, $\mathrm{p}(a_{\mathrm{H}}\gamma_{\mathrm{Cl}})$, and subsequently of the pH of buffer solutions.

The Harned cell is written as

$$\mathrm{Pt} \mid \mathrm{H_2} \mid \text{buffer S}, \mathrm{Cl^-} \mid \mathrm{AgCl} \mid \mathrm{Ag} \quad \text{(Cell I)}$$

and contains a standard buffer, S, with chloride ions, as potassium or sodium chloride, added in order to use the silver–silver chloride electrode as reference electrode. The application of the Nernst equation to the spontaneous cell reaction of Cell I:

$$\tfrac{1}{2}\mathrm{H_2} + \mathrm{AgCl} \rightarrow \mathrm{Ag(s)} + \mathrm{H^+} + \mathrm{Cl^-}$$

yields the potential difference E_{I} of the cell (corrected to 1 atm [101.325 kPa], the partial pressure of hydrogen gas used in electrochemistry in preference to 100 kPa) as

$$E_{\mathrm{I}} = E^{\mathrm{o}} - (RT/F)\ln 10 \lg [(m_{\mathrm{H}}\gamma_{\mathrm{H}}/m^{\mathrm{o}})(m_{\mathrm{Cl}}\gamma_{\mathrm{Cl}}/m^{\mathrm{o}})] \quad (3)$$

which can be rearranged, since $a_{\mathrm{H}} = m_{\mathrm{H}}\gamma_{\mathrm{H}}/m^{\mathrm{o}}$, to give the acidity function

$$\mathrm{p}(a_{\mathrm{H}}\gamma_{\mathrm{Cl}}) = -\lg(a_{\mathrm{H}}\gamma_{\mathrm{Cl}}) = (E_{\mathrm{I}} - E^{\mathrm{o}})/[(RT/F)\ln 10] + \lg(m_{\mathrm{Cl}}/m^{\mathrm{o}}) \quad (4)$$

where E^{o} is the standard potential difference of the cell, and hence of the silver–silver chloride electrode, and γ_{Cl} is the activity coefficient of the chloride ion.

The standard potential difference of the silver–silver chloride electrode, E^{o}, is determined from a Harned cell in which only HCl is present at a fixed molality (e.g., $m = 0.01\ \mathrm{mol\ kg}^{-1}$)

$$\mathrm{Pt} \mid \mathrm{H_2} \mid \mathrm{HCl}\ (m) \mid \mathrm{AgCl} \mid \mathrm{Ag} \quad \text{(Cell Ia)}$$

The application of the Nernst equation to the HCl cell (Ia) gives

$$E_{\mathrm{Ia}} = E^{\mathrm{o}} - (2RT/F)\ln 10 \lg[(m_{\mathrm{HCl}}/m^{\mathrm{o}})(\gamma_{\pm\mathrm{HCl}})] \quad (5)$$

where E_{Ia} has been corrected to 1 atmosphere partial pressure of hydrogen gas (101.325 kPa) and $\gamma_{\pm\mathrm{HCl}}$ is the mean ionic activity coefficient of HCl.

Values of the activity coefficient ($\gamma_{\pm\mathrm{HCl}}$) at molality 0.01 mol kg^{-1} and various temperatures were given by Bates and Robinson (Ref. 6). The standard potential difference depends on the method of preparation of the electrodes, but individual determinations of the activity coefficient of HCl at 0.01 mol kg^{-1} are more uniform than values of E^{o}. Hence the practical determination of the potential difference of the cell with HCl at 0.01 mol kg^{-1} is recommended at 298.15 K at which the mean ionic activity coefficient is 0.904. (It is unnecessary to repeat the measurement of E^{o} at other temperatures but simply to correct published smoothed values by the observed difference in E^{o} at 298.15 K.)

In national metrology institutes (NMIs), measurements of Cells I and Ia are often done simultaneously in a thermostat bath. Subtracting eqn. (5) from eqn. (3) gives

$$\Delta E = E_{\mathrm{I}} - E_{\mathrm{Ia}} = -(RT/F)\ln 10\{\lg[(m_{\mathrm{H}}\gamma_{\mathrm{H}}/m^{\mathrm{o}})(m_{\mathrm{Cl}}\gamma_{\mathrm{Cl}}/m^{\mathrm{o}})] - \lg[(m_{\mathrm{HCl}}/m^{\mathrm{o}})^2\gamma^2_{\pm\mathrm{HCl}}]\}$$

which is independent of the standard potential difference. Therefore, the subsequently calculated pH does not depend on the standard potential difference and hence does not depend on the assumption that the standard potential of the hydrogen electrode is zero at all temperatures. Therefore, the Harned cell gives an exact comparison between hydrogen ion activities at different temperatures.

The quantity $\mathrm{p}(a_{\mathrm{H}}\gamma_{\mathrm{Cl}}) = -\lg (a_{\mathrm{H}}\gamma_{\mathrm{Cl}})$, on the left-hand side of (4), is called the acidity function (5). To obtain the quantity pH according to eqn. (2) from the acidity function, it is necessary to evaluate $\lg \gamma_{\mathrm{Cl}}$ independently. This is done in two steps: (i) the value of $\lg (a_{\mathrm{H}}\gamma_{\mathrm{Cl}})$ at zero chloride molality, $\lg (a_{\mathrm{H}}\gamma_{\mathrm{Cl}})^{\mathrm{o}}$, is evaluated and (ii) a value for the activity of the chloride ion $\gamma^{\mathrm{o}}_{\mathrm{Cl}}$, at zero chloride molality (sometimes referred to as the limiting or 'trace' activity coefficient) is calculated using the Bates-Guggenheim convention (Ref. 7). The value of $\lg (a_{\mathrm{H}}\gamma_{\mathrm{Cl}})^{\mathrm{o}}$ corresponding to zero chloride molality is determined by linear extrapolation of measurements using Harned cells with at least three added molalities of sodium or potassium chloride ($I < 0.1\ \mathrm{mol\ kg}^{-1}$).

The value of $\lg (a_H\gamma_{Cl})^o$ corresponding to zero chloride molality is determined by linear extrapolation of measurements using Harned cells with at least three added molalities of sodium or potassium chloride ($I < 0.1$ mol kg^{-1}) in accord with eqn. (7):

$$-\lg (a_H\gamma_{Cl}) = -\lg (a_H\gamma_{Cl})^o + Sm_{Cl} \quad (7)$$

where S is an empirical, temperature dependent, constant.

The Bates-Guggenheim convention (Ref. 7) assumes that the trace activity coefficient of the chloride ion γ^o_{Cl} is given by

$$\lg \gamma^o_{Cl} = -A\, I^{1/2}/(1 + Ba\, I^{1/2}) \quad (8)$$

where A is the Debye-Hückel temperature dependent constant (limiting slope), a is the *mean* distance of closest approach of the ions (ion size parameter), Ba is set equal to 1.5 (mol kg^{-1})$^{-1/2}$ at all temperatures in the range 5–50 °C, and I is the ionic strength of the buffer (which for its evaluation requires knowledge of appropriate acid dissociation constants).

The various stages in the assignment of primary standard pH values are combined in eqn. (9), which is derived from eqns. (4), (5), and (8)

$$\text{pH(PS)} = \lim m_{Cl}\rightarrow_0 \{(E_I - E^o)/[(RT/F)\text{In } 10] + \lg (m_{Cl}/m^o)\} - A\, I^{1/2}/[1 + 1.5\,(I/m^o)^{1/2}] \quad (9)$$

In order for a particular buffer solution to be considered a primary buffer solution, it must be of the "highest metrological" quality (Ref. 4) in accordance with the definition of a primary standard. It is recommended that it have the following attributes (Ref. 9):

1. High buffer value in the range 0.016–0.07 (mol OH^-)/pH.
2. Small dilution value at half concentration (change in pH with change in buffer concentration) in the range 0.01–0.20.
3. Small dependence of pH on temperature less than ±0.01 K^{-1}.
4. Low residual liquid junction potential <0.01 in pH.
5. Ionic strength ≤0.1 mol kg^{-1} to permit applicability of Bates-Guggenheim convention.
6. NMI certificate for specific batch.
7. Reproducible purity of preparation (lot to lot differences of $|\Delta\text{pH(PS)}| < 0.003$).
8. Long-term stability of stored solid material.

Values for the above and other important parameters for the primary and secondary buffer materials are given in Table 1. Column definitions are as follows.

Column heading	Definition
Name	Name of salt or solid substance
Formula	Molecular formula of solid or substance
Molality	Molality of standard buffer solution, in mol kg^{-1}
Molar mass	Molar mass of salt or solid substance, g mol^{-1}
Density	Density of standard buffer solution, in g mL^{-1}
Amount conc. at 20 °C	Concentration amount at 20 °C, in mol dm^{-3}
Mass in g to make 1 dm³	Mass of salt or solid substance needed to make 1 dm^3; = molar mass × Amount conc. at 20 °C; in g
Dilution value $\Delta pH_{1/2}$	Change of pH value at half concentration, in pH units
Buffer value (β)	Strength of pH buffering, in units mol (OH^-) dm^{-3}
pH Temperature coefficient	Change of pH with temperature, in pH units K^{-1}

Primary Standard Buffers

As there can be significant variations in the purity of samples of a buffer of the same nominal chemical composition, it is essential that the primary buffer material used has been certified with values that have been measured with Cell I. The Harned cell is used by many national metrological institutes for accurate measurements of pH of buffer solutions.

Typical values of the pH(PS) of the seven solutions from the six accepted primary standard reference buffers, which meet the conditions stated above, are listed in Table 2. Batch-to-batch variations in purity can result in changes in the pH value of samples of at most 0.003. The typical values in Table 2 should not be used in place of the certified value (from a Harned cell measurement) for a specific batch of buffer material.

The required attributes listed above effectively limit the range of primary buffers available to between pH 3 and 10 (at 25 °C). Calcium hydroxide and potassium tetraoxalate are excluded because the contribution of hydroxide or hydrogen ions to the ionic strength is significant. Also excluded are the nitrogen bases of the type BH^+ (such as tris(hydroxymethyl)aminomethane and piperazine phosphate) and the zwitterionic buffers (e.g., HEPES and MOPS [Ref. 10]). These do not comply because either the Bates-Guggenheim convention is not applicable, or the liquid junction potentials are high. This means the choice of primary standards is restricted to buffers derived from oxy-carbon, -phosphorus, -boron and mono, di- and tri-protic carboxylic acids. The uncertainties (Ref. 11) associated with Harned cell measurements are calculated (Ref. 1) to be 0.004 in pH at NMIs, with typical variation between batches of primary standard buffers of 0.003.

Secondary Standards

Substances that do not fulfill all the criteria for primary standards, but to which pH values can be assigned using Cell I are considered to be secondary standards (Table 3). Reasons for their exclusion as primary standards include difficulties in achieving consistent and suitable chemical quality (e.g. acetic acid is a liquid), suspected high liquid junction potential, or inappropriateness of the Bates-Guggenheim convention (e.g., other charge-type buffers). The uncertainty is higher (e.g., 0.01) for biological buffers. Certain other substances, which cannot be used in cells containing hydrogen gas electrodes, are also classed as secondary standards.

Calibration Procedures

1. One-point calibration
2. A single-point calibration is insufficient to determine both slope and one-point parameters. The theoretical value for the slope can be assumed but the practical slope may be up to 5% lower. Alternatively, a value for the practical slope can be assumed from the manufacturer's prior calibration. The one-point calibration therefore yields only an estimate of pH(X). Since both parameters may change with age of the electrodes, this is not a reliable procedure.
3. Two-point calibration [target uncertainty: 0.02–0.03 at 25 °C]
4. In the majority of practical applications, glass electrodes cells are calibrated by a two-point calibration, or bracketing, procedure using two standard buffer solutions, with pH values, $\text{pH}(S_1)$ and $\text{pH}(S_2)$, bracketing the unknown pH(X). Bracketing is often taken to mean that the $\text{pH}(S_1)$ and $\text{pH}(S_2)$ buffers selected from Table 2 should be those that are immediately above and below pH(X). This may

not be appropriate in all situations and choice of a wider range may be better.

5. Multi-point calibration [target uncertainty: 0.01–0.03 at 25 °C].
6. Multi-point calibration is carried out using up to five standard buffers. The use of more than five points yields no significant improvement in the statistical information obtainable.
7. Details of uncertainty computations (Ref. 11) have been given (Ref. 1).

Measurement of pH and Choice of pH Standard Solutions

1. If pH is not required to better than ±0.05 any pH standard solution may be selected.
2. If pH is required to ±0.002 and interpretation in terms of hydrogen ion concentration or activity is desired, choose a standard solution, pH(PS), to match X as closely as possible in terms of pH, composition and ionic strength.
3. Alternatively, a bracketing procedure may be adopted whereby two standard solutions are chosen whose pH values, pH(S1), pH(S2) are on either side of pH(X). Then if the corresponding potential difference measurements are E(S1), E(S2), E(X), then pH(X) is obtained from

$$\mathrm{pH(X)} = \mathrm{pH(S1)} + [E(\mathrm{X}) - E(\mathrm{S1})]/\%k$$

where $\%k = 100[E(\mathrm{S2}) - E(\mathrm{S1})]/[\mathrm{pH(S2)} - \mathrm{pH(S1)}]$ is the apparent percentage slope. This procedure is very easily done on some pH meters simply by adjusting downwards the slope factor control with the electrodes in S2. The purpose of the bracketing procedure is to compensate for deficiencies in the electrodes and measuring system.

Information to Be Given about the Measurement of pH(X)

The standard solutions selected for calibration of the pH meter system should be reported with the measurement as follows:

System calibrated with pH(S) = ... at ... K
System calibrated with two primary standards, pH(PS1) = ... and pH(PS2) = ... at .... K
System calibrated with n standards, pH(S1) = ..., pH(S2) = ... etc. at .... K

Interpretation of pH(X) in Terms of Hydrogen Ion Concentration

The defined pH has no simple interpretation in terms of hydrogen ion concentration but the mean ionic activity coefficient of a typical 1:1 electrolyte can be used to obtain hydrogen ion concentration subject to an uncertainty of 3.9% in concentration, corresponding to 0.02 in pH.

References

1. Buck, R.P., Rondinini, S., Covington, A.K., Baucke, F.G.K., Brett, C.M.A., Camoes, M.F.C., Milton, M.J.T., Mussini, T., Naumann, R., Pratt, K.W., Spitzer, P., and Wilson, G.S. *Pure Appl. Chem.* 74, 2105, 2002. <https://doi.org/10.1351/pac200274112169>
2. Sørensen, S.P.L. *Comp. Rend. Trav. Lab. Carlsberg* 8, 1, 1909.
3. Sørensen, S.P.L., and Linderstrøm-Lang, K.L., *Comp. Rend. Trav. Lab. Carlsberg* 15, 1924.
4. BIPM, *Com. Cons. Quantité de Matière* 4, 1998. See also M.J.T. Milton and T.J. Quinn, *Metrologia* 38, 289, 2001. <https://doi.org/10.1088/0026-1394/38/4/1>
5. Harned H.S., and Owen, B.B., *The Physical Chemistry of Electrolytic Solutions*, Ch 14, Reinhold, New York, 1958.
6. Bates R.G., and Robinson, R.A., *J. Soln. Chem.* 9, 455, 1980.
7. Bates R.G., and Guggenheim, E.A., *Pure Appl. Chem.* 1, 163, 1960. <https://doi.org/10.1351/pac196001010163>
8. *International Vocabulary of Metrology – Basic and General Concepts and Associated Terms (VIM), Third Edition,* JCGM 200:2012, BIPM, 2012.
9. Bates, R.G. *Determination of pH,* Wiley, New York, 1973.
10. Good, N.E. et al., *Biochem. J.* 5, 467, 1966. <https://doi.org/10.1021/bi00866a011>
11. *Evaluation of Measurement Data — Guide to the Expression of Uncertainty in Measurement (GUM)*, JCGM 100:2008, BIPM, IEC, IFCC, ISO, IUPAC, IUPAP, OIML, 2008.

TABLE 1. Properties of Some Primary and Secondary Standard Buffer Substances and Solutions (Primary Standards in Bold Face)

Salt or solid substance	Formula	Molality/ mol kg^{-1}	Molar mass/ g mol^{-1}	Density/ g mL^{-1}	Amount conc. at 20 °C/ mol dm^{-3}	Mass/ g to make 1 dm^3	Dilution value $\Delta pH_{1/2}$	Buffer value (β)/mol OH$^-$ dm^{-3}	pH Temperature coefficient/K^{-1}
Potassium tetroxalate dihydrate	$KH_3C_4O_8{\cdot}2H_2O$	0.1	254.191	1.0091	0.09875	25.101			
Potassium tetraoxalate dihydrate	$KH_3C_4O_8{\cdot}2H_2O$	0.05	254.191	1.0032	0.04965	12.620	0.186	0.070	0.001
Potassium hydrogen tartrate (sat at 25 °C)	$KHC_4H_4O_6$	0.0341	188.18	1.0036	0.034	6.4	0.049	0.027	-0.0014
Potassium dihydrogen citrate	$KH_2C_6H_5O_7$	0.05	230.22	1.0029	0.04958	11.41	0.024	0.034	-0.022
Potassium hydrogen phthalate	$KHC_8H_4O_4$	0.05	204.44	1.0017	0.04958	10.12	0.052	0.016	0.00012
Disodium hydrogen orthophosphate (0.025 m) + potassium dihydrogen orthophosphate (0.025 m)	Na_2HPO_4 KH_2PO_4	0.025	141.958	1.0038	0.02492	3.5379	0.080	0.029	-0.0028
Disodium hydrogen orthophosphate (0.03043 m) + potassium dihydrogen orthophosphate (0.00869 m)	Na_2HPO_4	0.03043	141.959	1.0020	0.08665	4.302	0.07	0.016	-0.0028
Disodium tetraborate decahydrate	$Na_2B_4O_7{\cdot}10H_2O$	0.05	381.367	1.0075	0.04985	19.012			

Salt or solid substance	Formula	Molality/ mol kg^{-1}	Molar mass/ g mol^{-1}	Density/ g mL^{-1}	Amount conc. at 20 °C/ mol dm^{-3}	Mass/ g to make 1 dm^3	Dilution value $\Delta pH_{1/2}$	Buffer value (β)/mol OH$^-$ dm^{-3}	pH Temperature coefficient/K^{-1}
Disodium tetraborate decahydrate	$Na_2B_4O_7{\cdot}10H_2O$	0.01	381.367	1.0001	0.00998	3.806	0.01	0.020	-0.0082
Sodium hydrogen carbonate (0.025 m) + sodium carbonate (0.025 m)	$NaHCO_3$	0.025	84.01	1.0013	0.02492	2.092	0.079	0.029	-0.0096
Calcium hydroxide (sat. at 25 °C)	$Ca(OH)_2$	0.0203	74.09	0.9991	0.02025	1.5	-0.28	0.09	-0.033

TABLE 2. Typical Values of pH(PS) for Primary Standards at 0–50 °C

Primary standards (PS)	0 °C	5 °C	10 °C	15 °C	20 °C	25 °C	30 °C	35 °C	37 °C	40 °C	50 °C
Sat. potassium hydrogen tartrate (at 25 °C)						3.557	3.552	3.549	3.548	3.547	3.549
0.05 mol kg^{-1} potassium dihydrogen citrate	3.863	3.840	3.820	3.802	3.788	3.776	3.766	3.759	3.756	3.754	3.749
0.05 mol kg^{-1} potassium hydrogen phthalate	4.000	3.998	3.997	3.998	4.000	4.005	4.011	4.018	4.022	4.027	4.050
0.025 mol kg^{-1} disodium hydrogen phosphate + 0.025 mol kg^{-1} potassium dihydrogen phosphate	6.984	6.951	6.923	6.900	6.881	6.865	6.853	6.844	6.841	6.838	6.833
0.03043 mol kg^{-1} disodium hydrogen phosphate + 0.008695 mol kg–1 potassium dihydrogen phosphate	7.534	7.500	7.472	7.448	7.429	7.413	7.400	7.389	7.386	7.380	7.367
0.01 mol kg^{-1} disodium tetraborate	9.464	9.395	9.332	9.276	9.225	9.180	9.139	9.102	9.088	9.068	9.011
0.025 mol kg^{-1} sodium hydrogen carbonate + 0.025 mol kg^{-1} sodium carbonate	10.317	10.245	10.179	10.118	10.062	10.012	9.966	9.926	9.910	9.889	9.828

TABLE 3. Values of pH(SS) as a Function of Temperature for Some Secondary Standards from Harned Cell I Measurements

Secondary standards	0 °C	5 °C	10 °C	15 °C	20 °C	25 °C	30 °C	37 °C	40 °C	50 °C
0.05 mol kg^{-1} potassium tetroxalate[a]	1.67	1.67	1.67	1.67	1.68	1.68	1.68	1.69	1.69	1.71
0.05 mol kg^{-1} sodium hydrogen diglycolate[b]		3.47	3.47	3.48	3.48	3.49	3.50	3.52	3.53	3.56
0.1 mol dm^{-3} acetic acid + 0.1 mol dm^{-3} sodium acetate	4.68	4.67	4.67	4.66	4.66	4.65	4.65	4.66	4.66	4.68
0.01 mol dm^{-3} acetic acid + 0.1 mol dm^{-3} sodium acetate	4.74	4.73	4.73	4.72	4.72	4.72	4.72	4.73	4.73	4.75
0.02 mol kg^{-1} piperazine phosphate[c]	6.58	6.51	6.45	6.39	6.34	6.29	6.24	6.16	6.14	6.06
0.05 mol kg^{-1} tris hydrochloride + 0.01667 mol kg^{-1} tris[c]	8.47	8.30	8.14	7.99	7.84	7.70	7.56	7.38	7.31	7.07
0.05 mol kg^{-1} disodium tetraborate	9.51	9.43	9.36	9.30	9.25	9.19	9.15	9.09	9.07	9.01
Saturated (at 25 °C) calcium hydroxide	13.42	13.21	13.00	12.81	12.63	12.45	12.29	12.07	11.98	11.71

[a] Potassium trihydrogen dioxalate ($KH_3C_4O_8$).
[b] Sodium hydrogen 2,2-oxydiacetate.
[c] 2-Amino-2-(hydroxymethyl)-1,3 propanediol or tris(hydroxymethyl)aminomethane.

BUFFER SOLUTIONS GIVING ROUND VALUES OF pH AT 25 °C

This table gives a series of buffer solutions A, B, ··· J that produce round values of pH from 1.0 to 13.0. The value of *x* to be used for each pH value is given. The buffers are:

A: 25 mL of 0.2 molar KCl + *x* mL of 0.2 molar HCl.
B: 50 mL of 0.1 molar potassium hydrogen phthalate + *x* mL of 0.1 molar HCl.
C: 50 mL of 0.1 molar potassium hydrogen phthalate + *x* mL of 0.1 molar NaOH.
D: 50 mL of 0.1 molar potassium dihydrogen phosphate + *x* mL of 0.1 molar NaOH.
E: 50 mL of 0.1 molar tris(hydroxymethyl)aminomethane + *x* mL of 0.1 molar HCl.
F: 50 mL of 0.025 molar borax + *x* mL of 0.1 molar HCl.
G: 50 mL of 0.025 molar borax + *x* mL of 0.1 molar NaOH.
H: 50 mL of 0.05 molar sodium bicarbonate + *x* mL of 0.1 molar NaOH.
I: 50 mL of 0.05 molar disodium hydrogen phosphate + *x* mL of 0.1 molar NaOH.
J: 25 mL of 0.2 molar KCl + *x* mL of 0.2 molar NaOH.

Final volume of mixtures = 100 mL. Borax is sodium tetraborate decahydrate; see Ref 2 for its preparation using NIST Standard Reference Material 187f. The IUPAC name for potassium hydrogen phthalate is potassium 2-carboxybenzoate. Buffer solutions A-D are from Ref 1; E-J are from Ref. 2.

References

1. Bower, V. E., and Bates, R. G., *J. Res. Natl. Bur. Stand.*, 55, 197, 1955 (A–D). <https://doi.org/10.6028/jres.055.021>
2. Bates, R. G., and Bower, V. E., *Anal. Chem.*, 28, 1322, 1956 (E–J). <https://doi.org/10.1021/ac60116a029>

Preparation of Buffer Solutions Giving Round Values of pH at 25 °C (Value of Component *x* Needed for Each Buffer Solution)

A	*x*	B	*x*	C	*x*	D	*x*	E	*x*	F	*x*	G	*x*	H	*x*	I	*x*	J	*x*
1.0	67.0	**2.2**	49.5	**4.1**	1.3	**5.8**	3.6	**7.0**	46.6	**8.0**	20.5	**9.2**	0.9	**9.6**	5.0	**10.9**	3.3	**12.0**	6.0
1.1	52.8	**2.3**	45.8	**4.2**	3.0	**5.9**	4.6	**7.1**	45.7	**8.1**	19.7	**9.3**	3.6	**9.7**	6.2	**11.0**	4.1	**12.1**	8.0
1.2	42.5	**2.4**	42.2	**4.3**	4.7	**6.0**	5.6	**7.2**	44.7	**8.2**	18.8	**9.4**	6.2	**9.8**	7.6	**11.1**	5.1	**12.2**	10.2
1.3	33.6	**2.5**	38.8	**4.4**	6.6	**6.1**	6.8	**7.3**	43.4	**8.3**	17.7	**9.5**	8.8	**9.9**	9.1	**11.2**	6.3	**12.3**	12.8
1.4	26.6	**2.6**	35.4	**4.5**	8.7	**6.2**	8.1	**7.4**	42.0	**8.4**	16.6	**9.6**	11.1	**10.0**	10.7	**11.3**	7.6	**12.4**	16.2
1.5	20.7	**2.7**	32.1	**4.6**	11.1	**6.3**	9.7	**7.5**	40.3	**8.5**	15.2	**9.7**	13.1	**10.1**	12.2	**11.4**	9.1	**12.5**	20.4
1.6	16.2	**2.8**	28.9	**4.7**	13.6	**6.4**	11.6	**7.6**	38.5	**8.6**	13.5	**9.8**	15.0	**10.2**	13.8	**11.5**	11.1	**12.6**	25.6
1.7	13.0	**2.9**	25.7	**4.8**	16.5	**6.5**	13.9	**7.7**	36.6	**8.7**	11.6	**9.9**	16.7	**10.3**	15.2	**11.6**	13.5	**12.7**	32.2
1.8	10.2	**3.0**	22.3	**4.9**	19.4	**6.6**	16.4	**7.8**	34.5	**8.8**	9.6	**10.0**	18.3	**10.4**	16.5	**11.7**	16.2	**12.8**	41.2
1.9	8.1	**3.1**	18.8	**5.0**	22.6	**6.7**	19.3	**7.9**	32.0	**8.9**	7.1	**10.1**	19.5	**10.5**	17.8	**11.8**	19.4	**12.9**	53.0
2.0	6.5	**3.2**	15.7	**5.1**	25.5	**6.8**	22.4	**8.0**	29.2	**9.0**	4.6	**10.2**	20.5	**10.6**	19.1	**11.9**	23.0	**13.0**	66.0
2.1	5.10	**3.3**	12.9	**5.2**	28.8	**6.9**	25.9	**8.1**	26.2	**9.1**	2.0	**10.3**	21.3	**10.7**	20.2	**12.0**	26.9		
2.2	3.9	**3.4**	10.4	**5.3**	31.6	**7.0**	29.1	**8.2**	22.9			**10.4**	22.1	**10.8**	21.2				
		3.5	8.2	**5.4**	34.1	**7.1**	32.1	**8.3**	19.9			**10.5**	22.7	**10.9**	22.0				
		3.6	6.3	**5.5**	36.6	**7.2**	34.7	**8.4**	17.2			**10.6**	23.3	**11.0**	22.7				
		3.7	4.5	**5.6**	38.8	**7.3**	37.0	**8.5**	14.7			**10.7**	23.8						
		3.8	2.9	**5.7**	40.6	**7.4**	39.1	**8.6**	12.2			**10.8**	24.25						
		3.9	1.4	**5.8**	42.3	**7.5**	40.9	**8.7**	10.3										
		4.0	0.1	**5.9**	43.7	**7.6**	42.4	**8.8**	8.5										
						7.7	43.5	**8.9**	7.0										
						7.8	44.5	**9.0**	5.7										
						7.9	45.3												
						8.0	46.1												

CONCENTRATIVE PROPERTIES OF AQUEOUS SOLUTIONS

This table gives properties of aqueous solutions of 66 substances as a function of concentration. All data refer to a temperature of 20 °C. The column definitions are as follows.

Column heading	Definition
Mass %	Mass of solute divided by total mass of solution, expressed as percent
m	Molality, moles of solute per kg of water
c	Molarity, moles of solute per liter of solution
ρ	Density of solution, in g cm^{-3}
n	Index of refraction, relative to air, at a wavelength of 589 nm (sodium D line); the index of pure water at 20 °C is 1.3330
Δ	Freezing point depression relative to pure water, in °C
η	Absolute (dynamic) viscosity, in mPa s (equal to centipoise, cP); the viscosity of pure water at 20 °C is 1.002 mPa s

Density data for aqueous solutions over a wider range of temperatures and pressures (and for additional compounds not in this table) may be found in Ref. 2.

References

1. Wolf, A. V., *Aqueous Solutions and Body Fluids*, Hoeber, 1966.
2. Söhnel, O., and Novotny, P., *Densities of Aqueous Solutions of Inorganic Substances*, Elsevier, Amsterdam, 1985.

Concentrative Properties of Aqueous Solutions: Density, Refractive Index, Freezing Point Depression, and Viscosity

Mass %	m/mol kg^{-1}	c/mol L^{-1}	ρ/g cm^{-3}	n	Δ/°C	η/mPa s
Acetic acid						
0.5	0.084	0.083	0.9989	1.3334	0.16	1.012
1.0	0.168	0.166	0.9996	1.3337	0.32	1.022
2.0	0.340	0.333	1.0011	1.3345	0.63	1.042
3.0	0.515	0.501	1.0025	1.3352	0.94	1.063
4.0	0.694	0.669	1.0038	1.3359	1.26	1.084
5.0	0.876	0.837	1.0052	1.3366	1.58	1.105
6.0	1.063	1.006	1.0066	1.3373	1.90	1.125
7.0	1.253	1.175	1.0080	1.3381	2.23	1.143
8.0	1.448	1.345	1.0093	1.3388	2.56	1.162
9.0	1.647	1.515	1.0107	1.3395	2.89	1.186
10.0	1.850	1.685	1.0121	1.3402	3.23	1.210
11.0	2.058	1.856	1.0134	1.3409	3.57	1.231
12.0	2.271	2.028	1.0147	1.3416	3.91	1.253
13.0	2.488	2.200	1.0161	1.3423	4.26	1.275
14.0	2.711	2.372	1.0174	1.3430	4.61	1.298
15.0	2.939	2.545	1.0187	1.3437	4.97	1.320
16.0	3.172	2.718	1.0200	1.3444	5.33	1.341
17.0	3.411	2.891	1.0213	1.3451	5.69	1.360
18.0	3.655	3.065	1.0225	1.3458	6.06	1.380
19.0	3.906	3.239	1.0238	1.3465	6.43	1.405
20.0	4.163	3.414	1.0250	1.3472	6.81	1.431
22.0	4.697	3.764	1.0275	1.3485	7.57	1.478
24.0	5.259	4.116	1.0299	1.3498	8.36	1.525
26.0	5.851	4.470	1.0323	1.3512	9.17	1.572
28.0	6.476	4.824	1.0346	1.3525	10.00	1.613
30.0	7.137	5.180	1.0369	1.3537	10.84	1.669
32.0	7.837	5.537	1.0391	1.3550	11.70	1.715
34.0	8.579	5.896	1.0413	1.3562	12.55	1.762
36.0	9.367	6.255	1.0434	1.3574	13.38	1.812
38.0	10.207	6.615	1.0454	1.3586		1.852
40.0	11.102	6.977	1.0474	1.3598		1.912
42.0	12.059	7.339	1.0493	1.3610		1.960
44.0	13.084	7.701	1.0510	1.3621		2.007
46.0	14.186	8.065	1.0528	1.3632		2.052
48.0	15.372	8.429	1.0545	1.3642		2.110
50.0	16.653	8.794	1.0562	1.3653		2.158
52.0	18.041	9.159	1.0577	1.3663		2.212
54.0	19.549	9.525	1.0592	1.3673		2.265
56.0	21.194	9.890	1.0605	1.3682		2.308
58.0	22.997	10.256	1.0618	1.3691		2.360
60.0	24.979	10.620	1.0629	1.3700		2.409
62.0	27.170	10.986	1.0640	1.3708		2.456
64.0	29.605	11.351	1.0650	1.3716		2.502
66.0	32.326	11.715	1.0659	1.3724		2.553
68.0	35.387	12.080	1.0668	1.3732		2.594
70.0	38.857	12.441	1.0673	1.3738		2.629
72.0	42.821	12.801	1.0676	1.3745		2.662
74.0	47.396	13.159	1.0678	1.3751		2.687
76.0	52.734	13.517	1.0680	1.3757		2.714
78.0	59.042	13.874	1.0681	1.3762		2.720
80.0	66.611	14.228	1.0680	1.3767		2.720
82.0	75.863	14.580	1.0677	1.3770		2.696
84.0	87.427	14.930	1.0673	1.3773		2.658
86.0	102.296	15.275	1.0666	1.3774		2.596
88.0	122.120	15.619	1.0658	1.3774		2.511
90.0	149.875	15.953	1.0644	1.3771		2.386
92.0	191.507	16.284	1.0629	1.3766		2.240
94.0	260.894	16.602	1.0606	1.3759		2.036
96.0	399.667	16.911	1.0578	1.3748		1.813
98.0	815.987	17.198	1.0538	1.3734		1.535
100.0		17.447	1.0477	1.3716		1.223
Acetone						
0.5	0.087	0.086	0.9975	1.3334	0.16	1.013
1.0	0.174	0.172	0.9968	1.3337	0.32	1.024
1.5	0.262	0.257	0.9961	1.3341	0.48	1.035
2.0	0.351	0.343	0.9954	1.3344	0.65	1.047
2.5	0.441	0.428	0.9947	1.3348	0.81	1.059
3.0	0.533	0.513	0.9940	1.3352	0.97	1.072
3.5	0.624	0.599	0.9933	1.3355	1.13	1.085

Mass %	m/mol kg^{-1}	c/mol L^{-1}	ρ/g cm^{-3}	n	Δ/°C	η/mPa s
4.0	0.717	0.684	0.9926	1.3359	1.30	1.099
4.5	0.811	0.769	0.9919	1.3363	1.46	1.112
5.0	0.906	0.853	0.9912	1.3366	1.63	1.125
6.0	1.099	1.023	0.9899	1.3373	1.96	1.150
7.0	1.296	1.191	0.9886	1.3381	2.29	1.174
8.0	1.497	1.360	0.9874	1.3388	2.62	1.198
9.0	1.703	1.528	0.9861	1.3395	2.95	1.221
10.0	1.913	1.696	0.9849	1.3402	3.29	1.244

Ammonia

Mass %	m/mol kg^{-1}	c/mol L^{-1}	ρ/g cm^{-3}	n	Δ/°C	η/mPa s
0.5	0.295	0.292	0.9960	1.3332	0.55	1.009
1.0	0.593	0.584	0.9938	1.3335	1.14	1.015
2.0	1.198	1.162	0.9895	1.3339	2.32	1.029
3.0	1.816	1.736	0.9853	1.3344	3.53	1.043
4.0	2.447	2.304	0.9811	1.3349	4.78	1.057
5.0	3.090	2.868	0.9770	1.3354	6.08	1.071
6.0	3.748	3.428	0.9730	1.3359	7.43	1.085
7.0	4.420	3.983	0.9690	1.3365	8.95	1.099
8.0	5.106	4.533	0.9651	1.3370	10.34	1.113
9.0	5.807	5.080	0.9613	1.3376	11.90	1.127
10.0	6.524	5.622	0.9575	1.3381	13.55	1.141
11.0	7.257	6.160	0.9538	1.3387	15.29	1.155
12.0	8.007	6.695	0.9502	1.3393	17.13	1.169
13.0	8.774	7.226	0.9466	1.3398	19.07	1.182
14.0	9.558	7.753	0.9431	1.3404	21.13	1.195
15.0	10.362	8.275	0.9396	1.3410	23.32	1.207
16.0	11.184	8.794	0.9361	1.3416	25.63	1.218
17.0	12.026	9.310	0.9327	1.3422	28.09	1.228
18.0	12.889	9.823	0.9294	1.3428	30.70	1.237
19.0	13.773	10.332	0.9261	1.3434	33.47	1.245
20.0	14.679	10.837	0.9228	1.3440	36.42	1.254
22.0	16.561	11.838	0.9164	1.3453	43.36	1.268
24.0	18.542	12.826	0.9102	1.3465	51.38	1.280
26.0	20.630	13.801	0.9040	1.3477	60.77	1.288
28.0	22.834	14.764	0.8980	1.3490	71.66	
30.0	25.164	15.713	0.8920	1.3502	84.06	

Ammonium chloride

Mass %	m/mol kg^{-1}	c/mol L^{-1}	ρ/g cm^{-3}	n	Δ/°C	η/mPa s
0.5	0.094	0.093	0.9998	1.3340	0.32	0.999
1.0	0.189	0.187	1.0014	1.3349	0.64	0.996
2.0	0.382	0.376	1.0045	1.3369	1.27	0.992
3.0	0.578	0.565	1.0076	1.3388	1.91	0.988
4.0	0.779	0.756	1.0107	1.3407	2.57	0.985
5.0	0.984	0.948	1.0138	1.3426	3.25	0.982
6.0	1.193	1.141	1.0168	1.3445	3.94	0.979
7.0	1.407	1.335	1.0198	1.3464	4.66	0.976
8.0	1.626	1.529	1.0227	1.3483	5.40	0.974
9.0	1.849	1.726	1.0257	1.3502	6.16	0.972
10.0	2.077	1.923	1.0286	1.3521	6.95	0.970
11.0	2.311	2.121	1.0315	1.3540	7.76	0.969
12.0	2.549	2.320	1.0344	1.3559	8.60	0.969
13.0	2.793	2.521	1.0373	1.3578	9.47	0.969
14.0	3.043	2.722	1.0401	1.3596		0.969
15.0	3.299	2.924	1.0429	1.3615		0.970
16.0	3.561	3.128	1.0457	1.3634		0.971
17.0	3.829	3.332	1.0485	1.3652		0.972
18.0	4.104	3.537	1.0512	1.3671		0.973
19.0	4.385	3.744	1.0540	1.3689		0.975
20.0	4.674	3.951	1.0567	1.3708		0.978
22.0	5.273	4.368	1.0621	1.3745		0.986
24.0	5.903	4.789	1.0674	1.3782		0.996

Ammonium sulfate

Mass %	m/mol kg^{-1}	c/mol L^{-1}	ρ/g cm^{-3}	n	Δ/°C	η/mPa s
0.5	0.038	0.038	1.0012	1.3338	0.17	1.008
1.0	0.076	0.076	1.0042	1.3346	0.33	1.014
2.0	0.154	0.153	1.0101	1.3363	0.63	1.027
3.0	0.234	0.231	1.0160	1.3379	0.92	1.041
4.0	0.315	0.309	1.0220	1.3395	1.21	1.057
5.0	0.398	0.389	1.0279	1.3411	1.49	1.073
6.0	0.483	0.469	1.0338	1.3428	1.77	1.090
7.0	0.570	0.551	1.0397	1.3444	2.05	1.108
8.0	0.658	0.633	1.0456	1.3460	2.33	1.127
9.0	0.748	0.716	1.0515	1.3476	2.61	1.147
10.0	0.841	0.800	1.0574	1.3492	2.89	1.168
12.0	1.032	0.971	1.0691	1.3523	3.47	1.210
14.0	1.232	1.145	1.0808	1.3555	4.07	1.256
16.0	1.441	1.323	1.0924	1.3586	4.69	1.305
18.0	1.661	1.504	1.1039	1.3616		1.359
20.0	1.892	1.688	1.1154	1.3647		1.421
22.0	2.134	1.876	1.1269	1.3677		1.490
24.0	2.390	2.067	1.1383	1.3707		1.566
26.0	2.659	2.262	1.1496	1.3737		1.650
28.0	2.943	2.460	1.1609	1.3766		1.743
30.0	3.243	2.661	1.1721	1.3795		1.847
32.0	3.561	2.866	1.1833	1.3824		1.961
34.0	3.898	3.073	1.1945	1.3853		2.086
36.0	4.257	3.284	1.2056	1.3881		2.222
38.0	4.638	3.499	1.2166	1.3909		2.371
40.0	5.045	3.716	1.2277	1.3938		2.530

Barium chloride

Mass %	m/mol kg^{-1}	c/mol L^{-1}	ρ/g cm^{-3}	n	Δ/°C	η/mPa s
0.5	0.024	0.024	1.0026	1.3337	0.12	1.009
1.0	0.049	0.048	1.0070	1.3345	0.23	1.016
2.0	0.098	0.098	1.0159	1.3360	0.46	1.026
3.0	0.149	0.148	1.0249	1.3375	0.69	1.037
4.0	0.200	0.199	1.0341	1.3391	0.93	1.049
5.0	0.253	0.251	1.0434	1.3406	1.18	1.062
6.0	0.307	0.303	1.0528	1.3422	1.44	1.075
7.0	0.361	0.357	1.0624	1.3438	1.70	1.087
8.0	0.418	0.412	1.0721	1.3454	1.98	1.101
9.0	0.475	0.468	1.0820	1.3470	2.27	1.114
10.0	0.534	0.524	1.0921	1.3487	2.58	1.129
12.0	0.655	0.641	1.1128	1.3520	3.22	1.161
14.0	0.782	0.763	1.1342	1.3555	3.92	1.195
16.0	0.915	0.889	1.1564	1.3591	4.69	1.234
18.0	1.054	1.019	1.1793	1.3627		1.277
20.0	1.201	1.156	1.2031	1.3664		1.325
22.0	1.355	1.297	1.2277	1.3703		1.378
24.0	1.517	1.444	1.2531	1.3741		1.437
26.0	1.687	1.597	1.2793	1.3781		1.503

Calcium chloride

Mass %	m/mol kg^{-1}	c/mol L^{-1}	ρ/g cm^{-3}	n	Δ/°C	η/mPa s
0.5	0.045	0.045	1.0024	1.3342	0.22	1.015
1.0	0.091	0.091	1.0065	1.3354	0.44	1.028
2.0	0.184	0.183	1.0148	1.3378	0.88	1.050
3.0	0.279	0.277	1.0232	1.3402	1.33	1.078
4.0	0.375	0.372	1.0316	1.3426	1.82	1.110
5.0	0.474	0.469	1.0401	1.3451	2.35	1.143
6.0	0.575	0.567	1.0486	1.3475	2.93	1.175
7.0	0.678	0.667	1.0572	1.3500	3.57	1.208
8.0	0.784	0.768	1.0659	1.3525	4.28	1.242
9.0	0.891	0.872	1.0747	1.3549	5.04	1.279

Mass %	m/mol kg⁻¹	c/mol L⁻¹	ρ/g cm⁻³	n	Δ/°C	η/mPa s
10.0	1.001	0.976	1.0835	1.3575	5.86	1.319
11.0	1.114	1.083	1.0923	1.3600	6.74	1.362
12.0	1.229	1.191	1.1014	1.3625	7.70	1.408
13.0	1.346	1.301	1.1105	1.3651	8.72	1.457
14.0	1.467	1.413	1.1198	1.3677	9.83	1.508
15.0	1.590	1.526	1.1292	1.3704	11.01	1.564
16.0	1.716	1.641	1.1386	1.3730	12.28	1.625
17.0	1.846	1.759	1.1482	1.3757	13.65	1.691
18.0	1.978	1.878	1.1579	1.3784	15.11	1.764
19.0	2.114	1.999	1.1677	1.3812	16.70	1.843
20.0	2.253	2.122	1.1775	1.3839	18.30	1.930
22.0	2.541	2.374	1.1976	1.3895	21.70	2.127
24.0	2.845	2.634	1.2180	1.3951	25.30	2.356
26.0	3.166	2.902	1.2388	1.4008	29.70	2.645
28.0	3.504	3.179	1.2600	1.4066	34.70	3.000
30.0	3.862	3.464	1.2816	1.4124	41.00	3.467
32.0	4.240	3.759	1.3036	1.4183	49.70	4.035
34.0	4.642	4.062	1.3260	1.4242		4.820
36.0	5.068	4.375	1.3488	1.4301		5.807
38.0	5.522	4.698	1.3720	1.4361		7.321
40.0	6.007	5.030	1.3957	1.4420		8.997

Cesium chloride

Mass %	m/mol kg⁻¹	c/mol L⁻¹	ρ/g cm⁻³	n	Δ/°C	η/mPa s
0.5	0.030	0.030	1.0020	1.3334	0.10	1.000
1.0	0.060	0.060	1.0058	1.3337	0.20	0.997
1.5	0.090	0.090	1.0097	1.3341	0.30	0.994
2.0	0.121	0.120	1.0135	1.3345	0.40	0.992
2.5	0.152	0.151	1.0174	1.3349	0.51	0.990
3.0	0.184	0.182	1.0214	1.3353	0.61	0.988
3.5	0.215	0.213	1.0253	1.3357	0.71	0.986
4.0	0.247	0.245	1.0293	1.3361	0.81	0.984
4.5	0.280	0.276	1.0334	1.3365	0.92	0.982
5.0	0.313	0.308	1.0374	1.3369	1.02	0.980
5.5	0.346	0.340	1.0415	1.3373	1.12	0.978
6.0	0.379	0.373	1.0456	1.3377	1.22	0.977
6.5	0.413	0.405	1.0498	1.3382	1.33	0.975
7.0	0.447	0.438	1.0540	1.3386	1.43	0.974
7.5	0.482	0.471	1.0582	1.3390	1.53	0.973
8.0	0.516	0.505	1.0625	1.3394	1.64	0.971
8.5	0.552	0.539	1.0668	1.3399	1.74	0.970
9.0	0.587	0.573	1.0711	1.3403	1.85	0.969
9.5	0.624	0.607	1.0754	1.3407	1.95	0.968
10.0	0.660	0.641	1.0798	1.3412	2.06	0.966
11.0	0.734	0.711	1.0887	1.3421	2.28	0.963
12.0	0.810	0.782	1.0978	1.3430	2.51	0.961
13.0	0.888	0.855	1.1070	1.3439	2.74	0.958
14.0	0.967	0.928	1.1163	1.3448	2.97	0.955
15.0	1.048	1.003	1.1258	1.3458	3.21	0.953
16.0	1.131	1.079	1.1355	1.3468	3.46	0.950
17.0	1.217	1.156	1.1453	1.3477	3.71	0.947
18.0	1.304	1.235	1.1552	1.3487	3.96	0.945
19.0	1.393	1.315	1.1653	1.3497	4.22	0.942
20.0	1.485	1.397	1.1756	1.3507	4.49	0.939
22.0	1.675	1.564	1.1967	1.3528		0.934
24.0	1.876	1.737	1.2185	1.3550		0.930
26.0	2.087	1.917	1.2411	1.3572		0.926
28.0	2.310	2.103	1.2644	1.3594		0.924
30.0	2.546	2.296	1.2885	1.3617		0.922
32.0	2.795	2.497	1.3135	1.3641		0.922
34.0	3.060	2.705	1.3393	1.3666		0.924
36.0	3.341	2.921	1.3661	1.3691		0.926
38.0	3.640	3.146	1.3938	1.3717		0.930
40.0	3.960	3.380	1.4226	1.3744		0.934
42.0	4.301	3.624	1.4525	1.3771		0.940
44.0	4.667	3.877	1.4835	1.3800		0.947
46.0	5.060	4.142	1.5158	1.3829		0.956
48.0	5.483	4.418	1.5495	1.3860		0.967
50.0	5.940	4.706	1.5846	1.3892		0.981
52.0	6.435	5.007	1.6212	1.3925		1.000
54.0	6.973	5.323	1.6596	1.3960		1.023
56.0	7.560	5.654	1.6999	1.3996		1.050
58.0	8.202	6.002	1.7422	1.4035		1.080
60.0	8.910	6.368	1.7868	1.4076		1.120
62.0	9.691	6.754	1.8340	1.4120		1.172
64.0	10.560	7.163	1.8842	1.4167		1.238

Citric acid

Mass %	m/mol kg⁻¹	c/mol L⁻¹	ρ/g cm⁻³	n	Δ/°C	η/mPa s
0.5	0.026	0.026	1.0002	1.3336	0.05	1.013
1.0	0.053	0.052	1.0022	1.3343	0.11	1.024
2.0	0.106	0.105	1.0063	1.3356	0.21	1.048
3.0	0.161	0.158	1.0105	1.3368	0.32	1.073
4.0	0.217	0.211	1.0147	1.3381	0.43	1.098
5.0	0.274	0.265	1.0189	1.3394	0.54	1.125
6.0	0.332	0.320	1.0232	1.3407	0.65	1.153
7.0	0.392	0.374	1.0274	1.3420	0.76	1.183
8.0	0.453	0.430	1.0316	1.3433	0.88	1.214
9.0	0.515	0.485	1.0359	1.3446	1.00	1.247
10.0	0.578	0.541	1.0402	1.3459	1.12	1.283
12.0	0.710	0.655	1.0490	1.3486	1.38	1.357
14.0	0.847	0.771	1.0580	1.3514	1.66	1.436
16.0	0.991	0.889	1.0672	1.3541	1.95	1.525
18.0	1.143	1.008	1.0764	1.3569	2.26	1.625
20.0	1.301	1.130	1.0858	1.3598	2.57	1.740
22.0	1.468	1.254	1.0953	1.3626	2.88	1.872
24.0	1.644	1.380	1.1049	1.3655	3.21	2.017
26.0	1.829	1.508	1.1147	1.3684	3.55	2.178
28.0	2.024	1.639	1.1246	1.3714	3.89	2.356
30.0	2.231	1.772	1.1346	1.3744	4.25	2.549

Copper(II) sulfate

Mass %	m/mol kg⁻¹	c/mol L⁻¹	ρ/g cm⁻³	n	Δ/°C	η/mPa s
0.5	0.031	0.031	1.0033	1.3339	0.08	1.017
1.0	0.063	0.063	1.0085	1.3348	0.14	1.036
2.0	0.128	0.128	1.0190	1.3367	0.26	1.084
3.0	0.194	0.194	1.0296	1.3386	0.37	1.129
4.0	0.261	0.261	1.0403	1.3405	0.48	1.173
5.0	0.330	0.329	1.0511	1.3424	0.59	1.221
6.0	0.400	0.399	1.0620	1.3443	0.70	1.276
7.0	0.472	0.471	1.0730	1.3462	0.82	1.336
8.0	0.545	0.543	1.0842	1.3481	0.93	1.400
9.0	0.620	0.618	1.0955	1.3501	1.05	1.469
10.0	0.696	0.694	1.1070	1.3520	1.18	1.543
11.0	0.774	0.771	1.1186	1.3540	1.31	1.620
12.0	0.854	0.850	1.1304	1.3560	1.45	1.701
13.0	0.936	0.930	1.1424	1.3581	1.60	1.790
14.0	1.020	1.013	1.1545	1.3601	1.75	1.889
15.0	1.106	1.097	1.1669	1.3622		2.004
16.0	1.193	1.182	1.1796	1.3644		2.136
17.0	1.283	1.270	1.1926	1.3666		2.285
18.0	1.375	1.360	1.2059	1.3689		2.449

1,2-Ethanediol

Mass %	m/mol kg⁻¹	c/mol L⁻¹	ρ/g cm⁻³	n	Δ/°C	η/mPa s
0.5	0.081	0.080	0.9988	1.3335	0.15	1.010

Mass %	m/mol kg⁻¹	c/mol L⁻¹	ρ/g cm⁻³	n	Δ/°C	η/mPa s
1.0	0.163	0.161	0.9995	1.3339	0.30	1.020
2.0	0.329	0.322	1.0007	1.3348	0.61	1.048
3.0	0.498	0.484	1.0019	1.3358	0.92	1.074
4.0	0.671	0.646	1.0032	1.3367	1.24	1.099
5.0	0.848	0.809	1.0044	1.3377	1.58	1.125
6.0	1.028	0.972	1.0057	1.3386	1.91	1.153
7.0	1.213	1.136	1.0070	1.3396	2.26	1.182
8.0	1.401	1.299	1.0082	1.3405	2.62	1.212
9.0	1.593	1.464	1.0095	1.3415	2.99	1.243
10.0	1.790	1.628	1.0108	1.3425	3.37	1.277
12.0	2.197	1.959	1.0134	1.3444	4.16	1.348
14.0	2.623	2.292	1.0161	1.3464	5.01	1.424
16.0	3.069	2.626	1.0188	1.3484	5.91	1.500
18.0	3.537	2.962	1.0214	1.3503	6.89	1.578
20.0	4.028	3.300	1.0241	1.3523	7.93	1.661
24.0	5.088	3.981	1.0296	1.3564	10.28	1.843
28.0	6.265	4.669	1.0350	1.3605	13.03	2.047
32.0	7.582	5.364	1.0405	1.3646	16.23	2.280
36.0	9.062	6.067	1.0460	1.3687	19.82	2.537
40.0	10.741	6.776	1.0514	1.3728	23.84	2.832
44.0	12.659	7.491	1.0567	1.3769	28.32	3.166
48.0	14.872	8.212	1.0619	1.3811	33.30	3.544
52.0	17.453	8.939	1.0670	1.3851	38.81	3.981
56.0	20.505	9.671	1.0719	1.3892	44.83	4.475
60.0	24.166	10.406	1.0765	1.3931	51.23	5.026

Ethanol

Mass %	m/mol kg⁻¹	c/mol L⁻¹	ρ/g cm⁻³	n	Δ/°C	η/mPa s
0.5	0.109	0.108	0.9973	1.3333	0.20	1.023
1.0	0.219	0.216	0.9963	1.3336	0.40	1.046
2.0	0.443	0.432	0.9945	1.3342	0.81	1.095
3.0	0.671	0.646	0.9927	1.3348	1.23	1.140
4.0	0.904	0.860	0.9910	1.3354	1.65	1.183
5.0	1.142	1.074	0.9893	1.3360	2.09	1.228
6.0	1.385	1.286	0.9878	1.3367	2.54	1.279
7.0	1.634	1.498	0.9862	1.3374	2.99	1.331
8.0	1.887	1.710	0.9847	1.3381	3.47	1.385
9.0	2.147	1.921	0.9833	1.3388	3.96	1.442
10.0	2.412	2.131	0.9819	1.3395	4.47	1.501
11.0	2.683	2.341	0.9805	1.3403	5.00	1.563
12.0	2.960	2.551	0.9792	1.3410	5.56	1.627
13.0	3.243	2.759	0.9778	1.3417	6.13	1.694
14.0	3.534	2.967	0.9765	1.3425	6.73	1.761
15.0	3.830	3.175	0.9752	1.3432	7.36	1.826
16.0	4.134	3.382	0.9739	1.3440	8.01	1.890
17.0	4.446	3.589	0.9726	1.3447	8.69	1.955
18.0	4.765	3.795	0.9713	1.3455	9.40	2.019
19.0	5.092	4.000	0.9700	1.3462	10.14	2.081
20.0	5.427	4.205	0.9687	1.3469	10.92	2.142
22.0	6.122	4.613	0.9660	1.3484	12.60	2.259
24.0	6.855	5.018	0.9632	1.3498	14.47	2.370
26.0	7.626	5.419	0.9602	1.3511	16.41	2.476
28.0	8.441	5.817	0.9571	1.3524	18.43	2.581
30.0	9.303	6.212	0.9539	1.3535	20.47	2.667
32.0	10.215	6.601	0.9504	1.3546	22.44	2.726
34.0	11.182	6.987	0.9468	1.3557	24.27	2.768
36.0	12.210	7.370	0.9431	1.3566	25.98	2.803
38.0	13.304	7.747	0.9392	1.3575	27.62	2.829
40.0	14.471	8.120	0.9352	1.3583	29.26	2.846
42.0	15.718	8.488	0.9311	1.3590	30.98	2.852
44.0	17.055	8.853	0.9269	1.3598	32.68	2.850
46.0	18.490	9.213	0.9227	1.3604	34.36	2.843
48.0	20.036	9.568	0.9183	1.3610	36.04	2.832
50.0	21.706	9.919	0.9139	1.3616	37.67	2.813
52.0	23.515	10.266	0.9095	1.3621	39.20	2.789
54.0	25.481	10.607	0.9049	1.3626	40.65	2.754
56.0	27.626	10.945	0.9004	1.3630	42.06	2.701
58.0	29.975	11.278	0.8958	1.3634	43.49	2.632
60.0	32.559	11.605	0.8911	1.3638	44.93	2.547
62.0	35.415	11.930	0.8865	1.3641	46.28	2.479
64.0	38.589	12.250	0.8818	1.3644	47.52	2.415
66.0	42.135	12.565	0.8771	1.3647	48.64	2.347
68.0	46.125	12.877	0.8724	1.3650	49.52	2.281
70.0	50.648	13.183	0.8676	1.3652		2.214
72.0	55.816	13.486	0.8629	1.3654		2.148
74.0	61.779	13.783	0.8581	1.3655		2.082
76.0	68.736	14.077	0.8533	1.3657		2.015
78.0	76.958	14.366	0.8485	1.3657		1.948
80.0	86.824	14.649	0.8436	1.3658		1.881
82.0	98.883	14.928	0.8387	1.3657		1.808
84.0	113.957	15.197	0.8335	1.3656		1.741
86.0	133.337	15.464	0.8284	1.3655		1.674
88.0	159.178	15.724	0.8232	1.3653		1.606
90.0	195.355	15.980	0.8180	1.3650		1.542
92.0	249.620	16.225	0.8125	1.3646		1.475
94.0	340.062	16.466	0.8070	1.3642		1.407
96.0	520.946	16.697	0.8013	1.3636		1.342
98.0		16.920	0.7954	1.3630		1.273
100.0		17.133	0.7893	1.3614		1.203

Ethylenediaminetetraacetic acid, disodium salt, dihydrate

Mass %	m/mol kg⁻¹	c/mol L⁻¹	ρ/g cm⁻³	n	Δ/°C	η/mPa s
0.5	0.015	0.015	1.0009	1.3339	0.07	1.017
1.0	0.030	0.030	1.0036	1.3348	0.14	1.032
1.5	0.045	0.045	1.0062	1.3356	0.21	1.046
2.0	0.061	0.060	1.0089	1.3365	0.27	1.062
2.5	0.076	0.075	1.0115	1.3374	0.33	1.077
3.0	0.092	0.090	1.0142	1.3383	0.40	1.093
3.5	0.108	0.106	1.0169	1.3392	0.46	1.109
4.0	0.124	0.121	1.0196	1.3400	0.52	1.125
4.5	0.140	0.137	1.0223	1.3409	0.58	1.142
5.0	0.157	0.152	1.0250	1.3418	0.65	1.160
5.5	0.173	0.168	1.0277	1.3427	0.71	1.178
6.0	0.190	0.184	1.0305	1.3436	0.77	1.197

Formic acid

Mass %	m/mol kg⁻¹	c/mol L⁻¹	ρ/g cm⁻³	n	Δ/°C	η/mPa s
0.5	0.109	0.109	0.9994	1.3333	0.21	1.006
1.0	0.219	0.217	1.0006	1.3336	0.42	1.011
2.0	0.443	0.436	1.0029	1.3342	0.82	1.017
3.0	0.672	0.655	1.0053	1.3348	1.24	1.195
4.0	0.905	0.876	1.0077	1.3354	1.67	1.032
5.0	1.143	1.097	1.0102	1.3359	2.10	1.039
6.0	1.387	1.320	1.0126	1.3365	2.53	1.046
7.0	1.635	1.544	1.0150	1.3371	2.97	1.052
8.0	1.889	1.768	1.0175	1.3376	3.40	1.058
9.0	2.149	1.994	1.0199	1.3382	3.84	1.064
10.0	2.414	2.221	1.0224	1.3387	4.27	1.070
12.0	2.962	2.678	1.0273	1.3397	5.19	1.082
14.0	3.537	3.139	1.0322	1.3408	6.11	1.094
16.0	4.138	3.605	1.0371	1.3418	7.06	1.106
18.0	4.769	4.074	1.0419	1.3428	8.08	1.119
20.0	5.431	4.548	1.0467	1.3437	9.11	1.132
24.0	6.861	5.507	1.0562	1.3456	11.10	1.156
28.0	8.449	6.481	1.0654	1.3475	13.10	1.179

Mass %	m/mol kg^{-1}	c/mol L^{-1}	ρ/g cm^{-3}	n	Δ/°C	η/mPa s
32.0	10.224	7.471	1.0746	1.3493	15.28	1.203
36.0	12.220	8.477	1.0839	1.3511	17.65	1.227
40.0	14.483	9.502	1.0935	1.3529	20.18	1.254
44.0	17.070	10.529	1.1015	1.3547	22.93	1.281
48.0	20.054	11.572	1.1097	1.3565	26.06	1.309
52.0	23.535	12.633	1.1183	1.3581	29.69	1.340
56.0	27.650	13.715	1.1273	1.3597	33.81	1.374
60.0	32.587	14.813	1.1364	1.3612	38.26	1.410
64.0	38.622	15.928	1.1456	1.3626	43.02	1.449
68.0	46.166	17.054	1.1544	1.3641		1.490
70.0	50.692	17.619	1.1586	1.3648		1.511

β-D-Fructose

Mass %	m/mol kg^{-1}	c/mol L^{-1}	ρ/g cm^{-3}	n	Δ/°C	η/mPa s
0.5	0.028	0.028	1.0002	1.3337	0.05	1.015
1.0	0.056	0.056	1.0021	1.3344	0.10	1.028
2.0	0.113	0.112	1.0061	1.3358	0.21	1.054
3.0	0.172	0.168	1.0101	1.3373	0.32	1.080
4.0	0.231	0.225	1.0140	1.3387	0.43	1.106
5.0	0.292	0.283	1.0181	1.3402	0.54	1.134
6.0	0.354	0.340	1.0221	1.3417	0.66	1.165
7.0	0.418	0.399	1.0262	1.3431	0.78	1.198
8.0	0.483	0.458	1.0303	1.3446	0.90	1.232
9.0	0.549	0.517	1.0344	1.3461	1.03	1.270
10.0	0.617	0.576	1.0385	1.3476	1.16	1.309
11.0	0.686	0.637	1.0427	1.3492	1.29	1.349
12.0	0.757	0.697	1.0469	1.3507	1.43	1.391
13.0	0.829	0.759	1.0512	1.3522	1.57	1.435
14.0	0.904	0.820	1.0554	1.3538	1.71	1.483
15.0	0.980	0.882	1.0597	1.3554	1.86	1.533
16.0	1.057	0.945	1.0640	1.3569	2.01	1.587
17.0	1.137	1.008	1.0684	1.3585	2.16	1.643
18.0	1.218	1.072	1.0728	1.3601	2.32	1.703
19.0	1.302	1.136	1.0772	1.3618	2.48	1.768
20.0	1.388	1.201	1.0816	1.3634	2.64	1.837
22.0	1.566	1.332	1.0906	1.3667	3.05	1.986
24.0	1.753	1.465	1.0996	1.3700	3.43	2.154
26.0	1.950	1.600	1.1089	1.3734	3.82	2.348
28.0	2.159	1.738	1.1182	1.3768	4.20	2.562
30.0	2.379	1.878	1.1276	1.3803		2.817
32.0	2.612	2.020	1.1372	1.3839		3.112
34.0	2.859	2.164	1.1469	1.3874		3.462
36.0	3.122	2.312	1.1568	1.3911		3.899
38.0	3.402	2.461	1.1668	1.3948		4.418
40.0	3.700	2.613	1.1769	1.3985		5.046
42.0	4.019	2.767	1.1871	1.4023		5.773
44.0	4.361	2.925	1.1975	1.4062		6.644
46.0	4.728	3.084	1.2080	1.4101		7.753
48.0	5.124	3.247	1.2187	1.4141		9.060

α-D-Glucose

Mass %	m/mol kg^{-1}	c/mol L^{-1}	ρ/g cm^{-3}	n	Δ/°C	η/mPa s
0.5	0.028	0.028	1.0001	1.3337	0.05	1.010
1.0	0.056	0.056	1.0020	1.3344	0.11	1.021
2.0	0.113	0.112	1.0058	1.3358	0.21	1.052
3.0	0.172	0.168	1.0097	1.3373	0.32	1.083
4.0	0.231	0.225	1.0136	1.3387	0.43	1.113
5.0	0.292	0.282	1.0175	1.3402	0.55	1.145
6.0	0.354	0.340	1.0214	1.3417	0.67	1.179
7.0	0.418	0.398	1.0254	1.3432	0.79	1.214
8.0	0.483	0.457	1.0294	1.3447	0.91	1.250
9.0	0.549	0.516	1.0334	1.3462	1.04	1.289
10.0	0.617	0.576	1.0375	1.3477	1.17	1.330
11.0	0.686	0.636	1.0416	1.3492	1.30	1.372
12.0	0.757	0.697	1.0457	1.3508	1.44	1.416
13.0	0.829	0.758	1.0498	1.3523	1.59	1.462
14.0	0.904	0.819	1.0540	1.3539	1.73	1.512
15.0	0.980	0.881	1.0582	1.3555	1.88	1.566
16.0	1.057	0.944	1.0624	1.3571	2.03	1.625
17.0	1.137	1.007	1.0667	1.3587	2.19	1.688
18.0	1.218	1.070	1.0710	1.3603	2.35	1.757
19.0	1.302	1.134	1.0753	1.3619	2.52	1.829
20.0	1.388	1.199	1.0797	1.3635	2.70	1.904
22.0	1.566	1.329	1.0884	1.3668	3.07	2.063
24.0	1.753	1.462	1.0973	1.3702	3.48	2.242
26.0	1.950	1.597	1.1063	1.3736	3.90	2.458
28.0	2.159	1.734	1.1154	1.3770	4.34	2.707
30.0	2.379	1.873	1.1246	1.3805	4.79	2.998
32.0	2.612	2.014	1.1340	1.3840		3.324
34.0	2.859	2.158	1.1434	1.3876		3.704
36.0	3.122	2.304	1.1529	1.3912		4.193
38.0	3.402	2.452	1.1626	1.3949		4.786
40.0	3.700	2.603	1.1724	1.3986		5.493
42.0	4.019	2.756	1.1823	1.4024		6.288
44.0	4.361	2.912	1.1924	1.4062		7.235
46.0	4.728	3.071	1.2026	1.4101		8.454
48.0	5.124	3.232	1.2130	1.4141		9.883
50.0	5.551	3.396	1.2235	1.4181		11.884
52.0	6.013	3.562	1.2342	1.4222		14.489
54.0	6.516	3.732	1.2451	1.4263		17.916
56.0	7.064	3.905	1.2562	1.4306		22.886
58.0	7.665	4.081	1.2676	1.4349		29.389
60.0	8.326	4.261	1.2793	1.4394		37.445

Glycerol

Mass %	m/mol kg^{-1}	c/mol L^{-1}	ρ/g cm^{-3}	n	Δ/°C	η/mPa s
0.5	0.055	0.054	0.9994	1.3336	0.07	1.022
1.0	0.110	0.109	1.0005	1.3342	0.18	1.034
2.0	0.222	0.218	1.0028	1.3353	0.41	1.060
3.0	0.336	0.327	1.0051	1.3365	0.63	1.088
4.0	0.452	0.438	1.0074	1.3376	0.85	1.116
5.0	0.572	0.548	1.0097	1.3388	1.08	1.145
6.0	0.693	0.659	1.0120	1.3400	1.32	1.176
7.0	0.817	0.771	1.0144	1.3412	1.56	1.207
8.0	0.944	0.883	1.0167	1.3424	1.81	1.240
9.0	1.074	0.996	1.0191	1.3436	2.06	1.275
10.0	1.207	1.109	1.0215	1.3448	2.32	1.310
12.0	1.481	1.337	1.0262	1.3472	2.88	1.386
14.0	1.768	1.568	1.0311	1.3496	3.47	1.469
16.0	2.068	1.800	1.0360	1.3521	4.09	1.560
18.0	2.384	2.035	1.0409	1.3547	4.76	1.658
20.0	2.715	2.271	1.0459	1.3572	5.46	1.766
24.0	3.429	2.752	1.0561	1.3624	7.01	2.01
28.0	4.223	3.242	1.0664	1.3676	8.77	2.32
32.0	5.110	3.742	1.0770	1.3730	10.74	2.69
36.0	6.108	4.252	1.0876	1.3785	12.96	3.15
40.0	7.239	4.771	1.0984	1.3841	15.50	3.73
44.0	8.532	5.300	1.1092	1.3897		4.48
48.0	10.024	5.838	1.1200	1.3954		5.45
52.0	11.764	6.385	1.1308	1.4011		6.73
56.0	13.820	6.944	1.1419	1.4069		8.47
60.0	16.288	7.512	1.1530	1.4129		10.9
64.0	19.305	8.092	1.1643	1.4189		14.3
68.0	23.075	8.680	1.1755	1.4249		19.4
72.0	27.923	9.277	1.1866	1.4310		27.2

Mass %	m/mol kg^{-1}	c/mol L^{-1}	ρ/g cm^{-3}	n	Δ/°C	η/mPa s
76.0	34.387	9.884	1.1976	1.4370		39.6
80.0	43.436	10.498	1.2085	1.4431		60.6
84.0	57.009	11.121	1.2192	1.4492		98
88.0	79.632	11.753	1.2299	1.4553		170
92.0	124.878	12.392	1.2404	1.4613		319
96.0	260.615	13.039	1.2508	1.4674		648
100.0		13.694	1.2611	1.4735		1460

Hydrogen chloride

Mass %	m/mol kg^{-1}	c/mol L^{-1}	ρ/g cm^{-3}	n	Δ/°C	η/mPa s
0.5	0.138	0.137	1.0007	1.3341	0.49	1.008
1.0	0.277	0.275	1.0031	1.3353	0.99	1.015
2.0	0.560	0.553	1.0081	1.3376	2.08	1.029
3.0	0.848	0.833	1.0130	1.3399	3.28	1.044
4.0	1.143	1.117	1.0179	1.3422	4.58	1.059
5.0	1.444	1.403	1.0228	1.3445	5.98	1.075
6.0	1.751	1.691	1.0278	1.3468	7.52	1.091
7.0	2.064	1.983	1.0327	1.3491	9.22	1.108
8.0	2.385	2.277	1.0377	1.3515	11.10	1.125
9.0	2.713	2.574	1.0426	1.3538	13.15	1.143
10.0	3.047	2.873	1.0476	1.3561	15.40	1.161
11.0	3.390	3.176	1.0526	1.3584	17.85	1.180
12.0	3.740	3.481	1.0576	1.3607	20.51	1.199
13.0	4.098	3.789	1.0626	1.3630		1.219
14.0	4.465	4.099	1.0676	1.3653		1.239
15.0	4.840	4.413	1.0726	1.3676		1.261
16.0	5.224	4.729	1.0777	1.3700		1.282
17.0	5.617	5.049	1.0828	1.3723		1.304
18.0	6.020	5.370	1.0878	1.3746		1.326
19.0	6.433	5.695	1.0929	1.3769		1.350
20.0	6.857	6.023	1.0980	1.3792		1.374
22.0	7.736	6.687	1.1083	1.3838		1.426
24.0	8.661	7.362	1.1185	1.3884		1.483
26.0	9.636	8.049	1.1288	1.3930		1.547
28.0	10.666	8.748	1.1391	1.3976		1.620
30.0	11.754	9.456	1.1492	1.4020		1.705
32.0	12.907	10.175	1.1594	1.4066		1.799
34.0	14.129	10.904	1.1693	1.4112		1.900
36.0	15.427	11.642	1.1791	1.4158		2.002
38.0	16.810	12.388	1.1886	1.4204		2.105
40.0	18.284	13.140	1.1977	1.4250		

Iron(III) chloride

Mass %	m/mol kg^{-1}	c/mol L^{-1}	ρ/g cm^{-3}	n	Δ/°C	η/mPa s
0.5	0.031	0.031	1.0025	1.3344	0.21	1.024
1.0	0.062	0.062	1.0068	1.3358	0.39	1.047
2.0	0.126	0.125	1.0153	1.3386	0.75	1.093
3.0	0.191	0.189	1.0238	1.3413	1.15	1.139
4.0	0.257	0.255	1.0323	1.3441	1.56	1.187
5.0	0.324	0.321	1.0408	1.3468	2.00	1.238
6.0	0.394	0.388	1.0493	1.3496	2.48	1.292
7.0	0.464	0.457	1.0580	1.3524	2.99	1.350
8.0	0.536	0.526	1.0668	1.3552	3.57	1.412
9.0	0.610	0.597	1.0760	1.3581	4.19	1.480
10.0	0.685	0.669	1.0853	1.3611	4.85	1.553
12.0	0.841	0.817	1.1040	1.3670	6.38	1.707
14.0	1.004	0.969	1.1228	1.3730	8.22	1.879
16.0	1.174	1.126	1.1420		10.45	2.080
18.0	1.353	1.289	1.1615		13.08	2.311
20.0	1.541	1.457	1.1816		16.14	2.570
24.0	1.947	1.810	1.2234		23.79	3.178
28.0	2.398	2.189	1.2679		33.61	4.038
32.0	2.901	2.595	1.3153		49.16	5.274

Mass %	m/mol kg^{-1}	c/mol L^{-1}	ρ/g cm^{-3}	n	Δ/°C	η/mPa s
36.0	3.468	3.030	1.3654			7.130
40.0	4.110	3.496	1.4176			9.674

D-Lactic acid

Mass %	m/mol kg^{-1}	c/mol L^{-1}	ρ/g cm^{-3}	n	Δ/°C	η/mPa s
0.5	0.056	0.055	0.9992	1.3335	0.10	1.014
1.0	0.112	0.111	1.0002	1.3340	0.19	1.027
2.0	0.227	0.223	1.0023	1.3350	0.38	1.056
3.0	0.343	0.334	1.0043	1.3360	0.57	1.084
4.0	0.463	0.447	1.0065	1.3370	0.76	1.110
5.0	0.584	0.560	1.0086	1.3380	0.95	1.138
6.0	0.709	0.673	1.0108	1.3390	1.16	1.167
7.0	0.836	0.787	1.0131	1.3400	1.36	1.198
8.0	0.965	0.902	1.0153	1.3410	1.57	1.229
9.0	1.098	1.017	1.0176	1.3420	1.79	1.262
10.0	1.233	1.132	1.0199	1.3430	2.02	1.296
12.0	1.514	1.365	1.0246	1.3450	2.49	1.366
14.0	1.807	1.600	1.0294	1.3470	2.99	1.441
16.0	2.115	1.837	1.0342	1.3491	3.48	1.522
18.0	2.437	2.076	1.0390	1.3511	3.96	1.607
20.0	2.775	2.318	1.0439	1.3532	4.44	1.699
24.0	3.506	2.807	1.0536	1.3573		1.902
28.0	4.317	3.305	1.0632	1.3615		2.136
32.0	5.224	3.811	1.0728	1.3657		2.414
36.0	6.244	4.325	1.0822	1.3700		2.730
40.0	7.401	4.847	1.0915	1.3743		3.114
44.0	8.722	5.377	1.1008	1.3786		3.566
48.0	10.247	5.917	1.1105	1.3828		4.106
52.0	12.026	6.466	1.1201	1.3871		4.789
56.0	14.129	7.023	1.1297	1.3914		5.579
60.0	16.652	7.588	1.1392	1.3958		6.679
64.0	19.736	8.161	1.1486	1.4001		8.024
68.0	23.590	8.741	1.1579	1.4045		9.863
72.0	28.546	9.328	1.1670	1.4088		12.866
76.0	35.154	9.922	1.1760	1.4131		16.974
80.0	44.405	10.522	1.1848	1.4173		22.164

β-D-Lactose

Mass %	m/mol kg^{-1}	c/mol L^{-1}	ρ/g cm^{-3}	n	Δ/°C	η/mPa s
0.5	0.015	0.015	1.0002	1.3337	0.03	1.013
1.0	0.030	0.029	1.0021	1.3345	0.06	1.026
1.5	0.044	0.044	1.0041	1.3352	0.08	1.041
2.0	0.060	0.059	1.0061	1.3359	0.11	1.058
2.5	0.075	0.074	1.0081	1.3367	0.14	1.074
3.0	0.090	0.089	1.0102	1.3375	0.17	1.089
3.5	0.106	0.103	1.0122	1.3382	0.20	1.105
4.0	0.122	0.119	1.0143	1.3390	0.23	1.120
4.5	0.138	0.134	1.0163	1.3398	0.26	1.137
5.0	0.154	0.149	1.0184	1.3406	0.29	1.154
5.5	0.170	0.164	1.0204	1.3413	0.32	1.172
6.0	0.186	0.179	1.0225	1.3421	0.35	1.191
6.5	0.203	0.195	1.0246	1.3429	0.39	1.211
7.0	0.220	0.210	1.0267	1.3437	0.42	1.232
7.5	0.237	0.225	1.0287	1.3445	0.46	1.254
8.0	0.254	0.241	1.0308	1.3453	0.50	1.276
8.5	0.271	0.256	1.0329	1.3460		1.298
9.0	0.289	0.272	1.0349	1.3468		1.321
9.5	0.307	0.288	1.0370	1.3476		1.345
10.0	0.325	0.304	1.0390	1.3484		1.370
11.0	0.361	0.335	1.0432	1.3500		1.421
12.0	0.398	0.367	1.0473	1.3515		1.476
13.0	0.437	0.399	1.0515	1.3531		1.533
14.0	0.476	0.432	1.0558	1.3548		1.593

Mass %	m/mol kg^{-1}	c/mol L^{-1}	ρ/g cm^{-3}	n	Δ/°C	η/mPa s
15.0	0.516	0.465	1.0602	1.3564		1.657
16.0	0.556	0.498	1.0648	1.3582		1.724
17.0	0.598	0.531	1.0696	1.3600		1.795
18.0	0.641	0.565	1.0746	1.3619		1.869

Lithium chloride

Mass %	m/mol kg^{-1}	c/mol L^{-1}	ρ/g cm^{-3}	n	Δ/°C	η/mPa s
0.5	0.119	0.118	1.0012	1.3341	0.42	1.019
1.0	0.238	0.237	1.0041	1.3351	0.84	1.037
2.0	0.481	0.476	1.0099	1.3373	1.72	1.072
3.0	0.730	0.719	1.0157	1.3394	2.68	1.108
4.0	0.983	0.964	1.0215	1.3415	3.73	1.146
5.0	1.241	1.211	1.0272	1.3436	4.86	1.185
6.0	1.506	1.462	1.0330	1.3457	6.14	1.226
7.0	1.775	1.715	1.0387	1.3478	7.56	1.269
8.0	2.051	1.971	1.0444	1.3499	9.11	1.313
9.0	2.333	2.230	1.0502	1.3520	10.79	1.360
10.0	2.621	2.491	1.0560	1.3541	12.61	1.411
12.0	3.217	3.022	1.0675	1.3583	16.59	1.522
14.0	3.840	3.564	1.0792	1.3625	21.04	1.647
16.0	4.493	4.118	1.0910	1.3668		1.787
18.0	5.178	4.683	1.1029	1.3711		1.942
20.0	5.897	5.260	1.1150	1.3755		2.128
22.0	6.653	5.851	1.1274	1.3799		2.341
24.0	7.449	6.453	1.1399	1.3844		2.600
26.0	8.288	7.069	1.1527	1.3890		2.925
28.0	9.173	7.700	1.1658	1.3936		3.318
30.0	10.109	8.344	1.1791	1.3983		3.785

Magnesium chloride

Mass %	m/mol kg^{-1}	c/mol L^{-1}	ρ/g cm^{-3}	n	Δ/°C	η/mPa s
0.5	0.053	0.053	1.0022	1.3343	0.26	1.024
1.0	0.106	0.106	1.0062	1.3356	0.52	1.046
2.0	0.214	0.213	1.0144	1.3381	1.06	1.091
3.0	0.325	0.322	1.0226	1.3406	1.65	1.139
4.0	0.438	0.433	1.0309	1.3432	2.30	1.188
5.0	0.553	0.546	1.0394	1.3457	3.01	1.241
6.0	0.670	0.660	1.0479	1.3483		1.298
7.0	0.791	0.777	1.0564	1.3508		1.358
8.0	0.913	0.895	1.0651	1.3534		1.423
9.0	1.039	1.015	1.0738	1.3560		1.493
10.0	1.167	1.137	1.0826	1.3587		1.570
12.0	1.432	1.387	1.1005	1.3641		1.745
14.0	1.710	1.645	1.1189	1.3695		1.956
16.0	2.001	1.911	1.1372	1.3749		2.207
18.0	2.306	2.184	1.1553	1.3804		2.507
20.0	2.626	2.467	1.1742	1.3859		2.867
22.0	2.962	2.758	1.1938	1.3915		3.323
24.0	3.317	3.060	1.2140	1.3972		3.917
26.0	3.690	3.371	1.2346	1.4030		4.694
28.0	4.085	3.692	1.2555	1.4089		5.709
30.0	4.501	4.022	1.2763	1.4148		7.017

Magnesium sulfate

Mass %	m/mol kg^{-1}	c/mol L^{-1}	ρ/g cm^{-3}	n	Δ/°C	η/mPa s
0.5	0.042	0.042	1.0033	1.3340	0.10	1.027
1.0	0.084	0.084	1.0084	1.3350	0.19	1.054
2.0	0.170	0.169	1.0186	1.3371	0.36	1.112
3.0	0.257	0.256	1.0289	1.3391	0.52	1.177
4.0	0.346	0.345	1.0392	1.3411	0.69	1.249
5.0	0.437	0.436	1.0497	1.3431	0.87	1.328
6.0	0.530	0.528	1.0602	1.3451	1.05	1.411
7.0	0.625	0.623	1.0708	1.3471	1.24	1.498
8.0	0.722	0.719	1.0816	1.3492	1.43	1.593

Mass %	m/mol kg^{-1}	c/mol L^{-1}	ρ/g cm^{-3}	n	Δ/°C	η/mPa s
9.0	0.822	0.817	1.0924	1.3512	1.64	1.702
10.0	0.923	0.917	1.1034	1.3532	1.85	1.829
12.0	1.133	1.122	1.1257	1.3572	2.31	2.104
14.0	1.352	1.336	1.1484	1.3613	2.86	2.412
16.0	1.582	1.557	1.1717	1.3654	3.67	2.809
18.0	1.824	1.788	1.1955	1.3694		3.360
20.0	2.077	2.027	1.2198	1.3735		4.147
22.0	2.343	2.275	1.2447	1.3776		5.199
24.0	2.624	2.532	1.2701	1.3817		6.498
26.0	2.919	2.800	1.2961	1.3858		8.066

α-Maltose

Mass %	m/mol kg^{-1}	c/mol L^{-1}	ρ/g cm^{-3}	n	Δ/°C	η/mPa s
0.5	0.015	0.015	1.0003	1.3337	0.03	1.016
1.0	0.030	0.029	1.0023	1.3345	0.06	1.030
2.0	0.060	0.059	1.0063	1.3359	0.11	1.060
3.0	0.090	0.089	1.0104	1.3374	0.17	1.092
4.0	0.122	0.119	1.0144	1.3389	0.23	1.126
5.0	0.154	0.149	1.0184	1.3404	0.29	1.162
6.0	0.186	0.179	1.0224	1.3420	0.35	1.200
7.0	0.220	0.210	1.0265	1.3435	0.42	1.239
8.0	0.254	0.241	1.0305	1.3450	0.48	1.281
9.0	0.289	0.272	1.0345	1.3466	0.55	1.325
10.0	0.325	0.303	1.0385	1.3482	0.62	1.372
11.0	0.361	0.335	1.0425	1.3497	0.69	1.422
12.0	0.398	0.367	1.0465	1.3513	0.77	1.474
13.0	0.437	0.399	1.0505	1.3529	0.84	1.530
14.0	0.476	0.431	1.0545	1.3546	0.92	1.588
15.0	0.516	0.464	1.0585	1.3562	1.00	1.649
16.0	0.556	0.497	1.0629	1.3578	1.08	1.715
17.0	0.598	0.530	1.0672	1.3595	1.17	1.784
18.0	0.641	0.564	1.0716	1.3612	1.25	1.859
19.0	0.685	0.597	1.0759	1.3628	1.34	1.940
20.0	0.730	0.631	1.0801	1.3644	1.43	2.021
22.0	0.824	0.700	1.0894	1.3678	1.64	2.216
24.0	0.923	0.770	1.0984	1.3714	1.85	2.463
26.0	1.026	0.842	1.1080	1.3749	2.08	2.753
28.0	1.136	0.914	1.1171	1.3785	2.34	3.066
30.0	1.252	0.988	1.1269	1.3821	2.62	3.427
32.0	1.375	1.063	1.1367	1.3858	2.93	3.918
34.0	1.505	1.139	1.1463	1.3896	3.25	4.447
36.0	1.643	1.216	1.1561	1.3935	3.60	5.050
38.0	1.791	1.295	1.1663	1.3974	3.99	5.832
40.0	1.948	1.375	1.1769	1.4013	4.41	6.926
42.0	2.116	1.457	1.1878	1.4051	4.88	8.191
44.0	2.295	1.540	1.1979	1.4094	5.35	9.649
46.0	2.489	1.624	1.2084	1.4136		11.473
48.0	2.697	1.710	1.2194	1.4177		14.148
50.0	2.921	1.797	1.2304	1.4217		17.786
52.0	3.165	1.886	1.2416	1.4260		22.034
54.0	3.429	1.976	1.2528	1.4308		28.757
56.0	3.718	2.068	1.2638	1.4350		38.226
58.0	4.034	2.159	1.2740	1.4394		49.298
60.0	4.382	2.253	1.2855	1.4440		

Manganese(II) sulfate

Mass %	m/mol kg^{-1}	c/mol L^{-1}	ρ/g cm^{-3}	n	Δ/°C	η/mPa s
1.0	0.067	0.067	1.0080	1.3348	0.16	1.046
2.0	0.135	0.135	1.0178	1.3366	0.31	1.090
3.0	0.205	0.204	1.0277	1.3384	0.44	1.137
4.0	0.276	0.275	1.0378	1.3402	0.57	1.187
5.0	0.349	0.347	1.0480	1.3420	0.70	1.242
6.0	0.423	0.421	1.0583	1.3438	0.84	1.301

Mass %	m/mol kg^{-1}	c/mol L^{-1}	ρ/g cm^{-3}	n	Δ/°C	η/mPa s
7.0	0.498	0.495	1.0688	1.3457	0.98	1.363
8.0	0.576	0.572	1.0794	1.3475	1.12	1.431
9.0	0.655	0.650	1.0902	1.3494	1.28	1.505
10.0	0.736	0.729	1.1012	1.3513	1.44	1.587
11.0	0.819	0.810	1.1123	1.3532	1.61	1.678
12.0	0.903	0.893	1.1236	1.3551	1.80	1.779
13.0	0.990	0.977	1.1351	1.3570	2.00	1.887
14.0	1.078	1.063	1.1467	1.3589	2.21	2.005
15.0	1.169	1.151	1.1585	1.3609	2.43	2.133
16.0	1.261	1.240	1.1705	1.3629	2.67	2.272
17.0	1.356	1.331	1.1827	1.3648	2.92	2.420
18.0	1.454	1.424	1.1950	1.3668	3.19	2.580
19.0	1.553	1.519	1.2075	1.3688	3.49	2.752
20.0	1.656	1.616	1.2203	1.3708	3.80	2.938

D-Mannitol

Mass %	m/mol kg^{-1}	c/mol L^{-1}	ρ/g cm^{-3}	n	Δ/°C	η/mPa s
0.5	0.028	0.027	1.0000	1.3337	0.05	1.019
1.0	0.055	0.055	1.0017	1.3345	0.10	1.032
1.5	0.084	0.083	1.0035	1.3352	0.16	1.044
2.0	0.112	0.110	1.0053	1.3359	0.21	1.057
2.5	0.141	0.138	1.0070	1.3367	0.26	1.069
3.0	0.170	0.166	1.0088	1.3374	0.32	1.081
3.5	0.199	0.194	1.0106	1.3381	0.37	1.094
4.0	0.229	0.222	1.0124	1.3389	0.43	1.107
4.5	0.259	0.251	1.0141	1.3396	0.48	1.121
5.0	0.289	0.279	1.0159	1.3403	0.54	1.135
5.5	0.319	0.307	1.0177	1.3411	0.60	1.150
6.0	0.350	0.336	1.0195	1.3418	0.66	1.166
6.5	0.382	0.364	1.0212	1.3425	0.72	1.183
7.0	0.413	0.393	1.0230	1.3433	0.77	1.200
7.5	0.445	0.422	1.0248	1.3440	0.84	1.218
8.0	0.477	0.451	1.0266	1.3447	0.90	1.236
8.5	0.510	0.480	1.0284	1.3455	0.96	1.256
9.0	0.543	0.509	1.0302	1.3462	1.02	1.275
9.5	0.576	0.538	1.0320	1.3469	1.08	1.294
10.0	0.610	0.567	1.0338	1.3477	1.15	1.314
11.0	0.678	0.626	1.0375	1.3491	1.28	1.355
12.0	0.749	0.686	1.0412	1.3506	1.41	1.398
13.0	0.820	0.746	1.0450	1.3521	1.55	1.443
14.0	0.894	0.806	1.0489	1.3536	1.69	1.489
15.0	0.969	0.867	1.0529	1.3552	1.84	1.537

Methanol

Mass %	m/mol kg^{-1}	c/mol L^{-1}	ρ/g cm^{-3}	n	Δ/°C	η/mPa s
0.5	0.157	0.156	0.9973	1.3331	0.28	1.022
1.0	0.315	0.311	0.9964	1.3332	0.56	1.040
2.0	0.637	0.621	0.9947	1.3334	1.14	1.070
3.0	0.965	0.930	0.9930	1.3336	1.75	1.100
4.0	1.300	1.238	0.9913	1.3339	2.37	1.131
5.0	1.643	1.544	0.9896	1.3341	3.02	1.163
6.0	1.992	1.850	0.9880	1.3343	3.71	1.196
7.0	2.349	2.155	0.9864	1.3346	4.41	1.229
8.0	2.714	2.459	0.9848	1.3348	5.13	1.264
9.0	3.087	2.762	0.9832	1.3351	5.85	1.297
10.0	3.468	3.064	0.9816	1.3354	6.60	1.329
11.0	3.858	3.365	0.9801	1.3356	7.36	1.360
12.0	4.256	3.665	0.9785	1.3359	8.14	1.389
13.0	4.664	3.964	0.9770	1.3362	8.93	1.418
14.0	5.081	4.262	0.9755	1.3365	9.72	1.446
15.0	5.508	4.560	0.9740	1.3367	10.53	1.474
16.0	5.945	4.856	0.9725	1.3370	11.36	1.501
17.0	6.393	5.152	0.9710	1.3373	12.23	1.528
18.0	6.851	5.447	0.9695	1.3376	13.13	1.554
19.0	7.321	5.740	0.9680	1.3379	14.06	1.579
20.0	7.803	6.034	0.9666	1.3381	15.02	1.604
22.0	8.803	6.616	0.9636	1.3387	16.98	1.652
24.0	9.856	7.196	0.9606	1.3392	19.04	1.697
26.0	10.966	7.771	0.9576	1.3397	21.23	1.735
28.0	12.138	8.341	0.9545	1.3402	23.59	1.769
30.0	13.376	8.908	0.9514	1.3407	25.91	1.795
32.0	14.688	9.470	0.9482	1.3411	28.15	1.814
34.0	16.078	10.028	0.9450	1.3415	30.48	1.827
36.0	17.556	10.580	0.9416	1.3419	32.97	1.835
38.0	19.129	11.127	0.9382	1.3422	35.60	1.839
40.0	20.807	11.669	0.9347	1.3425	38.60	1.837
42.0	22.601	12.205	0.9311	1.3427	41.50	1.831
44.0	24.523	12.734	0.9273	1.3429	44.50	1.821
46.0	26.587	13.259	0.9235	1.3430	47.80	1.805
48.0	28.810	13.777	0.9196	1.3431	51.20	1.785
50.0	31.211	14.288	0.9156	1.3431	54.50	1.761
52.0	33.812	14.792	0.9114	1.3431	58.10	1.735
54.0	36.639	15.290	0.9072	1.3430	62.00	1.708
56.0	39.723	15.783	0.9030	1.3429	66.00	1.676
58.0	43.101	16.269	0.8987	1.3428	70.00	1.641
60.0	46.816	16.749	0.8944	1.3426	74.50	1.600
62.0	50.923	17.224	0.8901	1.3425	79.30	1.553
64.0	55.486	17.690	0.8856	1.3422	84.40	1.503
66.0	60.586	18.148	0.8810	1.3419	89.60	1.456
68.0	66.323	18.598	0.8763	1.3415	96.30	1.413
70.0	72.826	19.040	0.8715	1.3411		1.368
72.0	80.257	19.476	0.8667	1.3407		1.318
74.0	88.831	19.904	0.8618	1.3402		1.267
76.0	98.835	20.324	0.8568	1.3397		1.218
78.0	110.657	20.737	0.8518	1.3391		1.172
80.0	124.844	21.144	0.8468	1.3385		1.128
82.0	142.183	21.539	0.8416	1.3379		1.074
84.0	163.858	21.931	0.8365	1.3372		1.023
86.0	191.725	22.311	0.8312	1.3365		0.970
88.0	228.881	22.684	0.8259	1.3357		0.916
90.0	280.899	23.045	0.8204	1.3348		0.861
92.0	358.926	23.396	0.8148	1.3339		0.806
94.0	488.972	23.732	0.8089	1.3328		0.749
96.0	749.064	24.072	0.8034	1.3316		0.695
98.0		24.396	0.7976	1.3304		0.639
100.0		24.710	0.7917	1.3290		0.586

Nitric acid

Mass %	m/mol kg^{-1}	c/mol L^{-1}	ρ/g cm^{-3}	n	Δ/°C	η/mPa s
0.5	0.080	0.079	1.0009	1.3336	0.28	1.004
1.0	0.160	0.159	1.0037	1.3343	0.56	1.005
2.0	0.324	0.320	1.0091	1.3356	1.12	1.007
3.0	0.491	0.483	1.0146	1.3368	1.70	1.010
4.0	0.661	0.648	1.0202	1.3381	2.32	1.014
5.0	0.835	0.814	1.0257	1.3394	2.96	1.018
6.0	1.013	0.982	1.0314	1.3407	3.63	1.022
7.0	1.194	1.152	1.0370	1.3421	4.33	1.027
8.0	1.380	1.324	1.0427	1.3434	5.05	1.032
9.0	1.570	1.498	1.0485	1.3447	5.81	1.038
10.0	1.763	1.673	1.0543	1.3460	6.60	1.044
11.0	1.961	1.851	1.0602	1.3474	7.42	1.051
12.0	2.164	2.030	1.0660	1.3487	8.27	1.058
13.0	2.371	2.212	1.0720	1.3500	9.15	1.066
14.0	2.583	2.395	1.0780	1.3514	10.08	1.075
15.0	2.801	2.580	1.0840	1.3527	11.04	1.084

Mass %	m/mol kg^{-1}	c/mol L^{-1}	ρ/g cm^{-3}	n	Δ/°C	η/mPa s
16.0	3.023	2.768	1.0901	1.3541	12.04	1.094
17.0	3.250	2.958	1.0963	1.3555	13.08	1.105
18.0	3.484	3.149	1.1025	1.3569	14.16	1.116
19.0	3.723	3.343	1.1087	1.3582	15.30	1.128
20.0	3.967	3.539	1.1150	1.3596		1.141
22.0	4.476	3.937	1.1277	1.3624		1.169
24.0	5.011	4.344	1.1406	1.3652		1.199
26.0	5.576	4.760	1.1536	1.3680		1.233
28.0	6.172	5.185	1.1668	1.3708		1.271
30.0	6.801	5.618	1.1801	1.3736		1.311
32.0	7.468	6.060	1.1934	1.3763		1.354
34.0	8.175	6.512	1.2068	1.3790		1.400
36.0	8.927	6.971	1.2202	1.3817		1.450
38.0	9.727	7.439	1.2335	1.3842		1.504
40.0	10.580	7.913	1.2466	1.3867		1.561

Oxalic acid

Mass %	m/mol kg^{-1}	c/mol L^{-1}	ρ/g cm^{-3}	n	Δ/°C	η/mPa s
0.5	0.056	0.056	1.0006	1.3336	0.16	1.013
1.0	0.112	0.111	1.0030	1.3342	0.30	1.023
1.5	0.169	0.167	1.0054	1.3347	0.44	1.033
2.0	0.227	0.224	1.0079	1.3353	0.57	1.044
2.5	0.285	0.281	1.0103	1.3359	0.71	1.055
3.0	0.343	0.337	1.0126	1.3364	0.84	1.065
3.5	0.403	0.395	1.0150	1.3370	0.97	1.076
4.0	0.463	0.452	1.0174	1.3375	1.09	1.086
4.5	0.523	0.510	1.0197	1.3381		1.097
5.0	0.585	0.568	1.0220	1.3386		1.108
5.5	0.646	0.626	1.0244	1.3392		1.118
6.0	0.709	0.684	1.0265	1.3397		1.129
6.5	0.772	0.743	1.0288	1.3402		1.140
7.0	0.836	0.802	1.0310	1.3407		1.150
7.5	0.901	0.861	1.0332	1.3413		1.161
8.0	0.966	0.920	1.0355	1.3418		1.172

Phosphoric acid

Mass %	m/mol kg^{-1}	c/mol L^{-1}	ρ/g cm^{-3}	n	Δ/°C	η/mPa s
0.5	0.051	0.051	1.0010	1.3335	0.12	1.010
1.0	0.103	0.102	1.0038	1.3340	0.24	1.020
2.0	0.208	0.206	1.0092	1.3349	0.46	1.050
3.0	0.316	0.311	1.0146	1.3358	0.69	1.079
4.0	0.425	0.416	1.0200	1.3367	0.93	1.108
5.0	0.537	0.523	1.0254	1.3376	1.16	1.138
6.0	0.651	0.631	1.0309	1.3385	1.38	1.169
7.0	0.768	0.740	1.0363	1.3394	1.62	1.200
8.0	0.887	0.850	1.0418	1.3403	1.88	1.232
9.0	1.009	0.962	1.0474	1.3413	2.16	1.267
10.0	1.134	1.075	1.0531	1.3422	2.45	1.303
11.0	1.261	1.189	1.0589	1.3431	2.72	1.341
12.0	1.392	1.304	1.0647	1.3441	3.01	1.382
13.0	1.525	1.420	1.0705	1.3450	3.38	1.424
14.0	1.661	1.538	1.0765	1.3460	3.76	1.469
15.0	1.801	1.657	1.0825	1.3470	4.08	1.516
16.0	1.944	1.777	1.0885	1.3480	4.45	1.565
17.0	2.090	1.899	1.0947	1.3489	4.82	1.616
18.0	2.240	2.022	1.1009	1.3500	5.25	1.671
19.0	2.394	2.147	1.1071	1.3510	5.72	1.727
20.0	2.551	2.273	1.1135	1.3520	6.23	1.788
22.0	2.878	2.529	1.1263	1.3540	7.38	1.914
24.0	3.223	2.791	1.1395	1.3561	8.69	2.049
26.0	3.585	3.059	1.1528	1.3582	10.12	2.198
28.0	3.968	3.333	1.1665	1.3604	11.64	2.365
30.0	4.373	3.614	1.1804	1.3625	13.23	2.553

Mass %	m/mol kg^{-1}	c/mol L^{-1}	ρ/g cm^{-3}	n	Δ/°C	η/mPa s
32.0	4.802	3.901	1.1945	1.3647	14.94	2.766
34.0	5.257	4.194	1.2089	1.3669	16.81	3.001
36.0	5.740	4.495	1.2236	1.3691	18.85	3.260
38.0	6.254	4.803	1.2385	1.3713	21.09	3.544
40.0	6.803	5.117	1.2536	1.3735	23.58	3.856

Potassium bromide

Mass %	m/mol kg^{-1}	c/mol L^{-1}	ρ/g cm^{-3}	n	Δ/°C	η/mPa s
0.5	0.042	0.042	1.0018	1.3336	0.15	1.000
1.0	0.085	0.084	1.0054	1.3342	0.29	0.998
2.0	0.171	0.170	1.0127	1.3354	0.59	0.994
3.0	0.260	0.257	1.0200	1.3366	0.88	0.990
4.0	0.350	0.345	1.0275	1.3379	1.18	0.985
5.0	0.442	0.435	1.0350	1.3391	1.48	0.981
6.0	0.536	0.526	1.0426	1.3403	1.78	0.977
7.0	0.633	0.618	1.0503	1.3416	2.10	0.974
8.0	0.731	0.711	1.0581	1.3429	2.42	0.970
9.0	0.831	0.806	1.0660	1.3441	2.74	0.967
10.0	0.934	0.903	1.0740	1.3454	3.07	0.964
11.0	1.039	1.000	1.0821	1.3467	3.42	0.961
12.0	1.146	1.099	1.0903	1.3481	3.76	0.958
13.0	1.256	1.200	1.0986	1.3494	4.12	0.956
14.0	1.368	1.302	1.1070	1.3507	4.49	0.953
15.0	1.483	1.406	1.1155	1.3521	4.86	0.951
16.0	1.601	1.512	1.1242	1.3535	5.25	0.949
17.0	1.721	1.619	1.1330	1.3548	5.64	0.948
18.0	1.845	1.727	1.1419	1.3562	6.04	0.946
19.0	1.971	1.838	1.1509	1.3577	6.46	0.945
20.0	2.101	1.950	1.1601	1.3591	6.88	0.944
22.0	2.370	2.179	1.1788	1.3620	7.76	0.943
24.0	2.654	2.416	1.1980	1.3650	8.70	0.943
26.0	2.952	2.661	1.2179	1.3680	9.68	0.944
28.0	3.268	2.914	1.2383	1.3711	10.72	0.947
30.0	3.601	3.175	1.2593	1.3743	11.82	0.952
32.0	3.954	3.445	1.2810	1.3776	12.98	0.959
34.0	4.329	3.724	1.3033	1.3809		0.968
36.0	4.727	4.012	1.3263	1.3843		0.979
38.0	5.150	4.311	1.3501	1.3878		0.993
40.0	5.602	4.620	1.3746	1.3914		1.010

Potassium carbonate

Mass %	m/mol kg^{-1}	c/mol L^{-1}	ρ/g cm^{-3}	n	Δ/°C	η/mPa s
0.5	0.036	0.036	1.0027	1.3339	0.18	1.013
1.0	0.073	0.073	1.0072	1.3347	0.34	1.025
2.0	0.148	0.147	1.0163	1.3365	0.66	1.048
3.0	0.224	0.223	1.0254	1.3382	0.99	1.071
4.0	0.301	0.299	1.0345	1.3399	1.32	1.094
5.0	0.381	0.378	1.0437	1.3416	1.67	1.119
6.0	0.462	0.457	1.0529	1.3433	2.03	1.146
7.0	0.545	0.538	1.0622	1.3450	2.40	1.174
8.0	0.629	0.620	1.0715	1.3467	2.77	1.204
9.0	0.716	0.704	1.0809	1.3484	3.17	1.235
10.0	0.804	0.789	1.0904	1.3501	3.57	1.269
12.0	0.987	0.963	1.1095	1.3535	4.45	1.339
14.0	1.178	1.144	1.1291	1.3569	5.39	1.414
16.0	1.378	1.330	1.1490	1.3603	6.42	1.497
18.0	1.588	1.523	1.1692	1.3637	7.55	1.594
20.0	1.809	1.722	1.1898	1.3671	8.82	1.707
24.0	2.285	2.139	1.2320	1.3739	11.96	1.978
28.0	2.814	2.584	1.2755	1.3807	16.01	2.331
32.0	3.405	3.057	1.3204	1.3874	21.46	2.834
36.0	4.070	3.559	1.3665	1.3940	28.58	3.503
40.0	4.824	4.093	1.4142	1.4006	37.55	4.360

Mass %	m/mol kg^{-1}	c/mol L^{-1}	ρ/g cm^{-3}	n	Δ/°C	η/mPa s
44.0	5.685	4.659	1.4633	1.4071		5.720
48.0	6.679	5.259	1.5142	1.4136		7.764
50.0	7.236	5.573	1.5404	1.4168		9.369

Potassium chloride

Mass %	m/mol kg^{-1}	c/mol L^{-1}	ρ/g cm^{-3}	n	Δ/°C	η/mPa s
0.5	0.067	0.067	1.0014	1.3337	0.23	1.000
1.0	0.135	0.135	1.0046	1.3343	0.46	0.999
2.0	0.274	0.271	1.0110	1.3357	0.92	0.999
3.0	0.415	0.409	1.0174	1.3371	1.38	0.998
4.0	0.559	0.549	1.0239	1.3384	1.85	0.997
5.0	0.706	0.691	1.0304	1.3398	2.32	0.996
6.0	0.856	0.835	1.0369	1.3411	2.80	0.994
7.0	1.010	0.980	1.0434	1.3425	3.29	0.992
8.0	1.166	1.127	1.0500	1.3438	3.80	0.990
9.0	1.327	1.276	1.0566	1.3452	4.30	0.989
10.0	1.490	1.426	1.0633	1.3466	4.81	0.988
11.0	1.658	1.579	1.0700	1.3479	5.33	0.989
12.0	1.829	1.733	1.0768	1.3493	5.88	0.990
13.0	2.004	1.890	1.0836	1.3507	6.45	0.992
14.0	2.184	2.048	1.0905	1.3521		0.994
15.0	2.367	2.208	1.0974	1.3535		0.997
16.0	2.555	2.370	1.1043	1.3549		0.999
17.0	2.747	2.534	1.1114	1.3563		1.001
18.0	2.944	2.701	1.1185	1.3577		1.004
19.0	3.146	2.869	1.1256	1.3592		1.007
20.0	3.353	3.039	1.1328	1.3606		1.012
22.0	3.783	3.386	1.1474	1.3635		1.024
24.0	4.236	3.742	1.1623	1.3665		1.040

Potassium dihydrogen phosphate

Mass %	m/mol kg^{-1}	c/mol L^{-1}	ρ/g cm^{-3}	n	Δ/°C	η/mPa s
0.5	0.037	0.037	1.0018	1.3336	0.13	1.010
1.0	0.074	0.074	1.0053	1.3342	0.25	1.019
1.5	0.112	0.111	1.0089	1.3348	0.37	1.028
2.0	0.150	0.149	1.0125	1.3354	0.49	1.038
2.5	0.188	0.187	1.0161	1.3359	0.60	1.048
3.0	0.227	0.225	1.0197	1.3365	0.72	1.060
3.5	0.267	0.263	1.0233	1.3371	0.84	1.071
4.0	0.306	0.302	1.0269	1.3377	0.96	1.083
4.5	0.346	0.341	1.0306	1.3382	1.08	1.096
5.0	0.387	0.380	1.0342	1.3388	1.19	1.108
5.5	0.428	0.419	1.0378	1.3394	1.30	1.121
6.0	0.469	0.459	1.0414	1.3400	1.41	1.133
6.5	0.511	0.499	1.0450	1.3405	1.52	1.146
7.0	0.553	0.539	1.0486	1.3411	1.63	1.160
7.5	0.596	0.580	1.0522	1.3417	1.74	1.173
8.0	0.639	0.621	1.0558	1.3422	1.84	1.187
8.5	0.683	0.662	1.0594	1.3428	1.94	1.201
9.0	0.727	0.703	1.0630	1.3434	2.04	1.215
9.5	0.771	0.745	1.0667	1.3439	2.14	1.230
10.0	0.816	0.786	1.0703	1.3445	2.23	1.245

Potassium hydrogen carbonate

Mass %	m/mol kg^{-1}	c/mol L^{-1}	ρ/g cm^{-3}	n	Δ/°C	η/mPa s
0.5	0.050	0.050	1.0014	1.3335	0.18	1.009
1.0	0.101	0.100	1.0046	1.3341	0.34	1.015
2.0	0.204	0.202	1.0114	1.3353	0.67	1.027
3.0	0.309	0.305	1.0181	1.3365	0.98	1.040
4.0	0.416	0.409	1.0247	1.3376	1.29	1.053
5.0	0.526	0.515	1.0310	1.3386	1.60	1.067
6.0	0.638	0.622	1.0379	1.3397	1.91	1.081
7.0	0.752	0.730	1.0446	1.3409	2.22	1.096
8.0	0.869	0.840	1.0514	1.3419	2.53	1.112
9.0	0.988	0.951	1.0581	1.3430	2.84	1.128
10.0	1.110	1.064	1.0650	1.3441	3.16	1.145
12.0	1.362	1.293	1.0788	1.3462	3.79	1.183
14.0	1.626	1.528	1.0929	1.3484	4.41	1.224
16.0	1.903	1.770	1.1073	1.3506		1.270
18.0	2.193	2.017	1.1221	1.3528		1.319
20.0	2.497	2.272	1.1372	1.3550		1.373
22.0	2.817	2.533	1.1527	1.3572		1.432
24.0	3.154	2.801	1.1685	1.3595		1.497

Potassium hydrogen phosphate

Mass %	m/mol kg^{-1}	c/mol L^{-1}	ρ/g cm^{-3}	n	Δ/°C	η/mPa s
0.5	0.029	0.029	1.0025	1.3338	0.13	1.013
1.0	0.058	0.058	1.0068	1.3345	0.25	1.023
1.5	0.087	0.087	1.0110	1.3353	0.37	1.034
2.0	0.117	0.117	1.0153	1.3361	0.49	1.046
2.5	0.147	0.146	1.0195	1.3368	0.61	1.057
3.0	0.178	0.176	1.0238	1.3376	0.73	1.069
3.5	0.208	0.207	1.0281	1.3384	0.86	1.081
4.0	0.239	0.237	1.0324	1.3392	0.97	1.094
4.5	0.271	0.268	1.0368	1.3399	1.10	1.107
5.0	0.302	0.299	1.0412	1.3407	1.22	1.120
5.5	0.334	0.330	1.0456	1.3415	1.34	1.133
6.0	0.366	0.362	1.0500	1.3422	1.46	1.147
6.5	0.399	0.394	1.0545	1.3430	1.58	1.162
7.0	0.432	0.426	1.0590	1.3438	1.70	1.177
7.5	0.466	0.458	1.0635	1.3445	1.82	1.193
8.0	0.499	0.491	1.0680	1.3453	1.95	1.209

Potassium hydroxide

Mass %	m/mol kg^{-1}	c/mol L^{-1}	ρ/g cm^{-3}	n	Δ/°C	η/mPa s
0.5	0.090	0.089	1.0025	1.3340	0.30	1.010
1.0	0.180	0.179	1.0068	1.3350	0.61	1.019
2.0	0.364	0.362	1.0155	1.3369	1.24	1.038
3.0	0.551	0.548	1.0242	1.3388	1.89	1.058
4.0	0.743	0.736	1.0330	1.3408	2.57	1.079
5.0	0.938	0.929	1.0419	1.3427	3.36	1.102
6.0	1.138	1.124	1.0509	1.3445	4.14	1.126
7.0	1.342	1.322	1.0599	1.3464	4.92	1.151
8.0	1.550	1.524	1.0690	1.3483		1.177
9.0	1.763	1.729	1.0781	1.3502		1.205
10.0	1.980	1.938	1.0873	1.3520		1.233
11.0	2.203	2.150	1.0966	1.3539		1.264
12.0	2.431	2.365	1.1059	1.3558		1.294
13.0	2.663	2.584	1.1153	1.3576		1.327
14.0	2.902	2.806	1.1246	1.3595		1.361
15.0	3.145	3.032	1.1341	1.3614		1.397
16.0	3.395	3.261	1.1435	1.3632		1.436
17.0	3.651	3.494	1.1531	1.3651		1.477
18.0	3.913	3.730	1.1626	1.3670		1.521
19.0	4.181	3.970	1.1722	1.3688		1.568
20.0	4.456	4.213	1.1818	1.3707		1.619
22.0	5.027	4.711	1.2014	1.3744		1.732
24.0	5.629	5.223	1.2210	1.3781		1.861
26.0	6.262	5.750	1.2408	1.3818		2.006
28.0	6.931	6.293	1.2609	1.3854		2.170
30.0	7.639	6.851	1.2813	1.3889		2.357
32.0	8.388	7.426	1.3020	1.3923		2.570
34.0	9.182	8.017	1.3230	1.3957		2.820
36.0	10.026	8.626	1.3444	1.3993		3.111
38.0	10.924	9.253	1.3661	1.4030		3.460
40.0	11.882	9.896	1.3881	1.4068		3.879
42.0	12.907	10.558	1.4104	1.4106		4.389

Mass %	m/mol kg^{-1}	c/mol L^{-1}	ρ/g cm^{-3}	n	Δ/°C	η/mPa s
44.0	14.004	11.239	1.4331	1.4143		5.013
46.0	15.183	11.938	1.4560	1.4179		5.781
48.0	16.453	12.654	1.4791	1.4214		6.726
50.0	17.824	13.389	1.5024	1.4247		7.892

Potassium iodide

Mass %	m/mol kg^{-1}	c/mol L^{-1}	ρ/g cm^{-3}	n	Δ/°C	η/mPa s
0.5	0.030	0.030	1.0019	1.3337	0.11	0.999
1.0	0.061	0.061	1.0056	1.3343	0.22	0.997
2.0	0.123	0.122	1.0131	1.3357	0.43	0.991
3.0	0.186	0.184	1.0206	1.3370	0.64	0.986
4.0	0.251	0.248	1.0282	1.3384	0.86	0.981
5.0	0.317	0.312	1.0360	1.3397	1.08	0.976
6.0	0.385	0.377	1.0438	1.3411	1.30	0.969
7.0	0.453	0.443	1.0517	1.3425	1.53	0.963
8.0	0.524	0.511	1.0598	1.3440	1.77	0.957
9.0	0.596	0.579	1.0679	1.3454	2.01	0.951
10.0	0.669	0.648	1.0762	1.3469	2.26	0.946
12.0	0.821	0.790	1.0931	1.3498	2.77	0.937
14.0	0.981	0.937	1.1105	1.3529	3.30	0.929
16.0	1.147	1.088	1.1284	1.3560	3.87	0.921
18.0	1.322	1.244	1.1469	1.3593	4.46	0.915
20.0	1.506	1.405	1.1659	1.3626	5.09	0.910
22.0	1.699	1.571	1.1856	1.3661	5.76	0.905
24.0	1.902	1.744	1.2060	1.3696	6.46	0.901
26.0	2.117	1.922	1.2270	1.3733	7.21	0.898
28.0	2.343	2.106	1.2487	1.3771	8.01	0.895
30.0	2.582	2.297	1.2712	1.3810	8.86	0.892
32.0	2.835	2.495	1.2944	1.3851	9.76	0.891
34.0	3.103	2.700	1.3185	1.3893	10.72	0.890
36.0	3.388	2.913	1.3434	1.3936	11.73	0.890
38.0	3.692	3.134	1.3692	1.3981	12.81	0.893
40.0	4.016	3.364	1.3959	1.4027	13.97	0.897

Potassium nitrate

Mass %	m/mol kg^{-1}	c/mol L^{-1}	ρ/g cm^{-3}	n	Δ/°C	η/mPa s
0.5	0.050	0.050	1.0014	1.3335	0.17	0.999
1.0	0.100	0.099	1.0045	1.3339	0.33	0.996
2.0	0.202	0.200	1.0108	1.3349	0.64	0.990
3.0	0.306	0.302	1.0171	1.3358	0.94	0.986
4.0	0.412	0.405	1.0234	1.3368	1.22	0.983
5.0	0.521	0.509	1.0298	1.3377	1.50	0.980
6.0	0.631	0.615	1.0363	1.3386	1.76	0.977
7.0	0.744	0.722	1.0428	1.3396	2.02	0.975
8.0	0.860	0.830	1.0494	1.3405	2.27	0.973
9.0	0.978	0.940	1.0560	1.3415	2.52	0.971
10.0	1.099	1.051	1.0627	1.3425	2.75	0.970
12.0	1.349	1.277	1.0762	1.3444		0.970
14.0	1.610	1.509	1.0899	1.3463		0.972
16.0	1.884	1.747	1.1039	1.3482		0.976
18.0	2.171	1.991	1.1181	1.3502		0.982
20.0	2.473	2.240	1.1326	1.3521		0.990
22.0	2.790	2.497	1.1473	1.3541		0.999
24.0	3.123	2.759	1.1623	1.3561		1.010

Potassium permanganate

Mass %	m/mol kg^{-1}	c/mol L^{-1}	ρ/g cm^{-3}	n	Δ/°C	η/mPa s
0.5	0.032	0.032	1.0017		0.11	1.001
1.0	0.064	0.064	1.0051		0.22	1.000
1.5	0.096	0.096	1.0085		0.32	0.999
2.0	0.129	0.128	1.0118		0.43	0.998
2.5	0.162	0.161	1.0152			0.996
3.0	0.196	0.193	1.0186			0.995
3.5	0.230	0.226	1.0220			0.994
4.0	0.264	0.260	1.0254			0.992
4.5	0.298	0.293	1.0288			0.991
5.0	0.333	0.327	1.0322			0.989
5.5	0.368	0.360	1.0356			0.987
6.0	0.404	0.394	1.0390			0.985

Potassium sulfate

Mass %	m/mol kg^{-1}	c/mol L^{-1}	ρ/g cm^{-3}	n	Δ/°C	η/mPa s
0.5	0.029	0.029	1.0022	1.3336	0.14	1.006
1.0	0.058	0.058	1.0062	1.3343	0.26	1.011
2.0	0.117	0.116	1.0143	1.3355	0.50	1.021
2.5	0.147	0.146	1.0183	1.3362	0.61	1.027
3.0	0.177	0.176	1.0224	1.3368	0.73	1.033
3.5	0.208	0.206	1.0265	1.3374	0.84	1.039
4.0	0.239	0.237	1.0306	1.3380	0.95	1.045
4.5	0.270	0.267	1.0347	1.3386	1.06	1.051
5.0	0.302	0.298	1.0388	1.3393	1.17	1.058
6.0	0.366	0.360	1.0470	1.3405		1.072
7.0	0.432	0.424	1.0553	1.3417		1.087
8.0	0.499	0.488	1.0637	1.3428		1.102
9.0	0.568	0.554	1.0721	1.3440		1.117
10.0	0.638	0.620	1.0806	1.3452		1.132

1-Propanol

Mass %	m/mol kg^{-1}	c/mol L^{-1}	ρ/g cm^{-3}	n	Δ/°C	η/mPa s
1.0	0.168	0.166	0.9963	1.3339	0.31	1.051
2.0	0.340	0.331	0.9946	1.3348	0.61	1.100
3.0	0.515	0.496	0.9928	1.3357	0.93	1.152
4.0	0.693	0.660	0.9911	1.3366	1.24	1.208
5.0	0.876	0.823	0.9896	1.3376	1.57	1.267
6.0	1.062	0.987	0.9882	1.3385	1.91	1.325
7.0	1.252	1.149	0.9868	1.3394	2.26	1.387
8.0	1.447	1.312	0.9855	1.3404	2.61	1.449
9.0	1.646	1.474	0.9842	1.3414	2.99	1.514
10.0	1.849	1.635	0.9829	1.3423	3.36	1.577
12.0	2.269	1.958	0.9804	1.3442	4.09	1.710
14.0	2.709	2.278	0.9779	1.3460	4.91	1.849
16.0	3.169	2.595	0.9749	1.3477	5.78	1.986
18.0	3.652	2.911	0.9719	1.3494	6.67	2.106
20.0	4.160	3.223	0.9686	1.3510	7.76	2.218
24.0	5.254	3.838	0.9612	1.3539	9.12	2.432
28.0	6.471	4.441	0.9533	1.3566	10.17	2.612
32.0	7.830	5.033	0.9452	1.3592	10.66	2.765
36.0	9.359	5.613	0.9370	1.3614		2.900
40.0	11.093	6.182	0.9288	1.3635		3.010
60.0	24.958	8.860	0.8875	1.3734		3.186
80.0	66.556	11.275	0.8470	1.3812		2.822
100.0		13.368	0.8034	1.3852		2.227

2-Propanol

Mass %	m/mol kg^{-1}	c/mol L^{-1}	ρ/g cm^{-3}	n	Δ/°C	η/mPa s
1.0	0.168	0.166	0.9960	1.3338	0.30	1.056
2.0	0.340	0.331	0.9939	1.3346	0.60	1.112
3.0	0.515	0.495	0.9920	1.3355	0.93	1.166
4.0	0.693	0.659	0.9902	1.3364	1.26	1.225
5.0	0.876	0.822	0.9884	1.3373	1.61	1.287
6.0	1.062	0.985	0.9871	1.3382	1.96	1.352
7.0	1.252	1.148	0.9855	1.3392	2.32	1.417
8.0	1.447	1.310	0.9843	1.3400	2.68	1.485
9.0	1.646	1.472	0.9831	1.3410	3.06	1.553
10.0	1.849	1.633	0.9816	1.3420	3.48	1.629
12.0	2.269	1.955	0.9793	1.3439	4.43	1.794
14.0	2.709	2.276	0.9772	1.3459	5.29	1.970
16.0	3.169	2.596	0.9751	1.3478	6.36	2.160

Mass %	m/mol kg^{-1}	c/mol L^{-1}	ρ/g cm^{-3}	n	Δ/°C	η/mPa s
18.0	3.652	2.913	0.9725	1.3496	7.40	2.352
20.0	4.160	3.227	0.9696	1.3514	8.52	2.550
40.0	11.093	6.191	0.9302	1.3642		
60.0	24.958	8.809	0.8824	1.3717		
80.0	66.556	11.103	0.8341	1.3742		
100.0		13.058	0.7848	1.3776		

Silver nitrate

Mass %	m/mol kg^{-1}	c/mol L^{-1}	ρ/g cm^{-3}	n	Δ/°C	η/mPa s
0.5	0.030	0.030	1.0027	1.3336	0.10	1.003
1.0	0.059	0.059	1.0070	1.3342	0.20	1.005
2.0	0.120	0.120	1.0154	1.3352	0.40	1.009
3.0	0.182	0.181	1.0239	1.3363	0.59	1.013
4.0	0.245	0.243	1.0327	1.3374	0.78	1.016
5.0	0.310	0.307	1.0417	1.3385	0.96	1.020
6.0	0.376	0.371	1.0506	1.3396	1.15	1.024
7.0	0.443	0.437	1.0597	1.3407	1.33	1.027
8.0	0.512	0.503	1.0690	1.3419	1.51	1.031
9.0	0.582	0.571	1.0785	1.3431	1.69	1.035
10.0	0.654	0.641	1.0882	1.3443	1.87	1.039
12.0	0.803	0.783	1.1079	1.3467	2.21	1.049
14.0	0.958	0.930	1.1284	1.3493	2.55	1.060
16.0	1.121	1.083	1.1496	1.3519	2.86	1.072
18.0	1.292	1.241	1.1715	1.3546		1.086
20.0	1.472	1.406	1.1942	1.3574		1.101
22.0	1.660	1.577	1.2177	1.3602		1.117
24.0	1.859	1.755	1.2420	1.3632		1.135
26.0	2.068	1.940	1.2672	1.3662		1.154
28.0	2.289	2.132	1.2933	1.3694		1.176
30.0	2.523	2.332	1.3204	1.3726		1.200
32.0	2.770	2.541	1.3487	1.3760		1.227
34.0	3.033	2.758	1.3780	1.3795		1.257
36.0	3.311	2.985	1.4087	1.3832		1.290
38.0	3.608	3.223	1.4407	1.3871		1.326
40.0	3.925	3.472	1.4743	1.3911		1.366

Sodium acetate

Mass %	m/mol kg^{-1}	c/mol L^{-1}	ρ/g cm^{-3}	n	Δ/°C	η/mPa s
0.5	0.061	0.061	1.0008	1.3337	0.22	1.021
1.0	0.123	0.122	1.0034	1.3344	0.43	1.040
2.0	0.249	0.246	1.0085	1.3358	0.88	1.080
3.0	0.377	0.371	1.0135	1.3372	1.34	1.124
4.0	0.508	0.497	1.0184	1.3386	1.82	1.171
5.0	0.642	0.624	1.0234	1.3400	2.32	1.222
6.0	0.778	0.752	1.0283	1.3414	2.85	1.278
7.0	0.918	0.882	1.0334	1.3428	3.40	1.337
8.0	1.060	1.013	1.0386	1.3442	3.98	1.401
9.0	1.206	1.145	1.0440	1.3456	4.57	1.468
10.0	1.354	1.279	1.0495	1.3470		1.539
12.0	1.662	1.552	1.0607	1.3498		1.688
14.0	1.984	1.829	1.0718	1.3526		1.855
16.0	2.322	2.112	1.0830	1.3554		2.054
18.0	2.676	2.400	1.0940	1.3583		2.284
20.0	3.047	2.694	1.1050	1.3611		2.567
22.0	3.438	2.993	1.1159	1.3639		2.948
24.0	3.849	3.297	1.1268	1.3666		3.400
26.0	4.283	3.606	1.1377	1.3693		3.877
28.0	4.741	3.921	1.1488	1.3720		4.388
30.0	5.224	4.243	1.1602	1.3748		4.940

Sodium bromide

Mass %	m/mol kg^{-1}	c/mol L^{-1}	ρ/g cm^{-3}	n	Δ/°C	η/mPa s
0.5	0.049	0.049	1.0021	1.3337	0.17	1.004
1.0	0.098	0.098	1.0060	1.3344	0.34	1.007
2.0	0.198	0.197	1.0139	1.3358	0.69	1.012
3.0	0.301	0.298	1.0218	1.3372	1.04	1.017
4.0	0.405	0.400	1.0298	1.3386	1.39	1.022
5.0	0.512	0.504	1.0380	1.3401	1.76	1.028
6.0	0.620	0.610	1.0462	1.3415	2.14	1.034
7.0	0.732	0.717	1.0546	1.3430	2.53	1.040
8.0	0.845	0.826	1.0630	1.3445	2.93	1.046
9.0	0.961	0.937	1.0716	1.3460	3.34	1.053
10.0	1.080	1.050	1.0803	1.3475	3.77	1.060
11.0	1.201	1.164	1.0892	1.3491	4.21	1.068
12.0	1.325	1.281	1.0981	1.3506	4.67	1.077
13.0	1.452	1.399	1.1072	1.3522	5.15	1.086
14.0	1.582	1.519	1.1164	1.3538	5.65	1.096
15.0	1.715	1.641	1.1257	1.3554	6.18	1.107
16.0	1.851	1.765	1.1352	1.3570	6.74	1.119
17.0	1.991	1.891	1.1448	1.3587	7.32	1.131
18.0	2.133	2.020	1.1546	1.3604		1.144
19.0	2.280	2.150	1.1645	1.3621		1.159
20.0	2.430	2.283	1.1745	1.3638		1.174
22.0	2.741	2.555	1.1951	1.3673		1.207
24.0	3.069	2.837	1.2163	1.3708		1.244
26.0	3.415	3.129	1.2382	1.3745		1.287
28.0	3.780	3.431	1.2608	1.3783		1.336
30.0	4.165	3.744	1.2842	1.3822		1.395
32.0	4.574	4.069	1.3083	1.3862		1.465
34.0	5.007	4.406	1.3333	1.3903		1.546
36.0	5.467	4.755	1.3592	1.3946		1.639
38.0	5.957	5.119	1.3860	1.3990		1.745
40.0	6.479	5.496	1.4138	1.4035		1.866

Sodium carbonate

Mass %	m/mol kg^{-1}	c/mol L^{-1}	ρ/g cm^{-3}	n	Δ/°C	η/mPa s
0.5	0.047	0.047	1.0034	1.3341	0.22	1.025
1.0	0.095	0.095	1.0086	1.3352	0.43	1.049
2.0	0.193	0.192	1.0190	1.3375	0.75	1.102
3.0	0.292	0.291	1.0294	1.3397	1.08	1.159
4.0	0.393	0.392	1.0398	1.3419	1.42	1.222
5.0	0.497	0.495	1.0502	1.3440	1.77	1.292
6.0	0.602	0.600	1.0606	1.3462	2.13	1.367
7.0	0.710	0.707	1.0711	1.3483		1.448
8.0	0.820	0.816	1.0816	1.3504		1.538
9.0	0.933	0.927	1.0922	1.3525		1.638
10.0	1.048	1.041	1.1029	1.3547		1.754
11.0	1.166	1.156	1.1136	1.3568		1.884
12.0	1.287	1.273	1.1244	1.3589		2.028
13.0	1.410	1.392	1.1353	1.3610		2.186
14.0	1.536	1.514	1.1463	1.3631		2.361
15.0	1.665	1.638	1.1574	1.3652		2.551

Sodium chloride

Mass %	m/mol kg^{-1}	c/mol L^{-1}	ρ/g cm^{-3}	n	Δ/°C	η/mPa s
0.1	0.017	0.017	0.9989	1.3332	0.06	1.004
0.2	0.034	0.034	0.9997	1.3333	0.12	1.006
0.3	0.051	0.051	1.0004	1.3335	0.18	1.008
0.4	0.069	0.069	1.0011	1.3337	0.24	1.009
0.5	0.086	0.086	1.0018	1.3339	0.30	1.011
1.0	0.173	0.172	1.0053	1.3347	0.59	1.020
1.5	0.261	0.259	1.0089	1.3356	0.89	1.028
2.0	0.349	0.346	1.0125	1.3365	1.19	1.036
2.5	0.439	0.435	1.0160	1.3374	1.49	1.044
3.0	0.529	0.523	1.0196	1.3383	1.79	1.052
3.5	0.621	0.613	1.0232	1.3391	2.10	1.060
4.0	0.713	0.703	1.0268	1.3400	2.41	1.068

Mass %	m/mol kg^{-1}	c/mol L^{-1}	ρ/g cm^{-3}	n	Δ/°C	η/mPa s
4.5	0.806	0.793	1.0304	1.3409	2.73	1.076
5.0	0.901	0.885	1.0340	1.3418	3.05	1.085
6.0	1.092	1.069	1.0413	1.3435	3.70	1.104
7.0	1.288	1.256	1.0486	1.3453	4.38	1.124
8.0	1.488	1.445	1.0559	1.3470	5.08	1.145
9.0	1.692	1.637	1.0633	1.3488	5.81	1.168
10.0	1.901	1.832	1.0707	1.3505	6.56	1.193
11.0	2.115	2.029	1.0781	1.3523	7.35	1.220
12.0	2.333	2.229	1.0857	1.3541	8.18	1.250
13.0	2.557	2.432	1.0932	1.3558	9.04	1.283
14.0	2.785	2.637	1.1008	1.3576	9.94	1.317
15.0	3.020	2.845	1.1085	1.3594	10.89	1.352
16.0	3.259	3.056	1.1162	1.3612	11.89	1.388
17.0	3.505	3.270	1.1240	1.3630	12.94	1.424
18.0	3.756	3.486	1.1319	1.3648	14.04	1.463
19.0	4.014	3.706	1.1398	1.3666	15.22	1.507
20.0	4.278	3.928	1.1478	1.3684	16.46	1.557
21.0	4.548	4.153	1.1558	1.3702	17.78	1.614
22.0	4.826	4.382	1.1640	1.3721	19.18	1.676
23.0	5.111	4.613	1.1721	1.3739	20.67	1.745
24.0	5.403	4.847	1.1804	1.3757		1.821
25.0	5.704	5.085	1.1887	1.3776		1.902
26.0	6.012	5.326	1.1972	1.3795		1.990

Sodium citrate dihydrate

Mass %	m/mol kg^{-1}	c/mol L^{-1}	ρ/g cm^{-3}	n	Δ/°C	η/mPa s
1.0	0.039	0.039	1.0049	1.3348	0.20	1.043
2.0	0.079	0.078	1.0120	1.3366	0.39	1.081
3.0	0.120	0.118	1.0186	1.3383	0.59	1.122
4.0	0.161	0.159	1.0260	1.3401	0.79	1.166
5.0	0.204	0.200	1.0331	1.3419	0.97	1.210
6.0	0.247	0.242	1.0405	1.3437	1.17	1.263
7.0	0.292	0.284	1.0482	1.3455	1.36	1.314
8.0	0.337	0.327	1.0557	1.3473	1.57	1.371
9.0	0.383	0.371	1.0632	1.3491	1.77	1.427
10.0	0.431	0.415	1.0708	1.3509	1.96	1.499
12.0	0.528	0.505	1.0861	1.3546	2.38	1.649
14.0	0.631	0.598	1.1019	1.3583	2.82	1.832
16.0	0.738	0.693	1.1173	1.3618	3.27	2.045
18.0	0.851	0.790	1.1327	1.3656	3.82	2.290
20.0	0.969	0.891	1.1492	1.3693	4.39	2.596
24.0	1.224	1.099	1.1813	1.3767		3.409
28.0	1.507	1.318	1.2151	1.3845		4.586
32.0	1.823	1.548	1.2487	1.3923		6.541
36.0	2.180	1.792	1.2843	1.4001		9.788

Sodium dihydrogen phosphate

Mass %	m/mol kg^{-1}	c/mol L^{-1}	ρ/g cm^{-3}	n	Δ/°C	η/mPa s
0.5	0.042	0.042	1.0019	1.3336	0.14	1.018
1.0	0.084	0.084	1.0056	1.3343	0.28	1.035
1.5	0.127	0.126	1.0094	1.3349	0.42	1.051
2.0	0.170	0.169	1.0131	1.3356	0.56	1.068
2.5	0.214	0.212	1.0168	1.3362	0.70	1.085
3.0	0.258	0.255	1.0206	1.3369	0.84	1.103
3.5	0.302	0.299	1.0244	1.3375	0.98	1.121
4.0	0.347	0.343	1.0281	1.3382	1.12	1.140
4.5	0.393	0.387	1.0319	1.3388	1.25	1.160
5.0	0.439	0.432	1.0358	1.3395	1.39	1.180
5.5	0.485	0.477	1.0396	1.3401	1.52	1.201
6.0	0.532	0.522	1.0434	1.3408	1.65	1.223
6.5	0.579	0.567	1.0473	1.3414	1.77	1.245
7.0	0.627	0.613	1.0511	1.3421	1.89	1.270
7.5	0.676	0.660	1.0550	1.3427	2.00	1.294

Mass %	m/mol kg^{-1}	c/mol L^{-1}	ρ/g cm^{-3}	n	Δ/°C	η/mPa s
8.0	0.725	0.706	1.0589	1.3434	2.12	1.319
8.5	0.774	0.753	1.0628	1.3440	2.23	1.345
9.0	0.824	0.800	1.0668	1.3447	2.35	1.371
9.5	0.875	0.848	1.0707	1.3453	2.47	1.399
10.0	0.926	0.896	1.0747	1.3460	2.58	1.428
11.0	1.030	0.993	1.0826	1.3473	2.82	1.488
12.0	1.137	1.091	1.0907	1.3486	3.06	1.552
13.0	1.246	1.191	1.0988	1.3499	3.29	1.620
14.0	1.357	1.292	1.1070	1.3512	3.53	1.694
15.0	1.471	1.394	1.1152	1.3525	3.78	1.775
16.0	1.588	1.499	1.1236	1.3538	4.03	1.861
17.0	1.707	1.604	1.1320	1.3552	4.29	1.952
18.0	1.830	1.711	1.1404	1.3565	4.55	2.050
19.0	1.955	1.820	1.1490	1.3578	4.82	2.159
20.0	2.084	1.930	1.1576	1.3592	5.10	2.283
22.0	2.351	2.155	1.1752	1.3618		2.550
24.0	2.632	2.387	1.1931	1.3646		2.850
26.0	2.929	2.625	1.2113	1.3673		3.214
28.0	3.242	2.870	1.2299	1.3700		3.682
30.0	3.572	3.123	1.2488	1.3728		4.300
32.0	3.923	3.383	1.2682	1.3756		5.079
34.0	4.294	3.650	1.2879	1.3784		6.008
36.0	4.689	3.925	1.3080	1.3812		7.098
38.0	5.109	4.208	1.3285	1.3840		8.363
40.0	5.557	4.499	1.3493	1.3869		9.814

Sodium hydrogen carbonate

Mass %	m/mol kg^{-1}	c/mol L^{-1}	ρ/g cm^{-3}	n	Δ/°C	η/mPa s
0.5	0.060	0.060	1.0018	1.3337	0.20	1.015
1.0	0.120	0.120	1.0054	1.3344	0.40	1.028
1.5	0.181	0.180	1.0089	1.3351	0.59	1.042
2.0	0.243	0.241	1.0125	1.3357	0.78	1.057
2.5	0.305	0.302	1.0160	1.3364	0.98	1.071
3.0	0.368	0.364	1.0196	1.3370	1.16	1.086
3.5	0.432	0.426	1.0231	1.3377	1.35	1.102
4.0	0.496	0.489	1.0266	1.3383	1.54	1.118
4.5	0.561	0.552	1.0301	1.3390	1.72	1.134
5.0	0.627	0.615	1.0337	1.3396	1.90	1.151
5.5	0.693	0.679	1.0372	1.3403	2.08	1.168
6.0	0.760	0.743	1.0408	1.3409	2.26	1.185

Sodium hydrogen phosphate

Mass %	m/mol kg^{-1}	c/mol L^{-1}	ρ/g cm^{-3}	n	Δ/°C	η/mPa s
0.5	0.035	0.035	1.0032	1.3340	0.17	1.021
1.0	0.071	0.071	1.0082	1.3349	0.32	1.042
1.5	0.107	0.107	1.0131	1.3358	0.46	1.064
2.0	0.144	0.143	1.0180	1.3368		1.088
2.5	0.181	0.180	1.0229	1.3377		1.113
3.0	0.218	0.217	1.0279	1.3386		1.138
3.5	0.255	0.255	1.0328	1.3396		1.165
4.0	0.293	0.292	1.0378	1.3405		1.193
4.5	0.332	0.331	1.0428	1.3414		1.223
5.0	0.371	0.369	1.0478	1.3424		1.254
5.5	0.410	0.408	1.0528	1.3433		1.286

Sodium hydroxide

Mass %	m/mol kg^{-1}	c/mol L^{-1}	ρ/g cm^{-3}	n	Δ/°C	η/mPa s
0.5	0.126	0.125	1.0039	1.3344	0.43	1.027
1.0	0.253	0.252	1.0095	1.3358	0.86	1.054
2.0	0.510	0.510	1.0207	1.3386	1.74	1.112
3.0	0.773	0.774	1.0318	1.3414	2.64	1.176
4.0	1.042	1.043	1.0428	1.3441	3.59	1.248
5.0	1.316	1.317	1.0538	1.3467	4.57	1.329
6.0	1.596	1.597	1.0648	1.3494	5.60	1.416

Mass %	m/mol kg^{-1}	c/mol L^{-1}	ρ/g cm^{-3}	n	Δ/°C	η/mPa s
7.0	1.882	1.883	1.0758	1.3520	6.69	1.510
8.0	2.174	2.174	1.0869	1.3546	7.87	1.616
9.0	2.473	2.470	1.0979	1.3572	9.12	1.737
10.0	2.778	2.772	1.1089	1.3597	10.47	1.882
11.0	3.090	3.080	1.1199	1.3623	11.89	2.039
12.0	3.409	3.393	1.1309	1.3648	13.42	2.201
13.0	3.736	3.711	1.1419	1.3673	15.04	2.376
14.0	4.070	4.036	1.1530	1.3697	16.76	2.568
15.0	4.412	4.365	1.1640	1.3722		2.789
16.0	4.762	4.701	1.1751	1.3746		3.043
17.0	5.121	5.041	1.1861	1.3770		3.344
18.0	5.488	5.387	1.1971	1.3793		3.698
19.0	5.865	5.739	1.2082	1.3817		4.119
20.0	6.250	6.096	1.2192	1.3840		4.619
22.0	7.052	6.827	1.2412	1.3885		5.765
24.0	7.895	7.579	1.2631	1.3929		7.100
26.0	8.784	8.352	1.2848	1.3971		8.744
28.0	9.723	9.145	1.3064	1.4012		10.832
30.0	10.715	9.958	1.3277	1.4051		13.517
32.0	11.766	10.791	1.3488	1.4088		16.844
34.0	12.880	11.643	1.3697	1.4123		20.751
36.0	14.064	12.512	1.3901	1.4156		25.290
38.0	15.324	13.398	1.4102	1.4186		30.461
40.0	16.668	14.300	1.4299	1.4215		36.312

Sodium nitrate

Mass %	m/mol kg^{-1}	c/mol L^{-1}	ρ/g cm^{-3}	n	Δ/°C	η/mPa s
0.5	0.059	0.059	1.0016	1.3336	0.20	1.004
1.0	0.119	0.118	1.0050	1.3341	0.40	1.007
2.0	0.240	0.238	1.0117	1.3353	0.79	1.012
3.0	0.364	0.359	1.0185	1.3364	1.18	1.018
4.0	0.490	0.483	1.0254	1.3375	1.56	1.025
5.0	0.619	0.607	1.0322	1.3387	1.94	1.032
6.0	0.751	0.734	1.0392	1.3398	2.32	1.040
7.0	0.886	0.862	1.0462	1.3409	2.70	1.049
8.0	1.023	0.991	1.0532	1.3421	3.08	1.059
9.0	1.164	1.123	1.0603	1.3432	3.46	1.069
10.0	1.307	1.256	1.0674	1.3443	3.84	1.081
12.0	1.604	1.527	1.0819	1.3466	4.60	1.107
14.0	1.915	1.806	1.0967	1.3489	5.37	1.138
18.0	2.583	2.387	1.1272	1.3536	6.98	1.215
20.0	2.941	2.689	1.1429	1.3559	7.81	1.263
24.0	3.715	3.318	1.1752	1.3607	9.52	1.377
28.0	4.575	3.981	1.2085	1.3654	11.28	1.522
30.0	5.042	4.326	1.2256	1.3678		1.609
34.0	6.061	5.044	1.2610	1.3726		1.818
40.0	7.844	6.200	1.3175	1.3802		2.226

Sodium phosphate

Mass %	m/mol kg^{-1}	c/mol L^{-1}	ρ/g cm^{-3}	n	Δ/°C	η/mPa s
0.5	0.031	0.031	1.0042	1.3343	0.19	1.033
1.0	0.062	0.062	1.0100	1.3356	0.37	1.064
1.5	0.093	0.093	1.0158	1.3369	0.53	1.094
2.0	0.124	0.125	1.0216	1.3381	0.67	1.126
2.5	0.156	0.157	1.0275	1.3394	0.79	1.161
3.0	0.189	0.189	1.0335	1.3406		1.198
3.5	0.221	0.222	1.0395	1.3419		1.238
4.0	0.254	0.255	1.0456	1.3432		1.281
4.5	0.287	0.289	1.0517	1.3444		1.327
5.0	0.321	0.323	1.0579	1.3457		1.375
5.5	0.355	0.357	1.0642	1.3470		1.426
6.0	0.389	0.392	1.0705	1.3482		1.480
6.5	0.424	0.427	1.0768	1.3495		1.538
7.0	0.459	0.462	1.0832	1.3507		1.598
7.5	0.495	0.498	1.0896	1.3519		1.662
8.0	0.530	0.535	1.0961	1.3532		1.729

Sodium sulfate

Mass %	m/mol kg^{-1}	c/mol L^{-1}	ρ/g cm^{-3}	n	Δ/°C	η/mPa s
0.5	0.035	0.035	1.0027	1.3338	0.17	1.013
1.0	0.071	0.071	1.0071	1.3345	0.32	1.026
2.0	0.144	0.143	1.0161	1.3360	0.61	1.058
3.0	0.218	0.217	1.0252	1.3376	0.87	1.091
4.0	0.293	0.291	1.0343	1.3391	1.13	1.126
5.0	0.371	0.367	1.0436	1.3406	1.36	1.163
6.0	0.449	0.445	1.0526	1.3420	1.56	1.202
7.0	0.530	0.523	1.0619	1.3435		1.244
8.0	0.612	0.603	1.0713	1.3449		1.289
9.0	0.696	0.685	1.0808	1.3464		1.337
10.0	0.782	0.768	1.0905	1.3479		1.390
11.0	0.870	0.852	1.1002	1.3494		1.447
12.0	0.960	0.938	1.1101	1.3509		1.508
13.0	1.052	1.025	1.1201	1.3524		1.574
14.0	1.146	1.114	1.1301	1.3539		1.646
15.0	1.242	1.204	1.1402	1.3553		1.725
16.0	1.341	1.296	1.1503	1.3567		1.812
17.0	1.442	1.389	1.1604	1.3581		1.905
18.0	1.545	1.483	1.1705	1.3595		2.005
19.0	1.651	1.579	1.1806	1.3608		2.112
20.0	1.760	1.677	1.1907	1.3620		2.227
22.0	1.986	1.875	1.2106	1.3643		2.481

Sodium thiosulfate

Mass %	m/mol kg^{-1}	c/mol L^{-1}	ρ/g cm^{-3}	n	Δ/°C	η/mPa s
0.5	0.032	0.032	1.0024	1.3340	0.14	1.012
1.0	0.064	0.064	1.0065	1.3351	0.28	1.023
2.0	0.129	0.128	1.0148	1.3371	0.57	1.044
3.0	0.196	0.194	1.0231	1.3392	0.84	1.066
4.0	0.264	0.261	1.0315	1.3413	1.09	1.090
5.0	0.333	0.329	1.0399	1.3434	1.34	1.115
6.0	0.404	0.398	1.0483	1.3454	1.59	1.141
7.0	0.476	0.468	1.0568	1.3475	1.83	1.169
8.0	0.550	0.539	1.0654	1.3496	2.06	1.199
9.0	0.626	0.611	1.0740	1.3517	2.30	1.231
10.0	0.703	0.685	1.0827	1.3538	2.55	1.267
12.0	0.862	0.835	1.1003	1.3581	3.06	1.345
14.0	1.030	0.990	1.1182	1.3624	3.60	1.435
16.0	1.205	1.150	1.1365	1.3667	4.17	1.537
18.0	1.388	1.315	1.1551	1.3711	4.76	1.657
20.0	1.581	1.485	1.1740	1.3756	5.37	1.798
22.0	1.784	1.660	1.1932	1.3801		1.958
24.0	1.997	1.841	1.2128	1.3847		2.141
26.0	2.222	2.027	1.2328	1.3893		2.356
28.0	2.460	2.219	1.2532	1.3940		2.596
30.0	2.711	2.417	1.2739	1.3987		2.903
32.0	2.976	2.621	1.2950	1.4035		3.305
34.0	3.258	2.831	1.3164	1.4084		3.792
36.0	3.558	3.047	1.3382	1.4132		4.359
38.0	3.876	3.269	1.3603	1.4181		5.011
40.0	4.216	3.498	1.3827	1.4229		5.758

Strontium chloride

Mass %	m/mol kg^{-1}	c/mol L^{-1}	ρ/g cm^{-3}	n	Δ/°C	η/mPa s
0.5	0.032	0.032	1.0027	1.3339	0.16	1.012
1.0	0.064	0.064	1.0071	1.3348	0.31	1.021
2.0	0.129	0.128	1.0161	1.3366	0.62	1.039
3.0	0.195	0.194	1.0252	1.3384	0.93	1.057

Mass %	m/mol kg^{-1}	c/mol L^{-1}	ρ/g cm^{-3}	n	Δ/°C	η/mPa s
4.0	0.263	0.261	1.0344	1.3402	1.26	1.076
5.0	0.332	0.329	1.0437	1.3421	1.61	1.096
6.0	0.403	0.399	1.0532	1.3440	1.98	1.116
7.0	0.475	0.469	1.0628	1.3459	2.38	1.136
8.0	0.549	0.541	1.0726	1.3478	2.80	1.157
9.0	0.624	0.615	1.0825	1.3498	3.25	1.180
10.0	0.701	0.689	1.0925	1.3518	3.74	1.204
12.0	0.860	0.843	1.1131	1.3558	4.81	1.258
14.0	1.027	1.002	1.1342	1.3599	6.03	1.317
16.0	1.202	1.167	1.1558	1.3641	7.41	1.383
18.0	1.385	1.338	1.1780	1.3684	8.98	1.460
20.0	1.577	1.515	1.2008	1.3728	10.74	1.549
22.0	1.779	1.699	1.2241	1.3772	12.74	1.650
24.0	1.992	1.890	1.2481	1.3817	14.99	1.765
26.0	2.216	2.087	1.2728	1.3864		1.897
28.0	2.453	2.293	1.2983	1.3911		2.056
30.0	2.703	2.507	1.3248	1.3961		2.245
32.0	2.968	2.730	1.3523	1.4013		2.527
34.0	3.250	2.962	1.3811	1.4067		2.846
36.0	3.548	3.205	1.4114	1.4124		3.206
Sucrose						
0.5	0.015	0.015	1.0002	1.3337	0.03	1.015
1.0	0.030	0.029	1.0021	1.3344	0.06	1.028
2.0	0.060	0.059	1.0060	1.3359	0.11	1.055
3.0	0.090	0.089	1.0099	1.3373	0.17	1.084
4.0	0.122	0.118	1.0139	1.3388	0.23	1.114
5.0	0.154	0.149	1.0178	1.3403	0.29	1.146
6.0	0.186	0.179	1.0218	1.3418	0.35	1.179
7.0	0.220	0.210	1.0259	1.3433	0.42	1.215
8.0	0.254	0.241	1.0299	1.3448	0.49	1.254
9.0	0.289	0.272	1.0340	1.3463	0.55	1.294
10.0	0.325	0.303	1.0381	1.3478	0.63	1.336
11.0	0.361	0.335	1.0423	1.3494	0.70	1.381
12.0	0.398	0.367	1.0465	1.3509	0.77	1.429
13.0	0.437	0.399	1.0507	1.3525	0.85	1.480
14.0	0.476	0.431	1.0549	1.3541	0.93	1.534
15.0	0.516	0.464	1.0592	1.3557	1.01	1.592
16.0	0.556	0.497	1.0635	1.3573	1.10	1.653
17.0	0.598	0.530	1.0678	1.3589	1.19	1.719
18.0	0.641	0.564	1.0722	1.3606	1.27	1.790
19.0	0.685	0.598	1.0766	1.3622	1.37	1.865
20.0	0.730	0.632	1.0810	1.3639	1.47	1.945
22.0	0.824	0.700	1.0899	1.3672	1.67	2.124
24.0	0.923	0.771	1.0990	1.3706	1.89	2.331
26.0	1.026	0.842	1.1082	1.3741	2.12	2.573
28.0	1.136	0.914	1.1175	1.3776	2.37	2.855
30.0	1.252	0.988	1.1270	1.3812	2.64	3.187
32.0	1.375	1.063	1.1366	1.3848	2.94	3.762
34.0	1.505	1.139	1.1464	1.3885	3.27	4.052
36.0	1.643	1.216	1.1562	1.3922	3.63	4.621
38.0	1.791	1.295	1.1663	1.3960	4.02	5.315
40.0	1.948	1.375	1.1765	1.3999	4.45	6.162
42.0	2.116	1.456	1.1868	1.4038	4.93	7.234
44.0	2.295	1.539	1.1972	1.4078		8.596
46.0	2.489	1.623	1.2079	1.4118		10.301
48.0	2.697	1.709	1.2186	1.4159		12.515
50.0	2.921	1.796	1.2295	1.4201		15.431
60.0	4.382	2.255	1.2864	1.4419		58.487
70.0	6.817	2.755	1.3472	1.4654		481.561
80.0	11.686	3.299	1.4117	1.4906		

Mass %	m/mol kg^{-1}	c/mol L^{-1}	ρ/g cm^{-3}	n	Δ/°C	η/mPa s
84.0	15.337	3.530	1.4383	1.5010		
Sulfuric acid						
0.5	0.051	0.051	1.0016	1.3336	0.21	1.010
1.0	0.103	0.102	1.0049	1.3342	0.42	1.019
2.0	0.208	0.206	1.0116	1.3355	0.80	1.036
3.0	0.315	0.311	1.0183	1.3367	1.17	1.059
4.0	0.425	0.418	1.0250	1.3379	1.60	1.085
5.0	0.537	0.526	1.0318	1.3391	2.05	1.112
6.0	0.651	0.635	1.0385	1.3403	2.50	1.136
7.0	0.767	0.746	1.0453	1.3415	2.95	1.159
8.0	0.887	0.858	1.0522	1.3427	3.49	1.182
9.0	1.008	0.972	1.0591	1.3439	4.08	1.206
10.0	1.133	1.087	1.0661	1.3451	4.64	1.230
11.0	1.260	1.204	1.0731	1.3463	5.25	1.256
12.0	1.390	1.322	1.0802	1.3475	5.93	1.282
13.0	1.524	1.441	1.0874	1.3488	6.67	1.309
14.0	1.660	1.563	1.0947	1.3500	7.49	1.337
15.0	1.799	1.685	1.1020	1.3513	8.35	1.367
16.0	1.942	1.810	1.1094	1.3525	9.26	1.399
17.0	2.088	1.936	1.1169	1.3538	10.23	1.434
18.0	2.238	2.064	1.1245	1.3551	11.29	1.470
19.0	2.392	2.193	1.1321	1.3563	12.43	1.508
20.0	2.549	2.324	1.1398	1.3576	13.64	1.546
22.0	2.876	2.592	1.1554	1.3602	16.48	1.624
24.0	3.220	2.866	1.1714	1.3628	19.85	1.706
26.0	3.582	3.147	1.1872	1.3653	24.29	1.797
28.0	3.965	3.435	1.2031	1.3677	29.65	1.894
30.0	4.370	3.729	1.2191	1.3701	36.21	2.001
32.0	4.798	4.030	1.2353	1.3725	44.76	2.122
34.0	5.252	4.339	1.2518	1.3749	55.28	2.255
36.0	5.735	4.656	1.2685	1.3773		2.392
38.0	6.249	4.981	1.2855	1.3797		2.533
40.0	6.797	5.313	1.3028	1.3821		2.690
42.0	7.383	5.655	1.3205	1.3846		2.872
44.0	8.011	6.005	1.3386	1.3870		3.073
46.0	8.685	6.364	1.3570	1.3895		3.299
48.0	9.411	6.734	1.3759	1.3920		3.546
50.0	10.196	7.113	1.3952	1.3945		3.826
52.0	11.045	7.502	1.4149	1.3971		4.142
54.0	11.969	7.901	1.4351	1.3997		4.499
56.0	12.976	8.312	1.4558	1.4024		4.906
58.0	14.080	8.734	1.4770	1.4050		5.354
60.0	15.294	9.168	1.4987	1.4077		5.917
62.0	16.635	9.608	1.5200			
64.0	18.126	10.063	1.5421			
66.0	19.792	10.529	1.5646			
68.0	21.666	11.006	1.5874			
70.0	23.790	11.494	1.6105			
72.0	26.218	11.994	1.6338			
74.0	29.019	12.505	1.6574			
76.0	32.287	13.026	1.6810			
78.0	36.149	13.554	1.7043			
80.0	40.783	14.088	1.7272			
82.0	46.447	14.623	1.7491			
84.0	53.528	15.153	1.7693			
86.0	62.631	15.671	1.7872			
88.0	74.769	16.170	1.8022			
90.0	91.762	16.649	1.8144			
92.0	117.251	17.109	1.8240			
94.0	159.734	17.550	1.8312			

Mass %	m/mol kg^{-1}	c/mol L^{-1}	ρ/g cm^{-3}	n	Δ/°C	η/mPa s
96.0	244.698	17.966	1.8355			
98.0	499.592	18.346	1.8361			
100.0		18.663	1.8305			
Trichloroacetic acid						
0.5	0.031	0.031	1.0008	1.3337	0.11	1.011
1.0	0.062	0.061	1.0034	1.3343	0.21	1.021
2.0	0.125	0.123	1.0083	1.3356	0.42	1.044
3.0	0.189	0.186	1.0133	1.3369	0.64	1.069
4.0	0.255	0.249	1.0182	1.3381	0.86	1.096
5.0	0.322	0.313	1.0230	1.3394	1.08	1.123
6.0	0.391	0.377	1.0279	1.3406	1.30	1.150
7.0	0.461	0.442	1.0328	1.3418	1.53	1.177
8.0	0.532	0.508	1.0378	1.3431	1.76	1.204
9.0	0.605	0.574	1.0428	1.3444	1.99	1.233
10.0	0.680	0.641	1.0479	1.3456	2.23	1.263
12.0	0.835	0.777	1.0583	1.3483	2.73	1.326
14.0	0.996	0.916	1.0692	1.3510	3.26	1.393
16.0	1.166	1.058	1.0806	1.3539	3.82	1.462
18.0	1.343	1.203	1.0921	1.3568		1.533
20.0	1.530	1.351	1.1035	1.3597		1.608
24.0	1.933	1.654	1.1260	1.3652		1.768
28.0	2.380	1.968	1.1485	1.3705		1.935
32.0	2.880	2.294	1.1713	1.3759		2.118
36.0	3.443	2.632	1.1947	1.3813		2.320
40.0	4.080	2.984	1.2188	1.3868		1.543
44.0	4.809	3.349	1.2435	1.3923		2.797
48.0	5.650	3.726	1.2682	1.3977		3.076
50.0	6.120	3.918	1.2803	1.4003		3.225
Tris(hydroxymethyl)methylamine						
0.5	0.041	0.041	0.9994	1.3337	0.08	1.014
1.0	0.083	0.083	1.0006	1.3344	0.16	1.027
2.0	0.168	0.166	1.0030	1.3359	0.31	1.054
3.0	0.255	0.249	1.0054	1.3374	0.47	1.083
4.0	0.344	0.333	1.0078	1.3388	0.64	1.115
5.0	0.434	0.417	1.0103	1.3403	0.80	1.148
6.0	0.527	0.502	1.0128	1.3418	0.97	1.182
7.0	0.621	0.587	1.0153	1.3433	1.15	1.218
8.0	0.718	0.672	1.0179	1.3448	1.33	1.256
9.0	0.816	0.758	1.0204	1.3463	1.51	1.295
10.0	0.917	0.844	1.0230	1.3478	1.70	1.337
12.0	1.126	1.019	1.0282	1.3508	2.08	1.427
14.0	1.344	1.194	1.0335	1.3539	2.47	1.527
16.0	1.572	1.372	1.0389	1.3570	2.90	1.642
18.0	1.812	1.552	1.0443	1.3601	3.36	1.772
20.0	2.064	1.733	1.0498	1.3633	3.85	1.920
22.0	2.328	1.917	1.0554	1.3665		2.083
24.0	2.607	2.102	1.0610	1.3697		2.261
26.0	2.900	2.289	1.0666	1.3730		2.464
28.0	3.210	2.478	1.0723	1.3763		2.705
30.0	3.538	2.670	1.0781	1.3797		2.998
32.0	3.885	2.863	1.0839	1.3831		3.344
34.0	4.253	3.058	1.0897	1.3865		3.736
36.0	4.643	3.256	1.0956	1.3900		4.177
38.0	5.059	3.456	1.1016	1.3935		4.667
40.0	5.503	3.657	1.1076	1.3970		5.208
Urea						
0.5	0.084	0.083	0.9995	1.3337	0.16	1.007
1.0	0.168	0.167	1.0007	1.3344	0.31	1.010
2.0	0.340	0.334	1.0033	1.3358	0.62	1.012
3.0	0.515	0.502	1.0058	1.3372	0.93	1.017
4.0	0.694	0.672	1.0085	1.3387	1.24	1.025
5.0	0.876	0.842	1.0111	1.3401	1.55	1.033
6.0	1.063	1.013	1.0138	1.3416	1.88	1.041
7.0	1.253	1.185	1.0165	1.3431	2.22	1.049
8.0	1.448	1.358	1.0192	1.3446	2.56	1.057
9.0	1.647	1.531	1.0220	1.3461	2.91	1.065
10.0	1.850	1.706	1.0248	1.3476	3.26	1.074
11.0	2.058	1.882	1.0276	1.3491	3.61	1.083
12.0	2.270	2.059	1.0304	1.3506	3.95	1.091
13.0	2.488	2.236	1.0332	1.3521	4.30	1.100
14.0	2.710	2.415	1.0360	1.3537	4.66	1.109
15.0	2.938	2.594	1.0388	1.3552	5.02	1.119
16.0	3.171	2.775	1.0417	1.3568	5.40	1.130
17.0	3.410	2.956	1.0445	1.3583	5.79	1.141
18.0	3.655	3.139	1.0473	1.3599	6.19	1.153
19.0	3.906	3.322	1.0502	1.3614	6.59	1.165
20.0	4.163	3.506	1.0530	1.3629	7.00	1.178
22.0	4.696	3.878	1.0586	1.3661	7.81	1.205
24.0	5.258	4.253	1.0643	1.3692	8.64	1.235
26.0	5.850	4.632	1.0699	1.3723	9.52	1.266
28.0	6.475	5.014	1.0756	1.3754	10.45	1.298
30.0	7.136	5.401	1.0812	1.3785	11.40	1.332
32.0	7.835	5.791	1.0869	1.3817	12.34	1.371
34.0	8.577	6.185	1.0926	1.3848	13.27	1.413
36.0	9.366	6.584	1.0984	1.3881	14.20	1.459
38.0	10.205	6.988	1.1044	1.3913	15.11	1.509
40.0	11.100	7.397	1.1106	1.3947	15.99	1.565
42.0	12.057	7.812	1.1171	1.3982	16.83	1.629
44.0	13.082	8.234	1.1239	1.4018	17.62	1.700
46.0	14.183	8.665	1.1313	1.4056		1.780
Zinc sulfate						
0.5	0.031	0.031	1.0034	1.3339	0.08	1.021
1.0	0.063	0.062	1.0085	1.3348	0.15	1.040
2.0	0.126	0.126	1.0190	1.3366	0.28	1.081
3.0	0.192	0.191	1.0296	1.3384	0.41	1.126
4.0	0.258	0.258	1.0403	1.3403	0.53	1.175
5.0	0.326	0.326	1.0511	1.3421	0.65	1.227
6.0	0.395	0.395	1.0620	1.3439	0.77	1.283
7.0	0.466	0.465	1.0730	1.3457	0.89	1.341
8.0	0.539	0.537	1.0842	1.3475	1.01	1.403
9.0	0.613	0.611	1.0956	1.3494	1.14	1.470
10.0	0.688	0.686	1.1071	1.3513	1.27	1.545
11.0	0.766	0.762	1.1188	1.3532	1.41	1.627
12.0	0.845	0.840	1.1308	1.3551	1.55	1.716
13.0	0.926	0.920	1.1429	1.3570	1.71	1.814
14.0	1.008	1.002	1.1553	1.3590	1.89	1.918
15.0	1.093	1.085	1.1679	1.3610	2.09	2.031
16.0	1.180	1.170	1.1806	1.3630	2.31	2.152

SOLUBILITY OF SELECTED GASES IN WATER

L. H. Gevantman

The values in this table are taken almost exclusively from the International Union of Pure and Applied Chemistry "Solubility Data Series." Unless noted, they comprise evaluated data fitted to a smoothing equation. The data at each temperature are then derived from the smoothing equation which expresses the mole fraction solubility X_1 of the gas in solution as:

$$\ln X_1 = A + B/T^* + C \ln T^*$$

where

$$T^* = T/100\ \mathrm{K}$$

All values refer to a partial pressure of the gas of 101.325 kPa (one atmosphere).

The equation constants, the standard deviation for $\ln X_1$ (except where noted), and the temperature range over which the equation applies are given in the column headed Equation constants. There are two exceptions. The equation for methane has an added term, DT^*. The equation for H_2Se and H_2S takes the form,

$$\ln X_1 = A + B/T + C \ln T + DT$$

where T is the temperature in kelvin.

Solubilities given for those gases that react with water, namely ozone, nitrogen oxides, chlorine and its oxides, hydrogen sulfide, hydrogen selenide, and sulfur dioxide, are recorded as bulk solubilities; i.e., all chemical species of the gas and its reaction products with water are included. The solubility of carbon dioxide (CO_2) can be found in "Solubility of Carbon Dioxide in Water at Various Temperatures and Pressures" in this section.

Solubility of Gases in Water — Constants in the Solubility Equation and Solubility Values at Selected Temperatures

T/K	Solubility (X_1)	Equation constants	Ref.
Hydrogen (H_2) M_r = 2.01588			
288.15	1.510×10^{-5}	$A = -48.1611$	1
293.15	1.455×10^{-5}	$B = 55.2845$	1
298.15	1.411×10^{-5}	$C = 16.8893$	1
301.15	1.377×10^{-5}	Std. dev. = ±0.54%	1
308.15	1.350×10^{-5}	Temp. range = 273.15—353.15	1
Deuterium (D_2) M_r = 4.0282			
283.15	$1.675 \times 10^{-5} \pm 0.57\%$	Averaged experimental values	1
288.15	$1.595 \times 10^{-5} \pm 0.57\%$		1
293.15	$1.512 \times 10^{-5} \pm 0.78\%$	Temp. range = 278.15—303.15	1
298.15	$1.460 \times 10^{-5} \pm 0.52\%$		1
303.15	$1.395 \times 10^{-5} \pm 0.37\%$		1
Helium (He) A_r = 4.0026			
288.15	7.123×10^{-6}	$A = -41.4611$	2
293.15	7.044×10^{-6}	$B = 42.5962$	2
298.15	6.997×10^{-6}	$C = 14.0094$	2
303.15	6.978×10^{-6}	Std. dev. = ±0.54%	2
308.15	6.987×10^{-6}	Temp. range = 273.15—348.15	2
Neon (Ne) A_r = 20.1797			
288.15	8.702×10^{-6}	$A = -52.8573$	2
293.15	8.395×10^{-6}	$B = 61.0494$	2
298.15	8.152×10^{-6}	$C = 18.9157$	2
303.15	7.966×10^{-6}	Std. dev. = ±0.47%	2
308.15	7.829×10^{-6}	Temp. range = 273.15—348.15	2
Argon (Ar) A_r = 39.948			
288.15	3.025×10^{-5}	$A = -57.6661$	3
293.15	2.748×10^{-5}	$B = 74.7627$	3
298.15	2.519×10^{-5}	$C = 20.1398$	3
303.15	2.328×10^{-5}	Std. dev. = ±0.26%	3
308.15	2.169×10^{-5}	Temp. range = 273.15—348.15	3
Krypton (Kr) A_r = 83.80			
288.15	5.696×10^{-5}	$A = -66.9928$	4
293.15	5.041×10^{-5}	$B = 91.0166$	4
298.15	4.512×10^{-5}	$C = 24.2207$	4
303.15	4.079×10^{-5}	Std. dev. = ±0.32%	4
308.15	3.725×10^{-5}	Temp. range = 273.15—353.15	4
Xenon (Xe) A_r = 131.29			
288.15	10.519×10^{-5}	$A = -74.7398$	4
293.15	9.051×10^{-5}	$B = 105.210$	4
298.15	7.890×10^{-5}	$C = 27.4664$	4
303.15	6.961×10^{-5}	Std. dev. = ±0.35%	4
308.15	6.212×10^{-5}	Temp. range = 273.15—348.15	4
Radon (^{222}Rn) A_r = 222			
288.15	2.299×10^{-4}	$A = -90.5481$	4
293.15	1.945×10^{-4}	$B = 130.026$	4
298.15	1.671×10^{-4}	$C = 35.0047$	4
303.15	1.457×10^{-4}	Std. dev. = ±1.02%	4
308.15	1.288×10^{-4}	Temp. range = 273.15—373.15	4
Oxygen (O_2) M_r = 31.9988			
288.15	2.756×10^{-5}	$A = -66.7354$	5
293.15	2.501×10^{-5}	$B = 87.4755$	5
298.15	2.293×10^{-5}	$C = 24.4526$	5
303.15	2.122×10^{-5}	Std. dev. = ±0.36%	5
308.15	1.982×10^{-5}	Temp. range = 273.15—348.15	5
Ozone (O_3) M_r = 47.9982			
293.15	$1.885 \times 10^{-6} \pm 10\%$	Derived from Henry's Law Constant Equation	5
	pH = 7.0		5
Nitrogen (N_2) M_r = 28.0134			
288.15	1.386×10^{-5}	$A = -67.3877$	6
293.15	1.274×10^{-5}	$B = 86.3213$	6
298.15	1.183×10^{-5}	$C = 24.7981$	6
303.15	1.108×10^{-5}	Std. dev. = ±0.72%	6
308.15	1.047×10^{-5}	Temp. range = 273.15—348.15	6
Nitrous oxide (N_2O) M_r = 44.0129			
288.15	5.948×10^{-4}	$A = -60.7467$	7
293.15	5.068×10^{-4}	$B = 88.8280$	7
298.15	4.367×10^{-4}	$C = 21.2531$	7
303.15	3.805×10^{-4}	Std. dev. = ±1.2%	7
308.15	3.348×10^{-4}	Temp. range = 273.15—313.15	7

T/K	Solubility (X_1)	Equation constants	Ref.
Nitric oxide (NO) M_r = 30.0061			
288.15	4.163×10^{-5}	$A = -62.8086$	7
293.15	3.786×10^{-5}	$B = 82.3420$	7
298.15	3.477×10^{-5}	$C = 22.8155$	7
303.15	3.222×10^{-5}	Std. dev. = ±0.76%	7
308.15	3.012×10^{-5}	Temp. range = 273.15—358.15	7
Carbon monoxide (CO) M_r = 28.0104			
288.15	2.095×10^{-5}	Derived from Henry's Law Constant Equation	8
293.15	1.918×10^{-5}		8
298.15	1.774×10^{-5}	Std. dev. = ±0.043%	8
303.15	1.657×10^{-5}	Temp. range = 273.15—328.15	8
308.15	1.562×10^{-5}		8
Hydrogen selenide (H_2Se) M_r = 80.976			
288.15	1.80×10^{-3}	$A = 9.15$	9
298.15	1.49×10^{-3}	$B = 974$	9
308.15	1.24×10^{-3}	$C = -3.542$	9
		$D = 0.0042$	9
		Std. dev. = $\pm 2.3 \times 10^{-5}$	9
		Temp. range = 288.15—343.15	9
Hydrogen sulfide (H_2S) M_r = 34.082			
288.15	2.335×10^{-3}	$A = -24.912$	9
293.15	2.075×10^{-3}	$B = 3477$	9
298.15	1.85×10^{-3}	$C = 0.3993$	9
303.15	1.66×10^{-3}	$D = 0.0157$	9
308.15	1.51×10^{-3}	Std. dev. = $\pm 6.5 \times 10^{-5}$	9
		Temp. range = 283.15—603.15	9
Sulfur dioxide (SO_2) M_r = 64.0648			
288.15	3.45×10^{-2}	$A = -25.2629$	10
293.15	2.90×10^{-2}	$B = 45.7552$	10
298.15	2.46×10^{-2}	$C = 5.6855$	10
303.15	2.10×10^{-2}	Std. dev. = ±1.8%	10
308.15	1.80×10^{-2}	Temp. range = 278.15—328.15	10
Chlorine (Cl_2) M_r = 70.9054			
283.15	$2.48 \times 10^{-3} \pm 2\%$	Experimental data	10
293.15	$1.88 \times 10^{-3} \pm 2\%$	Temp. range = 283.15—333.15	10
303.15	$1.50 \times 10^{-3} \pm 2\%$		10
313.15	$1.23 \times 10^{-3} \pm 2\%$		10
Chlorine monoxide (Cl_2O) M_r = 86.9048			
273.15	$5.25 \times 10^{-1} \pm 1\%$	Experimental data	10
276.61	$4.54 \times 10^{-1} \pm 1\%$	Temp. range = 273.15—293.15	10
283.15	$4.273 \times 10^{-1} \pm 1\%$		10
293.15	$3.353 \times 10^{-1} \pm 1\%$		10

T/K	Solubility (X_1)	Equation constants	Ref.
Chlorine dioxide (ClO_2) M_r = 67.4515			
288.15	2.67×10^{-2}	$A = 7.9163$	10
293.15	2.20×10^{-2}	$B = 0.4791$	10
298.15	1.823×10^{-2}	$C = 11.0593$	10
303.15	1.513×10^{-2}	Std. dev. = ±4.6%	10
308.15	1.259×10^{-2}	Temp. range = 283.15—333.15	10
Methane (CH_4) M_r = 16.0428			
288.15	3.122×10^{-5}	$A = -115.6477$	11
293.15	2.806×10^{-5}	$B = 155.5756$	11
298.15	2.552×10^{-5}	$C = 65.2553$	11
303.15	2.346×10^{-5}	$D = -6.1698$	11
308.15	2.180×10^{-5}	Std. dev. = ±0.056%	11
		Temp. range = 273.15—328.15	11
Ethane (C_2H_6) M_r = 30.0696			
288.15	4.556×10^{-5}	$A = -90.8225$	12
293.15	3.907×10^{-5}	$B = 126.9559$	12
298.15	3.401×10^{-5}	$C = 34.7413$	12
303.15	3.002×10^{-5}	Std. dev. = ±0.13%	12
308.15	2.686×10^{-5}	Temp. range = 273.15—323.15	12
Ethylene (C_2H_4) M_r = 28.053			
288.15	11.01×10^{-5}	$A = -66.9156$	13
293.15	9.70×10^{-5}	$B = 92.2101$	13
298.15	8.65×10^{-5}	$C = 24.3792$	13
303.15	7.79×10^{-5}	Std. dev. = ±0.39%	13
308.15	7.08×10^{-5}	Temp. range = 273.15—323.15	13
Propane (C_3H_8) M_r = 44.097			
288.15	3.813×10^{-5}	$A = -102.044$	14
293.15	3.200×10^{-5}	$B = 144.345$	14
298.15	2.732×10^{-5}	$C = 39.4740$	14
303.15	2.370×10^{-5}	Std. dev. = ±0.012%	14
308.15	2.088×10^{-5}	Temp. range = 273.15—347.15	14
Butane (C_4H_{10}) M_r = 58.123			
288.15	3.274×10^{-5}	$A = -102.029$	14
293.15	2.687×10^{-5}	$B = 146.040$	14
298.15	2.244×10^{-5}	$C = 38.7599$	14
303.15	1.906×10^{-5}	Std. dev. = ±0.026%	14
308.15	1.645×10^{-5}	Temp. range = 273.15—349.15	14
Isobutane (C_4H_{10}) M_r = 58.123			
288.15	2.333×10^{-5}	$A = -129.714$	14
293.15	1.947×10^{-5}	$B = 183.044$	14
298.15	1.659×10^{-5}	$C = 53.4651$	14
303.15	1.443×10^{-5}	Std. dev. = ±0.034%	14
308.15	1.278×10^{-5}	Temp. range = 278.15—318.15	14

References

1. Young, C. L., Ed., *IUPAC Solubility Data Series*, Vol. 5/6, Hydrogen and Deuterium, Pergamon Press, Oxford, England, 1981.
2. Clever, H. L., Ed., *IUPAC Solubility Data Series*, Vol. 1, Helium and Neon, Pergamon Press, Oxford, England, 1979.
3. Clever, H. L., Ed., *IUPAC Solubility Data Series*, Vol. 4, Argon, Pergamon Press, Oxford, England, 1980.
4. Clever, H. L. Ed., *IUPAC Solubility Data Series*, Vol. 2, Krypton, Xenon and Radon, Pergamon Press, Oxford, England, 1979.
5. Battino, R., Ed., *IUPAC Solubility Data Series*, Vol. 7, Oxygen and Ozone, Pergamon Press, Oxford, England, 1981.
6. Battino, R., Ed., *IUPAC Solubility Data Series*, Vol. 10, Nitrogen and Air, Pergamon Press, Oxford, England, 1982.
7. Young, C. L., Ed., *IUPAC Solubility Data Series*, Vol. 8, Oxides of Nitrogen, Pergamon Press, Oxford, England, 1981.
8. Cargill, R. W., Ed., *IUPAC Solubility Data Series*, Vol. 43, Carbon Monoxide, Pergamon Press, Oxford, England, 1990.
9. Fogg, P. G. T. and Young, C. L., Eds., *IUPAC Solubility Data Series*, Vol. 32, Hydrogen Sulfide, Deuterium Sulfide, and Hydrogen Selenide, Pergamon Press, Oxford, England, 1988.
10. Young, C. L., Ed., *IUPAC Solubility Data Series*, Vol. 12, Sulfur Dioxide, Chlorine, Fluorine and Chlorine Oxides, Pergamon Press, Oxford, England, 1983.
11. Clever, H. L. and Young, C. L., Eds., *IUPAC Solubility Data Series*, Vol. 27/28, Methane, Pergamon Press, Oxford, England, 1987.
12. Hayduk, W., Ed., *IUPAC Solubility Data Series*, Vol. 9, Ethane, Pergamon Press, Oxford, England, 1982.
13. Hayduk, W., Ed., *IUPAC Solubility Data Series*, Vol. 57, Ethene, Oxford University Press, Oxford, England, 1994.
14. Hayduk, W., Ed., *IUPAC Solubility Data Series*, Vol. 24, Propane, Butane and 2-Methylpropane, Pergamon Press, Oxford, England, 1986.

SOLUBILITY OF CARBON DIOXIDE IN WATER AT VARIOUS TEMPERATURES AND PRESSURES

Given its importance, the solubility of CO_2 in water is presented in two tables. Table 1 contains the mole fraction solubility X_1 of CO_2 in water (column 2) for common temperatures (column 1) at a CO_2 partial pressure of 101.325 kPa. The values in this table are taken from Ref. 1, the International Union of Pure and Applied Chemistry "Solubility Data Series," and comprise evaluated data fitted to a Henry's Law Constant equation described in this reference. The right-hand column of Table 1 contains the standard deviation and temperature range of applicability.

Table 2 gives the solubility of CO_2 in water, expressed as a mole fraction of CO_2 in the liquid phase, for pressures up to 100 kPa and temperatures of 0 to 100 °C. (Note that 1 standard atmosphere equals 101.325 kPa.) Refs. 2-4 give data over a wider range of temperatures and pressures. The estimated uncertainty is about 2%.

CO_2 reacts with water and all solubility values are for bulk solubilities; i.e., all chemical species of CO_2 and its reaction products with water are included in both tables.

References

1. Scharin, P., Ed., *IUPAC Solubility Data Series*, Vol. 62, Carbon Dioxide in Water and Aqueous Electrolyte Solutions, Oxford University Press, Oxford, England, 1996.
2. Carroll, J. J., Slupsky, J. D., and Mather, A. E., *J. Phys. Chem. Ref. Data*, 20, 1201, 1991. <https://doi.org/10.1063/1.555900>
3. Fernandez-Prini, R. and Crovetto, R., *J. Phys. Chem. Ref. Data*, 18, 1231, 1989. <https://doi.org/10.1063/1.555834>
4. Crovetto, R., *J. Phys. Chem. Ref. Data*, 20, 575, 1991. <https://doi.org/10.1063/1.555905>

TABLE 1. Solubility of CO_2 in Water at Common Temperatures and 1 atm Partial Pressure of CO_2 – as Mole Fraction

T/K	Solubility (X_1)	Equation constants
288.15	8.21×10^{-4}	Derived from Henry's Law Constant Equation
293.15	7.07×10^{-4}	
298.15	6.15×10^{-4}	Std. dev. = ±1.1%
303.15	5.41×10^{-4}	Temp. range = 273.15–353.15
308.15	4.80×10^{-4}	

TABLE 2. Solubility of CO_2 in Water as a Function of CO_2 Partial Pressure - as 1000 × Mole Fraction of CO_2 in Liquid Phase

t/°C	5 kPa	10 kPa	20 kPa	30 kPa	40 kPa	50 kPa	100 kPa
0	0.067	0.135	0.269	0.404	0.538	0.671	1.337
5	0.056	0.113	0.226	0.338	0.451	0.564	1.123
10	0.048	0.096	0.191	0.287	0.382	0.477	0.950
15	0.041	0.082	0.164	0.245	0.327	0.409	0.814
20	0.035	0.071	0.141	0.212	0.283	0.353	0.704
25	0.031	0.062	0.123	0.185	0.247	0.308	0.614
30	0.027	0.054	0.109	0.163	0.218	0.271	0.541
35	0.024	0.048	0.097	0.145	0.193	0.242	0.481
40	0.022	0.043	0.087	0.130	0.173	0.216	0.431
45	0.020	0.039	0.078	0.117	0.156	0.196	0.389
50	0.018	0.036	0.071	0.107	0.142	0.178	0.354
55	0.016	0.033	0.065	0.098	0.131	0.163	0.325
60	0.015	0.030	0.060	0.090	0.121	0.150	0.300
65	0.014	0.028	0.056	0.084	0.112	0.140	0.279
70	0.013	0.026	0.052	0.079	0.105	0.131	0.261
75	0.012	0.025	0.049	0.074	0.099	0.123	0.245
80	0.012	0.023	0.047	0.070	0.093	0.116	0.232
85	0.011	0.022	0.044	0.067	0.089	0.111	0.221
90	0.011	0.021	0.042	0.064	0.085	0.106	0.211
95	0.010	0.020	0.041	0.061	0.082	0.102	0.203
100	0.010	0.020	0.039	0.059	0.079	0.098	0.196

AQUEOUS SOLUBILITY AND HENRY'S LAW CONSTANTS OF ORGANIC COMPOUNDS

The solubility in water of about 1300 organic compounds, including many compounds of environmental interest, is tabulated here. When data are available, values are given at several temperatures between 0 °C and 100 °C. Solids, liquids, and gases are included; additional data on gases can be found in the table "Solubility of Selected Gases in Water" in this section.

Solubility of solids is defined as the concentration of the compound in a solution that is in equilibrium with the solid phase at the specified temperature and one atmosphere pressure. For liquids whose water mixtures separate into two phases, the solubility given here is the concentration of the specified compound in the water-rich phase at equilibrium. In the case of gases (i.e., compounds whose vapor pressure at the specified temperature exceeds one atmosphere) the solubility is defined here as the concentration in the water phase when the partial pressure of the compound above the solution is 101.325 kPa (1 atm). Values for gases are marked with an asterisk.

The solubility values in this table are expressed as mass percent of solute, $s = 100w_2$, where the mass fraction w_2 is defined as

$$w_2 = m_2/(m_1 + m_2)$$

where m_2 is the mass of solute and m_1 the mass of water. For convenience, the solubility expressed in grams of solute that will dissolve in 1 kilogram of water is tabulated in the adjacent column to mass percent. For compounds with low solubility (e.g., $s < 1\%$), that column is, to a high approximation, numerically identical to the solubility expressed in grams of solute per liter of solution.

The mass fraction w_2 is related to other common measures of solubility as follows:

Molality: $m_2 = 1000\, w_2/M_2(1 - w_2)$
Molarity: $c_2 = 1000\rho w_2/M_2$
Mole fraction: $x_2 = (w_2/M_2)/\{(w_2/M_2) + (1 - w_2)/M_1\}$
Mass of solute per 100 g of H_2O: $100w_2/(1 - w_2)$
Mass of solute per liter of solution: $1000\rho w_2$

Here, M_2 is the molar mass of the solute, $M_1 = 18.015$ g/mol is the molar mass of water, and ρ is the density of the solution in g/mL.

Data have been selected from evaluated sources wherever possible, in particular the *IUPAC Solubility Data Series*. Many values come from experimental measurements reported in the *Journal of Chemical and Engineering Data* and the *Journal of Chemical Thermodynamics*, as well as critical review papers in the *Journal of Physical and Chemical Reference Data*. The primary source for each value is listed in the column following the solubility values; additional references of interest are sometimes given. Many of the references contain solubility data at other temperatures and pH values and in the presence of other compounds. The user is cautioned that wide variations of data are found in the literature for the lower solubility compounds. The references should be consulted for more information on these compounds.

The table also contains values of the Henry's Law constant k_H, which provides a measure of the partition of a substance between the atmosphere and the aqueous phase. Here, k_H is defined as the limit of p_2/c_2 as the concentration approaches zero, where p_2 is the partial pressure of the solute above the solution and c_2 is the concentration in the solution at equilibrium (other formulations of Henry's Law are often used (see Ref. 5, and especially Ref. 128). The values of k_H listed here are based on direct experimental measurement whenever available, but many of them are simply calculated as the ratio of the pure compound vapor pressure to the solubility. This approximation is reliable only for compounds of very low solubility. In fact, values of k_H found in the literature frequently differ by a factor of two or three, and variations over an order of magnitude are not unusual (Refs. 5, 128, and 129). Therefore, the data given here should be taken only as a rough indication of the true Henry's Law constant, which is difficult to measure precisely. Refs. 128 and 129 have a more complete compilation of measured and calculated Henry's Law constants.

All values of k_H refer to 25 °C. If the vapor pressure of the compound at 25 °C is greater than one atmosphere, it can be assumed that the k_H value has been calculated as $101.325/c_2$. The source of the Henry's Law data is given in the last column. The air-water partition coefficient (i.e., ratio of air concentration to water concentration when both are expressed in the same units) is equal to k_H/RT or $k_H/2.48$ in the units used here.

The definitions of the columns of the table are as follows.

Column heading	Definition
Name	Compound name, listed alphabetically by systematic name
Mol. form.	Molecular formula of compound in Hill Order
CAS Reg. No.	Chemical Abstracts Service Registry Number
Mol. wt.	Molecular weight of the compound; calculated with 2009 IUPAC values
***t*/°C**	Temperature of measurement
***s*/mass%**	Solubility of compound in water, in units of mass percent
***s*/g kg^{-1} H$_2$O**	Solubility of compound in water, in units of g kg^{-1} H_2O
Ref.	Reference number for solubility data
k_H/kPa m^3 mol^{-1}	Henry's Law constant, k_H, in units of kPa m^3 mol^{-1}
Ref.	Reference number for Henry's Law constant, k_H

We acknowledge Rolf Sander and Jennifer E. Searles for their help with updating the Henry's Law constant data.

References

1. *Solubility Data Series, International Union of Pure and Applied Chemistry, Vol. 15*, Pergamon Press, Oxford, 1982.
2. *Solubility Data Series, International Union of Pure and Applied Chemistry, Vol. 20*, Pergamon Press, Oxford, 1985.
3. *Solubility Data Series, International Union of Pure and Applied Chemistry, Vol. 37*, Pergamon Press, Oxford, 1988.
4. *Solubility Data Series, International Union of Pure and Applied Chemistry, Vol. 38*, Pergamon Press, Oxford, 1988.
5. Mackay, D., and Shiu, W. Y., *J. Phys. Chem. Ref. Data*, 10, 1175, 1981. <https://doi.org/10.1063/1.555654>
6. Pearlman, R. S., and Yalkowsky, S. H., *J. Phys. Chem. Ref. Data*, 13, 975, 1984. <https://doi.org/10.1063/1.555712>
7. Shiu, W. Y., and Mackay, D., *J. Phys. Chem. Ref. Data*, 15, 911, 1986. <https://doi.org/10.1063/1.555755>
8. Varhanickova, D., Lee, S. C., Shiu, W. Y., and Mackay, D., *J. Chem. Eng. Data*, 40, 620, 1995. <https://doi.org/10.1021/je00019a018>
9. Miller, M. M., Ghodbane, S., Wasik, S. P., Tewari, Y. B., and Martire, D. E., *J. Chem. Eng. Data*, 29, 184, 1984. <https://doi.org/10.1021/je00036a027>
10. Riddick, J. A., Bunger, W. B., and Sakano, T. K., *Organic Solvents, Fourth Edition*, John Wiley & Sons, New York, 1986.

11. Mackay, D., Shiu, W. Y., and Ma, K. C., *Illustrated Handbook of Physical-Chemical Properties and Environmental Fate for Organic Chemicals, Vol. I*, Lewis Publishers/CRC Press, Boca Raton, FL, 1992.
12. Mackay, D., Shiu, W. Y., and Ma, K. C., *Illustrated Handbook of Physical-Chemical Properties and Environmental Fate for Organic Chemicals, Vol. II*, Lewis Publishers/CRC Press, Boca Raton, FL, 1992.
13. Mackay, D., Shiu, W. Y., and Ma, K. C., *Illustrated Handbook of Physical-Chemical Properties and Environmental Fate for Organic Chemicals, Vol. III*, Lewis Publishers/CRC Press, Boca Raton, FL, 1993.
14. Horvath, A. L., *Halogenated Hydrocarbons*, Marcel Dekker, New York, 1982.
15. Howard, P. H., *Handbook of Environmental Fate and Exposure Data for Organic Chemicals, Vol. I*, Lewis Publishers/CRC Press, Boca Raton, FL, 1989.
16. Howard, P. H., *Handbook of Environmental Fate and Exposure Data for Organic Chemicals, Vol. II*, Lewis Publishers/CRC Press, Boca Raton, FL, 1990.
17. Banergee, S., Yalkowsky, S. H., and Valvani, S. C., *Environ. Sci. Technol.*, 14, 1227, 1980. <https://doi.org/10.1021/es60170a013>
18. Gevantman, L. H., in *CRC Handbook of Chemistry and Physics, 90th Edition*, pp. 8–80, CRC Press, Boca Raton, FL, 2009.
19. Wilhelm, E., Battino, R., and Wilcock, R. J., *Chem. Rev.* 77, 219, 1977. <https://doi.org/10.1021/cr60306a003>
20. Stephenson, R. M., *J. Chem. Eng. Data*, 37, 80, 1992. <https://doi.org/10.1021/je00005a024>
21. Stephenson, R. M., Stuart, J., and Tabak, M., *J. Chem. Eng. Data*, 29, 287, 1984. <https://doi.org/10.1021/je00037a019>
22. Shiu, W.-Y., and Ma, K.-C, *J. Phys. Chem. Ref. Data*, 29, 41, 2000. <https://doi.org/10.1063/1.556055>
23. Lun, R., Varhanickova, D., Shiu, W.-Y., and Mackay, D., *J. Chem. Eng. Data*, 42, 951 (1997). <https://doi.org/10.1021/je970069v>
24. Huang, G.-L., Xiao, H., Chi, J., Shiu, W.-Y., and Mackay, D., *J. Chem. Eng. Data*, 45, 411, 2000. <https://doi.org/10.1021/je990262k>
25. Horvath, A. L., Getzen, F. W., and Maczynska, Z., *J. Phys. Chem. Ref. Data*, 28, 395, 2000 [IUPAC No. 67]. <https://doi.org/10.1063/1.556039>
26. Dawson, R. M. C., Elliott, D. C., Elliott, W. H., and Jones, K. M., *Data for Biochemical Research, Third Edition*, Clarendon Press, Oxford, 1986.
27. Stephen, H., and Stephen, T., *Solubilities of Organic and Inorganic Compounds*, MacMillan, New York, 1963.
28. Shiu, W.-Y., and Mackay, D., *J. Chem. Eng. Data* 42, 27, 1997. <https://doi.org/10.1021/je960218u>
29. Hinz, H.-J., ed., *Thermodynamic Data for Biochemistry and Biotechnology*, Springer-Verlag, Berlin, 1986.<https://doi.org/10.1007/978-3-642-71114-5>
30. Budavari, S., ed., *The Merck Index, Twelfth Edition*, Merck & Co., Rahway, NJ, 1996.
31. Bamford, H. A., Poster, D. L., and Baker, J. E., *J. Chem. Eng. Data*, 45, 1069, 2000. <https://doi.org/10.1021/je0000266>
32. Lide, D. R., and Milne, G. W. A., *Handbook of Data on Organic Compounds, Third Edition*, CRC Press, Boca Raton, FL, 1994.
33. Apelblat, A., and Manzurola, E., *J. Chem. Thermodynamics* 21, 1005, 1989.
34. Apelblat, A., and Manzurola, E., *J. Chem. Thermodynamics* 22, 289, 1990. <https://doi.org/10.1016/0021-9614(90)90201-Z>
35. Horvath, A. L., and Getzen, F. W., *J. Phys. Chem. Ref. Data* 28, 649, 1999 [IUPAC No. 68]. <https://doi.org/10.1063/1.556051]
36. Sazonov, V. P., Marsh, K. N., and Hefter, G. T., *J. Phys. Chem. Ref. Data* 29, 1165, 2000 [IUPAC No. 71]. <https://doi.org/10.1063/1.1329911>
37. Verbruggen, E. M. J., Hermens, J. L. M., and Tolls, J., *J. Phys. Chem. Ref. Data* 29, 1435, 2000. <https://doi.org/10.1063/1.1347983>
38. Sazonov, V. P., Shaw, D. G., and Marsh, K. N., *J. Phys. Chem. Ref. Data* 31, 1, 2002 [IUPAC No. 77]. <https://doi.org/10.1063/1.1417522>
39. Sazonov, V. P., and Shaw, D. G., *J. Phys. Chem. Ref. Data* 31, 989, 2002 [IUPAC No. 78]. <https://doi.org/10.1063/1.1494086>
40. Yalkowsky, S. H., and He, Y., *Handbook of Aqueous Solubility Data*, CRC Press, Boca Raton, FL, 2003. <https://doi.org/10.1201/9780203490396>
41. Shiu, W.-Y., and Ma, K.-C., *J. Phys. Chem. Ref. Data* 29, 387, 2000. <https://doi.org/10.1063/1.1286267>
42. Shaw, D. G., and Maczynski, A., *J. Phys. Chem. Ref. Data* 35, 687, 2006 [IUPAC No. 81, Part 11]. <https://doi.org/10.1063/1.2132315>
43. Nordstrom, F. L., and Rasmuson, A. C., *J. Chem. Eng. Data* 51, 1668, 2006. <https://doi.org/10.1021/je060134d>
44. Nordstrom, F. L., and Rasmuson, A. C., *J. Chem. Eng. Data* 51, 1775, 2006. <https://doi.org/10.1021/je060178m>
45. Sapoundjiev, D., Lorenz, H,. and Seidel-Morgenstern, A., *J. Chem. Eng. Data* 51, 1562, 2006. <https://doi.org/10.1021/je060053h>
46. Marche, C., Ferronato, C., and Jose, J., *J. Chem. Eng. Data* 48, 967, 2003. <https://doi.org/10.1021/je025659u>
47. Lu, J., Wang, X., Yang, X., and Ching, C., *J. Chem. Eng. Data* 51, 1593, 2006. <https://doi.org/10.1021/je0600754>
48. Achard, C., Jaoui, M., Schwing, M., and Rogalski, M., *J. Chem. Eng. Data* 41, 504, 1996. <https://doi.org/10.1021/je950202o>
49. Shareef, A., et al., *J. Chem. Eng. Data* 51, 879, 2006. <https://doi.org/10.1021/je050318c>
50. Clever, H. L., et al., *J. Phys. Chem. Ref. Data* 34, 201, 2005 [IUPAC No. 80]. <https://doi.org/10.1063/1.2062308>
51. Jaoui, M., Achard, C., and Rogalski, M., *J. Chem. Eng. Data* 47, 297, 2002. <https://doi.org/10.1021/je0102309>
52. Fichan, I., Larroche, C., and Gros, J. B., *J. Chem. Eng. Data* 44, 56, 1999. <https://doi.org/10.1021/je980070+>
53. Freire, M. G., et al., *J. Chem. Eng. Data* 50, 237, 2005. <https://doi.org/10.1021/je049707h>
54. Domanska, U., and Kozlowska, M. K., *J. Chem. Eng. Data* 47, 456, 2002. <https://doi.org/10.1021/je0103014>
55. Phelan, J. M., and Barnett, J. L., *J. Chem. Eng. Data* 46, 375, 2001. <https://doi.org/10.1021/je000300w>
56. Long, B-W., Wang, L-S., and Wu, J-S., *J. Chem. Eng. Data* 50, 136, 2005. <https://doi.org/10.1021/je049784c>
57. Marche, C., Ferronato, C., and Jose, J., *J. Chem. Eng. Data* 49, 937, 2004. <https://doi.org/10.1021/je0342567>
58. Oleszek-Kudlak, S., Shibata, E., and Nakamura, T., *J. Chem. Eng. Data* 49, 570, 2004. <https://doi.org/10.1021/je034170d>
59. Lynch, J. C., et al., *J. Chem. Eng. Data* 46, 1549, 2001.
60. Xiao, H., Li, N., and Wania, F., *J. Chem. Eng. Data* 49, 173, 2004. <https://doi.org/10.1021/je034214i>
61. Ma, J. H. Y., Hung, H., Shiu, W-Y., and Mackay, D., *J. Chem. Eng. Data* 46, 619, 2001. <https://doi.org/10.1021/je000341s>
62. Carta, R., and Tola, G., *J. Chem. Eng. Data* 41, 414, 1996; 44, 563, 1999. <https://doi.org/10.1021/je9501853>
63. Kao, H. D., et al., *Pharm. Res.* 17, 978, 2000. <https://doi.org/10.1023/A:1007583422634>
64. Heric, E. L., and Langford, R. E., *J. Chem. Eng. Data* 17, 471, 1972. <https://doi.org/10.1021/je60055a013>
65. Marche, C., Delépine, H., Ferronato, C., and Jose, J., *J. Chem. Eng. Data* 48, 398, 2003. <https://doi.org/10.1021/je025609p>
66. Wang, L-C, and Wang, F-A, *J. Chem. Eng. Data* 49, 155, 2004. <https://doi.org/10.1021/je049968r>
67. Shen, L, and Wania, F., *J. Chem. Eng. Data* 50, 742, 2005. <https://doi.org/10.1021/je049693f>
68. Oleszek-Kudlak, S., Shibata, E., and Nakamura, T., *J. Chem. Eng. Data* 52, 1824, 2007. <https://doi.org/10.1021/je700185m>
69. Zhao, H-K., Li, R-R, Ji, H-Z, Zhang, D-S., Tang, C., and Yang, L-Q., *J. Chem. Eng. Data* 52, 2072, 2007. <https://doi.org/10.1021/je7002594>
70. Yang, X., Wang, X., and Ching, C. B., *J. Chem. Eng. Data* 53, 1133, 2008. <https://doi.org/10.1021/je7006988>
71. Liu, L., and Chen, J., *J. Chem. Eng. Data* 53, 1649, 2008. <https://doi.org/10.1021/je800078j>
72. Szterner, P., *J. Chem. Eng. Data* 53, 1738, 2008. <https://doi.org/10.1021/je800029c>
73. Kong, M-Z., Shi, X-H, Cao, Y-C., and Zhou, C-R., *J. Chem. Eng. Data* 53, 615, 2008. <https://doi.org/10.1021/je7004038>
74. Daneshfar, A., Ghaziaskar, H. S., and Homayoun, N., *J. Chem. Eng. Data* 53, 776, 2008. <https://doi.org/10.1021/je700633w>
75. Manzurola, E., and Apelblat, A., *J. Chem. Thermodynamics* 34, 1127, 2002. <https://doi.org/10.1006/jcht.2002.0975>

76. Apelblat, A., Manzurola, E., and Balal, N. A., *J. Chem. Thermodynamics* 38, 565, 2006. <https://doi.org/10.1016/j.jct.2005.07.007>
77. Apelblat, A., and Mishelevich, A., *J. Chem. Thermodynamics* 40, 897, 2008. <https://doi.org/10.1016/j.jct.2007.12.006>
78. Góral, M., Wiśniewska-Goclowska, B., and Mączyński, A., *J. Phys. Chem. Ref. Data* 35, 1391, 2006. <https://doi.org/10.1063/1.2203354>
79. Mączyński, A., Shaw, D. G., Góral, M., and Wiśniewska-Goclowska, B., *J. Phys. Chem. Ref. Data* 37, 1119, 2008. <https://doi.org/10.1063/1.2838022>
80. Mączyński, A., Shaw, D. G., Góral, M., and Wiśniewska-Goclowska, B., *J. Phys. Chem. Ref. Data* 37, 1147, 2008. <https://doi.org/10.1063/1.2839271>
81. Mączyński, A., Shaw, D. G., Góral, M., and Wiśniewska-Goclowska, B., *J. Phys. Chem. Ref. Data* 37, 1169, 2008. <https://doi.org/10.1063/1.2839272>
82. Mączyński, A., Shaw, D. G., Góral, M., and Wiśniewska-Goclowska, B., *J. Phys. Chem. Ref. Data* 37, 1517, 2008. <https://doi.org/10.1063/1.2945626>
83. Mączyński, A., Shaw, D. G., Góral, M., and Wiśniewska-Goclowska, B., *J. Phys. Chem. Ref. Data* 37, 1575, 2008. <https://doi.org/10.1063/1.2950300>
84. Mączyński, A., Shaw, D. G., Góral, M., and Wiśniewska-Goclowska, B., *J. Phys. Chem. Ref. Data* 37, 1611, 2008. <https://doi.org/10.1063/1.2953801>
85. Luning Prak, D. J., and O'Sullivam, D. W., *J. Chem. Eng. Data* 51, 448, 2006. <https://doi.org/10.1021/je050373l>
86. Shiu, W.-Y., Wania, F., Hung, H., and Mackay, D., *J. Chem. Eng. Data* 42, 293, 1997. <https://doi.org/10.1021/je960299u>
87. Mączyński, A., and Shaw, D. G., *J. Phys. Chem. Ref. Data* 36, 59, 2007. <https://doi.org/10.1063/1.2366707>
88. Mączyński, A., and Shaw, D. G., *J. Phys. Chem. Ref. Data* 36, 133, 2007. <https://doi.org/10.1063/1.2366719>
89. Góral, M., Wiśniewska-Goclowska, B., and Mączyński, A., *J. Phys. Chem. Ref. Data* 33, 1159, 2004. <https://doi.org/10.1063/1.1797038>
90. Heric, E. L., and Langford, R. E., *J. Chem. Eng. Data* 17, 209, 1972. <https://doi.org/10.1021/je60053a022>
91. Altschuh, J., Brüggemann, R., Santl, H., Eichinger, G., and Piringer, O. G., *Chemosphere* 39, 1871-1887, 1999. <doi:10.1016/S0045-6535(99)00082-X>.
92. Arp, H. P. H. and Schmidt, T. C., *Environ. Sci. Technol.* 38, 5405-5412, 2004. <doi:10.1021/ES049286O>.
93. Bakierowska, A.-M. and Trzeszczyński, J., *Fluid Phase Equilib.* 213, 139-146, 2003. <doi:10.1016/S0378-3812(03)00286-3>.
94. Brockbank, S. A., Russon, J. L., Giles, N. F., Rowley, R. L., and Wilding, W. V.,*Fluid Phase Equilib.*, 348, 45-51, 2013. <doi:10.1016/J.FLUID.2013.03.023>.
95. Brunner, S., Hornung, E., Santl, H., Wolff, E., Piringer, O. G., Altschuh, J., and Brüggemann, R., *Environ. Sci. Technol.* 24, 1751-1754, 1990. <doi:10.1021/ES00081A021>.
96. Dewulf, J., van Langenhove, H., and Everaert, P.,*J. Chromatogr. A* 830, 353-363, 1999. <doi:10.1016/S0021-9673(98)00877-2>.
97. Dohányosová, P., Sarraute, S., Dohnal, V., Majer, V., and Costa Gomes, M., *Ind. Eng. Chem. Res.* 43, 2805-2815, 2004. <doi:10.1021/IE030800T>.
98. Dohnal, V. and Hovorka, S., *Ind. Eng. Chem. Res.* 38, 2036-2043, 1999. <doi:10.1021/IE980743H>.
99. Dunnivant, F. M. and Elzerman, A. W.*Chemosphere* 17, 525-541, 1988. <doi:10.1016/0045-6535(88)90028-8>.
100. Dunnivant, F. M., Elzerman, A. W., Jurs, P. C., and Hasan, M. N., *Environ. Sci. Technol.* 26, 1567-1573, 1992. <doi:10.1021/ES00032A012>.
101. Fogg, P. and Sangster, J.,*Chemicals in the Atmosphere: Solubility, Sources and Reactivity*, John Wiley & Sons, Inc., New York, 2003.
102. Hiatt, M. H., *J. Chem. Eng. Data* 58, 902-908, 2013. <doi:10.1021/JE3010535>.
103. Hoff, J. T., Mackay, D., Gillham, R., and Shiu, W. Y., *Environ. Sci. Technol.* 27, 2174-2180, 1993. <doi:10.1021/ES00047A026>.
104. *Hazardous Substances Data Bank,TOXicology data NETwork (TOXNET)*, National Library of Medicine (US), 2015. <http://toxnet.nlm.nih.gov/newtoxnet/hsdb.htm>.
105. Ji, C. and Evans, E. M., *Environ. Toxicol. Chem.* 26, 231-236, 2007. <doi:10.1897/06-339R.1>.
106. Kim, B. R., Kalis, E. M., DeWulf, T., and Andrews, K. M., *Water Environ. Res.* 72, 65-74, 2000. <doi:10.2175/106143000X137121>.
107. Lau, F. K., Charles, M. J., and Cahill, T. M., *J. Chem. Eng. Data* 51, 871-878, 2006. <doi:10.1021/JE050308B>.
108. Lee, H., Kim, H.-J., and Kwon, J.-H., *J. Chem. Eng. Data* 57, 3296-3302, 2012. <doi:10.1021/JE300954S>.
109. Li, J. and Carr, P. W., *Anal. Chem.* 65, 1443-1450, 1993. <doi:10.1021/AC00058A023>.
110. Li, J., Dallas, A. J., Eikens, D. I., Carr, P. W., Bergmann, D. L., Hait, M. J., and Eckert, C. A., *Anal. Chem.* 65, 3212-3218, 1993. <doi:10.1021/AC00070A008>.
111. Li, N., Wania, F., Lei, Y. D., and Daly, G. L., *J. Phys. Chem. Ref. Data* 32, 1545-1590, 2003. <doi:10.1063/1.1562632>.
112. Ma, Y.-G., Lei, Y. D., Xiao, H., Wania, F., and Wang, W.-H., *J. Chem. Eng. Data*, 55, 819-825, 2010. <doi:10.1021/JE900477X>.
113. Mackay, D., Shiu, W. Y., Ma, K. C., and Lee, S. C., *Handbook of Physical-Chemical Properties and Environmental Fate for Organic Chemicals, Vol. I of Introduction and Hydrocarbons*, Taylor & Francis, Boca Raton, FL, 2006.
114. Mackay, D., Shiu, W. Y., Ma, K. C., and Lee, S. C., *Handbook of Physical-Chemical Properties and Environmental Fate for Organic Chemicals, Vol. II of Halogenated Hydrocarbons*, Taylor & Francis, Boca Raton, FL, 2006.
115. McPhedran, K. N., Seth, R., and Drouillard, K. G., *Chemosphere* 91, 1648-1652, 2013. <doi:10.1016/J.*CHEMOSPHERE*.2012.12.017>.
116. Nielsen, F., Olsen, E., and Fredenslund, A,, *Environ. Sci. Technol.* 28, 2133-2138, 1994. <doi:10.1021/ES00061A022>.
117. Oliver, B. G,, *Chemosphere* 14, 1087-1106, 1985. <doi:10.1016/0045-6535(85)90029-3>.
118. Reichl, A., *Messung und Korrelierung von Gaslöslichkeiten halogenierter Kohlenwasserstoffe, Ph.D. thesis*, Technische Universität Berlin, Germany, 1995.
119. Riederer, M., *Environ. Sci. Technol.* 24, 829-837, 1990. <doi:10.1021/ES00076A006>.
120. Ryu, S.-A. and Park, S.-J., *Fluid Phase Equilib.* 161, 295-304, 1999. <doi:10.1016/S0378-3812(99)00193-4>.
121. Sander, S. P., Friedl, R. R., Golden, D. M., Kurylo, M. J., Moortgat, G. K., Keller-Rudek, H., Wine, P. H., Ravishankara, A. R., Kolb, C. E., Molina, M. J., Finlayson-Pitts, B. J., Huie, R. E., and Orkin, V. L., *Chemical Kinetics and Photochemical Data for Use in Atmospheric Studies, Evaluation Number 15*, JPL Publication 06-2, Jet Propulsion Laboratory, Pasadena, CA, 2006. <http://jpldataeval.jpl.nasa.gov>.
122. Sander, S. P., Abbatt, J., Barker, J. R., Burkholder, J. B., Friedl, R. R., Golden, D. M., Huie, R. E., Kolb, C. E., Kurylo, M. J., Moortgat, G. K., Orkin, V. L., and Wine, P. H., *Chemical Kinetics and Photochemical Data for Use in Atmospheric Studies, Evaluation Number 17*, JPL Publication 10-6, Jet Propulsion Laboratory, Pasadena, CA, 2011. <http://jpldataeval.jpl.nasa.gov>.
123. Staudinger, J. and Roberts, P. V., *Crit. Rev. Environ. Sci. Technol.* 26, 205-297, 1996. <doi:10.1080/10643389609388492>.
124. Staudinger, J. and Roberts, P. V., *Chemosphere*, 44, 561-576, 2001. <doi:10.1016/S0045-6535(00)00505-1>.
125. Warneck, P., *Chemosphere* 69, 347-361, 2007. <doi:10.1016/J.*CHEMOSPHERE*.2007.04.088>.
126. Warneck, P. and Williams, J., *The Atmospheric Chemist's Companion: Numerical Data for Use in the Atmospheric Sciences*, Springer Verlag, Berlin, 2012. <doi:10.1007/978-94-007-2275-0>.
127. Moore, R. M., Geen, C. E., and Tait, V. K., *Chemosphere* 30, 1183-1191, 1995. <doi:10.1016/0045-6535(95)00009-W>.
128. Sander, R., *Atmos. Chem. Phys.* 15, 4399-4981, 2015. <doi:10.5194/acp-15-4399-2015>.
129. Sander, R., *Henry's Law Constants*, <satellite.mpic.de/henry>.

Aqueous Solubility and Henry's Law Constants for Organic Compounds

Name	Mol. form.	CAS Reg. No.	Mol. wt.	t/°C	s/mass%	s/g kg^{-1} H$_2$O	Ref.	k_H/kPa m^3 mol^{-1}	Ref.
Acenaphthene	$C_{12}H_{10}$	83-32-9	154.207	0	0.00015	0.0015	4		
				25	0.000380	0.00380	22	0.0143	112
				50	0.00092	0.0092	4		
Acenaphthylene	$C_{12}H_8$	208-96-8	152.192	20	0.0016	0.016	28	0.010	112
Acephate	$C_4H_{10}NO_3PS$	30560-19-1	183.166	20	≈28	≈390	40		
Acetamide	C_2H_5NO	60-35-5	59.067	20	40.8	689	10		
Acetanilide	C_8H_9NO	103-84-4	135.163	20	0.52	5.2	27		
				70	2.7	28	27		
Acetazolamide	$C_4H_6N_4O_3S_2$	59-66-5	222.246	30	0.10	1.0	40		
Acetohexamide	$C_{15}H_{20}N_2O_4S$	968-81-0	324.396	37	0.0013	0.013	40		
Acetonitrile	C_2H_3N	75-05-8	41.052	-10	31.7	464	39		
				-3	40.5	681	39		
Acetophenone	C_8H_8O	98-86-2	120.149	20	0.67	6.7	84	0.00091	124
				50	0.81	8.2	84		
				80	1.16	11.7	84		
Acetylene	C_2H_2	74-86-2	26.037	25	0.108	1.08	19		
2-(Acetyloxy)benzoic acid	$C_9H_8O_4$	50-78-2	180.158		0.25	2.5	27		
2-(Acetyloxy)-5-bromobenzoic acid	$C_9H_7BrO_4$	1503-53-3	259.054		0.07	0.7	30		
Acridine	$C_{13}H_9N$	260-94-6	179.217	25	0.00466	0.0466	6		
Acrolein	C_3H_4O	107-02-8	56.063	20	20.8	263	10		
Acrylamide	C_3H_5NO	79-06-1	71.078	20	≈27	≈370	40		
Acrylonitrile	C_3H_3N	107-13-1	53.063	20	7.35	79.3	10		
Adenine	$C_5H_5N_5$	73-24-5	135.128	25	0.104	1.04	29		
Adenosine	$C_{10}H_{13}N_5O_4$	58-61-7	267.242	25	0.51	5.1	29		
Alachlor	$C_{14}H_{20}ClNO_2$	15972-60-8	269.768	23	0.024	0.24	40		
L-Alanine	$C_3H_7NO_2$	56-41-7	89.094	25	14.30	167	26		
β-Alanine	$C_3H_7NO_2$	107-95-9	89.094	25	47.1	890	26		
Aldicarb	$C_7H_{14}N_2O_2S$	116-06-3	190.263	20	0.60	6.0	40		
Aldrin	$C_{12}H_8Cl_6$	309-00-2	364.910	25	0.00002	0.0002	67		
Allopurinol	$C_5H_4N_4O$	315-30-0	136.112	25	0.057	0.57	40		
Ametryn	$C_9H_{17}N_5S$	834-12-8	227.330	20	0.0190	0.190	40		
2-Amino-9,10-anthracenedione	$C_{14}H_9NO_2$	117-79-3	223.227	25	0.000016	0.00016	40		
4-Aminoazobenzene	$C_{12}H_{11}N_3$	60-09-3	197.235	25	0.0030	0.030	40		
				97	0.068	0.68	40		
4-Aminobenzenesulfonamide	$C_6H_8N_2O_2S$	63-74-1	172.205	20	0.71	7.2	40		
4-Aminobenzenesulfonic acid	$C_6H_7NO_3S$	121-57-3	173.190	7	0.59	5.9	27		
DL-2-Aminobutanoic acid	$C_4H_9NO_2$	2835-81-6	103.120	25	17.4	211	26		
DL-3-Aminobutanoic acid	$C_4H_9NO_2$	2835-82-7	103.120	25	55.6	1250	26		
4-Amino-*N*-[(butylamino)carbonyl] benzenesulfonamide	$C_{11}H_{17}N_3O_3S$	339-43-5	271.336	37	0.053	0.53	40		
3-Amino-2,5-dichlorobenzoic acid	$C_7H_5Cl_2NO_2$	133-90-4	206.027	25	0.070	0.70	40		
4-(2-Aminoethyl)phenol	$C_8H_{11}NO$	51-67-2	137.179	15	1.03	10.4	40		
6-Aminohexanoic acid	$C_6H_{13}NO_2$	60-32-2	131.173	25	46	852	29		
4-Amino-2-hydroxybenzoic acid	$C_7H_7NO_3$	65-49-6	153.136	20	0.20	2.0	40		
2-Amino-2-methylpropanoic acid	$C_4H_9NO_2$	62-57-7	103.120	25	12.1	138	26		
2-Aminophenol	C_6H_7NO	95-55-6	109.126	20	1.92	19.6	40		
3-Aminophenol	C_6H_7NO	591-27-5	109.126	20	2.56	26.3	40		
				70	≈24	≈320	40		
4-Aminophenol	C_6H_7NO	123-30-8	109.126	20	1.55	15.7	40		
Aminopyrine	$C_{13}H_{17}N_3O$	58-15-1	231.293	25	4.8	50	40		
Amitriptyline	$C_{20}H_{23}N$	50-48-6	277.404	24	0.00097	0.0097	40		
Amobarbital	$C_{11}H_{18}N_2O_3$	57-43-2	226.272	25	0.06	0.6	40		
Anilazine	$C_9H_5Cl_3N_4$	101-05-3	275.522	20	0.001	0.01	40		
Aniline	C_6H_7N	62-53-3	93.127	25	3.38	35.0	10	0.00019	91
Aniline-2-carboxylic acid	$C_7H_7NO_2$	118-92-3	137.137	20	0.349	3.49	40		
Aniline-4-carboxylic acid	$C_7H_7NO_2$	150-13-0	137.137	25	0.54	5.4	40		
Aniline hydrochloride	C_6H_8ClN	142-04-1	129.588	15	15.1	178	27		
Anisole	C_7H_8O	100-66-3	108.138	20	0.203	2.03	20	0.035	94
				40	0.184	1.84	20		

Name	Mol. form.	CAS Reg. No.	Mol. wt.	t/°C	s/mass%	s/g kg^{-1} H_2O	Ref.	k_H/kPa m^3 mol^{-1}	Ref.
				81	0.294	2.95	20		
Anthracene	$C_{14}H_{10}$	120-12-7	178.229	0	0.0000022	0.000022	42,4		
				25	0.0000044	0.000044	42,22	0.0050	112
				50	0.000029	0.00029	42		
9,10-Anthracenedione	$C_{14}H_8O_2$	84-65-1	208.213	25	0.00014	0.0014	40		
Apomorphine	$C_{17}H_{17}NO_2$	58-00-4	267.323	25	2.0	20	40		
L-Arginine	$C_6H_{14}N_4O_2$	74-79-3	174.201	25	15.44	183	26		
L-Ascorbic acid	$C_6H_8O_6$	50-81-7	176.124	25	25.2	337	33		
				50	41.0	695	33		
L-Asparagine	$C_4H_8N_2O_3$	70-47-3	132.118	25	2.45	25.1	26		
L-Aspartic acid	$C_4H_7NO_4$	56-84-8	133.104	10	0.29	2.9	77		
				25	0.49	4.9	77		
				50	1.31	13.3	77		
Atrazine	$C_8H_{14}ClN_5$	1912-24-9	215.684	25	0.007	0.07	26		
Atropine	$C_{17}H_{23}NO_3$	51-55-8	289.370	20	0.3	3	40		
Azinphos-methyl	$C_{10}H_{12}N_3O_3PS_2$	86-50-0	317.324	20	0.00209	0.0209	40		
trans-Azobenzene	$C_{12}H_{10}N_2$	17082-12-1	182.220	20	0.03	0.3	27		
Bayleton	$C_{14}H_{16}ClN_3O_2$	43121-43-3	293.749	20	0.026	0.26	40		
Bendiocarb	$C_{11}H_{13}NO_4$	22781-23-3	223.226	25	0.004	0.04	40		
Bentazon	$C_{10}H_{12}N_2O_3S$	25057-89-0	240.278	20	0.050	0.50	40		
Benzaldehyde	C_7H_6O	100-52-7	106.122	20	0.3	3	10		
Benzamide	C_7H_7NO	55-21-0	121.137	12	0.577	5.77	27		
Benz[*a*]anthracene	$C_{18}H_{12}$	56-55-3	228.288	10	0.00000038	0.0000038	42		
				25	0.00000093	0.0000093	42,22	0.00063	112
Benzene	C_6H_6	71-43-2	78.112	10	0.174	1.74	22		
				20	0.177	1.77	22		
				30	0.183	1.83	22		
				40	0.192	1.92	22		
				50	0.206	2.06	22		
				70	0.249	2.50	65		
				101	0.398	4.00	65		
Benzeneacetic acid	$C_8H_8O_2$	103-82-2	136.149	25	1.71	17.4	27		
1,2-Benzenediamine	$C_6H_8N_2$	95-54-5	108.141	20	3.02	31.1	40		
1,3-Benzenediamine	$C_6H_8N_2$	108-45-2	108.141	20	3.48	36.1	40		
1,4-Benzenediamine	$C_6H_8N_2$	106-50-3	108.141	24	3.45	35.7	40		
1,2-Benzenedicarboxamide	$C_8H_8N_2O_2$	88-96-0	164.162	30	0.59	5.9	40		
Benzeneethanol	$C_8H_{10}O$	60-12-8	122.164	25	1.72	17.5	40		
Benzenehexacarboxylic acid	$C_{12}H_6O_{12}$	517-60-2	342.169	25	49.3	972	76		
Benzenepentacarboxylic acid	$C_{11}H_6O_{10}$	1585-40-6	298.160	10	11.9	135	76		
				25	21.1	267	76		
				50	36.2	567	76		
1,2,3,4-Benzenetetracarboxylic acid	$C_{10}H_6O_8$	476-73-3	254.150	10	11.0	124	76		
				25	20.9	264	76		
				50	39.5	653	76		
1,2,3,5-Benzenetetracarboxylic acid	$C_{10}H_6O_8$	479-47-0	254.150	10	7.50	81.1	76		
				25	10.1	112	76		
				50	15.8	188	76		
1,2,4,5-Benzenetetracarboxylic acid	$C_{10}H_6O_8$	89-05-4	254.150	10	0.51	5.1	76		
				25	1.06	10.7	76		
				50	3.82	39.7	76		
1,2,3-Benzenetricarboxylic acid	$C_9H_6O_6$	569-51-7	210.140	10	2.39	24.5	76		
				25	4.78	50.2	76		
				50	17.4	211	76		
1,2,4-Benzenetricarboxylic acid	$C_9H_6O_6$	528-44-9	210.140	10	1.02	10.3	76		
				25	1.92	19.6	76		
				50	5.45	57.6	76		
1,3,5-Benzenetricarboxylic acid	$C_9H_6O_6$	554-95-0	210.140	10	0.110	1.10	76		
				25	0.207	2.07	76		
				50	0.598	6.02	76		
1,2,3-Benzenetriol	$C_6H_6O_3$	87-66-1	126.110	25	38.5	626	27		
1,3,5-Benzenetriol	$C_6H_6O_3$	108-73-6	126.110	20	1.12	11.3	27		

Name	Mol. form.	CAS Reg. No.	Mol. wt.	t/°C	s/mass%	s/g kg^{-1} H_2O	Ref.	k_H/kPa m^3 mol^{-1}	Ref.
p-Benzidine	$C_{12}H_{12}N_2$	92-87-5	184.236	24	0.0360	0.360	40		
1*H*-Benzimidazole	$C_7H_6N_2$	51-17-2	118.136	15	0.33	3.3	54		
				20	0.201	2.01	6		
1,3-Benzodioxole-5-carboxaldehyde	$C_8H_6O_3$	120-57-0	150.132	20	0.35	3.5	40		
Benzo[*b*]fluoranthene	$C_{20}H_{12}$	205-99-2	252.309	20	0.0000002	0.000002	40		
Benzo[*k*]fluoranthene	$C_{20}H_{12}$	207-08-9	252.309		0.00000008	0.0000008	40		
11*H*-Benzo[*a*]fluorene	$C_{17}H_{12}$	238-84-6	216.277	25	0.0000045	0.000045	42,4		
11*H*-Benzo[b]fluorene	$C_{17}H_{12}$	243-17-4	216.277	25	0.0000002	0.000002	42,4		
Benzoic acid	$C_7H_6O_2$	65-85-0	122.122	10	0.209	2.09	76		
				25	0.343	3.44	76		
				50	0.842	8.49	76		
Benzoin	$C_{14}H_{12}O_2$	579-44-2	212.244	25	0.03	0.3	40		
Benzonitrile	C_7H_5N	100-47-0	103.122	25	0.2	2	10		
Benzo[ghi]perylene	$C_{22}H_{12}$	191-24-2	276.330	25	0.000000026	0.00000026	42,4	0.000042	112
Benzophenone	$C_{13}H_{10}O$	119-61-9	182.217	20	0.0075	0.075	40		
2*H*-1-Benzopyran-2-one	$C_9H_6O_2$	91-64-5	146.143	20	0.190	1.90	40		
				60	0.69	6.9	40		
Benzo[*a*]pyrene	$C_{20}H_{12}$	50-32-8	252.309	25	0.00000043	0.0000043	42,22	0.000077	112
Benzo[e]pyrene	$C_{20}H_{12}$	192-97-2	252.309	8	0.00000032	0.0000032	42		
				17	0.00000044	0.0000044	42,22	0.000030	104
				25	0.00000048	0.0000048	42		
Benzo[f]quinoline	$C_{13}H_9N$	85-02-9	179.217	25	0.0079	0.079	6		
p-Benzoquinone	$C_6H_4O_2$	106-51-4	108.095	25	1.36	13.8	27		
Benzo[*b*]thiophene	C_8H_6S	95-15-8	134.199	20	0.0130	0.130	6		
Benzo[b]triphenylene	$C_{22}H_{14}$	215-58-7	278.346	25	0.0000027	0.000027	4		
Benzoxazole	C_7H_5NO	273-53-0	119.121	20	0.834	8.34	6		
N-Benzoylglycine	$C_9H_9NO_3$	495-69-2	179.172	25	0.37	3.7	29		
Benzoyl peroxide	$C_{14}H_{10}O_4$	94-36-0	242.227	20	0.000016	0.00016	40		
N-Benzoyl-L-phenylalanine	$C_{16}H_{15}NO_3$	2566-22-5	269.295	25	0.085	0.85	29		
Benzyl acetate	$C_9H_{10}O_2$	140-11-4	150.174	25	0.150	1.50	40		
Benzyl alcohol	C_7H_8O	100-51-6	108.138	20	0.08	0.8	10		
Benzyl formate	$C_8H_8O_2$	104-57-4	136.149	20	1.07	10.8	20		
				80	1.43	14.5	20		
Bifenthrin	$C_{23}H_{22}ClF_3O_2$	82657-04-3	422.868	25	0.00001	0.0001	32		
Biotin	$C_{10}H_{16}N_2O_3S$	58-85-5	244.310	25	0.035	0.35	40		
Biphenyl	$C_{12}H_{10}$	92-52-4	154.207	0	0.000272	0.00272	4		
				25	0.00054	0.0054	58,22	0.028	22
				50	0.0022	0.022	4		
2,2'-Bipyridine	$C_{10}H_8N_2$	366-18-7	156.184	25	0.61	6.1	40		
2,2'-Biquinoline	$C_{18}H_{12}N_2$	119-91-5	256.301	24	0.000102	0.00102	6		
Bis(4-aminophenyl) sulfone	$C_{12}H_{12}N_2O_2S$	80-08-0	248.300	25	0.016	0.16	40		
Bis(2-chloroethyl) ether	$C_4H_8Cl_2O$	111-44-4	143.012	20	1.04	10.5	20	0.0029	104
				81	1.26	12.8	20		
1,1-Bis(4-chlorophenyl)-2,2,2-trichloroethanol	$C_{14}H_9Cl_5O$	115-32-2	370.485	25	0.00013	0.0013	40		
Bis(2-ethylhexyl) phthalate	$C_{24}H_{38}O_4$	117-81-7	390.557	25	0.000027	0.00027	40		
2,2-Bis(4-hydroxyphenyl)propane	$C_{15}H_{16}O_2$	80-05-7	228.287	25	0.0300	0.30	49		
1,3-Bis(trifluoromethyl)benzene	$C_8H_4F_6$	402-31-3	214.108	25	0.0041	0.041	2		
Borneol, (-)-	$C_{10}H_{18}O$	464-45-9	154.249	25	0.046	0.46	52		
Bromacil	$C_9H_{13}BrN_2O_2$	314-40-9	261.115	25	0.082	0.82	40		
Bromobenzene	C_6H_5Br	108-86-1	157.008	10	0.0387	0.387	2		
				25	0.0445	0.445	2	0.21	5
				40	0.0516	0.516	2		
2-Bromobenzoic acid	$C_7H_5BrO_2$	88-65-3	201.018	25	0.185	1.85	27		
3-Bromobenzoic acid	$C_7H_5BrO_2$	585-76-2	201.018	25	0.040	0.40	27		
4-Bromobenzoic acid	$C_7H_5BrO_2$	586-76-5	201.018	25	0.0056	0.056	27		
1-Bromobutane	C_4H_9Br	109-65-9	137.018	25	0.087	0.87	35	2.17	103
4-Bromo-1-butene	C_4H_7Br	5162-44-7	135.003	25	0.076	0.76	35		
1-Bromo-2-chlorobenzene	C_6H_4BrCl	694-80-4	191.453	25	0.0124	0.124	2		
1-Bromo-3-chlorobenzene	C_6H_4BrCl	108-37-2	191.453	25	0.0118	0.118	2		
1-Bromo-4-chlorobenzene	C_6H_4BrCl	106-39-8	191.453	25	0.00442	0.0442	2		
1-Bromo-2-chloroethane	C_2H_4BrCl	107-04-0	143.410	30	0.683	6.83	25		

Name	Mol. form.	CAS Reg. No.	Mol. wt.	t/°C	s/mass%	s/g kg^{-1} H_2O	Ref.	k_H/kPa m^3 mol^{-1}	Ref.
Bromochloromethane	CH_2BrCl	74-97-5	129.384	25	1.7	17	10	0.15	102
1-Bromo-3-chloropropane	C_3H_6BrCl	109-70-6	157.437	25	0.223	2.23	35		
2-Bromo-2-chloro-1,1,1-trifluoroethane	$C_2HBrClF_3$	151-67-7	197.381	10	0.52	5.2	25		
				25	0.41	4.1	25		
				40	0.40	4.0	25		
Bromodichloromethane	$CHBrCl_2$	75-27-4	163.829	30	0.300	3.00	40		
Bromoethane	C_2H_5Br	74-96-4	108.965	0	1.05	10.6	25		
				25	0.90	9.0	25	0.77	110
1-Bromoheptane	$C_7H_{15}Br$	629-04-9	179.098	25	0.00067	0.0067	35		
1-Bromohexane	$C_6H_{13}Br$	111-25-1	165.071	25	0.00258	0.0258	35		
1-Bromo-4-iodobenzene	C_6H_4BrI	589-87-7	282.904	25	0.000794	0.00794	2		
Bromomethane	CH_3Br	74-83-9	94.939	20	1.80	18.3	5	0.59	121
1-Bromo-3-methylbutane	$C_5H_{11}Br$	107-82-4	151.045	16	0.020	0.20	35		
1-Bromo-2-methylpropane	C_4H_9Br	78-77-3	137.018	18	0.051	0.51	35		
1-Bromooctane	$C_8H_{17}Br$	111-83-1	193.125	25	0.000167	0.00167	35		
1-Bromopentane	$C_5H_{11}Br$	110-53-2	151.045	25	0.0127	0.127	35		
4-Bromophenol	C_6H_5BrO	106-41-2	173.007	25	1.86	19.0	2		
1-Bromopropane	C_3H_7Br	106-94-5	122.992	0	0.298	2.98	35		
				25	0.234	2.34	35	0.91	110
2-Bromopropane	C_3H_7Br	75-26-3	122.992	20	0.32	3.2	35	1.19	110
3-Bromopropene	C_3H_5Br	106-95-6	120.976	25	0.38	3.8	35		
4-Bromotoluene	C_7H_7Br	106-38-7	171.035	25	0.011	0.11	2		
Bromotrifluoromethane	$CBrF_3$	75-63-8	148.910	25	0.032*	0.32*	14		
5-Bromouracil	$C_4H_3BrN_2O_2$	51-20-7	190.983	25	0.288	2.89	72		
Brucine	$C_{23}H_{26}N_2O_4$	357-57-3	394.463	20	0.012	0.12	27		
1,3-Butadiene	C_4H_6	106-99-0	54.091	25	0.0735*	0.735*	5	7.69	5
Butanal	C_4H_8O	123-72-8	72.106	25	7.1	76	10		
Butanamide	C_4H_9NO	541-35-5	87.120	25	≈19	≈230	40		
Butane	C_4H_{10}	106-97-8	58.122	25	0.00724*	0.0724*	18	100.	5
2,3-Butanedione	$C_4H_6O_2$	431-03-8	86.090	20	31.7	464	20		
				80	21.8	279	20		
Butanenitrile	C_4H_7N	109-74-0	69.106	20	3.3	34	10		
1,2,3,4-Butanetetrol	$C_4H_{10}O_4$	149-32-6	122.120	20	38.0	613	27		
1-Butanethiol	$C_4H_{10}S$	109-79-5	90.187	20	0.0597	0.597	10		
1-Butanol	$C_4H_{10}O$	71-36-3	74.121	0	10.5	117	78,1		
				25	7.3	79	78,1		
				50	6.4	68	78,1		
				100	8.8	96	78		
2-Butanol	$C_4H_{10}O$	78-92-2	74.121	10	23.9	314	1,87		
				25	18.1	221	1,87		
				50	14.0	163	1,87		
2-Butanone	C_4H_8O	78-93-3	72.106	0	35.9	560	82		
				25	25.6	344	82		
				40	21.5	274	82		
				70	18.1	221	20		
				100	19.3	239	82		
trans-2-Butenal	C_4H_6O	123-73-9	70.090	20	15.6	185	10		
1-Butene	C_4H_8	106-98-9	56.107	25	0.0222*	0.222*	5	25.6	113
trans-2-Butenoic acid	$C_4H_6O_2$	107-93-7	86.090	20	7.1	76	26		
cis-2-Buten-1-ol	C_4H_8O	4088-60-2	72.106	20	16.6	199	10		
3-Buten-2-one	C_4H_6O	78-94-4	70.090	28	54.3	1190	82		
				50	35.6	553	82		
				80	37.6	603	82		
Butyl acetate	$C_6H_{12}O_2$	123-86-4	116.158	20	0.68	6.8	10		
sec-Butyl acetate	$C_6H_{12}O_2$	105-46-4	116.158	20	0.62	6.2	10		
Butyl 4-aminobenzoate	$C_{11}H_{15}NO_2$	94-25-7	193.243	25	0.018	0.18	40		
Butylbenzene	$C_{10}H_{14}$	104-51-8	134.218	25	0.00138	0.0138	22,89	1.30	5
sec-Butylbenzene, (±)-	$C_{10}H_{14}$	36383-15-0	134.218	25	0.0014	0.014	4,89	1.89	11
tert-Butylbenzene	$C_{10}H_{14}$	98-06-6	134.218	25	0.0032	0.032	4	1.20	5
Butyl ethyl ether	$C_6H_{14}O$	628-81-9	102.174	20	0.65	6.5	20		
				70	0.39	3.9	20		

Name	Mol. form.	CAS Reg. No.	Mol. wt.	t/°C	s/mass%	s/g kg⁻¹ H₂O	Ref.	k_H/kPa m³ mol⁻¹	Ref.
Butyl 4-hydroxybenzoate	$C_{11}H_{14}O_3$	94-26-8	194.227	25	0.020	0.20	40		
Butyl methyl ether	$C_5H_{12}O$	628-28-4	88.148	0	2.51	25.7	79		
				25	0.89	9.0	79		
4-*tert*-Butylphenol	$C_{10}H_{14}O$	98-54-4	150.217	25	0.058	0.58	40		
Butyl propanoate	$C_7H_{14}O_2$	590-01-2	130.185	22	0.572	5.72	27		
Butyl stearate	$C_{22}H_{44}O_2$	123-95-5	340.583	25	0.2	2	10		
Butyl vinyl ether	$C_6H_{12}O$	111-34-2	100.158	20	0.3	3	10		
1-Butyne	C_4H_6	107-00-6	54.091	25	0.287*	2.87*	5	1.92	5
Caffeine	$C_8H_{10}N_4O_2$	58-08-2	194.191	25	2.12	21.7	29		
Camphor, (+)	$C_{10}H_{16}O$	464-49-3	152.233	20	0.01	0.1	10		
trans-Camphoric acid, (±)-	$C_{10}H_{16}O_4$	560-08-7	200.232	25	0.8	8	27		
Cantharidin	$C_{10}H_{12}O_4$	56-25-7	196.200	20	0.003	0.03	40		
Caprolactam	$C_6H_{11}NO$	105-60-2	113.157	25	84.0	5250	10		
Captafol	$C_{10}H_9Cl_4NO_2S$	191906	349.061	20	0.000142	0.00142	40		
Captan	$C_9H_8Cl_3NO_2S$	133-06-2	300.590	20	0.00005	0.0005	40		
Carbaryl	$C_{12}H_{11}NO_2$	63-25-2	201.221	20	0.0102	0.102	40		
Carbazole	$C_{12}H_9N$	86-74-8	167.206	22	0.000120	0.00120	6		
Carbofuran	$C_{12}H_{15}NO_3$	1563-66-2	221.252	20	0.032	0.32	40		
Carbon dioxide	CO_2	124-38-9	44.010	25	0.150*	1.50*	18		
Carbon disulfide	CS_2	75-15-0	76.141	20	0.210	2.10	10		
Carbon monoxide	CO	630-08-0	28.010	25	0.00276*	0.0276*	18		
Carboxin	$C_{12}H_{13}NO_2S$	5234-68-4	235.302	25	0.017	0.17	40		
Carminic acid	$C_{22}H_{20}O_{13}$	1260-17-9	492.386	20	0.13	1.3	40		
Carnosine	$C_9H_{14}N_4O_3$	305-84-0	226.232	25	24.4	323	26		
Carvenol	$C_{10}H_{16}O$	99-48-9	152.233	25	0.29	2.9	52		
Carvenone, (*S*)-	$C_{10}H_{16}O$	10395-45-6	152.233	15	0.22	2.2	27		
Carvone, (±)-	$C_{10}H_{14}O$	22327-39-5	150.217	15	0.13	1.3	27		
(*S*)-Carvone	$C_{10}H_{14}O$	2244-16-8	150.217	25	0.13	1.3	52		
Cephalexin	$C_{16}H_{17}N_3O_4S$	15686-71-2	347.389	25	1.2	12	40		
Chloramphenicol	$C_{11}H_{12}Cl_2N_2O_5$	56-75-7	323.129	25	0.38	3.8	40		
Chlordane	$C_{10}H_6Cl_8$	57-74-9	409.779	25	0.00006	0.0006	67		
2-Chloroaniline	C_6H_6ClN	95-51-2	127.572	25	0.876	8.76	10		
3-Chloroaniline	C_6H_6ClN	108-42-9	127.572	20	0.54	5.4	40		
4-Chloroaniline	C_6H_6ClN	106-47-8	127.572	20	0.275	2.75	40		
Chlorobenzene	C_6H_5Cl	108-90-7	112.557	5	0.050	0.50	61		
				25	0.050	0.50	61		
				45	0.055	0.55	61,2		
Chlorobenzilate	$C_{16}H_{14}Cl_2O_3$	510-15-6	325.186	20	0.001	0.01	32		
2-Chlorobenzoic acid	$C_7H_5ClO_2$	118-91-2	156.567	25	0.209	2.09	27		
3-Chlorobenzoic acid	$C_7H_5ClO_2$	535-80-8	156.567	25	0.040	0.40	27		
4-Chlorobenzoic acid	$C_7H_5ClO_2$	74-11-3	156.567	25	0.072	0.72	27		
2-Chlorobiphenyl	$C_{12}H_9Cl$	2051-60-7	188.652	25	0.00055	0.0055	7	0.033	107
1-Chlorobutane	C_4H_9Cl	109-69-3	92.567	1	0.062	0.62	35		
				25	0.087	0.87	35	1.45	98
2-Chlorobutane	C_4H_9Cl	53178-20-4	92.567	0	0.107	1.07	35		
				25	0.092	0.92	35		
3-Chloro-2-butanone	C_4H_7ClO	4091-39-8	106.551	19	2.80	28.8	20		
				92	3.38	35.0	20		
Chlorodiazepoxide	$C_{16}H_{14}ClN_3O$	58-25-3	299.754	20	0.2	2	40		
Chlorodibromomethane	$CHBr_2Cl$	124-48-1	208.280	30	0.251	2.51	40		
Chlorodifluoromethane	$CHClF_2$	75-45-6	86.469	25	0.30*	3.0*	10	2.94	122
4-Chloro-2,5-dimethylphenol	C_8H_9ClO	1124-06-7	156.609	25	0.89	8.9	2		
4-Chloro-2,6-dimethylphenol	C_8H_9ClO	1123-63-3	156.609	25	0.52	5.2	2		
4-Chloro-3,5-dimethylphenol	C_8H_9ClO	88-04-0	156.609	25	0.34	3.4	2		
1-Chloro-2,4-dinitrobenzene	$C_6H_3ClN_2O_4$	97-00-7	202.552	25	0.00092	0.0092	40		
Chloroethane	C_2H_5Cl	75-00-3	64.514	0	0.45	4.5	25		
				25	0.67*	6.7*	25	1.20	125
Chloroethene	C_2H_3Cl	75-01-4	62.498	25	0.27*	2.7*	5	2.63	125
1-Chloro-2-fluorobenzene	C_6H_4ClF	348-51-6	130.547	25	0.0502	0.502	40		
Chlorofluoromethane	CH_2ClF	593-70-4	68.478	25	1.05*	10.6*	14		
1-Chloroheptane	$C_7H_{15}Cl$	629-06-1	134.647	25	0.00136	0.0136	35		

Thermochem

Name	Mol. form.	CAS Reg. No.	Mol. wt.	t/°C	s/mass%	s/g kg^{-1} H_2O	Ref.	k_H/kPa m^3 mol^{-1}	Ref.
1-Chlorohexane	$C_6H_{13}Cl$	544-10-5	120.620	5	0.0047	0.047	35		
				25	0.0064	0.064	35		
2-Chloro-4-hydroxy-5-methoxybenzaldehyde	$C_8H_7ClO_3$	18268-76-3	186.593	25	0.013	0.13	8		
3-Chloro-4-hydroxy-5-methoxybenzaldehyde	$C_8H_7ClO_3$	19463-48-0	186.593	25	0.093	0.93	8		
1-Chloro-2-iodobenzene	C_6H_4ClI	615-41-8	238.453	25	0.00689	0.0689	2		
1-Chloro-3-iodobenzene	C_6H_4ClI	625-99-0	238.453	25	0.00674	0.0674	2		
1-Chloro-4-iodobenzene	C_6H_4ClI	637-87-6	238.453	25	0.00311	0.0311	2		
Chloromethane	CH_3Cl	74-87-3	50.488	25	0.535*	5.35*	5	0.91	5
1-Chloro-2-methoxyethane	C_3H_7ClO	627-42-9	94.540	20	7.79	84.5	20		
				70	6.31	67.3	20		
(Chloromethyl)benzene	C_7H_7Cl	100-44-7	126.584	20	0.0493	0.493	10		
3-(Chloromethyl)heptane	$C_8H_{17}Cl$	123-04-6	148.674	20	0.01	0.1	10		
2-Chloro-6-methylphenol	C_7H_7ClO	87-64-9	142.583	25	0.36	3.6	2		
4-Chloro-2-methylphenol	C_7H_7ClO	1570-64-5	142.583	25	0.68	6.8	2		
4-Chloro-3-methylphenol	C_7H_7ClO	59-50-7	142.583	25	0.40	4.0	2		
(4-Chloro-2-methylphenoxy)acetic acid	$C_9H_9ClO_3$	94-74-6	200.618	25	0.117	1.17	40		
1-Chloro-2-methylpropane	C_4H_9Cl	513-36-0	92.567	25	0.92	9.2	35		
2-Chloro-2-methylpropane	C_4H_9Cl	507-20-0	92.567	15	0.29	2.9	35		
1-Chloro-2-methylpropene	C_4H_7Cl	513-37-1	90.552	25	0.916	9.16	5	0.192	104
1-Chloronaphthalene	$C_{10}H_7Cl$	90-13-1	162.616	25	0.00224	0.0224	5	0.0357	28
2-Chloronaphthalene	$C_{10}H_7Cl$	91-58-7	162.616	25	0.00117	0.0117	5	0.0333	28
1-Chloro-2-nitrobenzene	$C_6H_4ClNO_2$	88-73-3	157.555	20	0.0441	0.441	40		
1-Chloro-3-nitrobenzene	$C_6H_4ClNO_2$	121-73-3	157.555	20	0.0273	0.273	40		
1-Chloro-4-nitrobenzene	$C_6H_4ClNO_2$	100-00-5	157.555	20	0.0453	0.453	40		
3-Chloro-2-nitrobenzoic acid	$C_7H_4ClNO_4$	4771-47-5	201.565	25	0.047	0.47	27		
5-Chloro-2-nitrobenzoic acid	$C_7H_4ClNO_4$	2516-95-2	201.565	25	0.96	9.6	27		
1-Chlorooctane	$C_8H_{17}Cl$	111-85-3	148.674	25	0.0345	0.345	35		
Chloropentafluoroethane	C_2ClF_5	76-15-3	154.466	25	0.006*	0.06*	10	294	18
1-Chloropentane	$C_5H_{11}Cl$	543-59-9	106.594	5	0.020	0.20	35		
				25	0.0201	0.201	35	2.38	110
3-Chloropentane	$C_5H_{11}Cl$	616-20-6	106.594	25	0.025	0.25	35		
5-Chloro-2-pentanone	C_5H_9ClO	5891-21-4	120.577	22	4.7	49	20		
				71	13.5	156	20		
2-Chlorophenol	C_6H_5ClO	95-57-8	128.556	25	2.27	23.2	48,51,2		
3-Chlorophenol	C_6H_5ClO	108-43-0	128.556	25	2.2	22	2		
4-Chlorophenol	C_6H_5ClO	106-48-9	128.556	25	2.55	26.2	48,51,2		
N'-(4-Chlorophenyl)-*N,N*-dimethylurea	$C_9H_{11}ClN_2O$	150-68-5	198.648	25	0.023	0.23	26		
1-Chloropropane	C_3H_7Cl	540-54-5	78.541	25	0.250	2.50	35	1.45	110
2-Chloropropane	C_3H_7Cl	75-29-6	78.541	0	0.44	4.4	35		
				20	0.30	3.0	35		
3-Chloropropene	C_3H_5Cl	107-05-1	76.525	25	0.40	4.0	35	1.10	5
3-Chloropropene	C_3H_5Cl	107-05-1	76.525	50	0.13	1.3	35		
Chloropropham	$C_{10}H_{12}ClNO_2$	101-21-3	213.661	25	0.0080	0.080	40		
1-Chlorotetradecane	$C_{14}H_{29}Cl$	2425-54-9	232.833	25	0.0232	0.232	35		
Chlorothalonil	$C_8Cl_4N_2$	1897-45-6	265.911	25	0.00006	0.0006	40		
Chlorothiazide	$C_7H_6ClN_3O_4S_2$	58-94-6	295.724	25	0.0283	0.283	40		
2-Chlorotoluene	C_7H_7Cl	95-49-8	126.584	25	0.0117	0.117	61		
3-Chlorotoluene	C_7H_7Cl	108-41-8	126.584	25	0.0117	0.117	61		
4-Chlorotoluene	C_7H_7Cl	106-43-4	126.584	25	0.0123	0.123	61		
Chlorotrifluoromethane	$CClF_3$	75-72-9	104.459	25	0.009*	0.09*	10	107	18
3-Chloro-1,1,1-trifluoropropane	$C_3H_4ClF_3$	460-35-5	132.512	20	0.133	1.33	35		
2-Chloro-1,3,5-trinitrobenzene	$C_6H_2ClN_3O_6$	88-88-0	247.549	15	0.018	0.18	40		
5-Chlorouracil	$C_4H_3ClN_2O_2$	1820-81-1	146.532	25	0.250	2.51	72		
Chlorpyrifos	$C_9H_{11}Cl_3NO_3PS$	2921-88-2	350.586	20	0.000073	0.00073	40		
Chlorsulfuron	$C_{12}H_{12}ClN_5O_4S$	64902-72-3	357.773	25	2.71	27.9	32		
Cholic acid	$C_{24}H_{40}O_5$	81-25-4	408.572	20	0.028	0.28	26		
Chrysene	$C_{18}H_{12}$	218-01-9	228.288	7	0.00000007	0.0000007	42		
				25	0.00000019	0.0000019	42,22	0.000370	112
trans-Cinnamaldehyde	C_9H_8O	14371-10-9	132.159	25	0.135	1.35	40		
trans-Cinnamic acid	$C_9H_8O_2$	140-10-3	148.159	20	0.1	1	26		
				98	0.59	5.9	26		

Name	Mol. form.	CAS Reg. No.	Mol. wt.	t/°C	s/mass%	s/g kg⁻¹ H₂O	Ref.	k_H/kPa m³ mol⁻¹	Ref.
Citric acid	$C_6H_8O_7$	77-92-9	192.124	20	59	1440	26		
Clopyralid	$C_6H_3Cl_2NO_2$	1702-17-6	192.000	20	0.1	1	40		
Clorophene	$C_{13}H_{11}ClO$	120-32-1	218.678	20	0.42	4.2	40		
Cocaine	$C_{17}H_{21}NO_4$	50-36-2	303.354	25	0.17	1.7	27		
Codeine	$C_{18}H_{21}NO_3$	76-57-3	299.365	25	0.79	7.9	27		
Colchicine	$C_{22}H_{25}NO_6$	64-86-8	399.437	20	4	42	26		
Coronene	$C_{24}H_{12}$	191-07-1	300.352	25	0.000000014	0.00000014	42,4		
Creatine	$C_4H_9N_3O_2$	57-00-1	131.133	25	1.6	16	26		
o-Cresol	C_7H_8O	95-48-7	108.138	40	3.08	31.8	10		
m-Cresol	C_7H_8O	108-39-4	108.138	40	2.51	25.7	10		
p-Cresol	C_7H_8O	106-44-5	108.138	40	2.26	23.1	10		
Crufomate	$C_{12}H_{19}ClNO_3P$	299-86-5	291.711	20	0.50	5.0	40		
Cyanazine	$C_9H_{13}ClN_6$	21725-46-2	240.692	25	0.0171	0.171	40		
2-Cyanoacetamide	$C_3H_4N_2O$	107-91-5	84.076	20	11.5	130	40		
Cyanogen	C_2N_2	460-19-5	52.034	25	0.8*	8*	30		
Cyanogen chloride	CClN	506-77-4	61.471	0	5.7	60	40		
Cyanoguanidine	$C_2H_4N_4$	461-58-5	84.080	25	3.8	40	40		
Cyanuric acid	$C_3H_3N_3O_3$	108-80-5	129.074	25	0.259	2.59	40		
Cycloheptane	C_7H_{14}	291-64-5	98.186	25	0.0030	0.030	3	12.2	114
Cycloheptanone	$C_7H_{12}O$	502-42-1	112.169	20	3.61	37.5	20		
				92	2.82	29.0	20		
1,3,5-Cycloheptatriene	C_7H_8	544-25-2	92.139	25	0.064	0.64	3	0.476	13
Cycloheptene	C_7H_{12}	628-92-2	96.170	25	0.0066	0.066	3	3.85	113
1,4-Cyclohexadiene	C_6H_8	628-41-1	80.128	25	0.08	0.8	3	0.909	113
Cyclohexane	C_6H_{12}	110-82-7	84.159	25	0.0058	0.058	3	17.86	5
				70	0.0092	0.092	65		
				100	0.0163	0.163	65		
Cyclohexanecarboxylic acid	$C_7H_{12}O_2$	98-89-5	128.169	15	0.201	2.01	27		
Cyclohexanol	$C_6H_{12}O$	108-93-0	100.158	10	4.62	48.4	1		
				25	3.8	40	1		
				40	3.30	34.1	1		
Cyclohexanone	$C_6H_{10}O$	108-94-1	98.142	10	12.2	139	83		
				25	9.5	105	83		
				50	7.6	82	83		
				80	6.8	73	20		
Cyclohexanone oxime	$C_6H_{11}NO$	100-64-1	113.157	25	1.57	16.0	40		
Cyclohexene	C_6H_{10}	110-83-8	82.143	25	0.016	0.16	3	4.00	116
Cyclohexyl butanoate	$C_{10}H_{18}O_2$	1551-44-6	170.249	20	0.11	1.1	20		
				90	0.09	0.90	20		
Cyclooctane	C_8H_{16}	292-64-8	112.213	25	0.00079	0.0079	4	14.08	97
1,3-Cyclopentadiene	C_5H_6	542-92-7	66.102	25	0.068	0.68	3		
Cyclopentane	C_5H_{10}	287-92-3	70.133	25	0.0157	0.157	3	18.52	5
Cyclopentanol	$C_5H_{10}O$	96-41-3	86.132	19	10.6	119	88		
				50	8.3	91	88		
				90	9.2	101	88		
Cyclopentanone	C_5H_8O	120-92-3	84.117	0	37.7	605	20		
				20	31.0	449	20		
				80	24.8	330	20		
Cyclopentene	C_5H_8	142-29-0	68.118	25	0.054	0.54	3	4.35	93
Cyclopropane	C_3H_6	75-19-4	42.080	25	0.0484*	0.484*	19		
Cyfluthrin	$C_{22}H_{18}Cl_2FNO_3$	68359-37-5	434.287	20	0.0000002	0.000002	32		
Cygon	$C_5H_{12}NO_3PS_2$	60-51-5	229.258	20	2.6	27	40		
Cyhalothrin	$C_{23}H_{19}ClF_3NO_3$	91465-08-6	449.850	20	0.0000005	0.000005	32		
Cypermethrin	$C_{22}H_{19}Cl_2NO_3$	52315-07-8	416.297	20	0.000001	0.00001	32		
L-Cystine	$C_6H_{12}N_2O_4S_2$	56-89-3	240.300	25	0.0166	0.166	62		
Cytisine	$C_{11}H_{14}N_2O$	485-35-8	190.241	16	≈30	≈430	40		
Cytosine	$C_4H_5N_3O$	71-30-7	111.102	25	0.73	7.3	29		
Daminozide	$C_6H_{12}N_2O_3$	1596-84-5	160.170	25	9.1	100	40		
Dazomet	$C_5H_{10}N_2S_2$	533-74-4	162.276	25	0.12	1.2	40		
Decabromobiphenyl ether	$C_{12}Br_{10}O$	1163-19-5	959.167	25	0.0000025	0.000025	40		
Decachlorobiphenyl	$C_{12}Cl_{10}$	2051-24-3	498.658	25	0.00000000012	0.0000000012	7	0.0208	7

Name	Mol. form.	CAS Reg. No.	Mol. wt.	t/°C	s/mass%	s/g kg^{-1} H_2O	Ref.	k_H/kPa m^3 mol^{-1}	Ref.
cis-Decahydronaphthalene	$C_{10}H_{18}$	493-01-6	138.250	25	0.000089	0.00089	37		
trans-Decahydronaphthalene	$C_{10}H_{18}$	493-02-7	138.250	25	0.000089	0.00089	4	3.70	13
Decane	$C_{10}H_{22}$	124-18-5	142.282	0	0.0000015	0.000015	4	714	5
Decanedioic acid	$C_{10}H_{18}O_4$	111-20-6	202.248	20	0.10	1.0	40		
Decanoic acid	$C_{10}H_{20}O_2$	334-48-5	172.265	20	0.015	0.15	26		
1-Decanol	$C_{10}H_{22}O$	112-30-1	158.281	25	0.0037	0.037	1		
2-Decanone	$C_{10}H_{20}O$	693-54-9	156.265	25	0.0079	0.079	84		
4-Decanone	$C_{10}H_{20}O$	624-16-8	156.265	20	0.0238	0.238	20		
				80	0.0064	0.064	20		
1-Decene	$C_{10}H_{20}$	872-05-9	140.266	25	0.00057	0.0057	4		
2'-Deoxyadenosine	$C_{10}H_{13}N_5O_3$	958-09-8	251.242	25	0.67	6.7	29		
Deoxycholic acid	$C_{24}H_{40}O_4$	83-44-3	392.573	20	0.001	0.01	40		
Dexamethasone	$C_{22}H_{29}FO_5$	50-02-2	392.460	25	0.009	0.09	40		
Dibenz[a,j]acridine	$C_{21}H_{13}N$	224-42-0	279.335	25	0.000016	0.00016	6		
Dibenz[a,h]anthracene	$C_{22}H_{14}$	53-70-3	278.346	25	0.00000005	0.0000005	42,4		
Dibenz[a,j]anthracene	$C_{22}H_{14}$	224-41-9	278.346	27	0.0000012	0.000012	42,4		
13*H*-Dibenzo[a,i]carbazole	$C_{20}H_{13}N$	239-64-5	267.324	24	0.00000104	0.0000104	6		
Dibenzo[b,e][1,4]dioxin	$C_{12}H_8O_2$	262-12-4	184.191	25	0.000126	0.00126	68		
Dibenzofuran	$C_{12}H_8O$	132-64-9	168.191	25	0.000475	0.00475	41	0.0213	104
Dibenzothiophene	$C_{12}H_8S$	132-65-0	184.257	25	0.000103	0.00103	6		
Dibenzyl ether	$C_{14}H_{14}O$	103-50-4	198.260	35	0.0040	0.040	10		
o-Dibromobenzene	$C_6H_4Br_2$	583-53-9	235.904	25	0.00748	0.0748	2		
m-Dibromobenzene	$C_6H_4Br_2$	108-36-1	235.904	25	0.0064	0.064	2		
p-Dibromobenzene	$C_6H_4Br_2$	106-37-6	235.904	25	0.0020	0.020	2		
1,4-Dibromobutane	$C_4H_8Br_2$	110-52-1	215.915	25	0.035	0.35	35		
1,2-Dibromo-1-chloroethane	$C_2H_3Br_2Cl$	598-20-9	222.306	20	0.060	0.60	25		
1,2-Dibromo-3-chloropropane	$C_3H_5Br_2Cl$	96-12-8	236.333	20	0.123	1.23	35		
1,2-Dibromo-1,2-dichloroethane	$C_2H_2Br_2Cl_2$	683-68-1	256.751	20	0.070	0.70	25		
1,2-Dibromoethane	$C_2H_4Br_2$	106-93-4	187.861	20	0.412	4.14	20		
				50	0.493	4.95	20	0.833	102
				80	0.572	5.75	20		
1,2-Dibromo-1,1,2,3,3,3-hexafluoropropane	$C_3Br_2F_6$	661-95-0	309.830	21	0.0068	0.068	35		
3,5-Dibromo-4-hydroxybenzonitrile	$C_7H_3Br_2NO$	1689-84-5	276.913	25	0.013	0.13	40		
Dibromomethane	CH_2Br_2	74-95-3	173.835	20	1.28	13.0	20	0.086	13
				90	1.51	15.3	20		
2,4-Dibromophenol	$C_6H_4Br_2O$	615-58-7	251.903	25	0.2	2	2		
1,2-Dibromopropane	$C_3H_6Br_2$	78-75-1	201.888	25	0.143	1.43	10		
1,3-Dibromopropane	$C_3H_6Br_2$	109-64-8	201.888	25	0.169	1.69	35		
1,2-Dibromotetrafluoroethane	$C_2Br_2F_4$	124-73-2	259.823	25	0.00030	0.0030	25		
Dibutylamine	$C_8H_{19}N$	111-92-2	129.244	20	0.47	4.7	10		
Dibutyl ether	$C_8H_{18}O$	142-96-1	130.228	0	0.040	0.40	20	0.454	109
				20	0.023	0.23	20	0.48	13
				90	0.010	0.10	20		
Dibutyl phthalate	$C_{16}H_{22}O_4$	84-74-2	278.344	25	0.00112	0.0112	15		
Dibutyl sebacate	$C_{18}H_{34}O_4$	109-43-3	314.461	20	0.004	0.04	10		
o-Dichlorobenzene	$C_6H_4Cl_2$	95-50-1	147.002	5	0.012	0.12	61,58,2		
				25	0.015	0.15	61,58,2		
				45	0.020	0.20	61,58,2		
m-Dichlorobenzene	$C_6H_4Cl_2$	541-73-1	147.002	10	0.0103	0.103	41,2		
				25	0.0120	0.120	41,2	0.357	5
				45	0.0141	0.141	61,2		
p-Dichlorobenzene	$C_6H_4Cl_2$	106-46-7	147.002	10	0.00512	0.0512	2		
				25	0.0080	0.080	41	0.222	101
				50	0.0167	0.167	2		
3,5-Dichloro-1,2-benzenediol	$C_6H_4Cl_2O_2$	13673-92-2	179.001	25	0.78	7.8	8		
4,5-Dichloro-1,2-benzenediol	$C_6H_4Cl_2O_2$	3428-24-8	179.001	25	1.19	12.0	8		
3,3'-Dichloro-*p*-benzidine	$C_{12}H_{10}Cl_2N_2$	91-94-1	253.126	25	0.00031	0.0031	40		
2,5-Dichlorobiphenyl	$C_{12}H_8Cl_2$	34883-39-1	223.098	25	0.0002	0.002	7	0.040	100
2,6-Dichlorobiphenyl	$C_{12}H_8Cl_2$	33146-45-1	223.098	25	0.00014	0.0014	7		
1,1-Dichloro-2,2-bis(*p*-chlorophenyl)ethane	$C_{14}H_{10}Cl_4$	72-54-8	320.041	25	0.000009	0.00009	40		
				45	0.000024	0.00024	40		

Name	Mol. form.	CAS Reg. No.	Mol. wt.	t/°C	s/mass%	s/g kg⁻¹ H₂O	Ref.	k_H/kPa m³ mol⁻¹	Ref.
1,1-Dichlorobutane	$C_4H_8Cl_2$	541-33-3	127.013	25	0.050	0.50	35		
1,4-Dichlorobutane	$C_4H_8Cl_2$	110-56-5	127.013	25	0.16	1.6	35		
2,3-Dichlorobutane, (±)-	$C_4H_8Cl_2$	2211-67-8	127.013	20	0.056	0.56	35		
2,7-Dichlorodibenzo-*p*-dioxin	$C_{12}H_6Cl_2O_2$	33857-26-0	253.081	25	4.1E-07	0.0000041	68		
1,2-Dichloro-1,1-difluoroethane	$C_2H_2Cl_2F_2$	1649-08-7	134.940	24	0.49	4.9	25		
Dichlorodifluoromethane	CCl_2F_2	75-71-8	120.914	20	0.028*	0.28*	5	33.3	126
1,3-Dichloro-5,5-dimethyl hydantoin	$C_5H_6Cl_2N_2O_2$	118-52-5	197.019	20	0.050	0.50	40		
1,1-Dichloroethane	$C_2H_4Cl_2$	75-34-3	98.959	0	0.62	6.2	25		
				25	0.50	5.0	25	0.588	5
				50	0.50	5.0	25		
1,2-Dichloroethane	$C_2H_4Cl_2$	107-06-2	98.959	0	0.92	9.2	25		
				25	0.86	8.6	25	0.11	5
				50	1.05	10.6	25		
				100	2.17	22.2	25		
1,1-Dichloroethene	$C_2H_2Cl_2$	75-35-4	96.943	5	0.310	3.10	25		
				25	0.242	2.42	25	2.70	125
				50	0.225	2.25	25		
				90	0.355	3.55	25		
cis-1,2-Dichloroethene	$C_2H_2Cl_2$	156-59-2	96.943	10	0.76	7.6	25		
				25	0.64	6.4	25	0.38	125
				40	0.66	6.6	25		
trans-1,2-Dichloroethene	$C_2H_2Cl_2$	156-60-5	96.943	10	0.53	5.3	25		
				25	0.45	4.5	25	1.00	125
				40	0.41	4.1	25		
1,1-Dichloro-1-fluoroethane	$C_2H_3Cl_2F$	1717-00-6	116.949	25	0.042	0.42	25		
Dichlorofluoromethane	$CHCl_2F$	75-43-4	102.923	25	0.95*	9.5*	10		
1,2-Dichloro-1,1,2,3,3,3-hexafluoropropane	$C_3Cl_2F_6$	661-97-2	220.928	21	0.0096	0.096	35		
1,4-Dichloro-5-isopropyl-2-methylbenzene	$C_{10}H_{12}Cl_2$	81686-41-1	203.108	25	0.00049	0.0049	23		
Dichloromethane	CH_2Cl_2	75-09-2	84.933	25	1.73	17.6	20	0.278	121
3,6-Dichloro-2-methoxybenzoic acid	$C_8H_6Cl_2O_3$	1918-00-9	221.038	25	0.45	4.5	40		
(Dichloromethyl)benzene	$C_7H_6Cl_2$	98-87-3	161.029	30	0.025	0.25	10		
2,3-Dichloro-2-methylbutane	$C_5H_{10}Cl_2$	507-45-9	141.038	25	0.029	0.29	35		
2,4-Dichloro-6-methylphenol	$C_7H_6Cl_2O$	1570-65-6	177.028	25	0.0283	0.283	2		
2,6-Dichloro-4-methylphenol	$C_7H_6Cl_2O$	194649	177.028	25	0.0673	0.673	2		
2,3-Dichloro-1,4-naphthalenedione	$C_{10}H_4Cl_2O_2$	117-80-6	227.044	25	0.00001	0.0001	40		
1,2-Dichloro-4-nitrobenzene	$C_6H_3Cl_2NO_2$	99-54-7	192.000	20	0.0121	0.121	40		
1,2-Dichloropentane	$C_5H_{10}Cl_2$	1674-33-5	141.038	25	0.029	0.29	35		
1,5-Dichloropentane	$C_5H_{10}Cl_2$	628-76-2	141.038	19	0.02	0.2	35		
2,3-Dichloropentane	$C_5H_{10}Cl_2$	600-11-3	141.038	25	0.029	0.29	35		
Dichlorophene	$C_{13}H_{10}Cl_2O_2$	97-23-4	269.123	25	0.003	0.03	40		
2,3-Dichlorophenol	$C_6H_4Cl_2O$	576-24-9	163.001	25	0.82	8.3	40		
2,4-Dichlorophenol	$C_6H_4Cl_2O$	120-83-2	163.001	25	0.55	5.5	48,51,24		
2,6-Dichlorophenol	$C_6H_4Cl_2O$	87-65-0	163.001	25	0.262	2.62	40		
(2,4-Dichlorophenoxy)acetic acid	$C_8H_6Cl_2O_3$	94-75-7	221.038	25	0.07	0.7	40		
4-(2,4-Dichlorophenoxy)butanoic acid	$C_{10}H_{10}Cl_2O_3$	94-82-6	249.090	25	0.0046	0.046	40		
2-(2,4-Dichlorophenoxy)propanoic acid	$C_9H_8Cl_2O_3$	120-36-5	235.064	25	0.083	0.83	40		
1,2-Dichloropropane, (±)-	$C_3H_6Cl_2$	78-87-5	112.986	5	0.270	2.70	35		
				25	0.274	2.74	35	0.294	124
				40	0.297	2.97	35		
1,3-Dichloropropane	$C_3H_6Cl_2$	142-28-9	112.986	5	0.218	2.18	35		
				25	0.280	2.80	35		
cis-1,3-Dichloropropene	$C_3H_4Cl_2$	10061-01-5	110.970	20	0.27	2.7	5	0.238	5
trans-1,3-Dichloropropene	$C_3H_4Cl_2$	10061-02-6	110.970	20	0.28	2.8	5	0.179	5
2,3-Dichloropropene	$C_3H_4Cl_2$	78-88-6	110.970	25	0.215	2.15	5	0.357	5
1,2-Dichloro-1,1,2,2-tetrafluoroethane	$C_2Cl_2F_4$	76-14-2	170.921	25	0.013*	0.13*	10	111	118
2,4-Dichlorotoluene	$C_7H_6Cl_2$	95-73-8	161.029	25	0.00260	0.0260	61		
2,6-Dichlorotoluene	$C_7H_6Cl_2$	118-69-4	161.029	25	0.00233	0.0233	61		
2,2-Dichloro-1,1,1-trifluoroethane	$C_2HCl_2F_3$	306-83-2	152.930	25	0.46	4.6	25		
Diclofop-methyl	$C_{16}H_{14}Cl_2O_4$	51338-27-3	341.186	20	0.0003	0.003	32		
Dieldrin	$C_{12}H_8Cl_6O$	60-57-1	380.909	25	0.000020	0.00020	67		
Diethanolamine	$C_4H_{11}NO_2$	111-42-2	105.136	20	95.4	20700	10		

Name	Mol. form.	CAS Reg. No.	Mol. wt.	t/°C	s/mass%	s/g kg^{-1} H_2O	Ref.	k_H/kPa m^3 mol^{-1}	Ref.
1,1-Diethoxyethane	$C_6H_{14}O_2$	105-57-7	118.174	25	5	53	10		
1,2-Diethoxyethane	$C_6H_{14}O_2$	629-14-1	118.174	20	21.0	266	10		
2-(Diethylamino)-*N*-(2,6-dimethylphenyl) acetamide	$C_{14}H_{22}N_2O$	137-58-6	234.337	25	0.38	3.8	40		
o-Diethylbenzene	$C_{10}H_{14}$	135-01-3	134.218	20	0.0071	0.071	40		
p-Diethylbenzene	$C_{10}H_{14}$	105-05-5	134.218	20	0.0025	0.025	40		
Diethyl carbonate	$C_5H_{10}O_3$	105-58-8	118.131	20	1.8	18	40		
Diethyl ether	$C_4H_{10}O$	60-29-7	74.121	0	12.5	143	79		
				25	5.9	63	79	0.0909	102
				38	4.6	48	79		
				82	3.1	32	79	0.088	13
Diethyl glutarate	$C_9H_{16}O_4$	818-38-2	188.221	30	1.20	12.1	20		
				91	0.91	9.2	20		
Diethyl maleate	$C_8H_{12}O_4$	141-05-9	172.179	20	1.56	15.8	20		
				91	1.75	17.8	20		
Diethyl malonate	$C_7H_{12}O_4$	105-53-3	160.168	20	2.26	23.1	20		
				91	2.47	25.3	20		
Diethyl phthalate	$C_{12}H_{14}O_4$	84-66-2	222.237	25	0.12	1.2	40		
trans-Diethylstilbestrol	$C_{18}H_{20}O_2$	56-53-1	268.351	20	0.01	0.1	40		
Diethyl succinate	$C_8H_{14}O_4$	123-25-1	174.195	20	0.19	1.9	40		
Diethyl sulfide	$C_4H_{10}S$	352-93-2	90.187	25	0.307	3.07	40		
Diflubenzuron	$C_{14}H_9ClF_2N_2O_2$	35367-38-5	310.683	20	0.00002	0.0002	40		
o-Difluorobenzene	$C_6H_4F_2$	367-11-3	114.093	25	0.114	1.14	2		
m-Difluorobenzene	$C_6H_4F_2$	372-18-9	114.093	25	0.114	1.14	2		
p-Difluorobenzene	$C_6H_4F_2$	540-36-3	114.093	25	0.122	1.22	2		
1,1-Difluoroethane	$C_2H_4F_2$	75-37-6	66.050	20	0.29*	2.9*	50		
Digitoxin	$C_{41}H_{64}O_{13}$	71-63-6	764.939	25	0.0004	0.004	40		
Diglycolic acid	$C_4H_6O_5$	110-99-6	134.088	24	40.0	667	34		
				50	59.9	1490	34		
Digoxin	$C_{41}H_{64}O_{14}$	20830-75-5	780.939	25	0.0059	0.059	40		
Dihexyl ether	$C_{12}H_{26}O$	112-58-3	186.333	20	0.019	0.19	20		
				90	0.019	0.19	20		
1,2-Dihydrobenz[j]aceanthrylene	$C_{20}H_{14}$	479-23-2	254.325	27	0.00000035	0.0000035	42,6		
1,3-Dihydro-2*H*-benzimidazoL-2-one	$C_7H_6N_2O$	615-16-7	134.135	24	0.37	3.7	54		
1,2-Dihydro-3-methylbenz[j]aceanthrylene	$C_{21}H_{16}$	56-49-5	268.352	25	0.00000022	0.0000022	42,6		
				27	0.00000028	0.0000028	42		
2,3-Dihydro-6-propyl-2-thioxo-4(1H)-pyrimidinone	$C_7H_{10}N_2OS$	51-52-5	170.231	25	0.120	1.20	40		
1,7-Dihydro-6*H*-purine-6-thione	$C_5H_4N_4S$	50-44-2	152.178	25	0.0124	0.124	40		
3,4-Dihydro-2*H*-pyran	C_5H_8O	110-87-2	84.117	20	1.04	10.5	20		
				82	2.26	23.1	20		
1,4-Dihydroxy-9,10-anthracenedione	$C_{14}H_8O_4$	81-64-1	240.212	25	0.0000096	0.000096	40		
3,4-Dihydroxybenzoic acid	$C_7H_6O_4$	99-50-3	154.121	14	1.8	18	26		
				80	21.3	271	26		
17,21-Dihydroxypregna-1,4-diene-3,11,20-trione	$C_{21}H_{26}O_5$	53-03-2	358.428	25	0.012	0.12	40		
17,21-Dihydroxypregn-4-ene-3,11,20-trione	$C_{21}H_{28}O_5$	53-06-5	360.444	25	0.028	0.28	30		
o-Diiodobenzene	$C_6H_4I_2$	615-42-9	329.905	25	0.00192	0.0192	2		
m-Diiodobenzene	$C_6H_4I_2$	626-00-6	329.905	25	0.000185	0.00185	2		
p-Diiodobenzene	$C_6H_4I_2$	624-38-4	329.905	25	0.000893	0.00893	2		
cis-1,2-Diiodoethene	$C_2H_2I_2$	590-26-1	279.846	25	0.046	0.46	25		
trans-1,2-Diiodoethene	$C_2H_2I_2$	590-27-2	279.846	25	0.015	0.15	25		
Diiodomethane	CH_2I_2	75-11-6	267.836	30	0.124	1.24	10	0.043	127
3,5-Diiodo-L-tyrosine	$C_9H_9I_2NO_3$	300-39-0	432.981	25	0.062	0.62	26		
Diisopentyl ether	$C_{10}H_{22}O$	544-01-4	158.281	20	0.02	0.2	10		
Diisopropyl ether	$C_6H_{14}O$	108-20-3	102.174	20	0.79	8.0	20	0.256	92
				61	0.22	2.2	20		
1,2-Dimethoxybenzene	$C_8H_{10}O_2$	91-16-7	138.164	20	0.716	7.21	20		
				92	1.073	10.85	20		
3,3'-Dimethoxybenzidine	$C_{14}H_{16}N_2O_2$	119-90-4	244.289	25	0.006	0.06	40		
Dimethoxymethane	$C_3H_8O_2$	109-87-5	76.095	16	24.4	323	10		

Name	Mol. form.	CAS Reg. No.	Mol. wt.	t/°C	s/mass%	s/g kg⁻¹ H₂O	Ref.	k_H/kPa m³ mol⁻¹	Ref.
4-(Dimethylamino)azobenzene	$C_{14}H_{15}N_3$	60-11-7	225.289	20	0.00014	0.0014	40		
2′,3-Dimethyl-4-aminoazobenzene	$C_{14}H_{15}N_3$	97-56-3	225.289	37	0.0007	0.007	40		
2,5-Dimethylaniline	$C_8H_{11}N$	95-78-3	121.180	20	0.66	6.6	27		
N,N-Dimethylaniline	$C_8H_{11}N$	121-69-7	121.180	25	0.111	1.11	40		
9,10-Dimethylanthracene	$C_{16}H_{14}$	781-43-1	206.282	25	0.0000056	0.000056	42,4		
Dimethylarsinic acid	$C_2H_7AsO_2$	75-60-5	137.998	25	≈41	≈700	40		
7,12-Dimethylbenz[a]anthracene	$C_{20}H_{16}$	57-97-6	256.341	25	0.0000061	0.000061	42		
2,2-Dimethylbutane	C_6H_{14}	75-83-2	86.175	25	0.0021	0.021	3	172	5
2,3-Dimethylbutane	C_6H_{14}	79-29-8	86.175	25	0.0021	0.021	3	130	5
2,2-Dimethyl-1-butanol	$C_6H_{14}O$	1185-33-7	102.174	25	0.78	7.9	78,1		
2,3-Dimethyl-2-butanol	$C_6H_{14}O$	594-60-5	102.174	25	4.2	44	1		
3,3-Dimethyl-2-butanol, (±)-	$C_6H_{14}O$	20281-91-8	102.174	25	2.4	25	1		
3,3-Dimethyl-2-butanone	$C_6H_{12}O$	75-97-8	100.158	0	2.92	30.1	83		
				19	1.97	20.1	20		
				25	1.85	18.8	83		
				50	1.46	14.8	83		
				90	1.14	11.5	20		
2,3-Dimethyl-1-butene	C_6H_{12}	563-78-0	84.159	30	0.046	0.46	3		
cis-1,2-Dimethylcyclohexane	C_8H_{16}	112134	112.213	25	0.00060	0.0060	4	36	5
trans-1,2-Dimethylcyclohexane	C_8H_{16}	6876-23-9	112.213	30	0.00050	0.0050	57,4	58.8	97
				100	0.00293	0.0293	57,4		
Dimethyl ether	C_2H_6O	115-10-6	46.068	25	35.3*	546	79		
				50	29.2*	412	79		
Dimethylglyoxime	$C_4H_8N_2O_2$	95-45-4	116.119	20	0.06	0.6	40		
3,5-Dimethyl-4-heptanol	$C_9H_{20}O$	19549-79-2	144.254	15	0.072	0.72	1		
2,6-Dimethyl-4-heptanone	$C_9H_{18}O$	108-83-8	142.238	21	0.045	0.45	20		
				91	0.037	0.37	20		
1,2-Dimethyl-1*H*-imidazole	$C_5H_8N_2$	1739-84-0	96.131	19	94.3	16500	54		
Dimethyl maleate	$C_6H_8O_4$	624-48-6	144.126	25	8.0	87	10		
Dimethyl malonate	$C_5H_8O_4$	108-59-8	132.116	19	14.9	175	20		
				90	29.8	425	20		
1,3-Dimethylnaphthalene	$C_{12}H_{12}$	575-41-7	156.223	25	0.0008	0.008	4		
1,4-Dimethylnaphthalene	$C_{12}H_{12}$	571-58-4	156.223	25	0.00114	0.0114	4		
1,5-Dimethylnaphthalene	$C_{12}H_{12}$	571-61-9	156.223	25	0.00031	0.0031	4	0.0357	28
2,3-Dimethylnaphthalene	$C_{12}H_{12}$	581-40-8	156.223	25	0.00025	0.0025	4		
2,6-Dimethylnaphthalene	$C_{12}H_{12}$	581-42-0	156.223	25	0.00017	0.0017	4		
Dimethyl oxalate	$C_4H_6O_4$	553-90-2	118.089	20	5.82	61.8	27		
2,2-Dimethylpentane	C_7H_{16}	590-35-2	100.202	25	0.00044	0.0044	3	323	5
2,3-Dimethylpentane	C_7H_{16}	565-59-3	100.202	25	0.00052	0.0052	3	175	5
2,4-Dimethylpentane	C_7H_{16}	108-08-7	100.202	25	0.00042	0.0042	3	303	5
3,3-Dimethylpentane	C_7H_{16}	562-49-2	100.202	25	0.00059	0.0059	3	185	5
2,3-Dimethyl-2-pentanol	$C_7H_{16}O$	4911-70-0	116.201	25	1.5	15	1		
2,4-Dimethyl-2-pentanol	$C_7H_{16}O$	625-06-9	116.201	25	1.3	13	1		
2,2-Dimethyl-3-pentanol	$C_7H_{16}O$	3970-62-5	116.201	25	0.82	8.2	1		
2,3-Dimethyl-3-pentanol	$C_7H_{16}O$	595-41-5	116.201	25	1.6	16	1		
2,4-Dimethyl-3-pentanol	$C_7H_{16}O$	600-36-2	116.201	25	0.70	7.0	1		
2,4-Dimethyl-3-pentanone	$C_7H_{14}O$	565-80-0	114.185	20	0.52	5.2	20		
				90	0.30	3.0	20		
N,N-Dimethyl-*N*′-phenylurea	$C_9H_{12}N_2O$	101-42-8	164.203	25	0.32	3.2	40		
Dimethyl phthalate	$C_{10}H_{10}O_4$	131-11-3	194.184	25	0.40	4.0	15		
2,2-Dimethyl-1-propanol	$C_5H_{12}O$	75-84-3	88.148	12	3.87	40.3	78,1		
				25	3.26	33.7	78,1		
				80	2.84	29.2	78,1		
4-(1,1-Dimethylpropyl)phenol	$C_{11}H_{16}O$	80-46-6	164.244	25	0.017	0.17	40		
Dimethyl succinate	$C_6H_{10}O_4$	106-65-0	146.141	21	12.4	142	20		
				92	17.1	206	20		
Dimethyl sulfate	$C_2H_6O_4S$	77-78-1	126.132	18	2.7	28	27		
Dimethyl sulfide	C_2H_6S	75-18-3	62.134	25	2	20	10		
Dimethyl sulfoxide	C_2H_6OS	67-68-5	78.133	25	25.3	339	10		
Dimethyl terephthalate	$C_{10}H_{10}O_4$	120-61-6	194.184	25	0.00328	0.0328	40		
Dimethyl tetrachloroterephthalate	$C_{10}H_6Cl_4O_4$	1861-32-1	331.965	25	0.00005	0.0005	40		

Name	Mol. form.	CAS Reg. No.	Mol. wt.	t/°C	s/mass%	s/g kg^{-1} H_2O	Ref.	k_H/kPa m^3 mol^{-1}	Ref.
N,N-Dimethyl-N'-[3-(trifluoromethyl)phenyl] urea	$C_{10}H_{11}F_3N_2O$	2164-17-2	232.201	20	0.0105	0.105	40		
2,4-Dinitroaniline	$C_6H_5N_3O_4$	97-02-9	183.122	25	0.0078	0.078	40		
1,2-Dinitrobenzene	$C_6H_4N_2O_4$	528-29-0	168.107	20	0.21	2.1	27		
1,3-Dinitrobenzene	$C_6H_4N_2O_4$	99-65-0	168.107	20	2.09	21.3	27		
1,4-Dinitrobenzene	$C_6H_4N_2O_4$	100-25-4	168.107	20	1.30	13.2	27		
3,5-Dinitrobenzoic acid	$C_7H_4N_2O_6$	99-34-3	212.116	25	0.134	1.34	27		
2,4-Dinitrophenol	$C_6H_4N_2O_5$	51-28-5	184.106	25	0.069	0.69	48,51		
				35	0.098	0.98	48,51		
Dipentyl ether	$C_{10}H_{22}O$	693-65-2	158.281	25	0.11	1.1	81		
Diphenamid	$C_{16}H_{17}NO$	957-51-7	239.312	27	0.026	0.26	32		
Diphenylamine	$C_{12}H_{11}N$	122-39-4	169.222	20	0.0055	0.055	40		
				50	0.0058	0.058	40		
1,2-Diphenylethane	$C_{14}H_{14}$	103-29-7	182.261	25	0.00044	0.0044	6	0.0169	113
Diphenyl ether	$C_{12}H_{10}O$	101-84-8	170.206	25	0.0018	0.0180	6	0.0286	104
Diphenylmethane	$C_{13}H_{12}$	101-81-5	168.234	25	0.00014	0.0014	42,4	0.0132	104
Diphenyl phthalate	$C_{20}H_{14}O_4$	84-62-8	318.323	24	0.000008	0.00008	40		
1,3-Diphenyl-1-triazene	$C_{12}H_{11}N_3$	136-35-6	197.235	20	0.050	0.50	40		
N,N'-Diphenylurea	$C_{13}H_{12}N_2O$	102-07-8	212.246	20	0.015	0.15	40		
Dipropylamine	$C_6H_{15}N$	142-84-7	101.190	20	2.5	26	10		
Dipropyl ether	$C_6H_{14}O$	111-43-3	102.174	0	2.67	27.4	80		
				25	0.91	9.2	80	0.333	109
Diuron	$C_9H_{10}Cl_2N_2O$	330-54-1	233.093	25	0.0042	0.042	40		
Docosane	$C_{22}H_{46}$	629-97-0	310.600	22	0.0000006	0.000006	37		
Dodecane	$C_{12}H_{26}$	112-40-3	170.334	25	0.00000037	0.0000037	4	769	5
Dodecanedioic acid	$C_{12}H_{22}O_4$	693-23-2	230.301	20	0.004	0.04	40		
Dodecanoic acid	$C_{12}H_{24}O_2$	143-07-7	200.318	20	0.0055	0.055	26		
1-Dodecanol	$C_{12}H_{26}O$	112-53-8	186.333	25	0.0004	0.004	1		
Droperidol	$C_{22}H_{22}FN_3O_2$	548-73-2	379.427	30	0.00041	0.0041	40		
Eicosane	$C_{20}H_{42}$	112-95-8	282.547	25	0.00000019	0.0000019	42,4		
Emetine	$C_{29}H_{40}N_2O_4$	483-18-1	480.639	15	0.096	0.96	40		
Endrin	$C_{12}H_8Cl_6O$	72-20-8	380.909	25	0.000025	0.00025	67		
Ephedrine, (1R,2S)	$C_{10}H_{15}NO$	299-42-3	165.232	25	0.57	5.7	40		
Epichlorohydrin	C_3H_5ClO	13403-37-7	92.524	20	6.58	70.4	10	0.003	13
				65	7.2	78	40		
Epinephrine	$C_9H_{13}NO_3$	51-43-4	183.204	20	0.018	0.18	40		
1,2-Epoxy-4-(epoxyethyl)cyclohexane	$C_8H_{12}O_2$	106-87-6	140.180	20	13.4	155	40		
2,3-Epoxy-α-pinane	$C_{10}H_{16}O$	1686-14-2	152.233	25	0.039	0.39	52		
Erythromycin	$C_{37}H_{67}NO_{13}$	114-07-8	733.927	30	0.12	1.2	40		
				80	0.04	0.4	40		
Estra-1,3,5(10)-triene-3,17-diol (17β)	$C_{18}H_{24}O_2$	50-28-2	272.383	25	0.000151	0.00151	49		
Estrone	$C_{18}H_{22}O_2$	53-16-7	270.367	25	0.000130	0.00130	49		
Ethane	C_2H_6	74-84-0	30.069	25	0.00568*	0.0568*	18	50.0	5
1,2-Ethanediol, diacetate	$C_6H_{10}O_4$	111-55-7	146.141	25	13.3	153	40		
Ethinylestradiol	$C_{20}H_{24}O_2$	57-63-6	296.404	25	0.000921	0.00921	49		
Ethoxybenzene	$C_8H_{10}O$	103-73-1	122.164	25	0.12	1.2	10		
2-Ethoxyethyl acetate	$C_6H_{12}O_3$	111-15-9	132.157		14	163	30		
N-(4-Ethoxyphenyl)acetamide	$C_{10}H_{13}NO_2$	62-44-2	179.216	25	0.0502	0.502	40		
Ethyl acetate	$C_4H_8O_2$	141-78-6	88.106	25	8.08	87.9	10		
Ethyl acetoacetate	$C_6H_{10}O_3$	141-97-9	130.141	25	12	136	10		
Ethyl acrylate	$C_5H_8O_2$	140-88-5	100.117	25	1.50	15.2	10		
Ethylbenzene	C_8H_{10}	100-41-4	106.165	0	0.020	0.20	4,89		
				25	0.0161	0.161	22,89	0.769	5
				40	0.0200	0.200	4,89		
Ethyl benzoate	$C_9H_{10}O_2$	93-89-0	150.174	25	0.083	0.83	20		
Ethyl butanoate	$C_6H_{12}O_2$	105-54-4	116.158	20	0.49	4.9	10		
2-Ethyl-1-butanol	$C_6H_{14}O$	97-95-0	102.174	20	0.92	9.3	78		
				50	0.80	8.1	78		
Ethyl carbamate	$C_3H_7NO_2$	51-79-6	89.094	15	48	920	27		
Ethyl cyanoacetate	$C_5H_7NO_2$	105-56-6	113.116	20	25.9	350	10		
Ethylcyclohexane	C_8H_{16}	1678-91-7	112.213	30	0.00061	0.0061	57,4		

Name	Mol. form.	CAS Reg. No.	Mol. wt.	*t*/°C	*s*/mass%	*s*/g kg⁻¹ H_2O	Ref.	k_H/kPa m³ mol⁻¹	Ref.
				100	0.00212	0.0212	57,4		
Ethylcyclopentane	C_7H_{14}	1640-89-7	98.186	20	0.012	0.12	3		
Ethyl decanoate	$C_{12}H_{24}O_2$	110-38-3	200.318	20	0.0015	0.015	27		
Ethylene	C_2H_4	74-85-1	28.053	25	0.01336*	0.1336*	19	21.7	5
Ethyleneimine	C_2H_5N	151-56-4	43.068	20	0.90	9.1	40		
Ethyl formate	$C_3H_6O_2$	109-94-4	74.079	25	11.8	134	10		
Ethyl heptanoate	$C_9H_{18}O_2$	106-30-9	158.238	20	0.029	0.29	27		
Ethyl hexanoate	$C_8H_{16}O_2$	123-66-0	144.212	20	0.063	0.63	27		
2-Ethyl-1-hexanol	$C_8H_{18}O$	104-76-7	130.228	25	0.071	0.71	78		
				50	0.074	0.74	78		
2-Ethylhexylamine	$C_8H_{19}N$	104-75-6	129.244	20	0.25	2.5	10		
Ethyl 4-hydroxybenzoate	$C_9H_{10}O_3$	120-47-8	166.173	25	0.0080	0.080	40		
Ethyl isopropyl ether	$C_5H_{12}O$	625-54-7	88.148	25	0.52	5.2	79		
Ethyl 2-methylbutanoate, (+)	$C_7H_{14}O_2$	10307-61-6	130.185	19	0.257	2.58	20		
				91	0.151	1.51	20		
Ethyl 3-methylbutanoate	$C_7H_{14}O_2$	108-64-5	130.185	20	0.2	2	10		
Ethyl *N*-methylcarbamate	$C_4H_9NO_2$	105-40-8	103.120	15	69	2230	27		
1-Ethylnaphthalene	$C_{12}H_{12}$	1127-76-0	156.223	25	0.00101	0.0101	4	0.0385	5
2-Ethylnaphthalene	$C_{12}H_{12}$	939-27-5	156.223	25	0.00080	0.0080	4	0.083	5
O-Ethyl *O*-*p*-nitrophenyl benzenethiophosphonate	$C_{14}H_{14}NO_4PS$	2104-64-5	323.304	22	0.00031	0.0031	40		
N-Ethyl-*N*-nitrosourea	$C_3H_7N_3O_2$	759-73-9	117.107	20	1.3	13	40		
Ethyl nonanoate	$C_{11}H_{22}O_2$	123-29-5	186.292	20	0.003	0.03	27		
Ethyl octanoate	$C_{10}H_{20}O_2$	106-32-1	172.265	20	0.007	0.07	27		
Ethyl pentanoate	$C_7H_{14}O_2$	539-82-2	130.185	25	0.3	3	27		
3-Ethyl-3-pentanol	$C_7H_{16}O$	597-49-9	116.201	25	1.7	17	1		
4-Ethylphenol	$C_8H_{10}O$	123-07-9	122.164	20	0.59	5.9	40		
Ethyl propanoate	$C_5H_{10}O_2$	105-37-3	102.132	20	1.92	19.6	10		
Ethyl *N*-propylcarbamate	$C_6H_{13}NO_2$	623-85-8	131.173	15	7.70	83.4	27		
Ethyl propyl ether	$C_5H_{12}O$	628-32-0	88.148	25	1.87	19.2	79		
2-Ethyltoluene	C_9H_{12}	611-14-3	120.191	25	0.0075	0.075	89,5	0.435	5
4-Ethyltoluene	C_9H_{12}	622-96-8	120.191	25	0.0094	0.094	5	0.50	5
Ethyl vinyl ether	C_4H_8O	109-92-2	72.106	20	0.9	9	10		
Etoposide	$C_{29}H_{32}O_{13}$	33419-42-0	588.556	20	0.02	0.2	40		
Eucalyptol	$C_{10}H_{18}O$	470-82-6	154.249	21	0.379	3.79	40		
				50	0.170	1.70	40		
Fenamiphos	$C_{13}H_{22}NO_3PS$	22224-92-6	303.358	20	0.0329	0.329	40		
Fenbutatin oxide	$C_{60}H_{78}OSn_2$	13356-08-6	1052.68	23	0.0000005	0.000005	32		
α-Fenchol, (+)-	$C_{10}H_{18}O$	115823	154.249	25	0.083	0.83	52		
(±)-Fenchone	$C_{10}H_{16}O$	18492-37-0	152.233	20	0.2	2	84		
Fenoxycarb	$C_{17}H_{19}NO_4$	79127-80-3	301.338	20	0.0006	0.006	32		
Ferbam	$C_9H_{18}FeN_3S_6$	14484-64-1	416.494	20	0.013	0.13	40		
Fluoranthene	$C_{16}H_{10}$	206-44-0	202.250	20	0.000017	0.00017	42		
Fluoranthene	$C_{16}H_{10}$	206-44-0	202.250	25	0.000021	0.00021	42,22	0.00133	112
9*H*-Fluorene	$C_{13}H_{10}$	86-73-7	166.218	0	0.00007	0.0007	42,4		
				25	0.00019	0.0019	42,22	0.00909	112
				50	0.00063	0.0063	42,4		
Fluorescein	$C_{20}H_{12}O_5$	153954	332.306	20	0.005	0.05	27		
Fluorobenzene	C_6H_5F	462-06-6	96.102	19	0.170	1.70	20	0.625	5
				80	0.188	1.88	20		
2-Fluorobenzoic acid	$C_7H_5FO_2$	445-29-4	140.112	25	0.72	7.2	27		
3-Fluorobenzoic acid	$C_7H_5FO_2$	455-38-9	140.112	25	0.15	1.5	27		
4-Fluorobenzoic acid	$C_7H_5FO_2$	456-22-4	140.112	25	0.12	1.2	27		
Fluoroethane	C_2H_5F	353-36-6	48.059	25	0.216*	2.16*	14		
Fluoromethane	CH_3F	593-53-3	34.033	0	0.420*	4.20*	50		
				25	0.201*	2.01*	50		
				80	0.082*	0.82*	50		
1-Fluoropropane	C_3H_7F	460-13-9	62.086	14	0.386*	3.86*	14		
2-Fluoropropane	C_3H_7F	420-26-8	62.086	15	0.366*	3.66*	14		
5-Fluorouracil	$C_4H_3FN_2O_2$	51-21-8	130.077	25	1.77	18.0	72		
Folic acid	$C_{19}H_{19}N_7O_6$	59-30-3	441.397	0	0.001	0.01	26		

Name	Mol. form.	CAS Reg. No.	Mol. wt.	t/°C	s/mass%	s/g kg⁻¹ H₂O	Ref.	k_H/kPa m³ mol⁻¹	Ref.
				100	0.05	0.5	26		
Folpet	$C_9H_4Cl_3NO_2S$	133-07-3	296.558	20	0.00010	0.0010	40		
β-D-Fructose	$C_6H_{12}O_6$	53188-23-1	180.155	20	≈31	≈450	40		
Furan	C_4H_4O	110-00-9	68.074	25	1	10	10	0.556	104
2-Furancarboxylic acid	$C_5H_4O_3$	88-14-2	112.084	25	4.76	50.0	33		
				50	25.2	337	33		
Furfural	$C_5H_4O_2$	98-01-1	96.085	20	8.2	89	10		
Galactaric acid	$C_6H_{10}O_8$	526-99-8	210.138	14	0.33	3.3	40		
D-Galactose	$C_6H_{12}O_6$	59-23-4	180.155	20	40.6	684	27		
D-Glucitol	$C_6H_{14}O_6$	50-70-4	182.171	20	≈41	≈700	40		
α-D-Glucose	$C_6H_{12}O_6$	26655-34-5	180.155	15	45.0	818	27		
				30	54.6	1200	27		
				80	81.5	4400	27		
DL-Glutamic acid	$C_5H_9NO_4$	617-65-2	147.130	25	2.30	23.5	29		
L-Glutamic acid	$C_5H_9NO_4$	56-86-0	147.130	10	0.444	4.46	75		
				25	0.824	8.31	75		
				50	2.13	21.8	75		
L-Glutamine	$C_5H_{10}N_2O_3$	56-85-9	146.144	25	4.0	42	26		
Glycerol triacetate	$C_9H_{14}O_6$	102-76-1	218.203	25	5.8	62	10		
Glycine	$C_2H_5NO_2$	56-40-6	75.067	25	18.5	227	70		
				36	22.1	284	70		
				50	26.1	353	70		
Glycolic acid	$C_2H_4O_3$	79-14-1	76.051	25	71.2	2470	34		
				55	77.9	3520	34		
N-Glycylglycine	$C_4H_8N_2O_3$	556-50-3	132.118	25	18.8	232	47,29		
Glyphosate	$C_3H_8NO_5P$	1071-83-6	169.074	25	1.2	12	32		
Guanidinoacetic acid	$C_3H_7N_3O_2$	352-97-6	117.107	25	0.5	5	26		
Guanine	$C_5H_5N_5O$	73-40-5	151.127	25	0.0068	0.068	29		
Guanosine	$C_{10}H_{13}N_5O_5$	118-00-3	283.241	25	0.0500	0.500	29		
Haloperidol	$C_{21}H_{23}ClFNO_2$	52-86-8	375.865	30	0.0003	0.003	40		
Heptachlor	$C_{10}H_5Cl_7$	76-44-8	373.318	25	0.000018	0.00018	67		
2,2',3,3',4,4',6-Heptachlorobiphenyl	$C_{12}H_3Cl_7$	52663-71-5	395.323	25	0.0000002	0.000002	7	0.00526	114
Heptadecanoic acid	$C_{17}H_{34}O_2$	506-12-7	270.451	20	0.00042	0.0042	26		
1,6-Heptadiyne	C_7H_8	2396-63-6	92.139	25	0.125	1.25	3		
Heptanal	$C_7H_{14}O$	111-71-7	114.185	11	0.124	1.24	27		
Heptane	C_7H_{16}	142-82-5	100.202	25	0.000242	0.00242	46		
				50	0.000341	0.00341	46	181	104
				75	0.000570	0.00570	46		
				100	0.00108	0.0108	46		
Heptanedioic acid	$C_7H_{12}O_4$	111-16-0	160.168	25	6.347	67.77	33		
				50	42.80	748	33		
Heptanoic acid	$C_7H_{14}O_2$	111-14-8	130.185	15	0.24	2.4	27		
1-Heptanol	$C_7H_{16}O$	111-70-6	116.201	0	0.236	2.37	78		
				25	0.164	1.64	78,1	0.00263	127
				50	0.164	1.64	78,1		
				90	0.245	2.46	78		
2-Heptanol, (±)-	$C_7H_{16}O$	52390-72-4	116.201	30	0.33	3.3	1		
3-Heptanol, (*S*)-	$C_7H_{16}O$	26549-25-7	116.201	25	0.43	4.3	1		
4-Heptanol	$C_7H_{16}O$	589-55-9	116.201	25	0.47	4.7	1		
2-Heptanone	$C_7H_{14}O$	110-43-0	114.185	25	0.435	4.37	20	0.0161	106
				90	0.353	3.53	20		
3-Heptanone	$C_7H_{14}O$	106-35-4	114.185	20	0.479	4.81	20		
				90	0.309	3.10	20		
4-Heptanone	$C_7H_{14}O$	123-19-3	114.185	20	0.457	4.57	20		
				90	0.316	3.16	20		
1-Heptene	C_7H_{14}	592-76-7	98.186	25	0.032	0.32	3	43.5	104
trans-2-Heptene	C_7H_{14}	14686-13-6	98.186	25	0.015	0.15	3	41.7	5
Heptyl butanoate	$C_{11}H_{22}O_2$	5870-93-9	186.292	20	0.028	0.28	20		
				80	0.020	0.20	20		
1-Heptyne	C_7H_{12}	628-71-7	96.170	25	0.0094	0.094	3	7.69	113

Name	Mol. form.	CAS Reg. No.	Mol. wt.	t/°C	s/mass%	s/g kg⁻¹ H₂O	Ref.	k_H/kPa m³ mol⁻¹	Ref.
Hesperetin	$C_{16}H_{14}O_6$	520-33-2	302.278	15	0.00004	0.0004	71		
				25	0.00014	0.0014	71		
				35	0.00052	0.0052	71		
Hexachlorobenzene	C_6Cl_6	118-74-1	284.782	25	0.00000096	0.0000096	58	0.0667	67
				35	0.0000018	0.000018	58		
				55	0.0000038	0.000038	58		
2,2',3,3',4,4'-Hexachlorobiphenyl	$C_{12}H_4Cl_6$	38380-07-3	360.878	25	0.00000006	0.0000006	7	0.0357	31
2,2',4,4',6,6'-Hexachlorobiphenyl	$C_{12}H_4Cl_6$	33979-03-2	360.878	25	0.0000003	0.000003	41	0.0909	111
2,2',3,3',6,6'-Hexachlorobiphenyl	$C_{12}H_4Cl_6$	38411-22-2	360.878	25	0.0000004	0.000004	41		
Hexachloro-1,3-butadiene	C_4Cl_6	87-68-3	260.761	25	0.41	4.1	35		
1,2,3,4,5,6-Hexachlorocyclohexane, (1α,2α,3β,4α,5α,6β)	$C_6H_6Cl_6$	58-89-9	290.830	25	0.00078	0.0078	60		
				45	0.0015	0.015	60		
1,2,3,4,5,6-Hexachlorocyclohexane, (1α,2α,3β,4α,5β,6β)	$C_6H_6Cl_6$	319-84-6	290.830	25	0.00018	0.0018	60		
1,2,3,4,5,6-Hexachlorocyclohexane, (1α,2β,3α,4β,5α,6β)	$C_6H_6Cl_6$	319-85-7	290.830	25	0.00002	0.0002	60		
Hexachloroethane	C_2Cl_6	67-72-1	236.739	25	0.005	0.05	25	0.40	123
Hexachloropropene	C_3Cl_6	1888-71-7	248.750	20	0.00118	0.0118	35		
Hexacosafluorododecane	$C_{12}F_{26}$	307-59-5	638.086	20	0.00000096	0.0000096	35		
Hexacosane	$C_{26}H_{54}$	630-01-3	366.707	25	0.00000017	0.0000017	42,37		
Hexadecane	$C_{16}H_{34}$	544-76-3	226.441	25	0.0000004	0.000004	42,37		
Hexadecanoic acid	$C_{16}H_{32}O_2$	57-10-3	256.424	20	0.00072	0.0072	26		
1-Hexadecanol	$C_{16}H_{34}O$	36653-82-4	242.440	25	0.000003	0.00003	1		
1,5-Hexadiene	C_6H_{10}	592-42-7	82.143	25	0.017	0.17	3		
Hexafluorobenzene	C_6F_6	392-56-3	186.054	8	0.0778	0.778	53		
				28	0.0616	0.616	53		
				67	0.0636	0.636	53		
Hexahydro-1,3,5-trinitro-1,3,5-triazine	$C_3H_6N_6O_6$	121-82-4	222.116	3	0.0014	0.014	59		
				20	0.0037	0.037	59		
				25	0.0060	0.060	17		
				34	0.0086	0.086	59		
Hexamethylenetetramine	$C_6H_{12}N_4$	100-97-0	140.186	12	44.8	812	27		
Hexane	C_6H_{14}	110-54-3	86.175	25	0.00098	0.0098	46	169	5
				50	0.00114	0.0114	46		
				75	0.00167	0.0167	46		
				100	0.00291	0.0291	46		
1,6-Hexanediamine	$C_6H_{16}N_2$	124-09-4	116.204	5	≈42	≈720	40		
Hexanedinitrile	$C_6H_8N_2$	111-69-3	108.141	20	0.80	8.0	16		
1,6-Hexanedioic acid	$C_6H_{10}O_4$	124-04-9	146.141	15	1.48	15.0	26		
				100	61.5	1600	26		
Hexanoic acid	$C_6H_{12}O_2$	142-62-1	116.158	25	1.01	10.2	64		
				35	1.09	11.0	64		
				60	1.16	11.7	26		
1-Hexanol	$C_6H_{14}O$	111-27-3	102.174	0	0.79	7.9	1		
				10	0.70	7.0	78		
				25	0.59	5.9	78,1		
				50	0.55	5.5	78,1		
2-Hexanol	$C_6H_{14}O$	20281-86-1	102.174	25	1.4	14	1		
3-Hexanol	$C_6H_{14}O$	17015-11-1	102.174	25	1.6	16	1		
2-Hexanone	$C_6H_{12}O$	591-78-6	100.158	10	1.91	19.5	83		
				25	1.49	15.1	83		
				50	1.17	11.8	83		
3-Hexanone	$C_6H_{12}O$	589-38-8	100.158	25	1.47	14.9	83		
Hexatriacontane	$C_{36}H_{74}$	630-06-8	506.973	25	0.00000017	0.0000017	42,37		
Hexazinone	$C_{12}H_{20}N_4O_2$	51235-04-2	252.313	25	3.2	33	40		
1-Hexene	C_6H_{12}	592-41-6	84.159	25	0.0053	0.053	3	41.7	5
trans-2-Hexene	C_6H_{12}	4050-45-7	84.159	25	0.0067	0.067	3		
1-Hexen-3-ol	$C_6H_{12}O$	4798-44-1	100.158	25	2.52	25.9	1		
4-Hexen-2-ol	$C_6H_{12}O$	52387-50-5	100.158	25	3.81	39.6	1		
Hexyl acetate	$C_8H_{16}O_2$	142-92-7	144.212	20	0.02	0.2	10		

Name	Mol. form.	CAS Reg. No.	Mol. wt.	t/°C	s/mass%	s/g kg^{-1} H_2O	Ref.	k_H/kPa m^3 mol^{-1}	Ref.
sec-Hexyl acetate	$C_8H_{16}O_2$	108-84-9	144.212	20	0.13	1.3	10		
Hexylbenzene	$C_{12}H_{18}$	1077-16-3	162.271	25	0.00021	0.0021	4		
4-Hexyl-1,3-benzenediol	$C_{12}H_{18}O_2$	136-77-6	194.270	18	0.05	0.5	40		
Hexyl butanoate	$C_{10}H_{20}O_2$	2639-63-6	172.265	29	0.021	0.21	20		
1-Hexyne	C_6H_{10}	693-02-7	82.143	25	0.036	0.36	3	4.17	113
L-Histidine	$C_6H_9N_3O_2$	71-00-1	155.154	25	4.17	43.5	26		
Homocystine	$C_8H_{16}N_2O_4S_2$	870-93-9	268.354	25	0.02	0.2	26		
L-Homoserine	$C_4H_9NO_3$	672-15-1	119.119	25	52.4	1100	26		
Hydramethylnon	$C_{25}H_{24}F_6N_4$	67485-29-4	494.476	20	0.0000006	0.000006	32		
Hydrochlorothiazide	$C_7H_8ClN_3O_4S_2$	58-93-5	297.740	25	0.007	0.07	40		
Hydrocortisone	$C_{21}H_{30}O_5$	50-23-7	362.460	25	0.029	0.29	40		
Hydroflumethiazide	$C_8H_8F_3N_3O_4S_2$	135-09-1	331.293	37	0.068	0.68	40		
p-Hydroquinone	$C_6H_6O_2$	123-31-9	110.111	25	7.42	80.1	27		
17-Hydroxyandrost-4-en-3-one, (17β)	$C_{19}H_{28}O_2$	58-22-0	288.424	25	0.0024	0.024	40		
4-Hydroxybenzaldehyde	$C_7H_6O_2$	123-08-0	122.122	30	1.27	12.9	40		
2-Hydroxybenzamide	$C_7H_7NO_2$	65-45-2	137.137	10	0.122	1.22	44		
				25	0.241	2.42	44		
				50	0.737	7.42	44		
α-Hydroxybenzeneacetic acid, (±)-	$C_8H_8O_3$	611-72-3	152.148	25	11.3	127	27		
2-Hydroxybenzoic acid	$C_7H_6O_3$	69-72-7	138.121	10	0.119	1.19	43,33		
				25	0.189	1.89	43,33		
				50	0.521	5.24	43,33		
4-Hydroxybenzoic acid	$C_7H_6O_3$	99-96-7	138.121	15	0.8	8	26		
				75	2.5	26	27		
2-Hydroxybiphenyl	$C_{12}H_{10}O$	90-43-7	170.206	25	0.07	0.7	40		
4-Hydroxybiphenyl	$C_{12}H_{10}O$	92-69-3	170.206	25	0.0056	0.056	40		
4-Hydroxy-3-methoxybenzaldehyde	$C_8H_8O_3$	121-33-5	152.148	25	0.247	2.47	8		
3-Hydroxy-4-oxo-4*H*-pyran-2,6-dicarboxylic acid	$C_7H_4O_7$	497-59-6	200.103	25	0.84	8.4	27		
N-(4-Hydroxyphenyl)acetamide	$C_8H_9NO_2$	103-90-2	151.163	25	1.3	13	40		
trans-4-Hydroxy-L-proline	$C_5H_9NO_3$	51-35-4	131.130	25	26.5	361	26		
Hyoscyamine	$C_{17}H_{23}NO_3$	101-31-5	289.370	20	0.36	3.6	40		
Hypoxanthine	$C_5H_4N_4O$	68-94-0	136.112	25	0.070	0.70	29		
Ibuprofen	$C_{13}H_{18}O_2$	15687-27-1	206.281	25	0.0011	0.011	40		
				60	0.0048	0.048	40		
Imazaquin	$C_{17}H_{17}N_3O_3$	81335-37-7	311.335	20	0.009	0.09	32		
Imidacloprid	$C_9H_{10}ClN_5O_2$	105827-78-9	255.66	30	0.038	0.38	73		
				51	0.117	1.17	73		
Imidazole	$C_3H_4N_2$	288-32-4	68.077	19	67.3	2060	54		
2,4-Imidazolidinedione	$C_3H_4N_2O_2$	461-72-3	100.076	25	3.93	40.9	29		
Imidodicarbonic diamide	$C_2H_5N_3O_2$	108-19-0	103.080	15	1.5	15	40		
Iminodiacetic acid	$C_4H_7NO_4$	142-73-4	133.104	5	2.32	23.8	40		
Indan	C_9H_{10}	496-11-7	118.175	25	0.010	0.10	4		
1*H*-Indazole	$C_7H_6N_2$	271-44-3	118.136	20	0.0827	0.827	6		
Indeno[1,2,3-*cd*]pyrene	$C_{22}H_{12}$	193-39-5	276.330	20	0.00000002	0.0000002	40		
1*H*-Indole	C_8H_7N	120-72-9	117.149	20	0.187	1.87	6		
Indomethacin	$C_{19}H_{16}ClNO_4$	53-86-1	357.788	25	0.001	0.01	40		
Inosine	$C_{10}H_{12}N_4O_5$	58-63-9	268.226	20	1.6	16	29		
Iodobenzene	C_6H_5I	591-50-4	204.008	10	0.0193	0.193	2		
				25	0.0226	0.226	2	0.083	104
				45	0.0279	0.279	2		
2-Iodobenzoic acid	$C_7H_5IO_2$	88-67-5	248.018	25	0.095	0.95	27		
3-Iodobenzoic acid	$C_7H_5IO_2$	618-51-9	248.018	25	0.016	0.16	27		
4-Iodobenzoic acid	$C_7H_5IO_2$	619-58-9	248.018	25	0.0027	0.027	27		
1-Iodobutane	C_4H_9I	542-69-8	184.018	17	0.021	0.21	10	1.85	114
Iodoethane	C_2H_5I	75-03-6	155.965	0	0.44	4.4	25		
				25	0.40	4.0	25	0.67	101
1-Iodoheptane	$C_7H_{15}I$	4282-40-0	226.098	25	0.00035	0.0035	35		
Iodomethane	CH_3I	74-88-4	141.939	20	1.4	14	10	0.50	121
1-Iodopropane	C_3H_7I	107-08-4	169.992	0	0.114	1.14	35		
				20	0.100	1.00	35	0.909	110

Name	Mol. form.	CAS Reg. No.	Mol. wt.	t/°C	s/mass%	s/g kg⁻¹ H₂O	Ref.	k_H/kPa m³ mol⁻¹	Ref.
2-Iodopropane	C_3H_7I	75-30-9	169.992	0	0.167	1.67	35		
				20	0.140	1.40	35		
5-Iodouracil	$C_4H_3IN_2O_2$	696-07-1	237.983	25	0.49	4.9	72		
trans-β-Ionone	$C_{13}H_{20}O$	79-77-6	192.297	25	0.017	0.17	52		
Iopanoic acid	$C_{11}H_{12}I_3NO_2$	96-83-3	570.932	37	0.034	0.34	40		
Iprodione	$C_{13}H_{13}Cl_2N_3O_3$	36734-19-7	330.166	20	0.0013	0.013	40		
Isobutanal	C_4H_8O	78-84-2	72.106	20	9.1	100	10		
Isobutane	C_4H_{10}	75-28-5	58.122	25	0.00535*	0.0535*	18	110	121
Isobutene	C_4H_8	115-11-7	56.107	25	0.0263*	0.263*	5	17.9	18
Isobutyl acetate	$C_6H_{12}O_2$	110-19-0	116.158	20	0.63	6.3	10		
Isobutylbenzene	$C_{10}H_{14}$	538-93-2	134.218	25	0.0010	0.010	4	3.33	5
Isobutyl formate	$C_5H_{10}O_2$	542-55-2	102.132	22	1.0	10	10		
Isobutyl isobutanoate	$C_8H_{16}O_2$	97-85-8	144.212	20	0.5	5	10		
Isobutyl propanoate	$C_7H_{14}O_2$	540-42-1	130.185	19	0.225	2.26	20		
				91	0.142	1.42	20		
Isoguanine	$C_5H_5N_5O$	3373-53-3	151.127	25	0.006	0.06	26		
1*H*-Isoindole-1,3(2*H*)-dione	$C_8H_5NO_2$	85-41-6	147.132	25	0.036	0.36	40		
L-Isoleucine	$C_6H_{13}NO_2$	73-32-5	131.173	25	3.31	34.2	26		
Isoniazid	$C_6H_7N_3O$	54-85-3	137.139	25	11.0	124	40		
Isopentane	C_5H_{12}	78-78-4	72.149	25	0.00485	0.0485	3	139	5
Isopentyl acetate	$C_7H_{14}O_2$	123-92-2	130.185	20	0.2	2	10		
Isopentyl formate	$C_6H_{12}O_2$	110-45-2	116.158	22	0.3	3	27		
Isophorone	$C_9H_{14}O$	78-59-1	138.206	20	1.57	16.0	20		
				80	1.27	12.9	20		
Isophthalic acid	$C_8H_6O_4$	121-91-5	166.132	10	0.0062	0.062	76		
				25	0.0154	0.154	56		
				50	0.0395	0.395	56		
				80	0.123	1.23	56		
Isopropenylbenzene	C_9H_{10}	98-83-9	118.175	20	0.0116	0.116	40		
Isopropyl acetate	$C_5H_{10}O_2$	108-21-4	102.132	20	2.9	30	10		
Isopropylbenzene	C_9H_{12}	98-82-8	120.191	25	0.0050	0.050	22	0.833	124
1-Isopropyl-2-methylbenzene	$C_{10}H_{14}$	527-84-4	134.218	25	0.00482	0.0482	23		
1-Isopropyl-3-methylbenzene	$C_{10}H_{14}$	535-77-3	134.218	25	0.00425	0.0425	23		
1-Isopropyl-4-methylbenzene	$C_{10}H_{14}$	99-87-6	134.218	25	0.0051	0.051	23	0.77	5
Isopropyl phenylcarbamate	$C_{10}H_{13}NO_2$	122-42-9	179.216	20	0.01	0.1	40		
Isoquinoline	C_9H_7N	119-65-3	129.159	20	0.452	4.52	6		
Isosorbide dinitrate	$C_6H_8N_2O_8$	87-33-2	236.136	25	0.055	0.55	40		
Kepone	$C_{10}Cl_{10}O$	143-50-0	490.636	100	0.4	4	40		
L-Lanthionine	$C_6H_{12}N_2O_4S$	922-55-4	208.235	25	0.15	1.5	26		
Lasiocarpine	$C_{21}H_{33}NO_7$	303-34-4	411.490	20	0.67	6.7	40		
L-Leucine	$C_6H_{13}NO_2$	61-90-5	131.173	25	2.32	23.8	62		
Levodopa	$C_9H_{11}NO_4$	59-92-7	197.188	20	0.165	1.65	63		
d-Limonene	$C_{10}H_{16}$	5989-27-5	136.234	0	0.001	0.01	4		
				25	0.0020	0.020	52		
Linalol	$C_{10}H_{18}O$	22564-99-4	154.249	25	0.156	1.56	52		
Linuron	$C_9H_{10}Cl_2N_2O_2$	330-55-2	249.093	25	0.0075	0.075	40		
L-Lysine	$C_6H_{14}N_2O_2$	56-87-1	146.187	25	0.58	5.8	26		
Maleic acid	$C_4H_4O_4$	110-16-7	116.073	25	44.1	789	26		
Malic acid	$C_4H_6O_5$	617-48-1	134.088	26	59	1440	26		
Malonic acid	$C_3H_4O_4$	141-82-2	104.062	0	37.9	610	26		
				20	42.4	736	26		
				50	48.1	927	26		
Malononitrile	$C_3H_2N_2$	109-77-3	66.061	20	10.6	119	40		
α-Maltose	$C_{12}H_{22}O_{11}$	4482-75-1	342.296	20	51.9	1080	27		
D-Mannitol	$C_6H_{14}O_6$	69-65-8	182.171	25	17.7	215	27		
Mefenamic acid	$C_{15}H_{15}NO_2$	61-68-7	241.286	20	0.0026	0.026	40		
Melphalan	$C_{13}H_{18}Cl_2N_2O_2$	148-82-3	305.200	30	0.44	4.4	40		
Mercury(II) phenyl acetate	$C_8H_8HgO_2$	62-38-4	336.74	20	0.2	2	30		
Mesityl oxide	$C_6H_{10}O$	141-79-7	98.142	20	2.8	29	83		
Methacrylic acid	$C_4H_6O_2$	79-41-4	86.090	20	8.9	98	10		

Name	Mol. form.	CAS Reg. No.	Mol. wt.	t/°C	s/mass%	s/g kg^{-1} H_2O	Ref.	k_H/kPa m^3 mol^{-1}	Ref.
Methane	CH_4	74-82-8	16.043	25	0.00227*	0.0227*	18	71.4	126
Methazolamide	$C_5H_8N_4O_3S_2$	554-57-4	236.273	15	0.0472	0.472	40		
Methazole	$C_9H_6Cl_2N_2O_3$	20354-26-1	261.061	24	0.00015	0.0015	40		
Methidathion	$C_6H_{11}N_2O_4PS_3$	950-37-8	302.330	20	0.0187	0.187	40		
L-Methionine	$C_5H_{11}NO_2S$	63-68-3	149.212	25	5.3	56	26		
Methomyl	$C_5H_{10}N_2O_2S$	16752-77-5	162.210	25	5.5	58	40		
Methoxsalen	$C_{12}H_8O_4$	298-81-7	216.190	30	0.0048	0.048	40		
2-Methoxyaniline	C_7H_9NO	90-04-0	123.152	25	1.24	12.6	40		
4-Methoxyaniline	C_7H_9NO	104-94-9	123.152	20	1.14	11.5	40		
4-Methoxybenzaldehyde	$C_8H_8O_2$	123-11-5	136.149	25	0.429	4.29	40		
4-Methoxybenzoic acid	$C_8H_8O_3$	100-09-4	152.148	25	0.023	0.23	27		
Methoxychlor	$C_{16}H_{15}Cl_3O_2$	72-43-5	345.648	25	0.000005	0.00005	40		
2-Methoxy-2-methylbutane	$C_6H_{14}O$	994-05-8	102.174	20	1.10	11.1	20		
				79	0.36	3.6	20		
4-Methoxyphenol	$C_7H_8O_2$	150-76-5	124.138	20	2.51	25.7	40		
Methyclothiazide	$C_9H_{11}Cl_2N_3O_4S_2$	135-07-9	360.237	20	0.005	0.05	40		
Methyl acetate	$C_3H_6O_2$	79-20-9	74.079	20	24.5	325	10		
Methyl acrylate	$C_4H_6O_2$	96-33-3	86.090	25	4.94	52.0	10		
2-Methylacrylonitrile	C_4H_5N	126-98-7	67.090	20	2.57	26.4	10		
2-Methylaniline	C_7H_9N	95-53-4	107.153	20	1.66	16.9	10		
4-Methylaniline	C_7H_9N	106-49-0	107.153	21	7.35	79.3	10		
N-Methylaniline	C_7H_9N	100-61-8	107.153	25	0.56	5.6	40		
2-Methylanthracene	$C_{15}H_{12}$	613-12-7	192.256	6	0.0000007	0.000007	42		
				25	0.0000021	0.000021	42,22		
9-Methylanthracene	$C_{15}H_{12}$	779-02-2	192.256	25	0.000026	0.00026	42,4		
9-Methylbenz[*a*]anthracene	$C_{19}H_{14}$	2381-16-0	242.314	27	0.0000066	0.000066	42,4		
10-Methylbenz[*a*]anthracene	$C_{19}H_{14}$	2381-15-9	242.314	25	0.0000055	0.000055	42,4		
2-Methylbenzenesulfonamide	$C_7H_9NO_2S$	88-19-7	171.217	25	0.162	1.62	27		
3-Methylbenzenesulfonamide	$C_7H_9NO_2S$	1899-94-1	171.217	25	0.78	7.8	27		
4-Methylbenzenesulfonamide	$C_7H_9NO_2S$	70-55-3	171.217	25	0.316	3.16	27		
2-Methyl-1*H*-benzimidazole	$C_8H_8N_2$	615-15-6	132.163	20	0.145	1.45	6		
Methyl benzoate	$C_8H_8O_2$	93-58-3	136.149	20	0.21	2.1	10		
2-Methyl-1,3-butadiene	C_5H_8	78-79-5	68.118	25	0.061	0.61	3	7.769	5
				50	0.076*	0.76*	3		
Methyl butanoate	$C_5H_{10}O_2$	623-42-7	102.132		1.6	16	30		
3-Methylbutanoic acid	$C_5H_{10}O_2$	503-74-2	102.132	20	4.0	42	26		
2-Methyl-1-butanol, (±)-	$C_5H_{12}O$	34713-94-5	88.148	10	3.38	35.0	78		
				25	2.75	28.3	78		
				50	2.35	24.1	78		
3-Methyl-1-butanol	$C_5H_{12}O$	123-51-3	88.148	10	3.17	32.7	78,1		
				25	2.59	26.6	78,1		
				70	2.24	22.9	78,1		
2-Methyl-2-butanol	$C_5H_{12}O$	75-85-4	88.148	25	11.0	124	88,1		
				60	6.6	71	88,1		
3-Methyl-2-butanol, (±)-	$C_5H_{12}O$	70116-68-6	88.148	25	5.6	59	1		
3-Methyl-2-butanone	$C_5H_{10}O$	563-80-4	86.132	0	9.4	104	82		
				25	6.1	65	82		
				40	5.2	55	82		
3-Methyl-1-butene	C_5H_{10}	563-45-1	70.133	25	0.013*	0.13*	3	55.5	5
2-Methyl-2-butene	C_5H_{10}	513-35-9	70.133	25	0.041	0.41	3		
2-Methyl-3-buten-2-ol	$C_5H_{10}O$	115-18-4	86.132	18	27.4	377	88		
				29	18.4	225	88		
Methyl *tert*-butyl ether	$C_5H_{12}O$	1634-04-4	88.148	0	7.72	83.7	79		
				25	3.25	33.6	79		
				35	2.56	26.3	79		
				70	1.64	16.7	79		
Methyl carbamate	$C_2H_5NO_2$	598-55-0	75.067	15	69	2230	27		
5-Methylchrysene	$C_{19}H_{14}$	3697-24-3	242.314	27	0.0000062	0.000062	42,4		
Methylcyclohexane	C_7H_{14}	108-87-2	98.186	26	0.00161	0.0161	3	40.0	5
				100	0.00548	0.0548	3		
2-Methylcyclohexanone, (±)-	$C_7H_{12}O$	24965-84-2	112.169	0	2.93	30.2	84		

Name	Mol. form.	CAS Reg. No.	Mol. wt.	t/°C	s/mass%	s/g kg⁻¹ H₂O	Ref.	k_H/kPa m³ mol⁻¹	Ref.
				20	1.98	20.2	20		
				31	1.72	17.5	84		
				60	1.44	14.6	84		
				90	1.54	15.6	20		
4-Methylcyclohexanone	$C_7H_{12}O$	589-92-4	112.169	20	2.43	24.9	20		
				80	1.95	19.9	20		
1-Methylcyclohexene	C_7H_{12}	591-49-1	96.170	25	0.0052	0.052	3		
Methylcyclopentane	C_6H_{12}	96-37-7	84.159	25	0.0043	0.043	3	37.0	5
5-Methylcytosine	$C_5H_7N_3O$	554-01-8	125.129	25	0.45	4.5	26		
1-Methyl-2,4-dinitrobenzene	$C_7H_6N_2O_4$	121-14-2	182.134	12	0.0130	0.130	55		
				32	0.0270	0.270	85		
				62	0.098	0.98	85		
2-Methyl-4,6-dinitrophenol	$C_7H_6N_2O_5$	534-52-1	198.133		0.0130	0.130	40		
Methyl formate	$C_2H_4O_2$	107-31-3	60.052	25	23	300	10		
3-Methylheptane	C_8H_{18}	589-81-1	114.229	25	0.000079	0.00079	4	370	5
2-Methyl-2-heptanol	$C_8H_{18}O$	625-25-2	130.228	30	0.25	2.5	1		
5-Methyl-3-heptanone	$C_8H_{16}O$	541-85-5	128.212	20	0.192	1.92	20		
				90	0.131	1.31	20		
2-Methylhexane	C_7H_{16}	591-76-4	100.202	25	0.00025	0.0025	3	345	5
3-Methylhexane	C_7H_{16}	78918-91-9	100.202	25	0.00026	0.0026	3	249	13
2-Methyl-2-hexanol	$C_7H_{16}O$	625-23-0	116.201	25	1.0	10	1		
5-Methyl-2-hexanol	$C_7H_{16}O$	627-59-8	116.201	25	0.49	4.9	1		
3-Methyl-3-hexanol	$C_7H_{16}O$	597-96-6	116.201	25	1.2	12	1		
5-Methyl-2-hexanone	$C_7H_{14}O$	110-12-3	114.185	19	0.537	5.40	20		
				90	0.417	4.19	20		
5-Methyl-3-hexanone	$C_7H_{14}O$	623-56-3	114.185	20	0.47	4.7	20		
				81	0.32	3.2	20		
Methyl 4-hydroxybenzoate	$C_8H_8O_3$	99-76-3	152.148	25	0.24	2.4	40		
2-Methyl-1*H*-imidazole	$C_4H_6N_2$	693-98-1	82.104	18	23.2	302	54		
3-Methyl-1*H*-indole	C_9H_9N	83-34-1	131.174	20	0.050	0.50	6		
3-Methylisoquinoline	$C_{10}H_9N$	1125-80-0	143.185	20	0.092	0.92	6		
Methyl isothiocyanate	C_2H_3NS	556-61-6	73.117	20	0.75	7.6	40		
Methylmalonic acid	$C_4H_6O_4$	516-05-2	118.089	0	30.1	431	26		
				20	40	670	26		
Methyl methacrylate	$C_5H_8O_2$	80-62-6	100.117	20	1.56	15.8	10		
2-Methyl-3-(2-methylphenyl)-4(3*H*)-quinazolinone	$C_{16}H_{14}N_2O$	72-44-6	250.294	23	0.03	0.3	40		
1-Methylnaphthalene	$C_{11}H_{10}$	90-12-0	142.197	25	0.00281	0.0281	22	0.0455	101
2-Methylnaphthalene	$C_{11}H_{10}$	91-57-6	142.197	25	0.0025	0.025	4	0.0556	101
2-Methyl-1,4-naphthalenedione	$C_{11}H_8O_2$	58-27-5	172.181	25	0.016	0.16	40		
N-Methyl-*N*-nitrosourea	$C_2H_5N_3O_2$	684-93-5	103.080	14	2.3	24	40		
4-Methyloctane	C_9H_{20}	2216-34-4	128.255	25	0.0000115	0.000115	4	1000	5
Methyloxirane	C_3H_6O	75-56-9	58.079	20	40.5	681	10	0.0087	13
Methyl parathion	$C_8H_{10}NO_5PS$	298-00-0	263.208	10	0.00218	0.0218	40		
				20	0.00380	0.0380	40		
				30	0.0059	0.059	40		
2-Methylpentane	C_6H_{14}	107-83-5	86.175	25	0.00137	0.0137	3	169	5
3-Methylpentane	C_6H_{14}	96-14-0	86.175	25	0.00129	0.0129	3	172	5
2-Methyl-1-pentanol	$C_6H_{14}O$	105-30-6	102.174	25	0.76	7.7	78,1		
				50	0.70	7.0	78		
4-Methyl-1-pentanol	$C_6H_{14}O$	626-89-1	102.174	25	0.76	7.6	1		
2-Methyl-2-pentanol	$C_6H_{14}O$	590-36-3	102.174	25	3.2	33	1		
3-Methyl-2-pentanol	$C_6H_{14}O$	565-60-6	102.174	25	1.9	19	1		
4-Methyl-2-pentanol	$C_6H_{14}O$	108-11-2	102.174	27	1.5	15	1		
2-Methyl-3-pentanol	$C_6H_{14}O$	565-67-3	102.174	25	2.0	20	1		
3-Methyl-3-pentanol	$C_6H_{14}O$	77-74-7	102.174	25	4.3	45	1		
4-Methyl-2-pentanone	$C_6H_{12}O$	108-10-1	100.158	0	2.92	30.1	83		
				25	1.85	18.8	83		
				50	1.46	14.8	83		
2-Methyl-3-pentanone	$C_6H_{12}O$	565-69-5	100.158	25	1.5	15	83		
2-Methyl-1-pentene	C_6H_{12}	763-29-1	84.159	25	0.0078	0.078	3	27.8	5

Name	Mol. form.	CAS Reg. No.	Mol. wt.	t/°C	s/mass%	s/g kg^{-1} H_2O	Ref.	k_H/kPa m^3 mol^{-1}	Ref.
4-Methyl-1-pentene	C_6H_{12}	691-37-2	84.159	25	0.0048	0.048	3	62.5	5
1-Methylphenanthrene	$C_{15}H_{12}$	832-69-9	192.256	7	0.0000095	0.000095	42		
				25	0.0000269	0.000269	42,4		
Methylprednisolone	$C_{22}H_{30}O_5$	83-43-2	374.470	25	0.012	0.12	40		
Methyl propanoate	$C_4H_8O_2$	554-12-1	88.106		6	60	30		
2-Methylpropanoic acid	$C_4H_8O_2$	79-31-2	88.106	20	22.8	295	10		
2-Methyl-1-propanol	$C_4H_{10}O$	78-83-1	74.121	0	12.2	139	78,1		
				25	8.1	88	78,1	0.0010	121
				50	7.0	70	78,1		
Methyl propyl ether	$C_4H_{10}O$	557-17-5	74.121	0	5.4	57	79		
				25	3.0	31	79		
2-Methyl-2-propyl-1,3-propanediol dicarbamate	$C_9H_{18}N_2O_4$	57-53-4	218.250	25	0.33	3.3	40		
Methyl salicylate	$C_8H_8O_3$	119-36-8	152.148	30	0.74	7.4	10		
17-Methyltestosterone	$C_{20}H_{30}O_2$	58-18-4	302.451	25	0.0033	0.033	40		
2-Methyltetrahydrofuran	$C_5H_{10}O$	96-47-9	86.132	19	14.4	168	20	0.67	13
				71	6.0	64	20		
N-Methyl-*N*,2,4,6-tetranitroaniline	$C_7H_5N_5O_8$	479-45-8	287.144	20	0.0074	0.074	40		
Methylthiouracil	$C_5H_6N_2OS$	56-04-2	142.179	25	0.0533	0.533	40		
1-Methyl-2,3,4-trinitrobenzene	$C_7H_5N_3O_6$	602-29-9	227.131	14	0.0091	0.091	85,59		
				23	0.0116	0.116	85,59		
				61	0.0643	0.643	85,59		
Metronidazole	$C_6H_9N_3O_3$	443-48-1	171.153	20	0.93	9.4	40		
Mirex	$C_{10}Cl_{12}$	2385-85-5	545.543	25	0.0000085	0.000085	40		
Morphine	$C_{17}H_{19}NO_3$	57-27-2	285.338	20	0.015	0.15	27		
β-Myrcene	$C_{10}H_{16}$	123-35-3	136.234	25	0.030	0.30	52		
Naphthacene	$C_{18}H_{12}$	92-24-0	228.288	25	0.00000007	0.0000007	42,4	0.0000027	113
Naphthalene	$C_{10}H_8$	91-20-3	128.171	10	0.0019	0.019	4		
				25	0.00316	0.0316	22	0.0454	112
				50	0.0082	0.082	4		
1-Naphthaleneacetic acid	$C_{12}H_{10}O_2$	86-87-3	186.206	25	0.0415	0.415	40		
1-Naphthalenecarboxylic acid	$C_{11}H_8O_2$	86-55-5	172.181	25	0.0058	0.058	27		
1-Naphthalenylthiourea	$C_{11}H_{10}N_2S$	86-88-4	202.275	20	0.06	0.6	40		
1-Naphthol	$C_{10}H_8O$	90-15-3	144.170	20	0.111	1.11	40		
2-Naphthol	$C_{10}H_8O$	135-19-3	144.170	20	0.064	0.64	40		
				80	0.67	6.7	40		
1-Naphthylamine	$C_{10}H_9N$	134-32-7	143.185	20	0.17	1.7	40		
2-Naphthylamine	$C_{10}H_9N$	91-59-8	143.185	20	0.0189	0.189	40		
Narceine	$C_{23}H_{27}NO_8$	131-28-2	445.462	13	0.078	0.78	27		
Neopentane	C_5H_{12}	463-82-1	72.149	25	0.00332*	0.0332*	3	370	5
Nitrapyrin	$C_6H_3Cl_4N$	1929-82-4	230.907	20	0.0040	0.040	40		
2-Nitroaniline	$C_6H_6N_2O_2$	88-74-4	138.124	30	1.47	14.9	27		
3-Nitroaniline	$C_6H_6N_2O_2$	99-09-2	138.124	30	0.121	1.21	27		
4-Nitroaniline	$C_6H_6N_2O_2$	100-01-6	138.124	30	0.073	0.73	27		
2-Nitroanisole	$C_7H_7NO_3$	91-23-6	153.136	30	0.169	1.69	10		
4-Nitroanisole	$C_7H_7NO_3$	100-17-4	153.136	30	0.059	0.59	27		
3-Nitrobenzaldehyde	$C_7H_5NO_3$	99-61-6	151.120	25	0.16	1.6	27		
4-Nitrobenzaldehyde	$C_7H_5NO_3$	555-16-8	151.120	25	0.23	2.3	27		
Nitrobenzene	$C_6H_5NO_2$	98-95-3	123.110	25	0.21	2.1	17		
3-Nitro-1,2-benzenedicarboxylic acid	$C_8H_5NO_6$	603-11-2	211.129	25	1.63	16.6	69		
				40	2.97	30.6	69		
4-Nitro-1,2-benzenedicarboxylic acid	$C_8H_5NO_6$	610-27-5	211.129	25	60.9	1560	69		
				40	68.0	2125	69		
2-Nitrobenzoic acid	$C_7H_5NO_4$	552-16-9	167.120	25	0.55	5.5	40		
3-Nitrobenzoic acid	$C_7H_5NO_4$	121-92-6	167.120	10	0.197	1.97	75		
				25	0.313	3.14	75		
				50	0.90	9.1	75		
4-Nitrobenzoic acid	$C_7H_5NO_4$	62-23-7	167.120	25	0.0422	0.422	40		
Nitroethane	$C_2H_5NO_2$	79-24-3	75.067	25	4.4	46	38		
				50	5.3	56	38		
Nitrofen	$C_{12}H_7Cl_2NO_3$	1836-75-5	284.095	22	0.00095	0.0095	40		
Nitrofurantoin	$C_8H_6N_4O_5$	67-20-9	238.158	30	0.011	0.11	40		

Name	Mol. form.	CAS Reg. No.	Mol. wt.	t/°C	s/mass%	s/g kg^{-1} H_2O	Ref.	k_H/kPa m^3 mol^{-1}	Ref.
Nitrofurazone	$C_6H_6N_4O_4$	59-87-0	198.137	20	0.0238	0.238	40		
Nitroguanidine	$CH_4N_4O_2$	556-88-7	104.069	25	1.2	12	40		
Nitromethane	CH_3NO_2	75-52-5	61.041	0	9.2	101	36		
				25	11.0	124	36		
				50	14.8	174	36		
1-Nitronaphthalene	$C_{10}H_7NO_2$	86-57-7	173.169	18	0.005	0.05	40		
2-Nitrophenol	$C_6H_5NO_3$	88-75-5	139.109	25	0.170	1.70	48,51		
3-Nitrophenol	$C_6H_5NO_3$	554-84-7	139.109	20	2.14	21.9	27		
4-Nitrophenol	$C_6H_5NO_3$	100-02-7	139.109	20	1.56	15.8	48,51		
1-Nitropropane	$C_3H_7NO_2$	108-03-2	89.094	25	1.54	15.6	38		
				90	2.29	23.4	20		
2-Nitropropane	$C_3H_7NO_2$	79-46-9	89.094	25	1.75	17.8	38		
				90	2.36	24.2	20		
N-Nitrosodiethylamine	$C_4H_{10}N_2O$	55-18-5	102.134	24	9.6	106	40		
N-Nitrosodiphenylamine	$C_{12}H_{10}N_2O$	86-30-6	198.219	25	0.0035	0.035	17		
2-Nitrotoluene	$C_7H_7NO_2$	88-72-2	137.137	30	0.065	0.65	27		
3-Nitrotoluene	$C_7H_7NO_2$	99-08-1	137.137	30	0.050	0.50	27		
4-Nitrotoluene	$C_7H_7NO_2$	99-99-0	137.137	30	0.044	0.44	27		
2,2',3,3',4,5,5',6,6'-Nonachlorobiphenyl	$C_{12}HCl_9$	52663-77-1	464.213	25	0.0000000018	0.000000018	7		
1,8-Nonadiyne	C_9H_{12}	2396-65-8	120.191	25	0.0125	0.125	4		
Nonane	C_9H_{20}	111-84-2	128.255	25	0.000017	0.00017	4	454	120
				50	0.000022	0.00022	4		
Nonanedioic acid	$C_9H_{16}O_4$	123-99-9	188.221	25	0.1780	1.780	34		
				65	1.322	13.40	34		
Nonanoic acid	$C_9H_{18}O_2$	112-05-0	158.238	20	0.0284	0.284	26		
1-Nonanol	$C_9H_{20}O$	143-08-8	144.254	25	0.0129	0.129	78,1		
				90	0.0291	0.291	78		
2-Nonanol, (±)-	$C_9H_{20}O$	74683-66-2	144.254	15	0.026	0.26	1		
3-Nonanol, (±)-	$C_9H_{20}O$	74742-08-8	144.254	15	0.032	0.32	1		
4-Nonanol	$C_9H_{20}O$	52708-03-9	144.254	15	0.0026	0.026	1		
5-Nonanol	$C_9H_{20}O$	623-93-8	144.254	15	0.0032	0.032	1		
2-Nonanone	$C_9H_{18}O$	821-55-6	142.238	20	0.038	0.38	20		
				70	0.034	0.34	20		
3-Nonanone	$C_9H_{18}O$	925-78-0	142.238	30	0.056	0.56	20		
				80	0.046	0.46	20		
5-Nonanone	$C_9H_{18}O$	502-56-7	142.238	20	0.054	0.54	20		
				80	0.029	0.29	20		
1-Nonene	C_9H_{18}	124-11-8	126.239	25	0.000112	0.00112	40		
Nonyl formate	$C_{10}H_{20}O_2$	5451-92-3	172.265	10	0.012	0.12	20		
				90	0.039	0.39	20		
4-Nonylphenol	$C_{15}H_{24}O$	104-40-5	220.351	25	0.000636	0.00636	40		
1-Nonyne	C_9H_{16}	9/3/3452	124.223	25	0.00072	0.0072	4		
Norethisterone	$C_{20}H_{26}O_2$	68-22-4	298.419	25	0.00063	0.0063	40		
Norflurazon	$C_{12}H_9ClF_3N_3O$	27314-13-2	303.666	25	0.0028	0.028	40		
L-Norleucine	$C_6H_{13}NO_2$	327-57-1	131.173	25	1.5	15	26		
L-Norvaline	$C_5H_{11}NO_2$	6600-40-4	117.147	25	9.7	107	26		
Noscapine	$C_{22}H_{23}NO_7$	128-62-1	413.421	25	0.03	0.3	40		
2,2',3,3',5,5',6,6'-Octachlorobiphenyl	$C_{12}H_2Cl_8$	2136-99-4	429.768	25	0.00000015	0.0000015	41	0.0018	95
Octachlorodibenzo-*p*-dioxin	$C_{12}Cl_8O_2$	3268-87-9	459.751	25	2.3E-11	2.3E-10	68		
Octachloro-1,3-pentadiene	C_5Cl_8	1888-73-9	343.678	20	0.000020	0.00020	35		
Octacosane	$C_{28}H_{58}$	630-02-4	394.761	22	0.0000006	0.000006	37		
Octadecane	$C_{18}H_{38}$	593-45-3	254.495	25	0.00000021	0.0000021	42,37		
Octadecanoic acid	$C_{18}H_{36}O_2$	57-11-4	284.478	20	0.00029	0.0029	26		
1-Octadecanol	$C_{18}H_{38}O$	112-92-5	270.494	34	0.000011	0.00011	1		
Octane	C_8H_{18}	111-65-9	114.229	25	0.000073	0.00073	46	303	5
				50	0.000102	0.00102	47		
				75	0.000179	0.00179	46		
				100	0.000377	0.00377	46		
Octanedioic acid	$C_8H_{14}O_4$	505-48-6	174.195	25	0.242	2.43	34		
				50	0.557	5.570	34		
Octanoic acid	$C_8H_{16}O_2$	124-07-2	144.212	25	0.080	0.80	26		

Name	Mol. form.	CAS Reg. No.	Mol. wt.	t/°C	s/mass%	s/g kg⁻¹ H₂O	Ref.	k_H/kPa m³ mol⁻¹	Ref.
1-Octanol	$C_8H_{18}O$	111-87-5	130.228	25	0.0460	0.460	78		
				60	0.0536	0.536	78		
2-Octanol	$C_8H_{18}O$	4128-31-8	130.228	25	0.4	4	1		
2-Octanone	$C_8H_{16}O$	111-13-7	128.212	20	0.134	1.34	84		
				50	0.098	0.98	84		
				80	0.091	0.91	84		
3-Octanone	$C_8H_{16}O$	106-68-3	128.212	20	0.137	1.37	20		
				91	0.106	1.06	20		
1-Octene	C_8H_{16}	111-66-0	112.213	25	0.00027	0.0027	4	100	5
Octyl acetate	$C_{10}H_{20}O_2$	112-14-1	172.265	19	0.020	0.20	20		
				92	0.012	0.12	20		
1-Octyne	C_8H_{14}	629-05-0	110.197	25	0.0024	0.024	4	7.69	113
Orotic acid	$C_5H_4N_2O_4$	65-86-1	156.097	18	0.18	1.8	26		
Oryzalin	$C_{12}H_{18}N_4O_6S$	19044-88-3	346.359	25	0.00024	0.0024	40		
Ouabain	$C_{29}H_{44}O_{12}$	630-60-4	584.652	25	1.3	13	40		
Oxalic acid	$C_2H_2O_4$	144-62-7	90.035	20	8.69	95.2	27		
				80	45.8	845	27		
Oxamyl	$C_7H_{13}N_3O_3S$	23135-22-0	219.261	25	≈21	≈270	40		
4-Oxopentanoic acid	$C_5H_8O_3$	123-76-2	116.116	10	63.6	1750	34		
				25	84.0	5250	34		
4-Oxo-4*H*-pyran-2,6-dicarboxylic acid	$C_7H_4O_6$	99-32-1	184.103	25	1.45	14.7	27		
Papaverine	$C_{20}H_{21}NO_4$	58-74-2	339.386	37	0.0037	0.037	40		
Paraldehyde	$C_6H_{12}O_3$	123-63-7	132.157	25	11	124	30		
Parathion	$C_{10}H_{14}NO_5PS$	56-38-2	291.261	20	0.00129	0.0129	40		
Pendimethalin	$C_{13}H_{19}N_3O_4$	40487-42-1	281.308	20	0.00003	0.0003	40		
Pentachlorobenzene	C_6HCl_5	608-93-5	250.337	25	0.000050	0.00050	41	0.0714	67
2,3,4,5,6-Pentachlorobiphenyl	$C_{12}H_5Cl_5$	18259-05-7	326.433	25	0.0000008	0.000008	7		
2,2',4,5,5'-Pentachlorobiphenyl	$C_{12}H_5Cl_5$	37680-73-2	326.433	25	0.000001	0.00001	7	0.0244	111
Pentachloroethane	C_2HCl_5	76-01-7	202.294	25	0.049	0.49	25	0.222	5
Pentachloronitrobenzene	$C_6Cl_5NO_2$	82-68-8	295.335	20	0.000044	0.00044	40		
Pentachlorophenol	C_6HCl_5O	87-86-5	266.336	25	0.0021	0.021	48,51,24		
2,3,4,5,6-Pentachlorotoluene	$C_7H_3Cl_5$	877-11-2	264.364	25	0.0000028	0.000028	61		
Pentadecanoic acid	$C_{15}H_{30}O_2$	1002-84-2	242.398	20	0.0012	0.012	26		
1-Pentadecanol	$C_{15}H_{32}O$	629-76-5	228.414	25	0.000010	0.00010	1		
1,4-Pentadiene	C_5H_8	591-93-5	68.118	25	0.056	0.56	3	12	5
Pentaerythritol	$C_5H_{12}O_4$	115-77-5	136.147	15	5.3	56	30		
Pentaerythritol tetranitrate	$C_5H_8N_4O_{12}$	78-11-5	316.138	20	0.0002	0.002	40		
Pentanal	$C_5H_{10}O$	110-62-3	86.132	25	1.2	12	40		
Pentane	C_5H_{12}	109-66-0	72.149	25	0.0041	0.041	3	125	5
Pentanedioic acid	$C_5H_8O_4$	110-94-1	132.116	25	58.3	1400	33		
				50	78.1	3570	33		
2,4-Pentanedione	$C_5H_8O_2$	123-54-6	100.117	20	16.1	192	20		
				80	32.2	475	20		
Pentanoic acid	$C_5H_{10}O_2$	109-52-4	102.132	16	3.6	37	26		
				25	4.32	45.2	90		
				35	5.26	55.5	90		
1-Pentanol	$C_5H_{12}O$	71-41-0	88.148	0	3.23	33.4	78,1		
				25	2.14	21.9	78,1		
				50	1.83	18.6	78,1		
				90	2.12	21.7	78		
2-Pentanol	$C_5H_{12}O$	6032-29-7	88.148	25	4.3	45	21		
3-Pentanol	$C_5H_{12}O$	584-02-1	88.148	25	5.6	59	21		
2-Pentanone	$C_5H_{10}O$	107-87-9	86.132	0	8.7	95	20		
				25	5.5	58	20	0.00625	105
				80	3.8	40	20		
3-Pentanone	$C_5H_{10}O$	96-22-0	86.132	0	7.6	82	82		
				25	4.9	52	82		
3-Pentanone	$C_5H_{10}O$	96-22-0	86.132	80	3.6	37	82		
1-Pentene	C_5H_{10}	109-67-1	70.133	25	0.0148	0.148	3	40.0	5
cis-2-Pentene	C_5H_{10}	627-20-3	70.133	25	0.0203	0.203	3	22.7	5
Pentyl acetate	$C_7H_{14}O_2$	628-63-7	130.185	20	0.17	1.7	10		

Name	Mol. form.	CAS Reg. No.	Mol. wt.	t/°C	s/mass%	s/g kg⁻¹ H₂O	Ref.	k_H/kPa m³ mol⁻¹	Ref.
sec-Pentyl acetate (*S*)-	$C_7H_{14}O_2$	55621-90-4	130.185	25	0.2	2	27		
Pentylbenzene	$C_{11}H_{16}$	538-68-1	148.245	25	0.00043	0.0043	89,5	0.588	5
Pentylcyclopentane	$C_{10}H_{20}$	3741-00-2	140.266	25	0.0000115	0.000115	4	185	5
Pentyl propanoate	$C_8H_{16}O_2$	624-54-4	144.212	20	0.1	1	27		
1-Pentyne	C_5H_8	627-19-0	68.118	25	0.157	1.57	3	2.5	5
Perfluorocyclobutane	C_4F_8	115-25-3	200.030	5	0.00638*	0.0638*	50		
				25	0.00247*	0.0247*	50		
				45	0.00158*	0.0158*	50		
Perfluorodecane	$C_{10}F_{22}$	307-45-9	538.072	20	0.000031	0.00031	35		
Perfluoroheptane	C_7F_{16}	335-57-9	388.049	25	0.0000013	0.000013	35		
Perfluorohexane	C_6F_{14}	355-42-0	338.042	25	0.0000098	0.000098	35		
Perfluoro-2-methylpentane	C_6F_{14}	355-04-4	338.042	25	0.000017	0.00017	35		
Perfluorooctane	C_8F_{18}	307-34-6	438.057	25	0.00000017	0.0000017	35		
Perfluoropentane	C_5F_{12}	678-26-2	288.035	25	0.00012	0.0012	35		
Perfluoropropane	C_3F_8	76-19-7	188.019	15	0.0015*	0.015*	14		
Perfluoropropene	C_3F_6	116-15-4	150.022	25	0.0194*	0.194*	14		
Permethrin	$C_{21}H_{20}Cl_2O_3$	52645-53-1	391.288	20	0.00002	0.0002	32		
Perylene	$C_{20}H_{12}$	198-55-0	252.309	25	0.00000004	0.0000004	42,4	0.000435	119
Phenanthrene	$C_{14}H_{10}$	85-01-8	178.229	0	0.000039	0.00039	42		
				10	0.000047	0.00047	42,4		
				25	0.00012	0.0012	42,22	0.00435	112
				50	0.00042	0.0042	42,4		
Phenmedipham	$C_{16}H_{16}N_2O_4$	13684-63-4	300.309	25	0.00047	0.0047	32		
Phenobarbital	$C_{12}H_{12}N_2O_3$	50-06-6	232.234	25	0.12	1.2	40		
				45	0.26	2.6	40		
Phenol	C_6H_6O	108-95-2	94.111	15	7.60	82.3	48,51		
				25	8.40	91.7	48,51		
				35	9.31	102.7	48,51		
Phenolphthalein	$C_{20}H_{14}O_4$	77-09-8	318.323	20	0.018	0.18	27		
10*H*-Phenothiazine	$C_{12}H_9NS$	92-84-2	199.271	25	0.00016	0.0016	40		
2-Phenoxyethanol	$C_8H_{10}O_2$	122-99-6	138.164	20	2.53	26.0	40		
Phenyl acetate	$C_8H_8O_2$	122-79-2	136.149	20	0.59	5.9	20		
				91	0.91	9.2	20		
DL-Phenylalanine	$C_9H_{11}NO_2$	150-30-1	165.189	25	1.40	14.2	29		
L-Phenylalanine	$C_9H_{11}NO_2$	63-91-2	165.189	25	2.71	27.9	26		
Phenylbutazone	$C_{19}H_{20}N_2O_2$	50-33-9	308.374	25	0.0034	0.034	40		
1-Phenyl-1-propanone	$C_9H_{10}O$	93-55-0	134.174	19	0.32	3.2	20		
				80	0.24	2.4	20		
Phenylthiourea	$C_7H_8N_2S$	103-85-5	152.217	25	2.55	26.2	27		
Phenytoin	$C_{15}H_{12}N_2O_2$	57-41-0	252.268	37	0.0038	0.038	40		
Phosalone	$C_{12}H_{15}ClNO_4PS_2$	2310-17-0	367.808	20	0.00026	0.0026	40		
Phosmet	$C_{11}H_{12}NO_4PS_2$	732-11-6	317.321	25	0.0025	0.025	40		
Phthalic acid	$C_8H_6O_4$	88-99-3	166.132	10	0.464	4.66	76		
				25	0.719	7.24	76		
				50	1.76	17.9	76		
				65	3.57	37.0	33		
Phthalic anhydride	$C_8H_4O_3$	85-44-9	148.116	27	0.62	6.20	40		
Picene	$C_{22}H_{14}$	213-46-7	278.346	27	0.00000025	0.0000025	42,4		
α-Pinene, (-)	$C_{10}H_{16}$	7785-26-4	136.234	25	0.00050	0.0050	52		
β-Pinene, (1S)-	$C_{10}H_{16}$	18172-67-3	136.234	25	0.00110	0.0110	52		
2,5-Piperazinedione	$C_4H_6N_2O_2$	106-57-0	114.103	25	1.64	16.7	29		
2-Pivaloyl-1,3-indandione	$C_{14}H_{14}O_3$	83-26-1	230.259	25	0.0018	0.018	40		
Prednisolone	$C_{21}H_{28}O_5$	50-24-8	360.444	25	0.03	0.3	40		
Progesterone	$C_{21}H_{30}O_2$	57-83-0	314.462	25	0.00088	0.0088	40		
				41	0.00206	0.0206	40		
L-Proline	$C_5H_9NO_2$	147-85-3	115.131	25	61.9	1625	26		
Prometone	$C_{10}H_{19}N_5O$	1610-18-0	225.291	20	0.075	0.75	40		
Prometryn	$C_{10}H_{19}N_5S$	7287-19-6	241.357	20	0.0048	0.048	32		
Propachlor	$C_{11}H_{14}ClNO$	1918-16-7	211.688	20	0.07	0.7	40		
Propanal	C_3H_6O	123-38-6	58.079	25	30.6	441	10		

Name	Mol. form.	CAS Reg. No.	Mol. wt.	t/°C	s/mass%	s/g kg⁻¹ H₂O	Ref.	k_H/kPa m³ mol⁻¹	Ref.
Propane	C_3H_8	74-98-6	44.096	25	0.00669*	0.0669*	18	66.7	121
Propanenitrile	C_3H_5N	107-12-0	55.079	25	10.3	115	10		
Propanil	$C_9H_9Cl_2NO$	709-98-8	218.079	20	0.013	0.13	40		
Propazine	$C_9H_{16}ClN_5$	139-40-2	229.710	20	0.00086	0.0086	40		
Propene	C_3H_6	115-07-1	42.080	25	0.0200*	0.200*	5	21.3	5
1-Propene-2,3-dicarboxylic acid	$C_5H_6O_4$	97-65-4	130.100	20	7.7	83	26		
trans-1-Propene-1,2,3-tricarboxylic acid	$C_6H_6O_6$	4023-65-8	174.108	25	20.9	264	26		
				90	52.5	1105	26		
Propoxur	$C_{11}H_{15}NO_3$	114-26-1	209.242	20	0.193	1.93	40		
Propyl acetate	$C_5H_{10}O_2$	109-60-4	102.132	20	2.3	24	10		
Propylbenzene	C_9H_{12}	103-65-1	120.191	25	0.0052	0.052	22	0.714	5
Propyl butanoate	$C_7H_{14}O_2$	105-66-8	130.185	17	0.162	1.62	27		
Propylcyclopentane	C_8H_{16}	2040-96-2	112.213	25	0.00020	0.0020	4	90.9	5
Propyl formate	$C_4H_8O_2$	110-74-7	88.106	22	2.05	20.9	10		
Propyl 4-hydroxybenzoate	$C_{10}H_{12}O_3$	94-13-3	180.200	25	0.04	0.4	40		
Propyl propanoate	$C_6H_{12}O_2$	106-36-5	116.158	25	0.6	6	27		
Propyne	C_3H_4	74-99-7	40.064	25	0.364*	3.64*	5	1.11	5
Propyzamide	$C_{12}H_{11}Cl_2NO$	23950-58-5	256.127	25	0.0015	0.015	32		
Pyrene	$C_{16}H_{10}$	129-00-0	202.250	0	0.0000049	0.000049	42		
				15	0.0000069	0.000069	42		
				25	0.0000139	0.000139	42,22	0.00133	112
				50	0.000053	0.00053	42,4		
				75	0.000231	0.00231	42		
3-Pyridinecarboxamide	$C_6H_6N_2O$	98-92-0	122.124	20	≈33	≈490	40		
3-Pyridinecarboxylic acid	$C_6H_5NO_2$	59-67-6	123.110	24	1.63	16.6	66		
				52	3.40	35.2	66		
				72	5.20	54.9	66		
Pyrocatechol	$C_6H_6O_2$	120-80-9	110.111	20	31.1	451	27		
Pyrrole	C_4H_5N	109-97-7	67.090	25	4.5	47	10		
Quinic acid	$C_7H_{12}O_6$	77-95-2	192.166	9	29	410	26		
Quinidine	$C_{20}H_{24}N_2O_2$	56-54-2	324.417	20	0.020	0.20	27		
Quinine	$C_{20}H_{24}N_2O_2$	130-95-0	324.417	25	0.057	0.57	27		
Quinoline	C_9H_7N	91-22-5	129.159	20	0.633	6.33	6		
8-Quinolinol	C_9H_7NO	148-24-3	145.158	25	0.065	0.65	40		
Quinoxaline	$C_8H_6N_2$	91-19-0	130.147	50	54	1170	6		
Raffinose	$C_{18}H_{32}O_{16}$	512-69-6	504.437	20	12.5	143	27		
Reserpine	$C_{33}H_{40}N_2O_9$	50-55-5	608.679	30	0.0073	0.073	40		
Resorcinol	$C_6H_6O_2$	108-46-3	110.111	20	63.7	1750	27		
Riboflavin	$C_{17}H_{20}N_4O_6$	83-88-5	376.364	25	0.0075	0.075	40		
Ronnel	$C_8H_8Cl_3O_3PS$	299-84-3	321.546	20	0.00011	0.0011	40		
Rotenone	$C_{23}H_{22}O_6$	83-79-4	394.417	25	0.000017	0.00017	40		
Saccharin	$C_7H_5NO_3S$	81-07-2	183.185	25	0.40	4.0	27		
				100	4.0	42	27		
Salicylaldehyde	$C_7H_6O_2$	90-02-8	122.122	86	1.68	17.1	10		
Sarcosine	$C_3H_7NO_2$	107-97-1	89.094	25	30.0	429	26		
L-Serine	$C_3H_7NO_3$	56-45-1	105.093	25	20	250	26		
Shikimic acid	$C_7H_{10}O_5$	138-59-0	174.151		15	176	26		
Silvex	$C_9H_7Cl_3O_3$	93-72-1	269.509	25	0.014	0.14	40		
Solanine	$C_{45}H_{73}NO_{15}$	20562-02-1	868.060	15	0.0026	0.026	40		
L-Sorbose	$C_6H_{12}O_6$	87-79-6	180.155	17	≈26	≈350	40		
trans-Stilbene	$C_{14}H_{12}$	103-30-0	180.245	25	0.000029	0.00029	42,4	0.0714	104
Streptozotocin	$C_8H_{15}N_3O_7$	18883-66-4	265.221	25	0.50	5.0	40		
Strychnine	$C_{21}H_{22}N_2O_2$	57-24-9	334.412	20	0.013	0.13	27		
Styrene	C_8H_8	100-42-5	104.150	25	0.032	0.32	22	0.294	98
				50	0.046	0.46	4,89	0.3	13
Succinamide	$C_4H_8N_2O_2$	110-14-5	116.119	50	18.4	225	27		
Succinic acid	$C_4H_6O_4$	110-15-6	118.089	25	7.71	83.5	27		
				100	55	1220	27		
Succinonitrile	$C_4H_4N_2$	110-61-2	80.088	25	11.5	130	10		
Sucrose	$C_{12}H_{22}O_{11}$	57-50-1	342.296	20	67.1	2040	27		
			342.296	50	72.3	2610	27		

Name	Mol. form.	CAS Reg. No.	Mol. wt.	t/°C	s/mass%	s/g kg^{-1} H_2O	Ref.	k_H/kPa m^3 mol^{-1}	Ref.
			342.296	100	83.0	4880	27		
Sulfamethazine	$C_{12}H_{14}N_4O_2S$	57-68-1	278.330	20	0.053	0.53	40		
Sulfamethoxazole	$C_{10}H_{11}N_3O_3S$	723-46-6	253.277	25	0.0281	0.281	40		
Sulfathiazole	$C_9H_9N_3O_2S_2$	72-14-0	255.316	20	0.048	0.48	40		
Sulfisoxazole	$C_{11}H_{13}N_3O_3S$	127-69-5	267.304	37	0.03	0.3	40		
DL-Tartaric acid	$C_4H_6O_6$	133-37-9	150.087	0	8.95	98.3	26		
				20	17.1	206	26		
				100	65	1860	26		
L-Tartaric acid	$C_4H_6O_6$	87-69-4	150.087	20	58	1380	26		
				100	77	3350	26		
Tebuthiuron	$C_9H_{16}N_4OS$	34014-18-1	228.314	20	0.23	2.3	40		
Terbacil	$C_9H_{13}ClN_2O_2$	5902-51-2	216.664	25	0.071	0.71	40		
Terephthalic acid	$C_8H_6O_4$	100-21-0	166.132	10	0.0082	0.082	76		
				25	0.0065	0.065	76		
				50	0.0074	0.074	76		
o-Terphenyl	$C_{18}H_{14}$	84-15-1	230.304	25	0.000124	0.00124	42,40		
m-Terphenyl	$C_{18}H_{14}$	92-06-8	230.304	25	0.000152	0.00152	42,40		
p-Terphenyl	$C_{18}H_{14}$	92-94-4	230.304	25	0.00000180	0.000018	42,40		
α-Terpineol	$C_{10}H_{18}O$	12/2/2438	154.249	25	0.189	1.89	52		
1,2,4,5-Tetrabromobenzene	$C_6H_2Br_4$	636-28-2	393.696	25	0.00000434	0.0000434	2		
1,1,2,2-Tetrabromoethane	$C_2H_2Br_4$	79-27-6	345.653	0	0.052	0.52	25		
				25	0.068	0.68	25		
				50	0.106	1.06	25		
				100	0.307	3.07	25		
Tetrabromomethane	CBr_4	558-13-4	331.627	30	0.024	0.24	14		
1,2,3,4-Tetrachlorobenzene	$C_6H_2Cl_4$	634-66-2	215.892	25	0.0007	0.007	41	0.286	120
1,2,3,5-Tetrachlorobenzene	$C_6H_2Cl_4$	634-90-2	215.892	25	0.00035	0.0035	41	0.159	28
1,2,4,5-Tetrachlorobenzene	$C_6H_2Cl_4$	95-94-3	215.892	25	0.000060	0.00060	41	0.0556	115
3,4,5,6-Tetrachloro-1,2-benzenediol	$C_6H_2Cl_4O_2$	1198-55-6	247.891	25	0.071	0.71	8		
2,2',4',5-Tetrachlorobiphenyl	$C_{12}H_6Cl_4$	41464-40-8	291.988	25	0.0000016	0.000016	9		
2,3,4,5-Tetrachlorobiphenyl	$C_{12}H_6Cl_4$	33284-53-6	291.988	25	0.000002	0.00002	7		
2,3,5,6-Tetrachloro-2,5-cyclohexadiene-1,4-dione	$C_6Cl_4O_2$	118-75-2	245.875	20	0.025	0.25	40		
2,3,7,8-Tetrachlorodibenzo-*p*-dioxin	$C_{12}H_4Cl_4O_2$	1746-01-6	321.971	22	0.0000000019	0.000000019	40		
1,1,2,2-Tetrachloro-1,2-difluoroethane	$C_2Cl_4F_2$	76-12-0	203.830	27	0.016	0.16	25		
1,1,1,2-Tetrachloroethane	$C_2H_2Cl_4$	630-20-6	167.849	0	0.120	1.20	25		
				25	0.107	1.07	25	0.238	125
				50	0.123	1.23	25		
1,1,2,2-Tetrachloroethane	$C_2H_2Cl_4$	79-34-5	167.849	5	0.302	3.02	25		
				25	0.283	2.83	25	0.0417	125
				50	0.318	3.18	25		
Tetrachloroethene	C_2Cl_4	127-18-4	165.833	0	0.0273	0.273	20		
				20	0.0286	0.286	20	1.61	125
				80	0.0380	0.380	20		
Tetrachloromethane	CCl_4	56-23-5	153.823	25	0.065	0.65	20	2.94	122
				75	0.115	1.15	20		
2,3,4,6-Tetrachloro-5-methylphenol	$C_7H_4Cl_4O$	10460-33-0	245.918	25	0.00061	0.0061	2		
2,3,4,6-Tetrachlorophenol	$C_6H_2Cl_4O$	58-90-2	231.891	25	0.017	0.17	24		
1,1,1,3-Tetrachloro-2,2,3,3-tetrafluoropropane	$C_3Cl_4F_4$	2268-46-4	253.838	21	0.0052	0.052	35		
Tetracosane	$C_{24}H_{50}$	646-31-1	338.654	22	0.0000004	0.000004	37		
Tetradecane	$C_{14}H_{30}$	629-59-4	198.388	25	0.00000023	0.0000023	42,5		
Tetradecanoic acid	$C_{14}H_{28}O_2$	544-63-8	228.371	20	0.0020	0.020	26		
1-Tetradecanol	$C_{14}H_{30}O$	112-72-1	214.387	25	0.000031	0.00031	1		
Tetraethylsilane	$C_8H_{20}Si$	631-36-7	144.331	25	0.0000325	0.000325	10		
Tetrafluoroethene	C_2F_4	116-14-3	100.015	25	0.0158*	0.158*	19,50		
	C_2F_4	116-14-3	100.015	70	0.0090*	0.090*	50		
Tetrafluoromethane	CF_4	75-73-0	88.005	0	0.00390*	0.0390*	50		
				25	0.00185*	0.0185*	50,19		
				50	0.00134*	0.0134*	50		
Tetrahydro-2,5-dimethoxyfuran	$C_6H_{12}O_3$	696-59-3	132.157	21	32	470	20		
				90	19	235	20		

Name	Mol. form.	CAS Reg. No.	Mol. wt.	t/°C	s/mass%	s/g kg^{-1} H_2O	Ref.	k_H/kPa m^3 mol^{-1}	Ref.
1,2,3,4-Tetrahydronaphthalene	$C_{10}H_{12}$	119-64-2	132.202	20	0.0045	0.045	40		
Tetrahydropyran	$C_5H_{10}O$	142-68-7	86.132	20	8.57	93.7	20		
				81	4.29	44.8	20		
1,2,4,5-Tetramethylbenzene	$C_{10}H_{14}$	95-93-2	134.218	25	0.000348	0.00348	4	2.64	5
N,N,N',N'-Tetramethyl-4,4'-diaminobenzophenone	$C_{17}H_{20}N_2O$	90-94-8	268.353	20	0.04	0.4	40		
Tetramethylsilane	$C_4H_{12}Si$	75-76-3	88.224	25	0.00196	0.0196	10		
Theophylline	$C_7H_8N_4O_2$	58-55-9	180.165	20	0.52	5.2	29		
Thioacetamide	C_2H_5NS	62-55-5	75.133	25	12.3	140	40		
Thiourea	CH_4N_2S	62-56-6	76.121	20	10.6	119	40		
				80	≈37	≈590	40		
2-Thioxo-4-thiazolidinone	$C_3H_3NOS_2$	141-84-4	133.192	25	0.225	2.25	40		
Thiram	$C_6H_{12}N_2S_4$	137-26-8	240.432	20	0.003	0.03	40		
DL-Threonine	$C_4H_9NO_3$	80-68-2	119.119	10	14.34	167	45		
				20	15.69	186	45		
				40	19.84	248	45		
L-Threonine	$C_4H_9NO_3$	72-19-5	119.119	10	7.34	79.2	45		
				20	8.31	90.6	45		
				40	10.78	121	45		
Thymidine	$C_{10}H_{14}N_2O_5$	50-89-5	242.228	25	5.1	54	29		
Thymine	$C_5H_6N_2O_2$	65-71-4	126.114	25	0.35	3.5	29		
Thymol	$C_{10}H_{14}O$	89-83-8	150.217		0.1	1	30		
Tolazamide	$C_{14}H_{21}N_3O_3S$	1156-19-0	311.400	30	0.0065	0.065	40		
Tolbutamide	$C_{12}H_{18}N_2O_3S$	64-77-7	270.347	25	0.011	0.11	40		
o-Tolidine	$C_{14}H_{16}N_2$	119-93-7	212.290	25	0.13	1.3	40		
Toluene	C_7H_8	108-88-3	92.139	5	0.054	0.54	61		
				25	0.0519	0.519	61,22	0.667	124
				45	0.063	0.63	61		
				90	0.12	1.2	22		
p-Toluenesulfonic acid	$C_7H_8O_3S$	104-15-4	172.202	40	≈33	≈490	40		
o-Toluic acid	$C_8H_8O_2$	118-90-1	136.149	25	0.118	1.18	27		
m-Toluic acid	$C_8H_8O_2$	99-04-7	136.149	25	0.098	0.98	27		
p-Toluic acid	$C_8H_8O_2$	99-94-5	136.149	10	0.030	0.30	75		
				25	0.036	0.36	75		
				50	0.089	0.89	75		
1,3,5-Triazine-2,4,6-triamine	$C_3H_6N_6$	108-78-1	126.120	20	0.323	3.23	40		
				95	4.2	44	40		
1*H*-1,2,4-Triazol-3-amine	$C_2H_4N_4$	61-82-5	84.080	23	22	280	26		
1,2,4-Tribromobenzene	$C_6H_3Br_3$	615-54-3	314.800	25	0.0010	0.010	2		
1,3,5-Tribromobenzene	$C_6H_3Br_3$	626-39-1	314.800	25	0.0000789	0.000789	2		
1,1,2-Tribromoethane	$C_2H_3Br_3$	78-74-0	266.757	20	0.050	0.50	25		
Tribromofluoromethane	CBr_3F	353-54-8	270.721	25	0.040	0.40	14		
Tribromomethane	$CHBr_3$	75-25-2	252.731	25	0.30	3.0	5	0.0588	121
2,4,6-Tribromophenol	$C_6H_3Br_3O$	118-79-6	330.799	15	0.0007	0.007	2		
Tributylamine	$C_{12}H_{27}N$	102-82-9	185.349	25	0.0142	0.142	40		
Tributyl phosphate	$C_{12}H_{27}O_4P$	126-73-8	266.313	25	0.039	0.39	10		
Tributyrin	$C_{15}H_{26}O_6$	60-01-5	302.363	20	0.010	0.10	40		
Trichloroacetaldehyde	C_2HCl_3O	75-87-6	147.387	25	≈39	≈640	40		
Trichloroacetic acid	$C_2HCl_3O_2$	76-03-9	163.387	25	92.3	11990	27		
1,2,3-Trichlorobenzene	$C_6H_3Cl_3$	87-61-6	181.447	25	0.0021	0.021	41	0.159	94
1,2,4-Trichlorobenzene	$C_6H_3Cl_3$	120-82-1	181.447	15	0.0029	0.029	61		
				25	0.0037	0.037	61,41	0.172	108
				45	0.0047	0.047	61		
1,3,5-Trichlorobenzene	$C_6H_3Cl_3$	108-70-3	181.447	25	0.0008	0.008	41	0.556	117
3,4,5-Trichloro-1,2-benzenediol	$C_6H_3Cl_3O_2$	56961-20-7	213.446	25	0.051	0.51	8		
2,4,5-Trichlorobiphenyl	$C_{12}H_7Cl_3$	15862-07-4	257.543	25	0.000014	0.00014	7	0.0303	111
2,4,6-Trichlorobiphenyl	$C_{12}H_7Cl_3$	35693-92-6	257.543	25	0.00002	0.0002	7	0.0667	99
1,1,1-Trichloro-2,2-bis(4-chlorophenyl)ethane	$C_{14}H_9Cl_5$	50-29-3	354.486	25	0.0000004	0.000004	67		
2,4,6-Trichloro-3,5-dimethylphenol	$C_8H_7Cl_3O$	6972-47-0	225.500	25	0.00050	0.0050	2		
1,1,1-Trichloroethane	$C_2H_3Cl_3$	71-55-6	133.404	0	0.134	1.34	25		
				25	0.129	1.29	25	1.67	125

Name	Mol. form.	CAS Reg. No.	Mol. wt.	t/°C	s/mass%	s/g kg⁻¹ H₂O	Ref.	k_H/kPa m³ mol⁻¹	Ref.
				50	0.138	1.38	25		
1,1,2-Trichloroethane	$C_2H_3Cl_3$	79-00-5	133.404	0	0.425	4.25	25		
				25	0.459	4.59	25	0.0909	125
				50	0.536	5.36	25		
Trichloroethene	C_2HCl_3	79-01-6	131.388	0	0.145	1.45	25		
				25	0.128	1.28	25	0.909	125
				60	0.133	1.33	25		
Trichlorofluoromethane	CCl_3F	75-69-4	137.368	20	0.11	1.1	5	9.09	126
Trichloromethane	$CHCl_3$	67-66-3	119.378	25	0.80	8.0	20	0.40	121
				59	0.79	7.9	20		
1,2,4-Trichloro-5-methylbenzene	$C_7H_5Cl_3$	6639-30-1	195.474	25	0.00023	0.0023	61		
(Trichloromethyl)benzene	$C_7H_5Cl_3$	98-07-7	195.474	5	0.0053	0.053	10		
2,4,6-Trichloro-3-methylphenol	$C_7H_5Cl_3O$	551-76-8	211.473	25	0.0112	0.112	2		
Trichloronitromethane	CCl_3NO_2	76-06-2	164.376	0	0.227	2.27	40		
				25	0.162	1.62	40		
1,1,1-Trichloro-2,2,3,3,3-pentafluoropropane	$C_3Cl_3F_5$	4259-43-2	237.383	21	0.0058	0.058	35		
2,4,5-Trichlorophenol	$C_6H_3Cl_3O$	95-95-4	197.446	25	0.1	1	2		
2,4,6-Trichlorophenol	$C_6H_3Cl_3O$	88-06-2	197.446	25	0.069	0.69	48,51,24		
2,4,5-Trichlorophenoxyacetic acid	$C_8H_5Cl_3O_3$	93-76-5	255.483	25	0.028	0.28	40		
1,2,3-Trichloropropane	$C_3H_5Cl_3$	96-18-4	147.431	10	0.14	1.4	35		
				25	0.20	2.0	35	0.0277	124
1,1,2-Trichloro-1,2,2-trifluoroethane	$C_2Cl_3F_3$	76-13-1	187.375	25	0.017	0.17	25	34.5	96
Tri-*p*-cresyl phosphate	$C_{21}H_{21}O_4P$	78-32-0	368.363	25	0.00004	0.0004	40		
Tridecane	$C_{13}H_{28}$	629-50-5	184.361	25	0.000000033	0.00000033	37		
Tridecanoic acid	$C_{13}H_{26}O_2$	638-53-9	214.344	20	0.0033	0.033	26		
Triethylamine	$C_6H_{15}N$	121-44-8	101.190	20	5.5	58	10		
Triethylamine hydrochloride	$C_6H_{16}ClN$	554-68-7	137.651	25	57.8	1370	27		
Trifluoromethane	CHF_3	75-46-7	70.014	25	0.15*	1.5*	50,14		
3,4,5-Trihydroxybenzoic acid	$C_7H_6O_5$	149-91-7	170.120	25	1.52	15.4	74		
				50	3.82	39.7	74		
				100	25.0	333	27		
Triiodomethane	CHI_3	75-47-8	393.732	25	0.012	0.12	14		
Trimethoprim	$C_{14}H_{18}N_4O_3$	738-70-5	290.318	25	0.04	0.4	40		
1,2,3-Trimethylbenzene	C_9H_{12}	526-73-8	120.191	25	0.0070	0.070	22	0.323	5
1,2,4-Trimethylbenzene	C_9H_{12}	95-63-6	120.191	25	0.0057	0.057	22	0.588	5
1,3,5-Trimethylbenzene	C_9H_{12}	108-67-8	120.191	25	0.0050	0.050	22	0.588	5
2,3,3-Trimethyl-2-butanol	$C_7H_{16}O$	594-83-2	116.201	40	2.2	22	1		
1,1,3-Trimethylcyclohexane	C_9H_{18}	3073-66-3	126.239	25	0.000177	0.00177	4	105	113
1,1,3-Trimethylcyclopentane	C_8H_{16}	4516-69-2	112.213	25	0.00037	0.0037	4	159	5
2,2,5-Trimethylhexane	C_9H_{20}	3522-94-9	128.255	25	0.00008	0.0008	4	244	113
1,4,5-Trimethylnaphthalene	$C_{13}H_{14}$	2131-41-1	170.250	25	0.00021	0.0021	42,4		
2,6,8-Trimethyl-4-nonanone	$C_{12}H_{24}O$	123-18-2	184.318	10	0.012	0.12	20		
				80	0.014	0.14	20		
2,2,4-Trimethylpentane	C_8H_{18}	540-84-1	114.229	25	0.00022	0.0022	4	333	5
2,3,4-Trimethylpentane	C_8H_{18}	565-75-3	114.229	25	0.00018	0.0018	4	189	5
Trimethyl phosphate	$C_3H_9O_4P$	512-56-1	140.074	25	≈33	≈490	40		
1,3,5-Trinitrobenzene	$C_6H_3N_3O_6$	99-35-4	213.104	15	0.028	0.28	40		
2,4,6-Trinitrobenzoic acid	$C_7H_3N_3O_8$	129-66-8	257.114	23	1.97	20.1	40		
Trinitroglycerol	$C_3H_5N_3O_9$	55-63-0	227.087	25	0.13	1.3	40		
Trinitroglycerol	$C_3H_5N_3O_9$	55-63-0	227.087	80	0.34	3.4	40		
2,4,6-Trinitrophenol	$C_6H_3N_3O_7$	88-89-1	229.104	25	1.25	12.7	40		
				90	4.9	52	40		
2,4,6-Trinitrotoluene	$C_7H_5N_3O_6$	118-96-7	227.131	20	0.012	0.12	40		
				100	0.15	1.5	40		
2,4,6-Trinitro-*N*-(2,4,6-trinitrophenyl)aniline	$C_{12}H_5N_7O_{12}$	131-73-7	439.208	17	0.0060	0.060	40		
1,3,5-Trioxane	$C_3H_6O_3$	110-88-3	90.078	25	17.4	211	30		
Triphenylene	$C_{18}H_{12}$	217-59-4	228.288	25	0.0000043	0.000043	42,4	0.00001	12
Triphenyl phosphate	$C_{18}H_{15}O_4P$	115-86-6	326.283	24	0.000073	0.00073	40		
Triphenyltin hydroxide	$C_{18}H_{16}OSn$	76-87-9	367.029	20	0.0001	0.001	32		
Tris(hydroxymethyl)methylamine	$C_4H_{11}NO_3$	77-86-1	121.135	25	≈41	≈700	40		
L-Tryptophan	$C_{11}H_{12}N_2O_2$	73-22-3	204.225	25	1.30	13.2	26		

Name	Mol. form.	CAS Reg. No.	Mol. wt.	t/°C	s/mass%	s/g kg^{-1} H_2O	Ref.	k_H/kPa m^3 mol^{-1}	Ref.
DL-Tyrosine	$C_9H_{11}NO_3$	556-03-6	181.188	25	0.35	3.5	30		
L-Tyrosine	$C_9H_{11}NO_3$	60-18-4	181.188	25	0.0507	0.507	62		
Undecane	$C_{11}H_{24}$	1120-21-4	156.309	25	0.0000004	0.000004	37		
Uracil	$C_4H_4N_2O_2$	66-22-8	112.087	25	0.460	4.62	72		
Urea	CH_4N_2O	57-13-6	60.055	5	44	790	26		
				25	54.4	1200	26		
Uric acid	$C_5H_4N_4O_3$	69-93-2	168.111	20	0.002	0.02	26		
L-Valine	$C_5H_{11}NO_2$	72-18-4	117.147	25	8.13	88.5	26		
Valium	$C_{16}H_{13}ClN_2O$	439-14-5	284.739	25	0.005	0.05	40		
Vidarabine	$C_{10}H_{15}N_5O_5$	5536-17-4	285.257	20	0.051	0.51	40		
Vinclozolin	$C_{12}H_9Cl_2NO_3$	50471-44-8	286.110	20	0.1	1	32		
Vinyl acetate	$C_4H_6O_2$	108-05-4	86.090	20	2.0	20	10		
4-Vinylcyclohexene	C_8H_{12}	100-40-3	108.181	25	0.005	0.05	4		
Warfarin	$C_{19}H_{16}O_4$	81-81-2	308.328	20	0.004	0.04	40		
Xanthine	$C_5H_4N_4O_2$	69-89-6	152.112	20	0.05	0.5	26		
o-Xylene	C_8H_{10}	95-47-6	106.165	25	0.0171	0.171	22	0.417	101
				45	0.021	0.21	4		
m-Xylene	C_8H_{10}	108-38-3	106.165	0	0.0203	0.203	4		
				25	0.0161	0.161	22	0.714	124
				40	0.022	0.22	4		
p-Xylene	C_8H_{10}	106-42-3	106.165	0	0.0160	0.160	4		
				25	0.0181	0.181	22	0.526	101
				40	0.022	0.22	4		
2,3-Xylenol	$C_8H_{10}O$	526-75-0	122.164	25	0.457	4.57	40		
2,4-Xylenol	$C_8H_{10}O$	105-67-9	122.164	25	0.787	7.87	10		
2,5-Xylenol	$C_8H_{10}O$	95-87-4	122.164	25	0.354	3.54	40		
2,6-Xylenol	$C_8H_{10}O$	576-26-1	122.164	25	0.60	6.0	40		
3,4-Xylenol	$C_8H_{10}O$	95-65-8	122.164	25	0.477	4.77	40		
3,5-Xylenol	$C_8H_{10}O$	108-68-9	122.164	29	0.62	6.2	10		
D-Xylose	$C_5H_{10}O_5$	58-86-6	150.130	25	≈30	≈430	40		
Ziram	$C_6H_{12}N_2S_4Zn$	137-30-4	305.841	20	0.0065	0.065	40		

* Indicates a value of *s* for a gas at a partial pressure of 101.325 kPa (1 atm) in equilibrium with the solution.

AQUEOUS SOLUBILITY OF INORGANIC COMPOUNDS AT VARIOUS TEMPERATURES

The solubility of over 300 common inorganic compounds in water is tabulated here as a function of temperature. Solubility is defined as the concentration of the compound in a solution that is in equilibrium with a solid phase at the specified temperature. In this table the solid phase is generally the most stable crystalline phase at the temperature in question. An asterisk * on solubility values in adjacent columns indicates that the solid phase changes between those two temperatures (usually from one hydrated phase to another or from a hydrate to the anhydrous solid). In such cases the slope of the solubility vs. temperature curve may show a discontinuity.

All solubility values are expressed as mass percent of solute, $100 \times w_2$, where

$$w_2 = m_2/(m_1 + m_2)$$

and m_2 is the mass of solute and m_1 the mass of water. This quantity is related to other common measures of solubility as follows:

Molarity: $c_2 = 1000\rho w_2/M_2$
Molality: $m_2 = 1000w_2/M_2(1\text{-}w_2)$
Mole fraction: $x_2 = (w_2/M_2)/\{(w_2/M_2) + (1\text{-}w_2)/M_1\}$
Mass of solute per 100 g of H_2O: $r_2 = 100w_2/(1\text{-}w_2)$

Here M_2 is the molar mass of the solute and M_1 = 18.015 g/mol is the molar mass of water; ρ is the density of the solution in g cm^{-3}.

The data in the table have been derived from the references indicated; in many cases the data have been refitted or interpolated in order to present solubility at rounded values of temperature. Where available, values were taken from the IUPAC *Solubility Data Series* (Ref. 1) or related papers in the *Journal of Physical and Chemical Reference Data* (Refs. 2 to 5), which present carefully evaluated data.

The solubility of sparingly soluble compounds that do not appear in this table may be calculated from the data in the table "Solubility Product Constants of Inorganic Salts." Solubility of inorganic gases may be found in the table "Solubility of Selected Gases in Water." Both tables are in this section.

Compounds are listed alphabetically by formula.

References

1. *Solubility Data Series*, International Union of Pure and Applied Chemistry. Volumes 1 to 53 were published by Pergamon Press, Oxford, from 1979 to 1994; subsequent volumes were published by Oxford University Press, Oxford. The number following the colon is the volume number in the series. Current reports in the series appear in the *Journal of Physical and Chemical Reference Data*.
2. Clever, H. L., and Johnston, F. J., *J. Phys. Chem. Ref. Data*, 9, 751, 1980. <https://doi.org/10.1063/1.555628>
3. Marcus, Y., *J. Phys. Chem. Ref. Data*, 9, 1307, 1980. <https://doi.org/10.1063/1.555633>
4. Clever, H. L., Johnson, S. A., and Derrick, M. E., *J. Phys. Chem. Ref. Data*, 14, 631, 1985. <https://doi.org/10.1063/1.555732>
5. Clever, H. L., Johnson, S. A., and Derrick, M. E., *J. Phys. Chem. Ref. Data*, 21, 941, 1992. <https://doi.org/10.1063/1.555909>
6. Söhnel, O., and Novotny, P., *Densities of Aqueous Solutions of Inorganic Substances*, Elsevier, Amsterdam, 1985.
7. Krumgalz, B.S., *Mineral Solubility in Water at Various Temperatures*, Israel Oceanographic and Limnological Research Ltd., Haifa, 1994.
8. Potter, R. W., and Clynne, M. A., *J. Research U.S. Geological Survey*, 6, 701, 1978; Clynne, M. A., and Potter, R. W., *J. Chem. Eng. Data*, 24, 338, 1979.
9. Marshal, W. L., and Slusher, R., *J. Phys. Chem.*, 70, 4015, 1966; Knacke, O., and Gans, W., *Zeit. Phys. Chem.*, NF, 104, 41, 1977. <https://doi.org/10.1021/j100884a044>
10. Stephen, H., and Stephen, T., *Solubilities of Inorganic and Organic Compounds, Vol. 1*, Macmillan, New York, 1963.

Aqueous Solubility of Inorganic Compounds in Mass% as a Function of Temperature

Formula	0 °C	10 °C	20 °C	25 °C	30 °C	40 °C	50 °C	60 °C	70 °C	80 °C	90 °C	100 °C	Ref.
$AgBrO_3$				0.193							1.32		7
$AgClO_2$	0.17	0.31	0.47	0.55	0.64	0.82	1.02	1.22	1.44	1.66	1.88	2.11	7
$AgClO_3$				15									7
$AgClO_4$	81.6	83.0	84.2	84.8	85.3	86.3	86.9	87.5	87.9	88.3	88.6	88.8	6
$AgNO_2$	0.155			0.413									7
$AgNO_3$	55.9	62.3	67.8	70.1	72.3	76.1	79.2	81.7	83.8	85.4	86.7	87.8	6
Ag_2SO_4	0.56	0.67	0.78	0.83	0.88	0.97	1.05	1.13	1.20	1.26	1.32	1.39	7
$AlCl_3$	30.84	30.91	31.03	31.10	31.18	31.37	31.60	31.87	32.17	32.51	32.90	33.32	7
$Al(ClO_4)_3$	54.9										64.4		7
AlF_3	0.25	0.34	0.44	0.50	0.56	0.68	0.81	0.96	1.11	1.28	1.45	1.64	7
$Al(NO_3)_3$	37.0	38.2	39.9	40.8	42.0	44.5	47.3	50.4	53.8*			61.5*	6
$Al_2(SO_4)_3$	27.5			27.8	28.2	29.2	30.7	32.6	34.9	37.6	40.7	44.2	7
As_2O_3	1.19	1.48	1.80	2.01	2.27	2.86	3.43	4.11	4.89	5.77	6.72	7.71	10
$BaBr_2$	47.6	48.5	49.5	50.0	50.4	51.4	52.5	53.5	54.5	55.5	56.6	57.6	6
$Ba(BrO_3)_2$	0.285	0.442	0.656	0.788	0.935	1.30	1.74	2.27	2.90	3.61	4.40	5.25	1:14
$Ba(C_2H_3O_2)_2$	37.0			44.2									7
$BaCl_2$	23.30	24.88	26.33	27.03	27.70	29.00	30.27	31.53	32.81	34.14	35.54	37.05	8
$Ba(ClO_2)_2$	30.5			31.3								44.7	7
$Ba(ClO_3)_2$	16.90	21.23	23.66	27.50	29.43	33.16	36.69	40.05	43.04	45.90	48.70	51.17	1:14
$Ba(ClO_4)_2$	67.30	70.96	74.30	75.75	77.05	79.23	80.92	82.21	83.16	83.88	84.43	84.90	7
BaF_2		0.158		0.161									7

Formula	0 °C	10 °C	20 °C	25 °C	30 °C	40 °C	50 °C	60 °C	70 °C	80 °C	90 °C	100 °C	Ref.
BaI_2	62.5	64.7	67.3	68.8	69.1	69.5	70.1	70.7	71.3	72.0	72.7	73.4	6
$Ba(IO_3)_2$	0.0182	0.0262	0.0342	0.0396	0.045*	0.058*	0.073	0.090	0.109	0.131	0.156	0.182	1:14
$Ba(NO_2)_2$	31.1	36.6	41.8	44.3	46.8	51.6	56.2	60.5	64.6	68.5	72.1	75.6	10
$Ba(NO_3)_2$	4.7	6.3	8.2	9.3	10.2	12.4	14.7	17.0	19.3	21.5	23.5	25.5	6
$Ba(OH)_2$	1.67			4.68	8.4	19	33	52	74	100			7
BaS	2.79	4.78	6.97	8.21	9.58	12.67	16.18	20.05	24.19	28.55	33.04	37.61	7
$BaSO_3$				0.0011									1:26
$Ba(SCN)_2$				62.6									7
$BeCl_2$	40.5			41.7									7
$Be(ClO_4)_2$				59.5									7
$BeSO_4$	26.69	27.58	28.61	29.22	29.90	31.51	33.39	35.50	37.78	40.21	42.72	45.28	7
$CaBr_2$	55	56	59	61	63	68	71	73					10
$CaCl_2$	36.70	39.19	42.13	44.83*	49.12*	52.85*	56.05*	56.73	57.44	58.21	59.04	59.94	8
$Ca(ClO_3)_2$	63.2	64.2	65.5	66.3	67.2	69.0	71.0	73.2	75.5*	77.4*	77.7	78.0	1:14
$Ca(ClO_4)_2$				65.3									7
CaF_2	0.0013			0.0016									10
CaI_2	64.6	66.0	67.6	68.3	69.0	70.8	72.4	74.0	76.0	78.0	79.6	81.0	7
$Ca(IO_3)_2$	0.082	0.155	0.243	0.305	0.384*	0.517*	0.590	0.652	0.811*	0.665*	0.668		1:14
$Ca(NO_2)_2$	38.6	39.5	44.5	48.6									7
$Ca(NO_3)_2$	50.1	53.1	56.7	59.0	60.9	65.4	77.8	78.1	78.2	78.3	78.4	78.5	6
$CaSO_3$			0.0059	0.0054	0.0049	0.0041	0.0035	0.0030	0.0026	0.0023	0.0020	0.0019	1:26
$CaSO_4$	0.174	0.191	0.202	0.205	0.208	0.210	0.207	0.201	0.193	0.184	0.173	0.163	9
$CdBr_2$	36.0	43.0	49.9	53.4	56.4	60.3*	60.3*	60.5	60.7	60.9	61.3	61.6	6
CdC_2O_4				0.0060									5
$CdCl_2$	47.2	50.1	53.2	54.6	56.3*	57.3*	57.5	57.8	58.1	58.51	58.98	59.5	6
$Cd(ClO_4)_2$				58.7								66.9	7
CdF_2		5.82	4.65	4.18	3.76								5
CdI_2	44.1	44.9	45.8	46.3	46.8	47.9	49.0	50.2	51.5	52.7	54.1	55.4	6
$Cd(IO_3)_2$				0.091									5
$Cd(NO_3)_2$	55.4	57.1	59.6	61.0	62.8	66.5	70.6	86.1	86.5	86.8	87.1	87.4	6
$CdSO_4$	43.1	43.1	43.2	43.4	43.6	44.1	43.5	42.5	41.4	40.2	38.5	36.7	6
$CdSeO_4$	42.04	40.59	39.02	38.18	37.29	35.35	33.15	30.65	27.84	24.69	21.24	17.49	5
$Ce(NO_3)_3$	57.99	59.80	61.89	63.05	64.31*	67.0*	68.6	71.1*	74.9*	79.2	80.9	83.1	1:13
$CoCl_2$	30.30	32.60	34.87	35.99	37.10	39.27	41.38	43.46	45.50	47.51	49.51	51.50	7
$Co(ClO_4)_2$	50.0			53.0									7
CoF_2				1.4									7
CoI_2	58.00	61.78	65.35	66.99	68.51	71.17	73.41	75.29	76.89	78.28	79.52	80.70	7
$Co(NO_2)_2$	0.076			0.49									7
$Co(NO_3)_2$	45.5	47.0	49.4	50.8	52.4	56.0	60.1	62.6	64.9	67.7			6
$CoSO_4$	19.9	23.0	26.1	27.7	29.2	32.3	34.4	35.9	35.5	33.2	30.6	27.8	6
$Co(SCN)_2$				50.7									7
CrO_3	62.2	62.3	62.6	62.8	63.0	63.5	64.1	64.7	65.5	66.2	67.1	67.9	6
$CsBr$				55.2									7
$CsBrO_3$	1.16	1.93	3.01	3.69	4.46	6.32	8.60	11.32	14.45	17.96	21.83	25.98	1:30
$CsCl$	61.83	63.48	64.96	65.64	66.29	67.50	68.60	69.61	70.54	71.40	72.21	72.96	1:47
$CsClO_3$	2.40	3.87	5.94	7.22	8.69	12.15	16.33	21.14	26.45	32.10	37.89	43.42	1:30
$CsClO_4$	0.79	1.01	1.51	1.96	2.57	4.28	6.55	9.29	12.41	15.80	19.39	23.07	7
CsI	30.9	37.2	43.2	45.9	48.6	53.3	57.3	60.7	63.6	65.9	67.7	69.2	6
$CsIO_3$	1.08	1.58	2.21	2.59	3.02	3.96	5.06	6.29	7.70	9.20	10.79	12.45	1:30
$CsNO_3$	8.46	13.0	18.6	21.8	25.1	32.0	39.0	45.7	51.9	57.3	62.1	66.2	6
$CsOH$					75								7
Cs_2SO_4	62.6	63.4	64.1	64.5	64.8	65.5	66.1	66.7	67.3	67.8	68.3	68.8	6
$CuBr_2$				55.8									7
$CuCl_2$	40.8	41.7	42.6	43.1	43.7	44.8	46.0	47.2	48.5	49.9	51.3	52.7	6
$Cu(ClO_4)_2$	54.3				59.3								7
CuF_2				0.075									7
$Cu(NO_3)_2$	45.2	49.8	56.3	59.2	61.1	62.0	63.1	64.5	65.9	67.5	69.2	71.0	6
$CuSO_4$	12.4	14.4	16.7	18.0	19.3	22.2	25.4	28.8	32.4	36.3	40.3	43.5	6
$CuSeO_4$	10.6			16.0									7
$Dy(NO_3)_3$	58.79	59.99	61.49	62.35	63.29	65.43	68.04	71.58					1:13
$Er(NO_3)_3$	61.58	63.15	64.84	65.75	66.69	68.70	70.96	73.64	77.75				1:13

Formula	0 °C	10 °C	20 °C	25 °C	30 °C	40 °C	50 °C	60 °C	70 °C	80 °C	90 °C	100 °C	Ref.
$Eu(NO_3)_3$	55.2	56.7	58.5	59.4	60.4	62.5	64.6						1:13
$FeBr_2$				54.6*								64.8*	7
$FeCl_2$	33.2*			39.4*								48.7*	7
$Fe(ClO_4)_2$	63.39			67.76									7
$FeCl_3$	42.7	44.9	47.9	47.7	51.6	74.8	76.7	84.6	84.3	84.3	84.4	84.7	6
FeF_3				5.59									7
$Fe(NO_3)_2$	41.44			46.67									7
$Fe(NO_3)_3$	40.15			46.57									7
$FeSO_4$	13.5	17.0	20.8	22.8	24.8	28.8	32.8	35.5	33.6	30.4	27.1	24.0	6
$Gd(NO_3)_3$	56.3	57.7	59.2	60.1	61.0	62.9	65.2	67.9	71.5				1:13
HIO_3	73.45	74.10	74.98	75.48	76.03	77.20	78.46	79.78	81.13	82.48	83.82	85.14	1:30
H_3BO_3	2.61	3.57	4.77	5.48	6.27	8.10	10.3	12.9	15.9	19.3	23.1	27.3	6
$HgBr_2$	0.26	0.37	0.52	0.61	0.72	0.96	1.26	1.63	2.08	2.61	3.23	3.95	4
$Hg(CN)_2$	6.57	7.83	9.33	10.2	11.1	13.1	15.5	18.2	21.2	24.6	28.3	32.3	6
$HgCl_2$	4.24	5.05	6.17	6.81	7.62	9.53	12.02	15.18	19.16	24.06	29.90	36.62	4
HgI_2			0.0041	0.0055	0.0072	0.0122	0.0199						4
$Hg(SCN)_2$				0.070									4
Hg_2Cl_2				0.0004									3
$Hg_2(ClO_4)_2$	73.8			79.8*								85.3*	7
Hg_2SO_4	0.038	0.043	0.048	0.051	0.054	0.059	0.065	0.070	0.076	0.082	0.088	0.093	4
$Ho(NO_3)_3$				63.8									1:13
KBF_4	0.28	0.34	0.45	0.55	0.75	1.38	2.09	2.82	3.58	4.34	5.12	5.90	10
KBr	35.0	37.3	39.4	40.4	41.4	43.2	44.8	46.2	47.6	48.8	49.8	50.8	6
$KBrO_3$	2.97	4.48	6.42	7.55	8.79	11.57	14.71	18.14	21.79	25.57	29.42	33.28	1:30
$KC_2H_3O_2$	68.40	70.29	72.09	72.92	73.70	75.08	76.27	77.31	78.22	79.04	79.80	80.55	7
KCl	21.74	23.61	25.39	26.22	27.04	28.59	30.04	31.40	32.66	33.86	34.99	36.05	1:47
$KClO_3$	3.03	4.67	6.74	7.93	9.21	12.06	15.26	18.78	22.65	26.88	31.53	36.65	1:30
$KClO_4$	0.70	1.10	1.67	2.04	2.47	3.54	4.94	6.74	8.99	11.71	14.94	18.67	6
KF	30.90	39.8	47.3	50.41	53.2					60.0			7
$KHCO_3$	18.62	21.73	24.92	26.6	28.13	31.32	34.46	37.51	40.45				6
$KHSO_4$	27.1	29.7	32.3	33.6	35.0	37.8	40.5	43.4	46.2	49.02	51.82	54.6	6
KH_2PO_4	11.74	14.91	18.25	19.97	21.77	25.28	28.95	32.76	36.75	40.96	45.41	50.12	1:31
KI	56.0	57.6	59.0	59.7	60.4	61.6	62.8	63.8	64.8	65.7	66.6	67.4	6
KIO_3	4.53	5.96	7.57	8.44	9.34	11.09	13.22	15.29	17.41	19.58	21.78	24.03	1:30
KIO_4	0.16	0.22	0.37	0.51	0.70	1.24	1.96	2.83	3.82	4.89	6.02	7.17	7
$KMnO_4$	2.74	4.12	5.96	7.06	8.28	11.11	14.42	18.16					6
KNO_2	73.7	74.6	75.3	75.7	76.0	76.7	77.4	78.0	78.5	79.1	79.6	80.1	6
KNO_3	12.0	17.6	24.2	27.7	31.3	38.6	45.7	52.2	58.0	63.0	67.3	70.8	6
KOH	48.7	50.8	53.2	54.7	56.1	57.9	58.6	59.5	60.6	61.8	63.1	64.6	6
$KSCN$	63.8	66.4	69.1	70.4	71.6	74.1	76.5	78.9	81.1	83.3	85.3	87.3	6
K_2CO_3	51.3	51.7	52.3	52.7	53.1	54.0	54.9	56.0	57.2	58.4	59.6	61.0	6
K_2CrO_4	37.1	38.1	38.9	39.4	39.8	40.5	41.3	41.9	42.6	43.2	43.8	44.3	6
$K_2Cr_2O_7$	4.30	7.12	10.9	13.1	15.5	20.8	26.3	31.7	36.9	41.5	45.5	48.9	6
K_2HAsO_4	48.5*			63.6*								79.8*	7
K_2HPO_4	57.0	59.1	61.5	62.7	64.1	67.7*		72.7*					1:31
K_2MoO_4				64.7							66.5		7
K_2SO_3	51.30	51.39	51.49	51.55	51.62	51.76	51.93	52.11	52.32	52.54	52.79	53.06	1:26
K_2SO_4	7.11	8.46	9.95	10.7	11.4	12.9	14.2	15.5	16.7	17.7	18.6	19.3	6
$K_2S_2O_3$	49.0*			62.3*							75.7*		7
$K_2S_2O_5$	22.1	26.7	31.1	33.1	35.2	39.0	42.6	46.0	49.1	52.0	54.6		1:26
K_2SeO_3	68.4*			68.5*								68.5*	7
K_2SeO_4	52.70	52.93	53.17	53.30	53.43	53.70	53.99	54.30	54.61	54.94	55.26	55.60	7
K_3AsO_4	51.5*			55.6*								73*	7
$K_3Fe(CN)_6$	23.9	27.6	31.1	32.8	34.3	37.2	39.6	41.7	43.5	45.0	46.1	47.0	6
K_3PO_4	44.3			51.4									7
$K_4Fe(CN)_6$	12.5	17.3	22.0	23.9	25.6	29.2	32.5	35.5	38.2	40.6	41.4	43.1	6
$LaCl_3$	49.0	48.5	48.6	48.9	49.3	50.5	52.1	54.0	56.3	58.9	61.7		6
$La(NO_3)_3$	55.0	56.9	58.9	60.0	61.1	63.6	66.3	69.9*	74.1*				1:13
$LiBr$	58.4	60.1	62.7	64.4	65.9	67.8	68.3	69.0	69.8	70.7	71.7	72.8	6
$LiBrO_3$	61.03	62.62	64.44	65.44	66.51	68.90	71.68*	73.24*	74.43	75.66	76.93	78.32	1:30
$LiC_2H_3O_2$	23.76	26.49	29.42	31.02	32.72	36.48	40.65	45.15	49.93	54.91	60.04	65.26	7

Formula	0 °C	10 °C	20 °C	25 °C	30 °C	40 °C	50 °C	60 °C	70 °C	80 °C	90 °C	100 °C	Ref.
LiCl	40.45	42.46*	45.29*	45.81	46.25	47.30	48.47	49.78	51.27	52.98	54.98*	56.34*	1:47
$LiClO_3$	73.2	75.6*	80.8*	82.1	83.4	85.9*	87.1*	88.2	89.6	91.3	93.4	95.7	1:30
$LiClO_4$	30.1	32.6	35.5	37.0	38.6	41.9	45.5	49.2	53.2	57.2	61.3	71.4	6
LiF	0.120	0.126	0.131	0.134									7
LiH_2PO_4	55.8												7
LiI	59.4	60.5	61.7	62.3	63.0	64.3	65.8	67.3	68.8	81.3	81.7	82.6	6
$LiIO_3$				43.8									1:30
$LiNO_2$	41	45	49	51	53	56	60	63	66	68			10
$LiNO_3$	34.8	37.6	42.7	50.5	57.9	60.1	62.2	64.0	65.7	67.2	68.5	69.7	6
LiOH	10.8	10.8	11.0	11.1	11.3	11.7	12.2	12.7	13.4	14.2	15.1	16.1	6
LiSCN				54.5									7
Li_2CO_3	1.54	1.43	1.33	1.28	1.24	1.15	1.07	0.99	0.92	0.85	0.78	0.72	7
$Li_2C_2O_4$				5.87									7
Li_2HPO_3	9.07	8.40	7.77	7.47	7.18	6.64	6.16	5.71	5.30	4.91	4.53	4.16	7
Li_2SO_4	26.3	25.9	25.6	25.5	25.3	25.0	24.8	24.5	24.3	24.0	23.8	23.6	6
Li_3PO_4				0.027									1:31
$Lu(NO_3)_3$				71.1									1:13
$MgBr_2$	49.3	49.8	50.3	50.6	50.9	51.5	52.1	52.8	53.5	54.2	55.0	55.7	6
$Mg(BrO_3)_2$	43.0	45.2	48.0	49.4	51.0	54.3	57.9	61.6	65.3	69.0*	70.9*	71.7	1:14
MgC_2O_4				0.038									7
$Mg(C_2H_3O_2)_2$	36.18	37.55	38.92	39.61									7
$MgCl_2$	33.96	34.85	35.58	35.90	36.20	36.77	37.34	37.97	38.71	39.62	40.75	42.15	8
$Mg(ClO_3)_2$	53.35	54.40	56.81	58.66	60.91*	65.46*	67.33	69.27	71.01	72.44	73.48		1:14
$Mg(ClO_4)_2$	47.8	48.7	49.6	50.1	50.5	51.3	52.1						6
$MgCrO_4$	32.06*			35.39*									7
$MgCr_2O_7$				58.9						67.0			7
MgF_2				0.013									7
MgI_2	54.7	56.1	58.2	59.4	60.8	63.9	65.0	65.0	65.0	65.0	65.1	65.2	6
$Mg(IO_3)_2$	3.19*	6.70*	7.92	8.52	9.11	10.45	11.99	13.7	15.6	17.6	19.6		1:14
$Mg(NO_2)_2$				47									7
$Mg(NO_3)_2$	38.4	39.5	40.8	41.6	42.4	44.1	45.9	47.9	50.0	52.2	70.6	72.0	6
$MgSO_3$	0.32	0.37	0.46	0.52	0.61	0.87*	0.85*	0.76	0.69	0.64	0.62	0.60	1:26
$MgSO_4$	18.2	21.7	25.1	26.3	28.2	30.9	33.4	35.6	36.9	35.9	34.7	33.3	6
MgS_2O_3	30.7			34.1									7
$MgSeO_4$	31.4*			35.7*								47*	7
$MnBr_2$	56.00	57.72	59.39	60.19	60.96	62.41	63.75	65.01	66.19	67.32	68.42	69.50	7
$MnCl_2$	38.7	40.6	42.5	43.6	44.7	47.0	49.4	54.1	54.7	55.2	55.7	56.1	6
MnF_2	0.80*			1.01*								0.48	7
$Mn(IO_3)_2$				0.27							0.34		7
$Mn(NO_3)_2$	50.5			61.7									7
$MnSO_4$	34.6	37.3	38.6	38.9	38.9	37.7	36.3	34.6	32.8	30.8	28.8	26.7	6
NH_4Br	37.5	40.2	42.7	43.9	45.1	47.3	49.4	51.3	53.0	54.6	56.1	57.4	7
NH_4Cl	22.92	25.12	27.27	28.34	29.39	31.46	33.50	35.49	37.46	39.40	41.33	43.24	1:47
NH_4ClO_4	10.8	14.1	17.8	19.7	21.7	25.8	29.8	33.6	37.3	40.7	43.8	46.6	6
NH_4F	41.7	43.2	44.7	45.5	46.3	47.8	49.3	50.9	52.5	54.1			7
NH_4HCO_3	10.6	13.7	17.6	19.9	22.4	27.9	34.2	41.4	49.3	58.1	67.6	78.0	7
$NH_4H_2AsO_4$	25.2	29.0	32.7	34.5	36.3	39.7	43.1	46.2	49.3	52.2	55.0		7
$NH_4H_2PO_4$	17.8	22.0	26.4	28.8	31.2	36.2	41.6	47.2	53.0	59.2	65.7	72.4	7
NH_4I	60.7	62.1	63.4	64.0	64.6	65.8	66.8	67.8	68.7	69.6	70.4	71.1	6
NH_4IO_3				3.70	4.20	5.64	7.63						1:30
NH_4NO_2	55.7	59.0	64.9	68.8									7
NH_4NO_3	54.0	60.1	65.5	68.0	70.3	74.3	77.7	80.8	83.4	85.8	88.2	90.3	6
NH_4SCN				64.4					81.1				7
$(NH_4)_2C_2O_4$	2.31	3.11	4.25	4.94	5.73	7.56	9.73	12.2	15.1	18.3	21.8	25.7	7
$(NH_4)_2HPO_4$	36.4	38.2	40.0	41.0	42.0	44.1	46.2	48.5	50.9	53.3	55.9	58.6	7
$(NH_4)_2SO_3$	32.2	34.9	37.7	39.1	40.6	43.7	47.0	50.6	54.5	58.9			1:26
$(NH_4)_2SO_4$	41.3	42.1	42.9	43.3	43.8	44.7	45.6	46.6	47.5	48.5	49.5	50.5	6
$(NH_4)_2S_2O_5$	65.5	67.9	69.8	70.5	71.3	72.3	72.9	73.1					1:26
$(NH_4)_2S_2O_8$	37.00	40.45	43.84	45.49	47.11	50.25	53.28	56.23	59.13	62.00			7
$(NH_4)_2SeO_3$	49.0	51.1	53.4	54.7	56.0	58.9	62.0	65.4	69.1				7
$(NH_4)_2SeO_4$				54.02									7

Formula	0 °C	10 °C	20 °C	25 °C	30 °C	40 °C	50 °C	60 °C	70 °C	80 °C	90 °C	100 °C	Ref.
$(NH_4)_3PO_4$				15.5									7
$NaBr$	44.4	45.9	47.7	48.6	49.6	51.6	53.7	54.1	54.3	54.5	54.7	54.9	6
$NaBrO_3$	20.0	23.22	26.65	28.28	29.86	32.83	35.55	38.05	40.37	42.52			1:30
$NaCHO_2$	30.8	37.9	45.7	48.7	50.6	52.0	53.5	55.0					6
$NaC_2H_3O_2$	26.5	28.8	31.8	33.5	35.5	39.9	45.1	58.3	59.3	60.5	61.7	62.9	6
$NaCl$	26.28	26.32	26.41	26.45	26.52	26.67	26.84	27.03	27.25	27.50	27.78	28.05	1:47
$NaClO$	22.7			44.4									7
$NaClO_2$				97.0*				95.3*					7
$NaClO_3$	44.27	46.67	49.3	50.1	51.2	53.6	55.5	57.0	58.5	60.5	63.3	67.1	1:30
$NaClO_4$	61.9	64.1	66.2	67.2	68.3	70.4	72.5	74.1	74.7	75.4	76.1	76.7	6
NaF	3.52	3.72	3.89	3.97	4.05	4.20	4.34	4.46	4.57	4.66	4.75	4.82	6
$NaHCO_3$	6.48	7.59	8.73	9.32	9.91	11.13	12.40	13.70	15.02	16.37	17.73	19.10	7
$NaHSO_4$				22.2								33.3	10
NaH_2PO_4	36.54	41.07	46.00	48.68	51.54	57.89*	61.7*	62.3*	65.9	68.7			1:31
NaI	61.2	62.4	63.9	64.8	65.7	67.7	69.8	72.0	74.7	74.8	74.9	75.1	6
$NaIO_3$	2.43	4.40	7.78*	8.65*	9.60	11.67	13.99	16.52	19.25*	21.1*	22.9	24.7	1:30
$NaIO_4$				12.62									7
$NaNO_2$	41.9	43.4	45.1	45.9	46.8	48.7	50.7	52.8	55.0	57.2	59.5	61.8	6
$NaNO_3$	42.2	44.4	46.6	47.7	48.8	51.0	53.2	55.3	57.5	59.6	61.7	63.8	6
$NaOH$	30	39	46	50	53	58	63	67	71	74	76	79	10
$NaSCN$		52.9	57.1	60.2	62.7	63.5	64.2	65.0	65.9	66.9	67.9	69.0	6
$Na_2B_4O_7$	1.23	1.71	2.50	3.07	3.82	6.02	9.7	14.9	17.1	19.9	23.5	28.0	6
Na_2CO_3	6.44	10.8	17.9	23.5	28.7	32.8	32.2	31.7	31.3	31.1	30.9	30.9	6
$Na_2C_2O_4$	2.62	2.95	3.30	3.48	3.65	4.00	4.36	4.71	5.06	5.41	5.75	6.08	6
Na_2CrO_4	22.6	32.3	44.6	46.7	46.9	48.9	51.0	53.4	55.3	55.5	55.8	56.1	6
$Na_2Cr_2O_7$	62.1	63.1	64.4	65.2	66.1	68.0	70.1	72.3	74.6	77.0	79.6	80.7	6
Na_2HAsO_4	5.6*			29.3*								67*	7
Na_2HPO_4	1.66	4.19	7.51	10.55	16.34*	35.17*	44.64*	45.20	46.81	48.78	50.52	51.53	1:31
Na_2MoO_4	30.6	38.8	39.4	39.4	39.8	40.3	41.0	41.7	42.6	43.5	44.5	45.5	6
Na_2S	11.1	13.2	15.7	17.1	18.6	22.1	26.7	28.1	30.2	33.0	36.4	41.0	6
Na_2SO_3	12.0	16.1	20.9	23.5	26.3*	27.3*	25.9	24.8	23.7	22.8	22.1	21.5	1:26
Na_2SO_4			16.13	21.94	29.22*	32.35*	31.55	30.90	30.39	30.02	29.79	29.67	8
$Na_2S_2O_3$	33.1	36.3	40.6	43.3	45.9	52.0	62.3	65.7	68.8	69.4	70.1	71.0	6
$Na_2S_2O_5$		38.4	39.5	40.0	40.6	41.8	43.0	44.2	45.5	46.8	48.1	49.5	1:26
Na_2SeO_3				47.3*								45*	7
Na_2SeO_4	11.7			36.9*								42.1*	7
Na_2WO_4	41.6	41.9	42.3	42.6	42.9	43.6	44.4	45.3	46.2	47.3	48.4	49.5	6
Na_3PO_4	4.28	7.30	10.8	12.6	14.1	16.6	22.9	28.4	32.4	37.6	40.4	43.5	6
$Na_4P_2O_7$	2.23	3.28	4.81	6.62	7.00	10.10	14.38	20.07	27.31	36.03	32.37	30.67	6
$NdCl_3$	49.0	49.3	49.7	50.0	50.4	51.2	52.2	53.3	54.5	55.8	57.1	58.5	6
$Nd(NO_3)_3$	55.76	57.49	59.37	60.38	61.43	63.69	66.27	69.47					1:13
$NiCl_2$	34.7	36.1	38.5	40.3	41.7	42.1	43.2	45.0	46.1	46.2	46.4	46.6	6
$Ni(ClO_4)_2$	51.1			52.8									7
NiF_2				2.50							2.52		7
NiI_2	55.40	57.68	59.78	60.69	61.50	62.80	63.73	64.38	64.80	65.09	65.30		7
$Ni(NO_3)_2$	44.1	46.0	48.4	49.8	51.3	54.6	58.3	61.0	63.1	65.6	67.9	69.0	6
$NiSO_4$	21.4	24.4	27.4	28.8	30.3*	32.0*	34.1	35.8	37.7	39.9	42.3	44.8	6
$Ni(SCN)_2$				35.48									7
$NiSeO_4$	21.6		26.2*									45.6*	7
$PbBr_2$	0.449	0.620	0.841	0.966	1.118	1.46	1.89						2
$PbCl_2$	0.66	0.81	0.98	1.07	1.17	1.39	1.64	1.93	2.24	2.60	2.99	3.42	2
$Pb(ClO_4)_2$				81.5									7
PbF_2		0.0603	0.0649	0.0670	0.0693								2
PbI_2	0.041	0.052	0.067	0.076	0.086	0.112	0.144	0.187	0.243	0.315			2
$Pb(IO_3)_2$				0.0025									7
$Pb(NO_3)_2$	28.46	32.13	35.67	37.38	39.05	42.22	45.17	47.90	50.42	52.72	54.82	56.75	2
$PbSO_4$	0.0033	0.0038	0.0042	0.0044	0.0047	0.0052	0.0058						2
$PrCl_3$	48.0	48.1	48.6	49.0	49.5	50.8	52.3	54.1	56.1	58.3			6
$Pr(NO_3)_3$	57.50	59.20	61.16	62.24	63.40*	65.7*	67.8	70.2	73.4				1:13
$RbBr$	47.4	50.1	52.6	53.8	54.9	57.0	58.8	60.6	62.1	63.5	64.8	65.9	6
$RbBrO_3$	0.97	1.55	2.36	2.87	3.45	4.87	6.64	8.78	11.29	14.15	17.32	20.76	1:30

Formula	0 °C	10 °C	20 °C	25 °C	30 °C	40 °C	50 °C	60 °C	70 °C	80 °C	90 °C	100 °C	Ref.
$RbCl$	43.58	45.65	47.53	48.42	49.27	50.86	52.34	53.67	54.92	56.08	57.16	58.15	1:47
$RbClO_3$	2.10	3.38	5.14	6.22	7.45	10.35	13.85	17.93	22.53	27.57	32.96	38.60	1:30
$RbClO_4$	1			1.5								17	7
RbF			75										7
$RbHCO_3$			53.7										7
RbI	55.8	58.6	61.1	62.3	63.4	65.4	67.2	68.8	70.3	71.6	72.7	73.8	6
$RbIO_3$	1.09	1.53	2.07	2.38	2.74	3.52	4.41	5.42	6.52	7.74	9.00	10.36	1:30
$RbNO_3$	16.4	25.0	34.6	39.4	44.2	53.1	60.8	67.2	72.2	76.1	79.0	81.2	6
$RbOH$					63.4								7
Rb_2CrO_4	38.27			43.26									7
Rb_2SO_4	27.3	30.0	32.5	33.7	34.8	36.9	38.7	40.3	41.8	43.0	44.1	44.9	6
$SbCl_3$	85.7			90.8									7
SbF_3	79.4			83.1									7
$Sc(NO_3)_3$	57.0	59.3	61.6	62.8	63.9	66.2	68.5						1:13
$SmCl_3$		48.0	48.2	48.4	48.6	49.2	50.0						6
$Sm(NO_3)_3$	54.83	56.33	58.08	59.05	60.08	62.38	65.05*	68.1*	70.8	74.2			1:13
$SnCl_2$	46	64											7
SnI_2			0.97									3.87	7
$SrBr_2$	46.0	48.3	50.6	51.7	52.9	55.2	57.6	59.9	62.3	64.6	66.8	69.0	6
$Sr(BrO_3)_2$	18.53	22.00	25.39	27.02	28.59	31.55	34.21	36.57	38.64*	40.2*	40.8	41.0	1:14
$SrCl_2$	31.94	32.93	34.43	35.37	36.43	38.93	41.94	45.44*	46.81*	47.69	48.70	49.87	8
$Sr(ClO_2)_2$	13.0	13.6	14.1	14.3	14.5	14.9	15.3	15.6	15.9				7
$Sr(ClO_3)_2$	63.29	63.42	63.64	63.77	63.93	64.29	64.70	65.16	65.65	66.18	66.74	67.31	1:14
$Sr(ClO_4)_2$	70.04*			75.35*		78.44*							7
SrF_2	0.011			0.021									7
SrI_2	62.5	62.8	63.5	63.9	64.5	65.8	67.3	69.0	70.8	72.7	74.7	79.2	6
$Sr(IO_3)_2$	0.102	0.126	0.152	0.165	0.179	0.206	0.233	0.259	0.284	0.307	0.328	0.346	1:14
$Sr(MnO_4)_2$	2.5												7
$Sr(NO_2)_2$					41.9	44.3						58.6	7
$Sr(NO_3)_2$	28.2	34.6	41.0	44.5	47.0	47.4	47.9	48.4	48.9	49.5	50.1	50.7	6
$Sr(OH)_2$	0.9			2.2									7
$SrSO_3$				0.0015									1:26
$SrSO_4$				0.0135									7
SrS_2O_3	8.8	13.2	17.7	20.0	22.2	26.8							7
$Tb(NO_3)_3$			60.6	61.02									1:13
Tl_2SO_4	2.65	3.56	4.61	5.19	5.80	7.09	8.46	9.89	11.33	12.77	14.18	15.53	6
$Tm(NO_3)_3$				67.9									1:13
$UO_2(NO_3)_2$	49.52	51.82	54.42	55.85	57.55	61.59	67.07						1:??
$Y(NO_3)_3$	55.57	56.93	58.75	59.86	61.11*	63.3*	64.9	67.9	72.5				1:13
$Yb(NO_3)_3$				70.5									1:13
$ZnBr_2$	79.3	80.1	81.8	83.0	84.1	85.6	85.8	86.1	86.3	86.6	86.8	87.1	6
ZnC_2O_4		0.0010	0.0019	0.0026									5
$ZnCl_2$		76.6	79.0	80.3	81.4	81.8	82.4	83.0	83.7	84.4	85.2	86.0	6
$Zn(ClO_4)_2$	44.29*			46.27*			48.70						7
ZnF_2				1.53									5
ZnI_2	81.1	81.2	81.3	81.4	81.5	81.7	82.0	82.3	82.6	83.0	83.3	83.7	6
$Zn(IO_3)_2$			0.58	0.64	0.69	0.77	0.82						5
$Zn(NO_3)_2$	47.8	50.8	54.4	54.6	58.5	79.1	80.1	87.5	89.9				6
$ZnSO_3$			0.1786	0.1790	0.1794	0.1803	0.1812						5
$ZnSO_4$	29.1	32.0	35.0	36.6	38.2	41.3	43.0	42.1	41.0	39.9	38.8	37.6	6
$ZnSeO_4$	33.06	34.98	37.38	38.79	40.34								5

* Solid phase changes between these temperatures.

OCTANOL–WATER PARTITION COEFFICIENTS

The octanol–water partition coefficient, *P*, is a widely used parameter for correlating biological effects of organic substances. It is a property of the two-phase system in which water and 1-octanol are in equilibrium at a fixed temperature and the substance is distributed between the water-rich and octanol-rich phases. *P* is defined as the ratio of the equilibrium concentration of the substance in the octanol-rich phase to that in the water-rich phase, in the limit of zero concentration. In general, *P* tends to be large for compounds with extended nonpolar structures (such as long chain or multi-ring hydrocarbons) and small for compounds with highly polar groups. Thus *P* (or, in its more common form of expression, log *P*) provides a measure of the lipophilic vs. hydrophilic nature of a compound, which is an important consideration in assessing the potential toxicity. A discussion of methods of measurement and accuracy considerations for log *P* may be found in Ref. 1.

This table gives selected values of log *P* for about 450 organic compounds, including many of environmental importance. All values refer to a nominal temperature of 25 °C. The references contain data on many more compounds than are included here.

Column definitions are as follows.

Column heading	Definition
Name	Systematic name of compound; compounds are listed alphabetically
Mol. form.	Molecular formula of compound
log *P*	$\log_{10}$ of octanol-water partition coefficient
Ref.	Reference

References

1. Sangster, J., *J. Phys. Chem. Ref. Data*, 18, 1111, 1989. <https://doi.org/10.1063/1.555833>
2. Mackay, D., Shiu, W. Y., and Ma, K. C., *Illustrated Handbook of Physical-Chemical Properties and Environmental Fate for Organic Chemicals*, Lewis Publishers/CRC Press, Boca Raton, FL, 1992.
3. Shiu, W. Y., and Mackay, D., *J. Phys. Chem. Ref. Data*, 15, 911, 1986. <https://doi.org/10.1063/1.555755>
4. Pinsuwan, S., Li, L., and Yalkowsky, S. H., *J. Chem. Eng. Data*, 40, 623, 1995. <https://doi.org/10.1021/je00019a019>
5. *Solubility Data Series, International Union of Pure and Applied Chemistry, Vol. 20*, Pergamon Press, Oxford, 1985.
6. *Solubility Data Series, International Union of Pure and Applied Chemistry, Vol. 38*, Pergamon Press, Oxford, 1985.
7. Miller, M. M., Ghodbane, S., Wasik, S. P., Tewari, Y. B., and Martire, D. E., *J. Chem. Eng. Data*, 29, 184, 1984. <https://doi.org/10.1021/je00036a027>

Octanol–Water Partition Coefficients at 25 °C

Name	Mol. form.	log *P*	Ref.
Acenaphthene	$C_{12}H_{10}$	3.96	4
Acetaldehyde	C_2H_4O	0.45	1
Acetamide	C_2H_5NO	-1.26	1
Acetic acid	$C_2H_4O_2$	-0.17	1
Acetone	C_3H_6O	-0.24	1
Acetonitrile	C_2H_3N	-0.34	1
Acetophenone	C_8H_8O	1.63	1
Acridine	$C_{13}H_9N$	3.40	1
Acrolein	C_3H_4O	-0.01	1
Acrylamide	C_3H_5NO	-0.78	1
Acrylonitrile	C_3H_3N	0.25	1
Allyl alcohol	C_3H_6O	0.17	1
Allylamine	C_3H_7N	0.03	1
Aniline	C_6H_7N	0.90	1
Anisole	C_7H_8O	2.11	1
Anthracene	$C_{14}H_{10}$	4.56	4
trans-Azobenzene	$C_{12}H_{10}N_2$	3.82	1
Azulene	$C_{10}H_8$	3.22	1
Benzaldehyde	C_7H_6O	1.48	1
Benz[*a*]anthracene	$C_{18}H_{12}$	5.91	1
Benzene	C_6H_6	2.13	1
Benzeneacetaldehyde	C_8H_8O	1.78	1
Benzeneacetic acid	$C_8H_8O_2$	1.41	1
Benzeneacetonitrile	C_8H_7N	1.56	1
Benzeneethanamine	$C_8H_{11}N$	1.41	1
Benzeneethanol	$C_8H_{10}O$	1.36	1
Benzenepropanenitrile	C_9H_9N	1.72	1
Benzenepropanol	$C_9H_{12}O$	1.88	1
Benzenethiol	C_6H_6S	2.52	1
11*H*-Benzo[*a*]fluorene	$C_{17}H_{12}$	5.40	1
11*H*-Benzo[*b*]fluorene	$C_{17}H_{12}$	5.75	1
1-Benzofuran	C_8H_6O	2.67	1
Benzoic acid	$C_7H_6O_2$	1.88	4
Benzonitrile	C_7H_5N	1.56	1
Benzo[*ghi*]perylene	$C_{22}H_{12}$	6.90	1
Benzophenone	$C_{13}H_{10}O$	3.18	1
Benzo[*a*]pyrene	$C_{20}H_{12}$	6.20	4
Benzo[*b*]thiophene	C_8H_6S	3.12	1
Benzyl acetate	$C_9H_{10}O_2$	1.96	1
Benzyl alcohol	C_7H_8O	1.05	1
Benzylamine	C_7H_9N	1.09	1
Benzyl benzoate	$C_{14}H_{12}O_2$	3.97	1
Benzyl methyl ether	$C_8H_{10}O$	1.35	1
Benzyl phenyl ether	$C_{13}H_{12}O$	3.79	1
Biphenyl	$C_{12}H_{10}$	3.76	6
Bis(2-chloroethyl) ether	$C_4H_8Cl_2O$	1.12	2
Bromobenzene	C_6H_5Br	2.99	2
2-Bromobenzoic acid	$C_7H_5BrO_2$	2.20	4
3-Bromobenzoic acid	$C_7H_5BrO_2$	2.87	4
4-Bromobenzoic acid	$C_7H_5BrO_2$	2.86	4
1-Bromobutane	C_4H_9Br	2.75	1
Bromochloromethane	CH_2BrCl	1.41	2
Bromocyclohexane	$C_6H_{11}Br$	3.20	1
Bromoethane	C_2H_5Br	1.6	2
1-Bromoheptane	$C_7H_{15}Br$	4.36	1
1-Bromohexane	$C_6H_{13}Br$	3.80	1
Bromomethane	CH_3Br	1.19	2
(Bromomethyl)benzene	C_7H_7Br	2.92	1
1-Bromooctane	$C_8H_{17}Br$	4.89	1
1-Bromopentane	$C_5H_{11}Br$	3.37	1
1-Bromopropane	C_3H_7Br	2.1	2
2-Bromopropane	C_3H_7Br	1.9	2

Name	Mol. form.	log *P*	Ref.
3-Bromopropene	C_3H_5Br	1.79	1
1,3-Butadiene	C_4H_6	1.99	1
Butanal	C_4H_8O	0.88	1
Butanamide	C_4H_9NO	-0.21	1
Butanenitrile	C_4H_7N	0.60	1
1-Butanethiol	$C_4H_{10}S$	2.28	1
Butanoic acid	$C_4H_8O_2$	0.79	1
1-Butanol	$C_4H_{10}O$	0.84	1
2-Butanol	$C_4H_{10}O$	0.65	1
2-Butanone	C_4H_8O	0.29	1
cis-2-Butene	C_4H_8	2.33	1
trans-2-Butene	C_4H_8	2.31	1
Butyl acetate	$C_6H_{12}O_2$	1.82	1
Butylamine	$C_4H_{11}N$	0.86	1
tert-Butylamine	$C_4H_{11}N$	0.40	1
Butylbenzene	$C_{10}H_{14}$	4.26	1
tert-Butylbenzene	$C_{10}H_{14}$	4.11	1
Butyl methacrylate	$C_8H_{14}O_2$	2.88	1
4-Butylphenol	$C_{10}H_{14}O$	3.65	1
2-Butyne	C_4H_6	1.46	1
Carbazole	$C_{12}H_9N$	3.72	1
Chlorobenzene	C_6H_5Cl	2.84	1
2-Chlorobiphenyl	$C_{12}H_9Cl$	4.52	1
3-Chlorobiphenyl	$C_{12}H_9Cl$	4.58	1
4-Chlorobiphenyl	$C_{12}H_9Cl$	4.61	1
1-Chlorobutane	C_4H_9Cl	2.64	2
Chloroethane	C_2H_5Cl	1.43	2
Chloroethene	C_2H_3Cl	1.38	2
1-Chloroheptane	$C_7H_{15}Cl$	4.15	1
Chloromethane	CH_3Cl	0.91	2
(Chloromethyl)benzene	C_7H_7Cl	2.30	1
1-Chloronaphthalene	$C_{10}H_7Cl$	3.90	1
2-Chloronaphthalene	$C_{10}H_7Cl$	3.98	1
1-Chloropropane	C_3H_7Cl	2.04	1
2-Chloropropane	C_3H_7Cl	1.90	1
2-Chlorotoluene	C_7H_7Cl	3.42	1
3-Chlorotoluene	C_7H_7Cl	3.28	1
4-Chlorotoluene	C_7H_7Cl	3.33	1
Chrysene	$C_{18}H_{12}$	5.73	4
trans-Cinnamic acid	$C_9H_8O_2$	2.13	1
Coronene	$C_{24}H_{12}$	6.05	4
o-Cresol	C_7H_8O	1.98	1
m-Cresol	C_7H_8O	1.98	1
p-Cresol	C_7H_8O	1.97	1
Cyclododecanone	$C_{12}H_{22}O$	4.10	1
1,3,5-Cycloheptatriene	C_7H_8	2.63	2
1,4-Cyclohexadiene	C_6H_8	2.3	2
Cyclohexane	C_6H_{12}	3.44	1
Cyclohexanol	$C_6H_{12}O$	1.23	1
Cyclohexanone	$C_6H_{10}O$	0.81	1
Cyclohexene	C_6H_{10}	2.86	1
2-Cyclohexen-1-one	C_6H_8O	0.61	1
Cyclohexylamine	$C_6H_{13}N$	1.49	1
Cyclooctane	C_8H_{16}	4.45	2
Cyclopentane	C_5H_{10}	3.00	1
Decachlorobiphenyl	$C_{12}Cl_{10}$	8.26	3
Decane	$C_{10}H_{22}$	6.25	1
Decanoic acid	$C_{10}H_{20}O_2$	4.09	1
1-Decanol	$C_{10}H_{22}O$	4.57	1
2-Decanone	$C_{10}H_{20}O$	3.77	1
Dibenzofuran	$C_{12}H_8O$	4.12	1
Dibromomethane	CH_2Br_2	2.3	2
Dibutyl ether	$C_8H_{18}O$	3.21	1
o-Dichlorobenzene	$C_6H_4Cl_2$	3.38	5
m-Dichlorobenzene	$C_6H_4Cl_2$	3.48	5
p-Dichlorobenzene	$C_6H_4Cl_2$	3.38	5
2,5-Dichlorobiphenyl	$C_{12}H_8Cl_2$	5.10	3
2,6-Dichlorobiphenyl	$C_{12}H_8Cl_2$	5.00	3
Dichlorodifluoromethane	CCl_2F_2	2.16	2
1,1-Dichloroethane	$C_2H_4Cl_2$	1.79	2
1,2-Dichloroethane	$C_2H_4Cl_2$	1.48	2
1,1-Dichloroethene	$C_2H_2Cl_2$	2.13	2
cis-1,2-Dichloroethene	$C_2H_2Cl_2$	1.86	2
trans-1,2-Dichloroethene	$C_2H_2Cl_2$	1.93	2
Dichloromethane	CH_2Cl_2	1.25	2
2,4-Dichlorophenol	$C_6H_4Cl_2O$	3.23	4
1,2-Dichloropropane, (±)-	$C_3H_6Cl_2$	2.0	2
cis-1,3-Dichloropropene	$C_3H_4Cl_2$	2.03	2
Diethylamine	$C_4H_{11}N$	0.58	1
Diethyl carbonate	$C_5H_{10}O_3$	1.21	1
Diethyl ether	$C_4H_{10}O$	0.89	1
Diethyl sulfide	$C_4H_{10}S$	1.95	1
Difluoromethane	CH_2F_2	0.20	1
2,3-Dihydrobenzofuran	C_8H_8O	2.14	1
2,5-Dihydrofuran	C_4H_6O	0.46	1
1,2-Dihydro-3-methylbenz[*j*]aceanthrylene	$C_{21}H_{16}$	6.75	1
Diiodomethane	CH_2I_2	2.5	2
Diisopropyl ether	$C_6H_{14}O$	1.52	1
N,N-Dimethylacetamide	C_4H_9NO	-0.77	1
Dimethylamine	C_2H_7N	-0.38	1
N,N-Dimethylaniline	$C_8H_{11}N$	2.31	1
9,10-Dimethylanthracene	$C_{16}H_{14}$	5.69	1
N,N-Dimethylbenzylamine	$C_9H_{13}N$	1.98	1
4,4'-Dimethylbiphenyl	$C_{14}H_{14}$	5.09	1
2,2-Dimethylbutane	C_6H_{14}	3.82	1
2,3-Dimethylbutane	C_6H_{14}	3.85	2
3,3-Dimethyl-2-butanol, (±)-	$C_6H_{14}O$	1.48	1
Dimethyl ether	C_2H_6O	0.10	1
N,N-Dimethylformamide	C_3H_7NO	-1.01	1
1,2-Dimethylnaphthalene	$C_{12}H_{12}$	4.31	1
1,4-Dimethylnaphthalene	$C_{12}H_{12}$	4.37	1
2,2-Dimethyl-1-propanol	$C_5H_{12}O$	1.31	1
Dimethyl sulfone	$C_2H_6O_2S$	-1.41	1
Dimethyl sulfoxide	C_2H_6OS	-1.35	1
Diphenylamine	$C_{12}H_{11}N$	3.44	4
1,2-Diphenylethane	$C_{14}H_{14}$	4.70	1
Diphenyl ether	$C_{12}H_{10}O$	4.21	1
Diphenylmethane	$C_{13}H_{12}$	4.14	1
Diphenylmethanol	$C_{13}H_{12}O$	2.67	1
Diphenyl sulfide	$C_{12}H_{10}S$	4.45	1
Dipropylamine	$C_6H_{15}N$	1.67	1
Dipropyl ether	$C_6H_{14}O$	2.03	1
Dodecanoic acid	$C_{12}H_{24}O_2$	4.6	1
1-Dodecanol	$C_{12}H_{26}O$	5.13	1
Eicosanoic acid	$C_{20}H_{40}O_2$	9.29	1
5,8,11,14-Eicosatetraenoic acid, (all-*cis*)	$C_{20}H_{32}O_2$	6.98	1
Epichlorohydrin	C_3H_5ClO	0.30	2
Ethanol	C_2H_6O	-0.30	1
Ethoxybenzene	$C_8H_{10}O$	2.51	1
Ethyl acetate	$C_4H_8O_2$	0.73	1
Ethyl acrylate	$C_5H_8O_2$	1.32	1
Ethylamine	C_2H_7N	-0.13	1

Name	Mol. form.	log *P*	Ref.
4-Ethylaniline	$C_8H_{11}N$	1.96	1
Ethylbenzene	C_8H_{10}	3.15	1
Ethyl benzoate	$C_9H_{10}O_2$	2.64	1
2-Ethylfuran	C_6H_8O	2.40	1
Ethyl methacrylate	$C_6H_{10}O_2$	1.94	1
Ethylmethylamine	C_3H_9N	0.15	1
1-Ethylnaphthalene	$C_{12}H_{12}$	4.40	1
2-Ethylphenol	$C_8H_{10}O$	2.47	1
3-Ethylphenol	$C_8H_{10}O$	2.50	1
4-Ethylphenol	$C_8H_{10}O$	2.50	1
Ethyl propanoate	$C_5H_{10}O_2$	1.21	1
2-Ethyltoluene	C_9H_{12}	3.53	1
4-Ethyltoluene	C_9H_{12}	3.63	2
Ethyl vinyl ether	C_4H_8O	1.04	1
Fluoranthene	$C_{16}H_{10}$	5.07	4
9*H*-Fluorene	$C_{13}H_{10}$	4.20	4
9*H*-Fluoren-9-one	$C_{13}H_8O$	3.58	1
Fluorobenzene	C_6H_5F	2.27	2
1-Fluorobutane	C_4H_9F	2.58	1
Fluoromethane	CH_3F	0.51	1
1-Fluoropentane	$C_5H_{11}F$	2.33	1
Formaldehyde	CH_2O	0.35	1
Formamide	CH_3NO	-1.51	1
Formic acid	CH_2O_2	-0.54	1
Furan	C_4H_4O	1.34	1
2,2',3,3',4,4',6-Heptachlorobiphenyl	$C_{12}H_3Cl_7$	6.70	3
Heptane	C_7H_{16}	4.50	1
1-Heptanol	$C_7H_{16}O$	2.62	1
2-Heptanol, (±)-	$C_7H_{16}O$	2.31	1
3-Heptanol, (*S*)-	$C_7H_{16}O$	2.24	1
4-Heptanol	$C_7H_{16}O$	2.22	1
2-Heptanone	$C_7H_{14}O$	1.98	1
1-Heptene	C_7H_{14}	3.99	1
Heptylamine	$C_7H_{17}N$	2.57	1
Hexachlorobenzene	C_6Cl_6	5.47	5
2,2',3,3',4,4'-Hexachlorobiphenyl	$C_{12}H_4Cl_6$	7.00	3
2,2',4,4',6,6'-Hexachlorobiphenyl	$C_{12}H_4Cl_6$	7.00	3
2,2',3,3',6,6'-Hexachlorobiphenyl	$C_{12}H_4Cl_6$	6.70	3
Hexachloroethane	C_2Cl_6	4.00	4
Hexadecanoic acid	$C_{16}H_{32}O_2$	7.17	1
1,5-Hexadiene	C_6H_{10}	2.8	2
Hexamethylbenzene	$C_{12}H_{18}$	4.69	4
Hexanal	$C_6H_{12}O$	1.78	1
Hexane	C_6H_{14}	4.00	1
Hexanenitrile	$C_6H_{11}N$	1.66	1
Hexanoic acid	$C_6H_{12}O_2$	1.92	1
1-Hexanol	$C_6H_{14}O$	2.03	1
2-Hexanol	$C_6H_{14}O$	1.76	1
3-Hexanol	$C_6H_{14}O$	1.65	1
2-Hexanone	$C_6H_{12}O$	1.38	1
1-Hexene	C_6H_{12}	3.40	1
5-Hexen-2-one	$C_6H_{10}O$	1.02	1
Hexylamine	$C_6H_{15}N$	2.06	1
Hexylbenzene	$C_{12}H_{18}$	5.52	1
1-Hexyne	C_6H_{10}	2.73	2
5-Hexyn-2-one	C_6H_8O	0.58	1
2-Hydroxybenzoic acid	$C_7H_6O_3$	2.20	4
Indan	C_9H_{10}	3.33	1
Indene	C_9H_8	2.92	1
1*H*-Indole	C_8H_7N	2.14	1
Iodobenzene	C_6H_5I	3.28	2

Name	Mol. form.	log *P*	Ref.
1-Iodobutane	C_4H_9I	3	2
Iodoethane	C_2H_5I	2	2
1-Iodoheptane	$C_7H_{15}I$	4.70	1
Iodomethane	CH_3I	1.5	2
1-Iodopropane	C_3H_7I	2.5	2
Isobutane	C_4H_{10}	2.8	2
Isobutene	C_4H_8	2.35	1
Isobutylbenzene	$C_{10}H_{14}$	4.01	2
Isopropylamine	C_3H_9N	0.26	1
Isopropylbenzene	C_9H_{12}	3.66	1
Isopropyl benzoate	$C_{10}H_{12}O_2$	3.18	1
1-Isopropyl-4-methylbenzene	$C_{10}H_{14}$	4.10	1
Isoquinoline	C_9H_7N	2.08	1
Methacrylic acid	$C_4H_6O_2$	0.93	1
Methanol	CH_4O	-0.74	1
N-Methylacetamide	C_3H_7NO	-1.05	1
Methyl acetate	$C_3H_6O_2$	0.18	1
4-Methylacetophenone	$C_9H_{10}O$	2.19	1
Methyl acrylate	$C_4H_6O_2$	0.80	1
Methylamine	CH_5N	-0.57	1
2-Methylaniline	C_7H_9N	1.32	1
3-Methylaniline	C_7H_9N	1.40	1
4-Methylaniline	C_7H_9N	1.39	1
N-Methylaniline	C_7H_9N	1.66	1
2-Methylanisole	$C_8H_{10}O$	2.74	1
3-Methylanisole	$C_8H_{10}O$	2.66	1
4-Methylanisole	$C_8H_{10}O$	2.81	1
2-Methylanthracene	$C_{15}H_{12}$	5.15	2
9-Methylanthracene	$C_{15}H_{12}$	5.07	1
2-Methylbenzaldehyde	C_8H_8O	2.26	1
α-Methylbenzeneacetic acid, (±)-	$C_9H_{10}O_2$	1.80	1
α-Methylbenzenemethanol	$C_8H_{10}O$	1.42	1
3-Methylbenzenemethanol	$C_8H_{10}O$	1.60	1
4-Methylbenzenemethanol	$C_8H_{10}O$	1.58	1
Methyl benzoate	$C_8H_8O_2$	2.20	1
4-Methylbiphenyl	$C_{13}H_{12}$	4.63	1
3-Methyl-1-butanol	$C_5H_{12}O$	1.28	1
2-Methyl-2-butanol	$C_5H_{12}O$	0.89	1
3-Methyl-2-butanol, (±)-	$C_5H_{12}O$	1.28	1
3-Methyl-2-butanone	$C_5H_{10}O$	0.56	1
Methyl *tert*-butyl ether	$C_5H_{12}O$	0.94	1
Methylcyclohexane	C_7H_{14}	3.88	1
Methylcyclopentane	C_6H_{12}	3.37	2
Methyl decanoate	$C_{11}H_{22}O_2$	4.41	1
1-Methyl-9*H*-fluorene	$C_{14}H_{12}$	4.97	1
2-Methylfuran	C_5H_6O	1.85	1
5-Methyl-2-hexanone	$C_7H_{14}O$	1.88	1
Methyl methacrylate	$C_5H_8O_2$	1.38	1
1-Methylnaphthalene	$C_{11}H_{10}$	3.87	1
2-Methylnaphthalene	$C_{11}H_{10}$	4.00	1
5-Methyl-2-octanone	$C_9H_{18}O$	2.92	1
Methyloxirane	C_3H_6O	0.03	1
3-Methylpentane	C_6H_{14}	3.60	2
4-Methyl-2-pentanone	$C_6H_{12}O$	1.31	1
4-Methyl-1-pentene	C_6H_{12}	2.5	2
1-Methylphenanthrene	$C_{15}H_{12}$	5.14	2
4-Methylphenyl acetate	$C_9H_{10}O_2$	2.11	1
2-Methyl-1-propanol	$C_4H_{10}O$	0.76	1
2-Methyl-2-propanol	$C_4H_{10}O$	0.35	1
2-Methylpyridine	C_6H_7N	1.11	1
3-Methylpyridine	C_6H_7N	1.20	1

Name	Mol. form.	log *P*	Ref.
4-Methylpyridine	C_6H_7N	1.22	1
1-Methylpyrrole	C_5H_7N	1.21	1
2-Methyltetrahydrofuran	$C_5H_{10}O$	1.85	2
Naphthacene	$C_{18}H_{12}$	5.76	1
Naphthalene	$C_{10}H_8$	3.34	4
1-Naphthol	$C_{10}H_8O$	2.84	1
2-Naphthol	$C_{10}H_8O$	2.70	1
Neopentane	C_5H_{12}	3.11	1
Nitrobenzene	$C_6H_5NO_2$	1.85	1
1-Nitrobutane	$C_4H_9NO_2$	1.47	1
Nitroethane	$C_2H_5NO_2$	0.18	1
Nitromethane	CH_3NO_2	-0.33	1
1-Nitropentane	$C_5H_{11}NO_2$	2.01	1
1-Nitropropane	$C_3H_7NO_2$	0.87	1
4-Nitrotoluene	$C_7H_7NO_2$	2.42	1
2,2',3,3',4,5,5',6,6'-Nonachlorobiphenyl	$C_{12}HCl_9$	8.16	3
Nonane	C_9H_{20}	5.65	1
1-Nonanol	$C_9H_{20}O$	4.02	1
2-Nonanone	$C_9H_{18}O$	3.16	1
1-Nonene	C_9H_{18}	5.15	1
2,2',3,3',5,5',6,6'-Octachlorobiphenyl	$C_{12}H_2Cl_8$	7.10	3
cis,cis-9,12-Octadecadienoic acid	$C_{18}H_{32}O_2$	7.05	1
Octadecanoic acid	$C_{18}H_{36}O_2$	8.23	1
cis,cis,cis-9,12,15-Octadecatrienoic acid	$C_{18}H_{30}O_2$	6.46	1
cis-9-Octadecenoic acid	$C_{18}H_{34}O_2$	7.64	1
Octane	C_8H_{18}	5.15	1
Octanenitrile	$C_8H_{15}N$	2.75	1
Octanoic acid	$C_8H_{16}O_2$	3.05	1
1-Octanol	$C_8H_{18}O$	3.07	1
2-Octanol	$C_8H_{18}O$	2.90	1
4-Octanol	$C_8H_{18}O$	2.68	1
2-Octanone	$C_8H_{16}O$	2.37	1
1-Octene	C_8H_{16}	4.57	1
Octylbenzene	$C_{14}H_{22}$	6.30	1
Oxirane	C_2H_4O	-0.30	1
Pentachlorobenzene	C_6HCl_5	5.03	5
2,3,4,5,6-Pentachlorobiphenyl	$C_{12}H_5Cl_5$	6.30	3
2,2',4,5,5'-Pentachlorobiphenyl	$C_{12}H_5Cl_5$	6.40	3
Pentachloroethane	C_2HCl_5	2.89	2
Pentachlorophenol	C_6HCl_5O	5.07	4
1,4-Pentadiene	C_5H_8	2.48	1
Pentamethylbenzene	$C_{11}H_{16}$	4.56	1
Pentane	C_5H_{12}	3.45	1
Pentanenitrile	C_5H_9N	0.94	1
Pentanoic acid	$C_5H_{10}O_2$	1.39	1
1-Pentanol	$C_5H_{12}O$	1.51	1
2-Pentanol	$C_5H_{12}O$	1.25	1
3-Pentanol	$C_5H_{12}O$	1.21	1
2-Pentanone	$C_5H_{10}O$	0.84	1
3-Pentanone	$C_5H_{10}O$	0.82	1
1-Pentene	C_5H_{10}	2.2	2
Pentylamine	$C_5H_{13}N$	1.49	1
Pentylbenzene	$C_{11}H_{16}$	4.90	1
1-Pentyne	C_5H_8	1.98	1
Perylene	$C_{20}H_{12}$	6.25	1
Phenanthrene	$C_{14}H_{10}$	4.52	4
Phenol	C_6H_6O	1.48	4
Phenyl acetate	$C_8H_8O_2$	1.49	1
2-Phenylacetophenone	$C_{14}H_{12}O$	3.18	1
Phenylacetylene	C_8H_6	2.40	1
N-Phenylbenzamide	$C_{13}H_{11}NO$	2.62	1
Phenyl benzoate	$C_{13}H_{10}O_2$	3.59	1
4-Phenylcyclohexanone	$C_{12}H_{14}O$	2.45	1
Phenyl formate	$C_7H_6O_2$	1.26	1
Phenyloxirane	C_8H_8O	1.61	1
1-Phenyl-1-propanone	$C_9H_{10}O$	2.19	1
1-Phenyl-2-propanone	$C_9H_{10}O$	1.44	1
1-Phenyl-2-propylamine, (±)-	$C_9H_{13}N$	1.76	1
4-Phenylpyridine	$C_{11}H_9N$	2.59	1
Piperidine	$C_5H_{11}N$	0.84	1
Propanal	C_3H_6O	0.59	1
Propanenitrile	C_3H_5N	0.16	1
1-Propanethiol	C_3H_8S	1.81	1
Propanoic acid	$C_3H_6O_2$	0.33	1
1-Propanol	C_3H_8O	0.25	1
2-Propanol	C_3H_8O	0.05	1
Propargyl alcohol	C_3H_4O	-0.38	1
Propyl acetate	$C_5H_{10}O_2$	1.24	1
Propylamine	C_3H_9N	0.48	1
Propylbenzene	C_9H_{12}	3.69	1
Propyl formate	$C_4H_8O_2$	0.83	1
2-Propylphenol	$C_9H_{12}O$	2.93	1
4-Propylphenol	$C_9H_{12}O$	3.20	1
Pyrene	$C_{16}H_{10}$	5.08	4
Pyridine	C_5H_5N	0.65	1
Pyrrole	C_4H_5N	0.75	1
Pyrrolidine	C_4H_9N	0.46	1
Quinoline	C_9H_7N	2.03	1
trans-Stilbene	$C_{14}H_{12}$	4.81	1
Styrene	C_8H_8	3.05	1
1,2,3,4-Tetrachlorobenzene	$C_6H_2Cl_4$	4.55	5
1,2,3,5-Tetrachlorobenzene	$C_6H_2Cl_4$	4.65	5
1,2,4,5-Tetrachlorobenzene	$C_6H_2Cl_4$	4.51	5
2,2',4',5-Tetrachlorobiphenyl	$C_{12}H_6Cl_4$	5.73	7
2,3,4,5-Tetrachlorobiphenyl	$C_{12}H_6Cl_4$	5.72	3
1,1,2,2-Tetrachloroethane	$C_2H_2Cl_4$	2.39	2
Tetrachloroethene	C_2Cl_4	2.88	2
Tetrachloromethane	CCl_4	2.64	2
Tetradecanoic acid	$C_{14}H_{28}O_2$	6.1	1
Tetrahydrofuran	C_4H_8O	0.46	1
Tetrahydropyran	$C_5H_{10}O$	0.82	1
1,2,3,4-Tetramethylbenzene	$C_{10}H_{14}$	4.00	1
1,2,3,5-Tetramethylbenzene	$C_{10}H_{14}$	4.10	1
1,2,4,5-Tetramethylbenzene	$C_{10}H_{14}$	4.10	2
Thiophene	C_4H_4S	1.81	1
Toluene	C_7H_8	2.73	1
o-Toluic acid	$C_8H_8O_2$	2.32	4
m-Toluic acid	$C_8H_8O_2$	2.37	1
p-Toluic acid	$C_8H_8O_2$	2.34	1
Tribromomethane	$CHBr_3$	2.38	2
1,2,3-Trichlorobenzene	$C_6H_3Cl_3$	4.04	5
1,2,4-Trichlorobenzene	$C_6H_3Cl_3$	3.98	5
1,3,5-Trichlorobenzene	$C_6H_3Cl_3$	4.02	5
2,4,5-Trichlorobiphenyl	$C_{12}H_7Cl_3$	5.60	3
2,4,6-Trichlorobiphenyl	$C_{12}H_7Cl_3$	5.47	3
1,1,1-Trichloroethane	$C_2H_3Cl_3$	2.49	2
1,1,2-Trichloroethane	$C_2H_3Cl_3$	2.38	2
Trichloroethene	C_2HCl_3	2.53	2
Trichlorofluoromethane	CCl_3F	2.53	2
Trichloromethane	$CHCl_3$	1.97	2
1,2,3-Trichloropropane	$C_3H_5Cl_3$	2.63	2
1,1,2-Trichloro-1,2,2-trifluoroethane	$C_2Cl_3F_3$	3.16	2

Name	Mol. form.	log *P*	Ref.
Triethylamine	$C_6H_{15}N$	1.45	1
Trimethylamine	C_3H_9N	0.16	1
1,2,3-Trimethylbenzene	C_9H_{12}	3.60	1
1,2,4-Trimethylbenzene	C_9H_{12}	3.63	1
1,3,5-Trimethylbenzene	C_9H_{12}	3.42	1
2,3,6-Trimethylphenol	$C_9H_{12}O$	2.67	1
2,4,6-Trimethylphenol	$C_9H_{12}O$	2.46	1
Triphenylamine	$C_{18}H_{15}N$	5.74	1
Triphenylene	$C_{18}H_{12}$	5.49	4
Triphenylmethanol	$C_{19}H_{16}O$	3.68	1
Tripropylamine	$C_9H_{21}N$	2.79	1
2-Undecanone	$C_{11}H_{22}O$	4.09	1
Vinyl acetate	$C_4H_6O_2$	0.73	1
o-Xylene	C_8H_{10}	3.12	1
m-Xylene	C_8H_{10}	3.20	1
p-Xylene	C_8H_{10}	3.15	1
2,4-Xylenol	$C_8H_{10}O$	2.35	1
2,5-Xylenol	$C_8H_{10}O$	2.34	1
2,6-Xylenol	$C_8H_{10}O$	2.36	1
3,4-Xylenol	$C_8H_{10}O$	3.23	1
3,5-Xylenol	$C_8H_{10}O$	2.35	1

SOLUBILITY PRODUCT CONSTANTS OF INORGANIC SALTS

The solubility product constant K_{sp} is a useful parameter for calculating the aqueous solubility of sparingly soluble compounds under various conditions. It may be determined by direct measurement or calculated from the standard Gibbs energies of formation $\Delta_f G°$ of the species involved at their standard states. Thus, if $K_{sp} = [M^+]^m \cdot [A^-]^n$ is the equilibrium constant for the reaction

$$M_mA_n(s) \rightleftharpoons mM^+(aq) + nA^-(aq)$$

where M_mA_n is the slightly soluble substance and M^+ and A^- are the ions produced in solution by the dissociation of M_mA_n, then the Gibbs energy change is

$$\Delta G° = m\,\Delta_f G°\,(M^+,aq) + n\,\Delta_f G°\,(A^-,aq) - \Delta_f G°\,(M_mA_n,\,s)$$

The solubility product constant is calculated from the equation

$$\ln K_{sp} = -\Delta G°/RT$$

The table gives selected values of K_{sp} at 25 °C. Many of these have been calculated from standard state thermodynamic data in Refs. 1 and 2; other values are taken from publications of the IUPAC Solubility Data Project (Refs. 3 to 7).

The above formulation is not convenient for treating sulfides because the S^{-2} ion is usually not present in significant concentrations (see Ref. 8). This is due to the hydrolysis reaction

$$S^{-2} + H_2O \rightleftharpoons HS^- + OH^-$$

which is strongly shifted to the right except in very basic solutions. Furthermore, the equilibrium constant for this reaction, which depends on the second ionization constant of H_2S, is poorly known. Therefore, it is more useful in the case of sulfides to define a different solubility product K_{spa} based on the reaction

$$M_mS_n(s) + 2H^+ \rightleftharpoons mM^+ + nH_2S\ (aq)$$

Column definitions for the table are as follows.

Column heading	Definition
Name	Name of inorganic salt; listed alphabetically
Mol. form.	Molecular formula of inorganic salt
K_{sp}	Solubility product constant at 25 °C
K_{spa}	Solubility product constant for sulfides at 25 °C (see discussion above)

Values of K_{spa} taken from Ref. 8 are given for several sulfides in the last column. Additional discussion of sulfide equilibria may be found in Refs. 7 and 9.

References

1. Wagman, D. D., Evans, W. H., Parker, V. B., Schumm, R. H., Halow, I., Bailey, S. M., Churney, K. L., and Nuttall, R. L., *The NBS Tables of Chemical Thermodynamic Properties, J. Phys. Chem. Ref. Data*, Vol. 11, Suppl. 2, 1982.
2. Garvin, D., Parker, V. B., and White, H. J., *CODATA Thermodynamic Tables*, Hemisphere, New York, 1987.
3. *Solubility Data Series* (53 Volumes), International Union of Pure and Applied Chemistry, Pergamon Press, Oxford, 1979–1992.
4. Clever, H. L., and Johnston, F. J., *J. Phys. Chem. Ref. Data*, 9, 751, 1980. <https://doi.org/10.1063/1.555628>
5. Marcus, Y., *J. Phys. Chem. Ref. Data*, 9, 1307, 1980. <https://doi.org/10.1063/1.555633>
6. Clever, H. L., Johnson, S. A., and Derrick, M. E., *J. Phys. Chem. Ref. Data*, 14, 631, 1985. <https://doi.org/10.1063/1.555732>
7. Clever, H. L., Johnson, S. A., and Derrick, M. E., *J. Phys. Chem. Ref. Data*, 21, 941, 1992. <https://doi.org/10.1063/1.555909>
8. Myers, R. J., *J. Chem. Educ.*, 63, 687, 1986. <https://doi.org/10.1021/ed063p687>
9. Licht, S., *J. Electrochem. Soc.*, 135, 2971, 1988. <https://doi.org/10.1149/1.2095471>

Solubility Product Constants of Inorganic Salts

Name	Mol. form.	K_{sp}	K_{spa}
Aluminum phosphate	$AlPO_4$	$9.84 \cdot 10^{-21}$	
Barium bromate	$Ba(BrO_3)_2$	$2.43 \cdot 10^{-4}$	
Barium carbonate	$BaCO_3$	$2.58 \cdot 10^{-9}$	
Barium chromate(VI)	$BaCrO_4$	$1.17 \cdot 10^{-10}$	
Barium fluoride	BaF_2	$1.84 \cdot 10^{-7}$	
Barium hydroxide octahydrate	$Ba(OH)_2 \cdot 8H_2O$	$2.55 \cdot 10^{-4}$	
Barium iodate	$Ba(IO_3)_2$	$4.01 \cdot 10^{-9}$	
Barium iodate monohydrate	$Ba(IO_3)_2 \cdot H_2O$	$1.67 \cdot 10^{-9}$	
Barium molybdate	$BaMoO_4$	$3.54 \cdot 10^{-8}$	
Barium selenate	$BaSeO_4$	$3.40 \cdot 10^{-8}$	
Barium sulfate	$BaSO_4$	$1.08 \cdot 10^{-10}$	
Barium sulfite	$BaSO_3$	$5.0 \cdot 10^{-10}$	
Beryllium hydroxide (α)	$Be(OH)_2$	$6.92 \cdot 10^{-22}$	
Bismuth arsenate	$BiAsO_4$	$4.43 \cdot 10^{-10}$	
Bismuth triiodide	BiI_3	$7.71 \cdot 10^{-19}$	
Cadmium arsenate	$Cd_3(AsO_4)_2$	$2.2 \cdot 10^{-33}$	
Cadmium carbonate	$CdCO_3$	$1.0 \cdot 10^{-12}$	
Cadmium fluoride	CdF_2	$6.44 \cdot 10^{-3}$	
Cadmium hydroxide	$Cd(OH)_2$	$7.2 \cdot 10^{-15}$	
Cadmium iodate	$Cd(IO_3)_2$	$2.5 \cdot 10^{-8}$	
Cadmium oxalate trihydrate	$CdC_2O_4 \cdot 3H_2O$	$1.42 \cdot 10^{-8}$	
Cadmium phosphate	$Cd_3(PO_4)_2$	$2.53 \cdot 10^{-33}$	
Cadmium sulfide	CdS		$8 \cdot 10^{-7}$
Calcium carbonate (calcite)	$CaCO_3$	$3.36 \cdot 10^{-9}$	
Calcium fluoride	CaF_2	$3.45 \cdot 10^{-11}$	
Calcium hydroxide	$Ca(OH)_2$	$5.02 \cdot 10^{-6}$	
Calcium iodate	$Ca(IO_3)_2$	$6.47 \cdot 10^{-6}$	
Calcium iodate hexahydrate	$Ca(IO_3)_2 \cdot 6H_2O$	$7.10 \cdot 10^{-7}$	
Calcium molybdate	$CaMoO_4$	$1.46 \cdot 10^{-8}$	
Calcium oxalate monohydrate	$CaC_2O_4 \cdot H_2O$	$2.32 \cdot 10^{-9}$	
Calcium phosphate	$Ca_3(PO_4)_2$	$2.07 \cdot 10^{-33}$	
Calcium sulfate	$CaSO_4$	$4.93 \cdot 10^{-5}$	
Calcium sulfate dihydrate	$CaSO_4 \cdot 2H_2O$	$3.14 \cdot 10^{-5}$	
Calcium sulfite hemihydrate	$CaSO_3 \cdot 0.5H_2O$	$3.1 \cdot 10^{-7}$	
Cesium perchlorate	$CsClO_4$	$3.95 \cdot 10^{-3}$	
Cesium periodate	$CsIO_4$	$5.16 \cdot 10^{-6}$	
Cobalt(II) arsenate	$Co_3(AsO_4)_2$	$6.80 \cdot 10^{-29}$	
Cobalt(II) hydroxide	$Co(OH)_2$	$5.92 \cdot 10^{-15}$	
Cobalt(II) iodate dihydrate	$Co(IO_3)_2 \cdot 2H_2O$	$1.21 \cdot 10^{-2}$	
Cobalt(II) phosphate	$Co_3(PO_4)_2$	$2.05 \cdot 10^{-35}$	

Name	Mol. form.	K_{sp}	K_{spa}
Copper(II) arsenate	$Cu_3(AsO_4)_2$	$7.95 \cdot 10^{-36}$	
Copper(I) bromide	$CuBr$	$6.27 \cdot 10^{-9}$	
Copper(I) chloride	$CuCl$	$1.72 \cdot 10^{-7}$	
Copper(I) cyanide	$CuCN$	$3.47 \cdot 10^{-20}$	
Copper(II) iodate monohydrate	$Cu(IO_3)_2 \cdot H_2O$	$6.94 \cdot 10^{-8}$	
Copper(I) iodide	CuI	$1.27 \cdot 10^{-12}$	
Copper(II) oxalate	CuC_2O_4	$4.43 \cdot 10^{-10}$	
Copper(II) phosphate	$Cu_3(PO_4)_2$	$1.40 \cdot 10^{-37}$	
Copper(II) sulfide	CuS		$6 \cdot 10^{-16}$
Copper(I) thiocyanate	$CuSCN$	$1.77 \cdot 10^{-13}$	
Europium(III) hydroxide	$Eu(OH)_3$	$9.38 \cdot 10^{-27}$	
Gallium(III) hydroxide	$Ga(OH)_3$	$7.28 \cdot 10^{-36}$	
Iron(II) carbonate	$FeCO_3$	$3.13 \cdot 10^{-11}$	
Iron(II) fluoride	FeF_2	$2.36 \cdot 10^{-6}$	
Iron(II) hydroxide	$Fe(OH)_2$	$4.87 \cdot 10^{-17}$	
Iron(III) hydroxide	$Fe(OH)_3$	$2.79 \cdot 10^{-39}$	
Iron(III) phosphate dihydrate	$FePO_4 \cdot 2H_2O$	$9.91 \cdot 10^{-16}$	
Iron(II) sulfide	FeS		$6 \cdot 10^{2}$
Lanthanum iodate	$La(IO_3)_3$	$7.50 \cdot 10^{-12}$	
Lead(II) bromide	$PbBr_2$	$6.60 \cdot 10^{-6}$	
Lead(II) carbonate	$PbCO_3$	$7.40 \cdot 10^{-14}$	
Lead(II) chloride	$PbCl_2$	$1.70 \cdot 10^{-5}$	
Lead(II) fluoride	PbF_2	$3.3 \cdot 10^{-8}$	
Lead(II) hydroxide	$Pb(OH)_2$	$1.43 \cdot 10^{-20}$	
Lead(II) iodate	$Pb(IO_3)_2$	$3.69 \cdot 10^{-13}$	
Lead(II) iodide	PbI_2	$9.8 \cdot 10^{-9}$	
Lead(II) selenate	$PbSeO_4$	$1.37 \cdot 10^{-7}$	
Lead(II) sulfate	$PbSO_4$	$2.53 \cdot 10^{-8}$	
Lead(II) sulfide	PbS		$3 \cdot 10^{-7}$
Lithium carbonate	Li_2CO_3	$8.15 \cdot 10^{-4}$	
Lithium fluoride	LiF	$1.84 \cdot 10^{-3}$	
Lithium phosphate	Li_3PO_4	$2.37 \cdot 10^{-11}$	
Magnesium carbonate	$MgCO_3$	$6.82 \cdot 10^{-6}$	
Magnesium carbonate pentahydrate	$MgCO_3 \cdot 5H_2O$	$3.79 \cdot 10^{-6}$	
Magnesium carbonate trihydrate	$MgCO_3 \cdot 3H_2O$	$2.38 \cdot 10^{-6}$	
Magnesium fluoride	MgF_2	$5.16 \cdot 10^{-11}$	
Magnesium hydroxide	$Mg(OH)_2$	$5.61 \cdot 10^{-12}$	
Magnesium oxalate dihydrate	$MgC_2O_4 \cdot 2H_2O$	$4.83 \cdot 10^{-6}$	
Magnesium phosphate	$Mg_3(PO_4)_2$	$1.04 \cdot 10^{-24}$	
Manganese(II) carbonate	$MnCO_3$	$2.24 \cdot 10^{-11}$	
Manganese(II) iodate	$Mn(IO_3)_2$	$4.37 \cdot 10^{-7}$	
Manganese(II) oxalate dihydrate	$MnC_2O_4 \cdot 2H_2O$	$1.70 \cdot 10^{-7}$	
Manganese(II) sulfide (α form)	MnS		$3 \cdot 10^{7}$
Mercury(I) bromide	Hg_2Br_2	$6.40 \cdot 10^{-23}$	
Mercury(II) bromide	$HgBr_2$	$6.2 \cdot 10^{-20}$	
Mercury(I) carbonate	Hg_2CO_3	$3.6 \cdot 10^{-17}$	
Mercury(I) chloride	Hg_2Cl_2	$1.43 \cdot 10^{-18}$	
Mercury(I) fluoride	Hg_2F_2	$3.10 \cdot 10^{-6}$	
Mercury(I) iodide	Hg_2I_2	$5.2 \cdot 10^{-29}$	
Mercury(II) iodide (red)	HgI_2	$2.9 \cdot 10^{-29}$	
Mercury(I) oxalate	$Hg_2C_2O_4$	$1.75 \cdot 10^{-13}$	
Mercury(I) sulfate	Hg_2SO_4	$6.5 \cdot 10^{-7}$	
Mercury(II) sulfide (red)	HgS		$4 \cdot 10^{-33}$
Mercury(II) sulfide (black)	HgS		$2 \cdot 10^{-32}$
Mercury(I) thiocyanate	$Hg_2(SCN)_2$	$3.2 \cdot 10^{-20}$	
Neodymium carbonate	$Nd_2(CO_3)_3$	$1.08 \cdot 10^{-33}$	
Nickel(II) carbonate	$NiCO_3$	$1.42 \cdot 10^{-7}$	
Nickel(II) hydroxide	$Ni(OH)_2$	$5.48 \cdot 10^{-16}$	
Nickel(II) iodate	$Ni(IO_3)_2$	$4.71 \cdot 10^{-5}$	

Name	Mol. form.	K_{sp}	K_{spa}
Nickel(II) phosphate	$Ni_3(PO_4)_2$	$4.74 \cdot 10^{-32}$	
Palladium(II) thiocyanate	$Pd(SCN)_2$	$4.39 \cdot 10^{-23}$	
Potassium hexachloroplatinate	K_2PtCl_6	$7.48 \cdot 10^{-6}$	
Potassium perchlorate	$KClO_4$	$1.05 \cdot 10^{-2}$	
Potassium periodate	KIO_4	$3.71 \cdot 10^{-4}$	
Praseodymium(III) hydroxide	$Pr(OH)_3$	$3.39 \cdot 10^{-24}$	
Radium iodate	$Ra(IO_3)_2$	$1.16 \cdot 10^{-9}$	
Radium sulfate	$RaSO_4$	$3.66 \cdot 10^{-11}$	
Rubidium perchlorate	$RbClO_4$	$3.00 \cdot 10^{-3}$	
Scandium fluoride	ScF_3	$5.81 \cdot 10^{-24}$	
Scandium hydroxide	$Sc(OH)_3$	$2.22 \cdot 10^{-31}$	
Silver(I) acetate	$AgC_2H_3O_2$	$1.94 \cdot 10^{-3}$	
Silver(I) arsenate	Ag_3AsO_4	$1.03 \cdot 10^{-22}$	
Silver(I) bromate	$AgBrO_3$	$5.38 \cdot 10^{-5}$	
Silver(I) bromide	$AgBr$	$5.35 \cdot 10^{-13}$	
Silver(I) carbonate	Ag_2CO_3	$8.46 \cdot 10^{-12}$	
Silver(I) chloride	$AgCl$	$1.77 \cdot 10^{-10}$	
Silver(I) chromate	Ag_2CrO_4	$1.12 \cdot 10^{-12}$	
Silver(I) cyanide	$AgCN$	$5.97 \cdot 10^{-17}$	
Silver(I) iodate	$AgIO_3$	$3.17 \cdot 10^{-8}$	
Silver(I) iodide	AgI	$8.52 \cdot 10^{-17}$	
Silver(I) oxalate	$Ag_2C_2O_4$	$5.40 \cdot 10^{-12}$	
Silver(I) phosphate	Ag_3PO_4	$8.89 \cdot 10^{-17}$	
Silver(I) sulfate	Ag_2SO_4	$1.20 \cdot 10^{-5}$	
Silver(I) sulfide	Ag_2S		$6 \cdot 10^{-30}$
Silver(I) sulfite	Ag_2SO_3	$1.50 \cdot 10^{-14}$	
Silver(I) thiocyanate	$AgSCN$	$1.03 \cdot 10^{-12}$	
Strontium arsenate	$Sr_3(AsO_4)_2$	$4.29 \cdot 10^{-19}$	
Strontium carbonate	$SrCO_3$	$5.60 \cdot 10^{-10}$	
Strontium fluoride	SrF_2	$4.33 \cdot 10^{-9}$	
Strontium iodate	$Sr(IO_3)_2$	$1.14 \cdot 10^{-7}$	
Strontium iodate hexahydrate	$Sr(IO_3)_2 \cdot 6H_2O$	$4.55 \cdot 10^{-7}$	
Strontium iodate monohydrate	$Sr(IO_3)_2 \cdot H_2O$	$3.77 \cdot 10^{-7}$	
Strontium sulfate	$SrSO_4$	$3.44 \cdot 10^{-7}$	
Thallium(I) bromate	$TlBrO_3$	$1.10 \cdot 10^{-4}$	
Thallium(I) bromide	$TlBr$	$3.71 \cdot 10^{-6}$	
Thallium(I) chloride	$TlCl$	$1.86 \cdot 10^{-4}$	
Thallium(I) chromate	Tl_2CrO_4	$8.67 \cdot 10^{-13}$	
Thallium(III) hydroxide	$Tl(OH)_3$	$1.68 \cdot 10^{-44}$	
Thallium(I) iodate	$TlIO_3$	$3.12 \cdot 10^{-6}$	
Thallium(I) iodide	TlI	$5.54 \cdot 10^{-8}$	
Thallium(I) thiocyanate	$TlSCN$	$1.57 \cdot 10^{-4}$	
Tin(II) hydroxide	$Sn(OH)_2$	$5.45 \cdot 10^{-27}$	
Tin(II) sulfide	SnS		$1 \cdot 10^{-5}$
Yttrium carbonate	$Y_2(CO_3)_3$	$1.03 \cdot 10^{-31}$	
Yttrium fluoride	YF_3	$8.62 \cdot 10^{-21}$	
Yttrium hydroxide	$Y(OH)_3$	$1.00 \cdot 10^{-22}$	
Yttrium iodate	$Y(IO_3)_3$	$1.12 \cdot 10^{-10}$	
Zinc arsenate	$Zn_3(AsO_4)_2$	$2.8 \cdot 10^{-28}$	
Zinc carbonate	$ZnCO_3$	$1.46 \cdot 10^{-10}$	
Zinc carbonate monohydrate	$ZnCO_3 \cdot H_2O$	$5.42 \cdot 10^{-11}$	
Zinc fluoride	ZnF_2	$3.04 \cdot 10^{-2}$	
Zinc hydroxide	$Zn(OH)_2$	$3 \cdot 10^{-17}$	
Zinc iodate dihydrate	$Zn(IO_3)_2 \cdot 2H_2O$	$4.1 \cdot 10^{-6}$	
Zinc oxalate dihydrate	$ZnC_2O_4 \cdot 2H_2O$	$1.38 \cdot 10^{-9}$	
Zinc selenide	$ZnSe$	$3.6 \cdot 10^{-26}$	
Zinc selenite monohydrate	$ZnSeO_3 \cdot H_2O$	$1.59 \cdot 10^{-7}$	
Zinc sulfide (sphalerite)	ZnS		$2 \cdot 10^{-4}$
Zinc sulfide (wurtzite)	ZnS		$3 \cdot 10^{-2}$

SOLUBILITY OF COMMON SALTS AT VARIOUS TEMPERATURES

This table gives the aqueous solubility of selected salts at temperatures from 10 °C to 40 °C. Values are given in molality terms.

References

1. Apelblat, A., *J. Chem. Thermodynamics*, 24, 619, 1992. <https://doi.org/10.1016/S0021-9614(05)80033-3>
2. Apelblat, A., *J. Chem. Thermodynamics*, 25, 63, 1993. <https://doi.org/10.1006/jcht.1993.1008>
3. Apelblat, A., *J. Chem. Thermodynamics*, 25, 1513, 1993. <https://doi.org/10.1006/jcht.1993.1151>
4. Apelblat, A. and Korin, E., *J. Chem. Thermodynamics*, 30, 59, 1998. <https://doi.org/10.1006/jcht.1997.0275>

Solubility of Common Salts (in mol/kg) at the Indicated Temperature

Name	Mol. form.	Mol. wt.	10 °C	15 °C	20 °C	25 °C	30 °C	35 °C	40 °C	Ref.
Ammonium chloride	NH_4Cl	53.492	6.199	6.566	6.943	7.331				2
Ammonium nitrate	NH_4NO_3	80.043	18.809	21.163	23.721	26.496				2
Ammonium sulfate	$(NH_4)_2SO_4$	132.140	5.494	5.589	5.688	5.790	5.896	6.005		3
Barium chloride	$BaCl_2$	208.233	1.603	1.659	1.716	1.774	1.834	1.895	1.958	1
Calcium nitrate	$Ca(NO_3)_2$	164.087	6.896	7.398	7.986	8.675	9.480	10.421		1
Copper(II) sulfate	$CuSO_4$	159.609	1.055	1.153	1.260	1.376	1.502	1.639		3
Iron(II) sulfate	$FeSO_4$	151.908	1.352	1.533	1.729	1.940	2.165	2.405		3
Lithium chloride	$LiCl$	42.394	19.296	19.456	19.670	19.935				2
Magnesium nitrate	$Mg(NO_3)_2$	148.314	4.403	4.523	4.656	4.800	4.958	5.130	5.314	1
Manganese(II) chloride	$MnCl_2$	125.844	5.421	5.644	5.884	6.143	6.422	6.721		3
Potassium bromide	KBr	119.002	5.002	5.237	5.471	5.703	5.932	6.157		3
Potassium carbonate	K_2CO_3	138.206	7.756	7.846	7.948	8.063	8.191	8.331	8.483	1
Potassium iodate	KIO_3	214.001	0.291	0.333	0.378	0.426	0.478	0.534	0.593	4
Rubidium chloride	$RbCl$	120.921	6.911	7.180	7.449	7.717	7.986	8.253	8.520	4
Sodium bromide	$NaBr$	102.894	8.258	8.546	8.856	9.191	9.550	9.937	10.351	4
Sodium chloride	$NaCl$	58.443	6.110	6.121	6.136	6.153	6.174	6.197	6.222	4
Sodium nitrate	$NaNO_3$	84.995	9.395	9.819	10.261	10.723	11.204	11.706	12.230	4
Sodium nitrite	$NaNO_2$	68.996	11.111	11.484	11.883	12.310	12.766	13.253	13.772	4
Zinc sulfate	$ZnSO_4$	161.472	2.911	3.116	3.336	3.573	3.827	4.099	4.194	1

SOLUBILITY OF HYDROCARBONS IN SEAWATER

Concern about pollution of the oceans has stimulated measurements of the solubility of organic compounds in seawater. This table gives the solubility of several hydrocarbons in seawater. The data are derived from a review in the IUPAC Solubility Data Series (Ref. 1).

Solubility is expressed in this table as parts per million by mass, i.e.,

$$S/\text{ppm(mass)} = 10^6 \times w_2 = 10^6 \times m_2/(m_1 + m_2)$$

where m_1 and m_2 are the masses of solvent (seawater) and solute, respectively, under saturation conditions, and w_2 is the mass fraction. Because the solubilities in this table are very low, the value of S is effectively the mass of hydrocarbon in grams per 1000 kg of seawater. Column definitions are as follows.

Column heading	Definition
Name	Hydrocarbon name
Mol. form.	Molecular formula of hydrocarbon
Salinity	Salinity is a standardized measure of the concentration of dissolved salts, as explained in the table "Properties of Seawater" in Section 14; salinity values in the open oceans at mid-latitude typically fall between 34 and 36
t	Temperature of measurement, in °C
S	Solubility, as parts per million by mass

Ref. 1 gives details of the method of measurement and an indication of the reliability of the values.

Reference

1. Shaw, David G., and Maczynski, A., IUPAC-NIST Solubility Data Series 81. Hydrocarbons with Water and Seawater — Revised and Updated. Part 12. C5-C26 Hydrocarbons with Seawater, *J. Phys. Chem. Ref. Data* 35, 785, 2006. <https://doi.org/10.1063/1.2132316>

Solubility of Hydrocarbon Compounds in Seawater

Name	Mol. form.	Salinity	t/°C	S/ppm(mass)
Acenaphthene	$C_{12}H_{10}$	35	15	0.21
Acenaphthene	$C_{12}H_{10}$	35	25	1.8
Anthracene	$C_{14}H_{10}$	35	25	0.031
Benz[*a*]anthracene	$C_{18}H_{12}$	35	25	0.0056
Benzene	C_6H_6	34.4	0	1320
Benzene	C_6H_6	35	25	1360
Benzo[*ghi*]perylene	$C_{22}H_{12}$	6	25	0.00021
Benzo[*a*]pyrene	$C_{20}H_{12}$	6	25	0.00013
Benzo[*e*]pyrene	$C_{20}H_{12}$	30	25	0.0033
Benzo[*b*]triphenylene	$C_{22}H_{14}$	6	25	0.027
Biphenyl	$C_{12}H_{10}$	35	25	4.76
Butylbenzene	$C_{10}H_{14}$	34.5	25	7.1
sec-Butylbenzene, (±)-	$C_{10}H_{14}$	34.5	25	12
tert-Butylbenzene	$C_{10}H_{14}$	34.5	25	21
Chrysene	$C_{18}H_{12}$	35	25	0.0011
Dibenz[*a,h*]anthracene	$C_{22}H_{14}$	6	25	0.021
Dibenz[*a,j*]anthracene	$C_{22}H_{14}$	6	25	0.010
Dodecane	$C_{12}H_{26}$	35	25	0.0029
Eicosane	$C_{20}H_{42}$	35	25	0.0008
Ethylbenzene	C_8H_{10}	34.4	0	140
Ethylbenzene	C_8H_{10}	34.4	10	129
Ethylbenzene	C_8H_{10}	34.4	25	111
Fluoranthene	$C_{16}H_{10}$	35	25	0.124
9*H*-Fluorene	$C_{13}H_{10}$	35	25	1.2
Heptane	C_7H_{16}	6	25	10.3
Hexacosane	$C_{26}H_{54}$	35	25	0.0001
Hexadecane	$C_{16}H_{34}$	35	25	0.0004
Hexane	C_6H_{14}	35.3	25	7.9
Isopropylbenzene	C_9H_{12}	34.5	25	43
2-Methylanthracene	$C_{15}H_{12}$	35	25	0.013
Methylcyclopentane	C_6H_{12}	34.5	25	29
1-Methylnaphthalene	$C_{11}H_{10}$	30	25	23
1-Methylphenanthrene	$C_{15}H_{12}$	35	25	0.20
Naphthalene	$C_{10}H_8$	35	25	22.8
Nonane	C_9H_{20}	6	25	0.43
Octadecane	$C_{18}H_{38}$	35	25	0.0008
Pentane	C_5H_{12}	34.5	25	28
Phenanthrene	$C_{14}H_{10}$	34	25	0.69
Pyrene	$C_{16}H_{10}$	35	25	0.086
Tetradecane	$C_{14}H_{30}$	35	25	0.0017
Toluene	C_7H_8	34.4	0	450
Toluene	C_7H_8	35	25	387
1,2,3-Trimethylbenzene	C_9H_{12}	34.5	25	49
1,2,4-Trimethylbenzene	C_9H_{12}	34.5	25	40
1,3,5-Trimethylbenzene	C_9H_{12}	34.5	25	31
Undecane	$C_{11}H_{24}$	6	25	0.01
o-Xylene	C_8H_{10}	34.5	25	130
m-Xylene	C_8H_{10}	34.5	25	106
p-Xylene	C_8H_{10}	34.5	25	111

SOLUBILITY OF ORGANIC COMPOUNDS IN PRESSURIZED HOT WATER

Liquid water at elevated temperatures and pressures, but still in the subcritical region, is of interest as a solvent in various laboratory and industrial processes. In effect, this means water at a temperature between about 100 °C and 373 °C, the critical temperature, and at pressures up to 400 bar or greater. Because the dielectric constant of water decreases with increasing temperature, the solubility of many compounds, especially nonpolar compounds, increases dramatically at higher temperature. The fact that solubility can be fine-tuned by controlling temperature and pressure makes pressurized hot water a useful tool in various extraction and reaction processes.

This table gives a sample of the variations of solubility with temperature and pressure for several compounds, mostly hydrocarbons. More information is available in the references. Column definitions are as follows.

Column heading	Definition
Name	Name of compound; compounds are listed alphabetically by name
Mol. form.	Compound molecular formula
t	Temperature, in °C
p	Pressure, in bar
S_{mf}	Solute solubility as mole fraction x_2 in units $10^4 \times x_2$
$S_{m\%}$	Solute solubility as mass percent = 100 w_2 where w_2 is mass fraction
Ref.	Reference from which data were taken

References

1. *Solubility Data Series, International Union of Pure and Applied Chemistry, Vol. 38*, Pergamon Press, Oxford, 1988.
2. Shaw, D. G., and Maczynski, A., *J. Phys. Chem. Ref. Data* 35, 687, 2006.
3. Stephenson, R. M., *J. Chem. Eng. Data* 37, 80, 1992. <https://doi.org/10.1021/je00005a024>
4. Lun, R., Varhanickova, D., Shiu, W.-Y., and Mackay, D., *J. Chem. Eng. Data* 42, 951, 1997. <https://doi.org/10.1021/je970069v>
5. Miller, D. J., et al., *J. Chem. Eng. Data* 43, 1043, 1998. <https://doi.org/10.1021/je980094g>
6. Miller, D. J., and Hawthorne, S. B., *J. Chem. Eng. Data* 45, 78, 2000. <https://doi.org/10.1021/je990190x>
7. Ma, J. H. Y., Hung, H., Shiu, W-Y., and Mackay, D., *J. Chem. Eng. Data* 46, 619, 2001. <https://doi.org/10.1021/je000341s>
8. Marche, C., Ferronato, C., and Jose, J., *J. Chem. Eng. Data* 48, 967, 2003. <https://doi.org/10.1021/je025659u>
9. Oleszek-Kudlak, S., Shibata, E., and Nakamura, T., *J. Chem. Eng. Data* 49, 570, 2004. <https://doi.org/10.1021/je034170d>
10. Marche, C., Ferronato, C., and Jose, J., *J. Chem. Eng. Data* 49, 937, 2004. <https://doi.org/10.1021/je0342567>
11. Andersson, T. A., Hartonen, K. M., and Riekkola, M-L., *J. Chem. Eng. Data* 50, 1177, 2005. <https://doi.org/10.1021/je0495886>
12. Karasek, P., Planeta, J., and Roth, M., *J. Chem. Eng. Data* 51, 616, 2006. <https://doi.org/10.1021/je050427r>
13. Shiu, W.-Y., and Ma, K.-C, *J. Phys. Chem. Ref. Data* 29, 41, 2000. <https://doi.org/10.1063/1.556055>

Aqueous Solubility of Organic Compounds at Various Pressures and Temperatures

Name Mol. form.	*t*/°C	*p*/bar	$S_{mf}/10^3 x_2$	$S_{m\%}$/%	Ref.
Acenaphthene					
$C_{12}H_{10}$	25	1	0.000444	0.000380	13
	250	50	1.25	1.06	11
Anthracene					
$C_{14}H_{10}$	25	1	0.0000074	0.0000044	2
	50	50	0.000017	0.000017	5
	100	45	0.00032	0.00032	5
	100	39	0.000457	0.00045	12
	150	50	0.0102	0.0101	11
	200	77	0.13	0.13	12
	250	50	0.497	0.49	11
	300	100	3.78	3.62	11
Benz[a]anthracene					
$C_{18}H_{12}$	25	1	0.00000073	0.00000093	2
	60	50	0.00000846	0.0000107	12
	100	50	0.000113	0.000143	12
	120	52	0.000418	0.00053	12
	150	49	0.00296	0.00375	12
Benzene					
C_6H_6	25	1	0.40	0.178	13
	25	65	0.40	0.173	6
	25	400	0.33	0.143	6
	50	65	0.47	0.203	6
	100	65	0.89	0.38	6
	150	65	2.2	0.95	6
	200	65	5.0	2.13	6
	200	400	4.1	1.75	6
Carbazole					
$C_{12}H_9N$	25	1	0.00013	0.00012	5
	25	54	0.00011	0.000102	5
	50	56	0.00045	0.00042	5
	100	54	0.0099	0.0092	5
	150	54	0.162	0.150	5
	200	52	1.9	1.74	5
Chrysene					
$C_{18}H_{12}$	25	1	0.00000016	0.00000019	2
	25	32	0.00000063	0.0000008	5
	50	36	0.000001	0.0000013	5
	100	38	0.000013	0.000016	5
	150	43	0.00060	0.00076	5
	200	45	0.0158	0.020	5
	225	62	0.0758	0.096	5
***o*-Dichlorobenzene**					
$C_6H_4Cl_2$	25	1	0.018	0.0094	9
	50	65	0.023	0.019	6
	100	65	0.055	0.045	6
	150	65	0.18	0.15	6
	200	65	0.57	0.46	6
***trans*-1,2-Dimethylcyclohexane**					
C_8H_{16}	25	1	0.008	0.00050	10
	101	7	0.0047	0.0029	10

Name Mol. form.	t/°C	p/bar	$S_{mf}/10^3x_2$	$S_{m\%}$/%	Ref.
	131	7	0.0108	0.0067	10
	151	7	0.0223	0.0139	10
	170	7	0.0356	0.0222	10
Ethylcyclohexane					
C_8H_{16}	25	1	0.00098	0.00061	10
	100	7	0.00340	0.00212	10
	131	7	0.0085	0.0053	10
	151	7	0.01665	0.0104	10
	171	7	0.0334	0.0208	10
Heptane					
C_7H_{16}	25	1	0.0004352	0.000242	8
	50	7	0.000613	0.00034096	8
	100	7	0.001938	0.00108	8
	125	7	0.00400	0.00222	8
	150	7	0.00878	0.00488	8
	170	7	0.01701	0.00946	8
Hexane					
C_6H_{14}	25	1	0.002045	0.00098	8
	100	7	0.006074	0.0029	8
	125	7	0.01192	0.0057	8
	150	7	0.02555	0.0122	8
	170	7	0.04935	0.0236	8
1-Isopropyl-4-methylbenzene					
$C_{10}H_{14}$	25	1	0.0030	0.0051	4
	50	60	0.0040	0.0030	6
	100	60	0.011	0.0082	6
	150	60	0.043	0.032	6
	200	60	0.20	0.15	6
Methylcyclohexane					
C_7H_{14}	25	1	0.00293	0.00151	10
	100	7	0.01006	0.0055	10
	131	7	0.0244	0.0133	10
	151	7	0.0423	0.0231	10
	171	7	0.0708	0.0386	10
Naphthalene					
$C_{10}H_8$	25	1	0.00444	0.00316	13
	40	50	0.00692	0.0049	12
	50	50	0.0114	0.0081	12
	65	50	0.0264	0.0188	12
	75	50	0.0435	0.0309	12
Octane					
C_8H_{18}	25	1	0.0001158	0.000073	8
	100	7	0.0005943	0.000377	8
	125	7	0.0014163	0.000898	8
	150	7	0.0036957	0.00234	8
	170	7	0.0083483	0.00529	8
	200	65	0.029	0.018	6
Perylene					
$C_{20}H_{12}$	25	1	0.00000003	0.00000004	2

Name Mol. form.	t/°C	p/bar	$S_{mf}/10^3x_2$	$S_{m\%}$/%	Ref.
	50	50	0.00000029	0.0000004	5
	100	45	0.00000210	0.00000294	5
	150	47	0.000120	0.000168	5
	200	48	0.0050	0.0070	5
Pyrene					
$C_{16}H_{10}$	25	1	0.000012	0.0000139	2
	100	50	0.000637	0.00072	11
	100	200	0.00078	0.00087	5
	140	50	0.0054	0.0061	11
	200	50	0.0492	0.055	11
	250	50	0.205	0.23	11
	300	50	1.41	1.56	11
***p*-Terphenyl**					
$C_{18}H_{14}$	25	1	0.00000141	0.00000180	2
	100	49	0.0000219	0.000028	12
	140	51	0.000372	0.000476	12
	180	55	0.00626	0.0080	12
	200	53	0.0241	0.0308	12
	210	54	0.0393	0.0502	12
Tetrachloroethene					
C_2Cl_4	25	1	0.0285	0.0286	3
	50	65	0.027	0.025	6
	100	65	0.059	0.054	6
	150	65	0.18	0.17	6
	200	65	0.59	0.54	6
Toluene					
C_7H_8	25	1	0.107	0.0519	7,13
	50	50	0.125	0.064	6
	100	50	0.27	0.138	6
	150	50	0.66	0.337	6
	200	50	1.9	0.96	6
2,2,4-Trimethylpentane					
C_8H_{18}	25	1	0.00035	0.00022	1
	50	65	0.00052	0.00033	6
	100	65	0.0020	0.00127	6
	150	65	0.0102	0.0065	6
	200	65	0.061	0.0387	6
Triphenylene					
$C_{18}H_{12}$	25	1	0.0000034	0.0000043	2
	100	51	0.0000899	0.000114	12
	140	50	0.00126	0.00160	12
	180	64	0.0123	0.0156	12
	195	60	0.0283	0.0359	12
***m*-Xylene**					
C_8H_{10}	25	1	0.028	0.0161	13
	50	60	0.036	0.021	6
	100	60	0.085	0.050	6
	150	60	0.27	0.159	6
	200	60	0.88	0.516	6

SOLUBILITY CHART FOR INORGANIC SALTS

Abbreviations:

W: Soluble in water
A: Insoluble in water but soluble in acids
w: Sparingly soluble in water but soluble in acids
a: Insoluble in water and only sparingly soluble in acids
I: Insoluble in water and acids
d: Decomposes in water

Solubility Chart for Inorganic Salts

No.	Anion	Ag	Al	Au (I)	Au (II)	Ba	Bi	Ca	Cd	Co	Cr	Cu	Fe (II)	Fe (III)	H
1	Acetate —($C_2H_3O_2$)	w Ag(—)	W Al(—)$_3$	W	W	W Ba(—)$_2$	W Bi(—)$_3$	W Ca(—)$_2$	W Cd(—)$_2$	W Co(—)$_2$	W Cr(—)$_3$	W Cu(—)$_2$	Fe(—)$_2$	Fe_2(—)$_6$	W $C_2H_4O_2$
2	Arsenate —(AsO_4)	A Ag_3(—)	a Al(—)			w Ba_3(—)$_2$	A Bi(—)	w Ca_3(—)$_2$	A Cd_3(—)$_2$	A Co_3(—)$_2$		A Cu_3(—)$_2$	A Fe_3(—)$_2$	A Fe(—)	W H_3AsO_4
3	Arsenite —(AsO_3)	A Ag_3H(—)						w Ca_3(—)$_2$		A Co_3H_6(—)$_4$		A CuH(—)			
4	Benzoate —($C_7H_5O_2$)	w Ag(—)				W Ba(—)$_2$	A Bi(—)$_3$	W Ca(—)$_2$	W Cd(—)$_2$	W Co(—)$_2$		w Cu(—)$_2$	W Fe(—)$_2$	A Fe_2(—)$_6$	W $C_7H_6O_2$
5	Bromide	a AgBr	W $AlBr_2$	w AuBr	W $AuBr_3$	W $BaBr_2$	d $BiBr_3$	W $CaBr_2$	W $CdBr_2$	W $CoBr_2$	W(I) $CrBr_3$*	W $CuBr_2$	W $FeBr_2$	W $FeBr_3$	W HBr
6	Carbonate	A Ag_2CO_3				w $BaCO_3$		w $CaCO_3$	A $CdCO_3$	A $CoCO_3$	W $CrCO_3$		w $FeCO_3$		
7	Chlorate —(ClO_3)	W Ag(—)	W Al(—)$_3$			W Ba(—)$_2$	W Bi(—)$_3$	W Ca(—)$_2$	W Cd(—)$_2$	W Co(—)$_2$		W Cu(—)$_2$	W Fe(—)$_2$	W Fe(—)$_3$	W $HClO_3$
8	Chloride	a AgCl	W $AlCl_3$	w AuCl	W $AuCl_3$	W $BaCl_2$	d $BiCl_3$	W $CaCl_2$	W $CdCl_2$	W $CoCl_2$	I $CrCl_3$	W $CuCl_2$	W $FeCl_2$	W $FeCl_3$	W HCl
9	Chromate —(CrO_4)	w Ag_2(—)				A Ba(—)		W Ca(—)	A Cd(—)	A Co(—)				A Fe_2(—)$_3$	
10	Citrate —($C_6H_5O_7$)	w Ag_3(—)	W Al(—)			w Ba_3(—)$_2$	A Bi(—)	w Ca_3(—)$_2$	A Cd_3(—)$_2$	w Co_3(—)$_2$				W Fe(—)	W $C_6H_8O_7$
11	Cyanide	a AgCN		w AuCN	W $Au(CN)_3$	W $Ba(CN)_2$	w $Bi(CN)_3$	W $Ca(CN)_2$	W $Cd(CN)_2$	A $Co(CN)_2$	A $Cr(CN)_3$	A $Cu(CN)_2$	a $Fe(CN)_2$		W HCN
12	Ferricyanide —($Fe(CN)_6$)	I Ag_3(—)				w Ba_3(—)$_2$		W Ca_3(—)$_2$	A Cd_3(—)$_2$	I Co_3(—)$_2$		I Cu_3(—)$_2$	I Fe_3(—)$_2$		W H_3(—)
13	Ferrocyanide —($Fe(CN)_6$)	I Ag_4(—)	w Al_4(—)$_3$			W Ba_2(—)		W Ca_2(—)	A Cd_2(—)	I Co_2(—)		I Cu_2(—)	I Fe_2(—)	a Fe_4(—)$_3$	W H_4(—)
14	Fluoride	W AgF	W AlF_3			w BaF_2	W BiF_3	w CaF_2	W CdF_2	W CoF_2	W(a) CrF_3*	w CuF_2	w FeF_2	w FeF_3	W HF
15	Formate —(CHO_2)	W Ag(—)	W Al(—)$_3$			W Ba(—)$_2$	W Bi(—)$_3$	W Ca(—)$_2$	W Cd(—)$_2$	W Co(—)$_2$		W Cu(—)$_2$	W Fe(—)$_2$	W Fe(—)$_3$	W CH_2O_2
16	Hydroxide		A $Al(OH)_3$	W AuOH	A $Au(OH)_3$	W $Ba(OH)_2$	A $Bi(OH)_3$	W $Ca(OH)_2$	A $Cd(OH)_2$	A $Co(OH)_2$	A $Cr(OH)_3$	A $Cu(OH)_2$	A $Fe(OH)_2$	A $Fe(OH)_3$	
17	Iodide	I AgI	W AlI_3	a AuI	a AuI_3	W BaI_2	A BiI_3	W CaI_2	W CdI_2	W CoI_2	W CrI_3	a CuI	W FeI_2	W FeI_3	W HI
18	Nitrate	W $AgNO_3$	W $Al(NO_3)_3$			W $Ba(NO_3)_2$	d $Bi(NO_3)_3$	W $Ca(NO_3)_2$	W $Cd(NO_3)_2$	W $Co(NO_3)_2$	W $Cr(NO_3)_3$	W $Cu(NO_3)_2$	W $Fe(NO_3)_2$	W $Fe(NO)_3$	W HNO_3
19	Oxalate —(C_2O_4)	a Ag_2(—)	A Al_2(—)$_3$			w Ba(—)	A Bi_2(—)$_3$	A Ca(—)	w Cd(—)	A Co(—)	W Cr(—)	A Cu(—)	A Fe(—)	W Fe_2(—)$_3$	W $C_2H_2O_4$
20	Oxide	w Ag_2O	a Al_2O_3	Au_2O	A Au_2O_3	W BaO	A Bi_2O_3	w CaO	A CdO	A CoO	a Cr_2O_3	A CuO	A FeO	A Fe_2O_3	W H_2O_2
21	Phosphate	A Ag_3PO_4	A $AlPO_4$		H_3PO_4	A $Ba_3(PO_4)_2$	A $BiPO_4$	w $Ca_3(PO_4)_2$	A $Cd_3(PO_4)_2$	A $Co_3(PO_4)_2$	w $Cr_2(PO_4)_2$	A $Cu_3(PO_4)_2$	A $FePO_4$	w	W $Fe_3(PO_4)_2$
22	Silicate, —(SiO_3)		I Al_2(—)$_3$			W Ba(—)		w Ca(—)	A Cd(—)	A Co_2SiO_4		A Cu(—)			I H_2SiO_3
23	Sulfate	w Ag_2SO_4	W $Al_2(SO_4)_3$			a $BaSO_4$	d $Bi_2(SO_4)_3$	w $CaSO_4$	W $CdSO_4$	W $CoSO_4$	W(I) $Cr_2(SO_4)_3$*	W $CuSO_4$	W $FeSO_4$	w $Fe(SO_4)_3$	W H_2SO_4
24	Sulfide	A Ag_2S	d Al_2S_3	I Au_2S	I Au_2S_3	d BaS	A Bi_2S_3	w CaS	A CdS	A CoS	d Cr_2S_3	A CuS	A FeS	d Fe_2S_3	W H_2S
25	Tartrate —($C_4H_4O_6$)	A Ag_2(—)	w Al_2(—)$_3$	I	I	w Ba(—)	A Bi_2(—)$_3$	w Ca(—)	A Cd(—)	A Co(—)	d	A Cu(—)	A Fe(—)	d Fe_2(—)$_3$	W $C_4H_6O_6$
26	Thiocyanide	I AgCNS				W $Ba(CNS)_2$		W Ca(CNS)		W $Co(CNS)_2$		d CuCNS	W $Fe(CNS)_2$	W $Fe(CNS)_3$	W CNSH

* Indicates two modifications of the salt.

Thermochem

No.	Hg (I)	Hg (II)	K	Mg	Mn	NH_4	Na	Ni	Pb	Pt	Sb	Sn (II)	Sn (IV)	Sr	Zn
1	w $Hg(—)$	W $Hg(—)_2$	W $K(—)$	W $Mg(—)_2$	W $Mn(—)_2$	W $NH_4(—)$	w $Na(—)$	W $Ni(—)_2$	W $Pb(—)_2$			d $Sn(—)_2$	W $Sn(—)_4$	W $Sr(—)_2$	W $Zn(—)_2$
2	A $Hg_3(—)$	w $Hg_3(—)_2$	W $K_3(—)$	A $Mg_3(—)$	w $MnH(—)$	W $(NH_4)_3(—)$	W $Na_3(—)$	A $Ni_3(—)_2$	A $PbH(—)$		A $Sb(—)$			w $SrH(—)$	A $Zn_3(—)_2$
3	A $Hg_3(—)$	A $Hg_3(—)$	W K_3AsO_3	W $Mg_3(—)_2$	A $Mn_3H_6(—)_4$	W NH_4AsO_2	W $Na_2H(—)$	A $Ni_3H_6(—)_4$			A $Sb(—)$	A $Sn_3(—)_2$		w $Sr_3(—)_2$	
4	A $Hg_2(—)_2$	w $Hg(—)_2$	W $K(—)$	W $Mg(—)_2$	W $Mn(—)_2$	W $NH_4(—)$	W $Na(—)$	w $Ni(—)_2$	w $Pb(—)_2$						W $Zn(—)_2$
5	A $HgBr$	W $HgBr_2$	W KBr	W $MgBr_2$	W $MnBr_2$	W NH_4Br	W $NaBr$	W $NiBr_2$	W $PbBr_2$	w $PtBr_4$	d $SbBr_3$	W $SnBr_2$	W $SnBr_4$	W $SrBr_2$	W $ZnBr_2$
6	A Hg_2CO_3		W K_2CO_3	w $MgCO_3$	w $MnCO_3$	W $(NH_4)_2CO_3$	W Na_2CO_3	w $NiCO_3$	A $PbCO_3$					w $SrCO_3$	w $ZnCO_3$
7	W $Hg(—)$	W $Hg(—)_2$	W $K(—)$	W $Mg(—)_2$	W $Mn(—)_2$	W $NH_4(—)$	W $Na(—)$	W $Ni(—)_2$	W $Pb(—)_2$			W $Sn(—)_2$		W $Sr(—)_2$	W $Zn(—)_2$
8	a $HgCl$	W $HgCl_2$	W KCl	W $MgCl_2$	W $MnCl_2$	W NH_4Cl	W $NaCl$	W iCl_2	W $PbCl_2$	W $PtCl_4$	W $SbCl_3$	W $SnCl_2$	W $SnCl_4$	W $SrCl_2$	W $ZnCl_2$
9	w $Hg_2(—)$	w $Hg(—)$	W $K_2(—)$	W $Mg(—)$		W $(NH_4)_2(—)$	W $Na_2(—)$	A $Ni(—)$	A $Pb(—)$			A $Sn(—)$	W $Sn(—)_2$	w $Sr(—)$	w $Zn(—)$
10	w $Hg_3(—)$.		W $K_3(—)$	W $Mg_3(—)_2$	w $MnH(—)$	W $NH_4)_3(—)$	W $Na_3(—)$	W $Ni_3(—)_2$	W $Pb_3(—)_2$					A $SrH(—)$	w $Zn_3(—)_2$
11	A $HgCN$	W $Hg(CN)_2$	W KCN	W $Mg(CN)_2$		W NH_4CN	W $NaCN$	a $Ni(CN)_2$	w $Pb(CN)_2$	I $Pt(CN)_2$				W $Sr(CN)_2$	A $Zn(CN)_2$
12		A $Hg_3(—)_2$	W $K_3(—)$	W $Mg_3(—)_2$		W $NH_4)_3(—)$	W $Na_3(—)$	I $Ni_3(—)_2$	w $Pb_3(—)_2$			A $Sn_3(—)_2$		W $Sr_3(—)_2$	A $Zn_3(—)_2$
13		I $Hg_2(—)$	W $K_4(—)$	W $Mg_2(—)$	A $Mn_2(—)$	W $NH_4)_4(—)$	W $Na_4(—)$	I $Ni_2(—)$	a $Pb_2(—)$			a $Sn_2(—)$		W $Sr_2(—)$	I $Zn_2(—)$
14	d HgF	d HgF_2	W KF	w MgF_2	A MnF_2	W NH_4F	W NaF	w NiF_2	w PbF_2	W PtF_4	W SbF_3	W SnF_2	W SnF_4	w SrF_2	w ZnF_2
15	w $Hg(—)$	W $Hg(—)_2$	W $K(—)$	W $Mg(—)_2$	W $Mn(—)_2$	W $NH_4(—)$	W $Na(—)$	W $Ni(—)_2$	W $Pb(—)_2$					W $Sr(—)_2$	W $Zn(—)_2$
16		A $Hg(OH)_2$	W KOH	A $Mg(OH)_2$	A $Mn(OH)_2$	W NH_4OH	W $NaOH$	w $Ni(OH)_2$	w $Pb(OH)_2$	A $Pt(OH)_4$		A $Sn(OH)_2$	w $Sn(OH)_4$	W $Sr(OH)_2$	A $Zn(OH)_2$
17	A HgI	w HgI_2	W KI	W MgI_2	W MnI_2	W NH_4I	W NaI	W NiI_2	w PbI_2	I PtI_2	d SbI_3	W SnI_2	d SnI_4	W SrI_2	W ZnI_2
18	W $HgNO_3$	W $Hg(NO_3)_2$	W KNO_3	W $Mg(NO_3)_2$	W $Mn(NO_3)_2$	W NH_4NO_3	W $NaNO_3$	W $Ni(NO_3)_2$	W $Pb(NO_3)_2$	W $Pt(NO_3)_4$		d $Sn(NO_3)_2$		W $Sr(NO_3)_2$	W $Zn(NO_3)_2$
19	a $Hg_2(—)$	A $Hg(—)$	W $K_2(—)$	w $Mg(—)$	w $Mn(—)$	W $(NH_4)_2(—)$	W $Na_2(—)$	A $Ni(—)$	A $Pb(—)$			A $Sn(—)$		w $Sr(—)$	A $Zn(—)$
20	A Hg_2O	w HgO	W K_2O	A MgO	A MnO		d Na_2O	A NiO	w PbO	A PtO	w Sb_2O_3	A SnO	A SnO_2	W SrO	w ZnO
21	A Hg_3PO_4	A $Hg_3(PO_4)_2$	W K_3PO_4	w $Mg_2(PO_4)_2$	w $Mn_3(PO_4)_2$	W $NH_4H_2PO_4$	W Na_3PO_4	A $Ni_3(PO_4)_2$	A $Pb_3(PO_4)_2$			A $Sn_3(PO_4)$		A $Sr_3(PO_4)_2$	A $Zn_3(PO_4)_2$
22			W $K_2(—)$	A $Mg(—)$	I $Mn(—)$		W $Na_2(—)$		A $Pb(—)$					A $Sr(—)$	A $Zn(—)$
23	w Hg_2SO_4	d $HgSO_4$	W K_2SO_4	W $MgSO_4$	W $MnSO_4$	W $(NH_4)_2SO_4$	W Na_2SO_4	W $NiSO_4$	w $PbSO_4$	W $Pt(SO_4)_2$	A $Sb_2(SO_4)_3$	W $SnSO_4$	W $Sn(SO_4)_2$	w $SrSO_4$	W $ZnSO_4$
24	I Hg_2S	I HgS	W K_2S	d MgS	A MnS	W $(NH_4)_2S$	W Na_2S	A NiS	A PbS	I PtS	A Sb_2S_3	A SnS	A SnS_2	W SrS	A ZnS
25	I $Hg_2(—)$	I	W $K_2(—)$	w $Mg(—)$	w $Mn(—)$	W $(NH_4)_2(—)$	W $Na_2(—)$	A $Ni(—)$	A $Pb(—)$	I	W $Sb_2(—)_3$	A $Sn(—)$	A	W $Sr(—)$	A $Zn(—)$
26	A $HgCNS$	w $Hg(CNS)_2$	W $KCNS$	W $Mg(CNS)_2$	W $Mn(CNS)_2$	W NH_4CNS	W $NaCNS$		w $Pb(CNS)_2$					w $Sr(CNS)_2$	W $Zn(CNS)_2$

* Indicates two modifications of the salt.

Section 6
Fluid Properties

Thermophysical Properties of Water and Steam . . . 6-1
Vapor Pressure and Other Saturation Properties of Water . . . 6-5
Standard Density of Water . . . 6-7
Fixed-Point Properties of H_2O and D_2O . . . 6-9
Properties of Saturated Liquid D_2O . . . 6-10
Properties of Ice and Supercooled Water . . . 6-11
Vapor Pressure of Ice . . . 6-12
Melting Point of Ice as a Function of Pressure . . . 6-13
Permittivity (Dielectric Constant) of Water at Various Frequencies . . . 6-14
Thermophysical Properties of Air . . . 6-15
Thermophysical Properties of Fluids . . . 6-21
Thermophysical Properties of Selected Fluids at Saturation . . . 6-40
Virial Coefficients of Selected Gases . . . 6-51
Mean Free Path and Related Properties of Gases . . . 6-54
Influence of Pressure on Freezing Points . . . 6-55
Critical Constants of Organic Compounds . . . 6-56
Critical Constants of Inorganic Compounds . . . 6-82
Sublimation Pressure of Solids . . . 6-85
Vapor Pressure of Compounds and Elements . . . 6-87
Vapor Pressure of Fluids at Temperatures below 300 K . . . 6-117
Vapor Pressure of Saturated Salt Solutions . . . 6-126
Enthalpy of Vaporization . . . 6-127
Enthalpy of Fusion . . . 6-144
Compressibility and Expansion Coefficients of Liquids . . . 6-154
Temperature and Pressure Dependence of Liquid Density . . . 6-156
Properties of Cryogenic Fluids . . . 6-161
Properties of Liquid Helium . . . 6-163
Properties of Refrigerants . . . 6-164
Properties of Gas Clathrate Hydrates . . . 6-167
Properties of Ionic Liquids . . . 6-178
Surface Tension of Common Liquids . . . 6-182
Surface Tension of Aqueous Mixtures . . . 6-187
Surface Active Chemicals (Surfactants) . . . 6-188
Permittivity (Dielectric Constant) of Liquids . . . 6-193
Permittivity (Dielectric Constant) of Gases . . . 6-216
Azeotropic Data for Binary Mixtures . . . 6-217
Viscosity of Gases . . . 6-233
Viscosity of Liquids . . . 6-235
Thermal Conductivity of Gases and Refrigerants . . . 6-240
Thermal Conductivity of Liquids . . . 6-243
Diffusion in Gases . . . 6-249
Diffusion of Gases in Water . . . 6-251
Diffusion Coefficients in Liquids at Infinite Dilution . . . 6-252

THERMOPHYSICAL PROPERTIES OF WATER AND STEAM

Eric W. Lemmon and Allan H. Harvey

These tables summarize the thermophysical properties of water and steam at equilibrium as accepted by the International Association for the Properties of Water and Steam (http://www.iapws.org) for general and scientific use. The thermodynamic properties are calculated from the equation of state of Wagner and Pruß (Ref. 6). The reference state for these tables is the liquid at the triple point, at which the internal energy and entropy are taken as zero.

Table 1 contains the thermophysical properties of water at 1 bar (0.1 MPa) pressure with temperatures in °C. Table 2 has the same properties for liquid and gaseous states at equilibrium (saturation) as a function of temperature. Table 3 gives the properties along isobars. The column definitions for the three tables are as follows.

Column heading	Definition
t	Temperature, in °C
P	Pressure, in MPa
ρ	Density, in kg m^{-3}
H	Enthalpy, in units kJ kg^{-1}
S	Entropy, in units kJ kg^{-1} K^{-1}
C_v	Isochoric heat capacity, in units kJ kg^{-1} K^{-1}
C_p	Isobaric heat capacity, in units kJ kg^{-1} K^{-1}
μ	Speed of sound, in m s^{-1}
D	Static dielectric constant
η	Viscosity, in units μPa s
λ	Thermal conductivity, in units mW m^{-1} K^{-1}

In the saturation tables (Table 2), the first line of identical temperatures is for the liquid state and the second line is for the vapor state. A duplicate entry in the isobar table indicates a phase transition (liquid-vapor) at that temperature; property values are then given for both phases. These are identified by the high densities in the liquid and the low densities in the vapor. The temperature scale is ITS-90. Additional calculations at state points not listed below can be obtained by using the NIST Standard Reference Data program REFPROP (Ref. 5) or the water-specific program Steam (Ref. 2).

The uncertainty in density of the equation of state is 0.0001% at 1 atm in the liquid phase, and 0.001% at other liquid states at pressures up to 10 MPa and temperatures to 423 K. In the vapor phase, the uncertainty is 0.05% or less. The uncertainties rise at higher temperatures and/or pressures but are generally less than 0.1% in density except at extreme conditions. The uncertainty in pressure in the critical region is 0.1%. The uncertainty of the speed of sound is 0.15% in the vapor and 0.1% or less in the liquid and increases near the critical region and at high temperatures and pressures. The uncertainty in the isobaric heat capacity is 0.2% in the vapor and 0.1% in the liquid, with increasing values in the critical region and at high pressures. The uncertainties of saturation conditions are 0.025% in vapor pressure, 0.0025% in saturated liquid density, and 0.1% in saturated vapor density. The uncertainties in the saturated densities increase substantially as the critical region is approached.

References

1. Fernández, D.P., Goodwin, A.R.H., Lemmon, E.W., Levelt Sengers, J.M.H., and Williams, R.C., A Formulation for the Static Permittivity of Water and Steam at Temperatures from 238 K to 873 K at Pressures up to 1200 MPa, Including Derivatives and Debye-Hückel Coefficients, *J. Phys. Chem. Ref. Data* 26, 1125, 1997. <https://doi.org/10.1063/1.555997>
2. Harvey, A.H., and Lemmon, E.W., NIST Standard Reference Database 10: NIST/ASME Steam Properties, Version 3.0, National Institute of Standards and Technology, Standard Reference Data Program, Gaithersburg, MD, 2013 <http://www.nist.gov/srd/nist10.cfm>.
3. Huber, M.L., Perkins, R.A., Laesecke, A., Friend, D.G., Sengers, J.V., Assael, M.J., Metaxa, I.M., Vogel, E., Mareš, R., and Miyagawa, K., New International Formulation for the Viscosity of Water, *J. Phys. Chem. Ref. Data* 38, 101, 2009. <https://doi.org/10.1063/1.3088050>
4. Huber, M.L., Perkins, R.A., Friend, D.G., Sengers, J.V., Assael, M.J., Metaxa, I.N., Miyagawa, K., Hellmann, R., and Vogel, E., New International Formulation for the Thermal Conductivity of H_2O, *J. Phys. Chem. Ref. Data* 41, 033102, 2012. <https://doi.org/10.1063/1.4738955>
5. Lemmon, E.W., Huber, M.L., and McLinden, M.O., NIST Standard Reference Database 23: Reference Fluid Thermodynamic and Transport Properties-REFPROP, Version 9.1, National Institute of Standards and Technology, Standard Reference Data Program, Gaithersburg, MD, 2013 <http://www.nist.gov/srd/nist23.cfm>.
6. Wagner, W. and Pruß, A., The IAPWS Formulation 1995 for the Thermodynamic Properties of Ordinary Water Substance for General and Scientific Use, *J. Phys. Chem. Ref. Data* 31, 387, 2002. <https://doi.org/10.1063/1.1461829>

TABLE 1. Thermophysical Properties of H_2O from the Triple Point to 100 °C at P = 1 bar (0.100MPa)

t/°C	P/MPa	ρ/kg m^{-3}	H/kJ kg^{-1}	S/kJ kg^{-1} K^{-1}	C_v/kJ kg^{-1} K^{-1}	C_p/kJ kg^{-1} K^{-1}	u/m s^{-1}	D	η/μPa s	λ/mW m^{-1} K^{-1}
0.01	0.1	999.84	0.10186	0.000007	4.2170	4.2194	1402.4	87.899	1791.1	555.67
10	0.1	999.70	42.118	0.15108	4.1906	4.1952	1447.3	83.974	1305.9	578.78
20	0.1	998.21	84.006	0.29646	4.1567	4.1841	1482.3	80.223	1001.6	598.01
25	0.1	997.05	104.92	0.36720	4.1376	4.1813	1496.7	78.408	890.02	606.52
30	0.1	995.65	125.82	0.43673	4.1172	4.1798	1509.2	76.634	797.22	614.39
40	0.1	992.22	167.62	0.57237	4.0734	4.1794	1528.9	73.201	652.73	628.48
50	0.1	988.03	209.42	0.70377	4.0262	4.1813	1542.6	69.916	546.52	640.62
60	0.1	983.20	251.25	0.83125	3.9765	4.1850	1551.0	66.774	466.03	651.00
70	0.1	977.76	293.12	0.95509	3.9251	4.1901	1554.7	63.770	403.55	659.76
80	0.1	971.79	335.05	1.0755	3.8728	4.1968	1554.4	60.898	354.05	666.99
90	0.1	965.31	377.06	1.1928	3.8204	4.2052	1550.4	58.152	314.17	672.79

t/°C	P/MPa	ρ/kg m^{-3}	H/kJ kg^{-1}	S/kJ kg^{-1} K^{-1}	C_v/kJ kg^{-1} K^{-1}	C_p/kJ kg^{-1} K^{-1}	u/m s^{-1}	D	η/μPa s	λ/mW m^{-1} K^{-1}
99.606	0.1	958.63	417.50	1.3028	3.7702	4.2152	1543.5	55.628	282.75	677.06
99.606	0.1	0.59034	2674.9	7.3588	1.5548	2.0784	471.99	1.0058	12.218	24.532
100	0.1	0.58967	2675.8	7.3610	1.5535	2.0766	472.28	1.0058	12.234	24.564

TABLE 2. Thermophysical Properties of H_2O from the Triple Point to the Critical Point with Liquid and Gaseous States in Equilibrium (Saturation)

t/°C	P/MPa	ρ/kg m^{-3}	H/kJ kg^{-1}	S/kJ kg^{-1} K^{-1}	C_v/kJ kg^{-1} K^{-1}	C_p/kJ kg^{-1} K^{-1}	u/m s^{-1}	D	η/μPa s	λ/mW m^{-1} K^{-1}
0.01	0.000612	999.79	0.000612	0.0	4.2174	4.2199	1402.3	87.895	1791.4	555.60
0.01	0.000612	0.0048546	2500.9	9.1555	1.4184	1.8844	409.00	1.00006	8.9458	16.761
10	0.0012282	999.65	42.021	0.15109	4.1910	4.1955	1447.1	83.971	1306.0	578.71
10	0.0012282	0.0094071	2519.2	8.8998	1.4269	1.8947	416.17	1.00012	9.2384	17.412
20	0.0023393	998.16	83.914	0.29648	4.1570	4.1844	1482.2	80.219	1001.6	597.95
20	0.0023393	0.017314	2537.4	8.6660	1.4359	1.9059	423.18	1.00021	9.5441	18.087
25	0.0031699	997.00	104.83	0.36722	4.1379	4.1816	1496.5	78.405	890.04	606.46
25	0.0031699	0.023075	2546.5	8.5566	1.4405	1.9118	426.63	1.00028	9.7009	18.433
30	0.0042470	995.61	125.73	0.43675	4.1175	4.1801	1509.0	76.630	797.22	614.34
30	0.0042470	0.030415	2555.5	8.4520	1.4452	1.9180	430.03	1.00036	9.8602	18.786
40	0.0073849	992.18	167.53	0.57240	4.0737	4.1796	1528.7	73.197	652.72	628.44
40	0.0073849	0.051242	2573.5	8.2555	1.4552	1.9314	436.71	1.00059	10.185	19.509
50	0.012352	988.00	209.34	0.70381	4.0264	4.1815	1542.4	69.913	546.50	640.57
50	0.012352	0.083147	2591.3	8.0748	1.4663	1.9468	443.21	1.00094	10.516	20.261
60	0.019946	983.16	251.18	0.83129	3.9767	4.1851	1550.8	66.772	466.02	650.96
60	0.019946	0.13043	2608.8	7.9081	1.4789	1.9648	449.50	1.0014	10.854	21.043
70	0.031201	977.73	293.07	0.95513	3.9252	4.1902	1554.6	63.768	403.53	659.72
70	0.031201	0.19843	2626.1	7.7540	1.4937	1.9862	455.57	1.0021	11.195	21.860
80	0.047414	971.77	335.01	1.0756	3.8729	4.1969	1554.3	60.896	354.04	666.97
80	0.047414	0.29367	2643.0	7.6111	1.5111	2.0120	461.39	1.0030	11.539	22.717
90	0.070182	965.30	377.04	1.1929	3.8204	4.2053	1550.4	58.151	314.17	672.77
90	0.070182	0.42390	2659.5	7.4781	1.5316	2.0429	466.94	1.0043	11.885	23.618
100	0.10142	958.35	419.17	1.3072	3.7682	4.2157	1543.2	55.527	281.58	677.21
100	0.10142	0.59817	2675.6	7.3541	1.5558	2.0800	472.20	1.0059	12.232	24.570
110	0.14338	950.95	461.42	1.4188	3.7167	4.2283	1532.9	53.018	254.61	680.35
110	0.14338	0.82693	2691.1	7.2381	1.5843	2.1244	477.13	1.0080	12.580	25.579
120	0.19867	943.11	503.81	1.5279	3.6662	4.2435	1519.9	50.620	232.03	682.24
120	0.19867	1.1221	2705.9	7.1291	1.6177	2.1770	481.73	1.0105	12.927	26.652
140	0.36154	926.13	589.16	1.7392	3.5694	4.2826	1486.2	46.131	196.64	682.53
140	0.36154	1.9667	2733.4	6.9293	1.7002	2.3109	489.82	1.0177	13.618	29.016
160	0.61823	907.45	675.47	1.9426	3.4788	4.3354	1443.2	42.018	170.43	678.73
160	0.61823	3.2596	2757.4	6.7491	1.8044	2.4883	496.29	1.0282	14.304	31.721
180	1.0028	887.00	763.05	2.1392	3.3949	4.4050	1391.7	38.235	150.38	671.28
180	1.0028	5.1588	2777.2	6.5840	1.9279	2.7129	501.04	1.0431	14.985	34.832
200	1.5549	864.66	852.27	2.3305	3.3179	4.4958	1332.1	34.742	134.58	660.01
200	1.5549	7.8610	2792.0	6.4302	2.0666	2.9895	503.92	1.0636	15.666	38.426
220	2.3196	840.22	943.58	2.5177	3.2479	4.6146	1264.5	31.495	121.77	645.26
220	2.3196	11.615	2800.9	6.2840	2.2172	3.3289	504.77	1.0915	16.354	42.606
240	3.3469	813.37	1037.6	2.7020	3.1850	4.7719	1189.0	28.455	111.06	627.17
240	3.3469	16.749	2803.0	6.1423	2.3794	3.7537	503.32	1.1292	17.062	47.525
260	4.6923	783.63	1135.0	2.8849	3.1301	4.9856	1105.3	25.580	101.81	605.78
260	4.6923	23.712	2796.6	6.0016	2.5555	4.3075	499.21	1.1802	17.810	53.446
280	6.4166	750.28	1236.9	3.0685	3.0849	5.2889	1012.6	22.824	93.550	581.03
280	6.4166	33.165	2779.9	5.8579	2.7503	5.0731	491.93	1.2505	18.630	60.878
300	8.5879	712.14	1345.0	3.2552	3.0530	5.7504	909.40	20.135	85.855	552.65
300	8.5879	46.168	2749.6	5.7059	2.9708	6.2197	480.73	1.3504	19.580	70.900
320	11.284	667.09	1462.2	3.4494	3.0428	6.5373	793.16	17.440	78.310	520.02
320	11.284	64.638	2700.6	5.5372	3.2276	8.1589	464.43	1.5012	20.773	86.156
340	14.601	610.67	1594.5	3.6601	3.0781	8.2080	658.27	14.606	70.331	481.93
340	14.601	92.759	2621.8	5.3356	3.5430	12.236	440.72	1.7555	22.477	114.52
360	18.666	527.59	1761.7	3.9167	3.2972	15.004	479.74	11.225	60.306	439.16
360	18.666	143.90	2481.5	5.0536	4.0068	27.356	402.37	2.3096	25.638	191.44
373.946	22.064	322.00	2084.3	4.4070				5.3606		

TABLE 3. Thermophysical Properties of H_2O along Isobars from 0.1 MPa to 100 MPa

T/K	P/MPa	ρ/kg m^{-3}	H/kJ kg^{-1}	S/kJ kg^{-1} K^{-1}	C_v/kJ kg^{-1} K^{-1}	C_p/kJ kg^{-1} K^{-1}	u/m s^{-1}	D	η/μPa s	λ/mW m^{-1} K^{-1}
P = 0.1 MPa (1 bar)										
273.16	0.1	999.84	0.10186	0.000007	4.2170	4.2194	1402.4	87.899	1791.1	555.67
280	0.1	999.91	28.894	0.10411	4.1998	4.2009	1434.3	85.192	1433.6	571.98
300	0.1	996.56	112.65	0.39306	4.1302	4.1806	1501.5	77.747	853.74	609.50
320	0.1	989.43	196.25	0.66281	4.0414	4.1805	1538.9	70.935	576.73	637.00
340	0.1	979.54	279.93	0.91646	3.9414	4.1883	1554.0	64.702	421.63	657.17
360	0.1	967.40	363.82	1.1562	3.8369	4.2023	1552.1	59.004	325.86	671.11
372.756	0.1	958.63	417.50	1.3028	3.7702	4.2152	1543.5	55.628	282.75	677.06
372.756	0.1	0.59034	2674.9	7.3588	1.5548	2.0784	471.99	1.0058	12.218	24.532
380	0.1	0.57824	2689.9	7.3986	1.5356	2.0507	477.08	1.0056	12.498	25.134
400	0.1	0.54761	2730.4	7.5025	1.5082	2.0078	490.31	1.0051	13.278	26.825
450	0.1	0.48458	2829.7	7.7365	1.4943	1.9752	520.60	1.0040	15.267	31.269
500	0.1	0.43514	2928.6	7.9447	1.5082	1.9813	548.31	1.0033	17.299	36.032
550	0.1	0.39507	3028.1	8.1344	1.5319	2.0010	574.19	1.0027	19.356	41.093
600	0.1	0.36185	3128.8	8.3096	1.5600	2.0268	598.61	1.0023	21.425	46.424
650	0.1	0.33384	3230.8	8.4730	1.5903	2.0557	621.79	1.0020	23.496	51.996
700	0.1	0.30988	3334.4	8.6264	1.6222	2.0867	643.92	1.0017	25.562	57.782
750	0.1	0.28915	3439.5	8.7715	1.6553	2.1191	665.11	1.0015	27.617	63.759
800	0.1	0.27102	3546.3	8.9093	1.6892	2.1525	685.47	1.0013	29.657	69.908
900	0.1	0.24085	3765.0	9.1668	1.7589	2.2216	724.03	1.0011	33.680	82.647
1000	0.1	0.21673	3990.7	9.4045	1.8297	2.2921	760.17	1.00088	37.615	95.877
1100	0.1	0.19701	4223.4	9.6263	1.9000	2.3621	794.33	1.00074	41.453	109.50
1200	0.1	0.18058	4463.0	9.8348	1.9682	2.4302	826.85	1.00063	45.192	123.44
P = 1 MPa (10 bar)										
273.16	1.0	1000.3	1.0180	0.000066	4.2127	4.2150	1403.9	87.937	1789.1	556.36
280	1.0	1000.3	29.783	0.10407	4.1960	4.1973	1435.7	85.228	1432.5	572.59
300	1.0	996.96	113.48	0.39281	4.1272	4.1781	1503.0	77.781	853.66	610.00
320	1.0	989.82	197.03	0.66242	4.0390	4.1784	1540.5	70.967	576.89	637.47
340	1.0	979.93	280.67	0.91594	3.9395	4.1863	1555.7	64.734	421.86	657.64
360	1.0	967.81	364.52	1.1556	3.8353	4.2004	1553.9	59.035	326.10	671.61
380	1.0	953.74	448.73	1.3832	3.7315	4.2220	1538.4	53.827	262.82	680.00
400	1.0	937.87	533.47	1.6005	3.6315	4.2535	1511.3	49.065	218.82	683.32
450	1.0	890.39	749.20	2.1086	3.4076	4.3924	1400.6	38.814	153.23	672.77
453.028	1.0	887.13	762.52	2.1381	3.3954	4.4045	1392.0	38.258	150.49	671.33
453.028	1.0	5.1450	2777.1	6.5850	1.9271	2.7114	501.02	1.0430	14.981	34.812
500	1.0	4.5323	2891.2	6.8250	1.6699	2.2795	535.74	1.0344	17.054	38.473
550	1.0	4.0581	3001.8	7.0359	1.6159	2.1647	565.75	1.0282	19.215	42.908
600	1.0	3.6871	3109.0	7.2224	1.6098	2.1292	592.58	1.0236	21.349	47.802
650	1.0	3.3843	3215.2	7.3925	1.6227	2.1254	617.34	1.0202	23.462	53.075
700	1.0	3.1305	3321.7	7.5504	1.6447	2.1368	640.55	1.0174	25.555	58.664
750	1.0	2.9140	3429.0	7.6984	1.6715	2.1566	662.53	1.0153	27.628	64.516
800	1.0	2.7265	3537.5	7.8384	1.7014	2.1816	683.48	1.0135	29.680	70.591
900	1.0	2.4174	3758.5	8.0986	1.7663	2.2402	722.85	1.0108	33.716	83.281
1000	1.0	2.1723	3985.7	8.3380	1.8346	2.3048	759.50	1.0088	37.655	96.534
1100	1.0	1.9729	4219.5	8.5608	1.9034	2.3713	794.01	1.0074	41.494	110.21
1200	1.0	1.8074	4460.0	8.7699	1.9708	2.4371	826.77	1.0063	45.231	124.21
P = 10 MPa (100 bar)										
273.16	10.0	1004.8	10.111	0.00049	4.1721	4.1726	1418.4	88.311	1770.0	563.05
280	10.0	1004.7	38.613	0.10355	4.1593	4.1622	1450.3	85.588	1422.2	578.60
300	10.0	1001.0	121.73	0.39029	4.0984	4.1536	1518.2	78.113	852.99	614.97
320	10.0	993.70	204.84	0.65846	4.0157	4.1580	1556.4	71.284	578.57	642.13
340	10.0	983.84	288.08	0.91079	3.9205	4.1672	1572.6	65.044	424.17	662.33
360	10.0	971.85	371.56	1.1493	3.8198	4.1810	1572.0	59.345	328.53	676.49
380	10.0	957.99	455.37	1.3759	3.7188	4.2013	1558.0	54.140	265.22	685.19
400	10.0	942.42	539.67	1.5921	3.6210	4.2302	1532.7	49.385	221.17	688.88
450	10.0	896.16	753.94	2.0967	3.4010	4.3553	1428.4	39.172	155.48	679.41
500	10.0	838.02	977.18	2.5669	3.2211	4.6022	1271.3	30.794	119.83	646.41
550	10.0	761.82	1218.8	3.0270	3.0865	5.1407	1054.6	23.531	96.080	590.37

T/K	P/MPa	ρ/kg m^{-3}	H/kJ kg^{-1}	S/kJ kg^{-1} K^{-1}	C_v/kJ kg^{-1} K^{-1}	C_p/kJ kg^{-1} K^{-1}	u/m s^{-1}	D	η/μPa s	λ/mW m^{-1} K^{-1}
584.147	10.0	688.42	1408.1	3.3606	3.0438	6.1237	847.33	18.660	81.718	535.29
584.147	10.0	55.463	2725.5	5.6160	3.1065	7.1408	472.51	1.4248	20.194	78.344
600	10.0	49.773	2820.0	5.7756	2.6239	5.1365	503.34	1.3649	21.017	72.247
650	10.0	40.479	3022.6	6.1009	2.1103	3.3968	562.10	1.2672	23.472	68.284
700	10.0	35.355	3177.4	6.3305	1.9338	2.8741	602.20	1.2145	25.773	69.997
750	10.0	31.810	3314.6	6.5200	1.8625	2.6452	634.58	1.1793	27.973	73.764
800	10.0	29.107	3443.7	6.6867	1.8367	2.5313	662.61	1.1536	30.101	78.664
900	10.0	25.123	3691.6	6.9787	1.8439	2.4458	710.98	1.1182	34.201	90.420
1000	10.0	22.241	3935.5	7.2357	1.8843	2.4397	753.03	1.0948	38.144	103.67
1100	10.0	20.017	4180.6	7.4693	1.9377	2.4661	791.02	1.0782	41.960	117.74
1200	10.0	18.230	4429.2	7.6855	1.9957	2.5070	826.16	1.0659	45.663	132.27
P = 100 MPa (1000 bar)										
273.16	100.0	1045.3	95.444	-0.0083717	3.8761	3.9053	1575.5	91.834	1660.1	616.62
280	100.0	1043.6	122.26	0.088571	3.8869	3.9328	1603.8	88.976	1367.4	627.72
300	100.0	1037.2	201.44	0.36171	3.8751	3.9798	1667.9	81.216	859.19	657.72
320	100.0	1028.9	281.30	0.61941	3.8289	4.0043	1707.7	74.222	599.18	683.70
340	100.0	1019.0	361.55	0.86265	3.7637	4.0194	1728.7	67.891	448.03	705.08
360	100.0	1007.8	442.05	1.0927	3.6883	4.0309	1735.4	62.150	352.49	721.60
380	100.0	995.37	522.79	1.3110	3.6089	4.0427	1730.9	56.937	288.37	733.29
400	100.0	981.82	603.78	1.5187	3.5293	4.0569	1717.3	52.201	243.33	740.39
450	100.0	943.51	807.84	1.9993	3.3430	4.1105	1652.8	42.149	175.71	740.46
500	100.0	899.21	1015.4	2.4366	3.1820	4.1968	1555.7	34.149	139.57	720.25
550	100.0	848.78	1228.2	2.8421	3.0452	4.3234	1435.5	27.667	117.36	684.16
600	100.0	791.49	1448.6	3.2256	2.9295	4.5019	1300.4	22.290	101.85	634.92
650	100.0	726.21	1679.5	3.5952	2.8329	4.7503	1158.6	17.717	89.647	575.50
700	100.0	651.77	1925.0	3.9589	2.7538	5.0832	1020.0	13.754	79.123	508.87
750	100.0	568.52	2188.5	4.3223	2.6866	5.4492	898.84	10.336	69.732	438.73
800	100.0	482.23	2466.5	4.6811	2.6169	5.6108	813.97	7.5622	61.842	371.17
900	100.0	343.61	3000.1	5.3104	2.4386	4.8879	765.30	4.2835	52.771	275.39
1000	100.0	265.45	3440.1	5.7749	2.2950	3.9788	792.50	2.9559	50.506	237.77
1100	100.0	220.62	3809.8	6.1276	2.2276	3.4715	832.67	2.3472	51.089	232.34
1200	100.0	191.53	4142.5	6.4172	2.2098	3.2098	872.28	2.0111	52.802	238.54

VAPOR PRESSURE AND OTHER SATURATION PROPERTIES OF WATER

Eric W. Lemmon

This table summarizes the vapor pressure, enthalpy (heat) of vaporization, and surface tension γ of water as accepted by the International Association for the Properties of Water and Steam <www.iapws.org> for general and scientific use. The vapor pressure and heat of vaporization are calculated from the equation of state of Wagner and Pruss (Ref. 1). The temperature scale is ITS-90. Additional calculations at state points not listed below can be obtained by using the NIST Standard Reference Data program REFPROP <www.nist.gov/srd/nist23.cfm> or the water-specific program Steam Properties <www.nist.gov/srd/nist10.cfm>. Column definitions are as follows.

Column heading	Definition
t	Temperature, in °C
P	Vapor pressure, in kPa
$\Delta_{vap}H$	Enthalpy (heat) of vaporization, in units kJ kg^{-1}
γ	Surface tension, in units of mN m^{-1}

References

1. Wagner, W. and Pruss, A., The IAPWS Formulation 1995 for the Thermodynamic Properties of Ordinary Water Substance for General and Scientific Use, *J. Phys. Chem. Ref. Data*, 31, 387, 2002. <https://doi.org/10.1063/1.1461829>
2. International Association for the Properties of Water and Steam, Revised Release on the Surface Tension of Ordinary Water Substance, IAPWS R1-76 (2014) <www.iapws.org/relguide/Surf-H2O.html>.

Vapor Pressure, Enthalpy of Vaporization, and Surface Tension of Water

t/°C	P/kPa	$\Delta_{vap}H$/kJ kg^{-1}	γ/mN m^{-1}
0.01	0.61165	2500.9	75.65
2	0.70599	2496.2	75.37
4	0.81355	2491.4	75.08
6	0.93536	2486.7	74.80
8	1.0730	2481.9	74.51
10	1.2282	2477.2	74.22
12	1.4028	2472.5	73.93
14	1.5990	2467.7	73.63
16	1.8188	2463.0	73.34
18	2.0647	2458.3	73.04
20	2.3393	2453.5	72.74
22	2.6453	2448.8	72.43
24	2.9858	2444.0	72.13
25	3.1699	2441.7	71.97
26	3.3639	2439.3	71.82
28	3.7831	2434.6	71.51
30	4.2470	2429.8	71.19
32	4.7596	2425.1	70.88
34	5.3251	2420.3	70.56
36	5.9479	2415.5	70.24
38	6.6328	2410.8	69.92
40	7.3849	2406.0	69.60
42	8.2096	2401.2	69.27
44	9.1124	2396.4	68.94
46	10.099	2391.6	68.61
48	11.177	2386.8	68.28
50	12.352	2381.9	67.94
52	13.631	2377.1	67.61
54	15.022	2372.3	67.27
56	16.533	2367.4	66.93
58	18.171	2362.5	66.58
60	19.946	2357.7	66.24
62	21.867	2352.8	65.89
64	23.943	2347.8	65.54
66	26.183	2342.9	65.19
68	28.599	2338.0	64.84
70	31.201	2333.0	64.48
72	34.000	2328.1	64.12
74	37.009	2323.1	63.76
76	40.239	2318.1	63.40
78	43.703	2313.0	63.04
80	47.414	2308.0	62.67
82	51.387	2302.9	62.31
84	55.635	2297.9	61.94
86	60.173	2292.8	61.56
88	65.017	2287.6	61.19
90	70.182	2282.5	60.82
92	75.684	2277.3	60.44
94	81.541	2272.1	60.06
96	87.771	2266.9	59.68
98	94.390	2261.7	59.30
100	101.42	2256.4	58.91
102	108.87	2251.1	58.53
104	116.78	2245.8	58.14
106	125.15	2240.4	57.75
108	134.01	2235.1	57.36
110	143.38	2229.6	56.96
112	153.28	2224.2	56.57
114	163.74	2218.7	56.17
116	174.77	2213.2	55.77
118	186.41	2207.7	55.37
120	198.67	2202.1	54.97
122	211.59	2196.5	54.56
124	225.18	2190.9	54.16
126	239.47	2185.2	53.75
128	254.50	2179.5	53.34
130	270.28	2173.7	52.93
132	286.85	2167.9	52.52
134	304.23	2162.1	52.11
136	322.45	2156.2	51.69
138	341.54	2150.3	51.27
140	361.54	2144.3	50.86
142	382.47	2138.3	50.44
144	404.37	2132.2	50.01
146	427.26	2126.1	49.59
148	451.18	2119.9	49.17
150	476.16	2113.7	48.74
152	502.25	2107.5	48.31
154	529.46	2101.2	47.89
156	557.84	2094.8	47.46
158	587.42	2088.4	47.02
160	618.23	2082.0	46.59
162	650.33	2075.5	46.16
164	683.73	2068.9	45.72
166	718.48	2062.3	45.28
168	754.62	2055.6	44.85
170	792.19	2048.8	44.41
172	831.22	2042.0	43.97
174	871.76	2035.1	43.52
176	913.84	2028.2	43.08
178	957.51	2021.2	42.64
180	1002.8	2014.2	42.19
182	1049.8	2007.0	41.74
184	1098.5	1999.8	41.30
186	1148.9	1992.6	40.85
188	1201.1	1985.3	40.40
190	1255.2	1977.9	39.95
192	1311.2	1970.4	39.49
194	1369.1	1962.8	39.04
196	1429.0	1955.2	38.59
198	1490.9	1947.5	38.13

t/°C	P/kPa	$\Delta_{vap}H$/kJ kg^{-1}	γ/mN m^{-1}
200	1554.9	1939.7	37.67
202	1621.0	1931.9	37.22
204	1689.3	1923.9	36.76
206	1759.8	1915.9	36.30
208	1832.6	1907.8	35.84
210	1907.7	1899.6	35.38
212	1985.1	1891.4	34.92
214	2065.0	1883.0	34.46
216	2147.3	1874.6	33.99
218	2232.2	1866.0	33.53
220	2319.6	1857.4	33.07
222	2409.6	1848.6	32.60
224	2502.3	1839.8	32.14
226	2597.8	1830.9	31.67
228	2696.0	1821.8	31.20
230	2797.1	1812.7	30.74
232	2901.0	1803.5	30.27
234	3008.0	1794.1	29.80
236	3117.9	1784.7	29.33
238	3230.8	1775.1	28.86
240	3346.9	1765.4	28.39
242	3466.2	1755.6	27.92
244	3588.7	1745.7	27.45
246	3714.5	1735.6	26.98
248	3843.6	1725.5	26.51
250	3976.2	1715.2	26.04
252	4112.2	1704.7	25.57
254	4251.8	1694.2	25.10
256	4394.9	1683.5	24.63
258	4541.7	1672.6	24.16
260	4692.3	1661.6	23.69
262	4846.6	1650.5	23.22
264	5004.7	1639.2	22.75
266	5166.8	1627.8	22.28
268	5332.9	1616.2	21.81
270	5503.0	1604.4	21.34
272	5677.2	1592.5	20.87
274	5855.6	1580.4	20.40
276	6038.3	1568.1	19.93
278	6225.2	1555.6	19.46
280	6416.6	1543.0	18.99
282	6612.4	1530.1	18.53
284	6812.8	1517.1	18.06
286	7017.7	1503.8	17.59
288	7227.4	1490.4	17.13
290	7441.8	1476.7	16.66
292	7661.0	1462.7	16.20
294	7885.2	1448.6	15.74
296	8114.3	1434.2	15.28
298	8348.5	1419.5	14.82
300	8587.9	1404.6	14.36
302	8832.5	1389.4	13.90
304	9082.4	1374.0	13.45
306	9337.8	1358.2	12.99
308	9598.6	1342.1	12.54
310	9865.1	1325.7	12.09
312	10137	1309.0	11.64
314	10415	1291.9	11.19
316	10699	1274.5	10.75
318	10989	1256.6	10.30
320	11284	1238.4	9.86
322	11586	1219.7	9.43
324	11895	1200.6	8.99
326	12209	1180.9	8.56
328	12530	1160.8	8.13
330	12858	1140.2	7.70
332	13193	1118.9	7.28
334	13534	1097.1	6.86
336	13882	1074.6	6.44
338	14238	1051.3	6.03
340	14601	1027.3	5.63
342	14971	1002.5	5.22
344	15349	976.7	4.83
346	15734	949.9	4.43
348	16128	922.0	4.05
350	16529	892.7	3.67
352	16939	862.1	3.29
354	17358	829.8	2.93
356	17785	795.5	2.57
358	18221	759.0	2.22
360	18666	719.8	1.88
362	19121	677.3	1.55
364	19585	630.5	1.23
366	20060	578.2	0.93
368	20546	517.8	0.65
370	21044	443.8	0.39
372	21554	340.3	0.16
373.95	22064	0.0	0.0

STANDARD DENSITY OF WATER

This table gives the density ρ of water in the temperature range from 0 °C to 100 °C at a pressure of 101325 Pa (one standard atmosphere). Temperatures are given on the ITS-90 scale. From 0 °C to 40 °C the values are taken from the publication in Ref. 1 and refer to standard mean ocean water (SMOW), free from dissolved salts and gases. SMOW is a standard water sample of high purity and known isotopic composition. Methods of correcting for different isotopic compositions are discussed in Ref. 2. The remaining values are calculated from the NIST-REFPROP program, Ref. 3, which obtains thermodynamic properties from the equation of state of Wagner and Pruss given in Ref. 4.

References

1. Tanaka, M., Girard, G., Davis, R., Peuto, A., and Bignell, N., *Metrologia* 38, 301, 2001. <https://doi.org/10.1088/0026-1394/38/4/3>
2. Marsh, K. N., Ed., *Recommended Reference Materials for the Realization of Physicochemical Properties,* Blackwell Scientific Publications, Oxford, 1987.
3. Lemmon, E.W., Huber, M.L., and McLinden, M.O., NIST Standard Reference Database 23: Reference Fluid Thermodynamic and Transport Properties-REFPROP, Version 9.0, National Institute of Standards and Technology, Standard Reference Data Program, Gaithersburg, MD, 2010 <http://www.nist.gov/srd/nist23.cfmwww.nist.gov/srd/nist23.cfm>.
4. Wagner, W., and Pruss, A., *J. Phys. Chem. Ref. Data* 31, 387, 2002. <https://doi.org/10.1063/1.1461829>

Standard Density of Water as a Function of Temperature

t/°C	ρ/g cm^{-3}	t/°C	ρ/g cm^{-3}	t/°C	ρ/g cm^{-3}	t/°C	ρ/g cm^{-3}	t/°C	ρ/g cm^{-3}
0.1	0.9998495	4.2	0.9999746	8.3	0.9998325	12.4	0.9994539	16.5	0.9988633
0.2	0.9998560	4.3	0.9999742	8.4	0.9998260	12.5	0.9994419	16.6	0.9988464
0.3	0.9998624	4.4	0.9999736	8.5	0.9998193	12.6	0.9994298	16.7	0.9988294
0.4	0.9998685	4.5	0.9999728	8.6	0.9998125	12.7	0.9994176	16.8	0.9988123
0.5	0.9998745	4.6	0.9999719	8.7	0.9998056	12.8	0.9994052	16.9	0.9987951
0.6	0.9998803	4.7	0.9999709	8.8	0.9997985	12.9	0.9993927	17.0	0.9987778
0.7	0.9998859	4.8	0.9999697	8.9	0.9997912	13.0	0.9993801	17.1	0.9987603
0.8	0.9998913	4.9	0.9999683	9.0	0.9997839	13.1	0.9993674	17.2	0.9987428
0.9	0.9998966	5.0	0.9999668	9.1	0.9997764	13.2	0.9993546	17.3	0.9987251
1.0	0.9999017	5.1	0.9999651	9.2	0.9997687	13.3	0.9993416	17.4	0.9987073
1.1	0.9999066	5.2	0.9999633	9.3	0.9997610	13.4	0.9993285	17.5	0.9986895
1.2	0.9999113	5.3	0.9999613	9.4	0.9997530	13.5	0.9993153	17.6	0.9986715
1.3	0.9999158	5.4	0.9999592	9.5	0.9997450	13.6	0.9993020	17.7	0.9986534
1.4	0.9999202	5.5	0.9999569	9.6	0.9997368	13.7	0.9992885	17.8	0.9986351
1.5	0.9999244	5.6	0.9999544	9.7	0.9997285	13.8	0.9992749	17.9	0.9986168
1.6	0.9999285	5.7	0.9999518	9.8	0.9997200	13.9	0.9992612	18.0	0.9985984
1.7	0.9999323	5.8	0.9999491	9.9	0.9997114	14.0	0.9992474	18.1	0.9985798
1.8	0.9999360	5.9	0.9999462	10.0	0.9997027	14.1	0.9992335	18.2	0.9985611
1.9	0.9999396	6.0	0.9999431	10.1	0.9996938	14.2	0.9992194	18.3	0.9985424
2.0	0.9999429	6.1	0.9999400	10.2	0.9996848	14.3	0.9992052	18.4	0.9985235
2.1	0.9999461	6.2	0.9999366	10.3	0.9996757	14.4	0.9991909	18.5	0.9985045
2.2	0.9999491	6.3	0.9999331	10.4	0.9996665	14.5	0.9991765	18.6	0.9984854
2.3	0.9999519	6.4	0.9999295	10.5	0.9996571	14.6	0.9991619	18.7	0.9984662
2.4	0.9999546	6.5	0.9999257	10.6	0.9996475	14.7	0.9991473	18.8	0.9984469
2.5	0.9999571	6.6	0.9999217	10.7	0.9996379	14.8	0.9991325	18.9	0.9984275
2.6	0.9999595	6.7	0.9999176	10.8	0.9996281	14.9	0.9991176	19.0	0.9984079
2.7	0.9999616	6.8	0.9999134	10.9	0.9996182	15.0	0.9991026	19.1	0.9983883
2.8	0.9999636	6.9	0.9999090	11.0	0.9996081	15.1	0.9990874	19.2	0.9983686
2.9	0.9999655	7.0	0.9999045	11.1	0.9995979	15.2	0.9990722	19.3	0.9983487
3.0	0.9999672	7.1	0.9998998	11.2	0.9995876	15.3	0.9990568	19.4	0.9983287
3.1	0.9999687	7.2	0.9998950	11.3	0.9995772	15.4	0.9990413	19.5	0.9983087
3.2	0.9999700	7.3	0.9998900	11.4	0.9995666	15.5	0.9990257	19.6	0.9982885
3.3	0.9999712	7.4	0.9998849	11.5	0.9995559	15.6	0.9990100	19.7	0.9982682
3.4	0.9999722	7.5	0.9998797	11.6	0.9995451	15.7	0.9989942	19.8	0.9982478
3.5	0.9999731	7.6	0.9998743	11.7	0.9995341	15.8	0.9989782	19.9	0.9982273
3.6	0.9999738	7.7	0.9998687	11.8	0.9995230	15.9	0.9989621	20.0	0.9982067
3.7	0.9999743	7.8	0.9998631	11.9	0.9995118	16.0	0.9989459	20.1	0.9981860
3.8	0.9999747	7.9	0.9998572	12.0	0.9995005	16.1	0.9989296	20.2	0.9981652
3.9	0.9999749	8.0	0.9998513	12.1	0.9994890	16.2	0.9989132	20.3	0.9981443
4.0	0.9999749	8.1	0.9998452	12.2	0.9994774	16.3	0.9988967	20.4	0.9981233
4.1	0.9999748	8.2	0.9998389	12.3	0.9994657	16.4	0.9988800	20.5	0.9981022

t/°C	ρ/g cm^{-3}	t/°C	ρ/g cm^{-3}	t/°C	ρ/g cm^{-3}	t/°C	ρ/g cm^{-3}	t/°C	ρ/g cm^{-3}
20.6	0.9980810	25.7	0.9968651	30.8	0.9954044	35.9	0.9937199	50.0	0.98804
20.7	0.9980596	25.8	0.9968387	30.9	0.9953734	36.0	0.9936847	51.0	0.98758
20.8	0.9980382	25.9	0.9968123	31.0	0.9953424	36.1	0.9936495	52.0	0.98712
20.9	0.9980167	26.0	0.9967857	31.1	0.9953113	36.2	0.9936142	53.0	0.98665
21.0	0.9979950	26.1	0.9967591	31.2	0.9952801	36.3	0.9935788	54.0	0.98617
21.1	0.9979733	26.2	0.9967324	31.3	0.9952488	36.4	0.9935434	55.0	0.98569
21.2	0.9979514	26.3	0.9967055	31.4	0.9952175	36.5	0.9935078	56.0	0.98521
21.3	0.9979295	26.4	0.9966786	31.5	0.9951860	36.6	0.9934722	57.0	0.98471
21.4	0.9979074	26.5	0.9966516	31.6	0.9951545	36.7	0.9934365	58.0	0.98421
21.5	0.9978853	26.6	0.9966245	31.7	0.9951228	36.8	0.9934007	59.0	0.98371
21.6	0.9978630	26.7	0.9965973	31.8	0.9950911	36.9	0.9933649	60.0	0.98320
21.7	0.9978407	26.8	0.9965700	31.9	0.9950593	37.0	0.9933290	61.0	0.98268
21.8	0.9978182	26.9	0.9965426	32.0	0.9950275	37.1	0.9932929	62.0	0.98216
21.9	0.9977956	27.0	0.9965151	32.1	0.9949955	37.2	0.9932569	63.0	0.98163
22.0	0.9977730	27.1	0.9964875	32.2	0.9949635	37.3	0.9932207	64.0	0.98109
22.1	0.9977502	27.2	0.9964599	32.3	0.9949313	37.4	0.9931844	65.0	0.98055
22.2	0.9977273	27.3	0.9964321	32.4	0.9948991	37.5	0.9931481	66.0	0.98000
22.3	0.9977044	27.4	0.9964043	32.5	0.9948668	37.6	0.9931117	67.0	0.97945
22.4	0.9976813	27.5	0.9963763	32.6	0.9948344	37.7	0.9930753	68.0	0.97890
22.5	0.9976582	27.6	0.9963483	32.7	0.9948020	37.8	0.9930387	69.0	0.97833
22.6	0.9976349	27.7	0.9963202	32.8	0.9947694	37.9	0.9930021	70.0	0.97776
22.7	0.9976115	27.8	0.9962920	32.9	0.9947368	38.0	0.9929654	71.0	0.97719
22.8	0.9975881	27.9	0.9962637	33.0	0.9947041	38.1	0.9929286	72.0	0.97661
22.9	0.9975645	28.0	0.9962353	33.1	0.9946713	38.2	0.9928917	73.0	0.97603
23.0	0.9975408	28.1	0.9962068	33.2	0.9946384	38.3	0.9928548	74.0	0.97544
23.1	0.9975171	28.2	0.9961783	33.3	0.9946055	38.4	0.9928178	75.0	0.97484
23.2	0.9974932	28.3	0.9961496	33.4	0.9945724	38.5	0.9927807	76.0	0.97424
23.3	0.9974692	28.4	0.9961208	33.5	0.9945393	38.6	0.9927435	77.0	0.97364
23.4	0.9974452	28.5	0.9960920	33.6	0.9945061	38.7	0.9927063	78.0	0.97303
23.5	0.9974210	28.6	0.9960631	33.7	0.9944728	38.8	0.9926689	79.0	0.97241
23.6	0.9973968	28.7	0.9960341	33.8	0.9944394	38.9	0.9926316	80.0	0.97179
23.7	0.9973724	28.8	0.9960050	33.9	0.9944060	39.0	0.9925941	81.0	0.97116
23.8	0.9973480	28.9	0.9959758	34.0	0.9943724	39.1	0.9925565	82.0	0.97053
23.9	0.9973234	29.0	0.9959465	34.1	0.9943388	39.2	0.9925189	83.0	0.96990
24.0	0.9972988	29.1	0.9959171	34.2	0.9943051	39.3	0.9924812	84.0	0.96926
24.1	0.9972740	29.2	0.9958876	34.3	0.9942713	39.4	0.9924434	85.0	0.96861
24.2	0.9972492	29.3	0.9958581	34.4	0.9942375	39.5	0.9924056	86.0	0.96796
24.3	0.9972243	29.4	0.9958285	34.5	0.9942035	39.6	0.9923677	87.0	0.96731
24.4	0.9971992	29.5	0.9957987	34.6	0.9941695	39.7	0.9923297	88.0	0.96664
24.5	0.9971741	29.6	0.9957689	34.7	0.9941354	39.8	0.9922916	89.0	0.96598
24.6	0.9971489	29.7	0.9957390	34.8	0.9941012	39.9	0.9922534	90.0	0.96531
24.7	0.9971236	29.8	0.9957090	34.9	0.9940669	40.0	0.9922152	91.0	0.96463
24.8	0.9970981	29.9	0.9956790	35.0	0.9940326	41.0	0.99183	92.0	0.96396
24.9	0.9970726	30.0	0.9956488	35.1	0.9939982	42.0	0.99144	93.0	0.96327
25.0	0.9970470	30.1	0.9956185	35.2	0.9939637	43.0	0.99104	94.0	0.96258
25.1	0.9970213	30.2	0.9955882	35.3	0.9939291	44.0	0.99063	95.0	0.96189
25.2	0.9969955	30.3	0.9955578	35.4	0.9938944	45.0	0.99021	96.0	0.96119
25.3	0.9969696	30.4	0.9955273	35.5	0.9938597	46.0	0.98979	97.0	0.96049
25.4	0.9969436	30.5	0.9954967	35.6	0.9938248	47.0	0.98936	98.0	0.95978
25.5	0.9969176	30.6	0.9954660	35.7	0.9937899	48.0	0.98893	99.0	0.95907
25.6	0.9968914	30.7	0.9954352	35.8	0.9937549	49.0	0.98848	99.974	0.95837

FIXED-POINT PROPERTIES OF H_2O AND D_2O

Allan H. Harvey

The following table contains important fixed properties of water (H_2O) and heavy water (D_2O). The properties included are defined below.

Property	Definition
Molar mass	The average mass of the molecule with respect to the mass of a carbon-12 atom being assigned the value 12; in g mol^{-1}
Melting point (101.325 kPa)	The temperature at which the solid and liquid phases of a substance are in equilibrium at standard atmospheric pressure (101.325 kPa); in °C
Boiling point (101.325 kPa)	The temperature at which the liquid and gas phases of a substance are in equilibrium at standard atmospheric pressure (101.325 kPa); in °C
Triple-point temperature	The temperature at which the solid, liquid, and gas phases of a substance are in thermodynamic equilibrium; in °C
Triple-point pressure	The pressure at which the solid, liquid, and gas phases of a substance are in thermodynamic equilibrium; in Pa
Triple-point density (liq)	The density of the liquid at the triple point; in g cm^{-3}
Critical temperature	The temperature of the critical point, which is the point on the phase diagram of a two-phase system at which the two coexisting phases have identical properties and therefore represent a single phase; at the liquid-gas critical point of a pure substance, the distinction between liquid and gas vanishes; in °C
Critical pressure	The pressure of the critical point, which is the point on the phase diagram of a two-phase system at which the two coexisting phases have identical properties and therefore represent a single phase; at the liquid-gas critical point of a pure substance, the distinction between liquid and gas vanishes; in MPa
Critical density	The density of the substance at the critical point, which is the point on the phase diagram of a two-phase system at which the two coexisting phases have identical properties and therefore represent a single phase; at the liquid-gas critical point of a pure substance, the distinction between liquid and gas vanishes; in g cm^{-3}
Maximum density (101.325 kPa)	The maximum density of the substance at standard atmospheric pressure (101.325 kPa); in g cm^{-3}
Temperature of maximum density	The temperature at which the substance has its maximum density at standard atmospheric pressure (101.325 kPa); in °C

Temperatures are given on the ITS-90 scale.

References

1. International Association for the Properties of Water and Steam (IAPWS), IAPWS G5-01 (2016), *Guideline on the Use of Fundamental Physical Constants and Basic Constants of Water* (2001; 2016 update), available from <http://www.iapws.org>.
2. IAPWS, IAPWS R14-08(2011), *Revised Release on the Pressure along the Melting and Sublimation Curves of Ordinary Water Substance*, 2011, available from <http://www.iapws.org>.
3. IAPWS, IAPWS R2-83(1992), *IAPWS Release on the Values of Temperature, Pressure, and Density of Ordinary and Heavy Water Substances at Their Respective Critical Points*, 1992, available from <http://www.iapws.org>.
4. Wagner, W., and Pruß, A., The IAPWS Formulation 1995 for the Thermodynamic Properties of Ordinary Water Substance for General and Scientific Use, *J. Phys. Chem. Ref. Data* 31, 387, 2002. <https://doi.org/10.1063/1.1461829>
5. Herrig, S., Thol, M., Harvey, A. H., and Lemmon, E. W., A Reference Equation of State for Heavy Water, *J. Phys. Chem. Ref. Data* 47, 043102 (2018). <https://doi.org/10.1063/1.5053993>
6. Guildner, L. A., Johnson, D. P., and Jones, F. E., Vapor Pressure of Water at Its Triple Point, *J. Res. Nat. Bur. Stand.* 80A, 505, 1976. <https://doi.org/10.6028/jres.080A.054>
7. Harvey, A. H., and Lemmon, E.W., NIST Standard Reference Database 10: *NIST/ASME Steam Properties*, Version 3.0, National Institute of Standards and Technology, Standard Reference Data Program, Gaithersburg, MD, 2013, available from <http://www.nist.gov/srd/nist-10>.
8. Lemmon, E. W., Bell, I. H., Huber, M. L., and McLinden, M. O., NIST Standard Reference Database 23: *Reference Fluid Thermodynamic and Transport Properties-REFPROP*, Version 10.0, National Institute of Standards and Technology, Standard Reference Data Program, Gaithersburg, MD, 2018, available from <http://www.nist.gov/srd/refprop>.

Fixed-Point Properties of H_2O and D_2O

	Unit	H_2O	D_2O
Molar mass	g mol^{-1}	18.015268	20.027508
Melting point (101.325 kPa)	°C	0.0025	3.81
Boiling point (101.325 kPa)	°C	99.974	101.40
Triple-point temperature	°C	0.01[a]	3.82
Triple-point pressure	Pa	611.657	661.6
Triple-point density (liquid)	g cm^{-3}	0.99979	1.1053
Critical temperature	°C	373.946	370.697
Critical pressure	MPa	22.064	21.671
Critical density	g cm^{-3}	0.322	0.356
Maximum density (101.325 kPa)	g cm^{-3}	0.999975	1.1059
Temperature of maximum density	°C	3.98	11.60

[a] Exact.

Fluids

PROPERTIES OF SATURATED LIQUID D_2O

Allan H. Harvey

Properties of saturated liquid heavy water, D_2O, are given in this table as a function of temperature from the melting point to the critical point. Properties are calculated from formulations adopted for general and scientific use by the International Association for the Properties of Water and Steam (IAPWS). The background (including uncertainties) for the equation of state used for vapor pressure, density, and heat capacity is given by Herrig et al. (Ref. 1), and the background for the transport property correlations is given by Matsunaga and Nagashima (Ref. 2). The unpublished surface tension correlation and the other IAPWS formulations may be found on the IAPWS Web site <http://www.iapws.org>. The temperature scale is ITS-90. Additional calculations at state points not listed below can be obtained by using the NIST REFPROP program (Ref. 3) <http://www.nist.gov/srd/nist23.cfm>. The column definitions are as follows.

Column heading	Definition
t	Temperature, in °C
P	Vapor pressure, in kPa
ρ	Density, in kg m^{-3}
C_p	Isobaric heat capacity, in units kJ kg^{-1} K^{-1}
η	Viscosity, in units mPa s
λ	Thermal conductivity, in units W m^{-1} K^{-1}
σ	Surface tension, in units mN m^{-1}

References

1. Herrig, S., Thol, M., Harvey, A. H., and Lemmon, E. W., A Reference Equation of State for Heavy Water, *J. Phys. Chem. Ref. Data* 47, 043102, 2018. <https://doi.org/10.1063/1.5053993>
2. Matsunaga, N. and Nagashima, A., *J. Phys. Chem. Ref. Data* 12, 933, 1983. <https://doi.org/10.1063/1.555694>
3. Lemmon, E.W., Bell, I.H., Huber, M.L., and McLinden, M.O., NIST Standard Reference Database 23: *Reference Fluid Thermodynamic and Transport Properties-REFPROP*, Version 10.0, National Institute of Standards and Technology, Standard Reference Data Program, Gaithersburg, MD, 2018, available from <http://www.nist.gov/srd/nist23.cfm>.

Properties of Saturated Liquid D_2O

t/°C	P/kPa	ρ/kg m^{-3}	C_p/ kJ kg^{-1}K^{-1}	η/mPa s	λ/ W m^{-1} K^{-1}	σ/mN m^{-1}
3.82	0.6616	1105.3	4.247	2.087	0.564	74.93
10	1.027	1105.9	4.224	1.680	0.574	74.06
20	1.999	1105.3	4.200	1.247	0.589	72.61
30	3.701	1103.2	4.186	0.972	0.600	71.09
40	6.547	1100.0	4.177	0.785	0.610	69.52
50	11.117	1095.7	4.171	0.651	0.618	67.89
60	18.20	1090.6	4.168	0.552	0.625	66.21
70	28.80	1084.7	4.165	0.476	0.629	64.47
80	44.24	1078.2	4.163	0.416	0.633	62.67
90	66.09	1071.1	4.163	0.368	0.635	60.82
100	96.31	1063.4	4.163	0.329	0.636	58.93
110	137.16	1055.1	4.165	0.296	0.636	56.98
120	191.3	1046.4	4.170	0.269	0.635	54.99
130	261.8	1037.2	4.176	0.246	0.632	52.95
140	352.0	1027.4	4.185	0.227	0.629	50.87
150	465.7	1017.2	4.198	0.210	0.625	48.75
160	607.2	1006.5	4.214	0.195	0.620	46.59
170	781.0	995.2	4.234	0.182	0.614	44.39
180	992.0	983.5	4.259	0.171	0.607	42.16
190	1245.5	971.2	4.290	0.161	0.600	39.90
200	1547	958.3	4.327	0.152	0.592	37.61
210	1903	944.8	4.371	0.144	0.583	35.29
220	2319	930.6	4.423	0.136	0.574	32.95
230	2802	915.8	4.485	0.130	0.563	30.59
240	3360	900.1	4.558	0.123	0.553	28.22
250	3998	883.7	4.645	0.118	0.541	25.84
260	4726	866.3	4.750	0.112	0.529	23.45
270	5550	847.8	4.875	0.107	0.516	21.07
280	6480	828.2	5.028	0.103	0.502	18.69
290	7525	807.1	5.219	0.098	0.488	16.33
300	8693	784.5	5.463	0.094	0.473	13.99
310	9997	769.8	5.785	0.089	0.458	11.68
320	11447	732.6	6.234	0.085	0.442	9.428
330	13057	702.1	6.902	0.080	0.425	7.238
340	14843	666.7	8.006	0.075	0.408	5.141
350	16823	623.6	10.16	0.069	0.391	3.173
360	19020	565.7	16.22	0.062	0.382	1.405
370	21477	440.1	264.3	0.047	0.526	0.0467
370.697	21662	356.0				0

PROPERTIES OF ICE AND SUPERCOOLED WATER

Allan H. Harvey

The following tables contain the properties of ice and supercooled water. Table 1 contains properties at ambient pressure as a function of temperature for the common form of ice, hexagonal ice, designated as ice Ih. The data column definitions for Table 1 are as follows.

Column heading	Definition
t	Temperature, in °C
ρ	Mass density, in g cm^{-3} (Refs. 2, 3)
α_v	Cubic expansion coefficient, $\alpha_v = -(1/V)(\partial V/\partial T)_p$, in units 10^3 K^{-1} (Refs. 2, 3)
κ_S	Isentropic compressibility, $\kappa_s = -(1/V)(\partial V/\partial p)_s$, in units GPa^{-1} (Refs. 2, 3)
c_p	Specific heat capacity at constant pressure c_p, in units J g^{-1} K^{-1} (Refs. 2, 3)
λ	Thermal conductivity λ, in units W m^{-1} K^{-1} (Refs. 4, 5)
ε	Static dielectric constant ε (relative permittivity) (Ref. 6)

Density data for supercooled water as a function of temperature are given in Table 2; the data columns for Table 2 are defined as follows.

Column heading	Definition
t	Temperature, in °C
ρ	Mass density, in g cm^{-3} (Refs. 2, 3)

Table 3 has the enthalpy of fusion and sublimation for phase transitions of ice to water (fusion) and ice to gas (sublimation).

The data given here refer to that form at standard atmospheric pressure (101.325 kPa). Data have been taken from the references indicated, which in most cases are formulations based on critical evaluation of available experimental data. Most properties are sensitive to the method of preparation of the sample because air and other gases are sometimes occluded. For this reason, there is often disagreement among values in the literature. For all properties except the dielectric constant of ice, the cited reference contains information on the uncertainty of the property.

References

1. Holten, V., Sengers, J.V., and Anisimov, M.A., *J. Phys. Chem. Ref. Data* 43, 043101, 2014. <https://doi.org/10.1063/1.4895593>
2. Wagner, W., and Pruß, A., *J. Phys. Chem. Ref. Data* 31, 387, 2002. <https://doi.org/10.1063/1.1461829>
3. Feistel, R., and Wagner, W., *J. Phys. Chem. Ref. Data* 35, 1021, 2006. <https://doi.org/10.1063/1.2183324>
4. International Association for the Properties of Water and Steam (IAPWS), *Revised Release on the Equation of State 2006 for H_2O Ice Ih* (2009), available from <www.iapws.org>.
5. Andersson, O., and Inaba, A., *Phys. Chem. Chem. Phys.* 7, 1441, 2005. <https://doi.org/10.1039/b500373c>
6. Slack, G.A., *Phys. Rev. B* 22, 3065, 1980. <https://doi.org/10.1103/PhysRevB.22.3065>
7. Wörz, O., and Cole, R.H., *J. Chem. Phys.* 51, 1546, 1969. <https://doi.org/10.1063/1.1672209>

TABLE 1. Properties of Ice Ih at Atmospheric Pressure

t/°C	ρ/g cm^{-3}	$10^3 \times \alpha_v$/K^{-1}	κ_s/GPa^{-1}	c_p/J g^{-1} K^{-1}	λ/W m^{-1} K^{-1}	ε
0	0.9167	0.160	0.114	2.10	2.16	91.2
-10	0.9182	0.155	0.113	2.02	2.26	95.1
-20	0.9196	0.150	0.111	1.95	2.38	99.4
-30	0.9209	0.144	0.110	1.88	2.50	104.1
-40	0.9222	0.138	0.108	1.80	2.63	109.2
-50	0.9235	0.132	0.107	1.73	2.77	115.0
-60	0.9247	0.126	0.106	1.66	2.93	121.4
-80	0.9269	0.111	0.103	1.52	3.27	136.6
-100	0.9288	0.095	0.101	1.38	3.69	
-120	0.9304	0.078	0.099	1.25	4.2	
-140	0.9317	0.060	0.097	1.11	4.9	
-160	0.9326	0.041	0.096	0.97	5.7	
-180	0.9332	0.025	0.095	0.82	7.0	
-200	0.9336	0.013	0.095	0.65	8.9	
-220	0.9337	0.0050	0.095	0.47	12.2	
-240	0.9338	0.0012	0.095	0.27	20	
-260	0.9338	0.00008	0.095	0.036		

TABLE 2. Density of Supercooled Water at Atmospheric Pressure

t/°C	ρ/g cm^{-3}
0	0.99984
-5	0.9993
-10	0.9981
-15	0.9963
-20	0.9935
-25	0.9896
-30	0.9839
-35	0.9754

TABLE 3. Phase Transition Properties of Ice Ih (Ref. 3)

$\Delta_{fus}H(0\ °C) = 333.4$ J g^{-1}

$\Delta_{subl}H(0\ °C) = 2834$ J g^{-1}

VAPOR PRESSURE OF ICE

The values of the vapor (sublimation) pressure of ice Ih were calculated from the equation recommended by the International Association for the Properties of Water and Steam (IAPWS) in 2008. See Refs. 1 and 2 for details on the uncertainty in different temperature ranges. Table 1 is for temperatures in °C; Table 2 is for temperatures in K. The first entry in both tables is the triple point of water, whose pressure has an uncertainty of 0.010 Pa. Column definitions are as follows.

Column heading	Definition
t*, *T	Temperature, in °C (Table 1); in K (Table 2)
ρ	Vapor pressure of ice, in Pa

References

1. International Association for the Properties of Water and Steam (IAPWS), *Revised Release on the Pressure along the Melting and Sublimation Curves of Ordinary Water Substance, IAPWS R14-08* (2011), available from <http://www.iapws.org>.
2. Wagner, W., Riethmann, T., Feistel, R., and Harvey, A. H., New Equations for the Sublimation Pressure and Melting Pressure of H_2O Ice Ih, *J. Phys. Chem. Ref. Data* 40, 143103, 2011. <https://doi.org/10.1063/1.3657937>

TABLE 1. Vapor Pressure of Ice for Temperature in °C

t/°C	*p*/Pa	*t*/°C	*p*/Pa	*t*/°C	*p*/Pa	*t*/°C	*p*/Pa
0.01	611.657	-13	198.49	-34	24.89	-110	1.61×10^{-4}
0	611.15	-14	181.19	-36	20.04	-120	1.41×10^{-5}
-1	562.66	-15	165.27	-38	16.07	-130	8.75×10^{-7}
-2	517.70	-16	150.65	-40	12.84	-140	3.62×10^{-8}
-3	476.04	-17	137.22	-45	7.203	-150	9.00×10^{-10}
-4	437.45	-18	124.90	-50	3.938	-160	1.18×10^{-11}
-5	401.74	-19	113.60	-55	2.094	-170	6.80×10^{-14}
-6	368.71	-20	103.24	-60	1.081	-180	1.32×10^{-16}
-7	338.17	-22	85.08	-65	0.541	-190	5.89×10^{-20}
-8	309.95	-24	69.89	-70	0.262	-200	3.31×10^{-24}
-9	283.91	-26	57.24	-75	0.122	-210	8.86×10^{-30}
-10	259.87	-28	46.72	-80	5.48×10^{-2}	-220	2.06×10^{-37}
-11	237.71	-30	38.01	-90	9.68×10^{-3}		
-12	217.29	-32	30.81	-100	1.40×10^{-3}		

TABLE 2. Vapor Pressure of Ice for Temperature in K

T/K	*p*/Pa	*T*/K	*p*/Pa	*T*/K	*p*/Pa	*T*/K	*p*/Pa
273.16	611.657	262	234.54	235	15.81	140	3.37×10^{-7}
273.15	611.15	261	214.37	230	8.947	130	1.20×10^{-8}
273	603.65	260	195.80	225	4.939	120	2.48×10^{-10}
272	555.70	258	163.00	220	2.654	110	2.57×10^{-12}
271	511.25	256	135.30	215	1.386	100	1.09×10^{-14}
270	470.06	254	111.98	210	0.702	90	1.39×10^{-17}
269	431.92	252	92.40	205	0.344	80	3.50×10^{-21}
268	396.62	250	76.01	200	0.163	70	8.58×10^{-26}
267	363.97	248	62.33	190	3.24×10^{-2}	60	6.51×10^{-32}
266	333.79	246	50.95	180	5.39×10^{-3}	50	1.93×10^{-40}
265	305.91	244	41.51	170	7.30×10^{-4}		
264	280.18	242	33.70	160	7.73×10^{-5}		
263	256.43	240	27.27	150	6.10×10^{-6}		

MELTING POINT OF ICE AS A FUNCTION OF PRESSURE

Table 1 contains data on the phase transitions of water. Column definitions for Table 1 are as follows.

Column heading	Definition
Phase transition	Ice phase transition
p	Pressure needed for transition at specified temperature, in MPa
t	Temperature for transition, in °C

Table 2 gives the melting temperature of ice at various pressures, calculated from the equation for the ice Ih – liquid water phase boundary recommended by the International Association for the Properties of Water and Steam (IAPWS) in 2008. Column definitions for Table 2 are as follows.

Column heading	Definition
p	Specified pressure of melting, in MPa
t	Melting point of ice at specified pressure, in °C

See Refs. 1 and 2 for information on the solid/liquid transitions for high-pressure forms of ice. IAPWS gives the following locations for the triple points where equilibrium exists among two ice forms and liquid water.

References

1. International Association for the Properties of Water and Steam, *Revised Release on the Pressure along the Melting and Sublimation Curves of Ordinary Water Substance* IAPWS R14-08 (2011), available from <http://www.iapws.org>..
2. Wagner, W., Riethmann, T., Feistel, R., and Harvey, A. H., New Equations for the Sublimation Pressure and Melting Pressure of H_2O Ice Ih, *J. Phys. Chem. Ref. Data*, 40, 143103, 2011. <https://doi.org/10.1063/1.3657937>

TABLE 1. Phase Transitions between Ice Forms

Phase transition	*p*/MPa	*t*/°C
ice I – ice III	208.566	−21.985
ice III – ice V	350.1	−16.986
ice V – ice VI	632.4	0.16
ice VI – ice VII	2216	81.85

TABLE 2. Melting Temperature of Ice Ih as a Function of Pressure

p/MPa	*t*/°C
0.000 612	0.01
0.1	0.0026
1	-0.064
2	-0.14
5	-0.37
10	-0.75
15	-1.14
20	-1.54
30	-2.36
40	-3.21
50	-4.09
60	-5.00
70	-5.94
80	-6.91
90	-7.91
100	-8.94
120	-11.09
140	-13.35
160	-15.73
180	-18.22
200	-20.83

PERMITTIVITY (DIELECTRIC CONSTANT) OF WATER AT VARIOUS FREQUENCIES

The permittivity of liquid water in the radio frequency and microwave regions can be represented by the Debye equation (Refs. 1 and 2):

$$\varepsilon' = \varepsilon_\infty + \frac{\varepsilon_s - \varepsilon_\infty}{1+\omega^2\tau^2}$$

$$\varepsilon'' = \frac{(\varepsilon_s - \varepsilon_\infty)\omega\tau}{1+\omega^2\tau^2}$$

where $\varepsilon = \varepsilon' + i\,\varepsilon''$ is the (complex) relative permittivity (i.e., the absolute permittivity divided by the permittivity of free space $\varepsilon_0 = 8.854 \cdot 10^{-12}$ F m^{-1}). Here ε_s is the static permittivity (see Ref. 3 and the table "Thermophysical Properties of Water and Steam" in this section); ε_∞ is a parameter describing the permittivity in the high-frequency limit; τ is the relaxation time for molecular orientation; and $\omega = 2\pi f$ is the angular frequency. The values in Table 2 have been calculated from parameters given in Ref. 2, as shown in Table 1. Column definitions for Table 2 are as follows.

Column heading	Definition
Frequency	Frequency, in kHz, MHz, or GHz
ε′ (temperature)	Real part of the relative permittiviy at the indicated temperature in °C
ε″ (temperature)	Imaginary part of the relative permittivity at the indicated temperature in °C

Other useful quantities that can be calculated from the values in the table are the loss tangent:

$$\tan\delta = \varepsilon''/\varepsilon'$$

and the absorption coefficient α which describes the power attenuation per unit length ($P = P_0 e^{-\alpha l}$):

$$\alpha = \frac{\pi f \varepsilon''}{c\sqrt{\varepsilon'}}$$

and c is the speed of light. The last equation is valid when $\varepsilon''/\varepsilon' << 1$.

References

1. Fernández, D. P., Mulev, Y., Goodwin, A. R. H., and Levelt Sengers, J. M. H., *J. Phys. Chem. Ref. Data* 24, 33, 1995. <https://doi.org/10.1063/1.555977>
2. Kaatze, U., *J. Chem. Eng. Data* 34, 371, 1989. <https://doi.org/10.1021/je00058a001>
3. Archer, D. G., and Wang, P., *J. Phys. Chem. Ref. Data* 19, 371, 1990. <https://doi.org/10.1063/1.555853>

TABLE 1. Parameters for Calculation of Permittivity

	0 °C	25 °C	50 °C
ε_∞	5.7	5.2	4.0
τ/ps	17.67	8.27	4.75

TABLE 2. Permittivity for Liquid Water at 0, 25, and 50 °C and Various Frequencies

Frequency	ε′ (0 °C)	ε″ (0 °C)	ε′ (25 °C)	ε″ (25 °C)	ε′ (50 °C)	ε″ (50 °C)
0	87.90	0.00	78.36	0.00	69.88	0.00
1 kHz	87.90	0.00	78.36	0.00	69.88	0.00
1 MHz	87.90	0.01	78.36	0.00	69.88	0.00
10 MHz	87.90	0.09	78.36	0.04	69.88	0.02
100 MHz	87.89	0.91	78.36	0.38	69.88	0.20
200 MHz	87.86	1.82	78.35	0.76	69.88	0.39
500 MHz	87.65	4.55	78.31	1.90	69.87	0.98
1 GHz	86.90	9.01	78.16	3.79	69.82	1.96
2 GHz	84.04	17.39	77.58	7.52	69.65	3.92
3 GHz	79.69	24.64	76.62	11.13	69.36	5.85
4 GHz	74.36	30.49	75.33	14.58	68.95	7.75
5 GHz	68.54	34.88	73.73	17.81	68.45	9.62
10 GHz	42.52	40.88	62.81	29.93	64.49	18.05
20 GHz	19.56	30.78	40.37	36.55	52.57	28.99
30 GHz	12.50	22.64	26.53	33.25	40.57	32.74
40 GHz	9.67	17.62	18.95	28.58	31.17	32.43
50 GHz	8.28	14.34	14.64	24.53	24.42	30.47

THERMOPHYSICAL PROPERTIES OF AIR

Eric W. Lemmon

These tables summarize the thermophysical properties of air in the liquid and gaseous states as calculated from the pseudo-pure fluid equation of state of Lemmon et al. (Ref. 1). Table 1 refers to liquid and gaseous air at equilibrium as a function of temperature. The tabulated properties are the bubble-point pressure; the dew-point pressure; density; enthalpy; entropy; isochoric heat capacity; isobaric heat capacity; speed of sound; viscosity; and thermal conductivity. The normal boiling point of air, i.e., the temperature at which the bubble-point pressure reaches 1 standard atmosphere (1.01325 bar), is 78.90 K (-194.25 °C). The column definitions for Tables 1 and 2 are as follows.

Column heading	Definition
T	Temperature, in K
P	Bubble-point pressure or dew-point pressure, in MPa [Not relevant for Table 2]
ρ	Density, in kg m^{-3}
H	Enthalpy, in units kJ kg^{-1}
S	Entropy, in units kJ kg^{-1} K^{-1}
C_v	Isochoric heat capacity, in units kJ kg^{-1} K^{-1}
C_p	Isobaric heat capacity, in units kJ kg^{-1} K^{-1}
u	Speed of sound, in units m s^{-1}
η	Viscosity, in units μPa s
λ	Thermal conductivity, in units mW m^{-1} K^{-1}

In Table 1, the first line of identical temperatures is the bubble-point pressure, i.e., the pressure at which boiling begins as the pressure of the liquid is lowered; the second line is the dew-point pressure, i.e., the pressure at which condensation begins as the pressure of the gas is raised.

Table 2 gives the properties of air along various isobars. An entry with non-integer temperatures in the isobar section indicates a phase transition (liquid–vapor) at these temperatures; property values are then given for both phases. These are identified by the high densities in the liquid and the low densities in the vapor. Additional calculations at state points not listed below can be obtained by using the NIST program REFPROP <http://www.nist.gov/srd/nist23.cfm>.

In the range from the solidification point to 873 K at pressures to 70 MPa, the estimated uncertainty of density values calculated with the equation of state is 0.1%. The estimated uncertainty of calculated speed-of-sound values is 0.2%, and that for calculated heat capacities is 1%. At temperatures above 873 K and 70 MPa, the estimated uncertainty of calculated density values is 0.5%, increasing to 1.0% at 2000 K and 2000 MPa.

References

1. Lemmon, E.W., Jacobsen, R.T, Penoncello, S.G., and Friend, D.G., Thermodynamic Properties of Air and Mixtures of Nitrogen, Argon, and Oxygen from 60 to 2000 K at Pressures to 2000 MPa, *J. Phys. Chem. Ref. Data*, 29, 331, 2000. <https://doi.org/10.1063/1.1285884>
2. Lemmon, E.W. and Jacobsen, R.T, Viscosity and Thermal Conductivity Equations for Nitrogen, Oxygen, Argon, and Air, *Int. J. Thermophys.*, 25, 21, 2004. <https://doi.org/10.1023/B:IJOT.0000022327.04529.f3>
3. Lemmon, E.W., Huber, M.L., and McLinden, M.O., NIST Standard Reference Database 23: Reference Fluid Thermodynamic and Transport Properties-REFPROP, Version 9.0, National Institute of Standards and Technology, Standard Reference Data Program, Gaithersburg, MD, 2010. <http://www.nist.gov/srd/nist23.cfm>

TABLE 1. Thermophysical Properties of Air along the Boiling and Condensation Curves

T/ K	P/ MPa	ρ/ kg m^{-3}	H/ kJ kg^{-1}	S/ kJ kg^{-1} K^{-1}	C_v/ kJ kg^{-1} K^{-1}	C_p/ kJ kg^{-1} K^{-1}	u/ m s^{-1}	η/ μPa s	λ/ mW m^{-1} K^{-1}
59.75	0.005265	957.6	-36.66	-0.5306	1.174	1.901	1030	376.6	171.4
59.75	0.002432	0.1421	185.5	3.340	0.7184	1.009	154.8	4.220	5.294
60	0.005546	956.5	-36.19	-0.5226	1.173	1.901	1028	371.9	171.0
60	0.002584	0.1504	185.8	3.326	0.7186	1.009	155.1	4.238	5.320
62	0.008270	948.2	-32.38	-0.4603	1.157	1.901	1012	336.9	167.8
62	0.004111	0.2318	187.7	3.225	0.7198	1.012	157.6	4.386	5.529
64	0.01200	939.9	-28.58	-0.3999	1.143	1.902	995.8	306.3	164.5
64	0.006325	0.3460	189.6	3.132	0.7212	1.015	160.0	4.532	5.739
66	0.01699	931.5	-24.77	-0.3414	1.129	1.903	979.1	279.4	161.3
66	0.009442	0.5018	191.5	3.047	0.7230	1.019	162.3	4.679	5.950
68	0.02352	923.0	-20.95	-0.2846	1.115	1.906	962.2	255.7	158.0
68	0.01371	0.7089	193.4	2.968	0.7252	1.024	164.5	4.825	6.162
70	0.03191	914.4	-17.13	-0.2293	1.102	1.908	945.1	234.8	154.7
70	0.01943	0.9785	195.2	2.896	0.7277	1.030	166.7	4.970	6.376
72	0.04250	905.7	-13.31	-0.1756	1.090	1.912	927.7	216.3	151.4
72	0.02692	1.322	197.0	2.828	0.7305	1.037	168.7	5.115	6.592
74	0.05566	897.0	-9.468	-0.1232	1.078	1.917	910.0	199.9	148.1
74	0.03655	1.753	198.7	2.766	0.7338	1.046	170.6	5.260	6.810
76	0.07179	888.1	-5.617	-0.07209	1.067	1.923	892.1	185.2	144.8
76	0.04870	2.285	200.4	2.708	0.7375	1.055	172.5	5.405	7.031
78	0.09129	879.1	-1.751	-0.02217	1.056	1.930	873.9	172.1	141.5
78	0.06381	2.933	202.0	2.653	0.7416	1.066	174.2	5.549	7.256

T/ K	P/ MPa	ρ/ kg m^{-3}	H/ kJ kg^{-1}	S/ kJ kg^{-1} K^{-1}	C_v/ kJ kg^{-1} K^{-1}	C_p/ kJ kg^{-1} K^{-1}	u/ m s^{-1}	η/ μPa s	λ/ mW m^{-1} K^{-1}
80	0.1146	870.0	2.132	0.02665	1.045	1.938	855.4	160.4	138.2
80	0.08232	3.711	203.6	2.602	0.7460	1.078	175.8	5.694	7.485
82	0.1422	860.7	6.036	0.07444	1.035	1.948	836.7	149.8	134.8
82	0.1047	4.635	205.1	2.554	0.7510	1.092	177.4	5.839	7.719
84	0.1745	851.3	9.962	0.1213	1.025	1.959	817.6	140.2	131.4
84	0.1315	5.724	206.5	2.509	0.7563	1.108	178.8	5.984	7.959
86	0.2121	841.7	13.91	0.1673	1.016	1.972	798.2	131.5	128.1
86	0.1631	6.993	207.8	2.466	0.7620	1.125	180.0	6.131	8.206
88	0.2553	832.0	17.90	0.2125	1.007	1.986	778.6	123.6	124.8
88	0.2002	8.464	209.1	2.425	0.7682	1.144	181.2	6.278	8.461
90	0.3048	822.0	21.91	0.2569	0.9984	2.003	758.5	116.4	121.4
90	0.2432	10.16	210.3	2.386	0.7748	1.166	182.2	6.427	8.725
92	0.3609	811.8	25.97	0.3007	0.9902	2.022	738.2	109.7	118.0
92	0.2927	12.09	211.4	2.349	0.7817	1.190	183.1	6.578	9.001
94	0.4243	801.4	30.06	0.3439	0.9825	2.044	717.5	103.6	114.6
94	0.3493	14.29	212.3	2.313	0.7891	1.217	183.8	6.732	9.289
96	0.4954	790.7	34.21	0.3866	0.9752	2.069	696.5	97.88	111.2
96	0.4136	16.78	213.2	2.279	0.7969	1.248	184.5	6.889	9.593
98	0.5749	779.7	38.41	0.4288	0.9684	2.098	675.0	92.57	107.8
98	0.4861	19.60	214.0	2.246	0.8052	1.282	184.9	7.050	9.915
100	0.6631	768.4	42.66	0.4707	0.9619	2.131	653.3	87.61	104.4
100	0.5674	22.76	214.6	2.213	0.8138	1.320	185.3	7.215	10.26
102	0.7608	756.7	46.98	0.5122	0.9560	2.168	631.1	82.94	101.0
102	0.6582	26.32	215.1	2.182	0.8230	1.363	185.5	7.387	10.63
104	0.8684	744.6	51.38	0.5535	0.9505	2.212	608.5	78.53	97.62
104	0.7590	30.31	215.5	2.151	0.8326	1.413	185.6	7.566	11.02
106	0.9864	732.1	55.86	0.5947	0.9456	2.262	585.5	74.35	94.25
106	0.8706	34.78	215.7	2.120	0.8429	1.470	185.5	7.754	11.46
108	1.116	719.1	60.44	0.6358	0.9412	2.321	562.1	70.36	90.89
108	0.9934	39.79	215.8	2.089	0.8537	1.536	185.2	7.952	11.94
110	1.256	705.5	65.12	0.6769	0.9375	2.390	538.2	66.54	87.55
110	1.128	45.41	215.6	2.059	0.8653	1.614	184.9	8.163	12.47
112	1.409	691.2	69.93	0.7182	0.9345	2.472	513.9	62.87	84.24
112	1.276	51.73	215.2	2.028	0.8777	1.708	184.3	8.391	13.07
114	1.575	676.2	74.87	0.7598	0.9324	2.571	489.0	59.31	80.96
114	1.437	58.84	214.6	1.997	0.8912	1.821	183.6	8.637	13.76
116	1.755	660.3	79.98	0.8019	0.9312	2.693	463.5	55.85	77.72
116	1.612	66.88	213.8	1.965	0.9059	1.961	182.7	8.909	14.56
118	1.948	643.4	85.29	0.8447	0.9312	2.847	437.3	52.47	74.52
118	1.801	76.04	212.6	1.932	0.9220	2.139	181.7	9.210	15.50
120	2.156	625.1	90.83	0.8885	0.9327	3.048	410.2	49.13	71.36
120	2.007	86.55	211.0	1.898	0.9402	2.374	180.4	9.552	16.63
122	2.379	605.3	96.66	0.9338	0.9363	3.323	382.0	45.81	68.24
122	2.229	98.76	208.9	1.861	0.9608	2.694	179.1	9.946	18.04
124	2.617	583.3	102.9	0.9811	0.9427	3.723	352.3	42.46	65.17
124	2.468	113.2	206.3	1.821	0.9847	3.157	177.5	10.41	19.85
126	2.872	558.3	109.7	1.032	0.9537	4.367	320.4	39.01	62.18
126	2.727	130.6	202.9	1.777	1.013	3.882	175.8	10.98	22.29
128	3.143	528.3	117.3	1.088	0.9728	5.589	285.0	35.33	59.44
128	3.006	152.6	198.3	1.725	1.049	5.166	174.0	11.72	25.84
130	3.429	488.3	126.7	1.157	1.010	8.849	243.7	31.07	58.05
130	3.308	182.7	191.7	1.660	1.096	8.033	171.9	12.77	31.81
132	3.723	411.2	142.6	1.273	1.117	35.04	189.1	24.47	67.80
132	3.646	235.4	179.7	1.556	1.168	20.65	169.4	14.80	47.00
132.63	3.785	302.6	164.5	1.437				17.83	

Fluids

TABLE 2. Thermophysical Properties of Air along Various Isobars

T/ K	ρ/ kg m^{-3}	H/ kJ kg^{-1}	S/ kJ kg^{-1} K^{-1}	C_v/ kJ kg^{-1} K^{-1}	C_p/ kJ kg^{-1} K^{-1}	u/ m s^{-1}	η/ μPa s	λ/ mW m^{-1} K^{-1}
P = 0.1 MPa (1 bar)								
60	956.7	-36.11	-0.5230	1.173	1.901	1029	372.4	171.1
78.79	875.5	-0.2237	-0.002818	1.051	1.933	866.7	167.4	140.2
81.61	4.442	204.8	2.563	0.7500	1.089	177.1	5.811	7.673
100	3.557	224.3	2.779	0.7282	1.040	198.2	7.107	9.469
120	2.938	244.9	2.966	0.7211	1.022	218.3	8.457	11.38
140	2.507	265.2	3.123	0.7184	1.014	236.4	9.750	13.24
160	2.188	285.5	3.258	0.7172	1.011	253.2	10.99	15.05
180	1.942	305.6	3.377	0.7166	1.008	268.8	12.18	16.80
200	1.746	325.8	3.483	0.7163	1.007	283.5	13.33	18.50
220	1.586	345.9	3.579	0.7163	1.006	297.4	14.44	20.16
240	1.453	366.0	3.667	0.7164	1.006	310.7	15.51	21.77
260	1.341	386.2	3.747	0.7168	1.006	323.4	16.55	23.35
280	1.245	406.3	3.822	0.7173	1.006	335.6	17.56	24.88
300	1.161	426.4	3.891	0.7181	1.007	347.4	18.54	26.38
320	1.089	446.5	3.956	0.7192	1.007	358.7	19.49	27.85
340	1.024	466.7	4.018	0.7206	1.009	369.6	20.41	29.29
360	0.9674	486.9	4.075	0.7223	1.010	380.3	21.32	30.71
380	0.9164	507.1	4.130	0.7243	1.012	390.5	22.20	32.09
400	0.8706	527.4	4.182	0.7266	1.014	400.5	23.06	33.45
500	0.6964	629.5	4.410	0.7426	1.030	446.4	27.09	39.94
600	0.5803	733.6	4.599	0.7641	1.051	487.1	30.77	46.01
700	0.4974	839.9	4.763	0.7879	1.075	523.9	34.18	51.76
800	0.4352	948.6	4.908	0.8117	1.099	557.8	37.37	57.25
900	0.3869	1060	5.039	0.8340	1.121	589.6	40.39	62.54
1000	0.3482	1173	5.158	0.8540	1.141	619.6	43.28	67.68
P = 0.5 MPa (5 bar)								
60	957.3	-35.80	-0.5248	1.173	1.900	1031	374.6	171.4
80	870.9	2.387	0.02430	1.046	1.934	858.5	161.4	138.6
96.12	790.0	34.46	0.3892	0.9748	2.071	695.2	97.55	111.0
98.36	20.14	214.1	2.240	0.8067	1.288	185.0	7.079	9.974
100	19.65	216.2	2.261	0.7967	1.261	187.4	7.192	10.10
120	15.48	239.6	2.475	0.7461	1.115	212.4	8.542	11.81
140	12.94	261.4	2.643	0.7311	1.068	232.8	9.834	13.58
160	11.17	282.5	2.784	0.7245	1.045	250.9	11.07	15.32
180	9.842	303.3	2.906	0.7213	1.032	267.4	12.26	17.04
200	8.811	323.8	3.014	0.7195	1.025	282.6	13.41	18.71
220	7.981	344.3	3.112	0.7186	1.020	297.0	14.51	20.34
240	7.297	364.6	3.200	0.7182	1.017	310.6	15.58	21.94
260	6.724	385.0	3.282	0.7182	1.015	323.5	16.62	23.50
280	6.235	405.2	3.357	0.7185	1.014	335.9	17.62	25.02
300	5.813	425.5	3.427	0.7191	1.013	347.8	18.60	26.51
320	5.446	445.8	3.492	0.7200	1.013	359.3	19.54	27.97
340	5.123	466.0	3.553	0.7213	1.013	370.3	20.47	29.41
360	4.836	486.3	3.611	0.7229	1.014	381.0	21.37	30.81
380	4.580	506.6	3.666	0.7248	1.016	391.3	22.24	32.19
400	4.350	526.9	3.718	0.7271	1.018	401.3	23.10	33.55
500	3.477	629.3	3.947	0.7429	1.032	447.3	27.13	40.02
600	2.897	733.5	4.137	0.7643	1.053	488.0	30.80	46.07
700	2.483	840.0	4.301	0.7881	1.076	524.8	34.20	51.80
800	2.173	948.8	4.446	0.8119	1.100	558.7	37.40	57.29
900	1.932	1060	4.577	0.8341	1.122	590.5	40.42	62.58
1000	1.739	1173	4.696	0.8542	1.142	620.4	43.30	67.71
P = 1 MPa (10 bar)								
60	958.0	-35.42	-0.5271	1.174	1.898	1033	377.2	171.7
80	872.2	2.720	0.02129	1.047	1.930	862.5	162.8	139.1
100	770.1	42.76	0.4673	0.9623	2.119	658.2	88.33	105.0
106.22	730.7	56.36	0.5992	0.9451	2.268	583.0	73.90	93.88

T/ K	ρ/ kg m⁻³	H/ kJ kg⁻¹	S/ kJ kg⁻¹ K⁻¹	C_v/ kJ kg⁻¹ K⁻¹	C_p/ kJ kg⁻¹ K⁻¹	u/ m s⁻¹	η/ μPa s	λ/ mW m⁻¹ K⁻¹
108.10	40.07	215.8	2.088	0.8543	1.540	185.2	7.963	11.96
120	33.48	232.3	2.233	0.7844	1.285	204.0	8.718	12.59
140	27.02	256.3	2.419	0.7481	1.148	228.2	9.978	14.11
160	22.94	278.7	2.568	0.7341	1.093	248.1	11.20	15.74
180	20.03	300.2	2.695	0.7273	1.065	265.7	12.38	17.38
200	17.83	321.3	2.806	0.7236	1.048	281.7	13.51	19.00
220	16.09	342.2	2.906	0.7216	1.038	296.6	14.61	20.60
240	14.68	362.9	2.996	0.7204	1.031	310.6	15.67	22.17
260	13.50	383.5	3.078	0.7199	1.026	323.8	16.70	23.70
280	12.50	403.9	3.154	0.7199	1.023	336.4	17.70	25.21
300	11.64	424.4	3.224	0.7203	1.021	348.4	18.67	26.68
320	10.90	444.8	3.290	0.7211	1.020	360.0	19.62	28.13
340	10.25	465.2	3.352	0.7222	1.019	371.1	20.54	29.55
360	9.668	485.6	3.410	0.7237	1.019	381.9	21.43	30.95
380	9.153	506.0	3.465	0.7255	1.020	392.3	22.31	32.32
400	8.690	526.4	3.518	0.7277	1.022	402.3	23.16	33.67
500	6.943	629.1	3.747	0.7434	1.034	448.5	27.18	40.11
600	5.784	733.5	3.937	0.7646	1.054	489.2	30.84	46.15
700	4.957	840.0	4.101	0.7884	1.077	526.0	34.24	51.87
800	4.338	948.9	4.247	0.8121	1.100	559.9	37.43	57.35
900	3.857	1060	4.378	0.8343	1.122	591.5	40.45	62.63
1000	3.472	1173	4.497	0.8543	1.142	621.5	43.33	67.75
P = 2 MPa (20 bar)								
60.11[a]	959.1	-34.44	-0.5282	1.175	1.895	1037	380.5	172.2
80	874.6	3.390	0.01535	1.048	1.921	870.2	165.5	140.1
100	775.0	43.09	0.4576	0.9636	2.086	672.5	90.43	106.6
118.52	638.8	86.69	0.8559	0.9314	2.894	430.4	51.60	73.71
119.94	86.20	211.0	1.899	0.9396	2.365	180.5	9.540	16.59
120	86.03	211.2	1.900	0.9382	2.354	180.7	9.542	16.57
140	59.88	244.8	2.161	0.7883	1.387	218.3	10.44	15.68
160	48.61	270.5	2.333	0.7545	1.213	242.7	11.55	16.80
180	41.54	294.0	2.471	0.7396	1.139	262.8	12.67	18.21
200	36.52	316.3	2.589	0.7319	1.100	280.3	13.77	19.69
220	32.70	338.0	2.692	0.7275	1.076	296.2	14.84	21.19
240	29.66	359.4	2.785	0.7249	1.060	310.8	15.88	22.68
260	27.18	380.5	2.870	0.7235	1.050	324.6	16.89	24.16
280	25.11	401.4	2.947	0.7228	1.042	337.6	17.88	25.62
300	23.34	422.2	3.019	0.7227	1.037	349.9	18.84	27.06
320	21.82	442.9	3.086	0.7231	1.033	361.7	19.77	28.47
340	20.49	463.5	3.148	0.7240	1.031	373.0	20.68	29.87
360	19.32	484.1	3.207	0.7253	1.030	383.9	21.57	31.24
380	18.27	504.7	3.263	0.7269	1.029	394.4	22.44	32.59
400	17.34	525.3	3.315	0.7290	1.029	404.5	23.29	33.93
500	13.84	628.6	3.546	0.7442	1.039	450.8	27.28	40.31
600	11.52	733.4	3.737	0.7653	1.057	491.5	30.93	46.30
700	9.878	840.2	3.902	0.7889	1.079	528.3	34.32	52.00
800	8.646	949.3	4.047	0.8125	1.102	562.1	37.49	57.46
900	7.689	1061	4.178	0.8346	1.123	593.7	40.50	62.73
1000	6.923	1174	4.298	0.8546	1.143	623.6	43.38	67.84
P = 5 MPa (50 bar)								
60.64[a]	961.4	-31.10	-0.5246	1.176	1.886	1048	386.2	173.3
80	881.7	5.437	-0.001766	1.054	1.898	892.1	173.5	143.0
100	788.3	44.27	0.4311	0.9681	2.009	710.6	96.44	111.1
120	665.1	87.67	0.8256	0.9194	2.429	496.8	57.06	79.02
140	321.3	172.5	1.467	1.049	8.515	199.5	19.25	43.18
160	151.4	240.7	1.930	0.8269	1.916	231.8	13.80	22.96
180	116.7	273.4	2.123	0.7785	1.453	258.2	14.14	22.02
200	97.93	300.6	2.267	0.7569	1.288	279.4	14.90	22.54
220	85.43	325.5	2.385	0.7451	1.205	297.7	15.77	23.49
240	76.25	349.0	2.488	0.7381	1.155	314.0	16.68	24.62

T/ K	ρ/ kg m⁻³	H/ kJ kg⁻¹	S/ kJ kg⁻¹ K⁻¹	C_v/ kJ kg⁻¹ K⁻¹	C_p/ kJ kg⁻¹ K⁻¹	u/ m s⁻¹	η/ μPa s	λ/ mW m⁻¹ K⁻¹
260	69.12	371.8	2.579	0.7338	1.123	328.8	17.60	25.83
280	63.36	394.0	2.661	0.7312	1.101	342.7	18.52	27.09
300	58.59	415.8	2.736	0.7297	1.085	355.6	19.42	28.39
320	54.55	437.4	2.806	0.7291	1.074	367.9	20.31	29.69
340	51.07	458.8	2.871	0.7292	1.065	379.6	21.18	30.98
360	48.05	480.1	2.932	0.7298	1.059	390.7	22.04	32.27
380	45.38	501.2	2.989	0.7310	1.055	401.4	22.88	33.55
400	43.01	522.3	3.043	0.7327	1.052	411.7	23.70	34.81
500	34.21	627.4	3.277	0.7466	1.052	458.3	27.61	40.97
600	28.48	733.2	3.470	0.7671	1.066	498.9	31.20	46.83
700	24.42	840.8	3.636	0.7903	1.085	535.4	34.55	52.43
800	21.39	950.4	3.782	0.8136	1.106	569.0	37.69	57.83
900	19.03	1062	3.914	0.8356	1.127	600.3	40.68	63.04
1000	17.14	1176	4.034	0.8554	1.146	629.9	43.54	68.12
P = 10 MPa (100 bar)								
61.52[a]	965.2	-25.55	-0.5187	1.177	1.871	1064	395.1	175.0
80	892.4	8.950	-0.02831	1.063	1.868	925.2	186.8	147.5
100	806.9	46.73	0.3930	0.9767	1.924	763.5	105.8	117.8
120	706.1	86.65	0.7565	0.9192	2.094	591.1	67.00	89.01
140	573.7	132.1	1.106	0.8916	2.517	418.1	41.82	63.61
160	397.0	188.0	1.479	0.8787	2.882	297.6	24.99	43.38
180	273.3	238.0	1.774	0.8267	2.098	283.1	19.31	32.89
200	214.1	274.8	1.968	0.7908	1.646	296.3	18.16	29.63
220	179.9	305.4	2.114	0.7701	1.434	312.4	18.17	28.80
240	157.0	332.8	2.233	0.7574	1.317	328.0	18.60	28.89
260	140.2	358.3	2.336	0.7492	1.245	342.7	19.20	29.38
280	127.3	382.7	2.426	0.7439	1.196	356.5	19.90	30.15
300	116.9	406.3	2.507	0.7404	1.162	369.5	20.64	31.12
320	108.3	429.3	2.582	0.7383	1.138	381.8	21.40	32.15
340	101.0	451.8	2.650	0.7372	1.120	393.5	22.17	33.22
360	94.80	474.1	2.714	0.7369	1.106	404.6	22.95	34.33
380	89.36	496.1	2.773	0.7374	1.096	415.3	23.72	35.45
400	84.57	517.9	2.829	0.7384	1.088	425.6	24.49	36.58
500	67.06	625.7	3.070	0.7505	1.073	471.8	28.19	42.26
600	55.82	733.3	3.266	0.7699	1.080	511.9	31.67	47.84
700	47.90	842.0	3.433	0.7926	1.095	547.8	34.94	53.26
800	42.00	952.4	3.581	0.8155	1.113	580.8	38.04	58.52
900	37.42	1065	3.713	0.8372	1.132	611.6	40.99	63.64
1000	33.75	1179	3.833	0.8568	1.150	640.7	43.81	68.64
P = 20 MPa (200 bar)								
63.24[a]	972.3	-14.49	-0.5069	1.180	1.847	1094	411.4	178.4
80	911.3	16.25	-0.07568	1.081	1.825	982.1	213.1	155.4
100	836.2	52.71	0.3311	0.9948	1.827	846.0	123.1	128.7
120	755.9	89.57	0.6670	0.9334	1.864	711.9	82.28	103.9
140	668.7	127.4	0.9589	0.8906	1.926	589.6	58.52	82.08
160	575.8	166.5	1.220	0.8606	1.977	491.8	43.00	65.18
180	484.9	206.0	1.452	0.8365	1.946	428.0	33.22	53.43
200	407.6	243.6	1.651	0.8143	1.808	397.2	27.80	46.02
220	348.5	278.1	1.815	0.7956	1.642	387.6	25.12	41.70
240	304.6	309.5	1.952	0.7812	1.506	388.7	23.92	39.30
260	271.5	338.6	2.068	0.7705	1.405	394.7	23.48	37.98
280	245.7	365.9	2.169	0.7627	1.331	402.8	23.47	37.46
300	225.0	392.0	2.259	0.7572	1.277	412.0	23.70	37.57
320	208.0	417.1	2.340	0.7532	1.236	421.5	24.08	37.94
340	193.7	441.5	2.414	0.7506	1.204	431.1	24.56	38.47
360	181.5	465.3	2.482	0.7491	1.180	440.6	25.10	39.13
380	170.9	488.7	2.546	0.7485	1.161	450.0	25.68	39.88
400	161.7	511.7	2.605	0.7486	1.146	459.2	26.29	40.69
500	128.3	624.0	2.856	0.7575	1.108	501.8	29.49	45.27
600	107.0	734.4	3.057	0.7753	1.103	539.5	32.69	50.19

T/ K	*ρ*/ kg m⁻³	*H*/ kJ kg⁻¹	*S*/ kJ kg⁻¹ K⁻¹	C_v/ kJ kg⁻¹ K⁻¹	C_p/ kJ kg⁻¹ K⁻¹	*u*/ m s⁻¹	*η*/ μPa s	*λ*/ mW m⁻¹ K⁻¹
700	92.10	845.1	3.227	0.7969	1.112	573.7	35.78	55.18
800	80.99	957.0	3.377	0.8191	1.126	605.2	38.75	60.13
900	72.34	1070	3.510	0.8402	1.142	634.8	41.61	65.02
1000	65.41	1185	3.631	0.8594	1.157	662.8	44.36	69.84
***P* = 50 MPa (500 bar)**								
68.21[a]	991.5	18.30	-0.4728	1.192	1.793	1173	450.9	187.5
80	955.5	39.24	-0.1896	1.128	1.761	1112	292.7	174.1
100	895.9	73.98	0.1982	1.043	1.715	1012	171.6	152.8
120	837.8	107.9	0.5077	0.9811	1.680	920.9	118.9	133.2
140	781.1	141.2	0.7643	0.9342	1.649	839.4	90.60	115.9
160	726.2	173.9	0.9826	0.8983	1.619	770.2	72.86	101.0
180	673.9	205.9	1.171	0.8706	1.586	714.1	60.73	88.73
200	624.9	237.3	1.337	0.8488	1.549	670.6	52.14	79.05
220	579.8	267.9	1.482	0.8313	1.509	638.1	45.98	71.65
240	539.0	297.7	1.612	0.8171	1.467	614.7	41.60	66.11
260	502.4	326.6	1.728	0.8057	1.425	598.5	38.49	61.97
280	470.0	354.7	1.832	0.7964	1.385	588.0	36.32	58.99
300	441.2	382.0	1.926	0.7890	1.349	581.7	34.82	57.07
320	415.6	408.7	2.012	0.7832	1.316	578.4	33.83	55.77
340	392.9	434.7	2.091	0.7787	1.287	577.4	33.19	54.93
360	372.7	460.2	2.164	0.7754	1.262	578.1	32.82	54.41
380	354.6	485.2	2.232	0.7731	1.241	580.0	32.66	54.15
400	338.2	509.8	2.295	0.7717	1.223	582.9	32.64	54.07
500	276.1	629.0	2.561	0.7749	1.168	604.1	33.85	55.45
600	234.6	744.6	2.772	0.7890	1.148	629.5	36.00	58.34
700	204.6	859.2	2.948	0.8080	1.147	655.5	38.46	61.93
800	181.8	974.2	3.102	0.8285	1.154	681.0	41.00	65.86
900	163.8	1090	3.238	0.8483	1.164	705.9	43.55	69.98
1000	149.2	1207	3.362	0.8665	1.175	730.1	46.07	74.20

[a] Freezing point for the liquid state.

THERMOPHYSICAL PROPERTIES OF FLUIDS

Eric W. Lemmon and Ian H. Bell

These tables give thermodynamic and transport properties of a variety of fluids, as generated from the correlations presented in the references below. The properties tabulated are defined in the table below. All extensive properties are given on a mass basis. Not all properties are included for every substance. The references should be consulted for information on the uncertainties. Column definitions for the tables are as follows.

Column heading	Definition
T	Temperature, in K
P	Pressure, in units of MPa
ρ	Density, in units of kg m^{-3}
H	Enthalpy, in units of kJ kg^{-1}
S	Entropy, in units of kJ kg^{-1} K^{-1}
C_v	Isochoric heat capacity, in units of kJ kg^{-1} K^{-1}
C_p	Isobaric heat capacity, in units of kJ kg^{-1} K^{-1}
u	Speed of sound, in units of m s^{-1}
η	Viscosity, in units of μPa s
λ	Thermal conductivity, in units of mW m^{-1} K^{-1}
D	Static dielectric constant, dimensionless

Values are given first along the saturation line. The first two rows are the properties at the triple point. The final row gives the properties at the critical point. Two rows are given for each temperature (except at the critical point); the first row gives the values of the liquid phase (note the high density), and the second row gives the values of the vapor phase (lower densities). Following the saturation tables, values are given as a function of temperature for several isobars. A duplicate entry in the isobar section indicates a phase transition (liquid–vapor) at that temperature; property values are then given for both phases. The phase can be determined by noting the sharp decrease in density between two successive temperature entries; all rows above this point refer to the liquid phase, and all rows below refer to the gas phase. If there is no sharp discontinuity in density, all data in the table refer to the supercritical region (i.e., the isobar is above the critical pressure). If the first temperature in the isobars is not an integer, this state point refers to the properties of the liquid at the melting line. All temperatures are given on ITS-90, except those for oxygen, where the equation of state still uses the IPTS-68 temperature scale.

Properties for each fluids are given in individual tables as shown below. Reference state information is included.

Table no.	Fluid name	Fluid mol. form.	Reference state
1	Nitrogen	N_2	Zero enthalpy in the gas phase at 0 K
2	Oxygen	O_2	Zero enthalpy in the gas phase at 0 K
3	Argon	Ar	Zero enthalpy in the gas phase at 0 K
4	paraHydrogen	H_2 (para)	Zero enthalpy and entropy at the saturated liquid state at the normal boiling point
5	Helium	^4He	Zero enthalpy and entropy at the saturated liquid state at the normal boiling point
6	Methane	CH_4	Zero enthalpy and entropy at the saturated liquid state at the normal boiling point
7	Ethane	C_2H_6	Zero enthalpy and entropy at the saturated liquid state at the normal boiling point
8	Propane	C_3H_8	Enthalpy = 200 kJ kg^{-1} and entropy = 1 kJ kg^{-1} K^{-1} at −40 °C
9	Carbon dioxide	CO_2	Enthalpy = 200 kJ kg^{-1} and entropy = 1 kJ kg^{-1} K^{-1} at −40 °C

Additional calculations at state points not listed below and for fluids not contained here can be obtained with the use of the NIST Standard Reference Data program REFPROP <http://www.nist.gov/srd/refprop>.

References

1. Arp, V.D., McCarty, R.D., and Friend, D.G., Thermophysical Properties of Helium-4 from 0.8 to 1500 K with Pressures to 2000 MPa, *NIST Technical Note 1334 (revised)*, National Institute of Standardsand Technology, Boulder, CO,1998. <https://doi.org/10.6028/NIST.TN.1334>
2. Assael, M.J., Assael. J.-A.M., Huber, M.L., Perkins, R.A., and Takata, Y., Correlation of the Thermal Conductivity of Normal and Parahydrogen from the Triple Point to 1000 K and up to 100 MPa, *J. Phys. Chem. Ref. Data*, 40, 033101, 2011. <https://doi.org/10.1063/1.3606499>
3. Bücker, D. and Wagner, W., A Reference Equation of State for the Thermodynamic Properties of Ethane for Temperatures from the Melting Line to 675 K and Pressures up to 900 MPa, *J. Phys. Chem. Ref. Data*, 35(1):205-266, 2006. <https://doi.org/10.1063/1.1859286>
4. Friend, D.G., Ely, J.F., and Ingham, H., Tables for the Thermophysical Properties of Methane, *NIST Technical Note 1325*, National Institute of Standards and Technology, Boulder, CO, 1989. <https://doi.org/10.6028/NIST.TN.1325>
5. Friend, D.G., Ingham, H., and Ely, J.F., Thermophysical Properties of Ethane, *J. Phys. Chem. Ref. Data*, 20(2):275-347, 1991. <https://doi.org/10.1063/1.555881>
6. Hands, B.A. and Arp, V.D., A Correlation of Thermal Conductivity Data for Helium, *Cryogenics*, 21(12):697-703, 1981. <https://doi.org/10.1016/0011-2275(81)90211-3>
7. Harvey, A. H. and Lemmon, E. W., Method for Estimating the Dielectric Constant of Natural Gas Mixtures, *Int. J. Thermophys.*, 26, 31-46, 2005. <https://doi.org/10.1007/s10765-005-2351-5>
8. Huber, M.L., Sykioti, E.A., Assael, M.J., and Perkins, R.A., Reference Correlation of the Thermal Conductivity of Carbon Dioxide from the Triple Point to 1100 K and up to 200 MPa, *J. Phys. Chem. Ref. Data*, 45, 013102, 2016. <https://doi.org/10.1063/1.4940892>
9. Laesecke, A. and Muzny, C.D., Reference Correlation for the Viscosity of Carbon Dioxide, *J. Phys. Chem. Ref. Data*, 46, 013107, 2017. <https://doi.org/10.1063/1.4977429>
10. Leachman, J.W., Jacobsen, R.T, Penoncello, S.G., and Lemmon, E.W., Fundamental Equations of State for Parahydrogen, Normal Hydrogen, and Orthohydrogen, *J. Phys. Chem. Ref. Data*, 38(3):721-748, 2009. <https://doi.org/10.1063/1.3160306>
11. Lemmon, E.W. and Jacobsen, R.T, Viscosity and Thermal Conductivity Equations for Nitrogen, Oxygen, Argon, and Air, *Int. J. Thermophys.*, 25:21-69, 2004. <https://doi.org/10.1023/B:IJOT.0000022327.04529.f3>
12. Lemmon, E.W., McLinden, M.O., and Wagner, W., Thermodynamic Properties of Propane. III. A Reference Equation of State for Temperatures from the Melting Line to 650 K and Pressures up to 1000 MPa, *J. Chem. Eng. Data*, 54:3141-3180, 2009. <https://doi.org/10.1021/je900217v>

13. Marsh, K., Perkins, R., and Ramires, M.L.V., Measurement and Correlation of the Thermal Conductivity of Propane from 86 to 600 K at Pressures to 70 MPa, *J. Chem. Eng. Data*, 47(4):932-940, 2002. <https://doi.org/10.1021/je010001m>
14. Muzny, C.D., Huber, M.L., and Kazakov, A.F., Correlation for the Viscosity of Normal Hydrogen Obtained from Symbolic Regression, *J. Chem. Eng. Data*, 58:969-979, 2013. <https://doi.org/10.1021/je301273j>
15. Ortiz-Vega, D.O., Hall, K.R., Holste, J.C., Arp, V.D., Harvey, A.H., and Lemmon, E.W., to be submitted to *J. Phys. Chem. Ref. Data*, 2019.
16. Quinones-Cisneros, S.E., Huber, M.L., and Deiters, U.K., unpublished work, 2011.
17. Schmidt, R. and Wagner, W., A New Form of the Equation of State for Pure Substances and its Application to Oxygen, *Fluid Phase Equilib.*, 19:175-200, 1985. <https://doi.org/10.1016/0378-3812(85)87016-3>
18. Setzmann, U. and Wagner, W., A New Equation of State and Tables of Thermodynamic Properties for Methane Covering the Range from the Melting Line to 625 K at Pressures up to 1000 MPa, *J. Phys. Chem. Ref. Data*, 20(6):1061-1151, 1991. <https://doi.org/10.1063/1.555898>
19. Span, R. and Wagner, W., A New Equation of State for Carbon Dioxide Covering the Fluid Region from the Triple-Point Temperature to 1100 K at Pressures up to 800 MPa, *J. Phys. Chem. Ref. Data*, 25(6):1509-1596, 1996. <https://doi.org/10.1063/1.555991>
20. Span, R., Lemmon, E.W., Jacobsen, R.T, Wagner, W., and Yokozeki, A., A Reference Equation of State for the Thermodynamic Properties of Nitrogen for Temperatures from 63.151 to 1000 K and Pressures to 2200 MPa, *J. Phys. Chem. Ref. Data*, 29(6):1361-1433, 2000. <https://doi.org/10.1063/1.1349047>
21. Tegeler, Ch., Span, R., and Wagner, W., A New Equation of State for Argon Covering the Fluid Region for Temperatures from the Melting Line to 700 K at Pressures up to 1000 MPa, *J. Phys. Chem. Ref. Data*, 28(3):779-850, 1999. <https://doi.org/10.1063/1.556037>
22. Vogel, E. and Herrmann, S., New Formulation for the Viscosity of Propane, *J. Phys. Chem. Ref. Data*, 45, 043103, 2016. <https://doi.org/10.1063/1.4966928>
23. Vogel, E., Span, R., and Herrmann, S., Reference Correlation for the Viscosity of Ethane, *J. Phys. Chem. Ref. Data*, 44, 043101, 2015. <https://doi.org/10.1063/1.4930838>

TABLE 1. Thermophysical Properties of Nitrogen (N_2)

T/ K	P/ MPa	ρ/ kg m^{-3}	H/ kJ kg^{-1}	S/ kJ kg^{-1} K^{-1}	C_v/ kJ kg^{-1} K^{-1}	C_p/ kJ kg^{-1} K^{-1}	u/ m s^{-1}	η/ µPa s	λ/ mW m^{-1} K^{-1}	D
Saturation										
63.151	0.01252	867.2	-150.7	2.426	1.176	2.000	995.3	311.6	173.3	1.47003
63.151	0.01252	0.6743	64.78	5.838	0.7499	1.058	161.1	4.376	5.621	1.00032
70.0	0.03854	838.5	-137	2.632	1.130	2.014	925.7	220.2	159.5	1.45241
70.0	0.03854	1.896	71.10	5.605	0.7580	1.082	168.4	4.883	6.355	1.00089
80.0	0.1369	793.9	-116.6	2.903	1.069	2.056	824.4	145.1	139.6	1.42541
80.0	0.1369	6.089	79.10	5.349	0.7773	1.145	176.7	5.652	7.506	1.00286
90.0	0.3605	745.0	-95.52	3.147	1.020	2.141	719.0	102.8	119.8	1.39622
90.0	0.3605	15.08	84.97	5.153	0.8078	1.266	181.8	6.482	8.868	1.00710
100.0	0.7783	689.4	-73.21	3.376	0.9832	2.318	605.2	75.76	100.1	1.36351
100.0	0.7783	31.96	87.77	4.986	0.8548	1.503	183.3	7.429	10.73	1.01510
110.0	1.466	621.5	-48.49	3.601	0.9667	2.743	476.4	55.99	80.44	1.32430
110.0	1.466	62.58	85.84	4.823	0.9284	2.062	180.8	8.626	13.83	1.02974
120.0	2.511	523.4	-17.87	3.851	1.011	4.508	317.3	38.43	61.01	1.26895
120.0	2.511	125.1	74.17	4.618	1.099	4.631	172.6	10.62	21.71	1.06012
126.192	3.396	313.3	29.23	4.215				18.30		1.15559
P = 0.1 MPa (1 bar)										
63.170	0.1	867.3	-150.6	2.426	1.176	2.000	995.6	311.6	173.3	1.47007
77.244	0.1	806.6	-122.2	2.831	1.085	2.041	852.5	161.4	145.1	1.43304
77.244	0.1	4.556	77.07	5.412	0.7710	1.123	174.7	5.435	7.174	1.00214
80.0	0.1	4.379	80.15	5.451	0.7666	1.112	178.3	5.623	7.443	1.00206
100.0	0.1	3.437	101.9	5.694	0.7514	1.071	201.6	6.958	9.381	1.00162
120.0	0.1	2.840	123.2	5.888	0.7466	1.057	222.0	8.244	11.27	1.00133
140.0	0.1	2.424	144.2	6.050	0.7447	1.050	240.4	9.480	13.11	1.00114
160.0	0.1	2.116	165.2	6.190	0.7438	1.047	257.3	10.67	14.89	1.00099
180.0	0.1	1.878	186.1	6.313	0.7433	1.045	273.2	11.81	16.61	1.00088
200.0	0.1	1.688	207.0	6.423	0.7430	1.043	288.1	12.91	18.28	1.00079
220.0	0.1	1.534	227.9	6.523	0.7429	1.043	302.3	13.97	19.90	1.00072
240.0	0.1	1.405	248.7	6.613	0.7428	1.042	315.8	15.00	21.48	1.00066
260.0	0.1	1.297	269.5	6.697	0.7428	1.042	328.7	15.99	23.01	1.00061
280.0	0.1	1.204	290.4	6.774	0.7429	1.041	341.2	16.96	24.51	1.00057
300.0	0.1	1.123	311.2	6.846	0.7432	1.041	353.2	17.89	25.97	1.00053
320.0	0.1	1.053	332.0	6.913	0.7436	1.042	364.7	18.80	27.39	1.00049
340.0	0.1	0.9909	352.9	6.976	0.7441	1.042	375.9	19.68	28.79	1.00047
360.0	0.1	0.9357	373.7	7.036	0.7450	1.043	386.8	20.55	30.15	1.00044
380.0	0.1	0.8864	394.6	7.092	0.7461	1.044	397.3	21.39	31.49	1.00042
400.0	0.1	0.8421	415.5	7.146	0.7475	1.045	407.5	22.21	32.81	1.00040

T/ K	P/ MPa	ρ/ kg m⁻³	H/ kJ kg⁻¹	S/ kJ kg⁻¹ K⁻¹	C_v/ kJ kg⁻¹ K⁻¹	C_p/ kJ kg⁻¹ K⁻¹	u/ m s⁻¹	η/ μPa s	λ/ mW m⁻¹ K⁻¹	D
500.0	0.1	0.6736	520.5	7.380	0.7592	1.056	454.6	26.06	39.04	1.00032
600.0	0.1	0.5613	627.0	7.574	0.7781	1.075	496.3	29.58	44.84	1.00026
700.0	0.1	0.4811	735.6	7.741	0.8011	1.098	533.9	32.83	50.31	1.00023
800.0	0.1	0.4210	846.6	7.890	0.8254	1.122	568.4	35.89	55.51	1.00020
900.0	0.1	0.3742	960.0	8.023	0.8488	1.146	600.7	38.78	60.52	1.00018
1000.0	0.1	0.3368	1076.	8.145	0.8705	1.167	631.1	41.54	65.36	1.00016
P = 1.0 MPa (10 bar)										
63.368	1.0	868.0	-149.5	2.427	1.177	1.995	998.9	311.8	173.7	1.47049
80.0	1.0	796.3	-116	2.896	1.071	2.044	832.3	147.3	140.6	1.42683
100.0	1.0	690.8	-73.18	3.373	0.9833	2.305	609.4	76.26	100.6	1.36432
103.747	1.0	665.8	-64.33	3.460	0.9739	2.431	559.2	67.78	92.74	1.34984
103.747	1.0	41.33	87.73	4.926	0.8786	1.652	182.8	7.835	11.67	1.01956
120.0	1.0	32.10	110.9	5.134	0.7980	1.297	208.9	8.714	12.49	1.01517
140.0	1.0	26.01	135.5	5.324	0.7701	1.177	232.8	9.844	14.00	1.01227
160.0	1.0	22.11	158.5	5.477	0.7588	1.127	252.9	10.96	15.60	1.01043
180.0	1.0	19.33	180.7	5.608	0.7531	1.101	270.6	12.06	17.21	1.00911
200.0	1.0	17.21	202.6	5.723	0.7500	1.085	286.9	13.13	18.81	1.00811
220.0	1.0	15.54	224.2	5.826	0.7480	1.075	301.9	14.16	20.37	1.00732
240.0	1.0	14.17	245.6	5.919	0.7468	1.067	316.1	15.17	21.90	1.00667
260.0	1.0	13.04	266.9	6.005	0.7460	1.062	329.5	16.14	23.39	1.00614
280.0	1.0	12.07	288.1	6.083	0.7456	1.059	342.4	17.09	24.86	1.00569
300.0	1.0	11.25	309.2	6.156	0.7454	1.056	354.6	18.01	26.29	1.00530
320.0	1.0	10.53	330.3	6.224	0.7455	1.054	366.4	18.91	27.69	1.00496
340.0	1.0	9.901	351.4	6.288	0.7458	1.053	377.8	19.79	29.07	1.00466
360.0	1.0	9.343	372.4	6.348	0.7464	1.052	388.8	20.64	30.42	1.00440
380.0	1.0	8.846	393.5	6.405	0.7474	1.052	399.4	21.48	31.74	1.00416
380.0	1.0	8.846	393.5	6.405	0.7474	1.052	399.4	21.48	31.74	1.00416
480.0	1.0	6.992	498.9	6.651	0.7570	1.058	448.0	25.39	38.03	1.00329
580.0	1.0	5.784	605.4	6.853	0.7744	1.074	490.7	28.95	43.87	1.00272
680.0	1.0	4.934	713.9	7.025	0.7968	1.095	529.0	32.24	49.37	1.00232
780.0	1.0	4.302	824.6	7.177	0.8209	1.119	564.0	35.32	54.61	1.00202
880.0	1.0	3.814	937.7	7.314	0.8446	1.142	596.6	38.24	59.63	1.00180
980.0	1.0	3.426	1053.	7.438	0.8666	1.164	627.3	41.03	64.50	1.00161
1000.0	1.0	3.358	1076.	7.461	0.8708	1.168	633.2	41.57	65.45	1.00158
P = 10.0 MPa (100 bar)										
65.321	10.0	875.0	-138.3	2.441	1.186	1.955	1031.	314.0	177.1	1.47447
80.0	10.0	818.4	-109.6	2.837	1.093	1.960	904.1	169.0	150.4	1.43984
100.0	10.0	733.6	-69.89	3.280	0.9997	2.022	734.2	93.65	115.9	1.38918
120.0	10.0	632.9	-27.82	3.663	0.9401	2.218	559.6	58.57	84.11	1.33060
140.0	10.0	499.8	20.66	4.036	0.9127	2.676	392.0	36.11	57.60	1.25566
160.0	10.0	344.2	76.44	4.408	0.8849	2.668	300.8	22.40	40.20	1.17162
180.0	10.0	248.7	122.6	4.681	0.8407	1.990	294.7	18.50	32.06	1.12203
200.0	10.0	199.4	158.4	4.870	0.8107	1.628	307.7	17.70	29.38	1.09701
220.0	10.0	169.4	188.9	5.015	0.7927	1.445	323.2	17.74	28.72	1.08198
240.0	10.0	148.8	216.7	5.136	0.7813	1.340	338.4	18.13	28.87	1.07172
260.0	10.0	133.4	242.8	5.240	0.7737	1.273	352.9	18.67	29.39	1.06415
280.0	10.0	121.4	267.7	5.333	0.7685	1.227	366.6	19.29	30.21	1.05827
300.0	10.0	111.7	291.9	5.417	0.7649	1.195	379.5	19.96	31.14	1.05351
320.0	10.0	103.6	315.6	5.493	0.7623	1.171	391.8	20.66	32.13	1.04958
340.0	10.0	96.80	338.8	5.563	0.7606	1.153	403.6	21.37	33.16	1.04625
360.0	10.0	90.89	361.7	5.629	0.7596	1.138	414.8	22.08	34.21	1.04338
380.0	10.0	85.74	384.4	5.690	0.7592	1.127	425.6	22.80	35.28	1.04088
380.0	10.0	85.74	384.4	5.690	0.7592	1.127	425.6	22.80	35.28	1.04088
480.0	10.0	67.22	495.5	5.950	0.7648	1.102	474.2	26.31	40.66	1.03196
580.0	10.0	55.58	605.5	6.158	0.7802	1.102	516.1	29.64	45.95	1.02637
680.0	10.0	47.49	716.3	6.334	0.8013	1.115	553.4	32.79	51.09	1.02251
780.0	10.0	41.50	828.7	6.488	0.8247	1.133	587.3	35.77	56.07	1.01965
880.0	10.0	36.88	943.0	6.626	0.8478	1.153	618.9	38.62	60.90	1.01746
980.0	10.0	33.20	1059.	6.751	0.8694	1.173	648.6	41.35	65.61	1.01571
1000.0	10.0	32.56	1083.	6.775	0.8735	1.176	654.4	41.88	66.54	1.01540

TABLE 2. Thermophysical Properties of Oxygen (O_2)

T/ K	P/ MPa	ρ/ kg m^{-3}	H/ kJ kg^{-1}	S/ kJ kg^{-1} K^{-1}	C_v/ kJ kg^{-1} K^{-1}	C_p/ kJ kg^{-1} K^{-1}	u/ m s^{-1}	η/ µPa s	λ/ mW m^{-1} K^{-1}	D
Saturation										
54.361	0.0001463	1306.	-193.6	2.092	1.195	1.673	1123.	773.6	201.9	1.56799
54.361	0.0001463	0.01036	49.11	6.557	0.6638	0.9260	140.3	4.096	4.420	1.00000
60.0	0.0007258	1282.	-184.2	2.257	1.089	1.673	1127.	578.1	194.0	1.55615
60.0	0.0007258	0.04659	54.19	6.230	0.6817	0.9475	147.0	4.553	4.984	1.00002
70.0	0.006262	1237.	-167.4	2.516	1.017	1.678	1066.	371.8	179.7	1.53399
70.0	0.006262	0.3457	63.09	5.809	0.7052	0.9780	158.1	5.356	5.992	1.00013
80.0	0.03012	1190.	-150.6	2.740	0.9697	1.682	987.4	261.2	165.5	1.51120
80.0	0.03012	1.468	71.69	5.519	0.6950	0.9743	168.4	6.149	7.028	1.00054
90.0	0.09935	1142.	-133.7	2.938	0.9296	1.699	905.9	195.6	151.1	1.48766
90.0	0.09935	4.387	79.55	5.308	0.6758	0.9705	177.3	6.936	8.124	1.00163
100.0	0.2540	1091.	-116.4	3.118	0.8949	1.738	822.2	152.6	136.6	1.46295
100.0	0.2540	10.42	86.16	5.144	0.6752	1.006	184.1	7.728	9.336	1.00387
110.0	0.5434	1035.	-98.64	3.286	0.8658	1.807	734.8	121.5	121.9	1.43650
110.0	0.5434	21.28	91.05	5.010	0.6988	1.101	188.1	8.547	10.75	1.00791
120.0	1.022	973.9	-79.9	3.444	0.8430	1.927	641.5	97.43	107.2	1.40746
120.0	1.022	39.31	93.75	4.892	0.7415	1.276	189.4	9.427	12.51	1.01465
130.0	1.749	902.5	-59.66	3.600	0.8293	2.153	539.5	77.57	92.63	1.37432
130.0	1.749	68.37	93.47	4.778	0.8002	1.600	187.8	10.45	14.94	1.02558
140.0	2.788	813.2	-36.7	3.761	0.8323	2.691	423.1	60.22	78.22	1.33363
140.0	2.788	116.8	88.47	4.655	0.8834	2.370	182.8	11.82	18.98	1.04395
150.0	4.219	675.5	-6.671	3.955	0.9057	5.464	273.8	42.90	64.19	1.27248
150.0	4.219	214.9	72.56	4.483	1.049	6.625	172.8	14.72	29.67	1.08192
154.581	5.043	436.1	32.42	4.201				24.84		1.17084
***P* = 0.1 MPa (1 bar)**										
54.371	0.1	1306.	-193.5	2.092	1.195	1.673	1124.	773.8	202.0	1.56802
60.0	0.1	1282.	-184.1	2.257	1.089	1.673	1128.	578.5	194.0	1.55621
80.0	0.1	1191.	-150.6	2.739	0.9699	1.681	987.7	261.4	165.5	1.51126
90.062	0.1	1142.	-133.6	2.939	0.9293	1.699	905.4	195.3	151.0	1.48751
90.062	0.1	4.413	79.60	5.307	0.6757	0.9705	177.4	6.940	8.131	1.00164
100.0	0.1	3.941	88.99	5.405	0.6527	0.9352	188.4	7.712	9.085	1.00146
120.0	0.1	3.252	107.6	5.575	0.6543	0.9280	207.3	9.219	10.99	1.00121
140.0	0.1	2.774	126.1	5.718	0.6530	0.9218	224.6	10.67	12.86	1.00103
160.0	0.1	2.420	144.5	5.841	0.6519	0.9180	240.5	12.07	14.69	1.00090
180.0	0.1	2.147	162.8	5.949	0.6513	0.9158	255.4	13.41	16.48	1.00080
200.0	0.1	1.930	181.1	6.045	0.6512	0.9146	269.3	14.72	18.24	1.00072
220.0	0.1	1.753	199.4	6.132	0.6516	0.9142	282.6	15.98	19.95	1.00065
240.0	0.1	1.606	217.7	6.212	0.6525	0.9146	295.2	17.20	21.64	1.00060
260.0	0.1	1.482	236.0	6.285	0.6539	0.9156	307.2	18.38	23.28	1.00055
280.0	0.1	1.376	254.3	6.353	0.6560	0.9174	318.7	19.53	24.90	1.00051
300.0	0.1	1.284	272.7	6.416	0.6587	0.9199	329.7	20.65	26.49	1.00048
320.0	0.1	1.203	291.1	6.476	0.6621	0.9231	340.3	21.74	28.04	1.00045
340.0	0.1	1.132	309.6	6.532	0.6661	0.9269	350.5	22.80	29.58	1.00042
360.0	0.1	1.069	328.2	6.585	0.6707	0.9313	360.4	23.84	31.08	1.00040
380.0	0.1	1.013	346.9	6.635	0.6757	0.9363	369.9	24.85	32.57	1.00038
400.0	0.1	0.9622	365.7	6.684	0.6812	0.9416	379.0	25.84	34.03	1.00036
500.0	0.1	0.7696	461.3	6.897	0.7119	0.9722	421.3	30.49	41.05	1.00029
600.0	0.1	0.6413	560.1	7.077	0.7432	1.003	458.9	34.73	47.66	1.00024
700.0	0.1	0.5497	661.9	7.234	0.7710	1.031	493.3	38.65	53.97	1.00020
800.0	0.1	0.4809	766.2	7.373	0.7945	1.055	525.4	42.33	60.02	1.00018
900.0	0.1	0.4275	872.6	7.498	0.8140	1.074	555.6	45.81	65.87	1.00016
1000.0	0.1	0.3848	980.9	7.612	0.8301	1.090	584.3	49.12	71.55	1.00014
***P* = 1.0 MPa (10 bar)**										
54.474	1.0	1307.	-192.8	2.093	1.190	1.669	1127.	775.0	202.1	1.56831
60.0	1.0	1283.	-183.6	2.254	1.089	1.671	1130.	582.7	194.4	1.55674
80.0	1.0	1192.	-150.1	2.736	0.9716	1.679	991.1	263.6	166.1	1.51204
100.0	1.0	1093.	-116.1	3.115	0.8964	1.731	826.8	153.9	137.2	1.46397

T/ K	P/ MPa	ρ/ kg m⁻³	H/ kJ kg⁻¹	S/ kJ kg⁻¹ K⁻¹	C_v/ kJ kg⁻¹ K⁻¹	C_p/ kJ kg⁻¹ K⁻¹	u/ m s⁻¹	η/ μPa s	λ/ mW m⁻¹ K⁻¹	D
119.621	1.0	976.3	-80.64	3.438	0.8438	1.921	645.2	98.25	107.8	1.40862
119.621	1.0	38.46	93.70	4.896	0.7396	1.268	189.4	9.392	12.43	1.01433
120.0	1.0	38.25	94.18	4.900	0.7355	1.257	190.0	9.421	12.45	1.01425
140.0	1.0	30.39	116.8	5.075	0.6851	1.065	214.2	10.89	13.90	1.01131
160.0	1.0	25.65	137.5	5.213	0.6708	1.005	233.7	12.29	15.54	1.00954
180.0	1.0	22.33	157.2	5.329	0.6631	0.9743	250.8	13.64	17.21	1.00831
200.0	1.0	19.84	176.5	5.431	0.6590	0.9568	266.3	14.94	18.88	1.00738
220.0	1.0	17.88	195.5	5.521	0.6571	0.9462	280.6	16.20	20.53	1.00665
240.0	1.0	16.30	214.4	5.603	0.6566	0.9397	294.0	17.41	22.15	1.00606
260.0	1.0	14.98	233.1	5.678	0.6572	0.9359	306.6	18.59	23.76	1.00557
280.0	1.0	13.86	251.8	5.748	0.6586	0.9342	318.5	19.73	25.34	1.00515
300.0	1.0	12.91	270.5	5.812	0.6609	0.9340	329.9	20.85	26.89	1.00480
320.0	1.0	12.08	289.2	5.872	0.6640	0.9351	340.8	21.93	28.43	1.00449
340.0	1.0	11.35	307.9	5.929	0.6677	0.9373	351.2	22.99	29.93	1.00422
360.0	1.0	10.71	326.7	5.983	0.6720	0.9404	361.2	24.02	31.42	1.00398
380.0	1.0	10.14	345.6	6.034	0.6769	0.9442	370.8	25.02	32.89	1.00377
400.0	1.0	9.623	364.5	6.082	0.6823	0.9487	380.1	26.01	34.33	1.00358
500.0	1.0	7.683	460.7	6.297	0.7126	0.9763	422.7	30.63	41.29	1.00286
600.0	1.0	6.398	559.8	6.478	0.7436	1.006	460.4	34.85	47.86	1.00238
700.0	1.0	5.483	661.8	6.635	0.7713	1.033	494.9	38.77	54.14	1.00204
800.0	1.0	4.798	766.2	6.774	0.7948	1.056	526.9	42.43	60.17	1.00179
900.0	1.0	4.265	872.8	6.900	0.8142	1.075	557.1	45.90	66.00	1.00159
1000.0	1.0	3.839	981.1	7.014	0.8303	1.091	585.8	49.20	71.67	1.00143
P = 10.0 MPa (100 bar)										
55.498	10.0	1312.	-185.4	2.103	1.149	1.640	1158.	786.0	203.9	1.57109
60.0	10.0	1294.	-178	2.231	1.092	1.653	1155.	625.3	198.0	1.56193
80.0	10.0	1207.	-144.8	2.708	0.9885	1.654	1023.	285.2	171.5	1.51940
100.0	10.0	1116.	-111.6	3.078	0.9136	1.672	877.1	169.5	144.9	1.47515
120.0	10.0	1015.	-77.51	3.389	0.8571	1.755	727.0	112.4	118.1	1.42679
140.0	10.0	892.5	-40.64	3.673	0.8191	1.966	564.5	75.70	91.70	1.36967
160.0	10.0	716.0	4.057	3.970	0.8137	2.659	379.7	47.96	66.52	1.29022
180.0	10.0	423.3	70.35	4.360	0.8161	3.312	250.9	26.07	42.24	1.16550
200.0	10.0	277.6	119.7	4.621	0.7506	1.894	259.0	20.74	31.77	1.10662
220.0	10.0	219.9	152.6	4.778	0.7152	1.462	278.2	20.24	29.46	1.08384
240.0	10.0	186.7	179.8	4.896	0.6973	1.280	295.8	20.67	29.19	1.07090
260.0	10.0	164.2	204.3	4.995	0.6878	1.183	311.6	21.40	29.67	1.06220
280.0	10.0	147.6	227.3	5.080	0.6830	1.124	326.1	22.25	30.48	1.05578
300.0	10.0	134.6	249.4	5.156	0.6809	1.086	339.4	23.15	31.47	1.05078
320.0	10.0	124.0	270.8	5.225	0.6809	1.060	351.7	24.07	32.56	1.04675
340.0	10.0	115.2	291.9	5.289	0.6823	1.042	363.3	25.00	33.74	1.04339
360.0	10.0	107.8	312.6	5.348	0.6848	1.030	374.2	25.92	34.96	1.04054
380.0	10.0	101.3	333.1	5.404	0.6882	1.022	384.6	26.84	36.20	1.03809
400.0	10.0	95.67	353.5	5.456	0.6923	1.017	394.5	27.74	37.45	1.03595
500.0	10.0	75.32	454.8	5.682	0.7187	1.015	438.7	32.07	43.71	1.02825
600.0	10.0	62.44	557.1	5.869	0.7477	1.031	476.8	36.10	49.86	1.02339
700.0	10.0	53.46	661.2	6.029	0.7743	1.050	511.2	39.87	55.84	1.02002
800.0	10.0	46.79	767.1	6.170	0.7970	1.069	543.1	43.43	61.66	1.01752
900.0	10.0	41.63	874.8	6.297	0.8159	1.085	572.9	46.80	67.32	1.01558
1000.0	10.0	37.51	984.0	6.412	0.8316	1.098	601.2	50.03	72.85	1.01404

TABLE 3. Thermophysical Properties of Argon (Ar)

T/ K	P/ MPa	ρ/ kg m⁻³	H/ kJ kg⁻¹	S/ kJ kg⁻¹ K⁻¹	C_v/ kJ kg⁻¹ K⁻¹	C_p/ kJ kg⁻¹ K⁻¹	u/ m s⁻¹	η/ μPa s	λ/ mW m⁻¹ K⁻¹	D
Saturation										
83.8058	0.06889	1417.	-121.4	1.329	0.5496	1.116	862.4	290.2	133.7	1.51232
83.8058	0.06889	4.055	42.28	3.283	0.3247	0.5550	168.1	6.856	5.359	1.00126
90.0	0.1335	1379.	-114.5	1.409	0.5268	1.121	819.5	240.0	124.6	1.49650
90.0	0.1335	7.436	44.57	3.176	0.3309	0.5757	172.8	7.413	5.835	1.00231

T/ K	P/ MPa	ρ/ kg m⁻³	H/ kJ kg⁻¹	S/ kJ kg⁻¹ K⁻¹	C_v/ kJ kg⁻¹ K⁻¹	C_p/ kJ kg⁻¹ K⁻¹	u/ m s⁻¹	η/ μPa s	λ/ mW m⁻¹ K⁻¹	D
100.0	0.3238	1314.	-103.1	1.528	0.4976	1.154	746.9	181.3	110.2	1.46993
100.0	0.3238	16.86	47.40	3.032	0.3445	0.6269	178.9	8.349	6.689	1.00525
110.0	0.6653	1243.	-91.13	1.639	0.4747	1.218	669.2	140.4	96.41	1.44136
110.0	0.6653	33.29	48.84	2.911	0.3633	0.7122	183.0	9.366	7.732	1.01039
120.0	1.213	1163.	-78.35	1.746	0.4576	1.332	584.2	110.2	83.13	1.40965
120.0	1.213	60.14	48.41	2.802	0.3893	0.8627	185.1	10.54	9.154	1.01884
130.0	2.025	1068.	-64.16	1.854	0.4492	1.564	487.9	85.86	70.43	1.37273
130.0	2.025	103.6	45.30	2.696	0.4275	1.172	184.8	12.03	11.45	1.03259
140.0	3.168	943.7	-47.16	1.971	0.4598	2.225	371.6	63.62	58.06	1.32519
140.0	3.168	178.9	37.47	2.576	0.4940	2.104	181.5	14.32	16.39	1.05677
150.0	4.735	680.4	-17.88	2.159	0.7060	23.58	174.7	36.78	57.63	1.22816
150.0	4.735	394.5	11.52	2.355	0.8218	35.47	157.0	21.18	55.88	1.12827
150.687	4.863	535.6	-4.332	2.248				27.63		1.17684
P = 0.1 MPa (1 bar)										
83.814	0.1	1417.	-121.4	1.329	0.5496	1.116	862.5	290.2	133.7	1.51233
87.178	0.1	1396.	-117.7	1.373	0.5366	1.117	839.2	261.3	128.7	1.50375
87.178	0.1	5.704	43.57	3.223	0.3279	0.5654	170.8	7.157	5.614	1.00178
100.0	0.1	4.915	50.69	3.299	0.3206	0.5470	184.2	8.234	6.450	1.00153
120.0	0.1	4.058	61.49	3.398	0.3161	0.5347	202.8	9.878	7.735	1.00126
140.0	0.1	3.461	72.12	3.480	0.3144	0.5293	219.6	11.48	8.987	1.00108
160.0	0.1	3.020	82.68	3.550	0.3135	0.5264	235.1	13.03	10.21	1.00094
180.0	0.1	2.680	93.19	3.612	0.3131	0.5247	249.6	14.53	11.39	1.00083
200.0	0.1	2.409	103.7	3.667	0.3128	0.5236	263.2	16.00	12.54	1.00075
220.0	0.1	2.189	114.1	3.717	0.3127	0.5229	276.2	17.42	13.66	1.00068
240.0	0.1	2.005	124.6	3.762	0.3125	0.5224	288.5	18.80	14.74	1.00062
260.0	0.1	1.850	135.0	3.804	0.3125	0.5220	300.3	20.15	15.80	1.00058
280.0	0.1	1.717	145.5	3.843	0.3124	0.5217	311.7	21.46	16.83	1.00053
300.0	0.1	1.603	155.9	3.879	0.3124	0.5215	322.7	22.74	17.84	1.00050
320.0	0.1	1.502	166.3	3.913	0.3124	0.5214	333.3	23.99	18.82	1.00047
340.0	0.1	1.414	176.8	3.944	0.3123	0.5212	343.5	25.21	19.77	1.00044
360.0	0.1	1.335	187.2	3.974	0.3123	0.5211	353.5	26.40	20.71	1.00042
380.0	0.1	1.264	197.6	4.002	0.3123	0.5210	363.2	27.56	21.62	1.00039
400.0	0.1	1.201	208.0	4.029	0.3123	0.5209	372.7	28.70	22.52	1.00037
500.0	0.1	0.9608	260.1	4.145	0.3123	0.5207	416.6	34.08	26.73	1.00030
600.0	0.1	0.8006	312.2	4.240	0.3122	0.5206	456.4	39.00	30.57	1.00025
700.0	0.1	0.6862	364.2	4.320	0.3122	0.5205	493.0	43.56	34.13	1.00021
800.0	0.1	0.6004	416.3	4.390	0.3122	0.5205	527.0	47.82	37.46	1.00019
900.0	0.1	0.5337	468.3	4.451	0.3122	0.5204	558.9	51.85	40.60	1.00017
1000.0	0.1	0.4803	520.3	4.506	0.3122	0.5204	589.1	55.69	43.58	1.00015
P = 1.0 MPa (10 bar)										
84.039	1.0	1418.	-120.8	1.330	0.5498	1.113	865.3	291.1	134.0	1.51276
100.0	1.0	1316.	-102.8	1.525	0.4984	1.148	751.6	183.1	110.9	1.47104
116.598	1.0	1191.	-82.82	1.710	0.4627	1.285	614.1	119.7	87.58	1.42088
116.598	1.0	49.55	48.81	2.839	0.3795	0.8007	184.6	10.12	8.608	1.01550
120.0	1.0	47.20	51.45	2.861	0.3682	0.7559	189.3	10.37	8.732	1.01476
140.0	1.0	37.75	65.11	2.967	0.3375	0.6350	212.1	11.87	9.704	1.01179
160.0	1.0	31.93	77.31	3.048	0.3266	0.5907	230.5	13.36	10.80	1.00997
180.0	1.0	27.83	88.89	3.116	0.3214	0.5686	246.8	14.83	11.90	1.00868
200.0	1.0	24.74	100.1	3.176	0.3185	0.5556	261.6	16.26	13.00	1.00772
220.0	1.0	22.31	111.1	3.228	0.3168	0.5473	275.3	17.65	14.07	1.00696
240.0	1.0	20.33	122.0	3.275	0.3157	0.5417	288.3	19.01	15.12	1.00634
260.0	1.0	18.69	132.8	3.319	0.3149	0.5376	300.5	20.34	16.15	1.00583
280.0	1.0	17.30	143.5	3.358	0.3144	0.5346	312.2	21.64	17.16	1.00539
300.0	1.0	16.11	154.2	3.395	0.3140	0.5324	323.4	22.90	18.14	1.00502
320.0	1.0	15.08	164.8	3.429	0.3137	0.5306	334.2	24.14	19.10	1.00470
340.0	1.0	14.17	175.4	3.462	0.3135	0.5292	344.7	25.35	20.04	1.00441
360.0	1.0	13.37	186.0	3.492	0.3133	0.5280	354.7	26.53	20.96	1.00416
380.0	1.0	12.65	196.6	3.520	0.3132	0.5271	364.5	27.68	21.87	1.00394
400.0	1.0	12.01	207.1	3.547	0.3131	0.5263	374.1	28.82	22.75	1.00374

T/ K	P/ MPa	ρ/ kg m⁻³	H/ kJ kg⁻¹	S/ kJ kg⁻¹ K⁻¹	C_v/ kJ kg⁻¹ K⁻¹	C_p/ kJ kg⁻¹ K⁻¹	u/ m s⁻¹	η/ μPa s	λ/ mW m⁻¹ K⁻¹	D
500.0	1.0	9.593	259.6	3.664	0.3127	0.5239	418.3	34.16	26.92	1.00299
600.0	1.0	7.988	311.9	3.760	0.3126	0.5227	458.1	39.06	30.73	1.00249
700.0	1.0	6.846	364.1	3.840	0.3125	0.5220	494.6	43.61	34.27	1.00213
800.0	1.0	5.990	416.3	3.910	0.3124	0.5215	528.6	47.87	37.58	1.00186
900.0	1.0	5.325	468.4	3.971	0.3124	0.5212	560.5	51.90	40.71	1.00166
1000.0	1.0	4.793	520.6	4.026	0.3124	0.5210	590.7	55.72	43.68	1.00149
P = 10.0 MPa (100 bar)										
86.275	10.0	1428.	-114.2	1.333	0.5519	1.085	891.3	300.0	137.1	1.51685
100.0	10.0	1349.	-99.27	1.493	0.5089	1.093	805.9	205.8	118.7	1.48430
120.0	10.0	1222.	-76.82	1.698	0.4650	1.163	674.3	130.0	93.55	1.43296
140.0	10.0	1066.	-52.01	1.888	0.4374	1.349	527.4	85.36	70.81	1.37161
160.0	10.0	833.6	-20.15	2.100	0.4356	1.970	357.8	50.84	49.64	1.28395
180.0	10.0	491.9	26.15	2.373	0.4254	2.091	259.3	27.88	30.35	1.16159
200.0	10.0	337.7	57.55	2.539	0.3815	1.215	267.7	23.43	23.57	1.10910
220.0	10.0	270.9	78.52	2.639	0.3585	0.9257	283.7	22.90	21.77	1.08686
240.0	10.0	231.2	95.59	2.713	0.3457	0.7960	299.0	23.22	21.31	1.07382
260.0	10.0	204.0	110.7	2.774	0.3379	0.7240	313.1	23.87	21.39	1.06491
280.0	10.0	183.6	124.7	2.826	0.3327	0.6789	326.1	24.68	21.73	1.05829
300.0	10.0	167.6	138.0	2.872	0.3290	0.6481	338.4	25.58	22.19	1.05312
320.0	10.0	154.6	150.7	2.913	0.3264	0.6261	350.0	26.53	22.78	1.04892
340.0	10.0	143.7	163.1	2.950	0.3243	0.6095	361.0	27.51	23.43	1.04542
360.0	10.0	134.4	175.1	2.985	0.3228	0.5967	371.6	28.49	24.11	1.04245
380.0	10.0	126.4	186.9	3.017	0.3215	0.5866	381.7	29.49	24.80	1.03989
400.0	10.0	119.4	198.6	3.046	0.3205	0.5784	391.5	30.48	25.50	1.03765
500.0	10.0	94.09	255.0	3.173	0.3174	0.5540	436.1	35.35	29.04	1.02957
600.0	10.0	78.03	309.8	3.272	0.3159	0.5423	475.7	39.97	32.48	1.02448
700.0	10.0	66.80	363.7	3.355	0.3151	0.5357	511.7	44.34	35.76	1.02093
800.0	10.0	58.47	417.0	3.427	0.3145	0.5317	545.1	48.47	38.88	1.01830
900.0	10.0	52.03	470.0	3.489	0.3142	0.5290	576.4	52.40	41.86	1.01628
1000.0	10.0	46.88	522.8	3.545	0.3139	0.5271	606.0	56.16	44.71	1.01466

TABLE 4. Thermophysical Properties of Parahydrogen (H_2 para)

T/ K	P/ MPa	ρ/ kg m⁻³	H/ kJ kg⁻¹	S/ kJ kg⁻¹ K⁻¹	C_v/ kJ kg⁻¹ K⁻¹	C_p/ kJ kg⁻¹ K⁻¹	u/ m s⁻¹	η/ μPa s	λ/ mW m⁻¹ K⁻¹	D
Saturation										
13.8033	0.007041	76.98	-53.74	-3.084	5.131	6.924	1263.	25.77	87.33	1.25267
13.8033	0.007041	0.1255	396.3	29.52	6.226	10.53	305.6	0.6362	10.26	1.00038
14.0	0.007884	76.82	-52.36	-2.986	5.158	6.981	1257.	25.13	88.00	1.25208
14.0	0.007884	0.1388	398.1	29.19	6.236	10.56	307.6	0.6470	10.43	1.00042
16.0	0.02155	75.13	-37.6	-2.013	5.334	7.659	1210.	20.01	93.90	1.24585
16.0	0.02155	0.3377	415.8	26.33	6.323	10.87	325.4	0.7565	12.25	1.00102
18.0	0.04815	73.25	-21.18	-1.068	5.472	8.513	1168.	16.45	98.06	1.23903
18.0	0.04815	0.6880	431.5	24.08	6.387	11.29	340.7	0.8652	14.23	1.00207
20.0	0.09341	71.14	-2.692	-0.1281	5.637	9.569	1119.	13.80	100.5	1.23147
20.0	0.09341	1.244	444.5	22.23	6.450	11.92	353.5	0.9750	16.39	1.00375
22.0	0.1635	68.74	18.32	0.8244	5.811	10.86	1059.	11.75	101.3	1.22301
22.0	0.1635	2.071	454.2	20.64	6.541	12.88	363.7	1.089	18.80	1.00625
24.0	0.2648	66.01	42.37	1.805	5.974	12.52	989.9	10.09	100.5	1.21345
24.0	0.2648	3.255	459.7	19.19	6.681	14.40	371.3	1.210	21.56	1.00984
26.0	0.4038	62.83	70.23	2.832	6.120	14.83	909.3	8.691	98.30	1.20242
26.0	0.4038	4.924	460.0	17.82	6.889	16.90	376.1	1.346	24.85	1.01491
28.0	0.5875	58.98	103.2	3.941	6.266	18.57	813.3	7.447	94.68	1.18922
28.0	0.5875	7.300	453.4	16.45	7.188	21.53	378.2	1.509	29.13	1.02216
30.0	0.8232	53.98	144.2	5.211	6.472	26.65	693.0	6.248	89.56	1.17223
30.0	0.8232	10.87	435.7	14.93	7.625	32.58	377.2	1.728	35.78	1.03313
32.0	1.120	45.90	204.1	6.945	7.010	68.19	522.6	4.844	83.88	1.14520
32.0	1.120	17.49	392.3	12.83	8.336	92.39	372.0	2.126	53.22	1.05368
32.938	1.286	31.32	295.6	9.625				3.143		1.09756

T/ K	P/ MPa	ρ/ kg m^{-3}	H/ kJ kg^{-1}	S/ kJ kg^{-1} K^{-1}	C_v/ kJ kg^{-1} K^{-1}	C_p/ kJ kg^{-1} K^{-1}	u/ m s^{-1}	η/ µPa s	λ/ mW m^{-1} K^{-1}	D
P = 0.1 MPa (1 bar)										
13.834	0.1	77.03	-52.49	-3.081	5.142	6.924	1265.	25.84	87.48	1.25286
20.0	0.1	71.14	-2.629	-0.1296	5.637	9.566	1119.	13.81	100.5	1.23150
20.227	0.1	70.88	-0.4432	-0.021	5.657	9.702	1112.	13.55	100.7	1.23055
20.227	0.1	1.323	445.8	22.04	6.458	12.01	354.8	0.9877	16.65	1.00399
40.0	0.1	0.6157	661.5	29.52	6.218	10.57	521.4	1.988	31.19	1.00186
60.0	0.1	0.4059	873.1	33.81	6.503	10.72	635.7	2.790	44.47	1.00122
80.0	0.1	0.3035	1096.	37.02	7.598	11.77	713.9	3.489	60.71	1.00091
100.0	0.1	0.2425	1348.	39.82	9.277	13.43	772.5	4.120	80.52	1.00073
120.0	0.1	0.2020	1633.	42.41	10.83	14.98	827.4	4.704	101.6	1.00061
140.0	0.1	0.1731	1943.	44.80	11.82	15.96	883.4	5.252	120.7	1.00052
160.0	0.1	0.1514	2267.	46.96	12.21	16.35	940.4	5.772	135.8	1.00046
180.0	0.1	0.1346	2594.	48.89	12.19	16.32	997.7	6.270	147.3	1.00041
200.0	0.1	0.1211	2918.	50.60	11.95	16.08	1054.	6.748	156.3	1.00037
220.0	0.1	0.1101	3237.	52.11	11.63	15.76	1110.	7.210	164.1	1.00033
240.0	0.1	0.1010	3549.	53.47	11.33	15.46	1163.	7.659	171.4	1.00030
260.0	0.1	0.09319	3856.	54.70	11.07	15.20	1214.	8.096	178.5	1.00028
280.0	0.1	0.08654	4157.	55.82	10.87	15.00	1263.	8.522	185.7	1.00026
300.0	0.1	0.08077	4456.	56.85	10.72	14.85	1310.	8.938	193.1	1.00024
320.0	0.1	0.07572	4752.	57.80	10.61	14.74	1355.	9.347	200.6	1.00023
340.0	0.1	0.07127	5045.	58.69	10.53	14.66	1398.	9.747	208.2	1.00022
360.0	0.1	0.06731	5338.	59.53	10.48	14.61	1439.	10.14	216.0	1.00020
380.0	0.1	0.06377	5630.	60.32	10.45	14.57	1479.	10.53	223.8	1.00019
400.0	0.1	0.06058	5921.	61.06	10.43	14.55	1518.	10.91	231.6	1.00018
500.0	0.1	0.04847	7374.	64.31	10.40	14.53	1698.	12.74	270.7	1.00015
600.0	0.1	0.04040	8828.	66.96	10.42	14.55	1859.	14.47	308.9	1.00012
700.0	0.1	0.03463	10290.	69.20	10.48	14.60	2006.	16.12	346.2	1.00011
800.0	0.1	0.03030	11750.	71.16	10.57	14.70	2142.	17.70	383.4	1.00009
900.0	0.1	0.02693	13230.	72.90	10.71	14.83	2268.	19.24	421.2	1.00008
1000.0	0.1	0.02424	14720.	74.47	10.87	14.99	2386.	20.73	459.7	1.00007
P = 1.0 MPa (10 bar)										
14.129	1.0	77.54	-40.46	-3.053	5.218	6.918	1282.	26.49	88.92	1.25464
20.0	1.0	72.29	6.046	-0.3233	5.634	9.187	1164.	14.78	102.4	1.23552
31.244	1.0	49.65	177.4	6.181	6.718	40.77	596.5	5.437	85.64	1.15769
31.244	1.0	14.31	414.3	13.76	8.017	52.90	374.7	1.933	43.58	1.04376
40.0	1.0	7.270	590.6	18.85	6.455	14.12	497.2	2.159	37.00	1.02207
60.0	1.0	4.228	838.4	23.90	6.588	11.70	634.2	2.896	47.58	1.01279
80.0	1.0	3.072	1075.	27.31	7.645	12.25	717.9	3.562	62.89	1.00929
100.0	1.0	2.429	1335.	30.19	9.307	13.72	778.4	4.173	82.19	1.00734
120.0	1.0	2.015	1624.	32.83	10.85	15.17	834.2	4.744	102.9	1.00609
140.0	1.0	1.723	1938.	35.24	11.84	16.09	890.7	5.283	121.8	1.00520
160.0	1.0	1.506	2264.	37.42	12.23	16.45	948.1	5.796	136.7	1.00455
180.0	1.0	1.338	2593.	39.36	12.20	16.40	1006.	6.289	148.1	1.00404
200.0	1.0	1.204	2918.	41.07	11.96	16.14	1062.	6.763	157.0	1.00364
220.0	1.0	1.095	3238.	42.60	11.64	15.81	1117.	7.223	164.7	1.00331
240.0	1.0	1.004	3551.	43.96	11.34	15.50	1171.	7.669	171.9	1.00303
260.0	1.0	0.9267	3858.	45.19	11.08	15.24	1222.	8.104	179.0	1.00280
280.0	1.0	0.8607	4161.	46.31	10.88	15.03	1271.	8.529	186.2	1.00260
300.0	1.0	0.8035	4460.	47.34	10.73	14.87	1317.	8.944	193.5	1.00243
320.0	1.0	0.7534	4756.	48.30	10.62	14.76	1362.	9.351	201.0	1.00228
340.0	1.0	0.7093	5050.	49.19	10.54	14.68	1405.	9.751	208.6	1.00215
360.0	1.0	0.6700	5343.	50.03	10.49	14.62	1446.	10.14	216.3	1.00203
380.0	1.0	0.6349	5635.	50.82	10.45	14.59	1486.	10.53	224.0	1.00192
400.0	1.0	0.6033	5927.	51.56	10.43	14.56	1525.	10.91	231.8	1.00183
500.0	1.0	0.4830	7381.	54.81	10.41	14.53	1704.	12.74	270.9	1.00146
600.0	1.0	0.4027	8835.	57.46	10.43	14.55	1865.	14.47	309.0	1.00122
700.0	1.0	0.3454	10290.	59.71	10.48	14.61	2011.	16.12	346.2	1.00105
800.0	1.0	0.3023	11760.	61.66	10.58	14.70	2147.	17.70	383.4	1.00092
900.0	1.0	0.2688	13230.	63.40	10.71	14.83	2272.	19.23	421.1	1.00082

T/ K	P/ MPa	ρ/ kg m⁻³	H/ kJ kg⁻¹	S/ kJ kg⁻¹ K⁻¹	C_v/ kJ kg⁻¹ K⁻¹	C_p/ kJ kg⁻¹ K⁻¹	u/ m s⁻¹	η/ μPa s	λ/ mW m⁻¹ K⁻¹	D
1000.0	1.0	0.2420	14720.	64.97	10.87	15.00	2390.	20.72	459.7	1.00074
P = 10.0 MPa (100 bar)										
16.806	10.0	81.86	76.24	-2.8	5.215	6.676	1498.	31.52	106.0	1.26970
20.0	10.0	79.91	98.92	-1.567	5.529	7.566	1467.	23.62	118.1	1.26244
40.0	10.0	63.19	310.5	5.532	6.665	13.45	1154.	8.965	120.2	1.20345
60.0	10.0	42.86	617.0	11.71	6.960	16.19	923.2	5.753	94.88	1.13508
80.0	10.0	29.87	933.6	16.27	7.952	15.47	884.9	5.061	93.80	1.09289
100.0	10.0	23.01	1244.	19.73	9.546	15.76	904.3	5.132	105.1	1.07104
120.0	10.0	18.87	1567.	22.67	11.05	16.56	942.4	5.424	121.1	1.05803
140.0	10.0	16.08	1905.	25.27	12.00	17.11	990.1	5.795	136.8	1.04932
160.0	10.0	14.06	2249.	27.57	12.37	17.22	1042.	6.197	149.5	1.04303
180.0	10.0	12.51	2592.	29.59	12.32	17.01	1096.	6.611	159.1	1.03824
200.0	10.0	11.28	2928.	31.36	12.07	16.63	1150.	7.027	166.7	1.03445
220.0	10.0	10.28	3257.	32.93	11.74	16.22	1203.	7.442	173.3	1.03138
240.0	10.0	9.453	3577.	34.32	11.43	15.84	1254.	7.853	179.6	1.02883
260.0	10.0	8.751	3891.	35.58	11.17	15.52	1303.	8.260	185.9	1.02667
280.0	10.0	8.149	4199.	36.72	10.96	15.27	1350.	8.661	192.4	1.02483
300.0	10.0	7.626	4502.	37.77	10.80	15.08	1394.	9.058	199.2	1.02323
320.0	10.0	7.167	4802.	38.73	10.68	14.94	1437.	9.449	206.1	1.02183
340.0	10.0	6.762	5100.	39.64	10.60	14.84	1478.	9.835	213.3	1.02059
360.0	10.0	6.401	5396.	40.48	10.55	14.77	1518.	10.22	220.6	1.01948
380.0	10.0	6.076	5691.	41.28	10.51	14.71	1556.	10.59	228.0	1.01850
400.0	10.0	5.784	5985.	42.03	10.49	14.68	1593.	10.96	235.5	1.01761
500.0	10.0	4.665	7447.	45.30	10.45	14.60	1764.	12.76	273.3	1.01420
600.0	10.0	3.911	8907.	47.96	10.46	14.59	1919.	14.47	310.6	1.01191
700.0	10.0	3.368	10370.	50.21	10.51	14.63	2062.	16.11	347.3	1.01026
800.0	10.0	2.957	11830.	52.17	10.60	14.72	2193.	17.69	384.0	1.00902
900.0	10.0	2.636	13310.	53.91	10.73	14.84	2315.	19.22	421.4	1.00805
1000.0	10.0	2.378	14800.	55.48	10.89	15.00	2430.	20.70	459.6	1.00727

TABLE 5. Thermophysical Properties of Helium (^4He)

T/ K	P/ MPa	ρ/ kg m⁻³	H/ kJ kg⁻¹	S/ kJ kg⁻¹ K⁻¹	C_v/ kJ kg⁻¹ K⁻¹	C_p/ kJ kg⁻¹ K⁻¹	u/ m s⁻¹	η/ μPa s	λ/ mW m⁻¹ K⁻¹	D
Saturation										
2.1768	0.005039	146.0	-6.859	-1.926						
2.1768	0.005039	1.174	15.86	8.513	3.106	5.447	84.38	0.5376	3.978	1.00046
2.2	0.005332	146.0	-6.786	-1.893	3.003	3.016	220.9	3.595	13.61	1.07802
2.2	0.005332	1.232	15.96	8.446	3.106	5.458	84.74	0.5449	4.039	1.00048
2.4	0.008351	145.4	-6.237	-1.663	2.290	2.414	224.0	3.714	14.49	1.07592
2.4	0.008351	1.801	16.76	7.920	3.106	5.570	87.67	0.6072	4.534	1.00070
2.6	0.01238	144.3	-5.743	-1.476	2.062	2.325	225.7	3.746	15.25	1.07379
2.6	0.01238	2.516	17.51	7.467	3.108	5.709	90.28	0.6695	4.994	1.00098
2.8	0.01756	142.9	-5.239	-1.303	2.013	2.424	225.0	3.729	15.93	1.07171
2.8	0.01756	3.394	18.19	7.065	3.113	5.881	92.59	0.7323	5.443	1.00132
3.0	0.02406	141.2	-4.699	-1.133	2.034	2.608	222.1	3.682	16.53	1.06966
3.0	0.02406	4.454	18.81	6.703	3.120	6.093	94.60	0.7962	5.894	1.00174
3.2	0.03202	139.3	-4.112	-0.9614	2.083	2.843	217.7	3.618	17.06	1.06764
3.2	0.03202	5.719	19.35	6.370	3.129	6.355	96.33	0.8618	6.351	1.00224
3.4	0.04160	137.1	-3.467	-0.7871	2.143	3.123	212.2	3.542	17.52	1.06563
3.4	0.04160	7.218	19.81	6.058	3.140	6.682	97.77	0.9295	6.820	1.00283
3.6	0.05294	134.6	-2.758	-0.6082	2.206	3.457	205.6	3.458	17.91	1.06360
3.6	0.05294	8.986	20.18	5.762	3.155	7.098	98.93	1.000	7.308	1.00353
3.8	0.06619	131.9	-1.973	-0.4231	2.271	3.865	198.2	3.367	18.22	1.06151
3.8	0.06619	11.07	20.44	5.475	3.172	7.639	99.80	1.074	7.818	1.00436
4.0	0.08151	128.7	-1.101	-0.2297	2.336	4.383	189.6	3.271	18.45	1.05932
4.0	0.08151	13.55	20.58	5.191	3.194	8.368	100.4	1.152	8.365	1.00535
4.2	0.09908	125.1	-0.1244	-0.0253	2.404	5.080	179.9	3.168	18.61	1.05699
4.2	0.09908	16.51	20.58	4.904	3.221	9.404	100.6	1.236	8.972	1.00653

T/ K	P/ MPa	ρ/ kg m^{-3}	H/ kJ kg^{-1}	S/ kJ kg^{-1} K^{-1}	C_v/ kJ kg^{-1} K^{-1}	C_p/ kJ kg^{-1} K^{-1}	u/ m s^{-1}	η/ µPa s	λ/ mW m^{-1} K^{-1}	D
4.4	0.1191	120.9	0.9848	0.1948	2.478	6.095	168.6	3.058	18.70	1.05443
4.4	0.1191	20.12	20.38	4.604	3.257	10.99	100.6	1.328	9.684	1.00799
4.6	0.1417	115.8	2.272	0.4383	2.567	7.763	155.2	2.935	18.76	1.05151
4.6	0.1417	24.67	19.93	4.277	3.310	13.69	100.2	1.430	10.59	1.00984
4.8	0.1673	109.3	3.827	0.7205	2.689	11.13	139.0	2.793	18.85	1.04799
4.8	0.1673	30.71	19.09	3.901	3.389	19.22	99.48	1.550	11.87	1.01232
5.0	0.1962	99.84	5.880	1.083	2.928	21.97	117.7	2.611	19.13	1.04319
5.0	0.1962	39.71	17.54	3.414	3.620	37.20	97.94	1.704	14.11	1.01606
5.1953	0.2283	69.58	11.54	2.116				2.128		1.02901
P = 0.1 MPa (1 bar)										
3.0	0.1	143.1	-4.272	-1.168	2.002	2.510	230.6	3.837	16.81	1.07081
4.0	0.1	129.7	-1.049	-0.2523	2.328	4.254	193.8	3.317	18.59	1.05983
4.210	0.1	124.9	-0.0733	-0.0149	2.407	5.120	179.4	3.163	18.61	1.05687
4.210	0.1	16.67	20.57	4.889	3.222	9.466	100.6	1.240	9.004	1.00660
5.0	0.1	11.76	26.63	6.214	3.159	6.740	119.6	1.389	10.21	1.00461
6.0	0.1	9.015	32.93	7.365	3.133	6.003	136.8	1.579	11.74	1.00352
7.0	0.1	7.421	38.77	8.266	3.123	5.714	150.8	1.761	13.16	1.00289
8.0	0.1	6.345	44.40	9.018	3.120	5.563	163.1	1.935	14.49	1.00247
9.0	0.1	5.558	49.92	9.667	3.118	5.472	174.3	2.100	15.73	1.00216
10.0	0.1	4.953	55.36	10.24	3.117	5.412	184.6	2.258	16.89	1.00193
12.0	0.1	4.078	66.10	11.22	3.117	5.340	203.3	2.557	19.03	1.00158
14.0	0.1	3.473	76.74	12.04	3.116	5.299	220.2	2.835	20.98	1.00135
16.0	0.1	3.027	87.31	12.75	3.117	5.273	235.8	3.097	22.81	1.00117
18.0	0.1	2.684	97.84	13.37	3.117	5.256	250.3	3.345	24.54	1.00104
20.0	0.1	2.411	108.3	13.92	3.117	5.244	264.0	3.582	26.20	1.00094
25.0	0.1	1.925	134.5	15.09	3.117	5.226	295.2	4.131	30.09	1.00075
30.0	0.1	1.602	160.6	16.04	3.117	5.216	323.4	4.634	33.72	1.00062
35.0	0.1	1.373	186.7	16.84	3.117	5.210	349.3	5.101	37.15	1.00053
40.0	0.1	1.201	212.7	17.54	3.117	5.206	373.3	5.542	40.44	1.00047
60.0	0.1	0.8008	316.7	19.65	3.117	5.199	456.8	7.117	52.55	1.00031
80.0	0.1	0.6008	420.7	21.14	3.117	5.196	527.2	8.503	63.52	1.00023
100.0	0.1	0.4807	524.6	22.30	3.117	5.195	589.3	9.778	73.71	1.00019
120.0	0.1	0.4007	628.5	23.25	3.117	5.194	645.3	10.79	83.33	1.00016
140.0	0.1	0.3435	732.4	24.05	3.116	5.194	696.9	11.94	92.50	1.00013
160.0	0.1	0.3006	836.2	24.74	3.116	5.194	744.9	13.05	101.3	1.00012
180.0	0.1	0.2672	940.1	25.35	3.116	5.194	790.0	14.11	109.8	1.00010
200.0	0.1	0.2405	1044.	25.90	3.116	5.193	832.7	15.14	118.0	1.00009
220.0	0.1	0.2187	1148.	26.40	3.116	5.193	873.3	16.14	126.0	1.00008
240.0	0.1	0.2005	1252.	26.85	3.116	5.193	912.1	17.12	133.7	1.00008
260.0	0.1	0.1851	1356.	27.26	3.116	5.193	949.3	18.08	141.3	1.00007
280.0	0.1	0.1718	1459.	27.65	3.116	5.193	985.0	19.01	148.7	1.00007
300.0	0.1	0.1604	1563.	28.01	3.116	5.193	1020.	19.93	156.0	1.00006
320.0	0.1	0.1504	1667.	28.34	3.116	5.193	1053.	20.83	163.1	1.00006
340.0	0.1	0.1415	1771.	28.66	3.116	5.193	1085.	21.72	170.1	1.00005
360.0	0.1	0.1337	1875.	28.95	3.116	5.193	1117.	22.59	177.0	1.00005
380.0	0.1	0.1266	1979.	29.24	3.116	5.193	1147.	23.45	183.7	1.00005
400.0	0.1	0.1203	2083.	29.50	3.116	5.193	1177.	24.29	190.4	1.00005
500.0	0.1	0.09625	2602.	30.66	3.116	5.193	1316.	28.36	222.3	1.00004
600.0	0.1	0.08022	3121.	31.61	3.116	5.193	1442.	32.22	252.4	1.00003
700.0	0.1	0.06876	3641.	32.41	3.116	5.193	1557.	35.89	281.1	1.00003
800.0	0.1	0.06017	4160.	33.10	3.116	5.193	1664.	39.43	308.5	1.00002
900.0	0.1	0.05348	4679.	33.71	3.116	5.193	1765.	42.85	335.0	1.00002
1000.0	0.1	0.04813	5198.	34.26	3.116	5.193	1861.	46.16	360.6	1.00002
P = 1.0 MPa (10 bar)										
3.0	1.0	158.2	0.9125	-1.423	1.738	1.959	296.7	5.526	19.19	1.08031
4.0	1.0	152.0	3.252	-0.7557	2.122	2.732	285.4	4.922	22.62	1.07235
5.0	1.0	143.1	6.427	-0.0509	2.401	3.659	267.7	4.367	24.59	1.06514
6.0	1.0	130.7	10.69	0.7226	2.612	4.948	244.1	3.893	25.18	1.05745
7.0	1.0	113.7	16.51	1.617	2.776	6.776	218.2	3.494	24.57	1.04846

T/ K	P/ MPa	ρ/ kg m^{-3}	H/ kJ kg^{-1}	S/ kJ kg^{-1} K^{-1}	C_v/ kJ kg^{-1} K^{-1}	C_p/ kJ kg^{-1} K^{-1}	u/ m s^{-1}	η/ μPa s	λ/ mW m^{-1} K^{-1}	D
8.0	1.0	92.91	24.21	2.642	2.905	8.479	198.5	3.206	23.41	1.03855
9.0	1.0	73.97	32.85	3.659	2.986	8.442	192.8	3.072	22.78	1.03008
10.0	1.0	61.06	40.85	4.503	3.033	7.599	198.0	3.062	22.85	1.02452
12.0	1.0	45.91	55.05	5.800	3.086	6.734	213.6	3.177	23.79	1.01820
14.0	1.0	37.15	68.04	6.802	3.109	6.290	230.0	3.350	25.03	1.01463
16.0	1.0	31.41	80.32	7.622	3.118	6.006	245.8	3.542	26.36	1.01232
18.0	1.0	27.33	92.13	8.318	3.122	5.819	260.7	3.739	27.71	1.01069
20.0	1.0	24.25	103.6	8.924	3.124	5.690	274.7	3.937	29.08	1.00947
25.0	1.0	19.05	131.6	10.17	3.127	5.503	306.5	4.420	32.44	1.00742
30.0	1.0	15.76	158.8	11.17	3.128	5.406	334.7	4.881	35.72	1.00613
35.0	1.0	13.47	185.7	11.99	3.128	5.349	360.4	5.321	38.92	1.00524
40.0	1.0	11.78	212.3	12.71	3.128	5.313	384.2	5.741	42.02	1.00457
60.0	1.0	7.867	317.8	14.84	3.126	5.246	466.5	7.270	53.72	1.00305
80.0	1.0	5.919	422.4	16.35	3.124	5.222	535.8	8.633	64.51	1.00230
100.0	1.0	4.747	526.7	17.51	3.123	5.211	597.0	9.892	74.60	1.00184
120.0	1.0	3.964	630.9	18.46	3.122	5.205	652.4	10.90	84.16	1.00154
140.0	1.0	3.403	734.9	19.27	3.121	5.201	703.4	12.04	93.29	1.00132
160.0	1.0	2.981	838.9	19.96	3.121	5.199	750.9	13.13	102.1	1.00116
180.0	1.0	2.653	942.9	20.57	3.120	5.197	795.6	14.19	110.5	1.00103
200.0	1.0	2.390	1047.	21.12	3.120	5.196	837.9	15.21	118.7	1.00093
220.0	1.0	2.174	1151.	21.61	3.119	5.195	878.2	16.20	126.7	1.00084
240.0	1.0	1.994	1255.	22.07	3.119	5.195	916.7	17.17	134.4	1.00077
260.0	1.0	1.841	1359.	22.48	3.119	5.194	953.7	18.12	142.0	1.00071
280.0	1.0	1.711	1462.	22.87	3.119	5.194	989.2	19.05	149.4	1.00066
300.0	1.0	1.597	1566.	23.23	3.118	5.194	1024.	19.96	156.6	1.00062
320.0	1.0	1.498	1670.	23.56	3.118	5.193	1057.	20.86	163.8	1.00058
340.0	1.0	1.410	1774.	23.88	3.118	5.193	1089.	21.74	170.7	1.00055
360.0	1.0	1.332	1878.	24.17	3.118	5.193	1120.	22.61	177.6	1.00052
380.0	1.0	1.262	1982.	24.45	3.118	5.193	1151.	23.47	184.4	1.00049
400.0	1.0	1.199	2086.	24.72	3.118	5.193	1180.	24.32	191.0	1.00047
500.0	1.0	0.9603	2605.	25.88	3.117	5.193	1319.	28.38	222.9	1.00037
600.0	1.0	0.8006	3124.	26.82	3.117	5.193	1444.	32.23	253.0	1.00031
700.0	1.0	0.6865	3643.	27.63	3.117	5.193	1559.	35.91	281.6	1.00027
800.0	1.0	0.6008	4163.	28.32	3.117	5.193	1666.	39.44	309.1	1.00023
900.0	1.0	0.5342	4682.	28.93	3.117	5.193	1767.	42.86	335.5	1.00021
1000.0	1.0	0.4808	5201.	29.48	3.116	5.193	1863.	46.17	361.1	1.00019
P = 10.0 MPa (100 bar)										
3.467	10.0	209.7	47.14	-1.925	1.269	1.379	558.8	29.20	34.32	1.11188
4.0	10.0	208.5	47.93	-1.714	1.442	1.585	559.6	24.73	38.68	1.10745
5.0	10.0	206.0	49.71	-1.318	1.751	1.980	559.4	19.03	45.55	1.10126
6.0	10.0	203.2	51.89	-0.9228	2.023	2.364	557.7	15.51	50.87	1.09660
7.0	10.0	200.1	54.43	-0.5306	2.256	2.729	554.9	13.19	54.88	1.09277
8.0	10.0	196.7	57.34	-0.1437	2.451	3.069	551.3	11.58	57.80	1.08942
9.0	10.0	193.1	60.56	0.2362	2.609	3.383	546.9	10.43	59.84	1.08638
10.0	10.0	189.3	64.09	0.6077	2.738	3.670	542.1	9.576	61.17	1.08354
12.0	10.0	181.2	71.95	1.323	2.924	4.173	531.9	8.435	62.25	1.07830
14.0	10.0	172.8	80.73	1.998	3.043	4.591	521.9	7.751	61.99	1.07344
16.0	10.0	164.2	90.26	2.634	3.119	4.931	512.9	7.333	61.06	1.06889
18.0	10.0	155.7	100.4	3.231	3.168	5.196	505.6	7.082	59.93	1.06464
20.0	10.0	147.5	111.0	3.789	3.199	5.391	500.2	6.942	58.86	1.06069
25.0	10.0	129.0	138.7	5.025	3.234	5.644	494.6	6.870	57.28	1.05221
30.0	10.0	113.7	167.1	6.061	3.243	5.702	497.5	7.002	57.27	1.04554
35.0	10.0	101.3	195.6	6.939	3.243	5.685	505.5	7.222	58.29	1.04031
40.0	10.0	91.32	223.9	7.695	3.239	5.642	516.3	7.483	59.90	1.03614
60.0	10.0	65.49	334.9	9.948	3.217	5.470	569.2	8.641	68.16	1.02565
80.0	10.0	51.21	443.2	11.51	3.199	5.370	622.8	9.805	76.86	1.01998
100.0	10.0	42.12	550.0	12.70	3.185	5.312	673.5	10.93	85.44	1.01640
120.0	10.0	35.80	655.9	13.66	3.175	5.276	721.3	11.83	93.83	1.01392
140.0	10.0	31.15	761.1	14.47	3.167	5.252	766.4	12.88	102.0	1.01211
160.0	10.0	27.58	866.0	15.18	3.161	5.236	809.2	13.89	110.1	1.01071

T/ K	P/ MPa	ρ/ kg m⁻³	H/ kJ kg⁻¹	S/ kJ kg⁻¹ K⁻¹	C_v/ kJ kg⁻¹ K⁻¹	C_p/ kJ kg⁻¹ K⁻¹	u/ m s⁻¹	η/ μPa s	λ/ mW m⁻¹ K⁻¹	D
180.0	10.0	24.75	970.6	15.79	3.156	5.224	849.9	14.87	118.0	1.00961
200.0	10.0	22.44	1075.	16.34	3.152	5.216	888.8	15.82	125.7	1.00871
220.0	10.0	20.54	1179.	16.84	3.149	5.210	926.2	16.75	133.3	1.00797
240.0	10.0	18.93	1283.	17.29	3.146	5.205	962.1	17.65	140.7	1.00734
260.0	10.0	17.55	1387.	17.71	3.144	5.201	996.8	18.54	148.0	1.00681
280.0	10.0	16.36	1491.	18.09	3.142	5.199	1030.	19.40	155.2	1.00635
300.0	10.0	15.33	1595.	18.45	3.140	5.196	1063.	20.25	162.3	1.00595
320.0	10.0	14.41	1699.	18.79	3.138	5.195	1095.	21.14	169.2	1.00559
340.0	10.0	13.60	1803.	19.10	3.137	5.193	1125.	22.01	176.0	1.00528
360.0	10.0	12.88	1907.	19.40	3.135	5.192	1155.	22.86	182.8	1.00500
380.0	10.0	12.23	2011.	19.68	3.134	5.191	1184.	23.71	189.4	1.00474
400.0	10.0	11.64	2115.	19.95	3.133	5.190	1213.	24.54	196.0	1.00451
500.0	10.0	9.381	2634.	21.10	3.129	5.188	1347.	28.56	227.5	1.00364
600.0	10.0	7.857	3152.	22.05	3.127	5.188	1468.	32.38	257.4	1.00305
700.0	10.0	6.758	3671.	22.85	3.125	5.188	1581.	36.04	285.9	1.00262
800.0	10.0	5.928	4190.	23.54	3.124	5.188	1686.	39.56	313.2	1.00230
900.0	10.0	5.280	4709.	24.15	3.123	5.188	1785.	42.96	339.6	1.00205
1000.0	10.0	4.759	5228.	24.70	3.122	5.188	1879.	46.26	365.1	1.00185

TABLE 6. Thermophysical Properties of Methane (CH_4)

T/ K	P/ MPa	ρ/ kg m⁻³	H/ kJ kg⁻¹	S/ kJ kg⁻¹ K⁻¹	C_v/ kJ kg⁻¹ K⁻¹	C_p/ kJ kg⁻¹ K⁻¹	u/ m s⁻¹	η/ μPa s	λ/ mW m⁻¹ K⁻¹	D
Saturation										
90.6941	0.01170	451.5	-71.82	-0.7099	2.168	3.368	1539.	193.6	211.2	1.67721
90.6941	0.01170	0.2507	472.4	5.291	1.574	2.110	249.1	3.598	8.798	1.00031
100.0	0.03438	438.9	-40.27	-0.3793	2.114	3.408	1452.	151.0	199.6	1.65478
100.0	0.03438	0.6746	490.2	4.925	1.589	2.146	260.1	3.914	9.899	1.00082
110.0	0.08813	424.8	-5.813	-0.0522	2.064	3.469	1355.	120.9	186.0	1.62995
110.0	0.08813	1.598	508.0	4.619	1.611	2.205	270.0	4.265	11.19	1.00196
120.0	0.1914	409.9	29.41	0.2521	2.020	3.549	1253.	98.68	172.0	1.60408
120.0	0.1914	3.262	524.0	4.374	1.639	2.293	277.8	4.625	12.62	1.00399
130.0	0.3673	394.0	65.63	0.5385	1.980	3.658	1148.	81.30	157.7	1.57683
130.0	0.3673	5.980	537.7	4.170	1.674	2.421	283.1	4.997	14.23	1.00733
140.0	0.6412	376.9	103.2	0.8116	1.945	3.813	1038.	67.35	143.4	1.54771
140.0	0.6412	10.15	548.3	3.991	1.717	2.611	285.9	5.390	16.10	1.01248
150.0	1.040	357.9	142.6	1.076	1.919	4.047	920.8	55.98	129.2	1.51599
150.0	1.040	16.33	555.2	3.827	1.773	2.908	286.0	5.824	18.34	1.02013
160.0	1.592	336.3	184.8	1.338	1.904	4.435	795.4	46.54	114.9	1.48045
160.0	1.592	25.38	557.1	3.664	1.847	3.419	283.0	6.336	21.20	1.03144
170.0	2.328	310.5	231.2	1.605	1.910	5.187	657.5	38.43	100.5	1.43870
170.0	2.328	38.97	551.5	3.490	1.956	4.459	276.7	7.017	25.31	1.04861
180.0	3.285	276.2	285.9	1.899	1.967	7.292	497.0	30.89	85.48	1.38453
180.0	3.285	61.38	532.8	3.271	2.140	7.574	266.0	8.135	33.48	1.07739
190.0	4.519	200.8	378.3	2.369	2.602	94.01	250.3	20.29	94.15	1.27031
190.0	4.519	125.2	459.0	2.794	2.855	140.8	238.5	12.24	120.5	1.16274
190.564	4.599	162.7	415.6	2.562				15.88		1.21521
***P* = 0.1 MPa (1 bar)**										
90.717	0.1	451.5	-71.6	-0.7097	2.168	3.367	1539.	193.8	211.2	1.67726
100.0	0.1	438.9	-40.17	-0.3798	2.114	3.408	1453.	151.1	199.6	1.65487
111.508	0.1	422.6	-0.5573	-0.005	2.057	3.480	1340.	117.2	183.9	1.62612
111.508	0.1	1.795	510.6	4.579	1.615	2.216	271.3	4.318	11.39	1.00220
120.0	0.1	1.655	529.2	4.740	1.594	2.174	282.8	4.653	12.31	1.00202
140.0	0.1	1.403	572.1	5.071	1.574	2.129	307.6	5.433	14.60	1.00172
160.0	0.1	1.221	614.5	5.354	1.568	2.110	330.0	6.201	16.94	1.00149
180.0	0.1	1.081	656.6	5.602	1.568	2.104	350.7	6.957	19.30	1.00132
200.0	0.1	0.9709	698.7	5.824	1.574	2.106	370.0	7.697	21.65	1.00119
220.0	0.1	0.8812	740.9	6.025	1.587	2.115	388.0	8.420	24.00	1.00108
240.0	0.1	0.8069	783.4	6.209	1.607	2.133	404.9	9.125	26.41	1.00099

T/ K	P/ MPa	ρ/ kg m⁻³	H/ kJ kg⁻¹	S/ kJ kg⁻¹ K⁻¹	C_v/ kJ kg⁻¹ K⁻¹	C_p/ kJ kg⁻¹ K⁻¹	u/ m s⁻¹	η/ μPa s	λ/ mW m⁻¹ K⁻¹	D
260.0	0.1	0.7442	826.3	6.381	1.634	2.159	420.8	9.812	28.90	1.00091
280.0	0.1	0.6906	869.8	6.542	1.670	2.194	435.7	10.48	31.50	1.00085
300.0	0.1	0.6443	914.1	6.695	1.713	2.236	449.7	11.14	34.21	1.00079
320.0	0.1	0.6038	959.3	6.841	1.763	2.285	463.1	11.77	37.06	1.00074
340.0	0.1	0.5681	1006.	6.981	1.819	2.340	475.7	12.39	40.04	1.00070
360.0	0.1	0.5364	1053.	7.117	1.880	2.401	487.8	13.00	43.16	1.00066
380.0	0.1	0.5081	1102.	7.248	1.945	2.466	499.4	13.60	46.40	1.00062
400.0	0.1	0.4826	1152.	7.376	2.013	2.534	510.6	14.18	49.78	1.00059
500.0	0.1	0.3859	1423.	7.981	2.381	2.901	561.9	16.90	68.26	1.00047
600.0	0.1	0.3215	1732.	8.543	2.754	3.273	608.0	19.39	88.74	1.00039
700.0	0.1	0.2756	2077.	9.074	3.110	3.629	650.8	21.70	110.5	1.00034
800.0	0.1	0.2411	2457.	9.581	3.441	3.959	690.9	23.87	133.0	1.00030
900.0	0.1	0.2143	2868.	10.06	3.744	4.263	729.0	25.92	156.0	1.00026
1000.0	0.1	0.1929	3308.	10.53	4.020	4.538	765.2	27.87	173.7	1.00024
P = 1.0 MPa (10 bar)										
90.947	1.0	451.8	-69.37	-0.707	2.169	3.363	1543.	195.8	211.8	1.67772
100.0	1.0	439.6	-38.76	-0.3862	2.116	3.401	1460.	153.3	200.5	1.65604
120.0	1.0	410.8	30.49	0.2447	2.022	3.536	1262.	99.85	172.9	1.60560
140.0	1.0	377.5	103.5	0.8069	1.946	3.798	1044.	67.82	144.0	1.54879
149.139	1.0	359.6	139.2	1.054	1.920	4.023	931.2	56.88	130.4	1.51885
149.139	1.0	15.70	554.8	3.841	1.767	2.876	286.1	5.784	18.13	1.01935
160.0	1.0	13.97	584.2	4.031	1.682	2.588	305.2	6.280	18.76	1.01720
180.0	1.0	11.81	633.6	4.322	1.631	2.380	333.7	7.085	20.64	1.01453
200.0	1.0	10.33	680.2	4.568	1.613	2.289	357.8	7.838	22.74	1.01269
220.0	1.0	9.217	725.5	4.784	1.613	2.248	379.1	8.563	24.91	1.01133
240.0	1.0	8.344	770.3	4.978	1.626	2.234	398.2	9.265	27.21	1.01025
260.0	1.0	7.634	815.0	5.157	1.648	2.238	415.8	9.948	29.62	1.00938
280.0	1.0	7.043	859.9	5.324	1.681	2.258	432.0	10.61	32.15	1.00865
300.0	1.0	6.542	905.4	5.480	1.722	2.289	447.0	11.26	34.81	1.00803
320.0	1.0	6.110	951.5	5.629	1.770	2.330	461.1	11.89	37.62	1.00750
340.0	1.0	5.733	998.6	5.772	1.825	2.379	474.4	12.51	40.56	1.00704
360.0	1.0	5.402	1047.	5.910	1.885	2.434	487.0	13.11	43.64	1.00663
380.0	1.0	5.108	1096.	6.043	1.949	2.495	499.0	13.70	46.86	1.00627
400.0	1.0	4.846	1147.	6.172	2.017	2.559	510.6	14.28	50.21	1.00595
500.0	1.0	3.860	1420.	6.781	2.383	2.915	563.0	16.99	68.60	1.00474
600.0	1.0	3.210	1730.	7.345	2.755	3.282	609.7	19.47	89.02	1.00394
700.0	1.0	2.750	2076.	7.878	3.110	3.635	652.8	21.77	110.7	1.00338
800.0	1.0	2.405	2456.	8.385	3.441	3.964	693.1	23.92	133.2	1.00296
900.0	1.0	2.137	2868.	8.870	3.744	4.266	731.2	25.97	156.2	1.00263
1000.0	1.0	1.924	3308.	9.334	4.020	4.541	767.5	27.92	173.9	1.00237
P = 10.0 MPa (100 bar)										
93.222	10.0	454.5	-47.07	-0.6806	2.176	3.325	1582.	212.8	216.9	1.68223
100.0	10.0	446.0	-24.48	-0.4466	2.139	3.344	1526.	174.7	209.0	1.66701
120.0	10.0	419.8	43.14	0.1696	2.047	3.424	1352.	112.1	183.3	1.62085
140.0	10.0	391.2	112.8	0.7062	1.969	3.552	1171.	78.32	157.0	1.57167
160.0	10.0	358.8	185.8	1.194	1.909	3.777	981.8	56.70	131.4	1.51718
180.0	10.0	319.6	265.3	1.661	1.873	4.226	780.8	41.53	107.0	1.45298
200.0	10.0	266.2	358.9	2.153	1.878	5.304	567.9	29.55	83.93	1.36870
220.0	10.0	187.6	482.7	2.742	1.903	6.713	404.4	19.37	63.31	1.25087
240.0	10.0	128.4	599.6	3.252	1.847	4.853	383.6	14.88	49.77	1.16707
260.0	10.0	101.0	683.8	3.589	1.803	3.730	402.8	13.79	44.88	1.12974
280.0	10.0	85.51	752.9	3.846	1.793	3.242	424.5	13.61	43.84	1.10902
300.0	10.0	75.18	815.1	4.060	1.807	3.002	444.5	13.75	44.40	1.09539
320.0	10.0	67.61	873.8	4.250	1.838	2.878	462.8	14.04	45.82	1.08549
340.0	10.0	61.75	930.6	4.422	1.880	2.817	479.5	14.41	47.77	1.07786
360.0	10.0	57.01	986.7	4.582	1.931	2.795	494.8	14.82	50.10	1.07173
380.0	10.0	53.08	1043.	4.733	1.989	2.798	509.1	15.26	52.73	1.06666
400.0	10.0	49.74	1099.	4.877	2.051	2.819	522.6	15.71	55.59	1.06238
500.0	10.0	38.32	1391.	5.529	2.401	3.055	581.0	18.04	72.48	1.04783

T/ K	P/ MPa	ρ/ kg m^{-3}	H/ kJ kg^{-1}	S/ kJ kg^{-1} K^{-1}	C_v/ kJ kg^{-1} K^{-1}	C_p/ kJ kg^{-1} K^{-1}	u/ m s^{-1}	η/ μPa s	λ/ mW m^{-1} K^{-1}	D
600.0	10.0	31.47	1712.	6.113	2.766	3.370	630.6	20.31	92.09	1.03918
700.0	10.0	26.82	2066.	6.657	3.117	3.696	675.2	22.47	113.3	1.03334
800.0	10.0	23.41	2451.	7.171	3.445	4.008	716.3	24.53	135.4	1.02907
900.0	10.0	20.79	2866.	7.660	3.747	4.300	754.7	26.51	158.1	1.02581
1000.0	10.0	18.72	3310.	8.128	4.022	4.568	791.1	28.40	175.6	1.02323

TABLE 7. Thermophysical Properties of Ethane (C_2H_6)

T/ K	P/ MPa	ρ/ kg m^{-3}	H/ kJ kg^{-1}	S/ kJ kg^{-1} K^{-1}	C_v/ kJ kg^{-1} K^{-1}	C_p/ kJ kg^{-1} K^{-1}	u/ m s^{-1}	η/ μPa s	λ/ mW m^{-1} K^{-1}	D
Saturation										
90.368	0.000001142	651.5	-219.2	-1.655	1.605	2.326	2009.	1292.	255.4	1.94483
90.368	0.000001142	0.00004571	375.6	4.927	0.8916	1.168	180.9	2.854	2.728	1.00000
100.0	0.00001108	640.9	-197	-1.422	1.541	2.283	1938.	890.0	247.7	1.92579
100.0	0.00001108	0.0004007	386.9	4.418	0.9107	1.187	189.9	3.157	3.291	1.00000
120.0	0.0003523	618.9	-151.5	-1.007	1.478	2.280	1794.	489.9	229.8	1.88657
120.0	0.0003523	0.01062	411.0	3.681	0.9528	1.230	206.9	3.784	4.523	1.00001
140.0	0.003814	596.6	-105.6	-0.6532	1.450	2.311	1649.	319.7	210.5	1.84725
140.0	0.003814	0.09880	435.6	3.213	1.003	1.284	222.0	4.407	5.851	1.00011
160.0	0.02141	573.6	-58.95	-0.3418	1.436	2.357	1501.	231.1	190.8	1.80737
160.0	0.02141	0.4890	460.3	2.904	1.048	1.338	235.1	5.025	7.309	1.00054
180.0	0.07864	549.5	-11.13	-0.0608	1.434	2.421	1350.	176.5	171.3	1.76641
180.0	0.07864	1.625	484.2	2.691	1.098	1.409	245.5	5.640	8.955	1.00181
200.0	0.2172	524.0	38.30	0.1981	1.444	2.512	1196.	138.8	152.4	1.72364
200.0	0.2172	4.170	506.2	2.538	1.179	1.537	252.3	6.258	10.87	1.00465
220.0	0.4920	496.3	90.01	0.4419	1.468	2.645	1037.	110.4	134.4	1.67809
220.0	0.4920	9.017	525.5	2.421	1.280	1.720	254.6	6.897	13.15	1.01008
240.0	0.9668	465.3	145.0	0.6767	1.507	2.847	873.3	87.97	117.3	1.62822
240.0	0.9668	17.43	540.9	2.326	1.388	1.976	252.1	7.604	15.99	1.01957
260.0	1.712	429.1	204.9	0.9095	1.566	3.195	700.5	69.29	101.0	1.57121
260.0	1.712	31.58	550.2	2.238	1.516	2.418	243.8	8.484	19.84	1.03567
280.0	2.807	382.7	273.1	1.152	1.654	3.987	512.4	52.71	85.20	1.50037
280.0	2.807	56.37	548.6	2.136	1.696	3.522	228.1	9.850	26.22	1.06440
300.0	4.357	303.5	364.4	1.451	1.912	10.02	274.9	34.70	71.31	1.38458
300.0	4.357	114.5	514.1	1.950	2.089	13.30	200.5	13.39	47.92	1.13418
305.322	4.872	206.2	439.0	1.690				22.63		1.25115
***P* = 0.1 MPa (1 bar)**										
90.384	0.1	651.5	-219	-1.655	1.605	2.326	2009.	1292.	255.5	1.94485
100.0	0.1	641.0	-196.9	-1.422	1.541	2.283	1939.	890.7	247.7	1.92586
120.0	0.1	619.0	-151.4	-1.007	1.478	2.279	1795.	490.2	229.9	1.88665
140.0	0.1	596.6	-105.5	-0.6535	1.450	2.311	1650.	319.9	210.6	1.84735
160.0	0.1	573.6	-58.85	-0.342	1.436	2.357	1502.	231.2	190.8	1.80747
180.0	0.1	549.5	-11.1	-0.0609	1.434	2.421	1351.	176.6	171.3	1.76644
184.325	0.1	544.1	-0.5954	-0.0032	1.435	2.438	1317.	167.3	167.2	1.75734
184.325	0.1	2.030	489.1	2.654	1.113	1.432	247.3	5.773	9.343	1.00226
200.0	0.1	1.856	511.8	2.771	1.149	1.457	257.9	6.261	10.58	1.00207
220.0	0.1	1.676	541.3	2.912	1.200	1.499	270.3	6.879	12.30	1.00187
240.0	0.1	1.529	571.8	3.045	1.260	1.554	281.8	7.490	14.19	1.00170
260.0	0.1	1.407	603.5	3.172	1.328	1.617	292.5	8.095	16.27	1.00157
280.0	0.1	1.303	636.5	3.294	1.401	1.688	302.6	8.695	18.52	1.00145
300.0	0.1	1.214	671.1	3.413	1.479	1.764	312.2	9.288	20.97	1.00135
320.0	0.1	1.137	707.1	3.530	1.560	1.844	321.4	9.876	23.60	1.00127
340.0	0.1	1.069	744.8	3.644	1.644	1.927	330.3	10.46	26.41	1.00119
360.0	0.1	1.009	784.2	3.756	1.729	2.011	338.8	11.03	29.40	1.00112
380.0	0.1	0.9550	825.3	3.867	1.815	2.096	347.1	11.60	32.53	1.00106
400.0	0.1	0.9067	868.1	3.977	1.901	2.181	355.2	12.16	35.82	1.00101
500.0	0.1	0.7242	1107.	4.509	2.315	2.594	393.1	14.83	53.97	1.00081
600.0	0.1	0.6031	1386.	5.015	2.689	2.967	427.6	17.26	73.88	1.00067
700.0	0.1	0.5167	1699.	5.498	3.021	3.298	459.6	19.40	94.25	1.00058

T/ K	P/ MPa	ρ/ kg m^{-3}	H/ kJ kg^{-1}	S/ kJ kg^{-1} K^{-1}	C_v/ kJ kg^{-1} K^{-1}	C_p/ kJ kg^{-1} K^{-1}	u/ m s^{-1}	η/ μPa s	λ/ mW m^{-1} K^{-1}	D
800.0	0.1	0.4520	2044.	5.958	3.314	3.591	489.6	21.20	114.0	1.00050
900.0	0.1	0.4018	2416.	6.396	3.572	3.849	517.9	22.61	132.0	1.00045
1000.0	0.1	0.3616	2813.	6.814	3.799	4.076	544.8	25.15	138.4	1.00040
P = 1.0 MPa (10 bar)										
90.529	1.0	651.7	-217.5	-1.653	1.605	2.324	2012.	1293.	255.7	1.94510
100.0	1.0	641.3	-195.7	-1.424	1.542	2.282	1942.	896.6	248.1	1.92644
120.0	1.0	619.4	-150.2	-1.01	1.480	2.278	1799.	493.4	230.4	1.88737
140.0	1.0	597.2	-104.4	-0.6563	1.452	2.309	1654.	322.0	211.1	1.84822
160.0	1.0	574.3	-57.8	-0.3452	1.437	2.354	1508.	232.7	191.5	1.80855
180.0	1.0	550.4	-10.13	-0.0646	1.435	2.416	1358.	177.9	172.1	1.76780
200.0	1.0	524.9	39.03	0.1943	1.445	2.505	1204.	139.8	153.2	1.72518
220.0	1.0	497.1	90.36	0.4388	1.468	2.637	1044.	111.1	135.0	1.67944
240.0	1.0	465.4	145.0	0.6764	1.507	2.846	873.9	88.02	117.4	1.62834
241.098	1.0	463.5	148.1	0.6894	1.510	2.861	864.0	86.86	116.4	1.62531
241.098	1.0	18.04	541.6	2.321	1.395	1.994	251.8	7.646	16.17	1.02025
260.0	1.0	15.93	577.9	2.466	1.400	1.880	268.9	8.230	17.70	1.01786
280.0	1.0	14.32	615.3	2.605	1.449	1.869	283.8	8.838	19.68	1.01604
300.0	1.0	13.07	652.9	2.735	1.513	1.898	296.9	9.437	21.93	1.01463
320.0	1.0	12.05	691.3	2.859	1.585	1.948	308.7	10.03	24.42	1.01349
340.0	1.0	11.21	730.9	2.979	1.663	2.010	319.6	10.61	27.12	1.01254
360.0	1.0	10.49	771.8	3.095	1.745	2.079	329.9	11.18	30.02	1.01173
380.0	1.0	9.861	814.1	3.210	1.828	2.154	339.5	11.75	33.09	1.01103
400.0	1.0	9.312	857.9	3.322	1.911	2.230	348.8	12.31	36.31	1.01041
500.0	1.0	7.323	1101.	3.862	2.321	2.620	390.2	14.97	54.29	1.00818
600.0	1.0	6.058	1381.	4.372	2.693	2.983	426.6	17.38	74.11	1.00677
700.0	1.0	5.173	1696.	4.857	3.023	3.309	459.7	19.51	94.43	1.00578
800.0	1.0	4.518	2042.	5.318	3.315	3.598	490.3	21.30	114.1	1.00505
900.0	1.0	4.011	2415.	5.757	3.573	3.855	519.0	22.70	132.1	1.00448
1000.0	1.0	3.608	2812.	6.175	3.800	4.081	546.2	25.21	138.6	1.00403
P = 10.0 MPa (100 bar)										
91.964	10.0	653.3	-202.5	-1.64	1.605	2.310	2034.	1302.	258.5	1.94751
100.0	10.0	644.8	-184.1	-1.448	1.553	2.275	1976.	957.5	252.3	1.93215
120.0	10.0	623.6	-138.8	-1.035	1.490	2.266	1839.	525.4	235.3	1.89429
140.0	10.0	602.2	-93.25	-0.6839	1.463	2.290	1701.	342.4	216.8	1.85661
160.0	10.0	580.4	-47.1	-0.3758	1.449	2.326	1564.	248.0	197.9	1.81879
180.0	10.0	558.1	-0.1206	-0.0992	1.447	2.374	1425.	190.8	179.2	1.78049
200.0	10.0	534.8	47.98	0.1542	1.456	2.439	1287.	151.7	161.3	1.74125
220.0	10.0	510.2	97.60	0.3905	1.478	2.527	1147.	122.8	144.3	1.70050
240.0	10.0	483.8	149.3	0.6152	1.511	2.646	1008.	100.1	128.3	1.65744
260.0	10.0	454.6	203.7	0.8331	1.557	2.809	866.9	81.72	113.3	1.61088
280.0	10.0	421.4	262.1	1.049	1.616	3.044	724.4	66.19	99.34	1.55888
300.0	10.0	381.3	326.4	1.271	1.689	3.419	579.3	52.56	86.15	1.49786
320.0	10.0	328.8	400.9	1.511	1.783	4.109	434.8	39.92	73.59	1.42048
340.0	10.0	255.9	493.3	1.791	1.888	5.070	319.9	28.14	62.08	1.31781
360.0	10.0	187.0	591.3	2.071	1.941	4.452	286.8	20.93	53.28	1.22573
380.0	10.0	148.4	671.0	2.287	1.976	3.608	295.2	18.35	49.31	1.17628
400.0	10.0	126.2	738.6	2.460	2.025	3.205	310.4	17.45	48.61	1.14850
500.0	10.0	80.57	1035.	3.123	2.369	2.933	377.8	17.83	59.73	1.09293
600.0	10.0	62.41	1338.	3.675	2.722	3.150	427.4	19.55	77.43	1.07144
700.0	10.0	51.79	1666.	4.180	3.043	3.415	467.8	21.30	96.80	1.05903
800.0	10.0	44.59	2021.	4.653	3.330	3.672	502.9	22.83	116.0	1.05068
900.0	10.0	39.30	2400.	5.100	3.584	3.909	534.3	24.05	133.6	1.04457
1000.0	10.0	35.20	2802.	5.523	3.809	4.122	563.2	26.04	140.8	1.03988

TABLE 8. Thermophysical Properties of Propane (C_3H_8)

T/ K	P/ MPa	ρ/ kg m^{-3}	H/ kJ kg^{-1}	S/ kJ kg^{-1} K^{-1}	C_v/ kJ kg^{-1} K^{-1}	C_p/ kJ kg^{-1} K^{-1}	u/ m s^{-1}	η/ µPa s	λ/ mW m^{-1} K^{-1}	D
Saturation										
85.525	0.0000000001720	733.1	-196.6	-1.396	1.355	1.916	2136.	10960.	207.9	2.08838
85.525	0.0000000001720	0.00000001067	366.3	5.186	0.6907	0.8792	143.3	2.326	1.706	1.00000
100.0	0.00000002527	718.1	-168.8	-1.095	1.341	1.930	2038.	3858.	203.2	2.05908
100.0	0.00000002527	0.000001340	379.4	4.387	0.7475	0.9361	153.7	2.724	2.417	1.00000
120.0	0.000002964	697.8	-129.9	-0.7411	1.335	1.957	1901.	1519.	194.4	2.01930
120.0	0.000002964	0.0001310	398.9	3.666	0.8210	1.010	166.8	3.274	3.500	1.00000
140.0	0.00007899	677.6	-90.51	-0.4372	1.335	1.988	1767.	838.1	183.9	1.98025
140.0	0.00007899	0.002993	419.7	3.207	0.8878	1.076	178.9	3.824	4.699	1.00000
160.0	0.0008502	657.3	-50.38	-0.1693	1.342	2.025	1633.	542.6	172.4	1.94163
160.0	0.0008502	0.02821	441.8	2.907	0.9514	1.141	190.0	4.373	6.013	1.00003
180.0	0.005068	636.6	-9.433	0.0717	1.356	2.070	1499.	383.5	160.2	1.90310
180.0	0.005068	0.1499	464.9	2.707	1.017	1.209	200.1	4.920	7.439	1.00016
200.0	0.02019	615.4	32.53	0.2926	1.383	2.127	1366.	286.8	147.9	1.86434
200.0	0.02019	0.5417	488.6	2.573	1.088	1.287	208.7	5.460	8.971	1.00059
220.0	0.06057	593.4	75.80	0.4984	1.421	2.199	1233.	223.0	135.7	1.82499
220.0	0.06057	1.499	512.7	2.484	1.169	1.381	215.6	5.992	10.61	1.00164
240.0	0.1480	570.4	120.7	0.6932	1.469	2.289	1102.	177.9	124.0	1.78460
240.0	0.1480	3.438	536.6	2.426	1.259	1.494	220.1	6.518	12.38	1.00376
260.0	0.3107	545.8	167.7	0.8801	1.528	2.403	971.7	144.0	112.9	1.74256
260.0	0.3107	6.903	560.1	2.389	1.357	1.630	221.9	7.050	14.32	1.00755
280.0	0.5817	519.2	217.3	1.062	1.596	2.547	840.3	117.3	102.5	1.69800
280.0	0.5817	12.62	582.3	2.365	1.466	1.803	220.4	7.611	16.54	1.01383
300.0	0.9977	489.4	270.2	1.241	1.675	2.740	706.8	95.27	92.86	1.64961
300.0	0.9977	21.63	602.6	2.349	1.588	2.041	214.8	8.245	19.24	1.02381
320.0	1.599	454.9	327.3	1.421	1.764	3.028	569.1	76.41	83.95	1.59504
320.0	1.599	35.74	619.5	2.334	1.729	2.416	204.4	9.049	22.79	1.03957
340.0	2.431	411.8	390.9	1.608	1.872	3.585	422.5	59.34	75.58	1.52907
340.0	2.431	58.88	629.8	2.311	1.898	3.197	187.4	10.26	28.25	1.06579
360.0	3.555	345.6	468.2	1.820	2.049	5.984	251.0	41.61	67.78	1.43242
360.0	3.555	105.4	622.4	2.249	2.183	7.111	161.1	12.85	41.69	1.11995
369.890	4.251	220.5	555.2	2.052				25.70		1.26277
***P* = 0.1 MPa (1 bar)**										
85.534	0.1	733.1	-196.5	-1.396	1.355	1.916	2137.	10960.	207.9	2.08840
100.0	0.1	718.2	-168.7	-1.096	1.341	1.930	2038.	3861.	203.2	2.05912
120.0	0.1	697.8	-129.8	-0.7413	1.335	1.957	1902.	1520.	194.5	2.01936
140.0	0.1	677.6	-90.39	-0.4374	1.335	1.988	1767.	838.8	184.0	1.98032
160.0	0.1	657.3	-50.27	-0.1695	1.342	2.025	1633.	543.0	172.4	1.94172
180.0	0.1	636.7	-9.328	0.0715	1.357	2.070	1500.	383.8	160.2	1.90320
200.0	0.1	615.5	32.62	0.2924	1.383	2.127	1366.	287.0	147.9	1.86444
220.0	0.1	593.5	75.84	0.4983	1.421	2.198	1234.	223.1	135.8	1.82505
230.738	0.1	581.2	99.69	0.6042	1.446	2.245	1163.	197.0	129.4	1.80347
230.738	0.1	2.387	525.6	2.450	1.216	1.439	218.3	6.275	11.54	1.00261
240.0	0.1	2.284	539.0	2.507	1.249	1.467	222.9	6.532	12.40	1.00249
260.0	0.1	2.092	569.0	2.627	1.324	1.534	232.3	7.085	14.32	1.00228
280.0	0.1	1.932	600.5	2.744	1.405	1.610	241.1	7.631	16.36	1.00211
300.0	0.1	1.796	633.5	2.857	1.490	1.692	249.4	8.173	18.51	1.00196
320.0	0.1	1.679	668.2	2.969	1.578	1.777	257.3	8.708	20.78	1.00183
340.0	0.1	1.576	704.6	3.080	1.668	1.865	264.9	9.239	23.16	1.00172
360.0	0.1	1.486	742.8	3.189	1.758	1.954	272.3	9.763	25.66	1.00162
380.0	0.1	1.406	782.8	3.297	1.848	2.042	279.4	10.28	28.27	1.00153
400.0	0.1	1.334	824.5	3.404	1.937	2.130	286.2	10.80	30.99	1.00145
500.0	0.1	1.064	1059.	3.925	2.356	2.547	318.3	13.26	46.36	1.00116
600.0	0.1	0.8852	1332.	4.422	2.722	2.912	347.4	15.56	64.63	1.00096
700.0	0.1	0.7581	1640.	4.896	3.038	3.228	374.2	17.66	85.81	1.00082
800.0	0.1	0.6631	1976.	5.345	3.313	3.502	399.3	19.56	109.9	1.00072

T/ K	P/ MPa	ρ/ kg m^{-3}	H/ kJ kg^{-1}	S/ kJ kg^{-1} K^{-1}	C_v/ kJ kg^{-1} K^{-1}	C_p/ kJ kg^{-1} K^{-1}	u/ m s^{-1}	η/ μPa s	λ/ mW m^{-1} K^{-1}	D
900.0	0.1	0.5892	2339.	5.772	3.553	3.743	422.8	21.28	136.9	1.00064
1000.0	0.1	0.5302	2724.	6.177	3.764	3.953	445.1	23.14	143.3	1.00058
P = 1.0 MPa (10 bar)										
85.617	1.0	733.3	-195.3	-1.396	1.356	1.916	2139.	10990.	208.1	2.08856
100.0	1.0	718.5	-167.6	-1.097	1.342	1.930	2041.	3896.	203.5	2.05953
120.0	1.0	698.2	-128.8	-0.7432	1.336	1.956	1905.	1532.	194.8	2.01987
140.0	1.0	678.1	-89.34	-0.4393	1.336	1.987	1771.	844.9	184.3	1.98096
160.0	1.0	657.8	-49.24	-0.1717	1.343	2.024	1638.	547.0	172.8	1.94249
180.0	1.0	637.3	-8.331	0.0692	1.358	2.068	1505.	386.6	160.7	1.90415
200.0	1.0	616.2	33.57	0.2899	1.384	2.124	1372.	289.3	148.4	1.86560
220.0	1.0	594.4	76.72	0.4955	1.422	2.194	1241.	225.1	136.3	1.82648
240.0	1.0	571.5	121.5	0.6901	1.470	2.284	1111.	179.6	124.6	1.78630
260.0	1.0	547.0	168.2	0.8771	1.528	2.395	980.1	145.3	113.5	1.74433
280.0	1.0	520.1	217.5	1.060	1.596	2.539	846.7	118.1	102.9	1.69945
300.0	1.0	489.5	270.2	1.241	1.675	2.740	706.9	95.28	92.86	1.64962
300.092	1.0	489.3	270.4	1.242	1.675	2.741	706.2	95.18	92.82	1.64937
300.092	1.0	21.68	602.7	2.349	1.588	2.042	214.8	8.248	19.25	1.02386
320.0	1.0	19.34	642.9	2.479	1.646	2.009	229.9	8.809	21.49	1.02125
340.0	1.0	17.61	683.2	2.601	1.717	2.032	242.4	9.362	23.91	1.01931
360.0	1.0	16.25	724.3	2.719	1.795	2.081	253.5	9.905	26.46	1.01779
380.0	1.0	15.12	766.5	2.833	1.876	2.142	263.4	10.44	29.13	1.01654
400.0	1.0	14.17	810.0	2.944	1.958	2.211	272.6	10.96	31.92	1.01549
500.0	1.0	10.93	1050.	3.478	2.363	2.583	311.6	13.46	47.56	1.01191
600.0	1.0	8.964	1326.	3.980	2.725	2.933	344.1	15.76	66.04	1.00976
700.0	1.0	7.623	1635.	4.456	3.040	3.242	372.9	17.86	87.38	1.00830
800.0	1.0	6.642	1973.	4.907	3.315	3.512	399.1	19.76	111.6	1.00723
900.0	1.0	5.889	2336.	5.334	3.555	3.750	423.4	21.47	138.7	1.00641
1000.0	1.0	5.292	2722.	5.741	3.765	3.959	446.3	23.61	143.4	1.00576
P = 10.0 MPa (100 bar)										
86.440	10.0	735.2	-182.9	-1.395	1.362	1.914	2159.	11230.	209.8	2.09015
100.0	10.0	721.5	-156.9	-1.115	1.350	1.927	2070.	4252.	205.7	2.06350
120.0	10.0	701.8	-118.1	-0.7615	1.344	1.951	1939.	1655.	197.6	2.02490
140.0	10.0	682.2	-78.8	-0.4586	1.345	1.979	1809.	907.0	187.7	1.98713
160.0	10.0	662.7	-38.9	-0.1922	1.352	2.012	1682.	586.7	176.7	1.94997
180.0	10.0	643.1	1.726	0.0470	1.367	2.052	1556.	415.6	165.2	1.91318
200.0	10.0	623.2	43.23	0.2656	1.393	2.101	1432.	312.3	153.5	1.87653
220.0	10.0	602.8	85.84	0.4686	1.430	2.162	1310.	244.5	142.0	1.83978
240.0	10.0	581.9	129.8	0.6598	1.478	2.236	1192.	196.9	130.9	1.80267
260.0	10.0	560.1	175.4	0.8422	1.535	2.325	1077.	161.6	120.5	1.76484
280.0	10.0	537.1	222.9	1.018	1.600	2.429	963.4	134.1	110.7	1.72590
300.0	10.0	512.7	272.7	1.190	1.673	2.550	852.7	111.9	101.8	1.68529
320.0	10.0	486.1	325.0	1.359	1.751	2.692	744.0	93.42	93.68	1.64228
340.0	10.0	456.7	380.5	1.527	1.833	2.865	637.1	77.64	86.37	1.59580
360.0	10.0	423.2	439.9	1.697	1.921	3.085	532.1	63.84	79.80	1.54422
380.0	10.0	383.4	504.5	1.871	2.014	3.385	430.5	51.50	73.83	1.48493
400.0	10.0	334.5	576.0	2.054	2.111	3.790	339.0	40.34	68.29	1.41464
500.0	10.0	143.3	947.1	2.887	2.438	3.221	275.9	19.50	58.86	1.16393
600.0	10.0	99.02	1263.	3.462	2.755	3.183	330.8	19.46	73.95	1.11103
700.0	10.0	78.82	1591.	3.967	3.058	3.386	372.8	20.75	95.32	1.08759
800.0	10.0	66.46	1940.	4.434	3.328	3.610	407.1	22.22	120.1	1.07350
900.0	10.0	57.89	2312.	4.871	3.565	3.821	436.7	23.65	148.0	1.06382
1000.0	10.0	51.49	2704.	5.284	3.774	4.013	463.1	27.61	144.9	1.05665

TABLE 9. Thermophysical Properties of Carbon Dioxide (CO_2)

T/ K	P/ MPa	ρ/ kg m^{-3}	H/ kJ kg^{-1}	S/ kJ kg^{-1} K^{-1}	C_v/ kJ kg^{-1} K^{-1}	C_p/ kJ kg^{-1} K^{-1}	u/ m s^{-1}	η/ μPa s	λ/ mW m^{-1} K^{-1}	D
Saturation										
216.592	0.5180	1178.	80.04	0.5213	0.9747	1.953	975.8	253.4	177.2	1.75696
216.592	0.5180	13.76	430.4	2.139	0.6292	0.9087	222.8	10.89	11.12	1.00694
220.0	0.5991	1166.	86.73	0.5517	0.9698	1.962	951.2	239.3	172.9	1.74853
220.0	0.5991	15.82	431.6	2.119	0.6389	0.9303	223.1	11.06	11.42	1.00798
230.0	0.8929	1129.	106.6	0.6387	0.9567	1.997	879.1	202.8	160.6	1.72239
230.0	0.8929	23.27	434.6	2.065	0.6700	1.005	223.6	11.58	12.41	1.01178
240.0	1.282	1089.	126.8	0.7235	0.9454	2.051	806.4	172.5	148.5	1.69398
240.0	1.282	33.30	436.5	2.014	0.7053	1.103	223.0	12.12	13.57	1.01692
250.0	1.785	1046.	147.7	0.8068	0.9364	2.132	731.8	146.9	136.5	1.66298
250.0	1.785	46.64	437.0	1.964	0.7459	1.237	221.2	12.73	14.98	1.02382
260.0	2.419	998.9	169.4	0.8895	0.9323	2.255	652.6	125.1	124.7	1.62876
260.0	2.419	64.42	435.9	1.914	0.7943	1.429	218.2	13.42	16.80	1.03309
270.0	3.203	945.8	192.4	0.9732	0.9396	2.453	565.5	106.0	112.9	1.59026
270.0	3.203	88.37	432.6	1.863	0.8517	1.731	213.8	14.25	19.35	1.04574
280.0	4.161	883.6	217.3	1.060	0.9605	2.814	471.5	88.75	101.0	1.54547
280.0	4.161	121.7	425.9	1.805	0.9232	2.277	207.7	15.36	23.36	1.06363
290.0	5.318	804.7	245.6	1.154	0.9937	3.676	371.9	72.19	88.93	1.48961
290.0	5.318	172.0	413.8	1.734	1.026	3.614	199.4	17.05	31.05	1.09112
300.0	6.713	679.2	283.4	1.276	1.120	8.698	245.7	53.19	80.84	1.40346
300.0	6.713	268.6	387.1	1.622	1.248	11.92	185.3	20.81	55.71	1.14588
304.1282	7.377	467.6	332.2	1.434				32.35		1.26600
***P* = 0.1 MPa (1 bar)**										
220.0	0.1	2.439	442.2	2.492	0.5791	0.7807	233.4	11.11	10.96	1.00122
240.0	0.1	2.228	458.0	2.561	0.5981	0.7962	243.2	12.09	12.34	1.00112
260.0	0.1	2.052	474.1	2.625	0.6184	0.8142	252.3	13.07	13.77	1.00103
280.0	0.1	1.902	490.6	2.686	0.6390	0.8333	261.0	14.04	15.26	1.00095
300.0	0.1	1.773	507.4	2.745	0.6593	0.8525	269.4	15.00	16.77	1.00089
320.0	0.1	1.661	524.7	2.800	0.6791	0.8715	277.4	15.95	18.32	1.00083
340.0	0.1	1.562	542.3	2.854	0.6982	0.8900	285.2	16.89	19.90	1.00078
360.0	0.1	1.474	560.3	2.905	0.7165	0.9079	292.8	17.82	21.49	1.00074
380.0	0.1	1.396	578.6	2.955	0.7341	0.9251	300.1	18.74	23.10	1.00070
400.0	0.1	1.326	597.3	3.002	0.7510	0.9417	307.3	19.64	24.72	1.00066
500.0	0.1	1.059	695.2	3.221	0.8255	1.015	340.6	23.92	32.88	1.00053
600.0	0.1	0.8824	799.9	3.411	0.8867	1.076	370.8	27.88	40.96	1.00044
700.0	0.1	0.7562	910.2	3.581	0.9375	1.127	398.7	31.53	48.83	1.00038
800.0	0.1	0.6616	1025.	3.735	0.9800	1.169	424.7	34.94	56.44	1.00033
900.0	0.1	0.5880	1144.	3.874	1.015	1.205	449.2	38.15	63.75	1.00029
1000.0	0.1	0.5292	1266.	4.003	1.045	1.234	472.4	41.18	70.78	1.00027
***P* = 1.0 MPa (10 bar)**										
216.695	1.0	1179.	80.37	0.5210	0.9751	1.950	977.8	253.9	177.4	1.75735
220.0	1.0	1167.	86.83	0.5506	0.9703	1.959	953.6	240.1	173.2	1.74911
233.028	1.0	1117.	112.7	0.6646	0.9530	2.011	857.2	193.1	156.9	1.71404
233.028	1.0	26.01	435.3	2.049	0.6803	1.032	223.5	11.74	12.74	1.01318
240.0	1.0	24.86	442.4	2.079	0.6733	0.9991	228.5	12.10	13.16	1.01259
260.0	1.0	22.21	461.7	2.157	0.6656	0.9450	241.2	13.13	14.47	1.01123
280.0	1.0	20.20	480.4	2.226	0.6709	0.9252	252.3	14.13	15.87	1.01019
300.0	1.0	18.58	498.8	2.289	0.6822	0.9209	262.4	15.11	17.33	1.00937
320.0	1.0	17.23	517.3	2.349	0.6960	0.9243	271.8	16.07	18.84	1.00868
340.0	1.0	16.09	535.8	2.405	0.7111	0.9320	280.6	17.02	20.38	1.00810
360.0	1.0	15.11	554.6	2.459	0.7266	0.9421	289.0	17.95	21.95	1.00760
380.0	1.0	14.24	573.5	2.510	0.7422	0.9534	297.0	18.87	23.54	1.00716
400.0	1.0	13.48	592.7	2.559	0.7575	0.9655	304.7	19.77	25.14	1.00677
500.0	1.0	10.66	692.4	2.781	0.8282	1.027	339.8	24.05	33.23	1.00535
600.0	1.0	8.845	798.0	2.974	0.8881	1.083	370.9	27.98	41.27	1.00444
700.0	1.0	7.564	908.8	3.144	0.9384	1.132	399.2	31.62	49.12	1.00379
800.0	1.0	6.610	1024.	3.298	0.9805	1.173	425.5	35.02	56.70	1.00332

T/ K	P/ MPa	ρ/ kg m⁻³	H/ kJ kg⁻¹	S/ kJ kg⁻¹ K⁻¹	C_v/ kJ kg⁻¹ K⁻¹	C_p/ kJ kg⁻¹ K⁻¹	u/ m s⁻¹	η/ μPa s	λ/ mW m⁻¹ K⁻¹	D
900.0	1.0	5.872	1143.	3.438	1.016	1.207	450.2	38.21	64.00	1.00295
1000.0	1.0	5.283	1265.	3.567	1.046	1.236	473.6	41.24	71.02	1.00265
***P* = 10.0 MPa (100 bar)**										
218.600	10.0	1190.	86.78	0.5155	0.9827	1.902	1012.	264.1	182.1	1.76433
220.0	10.0	1186.	89.44	0.5277	0.9806	1.904	1003.	258.2	180.4	1.76113
240.0	10.0	1115.	127.9	0.6949	0.9536	1.949	870.9	189.2	156.9	1.71160
260.0	10.0	1035.	167.8	0.8545	0.9340	2.052	735.5	140.2	134.2	1.65372
280.0	10.0	938.2	210.8	1.014	0.9263	2.280	588.7	103.3	111.7	1.58337
300.0	10.0	801.6	261.8	1.189	0.9496	2.991	414.3	71.83	88.19	1.48612
320.0	10.0	448.3	362.9	1.514	1.058	7.617	219.1	31.76	61.92	1.25290
340.0	10.0	258.6	443.4	1.759	0.9025	2.402	238.1	22.60	36.69	1.13908
360.0	10.0	208.2	483.0	1.873	0.8526	1.707	257.8	21.86	32.55	1.11030
380.0	10.0	180.4	514.3	1.957	0.8346	1.457	273.5	22.02	31.67	1.09469
400.0	10.0	161.5	542.1	2.029	0.8287	1.333	286.9	22.47	31.87	1.08425
500.0	10.0	113.1	663.8	2.301	0.8552	1.162	337.4	25.77	37.14	1.05801
600.0	10.0	89.94	779.3	2.511	0.9017	1.156	375.4	29.29	44.26	1.04580
700.0	10.0	75.49	895.9	2.691	0.9465	1.178	407.4	32.66	51.60	1.03828
800.0	10.0	65.35	1015.	2.850	0.9859	1.205	435.8	35.87	58.88	1.03305
900.0	10.0	57.76	1137.	2.993	1.020	1.231	461.8	38.91	65.97	1.02916
1000.0	10.0	51.82	1261.	3.124	1.048	1.255	485.9	41.82	72.85	1.02613

THERMOPHYSICAL PROPERTIES OF SELECTED FLUIDS AT SATURATION

Eric W. Lemmon and Ian H. Bell

These tables give thermodynamic and transport properties, as defined in the table below, for a variety of fluids, as generated from the correlations presented in the references. All extensive properties are given on a mass basis. Not all properties are included for every substance. The references should be consulted for information on the uncertainties. Values are given along the saturation line.

Column heading	Definition
T	Temperature, in K
P	Pressure, in units of MPa
ρ	Density, in units of kg m^{-3}
H	Enthalpy, in units of kJ kg^{-1}
S	Entropy, in units of kJ kg^{-1} K^{-1}
C_v	Isochoric heat capacity, in units of kJ kg^{-1} K^{-1}
C_p	Isobaric heat capacity, in units of kJ kg^{-1} K^{-1}
u	Speed of sound, in units of m s^{-1}
η	Viscosity, in units of μPa s
λ	Thermal conductivity, in units of mW m^{-1} K^{-1}
D	Static dielectric constant, dimensionless [Not given for every fluid]

Two blocks (or sets) are given for each fluid; the first set gives the values of the liquid phase (note the high density), and the second set gives the values of the vapor phase (at low densities). The first line of each block gives the properties at the triple point (except for R-1234yf and R-1234ze(E)). The final line in the first block gives the properties at the critical point.

The liquids covered here are listed below. The right-hand column summarizes the reference state for each liquid.

Table no.	Liquid name	Liquid mol. form.	Reference state
1	Ammonia	NH_3	Zero energy and entropy at the triple point
2	Benzene	C_6H_6	Zero enthalpy and entropy at the saturated liquid state at the normal boiling point
3	Butane	C_4H_{10}	Enthalpy = 200 kJ kg^{-1} and entropy = 1 kJ kg^{-1} K^{-1} at −40 °C
4	Carbon Monoxide	CO	Zero enthalpy and entropy at the saturated liquid state at the normal boiling point
5	Ethanol	C_2H_5OH	Enthalpy = 200 kJ kg^{-1} and entropy = 1 kJ kg^{-1} K^{-1} at −40 °C
6	Ethylene	C_2H_4	Zero enthalpy and entropy at the saturated liquid state at the normal boiling point
7	Hydrogen Sulfide	H_2S	Zero enthalpy and entropy at the saturated liquid state at the normal boiling point
8	Isobutane	C_4H_{10}	Enthalpy = 200 kJ kg^{-1} and entropy = 1 kJ kg^{-1} K^{-1} at −40 °C
9	Propene	C_3H_6	Enthalpy = 200 kJ kg^{-1} and entropy = 1 kJ kg^{-1} K^{-1} at −40 °C
10	R-134a (1,1,1,2-Tetra-fluoroethane)	CF_3CH_2F	Enthalpy = 200 kJ kg^{-1} and entropy = 1 kJ kg^{-1} K^{-1} at −40 °C
11	R-1234yf (2,2,2,3-Tetrafluoro-propene)	$CF_3CF{=}CH_2$	Enthalpy = 200 kJ kg^{-1} and entropy = 1 kJ kg^{-1} K^{-1} at −40 °C
12	R-1234ze(E) (trans-1,3,3,3-Tetrafluoro-propene)	$CHF{=}CHCF_3$ (trans)	Enthalpy = 200 kJ kg^{-1} and entropy = 1 kJ kg^{-1} K^{-1} at −40 °C
13	Sulfur Hexafluoride	SF_6	Enthalpy = 200 kJ kg^{-1} and entropy = 1 kJ kg^{-1} K^{-1} at −40 °C
14	Sulfur Dioxide	SO_2	Zero enthalpy and entropy at the saturated liquid state at the normal boiling point
15	Toluene	$CH_3C_6H_5$	Zero enthalpy and entropy at the saturated liquid state at the normal boiling point

Additional calculations at state points not listed below and for fluids not contained herein can be obtained with the use of the NIST Standard Reference Data program REFPROP Version 10.0 <http://www.nist.gov/srd/refprop>.

References

1. Assael, M.J., Koini, I.A., Antoniadis, K.D., Huber, M.L., Abdulagatov, I.M., and Perkins, R.A., Reference Correlation of the Thermal Conductivity of Sulfur Hexafluoride from the Triple Point to 1000 K and up to 150 MPa, *J. Phys. Chem. Ref. Data*, 41, 023104, 2012. <https://doi.org/10.1063/1.4708620>
2. Assael, M.J., Koutian, A., Huber, M.L., and Perkins, R.A., Reference Correlations of the Thermal Conductivity of Ethylene and Propylene, *J. Phys. Chem. Ref. Data*, 45(3), 033104, 2016. <https://doi.org/10.1063/1.4958984>
3. Assael, M.J., Mihailidou, E., Huber, M.L., and Perkins, R.A., Reference Correlation of the Thermal Conductivity of Benzene from the Triple Point to 725 K and up to 500 MPa, *J. Phys. Chem. Ref. Data*, 41(4), 043102, 2012. <https://doi.org/10.1063/1.4755781>
4. Assael, M.J., Mylona, S.K., Huber, M.L., and Perkins, R.A., Reference Correlation of the Thermal Conductivity of Toluene from the Triple Point to 1000 K and up to 1000 MPa, *J. Phys. Chem. Ref. Data*, 41, 023101, 2012. <https://doi.org/10.1063/1.3700155>
5. Assael, M.J., Sykioti, E.A., Huber, M.L., and Perkins, R.A., Reference Correlation of the Thermal Conductivity of Ethanol from the Triple Point to 600 K and up to 245 MPa, *J. Phys. Chem. Ref. Data*, 42(2), 023102, 2013. <https://doi.org/10.1063/1.4797368>
6. Avgeri, S., Assael, M.J., Huber, M.L., and Perkins, R.A., Reference Correlation of the Viscosity of Benzene from the Triple Point to 675 K and up to 300 MPa, *J. Phys. Chem. Ref. Data*, 43(3), 033103, 2014. <https://doi.org/10.1063/1.4892935>
7. Avgeri, S., Assael, M.J., Huber, M.L., and Perkins, R.A., Reference Correlation of the Viscosity of Toluene from the Triple Point to 675 K and up to 500 MPa, *J. Phys. Chem. Ref. Data*, 44(3), 033101, 2015. <https://doi.org/10.1063/1.4926955>
8. Bücker, D. and Wagner, W., Reference Equations of State for the Thermodynamic Properties of Fluid Phase n-Butane and Isobutane, *J. Phys. Chem. Ref. Data*, 35(1):929-1019, 2006. <https://doi.org/10.1063/1.1901687>

9. Gao, K., Wu, J., Bell, I.H., and Lemmon, E.W., Thermodynamic Properties of Ammonia for Temperatures from the Melting Line to 725 K and Pressures to 1000 MPa, to be submitted to *J. Phys. Chem. Ref. Data*, 2019.
10. Gao, K., Wu, J., Zhang, P., and Lemmon, E.W., A Helmholtz Energy Equation of State for Sulfur Dioxide, *J. Chem. Eng. Data*, 61:2859-2872, 2016. <https://doi.org/10.1021/acs.jced.6b00195>
11. Guder, C. and Wagner, W., A Reference Equation of State for the Thermodynamic Properties of Sulfur Hexafluoride (SF6) for Temperatures from the Melting Line to 625 K and Pressures up to 150 MPa, *J. Phys. Chem. Ref. Data*, 38(1):33-94, 2009. <https://doi.org/10.1063/1.3037344>
12. Harvey, A. H., and Lemmon, E. W., Method for Estimating the Dielectric Constant of Natural Gas Mixtures, *Int. J. Thermophys.*, 26, 31-46, 2005. <https://doi.org/10.1007/s10765-005-2351-5>
13. Herrmann, S. and Vogel, E., New Formulation for the Viscosity of n-Butane, *J. Phys. Chem. Ref. Data*, 47, 013104, 2018. <https://doi.org/10.1063/1.5020802>
14. Holland, P.M., Eaton, B.E., and Hanley, H.J.M., A Correlation of the Viscosity and Thermal Conductivity Data of Gaseous and Liquid Ethylene, *J. Phys. Chem. Ref. Data*, 12(4):917-932, 1983. <https://doi.org/10.1063/1.555701>
15. Huber, M.L. and Assael, M.J., Correlations for the Viscosity of 2,3,3,3-Tetrafluoroprop-1-ene (R1234yf) and trans-1,2,2,2-Tetrafluoropropene (R1234ze(E)), *Int. J. Refrig.*, 71:39-45, 2016. <https://doi.org/10.1016/j.ijrefrig.2016.08.007>
16. Huber, M.L., Laesecke, A., and Perkins, R.A., Model for the Viscosity and Thermal Conductivity of Refrigerants, Including a New Correlation for the Viscosity of R134a, *Ind. Eng. Chem. Res.*, 42(13):3163-3178, 2003. <https://doi.org/10.1021/ie0300880>
17. Kiselev, S.B., Ely, J.F., Abdulagatov, I.M., and Huber, M.L., Generalized SAFT-DFT/DMT Model for the Thermodynamic, Interfacial, and Transport Properties of Associating Fluids: Application for n-Alkanols, *Ind. Eng. Chem. Res.*, 44:6916-6927, 2005. <https://doi.org/10.1021/ie050010e>
18. Lemmon, E.W. and Span, R., Short Fundamental Equations of State for 20 Industrial Fluids, *J. Chem. Eng. Data*, 51(3):785-850, 2006. <https://doi.org/10.1021/je050186n>
19. Lemmon, E.W., McLinden, M.O., Overhoff, U., and Wagner, W., A Reference Equation of State for Propylene for Temperatures from the Melting Line to 575 K and Pressures up to 1000 MPa, to be submitted to *J. Phys. Chem. Ref. Data*, 2019.
20. Monogenidou, S.A., Assael, M.J., and Huber, M.L. Reference Correlation of the Viscosity of Ammonia from the Triple Point to 700 K and up to 50 MPa, *J. Phys. Chem. Ref. Data*, 47(2), 023102, 2018. <https://doi.org/10.1063/1.5036724>
21. Monogenidou, S.A., Assael, M.J., and Huber, M.L., Reference Correlations for Thermal Conductivity of Ammonia from the Triple Point to 680 K and up to 80 MPa, *J. Phys. Chem. Ref. Data*, 47, 043101, 2018. <https://doi.org/10.1063/1.5053087>
22. Perkins, R.A, Ramires, M.L.V., Nieto de Castro, C.A., and Cusco, L., Measurement and Correlation of the Thermal Conductivity of Butane from 135 K to 600 K at Pressures to 70 MPa, *J. Chem. Eng. Data*, 47(5):1263-1271, 2002. <https://doi.org/10.1021/je0101202>
23. Perkins, R.A. and Huber, M.L., Measurement and Correlation of the Thermal Conductivity of 2,3,3,3-Tetrafluoroprop-1-ene (R1234yf) and trans-1,3,3,3-Tetrafluoropeopene (R1234ze), *J. Chem. Eng. Data*, 56(12):4868-4874, 2011. <https://doi.org/10.1021/je200811n>
24. Perkins, R.A., Laesecke, A., Howley, J., Ramires, M.L.V., Gurova, A.N., and Cusco, L., Experimental Thermal Conductivity Values for the IUPAC Round-Robin Sample of 1,1,1,2-Tetrafluoroethane (R134a), *NISTIR 6605*, NIST, Boulder, CO, 2000. <https://doi.org/10.6028/NIST.IR.6605>
25. Perkins, R.A., Measurement and Correlation of the Thermal Conductivity of Isobutane from 114 K to 600 K at Pressures to 70 MPa, *J. Chem. Eng. Data*, 47(5):1272-1279, 2002. <https://doi.org/10.1021/je010121u>
26. Quiñones-Cisneros, S.E., Huber, M.L., and Deiters, U.K., Correlation for the Viscosity of Sulfur Hexafluoride (SF6) from the Triple Point to 1000 K and Pressures to 50 MPa, *J. Phys. Chem. Ref. Data*, 41(2), 023102, 2012. <https://doi.org/10.1063/1.3702441>
27. Richter, M., McLinden, M.O., and Lemmon, E.W., Thermodynamic Properties of 2,3,3,3-Tetrafluoroprop-1-ene (R1234yf): Vapor Pressure and p-rho-T Measurements and an Equation of State, *J. Chem. Eng. Data*, 56(7):3254-3264, 2011. <https://doi.org/10.1021/je200369m>
28. Schmidt, K.A.G., Carroll, J.J., Quinones-Cisneros, S.E., and Kvamme, B., Hydrogen Sulphide Viscosity Model, *Proceedings of the 86th Annual GPA Convention*, March 11-14, San Antonio, TX, 2007.
29. Schroeder, J.A., Penoncello, S.G., and Schroeder, J.S., A New Fundamental Equation for Ethanol, *J. Phys. Chem. Ref. Data*, 43, 043102, 2014. <https://doi.org/10.1063/1.4895394>
30. Smukala, J., Span, R., and Wagner, W., A New Equation of State for Ethylene Covering the Fluid Region for Temperatures from the Melting Line to 450 K at Pressures up to 300 MPa, *J. Phys. Chem. Ref. Data*, 29(5):1053-1122, 2000. <https://doi.org/10.1063/1.1329318>
31. Thol, M. and Lemmon, E.W., Equation of State for the Thermodynamic Properties of trans-1,3,3,3-Tetrafluoropropene [R1234ze(E)], *Int. J. Thermophys.*, 37:28, 2016. <https://doi.org/10.1007/s10765-016-2040-6>
32. Thol, M., Lemmon, E.W., and Span, R., unpublished equation of state for benzene, 2018.
33. Tillner-Roth, R. and Baehr, H.D., An International Standard Formulation of the Thermodynamic Properties of 1,1,1,2-Tetrafluoroethane (HFC-134a) for Temperatures from 170 K to 455 K at Pressures up to 70 MPa, *J. Phys. Chem. Ref. Data*, 23:657-729, 1994. <https://doi.org/10.1063/1.555958>
34. Vogel, E., Kuechenmeister, C., and Bich, E., Viscosity Correlation for Isobutane over Wide Ranges of the Fluid Region, *Int. J. Thermophys.*, 21(2):343-356, 2000. <https://doi.org/10.1023/A:1006623310780>

Fluids

TABLE 1. Thermophysical Properties of Ammonia (NH_3) at Saturation

T/ K	P/ MPa	ρ/ kg m^{-3}	H/ kJ kg^{-1}	S/ kJ kg^{-1} K^{-1}	C_v/ kJ kg^{-1} K^{-1}	C_p/ kJ kg^{-1} K^{-1}	u/ m s^{-1}	η/ µPa s	λ/ mW m^{-1} K^{-1}
Saturated Liquid									
195.49	0.006053	733.9	0.008249	0.0	2.968	4.305	2011	570.6	610.9
200.0	0.008610	728.7	19.45	0.09833	2.969	4.318	1983	516.0	610.4
220.0	0.03373	705.5	106.5	0.5131	2.961	4.390	1859	347.3	600.1
240.0	0.1022	681.4	195.1	0.8981	2.921	4.466	1735	251.0	578.3
260.0	0.2552	656.1	285.4	1.258	2.866	4.546	1605	192.1	547.0
280.0	0.5507	629.2	377.5	1.598	2.815	4.649	1468	153.0	508.8
300.0	1.061	600.2	472.1	1.921	2.777	4.796	1323	125.0	465.9
320.0	1.872	568.3	570.2	2.233	2.754	5.023	1171	103.2	420.2
340.0	3.079	532.5	673.7	2.540	2.747	5.392	1009	85.03	373.5

T/ K	P/ MPa	ρ/ kg m⁻³	H/ kJ kg⁻¹	S/ kJ kg⁻¹ K⁻¹	C_v/ kJ kg⁻¹ K⁻¹	C_p/ kJ kg⁻¹ K⁻¹	u/ m s⁻¹	η/ μPa s	λ/ mW m⁻¹ K⁻¹
360.0	4.792	490.3	785.6	2.850	2.763	6.082	833.5	68.99	326.7
380.0	7.140	436.2	913.0	3.181	2.831	7.838	635.2	53.95	279.7
400.0	10.30	344.0	1089	3.610	3.158	22.39	384.0	37.32	238.1
405.56	11.36	233.3	1248	3.995				25.36	
Saturated Vapor									
195.49	0.006053	0.06371	1489	7.616	1.516	2.023	355.3	6.838	15.70
200.0	0.008610	0.08867	1497	7.487	1.527	2.039	359.0	6.953	16.03
220.0	0.03373	0.3184	1532	6.994	1.594	2.137	374.2	7.495	17.64
240.0	0.1022	0.8969	1564	6.603	1.695	2.293	386.9	8.071	19.54
260.0	0.2552	2.116	1592	6.284	1.829	2.513	396.3	8.669	21.83
280.0	0.5507	4.380	1614	6.016	1.987	2.805	402.3	9.281	24.63
300.0	1.061	8.244	1630	5.781	2.163	3.189	404.5	9.915	28.12
320.0	1.872	14.50	1637	5.567	2.357	3.723	402.3	10.60	32.60
340.0	3.079	24.39	1632	5.359	2.578	4.537	394.9	11.38	38.69
360.0	4.792	40.20	1610	5.142	2.837	5.978	381.1	12.41	48.01
380.0	7.140	67.33	1561	4.886	3.155	9.403	358.1	14.02	66.41
400.0	10.30	130.9	1435	4.475	3.821	37.95	314.4	17.94	142.3

TABLE 2. Thermophysical Properties of Benzene (C_6H_6) at Saturation

T/ K	P/ MPa	ρ/ kg m⁻³	H/ kJ kg⁻¹	S/ kJ kg⁻¹ K⁻¹	C_v/ kJ kg⁻¹ K⁻¹	C_p/ kJ kg⁻¹ K⁻¹	u/ m s⁻¹	η/ μPa s	λ/ mW m⁻¹ K⁻¹
Saturated Liquid									
278.674	0.004785	894.0	-133.4	-0.4227	1.173	1.685	1394	812.3	148.0
280.0	0.005140	892.6	-131.1	-0.4147	1.176	1.688	1387	794.2	147.5
300.0	0.01382	871.6	-96.88	-0.2966	1.223	1.739	1292	587.4	140.6
320.0	0.03207	850.4	-61.52	-0.1826	1.275	1.797	1199	456.6	134.1
340.0	0.06619	828.6	-24.92	-0.07177	1.330	1.861	1109	366.4	127.8
360.0	0.1243	806.2	13.02	0.03642	1.387	1.929	1022	300.4	121.7
380.0	0.2160	783.1	52.36	0.1424	1.445	2.000	936.5	250.0	115.7
400.0	0.3525	758.9	93.18	0.2467	1.504	2.076	853.1	210.3	109.8
420.0	0.5459	733.6	135.6	0.3494	1.563	2.156	771.1	178.2	103.9
440.0	0.8092	706.7	179.6	0.4510	1.622	2.244	689.9	151.8	98.07
460.0	1.157	677.8	225.5	0.5518	1.681	2.345	608.5	129.5	92.25
480.0	1.603	646.1	273.4	0.6524	1.740	2.470	525.8	110.3	86.46
500.0	2.165	610.1	324.0	0.7537	1.801	2.643	440.0	93.43	80.72
520.0	2.862	567.0	378.0	0.8573	1.864	2.931	348.2	77.87	75.10
540.0	3.721	508.9	437.9	0.9674	1.936	3.653	245.6	62.28	70.11
560.0	4.784	385.7	519.6	1.111	2.120	19.52	117.2	40.41	80.16
562.02	4.907	304.7	551.5	1.168				30.48	
Saturated Vapor									
278.674	0.004785	0.1618	314.3	1.184	0.8586	0.9667	182.1	7.082	9.130
280.0	0.005140	0.1731	315.6	1.181	0.8641	0.9722	182.5	7.114	9.222
300.0	0.01382	0.4361	335.3	1.144	0.9469	1.057	187.4	7.602	10.70
320.0	0.03207	0.9554	356.4	1.123	1.031	1.145	191.6	8.083	12.34
340.0	0.06619	1.876	378.5	1.115	1.116	1.236	195.1	8.556	14.12
360.0	0.1243	3.376	401.6	1.116	1.202	1.330	197.6	9.023	16.04
380.0	0.2160	5.671	425.5	1.124	1.287	1.427	199.1	9.487	18.08
400.0	0.3525	9.019	450.0	1.139	1.372	1.528	199.2	9.951	20.26
420.0	0.5459	13.74	474.8	1.157	1.455	1.636	197.9	10.42	22.60
440.0	0.8092	20.23	499.6	1.178	1.538	1.753	194.8	10.90	25.11
460.0	1.157	29.05	524.2	1.201	1.620	1.887	189.8	11.40	27.88
480.0	1.603	41.03	548.0	1.224	1.702	2.050	182.3	11.93	31.02

TABLE 3. Thermophysical Properties of Butane (C_4H_{10}) at Saturation

T/ K	P/ MPa	ρ/ kg m^{-3}	H/ kJ kg^{-1}	S/ kJ kg^{-1} K^{-1}	C_v/ kJ kg^{-1} K^{-1}	C_p/ kJ kg^{-1} K^{-1}	u/ m s^{-1}	η/ μPa s	λ/ mW m^{-1} K^{-1}	D
Saturated Liquid										
134.895	0.000000665710	735.0	-89.82	-0.4651	1.441	1.973	1827	2360	176.6	2.040
150.0	0.000008573	720.9	-59.93	-0.2550	1.442	1.986	1730	1387	171.2	2.015
170.0	0.0001162	702.2	-19.99	-0.005125	1.447	2.009	1610	826.9	163.0	1.982
190.0	0.0008498	683.5	20.50	0.2200	1.462	2.042	1495	564.5	154.0	1.949
210.0	0.004043	664.4	61.76	0.4264	1.488	2.086	1382	416.5	144.6	1.917
230.0	0.01411	645.0	104.0	0.6186	1.525	2.143	1271	322.0	135.2	1.883
250.0	0.03915	625.0	147.6	0.8000	1.573	2.213	1161	256.7	125.8	1.850
270.0	0.09148	604.1	192.7	0.9733	1.630	2.297	1052	208.8	116.7	1.815
290.0	0.1873	582.2	239.7	1.141	1.695	2.396	944.5	172.3	108.1	1.779
310.0	0.3463	558.8	288.9	1.303	1.765	2.511	837.3	143.2	99.94	1.741
330.0	0.5904	533.4	340.5	1.463	1.841	2.649	729.7	119.3	92.41	1.701
350.0	0.9442	505.2	395.1	1.622	1.922	2.820	620.4	99.02	85.49	1.657
370.0	1.434	472.8	453.4	1.781	2.010	3.055	506.8	81.11	79.17	1.607
390.0	2.092	433.2	516.7	1.944	2.112	3.454	384.7	64.63	73.36	1.548
410.0	2.958	377.1	589.1	2.119	2.247	4.677	246.5	47.96	67.96	1.467
425.125	3.796	228.0	693.9	2.363				24.84		1.266
Saturated Vapor										
134.895	0.0000006657	0.0000345	406.1	3.211	0.9634	1.106	148.9	3.342	4.855	1.000
150.0	0.000008573	0.0003995	423.2	2.966	1.015	1.158	156.5	3.726	5.579	1.000
170.0	0.0001162	0.004778	447.0	2.742	1.078	1.221	165.9	4.232	6.651	1.000
190.0	0.0008498	0.03130	471.9	2.596	1.141	1.285	174.7	4.733	7.851	1.000
210.0	0.004043	0.1351	497.9	2.503	1.208	1.354	182.8	5.228	9.175	1.000
230.0	0.01411	0.4330	524.9	2.448	1.282	1.433	189.8	5.715	10.62	1.000
250.0	0.03915	1.118	552.6	2.420	1.364	1.523	195.6	6.193	12.20	1.001
270.0	0.09148	2.463	580.8	2.411	1.454	1.626	199.7	6.664	13.91	1.003
290.0	0.1873	4.821	609.3	2.415	1.551	1.745	201.9	7.137	15.78	1.005
310.0	0.3463	8.645	637.8	2.429	1.654	1.883	201.8	7.623	17.85	1.009
330.0	0.5904	14.54	665.8	2.449	1.763	2.046	198.8	8.144	20.23	1.016
350.0	0.9442	23.39	692.8	2.473	1.877	2.250	192.5	8.739	23.06	1.025
370.0	1.434	36.69	717.7	2.496	1.994	2.538	182.0	9.482	26.62	1.040
390.0	2.092	57.55	738.1	2.512	2.129	3.075	165.7	10.56	31.56	1.063
410.0	2.958	95.37	748.0	2.507	2.306	4.840	141.4	12.61	40.62	1.106

TABLE 4. Thermophysical Properties of Carbon Monoxide (CO) at Saturation

T/ K	P/ MPa	ρ/ kg m^{-3}	H/ kJ kg^{-1}	S/ kJ kg^{-1} K^{-1}	C_v/ kJ kg^{-1} K^{-1}	C_p/ kJ kg^{-1} K^{-1}	u/ m s^{-1}	η/ μPa s	λ/ mW m^{-1} K^{-1}
Saturated Liquid									
68.16	0.01554	849.5	-28.96	-0.3863	1.262	2.157	998.2	291.5	165.3
70.0	0.02105	842.1	-24.99	-0.3290	1.243	2.150	980.5	267.0	161.8
80.0	0.08374	800.3	-3.524	-0.04334	1.154	2.142	883.4	175.7	143.3
90.0	0.2385	755.4	18.18	0.2099	1.088	2.188	783.6	124.8	125.0
100.0	0.5444	705.4	40.71	0.4428	1.037	2.306	678.7	92.77	106.9
110.0	1.067	647.4	64.89	0.6657	1.002	2.558	564.7	70.17	88.80
120.0	1.877	574.6	92.30	0.8924	0.9951	3.202	433.1	52.08	70.82
130.0	3.065	456.2	129.3	1.169	1.110	8.070	254.0	33.44	54.14
132.86	3.498	303.9	164.7	1.429					
Saturated Vapor									
68.16	0.01554	0.7761	203.0	3.017	0.7529	1.063	167.2	4.506	5.720
70.0	0.02105	1.027	204.7	2.953	0.7553	1.069	169.2	4.635	5.892
80.0	0.08374	3.658	213.4	2.668	0.7747	1.120	178.4	5.365	6.882
90.0	0.2385	9.660	220.4	2.457	0.8073	1.216	184.7	6.158	8.020
100.0	0.5444	21.20	224.9	2.285	0.8544	1.389	187.5	7.048	9.452
110.0	1.067	41.74	226.0	2.130	0.9190	1.725	186.7	8.108	11.52
120.0	1.877	78.77	221.2	1.967	1.016	2.601	182.0	9.553	15.29
130.0	3.065	164.8	201.1	1.721	1.235	9.853	171.9	12.64	29.33

TABLE 5. Thermophysical Properties of Ethanol (C_2H_5OH) at Saturation

T/ K	P/ MPa	ρ/ kg m^{-3}	H/ kJ kg^{-1}	S/ kJ kg^{-1} K^{-1}	C_v/ kJ kg^{-1} K^{-1}	C_p/ kJ kg^{-1} K^{-1}	u/ m s^{-1}	η/ μPa s	λ/ mW m^{-1} K^{-1}
Saturated Liquid									
159.0	0.0000000007185	909.0	-431.2	-1.725	1.306	1.794	1708	126600	220.9
160.0	0.0000000009015	907.9	-429.4	-1.713	1.323	1.804	1701	116200	220.0
180.0	0.00000004812	887.9	-392.0	-1.493	1.503	1.918	1608	33850	205.5
200.0	0.000001102	869.5	-353.2	-1.289	1.555	1.954	1528	14570	194.5
220.0	0.00001369	851.8	-313.7	-1.101	1.613	2.001	1443	7370	185.6
240.0	0.000108	834.5	-273.0	-0.9234	1.697	2.077	1358	4113	178.3
260.0	0.0006023	817.5	-230.5	-0.7533	1.802	2.179	1277	2465	172.3
280.0	0.002560	800.5	-185.7	-0.5875	1.922	2.302	1204	1565	167.3
300.0	0.008768	783.5	-138.2	-0.4238	2.056	2.449	1135	1044	163.1
320.0	0.02524	765.9	-87.56	-0.2604	2.201	2.620	1069	726.6	159.5
340.0	0.06303	747.6	-33.25	-0.09599	2.350	2.812	1005	524.3	156.2
360.0	0.1400	727.9	25.12	0.07045	2.497	3.023	938.0	389.6	153.0
380.0	0.2821	706.4	87.96	0.2397	2.639	3.257	867.4	296.0	149.9
400.0	0.5237	682.1	155.8	0.4127	2.771	3.516	790.5	228.1	146.6
420.0	0.9064	654.2	228.9	0.5896	2.880	3.790	706.3	177.1	143.3
440.0	1.478	621.8	307.2	0.7697	2.961	4.060	615.1	137.9	139.9
460.0	2.289	583.9	390.6	0.9519	3.027	4.365	517.8	107.6	136.8
480.0	3.398	537.8	480.3	1.139	3.113	4.959	408.1	83.55	134.0
500.0	4.872	467.4	586.1	1.348	3.273	7.648	266.0	60.98	133.3
514.71	6.268	273.2	758.7	1.680				30.57	
Saturated Vapor									
159.0	0.0000000007185	0.00000002504	610.9	4.830	0.8574	1.038	186.4	4.344	5.359
160.0	0.0000000009015	0.00000003122	612.0	4.795	0.8595	1.040	186.9	4.376	5.414
180.0	0.00000004812	0.000001481	633.2	4.203	0.9026	1.083	197.4	5.029	6.533
200.0	0.000001102	0.00003052	655.3	3.754	0.9487	1.129	207.3	5.676	7.715
220.0	0.00001369	0.0003449	678.4	3.409	0.9990	1.179	216.5	6.318	8.998
240.0	0.000108	0.002495	702.5	3.141	1.054	1.235	225.2	6.955	10.40
260.0	0.0006023	0.01285	727.6	2.932	1.115	1.296	233.4	7.586	11.93
280.0	0.002560	0.05078	753.6	2.767	1.181	1.365	241.1	8.211	13.60
300.0	0.008768	0.1629	780.4	2.638	1.254	1.444	248.1	8.829	15.40
320.0	0.02524	0.4426	807.5	2.537	1.336	1.535	254.3	9.439	17.34
340.0	0.06303	1.052	834.4	2.456	1.428	1.645	259.5	10.04	19.43
360.0	0.1400	2.245	860.5	2.391	1.532	1.778	263.3	10.63	21.71
380.0	0.2821	4.393	885.0	2.337	1.651	1.943	265.5	11.22	24.23
400.0	0.5237	8.010	907.4	2.292	1.786	2.150	265.6	11.82	27.10
420.0	0.9064	13.80	926.8	2.251	1.938	2.418	263.3	12.44	30.51
440.0	1.478	22.77	942.3	2.213	2.111	2.785	258.1	13.13	34.78
460.0	2.289	36.47	952.3	2.173	2.308	3.339	249.3	13.96	40.55
480.0	3.398	57.90	953.1	2.124	2.544	4.381	235.0	15.11	49.50
500.0	4.872	96.62	932.3	2.041	2.866	8.052	210.1	17.23	69.88

TABLE 6. Thermophysical Properties of Ethylene (C_2H_4) at Saturation

T/ K	P/ MPa	ρ/ kg m^{-3}	H/ kJ kg^{-1}	S/ kJ kg^{-1} K^{-1}	C_v/ kJ kg^{-1} K^{-1}	C_p/ kJ kg^{-1} K^{-1}	u/ m s^{-1}	η/ μPa s	λ/ mW m^{-1} K^{-1}	D
Saturated Liquid										
103.986	0.000122	654.6	-158.1	-1.179	1.622	2.429	1767	685.7	249.9	2.008
120.0	0.001368	634.2	-119.2	-0.8308	1.553	2.427	1660	427.5	232.6	1.966
140.0	0.01185	608.0	-70.81	-0.4581	1.465	2.408	1521	277.6	210.9	1.914
160.0	0.05623	580.9	-22.66	-0.1372	1.395	2.407	1377	200.1	189.3	1.863
180.0	0.1818	552.2	25.87	0.1473	1.346	2.441	1227	153.0	168.3	1.810
200.0	0.4555	521.2	75.68	0.4070	1.321	2.529	1070	120.5	147.9	1.755
220.0	0.9566	486.7	128.0	0.6516	1.320	2.700	903.5	95.68	128.3	1.695
240.0	1.773	446.1	184.9	0.8911	1.344	3.043	724.4	74.96	109.3	1.627
260.0	3.003	393.5	250.4	1.141	1.407	3.946	524.1	55.95	90.50	1.541
280.0	4.784	290.7	347.8	1.481	1.778	19.56	246.7	33.08	76.51	1.384

T/ K	P/ MPa	ρ/ kg m⁻³	H/ kJ kg⁻¹	S/ kJ kg⁻¹ K⁻¹	C_v/ kJ kg⁻¹ K⁻¹	C_p/ kJ kg⁻¹ K⁻¹	u/ m s⁻¹	η/ μPa s	λ/ mW m⁻¹ K⁻¹	D
282.35	5.042	214.2	399.4	1.661				22.88		1.274
Saturated Vapor										
103.986	0.000122	0.003958	409.4	4.279	0.8901	1.187	202.7	0.7727	4.075	1.000
120.0	0.001368	0.03852	428.3	3.731	0.8937	1.192	217.5	3.339	4.986	1.000
140.0	0.01185	0.2876	451.3	3.271	0.9064	1.212	233.9	4.841	6.195	1.000
160.0	0.05623	1.212	473.0	2.961	0.9336	1.260	247.4	5.700	7.556	1.001
180.0	0.1818	3.589	492.3	2.739	0.9783	1.347	257.0	6.401	9.168	1.004
200.0	0.4555	8.494	508.1	2.569	1.043	1.492	261.9	7.133	11.16	1.010
220.0	0.9566	17.45	519.1	2.429	1.132	1.738	261.5	8.007	13.71	1.020
240.0	1.773	33.07	522.7	2.299	1.254	2.211	254.9	9.162	17.31	1.039
260.0	3.003	61.54	513.5	2.153	1.439	3.512	240.5	10.96	23.83	1.073
280.0	4.784	140.7	457.5	1.873	1.981	29.26	208.9	16.18	59.20	1.174

TABLE 7. Thermophysical Properties of Hydrogen Sulfide (H_2S) at Saturation

T/ K	P/ MPa	ρ/ kg m⁻³	H/ kJ kg⁻¹	S/ kJ kg⁻¹ K⁻¹	C_v/ kJ kg⁻¹ K⁻¹	C_p/ kJ kg⁻¹ K⁻¹	u/ m s⁻¹	η/ μPa s	λ/ mW m⁻¹ K⁻¹	D
Saturated Liquid										
187.7	0.02326	992.3	-50.47	-0.2520	1.302	2.020	1438	543.8	272.1	9.484
200.0	0.05034	971.5	-25.72	-0.1244	1.263	2.003	1373	378.5	257.5	8.685
220.0	0.1437	936.5	14.28	0.06577	1.209	1.995	1268	265.4	234.0	7.597
240.0	0.3377	899.8	54.38	0.2393	1.164	2.009	1162	211.3	211.2	6.705
260.0	0.6875	860.6	95.09	0.4006	1.128	2.050	1054	175.0	189.2	5.955
280.0	1.256	818.0	137.0	0.5533	1.099	2.127	942.1	146.3	168.3	5.309
300.0	2.110	770.5	180.8	0.7008	1.078	2.262	825.0	121.9	148.5	4.735
320.0	3.323	715.7	227.8	0.8472	1.064	2.509	700.2	100.6	129.8	4.206
340.0	4.976	648.3	280.4	0.9989	1.064	3.055	562.6	81.06	111.7	3.688
360.0	7.171	551.1	345.6	1.175	1.099	5.271	398.9	61.01	94.05	3.103
373.1	8.999	347.3	450.5	1.449						2.182
Saturated Vapor										
187.7	0.02326	0.5120	524.5	2.811	0.7437	0.9976	245.8	7.673	8.192	1.002
200.0	0.05034	1.046	535.5	2.682	0.7507	1.012	252.8	8.188	8.779	1.004
220.0	0.1437	2.758	552.2	2.511	0.7671	1.047	262.6	9.055	9.748	1.009
240.0	0.3377	6.093	566.9	2.375	0.7898	1.102	270.1	9.954	10.77	1.019
260.0	0.6875	11.87	579.1	2.262	0.8190	1.183	275.0	10.88	11.95	1.037
280.0	1.256	21.16	588.0	2.164	0.8549	1.302	277.1	11.83	13.46	1.064
300.0	2.110	35.48	592.7	2.074	0.8981	1.488	276.1	12.80	15.66	1.106
320.0	3.323	57.40	591.4	1.983	0.9509	1.816	271.6	13.85	19.16	1.170
340.0	4.976	92.34	580.8	1.883	1.020	2.548	263.4	15.13	25.56	1.276
360.0	7.171	158.3	549.8	1.742	1.126	5.653	250.8	17.54	41.65	1.487

TABLE 8. Thermophysical Properties of Isobutane (C_4H_{10}) at Saturation

T/ K	P/ MPa	ρ/ kg m⁻³	H/ kJ kg⁻¹	S/ kJ kg⁻¹ K⁻¹	C_v/ kJ kg⁻¹ K⁻¹	C_p/ kJ kg⁻¹ K⁻¹	u/ m s⁻¹	η/ μPa s	λ/ mW m⁻¹ K⁻¹	D
Saturated Liquid										
113.73	0.00000002289	740.3	-112.4	-0.6798	1.174	1.689	2000	8767	157.9	2.109
150.0	0.00002388	706.0	-49.02	-0.1971	1.251	1.805	1714	1748	148.4	2.032
170.0	0.0002739	686.9	-12.30	0.03269	1.293	1.868	1579	1008	141.2	1.993
190.0	0.001763	667.5	25.72	0.2440	1.338	1.933	1451	657.3	133.2	1.954
210.0	0.007581	647.7	65.08	0.4409	1.388	2.004	1328	462.5	124.8	1.917
230.0	0.02439	627.4	105.9	0.6265	1.444	2.081	1210	342.1	116.3	1.879
250.0	0.06335	606.3	148.4	0.8034	1.507	2.168	1093	262.0	107.9	1.841
270.0	0.1402	584.2	192.8	0.9737	1.576	2.266	979.3	205.9	99.87	1.802
290.0	0.2744	560.7	239.3	1.139	1.651	2.379	866.5	164.8	92.23	1.762
310.0	0.4886	535.4	288.3	1.301	1.731	2.512	754.0	133.6	85.12	1.720
330.0	0.8076	507.4	340.2	1.461	1.817	2.676	640.5	109.0	78.60	1.674

T/ K	P/ MPa	ρ/ kg m⁻³	H/ kJ kg⁻¹	S/ kJ kg⁻¹ K⁻¹	C_v/ kJ kg⁻¹ K⁻¹	C_p/ kJ kg⁻¹ K⁻¹	u/ m s⁻¹	η/ μPa s	λ/ mW m⁻¹ K⁻¹	D
350.0	1.259	475.5	395.6	1.621	1.909	2.898	523.9	88.59	72.68	1.623
370.0	1.873	436.7	455.8	1.785	2.011	3.262	401.0	70.53	67.36	1.564
390.0	2.687	383.3	524.1	1.959	2.141	4.221	263.9	52.77	62.76	1.484
407.81	3.629	225.5	633.9	2.226				24.42		1.267
Saturated Vapor										
113.73	0.00000002289	0.000001407	368.3	3.547	0.7366	0.8796	139.4	2.848	2.272	1.000
150.0	0.00002388	0.001113	403.1	2.817	0.8944	1.037	157.8	3.799	4.461	1.000
170.0	0.0002739	0.01127	424.7	2.603	0.9747	1.118	166.9	4.317	5.813	1.000
190.0	0.001763	0.06499	447.6	2.465	1.055	1.199	175.4	4.827	7.265	1.000
210.0	0.007581	0.2539	471.9	2.378	1.137	1.284	183.1	5.329	8.813	1.000
230.0	0.02439	0.7523	497.3	2.328	1.224	1.377	189.6	5.821	10.45	1.001
250.0	0.06335	1.825	523.4	2.303	1.317	1.481	194.5	6.305	12.17	1.002
270.0	0.1402	3.828	550.1	2.297	1.417	1.599	197.6	6.787	14.00	1.004
290.0	0.2744	7.217	577.0	2.303	1.523	1.734	198.4	7.284	15.97	1.008
310.0	0.4886	12.59	603.7	2.318	1.635	1.893	196.4	7.822	18.15	1.014
330.0	0.8076	20.80	629.6	2.338	1.753	2.089	191.0	8.445	20.71	1.023
350.0	1.259	33.23	653.8	2.359	1.874	2.355	181.6	9.241	23.93	1.036
370.0	1.873	52.65	674.3	2.375	2.013	2.831	166.8	10.41	28.50	1.058
390.0	2.687	86.62	686.2	2.375	2.207	4.248	144.4	12.55	36.83	1.097

TABLE 9. Thermophysical Properties of Propene (Propylene) (C_3H_6) at Saturation

T/ K	P/ MPa	ρ/ kg m⁻³	H/ kJ kg⁻¹	S/ kJ kg⁻¹ K⁻¹	C_v/ kJ kg⁻¹ K⁻¹	C_p/ kJ kg⁻¹ K⁻¹	u/ m s⁻¹	η/ μPa s	λ/ mW m⁻¹ K⁻¹
Saturated Liquid									
87.953	0.0000000007472	768.1	-199.4	-1.420	1.569	2.182	2013	12100	321.8
100.0	0.00000004115	754.0	-173.4	-1.143	1.508	2.141	1952	4300	288.3
120.0	0.00000461	731.2	-131.0	-0.7565	1.432	2.095	1844	1470	244.6
140.0	0.0001171	708.8	-89.41	-0.4358	1.379	2.069	1728	760.5	211.9
160.0	0.001207	686.4	-48.11	-0.1600	1.346	2.064	1606	485.9	186.9
180.0	0.006928	663.9	-6.720	0.08366	1.333	2.078	1480	346.0	167.6
200.0	0.02675	640.8	35.18	0.3042	1.338	2.114	1352	261.6	152.3
220.0	0.07823	616.9	78.03	0.5080	1.359	2.170	1222	205.3	140.0
240.0	0.1872	591.8	122.3	0.6996	1.393	2.249	1091	165.3	129.7
260.0	0.3862	565.0	168.4	0.8828	1.437	2.354	959.8	135.4	120.8
280.0	0.7128	535.8	216.9	1.060	1.490	2.493	827.5	112.0	112.5
300.0	1.209	503.1	268.7	1.236	1.553	2.691	692.3	92.77	104.4
320.0	1.919	464.5	324.9	1.412	1.627	3.014	550.7	76.02	95.92
340.0	2.897	414.7	388.4	1.598	1.725	3.750	396.4	59.98	86.30
360.0	4.219	325.7	473.1	1.829	1.955	10.53	206.3	40.08	77.78
364.211	4.555	229.6	531.1	1.986					
Saturated Vapor									
87.953	0.0000000007472	0.00000004299	386.1	5.237	0.7037	0.9013	149.2	2.589	2.333
100.0	0.00000004115	0.000002082	397.2	4.563	0.7331	0.9307	158.4	2.900	2.882
120.0	0.00000461	0.0001944	416.2	3.804	0.7803	0.9779	172.4	3.429	3.857
140.0	0.0001171	0.004235	436.3	3.319	0.8288	1.027	185.1	3.974	4.922
160.0	0.001207	0.03822	457.2	2.998	0.8813	1.080	196.6	4.530	6.087
180.0	0.006928	0.1958	478.9	2.781	0.9389	1.141	206.9	5.091	7.365
200.0	0.02675	0.6864	501.0	2.633	1.003	1.213	215.6	5.652	8.772
220.0	0.07823	1.855	523.0	2.530	1.077	1.303	222.4	6.213	10.34
240.0	0.1872	4.179	544.5	2.459	1.162	1.416	226.7	6.782	12.09
260.0	0.3862	8.277	565.0	2.408	1.256	1.557	228.1	7.384	14.10
280.0	0.7128	14.99	583.8	2.371	1.360	1.740	225.9	8.063	16.46
300.0	1.209	25.57	600.0	2.340	1.478	2.002	219.5	8.895	19.42
320.0	1.919	42.31	611.7	2.308	1.611	2.439	208.1	10.04	23.52
340.0	2.897	70.66	614.7	2.263	1.772	3.485	190.2	11.90	30.56
360.0	4.219	138.1	588.4	2.149	2.090	14.37	161.9	16.82	54.26

TABLE 10. Thermophysical Properties of R-134a (1,1,1,2-Tetrafluoroethane, CF_3CH_2F) at Saturation

T/ K	P/ MPa	ρ/ kg m^{-3}	H/ kJ kg^{-1}	S/ kJ kg^{-1} K^{-1}	C_v/ kJ kg^{-1} K^{-1}	C_p/ kJ kg^{-1} K^{-1}	u/ m s^{-1}	η/ µPa s	λ/ mW m^{-1} K^{-1}
Saturated Liquid									
169.85	0.0003896	1591	71.46	0.4126	0.7922	1.184	1120	2154	145.3
180.0	0.001127	1564	83.48	0.4814	0.7912	1.187	1068	1479	139.1
200.0	0.006313	1510	107.4	0.6073	0.8016	1.206	967.6	867.3	127.8
220.0	0.02443	1455	131.8	0.7235	0.8193	1.233	869.9	582.2	117.2
240.0	0.07248	1398	156.8	0.8321	0.8403	1.267	775.0	420.2	107.3
260.0	0.1768	1337	182.6	0.9349	0.8631	1.308	682.1	316.6	97.92
280.0	0.3727	1272	209.3	1.033	0.8877	1.361	590.2	244.3	88.99
300.0	0.7028	1200	237.2	1.129	0.9144	1.432	497.9	190.5	80.34
320.0	1.217	1117	266.8	1.223	0.9443	1.543	404.0	147.8	71.78
340.0	1.972	1015	298.9	1.318	0.9802	1.751	306.4	111.8	63.08
360.0	3.040	870.1	336.1	1.421	1.039	2.437	196.0	78.15	54.06
374.21	4.059	511.9	389.6	1.562				34.69	
Saturated Vapor									
169.85	0.0003896	0.02817	334.9	1.964	0.5030	0.5853	126.8	6.829	3.080
180.0	0.001127	0.07701	340.9	1.911	0.5267	0.6097	130.1	7.232	3.893
200.0	0.006313	0.3898	353.1	1.836	0.5732	0.6586	136.0	8.015	5.498
220.0	0.02443	1.385	365.7	1.787	0.6204	0.7109	141.0	8.779	7.108
240.0	0.07248	3.837	378.3	1.755	0.6700	0.7705	144.7	9.521	8.732
260.0	0.1768	8.905	390.8	1.736	0.7234	0.8418	146.8	10.25	10.39
280.0	0.3727	18.23	402.5	1.724	0.7810	0.9296	146.6	10.98	12.12
300.0	0.7028	34.19	413.3	1.716	0.8426	1.044	143.9	11.77	14.01
320.0	1.217	60.71	422.2	1.709	0.9093	1.211	137.9	12.73	16.30
340.0	1.972	105.7	428.2	1.698	0.9852	1.524	127.6	14.16	19.71
360.0	3.040	193.6	427.1	1.674	1.085	2.606	111.2	17.14	27.37

TABLE 11. Thermophysical Properties of R-1234yf (2,3,3,3-Tetrafluoropropene, $CF_3CF{=}CH_2$) at Saturation

T/ K	P/ MPa	ρ/ kg m^{-3}	H/ kJ kg^{-1}	S/ kJ kg^{-1} K^{-1}	C_v/ kJ kg^{-1} K^{-1}	C_p/ kJ kg^{-1} K^{-1}	u/ m s^{-1}	η/ µPa s	λ/ mW m^{-1} K^{-1}
Saturated Liquid									
240.0	0.08603	1273	159.1	0.8412	0.7990	1.178	704.6	301.1	82.85
250.0	0.1327	1245	171.0	0.8898	0.8205	1.210	659.9	267.6	79.32
260.0	0.1972	1216	183.3	0.9378	0.8411	1.243	616.0	238.2	75.87
270.0	0.2834	1186	195.9	0.9852	0.8607	1.278	572.8	212.3	72.51
280.0	0.3959	1154	208.9	1.032	0.8794	1.315	529.7	189.2	69.24
290.0	0.5393	1121	222.3	1.079	0.8973	1.356	486.6	168.4	66.06
300.0	0.7187	1085	236.1	1.125	0.9147	1.401	443.1	149.6	62.97
310.0	0.9394	1047	250.3	1.171	0.9321	1.454	399.2	132.4	59.97
320.0	1.207	1005	265.1	1.217	0.9496	1.521	354.6	116.5	57.05
330.0	1.528	957.6	280.5	1.263	0.9667	1.617	308.2	101.6	54.18
340.0	1.910	902.8	296.8	1.311	0.9855	1.768	257.2	87.23	51.30
350.0	2.361	834.9	314.5	1.360	1.015	2.061	200.6	72.77	48.42
360.0	2.893	738.9	335.0	1.416	1.074	3.029	138.4	57.00	46.19
367.85	3.382	475.6	369.6	1.509				30.39	
Saturated Vapor									
240.0	0.08603	5.129	341.2	1.600	0.7096	0.8004	134.5	8.591	8.974
250.0	0.1327	7.713	347.9	1.597	0.7374	0.8352	135.4	9.098	9.765
260.0	0.1972	11.23	354.6	1.597	0.7656	0.8725	135.8	9.587	10.56
270.0	0.2834	15.89	361.2	1.597	0.7940	0.9127	135.6	10.06	11.37
280.0	0.3959	22.00	367.7	1.599	0.8227	0.9569	134.7	10.54	12.21
290.0	0.5393	29.88	374.0	1.602	0.8514	1.006	133.2	11.02	13.09
300.0	0.7187	39.99	379.9	1.604	0.8808	1.065	130.9	11.52	14.05
310.0	0.9394	52.95	385.5	1.607	0.9122	1.139	127.7	12.06	15.13
320.0	1.207	69.65	390.5	1.609	0.9467	1.238	123.6	12.68	16.41
330.0	1.528	91.48	394.8	1.610	0.9847	1.382	118.3	13.41	18.03

T/ K	P/ MPa	ρ/ kg m^{-3}	H/ kJ kg^{-1}	S/ kJ kg^{-1} K^{-1}	C_v/ kJ kg^{-1} K^{-1}	C_p/ kJ kg^{-1} K^{-1}	u/ m s^{-1}	η/ µPa s	λ/ mW m^{-1} K^{-1}
340.0	1.910	120.8	397.9	1.608	1.027	1.614	111.7	14.34	20.27
350.0	2.361	162.8	399.2	1.602	1.077	2.087	103.4	15.68	23.82
360.0	2.893	232.4	396.3	1.587	1.146	3.724	92.64	18.08	31.24

TABLE 12. Thermophysical Properties of R-1234ze(E) (trans-1,3,3,3-Tetrafluoropropene, $CHF{=}CHCF_3$ (trans)) at Saturation

T/ K	P/ MPa	ρ/ kg m^{-3}	H/ kJ kg^{-1}	S/ kJ kg^{-1} K^{-1}	C_v/ kJ kg^{-1} K^{-1}	C_p/ kJ kg^{-1} K^{-1}	u/ m s^{-1}	η/ µPa s	λ/ mW m^{-1} K^{-1}
Saturated Liquid									
240.0	0.05234	1332	157.3	0.8338	0.8580	1.260	791.3	409.0	95.70
250.0	0.08417	1305	170.0	0.8855	0.8665	1.275	746.7	357.2	91.78
260.0	0.1297	1278	182.8	0.9358	0.8756	1.292	702.3	312.4	87.96
270.0	0.1925	1249	195.9	0.9848	0.8853	1.312	658.0	273.7	84.22
280.0	0.2767	1220	209.1	1.033	0.8958	1.334	613.8	240.2	80.58
290.0	0.3866	1189	222.6	1.080	0.9070	1.361	569.8	211.2	77.03
300.0	0.5270	1157	236.4	1.126	0.9186	1.392	526.0	186.1	73.59
310.0	0.7029	1123	250.5	1.172	0.9301	1.430	481.9	164.2	70.24
320.0	0.9196	1086	265.0	1.217	0.9416	1.476	437.1	145.0	66.99
330.0	1.183	1046	280.0	1.263	0.9535	1.536	391.0	127.9	63.83
340.0	1.499	1001	295.5	1.308	0.9668	1.617	342.8	112.5	60.74
350.0	1.876	950.5	311.8	1.354	0.9833	1.738	292.0	98.21	57.72
360.0	2.320	889.6	329.3	1.402	1.006	1.953	237.6	84.28	54.79
370.0	2.843	809.9	348.7	1.454	1.043	2.476	177.8	69.50	52.15
380.0	3.460	670.2	374.1	1.519	1.142	6.942	105.8	49.40	54.24
382.51	3.635	489.2	395.5	1.574				31.29	
Saturated Vapor									
240.0	0.05234	3.071	360.8	1.682	0.7109	0.7935	136.1	8.714	9.021
250.0	0.08417	4.793	368.0	1.678	0.7312	0.8183	137.5	9.326	9.798
260.0	0.1297	7.201	375.0	1.675	0.7518	0.8449	138.5	9.932	10.57
270.0	0.1925	10.47	382.0	1.674	0.7729	0.8740	139.0	10.53	11.34
280.0	0.2767	14.82	388.8	1.675	0.7948	0.9061	139.0	11.13	12.12
290.0	0.3866	20.50	395.5	1.676	0.8174	0.9423	138.4	11.73	12.91
300.0	0.5270	27.82	401.9	1.678	0.8409	0.9840	137.1	12.34	13.75
310.0	0.7029	37.18	408.0	1.680	0.8654	1.033	135.2	12.96	14.65
320.0	0.9196	49.13	413.6	1.682	0.8909	1.094	132.4	13.59	15.65
330.0	1.183	64.41	418.8	1.683	0.9185	1.173	128.7	14.24	16.84
340.0	1.499	84.15	423.2	1.684	0.9502	1.286	124.0	14.93	18.34
350.0	1.876	110.2	426.6	1.682	0.9880	1.465	118.1	15.70	20.37
360.0	2.320	146.1	428.4	1.677	1.034	1.792	110.7	16.61	23.47
370.0	2.843	200.5	427.4	1.666	1.095	2.628	101.4	17.94	29.17
380.0	3.460	313.8	417.6	1.633	1.219	10.05	88.92	21.45	48.22

TABLE 13. Thermophysical Properties of Sulfur Hexafluoride (SF_6) at Saturation

T/ K	P/ MPa	ρ/ kg m^{-3}	H/ kJ kg^{-1}	S/ kJ kg^{-1} K^{-1}	C_v/ kJ kg^{-1} K^{-1}	C_p/ kJ kg^{-1} K^{-1}	u/ m s^{-1}	η/ µPa s	λ/ mW m^{-1} K^{-1}	D
Saturated Liquid										
223.555	0.2314	1845	154.1	0.8175	0.5275	0.8371	552.3	410.2	76.44	1.781
230.0	0.3013	1813	159.5	0.8414	0.5407	0.8573	521.5	373.3	74.78	1.765
240.0	0.4401	1761	168.3	0.8784	0.5604	0.8902	474.7	320.3	72.01	1.739
250.0	0.6220	1705	177.4	0.9151	0.5796	0.9264	428.6	272.8	69.06	1.712
260.0	0.8546	1646	186.9	0.9517	0.5986	0.9681	382.8	231.0	65.97	1.682
270.0	1.146	1580	196.8	0.9884	0.6176	1.019	336.9	195.0	62.76	1.650
280.0	1.506	1507	207.2	1.025	0.6372	1.087	290.2	163.9	59.40	1.615
290.0	1.942	1423	218.3	1.063	0.6589	1.190	241.6	136.5	55.86	1.575
300.0	2.468	1319	230.3	1.103	0.6863	1.380	189.8	111.3	52.06	1.526
310.0	3.098	1175	244.3	1.147	0.7308	1.961	131.5	85.19	48.20	1.460
318.7232	3.755	742.3	269.5	1.225				39.99		1.276

T/ K	P/ MPa	ρ/ kg m^{-3}	H/ kJ kg^{-1}	S/ kJ kg^{-1} K^{-1}	C_v/ kJ kg^{-1} K^{-1}	C_p/ kJ kg^{-1} K^{-1}	u/ m s^{-1}	η/ μPa s	λ/ mW m^{-1} K^{-1}	D
Saturated Vapor										
223.555	0.2314	19.56	264.7	1.312	0.4833	0.5631	112.8	11.91	7.628	1.007
230.0	0.3013	25.14	267.4	1.310	0.4996	0.5851	112.8	12.26	8.085	1.009
240.0	0.4401	36.21	271.6	1.309	0.5249	0.6219	112.3	12.82	8.855	1.012
250.0	0.6220	50.87	275.6	1.308	0.5506	0.6636	111.0	13.39	9.722	1.017
260.0	0.8546	70.10	279.5	1.308	0.5762	0.7121	109.0	14.01	10.73	1.024
270.0	1.146	95.32	283.0	1.308	0.6026	0.7739	106.1	14.69	11.94	1.033
280.0	1.506	128.7	286.1	1.307	0.6327	0.8623	102.1	15.50	13.49	1.044
290.0	1.942	173.9	288.4	1.305	0.6650	0.9997	96.92	16.56	15.59	1.060
300.0	2.468	238.4	289.6	1.300	0.7029	1.272	90.15	18.16	18.75	1.083
310.0	3.098	344.1	288.2	1.289	0.7618	2.195	81.33	21.28	24.78	1.122

TABLE 14. Thermophysical Properties of Sulfur Dioxide (SO_2) at Saturation

T/ K	P/ MPa	ρ/ kg m^{-3}	H/ kJ kg^{-1}	S/ kJ kg^{-1} K^{-1}	C_v/ kJ kg^{-1} K^{-1}	C_p/ kJ kg^{-1} K^{-1}	u/ m s^{-1}	η/ μPa s	λ/ mW m^{-1} K^{-1}	D
Saturated Liquid										
197.7	0.001666	1628	-89.40	-0.3905	0.8774	1.386	1301	999.9	214.0	25.30
200.0	0.002033	1622	-86.21	-0.3745	0.8745	1.384	1291	961.6	213.2	24.85
220.0	0.009353	1572	-58.72	-0.2435	0.8524	1.367	1208	707.8	205.9	21.41
240.0	0.03201	1521	-31.47	-0.1250	0.8358	1.359	1124	543.7	197.6	18.60
260.0	0.08790	1470	-4.273	-0.01630	0.8239	1.360	1040	429.4	188.4	16.26
280.0	0.2042	1417	23.07	0.08470	0.8159	1.372	954.9	345.7	178.3	14.28
300.0	0.4170	1362	50.78	0.1798	0.8104	1.395	868.4	282.1	167.5	12.57
320.0	0.7699	1303	79.12	0.2704	0.8071	1.434	779.4	232.2	155.9	11.08
340.0	1.313	1239	108.4	0.3579	0.8062	1.496	686.8	191.9	143.5	9.742
360.0	2.100	1167	139.2	0.4439	0.8089	1.596	588.9	158.3	130.4	8.517
380.0	3.193	1083	172.2	0.5305	0.8183	1.777	484.3	129.1	116.5	7.353
400.0	4.662	976.3	209.1	0.6215	0.8413	2.190	371.1	101.8	101.4	6.179
420.0	6.602	814.2	255.8	0.7298	0.8999	4.194	245.7	72.73	85.52	4.804
430.64	7.887	517.5	320.6	0.8773						2.989
Saturated Vapor										
197.7	0.001666	0.06503	354.9	1.857	0.4377	0.5686	182.3	8.464	5.775	1.000
200.0	0.002033	0.07845	356.2	1.837	0.4392	0.5702	183.3	8.562	5.854	1.000
220.0	0.009353	0.3292	367.2	1.693	0.4528	0.5866	191.3	9.418	6.556	1.001
240.0	0.03201	1.041	377.9	1.581	0.4701	0.6098	198.5	10.28	7.281	1.004
260.0	0.08790	2.672	388.0	1.493	0.4925	0.6429	204.5	11.14	8.030	1.009
280.0	0.2042	5.882	397.2	1.421	0.5207	0.6882	209.1	12.00	8.820	1.019
300.0	0.4170	11.54	405.2	1.361	0.5540	0.7479	212.1	12.87	9.701	1.036
320.0	0.7699	20.78	411.7	1.310	0.5916	0.8256	213.2	13.78	10.77	1.062
340.0	1.313	35.17	416.3	1.263	0.6327	0.9302	212.2	14.78	12.22	1.102
360.0	2.100	57.04	418.2	1.219	0.6777	1.084	208.9	15.94	14.35	1.164
380.0	3.193	90.49	416.2	1.173	0.7292	1.348	202.8	17.43	17.79	1.260
400.0	4.662	144.5	407.8	1.118	0.7939	1.951	193.7	19.63	24.16	1.426
420.0	6.602	251.8	383.8	1.035	0.8949	4.940	180.5	24.15	41.91	1.799

TABLE 15. Thermophysical Properties of Toluene ($CH_3C_6H_5$) at Saturation

T/ K	P/ MPa	ρ/ kg m^{-3}	H/ kJ kg^{-1}	S/ kJ kg^{-1} K^{-1}	C_v/ kJ kg^{-1} K^{-1}	C_p/ kJ kg^{-1} K^{-1}	u/ m s^{-1}	η/ μPa s	λ/ mW m^{-1} K^{-1}
Saturated Liquid									
178.0	0.00000003939	974.8	-344.9	-1.261	1.024	1.472	1888	36480	159.1
180.0	0.00000005534	972.9	-342.0	-1.244	1.024	1.472	1877	23770	158.8
200.0	0.000001083	953.5	-312.5	-1.089	1.039	1.479	1768	4583	155.1
220.0	0.00001148	934.6	-282.7	-0.9471	1.068	1.504	1665	2237	150.7
240.0	0.00007754	916.0	-252.2	-0.8147	1.108	1.542	1566	1375	145.9
260.0	0.0003731	897.5	-220.9	-0.6895	1.156	1.590	1472	946.1	140.8

T/ K	*P*/ MPa	*ρ*/ kg m^{-3}	*H*/ kJ kg^{-1}	*S*/ kJ kg^{-1} K^{-1}	C_v/ kJ kg^{-1} K^{-1}	C_p/ kJ kg^{-1} K^{-1}	*u*/ m s^{-1}	*η*/ μPa s	*λ*/ mW m^{-1} K^{-1}
280.0	0.001383	879.0	-188.6	-0.5697	1.210	1.646	1381	697.7	135.4
300.0	0.004177	860.4	-155.1	-0.4541	1.268	1.707	1294	539.7	129.9
320.0	0.01073	841.7	-120.3	-0.3418	1.329	1.773	1211	432.4	124.3
340.0	0.02417	822.6	-84.12	-0.2323	1.391	1.842	1130	355.9	118.8
360.0	0.04898	803.1	-46.54	-0.1250	1.455	1.915	1051	299.2	113.4
380.0	0.09099	783.0	-7.483	-0.01956	1.518	1.989	974.1	255.7	108.2
400.0	0.1573	762.2	33.10	0.08428	1.581	2.066	898.1	221.3	103.2
420.0	0.2562	740.5	75.25	0.1868	1.644	2.145	822.7	193.3	98.52
440.0	0.3970	717.6	119.0	0.2882	1.706	2.228	747.4	169.9	94.15
460.0	0.5897	693.3	164.5	0.3887	1.767	2.317	671.7	149.8	90.15
480.0	0.8456	667.0	211.9	0.4886	1.827	2.415	594.9	132.1	86.53
500.0	1.177	638.1	261.2	0.5882	1.887	2.529	516.5	116.0	83.34
520.0	1.596	605.5	312.8	0.6881	1.947	2.671	435.6	100.9	80.58
540.0	2.121	567.2	367.3	0.7891	2.010	2.876	350.8	86.07	78.27
591.75	4.126	292.0	565.8	1.130				28.32	
Saturated Vapor									
178.0	0.00000003939	0.000002453	147.1	1.503	0.5940	0.6843	136.0	4.470	4.130
180.0	0.00000005534	0.000003407	148.5	1.480	0.6004	0.6907	136.7	4.509	4.198
200.0	0.000001083	0.00006002	163.0	1.288	0.6668	0.7570	143.1	4.907	4.968
220.0	0.00001148	0.0005783	178.8	1.151	0.7373	0.8275	149.3	5.316	5.889
240.0	0.00007754	0.003581	196.1	1.053	0.8111	0.9014	155.1	5.733	6.950
260.0	0.0003731	0.01591	214.8	0.9865	0.8876	0.9781	160.7	6.156	8.141
280.0	0.001383	0.05483	235.0	0.9433	0.9660	1.057	166.0	6.584	9.452
300.0	0.004177	0.1549	256.7	0.9184	1.046	1.138	171.0	7.013	10.87
320.0	0.01073	0.3745	279.7	0.9080	1.126	1.221	175.5	7.443	12.39
340.0	0.02417	0.7995	303.9	0.9090	1.208	1.305	179.4	7.870	13.99
360.0	0.04898	1.545	329.3	0.9190	1.289	1.391	182.7	8.296	15.67
380.0	0.09099	2.757	355.7	0.9361	1.370	1.479	185.1	8.720	17.41
400.0	0.1573	4.612	382.8	0.9587	1.450	1.570	186.4	9.144	19.22
420.0	0.2562	7.327	410.7	0.9854	1.530	1.663	186.6	9.573	21.10
440.0	0.3970	11.17	439.0	1.015	1.610	1.762	185.4	10.01	23.07
460.0	0.5897	16.48	467.5	1.047	1.689	1.868	182.6	10.48	25.15
480.0	0.8456	23.74	495.9	1.080	1.767	1.988	177.9	10.98	27.43
500.0	1.177	33.64	523.7	1.113	1.846	2.132	170.9	11.55	30.02
520.0	1.596	47.33	550.4	1.145	1.927	2.323	161.2	12.22	33.16
540.0	2.121	66.88	574.7	1.173	2.012	2.625	148.1	13.10	37.37

VIRIAL COEFFICIENTS OF SELECTED GASES

Henry V. Kehiaian

This table gives second virial coefficients of about 110 inorganic and organic gases as a function of temperature. Selected data from the literature have been fitted by least squares to the equation

$$B / \text{cm}^3\text{mol}^{-1} = \sum_{i=1}^{n} a(i)[(T_0 / T) - 1]^{i-1}$$

where T_0 = 298.15 K. The table gives the minimum and maximum temperatures for which the equation is valid as well the the value of the second virial coefficient B at those temperature. The column definitions are as follows.

Column heading	Definition
Name	Compound name; listed alphabetically
Mol. form.	Molecular formula
T_{min}	Minimum temperature for which equation above is valid, in K
B_{Tmin}	Value of second virial coefficient B at T_{min}, in units $cm^3\ mol^{-1}$
T_{max}	Maximum temperature for which equation above is valid, in KB
B_{Tmax}	Value of second virial coefficient B at T_{max}, in units $cm^3\ mol^{-1}$
$a(1)$	First coefficient of smoothing equation
$a(2)$	Second coefficient of smoothing equation
$a(3)$	Third coefficient of smoothing equation
$a(4)$	Fourth coefficient of smoothing equation, when needed
$a(5)$	Fifth coefficient of smoothing equation, when needed

The smoothing equation may be used with the tabulated coefficients for interpolation within the indicated temperature range. It should not be used for extrapolation beyond this range. A useful compilation of virial coefficient data from the literature may be found in the reference.

Reference

Dymond, J. H., and Smith, E. B,, *The Virial Coefficients of Pure Gases and Mixtures, A Critical Compilation*, Oxford University Press, Oxford, 1980.

Virial Coefficients and Coefficients in the Equation Describing Their Temperature Dependence

Name	Mol. form.	T_{min}/ K	B_{Tmin}/ $cm^3 mol^{-1}$	T_{max}/ K	B_{Tmax}/ $cm^3 mol^{-1}$	$a(1)$	$a(2)$	$a(3)$	$a(4)$	$a(5)$
Acetaldehyde	CH_3CHO	290	-1352	470	-283	-1217	-4647	-5725		
Acetone	$(CH_3)_2CO$	300	-1996	480	-375	-2051	-8903	-18056	-16448	
Acetonitrile	CH_3CN	330	-3468	410	-1425	-5840	-29175	-47611		
Acetylene	$HC{\equiv}CH$	200	-573	270	-263	-216	-375	-716		
Ammonia	NH_3	290	-302	420	-101	-271	-1022	-2715	-4189	
Argon	Ar	100	-184	1000	22	-16	-60	-9.7	-1.5	
Benzene	C_6H_6	290	-1588	600	-291	-1477	-3851	-3683	-1423	
Boron trifluoride	BF_3	200	-338	440	-23	-106	-330	-251	-80	
Bromomethane	CH_3Br	280	-645	380	-274	-559	-1324			
Butane	C_4H_{10}	250	-1170	550	-171	-735	-1835	-1922	-1330	
1-Butanol	C_4H_9OH	350	-1693	440	-593	-2629	-6315			
2-Butanol	$C_4H_{10}O$	380	-1110	420	-721	-2232	-5209			
2-Butanone	C_4H_8O	310	-2056	370	-1135	-2282	-5907			
1-Butene	C_4H_8	300	-624	420	-294	-633	-1442	-932		
Carbon dioxide	CO_2	220	-244	1100	19	-127	-288	-118		
Carbon disulfide	CS_2	280	-932	430	-375	-807	-1829	-1371		
Carbon monoxide	CO	210	-36	480	11	-9	-58	-18		
Chlorine	Cl_2	210	-508	900	-12	-303	-555	9	329	68
1-Chlorobutane	C_4H_9Cl	330	-1224	570	-309	-1643	-4897	-6178	-3718	
Chlorodifluoromethane	$CHClF_2$	300	-343	425	-158	-347	-575	187		
Chloroethane	C_2H_5Cl	320	-634	600	-114	-777	-2205	-1764		
Chloromethane	CH_3Cl	280	-466	600	-58	-407	-887	-385		
1-Chloropropane	C_3H_7Cl	310	-1001	580	-198	-1121	-3271	-3786	-1974	
Chlorotrifluoromethane	CF_3Cl	240	-369	540	-39	-223	-504	-340	-291	
Cyclohexane	C_6H_{12}	300	-1698	560	-368	-1733	-5618	-9486	-7936	
Cyclopentane	C_5H_{10}	300	-1049	320	-918	-1062	-2116			
Cyclopropane	C_3H_6	300	-383	400	-204	-388	-861	-538		
Dichlorodifluoromethane	CF_2Cl_2	250	-769	460	-174	-486	-1217	-1188	-698	
1,2-Dichloroethane	CH_2ClCH_2Cl	370	-812	570	-295	-1362	-3240	-2100		
Dichlorofluoromethane	$CHCl_2F$	250	-728	450	-271	-562	-862			

Name	Mol. form.	T_{min}/ K	B_{Tmin}/ cm³ mol⁻¹	T_{max}/ K	B_{Tmax}/ cm³ mol⁻¹	$a(1)$	$a(2)$	$a(3)$	$a(4)$	$a(5)$
Dichloromethane	CH_2Cl_2	320	-706	420	-357	-913	-3371	-5013		
1,2-Dichloro-1,1,2,2-tetrafluoroethane	$C_2Cl_2F_4$	300	-801	500	-253	-812	-1773	-963		
Diethylamine	$(C_2H_5)_2NH$	320	-1228	400	-697	-1522	-5204	-15047	-28835	
Diethyl ether	$(C_2H_5)_2O$	280	-1550	420	-340	-1226	-4458	-7746	-10005	
Difluoromethane	CH_2F_2	280	-375	350	-238	-321	-754	-1300		
Dimethylamine	$(CH_3)_2NH$	310	-606	400	-322	-662	-1504	-667		
Dimethyl ether	CH_3OCH_3	275	-536	310	-418	-455	-965			
Ethane	C_2H_6	200	-409	600	-24	-184	-376	-143	-54	
Ethanol	C_2H_5OH	320	-2710	390	-622	-4475	-29719	-56716		
Ethyl acetate	$C_4H_8O_2$	330	-1543	400	-878	-2272	-8818	-13130		
Ethylamine	$C_2H_5NH_2$	300	-773	400	-363	-785	-2012	-1397		
Ethylene	$CH_2{=}CH_2$	240	-218	450	-52	-140	-296	-101		
Ethyl formate	C_2H_5OCHO	330	-1003	390	-614	-1371	-4231	-4312		
Fluorine	F_2	80	-378	260	-14	8.5	-163.2	84	-27.9	
Fluoromethane	CH_3F	280	-244	420	-87	-209	-525	-365		
Helium	He	2	-172	700	13	12.44	-1.25			
Heptane	C_7H_{16}	300	-2782	700	-304	-2834	-8523	-10068	-5051	
1-Heptene	$CH_2CH(CH_2)_4CH_3$	340	-1781	410	-1073	-2491	-6230	-3780		
Hexane	C_6H_{14}	300	-1920	450	-616	-1961	-6691	-13167	-15273	
Hydrogen	H_2	15	-230	400	18	15.4	-9	-0.21		
Hydrogen chloride	HCl	190	-451	470	-54	-144	-325	-277	-170	
Iodine pentafluoride	IF_5	320	-2540	410	-1443	-3077	-8474	-9116		
Iodomethane	CH_3I	310	-725	380	-427	-844	-3353	-6590		
Isobutane	$(CH_3)_3CH$	270	-900	510	-215	-707	-1719	-1282		
Isopentane	C_5H_{12}	280	-1263	450	-424	-1095	-2503	-1534		
Krypton	Kr	110	-363	700	8	-51	-118	-29	-5	
Methane	CH_4	110	-328	600	10	-43	-114	-19	-7	
Methanol	CH_3OH	320	-1431	400	-557	-1752	-4694			
Methyl acetate	$C_3H_6O_2$	320	-1320	390	-749	-1709	-6348	-9650		
Methylamine	CH_3NH_2	300	-451	550	-122	-459	-1191	-995		
Methylcyclopentane	$C_5H_9CH_3$	305	-1447	345	-1117	-1512	-2910			
Methyl formate	CH_3OCHO	320	-821	400	-435	-1035	-3425	-4203		
Methyl propanoate	$CH_3CH_2C(O)OCH_3$	330	-1588	400	-908	-2216	-7339	-8658		
2-Methyl-1-propanol	$C_4H_{10}O$	390	-1076	440	-636	-2269	-5065			
2-Methyl-2-propanol	$(CH_3)_3COH$	380	-924	420	-567	-1952	-4775			
2-Methylpyridine	C_6H_7N	360	-1656	430	-972	-2940	-8813	-7809		
3-Methylpyridine	C_6H_7N	380	-1819	430	-1166	-6304	-30415	-44549		
4-Methylpyridine	C_6H_7N	380	-1787	430	-1163	-6553	-32873	-49874		
Molybdenum(VI) fluoride	MoF_6	300	-896	390	-491	-914	-2922	-4778		
Neon	Ne	60	-25	600	15	10.8	-7.5	-0.4		
Neopentane	$C(CH_3)_4$	300	-916	550	-218	-931	-2387	-2641	-1810	
Nitric oxide	NO	120	-232	270	-24	-12	-119	89	-73	
Nitrogen	N_2	75	-274	700	24	-4.3	-55.7	-11.8		
Nitrous oxide	N_2O	240	-219	400	-68	-130	-307	-248		
Octane	C_8H_{18}	300	-4042	700	-375	-4123	-13120	-16408	-8580	
1-Octene	$CH_2CH(CH_2)_5CH_3$	360	-2147	410	-1485	-3273	-6557			
Oxygen	O_2	90	-241	400	-1	-16	-62	-8	-3	
Pentane	C_5H_{12}	300	-1234	550	-294	-1254	-3345	-2726		
2-Pentanone	$C_5H_{10}O$	330	-2850	390	-1332	-4962	-26372	-46537		
1-Pentene	$C_3H_7CH{=}CH_2$	310	-966	410	-495	-1055	-2377	-1189		
Phosphine	PH_3	190	-457	290	-166	-146	-733	1022	-1220	
Phosphorus(V) fluoride	PF_5	320	-162	460	-64	-186	-345			
Propane	C_3H_8	240	-641	560	-90	-386	-844	-720	-574	
1-Propanol	$CH_3CH_2CH_2OH$	380	-873	420	-606	-2690	-12040	-16738		
2-Propanol	$CH_3CHOHCH_3$	380	-821	420	-533	-3165	-16092	-24197		
Propene	$CH_3CH{=}CH_2$	280	-395	500	-106	-347	-727	-325		
Propyl formate	C_3H_7OCHO	330	-1496	400	-834	-2118	-7299	-8851		
Pyridine	C_5H_5N	350	-1257	440	-659	-1765	-3431			
Sulfur dioxide	SO_2	290	-465	470	-132	-430	-1193	-1029		
Sulfur hexafluoride	SF_6	200	-685	500	-68	-279	-647	-335	-72	

Name	Mol. form.	T_{min}/ K	B_{Tmin}/ $cm^3 mol^{-1}$	T_{max}/ K	B_{Tmax}/ $cm^3 mol^{-1}$	*a*(1)	*a*(2)	*a*(3)	*a*(4)	*a*(5)
Tetrachloromethane	CCl_4	320	-1345	420	-814	-1600	-4059	-4653		
Tetrafluoromethane	CF_4	250	-137	800	33	-88	-238	-70		
Tetrafluorosilane	SiF_4	210	-268	450	-32	-138	-312			
Toluene	$C_6H_5CH_3$	350	-1641	430	-903	-2620	-7548	-6349		
Trichlorofluoromethane	CCl_3F	240	-1140	480	-265	-786	-1428	-142		
Trichloromethane	$CHCl_3$	320	-1001	400	-559	-1193	-2936	-1751		
1,1,2-Trichloro-1,2,2-trifluoroethane	$CFCl_2CF_2Cl$	290	-1041	450	-500	-999	-1479			
Triethylamine	$(C_2H_5)_3N$	330	-1562	400	-983	-2061	-5735	-5899		
Trifluoromethane	CHF_3	200	-433	400	-91	-177	-399	-250		
Trimethylamine	$(CH_3)_3N$	310	-675	370	-450	-737	-1669	-986		
Tungsten(VI) fluoride	WF_6	320	-641	460	-317	-719	-1143			
Uranium(VI) fluoride	UF_6	320	-1030	440	-560	-1204	-2690	-2144		
Water	H_2O	300	-1126	1200	-11	-1158	-5157	-10301	-10597	-4415
Xenon	Xe	160	-421	650	-14	-130	-262	-87		
o-Xylene	$C_6H_4(CH_3)_2$	380	-2046	440	-1261	-5632	-22873	-28900		
m-Xylene	$C_6H_4(CH_3)_2$	380	-2082	440	-1184	-5808	-23244	-27607		
p-Xylene	$C_6H_4(CH_3)_2$	380	-2043	440	-1171	-4921	-16843	-16159		

MEAN FREE PATH AND RELATED PROPERTIES OF GASES

In the simplest version of the kinetic theory of gases, molecules are treated as hard spheres of diameter d which make binary collisions only. In this approximation the mean distance traveled by a molecule between successive collisions, the mean free path l, is related to the collision diameter by:

$$l = \frac{kT}{\pi\sqrt{2}Pd^2}$$

where P is the pressure, T the absolute temperature, and k the Boltzmann constant. At standard conditions (P = 100,000 Pa and T = 298.15 K), this relation becomes:

$$l = \frac{9.27 \cdot 10^{27}}{d^2}$$

where l and d are in meters.

Using the same model and the same standard pressure, the collision diameter can be calculated from the viscosity η by the kinetic theory relation:

$$\eta = \frac{2.67 \cdot 10^{-20}(MT)^{1/2}}{d^2}$$

where η is in units of μPa s and M is the molar mass in g mol^{-1}. Kinetic theory also gives a relation for the mean velocity $\bar{v}$ of molecules of mass m:

$$\bar{v} = \left(\frac{8kT}{\pi m}\right)^{1/2} = 145.5(T/M)^{1/2}\,\text{m/s}$$

Finally, the mean time τ between collisions can be calculated from the relation $\tau\bar{v} = l$, or $\tau = l/\bar{v}$; for argon, τ = 72.3 nm / 397 m s^{-1} = 0.182 × 10^{-9} s = 182 ps.

The table below gives values of the collision diameter, mean free path, mean velocity, and mean time betweeen collisons for some common gases at 25 °C and atmospheric pressure, all calculated from measured gas viscosities (see Refs. 2 and 3 and the table "Viscosity of Gases" in this section). Column definitions are as follows.

Column heading	Definition
Name	Name of gas
Mol. form.	Molecular formula of gas
d	Collision diameter, in nm
l	Mean free path, in nm
$\bar{v}$	Mean velocity, in m s^{-1}
τ	Mean time between collisions, in ps

It is seen from the above equations that the mean free path varies directly with T and inversely with P, while the mean velocity varies as the square root of T and, in this approximation, is independent of P.

A more accurate model, in which molecular interactions are described by a Lennard-Jones potential, gives mean free path values about 5% lower than this table (see Ref. 4).

References

1. Reid, R. C., Prausnitz, J. M., and Poling, B. E., *The Properties of Gases and Liquids, Fourth Edition*, McGraw-Hill, New York, 1987.
2. Lide, D. R., and Kehiaian, H. V., *CRC Handbook of Thermophysical and Thermochemical Data*, CRC Press, Boca Raton, FL, 1994.
3. Vargaftik, N. B., *Tables of Thermophysical Properties of Liquids and Gases, Second Edition*, John Wiley, New York, 1975. <https://doi.org/10.1007/978-3-642-52504-9_13>
4. Kaye, G. W. C., and Laby, T. H., *Tables of Physical and Chemical Constants, 15th Edition*, Longman, London, 1986.

Mean Free Path, Mean Velocity, and Time between Collisions for Common Gases

Name	Mol. form.	d/nm	l/nm	$\bar{v}$/m s^{-1}	τ/ps
Air		0.366	69.1	467	148
Ammonia	NH_3	0.432	49.9	609	82
Argon	Ar	0.358	72.3	397	182
Carbon dioxide	CO_2	0.453	45.1	379	119
Helium	He	0.215	200	1256	159
Hydrogen	H_2	0.271	126	1769	71
Krypton	Kr	0.408	55.6	274	203
Neon	Ne	0.254	143	559	256
Nitrogen	N_2	0.370	67.5	475	142
Oxygen	O_2	0.355	73.7	444	166
Xenon	Xe	0.478	40.5	219	185

INFLUENCE OF PRESSURE ON FREEZING POINTS

This table contains data on the variation of the freezing point of representative types of liquids with pressure. Substances are listed in alphabetical order. Note that 1 MPa = 0.01 kbar = 9.87 atm. The columns of the data table are defined as follows.

Column heading	Definition
Name	Chemical name of the substance
Mol. form.	Molecular formula written in the Hill format
0.1013 MPa	Freezing point at 0.1013 MPa (1 atmosphere), in °C
100 MPa	Freezing point at 100 MPa, in °C
1000 MPa	Freezing point at 1000 MPa, in °C

References

1. Isaacs, N. S., *Liquid Phase High Pressure Chemistry*, John Wiley, New York, 1981.
2. Merrill, L., *J. Phys. Chem. Ref. Data*, 6, 1205, 1977; 11, 1005, 1982. <https://doi.org/10.1063/1.555670>

Freezing Point in °C for Selected Gases at Different Pressures

Name	Mol. form.	0.1013 MPa	100 MPa	1000 MPa
Acetic acid	$C_2H_4O_2$	17	37	
Acetophenone	C_8H_8O	19.4	41.2	
Aniline	C_6H_7N	-6.0	13.5	140
Benzene	C_6H_6	5.538	33.4	
Benzonitrile	C_7H_5N	-12.82	7.6	
Benzyl alcohol	C_7H_8O	-15.5	0.2	
Bromobenzene	C_6H_5Br	-30.74	-12	108
Bromoethane	C_2H_5Br	-118.4	-108	
1-Bromonaphthalene	$C_{10}H_7Br$	6.1	6.1	
1-Bromopropane	C_3H_7Br	-110.1	-98	
4-Bromotoluene	C_7H_7Br	26.2	56.7	
Butanoic acid	$C_4H_8O_2$	-5.12	13.8	
1-Butanol	$C_4H_{10}O$	-88.60	-77.2	
Carbon disulfide	CS_2	-111.7	-98	
Chlorobenzene	C_6H_5Cl	-45.2	-28	84
4-Chlorotoluene	C_7H_7Cl	7.4	33.1	
o-Cresol	C_7H_8O	31.0	47.7	
m-Cresol	C_7H_8O	12.2	25.6	
p-Cresol	C_7H_8O	34.77	56.2	
Cyclohexane	C_6H_{12}	6.7	32.5	
Cyclohexanol	$C_6H_{12}O$	26	62.3	
1,2-Dibromoethane	$C_2H_4Br_2$	9.8	34.0	
p-Dichlorobenzene	$C_6H_4Cl_2$	53.1	79.1	
Dichloromethane	CH_2Cl_2	-94.9	-83	
N,N-Dimethylaniline	$C_8H_{11}N$	2.1	26.3	
1,4-Dioxane	$C_4H_8O_2$	11.75	23	
Ethanol	C_2H_6O	-114.14	-108	
Formamide	CH_3NO	2.57	10.8	
Formic acid	CH_2O_2	8.3	20.6	
Furan	C_4H_4O	-85.58	-73	
Hexamethyldisiloxane	$C_6H_{18}OSi_2$	-68.2	-37	
(-)-Menthol	$C_{10}H_{20}O$	42.1	60	
Methyl benzoate	$C_8H_8O_2$	-12.35	31.8	
2-Methyl-2-butanol	$C_5H_{12}O$	-8.7	13.4	
2-Methyl-2-propanol	$C_4H_{10}O$	25.81	58.1	
Naphthalene	$C_{10}H_8$	80.22	115.7	
Nitrobenzene	$C_6H_5NO_2$	5.65	13.5	
3-Nitrotoluene	$C_7H_7NO_2$	15.9	40.6	
Pentachloroethane	C_2HCl_5	-29.0	-6.3	
Potassium	K	63.5	78	170
Potassium chloride	ClK	771		945
Propanoic acid	$C_3H_6O_2$	-20.5	-1.2	
Silver(I) chloride	AgCl	455		545
Sodium	Na	97.794	106	167
Sodium chloride	ClNa	802.018		997
Sodium fluoride	FNa	996		1115
Tetrachloromethane	CCl_4	-22.8	14.2	
Tribromomethane	$CHBr_3$	8.69	31.5	
Trichloromethane	$CHCl_3$	-63.3	-45.2	
Water	H_2O	0.00	-9.0	
o-Xylene	C_8H_{10}	-25.16	-3.5	
m-Xylene	C_8H_{10}	-47.85	-25.2	
p-Xylene	C_8H_{10}	13.3	46.0	

CRITICAL CONSTANTS OF ORGANIC COMPOUNDS

Chris D. Muzny, Vladimir Diky, Andrei Kazakov, Robert D. Chirico, and Michael Frenkel

The parameters of the liquid-gas critical point are important constants in determining the behavior of fluids. This table lists the critical temperature, pressure, and molar volume, as well as the normal boiling point, for over 850 organic substances. The column definitions for the table are as follows.

Column heading	Definition
Name	Name of organic compound; compounds are listed alphabetically by name
Mol. form.	Molecular formula of compound
T_b	Normal boiling point in K at a pressure of 101.325 kPa (1 atmosphere); "sp" following the value indicates a sublimation point (temperature at which the solid is in equilibrium with the gas at a pressure of 101.325 kPa)
T_c	Critical temperature, in K
P_c	Critical pressure, in MPa
V_c	Critical molar volume, in units $cm^3\ mol^{-1}$
Ref.	Reference number; see text

The listed values of the critical constants are critically evaluated using the NIST ThermoData Engine, TDE (Ref. 1), designed to implement the dynamic data evaluation concept (Refs. 2–5). This concept requires large electronic databases capable of storing essentially all relevant experimental data known to date with detailed descriptions of metadata and uncertainties. The combination of these electronic databases with expert-system software, designed to automatically generate recommended property values based on available experimental and predicted data, leads to the ability to produce critically evaluated data dynamically or "to order." The evaluated data have been generated only for compounds for which experimental data for critical properties are available. Group contribution methods such as Joback-Reid (Ref. 6), Constantinou-Gani (Ref. 7), Marrero-Pardillo (Ref. 8), and Wilson-Jasperson (Ref. 9) as well as quantitative structure-property relationship (QSPR) methods (Ref. 5) were used within the TDE environment to validate available experimental data. Each recommended value in the table is characterized with a combined expanded uncertainty (Ref. 10) (level of confidence, approximately 95%) listed in parentheses. Only references to original experimental data actually used by TDE to generate critically evaluated data are indicated for each compound.

The values of the normal boiling temperatures provided in the table along with the combined expanded uncertainties listed in parentheses have also been critically evaluated using TDE. Additional details on the determination of the normal boiling temperatures using TDE can be found in the "Physical Constants of Organic Compounds" table in Section 3. The remaining values of the normal boiling temperatures (without uncertainties) are taken from the same table.

Critical Constants of Organic Compounds

Name	Mol. form.	T_b/K	T_c/K	P_c/MPa	$V_c/cm^3\ mol^{-1}$	Ref. (see text)
Acetaldehyde	C_2H_4O	294.0(6)	462(8)	7.5(10)	154(5)	11-13
Acetic acid	$C_2H_4O_2$	391.1(2)	593(2)	5.79(3)	171(2)	14-21
Acetic anhydride	$C_4H_6O_3$	412.7(3)	606(1)	4.00(8)	294(12)	22
Acetone	C_3H_6O	329.23(7)	508.1(2)	4.7(1)	221(20)	18, 23-31
Acetonitrile	C_2H_3N	354.8(2)	545.47(7)	4.88(5)	173(59)	32-40
Acetophenone	C_8H_8O	475.3(2)	709.5(7)	4.01(5)	373(40)	11, 41
Acetylene	C_2H_2	188.5 sp	308.4(4)	6.24(4)	119(11)	42-48
Acrylonitrile	C_3H_3N	350.4(2)	540(2)	4.6(1)	211(10)	49
Allene	C_3H_4	238.4(3)	394(4)	6.5(7)	167(8)	50
Allyl alcohol	C_3H_6O	370.1(5)	539.8(6)	5.76(4)	222(9)	51
Allylamine	C_3H_7N	327(2)	540.0(7)	4.83(3)	217(11)	52
Allyl ethyl ether	$C_5H_{10}O$	338(4)	518(10)		320(10)	21
2-Aminobiphenyl	$C_{12}H_{11}N$	571.5(2)	838(2)	3.52(3)	548(99)	53
2-(2-Aminoethoxy)ethanol	$C_4H_{11}NO_2$	496.3(1)	721(4)	4.88(10)	333(19)	54
N-(2-Aminoethyl)ethanolamine	$C_4H_{12}N_2O$	515(5)	739(2)	4.53(9)	340(17)	55
Aniline	C_6H_7N	457.3(4)	704(7)	5.3(1)	291(3)	39, 40, 58-60
Anisole	C_7H_8O	426.8(2)	646.1(2)	4.2(1)	355(12)	25, 39, 40, 49
Benzene	C_6H_6	353.23(7)	562.0(1)	4.90(2)	257(11)	19, 27, 28, 34, 61-97
Benzeneacetic acid	$C_8H_8O_2$	541(2)	766(8)	3.9(3)	372(16)	98
Benzenebutanoic acid	$C_{10}H_{12}O_2$	569(2)	783(8)	3.2(2)	493(18)	98
Benzenebutanol	$C_{10}H_{14}O$	537(3)	746(14)	3.1(2)	493(19)	584
Benzeneethanol	$C_8H_{10}O$	493(3)	724(4)	4.0(2)	390(15)	99
Benzeneheptanoic acid	$C_{13}H_{18}O_2$	585	798(8)	2.5(3)	662(21)	98
Benzenehexanoic acid	$C_{12}H_{16}O_2$	574	794(8)	2.6(3)	611(20)	98
Benzenepentanoic acid	$C_{11}H_{14}O_2$	583(1)	790(8)	3.1(2)	526(19)	98
Benzenepropanoic acid	$C_9H_{10}O_2$	557(2)	776(8)	3.5(2)	440(17)	98
Benzenepropanol	$C_9H_{12}O$	514(4)	732(14)	3.4(5)	450(14)	584
Benzonitrile	C_7H_5N	464(1)	691(9)	4.22(4)	348(7)	39, 40, 101, 102

Name	Mol. form.	T_b/K	T_c/K	P_c/MPa	V_c/cm³ mol⁻¹	Ref. (see text)
Benzophenone	$C_{13}H_{10}O$	579.1(2)	830(2)	3.0(1)	568(44)	101
Benzo[*b*]thiophene	C_8H_6S	494.1(4)	764(2)	4.68(4)	359(59)	100
Benzyl alcohol	C_7H_8O	478.5(2)	715(3)	4.3(2)	333(13)	103
[1,1'-Bicyclohexyl]-2-one	$C_{12}H_{20}O$	537	787(70)		584(20)	104
1,1'-Bicyclopentyl	$C_{10}H_{18}$	463.61(3)	690(2)	3.27(3)	497(41)	105
Biphenyl	$C_{12}H_{10}$	528.4(3)	773(5)	3.43(6)	481(69)	39, 106-109
Bis(2-aminoethyl)amine	$C_4H_{13}N_3$	479.7(3)	710(2)	4.43(7)	350(21)	55
1,1-Bis(difluoromethoxy)-1,2,2,2-tetrafluoroethane	$C_4H_2F_8O_2$	319.78	450(1)	2.40(8)	410(4)	110
Bis(difluoromethyl) ether	$C_2H_2F_4O$	278.7(4)	420.2(1)	4.16(6)	223(13)	111
Bis(2-ethylhexyl) phthalate	$C_{24}H_{38}O_4$	657	835(9)	1.1(2)	1495(27)	112
Bis(2-hydroxyethyl)methylamine	$C_5H_{13}NO_2$	518(1)	742(4)	4.2(4)	404(17)	54
Bis(2,2,2-trifluoroethyl) ether	$C_4H_4F_6O$	336.91	476.31(9)	2.78(5)	365(5)	113
Bis(trimethylsilyl)methane	$C_7H_{20}Si_2$	406	573.9(3)			114
Bromochlorodifluoromethane	$CBrClF_2$	269.3(7)	428(12)	4.31(5)	229(16)	115
Bromodifluoromethane	$CHBrF_2$	257.6(5)	412.0(3)	5.2(1)	173(18)	116
Bromoethane	C_2H_5Br	311.4(6)	503.9(4)	6.2(1)	214(10)	117, 118
1-Bromo-2-fluorobenzene	C_6H_4BrF	427	669.6(6)	4.3(6)	342(18)	119
1-Bromo-3-fluorobenzene	C_6H_4BrF	423	652.0(4)	4.2(6)	337(18)	119
1-Bromo-4-fluorobenzene	C_6H_4BrF	423(2)	654.8(4)	4.2(2)	338(18)	119
1-Bromopropane	C_3H_7Br	344.0(2)	536.9(1)	4.33(6)	271(6)	120
Bromotrifluoromethane	$CBrF_3$	215.4(4)	340.06(5)	3.96(1)	199(6)	121
1-Bromo-2-(trifluoromethyl)benzene	$C_7H_4BrF_3$	440.7	656.5(4)	3.3(8)	415(24)	119
1-Bromo-3-(trifluoromethyl)benzene	$C_7H_4BrF_3$	424.7	627.1(4)	3.2(7)	413(24)	119
1-Bromo-4-(trifluoromethyl)benzene	$C_7H_4BrF_3$	433	629.8(4)	3.2(8)	413(24)	119
1,3-Butadiene	C_4H_6	268.6(2)	425(1)	4.35(7)	221(23)	122, 123
Butanal	C_4H_8O	348.0(2)	537(2)		258(9)	104, 124
Butane	C_4H_{10}	272.7(5)	425.2(1)	3.79(1)	257(4)	29, 125-139
1,4-Butanediamine	$C_4H_{12}N_2$	429(10)	651(7)	4.5(5)	317(14)	140
1,2-Butanediol, (±)-	$C_4H_{10}O_2$	469.57(6)	680(2)	5.4(1)	298(12)	141
1,3-Butanediol	$C_4H_{10}O_2$	481.4(1)	679(17)	4.7(1)	302(67)	54, 141
1,4-Butanediol	$C_4H_{10}O_2$	502.7(4)	724(4)	5.5(2)	307(14)	49, 55
Butanenitrile	C_4H_7N	390.8(4)	585.40(7)	3.82(5)	265(6)	35, 39, 40, 142
1-Butanethiol	$C_4H_{10}S$	371.6(5)	570.1(6)	4.01(2)	324(12)	143, 144
Butanoic acid	$C_4H_8O_2$	436.9(1)	623(6)	4.0(3)	292(10)	145-147
1-Butanol	$C_4H_{10}O$	390.8(2)	563.0(4)	4.43(7)	280(14)	34, 80, 148-155
2-Butanol	$C_4H_{10}O$	372.6(2)	535(4)	4.2(1)	269(4)	148, 149, 153, 156
2-Butanone	C_4H_8O	352.8(2)	537(1)	4.18(2)	274(30)	25, 30, 33, 34, 38, 157
1-Butene	C_4H_8	266.9(2)	419.3(1)	4.00(5)	236(14)	123, 158-161
cis-2-Butene	C_4H_8	276.87(8)	435.7(2)	4.23(2)	235(4)	86, 123, 158
trans-2-Butene	C_4H_8	274.03(9)	428.6(1)	4.03(2)	238(4)	86, 123, 158
2-Butoxyethanol	$C_6H_{14}O_2$	444(2)	633.9(10)	3.3(1)	424(15)	11, 41
1-*tert*-Butoxy-2-ethoxyethane	$C_8H_{18}O_2$	421.3	585(3)	2.5(4)	546(14)	162
2-Butoxyethyl acetate	$C_8H_{16}O_3$	464.3(9)	640(2)	2.7(2)	551(21)	144, 163
1-*tert*-Butoxy-2-methoxyethane	$C_7H_{16}O_2$	404	574(1)	2.8(7)	480(13)	162
1-Butoxy-2-propanol	$C_7H_{16}O_2$	445(3)	625(1)	2.7(1)	479(20)	32
Butyl acetate	$C_6H_{12}O_2$	399.2(1)	578(10)	3.16(6)	403(6)	162, 164-167
sec-Butyl acetate	$C_6H_{12}O_2$	381(4)	571.1(5)	3.01(10)	398(14)	164, 165
tert-Butyl acetate	$C_6H_{12}O_2$	371.1(10)	541(4)	3.0(1)	399(7)	54
Butyl acrylate	$C_7H_{12}O_2$	419.8(6)	597.4(6)	2.76(3)	445(7)	51
Butylamine	$C_4H_{11}N$	350.2(2)	531.9(2)	4.20(4)	291(13)	168
sec-Butylamine	$C_4H_{11}N$	335.86(8)	514.3(2)	4.0(2)	284(11)	168
tert-Butylamine	$C_4H_{11}N$	317.17(7)	483.7(6)	3.85(6)	293(23)	75
Butylbenzene	$C_{10}H_{14}$	456.5(3)	660.5(1)	2.89(3)	498(18)	79, 86, 88
sec-Butylbenzene, (±)-	$C_{10}H_{14}$	446.5(4)	652(1)	2.94(3)	488(39)	58
tert-Butylbenzene	$C_{10}H_{14}$	442.3(3)	648(1)	3.00(3)	474(30)	58, 64
Butyl benzoate	$C_{11}H_{14}O_2$	522(3)	725(14)	2.4(3)	594(10)	169
Butyl butanoate	$C_8H_{16}O_2$	438.10(1)	612(3)	2.4(2)	550(9)	162
Butylcyclohexane	$C_{10}H_{20}$	454.1(6)	653.1(4)	2.57(7)	547(14)	170, 171
tert-Butylcyclohexane	$C_{10}H_{20}$	444.8(4)	652.0(4)	2.82(9)	537(15)	170
tert-Butyl ethyl ether	$C_6H_{14}O$	345.9(1)	509(2)	3.0(2)	394(4)	172
Butyl methyl ether	$C_5H_{12}O$	343.3(3)	512.7(1)	3.37(2)	340(2)	25, 173, 174

Name	Mol. form.	T_b/K	T_c/K	P_c/MPa	V_c/cm³ mol⁻¹	Ref. (see text)
Butyl propanoate	$C_7H_{14}O_2$	418.3(1)	594(1)	2.8(2)	464(10)	162
Butyl silicate	$C_{16}H_{36}O_4Si$	529	682(14)			57
Butyl vinyl ether	$C_6H_{12}O$	367(1)	540(1)	3.12(5)	379(10)	175
γ-Butyrolactone	$C_4H_6O_2$	477.8(4)	731(1)	5(1)	246(18)	49
Chlorobenzene	C_6H_5Cl	404.8(2)	632.4(1)	4.5(1)		19, 33, 34
1-Chlorobutane	C_4H_9Cl	351.6(2)	539.2(6)	4.1(2)	303(11)	119
2-Chlorobutane	C_4H_9Cl	344(8)	518.6(6)	3.4(2)	307(14)	119
1-Chloro-2,4-difluorobenzene	$C_6H_3ClF_2$	400	609.6(4)	4.0(7)	333(16)	119
1-Chloro-2,5-difluorobenzene	$C_6H_3ClF_2$	401	612.5(4)	4.0(7)	333(16)	119
1-Chloro-3,4-difluorobenzene	$C_6H_3ClF_2$	400	609.2(4)	4.0(6)	333(16)	119
1-Chloro-3,5-difluorobenzene	$C_6H_3ClF_2$	391.8	592.0(4)	3.9(7)	327(16)	119
1-Chloro-1,1-difluoroethane	$C_2H_3ClF_2$	264.03(7)	410.31(5)	4.06(3)	230(6)	113, 176-178
1-Chloro-2,2-difluoroethene	C_2HClF_2	254.4(5)	400.5(7)	4.54(7)	197(6)	179
Chlorodifluoromethane	$CHClF_2$	232.4(5)	369.30(5)	4.98(1)	165(2)	180-191
2-Chloro-2-(difluoromethoxy)-1,1,1-trifluoroethane	$C_3H_2ClF_5O$	322.5	467.8(6)	3.05(3)	316(24)	192
Chloroethane	C_2H_5Cl	285.5(2)	460.3(4)	5.24(4)	198(11)	193
Chloroethene	C_2H_3Cl	259.4(3)	425(5)	5.60(3)	171(9)	194
1-Chloro-2-fluorobenzene	C_6H_4ClF	410.8	633.8(4)	4.3(6)	319(21)	119
1-Chloro-3-fluorobenzene	C_6H_4ClF	401(25)	615.9(4)	4.2(6)	324(21)	119
1-Chloro-4-fluorobenzene	C_6H_4ClF	403	620.1(4)	4.2(4)	322(18)	119
1-Chloroheptane	$C_7H_{15}Cl$	432(2)	614(8)	3.1(6)	492(14)	119
1-Chlorohexane	$C_6H_{13}Cl$	408.2(5)	599(3)	3.3(3)	422(12)	119
Chloromethane	CH_3Cl	249.1(3)	416.24(4)	6.72(3)	136(2)	195, 196
2-Chloro-2-methylbutane	$C_5H_{11}Cl$	358(1)	509.1(6)	3.2(5)	397(15)	119
3-Chloro-3-methylpentane	$C_6H_{13}Cl$	389	528(3)		414(14)	119
2-Chloro-2-methylpropane	C_4H_9Cl	324.1(5)	497.8(1)	3.7(4)	308(13)	34
1-Chlorooctane	$C_8H_{17}Cl$	456(3)	643(2)	2.5(4)	543(13)	119
Chloropentafluoroacetone	C_3ClF_5O	281.0(9)	410.6(1)	2.89(1)	277(21)	197
Chloropentafluorobenzene	C_6ClF_5	391.11	570(1)	3.2(2)	367(25)	198
Chloropentafluoroethane	C_2ClF_5	234.0(2)	353.0(2)	3.141(10)	255(4)	199, 200
1-Chloropentane	$C_5H_{11}Cl$	381.1(3)	571.2(4)	3.3(2)	361(12)	119
1-Chloropropane	C_3H_7Cl	319.4(5)	503.3(4)	4.56(4)	268(24)	118, 193, 201, 202
2-Chloropropane	C_3H_7Cl	308.2(6)	482.4(4)	4.25(4)	245(16)	201, 202
1-Chloro-1,2,2,2-tetrafluoroethane	C_2HClF_4	261.19(9)	395.43(6)	3.62(1)	244(4)	203, 204
4-Chlorotoluene	C_7H_7Cl	435.0(2)	615.9(5)	2.33(9)	377(16)	88
1-Chloro-2,2,2-trifluoroethane	$C_2H_2ClF_3$	279.2(6)	425.0(2)	4.02(2)	232(6)	205
Chlorotrifluoroethene	C_2ClF_3	244.9(3)	380.1(1)	3.95(3)	214(12)	206, 207
2-Chloro-1,1,2-trifluoroethyl difluoromethyl ether	$C_3H_2ClF_5O$	330.0	475.0(6)	2.98(3)	343(25)	192
Chlorotrifluoromethane	$CClF_3$	191.67	301.9(2)	3.89(1)	180.3(1)	191, 208-216
o-Cresol	C_7H_8O	464.2(1)	697.6(2)	4.2(2)	336(12)	217, 218
m-Cresol	C_7H_8O	475.4(1)	705.8(4)	4.4(2)	337(12)	39, 40, 217, 218
p-Cresol	C_7H_8O	475.1(1)	704.6(3)	4.1(2)	349(13)	217, 218
Cyanogen	C_2N_2	252.1	397(3)	6.2(4)	149(8)	219
Cycloheptane	C_7H_{14}	392.0(2)	604.2(1)	3.85(4)	361(12)	34, 220, 221
Cyclohexane	C_6H_{12}	353.9(7)	553.4(3)	4.07(1)	307(12)	19, 34, 78, 82, 84, 88, 90, 144, 163, 170, 171, 220, 222-231
Cyclohexanol	$C_6H_{12}O$	434.1(2)	647.1(3)	4.3(1)	334(33)	49, 232, 233
Cyclohexanone	$C_6H_{10}O$	428.6(1)	665(1)	4.61(9)	354(12)	162, 166
Cyclohexene	C_6H_{10}	356.1(2)	560.45(5)	4.43(8)	290(20)	86, 201, 202
Cyclohexylamine	$C_6H_{13}N$	406.8(5)	626.8(9)	3.9(6)	349(15)	170
Cyclooctane	C_8H_{16}	424.3(1)	647.2(4)	3.55(6)	417(28)	34, 170, 220, 221
Cyclopentane	C_5H_{10}	322.4(1)	511.7(2)	4.51(7)	264(10)	34, 78, 90, 234-236
Cyclopentanol	$C_5H_{10}O$	413.6(2)	619(1)	4.9(1)	288(19)	233
Cyclopentanone	C_5H_8O	403.7(2)	624(2)	4.59(5)	276(17)	233
Cyclopentene	C_5H_8	317.4(2)	506.1(2)	4.78(5)	252(16)	90, 144, 163, 201, 202
Cyclopropane	C_3H_6	242(2)	398.2(4)	5.58(2)	164(9)	237, 238
2,2',3,3',4,4',5,5',6,6'-Decafluoro-1,1'-biphenyl	$C_{12}F_{10}$	480(2)	640(4)	2.3(3)	641(40)	239
cis-Decahydronaphthalene	$C_{10}H_{18}$	469.0(3)	702(1)	3.2(3)	492(19)	241
trans-Decahydronaphthalene	$C_{10}H_{18}$	460.5(2)	687(1)	3.1(1)	499(19)	241
Decamethylcyclopentasiloxane	$C_{10}H_{30}O_5Si_5$	486(3)	617.4(3)	1.04(2)	1201(8)	34, 114
Decanal	$C_{10}H_{20}O$	485(3)	674(1)	2.6(3)	601(14)	163, 242

Name	Mol. form.	T_b/K	T_c/K	P_c/MPa	V_c/cm³ mol⁻¹	Ref. (see text)
Decane	$C_{10}H_{22}$	447.3(1)	618.1(9)	2.10(3)	621(35)	16, 33, 65, 78, 86, 131, 243-252
1,10-Decanediamine	$C_{10}H_{24}N_2$	535(12)	736(8)	2.4(3)	654(28)	140
Decanedioic acid	$C_{10}H_{18}O_4$	647(5)	845(13)	2.5(1)	724(21)	253
Decanoic acid	$C_{10}H_{20}O_2$	543(1)	724(5)		638(24)	145, 146
1-Decanol	$C_{10}H_{22}O$	502(3)	690(10)	2.3(1)	624(87)	149, 152, 254, 255
2-Decanol	$C_{10}H_{22}O$	484	668.5(3)	2.3(5)	646(13)	255
3-Decanol	$C_{10}H_{22}O$	490(7)	666.1(3)	2.3(3)	643(13)	255
4-Decanol	$C_{10}H_{22}O$	487(3)	663.7(3)	2.3(1)	643(13)	255
5-Decanol	$C_{10}H_{22}O$	489(5)	663.2(4)	2.3(4)	646(13)	255
2-Decanone	$C_{10}H_{20}O$	484(3)	671.8(5)	2.2(3)	625(25)	256
3-Decanone	$C_{10}H_{20}O$	485(4)	668(1)	2.2(2)	628(15)	256
4-Decanone	$C_{10}H_{20}O$	479.7	662.9(5)	2.2(2)	636(18)	256
5-Decanone	$C_{10}H_{20}O$	477	661.0(4)	2.2(2)	628(25)	256
1-Decene	$C_{10}H_{20}$	444(1)	616.0(3)	2.16(5)	594(3)	257
Decylbenzene	$C_{16}H_{26}$	571(1)	752(8)	1.72(10)	879(29)	258
Decyl silicate	$C_{40}H_{84}O_4Si$		849(16)			57
Dibenzofuran	$C_{12}H_8O$	558.4(3)	824(2)	3.37(3)	494(32)	259
Dibenzothiophene	$C_{12}H_8S$	604.8(4)	897(2)	3.9(2)		260
1,2-Dibromo-1-chloro-1,2,2-trifluoroethane	$C_2Br_2ClF_3$	366.0(2)	560.6(2)	3.61(2)	368(4)	261
1,4-Dibromooctafluorobutane	$C_4Br_2F_8$	371(25)	532(2)	2.4(3)	452(29)	198
Dibutylamine	$C_8H_{19}N$	435(2)	607.5(2)	3.11(3)	532(21)	168
1,4-Di-*tert*-butylbenzene	$C_{14}H_{22}$	510.5(5)	708(2)	2.23(1)	732(70)	232
Dibutyl ether	$C_8H_{18}O$	414.8(3)	584.1(2)	2.4(2)	521(12)	262
Dibutyl phthalate	$C_{16}H_{22}O_4$	611(9)	797(9)	1.6(3)	954(18)	112
m-Dichlorobenzene	$C_6H_4Cl_2$	445(2)	685.7(4)	4.2(2)	366(22)	119
p-Dichlorobenzene	$C_6H_4Cl_2$	447.1(2)	669(5)	3.54(7)	364(22)	263
Dichlorodiethylsilane	$C_4H_{10}Cl_2Si$	403(2)	595.7(6)	3.06(3)	455(4)	264
Dichlorodifluoromethane	CCl_2F_2	243.4(1)	384.9(2)	4.12(1)	218(36)	191, 216, 265
Dichlorodimethylsilane	$C_2H_6Cl_2Si$	343.7(5)	520.3(6)	3.49(3)	350(5)	266
1,1-Dichloroethane	$C_2H_4Cl_2$	329.5(7)	523.4(1)	5.1(5)	248(12)	267
1,2-Dichloroethane	$C_2H_4Cl_2$	356.6(1)	561.5(4)	5.4(1)	225(8)	33, 34, 268, 269
cis-1,2-Dichloroethene	$C_2H_2Cl_2$	333(2)	535.8(4)	5.4(3)	220(15)	34
trans-1,2-Dichloroethene	$C_2H_2Cl_2$	320.79(8)	515.5(2)	5.3(2)	216(14)	33, 34
1,1-Dichloro-1-fluoroethane	$C_2H_3Cl_2F$	305.20(9)	477.3(1)	4.20(2)	253.7(6)	178, 270
Dichlorofluoromethane	$CHCl_2F$	282.1	451.6(4)	5.20(1)	196(1)	271
1,2-Dichloro-1,1,2,3,3,3-hexafluoropropane	$C_3Cl_2F_6$	307(2)	451.8(1)	2.63(7)	365(48)	272
Dichloromethane	CH_2Cl_2	313.0(3)	508.0(2)	6.35(5)	177(13)	273
1,2-Dichloropropane, (±)-	$C_3H_6Cl_2$	369.6	578(2)	4.63(6)	292(6)	119, 232
1,3-Dichloropropane	$C_3H_6Cl_2$	394.0(3)	615(3)	4.7(5)	299(16)	119
1,1-Dichloro-1,2,2,2-tetrafluoroethane	$C_2Cl_2F_4$	276(1)	418.6(8)	3.31(3)	294(8)	179
1,2-Dichloro-1,1,2,2-tetrafluoroethane	$C_2Cl_2F_4$	276.8(5)	418.74(6)	3.25(2)	295(3)	191, 274
1,2-Dichloro-1,1,2-trifluoroethane	$C_2HCl_2F_3$	303.2(1)	461.6(1)	3.77(8)	283.2(5)	178
2,2-Dichloro-1,1,1-trifluoroethane	$C_2HCl_2F_3$	301.0(6)	456.8(2)	3.67(1)	278(2)	182, 275-280
Didecyl phthalate	$C_{28}H_{46}O_4$	736(4)	870(10)	0.94(5)	1807(27)	112
1,1-Diethoxyethane	$C_6H_{14}O_2$	375(2)	539.7(4)	3.22(8)	426(12)	166
1,2-Diethoxyethane	$C_6H_{14}O_2$	393.8(7)	542(3)	2.14(2)	432(11)	162
Diethoxymethane	$C_5H_{12}O_2$	359(2)	532(1)	3.4(5)	370(10)	162
Diethylamine	$C_4H_{11}N$	328.6(1)	499.5(4)	3.75(2)	304(32)	38, 117, 133, 281, 282
p-Diethylbenzene	$C_{10}H_{14}$	457(1)	657.90(5)	2.80(8)	494(12)	79, 86
Diethylene glycol	$C_4H_{10}O_3$	518.7(2)	753(4)	4.8(2)	325(19)	283
Diethylene glycol diethyl ether	$C_8H_{18}O_3$	458(4)	612(10)	2.4(7)	587(18)	162
Diethylene glycol dimethyl ether	$C_6H_{14}O_3$	435(2)	617(4)	3.0(6)	450(16)	162
Diethylene glycol monobutyl ether	$C_8H_{18}O_3$	505(4)	692(3)	2.8(6)	546(26)	41
Diethylene glycol monobutyl ether acetate	$C_{10}H_{20}O_4$	521(2)	694(2)	2.15(5)	627(20)	55
Diethylene glycol monoethyl ether	$C_6H_{14}O_3$	475(3)	670(4)	3.2(1)	427(23)	49
Diethylene glycol monoethyl ether acetate	$C_8H_{16}O_4$	491(1)	670(12)	2.50(6)	524(18)	49, 55
Diethylene glycol monomethyl ether	$C_5H_{12}O_3$	467(2)	672(2)	3.7(2)	378(25)	49
Diethylene glycol monopropyl ether	$C_7H_{16}O_3$	488.0(4)	680(2)	3.05(7)	495(18)	41, 104
Diethyl ether	$C_4H_{10}O$	307.6(5)	466.8(3)	3.64(1)	280(5)	19, 77, 95, 155, 196, 284-301
Diethyl oxalate	$C_6H_{10}O_4$	459(1)	618(2)	2.14(2)	464(37)	101
Diethyl phthalate	$C_{12}H_{14}O_4$	571(2)	776(9)	2.2(2)	687(15)	112

Name	Mol. form.	T_b/K	T_c/K	P_c/MPa	V_c/cm³ mol⁻¹	Ref. (see text)
Diethyl succinate	$C_8H_{14}O_4$	490(1)	663(30)	2.26(2)	567(44)	101, 302
Diethyl sulfide	$C_4H_{10}S$	365.3(2)	557.5(10)	4.0(1)	322(8)	32, 303, 304
o-Difluorobenzene	$C_6H_4F_2$	367.1(5)	566.0(4)	4.28(7)	290(21)	119
m-Difluorobenzene	$C_6H_4F_2$	356.2(5)	548.4(4)	4.20(7)	289(21)	119
p-Difluorobenzene	$C_6H_4F_2$	362.1(3)	556.9(4)	4.28(7)	297(22)	119
1,1-Difluoroethane	$C_2H_4F_2$	249.13	386.4(1)	4.52(1)	178(2)	178, 180, 185, 305-307
1,1-Difluoroethene	$C_2H_2F_2$	187.7(8)	302.9(6)	4.48(5)	155(4)	179, 215
2,2-Difluoroethylbis(trifluoromethyl)amine	$C_4H_3F_8N$	324.6	460.20(9)	2.64(1)	375(1)	308
Difluoromethane	CH_2F_2	221.50(7)	351.28(3)	5.79(1)	121(4)	306, 309-316
3-Difluoromethoxy-1,1,1,2,2-pentafluoropropane	$C_4H_3F_7O$	319.09	455.1(1)	2.77(2)	363(1)	113
2-(Difluoromethoxy)-1,1,1-trifluoroethane	$C_3H_3F_5O$	302.4(2)	444.9(3)	3.43(1)	291(19)	113
2,4-Difluorotoluene	$C_7H_6F_2$	390	581.4(4)	3.7(4)	340(21)	119
2,5-Difluorotoluene	$C_7H_6F_2$	391	587.8(4)	3.8(5)	341(21)	119
2,6-Difluorotoluene	$C_7H_6F_2$	385	581.8(4)	3.7(4)	341(21)	119
3,4-Difluorotoluene	$C_7H_6F_2$	385	598.5(5)	3.8(6)	342(22)	119
Diheptyl phthalate	$C_{22}H_{34}O_4$	633	830(9)	1.24(8)	1153(22)	112
Dihexyl phthalate	$C_{20}H_{30}O_4$	652(5)	817(9)	1.3(1)	1061(22)	112
3,4-Dihydro-2*H*-pyran	C_5H_8O	358.7(2)	561(2)	4.63(8)	268(34)	75
Diisobutylamine	$C_8H_{19}N$	412.8	584.4(2)	3.20(6)	518(22)	168
Diisopropylamine	$C_6H_{15}N$	357(3)	523.1(2)	3.02(2)	407(18)	168
1,4-Diisopropylbenzene	$C_{12}H_{18}$	483.5(2)	675(1)	2.30(4)	610(65)	317
Diisopropyl ether	$C_6H_{14}O$	341.6(2)	500.2(7)	2.85(4)	386(5)	25, 290, 318, 319
1,2-Dimethoxyethane	$C_4H_{10}O_2$	358.2(1)	539(4)	3.91(5)	305(9)	162, 166, 175, 290
Dimethoxymethane	$C_3H_8O_2$	315.5(2)	488(11)	4.0(2)	259(10)	75, 162, 320
1,2-Dimethoxypropane	$C_5H_{12}O_2$	369	543(1)	3.4(6)	356(12)	162
2,2-Dimethoxypropane	$C_5H_{12}O_2$	350.6(7)	510(3)		360(13)	162
Dimethylamine	C_2H_7N	280.5(4)	437.5(4)	5.34(5)	188(13)	193, 322, 323
N,N-Dimethylaniline	$C_8H_{11}N$	466(1)	687.7(6)	3.63(9)	407(15)	39, 40
2,2-Dimethylbutane	C_6H_{14}	322.9(2)	489.1(5)	3.10(1)	364(1)	86, 324-328
2,3-Dimethylbutane	C_6H_{14}	331.2(3)	500.2(3)	3.13(1)	358(1)	19, 86, 252, 324-330
3,3-Dimethyl-2-butanone	$C_6H_{12}O$	379.3(2)	570.9(3)	3.67(3)	383(6)	164, 165
2,3-Dimethyl-1-butene	C_6H_{12}	328.74(4)	497.7(9)	3.31(1)	346(9)	170
3,3-Dimethyl-1-butene	C_6H_{12}	314.39(4)	477.4(9)	3.18(2)	348(10)	170
2,3-Dimethyl-2-butene	C_6H_{12}	346.34(6)	521.0(9)	3.4(1)	344(7)	170
Dimethyl carbonate	$C_3H_6O_3$	363.26(9)	557(1)	4.8(2)		331, 332
cis-1,3-Dimethylcyclohexane	C_8H_{16}	397.6(6)	587.7(5)	2.88(1)	429(10)	64
cis-1,4-Dimethylcyclohexane	C_8H_{16}	397.5(7)	603.2(3)	3.44(2)	434(7)	164, 165
trans-1,4-Dimethylcyclohexane	C_8H_{16}	392.5(5)	588(2)	3.04(1)	439(18)	74
Dimethyl disulfide	$C_2H_6S_2$	382.87(8)	608(4)	5.1(1)	266(8)	54
Dimethyl ether	C_2H_6O	248.4(2)	400.1(8)	5.31(3)	171(3)	125, 174, 186, 333-343
N,N-Dimethylformamide	C_3H_7NO	426.0(5)	649.6(8)	4.4(1)	262(9)	11, 344
Dimethyl glutarate	$C_7H_{12}O_4$	489(4)	682(14)	2.8(4)	488(14)	321
2,2-Dimethylheptane	C_9H_{20}	406(1)	576.7(5)	2.35(7)	546(12)	345
Dimethyl heptanedioate	$C_9H_{16}O_4$	518(2)	711(14)	2.4(1)	608(14)	321
2,2-Dimethylhexane	C_8H_{18}	380.0(4)	549.9(4)	2.53(3)	481(10)	346
2,3-Dimethylhexane	C_8H_{18}	388.8(5)	563.5(4)	2.63(2)	466(16)	346
2,4-Dimethylhexane	C_8H_{18}	382.6(4)	553(3)	2.55(2)	480(39)	330, 346
2,5-Dimethylhexane	C_8H_{18}	382.3(7)	550.0(3)	2.49(2)	485(20)	19, 346
3,3-Dimethylhexane	C_8H_{18}	385.1(6)	562.0(4)	2.65(2)	450(17)	346
3,4-Dimethylhexane	C_8H_{18}	390.9(4)	568.8(4)	2.69(2)	467(21)	346
Dimethyl 1,6-hexanedioate	$C_8H_{14}O_4$	504(3)	692(14)	2.5(5)	561(13)	321
Dimethyl malonate	$C_5H_8O_4$	454.3(6)	647(1)	3.5(1)	368(13)	317, 321
2,7-Dimethylnaphthalene	$C_{12}H_{12}$	535.6(3)	775(2)	3.02(3)	515(45)	347
Dimethyl octanedioate	$C_{10}H_{18}O_4$	532(8)	723(14)	2.3(3)	672(14)	321
Dimethyl oxalate	$C_4H_6O_4$	436.6(5)	632(14)	4.0(3)	315(12)	321, 348
2,2-Dimethyloxirane	C_4H_8O	324(2)	500(4)	4.4(1)	327(31)	54
2,2-Dimethylpentane	C_7H_{16}	352.4(3)	520.6(6)	2.77(4)	409(12)	346, 349
2,3-Dimethylpentane	C_7H_{16}	363.0(6)	537.5(6)	2.92(8)	394(23)	346, 349, 350
2,4-Dimethylpentane	C_7H_{16}	353.6(5)	520.0(7)	2.74(6)	415(19)	346, 349
3,3-Dimethylpentane	C_7H_{16}	359.2(6)	536.4(4)	2.94(2)	413(24)	346
2,3-Dimethyl-1-pentene	C_7H_{14}	357(1)	534(4)	2.9(8)	400(11)	170

Name	Mol. form.	T_b/K	T_c/K	P_c/MPa	V_c/cm³ mol⁻¹	Ref. (see text)
4,4-Dimethyl-1-pentene	C_7H_{14}	345.7(2)	516(4)	2.91(1)	406(11)	170
Dimethyl phthalate	$C_{10}H_{10}O_4$	555.9(2)	772(9)	2.76(8)	557(17)	112
2,3-Dimethylpyridine	C_7H_9N	434.3(4)	655.5(3)	4.03(1)	337(28)	86, 351
2,4-Dimethylpyridine	C_7H_9N	431.6(3)	647.1(9)	3.83(2)	363(34)	90, 352
2,5-Dimethylpyridine	C_7H_9N	430.15(5)	644.2(3)	4.11(4)	371(45)	86
2,6-Dimethylpyridine	C_7H_9N	417.2(1)	623.8(2)	3.80(4)	353(43)	90
3,4-Dimethylpyridine	C_7H_9N	452.3(3)	683.8(4)	4.06(1)	353(13)	86, 351, 352
3,5-Dimethylpyridine	C_7H_9N	445.1(1)	667.3(3)	3.84(2)	369(12)	86, 351
2,6-Dimethylquinoline	$C_{11}H_{11}N$	541.3(4)	786(2)	3.27(2)	505(24)	529
Dimethyl sebacate	$C_{12}H_{22}O_4$	562(3)	742(14)	2.1(2)	695(14)	321
Dimethyl succinate	$C_6H_{10}O_4$	470(1)	662(14)	3.5(2)	426(16)	321
Dimethyl sulfide	C_2H_6S	310.47(5)	503.0(3)	5.40(6)	201(9)	52, 117, 304
Dimethyl sulfoxide	C_2H_6OS	465.1(9)	707(1)	4.6(7)	228(7)	23
1,3-Dimethyl-1,1,3,3-tetraphenyldisiloxane	$C_{26}H_{26}OSi_2$	701	893(9)	1.38(10)	1300(89)	353
1,3-Dimethyltricyclo[3.3.1.1^{3,7}]decane	$C_{12}H_{20}$	476.53	708(2)	2.86(1)	595(115)	141
Dinonyl phthalate	$C_{26}H_{42}O_4$	686	858(9)	1.0(3)	1652(26)	112
Dioctyl phthalate	$C_{24}H_{38}O_4$	688(4)	840(9)	1.1(1)	1510(22)	112
1,4-Dioxane	$C_4H_8O_2$	374.4(3)	587.3(1)	5.2(2)	251(5)	34, 290, 354
Dipentyl phthalate	$C_{18}H_{26}O_4$	614(40)	811(9)		957(21)	112
Diphenyl ether	$C_{12}H_{10}O$	531.2(1)	766.9(8)	3.10(4)	526(23)	25, 355
Diphenylmethane	$C_{13}H_{12}$	537.4(3)	776(9)	3.02(7)		39, 302, 356, 357
Dipropylamine	$C_6H_{15}N$	380.7(9)	555.8(1)	3.6(1)	414(15)	168
Dipropylene glycol	$C_6H_{14}O_3$	504(2)	705(4)	3.4(1)	444(18)	54
Dipropyl ether	$C_6H_{14}O$	363.3(3)	531(2)	2.92(5)	402(8)	25, 52
Dipropyl phthalate	$C_{14}H_{18}O_4$	592(2)	784(9)	1.9(1)	816(20)	112
Diundecyl phthalate	$C_{30}H_{50}O_4$	711(25)	886(10)	0.89(10)	1590(25)	112
Docosane	$C_{22}H_{46}$	642(5)	786(6)	1.0(1)	1434(50)	62, 358, 359
Docosanoic acid	$C_{22}H_{44}O_2$	693(3)	837(8)	1.11(8)	1485(29)	360
1-Docosanol	$C_{22}H_{46}O$	680(12)	827(8)		1243(20)	361
1,2,2,3,3,4,4,5,5,6,6,7-Dodecafluoro-1-heptanol	$C_7H_4F_{12}O$	444.8	589(5)	2.0(2)	620(34)	362
Dodecane	$C_{12}H_{26}$	489.5(2)	658.8(9)	1.80(9)	747(18)	34, 86, 118, 243, 244, 363, 364
1,12-Dodecanediamine	$C_{12}H_{28}N_2$	572(13)	767(8)	2.0(3)	765(37)	140
Dodecanedioic acid	$C_{12}H_{22}O_4$	621(10)	859(13)	2.1(2)	730(19)	253
1-Dodecanethiol	$C_{12}H_{26}S$	550(3)	734(4)	1.81(10)	726(25)	99
Dodecanoic acid	$C_{12}H_{24}O_2$	572(1)	743(7)	1.9(2)	787(19)	360
1-Dodecanol	$C_{12}H_{26}O$	537.3(3)	719.4(6)	2.02(5)	805(15)	149
2-Dodecanone	$C_{12}H_{24}O$	520(6)	702(4)		742(20)	256
3-Dodecanone	$C_{12}H_{24}O$	523	701(2)	1.9(2)	680(19)	256
4-Dodecanone	$C_{12}H_{24}O$	528	697(2)	1.9(4)	672(19)	256
5-Dodecanone	$C_{12}H_{24}O$	521	695(5)	1.9(4)	678(19)	256
6-Dodecanone	$C_{12}H_{24}O$	522	694(2)	1.9(3)	677(19)	256
1-Dodecene	$C_{12}H_{24}$	486.6(9)	657.6(6)	1.88(1)	710(15)	257
Eicosane	$C_{20}H_{42}$	617.3(9)	768(6)	1.08(5)	1325(48)	358, 359
Eicosanoic acid	$C_{20}H_{40}O_2$	673(6)	820(8)	1.2(1)	1346(27)	360
1-Eicosanol	$C_{20}H_{42}O$	629	808(8)	1.1(2)	1130(18)	361
1-Eicosene	$C_{20}H_{40}$	614	772(15)	1.1(3)	1213(36)	365
Ethane	C_2H_6	184.6(4)	305.36(4)	4.88(1)	146(3)	44, 48, 92, 131, 132, 134, 214, 215, 229, 246, 366-393
1,2-Ethanediamine	$C_2H_8N_2$	390.1(5)	613.1(3)	6.71(4)	204(13)	49
1,2-Ethanediol	$C_2H_6O_2$	470.7(1)	719(5)	8.1(4)	180(11)	11, 32, 41, 394
1,1-Ethanediol, diacetate	$C_6H_{10}O_4$	441(3)	618(4)	2.9(1)	457(13)	54
Ethanethiol	C_2H_6S	308.2(1)	498.7(3)	5.53(8)	208(10)	304
Ethanol	C_2H_6O	351.39(9)	515(1)	6.25(4)	169(4)	19, 25, 34, 80, 85, 128, 148-150, 152, 222, 395-409
Ethanolamine	C_2H_7NO	443.5(4)	671(3)	8.0(5)	207(13)	41
Ethoxybenzene	$C_8H_{10}O$	443.0(2)	647(2)	3.45(5)	407(16)	39, 40
2-Ethoxyethyl acetate	$C_6H_{12}O_3$	429.8(4)	609(2)	3.07(3)	443(38)	141, 144, 163
2-Ethoxy-2-methylbutane	$C_7H_{16}O$	374.7(4)	546(2)	2.83(9)	448(29)	410
1-Ethoxy-1,1,2,2,3,3,4,4,4-nonafluorobutane	$C_6H_5F_9O$	350.04	482.0(1)	1.98(1)	518(2)	113
Ethyl acetate	$C_4H_8O_2$	350.3(2)	523.27(7)	3.88(2)	288(19)	19, 23, 400, 411-413
Ethylamine	C_2H_7N	289.8(2)	456.5(9)	5.6(1)	183(22)	193, 414

Name	Mol. form.	T_b/K	T_c/K	P_c/MPa	V_c/cm³ mol⁻¹	Ref. (see text)
Ethylbenzene	C_8H_{10}	409.4(4)	617.1(1)	3.61(1)	365(61)	68, 78, 79, 86, 88, 172, 345
Ethyl benzoate	$C_9H_{10}O_2$	485.7(2)	700(14)	3.01(5)	470(12)	169
Ethyl butanoate	$C_6H_{12}O_2$	394.3(4)	566.1(1)	3.2(3)	421(30)	415, 416
Ethyl *trans*-2-butenoate	$C_6H_{10}O_2$	413(5)	599(10)		382(7)	21
Ethylcyclohexane	C_8H_{16}	405.0(4)	606.9(4)	3.27(4)	431(12)	170, 171
Ethylcyclopentane	C_7H_{14}	376.7(6)	569.48(5)	3.40(7)	377(3)	236
Ethyl 2,2-dimethylpropanoate	$C_7H_{14}O_2$	391.5(4)	566(2)	2.88(2)	461(8)	417
Ethylene	C_2H_4	169.4(3)	282.35(3)	5.06(1)	130.9(2)	61, 246, 371, 418-436
Ethyl 3-ethoxypropanoate	$C_7H_{14}O_3$	441(2)	621(3)	2.7(1)	478(19)	11, 41
Ethyl formate	$C_3H_6O_2$	327.24(10)	508.5(5)	4.78(3)		413, 415
Ethyl heptanoate	$C_9H_{18}O_2$	461(2)	634(1)	2.3(2)	587(8)	162
3-Ethylhexane	C_8H_{18}	391.7(5)	565.5(4)	2.61(2)	450(15)	346
Ethyl hexanoate	$C_8H_{16}O_2$	438(1)	615(1)	2.6(2)	528(8)	162
2-Ethylhexanoic acid	$C_8H_{16}O_2$	500.7(1)	674(1)	2.75(6)	543(8)	144, 163
2-Ethyl-1-hexanol	$C_8H_{18}O$	459.4(2)	640.2(3)	3.0(2)	508(29)	437
2-Ethylhexyl acetate	$C_{10}H_{20}O_2$	473(1)	642(2)	2.02(1)	644(10)	410
Ethyl 3-methylbutanoate	$C_7H_{14}O_2$	408(3)	584(6)		463(9)	147, 162
Ethyl methyl ether	C_3H_8O	279(2)	437.8(2)	4.39(6)	219(5)	174, 304, 438
3-Ethyl-2-methylpentane	C_8H_{18}	388.8(6)	567.1(4)	2.70(2)	442(22)	346
3-Ethyl-3-methylpentane	C_8H_{18}	391.4(9)	576.5(4)	2.77(1)	463(13)	346
Ethyl 2-methylpropanoate	$C_6H_{12}O_2$	384(2)	554(4)	3.1(3)	421(76)	415
Ethyl methyl sulfide	C_3H_8S	339.8(3)	533(10)	4.62(3)	260(6)	303
Ethyl nonanoate	$C_{11}H_{22}O_2$	497(5)	664(1)	2.0(4)	715(8)	162
Ethyl octanoate	$C_{10}H_{20}O_2$	479(1)	652(12)		657(8)	147, 162
3-Ethylpentane	C_7H_{16}	366.6(4)	540.7(4)	2.90(3)	412(14)	346, 349
Ethyl pentanoate	$C_7H_{14}O_2$	415(3)	593(1)	2.8(4)	466(10)	162
2-Ethylphenol	$C_8H_{10}O$	477.7(1)	703(1)	3.7(3)	388(15)	218
3-Ethylphenol	$C_8H_{10}O$	491.6(1)	716(1)	3.8(3)	393(15)	218
4-Ethylphenol	$C_8H_{10}O$	491.12(6)	716(1)	3.1(6)	395(15)	218
Ethyl propanoate	$C_5H_{10}O_2$	372.1(2)	547(1)	3.37(5)	343(23)	19, 64, 162, 413, 415
Ethyl propyl ether	$C_5H_{12}O$	336(3)	500.2(4)	3.37(1)	343(44)	25, 304
Ethyl silicate	$C_8H_{20}O_4Si$	441(1)	587(12)	2.0(4)	701.3(2)	57
S-Ethyl thioacetate	C_4H_8OS	387(3)	590.5(2)	4.1(1)	320(10)	49
4-Ethyltoluene	C_9H_{12}	435.2(6)	640.2(5)	3.23(4)	446(12)	439, 440
Ethyl vinyl ether	C_4H_8O	309(2)	475(2)	4.06(4)	262(8)	290
Fluorobenzene	C_6H_5F	357.9(3)	560.10(7)	4.55(1)	272(9)	86, 441
Fluoroethane	C_2H_5F	235.5(3)	375.2(2)	5.02(1)	164(3)	206, 442, 443
Fluoromethane	CH_3F	194.84	317.42(1)	5.88(1)	112.4(1)	444, 445
2-Fluorotoluene	C_7H_7F	387(2)	591.2(4)	3.9(3)	323(47)	119
3-Fluorotoluene	C_7H_7F	389(2)	591.8(4)	3.9(3)	332(17)	119
4-Fluorotoluene	C_7H_7F	389.8(4)	592.1(8)	3.85(1)	332(17)	119
Formic acid	CH_2O_2	374	588(10)		115.9(1)	146
Furan	C_4H_4O	304.5(2)	490.2(2)	5.43(8)	218(3)	241, 290
Glycerol	$C_3H_8O_3$	562(3)	850(9)	7.6(8)	251(15)	394
Heneicosane	$C_{21}H_{44}$	632(6)	778(8)	1.0(1)	1366(48)	359
Heptadecane	$C_{17}H_{36}$	576(2)	736(1)	1.33(7)	1081(28)	359
Heptadecanoic acid	$C_{17}H_{34}O_2$	635(4)	792(8)	1.4(1)	1130(24)	360
1-Heptadecanol	$C_{17}H_{36}O$	597	780(8)	1.4(1)	1097(18)	361
1-Heptadecene	$C_{17}H_{34}$	574(3)	734(7)	1.34(8)	1053(35)	365
2,2,3,3,5,5,6-Heptafluoro-1,4-dioxane	$C_4HF_7O_2$	312.6	453(1)	2.86(6)	359(4)	110
1,1,1,2,2,3,3-Heptafluoropentan-4-one	$C_5H_3F_7O$	337.5	476.55(8)	2.57(1)	394.2(7)	308
1,1,1,2,3,3,3-Heptafluoropropane	C_3HF_7	256.81	375.0(1)	2.93(1)	299(7)	446-449
1,1,1,2,4,4,4-Heptafluoro-2-trifluoromethoxybutane	$C_5H_2F_{10}O$	322.73	447(1)	2.15(6)	465(5)	110
1,1,1,2,2,3,3-Heptafluoro-3-(trifluoromethoxy)propane	$C_4F_{10}O$	280.1	391.7(7)	1.89(5)	431(31)	450
2,2,4,4,6,8,8-Heptamethylnonane	$C_{16}H_{34}$	519(1)	692(4)	1.53(1)	957(27)	451
1,1,1,3,5,5,5-Heptamethyltrisiloxane	$C_7H_{22}O_2Si_3$	416(1)	553.4(6)	1.48(2)	828(3)	452
Heptanal	$C_7H_{14}O$	426(3)	616.8(4)	3.2(2)	434(7)	242
2-Heptanamine	$C_7H_{17}N$	414(4)	598(2)	2.9(3)	455(18)	170

Fluids

Name	Mol. form.	T_b/K	T_c/K	P_c/MPa	V_c/cm³ mol⁻¹	Ref. (see text)
Heptane	C_7H_{16}	371.53(7)	540.1(2)	2.74(1)	428(15)	34, 36, 62, 75, 78, 86, 131, 133-135, 157, 243, 244, 254, 286, 324, 346, 359, 394, 432, 453- 463
Heptanedioic acid	$C_7H_{12}O_4$	615.2	842(13)	3.3(2)	463(15)	253
Heptanoic acid	$C_7H_{14}O_2$	495(2)	678(2)	3.0(3)	476(5)	145, 146, 464
1-Heptanol	$C_7H_{16}O$	451(1)	632.4(6)	3.1(2)	430(9)	149, 254, 465
2-Heptanol, (±)-	$C_7H_{16}O$	432	608.4(6)	3.0(1)	442(2)	149, 465
3-Heptanol, (*S*)-	$C_7H_{16}O$	436(2)	605.4(3)	3.1(4)	451(3)	465
4-Heptanol	$C_7H_{16}O$	434(2)	602.6(3)	3.1(6)	455(4)	465
2-Heptanone	$C_7H_{14}O$	424.2(3)	611.4(2)	2.98(4)	436(4)	25, 172, 466
3-Heptanone	$C_7H_{14}O$	419(2)	606.6(2)	3.0(1)	433(5)	466
4-Heptanone	$C_7H_{14}O$	417(1)	602.0(2)	3.0(3)	434(5)	466
1-Heptene	C_7H_{14}	367(1)	537.3(3)	2.85(2)	409(2)	51, 52, 86, 124, 257, 467
cis-2-Heptene	C_7H_{14}	370(2)	548.5(6)	3.0(3)	410(7)	170
trans-2-Heptene	C_7H_{14}	371(2)	542.8(4)	3.0(2)	410(7)	170
trans-3-Heptene	C_7H_{14}	369(2)	538.6(7)	3.0(2)	411(7)	170
Heptylbenzene	$C_{13}H_{20}$	515(4)	708(7)	2.1(2)	680(16)	258
Heptyl silicate	$C_{28}H_{60}O_4Si$		778(16)			57
Hexacosane	$C_{26}H_{54}$	688(11)	816(8)	0.8(2)	1740(59)	358
Hexadecane	$C_{16}H_{34}$	560.1(7)	722.2(8)	1.4(2)	1009(53)	34, 244, 245
Hexadecanoic acid	$C_{16}H_{32}O_2$	624(6)	785(8)	1.5(2)	1059(23)	360
1-Hexadecanol	$C_{16}H_{34}O$	598(2)	770(8)	1.47(10)	1019(17)	361
1-Hexadecene	$C_{16}H_{32}$	558(1)	718(7)	1.4(2)	986(33)	365
Hexaethyldisiloxane	$C_{12}H_{30}OSi_2$	525(7)	692.9(1)		955(2)	34
Hexafluoroacetylacetone	$C_5H_2F_6O_2$	342(2)	485.1(5)	2.9(2)	313(49)	468
Hexafluorobenzene	C_6F_6	353.4(2)	516.4(5)	3.28(1)	337(4)	74, 198, 286, 469-473
2,2,4,4,5,5-Hexafluoro-1,3-dioxolane	$C_3F_6O_2$	251.0	368.1(7)	2.72(4)	293(30)	116
Hexafluoroethane	C_2F_6	195.1(1)	292.9(2)	3.03(1)	223(3)	172, 474-477
1,1,1,2,3,3-Hexafluoro-3-(2,2,3,3,3-pentafluoropropoxy)propane	$C_6H_3F_{11}O$	360.64	486.48(7)	1.95(1)	529(2)	113
1,1,1,2,3,3-Hexafluoropropane	$C_3H_2F_6$	279.35	412.40(6)	3.42(1)	270(5)	203, 447, 448, 478, 479
1,1,1,3,3,3-Hexafluoropropane	$C_3H_2F_6$	271.8(2)	398.07(6)	3.18(1)	262(18)	203
1,1,1,2,3,3-Hexafluoro-3-(2,2,3,3-tetrafluoropropoxy) propane	$C_6H_4F_{10}O$	379.07	516.2(3)	2.2(2)	543(34)	113
1,1,1,2,3,3-Hexafluoro-3-(2,2,2-trifluoroethoxy) propane	$C_5H_3F_9O$	345.87	475.74(9)	2.23(2)	455(2)	113
1,1,1,3,3,3-Hexafluoro-2-trifluoromethyl-2-methoxypropane	$C_5H_3F_9O$	327	463(1)	2.37(7)	448(5)	110
Hexamethylbenzene	$C_{12}H_{18}$	541(3)	758(2)	2.6(4)	581(15)	25
1,1,1,5,5,5-Hexamethyl-3,3-bis[(trimethylsilyl)oxy] trisiloxane	$C_{12}H_{36}O_4Si_5$	494.7	622.6(2)	1.03(2)	1323(90)	34
Hexamethyldisiloxane	$C_6H_{18}OSi_2$	373.7(3)	518.7(6)	1.95(2)	629(15)	480
2,6,10,15,19,23-Hexamethyltetracosane	$C_{30}H_{62}$	693(6)	796(2)	0.60(4)	2060(70)	481
Hexanal	$C_6H_{12}O$	402.8(4)	592(3)	3.4(2)	378(7)	163, 242
Hexane	C_6H_{14}	341.87(6)	507.5(1)	3.03(1)	366.0(8)	19, 29, 33, 34, 38, 52, 77, 78, 84, 86, 96, 106, 124, 131, 133, 134, 243, 244, 254, 281, 286, 318, 324-328, 401, 456, 458, 470, 482-488
1,6-Hexanediamine	$C_6H_{16}N_2$	470(2)	685(7)	3.6(5)	446(17)	140
1,6-Hexanedioic acid	$C_6H_{10}O_4$	610.7	841(13)	3.8(3)	449(17)	253
1,6-Hexanediol	$C_6H_{14}O_2$	481	741(10)	4.1(1)	404(13)	489
Hexanenitrile	$C_6H_{11}N$	436.7(3)	633.8(2)	2.99(6)	378(8)	35
Hexanoic acid	$C_6H_{12}O_2$	478.1(6)	661(7)		413(15)	145, 146, 360, 464
1-Hexanol	$C_6H_{14}O$	430.1(7)	611.0(4)	3.40(9)	381(30)	34, 149, 151, 152, 254, 255, 456
2-Hexanol	$C_6H_{14}O$	411(6)	585.9(5)	3.3(3)	406(8)	149, 255, 437
3-Hexanol	$C_6H_{14}O$	416(2)	582.4(4)	3.3(1)	378(14)	64, 254, 255
2-Hexanone	$C_6H_{12}O$	400.8(1)	586.7(5)	3.31(4)	377(4)	25, 172, 466
3-Hexanone	$C_6H_{12}O$	396.7(3)	583.1(5)	3.32(1)	378(4)	25, 466
Hexatriacontane	$C_{36}H_{74}$	777(7)	872(9)	0.47(7)	2711(2)	358
1-Hexene	C_6H_{12}	336.6(1)	504.1(9)	3.20(3)	381(9)	86, 201, 202, 252, 257, 490

Name	Mol. form.	T_b/K	T_c/K	P_c/MPa	V_c/cm³ mol⁻¹	Ref. (see text)
cis-2-Hexene	C_6H_{12}	342.1(5)	513.4(9)	3.34(6)	347(7)	170
trans-2-Hexene	C_6H_{12}	341.00(9)	509.0(7)	3.16(1)	353(7)	170
cis-3-Hexene	C_6H_{12}	339.6(5)	510(1)	3.29(1)	351(7)	170
trans-3-Hexene	C_6H_{12}	340.21(9)	507(2)	3.18(1)	352(7)	170
5-Hexen-2-one	$C_6H_{10}O$	402.3(5)	593.5(6)	3.51(4)	359(10)	51
Hexyl acetate	$C_8H_{16}O_2$	444.3(7)	618(1)	2.5(1)	526(8)	162
Hexylamine	$C_6H_{15}N$	405(1)	592.3(7)	3.4(3)	402(15)	170
Hexylbenzene	$C_{12}H_{18}$	499(2)	695(7)	2.4(2)	620(14)	258
Hexyl benzoate	$C_{13}H_{18}O_2$	545	748(14)	2.0(3)	658(14)	169
Hexyl silicate	$C_{24}H_{52}O_4Si$		757(16)			57
Indan	C_9H_{10}	451.0(4)	684.8(4)	3.95(3)	385(22)	25
Isobutanal	C_4H_8O	337.3(2)	543.6(6)	5.12(8)	283(10)	164, 165
Isobutane	C_4H_{10}	261.5(5)	407.84(7)	3.64(2)	256(7)	125, 132, 491, 492
Isobutene	C_4H_8	266.2(2)	418.0(3)	4.00(4)	240(3)	123, 158, 493, 494
Isobutyl acetate	$C_6H_{12}O_2$	390.1(6)	562(2)	2.97(6)	369(37)	166, 201, 202, 415
Isobutylbenzene	$C_{10}H_{14}$	445.9(4)	650(3)	3.0(2)	493(15)	96
Isobutyl butanoate	$C_8H_{16}O_2$	430(1)	611(6)	2.5(3)	524(9)	147
Isobutylcyclohexane	$C_{10}H_{20}$	444.5	642.1(6)	2.61(7)	550(14)	170
Isobutyl formate	$C_5H_{10}O_2$	371.6(3)	551(4)	3.9(4)	359(25)	415
Isobutyl isobutanoate	$C_8H_{16}O_2$	421(3)	602(6)		530(9)	147
Isobutyl 3-methylbutanoate	$C_9H_{18}O_2$	442(3)	621(6)		581(9)	147
Isobutyl propanoate	$C_7H_{14}O_2$	409(2)	586(8)		462(9)	162, 167
Isopentane	C_5H_{12}	300.98(6)	460.37(9)	3.35(6)	313(17)	19, 86, 96, 127, 495
Isopentyl acetate	$C_7H_{14}O_2$	414.8(7)	586.1(4)	2.76(7)	464(9)	166
Isopentyl butanoate	$C_9H_{18}O_2$	458.0(3)	619(6)		595(10)	147
Isopentyl nitrite	$C_5H_{11}NO_2$	372(3)	626(16)	5.07(4)	386.2(1)	52
Isopentyl propanoate	$C_8H_{16}O_2$	446(4)	611(6)		523(9)	147
Isopropyl acetate	$C_5H_{10}O_2$	361.8(2)	531.1(6)	3.31(4)	343(4)	64, 166, 172, 411, 490
Isopropylamine	C_3H_9N	305.0(2)	472.2(9)	4.55(7)	231(5)	75, 170
Isopropylbenzene	C_9H_{12}	425.6(2)	631(1)	3.2(1)	423(5)	79, 90, 96, 172
Isopropylcyclohexane	C_9H_{18}	427.6(4)	632.2(4)	3.1(2)	484(14)	170
Isopropyl formate	$C_4H_8O_2$	341(2)	534.6(5)	3.95(3)	294(11)	164, 165
1-Isopropyl-4-methylbenzene	$C_{10}H_{14}$	450(2)	654(8)	2.8(1)	495(16)	96, 147
Isopropyl methyl ether	$C_4H_{10}O$	304.0(5)	464.4(2)	3.76(1)	287(11)	25
Isoquinoline	C_9H_7N	516.4(6)	803(8)	5.07(3)	380(17)	218
d-Limonene	$C_{10}H_{16}$	450.8(5)	653(2)	2.81(2)	498(11)	496
(-)-Menthol	$C_{10}H_{20}O$	487(12)	694(5)	2.7(5)	539(14)	
Mesityl oxide	$C_6H_{10}O$	402.9(4)	605(2)	3.85(2)	353(27)	497
Methane	CH_4	111.7(2)	190.56(2)	4.60(1)	99(3)	132, 134, 498-508
Methane-*d*₄	CD_4		189.2(6)		98(3)	507
Methanethiol	CH_4S	279.2(1)	469.9(3)	7.24(9)	148(4)	304
Methanol	CH_4O	337.7(7)	512.7(6)	8.01(3)	117(4)	19, 38, 67, 80, 85, 87, 131, 148, 152, 155, 196, 405, 458, 482, 509-516
1-Methoxy-2,4-dimethylbenzene	$C_9H_{12}O$	465	682(4)	3.2(7)	451(18)	162
2-Methoxy-1,4-dimethylbenzene	$C_9H_{12}O$	467	677(1)	3.2(7)	451(15)	162
2-Methoxyethanol	$C_3H_8O_2$	397.5(1)	598(1)	5.28(8)	263(6)	49
2-Methoxyethyl acetate	$C_5H_{10}O_3$	415(3)	603(3)	3.6(5)	368(15)	162
4-Methoxy-1,1,1,2,2,3,3-heptafluorobutane	$C_5H_5F_7O$	344.13	481.5(2)	2.38(1)	431(2)	113
1-Methoxy-1,1,2,2,3,3-hexafluoropropane	$C_4H_4F_6O$	341.02	487.0(3)	2.9(1)	370(25)	113
2-Methoxy-2-methylbutane	$C_6H_{14}O$	359.6(1)	536(2)	3.23(9)	372(19)	172, 410, 517
5-Methoxy-1,1,2,2,3,3,4,4-octafluoropentane	$C_6H_6F_8O$	395.83	546.1(3)	2.40(7)	493(30)	113
1-Methoxy-2-propanol	$C_4H_{10}O_2$	393.2(6)	579.8(3)	4.11(4)	304(12)	49
2-Methoxy-1-propene	C_4H_8O	308.9(3)	478.5(6)	4.2(3)	257(13)	518
Methyl acetate	$C_3H_6O_2$	329.9(2)	506.7(4)	4.73(7)	227(22)	19, 411-413
Methylamine	CH_5N	266.8(3)	430.6(6)	7.61(9)	139(1)	130, 193, 322, 323
2-Methylaniline	C_7H_9N	473.2(4)	710(1)	3.6(1)	377(52)	519
3-Methylaniline	C_7H_9N	476.5(5)	709(10)	4.6(5)	346(17)	106
4-Methylaniline	C_7H_9N	474(1)	667(10)	3.3(7)	334(16)	106
N-Methylaniline	C_7H_9N	470(1)	702(5)	5.2(6)	347(16)	117
2-Methylanisole	$C_8H_{10}O$	446(2)	662(1)	3.6(3)	397(18)	162

Name	Mol. form.	T_b/K	T_c/K	P_c/MPa	V_c/cm³ mol⁻¹	Ref. (see text)
3-Methylanisole	$C_8H_{10}O$	450(2)	665(1)	3.6(3)	449(22)	162
4-Methylanisole	$C_8H_{10}O$	448(2)	667(1)	3.6(4)	396(15)	162
α-Methylbenzenemethanol	$C_8H_{10}O$	478(4)	699(5)	3.8(7)	399(16)	32
Methyl benzoate	$C_8H_8O_2$	472(2)	702(1)	3.8(1)	408(11)	517
2-Methylbutanal, (±)-	$C_5H_{10}O$	363	531.6(1)	4.04(3)	318(10)	124
Methyl butanoate	$C_5H_{10}O_2$	375.1(1)	554.4(1)	3.49(8)	341(20)	19, 413, 416
3-Methylbutanoic acid	$C_5H_{10}O_2$	449.7(2)	629(1)	3.4(2)	355(10)	146
2-Methyl-1-butanol, (±)-	$C_5H_{12}O$	402.2(4)	575.4(5)	3.9(1)	342(8)	254
3-Methyl-1-butanol	$C_5H_{12}O$	404.0(3)	579(2)	3.9(3)	335(7)	21, 91, 222, 254, 511, 520
2-Methyl-2-butanol	$C_5H_{12}O$	375.6	544(1)	3.71(5)	326(9)	147, 254
3-Methyl-2-butanol, (±)-	$C_5H_{12}O$	386.9(4)	556.1(5)	3.9(4)	336(9)	254
3-Methyl-2-butanone	$C_5H_{10}O$	367.4(2)	553.1(3)	3.83(10)	321(33)	157, 166
3-Methyl-1-butene	C_5H_{10}	293.3(2)	452.7(5)	3.51(4)	305(8)	490
2-Methyl-2-butene	C_5H_{10}	311.7(4)	470(1)	3.4(1)	299(7)	521
Methyl *tert*-butyl ether	$C_5H_{12}O$	328.3(1)	497.0(6)	3.41(5)	335(10)	25, 440
Methylcyclohexane	C_7H_{14}	374.1(1)	572.3(2)	3.48(9)	368(3)	34, 74, 78, 86, 88, 171, 235, 236
Methylcyclopentane	C_6H_{12}	345.0(2)	532.78(5)	3.79(5)	322(2)	78, 235, 236
2-Methylcyclopentanone	$C_6H_{10}O$	413(3)	631(2)	4.0(6)	328(17)	162
2-Methyl-*N,N*-dimethylaniline	$C_9H_{13}N$	458(2)	668.0(7)	3.12(8)	466(15)	39, 40
Methyl dodecanoate	$C_{13}H_{26}O_2$	541(2)	712(5)	1.4(4)	842(9)	218
1,1'-Methylenebis[(1-methylethyl)benzene]	$C_{19}H_{24}$	592	795(8)	1.6(1)	871(30)	394
Methyl formate	$C_2H_4O_2$	304.8(3)	487.16(10)	6.01(1)	172(6)	19, 412, 413
2-Methylfuran	C_5H_6O	337.1(2)	528(3)	4.77(8)	252(3)	290
2-Methylheptane	C_8H_{18}	390.8(9)	559.6(1)	2.50(2)	487(12)	86, 346, 522
3-Methylheptane	C_8H_{18}	392.1(6)	563.7(4)	2.54(2)	463(12)	346
4-Methylheptane	C_8H_{18}	390.9(5)	561.7(4)	2.54(2)	480(14)	346
Methyl heptanoate	$C_8H_{16}O_2$	442.9(4)	628(2)	2.6(4)	521(8)	162
4-Methyl-3-heptanol	$C_8H_{18}O$	430(2)	623.5(7)	2.8(4)	505(13)	437
5-Methyl-3-heptanol	$C_8H_{18}O$	427(2)	621.2(3)	2.8(3)	493(13)	437
2-Methyl-3-heptanone	$C_8H_{16}O$	431	615(1)	2.7(3)	487(11)	162
5-Methyl-3-heptanone	$C_8H_{16}O$	432(4)	619(4)	2.7(7)	484(11)	162
2-Methyl-1-heptene	C_8H_{16}	392(2)	567.5(9)	2.6(2)	466(10)	170
2-Methyl-2-heptene	C_8H_{16}	395(2)	569(1)	2.6(4)	465(11)	170
2-Methylhexane	C_7H_{16}	363.2(8)	530.4(1)	2.73(3)	420(15)	86, 346, 349, 522
3-Methylhexane	C_7H_{16}	365	535.4(5)	2.82(6)	405(19)	346, 349
5-Methyl-2-hexanone	$C_7H_{14}O$	412(2)	604(1)	2.9(3)	434(13)	162
2-Methyl-3-hexanone	$C_7H_{14}O$	407(3)	593(1)	2.9(4)	428(13)	162
2-Methyl-1-hexene	C_7H_{14}	365(2)	542(1)	2.9(3)	407(8)	170
5-Methyl-1-hexene	C_7H_{14}	358(1)	528.7(4)	2.9(2)	410(9)	170
N-Methylhexylamine	$C_7H_{17}N$	418	592(1)	2.8(8)	458(20)	170
Methyl isobutanoate	$C_5H_{10}O_2$	365(1)	540.7(5)	3.43(1)	341(42)	19, 413
Methyl methacrylate	$C_5H_8O_2$	373.8(2)	540.3(6)	2.97(6)	320(6)	51
1-Methylnaphthalene	$C_{11}H_{10}$	517.6(9)	771(5)	3.56(7)	479(22)	86, 218, 523
2-Methylnaphthalene	$C_{11}H_{10}$	514.3(3)	761(3)	3.37(6)	464(20)	218
2-Methyloctane	C_9H_{20}	416(1)	582.8(2)	2.30(2)	547(17)	345, 522
Methyloxirane	C_3H_6O	308	488.11(8)	5.44(2)		290, 524
Methyl pentafluoroethyl ether	$C_3H_3F_5O$	278.8(9)	406.81(5)	2.89(1)	301(5)	176, 448, 525
2-Methylpentane	C_6H_{14}	333.36(9)	497.9(2)	3.03(1)	371(2)	86, 127, 252, 325-328, 487, 522
3-Methylpentane	C_6H_{14}	336.5(5)	504.6(2)	3.12(1)	368.7(3)	252, 324-328, 526
Methyl pentanoate	$C_6H_{12}O_2$	400.51(6)	588.9(3)	3.20(5)	398(6)	164, 165
2-Methyl-1-pentanol	$C_6H_{14}O$	430(6)	604.4(5)	3.4(2)	410(8)	254
4-Methyl-1-pentanol	$C_6H_{14}O$	424(2)	603.5(7)	3.4(4)	406(7)	437
2-Methyl-2-pentanol	$C_6H_{14}O$	394(1)	559.5(7)	3.6(4)	410(11)	437
4-Methyl-2-pentanol	$C_6H_{14}O$	405.2(5)	574.4(5)		389(9)	437
2-Methyl-3-pentanol	$C_6H_{14}O$	401.1(2)	576(1)	3.5(1)	380(9)	254
3-Methyl-3-pentanol	$C_6H_{14}O$	402(4)	575.6(6)	3.5(2)	376(10)	254
4-Methyl-2-pentanone	$C_6H_{12}O$	388.9(2)	575.4(10)	3.4(1)	378(12)	166, 527
4-Methyl-1-pentene	C_6H_{12}	327(2)	493.1(5)	3.18(7)		170, 497
2-Methyl-2-pentene	C_6H_{12}	340.5(5)	509.3(5)	3.26(1)	348(7)	170
4-Methyl-*cis*-2-pentene	C_6H_{12}	329.6(1)	496.3(7)	3.24(1)	350(7)	170
Methyl pentyl ether	$C_6H_{14}O$	372(3)	546.5(2)	3.04(10)	395(14)	173, 174

Name	Mol. form.	T_b/K	T_c/K	P_c/MPa	V_c/cm³ mol⁻¹	Ref. (see text)
2-Methyl-1,3-propanediol	$C_4H_{10}O_2$	494(4)	708(2)	5.4(4)	300(12)	55
Methyl propanoate	$C_4H_8O_2$	351.8(2)	530.57(10)	4.0(2)		19, 412, 413, 415
2-Methylpropanoic acid	$C_4H_8O_2$	427.6(2)	605(2)	3.7(3)	296(10)	146
2-Methyl-1-propanol	$C_4H_{10}O$	380.99(7)	548(2)	4.30(4)	274(17)	25, 91, 148, 153, 155
2-Methyl-2-propanol	$C_4H_{10}O$	355.5(1)	506.2(1)	3.98(7)	283(4)	153
Methyl propyl ether	$C_4H_{10}O$	311.7(10)	476.2(2)	3.80(1)	281(7)	25
2-Methylpyridine	C_6H_7N	402.6(2)	622(1)	4.62(4)		75, 528
3-Methylpyridine	C_6H_7N	417.3(1)	644.8(6)	4.63(3)		90, 528
4-Methylpyridine	C_6H_7N	418.5(1)	645.8(5)	4.68(4)		75, 90
N-Methyl-2-pyrrolidinone	C_5H_9NO	477.4(3)	721.7(4)	4.5(4)	330(12)	144, 242
2-Methylquinoline	$C_{10}H_9N$	520.6(4)	778(2)	3.91(2)	447(49)	530
8-Methylquinoline	$C_{10}H_9N$	520.6(7)	787(2)	4.22(2)		530
Methyl salicylate	$C_8H_8O_3$	495.8(5)	709(30)	4.4(7)	436(17)	302
Methyl silicate	$C_4H_{12}O_4Si$	393.3(7)	558(12)	2.8(6)	464(22)	57
2-Methyltetrahydrofuran	$C_5H_{10}O$	353(1)	537(2)	3.74(6)	292(4)	290
(Methylthio)benzene	C_7H_8S	467.5(2)	706(4)	4.1(1)	374(13)	99
Methyl trifluoromethyl ether	$C_2H_3F_3O$	248.0(7)	377.92(6)	3.64(3)	219(2)	447, 525
Methyltris(trimethylsiloxy)silane	$C_{10}H_{30}O_3Si_4$	464.4	597.4(2)	1.23(2)	1089(74)	34
4-Morpholinecarboxaldehyde	$C_5H_9NO_2$	511(1)	779(4)	5.0(4)	326(14)	54
Naphthalene	$C_{10}H_8$	491.2(1)	748.3(4)	4.06(4)	408(21)	79, 82, 86, 241, 251, 355, 451, 531
Neopentane	C_5H_{12}	282.65(6)	433.71(1)	3.20(1)	311.6(7)	532
Nitromethane	CH_3NO_2	374.34(10)	588(3)	6.0(2)	175(2)	533, 534
Nonadecane	$C_{19}H_{40}$	603(3)	756(5)	1.16(7)	1216(43)	358, 359
1-Nonadecene	$C_{19}H_{38}$	604(17)	755(8)	1.2(2)	1196(36)	365
1,1,1,2,2,3,3,4,4-Nonafluorohexan-5-one	$C_6H_3F_9O$	360.47	498.97(8)	2.20(2)	504(2)	308
Nonanal	$C_9H_{18}O$	468(3)	658(2)	2.7(1)	546(10)	242
Nonane	C_9H_{20}	424.0(2)	594.2(5)	2.29(5)	547(23)	34, 36, 78, 86, 131, 243-245, 247, 249, 251, 454, 456, 535, 536
1,9-Nonanediamine	$C_9H_{22}N_2$	531.9	726(7)	2.6(3)	600(23)	140
Nonanedioic acid	$C_9H_{16}O_4$	630.3	844(13)	2.7(2)	586(17)	253
Nonanoic acid	$C_9H_{18}O_2$	529(1)	712(3)		592(16)	146
1-Nonanol	$C_9H_{20}O$	486.9(4)	670.6(5)	2.54(7)	555(75)	149, 152, 254, 255
2-Nonanol, (±)-	$C_9H_{20}O$	466.7	649(1)	2.53(10)	575(11)	149, 255
3-Nonanol, (±)-	$C_9H_{20}O$	468	648.0(3)	2.5(3)	577(12)	255
4-Nonanol	$C_9H_{20}O$	465.7	645.1(3)	2.5(3)	577(12)	255
2-Nonanone	$C_9H_{18}O$	467(1)	652.1(7)	2.5(1)	560(8)	49, 466
3-Nonanone	$C_9H_{18}O$	460(4)	648(4)	2.4(6)	560(7)	466
4-Nonanone	$C_9H_{18}O$	461(4)	643.7(3)	2.4(3)	560(7)	466
5-Nonanone	$C_9H_{18}O$	461.6(3)	641.4(3)	2.35(2)	560(7)	466
1-Nonene	C_9H_{18}	420.1(6)	594(1)	2.38(1)	529(2)	257
Nonyl silicate	$C_{36}H_{76}O_4Si$		830(16)			57
Octacosane	$C_{28}H_{58}$	705(6)	824(8)	0.8(1)	1916(65)	358
Octadecane	$C_{18}H_{38}$	589(2)	748(1)	1.3(1)	1167(41)	244
Octadecanoic acid	$C_{18}H_{36}O_2$	644(3)	803(8)	1.3(2)	1251(27)	360
1-Octadecanol	$C_{18}H_{38}O$	624(2)	790(8)	1.28(10)	1157(18)	361
1-Octadecene	$C_{18}H_{36}$	588.7(10)	748(8)	1.3(1)	1119(34)	365
1,1,1,2,2,3,3,4-Octafluorobutane	$C_4H_2F_8$	300.62	432.0(1)	2.80(2)	360(22)	537
1,2,2,3,3,4,4,5-Octafluoro-1-pentanol	$C_5H_4F_8O$	413.2	571(1)	2.9(1)	440(24)	362
Octafluorotetrahydrofuran	C_4F_8O	272.3	399.6(7)	2.68(9)	350(30)	450
Octamethylcyclotetrasiloxane	$C_8H_{24}O_4Si_4$	448.6(9)	585.8(9)	1.33(1)	1006(67)	114, 538
Octamethyltrisiloxane	$C_8H_{24}O_2Si_3$	425.7(8)	564.1(2)	1.42(1)	882(16)	539
Octanal	$C_8H_{16}O$	447(3)	639.3(3)	3.0(3)	489(7)	163, 242
Octane	C_8H_{18}	398.77(10)	568.7(1)	2.48(1)	490(22)	11, 19, 33, 34, 36, 78, 83, 86, 91, 96, 131, 133, 134, 243-245, 247, 249, 318, 319, 324, 346, 465, 469, 535, 540, 541
1,8-Octanediamine	$C_8H_{20}N_2$	498.8	712(7)	2.8(3)	547(20)	140
Octanedioic acid	$C_8H_{14}O_4$	618.7	843(13)	3.0(2)	520(16)	253
Octanenitrile	$C_8H_{15}N$	475(3)	674.4(4)	2.85(3)	494(10)	35

Name	Mol. form.	T_b/K	T_c/K	P_c/MPa	V_c/cm³ mol⁻¹	Ref. (see text)
Octanoic acid	$C_8H_{16}O_2$	513(1)	694(1)	2.9(3)	522(19)	145, 146, 360, 464
1-Octanol	$C_8H_{18}O$	467.9(8)	651(2)	2.80(7)	490(47)	25, 149, 152, 254, 255
2-Octanol	$C_8H_{18}O$	452(7)	629.5(9)	2.75(4)	519(10)	149, 254, 255
3-Octanol	$C_8H_{18}O$	457(6)	628.4(3)	2.8(4)	515(10)	255
4-Octanol	$C_8H_{18}O$	449.5	625.1(3)	2.8(3)	516(10)	255
2-Octanone	$C_8H_{16}O$	446(3)	632.7(2)	2.7(5)	497(6)	466
3-Octanone	$C_8H_{16}O$	439(4)	627.7(2)	2.7(3)	497(6)	466
4-Octanone	$C_8H_{16}O$	439(3)	623.8(2)	2.7(3)	497(6)	466
1-Octene	C_8H_{16}	394.5(2)	566.58(5)	2.68(2)	464(2)	86, 257
trans-2-Octene	C_8H_{16}	398.1(5)	569.8(4)	2.58(9)	471(9)	170
trans-4-Octene	C_8H_{16}	395.6(5)	566(1)	2.55(6)	472(9)	170
Octylamine	$C_8H_{19}N$	451.8(2)	641(1)	2.82(3)	494(41)	542
Octylbenzene	$C_{14}H_{22}$	536(2)	725(7)	2.0(2)	746(17)	258
Octyl silicate	$C_{32}H_{68}O_4Si$		812(16)			57
Oxazole	C_3H_3NO	342.7(2)	551(4)	6.8(2)	185(23)	54
Oxirane	C_2H_4O	283.6(1)	469(1)	7.2(2)	138(4)	543, 544
Paraldehyde	$C_6H_{12}O_3$	397(2)	563(10)		410(15)	12
Pentacene	$C_{22}H_{14}$		1115(47)		806(23)	545
1*H*-Pentadecafluoroheptane	C_7HF_{15}	368(2)	495.8(7)	1.7(5)	644(38)	546
Pentadecane	$C_{15}H_{32}$	543.8(4)	707(2)	1.54(9)	938(36)	243, 244, 245, 247, 249, 363, 547
Pentadecanoic acid	$C_{15}H_{30}O_2$	612(4)	777(8)	1.6(2)	1002(22)	360
1-Pentadecanol	$C_{15}H_{32}O$	591(2)	757(8)	1.6(2)	961(16)	361
1-Pentadecene	$C_{15}H_{30}$	541.6(4)	705(7)	1.56(5)	933(30)	365
Pentafluorobenzene	C_6HF_5	358(3)	530.93(5)	3.53(1)	322(22)	472, 548
3,3,4,4,4-Pentafluorobutan-2-one	$C_4H_3F_5O$	314.36	453(1)	2.90(6)	333(4)	110
Pentafluoroethane	C_2HF_5	225.06	339.2(2)	3.63(1)	210(3)	180, 309, 312, 315, 447, 448, 549-554
1,1,1,2,2-Pentafluoropentan-3-one	$C_5H_5F_5O$	335.24	475.5(1)	2.64(1)	356(1)	308
1,1,1,2,2-Pentafluoropropane	$C_3H_3F_5$	255.2(3)	380.1(4)	3.14(2)	273(3)	555
1,1,1,3,3-Pentafluoropropane	$C_3H_3F_5$	288.29	427.20(7)	3.66(2)	262(14)	203
1,1,2,2,3-Pentafluoropropane	$C_3H_3F_5$	298.28	447.57(6)	3.96(2)	258(13)	203
1,1,1,2,2-Pentafluoro-3-(1,1,2,2-tetrafluoroethoxy) propane	$C_5H_3F_9O$	343.5	473.0(1)	2.24(1)	457(2)	113
Pentafluoro(trifluoromethoxy)ethane	C_3F_8O	249.5	356.8(1)	2.4(5)	319(5)	442
Pentanal	$C_5H_{10}O$	376(2)	567(3)	3.1(3)	313(11)	104, 124
Pentane	C_5H_{12}	309.21(7)	469.7(1)	3.37(1)	310(1)	19, 34, 36, 52, 64, 77, 84-86, 127, 131, 133, 134, 201, 202, 243, 244, 254, 286, 324, 359, 394, 456, 490, 556-562
Pentanedioic acid	$C_5H_8O_4$	546(10)	840(13)	4.3(5)	343(13)	253
Pentanenitrile	C_5H_9N	413(1)	610.3(2)	3.58(5)	320(8)	35
Pentanoic acid	$C_5H_{10}O_2$	459.3(3)	639(2)	3.6(1)	347(15)	11, 41, 145, 146, 464
1-Pentanol	$C_5H_{12}O$	410.8(4)	587.9(4)	3.9(3)	331(9)	25, 34, 149, 329, 456, 465, 563
2-Pentanol	$C_5H_{12}O$	392.3(5)	560.4(2)	4.2(4)	340(3)	149, 254, 465
3-Pentanol	$C_5H_{12}O$	396(2)	559.6(3)	4.9(9)	325(2)	465
2-Pentanone	$C_5H_{10}O$	375.4(1)	561.0(2)	3.70(6)	324(4)	25, 172
3-Pentanone	$C_5H_{10}O$	375.1(1)	561.4(2)	3.73(7)	319(11)	25
1-Pentene	C_5H_{10}	303.2(3)	464.74(5)	3.55(2)	301.0(1)	86, 470, 559
cis-2-Pentene	C_5H_{10}	310.1(2)	474.9(4)	3.69(2)	301(7)	564
Pentyl acetate	$C_7H_{14}O_2$	422.6(3)	600(2)	2.79(3)		141, 162, 166
Pentylbenzene	$C_{11}H_{16}$	476(3)	675(7)	2.6(3)	559(12)	258
Pentyl benzoate	$C_{12}H_{16}O_2$	533(3)	736(14)	2.2(2)	661(14)	169
Pentyl formate	$C_6H_{12}O_2$	399(3)	576(4)	3.5(8)	453(35)	415
Pentyl silicate	$C_{20}H_{44}O_4Si$		714(14)			57
Perfluoroacetone	C_3F_6O	245.8(4)	357.2(1)	2.85(1)	329(7)	197, 470
Perfluorobutane	C_4F_{10}	271.1(8)	386.3(2)	2.33(2)	380(18)	565, 566
Perfluorocyclobutane	C_4F_8	267.24	388.4(1)	2.78(1)	316(8)	265, 470, 567, 568
Perfluorocyclohexane	C_6F_{12}	326.0 sp	457.1(5)	2.24(3)	424(32)	569
Perfluorocyclohexene	C_6F_{10}	324.9(1)	461.7(7)	2.6(1)	434(28)	546
Perfluorodecane	$C_{10}F_{22}$	408(3)	542.4(4)	1.45(3)	892(7)	240, 570
Perfluorodimethoxymethane	$C_3F_8O_2$	263.2(7)	372.4(2)	2.34(1)	370(30)	116, 571

Name	Mol. form.	T_b/K	T_c/K	P_c/MPa	V_c/cm³ mol⁻¹	Ref. (see text)
Perfluoro-2,3-dimethylbutane	C_6F_{14}	333.0(3)	463.0(1)	1.95(3)	523(18)	240, 572
Perfluoroethyl ethyl ether	$C_4H_5F_5O$	301(3)	431.23(8)	2.53(1)	366(3)	176
Perfluoroethyl 2,2,2-trifluoroethyl ether	$C_4H_2F_8O$	301.04	421.68(8)	2.33(1)	409(3)	176
Perfluoroheptane	C_7F_{16}	355.7(2)	477(3)	1.63(1)	603(14)	232, 566, 570, 573-576
Perfluoro-1-heptene	C_7F_{14}	354(4)	478.2(7)	1.7(3)	555(34)	546
Perfluorohexane	C_6F_{14}	330.4(2)	451(3)	1.88(2)	552(77)	198, 470, 546, 570, 572, 573, 577
1*H*-Perfluorohexane	C_6HF_{13}	345	471.8(7)	2.0(3)	504(32)	546
Perfluoro-1-hexene	C_6F_{12}	330.2	454.3(7)		462(31)	546
Perfluoroisobutane	C_4F_{10}	273	395.4(7)		396(22)	578
Perfluoroisopropyl methyl ether	$C_4H_3F_7O$	302(1)	433.30(8)	2.55(1)	369(3)	176
Perfluoromethylcyclohexane	C_7F_{14}	349.5(2)	486.5(10)	2.02(1)	561(4)	325, 326, 327, 566, 569
Perfluoromethylcyclopentane	C_6F_{12}	321.58	451.43(5)	2.17(1)	419(37)	579
Perfluoro-2-methylpentane	C_6F_{14}	330.8(3)	454.6(2)	1.87(2)	585(10)	572, 580
Perfluoro-3-methylpentane	C_6F_{14}	331(9)	450(1)	1.69(1)	511(36)	572
Perfluoronaphthalene	$C_{10}F_8$	473(8)	673(1)	2.9(6)	464(29)	546
Perfluorononane	C_9F_{20}	390(3)	524.0(1)	1.56(4)	846(48)	570
Perfluorooctane	C_8F_{18}	378(2)	502.3(1)	1.66(2)	738(17)	34, 240, 570, 573
Perfluorooxetane	C_3F_6O	244.6(6)	361.8(5)	3.10(2)	274(24)	116, 571
Perfluoropentane	C_5F_{12}	302.4(2)	421.8(1)	2.04(2)	463(7)	570, 573
1*H*-Perfluoropentane	C_5HF_{11}	319	443.9(7)	2.2(3)	413(28)	546
Perfluoropropane	C_3F_8	236.4(3)	345.03(8)	2.67(1)	301(12)	187, 305, 470, 581
Perfluoropropyl methyl ether	$C_4H_3F_7O$	307(1)	437.7(1)	2.48(1)	382(3)	176
Perfluorotoluene	C_7F_8	377.8(3)	534.4(2)	2.70(2)	423(26)	25
Perfluorovaleric acid	$C_5HF_9O_2$	415.4	545.5(2)	2.10(5)	496(29)	583
Phenol	C_6H_6O	455.0(1)	694.3(1)	5.5(2)	283(10)	218, 584
Phenyl acetate	$C_8H_8O_2$	468(1)	686(2)	3.60(6)	407(12)	55
Phenyl isocyanate	C_7H_5NO	439.5(4)	657(10)	3.6(1)	342(15)	54
α-Pinene, (-)	$C_{10}H_{16}$	429.2	644(2)	3.4(2)	472(10)	496
Piperazine	$C_4H_{10}N_2$	421.78(5)	660(5)	5.4(4)	283(16)	331, 489
Piperidine	$C_5H_{11}N$	379.34(9)	594.14(5)	4.7(1)	294(10)	39, 218
Propanal	C_3H_6O	321.2(2)	503.7(8)	5.04(3)	218(9)	104, 124, 163, 242
Propane	C_3H_8	231.04(9)	369.9(1)	4.25(1)	199(6)	29, 125, 127, 131-134, 252, 322, 323, 336, 371, 387, 470, 486, 509, 540, 585-596
1,3-Propanediamine	$C_3H_{10}N_2$	412.4(7)	632(7)	5.7(6)	257(13)	140
1,2-Propanediol	$C_3H_8O_2$	460.5(2)	676(1)	5.9(2)	237(10)	32
1,3-Propanediol	$C_3H_8O_2$	487.9(3)	718(2)	6.7(2)	255(12)	55
Propanenitrile	C_3H_5N	370.5(4)	561.3(2)	4.26(7)	211(7)	35
1-Propanethiol	C_3H_8S	340.9(1)	536.6(6)	4.7(1)	286(11)	143, 144
Propanoic acid	$C_3H_6O_2$	414.7(2)	603(3)	4.5(7)	232(12)	39, 145, 146, 493, 597, 598
1-Propanol	C_3H_8O	370.19(9)	536.8(2)	5.1(1)	220(22)	19, 34, 80, 148-150, 152, 153, 599, 600
2-Propanol	C_3H_8O	355.36(9)	508.3(2)	4.7(1)	226(1)	25, 91, 148, 149, 153, 324, 440, 520, 601, 602
Propene	C_3H_6	225.6(1)	364.9(5)	4.59(2)	184(11)	67, 123, 172, 181, 427, 603-610
2-Propoxyethanol	$C_5H_{12}O_2$	425(3)	614.7(7)	3.65(9)	364(13)	11, 41
1-Propoxy-2-propanol	$C_6H_{14}O_2$	423.4(7)	605(1)	3.1(1)	417(16)	32
Propyl acetate	$C_5H_{10}O_2$	374.2(2)	549.69(8)	3.37(7)	346(24)	19, 411, 413, 415
Propylamine	C_3H_9N	320.36(8)	499.2(4)	4.77(6)	230(12)	170, 193
Propylbenzene	C_9H_{12}	432.4(5)	638.3(1)	3.20(2)	441(6)	79, 86, 88
Propyl benzoate	$C_{10}H_{12}O_2$	504(2)	710(14)	2.6(3)	530(10)	169
Propyl butanoate	$C_7H_{14}O_2$	417(2)	593(1)	2.72(6)	463(10)	162, 166
Propylcyclohexane	C_9H_{18}	429.9(3)	630.8(9)	2.87(4)	489(13)	170, 171
Propylene carbonate	$C_4H_6O_3$	514.8(7)	763(2)	4.1(2)	256.5(5)	55
1,2-Propylene glycol 1-*tert*-butyl ether	$C_7H_{16}O_2$	425.3	601(4)	2.7(1)	468(15)	99
1,2-Propylene glycol monomethyl ether acetate	$C_6H_{12}O_3$	419.2(4)	598(1)	3.1(2)	432(16)	41, 144
Propyl formate	$C_4H_8O_2$	353.8(2)	538.1(1)	4.07(2)	281(31)	19, 412, 413, 415
Propyl isobutanoate	$C_7H_{14}O_2$	407(4)	582(11)		463(9)	162, 167
Propyl 3-methylbutanoate	$C_8H_{16}O_2$	428(3)	609(6)		523(9)	147
Propyl propanoate	$C_6H_{12}O_2$	395.4(1)	569(3)	3.1(1)	403(6)	162, 166

Name	Mol. form.	T_b/K	T_c/K	P_c/MPa	V_c/cm³ mol⁻¹	Ref. (see text)
Propyl silicate	$C_{12}H_{28}O_4Si$	499	649(12)			57
Propyne	C_3H_4	250.0	402(2)	5.63(6)	160(10)	605, 611
Pyrazine	$C_4H_4N_2$	389.5(1)	627(1)	6.49(3)	225(16)	612
Pyridine	C_5H_5N	388.4(1)	619(2)	5.63(7)	248(12)	16, 17, 60, 90, 246, 290, 520, 613
Pyrrole	C_4H_5N	402.89(4)	639.7(2)	8.0(2)	222(15)	241
Pyrrolidine	C_4H_9N	359.8(1)	568.6(2)	5.69(8)	259(3)	241, 290
Quinoline	C_9H_7N	510.3(5)	782(3)	4.75(10)	382(17)	218
Resorcinol	$C_6H_6O_2$	553(2)	836(10)	6.3(3)	292(10)	489
Styrene	C_8H_8	418.5(6)	635(2)	3.9(2)	357(15)	481
Succinic acid	$C_4H_6O_4$	507(3)	851(20)		308(21)	253
o-Terphenyl	$C_{18}H_{14}$	610(5)	857(6)	2.9(1)	737(37)	108
m-Terphenyl	$C_{18}H_{14}$	648(1)	883(7)	2.2(2)	747(37)	108, 358
p-Terphenyl	$C_{18}H_{14}$	649	913(22)	2.5(5)	713(37)	108, 614
Tetrachloromethane	CCl_4	349.9(2)	556.5(3)	4.57(7)	276(9)	19, 27, 469, 615-619
Tetracosane	$C_{24}H_{50}$	664(5)	800(5)	0.9(1)	1585(55)	358, 359
Tetradecamethylcycloheptasiloxane	$C_{14}H_{42}O_7Si_7$	548.5	683.2(2)	0.99(2)	1634(110)	34
Tetradecane	$C_{14}H_{30}$	526.7(4)	693(1)	1.56(8)	870(49)	34, 244
Tetradecanedioic acid	$C_{14}H_{26}O_4$	639(10)		1.9(2)		253
Tetradecanoic acid	$C_{14}H_{28}O_2$	599(1)	763(8)	1.6(2)	921(20)	360
1-Tetradecanol	$C_{14}H_{30}O$	569.0(4)	743(7)	1.70(4)	887(15)	361
2-Tetradecanone	$C_{14}H_{28}O$	562(6)	728(9)		896(26)	256
3-Tetradecanone	$C_{14}H_{28}O$	552	727(6)		896(26)	256
4-Tetradecanone	$C_{14}H_{28}O$	552	725(6)		900(27)	256
7-Tetradecanone	$C_{14}H_{28}O$	552	723(8)		904(27)	256
1-Tetradecene	$C_{14}H_{28}$	524.3(4)	691(7)	1.58(7)	851(24)	365
Tetraethylene glycol	$C_8H_{18}O_5$	588(7)	800(30)	2.8(7)	608(31)	283
Tetraethylsilane	$C_8H_{20}Si$	426.6(7)	606(2)	2.297(10)	596.4(2)	620
1,2,3,4-Tetrafluorobenzene	$C_6H_2F_4$	367.5(8)	550.8(2)	3.791(10)	312(22)	25
1,2,3,5-Tetrafluorobenzene	$C_6H_2F_4$	357.5(8)	535.2(2)	3.75(1)	311(22)	25
1,2,4,5-Tetrafluorobenzene	$C_6H_2F_4$	363.4(3)	543.3(2)	3.80(1)	309(22)	25
1,1,2,2-Tetrafluoro-2-(2,2-difluoroethoxy)ethane	$C_4H_4F_6O$	352.13	501.08(8)	3.09(2)	356(1)	113
1,1,1,2-Tetrafluoroethane	$C_2H_2F_4$	247.1(1)	374.2(2)	4.06(1)	200(2)	113, 182, 183, 188, 278-280, 305, 313, 315, 447, 448, 478, 621-631
1,1,2,2-Tetrafluoroethane	$C_2H_2F_4$	253(1)	391.75(8)	4.61(1)	192(1)	178, 632
Tetrafluoroethene	C_2F_4	197(1)	307(1)	3.94(5)	183(3)	633, 634
1,2,2,2-Tetrafluoroethyl difluoromethyl ether	$C_3H_2F_6O$	296(2)	428.95(8)	3.05(1)	315(2)	176
1,1,2,2-Tetrafluoroethyl 1,1,1-trifluoroethyl ether	$C_4H_3F_7O$	329.37	463.89(7)	2.71(1)	373(1)	113
Tetrafluoromethane	CF_4	145.3(1)	227.54(3)	3.73(3)	140(1)	506, 635
1,1,2,2-Tetrafluoro-3-methoxypropane	$C_4H_6F_4O$	347.5	505.4(1)	3.28(1)	331(1)	113
1,2,2,3-Tetrafluoro-1-propanol	$C_3H_4F_4O$	386.5	554(2)	3.3(2)	280(15)	362
1,1,2,2-Tetrafluoro-3-(1,1,2,2-tetrafluoroethoxy) propane	$C_5H_4F_8O$	366.32	510.07(8)	2.58(1)	440(2)	113
3,4,4,4-Tetrafluoro-3-trifluoromethylbutan-2-one	$C_5H_3F_7O$	328.76	468(1)	2.50(6)	409(5)	110
4,4,5,5-Tetrafluoro-2-trifluoromethyl-1,3-dioxolane	$C_4HF_7O_2$	304.7	435(1)	2.62(7)	376(4)	110
Tetrahydrofuran	C_4H_8O	339.2(1)	540(1)	5.29(6)	223(2)	23, 241, 290
1,2,3,4-Tetrahydronaphthalene	$C_{10}H_{12}$	480.4(3)	720(1)	3.6(1)	431(40)	144, 242, 636
Tetrahydropyran	$C_5H_{10}O$	361.2(4)	572.0(3)	4.8(2)	278(18)	75
Tetrahydrothiophene	C_4H_8S	394.3(2)	632.0(2)	5.4(6)	276(19)	143, 241
1,2,4,5-Tetraisopropylbenzene	$C_{18}H_{30}$	532	703(1)	1.65(2)	983(83)	317
1,2,4,5-Tetramethylbenzene	$C_{10}H_{14}$	470(1)	676(2)	2.9(3)	489(11)	39
1,1,3,3-Tetramethyl-1,3-diphenyldisiloxane	$C_{16}H_{22}OSi_2$	565	750(8)			353
2,2,3,3-Tetramethylhexane	$C_{10}H_{22}$	433(2)	623.0(5)	2.51(6)	574(14)	536
2,2,5,5-Tetramethylhexane	$C_{10}H_{22}$	410(2)	581.4(5)	2.19(1)	600(14)	536
2,2,3,3-Tetramethylpentane	C_9H_{20}	413.4(4)	607.5(5)	2.74(3)	514(15)	536
2,2,3,4-Tetramethylpentane	C_9H_{20}	406.2(8)	592.6(5)	2.60(3)	517(17)	536
2,2,4,4-Tetramethylpentane	C_9H_{20}	395.4(10)	574.6(5)	2.49(1)	532(16)	536
2,3,3,4-Tetramethylpentane	C_9H_{20}	414.7(7)	607.5(5)	2.72(4)	517(17)	536
Tetramethylsilane	$C_4H_{12}Si$	299.9(5)	449(2)	2.82(1)	362(7)	637-639
Tetramethylstannane	$C_4H_{12}Sn$	350.0(10)	521.77(5)	2.98(1)		640, 641
Thiacyclohexane	$C_5H_{10}S$	414.88(4)	684(44)	6.50(7)	284(11)	104

Name	Mol. form.	T_b/K	T_c/K	P_c/MPa	V_c/cm³ mol⁻¹	Ref. (see text)
Thiobis(trifluoromethane)	C_2F_6S	251.3	376.8(1)	3.2(5)	216(19)	442
Thiophene	C_4H_4S	357.3(1)	579.4(2)	5.7(2)	230(3)	241, 290
Thymol	$C_{10}H_{14}O$	506(3)	698(10)		528(21)	302
Toluene	C_7H_8	383.75(7)	591.9(2)	4.13(2)	314(7)	34, 63, 68, 74, 78, 79, 84, 86, 88, 90, 96, 163, 172, 258, 481, 636, 642-646
Triacontane	$C_{30}H_{62}$	724(7)	843(8)	0.6(1)	2055(69)	358
Tribromomethane	$CHBr_3$	422.4(5)	682(1)	5.8(2)	261(12)	647
Trichloroacetyl chloride	C_2Cl_4O	391.4(3)	604(2)	4.21(3)	331(54)	410
Trichloroethylsilane	$C_2H_5Cl_3Si$	371.9(7)	559.9(6)	3.34(4)	403(5)	266
Trichlorofluoromethane	CCl_3F	296.9(6)	471.1(2)	4.40(3)	248.0(9)	33, 34, 191, 271, 571
Trichloromethane	$CHCl_3$	334.4(1)	536.0(4)	5.5(2)	237(7)	18, 27, 28, 117
Trichloromethylsilane	CH_3Cl_3Si	339(2)	517.7(3)	3.52(3)	329(9)	264, 648
1,3,5-Trichloro-2,4,6-trifluorobenzene	$C_6Cl_3F_3$	472(27)	684.7(4)	3.3(1)	443(27)	25
1,1,2-Trichloro-1,2,2-trifluoroethane	$C_2Cl_3F_3$	320.8(2)	487.4(2)	3.40(2)	325(1)	191, 265, 271, 649
Tricosane	$C_{23}H_{48}$	654(9)	790(8)	0.9(1)	1527(53)	359
Tridecane	$C_{13}H_{28}$	508.6(4)	676(1)	1.68(4)	824(30)	34, 244, 249
1-Tridecanol	$C_{13}H_{28}O$	560(8)	732(7)	1.8(2)	828(14)	361
2-Tridecanone	$C_{13}H_{26}O$	541(1)	717(6)	1.8(2)	820(24)	256
3-Tridecanone	$C_{13}H_{26}O$	539	716(5)	1.7(5)	823(24)	256
4-Tridecanone	$C_{13}H_{26}O$	539	712(6)	1.7(5)	823(24)	256
5-Tridecanone	$C_{13}H_{26}O$	539	710(8)	1.7(5)	826(17)	256
6-Tridecanone	$C_{13}H_{26}O$	539	709(5)	1.7(5)	826(24)	256
7-Tridecanone	$C_{13}H_{26}O$	539(9)	708(5)	1.7(5)	830(24)	256
1-Tridecene	$C_{13}H_{26}$	506.0(7)	673(7)	1.74(5)	770(17)	365
Tridecylbenzene	$C_{19}H_{32}$	613(4)	790(8)	1.5(1)	1079(43)	258
Triethylamine	$C_6H_{15}N$	362.0(2)	535.6(3)	3.1(3)	392(25)	117, 286
1,3,5-Triethylbenzene	$C_{12}H_{18}$	489.0(9)	679(2)	2.32(1)	624(60)	332
Triethylene glycol	$C_6H_{14}O_4$	561.8(2)	775(30)	3.3(2)	454(25)	283
Trifluoroacetonitrile	C_2F_3N	204.4(8)	311.1(4)	3.61(4)	202(4)	470
1,2,3-Trifluorobenzene	$C_6H_3F_3$	368	560.3(4)	4.1(4)	296(20)	119
1,2,4-Trifluorobenzene	$C_6H_3F_3$	363	551.1(4)	4.1(6)	297(20)	119
1,3,5-Trifluorobenzene	$C_6H_3F_3$	350.2(5)	530.9(4)	3.8(2)	300(20)	119
1,1,1-Trifluoroethane	$C_2H_3F_3$	226.0(1)	345.89(7)	3.77(1)	195(15)	179, 203, 309, 478, 479, 549, 624, 650-653
2,2,2-Trifluoroethanol	$C_2H_3F_3O$	347.0(3)	498.57(5)	4.81(1)	211(12)	654, 655
2,2,2-Trifluoroethyl methyl ether	$C_3H_5F_3O$	304.77	448.98(8)	3.51(6)	277(3)	176
Trifluoroiodomethane	CF_3I	251.4(6)	396.44(6)	3.95(1)	231(3)	656-659
Trifluoromethane	CHF_3	191.2(1)	299.00(2)	4.82(1)	133(1)	310, 606, 660-664
Trifluoromethyl difluoromethyl ether	C_2HF_5O	238.2(2)	354.49(6)	3.36(2)	226(20)	203
(Trifluoromethyl)sulfur pentafluoride	CF_8S	252.4	381.2(1)	3.4(1)	284(4)	442
Trifluoromethyl 1,1,2,2-tetrafluoroethyl ether	C_3HF_7O	269.9(10)	387.8(5)	2.65(1)	341(22)	116, 571
Trifluoromethyl 1,2,2,2-tetrafluoroethyl ether	C_3HF_7O	263.6	377.26(6)	2.62(1)	321(2)	525
3,3,3-Trifluoro-1-propene	$C_3H_3F_3$	246(4)	378.6(5)	3.61(8)	229(14)	440
Trimethylamine	C_3H_9N	276.0(2)	433.0(6)	4.08(4)	254(6)	322, 323, 665
1,2,3-Trimethylbenzene	C_9H_{12}	449.2(4)	664.4(1)	3.45(3)	423(11)	79
1,2,4-Trimethylbenzene	C_9H_{12}	442.6(3)	649.1(1)	3.3(1)	436(12)	79, 86, 96, 251
1,3,5-Trimethylbenzene	C_9H_{12}	437.9(3)	637.31(10)	3.13(5)	435(12)	79
3,7,7-Trimethyl-bicyclo[4.1.0]hept-3-ene	$C_{10}H_{16}$	445	658(2)	2.9(5)	487(10)	496
2,2,3-Trimethylbutane	C_7H_{16}	354.0(1)	531.3(5)	2.96(3)	401(13)	346, 349
Trimethylchlorosilane	C_3H_9ClSi	330.8(4)	497.7(6)	3.20(3)	366(6)	266
1α,3α,5β-1,3,5-Trimethylcyclohexane	C_9H_{18}	414(2)	602(2)	2.6(3)	494(14)	74
3,3,5-Trimethylheptane	$C_{10}H_{22}$	430(3)	609.5(5)	2.32(5)	583(18)	536
2,2,5-Trimethylhexane	C_9H_{20}	397(2)	570(2)	2.46(3)	547(18)	350
2,2,3-Trimethylpentane	C_8H_{18}	383.0(4)	563.5(4)	2.73(2)	442(16)	346
2,2,4-Trimethylpentane	C_8H_{18}	372.4(2)	543.9(4)	2.57(2)	475(20)	33, 34, 86, 159, 346, 666, 667
2,3,3-Trimethylpentane	C_8H_{18}	387.9(3)	573.5(4)	2.82(3)	454(14)	346
2,3,4-Trimethylpentane	C_8H_{18}	386.6(3)	566.4(4)	2.72(2)	462(12)	346
cis-Tri(methylphenyl)trisiloxane	$C_{21}H_{24}O_3Si_3$		824(8)			353
2,4,6-Trimethyl-2,4,6-triphenylcyclotrisiloxane	$C_{21}H_{24}O_3Si_3$		839(8)			353
Tris(perfluorobutyl)amine	$C_{12}F_{27}N$	451(2)	566(4)	1.24(9)	1196(66)	582

Name	Mol. form.	T_b/K	T_c/K	P_c/MPa	V_c/cm^3 mol^{-1}	Ref. (see text)
Undecafluorocyclohexane	C_6HF_{11}	335.2	477.7(7)			546
Undecane	$C_{11}H_{24}$	469.1(3)	638.8(2)	2.01(3)	683(20)	34, 86, 131, 243, 249, 363, 535
Undecanoic acid	$C_{11}H_{22}O_2$	553	728(7)	2.1(2)	741(20)	360
1-Undecanol	$C_{11}H_{24}O$	519(2)	703.0(6)	2.15(7)	707(12)	149
2-Undecanone	$C_{11}H_{22}O$	506.3(3)	688(2)	2.08(1)	692(20)	256
3-Undecanone	$C_{11}H_{22}O$	500	685(2)	2.0(4)	692(20)	256
4-Undecanone	$C_{11}H_{22}O$	501	681(2)	2.0(2)	692(20)	256
5-Undecanone	$C_{11}H_{22}O$	500	679(2)	2.0(2)	692(20)	256
6-Undecanone	$C_{11}H_{22}O$	500.6(5)	678(2)	2.02(1)	692(20)	256
Undecylbenzene	$C_{17}H_{28}$	585(3)	763(8)	1.6(1)	946(35)	258
Vinyl acetate	$C_4H_6O_2$	345.8(3)	519.2(2)	4.17(3)	269(7)	156, 440
o-Xylene	C_8H_{10}	417.6(4)	630.26(10)	3.74(1)	372(40)	34, 68, 78, 79, 88, 669
m-Xylene	C_8H_{10}	412.3(4)	616.9(3)	3.54(1)	377(7)	34, 68, 79, 88, 90, 106, 668
p-Xylene	C_8H_{10}	411.5(5)	616.17(9)	3.55(2)	372(35)	34, 68, 74, 79, 88, 90, 669
2,3-Xylenol	$C_8H_{10}O$	490.03(5)	723(1)	4.1(3)	397(15)	218
2,4-Xylenol	$C_8H_{10}O$	484.09(3)	708(1)	3.5(3)	389(15)	218
2,5-Xylenol	$C_8H_{10}O$	484.29(8)	707(1)	3.9(1)	397(15)	25
2,6-Xylenol	$C_8H_{10}O$	474.18(5)	701(1)	3.8(1)	396(15)	218
3,4-Xylenol	$C_8H_{10}O$	500.46(5)	730(1)	4.9(5)	388(15)	218
3,5-Xylenol	$C_8H_{10}O$	494.86(5)	716(1)	3.8(2)	396(15)	218

References

1. Frenkel, M., Chirico, R. D., Diky, V. V., Kazakov, A., and Muzny, C. D., *ThermoData Engine*. NIST Standard Reference Database 103b, Version 4.0 (Pure Compounds, Binary Mixtures, and Chemical Reactions, TDE-SOURCE Version 4.3), National Institute of Standards and Technology, Gaithersburg, MD – Boulder, CO, 2009. <http://www.nist.gov/srd/nist103b.cfmhttp://www.nist.gov/srd/nist103b.cfm>
2. Frenkel, M., Chirico, R. D., Diky, V., Yan, X., Dong, Q., and Muzny, C., *J. Chem. Inf. Model.* 45, 816, 2005. <https://doi.org/10.1021/ci050067b>
3. Diky, V., Muzny, C. D., Lemmon, E. W., Chirico, R. D., and Frenkel, M., *J. Chem. Inf. Model.* 47, 1713, 2007. <https://doi.org/10.1021/ci700071t>
4. Diky, V., Chirico, R. D., Kazakov, A. F., Muzny, C., and Frenkel, M., *J. Chem. Inf. Model.* 49, 503, 2009. <https://doi.org/10.1021/ci800345e>
5. Diky, V., Chirico, R. D., Kazakov, A. F., Muzny, C., and Frenkel, M., *J. Chem. Inf. Model.* 49, 2883, 2009. <https://doi.org/10.1021/ci900340k>
6. Joback, K. G., and Reid, R. C., *Chem. Eng. Commun.* 57, 233, 1987. <https://doi.org/10.1080/00986448708960487>
7. Constantinou, L., and Gani, R., *AIChE J.* 40, 1697, 1994. <https://doi.org/10.1002/aic.690401011>
8. Marrero-Morejon, J., and Pardillo-Fontdevila, E., *AIChE J.* 45, 615, 1999. <https://doi.org/10.1002/aic.690450318>
9. Wilson, G. M., and Jasperson, L. V., AIChE Meeting, New Orleans, LA, 1996.
10. Chirico, R. D., Frenkel, M., Diky, V. V., Marsh, K. N., and Wilhoit, R. C., *J. Chem. Eng. Data* 48, 1344, 2003. <https://doi.org/10.1021/je034088i>
11. Teja, A. S., and Anselme, M. J., *AIChE Symp. Ser.* 86 (279), 115, 1990.
12. Hollmann, R., *Z. Phys. Chem., Stoechiom. Verwandtschaftsl.* 43, 129, 1903.
13. Van der Waals, J. D., *Continuity of Gas and Liquid Data*, 1st edition, Leipzig, p. 168, 1881.
14. Vandana, V., and Teja, A. S., *Fluid Phase Equilib.* 103, 113, 1995. <https://doi.org/10.1016/0378-3812(94)02591-N>
15. Ambrose, D., Ellender, J. H., Sprake, C. H. S., and Townsend, R., *J. Chem. Thermodyn.* 9, 735, 1977. <https://doi.org/10.1016/0021-9614(77)90017-9>
16. Kreglewski, A., *Rocz. Chem.* 31, 1001, 1957.
17. Swietoslawski, W., and Kreglewski, A., *Bull. Acad. Pol. Sci., Cl. 3* 2, 77, 1954.
18. Swietoslawski, W., and Kreglewski, A., *Bull. Acad. Pol. Sci., Cl. 3* 2, 187, 1954.
19. Young, S., *Sci. Proc. R. Dublin Soc.* 12, 374, 1910.
20. Young, S., *J. Chem. Soc.* 59, 903, 1891. <https://doi.org/10.1039/CT8915900903>
21. Pawlewski, B., *Ber. Dtsch. Chem. Ges.* 16, 2633, 1883. <https://doi.org/10.1002/cber.188301602207>
22. Ambrose, D., and Ghiassee, N. B., *J. Chem. Thermodyn.* 19, 911, 1987. <https://doi.org/10.1016/0021-9614(87)90037-1>
23. Sassa, Y., Konishi, R., and Katayama, T., *J. Chem. Eng. Data* 19, 44, 1974. <https://doi.org/10.1021/je60060a004>
24. Ambrose, D., Sprake, C. H. S., and Townsend, R., *J. Chem. Thermodyn.* 6, 693, 1974. <https://doi.org/10.1016/0021-9614(74)90119-0>
25. Ambrose, D., Broderick, B. E., and Townsend, R., *J. Appl. Chem. Biotechnol.* 24, 359, 1974. <https://doi.org/10.1002/jctb.5020240607>
26. Campbell, A. N., and Musbally, G. M., *Can. J. Chem.* 48, 3173, 1970. <https://doi.org/10.1139/v70-535>
27. Campbell, A. N., and Chatterjee, R. M., *Can. J. Chem.* 47, 3893, 1969. <https://doi.org/10.1139/v69-646>
28. Campbell, A. N., and Chatterjee, R. M., *Can. J. Chem.* 46, 575, 1968. <https://doi.org/10.1139/v68-095>
29. Kay, W. B., *J. Phys. Chem.* 68, 827, 1964. <https://doi.org/10.1021/j100786a021>
30. Rosenbaum, M., M.S. Thesis, Univ. Texas, Austin, TX, 1951.
31. Kuenen, J. P., and Robson, W. G., *Philos. Mag.* 3, 622, 1902. <https://doi.org/10.1080/14786440209462808>
32. VonNiederhausern, D. M., Wilson, L. C., Giles, N. F., and Wilson, G. M., *J. Chem. Eng. Data* 45, 154, 2000. <https://doi.org/10.1021/je990189y>
33. Christou, G., Young, C. L., and Svejda, P., *Ber. Bunsen-Ges. Phys. Chem.* 95, 510, 1991. <https://doi.org/10.1002/bbpc.19910950411>
34. Christou, G., Ph.D. Dissertation, Univ. Melbourne, 1988.
35. Castillo-Lopez, N., and Trejo Rodriguez, A., *J. Chem. Thermodyn.* 19, 671, 1987. <https://doi.org/10.1016/0021-9614(87)90073-5>
36. Trejo Rodriguez, A., and McLure, I. A., *Fluid Phase Equilib.* 12, 297, 1983. <https://doi.org/10.1016/0378-3812(83)80067-3>
37. Trejo Rodriguez, A., and McLure, I. A., *J. Chem. Thermodyn.* 11, 1113, 1979. <https://doi.org/10.1016/0021-9614(79)90143-5>

38. Khera, R., Ph.D. Thesis, Ohio State Univ., Columbus, OH, 1968.
39. Guye, P. A., and Mallet, E., *Arch. Sci. Phys. Nat.* 13, 274, 1902.
40. Guye, P. A., and Mallet, E., *C. R. Hebd. Seances Acad. Sci.* 133, 168, 1902.
41. Teja, A. S., and Rosenthal, D. J., *Experimental Results for Phase Equilibria and Pure Component Properties*, DIPPR Data Series No. 1, p. 96, 1991.
42. Goloborod'ko, N. P., and Khodeeva, S. M., *Russ. J. Phys. Chem.* (*Engl. Transl.*) 46, 235, 1972.
43. Mislavskaya, V. S., and Khodeeva, S. M., *Zh. Fiz. Khim.* 43, 2367, 1969.
44. Khodeeva, S. M., *Russ. J. Phys. Chem.* (*Engl. Transl.*) 40, 1061, 1966.
45. Ambrose, D., and Townsend, R., *Trans. Faraday Soc.* 60, 1025, 1964. <https://doi.org/10.1039/tf9646001025>
46. Ambrose, D., *Trans. Faraday Soc.* 52, 772, 1956. <https://doi.org/10.1039/tf9565200772>
47. McIntosh, D., *J. Phys. Chem.* 11, 306, 1907. <https://doi.org/10.1021/j150085a005>
48. Kuenen, J. P., *Philos. Mag.* 44, 174, 1897. <https://doi.org/10.1080/14786449708621051>
49. Wilson, L. C., Wilson, H. L., Wilding, W. V., and Wilson, G. M., *J. Chem. Eng. Data* 41, 1252, 1996. <https://doi.org/10.1021/je960052x>
50. Lespieau, R., and Chavanne, G., *C. R. Hebd. Seances Acad. Sci.* 140, 1035, 1905.
51. Wang, X., Jia, Q., Gao, J., Xia, S., and Ma, P. S., *J. Chem. Ind. Eng.* (*China*) 56, 1385, 2005.
52. Liang, Y.-H., Ma, P. S., and Zhang, H., *J. Chem. Ind. Eng.* (*China*) 51, 243, 2000.
53. Steele, W. V., Chirico, R. D., Knipmeyer, S. E., and Nguyen, A., *J. Chem. Thermodyn.* 23, 957, 1991. <https://doi.org/10.1016/S0021-9614(05)80177-6>
54. VonNiederhausern, D. M., Wilson, G. M., and Giles, N. F., *J. Chem. Eng. Data* 51, 1990, 2006. <https://doi.org/10.1021/je060269j>
55. Wilson, G. M., VonNiederhausern, D. M., and Giles, N. F., *J. Chem. Eng. Data* 47, 761, 2002. <https://doi.org/10.1021/je0100995>
56. Chashkin, Yu. R., Gorbunova, V. G., and Voronel, A. V., *Zh. Exp. Teor. Fiz.* 49, 432, 1965.
57. Nikitin, E. D., and Popov, A. P., *J. Chem. Eng. Data* 53, 1371, 2008. <https://doi.org/10.1021/je800086s>
58. Steele, W. V., Chirico, R. D., Knipmeyer, S. E., and Nguyen, A., *J. Chem. Eng. Data* 47, 648–666, 2002. <https://doi.org/10.1021/je0100847>
59. Lagutkin, O. D., and Kuropatkin, E. I., *Zh. Fiz. Khim.* 55, 1329, 1981.
60. Livingston, J., Morgan, R., and Higgins, E., *Z. Phys. Chem., Stoechiom. Verwandtschaftsl.* 64, 170, 1908. <https://doi.org/10.1515/zpch-1908-0108>
61. Liu, T., Fu, J., Wang, K., Gao, Y., and Yuan, W., *J. Chem. Eng. Data* 46, 809, 2001. <https://doi.org/10.1021/je000309y>
62. Nikitin, E. D., Pavlov, P. A., and Skutin, M., *Fluid Phase Equilib.* 161, 119, 1999. <https://doi.org/10.1016/S0378-3812(99)00169-7>
63. Chirico, R. D., and Steele, W. V., *Ind. Eng. Chem. Res.* 33, 157, 1994. <https://doi.org/10.1021/ie00025a021>
64. Zhang, J., Zhao, X., and Ma, P., *Huagong Xuebao* 43, 105, 1992.
65. Knipmeyer, S. E., Archer, D. G., Chirico, R. D., Gammon, B. E., Hossenlopp, I. A., Nguyen, A., Smith, N. K., Steele, W. V., and Strube, M. M., *Fluid Phase Equilib.* 52, 185, 1989. <https://doi.org/10.1016/0378-3812(89)80324-3>
66. Goodwin, R. D., *J. Phys. Chem. Ref. Data* 17, 1541, 1988. <https://doi.org/10.1063/1.555813>
67. Brunner, E., *J. Chem. Thermodyn.* 20, 1397, 1988. <https://doi.org/10.1016/0021-9614(88)90033-X>
68. Ambrose, D., *J. Chem. Thermodyn.* 19, 1007, 1987. <https://doi.org/10.1016/0021-9614(87)90048-6>
69. Kay, W. B., and Kreglewski, A., *Fluid Phase Equilib.* 11, 251, 1983. <https://doi.org/10.1016/0378-3812(83)85028-6>
70. Hales, J. L., and Gundry, H. A., *J. Phys. E* 16, 91, 1983. <https://doi.org/10.1088/0022-3735/16/1/018>
71. Hugill, J. A., and McGlashan, M. L., *J. Chem. Thermodyn.* 13, 429, 1981. <https://doi.org/10.1016/0021-9614(81)90049-5>
72. Ewing, M. B., McGlashan, M. L., and Tzias, P., *J. Chem. Thermodyn.* 13, 527, 1981. <https://doi.org/10.1016/0021-9614(81)90108-7>
73. Akhundov, T. S., and Abdullaev, F. G., *Izv. Vyssh. Uchebn. Zaved., Neft Gaz* 20, 73, 1977.
74. Powell, R. J., Swinton, F. L., and Young, C. L., *J. Chem. Thermodyn.* 2, 105, 1970. <https://doi.org/10.1016/0021-9614(70)90069-8>
75. Kobe, K. A., and Mathews, J. F., *J. Chem. Eng. Data* 15, 182, 1970. <https://doi.org/10.1021/je60044a012>
76. Artyukhovskaya, L. M., Shimanskaya, E. T., and Shimanskii, Yu. I., *Ukr. Fiz. Zh.* (*Ukr. Ed.*) 15, 1974, 1970.
77. Skripov, V. P., and Sinitsyn, E. N., *Zh. Fiz. Khim.* 42, 309, 1968.
78. Kay, W. B., and Hissong, D. W., *Proc. - Am. Pet. Inst., Div. Refin.* 47, 653, 1967. <https://doi.org/10.1093/ptj/47.7.653>
79. Ambrose, D., Broderick, B. E., and Townsend, R., *J. Chem. Soc. A* 633, 1967. <https://doi.org/10.1039/j19670000633>
80. Skaates, J. M., and Kay, W. B., *Chem. Eng. Sci.* 19, 431, 1964. <https://doi.org/10.1016/0009-2509(64)85070-3>
81. Makhan'ko, I. G., and Nozdrev, V. F., *Akust. Zh.* 10, 249, 1964.
82. Cheng, D. C. H., *Chem. Eng. Sci.* 18, 715, 1963. <https://doi.org/10.1016/0009-2509(63)85051-4>
83. Connolly, J. F., and Kandalic, G. A., *J. Chem. Eng. Data* 7, 137, 1962. <https://doi.org/10.1021/je60012a039>
84. Partington, E. J., Rowlinson, J. S., and Weston, J. F., *Trans. Faraday Soc.* 56, 479, 1960. <https://doi.org/10.1039/tf9605600479>
85. McCracken, P. G., Storvick, T. S., and Smith, J. M., *J. Chem. Eng. Data* 5, 130, 1960. <https://doi.org/10.1021/je60006a002>
86. Ambrose, D., Cox, J. D., and Townsend, R., *Trans. Faraday Soc.* 56, 1452, 1960. <https://doi.org/10.1039/tf9605601452>
87. Krichevskii, I. R., Khazanova, N. E., and Linshits, L. R., *Tr. GIAP* No. 9, 40, 1959.
88. Simon, M., *Bull. Soc. Chim. Belg.* 66, 375, 1957. <https://doi.org/10.1002/bscb.19570660129>
89. Krichevskii, I. R., Khazanova, N. E., and Linshits, L. R., *Zh.Fiz. Khim.* 31, 2711, 1957.
90. Ambrose, D., and Grant, D. G., *Trans. Faraday Soc.* 53, 771, 1957. <https://doi.org/10.1039/tf9575300771>
91. Kreglewski, A., *Rocz. Chem.* 29, 754, 1955.
92. Kay, W. B., and Nevens, T. D., *Chem. Eng. Prog., Symp. Ser.* 48, 108, 1952.
93. Bender, P., Furukawa, G. T., and Hyndman, J. R., *Ind. Eng. Chem.* 44, 387, 1952. <https://doi.org/10.1021/ie50506a050>
94. Gornowski, E. J., Amick, E. H., and Hixson, A. N., *Ind. Eng. Chem.* 39, 1348, 1947. <https://doi.org/10.1021/ie50454a022>
95. Schamhardt, H. O., Thesis, Amsterdam, the Netherlands, 1908.
96. Altschul, M., *Z. Phys. Chem., Stoechiom. Verwandtschaftsl.* 11, 577, 1893.
97. Young, S., *J. Chem. Soc., Trans.* 55, 486, 1889. <https://doi.org/10.1039/CT8895500486>
98. Nikitin, E. D., Popov, A. P., and Yatluk, Y. G., *J. Chem. Eng. Data* 51, 1335, 2006. <https://doi.org/10.1021/je060078g>
99. VonNiederhausern, D. M., Wilson, G. M., and Giles, N. F., *J. Chem. Eng. Data* 51, 1982, 2006. <https://doi.org/10.1021/je060088h>
100. Chirico, R. D., Knipmeyer, S. E., Nguyen, A., and Steele, W. V., *J. Chem. Thermodyn.* 23, 759, 1991. <https://doi.org/10.1016/S0021-9614(05)80194-6>
101. Steele, W. V., Chirico, R. D., Hossenlopp, I. A., Knipmeyer, S. E., Nguyen, A., and Smith, N. K., *Experimental Results for DIPPR 1990–91 Projects on Phase Equilibria and Pure Component Properties*, DIPPR Data Ser. No. 2, p. 188, 1994.
102. Guseinov, S. O., Naziev, Y. M., Farzaliev, B. I., and Movsunov, T. G., *Izv. Vyssh. Uchebn. Zaved., Neft Gaz* 21, 48, 1978.
103. Ambrose, D., and Ghiassee, N. B., *J. Chem. Thermodyn.* 22, 307, 1990. <https://doi.org/10.1016/0021-9614(90)90204-4>
104. Anselme, M. J., and Teja, A. S., *AIChE Symp. Ser.* 86 (279), 128, 1990.
105. Chirico, R. D., and Steele, W. V., *J. Chem. Thermodyn.* 36, 633, 2004. <https://doi.org/10.1016/j.jct.2004.04.003>

106. Glaser, F., and Ruland, H., *Chem.-Ing.-Tech.* 29, 772, 1957. <https://doi.org/10.1002/cite.330291204>
107. Chirico, R. D., Knipmeyer, S. E., Nguyen, A., and Steele, W. V., *J. Chem. Thermodyn.* 21, 1307, 1989. <https://doi.org/10.1016/0021-9614(89)90119-5>
108. Reiter, R. W., *NASA Document* N63, 1963.
109. Ellard, J. A., and Yanko, W. H., *U. S. A. E. C. Rep.* IDO-11008, 1963.
110. Sako, T., Yasumoto, M., Nakazawa, N., and Kamizawa, C., *J. Chem. Eng. Data* 46, 1078, 2001. <https://doi.org/10.1021/je000249w>
111. Defibaugh, D. R., Gillis, K. A., Moldover, M. R., Morrison, G., and Schmidt, J. W., *Fluid Phase Equilib.* 81, 285, 1992. <https://doi.org/10.1016/0378-3812(92)85158-5>
112. Nikitin, E. D., Popov, A. P., and Yatluk, Y. G., *J. Chem. Eng. Data* 51, 1326, 2006. <https://doi.org/10.1021/je060068f>
113. Yasumoto, M., Yamada, Y., Murata, J., Urata, S., and Otake, K., *J. Chem. Eng. Data* 48, 1368, 2003. <https://doi.org/10.1021/je0201976>
114. McLure, I. A., and Neville, J. F., *J. Chem. Thermodyn.* 14, 385, 1982. <https://doi.org/10.1016/0021-9614(82)90058-1>
115. Badylkes, S., *Kholod. Tekh.* 43, 18, 1966.
116. Salvi-Narkhede, M., Wang, B. -H., Adcock, J. I., and Van Hook, W. A., *J. Chem. Thermodyn.* 24, 1065, 1992. <https://doi.org/10.1016/S0021-9614(05)80017-5>
117. Herz, W., and Neukirch, E., *Z. Phys. Chem., Stoechiom. Verwandtschaftsl.* 104, 433, 1923. <https://doi.org/10.1515/zpch-1923-10429>
118. Adamenko, I. I., and Chernyavskaya, I. A., *Ukr. Fiz. Zh.* 11, 336, 1966.
119. Morton, D. W., Lui, M. P. W., Tran, C. A., and Young, C. L., *J. Chem. Eng. Data* 45, 437, 2000. <https://doi.org/10.1021/je9902490>
120. Li, Y., Ma, P., and Ruan, Y., *Shiyou Huagong* 22, 322, 1993.
121. Higashi, Y., Uematsu, M., and Watanabe, K., *Bull. JSME* 28, 2660, 1985. <https://doi.org/10.1299/jsme1958.28.2660>
122. Scott, R. B., Meyers, C. H., Rands, R. D., Brickwedde, F. G., and Bekkedahl, N., *J. Res. Natl. Bur. Stand.* (*U.S.*) 35, 39, 1945. <https://doi.org/10.6028/jres.035.017>
123. Cragoe, C. S., *Natl. Bur. Stand.* (*U.S.*) LC-736, 1943.
124. Ma, P. S., Gao, J., and Xia, S., *Chin. J. Chem. Eng.* 10, 473, 2002.
125. Yasumoto, M., Uchida, Y., Ochi, K., Furuya, T., and Otake, K., *J. Chem. Eng. Data* 50, 596, 2005. <https://doi.org/10.1021/je0496589>
126. Warowny, W., *J. Chem. Eng. Data* 41, 689, 1996. <https://doi.org/10.1021/je950242s>
127. Holcomb, C. D., Magee, J. W., and Haynes, W. M., *Research Report RR-147*, Gas Processors Association Project 916, Tulsa, OK, 1995.
128. Deak, A., Victorov, A. I., and De Loos, T. W., *Fluid Phase Equilib.* 107, 277, 1995. <https://doi.org/10.1016/0378-3812(94)02677-S>
129. Vasserman, A. A., Khasilev, I. P., and Cymarnyi, V. A., *Deposited Doc. VNIIKI*, Doc. No. 604-kk, 1989.
130. Li, L., and Kiran, E., *J. Chem. Eng. Data* 33, 342, 1988. <https://doi.org/10.1021/je00053a033>
131. Brunner, E., *J. Chem. Thermodyn.* 20, 273, 1988. <https://doi.org/10.1016/0021-9614(88)90124-3>
132. Younglove, B. A., and Ely, J. F., *J. Phys. Chem. Ref. Data* 16, 577, 1987. <https://doi.org/10.1063/1.555785>
133. Kreglewski, A., and Kay, W. B., *J. Phys. Chem.* 73, 3359, 1969. <https://doi.org/10.1021/j100844a035>
134. Golubev, I. F., and Agaev, N. A., *Dokl. Akad. Nauk SSSR* 151(4), 875, 1963.
135. Kay, W. B., *Ind. Eng. Chem.* 33, 590, 1941. <https://doi.org/10.1021/ie50377a009>
136. Sage, B. H., Hicks, B. L., and Lascey, W. N., *Ind. Eng. Chem.* 32, 1085, 1940. <https://doi.org/10.1021/ie50368a014>
137. Kay, W. B., *Ind. Eng. Chem.* 32, 353, 1940. <https://doi.org/10.1021/ie50363a015>
138. Beattie, J. A., Simard, G. L., and Su, G.-J., *J. Am. Chem. Soc.* 61, 24, 1939. <https://doi.org/10.1021/ja01870a007>
139. Kuenen, J. P., *Commun. Kamerlingh Onnes Lab., Univ. Leiden*, No. 125, 1, 1911.
140. Nikitin, E. D., Popov, A. P., and Yatluk, Y. G., *J. Chem. Eng. Data* 51, 609, 2006. <https://doi.org/10.1021/je050424e>
141. Steele, W. V., Chirico, R. D., Knipmeyer, S. E., and Nguyen, A., *J. Chem. Eng. Data* 41, 1255, 1996.
142. Eliosa, G., Murrieta-Guevara, F., Reza, J., and Trejo Rodriguez, A., *Fluid Phase Equilib.* 61, 99, 1990. <https://doi.org/10.1016/0378-3812(90)90007-A>
143. Tsonopoulos, C., and Ambrose, D., *J. Chem. Eng. Data* 46, 480, 2001. <https://doi.org/10.1021/je000210r>
144. Teja, A. S., and Anselme, M. J., *AIChE Symp. Ser.* 86 (279), 122, 1990.
145. Gude, M. T., Mendez-Santiago, J., and Teja, A. S., *J. Chem. Eng. Data* 42, 278, 1997. <https://doi.org/10.1021/je960231e>
146. Ambrose, D., and Ghiassee, N. B., *J. Chem. Thermodyn.* 19, 505, 1987. <https://doi.org/10.1016/0021-9614(87)90147-9>
147. Brown, J. C., *J. Chem. Soc., Trans.* 89, 311, 1906. <https://doi.org/10.1039/CT9068900311>
148. Lydersen, A. L., and Tsochev, V., *Chem. Eng. Technol.* 13, 125, 1990. <https://doi.org/10.1002/ceat.270130117>
149. Rosenthal, D. J., and Teja, A. S., *Ind. Eng. Chem. Res.* 28, 1693, 1989. <https://doi.org/10.1021/ie00095a020>
150. Christou, G., and Young, C. L., *Int. DATA Ser.*, Sel. Data Mixtures, Ser. A, 14(4), 245, 1986.
151. Naumova, A. A., Tyvina, T. N., and Fokina, V. V., *Zh. Prikl. Khim.* 1980, 1667, 1980.
152. Efremov, Yu. V., *Zh. Fiz. Khim.* 40, 1240, 1966. <https://doi.org/10.1121/1.1942972>
153. Ambrose, D., and Townsend, R., *J. Chem. Soc.* 54, 3614, 1963. <https://doi.org/10.1039/jr9630003614>
154. Singh, R., and Shemilt, L. W., *J. Chem. Phys.* 23, 1370, 1955. <https://doi.org/10.1063/1.1742307>
155. Kay, W. B., and Donham, W. E., *Chem. Eng. Sci.* 4, 1, 1955. <https://doi.org/10.1016/0009-2509(55)85001-4>
156. Stevens, R. M. M., Van Roermund, J. C., Jager, M. D., De Loos, T. W., and De Swaan Arons, J., *Fluid Phase Equilib.*, 138, 159, 1997. <https://doi.org/10.1016/S0378-3812(97)00163-5>
157. Kobe, K. A., Crawford, H. R., and Stephenson, R. W., *Ind. Eng. Chem.* 47, 1767, 1955. <https://doi.org/10.1021/ie50549a025>
158. Ihmels, E. C., Fischer, K., and Gmehling, J., *Fluid Phase Equilib.* 228–229, 155, 2005. <https://doi.org/10.1016/j.fluid.2004.09.013>
159. Li, J., Qin, Z., Wang, G., Dong, M., and Wang, J., *J. Chem. Eng. Data* 52, 1736, 2007. <https://doi.org/10.1021/je700132w>
160. Beattie, J. A., and Marple, S., *J. Am. Chem. Soc.* 72, 1449, 1950. <https://doi.org/10.1021/ja01160a006>
161. Olds, R. H., Sage, B. H., and Lacey, W. N., *Ind. Eng. Chem.* 38, 301, 1946. <https://doi.org/10.1021/ie50435a019>
162. Morton, D. W., Lui, M., and Young, C. L., *J. Chem. Thermodyn.* 31, 675, 1999. <https://doi.org/10.1006/jcht.1998.0480>
163. Teja, A. S., and Rosenthal, D. J., *AIChE Symp. Ser.* 86 (279), 133, 1990.
164. Ma, F., Wang, J., and Ruan, Y., *J. Chem. Eng. Chin. Univ.* 9, 62, 1995.
165. Ma, P., and Ruan, Y., *Gaoxiao Huaxue Gongcheng Xuebao* 9, 62, 1995.
166. Quadri, S. K., and Kudchadker, A. P., *J. Chem. Thermodyn.* 23, 129, 1991. <https://doi.org/10.1016/S0021-9614(05)80288-5>
167. Pawlewski, B., *Ber. Dtsch. Chem. Ges.* 15, 2460, 1882. <https://doi.org/10.1002/cber.188201502201>
168. Toczylkin. L. S., and Young, C. L., *J. Chem. Thermodyn.* 12, 365, 1980. <https://doi.org/10.1016/0021-9614(80)90149-4>
169. Nikitin, E. D., and Popov, A. P., *J. Chem. Eng. Data* 52, 1336, 2007. <https://doi.org/10.1021/je700049s>
170. Morton, D. W., Lui, M. P. W., Tran, C. A., and Young, C. L., *J. Chem. Eng. Data* 49, 283, 2004. <https://doi.org/10.1021/je0341357>
171. Nikitin, E. D., Popov, A. P., and Bogatishcheva, N. S., *J. Chem. Eng. Data* 48, 1137, 2003. <https://doi.org/10.1021/je0256535>
172. Wilson, L. C., Wilding, W. V., Wilson, H. L., and Wilson, G. M., *J. Chem. Eng. Data* 40, 765, 1995. <https://doi.org/10.1021/je00020a008>
173. Zawisza, A. C., and Glowka, S., *Bull. Acad. Pol. Sci., Ser. Sci. Chim.* 19, 191, 1971. <https://doi.org/10.1017/S0043174500048645>
174. Osipiuk, B., and Stryjek, R., *Bull. Acad. Pol. Sci., Ser. Sci. Chim.* 18, 289, 1970.

175. Steele, W. V., Chirico, R. D., Knipmeyer, S. E., Nguyen, A., and Smith, N. K., *J. Chem. Eng. Data* 41, 1285, 1996.
176. Sako, T., Sato, M., Nakazawa, N., Oowa, M., Yasumoto, M., Ito, H., And Yamashita, S., *J. Chem. Eng. Data* 41, 802, 1996. <https://doi.org/10.1021/je950327t>
177. Tanikawa, S., Tatoh, J., Maezawa, Y., Sato, H., and Watanabe, K., *J. Chem. Eng. Data* 37, 74, 1992. <https://doi.org/10.1021/je00005a022>
178. Chae, H. B., Schmidt, J. W., and Moldover, M. R., *J. Phys. Chem.* 94, 8840, 1990. <https://doi.org/10.1021/j100388a018>
179. Mears, W. H., Stahl, R. F., Orfeo, S. R., Shair, R. C., Kells, L. F., Thompson, W., and McCann, H., *Ind. Eng. Chem.* 47, 1449, 1955. <https://doi.org/10.1021/ie50547a052>
180. Yata, J., Hori, M., Kawakatsu, H., and Minamiyama, T., *Int. J. Thermophys.* 17, 65, 1996. <https://doi.org/10.1007/BF01448210>
181. Zhao, X., and Ma, P. S., *Chin. J. Chem. Eng.* 3, 233, 1995.
182. Nishiumi, H., Kohmatsu, S., Yokoyama, T., and Konda, A., *Fluid Phase Equilib.* 104, 131, 1995. <https://doi.org/10.1016/0378-3812(94)02644-G>
183. Economou, I. G., Peters, C. J., Florusse, L. J., and De Swaan Arons, J., *Fluid Phase Equilib.* 111, 239, 1995. <https://doi.org/10.1016/0378-3812(95)90814-7>
184. Nishiumi, H., Komatsu, M., Yokoyama, T., and Kohmatsu, S., *Fluid Phase Equilib.* 83, 109, 1993. <https://doi.org/10.1016/0378-3812(93)87013-Q>
185. Wang, J., Liu, Z. G., Tan, L. C., and Yin, J. M., *Fluid Phase Equilib.* 80, 203, 1992. <https://doi.org/10.1016/0378-3812(92)87068-X>
186. Noles, J. R., and Zollweg, J. A., *J. Chem. Eng. Data* 37, 306, 1992. <https://doi.org/10.1021/je00007a008>
187. Leu, A. D., and Robinson, D. B., *J. Chem. Eng. Data* 37, 7, 1992. <https://doi.org/10.1021/je00005a003>
188. Goodwin, A. R. H., Defibaugh, D. R., and Weber, L. A., *Int. J. Thermophys.* 13, 837, 1992. <https://doi.org/10.1007/BF00503911>
189. Zhimai, H., and Jianfen, H., *Gongcheng Rewuli Xuebao* 10, 233, 1989.
190. He, Z., Zhang, Y., and Hong, J., *The Second Asian Thermophysical Properties Conference*, Hunan University of Science and Technology, Guangzhou, China, p. 519, 1989.
191. *Chemicals and Plastics Physical Properties*, Union Carbide Corp. (Product Bulletin), 1968.
192. Ambrose, D., and Ghiassee, N. B., *J. Chem. Thermodyn.* 20, 765, 1988. <https://doi.org/10.1016/0021-9614(88)90029-8>
193. Berthoud, A., *J. Chim. Phys. Phys.-Chim. Biol.* 15, 3, 1917. <https://doi.org/10.1051/jcp/1917150003>
194. Cullick, A. S., and Ely, J. F., *J. Chem. Eng. Data* 27, 276, 1982. <https://doi.org/10.1021/je00029a014>
195. Mansoorian, H., Hall, K. R., Holste, J. C., and Eubank, P. T., *J. Chem. Thermodyn.* 13, 1001, 1981. <https://doi.org/10.1016/0021-9614(81)90001-X>
196. Centnerszwer, M., *Z. Phys. Chem., Stoechiom. Verwandtschaftsl.* 49, 199, 1904. <https://doi.org/10.1515/zpch-1904-0115>
197. Murphy, K. P., *J. Chem. Eng. Data* 9, 259, 1964. <https://doi.org/10.1021/je60021a041>
198. Skripov, V. P., and Muratov, G. N., *Russ. J. Phys. Chem.* (*Engl. Transl.*) 51, 806, 1977.
199. Yada, N., Uematsu, M., and Watanabe, K., *Nippon Kikai Gakkai Ronbunshu, B-hen* 55, 2426, 1989. <https://doi.org/10.1299/kikaib.55.2426>
200. Mears, W. H., Rosenthal. E., and Sinka, J. V., *J. Chem. Eng. Data* 11, 338, 1966. <https://doi.org/10.1021/je60030a014>
201. Ma, P., Fang, Z., Zhang, J., and Ruan, Y., *J. Chem. Eng. Chin. Univ.* 6, 112, 1992.
202. Ma, P., Fang, Z., Zhang, J., and Ruan, Y., *Gaoxiao Huaxue Gongcheng Xuebao* 6, 112, 1992.
203. Schmidt, J. W., Carrillo-Nava, E., and Moldover, M. R., *Fluid Phase Equilib.* 122, 187, 1996. <https://doi.org/10.1016/0378-3812(96)03044-0>
204. Fukushima, M., and Watanabe, N., *Nippon Reito Kyokai Ronbunshu* 10, 75, 1993.
205. Liu, Z., Liang, D., He, M., Ju, B., and Yin, J., *Gongchengrewuli Xuebao* 18, 261, 1997.
206. Booth, H. S., and Swinehart, C. F., *J. Am. Chem. Soc.* 57, 1337, 1935. <https://doi.org/10.1021/ja01310a051>
207. Oliver, G. D., Grisard, J. W., and Cunningham, C. W., *J. Am. Chem. Soc.* 73, 5719, 1951. <https://doi.org/10.1021/ja01156a067>
208. Weber, L. A., *J. Chem. Eng. Data* 34, 171, 1989. <https://doi.org/10.1021/je00056a007>
209. Shavandrin, A. M., and Li, S. A., *Inzh.-Fiz. Zh.* 37, 830, 1979. <https://doi.org/10.1007/BF01102225>
210. Vitkalov, V. S., Kolpakov, Y. D., and Skripov, V. P., *Zh. Fiz. Khim.* 50, 2336, 1976.
211. Oguchi, K., Tanishita, I., Watanabe, K., Yamaguchi, T., and Sasayama, A., *Bull. JSME* 18, 1456, 1975. <https://doi.org/10.1299/jsme1958.18.1456>
212. Muratov, G. N., and Skripov, V. P., *Zh. Fiz. Khim.* 49, 2148, 1975.
213. Levelt Sengers, J. M. H., Straub, J., and Vincentini-Missoni, M., *J. Chem. Phys.* 54, 5034, 1971. <https://doi.org/10.1063/1.1674794>
214. Tsiklis, D. S., and Prokhorov, V. M., *Dokl. Akad. Nauk SSSR* 174, 470, 1967.
215. Tsiklis, D. S., and Prokhorov, V. M., *Zh. Fiz. Khim.* 41, 2195, 1967.
216. Michels, A., Wassenaar, T., Wolkers, G. J., Prins, Chr., and van de Klundert, L., *J. Chem. Eng. Data* 11, 449, 1966.
217. Delaunois, C., *Ann. Mines Belg.* No. 1, 9, 1968.
218. Ambrose, D., *Trans. Faraday Soc.* 59, 1988, 1963. <https://doi.org/10.1039/tf9635901988>
219. Dewar, J., *Philos. Mag.* 18, 210, 1884. <https://doi.org/10.1080/14786448408627592>
220. Young, C. L., *Aust. J. Chem.* 25, 1625, 1972. <https://doi.org/10.1071/CH9721625>
221. Hicks, C. P., and Young, C. L., *Trans. Faraday Soc.* 67, 1605, 1971. <https://doi.org/10.1039/tf9716701605>
222. Fischer, R., and Reichel, T., *Mikrochem. Ver. Mikrochim. Acta* 31, 102, 1943. <https://doi.org/10.1007/BF01412990>
223. Zhang, R., Qin, Z., Wang, G., Dong, M., Hou, X., and Wang, J., *J. Chem. Eng. Data* 50, 1414, 2005. <https://doi.org/10.1021/je0500882>
224. Hugill, J. A., and McGlashan, M. L., *J. Chem. Thermodyn.* 10, 95, 1978. <https://doi.org/10.1016/0021-9614(78)90150-7>
225. Naziev, Y. M., Abasov, A. A., Nurberdiev, A. A., and Shakhverdiev, A. N., *Zh. Fiz. Khim.* 68, 434, 1974.
226. Krichevskii, I. R., and Sorina, G. A., *Zh. Fiz. Khim.* 34, 1420, 1960.
227. Richardson, M. J., and Rowlinson, J. S., *Trans. Faraday Soc.* 53, 1586, 1959.
228. Reamer, H. H., Sage, B. H., and Lacey, W. N., *Chem. Eng. Data Ser.* 3, 240, 1958. <https://doi.org/10.1021/i460004a014>
229. Kay, W. B., and Albert, R. E., *Ind. Eng. Chem.* 48, 422, 1956. <https://doi.org/10.1021/ie51398a024>
230. Rotinyantz, L., and Nagornov, N. N., *Z. Phys. Chem., Abt. A* 169, 20, 1934.
231. Young, S., and Fortey, E. C., *J. Chem. Soc., Trans.* 75, 873, 1899. <https://doi.org/10.1039/CT8997500873>
232. Steele, W. V., Chirico, R. D., Knipmeyer, S. E., and Nguyen, A., *J. Chem. Eng. Data* 42, 1021, 1997.
233. Ambrose, D., and Ghiassee, N. B., *J. Chem. Thermodyn.* 19, 903, 1987. <https://doi.org/10.1016/0021-9614(87)90036-X>
234. Alekhin, O. D., Krupskii, N. P., and Minchenko, Y. B., *Ukr. Fiz. Zh.* (*Ukr. Edc.*) 15, 509, 1970.
235. Kudchadker, A. P., Alani, G. H., and Zwolinski, B. J., *Chem. Rev.* 68, 659, 1968. <https://doi.org/10.1021/cr60256a002>
236. Kay, W. B., *J. Am. Chem. Soc.* 69, 1273, 1947. <https://doi.org/10.1021/ja01198a014>
237. Lin, D. C. K., Silberberg, I. H., and McKetta, J. J., *J. Chem. Eng. Data* 15, 483, 1970. <https://doi.org/10.1021/je60047a016>
238. Booth, H. S., and Morris, W. C., *J. Phys. Chem.* 62, 875, 1958. <https://doi.org/10.1021/j150565a033>
239. Grzyll, L. R., Ramos, C., and Back, D. D., *J. Chem. Eng. Data* 41, 446, 1996. <https://doi.org/10.1021/je950266z>
240. Ermakov, G. V., and Skripov, V. P., *Zh. Fiz. Khim.* 43, 1308, 1969.
241. Cheng, D. C. H., McCoubrey, J. C., and Phillips, D. G., *Trans. Faraday Soc.* 58, 224, 1962. <https://doi.org/10.1039/tf9625800224>

242. Gude, M. T., and Teja, A. S., *Experimental Results for DIPPR 1990-91 Projects on Phase Equilibria and Pure Component Properties*, DIPPR Data Series No. 2, p. 174, 1994.
243. Anselme, M. J., Gude, M., and Teja, A. S., *Fluid Phase Equilib.* 57, 317, 1990. <https://doi.org/10.1016/0378-3812(90)85130-3>
244. Rosenthal, D. J., and Teja, A. S., *AIChE J.* 35, 1829, 1989. <https://doi.org/10.1002/aic.690351109>
245. Smith, R. L., Teja, A. S., and Kay, W. B., *AIChE J.* 33, 232, 1987. <https://doi.org/10.1002/aic.690330209>
246. Brunner, E., *J. Chem. Thermodyn.* 19, 823, 1987. <https://doi.org/10.1016/0021-9614(87)90029-2>
247. Smith, R. L., Anselme, M., and Teja, A. S., *Proc. World Congress III Chem. Eng.*, Tokyo, Vol. II, p. 135, 1986.
248. Gehrig, M., and Lentz, H., *J. Chem. Thermodyn.* 15, 1159, 1983. <https://doi.org/10.1016/0021-9614(83)90007-1>
249. Mogollon, E., Kay, W. B., and Teja, A. S., *Ind. Eng. Chem. Fundam.* 21, 173, 1982. <https://doi.org/10.1021/i100006a012>
250. Cholpan, P. F., Sperkach, V. S., and Garkusha, L. N., *Fiz. Zhidk. Sostoyaniya* 9, 79, 1981.
251. Kay, W. B., and Pak, S. C., *J. Chem. Thermodyn.* 12, 673, 1980. <https://doi.org/10.1016/0021-9614(80)90089-0>
252. Chun, S. W., Ph.D. Thesis, Ohio State Univ., Columbus, OH, 1964.
253. Nikitin, E. D., Popov, A. P., Bogatishcheva, N. S., and Yatluk, Y. G., *J. Chem. Eng. Data* 49, 1515, 2004. <https://doi.org/10.1021/je0498356>
254. Quadri, S. K., Khilar, K. C., Kudchadker, A. P., and Patni, M. J., *J. Chem. Thermodyn.* 23, 67, 1991. <https://doi.org/10.1016/S0021-9614(05)80060-6>
255. Anselme, M. J., and Teja, A. S., *Fluid Phase Equilib.* 40, 127, 1988. <https://doi.org/10.1016/0378-3812(88)80025-6>
256. Pulliam, M. K., Gude, M. T., and Teja, A. S., *J. Chem. Eng. Data* 40, 455, 1995. <https://doi.org/10.1021/je00018a022>
257. Gude, M. T., Rosenthal, D. J., and Teja, A. S., *Fluid Phase Equilib.* 70, 55, 1991. <https://doi.org/10.1016/0378-3812(91)85004-E>
258. Nikitin, E. D., Popov, A. P., Bogatishcheva, N. S., and Yatluk, Y. G., *J. Chem. Eng. Data* 47, 1012, 2002. <https://doi.org/10.1021/je025514n>
259. Chirico, R. D., Gammon, B. E., Knipmeyer, S. E., Nguyen, A., Strube, M. M., Tsonopoulos, C., and Steele, W. V., *J. Chem. Thermodyn.* 22, 1075, 1990. <https://doi.org/10.1016/0021-9614(90)90157-L>
260. Chirico, R. D., Knipmeyer, S. E., Nguyen, A., and Steele, W. V., *J. Chem. Thermodyn.* 23, 431, 1991. <https://doi.org/10.1016/S0021-9614(05)80131-4>
261. Nisel'son, L. A., Tret'yakova, K. V., Yatko, M. E., Tsirut, E. K., and Antonova, N. P., *Thermophysical Properties of Matter and Substances*, Vol. 4, Rabinovich, V. A., Ed., Amerind Pub., New Delhi, p. 132, 1975.
262. Toczylkin. L. S., and Young, C. L., *J. Chem. Thermodyn.* 12, 355, 1980. <https://doi.org/10.1016/0021-9614(80)90148-2>
263. Golik, A. Z., and Ravikovich, S. D., *Zh. Fiz. Khim.* 23, 86, 1949.
264. Stepanov, N. G., *Russ. J. Phys. Chem.* (*Engl. Transl.*) 46, 464, 1972.
265. Krauss, R., and Stephan, K., *J. Phys. Chem. Ref. Data* 18, 43, 1989. <https://doi.org/10.1063/1.555842>
266. Stepanov, N. G., and Nozdrev, V. F., *Russ. J. Phys. Chem.* (*Engl. Transl.*) 42, 1300, 1968.
267. Garcia-Sanchez, F., and Trejo Rodriguez, A., *J. Chem. Thermodyn.* 19, 359, 1987. <https://doi.org/10.1016/0021-9614(87)90118-2>
268. Garcia-Sanchez, F., and Trejo Rodriguez, A., *J. Chem. Thermodyn.* 17, 981, 1985. <https://doi.org/10.1016/0021-9614(85)90012-6>
269. Hojendahl, K., *Mat.-Fys. Medd. - K. Dan. Vidensk. Selsk.* 24, 1, 1946.
270. Duarte-Garza, H. A., Hwang, C.-A., Kellerman, S. A., Miller, R. C., Hall, K. R., Holste, J. C., Marsh, K. N., and Gammon, B. E., *J. Chem. Eng. Data* 42, 497, 1997. <https://doi.org/10.1021/je9603584>
271. Benning, A. F., and McHarness, R. C., *Ind. Eng. Chem.* 32, 814, 1940. <https://doi.org/10.1021/ie50366a017>
272. Gorchakovskii, V. K., Zadov, V. E., and Podvezennyi, V. N., *Inzh.-Fiz. Zh.* 59, 122, 1990. <https://doi.org/10.1007/BF00871333>
273. Garcia-Sanchez, F., Romero-Martinez, A., and Trejo Rodriguez, A., *J. Chem. Thermodyn.* 21, 823, 1989. <https://doi.org/10.1016/0021-9614(89)90029-3>
274. Higashi, Y., Uematsu, M., and Watanabe, K., *Bull. JSME* 28, 2968, 1985. <https://doi.org/10.1299/jsme1958.28.2968>
275. Piao, C. C., Sato, H., and Watanabe, *K., J. Chem. Eng. Data* 36, 398, 1991. <https://doi.org/10.1021/je00004a016>
276. Weber, L. A., and Levelt Sengers, J. M. H., *Fluid Phase Equilib.* 55, 241, 1990. <https://doi.org/10.1016/0378-3812(90)85016-4>
277. Tanikawa, S., Kabata, Y., Sato, H., and Watanabe, K., *J. Chem. Eng. Data* 35, 381, 1990. <https://doi.org/10.1021/je00062a002>
278. Fukushima, M., Watanabe, N., and Kamimura, T., *Nippon Reito Kyokai Ronbunshu* 7, 243, 1990.
279. Fukushima, M., Watanabe, N., and Kamimura, T., *Nippon Reito Kyokai Ronbunshu* 7, 189, 1990.
280. Yamashita, T., Kubota, H., Tanaka, Y., Makita, T., and Kashiwagi, H., *Proc. 10th Symp. Thermophys. Prop.*, Japan, pp. 75–78, 1989.
281. Mandlekar, A. V., Kay, W. B., Smith, R. L., and Teja, A. S., *Fluid Phase Equilib.* 23, 79, 1985. <https://doi.org/10.1016/0378-3812(85)85029-9>
282. Herz, W., Z. *Anorg. Allg. Chem.* 149, 230, 1925. <https://doi.org/10.1002/zaac.19251490117>
283. Nikitin, E. D., Pavlov, P. A., and Popov, A. P., *J. Chem. Thermodyn.* 27, 43, 1995. <https://doi.org/10.1006/jcht.1995.0005>
284. Ratzsh, M. T., *Z. Phys. Chem. Leipzig* 243, 212, 1970.
285. Schmidt, G. C., *Justus Liebigs Ann. Chem.* 266, 266, 1891. <https://doi.org/10.1002/jlac.18912660304>
286. Young, C. L., *Int. DATA Ser., Sel. Data Mixtures, Ser. A*, No. 1, 66, 1975.
287. Ambrose, D., Sprake, C. H. S., and Townsend, R., *J. Chem. Thermodyn.* 4, 247, 1972. <https://doi.org/10.1016/0021-9614(72)90063-8>
288. Zawisza, A. C., *Bull. Acad. Pol. Sci., Ser. Sci. Chim.* 15, 291, 1967.
289. Stryjek, R., and Kreglewski, A., *Bull. Acad. Pol. Sci., Ser. Sci. Chim.* 13, 201, 1965.
290. Kobe, K. A., Ravicz, A. E., and Vohra, S. P., *J. Chem. Eng. Data* 1, 50, 1956. <https://doi.org/10.1021/i460001a010>
291. Schroeer, E., *Z. Phys. Chem., Abt. A* 140, 379, 1929.
292. Schroeer, E., *Z. Phys. Chem., Abt. A* 140, 241, 1929. <https://doi.org/10.1038/scientificamerican0329-241>
293. Wilip, J., *Eesti Vabariigi Tartu Ulik. Toim. A* 6 (2), 1924.
294. Audant, *C. R. Hebd. Seances Acad. Sci.* 170, 1573, 1920.
295. Prins, A., and Scheffer, F. E. C., *J. Phys. Chem.* 84, 827, 1913.
296. Travers, M. W., and Usher, F. L., *Z. Phys. Chem., Stoechiom. Verwandtschaftsl.* 57, 365, 1906.
297. Centerszwer, M., and Pakalneet, A., *Z. Phys. Chem., Stoechiom. Verwandtschaftsl.* 55, 303, 1906. <https://doi.org/10.1515/zpch-1906-5515>
298. Smits, A., *Z. Phys. Chem., Stoechiom. Verwandtschaftsl.* 52, 587, 1905.
299. Galitzine, B., and Wilip, J., *Bull. Acad. Pet.* 11, No. 3, 117, 1901.
300. De Vries, E. C., *Arch. Neerl. Sci. Exactes Nat.* 28, 215, 1895.
301. Ramsay, W., and Young, S., *Philos. Trans. R. Soc. London, A* 178, 57, 1887. <https://doi.org/10.1098/rsta.1887.0003>
302. Radice, G., Ph.D. Thesis, Univ. of Geneve, 1899.
303. Vespigniani, G. R., *Gazz. Chim. Ital.* 33, 73, 1903.
304. Berthoud, A., and Brum, R., *J. Chim. Phys. Phys.-Chim. Biol.* 21, 143, 1924. <https://doi.org/10.1051/jcp/19231924200143>
305. Grebenkov, A. J., Zhelezny, V. P., Klepatsky, P. M., Beljajeva, O. V., Chernjak, Y. A., Kotelevsky, Y. G., and Timofejev, B. D., *Int. J. Thermophys.* 17, 535, 1996. <https://doi.org/10.1007/BF01441501>
306. Holcomb, C. D., Niesen, V. G., Van Poolen, L. J., and Outcalt, S. L., *Fluid Phase Equilib.* 91, 145, 1993. <https://doi.org/10.1016/0378-3812(93)85085-Z>
307. Higashi, Y., Ashizawa, M., Kabata, Y., Majima, T., Uematsu, M., and Watanabe, K., *JSME Int. J.* 30, 1106, 1987. <https://doi.org/10.1299/jsme1987.30.1106>
308. Otake, K., Yasumoto, M., Yamada, Y., Murata, J., and Urata, S., *J. Chem. Eng. Data* 48, 1380, 2003. <https://doi.org/10.1021/je020210m>
309. Pitschmann, M., and Straub, J., *Int. J. Thermophys.* 23, 877, 2002. <https://doi.org/10.1023/A:1016511014722>
310. Diefenbacher, A., and Tuerk, M., *J. Chem. Thermodyn.* 31, 905, 1999. <https://doi.org/10.1006/jcht.1999.0500>

311. Shi, L., Zhu, M., Han, L., Duan, Y., Sun, L;, and Fu, Y.-D., *Science in China, Ser. E* 41, 435, 1998. <https://doi.org/10.1007/BF02917016>
312. Kuwabara, S., Aoyama, H., Sato, H., and Watanabe, K., *J. Chem. Eng. Data* 40, 112, 1995. <https://doi.org/10.1021/je00017a025>
313. Higashi, Y., *Int. J. Thermophys.* 16, 1175, 1995. <https://doi.org/10.1007/BF02081285>
314. Fu, Y. D., Han, L.-Z., and Zhu, M.-S., *Fluid Phase Equilib.* 111, 273, 1995. <https://doi.org/10.1016/0378-3812(95)02776-B>
315. Higashi, Y., *Int. J. Refrig.* 17, 524, 1994. <https://doi.org/10.1016/0140-7007(94)90028-0>
316. Malbrunot, P. F., Meunier, P. A., Scatena, G. M., Mears, W. H., Murphy, K. P., and Sinka, J. V., *J. Chem. Eng. Data* 13, 16, 1968. <https://doi.org/10.1021/je60036a006>
317. Steele, W. V., Chirico, R. D., Cowell, A. B., Knipmeyer, S. E., and Nguyen, A., *J. Chem. Eng. Data* 47, 725, 2002. <https://doi.org/10.1021/je0100847>
318. Young, C. L., *Int. DATA Ser., Sel. Data Mixtures, Ser. A*, No. 1, 159, 1975.
319. Durig, J. R., and Li, Y. S., *J. Chem. Phys.* 63, 4110, 1975. <https://doi.org/10.1063/1.431181>
320. Bourgou, A., *Bull. Soc. Chim. Belg.* 33, 101, 1924.
321. Verevkin, S. P., Kozlova, S. A., Emel'yanenko, V. N., Nikitin, E. D., Popov, A. P., and Krasnykh, E. L., *J. Chem. Eng. Data* 51, 1896, 2006. <https://doi.org/10.1021/je0602418>
322. Kay, W. B., and Young, C. L., *Int. DATA Ser., Sel. Data Mixtures, Ser. A*, No. 2, 154, 1974.
323. Weaver, D. L., M.S. Thesis, Ohio State Univ., Columbus, OH, 1973.
324. Young, C. L., *Int. DATA Ser., Sel. Data Mixtures, Ser. A*, No. 1, 47, 1974.
325. Genco, J. M., Teja, A. S., and Kay, W. B., *J. Chem. Eng. Data* 25, 350, 1980. <https://doi.org/10.1021/je60087a022>
326. Kay, W. B., and Young, C. L., *Int. DATA Ser., Sel. Data Mixtures, Ser. A*, No. 1, 52, 1975.
327. Genco, J. M., Ph.D. Thesis, Ohio State Univ., Columbus, OH, 1965.
328. Kay, W. B., *J. Am. Chem. Soc.* 68, 1336, 1946. <https://doi.org/10.1021/ja01211a074>
329. Quadri, S. K., and Kudchadker, A. P., *J. Chem. Thermodyn.* 24, 473, 1992. <https://doi.org/10.1016/S0021-9614(05)80119-3>
330. Young, S., and Fortey, E. C., *J. Chem. Soc.* 35, 1126, 1879.
331. Steele, W. V., Chirico, R. D., Knipmeyer, S. E., Nguyen, A., and Smith, N. K., *J. Chem. Eng. Data* 42, 1037, 1997.
332. Steele, W. V., Chirico, R. D., Knipmeyer, S. E., and Nguyen, A., *J. Chem. Eng. Data* 42, 1008, 1997. <https://doi.org/10.1021/je970102d>
333. Ihmels, E. C. C., and Lemmon, E. W., *Fluid Phase Equilib.* 260, 36, 2007. <https://doi.org/10.1016/j.fluid.2006.09.016>
334. Wu, J., Liu, Z., Wang, B., and Pan, J., *J. Chem. Eng. Data* 49, 704, 2004. <https://doi.org/10.1021/je034251+>
335. Noles, J. R., and Zollweg, J. A., *Fluid Phase Equilib.* 66, 275, 1991. <https://doi.org/10.1016/0378-3812(91)85061-X>
336. Glowka, S., *Bull. Acad. Pol. Sci., Ser. Sci. Chim.* 20, 163, 1972.
337. Zawisza, A. C., and Glowka, S., *Bull. Acad. Pol. Sci., Ser. Sci. Chim.* 18, 549, 1970.
338. Edwards, J., and Maass, O., *Can. J. Res., Sect. A* 12, 357, 1935. <https://doi.org/10.1139/cjr35-030>
339. Tapp, J. S., Steacie, E. W. R., and Maass, O., *Can. J. Res.* 9, 217, 1933. <https://doi.org/10.1139/cjr33-081>
340. Cardoso, E., and Coppola, A. A., J. *Chim. Phys. Phys.-Chim. Biol.* 20, 337, 1923. <https://doi.org/10.1051/jcp/1922200337>
341. Cardoso, E., and Bruno, A., *J. Chim. Phys. Phys.-Chim. Biol.* 20, 347, 1923. <https://doi.org/10.1051/jcp/1922200347>
342. Briner, E., and Cardoso, E., *J. Chim. Phys. Phys.-Chim. Biol.* 6, 641, 1908. <https://doi.org/10.1051/jcp/1908060641>
343. Briner, E., and Cardoso, E., *C. R. Hebd. Seances Acad. Sci.* 144, 911, 1907.
344. Bogoslovskii, V. E., Mikhalyuk, G. I., and Shamolin, A. I., *Zh. Prikl. Khim.* (*Leningrad*) 45, 1154, 1972.
345. Kay, W. B., and Hissong, D. W., *Proc. - Am. Pet. Inst., Div. Refin.* 49, 13, 1969.
346. McMicking, J. H., and Kay, W. B., *Proc., Am. Pet. Inst., Sect. 3* 45, 75, 1965.
347. Chirico, R. D., Knipmeyer, S. E., Nguyen, A., and Steele, W. V., *J. Chem. Thermodyn.* 25, 1461, 1993. <https://doi.org/10.1006/jcht.1993.1148>
348. Stern, S. A., and Kay, W. B., *J. Phys. Chem.* 61, 374, 1957. <https://doi.org/10.1021/j150549a027>
349. Edgar, G., and Calingaert, G., *J. Am. Chem. Soc.* 51, 1540, 1929. <https://doi.org/10.1021/ja01380a035>
350. Francis, A. W., *Ind. Eng. Chem.* 49, 1779, 1957. <https://doi.org/10.1021/ie50574a048>
351. Cox, J. D., *Trans. Faraday Soc.* 56, 959, 1960. <https://doi.org/10.1039/tf9605600959>
352. Steele, W. V., Chirico, R. D., Nguyen, A., and Knipmeyer, S. E., *J. Chem. Thermodyn.* 27, 311, 1995. <https://doi.org/10.1006/jcht.1995.0030>
353. Nikitin, E. D., Pavlov, P. A., and Popov, A. P., *J. Chem. Thermodyn.* 26, 1047, 1994. <https://doi.org/10.1006/jcht.1994.1122>
354. Cristou, G., Young, C. L., and Svejda, P., *Fluid Phase Equilib.* 67, 45, 1991. <https://doi.org/10.1016/0378-3812(91)90046-A>
355. Zhuravlev, D. I., *Zh. Fiz. Khim.* 9, 875, 1937.
356. Chirico, R. D., and Steele, W. V., *J. Chem. Eng. Data* 50, 1052, 2005. <https://doi.org/10.1021/je050034s>
357. Smith, R. L., Ph.D. Dissertation, Georgia Institute of Technology 1985.
358. Nikitin, E. D., Pavlov, P. A., and Popov, A. P., *Fluid Phase Equilib.* 141, 155, 1997. <https://doi.org/10.1016/S0378-3812(97)00202-1>
359. Nikitin, E. D., Pavlov, P. A., and Bessonova, N. V., *J. Chem. Thermodyn.* 26, 177, 1994. <https://doi.org/10.1006/jcht.1994.1036>
360. Nikitin, E. D., Pavlov, P. A., and Popov, A. P., *Fluid Phase Equilib.* 189, 151, 2001. <https://doi.org/10.1016/S0378-3812(01)00590-8>
361. Nikitin, E. D., Pavlov, P. A., and Popov, A. P., *Fluid Phase Equilib.* 149, 223, 1998. <https://doi.org/10.1016/S0378-3812(98)00265-9>
362. Sinicyn, E. N., Mikhalevich, L. A., and Yankovskaya, O. P., *Deposited Doc. VINITI*, Doc. No. 2510-V90, 1990.
363. Teja, A. S., Gude, M., and Rosenthal, D. J., *Fluid Phase Equilib.* 52, 193, 1989. <https://doi.org/10.1016/0378-3812(89)80325-5>
364. Beale, E. S., and Docksey, P., *J. Inst. Pet.* 21, 860, 1935.
365. Nikitin, E. D., and Popov, A. P., *Fluid Phase Equilib.* 166, 237, 1999. <https://doi.org/10.1016/S0378-3812(99)00301-5>
366. Horstman, S., Fischer, K., Gmehling, J., and Kolar, P., *J. Chem. Thermodyn.* 32, 451, 2000. <https://doi.org/10.1006/jcht.2000.0611>
367. Colgate, S. O., Sivaraman, A., and Dejsupa, C., *Fluid Phase Equilib.* 76, 175, 1992. <https://doi.org/10.1016/0378-3812(92)85086-N>
368. Friend, D. G., Ingham, H., and Ely, J. F., *J. Phys. Chem. Ref. Data* 20, 275, 1991. <https://doi.org/10.1063/1.555881>
369. Jangkamolkulchai, A., and Luks, K. D., *J. Chem. Eng. Data* 34, 92, 1989. <https://doi.org/10.1021/je00055a027>
370. Calado, J. C. G., Chang, E., Clancy, P., and Streett, W. B., *J. Phys. Chem.* 91, 3914, 1987. <https://doi.org/10.1021/j100298a037>
371. Brunner, E., *J. Chem. Thermodyn.* 17, 871, 1985. <https://doi.org/10.1016/0021-9614(85)90081-3>
372. Morrison, G., and Kincaid, J. M., *AIChE J.* 30, 257, 1984. <https://doi.org/10.1002/aic.690300213>
373. Sychev, V. V., Vasserman, A. A., Kozlov, A. D., Zagoruchenko, V. A., Spiridonov, G. A., and Tsymarny, V. A., *Thermodynamic Properties of Ethane*, Standards Publishing House, Moscow, 1982.
374. Morrison, G., *J. Phys. Chem.* 85, 759, 1981. <https://doi.org/10.1021/j150607a007>
375. Bulavin, L. A., and Shimanskii, Yu. I., *Zh. Eksp. Teor. Fiz.* 29, 482, 1979.
376. Strumpf, H. J., Collings, A. F., and Pings, C. J., *J. Chem. Phys.* 60, 3109, 1974. <https://doi.org/10.1063/1.1681497>
377. Burton, M., and Balzarini, D., *Can. J. Phys.* 52, 2011, 1974. <https://doi.org/10.1139/p74-266>
378. Douslin, D. R., and Harrison, R. H., *J. Chem. Thermodyn.* 5, 491, 1973. <https://doi.org/10.1016/S0021-9614(73)80097-7>
379. Berestov, A. T., Giterman, M. S., and Shmakov, N. G., *Sov. Phys. - JETP* (*Engl. Transl.*) 37, 1128, 1973.
380. Efremova, G. D., and Shvarts, A. V., *Russ. J. Phys. Chem.* (*Engl. Transl.*) 46, 237, 1972.

381. Miniovich, V. M., and Sorina, G. A., *Russ. J. Phys. Chem. (Engl. Transl.)* 45, 306, 1971.
382. Khazanova, N. E., and Sominskaya, E. E., *Russ. J. Phys. Chem. (Engl. Transl.)* 45, 88, 1971.
383. Bulavin, L. A., Ostanevich, Yu M., Simkina, A. P., and Stelkov, A. V., *Ukr. Fiz. Zh. (Ukr. Ed.)* 16, 90, 1971.
384. Chashkin, Yu. R., Smirnov, V. A., and Voronel, A. V., *Teplofiz. Svoistva Veshchestv Mater.* 2, 139, 1970.
385. Sliwinski, P., *Z. Phys. Chem. (Munich)* 63, 263, 1969. <https://doi.org/10.1524/zpch.1969.63.5_6.263>
386. Khazanova, N. E., Lesnevskaya, L. S., and Zakharova, A. V., *Khim. Prom-st. (Moscow)* 42, 364, 1966.
387. Matschke, D. E., and Thodos, G., *J. Chem. Eng. Data* 7, 232, 1962. <https://doi.org/10.1021/je60013a022>
388. Schmidt, E., and Thomas, W., *Forsch. Geb. Ingenieurw.* 20B, 161, 1954. <https://doi.org/10.1007/BF02558359>
389. Kay, W. B., and Brice, D. B., *Ind. Eng. Chem.* 45, 615, 1953. <https://doi.org/10.1021/ie50519a042>
390. Murray, F. E., and Mason, S. G., *Can. J. Chem.* 30, 550, 1952. <https://doi.org/10.1139/v52-067>
391. Mason, S. G., Naldrett, S. N., and Maass, O., *Can. J. Res., Sect.* B 18, 103, 1940. <https://doi.org/10.1139/cjr40b-015>
392. Beattie, J. A., Su, G.-J., and Simard, G. L., *J. Am. Chem. Soc.* 61, 924, 1939. <https://doi.org/10.1021/ja01873a045>
393. Price, T. W., *J. Chem. Soc.* 107, 188, 1915. <https://doi.org/10.1039/CT9150700188>
394. Nikitin, E. D., Pavlov, P. A., and Skripov, P. V., *J. Chem. Thermodyn.* 25, 869, 1993. <https://doi.org/10.1006/jcht.1993.1084>
395. Marshall, W. L., and Jones, E. V., *J. Inorg. Nucl. Chem.* 36, 2319, 1974. <https://doi.org/10.1016/0022-1902(74)80276-9>
396. Sajotschewsky, W., *Beibl. Ann. Phys.* 3, 741, 1879.
397. Mocharnyuk, R. F., *Zh. Obshch. Khim.* 30, 1098, 1960.
398. Golik, A. Z., Ravikovich, S. D., and Orishchenko, A. V., *Ukr. Khim. Zh. (Russ. Ed.)* 21, 167, 1955.
399. Polikhronidi, N. G., Abdulagatov, I. M., Stepanov, G. V., and Batyrova, R. G., *J. Supercrit. Fluids* 43, 1, 2007. <https://doi.org/10.1016/j.supflu.2007.05.004>
400. Hu, T., Qin, Z., Wang, G., Hou, X., and Wang, J., *J. Chem. Eng. Data* 49, 1809, 2004. <https://doi.org/10.1021/je049771z>
401. Sauermann, P., Holzapfel, K., Oprzynski, J., Kohler, F., Poot, W., and De Loos, T. W., *Fluid Phase Equilib.* 112, 249, 1995. <https://doi.org/10.1016/0378-3812(95)02798-J>
402. Mousa, A. H. N., *J. Chem. Eng. Jpn.* 20, 635, 1987. <https://doi.org/10.1252/jcej.20.635>
403. Wilson, K. S., Lindley, D. D., Kay, W. B., and Hershey, H. C., *J. Chem. Eng. Data* 29, 243, 1984. <https://doi.org/10.1021/je00037a003>
404. Hentze, G., *Thermochim. Acta* 20, 27, 1977. <https://doi.org/10.1016/0040-6031(77)85036-3>
405. Nozdrev, V. F., *Akust. Zh.* 2, 209, 1956.
406. Griswold, J., Haney, J. D., and Klein, V. A., *Ind. Eng. Chem.* 35, 701, 1943. <https://doi.org/10.1021/ie50402a015>
407. Battelli, A., *Mem. Torino, Ser.* 2 44, 57, 1893. <https://doi.org/10.1007/BF02709682>
408. Ramsay, W., and Young, S., *Philos. Trans. R. Soc. London* 177, 123, 1886. <https://doi.org/10.1098/rstl.1886.0004>
409. Strauss, O, *Beibl. Ann. Phys.* 6, 282, 1882.
410. Steele, W. V., Chirico, R. D., Knipmeyer, S. E., and Nguyen, A., *Experimental Results for DIPPR 1990–91 Projects on Phase Equilibria and Pure Component Properties*, DIPPR Data Series No. 2, 154, 1994.
411. Ambrose, D., Ellender, J. H., Gundry, H. A., Lee, D. A., and Townsend, R., *J. Chem. Thermodyn.* 13, 795, 1981. <https://doi.org/10.1016/0021-9614(81)90069-0>
412. Lambert, J. D., Clarke, J. S., Duke, J. F., Hicks, C. L., Lawrence, S. D., Morris, D. M., and Shone, M. G. T., *Proc. R. Soc. London, A* 249, 414, 1959. <https://doi.org/10.1098/rspa.1959.0033>
413. Young, S., and Thomas, G. L., *J. Chem. Soc.* 63, 1191, 1893. <https://doi.org/10.1039/CT8936301191>
414. Pohland, E., and Mehl, W., *Z. Phys. Chem., Abt.* A 164, 48, 1933. <https://doi.org/10.1515/zpch-1933-16406>
415. Nadezhdin, A., *Rep. Phys.* 23, 708, 1887.
416. Guseinov, K. D., and Zhabbarov, O., *Izv. Vyssh. Uchebn. Zaved. Neft Gaz* 2, 76, 1975.
417. Steele, W. V., Chirico, R. D., Cowell, A. B., Knipmeyer, S. E., and Nguyen, A., *J. Chem. Eng. Data* 47, 700, 2002. <https://doi.org/10.1021/je010086r>
418. Nowak, P., Kleinrahm, R., and Wagner, W., *J. Chem. Thermodyn.* 28, 1441, 1996. <https://doi.org/10.1006/jcht.1996.0126>
419. Hasch, B. M., and McHugh, M. A., *Fluid Phase Equilib.* 64, 251, 1991. <https://doi.org/10.1016/0378-3812(91)90017-2>
420. Jahangiri, M., Jacobsen, R. T., Stewart, R. B., and McCarty, R. D., *J. Phys. Chem. Ref. Data* 15, 593, 1986. <https://doi.org/10.1063/1.555753>
421. Younglove, B. A., *J. Phys. Chem. Ref. Data*, Vol. 11, suppl. No. 1, Am. Chem. Soc., Washington, DC, 1982.
422. McCarty, R. D., and Jacobsen, R. T., *NBS Tech. Note (U.S.)* 1045, 1981.
423. Thomas, W., and Zander, M., *Int. J. Thermophys* 1, 383, 1980. <https://doi.org/10.1007/BF00516565>
424. Hastings, J. R., Levelt Sengers, J. M. H., and Balfour, F. W., *J. Chem. Thermodyn.* 12, 1009, 1980. <https://doi.org/10.1016/0021-9614(80)90158-5>
425. Hastings, J. R., and Levelt Sengers, J. M. H., *Proc. 7th Symp. Thermophys. Prop.*, Cezairliyan, A., Ed., ASME, New York, p. 794, 1977.
426. Douslin, D. R., and Harrison, R. H., *J. Chem. Thermodyn.* 8, 301, 1976. <https://doi.org/10.1016/0021-9614(76)90072-0>
427. Bender, E., *Cryogenics* 15, 667, 1975. <https://doi.org/10.1016/0011-2275(75)90100-9>
428. Moldover, M. R., *J. Chem. Phys.* 61, 1766, 1974. <https://doi.org/10.1063/1.1682173>
429. Angus, S., Armstrong, B., and de Reuck, K. M., *International Thermodynamic Tables of the Fluid State, No. 2 Ethylene*, Butterworths, London, 1974.
430. Zernov, V. S., Kogan, V. B., and Lyubetskii, S. G., *Zh. Prikl. Khim. (Leningrad)* 44, 1819, 1971.
431. Shim, J., and Kohn, J. P., *J. Chem. Eng. Data* 9, 1, 1964. <https://doi.org/10.1021/je60020a001>
432. Kay, W. B., *Ind. Eng. Chem.* 40, 1459, 1948. <https://doi.org/10.1021/ie50464a026>
433. Diepen, G. A. M., and Scheffer, F. E. C., *J. Am. Chem. Soc.* 70, 4081, 1948. <https://doi.org/10.1021/ja01192a034>
434. Naldrett, S. N., and Maass, O., *Can. J. Res., Sect. B* 18, 118, 1940. <https://doi.org/10.1139/cjr40b-016>
435. Dacey, J. R., McIntosh, R. L., and Maass, O., *Can. J. Res., Sect. B* 17, 206, 1939. <https://doi.org/10.1139/cjr39b-031>
436. Maass, O., and Geddes, A. L., *Philos. Trans. R. Soc. London, A* 236, 303, 1937. <https://doi.org/10.1098/rsta.1937.0004>
437. Lawrenson, I. J., and Lee, D. A., *J. Chem. Thermodyn.* 10, 1111, 1978. <https://doi.org/10.1016/0021-9614(78)90086-1>
438. Zawisza, A. C., and Glowka, S., *Bull. Acad. Pol. Sci., Ser. Sci. Chim.* 18, 555, 1970.
439. Lyons, R. L., M.S. Thesis, Pennsylvania State Univ., University Park, PA, 1985.
440. Daubert, T. E., Jalowka, J. W., and Goren, V., *AIChE Symp. Ser.* 83 (256), 128, 1987.
441. Douslin, D. R., Moore, R. T., Dawson, J. P., and Waddington, G., *J. Am. Chem. Soc.* 80, 2031, 1958. <https://doi.org/10.1021/ja01542a001>
442. Beyerlein, A. L., Desmarteau, D. D., Kul, I., and Zhao, G., *Fluid Phase Equilib.* 150, 287, 1998. <https://doi.org/10.1016/S0378-3812(98)00328-8>
443. Parthasarathy, S., *Proc. - Indian Acad. Sci., Sect.* A 2, 497, 1935. <https://doi.org/10.1007/BF03046893>
444. Bominaar, S. A. R. C., Trappeniers, N. J., and Biswas, S. N., *J. Phys. Chem.* 94, 1097, 1990. <https://doi.org/10.1021/j100366a015>
445. Bominaar, S. A. R. C., Biswas, S. N., Trappeniers, N. J., and Ten Seldam, C. A., *J. Chem. Thermodyn.* 19, 959, 1987. <https://doi.org/10.1016/0021-9614(87)90043-7>
446. Froba, A. P., Botero, C., and Leipertz, A., *Int. J. Thermophys.* 27, 1609, 2006. <https://doi.org/10.1007/s10765-006-0122-6>
447. Uchida, Y., Yasumoto, M., Yamada, Y., Ochi, K., Furuya, T., and Otake, K., *J. Chem. Eng. Data* 49, 1615, 2004.

448. Otake, K., Uchida, Y., Yasumoto, M., Yamada, Y., Furuya, T., and Ochi, K., *J. Chem. Eng. Data* 49, 1643, 2004.
449. Hu, P., and Chen, Z. S., *Fluid Phase Equilib.* 221, 7, 2004. <https://doi.org/10.1016/j.fluid.2004.04.009>
450. Salvi-Narkhede, M., Adcock, J. L., Gakh, A., and Van Hook, W. A., *J. Chem. Thermodyn.* 25, 643, 1993. <https://doi.org/10.1006/jcht.1993.1060>
451. Ambrose, D., and Ghiassee, N. B., *J. Chem. Thermodyn.* 20, 1231, 1988. <https://doi.org/10.1016/0021-9614(88)90108-5>
452. Myers, J. E., Hershey, H. C., and Kay, W. B., *J. Chem. Thermodyn.* 11, 1019, 1979. <https://doi.org/10.1016/0021-9614(79)90132-0>
453. Golik, A. Z., and Adamenko, I. I., *Ukr. Fiz. Zh.* (*Ukr. Ed.*) 10, 443, 1965.
454. Golik, A. Z., and Ivanova, I. I., *Zh. Fiz. Khim.* 36, 1768, 1962.
455. Kay, W. B., *Ind. Eng. Chem.* 30, 459, 1938. <https://doi.org/10.1021/ie50340a023>
456. Christou, G., Sadus, R. J., and Young, C. L., *Fluid Phase Equilib.* 67, 259, 1991. <https://doi.org/10.1016/0378-3812(91)90060-K>
457. Kurumov, D. S., Grigor'ev, B. A., and Vasil'ev, Yu. L., *Teplofiz. Svoistva Veshchestv Mater.* No. 27, 101, 1989.
458. De Loos, T. W., Poot, W., and De Swaan Arons, J., *Fluid Phase Equilib.* 42, 209, 1988. <https://doi.org/10.1016/0378-3812(88)80060-8>
459. Artyukhovskaya, L. M., Shimanskaya, E. T., and Shimanskii, Yu I., *Opt. Spektrosk.* 37, 935, 1974.
460. Artyukhovskaya, L. M., Shimanskaya, E. T., and Shimanskii, Yu I., *Sov. Phys. - JETP* (*Engl. Transl.*) 37, 848, 1973.
461. Artyukhovskaya, L. M., Shimanskaya, E. T., and Shimanskii, Yu I., *Zh. Eksp. Teor. Fiz.* 63, 2159, 1972.
462. Smith, L. B., Beattie, J. A., and Kay, W. C., *J. Am. Chem. Soc.* 59, 1587, 1937. <https://doi.org/10.1021/ja01288a003>
463. Beattie, J. A., and Kay, W. C., *J. Am. Chem. Soc.* 59, 1586, 1937. <https://doi.org/10.1021/ja01288a002>
464. Rosenthal, D. J., Gude, M. T., Teja, A. S., and Mendez-Santiago, J., *Fluid Phase Equilib.* 135, 89, 1997. <https://doi.org/10.1016/S0378-3812(97)00053-8>
465. Smith, R. L., Anselme, M. J., and Teja, A. S., *Fluid Phase Equilib.* 31, 161, 1986. <https://doi.org/10.1016/0378-3812(86)90010-5>
466. Pulliam, M. K., Gude, M. T., and Teja, A. S., *Experimental Results for DIPPR 1990–91 Projects on Phase Equilibria and Pure Component Properties*, DIPPR Data Ser. No. 2, p. 184, 1994.
467. Naziev, Y. M., and Abasov, A. A., *Izv. Vyssh. Uchebn. Zaved., Neft Gaz* 12, 81, 1969.
468. Mousa, A. H. N., *J. Chem. Eng. Data* 26, 248, 1981. <https://doi.org/10.1021/je00025a006>
469. Hicks, C. P., and Young, C. L., *Chem. Rev.* 75, 119, 1975. <https://doi.org/10.1021/cr60294a001>
470. Mousa, A. H. N., Kay, W. B., and Kreglewski, A., *J. Chem. Thermodyn.* 4, 301, 1972. <https://doi.org/10.1016/0021-9614(72)90069-9>
471. Douslin, D. R., Harrison, R. H., and Moore, R. T., *J. Chem. Thermodyn.* 1, 305, 1969. <https://doi.org/10.1016/0021-9614(69)90050-0>
472. Evans, F. D., and Tiley, P. F., *J. Chem. Soc. B* 134, 1966. <https://doi.org/10.1039/j29660000134>
473. Counsell, J. F., Green, J. H. S., Hales, J. L., and Martin, J. F., *Trans. Faraday Soc.* 61, 212, 1965. <https://doi.org/10.1039/TF9656100212>
474. Saikawa, K., Kijima, J., Uematsu, M., and Watanabe, K., *J. Chem. Eng. Data* 24, 165, 1979. <https://doi.org/10.1021/je60082a002>
475. Kijima, J., Saikawa, K., Watanabe, K., Oguchi, K., and Tanishita, I., *Proc. 7th Symp. Thermophys. Prop.*, Cezairliyan, A., Ed., ASME, New York, p. 480, 1977.
476. Kim, K. Y., Ph.D. Dissertation, Univ. Michigan, Ann Arbor, MI, 1974.
477. Swarts, F., *Bull. Soc. Chim. Belg.* 42, 114, 1933.
478. Aoyama, H., Kishizawa, G., Sato, H., and Watanabe, K., *J. Chem. Eng. Data* 41, 1046, 1996.
479. Aoyama, H., Sato, H., and Watanabe, K., *Sixteenth Japan Symposium on Thermophysical Properties*, Hiroshima, p. 173, 1995.
480. McLure, I. A., and Dickinson, E., *J. Chem. Thermodyn.* 8, 93, 1976. <https://doi.org/10.1016/0021-9614(76)90156-7>
481. VonNiederhausern, D. M., Wilson, G. M., and Giles, N. F., *J. Chem. Eng. Data* 45, 157, 2000. <https://doi.org/10.1021/je990232h>
482. Liu, J., Qin, Z., Wang, G., Hou, X., and Wang, J., *J. Chem. Eng. Data* 48, 1610, 2003. <https://doi.org/10.1021/je034127q>
483. Gude, M. T., and Teja, A. S., *Fluid Phase Equilib.* 83, 139, 1993. <https://doi.org/10.1016/0378-3812(93)87016-T>
484. Grigor'ev, B. A., Rastorguev, Yu. L., Gerasimov, A. A., Kurumov, D. S., and Plotnikov, S. A., *Int. J. Thermophys.* 9, 439, 1988. <https://doi.org/10.1007/BF00513082>
485. Zawisza, A., *J. Chem. Thermodyn.* 17, 941, 1985. <https://doi.org/10.1016/0021-9614(85)90007-2>
486. Mousa, A. H. N., *J. Chem. Thermodyn.* 9, 1063, 1977. <https://doi.org/10.1016/0021-9614(77)90184-7>
487. Young, C. L., *Int. DATA Ser., Sel. Data Mixtures, Ser. A*, No. 1, 157, 1975.
488. Nichols, W. B., Reamer, H. H., and Sage, B. H., *AIChE J.* 3, 262, 1957. <https://doi.org/10.1002/aic.690030223>
489. VonNiederhausern, D. M., Wilson, G. M., and Giles, N. F., *J. Chem. Eng. Data* 51, 1986, 2006. <https://doi.org/10.1021/je0602465>
490. Ma, P., Ma, Y., and Zhang, J., *Gaoxiao Huaxue Gongcheng Xuebao* 5, 175, 1991.
491. Masui, G., Honda, Y., and Uematsu, M., *J. Chem. Thermodyn.* 38, 1711, 2006. <https://doi.org/10.1016/j.jct.2006.03.008>
492. Goodwin, R. D., and Haynes, W. M., *NBS Tech. Note* (*U.S.*) No. 1051, 1982.
493. Pryanikova, R. O., Plenkina, R. M., Kuzyakina, N. V., and Markina, I. A., *Khim. Prom-st.* (*Moscow*), 13. 1987.
494. Beattie, J. A., Ingersoll, H. G., and Stockmayer, W. H., *J. Am. Chem. Soc.* 64, 546, 1942. <https://doi.org/10.1021/ja01255a021>
495. Vohra, S. P., and Kobe, K. A., *J. Chem. Eng. Data* 4, 329, 1959. <https://doi.org/10.1021/je60004a012>
496. Smith, R. L., Negishi, E., Arai, K., and Saito, S., *J. Chem. Eng. Jpn.* 23, 99, 1990. <https://doi.org/10.1252/jcej.23.99>
497. Steele, W. V., Chirico, R. D., Cowell, A. B., Knipmeyer, S. E., and Nguyen, A., *J. Chem. Eng. Data* 42, 1053, 1997.
498. Setzmann, U., and Wagner, W., *J. Phys. Chem. Ref. Data* 20, 1061, 1991. <https://doi.org/10.1063/1.555898>
499. Friend, D. G., Ely, J. F., and Ingham, H., *J. Phys. Chem. Ref. Data* 18, 583, 1989. <https://doi.org/10.1063/1.555828>
500. Kleinrahm, R., and Wagner, W., *J. Chem. Thermodyn.* 18, 739, 1986. <https://doi.org/10.1016/0021-9614(86)90108-4>.
501. Calado, J. C. G., Dieters, U., and Strett, W. B., *J. Chem. Soc., Faraday Trans. 1 77*, 2503, 1981. <https://doi.org/10.1039/f19817702503>
502. Angus, S., Armstrong, B., and De Reuck, K. M., *International Thermodynamic Tables of the Fluid State, No. 5 Methane*, Pergamon, Oxford, 1978. <https://doi.org/10.1002/chin.197914365>
503. Goodwin, R. D., *NBS Tech. Note* (*U.S.*) No. 653, 1974.
504. Gielen, H., Jansoone, F., and Verbeke, O. B., *J. Chem. Phys.* 59, 5763, 1973. <https://doi.org/10.1063/1.1679940>
505. Jansoone, V., Gielen, H., De Boelpaep, J., and Verbeke, O. B., *Physica* (*Amsterdam*) 46, 213, 1970. <https://doi.org/10.1016/0031-8914(70)90223-5>
506. Terry, M. J., Lynch, J. T., Bunclark, M., Mansell, K. R., and Staveley, L. A. K., *J. Chem. Thermodyn.* 1, 413, 1969. <https://doi.org/10.1016/0021-9614(69)90072-X>
507. Grigor, A. F., and Steele, W. A., *J. Chem. Phys.* 48, 1032, 1968. <https://doi.org/10.1063/1.1668757>
508. Keyes, F. G., Taylor, R. S., and Smith, L. B., *J. Math. Phys.* (*Cambridge, Mass.*) 1, 211, 1922. <https://doi.org/10.1002/sapm192214211>
509. Kuenen, J. P., *Philos. Mag.* 6, 637, 1903. <https://doi.org/10.1080/14786440309463066>
510. Crismer, L., *Bull. Soc. Chim. Belg.* 18, 18, 1904.
511. Schmidt, G. C., *Z. Phys. Chem., Stoechiom. Verwandtschaftsl.* 8, 628, 1891.
512. Polikhronidi, N. G., Radzhabova, L. M., Rasulov, A. R., and Stepanov, G. V., *High Temp.* (*Engl. Transl.*) 44, 512, 2006. <https://doi.org/10.1007/s10740-006-0063-6>

513. Francesconi, Artur Zaghini, Lentz, H., and Franck, E. U., *J. Phys. Chem.* 85, 3303, 1981. <https://doi.org/10.1021/j150622a019>
514. Swami, D. R., Kumarkrishna Rao, V. N., and Narasinga Rao, N., *Trans., Indian Inst. Chem. Eng.* 9, 32, 1956.
515. Salzwedel, E., *Ann. Phys.* (*Leipzig*) 5, 853, 1930. <https://doi.org/10.1002/andp.19303970703>
516. Ramsay, W., and Young, S., *Philos. Trans. R. Soc. London, A* 178, 313, 1887. <https://doi.org/10.1098/rsta.1887.0011>
517. Steele, W. V., Chirico, R. D., Cowell, A. B., Knipmeyer, S. E., and Nguyen, A., *J. Chem. Eng. Data* 47, 667, 2002. <https://doi.org/10.1021/je0100847>
518. Gurarii, L. L., Kuleshov, G. G., Baglai, A. K., and Petrashkevich, R. I., *Khim.-Farm. Zh.* 21, 247, 1987.
519. Guseinov, S. O., Farzaliev, B. I., and Naziev, Y. M., *Izv. Vyssh. Uchebn. Zaved., Neft Gaz* 22, 52, 1979.
520. Kreglewski, A., *Bull. Acad. Pol. Sci., Cl.* 3 2, 191, 1954.
521. Kiyama, R., Suzuki, K., and Ikegami, T., *Rev. Phys. Chem. Jpn.* 21, 50, 1951.
522. Abara, J. A., Jennings, D. W., Kay, W. B., and Teja, A. S., *J. Chem. Eng. Data* 33, 242, 1988. <https://doi.org/10.1021/je00053a006>
523. Wilson, G. M., Johnston, R. H., Hwang, S.-C., and Tsonopoulos, C., *Ind. Eng. Chem. Process Des. Dev.* 20, 94, 1981. <https://doi.org/10.1021/i200012a015>
524. Rutenberg, O. L., and Shakhova, S. F., *Russ. J. Phys. Chem.* (*Engl. Transl.*) 47, 124, 1973.
525. Yasumoto, M., Uchida, Y., Ochi, K., Furuya, T., Shono, A., and Otake, K., *J. Chem. Eng. Data* 52, 1726, 2007. <https://doi.org/10.1021/je700123q>
526. Day, H. O., and Felsing, W. A., *J. Am. Chem. Soc.* 74, 1951, 1952. <https://doi.org/10.1021/ja01128a024>
527. Ambrose, D., and Ghiassee, N. B., *J. Chem. Thermodyn.* 20, 767, 1988. <https://doi.org/10.1016/0021-9614(88)90030-4>
528. Chirico, R. D., Knipmeyer, S. E., Nguyen, A., and Steele, W. V., *J. Chem. Thermodyn.* 31, 339, 1999. <https://doi.org/10.1006/jcht.1998.0451>
529. Chirico, R. D., Johnson, R. D. I., and Steele, W. V., *J. Chem. Thermodyn.* 39, 698, 2007. <https://doi.org/10.1016/j.jct.2006.10.012>
530. Chirico, R. D., and Steele, W. V., *J. Chem. Eng. Data* 50, 697, 2005. <https://doi.org/10.1021/je049595u>
531. Schroeer, E., *Z. Phys. Chem., Abt. B* 49, 271, 1941.
532. Dawson, P. P., Silberberg, I. H., and McKetta, J. J., *J. Chem. Eng. Data* 18, 7, 1973. <https://doi.org/10.1021/je60056a007>
533. Ambrose, D., Counsell, J. F., and Hicks, C. P., *J. Chem. Thermodyn.* 10, 771, 1978. <https://doi.org/10.1016/0021-9614(78)90135-0>
534. Griffin, D. N., *J. Am. Chem. Soc.* 71, 1423, 1949. <https://doi.org/10.1021/ja01172a079>
535. Matzik, I., and Schneider, G. M., *Ber. Bunsen-Ges. Phys. Chem.* 89, 551, 1985. <https://doi.org/10.1002/bbpc.19850890517>
536. Ambrose, D., and Townsend, R., *Trans. Faraday Soc.* 64, 2622, 1968. <https://doi.org/10.1039/tf9686402622>
537. Defibaugh, D. R., Carrillo-Nava, E., Hurly, J. J., Moldover, M. R., Schmidt, J. W., and Weber, L. A., *J. Chem. Eng. Data* 42, 488, 1997.
538. Young, C. L., *J. Chem. Thermodyn.* 4, 65, 1972. <https://doi.org/10.1016/S0021-9614(72)80009-0>
539. Lindley, D. D., and Hershey, H. C., *Fluid Phase Equilib.* 55, 109, 1990. <https://doi.org/10.1016/0378-3812(90)85007-W>
540. Kreglewski, A., *Bull. Acad. Pol. Sci., Cl.* 3 5, 323, 1957.
541. Young, S., *J. Chem. Soc.* 77, 1145, 1900. <https://doi.org/10.1039/CT9007701145>
542. Steele, W. V., Chirico, R. D., Knipmeyer, S. E., Nguyen, A., Smith, N. K., and Tasker, I. R., *J. Chem. Eng. Data* 41, 1269, 1996.
543. Post, R. G., *Unpublished Rep., Chem. Eng. No. 362*, Univ. Texas, Austin, TX, 1950.
544. Hess, L. G., and Tilton, V. V., *Ind. Eng. Chem.* 42, 1251, 1950. <https://doi.org/10.1021/ie50486a042>
545. Thodos, G., *AIChE J.* 3, 428, 1957. <https://doi.org/10.1002/aic.690030325>
546. Cheng, D. C. H., and McCoubrey, J. C., *J. Chem. Soc.* 4993, 1963.
547. Teja, A. S., and Smith, R. L., *AIChE J.* 33, 1560, 1987. <https://doi.org/10.1002/aic.690330917>
548. Ambrose, D., and Sprake, C. H. S., *J. Chem. Soc. A* 1263, 1971. <https://doi.org/10.1039/j19710001263>
549. Yata, J., Hori, M., Kohno, K., and Minamiyama, T., *High Temp. - High Press.* 29, 19, 1997. <https://doi.org/10.1068/htec283>
550. Duarte-Garza, H. A., Stouffer, C. E., Hall, K. R., Hall, K. R., Holste, J. C., Marsh, K. N., and Gammon, B. E., *J. Chem. Eng. Data* 42, 745, 1997. <https://doi.org/10.1021/je960362f>
551. Ye, F., Sato, H., and Watanabe, K., *J. Chem. Eng. Data* 40, 148, 1995. <https://doi.org/10.1021/je00017a033>
552. Tsvetkov, O. B., Kletskii, A. V., Laptev, Yu. A., Asambaev, A. J., and Zausaev, I. A., *Int. J. Thermophys.* 16, 1185, 1995. <https://doi.org/10.1007/BF02081286>
553. Sagawa, T., Sato, H., and Watanabe, K., *High Temp. - High Press.* 26, 193, 1994.
554. Wilson, L. C., Wilding, W. V., Wilson, G. M., Rowley, R. L., Felix, V. M., and Chisolm-Carter, T., *Fluid Phase Equilib.* 80, 167, 1992. <https://doi.org/10.1016/0378-3812(92)87065-U>
555. Shank, R. L., *J. Chem. Eng. Data* 12, 474, 1967. <https://doi.org/10.1021/je60035a004>
556. Gude, M. T., and Teja, A. S., *AIChE Symp. Ser.* 90 (298), 14, 1994.
557. Grigor'ev, B. A., Rastorguev, Yu. L., Kurumov, D. S., Gerasimov, A. A., Kharin, V. E., and Plotnikov, S. A., *Int. J. Thermophys.* 11, 487, 1990. <https://doi.org/10.1007/BF00500841>
558. Kratzke, H., *AIChE J.* 31, 693, 1985. <https://doi.org/10.1002/aic.690310421>
559. Wolfe, D., Kay, W. B., and Teja, A. S., *J. Chem. Eng. Data* 28, 319, 1983. <https://doi.org/10.1021/je00033a010>
560. Artyukhovskaya, L. M., Shimanskaya, E. T., and Shimanskii, Yu I., *Sov. Phys. - JETP* (*Engl. Transl.*) 59, 375, 1970.
561. Beattie, J. A., Levine, S. W., and Douslin, D. R., *J. Am. Chem. Soc.* 73, 4431, 1951. <https://doi.org/10.1021/ja01153a116>
562. Sage, B. H., and Lacey, W. N., *Ind. Eng. Chem.* 32, 992, 1940. <https://doi.org/10.1021/ie50367a030>
563. Quadri, S. K., and Kudchadker, A. P., *AIChE Symp. Ser.* 298, 1, 1994.
564. Lenoir, J. M., Rebert, C. J., and Hipkin, H. G., *J. Chem. Eng. Data* 16, 401, 1971. <https://doi.org/10.1021/je60051a026>
565. Brown, J. A., and Mears, W. H., *J. Phys. Chem.* 62, 960, 1958. <https://doi.org/10.1021/j150566a015>
566. Fowler, R. D., Hamilton, J. M., Kasper, J. S., Weber, C. E., Burford, W. B., and Anderson, H. C., *Ind. Eng. Chem.* 39, 375, 1947. <https://doi.org/10.1021/ie50447a628>
567. Martin, J. J., *J. Chem. Eng. Data* 7, 68, 1962. <https://doi.org/10.1021/je60012a020>
568. Douslin, D. R., Moore, R. T., and Waddington, G., *J. Phys. Chem.* 63, 1959, 1959. <https://doi.org/10.1021/j150581a037>
569. Rowlinson, J. S., and Thacker, R., *Trans. Faraday Soc.* 53, 1, 1957. <https://doi.org/10.1039/TF9575300001>
570. Ermakov, G. V., and Skripov, V. P., *Russ. J. Phys. Chem.* (*Engl. Transl.*) 41, 39, 1967.
571. Wang, B. -H., Adcock, J. L., Mathur, S. B., and Van Hook, W. A., *J. Chem. Thermodyn.* 23, 699, 1991. <https://doi.org/10.1016/S0021-9614(05)80208-3>
572. Taylor, Z. L., and Reed, T. M., *AIChE J.* 16, 738, 1970. <https://doi.org/10.1002/aic.690160510>
573. Vandana, V., Rosenthal, D. J., and Teja, A. S., *Fluid Phase Equilib.* 99, 209, 1994. <https://doi.org/10.1016/0378-3812(94)80032-4>
574. Milton, H. T., and Oliver, G. D., *J. Am. Chem. Soc.* 74, 3951, 1952. <https://doi.org/10.1021/ja01135a518>
575. Oliver, G. D., and Grisard, J. W., *J. Am. Chem. Soc.* 73, 1688, 1951. <https://doi.org/10.1021/ja01148a080>
576. Oliver, G. D., Blumkin, S., and Cunningham, C. W., *J. Am. Chem. Soc.* 73, 5722, 1951. <https://doi.org/10.1021/ja01156a068>
577. Mousa, A. H. N., *J. Chem. Eng. Data* 23, 133, 1978. <https://doi.org/10.1021/je60077a004>
578. McLure, I. A., Trejo Rodriguez, A., and Soares, V. A. M., *J. Chem. Thermodyn.* 14, 402, 1982. <https://doi.org/10.1016/0021-9614(82)90062-3>
579. Ewing, M. B., and Sanchez Ochoa, J. C., *J. Chem. Thermodyn.* 30, 189, 1998. <https://doi.org/10.1006/jcht.1997.0290>

580. Ernst, G., Gurtner, J., and Wirbser, H., *J. Chem. Thermodyn.* 29, 1125, 1997. <https://doi.org/10.1006/jcht.1997.0234>
581. Brown, J. A., *J. Chem. Eng. Data* 8, 106, 1963. <https://doi.org/10.1021/je60016a032>
582. Young, C. L., *Int. DATA Ser., Sel. Data Mixtures, Ser. A*, No. 4, 291, 1985.
583. Sinitsyn, E. N., Mikhalevich, L. A., Biryukova, L. V., Danilov, N. N., Muratov, G. N., and Fedorov, A. P., *Deposited Doc. VINITI*, Doc. No. 1516-80, 1980.
584. Nikitin, E. D., Popov, A. P., and Yatluk, Y. G., *J. Chem. Eng. Data* 52, 315, 2007. <https://doi.org/10.1021/je0604380>
585. Honda, Y., Sato, T., and Uematsu, M., *J. Chem. Thermodyn.* 40, 208, 2008. <https://doi.org/10.1016/j.jct.2007.06.021>
586. Jou, F.-Y., Carroll, J. J., and Mather, A. E., *Fluid Phase Equilib.* 109, 235, 1995. <https://doi.org/10.1016/0378-3812(95)02721-P>
587. Sychev, V. V., Vasserman, A. A., Kozlov, A. D., and Tsymarny, V. A., *Thermodynamic Properties Of Propane*, Standards Publishing House, Moscow, 1989.
588. Goodwin, R. D., and Haynes, W. M., *NBS Monogr.* (*U.S.*) No. 170, 1982.
589. Barber, J. R., Kay, W. B., and Teja, A. S., *AIChE J.* 28, 134, 1982. <https://doi.org/10.1002/aic.690280119>
590. Yesavage, V. F., Katz, D. L., and Powers, J. E., *J. Chem. Eng. Data* 14, 197, 1969. <https://doi.org/10.1021/je60041a032>
591. Clegg, H. P., and Rowlinson, J. S., *Trans. Faraday Soc.* 51, 1333, 1955. <https://doi.org/10.1039/TF9555101333>
592. Kay, W. B., and Rambosek, G. M., *Ind. Eng. Chem.* 45, 221, 1953. <https://doi.org/10.1021/ie50517a065>
593. Meyers, C. H., *J. Res. Natl. Bur. Stand.* (U.S.) 29, 157, 1942. <https://doi.org/10.6028/jres.029.006>
594. Meyer, R. E., Ph.D. Thesis, Pennsylvania State Univ., University Park, PA, 1941.
595. Deschner, W. W., and Brown, G. G., *Ind. Eng. Chem.* 32, 836, 1940. <https://doi.org/10.1021/ie50366a021>
596. Beattie, J. A., Poffenberger, N., and Hadlock, C., *J. Chem. Phys.* 3, 96, 1935. <https://doi.org/10.1063/1.1749615>
597. Efremova, G. D., and Sokolova, E. S., *Russ. J. Phys. Chem.* (*Engl. Transl.*) 46, 1084, 1972.
598. Anonymous, B., *International Critical Tables of Numerical Data, Phys., Chem. Technol.*, Vol. III, Washburn, E. W., Ed., McGraw-Hill, New York, 1928.
599. Kuenen, J. P., and Robson, W. G., *Philos. Mag.* 4, 116, 1902. <https://doi.org/10.1080/14786440209462823>
600. Ramsay, W., and Young, S., *Philos. Trans. R. Soc. London*, A 180, 137, 1889. <https://doi.org/10.1098/rsta.1889.0004>
601. Oh, B. C., Lee, S., Seo, J., and Kim, H., *J. Chem. Eng. Data* 49, 221, 2004. <https://doi.org/10.1021/je034066w>
602. Ambrose, D., Counsell, J. F., Lawrenson, I. J., and Lewis, G. B., *J. Chem. Thermodyn.* 10, 1033, 1978. <https://doi.org/10.1016/0021-9614(78)90078-2>
603. Seibert, F. M., and Burrell, G. A., *J. Am. Chem. Soc.* 37, 2683, 1915. <https://doi.org/10.1021/ja02177a011>
604. Lu, H., Newitt, D. M., and Ruhemann, M., *Proc. R. Soc. London, A* 178, 506, 1941. <https://doi.org/10.1098/rspa.1941.0072>
605. Maass, O., and Wright, C. H., *J. Am. Chem. Soc.* 43, 1098, 1921. <https://doi.org/10.1021/ja01438a013>
606. Ohgaki, K., Umezono, S., and Katayama, T., *J. Supercrit. Fluids* 3, 78, 1990. <https://doi.org/10.1016/0896-8446(90)90011-A>
607. Marchman, H., Prengle, H. W., and Motard, R. L., *Ind. Eng. Chem.* 41, 2658, 1949. <https://doi.org/10.1021/ie50479a063>
608. Farrington, P. S., and Sage, B. H., *Ind. Eng. Chem.* 41, 1734, 1949. <https://doi.org/10.1021/ie50476a050>
609. Vaughan, W. E., and Graves, N. R., *Ind. Eng. Chem.* 32, 1252, 1940. <https://doi.org/10.1021/ie50369a044>
610. Winkler, C. A., and Maass, O., *Can. J. Res.* 9, 613, 1933. <https://doi.org/10.1139/cjr33-113>
611. Vohra, S. P., Kang, T.-L., Kobe, K. A., and McKetta, J. J., *J. Chem. Eng. Data* 7, 150, 1962. <https://doi.org/10.1021/je60012a044>
612. Steele, W. V., Chirico, R. D., Knipmeyer, S. E., and Nguyen, A., *J. Chem. Eng. Data* 47, 689, 2002. <https://doi.org/10.1021/je010085z>
613. Chirico, R. D., Steele, W. V., Nguyen, A., Klots, T. D., and Knipmeyer, S. E., *J. Chem. Thermodyn.* 28, 797, 1996. <https://doi.org/10.1006/jcht.1996.0073>
614. Mandel, H., and Ewbank, N., *Atomics International* NAA-S-R-5129, 1960.
615. Gallant, R. W., *Hydrocarbon Process.* 45, 161, 1966.
616. Altunin, V. V., Geller, V. Z., Kremenvskaya, E. A., Perel'shtein, I. I., and Petrov, E. K., *Thermophysical Properties of Freons, Methane Ser.*, Part 2, Vol. 9, NSRDS-USSR, Selover, T. B., Ed., Hemisphere, New York, 1987. <https://doi.org/10.1007/978-3-662-30483-9>
617. Toczylkin. L. S., and Young, C. L., *Aust. J. Chem.* 30, 1591, 1977. <https://doi.org/10.1071/CH9771591>
618. Kordes, E., . *Elektrochem.* 58, 76, 1954. <https://doi.org/10.1017/S0368393100098291>
619. Lewis, D. T., *J. Appl. Chem.* 3, 154, 1953. <https://doi.org/10.1002/jctb.5010030404>
620. Steele, W. V., Chirico, R. D., Nguyen, A., Hossenlopp, I. A., and Smith, N. K., *DIPPR Data Ser.* 1, 101, 1991.
621. Poot, W., and De Loos, T. W., *Fluid Phase Equilib.* 222–223, 255, 2004. <https://doi.org/10.1016/j.fluid.2004.06.042>
622. Poot, W., and De Loos, T. W., *Fluid Phase Equilib.* 210, 69, 2003. <https://doi.org/10.1016/S0378-3812(03)00162-6>
623. Yata, J., Hori, M., Niki, M., Isono, Y., and Yanagitani, Y., *Fluid Phase Equilib.* 174, 221, 2000. <https://doi.org/10.1016/S0378-3812(00)00429-5>
624. Fujiwara, K., Nakamura, S., and Noguchi, M., *J. Chem. Eng. Data* 43, 55, 1998. <https://doi.org/10.1021/je970177h>
625. Morrison, G., and Ward, D., *Fluid Phase Equilib.* 62, 65, 1991. <https://doi.org/10.1016/0378-3812(91)87006-U>
626. Piao, C. C., Sato, H., and Watanabe, K., *ASHRAE Trans.* 96, 132, 1990.
627. Piao, C. C., Sato, H., and Watanabe, K., *ASHRAE Trans.* 41, 132, 1989.
628. Kubota, H., Yamashita, T., Tanaka, Y., and Makita, T., *Int. J. Thermophys.* 10, 629, 1989. <https://doi.org/10.1007/BF00507984>
629. Kabata, Y., Tanikawa, S., Uematsu, M., and Watanabe, K., *Int. J. Thermophys.* 10, 605, 1989. <https://doi.org/10.1007/BF00507982>
630. Basu, R. S., and Wilson, D. P., *Int. J. Thermophys.* 10, 591, 1989. <https://doi.org/10.1007/BF00507981>
631. Wilson, D. P., and Basu, R. S., *ASHRAE Trans.* 94, 2095, 1988.
632. Tatoh, J., Kuwabara, S., Sato, H., and Watanabe, K., *J. Chem. Eng. Data* 38, 116, 1993. <https://doi.org/10.1021/je00009a028>
633. Lebedeva, E. S., and Khodeeva, S. M., *Zh. Fiz. Khim.* 41, 2081, 1967.
634. Renfrew, M. M., and Lewis, E. E., *Ind. Eng. Chem.* 38, 870, 1946. <https://doi.org/10.1021/ie50441a009>
635. Chari, N. C., Ph.D. Dissertation, Univ. Michigan, Ann Arbor, MI, 1960.
636. Steele, W. V., Chirico, R. D., Knipmeyer, S. E., and Smith, N. K., *Report*, NIPPR-360, NTIS Order No. DE89000709, Dec. 1988.
637. McGlashan, M. L., and McKinnon, I. R., *J. Chem. Thermodyn.* 9, 1205, 1977. <https://doi.org/10.1016/0021-9614(77)90121-5>
638. Cipollint, N. E., and Allen, A. O., *J. Chem. Phys.* 67, 131, 1977. <https://doi.org/10.1063/1.434556>
639. Hicks, C. P., and Young, C. L., *J. Chem. Soc., Faraday Trans.* 1 72, 122, 1976. <https://doi.org/10.1039/f19767200122>
640. Hugill, J. A., and McGlashan, M. L., *J. Chem. Thermodyn.* 10, 85, 1978. <https://doi.org/10.1016/0021-9614(78)90149-0>
641. Bendtsen, J., *J. Raman Spectrosc.* 6, 306, 1977. <https://doi.org/10.1002/jrs.1250060609>
642. Abdulagatov, I. M., Polikhronidi, N. G., Bruno, T. J., Batyrova, R. G., and Stepanov, G. V., *Fluid Phase Equilib.* 263, 71, 2008. <https://doi.org/10.1016/j.fluid.2007.09.023>
643. Polikhronidi, N. G., Abdulagatov, I. M., Magee, J. W., and Batyrova, R. G., *J. Chem. Eng. Data* 46, 1064, 2001. <https://doi.org/10.1021/je000269y>
644. Goodwin, R. D., *J. Phys. Chem. Ref. Data* 18, 1565, 1989. <https://doi.org/10.1063/1.555837>
645. Akhundov, T. S., and Abdullaev, F. G., *Izv. Vyssh. Uchebn. Zaved., Neft Gaz* 12, 44, 1969.

646. Krase, N. W., and Goodman, J. B., *Ind. Eng. Chem.* 22, 13, 1930. <https://doi.org/10.1021/ie50241a004>
647. Buchowski, H., Janaszewski, B., and Teperek, J., *Bull. Acad. Pol. Sci., Ser. Sci. Chim.* 14, 403, 1966.
648. Sokolova, T. D., Prokof'eva, N. K., and Nisel'son, L. A., *Russ. J. Phys. Chem.* (*Engl. Transl.*) 47, 154, 1973.
649. Mastroianni, M. J., Stahl, R. F., and Sheldon, P. N., *J. Chem. Eng. Data* 23, 113, 1978. <https://doi.org/10.1021/je60077a005>
650. Weber, L. A., and Defibaugh, D. R., *J. Chem. Eng. Data* 41, 1477, 1996. <https://doi.org/10.1021/je9602071>
651. Higashi, Y., and Ikeda, T., *Fluid Phase Equilib.* 125, 139, 1996. <https://doi.org/10.1016/S0378-3812(96)03089-0>
652. Wang, H., Ma, Y., Lu, C., and Tian, Y., *J. Eng. Thermophys.* 14, 122, 1993.
653. Fukushima, M., *Nippon Reito Kyokai Ronbunshu* 10, 87, 1993.
654. Bier, K., Turk, M., and Zhai, J., *Vapour Pressure of Trifluoroethanol*, Insitut für Technische Thermodynamik und Kaltetechnik, Universität Karlsruhe (TH), D 7500 Karlsruhe I, FRG, 1991.
655. Bier, K., Tuerk, M., and Zhai, J., *Proc. Int. Inst. Ref., Comm. B1 Meet.*, Herzlia, Israel, pp. 129–139, 1990. <https://doi.org/10.1017/S0074180900241375>
656. Zhang, C., Duan, Y. -Y., Shi, L., Zhu, M. -S., and Han, L. -Z., *J. Tsinghua Univ.* (*Sci. & Technol.*) 40, 77, 2000.
657. Duan, Y. -Y., Shi, L., Zhu, M. -S., and Han, L. -Z., *J. Tsinghua Univ.* (*Sci. & Technol.*) 40, 60, 2000.
658. Duan, Y. -Y., Shi, L., Sun, L. -Q., Zhu, M. -S., and Han, L. -Z., *Int. J. Thermophys.* 21, 393, 2000. <https://doi.org/10.1023/A:1006683529436>
659. Duan, Y. -Y., Shi, L., Zhu, M. -S., and Han, L. -Z., *J. Chem. Eng. Data* 44, 501, 1999. <https://doi.org/10.1021/je980251b>
660. Khodeeva, S. M., and Gubochkin, I. V., *Russ. J. Phys. Chem.* (*Engl. Transl.*) 51, 998, 1977.
661. Diefenbacher, A., Crone, M., and Turk, M., *J. Chem. Thermodyn.* 30, 481, 1998. <https://doi.org/10.1006/jcht.1997.0320>
662. Hori, K., Okazaki, S., Uematsu, M., and Watanabe, K., *Proc. 8th Symp. Thermophys. Prop.*, Vol. II, Sengers, J. V., Ed., ASME, New York, pp. 370–376, 1982.
663. Wagner, W., *Kaeltetech.-Klim.* 20, 238, 1968. <https://doi.org/10.1007/BF00498351>
664. Hou, Y.-C., and Martin, J. J., *AIChE J.* 5, 125, 1959. <https://doi.org/10.1002/aic.690050126>
665. Day, H. O., and Felsing, W. A., *J. Am. Chem. Soc.* 72, 1698, 1950. <https://doi.org/10.1021/ja01160a077>
666. Kay, W. B., and Warzel, F. M., *Ind. Eng. Chem.* 43, 1150, 1951. <https://doi.org/10.1021/ie50497a044>
667. Beattie, J. A., and Edwards, D. G., *J. Am. Chem. Soc.* 70, 3382, 1948. <https://doi.org/10.1021/ja01190a049>
668. Akhundov, T. S., and Asadullaeva, N. N., *Izv. Vyssh. Uchebn. Zaved., Neft Gaz* 11, 83, 1968.
669. Akhundov, T. S., and Imanov, Sh. Yu., *Teplofiz. Svoistva Zhidk.* 48, 1970.

CRITICAL CONSTANTS OF INORGANIC COMPOUNDS

The parameters of the liquid-gas critical point are important constants in determining the behavior of fluids. This table lists the critical temperature, pressure, and molar volume, as well as the normal boiling point, for over 140 inorganic compounds. The column definitions for the table are as follows.

Column heading	Definition
Name	Name of inorganic compound; compounds are listed alphabetically by name
Mol. form.	Molecular formula of compound
T_b	Normal boiling point in K at a pressure of 101.325 kPa (1 atmosphere); an "sp" following the value indicates a sublimation point (temperature at which the solid is in equilibrium with the gas at a pressure of 101.325 kPa)
T_c	Critical temperature, in K
P_c	Critical pressure, in MPa
V_c	Critical molar volume, in cm³ mol⁻¹
Ref.	Reference number

The number of digits given for T_b, T_c, and P_c indicates the estimated accuracy of these quantities; however, values of T_c greater than 750 K may be in error by 10 K or more. Although most V_c values are given to three figures, they cannot be assumed accurate to better than a few percent.

All values are experimentally determined except for a few values, indicated by an asterisk*, that are based on extrapolations. Methods of measurement are described and critiqued in Ref. 1.

Critical Constants of Inorganic Compounds

Name	Mol. form.	T_b/K	T_c/K	P_c/MPa	V_c/cm³ mol⁻¹	Ref.
Aluminum	Al	2792	6700 *			2
Aluminum bromide	$AlBr_3$	528	763	2.89	310	3
Aluminum chloride	$AlCl_3$	453 sp	620	2.63	257	3
Aluminum iodide	AlI_3	655	983		408	3
Ammonia	NH_3	239.82	405.56	11.357	69.8	3,4
Ammonium chloride	NH_4Cl	611 sp	1155	163.5		3
Antimony(III) bromide	$SbBr_3$	561	904		300	3
Antimony(III) chloride	$SbCl_3$	493.5	794		272	3
Antimony(III) iodide	SbI_3	673	1102			3
Argon	Ar	87.302	150.687	4.863	75	3
Arsenic (gray)	As	889 sp	1673	22.3	35	3
Arsenic(III) chloride	$AsCl_3$	403	654		252	3
Arsine	AsH_3	210.7	373.1			3
Beryllium	Be	2741	5205 *			5
Bismuth	Bi	1837	4620 *			5
Bismuth tribromide	$BiBr_3$	735	1220		301	3
Bismuth trichloride	$BiCl_3$	714	1179	12.0	261	3
Boron tribromide	BBr_3	364.5	581		272	3
Boron trichloride	BCl_3	285.7	455	3.87	239	3
Boron trifluoride	BF_3	173.3	260.8	4.98	115	3
Boron triiodide	BI_3	482.7	773		356	3
Bromine	Br_2	332.0	588	10.34	127	3
Carbon dioxide	CO_2	194.686 sp	304.13	7.375	94	6
Carbon disulfide	CS_2	319.4	552	7.90	173	3
Carbon monoxide	CO	81.64	132.86	3.494	93	3,7,8
Carbon oxysulfide	OCS	223.0	375	5.88	137	3,7
Cesium	Cs	944	1938	9.4	341	9
Chlorine	Cl_2	239.11	417.0	7.991	123	3
Chlorine pentafluoride	ClF_5	260.1	416	5.27	233	3
Chlorotrifluorosilane	$SiClF_3$	203.2	307.7	3.46		3
Diborane	B_2H_6	180.66	289.8	4.05		3
Dichlorodifluorosilane	$SiCl_2F_2$	241	369.0	3.5		3
Difluoramine	NHF_2	250	403			3
cis-Difluorodiazine	N_2F_2	167.40	272	7.09		3
trans-Difluorodiazine	N_2F_2	161.70	260	5.57		3
Fluorine	F_2	85.04	144.41	5.1724	66	3
Fluorine monoxide	F_2O	128.9	215			3
Gallium(III) bromide	$GaBr_3$	552	806.7		303	3
Gallium(III) chloride	$GaCl_3$	474	694		263	3

Name	Mol. form.	T_b/K	T_c/K	P_c/MPa	V_c/cm³ mol⁻¹	Ref.
Gallium(III) iodide	GaI_3	613	951		395	3
Germane	GeH_4	185.1	312.2	4.95	147	3
Germanium	Ge	3106	9802 *			5
Germanium(IV) bromide	$GeBr_4$	459.50	718		392	3
Germanium(IV) chloride	$GeCl_4$	359.70	553.2	3.861	330	3
Germanium(IV) iodide	GeI_4	621	973		500	3
Hafnium(IV) bromide	$HfBr_4$	596 sp	746		415	3
Hafnium(IV) chloride	$HfCl_4$	590 sp	725.7	5.42	314	3
Hafnium(IV) iodide	HfI_4	667 sp	916		528	3
Helium	He	4.222	5.1953	0.22746	57	3
Hydrazine	N_2H_4	386.70	653	14.7		3
Hydrogen	H_2	20.271	32.938	1.2858	65	3
Hydrogen bromide	HBr	206.77	363.2	8.55		3
Hydrogen chloride	HCl	188	324.7	8.31	81	3
Hydrogen fluoride	HF	293	461	6.48	69	3
Hydrogen iodide	HI	237.60	424.0	8.31		3
Hydrogen peroxide	H_2O_2	423.4	728 *	22 *		10
Hydrogen selenide	H_2Se	231.90	411	8.92		3
Hydrogen sulfide	H_2S	213.60	373.1	9.00	99	3,7
Iodine	I_2	457.6	819		155	3
Iodine bromide	IBr	389 dec	719		139	3
Iron	Fe	3134	9340 *			5
Krypton	Kr	119.735	209.48	5.525	91	3,7
Lithium	Li	1615	3223 *	67 *	66	11
Manganese	Mn	2334	4325 *			5
Mercury	Hg	629.769	1764	167	43	3,12
Mercury(II) bromide	$HgBr_2$	591	1012			3
Mercury(II) chloride	$HgCl_2$	577	973		174	3
Mercury(II) iodide (yellow)	HgI_2	624	1072			3
Molybdenum(V) chloride	$MoCl_5$	541	850		369	3
Molybdenum(VI) fluoride	MoF_6	307.2	473	4.75	226	3
Neon	Ne	27.104	44.4918	2.6786	42	3
Niobium(V) chloride	$NbCl_5$	520.6	803.5	4.88	397	3
Niobium(V) fluoride	NbF_5	507	737	6.28	155	3
Nitric oxide	NO	121.41	180	6.48	58	3
Nitrogen	N_2	77.355	126.192	3.3958	90	3
Nitrogen chloride difluoride	$NClF_2$	206	337.5	5.15		3
Nitrogen tetroxide	N_2O_4	294.30	431	10.1	167	3
Nitrogen trifluoride	NF_3	144.40	234.0	4.46	126	3
Nitrosyl chloride	NOCl	267.7	440			3
Nitrous oxide	N_2O	184.67	309.52	7.245	97	3,7
Nitryl fluoride	NO_2F	200.8	349.5			3
Osmium(VIII) oxide	OsO_4	404.4	678			3
Oxygen	O_2	90.188	154.581	5.0430	73	3
Ozone	O_3	161.80	261.1	5.57	89	3
Perchloryl fluoride	ClO_3F	226.40	368.4	5.37	161	3
Phosphine	PH_3	185.40	324.5	6.54		3
Phosphonium chloride	PH_4Cl	246 sp	322.3	7.37		3
Phosphorothioc chloride difluoride	$PSClF_2$	279.5	439.2	4.14		3
Phosphorothioc trifluoride	PSF_3	220.90	346.0	3.82		3
Phosphorus (white)	P	553.7	994			3
Phosphorus(III) bromide	PBr_3	446.4	711		300	3
Phosphorus(III) chloride	PCl_3	349	563		264	3
Phosphorus(V) chloride	PCl_5	433 sp	646			3
Phosphorus(III) chloride difluoride	$PClF_2$	225.9	362.4	4.52		3
Phosphorus(III) dichloride fluoride	PCl_2F	287.00	463.0	4.96		3
Phosphorus(III) fluoride	PF_3	171.4	271.2	4.33		3
Potassium	K	1032	2223 *	16 *	209	11
Radon	Rn	211.5	377	6.28		3
Rhenium(VII) oxide	Re_2O_7	633	942		334	3
Rhenium(VI) oxytetrachloride	$ReOCl_4$	496	781		362	3
Rubidium	Rb	961	2093 *	16 *	247	11
Selenium (gray)	Se	958	1766	27.2		3

Name	Mol. form.	T_b/K	T_c/K	P_c/MPa	V_c/cm^3 mol^{-1}	Ref.
Selenium hexafluoride	SeF_6	226.6 sp	345.5			3
Selenium oxychloride	$SeOCl_2$	450	730	7.09	235	3
Silver	Ag	2435	6410 *			5
Sodium	Na	1156.090	2573 *	35 *	116	11
Sulfur (rhombic)	S	717.76	1314	20.7	57.0	3
Sulfur chloride pentafluoride	SF_5Cl	254.10	390.9			3
Sulfur dioxide	SO_2	263.13	430.64	7.884	122	3,7
Sulfur hexafluoride	SF_6	209.4 sp	318.723	3.77	197	3,14
Sulfur tetrafluoride	SF_4	232.70	364			3
Sulfur trioxide (β-form)	SO_3	317.7	491.0	8.2	127	3
Tantalum(V) bromide	$TaBr_5$	622.0	974		461	3
Tantalum(V) chloride	$TaCl_5$	512	767		402	3
Tellurium	Te	1261	2329 *			5
Tellurium hexafluoride	TeF_6	234.3 sp	356			3
Tellurium tetrachloride	$TeCl_4$	660	1002	8.56	310	3
Tetrabromosilane	$SiBr_4$	427	663		382	3
Tetrachlorosilane	$SiCl_4$	330.80	508.1	3.593	326	3
Tetrafluorohydrazine	N_2F_4	199	309	3.75		3
Tetrafluorosilane	SiF_4	187	259.0	3.72		3
Tetraiodosilane	SiI_4	560.50	944		558	3
Tin(IV) bromide	$SnBr_4$	478	744		417	3
Tin(IV) chloride	$SnCl_4$	387.30	591.9	3.75	351	3
Tin(IV) iodide	SnI_4	637.50	968		531	3
Titanium(IV) bromide	$TiBr_4$	506.7	795.7		391	3
Titanium(IV) chloride	$TiCl_4$	409.60	638	4.66	339	3
Titanium(IV) iodide	TiI_4	650	1040		505	3
Tribromosilane	$SiHBr_3$	382	610.0		305	3
Trichlorofluorosilane	$SiCl_3F$	285.40	438.6	3.58		3
Trichlorosilane	$SiHCl_3$	306	479		268	3
Trifluoramine oxide	NOF_3	185.7	303	6.43	147	3
Tungsten(VI) chloride	WCl_6	610	923		422	3
Tungsten(VI) fluoride	WF_6	290.3	444	4.34	233	3
Tungsten(VI) oxytetrachloride	$WOCl_4$	503	782		338	3
Uranium(VI) fluoride	UF_6	329.7 sp	505.8	4.66	250	3
Vanadyl chloride	VOCl	400	636		171	3
Water	H_2O	373.124	647.10	22.06	56	3
Xenon	Xe	165.051	289.733	5.8420	118	7,15
Xenon difluoride	XeF_2	387.50 sp	631	9.32	148	3
Xenon tetrafluoride	XeF_4	388.90 sp	612	7.04	188	3
Zirconium(IV) bromide	$ZrBr_4$	633 sp	805		424	3
Zirconium(IV) chloride	$ZrCl_4$	604 sp	778	5.77	319	3
Zirconium(IV) iodide	ZrI_4	704 sp	960		530	3

* Extrapolated.

References

1. Ambrose, D., and Young, C. L., *J. Chem. Eng. Data* 40, 345, 1995. <https://doi.org/10.1021/je00018a001>
2. Morel, V., Bultel, A., and Chéron, B. G., *Int. J. Thermophys.* 30, 1853, 2009. <https://doi.org/10.1007/s10765-009-0671-6>
3. Ambrose, D., "Vapor-Liquid Constants of Fluids," in *Handbook of the Thermodynamics of Organic Compounds*, Stevenson, R. M., and Malanowski, S., Eds., Elsevier, New York, 1987.
4. Sato, M., Masui, G., and Uematsu, M., *J. Chem. Thermodyn.* 37, 931, 2005. <https://doi.org/10.1016/j.jct.2004.12.016>
5. Velasco, S., Roman, F. L., White, J. A., and Mulero, A., *Fluid Phase Equilib.* 244, 11, 2006. <https://doi.org/10.1016/j.fluid.2006.03.017>
6. Nowak, P., Tielkes, T., Kleinrahm, R., and Wagner, W., *J. Chem. Thermodyn.* 29, 885, 1997. <https://doi.org/10.1006/jcht.1997.0208>
7. Lemmon, E. W., and Span, R., *J. Chem. Eng. Data* 51, 785, 2006. <https://doi.org/10.1021/je050186n>
8. Goodwin, R. D., *J. Phys. Chem. Ref. Data* 14, 849, 1985. <https://doi.org/10.1063/1.555742>
9. Vargaftik, N. B., *Int. J. Thermophys.* 11, 467, 1990. <https://doi.org/10.1007/BF00500839>
10. Nikitin, E. D., Pavlov, P. A., Popov, A. P., and Nikitina, H. E., *J. Chem. Thermodyn.* 27, 945, 1995. <https://doi.org/10.1006/jcht.1995.0100>
11. Dillon, I. G., Nelson, P. A., and Swanson, B. S., *J. Chem. Phys.* 44, 4229, 1966. <https://doi.org/10.1063/1.1726611>
12. Huber, M. L., Laesecke, A., and Friend, D. G., *The Vapor Pressure of Mercury*, NISTIR 6643, National Institute of Standards and Technology, Boulder, CO, March 2006; *Ind. Eng. Chem. Res.* 45, 7351, 2006. <https://doi.org/10.1021/ie060560s>
13. Rau, H., Kutty, T. R. N., and Guedes de Carvalho, J. R. F., *J. Chem. Thermodyn.* 5, 291, 1973. <https://doi.org/10.1016/S0021-9614(73)80089-8>
14. Funke, M., Kleinrahm, R., and Wagner, W., *J. Chem. Thermodyn.* 34, 717, 2002. <https://doi.org/10.1006/jcht.2001.0906>
15. Sifner, O., and Klomfar, J., *J. Phys. Chem. Ref. Data* 23, 63, 1994.

SUBLIMATION PRESSURE OF SOLIDS

This table gives the sublimation (vapor) pressure of some representative solids as a function of temperature. Entries include simple inorganic and organic substances in their solid phase below room temperature, as well as polycyclic organic compounds which show measurable sublimation pressure only at elevated temperatures. Substances are listed by molecular formula in the Hill order, with inorganic compounds preceding compounds containing carbon. Each substance has two rows of data. The top row has temperatures; the bottom row has the values of the sublimation pressure at that temperature.

Values marked by * represent the solid–liquid–gas triple point. Note that some pressure values are in pascals (Pa), and others are in kilopascals (kPa). For conversion, 1 kPa = 7.506 mmHg = 0.0098692 atm.

References

1. Lide, D. R. and Kehiaian, H. V., *CRC Handbook of Thermophysical and Thermochemical Data*, CRC Press, Boca Raton, FL, 1994.
2. *TRC Thermodynamic Tables*, Thermodynamic Research Center, Texas A&M University, College Station, TX.
3. Oja, V. and Suuberg, E. M., *J. Chem. Eng. Data* 43, 486, 1998. <https://doi.org/10.1021/je970222l>
4. Guder, C., and Wagner, W., *J. Phys. Chem. Ref. Data* 38, 33, 2009. [sulfur hexafluoride> <https://doi.org/10.1063/1.3037344>
5. Lemmon, E. W. and Span, R., *J. Chem. Eng. Data* 51 785, 2006. [krypton, xenon] <https://doi.org/10.1021/je050186n>
6. Setzmann, U., and Wagner, W., *J. Phys. Chem. Ref. Data* 20, 1061, 1991. [methane] <https://doi.org/10.1063/1.555898>
7. Span, R., and Wagner, W., *J. Phys. Chem. Ref. Data* 25, 1509, 1996. [carbon dioxide] <https://doi.org/10.1063/1.555991>
8. Katti, R. S., Jacobsen, R. T, Stewart, R. B., and Jahangiri, M., *Adv. Cryo. Eng.* 31, 1189, 1986. [neon] <https://doi.org/10.1007/978-1-4613-2213-9_132>
9. Span, R., Lemmon, E. W., Jacobsen, R. T, Wagner, W., and Yokozeki, A., *J. Phys. Chem. Ref. Data* 29, 1361, 2000. [nitrogen] <https://doi.org/10.1063/1.1349047>

Sublimation Pressure of Solids

Name	Mol. form.									
Inorganic										
Argon	Ar	*T*/K	55	60	65	70	75	80	83.81	
		p/kPa	0.2	0.8	2.8	7.7	18.7	40.7	68.89	
Hydrogen bromide	BrH	*T*/K	135	140	150	160	170	180	185.1	
		p/kPa	0.1	0.3	1.1	3.3	8.7	20.1	27.4	
Bromine	Br_2	*T*/K	170	180	190	200	210	220	230	240
		p/Pa	0.069	0.416	2.04	8.45	30.3	96.0	273	710
Hydrogen chloride	ClH	*T*/K	120	130	140	150	155	159.0		
		p/kPa	0.1	0.5	1.9	5.8	9.5	13.5		
Chlorine	Cl_2	*T*/K	120	130	140	150	160	172.17		
		p/Pa	0.144	1.52	11.2	63.1	283	1392		
Silicon tetrafluoride	F_4Si	*T*/K	130	140	150	160	170	175	180	186.3
		p/kPa	0.2	0.9	3.9	14.0	43.8	74.2	122.4	220.8
Sulfur hexafluoride	F_6S	*T*/K	150	165	180	190	200	210	220	223.554
		p/kPa	0.4566	2.596	10.94	25.07	52.71	103	189	232.7
Hydrogen iodide	HI	*T*/K	160	170	180	190	200	210	220	222.35
		p/kPa	0.2	0.8	2.2	5.3	11.7	23.6	44.1	49.3
Water	H_2O	*T*/K	190	210	225	240	250	260	270	273.16
		p/Pa	0.032	0.702	4.942	27.28	76.04	195.8	470.1	611.655
Hydrogen sulfide	H_2S	*T*/K	140	150	160	165	170	175	180	187.55
		p/kPa	0.2	0.6	1.9	3.2	5.2	8.3	12.7	23.18
Ammonia	H_3N	*T*/K	160	170	180	190	195	195.4		
		p/kPa	0.1	0.4	1.2	3.5	5.8	6.077		
Iodine	I_2	*T*/K	240	250	260	270	280	290	300	310
		p/Pa	0.081	0.297	0.971	2.89	7.92	20.1	47.9	107
Krypton	Kr	*T*/K	80	90	95	100	105	110	115.775	
		p/kPa	0.4160	2.670	5.840	11.81	22.35	39.39	73.53	
Nitrogen	N_2	*T*/K	50	55	60	63.151				
		p/kPa	0.4005	1.800	6.298	12.52				
Neon	Ne	*T*/K	13	15	17	19	21	22	23	24.5561
		p/kPa	0.003362	0.04915	0.3823	1.930	7.160	12.61	21.15	43.37
Nitric oxide	NO	*T*/K	85	90	95	100	105	109.5		
		p/kPa	0.1	0.4	1.3	3.8	10	21.916		
Xenon	Xe	*T*/K	110	120	130	140	150	155	160	161.405
		p/kPa	0.3433	1.298	4.343	12.83	33.32	50.96	74.58	81.77

Name	Mol. form.									
Organic										
Hydrogen cyanide	CHN	*T*/K	200	210	220	230	240	250	255	259.83
		p/kPa	0.2	0.4	1.0	2.2	4.8	9.7	13.6	18.62
Methane	CH_4	*T*/K	65	70	75	80	85	90	90.694	
		p/kPa	0.07307	0.2627	0.7965	2.102	4.949	10.59	11.696	
Carbon monoxide	CO	*T*/K	50	55	60	65	68.13			
		p/kPa	0.1	0.6	2.6	8.2	15.42			
Carbon dioxide	CO_2	*T*/K	130	140	155	170	185	194.7	205	216.592
		p/kPa	0.03087	0.1836	1.666	9.960	43.93	104.0	227.0	518.5
Hexachloroethane	C_2Cl_6	*T*/K	275	300	325	350	375	400	425	459.9
		p/Pa	0.004	0.056	0.383	1.62	5.30	14.8	36.4	107.4
Acetylene	C_2H_2	*T*/K	130	140	150	160	170	180	190	192.4
		p/kPa	0.2	0.7	2.6	7.8	20.6	49.0	106.3	126.0
Acetic acid	$C_2H_4O_2$	*T*/K	250	260	270	280	289.7			
		p/kPa	0.092	0.199	0.406	0.79	1.277			
Neopentane	C_5H_{12}	*T*/K	200	210	220	230	240	250	255	256.58
		p/kPa	0.7	1.6	3.6	7.3	13.9	24.8	32.4	35.8
1,2,3,4,5,6-Hexachlorocyclohexane, (1α,2α,3β,4α,5α,6β)	$C_6H_6Cl_6$	*T*/K	300	320	330	340	350	360	370	380
		p/Pa	0.01	0.13	0.39	1.04	2.66	6.42	14.8	32.7
Resorcinol	$C_6H_6O_2$	*T*/K	330	340	350	360	370	380		
		p/Pa	1.03	2.78	7.09	17.2	39.6	87.6		
p-Hydroquinone	$C_6H_6O_2$	*T*/K	350	360	370	380	390	400		
		p/Pa	1.20	3.18	7.96	19.0	43.4	95.1		
Naphthalene	$C_{10}H_8$	*T*/K	250	270	280	290	300	310	330	353.43
		p/Pa	0.036	0.514	1.662	4.918	13.43	34.15	182.9	999.6
Phenazine	$C_{12}H_8N_2$	*T*/K	290	300	310	320				
		p/Pa	0.0013	0.0046	0.0150	0.0448				
Dibenzofuran	$C_{12}H_8O$	*T*/K	300	310	320	330	340	350		
		p/Pa	0.408	1.21	3.35	8.71	21.4	50.0		
Carbazole	$C_{12}H_9N$	*T*/K	350	355	360					
		p/Pa	0.086	0.140	0.245					
Benz[*g*]isoquinoline-5,10-dione	$C_{13}H_7NO_2$	*T*/K	330	340	350	360	370	380		
		p/Pa	0.006	0.018	0.053	0.148	0.394	0.994		
1*H*-Phenalen-1-one	$C_{13}H_8O$	*T*/K	330	340	350					
		p/Pa	0.040	0.113	0.302					
3-Hydroxy-1*H*-phenalen-1-one	$C_{13}H_8O_2$	*T*/K	400	410	420	430				
		p/Pa	0.006	0.018	0.053	0.144				
Acridine	$C_{13}H_9N$	*T*/K	290	300	310	320				
		p/Pa	0.0024	0.0085	0.0278	0.0845				
Phenanthridine	$C_{13}H_9N$	*T*/K	310	320	330	340				
		p/Pa	0.020	0.066	0.206	0.603				
Anthracene	$C_{14}H_{10}$	*T*/K	320	330	340	350	360	370	380	390
		p/Pa	0.014	0.043	0.125	0.342	1.01	2.38	5.35	11.5
Phenanthrene	$C_{14}H_{10}$	*T*/K	300	310	320	330	340	350	360	
		p/Pa	0.025	0.085	0.270	0.796	2.02	4.89	11.2	
Pyrene	$C_{16}H_{10}$	*T*/K	320	330	340	350	360	370	380	390
		p/Pa	0.008	0.024	0.073	0.208	0.556	1.32	2.86	6.30
1-Pyrenol	$C_{16}H_{10}O$	*T*/K	360	370	380	390	400			
		p/Pa	0.005	0.016	0.047	0.135	0.364			
Benzo[*b*]naphtho[2,1-*d*]thiophene	$C_{16}H_{12}S$	*T*/K	330	340	350	360	370	380	390	
		p/Pa	0.001	0.004	0.012	0.036	0.098	0.255	0.631	
11*H*-Benzo[*b*]fluorene	$C_{17}H_{12}$	*T*/K	340	350	360	370	380	390	400	
		p/Pa	0.003	0.009	0.029	0.085	0.235	0.619	1.55	
6,11-Dihydroxy-5,12-naphthacenedione	$C_{18}H_{10}O_4$	*T*/K	420	430	440	450				
		p/Pa	0.008	0.022	0.055	0.131				
Chrysene	$C_{18}H_{12}$	*T*/K	390	400	410	420				
		p/Pa	0.087	0.221	0.539	1.26				
Naphthacene	$C_{18}H_{12}$	*T*/K	390	400	410	420	430	440	450	460
		p/Pa	0.005	0.014	0.035	0.084	0.194	0.432	0.928	1.929
Perylene	$C_{20}H_{12}$	*T*/K	390	400	410	420	430			
		p/Pa	0.006	0.015	0.040	0.102	0.246			
Pentacene	$C_{22}H_{14}$	*T*/K	450	460	470	480	490			
		p/Pa	0.002	0.006	0.013	0.031	0.069			
Coronene	$C_{24}H_{12}$	*T*/K	430	440	450	460	470	480	490	500
		p/Pa	0.004	0.01	0.021	0.046	0.097	0.197	0.389	0.747

VAPOR PRESSURE OF COMPOUNDS AND ELEMENTS

David R. Lide

This table gives vapor pressure data for about 1800 substances. In order to accommodate elements and compounds ranging from refractory to highly volatile in a single table, the temperature at which the vapor pressure reaches specified pressure values is listed.

Column heading	Definition
Name	Name of substance; listed alphabetically by name
Mol. form.	Molecular formula for substance
***t*/°C for 1 Pa**	Temperature at which vapor pressure reaches 1 Pa, in °C; symbol "*s*" following a value indicates that the substance is a solid at that temperature
***t*/°C for 10 Pa**	Temperature at which vapor pressure reaches 10 Pa, in °C; symbol "*s*" following a value indicates that the substance is a solid at that temperature
***t*/°C for 100 Pa**	Temperature at which vapor pressure reaches 100 Pa, in °C; symbol "*s*" following a value indicates that the substance is a solid at that temperature
***t*/°C for 1 kPa**	Temperature at which vapor pressure reaches 1 kPa, in °C; symbol "*s*" following a value indicates that the substance is a solid at that temperature
***t*/°C for 10 kPa**	Temperature at which vapor pressure reaches 10 kPa, in °C; symbol "*s*" following a value indicates that the substance is a solid at that temperature
***t*/°C for 100 kPa**	Temperature at which vapor pressure reaches 100 kPa, in °C; symbol "*s*" following a value indicates that the substance is a solid at that temperature
Ref.	Reference number

The data used in preparing the table came from a large number of sources; the main references used for each substance are indicated. Because the data were refit in most cases, values appearing in this table may not be identical with values in the source cited. The temperature entry in the 100 kPa column is close to, but not identical with, the normal boiling point (which is defined as the temperature at which the vapor pressure reaches 101.325 kPa). Although some temperatures are quoted to 0.1 °C, uncertainties of several degrees should generally be assumed.

More extensive and detailed vapor pressure data on selected important substances appear in other tables in this section of this *CRC Handbook* including "Vapor Pressure of Fluids at Temperatures below 300 K," "Vapor Pressure of Ice," and "Vapor Pressure and Other Saturation Properties of Water" in this section.

References

1. Lide, D.R., and Kehiaian, H.V., *CRC Handbook of Thermophysical and Thermochemical Data*, CRC Press, Boca Raton, FL, 1994.
2. Stull, D., in *American Institute of Physics Handbook, Third Edition*, Gray, D.E., Ed., McGraw-Hill, New York, 1972.
3. Hultgren, R., Desai, P.D., Hawkins, D.T., Gleiser, M., Kelley, K.K., and Wagman, D.D., *Selected Values of Thermodynamic Properties of the Elements*, American Society for Metals, Metals Park, OH, 1973.
4. Stull, D., *Ind. Eng. Chem.*, 39, 517, 1947. <https://doi.org/10.1021/ie50448a022>
5. *TRCVP, Vapor Pressure Database, Version 2.2P*, Thermodynamic Research Center, Texas A&M University, College Station, TX.
6. *TRC Thermodynamic Tables*, Thermodynamic Research Center, Texas A&M University, College Station, TX.
7. Ohe, S., *Computer Aided Data Book of Vapor Pressure*, Data Book Publishing Co., Tokyo, 1976.
8. Chase, M.W., Davies, C.A., Downey, J.R., Frurip, D.J., McDonald, R.A., and Syverud, A.N., *JANAF Thermochemical Tables, Third Edition, J. Phys. Chem. Ref. Data*, Vol. 14, Suppl. 1, 1985.
9. Barin, I., *Thermochemical Data of Pure Substances*, VCH Publishers, New York, 1993.
10. Jacobsen, R.T., et al, *International Thermodynamic Tables of the Fluid State, No. 10, Ethylene*, Blackwell Scientific Publications, Oxford, 1988.
11. Wakeham, W.A., *International Thermodynamic Tables of the Fluid State, No. 12, Methanol*, Blackwell Scientific Publications, Oxford, 1993.
12. Janz, G.J., *Molten Salts Handbook*, Academic Press, New York, 1967.
13. Ohse, R.W. *Handbook of Thermodynamic and Transport Properties of Alkali Metals*, Blackwell Scientific Publications, Oxford, 1994.
14. Gschneidner, K.A., in *CRC Handbook of Chemistry and Physics, 77th Edition*, p. 4–112, CRC Press, Boca Raton, FL, 1996.
15. Leider, H.R., Krikorian, O.H., and Young, D.A., *Carbon*, 11, 555, 1973. <https://doi.org/10.1016/0008-6223(73)90316-3>
16. Ruzicka, K., and Majer, V., *J. Phys. Chem. Ref. Data*, 23, 1, 1994. <https://doi.org/10.1063/1.555942>
17. Tillner-Roth, R., and Baehr, H.D., *J. Phys. Chem. Ref. Data*, 23, 657, 1994. <https://doi.org/10.1063/1.555958>
18. Younglove, B.A., and McLinden, M.O., *J. Phys. Chem. Ref. Data*, 23, 731, 1994. <https://doi.org/10.1063/1.555950>
19. Outcalt, S.L., and McLinden, M.O., *J. Phys. Chem. Ref. Data*, 25, 605, 1996. <https://doi.org/10.1063/1.555979>
20. Weber, L.A., and Defibaugh, D.R., *J. Chem. Eng. Data*, 41, 382, 1996. <https://doi.org/10.1021/je950217m>
21. Rodrigues, M.F., and Bernardo-Gil, M.G., *J. Chem. Eng. Data*, 41, 581, 1996. <https://doi.org/10.1021/je950324g>
22. Piacente, V., Gigli, G., Scardala, P., and Giustini, A., *J. Phys. Chem.*, 100, 9815, 1996. <https://doi.org/10.1021/jp9531669>
23. Barton, J.L., and Bloom, H., *J. Phys. Chem.*, 60, 1413, 1956. <https://doi.org/10.1021/j150544a018>
24. Sense, K.A., Alexander, C.A., Bowman, R.E., and Filbert, R.B., *J. Phys. Chem.*, 61, 337, 1957. <https://doi.org/10.1021/j150549a014>
25. Ewing, C.T., and Stern, K.H., *J. Phys. Chem.* 78, 1998, 1974. <https://doi.org/10.1021/j100613a005>
26. Cady, G.H., and Hargreaves, G.B., *J. Chem. Soc.*, 1563, 1961; 1568, 1961. <https://doi.org/10.1039/jr9610001563>
27. Skudlarski, K., Dudek, J., and Kapala, J., *J. Chem. Thermodynamics*, 19, 857, 1987. <https://doi.org/10.1016/0021-9614(87)90032-2>
28. Wagner, W., and de Reuck, K.M., *International Thermodynamic Tables of the Fluid State, No. 9, Oxygen*, Blackwell Scientific Publications, Oxford, 1987.
29. Marsh, K.N., Ed., *Recommended Reference Materials for the Realization of Physicochemical Properties*, Blackwell Scientific Publications, Oxford, 1987.
30. Alcock, C.B., Itkin, V.P., and Horrigan, M.K., *Canadian Metallurgical Quarterly*, 23, 309, 1984. <https://doi.org/10.1179/cmq.1984.23.3.309>
31. Stewart, R.B., and Jacobsen, R.T., *J. Phys. Chem. Ref. Data*, 18, 639, 1989. <https://doi.org/10.1063/1.555829>
32. Sifner, O., and Klomfar, J., *J. Phys. Chem. Ref. Data*, 23, 63, 1994. <https://doi.org/10.1063/1.555956>
33. Bah, A., and Dupont-Pavlovsky, N., *J. Chem. Eng. Data*, 40, 869, 1995. <https://doi.org/10.1021/je00020a028>
34. Behrens, R.G., and Rosenblatt, G., *J. Chem. Thermodynamics*, 4, 175, 1972. <https://doi.org/10.1016/0021-9614(72)90055-9>
35. Behrens, R.G., and Rosenblatt, G., *J. Chem. Thermodynamics*, 5, 173, 1973. <https://doi.org/10.1016/S0021-9614(73)80077-1>
36. Haar, L., Gallagher, J.S., and Kell, G.S., *NBS/NRC Steam Tables*, Hemisphere Publishing Corp., New York, 1984.
37. Wagner, W., Saul, A., and Pruss, A., *J. Phys. Chem. Ref. Data*, 23, 515. 1994. <https://doi.org/10.1063/1.555947>

38. Behrens, R.G., Lemons, R.S., and Rosenblatt, G., *J. Chem. Thermodynamics*, 6, 457, 1974. <https://doi.org/10.1016/0021-9614(74)90007-X>
39. Boublik, T., Fried, V., and Hala, E., *The Vapor Pressure of Pure Substances, Second Edition,* Elsevier, Amsterdam, 1984.
40. Goodwin, R.D., *J. Phys. Chem. Ref. Data*, 14, 849, 1985. <https://doi.org/10.1063/1.555742>
41. Younglove, B.A., and Ely, J.F., *J. Phys. Chem. Ref. Data*, 16, 577, 1987. <https://doi.org/10.1063/1.555785>

Vapor Pressure for Inorganic and Organic Substances at Various Temperatures

Name	Mol. form.	*t*/°C for 1 Pa	*t*/°C for 10 Pa	*t*/°C for 100 Pa	*t*/°C for 1 kPa	*t*/°C for 10 kPa	*t*/°C for 100 kPa	Ref.
Acenaphthene	$C_{12}H_{10}$				126.2	187[e]	276[e]	1
Acenaphthylene	$C_{12}H_8$	24 s	49.8 s	80.6 s				5
Acetaldehyde	C_2H_4O		-105[e]	-87[e]	-62.8	-29.4	20.0	5
Acetamide	C_2H_5NO	16.7 s	39.1 s	65.2 s	102.8	150.8	218.2	5
Acetic acid	$C_2H_4O_2$	-42.8 s	-26.7 s	-8 s	14.2 s	55.9	117.5	1,5
Acetic anhydride	$C_4H_6O_3$	-44[e]	-25[e]	-1[e]	31[e]	75.1	139.7	1
1-Acetonaphthone	$C_{12}H_{10}O$	37[e]	69[e]	107.0	154.6	215.2	294.9	5
2-Acetonaphthone	$C_{12}H_{10}O$	48.3 s		118.7	163.0	221.1	300.3	5
Acetone	C_3H_6O	-95	-81.8	-62.8	-35.6	1.3	55.7	1,5
Acetonitrile	C_2H_3N				-20[e]	21.4	81.2	1
Acetophenone	C_8H_8O			36[e]	73[e]	125.3	201.5	5
Acetyl bromide	C_2H_3BrO	-78[e]	-65[e]	-49[e]	-25[e]	13.9	84[e]	5
Acetyl chloride	C_2H_3ClO	-100[e]	-85[e]	-66[e]	-40[e]	-3.6	50.4	1
Acetylene[a]	C_2H_2			-146.6 s	-130.7 s	-110.6 s	-84.8 s	5
Acetyl fluoride	C_2H_3FO					-64.1	17.0	5
Acetyl iodide	C_2H_3IO				-0.6	47[e]	107.0	5
Acridine	$C_{13}H_9N$			124.4	176.2	246.0	345.4	5
Acrolein	C_3H_4O		-87[e]	-67[e]	-40[e]	-3.0	52.8	1
Acrylamide	C_3H_5NO			109.6	161[e]			5
Acrylic acid	$C_3H_4O_2$				35[e]	78.0	140.7	1
Acrylonitrile	C_3H_3N		-72[e]	-50[e]	-22[e]	17.7	77.0	1
Allene[a]	C_3H_4		-129[e]	-118[e]	-101.4	-76.7	-34.7	5
Allyl alcohol	C_3H_6O	-63[e]	-48[e]	-21.9	6.8	44.5	96.2	5
Allylamine	C_3H_7N		-88[e]	-65[e]	-37[e]	0.4	52[e]	5
Allyl ethyl ether	$C_5H_{10}O$			-56[e]	-28.7	9.8	67.2	5
Allyl glycidyl ether	$C_6H_{10}O_2$				40.1	85.7	152.8	5
Allyl isothiocyanate	C_4H_5NS	-45[e]	-27[e]	-3[e]	32.1	89[e]	198[e]	5
4-Allyl-2-methoxyphenol	$C_{10}H_{12}O_2$	9[e]	37[e]	72[e]	115.9	173.8	252.9	5
Allyltrichlorosilane	$C_3H_5Cl_3Si$					53.0	116.5	5
Aluminum	Al	1209	1359	1544	1781	2091	2517	2
Aluminum borohydride	AlB_3H_{12}				-46.8	-9.4	45.5	4
Aluminum chloride	$AlCl_3$	58.4 s	76.5 s	97.1 s	120.7 s	148.2 s	180.5 s	4
Aluminum fluoride	AlF_3	744 s	819 s	906 s	1008 s	1130 s	1276 s	8
Aluminum iodide	AlI_3				218	285	385	4
Aluminum oxide (α)	Al_2O_3			2122	2351	2629	2975	4
1-Amino-2-propanol	C_3H_9NO			18[e]	53.2	98.2	157.9	5
Ammonia[a]	H_3N	-139 s	-127 s	-112 s	-94.5 s	-71.3	-33.6	1,5,6
Ammonium bromide	BrH_4N	121 s	154 s	195 s	246 s	310.4 s	395.1 s	5
Ammonium chloride	ClH_4N	91 s	121 s	159 s	204.7 s	263.1 s	339.5 s	5
Ammonium iodide	H_4IN	125 s	159 s	201 s	253 s	318.4 s	405.2 s	5
Aniline	C_6H_7N		-2.5	26.7	63.5	112.5	183.5	1,5
Anisole	C_7H_8O		-21[e]	4[e]	38[e]	84[e]	153.2	1,5
Anthracene	$C_{14}H_{10}$	89.2 s	125.9 s	151.5 s	165 s	238.8	340.2	1,5
Antimony	Sb	534 s	603 s	738	946	1218	1585	2,3
Antimony(III) bromide	Br_3Sb				136.5	196.9	286.5	1
Antimony(III) iodide	I_3Sb				214.9	292.0	401.2	4
Antimony(III) oxide (valentinite)	O_3Sb_2	426.1 s	478 s	539 s	610 s	907	1420	4,35
Argon[a]	Ar		-226.4 s	-220.3 s	-212.4 s	-201.7 s	-186.0	1,5,31
Arsenic (gray)	As	280 s	323 s	373 s	433 s	508 s	601 s	3
Arsenic(III) chloride	$AsCl_3$			-8[e]	21.3	63.1	129.4	1
Arsenic(III) fluoride	AsF_3					8.1	56.0	4
Arsenic(III) iodide	AsI_3				187	261	367[e]	7
Arsenic(III) oxide (arsenolite)	As_2O_3	133.7 s	163.0 s	196.8 s	236.2 s	283.0		34
Astatine	At	88 s	119 s	156 s	202 s	258 s	334	2

Name	Mol. form.	t/°C for 1 Pa	t/°C for 10 Pa	t/°C for 100 Pa	t/°C for 1 kPa	t/°C for 10 kPa	t/°C for 100 kPa	Ref.
trans-Azobenzene	$C_{12}H_{10}N_2$			98.1	144.8	206.7	292.7	4
Azulene	$C_{10}H_8$	24.1 s	46 s	71.5 s	103.3	162.6	244.0	5
Barium	Ba	638 s	765	912	1115	1413	1897	9
Benzaldehyde	C_7H_6O		-9[e]	19[e]	54.6	104.6	178.3	1
Benzanthrone	$C_{17}H_{10}O$		184[e]	229.3	290.3	377.2	511[e]	5
Benzene[b]	C_6H_6			-40 s	-15.1 s	20.0	79.7	1,5
Benzeneacetonitrile	C_8H_7N	-3[e]	23[e]	55.3	97.4	153.7	233.1	5
1,3-Benzenediamine	$C_6H_8N_2$			94.5	140.2	200.8	285.0	5
Benzeneethanol	$C_8H_{10}O$	2[e]	25[e]	54[e]	92[e]	143.6	217.7	5
Benzenethiol	C_6H_6S		-15[e]	12[e]	47[e]	96.0	168.6	5
1,2,3-Benzenetriol	$C_6H_6O_3$				162.0	222.8	308.3	5
Benzil	$C_{14}H_{10}O_2$			123	175	246	346	4
1-Benzofuran	C_8H_6O		-16[e]	12[e]	47.9	97.7	170.7	5
Benzoin	$C_{14}H_{12}O_2$				181	248	342	4
Benzonitrile	C_7H_5N		-6[e]	23.9	63.1	115.7	190.0	5
p-Benzoquinone	$C_6H_4O_2$	-4.1 s	17.8 s	43.5 s	74.3 s	111.6 s		5
Benzoyl bromide	C_7H_5BrO	-15[e]	11[e]	42.6	83.9	139.5	218.0	5
Benzoyl chloride	C_7H_5ClO			27.5	67.0	120.4	196.7	5
Benzyl acetate	$C_9H_{10}O_2$	-11[e]	15[e]	46.6	86.9	139.5	211[e]	5
Benzyl alcohol	C_7H_8O	8[e]	28[e]	54[e]	88[e]	134.7	204.9	1
Benzylamine	C_7H_9N			25.6	62.6	112.7	183.9	5
Benzyl ethyl ether	$C_9H_{12}O$		-10[e]	20.4	59.3	111.3	184.5	5
Beryllium	Be	1189 s	1335	1518	1750	2054	2469	2
Beryllium bromide	$BeBr_2$	203 s	240 s	283 s	335 s	397 s	473 s	4
Beryllium chloride	$BeCl_2$	196 s	237 s	284 s	339 s	402 s	487	4
Beryllium fluoride	BeF_2		686[e]	767[e]	869	999	1172[e]	7
Beryllium iodide	BeI_2	188 s	229 s	276 s	333 s	402 s	487	4
Bicyclo[4.1.0]heptane	C_7H_{12}					49.9	116.3	5
Biphenyl	$C_{12}H_{10}$			69.0	111.1	169.5	254.7	1
Bis(2-aminoethyl)amine	$C_4H_{13}N_3$	-10[e]	13[e]	43[e]	80[e]	129.6	198[e]	5
Bis(2-chloroethyl) ether	$C_4H_8Cl_2O$	-32[e]	-9[e]	19.8	56.9	106.9	177.9	5
Bis(2-ethylhexyl) phthalate	$C_{24}H_{38}O_4$	122.0	153.2	189.2	231.3	281.1	341.1	5
Bis(2-hydroxyethyl) sulfide	$C_4H_{10}O_2S$			31[e]	114.2		282.0	5
Bismuth	Bi	668	768	892	1052	1265	1562	2
Bismuth tribromide	$BiBr_3$			217 s	273[d]	348[d]	455[d]	4,9
Bismuth trichloride	$BiCl_3$				248.9	328.6	438.7	1,4
Borane carbonyl	CH_3BO				-124	-99	-64	4
Boron	B	2075	2289	2549	2868	3272	3799	2
Boron tribromide	BBr_3			-45[e]	-15[e]	27.5	90.4	1
Boron trichloride[a]	BCl_3			-94.0	-70.5	-37.4	12.3	4
Boron trifluoride[a]	BF_3	-173.9 s	-166.0 s	-156.0 s	-143.0 s	-125.9	-101.1	4
Bromine[a]	Br_2	-87.7 s	-71.8 s	-52.7 s	-29.3 s	2.5	58.4	1
Bromobenzene	C_6H_5Br		-25[e]	1[e]	34.9	83.1	155.4	1
1-Bromobutane	C_4H_9Br	-68.4	-53.9	-34.1	-5.4	37.6	101.1	1,5
2-Bromobutane, (±)-	C_4H_9Br	-86[e]	-68[e]	-46[e]	-16[e]	26.6	90.7	5
trans-1-Bromo-1-butene	C_4H_7Br	-87[e]	-68[e]	-43.3	-11.4	31.9	94.4	5
2-Bromo-1-butene	C_4H_7Br	-87[e]	-70[e]	-48[e]	-20[e]	20.7	80.6	5
cis-2-Bromo-2-butene	C_4H_7Br	-90[e]	-72[e]	-49.0	-18.5	23.5	85.2	5
trans-2-Bromo-2-butene	C_4H_7Br	-86[e]	-67[e]	-43.4	-12.0	31.0	93.5	5
Bromochlorodifluoromethane	$CBrClF_2$	-136[e]	-123[e]	-106[e]	-83.4	-51.8	-4.3	1
1-Bromo-2-chloroethane	C_2H_4BrCl				-0.4	41.7	105.7	6
Bromochloromethane	CH_2BrCl	-83[e]	-69[e]	-50[e]	-25[e]	11.4	67.7	1
1-Bromo-3-chloropropane	C_3H_6BrCl	-51[e]	-31[e]	-6[e]	28[e]	74.1	142.9	5
2-Bromo-2-chloro-1,1,1-trifluoroethane	$C_2HBrClF_3$				-41.4	-4.8	49.8	1
1-Bromodecane	$C_{10}H_{21}Br$	9[e]	33[e]	63[e]	104[e]	159.2	240.0	5
Bromodifluoromethane	$CHBrF_2$		-128 s	-111.4 s	-89.7 s	-59.7 s	-16 s	5
1-Bromododecane	$C_{12}H_{25}Br$	31[e]	57[e]	90[e]	132[e]	190.8	275.3	5
Bromoethane	C_2H_5Br	-111[e]	-96[e]	-77[e]	-51.3	-15.5	38.0	5
Bromoethene	C_2H_3Br	-124[e]	-110[e]	-92[e]	-68[e]	-34.5	15.4	5
(2-Bromoethyl)cyclohexane	$C_8H_{15}Br$	-14[e]	8[e]	36.9	75.3	129.7	212.5	5

Name	Mol. form.	t/°C for 1 Pa	t/°C for 10 Pa	t/°C for 100 Pa	t/°C for 1 kPa	t/°C for 10 kPa	t/°C for 100 kPa	Ref.
1-Bromoheptane	$C_7H_{15}Br$	-30[e]	-9[e]	18[e]	54[e]	104.4	178.4	5
1-Bromohexane	$C_6H_{13}Br$	-45[e]	-25[e]	2[e]	36[e]	83.7	154.8	5
1-Bromo-4-isopropylbenzene	$C_9H_{11}Br$	-8[e]	15[e]	45[e]	84[e]	138.1	218.5	5
Bromomethane	CH_3Br				-77[e]	-44.3	3.3	1
(Bromomethyl)benzene	C_7H_7Br			25.4	66.8	121.7	198.3	5
1-Bromo-3-methylbutane	$C_5H_{11}Br$	-67[e]	-49[e]	-25[e]	8[e]	52.4	119.9	5
1-Bromo-2-methylpropane	C_4H_9Br	-85[e]	-68[e]	-46[e]	-16[e]	26.8	91.1	5
2-Bromo-2-methylpropane	C_4H_9Br					11.7	72.4	1,5
1-Bromonaphthalene	$C_{10}H_7Br$	17[e]	45[e]	80.3	126.7	189.8	280.5	5
1-Bromooctane	$C_8H_{17}Br$	-17[e]	6[e]	34[e]	72[e]	123.8	200.3	5
Bromopentafluorobenzene	C_6BrF_5			-10[e]	23[e]	68[e]	136.0	5
1-Bromopentane	$C_5H_{11}Br$	-60[e]	-41[e]	-16[e]	16[e]	61.5	129.1	5
2-Bromopentane	$C_5H_{11}Br$	-69[e]	-51[e]	-27[e]	5[e]	49.7	116.9	5
3-Bromopentane	$C_5H_{11}Br$	-68[e]	-50[e]	-26[e]	6[e]	50.8	118.1	5
1-Bromopropane	C_3H_7Br	-95[e]	-78[e]	-57[e]	-28[e]	11.6	70.6	1
2-Bromopropane	C_3H_7Br		-84[e]	-65[e]	-39.6	-1.7	59.1	1,5
cis-1-Bromopropene	C_3H_5Br	-100[e]	-84[e]	-64[e]	-37[e]	1.0	57.4	5
2-Bromopropene	C_3H_5Br	-112[e]	-95[e]	-75[e]	-47[e]	-9[e]	48.0	5
3-Bromopropene	C_3H_5Br	-98[e]	-80[e]	-58[e]	-28[e]	12[e]	69.6	5
Bromosilane	BrH_3Si				-81.0	-47.3	2.2	4
2-Bromotoluene	C_7H_7Br		-10[e]	17[e]	54[e]	104.8	181.1	5
3-Bromotoluene	C_7H_7Br	-34[e]	-11[e]	19.4	58.1	109.9	183.1	5
4-Bromotoluene	C_7H_7Br				57[e]	107.8	183.8	5
Bromotrichloromethane	$CBrCl_3$				-6[e]	38.9	104.4	5
Bromotrifluoromethane[a]	$CBrF_3$	-168[e]	-156[e]	-142[e]	-122.8	-96.6	-58.1	5
1,2-Butadiene	C_4H_6	-132[e]	-117[e]	-98[e]	-72.8	-38.9	10.5	5
1,3-Butadiene[a]	C_4H_6			-106[e]	-83[e]	-51.9	-4.7	1
Butanal	C_4H_8O	-88[e]	-72[e]	-50[e]	-22[e]	16.6	74.5	1,5
Butane[a]	C_4H_{10}	-134.3	-121.0	-103.9	-81.1	-49.1	-0.8	1,41
1,3-Butanediol	$C_4H_{10}O_2$	-4[e]	23[e]	55[e]	94[e]	142.9	206.1	5
1,4-Butanediol	$C_4H_{10}O_2$		45[e]	77[e]	116[e]	164.7	227.6	5
2,3-Butanediol	$C_4H_{10}O_2$		15[e]	43[e]	77[e]	121.2	180.3	5
2,3-Butanedione	$C_4H_6O_2$					30.7	84.8	5
1,4-Butanedithiol	$C_4H_{10}S_2$	-17[e]	5[e]	32[e]	69.1	119.9	195.1	5
Butanenitrile	C_4H_7N	-67[e]	-48[e]	-24[e]	8[e]	52.3	117.2	1
1-Butanethiol	$C_4H_{10}S$	-77[e]	-59[e]	-37[e]	-6[e]	35.4	98.0	5
2-Butanethiol	$C_4H_{10}S$	-86[e]	-69[e]	-47[e]	-17[e]	23.4	84.5	5
Butanoic acid	$C_4H_8O_2$			12.9	52.2	101.4	163.3	1,5
Butanoic anhydride	$C_8H_{14}O_3$	-28[e]	-2[e]	30[e]	71[e]	123.8	196.5	5
1-Butanol	$C_4H_{10}O$	-37[e]	-20[e]	0[e]	28[e]	64[e]	117.4	1
2-Butanol	$C_4H_{10}O$	-50[e]	-34[e]	-14[e]	12.6	48.2	99.2	1,5
2-Butanone	C_4H_8O	-85[e]	-68[e]	-46[e]	-18.1	21.2	79.2	1
2-Butanone oxime	C_4H_9NO		-18[e]	7[e]	38.9	81.9	142.9	5
trans-2-Butenal	C_4H_6O	-74[e]	-56[e]	-33[e]	-3[e]	39.7	102.4	5
1-Butene	C_4H_8	-139.0	-125.2	-107.8	-85.3	-53.7	-6.6	1,5
cis-2-Butene	C_4H_8	-131.2	-117.4	-99.8	-76.7	-44.8	3.4	1,5
trans-2-Butene	C_4H_8			-102[e]	-80[e]	-47.6	0.6	1
cis-2-Butene-1,4-diol	$C_4H_8O_2$	17[e]	44[e]	77[e]	117.4	168.5	234.9	5
trans-2-Butenedioyl dichloride	$C_4H_2Cl_2O_2$			8.0	45.6	94.3	159.8	5
3-Butenenitrile	C_4H_5N	-67[e]	-48[e]	-23.1	9.3	53.7	118.4	5
cis-2-Butenoic acid	$C_4H_6O_2$			30[e]	63[e]	106.7	168.9	5
trans-2-Butenoic acid	$C_4H_6O_2$				74[e]	120.8	184.9	5
3-Butenoic acid	$C_4H_6O_2$	-19[e]	2[e]	27[e]	61[e]	105.6	168.6	5
3-Buten-2-one	C_4H_6O					21[e]	81.0	5
1-Buten-3-yne	C_4H_4			-96.1	-73.4	-41.8	4.9	5
2-Butoxyethanol	$C_6H_{14}O_2$	-31[e]	-8[e]	20[e]	55[e]	103.2	170.2	5
Butyl acetate	$C_6H_{12}O_2$	-63[e]	-43[e]	-19[e]	14[e]	61.0	125.6	1,5
Butyl acrylate	$C_7H_{12}O_2$	-52[e]	-31[e]	-4.5	30.4	78.0	146.9	5
Butylamine	$C_4H_{11}N$			-46[e]	-18.1	20.0	75.9	5
sec-Butylamine	$C_4H_{11}N$			-55[e]	-29.1	7.5	62.3	5

Name	Mol. form.	t/°C for 1 Pa	t/°C for 10 Pa	t/°C for 100 Pa	t/°C for 1 kPa	t/°C for 10 kPa	t/°C for 100 kPa	Ref.
tert-Butylamine	$C_4H_{11}N$			-67[e]	-42.4	-8.1	43.7	5
N-Butylaniline	$C_{10}H_{15}N$	11[e]	35[e]	66[e]	106[e]	160.9	241.0	5
Butylbenzene	$C_{10}H_{14}$	-28[e]	-7[e]	21[e]	56.9	107.6	182.8	1,5
sec-Butylbenzene, (±)-	$C_{10}H_{14}$	-35[e]	-14[e]	13[e]	48[e]	98.3	172.8	5
tert-Butylbenzene	$C_{10}H_{14}$	-37[e]	-16[e]	10[e]	46[e]	94.9	168.6	5
Butyl benzoate	$C_{11}H_{14}O_2$	6[e]	34[e]	67.9	110.3	165[e]	237[e]	5
Butylcyclohexane	$C_{10}H_{20}$	-31[e]	-9[e]	18[e]	54[e]	104.7	180.4	5
tert-Butylcyclohexane	$C_{10}H_{20}$	-39[e]	-18[e]	9[e]	45[e]	95.3	171.1	5
Butylcyclopentane	C_9H_{18}	-45[e]	-24[e]	1[e]	36[e]	84[e]	156.1	5
Butylethylamine	$C_6H_{15}N$				6.1	47.7	107.0	5
Butyl ethyl ether	$C_6H_{14}O$	-78[e]	-61[e]	-39[e]	-10[e]	31.0	91.9	1
tert-Butyl ethyl ether	$C_6H_{14}O$	-90[e]	-74[e]	-53[e]	-24.6	14.4	72.6	5
Butyl ethyl sulfide	$C_6H_{14}S$	-49[e]	-30[e]	-5[e]	29[e]	74.8	143.8	5
Butyl formate	$C_5H_{10}O_2$			-29[e]	2[e]	44.4	105.7	5
Butyl methacrylate	$C_8H_{14}O_2$				47[e]	93.3	159.0	5
1-*tert*-Butyl-4-methylbenzene	$C_{11}H_{16}$	-24[e]	-2[e]	27[e]	64.1	115.5	190.8	5
Butyl methyl ether	$C_5H_{12}O$			-54[e]	-27[e]	12[e]	69.8	1
Butyl methyl sulfide	$C_5H_{12}S$		-43[e]	-19[e]	13[e]	57[e]	123.0	1
tert-Butyl methyl sulfide	$C_5H_{12}S$				-7.8	34.7	98.4	5
1-Butylnaphthalene	$C_{14}H_{16}$	67[e]	82[e]	103[e]	135[e]	186.7	288.6	5
2-Butylnaphthalene	$C_{14}H_{16}$	44[e]	67[e]	98[e]	139[e]	197.5	287.4	5
Butyl oleate	$C_{22}H_{42}O_2$	95.5	124.2	158[e]	198[e]	245[e]	304[e]	5
2-Butylphenol	$C_{10}H_{14}O$	7[e]	31[e]	61[e]	101[e]	155.2	234.4	5
Butyl phenyl ether	$C_{10}H_{14}O$	-16[e]	8[e]	38[e]	77[e]	131.3	209.7	5
Butyl stearate	$C_{22}H_{44}O_2$	99.6	128[e]	162[e]	201[e]	249[e]	307[e]	5
Butyltrichlorosilane	$C_4H_9Cl_3Si$					77.2	148.4	5
Butyl vinyl ether	$C_6H_{12}O$	-87[e]	-67[e]	-42[e]	-9.3	33.6	93.2	5
1-Butyne	C_4H_6	-125[e]	-111[e]	-94[e]	-71.2	-39.4	7.8	1
2-Butyne	C_4H_6		-89.2 s	-73.8 s	-53.5 s	-23.9	26.6	5
γ-Butyrolactone	$C_4H_6O_2$		-17[e]	24[e]	72[e]	130.2	203[e]	5
Cadmium	Cd	257 s	310 s	381	472	594	767	2
Cadmium bromide	Br_2Cd	373 s	435 s	509 s				27
Cadmium chloride	$CdCl_2$	412 s	471 s	541 s	634	768	959	12,23,27
Cadmium fluoride	CdF_2				1257	1461	1742	4
Cadmium iodide	CdI_2	296 s	344 s	406	498	622	795	4,27
Cadmium oxide	CdO	770 s	866 s	983 s	1128 s	1314 s	1558 s	4
Calcium	Ca	591 s	683 s	798 s	954	1170	1482	2
Camphene	$C_{10}H_{16}$					90.7	160.1	4
Camphor, (+)	$C_{10}H_{16}O$	-15.8 s	10 s	41.5 s	80.8 s	131.4 s	207.6	5
Caprolactam	$C_6H_{11}NO$	36.8 s	58.9 s	86.6 s			270	5
Carbazole	$C_{12}H_9N$					254.7	354.0	5
Carbon dioxide[a, b]	CO_2	-159.1 s	-148.9 s	-136.7 s	-121.6 s	-103.1 s	-78.6 s	5
Carbon diselenide	CSe_2			-24[e]	9.4	56.2	127[e]	1
Carbon disulfide	CS_2		-96[e]	-76[e]	-49[e]	-10.9	45.9	1
Carbon [fullerene-C_{70}]	C_{70}	598 s	662 s					22
Carbon (graphite)	C		2566 s	2775 s	3016 s	3299 s	3635 s	15
Carbon monoxide[a]	CO			-223 s	-216.5 s	-207.2 s	-191.7	40
Carbon oxyselenide	COSe			-120	-98	-67	-22	4
Carbon oxysulfide[a]	COS			-136[e]	-117[e]	-90.0	-50.4	1
Carbonyl chloride	CCl_2O	-127[e]	-113[e]	-96[e]	-73[e]	-40.6	7.2	5
Carbonyl dicyanide	C_3N_2O				-21.7	15.3	65.2	5
Cerium	Ce	1719	1921	2169	2481	2886	3432	14
Cesium	Cs	144.5	195.6	260.9	350.0	477.1	667.0	13,30
Cesium bromide	BrCs	531 s	601 s	701[d]	834[d]	1019[d]	1293[e]	9
Cesium chloride	ClCs			730	864	1043	1297	4
Cesium fluoride	CsF				825	999	1249	4
Cesium iodide	CsI	523 s	595 s	692	854	1029	1278	4,25
Chlorine[a]	Cl_2	-145 s	-133.7 s	-120.2 s	-103.6 s	-76.1	-34.2	1
Chlorine dioxide[a]	ClO_2					-34.3	10.5	5
Chlorine fluoride[a]	ClF				-144.4	-122.6	-90.2	5

Name	Mol. form.	t/°C for 1 Pa	t/°C for 10 Pa	t/°C for 100 Pa	t/°C for 1 kPa	t/°C for 10 kPa	t/°C for 100 kPa	Ref.
Chlorine pentafluoride	ClF_5				-88[e]	-59	-14	7
Chlorine trifluoride	ClF_3				-63.7	-33.0	11.4	5
Chloroacetic acid	$C_2H_3ClO_2$				78.4	123.9	188.9	1
Chloroacetyl chloride	$C_2H_2Cl_2O$			-23.7	5.6	46.1	105.6	5
2-Chloroaniline	C_6H_6ClN		10[e]	39.0	75.2	131.4	208.3	5
3-Chloroaniline	C_6H_6ClN	-5[e]	19.7	49.4	94.2	162[e]	1069[e]	5
2-Chloroanisole	C_7H_7ClO	-22[e]	2[e]	33[e]	72[e]	125.2	201[e]	5
Chlorobenzene	C_6H_5Cl		-43[e]	-17[e]	16.8	62.9	131.3	1,5
2-Chlorobenzoyl chloride	$C_7H_4Cl_2O$				93[e]	149[e]	237.0	5
3-Chlorobenzoyl chloride	$C_7H_4Cl_2O$				87.8	147[e]	225.0	5
2-Chloro-1,3-butadiene	C_4H_5Cl	-113[e]	-95[e]	-71[e]	-41[e]	0.3	59.0	5
1-Chlorobutane	C_4H_9Cl	-87[e]	-71[e]	-49[e]	-21[e]	18.4	78.1	1
2-Chlorobutane	C_4H_9Cl	-96[e]	-80[e]	-59[e]	-31.0	8.5	67.9	1
3-Chloro-1-butene	C_4H_7Cl			-64[e]	-36[e]	4[e]	63.6	5
cis-2-Chloro-2-butene	C_4H_7Cl	-100[e]	-83[e]	-62[e]	-34[e]	6[e]	66.4	5
trans-2-Chloro-2-butene	C_4H_7Cl	-102[e]	-86[e]	-65[e]	-37[e]	3[e]	62.2	5
Chlorocyclohexane	$C_6H_{11}Cl$		-35[e]	-9[e]	25[e]	71.6	142.1	5
1-Chlorodecane	$C_{10}H_{21}Cl$	2[e]	25[e]	54[e]	92[e]	145.7	225.3	5
1-Chloro-1,1-difluoroethane	$C_2H_3ClF_2$		-123[e]	-107[e]	-85.3	-55.4	-10.5	5
Chlorodifluoromethane[a]	$CHClF_2$	-152[e]	-141[e]	-126[e]	-107.1	-80.5	-41.1	5
1-Chloro-2,2-dimethylpropane	$C_5H_{11}Cl$				-17[e]	23.5	83.9	5
1-Chlorododecane	$C_{12}H_{25}Cl$	27[e]	51[e]	81[e]	122[e]	178.7	262.6	5
Chloroethane	C_2H_5Cl	-126[e]	-112[e]	-94[e]	-70[e]	-37.0	12.0	1
2-Chloroethanol	C_2H_5ClO	-61[e]	-39[e]	-12[e]	23[e]	67.1	127.3	5
Chloroethene	C_2H_3Cl	-139[e]	-127[e]	-110[e]	-89[e]	-59.0	-14.1	1
1-Chloro-2-ethylbenzene	C_8H_9Cl	-30[e]	-9[e]	18[e]	54[e]	103.7	177.9	5
1-Chloro-4-ethylbenzene	C_8H_9Cl	-27[e]	-6[e]	22[e]	58[e]	108.7	183.9	5
1-Chloro-1-fluoroethane	C_2H_4ClF				-69.9	-36.1	15.8	5
Chlorofluoromethane	CH_2ClF		-124[e]	-108[e]	-86.2	-55.7	-9.4	5
1-Chloroheptane	$C_7H_{15}Cl$	-39[e]	-19[e]	7[e]	41[e]	88.6	159.9	5
1-Chlorohexane	$C_6H_{13}Cl$	-55[e]	-36[e]	-11[e]	21[e]	66.7	134.6	5
1-Chloro-2-isopropylbenzene	$C_9H_{11}Cl$	-23[e]	-1[e]	27[e]	64[e]	114.6	190.5	5
1-Chloro-4-isopropylbenzene	$C_9H_{11}Cl$		3[e]	31[e]	69[e]	120.5	197.8	5
Chloromethane[a]	CH_3Cl	-140.2 s	-128.6 s	-114.7 s	-96[e]	-67.1	-24.4	1,33
(Chloromethyl)benzene	C_7H_7Cl	-34[e]	-11[e]	17.7	55.4	106.3	178.9	5
2-Chloro-2-methylbutane	$C_5H_{11}Cl$			-52[e]	-21[e]	21.8	85.2	5
3-(Chloromethyl)heptane	$C_8H_{17}Cl$					100.3	172.4	5
Chloromethyl methyl ether	C_2H_5ClO	-96[e]	-80[e]	-59[e]	-32[e]	6[e]	61[e]	5
1-Chloro-2-methylpropane	C_4H_9Cl	-94[e]	-78[e]	-56.6	-28.7	10.2	68.5	5
2-Chloro-2-methylpropane	C_4H_9Cl					-4.2	50.3	5
3-Chloro-2-methylpropene	C_4H_7Cl		-75[e]	-54[e]	-25[e]	13.8	71.5	5
Chloromethylsilane	CH_5ClSi	-129[e]	-115[e]	-97.9	-74.4	-41.5	8.3	5
1-Chloronaphthalene	$C_{10}H_7Cl$	14[e]	39[e]	70.5	112.8	171.6	258.6	5
1-Chloro-4-nitrobenzene	$C_6H_4ClNO_2$	15.4 s	35.8 s		97[e]	156.0	238[e]	5
1-Chloro-2-nitro-4-(trifluoromethyl)benzene	$C_7H_3ClF_3NO_2$	3[e]	26[e]	55[e]	92.8	145.2	222.0	5
1-Chlorononane	$C_9H_{19}Cl$	-11[e]	11[e]	39[e]	76[e]	127.8	204.7	5
1-Chlorooctane	$C_8H_{17}Cl$	-25[e]	-4[e]	23[e]	59[e]	108.8	182.9	5
Chloropentafluoroacetone	C_3ClF_5O	-122[e]	-109[e]	-93[e]	-71[e]	-39.4	7.4	5
Chloropentafluorobenzene	C_6ClF_5		-44[e]	-21[e]	11[e]	53.8	117.6	1
Chloropentafluoroethane	C_2ClF_5					-80.3	-39.4	1
1-Chloropentane	$C_5H_{11}Cl$	-73[e]	-55[e]	-32[e]	-1[e]	42.5	107.9	5
2-Chloropentane, (+)	$C_5H_{11}Cl$	-80[e]	-62[e]	-39[e]	-9[e]	33.2	96.1	5
3-Chloropentane	$C_5H_{11}Cl$	-77[e]	-60[e]	-37[e]	-7[e]	34.9	97.3	5
2-Chlorophenol	C_6H_5ClO				45.8	97.9	173.9	5
3-Chlorophenol	C_6H_5ClO			39.7	80.2	135.1	213.4	5
4-Chlorophenol	C_6H_5ClO			45.0	86.5	142.0	219.9	5
1-Chloropropane	C_3H_7Cl	-106[e]	-90[e]	-71[e]	-44.5	-8.1	46.2	1
2-Chloropropane	C_3H_7Cl		-91[e]	-74[e]	-51.1	-17.8	35.4	1,5
2-Chloro-1-propanol	C_3H_7ClO				23[e]	63.8	125.7	5
cis-1-Chloropropene	C_3H_5Cl	-114[e]	-100[e]	-81[e]	-55[e]	-20.1	32.4	5

Name	Mol. form.	t/°C for 1 Pa	t/°C for 10 Pa	t/°C for 100 Pa	t/°C for 1 kPa	t/°C for 10 kPa	t/°C for 100 kPa	Ref.
trans-1-Chloropropene	C_3H_5Cl		-97[e]	-77[e]	-52[e]	-16.2	37.0	5
2-Chloropropene	C_3H_5Cl	-120[e]	-106[e]	-87[e]	-63[e]	-28.7	22.3	5
3-Chloropropene	C_3H_5Cl	-107[e]	-92[e]	-72.4	-46.3	-9.8	44.6	5
2-Chloropyridine	C_5H_4ClN			7.4	45.8	97.3	169.9	5
2-Chlorostyrene	C_8H_7Cl	-33[e]	-10[e]	20[e]	58[e]	110.8	188[e]	5
Chlorosulfonic acid	$ClHO_3S$	-40[e]	-20[e]	5[e]	38.7	85.0	153.6	5
1-Chloro-1,1,2,2-tetrafluoroethane	C_2HClF_4			-110[e]	-87.6	-57.0	-12.1	5
2-Chlorothiophene	C_4H_3ClS		-62[e]	-35[e]	2[e]	51.8	123[e]	5
2-Chlorotoluene	C_7H_7Cl		-24[e]	3[e]	38[e]	86.3	158.7	1,5
3-Chlorotoluene	C_7H_7Cl	-41[e]	-21[e]	6[e]	41[e]	89[e]	161.8	5
4-Chlorotoluene	C_7H_7Cl				40[e]	88.9	161.5	1,5
Chlorotrifluoroethene	C_2ClF_3	-146[e]	-134[e]	-119[e]	-99[e]	-71[e]	-28.4	1
Chlorotrifluoromethane[a]	$CClF_3$	-176[e]	-167[e]	-155[e]	-139[e]	-116[e]	-81.7	5
1-Chloro-2-(trifluoromethyl)benzene	$C_7H_4ClF_3$			1[e]	34.5	81.8	151.8	5
1-Chloro-3-(trifluoromethyl)benzene	$C_7H_4ClF_3$	-53[e]	-34[e]	-9[e]	24.2	69.8	137.2	5
1-Chloro-4-(trifluoromethyl)benzene	$C_7H_4ClF_3$			-9[e]	24.2	70.4	138.1	5
3-Chloro-1,1,1-trifluoropropane	$C_3H_4ClF_3$	-102[e]	-87[e]	-68[e]	-43[e]	-8[e]	45.3	5
Chromium	Cr	1383 s	1534 s	1718 s	1950	2257	2669	2
Cobalt	Co	1517	1687	1892	2150	2482	2925	2
Cobalt(II) chloride	Cl_2Co					818	1048	4
Copper	Cu	1236	1388	1577	1816	2131	2563	2
Copper(I) chloride	ClCu		459	543	675	914	1477	4
Copper(I) iodide	CuI				636	864	1331	4
o-Cresol	C_7H_8O	-6.4 s	12.8 s	40.2	72.3	120.3	190.5	1,5
m-Cresol	C_7H_8O	20.8	33.6	52.4	82.6	130.6	201.8	1,5
p-Cresol	C_7H_8O	-0.2 s	20.7 s	52.7	83.1	130.7	201.5	1,5
Cyanic acid	CHNO			-81.1	-56.8	-23.9	23[e]	5
Cyanoacetylene	C_3HN			-58.7 s	-35.6 s	-7 s	42.0	5
Cyanogen	C_2N_2	-127 s	-114.1 s	-98.5 s	-79.2 s	-54.9 s	-21.4	5
Cyanogen bromide	CBrN				-13 s	17.7 s	61.0	1
Cyanogen chloride	CClN		-94.6 s	-78.1 s	-57 s	-29 s	13.0	5
Cyanogen fluoride	CFN		-135 s	-121.2 s	-104.1 s	-82.8 s	-46.2	1,5
Cyanogen iodide	CIN						153.8	5
Cyclobutane	C_4H_8				-71.8	-38.1	12.1	5
Cyclobutanone	C_4H_6O			-34[e]	-4[e]	37.1	97[e]	5
Cyclodecane	$C_{10}H_{20}$			29[e]	68[e]	121.3	201.8	1
1,5,9-Cyclododecatriene	$C_{12}H_{18}$	-14[e]	11[e]	44[e]	87[e]	145.0	229.8	5
Cycloheptane	C_7H_{14}				6[e]	51.1	118.4	1
Cycloheptanone	$C_7H_{12}O$			18[e]	53.7	104.0	178.7	5
Cycloheptene	C_7H_{12}			-30.0	3.4	47.5	108[e]	5
1,3-Cyclohexadiene	C_6H_8	-88[e]	-71[e]	-50[e]	-21[e]	19[e]	79.9	5
1,4-Cyclohexadiene	C_6H_8				-15[e]	27.3	85.0	5
Cyclohexane	C_6H_{12}	-85.6 s	-68.9 s	-47.6 s	-19.8 s	19.3	80.4	1,5
Cyclohexanethiol	$C_6H_{12}S$					84.8	158.3	5
Cyclohexanol	$C_6H_{12}O$			34[e]	61[e]	99.2	160.7	1
Cyclohexanone	$C_6H_{10}O$		-25[e]	1[e]	36[e]	84[e]	155.2	1
Cyclohexene	C_6H_{10}	-87[e]	-70[e]	-49[e]	-19[e]	21[e]	82.6	1
Cyclohexyl acetate	$C_8H_{14}O_2$					103.1	172.9	5
Cyclohexylamine	$C_6H_{13}N$			-9[e]	22[e]	66.6	133.5	1
Cyclohexylbenzene	$C_{12}H_{16}$		28[e]	58[e]	98[e]	154.7	239.5	5
Cyclohexylcyclohexane	$C_{12}H_{22}$		20[e]	53.1	96.0	154.1	237.2	5
cis,cis-1,5-Cyclooctadiene	C_8H_{12}		-37[e]	-8[e]	30[e]	80.2	150[e]	5
Cyclooctane	C_8H_{16}				30[e]	78[e]	150.7	1
1,3,5,7-Cyclooctatetraene	C_8H_8				24.3	71.0	140.1	5
1,3-Cyclopentadiene	C_5H_6			-77[e]	-51[e]	-14[e]	39.8	5
Cyclopentane	C_5H_{10}			-77.0	-45.4	-7.1	48.8	5
Cyclopentanethiol	$C_5H_{10}S$				18[e]	64[e]	131.7	5
Cyclopentanol	$C_5H_{10}O$		-13[e]	11.5	42.2	82.5	140.0	5
Cyclopentanone	C_5H_8O		-39[e]	-14[e]	19[e]	64[e]	130.3	1
Cyclopentene	C_5H_8	-109[e]	-94[e]	-74[e]	-48[e]	-11.1	43.8	5

Name	Mol. form.	t/°C for 1 Pa	t/°C for 10 Pa	t/°C for 100 Pa	t/°C for 1 kPa	t/°C for 10 kPa	t/°C for 100 kPa	Ref.
Cyclopentylamine	$C_5H_{11}N$	-66[e]	-48[e]	-26[e]	4[e]	45.8	108[e]	5
Cyclopropane	C_3H_6			-124[e]	-104[e]	-75.7	-33.1	1
Cyclopropyl methyl ketone	C_5H_8O		-57[e]	-31[e]	3[e]	49[e]	112[e]	5
cis-Decahydronaphthalene	$C_{10}H_{18}$	-26[e]	-4[e]	24[e]	62.4	115.5	195.3	1
trans-Decahydronaphthalene	$C_{10}H_{18}$		-10[e]	18[e]	55.3	107.9	186.8	1
Decamethylcyclopentasiloxane	$C_{10}H_{30}O_5Si_5$	-2[e]	19[e]	46[e]	82[e]	132.9	210.4	5
Decamethyltetrasiloxane	$C_{10}H_{30}O_3Si_4$	-31[e]	-6[e]	26[e]	66.8	118.8	193.9	5
Decanal	$C_{10}H_{20}O$		16[e]	47.2	86.3	137.7	208.0	5
Decane	$C_{10}H_{22}$		-10.6	16.7	52.3	101.1	173.7	16
Decanedioic acid	$C_{10}H_{18}O_4$	125.9 s						5
Decanenitrile	$C_{10}H_{19}N$	13[e]	36[e]	66[e]	105.8	160.6	241.6	5
1-Decanethiol	$C_{10}H_{22}S$	11[e]	34[e]	64[e]	103[e]	157.5	238.6	5
Decanoic acid	$C_{10}H_{20}O_2$	58[e]	80[e]	108[e]	145[e]	195.2	269.5	5
1-Decanol	$C_{10}H_{22}O$	30[e]	50[e]	75[e]	109[e]	157.3	230.6	1,39
4-Decanol	$C_{10}H_{22}O$	18[e]	37[e]	61[e]	93[e]	139[e]	210[e]	5
1-Decene	$C_{10}H_{20}$	-35.5	-13.7	13.7	49.0	97.9	170.1	1,5
Decyl acetate	$C_{12}H_{24}O_2$	12[e]	40[e]	74[e]	115.1	168.1	238[e]	5
Decylcyclopentane	$C_{15}H_{30}$	37[e]	61[e]	93[e]	134[e]	192.5	278.8	5
1-Decyne	$C_{10}H_{18}$	-34[e]	-13[e]	14[e]	51[e]	100.3	173.5	5
Diacetone alcohol	$C_6H_{12}O_2$	-41[e]	-17[e]	13[e]	50.1	98.5	164[e]	5
Diallyl sulfide	$C_6H_{10}S$	-58[e]	-38[e]	-12.4	21.7	68.8	138.1	5
Dibenzylamine	$C_{14}H_{15}N$	48[e]	77[e]	113.1	158.9	218.5	299.4	5
Diborane	B_2H_6			-162[e]	-147.0	-125.8	-92.6	1
m-Dibromobenzene	$C_6H_4Br_2$	-7[e]	16[e]	44[e]	83[e]	137.0	218.2	5
1,2-Dibromobutane	$C_4H_8Br_2$	-54[e]	-30[e]	0.4	39.6	92.1	166.1	5
1,4-Dibromobutane	$C_4H_8Br_2$	-13[e]	9[e]	37[e]	74[e]	124.0	196.5	5
1,2-Dibromo-1-chloro-1,2,2-trifluoroethane	$C_2Br_2ClF_3$						92.3	5
1,2-Dibromo-1,1-dichloroethane	$C_2H_2Br_2Cl_2$					103.6	177.8	5
1,2-Dibromo-1,2-dichloroethane	$C_2H_2Br_2Cl_2$		-11[e]	22[e]	64.1	119[e]	193[e]	5
Dibromodifluoromethane	CBr_2F_2		-110[e]	-91[e]	-66[e]	-30[e]	22.5	1
1,1-Dibromoethane	$C_2H_4Br_2$		-49[e]	-26[e]	5[e]	46.4	107.6	5
1,2-Dibromoethane	$C_2H_4Br_2$				18[e]	62.2	130.9	1
cis-1,2-Dibromoethene	$C_2H_2Br_2$		-45[e]	-21[e]	10[e]	52.2	114.8	1
trans-1,2-Dibromoethene	$C_2H_2Br_2$				-4[e]	42.2	107.4	5
Dibromomethane	CH_2Br_2			-37[e]	-7[e]	35.2	96.5	5
1,5-Dibromopentane	$C_5H_{10}Br_2$	1[e]	25[e]	54[e]	93[e]	145.6	221.8	5
1,2-Dibromopropane	$C_3H_6Br_2$	-46[e]	-26[e]	-2[e]	31[e]	75.3	139.5	5
1,3-Dibromopropane	$C_3H_6Br_2$	-30[e]	-9[e]	17[e]	52[e]	98.7	166.8	5
1,2-Dibromotetrafluoroethane	$C_2Br_2F_4$		-97[e]	-75[e]	-46[e]	-7.2	47.1	5
1,2-Dibutoxyethane	$C_{10}H_{22}O_2$	0[e]	20[e]	44[e]	78.4	127.1	202.9	5
Dibutylamine	$C_8H_{19}N$	-37[e]	-16[e]	10[e]	44[e]	90.8	159.1	5
Dibutyl ether	$C_8H_{18}O$	-55[e]	-35[e]	-8[e]	26[e]	73.0	141.2	5
Di-*sec*-butyl ether	$C_8H_{18}O$			-19[e]	12.1	55.4	120.6	5
Di-*tert*-butyl ether	$C_8H_{18}O$			-33[e]	-2[e]	41.7	106.8	1
Dibutyl maleate	$C_{12}H_{20}O_4$	12.3	50.4	94.0	144.2	203[e]	272[e]	5
Di-*tert*-butyl peroxide	$C_8H_{18}O_2$			-26[e]	4.3	46.6	110.5	5
Dibutyl phthalate	$C_{16}H_{22}O_4$		104.0	142.7	191.5	254.5	339.4	4
Dibutyl sulfide	$C_8H_{18}S$	-22[e]	0[e]	27[e]	63[e]	113.5	188.4	5
1,1-Dichloroacetone	$C_3H_4Cl_2O$				1[e]	47.8	118.0	5
o-Dichlorobenzene	$C_6H_4Cl_2$		-13[e]	16.3	53.9	104.6	180.0	1,5
m-Dichlorobenzene	$C_6H_4Cl_2$		-22[e]	8.0	46.7	97.8	172.5	1,5
p-Dichlorobenzene	$C_6H_4Cl_2$	-45.5 s	-21.8 s	8 s	46.7 s	99.0	173.6	1,5
1,1-Dichlorobutane	$C_4H_8Cl_2$			-25[e]	6[e]	49.3	113.4	5
1,2-Dichlorobutane	$C_4H_8Cl_2$			-28.4	5.8	53.1	123.1	5
1,4-Dichlorobutane	$C_4H_8Cl_2$		-26[e]	0[e]	35[e]	82.4	153.4	5
2,2-Dichlorobutane	$C_4H_8Cl_2$		-58[e]	-35[e]	-5[e]	37.8	102.1	5
1,1-Dichlorocyclohexane	$C_6H_{10}Cl_2$	-39[e]	-19[e]	8[e]	43[e]	93.5	170.5	5
cis-1,2-Dichlorocyclohexane	$C_6H_{10}Cl_2$			27[e]	69[e]	125.7	206.2	5
1,2-Dichloro-1,1-difluoroethane	$C_2H_2Cl_2F_2$	-101[e]	-87[e]	-68[e]	-42.2	-6.8	46.3	5
Dichlorodifluoromethane[a]	CCl_2F_2	-150[e]	-138[e]	-122[e]	-101.8	-73.1	-30.0	5

Name	Mol. form.	*t*/°C for 1 Pa	*t*/°C for 10 Pa	*t*/°C for 100 Pa	*t*/°C for 1 kPa	*t*/°C for 10 kPa	*t*/°C for 100 kPa	Ref.
2,2'-Dichlorodiisopropyl ether	$C_6H_{12}Cl_2O$		-1[e]	27.3	63.4	112.3	182.1	5
Dichlorodimethylsilane	$C_2H_6Cl_2Si$					11.1	70.1	5
1,1-Dichloroethane	$C_2H_4Cl_2$		-84[e]	-64[e]	-36.7	1.0	56.9	1
1,2-Dichloroethane	$C_2H_4Cl_2$				-16.4	23.7	83.1	1
1,1-Dichloroethene	$C_2H_2Cl_2$	-116[e]	-101[e]	-82[e]	-57[e]	-21.4	31.2	1
cis-1,2-Dichloroethene	$C_2H_2Cl_2$			-62[e]	-34[e]	3.8	60.3	1
trans-1,2-Dichloroethene	$C_2H_2Cl_2$				-44[e]	-7.5	47.3	1
1,1-Dichloro-1-fluoroethane	$C_2H_3Cl_2F$		-101[e]	-83[e]	-57.9	-22.7	31.4	5
1,2-Dichloro-1-fluoroethane	$C_2H_3Cl_2F$			-50[e]	-23.8	14.1	73.4	5
Dichlorofluoromethane	$CHCl_2F$	-76[e]	-70[e]	-61[e]	-49[e]	-28.7	8.6	1
1,5-Dichloro-1,1,3,3,5,5-hexamethyltrisiloxane	$C_6H_{18}Cl_2O_2Si_3$	-29[e]	-7[e]	22.2	59.7	110.5	183.4	5
1,2-Dichlorohexane	$C_6H_{12}Cl_2$				49[e]	98.1	171.7	5
Dichloromethane[a]	CH_2Cl_2		-92[e]	-73[e]	-48[e]	-12.5	39.3	1
(Dichloromethyl)benzene	$C_7H_6Cl_2$			31	72	130	213	4
Dichloromethylphenylsilane	$C_7H_8Cl_2Si$			32.4	71.8	126.0	205.0	5
Dichloromethylsilane	CH_4Cl_2Si			-77[e]	-51[e]	-14[e]	40.5	1
1,2-Dichloropentane	$C_5H_{10}Cl_2$				30[e]	77.4	147.8	5
1,5-Dichloropentane	$C_5H_{10}Cl_2$	-31[e]	-10[e]	17[e]	54[e]	104.1	178.9	5
Dichlorophenylarsine	$C_6H_5AsCl_2$	6.9	35.2	70[e]	113[e]	170[e]	245[e]	5
1,1-Dichloropropane	$C_3H_6Cl_2$				-14[e]	27.0	87.7	5
1,2-Dichloropropane, (±)-	$C_3H_6Cl_2$	-78[e]	-61[e]	-38.1	-8.1	33.7	95.9	5
1,3-Dichloropropane	$C_3H_6Cl_2$	-65[e]	-46[e]	-22[e]	10[e]	54.0	119.9	5
2,2-Dichloropropane	$C_3H_6Cl_2$				-28[e]	10.8	68.9	5
1,3-Dichloro-2-propanol	$C_3H_6Cl_2O$			21.8	59.0	107.6	173.9	5
1,1-Dichloro-1,2,2,2-tetrafluoroethane	$C_2Cl_2F_4$					-45.4	2.7	5
1,2-Dichloro-1,1,2,2-tetrafluoroethane	$C_2Cl_2F_4$				-76.8	-44.9	3.2	5
1,3-Dichloro-1,1,3,3-tetramethyldisiloxane	$C_4H_{12}Cl_2OSi_2$		-33[e]	-9[e]	23.8	69.1	136.5	5
2,5-Dichlorothiophene	$C_4H_2Cl_2S$			-20[e]	22[e]	81.4	171[e]	5
2,4-Dichlorotoluene	$C_7H_6Cl_2$		6[e]	33[e]	68.3	119.5	199.1	5
3,4-Dichlorotoluene	$C_7H_6Cl_2$	-13[e]	9[e]	38[e]	76[e]	129.3	208.4	5
2,2-Dichloro-1,1,1-trifluoroethane	$C_2HCl_2F_3$		-101.0	-82.2	-57.4	-23.3	26.7	18
Diethanolamine	$C_4H_{11}NO_2$	53[e]	77[e]	107[e]	146[e]	197.3	268[e]	5
Diethoxydimethylsilane	$C_6H_{16}O_2Si$	-62[e]	-44[e]	-21.2	9.1	51.0	113.0	5
1,1-Diethoxyethane	$C_6H_{14}O_2$	-68[e]	-49[e]	-26[e]	3.7	44.2	101.9	5
1,2-Diethoxyethane	$C_6H_{14}O_2$		-59[e]	-35.3	-2.8	44.4	118.8	5
Diethoxymethane	$C_5H_{12}O_2$		-65[e]	-43[e]	-14[e]	27.3	87.7	5
Diethylamine	$C_4H_{11}N$			-46[e]	-26[e]	5[e]	55.2	1
2-Diethylaminoethanol	$C_6H_{15}NO$					97[e]	160.6	5
N,N-Diethylaniline	$C_{10}H_{15}N$	-11[e]	14[e]	44.3	84.2	138.4	216.3	5
o-Diethylbenzene	$C_{10}H_{14}$	-28[e]	-6[e]	21[e]	58[e]	107.9	182.9	5
m-Diethylbenzene	$C_{10}H_{14}$	-28[e]	-7[e]	20[e]	56[e]	106.2	180.6	5
p-Diethylbenzene	$C_{10}H_{14}$	-28[e]	-6[e]	21[e]	57[e]	108.1	183.3	5
Diethyl carbonate	$C_5H_{10}O_3$		-42[e]	-17[e]	17[e]	61.6	125.9	5
Diethyl disulfide	$C_4H_{10}S_2$	-46[e]	-26[e]	0[e]	35[e]	82.4	153.5	5
Diethylene glycol	$C_4H_{10}O_3$	35[e]	58[e]	86[e]	123[e]	173.6	245.2	1
Diethylene glycol dibutyl ether	$C_{12}H_{26}O_3$	5[e]	34.4	70.2	115.3	174.1	253.8	5
Diethylene glycol diethyl ether	$C_8H_{18}O_3$	-32[e]	-7[e]	25[e]	64.9	117.1	189[e]	5
Diethylene glycol dimethyl ether	$C_6H_{14}O_3$	-42[e]	-20[e]	8.3	44.3	92.3	159.4	5
Diethylene glycol monobutyl ether	$C_8H_{18}O_3$	14[e]	37[e]	66.8	104.9	153[e]	230.4	5
Diethylene glycol monobutyl ether acetate	$C_{10}H_{20}O_4$	6[e]	34[e]	69[e]	112.6	169.2	245.4	5
Diethylene glycol monoethyl ether	$C_6H_{14}O_3$			40[e]	80.3	132.4	201.4	5
Diethylene glycol monoethyl ether acetate	$C_8H_{16}O_4$	-16[e]	10.6	43.9	86.2	141.3	216.6	5
Diethylene glycol monomethyl ether	$C_5H_{12}O_3$		12[e]	40[e]	76[e]	124.2	193.7	1
Diethyl ether	$C_4H_{10}O$	-111[e]	-96[e]	-77[e]	-52.6	-17.8	34.1	1
Diethyl glutarate	$C_9H_{16}O_4$	-1[e]	26[e]	60.2	103.3	159.6	236.5	5
Diethyl hexanedioate	$C_{10}H_{18}O_4$	4[e]	35[e]	72[e]	116.6	171.2	239.5	5
Diethyl maleate	$C_8H_{12}O_4$	-6[e]	20[e]	52.2	93.5	148.4	224.8	5
Diethyl malonate	$C_7H_{12}O_4$	-23[e]	4[e]	36.0	76.4	128.5	198.3	5
1,3-Diethyl-5-methylbenzene	$C_{11}H_{16}$	-26[e]	-1[e]	29.5	69.5	123.5	200.2	5
Diethyl oxalate	$C_6H_{10}O_4$	-5[e]	18[e]	44.9	79.4	124.3	185.2	5

Name	Mol. form.	t/°C for 1 Pa	t/°C for 10 Pa	t/°C for 100 Pa	t/°C for 1 kPa	t/°C for 10 kPa	t/°C for 100 kPa	Ref.
3,3-Diethylpentane	C_9H_{20}			-9[e]	26[e]	73.7	145.7	1
Diethylperoxide	$C_4H_{10}O_2$				-39[e]	3.6	65.0	5
Diethyl phthalate	$C_{12}H_{14}O_4$	12[e]	51[e]	96[e]	150.5	215.9	296.2	5
N,N-Diethyl-1,3-propanediamine	$C_7H_{18}N_2$				50.1	99.9	167.7	5
Diethyl sebacate	$C_{14}H_{26}O_4$		83[e]	120	166	225	305	4
Diethyl succinate	$C_8H_{14}O_4$	-6[e]	20[e]	51.0	91.1	143.7	216.1	5
Diethyl sulfate	$C_4H_{10}O_4S$		3[e]	36[e]	79[e]	134[e]	208.3	5
Diethyl sulfide	$C_4H_{10}S$	-80[e]	-62[e]	-40[e]	-10.8	30.3	91.7	1
1,1-Difluoroethane	$C_2H_4F_2$			-115.2	-94.6	-66.1	-24.3	19
Difluoromethane[a]	CH_2F_2	-156.7	-145.8	-131.9	-113.6	-88.6	-51.9	1
3,4-Dihydro-2*H*-pyran	C_5H_8O				-22[e]	22.0	84.9	5
Diiodomethane	CH_2I_2			17[e]	55[e]	106.1	181.6	5
Diiodosilane	H_2I_2Si				11.8	70.5	149.4	4
Diisobutylamine	$C_8H_{19}N$	-57[e]	-36[e]	-9.0	25.5	72.2	139.0	5
Diisopentyl ether	$C_{10}H_{22}O$			14.0	51.5	101.8	172.8	5
Diisopentyl sulfide	$C_{10}H_{22}S$			7[e]	82[e]	118[e]	139[e]	5
Diisopropylamine	$C_6H_{15}N$			-47[e]	-17.5	23.5	84.0	5
1,2-Diisopropylbenzene	$C_{12}H_{18}$	-14[e]	9[e]	37[e]	74[e]	125.9	203.2	5
1,3-Diisopropylbenzene	$C_{12}H_{18}$	-14[e]	8[e]	36[e]	74[e]	125.5	202.6	5
1,4-Diisopropylbenzene	$C_{12}H_{18}$	-6[e]	18[e]	49[e]	90[e]	148.8	238[e]	5
Diisopropyl ether	$C_6H_{14}O$		-76[e]	-55[e]	-28[e]	11[e]	68.1	1
Diisopropyl sulfide	$C_6H_{14}S$	-65[e]	-47[e]	-23[e]	9[e]	53.1	119.6	5
Diketene	$C_4H_4O_2$				19.3	63.3	126[e]	5
1,3-Dimethoxybenzene	$C_8H_{10}O_2$	18[e]	34[e]	56[e]	86.7	135.5	223[e]	5
Dimethoxyborane	$C_2H_7BO_2$	-116[e]	-101.9	-83.5	-59.2	-25.4	25[e]	5
1,2-Dimethoxyethane	$C_4H_{10}O_2$			-44[e]	-15[e]	25.2	85.2	1
Dimethoxymethane	$C_3H_8O_2$	-93[e]	-81[e]	-64[e]	-42[e]	-9.3	41.7	5
Dimethylacetal	$C_4H_{10}O_2$	-89[e]	-74[e]	-55[e]	-29[e]	7.7	64.1	5
N,N-Dimethylacetamide	C_4H_9NO	-8[e]	8[e]	28.0	56.4	98.2	165.7	1
Dimethylamine	C_2H_7N			-88[e]	-66.9	-37.2	6.6	1
(Dimethylamino)dimethylborane	$C_4H_{12}BN$		-81[e]	-60.1	-31.9	7.0	64.2	5
3-(Dimethylamino)propanenitrile	$C_5H_{10}N_2$				51.1	101.8	171.4	5
2,4-Dimethylaniline	$C_8H_{11}N$	-2[e]	21[e]	51[e]	88[e]	139.1	210.9	5
2,6-Dimethylaniline	$C_8H_{11}N$			37[e]	80[e]	137.7	217.7	5
N,N-Dimethylaniline	$C_8H_{11}N$			28[e]	66[e]	118.1	193.6	1
2,4-Dimethylbenzaldehyde	$C_9H_{10}O$	-3[e]	23[e]	54[e]	93.2	144.6	214.5	5
2,3-Dimethyl-1,3-butadiene	C_6H_{10}			-59[e]	-30[e]	9.7	68.1	5
2,2-Dimethylbutane	C_6H_{14}		-90[e]	-71.5	-45.5	-7.7	49.4	1
2,3-Dimethylbutane	C_6H_{14}	-103[e]	-87[e]	-66[e]	-39.0	-0.4	57.6	1
3,3-Dimethyl-1-butanol	$C_6H_{14}O$	-37[e]	-16[e]	9[e]	42[e]	84.3	142.5	5
2,3-Dimethyl-2-butanol	$C_6H_{14}O$			-5[e]	23[e]	61.3	118.2	5
3,3-Dimethyl-2-butanone	$C_6H_{12}O$			-30[e]	0[e]	42.5	105.7	1
2,3-Dimethyl-1-butene	C_6H_{12}	-103[e]	-87[e]	-67[e]	-39.9	-1.9	55.2	5
3,3-Dimethyl-1-butene	C_6H_{12}	-110[e]	-95[e]	-76[e]	-50.8	-14.5	40.8	5
2,3-Dimethyl-2-butene	C_6H_{12}		-75[e]	-54[e]	-25[e]	14[e]	72.9	1
1,1-Dimethylcyclohexane	C_8H_{16}			-27[e]	5[e]	50.6	119.1	5
cis-1,2-Dimethylcyclohexane	C_8H_{16}		-44[e]	-20[e]	14[e]	59.7	129.2	5
trans-1,2-Dimethylcyclohexane	C_8H_{16}	-68[e]	-49[e]	-25[e]	8[e]	53.9	122.9	5
cis-1,3-Dimethylcyclohexane	C_8H_{16}	-68[e]	-48[e]	-23[e]	10[e]	55.6	123.1	5
trans-1,3-Dimethylcyclohexane	C_8H_{16}	-62[e]	-45[e]	-23[e]	8[e]	51.5	120.9	5
cis-1,4-Dimethylcyclohexane	C_8H_{16}	-66[e]	-47[e]	-23[e]	10[e]	55.3	123.8	5
trans-1,4-Dimethylcyclohexane	C_8H_{16}			-27[e]	5[e]	50.6	118.9	5
1,1-Dimethylcyclopentane	C_7H_{14}		-69[e]	-47[e]	-17[e]	24.8	87.4	5
cis-1,2-Dimethylcyclopentane	C_7H_{14}			-38[e]	-8[e]	34.9	99.0	5
trans-1,2-Dimethylcyclopentane	C_7H_{14}	-83[e]	-66[e]	-43[e]	-13[e]	28.4	91.4	5
cis-1,3-Dimethylcyclopentane	C_7H_{14}	-84[e]	-66[e]	-44[e]	-14[e]	28.2	91.1	5
trans-1,3-Dimethylcyclopentane	C_7H_{14}	-84[e]	-67[e]	-44[e]	-14[e]	27.4	90.3	5
1,2-Dimethylcyclopentene	C_7H_{12}	-75[e]	-57[e]	-34[e]	-3[e]	40.2	105.3	5
1,5-Dimethylcyclopentene	C_7H_{12}	-77[e]	-59[e]	-36[e]	-5.5	37.3	101.5	5
cis-1,2-Dimethylcyclopropane	C_5H_{10}	-118[e]	-103[e]	-83[e]	-57[e]	-20[e]	36.6	5

Name	Mol. form.	t/°C for 1 Pa	t/°C for 10 Pa	t/°C for 100 Pa	t/°C for 1 kPa	t/°C for 10 kPa	t/°C for 100 kPa	Ref.
trans-1,2-Dimethylcyclopropane	C_5H_{10}	-122[e]	-108[e]	-89[e]	-63[e]	-27[e]	27.8	5
Dimethyl disulfide	$C_2H_6S_2$	-71[e]	-53[e]	-29[e]	1.7	45.0	109.3	5
N,N-Dimethylethanolamine	$C_4H_{11}NO$	-52[e]	-31[e]	-6[e]	27[e]	70.9	133[e]	5
Dimethyl ether[a]	C_2H_6O		-135[e]	-118[e]	-96.8	-67.6	-25.1	1,5
N,N-Dimethylformamide	C_3H_7NO	-39[e]	-20[e]	5[e]	38.0	83.9	152.6	1
Dimethyl glutarate	$C_7H_{12}O_4$	-11[e]	15[e]	47[e]	87.7	139.8	209.5	5
2,2-Dimethylheptane	C_9H_{20}	-58[e]	-39[e]	-15[e]	18[e]	63.6	132.3	5
2,3-Dimethylheptane	C_9H_{20}	-53[e]	-33[e]	-9[e]	25[e]	70.8	140.0	5
2,6-Dimethylheptane	C_9H_{20}	-55[e]	-36[e]	-12[e]	21[e]	66.4	134.7	5
2,6-Dimethyl-4-heptanone	$C_9H_{18}O$	-32[e]	-12[e]	14[e]	48[e]	96.2	167.7	5
2,5-Dimethyl-1,5-hexadiene	C_8H_{14}	-38[e]	-26[e]	-10[e]	14[e]	50.8	115.1	5
2,2-Dimethylhexane	C_8H_{18}	-73[e]	-55[e]	-32[e]	-1.5	41.6	106.4	5
2,3-Dimethylhexane	C_8H_{18}				5[e]	49.2	115.1	5
2,4-Dimethylhexane	C_8H_{18}				0.6	43.9	109.0	5
2,5-Dimethylhexane	C_8H_{18}	-71[e]	-53[e]	-30[e]	0.7	43.8	108.6	5
3,3-Dimethylhexane	C_8H_{18}	-72[e]	-54[e]	-30[e]	1.4	45.4	111.5	5
3,4-Dimethylhexane	C_8H_{18}				7[e]	50.9	117.3	5
Dimethyl 1,6-hexanedioate	$C_8H_{14}O_4$		28[e]	61[e]	103[e]	156.1	227.3	5
2,3-Dimethyl-2-hexene	C_8H_{16}	-65[e]	-47[e]	-23[e]	10[e]	54.3	121.3	5
cis-2,2-Dimethyl-3-hexene	C_8H_{16}	-74[e]	-56[e]	-33[e]	-3[e]	40.1	105.0	5
1,1-Dimethylhydrazine	$C_2H_8N_2$			-52[e]	-25.6	10.5	63[e]	5
1,2-Dimethylhydrazine	$C_2H_8N_2$		-49[e]	-33[e]	-9[e]	26.4	88[e]	1
Dimethyl isophthalate	$C_{10}H_{10}O_4$			85[e]	129.5	189.2	273[e]	5
2,4-Dimethyl-3-isopropylpentane	$C_{10}H_{22}$	-46[e]	-26[e]	0[e]	35[e]	83.2	156.5	5
Dimethyl maleate	$C_6H_8O_4$		5[e]	36[e]	76[e]	127.3	197[e]	5
Dimethyl malonate	$C_5H_8O_4$	-22[e]	1[e]	30.0	66.7	114.7	180.2	5
Dimethyl mercury	C_2H_6Hg				-13.5	29.0	92.1	5
1,2-Dimethylnaphthalene	$C_{12}H_{12}$	26[e]	51[e]	82[e]	123[e]	180.5	265.7	5
2,7-Dimethylnaphthalene	$C_{12}H_{12}$	31.5 s	53.1 s	78.8 s	115.9	175[e]	260[e]	5
2,4-Dimethyloctane	$C_{10}H_{22}$				38[e]	84.9	155.4	5
2,7-Dimethyloctane	$C_{10}H_{22}$	-39[e]	-19[e]	7[e]	41[e]	88.4	159.4	5
Dimethyl oxalate	$C_4H_6O_4$				50.5	98.1	163.0	5
2,2-Dimethylpentane	C_7H_{16}	-90[e]	-73[e]	-52[e]	-22.9	17.6	78.8	1
2,3-Dimethylpentane	C_7H_{16}	-87[e]	-68.4	-45.3	-14.9	26.8	89.3	5
2,4-Dimethylpentane	C_7H_{16}	-89[e]	-72[e]	-50[e]	-21.3	19.2	80.1	1
3,3-Dimethylpentane	C_7H_{16}	-88[e]	-71[e]	-49[e]	-18.8	22.9	85.6	1
2,2-Dimethyl-3-pentanol	$C_7H_{16}O$			9[e]	35[e]	73.1	135.5	5
2,4-Dimethyl-3-pentanone	$C_7H_{14}O$	-61[e]	-42[e]	-18[e]	14[e]	58.5	124.8	1
2,3-Dimethyl-1-pentene	C_7H_{14}	-85[e]	-68[e]	-46[e]	-17[e]	23.4	83.8	5
2,4-Dimethyl-1-pentene	C_7H_{14}	-88[e]	-71[e]	-50[e]	-21[e]	20.0	81.2	5
3,3-Dimethyl-1-pentene	C_7H_{14}	-87[e]	-71[e]	-50[e]	-21[e]	18.1	77.1	5
4,4-Dimethyl-1-pentene	C_7H_{14}	-94[e]	-78[e]	-57[e]	-28[e]	11.5	72.1	5
2,3-Dimethyl-2-pentene	C_7H_{14}	-79[e]	-62[e]	-39[e]	-9[e]	33.5	96.9	5
2,4-Dimethyl-2-pentene	C_7H_{14}	-84[e]	-68[e]	-46[e]	-18[e]	22.6	82.9	5
cis-3,4-Dimethyl-2-pentene	C_7H_{14}	-83[e]	-65[e]	-43[e]	-14[e]	27.2	88.8	5
trans-3,4-Dimethyl-2-pentene	C_7H_{14}	-82[e]	-64[e]	-42[e]	-13[e]	29.0	91.1	5
cis-4,4-Dimethyl-2-pentene	C_7H_{14}	-90[e]	-73[e]	-51[e]	-22[e]	18.6	80.0	5
trans-4,4-Dimethyl-2-pentene	C_7H_{14}	-90[e]	-73[e]	-52[e]	-23[e]	16.6	76.3	5
4,4-Dimethyl-1-pentyne	C_7H_{12}		-73[e]	-52[e]	-24[e]	15.9	75.6	5
4,4-Dimethyl-2-pentyne	C_7H_{12}		-70[e]	-48[e]	-19[e]	21.4	82.6	5
Dimethyl phthalate	$C_{10}H_{10}O_4$	27[e]	56[e]	92.7	137.8	195.8	272.7	5
2,2-Dimethylpropanenitrile	C_5H_9N					41.1	104.8	5
2,2-Dimethyl-1-propanol	$C_5H_{12}O$					59.2	112.7	5
2,3-Dimethylpyridine	C_7H_9N				42[e]	89.9	160.6	5
2,4-Dimethylpyridine	C_7H_9N		-25[e]	3.7	40.0	87.5	157.9	1,5
2,5-Dimethylpyridine	C_7H_9N			4[e]	39[e]	86.2	156.6	1
2,6-Dimethylpyridine	C_7H_9N			-3[e]	29.9	75.8	143.6	1
3,4-Dimethylpyridine	C_7H_9N		-9[e]	19[e]	55[e]	104.8	178.6	5
3,5-Dimethylpyridine	C_7H_9N			11[e]	48[e]	98[e]	171.5	1
Dimethyl sebacate	$C_{12}H_{22}O_4$		53[e]	97	150	214	293	4

Name	Mol. form.	*t*/°C for 1 Pa	*t*/°C for 10 Pa	*t*/°C for 100 Pa	*t*/°C for 1 kPa	*t*/°C for 10 kPa	*t*/°C for 100 kPa	Ref.
Dimethyl succinate	$C_6H_{10}O_4$			30[e]	70.4	123.3	195.4	5
Dimethyl sulfide	C_2H_6S		-96[e]	-77[e]	-51.2	-16.0	37.0	1,5
Dimethyl sulfone	$C_2H_6O_2S$				109[e]	166.8	248.9	5
Dimethyl sulfoxide	C_2H_6OS			27.4	65.0	115.9	188.6	1
Dimethyl terephthalate	$C_{10}H_{10}O_4$	56.6 s	79.4 s	106.1 s	137.9 s	197.9	282[e]	5
2,5-Dimethylthiophene	C_6H_8S		-43[e]	-16[e]	20[e]	67.5	134.8	5
1,1-Dinitropropane	$C_3H_6N_2O_4$	-9[e]	12[e]	39[e]	73.2	120[e]	187[e]	5
Dioctyl phthalate	$C_{24}H_{38}O_4$	130[e]	163.7	203.8	252[e]	311[e]	385[e]	5
1,3-Dioxane	$C_4H_8O_2$			-37[e]	-3[e]	43.4	106.0	5
1,4-Dioxane	$C_4H_8O_2$					39.6	101.0	1
1,3-Dioxolane	$C_3H_6O_2$		-72[e]	-50[e]	-22[e]	17.0	75.3	1
Dipentene	$C_{10}H_{16}$	-42[e]	-19[e]	10.6	48.7	100.2	173.9	5
Dipentylamine	$C_{10}H_{23}N$				77[e]	127.7	202.0	5
Dipentyl ether	$C_{10}H_{22}O$	-31[e]	-8[e]	22[e]	60[e]	111.6	186.2	5
Diphenylamine	$C_{12}H_{11}N$	48 s		102.8	150.5	213.7	301.4	5
1,1-Diphenylethane	$C_{14}H_{14}$	19[e]	47[e]	82.0	125.3	181[e]	254[e]	5
Diphenyl ether	$C_{12}H_{10}O$		44[e]	75[e]	116[e]	173[e]	257.4	5
Diphenylmethane	$C_{13}H_{12}$		45[e]	77[e]	119.3	177.7	263.6	1,5
Diphenyl sulfide	$C_{12}H_{10}S$	20[e]	51[e]	88.7	137.5	202.2	291.8	5
1,2-Dipropoxyethane	$C_8H_{18}O_2$			-44.2	-2.0	63.6	179.2	5
Dipropylamine	$C_6H_{15}N$		-48[e]	-25[e]	6[e]	47.5	108.8	5
Dipropylene glycol	$C_6H_{14}O_3$				110[e]	162.6	231.4	5
Dipropyl ether	$C_6H_{14}O$	-80[e]	-63[e]	-41[e]	-12[e]	28.8	89.7	1
Dipropyl oxalate	$C_8H_{14}O_4$	-4[e]	20[e]	49.9	88.6	140.4	213.0	5
Dipropyl succinate	$C_{10}H_{18}O_4$	11[e]	38[e]	72.1	115.4	172.3	250.4	5
Dipropyl sulfide	$C_6H_{14}S$	-50[e]	-30[e]	-6[e]	28[e]	73.6	142.4	5
m-Divinylbenzene	$C_{10}H_{10}$	-29[e]	-4[e]	27.1	67.6	122.1	199[e]	5
Divinyl ether	C_4H_6O		-99[e]	-80[e]	-56[e]	-22.1	28.0	5
Docosane	$C_{22}H_{46}$	83.5	115.0	154.0	203.6	274.8	368.0	5
Docosanoic acid	$C_{22}H_{44}O_2$	145.4	176.5	213.7	259.3	316.2	390[e]	5
cis-13-Docosenoic acid	$C_{22}H_{42}O_2$	126[e]	160[e]	199.4	247.4	306.5	381.1	5
trans-13-Docosenoic acid	$C_{22}H_{42}O_2$	134[e]	166[e]	203.6	249.8	307.6	382.0	5
Dodecamethylcyclohexasiloxane	$C_{12}H_{36}O_6Si_6$	18[e]	41[e]	69[e]	108[e]	162.2	244.7	5
Dodecanal	$C_{12}H_{24}O$			70[e]	116.2	175.9	256.6	5
Dodecane	$C_{12}H_{26}$	-5.4	18.2	47.6	85.8	138.2	215.8	16
Dodecanenitrile	$C_{12}H_{23}N$	36[e]	60[e]	92[e]	133[e]	190.5	275.5	5
Dodecanoic acid	$C_{12}H_{24}O_2$	78[e]	100[e]	128[e]	166[e]	219.1	298.1	5
1-Dodecanol	$C_{12}H_{26}O$				133[e]	185.0	264.1	1
1-Dodecene	$C_{12}H_{24}$	-8.3	15.2	44.8	82.9	135.4	212.8	5
1-Dodecyne	$C_{12}H_{22}$	-11[e]	13[e]	43[e]	82[e]	135.8	214.4	5
Dysprosium	Dy	1105 s	1250 s	1431[d]	1681[d]	2031[d]	2558[d]	3
Eicosamethylnonasiloxane	$C_{20}H_{60}O_8Si_9$			141[e]	183.1	236.7	307.1	5
Eicosane	$C_{20}H_{42}$	80.4	108.9	144.2	189.8	252.1	344[e]	16
1-Eicosanol	$C_{20}H_{42}O$	119[e]	143[e]	173[e]	213[e]	270.0	355.1	5
Epichlorohydrin	C_3H_5ClO			-21[e]	11[e]	53.8	115.5	5
1,2-Epoxybutane	C_4H_8O	-135[e]	-114[e]	-87[e]	-53[e]	-5.5	62.1	5
Erbium	Er	1231 s	1390 s	1612[d]	1890[d]	2279[d]	2859[d]	3
Ethane[a]	C_2H_6	-183.3 s	-173.2	-161.3	-145.3	-122.8	-88.8	41
1,2-Ethanediamine	$C_2H_8N_2$				17.0	57.5	116.6	1,5
1,2-Ethanediol	$C_2H_6O_2$	2[e]	24[e]	51.1	86.1	132.5	196.9	1
1,2-Ethanediol, diacetate	$C_6H_{10}O_4$	-17[e]	6[e]	35.0	71.9	121.1	190.0	5
1,2-Ethanediol, dinitrate	$C_2H_4N_2O_6$	4[e]	25.6	51.0	81[e]	117[e]	162[e]	5
Ethanethiol	C_2H_6S	-112[e]	-97[e]	-78[e]	-53[e]	-18[e]	34.7	1
Ethanol	C_2H_6O	-73[e]	-56[e]	-34[e]	-7[e]	29.2	78.0	1,5
Ethanolamine	C_2H_7NO		11[e]	35[e]	66.2	109.0	170.6	1
2-Ethoxyaniline	$C_8H_{11}NO$	0[e]	27[e]	60[e]	102.2	156.0	228.1	5
Ethoxybenzene	$C_8H_{10}O$		-9[e]	17[e]	51[e]	99[e]	169.3	5
2-Ethoxyethanol	$C_4H_{10}O_2$	-49[e]	-29[e]	-3[e]	30[e]	73.6	135.3	1
2-Ethoxyethyl acetate	$C_6H_{12}O_3$	-25[e]	-8[e]	14[e]	44.6	88.0	155.6	5
Ethyl acetate	$C_4H_8O_2$	-83[e]	-66[e]	-45[e]	-18[e]	20.4	76.8	1

Name	Mol. form.	t/°C for 1 Pa	t/°C for 10 Pa	t/°C for 100 Pa	t/°C for 1 kPa	t/°C for 10 kPa	t/°C for 100 kPa	Ref.
Ethyl acetoacetate	$C_6H_{10}O_3$	-25[e]	-3[e]	25.7	62.3	111.3	180.2	5
Ethyl acrylate	$C_5H_8O_2$		-55[e]	-32.7	-2.8	38.5	99.2	5
Ethylamine	C_2H_7N			-71[e]	-53[e]	-27[e]	16.4	1
4-Ethylaniline	$C_8H_{11}N$	-2[e]	21[e]	49[e]	87[e]	139.4	216.7	5
N-Ethylaniline	$C_8H_{11}N$	-15[e]	8[e]	38[e]	76.4	128.8	204.2	5
Ethylbenzene	C_8H_{10}	-56.2	-36.8	-12.0	21.1	67.1	135.7	1
Ethyl benzoate	$C_9H_{10}O_2$	-18[e]	8[e]	39[e]	80.1	135.1	212.8	5
Ethyl butanoate	$C_6H_{12}O_2$	-49[e]	-34[e]	-14[e]	14.3	55.2	121.1	5
2-Ethylbutanoic acid	$C_6H_{12}O_2$	-9[e]	16[e]	46[e]	83[e]	130.7	192.5	5
2-Ethyl-1-butanol	$C_6H_{14}O$		-5[e]	17[e]	46[e]	85.7	146.1	5
2-Ethyl-1-butene	C_6H_{12}	-98[e]	-81[e]	-60[e]	-32[e]	6.6	64.3	5
Ethyl chloroacetate	$C_4H_7ClO_2$			-2.6	32.6	79.1	143.8	5
Ethyl 2-chloropropanoate	$C_5H_9ClO_2$			1.4	36.4	82.5	146.0	5
Ethyl *trans*-cinnamate	$C_{11}H_{12}O_2$			79	125	187	271	4
Ethylcyanoacetate	$C_5H_7NO_2$	16[e]	39[e]	67.0	102.1	146.7	205.6	5
Ethylcyclobutane	C_6H_{12}	-99[e]	-82[e]	-61[e]	-32[e]	9[e]	70.2	5
Ethylcyclohexane	C_8H_{16}	-61[e]	-42[e]	-17[e]	15.8	61.9	131.3	5
1-Ethylcyclohexene	C_8H_{14}	-55[e]	-35[e]	-11[e]	22[e]	68[e]	136.5	5
Ethylcyclopentane	C_7H_{14}	-76[e]	-59[e]	-35[e]	-5[e]	38.4	103.0	5
1-Ethylcyclopentene	C_7H_{12}	-75[e]	-57[e]	-34[e]	-3[e]	40.7	105.8	5
Ethylcyclopropane	C_5H_{10}	-118[e]	-102[e]	-83[e]	-57[e]	-20[e]	35.5	5
Ethyl decanoate	$C_{12}H_{24}O_2$	8[e]	35[e]	69[e]	111.8	166.1	238[e]	5
Ethyl dichloroacetate	$C_4H_6Cl_2O_2$			2.6	40.1	89.1	156.3	5
Ethyl diethylmalonate	$C_{11}H_{20}O_4$			74[e]	105[e]	149.4	219[e]	5
Ethyldifluoroarsine	$C_2H_5AsF_2$			-36[e]	-6.0	35.0	93.1	5
1-Ethyl-2,4-dimethylbenzene	$C_{10}H_{14}$	-25[e]	-4[e]	24[e]	61[e]	112.2	187.9	5
1-Ethyl-3,5-dimethylbenzene	$C_{10}H_{14}$	-28[e]	-6[e]	21[e]	58[e]	108.3	183.2	5
2-Ethyl-1,3-dimethylbenzene	$C_{10}H_{14}$		-2[e]	26[e]	63[e]	113.7	189.5	5
2-Ethyl-1,4-dimethylbenzene	$C_{10}H_{14}$	-27[e]	-5[e]	23[e]	60[e]	110.6	186.4	5
3-Ethyl-1,2-dimethylbenzene	$C_{10}H_{14}$	-22[e]	0[e]	28[e]	66[e]	117.2	193.4	5
4-Ethyl-1,2-dimethylbenzene	$C_{10}H_{14}$	-24[e]	-2[e]	26[e]	63[e]	113.6	189.2	5
3-Ethyl-2,4-dimethylpentane	C_9H_{20}	-58[e]	-38[e]	-13[e]	20[e]	66.7	136.2	5
Ethylene[a]	C_2H_4				-155.6	-135.1	-104.0	1,10
Ethylene carbonate	$C_3H_4O_3$	12.7 s	37[e]				247	5
Ethyleneimine	C_2H_5N		-74[e]	-55[e]	-30[e]	4.1	55[e]	5
Ethyl formate	$C_3H_6O_2$		-80[e]	-61[e]	-35[e]	1[e]	54.0	1
3-Ethylhexane	C_8H_{18}				8[e]	52.1	118.1	5
Ethyl hexanoate	$C_8H_{16}O_2$	-31[e]	-9[e]	18.7	53.9	100.7	166.2	5
2-Ethylhexanoic acid	$C_8H_{16}O_2$				108[e]	159.6	226.6	5
2-Ethyl-1-hexanol	$C_8H_{18}O$			45[e]	75[e]	118.3	184.2	1
2-Ethylhexyl acetate	$C_{10}H_{20}O_2$	-11[e]	5[e]	26[e]	57.6	107.1	197.2	5
Ethyl hydroperoxide	$C_2H_6O_2$	-70[e]	-49[e]	-25[e]	6.8	47.0	101[e]	5
Ethyl isopropyl sulfide	$C_5H_{12}S$	-72[e]	-54[e]	-31[e]	0[e]	42.7	106.9	5
Ethyl isothiocyanate	C_3H_5NS				17.4	66[e]	136[e]	5
Ethyl levulinate	$C_7H_{12}O_3$		17[e]	45.3	82.6	133.2	205.7	5
Ethyl methacrylate	$C_6H_{10}O_2$				8[e]	53.2	116.8	5
Ethyl 3-methylbutanoate	$C_7H_{14}O_2$	-57[e]	-36[e]	-10[e]	23.9	69.5	134.4	5
trans-1-Ethyl-4-methylcyclohexane	C_9H_{18}	-53[e]	-33[e]	-8[e]	25[e]	71.8	141.5	5
1-Ethyl-1-methylcyclopentane	C_8H_{16}	-67[e]	-49[e]	-24[e]	8[e]	53.2	121.0	5
cis-1-Ethyl-2-methylcyclopentane	C_8H_{16}	-63[e]	-44[e]	-19[e]	13.3	59.1	127.6	5
1-Ethyl-1-methylcyclopropane	C_6H_{12}	-105[e]	-89[e]	-69[e]	-41[e]	-3[e]	56.3	5
Ethyl methyl ether	C_3H_8O	-98[e]	-89[e]	-77[e]	-60[e]	-34.8	7.0	5
3-Ethyl-4-methylhexane	C_9H_{20}			-9[e]	24[e]	70.6	139.9	5
3-Ethyl-2-methylpentane	C_8H_{18}	-69[e]	-50[e]	-27[e]	5[e]	48.9	115.2	5
3-Ethyl-3-methylpentane	C_8H_{18}	-70[e]	-51[e]	-27[e]	5[e]	50.2	117.8	5
Ethyl 2-methylpropanoate	$C_6H_{12}O_2$	-65[e]	-47[e]	-24.6	5.4	47.3	109.8	5
Ethyl methyl sulfide	C_3H_8S	-94[e]	-78[e]	-57[e]	-29.7	8.8	66.3	1
1-Ethylnaphthalene	$C_{12}H_{12}$	16[e]	41[e]	72[e]	114[e]	171.8	257.7	5
2-Ethylnaphthalene	$C_{12}H_{12}$	14[e]	39[e]	71[e]	113[e]	171.2	257.3	5
Ethyl nitrate	$C_2H_5NO_3$	-81[e]	-63[e]	-41[e]	-12[e]	28.2	87[e]	1

Name	Mol. form.	*t*/°C for 1 Pa	*t*/°C for 10 Pa	*t*/°C for 100 Pa	*t*/°C for 1 kPa	*t*/°C for 10 kPa	*t*/°C for 100 kPa	Ref.
1-Ethyl-4-nitrobenzene	$C_8H_9NO_2$	10[e]	36[e]	69[e]	111.6	168[e]	245[e]	5
Ethyl octanoate	$C_{10}H_{20}O_2$	-17[e]	9[e]	41[e]	81.4	133.2	203[e]	5
3-Ethylpentane	C_7H_{16}	-81[e]	-63[e]	-41[e]	-11[e]	30.5	93.1	1
3-Ethyl-1-pentene	C_7H_{14}	-85[e]	-68[e]	-46[e]	-17[e]	23.2	83.7	5
2-Ethylphenol	$C_8H_{10}O$		16.9	44.5	81.1	130.9	204.0	5
3-Ethylphenol	$C_8H_{10}O$	5.6	29.2	57.5	91.9	144.8	217.9	5
4-Ethylphenol	$C_8H_{10}O$			60[e]	95.5	144.6	217.5	5
Ethyl phenylacetate	$C_{10}H_{12}O_2$	-9[e]	19[e]	52[e]	95[e]	150.2	225[e]	5
5-Ethyl-2-picoline	$C_8H_{11}N$	-33[e]	-9.3	20[e]			178.0	5
Ethyl propanoate	$C_5H_{10}O_2$	-69[e]	-52[e]	-30[e]	-1[e]	38.9	98.7	1
Ethyl propyl ether	$C_5H_{12}O$	-92[e]	-77[e]	-57[e]	-30.5	6.7	63.4	1,5
Ethyl propyl sulfide	$C_5H_{12}S$	-64[e]	-46[e]	-23[e]	9[e]	52.7	118.0	5
2-Ethylpyridine	C_7H_9N	-46[e]	-26[e]	-1[e]	33[e]	79.3	149.0	5
3-Ethylpyridine	C_7H_9N	-38[e]	-17[e]	9[e]	44[e]	92.7	166.5	5
4-Ethylpyridine	C_7H_9N	-35[e]	-15[e]	11[e]	46[e]	94.4	168.6	5
Ethyl silicate	$C_8H_{20}O_4Si$	-77[e]	-52[e]	-21[e]	21.6	80.5	164.1	5
2-Ethylstyrene	$C_{10}H_{12}$	-31[e]	-8[e]	21[e]	60[e]	111.7	187[e]	5
3-Ethylstyrene	$C_{10}H_{12}$	-28[e]	-5.3	24.1	62.6	116[e]	193[e]	5
4-Ethylstyrene	$C_{10}H_{12}$	-31[e]	-8.2	21.3	60.5	115[e]	196[e]	5
Ethyl thiocyanate	C_3H_5NS	-39[e]	-20[e]	4[e]	35[e]	79.1	143.4	5
2-Ethyltoluene	C_9H_{12}	-40[e]	-19[e]	8[e]	43[e]	92.1	164.7	5
3-Ethyltoluene	C_9H_{12}	-42[e]	-21[e]	5[e]	40.4	88.9	160.8	5
4-Ethyltoluene	C_9H_{12}	-41[e]	-21[e]	6[e]	41[e]	89.2	161.5	5
Ethyl trichloroacetate	$C_4H_5Cl_3O_2$			15.3	51.9	100.1	166.6	5
1-Ethyl-2,4,5-trimethylbenzene	$C_{11}H_{16}$	-13[e]	11[e]	40[e]	79.4	132.1	207.7	5
2-Ethyl-1,3,5-trimethylbenzene	$C_{11}H_{16}$		6[e]	36[e]	75.7	129.6	207.6	5
Ethyl 10-undecenoate	$C_{13}H_{24}O_2$	32[e]	55[e]	86[e]	125.2	179.5	258.4	5
Ethyl vinyl ether	C_4H_8O		-102[e]	-81[e]	-53.1	-16.5	34.7	5
Eucalyptol	$C_{10}H_{18}O$			10.6	48.5	100.3	175.4	5
Europium	Eu	590 s	684 s	799 s	961	1179	1523	14
9*H*-Fluorene	$C_{13}H_{10}$	48.4 s			137.4	205.4	295[e]	5
Fluorine[a]	F_2	-235 s	-229.5 s	-222.9 s	-214.8	-204.3	-188.3	1,5
Fluorine monoxide[a]	F_2O	-211.7	-204.7	-195.9	-184.2	-168.2	-144.9	5
Fluorine nitrate	FNO_3	-160[e]	-149[e]	-135[e]	-115.1	-87.4	-45.0	5
Fluorobenzene	C_6H_5F				-16.9	24.2	84.4	1
1-Fluorobutane	C_4H_9F	-114[e]	-99[e]	-80[e]	-55[e]	-20.0	32.1	5
2-Fluorobutane	C_4H_9F	-117[e]	-103[e]	-85[e]	-60.7	-26.7	24.7	5
1-Fluorodecane	$C_{10}H_{21}F$	-22[e]	0[e]	27[e]	64[e]	113.3	185.7	5
Fluoroethane	C_2H_5F		-142[e]	-127[e]	-106.3	-78.7	-37.9	1
2-Fluoroethanol	C_2H_5FO			-22[e]	8.3	47.5	99[e]	5
Fluoroethene	C_2H_3F			-153.3	-135.2	-109.9	-72.2	5
1-Fluoroheptane	$C_7H_{15}F$	-64[e]	-45[e]	-22[e]	10[e]	53.3	117.4	5
1-Fluorohexane	$C_6H_{13}F$	-80[e]	-62[e]	-40[e]	-11[e]	30.4	91.1	5
Fluoromethane[a]	CH_3F				-130[e]	-111[e]	-78.6	1
1-Fluorooctane	$C_8H_{17}F$				29[e]	74.6	141.8	5
1-Fluoropentane	$C_5H_{11}F$	-97[e]	-80[e]	-60[e]	-32[e]	5.7	62.4	5
1-Fluoropropane	C_3H_7F	-133[e]	-120[e]	-103[e]	-80.7	-49.4	-2.8	5
Fluorosulfonic acid	FHO_3S	-14[e]	4[e]	28[e]	59.1	101.3	162.2	5
2-Fluorotoluene	C_7H_7F		-50[e]	-26[e]	5[e]	49.0	113.9	5
3-Fluorotoluene	C_7H_7F	-67[e]	-48[e]	-25[e]	7[e]	51.0	116.1	5
4-Fluorotoluene	C_7H_7F		-48[e]	-24[e]	7[e]	51[e]	116.2	5
1-Fluoro-4-(trifluoromethyl)benzene	$C_7H_4F_4$			-38[e]	-6[e]	38.6	102.3	5
Formaldehyde[a]	CH_2O				-91[e]	-61.7	-19.3	1
Formamide	CH_3NO		22[e]	53[e]	93[e]	145.0	218[e]	5
Formic acid	CH_2O_2	-56 s	-40.4 s	-22.3 s	-0.8 s	37.0	100.2	1,5
Francium	Fr	131[e]	181[e]	246[e]	335[e]	465[e]	673[e]	2
Fumaric acid	$C_4H_4O_4$	123.9 s	150 s	180 s				5
Furan	C_4H_4O			-78[e]	-54[e]	-20[e]	31.0	1
Furfural	$C_5H_4O_2$	-26[e]	-8[e]	16[e]	47[e]	92.4	161.4	1
Furfuryl alcohol	$C_5H_6O_2$	-30[e]	-5[e]	25[e]	62.6	109.3	169.7	5

Name	Mol. form.	t/°C for 1 Pa	t/°C for 10 Pa	t/°C for 100 Pa	t/°C for 1 kPa	t/°C for 10 kPa	t/°C for 100 kPa	Ref.
Gadolinium	Gd	1563[d]	1755[d]	1994[d]	2300[d]	2703[d]	3262[d]	3
Gallium	Ga	1037	1175	1347	1565	1852	2245	2
Geraniol	$C_{10}H_{18}O$	4[e]	31[e]	63.2	104.3	157.7	229.6	5
Geranyl acetate	$C_{12}H_{20}O_2$			67.7	110.8	166.9	242.9	5
Germanium	Ge	1371	1541	1750	2014	2360	2831	2
Germanium(IV) bromide	Br_4Ge				51	105	188	4
Glycerol	$C_3H_8O_3$	96[e]	113[e]	136[e]	168[e]	213.4	287[e]	1
Glycerol triacetate	$C_9H_{14}O_6$	37.6	62[e]	90[e]	124[e]	165[e]	214[e]	5
Glycolic acid	$C_2H_4O_3$						99.9	5
Gold	Au	1373	1541	1748	2008	2347	2805	2
Hafnium	Hf	2416	2681	3004	3406	3921	4603	9
Helium[a]	He					-270.6	-268.9	2
Heneicosane	$C_{21}H_{44}$	82.3	113.5	152.2	201.6	263.8	355.9	5
Heptacosane	$C_{27}H_{56}$	136.7	168.8	206.5	255.8	323.3	421.2	5
Heptadecane	$C_{17}H_{36}$	51.5	78.5	112.0	155.3	214.5	302[e]	16
1-Heptadecanol	$C_{17}H_{36}O$	94[e]	117[e]	146[e]	185[e]	240.1	323.3	5
Heptanal	$C_7H_{14}O$	-41[e]	-21[e]	4[e]	37[e]	83.7	152.3	5
Heptane	C_7H_{16}	-78.6	-60.2	-37.0	-6.6	35.4	98.0	16
1-Heptanethiol	$C_7H_{16}S$	-30[e]	-9[e]	18[e]	53[e]	102.7	176.4	5
Heptanoic acid	$C_7H_{14}O_2$	24[e]	46[e]	72[e]	107[e]	154.6	222.6	5
1-Heptanol	$C_7H_{16}O$		17[e]	40[e]	70.1	112.5	176[e]	1
2-Heptanol, (±)-	$C_7H_{16}O$	-9[e]	7[e]	27[e]	55.0	95.2	158.7	5
3-Heptanol, (*S*)-	$C_7H_{16}O$	-8[e]	7[e]	27[e]	54.5	93.9	156.3	5
4-Heptanol	$C_7H_{16}O$	-16[e]	1[e]	22[e]	51[e]	91.9	154.6	5
2-Heptanone	$C_7H_{14}O$		-22[e]	3[e]	36[e]	82.2	150.6	1
3-Heptanone	$C_7H_{14}O$		-28[e]	0[e]	36[e]	83.2	147.0	5
4-Heptanone	$C_7H_{14}O$	-27[e]	-6[e]	18.8	50.2	90.3	143.4	5
Heptanoyl chloride	$C_7H_{13}ClO$	-17[e]	4[e]	29.4	59.7	96.9	144.0	5
1-Heptene	C_7H_{14}	-82.1	-63.8	-40.6	-10.7	31.1	93.2	1,5
cis-2-Heptene	C_7H_{14}	-79[e]	-61[e]	-38[e]	-8[e]	34.3	98.0	5
trans-2-Heptene	C_7H_{14}	-79[e]	-61[e]	-39[e]	-8[e]	34.0	97.5	5
cis-3-Heptene	C_7H_{14}	-80[e]	-62[e]	-40[e]	-10[e]	32.3	95.3	5
trans-3-Heptene	C_7H_{14}	-80[e]	-62[e]	-40[e]	-10[e]	32.2	95.2	5
Heptyl acetate	$C_9H_{18}O_2$	-16[e]	6[e]	34[e]	70[e]	119.9	191.9	5
Heptylamine	$C_7H_{17}N$			5[e]	39[e]	86.7	156.4	5
Heptylbenzene	$C_{13}H_{20}$	12[e]	36[e]	66[e]	107[e]	162.7	246.2	5
Heptyl butanoate	$C_{11}H_{22}O_2$	2[e]	29[e]	62[e]	102.6	155.1	224.7	5
Heptylcyclohexane	$C_{13}H_{26}$	11[e]	34[e]	65[e]	105[e]	160.9	244.3	5
Heptylcyclopentane	$C_{12}H_{24}$	-1[e]	22[e]	51[e]	90[e]	143.5	223.5	5
1-Heptyne	C_7H_{12}	-75[e]	-57[e]	-35[e]	-5[e]	37.1	99.5	5
2-Heptyne	C_7H_{12}		-51[e]	-27[e]	4[e]	46.9	111.5	5
3-Heptyne	C_7H_{12}	-71[e]	-53[e]	-31[e]	0[e]	42.7	106.4	5
Hexachloro-1,3-butadiene	C_4Cl_6	-1[e]	22[e]	50[e]	86.7	137.0	209.7	5
Hexachloroethane	C_2Cl_6	-7.6 s	9.9 s	33.6 s	67.7 s	116.9 s	184.2 s	5
Hexachloropropene	C_3Cl_6	-12[e]	11[e]	40[e]	79[e]	132.8	213.6	5
Hexacosane	$C_{26}H_{54}$	125.1	158.8	200.1	252.1	314.3	411.3	5
Hexadecane	$C_{16}H_{34}$	41.1	67.4	100.3	142.7	200.7	286.3	16
Hexadecanoic acid	$C_{16}H_{32}O_2$		136[e]	165[e]	205[e]	261.9	350.2	5
1-Hexadecanol	$C_{16}H_{34}O$	99.5	130.6	171.9	175[e]	229.0	311.7	5
1-Hexadecene	$C_{16}H_{32}$	38.4	65.0	98.1	140.5	198.8	284.3	5
Hexadecylamine	$C_{16}H_{35}N$	63[e]	91[e]	126[e]	171[e]	232.6	320.5	5
trans-1,3-Hexadiene	C_6H_{10}	-86[e]	-70[e]	-51[e]	-24[e]	14[e]	72[e]	5
trans-1,4-Hexadiene	C_6H_{10}	-98[e]	-81[e]	-60[e]	-33[e]	7[e]	65[e]	5
1,5-Hexadiene	C_6H_{10}	-99[e]	-84[e]	-64[e]	-37[e]	0.9	59.2	5
cis,cis-2,4-Hexadiene	C_6H_{10}					18[e]	79.6	5
trans,cis-2,4-Hexadiene	C_6H_{10}	-89[e]	-73[e]	-52[e]	-23[e]	18[e]	79.6	5
trans,trans-2,4-Hexadiene	C_6H_{10}				-23[e]	18[e]	79.6	5
1,5-Hexadien-3-yne	C_6H_6	-82[e]	-66[e]	-44.3	-16.0	23.7	83.6	5
Hexaethylbenzene	$C_{18}H_{30}$				144.1	206.8	297.5	5
Hexafluorobenzene	C_6F_6		-56.9 s	-36 s	-11.5 s	22.6	79.9	1,5

Name	Mol. form.	t/°C for 1 Pa	t/°C for 10 Pa	t/°C for 100 Pa	t/°C for 1 kPa	t/°C for 10 kPa	t/°C for 100 kPa	Ref.
Hexafluoroethane[b]	C_2F_6			-155.2 s	-137.5 s	-113.4 s	-78.4 s	1,5
1,1,1,3,3,3-Hexafluoro-2-propanol	$C_3H_2F_6O$					12.7	57.1	5
Hexamethylbenzene	$C_{12}H_{18}$	46.3 s	72.5 s	81.7 s	121.8 s	178.3	263.7	5
Hexamethyldisiloxane	$C_6H_{18}OSi_2$		-56[e]	-34[e]	-5[e]	37.1	100.1	5
2,6,10,15,19,23-Hexamethyltetracosane	$C_{30}H_{62}$	66[e]	84[e]	105.8	131.9	163.7	203.2	5
Hexanal	$C_6H_{12}O$	-56[e]	-37[e]	-13[e]	19[e]	62.6	127.8	5
Hexane	C_6H_{14}	-96.4 s	-79.2	-57.6	-29.3	9.8	68.3	16
1,6-Hexanediamine	$C_6H_{16}N_2$				76.0	128.2	199.0	5
Hexanedinitrile	$C_6H_8N_2$	30[e]	61[e]	100[e]	148.6	211.8	297[e]	5
Hexanenitrile	$C_6H_{11}N$	-40[e]	-19[e]	8[e]	43[e]	91.5	163.2	1,5
1-Hexanethiol	$C_6H_{14}S$	-45[e]	-25[e]	1[e]	35[e]	81.7	152.2	5
2-Hexanethiol	$C_6H_{14}S$	-50[e]	-32[e]	-8[e]	25[e]	69.9	138.4	5
1,2,6-Hexanetriol	$C_6H_{14}O_3$	92[e]	114.8	146.0	191[e]			5
Hexanoic acid	$C_6H_{12}O_2$		33[e]	59[e]	93[e]	139.3	204.5	1
1-Hexanol	$C_6H_{14}O$		5[e]	28[e]	56.8	97.3	157.1	1
2-Hexanol	$C_6H_{14}O$	-28[e]	-10[e]	12[e]	41.4	81.5	139.6	1
3-Hexanol	$C_6H_{14}O$	-43[e]	-23[e]	1[e]	33[e]	75.4	135.1	1
2-Hexanone	$C_6H_{12}O$	-43[e]	-21[e]	4.2	34.5	61.9	127.2	1,5
3-Hexanone	$C_6H_{12}O$		-40[e]	-16[e]	15[e]	58.5	123.1	1
cis-1,3,5-Hexatriene	C_6H_8					21[e]	78[e]	5
1-Hexene	C_6H_{12}	-99.8	-82.8	-61.4	-33.7	5.2	63.1	1,5
cis-2-Hexene	C_6H_{12}	-97[e]	-80[e]	-58[e]	-30[e]	9.9	68.5	5
trans-2-Hexene	C_6H_{12}	-94[e]	-78[e]	-57[e]	-30[e]	9.3	67.5	5
cis-3-Hexene	C_6H_{12}	-96[e]	-79[e]	-59[e]	-30.8	7.9	66.0	5
trans-3-Hexene	C_6H_{12}	-95[e]	-79[e]	-58[e]	-30.0	8.8	66.7	5
Hexyl acetate	$C_8H_{16}O_2$	-37[e]	-13[e]	16[e]	52.8	100.4	164[e]	5
Hexylamine	$C_6H_{15}N$			-10[e]	22[e]	66.0	130.6	5
Hexylbenzene	$C_{12}H_{18}$	-2[e]	22[e]	51[e]	90[e]	144.5	225.5	5
Hexylcyclohexane	$C_{12}H_{24}$	-3[e]	20[e]	50[e]	89[e]	143.1	224.2	5
Hexylcyclopentane	$C_{11}H_{22}$	-15[e]	7[e]	36[e]	73[e]	125.0	202.5	5
2-(Hexyloxy)ethanol	$C_8H_{18}O_2$	-13[e]	14[e]	46[e]	86[e]	137.7	206.9	5
1-Hexyne	C_6H_{10}	-91[e]	-75[e]	-54[e]	-26[e]	12.8	71.0	5
2-Hexyne	C_6H_{10}	-84[e]	-67[e]	-46[e]	-17[e]	23.6	84.1	5
3-Hexyne	C_6H_{10}	-86[e]	-69[e]	-48[e]	-19.1	21.0	81.0	1,5
Holmium	Ho	1159 s	1311 s	1502[d]	1767[d]	2137[d]	2691[d]	3
Hydrazine	H_4N_2				14.7	55.6	113[e]	5
Hydrazoic acid	HN_3			-79[e]	-54[e]	-18.0	35.7	5
Hydrogen[a]	H_2					-258.6	-252.8	1
Hydrogen bromide[a]	BrH		-153.3 s	-140.4 s	-123.8 s	-101.5 s	-67.0	5
Hydrogen chloride[a]	ClH				-138.2 s	-118.0	-85.2	1,5
Hydrogen cyanide[a]	CHN			-77 s	-52.6 s	-22.7 s	25.4	1,5
Hydrogen disulfide	H_2S_2				-27[e]	12.2	70.7	5
Hydrogen fluoride[a]	FH				-71.1	-33.7	19.2	1,5
Hydrogen iodide[a]	HI	-146 s	-135.2 s	-120.8 s	-101.9 s	-75.9 s	-35.9	5
Hydrogen peroxide	H_2O_2			13[e]	45[e]	89.0	149.8	5
Hydrogen selenide	H_2Se	-145 s	-134 s	-120 s	-102.8 s	-78.9 s	-41.5	5
Hydrogen sulfide[a]	H_2S		-149 s	-136 s	-118.9 s	-95.9 s	-60.5	1,5
Hydrogen telluride	H_2Te					-46.6	-2.3	5
Hydroxylamine	H_3NO				43.7	73.3	109.8	4
3-Hydroxypropanenitrile	C_3H_5NO	-11[e]	18[e]	53[e]	96.1	150.3	220.8	5
Indan	C_9H_{10}	-33[e]	-12[e]	16[e]	52[e]	102.3	177.5	1
Indene	C_9H_8			12[e]	53.0	106.8	181.0	5
Indium	In	923	1052	1212	1417	1689	2067	2
Indium(III) bromide	Br_3In			304.6 s	328.7 s	364.8 s		1
1*H*-Indole	C_8H_7N	20.6 s	44.5 s				254.0	5
Iodine	I_2	-12.8 s	9.3 s	35.9 s	68.7 s	108 s	184.0	1,2
Iodobenzene	C_6H_5I	-30[e]	-7[e]	20.9	58.5	110.6	187.8	1
1-Iodobutane	C_4H_9I	-62[e]	-43[e]	-19[e]	14[e]	60.5	130.0	5
2-Iodobutane, (±)-	C_4H_9I	-70[e]	-51[e]	-27[e]	5[e]	50[e]	119.5	5
Iodoethane	C_2H_5I	-94[e]	-78[e]	-56[e]	-27.9	11.9	71.9	5

Name	Mol. form.	*t*/°C for 1 Pa	*t*/°C for 10 Pa	*t*/°C for 100 Pa	*t*/°C for 1 kPa	*t*/°C for 10 kPa	*t*/°C for 100 kPa	Ref.
Iodoethene	C_2H_3I				-41[e]	-3[e]	55.6	5
1-Iodoheptane	$C_7H_{15}I$	-19[e]	3[e]	32[e]	71[e]	123.8	203.4	5
1-Iodohexane	$C_6H_{13}I$	-33[e]	-11[e]	16[e]	53[e]	104.0	180.8	5
Iodomethane	CH_3I				-49[e]	-12.4	42.1	1
1-Iodo-3-methylbutane	$C_5H_{11}I$		-34[e]	-6.6	28.8	77.3	147.8	5
1-Iodo-2-methylpropane	C_4H_9I		-47[e]	-21.4	12.0	56.8	120.0	5
2-Iodo-2-methylpropane	C_4H_9I	-75.1 s	-58.8 s	-39.5 s	-5.2	41[e]	100.0	5
1-Iodooctane	$C_8H_{17}I$	-6[e]	18[e]	48[e]	87[e]	142.5	224.5	5
1-Iodopentane	$C_5H_{11}I$	-47[e]	-27[e]	-1[e]	34[e]	83.0	156.5	5
1-Iodopropane	C_3H_7I	-78[e]	-60[e]	-37[e]	-6[e]	36.9	102.0	5
2-Iodopropane	C_3H_7I	-89[e]	-71[e]	-47[e]	-16.3	26.5	89.2	5
3-Iodopropene	C_3H_5I	-80[e]	-62[e]	-39[e]	-8[e]	36[e]	101.5	5
Iodosilane	H_3ISi				-47.7	-10.1	45.2	4
2-Iodothiophene	C_4H_3IS			-25[e]	23[e]	94.9	181.0	5
Iridium	Ir	2440 s	2684	2979	3341	3796	4386	2
Iridium(VI) fluoride	F_6Ir	-88 s	-71 s	-51 s	-27 s	3.8 s	53.1	26
Iron	Fe	1455 s	1617	1818	2073	2406	2859	2
Iron(II) chloride	Cl_2Fe				685	821	1025	4
Iron(III) chloride	Cl_3Fe	118 s	153 s	190 s	229 s	268 s	319	4
Iron pentacarbonyl	C_5FeO_5				0	44	105	4
Isobutanal	C_4H_8O			-56[e]	-29[e]	8[e]	63.8	1
Isobutane[a]	C_4H_{10}		-129.0	-113.0	-90.9	-59.4	-12.0	1,41
Isobutene	C_4H_8	-139.1	-125.5	-108.2	-85.5	-54.5	-7.3	1,5
Isobutyl acetate	$C_6H_{12}O_2$	-63[e]	-45[e]	-21[e]	10[e]	53.4	116[e]	5
Isobutylamine	$C_4H_{11}N$	-85[e]	-70[e]	-50[e]	-24.5	12.0	67.3	5
Isobutylbenzene	$C_{10}H_{14}$	-36[e]	-15[e]	12[e]	47.9	97.8	172.3	5
Isobutylcyclohexane	$C_{10}H_{20}$	-37[e]	-16[e]	10[e]	46[e]	95.9	170.8	5
Isobutylcyclopentane	C_9H_{18}	-105[e]	-88[e]	-64[e]	-28[e]	31[e]	147.0	5
Isobutyl formate	$C_5H_{10}O_2$	-69[e]	-53[e]	-31[e]	-3[e]	37.4	97.6	5
Isobutyl isobutanoate	$C_8H_{16}O_2$	-47[e]	-26[e]	0.4	34.8	81.1	147.0	5
Isobutyl 3-methylbutanoate	$C_9H_{18}O_2$			11.3	48.3	97.9	168.3	5
Isobutyl nitrate	$C_4H_9NO_3$			-18[e]	15.1	59.2	123.0	5
Isobutyl propanoate	$C_7H_{14}O_2$	-35[e]	-19[e]	2[e]	31[e]	72.0	136.1	5
Isobutyl vinyl ether	$C_6H_{12}O$	-87[e]	-68[e]	-44[e]	-13[e]	26.5	80.7	5
Isoeugenol	$C_{10}H_{12}O_2$				125[e]	185.3	267.1	5
Isopentane	C_5H_{12}	-119[e]	-105[e]	-86[e]	-61[e]	-26[e]	27.5	1
Isopentyl acetate	$C_7H_{14}O_2$	-51[e]	-30[e]	-4[e]	30.3	76.2	141.4	5
Isopentyl butanoate	$C_9H_{18}O_2$				55[e]	105.6	178.4	5
Isopentyl formate	$C_6H_{12}O_2$	-60[e]	-41[e]	-17[e]	15[e]	59.1	124[e]	5
Isopentyl isopentanoate	$C_{10}H_{20}O_2$			22[e]	62.8	116.9	193.6	5
Isopentyl propanoate	$C_8H_{16}O_2$			3.1	40.7	90.6	159.8	5
Isophorone	$C_9H_{14}O$		1[e]	33.1	75.1	132.4	215.1	5
Isopropenylbenzene	C_9H_{10}			3.2	41.5	92.8	164.9	5
p-Isopropenylisopropylbenzene	$C_{12}H_{16}$	-11[e]	15[e]	46[e]	87[e]	142.4	221[e]	5
Isopropyl acetate	$C_5H_{10}O_2$		-61[e]	-40[e]	-11[e]	29.8	88.2	5
Isopropylamine	C_3H_9N		-91[e]	-74[e]	-50.4	-17.6	31.5	1,5
4-Isopropylbenzaldehyde	$C_{10}H_{12}O$			54.1	96.0	152.2	231.5	5
Isopropylbenzene	C_9H_{12}	-46[e]	-26[e]	-1[e]	33[e]	80.9	152.0	1
Isopropyl chloroacetate	$C_5H_9ClO_2$			-2[e]	35.0	83.3	148.1	5
Isopropylcyclohexane	C_9H_{18}	-48[e]	-28[e]	-2[e]	33[e]	81.3	154.0	5
Isopropylcyclopentane	C_8H_{16}	-65[e]	-46[e]	-21[e]	12[e]	57.3	125.9	5
Isopropylcyclopropane	C_6H_{12}	-104[e]	-88[e]	-68[e]	-40[e]	-1[e]	57.9	5
Isopropyl formate	$C_4H_8O_2$	-80[e]	-65[e]	-47[e]	-22.2	13.2	67.7	5
Isopropyl isobutanoate	$C_7H_{14}O_2$		-44[e]	-19.7	12.2	56.0	120.1	5
5-Isopropyl-2-methylaniline	$C_{10}H_{15}N$	19[e]	43[e]	72[e]	107.4	150[e]	204[e]	5
1-Isopropyl-2-methylbenzene	$C_{10}H_{14}$	-39[e]	-16[e]	13[e]	51[e]	103.1	177.8	5
1-Isopropyl-3-methylbenzene	$C_{10}H_{14}$	-34[e]	-13[e]	14[e]	50[e]	99.9	174.6	5
1-Isopropyl-4-methylbenzene	$C_{10}H_{14}$	-33[e]	-12[e]	16[e]	52[e]	102.2	176.6	5
Isopropyl methyl ether	$C_4H_{10}O$				-56[e]	-21.2	30.4	5
Isopropyl methyl sulfide	$C_4H_{10}S$	-85[e]	-68[e]	-46[e]	-17[e]	23.4	84.3	5

Name	Mol. form.	t/°C for 1 Pa	t/°C for 10 Pa	t/°C for 100 Pa	t/°C for 1 kPa	t/°C for 10 kPa	t/°C for 100 kPa	Ref.
1-Isopropylnaphthalene	$C_{13}H_{14}$	27[e]	51[e]	82[e]	123.2	180.8	267.3	5
Isopropyl propyl sulfide	$C_6H_{14}S$				18.5	63.8	131.6	5
4-Isopropylstyrene	$C_{11}H_{14}$	-25[e]	-1[e]	30.2	70.3	124.5	202.1	5
Isoquinoline	C_9H_7N		30.2	60.7	101.3	157.9	242.7	1,5
Ketene	C_2H_2O		-151[e]	-135[e]	-115[e]	-88.2	-50.0	1
Krypton[a]	Kr	-214.0 s	-208.0 s	-199.4 s	-188.9 s	-174.6 s	-153.6	5
Lanthanum	La	1732[d]	1935[d]	2185[d]	2499[d]	2905[d]	3453[d]	3
Lead	Pb	705	815	956	1139	1387	1754	2
Lead(II) bromide	Br_2Pb	374	431	502	597	726	914	4
Lead(II) chloride	Cl_2Pb			541[e]	637	765	949	23
Lead(II) fluoride	F_2Pb				865	1054	1292	4
Lead(II) iodide	I_2Pb			470	558	682	869	4
Lead(II) oxide (massicot)	OPb	724	816	928	1065	1241	1471	4
Lead(II) sulfide	PbS	656 s	741 s	838 s	953 s	1088 s	1280	4
d-Limonene	$C_{10}H_{16}$	-45[e]	-21[e]	9.1	48.0	100.4	174.5	5
l-Limonene	$C_{10}H_{16}$	-33[e]	-12[e]	16[e]	52.0	102.3	177.0	21
Lithium	Li	524.3	612.3	722.1	871.2	1064.3	1337.1	13,30
Lithium bromide	BrLi		630	733	868	1049	1308	4
Lithium chloride	ClLi		649[d]	761[d]	905[d]	1101[d]	1381[d]	8
Lithium fluoride	FLi	801 s	896	1024	1188	1395	1672	4,12,25
Lithium iodide	ILi	545	619	710	824	972	1170	4
Lutetium	Lu	1633 s	1829.8	2072.8	2380[d]	2799[d]	3390[d]	3
Magnesium	Mg	428 s	500 s	588 s	698	859	1088	2
Magnesium chloride	Cl_2Mg			762	908	1111	1414	4
Maleic anhydride	$C_4H_2O_3$				73.7	127.9	201.7	5
Manganese	Mn	955 s	1074 s	1220 s	1418	1682	2060	2
Manganese(II) chloride	Cl_2Mn				760	933	1189	4
Mercury[b]	Hg	42.0	76.6	120.0	175.6	250.3	355.9	29,30
Mercury(II) bromide	Br_2Hg	71 s	98 s	132 s	174 s	227 s	318	4
Mercury(II) chloride	Cl_2Hg	64.4 s	94.7 s	130.8 s	174.5 s	228.5 s	304.0	4
Mercury(II) iodide (red)	HgI_2	85.1 s	115.6 s	152.4 s	197.8 s	255.1 s	353.6	4
Mesityl oxide	$C_6H_{10}O$	-56[e]	-37[e]	-13[e]	19[e]	63.5	129.3	5
Methacrylic acid	$C_4H_6O_2$			22[e]	56[e]	99.9	161.5	5
Methane[a]	CH_4	-220 s	-214.2 s	-206.8 s	-197 s	-183.6 s	-161.7	5,41
Methanethiol	CH_4S		-115[e]	-97[e]	-74[e]	-41.7	5.7	1
Methanol[a]	CH_4O	-87[e]	-69[e]	-47.5	-20.4	15.2	64.2	11
4-Methoxybenzaldehyde	$C_8H_8O_2$	9[e]	35[e]	68.1	110.8	167.9	248.5	5
2-Methoxyethanol	$C_3H_8O_2$	-57[e]	-37[e]	-12[e]	21[e]	63.8	124.3	1
2-Methoxyethyl acetate	$C_5H_{10}O_3$	-47[e]	-26[e]	0[e]	34[e]	79.4	144.1	5
4-Methoxy-4-methyl-2-pentanone	$C_7H_{14}O_2$				43[e]	89.8	160[e]	5
1-Methoxy-4-(2-propenyl)benzene	$C_{10}H_{12}O$			48.5	88.0	140.7	214.6	5
N-Methylacetamide	C_3H_7NO	-13.3 s	13 s	43[e]	83.8	136.1	206.3	5
Methyl acetate	$C_3H_6O_2$	-95[e]	-79[e]	-59[e]	-33[e]	3.3	56.6	1
Methyl acetoacetate	$C_5H_8O_3$				50.1	101.1	171.3	5
Methyl acrylate	$C_4H_6O_2$		-71[e]	-48[e]	-18[e]	22[e]	79.9	5
2-Methylacrylonitrile	C_4H_5N				-12[e]	29.0	89.8	5
Methylamine	CH_5N				-76.7	-48.1	-6.6	1
2-Methylaniline	C_7H_9N	1.0	18.8	42.6	76.1	125.6	199.9	1,5
3-Methylaniline	C_7H_9N	3.8	22.0	46.2	80.1	128.8	202.9	1,5
4-Methylaniline	C_7H_9N				77.1	126.2	199.9	5
N-Methylaniline	C_7H_9N	-16[e]	6[e]	34[e]	70.3	121.1	195.8	1
4-Methyl-1,3-benzenediamine	$C_7H_{10}N_2$			100.4	145.3	202.9	279.5	5
3-Methylbenzenethiol	C_7H_8S		0[e]	29[e]	66[e]	117.9	194.6	5
Methyl benzoate	$C_8H_8O_2$		-1[e]	29[e]	68[e]	121.2	198.9	5
2-Methylbenzonitrile	C_8H_7N		1[e]	32.1	72.2	126.6	204.7	5
4-Methylbenzonitrile	C_8H_7N			40.1	78.7	134.3	221.3	5
1-Methylbicyclo[3.1.0]hexane	C_7H_{12}					29.8	92.6	5
3-Methyl-1,2-butadiene	C_5H_8	-111[e]	-95[e]	-75[e]	-49.2	-13.1	40.4	5
2-Methyl-1,3-butadiene	C_5H_8	-115[e]	-100[e]	-81[e]	-55.4	-19.7	33.7	1,5
2-Methyl-1-butanethiol, (+)	$C_5H_{12}S$				8.0	52.3	118.5	5

Name	Mol. form.	*t*/°C for 1 Pa	*t*/°C for 10 Pa	*t*/°C for 100 Pa	*t*/°C for 1 kPa	*t*/°C for 10 kPa	*t*/°C for 100 kPa	Ref.
3-Methyl-1-butanethiol	$C_5H_{12}S$				7.8	51.9	117.9	5
2-Methyl-2-butanethiol	$C_5H_{12}S$				-8.0	34.6	98.7	5
Methyl butanoate	$C_5H_{10}O_2$	-68[e]	-50[e]	-28[e]	0.9	41.7	102.3	5
2-Methylbutanoic acid	$C_5H_{10}O_2$	-10[e]	10[e]	36[e]	69[e]	112.8	175.2	5
3-Methylbutanoic acid	$C_5H_{10}O_2$	-15.8	4[e]	30.0	64.7	110.6	176.1	5
2-Methyl-1-butanol, (±)-	$C_5H_{12}O$	-27[e]	-11[e]	9[e]	36.2	73.4	128.3	1
3-Methyl-1-butanol	$C_5H_{12}O$	-22[e]	-7[e]	13[e]	39.1	75.7	130.1	5
2-Methyl-2-butanol	$C_5H_{12}O$			-5[e]	17.7	50.6	101.7	1,5
3-Methyl-2-butanol, (±)-	$C_5H_{12}O$			-3[e]	22.7	58.2	111.1	5
3-Methyl-2-butanone	$C_5H_{10}O$	-69[e]	-54[e]	-34[e]	-6.9	32.2	94.0	1,5
2-Methyl-1-butene	C_5H_{10}	-117.7	-102.2	-82.7	-57.2	-21.9	30.8	1,5
3-Methyl-1-butene	C_5H_{10}	-125.0	-110.1	-91.2	-66.7	-32.1	19.7	1,5
2-Methyl-2-butene	C_5H_{10}	-113.4	-97.6	-77.7	-51.6	-15.8	38.2	1,5
3-Methyl-3-buten-2-one	C_5H_8O			-35[e]	-5[e]	36.0	97.3	5
3-Methylbutyl benzoate	$C_{12}H_{16}O_2$			66[e]	115.0	177.7	261.4	5
Methyl *tert*-butyl ether	$C_5H_{12}O$			-66[e]	-39[e]	-2[e]	54.8	1
3-Methylbutyl nitrate	$C_5H_{11}NO_3$		-26[e]	1.0	35.5	81.7	147.0	5
3-Methyl-1-butyne	C_5H_8			-82[e]	-57.5	-23.1	28.6	5
Methyl chloroacetate	$C_3H_5ClO_2$		-28[e]	-5[e]	25[e]	66.9	129.1	5
Methyl cyanoacetate	$C_4H_5NO_2$	-3[e]	19[e]	48[e]	84[e]	134.0	204.6	5
Methylcyclohexane	C_7H_{14}	-79[e]	-62[e]	-39[e]	-7.9	35.5	100.5	1
1-Methylcyclohexene	C_7H_{12}	-72[e]	-53[e]	-30[e]	1[e]	45[e]	109.8	5
4-Methylcyclohexene	C_7H_{12}	-76[e]	-59[e]	-36[e]	-5[e]	37.9	102.3	5
Methylcyclopentane	C_6H_{12}	-97[e]	-80[e]	-58[e]	-28.8	11.6	71.4	1,5
Methylcyclopropane	C_4H_8	-130[e]	-116[e]	-99.3	-76.3	-44.2	4.2	5
2-Methyldecane	$C_{11}H_{24}$	-20[e]	1[e]	28[e]	64[e]	114.0	188.7	5
3-Methyldecane	$C_{11}H_{24}$	-35[e]	-10[e]	22[e]	61.9	115.6	190.4	5
4-Methyldecane	$C_{11}H_{24}$	-38[e]	-12[e]	20[e]	60.8	113.9	186.4	5
Methyl decanoate	$C_{11}H_{22}O_2$	10[e]	33[e]	62[e]	100.9	154.0	232[e]	5
Methyl dichloroacetate	$C_3H_4Cl_2O_2$	-44[e]	-25[e]	0[e]	33[e]	77.7	142.3	5
Methyldifluoroarsine	CH_3AsF_2				-15[e]	22.1	76.1	5
2-Methyl-*N,N*-dimethylaniline	$C_9H_{13}N$	-25[e]	-3[e]	24.4	60.6	110.7	184.5	5
Methyl dimethylthioborane	C_3H_9BS			-62[e]	-30.4	11.4	70.7	5
Methyldiphenylamine	$C_{13}H_{13}N$	35[e]	63[e]	98.4	143.1	201.6	281.6	5
Methyl dodecanoate	$C_{13}H_{26}O_2$	38[e]	61[e]	90[e]	130[e]	184.9	269[e]	5
Methylenecyclohexane	C_7H_{12}	-76[e]	-58[e]	-35[e]	-5[e]	38[e]	103.0	5
N-Methylformamide	C_2H_5NO		13[e]	41[e]	78[e]	127.9	199.1	1
Methyl formate	$C_2H_4O_2$		-95[e]	-76[e]	-51.8	-18.1	31.4	5
2-Methylfuran	C_5H_6O			-66[e]	-35[e]	6[e]	64.5	1
2-Methylheptane	C_8H_{18}	-69[e]	-49.1	-24.5	7.6	51.6	117.2	1,5
3-Methylheptane	C_8H_{18}	-67[e]	-48.1	-23.6	8.5	52.7	118.5	1,5
4-Methylheptane	C_8H_{18}	-65[e]	-47[e]	-24[e]	7.8	51.6	117.2	5
Methyl heptanoate	$C_8H_{16}O_2$	-30[e]	-9[e]	19[e]	54.2	102.4	172[e]	5
3-Methyl-3-heptanol	$C_8H_{18}O$	-13[e]	4[e]	26[e]	55[e]	96.3	160.3	5
4-Methyl-3-heptanol	$C_8H_{18}O$	-52[e]	-28[e]	1[e]	39[e]	87.6	155.0	5
5-Methyl-3-heptanol	$C_8H_{18}O$	-35[e]	-16[e]	8[e]	40[e]	84.8	153.0	5
4-Methyl-4-heptanol	$C_8H_{18}O$	-17[e]	1[e]	24[e]	55[e]	97.2	160.7	5
2-Methyl-1-heptene	C_8H_{16}	-66[e]	-48[e]	-24[e]	8[e]	52.3	118.7	5
Methyl hexadecanoate	$C_{17}H_{34}O_2$	65[e]	93	129	177			4
2-Methylhexane	C_7H_{16}	-82[e]	-65[e]	-43[e]	-13[e]	27.8	89.7	1
3-Methylhexane	C_7H_{16}	-81[e]	-64[e]	-42[e]	-12[e]	29.2	91.5	1
Methyl hexanoate	$C_7H_{14}O_2$	-47[e]	-26[e]	2[e]	36.6	83.3	149[e]	5
5-Methyl-2-hexanone	$C_7H_{14}O$		-27[e]	-2[e]	31.0	76.6	144.4	5
2-Methyl-1-hexene	C_7H_{14}	-81[e]	-64[e]	-42[e]	-12[e]	29.3	91.6	5
4-Methyl-1-hexene	C_7H_{14}	-84[e]	-67[e]	-45[e]	-16[e]	25.3	86.3	5
2-Methyl-2-hexene	C_7H_{14}	-80[e]	-63[e]	-40[e]	-10[e]	32.0	95.0	5
cis-3-Methyl-2-hexene	C_7H_{14}	-79[e]	-62[e]	-39[e]	-9[e]	33.4	96.8	5
trans-4-Methyl-2-hexene	C_7H_{14}	-83[e]	-66[e]	-44[e]	-15[e]	25.9	87.1	5
trans-5-Methyl-2-hexene	C_7H_{14}	-83[e]	-66[e]	-44[e]	-15[e]	26.3	87.7	5
trans-2-Methyl-3-hexene	C_7H_{14}	-84[e]	-67[e]	-45[e]	-16[e]	24.6	85.5	5

Name	Mol. form.	t/°C for 1 Pa	t/°C for 10 Pa	t/°C for 100 Pa	t/°C for 1 kPa	t/°C for 10 kPa	t/°C for 100 kPa	Ref.
5-Methyl-1-hexyne	C_7H_{12}	-80[e]	-62[e]	-40[e]	-11[e]	30.1	91.4	5
5-Methyl-2-hexyne	C_7H_{12}	-75[e]	-57[e]	-34[e]	-4[e]	38.6	102.0	5
2-Methyl-3-hexyne	C_7H_{12}	-78[e]	-61[e]	-39[e]	-9[e]	32.6	94.8	5
Methylhydrazine	CH_6N_2			-31[e]	-4.7	32.9	91[e]	1
Methyl isobutanoate	$C_5H_{10}O_2$	-83[e]	-65[e]	-41[e]	-11[e]	31[e]	92.1	5
Methyl isocyanate	C_2H_3NO				-43.5	-10.2	38.8	1
Methyl isopentanoate	$C_6H_{12}O_2$					53.3	116.3	5
Methyl methacrylate	$C_5H_8O_2$			-31[e]	-1[e]	39.7	100.0	1
1-Methylnaphthalene	$C_{11}H_{10}$	5[e]	29[e]	60[e]	102[e]	159.1	244.1	1
2-Methylnaphthalene	$C_{11}H_{10}$			57[e]	99[e]	156.0	240.5	1
Methyl nitrate	CH_3NO_3		-75[e]	-55[e]	-27[e]	9.8	63[e]	5
Methyl 2-nitrobenzoate	$C_8H_7NO_4$	17[e]	49[e]	89[e]	140[e]	208[e]	302[e]	5
2-Methylnonane	$C_{10}H_{22}$	-34[e]	-14[e]	12[e]	47[e]	94.8	166.5	5
3-Methylnonane	$C_{10}H_{22}$	-34[e]	-14[e]	12[e]	47[e]	95.1	167.3	5
4-Methylnonane	$C_{10}H_{22}$	-36[e]	-16[e]	10[e]	45[e]	93.1	165.2	5
5-Methylnonane	$C_{10}H_{22}$	-36[e]	-16[e]	10[e]	45[e]	92.6	164.6	5
Methyl *cis*-9-octadecenoate	$C_{19}H_{36}O_2$	85[e]	114[e]	149.7	195.6	256[e]	340[e]	5
2-Methyloctane	C_9H_{20}	-49[e]	-30[e]	-5[e]	28[e]	73.9	142.8	5
3-Methyloctane	C_9H_{20}	-49[e]	-29[e]	-5[e]	29[e]	74.7	143.7	5
4-Methyloctane	C_9H_{20}	-50[e]	-30[e]	-6[e]	27[e]	73.2	141.9	5
Methyl octanoate	$C_9H_{18}O_2$	-26[e]	-9[e]	13[e]	40[e]	76[e]	127.9	5
2-Methyl-1-octene	C_9H_{18}	-53[e]	-34[e]	-9[e]	25[e]	72[e]	144.1	5
Methyloxirane	C_3H_6O	-109[e]	-95[e]	-76[e]	-51.5	-17.2	33.9	5
cis-2-Methyl-1,3-pentadiene	C_6H_{10}	-92[e]	-75[e]	-54[e]	-26[e]	14[e]	75.6	5
2-Methylpentane	C_6H_{14}	-100[e]	-84[e]	-64[e]	-36[e]	2[e]	59.9	1
3-Methylpentane	C_6H_{14}	-99[e]	-83[e]	-62[e]	-34.3	4.6	62.9	1
2-Methyl-2,4-pentanediol	$C_6H_{14}O_2$	-8[e]	17[e]	48[e]	86[e]	134.4	197.5	5
4-Methylpentanenitrile	$C_6H_{11}N$		-50[e]	-20[e]	20[e]	75.2	155.2	5
Methyl pentanoate	$C_6H_{12}O_2$				19.2	63.7	127.4	5
4-Methylpentanoic acid	$C_6H_{12}O_2$	36[e]	49[e]	67.1	92.9	133.6	206.8	5
2-Methyl-1-pentanol	$C_6H_{14}O$			14[e]	45.9	88.3	147.6	5
4-Methyl-1-pentanol	$C_6H_{14}O$			24[e]	53[e]	92.4	151.4	5
2-Methyl-2-pentanol	$C_6H_{14}O$	-29[e]	-15[e]	3[e]	27.1	63.0	120.9	5
3-Methyl-2-pentanol	$C_6H_{14}O$				36.5	76.1	133.8	5
4-Methyl-2-pentanol	$C_6H_{14}O$	-43[e]	-24[e]	0[e]	30[e]	71.9	131.3	5
2-Methyl-3-pentanol	$C_6H_{14}O$				29.8	68.8	126.0	5
3-Methyl-3-pentanol	$C_6H_{14}O$		-23[e]	-4[e]	22.9	61.1	121.1	5
3-Methyl-2-pentanone, (±)-	$C_6H_{12}O$				8.5	52.7	117.0	5
4-Methyl-2-pentanone	$C_6H_{12}O$	-61[e]	-43[e]	-21[e]	9[e]	51.5	116.1	5
2-Methyl-3-pentanone	$C_6H_{12}O$					50.2	113.0	5
2-Methyl-1-pentene	C_6H_{12}	-98[e]	-82[e]	-62[e]	-34.2	4.1	61.7	5
3-Methyl-1-pentene	C_6H_{12}	-104[e]	-88[e]	-68[e]	-41.5	-3.6	53.8	5
4-Methyl-1-pentene	C_6H_{12}	-105[e]	-89[e]	-69[e]	-41.6	-3.6	53.5	5
2-Methyl-2-pentene	C_6H_{12}	-95[e]	-78[e]	-58[e]	-30[e]	9.0	66.9	5
3-Methyl-*cis*-2-pentene	C_6H_{12}	-95[e]	-79[e]	-58[e]	-30[e]	8.9	67.3	5
3-Methyl-*trans*-2-pentene	C_6H_{12}	-93[e]	-77[e]	-55[e]	-27.4	11.7	70.0	5
4-Methyl-*cis*-2-pentene	C_6H_{12}	-102[e]	-86[e]	-66[e]	-38.7	-0.9	56.0	5
4-Methyl-*trans*-2-pentene	C_6H_{12}	-100[e]	-84[e]	-64[e]	-36.8	1.2	58.2	5
4-Methyl-4-penten-2-one	$C_6H_{10}O$	-59[e]	-41[e]	-17[e]	14[e]	57.0	121.0	5
4-Methyl-1-pentyne	C_6H_{10}	-97[e]	-81[e]	-61[e]	-34[e]	4.1	60.7	5
4-Methyl-2-pentyne	C_6H_{10}	-91[e]	-74[e]	-54[e]	-26[e]	13.8	72.7	5
N-Methylpropanamide	C_4H_9NO				81.1	105[e]		5
2-Methyl-1-propanethiol	$C_4H_{10}S$		-66[e]	-44[e]	-15[e]	26.5	88.1	5
2-Methyl-2-propanethiol	$C_4H_{10}S$					5.8	63.8	5
Methyl propanoate	$C_4H_8O_2$	-80[e]	-64[e]	-43[e]	-15.8	22.2	79.0	1
2-Methylpropanoic acid	$C_4H_8O_2$	-30.1	-8.2	18.1	50.5	92.9	154.0	5
2-Methyl-1-propanol	$C_4H_{10}O$	-39[e]	-24[e]	-5[e]	20.9	56.0	107.6	1,5
2-Methyl-2-propanol	$C_4H_{10}O$					34.4	82.1	1,5
2-Methyl-2-propenoyl chloride	C_4H_5ClO		-57[e]	-35[e]	-5[e]	36.4	98.2	5
1-Methyl-2-propylbenzene	$C_{10}H_{14}$	-27[e]	-6[e]	22[e]	58.2	108.9	184.3	5

Name	Mol. form.	t/°C for 1 Pa	t/°C for 10 Pa	t/°C for 100 Pa	t/°C for 1 kPa	t/°C for 10 kPa	t/°C for 100 kPa	Ref.
1-Methyl-3-propylbenzene	$C_{10}H_{14}$	-29[e]	-8[e]	20[e]	56.1	106.5	181.3	5
1-Methyl-4-propylbenzene	$C_{10}H_{14}$	-29[e]	-7[e]	20[e]	56.6	107.4	182.8	5
cis-1-Methyl-2-propylcyclopentane	C_9H_{18}	-52[e]	-33[e]	-7[e]	28[e]	77[e]	152.0	5
trans-1-Methyl-2-propylcyclopentane	C_9H_{18}	-56[e]	-36[e]	-11[e]	23[e]	72[e]	145.8	5
Methyl propyl ether	$C_4H_{10}O$				-40[e]	-11.3	38.7	5
Methyl propyl sulfide	$C_4H_{10}S$	-78[e]	-61[e]	-38[e]	-8[e]	33.1	95.1	5
2-Methylpyridine	C_6H_7N	-56.5	-37.8	-13.9	18.3	62.9	129.0	1,5
3-Methylpyridine	C_6H_7N			-5[e]	28.8	75.2	143.7	1
4-Methylpyridine	C_6H_7N	-58.2 s	-43.1 s	-3.9 s	29.6	76.1	144.9	1,5
1-Methylpyrrole	C_5H_7N				8[e]	49.9	112.3	5
N-Methylpyrrolidine	$C_5H_{11}N$				-23[e]	18.5	78[e]	5
N-Methyl-2-pyrrolidinone	C_5H_9NO	1[e]	24[e]	53.1	92.3	147.2	229[e]	5
2-Methylquinoline	$C_{10}H_9N$	5.3	31.9	63.8	102.9	165.8	247.2	5
4-Methylquinoline	$C_{10}H_9N$	29[e]	54[e]	85[e]	127[e]	183.0	265.1	5
6-Methylquinoline	$C_{10}H_9N$	27[e]	51[e]	81[e]	122[e]	179.2	264.5	5
8-Methylquinoline	$C_{10}H_9N$	15[e]	40[e]	70[e]	111[e]	166.1	247.3	5
Methyl salicylate	$C_8H_8O_3$	-1[e]	22[e]	51[e]	88.8	141.8	219.9	5
Methylsilane	CH_6Si			-144[e]	-124.6	-97.5	-57.5	5
Methyl silicate	$C_4H_{12}O_4Si$				14.4	59.3	119.7	5
Methyl silyl ether	CH_6OSi				-90.2	-61.8	-18[e]	1
Methyl tetradecanoate	$C_{15}H_{30}O_2$		75[e]	110	155	214	295	4
2-Methyltetrahydrofuran	$C_5H_{10}O$				-20[e]	19.7	79.8	5
4-Methylthiazole	C_4H_5NS						67.0	5
Methyl thiocyanate	C_2H_3NS			-18.4	16.2	63.5	132.5	5
2-Methylthiophene	C_5H_6S		-58[e]	-32[e]	2[e]	47.9	112.2	1
3-Methylthiophene	C_5H_6S		-53[e]	-28[e]	6[e]	50.6	115.1	1
Methyl 10-undecenoate	$C_{12}H_{22}O_2$	10[e]	38[e]	73[e]	116[e]	172.2	247.1	5
Methyl vinyl ether	C_3H_6O			-114[e]	-89[e]	-52.7	4.6	1
Molybdenum	Mo	2469 s	2721	3039	3434	3939	4606	2
Molybdenum carbonyl	C_6MoO_6		17.4 s	42.8 s	73.1 s	109.9 s	155.4 s	5
Molybdenum(V) fluoride	F_5Mo				86.6	140.3	213[e]	26
Molybdenum(VI) fluoride	F_6Mo	-98 s	-82 s	-64 s	-41.2 s	-13.4 s	33.5	26
Molybdenum(VI) oxide	MoO_3				801	935	1151	4
Molybdenum(VI) oxytetrafluoride	F_4MoO	-21 s	3 s	33 s	69.3 s	117.3	184.1	26
Morpholine	C_4H_9NO				21[e]	64.5	128.5	1
β-Myrcene	$C_{10}H_{16}$			9.4	47.3	98.3	171.0	5
Myristicin	$C_{11}H_{12}O_3$	23[e]	53[e]	88.9	135.2	196.0	279.4	5
Naphthalene[b]	$C_{10}H_8$	3.2 s	24.1 s	49.3 s	80.7	135.6	217.5	1,5
1-Naphthalenecarboxylic acid	$C_{11}H_8O_2$				191.9	239.3	299.6	5
2-Naphthalenecarboxylic acid	$C_{11}H_8O_2$				197.9	246.0	308.1	5
1-Naphthol	$C_{10}H_8O$				137.2	196.7	281.8	5
2-Naphthol	$C_{10}H_8O$				140.7	200.5	286.8	5
1-Naphthylamine	$C_{10}H_9N$		62[e]	99.0	146.9	210.7	300.1	5
2-Naphthylamine	$C_{10}H_9N$	36.3 s	65.9 s	103 s	150.9	215.1	305.5	5
Neodymium	Nd	1322.3	1501.2	1725.3	2023[d]	2442[d]	3063[d]	3
Neon[a]	Ne	-261 s	-260 s	-258 s	-255 s	-252 s	-246.1	2
Neopentane[a]	C_5H_{12}		-107.5 s	-90.8 s	-68.8 s	-38.5 s	9.2	1,5
Nickel	Ni	1510	1677	1881	2137	2468	2911	2
Nickel carbonyl [$Ni(CO)_4$]	C_4NiO_4					-12	42	4
Nickel(II) chloride	Cl_2Ni	534 s	592 s	662 s	747 s	852 s	985 s	4
Niobium	Nb	2669	2934	3251	3637	4120	4740	2
Niobium(V) fluoride	F_5Nb				80	140	224	4
Nitric acid	HNO_3			-37[e]	-9[e]	28.4	82.2	5
Nitric oxide[a]	NO	-201 s	-195 s	-188 s	-179.3 s	-168.1 s	-151.9	5
4-Nitroaniline	$C_6H_6N_2O_2$	87.8 s			192.0	252.6	331.2	5
2-Nitroanisole	$C_7H_7NO_3$	15[e]	45[e]	82[e]	129[e]	189.4	271.8	5
Nitrobenzene	$C_6H_5NO_2$		10[e]	40[e]	78[e]	132[e]	210.3	1
Nitroethane	$C_2H_5NO_2$	-61[e]	-44[e]	-21[e]	8.3	50.1	113.5	5
Nitrogen[a]	N_2	-236 s	-232 s	-226.8 s	-220.2 s	-211.1 s	-195.9	1,5
Nitrogen pentoxide	N_2O_5	-71 s	-56 s	-40 s	-19.9 s	3.9 s	33.2	5

Name	Mol. form.	t/°C for 1 Pa	t/°C for 10 Pa	t/°C for 100 Pa	t/°C for 1 kPa	t/°C for 10 kPa	t/°C for 100 kPa	Ref.
Nitrogen tetroxide	N_2O_4	-92 s	-78 s	-61 s	-41.1 s	-16.6 s	28.7	5
Nitrogen trichloride	Cl_3N				-25[e]	13.2	70.6	5
Nitrogen trifluoride[a]	F_3N	-201[e]	-194[e]	-185[e]	-172.8	-155.5	-129.2	5
Nitromethane	CH_3NO_2				-2[e]	40[e]	100.8	1
4-Nitrophenol	$C_6H_5NO_3$	72.6 s	97.4 s					5
1-Nitropropane	$C_3H_7NO_2$	-56[e]	-37[e]	-13[e]	20[e]	64.8	130.8	1
2-Nitropropane	$C_3H_7NO_2$		-48[e]	-22[e]	10.7	55.6	119.8	1
N-Nitrosodimethylamine	$C_2H_6N_2O$				30.7	80.5	149.8	5
Nitrosyl chloride	ClNO		-116 s	-100 s	-78.7 s	-50.2	-5.7	5
Nitrosyl fluoride	FNO			-131[e]	-116.1	-94.3	-60.1	5
2-Nitrotoluene	$C_7H_7NO_2$	23[e]	40[e]	62[e]	94[e]	141.9	221.9	5
3-Nitrotoluene	$C_7H_7NO_2$			45[e]	89.7	148.7	231.3	5
1-Nitro-3-(trifluoromethyl)benzene	$C_7H_4F_3NO_2$		11[e]	39[e]	76.2	127.3	202.2	5
Nitrous oxide[a]	N_2O	-167 s	-157 s	-145.4 s	-131.1 s	-112.9 s	-88.7	5
Nitryl chloride	$ClNO_2$	-121[e]	-113[e]	-102[e]	-86.1	-60.9	-15.7	5
Nitryl fluoride	FNO_2		-156[e]	-144[e]	-128.1	-106.0	-72.6	5
Nonacosane	$C_{29}H_{60}$	148.2	182.8	221.2	271.5	340.2	439.7	5
Nonadecane	$C_{19}H_{40}$	71.1	99.1	133.8	178.8	240.1	330[e]	16
Nonanal	$C_9H_{18}O$		-3[e]	27.4	65.5	115.6	184.6	5
Nonane	C_9H_{20}	-46.8	-26.0	0.0	34.0	80.8	150.3	16
Nonanenitrile	$C_9H_{17}N$	-3[e]	21[e]	50.9	90.7	145.4	225.1	5
1-Nonanethiol	$C_9H_{20}S$	-2[e]	21[e]	49[e]	87[e]	140.4	219.2	5
Nonanoic acid	$C_9H_{18}O_2$	48[e]	69[e]	97[e]	133[e]	182.7	255.1	5
1-Nonanol	$C_9H_{20}O$		40[e]	64[e]	96.9	141.0	213.0	5,39
3-Nonanol, (±)-	$C_9H_{20}O$		24[e]	47[e]	78[e]	123.0	194.2	5
4-Nonanol	$C_9H_{20}O$			45[e]	76.4	121.3	192.0	5
5-Nonanol	$C_9H_{20}O$	13[e]	31[e]	54[e]	84.5	128.1	194.7	5
2-Nonanone	$C_9H_{18}O$		8[e]	35[e]	71[e]	121.0	194.0	5
5-Nonanone	$C_9H_{18}O$			-1[e]	39.1	94[e]	188[e]	5
1-Nonene	C_9H_{18}	-50.1	-29.4	-3.3	30.4	77.1	146.4	1,5
Nonylamine	$C_9H_{21}N$		9[e]	37[e]	75[e]	126.2	202.1	5
Nonylbenzene	$C_{15}H_{24}$	33.0	58.9	92.0	135.4	193.7	281.4	5
Nonylcyclohexane	$C_{15}H_{30}$	35[e]	60[e]	92[e]	134[e]	193.4	280.9	5
Nonylcyclopentane	$C_{14}H_{28}$	25[e]	49[e]	80[e]	120[e]	177.2	261.5	5
2,5-Norbornadiene	C_7H_8				-15[e]	27.4	91[e]	5
Octacosane	$C_{28}H_{58}$	136.5	169.8	210.9	263.1	332.0	430.6	5
Octadecane	$C_{18}H_{38}$	61.5	89.0	123.1	167.3	227.6	316[e]	16
Octadecanoic acid	$C_{18}H_{36}O_2$		153[e]	183[e]	223[e]	281.6	374.5	5
1-Octadecanol	$C_{18}H_{38}O$	106[e]	130[e]	160[e]	200.5	257.3	343.0	5
cis-9-Octadecenoic acid	$C_{18}H_{34}O_2$	94[e]	126[e]	165.5	214.5	277.0	359.7	5
trans-9-Octadecenoic acid	$C_{18}H_{34}O_2$		124[e]	166	216	280	361	4
Octanal	$C_8H_{16}O$			6[e]	45.7	97.8	170.2	5
Octane	C_8H_{18}		-42.6	-17.9	14.4	58.9	125.3	16
Octanenitrile	$C_8H_{15}N$	-15[e]	8[e]	37[e]	75[e]	127.7	204.4	5
1-Octanethiol	$C_8H_{18}S$	-15[e]	6[e]	34[e]	71[e]	122.1	198.5	5
Octanoic acid	$C_8H_{16}O_2$	37[e]	58[e]	85[e]	120[e]	165.5	238.4	1,5
1-Octanol	$C_8H_{18}O$	12[e]	30[e]	53[e]	84[e]	128.2	194.8	1,39
2-Octanol	$C_8H_{18}O$			40[e]	69.9	112.5	179.4	1,39
3-Octanol	$C_8H_{18}O$	12[e]	24[e]	40[e]	64[e]	102.8	174.1	1
4-Octanol	$C_8H_{18}O$			40[e]	66.9	107.3	176.0	1,39
2-Octanone	$C_8H_{16}O$		-3[e]	23[e]	57[e]	103.8	172.1	5
3-Octanone	$C_8H_{16}O$			8[e]	47.7	97[e]	161[e]	5
Octanoyl chloride	$C_8H_{15}ClO$	1[e]	22[e]	46[e]	74.7	109[e]	150[e]	5
1-Octene	C_8H_{16}	-65.7	-46.1	-21.4	10.5	54.9	120.9	1,5
cis-2-Octene	C_8H_{16}	-59[e]	-41[e]	-17[e]	15[e]	59[e]	125.2	5
trans-2-Octene	C_8H_{16}	-59[e]	-41[e]	-17[e]	14[e]	59[e]	124.5	5
cis-3-Octene	C_8H_{16}	-65[e]	-46[e]	-22[e]	10[e]	55.1	122.4	5
trans-3-Octene	C_8H_{16}	-61[e]	-43[e]	-19[e]	13[e]	57[e]	122.8	5
cis-4-Octene	C_8H_{16}	-63[e]	-44[e]	-20[e]	11[e]	56[e]	122.1	5
trans-4-Octene	C_8H_{16}	-65[e]	-46[e]	-22[e]	10[e]	54.6	121.8	5

Name	Mol. form.	t/°C for 1 Pa	t/°C for 10 Pa	t/°C for 100 Pa	t/°C for 1 kPa	t/°C for 10 kPa	t/°C for 100 kPa	Ref.
Octyl acetate	$C_{10}H_{20}O_2$	-26[e]	-3[e]	27[e]	66.3	120.0	198.2	5
Octylbenzene	$C_{14}H_{22}$	20.1	46.2	79.1	121.9	178.1	263.8	5
Octylcyclohexane	$C_{14}H_{28}$	16.9	44.3	77.8	120.0	177.6	263.2	5
Octylcyclopentane	$C_{13}H_{26}$	13[e]	36[e]	66[e]	106[e]	160.9	243.1	5
1-Octyne	C_8H_{14}	-59[e]	-40[e]	-16[e]	16[e]	60.3	125.8	1
2-Octyne	C_8H_{14}	-52[e]	-33[e]	-8[e]	25[e]	70.6	137.8	1
3-Octyne	C_8H_{14}	-55[e]	-35[e]	-11[e]	22[e]	66.8	132.8	1
4-Octyne	C_8H_{14}	-56[e]	-36[e]	-12[e]	21[e]	65.6	131.4	1
Osmium	Os	2887 s	3150	3478	3875	4365	4983	2
Osmium(V) fluoride	F_5Os			74.1	113.2	162.3	226[e]	26
Osmium(VI) fluoride	F_6Os	-89 s	-73 s	-54 s	-30.6 s	-1.7 s	47.4	26
2-Oxetanone	$C_3H_4O_2$		-21[e]	8[e]	45.5	93.8	159.3	5
Oxirane	C_2H_4O		-111[e]	-93[e]	-70[e]	-37.0	10.2	1
Oxygen[a]	O_2				-211.9	-200.5	-183.1	1,28
Ozone[a]	O_3	-189[e]	-182[e]	-172[e]	-158[e]	-139.7	-111.5	5
Palladium	Pd	1448 s	1624	1844	2122	2480	2961	2
Paraldehyde	$C_6H_{12}O_3$				17[e]	62.2	124[e]	5
Pentaborane(9)	B_5H_9				-34.8	3.8	57.6	4
Pentachloroethane	C_2HCl_5		-23[e]	3[e]	37.4	86.0	159.4	1
Pentacosane	$C_{25}H_{52}$	119.7	152.7	193.2	244.4	305.0	401.1	5
1*H*-Pentadecafluoroheptane	C_7HF_{15}				-7[e]	35.9	96.0	5
Pentadecane	$C_{15}H_{32}$	30.5	56.1	88.1	129.6	186.3	270.1	16
1,2-Pentadiene	C_5H_8	-109[e]	-93[e]	-73[e]	-46.1	-9.7	44.5	5
cis-1,3-Pentadiene	C_5H_8	-109[e]	-93[e]	-73[e]	-47.0	-10.5	43.7	1,5
trans-1,3-Pentadiene	C_5H_8			-75[e]	-49.0	-13[e]	42[e]	1
1,4-Pentadiene	C_5H_8	-120[e]	-105[e]	-86[e]	-60.9	-26.2	25.6	5
2,3-Pentadiene	C_5H_8	-106[e]	-90[e]	-70[e]	-42.9	-6.3	47.9	5
Pentafluorobenzene	C_6HF_5			-41[e]	-13[e]	27[e]	85.3	5
Pentafluorophenol	C_6HF_5O				39[e]	82[e]	145.2	5
1,1,1,2,2-Pentafluoropropane	$C_3H_3F_5$					-60[e]	-17.9	5
2,3,4,5,6-Pentafluorotoluene	$C_7H_3F_5$			-20[e]	11[e]	53.6	117.0	5
2,2,3,3,4-Pentamethylpentane	$C_{10}H_{22}$		-24[e]	3[e]	39[e]	89.1	165.5	5
2,2,3,4,4-Pentamethylpentane	$C_{10}H_{22}$		-29[e]	-3[e]	33[e]	82.8	158.7	5
Pentanal	$C_5H_{10}O$	-71[e]	-53[e]	-31[e]	-1[e]	40.8	102.6	5
Pentane[b]	C_5H_{12}	-115.5	-99.8	-80.0	-54.0	-18.1	35.7	16
Pentanedinitrile	$C_5H_6N_2$	24.1	52[e]	85[e]	126[e]	178[e]	245[e]	5
Pentanedioic acid	$C_5H_8O_4$		121[e]	153.2	191.9	240.3	302.5	5
1,5-Pentanediol	$C_5H_{12}O_2$	25[e]	52[e]	85[e]	125[e]	175.1	238.9	5
2,4-Pentanedione	$C_5H_8O_2$			-5[e]	24.7	67.8	137.4	1
Pentanenitrile	C_5H_9N	-54[e]	-34[e]	-8[e]	26[e]	72.2	140.9	1
1-Pentanethiol	$C_5H_{12}S$	-60[e]	-41[e]	-17[e]	15[e]	60[e]	126.2	1
2-Pentanethiol	$C_5H_{12}S$	-70[e]	-52[e]	-28[e]	3[e]	46.6	111.9	5
3-Pentanethiol	$C_5H_{12}S$	-70[e]	-51[e]	-28[e]	4[e]	47.7	113.4	5
Pentanoic acid	$C_5H_{10}O_2$	-7.4	15.3	42.7	76.3	122.1	185.7	5
1-Pentanol	$C_5H_{12}O$	-27[e]	-10[e]	12[e]	41[e]	79.8	137.4	5
2-Pentanol	$C_5H_{12}O$	-35[e]	-19[e]	1[e]	28.0	64.9	118.7	1
3-Pentanol	$C_5H_{12}O$	-41[e]	-25[e]	-4[e]	24[e]	61.1	114.9	5
2-Pentanone	$C_5H_{10}O$				-1[e]	40.3	101.9	1,5
3-Pentanone	$C_5H_{10}O$			-31[e]	-1[e]	40[e]	101.6	1
1-Pentene	C_5H_{10}	-118.9	-103.4	-84.0	-58.8	-23.3	29.6	1,5
cis-2-Pentene	C_5H_{10}	-113.8	-98.1	-78.4	-52.7	-16.8	36.6	1,5
trans-2-Pentene	C_5H_{10}	-114.5	-98.9	-79.1	-53.3	-17.5	36.0	1,5
4-Pentenoic acid	$C_5H_8O_2$	0[e]	19[e]	44[e]	77[e]	122.0	187.5	5
Pentyl acetate	$C_7H_{14}O_2$	-58[e]	-39[e]	-14[e]	20[e]	70.1	149[e]	5
Pentylamine	$C_5H_{13}N$		-52[e]	-29[e]	1[e]	42.8	104.0	5
Pentylbenzene	$C_{11}H_{16}$	-14[e]	8[e]	37[e]	74[e]	126.7	204.9	5
Pentylcyclohexane	$C_{11}H_{22}$	-17[e]	6[e]	34[e]	72[e]	124.2	202.7	5
1-Pentylnaphthalene	$C_{15}H_{18}$	34[e]	62[e]	96[e]	141.3	202.2	289[e]	5
1-Pentyne	C_5H_8			-75[e]	-49.1	-13.5	39.9	5
2-Pentyne	C_5H_8	-100[e]	-85[e]	-65[e]	-37.9	-0.5	55.7	5

Name	Mol. form.	t/°C for 1 Pa	t/°C for 10 Pa	t/°C for 100 Pa	t/°C for 1 kPa	t/°C for 10 kPa	t/°C for 100 kPa	Ref.
Perfluoroacetone	C_3F_6O			-113[e]	-94[e]	-67.8	-27.6	5
Perfluorobutane	C_4F_{10}		-122[e]	-105[e]	-82[e]	-49.8	-2.5	1,5
Perfluorocyclobutane	C_4F_8						-6.2	1
Perfluorocyclohexane	C_6F_{12}				-46.2 s	-7.6 s	48.9 s	5
Perfluorodecane	$C_{10}F_{22}$					52[e]	132.9	5
Perfluoro-2,3-dimethylbutane	C_6F_{14}					4.3	59.3	5
Perfluoroheptane	C_7F_{16}		-62[e]	-41[e]	-14[e]	24.7	82.1	1
Perfluorohexane	C_6F_{14}		-75[e]	-57[e]	-32[e]	2.8	56.8	5
Perfluoromethylcyclohexane	C_7F_{14}				-21[e]	18[e]	75.9	1
Perfluoro-2-methylpentane	C_6F_{14}				-33[e]	2.9	57.1	5
Perfluoro-3-methylpentane	C_6F_{14}	-95[e]	-80[e]	-60[e]	-34[e]	2.8	57.9	5
Perfluoronaphthalene	$C_{10}F_8$	5.2 s	25.1 s	48.1 s				5
Perfluorononane	C_9F_{20}					40[e]	114.7	5
Perfluorooctane	C_8F_{18}				5[e]	45.0	105.6	5
Perfluoropentane	C_5F_{12}				-54.7	-20.9	28.6	5
Perfluoropropane	C_3F_8		-139[e]	-124[e]	-105[e]	-77.5	-37.0	1
Perfluoropropene	C_3F_6	-150[e]	-138[e]	-122[e]	-101[e]	-72[e]	-30.6	5
Peroxyacetic acid	$C_2H_4O_3$				14.4	55.3	109.7	5
β-Phellandrene	$C_{10}H_{16}$			16[e]	53.2	104[e]	171.0	5
Phenanthrene	$C_{14}H_{10}$	53 s	83 s	120.8	170.4	238.4	337.7	5
Phenanthridine	$C_{13}H_9N$	79 s						5
Phenol	C_6H_6O	-9.7 s	9.6 s	34.1 s	68.9	113.7	181.4	1,5
2-Phenoxyethanol	$C_8H_{10}O_2$	21[e]	46[e]	75.9	115.4	168.7	244.8	5
Phenyl acetate	$C_8H_8O_2$		3[e]	33.1	72.2	123.9	195.5	5
Phenyl benzoate	$C_{13}H_{10}O_2$			102.3	151.4	217.9	313.3	5
2-Phenylethyl acetate	$C_{10}H_{12}O_2$	-4[e]	22[e]	54[e]	96[e]	152.3	232.0	5
Phenylhydrazine	$C_6H_8N_2$		38[e]	69[e]	109[e]	163.9	242.5	5
Phenyl isopropyl ether	$C_9H_{12}O$	-20[e]	-1[e]	23[e]	56[e]	103.7	176.9	5
Phenyl isothiocyanate	C_7H_5NS				79.4	105[e]	117[e]	5
1-Phenyl-2-propylamine, (±)-	$C_9H_{13}N$			33[e]	70.1	118[e]	202.0	5
Phenyl propyl ether	$C_9H_{12}O$		-10[e]	21[e]	61[e]	113.9	189.3	5
Phenyl salicylate	$C_{13}H_{10}O_3$				166.0	224.8	312.4	5
Phosphine[a]	H_3P	-182 s	-173 s	-161 s	-145 s	-122.7	-88.0	5
Phosphorus (white)	P	6 s	34 s	69	115	180	276	3,9
Phosphorus (red)	P	182 s	216 s	256 s	303 s	362 s	431 s	2,3
Phosphorus(III) bromide	Br_3P		-23[e]	5[e]	42.3	94.6	172.6	5
Phosphorus(V) bromide	Br_5P		-19 s	4 s	31 s	65.5 s	110.1	5
Phosphorus(III) chloride	Cl_3P	-93[e]	-77[e]	-55[e]	-26.0	14.5	75.7	5
Phosphorus(V) chloride	Cl_5P	-2 s	19 s	44 s	74 s	111.4 s	158.9 s	5
Phosphorus(III) chloride difluoride	ClF_2P				-119.5	-91.1	-47.6	5
Phosphorus(III) dichloride fluoride	Cl_2FP				-71.1	-37.4	13.5	5
Phosphorus(V) dichloride trifluoride	Cl_2F_3P		-120[e]	-101[e]	-77.1	-44.3	3[e]	7
Phosphorus(III) fluoride[a]	F_3P				-152[e]	-132.6	-101.4	5
Phosphorus(V) fluoride	F_5P	-157 s	-148 s	-137 s	-124.5 s	-108.6 s	-84.8	5
Phosphorus(III) oxide	O_3P_2				47.3	100.3	172.8	4
Phosphorus(V) oxide	O_5P_2	285 s	328 s	377.5 s	434.4 s	500.5 s	591	4
Phosphoryl bromide	Br_3OP				64[e]	115.5	191.4	5
Phosphoryl chloride	Cl_3OP					39.9	105.0	5
Phosphoryl fluoride	F_3OP	-124 s	-113 s	-100 s	-83.7 s	-64.1 s	-39.7 s	5
Phthalic anhydride	$C_8H_4O_3$	48.2 s	72.4 s			192.7	284.2	5
α-Pinene	$C_{10}H_{16}$	-48[e]	-27[e]	-1[e]	33.6	82.2	155.1	21
β-Pinene	$C_{10}H_{16}$	-43[e]	-22[e]	5.0	40.6	90.5	165.5	21
Piperidine	$C_5H_{11}N$				2[e]	43.3	105.8	5
Platinum	Pt	2057	2277[e]	2542	2870	3283	3821	2
Plutonium	Pu	1483	1680	1925	2238	2653	3226	2
Polonium	Po				573[e]	730.2	963.3	5
Polonium(IV) chloride	Cl_4Po					300.6	389.4	5
Potassium	K	200.2	256.5	328	424	559	756.2	13,30
Potassium bromide	BrK	597 s	674 s	773				25
Potassium chloride	ClK	625 s	704 s	804	945	1137	1411	12,23,25

Fluids

Name	Mol. form.	*t*/°C for 1 Pa	*t*/°C for 10 Pa	*t*/°C for 100 Pa	*t*/°C for 1 kPa	*t*/°C for 10 kPa	*t*/°C for 100 kPa	Ref.
Potassium fluoride	FK			869	1017	1216	1499	4
Potassium hydroxide	HKO	520[e]	601[e]	704	842	1035	1325	4
Potassium iodide	IK			731	866	1052	1322	4
Praseodymium	Pr	1497.7	1699.4	1954[d]	2298[d]	2781[d]	3506[d]	3
Propanal	C_3H_6O			-69[e]	-42[e]	-6[e]	47.7	1
Propane[a]	C_3H_8	-156.9	-145.6	-130.9	-111.4	-83.8	-42.3	1,41
1,2-Propanediamine, (±)-	$C_3H_{10}N_2$		-35.4	-12.0	18.8	61[e]	119[e]	5
1,2-Propanediol	$C_3H_8O_2$	-11[e]	13[e]	42[e]	78[e]	125.0	187.2	5
1,3-Propanediol	$C_3H_8O_2$	4[e]	30[e]	62[e]	101[e]	149.9	214.0	5
1,3-Propanedithiol	$C_3H_8S_2$	-53[e]	-28[e]	3[e]	43[e]	97[e]	172.4	5
Propanenitrile	C_3H_5N	-69.4	-55.3	-36.0	-7.9	35.2	97.4	1,5
1-Propanethiol	C_3H_8S	-94[e]	-78[e]	-57[e]	-29.1	9.6	67.4	1,5
2-Propanethiol	C_3H_8S	-102[e]	-87[e]	-67[e]	-41[e]	-3[e]	52.2	1
Propanoic acid	$C_3H_6O_2$			0[e]	35.1	79.9	140.8	1,5
Propanoic anhydride	$C_6H_{10}O_3$	-32[e]	-15[e]	6[e]	36[e]	77.6	142.9	5
1-Propanol	C_3H_8O	-54[e]	-38[e]	-16[e]	10[e]	47[e]	96.9	1,5
2-Propanol	C_3H_8O	-65[e]	-49[e]	-28[e]	-1.3	33.6	82.0	1,5
Propene[a]	C_3H_6	-160.6	-149.0	-134.3	-114.9	-88.2	-47.9	1,5
cis-1-Propenylbenzene	C_9H_{10}	-38[e]	-15.4	13.3	51.4	103.7	178.4	5
trans-1-Propenylbenzene	C_9H_{10}		-16[e]	13.3	51.6	103.7	178.4	5
2-Propoxyethanol	$C_5H_{12}O_2$				40[e]	85.6	149.3	5
Propyl acetate	$C_5H_{10}O_2$	-69[e]	-51[e]	-29[e]	0[e]	40.9	101.2	1
Propylamine	C_3H_9N		-81[e]	-63[e]	-38.3	-4.1	46.9	1,5
Propylbenzene	C_9H_{12}	-43[e]	-23[e]	4[e]	38[e]	86.7	158.8	1
Propyl benzoate	$C_{10}H_{12}O_2$	-8[e]	18[e]	50.2	92.3	149.2	230.5	5
Propyl butanoate	$C_7H_{14}O_2$	-35[e]	-19[e]	3[e]	32.0	74.9	142.8	5
Propylcyclohexane	C_9H_{18}	-46[e]	-26[e]	0[e]	35.1	83.6	156.2	5
Propylcyclopentane	C_8H_{16}	-60[e]	-41[e]	-16[e]	16.5	62.1	130.5	5
1-Propylcyclopentanol	$C_8H_{16}O$	9[e]	24[e]	43[e]	69.0	108.4	173.5	5
Propylene carbonate	$C_4H_6O_3$	-40[e]	-5[e]	43[e]	112[e]	220[e]	410[e]	5
Propyl formate	$C_4H_8O_2$	-78[e]	-62[e]	-42[e]	-15.1	23.0	80.4	1,5
Propyl hexanoate	$C_9H_{18}O_2$	-26[e]	-2[e]	28[e]	65.1	113.4	178[e]	5
Propyl isobutanoate	$C_7H_{14}O_2$		-28[e]	-5.7	24.5	67.5	133.3	5
Propyl methacrylate	$C_7H_{12}O_2$				26[e]	73.8	139.7	5
Propyl 3-methylbutanoate	$C_8H_{16}O_2$			1.8	38.9	87.9	155.6	5
Propyl nitrate	$C_3H_7NO_3$			-23.9	6.1	48.1	111[e]	5
Propyl octanoate	$C_{11}H_{22}O_2$	-2[e]	23[e]	55[e]	94.0	145.2	215[e]	5
Propyl propanoate	$C_6H_{12}O_2$	-62[e]	-42[e]	-18[e]	14[e]	58.3	122.0	5
Propyne	C_3H_4				-94[e]	-65.3	-23.2	1
Pulegone	$C_{10}H_{16}O$	37[e]	49.1	66.4	92.2	135.1	220.2	5
Pyridine	C_5H_5N			-23[e]	8[e]	51.0	114.9	1
Pyrrole	C_4H_5N			-8[e]	24[e]	66.7	129.4	1
Pyrrolidine	C_4H_9N		-59[e]	-38[e]	-10[e]	28.5	86.2	1
Quinoline	C_9H_7N	-1.3	23.7	55.4	96.8	153.4	236.5	1,5
Radium	Ra	546 s	633 s	764	936	1173	1526	2
Radon[a]	Rn	-163 s	-152 s	-139 s	-121.4 s	-97.6 s	-62.3	5
Rhenium	Re	3030 s	3341	3736	4227	4854	5681	2
Rhenium(VI) dioxydifluoride	F_2O_2Re				89.2	131.9	185[e]	26
Rhenium(V) fluoride	F_5Re			58.8	99.5	152[e]	221[e]	26
Rhenium(VI) fluoride	F_6Re	-97 s	-82 s	-63 s	-40.2 s	-11.9 s	33.4	26
Rhenium(VII) oxide	O_7Re_2	147 s	176 s	208 s	244 s	284 s	362	4
Rhenium(VII) oxypentafluoride	F_5ORe	-103 s	-84 s	-59 s	-28 s	13.7 s	72.8	26
Rhenium(VI) oxytetrafluoride	F_4ORe	5 s	26 s	50.7 s	80.1 s	117.1	171.2	26
Rhodium	Rh	2015	2223	2476	2790	3132	3724	2
Rubidium	Rb	160.4	212.5	278.9	368	496.1	685.3	13,30
Rubidium bromide	BrRb			766	903	1087	1350	4
Rubidium chloride	ClRb			777	916	1105	1379	4
Rubidium fluoride	FRb			910	1001	1145	1409	4,12
Rubidium iodide	IRb			733	866	1045	1302	4
Ruthenium	Ru	2315 s	2538	2814	3151	3572	4115	2

Name	Mol. form.	t/°C for 1 Pa	t/°C for 10 Pa	t/°C for 100 Pa	t/°C for 1 kPa	t/°C for 10 kPa	t/°C for 100 kPa	Ref.
Salicylaldehyde	$C_7H_6O_2$		-1[e]	29[e]	68[e]	120.7	196.2	5
Samarium	Sm	728 s	833 s	967 s	1148[d]	1402[d]	1788[d]	3
Scandium	Sc	1372 s	1531 s	1733[d]	1993[d]	2340[d]	2828[d]	3
Selenium	Se	227	279	344	431	540	685	3
Selenium dioxide	O_2Se	124.5 s	153.9 s	188 s	228 s	275 s	315 s	38
Selenium hexafluoride	F_6Se	-143 s	-132 s	-118 s	-100.7 s	-77.8 s	-46.5 s	5
Selenium tetrachloride	Cl_4Se	23 s	45 s	71 s	102 s	141.4 s	191.1 s	5
Selenium tetrafluoride	F_4Se				13.6	51.6	104.7	5
Silane[a]	H_4Si			-181	-165.4	-143.7	-111.8	4
Silicon	Si	1635	1829	2066	2363	2748	3264	2
Silicon dioxide (α-quartz)	O_2Si	1966[d]	2149[d]	2368[d]				8
Silver	Ag	1010	1140	1302	1509	1782	2160	2
Silver(I) bromide	AgBr	569[d]	656[d]	765[d]	905[d]	1093[d]	1359[d]	9
Silver(I) chloride	AgCl	670	769	873	1052	1264	1561	4
Silver(I) iodide	AgI	594	686	803	959	1177	1503	4
Sodium	Na	280.6	344.2	424.3	529	673	880.2	13,30
Sodium bromide	BrNa			791	931	1120	1389	4
Sodium chloride	ClNa	653 s	733 s	835	987	1182	1461	12,23,25
Sodium cyanide	CNNa		672[e]	798	961	1182	1497	4
Sodium fluoride	FNa		920 s	1058	1218	1426	1702	4,12,24
Sodium hydroxide	HNaO	513	605	722	874	1080	1377	4
Sodium iodide	INa			753	883	1058	1301	4
Spiro[2.2]pentane	C_5H_8	-110[e]	-95[e]	-76[e]	-51[e]	-15[e]	38.6	5
Stearaldehyde	$C_{18}H_{36}O$			142[e]	186[e]	246.9	336.7	5
cis-Stilbene	$C_{14}H_{12}$	26[e]	54[e]	88[e]	130.4	183[e]	253[e]	5
trans-Stilbene	$C_{14}H_{12}$				155.6	218.1	305.8	5
Strontium	Sr	523 s	609 s	717 s	866	1072	1373	2
Strontium oxide	OSr	1789 s	1903 s	2047 s	2235 s	2488 s		4
Styrene	C_8H_8		-31[e]	-5[e]	28.6	75.4	144.7	1
Succinic anhydride	$C_4H_4O_3$				121[e]	180.8	260.8	5
Succinonitrile	$C_4H_4N_2$	24.8 s					266.0	5
Sulfolane	$C_4H_8O_2S$		49[e]	87[e]	135[e]	198.0	283.5	5
Sulfur (rhombic)	S	102 s	135	176	235	318	444	3
Sulfur bromide [BrSSBr]	Br_2S_2	-7[e]	15[e]	42[e]	78.4	128.1	200.9	5
Sulfur chloride [ClSSCl]	Cl_2S_2	-55[e]	-36[e]	-12[e]	21.0	67.2	137.1	5
Sulfur decafluoride	$F_{10}S_2$					-22.0	28.5	5
Sulfur dichloride	Cl_2S	-76[e]	-61[e]	-41[e]	-16.7	15.3	58.7	5
Sulfur dioxide[a]	O_2S			-98 s	-80 s	-52.2	-10.3	1,5
Sulfur hexafluoride[a]	F_6S	-158 s	-147 s	-133.6 s	-116.6 s	-94.4 s	-64.1 s	5
Sulfuric acid	H_2O_4S	72	103	140	187	248	330	4
Sulfur tetrafluoride	F_4S				-110.0	-82.1	-40.3	5
Sulfur trioxide (α-form)	O_3S				-20 s	6.6 s	44.5	5
Sulfuryl chloride	Cl_2O_2S				-27[e]	11.8	69.0	5
Tantalum	Ta	3024	3324	3684	4122	4666	5361	2
Tantalum(V) fluoride	F_5Ta					119	229	4
Technetium	Tc	2454[e]	2725[e]	3051[e]	3453[e]	3961[e]	4621[e]	2
Tellurium	Te			502[e]	615[e]	768.8	992.4	5
Tellurium hexafluoride	F_6Te	-142 s	-130 s	-115 s	-96 s	-71.8 s	-39.1 s	5
Tellurium tetrachloride	Cl_4Te				237[e]	299.4	387.8	5
Terbium	Tb	1516.1	1706.1	1928[d]	2232[d]	2640[d]	3218[d]	3
o-Terphenyl	$C_{18}H_{14}$	66[e]	94[e]	129[e]	176[e]	241.3	336.3	5
m-Terphenyl	$C_{18}H_{14}$	87[e]	118[e]	156[e]	206.6	275.3	374.6	5
p-Terphenyl	$C_{18}H_{14}$	127.1 s	154.7 s		217.2	284.0	383.0	5
α-Terpineol	$C_{10}H_{18}O$			48	89	142	217	4
Terpinolene	$C_{10}H_{16}$			26.5	64.9	115.4	184.6	5
1,1,2,2-Tetrabromoethane	$C_2H_2Br_4$	14[e]	38[e]	69[e]	109[e]	163.7	242.9	5
Tetrabromoethene	C_2Br_4		-54.5 s	-31.7 s	-3.5 s	32.2 s	226.0	5
Tetrabromomethane	CBr_4			25.6 s	65.8 s	111.6	188.9	5
1,1,1,2-Tetrachloro-2,2-difluoroethane	$C_2Cl_4F_2$				-7[e]	31.0	91.1	5
1,1,2,2-Tetrachloro-1,2-difluoroethane	$C_2Cl_4F_2$					32.3	92.5	1

Name	Mol. form.	t/°C for 1 Pa	t/°C for 10 Pa	t/°C for 100 Pa	t/°C for 1 kPa	t/°C for 10 kPa	t/°C for 100 kPa	Ref.
1,1,1,2-Tetrachloroethane	$C_2H_2Cl_4$	-58[e]	-40[e]	-15[e]	17[e]	62.2	129.7	1
1,1,2,2-Tetrachloroethane	$C_2H_2Cl_4$		-22[e]	1[e]	32.4	76.9	144.7	1
Tetrachloroethene	C_2Cl_4			-22[e]	10[e]	54.4	120.7	1
Tetrachloromethane[a]	CCl_4	-79.4 s	-70.8 s	-53.5 s	-24.4 s	15.8	76.2	1,5
1,1,1,2-Tetrachloropropane	$C_3H_4Cl_4$	-48[e]	-28[e]	-2[e]	32[e]	79.1	149.5	5
Tetrachlorosilane[a]	Cl_4Si				-39[e]	0[e]	57.3	1
Tetracosane	$C_{24}H_{50}$	115.0	148.1	188.5	239.1	295.4	390.6	5
Tetradecamethylhexasiloxane	$C_{14}H_{42}O_5Si_6$	6[e]	36[e]	72[e]	117[e]	176.0	259.1	5
Tetradecane	$C_{14}H_{30}$	19.1	44.1	75.3	115.7	171.1	253.0	16
Tetradecanenitrile	$C_{14}H_{27}N$	52[e]	79[e]	114.0	159.0	219.7	306.3	5
Tetradecanoic acid	$C_{14}H_{28}O_2$	96[e]	118[e]	147[e]	186[e]	241.3	325.6	5
1-Tetradecanol	$C_{14}H_{30}O$	80.0	110.5	149.6	152[e]	205.3	286.7	5
1-Tetradecene	$C_{14}H_{28}$	16.1	41.3	72.7	113.2	168.7	250.6	5
Tetradecylamine	$C_{14}H_{31}N$			104[e]	147[e]	206.1	290.9	5
Tetraethylene glycol	$C_8H_{18}O_5$	89[e]	117[e]	151.1	192.2	242.9	307.3	5
Tetraethylene glycol dimethyl ether	$C_{10}H_{22}O_5$				138[e]	200.9	275.3	5
Tetraethylsilane	$C_8H_{20}Si$			-6.5	30.5	80.6	152.6	5
1,2,3,4-Tetrafluorobenzene	$C_6H_2F_4$			-36[e]	-7[e]	33.8	94.0	1
1,2,3,5-Tetrafluorobenzene	$C_6H_2F_4$			-43[e]	-14[e]	25.5	84.1	1
1,2,4,5-Tetrafluorobenzene	$C_6H_2F_4$					30.7	89.9	1
Tetrafluorodiborane	B_2F_4						-34	1
1,1,2,2-Tetrafluoro-1,2-dinitroethane	$C_2F_4N_2O_4$				-30[e]	6.4	59.5	5
1,1,1,2-Tetrafluoroethane	$C_2H_2F_4$				-94.3	-66.8	-26.4	17
1,1,2,2-Tetrafluoroethane	$C_2H_2F_4$				-96.0	-66.9	-23.3	5
Tetrafluoroethene	C_2F_4				-132.3	-109.7	-75.8	1
Tetrafluoromethane[a]	CF_4	-199.9 s	-193 s	-183.9 s	-171.6	-153.9	-128.3	1,5
2,2,3,3-Tetrafluoro-1-propanol	$C_3H_4F_4O$			-10[e]	17[e]	53.9	107.2	5
Tetrafluorosilane[a]	F_4Si	-166 s	-157 s	-145.6 s	-132.3 s	-115.7 s	-94.9 s	4,7
cis-Tetrahydro-2,5-dimethylthiophene	$C_6H_{12}S$	-53[e]	-34[e]	-8[e]	25[e]	72.0	142.1	5
Tetrahydrofuran	C_4H_8O	-94[e]	-78[e]	-57.3	-29.8	9[e]	65.6	1
Tetrahydrofurfuryl alcohol	$C_5H_{10}O_2$	-40[e]	-16[e]	15[e]	55[e]	106[e]	176.8	5
1,2,3,4-Tetrahydro-5-methylnaphthalene	$C_{11}H_{14}$	9[e]	31[e]	60[e]	99[e]	153.1	233.8	5
1,2,3,4-Tetrahydro-6-methylnaphthalene	$C_{11}H_{14}$	17[e]	36[e]	62[e]	97[e]	147.8	228.5	5
Tetrahydro-3-methyl-2*H*-thiopyran	$C_6H_{12}S$	-48[e]	-27[e]	0[e]	35[e]	84.1	157.5	5
1,2,3,4-Tetrahydronaphthalene	$C_{10}H_{12}$	-21[e]	3[e]	33.2	74.1	127.4	207.8	5
Tetrahydropyran	$C_5H_{10}O$				-15[e]	26.0	88[e]	5
Tetrahydro-2*H*-pyran-2-one	$C_5H_8O_2$		5[e]	35.1	74.4	128.3	207.0	5
Tetrahydrothiophene	C_4H_8S	-66[e]	-47[e]	-23[e]	9.4	54.1	120.5	1
1,2,3,4-Tetramethylbenzene	$C_{10}H_{14}$		7[e]	36[e]	74[e]	126.6	204.5	5
1,2,3,5-Tetramethylbenzene	$C_{10}H_{14}$	-19[e]	3[e]	32[e]	69[e]	120.9	197.5	5
1,2,4,5-Tetramethylbenzene	$C_{10}H_{14}$					119.9	196.3	5
2,2,3,3-Tetramethylbutane	C_8H_{18}	-62.5 s	-44 s	-20.9 s	8.9 s	48.8 s	105.8	5
1,1,3,3-Tetramethylcyclopentane	C_9H_{18}	-72[e]	-54[e]	-30[e]	2[e]	47[e]	117.4	5
2,2,3,3-Tetramethylhexane	$C_{10}H_{22}$	-46[e]	-25[e]	1[e]	36[e]	85.6	159.8	5
2,2,5,5-Tetramethylhexane	$C_{10}H_{22}$			-10[e]	22[e]	68.3	137.0	5
2,2,3,3-Tetramethylpentane	C_9H_{20}				21[e]	68.5	139.8	1
2,2,3,4-Tetramethylpentane	C_9H_{20}	-61[e]	-42[e]	-17[e]	16[e]	62.5	132.6	1
2,2,4,4-Tetramethylpentane	C_9H_{20}		-49[e]	-25[e]	8[e]	53.2	121.8	1
2,3,3,4-Tetramethylpentane	C_9H_{20}	-57[e]	-37[e]	-12[e]	22[e]	69.7	141.1	1
2,2,4,4-Tetramethyl-3-pentanol	$C_9H_{20}O$				58	100	167	5
Tetramethylsilane	$C_4H_{12}Si$			-83[e]	-59[e]	-25[e]	26.7	5
Tetramethylstannane	$C_4H_{12}Sn$			-55.0	-25.6	16.6	77.7	5
Tetramethylurea	$C_5H_{12}N_2O$			20.7	58.0	106.7	179.5	5
Tetranitromethane	CN_4O_8				18.0	61.8	124[e]	5
Thallium	Tl	609	704	824	979	1188	1485	2
Thallium(I) bromide	BrTl				509	635	817	4
Thallium(I) chloride	ClTl				504	626	806	4
Thallium(I) iodide	ITl				520	644	821	4
Thiacyclohexane	$C_5H_{10}S$				24[e]	71.1	141.2	5
Thiazole	C_3H_3NS					54.4	117.8	5

Name	Mol. form.	t/°C for 1 Pa	t/°C for 10 Pa	t/°C for 100 Pa	t/°C for 1 kPa	t/°C for 10 kPa	t/°C for 100 kPa	Ref.
Thietane	C_3H_6S		-62[e]	-40[e]	-9[e]	32.5	94.5	5
Thionyl bromide	Br_2OS	-49[e]	-29[e]	-5[e]	27.8	72.9	139.6	5
Thionyl chloride	Cl_2OS	-99[e]	-81[e]	-58[e]	-27.1	14.6	75.2	5
Thionyl fluoride	F_2OS			-124[e]	-106.5	-81.5	-44.1	5
Thiophene	C_4H_4S				-17[e]	23.7	83.7	5
Thorium	Th	2360	2634	2975	3410	3986	4782	2
Thulium	Tm	844 s	962 s	1108 s	1297 s	1548[d]	1944[d]	3
Thymol	$C_{10}H_{14}O$	18.9 s	37.9 s	59.5	101.2	155.0	230.4	5
Tin	Sn	1224	1384	1582	1834	2165	2620	2
Tin(IV) bromide	Br_4Sn				67	122	204	4
Tin(II) chloride	Cl_2Sn		253	308	381	479	622	4
Tin(IV) iodide	I_4Sn				167.1	242.7	347.7	4
Titanium	Ti	1709	1898	2130[e]	2419	2791	3285	2
Toluene	C_7H_8	-78.1	-57.1	-31.3	1.5	45.2	110.1	5
Toluene-2,4-diisocyanate	$C_9H_6N_2O_2$		39[e]	72[e]	113.9	169.7	247[e]	5
Tribromoacetaldehyde	C_2HBr_3O			15.0	52.7	103.0	173.5	5
1,2,3-Tribromobutane	$C_4H_7Br_3$	0[e]	23[e]	53[e]	91[e]	143.7	219.5	5
1,2,4-Tribromobutane	$C_4H_7Br_3$	-3[e]	20[e]	49[e]	87[e]	139.4	214.5	5
1,1,2-Tribromoethane	$C_2H_3Br_3$	-18[e]	4[e]	32[e]	68[e]	117.1	188.4	5
Tribromomethane	$CHBr_3$				30.5	78.3	148.8	1
Tributylamine	$C_{12}H_{27}N$	-26[e]	1[e]	35[e]	77.7	134.5	213.4	5
Tributyl phosphate	$C_{12}H_{27}O_4P$					205[e]	288.3	5
Trichloroacetaldehyde	C_2HCl_3O			-41.6	-9.8	33.8	97.4	5
Trichloroacetic acid	$C_2HCl_3O_2$				83.8	130.0	197.2	1,5
Trichloroacetonitrile	C_2Cl_3N				-16[e]	25.3	85.1	1
Trichloroacetyl chloride	C_2Cl_4O			-25[e]	7[e]	51.7	117.8	1,5
1,1,1-Trichloroethane	$C_2H_3Cl_3$				-25.3	14.2	73.7	5
1,1,2-Trichloroethane	$C_2H_3Cl_3$			-23[e]	7[e]	49.9	113.4	1
Trichloroethene	C_2HCl_3	-74[e]	-59[e]	-39[e]	-12[e]	26.7	86.8	1
Trichloroethoxysilane	$C_2H_5Cl_3OSi$	-78[e]	-60[e]	-36.0	-4.6	38.7	102.0	5
Trichloroethylsilane	$C_2H_5Cl_3Si$	-79[e]	-61[e]	-38[e]	-8[e]	34.9	98.7	5
Trichlorofluoromethane[a]	CCl_3F		-107[e]	-89[e]	-63[e]	-28.5	23.3	1,5
Trichloromethane[a]	$CHCl_3$			-61[e]	-34[e]	4.3	60.8	1
(Trichloromethyl)benzene	$C_7H_5Cl_3$		9[e]	40.6	81.5	136.2	213.0	5
Trichloromethylsilane	CH_3Cl_3Si		-83[e]	-61[e]	-33[e]	7[e]	65.7	1
Trichloronitromethane	CCl_3NO_2		-59[e]	-30[e]	4.4	47.8	112.0	5
2,4,6-Trichlorophenol	$C_6H_3Cl_3O$			71.8	114.0	169.5	245.7	5
Trichlorophenylsilane	$C_6H_5Cl_3Si$			33[e]	70.2	122.6	201[e]	5
1,1,3-Trichloropropane	$C_3H_5Cl_3$	-51[e]	-31[e]	-5[e]	28[e]	75.3	145.1	5
1,2,3-Trichloropropane	$C_3H_5Cl_3$			2[e]	37[e]	84.9	156.3	5
Trichlorosilane	Cl_3HSi			-81[e]	-56[e]	-21[e]	31.6	7
1,3,5-Trichloro-2,4,6-trifluorobenzene	$C_6Cl_3F_3$	-19[e]	4[e]	32[e]	70[e]	121.7	197.9	1
1,1,1-Trichloro-2,2,2-trifluoroethane	$C_2Cl_3F_3$						45.6	1,5
1,1,2-Trichloro-1,2,2-trifluoroethane	$C_2Cl_3F_3$					-8.2	47.3	1,5
Tricosane	$C_{23}H_{48}$	102.9	135.1	174.8	221[e]	285.3	379.5	5
Tri-*o*-cresyl phosphate	$C_{21}H_{21}O_4P$	119.0	156.1	201.0	256.3	326.3	418[e]	5
Tri-*m*-cresyl phosphate	$C_{21}H_{21}O_4P$	147.8	177.3	211.4	251.3	298[e]	355[e]	5
Tri-*p*-cresyl phosphate	$C_{21}H_{21}O_4P$	140.6	174[e]	214[e]	262[e]	320[e]	392[e]	5
Tridecane	$C_{13}H_{28}$	7.2	31.5	61.8	101.1	155.1	234.9	16
Tridecanoic acid	$C_{13}H_{26}O_2$	87[e]	109[e]	138[e]	176[e]	230.3	311.5	5
1-Tridecanol	$C_{13}H_{28}O$	71.6	101.0	103[e]	140[e]	192.3	273.1	5
1-Tridecene	$C_{13}H_{26}$	4.1	28.5	59.0	98.3	152.5	232.3	5
Tris(2-hydroxyethyl)amine	$C_6H_{15}NO_3$	75[e]	108[e]	148[e]	196[e]	256.7	334[e]	5
Triethylamine	$C_6H_{15}N$	-58[e]	-45[e]	-29[e]	-5[e]	29.9	88.5	1
Triethylene glycol	$C_6H_{14}O_4$	44[e]	74[e]	109.0	152.6	207.2	277.9	5
Triethyl phosphate	$C_6H_{15}O_4P$			34	76	132	211	4
Trifluoroacetic acid	$C_2HF_3O_2$					16.8	71.4	1,5
Trifluoroacetic acid anhydride	$C_4F_6O_3$			-63[e]	-39[e]	-7.1	38.8	5
Trifluoroacetonitrile	C_2F_3N				-126.1	-102.5	-67.8	1
1,3,5-Trifluorobenzene	$C_6H_3F_3$					18.2	75.0	5

Name	Mol. form.	t/°C for 1 Pa	t/°C for 10 Pa	t/°C for 100 Pa	t/°C for 1 kPa	t/°C for 10 kPa	t/°C for 100 kPa	Ref.
1,1,1-Trifluoroethane	$C_2H_3F_3$				-113[e]	-86.6	-47.8	1
2,2,2-Trifluoroethanol	$C_2H_3F_3O$			-33[e]	-8[e]	26.0	74[e]	5
Trifluoromethane[a]	CHF_3			-152[e]	-136[e]	-114.4	-82.3	1
(Trifluoromethyl)benzene	$C_7H_5F_3$				-3[e]	39[e]	101.6	5
Trifluoromethyl difluoromethyl ether	C_2HF_5O	-147[e]	-136[e]	-121[e]	-102[e]	-75.0	-35.4	20
Triiodomethane	CHI_3	51.1 s	82.7 s	121[e]			218.0	5
Triisobutylamine	$C_{12}H_{27}N$		1[e]	28.9	64.9	112.5	178.5	5
Triisopropyl borate	$C_9H_{21}BO_3$					73.1	139.0	5
Trimethylamine	C_3H_9N		-114[e]	-97[e]	-75.0	-43.8	2.6	1,5
2,4,6-Trimethylaniline	$C_9H_{13}N$	12[e]	36[e]	66[e]	104.1	154.9	226[e]	5
Trimethylarsine	C_3H_9As			-74[e]	-45[e]	-5.4	52.0	5
1,2,3-Trimethylbenzene	C_9H_{12}		-12[e]	15[e]	52[e]	101.5	175.6	1
1,2,4-Trimethylbenzene	C_9H_{12}	-37[e]	-16[e]	11[e]	47[e]	95.9	168.9	1
1,3,5-Trimethylbenzene	C_9H_{12}	-39[e]	-18[e]	9[e]	43.7	92.4	164.3	1
Trimethyl borate	$C_3H_9BO_3$				-14[e]	15.6	67.9	5
2,2,3-Trimethylbutane	C_7H_{16}				-23.2	18.1	80.4	5
2,3,3-Trimethyl-1-butene	C_7H_{14}	-91[e]	-75[e]	-53[e]	-24.2	16.3	77.5	5
Trimethylchlorosilane	C_3H_9ClSi				-37.8	0.4	57.3	5
1,1,2-Trimethylcyclohexane	C_9H_{18}			-12[e]	23[e]	71.5	145.5	5
1,1,3-Trimethylcyclohexane	C_9H_{18}	-60[e]	-41[e]	-16[e]	18[e]	65.2	136.1	5
1α,2β,4β-1,2,4-Trimethylcyclohexane	C_9H_{18}	-71[e]	-50[e]	-22[e]	15[e]	65.7	140.7	5
1α,3α,5β-1,3,5-Trimethylcyclohexane	C_9H_{18}	-72[e]	-50[e]	-22[e]	14[e]	65.1	140.0	5
1,1,2-Trimethylcyclopentane	C_8H_{16}				2[e]	46.2	113.2	5
1,1,3-Trimethylcyclopentane	C_8H_{16}	-77[e]	-59[e]	-36[e]	-5[e]	38.7	104.4	5
1α,2α,4β-1,2,4-Trimethylcyclopentane	C_8H_{16}	-70[e]	-52[e]	-28[e]	4[e]	48.9	116.2	5
1α,2β,4α-1,2,4-Trimethylcyclopentane	C_8H_{16}	-74[e]	-56[e]	-33[e]	-1[e]	42.8	108.8	5
1,1,2-Trimethylcyclopropane	C_6H_{12}	-109[e]	-94[e]	-73[e]	-46[e]	-7[e]	52.0	5
2,2,6-Trimethylheptane	$C_{10}H_{22}$	-46[e]	-27[e]	-2[e]	32[e]	78.5	148.4	5
3,3,5-Trimethylheptane	$C_{10}H_{22}$			0[e]	35[e]	82.7	155.2	5
2,2,4-Trimethylhexane	C_9H_{20}	-66.1	-46.4	-21.3	11.8	57.7	126.0	5
2,2,5-Trimethylhexane	C_9H_{20}	-65.1	-45.8	-21.2	11.2	56.2	123.7	1,5
2,3,3-Trimethylhexane	C_9H_{20}	-58[e]	-38[e]	-13[e]	20[e]	66.7	137.2	5
2,3,5-Trimethylhexane	C_9H_{20}	-60[e]	-41[e]	-16[e]	17[e]	62.3	130.9	5
2,4,4-Trimethylhexane	C_9H_{20}	-62[e]	-43[e]	-18[e]	15[e]	61.0	130.2	5
3,3,4-Trimethylhexane	C_9H_{20}	-53[e]	-33[e]	-7[e]	28[e]	76.3	148.9	5
2,4,7-Trimethyloctane	$C_{11}H_{24}$				43[e]	94[e]	170.4	5
Trimethylolpropane	$C_6H_{14}O_3$	73[e]	98[e]	128[e]	167.8	220.5	295[e]	5
2,2,3-Trimethylpentane	C_8H_{18}	-74[e]	-56[e]	-32[e]	-0.8	43.1	109.4	5
2,2,4-Trimethylpentane	C_8H_{18}	-81.9	-63.4	-39.8	-8.9	34.0	98.8	5
2,3,3-Trimethylpentane	C_8H_{18}	-72[e]	-54[e]	-30[e]	2.1	46.9	114.3	5
2,3,4-Trimethylpentane	C_8H_{18}	-74[e]	-54.5	-30.0	2.2	46.7	113.1	1,5
2,4,4-Trimethyl-2-pentanol	$C_8H_{18}O$		-7[e]	13[e]	40[e]	79.8	146.1	5
2,2,4-Trimethyl-3-pentanol	$C_8H_{18}O$	-2[e]	9[e]	24[e]	47[e]	82.6	150.4	5
2,2,4-Trimethyl-3-pentanone	$C_8H_{16}O$			11.3	42.1	81.7	134.6	5
2,3,3-Trimethyl-1-pentene	C_8H_{16}		-53[e]	-30[e]	1[e]	43.8	107.9	5
2,4,4-Trimethyl-1-pentene	C_8H_{16}	-79[e]	-61[e]	-38[e]	-7[e]	36.2	101.0	5
2,3,4-Trimethyl-2-pentene	C_8H_{16}	-68[e]	-49[e]	-26[e]	6[e]	50.0	115.8	5
2,4,4-Trimethyl-2-pentene	C_8H_{16}	-73[e]	-56[e]	-33[e]	-2[e]	40.4	104.5	5
Trimethyl phosphate	$C_3H_9O_4P$	-31[e]	-7[e]	23.6	62.8	116.0	192.0	5
Trimethylphosphine	C_3H_9P			-81[e]	-53[e]	-15.0	37.1	5
Trimethylstibine	C_3H_9Sb			-56[e]	-23.8	19[e]	80[e]	5
Trinitroglycerol	$C_3H_5N_3O_9$	48.6	75.7	118[e]	191[e]	353[e]	1007[e]	5
1,3,5-Trioxane	$C_3H_6O_3$					53[e]	113.7	1
Triphenylmethane	$C_{19}H_{16}$	81 s		112[e]	175[e]	254.6	360.0	5
Tripropylamine	$C_9H_{21}N$	-39[e]	-18[e]	8[e]	42[e]	88.2	156.0	5
Tris(perfluorobutyl)amine	$C_{12}F_{27}N$		3[e]	29.0	63.3	109.9	176.8	5
Tungsten	W	3204 s	3500	3864	4306	4854	5550	2
Tungsten(VI) fluoride	F_6W	-107 s	-92 s	-74 s	-52.1 s	-24.8 s	16.9	26
Tungsten(VI) oxytetrafluoride	F_4OW	2 s	25 s	52.1 s	84.3 s	126.7	185.4	26
Undecane	$C_{11}H_{24}$	-18.4	4.3	32.6	69.5	120.2	195.4	16

Name	Mol. form.	*t*/°C for 1 Pa	*t*/°C for 10 Pa	*t*/°C for 100 Pa	*t*/°C for 1 kPa	*t*/°C for 10 kPa	*t*/°C for 100 kPa	Ref.
Undecanenitrile	$C_{11}H_{21}N$			78.6	120.3	177.3	259.9	5
1-Undecanethiol	$C_{11}H_{24}S$	23[e]	47[e]	77[e]	118[e]	173.6	256.8	5
Undecanoic acid	$C_{11}H_{22}O_2$	68[e]	90[e]	118[e]	156[e]	207.2	283.6	5
1-Undecanol	$C_{11}H_{24}O$	52.2	80.0	82[e]	118[e]	167.6	244.1	5
2-Undecanone	$C_{11}H_{22}O$	17[e]	37[e]	64.3	103.0	153.6	232.6	1,5
6-Undecanone	$C_{11}H_{22}O$		28[e]	57[e]	95[e]	148.4	226.9	1
1-Undecene	$C_{11}H_{22}$	-21.6	1.2	29.7	66.4	117.1	192.2	5
cis-2-Undecene	$C_{11}H_{22}$	-14[e]	7[e]	34[e]	70.2	120.6	196[e]	5
trans-2-Undecene	$C_{11}H_{22}$	-14[e]	7[e]	33[e]	69.3	119.6	195[e]	5
cis-4-Undecene	$C_{11}H_{22}$	-19[e]	3[e]	30[e]	66.6	117.1	192[e]	5
trans-4-Undecene	$C_{11}H_{22}$	-17[e]	4[e]	31[e]	67.1	117.4	193[e]	5
cis-5-Undecene	$C_{11}H_{22}$	-19[e]	2[e]	30[e]	66.2	116.7	191[e]	5
trans-5-Undecene	$C_{11}H_{22}$	-18[e]	3[e]	31[e]	67.0	117.4	192[e]	5
10-Undecenoic acid	$C_{11}H_{20}O_2$	35[e]	67[e]	105[e]	150.0	205.4	274.5	5
1-Undecyne	$C_{11}H_{20}$	-22[e]	0[e]	29[e]	67[e]	118.5	194.5	5
2-Undecyne	$C_{11}H_{20}$	-17[e]	6[e]	35[e]	74[e]	127.4	205.4	5
Uranium	U	2052	2291	2586	2961	3454	4129	2
Vanadium	V	1828 s	2016	2250	2541	2914	3406	2
Vinyl acetate	$C_4H_6O_2$	-88[e]	-71[e]	-50[e]	-22[e]	16.2	72.2	1
Vinyl butanoate	$C_6H_{10}O_2$					53[e]	114.5	5
4-Vinylcyclohexene	C_8H_{12}	-62[e]	-43[e]	-19[e]	14.1	59.9	129[e]	5
Vinyl formate	$C_3H_4O_2$			-58[e]	-34[e]	-1.6	46.2	1
Vinyl propanoate	$C_5H_8O_2$					31.2	94[e]	5
Water[b, c]	H_2O	-60.7 s	-42.2 s	-20.3 s	7.0	45.8	99.6	36,37
Xenon[a]	Xe	-190 s	-181 s	-170 s	-155.8 s	-136.6 s	-108.4	5,32
Xenon difluoride	F_2Xe			2.9 s	31.8 s	67.9 s	114 s	1,5
o-Xylene	C_8H_{10}			-7[e]	27[e]	74.2	143.9	1
m-Xylene	C_8H_{10}		-35[e]	-10[e]	23.4	69.8	138.7	1
p-Xylene	C_8H_{10}				22.4	68.9	137.9	1
2,3-Xylenol	$C_8H_{10}O$	14.3 s	34.3 s	57.2 s	91.4	141.7	216.4	1,5
2,4-Xylenol	$C_8H_{10}O$			50.2	85.5	137.2	210.5	1,5
2,5-Xylenol	$C_8H_{10}O$	13.4 s	33.2 s	55.9 s	87.4	137.0	210.6	5
2,6-Xylenol	$C_8H_{10}O$	-3.1 s	16.7 s	39.6 s	75.3	125.9	200.6	1,5
3,4-Xylenol	$C_8H_{10}O$	19.7 s	40.2 s	63.7 s	102.1	152.3	226.4	1,5
3,5-Xylenol	$C_8H_{10}O$	16.5 s	37.2 s	61.1 s	98.0	147.9	221.3	1,5
Ytterbium	Yb	463 s	540 s	637 s	774 s	993[d]	1192[d]	3
Yttrium	Y	1610.1	1802.3	2047[d]	2354[d]	2763[d]	3334[d]	3
Zinc	Zn	337 s	397 s	477	579	717	912[e]	2
Zinc chloride	Cl_2Zn	305[d]	356[d]	419[d]	497[d]	596[d]	726[d]	4,9,12
Zinc fluoride	F_2Zn	731 s	813 s	911[d]	1048[d]	1237[d]	1503[d]	9
Zinc iodide	I_2Zn	301 s	351 s	409 s	488[d]	598[d]	750[d]	9
Zirconium	Zr	2366	2618	2924	3302	3780	4405	2
Zirconium(IV) bromide	Br_4Zr	136 s	167 s	203 s	245 s	295 s	356 s	4
Zirconium(IV) chloride	Cl_4Zr	117 s	146 s	181 s	222 s	272 s	336 s	9
Zirconium(IV) iodide	I_4Zr	187 s	220 s	259 s	305 s	361 s	430 s	4

[a] More detailed data on this compound can be found in "Vapor Pressure of Fluids at Temperatures below 300 K" in this section.
[b] See also "Recommended Data for Vapor-Pressure Calibration" in this section of the Online Edition of the *CRC Handbook*.
[c] See also "Vapor Pressure of Ice" and "Vapor Pressure and Other Saturation Properties of Water" in this section.
[d] Value was calculated from ideal gas thermodynamic functions (Ref. 8).
[e] Value obtained by extrapolation beyond the region where experimental values exist.
[s] Solid phase value.

VAPOR PRESSURE OF FLUIDS AT TEMPERATURES BELOW 300 K

This table gives vapor pressures of 65 important fluids in the temperature range 2 K to 300 K. The data have been taken from evaluated sources, as indicated in the references. The symbol *"s"* appearing as a superscript on a vapor pressure value indicates that the substance is solid at that temperature. Pressures are given in kilopascals (kPa).

References

1. Ortiz-Vega, D. O., Hall, K. R., Holste, J. C., Arp, V. D., and Lemmon, E. W., to be published in *J. Phys. Chem. Ref. Data* 2016. [helium]
2. Leachman, J. W., Jacobsen, R. T, Penoncello, S. G., and Lemmon, E. W., *J. Phys. Chem. Ref. Data* 38, 721, 2009. [parahydrogen] <https://doi.org/10.1063/1.3160306>
3. Katti, R. S., Jacobsen, R. T, Stewart, R. B., and Jahangiri, M., *Adv. Cryo. Eng.* 31, 1189, 1986. [neon] <https://doi.org/10.1007/978-1-4613-2213-9_132>
4. Tegeler, Ch., Span, R., and Wagner, W., *J. Phys. Chem. Ref. Data* 28, 779, 1999. [argon] <https://doi.org/10.1063/1.556037>
5. DIPPR Data Compilation of Pure Compounds Properties, Design Institute for Physical Properties Data, American Institute of Chemical Engineers, 1987. [boron trichloride, boron trifluoride, hydrogen bromide, bromine, chlorine fluoride, hydrogen chloride, chlorine dioxide, silicon tetrachloride, hydrogen fluoride, difluorine oxide, phosphorus trifluoride, silicon tetrafluoride, hydrogen iodide, phosphine, silane, nitric oxide, ozone, bromotrifluoromethane, tetrachloromethane, trichloromethane, hydrogen cyanide, dichloromethane, formaldehyde, chloromethane, acetylene, propadiene, buta-1,3-diene]
6. Angus, S., Armstrong, B., de Reuck, K.M., and Craven, W., *International Tables for the Fluid State, No. 8, Chlorine*, Pergamon Press, Oxford, 1985. [chlorine]
7. de Reuck, K. M., *International Thermodynamic Tables of the Fluid State, No. 11, Fluorine*, International Union of Pure and Applied Chemistry, Pergamon Press, Oxford, 1990. [fluorine]
8. Younglove, B. A., *J. Phys. Chem. Ref. Data* Vol. 11, Suppl. 1, 1, 1982. [nitrogen trifluoride]
9. Guder, C., and Wagner, W., *J. Phys. Chem. Ref. Data* 38, 33, 2009. [sulfur hexafluoride] <https://doi.org/10.1063/1.3037344>
10. Lemmon, E. W. and Span, R., *J. Chem. Eng. Data*, 51 785, 2006. [hydrogen sulfide, krypton, nitrous oxide, sulfur dioxide, xenon, carbon monoxide, carbon oxysulfide, R41-fluoromethane, neopentane]
11. Tillner-Roth, R., Harms-Watzenberg, F., and Baehr, H. D., Eine neue Fundamentalgleichung fuer Ammoniak, *DKV-Tagungsbericht* 20, 167, 1993. [ammonia]
12. Span, R., Lemmon, E. W., Jacobsen, R. T, Wagner, W., and Yokozeki, A., *J. Phys. Chem. Ref. Data* 29, 1361, 2000. [nitrogen] <https://doi.org/10.1063/1.1349047>
13. Schmidt, R., and Wagner, W., *Fluid Phase Equilib.* 19, 175, 1985. [oxygen] <https://doi.org/10.1016/0378-3812(85)87016-3>
14. Magee, J. W., Outcalt, S. L., and Ely, J. F., *Int. J. Thermophys.* 21, 1097, 2000. [R13-chlorotrifluoromethane] <https://doi.org/10.1023/A:1026446004383>
15. Marx, V., Pruss, A., and Wagner, W., Duesseldorf, VDI Verlag, Series 19 (Waermetechnik/Kaeltetechnik), No. 57, 1992. [R12-dichlorodifluoromethane]
16. Jacobsen, R. T, Penoncello, S. G., and Lemmon, E. W., *Fluid Phase Equilib.* 80, 45, 1992. [R11-trichlorofluoromethane] <https://doi.org/10.1016/0378-3812(92)87054-Q>
17. Platzer, B., Polt, A., and Maurer, G., *Thermophysical Properties of Refrigerants*, Springer-Verlag, Berlin, 1990. [R14-tetrafluoromethane] <https://doi.org/10.1007/978-3-662-02608-3_1>
18. Span, R., and Wagner, W., *J. Phys. Chem. Ref. Data* 25, 1509, 1996. [carbon dioxide] <https://doi.org/10.1063/1.555991>
19. Kamei, A., Beyerlein, S. W., and Jacobsen, R. T, *Int. J. Thermophys.* 16, 1155, 1995. [R22- chlorodifluoromethane] <https://doi.org/10.1007/BF02081283>
20. Penoncello, S. G., Lemmon, E. W., Jacobsen, R. T, and Shan, Z., *J. Phys. Chem. Ref. Data* 32, 1473, 2003. [R23-trifluoromethane] <https://doi.org/10.1063/1.1559671>
21. Tillner-Roth, R., and Yokozeki, A., *J. Phys. Chem. Ref. Data* 26, 1273, 1997. [R32-difluoromethane] <https://doi.org/10.1063/1.556002>
22. Setzmann, U., and Wagner, W., *J. Phys. Chem. Ref. Data* 20, 1061, 1991. [methane] <https://doi.org/10.1063/1.555898>
23. de Reuck, K. M., and Craven, R. J. B., *International Thermodynamic Tables of the Fluid State, No. 12, Methanol*, IUPAC, Blackwell Scientific Publications, London, 1993. [methanol]
24. Smukala, J., Span, R., and Wagner, W., *J. Phys. Chem. Ref. Data* 29, 1053, 2000. [ethylene] <https://doi.org/10.1063/1.1329318>
25. Buecker, D., and Wagner, W., *J. Phys. Chem. Ref. Data* 35, 205, 2006. [ethane] <https://doi.org/10.1063/1.1859286>
26. Wu, J., Zhou, Y., and Lemmon, E. W., *J. Phys. Chem. Ref. Data* 40, 1, 2011. [dimethyl ether]
27. Lemmon, E. W., Overhoff, U., McLinden, M. O., and Wagner, W., to be published in *J. Phys. Chem. Ref. Data* 2015. [propylene]
28. Lemmon, E. W., McLinden, M. O., and Wagner, W., *J. Chem. Eng. Data* 54, 3141, 2009. [propane] <https://doi.org/10.1021/je900217v>
29. Buecker, D., and Wagner, W., *J. Phys. Chem. Ref. Data* 35, 929, 2006. [butane, isobutane] <https://doi.org/10.1063/1.1901687>
30. Span, R., and Wagner, W., *Int. J. Thermophys.* 24, 41, 2003. [pentane] <https://doi.org/10.1023/A:1022310214958>
31. TRC Thermodynamic Tables: Non-Hydrocarbons, Thermodynamic Research Center, Texas A&M University, College Station, TX, 1985. [nitric oxide, radon]
32. Stevenson, R.M., and Malanowski, S., *Handbook of the Thermodynamics of Organic Compounds*, Elsevier, New York, 1987. [hydrogen cyanide, acetylene]

Vapor Pressure of Fluids at Low Temperatures

T/K	*p*/kPa
Acetylene	**HC≡CH**
130	0.1^s
135	0.3^s
140	0.7^s
145	1.3^s
150	2.6^s
155	4.6^s
160	7.8^s
165	12.8^s
170	20.6^s
175	32.2^s
180	49.0^s
185	72.9^s
190	106^s
195	146
200	190
205	244
210	309
215	385
220	475
225	579
230	699
235	837
240	993
245	1170
250	1370
255	1593
260	1843
265	2121
270	2429
275	2771
280	3150
285	3567
290	4028
295	4535
300	5093
Allene	**CH_2=C=CH_2**
150	0.1
155	0.2
160	0.3
165	0.6
170	1.0
175	1.7
180	2.7

T/K	p/kPa
185	4.1
190	6.1
195	8.9
200	12.5
205	17.4
210	23.7
215	31.6
220	41.4
225	53.5
230	68.2
235	85.8
240	107
245	131
250	160
255	193
260	230
265	273
270	322
275	376
280	438
285	506
290	582
295	666
300	759

Ammonia NH_3

T/K	p/kPa
160	0.1[s]
165	0.2[s]
170	0.3[s]
175	0.6[s]
180	1.2[s]
185	2.1[s]
190	3.5[s]
195	5.8[s]
200	8.651
205	12.51
210	17.74
215	24.69
220	33.79
225	45.51
230	60.41
235	79.09
240	102.2
245	130.6
250	164.9
255	206.2
260	255.3
265	313.2
270	381.1
275	459.9
280	550.9
285	655.3
290	774.4
295	909.4
300	1062

Argon Ar

T/K	p/kPa
50	0.03714[s]
55	0.2025[s]
60	0.8322[s]
65	2.752[s]
70	7.670[s]
75	18.65[s]
80	40.57[s]
85	78.90
90	133.5
95	213.0
100	323.8
105	472.2
110	665.3
115	909.8
120	1213
125	1582
130	2025
135	2551
140	3168
145	3890
150	4735

Boron trichloride BCl_3

T/K	p/kPa
180	0.1
185	0.2
190	0.3
195	0.5
200	0.8
205	1.2
210	1.8
215	2.6
220	3.8
225	5.2
230	7.2
235	9.7
240	12.9
245	17.0
250	22.0
255	28.1
260	35.6
265	44.5
270	55.1
275	67.6
280	82.2
285	99.1
290	119
295	141
300	166

Boron trifluoride BF_3

T/K	p/kPa
145	7.7
150	13.4
155	22.3
160	35.2
165	53.7
170	79.1
175	113
180	157
185	214
190	285
195	372
200	479
205	608
210	762
215	944
220	1160
225	1413
230	1709
235	2056
240	2460
245	2913
250	3481
255	4123
260	4874

Bromine Br_2

T/K	p/kPa
220	0.1[s]
225	0.2[s]
230	0.3[s]
235	0.4[s]
240	0.7[s]
245	1.1[s]
250	1.7[s]
255	2.6[s]
260	3.8[s]
265	5.5[s]
270	7.3
275	9.5
280	12.3
285	15.6
290	19.7
295	24.6
300	30.5

Bromotrifluoromethane CF_3Br

T/K	p/kPa
135	0.1
140	0.3
145	0.5
150	0.9
155	1.5
160	2.5
165	3.9
170	5.9
175	8.8
180	12.8
185	18.1
190	25.1
195	34.1
200	45.6
205	60.0
210	77.8
215	99.5
220	126
225	157
230	194
235	237
240	287
245	344
250	410
255	485
260	570
265	665
270	771
275	889
280	1021
285	1166
290	1325
295	1501
300	1692

1,3-Butadiene C_4H_6

T/K	p/kPa
170	0.1
175	0.2
180	0.4
185	0.6
190	1.0
195	1.5
200	2.3
205	3.4
210	4.8
215	6.7
220	9.2
225	12.5
230	16.7
235	21.9
240	28.4
245	36.3
250	46.0
255	57.6
260	71.3
265	87.6
270	107
275	129
280	154
285	184
290	217
295	255
300	297

Butane C_4H_{10}

T/K	p/kPa
170	0.1
175	0.2
180	0.3
185	0.5
190	0.8
195	1.3
200	1.9
205	2.8
210	4.0
215	5.7
220	7.8
225	10.6
230	14.1
235	18.5
240	24.1
245	30.9
250	39.1
255	49.1
260	61.0
265	75.0
270	91.5
275	111
280	133
285	159
290	188
295	221
300	258

Carbon dioxide CO_2

T/K	p/kPa
135	0.1[s]
140	0.2[s]

T/K	p/kPa
145	0.4[s]
150	0.8[s]
155	1.7[s]
160	3.1[s]
165	5.7[s]
170	9.9[s]
175	16.8[s]
180	27.6[s]
185	44.0[s]
190	68.4[s]
195	104[s]
200	155[s]
205	227[s]
210	327[s]
215	465[s]
220	599
225	734
230	893
235	1075
240	1283
245	1519
250	1785
255	2085
260	2419
265	2790
270	3203
275	3658
280	4161
285	4714
290	5318
295	5984
300	6713

Carbon monoxide CO

T/K	p/kPa
50	0.1[s]
55	0.6[s]
60	2.6[s]
65	8.2[s]
70	21.0
75	44.4
80	83.7
85	147
90	239
95	371
100	545
105	771
110	1067
115	1428
120	1877
125	2400
130	3064

Carbon oxysulfide OCS

T/K	p/kPa
140	0.1
145	0.2
150	0.4
155	0.8
160	1.3
165	2.2
170	3.4
175	5.2
180	7.8
185	11.3
190	15.9
195	22.1
200	30.0
205	40.1
210	52.7
215	68.2
220	87.2
225	110
230	137
235	169
240	207
245	250
250	301
255	358
260	423
265	497
270	580
275	673
280	777
285	892
290	1019
295	1159
300	1313

Chlorine Cl_2

T/K	p/kPa
175	1.8
180	2.8
185	4.2
190	6.1
195	8.7
200	12.3
205	16.9
210	22.9
215	30.5
220	40.1
225	51.9
230	66.4
235	84.0
240	105
245	130
250	160
255	194
260	234
265	280
270	332
275	392
280	459
285	535
290	619
295	714
300	818

Chlorine dioxide ClO_2

T/K	p/kPa
195	0.1
200	0.3
205	0.5
210	0.9
215	1.4
220	2.3
225	3.5
230	5.3
235	7.6
240	10.8
245	14.9
250	20.1
255	26.6
260	34.6
265	44.4
270	56.1
275	69.9
280	86.2
285	105
290	127
295	151
300	179

Chlorine flouride ClF

T/K	p/kPa
115	0.1
120	0.3
125	0.6
130	1.2
135	2.1
140	3.6
145	6.0
150	9.5
155	14.6
160	21.8
165	31.7
170	44.8
175	62.0
180	84.2
185	112
190	147
195	190
200	242
205	304
210	378
215	464
220	564
225	680
230	812
235	961
240	1130
245	1319
250	1529
255	1762
260	2019
265	2301
270	2608
275	2941
280	3303
285	3693
290	4111
295	4560
300	5039

Chlorodifluoromethane $CHClF_2$

T/K	p/kPa
150	0.1
155	0.3
160	0.5
165	0.8
170	1.4
175	2.3
180	3.6
185	5.5
190	8.1
195	11.8
200	16.7
205	23.1
210	31.5
215	42.1
220	55.3
225	71.7
230	91.6
235	116
240	144
245	178
250	218
255	264
260	317
265	377
270	446
275	525
280	613
285	711
290	821
295	944
300	1080

Chloromethane CH_3Cl

T/K	p/kPa
185	2.1
190	3.1
195	4.6
200	6.7
205	9.5
210	13.1
215	17.9
220	24.0
225	31.8
230	41.4
235	53.3
240	67.7
245	85.1
250	106
255	131
260	159
265	193
270	232
275	277
280	327
285	385
290	450
295	524
300	606

Chlorotrifluoromethane CF_3Cl

T/K	p/kPa
100	0.002689
105	0.008350
110	0.02310
115	0.05783
120	0.1327
125	0.2823
130	0.5620
135	1.055
140	1.880

T/K	p/kPa
145	3.199
150	5.226
155	8.231
160	12.54
165	18.56
170	26.74
175	37.61
180	51.76
185	69.84
190	92.56
195	120.7
200	155.0
205	196.3
210	245.5
215	303.6
220	371.4
225	450.0
230	540.3
235	643.4
240	760.2
245	892.0
250	1040
255	1205
260	1388
265	1591
270	1815
275	2061
280	2332
285	2628
290	2954
295	3311
300	3706

Dichlorodifluoromethane CF_2Cl_2

T/K	p/kPa
125	0.001647
130	0.004242
135	0.01009
140	0.02238
145	0.04661
150	0.09183
155	0.1721
160	0.3083
165	0.5303
170	0.8793
175	1.410
180	2.194
185	3.322
190	4.902
195	7.069
200	9.981
205	13.82
210	18.79
215	25.14
220	33.12
225	43.02
230	55.15
235	69.85
240	87.48
245	108.4
250	133.1
255	161.9
260	195.2
265	233.6
270	277.5
275	327.3
280	383.6
285	446.9
290	517.7
295	596.6
300	683.9

Difluoromethane CH_2F_2

T/K	p/kPa
200	0.1
205	0.2
210	0.3
215	0.4
220	0.6
225	0.9
230	1.4
235	2.0
240	2.8
245	3.8
250	5.3
255	7.1
260	9.5
265	12.4
270	16.1
275	20.7
280	26.3
285	33.0
290	41.1
295	50.8
300	62.1
140	0.1
145	0.2
150	0.3
155	0.6
160	1.0
165	1.7
170	2.8
175	4.4
180	6.8
185	10.2
190	14.8
195	21.2
200	29.5
205	40.5
210	54.5
215	72.1
220	94.1
225	121
230	154
235	193
240	240
245	295
250	360
255	434
260	521
265	620
270	732
275	860
280	1004
285	1165
290	1346
295	1547
300	1770

Dimethyl ether CH_3OCH_3

T/K	p/kPa
155	0.1
160	0.2
165	0.3
170	0.5
175	0.9
180	1.4
185	2.1
190	3.2
195	4.7
200	6.8
205	9.6
210	13.3
215	18.1
220	24.3
225	32.1
230	41.9
235	53.9
240	68.6
245	86.3
250	108
255	133
260	162
265	197
270	237
275	283
280	335
285	395
290	463
295	538
300	623

Ethane C_2H_6

T/K	p/kPa
115	0.1
120	0.4
125	0.7
130	1.3
135	2.2
140	3.8
145	6.0
150	9.7
155	15.0
160	21.5
165	31.0
170	42.9
175	59.0
180	78.7
185	104
190	135
195	172
200	217
205	271
210	334
215	407
220	492
225	590
230	700
235	826
240	967
245	1125
250	1301
255	1496
260	1712
265	1949
270	2210
275	2495
280	2806
285	3146
290	3515
295	3917
300	4355

Ethylene $CH_2{=}CH_2$

T/K	p/kPa
110	0.33
115	0.70
120	1.38
125	2.7
130	4.5
135	7.7
140	11.9
145	18.3
150	27.5
155	39.9
160	56.4
165	77.9
170	105
175	140
180	182
185	234
190	296
195	369
200	456
205	557
210	673
215	806
220	958
225	1128
230	1321
235	1535
240	1774
245	2039
250	2331
255	2652
260	3005
265	3391
270	3813
275	4275

Fluorine F_2

T/K	p/kPa
55	0.382
60	1.49
65	4.61
70	12.0
75	27.1
80	54.7
85	101
90	173
95	279
100	428

T/K	p/kPa
105	628
110	889
115	1220
120	1630
125	2140
130	2750
135	3470
140	4340

Fluorine monoxide F_2O

T/K	p/kPa
75	0.1
80	0.2
85	0.5
90	1.2
95	2.6
100	5.3
105	10.1
110	18.0
115	30.5
120	49.3
125	76.7
130	115
135	168
140	237
145	328
150	444
155	588
160	766
165	981
170	1238
175	1541
180	1895
185	2303
190	2771
195	3302
200	3899
205	4567
210	5308

Fluoromethane CH_3F

T/K	p/kPa
135	0.6
140	1.2
145	2.1
150	3.6
155	5.9
160	9.3
165	14.1
170	20.9
175	29.9
180	42.0
185	57.6
190	77.4
195	102
200	133
205	171
210	216
215	270
220	333
225	408
230	495
235	595
240	711
245	843
250	993
255	1163
260	1355
265	1571
270	1813
275	2084
280	2387
285	2724
290	3099
295	3516
300	3978

Formaldehyde HCHO

T/K	p/kPa
185	1.3
190	2.0
195	3.0
200	4.4
205	6.4
210	9.1
215	12.7
220	17.4
225	23.4
230	31.0
235	40.6
240	52.5
245	67.0
250	84.6
255	106
260	131
265	161
270	196
275	236
280	283
285	337
290	399
295	470
300	549

Helium He

T/K	p/kPa
2.2	5.332
2.3	6.725
2.4	8.351
2.5	10.23
2.6	12.38
2.7	14.81
2.8	17.56
2.9	20.64
3.0	24.06
3.1	27.85
3.2	32.02
3.3	36.60
3.4	41.60
3.5	47.04
3.6	52.94
3.7	59.31
3.8	66.19
3.9	73.58
4.0	81.51
4.1	90.00
4.2	99.08
4.3	108.8
4.4	119.1
4.5	130.1
4.6	141.7
4.7	154.1
4.8	167.3
4.9	181.3
5.0	196.2
5.1	212.1

Hydrogen H_2

T/K	p/kPa
14.0	7.884
14.5	10.38
15.0	13.43
15.5	17.13
16.0	21.55
16.5	26.77
17.0	32.89
17.5	39.98
18.0	48.15
18.5	57.48
19.0	68.07
19.5	80.02
20.0	93.41
20.5	108.4
21.0	125.0
21.5	143.3
22.0	163.5
22.5	185.6
23.0	209.8
23.5	236.2
24.0	264.8
24.5	295.7
25.0	329.2
25.5	365.2
26.0	403.8
26.5	445.3
27.0	489.6
27.5	537.0
28.0	587.5
28.5	641.2
29.0	698.3
29.5	758.9
30.0	823.2
30.5	891.2
31.0	963.3
31.5	1040
32.0	1120
32.5	1206

Hydrogen bromide HBr

T/K	p/kPa
135	0.1^{s}
140	0.3^{s}
145	0.6^{s}
150	1.1
155	1.9
160	3.3
165	5.4
170	8.7^{s}
175	13.4^{s}
180	20.1^{s}
185	29.5^{s}
190	37.9
195	51.8
200	69.5
205	91.8
210	119
215	153
220	194
225	242
230	299
235	366
240	443
245	532
250	633
255	748
260	878
265	1023
270	1185
275	1364
280	1562
285	1780
290	2018
295	2278
300	2561

Hydrogen chloride HCl

T/K	p/kPa
120	0.1
125	0.3
130	0.5
135	1.0
140	1.9
145	3.4
150	5.8
155	9.5
160	14.7
165	22.0
170	31.9
175	45.1
180	62.5
185	84.7
190	113
195	148
200	190
205	242
210	304
215	377
220	463
225	563
230	678
235	811
240	961
245	1132
250	1325
255	1542
260	1784
265	2054
270	2354
275	2686
280	3053
285	3457
290	3901
295	4388
300	4921

Hydrogen cyanide HCN

T/K	*p*/kPa
200	0.1^s
205	0.2^s
210	0.4^s
215	0.6^s
220	1.0^s
225	1.5^s
230	2.2^s
235	3.3^s
240	4.7^s
245	6.8^s
250	9.7^s
255	13.6^s
260	18.8
265	24.1
270	30.5
275	38.3
280	47.7
285	58.8
290	72.1
295	87.6
300	105.9

Hydrogen fluoride HF

T/K	*p*/kPa
190	0.3
195	0.5
200	0.8
205	1.2
210	1.7
215	2.3
220	3.2
225	4.4
230	5.9
235	7.9
240	10.3
245	13.4
250	17.2
255	21.8
260	27.4
265	34.2
270	42.2
275	51.8
280	63.1
285	76.3
290	91.7
295	110
300	130

Hydrogen iodide HI

T/K	*p*/kPa
155	0.1
160	0.2
165	0.4
170	0.8
175	1.3^s
180	2.2^s
185	3.4^s
190	5.3^s
195	8.0^s
200	11.7^s
205	16.8^s
210	23.6^s
215	32.5^s
220	44.0^s
225	56.2
230	71.4
235	89.7
240	112
245	137
250	168
255	203
260	244
265	290
270	343
275	404
280	472
285	548
290	633
295	727
300	831

Hydrogen sulfide H_2S

T/K	*p*/kPa
135	0.1^s
140	0.2^s
145	0.3^s
150	0.6^s
155	1.1^s
160	1.9^s
165	3.2
170	5.2
175	8.3
180	12.7
185	18.9^s
190	27.11
195	37.27
200	50.34
205	66.87
210	87.47
215	112.8
220	143.7
225	180.7
230	224.9
235	276.9
240	337.7
245	408.2
250	489.3
255	582.1
260	687.5
265	806.5
270	940.2
275	1090
280	1256
285	1440
290	1643
295	1866
300	2110

Isobutane C_4H_{10}

T/K	*p*/kPa
160	0.1
165	0.1
170	0.3
175	0.4
180	0.7
185	1.1
190	1.7
195	2.5
200	3.7
205	5.3
210	7.4
215	10.2
220	13.8
225	18.3
230	24.0
235	31.1
240	39.8
245	50.3
250	62.9
255	77.8
260	95.4
265	116
270	140
275	167
280	198
285	234
290	274
295	319
300	370

Krypton Kr

T/K	*p*/kPa
60	0.001574^s
65	0.008754^s
70	0.03811^s
75	0.1364^s
80	0.4160^s
85	1.113^s
90	2.670^s
95	5.840^s
100	11.81^s
105	22.35^s
110	39.89^s
115	67.71^s
120	103.4
125	150.1
130	211.2
135	289.6
140	387.8
145	508.7
150	655.1
155	830.1
160	1036
165	1277
170	1556
175	1875
180	2239
185	2651
190	3114
195	3634
200	4216
205	4869

Methane CH_4

T/K	*p*/kPa
65	0.1
70	0.3
75	0.8
80	2.1
85	4.9
90	10.6
95	20.0
100	34.5
105	57.0
110	88.4
115	133
120	192
125	269
130	368
135	491
140	642
145	824
150	1041
155	1297
160	1594
165	1937
170	2331
175	2779
180	3288
185	3865
190	4520

Methanol CH_3OH

T/K	*p*/kPa
230	0.1
235	0.2
240	0.4
245	0.5
250	0.8
255	1.2
260	1.7
265	2.4
270	3.3
275	4.5
280	6.2
285	8.3
290	11.0
295	14.4
300	18.7

Neon Ne

T/K	*p*/kPa
25	50.92
26	71.61
27	98.17
28	131.6
29	172.9
30	223.1
31	283.2
32	354.5
33	437.8
34	534.4
35	645.4
36	772.0
37	915.4
38	1077
39	1258
40	1460
41	1685
42	1935
43	2212

T/K	p/kPa
44	2517

Neopentane $C(CH_3)_4$

T/K	p/kPa
185	0.1[s]
190	0.2[s]
195	0.4[s]
200	0.7[s]
205	1.1[s]
210	1.6[s]
215	2.4[s]
220	3.6[s]
225	5.2[s]
230	7.3[s]
235	10.2[s]
240	13.9[s]
245	18.7[s]
250	24.8[s]
255	32.4[s]
260	41.6
265	51.4
270	63.0
275	76.6
280	92.3
285	111
290	131
295	155
300	182

Nitric oxide NO

T/K	p/kPa
85	0.1[s]
90	0.4[s]
95	1.3[s]
100	3.8[s]
105	10.0[s]
110	23.5
115	46.8
120	86.5
125	151
130	248
135	391
140	592
145	867
150	1231
155	1703
160	2302
165	3050
170	3971
175	5089
180	6433

Nitrogen N_2

T/K	p/kPa
50	0.4005[s]
55	1.800[s]
60	6.298[s]
65	17.40
70	38.54
75	76.04
80	136.9
85	228.9
90	360.5
95	540.5
100	778.3
105	1083
110	1466
115	1937
120	2511
125	3207

Nitrogen trifluoride NF_3

T/K	p/kPa
85	0.0530
90	0.156
95	0.406
100	0.946
105	2.02
110	3.98
115	7.33
120	12.8
125	21.1
130	33.5
135	51.0
140	75.2
145	108
150	150
155	204
160	272
165	356
170	458
175	580
180	725
185	895
190	1090
195	1320
200	1580
205	1870
210	2210
215	2580
220	3000
225	3470
230	3990

Nitrous oxide N_2O

T/K	p/kPa
115	0.1
120	0.1
125	0.3
130	0.7
135	1.3
140	2.5
145	4.3
150	7.1
155	11.4
160	17.6
165	26.4
170	38.5
175	54.7
180	75.9
185	103.3
190	137.8
195	180.9
200	233.7
205	297.7
210	374.3
215	465.1
220	571.6
225	695.4
230	838.3
235	1002
240	1188
245	1398
250	1634
255	1898
260	2192
265	2518
270	2877
275	3273
280	3707
285	4182
290	4701
295	5268
300	5887

Oxygen O_2

T/K	p/kPa
50	0.03096[s]
55	0.1786
60	0.7258
65	2.335
70	6.262
75	14.55
80	30.12
85	56.83
90	99.35
95	163.1
100	254.0
105	378.5
110	543.4
115	755.6
120	1022
125	1351
130	1749
135	2225
140	2788
145	3448
150	4219

Ozone O_3

T/K	p/kPa
100	0.1
105	0.2
110	0.4
115	1.0
120	2.0
125	3.8
130	6.8
135	11.5
140	18.7
145	29.1
150	43.7
155	63.6
160	89.9
165	124
170	168
175	222
180	289
185	367
190	468
195	584
200	721
205	881
210	1068
215	1285
220	1536
225	1824
230	2155
235	2534
240	2968
245	3464
250	4031
255	4678
260	5417

Pentane C_5H_{12}

T/K	p/kPa
220	1.0
225	1.5
230	2.1
235	3.0
240	4.2
245	5.7
250	7.6
255	10.0
260	13.0
265	16.6
270	21.1
275	26.6
280	33.1
285	40.8
290	50.0
295	60.7
300	73.2

Phosphine PH_3

T/K	p/kPa
110	0.1
115	0.2
120	0.4
125	0.7
130	1.3
135	2.3
140	3.9
145	6.2
150	9.6
155	14.5
160	21.1
165	30.0
170	41.6
175	56.6
180	75.6
185	99.2
190	128
195	163
200	205
205	254
210	312
215	379
220	456
225	544
230	644
235	756
240	881
245	1019
250	1172
255	1341

T/K	p/kPa
260	1525
265	1725
270	1942
275	2176
280	2428
285	2699
290	2987
295	3295
300	3621

Phosphorus trifluoride PF_3

T/K	p/kPa
105	0.1
110	0.2
115	0.5
120	1.0
125	1.9
130	3.5
135	5.9
140	9.5
145	14.9
150	22.5
155	33.1
160	47.3
165	66.0
170	90.1
175	121
180	159
185	206
190	262
195	330
200	410
205	503
210	611
215	736
220	877
225	1037
230	1217
235	1418
240	1640
245	1885
250	2154
255	2448
260	2767
265	3112

Propane C_3H_8

T/K	p/kPa
160	0.8
165	1.4
170	2.2
175	3.3
180	5.0
185	7.3
190	10.5
195	15.0
200	20.1
205	27.0
210	36.0
215	47.0
220	60.0
225	77.0
230	97.0
235	120
240	148
245	180
250	218
255	261
260	311
265	367
270	431
275	502
280	582
285	671
290	769
295	878
300	998

Propene $CH_3CH_2{=}CH_2$

T/K	p/kPa
140	0.1
145	0.2
150	0.4
155	0.7
160	1.2
165	2.0
170	3.1
175	4.7
180	7.0
185	10.1
190	14.2
195	19.7
200	26.9
205	35.9
210	47.3
215	61.3
220	78.5
225	99.2
230	124
235	153
240	188
245	228
250	274
255	327
260	387
265	456
270	533
275	619
280	715
285	822
290	940
295	1069
300	1212

Radon Rn

T/K	p/kPa
130	0.1
135	0.3
140	0.5
145	0.9
150	1.5
155	2.4
160	3.8
165	5.8
170	8.6
175	12.5
180	17.7
185	24.5
190	33.2
195	44.4
200	58.2
205	75.3
210	96.0
215	121
220	151
225	185

Silane SiH_3

T/K	p/kPa
95	0.1
100	0.2
105	0.4
110	1.0
115	1.9
120	3.5
125	6.1
130	10.0
135	15.8
140	24.1
145	35.3
150	50.3
155	69.8
160	94.6
165	126
170	164
175	210
180	265
185	331
190	408
195	498
200	602
205	722
210	859
215	1017
220	1196
225	1398
230	1628
235	1888
240	2180
245	2509
250	2880
255	3296
260	3763
265	4288

Sulfur dioxide SO_2

T/K	p/kPa
170	0.1
175	0.2
180	0.3
185	0.5
190	0.8
195	1.3
200	2.026
205	3.068
210	4.539
215	6.574
220	9.334
225	13.01
230	17.84
235	24.06
240	31.99
245	41.95
250	54.31
255	69.48
260	87.91
265	110.1
270	136.5
275	167.7
280	204.3
285	246.9
290	296.1
295	352.7
300	417.2

Sulfur hexafluoride SF_6

T/K	p/kPa
115	0.001259^{s}
120	0.003635^{s}
125	0.009614^{s}
130	0.02353^{s}
135	0.05379^{s}
140	0.1157^{s}
145	0.2355^{s}
150	0.4566^{s}
155	0.8471^{s}
160	1.510^{s}
165	2.596^{s}
170	4.317^{s}
175	6.968^{s}
180	10.94^{s}
185	16.75^{s}
190	25.07^{s}
195	36.71^{s}
200	52.71^{s}
205	74.31^{s}
210	103.0
215	140.6
220	189.0
225	245.9
230	301.3
235	365.8
240	440.1
245	525.2
250	622.0
255	731.5
260	854.6
265	992.6
270	1146
275	1317
280	1506
285	1714
290	1942
295	2193
300	2468

Tetrachloromethane CCl_4

T/K	p/kPa
255	1.5
260	2.1
265	2.8
270	3.7
275	4.9
280	6.4
285	8.2
290	10.5

T/K	p/kPa
295	13.2
300	16.5

Tetrachlorosilane $SiCl_4$

T/K	p/kPa
210	0.1
215	0.2
220	0.3
225	0.5
230	0.7
235	1.0
240	1.5
245	2.0
250	2.8
255	3.8
260	5.0
265	6.6
270	8.6
275	11.1
280	14.2
285	17.9
290	22.3
295	27.7
300	34.0

Tetrafluoromethane CF_4

T/K	p/kPa
90	0.1
95	0.3
100	0.8
105	1.7
110	3.4
115	6.5
120	11.5
125	19.3
130	30.8
135	47.4
140	70.2
145	101
150	141
155	191
160	254
165	332
170	425
175	537
180	669
185	824
190	1005
195	1216
200	1460
205	1743
210	2073
215	2457
220	2907
225	3438

Tetrafluorosilane SiF_4

T/K	p/kPa
125	0.1
130	0.2^s
135	0.4^s
140	0.9^s
145	1.9^s
150	3.8^s
155	7.5^s
160	14.0^s
165	25.2^s
170	43.8^s
175	74.2^s
180	122^s
185	197^s
190	280
195	376
200	488
205	618
210	766
215	932
220	1117
225	1324
230	1555
235	1816
240	2111
245	2449
250	2841
255	3301

Trichlorofluoromethane CCl_3F

T/K	p/kPa
190	0.2
195	0.3
200	0.4
205	0.6
210	1.0
215	1.4
220	2.0
225	2.9
230	4.1
235	5.6
240	7.6
245	10.1
250	13.3
255	17.2
260	22.1
265	28.0
270	35.1
275	43.7
280	53.8
285	65.7
290	79.6
295	95.6
300	114.1

Trichloromethane $CHCl_3$

T/K	p/kPa
215	0.1
220	0.2
225	0.3
230	0.4
235	0.7
240	1.0
245	1.4
250	2.0
255	2.7
260	3.7
265	5.0
270	6.6
275	8.7
280	11.3
285	14.4
290	18.3
295	22.9
300	28.5

Trifluoromethane CHF_3

T/K	p/kPa
120	0.1
125	0.2
130	0.4
135	0.7
140	1.4
145	2.5
150	4.3
155	7.1
160	11.1
165	17.0
170	25.3
175	36.5
180	51.4
185	70.9
190	95.8
195	127
200	166
205	214
210	271
215	340
220	421
225	516
230	626
235	754
240	900
245	1067
250	1257
255	1472
260	1713
265	1984
270	2287
275	2624
280	3000
285	3418
290	3881
295	4393

Xenon Xe

T/K	p/kPa
100	0.08037^s
105	0.1686^s
110	0.3433^s
115	0.6778^s
120	1.298^s
125	2.411^s
130	4.343^s
135	7.583^s
140	12.83^s
145	21.02^s
150	33.32^s
155	50.96^s
160	74.58^s
165	101.0
170	133.4
175	173.3
180	221.5
185	279.3
190	347.7
195	427.9
200	520.9
205	628.0
210	750.2
215	888.9
220	1045
225	1220
230	1416
235	1632
240	1871
245	2134
250	2423
255	2738
260	3082
265	3456
270	3862
275	4303
280	4782
285	5303

^{s} Solid.

VAPOR PRESSURE OF SATURATED SALT SOLUTIONS

This table gives the vapor pressure of water above saturated solutions of some common salts at temperature intervals from 10 °C to 40 °C. Data on pure water are given on the last line for comparison. Column definitions are as follows.

Column heading	Definition
Name	Name of salt forming saturated solution; salts are listed alphabetically by name
Mol. form.	Formula of salt forming saturated solution
P_{vap}(temperature)	Vapor pressure of saturated solution at temperature specified in parentheses in °C, in kPa
Ref.	Reference for data values

The references provide additional information on activity coefficients, osmotic coefficients, and enthalpies of vaporization.

References

1. Apelblat, A., *J. Chem. Thermodynamics*, 24, 619, 1992. <https://doi.org/10.1016/S0021-9614(05)80033-3>
2. Apelblat, A., *J. Chem. Thermodynamics*, 25, 63, 1993. <https://doi.org/10.1006/jcht.1993.1008>
3. Apelblat, A., *J. Chem. Thermodynamics*, 25, 1513, 1993. <https://doi.org/10.1006/jcht.1993.1151>
4. Apelblat, A. and Korin, E., *J. Chem. Thermodynamics*, 30, 59, 1998. <https://doi.org/10.1006/jcht.1997.0275>

Vapor Pressure of Saturated Solutions at the Indicated Temperature

Name	Mol. form.	P_{vap}(10 °C)/kPa	P_{vap}(15 °C)/kPa	P_{vap}(20 °C)/kPa	P_{vap}(25 °C)/kPa	P_{vap}(30 °C)/kPa	P_{vap}(35 °C)/kPa	P_{vap}(40 °C)/kPa	Ref.
Ammonium chloride	NH_4Cl	0.971	1.328	1.836	2.481				2
Ammonium nitrate	NH_4NO_3	0.853	1.152	1.524	1.972				2
Ammonium sulfate	$(NH_4)_2SO_4$	0.901	1.319	1.871	2.573	3.439	4.474		3
Barium chloride	$BaCl_2$	0.971	1.443	2.073	2.887	3.903	5.133	6.576	1
Calcium nitrate	$Ca(NO_3)_2$	0.701	1.015	1.381	1.772	2.154	2.487		1
Copper(II) sulfate	$CuSO_4$	1.113	1.574	2.189	2.996	4.037	5.363		3
Iron(II) sulfate	$FeSO_4$	0.978	1.516	2.208	3.035	3.950	4.884		3
Lithium chloride	LiCl	0.128	0.193	0.279	0.384				2
Magnesium nitrate	$Mg(NO_3)_2$	0.726	0.999	1.339	1.749	2.231	2.782	3.397	1
Manganese(II) chloride	$MnCl_2$	0.697	1.064	1.515	2.020	2.535	3.002		3
Potassium bromide	KBr	0.953	1.338	1.853	2.533	3.419	4.563		3
Potassium carbonate	K_2CO_3	0.541	0.802	1.134	1.536	1.997	2.499	3.016	1
Potassium iodate	KIO_3	1.100	1.564	2.177	2.970	3.979	5.236	6.778	4
Rubidium chloride	RbCl	0.862	1.215	1.684	2.298	3.088	4.089	5.343	4
Sodium bromide	NaBr	0.722	1.004	1.376	1.858	2.475	3.255	4.229	4
Sodium chloride	NaCl	0.921	1.285	1.768	2.401	3.218	4.262	5.581	4
Sodium nitrate	$NaNO_3$	0.884	1.244	1.719	2.335	3.121	4.109	5.333	4
Sodium nitrite	$NaNO_2$	0.703	0.994	1.381	1.888	2.540	3.368	4.403	4
Zinc sulfate	$ZnSO_4$	0.945	1.401	1.986	2.698	3.523	4.431	5.382	1
Water	H_2O	1.228	1.706	2.339	3.169	4.246	5.627	7.381	

ENTHALPY OF VAPORIZATION

The molar enthalpy (heat) of vaporization $\Delta_{vap}H$, which is defined as the enthalpy change in the conversion of one mole of liquid to gas at constant temperature, is tabulated here for about 950 inorganic and organic compounds. Values are given, when available, both at the normal boiling point t_b, referred to a pressure of 101.325 kPa (760 mmHg), and at 25 °C.

The values in this table were measured either by calorimetric techniques or by application of the Claperyon equation to the variation of vapor pressure with temperature. See Ref. 1 for a discussion of the accuracy of different experimental techniques and methods of estimating enthalpy of vaporization at other temperatures. Several of the references present empirical techniques for correlating enthalpy of vaporization with molecular structure.

Column definitions are as follows.

Column heading	Column
Name	Name of compound; compounds are listed alphabetically by systematic name
Mol. form.	Compound formula
t_b	Boiling point, in °C
$\Delta_{vap}H(t_b)$	Molar enthalpy of vaporization at normal boiling point, in units kJ mol^{-1}
$\Delta_{vap}H$**(25 °C)**	Molar enthalpy of vaporization at 25 °C, in units kJ mol^{-1}
Ref.	Reference for data values

References

1. Majer, V., and Svoboda, V., *Enthalpies of Vaporization of Organic Compounds*, Blackwell Scientific Publications, Oxford, 1985.
2. Chase, M. W., Davies, C. A., Downey, J. R., Frurip, D. J., McDonald, R. A., and Syverud, A. N., *JANAF Thermochemical Tables, Third Edition, J. Phys. Chem. Ref. Data* 14, Suppl. 1, 1985.
3. *Landolt-Börnstein, Numerical Data and Functional Relationships in Science and Technology, Sixth Edition*, II/4, *Caloric Quantities of State*, Springer-Verlag, Heidelberg, 1961.
4. Daubert, T. E., Danner, R. P., Sibul, H. M., and Stebbins, C. C., *Physical and Thermodynamic Properties of Pure Compounds: Data Compilation*, extant 1994 (core with 4 supplements), Taylor & Francis, Bristol, PA.
5. Ruzicka, K., and Majer, V., Simultaneous Treatment of Vapor Pressures and Related Thermal Data between the Triple and Normal Boiling Temperatures for *n*-Alkanes C5–C20, *J. Phys. Chem. Ref. Data* 23, 1, 1994. <https://doi.org/10.1063/1.555942>
6. Verevkin, S. P., Thermochemistry of Amines: Experimental Standard Molar Enthalpies of Formation of Some Aliphatic and Aromatic Amines, *J. Chem. Thermodynamics* 29, 891, 1997. <https://doi.org/10.1006/jcht.1997.0212>
7. Cady, G. H., and Hargreaves, G. B., The Vapor Pressure of Some Heavy Transition Metal Hexafluorides, *J. Chem. Soc.* 1563, 1578, 1961. <https://doi.org/10.1039/jr9610001563>
8. Steele, W. V., Chirico, R. D., Knipmeyer, S. E., and Nguyen, A., *J. Chem. Eng. Data* 41, 1255, 1996.
9. Nichols, G., et al., *J. Chem. Eng. Data* 51, 475, 2006.
10. Dias, A. M. A., et al., *J. Chem. Eng. Data* 50, 1328, 2005. <https://doi.org/10.1021/je050056e>
11. Umnahanant, p., et al., *J. Chem. Eng. Data* 51, 2246, 2006. <https://doi.org/10.1021/je060333x>
12. Raganov, G. N., Pisarev, P. N., and Emel'yanenko, V. N., *J. Chem. Eng. Data* 50, 1114, 2005.
13. Verevkin, S. P., et al., *J. Chem. Eng. Data* 51, 1896, 2006. <https://doi.org/10.1021/je0602418>
14. Verevkin, S. P., *J. Chem. Thermodynamics* 38, 1111, 2006. <https://doi.org/10.1016/j.jct.2005.11.009>
15. Lide, D. R., *Handbook of Organic Solvents*, CRC Press, Boca Raton, FL, 1995.
16. Steele, W. V., Chirico, R. D., Nguyen, A., and Knipmeyer, S. E., *J. Chem. Thermodynamics* 27, 322, 1995. <https://doi.org/10.1006/jcht.1995.0149>
17. Ribeiro da Silva, M., et al., *J. Chem. Thermodynamics* 27, 565, 1995. <https://doi.org/10.1006/jcht.1995.0057>
18. Kabo, G. J., et al., *J. Chem. Thermodynamics* 27, 953, 1995. <https://doi.org/10.1006/jcht.1995.0101>
19. Duarte-Garza, H. A., et al., *J. Chem. Eng. Data* 42, 497, 1997. <https://doi.org/10.1021/je9603584>
20. Steele, W. V., and Chirico, R. D., *J. Phys. Chem. Ref. Data* 22, 377, 1993. <https://doi.org/10.1063/1.555937>
21. Linstrom, P. J., and Mallard, W. G., Eds., *NIST Chemistry WebBook*, NIST Standard Reference Database No. 69, June 2005, National Institute of Standards and Technology, Gaithersburg, MD, <http://webbook.nist.gov>.
22. Weber, L. A. and Defibaugh, D. R., *J. Chem. Eng. Data* 41, 1477, 1996. <https://doi.org/10.1021/je9602071>
23. Riddick, J. A., Bunger, W. B., and Sakano, T. K., *Organic Solvents, Fourth Edition*, John Wiley & Sons, New York, 1986.

Molar Enthalpy of Vaporization at the Boiling Point and at 25 °C

Name	Mol. form.	t_b/°C	$\Delta_{vap}H(t_b)$/ kJ mol^{-1}	$\Delta_{vap}H$(25 °C)/ kJ mol^{-1}	Ref.
Acetaldehyde	C_2H_4O	20.8	25.76	25.47	1
Acetic acid	$C_2H_4O_2$	117.9	23.70	23.36	1
Acetic anhydride	$C_4H_6O_3$	139.5	38.2		
Acetone	C_3H_6O	56.08	29.10	30.99	1
Acetonitrile	C_2H_3N	81.6	29.75	32.94	1
Acetophenone	C_8H_8O	202.1	43.98	55.40	8
Acrolein	C_3H_4O	52.3	28.3		
Acrylonitrile	C_3H_3N	77.2	32.6		
Allyl acetate	$C_5H_8O_2$	104	36.3		
Allyl alcohol	C_3H_6O	96.9	40.0		
Aluminum	Al	2519	294		
Aluminum borohydride	AlB_3H_{12}	44.5	30		
Aluminum bromide	$AlBr_3$	255	23.5		
Aluminum iodide	AlI_3	382	32.2		

Name	Mol. form.	t_b/°C	$\Delta_{vap}H(t_b)$/ kJ mol^{-1}	$\Delta_{vap}H$(25 °C)/ kJ mol^{-1}	Ref.
2-Amino-2-methyl-1-propanol	$C_4H_{11}NO$	163.8	50.6		
Ammonia	H_3N	-33.33	23.33	19.86	
Aniline	C_6H_7N	184.1	42.44	55.83	1
Anisole	C_7H_8O	153.6	38.97	46.90	1
Antimony(III) bromide	Br_3Sb	288	59		
Antimony(III) chloride	Cl_3Sb	220.3	45.19		
Antimony(III) iodide	I_3Sb	400	68.6		
Argon	Ar	-185.848	6.43		
Arsenic(III) bromide	$AsBr_3$	221	41.8		
Arsenic(III) chloride	$AsCl_3$	130	35.01		
Arsenic(III) fluoride	AsF_3	57.13	29.7		
Arsenic(V) fluoride	AsF_5	-52.8	20.8		
Arsenic(III) iodide	AsI_3	424	59.3		
Arsine	AsH_3	-62.5	16.69		
Azobutane	$C_8H_{18}N_2$			49.31	1
Azopropane	$C_6H_{14}N_2$	114		39.88	1
Barium	Ba	≈1845	140		
Benzaldehyde	C_7H_6O	178.7	42.5		
Benzene	C_6H_6	80.08	30.72	33.83	1
Benzenethiol	C_6H_6S	169.1	39.93	47.56	1
Benzonitrile	C_7H_5N	191	45.9		
Benzyl acetate	$C_9H_{10}O_2$	215	49.4		
Benzyl alcohol	C_7H_8O	205.3	50.48		1
Benzylamine	C_7H_9N	185		60.16	1
N-Benzylaniline	$C_{13}H_{13}N$	306.5		79.6	6
Benzyl benzoate	$C_{14}H_{12}O_2$	321.3	53.6		
Beryllium chloride	$BeCl_2$	482	105		
Beryllium iodide	BeI_2	590	70.5		
Bis(2-chloroethyl) ether	$C_4H_8Cl_2O$	178	45.2		
Bis(ethoxymethyl) ether	$C_6H_{14}O_3$	136	36.17	44.69	1
Bismuth	Bi	1564	151		
Bismuth tribromide	$BiBr_3$	462	75.4		
Bismuth trichloride	$BiCl_3$	441	72.61		
Boron	B	4000	480		
Boron tribromide	BBr_3	91.3	30.5		
Boron trichloride	BCl_3	12.5	23.77	23.1	
Boron trifluoride	BF_3	-99.9	19.33		
Boron triiodide	BI_3	209.5	40.5		
Bromine	Br_2	58.8	29.96	30.91	
Bromine fluoride	BrF	≈20 dec	25.1		
Bromine pentafluoride	BrF_5	41.3	30.6		
Bromine trifluoride	BrF_3	125.8	47.57		
Bromobenzene	C_6H_5Br	155.9		44.54	1
1-Bromobutane	C_4H_9Br	101.4	32.51	36.64	1
2-Bromobutane, (±)-	C_4H_9Br	91	30.77	34.41	1
Bromochloromethane	CH_2BrCl	67.9	30.0		
2-Bromo-2-chloro-1,1,1-trifluoroethane	$C_2HBrClF_3$	50	28.08	29.61	1
Bromoethane	C_2H_5Br	38.2	27.04	28.03	1
Bromoethene	C_2H_3Br	16	23.4		
1-Bromoheptane	$C_7H_{15}Br$	179		50.60	1
1-Bromohexane	$C_6H_{13}Br$	156		45.89	1
Bromomethane	CH_3Br	3.4	23.91	22.81	1
1-Bromo-2-methylpropane	C_4H_9Br	91.3	31.33	34.82	1
2-Bromo-2-methylpropane	C_4H_9Br	73.3	29.23	31.81	1
1-Bromonaphthalene	$C_{10}H_7Br$	280	39.3		
1-Bromooctane	$C_8H_{17}Br$	199		55.77	1
1-Bromopentane	$C_5H_{11}Br$	126	35.01	41.28	1
1-Bromopropane	C_3H_7Br	70.8	29.84	32.01	1
2-Bromopropane	C_3H_7Br	59.34	28.33	30.17	1
3-Bromopropene	C_3H_5Br	70.1	30.24	32.73	1

Name	Mol. form.	t_b/°C	$\Delta_{vap}H(t_b)$/ kJ mol^{-1}	$\Delta_{vap}H$(25 °C)/ kJ mol^{-1}	Ref.
Bromosilane	BrH_3Si	1.9	24.4		
1,2-Butadiene	C_4H_6	11.0	24.02	23.21	1
1,3-Butadiene	C_4H_6	-4.6	22.47	20.86	1
Butanal	C_4H_8O	74.8	31.5		
Butane	C_4H_{10}	-0.5	22.44	21.02	1
1,2-Butanediol, (±)-	$C_4H_{10}O_2$	196.42	52.84	71.55	8
1,3-Butanediol	$C_4H_{10}O_2$	208.2	54.31	74.46	8
1,4-Butanediol	$C_4H_{10}O_2$	229.5		77.1	11
1,4-Butanedithiol	$C_4H_{10}S_2$	195.5		55.10	
Butanenitrile	C_4H_7N	117.6	33.68	39.33	1
1-Butanethiol	$C_4H_{10}S$	98.4	32.23	36.63	1
2-Butanethiol	$C_4H_{10}S$	85.0	30.59	33.99	1
Butanoic acid	$C_4H_8O_2$	163.7		40.45	1
Butanoic anhydride	$C_8H_{14}O_3$	195	50.0		
1-Butanol	$C_4H_{10}O$	117.6	43.29	52.35	1
2-Butanol	$C_4H_{10}O$	99.4	40.75	49.72	1
2-Butanone	C_4H_8O	79.6	31.30	34.79	1
1-Butene	C_4H_8	-6.3	22.07	20.22	1
cis-2-Butene	C_4H_8	3.72	23.34	22.16	1
trans-2-Butene	C_4H_8	0.88	22.72	21.40	1
2-Butoxyethanol	$C_6H_{14}O_2$	171		56.59	1
Butyl acetate	$C_6H_{12}O_2$	126.0	36.28	43.86	1
tert-Butyl acetate	$C_6H_{12}O_2$	97.9	33.07	38.03	1
Butylamine	$C_4H_{11}N$	77.0	31.81	35.72	1
sec-Butylamine	$C_4H_{11}N$	62.71	29.92	32.85	1
tert-Butylamine	$C_4H_{11}N$	44.02	28.27	29.64	1
Butylbenzene	$C_{10}H_{14}$	183.3	38.87	50.8	14
sec-Butylbenzene, (±)-	$C_{10}H_{14}$	173.3		48.1	14
tert-Butylbenzene	$C_{10}H_{14}$	169.1		47.6	14
Butylcyclohexane	$C_{10}H_{20}$	180.9		49.36	1
Butylcyclopentane	C_9H_{18}	156	36.16	45.89	1
Butylethylamine	$C_6H_{15}N$	104.8	33.97	40.15	1
Butyl ethyl ether	$C_6H_{14}O$	89	31.63	36.32	1
Butyl ethyl sulfide	$C_6H_{14}S$	144.2	37.01	44.51	1
Butyl formate	$C_5H_{10}O_2$	106.1	36.58	41.11	1
tert-Butyl isobutyl ether	$C_8H_{18}O$	112.9	33.11	40.5	
Butyl methyl ether	$C_5H_{12}O$	70.1	29.55	32.37	1
sec-Butyl methyl ether	$C_5H_{12}O$	59.1	28.09	30.23	1
Butyl methyl sulfide	$C_5H_{12}S$	123.4	34.47	40.46	1
tert-Butyl methyl sulfide	$C_5H_{12}S$	98.9	31.47	35.84	1
Butyl propyl ether	$C_7H_{16}O$	117	33.72	40.22	1
Butyl vinyl ether	$C_6H_{12}O$	94	31.58	36.17	1
1-Butyne	C_4H_6	8.1	24.52	23.35	1
2-Butyne-1,4-diol	$C_4H_6O_2$	238		81.5	11
γ-Butyrolactone	$C_4H_6O_2$	204.6	52.2		
Cadmium	Cd	767	99.87		
Cadmium bromide	Br_2Cd	863	115		
Cadmium chloride	$CdCl_2$	964	124.3		
Cadmium fluoride	CdF_2	1750	214		
Cadmium iodide	CdI_2	744	115		
Camphor, (+)	$C_{10}H_{16}O$	209	59.5		
Carbon disulfide	CS_2	46.2	26.74	27.51	1
Carbon monoxide	CO	-191.51	6.04		1
Chlorine	Cl_2	-34.04	20.41	17.65	
Chlorine dioxide	ClO_2	11	30		
Chlorine fluoride	ClF	-101.1	24		
Chlorine monoxide	Cl_2O	2.2	25.9		
Chlorine trifluoride	ClF_3	11.75	27.53		
2-Chloroaniline	C_6H_6ClN	209	44.4		
Chlorobenzene	C_6H_5Cl	131.6	35.19	40.97	1

Name	Mol. form.	t_b/°C	$\Delta_{vap}H(t_b)$/ kJ mol^{-1}	$\Delta_{vap}H$(25 °C)/ kJ mol^{-1}	Ref.
1-Chlorobutane	C_4H_9Cl	78.4	30.39	33.51	1
2-Chlorobutane	C_4H_9Cl	71	29.17	31.53	1
Chlorodifluoromethane	$CHClF_2$	-40.8	20.2		
Chloroethane	C_2H_5Cl	12.3	24.65		1
2-Chloroethanol	C_2H_5ClO	126	41.4		
Chloroethene	C_2H_3Cl	-13.8	20.8		
1-Chloroheptane	$C_7H_{15}Cl$	159		47.66	1
1-Chlorohexane	$C_6H_{13}Cl$	135.0	35.67	42.83	1
Chloromethane	CH_3Cl	-24.1	21.40	18.92	1
1-Chloro-3-methylbutane	$C_5H_{11}Cl$	99	32.02	36.24	1
1-Chloro-2-methylpropane	C_4H_9Cl	69	29.22	31.67	1
2-Chloro-2-methylpropane	C_4H_9Cl	50.9	27.55	28.98	1
1-Chloronaphthalene	$C_{10}H_7Cl$	259	52.1		
1-Chlorooctane	$C_8H_{17}Cl$	183		52.42	1
Chloropentafluorobenzene	C_6ClF_5	117.96	34.76	41.07	1
Chloropentafluoroethane	C_2ClF_5	-39.2	19.41		1
1-Chloropentane	$C_5H_{11}Cl$	107.9	33.15	38.24	1
2-Chloropentane, (+)	$C_5H_{11}Cl$	96.3	31.79	36.03	1
1-Chloropropane	C_3H_7Cl	46.2	27.18	28.35	1
2-Chloropropane	C_3H_7Cl	35.0	26.30	26.90	1
3-Chloropropene	C_3H_5Cl	44.8	29.0		
Chlorosilane	ClH_3Si	-30.4	21		
2-Chlorotoluene	C_7H_7Cl	158.8	37.5		
4-Chlorotoluene	C_7H_7Cl	161.8	38.7		
Chlorotrifluoromethane	$CClF_3$	-81.48	15.8		
Chlorotrifluorosilane	ClF_3Si	-70.0	18.7		
Cholesterol	$C_{27}H_{46}O$	459		148.0	9
Chromium(II) chloride	Cl_2Cr	1120	197		
Chromium(VI) dichloride dioxide	Cl_2CrO_2	117	35.1		
o-Cresol	C_7H_8O	191.0	45.19		1
m-Cresol	C_7H_8O	202.2	47.40	61.71	1
p-Cresol	C_7H_8O	201.9	47.45		1
Cyanogen	C_2N_2	-21.1	23.33	19.75	1
Cyclobutane	C_4H_8	12.5	24.19	23.51	1
Cyclobutanecarbonitrile	C_5H_7N	148	36.88	44.34	1
Cyclohexane	C_6H_{12}	80.7	29.97	33.01	1
Cyclohexanecarbonitrile	$C_7H_{11}N$	188		51.92	1
Cyclohexanethiol	$C_6H_{12}S$	158.8	37.06	44.57	1
Cyclohexanol	$C_6H_{12}O$	160.9		62.01	1
Cyclohexanone	$C_6H_{10}O$	155.4		45.06	1
Cyclohexene	C_6H_{10}	82.9	30.46	33.47	1
1-Cyclohexenecarbonitrile	C_7H_9N			53.55	1
Cyclohexylamine	$C_6H_{13}N$	133.6	36.14	43.67	1
Cyclohexylbenzene	$C_{12}H_{16}$	239		60.8	14
Cyclohexylcyclohexane	$C_{12}H_{22}$	239		57.98	1
Cyclopentane	C_5H_{10}	49.2	27.30	28.52	1
Cyclopentanecarbonitrile	C_6H_9N	169		43.43	1
Cyclopentanethiol	$C_5H_{10}S$	132.2	35.32	41.42	1
Cyclopentanol	$C_5H_{10}O$	140.4		57.05	
Cyclopentanone	C_5H_8O	130.5	36.35	42.72	1
1-Cyclopentenecarbonitrile	C_6H_7N			44.98	1
Cyclopropane	C_3H_6	-31	20.05	16.93	1
Cyclopropanecarbonitrile	C_4H_5N	135	35.55	41.94	1
Cyclopropylbenzene	C_9H_{10}	172		50.22	1
Cyclopropyl methyl ketone	C_5H_8O	111	34.07	39.41	1
cis-Decahydronaphthalene	$C_{10}H_{18}$	195.8	41.0		
trans-Decahydronaphthalene	$C_{10}H_{18}$	187.3	40.2		
Decane	$C_{10}H_{22}$	174.1	39.58	51.42	
1,10-Decanediol	$C_{10}H_{22}O_2$			120.0	11
Decanenitrile	$C_{10}H_{19}N$	241		66.84	1

Name	Mol. form.	t_b/°C	$\Delta_{vap}H(t_b)$/ kJ mol^{-1}	$\Delta_{vap}H$(25 °C)/ kJ mol^{-1}	Ref.
1-Decanethiol	$C_{10}H_{22}S$	240		65.48	1
1-Decanol	$C_{10}H_{22}O$	229		81.50	1
1-Decene	$C_{10}H_{20}$	171		50.43	1
Decylbenzene	$C_{16}H_{26}$	298		78.2	14
Diborane	B_2H_6	-92.49	14.28		
1,4-Dibromobutane	$C_4H_8Br_2$	197		53.09	1
1,2-Dibromo-1-chloro-1,2,2-trifluoroethane	$C_2Br_2ClF_3$	92.8	31.17	35.04	1
1,2-Dibromoethane	$C_2H_4Br_2$	131.3	34.77	41.73	1
Dibromomethane	CH_2Br_2	97.0	32.92	36.97	1
1,2-Dibromopropane	$C_3H_6Br_2$	140	35.61	41.67	1
1,3-Dibromopropane	$C_3H_6Br_2$	164		47.45	1
Dibromosilane	Br_2H_2Si	66	31		
1,2-Dibromotetrafluoroethane	$C_2Br_2F_4$	47.1	27.03	28.39	1
Dibutylamine	$C_8H_{19}N$	162	38.44	49.45	1
Dibutyl ether	$C_8H_{18}O$	141.6	36.49	44.97	1
Di-*sec*-butyl ether	$C_8H_{18}O$	121.9	34.06	40.84	1
Di-*tert*-butyl ether	$C_8H_{18}O$	107.1	32.15	37.61	1
Dibutyl phthalate	$C_{16}H_{22}O_4$	338	79.2		
Dibutyl sulfide	$C_8H_{18}S$	168		52.96	1
Di-*tert*-butyl sulfide	$C_8H_{18}S$	152.3	33.26	43.76	1
o-Dichlorobenzene	$C_6H_4Cl_2$	180.2	39.66	50.21	1
m-Dichlorobenzene	$C_6H_4Cl_2$	172	38.62	48.58	1
p-Dichlorobenzene	$C_6H_4Cl_2$	173.9	38.79	49.0	1
1,2-Dichlorobutane	$C_4H_8Cl_2$	123.9	33.90	39.58	1
1,4-Dichlorobutane	$C_4H_8Cl_2$	155		46.36	1
Dichlorodifluoromethane	CCl_2F_2	-29.8	20.1		
Dichlorodifluorosilane	Cl_2F_2Si	-32	21.2		
1,1-Dichloroethane	$C_2H_4Cl_2$	56.3	28.85	30.62	1
1,2-Dichloroethane	$C_2H_4Cl_2$	83.4	31.98	35.16	1
1,1-Dichloroethene	$C_2H_2Cl_2$	31.6	26.14	26.48	1
cis-1,2-Dichloroethene	$C_2H_2Cl_2$	60	30.2		
trans-1,2-Dichloroethene	$C_2H_2Cl_2$	47.64	28.9		
1,1-Dichloro-1-fluoroethane	$C_2H_3Cl_2F$	32.05	26.06	26.48	
Dichlorofluoromethane	$CHCl_2F$	8.9	25.2		
1,2-Dichloro-1,1,2,3,3,3-hexafluoropropane	$C_3Cl_2F_6$	34	26.28	26.93	1
1,2-Dichlorohexane	$C_6H_{12}Cl_2$	172.2		48.16	1
Dichloromethane	CH_2Cl_2	39.8	28.06	28.82	1
1,2-Dichloropentane	$C_5H_{10}Cl_2$	148.2	36.45	43.89	1
1,5-Dichloropentane	$C_5H_{10}Cl_2$	182.9		50.71	1
1,3-Dichloropropane	$C_3H_6Cl_2$	120.8	35.18	40.75	1
Dichlorosilane	Cl_2H_2Si	8.3	25	24.2	
1,2-Dichloro-1,1,2,2-tetrafluoroethane	$C_2Cl_2F_4$	3.6	23.3		
Dicyclopropyl ketone	$C_7H_{10}O$	161		53.70	1
Diethanolamine	$C_4H_{11}NO_2$	271.2	65.2		
1,1-Diethoxyethane	$C_6H_{14}O_2$	102	36.28	43.20	1
1,2-Diethoxyethane	$C_6H_{14}O_2$	120.6	36.28	43.20	1
Diethoxymethane	$C_5H_{12}O_2$	86	31.33	35.65	1
Diethylamine	$C_4H_{11}N$	55.4	29.06	31.31	1
Diethyl carbonate	$C_5H_{10}O_3$	125.9		43.60	1
Diethyl disulfide	$C_4H_{10}S_2$	154.0	37.58	45.18	1
Diethylene glycol	$C_4H_{10}O_3$	245.5	52.3		
Diethylene glycol diethyl ether	$C_8H_{18}O_3$	185		58.40	1
Diethylene glycol dimethyl ether	$C_6H_{14}O_3$	162	36.17	44.69	1
Diethylene glycol monoethyl ether	$C_6H_{14}O_3$	202	47.5		
Diethylene glycol monomethyl ether	$C_5H_{12}O_3$	194	46.6		
Diethyl ether	$C_4H_{10}O$	34.4	26.52	27.10	1
Diethyl malonate	$C_7H_{12}O_4$	200	54.8		
Diethyl oxalate	$C_6H_{10}O_4$	186	42.0		
3,3-Diethylpentane	C_9H_{20}	146.2	34.61	42.0	1
Diethyl sulfide	$C_4H_{10}S$	92.1	31.77	35.80	1

Name	Mol. form.	t_b/°C	$\Delta_{vap}H(t_b)$/ kJ mol^{-1}	$\Delta_{vap}H$(25 °C)/ kJ mol^{-1}	Ref.
Difluorine dioxide	F_2O_2	-57 extrap	19.1		
o-Difluorobenzene	$C_6H_4F_2$	93.9	32.21	36.18	1
m-Difluorobenzene	$C_6H_4F_2$	83.0	31.10	34.59	1
p-Difluorobenzene	$C_6H_4F_2$	88.9	31.77	35.54	1
1,1-Difluoroethane	$C_2H_4F_2$	-24.02	21.56	19.08	1
Difluorosilane	F_2H_2Si	-77.8	16.3		
Digermane	Ge_2H_6	29	25.1		
2,3-Dihydrothiophene	C_4H_6S	112	33.24	37.74	1
2,5-Dihydrothiophene	C_4H_6S	122.4	34.83	39.95	1
Diiodomethane	CH_2I_2	182	42.5		
Diisobutyl sulfide	$C_8H_{18}S$	173		48.71	1
Diisopentyl ether	$C_{10}H_{22}O$	172	35.1		
Diisopropylamine	$C_6H_{15}N$	84	30.40	34.61	1
Diisopropyl ether	$C_6H_{14}O$	68.4	29.10	32.12	1
Diisopropyl sulfide	$C_6H_{14}S$	120.0	33.80	39.60	1
Diketene	$C_4H_4O_2$	127.0	36.80	42.89	1
1,2-Dimethoxyethane	$C_4H_{10}O_2$	85.0	32.42	36.39	1
N,N-Dimethylacetamide	C_4H_9NO	165.9		50.24	1
Dimethylamine	C_2H_7N	7.3	26.40	25.05	1
2,4-Dimethylaniline	$C_8H_{11}N$	215		61.3	6
2,5-Dimethylaniline	$C_8H_{11}N$	214		61.7	6
N,N-Dimethylaniline	$C_8H_{11}N$	193		52.83	1
2,2-Dimethylbutane	C_6H_{14}	49.7	26.31	27.68	1
2,3-Dimethylbutane	C_6H_{14}	58.0	27.38	29.12	1
2,3-Dimethyl-2-butanethiol	$C_6H_{14}S$	126.1		39.3	
3,3-Dimethyl-2-butanone	$C_6H_{12}O$	106.1	33.39	37.91	1
2,3-Dimethyl-1-butene	C_6H_{12}	55.59		29.18	1
3,3-Dimethyl-1-butene	C_6H_{12}	41.24		26.61	
2,3-Dimethyl-2-butene	C_6H_{12}	73.19	29.64	32.51	
1,1-Dimethylcyclohexane	C_8H_{16}	119.5	32.51	37.92	1
cis-1,2-Dimethylcyclohexane	C_8H_{16}	129.7	33.47	39.70	1
trans-1,2-Dimethylcyclohexane	C_8H_{16}	123.4	32.96	38.36	1
cis-1,3-Dimethylcyclohexane	C_8H_{16}	124.4	32.91	38.26	1
trans-1,3-Dimethylcyclohexane	C_8H_{16}	120.1	33.39	39.16	1
cis-1,4-Dimethylcyclohexane	C_8H_{16}	124.3	33.28	39.02	1
trans-1,4-Dimethylcyclohexane	C_8H_{16}	119.3	32.56	37.90	1
cis-1,3-Dimethylcyclopentane	C_7H_{14}	91.7	30.40	34.20	1
Dimethyl disulfide	$C_2H_6S_2$	109.72	33.78	37.86	1
Dimethyl ether	C_2H_6O	-24.8	21.51	18.51	1
N,N-Dimethylformamide	C_3H_7NO	152.8		46.89	1
Dimethyl glutarate	$C_7H_{12}O_4$	216		65.7	13
Dimethyl heptanedioate	$C_9H_{16}O_4$	245		73.5	13
2,6-Dimethyl-4-heptanol	$C_9H_{20}O$	193		65.17	12
2,6-Dimethyl-4-heptanone	$C_9H_{18}O$	157		50.92	1
2,2-Dimethylhexane	C_8H_{18}	106.8	32.07	37.28	1
2,3-Dimethylhexane	C_8H_{18}	115.6	33.17	38.78	1
2,4-Dimethylhexane	C_8H_{18}	109.4	32.51	37.76	1
2,5-Dimethylhexane	C_8H_{18}	109.1	32.54	37.85	1
3,3-Dimethylhexane	C_8H_{18}	111.9	32.31	37.53	1
3,4-Dimethylhexane	C_8H_{18}	117.7	33.24	38.97	1
Dimethyl 1,6-hexanedioate	$C_8H_{14}O_4$	231		69.0	13
2,5-Dimethyl-2,5-hexanediol	$C_8H_{18}O_2$	233		85.2	11
cis-2,2-Dimethyl-3-hexene	C_8H_{16}	105.4		36.86	
trans-2,2-Dimethyl-3-hexene	C_8H_{16}	100.9		37.03	
2,5-Dimethyl-3-hexyne-2,5-diol	$C_8H_{14}O_2$	206		82.8	11
1,1-Dimethylhydrazine	$C_2H_8N_2$	62.4	32.55	35.0	1
Dimethyl malonate	$C_5H_8O_4$	181.1		57.5	13
Dimethyl nonanedioate	$C_{11}H_{20}O_4$			82.3	13
2,4-Dimethyloctane	$C_{10}H_{22}$	154	36.47	47.13	1
Dimethyl octanedioate	$C_{10}H_{18}O_4$	259		78.1	13

Fluids

Name	Mol. form.	t_b/°C	$\Delta_{vap}H(t_b)$/ kJ mol⁻¹	$\Delta_{vap}H$(25 °C)/ kJ mol⁻¹	Ref.
Dimethyl oxalate	$C_4H_6O_4$	163.4		54.7	13
3,3-Dimethyloxetane	$C_5H_{10}O$	78	30.85	33.94	1
2,2-Dimethylpentane	C_7H_{16}	79.2	29.23	32.42	1
2,3-Dimethylpentane	C_7H_{16}	89.8	30.46	34.26	1
2,4-Dimethylpentane	C_7H_{16}	80.4	29.55	32.88	1
3,3-Dimethylpentane	C_7H_{16}	86.0	29.62	33.03	1
2,2-Dimethyl-3-pentanone	$C_7H_{14}O$	125	36.09	42.34	1
2,4-Dimethyl-3-pentanone	$C_7H_{14}O$	125.2	34.64	41.51	1
2,4-Dimethyl-1-pentene	C_7H_{14}	81.6		33.03	
4,4-Dimethyl-1-pentene	C_7H_{14}	72.5		31.13	
2,4-Dimethyl-2-pentene	C_7H_{14}	83.3		34.19	
cis-4,4-Dimethyl-2-pentene	C_7H_{14}	80.4		32.56	
trans-4,4-Dimethyl-2-pentene	C_7H_{14}	76.7		32.81	
2,2-Dimethylpropanenitrile	C_5H_9N	105.2	32.40	37.35	1
2,2-Dimethyl-1-propanethiol	$C_5H_{12}S$	103.6		36.4	
2,3-Dimethylpyridine	C_7H_9N	161.1	39.08	47.82	
2,4-Dimethylpyridine	C_7H_9N	158.4	38.53	47.49	
2,5-Dimethylpyridine	C_7H_9N	157.00	38.68	47.04	
2,6-Dimethylpyridine	C_7H_9N	144.0	37.46	45.34	
3,4-Dimethylpyridine	C_7H_9N	179.1	39.99	50.50	
3,5-Dimethylpyridine	C_7H_9N	171.9	39.46	49.33	
Dimethyl sebacate	$C_{12}H_{22}O_4$	289		86.4	13
Dimethyl succinate	$C_6H_{10}O_4$	197		61.0	13
Dimethyl sulfide	C_2H_6S	37.32	27.0	27.65	1
Dimethyl sulfoxide	C_2H_6OS	191.9	43.1		
1,3-Dioxane	$C_4H_8O_2$	105	34.37	39.09	1
1,4-Dioxane	$C_4H_8O_2$	101.2	34.16	38.60	1
Diphenyl ether	$C_{12}H_{10}O$	258.0	48.2		
Diphosphine	H_4P_2	63.5 dec	28.8		
1,2-Dipropoxyethane	$C_8H_{18}O_2$	161		50.62	1
Dipropylamine	$C_6H_{15}N$	107.5	33.47	40.04	1
Dipropyl ether	$C_6H_{14}O$	90.1	31.31	35.69	1
Dipropyl sulfide	$C_6H_{14}S$	142.8	36.60	44.21	1
Disilane	H_6Si_2	-14.8	21.2		
1-Docosanol	$C_{22}H_{46}O$	407		135.9	9
Dodecane	$C_{12}H_{26}$	216.3	44.09	61.52	
1,12-Dodecanediol	$C_{12}H_{26}O_2$			135	11
Dodecanenitrile	$C_{12}H_{23}N$	276.1		76.12	1
1-Dodecanol	$C_{12}H_{26}O$	264.1		90.8	9
1-Dodecene	$C_{12}H_{24}$	213.4		60.78	1
Dodecylbenzene	$C_{18}H_{30}$	329		86.6	14
Eicosane	$C_{20}H_{42}$	344.1	58.49	101.81	5
1-Eicosanol	$C_{20}H_{42}O$	356		125.9	9
1,2-Epoxybutane	C_4H_8O	63.4	30.3		
Ethane	C_2H_6	-88.6	14.69	5.16	1
1,2-Ethanediamine	$C_2H_8N_2$	116.9	37.98	44.98	1
1,2-Ethanediol	$C_2H_6O_2$	197.5	50.5	63.9	11
1,2-Ethanediol, diacetate	$C_6H_{10}O_4$	184		61.44	1
1,2-Ethanedithiol	$C_2H_6S_2$	144	37.93	44.68	
Ethanethiol	C_2H_6S	35.0	26.79	27.30	1
Ethanol	C_2H_6O	78.24	38.56	42.32	1
Ethanolamine	C_2H_7NO	170.3	49.83		1
Ethoxybenzene	$C_8H_{10}O$	169.8		51.04	1
2-Ethoxyethanol	$C_4H_{10}O_2$	134.7	39.22	48.21	1
2-Ethoxyethyl acetate	$C_6H_{12}O_3$	156.6	40.76	52.61	8
1-Ethoxy-2-methoxyethane	$C_5H_{12}O_2$	88	34.33	39.83	1
N-Ethylacetamide	C_4H_9NO	205		64.89	1
Ethyl acetate	$C_4H_8O_2$	77.1	31.94	35.60	1
Ethyl acrylate	$C_5H_8O_2$	98.9	34.7		
N-Ethylaniline	$C_8H_{11}N$	204		58.3	6

Name	Mol. form.	t_b/°C	$\Delta_{vap}H(t_b)$/ kJ mol^{-1}	$\Delta_{vap}H$(25 °C)/ kJ mol^{-1}	Ref.
Ethylbenzene	C_8H_{10}	136.2	35.57	42.24	1
Ethyl butanoate	$C_6H_{12}O_2$	121.1	35.47	42.68	1
2-Ethyl-1-butanol	$C_6H_{14}O$	155	43.2		
2-Ethyl-1-butene	C_6H_{12}	64.7		31.13	1
Ethyl chloroacetate	$C_4H_7ClO_2$	144	40.43	49.47	1
Ethylcyclobutane	C_6H_{12}	70	28.67	31.24	1
Ethylcyclohexane	C_8H_{16}	131.8	34.04	40.56	1
Ethylcyclopentane	C_7H_{14}	103.5	31.96	36.40	1
Ethyl dichloroacetate	$C_4H_6Cl_2O_2$	155		50.60	1
Ethyl 2,2-dimethylpropanoate	$C_7H_{14}O_2$	118.3	34.51	41.25	1
Ethylene	C_2H_4	-103.8	13.53		1
N-Ethylformamide	C_3H_7NO	198		58.44	1
Ethyl formate	$C_3H_6O_2$	54.09	29.91	31.96	1
3-Ethylhexane	C_8H_{18}	118.5	33.59	39.64	1
Ethyl hexanoate	$C_8H_{16}O_2$	165		51.72	1
2-Ethylhexanoic acid	$C_8H_{16}O_2$	227.5		75.60	1
2-Ethyl-1-hexanol	$C_8H_{18}O$	186.2	54.2	68.51	12
2-Ethylhexyl acetate	$C_{10}H_{20}O_2$	200	43.5		
2-Ethylhexylamine	$C_8H_{19}N$	172	40.0		
Ethylisopropylamine	$C_5H_{13}N$	68	29.94	33.13	1
Ethyl isopropyl ether	$C_5H_{12}O$	54	28.21	30.08	1
Ethyl isopropyl sulfide	$C_5H_{12}S$	107.3	32.74	37.78	1
Ethyl 3-methylbutanoate	$C_7H_{14}O_2$	135	37.0		
2-Ethyl-3-methyl-1-butene	C_7H_{14}	86.3		34.35	
1-Ethyl-1-methylcyclopentane	C_8H_{16}	121.5	33.20	38.85	1
3-Ethyl-2-methylpentane	C_8H_{18}	115.6	32.93	38.52	1
3-Ethyl-3-methylpentane	C_8H_{18}	118.2	32.78	37.99	1
3-Ethyl-2-methyl-1-pentene	C_8H_{16}	109.2		37.27	
Ethyl 2-methylpropanoate	$C_6H_{12}O_2$	111	33.67	39.83	1
Ethyl methyl sulfide	C_3H_8S	66.6	29.53	31.85	1
3-Ethylpentane	C_7H_{16}	93.4	31.12	35.22	1
Ethyl pentanoate	$C_7H_{14}O_2$	142	36.96	47.01	1
3-Ethyl-3-pentanol	$C_7H_{16}O$	142		57.34	12
Ethyl pentyl ether	$C_7H_{16}O$	118	34.41	41.01	1
Ethyl propanoate	$C_5H_{10}O_2$	98.9	33.88	39.21	1
Ethyl propyl ether	$C_5H_{12}O$	63	28.94	31.43	1
Ethyl propyl sulfide	$C_5H_{12}S$	118.5	34.24	39.97	1
Ethyl trichloroacetate	$C_4H_5Cl_3O_2$	167		50.97	1
Ethyl vinyl ether	C_4H_8O	36	26.2		
Fluorine	F_2	-188.11	6.62		
Fluorine monoxide	F_2O	-144.3	11.09		
Fluorobenzene	C_6H_5F	84.7	31.19	34.58	1
1-Fluorooctane	$C_8H_{17}F$	146	40.43	49.65	1
Fluorosilane	FH_3Si	-98.6	18.8		
2-Fluorotoluene	C_7H_7F	114	35.4		
4-Fluorotoluene	C_7H_7F	116.6	34.08	39.42	1
Formamide	CH_3NO	217		60.15	1
Formic acid	CH_2O_2	101	22.69	20.10	1
Furan	C_4H_4O	31.3	27.10	27.45	1
Furfural	$C_5H_4O_2$	161.5	43.2		
Furfuryl alcohol	$C_5H_6O_2$	168	53.6		
Gallium	Ga	2229	254		
Gallium(III) bromide	Br_3Ga	279	38.9		
Gallium(III) chloride	Cl_3Ga	201	23.9		
Gallium(III) iodide	GaI_3	340	56.5		
Germane	GeH_4	-88.1	14.06		
Germanium	Ge	2833	334		
Germanium(IV) bromide	Br_4Ge	186.35	41.4		
Germanium(IV) chloride	Cl_4Ge	86.55	27.9		
Glycerol	$C_3H_8O_3$	289	61.0		

Name	Mol. form.	t_b/°C	$\Delta_{vap}H(t_b)$/ kJ mol⁻¹	$\Delta_{vap}H$(25 °C)/ kJ mol⁻¹	Ref.
Glycerol triacetate	$C_9H_{14}O_6$	259		85.74	1
Gold	Au	2836	324		
Helium	He	-268.928	0.08		
Heptadecane	$C_{17}H_{36}$	303	53.58	86.47	
1-Heptadecanol	$C_{17}H_{36}O$	324		112.5	9
6-Heptadecanol	$C_{17}H_{36}O$			108.6	9
7-Heptadecanol	$C_{17}H_{36}O$			108.2	9
9-Heptadecanol	$C_{17}H_{36}O$			108.5	9
Heptane	C_7H_{16}	98.38	31.77	36.57	1
1,7-Heptanediol	$C_7H_{16}O_2$	262		97.9	11
1-Heptanol	$C_7H_{16}O$	178		66.81	1
3-Heptanol, (*S*)-	$C_7H_{16}O$	163	42.5		
2-Heptanone	$C_7H_{14}O$	151.0		47.24	1
1-Heptene	C_7H_{14}	94		35.49	1
cis-2-Heptene	C_7H_{14}	97		36.26	
trans-2-Heptene	C_7H_{14}	98		36.27	
cis-3-Heptene	C_7H_{14}	96		35.81	
trans-3-Heptene	C_7H_{14}	96		35.84	
Heptylamine	$C_7H_{17}N$	153		49.96	1
Heptylbenzene	$C_{13}H_{20}$	242		64.2	14
1-Hexacosanol	$C_{26}H_{54}O$			153.7	9
Hexadecane	$C_{16}H_{34}$	286.9	51.84	81.35	
1,16-Hexadecanediol	$C_{16}H_{34}O_2$			163	11
1-Hexadecanol	$C_{16}H_{34}O$	325		107.7	9
1-Hexadecene	$C_{16}H_{32}$	285		80.25	1
Hexadecylbenzene	$C_{22}H_{38}$	385		104.8	14
Hexafluoroacetylacetone	$C_5H_2F_6O_2$	69	27.05	30.58	1
Hexafluorobenzene	C_6F_6	80.2	31.66	35.71	1
Hexafluoroethane	C_2F_6	-78.1	16.15		1
Hexane	C_6H_{14}	68.72	28.85	31.56	1
1,6-Hexanediol	$C_6H_{14}O_2$	208		90.2	10,11
Hexanenitrile	$C_6H_{11}N$	163.5		47.91	1
1-Hexanol	$C_6H_{14}O$	156.9	44.50	61.61	1
2-Hexanol	$C_6H_{14}O$	138	41.01	58.46	1
2-Hexanone	$C_6H_{12}O$	127.6	36.35	43.14	1
3-Hexanone	$C_6H_{12}O$	123.5	35.36	42.47	1
1-Hexene	C_6H_{12}	63.4		30.61	1
cis-2-Hexene	C_6H_{12}	68.9		32.19	1
trans-2-Hexene	C_6H_{12}	67.85		31.60	1
cis-3-Hexene	C_6H_{12}	66.4		31.23	
trans-3-Hexene	C_6H_{12}	67.06		31.55	
Hexylamine	$C_6H_{15}N$	132	36.54	45.10	1
Hexylbenzene	$C_{12}H_{18}$	226		60.4	14
Hexyl methyl ether	$C_7H_{16}O$	125	34.93	42.07	1
Hydrazine	H_4N_2	113.55	41.8	44.7	
Hydrazoic acid	HN_3	35.7	30.5		
Hydrogen	H_2	-252.879	0.90		
Hydrogen bromide	BrH	-66.38		12.69	
Hydrogen chloride	ClH	-85	16.15	9.08	
Hydrogen disulfide	H_2S_2	70.7		33.78	
Hydrogen iodide	HI	-35.55	19.76	17.36	
Hydrogen peroxide	H_2O_2	150.2		51.6	
Hydrogen selenide	H_2Se	-41.25	19.7		
Hydrogen sulfide	H_2S	-59.55	18.67	14.08	
Hydrogen telluride	H_2Te	-2	19.2		
Indan	C_9H_{10}	177.8	39.63	48.79	1
Indium(I) bromide	BrIn	656	92		
Indium(I) iodide	IIn	712	90.8		
Iodine	I_2	184.4	41.57		
Iodine pentafluoride	F_5I	100.5	41.3		

Name	Mol. form.	t_b/°C	$\Delta_{vap}H(t_b)$/ kJ mol^{-1}	$\Delta_{vap}H$(25 °C)/ kJ mol^{-1}	Ref.
Iodobenzene	C_6H_5I	188.5	39.5		
1-Iodobutane	C_4H_9I	130	34.66	40.63	1
2-Iodobutane, (±)-	C_4H_9I	118	33.27	38.46	1
Iodoethane	C_2H_5I	72	29.44	31.93	1
1-Iodohexane	$C_6H_{13}I$	182		49.75	1
Iodomethane	CH_3I	42.4	27.34	27.97	1
1-Iodo-2-methylpropane	C_4H_9I	121	33.54	38.83	1
2-Iodo-2-methylpropane	C_4H_9I	98	31.43	35.41	1
1-Iodopentane	$C_5H_{11}I$	156		45.27	1
1-Iodopropane	C_3H_7I	102	32.08	36.25	1
2-Iodopropane	C_3H_7I	89	30.68	34.06	1
Iridium(VI) fluoride	F_6Ir	53.6	30.9		7
Isobutane	C_4H_{10}	-11.7	21.30	19.23	1
Isobutyl acetate	$C_6H_{12}O_2$	116.9	35.9		
Isobutylamine	$C_4H_{11}N$	68.8	30.61	33.85	1
Isobutylbenzene	$C_{10}H_{14}$	172.7		48.0	14
Isobutyl formate	$C_5H_{10}O_2$	98.4	33.6		
Isobutyl isobutanoate	$C_8H_{16}O_2$	148	38.2		
Isobutyl methyl ether	$C_5H_{12}O$	58.6	28.02	30.13	1
Isopentane	C_5H_{12}	27.83	24.69	24.85	1
Isopentyl acetate	$C_7H_{14}O_2$	141.6	37.5		
Isopentyl isopentanoate	$C_{10}H_{20}O_2$	190	45.9		
Isopropyl acetate	$C_5H_{10}O_2$	88.6	32.93	37.20	1
Isopropylamine	C_3H_9N	31.8	27.83	28.36	1
Isopropylbenzene	C_9H_{12}	152.4		45.13	1
Isopropylcyclohexane	C_9H_{18}	154.4		44.02	1
Isopropylcyclopentane	C_8H_{16}	126.4	33.56	39.44	1
Isopropylmethylamine	$C_4H_{11}N$	53.3	28.71	30.69	1
1-Isopropyl-4-methylbenzene	$C_{10}H_{14}$	177	38.2		
Isopropyl methyl ether	$C_4H_{10}O$	30.8	26.05	26.41	1
Isopropyl methyl sulfide	$C_4H_{10}S$	84.7	30.71	34.15	1
Isopropylpropylamine	$C_6H_{15}N$	96.2	32.14	37.23	1
Isopropyl propyl sulfide	$C_6H_{14}S$	132.0	35.11	41.78	1
Isoquinoline	C_9H_7N	243.2	49.0	60.26	
Krypton	Kr	-153.415	9.08		
Lead	Pb	1749	179.5		
Lead(II) bromide	Br_2Pb	892	133		
Lead(II) chloride	Cl_2Pb	951	127		
Lead(II) fluoride	F_2Pb	1293	160.4		
Lead(II) iodide	I_2Pb	872 dec	104		
Lithium fluoride	FLi	1673	147		
Lithium hydroxide	HLiO	1626	188		
Mercury	Hg	356.619	59.11		
Mercury(II) bromide	Br_2Hg	318	58.89		
Mercury(II) chloride	Cl_2Hg	304	58.9		
Mercury(II) iodide (yellow)	HgI_2	351	59.2		
Mesityl oxide	$C_6H_{10}O$	129.7	36.1		
Methane	CH_4	-161.5	8.19		1
Methanol	CH_4O	64.5	35.21	37.43	1
2-Methoxyethanol	$C_3H_8O_2$	124.3	37.54	45.17	1
2-Methoxyethyl acetate	$C_5H_{10}O_3$	142	43.9		
Methyl acetate	$C_3H_6O_2$	56.7	30.32	32.29	1
Methyl acrylate	$C_4H_6O_2$	80.1	33.1		
2-Methylacrylonitrile	C_4H_5N	90	31.8		
Methylamine	CH_5N	-6.4	25.60	23.37	1
2-Methylaniline	C_7H_9N	200.0	44.6		
3-Methylaniline	C_7H_9N	203.3	44.9		
4-Methylaniline	C_7H_9N	201	44.3		
Methyl benzoate	$C_8H_8O_2$	199		55.57	1
1-Methylbicyclo[3.1.0]hexane	C_7H_{12}	88	31.07	34.77	1

Name	Mol. form.	t_b/°C	$\Delta_{vap}H(t_b)$/ kJ mol⁻¹	$\Delta_{vap}H$(25 °C)/ kJ mol⁻¹	Ref.
3-Methylbutanenitrile	C_5H_9N	129	35.10	41.64	1
2-Methyl-1-butanethiol, (+)	$C_5H_{12}S$	119.0	33.79	39.45	1
2-Methyl-2-butanethiol	$C_5H_{12}S$	99.1	31.37	35.67	1
3-Methyl-2-butanethiol	$C_5H_{12}S$	109.8		37.5	
Methyl butanoate	$C_5H_{10}O_2$	101.9	33.79	39.28	1
2-Methylbutanoic acid	$C_5H_{10}O_2$	177		46.91	1
2-Methyl-1-butanol, (±)-	$C_5H_{12}O$	129.0		55.16	1
3-Methyl-1-butanol	$C_5H_{12}O$	130.8	44.07	55.61	1
2-Methyl-2-butanol	$C_5H_{12}O$	102.4	39.04	50.10	1
3-Methyl-2-butanol, (±)-	$C_5H_{12}O$	113.7		53.0	
3-Methyl-2-butanone	$C_5H_{10}O$	94.2	32.35	36.78	1
2-Methyl-1-butene	C_5H_{10}	31.1	25.50	25.92	
3-Methyl-1-butene	C_5H_{10}	20.1		23.77	
2-Methyl-2-butene	C_5H_{10}	38.5	26.31	27.06	1
(1-Methylbutyl)benzene	$C_{11}H_{16}$	199		53.0	14
Methyl *tert*-butyl ether	$C_5H_{12}O$	55.1	27.94	29.82	1
Methyl chloroacetate	$C_3H_5ClO_2$	130	39.23	46.73	1
Methyl cyanoacetate	$C_4H_5NO_2$	203	48.2		
Methyl cyclobutanecarboxylate	$C_6H_{10}O_2$	135.5	37.13	44.72	1
Methylcyclohexane	C_7H_{14}	100.9	31.27	35.36	1
1-Methylcyclohexanol	$C_7H_{14}O$	155	79.0		
cis-2-Methylcyclohexanol	$C_7H_{14}O$	165	48.5		
trans-2-Methylcyclohexanol, (±)-	$C_7H_{14}O$	168.4	53.0		
Methylcyclopentane	C_6H_{12}	71.8	29.08	31.64	1
Methyl cyclopropanecarboxylate	$C_5H_8O_2$	112	35.25	41.27	1
2-Methyldecane	$C_{11}H_{24}$	189.2	40.25	54.28	1
4-Methyldecane	$C_{11}H_{24}$	188	40.70	53.76	1
Methyl dichloroacetate	$C_3H_4Cl_2O_2$	127	39.28	47.72	1
Methyl 2,2-dimethylpropanoate	$C_6H_{12}O_2$	102	33.42	38.76	1
Methyl dodecanoate	$C_{13}H_{26}O_2$	268		77.17	1
N-Methylformamide	C_2H_5NO	186		56.19	1
Methyl formate	$C_2H_4O_2$	31.6	27.92	28.35	1
2-Methylheptane	C_8H_{18}	117.6	33.26	39.67	1
3-Methylheptane	C_8H_{18}	118.9	33.66	39.83	1
4-Methylheptane	C_8H_{18}	117.7	33.35	39.69	1
Methyl heptanoate	$C_8H_{16}O_2$	169.7		51.62	1
2-Methyl-2-heptanol	$C_8H_{18}O$	173		62.87	12
2-Methylhexane	C_7H_{16}	90.0	30.62	34.87	1
3-Methylhexane	C_7H_{16}	92	30.9		
Methyl hexanoate	$C_7H_{14}O_2$	151	38.55	48.04	1
2-Methyl-2-hexanol	$C_7H_{16}O$	142		58.57	12
5-Methyl-3-hexanol, (±)-	$C_7H_{16}O$	147		59.82	12
cis-3-Methyl-3-hexene	C_7H_{14}	95.3		36.31	
trans-3-Methyl-3-hexene	C_7H_{14}	94		35.70	
Methylhydrazine	CH_6N_2	83	36.12	40.37	1
Methyl isobutanoate	$C_5H_{10}O_2$	92	32.61	37.32	1
Methyl methacrylate	$C_5H_8O_2$	100.6	36.0		
1-Methylnaphthalene	$C_{11}H_{10}$	244.4	45.5		
2-Methylnonane	$C_{10}H_{22}$	167	38.23	49.63	1
3-Methylnonane	$C_{10}H_{22}$	168	38.26	49.71	1
5-Methylnonane	$C_{10}H_{22}$	164	38.14	49.34	1
Methyl octanoate	$C_9H_{18}O_2$	194.1		56.41	1
Methyloxirane	C_3H_6O	35	27.35	27.89	1
2-Methylpentane	C_6H_{14}	60.21	27.79	29.89	1
3-Methylpentane	C_6H_{14}	63.3	28.06	30.28	1
2-Methyl-2,4-pentanediol	$C_6H_{14}O_2$	197.9	57.3		
Methyl pentanoate	$C_6H_{12}O_2$	127.36	35.36	43.10	1
2-Methyl-1-pentanol	$C_6H_{14}O$	157	50.2		
4-Methyl-1-pentanol	$C_6H_{14}O$	151	44.46	60.47	1
2-Methyl-2-pentanol	$C_6H_{14}O$	121	39.59	54.77	1

Name	Mol. form.	t_b/°C	$\Delta_{vap}H(t_b)$/ kJ mol^{-1}	$\Delta_{vap}H(25\ °C)$/ kJ mol^{-1}	Ref.
4-Methyl-2-pentanol	$C_6H_{14}O$	132.0	44.2		
3-Methyl-2-pentanone, (±)-	$C_6H_{12}O$	117	34.16	40.53	1
4-Methyl-2-pentanone	$C_6H_{12}O$	115.7	34.49	40.61	1
2-Methyl-3-pentanone	$C_6H_{12}O$	118	33.84	39.79	1
2-Methyl-1-pentene	C_6H_{12}	62.1		30.48	1
3-Methyl-1-pentene	C_6H_{12}	54		28.62	
4-Methyl-1-pentene	C_6H_{12}	54		28.71	1
2-Methyl-2-pentene	C_6H_{12}	67.3		31.60	1
3-Methyl-*cis*-2-pentene	C_6H_{12}	70.4		32.09	
3-Methyl-*trans*-2-pentene	C_6H_{12}	67.67		31.35	
4-Methyl-*cis*-2-pentene	C_6H_{12}	56.4		29.48	1
4-Methyl-*trans*-2-pentene	C_6H_{12}	58.58		29.97	1
Methyl pentyl ether	$C_6H_{14}O$	99	32.02	36.85	1
Methyl pentyl sulfide	$C_6H_{14}S$	148	37.41	45.24	1
2-Methylpropanenitrile	C_4H_7N	102	32.39	37.13	1
2-Methyl-1-propanethiol	$C_4H_{10}S$	88.5	31.01	34.63	1
2-Methyl-2-propanethiol	$C_4H_{10}S$	64.2	28.45	30.78	1
Methyl propanoate	$C_4H_8O_2$	78.6	32.24	35.85	1
2-Methylpropanoic acid	$C_4H_8O_2$	154.4		35.30	1
2-Methyl-1-propanol	$C_4H_{10}O$	107.84	41.82	50.82	1
2-Methyl-2-propanol	$C_4H_{10}O$	82.3	39.07	46.69	1
Methyl propyl ether	$C_4H_{10}O$	38.5	26.75	27.60	1
Methyl propyl sulfide	$C_4H_{10}S$	95.5	32.08	36.24	1
2-Methylpyridine	C_6H_7N	129.4	36.17	42.48	1
3-Methylpyridine	C_6H_7N	144.1	37.35	44.44	1
4-Methylpyridine	C_6H_7N	145.3	37.51	44.56	1
2-Methylquinoline	$C_{10}H_9N$	247.4		66.1	
4-Methylquinoline	$C_{10}H_9N$	266		67.6	
6-Methylquinoline	$C_{10}H_9N$	265		67.7	
8-Methylquinoline	$C_{10}H_9N$	247.4		65.7	
Methyl salicylate	$C_8H_8O_3$	222.6	46.7		
4-Methylthiazole	C_4H_5NS	133.2	37.58	43.85	1
2-Methylthiophene	C_5H_6S	112.5	33.90	38.87	1
3-Methylthiophene	C_5H_6S	115.4	34.24	39.43	1
Methyl trichloroacetate	$C_3H_3Cl_3O_2$	152		48.33	1
Molybdenum(V) chloride	Cl_5Mo	268	62.8		
Molybdenum(V) fluoride	F_5Mo	213.6	51.8		7
Molybdenum(VI) fluoride	F_6Mo	34.0	29.0		7
Molybdenum(VI) oxide	MoO_3	1155	138		
Molybdenum(VI) oxytetrafluoride	F_4MoO	186.0	50.6		7
Morpholine	C_4H_9NO	128.2	37.1		
Naphthalene	$C_{10}H_8$	218.0	43.2		
Neon	Ne	-246.046	1.71		
Neopentane	C_5H_{12}	9.50	22.74	21.84	1
Niobium(V) chloride	Cl_5Nb	247.4	52.7		
Niobium(V) fluoride	F_5Nb	234	52.3		
Nitric acid	HNO_3	83		39.1	
Nitric oxide	NO	-151.74	13.83		
Nitrobenzene	$C_6H_5NO_2$	210.7		55.01	1
Nitroethane	$C_2H_5NO_2$	114.1	38.0		
Nitrogen	N_2	-195.795	5.57		
Nitrogen tetroxide	N_2O_4	21.15	38.12		
Nitrogen trifluoride	F_3N	-128.75	11.56		
Nitromethane	CH_3NO_2	101.19	33.99	38.27	1
1-Nitropropane	$C_3H_7NO_2$	131.2	38.5		
2-Nitropropane	$C_3H_7NO_2$	120.2	36.8		
Nitrosyl chloride	ClNO	-5.5	25.78		
Nitrosyl fluoride	FNO	-59.9	19.28		
Nitrous oxide	N_2O	-88.48	16.53		
Nitryl chloride	$ClNO_2$	-15	25.7		

Name	Mol. form.	t_b/°C	$\Delta_{vap}H(t_b)$/ kJ mol^{-1}	$\Delta_{vap}H$(25 °C)/ kJ mol^{-1}	Ref.
Nitryl fluoride	FNO_2	-72.4	18.05		
Nonadecane	$C_{19}H_{40}$	330	56.93	96.4	5
Nonane	C_9H_{20}	150.8	37.18	46.55	
1,9-Nonanediol	$C_9H_{20}O_2$			112.5	11
1-Nonanol	$C_9H_{20}O$	213.7		76.86	1
2-Nonanone	$C_9H_{18}O$	194		56.44	1
5-Nonanone	$C_9H_{18}O$	188.4		53.30	1
Nonylbenzene	$C_{15}H_{24}$	280		74.1	14
Octadecane	$C_{18}H_{38}$	316	55.23	91.44	5
1-Octadecanol	$C_{18}H_{38}O$	351		116.8	9
cis-9-Octadecenoic acid	$C_{18}H_{34}O_2$	360	67.4		
Octane	C_8H_{18}	125.62	34.41	41.49	1
1,8-Octanediol	$C_8H_{18}O_2$			104.9	11
Octanenitrile	$C_8H_{15}N$	202		56.80	1
Octanoic acid	$C_8H_{16}O_2$	240	58.5		
1-Octanol	$C_8H_{18}O$	194.7		70.98	1
2-Octanol	$C_8H_{18}O$	179	44.4		
1-Octene	C_8H_{16}	121.3	34.07	40.34	
Octylbenzene	$C_{14}H_{22}$	263		69.1	14
1-Octyne	C_8H_{14}	126.2	35.83	42.30	1
2-Octyne	C_8H_{14}	138.0	37.26	44.49	1
3-Octyne	C_8H_{14}	135.6	36.94	43.92	1
4-Octyne	C_8H_{14}	133.5	36.0	42.73	1
Osmium(V) fluoride	F_5Os	233	65.6		7
Osmium(VI) fluoride	F_6Os	47.5	28.1		7
Oxetane	C_3H_6O	47.6	28.67	29.85	1
2-Oxetanone	$C_3H_4O_2$	161		47.03	1
Oxirane	C_2H_4O	10.4	25.54	24.75	1
Oxygen	O_2	-182.962	6.82		
Pentaborane(11)	B_5H_{11}	65	31.8		
Pentachloroethane	C_2HCl_5	161	36.9		
Pentadecane	$C_{15}H_{32}$	270.6	50.08	76.77	
1,15-Pentadecanediol	$C_{15}H_{32}O_2$			139	11
1-Pentadecanol	$C_{15}H_{32}O$	318		103.5	9
Pentadecylbenzene	$C_{21}H_{36}$	373		100.3	14
Pentafluorobenzene	C_6HF_5	85	32.15	36.27	1
2,3,4,5,6-Pentafluorotoluene	$C_7H_3F_5$	117	34.75	41.12	1
2,2,4,6,6-Pentamethylheptane	$C_{12}H_{26}$	178		48.97	1
Pentane	C_5H_{12}	36.06	25.79	26.43	1
1,5-Pentanediol	$C_5H_{12}O_2$	241	60.7	83.0	11
2,4-Pentanedione	$C_5H_8O_2$	140.7	34.30	41.77	1
Pentanenitrile	C_5H_9N	140	36.09	43.60	1
1-Pentanethiol	$C_5H_{12}S$	126.6	34.88	41.24	1
Pentanoic acid	$C_5H_{10}O_2$	186.1	44.1		
1-Pentanol	$C_5H_{12}O$	137.6	44.36	57.02	1
2-Pentanol	$C_5H_{12}O$	119.1	41.40	54.21	1
3-Pentanol	$C_5H_{12}O$	123		54.0	1
2-Pentanone	$C_5H_{10}O$	102.2	33.44	38.40	1
3-Pentanone	$C_5H_{10}O$	101.9	33.45	38.52	1
1-Pentene	C_5H_{10}	30.0	25.20	25.47	1
cis-2-Pentene	C_5H_{10}	36.9		26.86	1
trans-2-Pentene	C_5H_{10}	36.3		26.76	1
trans-3-Pentenenitrile	C_5H_7N	144	37.09	44.77	1
Pentyl acetate	$C_7H_{14}O_2$	149.4	38.42	48.56	8
Pentylamine	$C_5H_{13}N$	104.7	34.01	40.08	1
Pentylbenzene	$C_{11}H_{16}$	203		55.1	14
Pentylcyclohexane	$C_{11}H_{22}$	204		53.88	1
Perchloryl fluoride	$ClFO_3$	-46.75	19.33		
Perfluorobutane	C_4F_{10}	-2.1	22.9		
Perfluorocyclobutane	C_4F_8	-5.91	23.2		

Name	Mol. form.	t_b/°C	$\Delta_{vap}H(t_b)$/ kJ mol^{-1}	$\Delta_{vap}H$(25 °C)/ kJ mol^{-1}	Ref.
Perfluorodecalin	$C_{10}F_{18}$	143		41.54	10
Perfluorohexane	C_6F_{14}	57.2		32.47	10
Perfluorononane	C_9F_{20}	117		45.27	10
Perfluorooctane	C_8F_{18}	105	33.38	41.13	1
Perfluorotoluene	C_7F_8	104.6		40.52	10
Phenanthrene	$C_{14}H_{10}$	338.4		75.50	1
Phenol	C_6H_6O	181.8	45.69	57.82	1
Phosphine	H_3P	-87.75	14.6		
Phosphorothioc trifluoride	F_3PS	-52.25	19.6		
Phosphorus (white)	P	280.5	12.4	14.2	
Phosphorus(III) bromide	Br_3P	173.2	38.8		
Phosphorus(III) chloride	Cl_3P	76	30.5	32.1	
Phosphorus(III) chloride difluoride	ClF_2P	-47.3	17.6		
Phosphorus(III) dichloride fluoride	Cl_2FP	13.85	24.9		
Phosphorus(III) fluoride	F_3P	-101.8	16.5		
Phosphorus(V) fluoride	F_5P	-84.6	17.2		
Phosphorus(III) iodide	I_3P	227 dec	43.9		
Phosphoryl bromide	Br_3OP	191.7	38		
Phosphoryl chloride	Cl_3OP	105.5	34.35	38.6	
Piperidine	$C_5H_{11}N$	106.19		39.29	1
Propanal	C_3H_6O	48.0	28.31	29.62	1
Propane	C_3H_8	-42.11	19.04	14.79	1
1,3-Propanediamine	$C_3H_{10}N_2$	139.2	40.85	50.16	1
1,2-Propanediol	$C_3H_8O_2$	187.3	52.4		
1,3-Propanediol	$C_3H_8O_2$	214.7	57.9	69.8	11
1,3-Propanedithiol	$C_3H_8S_2$	172.9		49.66	1
Propanenitrile	C_3H_5N	97.3	31.81	36.03	1
1-Propanethiol	C_3H_8S	67.7	29.54	31.89	1
2-Propanethiol	C_3H_8S	52.6	27.91	29.45	1
Propanoic acid	$C_3H_6O_2$	141.5		32.14	1
Propanoic anhydride	$C_6H_{10}O_3$	168	41.7		
1-Propanol	C_3H_8O	97.04	41.44	47.45	1
2-Propanol	C_3H_8O	82.21	39.85	45.39	1
Propene	C_3H_6	-47.6	18.42	14.24	
2-Propoxyethanol	$C_5H_{12}O_2$	152	41.40	52.12	1
Propyl acetate	$C_5H_{10}O_2$	101.0	33.92	39.72	1
Propylamine	C_3H_9N	47.21	29.55	31.27	1
Propylbenzene	C_9H_{12}	159.2		46.22	1
Propylcyclohexane	C_9H_{18}	156.7		45.08	1
Propylcyclopentane	C_8H_{16}	130.9	34.70	41.08	1
Propyl formate	$C_4H_8O_2$	80.6	33.61	37.53	1
Propyl propanoate	$C_6H_{12}O_2$	122.2	35.54	43.45	1
Pyridazine	$C_4H_4N_2$	208		53.47	1
Pyridine	C_5H_5N	115.2	35.09	40.21	1
Pyrimidine	$C_4H_4N_2$	123.8	43.09	49.79	1
Pyrrole	C_4H_5N	129.74	38.75	45.09	1
Pyrrolidine	C_4H_9N	86.6	33.01	37.52	1
Quinoline	C_9H_7N	237.1	49.7	59.30	
Rhenium(VII) dioxytrifluoride	F_3O_2Re	185.4	65.7		7
Rhenium(V) fluoride	F_5Re	221.3	58.1		7
Rhenium(VI) fluoride	F_6Re	33.8	28.7		7
Rhenium(VI) oxytetrafluoride	F_4ORe	171.7	61.0		7
Salicylaldehyde	$C_7H_6O_2$	208	38.2		
Selenium (α form)	Se	685	95.48		
Selenium tetrafluoride	F_4Se	101.6	47.2		
Silane	H_4Si	-111.9	12.1		
Silver(I) bromide	AgBr	1502	198		
Silver(I) chloride	AgCl	1547	199		
Silver(I) iodide	AgI	1506	143.9		
Sodium hydroxide	HNaO	1388	175		

Name	Mol. form.	t_b/°C	$\Delta_{vap}H(t_b)$/ kJ mol^{-1}	$\Delta_{vap}H$(25 °C)/ kJ mol^{-1}	Ref.
Spiro[2.2]pentane	C_5H_8	39.0	26.76	27.49	1
Stannane	H_4Sn	-51.8	19.05		
Stibine	H_3Sb	-17	21.3		
Styrene	C_8H_8	145.3	38.7		
Succinonitrile	$C_4H_4N_2$	266	48.5		
Sulfur (rhombic)	S	444.61	45		
Sulfur dioxide	O_2S	-10.02	24.94	22.92	
Sulfur hexafluoride	F_6S	-63.8 sp		8.99	
Sulfur tetrafluoride	F_4S	-40.45	26.44		
Sulfur trioxide (α-form)	O_3S	subl		43.14	
Sulfuryl chloride	Cl_2O_2S	69.4	31.4	30.1	
Tantalum(V) bromide	Br_5Ta	348.8	62.3		
Tantalum(V) chloride	Cl_5Ta	239	54.8		
Tantalum(V) fluoride	F_5Ta	229.5	56.9		
Tellurium	Te	988	114.1		
Tellurium tetrachloride	Cl_4Te	387	77		
Tetraborane(10)	B_4H_{10}	18	27.1		
1,1,2,2-Tetrabromoethane	$C_2H_2Br_4$	248	48.7		
Tetrabromosilane	Br_4Si	154	37.9		
1,1,2,2-Tetrachloroethane	$C_2H_2Cl_4$	146.0	37.64	45.71	1
Tetrachloroethene	C_2Cl_4	121.2	34.68	39.68	1
Tetrachloromethane	CCl_4	76.7	29.82	32.43	1
Tetrachlorosilane	Cl_4Si	57.65	28.7	29.7	
Tetradecane	$C_{14}H_{30}$	253.5	48.16	71.73	
1,14-Tetradecanediol	$C_{14}H_{30}O_2$			149.7	11
Tetradecanenitrile	$C_{14}H_{27}N$			85.29	1
1-Tetradecanol	$C_{14}H_{30}O$	295.8		98.9	9
Tetradecylbenzene	$C_{20}H_{34}$	347		95.8	14
Tetrafluorodiborane	B_2F_4	-34.0	28		
Tetrafluorohydrazine	F_4N_2	-74	13.27		
Tetrahydrofuran	C_4H_8O	66.0	29.81	31.99	1
Tetrahydrofurfuryl alcohol	$C_5H_{10}O_2$	176.3	45.2		
1,2,3,4-Tetrahydronaphthalene	$C_{10}H_{12}$	207.2	43.9		
Tetrahydropyran	$C_5H_{10}O$	88.0	31.17	34.58	1
Tetrahydrothiophene	C_4H_8S	121.1	34.66	39.43	1
Tetraiodosilane	I_4Si	287.35	50.2		
2,2,3,3-Tetramethylbutane	C_8H_{18}	106.32		42.90	1
2,2,4,4-Tetramethylpentane	C_9H_{20}	122.2	32.51	38.49	1
Tetranitromethane	CN_4O_8	125.6	40.74	49.93	1
Thallium(I) bromide	BrTl	819	99.56		
Thallium(I) chloride	ClTl	720	102.2		
Thallium(I) iodide	ITl	824	104.7		
Thallium(I) sulfide	STl_2	1367	154		
Thiacyclohexane	$C_5H_{10}S$	141.73	35.96	42.58	1
Thietane	C_3H_6S	95.0	32.32	35.97	1
Thionitrosyl fluoride (NSF)	FNS	4.8	22.2		
Thionyl chloride	Cl_2OS	75.6	31.7	31	
Thionyl fluoride	F_2OS	-43.8	21.8		
Thiophene	C_4H_4S	84.1	31.48	34.70	1
Thorium(IV) chloride	Cl_4Th	921	146.4		
Thorium(IV) fluoride	F_4Th	1680	258		
Tin(II) bromide	Br_2Sn	639	102		
Tin(IV) bromide	Br_4Sn	205	43.5		
Tin(II) chloride	Cl_2Sn	623	86.8		
Tin(IV) chloride	Cl_4Sn	114.15	34.9		
Tin(II) iodide	I_2Sn	714	105		
Tin(IV) iodide	I_4Sn	364.35	56.9		
Titanium(IV) bromide	Br_4Ti	233.5	44.37		
Titanium(II) chloride	Cl_2Ti	1500	232		
Titanium(III) chloride	Cl_3Ti	960	124		

Name	Mol. form.	t_b/°C	$\Delta_{vap}H(t_b)$/ kJ mol^{-1}	$\Delta_{vap}H$(25 °C)/ kJ mol^{-1}	Ref.
Titanium(IV) chloride	Cl_4Ti	136.45	36.2		
Titanium(IV) iodide	I_4Ti	377	58.4		
Toluene	C_7H_8	110.60	33.18	38.01	1
Triacetamide	$C_6H_9NO_3$			60.41	1
Tribromomethane	$CHBr_3$	149.2	39.66	46.05	1
Tribromosilane	Br_3HSi	109	34.8		
Tributylamine	$C_{12}H_{27}N$	207	46.9		
Tributyl borate	$C_{12}H_{27}BO_3$	233.8	56.1		
1,1,1-Trichloroethane	$C_2H_3Cl_3$	74.02	29.86	32.50	1
1,1,2-Trichloroethane	$C_2H_3Cl_3$	113	34.82	40.24	1
Trichloroethene	C_2HCl_3	86.8	31.40	34.54	1
Trichlorofluoromethane	CCl_3F	23.7	25.1		
Trichloromethane	$CHCl_3$	61.2	29.24	31.28	1
1,2,3-Trichloropropane	$C_3H_5Cl_3$	158	37.1		
Trichlorosilane	Cl_3HSi	33		25.7	
1,1,1-Trichloro-2,2,2-trifluoroethane	$C_2Cl_3F_3$	46	26.85	28.08	1
1,1,2-Trichloro-1,2,2-trifluoroethane	$C_2Cl_3F_3$	47.6	27.04	28.40	1
Tridecane	$C_{13}H_{28}$	235.4	46.20	66.68	
1,13-Tridecanediol	$C_{13}H_{28}O_2$			133	11
1-Tridecanol	$C_{13}H_{28}O$	287		94.7	9
Tridecylbenzene	$C_{19}H_{32}$	340		91.8	14
Triethylamine	$C_6H_{15}N$	88.8	31.01	34.84	1
Triethylene glycol	$C_6H_{14}O_4$	288.6	71.4		
Trifluoroacetic acid	$C_2HF_3O_2$	72	33.3		
1,1,1-Trifluoroethane	$C_2H_3F_3$	-47.2	18.99		
(Trifluoromethyl)benzene	$C_7H_5F_3$	102.0	32.63	37.60	1
Trifluorosilane	F_3HSi	-95	16.2		
Trigermane	Ge_3H_8	110.5	32.2		
Trimethylamine	C_3H_9N	2.8	22.94	21.66	1
1,2,3-Trimethylbenzene	C_9H_{12}	176.0		49.05	1
1,2,4-Trimethylbenzene	C_9H_{12}	169.4		47.93	1
1,3,5-Trimethylbenzene	C_9H_{12}	164.7		47.50	1
2,2,3-Trimethylbutane	C_7H_{16}	80.8	28.90	32.05	1
2,3,3-Trimethyl-1-butene	C_7H_{14}	77.8		32.09	
2,2,5-Trimethylhexane	C_9H_{20}	124	33.65	40.16	1
2,3,5-Trimethylhexane	C_9H_{20}	131	34.43	41.41	1
3,5,5-Trimethyl-1-hexanol	$C_9H_{20}O$	193		67.86	12
2,4,7-Trimethyloctane	$C_{11}H_{24}$	170	38.22	49.91	1
2,2,3-Trimethylpentane	C_8H_{18}	109.8	31.94	36.91	1
2,2,4-Trimethylpentane	C_8H_{18}	99.2	30.79	35.14	1
2,3,3-Trimethylpentane	C_8H_{18}	114.7	32.12	37.27	1
2,3,4-Trimethylpentane	C_8H_{18}	113.4	32.36	37.75	1
2,2,4-Trimethyl-3-pentanone	$C_8H_{16}O$	146	35.64	43.30	1
2,4,4-Trimethyl-1-pentene	C_8H_{16}	101.3		35.59	
2,4,4-Trimethyl-2-pentene	C_8H_{16}	104.9		37.23	
2,3,6-Trimethylpyridine	$C_8H_{11}N$	170	39.95	50.61	1
2,4,6-Trimethylpyridine	$C_8H_{11}N$	170	39.87	50.33	1
Trisilane	H_8Si_3	52.9	28.5		
Tris(perfluorobutyl)amine	$C_{12}F_{27}N$	178	46.4		
Tungsten(VI) chloride	Cl_6W	337	52.7		
Tungsten(VI) fluoride	F_6W	17.1	26.5		7
Tungsten(VI) oxytetrachloride	Cl_4OW	230	67.8		
Tungsten(VI) oxytetrafluoride	F_4OW	185.9	59.5		7
Undecane	$C_{11}H_{24}$	195.9	41.91	56.58	
1,11-Undecanediol	$C_{11}H_{24}O_2$			132	11
Undecanenitrile	$C_{11}H_{21}N$	265		71.14	1
1-Undecanol	$C_{11}H_{24}O$	246		85.8	9
Undecylbenzene	$C_{17}H_{28}$	312		82.4	14
Vanadium(IV) chloride	Cl_4V	151	41.4	42.5	
Vanadium(V) fluoride	F_5V	48.3	44.52		

Name	Mol. form.	t_b/°C	$\Delta_{vap}H(t_b)$/ kJ mol^{-1}	$\Delta_{vap}H$(25 °C)/ kJ mol^{-1}	Ref.
Vanadyl trichloride	Cl_3OV	127	36.78		
Vinyl acetate	$C_4H_6O_2$	72.6	34.6		
Water	H_2O	99.974	40.65	43.98	
Xenon	Xe	-108.099	12.57		
o-Xylene	C_8H_{10}	144.4	36.24	43.43	1
m-Xylene	C_8H_{10}	139.1	35.66	42.65	1
p-Xylene	C_8H_{10}	138.3	35.67	42.40	1
2,4-Xylenol	$C_8H_{10}O$	210.94		64.96	1
2,5-Xylenol	$C_8H_{10}O$	211.14	46.9		
2,6-Xylenol	$C_8H_{10}O$	201.03		75.31	1
3,4-Xylenol	$C_8H_{10}O$	227.31		85.03	1
3,5-Xylenol	$C_8H_{10}O$	221.71		82.01	1
Zinc bromide	Br_2Zn	≈670	118		
Zinc chloride	Cl_2Zn	732	126		
Zinc fluoride	F_2Zn	1500	190.1		

ENTHALPY OF FUSION

This table lists the molar enthalpy (heat) of fusion, $\Delta_{fus}H$, for over 1100 inorganic and organic compounds. All values refer to the enthalpy change at equilibrium between the liquid phase and the most stable solid phase at the phase transition temperature. Most values of $\Delta_{fus}H$ are given at the normal melting point t_m; see column definition table for details. Temperatures are given on the ITS–90 scale.

Column definitions are as follows.

Column heading	Definition
Name	Name of substance, either an IUPAC systematic name or, in the case of drugs and other complex compounds, a common synonym; listed alphabetically
Mol. form.	Molecular formula in the Hill convention
t_m	Normal meting point; a "tp" following the entry in the melting point column indicates a triple-point temperature, where the solid, liquid, and gas phases are in equilibrium; all temperatures are in °C; a few entries have the triple-point pressure indicated; (dec) = decomposes
$\Delta_{fus}H$	Molar enthalpy of fusion, in units kJ mol^{-1}

References

1. Chase, M. W., Davies, C. A., Downey, J. R., Frurip, D. J., McDonald, R. A., and Syverud, A. N., *JANAF Thermochemical Tables, Third Edition, J. Phys. Chem. Ref. Data* Vol. 14, Suppl. 1, 1985.
2. Chase, M. W., *NIST–JANAF Thermochemical Tables, Fourth Edition, J. Phys. Chem. Ref. Data* Monograph No. 9, 1998.
3. Gurvich, L. V., Veyts, I. V., and Alcock, C. B., *Thermodynamic Properties of Individual Substances, Fourth Edition;* Vol. 2, Hemisphere Publishing Corp., New York, 1991; Vol. 3, CRC Press, Boca Raton, FL, 1994.
4. Dinsdale, A. T., "SGTE Data for Pure Elements," *CALPHAD,* 15, 317–425, 1991. <https://doi.org/10.1016/0364-5916(91)90030-N>
5. *Landolt–Börnstein, Numerical Data and Functional Relationships in Science and Technology, New Series,* IV/8A, "Enthalpies of Fusion and Transition of Organic Compounds," Springer–Verlag, Heidelberg, 1995.
6. *Landolt–Börnstein, Numerical Data and Functional Relationships in Science and Technology, New Series,* IV/19A, "Thermodynamic Properties of Inorganic Materials compiled by SGTE," Springer–Verlag, Heidelberg; Part 1, 1999; Part 2; 1999; Part 3, 2000; Part 4, 2001.
7. Janz, G. J., et al., *Physical Properties Data Compilations Relevant to Energy Storage. II. Molten Salts,* Nat. Stand. Ref. Data Sys.– Nat. Bur. Standards (U.S.), No. 61, Part 2, 1979. <https://doi.org/10.6028/NBS.NSRDS.61p2>
8. Dirand, M., Bouroukba, M., Chevallier, V., Petitjean, D., Behar, E., and Ruffier-Meray, V., "Normal Alkanes, Multialkane Synthetic Model Mixtures, and Real Petroleum Waxes: Crystallographic Structures, Thermodynamic Properties, and Crystallization," *J. Chem. Eng. Data* 47, 115–143, 2002. <https://doi.org/10.1021/je0100084>
9. Linstrom, P. J., and Mallard, W. G., Editors, *NIST Chemistry WebBook,* NIST Standard Reference Database No. 69, June 2005, National Institute of Standards and Technology, Gaithersburg, MD, <http://webbook.nist.gov>.
10. Thermodynamic Research Center, National Institute of Standards and Technology, *TRC Thermodynamic Tables,* <http://trc.nist.gov>.
11. Sangster, J., "Phase Diagrams and Thermodynamic Properties of Binary Systems of Drugs," *J. Phys. Chem. Ref. Data* 28, 889, 1999. <https://doi.org/10.1063/1.556040>
12. Chevallier, V., et al., *J. Chem. Eng. Data* 46, 1114, 2001. <https://doi.org/10.1021/je0003501>
13. Dirand, M., et al., *J. Chem. Eng. Data* 47, 115, 2002. <https://doi.org/10.1021/je0100084>
14. Rycerz, L., and Gaune-Escard, M., *J. Chem. Eng. Data* 49, 1078, 2004. <https://doi.org/10.1021/je049910c>
15. Ott, J. B., and Goates, J. R., *J. Chem. Eng. Data* 41, 669, 1996. <https://doi.org/10.1021/je9601063>

Molar Enthalpy of Fusion

Name	Mol. form.	t_m/°C	$\Delta_{fus}H$/ kJ mol^{-1}
Acenaphthene	$C_{12}H_{10}$	93	21.49
Acenaphthylene	$C_{12}H_8$	89.4	6.9
Acetaldehyde	C_2H_4O	-123.4	2.31
Acetamide	C_2H_5NO	80.16	15.59
Acetanilide	C_8H_9NO	114.35	21.3
Acetic acid	$C_2H_4O_2$	17	11.73
Acetic anhydride	$C_4H_6O_3$	-73.4	10.5
Acetone	C_3H_6O	-94.9	5.77
Acetonitrile	C_2H_3N	-44	8.16
Acrylic acid	$C_3H_4O_2$	13.56	9.51
Acrylonitrile	C_3H_3N	-83.51	6.23
Actinium	Ac	1050	12.0
Allene	C_3H_4	-136.4	4.40
Aluminum	Al	660.323	10.71
Aluminum bromide	$AlBr_3$	97.5	11.25
Aluminum chloride	$AlCl_3$	192.6	35.35
Aluminum fluoride	AlF_3	2250 tp (220 MPa)	98
Aluminum iodide	AlI_3	188.28	15.90
Aluminum oxide (α)	Al_2O_3	2053	111.1
Aluminum sulfide	Al_2S_3	1100	66
Americium	Am	1176	14.39
N-[(4-Aminophenyl)sulfonyl] acetamide	$C_8H_{10}N_2O_3S$	182.0	22.4
3-Amino-1-propanol	C_3H_9NO	12.1	19.7
4-Amino-*N*-2-pyridinylbenzenesulfonamide	$C_{11}H_{11}N_3O_2S$	190	34.4
4-Amino-*N*-2-pyrimidinylbenzenesulfonamide	$C_{10}H_{10}N_4O_2S$	261	42.6
Aminopyrine	$C_{13}H_{17}N_3O$	107.5	27.6
Ammonia	H_3N	-77.65	5.66
Ammonium chloride	ClH_4N	520.1 tp (dec)	10.6
Ammonium fluoride	FH_4N	238	12.6
Ammonium iodide	H_4IN	551 dec	21
Ammonium nitrate	$H_4N_2O_3$	169.7	5.86
Ampyrone	$C_{11}H_{13}N_3O$	109	24.9
Aniline	C_6H_7N	-6.0	10.54
Aniline-2-carboxylic acid	$C_7H_7NO_2$	144.6	20.5
Aniline-4-carboxylic acid	$C_7H_7NO_2$	188.2	22.5
Anisole	C_7H_8O	-37.3	12.9
Anthracene	$C_{14}H_{10}$	216	29.4
Antimony (gray)	Sb	630.628	19.79

Name	Mol. form.	t_m/°C	$\Delta_{fus}H$/ kJ mol^{-1}
Antimony(III) bromide	Br_3Sb	97	14.6
Antimony(III) chloride	Cl_3Sb	73.4	12.97
Antimony(III) fluoride	F_3Sb	287	22.8
Antimony(III) iodide	I_3Sb	171	22.8
Antimony(III) oxide (valentinite)	O_3Sb_2	655	54
Antimony(III) sulfide	S_3Sb_2	550	47.9
Argon	Ar	-189.34	1.18
Arsenic (gray)	As	817	24.44
Arsenic(III) bromide	$AsBr_3$	31.1	11.7
Arsenic(III) chloride	$AsCl_3$	-16	10.1
Arsenic(III) fluoride	AsF_3	-5.9	10.4
Arsenic(III) iodide	AsI_3	141	21.8
Arsenic(V) oxide	As_2O_5	730	60
Arsenic(III) oxide (claudetite)	As_2O_3	314	18
Arsenic(III) selenide	As_2Se_3	377	40.8
Arsenic(III) sulfide	As_2S_3	312	28.7
Arsenic sulfide	As_4S_4	307	25.4
Arsenic(III) telluride	As_2Te_3	375	46.0
trans-Azobenzene	$C_{12}H_{10}N_2$	67.88	22.52
trans-Azoxybenzene	$C_{12}H_{10}N_2O$	34.6	17.9
Barbital	$C_8H_{12}N_2O_3$	189	24.7
Barium	Ba	727	7.12
Barium bromide	$BaBr_2$	857	32.2
Barium carbonate	$CBaO_3$	1555	40
Barium chloride	$BaCl_2$	961	15.85
Barium fluoride	BaF_2	1368	23.36
Barium hydride	BaH_2	1200	25
Barium hydroxide	BaH_2O_2	408	16
Barium iodide	BaI_2	711	26.5
Barium oxide	BaO	1973	46
Barium sulfate	BaO_4S	1580	40
Barium sulfide	BaS	2227	63
Benzaldehyde	C_7H_6O	-57.12	9.32
Benzamide	C_7H_7NO	128	19.5
Benz[*a*]anthracene	$C_{18}H_{12}$	160	21.4
Benzene	C_6H_6	5.538	9.87
Benzeneacetic acid	$C_8H_8O_2$	76.7	16.3
1,2-Benzenediamine	$C_6H_8N_2$	103	23.1
1,3-Benzenediamine	$C_6H_8N_2$	65.5	15.57
1,4-Benzenediamine	$C_6H_8N_2$	140.3	23.8
Benzenethiol	C_6H_6S	-14.87	11.48
p-Benzidine	$C_{12}H_{12}N_2$	127.0	19.1
Benzil	$C_{14}H_{10}O_2$	94.84	23.5
Benzoic acid	$C_7H_6O_2$	122.352	18.02
Benzonitrile	C_7H_5N	-12.82	9.1
Benzo[*c*]phenanthrene	$C_{18}H_{12}$	67	16.3
Benzophenone	$C_{13}H_{10}O$	48.0	18.19
Benzo[*a*]pyrene	$C_{20}H_{12}$	179	17.3
Benzo[*e*]pyrene	$C_{20}H_{12}$	180	16.6
p-Benzoquinone	$C_6H_4O_2$	113	18.5
Benzoyl chloride	C_7H_5ClO	-0.5	19.2
Benzyl alcohol	C_7H_8O	-15.5	8.97
Beryllium	Be	1287	7.895
Beryllium bromide	$BeBr_2$	508	18
Beryllium carbide	CBe_2	2127	75.3
Beryllium chloride	$BeCl_2$	415	8.66
Beryllium fluoride	BeF_2	552	4.77
Beryllium iodide	BeI_2	480	20.92
Beryllium nitride	Be_3N_2	2200	111
Beryllium oxide	BeO	2578	86
Beryllium sulfate	BeO_4S	1127	6
2,2'-Binaphthalene	$C_{20}H_{14}$	188.1	38.9
Biphenyl	$C_{12}H_{10}$	68.93	18.57
Bismuth	Bi	271.402	11.106
Bismuth oxide	Bi_2O_3	825	14.7
Bismuth sulfide	Bi_2S_3	777	78.2
Bismuth tribromide	$BiBr_3$	219	21.7
Bismuth trichloride	$BiCl_3$	234	23.6
Bismuth trifluoride	BiF_3	649	21.6
Bismuth triiodide	BiI_3	408.6	39.1
Boric acid	BH_3O_3	170.9	22.3
Boron	B	2077	50.2
Boron nitride	BN	2967	81
Boron oxide	B_2O_3	450	24.56
Boron sulfide	B_2S_3	563	48.12
Boron trichloride	BCl_3	-107.3	2.10
Boron trifluoride	BF_3	-126.8	4.20
Bromine	Br_2	-7.2	10.57
Bromine pentafluoride	BrF_5	-60.5	5.67
Bromobenzene	C_6H_5Br	-30.74	10.70
1-Bromobutane	C_4H_9Br	-112.5	9.23
2-Bromobutane, (±)-	C_4H_9Br	-112.6	6.89
Bromoethane	C_2H_5Br	-118.4	7.47
Bromoethene	C_2H_3Br	-139.5	5.12
1-Bromoheptane	$C_7H_{15}Br$	-56.1	21.8
1-Bromohexane	$C_6H_{13}Br$	-84.9	18.1
Bromomethane	CH_3Br	-93.7	5.98
1-Bromonaphthalene	$C_{10}H_7Br$	6.1	15.2
2-Bromonaphthalene	$C_{10}H_7Br$	58	14.4
1-Bromooctane	$C_8H_{17}Br$	-55.0	24.7
1-Bromopentane	$C_5H_{11}Br$	-88.0	14.37
1-Bromopropane	C_3H_7Br	-110.1	6.44
2-Bromopropane	C_3H_7Br	-88.9	6.53
Bromotrichloromethane	$CBrCl_3$	-5.6	2.53
1,2-Butadiene	C_4H_6	-136.20	6.96
1,3-Butadiene	C_4H_6	-108.9	7.98
Butanal	C_4H_8O	-96.86	10.77
Butane	C_4H_{10}	-138.2	4.66
1,4-Butanediol	$C_4H_{10}O_2$	20.43	18.70
1-Butanethiol	$C_4H_{10}S$	-115.66	10.46
Butanoic acid	$C_4H_8O_2$	-5.12	11.59
1-Butanol	$C_4H_{10}O$	-88.60	9.37
2-Butanol	$C_4H_{10}O$	-88.44	5.97
2-Butanone	C_4H_8O	-86.67	8.39
1-Butene	C_4H_8	-185.33	3.96
cis-2-Butene	C_4H_8	-138.89	7.31
trans-2-Butene	C_4H_8	-105.52	9.76
cis-2-Butenoic acid	$C_4H_6O_2$	15	12.6
trans-2-Butenoic acid	$C_4H_6O_2$	71.3	13.0
tert-Butylamine	$C_4H_{11}N$	-66.92	0.882
Butylbenzene	$C_{10}H_{14}$	-87.81	11.22
Butylcyclohexane	$C_{10}H_{20}$	-74.68	14.16
Butyl methyl ether	$C_5H_{12}O$	-115.7	10.85
1-Butyne	C_4H_6	-125.7	6.03
2-Butyne	C_4H_6	-32.2	9.23
γ-Butyrolactone	$C_4H_6O_2$	-43.36	9.57
Cadmium	Cd	321.069	6.21
Cadmium bromide	Br_2Cd	568	33.35
Cadmium chloride	$CdCl_2$	568	48.58
Cadmium fluoride	CdF_2	1075	22.6

Name	Mol. form.	t_m/°C	$\Delta_{fus}H$/ kJ mol^{-1}
Cadmium iodide	CdI_2	388	15.3
Cadmium nitrate	CdN_2O_6	360	18.3
Caffeine	$C_8H_{10}N_4O_2$	236.1	22.0
Calcium	Ca	842	8.54
Calcium bromide	Br_2Ca	742	29.1
Calcium carbonate (calcite)	$CCaO_3$	800	36
Calcium chloride	$CaCl_2$	775	28.05
Calcium fluoride	CaF_2	1418	30
Calcium hydride	CaH_2	1000	6.7
Calcium iodide	CaI_2	783	41.8
Calcium nitrate	CaN_2O_6	561	23.4
Calcium oxide	CaO	2613	80
Calcium sulfate	CaO_4S	1460	28
Calcium sulfide	CaS	2524	70
Carbazole	$C_{12}H_9N$	245	24.1
Carbon dioxide	CO_2	-56.561 tp	9.02
Carbon diselenide	CSe_2	-43.6	6.36
Carbon disulfide	CS_2	-111.7	4.39
Carbon (graphite)	C	4489 tp (10.3 MPa)	117.4
Carbon monoxide	CO	-205.1	0.833
Carbon oxysulfide	COS	-138.8	4.73
Carbonyl chloride	CCl_2O	-127.77	5.74
Cerium	Ce	799	5.460
Cerium(III) bromide	Br_3Ce	732	51.9
Cerium(III) chloride	$CeCl_3$	807	53.1
Cerium(III) fluoride	CeF_3	1430	55.6
Cerium(III) iodide	CeI_3	760	51.0
Cerium(III) oxide	Ce_2O_3	2250	120
Cerium(IV) oxide	CeO_2	2480	80
Cesium	Cs	28.5	2.09
Cesium carbonate	CCs_2O_3	793	31
Cesium chloride	ClCs	646	20.4
Cesium chromate	$CrCs_2O_4$	963	35.3
Cesium fluoride	CsF	703	21.7
Cesium hydride	CsH	528	15
Cesium hydroxide	CsHO	342.3	7.78
Cesium iodide	CsI	632	25.7
Cesium metaborate	$BCsO_2$	732	27
Cesium molybdate	Cs_2MoO_4	956.3	31.8
Cesium nitrate	$CsNO_3$	409	13.8
Cesium nitrite	$CsNO_2$	406	10.9
Cesium oxide	Cs_2O	495	20
Cesium peroxide	Cs_2O_2	594	22
Cesium sulfate	Cs_2O_4S	1005	35.7
Chlorine	Cl_2	-101.5	6.40
Chloroacetic acid	$C_2H_3ClO_2$	62.0	12.28
2-Chloroaniline	C_6H_6ClN	-2.3	11.9
3-Chloroaniline	C_6H_6ClN	-10.3	10.15
4-Chloroaniline	C_6H_6ClN	70.4	20.0
Chlorobenzene	C_6H_5Cl	-45.2	9.6
2-Chlorobenzoic acid	$C_7H_5ClO_2$	140.4	25.6
Chlorocyclohexane	$C_6H_{11}Cl$	-45	2.043
Chlorodifluoromethane	$CHClF_2$	-157.41	4.12
Chloroethane	C_2H_5Cl	-138.4	4.45
Chloroethene	C_2H_3Cl	-153.84	4.92
Chloromethane	CH_3Cl	-97.6	6.43
2-Chloro-2-methylpropane	C_4H_9Cl	-25.60	2.07
1-Chloronaphthalene	$C_{10}H_7Cl$	-6.0	12.9
2-Chloronaphthalene	$C_{10}H_7Cl$	58.02	14.0
1-Chloro-2-nitrobenzene	$C_6H_4ClNO_2$	32.1	17.9
1-Chloro-3-nitrobenzene	$C_6H_4ClNO_2$	43.6	19.4
1-Chloro-4-nitrobenzene	$C_6H_4ClNO_2$	82.2	14.1
Chloropentafluoroethane	C_2ClF_5	-99.4	1.86
2-Chlorophenol	C_6H_5ClO	8	13.0
3-Chlorophenol	C_6H_5ClO	32.5	14.9
4-Chlorophenol	C_6H_5ClO	43.1	14.1
1-Chloropropane	C_3H_7Cl	-122.9	5.54
2-Chloropropane	C_3H_7Cl	-117.1	7.39
2-Chlorotoluene	C_7H_7Cl	-35.9	9.6
Chlorotrifluoroethene	C_2ClF_3	-158.14	5.55
Chromium	Cr	1907	21.00
Chromium(II) bromide	Br_2Cr	842	45
Chromium(III) bromide	Br_3Cr	812	60
Chromium(II) chloride	Cl_2Cr	824	45.0
Chromium(III) chloride	Cl_3Cr	827	60
Chromium(II) fluoride	CrF_2	894	34
Chromium(III) fluoride	CrF_3	1425	66
Chromium(II) iodide	CrI_2	867	46
Chromium(III) iodide	CrI_3	857	61
Chromium(III) oxide	Cr_2O_3	2432	125
Chromium(VI) oxide	CrO_3	197	14.2
Chromium(II) sulfide	CrS	1567	25.5
Chrysene	$C_{18}H_{12}$	255.0	26.2
Cobalt	Co	1495	16.20
Cobalt(II) bromide	Br_2Co	678	43
Cobalt(II) chloride	Cl_2Co	737	46.0
Cobalt(II) fluoride	CoF_2	1127	58.1
Cobalt(II) iodide	CoI_2	520	35
Cobalt(II) selenite	CoO_3Se	659	16.3
Cobalt(II) sulfide	CoS	1117	30
Copper	Cu	1084.62	13.26
Copper(I) bromide	BrCu	483	5.1
Copper(I) chloride	ClCu	423	7.08
Copper(II) chloride	Cl_2Cu	598	15.0
Copper(II) fluoride	CuF_2	836	55
Copper(I) iodide	CuI	591	7.93
Copper(I) oxide	Cu_2O	1244	65.6
Copper(II) oxide	CuO	1227	49
Copper(I) sulfide	Cu_2S	1129	9.62
Coronene	$C_{24}H_{12}$	437.3	19.2
o-Cresol	C_7H_8O	31.0	15.82
m-Cresol	C_7H_8O	12.2	10.71
p-Cresol	C_7H_8O	34.77	12.71
Curium	Cm	1345	14.64
Cyanamide	CH_2N_2	45.55	7.27
Cyanogen	C_2N_2	-27.83	8.11
Cyclobutane	C_4H_8	-90.7	1.09
Cycloheptane	C_7H_{14}	-8.0	1.88
Cycloheptanol	$C_7H_{14}O$	7.15	1.60
Cyclohexane	C_6H_{12}	6.7	2.68
Cyclohexanol	$C_6H_{12}O$	26	1.78
Cyclohexanone	$C_6H_{10}O$	-27.93	1.328
Cyclohexene	C_6H_{10}	-103.5	3.29
Cyclohexylamine	$C_6H_{13}N$	-17.7	17.5
Cyclohexylbenzene	$C_{12}H_{16}$	7.02	15.6
Cyclooctane	C_8H_{16}	14.82	2.41
Cyclopentane	C_5H_{10}	-93.4	0.61
Cyclopentanol	$C_5H_{10}O$	-17	1.535
Cyclopentene	C_5H_8	-135.02	3.36

Name	Mol. form.	t_m/°C	$\Delta_{fus}H$/ kJ mol⁻¹
Cyclopentylamine	$C_5H_{11}N$	-82.69	8.31
Cyclopropane	C_3H_6	-127.6	5.44
Cyclopropylamine	C_3H_7N	-35.38	13.18
Decaborane(14)	$B_{10}H_{14}$	98.78	21.97
cis-Decahydronaphthalene	$C_{10}H_{18}$	-42.9	9.49
trans-Decahydronaphthalene	$C_{10}H_{18}$	-30.35	14.41
Decanal	$C_{10}H_{20}O$	-3.9	34.5
Decane	$C_{10}H_{22}$	-29.61	28.72
Decanedioic acid	$C_{10}H_{18}O_4$	131	40.8
Decanoic acid	$C_{10}H_{20}O_2$	31.39	27.8
1-Decanol	$C_{10}H_{22}O$	7	43
1-Decene	$C_{10}H_{20}$	-66.21	13.81
5,5-Diallyl-2,4,6(1*H*,3*H*,5*H*)-pyrimidinetrione	$C_{10}H_{12}N_2O_3$	172	32.3
1,2-Dibromoethane	$C_2H_4Br_2$	9.8	10.89
1,2-Dibromopropane	$C_3H_6Br_2$	-55.4	8.94
1,3-Dibromopropane	$C_3H_6Br_2$	-35	14.6
1,2-Dibromotetrafluoroethane	$C_2Br_2F_4$	-110.1	7.04
o-Dichlorobenzene	$C_6H_4Cl_2$	-17.0	12.4
m-Dichlorobenzene	$C_6H_4Cl_2$	-24.8	12.6
p-Dichlorobenzene	$C_6H_4Cl_2$	53.1	18.19
1,1-Dichloroethane	$C_2H_4Cl_2$	-96.93	7.87
1,2-Dichloroethane	$C_2H_4Cl_2$	-35.6	8.84
1,1-Dichloroethene	$C_2H_2Cl_2$	-122.5	6.51
cis-1,2-Dichloroethene	$C_2H_2Cl_2$	-80.0	7.2
Dichloromethane	CH_2Cl_2	-94.9	4.60
1,2-Dichloropropane, (±)-	$C_3H_6Cl_2$	-100.53	6.40
2,2-Dichloropropane	$C_3H_6Cl_2$	-33.9	2.30
1,2-Dichloro-1,1,2,2-tetrafluoroethane	$C_2Cl_2F_4$	-92.52	1.51
Diethyl ether	$C_4H_{10}O$	-116.22	7.19
3,3-Diethylpentane	C_9H_{20}	-33.04	10.09
Diethyl sulfide	$C_4H_{10}S$	-103.9	10.90
o-Difluorobenzene	$C_6H_4F_2$	-47.1	11.05
m-Difluorobenzene	$C_6H_4F_2$	-69.11	8.58
1,2-Dihydro-1,5-dimethyl-2-phenyl-3*H*-pyrazol-3-one	$C_{11}H_{12}N_2O$	108.0	27.3
Diisopropyl ether	$C_6H_{14}O$	-85.37	12.04
1,2-Dimethoxyethane	$C_4H_{10}O_2$	-69.0	12.6
Dimethoxymethane	$C_3H_8O_2$	-105.11	8.33
Dimethylamine	C_2H_7N	-93	5.94
2,2-Dimethylbutane	C_6H_{14}	-99.0	0.58
2,3-Dimethylbutane	C_6H_{14}	-128.1	0.79
2,3-Dimethyl-2-butene	C_6H_{12}	-74.3	6.45
1,1-Dimethylcyclohexane	C_8H_{16}	-33.31	2.07
cis-1,2-Dimethylcyclohexane	C_8H_{16}	-49.83	1.64
trans-1,2-Dimethylcyclohexane	C_8H_{16}	-88.12	10.49
cis-1,3-Dimethylcyclohexane	C_8H_{16}	-75.51	10.82
trans-1,3-Dimethylcyclohexane	C_8H_{16}	-90.05	9.87
cis-1,4-Dimethylcyclohexane	C_8H_{16}	-87.4	9.31
trans-1,4-Dimethylcyclohexane	C_8H_{16}	-36.9	12.33
Dimethyl disulfide	$C_2H_6S_2$	-84.67	9.19
Dimethyl ether	C_2H_6O	-141.49	4.94
N,N-Dimethylformamide	C_3H_7NO	-60.3	7.90
1,1-Dimethylhydrazine	$C_2H_8N_2$	-57.15	10.07
1,2-Dimethylhydrazine	$C_2H_8N_2$	-8.86	13.64
Dimethyl oxalate	$C_4H_6O_4$	51	21.1
2,2-Dimethylpentane	C_7H_{16}	-123.71	5.82
2,4-Dimethylpentane	C_7H_{16}	-119.16	6.85
3,3-Dimethylpentane	C_7H_{16}	-134.4	6.85
Dimethyl sulfide	C_2H_6S	-98.26	7.99
Dimethyl sulfone	$C_2H_6O_2S$	108.83	18.30
Dimethyl sulfoxide	C_2H_6OS	18.52	14.37
N,N-Dimethylurea	$C_3H_8N_2O$	181.2	23.0
N,N'-Dimethylurea	$C_3H_8N_2O$	106	13.0
Dimethyl zinc	C_2H_6Zn	-43.01	6.83
1,4-Dioxane	$C_4H_8O_2$	11.75	12.84
1,3-Dioxolane	$C_3H_6O_2$	-97.21	6.57
Diphenylamine	$C_{12}H_{11}N$	53.2	18.5
Diphenyl ether	$C_{12}H_{10}O$	26.865	17.22
Diphenylmethane	$C_{13}H_{12}$	25.22	18.6
Dipropyl ether	$C_6H_{14}O$	-114.8	10.8
Divinyl ether	C_4H_6O	-101	7.9
Docosane	$C_{22}H_{46}$	43.8	48.8
Dodecane	$C_{12}H_{26}$	-9.55	36.8
Dodecanoic acid	$C_{12}H_{24}O_2$	43.82	36.3
1-Dodecanol	$C_{12}H_{26}O$	24.2	40.2
1-Dodecene	$C_{12}H_{24}$	-35.19	19.9
Dotriacontane	$C_{32}H_{66}$	69.7	75.8
Dysprosium	Dy	1412	11.35
Dysprosium(III) fluoride	DyF_3	1157	58.6
Dysprosium(III) oxide	Dy_2O_3	2408	120
Eicosane	$C_{20}H_{42}$	36.48	69.9
1-Eicosanol	$C_{20}H_{42}O$	63.9	42
Einsteinium	Es	860	9.41
Erbium	Er	1529	19.90
Erbium chloride	Cl_3Er	776	32.6
Erbium fluoride	ErF_3	1146	28.2
Erbium oxide	Er_2O_3	2418	130
Estra-1,3,5(10)-triene-3,17-diol 3-benzoate, (17β)	$C_{25}H_{28}O_3$	198	41.8
Ethane	C_2H_6	-182.77	2.72 [a]
1,2-Ethanediamine	$C_2H_8N_2$	11.14	22.58
1,2-Ethanediol	$C_2H_6O_2$	-13	9.96
Ethanethiol	C_2H_6S	-147.89	4.98
Ethanol	C_2H_6O	-114.14	4.931
Ethinylestradiol	$C_{20}H_{24}O_2$	186	27.9
N-(4-Ethoxyphenyl)acetamide	$C_{10}H_{13}NO_2$	135	33.0
Ethyl acetate	$C_4H_8O_2$	-83.8	10.48
Ethyl 4-aminobenzoate	$C_9H_{11}NO_2$	89.6	22.3
Ethylbenzene	C_8H_{10}	-94.95	9.18
Ethylcyclohexane	C_8H_{16}	-111.28	8.33
Ethylene	C_2H_4	-169.15	3.35
Ethyl methyl sulfide	C_3H_8S	-105.89	9.76
3-Ethylpentane	C_7H_{16}	-118.55	9.55
2-Ethyltoluene	C_9H_{12}	-80.7	9.96
3-Ethyltoluene	C_9H_{12}	-95.7	7.6
4-Ethyltoluene	C_9H_{12}	-62.7	12.7
Europium	Eu	822	9.21
Europium(II) bromide	Br_2Eu	683	25.1
Europium(III) chloride	Cl_3Eu	623	33.1
Europium(III) fluoride	EuF_3	647	6.40
Europium(II) oxide	EuO	1967	40
Europium(III) oxide	Eu_2O_3	2350	117
Fluoranthene	$C_{16}H_{10}$	110.2	18.69
9*H*-Fluorene	$C_{13}H_{10}$	114.76	19.58
Fluorine	F_2	-219.67	0.51
Fluorobenzene	C_6H_5F	-42.18	11.31
Formamide	CH_3NO	2.57	8.44
Formic acid	CH_2O_2	8.3	12.68
Furan	C_4H_4O	-85.58	3.80

Name	Mol. form.	t_m/°C	$\Delta_{fus}H$/ kJ mol^{-1}
Furfural	$C_5H_4O_2$	-38.3	14.37
Furfuryl alcohol	$C_5H_6O_2$	-14.5	13.13
Gadolinium	Gd	1313	9.67
Gadolinium(III) bromide	Br_3Gd	785	38.1
Gadolinium(III) chloride	Cl_3Gd	602	40.6
Gadolinium(III) fluoride	F_3Gd	1232	52.4
Gadolinium(III) iodide	GdI_3	930	54.0
Gadolinium(III) oxide	Gd_2O_3	2425	60
Gallium	Ga	29.7646	5.585
Gallium antimonide	GaSb	712	25.1
Gallium arsenide	AsGa	1238	87.64
Gallium(III) bromide	Br_3Ga	123	11.7
Gallium(III) chloride	Cl_3Ga	77.9	11.51
Gallium(III) iodide	GaI_3	212	12.9
Gallium(III) oxide	Ga_2O_3	1807	100
Germanium	Ge	938.25	36.94
Germanium(IV) bromide	Br_4Ge	26.1	12
Germanium(II) iodide	GeI_2	428	33.3
Germanium(IV) iodide	GeI_4	146	19.1
Germanium(IV) oxide	GeO_2	1116	12.6
Germanium(II) selenide	GeSe	675	24.7
Germanium(II) sulfide	GeS	658	21.3
Germanium(IV) sulfide	GeS_2	840	16.3
Germanium(II) telluride	GeTe	724	47.3
Glycerol	$C_3H_8O_3$	18.2	18.3
Gold	Au	1064.18	12.55
Hafnium	Hf	2233	27.20
Hafnium nitride	HfN	3310	62.8
Hafnium(IV) oxide	HfO_2	2800	96
Heneicosane	$C_{21}H_{44}$	40.4	45.21
Heptacosane	$C_{27}H_{56}$	58.8	61.9
Heptadecane	$C_{17}H_{36}$	21.97	40.16
Heptanal	$C_7H_{14}O$	-43.94	23.2
Heptane	C_7H_{16}	-90.549	14.03
Heptanoic acid	$C_7H_{14}O_2$	-7.17	15.13
1-Heptanol	$C_7H_{16}O$	-33.2	18.17
1-Heptene	C_7H_{14}	-118.83	12.41
Hexachlorobenzene	C_6Cl_6	230	25.2
Hexachloroethane	C_2Cl_6	186.8	9.75
Hexacontane	$C_{60}H_{122}$	100	193.2
Hexacosane	$C_{26}H_{54}$	56.09	60.0
Hexadecane	$C_{16}H_{34}$	18.18	53.36
Hexadecanoic acid	$C_{16}H_{32}O_2$	62.49	53.7
1-Hexadecanol	$C_{16}H_{34}O$	49.30	33.6
Hexafluorobenzene	C_6F_6	5.10	11.59
Hexafluoroethane	C_2F_6	-100.02	2.69
Hexamethylbenzene	$C_{12}H_{18}$	165.6	20.6
Hexanal	$C_6H_{12}O$	-58.2	13.3
Hexane	C_6H_{14}	-95.27	13.08
1,6-Hexanedioic acid	$C_6H_{10}O_4$	151.5	36.3
1,6-Hexanediol	$C_6H_{14}O_2$	41.5	22.2
1-Hexanol	$C_6H_{14}O$	-46.4	15.38
2-Hexanone	$C_6H_{12}O$	-55.45	14.9
3-Hexanone	$C_6H_{12}O$	-55.4	13.49
Hexatetracontane	$C_{46}H_{94}$	87.6	176.0
Hexatriacontane	$C_{36}H_{74}$	75.81	87.7
1-Hexene	C_6H_{12}	-139.76	9.35
cis-2-Hexene	C_6H_{12}	-141.12	8.88
Holmium	Ho	1472	11.76
Holmium bromide	Br_3Ho	919	50.1
Holmium chloride	Cl_3Ho	720	30.5
Holmium fluoride	F_3Ho	1143	56.3
Holmium oxide	Ho_2O_3	2415	130
Hydrazine	H_4N_2	1.54	12.66
Hydrogen	H_2	-259.16	0.12
Hydrogen bromide	BrH	-86.80	2.41
Hydrogen chloride	ClH	-114.17	2.00
Hydrogen cyanide	CHN	-13.28	8.41
Hydrogen fluoride	FH	-83.36	4.58
Hydrogen iodide	HI	-50.76	2.87
Hydrogen peroxide	H_2O_2	-0.43	12.50
Hydrogen sulfide	H_2S	-85.5	2.38
p-Hydroquinone	$C_6H_6O_2$	173	26.8
2-Hydroxybenzoic acid	$C_7H_6O_3$	158.6	14.2
N-(4-Hydroxyphenyl)acetamide	$C_8H_9NO_2$	168.0	30.5
Imidazole	$C_3H_4N_2$	89.52	12.82
Indan	C_9H_{10}	-51.34	8.60
Indene	C_9H_8	-1.45	10.20
Indium	In	156.5985	3.291
Indium antimonide	InSb	524	47.7
Indium arsenide	AsIn	942	77.0
Indium(I) bromide	BrIn	285	24.3
Indium(III) bromide	Br_3In	420	26
Indium(I) chloride	ClIn	225	9.20
Indium(III) chloride	Cl_3In	583	27
Indium(III) fluoride	F_3In	1172	64
Indium(I) iodide	IIn	364.4	17.26
Indium(II) iodide	I_2In	155	1.29
Indium(III) iodide	I_3In	207	18.48
Indium(III) oxide	In_2O_3	1912	105
Indium(II) sulfide	InS	692	36.0
Indomethacin	$C_{19}H_{16}ClNO_4$	160	36.9
Iodine	I_2	113.7	15.52
Iodine chloride	ClI	27.38	11.6
Iodobenzene	C_6H_5I	-30.7	9.75
Iridium	Ir	2446	41.12
Iridium(VI) fluoride	F_6Ir	44	8.40
Iron	Fe	1538	13.81
Iron boride (FeB)	BFe	1658	62.66
Iron(II) bromide	Br_2Fe	691	43.0
Iron(II) chloride	Cl_2Fe	677	42.83
Iron(III) chloride	Cl_3Fe	307.6	40
Iron(II) fluoride	F_2Fe	1100	50
Iron(III) fluoride	F_3Fe	367	0.58
Iron(II) iodide	FeI_2	594	39
Iron(II) oxide	FeO	1377	24.1
Iron(II,III) oxide	Fe_3O_4	1597	138
Iron(III) oxide	Fe_2O_3	1539	87
Iron sodium oxide	$FeNaO_2$	1347	49.4
Iron(II) sulfide	FeS	1188	31.5
Isobutane	C_4H_{10}	-159.38	4.54
Isobutene	C_4H_8	-140.7	5.92
Isopentane	C_5H_{12}	-159.8	5.15
Isopropylamine	C_3H_9N	-95.119	7.33
Isopropylbenzene	C_9H_{12}	-96.01	7.33
1-Isopropyl-4-methylbenzene	$C_{10}H_{14}$	-68.1	9.66
Isoquinoline	C_9H_7N	26.46	13.54
Khellin	$C_{14}H_{12}O_5$	154 dec	32.3
Krypton	Kr	-157.37	1.64
Lanthanum	La	920	6.20

Name	Mol. form.	t_m/°C	$\Delta_{fus}H$/ kJ mol^{-1}
Lanthanum bromide	Br_3La	788	54.0
Lanthanum chloride	Cl_3La	858	54.4
Lanthanum fluoride	F_3La	1493	50.2
Lanthanum iodide	I_3La	778	56.1
Lead	Pb	327.462	4.774
Lead(II) bromide	Br_2Pb	371	16.44
Lead(II) chloride	Cl_2Pb	501	21.88
Lead(II) fluoride	F_2Pb	830	14.7
Lead(II) iodide	I_2Pb	410	23.4
Lead(II) oxide (massicot)	OPb	887	25.6
Lead(II) sulfate	O_4PbS	1087	40.2
Lead(II) sulfide	PbS	1113	49.4
Lithium	Li	180.50	3.00
Lithium aluminate	$AlLiO_2$	1610	87.9
Lithium bromide	BrLi	550	17.66
Lithium carbonate	CLi_2O_3	732	44.8
Lithium chloride	ClLi	610	19.8
Lithium chromate	$CrLi_2O_4$	482	30.5
Lithium fluoride	FLi	848.2	27.09
Lithium hexafluoroaluminate	AlF_6Li_3	785	86.19
Lithium hydride	HLi	692	21.8
Lithium hydride-*d*	DLi	694	22
Lithium hydroxide	HLiO	473	20.9
Lithium iodide	ILi	469	14.6
Lithium metasilicate	Li_2O_3Si	1201	28
Lithium nitrate	$LiNO_3$	253	26.7
Lithium nitrite	$LiNO_2$	222	9.2
Lithium perchlorate	$ClLiO_4$	236	29.3
Lithium sulfate	Li_2O_4S	860	9.00
Lutetium	Lu	1663	18.65
Lutetium oxide	Lu_2O_3	2490	133
Magnesium	Mg	650	8.48
Magnesium bromide	Br_2Mg	711	39.3
Magnesium carbonate	$CMgO_3$	990	59
Magnesium chloride	Cl_2Mg	714	43.1
Magnesium fluoride	F_2Mg	1263	58.7
Magnesium hydride	H_2Mg	327	14
Magnesium iodide	I_2Mg	634	26
Magnesium orthosilicate	Mg_2O_4Si	1897	71
Magnesium oxide	MgO	2825	77
Magnesium phosphate	$Mg_3O_8P_2$	1348	121
Magnesium sulfate	MgO_4S	1137	14.6
Magnesium sulfide	MgS	2226	63
Magnesium tetraboride	B_4Mg	727	0.0
Maleic anhydride	$C_4H_2O_3$	52.56	13.60
Manganese	Mn	1246	12.91
Manganese(II) bromide	Br_2Mn	698	33.5
Manganese(II) chloride	Cl_2Mn	650	30.7
Manganese(II) fluoride	F_2Mn	900	30
Manganese(II) iodide	I_2Mn	638	41.8
Manganese(II) oxide	MnO	1842	43.9
Manganese(II) sulfide (α form)	MnS	1530	26.1
Mercury	Hg	-38.8290	2.295
Mercury(II) bromide	Br_2Hg	241	17.9
Mercury(II) chloride	Cl_2Hg	277	19.41
Mercury(II) fluoride	F_2Hg	645 dec	23.0
Mercury(I) iodide	Hg_2I_2	290	31.4
Mercury(II) iodide (yellow)	HgI_2	256	15.6
Mercury(II) sulfide (black)	HgS	820	40
Metaboric acid (γ form)	BHO_2	236	14.3
Methane	CH_4	-182.475	0.94
Methanethiol	CH_4S	-122.98	5.91
Methanol	CH_4O	-97.5	3.215
Methyl acetate	$C_3H_6O_2$	-98.2	7.49
Methylamine	CH_5N	-93.42	6.13
2-Methylaniline	C_7H_9N	-14.41	11.66
3-Methylaniline	C_7H_9N	-30.8	7.9
4-Methylaniline	C_7H_9N	43.3	18.9
Methyl benzoate	$C_8H_8O_2$	-12.35	9.74
2-Methyl-1,3-butadiene	C_5H_8	-146.1	4.93
2-Methyl-2-butanol	$C_5H_{12}O$	-8.7	4.46
3-Methyl-2-butanone	$C_5H_{10}O$	-93.13	9.34
2-Methyl-1-butene	C_5H_{10}	-137.53	7.91
3-Methyl-1-butene	C_5H_{10}	-168.41	5.36
2-Methyl-2-butene	C_5H_{10}	-133.72	7.60
Methyl *tert*-butyl ether	$C_5H_{12}O$	-108.6	7.60
Methylcyclohexane	C_7H_{14}	-126.6	6.75
Methylcyclopentane	C_6H_{12}	-142.419	6.93
Methylcyclopropane	C_4H_8	-177.2	2.8
2-Methylfuran	C_5H_6O	-91.2	8.55
2-Methylheptane	C_8H_{18}	-109	11.92
3-Methylheptane	C_8H_{18}	-120.48	11.69
4-Methylheptane	C_8H_{18}	-121.0	10.8
2-Methylhexane	C_7H_{16}	-118.23	9.19
Methylhydrazine	CH_6N_2	-52.3	10.42
Methyl methacrylate	$C_5H_8O_2$	-47.55	14.4
1-Methylnaphthalene	$C_{11}H_{10}$	-30.43	6.95
2-Methylnaphthalene	$C_{11}H_{10}$	34.6	12.13
Methyl nitrate	CH_3NO_3	-82.9	8.24
Methyloxirane	C_3H_6O	-111.9	6.53
2-Methylpentane	C_6H_{14}	-153.60	6.27
3-Methylpentane	C_6H_{14}	-162.89	5.30
2-Methyl-1-propanol	$C_4H_{10}O$	-101.96	6.32
2-Methyl-2-propanol	$C_4H_{10}O$	25.81	6.70
2-Methylpyridine	C_6H_7N	-66.65	9.72
3-Methylpyridine	C_6H_7N	-18.1	14.18
4-Methylpyridine	C_6H_7N	3.68	12.58
N-Methylurea	$C_2H_6N_2O$	101.3	14.0
Molybdenum	Mo	2622	37.48
Molybdenum boride [Mo_2B_5]	B_5Mo_2	2210	226
Molybdenum(IV) chloride	Cl_4Mo	317	16.7
Molybdenum(V) chloride	Cl_5Mo	194	19
Molybdenum(VI) dioxydichloride	Cl_2MoO_2	176	17.0
Molybdenum(V) fluoride	F_5Mo	45.67	6.1
Molybdenum(VI) fluoride	F_6Mo	17.5	4.33
Molybdenum monoboride	BMo	2600	55.23
Molybdenum(VI) oxide	MoO_3	802	48.7
Molybdenum(VI) oxytetrachloride	Cl_4MoO	105	14.3
Molybdenum(VI) oxytetrafluoride	F_4MoO	97.2	4
Molybdenum(V) oxytrichloride	Cl_3MoO	310	22
Molybdenum(III) sulfide	Mo_2S_3	1807	0.13
Morpholine	C_4H_9NO	-4.8	14.5
Naphthalene	$C_{10}H_8$	80.22	19.01
1-Naphthol	$C_{10}H_8O$	95.1	23.1
2-Naphthol	$C_{10}H_8O$	122	18.1
Neodymium	Nd	1016	7.14
Neodymium(III) bromide	Br_3Nd	682	45.3
Neodymium(III) chloride	Cl_3Nd	759	48.5
Neodymium(III) fluoride	F_3Nd	1377	54.8

Name	Mol. form.	t_m/°C	$\Delta_{fus}H$/ kJ mol^{-1}
Neodymium(III) iodide	I_3Nd	787	41.5
Neon	Ne	-248.59	0.328
Neopentane	C_5H_{12}	-16.37	3.10
Neptunium	Np	644	3.20
Nickel	Ni	1455	17.48
Nickel boride (Ni_2B)	BNi_2	1125	42.15
Nickel boride (Ni_3B)	BNi_3	1166	72.28
Nickel(II) bromide	Br_2Ni	963	56
Nickel(II) chloride	Cl_2Ni	1031	77.9
Nickel disulfide	NiS_2	1007	65.7
Nickel(II) fluoride	F_2Ni	1380	69
Nickel(II) iodide	I_2Ni	800	48
Nickel(II) oxide	NiO	1957	50.7
Nickel subsulfide	Ni_3S_2	789	19.7
Nickel(II) sulfide	NiS	976	30.1
Niobium	Nb	2477	30
Niobium(V) bromide	Br_5Nb	254	24.0
Niobium(V) chloride	Cl_5Nb	205.8	33.9
Niobium(V) fluoride	F_5Nb	80	12.2
Niobium(V) iodide	I_5Nb	327	37.7
Niobium nitride	NNb	2050	46.0
Niobium(II) oxide	NbO	1937	85.4
Niobium(IV) oxide	NbO_2	1901	92
Niobium(V) oxide	Nb_2O_5	1512	104.3
Nitric acid	HNO_3	-41.6	10.5
Nitric oxide	NO	-163.6	2.30
2-Nitroaniline	$C_6H_6N_2O_2$	71	16.1
3-Nitroaniline	$C_6H_6N_2O_2$	112	23.6
4-Nitroaniline	$C_6H_6N_2O_2$	147.7	21.2
Nitrobenzene	$C_6H_5NO_2$	5.65	12.12
Nitroethane	$C_2H_5NO_2$	-89.42	9.85
Nitrogen	N_2	-210.0	0.71
Nitrogen tetroxide	N_2O_4	-9.3	14.65
Nitromethane	CH_3NO_2	-28.7	9.70
2-Nitrophenol	$C_6H_5NO_3$	44.9	17.7
3-Nitrophenol	$C_6H_5NO_3$	95	20.6
4-Nitrophenol	$C_6H_5NO_3$	113.8	18.8
Nitrosobenzene	C_6H_5NO	67.8	31.0
4-Nitrotoluene	$C_7H_7NO_2$	51.7	16.81
Nitrous oxide	N_2O	-90.8	6.54
Nonacosane	$C_{29}H_{60}$	63.7	66.9
Nonadecane	$C_{19}H_{40}$	31.5	45.8
Nonanal	$C_9H_{18}O$	-19.3	30.5
Nonane	C_9H_{20}	-53.47	15.47
Nonanoic acid	$C_9H_{18}O_2$	12.38	19.82
5-Nonanone	$C_9H_{18}O$	-3.84	24.93
Octacosane	$C_{28}H_{58}$	61.3	65.1
Octadecane	$C_{18}H_{38}$	28.17	61.7
Octadecanoic acid	$C_{18}H_{36}O_2$	69.3	61.2
1-Octadecanol	$C_{18}H_{38}O$	58.0	45
Octane	C_8H_{18}	-56.73	20.73
Octanoic acid	$C_8H_{16}O_2$	16.51	21.35
1-Octanol	$C_8H_{18}O$	-14.7	23.7
Octatriacontane	$C_{38}H_{78}$	78.6	133.2
1-Octene	C_8H_{16}	-101.66	15.31
Osmium	Os	3033	57.85
Osmium(VIII) oxide	O_4Os	40.6	14.3
2-Oxepanone	$C_6H_{10}O_2$	-1.02	13.83
Oxetane	C_3H_6O	-97	6.5
Oxirane	C_2H_4O	-112.46	5.17
4-Oxopentanoic acid	$C_5H_8O_3$	33.0	9.22
Oxygen	O_2	-218.79	0.44
Palladium	Pd	1554.8	16.74
Palladium(II) chloride	Cl_2Pd	679	18.41
Paraldehyde	$C_6H_{12}O_3$	12	13.5
Pentachloroethane	C_2HCl_5	-29.0	11.3
Pentacontane	$C_{50}H_{102}$	91.7	162.4
Pentacosane	$C_{25}H_{52}$	53.3	56.9
Pentadecane	$C_{15}H_{32}$	9.95	34.6
cis-1,3-Pentadiene	C_5H_8	-140.81	5.64
trans-1,3-Pentadiene	C_5H_8	-87.5	7.14
1,4-Pentadiene	C_5H_8	-148.3	6.12
Pentaerythritol	$C_5H_{12}O_4$	258	4.8
Pentafluorobenzene	C_6HF_5	-47.3	10.87
Pentafluorophenol	C_6HF_5O	32.80	16.41
2,3,4,5,6-Pentafluorotoluene	$C_7H_3F_5$	-29.79	13.1
Pentane	C_5H_{12}	-129.67	8.40
Pentanedioic acid	$C_5H_8O_4$	97.9	20.3
Pentanenitrile	C_5H_9N	-96.2	9
1-Pentanethiol	$C_5H_{12}S$	-75.69	17.53
Pentanoic acid	$C_5H_{10}O_2$	-33.63	14.16
1-Pentanol	$C_5H_{12}O$	-77.58	10.50
2-Pentanone	$C_5H_{10}O$	-76.83	10.63
3-Pentanone	$C_5H_{10}O$	-38.98	11.59
Pentatriacontane	$C_{35}H_{72}$	74.4	86.3
1-Pentene	C_5H_{10}	-165.13	5.94
cis-2-Pentene	C_5H_{10}	-151.35	7.11
trans-2-Pentene	C_5H_{10}	-140.20	8.35
Perfluoroacetone	C_3F_6O	-125.45	8.38
Perfluorobutane	C_4F_{10}	-129	7.66
Perfluorocyclobutane	C_4F_8	-40.16	2.77
Perfluoroheptane	C_7F_{16}	-51.3	6.95
Perfluorohexane	C_6F_{14}	-86.1	6.84
Perfluoropropane	C_3F_8	-147.7	0.477
Perfluorotoluene	C_7F_8	-65.48	11.54
Perylene	$C_{20}H_{12}$	276	31.9
Phenanthrene	$C_{14}H_{10}$	99	16.46
Phenobarbital	$C_{12}H_{12}N_2O_3$	176	27.8
Phenol	C_6H_6O	40.89	11.51
α-Phenylbenzeneacetic acid	$C_{14}H_{12}O_2$	147.26	31.3
Phenylbutazone	$C_{19}H_{20}N_2O_2$	104	27.7
Phenylhydrazine	$C_6H_8N_2$	20	14.05
Phosphinic acid	H_3O_2P	26.5	9.7
Phosphonic acid	H_3O_3P	74.4	12.8
Phosphoric acid	H_3O_4P	42.4	13.4
Phosphorus (white)	P	44.15	0.659
Phosphorus (red)	P	579.2	18.54
Phosphorus(III) chloride	Cl_3P	-93	7.10
Phosphorus heptasulfide	P_4S_7	308	36.6
Phosphorus(V) oxide	O_5P_2	562	27.2
Phosphorus sesquisulfide	P_4S_3	173	20.1
Phosphoryl chloride	Cl_3OP	1.18	13.1
Piperidine	$C_5H_{11}N$	-11.05	14.85
Platinum	Pt	1768.2	22.175
Plutonium	Pu	640	2.824
Plutonium(III) bromide	Br_3Pu	681	58.6
Plutonium(III) chloride	Cl_3Pu	760	63.6
Plutonium(III) fluoride	F_3Pu	1396	59.8
Plutonium(IV) fluoride	F_4Pu	1037	42.7
Plutonium(VI) fluoride	F_6Pu	51.6	18.6

Name	Mol. form.	t_m/°C	$\Delta_{fus}H$/ kJ mol^{-1}
Plutonium(III) iodide	I_3Pu	777	50.2
Plutonium(III) oxide	O_3Pu_2	2085	113
Plutonium(IV) oxide	O_2Pu	2390	67
Polonium	Po	254	10.0
Potassium	K	63.5	2.335
Potassium acetate	$C_2H_3KO_2$	309	7.65
Potassium aluminate	$AlKO_2$	1713	82
Potassium bromide	BrK	734	25.52
Potassium carbonate	CK_2O_3	899	27.6
Potassium chloride	ClK	771	26.28
Potassium chromate	CrK_2O_4	974	33.0
Potassium cyanide	CKN	622	14.6
Potassium fluoride	FK	858	27.2
Potassium fluoroborate	BF_4K	570	17.66
Potassium hydride	HK	619	21
Potassium hydrogen fluoride	F_2HK	238.8	6.62
Potassium hydroxide	HKO	406	7.90
Potassium iodide	IK	681	24.0
Potassium metaborate	BKO_2	947	31.38
Potassium nitrate	KNO_3	334	9.6
Potassium nitrite	KNO_2	438	16.7
Potassium oxide	K_2O	740	27
Potassium peroxide	K_2O_2	545	20.5
Potassium sulfate	K_2O_4S	1069	36.6
Potassium sulfide	K_2S	948	16.15
Potassium superoxide	KO_2	535	20.6
Praseodymium	Pr	931	6.89
Praseodymium(III) bromide	Br_3Pr	693	47.3
Praseodymium(III) chloride	Cl_3Pr	786	50.6
Praseodymium(III) fluoride	F_3Pr	1399	57.3
Praseodymium(III) iodide	I_3Pr	738	53.1
Propane	C_3H_8	-187.62	3.50
1,3-Propanediol	$C_3H_8O_2$	-27.6	7.1
Propanenitrile	C_3H_5N	-93	5.03
1-Propanethiol	C_3H_8S	-113.12	5.48
2-Propanethiol	C_3H_8S	-130.50	5.74
Propanoic acid	$C_3H_6O_2$	-20.5	10.66
1-Propanol	C_3H_8O	-124.39	5.37
2-Propanol	C_3H_8O	-87.91	5.41
Propene	C_3H_6	-185.19	3.003
Propylamine	C_3H_9N	-84.78	10.97
Propylbenzene	C_9H_{12}	-99.52	9.27
Propylcyclohexane	C_9H_{18}	-94.86	10.37
Protactinium	Pa	1572	12.34
Pyrazine	$C_4H_4N_2$	52.30	12.9
1*H*-Pyrazole	$C_3H_4N_2$	59.9	14.0
Pyrene	$C_{16}H_{10}$	150.62	17.36
Pyridine	C_5H_5N	-41.63	8.28
3-Pyridinecarboxamide	$C_6H_6N_2O$	128.8	23.2
Pyrocatechol	$C_6H_6O_2$	104.6	22.8
Pyrrole	C_4H_5N	-23.39	7.91
Pyrrolidine	C_4H_9N	-57.79	8.58
Quinoline	C_9H_7N	-14.78	10.66
Radium	Ra	696	7.7
Resorcinol	$C_6H_6O_2$	109.8	20.4
Rhenium	Re	3185	34.08
Rhenium(VII) oxide	O_7Re_2	327	65.7
Rhodium	Rh	1963	26.59
Rubidium	Rb	39.30	2.19
Rubidium bromide	BrRb	692	23.3
Rubidium carbonate	CO_3Rb_2	873	30
Rubidium chloride	ClRb	724	24.4

Name	Mol. form.	t_m/°C	$\Delta_{fus}H$/ kJ mol^{-1}
Rubidium fluoride	FRb	795	25.8
Rubidium hydride	HRb	585	22
Rubidium hydroxide	HORb	385	8.0
Rubidium iodide	IRb	656	22.1
Rubidium metaborate	BO_2Rb	860	31
Rubidium nitrate	NO_3Rb	310	4.6
Rubidium nitrite	NO_2Rb	422	12.1
Rubidium oxide	ORb_2	505	20
Rubidium peroxide	O_2Rb_2	570	21
Rubidium sulfate	O_4Rb_2S	1066	37.3
Rubidium superoxide	O_2Rb	540	21
Ruthenium	Ru	2333	38.59
Ruthenium(V) fluoride	F_5Ru	101	74.5
Samarium	Sm	1072	8.62
Samarium(III) oxide	O_3Sm_2	2335	119
Scandium	Sc	1541	14.10
Scandium chloride	Cl_3Sc	967	67.4
Scandium fluoride	F_3Sc	1552	62.6
Scandium oxide	O_3Sc_2	2489	127
Selenium (gray)	Se	220.8	6.69
Selenium dioxide	O_2Se	360	17.6
Silicon	Si	1414	50.21
Silicon dioxide (cristobalite)	O_2Si	1722	9.6
Silicon monosulfide	SSi	1090	31
Silver	Ag	961.78	11.30
Silver(I) bromide	AgBr	430	9.163
Silver(I) chloride	AgCl	455	13.054
Silver(I) iodide	AgI	558	9.414
Silver(I) nitrate	$AgNO_3$	210	11.72
Silver(I) oxide	Ag_2O	827	15
Silver(I) sulfate	Ag_2O_4S	660	17.99
Silver(I) sulfide	Ag_2S	836	7.9
Sodium	Na	97.794	2.60
Sodium acetate	$C_2H_3NaO_2$	328.2	17.9
Sodium bromate	$BrNaO_3$	381	28.11
Sodium bromide	BrNa	747	26.23
Sodium carbonate	CNa_2O_3	856	29.7
Sodium chlorate	$ClNaO_3$	248	22.6
Sodium chloride	ClNa	802.018	28.16
Sodium chromate	$CrNa_2O_4$	794	24.7
Sodium cyanide	CNNa	562	8.79
Sodium fluoride	FNa	996	33.35
Sodium formate	$CHNaO_2$	257.3	17.7
Sodium hexafluoroaluminate	AlF_6Na_3	1013	114.4
Sodium hexafluorosilicate	F_6Na_2Si	847	99.6
Sodium hydride	HNa	638	26
Sodium hydrogen carbonate	$CHNaO_3$	527	25
Sodium hydroxide	HNaO	323	6.60
Sodium iodate	$INaO_3$	422	35.1
Sodium iodide	INa	661	23.7
Sodium metaborate	$BNaO_2$	966	36.2
Sodium metasilicate	Na_2O_3Si	1089	51.8
Sodium nitrate	$NNaO_3$	306.5	15.5
Sodium nitrite	$NNaO_2$	284	14.9
Sodium oxide	Na_2O	1134	47.7
Sodium peroxide	Na_2O_2	675	24.5
Sodium sulfate	Na_2O_4S	884	23.85
Sodium sulfide	Na_2S	1172	19
Sodium sulfite	Na_2O_3S	911	25.9
Spiro[2.2]pentane	C_5H_8	-106.98	6.43
trans-Stilbene	$C_{14}H_{12}$	124.82	27.7
Strontium	Sr	777	7.43

Name	Mol. form.	t_m/°C	$\Delta_{fus}H$/ kJ mol^{-1}
Strontium bromide	Br_2Sr	657	10.5
Strontium carbonate	CO_3Sr	1494	40
Strontium chloride	Cl_2Sr	874	16.22
Strontium fluoride	F_2Sr	1477	29.7
Strontium hydride	H_2Sr	1050	23
Strontium hydroxide	H_2O_2Sr	535	23
Strontium iodide	I_2Sr	538	19.7
Strontium nitrate	N_2O_6Sr	570	44.6
Strontium oxide	OSr	2531	81
Strontium sulfate	O_4SSr	1606	36
Strontium sulfide	SSr	2226	63
Styrene	C_8H_8	-30.65	10.9
Succinic acid	$C_4H_6O_4$	185	32.4
Succinic anhydride	$C_4H_4O_3$	119.5	20.4
Succinonitrile	$C_4H_4N_2$	57.985	3.70
Sulfamerazine	$C_{11}H_{12}N_4O_2S$	238	38.7
Sulfamethoxazole	$C_{10}H_{11}N_3O_3S$	167	32.2
Sulfamethoxypyridazine	$C_{11}H_{12}N_4O_3S$	182.5	31.3
Sulfathiazole	$C_9H_9N_3O_2S_2$	200	26.4
Sulfisoxazole	$C_{11}H_{13}N_3O_3S$	195.0	30.2
Sulfur (monoclinic)	S	115.21	1.721
Sulfur hexafluoride	F_6S	-49.596 tp	5.02
Sulfuric acid	H_2O_4S	10.31	10.71
Sulfur trioxide (γ-form)	O_3S	16.8	8.60
Tantalum	Ta	3017	36.57
Tantalum boride [TaB_2]	B_2Ta	3100	83.68
Tantalum(V) bromide	Br_5Ta	240	37.7
Tantalum(V) chloride	Cl_5Ta	216.6	35.1
Tantalum(V) fluoride	F_5Ta	96.9	12
Tantalum(V) iodide	I_5Ta	496	7.74
Tantalum nitride	NTa	3090	6.7
Tantalum nitride (Ta2N)	NTa_2	2727	92.0
Tantalum(V) oxide	O_5Ta_2	1875	120
Technetium	Tc	2157	33.29
Tellurium	Te	449.51	17.38
Tellurium dioxide	O_2Te	733	28.9
Tellurium tetrabromide	Br_4Te	380	24.7
Tellurium tetrachloride	Cl_4Te	224	18.9
Terbium	Tb	1359	10.15
Terbium(III) bromide	Br_3Tb	830	31.5
Terbium(III) chloride	Cl_3Tb	582	19.5
o-Terphenyl	$C_{18}H_{14}$	56.19	17.19
p-Terphenyl	$C_{18}H_{14}$	213.8	35.3
Tetrabromomethane	CBr_4	90	3.76
1,1,2,2-Tetrachloro-1,2-difluoroethane	$C_2Cl_4F_2$	26.54	3.67
1,1,2,2-Tetrachloroethane	$C_2H_2Cl_4$	-42.4	9.17
Tetrachloroethene	C_2Cl_4	-22.2	10.88
Tetrachloromethane	CCl_4	-22.8	2.56
Tetrachlorosilane	Cl_4Si	-68.74	7.60
Tetracontane	$C_{40}H_{82}$	81.4	135.5
Tetracosane	$C_{24}H_{50}$	50.3	54.4
Tetradecane	$C_{14}H_{30}$	5.87	45.07
Tetradecanoic acid	$C_{14}H_{28}O_2$	54.16	45.1
1-Tetradecanol	$C_{14}H_{30}O$	37.7	25.1 [a]
1,2,3,5-Tetrafluorobenzene	$C_6H_2F_4$	-46.2	6.36
1,2,4,5-Tetrafluorobenzene	$C_6H_2F_4$	3.88	15.05
Tetrafluoroethene	C_2F_4	-131.14	7.72
Tetrafluoromethane	CF_4	-183.58	0.704
Tetrahydrofuran	C_4H_8O	-108.38	8.54
Tetrahydropyran	$C_5H_{10}O$	-49.1	1.8
Tetrahydrothiophene	C_4H_8S	-96.13	7.35

Name	Mol. form.	t_m/°C	$\Delta_{fus}H$/ kJ mol^{-1}
Tetraiodosilane	I_4Si	120.5	19.7
1,2,4,5-Tetramethylbenzene	$C_{10}H_{14}$	79.2	21
Tetramethyl lead	$C_4H_{12}Pb$	-30.2	10.80
2,2,3,3-Tetramethylpentane	C_9H_{20}	-9.75	2.33
2,2,4,4-Tetramethylpentane	C_9H_{20}	-66.53	9.74
Tetramethylsilane	$C_4H_{12}Si$	-99.063	6.87
Tetramethylstannane	$C_4H_{12}Sn$	-55.09	9.30
Tetratetracontane	$C_{44}H_{90}$	86.0	149.6
Tetratriacontane	$C_{34}H_{70}$	72.8	79.4
1*H*-Tetrazole	CH_2N_4	156.9	18.2
Thallium	Tl	304	4.142
Thallium(I) bromide	BrTl	460	16.4
Thallium(I) carbonate	CO_3Tl_2	273	18
Thallium(I) chloride	ClTl	431	15.56
Thallium(I) fluoride	FTl	326	13.87
Thallium(I) formate	CHO_2Tl	101	10.9
Thallium(I) iodide	ITl	441.7	14.7
Thallium(I) nitrate	NO_3Tl	206	9.6
Thallium(I) oxide	OTl_2	579	30.3
Thallium(III) oxide	O_3Tl_2	834	53
Thallium(I) sulfate	O_4STl_2	632	23.8
Thallium(I) sulfide	STl_2	457	23.0
Thiazole	C_3H_3NS	-33.61	9.57
Thietane	C_3H_6S	-73.19	8.25
Thiophene	C_4H_4S	-38.12	5.07
Thiourea	CH_4N_2S	176	14.0
Thorium	Th	1750	13.81
Thorium(IV) bromide	Br_4Th	679	54.4
Thorium(IV) chloride	Cl_4Th	770	43.9
Thorium(IV) fluoride	F_4Th	1110	41.8
Thorium(IV) iodide	I_4Th	566	48.1
Thorium(IV) oxide	O_2Th	3350	90
Thulium	Tm	1545	16.84
Thulium(III) chloride	Cl_3Tm	845	34.9
Thulium(III) fluoride	F_3Tm	1158	28.9
Thymol	$C_{10}H_{14}O$	49.6	21.3
Tin (white)	Sn	231.928	7.15
Tin(II) bromide	Br_2Sn	232	18.0
Tin(IV) bromide	Br_4Sn	29.1	12.2
Tin(II) chloride	Cl_2Sn	247.0	14.52
Tin(IV) chloride	Cl_4Sn	-34.07	9.20
Tin(II) fluoride	F_2Sn	215	10.5
Tin(IV) fluoride	F_4Sn	442	27.6
Tin(II) iodide	I_2Sn	320	18.0
Tin(IV) iodide	I_4Sn	402	0.16
Tin(II) oxide	OSn	977	27.7
Tin(IV) oxide	O_2Sn	1630	23.4
Tin(II) sulfide	SSn	881	31.6
Tin(II) telluride	SnTe	806	45.2
Titanium	Ti	1670	14.15
Titanium boride	B_2Ti	2920	100.4
Titanium(IV) bromide	Br_4Ti	38.3	12.9
Titanium(II) chloride	Cl_2Ti	1035	34.3
Titanium(IV) chloride	Cl_4Ti	-24.12	9.97
Titanium(IV) fluoride	F_4Ti	377	41
Titanium(IV) iodide	I_4Ti	155	19.8
Titanium nitride	NTi	2947	66.9
Titanium(III) oxide	O_3Ti_2	1842	104.6
Titanium(IV) oxide (rutile)	O_2Ti	1912	68
Titanium(II) sulfide	STi	1927	32
Toluene	C_7H_8	-95.0	6.64
o-Toluic acid	$C_8H_8O_2$	103.4	19.5

Name	Mol. form.	t_m/°C	$\Delta_{fus}H$/ kJ mol⁻¹
m-Toluic acid	$C_8H_8O_2$	109.3	15.7
p-Toluic acid	$C_8H_8O_2$	180	22.7
Triacontane	$C_{30}H_{62}$	65.9	68.3
1,3,5-Triazine	$C_3H_3N_3$	80.3	14.56
Tribromomethane	$CHBr_3$	8.69	11.05
Trichloroacetic acid	$C_2HCl_3O_2$	59.1	5.90
1,2,3-Trichlorobenzene	$C_6H_3Cl_3$	53	17.9
1,2,4-Trichlorobenzene	$C_6H_3Cl_3$	17.0	16.4
1,3,5-Trichlorobenzene	$C_6H_3Cl_3$	62.8	18.1
1,1,1-Trichloroethane	$C_2H_3Cl_3$	-30	2.35
1,1,2-Trichloroethane	$C_2H_3Cl_3$	-36.3	11.46
Trichloroethene	C_2HCl_3	-84.7	8.45
Trichlorofluoromethane	CCl_3F	-110.44	6.89
Trichloromethane	$CHCl_3$	-63.3	9.5
1,1,2-Trichloro-1,2,2-trifluoroethane	$C_2Cl_3F_3$		2.47
Tricosane	$C_{23}H_{48}$	47.4	50.86
Tridecane	$C_{13}H_{28}$	-5.35	28.50
1-Tridecanol	$C_{13}H_{28}O$	31	41.4
1,1,1-Trifluoroethane	$C_2H_3F_3$	-111.6	6.19
Trifluoromethane	CHF_3	-155.18	4.06
Triiodomethane	CHI_3	120	16.44
Trimethoprim	$C_{14}H_{18}N_4O_3$	199	49.4
Trimethylamine	C_3H_9N	-117.1	7
1,2,3-Trimethylbenzene	C_9H_{12}	-25.32	8.18
1,2,4-Trimethylbenzene	C_9H_{12}	-43.8	13.19
1,3,5-Trimethylbenzene	C_9H_{12}	-44.69	9.51
2,2,3-Trimethylbutane	C_7H_{16}	-24.56	2.26
2,2,4-Trimethylpentane	C_8H_{18}	-107.36	9.20
1,3,5-Trinitrobenzene	$C_6H_3N_3O_6$	121.3	15.4
Trinitroglycerol	$C_3H_5N_3O_9$	12.8	21.87
2,4,6-Trinitrotoluene	$C_7H_5N_3O_6$	80.9	22.9
1,3,5-Trioxane	$C_3H_6O_3$	60	15.11
Triphenylamine	$C_{18}H_{15}N$	126.5	24.9
Triphenylene	$C_{18}H_{12}$	197.82	24.74
Tritriacontane	$C_{33}H_{68}$	71.2	79.5
Tungsten	W	3414	52.31
Tungsten boride	BW	2800	80
Tungsten boride (W_2B)	BW_2	2740	117
Tungsten boride [W_2B_5]	B_5W_2	2370	240
Tungsten(V) bromide	Br_5W	286	17.2
Tungsten(V) chloride	Cl_5W	253	20.6
Tungsten(VI) chloride	Cl_6W	282	6.69
Tungsten(VI) fluoride	F_6W	1.9	4.10
Tungsten(VI) oxide	O_3W	1473	73
Tungsten(VI) oxytetrachloride	Cl_4OW	210	18.8
Tungsten(VI) oxytetrafluoride	F_4OW	105	6
Undecane	$C_{11}H_{24}$	-25.54	22.2
Uranium	U	1135	9.14
Uranium(III) bromide	Br_3U	727	43.9
Uranium(IV) bromide	Br_4U	519	55.2
Uranium(IV) chloride	Cl_4U	590	44.8
Uranium(III) fluoride	F_3U	1495	36.8
Uranium(IV) fluoride	F_4U	1036	47
Uranium(V) fluoride	F_5U	348	35
Uranium(VI) fluoride	F_6U	64.06 tp	19.2
Uranium(IV) iodide	I_4U	506	42.1
Uranium(IV) oxide	O_2U	2847	74.2
Uranyl chloride	Cl_2O_2U	577	44.06
Urea	CH_4N_2O	132.4	13.9
Vanadium	V	1910	21.5
Vanadium(II) chloride	Cl_2V	1350	35.0
Vanadium(IV) chloride	Cl_4V	-28	2.30
Vanadium(II) fluoride	F_2V	1490	44
Vanadium(III) fluoride	F_3V	1395	57
Vanadium(V) fluoride	F_5V	19.5	49.96
Vanadium(II) oxide	OV	1790	50
Vanadium(III) oxide	O_3V_2	1957	140
Vanadium(IV) oxide	O_2V	1545	56.0
Vanadium(V) oxide	O_5V_2	681	64
Water	H_2O	0.00	6.01
Xenon	Xe	-111.75	2.27
Xenon difluoride	F_2Xe	129.03 tp	16.8
Xenon hexafluoride	F_6Xe	49.48	5.74
Xenon tetrafluoride	F_4Xe	117.10 tp	16.3
o-Xylene	C_8H_{10}	-25.16	13.6
m-Xylene	C_8H_{10}	-47.85	11.6
p-Xylene	C_8H_{10}	13.3	17.12
2,3-Xylenol	$C_8H_{10}O$	72.7	21.0
2,5-Xylenol	$C_8H_{10}O$	74.9	23.4
2,6-Xylenol	$C_8H_{10}O$	45.4	18.9
3,4-Xylenol	$C_8H_{10}O$	65.1	18.1
3,5-Xylenol	$C_8H_{10}O$	63.4	17.4
Ytterbium	Yb	824	7.66
Ytterbium(III) chloride	Cl_3Yb	854	35.4
Yttrium	Y	1522	11.39
Yttrium chloride	Cl_3Y	721	31.5
Yttrium fluoride	F_3Y	1155	27.9
Yttrium oxide	O_3Y_2	2439	81
Zinc	Zn	419.527	7.068
Zinc bromide	Br_2Zn	402	15.7
Zinc chloride	Cl_2Zn	325	10.30
Zinc fluoride	F_2Zn	872	40
Zinc iodide	I_2Zn	450	17
Zinc oxide	OZn	1974	70
Zinc phosphide [ZnP_2]	P_2Zn	980	92.9
Zinc selenite	O_3SeZn	621	46.4
Zinc sulfide (wurtzite)	SZn	1827	30
Zinc telluride	TeZn	1295	63
Zirconium	Zr	1854	21.00
Zirconium boride	B_2Zr	3050	104.6
Zirconium(II) bromide	Br_2Zr	827	28
Zirconium(III) bromide	Br_3Zr	727	33
Zirconium(IV) bromide	Br_4Zr	450 tp	30
Zirconium(II) chloride	Cl_2Zr	722	27.0
Zirconium(III) chloride	Cl_3Zr	627	30
Zirconium(IV) chloride	Cl_4Zr	437 tp	29
Zirconium(II) fluoride	F_2Zr	902	37.7
Zirconium(III) fluoride	F_3Zr	927	50
Zirconium(IV) fluoride	F_4Zr	910	61
Zirconium(II) iodide	I_2Zr	827	28
Zirconium(III) iodide	I_3Zr	727	33
Zirconium(IV) iodide	I_4Zr	500	32
Zirconium nitride	NZr	2952	67.4
Zirconium(IV) oxide	O_2Zr	2710	90
Zirconium(IV) sulfide	S_2Zr	1550	45

[a] The value of $\Delta_{fus}H$ includes the enthalpy of transition between two crystalline phases whose transformation occurs within 1 °C of the melting point.

COMPRESSIBILITY AND EXPANSION COEFFICIENTS OF LIQUIDS

This table gives data on the variation of the density of some common liquids with pressure and temperature. The pressure dependence is described to first order by the isothermal compressibility coefficient κ defined as

$$\kappa T = -(1/V)\ (\partial V/\partial P)T$$

where V is the volume and P the pressure, and the temperature dependence by the cubic expansion coefficient α,

$$\alpha V = (1/V)\ (\partial V/\partial T)P$$

More precise data on the variation of density with temperature over a wide temperature range can be found in Ref. 1. Column definitions are as follows.

Column heading	Definition
Name	Name of liquid; listed alphabetically by name
Mol. form.	Molecular formula of liquid
t	Temperature at which properties are given, in °C
$\kappa_T \times 10^4$	Isothermal compressibiliy coefficient in units MPa^{-1}; value as written is 10^4 the true value
$\alpha_V \times 10^3$	Cubic expansion coefficient, in units $°C^{-1}$; value as written is 10^3 the true value

References

1. Lide, D. R., and Kehiaian, H. V., *CRC Handbook of Thermophysical and Thermochemical Data*, CRC Press, Boca Raton, FL, 1994.
2. Le Neindre, B., *Effets des Hautes et Très Hautes Pressions*, in *Techniques de l'Ingénieur*, Paris, 1991.
3. *Landolt-Börnstein, Numerical Data and Functional Relationships in Science and Technology, New Series*, IV/4, *High-Pressure Properties of Matter*, Springer-Verlag, Heidelberg, 1980.
4. Riddick, J.A., Bunger, W.B., and Sakano, T.K., *Organic Solvents, Fourth Edition*, John Wiley & Sons, New York, 1986.
5. Isaacs, N. S., *Liquid Phase High Pressure Chemistry*, John Wiley, New York, 1981.

Compressibility and Expansion Coefficients of Liquids

Name	Mol. form.	t/°C	$\kappa_T \times 10^4$/ MPa^{-1}	$\alpha_V \times 10^3$/ $°C^{-1}$
Acetic acid	$C_2H_4O_2$	20	9.08	1.08
Acetic acid	$C_2H_4O_2$	80	13.7	1.38
Acetone	C_3H_6O	20	12.62	1.46
Acetone	C_3H_6O	40	15.6	1.57
Aniline	C_6H_7N	20	4.53	0.81
Aniline	C_6H_7N	80	6.32	0.91
Anisole	C_7H_8O	20	6.60	0.951
Benzene	C_6H_6	25	9.66	1.14
Benzene	C_6H_6	45	11.28	1.21
Bromobenzene	C_6H_5Br	20	6.46	0.86
1-Bromobutane	C_4H_9Br	20		1.13
1-Bromobutane	C_4H_9Br	25	10.26	
Bromoethane	C_2H_5Br	20	11.53	1.31
1-Bromopentane	$C_5H_{11}Br$	0	8.42	
1-Bromopentane	$C_5H_{11}Br$	25		1.04
1-Bromopropane	C_3H_7Br	0	10.22	
1-Bromopropane	C_3H_7Br	25		1.2
1-Butanol	$C_4H_{10}O$	0	8.10	1.12
Carbon disulfide	CS_2	20	9.38	1.12
Carbon disulfide	CS_2	35		1.16
Carbon disulfide	CS_2	40	10.6	
Chlorobenzene	C_6H_5Cl	20	7.45	0.94
1-Chloropropane	C_3H_7Cl	0	12.09	
1-Chloropropane	C_3H_7Cl	20		1.4
Cycloheptane	C_7H_{14}	20	9.22	
Cyclohexane	C_6H_{12}	20	11.30	1.15
Cyclohexane	C_6H_{12}	60	15.2	1.29
Cyclooctane	C_8H_{16}	20	8.03	
Cyclopentane	C_5H_{10}	20	13.31	1.35
Decane	$C_{10}H_{22}$	25	10.94	1.02
Dibromomethane	CH_2Br_2	27	6.85	
Dibutyl phthalate	$C_{16}H_{22}O_4$	0	5.0	
Dibutyl phthalate	$C_{16}H_{22}O_4$	25		0.86
1,1-Dichloroethane	$C_2H_4Cl_2$	20	7.97	
1,1-Dichloroethane	$C_2H_4Cl_2$	25		0.93
1,2-Dichloroethane	$C_2H_4Cl_2$	20		1.14
1,2-Dichloroethane	$C_2H_4Cl_2$	30	8.46	
trans-1,2-Dichloroethene	$C_2H_2Cl_2$	25	11.2	1.36
Dichloromethane	CH_2Cl_2	25	10.3	1.39
Diethylene glycol	$C_4H_{10}O_3$	0	3.34	
Diethylene glycol	$C_4H_{10}O_3$	20		0.635
Diethyl ether	$C_4H_{10}O$	20	18.65	1.65
Diethyl ether	$C_4H_{10}O$	30	20.85	1.72
2,3-Dimethylbutane	C_6H_{14}	20	17.97	
2,3-Dimethylbutane	C_6H_{14}	25		1.39
Dodecane	$C_{12}H_{26}$	25	9.88	0.93
1,2-Ethanediol	$C_2H_6O_2$	20	3.64	0.626
Ethanol	C_2H_6O	20	11.19	1.40
Ethanol	C_2H_6O	70	15.93	1.67
Ethyl acetate	$C_4H_8O_2$	20	11.32	1.35
Ethyl acetate	$C_4H_8O_2$	60	16.2	1.54
Glycerol	$C_3H_8O_3$	0	2.54	
Glycerol	$C_3H_8O_3$	20		0.520
Glycerol triacetate	$C_9H_{14}O_6$	0	4.49	
Glycerol triacetate	$C_9H_{14}O_6$	25		0.94
Heptane	C_7H_{16}	25	14.38	1.26
Hexadecane	$C_{16}H_{34}$	25	8.57	
Hexadecane	$C_{16}H_{34}$	45	9.78	
Hexane	C_6H_{14}	25	16.69	1.41
Hexane	C_6H_{14}	45	20.27	1.52
1-Hexanol	$C_6H_{14}O$	25	8.24	1.03
1-Iodobutane	C_4H_9I	0	7.73	
1-Iodobutane	C_4H_9I	25		1.02
Iodoethane	C_2H_5I	20	9.82	

Name	Mol. form.	t/°C	$\kappa_T \times 10^4$/ MPa^{-1}	$\alpha_V \times 10^3$/ °C^{-1}
Iodoethane	C_2H_5I	25		1.17
Iodomethane	CH_3I	25		1.26
Iodomethane	CH_3I	27	10.3	
1-Iodopentane	$C_5H_{11}I$	0	7.56	
1-Iodopropane	C_3H_7I	0	10.22	
1-Iodopropane	C_3H_7I	25		1.09
Mercury	Hg	20	0.401	0.1811
Methanol	CH_4O	20	12.14	1.49
Methanol	CH_4O	40	13.83	1.59
Methyl *cis*-9-octadecenoate	$C_{19}H_{36}O_2$	0	6.18	
Methyl *cis*-9-octadecenoate	$C_{19}H_{36}O_2$	60		0.85
2-Methylpentane	C_6H_{14}	0	13.97	
2-Methylpentane	C_6H_{14}	25		1.43
3-Methylpentane	C_6H_{14}	0	14.57	
3-Methylpentane	C_6H_{14}	25		1.40
Nitrobenzene	$C_6H_5NO_2$	20	4.93	
Nitrobenzene	$C_6H_5NO_2$	25		0.833
Nonane	C_9H_{20}	25	11.75	1.08
Octane	C_8H_{18}	25	12.82	1.16
Octane	C_8H_{18}	45	15.06	1.23
1-Octanol	$C_8H_{18}O$	25	7.64	0.827
Pentadecane	$C_{15}H_{32}$	25	8.82	
Pentane	C_5H_{12}	25	21.80	1.64
1-Pentanol	$C_5H_{12}O$	0	7.71	1.02
Phenol	C_6H_6O	60	6.05	0.82
Phosphorus(III) chloride	Cl_3P	20	9.45	1.9
1,2-Propanediol	$C_3H_8O_2$	0	4.45	
1,2-Propanediol	$C_3H_8O_2$	20		0.695
1,3-Propanediol	$C_3H_8O_2$	0	4.09	
1,3-Propanediol	$C_3H_8O_2$	20		0.61
1-Propanol	C_3H_8O	0	8.43	1.22
2-Propanol	C_3H_8O	40	13.32	1.55
Tetrachloroethene	C_2Cl_4	25	7.56	1.02
Tetrachloromethane	CCl_4	20	10.50	1.14
Tetrachloromethane	CCl_4	40	12.20	1.21
Tetrachloromethane	CCl_4	70	15.6	1.33
Tetradecane	$C_{14}H_{30}$	25	9.10	0.87
Toluene	C_7H_8	20	8.96	1.05
Toluene	C_7H_8	50	11.0	1.13
Tribromomethane	$CHBr_3$	25		0.91
Tribromomethane	$CHBr_3$	50	8.76	
Trichloroethene	C_2HCl_3	25	8.57	1.17
Trichloromethane	$CHCl_3$	20	9.96	1.21
Trichloromethane	$CHCl_3$	50	12.9	1.33
Tridecane	$C_{13}H_{28}$	25	9.48	0.90
Tris(2-hydroxyethyl)amine	$C_6H_{15}NO_3$	0	3.61	
Tris(2-hydroxyethyl)amine	$C_6H_{15}NO_3$	55		0.53
1,3,5-Trimethylbenzene	C_9H_{12}	25	8.14	0.94
Undecane	$C_{11}H_{24}$	25	10.31	0.97
Water	H_2O	20	4.591	0.206
Water	H_2O	25	4.524	0.256
Water	H_2O	30	4.475	0.302
o-Xylene	C_8H_{10}	25	8.10	0.96
m-Xylene	C_8H_{10}	20	8.46	0.99
p-Xylene	C_8H_{10}	25	8.59	1.00

TEMPERATURE AND PRESSURE DEPENDENCE OF LIQUID DENSITY

Ivan Cibulka

This table provides the data to calculate the temperature and pressure dependence of the denisty of 61 organic liquids using two different methods: the Tait equation and the Wagner function.

Tait Equation: The Tait equation (Refs. 1,2) gives the ratio between the density at pressure P, $\rho(T,P)$, relative to the density at a reference pressure, $\rho(T,P_{ref})$, at the same temperature T.

$$\frac{\rho(T,P)}{\rho(T,P_{\mathrm{ref}})}=\frac{1}{1-C(T)\ln\dfrac{B(T)+P}{B(T)+P_{\mathrm{ref}}}}$$
$$C(T)=a_1+b_1(T/K)+c_1(T/K)^2 \qquad (1)$$
$$B(T)/\mathrm{MPa}=a_2+a_3(T/K)+a_4(T/K)^2+a_5(T/K)^3+a_6(T/K)^4$$

Parameters a_i (i = 1,...,6) and b_1 are given in the second row of each entry in the table below. Parameter c_1 is zero for most substances, and therefore its value for heptane, the only organic liquid included in the table for which it is relevant, is given in a footnote.

The reference pressure is P_{ref} = 0.101325 MPa at temperatures either at or below the normal boiling point temperature (T_{nbp}) and $P_{ref} = P_{sat}(T)$ (saturated vapor pressure) at temperatures $T > T_{nbp}$. Ranges of validity of the equation (T_{min}, T_{max}, P_{max}) are derived from ranges of experimental data; the minimal pressure of validity is taken as P_{ref}, i.e., interpolation between P_{ref} and lowest experimental pressure is allowed. The upper limit of application is the freezing line (if not limited by the ranges of validity). To avoid any large-scale extrapolation, the validity ranges are rectangular areas $(T_{max} - T_{min})P_{max}$. If in a specific temperature interval(s) the maximum experimental pressure exceeded the given value of P_{max}, then the maximum pressure given in the table is denoted by (r) which means that the validity range given in the table is a rectangular subset of the non-rectangular experimental T, P range. In a few cases P_{max} is given as a ratio where the first value corresponds to T_{min} and the second one to T_{max}, i.e., the validity range has approximately a trapezoidal shape.

Values of parameters were taken from the papers (Refs. 3–10) where detailed information on the fits, experimental data, and application ranges is available. The numerical values of the parameters are different from those reported in papers (Refs. 3–10) because the forms of polynomials $C(T)$ and $B(T)$ differ. Also, the parameters recorded in the table below do not necessarily correspond to those in Refs. 3–10 as some fits were updated using newly published experimental data.

Smoothing function: To determine the density at the reference pressure $\rho(T,P_{ref})$ for use in Eq. 1, one of two smoothing functions is used, both polynomial expansions.

$$\rho(T,P_{\mathrm{ref}})/(\mathrm{kg\,m^{-3}})=\sum_{i=1}^{N_p}a_i(T/\mathrm{K})^{(i-1)} \qquad (2)$$

$$\rho(T,P_{\mathrm{ref}})=\rho_c\left[1+\sum_{i=1}^{N_p}a_i(1-T_r)^{(i/3)}\right],\quad T_r=T/T_c \qquad (3)$$

where N_p is the number of adjustable parameters, a_i, whose values are given in the third row of each entry, as prefaced by the appropriate equation number. Values of the critical density ρ_c and the critical temperature T_c used for the fits using Eq. 3 are also recorded in the table. Parameters were mostly taken from Refs. 3–10, those for 1-alkanols C_1 to C_{10} and n-alkanes C_5 to C_{16} are from Ref. 11. Data used for the fits were predominantly recommended values published in the TRC Thermodynamic Tables (Refs. 12,13), sometimes combined with the original experimental data or, in a few cases, the original experimental data were correlated.

RMSD is a relative root-mean-square deviation (in percent) between experimental values of density and those calculated from the particular function (Tait Eq. 1 or Eqs. 2,3).

$$RMSD/\%=100\left\{\frac{1}{N}\sum_{i=1}^{N}\left(\frac{\rho_{\mathrm{exp}}-\rho_{\mathrm{calc}}}{\rho_{\mathrm{exp}}}\right)^2\right\}^{1/2}$$

where N is the number of experimental values included in the fit.

Wagner equation: If the maximum temperature T_{max} of validity of the Tait equation is greater than the normal boiling point temperature T_{nbp}, then the Wagner equation is applicable in the form of either

$$P_{\mathrm{sat}}(T)=P_c\exp\left[\frac{a_1(1-T_r)+a_2(1-T_r)^{1.5}+a_3(1-T_r)^{2.5}+a_4(1-T_r)^5}{T_r}\right] \qquad (4)$$

or

$$P_{\mathrm{sat}}(T)=P_c\exp\left[\frac{a_1(1-T_r)+a_2(1-T_r)^{1.5}+a_3(1-T_r)^3+a_4(1-T_r)^6}{T_r}\right] \qquad (5)$$

where $T_r = T/T_c$ are recorded in the third line for each substance. Values of the critical pressure P_c and critical temperature T_c used in Eqs. 4 and 5 are also recorded in the table as prefaced by the equation number. Values of the critical temperature may differ a little from those recorded for the function, Eq. 3. Parameters of Eqs. 4 and 5 were taken mostly from the papers by McGarry (Ref. 14) and Ambrose and Walton (Ref. 15); in a few cases, the fits were performed using original experimental data or in combination with the recommended values from the TRC Thermodynamic Tables (Refs. 12,13).

The two right-hand-most columns gives values of the isothermal compressibility coefficient, $\kappa_T = -(1/V)(\partial V/\partial P)_T = (1/\rho)(\partial\rho/\partial P)_T$, and the isobaric cubic expansion coefficient, $\alpha_p = (1/V)(\partial V/\partial T)_p = -(1/\rho)(\partial\rho/\partial T)_p$, calculated for T = 298.15 K and P = 0.101325 MPa using Tait Eq. 1 and from the $\rho(T, P_{ref})$ equation, respectively. In a very few cases when the lower temperature limit of the Tait equation T_{min} is greater than 298.15 K, the extrapolated values of isothermal compressibility are given.

The column and row definitions are as follows.

Column heading	Definition
Row 1	**Entry information**
Mol. form.	Molecular formula of liquid; liquids listed by Hill order
Name	Liquid name
T_{nbp}	Normal boiling point, in K
κ_T	Isothermal compressibility coefficient, in units Pa^{-1}
α_p	Isobaric cubic expansion coefficient, in units K^{-1}
Ref.	Reference
Row 2	**Tait equations**
Eq.	Tait equation number; see text
$a_1, a_2, a_3, a_4, a_5, b_1$	Tait equation coefficients; for heptane, coefficient c_1 is required and given in Footnote b
T_{min}/T_{max}	Temperature range of validity for Tait equation, in K
P_{max}	Maximum pressure for validity of Tait equation, in MPa
RMSD	Relative root-mean-square deviation (in percent) between experimental and calculated (Eqs. 1-3) values of density
Rows 3 and 4	**Smoothing and Wagner equations**
Eq.	Equation number
a_1, a_2, a_3, a_4, a_5	Equation coefficients; see Footnote a for extra term needed for ethanol calculations
T_c	Critical temperature, in K
P_c	Critical pressure, in MPa
ρ_c	Density at critical temperature and pressure, in kg m^{-3}

References

1. Tait, P. G., in *Physics and Chemistry of the Voyage of H.M.S. Challenger*, Vol. II, Part IV, Thomson, C. W., and Murray, J., Eds., H.M.S.O., London, 1889.
2. Tamman, G., *Z. Phys. Chem.* 17, 620, 1895. <https://doi.org/10.1007/BF01841600>
3. Cibulka, I., and Ziková, M., *J. Chem. Eng. Data* 39, 876, 1994. <https://doi.org/10.1021/je00016a055>
4. Cibulka, I., and Hnědkovský, L., *J. Chem. Eng. Data* 41, 657, 1996. <https://doi.org/10.1021/je960058m>
5. Cibulka, I., Hnědkovský, L., and Takagi, T., *J. Chem. Eng. Data* 42, 2, 1997. <https://doi.org/10.1021/je960199o>
6. Cibulka, I., Hnědkovský, L., and Takagi, T., *J. Chem. Eng. Data* 42, 415, 1997. <https://doi.org/10.1021/je960199o>
7. Cibulka, I., and Takagi, T., *J. Chem. Eng. Data* 44, 411, 1999. <https://doi.org/10.1021/je980278v>
8. Cibulka, I., and Takagi, T., *J. Chem. Eng. Data* 44, 1105, 1999. <https://doi.org/10.1021/je990140s>
9. Cibulka, I., Takagi, T., and Růžička, K., *J. Chem. Eng. Data* 46, 2, 2001. <https://doi.org/10.1021/je0002383>
10. Cibulka, I., and Takagi, T., *J. Chem. Eng. Data* 47, 1037, 2002. <https://doi.org/10.1021/je0200463>
11. Cibulka, I., *Fluid Phase Equilib.* 89, 1, 1993. <https://doi.org/10.1016/0378-3812(93)85042-K>
12. TRC Thermodynamic Tables, Hydrocarbons, Thermodynamics Research Center (TRC), NIST, Thermophysical Properties Division, Boulder, CO.
13. TRC Thermodynamic Tables, Non-Hydrocarbons, Thermodynamics Research Center (TRC), NIST, Thermophysical Properties Division, Boulder, CO.
14. McGarry, J., *Ind. Eng. Chem., Process Des. Develop.* 22, 313, 1983. <https://doi.org/10.1021/i200021a023>
15. Ambrose, D., and Walton, J., *Pure Appl. Chem.* 61, 1395, 1989. <https://doi.org/10.1351/pac198961081395>

Parameters to Calculate Temperature and Pressure Dependence of Density and Calculated Values of Physical Constants for Selected Liquids

Mol. form.	Eq.	a_1	a_2	a_3	a_4	a_5	a_6	b_1	T_{min}/T_{max} K	P_{max} MPa	T_c K	P_c MPa	ρ_c kg m^{-3}	RMSD %
CCl_4	T_{nbp} = 349.9 K	κ_T = 1.074 GPa^{-1}	α_P = 1.209 kK^{-1}	Ref. 9										
Tetrachloromethane	1	$9.33340\cdot10^{-2}$	$1.11363\cdot10^{3}$	-8.68453	$2.80698\cdot10^{-2}$	$-4.22880\cdot10^{-5}$	$2.37923\cdot10^{-8}$		273/413	51/388				0.04
	3	1.58994	2.51946	-5.82313	6.96793	-2.51359			253/554		556.4		557.33	0.042
	5	-7.07139	1.71497	-2.8993	-2.49466				250/556		556.4	4.551		
$CHBr_3$	T_{nbp} = 422.3 K	κ_T = 0.809 GPa^{-1}	α_P = 0.907 kK^{-1}	Ref. 9										
Tribromomethane	1	$1.03492\cdot10^{-1}$	$2.64208\cdot10^{2}$	$-4.57399\cdot10^{-1}$					323/368	150/343				0.058
	2	$3.55953\cdot10^{3}$	-1.96212	$-1.08712\cdot10^{-3}$					283/403					0.001
$CHCl_3$	T_{nbp} = 334.4 K	κ_T = 1.037 GPa^{-1}	α_P = 1.274 kK^{-1}	Ref. 9										
Trichloromethane	1	$9.57210\cdot10^{-2}$	$4.79593\cdot10^{2}$	-1.84011	$1.81340\cdot10^{-3}$				273/348	100(r)				0.031
	3	3.56339	-3.86051	3.35636					213/333		536.4		499.49	0.043
	5	-6.95546	1.16625	-2.1397	-3.44421				215/536		536.4	5.366		
CH_2Cl_2	T_{nbp} = 313.4 K	κ_T = 1.032 GPa^{-1}	α_P = 1.428 kK^{-1}	Ref. 9										
Dichloromethane	1	$9.76370\cdot10^{-2}$	$5.24365\cdot10^{2}$	-2.06633	$2.09494\cdot10^{-3}$				293/423	100(r)				0.091
	3	3.00368	-2.19763	2.34269					178/383		510		440.07	0.014
	5	-7.35739	2.17546	-4.07038	3.50701				233/510		510	6.3		
CH_3I	T_{nbp} = 315.6 K	κ_T = 1.052 GPa^{-1}	α_P = 1.255 kK^{-1}	Ref. 9										
Iodomethane	1	$9.54770\cdot10^{-2}$	$5.36810\cdot10^{2}$	-2.25115	$2.53188\cdot10^{-3}$				253/313	160				0.038
	2	$3.48981\cdot10^{3}$	-7.47709	$1.83592\cdot10^{-2}$	$-2.36742\cdot10^{-5}$				213/313					0.011
CH_3OH	T_{nbp} = 337.7 K	κ_T = 1.231 GPa^{-1}	α_P = 1.201 kK^{-1}	Ref. 3				$-5.322\cdot10^{-5}$						
Methanol	1	$1.15068\cdot10^{-1}$	$6.49718\cdot10^{2}$	-4.34583	$1.30722\cdot10^{-2}$	$-2.00292\cdot10^{-5}$	$1.20566\cdot10^{-8}$		183/483	104(r)				0.059
	3	2.62781	-4.04742	$1.58343\cdot10$	$-2.25066\cdot10$	$1.09160\cdot10$	$3.04774\cdot10^{-1}$		175/509		512.6		272	0.18
	4	-8.63571	1.17982	-2.479	-1.024				175/513		512.64	8.092		
C_2Cl_4	T_{nbp} = 394.5 K	κ_T = 0.744 GPa^{-1}	α_P = 1.012 kK^{-1}	Ref. 9										
Tetrachloroethene	1	$1.01727\cdot10^{-1}$	$1.36610\cdot10^{2}$						298/298	101				0.007
	3	1.10232	1.45142						298/313		620.2		571.84	0.019

Mol. form.	Eq.	a_1	a_2	a_3	a_4	a_5	a_6	b_1	T_{min}/T_{max} K	P_{max} MPa	T_c K	P_c MPa	ρ_c kg m^{-3}	RMSD %
C_2HCl_3	T_{nbp} = 360.4 K	κ_T = 0.870 GPa^{-1}	α_P = 1.142 kK^{-1}	9										
Trichloroethene	1	$1.00987\cdot10^{-1}$	$1.16005\cdot10^{2}$						298/298	101				0.01
	3	1.30631	1.33512						291/315		571		513.24	0.008
CH_3CHCl_2	T_{nbp} = 330.4 K	κ_T = 1.144 GPa^{-1}	α_P = 1.320 kK^{-1}	9										
1,1-Dichloroethane	1	$9.79930\cdot10^{-2}$	$4.40906\cdot10^{2}$	-1.67041	$1.60492\cdot10^{-3}$				298/398	101(r)				0.028
	3	2.81773	-2.41755	2.40906					263/398		523		419.32	0.084
	4	-9.51254	7.49332	-7.87028	$6.32540\cdot10^{-1}$				320/450		523	5.07		
CH_2ClCH_2Cl	T_{nbp} = 356.7 K	κ_T = 0.801 GPa^{-1}	α_P = 1.158 kK^{-1}	Ref. 8										
1,2-Dichloroethane	1	$9.60440\cdot10^{-2}$	$5.39476\cdot10^{2}$	-1.94358	$1.79758\cdot10^{-3}$				278/398	101(r)				0.027
	3	2.7766	-2.33774	2.3153					263/398		561.6		439.82	0.063
	5	-7.36864	1.76727	-3.34295	-1.4353				260/566		566	5.362		
CH_3COOH	T_{nbp} = 391.1 K	κ_T = 0.897 GPa^{-1}	α_P = 1.085 kK^{-1}	Ref. 5										
Acetic acid	1	$9.44550\cdot10^{-2}$	$7.22131\cdot10^{2}$	-3.33202	$4.23588\cdot10^{-3}$				293/328	25/253				0.04
	3	4.74861	$-1.76644\cdot10$	$4.44807\cdot10$	$-4.82535\cdot10$	$1.98308\cdot10$			293/493		592.71		351.19	0.003
C_2H_5Br	T_{nbp} = 311.6 K	κ_T = 1.344 GPa^{-1}	α_P = 1.475 kK^{-1}	Ref. 9										
Bromoethane	1	$9.43490\cdot10^{-2}$	$7.70215\cdot10^{2}$	-4.0647	$5.75736\cdot10^{-3}$				253/313	157(r)				0.055
	3	1.50516	1.35063						253/313		503.9		506.82	0.058
	5	-9.14807	5.49831	-6.68657	6.27287				301/504		503.8	6.232		
C_2H_5OH	T_{nbp} = 351.4 K	κ_T = 1.117 GPa^{-1}	α_P = 1.097 kK^{-1}	Ref. 3				$-4.588\cdot10^{-5}$						
Ethanol[a]	1	$1.09012\cdot10^{-1}$	$5.22523\cdot10^{2}$	-2.99367	$8.25136\cdot10^{-3}$	$-1.31014\cdot10^{-5}$	$8.74008\cdot10^{-9}$		193/480	200(r)				0.055
	3	$-9.92640\cdot10^{-1}$	$3.80287\cdot10$	$-1.81117\cdot10^{2}$	$4.45905\cdot10^{2}$	$-5.88518\cdot10^{2}$	$3.93920\cdot10^{2}$		159/508		513.88		276	0.16
	4	-8.68587	1.17831	-4.8762	1.588				159/514		513.92	6.132		
$(CH_2OH)_2$	T_{nbp} = 470.5 K	κ_T = 0.368 GPa^{-1}	α_P = 0.639 kK^{-1}	Ref. 6										
1,2-Ethanediol	1	$9.50140\cdot10^{-2}$	$6.74427\cdot10^{2}$	-1.77594	$1.27583\cdot10^{-3}$				298/378	100(r)				0.031
	3	1.77482	1.11208						298/378		790		333.7	0.022
$(CH_3)_2CO$	T_{nbp} = 329.2 K	κ_T = 1.293 GPa^{-1}	α_P = 1.457 kK^{-1}	Ref. 5										
Acetone	1	$9.92390\cdot10^{-2}$	$6.22264\cdot10^{2}$	-2.94315	$3.73329\cdot10^{-3}$				278/323	392				0.032
	3	1.83805	$5.82830\cdot10^{-1}$	$-1.00291\cdot10^{-1}$	$5.70410\cdot10^{-1}$				179/506		508.1		277.9	0.025
C_3H_7Br	T_{nbp} = 344.0 K	κ_T = 1.137 GPa^{-1}	α_P = 1.251 kK^{-1}	Ref. 9										
1-Bromopropane	1	$8.88710\cdot10^{-2}$	$4.24716\cdot10^{2}$	-1.81352	$2.55049\cdot10^{-3}$	$-1.23227\cdot10^{-6}$			280/500	100				0.094
	3	1.65835	1.17536						280/340		536		455.53	0.009
	4	-9.68919	7.16793	-7.05282	1.00727				300/536		536	4.75		
C_3H_7Cl	T_{nbp} = 319.6 K	κ_T = 1.382 GPa^{-1}	α_P = 1.462 kK^{-1}	Ref. 9										
1-Chloropropane	1	$1.07245\cdot10^{-1}$	$2.78606\cdot10^{2}$	$-5.17553\cdot10^{-2}$	$-3.83546\cdot10^{-3}$	$5.85824\cdot10^{-6}$			273/368	98(r)				0.078
	3	4.15272	-5.73954	4.74431					253/343		503		309.22	0.015
	5	-7.55764	2.60153	-5.06041	3.31163				248/503		503	4.58		
C_3H_7I	T_{nbp} = 375.7 K	κ_T = 0.904 GPa^{-1}	α_P = 1.072 kK^{-1}	Ref. 9										
1-Iodopropane	1	$1.01044\cdot10^{-1}$	$3.01114\cdot10^{2}$	$-6.61560\cdot10^{-1}$	$8.74342\cdot10^{-5}$				273/368	1177				0.07
	2	$2.22323\cdot10^{3}$	-1.38867	$-7.96985\cdot10^{-4}$					288/361					0.196
$CH_3CH_2CH_2OH$	T_{nbp} = 370.3 K	κ_T = 1.007 GPa^{-1}	α_P = 0.999 kK^{-1}	Ref. 3				$-3.058\cdot10^{-5}$						
1-Propanol	1	$9.60290\cdot10^{-2}$	$3.10861\cdot10^{2}$	$-8.35531\cdot10^{-1}$	$4.27987\cdot10^{-5}$	$7.78234\cdot10^{-7}$			170/524	49(r)				0.109
	3	$9.40534\cdot10^{-1}$	$1.29442\cdot10$	$-5.39519\cdot10$	$1.13639\cdot10^{2}$	$-1.13866\cdot10^{2}$	$4.33832\cdot10$		153/530		536.74		274	0.16
	4	-8.53706	1.96214	-7.6918	2.945				147/537		536.78	5.168		
$CH_3CHOHCH_3$	T_{nbp} = 355.4 K	κ_T = 1.123 GPa^{-1}	α_P = 1.099 kK^{-1}	Ref. 6										
2-Propanol	1	$8.90020\cdot10^{-2}$	$1.87411\cdot10^{2}$	$-2.36391\cdot10^{-1}$	$-4.24787\cdot10^{-4}$				273/400	50(r)				0.054
	3	$6.24542\cdot10^{-1}$	6.18938	-6.79039	2.49009				243/430		508.3		273.16	0.093
	5	-8.16927	$-9.43213\cdot10^{-2}$	-8.1004	7.85				250/508		508.3	4.742		
$C_3H_8O_2$	T_{nbp} = 460.2 K	κ_T = 0.481 GPa^{-1}	α_P = 0.714 kK^{-1}	Ref. 6										
1,2-Propanediol	1	$8.38940\cdot10^{-2}$	$9.09931\cdot10^{2}$	-4.03075	$5.24295\cdot10^{-3}$				273/368	200				0.019
	2	$1.18071\cdot10^{3}$	$-2.56645\cdot10^{-1}$	$-8.05917\cdot10^{-4}$					283/363					0.017
$C_3H_8O_2$	T_{nbp} = 487.6 K	κ_T = 0.403 GPa^{-1}	α_P = 0.593 kK^{-1}	Ref. 6										
1,3-Propanediol	1	$9.10310\cdot10^{-2}$	$5.95160\cdot10^{2}$	-1.64714	$1.36994\cdot10^{-3}$				273/368	200				0.067
	3	3.64841	-2.06696	1.22738					283/363		685.7		318.39	0.002
$CH_2OHCHOHCH_2OH$	T_{nbp} = 563.1 K	κ_T = 0.231 GPa^{-1}	α_P = 0.486 kK^{-1}	Ref. 6										
Glycerol	1	$1.14255\cdot10^{-1}$	$9.25959\cdot10^{2}$	-1.5814	$4.48909\cdot10^{-4}$				223/368	686				0.061
	3	6.9496	-9.25236	5.43535					214/364		800		351.51	0.069
$C_4H_8O_2$	T_{nbp} = 350.3 K	κ_T = 1.204 GPa^{-1}	α_P = 1.425 kK^{-1}	Ref. 5										
Ethyl acetate	1	$8.94090\cdot10^{-2}$	$1.05394\cdot10^{3}$	-7.587	$1.98689\cdot10^{-2}$	$-1.82587\cdot10^{-5}$			253/343	49(r)				0.106
	3	1.69043	2.6415	-4.70476	3.52368				253/473		523.2		308.06	0.015
C_4H_9Br	T_{nbp} = 374.8 K	κ_T = 1.046 GPa^{-1}	α_P = 1.119 kK^{-1}	Ref. 9										
1-Bromobutane	1	$9.76600\cdot10^{-2}$	$4.97301\cdot10^{2}$	-2.02541	$2.24791\cdot10^{-3}$				273/368	1177				0.074
	3	1.59525	1.23746						293/323		572		421.6	0.003
C_4H_9OH	T_{nbp} = 390.9 K	κ_T = 0.914 GPa^{-1}	α_P = 0.949 kK^{-1}	Ref. 3				$-3.268\cdot10^{-5}$						
1-Butanol	1	$1.01542\cdot10^{-1}$	$5.01888\cdot10^{2}$	-2.41576	$5.35877\cdot10^{-3}$	$-7.31982\cdot10^{-6}$	$4.60628\cdot10^{-9}$		195/524	49(r)				0.053
	3	-3.57379	$5.02450\cdot10$	$-1.75934\cdot10^{2}$	$3.08588\cdot10^{2}$	$-2.65628\cdot10^{2}$	$8.94997\cdot10$		186/559		563.01		271	0.352
	4	-8.40615	2.2301	-8.2486	$-7.11000\cdot10^{-1}$				185/563		563.05	4.424		

Mol. form.	Eq.	a_1	a_2	a_3	a_4	a_5	a_6	b_1	T_{min}/T_{max} K	P_{max} MPa	T_c K	P_c MPa	ρ_c kg m^{-3}	RMSD %
$(C_2H_5)_2O$	T_{nbp} = 307.6 K	κ_T = 1.740 GPa^{-1}	α_P = 1.657 kK^{-1}	Ref. 5										
Diethyl ether	1	$9.70250\cdot10^{-2}$	$3.95413\cdot10^{2}$	-1.68339	$1.82398\cdot10^{-3}$				293/353	981				0.239
	3	3.31808	-8.55179	$2.09100\cdot10$	$-2.10672\cdot10$	8.26525			140/430		466.74		264.72	0.027
	5	-7.29916	1.24828	-2.91931	-3.3674				250/467		466.74	3.646		
C_5H_{10}	T_{nbp} = 322.4 K	κ_T = 1.308 GPa^{-1}	α_P = 1.325 kK^{-1}	Ref. 8										
Cyclopentane	1	$8.85580\cdot10^{-2}$	$8.94968\cdot10^{2}$	-7.76576	$3.01636\cdot10^{-2}$	$-5.82645\cdot10^{-5}$	$4.43996\cdot10^{-8}$		193/353	48/196				0.05
	3	1.68431	$6.19856\cdot10^{-1}$	$4.90670\cdot10^{-2}$	$2.19121\cdot10^{-1}$				179/508		511.7		275.04	0.031
	5	-6.51809	$3.84422\cdot10^{-1}$	-1.11706	-4.50275				289/512		511.6	4.509		
$C_5H_{11}Br$	T_{nbp} = 402.7 K	κ_T = 0.943 GPa^{-1}	α_P = 1.076 kK^{-1}	Ref. 9										
1-Bromopentane	1	$8.98660\cdot10^{-2}$	$5.28042\cdot10^{2}$	-2.4162	$3.92466\cdot10^{-3}$	$-2.31231\cdot10^{-6}$			283/523	98				0.061
	2	$1.60209\cdot10^{3}$	-1.30486						283/363					0.006
	4	-8.59341	3.46288	-3.06661	-4.07997				294/606		605.9	3.9		
C_5H_{12}	T_{nbp} = 309.2 K	κ_T = 2.133 GPa^{-1}	α_P = 1.610 kK^{-1}	Ref. 4				$-3.788\cdot10^{-5}$						
Pentane	1	$1.00620\cdot10^{-1}$	$6.52014\cdot10^{2}$	-4.94167	$1.54854\cdot10^{-2}$	$-2.36811\cdot10^{-5}$	$1.44530\cdot10^{-8}$		173/373	87/285				0.119
	3	1.17756	3.89157	-5.50896	3.29181				143/444		469.8		232	0.154
	4	-7.30698	1.75845	-2.1629	-2.913				143/470		469.8	3.375		
$C_5H_{11}OH$	T_{nbp} = 411.2 K	κ_T = 0.866 GPa^{-1}	α_P = 0.905 kK^{-1}	Ref. 3										
1-Pentanol	1	$9.52000\cdot10^{-2}$	$3.00019\cdot10^{2}$	$-5.62481\cdot10^{-1}$	$-5.98193\cdot10^{-4}$	$1.15744\cdot10^{-6}$			233/550	59(r)				0.149
	3	-2.61129	$4.23375\cdot10$	$-1.49640\cdot10^{2}$	$2.62210\cdot10^{2}$	$-2.23090\cdot10^{2}$	$7.37879\cdot10$		195/585		588.11		270	0.353
	4	-8.98005	3.91624	-9.9081	-2.191				196/588		588.15	3.909		
C_6H_5Br	T_{nbp} = 429.2 K	κ_T = 0.668 GPa^{-1}	α_P = 0.901 kK^{-1}	Ref. 9										
Bromobenzene	1	$9.73880\cdot10^{-2}$	$4.87590\cdot10^{2}$	-1.49313	$1.16283\cdot10^{-3}$				278/358	100(r)				0.043
	3	2.84948	-2.30055	2.31394					298/358		670		484.6	0.002
C_6H_5Cl	T_{nbp} = 404.9 K	κ_T = 0.752 GPa^{-1}	α_P = 0.969 kK^{-1}	Ref. 9				$-3.690\cdot10^{-5}$						
Chlorobenzene	1	1.06896·10–1	$5.95178\cdot10^{2}$	-2.52143	$4.02563\cdot10^{-3}$	$-3.10524\cdot10^{-6}$	$1.06813\cdot10^{-9}$		278/583	50(r)				0.057
	3	$1.17137\cdot10^{-1}$	$1.23712\cdot10$	$-2.84464\cdot10$	$2.94048\cdot10$	$-1.07075\cdot10$			253/630		632.4		365.45	0.008
	5	-7.587	2.26551	-4.09118	$1.70377\cdot10^{-1}$				335/632		632.4	4.519		
$C_6H_5NO_2$	T_{nbp} = 484.0 K	κ_T = 0.503 GPa^{-1}	α_P = 0.826 kK^{-1}	Ref. 10										
Nitrobenzene	1	$9.31940\cdot10^{-2}$	$6.10698\cdot10^{2}$	-1.89353	$1.56305\cdot10^{-3}$				293/358	100				0.005
	3	4.47929	-5.72846	4.39478					273/373		718		362.09	0.027
C_6H_6	T_{nbp} = 353.2 K	κ_T = 0.965 GPa^{-1}	α_P = 1.222 kK^{-1}	Ref. 7										
Benzene	1	$9.36560\cdot10^{-2}$	$3.28066\cdot10^{2}$	$2.79270\cdot10^{-2}$	$-6.54288\cdot10^{-3}$	$1.66249\cdot10^{-5}$	$-1.24527\cdot10^{-8}$		283/499	58(r)				0.079
	3	1.81679	$-1.44828\cdot10^{-1}$	3.65168	-5.85231	3.50125			279/561		562.16		301.6	0.024
	4	-7.01433	1.55256	-1.8479	-3.713				288/562		562.16	4.898		
$C_6H_5NH_2$	T_{nbp} = 457.3 K	κ_T = 0.467 GPa^{-1}	α_P = 0.830 kK^{-1}	Ref. 10										
Aniline	1	$9.43630\cdot10^{-2}$	$7.27479\cdot10^{2}$	-2.43017	$2.23750\cdot10^{-3}$				298/358	100(r)				0.043
	3	2.71043	-3.14236	$1.48381\cdot10$	$-3.89610\cdot10$	$4.62777\cdot10$	$-1.91216\cdot10$		263/699		699		332.6	0.093
C_6H_{12}	T_{nbp} = 353.9 K	κ_T = 1.135 GPa^{-1}	α_P = 1.207 kK^{-1}	Ref. 8										
Cyclohexane	1	$8.51590\cdot10^{-2}$	$6.06614\cdot10^{2}$	-3.64438	$9.15077\cdot10^{-3}$	$-1.15323\cdot10^{-5}$	$5.95736\cdot10^{-9}$		287/523	81(r)				0.087
	3	1.62642	$9.95043\cdot10^{-1}$	$-2.55766\cdot10^{-1}$	$-8.79980\cdot10^{-2}$	$4.74534\cdot10^{-1}$			273/553		553.5		273.25	0.038
	5	-6.96009	1.31328	-2.75683	-2.45491				293/554		553.64	4.075		
C_6H_{14}	T_{nbp} = 341.9 K	κ_T = 1.645 GPa^{-1}	α_P = 1.387 kK^{-1}	Ref. 4				$-4.522\cdot10^{-5}$						
Hexane	1	$1.05863\cdot10^{-1}$	$4.87649\cdot10^{2}$	-2.71631	$5.75951\cdot10^{-3}$	$-5.74075\cdot10^{-6}$	$2.33450\cdot10^{-9}$		223/498	200(r)				0.112
	3	1.59756	1.84266	-1.72631	$4.94308\cdot10^{-1}$	$6.46314\cdot10^{-1}$			183/507		507.9		234	0.124
	4	-7.53998	1.83759	-2.5438	-3.163				178/508		507.9	3.035		
C_6H_{14}	T_{nbp} = 333.4 K	κ_T = 1.793 GPa^{-1}	α_P = 1.430 kK^{-1}	Ref. 8										
2-Methylpentane	1	$8.95370\cdot10^{-2}$	$3.60920\cdot10^{2}$	-1.7065	$2.61411\cdot10^{-3}$	$-1.30788\cdot10^{-6}$			273/473	32(r)				0.272
	3	5.72227	$-2.82931\cdot10$	$7.80587\cdot10$	$-9.08956\cdot10$	$3.90010\cdot10$			273/473		497.5		234.82	0.09
	5	-7.2875	1.29015	-2.97853	-2.17234				240/498		498.1	3.033		
C_6H_{14}	T_{nbp} = 336.4 K	κ_T = 1.714 GPa^{-1}	α_P = 1.350 kK^{-1}	Ref. 8										
3-Methylpentane	1	$8.79150\cdot10^{-2}$	$2.68648\cdot10^{2}$	-1.00644	$9.29350\cdot10^{-4}$				293/473	32(r)				0.086
	3	4.03149	$-1.18248\cdot10$	$2.42559\cdot10$	$-1.83295\cdot10$	4.33405			293/473		504.5		234.82	0.143
	5	-7.27084	1.26113	-2.81741	-2.17642				235/504		504.4	3.122		
C_6H_{14}	T_{nbp} = 331.2 K	κ_T = 1.831 GPa^{-1}	α_P = 1.364 kK^{-1}	Ref. 8				$-4.085\cdot10^{-5}$						
2,3-Dimethylbutane	1	$6.09940\cdot10^{-2}$	$2.88312\cdot10^{2}$	-1.3888	$2.23422\cdot10^{-3}$	$-1.24468\cdot10^{-6}$			208/473	32(r)				0.072
	3	4.23996	$-1.47401\cdot10$	$3.56742\cdot10$	$-3.61641\cdot10$	$1.37668\cdot10$			208/463		499.98		240.72	0.078
	5	-7.2787	1.56349	-3.05387	-1.57752				235/500		500.3	3.146		
1-Hexanol	T_{nbp} = 430.5 K	κ_T = 0.828 GPa^{-1}	α_P = 0.878 kK^{-1}	Ref. 3										
	1	$9.45430\cdot10^{-2}$	$3.56654\cdot10^{2}$	$-9.63938\cdot10^{-1}$	$3.19317\cdot10^{-4}$	$7.27932\cdot10^{-7}$	$-3.68450\cdot10^{-10}$		298/503	400				0.045
	3	$-1.55309\cdot10^{-1}$	$1.77262\cdot10$	$-5.76007\cdot10$	$9.99026\cdot10$	$-8.65883\cdot10$	$2.96281\cdot10$		223/607		610.7		268	0.191
	4	-9.49034	5.13288	$-1.05817\cdot10$	-5.154				226/611		610.7	3.47		
$C_6H_5CH_3$	T_{nbp} = 383.8 K	κ_T = 0.900 GPa^{-1}	α_P = 1.080 kK^{-1}	Ref. 7				$-5.004\cdot10^{-5}$						
Toluene	1	$1.08655\cdot10^{-1}$	$6.58594\cdot10^{2}$	-3.36063	$6.79417\cdot10^{-3}$	$-6.73894\cdot10^{-6}$	$2.80015\cdot10^{-9}$		179/583	50(r)				0.052
	3	2.33057	-3.10784	$1.04839\cdot10$	$-1.28168\cdot10$	6.29458	$-2.64550\cdot10^{-1}$		178/588		591.79		291.59	0.041
	5	-7.28607	1.38091	-2.83433	-2.79168				309/592		591.72	4.106		

Fluids

Mol. form.	Eq.	a_1	a_2	a_3	a_4	a_5	a_6	b_1	T_{min}/T_{max} K	P_{max} MPa	T_c K	P_c MPa	ρ_c kg m^{-3}	RMSD %
C_7H_8O	T_{nbp} = 426.9 K	κ_T = 0.657 GPa^{-1}	α_P = 0.949 kK^{-1}	Ref. 5										
Anisole	1	$9.57370\cdot10^{-2}$	$3.76997\cdot10^{2}$	$-7.76199\cdot10^{-1}$					298/353	196				0.032
	3	2.32023	-1.28589	1.68888					273/353		645.6		335.84	0.031
Cycloheptane	T_{nbp} = 391.6 K	κ_T = 0.904 GPa^{-1}	α_P = 1.062 kK^{-1}	Ref. 8										
	1	$9.09350\cdot10^{-2}$	$3.25145\cdot10^{2}$	$-8.81409\cdot10^{-1}$	$4.29576\cdot10^{-4}$				294/393	40(r)				0.041
	2	$1.06207\cdot10^{3}$	$-8.56800\cdot10^{-1}$						298/353					0.003
C_7H_{16}	T_{nbp} = 371.6 K	κ_T = 1.429 GPa^{-1}	α_P = 1.251 kK^{-1}	Ref. 4				$-7.924\cdot10^{-5}$						
Heptane[b]	1	$8.14360\cdot10^{-2}$	$5.31891\cdot10^{2}$	-3.3849	$9.42976\cdot10^{-3}$	$-1.36183\cdot10^{-5}$	$8.10257\cdot10^{-9}$		198/511	150(r)				0.153
	3	1.33159	3.30092	-4.50961	2.76549				183/538		540.11		236	0.377
	4	-7.77404	1.85614	-2.8298	-3.507				183/540		540.15	2.735		
$C_6H_4(CH_3)_2$	T_{nbp} = 417.6 K	κ_T = 0.806 GPa^{-1}	α_P = 0.944 kK^{-1}	Ref. 7										
o-Xylene	1	$8.02720\cdot10^{-2}$	$4.52200\cdot10^{2}$	-1.8115	$2.45174\cdot10^{-3}$	$-1.15349\cdot10^{-6}$			257/598	51(r)				0.082
	3	1.93903	$6.93214\cdot10^{-1}$	$-5.52518\cdot10^{-1}$	$7.42481\cdot10^{-1}$				248/628		630.3		287.72	0.006
	5	-7.53357	1.40968	-3.10985	-2.85992				337/630		630.25	3.733		
$C_6H_4(CH_3)_2$	T_{nbp} = 412.3 K	κ_T = 0.857 GPa^{-1}	α_P = 0.987 kK^{-1}	Ref. 7										
m-Xylene	1	$8.13550\cdot10^{-2}$	$5.95204\cdot10^{2}$	-3.17132	$6.99286\cdot10^{-3}$	$-7.69418\cdot10^{-6}$	$3.47267\cdot10^{-9}$		230/598	20(r)				0.054
	3	1.96653	$7.70424\cdot10^{-1}$	$-8.14622\cdot10^{-1}$	$9.44937\cdot10^{-1}$				225/613		617.05		282.36	0.007
	5	-7.59222	1.39441	-3.22746	-2.40376				332/617		616.97	3.537		
$C_6H_4(CH_3)_2$	T_{nbp} = 411.5 K	κ_T = 0.894 GPa^{-1}	α_P = 1.003 kK^{-1}	Ref. 7										
p-Xylene	1	$8.45420\cdot10^{-2}$	$4.38870\cdot10^{2}$	-1.77395	$2.41453\cdot10^{-3}$	$-1.13719\cdot10^{-6}$			288/598	51(r)				0.076
	3	2.0168	$6.60504\cdot10^{-1}$	$-7.70676\cdot10^{-1}$	$9.97453\cdot10^{-1}$				286/603		616.2		280.13	0.005
	5	-7.63495	1.50724	-3.19678	-2.7871				331/616		616.15	3.513		
Cyclooctane	T_{nbp} = 422.2 K	κ_T = 0.798 GPa^{-1}	α_P = 0.960 kK^{-1}	Ref. 8										
	1	$8.66040\cdot10^{-2}$	$3.85135\cdot10^{2}$	-1.19668	$9.00890\cdot10^{-4}$				314/394	40				0.026
	3	1.25843	1.53566						293/394		647.2		273.7	0.04
C_8H_{18}	T_{nbp} = 398.8 K	κ_T = 1.255 GPa^{-1}	α_P = 1.158 kK^{-1}	Ref. 4				$-5.741\cdot10^{-5}$						
Octane	1	$1.09943\cdot10^{-1}$	$1.17639\cdot10^{3}$	$-1.02441\cdot10$	$3.73444\cdot10^{-2}$	$-6.42362\cdot10^{-5}$	$4.23381\cdot10^{-8}$		248/393	108(r)				0.069
	3	1.96977	-1.10062	6.36417	-8.69348	4.42005			216/563		568.91		237	0.221
	4	-8.04937	2.03865	-3.312	-3.648				216/569		568.95	2.49		
$C_7H_{15}CH_2OH$	T_{nbp} = 468.3 K	κ_T = 0.743 GPa^{-1}	α_P = 0.844 kK^{-1}	Ref. 3										
1-Octanol	1	$9.39730\cdot10^{-2}$	$5.22871\cdot10^{2}$	-1.99322	$2.56178\cdot10^{-3}$	$-1.13255\cdot10^{-6}$			283/623	79(r)				0.177
	3	-3.53373	$5.48581\cdot10$	$-1.99653\cdot10^{2}$	$3.45797\cdot10^{2}$	$-2.85030\cdot10^{2}$	$9.05445\cdot10$		258/643		652.5		266	0.262
	4	$-1.00144\cdot10$	5.90629	$-1.04026\cdot10$	-9.048				258/653		652.5	2.86		
C_9H_{12}	T_{nbp} = 437.9 K	κ_T = 0.786 GPa^{-1}	α_P = 0.944 kK^{-1}	Ref. 7										
1,3,5-Trimethylbenzene	1	$8.84360\cdot10^{-2}$	$4.00368\cdot10^{2}$	-1.18616	$7.38886\cdot10^{-4}$				238/362	200				0.056
	3	$1.23001\cdot10$	$-3.23171\cdot10$	$3.37700\cdot10$	$-1.06994\cdot10$				238/353		637.25		277.59	0.012
C_9H_{20}	T_{nbp} = 424.0 K	κ_T = 1.125 GPa^{-1}	α_P = 1.088 kK^{-1}	Ref. Ref. 4				$-2.960\cdot10^{-5}$						
Nonane	1	$1.04176\cdot10^{-1}$	$6.60874\cdot10^{2}$	-3.94793	$9.86964\cdot10^{-3}$	$-1.23014\cdot10^{-5}$	$6.27200\cdot10^{-9}$		248/511	65(r)				0.064
	3	1.92778	$9.30219\cdot10^{-1}$	-1.33413	1.39282				223/511		594.9		238	0.078
	4	-8.32886	2.25707	-3.8257	-3.732				220/595		594.9	2.29		
Decane	T_{nbp} = 447.3 K	κ_T = 1.096 GPa^{-1}	α_P = 1.042 kK^{-1}	Ref. 4				$-2.778\cdot10^{-5}$						
	1	$7.76110\cdot10^{-2}$	$5.12895\cdot10^{2}$	-2.98749	$7.58178\cdot10^{-3}$	$-9.71758\cdot10^{-6}$	$5.01670\cdot10^{-9}$		248/503	200(r)				0.079
	3	$3.29139\cdot10^{-1}$	7.36434	-9.9851	5.28361				243/511		617.61		239	0.108
	4	-8.60643	2.44659	-4.2925	-3.908				244/618		617.65	2.105		
$C_{11}H_{24}$	T_{nbp} = 469.1 K	κ_T = 1.078 GPa^{-1}	α_P = 0.998 kK^{-1}	Ref. 4				$-1.105\cdot10^{-5}$						
Undecane	1	$9.36240\cdot10^{-2}$	$9.78153\cdot10^{3}$	$-1.13416\cdot10^{2}$	$4.97036\cdot10^{-1}$	$-9.65143\cdot10^{-4}$	$6.97765\cdot10^{-7}$		258/423	50/500				0.044
	3	$-2.75532\cdot10^{-1}$	9.28503	$-1.19923\cdot10$	5.96957				258/473		638.81		240	0.069
Dodecane	T_{nbp} = 489.5 K	κ_T = 0.998 GPa^{-1}	α_P = 0.977 kK^{-1}	Ref. 4										
	1	9.0545010^{-2}	$9.00366\cdot10^{3}$	$-1.03134\cdot10^{2}$	$4.49084\cdot10^{-1}$	$-8.70464\cdot10^{-4}$	$6.31007\cdot10^{-7}$		268/393	20/442				0.083
	3	$-3.04946\cdot10^{-2}$	8.32535	$-1.08268\cdot10$	5.55178				263/483		658.6		240	0.055
$C_{13}H_{28}$	T_{nbp} = 508.6 K	κ_T = 0.951 GPa^{-1}	α_P = 0.947 kK^{-1}	Ref. 4										
Tridecane	1	$8.79880\cdot10^{-2}$	$3.53129\cdot10^{2}$	-1.17372	$1.00334\cdot10^{-3}$				303/473	500				0.08
	3	2.52779	-1.64864	2.10099					265/473		676		240	0.139
Tetradecane	T_{nbp} = 526.7 K	κ_T = 0.918 GPa^{-1}	α_P = 0.924 kK^{-1}	Ref. 4										
	1	$9.01310\cdot10^{-2}$	$2.44902\cdot10^{2}$	$-4.92510\cdot10^{-1}$					293/358	50/367				0.04
	3	3.50992	-4.09768	3.61064					279/372		693		241	0.05
Pentadecanenitrile	T_{nbp} = 543.8 K	κ_T = 0.835 GPa^{-1}	α_P = 0.882 kK^{-1}	Ref. 4										
	1	$8.85030\cdot10^{-2}$	$4.04942\cdot10^{2}$	-1.37693	$1.25370\cdot10^{-3}$				311/408	69/655				0.048
	3	1.08801	1.82524						288/408		708		241	0.038
$C_{16}H_{34}$	T_{nbp} = 560.0 K	κ_T = 0.866 GPa^{-1}	α_P = 0.889 kK^{-1}	Ref. 4										
Hexadecane	1	$9.05120\cdot10^{-2}$	$-6.87292\cdot10^{2}$	8.07128	$-2.59370\cdot10^{-2}$	$2.60661\cdot10^{-5}$			293/393	10/451				0.047
	3	2.61306	-1.99964	2.40011					293/490		722		241	0.035

[a] Additional term $a_7(1 - T_r)^{(7/3)}$ is included in Eq. 3 with $a_7 = -1.04344\cdot10^{2}$.
[b] Coefficient for additional term in Eq. 1 is $c_1 = -1.513\cdot10^{-7}$.

Fluids

PROPERTIES OF CRYOGENIC FLUIDS

Eric W. Lemmon

These three tables give physical and thermodynamic properties of ten cryogenic fluids, with the use of the following nomenclature. Table 1 has properties at the triple point of the fluids, indicated by a subscript t. Table 2 properties are at the normal boiling point (101.325 kPa), indicated by a subscript b, and Table 3 has properties at the critical point, indicated by the subscript c. The column headings in the tables are as follows.

Column heading	Definition
Name	Name of cryogenic fluid
Mol. form.	Molecular formula of cryogenic fluid
M	Molar mass, in g mol^{-1}
T	Temperature, in K; subscript t indicates at triple point, b indicates at boiling point, c indicates at critical point
P	Pressure, in kPa or MPa as shown; subscript t indicates at triple point, b indicates at boiling point, c indicates at critical point
ρ	Density, in kg m^{-3}; subscript t indicates at triple point, b indicates at boiling point, c indicates at critical point
$\Delta_{vap}H$	Enthalpy of vaporization at the normal boiling point, in units kJ kg^{-1}
$\Delta_{fus}H$	Enthalpy of fusion at the triple point, in units kJ kg^{-1}
C_p	Heat capacity at constant pressure, in units kJ kg^{-1} K^{-1}
u	Speed of sound, in units m s^{-1}

All properties except the heat of fusion are calculated from the REFPROP 10 program, see Ref. 1. *Not all of the listed molar masses match the current values but are those reported along with the equation of state in the references below (mainly those published before 2009).* Temperatures are listed on the ITS-90 scale, except values for oxygen that were obtained from an older equation of state based on the IPTS-68 temperature scale. The properties of hydrogen are given for the para form of the molecule. The triple-point temperature of air is the solidification temperature of the liquid (see Ref. 3), and the boiling-point temperature is given at the bubble point (i.e., the liquid phase temperature at which boiling begins). The dew-point (vapor) properties of air at 101.325 kPa are calculated at a temperature of 81.72 K; the liquid and vapor properties of these two state points are not in equilibrium. The triple-point properties of helium are given at the temperature of the lambda line (where the fluid changes from normal to superfluid helium).

References

1. Lemmon, E.W., Bell, I.H., Huber, M.L., McLinden, M.O. NIST Standard Reference Database 23: Reference Fluid Thermodynamic and Transport Properties-REFPROP, Version 10.0, National Institute of Standards and Technology, Standard Reference Data Program, Gaithersburg, MD, 2018 <http://www.nist.gov/srd/refprop>.
2. Leachman, J.W., Jacobsen, R.T, Penoncello, S.G., and Lemmon, E.W., Fundamental Equations of State for Parahydrogen, Normal Hydrogen, and Orthohydrogen, *J. Phys. Chem. Ref. Data* 38, 721, 2009. <https://doi.org/10.1063/1.3160306>
3. Lemmon, E.W., Jacobsen, R.T, Penoncello, S.G., and Friend, D.G., Thermodynamic Properties of Air and Mixtures of Nitrogen, Argon, and Oxygen from 60 to 2000 K at Pressures to 2000 MPa, *J. Phys. Chem. Ref. Data* 29, 331, 2000. <https://doi.org/10.1063/1.1285884>
4. Lemmon, E.W. and Span, R., Short Fundamental Equations of State for 20 Industrial Fluids, *J. Chem. Eng. Data* 51, 785, 2006. <https://doi.org/10.1021/je050186n>
5. Ortiz-Vega, D.O., Hall, K.R., Holste, J.C., Arp, V.D., Harvey, A.H., and Lemmon, E.W., Final Equation of State, to be submitted to *J. Phys. Chem. Ref. Data* 2019.
6. Schmidt, R. and Wagner, W., A New Form of the Equation of State for Pure Substances and Its Application to Oxygen, *Fluid Phase Equilib.* 19, 175, 1985. <https://doi.org/10.1016/0378-3812(85)87016-3>
7. Setzmann, U. and Wagner, W., A New Equation of State and Tables of Thermodynamic Properties for Methane Covering the Range from the Melting Line to 625 K at Pressures up to 1000 MPa, *J. Phys. Chem. Ref. Data* 20, 1061, 1991. <https://doi.org/10.1063/1.555898>
8. Span, R., Lemmon, E.W., Jacobsen, R.T, Wagner, W., and Yokozeki, A., A Reference Equation of State for the Thermodynamic Properties of Nitrogen for Temperatures from 63.151 to 1000 K and Pressures to 2200 MPa, *J. Phys. Chem. Ref. Data* 29, 1361, 2000. <https://doi.org/10.1063/1.1349047>
9. Tegeler, Ch., Span, R., and Wagner, W., A New Equation of State for Argon Covering the Fluid Region for Temperatures from the Melting Line to 700 K at Pressures up to 1000 MPa, *J. Phys. Chem. Ref. Data* 28, 779, 1999. <https://doi.org/10.1063/1.556037>
10. Thol, M., Beckmüller, R., Weiss, R., Harvey, A.H., Lemmon, E.W., Jacobsen, R.T, and Span, R., Thermodynamic Properties for Neon for Temperatures from the Triple Point to 700 K at Pressures to 700 MPa, to be submitted to *J. Phys. Chem. Ref. Data* 2019.

TABLE 1. Properties of Cryogenic Fluids at Their Triple Point

Name	Mol. form.	M	T_t/K	P_t/kPa	ρ_t(l)/kg m^{-3}	$\Delta_{fus}H$/kJ kg^{-1}
Air	Air	28.9655	59.75	5.265	957.8	
Nitrogen	N_2	28.01348	63.151	12.52	867.2	25.3
Oxygen	O_2	31.9988	54.361	0.1463	1306	13.7
Hydrogen	para-H_2	2.01588	13.8033	7.041	76.98	59.5
Helium	He	4.002602	2.1768	5.039	146.0	
Neon	Ne	20.179	24.5561	43.36	1250	16.8
Argon	Ar	39.948	83.8058	68.89	1417	28.0
Krypton	Kr	83.798	115.775	73.53	2447	16.3
Xenon	Xe	131.293	161.405	81.77	2966	13.8
Methane	CH_4	16.0428	90.6941	11.70	451.5	58.41

TABLE 2. Properties of Cryogenic Fluids at Their Normal Boiling Point (101.325 kPa, 760 mmHg)

Name	Mol. form.	T_b /K	$\Delta_{vap}H$/kJ kg^{-1}	ρ_{liq}/kg m^{-3}	ρ_{vap}/kg m^{-3}	$C_{p,liq}$/kJ kg^{-1} K^{-1}	$C_{p,vap}$/kJ kg^{-1} K^{-1}	u_{liq}/m s^{-1}	u_{vap}/m s^{-1}
Air	Air	78.903	204.8	875.2	4.497	1.933	1.090	865.5	177.1
Nitrogen	N_2	77.355	199.2	806.1	4.612	2.041	1.124	851.4	174.8
Oxygen	O_2	90.188	213.1	1141	4.467	1.699	0.9707	904.3	177.5
Hydrogen	para-H_2	20.271	446.1	70.83	1.339	9.729	12.03	1111	355.0
Helium	He	4.2238	20.56	124.7	16.90		9.556		100.6
Neon	Ne	27.100	85.79	1206	9.582	1.893	1.154	595.4	132.9
Argon	Ar	87.302	161.1	1395	5.773	1.117	0.5658	838.3	170.9
Krypton	Kr	119.730	107.1	2417	8.815	0.5198	0.2748	682.6	137.9
Xenon	Xe	165.050	95.59	2942	10.01	0.3393	0.1751	643.1	129.4
Methane	CH_4	111.667	510.8	422.4	1.816	3.481	2.218	1338	271.5

TABLE 3. Properties of Cryogenic Fluids at Their Critical Point

Name	Mol. form.	T_c/K	P_c/Mpa	ρ_c/kg m^{-3}
Air	Air	132.5306	3.7860	342.6
Nitrogen	N_2	126.192	3.3958	313.3
Oxygen	O_2	154.581	5.0430	436.1
Hydrogen	para-H_2	32.938	1.2858	31.32
Helium	He	5.1953	0.22832	69.58
Neon	Ne	44.400	2.6616	486.3
Argon	Ar	150.687	4.8630	535.6
Krypton	Kr	209.480	5.5250	909.2
Xenon	Xe	289.733	5.8420	1103
Methane	CH_4	190.564	4.5992	162.7

PROPERTIES OF LIQUID HELIUM

The following helium property data were obtained by a critical evaluation of all existing experimental measurements on liquid helium, using a fitting procedure described in the reference. All values refer to liquid helium at saturated vapor pressure; temperatures are on the ITS-90 scale. Several properties show a singularity at the lambda point (2.1768 K). The column definitions are as follows.

Column heading	Definition
T	Temperature, in K
p	Vapor pressure, in kPa
ρ	Density, in g cm^{-3}
C_s	Molar heat capacity, in units J mol^{-1} K^{-1}
$\Delta_{vap}H$	Molar enthalpy of vaporization, in units J mol^{-1}
ε	Relative permittivity (dielectric constant), dimensionless
σ	Surface tension, in units mN m^{-1}
α_V	Cubic expansion coefficient, in units 10^{3}·K^{-1}
η	Viscosity, in units µPa s
λ	Thermal conductivity, in units W cm^{-1} K^{-1}

Reference

Donnelly, R. J., and Barenghi, C. F., *J. Phys. Chem. Ref. Data* 27, 1217, 1998. <https://doi.org/10.1063/1.556028>

Properties of Liquid Helium

T/K	p/kPa	ρ/g cm^{-3}	C_s/J mol^{-1} K^{-1}	$\Delta_{vap}H$/J mol^{-1}	ε	σ/mN m^{-1}	$10^3\alpha_V$/K^{-1}	η/µPa s	λ/W cm^{-1} K^{-1}
0.0		0.1451397	0	59.83	1.057255		0.000		
0.5		0.1451377	0.010	70.24	1.057254	0.3530	0.107		
1.0	0.01558	0.1451183	0.415	80.33	1.057246	0.3471	0.309	3.873	
1.5	0.4715	0.1451646	4.468	89.35	1.057265	0.3322	-2.36	1.346	
2.0	3.130	0.1456217	21.28	93.07	1.057449	0.3021	-12.2	1.468	
2.5	10.23	0.1448402	9.083	92.50	1.057135	0.2623	39.4	3.259	0.1497
3.0	24.05	0.1412269	9.944	94.11	1.055683	0.2161	61.5	3.517	0.1717
3.5	47.05	0.1360736	12.37	92.84	1.053615	0.1626	88.7	3.509	0.1868
4.0	81.62	0.1289745	15.96	87.00	1.050770	0.1095	129	3.319	0.1965
4.5	130.3	0.1188552	21.8	75.86	1.046725	0.0609	211		
5.0	196.0		44.7	47.67		0.0157			

PROPERTIES OF REFRIGERANTS

This table gives physical properties and safe exposure limits for compounds that have been used as working fluids in traditional refrigeration systems or are under consideration as replacements in newer systems. Some are also used as solvents and blowing agents. Many of the compounds listed are believed to be less harmful to the environment than the traditional halocarbon refrigerants.

Column definitions for the table are as follows.

Column heading	Definition
Name	Chemical name of refrigerant; listed alphabetically by name
ASHRAE code	ASHRAE standard refrigerant designation (Ref. 1); see text for discussion
Mol. form.	Molecular formula for refrigerant
CAS Reg. No.	Chemical Abstracts Service Registry Number
t_m	Normal melting point, in °C ("tp" indicates a triple point)
t_b	Normal boiling point, in °C at 101.325 kPa or 760 mmHg ("sp" indicates a sublimation point)
t_c	Critical temperature, in °C
p_c	Critical pressure, in MPa
TWA	Threshold limit value, expressed as the time-weighted average over an 8-hr workday and 40-hr workweek, for safe exposure to the vapor; in units of parts per million (ppm) by volume
STEL	Short-term exposure limit, which should not be exceeded for more than 15 min, in units of parts per million (ppm)

The ASHRAE codes are often prefixed by symbols such as CFC- (for chlorofluorocarbon), HCFC- (for hydrochlorofluorocarbon), or simply R- (for refrigerant). The "R" number assigned to refrigerants is specified by ANSI/ASHRAE Standard 34. This system is most useful for the hydrocarbons and halocarbons with one to three carbons; for such molecules the chemical composition can be determined from the number and vice versa. The first digit on the far right is the number of fluorine atoms in the compound. The second digit from the right is one more than the number of hydrogen atoms. The third digit from the right is one less than the number of carbon atoms; for single-carbon compounds, this digit is omitted. The fourth digit from the right is equal to the number of unsaturated carbon–carbon bonds; for saturated compounds, this digit is omitted. The number of bromine and iodine atoms is indicated, if needed, by appending "Bn" or "In" to the digits specified by the above rules, where "n" is the number of bromine or iodine atoms. All atoms not specified by the above are assumed to be chlorine. Appended lowercase letter(s) designate different isomers. Additional rules are used to specify cyclic compounds, ethers, inorganic fluids (R700- and R7000-series), miscellaneous organic compounds (R600-series), and blends (R400- and R500-series).

Further references and additional data on the critical properties may be found in the tables "Critical Constants of Organic Compounds" and "Critical Constants of Inorganic Compounds" in this section. Details on threshold limits are given in Section 16.

References

1. ASHRAE (2007). ANSI/ASHRAE Standard 34-2007 Designation and Safety Classification of Refrigerants, American Society of Heating, Refrigerating and Air-Conditioning Engineers, Atlanta, GA.
2. *ASHRAE Fundamentals Handbook 2001*, Chapter 19, Refrigerants, American Society of Heating, Refrigerating, and Air-Conditioning Engineers, Atlanta, GA, 2001.
3. Platzer, B., Polt, A., and Mauer, G., *Thermophysical Properties of Refrigerants*, Springer, Berlin, 1990. <https://doi.org/10.1007/978-3-662-02608-3>
4. Sako, T., Sato, M., Nakazawa, N., Oowa, M., Yasumoto, M., Ito, H., and Yamashita, S., *J. Chem. Eng. Data* 41, 802, 1996. <https://doi.org/10.1021/je950327t>
5. Schmidt, J. W., Carrillo-Nava, E., and Moldover, M. R., *Fluid Phase Equilib.* 122, 187, 1996. <https://doi.org/10.1016/0378-3812(96)03044-0>
6. Salvi-Narkhede, M., Wang, B-H., Adcock, J. L., and Van Hook, W. A., *J. Chem. Thermodyn.* 24, 1065, 1992. <https://doi.org/10.1016/S0021-9614(05)80017-5>
7. Fialho, P. S., and Nieto de Castro, C. A., *Int. J. Thermophys.* 21, 385, 2000. <https://doi.org/10.1023/A:1006631512597>
8. Daubert, T. E., Danner, R. P., Sibul, H. M., and Stebbins, C. C., *Physical and Thermodynamic Properties of Pure Compounds: Data Compilation,* extant 2002 (core with supplements), Taylor & Francis, Bristol, PA.
9. McLinden, M.O., Lemmon, E.W., and Huber, M.L., The REFPROP Database for the Thermophysical Properties of Refrigerants, 21st International Congress of Refrigeration, International Institute of Refrigeration, Paper ICR0443, Washington, DC, 2003.
10. Lemmon, E.W., Huber, M.L., and McLinden, M.O., NIST Standard Reference Database 23: Reference Fluid Thermodynamic and Transport Properties - REFPROP, Version 9.0, National Institute of Standards and Technology, Standard Reference Data Program, Gaithersburg, MD, 2010, <http://www.nist.gov/srd/nist23.cfm>.

Properties of Refrigerants

Name	ASHRAE code	Mol. form.	CAS Reg. No.	t_m/°C	t_b/°C	t_c/°C	p_c/MPa	TWA/ ppm	STEL/ ppm
Ammonia	717	H_3N	7664-41-7	-77.65	-33.33	132.41	11.357	25	35
Bis(difluoromethyl) ether	E134	$C_2H_2F_4O$	1691-17-4		5.5	147.1	4.16		
Bromochlorodifluoromethane	12B1	$CBrClF_2$	353-59-3	-159.5	-3.9	155	4.31		
Bromodifluoromethane	22B1	$CHBrF_2$	1511-62-2	-145	-15.6	138.9	5.2		
Bromotrifluoromethane	13B1	$CBrF_3$	75-63-8	-174.4	-57.8	66.91	3.96	1000	
Butane	600	C_4H_{10}	106-97-8	-138.2	-0.5	152.1	3.79		1000
Carbon dioxide	744	CO_2	124-38-9	-56.561 tp	-78.464 sp	30.98	7.375	5000	30000
1-Chloro-1,1-difluoroethane	142b	$C_2H_3ClF_2$	75-68-3	-130.43	-9.12	137.16	4.06		
1-Chloro-2,2-difluoroethane	142	$C_2H_3ClF_2$	338-65-8		35				
Chlorodifluoromethane	22	$CHClF_2$	75-45-6	-157.41	-40.8	96.15	4.98	1000	

Name	ASHRAE code	Mol. form.	CAS Reg. No.	t_m/°C	t_b/°C	t_c/°C	p_c/MPa	TWA/ ppm	STEL/ ppm
Chloroethane	160	C_2H_5Cl	75-00-3	-138.4	12.3	187.2	5.24	100	
Chloroethene	1140	C_2H_3Cl	75-01-4	-153.84	-13.8	152	5.60	1	
1-Chloro-1-fluoroethane	151a	C_2H_4ClF	1615-75-4		16				
1-Chloro-2-fluoroethane	151	C_2H_4ClF	762-50-5		53.1				
Chlorofluoromethane	31	CH_2ClF	593-70-4	-135.1	-9.1	154			
1-Chloro-1,2,2,3,3,4,4-heptafluorocyclobutane	C317	C_4ClF_7	377-41-3	-39.1	25				
Chloromethane	40	CH_3Cl	74-87-3	-97.6	-24.1	143.09	6.72	50	100
Chloropentafluoroethane	115	C_2ClF_5	76-15-3	-99.4	-39.2	79.9	3.141	1000	
1-Chloro-1,1,2,2-tetrafluoroethane	124a	C_2HClF_4	354-25-6	-117	-13	126.8			
1-Chloro-1,2,2,2-tetrafluoroethane	124	C_2HClF_4	2837-89-0	-199.15	-11.96	122.28	3.62		
1-Chloro-1,1,2-trifluoroethane	133b	$C_2H_2ClF_3$	421-04-5		16				
1-Chloro-1,2,2-trifluoroethane	133	$C_2H_2ClF_3$	431-07-2		17.3				
1-Chloro-2,2,2-trifluoroethane	133a	$C_2H_2ClF_3$	75-88-7	-105.5	6.0	151.9	4.02		
Chlorotrifluoroethene	1113	C_2ClF_3	79-38-9	-158.14	-28.3	107	3.95		
Chlorotrifluoromethane	13	$CClF_3$	75-72-9	-181.2	-81.48	28.8	3.89		
Dibromodifluoromethane	12B2	CBr_2F_2	75-61-6	-110.1	22.79	198.2		100	
1,2-Dibromotetrafluoroethane	114B2	$C_2Br_2F_4$	124-73-2	-110.1	47.1	214.7			
1,2-Dichloro-1,1-difluoroethane	132b	$C_2H_2Cl_2F_2$	1649-08-7	-101.2	47				
1,2-Dichloro-1,2-difluoroethane	132	$C_2H_2Cl_2F_2$	431-06-1	-101.2	59.6				
1,1-Dichloro-2,2-difluoroethene	1112a	$C_2Cl_2F_2$	79-35-6	-116	19				
Dichlorodifluoromethane	12	CCl_2F_2	75-71-8	-157.05	-29.8	111.8	4.12	1000	
1,1-Dichloroethane	150a	$C_2H_4Cl_2$	75-34-3	-96.93	56.3	250.3	5.1	100	
1,2-Dichloroethane	150	$C_2H_4Cl_2$	107-06-2	-35.6	83.4	288.4	5.4	10	
trans-1,2-Dichloroethene	1130	$C_2H_2Cl_2$	156-60-5	-49.8	47.64	242.4	5.3	200	
1,1-Dichloro-1-fluoroethane	141b	$C_2H_3Cl_2F$	1717-00-6	-103.5	32.05	204.2	4.20		
1,2-Dichloro-1-fluoroethane	141	$C_2H_3Cl_2F$	430-57-9	-60	74				
Dichlorofluoromethane	21	$CHCl_2F$	75-43-4	-130.35	8.9	178.5	5.20	10	
1,2-Dichloro-1,2,3,3,4,4-hexafluorocyclobutane	C316	$C_4Cl_2F_6$	356-18-3	-24.2	59.5	224			
1,3-Dichloro-1,1,2,2,3,3-hexafluoropropane	216ca	$C_3Cl_2F_6$	662-01-1	-125.4	35.7	180			
Dichloromethane	30	CH_2Cl_2	75-09-2	-94.9	39.8	234.9	6.35	50	
1,1-Dichloro-1,2,2,2-tetrafluoroethane	114a	$C_2Cl_2F_4$	374-07-2	-56.6	3	145.5	3.31		
1,2-Dichloro-1,1,2,2-tetrafluoroethane	114	$C_2Cl_2F_4$	76-14-2	-92.52	3.6	145.59	3.25	1000	
1,2-Dichloro-1,1,2-trifluoroethane	123a	$C_2HCl_2F_3$	354-23-4	-78	30.0	188.5	3.77		
2,2-Dichloro-1,1,1-trifluoroethane	123	$C_2HCl_2F_3$	306-83-2	-107.15	27.8	183.7	3.67		
Diethyl ether	610	$C_4H_{10}O$	60-29-7	-116.22	34.4	193.7	3.64	400	500
1,1-Difluoroethane	152a	$C_2H_4F_2$	75-37-6	-118.59	-24.02	113.3	4.52		
1,2-Difluoroethane	152	$C_2H_4F_2$	624-72-6		26				
1,1-Difluoroethene	1132a	$C_2H_2F_2$	75-38-7	-144	-85.5	29.8	4.48	500	
Difluoromethane	32	CH_2F_2	75-10-5	-136.8	-51.65	78.13	5.79		
2-(Difluoromethoxy)-1,1,1-trifluoroethane	245mf	$C_3H_3F_5O$	1885-48-9		29.2	171.8	3.43		
Difluoromethyl 1,1,2-trifluoroethyl ether	245qc	$C_3H_3F_5O$	69948-24-9		43.1				
Dimethyl ether	E170	C_2H_6O	115-10-6	-141.49	-24.8	127	5.31		
Ethane	170	C_2H_6	74-84-0	-182.77	-88.6	32.21	4.88		
Ethylene	1150	C_2H_4	74-85-1	-169.15	-103.8	9.20	5.06	200	
Fluoroethane	161	C_2H_5F	353-36-6	-143.2	-37.7	102.1	5.02		
Fluoroethene	1141	C_2H_3F	75-02-5	-160.5	-72	54.8		1	
Fluoromethane	41	CH_3F	593-53-3	-143.33	-78.31	44.27	5.88		
1,1,1,2,3,3,3-Heptafluoropropane	227ea	C_3HF_7	431-89-0	-126.80	-16.34	101.9	2.93		
Hexachloroethane	110	C_2Cl_6	67-72-1	186.8	184.7 sp	422		1	
Hexafluoroethane	116	C_2F_6	76-16-4	-100.02	-78.1	19.8	3.03		
1,1,1,2,3,3-Hexafluoropropane	236ea	$C_3H_2F_6$	431-63-0		6.20	139.25	3.42		
1,1,1,3,3,3-Hexafluoropropane	236fa	$C_3H_2F_6$	690-39-1	-93.63	-1.4	124.92	3.18		
Isobutane	600a	C_4H_{10}	75-28-5	-159.38	-11.7	134.69	3.64		1000
Methane	50	CH_4	74-82-8	-182.475	-161.5	-82.59	4.60		
Methyl formate	611	$C_2H_4O_2$	107-31-3	-99.7	31.6	214.01	6.01	50	100
Methyl pentafluoroethyl ether	245mc	$C_3H_3F_5O$	22410-44-2		5.6	133.66	2.89		
Methyl 1,1,2,2-tetrafluoroethyl ether	254pc	$C_3H_4F_4O$	425-88-7	-107	37.1				
Methyl trifluoromethyl ether	143m	$C_2H_3F_3O$	421-14-7	-149.2	-25.2	104.77	3.64		
Pentachloroethane	120	C_2HCl_5	76-01-7	-29.0	161				
Pentachlorofluoroethane	111	C_2Cl_5F	354-56-3	101	138				

Name	ASHRAE code	Mol. form.	CAS Reg. No.	t_m/°C	t_b/°C	t_c/°C	p_c/MPa	TWA/ ppm	STEL/ ppm
Pentafluoroethane	125	C_2HF_5	354-33-6	-100.63	-48.09	66.1	3.63		
1,1,1,2,2-Pentafluoropropane	245cb	$C_3H_3F_5$	1814-88-6		-18.0	107	3.14		
1,1,1,3,3-Pentafluoropropane	245fa	$C_3H_3F_5$	460-73-1	-102.10	15.14	154.05	3.66		
1,1,2,2,3-Pentafluoropropane	245ca	$C_3H_3F_5$	679-86-7		25.13	174.42	3.96		
Perfluorocyclobutane	C318	C_4F_8	115-25-3	-40.16	-5.91	115.3	2.78		
Perfluoroisopropyl methyl ether	347mmy	$C_4H_3F_7O$	22052-84-2		29	160.15	2.55		
Perfluoropropane	218	C_3F_8	76-19-7	-147.7	-36.8	71.88	2.67		
Perfluoropropyl methyl ether	347mcc	$C_4H_3F_7O$	375-03-1		34	164.6	2.48		
Propane	290	C_3H_8	74-98-6	-187.62	-42.11	96.8	4.25		
Propene	1270	C_3H_6	115-07-1	-185.19	-47.6	91.8	4.59	500	
Sulfur dioxide	764	O_2S	7446-09-5	-75.45	-10.02	157.49	7.884		0.25
1,1,1,2-Tetrachloro-2,2-difluoroethane	112a	$C_2Cl_4F_2$	76-11-9	41.0	96			100	
1,1,2,2-Tetrachloro-1,2-difluoroethane	112	$C_2Cl_4F_2$	76-12-0	26.54	92.83	278		50	
1,1,2,2-Tetrachloroethane	130	$C_2H_2Cl_4$	79-34-5	-42.4	146.0	388		1	
1,1,1,2-Tetrachloro-2-fluoroethane	121a	C_2HCl_4F	354-11-0	-95.3	117.1				
1,1,2,2-Tetrachloro-1-fluoroethane	121	C_2HCl_4F	354-14-3	-82.6	116.7				
Tetrachloromethane	10	CCl_4	56-23-5	-22.8	76.7	283.4	4.57	5	10
1,1,1,2-Tetrafluoroethane	134a	$C_2H_2F_4$	811-97-2	-103.30	-26.1	101.1	4.06		
1,1,2,2-Tetrafluoroethane	134	$C_2H_2F_4$	359-35-3	-89	-20	118.60	4.61		
Tetrafluoroethene	1114	C_2F_4	116-14-3	-131.14	-76	34	3.94	2	
1,2,2,2-Tetrafluoroethyl difluoromethyl ether	236me	$C_3H_2F_6O$	57041-67-5		23	155.80	3.05		
Tetrafluoromethane	14	CF_4	75-73-0	-183.58	-127.9	-45.61	3.73		
2,3,3,3-Tetrafluoropropene	1234yf	$C_3H_2F_4$	754-12-1		-29.5	94.8			
1,1,1-Trichloro-2,2-difluoroethane	122b	$C_2HCl_3F_2$	354-12-1		75				
1,2,2-Trichloro-1,1-difluoroethane	122	$C_2HCl_3F_2$	354-21-2	-150.1	71.9				
1,2,2-Trichloro-1,2-difluoroethane	122a	$C_2HCl_3F_2$	354-15-4	-174	73.2				
1,1,1-Trichloroethane	140a	$C_2H_3Cl_3$	71-55-6	-30	74.02	272		350	450
1,1,2-Trichloroethane	140	$C_2H_3Cl_3$	79-00-5	-36.3	113	329		10	
Trichloroethene	1120	C_2HCl_3	79-01-6	-84.7	86.8	271.1		10	25
1,1,2-Trichloro-2-fluoroethane	131	$C_2H_2Cl_3F$	359-28-4		102.4				
Trichlorofluoromethane	11	CCl_3F	75-69-4	-110.44	23.7	198	4.40		1000
Trichloromethane	20	$CHCl_3$	67-66-3	-63.3	61.2	262.9	5.5	10	
1,1,1-Trichloro-2,2,2-trifluoroethane	113a	$C_2Cl_3F_3$	354-58-5	14.37	46	209.8			
1,1,2-Trichloro-1,2,2-trifluoroethane	113	$C_2Cl_3F_3$	76-13-1		47.6	214.3	3.40	1000	1250
1,1,1-Trifluoroethane	143a	$C_2H_3F_3$	420-46-2	-111.6	-47.2	72.74	3.77		
1,1,2-Trifluoroethane	143	$C_2H_3F_3$	430-66-0	-84	3.5	156.7			
2,2,2-Trifluoroethyl methyl ether	E143a	$C_3H_5F_3O$	460-43-5		31.62	175.83	3.51		
Trifluoromethane	23	CHF_3	75-46-7	-155.18	-82.0	25.85	4.82		
Trifluoromethyl difluoromethyl ether	E125	C_2HF_5O	3822-68-2	-157	-35.0	81.34	3.36		
Trifluoromethyl 1,1,2,2-tetrafluoroethyl ether	227ca2	C_3HF_7O	2356-61-8	-141	-3.3	114.7	2.65		
Trifluoromethyl 1,2,2,2-tetrafluoroethyl ether	227me	C_3HF_7O	2356-62-9		-9.6	104.11	2.62		

PROPERTIES OF GAS CLATHRATE HYDRATES

Carolyn A. Koh, M. Naveed Khan, and E. Dendy Sloan

Gas clathrate hydrates (also known as gas hydrates) are crystalline inclusion compounds composed of hydrogen-bonded water cavities (host) that encage small gas (guest) molecules in cavities. Generally, a maximum of one guest molecule occupies each water cavity. Typical guest molecules that form gas hydrates are methane, ethane, carbon dioxide, and propane (see gas hydrate phase equilibria data in Table 2). The three important structures of gas hydrates are: cubic structure I (sI) with two small and six large cavities (cages); cubic structure II (sII) with sixteen small and eight large cavities; and hexagonal structure (sH) with three small. two medium, and one large cavities in the unit cell. The structural and physical properties of each type are given in Tables 1a and 1b. Data have been taken from the references indicated.

TABLE 1a. Gas Hydrate Structural Properties (All values are from Ref. 1 unless noted otherwise)

sI Small	sI Large	sII Small	sII Large	sH Small	sH Medium	sH Large
Crystal system						
Cubic	Cubic	Cubic	Cubic	Hexagonal	Hexagonal	Hexagonal
Space group						
Pm3n (No. 223)[b]	Pm3n (No. 223)[b]	Fd3m (No. 227)[b]	Fd3m (No. 227)[b]	P6/mmm (No. 191)[b]	P6/mmm (No. 191)[b]	P6/mmm (No. 191)[b]
Lattice description						
Primitive	Primitive	Face centered	Face centered	Hexagonal	Hexagonal	Hexagonal
Lattice parameters[a]						
a = 12 Å $\alpha = \beta = \gamma = 90°$	a = 12 Å $\alpha = \beta = \gamma = 90°$	a = 17.3 Å $\alpha = \beta = \gamma = 90°$	a = 17.3 Å $\alpha = \beta = \gamma = 90°$	a = 12.2 Å, c = 10.1 Å $\alpha = \beta = 90°$, $\gamma = 120°$	a = 12.2 Å, c = 10.1 Å $\alpha = \beta = 90°$, $\gamma = 120°$	a = 12.2 Å, c = 10.1 Å $\alpha = \beta = 90°$, $\gamma = 120°$
Ideal unit cell formula						
$6(5^{12}6^2)\cdot 2(5^{12})\cdot 46H_2O$	$6(5^{12}6^2)\cdot 2(5^{12})\cdot 46H_2O$	$8(5^{12}6^4)\cdot 16(5^{12})\cdot 136H_2O$	$8(5^{12}6^4)\cdot 16(5^{12})\cdot 136H_2O$	$1(5^{12}6^8)\cdot 3(5^{12})\cdot 2(4^35^66^3)\cdot 34H_2O$	$1(5^{12}6^8)\cdot 3(5^{12})\cdot 2(4^35^66^3)\cdot 34H_2O$	$1(5^{12}6^8)\cdot 3(5^{12})\cdot 2(4^35^66^3)\cdot 34H_2O$
Cavity						
Small	Large	Small	Large	Small	Medium	Large
Description						
5^{12}	$5^{12}6^2$	5^{12}	$5^{12}6^4$	5^{12}	$4^35^66^3$	$5^{12}6^8$
Number of cavities/unit cell						
2	6	16	8	3	2	1
Average cavity radius (Å)[c]						
3.95	4.33	3.91	4.73	3.94[d]	4.04[d]	5.79[d]
H_2O molecules/cavity[e]						
20	24	20	28	20	20	36

[a] Lattice parameters are a function of temperature, pressure, and guest composition. Typical average values given.
[b] Space group reference numbers from the International Tables of Crystallography.
[c] The average cavity radius will vary with temperature, pressure, and guest composition.
[d] From the atomic coordinates measured using single crystal x-ray diffraction on 2,2-dimethylpentane·$5(Xe,H_2S)\cdot 34H_2O$ at 173 K (Ref. 2). The Rietveld refinement package, GSAS was used to determine the atomic distances for each cage oxygen to the cage center.
[e] Number of oxygen atoms at the periphery of each cavity.

TABLE 1b. Physical Properties of sI, sII Hydrates Compared to Ice Ih (All values are from Refs. 1, 3, 4, and 5 unless noted otherwise)

Property	Ice Ih	sI	sII
Dielectric constant at 273 K	94	~58	~58
H_2O reorientation time at 273 K (μs)	21	~10	~10
H_2O diffusion jump time (μs)	2.7	>200	>200
Isothermal Young's modulus at 268 K (10^9 Pa)	9.5	8.4[est]	8.2[est]
Poisson's ratio	0.3301[f]	0.31403[f]	0.31119[f]
Bulk modulus (GPa)	9.097[f]	8.762[f]	8.482[f]
Shear modulus (GPa)	3.488[f]	3.574[f]	3.6663[f]
Compressional velocity, V_p(m/s)	3870.1[f]	3778[f]	3821.8[f]
Shear velocity, V_s(m/s)	1949[f]	1963.6	2001.14[g]
Linear thermal expansion at 200 K (K^{-1})	56×10^{-6}	77×10^{-6}	52×10^{-6}
Thermal conductivity (W m^{-1} K^{-1}) at 263 K	2.18±0.01[h]	0.51±0.01[h]	0.50±0.01[h]
Adiabatic bulk compression at 273 K (GPa)	12	14[est]	14[est]
Heat capacity (J kg^{-1} K^{-1})	1700±200[h]	2080	2130±40[h]
Refractive index (632.8 nm, −3 °C)	1.3082 (Ref. 9)	1.346 (Ref. 9)	1.350 (Ref. 9)
Density (g/cm^3)	0.91[j]	0.94	1.291[k]

[f] At 253–268 K, 22.4–32.8 MPa (ice, Ih), 258–288 K, 27.1–62.1 MPa (CH_4, sI), 258–288 K, 30.5–91.6 MPa (CH_4–C_2H_6, sII), Ref. 6.
[g] At 258–288 K, 26.6–62.1 MPa, Ref. 7.
[h] At 248–268 K (ice, Ih), 253–288 K (CH_4, sI), 248–265.5 K (THF, sII), Ref. 8.
[j] Fractional occupancy (calculated from a theoretical model) in small (S) and large (L) cavities: sI = CH_4: 0.87 (S) and CH_4: 0.973 (L); sII = CH_4: 0.672 (S), 0.057 (L); C_2H_6: 0.096 (L) only; C_3H_8: 0.84 (L) only.
[k] Calculated for 2,2-dimethylpentane·$5(Xe,H_2S)\cdot 34H_2O$, Ref. 2; est = estimated.

References for Table 1

1. Sloan, E .D. and Koh, C. A., *Clathrate Hydrates of Natural Gases, Third Edition*, CRC Press, 2008.
2. Udachin, K. A., Ratcliffe, C. I., Enright, G. D., and Ripmeester, J. A., *Supramol. Chem.*, 8, 173, 1997. <https://doi.org/10.1080/10610279708034933>
3. Davidson, D. W., *Natural Gas Hydrates* (Cox, J. L., Ed.) Butterworths, Boston, 1, 1983.
4. Davidson, D. W., Handa, Y. P., and Ripmeester, J. A., *J. Phys. Chem.*, 90, 6549, 1986. <https://doi.org/10.1021/j100282a026>
5. Ripmeester, J. A., Ratcliffe, C. I., Klug, D. D., and Tse, J. S., in *Proc. First International Conference on Natural Gas Hydrates*, (Sloan, E.D., Happel, J., and Hnatow, M.A., Eds.) *Annals of the New York Academy of Sciences*, 715, 161, 1994. <https://doi.org/10.1111/j.1749-6632.1994.tb38832.x>
6. Helgerud, M. B., Circone, S., Stern, L., Kirby, S., and Lorenson, T. D., in *Proc. Fourth International Conference on Gas Hydrates*, p. 716, Yokohama, May 19–23, 2002.
7. Helgerud, M .B., Waite, W. F., Kirby, S. H., and Nur, A., *Can. J. Phys.*, 81, 47, 2003. <https://doi.org/10.1139/p03-016>
8. Waite, W. F., Gilbert, L. Y., Winters, W. J., and Mason, D. H., in *Proc. Fifth International Conference on Gas Hydrates*, Trondheim, Norway, June 13–16, Paper 5042, 2005.
9. Bylov, M. and Rasmussen, P., *Chem. Eng. Sci.*, 52, 3295, 1997. <https://doi.org/10.1016/S0009-2509(97)00144-9>

Table 2: Phase Equilibria Data of Gas Clathrate Hydrates

This table gives measured phase equilibria data of pure sI and sII gas clathrate hydrates (see Table 1 for gas hydrate structure and physical property data). The temperature and pressure conditions at which gas hydrates are stable are listed here for typical guest molecules (Tables 2a–2g). For example, data for methane hydrate (Table 2a) show that at 277.1 K, methane hydrate will dissociate at pressures below 3.81 MPa. In addition to small hydrocarbons (methane, ethane, and propane) and carbon dioxide, the nitrogen (N_2), xenon, and hydrogen sulfide hydrate systems have been measured extensively, and the hydrate phase equilibria data are available at temperatures and pressures as low as about 211.2 K and 6.5 kPa to as high as about 340.15 K and 1500 kPa.

TABLE 2a. Methane Hydrate (Ref. 1)

T/K	*P*/MPa
Ice-Hydrate-Vapor **(Ref. 2)**	
262.4	1.79
264.2	1.9
266.5	2.08
268.6	2.22
270.9	2.39
Liquid Water-Hydrate-Vapor **(Ref. 2)**	
273.7	2.77
274.3	2.9
275.4	3.24
275.9	3.42
275.9	3.43
277.1	3.81
279.3	4.77
280.4	5.35
280.9	5.71
281.5	6.06
282.6	6.77
284.3	8.12
285.9	9.78
Liquid Water-Hydrate-Vapor **(Ref. 3)**	
295.7	33.99
295.9	35.3
301.0	64.81
302.0	77.5
Liquid Water-Hydrate-Vapor **(Ref. 4)**	
285.7	9.62
286.3	10.31
286.1	10.1
285.7	9.62
289.0	13.96
292.1	21.13
295.9	34.75
298.7	48.68
300.9	62.4
301.6	68.09
Liquid Water-Hydrate-Vapor **(Ref. 5)**	
275.4	2.87
276.2	3.37
277.2	3.9
278.2	4.5
279.2	4.9
281.2	6.1
Ice-Hydrate-Vapor **(Ref. 6)**	
190.2	0.08251
198.2	0.1314
208.2	0.222
218.2	0.3571
243.2	0.955
262.4	1.798

TABLE 2b. Ethane Hydrate

T/K	*P*/kPa
Ice-Hydrate-Vapor **(Ref. 7)**	
260.8	294
260.9	290
269.3	441
Liquid Water-Hydrate-Vapor **(Ref. 7)**	
273.4	545
275.4	669
277.6	876
279.1	1048
281.1	1317
282.8	1641
284.4	2137
284.6	2055
285.8	2537
287.0	3054
Liquid Water-Hydrate-Liquid Ethane **(Ref. 7)**	
287.7	4909
287.8	3413
287.8	4289
288.1	3716
288.1	6840
288.2	4944
288.2	5082
288.3	4358
288.4	6840
Ice-Hydrate-Vapor **(Ref. 1)**	
263.6	313
266.5	357
269.3	405
272.0	457
Liquid Water-Hydrate-Vapor **(Ref. 2)**	
273.7	510
273.7	503
274.8	579
275.9	662
277.6	814
278.7	931
279.3	1007
279.8	1083
280.4	1165
280.4	1165
280.9	1255
281.5	1345
282.1	1448
282.6	1558
283.2	1689
284.3	1986
285.4	2303
285.4	2310
286.5	2730
Liquid Water-Hydrate-Vapor **(Ref. 8)**	
277.5	780
278.1	840
279.9	1040
281.5	1380
283.3	1660
284.5	2100
286.5	2620

TABLE 2c. Propane Hydrate

T/K	P/kPa
***Ice-Hydrate-Vapor* (Ref. 1)**	
261.2	100
264.2	115
267.4	132
267.6	135
269.8	149
272.2	167
272.9	172
***Liquid Water-Hydrate-Vapor* (Ref. 2)**	
273.7	183
274.8	232
275.4	270
275.9	301
277.1	386
***Ice-Hydrate-Vapor* (Ref. 9)**	
247.9	48.2
251.4	58.3
251.6	58.3
255.4	69.6
258.2	81.1
260.8	90.5
260.9	94.5
262.1	99.4
***Liquid Water-Hydrate-Vapor* (Ref. 10)**	
276.77	0.368
277.01	0.377
277.22	0.405
277.36	0.425
277.44	0.433
277.87	0.473
278.01	0.527
278.55	0.547
***Liquid Water-Hydrate-Liquid Propane* (Ref. 10)**	
278.71	0.643
278.75	0.893
278.75	1.393
278.75	1.891
278.78	1.893
278.80	2.391
278.80	2.891
278.79	2.893
278.75	3.891
278.77	3.391
278.81	4.391
278.79	5.892
278.86	6.392
278.88	6.892
278.80	8.393
278.84	8.893
278.89	9.893

TABLE 2d. Carbon Dioxide Hydrate (Ref. 1)

T/K	P/MPa
***Liquid Water-Hydrate-Vapor* (Ref. 11)**	
279.6	2.74
282.1	4.01
282.8	4.36
***Liquid Water-Hydrate-Liquid Carbon Dioxide* (Ref. 11)**	
282.9	5.03
282.9	5.62
283.1	6.47
283.2	9.01
283.6	11.98
283.9	14.36
***Liquid Water-Hydrate-Vapor* (Ref. 12)**	
276.52	1.82
277.85	1.95
278.52	2.21
279.49	2.62
280.44	2.88
281.49	3.35
281.97	3.68
282.00	3.69
282.45	3.85
282.50	4.01
***Liquid Water-Hydrate-Liquid Carbon Dioxide* (Ref. 12)**	
283.33	5.97
283.36	7.35

TABLE 2e. Nitrogen (N_2) Hydrate

T/K	P/kPa
***Liquid Water-Hydrate-Vapor* (Ref. 13)**	
272.0	14480
272.6	15300
272.8	15910
273.8	15910
273.2	16010
273.2	16310
273.4	16620
274.0	17530
274.2	17730
274.8	19150
274.8	19250
275.2	19660
275.6	20670
275.8	21580
276.2	22390
276.6	23100
277.2	24830
278.2	27360
278.2	27970
278.6	28270
279.2	29890
279.2	30300
280.2	33940
281.2	37490
281.6	38610
282.2	41440
283.2	45900
284.2	50660
284.6	52290
285.2	55430
286.2	61400
287.2	67790
287.8	71230
288.4	74580
289.2	81470
290.2	89370
290.6	92210
291.0	95860
***Liquid Water-Hydrate-Vapor* (Ref. 14)**	
277.6	24930
282.2	36820
286.7	63710
291.6	101980
293.0	115490
294.3	128800
296.6	153480
297.7	169270
298.8	192370
299.7	207780
300.6	219600
302.6	268320
304.7	317650
305.5	328890
***Liquid Water-Hydrate-Vapor* (Ref. 15)**	
273.2	16270
273.7	17130
274.9	19130
276.5	23690
277.4	25200
278.6	28610
279.3	30270
281.1	35160

T/K	*P*/kPa
***Liquid Water-Hydrate-Vapor* (Ref. 16)**	
285.63	55000
287.87	59000
289.47	78000
290.80	88000
291.96	101000
292.90	113000
294.60	127000
295.61	139000
296.62	152000
297.32	162000
297.86	169000
298.47	180000
299.31	195000
299.92	206000
300.49	219000
301.86	240000
302.64	256000
303.08	265000
303.82	280000
304.23	294000
304.43	304000
304.56	306000
305.46	324000
305.85	331000
306.26	342000
306.74	354000
307.21	373000
307.50	383000
308.09	398000
308.57	412000
308.82	420000
309.21	431000
309.43	439000

TABLE 2f. Xenon Hydrate

T/K	*P*/kPa
***Liquid Water-Hydrate-Vapor* (Ref. 17)**	
211.2	6.5
223.0	13.5
229.4	19.4
238.2	30.8
253.3	64.8
268.2	121.3
***Liquid Water-Hydrate-Vapor* (Ref. 18)**	
273.15	153.976
275.15	179.301
277.15	220.834
278.15	252.237
279.15	283.64
281.15	343.407
283.15	418.369
285.15	497.383
***Liquid Water-Hydrate-Vapor* (Ref. 19)**	
286.65	675
295.35	1780
300.15	2380
300.65	2680
302.15	3060
303.55	3560
305.45	4060
307.15	5100
307.35	5400
308.45	6100
309.05	6600
309.75	6900
310.15	7800
310.35	8620
310.55	9060
311.25	11200
311.75	13400
311.95	13900
312.15	16200
312.55	17700
312.95	20000
313.15	17000
316.65	37500
319.15	60000
320.45	63000
323.75	90000
327.15	125000
328.75	145000
330.15	159000
331.85	184000
333.15	195000
333.75	215000
334.95	220000
335.65	234000
336.85	250000
337.15	258000
349.45	549000
349.15	548000
348.95	555800
349.15	560000
349.65	579600
349.65	587100
350.95	600000
313.45	21900
313.95	25000
314.65	28000
315.15	30000
316.75	35000
317.75	45000
317.65	50000
319.35	57500
322.25	75000
324.95	100000
326.15	113000
327.05	118000
327.35	125000
328.05	130000
328.35	137000
329.65	140000
329.65	150000
331.35	159300
331.15	165000
331.15	170500
331.45	175000
331.65	180000
332.55	182500
333.35	190600
334.15	200000
333.95	207000
334.55	210000
334.85	216000
335.35	217400
335.35	220000
335.55	228000
335.95	230000
335.65	234000
336.95	240000
336.65	241300
337.65	250000
336.95	251700
337.35	255000
337.95	264400
338.95	287500
338.65	303000
340.15	308000
339.65	310000
351.65	645000
351.85	660000
352.15	684000
352.35	700000
352.35	725000
352.65	730000
352.65	750000
352.25	760000
352.75	780000
352.65	800000
352.65	820000
352.65	850000
351.85	875000
352.65	880000
352.35	900000
352.65	920000
351.65	937000
352.15	950000
351.35	965000
350.95	970000
351.95	980000
350.65	1000000
351.65	1000000
350.35	1016000
351.25	1025000
350.65	1070000
349.85	1080000
349.05	1100000
348.75	1133000
348.05	1162000

T/K	P/kPa
348.45	1200000
347.25	1230000
346.15	1250000
345.85	1270000
345.55	1290000
344.95	1300000
344.35	1325000
344.45	1340000
343.85	1355000
343.25	1377000
343.15	1390000
342.95	1400000
342.55	1408000
297.90	426000
314.00	479000
316.40	493000
320.20	502000
324.40	520000
331.60	544000
331.70	548000
334.60	561000
339.65	315000
340.65	333000
341.65	347000
341.95	355400
342.95	376200
343.35	392000
343.65	395600
343.15	1420000
342.15	1425000
341.85	1445000
340.85	1463000
340.95	1475000
341.15	1490000
340.15	1500000
339.85	1500000
343.15	398000
344.65	409000
344.35	415700
344.95	421700
344.95	448500
346.15	453000
346.35	474500
345.95	488000
346.95	488700
347.65	498000
346.65	499000
346.65	503000
347.95	506600
348.15	523000
348.15	529700
348.65	540000
344.7	591000
353.2	624000
354.4	629000
356.6	641000

Liquid Water-Hydrate-Vapor **(Ref. 20)**

T/K	P/kPa
283.36	5140
285.37	5360
287.37	5580
288.85	5750
289.36	5810
289.66	5850
289.76	5860
290.46	930
292.47	1110
294.50	1340
295.91	1540
296.91	1710
298.92	2090
300.92	2560
302.92	3160
304.90	3960
306.13	4520
307.93	5710
308.76	6510
309.63	7560
310.09	8510
310.99	11560
311.40	13100
312.11	16180
313.23	21530
314.02	25580
314.87	30360
315.84	36230
317.60	47400
318.96	56930
320.56	69020

TABLE 2g. Hydrogen Sulfide Hydrate

Ice-Hydrate-Vapor **(Ref. 21)**

T/K	P/kPa
250.5	34
255.4	44
258.2	50
260.9	57
263.7	64
265.3	69
266.5	72
269.3	81
272.1	90
272.8	93

Liquid Water-Hydrate-Vapor **(Ref. 21)**

T/K	P/kPa
272.8	93[Q1]
277.6	157
283.2	280
285.2	345
288.7	499
291.8	689
294.3	890
295.7	1034
298.5	1379
299.8	1596
300.5	1724
302.1	2068
302.7	2239[Q2]

Liquid Water-Hydrate-Liquid Hydrogen Sulfide **(Ref. 21)**

T/K	P/kPa
302.7	2239
302.8	3447
303.1	6895
303.2	7826
303.4	10342
303.7	13790
303.7	14190
304.0	17237
304.3	20685
304.3	20954
304.6	24132
304.8	27580
304.8	27842
305.1	31027
305.3	34475
305.4	35068

Liquid Water-Hydrate-Vapor **(Ref. 22)**

T/K	P/kPa
283.2	310
291.2	710
299.7	1496
302.7	2241

Liquid Water-Hydrate-Liquid Hydrogen Sulfide **(Ref. 23)**

T/K	P/kPa
298.0	2030
299.0	2050
299.1	2060
299.4	2080
299.6	2080
299.7	2090
299.8	2090
300.1	2090
300.4	2120
300.8	2110
301.0	2130
301.1	2150
301.2	2170
301.2	2150
301.4	2170
301.4	2180
301.6	2200
301.6	2220
302.6	2240

Liquid Water-Hydrate-Vapor **(Ref. 23)**

T/K	P/kPa
298.6	1610
298.8	1620
299.0	1710

T/K	P/kPa
299.1	1680
299.2	1700
299.4	1700
299.8	1750
299.8	1770
300.1	1810
300.2	1850
300.4	1870
300.7	1970
300.8	2070
Liquid Water-Hydrate-Vapor **(Ref. 24)**	
277.7	164
279.0	188
280.5	218
281.7	247
283.1	283
284.6	332
286.1	380
287.4	458
288.6	514
290.0	578
291.3	672
292.5	764
294.0	865
295.3	988
296.6	1131
297.5	1237
298.8	1425
300.4	1692
301.3	1861

T/K	P/kPa
Liquid Water-Hydrate-Vapor **(Ref. 25)**	
273.6	107.89
274.1	112.08
274.6	117.72
275.1	123.81
275.6	132.39
276.1	137.31
276.6	146.44
277.1	151.72
277.6	161.49
278.1	167.42
278.6	177.48
279.1	185.06
279.6	195.91
280.1	204.31
280.6	216.08
281.1	226.05
281.6	238.58
282.1	249.94
282.6	264.69
283.1	276.82
283.6	291.62
284.1	305.08
284.6	322.07
285.1	338.16
285.6	356.2
286.1	372.71
286.6	393.14
287.1	412.55
287.6	432.16

T/K	P/kPa
288.1	453.87
288.6	478.05
289.0	503.19
289.6	529.25
290.0	560.39
290.5	585.85
291.0	620.23
291.5	650.69
292.0	681.87
292.5	723.21
293.0	762.36
293.5	802.82
294.0	846.56
294.5	892.44
295.0	940.69
295.5	991.56
296.0	1046.1
296.5	1104.06
297.0	1167.65
297.5	1236.72
298.0	1306.68
298.5	1381.8
299.0	1460.44
299.5	1548.55
300.0	1639.3
300.5	1738.63
301.0	1845.91
301.5	1961.13

Q1 Lower Quadruple Point.
Q2 Upper Quadruple Point.

References for Table 2

1. Sloan, E. D. and Koh, C. A., *Clathrate Hydrates of Natural Gases, Third Edition*, CRC Press, 2008. <https://doi.org/10.1201/9781420008494>
2. Deaton, W. M. and Frost, E. M., Jr., *Gas Hydrates and Their Relation to the Operation of Natural-Gas Pipe Lines*, U.S. Bureau of Mines Monograph 8, p. 101, 1946.
3. Kobayashi, R. and Katz, D. L., *Trans AIME*, 186, 66, 1949. <https://doi.org/10.2118/949066-G>
4. McLeod, H. O. and Campbell, J. M., *J. Petl Tech.*, 222, 590, 1961. <https://doi.org/10.2118/1566-G-PA>
5. Thakore, J. L. and Holder, G. D., *Ind. Eng Chem. Res.*, 26, 462, 1987. <https://doi.org/10.1021/ie00063a011>
6. Makogon, T. Y. and Sloan, E. D., *J. Chem. Eng. Data*, 39, 351, 1994. <https://doi.org/10.1021/je00014a035>
7. Roberts, O. L., Brownscombe, E. R., and Howe, L. S., *Oil Gas J.*, 39, 37, 1940.
8. Holder, G. D. and Grigoriou, G. C., *J. Chem. Thermodyn.*, 12, 1093, 1980. <https://doi.org/10.1016/0021-9614(80)90166-4>
9. Holder, G. D. and Godbole, S. P., *AIChE J.*, 28, 930, 1982. <https://doi.org/10.1002/aic.690280607>
10. Mooijer-van den Heuvel, M. M., Peters, C. J., and de Swaan Arons, J., *Fluid Phase Equilib.*, 193, 245, 2002. <https://doi.org/10.1016/S0378-3812(01)00757-9>
11. Ng, H.-J. and Robinson, D. B., *Fluid Phase Equilib.*, 21, 145, 1985. <https://doi.org/10.1016/0378-3812(85)90065-2>
12. Mooijer-van den Heuvel, M.M., Witteman, R., and Peters, C.J., *Fluid Phase Equilib.*, 182, 97, 2001. <https://doi.org/10.1016/S0378-3812(01)00384-3>
13. Van Cleeff, A., and Diepen, G. A. M., *Recueil des Travaux Chimiques des Pays-Bas*, 79, 582, 1960. <https://doi.org/10.1002/recl.19600790606>
14. Marshall, D. R., Saito, S., and Kobayashi, R., *AIChE Journal*, 10, 202, 1964. <https://doi.org/10.1002/aic.690100214>
15. Jhaveri, J. and Robinson, D. B., *Canadian Journal of Chemical Engineering*, 43, 75, 1965. <https://doi.org/10.1002/cjce.5450430207>
16. Sugahara, K. et al., *Journal of Supramolecular Chemistry*, 2, 365, 2002. <https://doi.org/10.1016/S1472-7862(03)00060-1>
17. Barrer, R. M. and Edge, A. V. J., *Proceedings of the Royal Society of London A: Mathematical, Physical and Engineering Sciences*, 300, 1, 1967. <https://doi.org/10.1098/rspa.1967.0154>
18. Ewing, G. J., and Ionescu, L. G., *Journal of Chemical and Engineering Data*, 19, 367, 1974. <https://doi.org/10.1021/je60063a006>
19. Dyadin, Y. A. et al., *Journal of Inclusion Phenomena and Molecular Recognition in Chemistry*, 28, 271, 1997. <https://doi.org/10.1023/A:1007911123739>
20. Ohgaki, K. et al., *Fluid Phase Equilibria*, 175, 1, 2000. <https://doi.org/10.1016/S0378-3812(00)00374-5>
21. Selleck, F. T., Carmichael, L.T., and Sage, B.H., *Industrial and Engineering Chemistry*, 44, 2219, 1952. <https://doi.org/10.1021/ie50513a064>
22. Bond, D. C. and Russell, N.B., *Transactions of the AIME*, 179, 192, 1949. <https://doi.org/10.2118/949192-G>
23. Carroll, J. J. and Mather, A. E., *Canadian Journal of Chemical Engineering*, 69, 1206, 1991. <https://doi.org/10.1002/cjce.5450690522>
24. Mohammadi, A. H. and Richon, D., *Journal of Chemical and Engineering Data*, 54, 2338, 2009. <https://doi.org/10.1021/je900209y>
25. Ward, Z. T. et al., *Journal of Chemical and Engineering Data*, 60, 403, 2014. <https://doi.org/10.1021/je500657f>

Table 3: Phase Equilibria Data for Ternary Gas Clathrate Hydrates

Structure H Hydrate phase equilibria data for multi-component systems are reported in Table 3 for various ternary systems involving methane as a co-guest and methylcyclohexane as an sH hydrate former. The structure H (CH_4+MCH+H_2O) ternary system has been measured extensively, and experimental data are available at temperature and pressure ranges of 251.50-290.62 K and 519-11933 kPa, respectively.

TABLE 3. Ternary Methane+MCH (Methylcyclohexane)+Water *Structure H* Hydrate

T/K	*P*/kPa
***Liquid Water-Methane-MCH Hydrate-Vapor* (Ref. 1)**	
275.9	2078
277.4	2482
279.2	3088
280.8	2795
***Liquid Water-Methane-MCH Hydrate-Vapor* (Ref. 2)**	
275.65	1599
277.65	2137
279.45	2688
281.15	3357
***Liquid Water-Methane-MCH Hydrate-Vapor* (Ref. 3)**	
282.65	3990
284.15	4860
286.45	6470
287.65	7610
289.15	8820
290.25	10500
***Liquid Water-Methane-MCH Hydrate-Vapor* (Ref. 4)**	
277.1	2041
279.9	2951
282.2	3937
283.4	4606
287.1	7391
***Liquid Water-Methane-MCH Hydrate-Vapor* (Ref. 5)**	
279.48	2650
280.49	2930
281.43	3170
282.42	3870
283.39	4370
284.44	4810
284.95	5200
285.44	5500
285.95	5960
286.49	6340
286.96	6750
287.46	7340
288.40	8330
288.97	9130
***Liquid Water-Methane-MCH Hydrate-Vapor* (Ref. 6)**	
273.60	1354
274.99	1639
277.48	2156
279.95	3086
282.29	4084
285.19	5862
287.59	8076
289.48	10228
290.62	11933
***Liquid Water-Methane-MCH Hydrate-Vapor* (Ref. 7)**	
274.09	1420
274.78	1540
275.28	1660
275.79	1750
276.26	1870
276.79	1990
277.26	2110
277.8	2250
278.30	2390
279.25	2700
280.26	3050
281.27	3450
282.28	3900
283.29	4430
284.30	5030
285.29	5720
286.28	6500
287.25	7420
288.29	8480
***Liquid Water-Methane-MCH Hydrate-Vapor* (Ref. 8)**	
279.48	2650
280.49	2930
281.43	3170
282.42	3870
283.39	4370
284.44	4810
284.95	5200
285.44	5500
285.95	5960
286.49	6340
286.96	6750
287.46	7340
288.40	8330
288.97	9130
***Liquid Water-Methane-MCH Hydrate-Vapor* (Ref. 9)**	
251.50	519
253.15	559
255.70	619
258.10	686
261.00	774
264.00	873
267.00	984
269.05	1063
271.00	1145
272.60	1213

References for Table 3

1. Becke, P. Kessel, D., and Rahimian, I., *European Petroleum Conference, Society of Petroleum Engineers*, 1992.
2. Mehta, A. P., and Sloan Jr., E. D., *Journal of Chemical and Engineering Data*, 38, 580, 1993. <https://doi.org/10.1021/je00012a027>
3. Thomas, M. and Behar, E., *Proceedings of the Annual Convention-Gas Processors Association*, 1994.
4. Tohidi, B. et al., *International Conference on Natural Gas Hydrates*, 1996.
5. Mooijer-van den Heuvel, M. M., Peters, C. J., and de Swaan A. J., *Fluid Phase Equilibria*, 172, 73, 2000. <https://doi.org/10.1016/S0378-3812(00)00367-8>
6. Sun, Z.-G. et al., *Fluid Phase Equilibria*, 198, 293, 2002. <https://doi.org/10.1016/S0378-3812(01)00806-8>
7. Nakamura, T. et al., *Chemical Engineering Science*, 58, 269, 2003. <https://doi.org/10.1016/S0009-2509(02)00518-3>
8. Mooijer-van den Heuvel, M. M. "Phase behavior and structural aspects of ternary clathrate hydrate systems: the role of additives," *Diss. Ph. D. Thesis, Technische Universiteit Delft*, 2004.
9. Ohmura, R. et al., *Journal of Chemical and Engineering Data*, 50, 993, 2005. <https://doi.org/10.1021/je0495381>

Table 4: Phase Equilibria Data of Multi-Component Gas Clathrate Hydrates

Gas hydrate phase equilibria data for multi-component systems, which includes CH_4+C_3H_8+Water, CH_4+CO_2+Water, CH_4+H_2S+Water, and CH_4+C_2H_6+Water, are reported in Tables 4a–4d for various concentrations, temperatures, and pressure ranges.

TABLE 4a. Ternary CH_4+C_3H_8+Water Hydrate Phase Equilibria

T/K	*P*/kPa	x_{CH4}
***Liquid Water-Hydrate-Vapor* (Ref. 1)**		
274.8	1627	0.99
277.6	2248	0.99
277.6	2255	0.99
280.4	3123	0.99
283.0	4357	0.99
274.8	1151	0.974
277.6	1593	0.974
280.4	2193	0.974
283.2	3013	0.974
274.8	814	0.952
277.6	1138	0.952
280.4	1586	0.952
283.2	2227	0.952
274.8	552	0.883
277.6	779	0.883
280.4	1110	0.883
283.2	1558	0.883
274.8	365	0.712
277.6	538	0.712
280.4	800	0.712
280.4	800	0.712
283.2	1151	0.712
274.8	272	0.362
277.6	436	0.362
280.4	687	0.362
***Liquid Water-Hydrate-Vapor* (Ref. 2)**		
290.5	6929	0.965
290.7	6929	0.965
293.3	10446	0.965
294.5	13893	0.965
296.6	20857	0.965
299.1	34508	0.965
301.6	48367	0.965
303.7	62294	0.965
303.7	62467	0.965
304.4	68982	0.965
293.1	7412	0.945
292.8	7412	0.945
296.2	13893	0.945
298.5	23615	0.945
300.6	34577	0.945
302.7	48367	0.945
304.9	62225	0.945
***Liquid Water-Hydrate-Vapor* (Ref. 3)**		
274.88	263	0.2375
276.45	350	0.2375
277.74	443	0.2375
279.11	560	0.2375
280.23	689	0.2375
274.45	270	0.371
275.91	343	0.371
277.08	419	0.371
278.65	536	0.371
280.16	691	0.371
282.31	945	0.371
279.80	3032	0.0072
279.91	6996	0.0072
280.04	10271	0.0072
280.21	15250	0.0072
280.49	1834	0.0092
280.59	3166	0.0092
280.71	6989	0.0092
280.91	11002	0.0092
281.11	15394	0.0092
282.35	1756	0.022
282.57	3170	0.022
282.88	6003	0.022
285.09	2060	0.0446
285.30	3611	0.0446
285.69	6590	0.0446
286.25	10140	0.0446
286.67	14326	0.0446
288.01	2811	0.078
288.27	4804	0.078
288.72	7362	0.078
289.27	11496	0.078
289.84	15339	0.078
290.37	3363	0.137
290.66	5507	0.137
290.91	6362	0.137
291.56	9375	0.137
291.66	10389	0.137
292.10	13133	0.137
292.73	16939	0.137
294.24	5259	0.257
294.85	7810	0.257
295.62	10816	0.257
296.39	13795	0.257
297.15	17063	0.257
297.14	8431	0.4823
297.93	11078	0.4823
298.66	13988	0.4823
299.67	17491	0.4823
298.09	10430	0.5635
298.98	13181	0.5635
299.69	15381	0.5635
300.22	17408	0.5635
298.25	10706	0.594
298.94	12623	0.594
299.40	14526	0.594
300.20	17291	0.594
298.02	10600	0.651
298.93	13891	0.651
300.12	17125	0.651
***Liquid Water-Hydrate-Vapor* (Ref. 4)**		
278.15	1306	0.956
278.15	1144	0.947
278.15	848	0.894
278.15	630	0.768
278.15	496	0.53
278.15	489	0.51
278.15	479	0.502
278.15	475	0.468
278.15	479	0.412
278.15	458	0.394
278.15	458	0.39
278.15	479	0.03
278.15	480	0.026
278.15	509	0
***Liquid Water-Hydrate-Vapor* (Ref. 5)**		
274.55	443	0.8813
277.20	645	0.8813
280.45	956	0.8813
283.65	1452	0.8813
289.25	2638	0.8813
291.05	3607	0.8813
291.32	3799	0.8813
295.15	6630	0.8813
295.71	7306	0.8813
296.91	8276	0.8813
299.35	15216	0.8813
***Liquid Water-Hydrate-Vapor* (Ref. 6)**		
278.09	1416	0.9707
283.52	2640	0.9707
288.80	5080	0.9707
292.87	9196	0.9707
295.33	14966	0.9707
296.59	19814	0.9707
297.53	24363	0.9707

TABLE 4b. Ternary CH_4+CO_2+Water Hydrate

T/K	*P*/kPa	x_{CO2}
***Liquid Water-Hydrate-Vapor* (Ref. 7)**		
277.0	2841	0.539
278.9	3461	0.539
278.9	3434	0.545
280.9	4240	0.545
282.9	5171	0.545
284.7	6467	0.545
275.5	1993	0.777
279.2	3082	0.777
276.4	3199	0.274
278.4	3951	0.274
281.0	5102	0.274
283.8	6895	0.274
279.5	2999	0.824
282.2	4275	0.824
283.8	5274	0.824
285.5	6895	0.824
285.7	6998	0.824
***Liquid Water-Hydrate-Vapor* (Ref. 8)**		
273.7	2520	0.1
275.8	3100	0.09
277.8	3830	0.08
280.2	4910	0.08
283.2	6800	0.08
285.1	8400	0.08
287.2	10760	0.09
274.6	2590	0.14
276.9	3240	0.13
279.1	4180	0.13
281.6	5380	0.13
284.0	7170	0.13
286.1	9240	0.12
287.4	10950	0.13
273.8	2120	0.25
279.4	3960	0.22
283.4	6230	0.22
285.2	7750	0.21
287.6	10440	0.25
273.7	1810	0.44
276.9	2630	0.42
280.7	4030	0.40
283.1	5430	0.39
285.1	6940	0.39
287.4	9780	0.39
275.6	1990	0.50
278.5	2980	0.47
280.9	4140	0.40
281.8	4470	0.41
285.1	6840	0.44
287.4	9590	0.45
274.6	1660	0.73
276.4	2080	0.70
278.2	2580	0.68
280.2	3280	0.68
282.0	4120	0.67
273.7	1450	0.79
275.9	1880	0.78
277.8	2370	0.76
279.6	2970	0.75
281.6	3790	0.74
282.7	4370	0.85
***Liquid Water-Hydrate-Vapor* (Ref. 9)**		
271.25	1271	0.77
271.41	1434	0.50
271.37	2022	0.25
271.29	1294	0.77
271.44	1450	0.50
271.37	2033	0.25

TABLE 4c. Ternary CH_4+H_2S+Water Hydrate

T/K	*P*/kPa	x_{CH4}
***Liquid Water-Hydrate-Vapor* (Ref. 10)**		
288.7	4830	0.93
284.3	2590	0.91
282.3	3030	0.937
287.1	4790	0.935
290.1	6790	0.93
279.3	2210	0.935
290.1	6380	0.930
278.7	2830	0.970
282.9	4270	0.969
287.6	6650	0.9705
276.5	2030	0.961
278.4	3240	0.990
282.3	4620	0.9896
284.8	6690	0.9894
287.6	2100	0.780
295.4	5070	0.790
279.8	1030	0.780
281.5	2070	0.905
287.3	3590	0.890
292.1	6000	0.885
***Liquid Water-Hydrate-Vapor* (Ref. 11)**		
283.31	1998.734	0.8866
289.726	4110.473	0.8863
292.068	5990.616	0.8985
288.62	3586.468	0.8833
290.932	4763.755	0.8914
293.224	6869.082	0.8916

TABLE 4d. Ternary CH_4+C_2H_6+Water Hydrate

T/K	*P*/kPa	x_{C2H6}
***Liquid Water-Hydrate-Vapor* (Ref. 1)**		
274.8	2861	0.012
277.6	3806	0.012
280.4	5088	0.012
274.8	2365	0.022
277.6	3227	0.022
280.4	4413	0.022
282.6	5667	0.022
283.2	6088	0.022
274.8	2158	0.029
277.6	2958	0.029
280.4	4033	0.029
274.8	1841	0.05
274.8	1841	0.05
277.6	2530	0.05
280.4	3447	0.05
283.2	4771	0.05
274.8	1524	0.096
277.6	2096	0.096
280.4	2889	0.096
283.2	3964	0.096
274.8	945	0.436
277.6	1289	0.436
280.4	1758	0.436
283.2	2434	0.436
***Liquid Water-Hydrate-Vapor* (Ref. 12)**		
302	68430	0.054
301.2	62225	0.054
299.1	48229	0.054
296.6	34439	0.054
293.6	24235	0.054
289.7	13893	0.054
287.9	10446	0.054
284.9	6929	0.054
304.1	68568	0.191
303.1	61949	0.191
301.3	48643	0.191
299	35611	0.191
296.4	23477	0.191
293.3	13893	0.191
291.7	10446	0.191
288.8	6998	0.191
***Liquid Water-Hydrate-Vapor* (Ref. 13)**		
274.2	883	0.419
274.2	958	0.31
274.2	972	0.292
274.2	986	0.27
274.2	1165	0.156
274.2	1448	0.082

References for Table 4

1. Deaton, W. M., and Frost Jr. E. M., Gas hydrates and their relation to the operation of natural-gas pipe lines, *No. BM-Mon-8. Bureau of Mines*, Amarillo, TX (USA), 1946.
2. McLeod Jr, H. O., and Campbell J. M., *Journal of Petroleum Technology*, 13, 590, 1961. <https://doi.org/10.2118/1566-G-PA>
3. Verma, V. K., Hand, J. H., and Katz D. L.. *GVC/AIChE Joint Meeting*, 10, 1975.
4. Thakore, J. L. and Holder G. D., *Industrial and Engineering Chemistry Research*, 26, 462, 1987. <https://doi.org/10.1021/ie00063a011>
5. Song, K. Y. and Kobayashi, R., *Fluid Phase Equilibria*, 47, 295, 1989. <https://doi.org/10.1016/0378-3812(89)80181-5>
6. Nixdorf, J. and Oellrich L. R., *Fluid Phase Equilibria*, 139, 325, 1997. <https://doi.org/10.1016/S0378-3812(97)00141-6>
7. Unruh, C. H. and Katz D. L., *Journal of Petroleum Technology*, 1, 83, 1949. <https://doi.org/10.2118/949983-G>
8. Adisasmito, S., Frank III, R. J., and Sloan Jr., E. D., *Journal of Chemical and Engineering Data*, 36, 68, 1991. <https://doi.org/10.1021/je00001a020>
9. Hachikubo, A. et al., *Fourth International Conference on Gas Hydrates*, Japan, 2002.
10. Bond, D. C. and Russell, N. B., *Transactions of the AIME*, 179, 192, 1949. <https://doi.org/10.2118/949192-G>
11. Ward, Z. T., "Phase equilibria of gas hydrates containing hydrogen sulfide and carbon dioxide," *Diss. Colorado School of Mines*, 2015.
12. McLeod Jr., H. O. and Campbell J. M.. *Journal of Petroleum Technology*, 13, 590, 1961. <https://doi.org/10.2118/1566-G-PA>
13. Subramanian, S. et al., *Chemical Engineering Science,*, 55, 1981, 2000. <https://doi.org/10.1016/S0009-2509(99)00389-9>

Table 5: Hydrate Phase Equilibria for Inhibited Gas Hydrate Systems

Tables 5a-5c provide a brief review of hydrate phase equilibria for inhibited gas hydrate systems. The methane hydrate phase equilibria data in the presence of sodium chloride (NaCl), methanol (MeOH), and 1,2-ethanediol (monoethylene glycol (MEG)) are included.

TABLE 5a. Methane Hydrate Phase Equilibria in the Presence of Sodium Chloride

T/K	P/kPa	x_{NaCl}
***Liquid Water-Hydrate-Vapor* (Ref. 1)**		
272.7	1416	0.03986
275.3	2640	0.03986
277.7	5080	0.03986
269.2	9196	0.05978
271.4	14966	0.05978
275.1	19814	0.05978
***Liquid Water-Hydrate-Vapor* (Ref. 2)**		
265.9	3778	0.2 (wt)
267.8	4626	0.2 (wt)
272.3	6178	0.2 (wt)
272.3	7191	0.2 (wt)
275.7	11094	0.2 (wt)
276.4	10818	0.2 (wt)
276.3	13659	0.2 (wt)
***Liquid Water-Hydrate-Vapor* (Ref. 3)**		
280.66	6600	0.02001
286.00	13710	0.02001
289.42	21380	0.02001
291.71	28180	0.02001
293.42	35940	0.02001
293.51	34840	0.02001
295.62	45520	0.02001
296.30	48750	0.02001
297.42	56200	0.02001
298.42	63250	0.02001
299.06	67810	0.02001
279.16	7510	0.03611
284.53	14070	0.03611
287.50	23380	0.03611
288.30	23880	0.03611
289.23	27190	0.03611
290.55	34690	0.03611
292.15	41850	0.03611
293.37	49000	0.03611
294.58	57630	0.03611
295.47	64600	0.03611
296.03	71560	0.03611
274.40	7920	0.05994
280.25	14320	0.05994
282.33	20740	0.05994
284.67	29120	0.05994
286.23	35930	0.05994
287.47	42930	0.05994
288.42	49530	0.05994
289.21	56470	0.05994
290.37	64140	0.05994
291.00	70560	0.05994
270.66	7850	0.08014
275.22	14940	0.08014
278.04	22980	0.08014
279.20	28180	0.08014
281.29	37860	0.08014
282.39	42260	0.08014
283.49	50570	0.08014
284.36	58850	0.08014
284.92	64030	0.08014
285.76	71300	0.08014

TABLE 5b. Methane Hydrate Phase Equilibria in the Presence of Methanol (MeOH)

T/K	P/kPa	f_{MeOH}
***Liquid Water-Hydrate-Vapor* (Ref. 2)**		
266.23	2140	0.1
271.24	3410	0.1
275.87	5630	0.1
280.31	9070	0.1
283.67	13320	0.1
286.40	18820	0.1
263.34	2830	0.2
267.51	4200	0.2
270.08	5610	0.2
273.55	8410	0.2
277.56	13300	0.2
280.17	18750	0.2
***Liquid Water-Hydrate-Vapor* (Ref. 4)**		
233.05	1470	0.50
240.05	2950	0.50
247.35	7240	0.50
250.35	10540	0.50
255.25	16980	0.50
250.85	2380	0.35
256.25	3690	0.35
260.25	6810	0.35
264.55	10160	0.35
267.75	13680	0.35
268.45	17220	0.35
270.05	20510	0.35
***Liquid Water-Hydrate-Vapor* (Ref. 5)**		
277.85	4200	0.2
279.65	5000	0
278.65	16800	0.2
291.45	18000	0
282.15	24000	0.2
293.85	25500	0
296.75	35000	0
285.25	35900	0.2
296.25	36100	0

T/K	P/kPa	f_{MeOH}
297.25	40500	0
297.85	43200	0
286.95	50500	0.2
299.55	54000	0
288.45	57000	0.2
289.00	57600	0.2
289.05	62000	0.2
301.35	62500	0
288.95	62500	0.2
301.55	65000	0
291.35	75700	0.2
303.35	76500	0
305.15	92000	0
294.05	92500	0.2
294.45	96200	0.2
295.15	100000	0.2

TABLE 5c. Methane Hydrate Phase Equilibria in the Presence of 1,2-Ethanediol (MEG)

T/K	P/kPa	f_{MEG}
***Liquid Water-Hydrate-Vapor* (Ref. 4)**		
270.24	2420	0.1
273.49	3400	0.1
280.19	6530	0.1
287.10	15610	0.1
267.59	3770	0.3
269.73	4930	0.3
274.36	7860	0.3
280.14	16140	0.3
279.89	16380	0.3
263.43	9890	0.5
266.32	14080	0.5
266.48	15240	0.5

References for Table 5

1. Kharrat, M. and Dalmazzone, D., *Journal of Chemical Thermodynamics*, 35, 1489, 2003. <https://doi.org/10.1016/S0021-9614(03)00121-6>
2. Sloan Jr., E. D. and Koh, C., *Clathrate Hydrates of Natural Gases*, CRC Press, 2007. <https://doi.org/10.1201/9781420008494>
3. Jager, M. D., and Sloan Jr, E. D., *Fluid Phase Equilibria*, 185, 89, 2001. <https://doi.org/10.1016/S0378-3812(01)00459-9>
4. Robinson, D. B., *Journal of Canadian Petroleum Technology*, 25 (04), 1986. <https://doi.org/10.2118/86-04-01>
5. Blanc, C. and Tornier-Lasserve, J., *World Oil (USA)*, 211 (5), 1990.

PROPERTIES OF IONIC LIQUIDS

Eugene Paulechka and Chris Muzny

Ionic liquids are a class of organic salts with melting points below 100 °C. Many of these salts are liquid at room temperature. This table lists four thermophysical properties for 153 ionic liquids. Column definitions for the table are as follows.

Column heading	Definition
Name	Systematic name of organic salt
Mol. form.	Molecular formula in the Hill convention
CAS Reg. No.	Chemical Abstracts Service Registry Number
Mol. wt.	Molecular weight (relative molar mass)
t_m	Normal melting point, in °C; the notation "gl" indicates a glass–liquid transition
ρ	Density at 25 °C, in g cm^{-3}
η	Viscosity at 25 °C, in mPa s
C_p	Isobaric heat capacity at 25 °C in J K^{-1} mol^{-1}; the superscripts indicate the temperature in °C, if different from 25 °C

The original experimental data used in this table can be found in the ILThermo database (Ref. 1). The melting points, densities, and viscosities are provided if at least three independent and consistent data sources are available for a compound. A few exceptions were made if the melting point (triple-point temperature) was determined by adiabatic calorimetry. If several polymorphs that melt at different temperatures are known for a compound, the highest melting point was specified. The glass-liquid transition temperatures are those determined by differential scanning calorimetry at a typical heating rate of 10 K·min^{-1}.

The listed values of the thermophysical properties were critically evaluated using the NIST ThermoData Engine (Ref. 2). For the heat capacity, the procedures described in Ref. 3 were additionally used to validate experimental data. Each recommended value in the table is characterized with an expanded uncertainty (level of confidence approximately 95%, k = 2) listed in parentheses. Compounds are listed alphabetically by name.

References

1. Kazakov, A., Magee, J.W., Chirico, R.D., Paulechka, E., Diky, V., Muzny, C.D., Kroenlein, K., and Frenkel, M., *NIST Standard Reference Database 147: NIST Ionic Liquids Database - (ILThermo), Version 2.0*, National Institute of Standards and Technology, Gaithersburg, MD, 2016, <ilthermo.boulder.nist.gov/>.
2. Diky, V., Chirico, R. D., Frenkel, M., Bazyleva, A., Magee, J. W., Paulechka, E., Kazakov, A. F., Lemmon, E. W., Muzny, C. D., Smolyanitsky, A. Y., Townsend, S., and Kroenlein, K., *NIST Standard Reference Database 103b: ThermoData Engine - Pure Compounds, Binary Mixtures, Ternary Mixtures, and Chemical Reactions, Version 10.1,*
3. Paulechka, E., *J. Phys. Chem. Ref. Data*, 39, 033108, 2010. <https://doi.org/10.1063/1.3463478>

Properties of Ionic Liquids

Name	Mol. form.	CAS Reg. No.	Mol. wt.	t_m/°C	ρ/g cm^{-3}	η/mPa s	C_p/J K^{-1} mol^{-1}
1-Benzyl-3-methylimidazolium bis(trifluoromethylsulfonyl)imide	$C_{13}H_{13}F_6N_3O_4S_2$	433337-24-7	453.37				599(2)
1-Benzyl-3-methylimidazolium chloride	$C_{11}H_{13}ClN_2$	36443-80-8	208.69				338(1)
1-Benzyl-3-methylimidazolium tetrafluoroborate	$C_{11}H_{13}BF_4N_2$	500996-04-3	260.04				$417(4)^{72}$
1-Benzyl-3-methylimidazolium 1,1,2,2-tetrafluoroethanesulfonate	$C_{13}H_{14}F_4N_2O_3S$		354.32				510(2)
1-Butyl-2,3-dimethylimidazolium bis(trifluoromethylsulfonyl) imide	$C_{11}H_{17}F_6N_3O_4S_2$	350493-08-2	433.38		1.4188(4)	102(2)	
1-Butyl-2,3-dimethylimidazolium tetrafluoroborate	$C_9H_{17}N_2BF_4$	402846-78-0	240.05		1.1932(1)		
2-Butyl-1-ethylpyridinium bis(trifluoromethylsulfonyl)imide	$C_{13}H_{18}F_6N_2O_4S_2$	1642855-33-1	444.41				624(2)
1-Butyl-3-methylimidazolium acetate	$C_{10}H_{18}N_2O_2$	284049-75-8	198.27		1.0526(5)		383(2)
1-Butyl-3-methylimidazolium bis(pentafluoroethylsulfonyl)imide	$C_{12}H_{15}F_{10}N_3O_4S_2$	254731-29-8	519.37		1.514(1)	116(5)	
1-Butyl-3-methylimidazolium bis(trifluoromethylsulfonyl)imide	$C_{10}H_{15}F_6N_3O_4S_2$	174899-83-3	419.36	-2.93(4)	1.4357(9)	50.9(0.9)	566(1)
1-Butyl-3-methylimidazolium bromide	$C_8H_{15}BrN_2$	85100-77-2	219.13	78.2(2)	1.294(3)		311(1)
1-Butyl-3-methylimidazolium chloride	$C_8H_{15}ClN_2$	79917-90-1	174.67	67.85(1)	1.082(2)		317(5)
1-Butyl-3-methylimidazolium dibutyl phosphate	$C_{16}H_{33}N_2O_4P$	663199-28-8	348.42		1.045(2)		
1-Butyl-3-methylimidazolium dicyanamide	$C_{10}H_{15}N_5$	448245-52-1	205.27	-2.3(1)	1.0605(9)	30.1(0.6)	376(2)
1-Butyl-3-methylimidazolium dimethyl phosphate	$C_{10}H_{21}N_2O_4P$	891772-94-4	264.26		1.160(2)		
1-Butyl-3-methylimidazolium hexafluorophosphate	$C_8H_{15}F_6N_2P$	174501-64-5	284.18	11.40(6)	1.3654(9)	249(15)	408(2)
1-Butyl-3-methylimidazolium iodide	$C_8H_{15}IN_2$	65039-05-6	266.13	18.77(4)	1.4809(9)		310(1)
1-Butyl-3-methylimidazolium methyl sulfate	$C_9H_{18}N_2O_4S$	401788-98-5	250.31		1.2081(8)	216(11)	
1-Butyl-3-methylimidazolium nitrate	$C_8H_{15}N_3O_3$	179075-88-8	201.23	36.01(7)	1.157(1)		354(1)
1-Butyl-3-methylimidazolium octyl sulfate	$C_{16}H_{32}N_2O_4S$	445473-58-5	348.50		1.067(1)		
1-Butyl-3-methylimidazolium tetrafluoroborate	$C_8H_{15}BF_4N_2$	174501-65-6	226.02	-85(2) gl	1.2013(5)	104(4)	367(1)
1-Butyl-3-methylimidazolium thiocyanate	$C_9H_{15}N_3S$	344790-87-0	197.30		1.0699(6)	56(2)	
1-Butyl-3-methylimidazolium tosylate	$C_{15}H_{22}N_2O_3S$	410522-18-8	310.41	71(1)			$548(5)^{77}$
1-Butyl-3-methylimidazolium tricyanomethide	$C_{12}H_{15}N_5$	878027-73-7	229.29		1.047(5)	27.8(0.4)	
1-Butyl-3-methylimidazolium trifluoroacetate	$C_{10}H_{15}F_3N_2O_2$	174899-94-6	252.24	23.28(9)	1.216(1)		408(2)

Name	Mol. form.	CAS Reg. No.	Mol. wt.	t_m/°C	ρ/g cm^{-3}	η/mPa s	C_p/J K^{-1} mol^{-1}
1-Butyl-3-methylimidazolium trifluoromethanesulfonate	$C_9H_{15}F_3N_2O_3S$	174899-66-2	288.28	17.83(4)	1.298(1)	89(3)	428(2)
1-Butyl-3-methylimidazolium tris(pentafluoroethyl) trifluorophosphate	$C_{14}H_{15}F_{18}N_2P$	917762-91-5	584.23		1.6245(7)		
1-Butyl-3-methylpyridinium bis(trifluoromethylsulfonyl)imide	$C_{12}H_{16}F_6N_2O_4S_2$	344790-86-9	430.38	-84(3) gl	1.4145(7)	64.5(0.6)	578(2)
1-Butyl-4-methylpyridinium bis(trifluoromethylsulfonyl)imide	$C_{12}H_{16}F_6N_2O_4S_2$	475681-62-0	430.38		1.417(2)	56(1)	
1-Butyl-3-methylpyridinium dicyanamide	$C_{12}H_{16}N_4$	712355-12-9	216.29		1.0494(5)		
1-Butyl-2-methylpyridinium tetrafluoroborate	$C_{10}H_{16}BF_4N$	286453-46-1	237.05		1.2012(5)		
1-Butyl-3-methylpyridinium tetrafluoroborate	$C_{10}H_{16}BF_4N$	597581-48-1	237.05		1.1825(3)	177(5)	388(12)
1-Butyl-4-methylpyridinium tetrafluoroborate	$C_{10}H_{16}BF_4N$	343952-33-0	237.05		1.1825(5)	201(9)	
1-Butyl-3-methylpyridinium trifluoromethanesulfonate	$C_{11}H_{16}F_3NO_3S$	857841-32-8	299.31		1.279(1)		
1-Butyl-1-methylpyrrolidinium bis(trifluoromethylsulfonyl)imide	$C_{11}H_{20}F_6N_2O_4S_2$	223437-11-4	422.40	-7.4(1)	1.3945(4)	77(1)	586(2)
1-Butyl-1-methylpyrrolidinium dicyanamide	$C_{11}H_{20}N_4$	370865-80-8	208.31		1.0135(9)	33.8(0.7)	
1-Butyl-1-methylpyrrolidinium tetracyanoborate	$C_{13}H_{20}BN_5$	1266721-18-9	257.15		0.9761(4)		
1-Butyl-1-methylpyrrolidinium trifluoromethanesulfonate	$C_{10}H_{20}F_3NO_3S$	367522-96-1	291.33		1.2527(3)	166(5)	
1-Butyl-1-methylpyrrolidinium tris(pentafluoroethyl) trifluorophosphate	$C_{15}H_{20}F_{18}NP$	851856-47-8	587.27		1.5833(9)	219(8)	
1-Butylpyridinium bis(trifluoromethylsulfonyl)imide	$C_{11}H_{14}F_6N_2O_4S_2$	187863-42-9	416.35		1.448(1)	60(1)	
1-Butylpyridinium tetrafluoroborate	$C_9H_{14}BF_4N$	203389-28-0	223.02		1.2137(3)	167(8)	
Butyltrimethylammonium bis(trifluoromethylsulfonyl)imide	$C_9H_{18}F_6N_2O_4S_2$	258273-75-5	396.36	17.08(5)	1.3922(7)	105(3)	559(2)
2-Decyl-1-ethylpyridinium bis(trifluoromethylsulfonyl)imide	$C_{19}H_{30}F_6N_2O_4S_2$	1642855-44-4	528.57				811(2)
1-Decyl-3-methylimidazolium bis(trifluoromethylsulfonyl)imide	$C_{16}H_{27}F_6N_3O_4S_2$	433337-23-6	503.52	4.18(5)	1.278(1)	119(4)	755(2)
1-Decyl-3-methylimidazolium hexafluorophosphate	$C_{14}H_{27}F_6N_2P$	362043-46-7	368.35				616(6)[42]
1-Decyl-3-methylimidazolium tetracyanoborate	$C_{18}H_{27}BN_6$	1201894-90-7	338.27		0.9625(2)		
Decyltriethylammonium bis(trifluoromethylsulfonyl)imide	$C_{18}H_{36}F_6N_2O_4S_2$	1190369-93-7	522.60				842(5)
1,3-Dibutylimidazolium bis(trifluoromethylsulfonyl)imide	$C_{13}H_{21}F_6N_3O_4S_2$	749921-07-1	461.44				657(2)
1,3-Didecylimidazolium bis(trifluoromethylsulfonyl)imide	$C_{25}H_{45}F_6N_3O_4S_2$	1453194-51-8	629.76				1027(3)
1,3-Diethylimidazolium bis(trifluoromethylsulfonyl)imide	$C_9H_{13}F_6N_3O_4S_2$	174899-88-8	405.33				533(2)
1,2-Diethylpyridinium bis(trifluoromethylsulfonyl)imide	$C_{11}H_{14}F_6N_2O_4S_2$	1642855-30-8	416.35				566(2)
1,3-Diheptylimidazolium bis(trifluoromethylsulfonyl)imide	$C_{19}H_{33}F_6N_3O_4S_2$	1453194-49-4	545.60				839(3)
1,3-Dihexylimidazolium bis(trifluoromethylsulfonyl)imide	$C_{17}H_{29}F_6N_3O_4S_2$	1394227-32-7	517.54				779(2)
1,3-Dimethylimidazolium bis(trifluoromethylsulfonyl)imide	$C_7H_9F_6N_3O_4S_2$	174899-81-1	377.27		1.569(1)		472(1)
1,3-Dimethylimidazolium dimethyl phosphate	$C_7H_{15}N_2O_4P$	654058-04-5	222.18		1.260(1)	276(19)	
1,3-Dimethylimidazolium methyl sulfate	$C_6H_{12}N_2O_4S$	97345-90-9	208.23		1.329(3)	72(1)	
1,3-Dinonylimidazolium bis(trifluoromethylsulfonyl)imide	$C_{23}H_{41}F_6N_3O_4S_2$	1453194-50-7	601.71				966(3)
1,3-Dioctylimidazolium bis(trifluoromethylsulfonyl)imide	$C_{21}H_{37}F_6N_3O_4S_2$	220749-78-0	573.65				906(3)
1,3-Dipentylimidazolium bis(trifluoromethylsulfonyl)imide	$C_{15}H_{25}F_6N_3O_4S_2$	1138216-85-9	489.49				718(2)
1,3-Dipropylimidazolium bis(trifluoromethylsulfonyl)imide	$C_{11}H_{17}F_6N_3O_4S_2$	1138216-84-8	433.38				595(2)
1-Dodecyl-3-methylimidazolium bis(trifluoromethylsulfonyl) imide	$C_{18}H_{31}F_6N_3O_4S_2$	404001-48-5	531.57				820(2)
1-Dodecyl-3-methylimidazolium hexafluorophosphate	$C_{16}H_{31}F_6N_2P$	219947-93-0	396.40				681(7)[62]
Dodecyltriethylammonium bis(trifluoromethylsulfonyl)imide	$C_{20}H_{40}F_6N_2O_4S_2$	876048-22-5	550.66				905(5)
1-Ethyl-2-heptylpyridium bis(trifluoromethylsulfonyl)imide	$C_{16}H_{24}F_6N_2O_4S_2$	1642855-39-7	486.49				718(2)
1-Ethyl-2-hexylpyridinium bis(trifluoromethylsulfonyl)imide	$C_{15}H_{22}F_6N_2O_4S_2$	1642855-37-5	472.46				686(2)
1-Ethyl-3-methylimidazolium acetate	$C_8H_{14}N_2O_2$	143314-17-4	170.21		1.098(1)	137(5)	322(1)
1-Ethyl-3-methylimidazolium bis(pentafluoroethylsulfonyl)imide	$C_{10}H_{11}F_{10}N_3O_4S_2$	216299-76-2	491.32		1.592(2)		
1-Ethyl-3-methylimidazolium bis(trifluoromethylsulfonyl)imide	$C_8H_{11}F_6N_3O_4S_2$	174899-82-2	391.30	-1.71(7)	1.5183(4)	32.2(0.6)	506(2)
1-Ethyl-3-methylimidazolium bromide	$C_6H_{11}BrN_2$	65039-08-9	191.07	76.8(2)			265(1)[77]
1-Ethyl-3-methylimidazolium dicyanamide	$C_8H_{11}N_5$	370865-89-7	177.21		1.105(4)	16(2)	315(1)
1-Ethyl-3-methylimidazolium diethyl phosphate	$C_{10}H_{21}N_2O_4P$	848641-69-0	264.26		1.146(1)		
1-Ethyl-3-methylimidazolium dimethyl phosphate	$C_8H_{17}N_2O_4P$	945611-27-8	236.21		1.219(1)		412(1)
1-Ethyl-3-methylimidazolium ethyl sulfate	$C_8H_{16}N_2O_4S$	342573-75-5	236.29		1.2368(4)	97(2)	383(2)
1-Ethyl-3-methylimidazolium hexafluorophosphate	$C_6H_{11}F_6N_2P$	155371-19-0	256.13	60(1)			362(4)[67]
1-Ethyl-3-methylimidazolium hexyl sulfate	$C_{12}H_{24}N_2O_4S$	942916-86-1	292.39		1.1303(6)		
1-Ethyl-3-methylimidazolium hydrogen sulfate	$C_6H_{12}N_2O_4S$	412009-61-1	208.23		1.3658(4)		
1-Ethyl-3-methylimidazolium iodide	$C_6H_{11}IN_2$	35935-34-3	238.07	77(3)			
1-Ethyl-3-methylimidazolium methanesulfonate	$C_7H_{14}N_2O_3S$	145022-45-3	206.26		1.2417(7)	160(5)	346(1)
1-Ethyl-3-methylimidazolium 2-(2-methoxyethoxy)ethyl sulfate	$C_{11}H_{22}N_2O_6S$	790663-77-3	310.37		1.236(1)		
1-Ethyl-3-methylimidazolium methyl phosphonate	$C_7H_{15}N_2O_3P$	81994-80-1	206.18		1.192(1)		
1-Ethyl-3-methylimidazolium methyl sulfate	$C_7H_{14}N_2O_4S$	516474-01-4	222.26		1.284(1)	78(7)	
1-Ethyl-3-methylimidazolium octyl sulfate	$C_{14}H_{28}N_2O_4S$	790663-79-5	320.45		1.0948(7)	568(21)	
1-Ethyl-3-methylimidazolium tetracyanoborate	$C_{10}H_{11}BN_6$	742099-80-5	226.05		1.0364(2)	18.4(0.3)	407(1)

Name	Mol. form.	CAS Reg. No.	Mol. wt.	t_m/°C	ρ/g cm^{-3}	η/mPa s	C_p/J K^{-1} mol^{-1}
1-Ethyl-3-methylimidazolium tetrafluoroborate	$C_6H_{11}BF_4N_2$	143314-16-3	197.97	14(1)	1.281(1)	38(1)	305(2)
1-Ethyl-3-methylimidazolium thiocyanate	$C_7H_{11}N_3S$	331717-63-6	169.25	-95(3) gl	1.1164(7)	24.4(0.3)	281(1)
1-Ethyl-3-methylimidazolium tosylate	$C_{13}H_{18}N_2O_3S$	328090-25-1	282.36		1.221(2)		
1-Ethyl-3-methylimidazolium tricyanomethanide	$C_{10}H_{11}N_5$	666823-18-3	201.23		1.0814(3)	14.5(0.2)	
1-Ethyl-3-methylimidazolium trifluoromethanesulfonate	$C_7H_{11}F_3N_2O_3S$	145022-44-2	260.23				363(1)
1-Ethyl-3-methylimidazolium tris(pentafluoroethyl) trifluorophosphate	$C_{12}H_{11}F_{18}N_2P$	377739-43-0	556.17		1.7087(6)	59(1)	
1-Ethyl-3-methylpyridinium ethyl sulfate	$C_{10}H_{17}NO_4S$	872672-50-9	247.31		1.2211(5)		
1-Ethyl-1-methylpyrrolidinium bis(trifluoromethylsulfonyl)imide	$C_9H_{16}F_6N_2O_4S_2$	223436-99-5	394.35	90(1)			
1-Ethyl-2-nonylpyridinium bis(trifluoromethylsulfonyl)imide	$C_{18}H_{28}F_6N_2O_4S_2$	1642855-43-3	514.54				779(2)
1-Ethyl-2-octylpyridinium bis(trifluoromethylsulfonyl)imide	$C_{17}H_{26}F_6N_2O_4S_2$	1642855-41-1	500.51				749(2)
1-Ethyl-2-pentylpyridinium bis(trifluoromethylsulfonyl)imide	$C_{14}H_{20}F_6N_2O_4S_2$	1642855-35-3	458.43				653(2)
1-Ethyl-3-propylimidazolium bis(trifluoromethylsulfonyl)imide	$C_{10}H_{15}F_6N_3O_4S_2$	347882-21-7	419.36				565(2)
1-Ethyl-2-propylpyridinium bis(trifluoromethylsulfonyl)imide	$C_{12}H_{16}F_6N_2O_4S_2$	1642855-31-9	430.38				594(2)
1-Ethylpyridinium bis(trifluoromethylsulfonyl)imide	$C_9H_{10}F_6N_2O_4S_2$	712354-97-7	388.30		1.5359(4)	40.3(0.7)	
1-Heptyl-3-methylimidazolium bis(trifluoromethylsulfonyl)imide	$C_{13}H_{21}F_6N_3O_4S_2$	425382-14-5	461.44		1.3449(6)		659(2)
1-Heptyl-3-methylimidazolium hexafluorophosphate	$C_{11}H_{21}F_6N_2P$	357915-04-9	326.27		1.2623(4)		500(5)
1-Hexadecyl-3-methylimidazolium bis(trifluoromethylsulfonyl) imide	$C_{22}H_{39}F_6N_3O_4S_2$	404001-50-9	587.68	46.10(5)			972(4)[47]
1-Hexyl-3-methylimidazolium acetate	$C_{12}H_{22}N_2O_2$	888320-05-6	226.32				441(4)
1-Hexyl-3-methylimidazolium bis(trifluoromethylsulfonyl)imide	$C_{12}H_{19}F_6N_3O_4S_2$	382150-50-7	447.41	-1.1(1)	1.3719(3)	69.5(0.9)	629(2)
1-Hexyl-3-methylimidazolium bromide	$C_{10}H_{19}BrN_2$	85100-78-3	247.18		1.2239(7)		
1-Hexyl-3-methylimidazolium chloride	$C_{10}H_{19}ClN_2$	171058-17-6	202.73		1.0408(9)		
1-Hexyl-3-methylimidazolium dicyanamide	$C_{12}H_{19}N_5$	927902-57-6	233.32		1.0288(7)		
1-Hexyl-3-methylimidazolium hexafluorophosphate	$C_{10}H_{19}F_6N_2P$	304680-35-1	312.24		1.293(7)	470(23)	469(5)
1-Hexyl-3-methylimidazolium iodide	$C_{10}H_{19}IN_2$	178631-05-5	294.18		1.384(2)		
1-Hexyl-3-methylimidazolium tetracyanoborate	$C_{14}H_{19}BN_6$	1240857-50-4	282.16		0.9900(3)	49.8(0.5)	
1-Hexyl-3-methylimidazolium tetrafluoroborate	$C_{10}H_{19}BF_4N_2$	244193-50-8	254.08		1.1457(7)		431(1)
1-Hexyl-3-methylimidazolium trifluoromethanesulfonate	$C_{11}H_{19}F_3N_2O_3S$	460345-16-8	316.34		1.238(1)		
1-Hexyl-3-methylimidazolium bis(pentafluoroethyl) trifluorophosphate	$C_{16}H_{19}F_{18}N_2P$	713512-19-7	612.28		1.5506(6)		
1-Hexyl-3-methylpyridinium bis(trifluoromethylsulfonyl)imide	$C_{14}H_{20}F_6N_2O_4S_2$	547718-92-3	458.43			85(2)	
1-Hexylpyridinium bis(trifluoromethylsulfonyl)imide	$C_{13}H_{18}F_6N_2O_4S_2$	460983-97-5	444.41		1.388(2)	91(6)	
1-Hexylpyridinium tetrafluoroborate	$C_{11}H_{18}BF_4N$	474368-70-2	251.07		1.155(1)		
1-(2-Hydroxyethyl)-3-methylimidazolium tris(pentafluoroethyl) trifluorophosphate	$C_{12}H_{11}F_{18}N_2OP$	1187194-46-2	572.17		1.7652(5)		
1-Isobutyl-3-methylimidazolium bis(trifluoromethylsulfonyl) imide	$C_{10}H_{15}F_6N_3O_4S_2$	174899-85-5	419.36				565(2)
1-Isobutyl-1-methylpiperidinium bis(trifluoromethylsulfonyl) imide	$C_{12}H_{22}F_6N_2O_4S_2$	1566595-48-9	436.43				608(2)
1-Isobutyl-3-methylpyridinium bis(trifluoromethylsulfonyl)imide	$C_{12}H_{16}F_6N_2O_4S_2$	1621008-58-9	430.38				579(2)
1-Isobutyl-1-methylpyrrolidinium bis(trifluoromethylsulfonyl) imide	$C_{11}H_{20}F_6N_2O_4S_2$	1111670-82-6	422.40				582(2)
1-(2-Methoxyethyl)-1-methylpyrrolidinium bis(trifluoromethylsulfonyl)imide	$C_{10}H_{18}F_6N_2O_5S_2$	757240-24-7	424.37		1.4548(8)		
1-(2-Methoxyethyl)-1-methylpyrrolidinium tris(pentafluoroethyl)trifluorophosphate	$C_{14}H_{18}F_{18}NOP$	1195983-48-2	589.24		1.6306(6)		
1-Methyl-3-nonylimidazolium bis(trifluoromethylsulfonyl)imide	$C_{15}H_{25}F_6N_3O_4S_2$	433337-21-4	489.49		1.2985(7)		
1-Methyl-3-nonylimidazolium hexafluorophosphate	$C_{13}H_{25}F_6N_2P$	343952-29-4	354.32		1.2127(5)		569(6)
1-Methyl-3-octylimidazolium bis(trifluoromethylsulfonyl)imide	$C_{14}H_{23}F_6N_3O_4S_2$	178631-04-4	475.46	-9.19(4)	1.3206(6)	90(3)	691(2)
1-Methyl-3-octylimidazolium bromide	$C_{12}H_{23}BrN_2$	61545-99-1	275.23		1.175(1)		
1-Methyl-3-octylimidazolium chloride	$C_{12}H_{23}ClN_2$	64697-40-1	230.78		1.010(1)		
1-Methyl-3-octylimidazolium hexafluorophosphate	$C_{12}H_{23}F_6N_2P$	304680-36-2	340.29		1.236(1)	723(19)	536(5)
1-Methyl-3-octylimidazolium tetrafluoroborate	$C_{12}H_{23}BF_4N_2$	244193-52-0	282.13	-27.34(4)	1.102(1)	340(8)	498(2)
1-Methyl-3-octylimidazolium trifluoromethanesulfonate	$C_{13}H_{23}F_3N_2O_3S$	403842-84-2	344.39		1.1932(2)		
1-Methyl-3-pentylimidazolium bis(trifluoromethylsulfonyl)imide	$C_{11}H_{17}F_6N_3O_4S_2$	280779-53-5	433.38		1.4036(6)		596(2)
1-Methyl-3-pentylimidazolium hexafluorophosphate	$C_9H_{17}F_6N_2P$	280779-52-4	298.21		1.3276(5)		437(4)
1-Methyl-3-pentylimidazolium tetrafluoroborate	$C_9H_{17}BF_4N_2$	244193-49-5	240.05		1.1768(9)		
1-Methyl-3-propylimidazolium bis(trifluoromethylsulfonyl)imide	$C_9H_{13}F_6N_3O_4S_2$	216299-72-8	405.33	-87(1) gl	1.4747(6)	44(1)	535(2)
1-Methyl-3-propylimidazolium bromide	$C_7H_{13}BrN_2$	85100-76-1	205.10	36.39(5)			281(1)

Name	Mol. form.	CAS Reg. No.	Mol. wt.	t_m/°C	ρ/g cm^{-3}	η/mPa s	C_p/J K^{-1} mol^{-1}
1-Methyl-1-propylpiperidinium bis(trifluoromethylsulfonyl) imide	$C_{11}H_{20}F_6N_2O_4S_2$	608140-12-1	422.40		1.4101(8)		
3-Methyl-1-propylpyridinium bis(trifluoromethylsulfonyl)imide	$C_{11}H_{14}F_6N_2O_4S_2$	817575-06-7	416.35		1.4481(6)	54.4(0.9)	
1-Methyl-1-propylpyrrolidinium bis(trifluoromethylsulfonyl) imide	$C_{10}H_{18}F_6N_2O_4S_2$	223437-05-6	408.37	11(1)	1.4276(5)	59(4)	554(6)
1-Methyl-3-tetradecylimidazolium bis(trifluoromethylsulfonyl) imide	$C_{20}H_{35}F_6N_3O_4S_2$	404001-49-6	559.63	35.57(5)			897(4)[37]
Methyltrioctylammonium bis(trifluoromethylsulfonyl)imide	$C_{27}H_{54}F_6N_2O_4S_2$	375395-33-8	648.85		1.103(2)		
Methyltrioctylammonium chloride	$C_{25}H_{54}ClN$	5137-55-3	404.16		0.887(2)		
1-Methyl-3-propylimidazolium tetrafluoroborate	$C_7H_{13}BF_4N_2$	244193-48-4	212.00		1.2356(9)		
Tributylethylphosphonium diethyl phosphate	$C_{18}H_{42}O_4P_2$	20445-94-7	384.48		1.0086(9)		
Triethylhexylammonium bis(trifluoromethylsulfonyl)imide	$C_{14}H_{28}F_6N_2O_4S_2$	210230-46-9	466.50		1.2887(3)	182(8)	713(4)
Triethyloctylammonium bis(trifluoromethylsulfonyl)imide	$C_{16}H_{32}F_6N_2O_4S_2$	210230-48-1	494.55	-77(3) gl	1.2506(6)	226(12)	775(5)
Triethyltetradecylammonium bis(trifluoromethylsulfonyl)imide	$C_{22}H_{44}F_6N_2O_4S_2$	1190369-94-8	578.71				972(6)
Tributylmethylammonium bis(trifluoromethylsulfonyl)imide	$C_{15}H_{30}F_6N_2O_4S_2$	405514-94-5	480.52		1.2626(6)		
Trihexyl(tetradecyl)phosphonium bis(trifluoromethylsulfonyl) imide	$C_{34}H_{68}F_6NO_4PS_2$	460092-03-9	764.00		1.0672(8)		
Trihexyl(tetradecyl)phosphonium bis(2,4,4-trimethylpentyl) phosphinate	$C_{48}H_{102}O_2P_2$	465527-59-7	773.29		0.8856(8)		
Trihexyl(tetradecyl)phosphonium bromide	$C_{32}H_{68}BrP$	654057-97-3	563.77		0.956(2)		
Trihexyl(tetradecyl)phosphonium chloride	$C_{32}H_{68}ClP$	258864-54-9	519.32		0.8906(6)		
Trihexyl(tetradecyl)phosphonium dicyanamide	$C_{34}H_{68}N_3P$	701921-71-3	549.91		0.8987(6)	405(24)	
Trihexyl(tetradecyl)phosphonium tris(pentafluoroethyl) trifluorophosphate	$C_{38}H_{68}F_{18}P_2$	883860-35-3	928.87		1.1826(8)	346(12)	
Triisobutylmethylphosphonium *p*-toluenesulfonate	$C_{20}H_{37}O_3PS$	344774-05-6	388.55		1.0731(6)		

SURFACE TENSION OF COMMON LIQUIDS

Ian H. Bell

The surface tension γ of about 230 liquids is tabulated here as a function of temperature. Values of γ are given in units of millinewtons per meter (mN m^{-1}), which is equivalent to dyn cm^{-1} in cgs units. The uncertainty of the values is 0.1 to 0.2 mN m^{-1} or less in most cases. Values at temperatures between the points tabulated can be obtained by linear interpolation to a good approximation. A more extensive compilation of surface tension may be found in Ref. 1. Column definitions for the table are as follows.

Column heading	Definition
Name	Liquid name; liquids are listed alphabetically by name
Synonym	Common synonym
Mol. form.	Molecular formula of liquid
γ (temp)	Surface tension at temperature (in °C) within parentheses, in units mN m^{-1}
Ref.	Source of data

Most of the values from Ref. 1 were obtained by capillary measurements in which the liquid phase is in contact with its vapor plus air at ambient pressure. Values from Refs. 3-6 refer to measurements in which the liquid and vapor phases are in equilibrium, in the absence of other gases. Details may be found in the references.

References

1. Jasper, J. J., *J. Phys. Chem. Ref. Data*, 1, 841, 1972. <https://doi.org/10.1063/1.3253106>
2. Kahl, H., Wadewitz, T., and Winkelmann, J., *J. Chem. Eng. Data*, 48, 580, 2003. <https://doi.org/10.1021/je0201323>
3. Mulero, A., Cachadiña., I, and Parra, M. I., *J. Phys. Chem. Ref. Data*, 41, 043105, 2012. <https://doi.org/10.1063/1.4768782>
4. Mulero, A. and Cachadiña, I., *J. Phys. Chem. Ref. Data*, 43, 023104, 2014. <https://doi.org/10.1063/1.4878755>
5. Mulero, A., Cachadiña, I., and Sanjuán, E. L., *J. Phys. Chem. Ref. Data*, 44, 033104, 2015. <https://doi.org/10.1063/1.4927858>
6. Cachadiña, I., Mulero, A., and Tian, J., *J. Phys. Chem. Ref. Data*, 44, 023104, 2015. <https://doi.org/10.1063/1.4921749>

Surface Tension of Common Liquids at Various Temperatures

Name	Synonym	Mol. form.	γ(10 °C)/ mN m^{-1}	γ(25 °C)/ mN m^{-1}	γ(50 °C)/ mN m^{-1}	γ(75 °C)/ mN m^{-1}	γ(100 °C)/ mN m^{-1}	Ref.
Acetaldehyde	Ethanal	C_2H_4O	22.54	20.50	17.10			1
Acetic acid	Ethanoic acid	$C_2H_4O_2$		27.10	24.61	22.13		1
Acetic anhydride	Acetyl acetate	$C_4H_6O_3$	34.08	31.93	28.34	24.75	21.16	1
Acetone	2-Propanone	C_3H_6O	24.60	22.71	19.60			2,3
Acetonitrile	Methyl cyanide	C_2H_3N		28.66	25.51			1
Acetophenone	Methyl phenyl ketone	C_8H_8O		39.04	36.15	33.27		1
Allyl alcohol	2-Propen-1-ol	C_3H_6O	26.63	25.30	23.06	20.79		5
Ammonia	R-717	H_3N	24.04	20.58	14.94	9.61	4.76	6
Aniline	Benzenamine	C_6H_7N		42.12	39.41	36.69		1
Anisole	Methoxybenzene	C_7H_8O		35.10	32.09	29.08		1
Benzaldehyde	Benzenecarboxaldehyde	C_7H_6O	39.63	38.00	35.27	32.55	29.82	1
Benzene	[6]Annulene	C_6H_6		28.21	24.93	21.73		3
Benzonitrile	Phenyl cyanide	C_7H_5N		38.79	35.90	33.00		1
Benzyl alcohol	Benzenemethanol	C_7H_8O				31.37	27.67	5
Benzylamine	Benzenemethanamine	C_7H_9N		39.30	36.27	33.23		1
Benzyl benzoate	Benzyl benzenecarboxylate	$C_{14}H_{12}O_2$	44.47	42.82	40.06	37.31	34.55	1
Bromine		Br_2	43.68	40.95	36.40			1
Bromobenzene	Phenyl bromide	C_6H_5Br	36.98	35.24	32.34	29.44	26.54	1
1-Bromobutane	Butyl bromide	C_4H_9Br	27.58	25.90	23.08	20.27	17.45	1
Bromoethane	Ethyl bromide	C_2H_5Br	25.36	23.62				1
1-Bromopropane	Propyl bromide	C_3H_7Br	27.08	25.26	22.21			1
2-Bromopropane	Isopropyl bromide	C_3H_7Br	25.03	23.25	20.30			1
3-Bromopropene	Allyl bromide	C_3H_5Br		26.31	23.17			1
Bromotrifluoromethane	Halon-1301	$CBrF_3$	5.51	3.75	1.19			6
Butanenitrile	Propyl cyanide	C_4H_7N		26.92	24.33	21.73		1
Butanoic acid	Butyric acid	$C_4H_8O_2$		26.05	23.75	21.45		1
1-Butanol	Butyl alcohol	$C_4H_{10}O$	25.28	24.13	22.13	20.03	17.83	5
2-Butanol	*sec*-Butyl alcohol	$C_4H_{10}O$	24.24	23.00	20.91	18.79	16.62	5
2-Butanone	Methyl ethyl ketone	C_4H_8O		23.97	21.16			1
2-Butoxyethanol	Ethylene glycol monobutyl ether	$C_6H_{14}O_2$	27.36	26.14	24.10	22.06	20.02	1
Butyl acetate		$C_6H_{12}O_2$	26.48	24.88	22.21	19.54	16.87	1
Butylamine	1-Butanamine	$C_4H_{11}N$		23.44	20.63			1
tert-Butylamine	2-Methyl-2-propanamine	$C_4H_{11}N$		16.87				1

Name	Synonym	Mol. form.	γ(10 °C)/ mN m⁻¹	γ(25 °C)/ mN m⁻¹	γ(50 °C)/ mN m⁻¹	γ(75 °C)/ mN m⁻¹	γ(100 °C)/ mN m⁻¹	Ref.
Butyl formate		$C_5H_{10}O_2$	26.05	24.52	21.95	19.39	16.82	1
Carbon dioxide	Carbonic anhydride	CO_2	2.76	0.56				6
Carbon disulfide	Carbon bisulfide	CS_2	33.81	31.58	27.87			1
Chlorobenzene	Phenyl chloride	C_6H_5Cl	34.78	32.99	30.02	27.04	24.06	1
1-Chlorobutane	Butyl chloride	C_4H_9Cl	24.85	23.18	20.39			1
Chlorodifluoromethane	HCFC-22	$CHClF_2$	10.29	8.11	4.72	1.75		6
1-Chlorohexane	Hexyl chloride	$C_6H_{13}Cl$	27.28	25.73	23.13	20.54	17.94	1
Chloropentafluoroethane	CFC-115	C_2ClF_5	6.37	4.74	2.23	0.21		6
1-Chloropentane	Pentyl chloride	$C_5H_{11}Cl$	26.01	24.40	21.71	19.02	16.33	1
2-Chlorophenol		C_6H_5ClO		39.70	36.89	34.09	31.28	1
3-Chlorophenol		C_6H_5ClO		41.18	38.66	36.13	33.61	1
1-Chloropropane	Propyl chloride	C_3H_7Cl	23.16	21.30				1
2-Chloropropane	Isopropyl chloride	C_3H_7Cl	20.49	19.16				1
3-Chloropropene	Allyl chloride	C_3H_5Cl		23.14				1
o-Cresol	2-Methylphenol	C_7H_8O		36.83	34.39	31.92	29.40	5
m-Cresol	3-Methylphenol	C_7H_8O		35.77	33.42	31.08	28.76	5
Cyclohexane	Hexahydrobenzene	C_6H_{12}	26.24	24.42	21.44	18.55		2,3
Cyclohexanol	Cyclohexyl alcohol	$C_6H_{12}O$		33.25	30.56	28.06	25.71	5
Cyclohexanone	Pimelic ketone	$C_6H_{10}O$	36.43	34.57	31.46	28.36	25.25	1
Cyclohexene	Tetrahydrobenzene	C_6H_{10}	28.01	26.17	23.12			1
Cyclohexylamine	Cyclohexanamine	$C_6H_{13}N$		31.22	28.25	25.28		1
Cyclopentane	Pentamethylene	C_5H_{10}	24.01	21.85	18.38			4
Cyclopentanone	Adipic ketone	C_5H_8O	34.45	32.80	30.05	27.30	24.55	1
Cyclopentene		C_5H_8	24.45	22.20				1
Decane		$C_{10}H_{22}$	24.81	23.39	21.05	18.78	16.56	3
1-Decanol	Capric alcohol	$C_{10}H_{22}O$	28.96	28.02	26.41	24.73	23.01	5
1,2-Dibromoethane	Ethylene dibromide	$C_2H_4Br_2$		39.55	36.25	32.95		1
Dibromomethane	Methylene bromide	CH_2Br_2		39.38	35.43	31.55		6
Dibutylamine	*N*-Butylbutanamine	$C_8H_{19}N$		24.12	21.74	19.36		1
m-Dichlorobenzene	1,3-Dichlorobenzene	$C_6H_4Cl_2$	37.15	35.43	32.57	29.70	26.83	1
Dichlorodifluoromethane	CFC-12	CCl_2F_2	10.50	8.58	5.58	2.89	0.68	3
1,1-Dichloroethane	Ethylidene dichloride	$C_2H_4Cl_2$		24.06				6
1,2-Dichloroethane	Ethylene dichloride	$C_2H_4Cl_2$		31.86	28.29	24.72		1
Dichloromethane	Methylene chloride	CH_2Cl_2		27.20				1,6
1,2-Dichloropropane, (±)-	Propylene dichloride	$C_3H_6Cl_2$		28.32	25.22	22.12		1
1,2-Dichloro-1,1,2,2-tetrafluoroethane	CFC-114	$C_2Cl_2F_4$	12.65	10.94	8.20	5.61	3.23	6
1,1-Diethoxyethane	Acetal	$C_6H_{14}O_2$		20.89	18.31	15.74		1
Diethylamine	*N*-Ethylethanamine	$C_4H_{11}N$		19.85				1
Diethylene glycol	Diglycol	$C_4H_{10}O_3$		44.82	42.78	40.72	38.63	5
Diethyl ether	Ethyl ether	$C_4H_{10}O$		16.5				1,4
Diethyl sulfide	Ethyl sulfide	$C_4H_{10}S$	26.22	24.57	21.8			1
1,1-Difluoroethane	HFC-152a	$C_2H_4F_2$	11.91	9.89	6.64	3.59	0.95	6
Difluoromethane	HFC-32	CH_2F_2	9.29	6.85	3.10	0.04		6
Diisobutylamine	2-Methyl-*N*-(2-methylpropyl)-1-propanamine	$C_8H_{19}N$		21.72	19.44	17.16		1
Diisopropylamine	*N*-Isopropyl-2-propanamine	$C_6H_{15}N$		19.14	16.45			1
Diisopropyl ether	Isopropyl ether	$C_6H_{14}O$		17.27	14.65			1
Dimethylamine	*N*-Methylmethanamine	C_2H_7N		26.34				1
N,N-Dimethylaniline	*N,N*-Dimethylbenzenamine	$C_8H_{11}N$		35.52	32.9	30.27		1
Dimethyl disulfide	Methyl disulfide	$C_2H_6S_2$		33.39	30.04			1
N,N-Dimethylformamide	DMF	C_3H_7NO	37.56	35.74	32.70	29.66	26.62	2
2,5-Dimethylhexane	Biisobutyl	C_8H_{18}	20.77	19.40	17.12	14.84	12.56	1
2,3-Dimethylpyridine	2,3-Lutidine	C_7H_9N		32.71	30.04	27.36		1
Dimethyl sulfide	2-Thiapropane	C_2H_6S	25.27	24.06				1
Dimethyl sulfoxide	DMSO	C_2H_6OS		42.92	40.06			1
1,4-Dioxane	1,4-Dioxacyclohexane	$C_4H_8O_2$		32.75	29.28	25.80	22.32	1
Diphenyl ether	Oxybisbenzene	$C_{12}H_{10}O$		26.75	24.80			1
Dipropylamine	*N*-Propyl-1-propanamine	$C_6H_{15}N$		22.31	19.75	17.20		1
Dodecane		$C_{12}H_{26}$	26.32	24.93	22.70	20.55	18.49	3

Name	Synonym	Mol. form.	γ(10 °C)/ mN m⁻¹	γ(25 °C)/ mN m⁻¹	γ(50 °C)/ mN m⁻¹	γ(75 °C)/ mN m⁻¹	γ(100 °C)/ mN m⁻¹	Ref.
Epichlorohydrin	(Chloromethyl)oxirane	C_3H_5ClO	38.4	36.36	32.96	29.56	26.16	1
Ethane		C_2H_6	2.03	0.48				3
1,2-Ethanediol	Ethylene glycol	$C_2H_6O_2$		48.02	45.86	43.66	41.42	5
Ethanethiol	Ethyl mercaptan	C_2H_6S		23.08				1
Ethanol	Ethyl alcohol	C_2H_6O	23.12	21.91	19.85			1,5
Ethanolamine	Glycinol	C_2H_7NO		48.32	45.53	42.73		1
Ethoxybenzene	Phenetole	$C_8H_{10}O$		32.41	29.65	26.89		1
2-Ethoxyethanol	Ethylene glycol monoethyl ether	$C_4H_{10}O_2$		28.35	26.11	23.86	21.62	1
Ethyl acetate		$C_4H_8O_2$	25.13	23.39	20.49	17.58	14.68	1
Ethylamine	Ethanamine	C_2H_7N		19.20				1,6
N-Ethylaniline		$C_8H_{11}N$		36.33	33.65	30.98		1
Ethylbenzene	Phenylethane	C_8H_{10}	30.16	28.52	25.82	23.16	20.56	4
Ethyl butanoate	Ethyl butyrate	$C_6H_{12}O_2$	25.51	23.94	21.33	18.71	16.10	1
Ethylcyclohexane		C_8H_{16}	26.73	25.15	22.51			1
Ethyl formate		$C_3H_6O_2$	25.16	23.18				1
Ethyl propanoate	Ethyl propionate	$C_5H_{10}O_2$	25.55	23.80	20.88	17.96		1
Fluorobenzene	Phenyl fluoride	C_6H_5F	28.47	26.66	23.65	20.64		1
Formamide	Methanamide	CH_3NO		57.03	54.92	52.82	50.71	1
Formic acid	Methanoic acid	CH_2O_2		37.13	34.38	31.64		1
Furfural	2-Furaldehyde	$C_5H_4O_2$	45.08	43.09	39.78	36.46	33.14	1
1,1,1,2,3,3,3-Heptafluoropropane	HFC-227ea	C_3HF_7	8.79	7.04	4.30	1.87	0.05	6
Heptane		C_7H_{16}	21.23	19.73	17.28	14.88		2,3
Heptanoic acid	Enanthic acid	$C_7H_{14}O_2$		27.76	25.64			1
1-Heptanol	Heptyl alcohol	$C_7H_{16}O$	27.57	26.44	24.52	22.58	20.60	5
2-Heptanone	Methyl pentyl ketone	$C_7H_{14}O$		26.12	23.48			1
1-Heptene		C_7H_{14}	21.29	19.80	17.33	14.85		1
Hexadecane	Cetane	$C_{16}H_{34}$		27.05	24.91	22.78	20.64	1
Hexafluoroethane	Perfluoroethane	C_2F_6	0.64					6
1,1,1,3,3,3-Hexafluoropropane	HFC-236fa	$C_3H_2F_6$	11.43	9.61	6.71	4.05	1.70	6
Hexane		C_6H_{14}	19.46	17.88	15.30			3
Hexanedinitrile	Adiponitrile	$C_6H_8N_2$		45.45	43.02	40.58		1
Hexanenitrile	Capronitrile	$C_6H_{11}N$		27.37	25.11	22.84		1
1-Hexanol	Caproyl alcohol	$C_6H_{14}O$		25.77	23.77	21.79	19.80	5
2-Hexanone	Butyl methyl ketone	$C_6H_{12}O$		25.45	22.72			1
1-Hexene		C_6H_{12}	19.44	17.90	15.33			1
Hydrazine		H_4N_2		66.39				1
Iodobenzene	Phenyl iodide	C_6H_5I	40.40	38.71	35.91	33.10	30.29	1
1-Iodobutane	Butyl iodide	C_4H_9I	29.79	28.24	25.67	23.09	20.51	1
Iodoethane	Ethyl iodide	C_2H_5I	30.38	28.46	25.24			1
Iodomethane	Methyl iodide	CH_3I	32.19	30.34				1
Isobutane	2-Methylpropane	C_4H_{10}	11.69	10.00	7.28	4.70	2.37	3
Isobutyl acetate	2-Methylpropyl acetate	$C_6H_{12}O_2$	24.58	23.06	20.53	17.99	15.46	1
Isobutylamine	2-Methyl-1-propanamine	$C_4H_{11}N$		21.75	19.02			1
Isopropyl acetate	1-Methylethyl acetate	$C_5H_{10}O_2$	23.37	21.76	19.08	16.40		1
Isopropylbenzene	Cumene	C_9H_{12}	29.27	27.69	25.05	22.42	19.78	1
Mercury	Quicksilver	Hg	488.55	485.48	480.36	475.23	470.11	1
Methanol	Methyl alcohol	CH_4O	23.47	22.17	19.98			1,5
2-Methoxyethanol	Ethylene glycol monomethyl ether	$C_3H_8O_2$	32.32	30.84	28.38	25.92	23.46	1
Methyl acetate		$C_3H_6O_2$	26.66	24.73	21.51			1
Methylamine	Methanamine	CH_5N		19.09				6
N-Methylaniline	*N*-Methylbenzenamine	C_7H_9N		36.90	34.47	32.05		1
Methyl benzoate	Methyl benzenecarboxylate	$C_8H_8O_2$		37.17	34.25	31.32		1
Methyl butanoate		$C_5H_{10}O_2$	26.34	24.62	21.76	18.89	16.03	1
3-Methyl-1-butanol	Isopentyl alcohol	$C_5H_{12}O$	24.95	23.73	21.69	19.64	17.58	5
2-Methyl-2-butene		C_5H_{10}	18.61	17.15				1
Methylcyclohexane		C_7H_{14}	24.94	23.27	20.55			4
Methylcyclopentane		C_6H_{12}	23.47	21.72	18.82			1
Methyl formate		$C_2H_4O_2$	26.72	24.36	20.43	16.50	12.57	1
Methyl hexadecanoate	Methyl palmitate	$C_{17}H_{34}O_2$			27.76	25.44	23.23	4
3-Methylhexane		C_7H_{16}	20.76	19.31	16.88	14.46		1

Name	Synonym	Mol. form.	γ(10 °C)/ mN m⁻¹	γ(25 °C)/ mN m⁻¹	γ(50 °C)/ mN m⁻¹	γ(75 °C)/ mN m⁻¹	γ(100 °C)/ mN m⁻¹	Ref.
Methyl octadecanoate	Methyl stearate	$C_{19}H_{38}O_2$			28.41	26.17	24.03	4
2-Methylpentane	Isohexane	C_6H_{14}	18.4	16.87	14.38			3
3-Methylpentane		C_6H_{14}	19.2	17.61	14.96			1
Methyl propanoate	Methyl propionate	$C_4H_8O_2$	26.32	24.44	21.29			1
2-Methyl-2-propanol	*tert*-Butyl alcohol	$C_4H_{10}O$		20.01	17.75			5
2-Methylpyridine	2-Picoline	C_6H_7N		33.00	29.90	26.79		1
N-Methyl-2-pyrrolidinone	1-Methyl-2-pyrrolidinone	C_5H_9NO	41.94	40.21	37.33	34.45	31.57	2
Methyl salicylate	Methyl 2-hydroxybenzoate	$C_8H_8O_3$	40.98	39.22	36.28	33.35	30.41	1
Nitrobenzene		$C_6H_5NO_2$			40.56	37.66	34.77	1
Nitroethane		$C_2H_5NO_2$	34.02	32.13	29.00			1
Nitromethane	Nitrocarbol	CH_3NO_2	39.04	36.53	32.33			1
2-Nitropropane	Isonitropropane	$C_3H_7NO_2$	31.02	29.29	26.39			1
Nonane		C_9H_{20}	23.82	22.38	20.03	17.73	15.49	3
1-Nonanol	Nonyl alcohol	$C_9H_{20}O$	28.60	27.59	25.86	24.08	22.25	5
5-Nonanone	Dibutyl ketone	$C_9H_{18}O$		26.28	23.85			1
Octadecane		$C_{18}H_{38}$		27.87	25.77	23.66	21.55	1
Octane		C_8H_{18}	22.66	21.17	18.77	16.45	14.21	3
1-Octanol	Capryl alcohol	$C_8H_{18}O$	28.11	26.96	25.07			5
Paraldehyde	2,4,6-Trimethyl-1,3,5-trioxane	$C_6H_{12}O_3$	27.22	25.63	22.97	20.32	17.66	1
Pentachloroethane	Refrigerant 120	C_2HCl_5		34.11	31.21	28.24		6
Pentanal	Valeraldehyde	$C_5H_{10}O$	26.95	25.44	22.91			1
Pentane		C_5H_{12}	17.12	15.45				3
1-Pentanol	Amyl alcohol	$C_5H_{12}O$	26.19	25.01	23.01	20.98	18.89	5
2-Pentanol	*sec*-Amyl alcohol	$C_5H_{12}O$	25.01	23.46	20.91	18.41	15.94	5
2-Pentanone	Methyl propyl ketone	$C_5H_{10}O$		23.25	21.62			1
3-Pentanone	Diethyl ketone	$C_5H_{10}O$		24.74	22.13			1
1-Pentene	α-Amylene	C_5H_{10}	17.10	15.45				1
Pentyl acetate	Amyl acetate	$C_7H_{14}O_2$	26.67	25.17	22.69	20.20	17.72	1
Pentylamine	Amylamine	$C_5H_{13}N$		24.69	22.14	19.58		1
Perfluoropropane	Octafluoropropane	C_3F_8	5.32	3.77	1.45			6
Phenol	Hydroxybenzene	C_6H_6O			37.69	35.05	32.42	5
Phosphorus(III) chloride	Phosphorus trichloride	Cl_3P		27.98	24.81			1
Phosphoryl chloride	Phosphorus oxychloride	Cl_3OP		32.03	28.85	25.66		1
Piperidine	Azacyclohexane	$C_5H_{11}N$	30.64	28.91	26.03	23.14	20.26	1
Propane	LPG	C_3H_8	8.87	7.02	4.14	1.61		3
1,2-Propanediol	1,2-Propylene glycol	$C_3H_8O_2$	36.85	35.99	34.41	32.65	30.69	5
Propanenitrile	Ethyl cyanide	C_3H_5N		26.75	23.87			1
1-Propanethiol	Propyl mercaptan	C_3H_8S		24.20	21.02			1
2-Propanethiol	Isopropyl mercaptan	C_3H_8S		21.33	18.39			1
Propanoic acid	Propionic acid	$C_3H_6O_2$		26.20	23.72	21.23		1
1-Propanol	Propyl alcohol	C_3H_8O	24.59	23.37	21.32	19.20		5
2-Propanol	Isopropyl alcohol	C_3H_8O	22.06	20.92	18.97	16.95		5
Propyl acetate		$C_5H_{10}O_2$	25.48	23.80	21.00	18.20	15.40	1
Propylamine	1-Propanamine	C_3H_9N		21.75				1
Pyridazine	1,2-Diazabenzene	$C_4H_4N_2$	49.51	47.96	45.37	42.78	40.19	1
Pyridine	Azine	C_5H_5N		36.56	33.29	30.03		1
Pyrimidine	1,3-Diazine	$C_4H_4N_2$		30.33	27.80	25.28	22.75	1
Pyrrole	Imidole	C_4H_5N	38.71	37.06	34.31			1
Pyrrolidine	Azacyclopentane	C_4H_9N	30.58	29.23	26.98			1
Quinoline	1-Azanaphthalene	C_9H_7N	44.19	42.59	39.94	37.28	34.62	1
Sulfur hexafluoride		F_6S	3.29	1.63				3
Sulfuryl chloride		Cl_2O_2S		28.78				1
1,1,2,2-Tetrachloroethane	Acetylene tetrachloride	$C_2H_2Cl_4$		35.58	32.41	29.24	26.07	1
Tetrachloromethane	Carbon tetrachloride	CCl_4		26.18	23.23	20.18	17.08	6
Tetrachlorosilane	Silicon tetrachloride	Cl_4Si	19.78	18.29	15.80			1
Tetradecane		$C_{14}H_{30}$	27.43	26.13	23.96	21.78	19.61	1
1,1,1,2-Tetrafluoroethane	HFC-134a	$C_2H_2F_4$	10.08	8.07	4.92	2.14	0.04	6
1,2,3,4-Tetrahydronaphthalene	Tetralin	$C_{10}H_{12}$		33.17	30.78	28.40		1
Thiophene	Thiofuran	C_4H_4S		30.68	27.36			1
Toluene	Methylbenzene	C_7H_8	29.76	27.91	24.88	21.93	19.07	2,3

Name	Synonym	Mol. form.	γ(10 °C)/ mN m⁻¹	γ(25 °C)/ mN m⁻¹	γ(50 °C)/ mN m⁻¹	γ(75 °C)/ mN m⁻¹	γ(100 °C)/ mN m⁻¹	Ref.
Tribromomethane	Bromoform	$CHBr_3$		44.84	41.61	38.33		6
Tributylamine	*N,N*-Dibutyl-1-butanamine	$C_{12}H_{27}N$		24.39	22.32	20.24		1
1,1,1-Trichloroethane	Methyl chloroform	$C_2H_3Cl_3$		25.18	22.07			1
1,1,2-Trichloroethane	Vinyl trichloride	$C_2H_3Cl_3$		34.02	30.65	27.27	23.89	1
Trichloromethane	Chloroform	$CHCl_3$		26.65	23.42	20.15		6
Tridecane		$C_{13}H_{28}$	26.86	25.55	23.37	21.19	19.01	1
Triethylamine	*N,N*-Diethylethanamine	$C_6H_{15}N$		20.22	17.74			1
Trifluoroacetic acid		$C_2HF_3O_2$		13.53	11.42			1
Trifluoromethane	HFC-23	CHF_3	1.44	0.03				6
Trimethylamine	*N,N*-Dimethylmethanamine	C_3H_9N		13.41				1
1,2,4-Trimethylbenzene	Pseudocumene	C_9H_{12}	30.74	29.20	26.64	24.07	21.51	1
1,3,5-Trimethylbenzene	Mesitylene	C_9H_{12}	28.89	27.55	25.31	23.07	20.82	1
Undecane	Hendecane	$C_{11}H_{24}$	25.56	24.21	21.96	19.70	17.45	1
Water		H_2O	74.29	72.06	68.02	63.63	58.92	3
Water-d_2	Heavy water	D_2O	72.90	70.95	67.33	63.25	58.75	3
Xenon		Xe	0.44					3
o-Xylene	1,2-Dimethylbenzene	C_8H_{10}	31.15	29.51	26.81	24.16	21.56	4
m-Xylene	1,3-Dimethylbenzene	C_8H_{10}	29.79	28.12	25.38	22.70	20.08	4
p-Xylene	1,4-Dimethylbenzene	C_8H_{10}		27.80	25.18	22.61	20.08	4

SURFACE TENSION OF AQUEOUS MIXTURES

The composition dependence of the surface tension of binary mixtures of several compounds with water is given in this table. The data are tabulated as a function of the mass percent of the non-aqueous component. Data for methanol, ethanol, 1-propanol, and 2-propanol are taken from Ref. 1, which also gives values at other temperatures. Column definitions are as follows.

Column heading	**Definition**
Name	Name of compound in aqueous mixture; compounds are listed alphabetically by name
t	Temperature for surface tension data, in °C
***γ* (mass %)**	Surface tension of aqueous mixture with specified mass % of compound, in mN m^{-1}

References

1. Vazquez, G., Alvarez, E., and Navaza, J. M., *J. Chem. Eng. Data*, 40, 611, 1995.<https://doi.org/10.1021/je00019a016>
2. *Landolt-Börnstein, Numerical Data and Functional Relationships in Science and Technology, New Series*, IV/16, *Surface Tension*, Springer-Verlag, Heidelberg, 1997.

Surface Tension in mN m^{-1} for the Specified Mass % of the Substance in Water

Name	*t*/°C	*γ*(0%)	*γ*(10%)	*γ*(20%)	*γ*(30%)	*γ*(40%)	*γ*(50%)	*γ*(60%)	*γ*(70%)	*γ*(80%)	*γ*(90%)	*γ*(100%)
Acetic acid	30	71.2	51.4	43.3	41.2	38.2	37.4	36.1	33.5	31.5	30.2	26.3
Acetone	25	72.0	44.9	40.5	36.7	33.0	30.1	29.4	29.4	27.6	24.5	23.1
Acetonitrile	20	72.8	48.5	40.2	34.1	31.6	30.6	30.0	29.6	29.1	28.7	28.4
1,2-Butanediol, (±)-	25	72.0	66.1	60.4	55.1	50.1	45.6	43.3	41.9	40.8	39.2	35.8
1,3-Butanediol, (±)-	30	71.2	58.1	51.6	48.7	45.8	43.9	42.4	41.2	40.0	39.0	37.0
1,4-Butanediol	30	71.2	61.2	56.9	54.2	52.0	50.7	49.5	47.9	46.6	45.2	43.8
Butanoic acid	30	71.2	42.4	37.5	35.5	34.8	32.2	30.8	29.2	27.4	26.3	25.5
2-Butanone	20	72.8	41.6	32.2				25.2				24.6
γ-Butyrolactone	30	71.2	64	58	53	50	48	46	45	44	42.8	42.7
Chloroacetic acid	25	72.0	59.8	53.6	51.3	49.7	48.3	47.5	46.1			
Diethanolamine	25	72.0	66.8	63.2	60.7	58.8	57.2	55.7	54.3	52.7	50.6	47.2
N,N-Dimethylacetamide	25	72.0	72.0	72.0	72.4	73.5	74.9	75.4	73.0	65.7	54.7	36.4
N,N-Dimethylformamide	25	72.0	65.4	59.2	53.8	49.6	47.3	46.9	44.9	42.3	38.4	35.2
1,4-Dioxane	25	72.0					41.2	39.6	37.9	36.2	34.5	33.7
1,2-Ethanediol	20	72.8	68.5	64.9	61.9		57.0					48.2
Ethanol	25	72.01	47.53	37.97	32.98	30.16	27.96	26.23	25.01	23.82	22.72	21.82
Formic acid	20	72.8	66	60	55.7	52.2	50.3	48.8	47.1	44.7	40.9	38.0
Glycerol	25	72.0	70.5	69.5	68.5	67.9	67.4	66.9	66.5	65.7	64.5	62.5
Methanol	25	72.01	56.18	47.21	41.09	36.51	32.86	29.83	27.48	25.54	23.93	22.51
Morpholine	20	72.8	65.1	60.7	58.9	56.7	53.0	49.6	47.0	43.7	41.8	38.7
Nitric acid	20	72.8	71.9	70.7	68.9	66.6	63.8	60.6	56.8	52.6	47.9	42.6
1,2-Propanediol	30	71.2	60.5	54.9	50.7	47.2	44.5	41.5	38.6	37.6	36.3	35.5
1,3-Propanediol	30	71.2	62.6	58.8	55.7	53.8	52.8	51.7	50.8	49.6	48.2	47.0
Propanoic acid	30	71.2	46.6	42.2	37.7	35.6	33.1	31.7	30.2	28.2	27.4	25.8
1-Propanol	25	72.01	34.32	27.84	25.98	25.26	24.80	24.49	24.08	23.86	23.59	23.28
2-Propanol	25	72.01	40.42	30.57	26.82	25.27	24.26	23.51	22.68	22.14	21.69	21.22
Pyridine	25	72.0	52.8	51.2	48.0	46.8	46.6	45.8	45.0	43.6	40.9	37.0
Sulfolane	20	72.8					62.5	61.6	59.6	57.1	54.9	50.9
Sulfuric acid	50	67.9	73.5	75.1	73.6	71.2	68.0	64.1	60.0	56.4	53.6	51.7
Trichloroacetaldehyde	25	72.0	56.7	51.0	46.7	44.1	43.0	42.5	41.5	38.9	34.7	29.4
Trichloroacetic acid	25	72.0	55.8	46.5	42.8	41.6	40.6	39.4	38.3	37.4	36.5	

SURFACE ACTIVE CHEMICALS (SURFACTANTS)

Thomas J. Bruno and Paris D. N. Svoronos

The following table provides the structure and properties of common surface active chemicals or surfactants (Refs. 1-2). These reagents are used industrially to decrease surface tension in many different applications, and they are used in the laboratory in many analytical and biochemical procedures, as well as consumer applications. In this way, the table complements many of the tables presented in Sections 7 and 8, in which these surfactants are incorporated in reagents. The table is arranged alphabetically according to chemical name, although the most common generic name or abbreviation is provided in bold type (unless the chemical name is also the most common name). The surfactant class is provided (anionic, cationic, or non-ionic). Anionic surfactants are molecules in which the hydrophilic moiety is a negatively charged group such as a sulfonate, sulfate, or carboxylate. Cationic surfactants are molecules in which the hydrophilic moiety is a positively charged group such as a quaternary ammonium ion. Non-ionic surfactants are molecules in which the hydrophilic moiety is uncharged, such as an ethoxylate. Another classification exists, though not represented here, called amphoteric surfactants, in which the ionic character is pH dependent.

When available, the melting and normal boiling temperatures are provided; if decomposition occurs before this state point is reached, this is indicated by the notation "dec." The measured density of either the liquid or a solution is provided, along with temperature, whenever possible. The predicted density, while available for some surfactants in which measurements are unavailable, is not provided here.

When available, the critical micelle concentration (CMC), the concentration of the surfactant above which micelles spontaneously form, is provided. The preferred unit is mmol per liter (mmol L^{-1}), but in the case of mixtures, this is provided as a percent or ppm, usually on the basis of mass (mass/mass). The CMC is dependent on temperature, the ionic strength of the solution, and, to some extent, the measurement technique. For anionic and cationic surfactants, the CMC is reduced by increasing ionic strength, but temperature has a minor effect. For non-ionic surfactants, the CMC is relatively insensitive to ionic strength, but increases as temperature increases. For the CMC values given here, the temperature is provided whenever possible. If not specified, the temperature is to be regarded as ambient.

When available, the aggregation number is provided (Refs. 3 and 4). This is the mean number of surfactant molecules present in a micelle after the CMC has been reached. It is measured using luminescent probes by varying surfactant concentration and is dependent on temperature and the concentration of organic or ionic species present. The hydrophilic lipophilic balance (HLB) is provided when available. This is a fit-for-purpose property rather than a fundamental property, which has evolved in definition and determination since being devised, a matter beyond the scope of this entry. The HLB is a numerical scale between 0 and 20 descriptive of the tendency of a surfactant to be either hydrophilic or hydrophobic. A relatively high HLB (greater than 10) indicates hydrophilicity, or good water or polar solvent solubility. A relatively low HLB (lower than 10) indicates lipophilicity, or good solubility in nonpolar solvents such as oils or organics. The HLB changes with concentration and mixture composition.

Column definitions for the table are as follows.

Column heading	Definition
Name	Chemical name of surfactant; listed alphabetically; structure diagram is shown below name
Synonym	Common synonyms
Mol. form.	Molecular formula of surfactant, in Hill order
Mol. wt.	Molecular weight
t_b	Normal boiling point temperature (at 101.325 kPa), in °C
t_m	Melting point, in °C
ρ	Denisty, g cm^{-3}
CMC	Critical micelle concentration, in units mmol L^{-1} (see discussion above)
Aggr. no.	Aggregation number (see discussion above)
HLB	Hydrophilic/hydrophobic tendency (see discussion above)
Sol. H_2O	Solubility in water

References

1. Lange, R. K., *Surfactants: A Practical Handbook*, Hanser Publishers, Munich, 1999.
2. Porter, M. R., *Handbook of Surfactants*, Blackie Academic and Professional, London, 1994.
3. Alargova, R. G., Kochijashky, I. I., Sierra, M. L., and Zana, R., Micelle aggregarion numbers and surfactants in aqueous solutions: a comparison between the results from steady state and time resolved fluorescence quenching, *Langmuir* 14, 5412-5418, 1998. <https://doi.org/10.1021/la980565x>
4. *Technical Information Surface Chemistry*, Akso Nobel, 2015. <http://surface.akzonobel.com>

Properties of Surfactants

Name	Synonym	Mol. form.	Mol. wt.	t_b/°C	t_m/°C	ρ/g cm^{-3}	CMC/ mmol L^{-1}	Aggr. no.	HLB	Sol. H_2O
Anionic surfactants										
Abietic acid	abietinic acid; sylvic acid; abieta-7,13-dien-18-oic acid (1*R*,4a*R*,4b*R*,10a*R*)-7-isopropyl-1,4a- dimethyl-1,2,3,4,4a,4b,5,6,10,10a-decahydrophenanthrene-1-carboxylic acid	$C_{20}H_{30}O_2$	302.451	439.5	173.5	1.06^{25}	2 at 25 °C			Insoluble
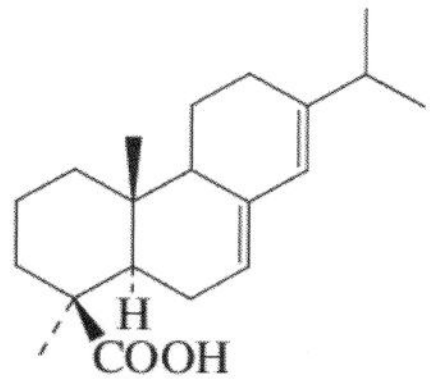							**Soluble in:** ethanol, acetone, ether, chloroform, and benzene			**Notes:** Natural product is primary component of resin acid of pine wood; used as bow rosin; used at high pH.
Bis(2-ethylhexyl) sodium sulfosuccinate	sodium dioctyl sulfosuccinate; dioctyl sulfosuccinate sodium salt; sodium 1,4-bis(2-ethylhexoxy)-1,4-dioxobutane-2-sulfonate; 1,4-bis (2-ethylhexyl) sodium sulfosuccinate; sulfosuccinic acid 1,4-bis (2-ethylhexyl) ester sodium salt; sulfobutanedioic acid 1,4-bis (2-ethylhexyl) ester sodium salt; dioctyl sodium sulfosuccinate; colace; ex-lax; senokot; comfolax	$C_{20}H_{37}NaO_7S$	444.559	dec	155	1.1^{25}				15 g/L at 25 °C; 23 g/L at 40 °C; 30 g/L at 50 °C; 55 g/L at 70 °C
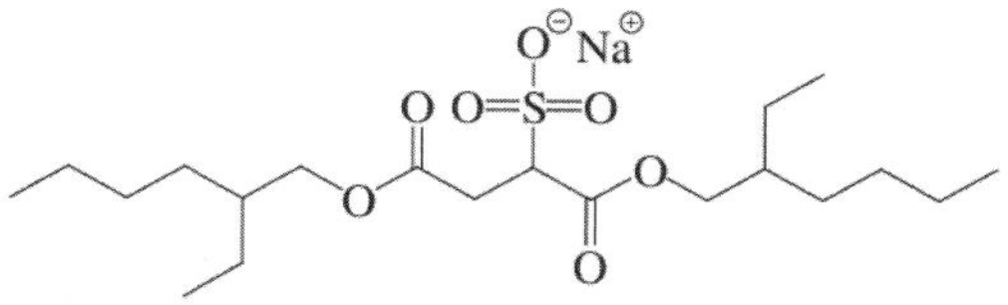							**Soluble in:** ethanol (20 g/L), chloroform (300 g/L), diethyl ether (300 g/L), petroleum ether (unlimited), glycerol, carbon tetrachloride, xylene, petroleum ether, acetone, and vegetable oils			**Notes:** Stable in acid and neutral solutions; hydrolyzes in basic media; used as emulsifier; dispersant, emulsifier, wetting agent, and laxative to treat constipation and earwax removal.
Deoxycholic acid	(3α,5β,12α,20*R*)-3,12-dihydroxycholan-24-oic acid; cholanoic acid; deoxycholate; deoxycholic; Pyrochol; Septochol(l)	$C_{24}H_{40}O_4$	392.573	dec	177		5 at 25 °C	22	17.6	0.24 mg/mL at 15 °C
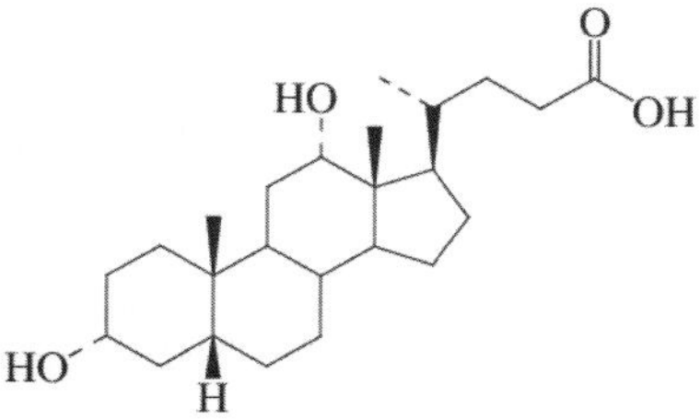							**Soluble in:** ethanol (1 g/100 mL at 20 °C), acetone, ether, and chloroform			**Notes:** Readily dialyzable; used in a modified procedure to recover 40% - 80% of a protein; pK_a = 6.58; λ_{max} = 310 nm; specific rotation +55°; forms molecular coordination compounds; complexes with fatty acids; sodium salt is much more soluble (333 mg/mL in water at 15 °C), but its aqueous solutions precipitate at pH < 5.
Perfluorobutanesulfonic acid	FC-98; PFBS; nonaflate; 1,1,2,2,3,3,4,4,4-nonafluorobutanesulfonic acid; 1,1,2,2,3,3,4,4,4- nonafluorobutane-1-sulfonic acid; nonafluorobutanesulfonic acid; nonafluoro-1-butanesulfonic acid; nonafluorobutane-1-sulfonic acid; nonafluorobutanesulfonic acid; perfluorobutane sulfonate	$C_4HF_9O_3S$	300.1	211	≈80	1.811^{25}				reacts violently
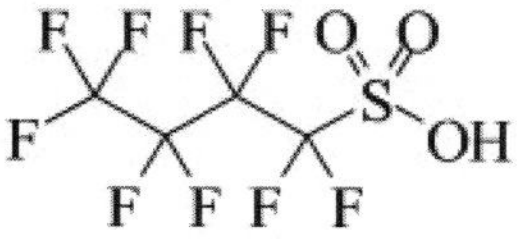										**Notes:** Used as a replacement for perfluoroctanesulfonic acid stain repellent.

Fluids

Name	Synonym	Mol. form.	Mol. wt.	t_b/°C	t_m/°C	ρ/g cm^{-3}	CMC/ mmol L^{-1}	Aggr. no.	HLB	Sol. H_2O
Sodium dodecyl sulfate	sodium lauryl sulfate, **SDS**, sodium monododecyl sulfate; sodium lauryl sulfate; sodium monolauryl sulfate; sodium dodecanesulfate; sodium coco-sulfate; dodecyl alcohol, hydrogen sulfate, sodium salt; n-dodecyl sulfate sodium; sulfuric acid monododecyl ester sodium salt	$C_{12}H_{25}NaO_4S$	288.380	dec	205	1.06[25]	8.2 at 25 °C	62	40	200 mg/mL at 20 °C
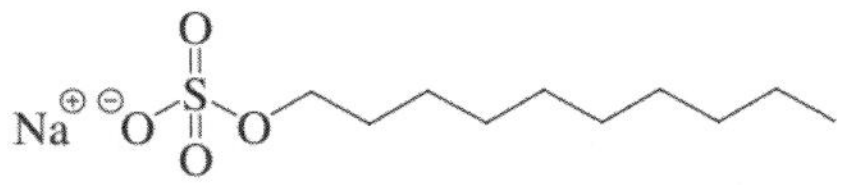							**Soluble in:** ethanol			**Notes:** Aids in lysing cells during DNA extraction and for unraveling proteins in SDS-PAGE; low pK_a; can be used in pH as low as 4; can be prone to foaming.
Cationic surfactant										
Cetyltrimethylammonium bromide	**CTAB**, CTABr, alkyltrimethylammonium bromide; HTAB; hexadecyl palmityltrimethylammonium bromide; palmityltrimethylammonium bromide	$C_{19}H_{42}N$	364.45	dec	240		0.92 – 1.00 at 25 °C	75-170	≈10	0.3 g/100 mL at 20 °C
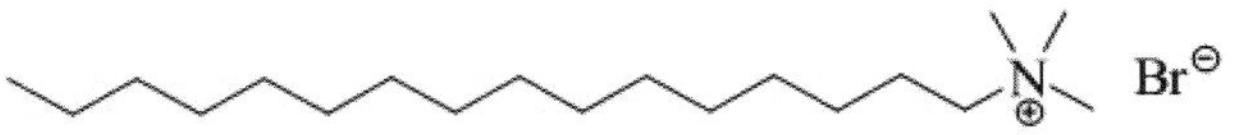							**Soluble in:** alcohols, slightly soluble in acetone; insoluble in diethyl ether and benzene			**Notes:** Aids in high molecular mass DNA isolation and PCR analysis; titrant for perchlorate; phase transfer catalyst in arene and heterocycle reductions.
Ionic (amphiphilic) surfactant										
Sodium lauroyl sarcosinate	**sarkosyl**; sarkosyl NL; sodium *N*-lauroylsarcosinate; sarcosyl; sarcosyl NL; maprosyl 30; sodium [dodecanoyl(methyl)amino]acetate; n-lauroylsarcosine, sodium salt; *N*-methyl-*N*-(1-oxododecyl)glycine, sodium salt; sodium-n-lauriyl sarcosinate	$C_{15}H_{28}NNaO_3$	293.378	dec	140					7.0% - 8.6% mass/mass
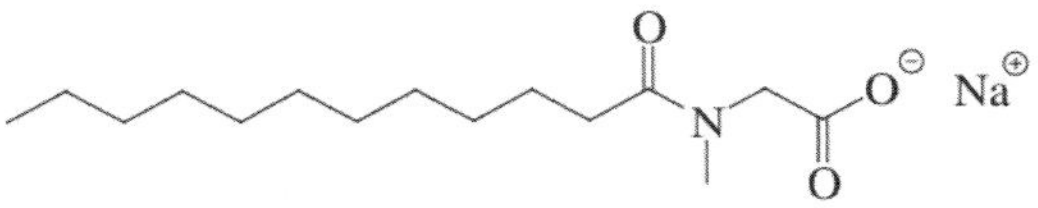										**Notes:** White powder (94%); pH 7.5-8.5 (10% sol); foam stabilizer; lubricant; corrosion inhibitor; bacteriostat; used in emulsion polymerization.
Non-ionic surfactants										
Dodecyl-β-D-maltoside	n-dodecyl-beta-D-maltopyranoside; dodecyl 4-O-alpha-D-glucopyranosyl-beta-D-glucopyranoside; dodecyl B-D-maltopyranoside	$C_{24}H_{46}O_{11}$	510.62	dec	225	1.26[25]	0.15 - 0.17 at 25 °C	70-140	≈10	soluble
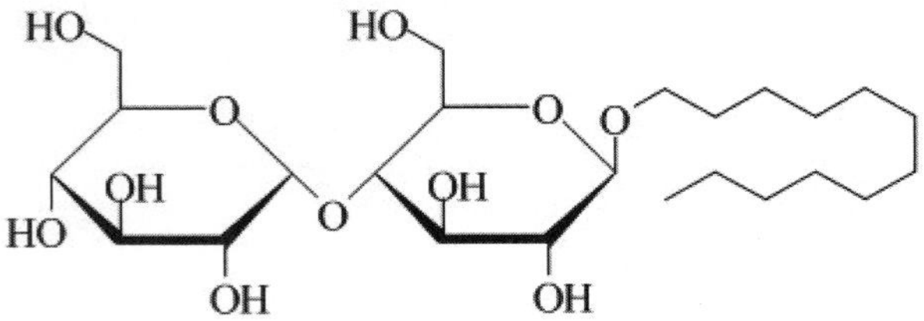										**Notes:** Used to aid in the solubilization and isolation of hydrophobic membrane proteins to preserve their activity.
Ethoxylated tall oil	**Ethofat242/25**, Renex; Industrol TO 16HR; OKM; OKM (surfactant); OKM 10; OKM 12; OKM 50; OKM 60; OKM75; T 13; T 13 (emulsifier); Teric T 2	Mixture	945	dec		1.08[25a]	4[a] at 25 °C		12.2	soluble
										Notes: Derived from tall oil obtained from paper processing, can be used at wide range of pH.

Name	Synonym	Mol. form.	Mol. wt.	t_b/°C	t_m/°C	ρ/g cm^{-3}	CMC/ mmol L^{-1}	Aggr. no.	HLB	Sol. H_2O
Glycerol1-dodecanoate,(±)-	glyceryl laurate, glycerol monolaurate; glycerol α-monolaurate; glycerin monolaurate; glycerol 1-laurate; glycerol 1-monolaurate; glycerol α- monolaurate; dodecanoic acid 2,3-dihydroxypropyl ester; 1-lauroyl-glycerol; monolauroylglycerin; 2,3-dihydroxypropyl dodecanoate; 2,3-dihydroxypropyl laurate; dodecanoic acid α-monoglyceride; 1-mono-dodecanoyl glycerol; lauricidin R; 1-monomyristin; imwitor 312; lauricidin 802; dimodan ML 90; monomuls L 90; monomuls 90L12	$C_{15}H_{30}O_4$	274.397	397	63.2					12.67 mg/L at 25 °C
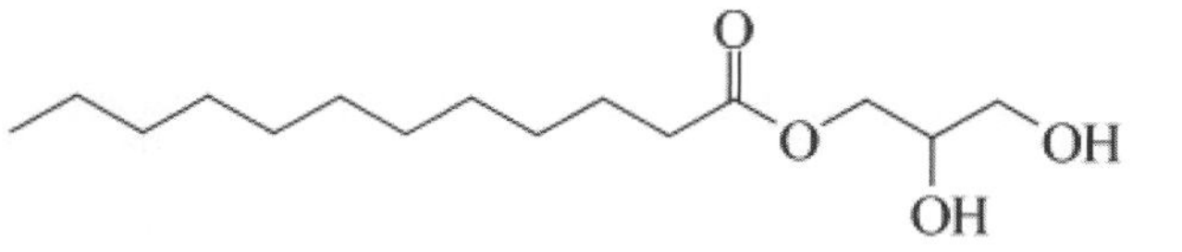							**Soluble in:** methanol and chloroform (50 mg/L)			**Notes:** White fluffy semisolid; has shown both antibacterial and antiviral activity *in vitro*; used against some infections but clinical studies are incomplete; used in deodorants and in some cases as an emulsifier in food additives; used as a methane mitigation agent in ruminants.
Octyl-β-D-1-thioglucopyranoside	(1*S*)-octyl-β-D-thioglucoside; n-octyl-β-D-thioglucoside; OTG; (2*R*,3*S*,4*S*,5*R*,6*R*)-2-(hydroxymethyl)-6-octylsulfanyl-oxane-3,4,5-triol; (2*R*,3*S*,4*S*,5*R*,6*R*)-2-(hydroxymethyl)-6-(octylthio) tetrahydro-2H-pyran-3,4,5-triol; octylthioglycoside; octyl thioglycoside; octyl-β-D-thioglucopyranoside	$C_{14}H_{28}O_5S$	308.43	dec	128		9 at 25 °C			slightly soluble
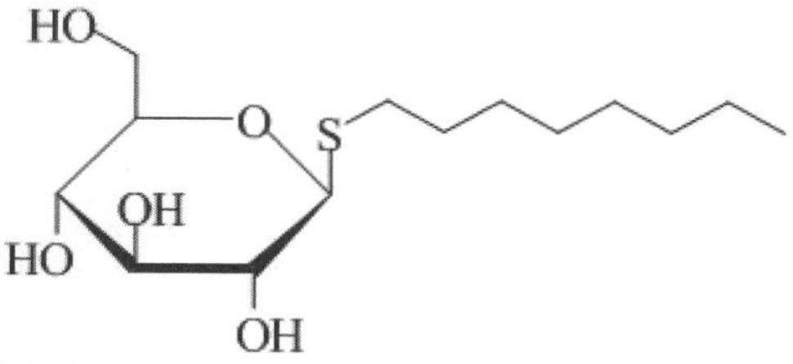							**Soluble in:** ethanol (50 mg/mL)			**Notes:** Colorless powder; used for cell lysis and for solubilizing proteins without denaturing them; dialyzable.
Polyethylene glycol *tert*-octylphenyl ether	Octoxynol-9, **Triton X-100**, TX-100		647	270	6	1.07[25]	0.24 - 0.27 at 25 °C	140	13.5	soluble
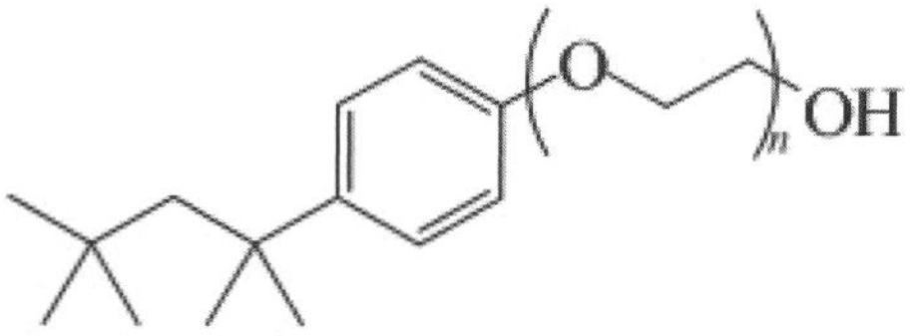							**Soluble in:** benzene, toluene, xylene, trichloroethylene, ethylene glycol, ethyl ether, ethanol, isopropanol, and ethylene dichloride			**Notes:** Non-ionic surfactant and emulsifier widely used for solubilizing membrane proteins and lysing cells; no antimicrobial properties; absorbs in the UV, so can interfere with protein quantitation.
Polyoxyethylene (20) sorbitan monolaurate	Polysorbate 20; PEG(20) sorbitan monolaurate; Alkest TW 20; **Tween 20**	$C_{58}H_{114}O_{26}$	1227.54	dec		1.06[25]	0.06 at 25 °C	62	16.7	soluble
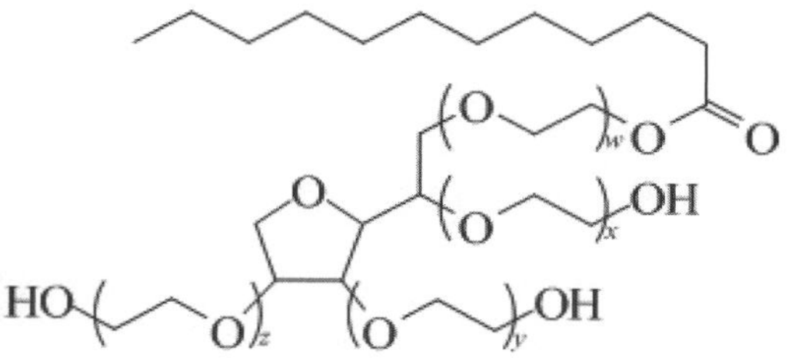 $w+x+y+z=20$										**Notes:** Washing agent in immunoassay; lysing mammalian cells; used industrially in cleaning applications.

Name	Synonym	Mol. form.	Mol. wt.	t_b/°C	t_m/°C	ρ/g cm^{-3}	CMC/ mmol L^{-1}	Aggr. no.	HLB	Sol. H_2O
Polyoxyethylene (20) sorbitan monooleate 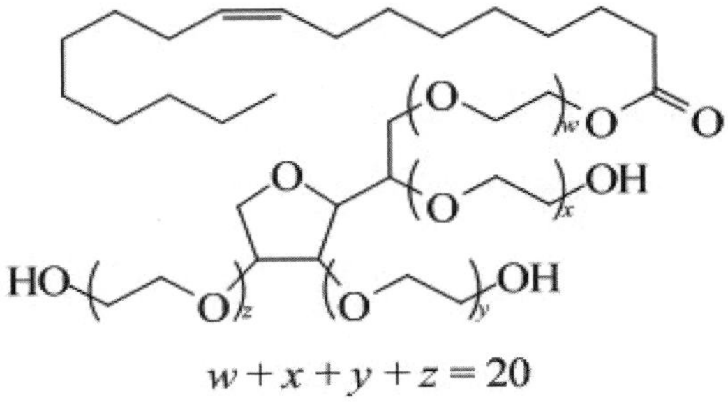	Polysorbate 80; Alkest TW 80; **Tween 80**	$C_{64}H_{124}O_{26}$	1310	dec		1.07^{25}	0.01^{25}		13.4-15	soluble **Notes:** Used to stabilize aqueous drug formulations; emulsifier for drugs; surfactant used in cosmetics and soaps, and in mouthwash, oily liquid.
Tergitol 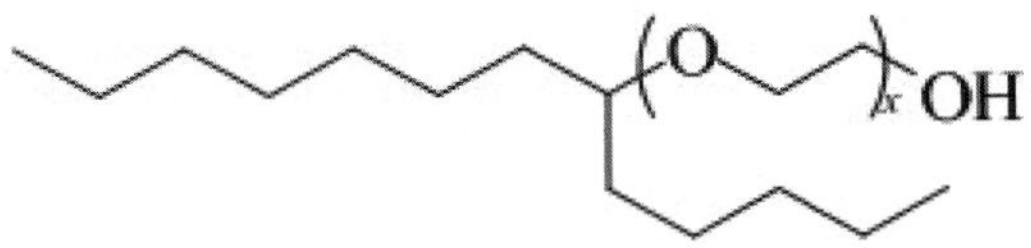	secondary alcohol ethoxylate			dec		1.006^{20}	52 ppm (mass/ mass) **Soluble in:** chlorinated solvents and most polar organics (ethanol, acetone)			soluble **Notes:** Pale yellow liquid; pH (1% aqueous solution) 7.1; chemically stable in the presence of dilute acids, bases, and salts; compatible with anionic, cationic, and other non-ionic surfactants; used in chemical analysis, high-performance cleaners and agrochemicals; member of the large Tergitol family of surfactants.

[a] Typical value.

PERMITTIVITY (DIELECTRIC CONSTANT) OF LIQUIDS

Christian Wohlfarth

The permittivity of a substance (often called the dielectric constant) is the ratio of the electric displacement D to the electric field strength E when an external field is applied to the substance. The quantity tabulated here is the relative permittivity, which is the ratio of the actual permittivity to the permittivity of a vacuum; it is a dimensionless number.

The table gives the static relative permittivity ε, i.e., the relative permittivity measured in static fields or at low frequencies where no relaxation effects occur. When available, a temperature close to 20 °C was chosen, or (as it is the case for many of the substances included here) ε is given at the only temperature for which data are available. The static permittivity refers to nominal atmospheric pressure as long as the corresponding temperature is below the normal boiling point. Otherwise, at temperatures above the normal boiling point, the pressure is understood to be the saturated vapor pressure of the substance.

For substances where information on the temperature dependence of the permittivity is available, the table gives the coefficients of a simple polynomial fitting of permittivity to temperature with an equation of the form

$$\varepsilon(T) = a + bT + cT^2 + dT^3$$

where T is the absolute temperature in K.

The coefficients of the fitting equation can be used to calculate dielectric constants within the fitted temperature range but should not be used for extrapolation outside the stated temperature range. The user who needs dielectric constant data with more accuracy than can be provided by this equation is referred to Ref. 1, which gives the original data together with their literature source. Column definitions for the table are as follows.

The notation E-xy means 10^{-xy}.

Column heading	Definition
Name	Name of liquid; substances are listed alphabetically by systematic name
Mol. form.	Molecular formula of liquid
ε	Static relative permittivity (dielectric constant); the value is at the temperature in °C specified by the superscript
a	Constant in temperature fitting equation
b	Coefficient of T in the temperature fitting equation
c	Coefficient of T^2 in the temperature fitting equation
d	Coefficient of T^3 in the temperature fitting equation
Temp. range	Temperature range for which the fitting equation is valid, in K

References

1. Wohlfarth, Ch., Static Dielectric Constants of Pure Liquids and Binary Liquid Mixtures, *Landolt-Börnstein, Numerical Data and Functional Relationships in Science and Technology, New Series*, Editor in Chief, O. Madelung, Group IV, Macroscopic and Technical Properties of Matter, Volume 6, Springer-Verlag, Berlin, 1991.
2. Marsh, K. N., Ed., *Recommended Reference Materials for the Realization of Physicochemical Properties*, Blackwell Scientific Publications, Oxford, 1987.

Permittivity (Dielectric Constants) of Liquids

Name	Mol. form.	ε	a	b	c	d	Temp. range/K
Acetaldehyde	C_2H_4O	21.0^{18}					
Acetaldoxime	C_2H_5NO	4.70^{25}					
Acetamide	C_2H_5NO	67.6^{91}	-0.20055E+03	0.15515E+01	-0.22392E-02		364 - 448
Acetic acid	$C_2H_4O_2$	6.20^{20}	-0.15731E+02	0.12662E+00	-0.17738E-03		293 - 363
Acetic anhydride	$C_4H_6O_3$	22.45^{20}					
2-Acetonaphthone	$C_{12}H_{10}O$	13.03^{60}	0.14538E+03	-0.73040E+00	0.10000E-02		333 - 363
Acetone	C_3H_6O	21.01^{20}	0.88157E+02	-0.34300E+00	0.38925E-03		273 - 323
Acetonitrile	C_2H_3N	36.64^{20}	0.29724E+03	-0.15508E+01	0.22591E-02		288 - 333
Acetophenone	C_8H_8O	17.44^{25}	0.26099E+02	0.64048E-02	-0.11905E-03		298 - 333
4-Acetylanisole	$C_9H_{10}O_2$	17.3^{40}					
Acetyl chloride	C_2H_3ClO	15.8^{22}					
Acetylene	C_2H_2	2.4841^{-78}					
N-Acetylethanolamine	$C_4H_9NO_2$	96.6^{25}	0.37016E+03	-0.13113E+01	0.13214E-02		298 - 348
2-(Acetyloxy)benzoic acid	$C_9H_8O_4$	6.55^{60}	0.69994E+01	-0.14553E-02			333 - 416
4-Acetylthioanisole	$C_9H_{10}OS$	11.34^{82}					
Acrylonitrile	C_3H_3N	33.0^{20}	0.11109E+03	-0.36806E+00	0.34879E-03		233 - 413
Allene	C_3H_4	2.025^{-4}	0.26049E+01	-0.44147E-03	-0.63420E-05		156 - 269
Allyl alcohol	C_3H_6O	19.7^{20}	0.62714E+02	-0.14771E+00	0.37879E-05		213 - 303
Allylbenzene	C_9H_{10}	2.63^{20}					
Allyl isocyanate	C_4H_5NO	15.15^{15}	0.34299E+02	-0.66444E-01			288 - 333
4-Allyl-2-methoxyphenol	$C_{10}H_{12}O_2$	9.55^{20}	0.52377E+02	-0.24380E+00	0.33333E-03		273 - 323
Aluminum bromide	$AlBr_3$	3.38^{100}					
Ammonia	H_3N	16.61^{20}	0.66756E+02	-0.24696E+00	0.25913E-03		238 - 323
Aniline	C_6H_7N	7.06^{20}	0.89534E+01	0.38990E-02	-0.36310E-04		293 - 413

Name	Mol. form.	ε	a	b	c	d	Temp. range/K
Anisole	C_7H_8O	4.30^{21}	0.10887E+02	-0.32372E-01	0.33629E-04		294 - 413
Anthracene	$C_{14}H_{10}$	2.649^{229}	0.20571E+02	-0.69169E-01	0.66667E-04		502 - 516
Antimony(V) chloride	Cl_5Sb	3.222^{20}	0.45413E+01	-0.45078E-02			276 - 320
Argon	Ar	1.3247^{-133}	0.12408E+01	0.68755E-02	-0.45344E-04		87 - 149
Arsine	AsH_3	2.40^{-72}	0.37674E+01	-0.97454E-02	0.14537E-04		157 - 201
trans-Azoxybenzene	$C_{12}H_{10}N_2O$	5.2^{38}					
Benzaldehyde	C_7H_6O	17.85^{20}	0.35046E+02	-0.61271E-01	0.16222E-04		301 - 346
Benzene	C_6H_6	2.2825^{20}	0.26706E+01	-0.91648E-03	-0.14257E-05		293 - 513
Benzeneacetic acid	$C_8H_8O_2$	3.47^{80}	0.24104E+01	0.30000E-02			353 - 393
Benzeneacetonitrile	C_8H_7N	17.87^{26}	0.82175E+02	-0.37416E+00	0.53220E-03		299 - 343
Benzeneethanol	$C_8H_{10}O$	12.31^{20}	0.12170E+03	-0.63124E+00	0.87776E-03		278 - 333
Benzenemethanethiol	C_7H_8S	4.705^{25}	0.16628E+02	-0.68276E-01	0.94636E-04		298 - 358
Benzenepropanethiol	$C_9H_{12}S$	4.36^{30}	0.82411E+01	-0.15034E-01	0.73617E-05		303 - 358
Benzenepropanol	$C_9H_{12}O$	11.97^{20}	0.94482E+02	-0.45540E+00	0.59307E-03		213 - 303
Benzenesulfonyl chloride	$C_6H_5ClO_2S$	28.90^{50}	0.83886E+02	-0.23405E+00	0.19713E-03		323 - 473
Benzenethiol	C_6H_6S	4.26^{30}	0.57155E+01	-0.70336E-02	0.73617E-05		303 - 358
Benzil	$C_{14}H_{10}O_2$	13.04^{95}	-0.23599E+02	0.22715E+00	-0.34667E-03		368 - 393
Benzonitrile	C_7H_5N	25.9^{20}	0.57605E+02	-0.13354E+00	0.87767E-04		273 - 453
Benzophenone	$C_{13}H_{10}O$	12.62^{27}	0.34130E+02	-0.10249E+00	0.10268E-03		300 - 420
2*H*-1-Benzopyran-2-one	$C_9H_6O_2$	34.04^{70}	0.11311E+03	-0.33804E+00	0.31324E-03		343 - 423
Benzoyl bromide	C_7H_5BrO	21.33^{20}	0.84231E+02	-0.31089E+00	0.32857E-03		283 - 313
Benzoyl chloride	C_7H_5ClO	23.0^{20}					
Benzoyl fluoride	C_7H_5FO	22.7^{20}					
Benzyl acetate	$C_9H_{10}O_2$	5.34^{30}	0.11727E+02	-0.30869E-01	0.32340E-04		303 - 358
Benzyl alcohol	C_7H_8O	11.916^{30}	0.13661E+03	-0.72127E+00	0.10225E-02		303 - 333
Benzylamine	C_7H_9N	5.18^{20}					
Benzyl benzoate	$C_{14}H_{12}O_2$	5.26^{30}	0.76856E+01	-0.80000E-02	-0.80361E-15		303 - 358
Benzyl butanoate	$C_{11}H_{14}O_2$	4.55^{28}					
Benzylethylamine	$C_9H_{13}N$	4.3^{20}					
Benzyl ethyl ether	$C_9H_{12}O$	3.90^{25}					
Benzyl formate	$C_8H_8O_2$	6.34^{30}	0.26162E+02	-0.11026E+00	0.14787E-03		303 - 358
Benzyl nitrite	$C_7H_7NO_2$	7.78^{25}					
Benzyl phenyl ether	$C_{13}H_{12}O$	3.748^{40}					
Benzyl propanoate	$C_{10}H_{12}O_2$	5.11^{30}	0.42301E+01	0.13962E-01	-0.36426E-04		303 - 358
Benzyl salicylate	$C_{14}H_{12}O_3$	4.12^{28}					
Biphenyl	$C_{12}H_{10}$	2.53^{75}	0.26869E+01	0.63072E-03	-0.30995E-05		348 - 428
Bis(2-aminoethyl)amine	$C_4H_{13}N_3$	12.62^{20}	0.57840E+02	-0.23873E+00	0.28841E-03		213 - 333
N,N'-Bis(2-aminoethyl)-1,2-ethanediamine	$C_6H_{18}N_4$	10.76^{20}	0.50699E+02	-0.21730E+00	0.27582E-03		213 - 333
Bis(2-chloroethyl) ether	$C_4H_8Cl_2O$	21.20^{20}					
Bis(chloromethyl) ether	$C_2H_4Cl_2O$	3.51^{20}					
Bis(2-hydroxyethyl) sulfide	$C_4H_{10}O_2S$	28.61^{20}	0.13128E+03	-0.52719E+00	0.60465E-03		253 - 333
1,3-Bis(trifluoromethyl)benzene	$C_8H_4F_6$	5.98^{30}					
l-Bornyl acetate	$C_{12}H_{20}O_2$	4.46^{30}	0.60791E+01	0.98200E-02	-0.50000E-04		303 - 323
Boron tribromide	BBr_3	2.58^{0}					
Bromine	Br_2	3.1484^{25}	0.32701E+01	-0.12535E-03			273 - 327
Bromine pentafluoride	BrF_5	7.91^{25}	0.11428E+02	-0.11822E-01			262 - 298
Bromine trifluoride	BrF_3	106.8^{25}					
3-Bromoaniline	C_6H_6BrN	13.0^{20}					
2-Bromoanisole	C_7H_7BrO	8.96^{30}	0.12023E+02	-0.59116E-02	-0.13787E-04		303 - 358
4-Bromoanisole	C_7H_7BrO	7.40^{30}	0.74367E+01	0.12648E-01	-0.42128E-04		303 - 358
Bromobenzene	C_6H_5Br	5.45^{20}	0.94100E+01	-0.12537E-01	-0.31127E-05		234 - 333
1-Bromobutane	C_4H_9Br	7.315^{10}	0.22542E+02	-0.79306E-01	0.89867E-04		183 - 363
2-Bromobutane, (±)-	C_4H_9Br	8.64^{25}	0.18461E+02	-0.32933E-01			274 - 328
2-Bromobutanoic acid, (±)-	$C_4H_7BrO_2$	7.2^{20}					
cis-2-Bromo-2-butene	C_4H_7Br	5.38^{20}					
trans-2-Bromo-2-butene	C_4H_7Br	6.76^{20}					
Bromochlorodifluoromethane	$CBrClF_2$	3.920^{-150}	0.52442E+01	-0.11000E-01			123 - 223
1-Bromo-2-chloroethane	C_2H_4BrCl	7.41^{10}	0.19493E+02	-0.59054E-01	0.58036E-04		263 - 363
Bromocyclohexane	$C_6H_{11}Br$	8.0026^{30}					
1-Bromodecane	$C_{10}H_{21}Br$	4.44^{25}	0.11202E+02	-0.33491E-01	0.36314E-04		274 - 328

Name	Mol. form.	ε	a	b	c	d	Temp. range/K
1-Bromododecane	$C_{12}H_{25}Br$	4.07^{25}	0.86103E+01	-0.20891E-01	0.18994E-04		274 - 328
Bromoethane	C_2H_5Br	9.01^{25}	0.28473E+02	-0.85495E-01	0.67971E-04		243 - 308
1-Bromo-2-ethoxybenzene	C_8H_9BrO	7.04^{40}	0.23146E+02	-0.75753E-01	0.77778E-04		313 - 358
1-Bromo-2-ethylbenzene	C_8H_9Br	5.55^{25}					
1-Bromo-3-ethylbenzene	C_8H_9Br	5.56^{25}					
1-Bromo-4-ethylbenzene	C_8H_9Br	5.42^{25}					
1-Bromo-2-fluorobenzene	C_6H_4BrF	4.72^{25}					
1-Bromo-3-fluorobenzene	C_6H_4BrF	4.85^{25}					
1-Bromo-4-fluorobenzene	C_6H_4BrF	2.60^{25}					
1-Bromoheptane	$C_7H_{15}Br$	5.255^{30}	0.15289E+02	-0.50621E-01	0.57753E-04		203 - 343
2-Bromoheptane	$C_7H_{15}Br$	6.46^{22}					
4-Bromoheptane	$C_7H_{15}Br$	6.81^{22}					
1-Bromohexadecane	$C_{16}H_{33}Br$	3.68^{25}	0.58668E+01	-0.73333E-02	-0.52666E-14		298 - 328
1-Bromohexane	$C_6H_{13}Br$	5.82^{25}	0.15233E+02	-0.44385E-01	0.43039E-04		274 - 328
Bromomethane	CH_3Br	9.71^{3}	0.40580E+02	-0.18418E+00	0.26219E-03		195 - 276
(Bromomethyl)benzene	C_7H_7Br	6.658^{20}	0.18482E+02	-0.57207E-01	0.57321E-04		273 - 323
1-Bromo-3-methylbutane	$C_5H_{11}Br$	6.33^{18}	0.27743E+02	-0.13927E+00	0.22627E-03		123 - 292
2-Bromo-2-methylbutane	$C_5H_{11}Br$	9.21^{25}					
1-Bromo-2-methylpropane	C_4H_9Br	7.70^{0}	0.37558E+02	-0.20571E+00	0.35496E-03		112 - 273
2-Bromo-2-methylpropane	C_4H_9Br	10.98^{20}	0.35085E+02	-0.14075E+00	0.19960E-03		258 - 293
1-Bromonaphthalene	$C_{10}H_7Br$	4.768^{25}	0.10561E+02	-0.27671E-01	0.27655E-04		293 - 323
1-Bromo-3-nitrobenzene	$C_6H_4BrNO_2$	20.2^{55}	0.81413E+02	-0.27645E+00	0.27367E-03		328 - 413
1-Bromononane	$C_9H_{19}Br$	4.74^{25}	0.79870E+01	-0.10488E-01	-0.13450E-05		274 - 328
1-Bromooctadecane	$C_{18}H_{37}Br$	3.53^{30}	0.46790E+01	-0.30355E-02	-0.24798E-05		303 - 332
1-Bromooctane	$C_8H_{17}Br$	5.0957^{20}	0.12404E+02	-0.35050E-01	0.34542E-04		283 - 353
2-Bromooctane, (±)-	$C_8H_{17}Br$	5.44^{20}					
1-Bromopentadecane	$C_{15}H_{31}Br$	3.88^{20}					
1-Bromopentane	$C_5H_{11}Br$	6.31^{26}	0.20954E+02	-0.78743E-01	0.98908E-04		183 - 328
3-Bromopentane	$C_5H_{11}Br$	8.37^{25}					
1-Bromopropane	C_3H_7Br	8.09^{20}	0.17769E+02	-0.32599E-01			274 - 328
2-Bromopropane	C_3H_7Br	9.46^{20}	0.26195E+02	-0.72995E-01	0.55454E-04		186 - 328
2-Bromopropanoic acid, (±)-	$C_3H_5BrO_2$	11.0^{21}					
3-Bromopropene	C_3H_5Br	7.0^{20}					
(3-Bromopropyl)benzene	$C_9H_{11}Br$	5.41^{29}	0.11360E+02	-0.27471E-01	0.25775E-04		302 - 358
2-Bromopyridine	C_5H_4BrN	23.18^{25}	0.73391E+02	-0.23678E+00	0.22930E-03		298 - 398
1-Bromotetradecane	$C_{14}H_{29}Br$	3.84^{20}	0.10058E+02	-0.33905E-01	0.43528E-04		274 - 328
2-Bromotoluene	C_7H_7Br	4.641^{20}	0.10229E+02	-0.25050E-01	0.20357E-04		273 - 323
3-Bromotoluene	C_7H_7Br	5.566^{20}	0.11522E+02	-0.24946E-01	0.15714E-04		273 - 323
4-Bromotoluene	C_7H_7Br	5.503^{20}	0.10014E+02	-0.13918E-01	-0.50000E-05		273 - 293
Bromotrichloromethane	$CBrCl_3$	2.405^{20}	0.29249E+01	-0.17650E-02			273 - 333
1-Bromotridecane	$C_{13}H_{27}Br$	4.19^{8}					
Bromotrifluoromethane	$CBrF_3$	3.730^{-150}	0.54154E+01	-0.13680E-01			123 - 173
1-Bromoundecane	$C_{11}H_{23}Br$	4.61^{-1}					
1,3-Butadiene	C_4H_6	2.050^{-8}	0.27674E+01	-0.26738E-02			185 - 265
Butanal	C_4H_8O	13.45^{25}					
Butane	C_4H_{10}	1.7697^{22}	0.22379E+01	-0.13884E-02	-0.66711E-06		135 - 303
1,2-Butanediol, (±)-	$C_4H_{10}O_2$	22.4^{25}	0.63702E+02	-0.13807E+00			278 - 323
1,3-Butanediol	$C_4H_{10}O_2$	28.8^{25}	0.72883E+02	-0.14770E+00			278 - 323
1,4-Butanediol	$C_4H_{10}O_2$	31.9^{25}	0.13079E+03	-0.46985E+00	0.46320E-03		288 - 328
2,3-Butanedione	$C_4H_6O_2$	4.04^{25}	0.46907E+01	-0.22302E-02			278 - 348
Butanenitrile	C_4H_7N	24.83^{20}	0.53884E+02	-0.99257E-01			293 - 333
1,2,3,4-Butanetetrol	$C_4H_{10}O_4$	28.2^{120}					
1-Butanethiol	$C_4H_{10}S$	5.204^{15}	0.11201E+02	-0.20767E-01			273 - 318
2-Butanethiol	$C_4H_{10}S$	5.645^{15}	0.10866E+02	-0.17993E-01			273 - 318
Butanoic acid	$C_4H_8O_2$	2.98^{14}	0.15010E+01	0.50046E-02			287 - 403
Butanoic anhydride	$C_8H_{14}O_3$	12.8^{20}					
1-Butanol	$C_4H_{10}O$	17.84^{20}	0.10578E+03	-0.50587E+00	0.84733E-03	-0.48841E-06	193 - 553
2-Butanol	$C_4H_{10}O$	17.26^{20}	0.13850E+03	-0.75146E+00	0.14086E-02	-0.89512E-06	172 - 533
2-Butanone	C_4H_8O	18.56^{20}	0.15457E+02	0.90152E-01	-0.27100E-03		293 - 333
2-Butanone oxime	C_4H_9NO	3.4^{20}					

Name	Mol. form.	*ε*	*a*	*b*	*c*	*d*	Temp. range/K
1-Butene	C_4H_8	2.2195^{-53}	0.29354E+01	-0.32580E-02			220 - 250
cis-2-Butene	C_4H_8	1.960^{23}	0.28802E+01	-0.31064E-02			197 - 296
Butoxyacetylene	$C_6H_{10}O$	6.62^{25}					
N-Butylacetamide	$C_6H_{13}NO$	104.0^{20}	0.70739E+03	-0.37369E+01	0.71585E-02	-0.48716E-05	253 - 493
Butyl acetate	$C_6H_{12}O_2$	5.07^{20}	0.13825E+02	-0.43994E-01	0.48214E-04		253 - 353
sec-Butyl acetate	$C_6H_{12}O_2$	5.135^{20}	0.12427E+02	-0.32035E-01	0.24286E-04		273 - 323
tert-Butyl acetate	$C_6H_{12}O_2$	5.672^{20}	0.55435E+02	-0.30494E+00	0.46107E-03		273 - 323
tert-Butylacetic acid	$C_6H_{12}O_2$	2.85^{23}					
Butyl acrylate	$C_7H_{12}O_2$	5.25^{28}	0.38296E+02	-0.19109E+00	0.27006E-03		301 - 343
Butylamine	$C_4H_{11}N$	4.71^{20}	0.13322E+02	-0.44176E-01	0.50250E-04		223 - 333
Butylbenzene	$C_{10}H_{14}$	2.359^{20}					
sec-Butylbenzene, (±)-	$C_{10}H_{14}$	2.357^{20}	0.28348E+01	-0.68586E-03	-0.32143E-05		273 - 323
tert-Butylbenzene	$C_{10}H_{14}$	2.359^{20}	0.27924E+01	-0.38350E-03	-0.37500E-05		273 - 323
Butyl benzoate	$C_{11}H_{14}O_2$	5.52^{30}	0.77854E+01	-0.34972E-02	-0.13149E-04		303 - 358
Butyl butanoate	$C_8H_{16}O_2$	4.39^{25}	0.79684E+01	-0.12000E-01	0.15266E-13		298 - 318
Butyl formate	$C_5H_{10}O_2$	6.10^{30}	0.21532E+02	-0.84106E-01	0.10952E-03		288 - 323
Butyl nitrate	$C_4H_9NO_3$	13.10^{20}					
tert-Butyl nitrite	$C_4H_9NO_2$	11.47^{25}					
5-Butylnonane	$C_{13}H_{28}$	2.0319^{20}					
2-Butyl-1-octanol	$C_{12}H_{26}O$	3.28^{90}					
Butyl oleate	$C_{22}H_{42}O_2$	4.00^{25}					
N-Butyl-*N*-phenylacetamide	$C_{12}H_{17}NO$	11.66^{25}					
Butyl phenyl ether	$C_{10}H_{14}O$	3.734^{20}					
Butyl propanoate	$C_7H_{14}O_2$	4.838^{20}					
Butyl stearate	$C_{22}H_{44}O_2$	3.120^{25}	0.73894E+02	-0.46261E+00	0.75500E-03		298 - 343
Butyl thiophene-2-carboxylate	$C_9H_{12}O_2S$	6.40^{20}					
Butyl trichloroacetate	$C_6H_9Cl_3O_2$	7.480^{20}					
γ-Butyrolactone	$C_4H_6O_2$	39.0^{20}					
Carbon dioxide	CO_2	1.4492^{22}	0.79062E+00	0.10639E-01	-0.28510E-04		220 - 300
Carbon disulfide	CS_2	2.6320^{20}	0.45024E+01	-0.12054E-01	0.19147E-04		154 - 319
Carbon oxyselenide	COSe	3.47^{10}	0.48740E+01	-0.49425E-02			219 - 283
Carbon oxysulfide	COS	4.47^{-88}	0.84702E+01	-0.21488E-01			143 - 185
Carbonyl chloride	CCl_2O	4.30^{22}					
Carvenone, (*S*)-	$C_{10}H_{16}O$	18.8^{20}					
Chlorine	Cl_2	2.147^{-65}	0.29440E+01	-0.44649E-02	0.30388E-05		208 - 240
Chlorine pentafluoride	ClF_5	4.28^{-80}	0.78192E+01	-0.20860E-01	0.13132E-04		193 - 256
Chlorine trifluoride	ClF_3	4.394^{20}	0.96716E+01	-0.18000E-01			273 - 313
Chloroacetic acid	$C_2H_3ClO_2$	12.35^{65}	0.17310E+02	-0.14674E-01			338 - 393
2-Chloroaniline	C_6H_6ClN	13.40^{20}					
3-Chloroaniline	C_6H_6ClN	13.3^{20}					
4-Chloroanisole	C_7H_7ClO	7.84^{20}	0.64019E+01	0.30560E-01	-0.87500E-04		293 - 333
Chlorobenzene	C_6H_5Cl	5.6895^{20}	0.19471E+02	-0.70786E-01	0.82466E-04		293 - 430
4-Chlorobenzenethiol	C_6H_5ClS	3.59^{65}					
2-Chlorobornane	$C_{10}H_{17}Cl$	5.21^{95}					
2-Chloro-1,3-butadiene	C_4H_5Cl	4.914^{20}					
1-Chlorobutane	C_4H_9Cl	7.276^{20}	0.13565E+02	-0.10161E-01	-0.38750E-04		273 - 323
2-Chlorobutane	C_4H_9Cl	8.564^{20}	0.30376E+02	-0.11377E+00	0.13429E-03		273 - 323
Chlorocyclohexane	$C_6H_{11}Cl$	7.9505^{30}					
1-Chlorodecane	$C_{10}H_{21}Cl$	4.581^{20}	0.68741E+01	-0.12210E-02	-0.22500E-04		293 - 323
4-Chloro-1,3-dioxolan-2-one	$C_3H_3ClO_3$	62.0^{40}					
1-Chlorododecane	$C_{12}H_{25}Cl$	4.17^{25}	0.10002E+02	-0.27798E-01	0.27559E-04		274 - 328
Chloroethane	C_2H_5Cl	9.45^{20}	0.60693E+02	-0.31290E+00	0.47154E-03		237 - 293
2-Chloroethanol	C_2H_5ClO	25.80^{20}	0.11155E+03	-0.30149E+00			140 - 175
1-Chloro-2-ethylbenzene	C_8H_9Cl	4.36^{25}					
1-Chloro-3-ethylbenzene	C_8H_9Cl	5.18^{25}					
1-Chloro-4-ethylbenzene	C_8H_9Cl	5.16^{25}					
2-Chloroethyl isocyanate	C_3H_4ClNO	29.1^{15}	0.64311E+02	-0.12217E+00			288 - 403
1-Chloro-2-fluorobenzene	C_6H_4ClF	6.10^{25}					
1-Chloro-3-fluorobenzene	C_6H_4ClF	4.96^{25}					
1-Chloro-4-fluorobenzene	C_6H_4ClF	3.34^{25}					

Name	Mol. form.	ε	*a*	*b*	*c*	*d*	Temp. range/K
1-Chloroheptane	$C_7H_{15}Cl$	5.521^{20}	0.14279E+02	-0.39431E-01	0.32321E-04		273 - 323
2-Chloroheptane	$C_7H_{15}Cl$	6.52^{22}					
3-Chloroheptane	$C_7H_{15}Cl$	6.70^{22}					
4-Chloroheptane	$C_7H_{15}Cl$	6.54^{22}					
1-Chlorohexane	$C_6H_{13}Cl$	6.104^{20}	0.15994E+02	-0.43647E-01	0.33393E-04		273 - 323
6-Chloro-1-hexanol	$C_6H_{13}ClO$	21.6^{-31}	-0.73364E+01	0.46377E+00	-0.14202E-02		195 - 242
Chloromethane	CH_3Cl	10.0^{22}	0.42775E+02	-0.16175E+00	0.17108E-03		190 - 392
(Chloromethyl)benzene	C_7H_7Cl	6.854^{20}	0.17108E+02	-0.45285E-01	0.35000E-04		273 - 323
1-Chloro-3-methylbutane	$C_5H_{11}Cl$	6.10^{19}	0.22228E+02	-0.93189E-01	0.12991E-03		171 - 297
2-Chloro-2-methylbutane	$C_5H_{11}Cl$	12.31^{-50}	0.55104E+02	-0.29866E+00	0.47840E-03		201 - 223
1-Chloro-4-methyl-2-nitrobenzene	$C_7H_6ClNO_2$	28.07^{28}					
1-Chloro-2-methylpropane	C_4H_9Cl	7.027^{20}	0.14945E+02	-0.33747E-01	0.23036E-04		273 - 323
2-Chloro-2-methylpropane	C_4H_9Cl	9.663^{20}	0.35077E+02	-0.12867E+00	0.14304E-03		273 - 323
1-Chloronaphthalene	$C_{10}H_7Cl$	5.04^{25}	0.84861E+01	-0.12357E-01	0.26899E-05		274 - 328
1-Chloro-2-nitrobenzene	$C_6H_4ClNO_2$	37.7^{50}	0.16800E+03	-0.59708E+00	0.59957E-03		323 - 436
1-Chloro-3-nitrobenzene	$C_6H_4ClNO_2$	20.9^{50}	0.77193E+02	-0.25118E+00	0.23798E-03		323 - 433
1-Chloro-4-nitrobenzene	$C_6H_4ClNO_2$	8.09^{120}					
2-Chloro-2-nitropropane	$C_3H_6ClNO_2$	31.90^{-23}					
1-Chlorononane	$C_9H_{19}Cl$	4.803^{20}	0.95528E+01	-0.16200E-01	-0.16365E-13		293 - 323
1-Chlorooctane	$C_8H_{17}Cl$	5.05^{25}	0.11346E+02	-0.25120E-01	0.13450E-04		274 - 328
2-Chlorooctane	$C_8H_{17}Cl$	5.42^{20}					
1-Chloropentane	$C_5H_{11}Cl$	6.654^{20}	0.18626E+02	-0.54719E-01	0.47143E-04		273 - 323
2-Chlorophenol	C_6H_5ClO	7.40^{23}	0.29755E+02	-0.11256E+00	0.12390E-03		296 - 448
3-Chlorophenol	C_6H_5ClO	6.255^{20}					
4-Chlorophenol	C_6H_5ClO	11.18^{41}	0.31997E+02	-0.94241E-01	0.88392E-04		314 - 453
4-Chlorophenyl isocyanate	C_7H_4ClNO	3.177^{15}	0.40896E+01	-0.31667E-02			288 - 348
1-Chloropropane	C_3H_7Cl	8.588^{20}	0.21214E+02	-0.43130E-01			273 - 313
3-Chloro-1,2-propanediol	$C_3H_7ClO_2$	31.0^{20}					
3-Chloro-1,2-propanediol dinitrate	$C_3H_5ClN_2O_6$	17.50^{20}					
3-Chloro-1-propanol	C_3H_7ClO	36.0^{-58}	0.12436E+03	-0.60841E+00	0.92060E-03		145 - 215
1-Chloro-2-propanol	C_3H_7ClO	59.0^{-120}	-0.19169E+02	0.13605E+01	-0.55567E-02		153 - 177
2-Chloropropene	C_3H_5Cl	8.92^{26}					
3-Chloropropene	C_3H_5Cl	8.2^{20}					
2-Chloropyridine	C_5H_4ClN	27.32^{25}	0.98702E+02	-0.34237E+00	0.34502E-03		298 - 398
2-Chlorotoluene	C_7H_7Cl	4.721^{20}	0.11507E+02	-0.31148E-01	0.27143E-04		273 - 323
3-Chlorotoluene	C_7H_7Cl	5.763^{20}	0.13921E+02	-0.37186E-01	0.31786E-04		273 - 323
4-Chlorotoluene	C_7H_7Cl	6.25^{20}	0.20265E+01	0.40060E-01	-0.87500E-04		293 - 333
Chlorotrifluoromethane	$CClF_3$	3.010^{-150}	0.43677E+01	-0.11020E-01			123 - 173
3-Chloro-1,1,1-trifluoropropane	$C_3H_4ClF_3$	7.32^{22}	0.22361E+02	-0.68840E-01	0.60594E-04		275 - 313
trans-Cinnamaldehyde	C_9H_8O	17.72^{33}	0.41837E+02	-0.11060E+00	0.10401E-03		306 - 354
o-Cresol	C_7H_8O	6.76^{25}	0.21633E+02	-0.71069E-01	0.70590E-04		298 - 453
m-Cresol	C_7H_8O	12.44^{25}	0.81716E+02	-0.35039E+00	0.39878E-03		274 - 463
p-Cresol	C_7H_8O	13.05^{25}	0.70253E+02	-0.28870E+00	0.31979E-03		298 - 453
Cyanoacetic acid	$C_3H_3NO_2$	33.4^{4}					
Cyanoacetylene	C_3HN	72.3^{19}	0.91803E+03	-0.49149E+01	0.69104E-02		281 - 314
Cyclobutanone	C_4H_6O	14.27^{25}	0.43974E+02	-0.15712E+00	0.19264E-03		220 - 317
Cyclododecanone	$C_{12}H_{22}O$	11.4^{30}	0.39327E+02	-0.13248E+00	0.13298E-03		303 - 423
Cycloheptane	C_7H_{14}	2.0784^{20}	0.25136E+01	-0.15089E-02	0.84915E-07		278 - 333
Cycloheptanone	$C_7H_{12}O$	13.16^{25}	0.17511E+03	-0.11221E+01	0.19417E-02		258 - 298
Cycloheptene	C_7H_{12}	2.265^{22}	0.32309E+01	-0.42373E-02	0.32572E-05		227 - 363
1,3-Cyclohexadiene	C_6H_8	2.68^{-89}					
1,4-Cyclohexadiene	C_6H_8	2.211^{23}	0.27459E+01	-0.16975E-02	-0.36461E-06		232 - 356
Cyclohexane	C_6H_{12}	2.0243^{20}	0.24293E+01	-0.12095E-02	-0.58741E-06		283 - 333
Cyclohexanecarboxylic acid	$C_7H_{12}O_2$	2.67^{31}					
1,4-Cyclohexanedione	$C_6H_8O_2$	4.40^{78}					
Cyclohexanemethanol	$C_7H_{14}O$	9.70^{60}	0.10164E+03	-0.45839E+00	0.54762E-03		333 - 368
Cyclohexanethiol	$C_6H_{12}S$	5.420^{25}					
Cyclohexanol	$C_6H_{12}O$	16.40^{20}	0.10173E+03	-0.43072E+00	0.47926E-03		293 - 423
Cyclohexanone	$C_6H_{10}O$	16.1^{20}	0.41577E+02	-0.11463E+00	0.92454E-04		253 - 423
Cyclohexanone oxime	$C_6H_{11}NO$	3.04^{89}					

Name	Mol. form.	ε	*a*	*b*	*c*	*d*	Temp. range/K
Cyclohexene	C_6H_{10}	2.2176^{20}	0.30598E+01	-0.39841E-02	0.37554E-05		141 - 313
Cyclohexyl acetate	$C_8H_{14}O_2$	5.08^{20}					
Cyclohexylamine	$C_6H_{13}N$	4.547^{20}					
Cyclohexyl butanoate	$C_{10}H_{18}O_2$	4.58^{20}					
Cyclohexyl formate	$C_7H_{12}O_2$	6.47^{20}					
2-Cyclohexylphenol	$C_{12}H_{16}O$	3.97^{55}					
4-Cyclohexylphenol	$C_{12}H_{16}O$	4.42^{131}					
Cyclohexyl propanoate	$C_9H_{16}O_2$	4.82^{20}					
Cyclooctane	C_8H_{16}	2.116^{22}	0.25036E+01	-0.12460E-02	-0.23175E-06		295 - 411
cis-Cyclooctene	C_8H_{14}	2.306^{23}	0.31115E+01	-0.32058E-02	0.16713E-05		269 - 406
Cyclopentane	C_5H_{10}	1.9687^{20}	0.24287E+01	-0.15304E-02	-0.13095E-06		278 - 313
Cyclopentanecarbonitrile	C_6H_9N	22.68^{20}	0.69830E+02	-0.25303E+00	0.31491E-03		201 - 293
Cyclopentanol	$C_5H_{10}O$	18.5^{15}	0.10565E+03	-0.44244E+00	0.48657E-03		258 - 323
Cyclopentanone	C_5H_8O	13.58^{25}	0.24083E+02	-0.30286E-01	-0.16802E-04		219 - 298
Cyclopentene	C_5H_8	2.083^{22}	0.28177E+01	-0.27597E-02	0.89346E-06		171 - 319
cis-Decahydronaphthalene	$C_{10}H_{18}$	2.219^{20}	0.25410E+01	-0.11420E-02	0.15092E-06		293 - 373
trans-Decahydronaphthalene	$C_{10}H_{18}$	2.184^{20}	0.26615E+01	-0.21241E-02	0.16864E-05		293 - 373
Decamethylcyclopentasiloxane	$C_{10}H_{30}O_5Si_5$	2.50^{20}					
Decamethyltetrasiloxane	$C_{10}H_{30}O_3Si_4$	2.370^{20}					
Decane	$C_{10}H_{22}$	1.9853^{20}	0.24054E+01	-0.15445E-02	0.44643E-06		253 - 393
1-Decanol	$C_{10}H_{22}O$	7.93^{20}	0.47195E+02	-0.20740E+00	0.24942E-03		293 - 343
2-Decanol	$C_{10}H_{22}O$	5.82^{25}	0.13621E+03	-0.81000E+00	0.12500E-02		288 - 308
3-Decanol	$C_{10}H_{22}O$	4.05^{25}	0.52090E+02	-0.31020E+00	0.50000E-03		288 - 308
4-Decanol	$C_{10}H_{22}O$	3.42^{25}	-0.11260E+02	0.93960E-01	-0.15000E-03		288 - 308
5-Decanol	$C_{10}H_{22}O$	3.24^{25}	-0.25832E+01	0.31456E-01	-0.40000E-04		288 - 308
2-Decanone	$C_{10}H_{20}O$	8.3^{14}					
1-Decene	$C_{10}H_{20}$	2.136^{20}	0.19091E+01	0.33442E-02	-0.87500E-05		273 - 323
cis-5-Decene	$C_{10}H_{20}$	2.071^{25}					
trans-5-Decene	$C_{10}H_{20}$	2.030^{25}					
Decyl acetate	$C_{12}H_{24}O_2$	3.75^{20}					
Decylamine	$C_{10}H_{23}N$	3.31^{20}	0.61497E+01	-0.12801E-01	0.10606E-04		293 - 373
Diacetone alcohol	$C_6H_{12}O_2$	18.2^{25}					
Dibenzofuran	$C_{12}H_8O$	3.00^{100}					
Dibenzylamine	$C_{14}H_{15}N$	3.446^{20}					
Dibenzyl ether	$C_{14}H_{14}O$	3.821^{20}	0.80154E+01	-0.20536E-01	0.21250E-04		293 - 333
Diborane	B_2H_6	1.8725^{-93}	0.23848E+01	-0.29501E-02	0.64189E-06		108 - 181
o-Dibromobenzene	$C_6H_4Br_2$	7.86^{20}	-0.81849E-02	0.62671E-01	-0.12222E-03		293 - 353
m-Dibromobenzene	$C_6H_4Br_2$	4.81^{20}	0.93214E+01	-0.20273E-01	0.16667E-04		293 - 353
p-Dibromobenzene	$C_6H_4Br_2$	2.57^{95}					
1,2-Dibromobutane	$C_4H_8Br_2$	4.74^{20}	0.11199E+03	-0.63334E+00	0.91250E-03		293 - 333
1,3-Dibromobutane	$C_4H_8Br_2$	9.14^{20}	0.34031E+02	-0.13254E+00	0.16250E-03		293 - 333
1,4-Dibromobutane	$C_4H_8Br_2$	8.68^{30}	0.20944E+02	-0.55620E-01	0.50000E-04		303 - 333
2,3-Dibromobutane	$C_4H_8Br_2$	6.245^{25}	0.23849E+02	-0.96300E-01	0.12500E-03		293 - 333
1,10-Dibromodecane	$C_{10}H_{20}Br_2$	6.56^{30}	0.17350E+02	-0.50328E-01	0.48633E-04		303 - 368
Dibromodichloromethane	CBr_2Cl_2	2.542^{25}	0.32330E+01	-0.23162E-02			298 - 333
Dibromodifluoromethane	CBr_2F_2	2.939^{0}	0.67296E+01	-0.22133E-01	0.30213E-04		139 - 273
1,2-Dibromoethane	$C_2H_4Br_2$	4.9612^{20}	0.67142E+01	-0.59800E-02			293 - 313
cis-1,2-Dibromoethene	$C_2H_2Br_2$	7.08^{25}					
trans-1,2-Dibromoethene	$C_2H_2Br_2$	2.88^{25}					
1,2-Dibromoheptane	$C_7H_{14}Br_2$	3.77^{25}					
2,3-Dibromoheptane	$C_7H_{14}Br_2$	5.08^{25}					
3,4-Dibromoheptane	$C_7H_{14}Br_2$	4.70^{25}					
1,6-Dibromohexane	$C_6H_{12}Br_2$	8.52^{25}	-0.55185E+01	0.11746E+00	-0.23658E-03		274 - 328
3,4-Dibromohexane	$C_6H_{12}Br_2$	6.732^{25}					
Dibromomethane	CH_2Br_2	7.77^{10}	0.18060E+02	-0.36333E-01			283 - 313
1,2-Dibromo-2-methylpropane	$C_4H_8Br_2$	4.1^{20}					
1,9-Dibromononane	$C_9H_{18}Br_2$	7.153^{20}	0.18931E+02	-0.57764E-01	0.60000E-04		293 - 343
1,8-Dibromooctane	$C_8H_{16}Br_2$	7.43^{25}	0.94117E+00	0.61520E-01	-0.13333E-03		298 - 328
1,2-Dibromopentane	$C_5H_{10}Br_2$	4.39^{25}					
1,4-Dibromopentane	$C_5H_{10}Br_2$	9.05^{20}	0.26443E+02	-0.88640E-01	0.10000E-03		293 - 333

Name	Mol. form.	ε	*a*	*b*	*c*	*d*	Temp. range/K
1,5-Dibromopentane	$C_5H_{10}Br_2$	9.14^{30}	0.38192E+02	-0.15648E+00	0.20000E-03		303 - 333
1,2-Dibromopropane	$C_3H_6Br_2$	4.60^{10}	0.54973E+01	-0.31695E-02			283 - 333
1,3-Dibromopropane	$C_3H_6Br_2$	9.482^{20}	0.29193E+02	-0.94450E-01	0.92800E-04		293 - 368
1,2-Dibromotetrafluoroethane	$C_2Br_2F_4$	2.34^{25}					
N,N-Dibutylacetamide	$C_{10}H_{21}NO$	19.1^{20}					
Dibutylamine	$C_8H_{19}N$	2.765^{20}	0.52504E+01	-0.10538E-01	0.71485E-05		243 - 323
Dibutyl ether	$C_8H_{18}O$	3.0830^{20}	0.65383E+01	-0.16172E-01	0.14969E-04		293 - 314
N,N-Dibutylformamide	$C_9H_{19}NO$	18.4^{20}					
Di-*tert*-butyl ketone	$C_9H_{18}O$	10.0^{14}					
Dibutyl phthalate	$C_{16}H_{22}O_4$	6.58^{20}	0.12444E+02	-0.20000E-01			293 - 333
Dibutyl sebacate	$C_{18}H_{34}O_4$	4.54^{20}					
Dibutyl sulfide	$C_8H_{18}S$	4.29^{25}					
Dibutyl sulfone	$C_8H_{18}O_2S$	25.72^{50}	0.66248E+02	-0.16417E+00	0.12001E-03		323 - 398
Dibutyl sulfoxide	$C_8H_{18}OS$	24.73^{40}	0.67156E+02	-0.16448E+00	0.92275E-04		313 - 393
Dibutyl tartrate	$C_{12}H_{22}O_6$	9.4^{41}					
Dichloroacetic acid	$C_2H_2Cl_2O_2$	8.33^{20}	0.11014E+02	-0.10859E-01	0.49242E-05		284 - 363
Dichloroacetic anhydride	$C_4H_2Cl_4O_3$	15.8^{25}					
1,1-Dichloroacetone	$C_3H_4Cl_2O$	14.6^{20}					
o-Dichlorobenzene	$C_6H_4Cl_2$	10.12^{20}	0.13629E+02	0.10622E-02	-0.44444E-04		293 - 353
m-Dichlorobenzene	$C_6H_4Cl_2$	5.02^{20}	0.77565E+01	-0.93333E-02	-0.26880E-14		293 - 353
p-Dichlorobenzene	$C_6H_4Cl_2$	2.3943^{55}	0.26999E+01	-0.35325E-03	-0.17619E-05		328 - 363
1,2-Dichlorobutane	$C_4H_8Cl_2$	7.74^{20}	0.31925E+02	-0.13232E+00	0.17007E-03		293 - 356
1,4-Dichlorobutane	$C_4H_8Cl_2$	9.30^{35}	0.59766E+01	0.49300E-01	-0.12500E-03		308 - 338
1,10-Dichlorodecane	$C_{10}H_{20}Cl_2$	6.68^{35}	-0.57423E+01	0.94220E-01	-0.17500E-03		308 - 338
Dichlorodifluoromethane	CCl_2F_2	3.500^{-150}	0.46984E+01	-0.97600E-02			123 - 223
1,1-Dichloroethane	$C_2H_4Cl_2$	10.10^{25}	0.24429E+02	-0.48000E-01			288 - 318
1,2-Dichloroethane	$C_2H_4Cl_2$	10.42^{20}	0.24404E+02	-0.47892E-01			293 - 343
1,1-Dichloroethene	$C_2H_2Cl_2$	4.60^{20}					
cis-1,2-Dichloroethene	$C_2H_2Cl_2$	9.20^{25}					
trans-1,2-Dichloroethene	$C_2H_2Cl_2$	2.14^{20}					
1,7-Dichloroheptane	$C_7H_{14}Cl_2$	8.34^{25}					
1,6-Dichlorohexane	$C_6H_{12}Cl_2$	8.60^{35}	0.11277E+02	0.67200E-02	-0.50000E-04		308 - 338
Dichloromethane	CH_2Cl_2	8.93^{25}	0.40452E+02	-0.17748E+00	0.23942E-03		184 - 306
(Dichloromethyl)benzene	$C_7H_6Cl_2$	6.9^{20}					
1,2-Dichloro-2-methylpropane	$C_4H_8Cl_2$	7.15^{23}	0.39429E+02	-0.20028E+00	0.30917E-03		165 - 296
1,1-Dichloro-1-nitroethane	$C_2H_3Cl_2NO_2$	16.3^{30}	0.37576E+02	-0.70400E-01			303 - 333
1,8-Dichlorooctane	$C_8H_{16}Cl_2$	7.64^{25}					
1,2-Dichloropentane	$C_5H_{10}Cl_2$	6.89^{20}	0.19016E+02	-0.57954E-01	0.56801E-04		293 - 356
1,5-Dichloropentane	$C_5H_{10}Cl_2$	9.92^{25}					
1,2-Dichloropropane, (±)-	$C_3H_6Cl_2$	8.37^{20}	0.18915E+02	-0.35907E-01			281 - 323
1,3-Dichloropropane	$C_3H_6Cl_2$	10.27^{30}	0.21609E+02	-0.37333E-01			303 - 333
2,2-Dichloropropane	$C_3H_6Cl_2$	11.37^{20}	0.32421E+02	-0.72188E-01			245 - 293
2,5-Dichlorostyrene	$C_8H_6Cl_2$	2.58^{25}					
1,2-Dichloro-1,1,2,2-tetrafluoroethane	$C_2Cl_2F_4$	2.4842^{0}	0.36663E+01	-0.42271E-02	-0.36255E-06		193 - 273
2,4-Dichlorotoluene	$C_7H_6Cl_2$	5.68^{28}					
2,6-Dichlorotoluene	$C_7H_6Cl_2$	3.36^{28}					
3,4-Dichlorotoluene	$C_7H_6Cl_2$	9.39^{28}					
Dicyclohexyl ether	$C_{12}H_{22}O$	3.45^{20}	0.95324E+01	-0.31740E-01	0.37500E-04		293 - 333
Dicyclohexyl hexanedioate	$C_{18}H_{30}O_4$	4.84^{35}					
Dicyclopentadiene	$C_{10}H_{12}$	2.43^{40}	0.30564E+01	-0.20000E-02	0.82443E-15		313 - 373
Didecyl ether	$C_{20}H_{42}O$	2.644^{20}	0.41465E+01	-0.62240E-02	0.37500E-05		293 - 333
Diethanolamine	$C_4H_{11}NO_2$	25.75^{20}	0.73435E+02	-0.21377E+00	0.17500E-03		273 - 323
Diethoxydimethylsilane	$C_6H_{16}O_2Si$	3.216^{25}					
1,2-Diethoxyethane	$C_6H_{14}O_2$	3.90^{20}	0.99099E+01	-0.33403E-01	0.44048E-04		223 - 303
Diethoxymethane	$C_5H_{12}O_2$	2.527^{20}	0.25294E+01	0.73988E-04	-0.28331E-06		227 - 293
N,N-Diethylacetamide	$C_6H_{13}NO$	32.1^{20}					
Diethylamine	$C_4H_{11}N$	3.680^{20}	0.26462E+02	-0.13750E+00	0.20373E-03		243 - 323
N,N-Diethylaniline	$C_{10}H_{15}N$	5.15^{30}	0.50773E+01	0.15399E-01	-0.50000E-04		303 - 328
o-Diethylbenzene	$C_{10}H_{14}$	2.594^{20}					
m-Diethylbenzene	$C_{10}H_{14}$	2.369^{20}					

Name	Mol. form.	ε	a	b	c	d	Temp. range/K
p-Diethylbenzene	$C_{10}H_{14}$	2.259^{20}					
Diethyl carbonate	$C_5H_{10}O_3$	2.820^{24}					
Diethylene glycol	$C_4H_{10}O_3$	31.82^{20}	0.13973E+03	-0.54725E+00	0.61149E-03		288 - 343
Diethylene glycol dimethyl ether	$C_6H_{14}O_3$	7.23^{25}	0.28291E+02	-0.11236E+00	0.14000E-03		298 - 333
Diethyl ether	$C_4H_{10}O$	4.2666^{20}	0.79725E+01	-0.12519E-01			283 - 301
N,N-Diethylformamide	$C_5H_{11}NO$	29.6^{20}					
Diethyl fumarate	$C_8H_{12}O_4$	6.56^{23}					
Diethyl glutarate	$C_9H_{16}O_4$	6.659^{30}					
Diethyl hexanedioate	$C_{10}H_{18}O_4$	6.109^{20}	0.14824E+02	-0.40749E-01	0.37600E-04		293 - 343
Diethyl maleate	$C_8H_{12}O_4$	7.560^{25}	0.13953E+02	-0.21969E-01	0.17817E-05		298 - 343
Diethyl malonate	$C_7H_{12}O_4$	7.550^{31}	0.14809E+02	-0.31207E-01	0.24066E-04		304 - 393
1,3-Diethyl-5-methylbenzene	$C_{11}H_{16}$	2.264^{20}					
Diethyl nonanedioate	$C_{13}H_{24}O_4$	5.133^{30}					
Diethyl oxalate	$C_6H_{10}O_4$	8.266^{20}	0.21938E+02	-0.66226E-01	0.66800E-04		293 - 368
Diethyl phthalate	$C_{12}H_{14}O_4$	7.86^{20}					
Diethyl sebacate	$C_{14}H_{26}O_4$	4.995^{30}	0.39143E+02	-0.20965E+00	0.32000E-03		303 - 313
Diethylsilane	$C_4H_{12}Si$	2.544^{20}					
Diethyl succinate	$C_8H_{14}O_4$	6.098^{20}	0.80213E+01	0.11810E-02	-0.26400E-04		293 - 343
Diethyl sulfate	$C_4H_{10}O_4S$	29.2^{20}					
Diethyl sulfide	$C_4H_{10}S$	5.723^{25}					
Diethyl sulfite	$C_4H_{10}O_3S$	15.6^{20}					
o-Difluorobenzene	$C_6H_4F_2$	13.38^{28}	0.59107E+02	-0.23611E+00	0.27987E-03		273 - 323
m-Difluorobenzene	$C_6H_4F_2$	5.01^{28}	0.14448E+02	-0.46982E-01	0.51948E-04		273 - 323
Difluoromethane	CH_2F_2	53.74^{-121}	0.19428E+03	-0.12939E+01	0.24280E-02		152 - 224
Dihexyl phthalate	$C_{20}H_{30}O_4$	5.62^{20}					
o-Diiodobenzene	$C_6H_4I_2$	5.41^{50}	0.31150E+02	-0.14428E+00	0.20000E-03		323 - 353
m-Diiodobenzene	$C_6H_4I_2$	4.11^{50}					
p-Diiodobenzene	$C_6H_4I_2$	2.88^{120}					
cis-1,2-Diiodoethene	$C_2H_2I_2$	4.46^{72}					
Diiodomethane	CH_2I_2	5.32^{25}					
Diisobutyl hexanedioate	$C_{14}H_{26}O_4$	5.19^{20}					
Diisopentyl ether	$C_{10}H_{22}O$	2.817^{20}	0.44690E+01	-0.63710E-02	0.25000E-05		293 - 323
Diisopropyl ether	$C_6H_{14}O$	3.805^{30}					
Diisopropyl oxalate	$C_8H_{14}O_4$	6.403^{20}	0.10709E+02	-0.16328E-01	0.56000E-05		293 - 368
1,2-Dimethoxybenzene	$C_8H_{10}O_2$	4.45^{20}	0.74604E+01	-0.13445E-01	0.10737E-04		293 - 443
1,3-Dimethoxybenzene	$C_8H_{10}O_2$	5.363^{25}	0.11911E+02	-0.30804E-01	0.29643E-04		298 - 358
1,4-Dimethoxybenzene	$C_8H_{10}O_2$	5.60^{61}	0.11289E+02	-0.20765E-01	0.11987E-04		334 - 463
Dimethoxydimethylsilane	$C_4H_{12}O_2Si$	3.663^{25}					
1,2-Dimethoxyethane	$C_4H_{10}O_2$	7.30^{24}	0.48832E+02	-0.24218E+00	0.34413E-03		256 - 318
Dimethoxymethane	$C_3H_8O_2$	2.644^{20}	0.25877E+01	-0.93019E-03	0.38472E-05		171 - 293
N,N-Dimethylacetamide	C_4H_9NO	38.85^{21}	0.15420E+03	-0.57506E+00	0.61911E-03		294 - 433
N,N-Dimethylaniline	$C_8H_{11}N$	4.90^{25}	0.84052E+01	-0.13549E-01	0.62835E-05		289 - 453
2,6-Dimethylanisole	$C_9H_{12}O$	3.780^{20}	0.76700E+01	-0.18298E-01	0.17143E-04		293 - 333
3,5-Dimethylanisole	$C_9H_{12}O$	3.711^{20}	0.54981E+01	-0.56651E-02	-0.14286E-05		293 - 333
N,N-Dimethylbenzamide	$C_9H_{11}NO$	20.77^{45}	0.76725E+02	-0.26908E+00	0.29409E-03		318 - 443
α,α-Dimethylbenzenemethanol	$C_9H_{12}O$	5.61^{30}	0.57072E+01	0.86568E-02	-0.29580E-04		303 - 373
2,3-Dimethyl-1,3-butadiene	C_6H_{10}	2.102^{20}	0.26258E+01	-0.17990E-02	0.12035E-06		223 - 323
2,2-Dimethylbutane	C_6H_{14}	1.869^{20}	0.22740E+01	-0.96229E-03	-0.14286E-05		273 - 313
2,3-Dimethylbutane	C_6H_{14}	1.889^{20}	0.24305E+01	-0.20081E-02	0.53571E-06		273 - 323
2,2-Dimethyl-1-butanol	$C_6H_{14}O$	10.5^{20}	0.14054E+03	-0.72925E+00	0.97821E-03		243 - 393
3,3-Dimethyl-2-butanone	$C_6H_{12}O$	12.73^{20}	0.66857E+02	-0.28552E+00	0.34422E-03		243 - 293
Dimethyl carbonate	$C_3H_6O_3$	3.087^{25}					
1,2-Dimethylcyclohexene	C_8H_{14}	2.144^{23}	0.26443E+01	-0.17973E-02	0.35815E-06		211 - 374
1,3-Dimethylcyclohexene	C_8H_{14}	2.182^{23}	0.29951E+01	-0.34615E-02	0.24026E-05		213 - 373
Dimethyldiphenoxysilane	$C_{14}H_{16}O_2Si$	3.500^{25}	0.51669E+01	-0.77001E-02	0.70156E-05		283 - 353
Dimethyl disulfide	$C_2H_6S_2$	9.6^{25}	0.19109E+02	-0.32000E-01			298 - 323
Dimethyl ether	C_2H_6O	6.18^{-15}	0.22389E+02	-0.86524E-01	0.91291E-04		155 - 258
N,N-Dimethylformamide	C_3H_7NO	38.25^{20}	0.15364E+03	-0.60367E+00	0.71505E-03		213 - 353
Dimethyl glutarate	$C_7H_{12}O_4$	7.87^{20}	0.20697E+02	-0.57794E-01	0.48405E-04		293 - 433
2,4-Dimethylheptane	C_9H_{20}	1.89^{20}					

Name	Mol. form.	*ε*	*a*	*b*	*c*	*d*	Temp. range/K
2,5-Dimethylheptane	C_9H_{20}	1.89^{20}					
2,6-Dimethylheptane	C_9H_{20}	1.987^{20}					
2,6-Dimethyl-4-heptanone	$C_9H_{18}O$	9.91^{20}	0.33178E+02	-0.11290E+00	0.11454E-03		273 - 393
2,2-Dimethylhexane	C_8H_{18}	1.9498^{20}					
2,5-Dimethylhexane	C_8H_{18}	1.9619^{21}	0.25821E+01	-0.26804E-02	0.19404E-05		294 - 324
3,3-Dimethylhexane	C_8H_{18}	1.9645^{20}					
3,4-Dimethylhexane	C_8H_{18}	1.9814^{19}	0.26849E+01	-0.33712E-02	0.32949E-05		292 - 324
Dimethyl 1,6-hexanedioate	$C_8H_{14}O_4$	6.84^{20}	0.11739E+02	-0.17281E-01	0.11447E-05		293 - 433
2,2-Dimethyl-1-hexanol	$C_8H_{18}O$	4.50^{20}	0.91244E+01	-0.21785E-01	0.21018E-04		283 - 393
2,5-Dimethyl-2-hexene	C_8H_{16}	2.431^{20}					
Dimethyl malonate	$C_5H_8O_4$	9.82^{20}	0.26470E+02	-0.76656E-01	0.67888E-04		293 - 433
1,6-Dimethylnaphthalene	$C_{12}H_{12}$	2.7250^{20}					
2,7-Dimethyloctane	$C_{10}H_{22}$	1.98^{20}					
2,2-Dimethyloctanoic acid	$C_{10}H_{20}O_2$	2.8^{23}					
2,2-Dimethyl-1-octanol	$C_{10}H_{22}O$	7.86^{20}	0.69536E+02	-0.34596E+00	0.46250E-03		293 - 333
2,2-Dimethylpentane	C_7H_{16}	1.915^{20}	0.23414E+01	-0.14362E-02	-0.51322E-07		153 - 353
2,3-Dimethylpentane	C_7H_{16}	1.929^{20}	0.25637E+01	-0.26328E-02	0.16071E-05		273 - 323
2,4-Dimethylpentane	C_7H_{16}	1.902^{20}	0.23979E+01	-0.17436E-02	0.17857E-06		273 - 323
3,3-Dimethylpentane	C_7H_{16}	1.9419^{18}	0.24007E+01	-0.16802E-02	0.36069E-06		291 - 322
2,2-Dimethyl-1-pentanol	$C_7H_{16}O$	6.020^{20}	0.37318E+02	-0.17095E+00	0.22022E-03		283 - 393
Dimethyl phthalate	$C_{10}H_{10}O_4$	8.66^{20}					
2,2-Dimethylpropanal	$C_5H_{10}O$	9.051^{20}	0.18645E+02	-0.32395E-01	-0.16157E-05		280 - 333
2,2-Dimethylpropanamide	$C_5H_{11}NO$	20.13^{25}	0.10400E+03	-0.46017E+00	0.60000E-03		298 - 328
2,2-Dimethylpropanenitrile	C_5H_9N	21.1^{20}	0.58418E+02	-0.16884E+00	0.14131E-03		293 - 453
2,2-Dimethyl-1-propanol	$C_5H_{12}O$	8.35^{60}	0.92350E+02	-0.41870E+00	0.50000E-03		333 - 373
2,5-Dimethylpyrazine	$C_6H_8N_2$	2.436^{20}					
2,6-Dimethylpyrazine	$C_6H_8N_2$	2.653^{35}					
2,4-Dimethylpyridine	C_7H_9N	9.60^{20}	0.25895E+02	-0.73900E-01	0.62500E-04		293 - 333
2,6-Dimethylpyridine	C_7H_9N	7.33^{20}	0.17714E+02	-0.39080E-01	0.12500E-04		293 - 333
2,6-Dimethylpyridine-1-oxide	C_7H_9NO	46.11^{25}	0.22765E+03	-0.90760E+00	0.10011E-02		298 - 398
Dimethyl succinate	$C_6H_{10}O_4$	7.19^{20}	0.13551E+02	-0.23109E-01	0.55440E-05		293 - 433
Dimethyl sulfate	$C_2H_6O_4S$	55.0^{25}					
Dimethyl sulfide	C_2H_6S	6.70^{21}					
Dimethyl sulfone	$C_2H_6O_2S$	47.39^{110}	0.10830E+03	-0.15900E+00			383 - 398
Dimethyl sulfoxide	C_2H_6OS	47.24^{20}	0.38478E+02	0.16939E+00	-0.47423E-03		288 - 343
1,3-Dinitrobenzene	$C_6H_4N_2O_4$	22.9^{92}	0.10406E+03	-0.34133E+00	0.32609E-03		365 - 413
2,2-Dinitropropane	$C_3H_6N_2O_4$	42.4^{52}					
Dioctyl phthalate	$C_{24}H_{38}O_4$	5.22^{20}					
Dioctyl sebacate	$C_{26}H_{50}O_4$	4.01^{26}					
1,4-Dioxane	$C_4H_8O_2$	2.2189^{20}	0.27299E+01	-0.17440E-02			293 - 313
Dipentyl ether	$C_{10}H_{22}O$	2.798^{25}					
Dipentyl phthalate	$C_{18}H_{26}O_4$	6.00^{20}					
Dipentyl sulfide	$C_{10}H_{22}S$	3.826^{25}					
Dipentyl sulfoxide	$C_{10}H_{22}OS$	18.8^{75}					
Diphenylamine	$C_{12}H_{11}N$	3.73^{50}					
1,2-Diphenylethane	$C_{14}H_{14}$	2.47^{58}	0.31178E+01	-0.21572E-02	0.59800E-06		331 - 451
Diphenyl ether	$C_{12}H_{10}O$	3.726^{10}					
Diphenylmethane	$C_{13}H_{12}$	2.540^{30}	0.30638E+01	-0.17286E-02			303 - 333
Diphenyl sulfide	$C_{12}H_{10}S$	5.43^{25}					
Diphenyl sulfone	$C_{12}H_{10}O_2S$	21.1^{133}					
Diphenyl sulfoxide	$C_{12}H_{10}OS$	16.6^{72}					
Dipropylamine	$C_6H_{15}N$	2.923^{20}	0.11376E+02	-0.49796E-01	0.71792E-04		243 - 323
Dipropyl ether	$C_6H_{14}O$	3.38^{24}	0.14600E+02	-0.72670E-01	0.11742E-03		161 - 297
Dipropyl sulfone	$C_6H_{14}O_2S$	32.62^{30}	0.70195E+02	-0.15008E+00	0.86506E-04		303 - 398
Dipropyl sulfoxide	$C_6H_{14}OS$	30.37^{30}	0.84868E+02	-0.23486E+00	0.18198E-03		303 - 373
Divinyl ether	C_4H_6O	3.94^{15}					
Docosane	$C_{22}H_{46}$	2.0840^{20}					
1-Docosanol	$C_{22}H_{46}O$	2.94^{75}	0.82062E+01	-0.25069E-01	0.28571E-04		348 - 373
Dodecane	$C_{12}H_{26}$	2.0120^{20}	0.23697E+01	-0.12200E-02	-0.36375E-16		283 - 363
1-Dodecanol	$C_{12}H_{26}O$	5.82^{30}	0.18518E+02	-0.44859E-01	0.99900E-05		303 - 358

Name	Mol. form.	ε	*a*	*b*	*c*	*d*	Temp. range/K
1-Dodecene	$C_{12}H_{24}$	2.152^{20}	0.22581E+01	0.11106E-02	-0.50000E-05		273 - 323
Dodecyl acetate	$C_{14}H_{28}O_2$	3.6^{20}					
Dodecylamine	$C_{12}H_{27}N$	3.07^{30}	0.27999E+01	0.44810E-02	-0.11905E-04		303 - 373
Eicosamethylnonasiloxane	$C_{20}H_{60}O_8Si_9$	2.645^{20}	0.57840E+01	-0.16568E-01	0.20000E-04		293 - 323
1-Eicosanol	$C_{20}H_{42}O$	3.13^{65}	0.21700E+01	0.12497E-01	-0.28571E-04		338 - 363
Epichlorohydrin	C_3H_5ClO	22.6^{20}					
Ethane	C_2H_6	1.9356^{-178}	0.20185E+01	-0.51493E-03	-0.48148E-05		95 - 295
1,2-Ethanediamine	$C_2H_8N_2$	13.82^{20}	0.48922E+02	-0.17021E+00	0.17262E-03		273 - 333
1,2-Ethanediol	$C_2H_6O_2$	41.4^{20}	0.14355E+03	-0.48573E+00	0.46703E-03		293 - 423
1,2-Ethanediol, diacetate	$C_6H_{10}O_4$	7.7^{17}	0.25093E+02	-0.95171E-01	0.12224E-03		223 - 290
1,2-Ethanediol, dihexadecanoate	$C_{34}H_{66}O_4$	2.89^{75}					
1,2-Ethanediol, dinitrate	$C_2H_4N_2O_6$	28.26^{20}					
1,2-Ethanediol, distearate	$C_{38}H_{74}O_4$	2.79^{80}					
1,2-Ethanediol, ditetradecanoate	$C_{30}H_{58}O_4$	2.98^{70}					
1,2-Ethanediol, monoacetate	$C_4H_8O_3$	12.95^{30}					
1,2-Ethanediol, monosulfite	$C_2H_4O_3S$	39.6^{25}	0.85483E+02	-0.15400E+00			298 - 328
1,2-Ethanedithiol	$C_2H_6S_2$	7.26^{20}	0.11228E+02	-0.13500E-01			293 - 333
Ethanethiol	C_2H_6S	6.667^{25}					
Ethanol	C_2H_6O	25.3^{20}	0.15145E+03	-0.87020E+00	0.19570E-02	-0.15512E-05	163 - 523
Ethanolamine	C_2H_7NO	31.94^{20}	0.14890E+03	-0.62491E+00	0.77143E-03		253 - 293
Ethoxyacetylene	C_4H_6O	8.05^{25}					
4-Ethoxyaniline	$C_8H_{11}NO$	7.43^{25}					
Ethoxybenzene	$C_8H_{10}O$	4.216^{20}	-0.15043E+02	0.13752E+00	-0.24500E-03		293 - 313
2-Ethoxyethanol	$C_4H_{10}O_2$	13.38^{25}					
2-Ethoxyethyl acetate	$C_6H_{12}O_3$	7.567^{30}	0.23290E+02	-0.71566E-01	0.65000E-04		303 - 323
1-Ethoxynaphthalene	$C_{12}H_{12}O$	3.3^{19}					
Ethoxytrimethylsilane	$C_5H_{14}OSi$	3.013^{25}					
N-Ethylacetamide	C_4H_9NO	135.0^{20}	0.74494E+03	-0.31400E+01	0.36131E-02		213 - 353
Ethyl acetate	$C_4H_8O_2$	6.0814^{20}	0.15646E+02	-0.44066E-01	0.39137E-04		293 - 433
Ethyl acetoacetate	$C_6H_{10}O_3$	14.0^{20}					
Ethyl acrylate	$C_5H_8O_2$	6.05^{30}	0.47827E+02	-0.24394E+00	0.35000E-03		303 - 343
Ethylamine	C_2H_7N	8.7^{0}	0.30163E+02	-0.79000E-01			233 - 273
Ethyl 2-aminobenzoate	$C_9H_{11}NO_2$	4.14^{25}					
4-Ethylaniline	$C_8H_{11}N$	4.84^{25}					
N-Ethylaniline	$C_8H_{11}N$	5.87^{20}					
N-Ethylbenzamide	$C_9H_{11}NO$	42.6^{80}	-0.20109E+03	0.17866E+01	-0.31065E-02		353 - 389
Ethylbenzene	C_8H_{10}	2.4463^{20}	0.35969E+01	-0.53169E-02	0.47500E-05		293 - 323
α-Ethylbenzenemethanol	$C_9H_{12}O$	6.68^{20}	0.44520E+02	-0.21505E+00	0.29443E-03		233 - 373
Ethyl benzoate	$C_9H_{10}O_2$	6.20^{20}	0.18216E+02	-0.62361E-01	0.72884E-04		288 - 343
Ethyl benzoylacetate	$C_{11}H_{12}O_3$	13.50^{30}	0.93644E+01	0.74280E-01	-0.20000E-03		303 - 323
Ethyl bromoacetate	$C_4H_7BrO_2$	9.75^{30}	0.15627E+02	-0.19600E-01			303 - 333
Ethyl 2-bromobutanoate	$C_6H_{11}BrO_2$	8.57^{30}	0.49005E+02	-0.23193E+00	0.32500E-03		303 - 333
Ethyl 2-bromo-2-methylpropanoate	$C_6H_{11}BrO_2$	8.55^{30}	0.77044E+02	-0.40784E+00	0.60000E-03		303 - 333
Ethyl 2-bromopropanoate	$C_5H_9BrO_2$	9.4^{20}					
Ethyl butanoate	$C_6H_{12}O_2$	5.18^{28}	0.48698E+02	-0.25660E+00	0.37237E-03		301 - 343
2-Ethylbutanoic acid	$C_6H_{12}O_2$	2.72^{23}					
2-Ethyl-1-butanol	$C_6H_{14}O$	6.19^{89}					
Ethyl 2-butenoate	$C_6H_{10}O_2$	5.4^{20}					
Ethyl carbamate	$C_3H_7NO_2$	14.14^{55}	0.32431E+02	-0.65097E-01	0.28571E-04		328 - 368
Ethyl chloroformate	$C_3H_5ClO_2$	9.736^{36}	0.15356E+02	-0.18250E-01			309 - 349
Ethyl 2-chloropropanoate	$C_5H_9ClO_2$	11.95^{30}	0.25965E+02	-0.46250E-01			303 - 343
Ethyl 3-chloropropanoate	$C_5H_9ClO_2$	10.19^{30}	0.21951E+02	-0.38750E-01			303 - 343
Ethyl *trans*-cinnamate	$C_{11}H_{12}O_2$	5.63^{20}					
Ethylcyanoacetate	$C_5H_7NO_2$	31.62^{-10}					
Ethylcyclobutane	C_6H_{12}	1.965^{20}					
Ethyl cyclohexanecarboxylate	$C_9H_{16}O_2$	4.64^{20}					
Ethylcyclopropane	C_5H_{10}	1.933^{20}					
Ethyl decanoate	$C_{12}H_{24}O_2$	3.75^{20}	0.70969E+01	-0.15080E-01	0.12500E-04		293 - 353
1-Ethyl-3,5-dimethylbenzene	$C_{10}H_{14}$	2.275^{20}					
Ethyl dodecanoate	$C_{14}H_{28}O_2$	3.94^{0}					

Name	Mol. form.	ε	a	b	c	d	Temp. range/K
Ethylene	C_2H_4	1.4833^{-3}	0.13546E+01	0.62614E-02	-0.21374E-04		200 - 270
Ethylene carbonate	$C_3H_4O_3$	89.78^{40}	0.20746E+03	-0.37610E+00			313 - 343
Ethyleneimine	C_2H_5N	18.3^{25}	0.61405E+02	-0.14474E+00			273 - 298
N-Ethylformamide	C_3H_7NO	102.7^{25}	0.64764E+03	-0.28499E+01	0.34286E-02		298 - 338
Ethyl formate	$C_3H_6O_2$	8.57^{15}	0.15884E+02	-0.25333E-01			288 - 318
2-Ethylheptanoic acid	$C_9H_{18}O_2$	1.98^{20}					
Ethyl hexadecanoate	$C_{18}H_{36}O_2$	3.07^{30}	0.57938E+01	-0.12294E-01	0.10919E-04		303 - 455
3-Ethylhexane	C_8H_{18}	1.9617^{20}					
2-Ethyl-1,3-hexanediol	$C_8H_{18}O_2$	18.73^{20}	0.57919E+02	-0.17128E+00	0.12949E-03		233 - 333
Ethyl hexanoate	$C_8H_{16}O_2$	4.45^{20}	0.11007E+02	-0.32800E-01	0.35714E-04		253 - 353
2-Ethylhexanoic acid	$C_8H_{16}O_2$	2.64^{23}					
2-Ethyl-1-hexanol	$C_8H_{18}O$	7.58^{25}	0.86074E+02	-0.42636E+00	0.55078E-03		208 - 318
Ethyl isocyanate	C_3H_5NO	19.7^{20}					
Ethyl isopentyl ether	$C_7H_{16}O$	3.955^{20}	0.66541E+01	-0.55450E-02	-0.12500E-04		293 - 323
Ethyl isothiocyanate	C_3H_5NS	19.6^{20}					
Ethyl lactate	$C_5H_{10}O_3$	15.4^{30}	0.31225E+02	-0.43531E-01	-0.28571E-04		273 - 373
Ethyl methacrylate	$C_6H_{10}O_2$	5.68^{30}	0.40962E+02	-0.20520E+00	0.29286E-03		303 - 343
Ethyl 3-methylbutanoate	$C_7H_{14}O_2$	4.71^{20}					
Ethyl *N*-methylcarbamate	$C_4H_9NO_2$	21.10^{25}	0.11477E+03	-0.47568E+00	0.54127E-03		298 - 373
Ethyl methyl carbonate	$C_4H_8O_3$	2.985^{20}					
3-Ethyl-3-methylpentane	C_8H_{18}	1.9869^{18}	0.25983E+01	-0.28027E-02	0.24195E-05		292 - 324
Ethyl nitrate	$C_2H_5NO_3$	19.7^{20}					
1-Ethyl-2-nitrobenzene	$C_8H_9NO_2$	21.9^{0}					
Ethyl octadecanoate	$C_{20}H_{40}O_2$	2.958^{40}	0.70930E+01	-0.19081E-01	0.19555E-04		331 - 440
Ethyl *cis*-9-octadecenoate	$C_{20}H_{38}O_2$	3.17^{28}	0.57033E+01	-0.11223E-01	0.93447E-05		301 - 423
3-Ethylpentane	C_7H_{16}	1.942^{20}	0.23771E+01	-0.15140E-02	0.10093E-06		163 - 363
Ethyl pentanoate	$C_7H_{14}O_2$	4.71^{18}					
3-Ethyl-3-pentanol	$C_7H_{16}O$	3.158^{20}					
3-Ethyl-2-pentene	C_7H_{14}	2.051^{20}					
Ethyl pentyl ether	$C_7H_{16}O$	3.6^{23}					
Ethyl phenylacetate	$C_{10}H_{12}O_2$	5.320^{20}					
Ethyl phenyl sulfone	$C_8H_{10}O_2S$	39.0^{75}					
Ethyl propanoate	$C_5H_{10}O_2$	5.76^{20}					
2-Ethylpyridine	C_7H_9N	8.33^{20}	0.36397E+02	-0.15070E+00	0.18750E-03		293 - 333
4-Ethylpyridine	C_7H_9N	10.98^{20}	-0.73831E+01	0.14326E+00	-0.27500E-03		293 - 333
Ethyl 4-pyridinecarboxylate	$C_8H_9NO_2$	8.95^{20}					
Ethyl salicylate	$C_9H_{10}O_3$	8.48^{35}	0.18910E+02	-0.35623E-01	0.46529E-05		225 - 321
Ethyl silicate	$C_8H_{20}O_4Si$	2.50^{20}					
4-Ethylstyrene	$C_{10}H_{12}$	3.350^{25}					
Ethyl tetradecanoate	$C_{16}H_{32}O_2$	3.50^{20}	0.52642E+01	-0.60000E-02	-0.47358E-15		293 - 353
(Ethylthio)benzene	$C_8H_{10}S$	4.95^{25}					
Ethyl thiophene-2-carboxylate	$C_7H_8O_2S$	6.18^{20}					
2-Ethyltoluene	C_9H_{12}	2.595^{20}					
3-Ethyltoluene	C_9H_{12}	2.365^{20}					
4-Ethyltoluene	C_9H_{12}	2.265^{20}					
Ethyl trichloroacetate	$C_4H_5Cl_3O_2$	8.428^{20}					
Ethyl undecanoate	$C_{13}H_{26}O_2$	3.55^{20}					
Eucalyptol	$C_{10}H_{18}O$	4.57^{25}					
(±)-Fenchone	$C_{10}H_{16}O$	12.8^{21}					
Fluorine	F_2	1.4913^{-220}	0.14144E+01	0.26387E-02	-0.28356E-04		54 - 144
Fluorobenzene	C_6H_5F	5.465^{20}					
1-Fluoro-2-iodobenzene	C_6H_4FI	8.22^{25}					
1-Fluoro-4-iodobenzene	C_6H_4FI	3.12^{25}					
Fluoromethane	CH_3F	51.0^{-142}	0.11338E+03	-0.63979E+00	0.96983E-03		150 - 299
1-Fluorooctane	$C_8H_{17}F$	3.89^{20}					
1-Fluoropentane	$C_5H_{11}F$	3.931^{20}					
2-Fluorotoluene	C_7H_7F	4.23^{25}					
3-Fluorotoluene	C_7H_7F	5.41^{25}					
4-Fluorotoluene	C_7H_7F	5.88^{25}					
Formamide	CH_3NO	111.0^{20}	0.26076E+03	-0.61145E+00	0.34296E-03		278 - 333

Name	Mol. form.	ε	*a*	*b*	*c*	*d*	Temp. range/K
Formic acid	CH_2O_2	51.1^{25}	0.14040E+03	-0.24673E+00	-0.17151E-03		287 - 358
Furan	C_4H_4O	2.94^{25}	0.13636E+01	0.12864E-01	-0.22701E-04		188 - 277
Furfural	$C_5H_4O_2$	42.1^{20}					
Furfuryl alcohol	$C_5H_6O_2$	16.85^{25}					
Germanium(IV) bromide	Br_4Ge	2.955^{27}	0.34450E+01	-0.16083E-02			300 - 316
Germanium(IV) chloride	Cl_4Ge	2.463^{0}	-0.55078E+01	0.64881E-01	-0.13091E-03		246 - 273
D-Glucitol	$C_6H_{14}O_6$	35.5^{80}					
Glycerol	$C_3H_8O_3$	46.53^{20}	0.77503E+02	-0.37984E-01	-0.23107E-03		288 - 343
Glycerol 1-acetate	$C_5H_{10}O_4$	38.57^{-31}	0.10653E+03	-0.26439E+00	-0.62371E-04		215 - 242
Glycerol 1,3-diacetate	$C_7H_{12}O_5$	9.80^{15}	0.28321E+02	-0.89073E-01	0.86891E-04		258 - 374
Glycerol triacetate	$C_9H_{14}O_6$	7.11^{20}	0.17819E+02	-0.53656E-01	0.57759E-04		219 - 304
Glycerol trielaidate	$C_{57}H_{104}O_6$	2.980^{40}					
Glycerol trihexanoate	$C_{21}H_{38}O_6$	4.476^{20}					
Glycerol trilaurate	$C_{39}H_{74}O_6$	3.287^{40}					
Glycerol trioctanoate	$C_{27}H_{50}O_6$	3.931^{20}					
Glycerol trioleate	$C_{57}H_{104}O_6$	3.109^{20}					
Glycerol tripalmitate	$C_{51}H_{98}O_6$	2.901^{55}	-0.29131E+01	0.32206E-01	-0.44154E-04		328 - 393
Glycerol tristearate	$C_{57}H_{110}O_6$	2.740^{80}					
Helium	He	1.0555^{-271}	0.10640E+01	-0.35584E-02			2 - 4
Heptadecane	$C_{17}H_{36}$	2.0578^{20}	0.23627E+01	-0.10400E-02	-0.10397E-12		293 - 308
1-Heptadecanol	$C_{17}H_{36}O$	3.41^{60}					
9-Heptadecanone	$C_{17}H_{34}O$	5.43^{55}	0.44176E+02	-0.21183E+00	0.28571E-03		328 - 363
1,6-Heptadiene	C_7H_{12}	2.161^{20}	0.30815E+01	-0.36095E-02	0.16354E-05		184 - 293
2,2,3,3,4,4,4-Heptafluoro-1-butanol	$C_4H_3F_7O$	14.4^{25}					
Heptanal	$C_7H_{14}O$	9.07^{22}					
Heptane	C_7H_{16}	1.9209^{20}	0.24740E+01	-0.22577E-02	0.12428E-05		273 - 373
1-Heptanethiol	$C_7H_{16}S$	4.194^{20}	0.71333E+01	-0.97320E-02	-0.12500E-05		273 - 333
Heptanoic acid	$C_7H_{14}O_2$	3.04^{15}	0.36423E+01	-0.31996E-02	0.39362E-05		288 - 423
1-Heptanol	$C_7H_{16}O$	11.75^{20}	0.60662E+02	-0.24049E+00	0.25155E-03		239 - 513
2-Heptanol, (±)-	$C_7H_{16}O$	9.72^{21}	0.10050E+03	-0.49793E+00	0.64504E-03		207 - 365
3-Heptanol, (*S*)-	$C_7H_{16}O$	7.07^{23}	0.19586E+03	-0.11465E+01	0.17175E-02		248 - 349
4-Heptanol	$C_7H_{16}O$	6.18^{23}	0.28995E+03	-0.18499E+01	0.30109E-02		270 - 301
2-Heptanone	$C_7H_{14}O$	11.95^{20}	0.38348E+02	-0.12531E+00	0.12005E-03		253 - 413
3-Heptanone	$C_7H_{14}O$	12.7^{20}					
4-Heptanone	$C_7H_{14}O$	12.60^{20}	0.41520E+02	-0.13839E+00	0.13497E-03		253 - 393
1-Heptene	C_7H_{14}	2.092^{20}	0.21755E+01	0.13896E-02	-0.57049E-05		273 - 323
Heptyl acetate	$C_9H_{18}O_2$	4.2^{20}					
Heptylamine	$C_7H_{17}N$	3.81^{20}	0.87794E+01	-0.24363E-01	0.25325E-04		253 - 373
Heptylbenzene	$C_{13}H_{20}$	2.26^{20}					
Hexachloroacetone	C_3Cl_6O	3.925^{19}	0.76423E+01	-0.15838E-01	0.10618E-04		269 - 303
Hexachloro-1,3-butadiene	C_4Cl_6	2.55^{20}					
Hexadecane	$C_{16}H_{34}$	2.0460^{20}	0.23861E+01	-0.11600E-02	0.25555E-15		293 - 363
Hexadecanoic acid	$C_{16}H_{32}O_2$	2.417^{65}					
1-Hexadecanol	$C_{16}H_{34}O$	3.69^{60}	0.85935E+01	-0.14714E-01	-0.45533E-13		333 - 363
Hexadecyl acetate	$C_{18}H_{36}O_2$	3.19^{35}	0.47310E+01	-0.50000E-02	0.41338E-14		308 - 348
Hexadecylamine	$C_{16}H_{35}N$	2.71^{55}					
Hexadecyl stearate	$C_{34}H_{68}O_2$	2.61^{60}					
1,5-Hexadiene	C_6H_{10}	2.125^{21}	0.30014E+01	-0.28668E-02	-0.31026E-06		151 - 294
cis,cis-2,4-Hexadiene	C_6H_{10}	2.163^{24}	0.27284E+01	-0.17178E-02	-0.62926E-06		234 - 351
trans,trans-2,4-Hexadiene	C_6H_{10}	2.123^{24}	0.26774E+01	-0.16977E-02	-0.55637E-06		232 - 353
Hexaethyldisiloxane	$C_{12}H_{30}OSi_2$	2.259^{25}	0.36559E+01	-0.72406E-02	0.85714E-05		298 - 333
Hexafluorobenzene	C_6F_6	2.029^{25}	0.24041E+01	-0.83086E-03	-0.14286E-05		298 - 338
1,1,1,3,3,3-Hexafluoro-2-propanol	$C_3H_2F_6O$	16.70^{20}					
Hexamethylbenzene	$C_{12}H_{18}$	2.172^{176}	0.35710E+01	-0.46912E-02	0.35088E-05		449 - 489
Hexamethylcyclotrisiloxane	$C_6H_{18}O_3Si_3$	2.139^{70}					
Hexamethyldisilazane	$C_6H_{19}NSi_2$	2.273^{21}	0.23358E+01	0.16127E-02	-0.62078E-05		294 - 333
Hexamethyldisiloxane	$C_6H_{18}OSi_2$	2.179^{20}	0.34537E+01	-0.61530E-02	0.61544E-05		213 - 313
Hexamethylene diisocyanate	$C_8H_{12}N_2O_2$	14.41^{15}	0.26715E+02	-0.42696E-01			288 - 403
Hexamethylphosphoric triamide	$C_6H_{18}N_3OP$	31.3^{20}	0.95666E+02	-0.29769E+00	0.26407E-03		283 - 363
2,6,10,15,19,23-Hexamethyltetracosane	$C_{30}H_{62}$	1.9106^{100}					

Name	Mol. form.	ε	a	b	c	d	Temp. range/K
Hexane	C_6H_{14}	1.8865^{20}	0.19768E+01	0.70933E-03	-0.34470E-05		293 - 473
Hexanenitrile	$C_6H_{11}N$	17.26^{25}					
1-Hexanethiol	$C_6H_{14}S$	4.436^{20}	0.11774E+02	-0.37298E-01	0.41875E-04		273 - 333
1,2,6-Hexanetriol	$C_6H_{14}O_3$	31.5^{12}	0.26127E+03	-0.14552E+01	0.22765E-02		261 - 285
Hexanoic acid	$C_6H_{12}O_2$	2.600^{25}	0.21730E+01	0.14840E-02	-0.16526E-06		298 - 433
1-Hexanol	$C_6H_{14}O$	13.03^{20}	0.62744E+02	-0.24214E+00	0.24704E-03		233 - 513
2-Hexanol	$C_6H_{14}O$	11.06^{25}					
3-Hexanol	$C_6H_{14}O$	9.66^{25}					
2-Hexanone	$C_6H_{12}O$	14.56^{20}	0.70378E+02	-0.29385E+00	0.35289E-03		243 - 293
1-Hexene	C_6H_{12}	2.077^{21}	0.31476E+01	-0.50003E-02	0.46673E-05		149 - 294
trans-2-Hexene	C_6H_{12}	1.978^{22}	0.24338E+01	-0.11323E-02	-0.13720E-05		157 - 295
cis-3-Hexene	C_6H_{12}	2.069^{23}	0.30691E+01	-0.45458E-02	0.39898E-05		155 - 296
trans-3-Hexene	C_6H_{12}	1.954^{20}					
Hexyl acetate	$C_8H_{16}O_2$	4.42^{20}					
Hexylamine	$C_6H_{15}N$	4.08^{20}	0.80244E+01	-0.16627E-01	0.10874E-04		253 - 373
Hexylbenzene	$C_{12}H_{18}$	2.3^{20}					
Hexyl benzoate	$C_{13}H_{18}O_2$	4.80^{20}					
1-Hexyne	C_6H_{10}	2.621^{23}	0.58591E+01	-0.17099E-01	0.20856E-04		184 - 296
Hydrazine	H_4N_2	51.7^{25}	0.22061E+03	-0.89633E+00	0.11066E-02		278 - 323
Hydrogen	H_2	1.2792^{-260}	0.13327E+01	-0.51946E-02			14 - 19
Hydrogen bromide	BrH	8.23^{-86}					
Hydrogen chloride	ClH	14.3^{-114}	0.47316E+02	-0.28455E+00	0.48650E-03		159 - 258
Hydrogen cyanide	CHN	114.9^{20}	0.37331E+04	-0.23180E+02	0.36963E-01		258 - 299
Hydrogen fluoride	FH	83.6^{0}	0.50352E+03	-0.19297E+01	0.14372E-02		200 - 273
Hydrogen iodide	HI	3.87^{-53}	0.51557E+03	-0.44552E+01	0.96795E-02		220 - 236
Hydrogen peroxide	H_2O_2	74.6^{17}	0.48511E+03	-0.23145E+01	0.31020E-02		233 - 303
Hydrogen sulfide	H_2S	5.93^{10}	0.14736E+02	-0.33675E-01	0.96740E-05		212 - 363
(Hydroxyacetyl)benzene	$C_8H_8O_2$	21.33^{25}	0.42286E+02	-0.69215E-01	-0.35714E-05		298 - 368
2-Hydroxybutanoic acid, (±)-	$C_4H_8O_3$	37.7^{23}					
3-Hydroxybutanoic acid, (±)-	$C_4H_8O_3$	31.5^{23}					
3-Hydroxypropanoic acid	$C_3H_6O_3$	30.0^{23}					
Iodine	I_2	11.08^{118}	0.64730E+02	-0.29266E+00	0.39759E-03		391 - 441
Iodine heptafluoride	F_7I	1.75^{25}					
Iodine pentafluoride	F_5I	37.13^{20}	0.95184E+02	-0.19800E+00			273 - 313
Iodobenzene	C_6H_5I	4.59^{20}	0.89442E+01	-0.20008E-01	0.17641E-04		243 - 323
1-Iodobutane	C_4H_9I	6.27^{20}	0.16493E+02	-0.50262E-01	0.52485E-04		293 - 323
2-Iodobutane, (±)-	C_4H_9I	7.873^{20}	0.10883E+02	-0.14680E-02	-0.30000E-04		293 - 323
1-Iodododecane	$C_{12}H_{25}I$	3.91^{25}	0.34641E+01	0.97404E-02	-0.27602E-04		293 - 323
Iodoethane	C_2H_5I	7.82^{20}	0.25598E+02	-0.94367E-01	0.11424E-03		183 - 343
1-Iodoheptane	$C_7H_{15}I$	4.92^{25}	0.11856E+02	-0.33493E-01	0.34368E-04		294 - 323
3-Iodoheptane	$C_7H_{15}I$	6.39^{22}					
1-Iodohexadecane	$C_{16}H_{33}I$	3.57^{20}	0.79531E+01	-0.22859E-01	0.26955E-04		293 - 323
1-Iodohexane	$C_6H_{13}I$	5.35^{20}	0.16685E+02	-0.61309E-01	0.77262E-04		293 - 323
Iodomethane	CH_3I	6.97^{20}	0.24264E+02	-0.93914E-01	0.11926E-03		223 - 303
1-Iodo-3-methylbutane	$C_5H_{11}I$	5.6^{19}					
2-Iodo-2-methylbutane	$C_5H_{11}I$	8.192^{20}					
2-Iodo-2-methylpropane	C_4H_9I	6.65^{10}	0.76780E+01	0.69900E-02	-0.37500E-04		283 - 323
1-Iodooctane	$C_8H_{17}I$	4.67^{20}	0.12452E+02	-0.41229E-01	0.50108E-04		233 - 313
1-Iodopentane	$C_5H_{11}I$	5.78^{20}	0.15753E+02	-0.50543E-01	0.56401E-04		293 - 323
3-Iodopentane	$C_5H_{11}I$	7.432^{20}					
1-Iodopropane	C_3H_7I	7.07^{20}	0.13744E+02	-0.22745E-01			293 - 323
2-Iodopropane	C_3H_7I	8.19^{25}					
3-Iodopropene	C_3H_5I	6.1^{19}					
4-Iodotoluene	C_7H_7I	4.4^{35}					
trans-α-Ionone, (±)-	$C_{13}H_{20}O$	10.78^{19}					
trans-β-Ionone	$C_{13}H_{20}O$	11.66^{24}					
Iron pentacarbonyl	C_5FeO_5	2.602^{20}					
Isobutane	C_4H_{10}	1.7518^{22}	0.23295E+01	-0.19953E-02	0.14197E-06		115 - 303
Isobutene	C_4H_8	2.1225^{16}	0.33701E+01	-0.43295E-02			220 - 289
Isobutyl acetate	$C_6H_{12}O_2$	5.068^{20}	0.14323E+02	-0.46048E-01	0.49286E-04		273 - 323

Name	Mol. form.	ε	a	b	c	d	Temp. range/K
Isobutylbenzene	$C_{10}H_{14}$	2.318^{20}	0.28055E+01	-0.92614E-03	-0.25000E-05		273 - 323
Isobutyl benzoate	$C_{11}H_{14}O_2$	5.39^{18}					
Isobutyl chlorocarbonate	$C_5H_9ClO_2$	9.1^{20}					
Isobutyl formate	$C_5H_{10}O_2$	6.41^{20}					
Isobutyl isocyanate	C_5H_9NO	11.638^{20}	0.38026E+02	-0.12714E+00	0.12679E-03		293 - 353
Isobutyl pentanoate	$C_9H_{18}O_2$	3.8^{19}					
Isobutyl trichloroacetate	$C_6H_9Cl_3O_2$	7.667^{20}					
Isobutyl vinyl ether	$C_6H_{12}O$	3.34^{20}	0.48060E+01	-0.50000E-02	-0.41495E-14		293 - 323
Isopentane	C_5H_{12}	1.845^{20}	0.22384E+01	-0.12985E-02	-0.16182E-06		143 - 293
Isopentyl acetate	$C_7H_{14}O_2$	4.72^{20}					
Isopentyl butanoate	$C_9H_{18}O_2$	4.0^{20}					
Isopentyl formate	$C_6H_{12}O_2$	5.44^{15}	0.29257E+02	-0.14028E+00	0.20000E-03		288 - 323
Isopentyl isopentanoate	$C_{10}H_{20}O_2$	4.39^{15}	0.14698E+02	-0.57726E-01	0.76190E-04		288 - 323
Isopentyl lactate	$C_8H_{16}O_3$	11.2^{0}	0.48649E+02	-0.21253E+00	0.27619E-03		273 - 373
Isopentyl pentanoate	$C_{10}H_{20}O_2$	3.6^{19}					
Isopentyl propanoate	$C_8H_{16}O_2$	5.21^{0}	0.17665E+02	-0.71718E-01	0.95635E-04		273 - 373
Isopentyl salicylate	$C_{12}H_{16}O_3$	7.26^{20}	0.13129E+02	-0.19190E-01	-0.36060E-05		225 - 397
Isopentyl trichloroacetate	$C_7H_{11}Cl_3O_2$	7.287^{20}					
Isopropenylbenzene	C_9H_{10}	2.28^{20}					
Isopropylamine	C_3H_9N	5.6268^{20}	0.40429E+02	-0.21441E+00	0.32634E-03		213 - 298
4-Isopropylbenzaldehyde	$C_{10}H_{12}O$	10.68^{15}					
Isopropylbenzene	C_9H_{12}	2.381^{20}	0.31149E+01	-0.30801E-02	0.19643E-05		273 - 323
1-Isopropyl-4-methylbenzene	$C_{10}H_{14}$	2.2322^{25}	0.25266E+01	-0.25121E-03	-0.24867E-05		277 - 333
Isopropyl nitrite	$C_3H_7NO_2$	13.92^{-13}	0.74578E+02	-0.38283E+00	0.57071E-03		150 - 300
Isoquinoline	C_9H_7N	11.0^{25}	0.14412E+03	-0.79935E+00	0.11839E-02		298 - 323
Krypton	Kr	1.664^{-153}					
Lead(IV) chloride	Cl_4Pb	2.78^{20}					
d-Limonene	$C_{10}H_{16}$	2.3746^{25}					
l-Limonene	$C_{10}H_{16}$	2.3738^{25}					
Maleic anhydride	$C_4H_2O_3$	52.75^{53}					
Manganese(VII) oxide	Mn_2O_7	3.28^{20}	0.37655E+01	-0.16463E-02			283 - 312
D-Mannitol	$C_6H_{14}O_6$	24.6^{170}					
Menthol	$C_{10}H_{20}O$	3.90^{36}	0.68202E+01	-0.15894E-01	0.20837E-04		309 - 358
Mesityl oxide	$C_6H_{10}O$	15.6^{0}					
Methane	CH_4	1.6761^{-182}	0.15996E+01	0.27434E-02	-0.22086E-04		91 - 184
Methanesulfonyl chloride	CH_3ClO_2S	34.0^{20}	0.10384E+03	-0.33838E+00	0.34156E-03		293 - 373
Methanol	CH_4O	33.0^{20}	0.19341E+03	-0.92211E+00	0.12839E-02		177 - 293
Methan-d_1-ol	CH_3DO	31.68^{24}	0.20839E+03	-0.10318E+01	0.14740E-02		176 - 298
2-Methoxyaniline	C_7H_9NO	5.230^{30}	0.79911E+01	-0.92183E-02	0.37879E-06		303 - 393
3-Methoxyaniline	C_7H_9NO	8.76^{25}	0.28179E+02	-0.97840E-01	0.11027E-03		289 - 393
4-Methoxyaniline	C_7H_9NO	7.85^{60}	0.30149E+02	-0.10523E+00	0.11467E-03		333 - 453
4-Methoxybenzaldehyde	$C_8H_8O_2$	22.0^{30}					
4-Methoxybenzenesulfonyl chloride	$C_7H_7ClO_3S$	27.2^{41}					
2-Methoxyethanol	$C_3H_8O_2$	17.2^{25}	0.11803E+03	-0.58000E+00	0.81001E-03		254 - 318
1-Methoxynaphthalene	$C_{11}H_{10}O$	4.020^{20}	0.71885E+01	-0.14838E-01	0.13750E-04		293 - 333
2-Methoxynaphthalene	$C_{11}H_{10}O$	3.563^{80}	0.56702E+01	-0.69754E-02	0.28571E-05		353 - 373
2-Methoxyphenol	$C_7H_8O_2$	11.95^{25}	0.31751E+02	-0.88173E-01	0.72953E-04		291 - 448
3-Methoxyphenol	$C_7H_8O_2$	11.59^{25}	0.37279E+02	-0.12113E+00	0.11698E-03		298 - 433
4-Methoxyphenol	$C_7H_8O_2$	11.05^{61}	0.39483E+02	-0.12142E+00	0.10841E-03		334 - 453
4-Methoxyphenyl isocyanate	$C_8H_7NO_2$	10.26^{60}	0.20780E+02	-0.31571E-01			333 - 403
N-Methylacetamide	C_3H_7NO	179.0^{30}	0.15975E+04	-0.90451E+01	0.18345E-01	-0.12998E-04	303 - 473
Methyl acetate	$C_3H_6O_2$	7.07^{15}	0.13190E+02	-0.21226E-01			276 - 318
Methyl 2-(acetyloxy)benzoate	$C_{10}H_{10}O_4$	5.31^{56}	0.19579E+02	-0.69970E-01	0.80889E-04		329 - 371
Methyl acrylate	$C_4H_6O_2$	7.03^{30}	0.11968E+02	-0.16500E-01			303 - 333
Methylamine	CH_5N	16.7^{-58}	0.34398E+02	-0.73630E-01	-0.41279E-04		198 - 258
Methyl 2-aminobenzoate	$C_8H_9NO_2$	21.9^{25}					
2-Methylaniline	C_7H_9N	6.138^{25}	0.10988E+02	-0.18976E-01	0.91958E-05		298 - 398
3-Methylaniline	C_7H_9N	5.816^{25}	0.13477E+02	-0.35551E-01	0.33135E-04		298 - 398
4-Methylaniline	C_7H_9N	5.058^{60}	0.78897E+01	-0.10196E-01	0.51190E-05		333 - 403
N-Methylaniline	C_7H_9N	5.96^{20}					

Name	Mol. form.	ε	a	b	c	d	Temp. range/K
2-Methylanisole	$C_8H_{10}O$	3.502[20]	0.50825E+01	-0.62297E-02	0.28571E-05		293 - 333
3-Methylanisole	$C_8H_{10}O$	3.967[20]	0.12830E+02	-0.49701E-01	0.66429E-04		293 - 333
4-Methylanisole	$C_8H_{10}O$	3.914[20]	0.86608E+01	-0.23510E-01	0.25000E-04		293 - 333
α-Methylbenzenemethanol	$C_8H_{10}O$	8.77[20]	0.32971E+02	-0.12042E+00	0.12809E-03		293 - 423
4-Methylbenzenethiol	C_7H_8S	4.74[50]	0.87052E+01	-0.15347E-01	0.95238E-05		323 - 358
Methyl benzoate	$C_8H_8O_2$	6.642[30]	0.17486E+02	-0.51027E-01	0.50222E-04		303 - 393
Methyl 3-bromopropanoate	$C_4H_7BrO_2$	5.81[30]	0.36001E+01	0.72500E-02			303 - 343
2-Methyl-1,3-butadiene	C_5H_8	2.098[20]	0.28170E+01	-0.23147E-02	-0.43975E-06		198 - 293
2-Methyl-2-butanethiol	$C_5H_{12}S$	5.087[20]	0.15116E+02	-0.50700E-01	0.56250E-04		273 - 333
Methyl butanoate	$C_5H_{10}O_2$	5.48[28]	0.38604E+02	-0.19171E+00	0.27128E-03		301 - 343
2-Methyl-1-butanol, (±)-	$C_5H_{12}O$	15.63[25]	0.14020E+02	0.13948E+00	-0.45000E-03		288 - 318
3-Methyl-1-butanol	$C_5H_{12}O$	15.63[20]	0.79733E+02	-0.31272E+00	0.32014E-03		173 - 513
2-Methyl-2-butanol	$C_5H_{12}O$	5.78[25]	0.11662E+03	-0.69756E+00	0.10920E-02		268 - 318
3-Methyl-2-butanol, (±)-	$C_5H_{12}O$	12.1[25]					
3-Methyl-2-butanone	$C_5H_{10}O$	10.37[20]	0.30695E+02	-0.10962E+00	0.13810E-03		293 - 328
2-Methyl-1-butene	C_5H_{10}	2.180[20]					
2-Methyl-2-butene	C_5H_{10}	1.979[23]	0.26064E+01	-0.19578E-02	-0.53908E-06		225 - 296
Methyl *trans*-2-butenoate	$C_5H_8O_2$	6.6645[20]					
Methyl carbamate	$C_2H_5NO_2$	18.48[55]	0.36773E+02	-0.55700E-01			328 - 368
Methyl chloroacetate	$C_3H_5ClO_2$	12.0[20]					
Methyl 4-chlorobutanoate	$C_5H_9ClO_2$	9.51[30]	0.17127E+02	-0.25000E-01			303 - 343
Methyl 2-chloropropanoate	$C_4H_7ClO_2$	11.45[30]	0.22449E+02	-0.36250E-01			303 - 343
Methylcyclohexane	C_7H_{14}	2.024[20]					
Methyl cyclohexanecarboxylate	$C_8H_{14}O_2$	4.87[20]					
cis-2-Methylcyclohexanol	$C_7H_{14}O$	9.375[20]	0.17315E+03	-0.98794E+00	0.14634E-02		273 - 323
cis-3-Methylcyclohexanol, (±)-	$C_7H_{14}O$	13.79[20]	0.65896E+02	-0.21954E+00	0.14107E-03		273 - 323
cis-4-Methylcyclohexanol	$C_7H_{14}O$	13.45[20]	0.65021E+02	-0.22896E+00	0.17946E-03		273 - 323
2-Methylcyclohexanone, (±)-	$C_7H_{12}O$	14.0[20]					
3-Methylcyclohexanone, (±)-	$C_7H_{12}O$	12.4[20]					
4-Methylcyclohexanone	$C_7H_{12}O$	12.35[20]					
Methylcyclopentane	C_6H_{12}	1.9853[20]	0.21587E+01	-0.22450E-03	-0.12500E-05		293 - 323
1-Methylcyclopentanol	$C_6H_{12}O$	7.11[37]	0.75444E+02	-0.36617E+00	0.47021E-03		310 - 333
2-Methyl-*N,N*-dimethylaniline	$C_9H_{13}N$	3.4[20]					
4-Methyl-*N,N*-dimethylaniline	$C_9H_{13}N$	3.9[20]					
Methyl dodecanoate	$C_{13}H_{26}O_2$	3.539[20]					
N-Methylformamide	C_2H_5NO	189.0[20]	0.10383E+04	-0.43165E+01	0.48398E-02		276 - 353
Methyl formate	$C_2H_4O_2$	9.20[15]	0.19699E+02	-0.36429E-01			288 - 302
2-Methylfuran	C_5H_6O	2.76[20]					
Methyl heptadecanoate	$C_{18}H_{36}O_2$	3.07[40]					
2-Methylheptane	C_8H_{18}	1.9519[20]					
Methyl heptanoate	$C_8H_{16}O_2$	4.355[20]					
2-Methyl-1-heptanol, (±)-	$C_8H_{18}O$	5.16[20]	0.61698E+02	-0.33647E+00	0.49066E-03		236 - 328
3-Methyl-1-heptanol	$C_8H_{18}O$	2.884[17]	0.84687E+01	-0.33712E-01	0.49793E-04		241 - 316
4-Methyl-1-heptanol	$C_8H_{18}O$	4.63[17]	0.48612E+02	-0.26773E+00	0.39972E-03		237 - 332
5-Methyl-1-heptanol, (±)-	$C_8H_{18}O$	7.68[17]	0.54581E+02	-0.24772E+00	0.29734E-03		235 - 328
6-Methyl-1-heptanol	$C_8H_{18}O$	10.54[17]	0.57997E+02	-0.23517E+00	0.24663E-03		265 - 328
2-Methyl-2-heptanol	$C_8H_{18}O$	3.43[19]					
3-Methyl-2-heptanol	$C_8H_{18}O$	7.47[16]	0.39178E+02	-0.17976E+00	0.24218E-03		229 - 329
4-Methyl-2-heptanol	$C_8H_{18}O$	3.59[17]	0.39715E+02	-0.23115E+00	0.36771E-03		240 - 333
5-Methyl-2-heptanol	$C_8H_{18}O$	7.5[5]	0.68568E+02	-0.40706E+00	0.67433E-03		230 - 279
6-Methyl-2-heptanol	$C_8H_{18}O$	6.41[17]	0.77520E+02	-0.41724E+00	0.59448E-03		239 - 329
2-Methyl-3-heptanol, (±)-	$C_8H_{18}O$	3.260[20]	-0.59739E+01	0.56700E-01	-0.83125E-04		343 - 403
3-Methyl-3-heptanol	$C_8H_{18}O$	3.013[20]	-0.38440E+01	0.42327E-01	-0.61250E-04		343 - 403
4-Methyl-3-heptanol	$C_8H_{18}O$	3.312[20]	-0.48003E+01	0.50740E-01	-0.75000E-04		343 - 403
5-Methyl-3-heptanol	$C_8H_{18}O$	3.832[20]	0.61967E+01	-0.63750E-02			343 - 383
6-Methyl-3-heptanol, (±)-	$C_8H_{18}O$	4.992[20]	0.23037E+02	-0.98029E-01	0.12479E-03		283 - 383
2-Methyl-4-heptanol	$C_8H_{18}O$	3.338[23]	0.42102E+00	0.10427E-01	-0.20438E-05		230 - 333
3-Methyl-4-heptanol	$C_8H_{18}O$	7.46[17]	0.33354E+02	-0.14077E+00	0.17750E-03		230 - 330
4-Methyl-4-heptanol	$C_8H_{18}O$	2.902[23]					
cis-3-Methyl-2-heptene	C_8H_{16}	2.436[20]					

Name	Mol. form.	ε	a	b	c	d	Temp. range/K
2-Methylheptyl acetate, (±)-	$C_{10}H_{20}O_2$	4.27^{15}	0.23285E+02	-0.11538E+00	0.17143E-03		288 - 323
Methyl hexadecanoate	$C_{17}H_{34}O_2$	3.124^{40}					
2-Methylhexane	C_7H_{16}	1.9221^{20}	0.24759E+01	-0.22535E-02	0.12500E-05		293 - 323
3-Methylhexane	C_7H_{16}	1.920^{20}	0.27089E+01	-0.37908E-02	0.37500E-05		273 - 323
Methyl hexanedioate	$C_7H_{12}O_4$	6.69^{20}	0.11962E+02	-0.23973E-01	0.20608E-04		293 - 433
Methyl hexanoate	$C_7H_{14}O_2$	4.615^{20}					
2-Methyl-2-hexanol	$C_7H_{16}O$	3.257^{24}					
3-Methyl-2-hexanol	$C_7H_{16}O$	4.990^{24}	0.59724E+02	-0.32417E+00	0.47058E-03		244 - 372
3-Methyl-3-hexanol	$C_7H_{16}O$	3.248^{25}					
5-Methyl-2-hexanone	$C_7H_{14}O$	13.53^{20}	0.52353E+02	-0.17695E+00	0.15195E-03		293 - 333
2-Methyl-2-hexene	C_7H_{14}	2.962^{20}					
Methyl isocyanate	C_2H_3NO	21.75^{16}					
Methyl methacrylate	$C_5H_8O_2$	6.32^{30}	0.32098E+02	-0.14568E+00	0.20000E-03		303 - 343
Methyl 2-methoxybenzoate	$C_9H_{10}O_3$	7.7^{21}					
Methyl 4-methylbenzoate	$C_9H_{10}O_2$	4.3^{33}					
1-Methylnaphthalene	$C_{11}H_{10}$	2.915^{20}	0.45126E+01	-0.76480E-02	0.75000E-05		293 - 333
2-Methylnaphthalene	$C_{11}H_{10}$	2.747^{40}					
Methyl nitrate	CH_3NO_3	23.9^{20}					
Methyl nitrite	CH_3NO_2	20.77^{-73}	0.11071E+03	-0.73428E+00	0.14054E-02		110 - 260
Methyl 2-nitrobenzoate	$C_8H_7NO_4$	27.76^{27}					
Methyl nonadecanoate	$C_{20}H_{40}O_2$	2.982^{40}					
Methyl nonanoate	$C_{10}H_{20}O_2$	3.943^{20}					
5-Methyl-4-nonene	$C_{10}H_{20}$	2.175^{20}					
Methyl *cis,cis*-9,12-octadecadienoate	$C_{19}H_{34}O_2$	3.466^{20}					
Methyl octadecanoate	$C_{19}H_{38}O_2$	3.021^{40}					
Methyl *cis,cis,cis*-9,12,15-octadecatrienoate	$C_{19}H_{32}O_2$	3.355^{20}					
Methyl *cis*-9-octadecenoate	$C_{19}H_{36}O_2$	3.211^{20}					
2-Methyloctane	C_9H_{20}	1.967^{20}					
4-Methyloctane	C_9H_{20}	1.967^{20}					
Methyl octanoate	$C_9H_{18}O_2$	4.101^{20}					
2-Methyloctanoic acid	$C_9H_{18}O_2$	2.39^{20}					
Methyl pentadecanoate	$C_{16}H_{32}O_2$	3.296^{20}					
cis-2-Methyl-1,3-pentadiene	C_6H_{10}	2.422^{25}					
3-Methyl-1,3-pentadiene	C_6H_{10}	2.426^{25}					
4-Methyl-1,3-pentadiene	C_6H_{10}	2.599^{20}	0.51328E+01	-0.12774E-01	0.14215E-04		198 - 323
2-Methylpentane	C_6H_{14}	1.886^{20}	0.20745E+01	0.50871E-03	-0.39286E-05		273 - 323
3-Methylpentane	C_6H_{14}	1.886^{20}	0.24739E+01	-0.23190E-02	0.10714E-05		273 - 323
2-Methyl-2,4-pentanediol	$C_6H_{14}O_2$	25.86^{20}	0.14531E+03	-0.65285E+00	0.83503E-03		203 - 333
4-Methylpentanenitrile	$C_6H_{11}N$	17.5^{22}					
Methyl pentanoate	$C_6H_{12}O_2$	4.992^{20}					
3-Methyl-1-pentanol, (±)-	$C_6H_{14}O$	15.2^{25}					
3-Methyl-3-pentanol	$C_6H_{14}O$	4.322^{20}					
4-Methyl-2-pentanone	$C_6H_{12}O$	13.11^{20}	0.36341E+02	-0.97119E-01	0.61896E-04		204 - 373
1-Methyl-1-phenylhydrazine	$C_7H_{10}N_2$	7.3^{19}					
Methyl phenyl sulfone	$C_7H_8O_2S$	37.9^{100}					
N-Methylpropanamide	C_4H_9NO	170.0^{20}					
2-Methylpropanenitrile	C_4H_7N	24.42^{20}	0.52554E+02	-0.96000E-01			293 - 313
2-Methyl-1-propanethiol	$C_4H_{10}S$	4.961^{25}					
2-Methyl-2-propanethiol	$C_4H_{10}S$	5.475^{20}	0.10597E+02	-0.17500E-01			283 - 313
Methyl propanoate	$C_4H_8O_2$	6.200^{20}	0.12798E+02	-0.22540E-01			293 - 333
2-Methylpropanoic acid	$C_4H_8O_2$	2.58^{20}					
2-Methylpropanoic anhydride	$C_8H_{14}O_3$	13.6^{19}					
2-Methyl-1-propanol	$C_4H_{10}O$	17.93^{20}	0.10762E+03	-0.51398E+00	0.83702E-03	-0.45299E-06	173 - 533
2-Methyl-2-propanol	$C_4H_{10}O$	12.47^{25}	0.22541E+03	-0.14990E+01	0.34050E-02	-0.25968E-05	298 - 503
2-Methylpyridine	C_6H_7N	10.18^{20}	0.34560E+02	-0.11980E+00	0.12500E-03		293 - 333
3-Methylpyridine	C_6H_7N	11.10^{30}	0.19643E+03	-0.11167E+01	0.16667E-02		303 - 333
4-Methylpyridine	C_6H_7N	12.2^{20}	0.33765E+02	-0.10113E+00	0.93860E-04		274 - 333
2-Methylpyridine-1-oxide	C_6H_7NO	36.4^{50}	0.11705E+03	-0.35301E+00	0.32000E-03		323 - 398
3-Methylpyridine-1-oxide	C_6H_7NO	28.26^{45}	0.59851E+02	-0.12682E+00	0.86622E-04		318 - 398
N-Methylpyrrolidine	$C_5H_{11}N$	32.2^{25}					

Name	Mol. form.	ε	a	b	c	d	Temp. range/K
N-Methyl-2-pyrrolidinone	C_5H_9NO	32.55^{20}					
2-Methylquinoline	$C_{10}H_9N$	7.24^{20}	0.11688E+02	-0.78400E-02	-0.25000E-04		293 - 333
4-Methylquinoline	$C_{10}H_9N$	9.31^{20}	0.17788E+02	-0.32580E-01	0.12500E-04		293 - 333
6-Methylquinoline	$C_{10}H_9N$	8.48^{20}	0.21696E+02	-0.63400E-01	0.62500E-04		293 - 333
8-Methylquinoline	$C_{10}H_9N$	6.58^{20}	0.19356E+02	-0.61900E-01	0.62500E-04		293 - 333
Methyl salicylate	$C_8H_8O_3$	8.80^{41}	0.20501E+02	-0.39045E-01	0.68298E-05		223 - 398
Methyl silicate	$C_4H_{12}O_4Si$	6.0^{20}					
3-Methyl sulfolane	$C_5H_{10}O_2S$	29.4^{25}	0.53158E+02	-0.93730E-01	0.47275E-04		298 - 398
Methyl tetradecanoate	$C_{15}H_{30}O_2$	3.352^{20}					
2-Methyltetrahydrofuran	$C_5H_{10}O$	6.97^{25}					
(Methylthio)benzene	C_7H_8S	4.88^{30}	0.21841E+02	-0.97630E-01	0.13750E-03		303 - 343
Methyl tridecanoate	$C_{14}H_{28}O_2$	3.442^{20}					
Methyltriphenoxysilane	$C_{19}H_{18}O_3Si$	3.628^{25}					
Methyl undecanoate	$C_{12}H_{24}O_2$	3.671^{20}					
Monomethyl glutarate	$C_6H_{10}O_4$	8.37^{20}	0.16779E+02	-0.39839E-01	0.38095E-04		293 - 363
Morpholine	C_4H_9NO	7.42^{25}					
β-Myrcene	$C_{10}H_{16}$	2.3^{25}					
Naphthalene	$C_{10}H_8$	2.54^{90}					
N-1-Naphthalenylacetamide	$C_{12}H_{11}NO$	24.3^{160}	0.84739E+02	-0.12391E+00	-0.35714E-04		433 - 533
1-Naphthol	$C_{10}H_8O$	5.03^{100}	0.16489E+02	-0.46700E-01	0.42857E-04		373 - 453
2-Naphthol	$C_{10}H_8O$	4.95^{140}	0.92865E+01	-0.10500E-01	0.42501E-15		413 - 453
1-Naphthylamine	$C_{10}H_9N$	5.20^{60}	0.10577E+02	-0.22114E-01	0.17857E-04		333 - 453
2-Naphthylamine	$C_{10}H_9N$	5.26^{120}	0.19722E+02	-0.60679E-01	0.60714E-04		393 - 473
1-Naphthyl 2-hydroxybenzoate	$C_{17}H_{12}O_3$	6.30^{20}	0.11229E+02	-0.18857E-01	0.70332E-05		293 - 353
Neon	Ne	1.1907^{-247}	0.12667E+01	-0.29064E-02			26 - 29
Neopentane	C_5H_{12}	1.769^{23}	0.10949E+02	-0.63057E-01	0.10835E-03		251 - 296
L-Nicotine	$C_{10}H_{14}N_2$	8.937^{20}	0.21347E+02	-0.57177E-01	0.50655E-04		293 - 363
Nitric oxide	NO	2.00^{-149}					
2-Nitroaniline	$C_6H_6N_2O_2$	47.3^{80}	0.18900E+03	-0.56977E+00	0.47484E-03		353 - 468
3-Nitroaniline	$C_6H_6N_2O_2$	35.6^{125}	0.20352E+03	-0.66582E+00	0.61310E-03		398 - 468
4-Nitroaniline	$C_6H_6N_2O_2$	78.5^{155}	0.48673E+03	-0.15040E+01	0.12857E-02		428 - 468
2-Nitroanisole	$C_7H_7NO_3$	45.75^{20}	0.16684E+03	-0.58196E+00	0.57382E-03		293 - 423
3-Nitroanisole	$C_7H_7NO_3$	25.7^{45}	0.65402E+02	-0.16460E+00	0.12560E-03		318 - 443
4-Nitroanisole	$C_7H_7NO_3$	26.95^{65}	0.59811E+02	-0.10955E+00	0.36042E-04		338 - 443
Nitrobenzene	$C_6H_5NO_2$	35.6^{20}	0.11212E+03	-0.35211E+00	0.31128E-03		279 - 533
Nitroethane	$C_2H_5NO_2$	29.11^{15}	0.57406E+02	-0.97657E-01			276 - 333
Nitrogen	N_2	1.4680^{-210}	0.12550E+01	0.67949E-02	-0.56704E-04		63 - 126
Nitrogen tetroxide	N_2O_4	2.44^{20}	0.28212E+01	-0.13000E-02			253 - 293
Nitrogen trioxide	N_2O_3	31.13^{-70}	0.92287E+02	-0.43306E+00	0.65000E-03		203 - 243
Nitromethane	CH_3NO_2	37.27^{20}	0.11227E+03	-0.35591E+00	0.34206E-03		288 - 343
1-Nitronaphthalene	$C_{10}H_7NO_2$	19.68^{60}	0.36267E+02	-0.41283E-01	-0.25595E-04		333 - 403
1-Nitrooctane	$C_8H_{17}NO_2$	11.46^{20}					
2-Nitrophenol	$C_6H_5NO_3$	16.50^{50}	0.33827E+02	-0.62123E-01	0.26774E-04		323 - 453
3-Nitrophenol	$C_6H_5NO_3$	35.45^{100}	0.18967E+03	-0.66144E+00	0.66532E-03		373 - 458
4-Nitrophenol	$C_6H_5NO_3$	42.20^{120}	0.22901E+03	-0.74264E+00	0.68006E-03		393 - 463
1-Nitropropane	$C_3H_7NO_2$	24.70^{15}	0.94999E+02	-0.38358E+00	0.48480E-03		276 - 333
2-Nitropropane	$C_3H_7NO_2$	26.74^{15}	0.60138E+02	-0.11566E+00			276 - 303
Nitrosyl bromide	BrNO	13.4^{15}					
Nitrosyl chloride	ClNO	18.2^{12}					
4-Nitrothioanisole	$C_7H_7NO_2S$	21.7^{73}					
2-Nitrotoluene	$C_7H_7NO_2$	26.26^{20}	0.10420E+03	-0.41726E+00	0.51607E-03		273 - 323
3-Nitrotoluene	$C_7H_7NO_2$	24.95^{30}	0.62492E+02	-0.16235E+00	0.12844E-03		303 - 403
4-Nitrotoluene	$C_7H_7NO_2$	22.2^{58}					
Nonadecane	$C_{19}H_{40}$	2.0706^{20}					
10-Nonadecanone	$C_{19}H_{38}O$	5.37^{80}					
Nonane	C_9H_{20}	1.9722^{20}	0.23894E+01	-0.14830E-02	0.14881E-06		253 - 393
Nonanenitrile	$C_9H_{17}N$	12.08^{20}					
Nonanoic acid	$C_9H_{18}O_2$	2.475^{22}	0.25039E+01	0.67274E-03	-0.24180E-05		295 - 365
1-Nonanol	$C_9H_{20}O$	8.83^{20}	0.97467E+02	-0.51103E+00	0.71429E-03		288 - 343
2-Nonanol, (±)-	$C_9H_{20}O$	6.66^{25}	0.10136E+03	-0.55612E+00	0.80000E-03		288 - 308

Name	Mol. form.	ε	a	b	c	d	Temp. range/K
3-Nonanol, (±)-	$C_9H_{20}O$	4.49[25]	0.55214E+02	-0.31920E+00	0.50000E-03		288 - 308
4-Nonanol	$C_9H_{20}O$	3.69[25]	0.27954E+01	0.30000E-02	-0.52375E-13		288 - 308
5-Nonanol	$C_9H_{20}O$	3.54[25]	-0.25463E+01	0.35320E-01	-0.50000E-04		288 - 308
2-Nonanone	$C_9H_{18}O$	9.14[22]					
5-Nonanone	$C_9H_{18}O$	10.6[20]					
1-Nonene	C_9H_{18}	2.180[20]	0.22710E+01	0.15797E-02	-0.64286E-05		273 - 323
2-Nonenoic acid	$C_9H_{16}O_2$	2.5[23]					
Nonyl acetate	$C_{11}H_{22}O_2$	3.87[20]					
Nonylamine	$C_9H_{21}N$	3.42[20]	0.53575E+01	-0.71982E-02	0.19481E-05		293 - 373
cis,cis-9,12-Octadecadienoic acid	$C_{18}H_{32}O_2$	2.754[20]	0.32073E+01	-0.15477E-02			275 - 368
Octadecanoic acid	$C_{18}H_{36}O_2$	2.314[20]	0.27159E+01	-0.13300E-02			293 - 373
1-Octadecanol	$C_{18}H_{38}O$	3.38[60]	0.73784E+01	-0.12000E-01	-0.22871E-13		333 - 363
cis,cis,cis-9,12,15-Octadecatrienoic acid	$C_{18}H_{30}O_2$	2.825[20]	0.33867E+01	-0.19181E-02			274 - 368
cis-9-Octadecenoic acid	$C_{18}H_{34}O_2$	2.336[20]	0.25385E+01	-0.69448E-03			275 - 368
Octadecyl acetate	$C_{20}H_{40}O_2$	3.07[35]	0.44569E+01	-0.45000E-02	0.33923E-14		308 - 348
Octadecylamine	$C_{18}H_{39}N$	2.67[53]					
1,7-Octadiene	C_8H_{14}	2.186[20]	0.28376E+01	-0.17442E-02	-0.16141E-05		214 - 293
2,2,3,3,4,4,5,5-Octafluoro-1-pentanol	$C_5H_4F_8O$	15.30[25]					
Octamethylcyclotetrasiloxane	$C_8H_{24}O_4Si_4$	2.390[23]	0.36286E+01	-0.56885E-02	0.50874E-05		296 - 333
Octane	C_8H_{18}	1.948[20]	0.22590E+01	-0.84212E-03	-0.75758E-06		233 - 393
Octanenitrile	$C_8H_{15}N$	13.90[20]					
1-Octanethiol	$C_8H_{18}S$	3.949[20]	0.63667E+01	-0.87920E-02	0.18750E-05		273 - 333
Octanoic acid	$C_8H_{16}O_2$	2.85[15]	0.29391E+01	-0.38721E-03			288 - 423
1-Octanol	$C_8H_{18}O$	10.30[20]	0.51647E+02	-0.20371E+00	0.21320E-03		258 - 513
2-Octanol	$C_8H_{18}O$	8.13[20]	0.63760E+02	-0.27643E+00	0.31075E-03		213 - 513
3-Octanol	$C_8H_{18}O$	5.55[20]	0.12505E+03	-0.70646E+00	0.10245E-02		223 - 383
4-Octanol	$C_8H_{18}O$	4.48[20]	0.51049E+02	-0.26664E+00	0.37280E-03		243 - 403
2-Octanone	$C_8H_{16}O$	9.51[20]	-0.16219E+02	0.18799E+00	-0.34156E-03		293 - 333
3-Octanone	$C_8H_{16}O$	10.50[30]					
1-Octene	C_8H_{16}	2.113[20]	0.24348E+01	0.34200E-03	-0.50000E-05		273 - 323
cis-3-Octene	C_8H_{16}	2.062[25]					
trans-3-Octene	C_8H_{16}	2.002[25]					
cis-4-Octene	C_8H_{16}	2.053[25]					
trans-4-Octene	C_8H_{16}	2.004[25]					
Octyl acetate	$C_{10}H_{20}O_2$	4.18[15]	-0.34691E+01	0.58106E-01	-0.10952E-03		288 - 323
Octylamine	$C_8H_{19}N$	3.58[20]	0.77931E+01	-0.20015E-01	0.19347E-04		273 - 373
Octylbenzene	$C_{14}H_{22}$	2.26[20]					
Oxalyl chloride	$C_2Cl_2O_2$	3.470[21]					
Oxirane	C_2H_4O	12.42[20]	0.52661E+02	-0.21337E+00	0.25947E-03		293 - 243
Oxygen	O_2	1.5684[-219]	0.15434E+01	0.14615E-02	-0.21964E-04		55 - 154
Ozone	O_3	4.75[-183]	0.86344E+01	-0.54807E-01	0.12596E-03		90 - 185
Pentaborane(9)	B_5H_9	21.1[25]	0.40952E+03	-0.24414E+01	0.38225E-02		226 - 298
Pentachloroethane	C_2HCl_5	3.716[25]	0.65972E+01	-0.96800E-02			298 - 338
2,3,4,5,6-Pentachlorotoluene	$C_7H_3Cl_5$	4.8[20]					
Pentadecane	$C_{15}H_{32}$	2.0391[20]	0.23792E+01	-0.11600E-02	-0.71069E-16		283 - 363
1-Pentadecanol	$C_{15}H_{32}O$	3.70[60]					
Pentadecylamine	$C_{15}H_{33}N$	2.85[40]					
cis-1,3-Pentadiene	C_5H_8	2.319[25]					
1,4-Pentadiene	C_5H_8	2.054[21]	0.29994E+01	-0.34578E-02	0.85300E-06		178 - 294
Pentamethylbenzene	$C_{11}H_{16}$	2.358[61]	0.30196E+01	-0.22619E-02	0.83831E-06		334 - 413
Pentanal	$C_5H_{10}O$	10.00[20]					
Pentane	C_5H_{12}	1.8371[20]					
1,2-Pentanediol, (±)-	$C_5H_{12}O_2$	17.31[24]	0.18436E+03	-0.10682E+01	0.17037E-02		197 - 297
1,4-Pentanediol	$C_5H_{12}O_2$	26.74[23]	0.13568E+03	-0.59198E+00	0.75398E-03		193 - 318
1,5-Pentanediol	$C_5H_{12}O_2$	26.2[20]	0.11858E+03	-0.45920E+00	0.49341E-03		243 - 343
2,3-Pentanediol	$C_5H_{12}O_2$	17.37[24]	0.95876E+02	-0.46463E+00	0.67434E-03		238 - 297
2,4-Pentanediol	$C_5H_{12}O_2$	24.69[21]	0.11914E+03	-0.52569E+00	0.69607E-03		224 - 294
2,4-Pentanedione	$C_5H_8O_2$	26.524[30]					
Pentanenitrile	C_5H_9N	20.04[20]	0.55793E+02	-0.15750E+00	0.12432E-03		183 - 333
1-Pentanethiol	$C_5H_{12}S$	4.847[20]	0.71131E+01	-0.30228E-02	-0.16414E-04		273 - 333

Name	Mol. form.	*ε*	*a*	*b*	*c*	*d*	Temp. range/K
Pentanoic acid	$C_5H_{10}O_2$	2.661^{21}	0.33491E+01	-0.75156E-02	0.17820E-04		250 - 344
1-Pentanol	$C_5H_{12}O$	15.13^{25}	0.73397E+02	-0.28165E+00	0.28427E-03		213 - 513
2-Pentanol	$C_5H_{12}O$	13.71^{25}	0.16437E+03	-0.86506E+00	0.11955E-02		273 - 323
3-Pentanol	$C_5H_{12}O$	13.35^{25}	0.12838E+03	-0.60980E+00	0.75000E-03		288 - 318
2-Pentanone	$C_5H_{10}O$	15.45^{20}	0.40893E+02	-0.10423E+00	0.60557E-04		204 - 353
3-Pentanone	$C_5H_{10}O$	17.00^{20}	0.12690E+02	0.95177E-01	-0.27321E-03		233 - 353
2-Pentanone oxime	$C_5H_{11}NO$	3.3^{20}					
1-Pentene	C_5H_{10}	2.011^{20}	-0.11438E+01	0.25420E-01	-0.50000E-04		273 - 293
Pentyl acetate	$C_7H_{14}O_2$	4.79^{20}	0.12091E+02	-0.36536E-01	0.39732E-04		253 - 353
Pentylamine	$C_5H_{13}N$	4.27^{20}	0.11274E+02	-0.34965E-01	0.37706E-04		223 - 353
Pentyl benzoate	$C_{12}H_{16}O_2$	5.07^{20}					
Pentyl butanoate	$C_9H_{18}O_2$	4.08^{28}	0.59029E+01	-0.49905E-02	-0.34292E-05		301 - 343
Pentyl cinnamate	$C_{14}H_{18}O_2$	4.89^{20}					
Pentyl formate	$C_6H_{12}O_2$	5.7^{19}					
Pentyl hexanoate	$C_{11}H_{22}O_2$	4.22^{15}	0.83503E+01	-0.18449E-01	0.14286E-04		288 - 323
Pentyl nitrite	$C_5H_{11}NO_2$	7.21^{25}					
Pentyl pentanoate	$C_{10}H_{20}O_2$	4.076^{32}	0.77641E+01	-0.14335E-01	0.73740E-05		306 - 393
Pentyl propanoate	$C_8H_{16}O_2$	4.552^{20}					
Pentyl salicylate	$C_{12}H_{16}O_3$	6.25^{28}					
Perchloryl fluoride	$ClFO_3$	2.194^{-123}	0.23808E+01	-0.38629E-03	-0.57143E-05		125 - 150
Perfluoroacetone	C_3F_6O	2.104^{-71}	0.34809E+01	-0.92883E-02	0.12282E-04		151 - 238
Perfluoroheptane	C_7F_{16}	1.847^{16}					
Perfluorohexane	C_6F_{14}	1.76^{25}					
Perfluoromethylcyclohexane	C_7F_{14}	1.82^{25}					
Phenanthrene	$C_{14}H_{10}$	2.72^{110}					
Phenol	C_6H_6O	12.40^{30}	0.63391E+02	-0.24988E+00	0.26930E-03		303 - 433
Phenoxyacetylene	C_8H_6O	4.76^{25}					
Phenyl acetate	$C_8H_8O_2$	5.403^{25}	0.11327E+02	-0.26707E-01	0.22938E-04		298 - 404
Phenylacetylene	C_8H_6	2.98^{25}					
Phenyl 2-(acetyloxy)benzoate	$C_{15}H_{12}O_4$	4.33^{111}					
Phenyl butanoate	$C_{10}H_{12}O_2$	4.48^{20}					
2-Phenylethyl acetate	$C_{10}H_{12}O_2$	4.93^{24}					
Phenylhydrazine	$C_6H_8N_2$	7.15^{20}					
Phenyl isocyanate	C_7H_5NO	8.940^{20}	0.17541E+02	-0.29790E-01	0.15476E-05		293 - 353
Phenyl laurate	$C_{18}H_{28}O_2$	3.28^{20}					
1-Phenyl-2-methyl-2-propanol	$C_{10}H_{14}O$	5.71^{25}	0.21922E+02	-0.84231E-01	0.99475E-04		298 - 423
Phenyl pentanoate	$C_{11}H_{14}O_2$	4.30^{20}					
Phenyl propanoate	$C_9H_{10}O_2$	4.77^{20}					
1-Phenyl-2-propanol	$C_9H_{12}O$	9.35^{20}	0.10762E+03	-0.56026E+00	0.76915E-03		233 - 373
Phenyl salicylate	$C_{13}H_{10}O_3$	6.92^{17}	0.26545E+02	-0.11180E+00	0.15220E-03		290 - 358
Phenyl silicate	$C_{24}H_{20}O_4Si$	3.4915^{60}					
Phosphorothioc trichloride	Cl_3PS	4.94^{25}					
Phosphorus (white)	P	4.096^{34}	0.79018E+00	0.23911E-01	-0.42826E-04		307 - 358
Phosphorus(III) chloride	Cl_3P	3.498^{17}	0.59098E+01	-0.83322E-02			290 - 333
Phosphorus(V) chloride	Cl_5P	2.85^{160}					
Phosphorus(V) dichloride trifluoride	Cl_2F_3P	2.8129^{-45}	0.46501E+01	-0.80358E-02			172 - 229
Phosphorus(V) tetrachloride fluoride	Cl_4FP	2.6499^{-1}	0.33503E+01	-0.29651E-02			244 - 273
Phosphorus(V) trichloride difluoride	Cl_3F_2P	2.3752^{-5}	0.28905E+01	-0.19228E-02			215 - 268
Phosphoryl chloride	Cl_3OP	14.1^{20}					
Pinane	$C_{10}H_{18}$	2.1456^{25}					
α-Pinene	$C_{10}H_{16}$	2.1787^{25}					
β-Pinene	$C_{10}H_{16}$	2.4970^{25}					
Piperidine	$C_5H_{11}N$	4.33^{20}	0.82317E+01	-0.11229E-01	-0.71429E-05		293 - 333
Propanal	C_3H_6O	18.5^{17}					
Propane	C_3H_8	1.6678^{20}	0.22883E+01	-0.23276E-02	0.84710E-06		90 - 300
1,2-Propanediol	$C_3H_8O_2$	27.5^{30}	0.24546E+03	-0.15738E+01	0.38068E-02	-0.32544E-05	193 - 403
1,3-Propanediol	$C_3H_8O_2$	35.1^{20}	0.11365E+03	-0.36680E+00	0.33766E-03		288 - 328
1,2-Propanedithiol	$C_3H_8S_2$	7.24^{20}	0.14667E+02	-0.32660E-01	0.25000E-04		293 - 333
1,3-Propanedithiol	$C_3H_8S_2$	8.11^{30}	0.66607E+01	0.31310E-01	-0.87500E-04		303 - 343
Propanenitrile	C_3H_5N	29.7^{20}	0.82222E+02	-0.22937E+00	0.17424E-03		213 - 473

Name	Mol. form.	ε	*a*	*b*	*c*	*d*	Temp. range/K
1-Propanethiol	C_3H_8S	5.937^{15}	0.11602E+02	-0.19580E-01			273 - 318
2-Propanethiol	C_3H_8S	5.952^{25}					
Propanoic acid	$C_3H_6O_2$	3.44^{25}	0.18793E+01	0.46841E-02	0.19983E-05		289 - 408
Propanoic anhydride	$C_6H_{10}O_3$	18.30^{20}					
1-Propanol	C_3H_8O	20.8^{20}	0.98045E+02	-0.36860E+00	0.36422E-03		193 - 493
2-Propanol	C_3H_8O	20.18^{20}	0.10416E+03	-0.41011E+00	0.42049E-03		193 - 493
Propargyl alcohol	C_3H_4O	20.8^{20}	0.99895E+02	-0.38911E+00	0.40776E-03		213 - 293
Propene	C_3H_6	2.1365^{-53}	0.29623E+01	-0.37564E-02			220 - 250
1-Propenylbenzene (unspecified isomer)	C_9H_{10}	2.73^{20}					
Propyl acetate	$C_5H_{10}O_2$	5.62^{20}	0.17677E+02	-0.61404E-01	0.69196E-04		253 - 353
Propylamine	C_3H_9N	5.08^{23}	0.17719E+02	-0.59022E-01	0.54780E-04		204 - 296
N-Propylaniline	$C_9H_{13}N$	5.48^{20}					
Propylbenzene	C_9H_{12}	2.370^{20}	0.26933E+01	0.21679E-03	-0.44643E-05		273 - 323
Propyl benzoate	$C_{10}H_{12}O_2$	5.78^{30}	0.10927E+02	-0.20535E-01	0.11745E-04		303 - 358
Propyl butanoate	$C_7H_{14}O_2$	4.3^{20}					
Propyl carbamate	$C_4H_9NO_2$	12.06^{65}	0.24356E+02	-0.36400E-01			338 - 378
Propyl chlorocarbonate	$C_4H_7ClO_2$	11.2^{20}					
Propyl *trans*-cinnamate	$C_{12}H_{14}O_2$	5.45^{20}					
Propylene carbonate	$C_4H_6O_3$	66.14^{20}	0.15940E+03	-0.39530E+00	0.26284E-03		273 - 333
Propyl formate	$C_4H_8O_2$	6.92^{30}					
4-Propylheptane	$C_{10}H_{22}$	1.9955^{20}					
Propyl nitrite	$C_3H_7NO_2$	12.35^{-23}	0.70552E+02	-0.40362E+00	0.66687E-03		110 - 310
Propyl pentanoate	$C_8H_{16}O_2$	4.0^{19}					
N-Propylpropanamide	$C_6H_{13}NO$	118.1^{25}	0.58846E+03	-0.22012E+01	0.20870E-02		298 - 328
Propyl propanoate	$C_6H_{12}O_2$	5.249^{20}					
Propyl silicate	$C_{12}H_{28}O_4Si$	3.21^{25}					
Propyl trichloroacetate	$C_5H_7Cl_3O_2$	8.32^{25}					
Propyne	C_3H_4	3.218^{-27}	0.60871E+01	-0.11730E-01			185 - 246
Pyrazine	$C_4H_4N_2$	2.80^{50}					
Pyridine	C_5H_5N	13.260^{20}	0.43991E+02	-0.15150E+00	0.15925E-03		293 - 323
2-Pyridinecarbonitrile	$C_6H_4N_2$	93.77^{30}	0.45596E+03	-0.17746E+01	0.19105E-02		303 - 398
3-Pyridinecarbonitrile	$C_6H_4N_2$	20.54^{50}	0.60484E+02	-0.17280E+00	0.15218E-03		323 - 398
4-Pyridinecarbonitrile	$C_6H_4N_2$	5.23^{80}	0.12533E+02	-0.30115E-01	0.26674E-04		353 - 398
Pyridine-1-oxide	C_5H_5NO	35.94^{70}	0.20878E+02	0.16450E+00	-0.35269E-03		343 - 398
Pyrocatechol	$C_6H_6O_2$	17.57^{115}	0.74930E+02	-0.22142E+00	0.18919E-03		388 - 463
Pyrrole	C_4H_5N	8.00^{20}	0.12672E+02	-0.14075E-01	-0.62671E-05		293 - 357
Pyrrolidine	C_4H_9N	8.30^{20}	0.38191E+02	-0.15462E+00	0.17941E-03		274 - 333
2-Pyrrolidone	C_4H_7NO	28.18^{25}	0.11054E+03	-0.47945E+00	0.68182E-03		298 - 338
Quinoline	C_9H_7N	9.16^{20}	0.33432E+02	-0.13497E+00	0.17788E-03		258 - 323
Resorcinol	$C_6H_6O_2$	13.55^{120}	0.30252E+02	-0.56443E-01	0.35578E-04		393 - 463
Salicylaldehyde	$C_7H_6O_2$	18.35^{20}	0.51315E+02	-0.15379E+00	0.14111E-03		289 - 453
Selenium (gray)	Se	5.44^{237}	0.67569E+01	-0.25829E-02			511 - 575
Selenium oxychloride	Cl_2OSe	46.2^{20}					
Styrene	C_8H_8	2.4737^{20}	0.44473E+01	-0.11422E-01	0.16000E-04		293 - 313
Succinonitrile	$C_4H_4N_2$	62.6^{25}	0.17724E+03	-0.54654E+00	0.54046E-03		236 - 351
N-Sulfinylaniline	C_6H_5NOS	6.97^{25}					
Sulfur (rhombic)	S	3.4991^{134}	0.51651E+01	-0.77381E-02	0.89120E-05		407 - 479
Sulfur chloride [ClSSCl]	Cl_2S_2	4.79^{15}					
Sulfur decafluoride	$F_{10}S_2$	2.0202^{20}					
Sulfur dichloride	Cl_2S	2.915^{25}					
Sulfur dioxide	O_2S	14.3^{20}	0.52045E+02	-0.16125E+00	0.11042E-03		213 - 449
Sulfur hexafluoride	F_6S	1.81^{-50}					
Sulfur trioxide (α-form)	O_3S	3.11^{18}					
Sulfuryl chloride	Cl_2O_2S	9.1^{20}					
α-Terpinene	$C_{10}H_{16}$	2.4526^{25}					
γ-Terpinene	$C_{10}H_{16}$	2.2738^{25}					
Terpinolene	$C_{10}H_{16}$	2.2918^{25}					
1,1,2,2-Tetrabromoethane	$C_2H_2Br_4$	6.72^{30}	0.16246E+02	-0.31500E-01			303 - 333
Tetrabutylstannane	$C_{16}H_{36}Sn$	9.74^{20}	0.56115E+02	-0.24812E+00	0.30682E-03		293 - 313
1,1,2,2-Tetrachloro-1,2-difluoroethane	$C_2Cl_4F_2$	2.52^{35}					

Name	Mol. form.	*ε*	*a*	*b*	*c*	*d*	Temp. range/K
1,2,3,4-Tetrachloro-5,6-dimethylbenzene	$C_8H_6Cl_4$	8.0^{20}					
1,2,3,5-Tetrachloro-4,6-dimethylbenzene	$C_8H_6Cl_4$	5.4^{20}					
1,1,1,2-Tetrachloroethane	$C_2H_2Cl_4$	9.22^{-66}	0.19606E+02	-0.49847E-01			207 - 233
1,1,2,2-Tetrachloroethane	$C_2H_2Cl_4$	8.50^{20}					
Tetrachloroethene	C_2Cl_4	2.268^{30}					
Tetrachloromethane	CCl_4	2.2379^{20}	0.28280E+01	-0.20339E-02	0.71795E-07		283 - 333
Tetrachlorosilane	Cl_4Si	2.248^{0}	0.58041E+01	-0.27129E-01	0.51678E-04		207 - 273
Tetradecane	$C_{14}H_{30}$	2.0343^{20}	0.23832E+01	-0.11900E-02	-0.51229E-16		283 - 363
1-Tetradecanol	$C_{14}H_{30}O$	4.42^{45}	0.12272E+02	-0.24667E-01	-0.13168E-13		318 - 358
Tetradecylamine	$C_{14}H_{31}N$	2.90^{39}					
Tetraethylene glycol	$C_8H_{18}O_5$	20.44^{20}	0.83547E+02	-0.31691E+00	0.34689E-03		253 - 333
Tetraethylene glycol dimethyl ether	$C_{10}H_{22}O_5$	7.68^{25}					
Tetraethylenepentamine	$C_8H_{23}N_5$	9.40^{20}	0.40553E+02	-0.16681E+00	0.20659E-03		213 - 333
Tetraethylsilane	$C_8H_{20}Si$	2.090^{20}					
Tetraethylstannane	$C_8H_{20}Sn$	2.241^{20}					
Tetraethylurea	$C_9H_{20}N_2O$	14.29^{24}	0.52820E+02	-0.18790E+00	0.19580E-03		205 - 411
Tetrafluoromethane	CF_4	1.685^{-147}	0.20350E+01	-0.27616E-02			126 - 142
2,2,3,3-Tetrafluoro-1-propanol	$C_3H_4F_4O$	21.03^{25}					
Tetrahydrofuran	C_4H_8O	7.52^{22}	0.30739E+02	-0.12946E+00	0.17195E-03		224 - 295
Tetrahydrofurfuryl alcohol	$C_5H_{10}O_2$	13.48^{30}					
1,2,3,4-Tetrahydronaphthalene	$C_{10}H_{12}$	2.771^{25}	0.29172E+01	0.12832E-02	-0.59453E-05		298 - 343
1,2,3,4-Tetrahydro-2-naphthol	$C_{10}H_{12}O$	11.70^{20}	0.98978E+02	-0.48267E+00	0.63008E-03		293 - 363
Tetrahydropyran	$C_5H_{10}O$	5.66^{20}	0.19793E+02	-0.76071E-01	0.94852E-04		234 - 333
Tetrakis(methylthio)methane	$C_5H_{12}S_4$	2.818^{70}					
Tetramethoxymethane	$C_5H_{12}O_4$	2.40^{20}					
1,2,3,4-Tetramethylbenzene	$C_{10}H_{14}$	2.538^{23}	0.33822E+01	-0.33630E-02	0.17475E-05		273 - 412
1,2,4,5-Tetramethylbenzene	$C_{10}H_{14}$	2.223^{83}	0.26834E+01	-0.10327E-02	-0.73533E-06		356 - 430
1,1,3,3-Tetramethylguanidine	$C_5H_{13}N_3$	11.5^{25}					
Tetramethylsilane	$C_4H_{12}Si$	1.921^{20}					
Tetramethylurea	$C_5H_{12}N_2O$	23.10^{20}					
Tetranitromethane	CN_4O_8	2.317^{20}					
Tetrapropylstannane	$C_{12}H_{28}Sn$	2.267^{20}					
Thioacetic acid	C_2H_4OS	14.30^{25}					
Thionyl bromide	Br_2OS	9.06^{20}					
Thionyl chloride	Cl_2OS	8.675^{25}					
Thiophene	C_4H_4S	2.739^{20}	0.32941E+01	-0.19019E-02			253 - 293
Thymol	$C_{10}H_{14}O$	4.259^{60}					
Tin(IV) bromide	Br_4Sn	3.169^{30}	0.50001E+01	-0.60383E-02			304 - 316
Tin(IV) chloride	Cl_4Sn	3.014^{0}	0.43951E+01	-0.48805E-02			234 - 273
Titanium(IV) chloride	Cl_4Ti	2.843^{-16}	0.33668E+01	-0.19675E-02			237 - 257
Toluene	C_7H_8	2.379^{23}	0.32584E+01	-0.34410E-02	0.15937E-05		207 - 316
Toluene-2,4-diisocyanate	$C_9H_6N_2O_2$	8.433^{20}	0.22174E+02	-0.66982E-01	0.68571E-04		293 - 353
p-Toluenesulfonyl chloride	$C_7H_7ClO_2S$	22.6^{70}					
Triacontane	$C_{30}H_{62}$	1.9112^{100}					
Tribromoacetaldehyde	C_2HBr_3O	7.6^{20}					
Tribromochloromethane	CBr_3Cl	2.601^{60}					
Tribromofluoromethane	CBr_3F	3.00^{20}	0.53203E+01	-0.11061E-01	0.10688E-04		206 - 323
Tribromomethane	$CHBr_3$	4.404^{10}	0.71707E+01	-0.98000E-02			283 - 343
Tribromonitromethane	CBr_3NO_2	9.034^{25}	0.16079E+02	-0.23630E-01			298 - 328
1,2,3-Tribromopropane	$C_3H_5Br_3$	6.00^{30}	0.11024E+02	-0.16596E-01			303 - 358
Tributylamine	$C_{12}H_{27}N$	2.340^{20}	0.19846E+01	0.28108E-02	-0.54545E-05		233 - 293
Tributyl borate	$C_{12}H_{27}BO_3$	2.23^{20}					
Tributyl phosphate	$C_{12}H_{27}O_4P$	8.34^{20}	0.26304E+02	-0.88480E-01	0.92857E-04		293 - 373
Tributyrin	$C_{15}H_{26}O_6$	5.72^{10}	0.13152E+02	-0.36684E-01	0.36795E-04		199 - 283
Trichloroacetaldehyde	C_2HCl_3O	6.8^{25}					
Trichloroacetic acid	$C_2HCl_3O_2$	4.34^{60}	0.13412E+01	0.90000E-02	-0.24130E-14		333 - 393
Trichloroacetic anhydride	$C_4Cl_6O_3$	5.0^{25}					
Trichloroacetonitrile	C_2Cl_3N	7.85^{19}					
1,2,2-Trichloro-1,1-difluoroethane	$C_2HCl_3F_2$	4.01^{30}	0.75423E+01	-0.11667E-01			303 - 333
1,1,1-Trichloroethane	$C_2H_3Cl_3$	7.243^{20}	0.27705E+02	-0.10621E+00	0.12424E-03		258 - 318

Name	Mol. form.	*ε*	*a*	*b*	*c*	*d*	Temp. range/K
1,1,2-Trichloroethane	$C_2H_3Cl_3$	7.1937^{25}	0.17147E+02	-0.33371E-01			288 - 318
Trichloroethene	C_2HCl_3	3.390^{28}	0.58319E+01	-0.80828E-02			302 - 338
Trichlorofluoromethane	CCl_3F	3.00^{20}	0.53203E+01	-0.11061E-01	0.10688E-04		206 - 323
Trichloromethane	$CHCl_3$	4.8069^{20}	0.15115E+02	-0.51830E-01	0.56803E-04		218 - 323
Trichloromethane-*d*	$CDCl_3$	4.67^{25}					
Trichloronitromethane	CCl_3NO_2	7.319^{20}	0.14403E+02	-0.24178E-01			276 - 333
1,2,3-Trichloropropane	$C_3H_5Cl_3$	7.5^{20}					
Tri-*o*-cresyl phosphate	$C_{21}H_{21}O_4P$	6.7^{25}					
Tridecane	$C_{13}H_{28}$	2.0213^{20}	0.23731E+01	-0.12000E-02	-0.21841E-15		283 - 363
1-Tridecanol	$C_{13}H_{28}O$	4.02^{60}					
7-Tridecanone	$C_{13}H_{26}O$	7.6^{30}					
1-Tridecene	$C_{13}H_{26}$	2.139^{20}	0.14154E+01	0.66514E-02	-0.14286E-04		273 - 323
Triethoxymethane	$C_7H_{16}O_3$	4.779^{20}					
Triethoxymethylsilane	$C_7H_{18}O_3Si$	3.845^{25}					
Triethylamine	$C_6H_{15}N$	2.418^{20}	0.29205E+01	-0.14007E-02	-0.13469E-05		233 - 323
1,3,5-Triethylbenzene	$C_{12}H_{18}$	2.256^{20}					
Triethylborane	$C_6H_{15}B$	1.974^{20}					
Triethylene glycol	$C_6H_{14}O_4$	23.69^{20}	0.91845E+02	-0.33827E+00	0.36062E-03		253 - 333
Triethylene glycol dimethyl ether	$C_8H_{18}O_4$	7.62^{25}					
Triethyl phosphate	$C_6H_{15}O_4P$	13.20^{25}	0.61230E+02	-0.26047E+00	0.33333E-03		298 - 333
Triethylphosphine oxide	$C_6H_{15}OP$	35.5^{50}					
Triethylphosphine sulfide	$C_6H_{15}PS$	39.0^{98}					
Triethylsilane	$C_6H_{16}Si$	2.323^{20}					
Trifluoroacetic acid	$C_2HF_3O_2$	8.42^{20}	0.21652E+02	-0.68146E-01	0.78571E-04		263 - 323
Trifluoroacetic acid anhydride	$C_4F_6O_3$	2.7^{25}					
2,2,2-Trifluoroethanol	$C_2H_3F_3O$	27.68^{20}	0.90593E+02	-0.21421E+00			293 - 318
Trifluoromethane	CHF_3	5.2^{21}	0.11442E+03	-0.75600E+00	0.13562E-02		130 - 263
(Trifluoromethyl)benzene	$C_7H_5F_3$	9.22^{25}					
Trihexyl borate	$C_{18}H_{39}BO_3$	2.22^{20}					
Triisopentylamine	$C_{15}H_{33}N$	2.29^{21}					
Trimethoxymethylsilane	$C_4H_{12}O_3Si$	4.9^{25}					
Trimethylamine	C_3H_9N	2.440^{25}	0.39745E+01	-0.51331E-02			273 - 298
1,2,3-Trimethylbenzene	C_9H_{12}	2.656^{20}	0.76006E+01	-0.29118E-01	0.41786E-04		273 - 323
1,2,4-Trimethylbenzene	C_9H_{12}	2.377^{20}	0.31517E+01	-0.30634E-02	0.14286E-05		273 - 323
1,3,5-Trimethylbenzene	C_9H_{12}	2.279^{20}	0.38998E+01	-0.88072E-02	0.11149E-04		288 - 358
Trimethyl borate	$C_3H_9BO_3$	2.2762^{20}					
2,2,3-Trimethylbutane	C_7H_{16}	1.930^{20}					
Trimethylchlorosilane	C_3H_9ClSi	10.21^{0}	-0.19492E+02	0.29806E+00	-0.69284E-03		223 - 273
2,4,6-Trimethyl-3-heptene (unspecified isomer)	$C_{10}H_{20}$	2.293^{20}					
2,2,3-Trimethylpentane	C_8H_{18}	1.960^{20}					
2,2,4-Trimethylpentane	C_8H_{18}	1.943^{20}	0.23677E+01	-0.14768E-02	0.94261E-07		173 - 373
2,3,3-Trimethylpentane	C_8H_{18}	1.9780^{20}					
2,3,4-Trimethylpentane	C_8H_{18}	1.9738^{20}					
2,4,4-Trimethyl-1-pentene	C_8H_{16}	2.0908^{25}					
Trimethylphenoxysilane	$C_9H_{14}OSi$	3.3953^{25}					
Trimethylphenylsilane	$C_9H_{14}Si$	2.3533^{25}	0.21463E+01	0.32711E-02	-0.86264E-05		288 - 323
Trimethyl phosphate	$C_3H_9O_4P$	20.6^{20}					
2,4,6-Trimethylpyridine	$C_8H_{11}N$	7.807^{25}	0.20990E+02	-0.57419E-01	0.44286E-04		298 - 358
Trinitroglycerol	$C_3H_5N_3O_9$	19.25^{20}					
2,4,6-Trinitrophenol	$C_6H_3N_3O_7$	4.0^{21}					
1,3,5-Trioxane	$C_3H_6O_3$	15.55^{65}					
Triphenylmethane	$C_{19}H_{16}$	2.46^{94}	0.40201E+01	-0.66507E-02	0.65329E-05		367 - 448
Tripropylamine	$C_9H_{21}N$	2.380^{20}	0.33380E+01	-0.86332E-02	0.18322E-04		243 - 293
Tripropylborane	$C_9H_{21}B$	2.026^{20}					
Tripropyl phosphate	$C_9H_{21}O_4P$	10.93^{20}	0.33166E+02	-0.10514E+00	0.10000E-03		293 - 373
Tris(perfluorobutyl)amine	$C_{12}F_{27}N$	2.15^{20}					
Undecane	$C_{11}H_{24}$	1.9972^{20}	0.23637E+01	-0.12500E-02	-0.85869E-16		283 - 363
1-Undecanol	$C_{11}H_{24}O$	5.98^{40}					
2-Undecanone	$C_{11}H_{22}O$	8.3^{12}					

Name	Mol. form.	*ε*	*a*	*b*	*c*	*d*	Temp. range/K
1-Undecene	$C_{11}H_{22}$	2.137^{20}	0.22132E+01	0.13121E-02	-0.53571E-05		273 - 323
Undecylamine	$C_{11}H_{25}N$	3.25^{20}	0.54945E+01	-0.96161E-02	0.66017E-05		293 - 373
Vanadium(IV) chloride	Cl_4V	3.05^{25}					
Vanadyl tribromide	Br_3OV	3.6^{25}	0.61112E+01	-0.84211E-02			203 - 298
Vanadyl trichloride	Cl_3OV	3.4^{25}					
2-Vinylpyridine	C_7H_7N	9.126^{20}					
4-Vinylpyridine	C_7H_7N	10.50^{20}					
Water	H_2O	80.1^{20}	0.24921E+03	-0.79069E+00	0.72997E-03		273 - 372
Xenon	Xe	1.880^{-112}					
Xenon hexafluoride	F_6Xe	4.10^{55}					
o-Xylene	C_8H_{10}	2.562^{20}	0.36163E+01	-0.40177E-02	0.14286E-05		273 - 323
m-Xylene	C_8H_{10}	2.359^{20}	0.28421E+01	-0.10191E-02	-0.21429E-05		273 - 323
p-Xylene	C_8H_{10}	2.2735^{20}	0.23140E+01	0.97221E-03	-0.37500E-05		293 - 363
2,3-Xylenol	$C_8H_{10}O$	4.81^{70}	0.14399E+02	-0.41438E-01	0.39244E-04		343 - 433
2,4-Xylenol	$C_8H_{10}O$	5.060^{30}	0.22125E+02	-0.85543E-01	0.96548E-04		303 - 363
2,5-Xylenol	$C_8H_{10}O$	5.36^{65}	0.18049E+02	-0.54991E-01	0.51656E-04		338 - 455
2,6-Xylenol	$C_8H_{10}O$	4.90^{40}	0.12284E+02	-0.32996E-01	0.29867E-04		313 - 453
3,4-Xylenol	$C_8H_{10}O$	9.02^{60}	0.54423E+02	-0.21153E+00	0.22508E-03		333 - 453
3,5-Xylenol	$C_8H_{10}O$	9.06^{50}	0.54251E+02	-0.21647E+00	0.23542E-03		323 - 453
Xylitol	$C_5H_{12}O_5$	40.0^{20}					

PERMITTIVITY (DIELECTRIC CONSTANT) OF GASES

Table 1 gives the relative permittivity ε (often called the dielectric constant) of some common gases at a temperature of 20 °C and pressure of one atmosphere (101.325 kPa). Values of the permanent dipole moment μ are also included. The density dependence of the permittivity is given by the equation

$$\frac{\varepsilon-1}{\varepsilon+2}=\rho_{\mathrm{m}}\left(\frac{4\pi N\alpha}{3}+\frac{4\pi N\mu^2}{9kT}\right)$$

where ρ_m is the molar density, N is Avogadro's number, k is the Boltzmann constant, T is the temperature, and α is the molecular polarizability. Therefore, in regions where the gas can be considered ideal, $\varepsilon - 1$ is approximately proportional to the pressure at constant temperature. For nonpolar gases ($\mu = 0$), $\varepsilon - 1$ is inversely proportional to temperature at constant pressure.

The number of significant figures indicates the accuracy of the values given. The values of ε for air, Ar, H_2, He, N_2, O_2, and CO_2 are recommended as reference values; these are accurate to 1 ppm or better. Column definitions for Table 1 are as follows.

Column heading	Definition
Name	Name of gas; listed alphabetically by name
Mol. form.	Molecular formula for gas
ε	Relative permittivity (dielectric constant) at 20 °C and 1 atmosphere pressure (101.325 kPa)
μ	Permanent dipole moment, in Debye units (1 D = 3.33564 × 10^{-30} C m)

Table 2 gives the permittivity of water vapor in equilibrium with liquid water as a function of temperature (derived from Ref. 4).

References

1. Maryott, A. A., and Buckley, F., *Table of Dielectric Constants and Electric Dipole Moments of Substances in the Gaseous State*, National Bureau of Standards Circular 537, 1953.
2. Harvey, A. H., and Lemmon, E. W., *Int. J. Thermophys.* 26, 31, 2005 [for nonpolar gases and light hydrocarbons] <https://doi.org/10.1007/s10765-005-2351-5>
3. *Landolt-Börnstein, Numerical Data and Functional Relationships in Science and Technology*, New Series, Group IV, Vol. 4, Springer-Verlag, Heidelberg, 1980 [for data at high pressures].
4. Fernández, D. P., Goodwin, A. R. H., Lemmon, E. W., Levelt Sengers, J. M. H., and Williams, R. C., *J. Phys. Chem. Ref. Data* 26, 1125, 1997 [for water vapor] <https://doi.org/10.1063/1.555997>

TABLE 1. Permittivity of Selected Gases at 20 °C and 1 Atmosphere Pressure

Name	Mol. form.	ε	μ/D
Acetylene	HC≡CH	1.00124	0
Air[a]		1.0005360	0
Ammonia	NH_3	1.00622	1.4718
Argon	Ar	1.0005169	0
Boron trifluoride	BF_3	1.0011	0
Bromomethane	CH_3Br	1.01028	1.8203
Butane	C_4H_{10}	1.00266	0.05
Carbon dioxide	CO_2	1.0009217	0
Carbon monoxide	CO	1.00065	0.10980
Chloroethane	C_2H_5Cl	1.01325	2.05
Chloroethene	CH_2=CHCl	1.0075	1.45
Chloromethane	CH_3Cl	1.01080	1.8963
Cyclopropane	C_3H_6	1.00178	0
Dimethyl ether	CH_3OCH_3	1.0062	1.30
Ethane	C_2H_6	1.001403	0
Ethylene	CH_2=CH_2	1.00135	0
Fluoromethane	CH_3F	1.00973	1.858
Helium	He	1.0000645	0
Hydrogen	H_2	1.0002532	0
Hydrogen bromide	HBr	1.00279	0.8272
Hydrogen chloride	HCl	1.00390	1.1086
Hydrogen iodide	HI	1.00214	0.448
Hydrogen sulfide	H_2S	1.00344	0.97833
Iodomethane	CH_3I	1.00914	1.6406
Isobutane	$(CH_3)_3CH$	1.00268	0.132
Krypton	Kr	1.000784	0
Methane	CH_4	1.0008181	0
Neon	Ne	1.000124	0
Nitric oxide	NO	1.00060	0.15872
Nitrogen	N_2	1.0005474	0
Nitrogen trifluoride	NF_3	1.0013	0.235
Nitrous oxide	N_2O	1.00104	0.16083
Oxygen	O_2	1.0004941	0
Ozone	O_3	1.0017	0.53373
Propane	C_3H_8	1.002032	0.084
Propene	CH_3CH=CH_2	1.00228	0.366
Sulfur dioxide	SO_2	1.00825	1.63305
Sulfur hexafluoride	SF_6	1.00200	0
Tetrafluoromethane	CF_4	1.00121	0
Xenon	Xe	1.00127	0

[a] Dry, CO_2 free.

TABLE 2. Permittivity of Saturated Water Vapor at Various Temperatures

t/°C	ε
0	1.000064
10	1.000120
20	1.000214
30	1.000363
40	1.000594
50	1.000935
60	1.00143
70	1.00211
80	1.00304
90	1.00428
100	1.00589

AZEOTROPIC DATA FOR BINARY MIXTURES

J. Gmehling, J. Menke, J. Krafczyk, K. Fischer, J.-C. Fontaine, and H. V. Kehiaian

In this table, important parameters are given for 808 azeotropic binary mixtures.

Binary homogeneous (single-phase) liquid mixtures having an extremum (maximum or minimum) vapor pressure P at constant temperature T, as a function of composition, are called azeotropic mixtures, or simply azeotropes. The composition is usually expressed as mole fractions, where x_1 for component 1 in the liquid phase and y_1 for component 1 in the vapor phase are identical. Mixtures that do not show a maximum or minimum are called zeotropic. A maximum (minimum) of the $P(x_1)$ or $P(y_1)$ curves corresponds to a minimum (maximum) of the boiling temperature T at constant P, plotted as a function of x_1 or y_1 [see $T(x_1)$ and $T(y_1)$ curves, Types I and III, in Figure 1]. Azeotropes in which the pressure is a maximum (temperature is a minimum) are often called positive azeotropes, while pressure-minimum (temperature-maximum) azeotropes are called negative azeotropes. The coordinates of an azeotropic point are the azeotropic temperature T_{Az}, pressure P_{Az}, and the vapor-phase composition $y_{1,Az}$, which is the same as the liquid-phase composition $x_{1,Az}$.

In the two-phase liquid-liquid region of partially miscible (heterogeneous) mixtures, the vapor pressure at constant T (or the boiling temperature at constant P) is independent of the global composition x_1 of the two coexisting liquid phases between the equilibrium compositions x_1' and x_1'' ($x_1' < x_1''$).

The constant vapor pressure (boiling temperature) above the two-phase region of certain partially miscible mixtures is usually larger (smaller) than the vapor pressure (boiling temperature) at any other liquid-phase composition in the homogeneous region. In this case, the vapor-phase composition is inside the miscibility gap. Mixtures of this type are called heteroazeotropic mixtures, or simply heteroazeotropes. (Figure 1, Type II), as opposed to the other types of azeotropes, called homoazeotropes.

Only in a few cases partially miscible mixtures present a positive or negative azeotropic point in the single-phase region, outside the miscibility gap, similar to the azeotropic points of homogeneous mixtures (Figure 1, Types IV and VI).

A few binary mixtures, for example the system perfluorobenzene + benzene, may present two azeotropic points at constant temperature (pressure), a positive and a negative one. They are called double azeotropic mixtures, or simply double azeotropes. (Figure 1, Type V).

The knowledge of the occurrence of azeotropic points in binary and higher systems is of special importance for the design of distillation processes. The number of theoretical stages of a distillation column required for the separation depends on the separation factor α_{12}, i.e., the ratio of the K_i-factors ($K_i = y_i/x_i$) of the components i ($i = 1, 2$). The required separation factor can be calculated with the following simplified relation (Ref. 1)

$$\alpha_{12} = K_1/K_2 = (y_1/x_1)/(y_2/x_2) = (\gamma_1 P_1^s)/(\gamma_2 P_2^s) \qquad (1)$$

where γ_i is the activity coefficient of component i in the liquid phase and P_i^s is the vapor pressure of the pure component i.

In distillation processes, only the difference between the separation factor and unity ($\alpha_{12} - 1$) can be exploited for the separation. If the separation factor is close to unity, a large number of theoretical stages is required for the separation. If the binary system to be separated shows an azeotropic point ($\alpha_{12} = 1$), the separation is impossible by ordinary distillation, even with an infinitely large number of stages.

Following eq. (1) azeotropic behavior will always occur in homogeneous binary systems when the vapor pressure ratio P_1^s/P_2^s is equal to the ratio of the activity coefficients γ_2/γ_1.

Various thermodynamic methods based on g^E–models (Wilson, NRTL, UNIQUAC) or group contribution methods (UNIFAC, modified UNIFAC, ASOG, PSRK) can be used for either calculating or predicting the required activity coefficients for the components under given conditions of temperature and composition (Ref. 2).

Because of the importance of azeotropic data for the design of distillation processes, compilations have been available in book form for quite some time (Refs. 3-7). The most recent printed data collection was published in 1994 (Ref. 8). A revised and extended version appeared in 2004 (Ref. 9).

A collection of approximately 47,400 zeotropic and azeotropic data sets, compiled from 6600 references, are stored in a comprehensive computerized data bank (Ref. 10). The references from the above-mentioned compilations and from the vapor-liquid equilibrium part of the Dortmund Data Bank (Ref. 11) were supplemented by references found from CAS online searches, private communications, data from industry, etc. Over 24,000 zeotropic data and over 20,000 azeotropic data are available for binary systems. Nearly 90% of the binary azeotropic data show a pressure maximum. In most cases (ca. 90%) these are homogeneous azeotropes, and in approximately 7–8% of the cases heterogeneous azeotropes are reported. Less than 10% of the data stored show a pressure minimum. Approximately 21,000 of the data sets stored were published after 1970.

Table 1 provides information about azeotropes for selected binary systems. Mixtures are listed alphabetically by name of the first component followed by the name of the second component. ***For convenience in search, in the Online Edition, each row is duplicated with the components in reverse order.***

Column headings for the table are as follows.

Column heading	Definition
Component 1	Name of first component; mixtures are listed alphabetically by the name of the first component
Mol. form. 1	Molecular formula of first component; in Hill order
Component 2	Name of second component
Mol. form. 2	Molecular formula of second component; in Hill order
T_{Az}	Azeotropic temperature, in K
$y_{1,Az}$	Vapor-phase composition of first component
P_{Az}	Azeotropic pressure, in kPa
Type	Azeotropic type, see definition of symbols below

The explanation of the type of azeotrope is given by the following codes:

O: Homogeneous azeotrope in a completely miscible system
L: Homogeneous azeotrope in a partially miscible system
E: Heterogeneous azeotrope
X: Pressure maximum
N: Pressure minimum
D: Double azeotrope
C: System contains a supercritical compound

Fluids

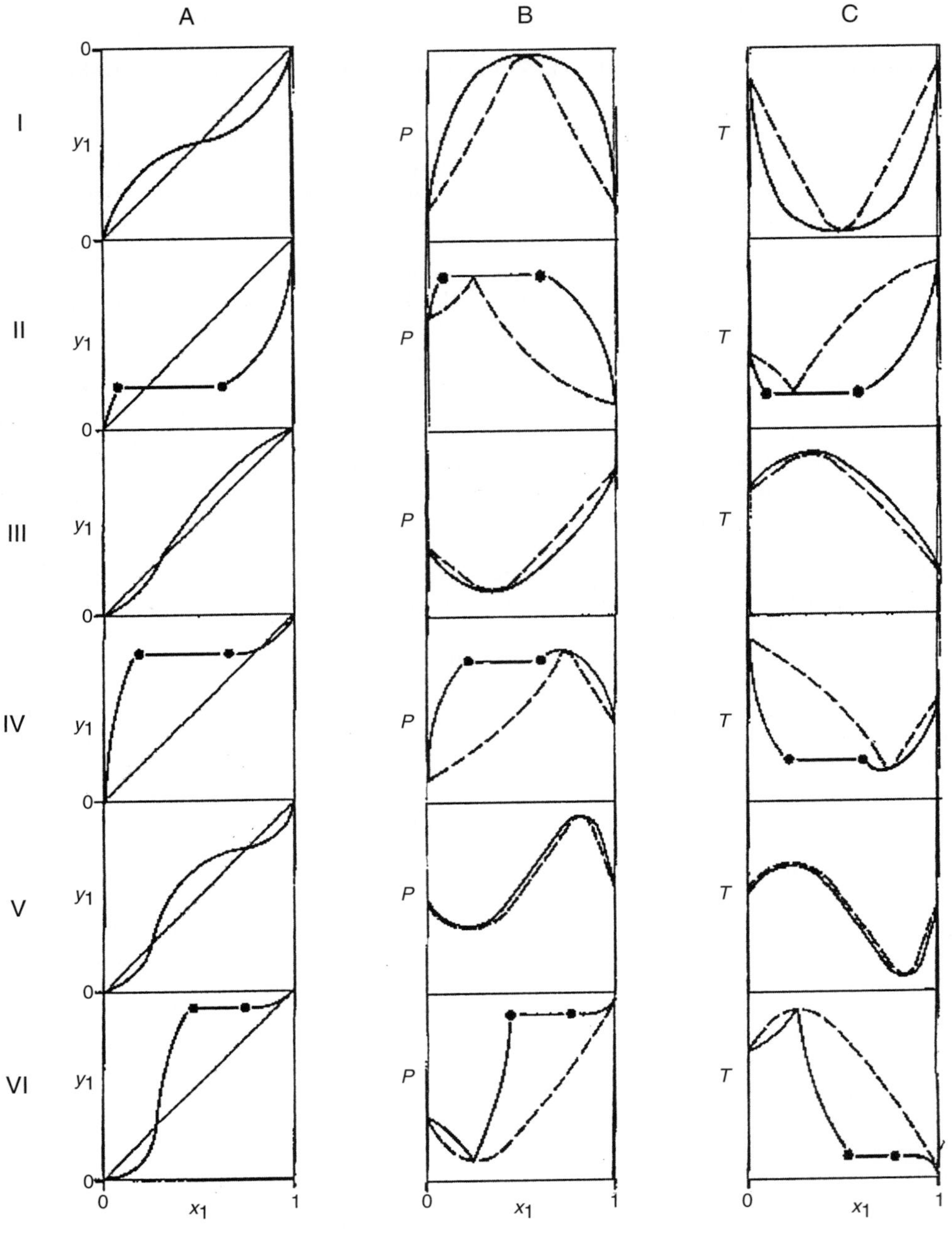

FIGURE 1. Different types of binary azeotropic systems: I — homogeneous pressure-maximum azeotrope in a completely miscible system (OX); II — heterogeneous pressure-maximum azeotrope (EX); III — homogeneous pressure-minimum azeotrope in a completely miscible system (ON); IV — homogeneous pressure-maximum azeotrope in a partially miscible system (LX); V–D: double azeotrope (OND, OXD); VI — homogeneous pressure-minimum azeotrope in a partially miscible system (LN). A — $y_1(x_1)$; B — $P(x_1)$ and $P(y_1)$; C — $T(x_1)$ and $T(y_1)$. Continuous line — (x_1); Dashed line — (y_1).

References

1. Gmehling, J. and Brehm, A., *Grundoperationen*, Thieme-Verlag, Stuttgart, 1996.
2. Gmehling, J. and Kolbe, B., *Thermodynamik*, VCH-Verlag, Weinheim, 1992.
3. Lecat, M., *Doctoral Dissertation*, 1908.
4. Lecat, M., *L'Azeotropisme*, Monograph, L'Auteur, Brussel, 1918.
5. Lecat, M., *Tables Azeotropiques*, Monograph, Lamertin, Brussel, 1949.
6. Ogorodnikov, S. K., Lesteva, T. M., and Kogan V. B., *Azeotropic Mixtures*, Khimia, Leningrad, 1971.
7. Horsley, L. H., *Azeotropic Data III*, American Chemical Society, Washington, DC, 1973. <https://doi.org/10.1021/ba-1973-0116]
8. Gmehling, J., Menke, J., Krafczyk, J., and Fischer, K., *Azeotropic Data*, 2 Volumes, VCH Verlag, Weinheim, 1994.
9. Gmehling, J., Menke, J., Krafczyk, J., and Fischer, K., *Azeotropic Data, Second Edition*, 3 Volumes, VCH Verlag, Weinheim, 2004.
10. Gmehling, J., Menke, J., Krafczyk, J., and Fischer, K., A Data Bank for Azeotropic Data, Status and Applications, *Fluid Phase Equilib.* 103, 51, 1995. <https://doi.org/10.1016/0378-3812(94)02569-M]
11. Dortmund Data Bank, < http://www.ddbst.de>.

Azeotropic Binary Mixtures: Temperature, Pressure, and Composition

Component 1	Mol. form. 1	Component 2	Mol. form. 2	T_{Az}/K	$y_{1,Az}$	P_{Az}/kPa	Type
Acetaldehyde	C_2H_4O	1,3-Butadiene	C_4H_6	268.15	0.0520	101.33	OX
Acetaldehyde	C_2H_4O	2-Methyl-1,3-butadiene	C_5H_8	292.23	0.8140	101.33	OX
Acetic acid	$C_2H_4O_2$	Decane	$C_{10}H_{22}$	390.05	0.9250	101.33	OX
Acetic acid	$C_2H_4O_2$	2,4-Dimethylpyridine	C_7H_9N	435.45	0.3022	101.33	ON
Acetic acid	$C_2H_4O_2$	Heptane	C_7H_{16}	364.95	0.4490	101.33	OX
Acetic acid	$C_2H_4O_2$	Hexane	C_6H_{14}	341.40	0.0839	101.33	OX
Acetic acid	$C_2H_4O_2$	3-Methyl-2-butanol, (±)-	$C_5H_{12}O$	392.65	0.7210	101.33	ON
Acetic acid	$C_2H_4O_2$	2-Methylpyridine	C_6H_7N	417.27	0.5120	101.33	ON
Acetic acid	$C_2H_4O_2$	Nonane	C_9H_{20}	386.05	0.8250	101.33	OX
Acetic acid	$C_2H_4O_2$	Octane	C_8H_{18}	378.85	0.6870	101.33	OX
Acetic acid	$C_2H_4O_2$	Pyridine	C_5H_5N	411.25	0.5780	101.33	ON
Acetic acid	$C_2H_4O_2$	Undecane	$C_{11}H_{24}$	391.15	0.9720	101.33	OX
Acetic acid	$C_2H_4O_2$	Vinyl butanoate	$C_6H_{10}O_2$	386.45	0.5750	101.33	OX
Acetic acid	$C_2H_4O_2$	*o*-Xylene	C_8H_{10}	389.75	0.8640	101.33	OX
Acetic acid	$C_2H_4O_2$	*p*-Xylene	C_8H_{10}	388.40	0.8200	101.33	OX
Acetic anhydride	$C_4H_6O_3$	Octane	C_8H_{18}	397.65	0.3500	129.80	OX
Acetic anhydride	$C_4H_6O_3$	1-Octene	C_8H_{16}	367.53	0.2840	53.88	OX
Acetone	C_3H_6O	1-Bromopropane	C_3H_7Br	328.75	0.9915	99.75	OX
Acetone	C_3H_6O	2-Chloro-2-methylpropane	C_4H_9Cl	322.05	0.1944	102.11	OX
Acetone	C_3H_6O	Cyclohexane	C_6H_{12}	330.05	0.7590	109.32	OX
Acetone	C_3H_6O	Diisopropyl ether	$C_6H_{14}O$	327.10	0.7424	100.17	OX
Acetone	C_3H_6O	Hexane	C_6H_{14}	322.95	0.6480	101.33	OX
Acetone	C_3H_6O	1-Hexene	C_6H_{12}	323.35	0.5973	101.40	OX
Acetone	C_3H_6O	Isopentane	C_5H_{12}	298.75	0.1730	101.33	OX
Acetone	C_3H_6O	Methanol	CH_4O	328.29	0.7600	101.33	OX
Acetone	C_3H_6O	Methyl acetate	$C_3H_6O_2$	328.85	0.6470	101.33	OX
Acetone	C_3H_6O	2-Methyl-1,3-butadiene	C_5H_8	306.95	0.0610	101.33	OX
Acetone	C_3H_6O	2-Methyl-1-butene	C_5H_{10}	303.25	0.1400	101.33	OX
Acetone	C_3H_6O	2-Methyl-2-butene	C_5H_{10}	308.75	0.2440	101.33	OX
Acetone	C_3H_6O	Methyl *tert*-butyl ether	$C_5H_{12}O$	324.35	0.4824	102.19	OX
Acetone	C_3H_6O	1-Methylcyclobutene	C_5H_8	307.75	0.2220	101.33	OX
Acetone	C_3H_6O	Methylcyclohexane	C_7H_{14}	318.15	0.9500	68.66	OX
Acetone	C_3H_6O	Methylenecyclobutane	C_5H_8	311.25	0.2800	101.33	OX
Acetone	C_3H_6O	2-Methyl-1-pentene	C_6H_{12}	333.40	0.5793	140.60	OX
Acetone	C_3H_6O	Tetrachloromethane	CCl_4	341.25	0.9663	149.93	OX
Acetone	C_3H_6O	Tetrahydrofuran	C_4H_8O	328.85	0.9603	100.35	OX
Acetone	C_3H_6O	Trichloromethane	$CHCl_3$	337.58	0.3602	101.33	ON
Acetone	C_3H_6O	Triethylamine	$C_6H_{15}N$	318.15	0.9800	68.13	OX
Acetonitrile	C_2H_3N	Benzene	C_6H_6	328.15	0.4560	54.65	OX
Acetonitrile	C_2H_3N	2-Butanone	C_4H_8O	352.15	0.3195	101.15	OX
Acetonitrile	C_2H_3N	1-Decene	$C_{10}H_{20}$	354.55	0.9924	100.51	OX
Acetonitrile	C_2H_3N	2-Ethoxy-2-methylbutane	$C_7H_{16}O$	348.85	0.7219	98.99	OX
Acetonitrile	C_2H_3N	Isopentane	C_5H_{12}	298.45	0.1040	101.33	EX
Acetonitrile	C_2H_3N	2-Methoxy-2-methylbutane	$C_6H_{14}O$	346.13	0.5835	100.56	OX
Acetonitrile	C_2H_3N	2-Methyl-1,3-butadiene	C_5H_8	306.75	0.0410	101.33	OX
Acetonitrile	C_2H_3N	2-Methyl-2-butene	C_5H_{10}	308.95	0.1320	101.33	OX
Acetonitrile	C_2H_3N	Methylenecyclobutane	C_5H_8	312.45	0.1450	101.33	OX
Acetonitrile	C_2H_3N	Methyl methacrylate	$C_5H_8O_2$	355.25	0.9866	102.07	OX
Acetonitrile	C_2H_3N	2-Methyl-2-propanol	$C_4H_{10}O$	333.15	0.6200	56.93	OX
Acetonitrile	C_2H_3N	1-Pentene	C_5H_{10}	301.85	0.0830	101.33	OX
Acetonitrile	C_2H_3N	2-Propanol	C_3H_8O	348.15	0.5287	100.81	OX
Acetonitrile	C_2H_3N	Tetrachlorosilane	Cl_4Si	321.05	0.3100	101.33	EX
Acetonitrile	C_2H_3N	Tetrahydrofuran	C_4H_8O	338.95	0.0784	101.13	OX
Acetonitrile	C_2H_3N	Vinyl acetate	$C_4H_6O_2$	344.65	0.1948	98.33	OX
Acetonitrile	C_2H_3N	Water	H_2O	349.95	0.6900	101.33	OX
Acetophenone	C_8H_8O	Water	H_2O	371.15	0.0325	101.19	EX
Acrolein	C_3H_4O	2-Methyl-1,3-butadiene	C_5H_8	306.45	0.1980	101.33	OX
Acrolein	C_3H_4O	Water	H_2O	325.45	0.9270	101.33	LX
Acrylonitrile	C_3H_3N	Benzene	C_6H_6	347.45	0.5575	101.46	OX
Acrylonitrile	C_3H_3N	Cyclohexane	C_6H_{12}	337.75	0.4836	101.94	OX

Component 1	Mol. form. 1	Component 2	Mol. form. 2	T_{Az}/K	$y_{1,Az}$	P_{Az}/kPa	Type
Acrylonitrile	C_3H_3N	Ethanol	C_2H_6O	343.95	0.5560	101.33	OX
Acrylonitrile	C_3H_3N	Hexane	C_6H_{14}	330.90	0.4048	101.05	OX
Acrylonitrile	C_3H_3N	Methylenecyclobutane	C_5H_8	313.80	0.1275	101.33	OX
Acrylonitrile	C_3H_3N	Water	H_2O	344.05	0.7150	101.33	EX
Allyl alcohol	C_3H_6O	Benzene	C_6H_6	349.90	0.2203	101.33	OX
Allyl alcohol	C_3H_6O	Cyclohexane	C_6H_{12}	333.15	0.2790	63.98	OX
Allyl alcohol	C_3H_6O	Ethyl propanoate	$C_5H_{10}O_2$	367.65	0.5597	99.79	OX
Allyl alcohol	C_3H_6O	Water	H_2O	361.15	0.4438	101.33	OX
Aluminum chloride	$AlCl_3$	Phosphoryl chloride	Cl_3OP	660.15	0.5150	101.33	ONC
Aniline	C_6H_7N	Butylbenzene	$C_{10}H_{14}$	448.65	0.4993	101.33	OX
Aniline	C_6H_7N	*sec*-Butylbenzene, (±)-	$C_{10}H_{14}$	443.15	0.3021	101.33	OX
Aniline	C_6H_7N	*tert*-Butylbenzene	$C_{10}H_{14}$	438.25	0.2104	101.33	OX
Aniline	C_6H_7N	1-Butyl-2-methylbenzene	$C_{11}H_{16}$	458.15	0.8091	101.33	OX
Aniline	C_6H_7N	1-Butyl-3-methylbenzene	$C_{11}H_{16}$	456.55	0.7661	101.33	OX
Aniline	C_6H_7N	1-Butyl-4-methylbenzene	$C_{11}H_{16}$	457.05	0.7807	101.33	OX
Aniline	C_6H_7N	*o*-Cresol	C_7H_8O	464.29	0.0953	101.33	ON
Aniline	C_6H_7N	Decane	$C_{10}H_{22}$	440.43	0.4660	101.33	OX
Aniline	C_6H_7N	*o*-Diethylbenzene	$C_{10}H_{14}$	448.75	0.5024	101.33	OX
Aniline	C_6H_7N	*m*-Diethylbenzene	$C_{10}H_{14}$	447.45	0.4584	101.33	OX
Aniline	C_6H_7N	*p*-Diethylbenzene	$C_{10}H_{14}$	448.85	0.5086	101.33	OX
Aniline	C_6H_7N	Dodecane	$C_{12}H_{26}$	453.52	0.8220	101.33	OX
Aniline	C_6H_7N	1-Ethyl-3,5-dimethylbenzene	$C_{10}H_{14}$	449.65	0.5310	101.33	OX
Aniline	C_6H_7N	2-Ethyl-1,4-dimethylbenzene	$C_{10}H_{14}$	449.65	0.5310	101.33	OX
Aniline	C_6H_7N	Isobutylbenzene	$C_{10}H_{14}$	442.75	0.2890	101.33	OX
Aniline	C_6H_7N	1-Isopropyl-2-methylbenzene	$C_{10}H_{14}$	445.95	0.4052	101.33	OX
Aniline	C_6H_7N	1-Isopropyl-3-methylbenzene	$C_{10}H_{14}$	444.15	0.3419	101.33	OX
Aniline	C_6H_7N	1-Isopropyl-4-methylbenzene	$C_{10}H_{14}$	445.35	0.3829	101.33	OX
Aniline	C_6H_7N	1-Methyl-2-propylbenzene	$C_{10}H_{14}$	449.45	0.5270	101.33	OX
Aniline	C_6H_7N	1-Methyl-3-propylbenzene	$C_{10}H_{14}$	447.85	0.4711	101.33	OX
Aniline	C_6H_7N	1-Methyl-4-propylbenzene	$C_{10}H_{14}$	448.75	0.5035	101.33	OX
Aniline	C_6H_7N	Nonane	C_9H_{20}	422.35	0.1770	101.33	OX
Aniline	C_6H_7N	Phenol	C_6H_6O	459.09	0.6116	101.33	ON
Aniline	C_6H_7N	Tetradecane	$C_{14}H_{30}$	457.05	0.9770	101.33	OX
Aniline	C_6H_7N	1,2,3,5-Tetramethylbenzene	$C_{10}H_{14}$	456.55	0.7504	101.33	OX
Aniline	C_6H_7N	1,2,4,5-Tetramethylbenzene	$C_{10}H_{14}$	455.36	0.7349	101.33	OX
Aniline	C_6H_7N	Tridecane	$C_{13}H_{28}$	456.22	0.9300	101.33	OX
Aniline	C_6H_7N	1,2,3-Trimethylbenzene	C_9H_{12}	444.65	0.3331	101.33	OX
Aniline	C_6H_7N	1,2,4-Trimethylbenzene	C_9H_{12}	441.80	0.1850	101.33	OX
Aniline	C_6H_7N	1,3,5-Trimethylbenzene	C_9H_{12}	437.68	0.1071	101.33	OX
Aniline	C_6H_7N	Undecane	$C_{11}H_{24}$	449.05	0.6970	101.33	OX
Aniline	C_6H_7N	Water	H_2O	372.55	0.0420	101.33	EX
Benzaldehyde	C_7H_6O	Phenol	C_6H_6O	447.00	0.3999	73.00	ON
Benzene	C_6H_6	2-Butanone	C_4H_8O	351.53	0.5210	101.33	OX
Benzene	C_6H_6	Butylamine	$C_4H_{11}N$	343.15	0.3000	80.89	OX
Benzene	C_6H_6	Cyclohexane	C_6H_{12}	353.15	0.5460	109.18	OX
Benzene	C_6H_6	Dimethyl carbonate	$C_3H_6O_3$	353.50	0.8634	100.48	OX
Benzene	C_6H_6	Ethanol	C_2H_6O	341.25	0.5400	101.33	OX
Benzene	C_6H_6	Ethyl acetate	$C_4H_8O_2$	350.55	0.0547	102.45	OX
Benzene	C_6H_6	Heptane	C_7H_{16}	353.25	0.9922	101.32	OX
Benzene	C_6H_6	Hexafluorobenzene	C_6F_6	353.60	0.2400	101.33	OND
Benzene	C_6H_6	Hexafluorobenzene	C_6F_6	352.50	0.8168	101.33	OXD
Benzene	C_6H_6	Hexane	C_6H_{14}	341.45	0.0500	101.33	OX
Benzene	C_6H_6	Methanol	CH_4O	331.56	0.3910	101.33	OX
Benzene	C_6H_6	2-Methyl-2-butanol	$C_5H_{12}O$	352.35	0.8500	101.33	OX
Benzene	C_6H_6	Methylcyclopentane	C_6H_{12}	333.15	0.1390	69.93	OX
Benzene	C_6H_6	2-Methyl-1-propanol	$C_4H_{10}O$	352.45	0.9220	101.33	OX
Benzene	C_6H_6	1-Propanol	C_3H_8O	350.20	0.7940	101.33	OX
Benzene	C_6H_6	2-Propanol	C_3H_8O	345.03	0.6040	101.33	OX
Benzene	C_6H_6	Propyl formate	$C_4H_8O_2$	343.15	0.6230	76.08	OX
Benzene	C_6H_6	2,2,4-Trimethylpentane	C_8H_{18}	353.25	0.9751	101.32	OX
Benzene	C_6H_6	Tris(perfluoroethyl)amine	$C_6F_{15}N$	329.95	0.4100	101.33	EX

Component 1	Mol. form. 1	Component 2	Mol. form. 2	T_{Az}/K	$y_{1,Az}$	P_{Az}/kPa	Type
Benzene	C_6H_6	Water	H_2O	342.35	0.7020	101.33	EX
Benzonitrile	C_7H_5N	*o*-Cresol	C_7H_8O	468.91	0.5100	101.33	ON
Benzonitrile	C_7H_5N	*m*-Cresol	C_7H_8O	476.10	0.1441	101.33	ON
Benzonitrile	C_7H_5N	*p*-Cresol	C_7H_8O	476.95	0.0898	101.33	ON
Benzonitrile	C_7H_5N	Phenol	C_6H_6O	465.11	0.7655	101.33	ON
Benzonitrile	C_7H_5N	2,6-Xylenol	$C_8H_{10}O$	477.15	0.0807	101.33	ON
Benzyl alcohol	C_7H_8O	Decane	$C_{10}H_{22}$	445.75	0.2490	101.33	OX
Benzyl alcohol	C_7H_8O	Water	H_2O	373.05	0.0160	101.33	EX
Bis(2,2,2-trifluoroethyl) ether	$C_4H_4F_6O$	Ethanol	C_2H_6O	331.90	0.7160	101.30	OX
Bis(2,2,2-trifluoroethyl) ether	$C_4H_4F_6O$	Methanol	CH_4O	326.28	0.5550	101.30	OX
Bis(2,2,2-trifluoroethyl) ether	$C_4H_4F_6O$	1-Propanol	C_3H_8O	336.22	0.8900	101.30	OX
Bis(2,2,2-trifluoroethyl) ether	$C_4H_4F_6O$	2-Propanol	C_3H_8O	334.16	0.7770	101.30	OX
Bromobenzene	C_6H_5Br	Cyclohexanol	$C_6H_{12}O$	403.15	0.7390	52.45	OX
1-Bromobutane	C_4H_9Br	Water	H_2O	353.95	0.5050	101.33	EX
2-Bromo-2-chloro-1,1,1-trifluoroethane	$C_2HBrClF_3$	Diethyl ether	$C_4H_{10}O$	323.65	0.7200	93.33	ON
2-Bromo-2-chloro-1,1,1-trifluoroethane	$C_2HBrClF_3$	Methanol	CH_4O	317.25	0.8110	93.33	OX
Bromoethane	C_2H_5Br	Isopentane	C_5H_{12}	300.55	0.2180	101.33	OX
Bromoethane	C_2H_5Br	Methanol	CH_4O	308.05	0.8390	101.33	OX
Bromoethane	C_2H_5Br	2-Methyl-2-butene	C_5H_{10}	308.55	0.5110	101.33	OX
Bromoethane	C_2H_5Br	Methyl formate	$C_2H_4O_2$	303.05	0.2640	101.33	OX
1-Bromo-2-methylpropane	C_4H_9Br	Water	H_2O	348.45	0.6270	101.33	EX
1-Bromopropane	C_3H_7Br	Cyclohexane	C_6H_{12}	343.35	0.9219	98.84	OX
1-Bromopropane	C_3H_7Br	Methyl acetate	$C_3H_6O_2$	329.60	0.0273	99.56	OX
1-Bromopropane	C_3H_7Br	Nitromethane	CH_3NO_2	343.25	0.8980	99.82	OX
1-Bromopropane	C_3H_7Br	2-Propanol	C_3H_8O	339.15	0.7349	99.97	OX
1-Bromopropane	C_3H_7Br	Water	H_2O	336.35	0.7790	101.33	EX
2-Bromopropane	C_3H_7Br	Ethyl formate	$C_3H_6O_2$	326.15	0.2910	101.33	OX
1,3-Butadiene	C_4H_6	*cis*-2-Butene	C_4H_8	267.59	0.7650	101.33	OX
Butanal	C_4H_8O	Ethanol	C_2H_6O	345.45	0.6310	101.33	OX
Butanal	C_4H_8O	2-Methyl-1-pentene	C_6H_{12}	334.15	0.2293	101.48	OX
1,4-Butanediol	$C_4H_{10}O_2$	1-Pentadecanol	$C_{15}H_{32}O$	502.75	0.9980	101.33	OX
2,3-Butanedione	$C_4H_6O_2$	1-Propanol	C_3H_8O	359.30	0.6400	100.67	OX
2,3-Butanedione	$C_4H_6O_2$	2-Propanol	C_3H_8O	350.85	0.3546	100.95	OX
2,3-Butanedione	$C_4H_6O_2$	Toluene	C_7H_8	362.70	0.9513	101.34	OX
Butanoic acid	$C_4H_8O_2$	Butyl butanoate	$C_8H_{16}O_2$	434.60	0.6532	93.33	OXD
Butanoic acid	$C_4H_8O_2$	Butyl butanoate	$C_8H_{16}O_2$	434.78	0.8639	93.33	OND
Butanoic acid	$C_4H_8O_2$	Pyridine	C_5H_5N	436.35	0.9117	101.33	ON
Butanoic acid	$C_4H_8O_2$	Undecane	$C_{11}H_{24}$	435.55	0.9060	101.33	OX
Butanoic acid	$C_4H_8O_2$	Water	H_2O	372.95	0.0441	101.33	OX
1-Butanol	$C_4H_{10}O$	Butyl acetate	$C_6H_{12}O_2$	389.97	0.7700	101.33	OX
1-Butanol	$C_4H_{10}O$	1-Butylcyclopentene	C_9H_{16}	356.70	0.8450	79.99	OX
1-Butanol	$C_4H_{10}O$	Chlorobenzene	C_6H_5Cl	388.25	0.6950	101.33	OX
1-Butanol	$C_4H_{10}O$	Cyclohexane	C_6H_{12}	352.68	0.0787	101.33	OX
1-Butanol	$C_4H_{10}O$	Dibutyl ether	$C_8H_{18}O$	390.59	0.8754	101.33	OX
1-Butanol	$C_4H_{10}O$	Diethyl carbonate	$C_5H_{10}O_3$	370.85	0.6346	53.20	OX
1-Butanol	$C_4H_{10}O$	3-Ethylcyclopentene	C_7H_{12}	367.65	0.1900	101.33	OX
1-Butanol	$C_4H_{10}O$	Heptane	C_7H_{16}	366.55	0.2272	101.38	OX
1-Butanol	$C_4H_{10}O$	Hexane	C_6H_{14}	341.35	0.0370	101.33	OX
1-Butanol	$C_4H_{10}O$	Isobutyl acetate	$C_6H_{12}O_2$	387.15	0.5980	101.33	OX
1-Butanol	$C_4H_{10}O$	Nonane	C_9H_{20}	389.05	0.8128	101.33	OX
1-Butanol	$C_4H_{10}O$	1-Nonyne	C_9H_{16}	390.60	0.9400	101.33	OX
1-Butanol	$C_4H_{10}O$	Octane	C_8H_{18}	383.15	0.5500	102.79	OX
1-Butanol	$C_4H_{10}O$	1-Octene	C_8H_{16}	363.45	0.4530	53.33	OX
1-Butanol	$C_4H_{10}O$	*cis*-4-Octene	C_8H_{16}	382.35	0.5300	101.33	OX
1-Butanol	$C_4H_{10}O$	*trans*-4-Octene	C_8H_{16}	382.15	0.5310	101.33	OX
1-Butanol	$C_4H_{10}O$	1-Octyne	C_8H_{14}	386.50	0.6200	101.33	OX
1-Butanol	$C_4H_{10}O$	2-Octyne	C_8H_{14}	398.30	0.7910	101.33	OX
1-Butanol	$C_4H_{10}O$	Pyridine	C_5H_5N	392.00	0.7050	101.33	ON
1-Butanol	$C_4H_{10}O$	Styrene	C_8H_8	388.71	0.8923	98.39	OX
1-Butanol	$C_4H_{10}O$	Tetrachloromethane	CCl_4	349.71	0.0500	101.33	OX
1-Butanol	$C_4H_{10}O$	Toluene	C_7H_8	378.85	0.3320	101.33	OX

Component 1	Mol. form. 1	Component 2	Mol. form. 2	T_{Az}/K	$y_{1,Az}$	P_{Az}/kPa	Type
1-Butanol	$C_4H_{10}O$	Water	H_2O	365.45	0.2460	101.33	EX
1-Butanol	$C_4H_{10}O$	*o*-Xylene	C_8H_{10}	388.05	0.8671	100.13	OX
1-Butanol	$C_4H_{10}O$	*m*-Xylene	C_8H_{10}	387.75	0.7865	101.46	OX
1-Butanol	$C_4H_{10}O$	*p*-Xylene	C_8H_{10}	387.85	0.7823	99.73	OX
2-Butanol	$C_4H_{10}O$	Cyclohexane	C_6H_{12}	349.90	0.1892	101.02	OX
2-Butanol	$C_4H_{10}O$	Cyclohexene	C_6H_{10}	352.75	0.2046	101.25	OX
2-Butanol	$C_4H_{10}O$	1,4-Dioxane	$C_4H_8O_2$	371.75	0.5268	100.77	OX
2-Butanol	$C_4H_{10}O$	2-Ethoxy-2-methylbutane	$C_7H_{16}O$	367.75	0.4931	102.89	OX
2-Butanol	$C_4H_{10}O$	Heptane	C_7H_{16}	361.95	0.4116	102.70	OX
2-Butanol	$C_4H_{10}O$	Hexane	C_6H_{14}	348.15	0.1010	128.66	OX
2-Butanol	$C_4H_{10}O$	2-Methoxy-2-methylbutane	$C_6H_{14}O$	359.15	0.0991	102.12	OX
2-Butanol	$C_4H_{10}O$	Octane	C_8H_{18}	371.05	0.8001	101.30	OX
2-Butanol	$C_4H_{10}O$	3-Pentanone	$C_5H_{10}O$	370.50	0.6075	99.98	OX
2-Butanol	$C_4H_{10}O$	Toluene	C_7H_8	353.44	0.5550	56.67	OX
2-Butanol	$C_4H_{10}O$	Water	H_2O	360.50	0.3800	101.33	LX
2-Butanol	$C_4H_{10}O$	*m*-Xylene	C_8H_{10}	369.85	0.9717	101.06	OX
2-Butanol	$C_4H_{10}O$	*p*-Xylene	C_8H_{10}	369.55	0.9646	101.46	OX
2-Butanone	C_4H_8O	Cyclohexene	C_6H_{10}	343.29	0.5110	89.35	OX
2-Butanone	C_4H_8O	Diisopropyl ether	$C_6H_{14}O$	340.55	0.1938	101.56	OX
2-Butanone	C_4H_8O	Dipropyl ether	$C_6H_{14}O$	351.40	0.7785	100.88	OX
2-Butanone	C_4H_8O	Ethanol	C_2H_6O	347.15	0.4920	101.33	OX
2-Butanone	C_4H_8O	Ethyl acetate	$C_4H_8O_2$	349.55	0.1700	101.33	OX
2-Butanone	C_4H_8O	Heptane	C_7H_{16}	350.15	0.7670	101.33	OX
2-Butanone	C_4H_8O	Hexane	C_6H_{14}	337.15	0.3280	101.33	OX
2-Butanone	C_4H_8O	1-Hexene	C_6H_{12}	334.75	0.1760	100.58	OX
2-Butanone	C_4H_8O	Methanol	CH_4O	323.15	0.1980	58.80	OX
2-Butanone	C_4H_8O	Methylcyclohexane	C_7H_{14}	350.50	0.7984	98.93	OX
2-Butanone	C_4H_8O	2-Propanol	C_3H_8O	350.55	0.6170	101.33	OX
2-Butanone	C_4H_8O	Tetrachloromethane	CCl_4	346.99	0.3370	101.33	OX
2-Butanone	C_4H_8O	Water	H_2O	346.54	0.6520	101.33	LX
trans-2-Butenal	C_4H_6O	Octane	C_8H_{18}	353.15	0.4950	97.86	OX
trans-2-Butenal	C_4H_6O	Tetrachloromethane	CCl_4	348.15	0.3500	97.86	OX
trans-2-Butenal	C_4H_6O	Toluene	C_7H_8	374.15	0.5950	97.86	OX
trans-2-Butenal	C_4H_6O	Trichloroethene	C_2HCl_3	360.15	0.1000	97.86	OX
trans-2-Butenal	C_4H_6O	Trichloromethane	$CHCl_3$	329.15	0.0050	97.86	OX
cis-2-Butenenitrile	C_4H_5N	Water	H_2O	358.45	0.6168	101.33	EX
trans-2-Butenenitrile	C_4H_5N	Water	H_2O	363.05	0.3157	101.33	EX
Butyl acetate	$C_6H_{12}O_2$	1-Octene	C_8H_{16}	393.00	0.3030	101.33	OX
Butyl acetate	$C_6H_{12}O_2$	Water	H_2O	363.35	0.2987	101.33	EX
Butylamine	$C_4H_{11}N$	Ethanol	C_2H_6O	354.99	0.4100	101.33	ON
Butylamine	$C_4H_{11}N$	Water	H_2O	349.85	0.9300	101.33	OX
Butylbenzene	$C_{10}H_{14}$	Phenol	C_6H_6O	447.05	0.4465	101.33	OX
sec-Butylbenzene, (±)-	$C_{10}H_{14}$	*o*-Cresol	C_7H_8O	444.65	0.9062	101.33	OX
sec-Butylbenzene, (±)-	$C_{10}H_{14}$	*m*-Cresol	C_7H_8O	445.85	0.9864	101.33	OX
sec-Butylbenzene, (±)-	$C_{10}H_{14}$	*p*-Cresol	C_7H_8O	446.05	0.9814	101.33	OX
sec-Butylbenzene, (±)-	$C_{10}H_{14}$	Phenol	C_6H_6O	441.15	0.6871	101.33	OX
tert-Butylbenzene	$C_{10}H_{14}$	Phenol	C_6H_6O	439.95	0.7227	101.33	OX
Butyl butanoate	$C_8H_{16}O_2$	Water	H_2O	369.85	0.0890	101.33	EX
Butyl ethyl ether	$C_6H_{14}O$	Methanol	CH_4O	335.00	0.1990	98.84	OX
Butyl ethyl ether	$C_6H_{14}O$	Water	H_2O	349.85	0.5930	101.33	EX
tert-Butyl ethyl ether	$C_6H_{14}O$	Ethanol	C_2H_6O	339.95	0.6272	101.72	OX
tert-Butyl ethyl ether	$C_6H_{14}O$	Methanol	CH_4O	330.95	0.3998	101.54	OX
tert-Butyl ethyl ether	$C_6H_{14}O$	2-Methyl-2-propanol	$C_4H_{10}O$	342.85	0.7488	101.44	OX
Butyl formate	$C_5H_{10}O_2$	Formic acid	CH_2O_2	372.15	0.1300	101.33	OX
Butyl formate	$C_5H_{10}O_2$	Water	H_2O	356.95	0.4640	101.33	EX
tert-Butyl isopropyl ether	$C_7H_{16}O$	2-Methyl-2-propanol	$C_4H_{10}O$	350.90	0.4610	102.94	OX
tert-Butyl isopropyl ether	$C_7H_{16}O$	2-Propanol	C_3H_8O	349.95	0.4694	102.70	OX
1-Butyl-2-methylbenzene	$C_{11}H_{16}$	Phenol	C_6H_6O	455.55	0.1496	101.33	OX
1-Butyl-3-methylbenzene	$C_{11}H_{16}$	Phenol	C_6H_6O	454.15	0.1901	101.33	OX
1-Butyl-4-methylbenzene	$C_{11}H_{16}$	Phenol	C_6H_6O	454.55	0.1770	101.33	OX
Butyl methyl ether	$C_5H_{12}O$	Methanol	CH_4O	330.00	0.4485	100.08	OX

Component 1	Mol. form. 1	Component 2	Mol. form. 2	T_{Az}/K	$y_{1,Az}$	P_{Az}/kPa	Type
2-*sec*-Butyl-4-methylphenol	$C_{11}H_{16}O$	Quinoline	C_9H_7N	516.10	0.4861	101.33	ON
2-*tert*-Butyl-5-methylphenol	$C_{11}H_{16}O$	Quinoline	C_9H_7N	515.45	0.4146	101.33	ON
2-Butylphenol	$C_{10}H_{14}O$	Quinoline	C_9H_7N	515.70	0.4650	101.33	ON
2-*sec*-Butylphenol	$C_{10}H_{14}O$	Quinoline	C_9H_7N	514.70	0.3661	101.33	ON
4-*sec*-Butylphenol	$C_{10}H_{14}O$	Quinoline	C_9H_7N	516.45	0.5449	101.33	ON
2-*tert*-Butylphenol	$C_{10}H_{14}O$	Quinoline	C_9H_7N	513.70	0.2701	101.33	ON
3-*tert*-Butylphenol	$C_{10}H_{14}O$	Quinoline	C_9H_7N	517.05	0.5685	101.33	ON
Butyl propanoate	$C_7H_{14}O_2$	Water	H_2O	367.95	0.1660	101.33	EX
Carbon disulfide	CS_2	Methanol	CH_4O	310.65	0.7000	101.33	LX
1-Chlorobutane	C_4H_9Cl	Cyclohexane	C_6H_{12}	348.31	0.5800	95.85	OX
1-Chloro-2-methylpropane	C_4H_9Cl	Water	H_2O	333.95	0.8030	101.33	LX
1-Chloropropane	C_3H_7Cl	Methanol	CH_4O	313.35	0.7500	101.59	OX
3-Chloropropene	C_3H_5Cl	Hydrogen cyanide	CHN	296.45	0.1984	101.33	OX
3-Chloropropene	C_3H_5Cl	Methanol	CH_4O	312.15	0.7430	100.39	OX
o-Cresol	C_7H_8O	Decane	$C_{10}H_{22}$	433.15	0.3100	78.71	OX
o-Cresol	C_7H_8O	*p*-Diethylbenzene	$C_{10}H_{14}$	453.10	0.2694	101.33	OX
o-Cresol	C_7H_8O	Dodecane	$C_{12}H_{26}$	458.15	0.8466	93.55	OX
o-Cresol	C_7H_8O	1,2-Ethanediol	$C_2H_6O_2$	462.67	0.6203	101.33	OX
o-Cresol	C_7H_8O	1,2,4,5-Tetramethylbenzene	$C_{10}H_{14}$	462.37	0.6273	101.33	OX
o-Cresol	C_7H_8O	2,4,6-Trimethylpyridine	$C_8H_{11}N$	470.35	0.6561	101.33	ON
o-Cresol	C_7H_8O	Undecane	$C_{11}H_{24}$	433.15	0.5800	56.40	OX
o-Cresol	C_7H_8O	1-Undecene	$C_{11}H_{22}$	448.15	0.5516	83.07	OX
m-Cresol	C_7H_8O	Decane	$C_{10}H_{22}$	433.15	0.2170	75.85	OX
m-Cresol	C_7H_8O	*p*-Diethylbenzene	$C_{10}H_{14}$	454.10	0.1010	101.33	OX
m-Cresol	C_7H_8O	2,6-Dimethylpyridine	C_7H_9N	475.66	0.9869	101.32	ON
m-Cresol	C_7H_8O	3-Methylpyridine	C_6H_7N	477.01	0.8444	101.32	ON
m-Cresol	C_7H_8O	4-Methylpyridine	C_6H_7N	477.74	0.8178	101.32	ON
m-Cresol	C_7H_8O	Naphthalene	$C_{10}H_8$	474.65	0.9680	101.33	OX
m-Cresol	C_7H_8O	Nonane	C_9H_{20}	413.15	0.0400	76.54	OX
m-Cresol	C_7H_8O	Quinoline	C_9H_7N	511.20	0.0356	101.33	ON
m-Cresol	C_7H_8O	1,2,3,4-Tetrahydronaphthalene	$C_{10}H_{12}$	468.45	0.5900	93.10	OX
m-Cresol	C_7H_8O	1,2,4,5-Tetramethylbenzene	$C_{10}H_{14}$	466.87	0.3591	101.33	OX
p-Cresol	C_7H_8O	*p*-Diethylbenzene	$C_{10}H_{14}$	454.50	0.1105	101.33	OX
p-Cresol	C_7H_8O	Naphthalene	$C_{10}H_8$	474.55	0.9414	101.33	OX
Cyclohexane	C_6H_{12}	Diethoxymethane	$C_5H_{12}O_2$	353.21	0.8226	101.39	OX
Cyclohexane	C_6H_{12}	Dimethyl carbonate	$C_3H_6O_3$	346.95	0.6220	101.49	OX
Cyclohexane	C_6H_{12}	Ethanol	C_2H_6O	337.95	0.5460	102.26	OX
Cyclohexane	C_6H_{12}	Ethyl acetate	$C_4H_8O_2$	345.00	0.4610	102.45	OX
Cyclohexane	C_6H_{12}	Ethyl formate	$C_3H_6O_2$	323.15	0.1790	91.46	OX
Cyclohexane	C_6H_{12}	Methanol	CH_4O	328.75	0.3910	106.66	OX
Cyclohexane	C_6H_{12}	Methyl acetate	$C_3H_6O_2$	328.65	0.2000	101.33	OX
Cyclohexane	C_6H_{12}	2-Methyl-2-butanol	$C_5H_{12}O$	351.95	0.8900	101.33	OX
Cyclohexane	C_6H_{12}	3-Methyl-3-buten-1-ol	$C_5H_{10}O$	352.65	0.9785	101.10	OX
Cyclohexane	C_6H_{12}	2-Methyl-3-buten-2-ol	$C_5H_{10}O$	350.15	0.8096	101.20	OX
Cyclohexane	C_6H_{12}	2-Methyl-1-propanol	$C_4H_{10}O$	351.35	0.8675	101.45	OX
Cyclohexane	C_6H_{12}	1-Propanol	C_3H_8O	347.68	0.7510	101.33	OX
Cyclohexane	C_6H_{12}	2-Propanol	C_3H_8O	342.75	0.5950	101.33	OX
Cyclohexane	C_6H_{12}	Propyl acetate	$C_5H_{10}O_2$	353.15	0.9402	100.43	OX
Cyclohexane	C_6H_{12}	Trichloroethene	C_2HCl_3	353.40	0.9025	101.32	OX
Cyclohexane	C_6H_{12}	Tris(perfluoroethyl)amine	$C_6F_{15}N$	329.35	0.4310	101.33	EX
Cyclohexane	C_6H_{12}	Vinyl acetate	$C_4H_6O_2$	340.45	0.3800	101.33	OX
Cyclohexanol	$C_6H_{12}O$	Nonane	C_9H_{20}	410.20	0.3350	79.99	OX
Cyclohexanol	$C_6H_{12}O$	*o*-Xylene	C_8H_{10}	415.95	0.1426	101.33	OX
Cyclohexanol	$C_6H_{12}O$	*m*-Xylene	C_8H_{10}	411.85	0.0503	101.33	OX
Cyclohexanol	$C_6H_{12}O$	*p*-Xylene	C_8H_{10}	410.95	0.0505	101.33	OX
Cyclohexanone	$C_6H_{10}O$	3-Methyl-1-butanol	$C_5H_{12}O$	404.87	0.0906	101.33	OX
Cyclohexanone	$C_6H_{10}O$	1-Pentanol	$C_5H_{12}O$	392.37	0.0252	53.32	OX
Cyclohexanone	$C_6H_{10}O$	Water	H_2O	369.45	0.1306	101.33	EX
Cyclohexene	C_6H_{10}	1,4-Dioxane	$C_4H_8O_2$	355.75	0.8935	101.44	OX
Cyclohexene	C_6H_{10}	Ethyl acetate	$C_4H_8O_2$	347.45	0.3817	100.87	OX
Cyclohexene	C_6H_{10}	Methyl acetate	$C_3H_6O_2$	330.35	0.0879	102.87	OX

Component 1	Mol. form. 1	Component 2	Mol. form. 2	T_{Az}/K	$y_{1,Az}$	P_{Az}/kPa	Type
Cyclohexene	C_6H_{10}	2-Methyl-1-propanol	$C_4H_{10}O$	353.75	0.8637	100.31	OX
Cyclohexene	C_6H_{10}	2-Methyl-2-propanol	$C_4H_{10}O$	346.00	0.5828	99.61	OX
Cyclohexene	C_6H_{10}	2-Propanol	C_3H_8O	344.65	0.5729	101.40	OX
Cyclohexene	C_6H_{10}	Water	H_2O	343.95	0.6910	101.33	EX
Cyclohexylamine	$C_6H_{13}N$	Water	H_2O	369.55	0.1308	101.33	OX
1,3-Cyclopentadiene	C_5H_6	Dimethoxymethane	$C_3H_8O_2$	313.65	0.6650	101.33	OX
1,3-Cyclopentadiene	C_5H_6	Methanol	CH_4O	309.05	0.7880	101.33	OX
1,3-Cyclopentadiene	C_5H_6	2-Methyl-2-butene	C_5H_{10}	310.85	0.3000	101.33	OX
Cyclopentane	C_5H_{10}	Ethanol	C_2H_6O	323.44	0.8200	121.00	OX
Cyclopentanone	C_5H_8O	3-Methyl-1-butanol	$C_5H_{12}O$	402.02	0.5944	101.33	OX
Cyclopentanone	C_5H_8O	1-Pentanol	$C_5H_{12}O$	403.84	0.9196	101.33	OX
Cyclopentene	C_5H_8	Ethanol	C_2H_6O	323.40	0.8560	134.00	OX
Cyclopropyl methyl ketone	C_5H_8O	Water	H_2O	361.65	0.2940	101.19	EX
trans-Decahydronaphthalene	$C_{10}H_{18}$	Phenol	C_6H_6O	443.15	0.4581	99.85	OX
Decane	$C_{10}H_{22}$	4,4-Dimethyl-1,3-dioxane	$C_6H_{12}O_2$	405.35	0.0001	100.60	OX
Decane	$C_{10}H_{22}$	1-Heptanol	$C_7H_{16}O$	438.75	0.5692	101.33	OX
Decane	$C_{10}H_{22}$	1-Hexanol	$C_6H_{14}O$	427.05	0.2877	101.33	OX
Decane	$C_{10}H_{22}$	2-Methylaniline	C_7H_9N	446.91	0.8230	101.33	OX
Decane	$C_{10}H_{22}$	Nitromethane	CH_3NO_2	371.96	0.0761	99.73	EX
Decane	$C_{10}H_{22}$	1-Octanol	$C_8H_{18}O$	446.45	0.8971	101.33	OX
Decane	$C_{10}H_{22}$	1-Pentanol	$C_5H_{12}O$	410.65	0.0779	101.33	OX
Decane	$C_{10}H_{22}$	Phenol	C_6H_6O	434.15	0.5550	101.32	OX
Decane	$C_{10}H_{22}$	1,2,3-Trimethylbenzene	C_9H_{12}	433.35	0.5990	72.54	OX
Decane	$C_{10}H_{22}$	1,2,4-Trimethylbenzene	C_9H_{12}	433.35	0.1400	80.25	OX
Decane	$C_{10}H_{22}$	Water	H_2O	370.75	0.0820	101.33	EX
1-Decanol	$C_{10}H_{22}O$	Dodecane	$C_{12}H_{26}$	489.25	0.1068	101.33	OX
1-Decanol	$C_{10}H_{22}O$	Water	H_2O	373.13	0.0135	101.33	EX
Diacetone alcohol	$C_6H_{12}O_2$	Water	H_2O	370.00	0.0100	90.79	OX
Dibutylamine	$C_8H_{19}N$	Water	H_2O	370.05	0.1150	101.33	EX
Dibutyl ether	$C_8H_{18}O$	Water	H_2O	368.65	0.2372	101.33	EX
1,1-Dichloroethane	$C_2H_4Cl_2$	Hexane	C_6H_{14}	329.30	0.8025	101.21	OX
1,1-Dichloroethane	$C_2H_4Cl_2$	2-Propanol	C_3H_8O	329.55	0.8928	101.60	OX
1,2-Dichloroethane	$C_2H_4Cl_2$	Formic acid	CH_2O_2	350.17	0.5725	101.33	OX
1,2-Dichloroethane	$C_2H_4Cl_2$	Germanium(IV) chloride	Cl_4Ge	350.75	0.5370	101.33	OX
1,2-Dichloroethane	$C_2H_4Cl_2$	Methylcyclohexane	C_7H_{14}	354.65	0.8036	101.21	OX
1,2-Dichloroethane	$C_2H_4Cl_2$	2-Methyl-1-propanol	$C_4H_{10}O$	356.05	0.9173	101.26	OX
1,2-Dichloroethane	$C_2H_4Cl_2$	2-Methyl-2-propanol	$C_4H_{10}O$	349.45	0.5336	101.43	OX
1,2-Dichloroethane	$C_2H_4Cl_2$	2-Propanol	C_3H_8O	347.25	0.5258	100.32	OX
1,2-Dichloroethane	$C_2H_4Cl_2$	1,1,2,2-Tetrachloro-1,2-difluoroethane	$C_2Cl_4F_2$	353.80	0.7300	101.33	OX
1,2-Dichloroethane	$C_2H_4Cl_2$	Trichloroethene	C_2HCl_3	355.35	0.6676	101.36	OX
1,2-Dichloroethane	$C_2H_4Cl_2$	2,2,4-Trimethylpentane	C_8H_{18}	343.15	0.7600	73.13	OX
1,2-Dichloroethane	$C_2H_4Cl_2$	Water	H_2O	345.43	0.6430	101.33	EX
trans-1,2-Dichloroethene	$C_2H_2Cl_2$	1,1,1,2,3,3-Hexafluoro-3-(2,2,2-trifluoroethoxy)propane	$C_5H_3F_9O$	318.50	0.8390	101.30	OX
Dichloromethane	CH_2Cl_2	Ethanol	C_2H_6O	312.05	0.9600	101.33	OX
Diethoxymethane	$C_5H_{12}O_2$	Dimethyl carbonate	$C_3H_6O_3$	358.71	0.5563	100.42	OX
Diethoxymethane	$C_5H_{12}O_2$	Ethanol	C_2H_6O	348.30	0.3503	102.35	OX
Diethoxymethane	$C_5H_{12}O_2$	Hexane	C_6H_{14}	361.27	0.9101	102.30	OX
Diethoxymethane	$C_5H_{12}O_2$	Methanol	CH_4O	336.03	0.1873	101.52	OX
Diethoxymethane	$C_5H_{12}O_2$	1-Propanol	C_3H_8O	359.01	0.7680	99.43	OX
Diethoxymethane	$C_5H_{12}O_2$	2-Propanol	C_3H_8O	351.45	0.3893	98.61	OX
o-Diethylbenzene	$C_{10}H_{14}$	Phenol	C_6H_6O	447.15	0.4435	101.33	OX
m-Diethylbenzene	$C_{10}H_{14}$	Phenol	C_6H_6O	445.95	0.4848	101.33	OX
p-Diethylbenzene	$C_{10}H_{14}$	Phenol	C_6H_6O	446.45	0.5295	101.33	OX
Diethylene glycol	$C_4H_{10}O_3$	1-Nonanol	$C_9H_{20}O$	486.65	0.0095	101.33	OX
Diethylene glycol dimethyl ether	$C_6H_{14}O_3$	Water	H_2O	372.70	0.0321	101.33	OX
Diethylene glycol monobutyl ether	$C_8H_{18}O_3$	1,2-Ethanediol	$C_2H_6O_2$	469.15	0.0898	101.33	OX
Diethylene glycol monoethyl ether	$C_6H_{14}O_3$	1,2-Ethanediol	$C_2H_6O_2$	467.15	0.3520	101.33	OX
Diethylene glycol monoethyl ether	$C_6H_{14}O_3$	1,3-Propanediol	$C_3H_8O_2$	459.25	0.0650	101.33	OX
Diethylene glycol monoisobutyl ether	$C_8H_{18}O_3$	1,2-Ethanediol	$C_2H_6O_2$	467.55	0.1645	101.33	OX
Diethylene glycol monoisopropyl ether	$C_7H_{16}O_3$	1,2-Ethanediol	$C_2H_6O_2$	466.35	0.3036	101.33	OX

Component 1	Mol. form. 1	Component 2	Mol. form. 2	T_{Az}/K	$y_{1,Az}$	P_{Az}/kPa	Type
Diethylene glycol monomethyl ether	$C_5H_{12}O_3$	1,2-Ethanediol	$C_2H_6O_2$	463.95	0.5612	101.33	OX
Diethylene glycol monomethyl ether	$C_5H_{12}O_3$	1,3-Propanediol	$C_3H_8O_2$	455.25	0.3700	101.33	OX
Diethylene glycol monopropyl ether	$C_7H_{16}O_3$	1,2-Ethanediol	$C_2H_6O_2$	468.55	0.1552	101.33	OX
Diethyl ether	$C_4H_{10}O$	Methanol	CH_4O	305.15	0.9500	93.33	OX
Diethyl ether	$C_4H_{10}O$	Methyl formate	$C_2H_4O_2$	301.55	0.3970	101.33	OX
Diethyl ether	$C_4H_{10}O$	Pentane	C_5H_{12}	306.85	0.5500	101.33	OX
Dihydromethylpyran (unspecified isomer)	$C_6H_{10}O$	Toluene	C_7H_8	381.85	0.0207	101.30	OX
Dihydromethylpyran (unspecified isomer)	$C_6H_{10}O$	Water	H_2O	360.75	0.4159	100.93	EX
Diisopropylamine	$C_6H_{15}N$	2-Propanol	C_3H_8O	352.94	0.5110	101.33	OX
Diisopropylamine	$C_6H_{15}N$	Water	H_2O	347.25	0.6346	101.33	EX
Diisopropyl ether	$C_6H_{14}O$	Methanol	CH_4O	330.00	0.4610	101.61	OX
Diisopropyl ether	$C_6H_{14}O$	2-Methyl-2-propanol	$C_4H_{10}O$	340.45	0.8942	101.72	OX
Diisopropyl ether	$C_6H_{14}O$	2-Propanol	C_3H_8O	340.00	0.7950	103.36	OX
2,2-Dimethoxybutane	$C_6H_{14}O_2$	Toluene	C_7H_8	380.15	0.9180	101.44	OX
1,2-Dimethoxyethane	$C_4H_{10}O_2$	Methylcyclohexane	C_7H_{14}	350.00	0.8190	79.42	OX
Dimethoxymethane	$C_3H_8O_2$	2-Methyl-1,3-butadiene	C_5H_8	306.80	0.0160	101.33	OX
Dimethoxymethane	$C_3H_8O_2$	Methylenecyclobutane	C_5H_8	310.35	0.4630	101.33	OX
Dimethoxymethane	$C_3H_8O_2$	Water	H_2O	315.05	0.9731	101.38	LX
2,2-Dimethoxypropane	$C_5H_{12}O_2$	Methanol	CH_4O	334.15	0.2750	100.00	OX
Dimethylacetal	$C_4H_{10}O_2$	Methanol	CH_4O	330.35	0.5300	101.33	OX
N,N-Dimethylacetamide	C_4H_9NO	Ethylbenzene	C_8H_{10}	408.95	0.0037	101.70	OX
N,N-Dimethylacetamide	C_4H_9NO	*o*-Xylene	C_8H_{10}	416.95	0.0591	103.40	OX
2,4-Dimethylaniline	$C_8H_{11}N$	Dodecane	$C_{12}H_{26}$	482.95	0.4520	101.33	OX
2,4-Dimethylaniline	$C_8H_{11}N$	Tetradecane	$C_{14}H_{30}$	490.53	0.9840	101.33	OX
2,4-Dimethylaniline	$C_8H_{11}N$	Tridecane	$C_{13}H_{28}$	488.43	0.7880	101.33	OX
2,4-Dimethylaniline	$C_8H_{11}N$	Undecane	$C_{11}H_{24}$	468.13	0.1490	101.33	OX
α,α-Dimethylbenzenemethanol	$C_9H_{12}O$	Water	H_2O	371.25	0.0282	101.33	EX
2,3-Dimethylbutane	C_6H_{14}	Methanol	CH_4O	313.15	0.6380	85.50	OX
Dimethyl carbonate	$C_3H_6O_3$	Dipropyl ether	$C_6H_{14}O$	356.45	0.5044	100.73	OX
Dimethyl carbonate	$C_3H_6O_3$	Heptane	C_7H_{16}	355.15	0.5930	99.67	OX
Dimethyl carbonate	$C_3H_6O_3$	Hexane	C_6H_{14}	338.15	0.2540	98.46	OX
Dimethyl carbonate	$C_3H_6O_3$	Methanol	CH_4O	337.25	0.1496	102.52	OX
Dimethyl carbonate	$C_3H_6O_3$	Methylcyclopentane	C_6H_{12}	342.35	0.2680	103.46	OX
4,4-Dimethyl-1,3-dioxane	$C_6H_{12}O_2$	3-Methyl-3-buten-1-ol	$C_5H_{10}O$	403.05	0.2410	102.26	OX
4,4-Dimethyl-1,3-dioxane	$C_6H_{12}O_2$	Nonane	C_9H_{20}	402.15	0.8864	101.30	OX
4,4-Dimethyl-1,3-dioxane	$C_6H_{12}O_2$	Octane	C_8H_{18}	393.95	0.3343	101.20	OX
4,4-Dimethyl-1,3-dioxane	$C_6H_{12}O_2$	1-Propanol	C_3H_8O	368.20	0.0403	101.30	OX
4,4-Dimethyl-1,3-dioxane	$C_6H_{12}O_2$	Water	H_2O	366.00	0.2221	101.33	EX
4,4-Dimethyl-1,3-dioxane	$C_6H_{12}O_2$	*o*-Xylene	C_8H_{10}	404.65	0.9662	101.30	OX
cis-4,5-Dimethyl-1,3-dioxane	$C_6H_{12}O_2$	Water	H_2O	365.05	0.2034	101.50	EX
N,N-Dimethylformamide	C_3H_7NO	1,4-Dimethyl-4-vinylcyclohexene	$C_{10}H_{16}$	415.65	0.5880	101.33	OX
N,N-Dimethylformamide	C_3H_7NO	Heptane	C_7H_{16}	370.15	0.0800	101.33	OX
N,N-Dimethylformamide	C_3H_7NO	1-Methyl-3-(1-methylethylidene)cyclohexene	$C_{10}H_{16}$	419.05	0.7250	101.33	OX
2,6-Dimethylpyridine	C_7H_9N	3-Methylpyridine	C_6H_7N	416.64	0.7060	101.33	OX
2,6-Dimethylpyridine	C_7H_9N	4-Methylpyridine	C_6H_7N	417.08	0.8000	101.32	OX
2,6-Dimethylpyridine	C_7H_9N	Phenol	C_6H_6O	459.32	0.2461	101.32	ON
2,6-Dimethylpyridine	C_7H_9N	Water	H_2O	369.17	0.1353	101.33	EX
2,3-Dimethylquinoline	$C_{11}H_{11}N$	3,4-Xylenol	$C_8H_{10}O$	521.60	0.7887	101.33	ON
2,3-Dimethylquinoline	$C_{11}H_{11}N$	3,5-Xylenol	$C_8H_{10}O$	520.70	0.9470	101.33	ON
1,4-Dioxane	$C_4H_8O_2$	Ethanol	C_2H_6O	351.33	0.0520	101.33	OX
1,4-Dioxane	$C_4H_8O_2$	2-Ethoxy-2-methylbutane	$C_7H_{16}O$	369.15	0.5452	100.27	OX
1,4-Dioxane	$C_4H_8O_2$	Heptane	C_7H_{16}	364.30	0.4868	101.06	OX
1,4-Dioxane	$C_4H_8O_2$	2-Methyl-2-butanol	$C_5H_{12}O$	373.75	0.8119	99.62	OX
1,4-Dioxane	$C_4H_8O_2$	Methylcyclopentane	C_6H_{12}	343.85	0.0538	99.79	OX
1,4-Dioxane	$C_4H_8O_2$	Nitromethane	CH_3NO_2	373.25	0.5899	101.48	OX
1,4-Dioxane	$C_4H_8O_2$	1-Propanol	C_3H_8O	365.30	0.3582	101.30	OX
1,4-Dioxane	$C_4H_8O_2$	Propyl acetate	$C_5H_{10}O_2$	373.35	0.6334	101.13	OX
1,4-Dioxane	$C_4H_8O_2$	Triethylamine	$C_6H_{15}N$	343.15	0.2500	56.80	OX
1,4-Dioxane	$C_4H_8O_2$	Water	H_2O	360.65	0.4720	101.33	OX

Component 1	Mol. form. 1	Component 2	Mol. form. 2	T_{Az}/K	$y_{1,Az}$	P_{Az}/kPa	Type
1,3-Dioxolane	$C_3H_6O_2$	Methanol	CH_4O	334.66	0.3090	101.30	OX
1,3-Dioxolane	$C_3H_6O_2$	Water	H_2O	344.95	0.7480	101.30	OX
Dipropylamine	$C_6H_{15}N$	Water	H_2O	359.00	0.3954	101.33	EX
Dipropylene glycol monobutyl ether	$C_{10}H_{22}O_3$	1,2-Ethanediol	$C_2H_6O_2$	465.75	0.1870	101.33	OX
Dipropylene glycol monobutyl ether	$C_{10}H_{22}O_3$	1,2-Propanediol	$C_3H_8O_2$	459.65	0.0279	101.33	OX
Dipropylene glycol monoethyl ether	$C_8H_{18}O_3$	1,2-Ethanediol	$C_2H_6O_2$	458.65	0.5200	101.33	OX
Dipropylene glycol monoethyl ether	$C_8H_{18}O_3$	1,2-Propanediol	$C_3H_8O_2$	458.75	0.2222	101.33	OX
Dipropylene glycol monoisobutyl ether	$C_{10}H_{22}O_3$	1,2-Propanediol	$C_3H_8O_2$	459.05	0.0745	101.33	OX
Dipropylene glycol monoisopropyl ether	$C_9H_{20}O_3$	1,2-Propanediol	$C_3H_8O_2$	458.95	0.1870	101.33	OX
Dipropylene glycol monopropyl ether	$C_9H_{20}O_3$	1,2-Ethanediol	$C_2H_6O_2$	463.15	0.3410	101.33	OX
Dipropylene glycol monopropyl ether	$C_9H_{20}O_3$	1,2-Propanediol	$C_3H_8O_2$	458.95	0.0990	101.33	OX
Dodecane	$C_{12}H_{26}$	2-Methylaniline	C_7H_9N	468.90	0.2350	101.33	OX
Dodecane	$C_{12}H_{26}$	Nitromethane	CH_3NO_2	373.75	0.0154	99.73	EX
Dodecane	$C_{12}H_{26}$	1-Nonanol	$C_9H_{20}O$	480.65	0.4765	101.33	OX
Dodecane	$C_{12}H_{26}$	1-Octanol	$C_8H_{18}O$	466.95	0.1164	101.33	OX
Dodecane	$C_{12}H_{26}$	Phenol	C_6H_6O	450.73	0.2100	101.32	OX
1,2-Ethanediamine	$C_2H_8N_2$	Water	H_2O	391.85	0.5550	101.33	ON
1,2-Ethanediol	$C_2H_6O_2$	5-Methyl-1-heptanol, (±)-	$C_8H_{18}O$	457.65	0.3500	101.33	OX
1,2-Ethanediol	$C_2H_6O_2$	2,4,6-Trimethylpyridine	$C_8H_{11}N$	443.65	0.1734	101.33	OX
Ethanol	C_2H_6O	2-Ethoxy-2-methylbutane	$C_7H_{16}O$	349.35	0.7644	101.54	OX
Ethanol	C_2H_6O	Ethyl acetate	$C_4H_8O_2$	344.85	0.4590	101.33	OX
Ethanol	C_2H_6O	Fluorobenzene	C_6H_5F	343.85	0.4752	101.54	OX
Ethanol	C_2H_6O	1,1,1,2,3,3-Hexafluoro-3-(2,2,2-trifluoroethoxy)propane	$C_5H_3F_9O$	337.88	0.3980	101.30	OX
Ethanol	C_2H_6O	Hexane	C_6H_{14}	331.65	0.3410	101.33	OX
Ethanol	C_2H_6O	Isopentane	C_5H_{12}	299.95	0.0540	101.33	OX
Ethanol	C_2H_6O	Isopropyl acetate	$C_5H_{10}O_2$	349.85	0.7010	101.33	OX
Ethanol	C_2H_6O	2-Methoxy-2-methylbutane	$C_6H_{14}O$	346.81	0.5820	101.32	OX
Ethanol	C_2H_6O	Methyl acetate	$C_3H_6O_2$	329.79	0.0362	101.33	OX
Ethanol	C_2H_6O	2-Methyl-1,3-butadiene	C_5H_8	305.95	0.1500	101.33	OX
Ethanol	C_2H_6O	Methyl butanoate	$C_5H_{10}O_2$	346.30	0.8800	83.88	OX
Ethanol	C_2H_6O	3-Methyl-2-butanone	$C_5H_{10}O$	350.85	0.8250	101.33	OX
Ethanol	C_2H_6O	2-Methyl-2-butene	C_5H_{10}	309.79	0.0795	101.33	OX
Ethanol	C_2H_6O	Methyl *tert*-butyl ether	$C_5H_{12}O$	327.75	0.0380	101.33	OX
Ethanol	C_2H_6O	Methyl propanoate	$C_4H_8O_2$	346.30	0.5140	103.91	OX
Ethanol	C_2H_6O	Nitromethane	CH_3NO_2	333.15	0.7150	53.61	OX
Ethanol	C_2H_6O	Nonane	C_9H_{20}	351.35	0.9400	101.33	OX
Ethanol	C_2H_6O	Octane	C_8H_{18}	349.85	0.8250	101.33	OX
Ethanol	C_2H_6O	Pentane	C_5H_{12}	307.15	0.0537	101.33	OX
Ethanol	C_2H_6O	2-Pentanone	$C_5H_{10}O$	351.15	0.9779	100.50	OX
Ethanol	C_2H_6O	3-Pentanone	$C_5H_{10}O$	351.33	0.9590	101.33	OX
Ethanol	C_2H_6O	Tetrachloromethane	CCl_4	338.19	0.3860	101.33	OX
Ethanol	C_2H_6O	1,1,2,2-Tetrafluoroethyl 1,1,1-trifluoroethyl ether	$C_4H_3F_7O$	326.67	0.2000	101.30	OX
Ethanol	C_2H_6O	Tetrahydrofuran	C_4H_8O	344.95	0.1290	125.00	OX
Ethanol	C_2H_6O	Toluene	C_7H_8	349.75	0.8152	101.33	OX
Ethanol	C_2H_6O	Trichloroethene	C_2HCl_3	343.85	0.5259	101.33	OX
Ethanol	C_2H_6O	Trichloromethane	$CHCl_3$	332.45	0.1590	101.33	OX
Ethanol	C_2H_6O	1,1,2-Trichloro-1,2,2-trifluoroethane	$C_2Cl_3F_3$	317.75	0.1544	101.42	OX
Ethanol	C_2H_6O	2,2,3-Trimethyloxirane	$C_5H_{10}O$	343.45	0.2930	101.33	OX
Ethanol	C_2H_6O	2,2,4-Trimethylpentane	C_8H_{18}	344.42	0.6450	101.33	OX
Ethanol	C_2H_6O	Water	H_2O	351.25	0.8970	101.33	OX
2-Ethoxyethanol	$C_4H_{10}O_2$	Ethylbenzene	C_8H_{10}	401.05	0.4632	100.94	OX
2-Ethoxyethanol	$C_4H_{10}O_2$	Styrene	C_8H_8	405.75	0.6438	101.33	OX
2-Ethoxyethanol	$C_4H_{10}O_2$	*o*-Xylene	C_8H_{10}	404.95	0.5965	101.36	OX
2-Ethoxyethanol	$C_4H_{10}O_2$	*m*-Xylene	C_8H_{10}	401.75	0.5159	101.33	OX
2-Ethoxyethanol	$C_4H_{10}O_2$	*p*-Xylene	C_8H_{10}	402.55	0.5042	102.19	OX
2-Ethoxy-2-methylbutane	$C_7H_{16}O$	1,4-Dioxane	$C_4H_8O_2$	369.15	0.4548	100.27	OX
2-Ethoxy-2-methylbutane	$C_7H_{16}O$	Methanol	CH_4O	335.15	0.1264	97.28	OX
2-Ethoxy-2-methylbutane	$C_7H_{16}O$	2-Methyl-2-butanol	$C_5H_{12}O$	369.85	0.6096	100.52	OX
2-Ethoxy-2-methylbutane	$C_7H_{16}O$	3-Pentanone	$C_5H_{10}O$	371.15	0.5236	100.21	OX

Component 1	Mol. form. 1	Component 2	Mol. form. 2	T_{Az}/K	$y_{1,Az}$	P_{Az}/kPa	Type
2-Ethoxy-2-methylbutane	$C_7H_{16}O$	Propyl acetate	$C_5H_{10}O_2$	370.95	0.3471	100.03	OX
Ethyl acetate	$C_4H_8O_2$	Hexane	C_6H_{14}	338.00	0.3430	101.32	OX
Ethyl acetate	$C_4H_8O_2$	1-Hexene	C_6H_{12}	333.15	0.1230	91.47	OX
Ethyl acetate	$C_4H_8O_2$	Methanol	CH_4O	335.66	0.2880	101.33	OX
Ethyl acetate	$C_4H_8O_2$	Methylcyclohexane	C_7H_{14}	349.90	0.9001	101.83	OX
Ethyl acetate	$C_4H_8O_2$	2-Methyl-2-propanol	$C_4H_{10}O$	349.75	0.7778	101.28	OX
Ethyl acetate	$C_4H_8O_2$	Tetrachloromethane	CCl_4	347.95	0.4300	101.33	OX
Ethyl acetate	$C_4H_8O_2$	Water	H_2O	343.55	0.7010	101.33	EX
Ethylbenzene	C_8H_{10}	2-Methyl-1-butanol, (±)-	$C_5H_{12}O$	398.75	0.4343	99.46	OX
Ethylbenzene	C_8H_{10}	Water	H_2O	364.15	0.2779	101.33	EX
1-Ethyl-3,5-dimethylbenzene	$C_{10}H_{14}$	Phenol	C_6H_6O	447.95	0.4160	101.33	OX
2-Ethyl-1,4-dimethylbenzene	$C_{10}H_{14}$	Phenol	C_6H_6O	447.95	0.4160	101.33	OX
4-Ethyl-1,3-dioxane	$C_6H_{12}O_2$	Water	H_2O	365.75	0.2743	101.30	EX
Ethyl formate	$C_3H_6O_2$	Methanol	CH_4O	318.15	0.7000	81.34	OX
Ethyl formate	$C_3H_6O_2$	Water	H_2O	325.75	0.9300	101.33	EX
2-Ethyl-4-methylphenol	$C_9H_{12}O$	Naphthalene	$C_{10}H_8$	488.20	0.2218	101.33	OX
2-Ethyl-5-methylphenol	$C_9H_{12}O$	Naphthalene	$C_{10}H_8$	489.45	0.1710	101.33	OX
2-Ethylphenol	$C_8H_{10}O$	Naphthalene	$C_{10}H_8$	478.35	0.8005	101.33	OX
2-Ethylphenol	$C_8H_{10}O$	Quinoline	C_9H_7N	511.75	0.1041	101.33	ON
2-Ethylphenol	$C_8H_{10}O$	1,2,4,5-Tetramethylbenzene	$C_{10}H_{14}$	471.45	0.3707	101.33	OX
3-Ethylphenol	$C_8H_{10}O$	Naphthalene	$C_{10}H_8$	483.45	0.5551	101.33	OX
3-Ethylphenol	$C_8H_{10}O$	Quinoline	C_9H_7N	512.70	0.2089	101.33	ON
3-Ethylphenol	$C_8H_{10}O$	1,2,4,5-Tetramethylbenzene	$C_{10}H_{14}$	475.95	0.1249	101.33	OX
4-Ethylphenol	$C_8H_{10}O$	Naphthalene	$C_{10}H_8$	486.10	0.3762	101.33	OX
4-Ethylphenol	$C_8H_{10}O$	Quinoline	C_9H_7N	513.45	0.2832	101.33	ON
Ethyl propyl ether	$C_5H_{12}O$	Methanol	CH_4O	330.00	0.5950	112.25	OX
Fluorobenzene	C_6H_5F	Methanol	CH_4O	333.35	0.3375	101.62	OX
Fluorobenzene	C_6H_5F	2-Propanol	C_3H_8O	347.75	0.5334	101.25	OX
Formaldehyde	CH_2O	Water	H_2O	355.75	0.0700	53.33	OX
Formic acid	CH_2O_2	Water	H_2O	380.35	0.5728	101.33	ON
Formic acid	CH_2O_2	*m*-Xylene	C_8H_{10}	365.95	0.8545	101.33	EX
Heptane	C_7H_{16}	Methanol	CH_4O	331.95	0.2721	101.33	OX
Heptane	C_7H_{16}	Methyl acetate	$C_3H_6O_2$	323.15	0.0430	79.48	OX
Heptane	C_7H_{16}	3-Methyl-1-butanol	$C_5H_{12}O$	368.15	0.8984	95.06	OX
Heptane	C_7H_{16}	2-Methyl-2-butanol	$C_5H_{12}O$	348.15	0.6860	56.83	OX
Heptane	C_7H_{16}	3-Methyl-3-buten-1-ol	$C_5H_{10}O$	370.00	0.7900	101.30	OX
Heptane	C_7H_{16}	Methyl methacrylate	$C_5H_8O_2$	366.35	0.5403	99.94	OX
Heptane	C_7H_{16}	Nitroethane	$C_2H_5NO_2$	362.95	0.6480	101.33	OX
Heptane	C_7H_{16}	Nitromethane	CH_3NO_2	353.25	0.5210	101.33	EX
Heptane	C_7H_{16}	1-Nitropropane	$C_3H_7NO_2$	369.25	0.8370	101.33	OX
Heptane	C_7H_{16}	2-Nitropropane	$C_3H_7NO_2$	367.55	0.7080	101.33	OX
Heptane	C_7H_{16}	1-Pentanol	$C_5H_{12}O$	371.45	0.9424	101.33	OX
Heptane	C_7H_{16}	3-Pentanol	$C_5H_{12}O$	368.15	0.7999	98.62	OX
Heptane	C_7H_{16}	Perfluoroheptane	C_7F_{16}	328.16	0.3900	53.60	OX
Heptane	C_7H_{16}	1-Propanol	C_3H_8O	357.65	0.5170	101.33	OX
Heptane	C_7H_{16}	2-Propanol	C_3H_8O	349.55	0.3977	101.33	OX
Heptane	C_7H_{16}	Propyl acetate	$C_5H_{10}O_2$	366.75	0.5785	101.38	OX
Heptane	C_7H_{16}	Pyridine	C_5H_5N	368.61	0.6998	101.33	OX
Heptane	C_7H_{16}	Water	H_2O	352.35	0.5490	101.33	EX
1-Heptanol	$C_7H_{16}O$	Nonane	C_9H_{20}	423.45	0.1071	101.33	OX
1-Heptanol	$C_7H_{16}O$	Undecane	$C_{11}H_{24}$	447.85	0.8014	101.33	OX
1-Heptanol	$C_7H_{16}O$	Water	H_2O	371.99	0.0297	101.33	EX
1-Heptene	C_7H_{14}	Water	H_2O	350.20	0.5900	101.33	EX
Hexafluorobenzene	C_6F_6	Methanol	CH_4O	318.15	0.3900	61.73	OX
1,1,1,2,3,3-Hexafluoro-3-(2,2,2-trifluoroethoxy)propane	$C_5H_3F_9O$	Methanol	CH_4O	330.67	0.4400	101.30	OX
1,1,1,2,3,3-Hexafluoro-3-(2,2,2-trifluoroethoxy)propane	$C_5H_3F_9O$	2-Propanol	C_3H_8O	341.23	0.6580	101.30	OX
Hexane	C_6H_{14}	Methanol	CH_4O	333.15	0.4840	149.64	OX
Hexane	C_6H_{14}	Methyl acetate	$C_3H_6O_2$	326.65	0.3410	106.66	OX
Hexane	C_6H_{14}	2-Methyl-2-butanol	$C_5H_{12}O$	339.06	0.9564	93.55	OX

Component 1	Mol. form. 1	Component 2	Mol. form. 2	T_{Az}/K	$y_{1,Az}$	P_{Az}/kPa	Type
Hexane	C_6H_{14}	Methyl formate	$C_2H_4O_2$	302.65	0.1510	101.33	OX
Hexane	C_6H_{14}	2-Methyl-2-propanol	$C_4H_{10}O$	337.70	0.7498	101.30	OX
Hexane	C_6H_{14}	1-Propanol	C_3H_8O	348.15	0.8100	137.23	OX
Hexane	C_6H_{14}	2-Propanol	C_3H_8O	338.15	0.7100	112.66	OX
Hexane	C_6H_{14}	Tetrahydrofuran	C_4H_8O	323.15	0.4100	65.83	OX
Hexane	C_6H_{14}	Trichloromethane	$CHCl_3$	333.45	0.2160	101.33	OX
Hexane	C_6H_{14}	Tris(perfluoroethyl)amine	$C_6F_{15}N$	327.65	0.5160	101.33	OX
Hexane	C_6H_{14}	Vinyl acetate	$C_4H_6O_2$	335.25	0.5550	101.33	OX
Hexane	C_6H_{14}	Water	H_2O	334.75	0.7890	101.33	EX
1-Hexanol	$C_6H_{14}O$	Nonane	C_9H_{20}	416.95	0.3649	101.33	OX
1-Hexanol	$C_6H_{14}O$	Octane	C_8H_{18}	398.55	0.0886	101.33	OX
1-Hexanol	$C_6H_{14}O$	Water	H_2O	367.89	0.0568	101.33	EX
1-Hexene	C_6H_{12}	Methyl acetate	$C_3H_6O_2$	323.15	0.3660	92.08	OX
1-Hexene	C_6H_{12}	2-Methyl-3-buten-2-ol	$C_5H_{10}O$	336.55	0.9521	101.30	OX
1-Hexene	C_6H_{12}	2-Methyl-2-propanol	$C_4H_{10}O$	333.25	0.7350	101.30	OX
1-Hexene	C_6H_{12}	Water	H_2O	318.15	0.8490	63.35	EX
Hydrogen chloride	ClH	Water	H_2O	389.34	0.1083	133.32	ONC
Hydrogen fluoride	FH	Trichlorofluoromethane	CCl_3F	283.15	0.7840	129.45	EX
Hydrogen fluoride	FH	Water	H_2O	382.15	0.3508	101.33	ON
Isobutanal	C_4H_8O	Water	H_2O	332.80	0.8302	100.99	EX
Isobutyl acetate	$C_6H_{12}O_2$	Water	H_2O	361.05	0.3560	101.33	EX
Isobutylbenzene	$C_{10}H_{14}$	Phenol	C_6H_6O	441.75	0.6478	101.33	OX
Isobutyl formate	$C_5H_{10}O_2$	2-Methyl-1-propanol	$C_4H_{10}O$	370.90	0.8070	101.33	OX
Isobutyl formate	$C_5H_{10}O_2$	Water	H_2O	352.75	0.5540	101.33	EX
4-Isobutylphenol	$C_{10}H_{14}O$	Quinoline	C_9H_7N	515.95	0.4939	101.33	ON
Isopentane	C_5H_{12}	Methanol	CH_4O	297.05	0.9070	101.33	OX
Isopentane	C_5H_{12}	2-Methyl-1-buten-3-yne	C_5H_6	299.35	0.6380	101.33	OX
Isopentane	C_5H_{12}	3-Methyl-1-butyne	C_5H_8	297.15	0.4350	101.33	OX
Isopentane	C_5H_{12}	Methyl formate	$C_2H_4O_2$	291.55	0.5080	101.33	OX
Isopentane	C_5H_{12}	2-Propanol	C_3H_8O	298.15	0.8630	101.33	OX
Isopentane	C_5H_{12}	Tris(perfluoroethyl)amine	$C_6F_{15}N$	299.65	0.9020	101.33	OX
Isopentyl acetate	$C_7H_{14}O_2$	3-Methyl-1-butanol	$C_5H_{12}O$	403.95	0.0100	101.33	OX
Isopentyl acetate	$C_7H_{14}O_2$	1-Pentanol	$C_5H_{12}O$	407.45	0.4000	101.33	OX
Isopentyl acetate	$C_7H_{14}O_2$	Water	H_2O	367.05	0.2010	101.46	EX
Isopropenylbenzene	C_9H_{10}	Water	H_2O	369.95	0.1120	101.33	EX
Isopropyl acetate	$C_5H_{10}O_2$	Water	H_2O	349.75	0.6040	101.33	EX
Isopropylbenzene	C_9H_{12}	Water	H_2O	368.15	0.1660	101.33	EX
1-Isopropyl-2-methylbenzene	$C_{10}H_{14}$	Phenol	C_6H_6O	443.75	0.5357	101.33	OX
1-Isopropyl-3-methylbenzene	$C_{10}H_{14}$	Phenol	C_6H_6O	443.05	0.5973	101.33	OX
1-Isopropyl-4-methylbenzene	$C_{10}H_{14}$	Phenol	C_6H_6O	443.65	0.5570	101.33	OX
2-Isopropylphenol	$C_9H_{12}O$	Naphthalene	$C_{10}H_8$	483.15	0.5102	101.33	OX
2-Isopropylphenol	$C_9H_{12}O$	Quinoline	C_9H_7N	512.75	0.1985	101.33	ON
2-Isopropylphenol	$C_9H_{12}O$	1,2,4,5-Tetramethylbenzene	$C_{10}H_{14}$	476.25	0.1036	101.33	OX
3-Isopropylphenol	$C_9H_{12}O$	Quinoline	C_9H_7N	514.70	0.3891	101.33	ON
Isoquinoline	C_9H_7N	2-Methylnaphthalene	$C_{11}H_{10}$	513.90	0.2074	101.33	OX
Isoquinoline	C_9H_7N	3,4-Xylenol	$C_8H_{10}O$	519.75	0.7045	101.33	ON
Isoquinoline	C_9H_7N	3,5-Xylenol	$C_8H_{10}O$	518.05	0.8085	101.33	ON
Methacrylic acid	$C_4H_6O_2$	Water	H_2O	372.25	0.0536	98.93	OX
Methanol	CH_4O	1,3-Cyclopentadiene	C_5H_6	309.05	0.2120	101.33	OX
Methanol	CH_4O	2-Methoxy-2-methylbutane	$C_6H_{14}O$	335.55	0.7735	101.69	OX
Methanol	CH_4O	Methyl acetate	$C_3H_6O_2$	328.15	0.3480	107.19	OX
Methanol	CH_4O	2-Methyl-1,3-butadiene	C_5H_8	303.55	0.1670	101.33	OX
Methanol	CH_4O	2-Methyl-1-butene	C_5H_{10}	300.55	0.1720	101.33	OX
Methanol	CH_4O	3-Methyl-1-butene	C_5H_{10}	291.05	0.0890	101.33	OX
Methanol	CH_4O	2-Methyl-2-butene	C_5H_{10}	306.25	0.2160	101.33	OX
Methanol	CH_4O	Methyl *tert*-butyl ether	$C_5H_{12}O$	325.00	0.3140	103.15	OX
Methanol	CH_4O	1-Methylcyclobutene	C_5H_8	304.85	0.1900	101.33	OX
Methanol	CH_4O	Methylcyclohexane	C_7H_{14}	333.15	0.7520	102.87	EX
Methanol	CH_4O	2-Methyl-1-pentene	C_6H_{12}	330.00	0.4517	141.80	OX
Methanol	CH_4O	Nonane	C_9H_{20}	337.25	0.9526	101.33	OX
Methanol	CH_4O	Octane	C_8H_{18}	335.55	0.8830	101.33	LX

Component 1	Mol. form. 1	Component 2	Mol. form. 2	T_{Az}/K	$y_{1,Az}$	P_{Az}/kPa	Type
Methanol	CH_4O	*cis*-1,3-Pentadiene	C_5H_8	311.10	0.2300	101.33	OX
Methanol	CH_4O	*trans*-1,3-Pentadiene	C_5H_8	309.65	0.2110	101.33	OX
Methanol	CH_4O	Pentane	C_5H_{12}	303.20	0.1930	101.30	OX
Methanol	CH_4O	1-Pentene	C_5H_{10}	300.05	0.1469	102.47	OX
Methanol	CH_4O	Tetrahydrofuran	C_4H_8O	332.75	0.5040	101.33	OX
Methanol	CH_4O	*N,N,N',N'*-Tetramethylmethanediamine	$C_5H_{14}N_2$	335.15	0.7670	101.33	OX
Methanol	CH_4O	Toluene	C_7H_8	336.65	0.8820	101.33	OX
Methanol	CH_4O	Trichloromethane	$CHCl_3$	328.15	0.3520	107.99	OX
Methanol	CH_4O	2,2,3-Trimethyloxirane	$C_5H_{10}O$	334.95	0.6590	101.33	OX
Methanol	CH_4O	Vinyl acetate	$C_4H_6O_2$	332.05	0.6182	101.33	OX
2-Methoxyethanol	$C_3H_8O_2$	1-Octene	C_8H_{16}	380.75	0.4700	101.33	OX
2-Methoxyethanol	$C_3H_8O_2$	*cis*-4-Octene	C_8H_{16}	381.25	0.4900	101.33	OX
2-Methoxyethanol	$C_3H_8O_2$	*trans*-4-Octene	C_8H_{16}	381.05	0.4900	101.33	OX
2-Methoxyethanol	$C_3H_8O_2$	Styrene	C_8H_8	393.95	0.7787	98.93	OX
2-Methoxyethanol	$C_3H_8O_2$	Water	H_2O	372.65	0.0559	99.99	OX
2-Methoxyethanol	$C_3H_8O_2$	*o*-Xylene	C_8H_{10}	392.65	0.7127	98.79	OX
2-Methoxyethanol	$C_3H_8O_2$	*m*-Xylene	C_8H_{10}	392.15	0.6397	99.73	OX
2-Methoxyethanol	$C_3H_8O_2$	*p*-Xylene	C_8H_{10}	392.65	0.6303	99.99	OX
2-Methoxy-2-methylbutane	$C_6H_{14}O$	2-Methyl-2-propanol	$C_4H_{10}O$	353.20	0.4383	101.80	OX
Methyl acetate	$C_3H_6O_2$	Methylcyclopentane	C_6H_{12}	325.85	0.6917	99.50	OX
Methyl acetate	$C_3H_6O_2$	2-Methyl-1-pentene	C_6H_{12}	325.15	0.5931	100.38	OX
Methyl acetate	$C_3H_6O_2$	Trichloromethane	$CHCl_3$	337.51	0.3240	101.33	ON
Methyl acetate	$C_3H_6O_2$	Water	H_2O	330.05	0.8940	103.62	LX
2-Methylaniline	C_7H_9N	Tridecane	$C_{13}H_{28}$	472.55	0.9070	101.33	OX
2-Methylaniline	C_7H_9N	Undecane	$C_{11}H_{24}$	461.40	0.4930	101.33	OX
2-Methyl-1,3-butadiene	C_5H_8	2-Methyl-1-buten-3-yne	C_5H_6	305.88	0.2790	101.33	OX
2-Methyl-1,3-butadiene	C_5H_8	Methyl formate	$C_2H_4O_2$	298.90	0.4850	101.33	OX
2-Methyl-1,3-butadiene	C_5H_8	Pentane	C_5H_{12}	310.55	0.7421	114.66	OX
2-Methyl-1,3-butadiene	C_5H_8	Propanal	C_3H_6O	306.35	0.8300	101.33	OX
2-Methyl-1,3-butadiene	C_5H_8	2-Propanol	C_3H_8O	307.05	0.9850	101.33	OX
2-Methyl-1,3-butadiene	C_5H_8	Tris(perfluoroethyl)amine	$C_6F_{15}N$	303.35	0.8200	101.33	EX
2-Methyl-1,3-butadiene	C_5H_8	Water	H_2O	305.85	0.9480	101.33	EX
2-Methyl-1-butanol, (±)-	$C_5H_{12}O$	*o*-Xylene	C_8H_{10}	402.05	0.7417	101.87	OX
2-Methyl-1-butanol, (±)-	$C_5H_{12}O$	*m*-Xylene	C_8H_{10}	400.65	0.6316	101.85	OX
2-Methyl-1-butanol, (±)-	$C_5H_{12}O$	*p*-Xylene	C_8H_{10}	400.15	0.6273	101.07	OX
3-Methyl-1-butanol	$C_5H_{12}O$	Toluene	C_7H_8	383.15	0.1250	101.33	OX
3-Methyl-1-butanol	$C_5H_{12}O$	Water	H_2O	367.97	0.1735	101.33	EX
2-Methyl-2-butanol	$C_5H_{12}O$	Methylcyclohexane	C_7H_{14}	366.60	0.3965	99.87	OX
2-Methyl-2-butanol	$C_5H_{12}O$	Methylcyclopentane	C_6H_{12}	344.75	0.0551	101.80	OX
2-Methyl-2-butanol	$C_5H_{12}O$	Water	H_2O	360.85	0.3645	101.75	EX
3-Methyl-2-butanone	$C_5H_{10}O$	2-Propanol	C_3H_8O	354.75	0.1500	101.33	OX
2-Methyl-1-butene	C_5H_{10}	2-Methyl-1-buten-3-yne	C_5H_6	303.15	0.6550	101.33	OX
2-Methyl-1-butene	C_5H_{10}	Tris(perfluoroethyl)amine	$C_6F_{15}N$	301.95	0.8450	101.33	OX
2-Methyl-2-butene	C_5H_{10}	Methyl formate	$C_2H_4O_2$	297.75	0.4240	101.33	OX
2-Methyl-2-butene	C_5H_{10}	Nitromethane	CH_3NO_2	311.15	0.9430	101.33	LX
2-Methyl-2-butene	C_5H_{10}	1-Pentyne	C_5H_8	310.95	0.6700	101.33	OX
2-Methyl-2-butene	C_5H_{10}	2-Propanol	C_3H_8O	310.95	0.9540	101.33	OX
2-Methyl-2-butene	C_5H_{10}	Tris(perfluoroethyl)amine	$C_6F_{15}N$	307.65	0.8170	101.33	OX
2-Methyl-2-butene	C_5H_{10}	Water	H_2O	309.75	0.9350	101.33	EX
3-Methyl-3-buten-1-ol	$C_5H_{10}O$	Toluene	C_7H_8	381.55	0.2391	101.60	OX
3-Methyl-2-buten-1-ol	$C_5H_{10}O$	Water	H_2O	369.55	0.0859	101.33	EX
3-Methyl-3-buten-1-ol	$C_5H_{10}O$	Water	H_2O	333.15	0.1320	101.33	EX
2-Methyl-3-buten-2-ol	$C_5H_{10}O$	Tetrachloromethane	CCl_4	348.45	0.0991	101.06	OX
2-Methyl-3-buten-2-ol	$C_5H_{10}O$	Toluene	C_7H_8	366.55	0.7788	101.20	OX
2-Methyl-3-buten-2-ol	$C_5H_{10}O$	Water	H_2O	359.25	0.4230	101.33	LX
1-Methylcyclobutene	C_5H_8	Acetone	C_3H_6O	307.75	0.7780	101.33	OX
1-Methylcyclobutene	C_5H_8	Dimethoxymethane	$C_3H_8O_2$	309.05	0.7100	101.33	OX
Methylcyclohexane	C_7H_{14}	Methyl propanoate	$C_4H_8O_2$	352.45	0.1044	101.33	OX
Methylcyclohexane	C_7H_{14}	Nitromethane	CH_3NO_2	354.85	0.4877	101.33	EX
Methylcyclohexane	C_7H_{14}	3-Pentanone	$C_5H_{10}O$	366.95	0.5559	99.82	OX
Methylcyclohexane	C_7H_{14}	2-Propanol	C_3H_8O	350.85	0.3470	101.33	OX

Component 1	Mol. form. 1	Component 2	Mol. form. 2	T_{Az}/K	$y_{1,Az}$	P_{Az}/kPa	Type
Methylcyclohexane	C_7H_{14}	Propyl acetate	$C_5H_{10}O_2$	368.40	0.5254	100.90	OX
Methylcyclopentane	C_6H_{12}	2-Methyl-1-propanol	$C_4H_{10}O$	343.15	0.9433	100.35	OX
Methylcyclopentane	C_6H_{12}	2-Methyl-2-propanol	$C_4H_{10}O$	339.35	0.7441	99.93	OX
Methylcyclopentane	C_6H_{12}	1-Propanol	C_3H_8O	340.85	0.8271	101.19	OX
Methylcyclopentane	C_6H_{12}	2-Propanol	C_3H_8O	336.45	0.7100	98.14	OX
Methylenecyclobutane	C_5H_8	2-Methyl-2-propanol	$C_4H_{10}O$	314.65	0.9850	101.33	OX
Methylenecyclobutane	C_5H_8	Propanal	C_3H_6O	311.30	0.6400	101.33	OX
Methylenecyclobutane	C_5H_8	Water	H_2O	313.15	0.9788	101.30	EX
4-Methylenetetrahydropyran	$C_6H_{10}O$	Toluene	C_7H_8	381.15	0.5253	101.20	OX
Methyl formate	$C_2H_4O_2$	Pentane	C_5H_{12}	294.85	0.5740	101.33	OX
5-Methyl-1-heptanol, (±)-	$C_8H_{18}O$	1,2-Propanediol	$C_3H_8O_2$	456.85	0.4309	101.33	OX
3-Methylisoquinoline	$C_{10}H_9N$	3,4-Xylenol	$C_8H_{10}O$	524.35	0.9189	101.33	ON
Methyl methacrylate	$C_5H_8O_2$	Octane	C_8H_{18}	373.70	0.9651	100.16	OX
Methyl methacrylate	$C_5H_8O_2$	Water	H_2O	354.45	0.5004	101.33	EX
2-Methylnaphthalene	$C_{11}H_{10}$	Quinoline	C_9H_7N	511.05	0.0787	101.33	OX
2-Methyl-1-pentene	C_6H_{12}	Tetrahydrofuran	C_4H_8O	334.65	0.7133	101.29	OX
2-Methyl-1-pentene	C_6H_{12}	Trichloromethane	$CHCl_3$	333.95	0.3765	101.19	OX
Methyl propanoate	$C_4H_8O_2$	Water	H_2O	344.75	0.6950	101.33	EX
2-Methyl-1-propanol	$C_4H_{10}O$	Nitroethane	$C_2H_5NO_2$	375.81	0.5920	101.33	OX
2-Methyl-1-propanol	$C_4H_{10}O$	Octane	C_8H_{18}	376.58	0.6700	101.30	OX
2-Methyl-1-propanol	$C_4H_{10}O$	Tetrachloromethane	CCl_4	348.95	0.0920	101.33	OX
2-Methyl-1-propanol	$C_4H_{10}O$	Toluene	C_7H_8	374.35	0.4941	101.33	OX
2-Methyl-1-propanol	$C_4H_{10}O$	*m*-Xylene	C_8H_{10}	380.35	0.9300	101.33	OX
2-Methyl-1-propanol	$C_4H_{10}O$	*p*-Xylene	C_8H_{10}	380.30	0.9200	101.33	OX
2-Methyl-2-propanol	$C_4H_{10}O$	Octane	C_8H_{18}	343.15	0.9680	61.18	OX
2-Methyl-2-propanol	$C_4H_{10}O$	2-Propanol	C_3H_8O	343.05	0.4449	60.27	ON
2-Methyl-2-propanol	$C_4H_{10}O$	Toluene	C_7H_8	353.44	0.9200	93.61	OX
2-Methyl-2-propanol	$C_4H_{10}O$	1,1,2-Trichloro-1,2,2-trifluoroethane	$C_2Cl_3F_3$	319.95	0.0574	101.09	OX
2-Methyl-2-propanol	$C_4H_{10}O$	2,2,4-Trimethylpentane	C_8H_{18}	339.28	0.6040	59.49	OX
2-Methyl-2-propanol	$C_4H_{10}O$	Water	H_2O	353.00	0.5989	101.33	OX
2-Methylpropenal	C_4H_6O	Tetrachloromethane	CCl_4	339.15	0.4000	97.86	OX
1-Methyl-2-propylbenzene	$C_{10}H_{14}$	Phenol	C_6H_6O	447.75	0.4199	101.33	OX
1-Methyl-3-propylbenzene	$C_{10}H_{14}$	Phenol	C_6H_6O	446.35	0.4736	101.33	OX
1-Methyl-4-propylbenzene	$C_{10}H_{14}$	Phenol	C_6H_6O	447.15	0.4425	101.33	OX
2-Methylpyridine	C_6H_7N	Nonane	C_9H_{20}	402.35	0.8790	101.33	OX
2-Methylpyridine	C_6H_7N	Octane	C_8H_{18}	394.27	0.4610	101.33	OX
2-Methylpyridine	C_6H_7N	Phenol	C_6H_6O	458.33	0.2148	101.32	ON
3-Methylpyridine	C_6H_7N	Phenol	C_6H_6O	462.93	0.3082	101.32	ON
4-Methylpyridine	C_6H_7N	Water	H_2O	370.50	0.1028	101.33	OX
2-Methylquinoline	$C_{10}H_9N$	3,4-Xylenol	$C_8H_{10}O$	521.17	0.8353	101.33	ON
3-Methylquinoline	$C_{10}H_9N$	3,4-Xylenol	$C_8H_{10}O$	523.60	0.8848	101.33	ON
7-Methylquinoline	$C_{10}H_9N$	3,4-Xylenol	$C_8H_{10}O$	525.85	0.9534	101.33	ON
2-Methylquinoline	$C_{10}H_9N$	3,5-Xylenol	$C_8H_{10}O$	520.65	0.9906	101.33	ON
Molybdenum(V) chloride	Cl_5Mo	Tungsten(VI) chloride	Cl_6W	274.70	0.9750	101.33	OX
Naphthalene	$C_{10}H_8$	2,4,6-Trimethylphenol	$C_9H_{12}O$	486.70	0.6839	101.33	OX
Naphthalene	$C_{10}H_8$	2,3-Xylenol	$C_8H_{10}O$	485.45	0.5877	101.33	OX
Naphthalene	$C_{10}H_8$	2,4-Xylenol	$C_8H_{10}O$	481.25	0.3565	101.33	OX
Naphthalene	$C_{10}H_8$	2,5-Xylenol	$C_8H_{10}O$	481.25	0.3565	101.33	OX
Naphthalene	$C_{10}H_8$	2,6-Xylenol	$C_8H_{10}O$	475.70	0.0619	101.33	OX
Naphthalene	$C_{10}H_8$	3,4-Xylenol	$C_8H_{10}O$	490.95	0.8842	101.33	OX
Naphthalene	$C_{10}H_8$	3,5-Xylenol	$C_8H_{10}O$	489.33	0.7399	101.33	OX
Niobium(V) chloride	Cl_5Nb	Phosphoryl chloride	Cl_3OP	536.15	0.5980	101.33	ON
Nitric acid	HNO_3	Water	H_2O	393.20	0.3820	101.33	ON
Nitromethane	CH_3NO_2	Nonane	C_9H_{20}	369.29	0.8403	99.73	EX
Nitromethane	CH_3NO_2	Octane	C_8H_{18}	363.38	0.6964	99.73	EX
Nitromethane	CH_3NO_2	Undecane	$C_{11}H_{24}$	373.16	0.9619	99.73	EX
Nitromethane	CH_3NO_2	Water	H_2O	356.90	0.4840	101.33	EX
Nonane	C_9H_{20}	1-Pentanol	$C_5H_{12}O$	404.45	0.3758	101.33	OX
Nonane	C_9H_{20}	Pentyl acetate	$C_7H_{14}O_2$	419.20	0.4620	101.32	OX
Nonane	C_9H_{20}	Phenol	C_6H_6O	419.18	0.7820	101.32	OX
Nonane	C_9H_{20}	1-Propanol	C_3H_8O	369.95	0.0775	101.33	OX

Component 1	Mol. form. 1	Component 2	Mol. form. 2	T_{Az}/K	$y_{1,Az}$	P_{Az}/kPa	Type
Nonane	C_9H_{20}	Pyridine	C_5H_5N	388.15	0.0650	101.33	OX
Nonane	C_9H_{20}	Water	H_2O	367.95	0.1720	101.33	EX
Nonane	C_9H_{20}	*o*-Xylene	C_8H_{10}	417.40	0.1502	101.33	OX
1-Nonanol	$C_9H_{20}O$	Undecane	$C_{11}H_{24}$	468.45	0.0925	101.33	OX
1-Nonanol	$C_9H_{20}O$	Water	H_2O	373.00	0.0154	101.33	EX
1-Nonene	C_9H_{18}	Phenol	C_6H_6O	413.15	0.8703	86.57	OX
Octane	C_8H_{18}	1-Pentanol	$C_5H_{12}O$	393.15	0.7153	101.33	OX
Octane	C_8H_{18}	Phenol	C_6H_6O	398.17	0.9310	101.32	OX
Octane	C_8H_{18}	1-Propanol	C_3H_8O	366.85	0.2517	101.33	OX
Octane	C_8H_{18}	2-Propanol	C_3H_8O	354.63	0.1010	101.33	OX
Octane	C_8H_{18}	Tetrachloroethene	C_2Cl_4	371.90	0.1219	53.44	OX
Octane	C_8H_{18}	Water	H_2O	362.75	0.3150	101.33	EX
1-Octanol	$C_8H_{18}O$	Undecane	$C_{11}H_{24}$	460.05	0.4772	101.33	OX
1-Octanol	$C_8H_{18}O$	Water	H_2O	372.75	0.0180	101.33	EX
1-Octene	C_8H_{16}	Tetrachloroethene	C_2Cl_4	393.15	0.4100	101.33	OX
cis-4-Octene	C_8H_{16}	Tetrachloroethene	C_2Cl_4	393.65	0.2900	101.33	OX
trans-4-Octene	C_8H_{16}	Tetrachloroethene	C_2Cl_4	393.45	0.3300	101.33	OX
1-Octyne	C_8H_{14}	1-Propanol	C_3H_8O	369.00	0.1400	101.33	OX
Pentane	C_5H_{12}	1-Pentyne	C_5H_8	307.55	0.6950	101.33	OX
1-Pentanol	$C_5H_{12}O$	Water	H_2O	369.08	0.1367	101.33	EX
2-Pentanol	$C_5H_{12}O$	Water	H_2O	363.15	0.2450	92.49	EX
3-Pentanone	$C_5H_{10}O$	Water	H_2O	356.05	0.5250	101.33	EX
Phenol	C_6H_6O	Propylbenzene	C_9H_{12}	428.15	0.1150	91.85	OX
Phenol	C_6H_6O	Tetradecane	$C_{14}H_{30}$	452.48	0.9650	101.32	OX
Phenol	C_6H_6O	1,2,3,4-Tetrahydronaphthalene	$C_{10}H_{12}$	448.15	0.9031	84.25	OX
Phenol	C_6H_6O	1,2,3,5-Tetramethylbenzene	$C_{10}H_{14}$	454.25	0.7957	101.33	OX
Phenol	C_6H_6O	1,2,4,5-Tetramethylbenzene	$C_{10}H_{14}$	453.36	0.7857	101.33	OX
Phenol	C_6H_6O	1,2,3-Trimethylbenzene	C_9H_{12}	443.45	0.3936	101.33	OX
Phenol	C_6H_6O	1,2,4-Trimethylbenzene	C_9H_{12}	440.65	0.2409	101.33	OX
Phenol	C_6H_6O	1,3,5-Trimethylbenzene	C_9H_{12}	436.95	0.1828	101.33	OX
Phenol	C_6H_6O	1-Undecene	$C_{11}H_{22}$	443.15	0.6426	92.31	OX
Phosphorus(III) chloride	Cl_3P	Sulfuryl chloride	Cl_2O_2S	364.15	0.5000	101.33	ON
Phosphorus(III) chloride	Cl_3P	Thionyl chloride	Cl_2OS	345.85	0.5800	101.33	OX
Phosphoryl chloride	Cl_3OP	Tantalum (V) chloride	Cl_5Ta	558.85	0.4650	101.33	ON
Propanal	C_3H_6O	Water	H_2O	320.65	0.9400	101.33	LX
1,2-Propanediol diacetate	$C_7H_{12}O_4$	Water	H_2O	358.15	0.0260	59.41	EX
Propanoic acid	$C_3H_6O_2$	Pyridine	C_5H_5N	421.75	0.6860	101.33	ON
Propanoic acid	$C_3H_6O_2$	Water	H_2O	373.05	0.0500	101.33	OX
1-Propanol	C_3H_8O	Propyl acetate	$C_5H_{10}O_2$	367.88	0.6190	101.33	OX
1-Propanol	C_3H_8O	Styrene	C_8H_8	369.08	0.9884	98.13	OX
1-Propanol	C_3H_8O	Tetrachloromethane	CCl_4	346.28	0.1968	101.33	OX
1-Propanol	C_3H_8O	1,1,2,2-Tetrafluoroethyl 1,1,1-trifluoroethyl ether	$C_4H_3F_7O$	329.23	0.0350	101.30	OX
1-Propanol	C_3H_8O	Toluene	C_7H_8	365.35	0.6770	101.33	OX
1-Propanol	C_3H_8O	2,2,4-Trimethylpentane	C_8H_{18}	357.89	0.4580	101.30	OX
1-Propanol	C_3H_8O	Water	H_2O	360.80	0.4320	101.33	OX
1-Propanol	C_3H_8O	*o*-Xylene	C_8H_{10}	369.85	0.9886	98.66	OX
1-Propanol	C_3H_8O	*m*-Xylene	C_8H_{10}	369.90	0.9531	99.06	OX
1-Propanol	C_3H_8O	*p*-Xylene	C_8H_{10}	369.60	0.9531	99.99	OX
2-Propanol	C_3H_8O	Tetrachloromethane	CCl_4	341.83	0.3314	101.33	OX
2-Propanol	C_3H_8O	Toluene	C_7H_8	354.65	0.8370	101.33	OX
2-Propanol	C_3H_8O	Trichloromethane	$CHCl_3$	334.15	0.0500	101.33	OX
2-Propanol	C_3H_8O	1,1,2-Trichloro-1,2,2-trifluoroethane	$C_2Cl_3F_3$	319.35	0.0841	100.95	OX
2-Propanol	C_3H_8O	2,2,3-Trimethyloxirane	$C_5H_{10}O$	346.10	0.1400	101.33	OX
2-Propanol	C_3H_8O	2,2,4-Trimethylpentane	C_8H_{18}	349.58	0.6350	101.30	OX
2-Propanol	C_3H_8O	Water	H_2O	353.70	0.6740	101.33	OX
Propyl acetate	$C_5H_{10}O_2$	Water	H_2O	355.91	0.4772	101.33	EX
Propyl formate	$C_4H_8O_2$	Water	H_2O	344.85	0.6910	101.33	EX
2-Propylphenol	$C_9H_{12}O$	Quinoline	C_9H_7N	513.60	0.2757	101.33	ON
3-Propylphenol	$C_9H_{12}O$	Quinoline	C_9H_7N	514.70	0.3891	101.33	ON
4-Propylphenol	$C_9H_{12}O$	Quinoline	C_9H_7N	515.35	0.4549	101.33	ON

Component 1	Mol. form. 1	Component 2	Mol. form. 2	T_{Az}/K	$y_{1,Az}$	P_{Az}/kPa	Type
Propyl propanoate	$C_6H_{12}O_2$	Water	H_2O	362.05	0.3400	101.33	EX
Pyridine	C_5H_5N	Toluene	C_7H_8	383.19	0.2250	101.33	OX
Pyridine	C_5H_5N	Water	H_2O	367.30	0.2500	101.33	OX
Pyrrole	C_4H_5N	Water	H_2O	348.15	0.2486	50.13	EX
Quinoline	C_9H_7N	2,3-Xylenol	$C_8H_{10}O$	513.30	0.7316	101.33	ON
Quinoline	C_9H_7N	2,4-Xylenol	$C_8H_{10}O$	512.30	0.8283	101.33	ON
Quinoline	C_9H_7N	2,5-Xylenol	$C_8H_{10}O$	512.30	0.8283	101.33	ON
Quinoline	C_9H_7N	2,6-Xylenol	$C_8H_{10}O$	511.00	0.9110	101.33	ON
Quinoline	C_9H_7N	3,4-Xylenol	$C_8H_{10}O$	514.77	0.6093	101.33	ON
Quinoline	C_9H_7N	3,5-Xylenol	$C_8H_{10}O$	513.58	0.6713	101.33	ON
Styrene	C_8H_8	Water	H_2O	367.15	0.2000	101.33	EX
Tetrachloroethene	C_2Cl_4	1,1,2-Trichloroethane	$C_2H_3Cl_3$	385.95	0.2115	101.33	OX
Tetrahydrofuran	C_4H_8O	Water	H_2O	336.67	0.8172	101.33	OX
1,2,4,5-Tetramethylbenzene	$C_{10}H_{14}$	2,4-Xylenol	$C_8H_{10}O$	474.05	0.8131	101.33	OX
1,2,4,5-Tetramethylbenzene	$C_{10}H_{14}$	2,5-Xylenol	$C_8H_{10}O$	474.35	0.8237	101.33	OX
1,2,4,5-Tetramethylbenzene	$C_{10}H_{14}$	2,6-Xylenol	$C_8H_{10}O$	468.85	0.6520	101.33	OX
Toluene	C_7H_8	Water	H_2O	357.25	0.4770	101.33	EX
Tributylamine	$C_{12}H_{27}N$	Water	H_2O	372.80	0.0238	101.46	EX
Trichloroethene	C_2HCl_3	Water	H_2O	346.55	0.6440	101.33	EX
Trichloromethane	$CHCl_3$	Water	H_2O	329.27	0.8397	101.33	EX
1,1,2-Trichloro-1,2,2-trifluoroethane	$C_2Cl_3F_3$	2,2,2-Trifluoroethanol	$C_2H_3F_3O$	316.58	0.7770	101.33	EX
2,2,4-Trimethylpentane	C_8H_{18}	Water	H_2O	351.95	0.5580	101.33	EX
4-Vinyl-1,3-dioxane	$C_6H_{10}O_2$	Water	H_2O	367.65	0.1045	101.33	EX
Water	H_2O	*m*-Xylene	C_8H_{10}	365.15	0.7667	101.33	EX
Water	H_2O	*p*-Xylene	C_8H_{10}	365.15	0.7450	101.33	EX

VISCOSITY OF GASES

Marcia L. Huber

The following table gives the viscosity of some common gases as a function of temperature. Unless otherwise noted, the viscosity values refer to a pressure of 100 kPa (1 bar) or to the saturation vapor pressure if that is less than 100 kPa. The notation "Yes" in the column "Low-pressure limiting value" indicates that the low-pressure limiting value is given. The difference between the viscosity at 100 kPa and the limiting value at zero pressure is generally less than 2%. Uncertainties for the viscosities of gases in this table are usually less than 3%; uncertainty information on specific fluids can be found in the references. Viscosity is given in units of μPa s; note that 1 μPa s = 10^{-5} poise. Substances are listed in the modified Hill order. Definitions of the data column headings are as follows.

Column heading	Definition
Mol. form.	Molecular formula of gas; order is in modified Hill order
Name	Chemical name of gas
Low-pressure limiting value	"Yes" indicates viscosity value η is low-pressure limiting value
η(100 K), η(200 K), etc.	Viscosity η (μPa s) at the temperature (in K) indicated within parentheses
Ref.	References

References

1. Lemmon, E. W., and Jacobsen, R. T, Viscosity and Thermal Conductivity Equations for Nitrogen, Oxygen, Argon, and Air, *Int. J. Thermophys.* 25, 21, 2004. <https://doi.org/10.1023/B:IJOT.0000022327.04529.f3>
2. Vogel, E., Jäger, B., Hellmann, R., and Bich, E., *Ab initio* Pair Potential Energy Curve for the Argon Atom Pair and Thermophysical Properties for the Dilute Argon Gas. II. Thermophysical Properties for Low-Density Argon, *Mol. Phys.* 108, 3335, 2010. <https://doi.org/10.1080/00268976.2010.507557>
3. May, E. F., Berg, R. F., and Moldover, M. R., Reference Viscosities of H_2, CH_4, Ar, and Xe at Low Densities, *Int. J. Thermophys.* 28, 1085, 2007. <https://doi.org/10.1007/s10765-007-0198-7>
4. Vogel, E., Reference Viscosity of Argon at Low Density in the Temperature Range from 290 K to 680 K, *Int. J. Thermophys.* 31, 447, 2010. <https://doi.org/10.1007/s10765-010-0760-6>
5. Ho, C. Y., Ed., *Properties of Inorganic and Organic Fluids, CINDAS Data Series on Materials Properties*, Vol. V-1, Hemisphere Publishing Corp., New York, 1988.
6. Quiñones-Cisneros, S. E., Huber, M. L., and Deiters, U. K., Correlation for the Viscosity of Sulfur Hexafluoride (SF_6) from the Triple Point to 1000 K and Pressures to 50 MPa, *J. Phys. Chem. Ref. Data* 41, 023102 (2012). <https://doi.org/10.1063/1.3702441>
7. Mehl, J. B., Huber, M. L., and Harvey, A. H., *Ab Initio* Transport Coefficients of Gaseous Hydrogen, *Int. J. Thermophys.* 31, 740, 2010. <https://doi.org/10.1007/s10765-009-0697-9>
8. Assael, M. J., Mixafendi, M., and Wakeham, W. A., The Viscosity of Normal Deuterium in the Limit of Zero Density, *J. Phys. Chem. Ref. Data* 16, 189, 1987. <https://doi.org/10.1063/1.555778>
9. Huber, M. L., R. A. Perkins, R. A., Laesecke, A., Friend, D. G., Sengers, J. V., Assael, M. J., Metaxa, I. M., Vogel, E., Mares, R., and Miyagawa, K., New International Formulation for the Viscosity of Water, *J. Phys. Chem. Ref. Data* 38, 101, 2009. <https://doi.org/10.1063/1.3088050>
10. Hellmann, R. and Bich, E., Transport Properties of Dilute D_2O Vapour from First Principles, *Mol. Phys.* 115, 1057, 2017. <https://doi.org/10.1080/00268976.2016.1226443>
11. Hellmann, R., Bich, E., Vogel, E., and Vesovic, V., Thermophysical Properties of Dilute Hydrogen Sulfide Gas, *J. Chem. Eng. Data* 57, 1312, 2012. <https://doi.org/10.1021/je3000926>
12. Monogenidou, S.A., Assael, M.J., and Huber, M.L., Reference Correlation of the Viscosity of Ammonia from the Triple Point to 700 K and up to 50 MPa, *J. Phys. Chem. Ref. Data* 47, 023102, 2018. <https://doi.org/10.1063/1.5036724>
13. Cencek, W., Przybytek, M., Komasa, J., Mehl, J.B., Jeziorski, B., and Szalewicz, K., Effects of Adiabatic, Relativistic, and Quantum Electrodynamics Interactions on the Pair Potential and Thermophysical Properties of Helium, *J. Chem. Phys.* 136, 224303, 2012. <https://doi.org/10.1063/1.4712218>
14. Jäger, B., Hellmann, R., Bich, E., and Vogel, E., State-of-the-Art *Ab Initio* Potential Energy Curve for the Krypton Atom Pair and Thermophysical Properties of Dilute Krypton Gas, *J. Chem. Phys.* 144, 114304, 2016. <https://doi.org/10.1063/1.4943959>
15. Vogel, E., The Viscosities of Dilute Kr, Xe, and CO_2 Revisited: New Experimental Reference Data at Temperatures from 295 K to 690 K, *Int. J. Thermophys.* 37, 63, 2016. <https://doi.org/10.1007/s10765-016-2068-7>
16. Hellmann, R., *Ab Initio* Potential Energy Surface for the Nitrogen Molecule Pair and Thermophysical Properties of Nitrogen Gas, *Mol. Phys.* 111, 387, 2013. <https://doi.org/10.1080/00268976.2012.726379>
17. Vogel, E., Towards Reference Viscosities of Carbon Monoxide and Nitrogen at Low Density Using Measurements between 290 K and 680 K as Well as Theoretically Calculated Viscosities, *Int. J. Thermophys.* 33, 741, 2012. <https://doi.org/10.1007/s10765-012-1185-1>
18. Crusius, J.-P., Hellmann, R., Hassel, E. and Bich, E., *Ab Initio* Intermolecular Potential Energy Surface and Thermophysical Properties of Nitrous Oxide, *J. Chem. Phys.* 142, 244307, 2015. <https://doi.org/10.1063/1.4922830>
19. Bich, E., Hellmann, R., and Vogel, E., *Ab Initio* Potential Energy Curve for the Neon Atom Pair and Thermophysical Properties for the Dilute Neon Gas. II. Thermophysical Properties for Low-Density Neon, *Mol. Phys.* 106, 1107, 2008. <https://doi.org/10.1080/00268970801964207>
20. Hellmann, R., Jäger, B., and Bich, E., State-of-the-Art *Ab Initio* Potential Energy Curve for the Xenon Atom Pair and Related Spectroscopic and Thermodynamic Properties, *J. Chem. Phys.* 147, 034304, 2017. <https://doi.org/10.1063/1.4994267>
21. Laesecke, A., and Muzny, C. D., Reference Correlation for the Viscosity of Carbon Dioxide, *J. Phys. Chem. Ref. Data* 46, 013107, 2017. <https://doi.org/10.1063/1.4977429>
22. Laesecke, A., and Muzny, C.D., *Ab Initio* Calculated Results Require New Formulations for Properties in the Limit of Zero Density: The Viscosity of Methane (CH_4), *Int. J. Thermophys.*, 38, 181, 2017, Erratum: *Int. J. Thermophys.* 39, 52, 2018. <https://doi.org/10.1007/s10765-017-2305-8>
23. Xiang, H.-W., Huber, M. L., and Laesecke, A., A New Reference Correlation for the Viscosity of Methanol, *J. Phys. Chem. Ref. Data* 35, 1597, 2006. <https://doi.org/10.1063/1.2360605>
24. Holland, P. M., Eaton, B. E., and Hanley, H. J. M., A Correlation of the Viscosity and Thermal Conductivity Data of Gaseous and Liquid Ethylene, *J. Phys. Chem. Ref. Data* 12, 917, 1983. <https://doi.org/10.1063/1.555701>
25. Crusius, J.-P., Hellmann, R., Hassel, E. and Bich, E., Intermolecular Potential Energy Surface and Thermophysical Properties of Ethylene Oxide, *J. Chem. Phys.* 141, 164322, 2014. <https://doi.org/10.1063/1.4899074>
26. Herrmann, S., Hellmann, R., and Vogel, E., Update: Reference Correlation for the Viscosity of Ethane, *J. Phys. Chem. Ref. Data* 47, 023103, 2018. <https://doi.org/10.1063/1.5037239>
27. Vogel, E., and Herrmann, S., New Formulation for the Viscosity of Propane, *J. Phys. Chem. Ref. Data* 45, 043103, 2016. <https://doi.org/10.1063/1.4966928>

28. Herrmann, S., and Vogel, E., New Formulation for the Viscosity of n-Butane, *J. Phys. Chem. Ref. Data* 47, 013104, 2018. <https://doi.org/10.1063/1.5020802>
29. Vogel, E., Kuechenmeister, C., and Bich, E., Viscosity Correlation for Isobutane over Wide Ranges of the Fluid Region, *Int. J. Thermophys.* 21, 343, 2000. <https://doi.org/10.1023/A:1006623310780>
30. Michailidou, E.K., Assael, M.J., Huber, M.L., and Perkins, R.A., Reference Correlation of the Viscosity of *n*-Hexane from the Triple Point to 600 K and up to 100 MPa, *J. Phys. Chem. Ref. Data* 42, 033104, 2013. <https://doi.org/10.1063/1.4818980>

Viscosity of Gases as a Function of Temperature

Mol. form.	Name	Low-pressure limiting value	η(100 K)	η(200 K)	η(300 K)	η(400 K)	η(500 K)	η(600 K)	Ref.
	Air		7.1	13.3	18.5	23.1	27.1	30.8	1
Ar	Argon	Yes	8.1	15.9	22.7	28.6	33.9	38.8	2, 3*, 4*
BF_3	Boron trifluoride			12.3	17.1	21.7	26.1	30.2	5
ClH	Hydrogen chloride				14.6	19.7	24.3		5
F_6S	Sulfur hexafluoride				15.3	19.7	23.8	27.6	6
H_2	Normal hydrogen	Yes	4.1	6.8	8.9	10.9	12.8	14.5	3*,7
D_2	Deuterium	Yes	5.9	9.6	12.6	15.4	17.9	20.3	8
H_2O	Water	Yes			9.8	13.4	17.3	21.4	9
D_2O	Deuterium oxide	Yes			10.1	13.8	17.9	22.1	10
H_2S	Hydrogen sulfide	Yes		8.1	12.2	16.4	20.3	24.0	11
H_3N	Ammonia				10.2	13.9	17.7	21.4	12
He	Helium	Yes	9.6	15.1	19.9	24.3	28.3	32.2	13
Kr	Krypton	Yes	8.9	17.3	25.4	32.8	39.4	45.5	14,15*
NO	Nitric oxide			13.8	19.2	23.8	28.0	31.9	5
N_2	Nitrogen	Yes	6.7	12.8	17.8	22.1	25.9	29.4	16,17*
N_2O	Nitrous oxide	Yes		10	14.9	19.6	23.9	27.8	18
Ne	Neon	Yes	14.4	24.1	31.9	38.6	44.8	50.6	19
O_2	Oxygen		7.7	14.7	20.7	25.8	30.5	34.7	1
O_2S	Sulfur dioxide			8.6	12.9	17.5	21.7		5
Xe	Xenon	Yes	8.3	15.6	23.2	30.4	37.1	43.4	3*,15*,20
CO	Carbon monoxide		6.7	12.9	17.8	22.1	25.8	29.1	5,17*
CO_2	Carbon dioxide			10.1	15.0	19.6	23.9	27.9	21
$CHCl_3$	Chloroform				10.2	13.7	16.9	20.1	5
CH_4	Methane	Yes	3.9	7.7	11.1	14.2	16.9	19.3	3*,22
CH_4O	Methanol	Yes		6.6	9.7	13.0	16.4	19.8	23
C_2H_2	Acetylene				10.4	13.5	16.5		5
C_2H_4	Ethylene			7.0	10.4	13.6	16.5	19.2	24
C_2H_4O	Ethylene oxide	Yes		6.4	9.4	12.5	15.7	18.8	25
C_2H_6	Ethane	Yes	3.3	6.2	9.3	12.2	14.8	17.3	26
C_2H_6O	Ethanol					11.6	14.5	17.0	5
C_3H_8	Propane				8.2	10.8	13.3	15.6	27
C_4H_{10}	Butane				7.4	9.9	12.2	14.5	28
C_4H_{10}	Isobutane				7.5	9.9	12.2	14.4	29
$C_4H_{10}O$	Diethyl ether				7.6	10.1	12.4		5
C_5H_{12}	Pentane				6.7	9.2	11.4	13.4	5
C_6H_{14}	Hexane					8.4	10.4	12.4	30

* Reference contains more accurate data covering a restricted temperature range.

VISCOSITY OF LIQUIDS

The absolute viscosity of some common liquids at temperatures between –25 °C and 100 °C is given in this table. Values were derived by fitting experimental data to suitable expressions for the temperature dependence. All values are given in units of millipascal seconds (mPa s); this unit is identical to centipoise (cp).

Viscosity values correspond to a nominal pressure of 1 atmosphere. However, if a value is given at a temperature above the normal boiling point, the applicable pressure is understood to be the vapor pressure of the liquid at that temperature. A few values are given at a temperature slightly below the normal freezing point; these refer to the supercooled liquid.

The accuracy ranges from 1% in the best cases to 5 to 10% in the worst cases. Additional significant figures are included in the table to facilitate interpolation. Column definitions for the table are as follows.

Column heading	**Definition**
Name	Name of liquid; liquids are listed alphabetically by name
Mol. form.	Molecular formula of liquid
η(-25 °C), η(0 °C), etc.	Viscosity at the temperature (in °C) indicated within parentheses, in units mPa s

References

1. Viswanath, D. S. and Natarajan, G., *Data Book on the Viscosity of Liquids*, Hemisphere Publishing Corp., New York, 1989.
2. Daubert, T. E., Danner, R. P., Sibul, H. M., and Stebbins, C. C., *Physical and Thermodynamic Properties of Pure Compounds: Data Compilation*, extant 1994 (core with 4 supplements), Taylor & Francis, Bristol, PA (also available as database).
3. Ho, C. Y., Ed., *CINDAS Data Series on Material Properties*, Vol. V-1, *Properties of Inorganic and Organic Fluids*, Hemisphere Publishing Corp., New York, 1988.
4. Stephan, K. and Lucas, K., *Viscosity of Dense Fluids*, Plenum Press, New York, 1979. <https://doi.org/10.1007/978-1-4757-6931-9>
5. Vargaftik, N. B., *Tables of Thermophysical Properties of Liquids and Gases, Second Edition*, John Wiley, New York, 1975. <https://doi.org/10.1007/978-3-642-52504-9_13>

Viscosity of Liquids as a Function of Temperature

Name	Mol. form.	η(-25 °C)/ mPa s	η(0 °C)/ mPa s	η(25 °C)/ mPa s	η(50 °C)/ mPa s	η(75 °C)/ mPa s	η(100 °C)/ mPa s
Acetic acid	$C_2H_4O_2$			1.056	0.786	0.599	0.464
Acetic anhydride	$C_4H_6O_3$		1.241	0.843	0.614	0.472	0.377
Acetone	C_3H_6O	0.540	0.395	0.306	0.247		
Acetonitrile	C_2H_3N		0.400	0.369	0.284	0.234	
Acetophenone	C_8H_8O			1.681			0.634
Acetyl chloride	C_2H_3ClO			0.368	0.294		
Allyl alcohol	C_3H_6O			1.218	0.759	0.505	
Aniline	C_6H_7N			3.85	2.03	1.247	0.850
Anisole	C_7H_8O			1.056	0.747	0.554	0.427
Benzene	C_6H_6			0.604	0.436	0.335	
Benzonitrile	C_7H_5N			1.267	0.883	0.662	0.524
Benzyl alcohol	C_7H_8O			5.47	2.76	1.618	1.055
Benzylamine	C_7H_9N			1.624	1.080	0.769	0.577
Bromine	Br_2		1.252	0.944	0.746		
Bromobenzene	C_6H_5Br		1.560	1.074	0.798	0.627	0.512
1-Bromobutane	C_4H_9Br		0.815	0.606	0.471	0.379	
Bromoethane	C_2H_5Br	0.635	0.477	0.374			
1-Bromopropane	C_3H_7Br		0.645	0.489	0.387		
2-Bromopropane	C_3H_7Br		0.612	0.458	0.359		
3-Bromopropene	C_3H_5Br		0.620	0.471	0.373		
Butanenitrile	C_4H_7N			0.553	0.418	0.330	0.268
Butanoic acid	$C_4H_8O_2$		2.22	1.426	0.982	0.714	0.542
1-Butanol	$C_4H_{10}O$	12.19	5.18	2.54	1.394	0.833	0.533
2-Butanol	$C_4H_{10}O$			3.10	1.332	0.698	0.419
2-Butanone	C_4H_8O	0.720	0.533	0.405	0.315	0.249	
Butyl acetate	$C_6H_{12}O_2$		1.002	0.685	0.500	0.383	0.305
Butylamine	$C_4H_{11}N$		0.830	0.574	0.409	0.298	
Butylbenzene	$C_{10}H_{14}$			0.950	0.683	0.515	
Butyl formate	$C_5H_{10}O_2$		0.937	0.644	0.472	0.362	0.289
Carbon disulfide	CS_2		0.429	0.352			
2-Chloroaniline	C_6H_6ClN			3.32	1.913	1.248	0.887
Chlorobenzene	C_6H_5Cl	1.703	1.058	0.753	0.575	0.456	0.369

Name	Mol. form.	η(-25 °C)/ mPa s	η(0 °C)/ mPa s	η(25 °C)/ mPa s	η(50 °C)/ mPa s	η(75 °C)/ mPa s	η(100 °C)/ mPa s
1-Chlorobutane	C_4H_9Cl		0.556	0.422	0.329	0.261	
1-Chloro-1,1-difluoroethane	$C_2H_3ClF_2$	0.477	0.376				
Chloroethane	C_2H_5Cl	0.416	0.319				
2-Chlorophenol	C_6H_5ClO			3.59	1.835	1.131	0.786
3-Chlorophenol	C_6H_5ClO				4.04		
1-Chloropropane	C_3H_7Cl		0.436	0.334			
2-Chloropropane	C_3H_7Cl		0.401	0.303			
3-Chloropropene	C_3H_5Cl		0.408	0.314			
2-Chlorotoluene	C_7H_7Cl		1.390	0.964	0.710	0.547	0.437
3-Chlorotoluene	C_7H_7Cl		1.165	0.823	0.616	0.482	0.391
4-Chlorotoluene	C_7H_7Cl			0.837	0.621	0.483	0.390
o-Cresol	C_7H_8O				3.03	1.562	0.961
m-Cresol	C_7H_8O			12.91	4.42	2.09	1.207
Cyclohexane	C_6H_{12}			0.894	0.615	0.447	
Cyclohexanol	$C_6H_{12}O$			57.5	12.29	4.27	1.98
Cyclohexanone	$C_6H_{10}O$			2.02	1.321	0.919	0.671
Cyclohexene	C_6H_{10}		0.882	0.625	0.467	0.364	
Cyclohexylamine	$C_6H_{13}N$			1.944	1.169	0.782	0.565
Cyclopentane	C_5H_{10}		0.555	0.413	0.321		
cis-Decahydronaphthalene	$C_{10}H_{18}$	12.79	5.64	3.04	1.875	1.271	0.924
trans-Decahydronaphthalene	$C_{10}H_{18}$	6.19	3.24	1.948	1.289	0.917	0.689
Decane	$C_{10}H_{22}$	2.19	1.277	0.838	0.598	0.453	0.359
Decanoic acid	$C_{10}H_{20}O_2$				4.33	2.65	
1-Decanol	$C_{10}H_{22}O$			10.91	4.59		
Diacetone alcohol	$C_6H_{12}O_2$	28.7	6.62	2.80	1.829	1.648	
1,2-Dibromoethane	$C_2H_4Br_2$			1.595	1.116	0.837	0.661
Dibromomethane	CH_2Br_2	1.948	1.320	0.980	0.779	0.652	
Dibutylamine	$C_8H_{19}N$		1.509	0.918	0.619	0.449	0.345
Dibutyl ether	$C_8H_{18}O$	1.417	0.918	0.637	0.466	0.356	0.281
Dibutyl phthalate	$C_{16}H_{22}O_4$	483	66.4	16.63	6.47	3.50	2.43
o-Dichlorobenzene	$C_6H_4Cl_2$		1.958	1.324	0.962	0.739	0.593
m-Dichlorobenzene	$C_6H_4Cl_2$		1.492	1.044	0.787	0.628	0.525
1,1-Dichloroethane	$C_2H_4Cl_2$			0.464	0.362		
1,2-Dichloroethane	$C_2H_4Cl_2$		1.125	0.779	0.576	0.447	
cis-1,2-Dichloroethene	$C_2H_2Cl_2$	0.786	0.575	0.445			
trans-1,2-Dichloroethene	$C_2H_2Cl_2$	0.522	0.398	0.317	0.261		
Dichloromethane	CH_2Cl_2	0.727	0.533	0.413			
Diethanolamine	$C_4H_{11}NO_2$				109.5	28.700	9.10
Diethylamine	$C_4H_{11}N$			0.319	0.239		
Diethylene glycol	$C_4H_{10}O_3$			30.2	11.13	4.92	2.51
Diethyl ether	$C_4H_{10}O$		0.283	0.224			
Diethyl sulfide	$C_4H_{10}S$		0.558	0.422	0.331	0.267	
Diisobutylamine	$C_8H_{19}N$		1.115	0.723	0.511	0.384	0.303
Diisopropylamine	$C_6H_{15}N$			0.393	0.300	0.237	
N,N-Dimethylacetamide	C_4H_9NO			1.927			
Dimethylamine	C_2H_7N	0.300	0.232				
N,N-Dimethylaniline	$C_8H_{11}N$		1.996	1.300	0.911	0.675	0.523
N,N-Dimethylformamide	C_3H_7NO		1.176	0.794	0.624		
Dimethyl phthalate	$C_{10}H_{10}O_4$		63.2	14.36	5.31	2.82	1.98
Dimethyl sulfide	C_2H_6S		0.356	0.284			
Dimethyl sulfoxide	C_2H_6OS			1.987	1.290		
1,4-Dioxane	$C_4H_8O_2$			1.177	0.787	0.569	
Diphenyl ether	$C_{12}H_{10}O$				2.13	1.407	1.023
Diphenylmethane	$C_{13}H_{12}$					1.265	0.929
Dipropylamine	$C_6H_{15}N$		0.751	0.517	0.377	0.288	0.228
Dipropyl ether	$C_6H_{14}O$		0.542	0.396	0.304	0.242	
Dodecane	$C_{12}H_{26}$		2.28	1.383	0.930	0.673	0.514
Epichlorohydrin	C_3H_5ClO	2.49	1.570	1.073	0.781	0.597	0.474
1,2-Ethanediol	$C_2H_6O_2$			16.06	6.55	3.34	1.975
Ethanethiol	C_2H_6S		0.364	0.287			

Name	Mol. form.	η(-25 °C)/ mPa s	η(0 °C)/ mPa s	η(25 °C)/ mPa s	η(50 °C)/ mPa s	η(75 °C)/ mPa s	η(100 °C)/ mPa s
Ethanol	C_2H_6O	3.26	1.786	1.074	0.694	0.476	
Ethanolamine	C_2H_7NO			21.1	8.56	3.93	2.00
Ethoxybenzene	$C_8H_{10}O$			1.197	0.817	0.594	0.453
Ethyl acetate	$C_4H_8O_2$		0.578	0.423	0.325	0.259	
N-Ethylaniline	$C_8H_{11}N$		3.98	2.05	1.231	0.825	0.596
Ethylbenzene	C_8H_{10}		0.872	0.631	0.482	0.380	0.304
Ethyl butanoate	$C_6H_{12}O_2$			0.639	0.453		
Ethylcyclohexane	C_8H_{16}		1.139	0.784	0.579		
Ethyl formate	$C_3H_6O_2$		0.506	0.380	0.300		
2-Ethyl-1-hexanol	$C_8H_{18}O$		20.7	6.27	2.63	1.360	0.810
Ethyl propanoate	$C_5H_{10}O_2$		0.691	0.501	0.380	0.299	0.242
Fluorobenzene	C_6H_5F		0.749	0.550	0.423	0.338	
Formamide	CH_3NO		7.11	3.34	1.833		
Formic acid	CH_2O_2			1.607	1.030	0.724	0.545
Furan	C_4H_4O	0.661	0.475	0.361			
Furfural	$C_5H_4O_2$		2.50	1.587	1.143	0.906	0.772
Glycerol	$C_3H_8O_3$			934	152	39.8	14.76
Heptane	C_7H_{16}	0.757	0.523	0.387	0.301	0.243	
Heptanoic acid	$C_7H_{14}O_2$			3.84	2.28	1.488	1.041
1-Heptanol	$C_7H_{16}O$			5.81	2.60	1.389	0.849
2-Heptanol, (±)-	$C_7H_{16}O$			3.95	1.799	0.987	0.615
3-Heptanol, (*S*)-	$C_7H_{16}O$				1.957	0.976	0.584
4-Heptanol	$C_7H_{16}O$			4.21	1.695	0.882	0.539
2-Heptanone	$C_7H_{14}O$			0.714	0.407	0.297	
1-Heptene	C_7H_{14}		0.441	0.340	0.273	0.226	
Heptylamine	$C_7H_{17}N$			1.314	0.865	0.600	0.434
Hexadecane	$C_{16}H_{34}$			3.03	1.879	1.260	0.899
Hexafluorobenzene	C_6F_6			2.79	1.730	1.151	
Hexane	C_6H_{14}		0.405	0.300	0.240		
Hexanenitrile	$C_6H_{11}N$			0.912	0.650	0.488	0.382
1-Hexanol	$C_6H_{14}O$			4.58	2.27	1.270	0.781
2-Hexanone	$C_6H_{12}O$	1.300	0.840	0.583	0.429	0.329	0.262
1-Hexene	C_6H_{12}	0.441	0.326	0.252	0.202		
Hydrazine	H_4N_2			0.876	0.628	0.480	0.384
Hydrogen cyanide	CHN		0.235	0.183			
Indan	C_9H_{10}		2.23	1.357	0.931	0.692	0.545
Iodobenzene	C_6H_5I		2.35	1.554	1.117	0.854	0.683
Iodoethane	C_2H_5I		0.723	0.556	0.444	0.365	
Iodomethane	CH_3I		0.594	0.469			
1-Iodopropane	C_3H_7I		0.970	0.703	0.541	0.436	0.363
2-Iodopropane	C_3H_7I		0.883	0.653	0.506	0.407	
Isobutyl acetate	$C_6H_{12}O_2$			0.676	0.493	0.370	0.286
Isobutylamine	$C_4H_{11}N$		0.770	0.571	0.367		
Isopentane	C_5H_{12}	0.376	0.277	0.214			
Isophorone	$C_9H_{14}O$		4.20	2.33	1.415	0.923	0.638
Isopropylamine	C_3H_9N		0.454	0.325			
Isopropylbenzene	C_9H_{12}		1.075	0.737	0.547		
Mercury	Hg			1.526	1.402	1.312	1.245
Mesityl oxide	$C_6H_{10}O$	1.291	0.838	0.602	0.465	0.381	0.326
Methanol	CH_4O	1.258	0.793	0.544			
Methyl acetate	$C_3H_6O_2$		0.477	0.364	0.284		
Methylamine	CH_5N	0.319	0.231				
2-Methylaniline	C_7H_9N		10.33	3.82	1.936	1.198	0.839
3-Methylaniline	C_7H_9N		8.18	3.31	1.679	1.014	0.699
N-Methylaniline	C_7H_9N		4.12	2.04	1.222	0.825	0.606
Methyl benzoate	$C_8H_8O_2$			1.857			
Methyl butanoate	$C_5H_{10}O_2$		0.759	0.541	0.406	0.318	0.257
2-Methyl-1-butanol, (±)-	$C_5H_{12}O$			4.45	1.963	1.031	0.612
3-Methyl-1-butanol	$C_5H_{12}O$		8.63	3.69	1.842	1.031	0.631
2-Methyl-2-butene	C_5H_{10}		0.255	0.203			

Name	Mol. form.	η(-25 °C)/ mPa s	η(0 °C)/ mPa s	η(25 °C)/ mPa s	η(50 °C)/ mPa s	η(75 °C)/ mPa s	η(100 °C)/ mPa s
Methylcyclohexane	C_7H_{14}		0.991	0.679	0.501	0.390	0.316
Methylcyclopentane	C_6H_{12}	0.927	0.653	0.479	0.364		
N-Methylformamide	C_2H_5NO		2.55	1.678	1.155	0.824	0.606
Methyl formate	$C_2H_4O_2$		0.424	0.325			
4-Methyl-3-heptanol	$C_8H_{18}O$		1.904	1.085	0.702	0.497	0.375
5-Methyl-3-heptanol	$C_8H_{18}O$		2.05	1.178	0.762	0.536	0.401
3-Methylhexane	C_7H_{16}			0.350			
Methyl isobutanoate	$C_5H_{10}O_2$		0.672	0.488	0.373	0.296	
2-Methylpentane	C_6H_{14}		0.372	0.286	0.226		
3-Methylpentane	C_6H_{14}		0.395	0.306			
4-Methyl-2-pentanone	$C_6H_{12}O$			0.545	0.406		
Methyl propanoate	$C_4H_8O_2$		0.581	0.431	0.333	0.266	
2-Methylpropanoic acid	$C_4H_8O_2$		1.857	1.226	0.863	0.639	0.492
2-Methyl-2-propanol	$C_4H_{10}O$			4.31	1.421	0.678	
Methyl salicylate	$C_8H_8O_3$					1.102	0.815
Morpholine	C_4H_9NO			2.02	1.247	0.850	0.627
Nitrobenzene	$C_6H_5NO_2$		3.04	1.863	1.262	0.918	0.704
Nitroethane	$C_2H_5NO_2$	1.354	0.940	0.688	0.526	0.415	0.337
Nitrogen dioxide	NO_2		0.532	0.402			
Nitromethane	CH_3NO_2	1.311	0.875	0.630	0.481	0.383	0.317
1-Nitropropane	$C_3H_7NO_2$	1.851	1.160	0.798	0.589	0.460	0.374
Nonane	C_9H_{20}		0.964	0.665	0.488	0.375	0.300
Nonanoic acid	$C_9H_{18}O_2$			7.01	3.71	2.23	1.475
1-Nonanol	$C_9H_{20}O$			9.12	4.03		
5-Nonanone	$C_9H_{18}O$			1.199	0.834	0.619	0.484
Octadecane	$C_{18}H_{38}$				2.49	1.609	1.132
Octane	C_8H_{18}		0.700	0.508	0.385	0.302	0.243
Octanoic acid	$C_8H_{16}O_2$			5.02	2.66	1.654	1.147
1-Octanol	$C_8H_{18}O$			7.29	3.23	1.681	0.991
Paraldehyde	$C_6H_{12}O_3$			1.079	0.692	0.485	0.362
Pentachloroethane	C_2HCl_5		3.76	2.25	1.491	1.061	
Pentane	C_5H_{12}	0.351	0.274	0.224			
1-Pentanol	$C_5H_{12}O$	25.4	8.51	3.62	1.820	1.035	0.646
2-Pentanol	$C_5H_{12}O$			3.47	1.447	0.761	0.465
3-Pentanol	$C_5H_{12}O$			4.15	1.473	0.727	0.436
2-Pentanone	$C_5H_{10}O$		0.641	0.470	0.362	0.289	0.238
3-Pentanone	$C_5H_{10}O$		0.592	0.444	0.345	0.276	0.227
1-Pentene	C_5H_{10}	0.313	0.241	0.195			
Pentylamine	$C_5H_{13}N$		1.030	0.702	0.493	0.356	
Phenol	C_6H_6O				3.44	1.784	1.099
Phenylhydrazine	$C_6H_8N_2$			13.03	4.55	1.850	0.848
Phosphorus(III) chloride	Cl_3P	0.870	0.662	0.529	0.439		
Piperidine	$C_5H_{11}N$			1.573	0.958	0.649	0.474
Propanal	C_3H_6O			0.321	0.249		
1,2-Propanediol	$C_3H_8O_2$		248	40.4	11.30	4.77	2.75
Propanenitrile	C_3H_5N			0.294	0.240	0.202	
1-Propanethiol	C_3H_8S		0.503	0.385			
2-Propanethiol	C_3H_8S		0.477	0.357	0.280		
Propanoic acid	$C_3H_6O_2$		1.499	1.030	0.749	0.569	0.449
1-Propanol	C_3H_8O	8.64	3.81	1.945	1.107	0.685	
2-Propanol	C_3H_8O		4.62	2.04	1.028	0.576	
Propyl acetate	$C_5H_{10}O_2$		0.768	0.544	0.406	0.316	0.255
Propylamine	C_3H_9N			0.376			
Propyl formate	$C_4H_8O_2$		0.669	0.485	0.370	0.293	
Pyridine	C_5H_5N		1.361	0.879	0.637	0.497	0.409
Pyrrole	C_4H_5N		2.08	1.225	0.828	0.612	
Pyrrolidine	C_4H_9N	1.914	1.071	0.704	0.512		
Quinoline	C_9H_7N			3.34	1.892	1.201	0.833
Styrene	C_8H_8		1.050	0.695	0.507	0.390	0.310
Sulfolane	$C_4H_8O_2S$				6.28	3.82	2.56

Name	Mol. form.	η(-25 °C)/ mPa s	η(0 °C)/ mPa s	η(25 °C)/ mPa s	η(50 °C)/ mPa s	η(75 °C)/ mPa s	η(100 °C)/ mPa s
1,1,1,2-Tetrachloroethane	$C_2H_2Cl_4$	3.66	2.20	1.437	1.006	0.741	0.570
Tetrachloroethene	C_2Cl_4		1.114	0.844	0.663	0.535	0.442
Tetrachloromethane	CCl_4		1.321	0.908	0.656	0.494	
Tetrachlorosilane	Cl_4Si			99.4	96.2		
Tetradecane	$C_{14}H_{30}$			2.13	1.376	0.953	0.697
Tetrahydrofuran	C_4H_8O	0.849	0.605	0.456	0.359		
Tetrahydrothiophene	C_4H_8S			0.973	0.912		
Toluene	C_7H_8	1.165	0.778	0.560	0.424	0.333	0.270
Tribromomethane	$CHBr_3$			1.857	1.367	1.029	
1,1,1-Trichloroethane	$C_2H_3Cl_3$	1.847	1.161	0.793	0.578	0.428	
Trichloroethene	C_2HCl_3		0.703	0.545	0.444	0.376	
Trichlorofluoromethane	CCl_3F	0.740	0.539	0.421			
Trichloromethane	$CHCl_3$	0.988	0.706	0.537	0.427		
Trichlorosilane	Cl_3HSi		0.415	0.326			
1,1,2-Trichloro-1,2,2-trifluoroethane	$C_2Cl_3F_3$	1.465	0.945	0.656	0.481		
Tridecane	$C_{13}H_{28}$		2.91	1.724	1.129	0.796	0.594
Tris(2-hydroxyethyl)amine	$C_6H_{15}NO_3$			609	114	31.5	11.70
Triethylamine	$C_6H_{15}N$		0.455	0.347	0.273	0.221	
Trifluoroacetic acid	$C_2HF_3O_2$			0.808	0.571		
Undecane	$C_{11}H_{24}$		1.707	1.098	0.763	0.562	0.433
Water	H_2O		1.793	0.890	0.547	0.378	0.282
o-Xylene	C_8H_{10}		1.084	0.760	0.561	0.432	0.345
m-Xylene	C_8H_{10}		0.795	0.581	0.445	0.353	0.289
p-Xylene	C_8H_{10}			0.603	0.457	0.359	0.290

THERMAL CONDUCTIVITY OF GASES AND REFRIGERANTS

Marcia L. Huber

The following tables give the thermal conductivity of some common gases (Table 1) and selected refrigerants (Table 2) as a function of temperature. In Table 1, the thermal conductivity values refer to a pressure of 100 kPa (1 bar) or to the saturation vapor pressure if that is less than 100 kPa. The difference between the thermal conductivity at 100 kPa and the limiting value at zero pressure is usually less than 1%. Uncertainties for the thermal conductivities of gases in this table are generally less than 3%; uncertainty information on specific fluids can be found in the references. Thermal conductivity is given in units of mW m^{-1} K^{-1}. Definitions of the data columns for Table 1 are as follows.

Table 1 Column heading	Definition
Mol. form.	Molecular formula; gases are listed in modified Hill order
Gas	Gas name
Low-pressure limiting value	"Yes" indicates data are low-pressure limiting values
λ(100 K), λ(200 K), etc.	Thermal conductivity λ in mW m^{-1} K^{-1} at 100 K, 200 K, and other indicated temperatures
Ref.	References

In Table 2, thermal conductivity values are given for a few select refrigerants. Uncertainties for the thermal conductivities of refrigerants in this table are generally less than 3%; uncertainty information on specific fluids can be found in the references. Column definitions for Table 2 are as follows.

Table 2 Column heading	Definition
Mol. form.	Molecular formula; gases are listed in modified Hill order
Refrigerant	Refrigerant name
Refrig. no.	Refrigerant code
λ(300 K), λ(400 K), etc.	Thermal conductivity λ in mW m^{-1} K^{-1} at 300 K, 400 K, and other indicated temperatures
Ref.	Reference

References

1. Lemmon, E. W., and Jacobsen, R. T, Viscosity and Thermal Conductivity Equations for Nitrogen, Oxygen, Argon, and Air, *Int. J. Thermophys.* 25, 21, 2004. <https://doi.org/10.1023/B:IJOT.0000022327.04529.f3>
2. Vogel, E., Jäger, B., Hellmann, R., and Bich, E., *Ab Initio* Pair Potential Energy Curve for the Argon Atom Pair and Thermophysical Properties for the Dilute Argon Gas. II. Thermophysical Properties for Low-Density Argon, *Mol. Phys.* 108, 3335, 2010. <https://doi.org/10.1080/00268976.2010.507557>
3. May, E. F., Berg, R. F., and Moldover, M. R., Reference Viscosities of H_2, CH_4, Ar, and Xe at Low Densities, *Int. J. Thermophys.* 28, 1085, 2007. <https://doi.org/10.1007/s10765-007-0198-7>
4. Ho, C. Y., Ed., *Properties of Inorganic and Organic Fluids, CINDAS Data Series on Materials Properties,* Vol. V-1, Hemisphere Publishing Corp., New York, 1988.
5. Huber, M. L., Models for the Viscosity, Thermal Conductivity, and Surface Tension of Selected Pure Fluids as Implemented in REFPROP v10.0, *NISTIR 8209*, U.S. Department of Commerce, Gaithersburg, MD, 2018. <https://doi.org/10.6028/NIST.IR.8209>
6. Assael, M. J., Koini, I. A., Antoniadis, K. D., Huber, M. L., Abdulagatov, I. M., and Perkins, R. A., Reference Correlation of the Thermal Conductivity of Sulfur Hexafluoride from the Triple Point to 1000 K and up to 150 MPa, *J. Phys. Chem. Ref. Data*, 41, 023104, 2012. <https://doi.org/10.1063/1.4708620>
7. Mehl, J. B., Huber, M. L., and Harvey, A. H., *Ab Initio* Transport Coefficients of Gaseous Hydrogen, *Int. J. Thermophys.* 31, 740, 2010. <https://doi.org/10.1007/s10765-009-0697-9>
8. Huber, M. L., Perkins, R. A., Friend, D. G., and Sengers, J. V., Assael, M. J., Metaxa, I. N., Miyagawa, K., Hellmann, R., and Vogel, E., New International Formulation for the Thermal Conductivity of H_2O, *J. Phys. Chem. Ref. Data*, 41, 033102, 2012. <https://doi.org/10.1063/1.4738955>
9. Hellmann, R. and Bich, E., Transport Properties of Dilute D_2O Vapour from First Principles, *Mol. Phys.*, 115, 1057, 2017. <https://doi.org/10.1080/00268976.2016.1226443>
10. Hellmann, R., Bich, E., Vogel, E., and Vesovic, V., Thermophysical Properties of Dilute Hydrogen Sulfide Gas, *J. Chem. Eng. Data*, 57, 1312, 2012. <https://doi.org/10.1021/je3000926>
11. Monogenidou, S. A., Assael, M. J., and Huber, M. L., Reference Correlations for Thermal Conductivity of Ammonia from the Triple Point to 680 K and up to 80 MPa, *J. Phys. Chem. Ref. Data 47*, Article 043101, 2018. <https://doi.org/10.1063/1.5053087>
12. Cencek, W., Przybytek, M., Komasa, J., Mehl, J. B., Jeziorski, B., and Szalewicz, K., Effects of Adiabatic, Relativistic, and Quantum Electrodynamics Interactions in Helium Dimer on Thermophysical Properties of Helium, *J. Chem. Phys.* 136, 224303, 2012. <https://doi.org/10.1063/1.4712218>
13. Jäger, B., Hellmann, R., Bich, E., and Vogel, E., State-of-the-Art *Ab Initio* Potential Energy Curve for the Krypton Atom Pair and Thermophysical Properties of Dilute Krypton Gas, *J. Chem. Phys.*, 144, 114304, 2016. <https://doi.org/10.1063/1.4943959>
14. Hellmann, R., Ab Initio Potential Energy Surface for the Nitrogen Molecule Pair and Thermophysical Properties of Nitrogen Gas, *Mol. Phys.*, 111, 387, 2013. <https://doi.org/10.1080/00268976.2012.726379>
15. Crusius, J.-P., Hellmann, R., Hassel, E. and Bich, E., *Ab Initio* Intermolecular Potential Energy Surface and Thermophysical Properties of Nitrous Oxide, *J. Chem. Phys.*, 142, 244307, 2015. <https://doi.org/10.1063/1.4922830>
16. Bich, E., Hellmann, R., and Vogel, E., *Ab Initio* Potential Energy Curve for the Neon Atom Pair and Thermophysical Properties for the Dilute Neon Gas. II. Thermophysical Properties for Low-Density Neon, *Mol. Phys* 106, 813, 2008. <https://doi.org/10.1080/00268970801964207>
17. Hellmann, R., Jäger, B., and Bich, E., State-of-the-Art *Ab Initio* Potential Energy Curve for the Xenon Atom Pair and Related Spectroscopic and Thermodynamic Properties, *J. Chem. Phys.*, 147, 034304, 2017. <https://doi.org/10.1063/1.4994267>
18. Krauss, R., and Stephan, K., Thermal Conductivity of Refrigerants in a Wide Range of Temperature and Pressure, *J. Phys. Chem. Ref. Data* 18, 43, 1989. <https://doi.org/10.1063/1.555842>
19. Uribe, F. J., Mason, E. A., and Kestin, J., Thermal Conductivity of Nine Polyatomic Gases at Low Density, *J. Phys. Chem. Ref. Data* 19, 1123, 1990. <https://doi.org/10.1063/1.555864>
20. Millat J., and Wakeham, W. A., The Thermal Conductivity of Nitrogen and Carbon Monoxide in the Limit of Zero Density, *J. Phys. Chem. Ref. Data* 18, 565, 1989. <https://doi.org/10.1063/1.555827>
21. Huber, M. L., Sykioti, E. A., Assael, M. J., and Perkins, R. A., Reference Correlation of the Thermal Conductivity of Carbon Dioxide from the Triple Point to 1100 K and up to 200 MPa, *J. Phys. Chem. Ref. Data*, 45, 013102, 2016. <https://doi.org/10.1063/1.4940892>

22. Hellmann, R., Bich, E., Vogel, E., Dickinson, A. S., and Vesovic, V., Calculation of the Transport and Relaxation Properties of Methane. II. Thermal Conductivity, Thermomagnetic Effects, Volume Viscosity, and Nuclear-Spin Relaxation, *J. Chem. Phys.* 130, 124309, 2009. <https://doi.org/10.1063/1.3098317>
23. Sykioti, E. A., Assael, M. J., Huber, M. L., and Perkins, R. A., Reference Correlation of the Thermal Conductivity of Methanol from the Triple Point to 660 K and up to 245 MPa, *J. Phys. Chem. Ref.Data*, 42, 043101, 2013. <https://doi.org/10.1063/1.4829449>
24. Assael, M. J., Koutian, A., Huber, M. L., and Perkins, R. A., Reference Correlations of the Thermal Conductivity of Ethylene and Propylene, *J. Phys. Chem. Ref. Data*, 45, 033104, 2016. <https://doi.org/10.1063/1.4958984>
25. Crusius, J.-P., Hellmann, R., Hassel, E. and Bich, E., Intermolecular Potential Energy Surface and Thermophysical Properties of Ethylene Oxide, *J. Chem. Phys.*, 141, 164322, 2014. <https://doi.org/10.1063/1.4899074>
26. Hellmann, R., Reference Values for the Second Virial Coefficient and Three Dilute Gas Transport Properties of Ethane for a State-of-the-Art Intermolecular Potential Energy Surface, *J. Chem. Eng. Data*, 63, 470, 2018. <https://doi.org/10.1021/acs.jced.7b01069>
27. Perkins, R. A., Huber, M. L., and Assael, M. J., Measurement and Correlation of the Thermal Conductivity of *trans*-1-Chloro-3,3,3-Trifluoropropene (R1233zd(E)), *J. Chem. Eng. Data*, 62, 2659, 2017. <https://doi.org/10.1021/acs.jced.7b00106>
28. Perkins, R. A., Huber, M. L., and Assael, M. J., Measurements of the Thermal Conductivity of 1,1,1,3-3-Pentafluoropropane (R-245fa) and Correlations for the Viscosity and Thermal Conductivity Surfaces, *J. Chem. Eng. Data*, 61, 3286, 2016. <https://doi.org/10.1021/acs.jced.6b00350>
29. Hellmann, R., Intermolecular Potential Energy Surface and Thermophysical Properties of Propane, *J. Chem. Phys.*, 146, 114304, 2017. <https://doi.org/10.1063/1.4978412>
30. Perkins, R. A., Ramires, M. L. V., Nieto de Castro, C. A., and Cusco, L., Measurement and Correlation of the Thermal Conductivity of Butane from 135 K to 600 K at Pressures to 70 MPa, *J. Chem. Eng. Data* 47, 1263, 2002. <https://doi.org/10.1021/je0101202>
31. Perkins, R. A., Measurement and Correlation of the Thermal Conductivity of Isobutane from 114 K to 600 K at Pressures to 70 MPa, *J. Chem. Eng. Data* 47, 1272, 2002. <https://doi.org/10.1021/je010121u>
32. Vassiliou, C.-M., Assael, M. J., Huber, M. L., and Perkins, R. A., Reference Correlations of the Thermal Conductivity of Cyclopentane, iso-Pentane, and n-Pentane, *J. Phys. Chem. Ref. Data*, 44, 033102, 2015. <https://doi.org/10.1063/1.4927095>
33. Assael, M. J., Mylona, S. K., Tsiglifisi, Ch. A., Huber, M. L., and Perkins, R. A., Reference Correlation of the Thermal Conductivity of *n*-Hexane from the Triple Point to 600 K and up to 500 MPa, *J. Phys. Chem. Ref. Data*, 42, 013106, 2013. <https://doi.org/10.1063/1.4793335>
34. Vesovic, V., Wakeham, W. A., Olchowy, G. A., Sengers, J. V., Watson, J. T. R., and Millat, J., The Transport Properties of Carbon Dioxide, *J. Phys. Chem. Ref. Data*, 19, 763, 1990. <https://doi.org/10.1063/1.555875>
35. Holland, P. M., Eaton, B. E., and Hanley, H. J. M., A Correlation of the Viscosity and Thermal Conductivity Data of Gaseous and Liquid Ethylene, *J. Phys. Chem. Ref. Data* 12, 917, 1983. <https://doi.org/10.1063/1.555701>
36. Friend, D. G., Ingham, H., and Ely, J. F., Thermophysical Properties of Ethane, *J. Phys. Chem. Ref. Data* 20, 275, 1991. <https://doi.org/10.1063/1.555881>
37. Marsh, K., Perkins, R. A., and Ramires, M. L. V., Measurement and Correlation of the Thermal Conductivity of Propane from 86 to 600 K at Pressures to 70 MPa, *J. Chem. Eng. Data* 47, 932, 2002. <https://doi.org/10.1021/je010001m>
38. Hellmann, R., Bich, E., Vogel, E., and Vesovic, V., Thermophysical Properties of Dilute Hydrogen Sulfide Gas, *J. Chem. Eng. Data*, 57, 1312, 2012. <https://doi.org/10.1021/je3000926>
39. Sengers, J. V., and Watson, J. T. R., Improved International Formulations for the Viscosity and Thermal Conductivity of Water Substance, *J. Phys. Chem. Ref. Data* 15, 1291, 1986. <https://doi.org/10.1063/1.555763>
40. Matsunaga, N., and Nagashima, A., Transport Properties of Liquid and Gaseous D_2O over a Wide Range of Temperature and Pressure, *J. Phys. Chem. Ref. Data* 12, 933, 1983. <https://doi.org/10.1063/1.555694>
41. Tufeu, R., Ivanov, D. Y., Garrabos, Y., and Le Neindre, B., Thermal Conductivity of Ammonia in a Large Temperature and Pressure Range Including the Critical Region, *Ber. Bunsenges. Phys. Chem.* 88, 422, 1984. <https://doi.org/10.1002/bbpc.19840880421>
42. Bich, E., Millat, J., and Vogel, E., The Viscosity and Thermal Conductivity of Pure Monatomic Gases from Their Normal Boiling Point up to 5000 K in the Limit of Zero Density and at 0.101325 MPa, *J. Phys. Chem. Ref. Data* 19, 1289, 1990. <https://doi.org/10.1063/1.555846>

TABLE 1. Thermal Conductivity λ (mW m^{-1} K^{-1}) of Common Gases as a Function of Temperature

Mol. form.	Gas	Low-pressure limiting value	λ(100 K)	λ(200 K)	λ(300 K)	λ(400 K)	λ(500 K)	λ(600 K)	Ref.
	Air		9.5	18.5	26.4	33.5	39.9	46	1
Ar	Argon	Yes	6.3	12.4	17.7	22.4	26.5	30.3	2,3*
BF_3	Boron trifluoride				19.0	24.6			4
ClH	Hydrogen chloride			9.8	15.0	19.8	24.2	28.3	5
F_6S	Sulfur hexafluoride				13.1	20.6	27.6	34.1	6
H_2	Normal hydrogen	Yes	68.2	132.8	186.6	230.9	270.9	309.1	7
H_2O	Water	Yes			18.6	26.4	35.8	46.3	8
D_2O	Deuterium oxide	Yes			17.9	26	35.6	46.7	9
H_2S	Hydrogen sulfide	Yes		9.0	14.2	20.2	26.8	33.8	10
H_3N	Ammonia				25.4	37.5	52.0	68.6	11
He	Helium	Yes	74.7	118.3	155.7	189.6	221.4	251.6	12
Kr	Krypton	Yes	3.3	6.4	9.5	12.2	14.7	17.0	13
NO	Nitric oxide			17.8	25.9	33.1	39.6	46.2	4
N_2	Nitrogen	Yes	9.2	18.3	25.9	32.6	38.9	45.1	14
N_2O	Nitrous oxide	Yes		9.7	17.1	25.3	33.6	41.8	15
Ne	Neon	Yes	22.3	37.4	49.4	59.9	69.5	78.5	16
O_2	Oxygen		9.1	18.2	26.5	34.0	41.0	47.7	1
O_2S	Sulfur dioxide				9.6	14.3	20.0	25.6	4
Xe	Xenon	Yes	2.0	3.7	5.5	7.2	8.8	10.3	3*,17
CF_4	Tetrafluoromethane	Yes			16.0	24.1	32.2	39.9	19

Mol. form.	Gas	Low-pressure limiting value	λ(100 K)	λ(200 K)	λ(300 K)	λ(400 K)	λ(500 K)	λ(600 K)	Ref.
CO	Carbon monoxide	Yes			25.0	31.7	37.9	43.2	20
CO_2	Carbon dioxide			9.7	16.8	24.7	32.9	41.0	21
$CHCl_3$	Trichloromethane				7.5	11.1	15.1		4
CH_4	Methane	Yes	10.4	21.8	34.4	50.0	68.4	88.6	22
CH_4O	Methanol					25.6	38.2	52.7	23
C_2H_2	Acetylene				21.4	33.3	45.4	56.8	4
C_2H_4	Ethylene			10.5	21.1	36.4	55.1	75.8	24
C_2H_4O	Ethylene oxide	Yes		6.0	12.2	22.2	34.8	48.5	25
C_2H_6	Ethane	Yes	4.8	11.2	21.5	35.7	52.3	70.1	26
C_2H_6O	Ethanol				14.4	25.8	38.4	53.2	4
C_3H_6O	Acetone				11.5	20.2	30.6	42.7	4
C_3H_8	Propane	Yes		8.9	18.0	30.9	46.0	62.1	29
C_4F_8	Perfluorocyclobutane				12.1	19.1	26.3	33.5	5
C_4H_{10}	Butane				16.7	28.3	43.0	60.9	30
C_4H_{10}	Isobutane				17.1	28.9	43.2	60.2	31
$C_4H_{10}O$	Diethyl ether				15.1	25.0	37.1		4
C_5H_{10}	Cyclopentane					22.4	35.6	50.4	32
C_5H_{12}	Pentane					26.0	39.9	55.5	32
C_6H_{14}	Hexane					23.5	35.5	48.8	33

* References contain more accurate data covering a restricted temperature range.

TABLE 2. Thermal Conductivity λ (mW m^{-1} K^{-1}) of Selected Refrigerants as a Function of Temperature

Mol. form.	Refrigerant	Refrig. no.	λ(300 K)	λ(400 K)	λ(500 K)	λ(600 K)	Ref.
CCl_2F_2	Dichlorodifluoromethane	CFC-12	9.9	15.0	20.1	25.2	18
$C_2Cl_2F_4$	1,2-Dichlorotetrafluoroethane	CFC-114	10.8	15.8	20.6	25.1	5
$C_2Cl_3F_3$	1,1,2-Trichlorotrifluoroethane	CFC-113	9.0	13.6	18.3		18
$C_3H_3ClF_3$	*trans*-1-Chloro-3,3,3-trifluoropropene	R1233zd(E)	10.8	18.4	25.9	33.1	27
$C_3H_3F_5$	Pentafluoropropane	HFC-245fa	12.8	21.9	31	40.1	28

THERMAL CONDUCTIVITY OF LIQUIDS

Marcia L. Huber

This table gives the thermal conductivity λ of 277 liquids at temperatures between −25 °C and 100 °C. Values refer to nominal atmospheric pressure; however, when an entry is given for a temperature above the normal boiling point of the liquid, the pressure is understood to be the saturation vapor pressure at that temperature. Additional data at pressures above saturation can be found in most of the cited references. Values given to three decimal places (i.e., to 0.001 W m^{-1} K^{-1}) have an uncertainty of 2% to 5%. Values given to 0.0001 W m^{-1} K^{-1} are accurate to 1% or better. Entries reported with two digits should be considered as predictions with uncertainties in the range of 10-25%.

Column definitions for the table are as follows.

Column heading	Definition
Name	Name of liquid; liquids are listed alphabetically by name
Mol. form.	Molecular formula of liquid
λ(-25 °C), λ(0 °C), etc.	Thermal conductivity at temperature (in °C) indicated within parentheses, in units W m^{-1} K^{-1}
Ref.	Reference number

References

1. AIChE, *The DIPPR Information and Data Evaluation Manager for the Design Institute for Physical Properties*, DIADEM Pro, v14.0.0. 2019.
2. Huber, M. L., *NISTIR 8209*, 2018. https://doi.org/10.6028/NIST.IR.8209
3. Diky, V., Chirico, R. D., Frenkel, M., Bazyleva, A., Magee, J. W., Paulechka, E., Kazakov, A., Lemmon, E. W., Muzny, C. D., Smolyanitsky, A. Y., Townsend, S., and Kroenlein, K., *NIST ThermoData Engine - Pure Compounds, Binary Mixtures and Reactions - SRD 103b*, National Institute of Standards and Technology,https://doi.org/10.18434/T4Q30W (Accessed 2020-11-22). 2020.
4. Vargaftik, N. B., Filippov, L. P., Tarzimanov, A. A., and Totskii, E. E., *Handbook of Thermal Conductivity of Liquids and Gases*, CRC Press, Boca Raton FL, 1994.
5. Assael, M. J., Mihailidou, E. K., Huber, M. L., and Perkins, R. A., *J. Phys. Chem. Ref. Data* 41, 043102, 2012.
6. Assael, M. J., Charitidou, E., and Nieto de Castro, C. A., *Int. J. Thermophys.* 9, 813, 1988.
7. Koutian, A., Assael, M. J., Huber, M. L., and Perkins, R. A., *J. Phys. Chem. Ref. Data* 46, 013102, 2017.
8. Vassiliou, C. M., Assael, M. J., Huber, M. L., and Perkins, R. A., *J. Phys. Chem. Ref. Data* 44, 033102, 2015.
9. Watanabe, H., and Seong, D. J., *Int. J. Thermophys.* 23, 337, 2002. <https://doi.org/10.1023/A:1015158401299>
10. Watanabe, H., *J. Chem. Eng. Data* 48, 124, 2003. <https://doi.org/10.1021/je020125e>
11. Marsh, K. N., Ed., *Recommended Reference Materials for the Realization of Physicochemical Properties*, Blackwell Scientific Publications, Oxford, 1987.
12. Huber, M. L., Laesecke, A., and Perkins, R. A., *Energy & Fuels* 18, 968, 2004.
13. Assael, M. J., Sykioti, E. A., Huber, M. L., and Perkins, R. A., *J. Phys. Chem. Ref. Data* 42, 023102, 2013.
14. Mylona, S. K., Antoniadis, K. D., Assael, M. J., Huber, M. L., and Perkins, R. A., *J. Phys. Chem. Ref. Data* 43, 043104, 2014.
15. Assael, M. J., Bogdanou, I., Mylona, S. K., Huber, M. L., Perkins, R. A., and Vesovic V., *J. Phys. Chem. Ref. Data* 42, 023101, 2013.
16. Monogenidou, S. A., Assael, M. J., and Huber, M. L., *J. Phys. Chem. Ref. Data* 47, 013103, 2018.
17. Assael, M. J., Mylona, S. K., Tsiglifisi, C. A., Huber, M. L., and Perkins, R. A., *J. Phys. Chem. Ref. Data* 42, 033104, 2013.
18. Peralta-Martinez, M. V., Assael, M. J., Dix, M. J., Karagiannidis, L., and Wakeham, W. A., *Int. J. Thermophys.* 27, 681, 2006.
19. Sykioti, E. A., Assael, M. J., Huber, M. L., and Perkins, R. A., *J. Phys. Chem. Ref. Data* 42, 043101, 2013.
20. Assael, M. J., Mylona, S. K. , Huber, M. L., and Perkins, R. A., *J. Phys. Chem. Ref. Data* 41, 023101, 2012.
21. Krauss, R., and Stephan, K., *J. Phys. Chem. Ref. Data* 18, 43, 1989. <https://doi.org/10.1063/1.555842>
22. Huber, M. L., Perkins, R. A., Friend, D. G., Sengers, J. V., Assael, M. J., Metaxa, I.N., Miyagawa, K., Hellmann, R., and Vogel, E., *J. Phys. Chem. Ref. Data* 41, 033102, 2012.

Thermal Conductivity of Liquids as a Function of Temperature

Name	Mol. form.	λ(-25 °C)/ W m^{-1} K^{-1}	λ(0 °C)/ W m^{-1} K^{-1}	λ(25 °C)/ W m^{-1} K^{-1}	λ(50 °C)/ W m^{-1} K^{-1}	λ(75 °C)/ W m^{-1} K^{-1}	λ(100 °C)/ W m^{-1} K^{-1}	Ref.
Acetic acid	$C_2H_4O_2$			0.159	0.155	0.150	0.146	1
Acetic anhydride	$C_4H_6O_3$		0.176	0.170	0.164	0.158	0.152	1
Acetone	C_3H_6O	0.176	0.165	0.155	0.145	0.135	0.126	2
Acetonitrile	C_2H_3N	0.235	0.215	0.199	0.186			3
Acetophenone	C_8H_8O			0.147	0.146	0.144	0.142	4
Acrylonitrile	C_3H_3N	0.184	0.174	0.166	0.160			3
Allyl alcohol	C_3H_6O			0.162				4
Aniline	C_6H_7N			0.173	0.171	0.169	0.167	4
Anisole	C_7H_8O			0.145	0.142	0.139	0.136	4
Benzaldehyde	C_7H_6O			0.153	0.148	0.143	0.139	4
Benzene	C_6H_6			0.1412	0.1331	0.1253	0.1178	5
Benzonitrile	C_7H_5N			0.148	0.142	0.136	0.130	4
Benzyl alcohol	C_7H_8O			0.159	0.158	0.156	0.154	4
Bromobenzene	C_6H_5Br	0.119	0.115	0.111	0.107	0.103	0.099	4
1-Bromobutane	C_4H_9Br	0.112	0.107	0.103	0.098	0.093	0.088	4
Bromoethane	C_2H_5Br	0.107	0.104	0.101				4
1-Bromohexane	$C_6H_{13}Br$	0.115	0.111	0.108	0.104	0.101	0.097	4
1-Bromonaphthalene	$C_{10}H_7Br$			0.110	0.109	0.108	0.106	4

Name	Mol. form.	λ(-25 °C)/ W m⁻¹ K⁻¹	λ(0 °C)/ W m⁻¹ K⁻¹	λ(25 °C)/ W m⁻¹ K⁻¹	λ(50 °C)/ W m⁻¹ K⁻¹	λ(75 °C)/ W m⁻¹ K⁻¹	λ(100 °C)/ W m⁻¹ K⁻¹	Ref.
1-Bromononane	$C_9H_{19}Br$		0.116	0.112	0.109	0.106	0.103	4
1-Bromopentane	$C_5H_{11}Br$	0.113	0.109	0.105	0.101	0.097	0.093	4
1-Bromopropane	C_3H_7Br	0.108	0.104	0.099	0.094			4
1,2-Butadiene	C_4H_6	0.147	0.134					4
Butanal	C_4H_8O		0.145	0.140	0.134	0.128	0.122	3
1-Butanol	$C_4H_{10}O$		0.150	0.1476	0.1449	0.1423	0.140	6
2-Butanol	$C_4H_{10}O$		0.139	0.134	0.129	0.123	0.117	3
2-Butanone	C_4H_8O	0.158	0.152	0.146	0.140	0.134	0.128	1
Butyl acetate	$C_6H_{12}O_2$		0.143	0.136	0.130	0.123	0.116	4
Butylbenzene	$C_{10}H_{14}$	0.140	0.135	0.130	0.125	0.119	0.114	3
sec-Butylbenzene, (±)-	$C_{10}H_{14}$		0.129	0.124	0.119	0.114	0.108	4
tert-Butylbenzene	$C_{10}H_{14}$			0.117	0.114	0.110	0.106	4
Butyl formate	$C_5H_{10}O_2$			0.136	0.130	0.123	0.117	4
Butyl oleate	$C_{22}H_{42}O_2$			0.18	0.17	0.17	0.16	3
Butyl palmitate	$C_{20}H_{40}O_2$			0.151	0.148	0.144	0.140	4
Butyl propanoate	$C_7H_{14}O_2$			0.139	0.133	0.126	0.121	4
Butyl stearate	$C_{22}H_{44}O_2$			0.157	0.153	0.149	0.145	4
2-Butyne	C_4H_6	0.14	0.13	0.12				1
Carbon disulfide	CS_2	0.17	0.17	0.16				1
2-Chloroaniline	C_6H_6ClN			0.148				4
Chlorobenzene	C_6H_5Cl	0.138	0.131	0.124	0.118	0.113	0.109	2
Chloroethane	C_2H_5Cl	0.140	0.130	0.120	0.110	0.100		1
1-Chloronaphthalene	$C_{10}H_7Cl$			0.126				4
1-Chlorononane	$C_9H_{19}Cl$		0.132	0.128	0.124	0.120	0.115	4
1-Chlorooctane	$C_8H_{17}Cl$		0.130	0.127	0.124	0.121	0.119	4
1-Chloropentane	$C_5H_{11}Cl$		0.125	0.120	0.115	0.109		4
1-Chloropropane	C_3H_7Cl	0.129	0.123	0.116	0.110	0.104	0.098	4
o-Cresol	C_7H_8O			0.153				4
m-Cresol	C_7H_8O			0.149	0.147	0.145		4
Cycloheptane	C_7H_{14}			0.118	0.113	0.109	0.105	3
Cyclohexane	C_6H_{12}			0.1181	0.1115	0.106	0.102	7
Cyclohexanol	$C_6H_{12}O$			0.134	0.131	0.127	0.123	3
Cyclohexanone	$C_6H_{10}O$			0.144	0.135	0.129	0.124	3
Cyclohexene	C_6H_{10}	0.143	0.135	0.127	0.120	0.113	0.106	3
Cyclohexylbenzene	$C_{12}H_{16}$				0.12	0.12	0.12	3
Cyclopentane	C_5H_{10}	0.1463	0.1363	0.1266	0.117	0.108	0.100	8
Cyclopentene	C_5H_8	0.152	0.142	0.132	0.122	0.111	0.100	3
trans-Decahydronaphthalene	$C_{10}H_{18}$	0.117	0.115	0.113	0.111	0.109	0.106	1
Decanal	$C_{10}H_{20}O$		0.149	0.144	0.139	0.134	0.129	4
Decane	$C_{10}H_{22}$	0.142	0.1360	0.1296	0.1232	0.1167	0.110	9
Decanoic acid	$C_{10}H_{20}O_2$				0.150	0.145	0.140	3
1-Decanol	$C_{10}H_{22}O$			0.162	0.159	0.155	0.151	4
1-Decene	$C_{10}H_{20}$	0.142	0.135	0.129	0.122	0.116	0.110	3
Decyl acetate	$C_{12}H_{24}O_2$			0.148	0.136	0.128	0.123	3
1,2-Dibromoethane	$C_2H_4Br_2$			0.100	0.096	0.092	0.088	4
Dibromomethane	CH_2Br_2	0.120	0.114	0.108	0.103	0.097		1
1,2-Dibromotetrafluoroethane	$C_2Br_2F_4$	0.071	0.066	0.061	0.057	0.053	0.049	4
1,2-Dibutoxyethane	$C_{10}H_{22}O_2$			0.140	0.134	0.127	0.120	4
Dibutyl ether	$C_8H_{18}O$		0.139	0.132	0.125	0.118	0.112	4
Dibutyl phthalate	$C_{16}H_{22}O_4$	0.143	0.139	0.136	0.132	0.128	0.125	1
o-Dichlorobenzene	$C_6H_4Cl_2$		0.125	0.121	0.117	0.113	0.109	4
m-Dichlorobenzene	$C_6H_4Cl_2$		0.120	0.116	0.113	0.109		4
p-Dichlorobenzene	$C_6H_4Cl_2$				0.112	0.108	0.105	4
1,2-Dichloroethane	$C_2H_4Cl_2$	0.144	0.139	0.133	0.128	0.122	0.117	4
Dichloromethane	CH_2Cl_2	0.158	0.149	0.140	0.133	0.128	0.127	4
1,2-Diethoxyethane	$C_6H_{14}O_2$	0.164	0.156	0.148	0.140	0.132	0.125	1
o-Diethylbenzene	$C_{10}H_{14}$		0.133	0.127	0.122	0.116	0.111	4
Diethylene glycol dibutyl ether	$C_{12}H_{26}O_3$	0.154	0.150	0.146	0.142	0.137	0.133	1
Diethylene glycol monobutyl ether	$C_8H_{18}O_3$			0.163	0.158	0.153	0.148	4
Diethylene glycol monoethyl ether	$C_6H_{14}O_3$				0.188	0.184	0.180	4

Name	Mol. form.	λ(-25 °C)/ W m^-1 K^-1	λ(0 °C)/ W m^-1 K^-1	λ(25 °C)/ W m^-1 K^-1	λ(50 °C)/ W m^-1 K^-1	λ(75 °C)/ W m^-1 K^-1	λ(100 °C)/ W m^-1 K^-1	Ref.
Diethylene glycol monomethyl ether	$C_5H_{12}O_3$			0.190	0.185	0.180	0.175	4
Diethyl ether	$C_4H_{10}O$	0.142	0.134	0.126	0.117	0.107	0.095	3
Diethyl oxalate	$C_6H_{10}O_4$	0.163	0.158	0.153	0.148	0.142	0.137	1
Diethyl phthalate	$C_{12}H_{14}O_4$		0.147	0.143	0.140	0.136	0.132	1
Diiodomethane	CH_2I_2			0.098	0.093	0.088	0.083	4
N,N-Dimethylacetamide	C_4H_9NO		0.179	0.168	0.157	0.147	0.138	3
N,N-Dimethylaniline	$C_8H_{11}N$			0.142	0.138	0.133	0.128	3
2,2-Dimethylbutane	C_6H_{14}	0.108	0.1006	0.0934	0.0861	0.0788	0.072	10
2,3-Dimethylbutane	C_6H_{14}	0.115	0.1076	0.1003	0.0930	0.0857	0.078	10
N,N-Dimethylformamide	C_3H_7NO			0.184	0.178	0.173	0.167	3
2,2-Dimethylpentane	C_7H_{16}	0.111	0.1046	0.0980	0.0913	0.0847	0.078	10
2,3-Dimethylpentane	C_7H_{16}	0.120	0.1127	0.1059	0.0990	0.0922	0.085	10
2,4-Dimethylpentane	C_7H_{16}	0.116	0.1089	0.1020	0.0951	0.0882	0.081	10
3,3-Dimethylpentane	C_7H_{16}	0.113	0.1068	0.1001	0.0934	0.0867	0.080	10
Dimethyl phthalate	$C_{10}H_{10}O_4$			0.1473	0.1443	0.1409	0.1373	11
Dioctyl hexanedioate	$C_{22}H_{42}O_4$			0.157	0.153	0.149	0.145	4
1,4-Dioxane	$C_4H_8O_2$			0.162	0.157	0.150	0.141	3
Dipentyl ether	$C_{10}H_{22}O$			0.135	0.131	0.128	0.125	3
Diphenyl ether	$C_{12}H_{10}O$				0.138	0.134	0.130	1
Dipropyl ether	$C_6H_{14}O$		0.137	0.130	0.123	0.117		4
Dodecane	$C_{12}H_{26}$	0.148	0.141	0.135	0.129	0.124	0.118	12
1-Dodecanol	$C_{12}H_{26}O$				0.167	0.163	0.159	4
Epichlorohydrin	C_3H_5ClO	0.17	0.16	0.15	0.15	0.14	0.13	1
1,2-Ethanediol	$C_2H_6O_2$		0.258	0.256	0.255	0.254	0.253	3
Ethanol	C_2H_6O	0.176	0.169	0.163	0.1589	0.155	0.151	13
Ethanolamine	C_2H_7NO			0.240	0.238	0.236		4
Ethoxybenzene	$C_8H_{10}O$	0.148	0.144	0.140	0.136	0.132	0.128	1
2-Ethoxyethanol	$C_4H_{10}O_2$			0.169	0.164	0.158	0.152	3
Ethyl acetate	$C_4H_8O_2$	0.162	0.152	0.143	0.135	0.127	0.120	3
Ethyl acetoacetate	$C_6H_{10}O_3$			0.155	0.152	0.148	0.144	4
N-Ethylaniline	$C_8H_{11}N$			0.150				4
Ethylbenzene	C_8H_{10}	0.140	0.1341	0.1278	0.1215	0.115	0.110	14
Ethyl benzoate	$C_9H_{10}O_2$			0.138	0.132	0.126	0.122	3
Ethyl butanoate	$C_6H_{12}O_2$		0.143	0.137	0.131	0.126		4
Ethyl formate	$C_3H_6O_2$	0.181	0.171	0.160	0.149	0.138		4
Ethyl hexanoate	$C_8H_{16}O_2$		0.142	0.137	0.133	0.128	0.123	4
Ethyl hexyl ether	$C_8H_{18}O$		0.131	0.126	0.120	0.114	0.109	4
3-Ethylpentane	C_7H_{16}	0.128	0.1203	0.1128	0.1053	0.0978	0.090	10
Ethyl pentanoate	$C_7H_{14}O_2$			0.132				4
Ethyl propanoate	$C_5H_{10}O_2$				0.133	0.121		4
Fluorobenzene	C_6H_5F	0.141	0.132	0.124	0.116	0.109	0.102	3
Formic acid	CH_2O_2			0.267	0.265	0.263	0.261	4
Furan	C_4H_4O	0.14	0.13	0.13				1
Furfuryl alcohol	$C_5H_6O_2$			0.179				4
Germanium(IV) chloride	Cl_4Ge	0.111	0.105	0.100	0.095	0.090	0.084	4
Glycerol	$C_3H_8O_3$			0.285	0.288	0.292	0.296	4
Heptanal	$C_7H_{14}O$		0.146	0.139	0.132	0.126	0.121	3
Heptane	C_7H_{16}	0.138	0.1298	0.1222	0.1147	0.107	0.101	15
Heptanoic acid	$C_7H_{14}O_2$		0.147	0.142	0.137	0.132	0.127	1
1-Heptanol	$C_7H_{16}O$	0.170	0.164	0.158	0.152	0.146	0.140	3
3-Heptanone	$C_7H_{14}O$		0.143	0.137	0.131	0.125	0.119	4
4-Heptanone	$C_7H_{14}O$			0.136	0.131	0.125	0.120	4
1-Heptene	C_7H_{14}	0.137	0.1293	0.1216	0.1143	0.108	0.101	3
Heptyl acetate	$C_9H_{18}O_2$			0.135	0.128	0.122	0.116	4
Heptyl butanoate	$C_{11}H_{22}O_2$			0.139	0.134	0.129	0.123	4
Heptyl formate	$C_8H_{16}O_2$		0.141	0.137	0.132	0.128	0.123	4
Heptyl propanoate	$C_{10}H_{20}O_2$			0.137	0.132	0.127	0.122	4
Hexadecane	$C_{16}H_{34}$			0.144	0.139	0.135	0.131	16
Hexafluorobenzene	C_6F_6				0.083			4
Hexane	C_6H_{14}	0.137	0.128	0.1200	0.1127	0.106	0.100	17

Name	Mol. form.	λ(-25 °C)/ W m⁻¹ K⁻¹	λ(0 °C)/ W m⁻¹ K⁻¹	λ(25 °C)/ W m⁻¹ K⁻¹	λ(50 °C)/ W m⁻¹ K⁻¹	λ(75 °C)/ W m⁻¹ K⁻¹	λ(100 °C)/ W m⁻¹ K⁻¹	Ref.
Hexanedinitrile	$C_6H_8N_2$			0.174	0.158			4
Hexanoic acid	$C_6H_{12}O_2$			0.145	0.140	0.136	0.132	3
1-Hexanol	$C_6H_{14}O$	0.154	0.149	0.1447	0.1399	0.1351	0.130	6
2-Hexanone	$C_6H_{12}O$		0.145	0.140	0.134	0.128	0.122	3
1-Hexene	C_6H_{12}	0.135	0.1261	0.1173	0.1090	0.101	0.095	3
Hexyl acetate	$C_8H_{16}O_2$			0.135	0.129	0.123	0.118	4
Hexylbenzene	$C_{12}H_{18}$	0.137	0.133	0.129	0.126	0.123	0.120	3
Hexyl butanoate	$C_{10}H_{20}O_2$			0.137	0.132	0.127	0.121	4
Hexyl formate	$C_7H_{14}O_2$			0.143	0.133	0.126	0.119	3
Indan	C_9H_{10}			0.135				4
Iodobenzene	C_6H_5I	0.104	0.102	0.099	0.097	0.096	0.094	3
1-Iodobutane	C_4H_9I		0.094	0.090	0.085	0.081	0.077	4
Iodoethane	C_2H_5I		0.091	0.087	0.083	0.079		4
1-Iodohexane	$C_6H_{13}I$		0.098	0.095	0.091	0.088	0.084	4
1-Iodononane	$C_9H_{19}I$		0.105	0.102	0.099	0.095	0.092	4
1-Iodopentane	$C_5H_{11}I$		0.096	0.092	0.088	0.084	0.081	4
1-Iodopropane	C_3H_7I	0.096	0.092	0.087	0.083	0.078	0.074	4
2-Iodopropane	C_3H_7I	0.089	0.085	0.082	0.078	0.074	0.071	4
Isopentane	C_5H_{12}	0.118	0.109	0.1017	0.0942	0.0869	0.079	8
Isopropylbenzene	C_9H_{12}	0.128	0.124	0.120	0.115	0.111	0.106	3
1-Isopropyl-4-methylbenzene	$C_{10}H_{14}$	0.131	0.126	0.122	0.117	0.112	0.107	1
Mercury	Hg			8.171	8.795	9.403	9.996	18
Mesityl oxide	$C_6H_{10}O$	0.15	0.15	0.14	0.14	0.13	0.13	3
Methanol	CH_4O	0.210	0.205	0.2002	0.1954	0.191	0.186	19
2-Methoxyethanol	$C_3H_8O_2$				0.190	0.180	0.170	4
Methyl acetate	$C_3H_6O_2$	0.170	0.160	0.150	0.140	0.131	0.122	1
2-Methylaniline	C_7H_9N			0.162				4
3-Methylaniline	C_7H_9N			0.161				4
Methyl benzoate	$C_8H_8O_2$				0.137	0.131	0.126	3
2-Methyl-1,3-butadiene	C_5H_8	0.141	0.131	0.121	0.111			3
Methyl butanoate	$C_5H_{10}O_2$		0.146	0.139	0.132	0.124	0.117	3
2-Methyl-2-butanol	$C_5H_{12}O$		0.119	0.116	0.113	0.109	0.106	4
N-Methylformamide	C_2H_5NO			0.21	0.21	0.20	0.19	1
Methyl formate	$C_2H_4O_2$		0.194	0.187				4
2-Methylheptane	C_8H_{18}	0.127	0.1206	0.1139	0.1072	0.1005	0.094	10
3-Methylheptane	C_8H_{18}	0.128	0.1216	0.1149	0.1081	0.1014	0.095	10
2-Methylhexane	C_7H_{16}	0.125	0.1177	0.1105	0.1033	0.0961	0.089	10
3-Methylhexane	C_7H_{16}	0.126	0.1184	0.1112	0.1040	0.0968	0.090	10
Methyl hexanoate	$C_7H_{14}O_2$		0.145	0.138	0.131	0.124	0.118	3
Methyl methacrylate	$C_5H_8O_2$		0.156	0.147	0.137	0.127	0.117	4
Methyloxirane	C_3H_6O		0.181	0.171				4
2-Methylpentane	C_6H_{14}	0.120	0.1127	0.1050	0.0972	0.0894	0.082	10
3-Methylpentane	C_6H_{14}	0.122	0.1142	0.1064	0.0986	0.0909	0.083	10
Methyl pentanoate	$C_6H_{12}O_2$		0.143	0.137	0.131	0.124	0.117	3
Methyl propanoate	$C_4H_8O_2$		0.143	0.140	0.136	0.131	0.125	3
2-Methyl-2-propanol	$C_4H_{10}O$			0.112	0.110	0.109	0.108	4
N-Methyl-2-pyrrolidinone	C_5H_9NO			0.167	0.162	0.157		4
Nitrobenzene	$C_6H_5NO_2$			0.149	0.145	0.142	0.139	4
Nitroethane	$C_2H_5NO_2$			0.173	0.161	0.149		4
Nitromethane	CH_3NO_2	0.226	0.215	0.204	0.193	0.182	0.171	4
1-Nitropropane	$C_3H_7NO_2$	0.167	0.161	0.154	0.148	0.141	0.135	1
Nonane	C_9H_{20}	0.141	0.1337	0.1269	0.1201	0.1133	0.106	9
Nonanoic acid	$C_9H_{18}O_2$			0.150	0.145	0.140	0.136	3
1-Nonanol	$C_9H_{20}O$		0.173	0.166	0.159	0.154	0.150	3
1-Nonene	C_9H_{18}	0.139	0.132	0.124	0.118	0.111	0.105	3
Octadecane	$C_{18}H_{38}$				0.144	0.138	0.132	3
Octane	C_8H_{18}	0.139	0.1317	0.1244	0.1171	0.1097	0.102	9
Octanoic acid	$C_8H_{16}O_2$			0.147	0.142	0.137	0.132	3
1-Octanol	$C_8H_{18}O$		0.166	0.160	0.154	0.148	0.142	3
2-Octanone	$C_8H_{16}O$		0.144	0.137	0.131	0.125	0.120	3

Name	Mol. form.	λ(-25 °C)/ W m⁻¹ K⁻¹	λ(0 °C)/ W m⁻¹ K⁻¹	λ(25 °C)/ W m⁻¹ K⁻¹	λ(50 °C)/ W m⁻¹ K⁻¹	λ(75 °C)/ W m⁻¹ K⁻¹	λ(100 °C)/ W m⁻¹ K⁻¹	Ref.
1-Octene	C_8H_{16}	0.139	0.1312	0.1238	0.1169	0.111	0.105	3
Octyl butanoate	$C_{12}H_{24}O_2$			0.139	0.134	0.129	0.125	4
Octyl propanoate	$C_{11}H_{22}O_2$			0.135	0.130	0.125	0.120	4
Paraldehyde	$C_6H_{12}O_3$				0.143	0.139	0.134	3
Pentanal	$C_5H_{10}O$		0.146	0.139	0.133	0.127	0.121	4
Pentane	C_5H_{12}	0.131	0.1214	0.1119	0.103	0.095	0.087	8
1,5-Pentanediol	$C_5H_{12}O_2$		0.202	0.201	0.200	0.198	0.196	4
2,4-Pentanedione	$C_5H_8O_2$			0.154	0.150	0.146	0.143	4
Pentanoic acid	$C_5H_{10}O_2$			0.141	0.136	0.132	0.128	3
1-Pentanol	$C_5H_{12}O$		0.149	0.1460	0.1431	0.1401	0.137	6
2-Pentanone	$C_5H_{10}O$		0.149	0.142	0.135	0.128	0.121	4
3-Pentanone	$C_5H_{10}O$		0.151	0.144	0.137	0.129	0.122	4
1-Pentene	C_5H_{10}	0.131	0.1222	0.1127	0.102	0.091		3
Pentyl acetate	$C_7H_{14}O_2$		0.141	0.134	0.126	0.120	0.113	4
Pentylbenzene	$C_{11}H_{16}$	0.137	0.133	0.130	0.126	0.123	0.120	3
Pentyl propanoate	$C_8H_{16}O_2$			0.139	0.133	0.128	0.122	3
1-Pentyne	C_5H_8	0.144	0.136	0.127	0.119			4
Perfluorocyclobutane	C_4F_8	0.077	0.071	0.064	0.056			2
Perfluoroheptane	C_7F_{16}	0.068	0.064	0.060	0.056	0.053		4
Perfluorohexane	C_6F_{14}		0.067	0.065	0.064			4
Perfluorooctane	C_8F_{18}		0.066	0.062	0.059	0.055	0.052	4
Perfluoropropane	C_3F_8	0.062	0.056	0.051	0.046	0.041	0.035	4
Phenol	C_6H_6O				0.153	0.149	0.147	4
2-Phenoxyethanol	$C_8H_{10}O_2$			0.169	0.168	0.166	0.165	4
1,2-Propanediol	$C_3H_8O_2$	0.203	0.202	0.200	0.199	0.198	0.197	1
Propanoic acid	$C_3H_6O_2$		0.149	0.145	0.142	0.139	0.135	3
1-Propanol	C_3H_8O		0.152	0.1494	0.1469	0.1444	0.142	6
2-Propanol	C_3H_8O		0.139	0.135	0.130	0.125	0.119	3
Propyl acetate	$C_5H_{10}O_2$		0.146	0.140	0.135	0.130	0.124	4
Propylbenzene	C_9H_{12}	0.138	0.1333	0.1279	0.1223	0.117	0.111	3
Propyl formate	$C_4H_8O_2$		0.151	0.144	0.137	0.130		4
Propyl propanoate	$C_6H_{12}O_2$				0.133			4
Pyridine	C_5H_5N		0.171	0.166	0.162	0.157	0.153	4
Quinoline	C_9H_7N			0.147	0.144	0.141	0.138	4
Styrene	C_8H_8	0.15	0.14	0.14	0.13	0.13	0.13	3
1,1,2,2-Tetrachloro-1,2-difluoroethane	$C_2Cl_4F_2$			0.082	0.078	0.074	0.069	4
1,1,2,2-Tetrachloroethane	$C_2H_2Cl_4$	0.126	0.120	0.113	0.107	0.101	0.095	3
Tetrachloroethene	C_2Cl_4		0.117	0.109	0.102	0.097	0.093	3
Tetrachloromethane	CCl_4		0.106	0.102	0.096	0.091	0.085	3
Tetrachlorosilane	Cl_4Si			0.099	0.096			1
Tetradecane	$C_{14}H_{30}$			0.139	0.134	0.129	0.124	4
1-Tetradecanol	$C_{14}H_{30}O$				0.167	0.162	0.157	1
1-Tetradecene	$C_{14}H_{28}$			0.136	0.131	0.126	0.121	4
Tetraethylene glycol	$C_8H_{18}O_5$			0.188	0.187	0.185	0.184	3
Tetrahydrofuran	C_4H_8O	0.167	0.161	0.153	0.145			3
1,2,3,4-Tetrahydronaphthalene	$C_{10}H_{12}$			0.131	0.129	0.128	0.126	4
Thiophene	C_4H_4S	0.156	0.151	0.146	0.141	0.136		1
Tin(IV) chloride	Cl_4Sn	0.123	0.117	0.112	0.106	0.101	0.095	4
Titanium(IV) chloride	Cl_4Ti		0.143	0.138	0.134	0.129	0.124	4
Toluene	C_7H_8	0.144	0.1372	0.1304	0.1234	0.1166	0.1100	20
Tributylamine	$C_{12}H_{27}N$	0.127	0.124	0.120	0.117	0.114	0.111	3
1,2,3-Trichlorobenzene	$C_6H_3Cl_3$				0.110	0.108	0.106	4
1,2,4-Trichlorobenzene	$C_6H_3Cl_3$			0.112	0.109	0.106		4
1,1,1-Trichloroethane	$C_2H_3Cl_3$		0.106	0.101	0.096			1
Trichloroethene	C_2HCl_3	0.128	0.121	0.114	0.106	0.098	0.090	4
Trichlorofluoromethane	CCl_3F	0.102	0.094	0.087	0.079	0.072	0.066	3
Trichloromethane	$CHCl_3$	0.135	0.123	0.115	0.110			3
1,1,2-Trichloro-1,2,2-trifluoroethane	$C_2Cl_3F_3$	0.0847	0.0790	0.0736	0.0683			21
Tridecane	$C_{13}H_{28}$		0.143	0.136	0.131	0.125	0.120	3
1-Tridecene	$C_{13}H_{26}$			0.130	0.125	0.120	0.115	4

Fluids

Name	Mol. form.	λ(-25 °C)/ W m⁻¹ K⁻¹	λ(0 °C)/ W m⁻¹ K⁻¹	λ(25 °C)/ W m⁻¹ K⁻¹	λ(50 °C)/ W m⁻¹ K⁻¹	λ(75 °C)/ W m⁻¹ K⁻¹	λ(100 °C)/ W m⁻¹ K⁻¹	Ref.
Triethylamine	$C_6H_{15}N$	0.130	0.123	0.117	0.112			3
Triethylene glycol	$C_6H_{14}O_4$		0.193	0.195	0.196	0.196	0.196	4
Triethylene glycol dimethyl ether	$C_8H_{18}O_4$		0.164	0.160	0.156	0.152	0.148	3
Trimethylamine	C_3H_9N	0.14	0.13					1
1,2,4-Trimethylbenzene	C_9H_{12}	0.137	0.1320	0.1273	0.1225	0.117	0.112	3
1,3,5-Trimethylbenzene	C_9H_{12}	0.146	0.1407	0.1354	0.1298	0.124	0.118	3
2,2,3-Trimethylbutane	C_7H_{16}	0.107	0.1011	0.0950	0.0889	0.0828	0.077	10
2,2,4-Trimethylpentane	C_8H_{18}	0.107	0.1007	0.0948	0.0888	0.0829	0.077	10
2,3,4-Trimethylpentane	C_8H_{18}	0.115	0.1093	0.1035	0.0976	0.0918	0.086	10
Undecane	$C_{11}H_{24}$		0.141	0.134	0.128	0.122	0.117	3
Undecanoic acid	$C_{11}H_{22}O_2$				0.153	0.149		4
1-Undecanol	$C_{11}H_{24}O$			0.169	0.165	0.161	0.158	4
6-Undecanone	$C_{11}H_{22}O$			0.139	0.132	0.126	0.120	3
1-Undecene	$C_{11}H_{22}$	0.137	0.131	0.125	0.119	0.114	0.108	1
Vinyl acetate	$C_4H_6O_2$	0.168	0.159	0.150	0.142	0.133	0.124	1
Water	H_2O		0.5557	0.6065	0.6406	0.6635	0.6772	22
o-Xylene	C_8H_{10}	0.142	0.137	0.131	0.1244	0.1180	0.112	14
m-Xylene	C_8H_{10}	0.142	0.1360	0.1299	0.1234	0.1169	0.110	14
p-Xylene	C_8H_{10}			0.1266	0.1202	0.1140	0.108	14

DIFFUSION IN GASES

This table gives diffusion coefficients D_{12} for binary gas mixtures as a function of temperature. Specifically, for molecule 1 diffusing in a binary mixture of molecules 1 and 2 in the absence of temperature and pressure gradients, the flux equation is

$$J_1 = -nD_{12}\Delta x_1$$

where J_1 is the flux density of molecule 1, n is the total number density, and Δx_1 is the composition gradient in terms of mole fraction x_1. A similar equation holds for molecule 2. These equations hold only in the case of zero net flux, $J_1 + J_2 = 0$; more complicated equations are required in other cases (Refs. 1 and 2). It can be shown $D_{12} = D_{21}$ because there is no net flux and $x_1 + x_2 = 1$.

Values refer to atmospheric pressure. The diffusion coefficient is inversely proportional to pressure as long as the gas is in a regime where binary collisions dominate. See Ref. 1 for a discussion of the dependence of D_{12} on temperature and composition. Column definitions for the table are as follows.

Column heading	Definition
Mixture	Names of mixture components
D_{12}(-73 °C), D_{12}(0 °C), etc.	Diffusion coefficient, in units cm^2 s^{-1}, at temperature (in °C) indicated within the parentheses

The top portion of the table contains diffusion coefficients for eight gases in air; the lower portion of the table contains a selection of common binary gas mixtures.

References

1. Marrero, T. R., and Mason, E. A., *J. Phys. Chem. Ref. Data* 1, 1, 1972. <https://doi.org/10.1063/1.3253094>
2. Kestin, J., et al., *J. Phys. Chem. Ref. Data* 13, 229, 1984. <https://doi.org/10.1063/1.555703>

Diffusion Coefficients for Binary Gas Mixtures at 101.325 kPa and the Specified *t* /°C

Mixture	D_{12}(-73 °C)/ cm^2 s^{-1}	D_{12}(0 °C)/ cm^2 s^{-1}	D_{12}(20 °C)/ cm^2 s^{-1}	D_{12}(100 °C)/ cm^2 s^{-1}	D_{12}(200 °C)/ cm^2 s^{-1}	D_{12}(300 °C)/ cm^2 s^{-1}	D_{12}(400 °C)/ cm^2 s^{-1}
Large Excess of Air							
Ar-air		0.167	0.189	0.289	0.437	0.612	0.810
CH_4-air			0.210	0.321	0.485	0.678	0.899
CO-air			0.208	0.315	0.475	0.662	0.875
CO_2-air			0.160	0.252	0.390	0.549	0.728
H_2-air		0.668	0.756	1.153	1.747	2.444	3.238
H_2O-air			0.242	0.399	0.638	0.873	1.135
He-air		0.617	0.697	1.057	1.594	2.221	2.933
SF_6-air				0.150	0.233	0.329	0.438
Equimolar Mixture							
Ar-CH_4				0.306	0.467	0.657	0.876
Ar-CO		0.168	0.190	0.290	0.439	0.615	0.815
Ar-CO_2		0.129	0.148	0.235	0.365	0.517	0.689
Ar-H_2		0.698	0.794	1.228	1.876	2.634	3.496
Ar-He	0.381	0.645	0.726	1.088	1.617	2.226	2.911
Ar-Kr	0.064	0.117	0.134	0.210	0.323	0.456	0.605
Ar-N_2		0.168	0.190	0.290	0.439	0.615	0.815
Ar-Ne	0.160	0.277	0.313	0.475	0.710	0.979	1.283
Ar-O_2		0.166	0.187	0.285	0.430	0.600	0.793
Ar-SF_6				0.128	0.202	0.290	0.389
Ar-Xe	0.052	0.095	0.108	0.171	0.264	0.374	0.498
CH_4-H_2			0.708	1.084	1.648	2.311	3.070
CH_4-He			0.650	0.992	1.502	2.101	2.784
CH_4-N_2			0.208	0.317	0.480	0.671	0.890
CH_4-O_2			0.220	0.341	0.523	0.736	0.978
CH_4-SF_6				0.167	0.257	0.363	0.482
CO_2-C_3H_8			0.084	0.133	0.209		
CO_2-H_2	0.315	0.552	0.627	0.964	1.470	2.066	2.745
CO_2-H_2O			0.162	0.292	0.496	0.741	1.021
CO_2-He	0.300	0.513	0.580	0.878	1.321		
CO_2-N_2			0.160	0.253	0.392	0.553	0.733
CO_2-N_2O	0.055	0.099	0.113	0.177	0.276		
CO_2-Ne	0.131	0.227	0.258	0.395	0.603	0.847	
CO_2-O_2			0.159	0.248	0.380	0.535	0.710
CO_2-SF_6				0.099	0.155		

Mixture	D_{12}(-73 °C)/ cm² s⁻¹	D_{12}(0 °C)/ cm² s⁻¹	D_{12}(20 °C)/ cm² s⁻¹	D_{12}(100 °C)/ cm² s⁻¹	D_{12}(200 °C)/ cm² s⁻¹	D_{12}(300 °C)/ cm² s⁻¹	D_{12}(400 °C)/ cm² s⁻¹
$CO-CO_2$			0.162	0.250	0.384		
$CO-H_2$	0.408	0.686	0.772	1.162	1.743	2.423	3.196
CO-He	0.365	0.619	0.698	1.052	1.577	2.188	2.882
CO-Kr		0.131	0.149	0.227	0.346	0.485	0.645
$CO-N_2$	0.133	0.208	0.231	0.336	0.491	0.673	0.878
$CO-O_2$			0.202	0.307	0.462	0.643	0.849
$CO-SF_6$				0.144	0.226	0.323	0.432
D_2-H_2	0.631	1.079	1.219	1.846	2.778	3.866	5.103
H_2-He	0.775	1.320	1.490	2.255	3.394	4.726	6.243
H_2-Kr	0.340	0.601	0.682	1.053	1.607	2.258	2.999
H_2-N_2	0.408	0.686	0.772	1.162	1.743	2.423	3.196
H_2-Ne	0.572	0.982	1.109	1.684	2.541	3.541	4.677
H_2-O_2		0.692	0.782	1.188	1.792	2.497	3.299
H_2O-N_2			0.242	0.399			
H_2O-O_2			0.244	0.403	0.645	0.882	1.147
H_2-SF_6			0.412	0.649	0.998	1.400	1.851
H_2-Xe		0.513	0.581	0.890	1.349	1.885	2.493
He-Kr	0.330	0.559	0.629	0.942	1.404	1.942	2.550
$He-N_2$	0.365	0.619	0.698	1.052	1.577	2.188	2.882
He-Ne	0.563	0.948	1.066	1.592	2.362	3.254	4.262
$He-O_2$		0.641	0.723	1.092	1.640	2.276	2.996
$He-SF_6$			0.400	0.592	0.871	1.190	1.545
He-Xe	0.282	0.478	0.538	0.807	1.201	1.655	2.168
$Kr-N_2$		0.131	0.149	0.227	0.346	0.485	0.645
Kr-Ne	0.131	0.228	0.258	0.392	0.587	0.812	1.063
Kr-Xe	0.035	0.064	0.073	0.116	0.181	0.257	0.344
N_2-Ne			0.317	0.483	0.731	1.021	1.351
N_2-O_2			0.202	0.307	0.462	0.643	0.849
N_2-SF_6				0.148	0.231	0.328	0.436
N_2-Xe		0.107	0.122	0.188	0.287	0.404	0.539
Ne-Xe	0.111	0.193	0.219	0.332	0.498	0.688	0.901
O_2-SF_6			0.097	0.154	0.238	0.334	0.441

DIFFUSION OF GASES IN WATER

This table gives values of the diffusion coefficient, D, for diffusion of several common gases in water at various temperatures. For simple one-dimensional transport, the diffusion coefficient describes the time–rate of change of concentration, dc/dt, through the equation

$$dc/dt = Dd^2c/dx^2$$

where x is, for example, the perpendicular distance from a gas–liquid interface. The values below have been selected from the references indicated; in some cases, data have been refitted to permit interpolation in temperature. Column definitions for the table are as follows.

Column heading	Definition
Name	Name of gas; gases are listed in alphabetical order of name
Mol. form.	Molecular formula of gas
***D*(10 °C), *D*(15 °C), etc.**	Diffusion coefficient, in units 10^{-5} cm^2 s^{-1}, at temperature (in °C) indicated within parentheses
Ref.	Reference

Gas–liquid diffusion coefficients are difficult to measure, and large differences are found between values obtained by different authors and through different experimental methods. See Refs. 1 and 2 for a discussion of measurement techniques.

References

1. Jähne, B., Heinz, G., and Dietrich, W., *J. Geophys. Res.* 92, 10767, 1987. <https://doi.org/10.1029/JC092iC10p10767]
2. Himmelblau, D. M., *Chem. Rev.* 64, 527, 1964. <https://doi.org/10.1021/cr60231a002]
3. Boerboom, A. J. H., and Kleyn, G., *J. Chem. Phys.* 50, 1086, 1969. <https://doi.org/10.1063/1.1671161]
4. O'Brien, R. N., and Hyslop, W. F., *Can. J. Chem.* 55, 1415, 1977. <https://doi.org/10.1139/v77-196]
5. Maharajh, D. M., and Walkley, J., *Can. J. Chem.* 51, 944, 1973. <https://doi.org/10.1139/v73-140]
6. *Landolt-Börnstein, Numerical Data and Functional Relationships in Science and Technology, Sixth Edition*, II/5a, *Transport Phenomena I (Viscosity and Diffusion)*, Springer-Verlag, Heidelberg, 1969.

Diffusion Coefficients for Gases in Water at the Indicated Temperature

Name	Mol. form.	*D*(10 °C)/ 10^{-5} cm^2 s^{-1}	*D*(15 °C)/ 10^{-5} cm^2 s^{-1}	*D*(20 °C)/ 10^{-5} cm^2 s^{-1}	*D*(25 °C)/ 10^{-5} cm^2 s^{-1}	*D*(30 °C)/ 10^{-5} cm^2 s^{-1}	*D*>(35 °C)/ 10^{-5} cm^2 s^{-1}	Ref.
Acetylene	HC≡CH	1.43	1.59	1.78	1.99	2.23		2
Ammonia	NH_3		1.3	1.5				2
Argon	Ar				2.5			4
Bromomethane	CH_3Br				1.35			5
Carbon dioxide	CO_2	1.26	1.45	1.67	1.91	2.17	2.47	1
Chlorine	Cl_2		1.13	1.5	1.89			2,6
Chloromethane	CH_3Cl				1.40			5
Dichlorofluoromethane	$CHCl_2F$				1.80			5
Helium	He	5.67	6.18	6.71	7.28	7.87	8.48	1
Hydrogen	H_2	3.62	4.08	4.58	5.11	5.69	6.31	1
Hydrogen bromide	HBr				3.15			6
Hydrogen chloride	HCl				3.07			6
Hydrogen sulfide	H_2S				1.36			2,6
Krypton	Kr	1.20	1.39	1.60	1.84	2.11	2.40	1
Methane	CH_4	1.24	1.43	1.62	1.84	2.08	2.35	1
Neon	Ne	2.93	3.27	3.64	4.03	4.45	4.89	1
Nitrogen	N_2				2.0			2
Nitrogen dioxide	NO_2			1.23	1.4	1.59		2,6
Nitrous oxide	N_2O		1.62	2.11	2.57			2,6
Oxygen	O_2		1.67	2.01	2.42			2,6
Radon	Rn	0.81	0.96	1.13	1.33	1.55	1.80	1
Sulfur dioxide	SO_2			1.62	1.83	2.07	2.32	2
Xenon	Xe	0.93	1.08	1.27	1.47	1.70	1.95	1

DIFFUSION COEFFICIENTS IN LIQUIDS AT INFINITE DILUTION

This table lists diffusion coefficients D_{AB} at infinite dilution for some binary liquid mixtures. Values are given at 25 °C when available; it should be noted that the diffusion coefficient generally increases by 10% to 20% for a 10 °C increase above ambient temperature. The diffusion coefficient is defined by:

$$dc/dt = D_{AB}d^2c/dx^2$$

where x is the perpendicular distance from the point at which the solute is introduced. Column definitions are as follows.

Column heading	Definition
Solvent/Solute	Name of solvent; listed in alphabetical order; name of solute; listed alphabetically by name within each solvent group
t	Temperature, in °C
D_{AB}	Diffusion coefficient, in units 10^{-5} cm^2 s^{-1}
Ref.	Reference

References

1. Safi, A., Nicolas, C., Neau, E., and Chevalier, J.-L., *J. Chem. Eng. Data* 52, 977, 2007. <https://doi.org/10.1021/je6005604>
2. Safi, A., Nicolas, C., Neau, E., and Chevalier, J.-L., *J. Chem. Eng. Data* 52, 126, 2007. <https://doi.org/10.1021/je060289l>
3. Sanni, S. A., Fell, C. J. D., and Hutchison, H. P., *J. Chem. Eng. Data* 16, 424, 1971. <https://doi.org/10.1021/je60051a009>
4. Fan, Y., Qian, R., Shi, M., and Shi, J., *J. Chem. Eng. Data* 40, 1053, 1995. <https://doi.org/10.1021/je00021a004>
5. *Landolt-Börnstein Numerical Data and Functional Relationships in Science and Technology, Sixth Edition*, Vol. II/5a, Springer-Verlag, Heidelberg, 1969.

Diffusion Coefficients in Liquids at Infinite Dilution

Solvent/Solute	t/°C	D_{AB}/ 10^{-5} cm^2 s^{-1}	Ref.
Acetone			
Acetic acid	25	3.31	5
Benzene	25	4.25	2
Benzoic acid	25	2.62	5
Formic acid	25	3.77	5
Nitrobenzene	20	2.94	5
Tetrachloromethane	25	3.29	5
Trichloromethane	25	3.64	5
Water	25	4.56	5
Benzene			
Acetic acid	25	2.09	5
Aniline	25	1.96	5
Benzoic acid	25	1.38	5
Bromobenzene	8	1.45	5
2-Butanone	30	2.09	5
Chloroethene	8	1.77	5
Cyclohexane	25	2.09	3
Ethanol	25	3.02	5
Formic acid	25	2.28	5
Heptane	25	1.79	3
Methanol	25	3.80	5
Toluene	25	1.85	3
1,2,4-Trichlorobenzene	8	1.34	5
Trichloromethane	25	2.26	5
1-Butanol			
Benzene	25	1.00	5
Biphenyl	25	0.63	5
Butanoic acid	30	0.51	5
p-Dichlorobenzene	25	0.82	5
1,6-Hexanedioic acid	30	0.40	5
Methanol	30	0.59	5
cis-9-Octadecenoic acid	30	0.25	5
Propane	25	1.57	5
Water	25	0.56	5
Cyclohexane			
Benzene	25	1.92	1
Chlorobenzene	25	1.34	1
4-Chlorotoluene	25	1.28	1
Ethylbenzene	25	1.36	1
Naphthalene	25	1.18	1
Perylene	25	0.79	1
Pyrene	25	0.95	1
Tetrachloromethane	25	1.49	3
Toluene	25	1.66	1
Decane			
Benzene	25	2.16	1
Chlorobenzene	25	1.98	1
4-Chlorotoluene	25	1.80	1
Ethylbenzene	25	1.79	1
Naphthalene	25	1.65	1
Perylene	25	1.08	1
Pyrene	25	1.23	1
Toluene	25	1.93	1
Diethyl ether			
Trichloromethane	25	4.48	3
Ethanol			
Allyl alcohol	20	0.98	5
Benzene	25	1.88	2
Iodine	25	1.32	5
Iodobenzene	20	1.00	5
3-Methyl-1-butanol	20	0.81	5
Pyridine	20	1.10	5
Tetrachloromethane	25	1.50	5
Water	25	1.24	5
Ethyl acetate			
Acetic acid	20	2.18	5
Acetone	20	3.18	5

Solvent/Solute	t/°C	D_{AB}/ 10^{-5} cm² s⁻¹	Ref.
2-Butanone	30	2.93	5
Ethyl benzoate	20	1.85	5
Nitrobenzene	20	2.25	5
Water	25	3.20	5
Heptane			
Benzene	25	3.75	1
Chlorobenzene	25	3.42	1
4-Chlorotoluene	25	3.11	1
Ethylbenzene	25	3.15	1
Naphthalene	25	2.81	1
Perylene	25	1.89	1
Pyrene	25	2.16	1
Toluene	25	3.42	1
Hexane			
Benzene	25	4.70	1
Bromobenzene	8	2.60	5
2-Butanone	30	3.74	5
Chlorobenzene	25	4.16	1
4-Chlorotoluene	25	3.74	1
Dodecane	25	2.73	5
Ethylbenzene	25	3.73	1
Iodine	25	4.45	5
Methane	25	0.09	5
Naphthalene	25	3.55	1
Perylene	25	2.30	1
Propane	25	4.87	5
Pyrene	25	2.58	1
Tetrachloromethane	25	3.70	5
Toluene	25	4.12	1
Octane			
Benzene	25	3.19	1
Chlorobenzene	25	2.89	1
4-Chlorotoluene	25	2.62	1
Ethylbenzene	25	2.58	1
Naphthalene	25	2.35	1
Perylene	25	1.58	1
Pyrene	25	1.81	1
Toluene	25	2.83	1
1,3,5-Trimethylbenzene	30	2.21	4
o-Xylene	30	2.65	4
p-Xylene	30	2.82	4
Tetrachloromethane			
Acetone	25	1.75	5
Benzene	25	1.42	5
Cyclohexane	25	1.30	5
Ethanol	25	1.90	5
Iodine	30	1.63	5
Trichloromethane	25	1.66	5
Toluene			
Acetic acid	25	2.26	5
Benzene	25	2.54	3
Benzoic acid	25	1.49	5
Cyclohexane	25	2.42	3
Formic acid	25	2.65	5

Solvent/Solute	t/°C	D_{AB}/ 10^{-5} cm² s⁻¹	Ref.
Water	25	6.19	5
Trichloromethane			
Acetone	25	2.55	5
Benzene	25	2.89	5
2-Butanone	25	2.13	5
Butyl acetate	25	1.71	5
Cyclohexane	25	1.28	3
Diethyl ether	25	2.13	3
Ethanol	15	2.20	5
Ethyl acetate	25	2.02	5
2,2,4-Tirmethylpentane			
Benzene	30	3.46	4
Ethylbenzene	30	2.63	4
Toluene	30	3.15	4
1,3,5-Trimethylbenzene	30	2.26	4
o-Xylene	30	2.62	4
p-Xylene	30	3.03	4
Water			
Acetic acid	25	1.29	5
Acetone	25	1.28	5
Acetonitrile	15	1.26	5
L-Alanine	25	0.91	5
Allyl alcohol	15	0.90	5
Aniline	20	0.92	5
DL-Arabinose	20	0.69	5
Benzene	20	1.02	5
1-Butanol	25	0.56	5
Caprolactam	25	0.87	5
Chloroethene	25	1.34	5
Cyclohexane	20	0.84	5
Diethylamine	20	0.97	5
1,2-Ethanediol	25	1.16	5
Ethanol	25	1.24	5
Ethanolamine	25	1.08	5
Ethyl acetate	20	1.00	5
Ethylbenzene	20	0.81	5
Ethyl carbamate	15	0.80	5
α-D-Glucose	25	0.67	5
Glycerol	25	1.06	5
Glycine	25	1.05	5
β-D-Lactose	15	0.38	5
α-Maltose	15	0.38	5
D-Mannitol	15	0.50	5
Methane	25	1.49	5
Methanol	15	1.28	5
3-Methyl-1-butanol	10	0.69	5
Methylcyclopentane	20	0.85	5
Phenol	20	0.89	5
1-Propanol	15	0.87	5
Propene	25	1.44	5
Pyridine	25	0.58	5
Raffinose	15	0.33	5
Sucrose	25	0.52	5
Toluene	20	0.85	5
Urea	25	1.38	5

Section 7
Biochemistry

Properties of Amino Acids 7-1
Structures of Common Amino Acids 7-3
Properties of Purine and Pyrimidine Bases 7-5
The Genetic Code 7-6
Properties of Fatty Acids and Their Methyl Esters 7-7
Composition and Properties of Common Oils and Fats 7-9
Carbohydrate Names and Symbols 7-12
Standard Transformed Gibbs Energies of Formation for Biochemical Reactants 7-14
Apparent Equilibrium Constants for Enzyme-Catalyzed Reactions 7-17
Apparent Equilibrium Thermodynamics of Protein-Ligand Binding Reactions 7-21
Thermodynamic Quantities for the Ionization Reactions of Buffers in Water 7-30
Biological Buffers 7-33
Typical pH Values of Biological Materials and Foods 7-34
Properties and Functions of Common Drugs 7-35
Properties of Controlled Substances 7-57
Chemical Constituents of Human Blood 7-59
Chemical Composition of the Human Body 7-61

PROPERTIES OF AMINO ACIDS

This table gives selected properties of some important amino acids and closely related compounds. Compounds are listed by name in alphabetical order. The three-letter symbol is included for the 20 "standard" amino acids that are the basic constituents of proteins. Structures are given in the following table.

Column heading	Definition
Symbol	Three-letter symbol for standard amino acids; table lists amino acids alphabetically by this symbol
Name	Amino acid name
Mol. form.	Molecular formula of amino acid
Mol. wt.	Molecular weight
t_m	Melting point, in °C
pK_a, pK_b, pK_c, pK_d	Negative of the logarithm of the acid dissociation constants for the COOH and NH_2 groups (and, in some cases, other groups) in the molecule (at 25 °C)
pI	pH at the isoelectric point
S	Solubility in water, in units of grams of compound per kilogram of water; a temperature of 25 °C is understood unless otherwise stated; when quantitative data are not available, a qualitative indication is given
$\overline{V}_2$**/cm³ mol⁻¹**	Partial molar volume in aqueous solution at infinite dilution at 25 °C

Data on the enthalpy of formation of many of these compounds are included in the table "Standard Thermodynamic Properties of Chemical Substances" in Section 5 of this *CRC Handbook*. Absorption spectra and optical rotation data can be found in Ref. 3. Partial molar volume is taken from Ref. 5; other thermodynamic properties, including solubility as a function of temperature, are given in Refs. 3 and 5. Most of the pK values come from Refs. 1, 6, and 7.

References

1. Dawson, R. M. C., Elliott, D. C., Elliott, W. H., and Jones, K. M., *Data for Biochemical Research, Third Edition*, Clarendon Press, Oxford, 1986.
2. O'Neil, Maryadele J., Ed., *The Merck Index, 15th Edition*, Royal Society of Chemistry, Cambridge, 2013.
3. Sober, H. A., Ed., *CRC Handbook of Biochemistry. Selected Data for Molecular Biology*, CRC Press, Boca Raton, FL, 1968.
4. Voet, D., and Voet, J. G., *Biochemistry, Second Edition*, John Wiley & Sons, New York, 1995.
5. Hinz, H. J., Ed., *Thermodynamic Data for Biochemistry and Biotechnology*, Springer-Verlag, Heidelberg, 1986. <https://doi.org/10.1007/978-3-642-71114-5>
6. Fasman, G. D., Ed. *Practical Handbook of Biochemistry and Molecular Biology*, CRC Press, Boca Raton, FL, 1989.
7. Smith, R. M., and Martell, A. E., *NIST Standard Reference Database 46: Critically Selected Stability Constants of Metal Complexes Database*, Version 3.0, National Institute of Standards and Technology, Gaithersburg, MD, 1997.
8. Ramasami, P., *J. Chem. Eng. Data*, 47, 1164, 2002. <https://doi.org/10.1021/je025503u>

Properties of Amino Acids

Symbol	Name	Mol. form.	Mol. wt.	t_m/°C	pK_a	pK_b	pK_c	pK_d	pI	S/g kg⁻¹	$\overline{V}_2$/ cm³ mol⁻¹
Standard amino acids											
Ala	L-Alanine	$C_3H_7NO_2$	89.094	297 dec	2.33	9.71			6.00	166.9	60.54
Arg	L-Arginine	$C_6H_{14}N_4O_2$	174.201	260	2.03	9.00	12.10		10.76	182.6	127.42
Asn	L-Asparagine	$C_4H_8N_2O_3$	132.118	235	2.16	8.73			5.41	25.1	78.0
Asp	L-Aspartic acid	$C_4H_7NO_4$	133.104	270	1.95	9.66	3.71		2.77	5.04	74.8
Cys	L-Cysteine	$C_3H_7NO_2S$	121.159	240 dec	1.91	10.28	8.14		5.07	vs	73.45
Gln	L-Glutamine	$C_5H_{10}N_2O_3$	146.144	182	2.18	9.00			5.65	42	
Glu	L-Glutamic acid	$C_5H_9NO_4$	147.130	160 dec	2.16	9.58	4.15		3.22	8.6	89.85
Gly	Glycine	$C_2H_5NO_2$	75.067	290 dec	2.34	9.58			5.97	239	43.26
His	L-Histidine	$C_6H_9N_3O_2$	155.154	287 dec	1.70	9.09	6.04		7.59	43.5	98.3
Ile	L-Isoleucine	$C_6H_{13}NO_2$	131.173	284 dec	2.26	9.60			6.02	34.2	105.80
Leu	L-Leucine	$C_6H_{13}NO_2$	131.173	293	2.32	9.58			5.98	23.8	107.77
Lys	L-Lysine	$C_6H_{14}N_2O_2$	146.187	224 dec	2.15	9.16	10.67		9.74	5.8	108.5
Met	L-Methionine	$C_5H_{11}NO_2S$	149.212	281 dec	2.16	9.08			5.74	56	105.57
Phe	L-Phenylalanine	$C_9H_{11}NO_2$	165.189	283 dec	2.18	9.09			5.48	27.9	121.5
Pro	L-Proline	$C_5H_9NO_2$	115.131	221 dec	1.95	10.47			6.30	1625	82.76
Ser	L-Serine	$C_3H_7NO_3$	105.093	228 dec	2.13	9.05			5.68	250	60.62
Thr	L-Threonine	$C_4H_9NO_3$	119.119	256 dec	2.20	8.96			5.60	90.6	76.90
Trp	L-Tryptophan	$C_{11}H_{12}N_2O_2$	204.225	289 dec	2.38	9.34			5.89	13.2	143.8
Tyr	L-Tyrosine	$C_9H_{11}NO_3$	181.188	343 dec	2.24	9.04	10.10		5.66	0.51	
Val	L-Valine	$C_5H_{11}NO_2$	117.147	315	2.27	9.52			5.96	88	90.75
Other important amino acids and related compounds											
	N-Acetylglutamic acid	$C_7H_{11}NO_5$	189.166	199						s	
	*N*⁶-Acetyl-L-lysine	$C_8H_{16}N_2O_3$	188.224	265 dec	2.12	9.51					
	β-Alanine	$C_3H_7NO_2$	89.094	200 dec	3.51	10.08				890	58.28
	2-Aminoadipic acid	$C_6H_{11}NO_4$	161.156	207.0	2.14	4.21	9.77		3.18	2.2^{40}	

Symbol	Name	Mol. form.	Mol. wt.	t_m/°C	pK_a	pK_b	pK_c	pK_d	pI	S/g kg^{-1}	$\overline{V}_2$/ cm^3 mol^{-1}
	DL-2-Aminobutanoic acid	$C_4H_9NO_2$	103.120	304 dec	2.30	9.63			6.06	210	75.6
	DL-3-Aminobutanoic acid	$C_4H_9NO_2$	103.120	194.3	3.43	10.05			7.30	1250	76.3
	4-Aminobutanoic acid	$C_4H_9NO_2$	103.120	203 dec	4.02	10.35				971	73.2
	10-Aminodecanoic acid	$C_{10}H_{21}NO_2$	187.280								167.3
	7-Aminoheptanoic acid	$C_7H_{15}NO_2$	145.200							[vs]	120.0
	6-Aminohexanoic acid	$C_6H_{13}NO_2$	131.173	205					7.29	850	104.2
	2-Amino-2-methylpropanoic acid	$C_4H_9NO_2$	103.120	335	2.36	10.21			5.72	137	77.55
	L-3-Amino-2-methylpropanoic acid	$C_4H_9NO_2$	103.120	185						[s]	
	9-Aminononanoic acid	$C_9H_{19}NO_2$	173.253								151.3
	8-Aminooctanoic acid	$C_8H_{17}NO_2$	159.227								136.1
	5-Amino-4-oxopentanoic acid	$C_5H_9NO_3$	131.130	118	4.05	8.90					
	5-Aminopentanoic acid	$C_5H_{11}NO_2$	117.147	157 dec						[s]	87.6
	Aniline-2-carboxylic acid	$C_7H_7NO_2$	137.137	144.6	2.05	4.95				3.5[14]	
	Azaserine	$C_5H_7N_3O_4$	173.128	150 dec		8.55				[vs]	
	Canavanine	$C_5H_{12}N_4O_3$	176.174	172	2.50	6.60	9.25		7.93	[vs]	
	L-γ-Carboxyglutamic acid	$C_6H_9NO_6$	191.138	167	1.70	9.90	4.75	3.20			
	Carnosine	$C_9H_{14}N_4O_3$	226.232	260	2.51	9.35	6.76			322	
	Citrulline	$C_6H_{13}N_3O_3$	175.185	222	2.32	9.30			5.92	[s]	
	Creatine	$C_4H_9N_3O_2$	131.133	303 dec	2.63	14.30				16	
	L-Cysteic acid	$C_3H_7NO_5S$	169.157	260 dec	1.89	8.70	1.30			[vs]	
	L-Cystine	$C_6H_{12}N_2O_4S_2$	240.300	260 dec	1.50	8.80	2.05	8.03		0.17	
	2,4-Diaminobutanoic acid	$C_4H_{10}N_2O_2$	118.134	118.1	1.85	8.24	10.44		9.27	[s]	
	3,5-Dibromo-L-tyrosine	$C_9H_9Br_2NO_3$	338.980	245						2.72	
	3,5-Dichloro-L-tyrosine	$C_9H_9Cl_2NO_3$	250.078							1.97	
	3,5-Diiodo-L-tyrosine	$C_9H_9I_2NO_3$	432.981	213	2.12	9.10	6.16			0.62	
	Dopamine	$C_8H_{11}NO_2$	153.179			10.36	8.88			[s]	
	L-Ethionine	$C_6H_{13}NO_2S$	163.238	273 dec	2.18	9.05	13.10				
	N-Glycylglycine	$C_4H_8N_2O_3$	132.118	220	3.13	8.10				231	
	Guanidinoacetic acid	$C_3H_7N_3O_2$	117.107	282	2.82					5	
	Histamine	$C_5H_9N_3$	111.145	83		9.83	6.11			[vs]	
	L-Homocysteine	$C_4H_9NO_2S$	135.185	232	2.15	8.57	10.38		5.55	[s]	
	Homocystine	$C_8H_{16}N_2O_4S_2$	268.354	264	1.59	9.44	2.54	8.52		0.20	
	L-Homoserine	$C_4H_9NO_3$	119.119	203 dec	2.27	9.28			6.17	1100	
	3-Hydroxy-DL-glutamic acid	$C_5H_9NO_5$	163.129						3.28		
	5-Hydroxylysine	$C_6H_{14}N_2O_3$	162.186		2.13	8.85	9.83		9.15		
	trans-4-Hydroxy-L-proline	$C_5H_9NO_3$	131.130	274	1.82	9.47			5.74	361	84.49
	L-3-Iodotyrosine	$C_9H_{10}INO_3$	307.084	205 dec	2.20	9.10	8.70			[sls]	
	L-Kynurenine	$C_{10}H_{12}N_2O_3$	208.213	194 dec						[sls]	
	L-Lanthionine	$C_6H_{12}N_2O_4S$	208.235	294 dec						1.5	
	Levodopa	$C_9H_{11}NO_4$	197.188	277	2.32	8.72	9.96	11.79		1.65[20]	
	L-1-Methylhistidine	$C_7H_{11}N_3O_2$	169.181	249	1.69	8.85	6.48			200	
	L-Norleucine	$C_6H_{13}NO_2$	131.173	301 dec	2.31	9.68			6.09	15	107.7
	L-Norvaline	$C_5H_{11}NO_2$	117.147	307	2.31	9.65				107	91.8
	L-Ornithine	$C_5H_{12}N_2O_2$	132.161	140	1.94	8.78	10.52		9.73	[vs]	
	O-Phosphoserine	$C_3H_8NO_6P$	185.073	166 dec	2.14	9.80	5.70				
	L-Pyroglutamic acid	$C_5H_7NO_3$	129.115	162	3.32						
	Sarcosine	$C_3H_7NO_2$	89.094	212 dec	2.18	9.97				428	
	Taurine	$C_2H_7NO_3S$	125.147	328	-0.3	9.06				105	
	L-Thyroxine	$C_{15}H_{11}I_4NO_4$	776.871	235	2.20	10.01	6.45			[sls]	

[sls] Slightly soluble.
[s] Soluble.
[vs] Very soluble.
[14] At 14 °C.
[20] At 20 °C.
[40] At 40 °C.

STRUCTURES OF COMMON AMINO ACIDS

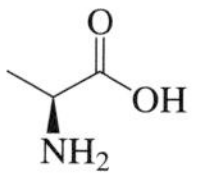

L-Alanine (Ala)

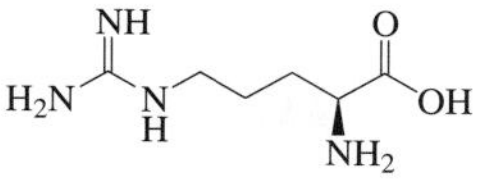

L-Arginine (Arg)

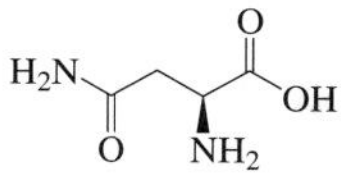

L-Asparagine (Asn)

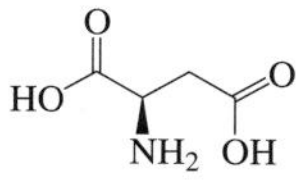

L-Aspartic acid (Asp)

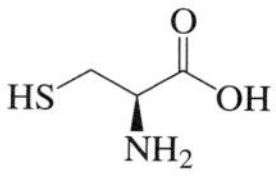

L-Cysteine (Cys)

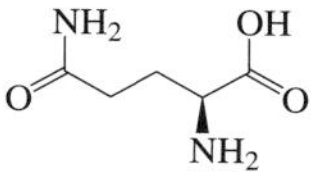

L-Glutamine (Gln)

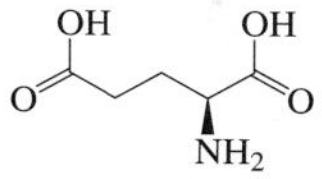

L-Glutamic acid (Glu)

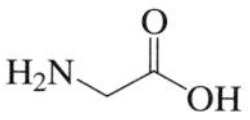

Glycine (Gly)

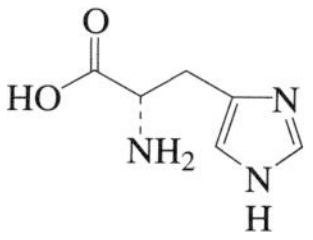

L-Histidine (His)

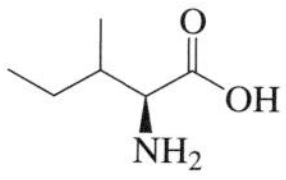

L-Isoleucine (Ile)

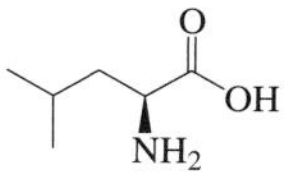

L-Leucine (Leu)

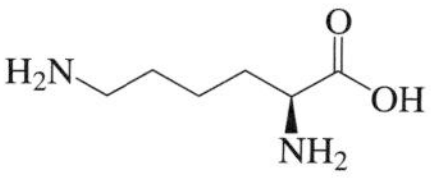

L-Lysine (Lys)

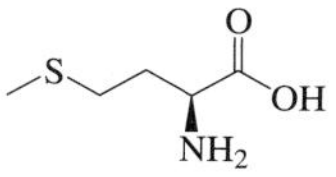

L-Methionine (Met)

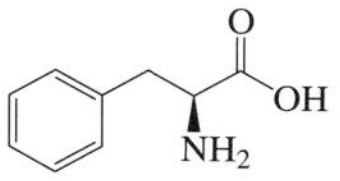

L-Phenylalanine (Phe)

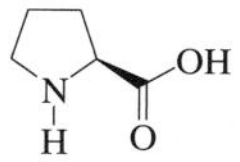

L-Proline (Pro)

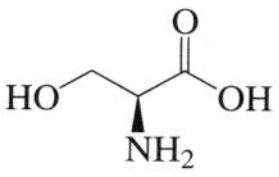

L-Serine (Ser)

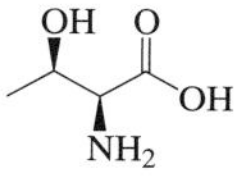

L-Threonine (Thr)

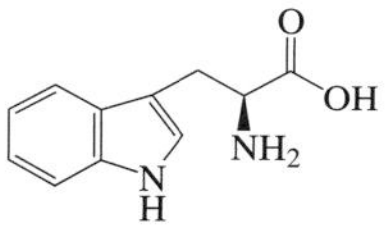

L-Tryptophan (Trp)

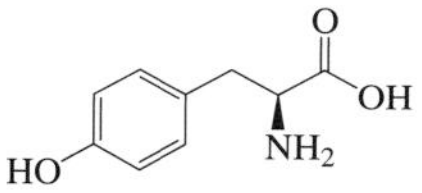

L-Tyrosine (Tyr)

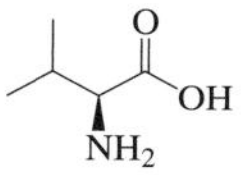

L-Valine (Val)

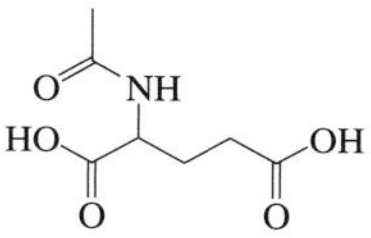

N-Acetylglutamic acid

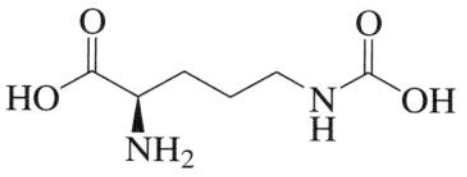

*N*6-Acetyl-L-lysine

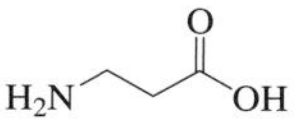

β-Alanine

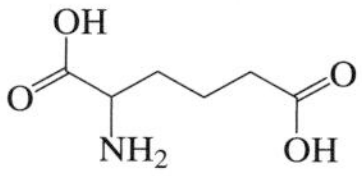

2-Aminoadipic acid

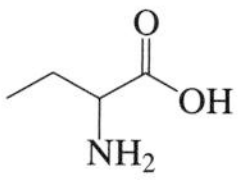

DL-2-Aminobutanoic acid

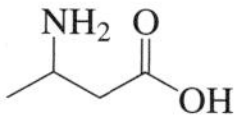

DL-3-Aminobutanoic acid

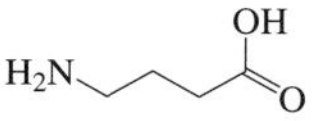

4-Aminobutanoic acid

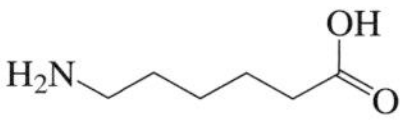

6-Aminohexanoic acid

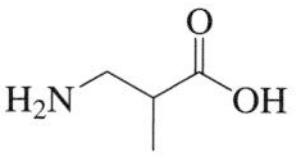

L-3-Amino-2-methylpropanoic acid

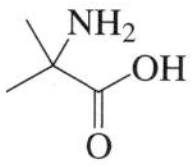

2-Amino-2-methylpropanoic acid

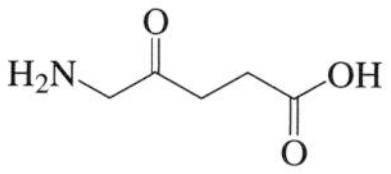

5-Amino-4-oxopentanoic acid

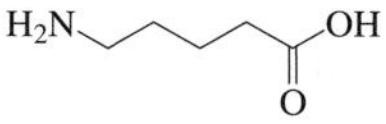

5-Aminopentanoic acid

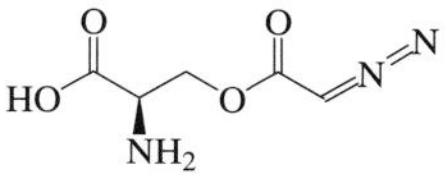

Azaserine

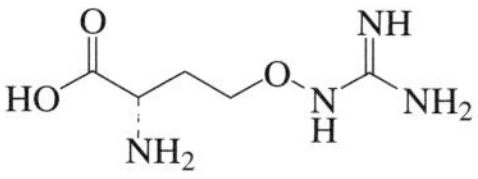

Canavanine

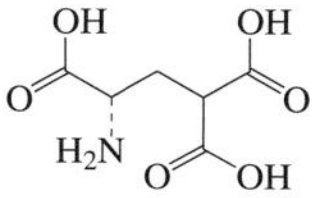

L-γ-Carboxyglutamic acid

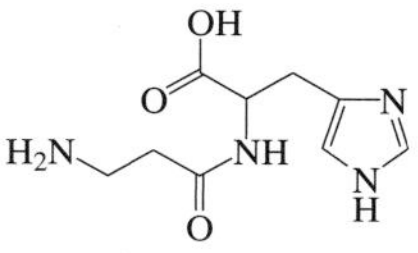

Carnosine

Biochem

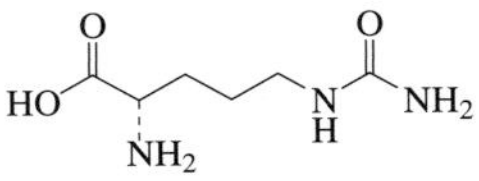

Citrulline

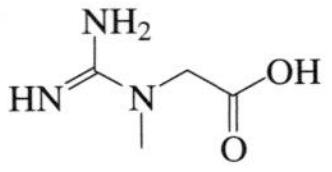

Creatine

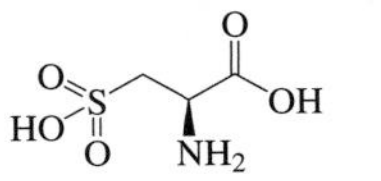

L-Cysteic acid

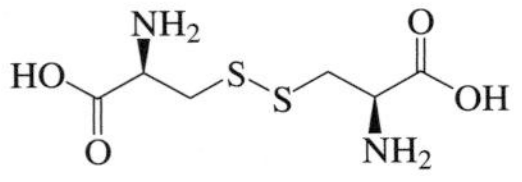

L-Cystine

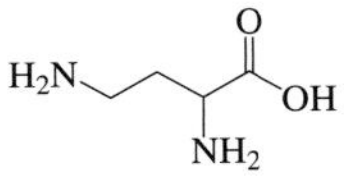

2,4-Diaminobutanoic acid

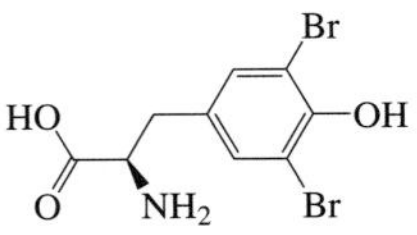

3,5-Dibromo-L-tyrosine

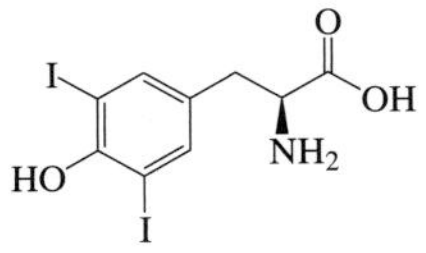

3,5-Diiodo-L-tyrosine

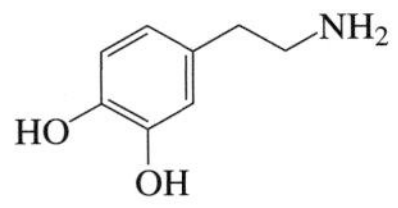

Dopamine

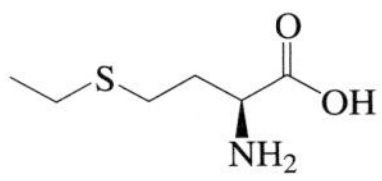

L-Ethionine

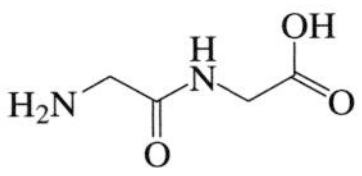

N-Glycylglycine

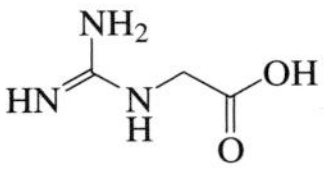

Guanidinoacetic acid

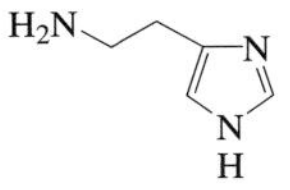

Histamine

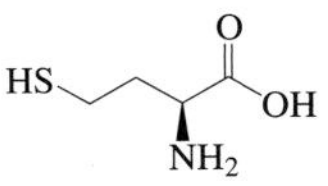

L-Homocysteine

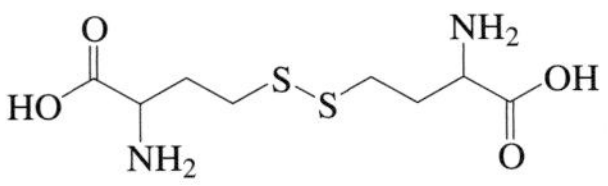

Homocystine

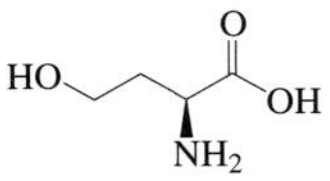

L-Homoserine

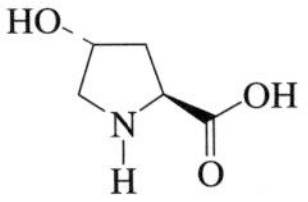

trans-4-Hydroxy-L-proline

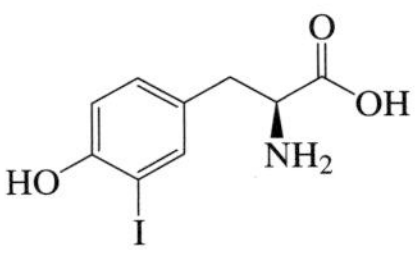

L-3-Iodotyrosine

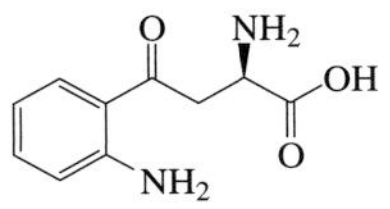

L-Kynurenine

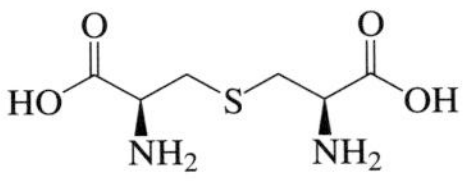

L-Lanthionine

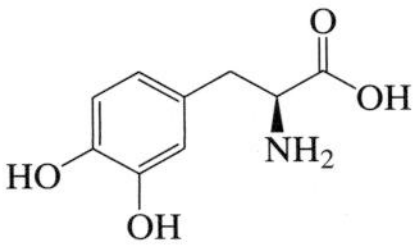

Levodopa

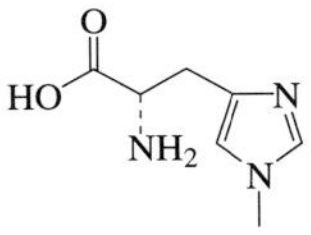

L-1-Methylhistidine

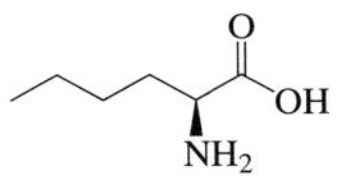

L-Norleucine

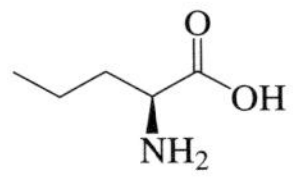

L-Norvaline

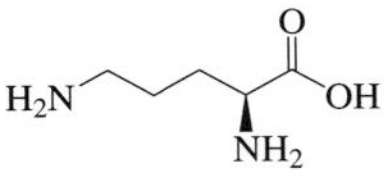

L-Ornithine

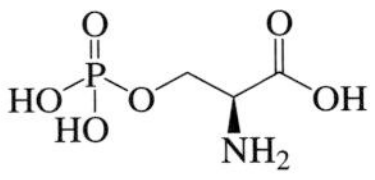

O-Phosphoserine

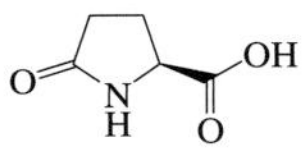

L-Pyroglutamic acid

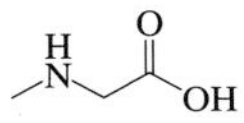

Sarcosine

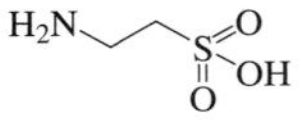

Taurine

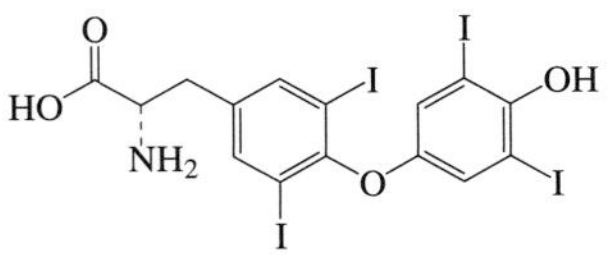

L-Thyroxine

Biochem

PROPERTIES OF PURINE AND PYRIMIDINE BASES

This table lists acid dissociation constants and solubilities for some important purine and pyrimidine bases that occur in nucleic acids. The columns are defined as follows.

Column heading	Definition
Common name	Common chemical name
Systematic name	IUPAC systematic chemical name
Mol. form.	Molecular formula
Mol. wt.	Molecular weight
pK_a values	pK_a values (negative logarithm of the acid dissociation constant) are given for each ionization stage
***S* (temp.)**	Aqueous solubility at the indicated temperature in °C (in parentheses), in units of grams per 100 grams of solution

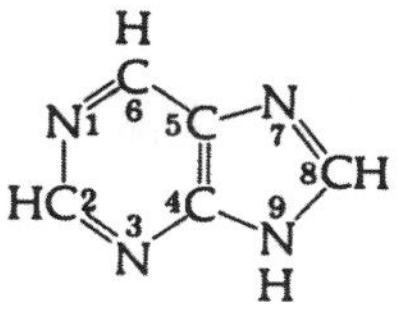

Purine

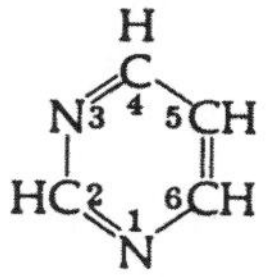

Pyrimidine

References

1. Dawson, R. M. C., et al., *Data for Biochemical Research, Third Edition*, Clarendon Press, Oxford, 1986.
2. O'Neil, M. J., Ed., *The Merck Index, 15th Edition*, Royal Society of Chemistry, 2013.

Acid Dissociation Constants and Solubilities of Purine and Pyrimidine Bases

Common name	Systematic name	Mol. form	Mol. wt.	pK_a			*S*/mass % (temp.)
Pyrimidines							
Cytosine	4-Amino-2-hydroxypyrimidine	$C_4H_5N_3O$	111.102	4.60	12.16		0.73 (25 °C)
5-Methylcytosine	4-Amino-5-methyl-3*H*-pyrimidin-2-one	$C_5H_7N_3O$	125.129	4.6	12.4		0.45 (25 °C)
5-Hydroxymethylcytosine	6-Amino-5-(hydroxymethyl)-1*H*-pyrimidin-2-one	$C_5H_7N_3O_2$	141.129	4.3	13		
Uracil	2,4-Dihydroxypyrimidine	$C_4H_4N_2O_2$	112.087	0.5	9.5	>13	0.27 (25 °C)
Thymine	5-Methyluracil	$C_5H_6N_2O_2$	126.114	9.94	>13		0.35 (25 °C)
Orotic acid	Uracil-6-carboxylic acid	$C_5H_4N_2O_4$	156.097	2.4	9.5	>13	0.18 (18 °C)
Purines							
Adenine	1*H*-Purin-6-amine	$C_5H_5N_5$	135.128	<1	4.3	9.83	0.104 (25 °C)
Guanine	2-Amino-6-hydroxypurine	$C_5H_5N_5O$	151.127	3.3	9.2	12.3	0.0068 (40 °C)
7-Methylguanine	2-Amino-6-hydroxy-7-methylpurine	$C_6H_7N_5O$	165.153	3.5	9.9		
Isoguanine	2-Hydroxy-6-aminopurine	$C_5H_5N_5O$	151.127	4.5	9.0		0.006 (25 °C)
Xanthine	2,6-Dioxopurine	$C_5H_4N_4O_2$	152.112	0.8	7.4	11.1	0.05 (20 °C)
Hypoxanthine	1*H*-Purin-6(9*H*)-one	$C_5H_4N_4O$	136.112	2.0	8.9	12.1	0.07 (19 °C)
Uric acid	2,6,8-Trihydroxypurine	$C_5H_4N_4O_3$	168.111	5.4	11.3		0.002 (20 °C)

THE GENETIC CODE

This table gives the correspondence between a messenger RNA codon and the amino acid that it specifies. The symbols for bases in the codon are:

Symbol: Amino acid
U: Uracil
C: Cytosine
A: Adenine
G: Guanine

The amino acid symbols are given in the table entitled "Structures of Common Amino Acids" in this section. A chain-initiating codon is indicated by init and a chain-terminating codon by term.

Example: UCA codes for Ser, UAC codes for Tyr, etc.

The Genetic Code

1st position	2nd position = U	2nd position = C	2nd position = A	2nd position = G	3rd position
U	Phe	Ser	Tyr	Cys	U
U	Phe	Ser	Tyr	Cys	C
U	Leu	Ser	**term**	**term**	A
U	Leu	Ser	**term**	Trp	G
C	Leu	Pro	His	Arg	U
C	Leu	Pro	His	Arg	C
C	Leu	Pro	Gln	Arg	A
C	Leu	Pro	Gln	Arg	G
A	Ile	Thr	Asn	Ser	U
A	Ile	Thr	Asn	Ser	C
A	Ile	Thr	Lys	Arg	A
A	Met (**init**)	Thr	Lys	Arg	G
GG	Val	Ala	Asp	Gly	U
GG	Val	Ala	Asp	Gly	C
GG	Val	Ala	Glu	Gly	A
GG	Val (**init**)	Ala	Glu	Gly	G

PROPERTIES OF FATTY ACIDS AND THEIR METHYL ESTERS

This table gives the names and selected properties of some important fatty acids and their methyl esters. It includes most of the acids that are significant constituents of naturally occurring oils and fats. Compounds are listed first by number of carbon atoms and, second, by the degree of unsaturation.

Column heading	Definition
Systematic name	IUPAC systematic chemical name
Common name	Common or trivial name
Mol. form.	The molecular formula written in the Hill convention
Acid code	Shorthand acid code commonly used; the first number in this code gives the number of carbon atoms; the number following the colon is the number of unsaturated centers (mainly double bonds); the location and orientation of the unsaturated centers follow. The symbols used are: c = *cis*; t = *trans*; a = acetylenic center; e = ethylenic center at end of chain; ep = epoxy; thus 9c,11t indicates a double bond with *cis* orientation at the No. 9 carbon and another with *trans* orientation at the No. 11 carbon. Additional details on the codes can be found in Ref. 1
CAS Reg. No.	Chemical Abstracts Service Registry Number for the compound
Mol. wt.	Molecular weight (relative molar mass) as calculated with the 2016 IUPAC Standard Atomic Weights
mp	Normal melting point, in °C
Me ester mp	Normal melting point, in °C of the methyl ester of the acid when available
Me ester bp	Normal boiling point in °C of the methyl ester of the acid when available; a superscript on the boiling point indicates the pressure in mmHg (torr); if there is no superscript, the value refers to one atmosphere (101.325 kPa or 760 mmHg)

The references cover many other fatty acids beyond those listed here and give additional properties.

We are indebted to Frank D. Gunstone for advice on the content of the table.

References

1. Gunstone, F. D., Harwood, J. L., and Dijkstra, A. J., eds., *The Lipid Handbook, Third Edition,* CRC Press, Boca Raton, FL, 2006.
2. Adlof, R. O., and Gunstone, F. D., *Common (non-systematic) Names for Fatty Acids,* <www.aocs.org/Documents/Membership/Divisions/Comman_Names_FattyAcids.pdf>.
3. Firestone, D., *Physical and Chemical Characteristics of Oils, Fats, and Waxes, Second Edition,* AOCS Press, Urbana, IL, 2006.
4. Dawson, R. M. C., Elliott, D. C., Elliott, W. H., and Jones, K. M., *Data for Biochemical Research, Third Edition,* Clarendon Press, Oxford, 1986.
5. Altman, P. L., and Dittmer, D. S., eds., *Biology Data Book, Second Edition,* Vol. 1, Federation of American Societies for Experimental Biology, Bethesda, MD, 1972.
6. Fasman, G. D., Ed., *Practical Handbook of Biochemistry and Molecular Biology,* CRC Press, Boca Raton, FL, 1989.

Properties of Fatty Acids and Their Methyl Esters

Systematic name	Common name	Mol. form.	Acid code	CAS Reg. No.	Mol. wt.	mp/°C	Me ester mp/°C	Me ester bp/°C
Butanoic acid	Butyric acid	$C_4H_8O_2$	4:0	107-92-6	88.106	-5.12	-85.8	102.8
Pentanoic acid	Valeric acid	$C_5H_{10}O_2$	5:0	109-52-4	102.132	-33.63		127.4
3-Methylbutanoic acid	Isovaleric acid	$C_5H_{10}O_2$	4:0 3-Me	503-74-2	102.132	-29.6		116.5
Hexanoic acid	Caproic acid	$C_6H_{12}O_2$	6:0	142-62-1	116.158	-4.1	-71	149.5
Heptanoic acid	Enanthic acid	$C_7H_{14}O_2$	7:0	111-14-8	130.185	-7.17	-56	174
Octanoic acid	Caprylic acid	$C_8H_{16}O_2$	8:0	124-07-2	144.212	16.51	-40	192.9
Nonanoic acid	Pelargonic acid	$C_9H_{18}O_2$	9:0	112-05-0	158.238	12.38		213.5
Decanoic acid	Capric acid	$C_{10}H_{20}O_2$	10:0	334-48-5	172.265	31.39	-18	224
9-Decenoic acid	Caproleic acid	$C_{10}H_{18}O_2$	10:1	14436-32-9	170.249	26.5		120^{20}
Undecanoic acid	Hendecanoic acid	$C_{11}H_{22}O_2$	11:0	112-37-8	186.292	28.47		123^{10}
Dodecanoic acid	Lauric acid	$C_{12}H_{24}O_2$	12:0	143-07-7	200.318	43.82	5.1	268
cis-9-Dodecenoic acid	Lauroleic acid	$C_{12}H_{22}O_2$	12:1 9c	2382-40-3	198.302			
Tridecanoic acid	Tridecylic acid	$C_{13}H_{26}O_2$	13:0	638-53-9	214.344	41.85	6.5	92^{1}
Tetradecanoic acid	Myristic acid	$C_{14}H_{28}O_2$	14:0	544-63-8	228.371	54.16	19.0	295; 155^{7}
cis-9-Tetradecenoic acid	Myristoleic acid	$C_{14}H_{26}O_2$	14:1 9c	13147-06-3	226.355	-4		
Pentadecanoic acid	Pentadecyclic acid	$C_{15}H_{30}O_2$	15:0	1002-84-2	242.398	52.52	18.5	153.5
Hexadecanoic acid	Palmitic acid	$C_{16}H_{32}O_2$	16:0	57-10-3	256.424	62.49	29.6	324
cis-9-Hexadecenoic acid	Palmitoleic acid	$C_{16}H_{30}O_2$	16:1 9c	373-49-9	254.408	2	33.7	325
Heptadecanoic acid	Margaric acid	$C_{17}H_{34}O_2$	17:0	506-12-7	270.451	61.08	30	185^{9}
Octadecanoic acid	Stearic acid	$C_{18}H_{36}O_2$	18:0	57-11-4	284.478	69.3	38.7	353
cis-6-Octadecenoic acid	Petroselinic acid	$C_{18}H_{34}O_2$	18:1 6c	593-39-5	282.462	30.0		
cis-9-Octadecenoic acid	Oleic acid	$C_{18}H_{34}O_2$	18:1 9c	112-80-1	282.462	14	-19.7	347
trans-9-Octadecenoic acid	Elaidic acid	$C_{18}H_{34}O_2$	18:1 9t	112-79-8	282.462	44.0	13.5	218^{24}
cis-11-Octadecenoic acid	Asclepic acid	$C_{18}H_{34}O_2$	18:1 11c	506-17-2	282.462	15		$163^{0.1}$
trans-11-Octadecenoic acid	Vaccenic acid	$C_{18}H_{34}O_2$	18:1 11t	693-72-1	282.462	43		172^{3}
cis-12,13-Epoxy-*cis*-9-octadecenoic acid	Vernolic acid	$C_{18}H_{32}O_3$	18:1 12,13-ep,9c	503-07-1	296.445	32.5		

Systematic name	Common name	Mol. form.	Acid code	CAS Reg. No.	Mol. wt.	mp/°C	Me ester mp/°C	Me ester bp/°C
cis-12-Hydroxy-9-octadecenoicacid,(*R*)-	Ricinoleic acid	$C_{18}H_{34}O_3$	18:1 12-OH,9c	141-22-0	298.461	-8.28		226^{15}
cis,trans-9,11-Octadecadienoic acid	Rumenic acid	$C_{18}H_{32}O_2$	18:2 9c,11t	1839-11-8	280.446	20		
cis-9-Octadecen-12-ynoic acid	Crepenynic acid	$C_{18}H_{30}O_2$	18:2 9c,12a	2277-31-8	278.430			
cis,cis-9,12-Octadecadienoic acid	Linoleic acid	$C_{18}H_{32}O_2$	18:2 9c,12c	60-33-3	280.446	-6.9	-36.6	347
trans,cis-10,12-Octadecadienoic acid		$C_{18}H_{32}O_2$	18:2 10t,12c	22880-03-1	280.446	23	-12	
cis,cis,cis-5,9,12-Octadecatrienoic acid	Pinolenic acid	$C_{18}H_{30}O_2$	18:3 5c,9c,12c	27213-43-0	278.430			
trans,cis,cis-5,9,12-Octadecatrienoic acid	Columbinic acid	$C_{18}H_{30}O_2$	18:3 5t,9c,12c	2441-53-4	278.430			
cis,cis,cis-6,9,12-Octadecatrienoic acid	γ-Linolenic acid	$C_{18}H_{30}O_2$	18:3 6c,9c,12c	506-26-3	278.430			$162^{0.5}$
trans,trans,cis-8,10,12-Octadecatrienoic acid	Calendic acid	$C_{18}H_{30}O_2$	18:3 8t,10t,12c	28872-28-8	278.430	40		
cis,trans,cis-9,11,13-Octadecatrienoic acid	Punicic acid	$C_{18}H_{30}O_2$	18:3 9c,11t,13c	544-72-9	278.430	62		
cis,trans,trans-9,11,13-Octadecatrienoic acid	*cis*-Eleostearic acid	$C_{18}H_{30}O_2$	18:3 9c,11t,13t	506-23-0	278.430	49		148^{1}
cis,cis,cis-9,12,15-Octadecatrienoic acid	α-Linolenic acid	$C_{18}H_{30}O_2$	18:3 9c,12c,15c	463-40-1	278.430	-10	-49	348
trans,trans,cis-9,11,13-Octadecatrienoic acid	Catalpic acid	$C_{18}H_{30}O_2$	18:3 9t,11t,13c	4337-71-7	278.430	32		
trans,trans,trans-9,11,13-Octadecatrienoic acid	*trans*-Eleostearic acid	$C_{18}H_{30}O_2$	18:3 9t,11t,13t	544-73-0	278.430	71.5	13	162^{1}
4,8,12,15-Octadecatetraenoic acid, all *cis*	Moroctic acid	$C_{18}H_{28}O_2$	18:4 4c,8c,12c,15c	67329-10-6	276.414			
6,9,12,15-Octadecatetraenoic acid (all *cis*)	Stearidonic acid	$C_{18}H_{28}O_2$	18:4 6c,9c,12c,15c	20290-75-9	276.414	-57		
cis,trans,trans,cis-9,11,13,15-Octadecatetraenoic acid	Parinaric acid	$C_{18}H_{28}O_2$	18:4 9c,11t,13t,15c	593-38-4	276.414	86		
Nonadecanoic acid		$C_{19}H_{38}O_2$	19:0	646-30-0	298.504	68.06	41.3	190^{4}
Eicosanoic acid	Arachidic acid	$C_{20}H_{40}O_2$	20:0	506-30-9	312.531	75.06	46.4	371
3,7,11,15-Tetramethylhexadecanoic acid	Phytanic acid	$C_{20}H_{40}O_2$	16:0 3,7,11,15-tetraMe	14721-66-5	312.531	-65		
cis-5-Eicosenoic acid	Icos-5-enoic acid	$C_{20}H_{38}O_2$	20:1 5c	7050-07-9	310.515	27		
cis-9-Eicosenoic acid	Gadoleic acid	$C_{20}H_{38}O_2$	20:1 9c	29204-02-2	310.515	24.5	-34	378
cis-11-Eicosenoic acid	Gondoic acid	$C_{20}H_{38}O_2$	20:1 11c	2462-94-4	310.515	24		
cis,cis,cis-8,11,14-Eicosatrienoic acid	Dihomo-γ-linolenic acid	$C_{20}H_{34}O_2$	20:3 8c,11c,14c	1783-84-2	306.483			
5,8,11,14-Eicosatetraenoic acid, (all-*cis*)	Arachidonic acid	$C_{20}H_{32}O_2$	20:4 5c,8c,11c,14c	506-32-1	304.467	-38		$195^{0.7}$
5,8,11,14,17-Eicosapentaenoic acid (all *cis*)	Timnodonic acid	$C_{20}H_{30}O_2$	20:5 5c,8c,11c,14c,17c	10417-94-4	302.451	-54		
Heneicosanoic acid		$C_{21}H_{42}O_2$	21:0	2363-71-5	326.557	82	49	207^{4}
Docosanoic acid	Behenic acid	$C_{22}H_{44}O_2$	22:0	112-85-6	340.583	80.6	53.3	402
cis-11-Docosenoic acid	Cetolic acid	$C_{22}H_{42}O_2$	22:1 11c	506-36-5	338.567	33		
cis-13-Docosenoic acid	Erucic acid	$C_{22}H_{42}O_2$	22:1 13c	112-86-7	338.567	33.0	-1.1	400
trans-13-Docosenoic acid	Brassidic acid	$C_{22}H_{42}O_2$	22:1 13t	506-33-2	338.567	61.9	35	
cis,cis-5,13-Docosadienoic acid		$C_{22}H_{40}O_2$	22:2 5c,13c	676-39-1	336.552	-4		
7,10,13,16,19-Docosapentaenoic acid, all (*cis*)		$C_{22}H_{34}O_2$	22:5 7c,10c,13c,16c,19c		330.504			
4,7,10,13,16,19-Docosahexaenoic acid (all *cis*)	Cervonic acid	$C_{22}H_{32}O_2$	22:6 4c,7c,10c,13c,16c,19c	2091-24-9	328.49	-45		
Tricosanoic acid		$C_{23}H_{46}O_2$	23:0	2433-96-7	354.610	79.6	53.4	
Tetracosanoic acid	Lignoceric acid	$C_{24}H_{48}O_2$	24:0	557-59-5	368.637	82	60	
cis-15-Tetracosenoic acid	Nervonic acid	$C_{24}H_{46}O_2$	24:1 15c	506-37-6	366.621	43	15	$165^{0.02}$
Pentacosanoic acid		$C_{25}H_{50}O_2$	25:0	506-38-7	382.664	77.5	62	
Hexacosanoic acid	Cerotic acid	$C_{26}H_{52}O_2$	26:0	506-46-7	396.690	87.7	63.8	286^{15}
Heptacosanoic acid		$C_{27}H_{54}O_2$	27:0	7138-40-1	410.717	87.6	64	
Octacosanoic acid	Montanic acid	$C_{28}H_{56}O_2$	28:0	506-48-9	424.744	90.9	67	
Nonacosanoic acid		$C_{29}H_{58}O_2$	29:0	4250-38-8	438.770	90.3	69	
Triacontanoic acid	Melissic acid	$C_{30}H_{60}O_2$	30:0	506-50-3	452.796	93.6	72	
Hentriacontanoic acid		$C_{31}H_{62}O_2$	31:0	38232-01-8	466.823	93.1		
Dotriacontanoic acid	Lacceric acid	$C_{32}H_{64}O_2$	32:0	3625-52-3	480.849	96.2		$192^{0.01}$

COMPOSITION AND PROPERTIES OF COMMON OILS AND FATS

This table lists some of the most common naturally occurring oils and fats. The oils and fats consist mainly of esters of glycerol (i.e., triglycerides) with fatty acids of 10 to 22 carbon atoms. The four fatty acids with the highest concentration are given for each oil; concentrations are given in weight percent. Because there is often a wide variation in composition depending on the source of the oil sample, a range (or sometimes an average) is generally given. More complete data on composition, including minor fatty acids, sterols, and tocopherols, can be found in the references. Column definitions for the table are as follows.

Column heading	Definition
Name	Name of oil or fat; the oils and fats are listed alphabetically by common name
Source	Type of source: plant, fish and other marine life, or land animal
Acid 1	Fatty acid with highest oil or fat. Acids are designated by the codes described in the table "Properties of Fatty Acids and Their Methyl Esters" in this section; that table gives the systematic and common names of the acids. Thus 18:2 9c,12c indicates a C18 acid with two double bonds in the 9 and 12 positions, both with a *cis* configuration (*cis,cis*-9,12-octadecadienoic acid, or linoleic acid)
Conc. 1	Concentration of Acid 1, in weight %
Acid 2	Fatty acid with second highest oil or fat concentration; see definition of Acid 1 for code explanation
Conc. 2	Concentration of Acid 2, in weight %
Acid 3	Fatty acid with third highest oil or fat concentration; see definition of Acid 1 for code explanation
Conc. 3	Concentration of Acid 3, in weight %
Acid 4	Fatty acid with fourth highest oil or fat concentration; see definition of Acid 1 for code explanation
Conc. 4	Concentration of Acid 4, in weight %
mp	The normal melting point, in °C
ρ	Typical value of the density of the oil or fat, in g cm^{-3}; superscript indicates the temperature in °C
n_D	Typical value of the refractive index; superscript indicates the temperature in °C
Sapon. value	Saponification value or number; the number of milligrams (mg) of potassium hydroxide (KOH) required to completely saponify 1 g of the oil or fat
Iodine value	Iodine value or number; the number of grams of iodine taken up by 100 gm of the oil or fat; a measure of the degree of unsaturation of the oil or fat

Notes

- The composition figure given for oleic acid (18:1 9c) often includes low levels of other 18:1 isomers.
- In some oils where a concentration is given for 18:2 9c,12c (linoleic acid), other isomers of 18:2 may be included.
- Likewise, where a concentration is given for 18:3 9c,12c,15c (α-linolenic acid), other isomers of 18:3 may be included.
- The acid 20:5 6c,9c,12c,15c,17c, which is prevalent in many fish oils, is often abbreviated as 20:5 ω-3 or 20:5 n-3.
- Jojoba oil consists primarily of wax esters of the acids listed here and long-chain alcohols.
- Kapok oil also contains up to 15% cyclopropene acids.
- Stillingia oil also contains 5-10% *trans,cis*-2,4-decadienoic acid (stillingic acid, 10:2 2t,4c).
- Capelin oil also contains about 10% 16:1.

The assistance of Frank D. Gunstone in preparing this table is gratefully acknowledged.

References

1. Firestone, D., *Physical and Chemical Characteristics of Oils, Fats, and Waxes, Second Edition*, AOCS Press, Urbana, IL, 2006.
2. Gunstone, F. D., Harwood, J. L., and Dijkstra, A. J., eds., *The Lipid Handbook, Third Edition*, CRC Press, Boca Raton, FL, 2006.
3. Dawson, R. M. C., Elliott, D. C., Elliott, W. H., and Jones, K. M., *Data for Biochemical Research, Third Edition*, Clarendon Press, Oxford, 1986.
4. Altman, P. L., and Dittmer, D. S., eds., *Biology Data Book, Second Edition*, Vol. 1, Federation of American Societies for Experimental Biology, Bethesda, MD, 1972.

Biochem

Composition and Properties of Common Oils and Fats

Name	Source	Acid 1	Conc. 1/ wt%	Acid 2	Conc. 2/ wt%	Acid 3	Conc. 3/ wt%	Acid 4	Conc. 4/ wt%	mp/ °C	ρ/g cm^{-3}	n_D	Sapon. value	Iodine value
Almond kernel oil	Plants	18:1 9c	43-70%	18:2 9c,12c	24-30%	16:0	4-13%	18:0	1-10%		0.910^{25}	1.467^{26}	188-200	89-101
Anchovy oil	Marine animals	16:0	16-20%	20:5 6c,9c, 12c,15c,17c	10-20%	18:1 9c	9-14%	16:1 undefined	8-12%				191-194	163-169
Apricot kernel oil	Plants	18:1 9c	58-66%	18:2 9c,12c	29-33%	16:0	4.6-6%	18:0	1%		0.910^{25}	1.469^{25}	185-199	97-110
Argan seed oil	Plants	18:1 9c	42-55%	18:2 9c,12c	30-34%	16:0	12-16%	18:0	2-7%		0.912^{20}	1.467^{20}	189-195	92-102
Avocado pulp oil	Plants	18:1 9c	56-74%	18:2 9c,12c	10-17%	16:0	9-18%	16:1 9c	3-9%		0.912^{25}	1.466^{25}	177-198	85-90
Babassu palm oil	Plants	12:0	40-55%	14:0	11-27%	18:1 9c	9-20%	16:0	5.2-11%	24	0.914^{25}	1.450^{40}	245-256	10-18
Beef tallow	Land animals	18:1 undefined	31-50%	18:0	25-40%	16:0	20-37%	14:0	1-6%	47	0.902^{25}	1.454^{40}	190-200	33-47

Name	Source	Acid 1	Conc. 1/ wt%	Acid 2	Conc. 2/ wt%	Acid 3	Conc. 3/ wt%	Acid 4	Conc. 4/ wt%	mp/ °C	ρ/g cm^{-3}	n_D	Sapon. value	Iodine value
Blackcurrant oil	Plants	18:2 9c,12c	45-50%	18:3 6c,9c,12c	14-20%	18:3 9c,12c,15c	12-15%	18:1 9c	9-13%		0.923^{20}	1.480^{20}	185-195	173-182
Borage (starflower) oil	Plants	18:2 9c,12c	36-40%	18:3 6c,9c,12c	17-25%	18:1 9c	14-21%	16:0	9.4-12%				189-192	141-160
Borneo tallow	Plants	18:0	39-43%	18:1 9c	34-37%	16:0	18-21%	20:0	1.0%	38	0.855^{100}	1.456^{40}	189-200	29-38
Butterfat	Land animals	16:0	28.1% (av.)	18:1 9c	20.8% (av.)	14:0	10.8% (av.)	18:0	10.6% (av.)	32	0.934^{15}	1.455^{40}	210-232	26-40
Cameline oil	Plants	18:3 9c,12c,15c	33-38%	18:2 9c,12c	15-16%	20:1 total	14-16%	18:1 9c	12-24%		0.924^{15}	1.477^{20}	180-190	127-155
Canola (rapeseed) oil (low erucic)	Plants	18:1 9c	52-67%	18:2 9c,12c	16-25%	18:3 9c,12c,15c	6-14%	16:0	3.3-6.0%	-10	0.915^{20}	1.466^{40}	182-193	110-126
Canola (rapeseed) oil (low linolenic)	Plants	18:1 9c	59-66%	18:2 9c,12c	24-29%	16:0	4-5%	18:3 9c,12c,15c	2-3%	-10				91
Capelin oil	Marine animals	18:1 9c	12-18%	22:1 undefined	9-25%	20:1 undefined	9-27%	16:0	8-12%			1.463^{50}	185-202	94-164
Caraway seed oil	Plants	18:1 9c	40%	18:2 9c,12c	30%	18:1 6c	26%	16:0	3%			1.471^{35}	178	128
Cashew nut oil	Plants	18:1 9c	57-80%	18:2 9c,12c	16-22%	16:0	4-17%	18:0	2-12%		0.914^{15}	1.463^{40}	180-196	79-89
Castor oil	Plants	18:1 12-OH,9c	88%	18:2 9c,12c	3-5%	18:1 undefined	2.9-6%	22:0	2.1%	-18	0.952^{25}	1.475^{25}	176-187	81-91
Cherry kernel oil	Plants	18:2 9c,12c	42-45%	18:1 9c	35-49%	16:0	4-9%	18:3 9c,11t,13t	3-10		0.918^{25}	1.468^{40}	190-198	110-118
Chicken egg lipids, yolk	Land animals	16:0	28%	18.1 9c	25%	18:0	17%	18:2 9c,12c	16%					
Chicken fat	Land animals	18:1 undefined	37%	16:0	22%	18:2	20%	18:0	6%		0.918^{15}	1.456^{40}		76-80
Chinese vegetable tallow	Plants	16:0	58-72%	18:1 9c	20-35%	18:0	1-8%	14:0	0.5-3.7%	44	0.887^{25}	1.456^{40}	200-218	16-29
Cocoa butter	Plants	18:0	31-37%	18:1 9c	31-35%	16:0	25-27%	18:2 9c,12c	2.8-4.0%	34	0.974^{25}	1.457^{40}	192-200	32-40
Coconut oil	Plants	12:0	45-51%	14:0	17-21%	16:0	7.7-10.2%	18:1 9c	5.4-9.9%	25	0.913^{40}	1.449^{40}	248-265	5-13
Cod liver oil	Marine animals	18:1 undefined	19-27%	18:1 9c	14-20%	16:0	10-14%	20:5 6c,9c,12c,15c,17c	8-14%		0.924^{15}	1.482^{25}	180-192	142-176
Cohune nut oil	Plants	12:0	44-48%	14:0	16-17%	18:1 9c	8-10%	16:0	7-10%		0.914^{25}	1.450^{40}	251-260	9-14
Coriander seed oil	Plants	18:1 6c	53%	18:1 9c	32%	18:2 9c,12c	7-14%	16:0	3-8%		0.908^{25}	1.464^{25}	182-191	86-100
Corn oil	Plants	18:2 9c,12c	40-66%	18:1 9c	20-42%	16:0	9-16%	18:0	0-3%	-20	0.919^{20}	1.472^{25}	187-195	107-135
Cottonseed oil	Plants	18:2 9c,12c	47-58%	16:0	18-26%	18:1 9c	14-22%	18:0	2.1-3.3%	-1	0.920^{20}	1.462^{40}	189-198	96-115
Crambe oil	Plants	22:1 13c	55-60%	18:1 9c	12-15%	18:2 9c,12c	8-10%	18:3 9c,12c,15c	6-7%		0.906^{25}	1.470^{25}		87-113
Cuphea oil	Plants	18:2 9c,12c	62-67%	16:0	17-18%	18:1 9c	12-14%	18:0	2%					
Euphorbia lagascae seed oil	Plants	18:1 12,13-ep,9c	64%	18:1 other	19%	18:2 9c,12c	9%	16:0	4%		0.952^{25}	1.473^{25}		102
Evening primrose oil	Plants	18:2 9c,12c	65-80%	18:3 6c,9c,12c	8-14%	16:0	6-10%	18:1 9c	5-12%			1.479^{20}	193-198	147-155
Grape seed oil	Plants	18:2 9c,12c	58-78%	18:1 9c	12-28%	16:0	5.5-11%	18:0	3-6%		0.923^{20}	1.475^{40}	188-194	130-138
Hazelnut oil (Chilean)	Plants	18:1 9c	39%	16:1 11c	22.7%	20:1 total	9.7%	22:1 total	9.5%					
Hazelnut oil (Filbert)	Plants	18:1 9c	72-84%	18:2 9c,12c	5.7-22%	16:0	4.1-7.2%	18:0	1.5-2.4%		0.909^{25}	1.473^{25}	188-197	83-90
Hempseed oil	Plants	18:2 9c,12c	45-60%	18:3 9c,12c,15c	15-30%	18:1 9c	11-16%	16:0	6-12%		0.921^{25}	1.472^{40}	190-195	145-166
Herring oil	Marine animals	16:0	8-25%	20:1	6-20%	18:1 undefined	5-22%	22:1	4-31%		0.914^{20}	1.474^{25}	161-192	115-160
Illipe (mowrah) butter	Plants	18:1 9c	34%	16:0	23%	18:0	23%	18:2 9c,12c	14%	27	0.862^{100}	1.460^{40}	188-207	53-70
Jojoba oil	Plants	20:1 total	66-74%	22:1 undefined	9-19%	18:1 9c	5-12%	24:1 15c	1-5%					
Kapok seed oil	Plants	18:1 9c	45-65%	16:0	10-28%	18:2 9c,12c	7-35%	18:0	2-9%	30	0.926^{15}	1.469^{25}	189-197	86-110
Kokum butter	Plants	18:0	49-56%	18:1 9c	39-49%	16:0	2-5%	18:2 9c,12c	1-2%	41		1.456^{40}	192	33-37
Kusum oil	Plants	18:1 9c	57-62%	20:0	20-25%	16:0	5-8%	18:0	2-6%			1.461^{40}	220-230	48-58
Linola oil	Plants	18:2 9c,12c	72%	18:1 9c	16%	16:0	5.6%	18:0	4.0%					142
Linseed oil	Plants	18:3 9c,12c,15c	52-58%	18:1 9c	18-20%	18:2 9c,12c	17%	18:2 undefined	16%	-24	0.924^{25}	1.480^{25}	188-196	170-203
Macadamia nut oil	Plants	18:1 9c	56-59%	16:1 9c	21-22%	16:0	8-9%	18:0	2-4%					
Mango seed oil	Plants	18:1 9c	38-50%	18:0	31-49%	18:2 9c,12c	3-6%	20:0	2-6%		0.912^{15}	1.461^{25}	188-195	39-48
Meadowfoam seed oil	Plants	20:1 5c	58-77%	22:1 total	8-24%	22:2 5c,13c	7-15%	18:1 9c	1-3%			1.464^{40}	168	86-91
Melon oil	Plants	18:2 9c,12c	67% (av.)	18:1 9c	12% (av.)	16:0	11% (av.)	18:0	9% (av.)					
Menhaden oil	Marine animals	16:0	14-23%	20:5 6c,9c,12c, 15c,17c	12-18%	16:1	7-15%	18:1 undefined	6-16%		0.920^{15}		192-199	150-200
Milk fats, cow	Land animals	16:0	28.2% (av.)	18:1 9c	21.4% (av.)	18:0	12.6% (av.)	14:0	10.6% (av.)					

Name	Source	Acid 1	Conc. 1/ wt%	Acid 2	Conc. 2/ wt%	Acid 3	Conc. 3/ wt%	Acid 4	Conc. 4/ wt%	mp/ °C	ρ/g cm^{-3}	n_D	Sapon. value	Iodine value
Milk fats, human	Land animals	18:1 9c	31.1% (av.)	16:0	21.6% (av.)	18:2 9c,12c	11.7% (av.)	14:0	6.6% (av.)					
Moringa peregrina seed oil	Plants	18:1 9c	70%	16:0	9%	18:0	3.8%	22:0	2.4%		0.903^{24}	1.460^{40}	185	70
Mustard seed oil	Plants	22:1 13c	43%	22:1 undefined	22-50%	18:3 9c,12c,15c	12%	18:2 9c,12c	10-24%		0.913^{20}	1.465^{40}	170-184	92-125
Mutton tallow	Land animals	18:1 undefined	30-42%	18:0	22-34%	16:0	20-27%	14:0	2-4%	48	0.946^{15}	1.455^{40}		35-46
Neem oil	Plants	18:1 9c	49-62%	18:0	14-24%	16:0	13-18%	18:2 9c,12c	7-15%	-3	0.912^{30}	1.462^{40}	195-205	68-71
Niger seed oil	Plants	18:2 9c,12c	52-78%	16:0	5-12%	18:1 9c	4-10%	18:0	2-12%		0.924^{15}	1.468^{40}	188-193	126-135
Nutmeg butter	Plants	14:0	76-83%	18:1 9c	5-11%	16:0	4-10%	12:0	3-6%	45		1.468^{40}	170-190	48-85
Oat bean oil	Plants	18:2 9c,12c	36%	18:1 9c	31%	16:0	3.8%	20:0	2.4%		0.904^{25}	1.4723^{30}	181-187	86-96
Oat oil	Plants	18:2 9c,12c	24-48%	18:1 9c	18-53%	16:0	13-39%	18:0	0.5-4%		0.917^{25}	1.467^{40}	190-199	105-116
Oiticica oil	Plants	18:3 9c,11t,13t	70-80%	16:0	7%	18:0	5%	18:1 9c	4-7%		0.972^{20}	1.514^{25}	188-193	140-150
Olive oil	Plants	18:1 9c	55-83%	18:2 9c,12c	9%	16:0	7.5-20%	18:2 9c,12c	3.5-21%	-6	0.911^{20}	1.469^{20}	184-196	75-94
Palm kernel oil	Plants	12:0	40-55%	14:0	14-18%	18:1 9c	12-21%	16:0	6.5-10%	24	0.922^{15}	1.450^{40}	230-250	14-21
Palm oil	Plants	16:0	40-48%	18:1 9c	36-44%	18:2 9c,12c	6.5-12%	18:0	3.5-6.5%	35	0.914^{15}	1.455^{40}	190-209	49-55
Palm olein	Plants	18:1 9c	40-44%	16:0	38-43%	18:2 9c,12c	10-13%	18:0	3.7-4.8%		0.914^{40}	1.459^{40}	194-202	>56
Palm stearin	Plants	16:0	48-74%	18:1 9c	16-36%	18:0	3.9-5.6%	18:2 9c,12c	3.2-9.8%		0.884^{60}	1.449^{40}	193-205	<48
Parsley seed oil	Plants	18:1 6c	69-76%	18:1 9c	12-15%	18:2 9c,12c	6-14%	16:0	2%			1.4800^{40}		110-120
Peanut oil	Plants	18:1 9c	36-67%	18:2 9c,12c	14-43%	16:0	8.3-14%	22:0	2.1-4.4%	3	0.914^{20}	1.463^{40}	187-196	86-107
Perilla oil	Plants	18:3 9c,12c,15c	59%	18:2 9c,12c	14-18%	18:1 9c	11-13%	16:0	6-9%		0.924^{25}	1.477^{25}	188-197	192-208
Phulwara butter	Plants	16:0	57-61%	18:1 9c	30-36%	18:2 9c,12c	3-4%	18:0	3-4%	43	0.862^{100}	1.458^{40}	188-200	40-51
Pine nut oil	Plants	18:2 9c,12c	47-51%	18:1 9c	36-39%	16:0	6-8%	18:0	2-3%		0.919^{15}		193-197	118-121
Poppy seed oil	Plants	18:2 9c,12c	62-73%	18:1 9c	16-30%	16:0	7-11%	18:0	1-4%	-15	0.916^{25}	1.469^{40}	188-196	132-146
Pork lard	Land animals	18:1 undefined	35-62%	16:0	20-32%	18:0	5-24%	18:2	3-16%	30	0.898^{20}			
Rice bran oil	Plants	18:1 9c	38-48%	18:2 9c,12c	16-36%	16:0	16-28%	18:0	2-4%		0.916^{25}	1.472^{25}	181-189	92-108
Safflower seed oil	Plants	18:2 9c,12c	68-83%	18:1 9c	8.4-30%	16:0	5.3-8.0%	18:0	1.9-2.9%		0.924^{15}	1.474^{25}	186-198	136-148
Safflower seed oil (high oleic)	Plants	18:1 9c	74-80%	18:2 9c,12c	13-18%	16:0	5-6%	18:0	1.5-2.0%		0.921^{20}	1.470^{25}		91-95
Sal fat	Plants	18:0	33-57%	18:1 9c	31-52%	16:0	6-23%	20:0	1-8%	33		1.456^{40}	175-192	31-45
Salmon oil	Marine animals	22:6 4c,7c,10c, 13c,16c,19	18%	20:5 6c,9c, 12c,15c,17c	13%	16:0	9.8%	16:1	4.8%		0.924^{15}	1.475^{25}	183-186	130-160
Sardine oil	Marine animals	20:5 6c,9c,12c, 15c,17c	9-35%	16:0	9-22%	18:1 undefined	7-17%	16:1	6-13%		0.915^{25}	1.464^{65}	188-199	159-192
Seal blubber oil, harp	Marine animals	18:1 9c	21%	20:1	12%	22:6 4c,7c,10c, 13c,16c,19c	7.6%	20:5 6c,9c, 12c,15c,17c	6.4%					
Sesame seed oil	Plants	18:2 9c,12c	40-51%	18:1 9c	33-44%	16:0	7.9-10.2%	18:0	4.4-6.7%	-6	0.917^{20}	1.467^{40}	187-195	104-120
Shark liver oil	Marine animals	18:1 undefined	45%	16:0	21%	20:1	12%	22:1	9%		0.917^{25}	1.476^{25}	170-190	150-300
Sheanut butter	Plants	18:1 9c	45-50%	18:0	36-41%	16:0	4-8%	18:2 9c,12c	4-8%	38	0.863^{100}	1.465^{40}	178-198	52-66
Soybean oil	Plants	18:2 9c,12c	50-57%	18:1 9c	18-28%	16:0	9-13%	18:3 9c,12c,15c	5.5-9.5%	-16	0.920^{20}	1.468^{40}	189-195	118-139
Stillingia seed kernel oil	Plants	18:3 9c,12c,15c	41-54%	18:2 total	24-30%	18:1 9c	7-10%	16:0	6-9%		0.937^{25}	1.483^{25}	202-212	169-191
Sunflower seed oil	Plants	18:2 9c,12c	48-74%	18:1 9c	13-40%	16:0	5-8%	18:0	2.5-7.0%	-17	0.919^{20}	1.474^{25}	188-194	118-145
Tall oil	Plants	18:2 9c,12c	41-52%	18:1 9c	41-48%	16:0	5-6%	18:0	2-3%		0.969^{25}	1.494^{25}	154-180	140-180
Trout lipids	Marine animals	16:0	21-24%	18:1 undefined	18-31%	18:2	7-16%	16:1	4-10%					
Tung oil	Plants	18:3 9c,11t,13t	71-82%	18:2 9c,12c	8-15%	18:1 9c	4-10%	18:0	3%	-2	0.912^{25}	1.517^{25}	189-195	160-175
Ucuhuba butter oil	Plants	14:0	64-73%	12:0	13-15%	18:1 9c	6-8%	16:0	3-9%		0.870^{100}	1.451^{50}	221-229	11-17
Vernonia seed oil	Plants	18:1 12,13-ep,9c	62-72%	18:2 9c,12c	9-17%	16:0	3-7%	18:0	2-6%		0.901^{30}	1.486^{32}	176	55
Walnut oil	Plants	18:2 9c,12c	56-60%	18:1 9c	17-19%	18:3 9c,12c,15c	13-14%	16:0	6-8%		0.921^{25}	1.474^{25}	189-197	138-162
Whale oil, minke	Marine animals	18:1 undefined	18%	20:1	17%	22:1	11%	16:1	9%					
Wheatgerm oil	Plants	18:2 9c,12c	50-59%	18:1 9c	13-23%	16:0	12-20%	18:3 9c,12c,15c	2-9%		0.926^{25}	1.479^{25}	179-217	100-128

CARBOHYDRATE NAMES AND SYMBOLS

The following table lists the systematic names and symbols for selected carbohydrates and some of their derivatives. The symbols for monosaccharide residues and derivatives are recommended by IUPAC for use in describing the structures of oligosaccharide chains. A more complete list can be found in the reference.

Reference

McNaught, A. D., *Pure Appl. Chem.*, 68, 1919–2008, 1996 <https://doi.org/10.1351/pac199668101919>.

Carbohydrate Names and Symbols

Common name	Symbol	Systematic name
Abequose	Abe	3,6-Dideoxy-D-*xylo*-hexose
N-Acetyl-2-deoxyneur-2-enaminic acid	Neu2en5Ac	
N-Acetylgalactosamine	GalNAc	
N-Acetylglucosamine	GlcNAc	
N-Acetylneuraminic acid	Neu5Ac	
Allose	All	*allo*-Hexose
Altrose	Alt	*altro*-Hexose
Apiose	Api	3-*C*-(Hydroxymethyl)-*glycero*-tetrose
Arabinitol	Ara-ol	Arabinitol
Arabinose	Ara	*arabino*-Pentose
Arcanose		2,6-Dideoxy-3-*C*-methyl-3-*O*-methyl-*xylo*-hexose
Ascarylose		3,6-Dideoxy-L-*arabino*-hexose
Boivinose		2,6-Dideoxy-D-gulose
Chalcose		4,6-Dideoxy-3-*O*-methyl-D-*xylo*-hexose
Cladinose		2,6-Dideoxy-3-*C*-methyl-3-*O*-methyl-L-*ribo*-hexose
Colitose		3,6-Dideoxy-L-*xylo*-hexose
Cymarose		6-Deoxy-3-*O*-methyl-*ribo*-hexose
3-Deoxy-D-*manno*-oct-2-ulosonic acid	Kdo	
2-Deoxyribose	dRib	2-Deoxy-*erythro*-pentose
2,3-Diamino-2,3-dideoxy-D-glucose	GlcN3N	
Diginose		2,6-Dideoxy-3-*O*-methyl-*lyxo*-hexose
Digitalose		6-Deoxy-3-*O*-methyl-D-galactose
Digitoxose		2,6-Dideoxy-D-*ribo*-hexose
3,4-Di-*O*-methylrhamnose	$Rha3,4Me_2$	
Ethyl glucopyranuronate	Glc*p*A6Et	
Evalose		6-Deoxy-3-*C*-methyl-D-mannose
Fructose	Fru	*arabino*-Hex-2-ulose
Fucitol	Fuc-ol	6-Deoxy-D-galactitol
Fucose	Fuc	6-Deoxygalactose
β-D-Galactopyranose 4-sulfate	β-D-Gal*p*4*S*	
Galactosamine	GalN	2-Amino-2-deoxygalactose
Galactose	Gal	*galacto*-Hexose
Glucitol	Glc-ol	
Glucosamine	GlcN	2-Amino-2-deoxyglucose
Glucose	Glc	*gluco*-Hexose
Glucuronic acid	GlcA	
N-Glycoloylneuraminic acid	Neu5Gc	
Gulose	Gul	*gulo*-Hexose
Hamamelose		2-*C*-(Hydroxymethyl)-D-ribose
Idose	Ido	*ido*-Hexose
Iduronic acid	IdoA	
Lactose	Lac	β-D-Galactopyranosyl-(1→4)-D-glucose
Lyxose	Lyx	*lyxo*-Pentose
Maltose		α-D-Glucopyranosyl-(1→4)-D-glucose
Mannose	Man	*manno*-Hexose
2-*C*-Methylxylose	Xyl2*C*Me	
Muramic acid	Mur	2-Amino-3-*O*-[(R)-1-carboxyethyl]-2-deoxy-D-glucose
Mycarose		2,6-Dideoxy-3-*C*-methyl-L-*ribo*-hexose
Mycinose		6-Deoxy-2,3-di-*O*-methyl-D-allose
Neuraminic acid	Neu	5-Amino-3,5-dideoxy-D-*glycero*-D-*galacto*-non-2-ulosonic acid

Biochem

Common name	Symbol	Systematic name
Panose		α-D-Glucopyranosyl-(1→6)-α-D-glucopyranosyl-(1→4)-D-glucose
Paratose		3,6-Dideoxy-D-*ribo*-hexose
Primeverose		β-D-Xylopyranosyl-(1→6)-D-glucose
Psicose	Psi	*ribo*-Hex-2-ulose
Quinovose	Qui	6-Deoxyglucose
Raffinose		β-D-Fructofuranosyl-α-D-galactopyranosyl-(1→6)-α-D-glucopyranoside
Rhamnose	Rha	6-Deoxymannose
Rhodinose		2,3,6-Trideoxy-L-*threo*-hexose
Ribose	Rib	*ribo*-Pentose
Ribose 5-phosphate	Rib5*P*	
Ribulose	Ribulo (Rul)	*erythro*-Pent-2-ulose
Rutinose		α-L-Rhamnopyranosyl-(1→6)-D-glucose
Sarmentose		2,6-Dideoxy-3-*O*-methyl-D-*xylo*-hexose
Sedoheptulose		D-*altro*-Hept-2-ulose
Sorbose	Sor	*xylo*-Hex-2-ulose
Streptose		5-Deoxy-3-*C*-formyl-L-lyxose
Sucrose		β-D-Fructofuranosyl-α-D-glucopyranoside
Tagatose	Tag	*lyxo*-Hex-2-ulose
Talose	Tal	*talo*-Hexose
Turanose		α-D-Glucopyranosyl-(1→3)-D-fructose
Tyvelose	Tyv	3,6-Dideoxy-D-*arabino*-hexose
Xylose	Xyl	*xylo*-Pentose
Xylulose	Xylulo (Xul)	*threo*-Pent-2-ulose

STANDARD TRANSFORMED GIBBS ENERGIES OF FORMATION FOR BIOCHEMICAL REACTANTS

Robert N. Goldberg and Robert A. Alberty

This table contains values of the standard transformed Gibbs energies of formation $\Delta_f G'^\circ$ for 130 biochemical reactants. Values of $\Delta_f G'^\circ$ are given at pH 7.0, the temperature 298.15 K, and the pressure 100 kPa for three ionic strengths: I = 0, I = 0.1 mol/L, and I = 0.25 mol/L. The table can be used for calculating apparent equilibrium constants K' and standard apparent reduction potentials E'° for biochemical reactions. Such a listing is more compact than tabulating the actual apparent equilibrium constants or standard apparent reduction potentials, which would require a very large number of reactant–product combinations. In the table, all reactants are in aqueous solution unless indicated otherwise.

A biochemical reactant is a sum of species. For example, ATP consists of an equilibrium mixture of the aqueous species ATP^{4-}, $HATP^{3-}$, H_2ATP^{2-}, $MgATP^{2-}$, etc. Similarly, phosphate refers to the equilibrium mixture of the aqueous species PO_4^{3-}, HPO_4^{2-}, $H_2PO_4^-$, H_3PO_4, $MgHPO_4$, etc. Biochemical reactions are written using biochemical reactants in terms of an apparent equilibrium constant K', which is distinct from the standard equilibrium constant K. This subject is discussed in an IUPAC report (see Ref. 1).

The apparent equilibrium constant K' and the standard transformed Gibbs energy change $\Delta_r G'^\circ$ for a biochemical reaction can be calculated from the $\Delta_f G'^\circ$ values by using the relationship

$$-RT \ln K' = \Delta_r G'^\circ = \Sigma \nu'_i \, \Delta_f G'^\circ$$

where the summation is over all of the biochemical reactants. The quantity ν'_i is the stoichiometric number of reactant i (ν'_i is positive for reactants on the right side of the equation and negative for reactants on the left side); R is the gas constant. As an example, the hydrolysis reaction of ATP is

$$\text{ATP} + H_2O(l) = \text{ADP} + \text{phosphate}$$

At pH 7.00 and I = 0.25 M, $\Delta_r G'^\circ$ and K' are calculated as follows:

$$\Delta_r G'^\circ = \{-1424.70 - 1059.49 - (-2292.50 - 155.66)\}\cdot(\text{kJ mol}^{-1}) = -36.03 \text{ kJ mol}^{-1}$$

$$K' = \exp[-(-36030 \text{ J mol}^{-1})/\{(8.3145 \text{ J mol}^{-1} \text{ K}^{-1})\cdot(298.15 \text{ K})\} = 2.05\cdot10^{6}$$

An example involving a biochemical half-cell reaction is

$$\text{acetaldehyde(aq)} + 2\, e^- = \text{ethanol(aq)}$$

At 298.15 K, pH 7.00, and I = 0, the standard apparent reduction potential E'° can be calculated as follows

$$E'^\circ = -(1/nF)\cdot\{\Delta_f G'^\circ(\text{ethanol}) - \Delta_f G'^\circ(\text{acetaldehyde})\}$$

where n is the number of electrons in the half-cell reaction and F is the Faraday constant. Then,

$$E'^\circ = [-1/(2\cdot9.6485\cdot10^4 \text{ C mol}^{-1})]\cdot(58.10\cdot10^3 \text{ J mol}^{-1} - 20.83\cdot10^3 \text{ J mol}^{-1}) = -0.193 \text{ V}$$

The columns are defined as follows.

Column heading	Definition
Name	Name of biochemical reactant; see discussion above; listed in alphabetical order
Note	Common abbreviation or note about physical or oxidation state
$\Delta_f G'^\circ(I=0)$	Standard transformed Gibbs energy of formation, in units kJ mol^{-1}, at pH 7.0, temperature 298.15 K, and pressure 100 kPa at ionic strength I = 0
$\Delta_f G'^\circ(I=0.1)$	Standard transformed Gibbs energy of formation, in units kJ mol^{-1}, at pH 7.0, temperature 298.15 K, and pressure 100 kPa at ionic strength I = 0.1
$\Delta_f G'^\circ(I=0.25)$	Standard transformed Gibbs energy of formation, in units kJ mol^{-1}, at pH 7.0, temperature 298.15 K, and pressure 100 kPa at ionic strength I = 0.25

References

1. Alberty, R. A., Cornish-Bowden, A., Gibson, Q. H., Goldberg, R. N., Hammes, G., Jencks, W., Tipton, K. F, Veech, R., Westerhoff, H. V., and Webb, E. C. *Pure Appl. Chem.* 66, 1641-1666, 1994.
2. Alberty, R. A., *Arch. Biochem. Biophys.*, 353, 116-130, 1998; 358, 25-39, 1998. <https://doi.org/10.1006/abbi.1998.0638>
3. Alberty, R. A., *Thermodynamics of Biochemical Reactions*, Wiley-Interscience, New York, 2003. <https://doi.org/10.1002/0471332607>
4. Alberty, R. A., *BasicBiochemData2: Data and Programs for Biochemical Thermodynamics*, <http://library.wolfram.com/infocenter/MathSource/797>.

Standard Transformed Gibbs Energies of Formation for Biochemical Reactants

Name	Note	$\Delta_f G'^\circ(I=0)$/ kJ mol^{-1}	$\Delta_f G'^\circ(I=0.1)$/ kJ mol^{-1}	$\Delta_f G'^\circ(I=0.25)$/ kJ mol^{-1}
Acetaldehyde		20.83	23.27	24.06
Acetate ion [CH_3COO^-]		-249.46	-248.23	-247.83
Acetone		80.04	83.71	84.90
Acetyl coenzyme A		-60.49	-58.65	-58.06
Acetylphosphate		-1109.34	-1107.57	-1107.02
cis-Aconitate		-797.26	-800.93	-802.12
Adenine		510.45	513.51	514.50
Adenosine		324.93	332.89	335.46

Biochem

Name	Note	$\Delta_f G'^\circ(I=0)$/ kJ mol^{-1}	$\Delta_f G'^\circ(I=0.1)$/ kJ mol^{-1}	$\Delta_f G'^\circ(I=0.25)$/ kJ mol^{-1}
Adenosine 5'-(trihydrogen diphosphate)	ADP	-1428.93	-1425.55	-1424.70
Adenosine 5'-monophosphate	AMP	-562.04	-556.53	-554.83
Adenosine 5'-triphosphate	ATP	-2292.61	-2292.16	-2292.50
D-Alanine		-91.31	-87.02	-85.64
Ammonia		80.50	82.34	82.93
α-D-Arabinose		-342.67	-336.55	-334.57
L-Asparagine		-206.28	-201.38	-199.80
L-Aspartate		-456.14	-453.08	-452.09
1,3-Biphosphoglycerate		-2202.06	-2205.69	-2207.30
Butanoate		-72.94	-69.26	-68.08
1-Butanol		227.72	233.84	235.82
Citrate		-963.46	-965.49	-966.23
Isocitrate		-956.82	-958.84	-959.58
Coenzyme A	CoA	-7.98	-7.43	-7.26
Carbon monoxide	aqueous	-119.90	-119.90	-119.90
Carbon monoxide	gas	-137.17	-137.17	-137.17
Carbon dioxide	aqueous (total)	-547.33	-547.15	-547.10
Carbon dioxide	gas	-394.36	-394.36	-394.36
Creatine		100.41	105.92	107.69
Creatinine		256.55	260.84	262.22
L-Cysteine		-59.23	-55.01	-53.65
L-Cystine		-187.03	-179.69	-177.32
Cytochrome c [oxidized]		0.00	-5.51	-7.29
Cytochrome c [reduced]		-24.51	-26.96	-27.75
Dihydroxyacetone phosphate		-1096.60	-1095.91	-1095.70
Ethanol		58.10	61.77	62.96
Ethyl acetate		-18.00	-13.10	-11.52
Ferredoxin [oxidized]		0.00	-0.61	-0.81
Ferredoxin [reduced]		38.07	38.07	38.07
Flavine adenine dinucleotide	oxidized	1238.65	1255.17	1260.51
Flavine adenine dinucleotide	reduced	1279.68	1297.43	1303.16
Flavin adenine dinucleotide-enzyme (FADenz) [oxidized]		1238.65	1255.17	1260.51
Flavin adenine dinucleotide-enzyme (FADenz) [reduced]		1229.96	1247.71	1253.44
Flavin mononucleotide (FMN) [oxidized]		759.17	768.35	771.32
Flavin mononucleotide (FMN) [reduced]		800.20	810.61	813.97
Formate ion [$CHOO^-$]		-311.04	-311.04	-311.04
D-Fructose		-436.03	-428.69	-426.32
D-Fructose 1,6-diphosphate		-2202.84	-2205.66	-2206.78
D-Fructose 6-phosphate		-1321.71	-1317.16	-1315.74
Fumarate		-521.97	-523.19	-523.58
D-Galactose		-429.45	-422.11	-419.74
Galactose 1-dihydrogen phosphate		-1317.50	-1313.01	-1311.60
α-D-Glucose		-436.42	-429.08	-426.71
α-D-Glucose 1-phosphate		-1318.03	-1313.34	-1311.89
D-Glucose 6-phosphate		-1325.00	-1320.37	-1318.92
Glutamate		-377.82	-373.54	-372.16
D-Glutamine		-128.46	-122.34	-120.36
Glutathione	oxidized	1198.69	1214.60	1219.74
Glutathione	reduced	625.75	634.76	637.62
Glutathione-coenzyme A		563.49	572.06	574.83
D-Glyceraldehyde 3-phosphate		-1088.94	-1088.25	-1088.04
Glycerol		-177.83	-172.93	-171.35
L-Glycerol 1-phosphate		-1080.22	-1077.83	-1077.13
Glycine		-180.13	-177.07	-176.08
Glycolate		-411.08	-409.86	-409.46
N-Glycylglycine		-200.55	-195.65	-194.07
Glyoxylate		-428.64	-428.64	-428.64
Hydrogen	aqueous	97.51	98.74	99.13
Hydrogen	gas	79.91	81.14	81.53
Hydrogen peroxide	aqueous	-54.12	-52.89	-52.50

Name	Note	$\Delta_f G'^\circ(I=0)$/ kJ mol^{-1}	$\Delta_f G'^\circ(I=0.1)$/ kJ mol^{-1}	$\Delta_f G'^\circ(I=0.25)$/ kJ mol^{-1}
3-Hydroxypropanoate		-318.62	-316.17	-315.38
Hypoxanthine		249.33	251.77	252.56
1*H*-Indole		503.49	507.78	509.16
Lactate		-316.94	-314.49	-313.70
α-Lactose		-688.29	-674.83	-670.48
L-Leucine		167.18	175.14	177.71
L-Isoleucine		175.53	183.49	186.06
D-Lyxose		-349.58	-343.46	-341.48
Malate		-682.88	-682.85	-682.85
α-Maltose		-695.65	-682.19	-677.84
D-Mannitol		-383.22	-374.65	-371.89
DL-Mannose		-430.52	-423.18	-420.81
Methane	aqueous	125.50	127.94	128.73
Methane	gas	109.11	111.55	112.34
Methanol		-15.48	-13.04	-12.25
L-Methionine		-63.40	-56.67	-54.49
Nicotinamide adenine dinucleotide (NAD) [oxidized]		1038.86	1054.17	1059.11
Nicotinamide adenine dinucleotide (NAD) [reduced]		1101.47	1115.55	1120.09
Nicotinamide adenine dinucleotide phosphate (NADP) [oxidized]		163.73	173.52	176.68
Nicotinamide adenine dinucleotide phosphate (NADP) [reduced]		229.67	235.79	237.77
Nitrogen	aqueous	18.70	18.70	18.70
Nitrogen	gas	0.00	0.00	0.00
Oxalate ion [$C_2O_4^{-2}$]		-673.90	-676.35	-677.14
Oxaloacetate		-713.38	-714.60	-715.00
Oxalosuccinate		-979.05	-979.05	-979.05
2-Oxoglutarate		-633.58	-633.58	-633.58
Oxygen	aqueous	16.40	16.40	16.40
Oxygen	gas	0.00	0.00	0.00
Palmitate		979.25	997.61	1003.54
L-Phenylalanine		232.42	239.15	241.33
Phosphate ion [PO_4^{-3}]		-1058.56	-1059.17	-1059.49
2-Phospho-D-glycerate		-1340.72	-1341.32	-1341.79
3-Phospho-D-glycerate		-1346.38	-1347.19	-1347.73
Phospho*enol*pyruvate		-1185.46	-1188.53	-1189.73
1-Propanol		143.84	148.74	150.32
2-Propanol		134.42	139.32	140.90
Pyrophosphate ion [$P_2O_7^{-4}$]		-1934.95	-1939.13	-1940.66
Pyruvate		-352.40	-351.18	-350.78
Retinal (all *trans*)		1118.78	1135.91	1141.45
Retinol		1170.78	1189.14	1195.07
D-Ribose		-339.23	-333.11	-331.13
D-Ribose 1-phosphate		-1215.87	-1212.24	-1211.14
D-Ribose 5-phosphate		-1223.95	-1220.32	-1219.22
D-Ribulose		-336.38	-330.26	-328.28
L-Serine		-231.18	-226.89	-225.51
D-Sorbose		-432.47	-425.13	-422.76
Succinate		-530.72	-530.65	-530.64
Succinyl coenzyme A		-349.90	-348.06	-347.47
Sucrose		-685.66	-672.20	-667.85
Thioredoxin [oxidized]		0.00	0.00	0.00
Thioredoxin [reduced]		54.32	55.41	55.74
DL-Tryptophan		364.78	372.12	374.49
L-Tyrosine		68.82	75.55	77.73
Ubiquinone [oxidized]		3596.07	3651.15	3668.94
Ubiquinone [reduced]		3586.06	3642.37	3660.55
Urate		-206.03	-204.81	-204.41
Urea		-42.97	-40.53	-39.74
Uric acid		-197.07	-194.63	-193.84
L-Valine		80.87	87.60	89.78
Water	liquid	-157.28	-156.05	-155.66
D-Xylose		-350.93	-344.81	-342.83
D-Xylulose		-346.59	-340.47	-338.49

APPARENT EQUILIBRIUM CONSTANTS FOR ENZYME-CATALYZED REACTIONS

Robert N. Goldberg

This table contains values of apparent equilibrium constants K' for selected enzyme-catalyzed reactions at specified temperatures T and pHs. In those cases where the ionic strength I and/or the pMg (pMg = $-\log_{10}[Mg^{2+}]$) have been reported, the values of these quantities are given.

There are two fundamentally different types of equilibrium constants. This is illustrated by the following example for the hydrolysis of adenosine 5′-triphosphate (ATP) to adenosine 5′-diphosphate (ADP) and phosphate:

$$ATP + H_2O = ADP + \text{phosphate} \quad (1)$$

The apparent equilibrium constant for the overall biochemical reaction (1) is

$$K' = [ADP][\text{phosphate}]/([ATP]c^{\circ}) \quad (2)$$

The biochemical reactants ATP, ADP, and phosphate each exist in several different ionized and metal bound forms. For example, ATP is an equilibrium mixture of the species ATP^{4-}, $HATP^{3-}$, H_2ATP^{2-}, $MgATP^{2-}$, $MgHATP^{-}$, Mg_2ATP^{0}. Additional species would also have to be considered if Ca^{2+} were present. Thus, ATP has often been denoted in the literature as ΣATP or as $(ATP)_{tot}$. When it is clear that one is dealing with total amounts of substances, it is not necessary to use either the Σ or "tot." Thus, these designations are not used in this table. In the above equation, c°=1 mol dm^{-3}; it is included to make K' dimensionless. The standard transformed Gibbs energy of reaction $\Delta_r G'^{\circ}$ at specified conditions of temperature T, pressure P, ionic strength I, pH, and pMg can be calculated from K':

$$\Delta_r G'^{\circ} = -RT\ln K' \quad (3)$$

The molar gas constant, R, is equal to 8.314462 J K^{-1} mol^{-1}. $\Delta_r G'^{\circ}$ and the apparent equilibrium constant, K', can be used to calculate the position of equilibrium of overall biochemical reactions.

It is also possible to choose a chemical reference reaction that involves selected solute species:

$$ATP^{4-} + H_2O = ADP^{3-} + HPO_4^{2-} + H^{+} \quad (4)$$

The equilibrium constant for this reference reaction is

$$K = [ADP^{3-}][HPO_4^{2-}][H+]/\{[ATP^{4-}](c^{\circ})^2\} \quad (5)$$

Equations and algorithms that relate these two different types of equilibrium constants have been published in Refs. 2–4. To calculate the equilibrium constant K for the reference reaction from the apparent equilibrium constant K', or vice versa, one needs the equilibrium constants for the binding of H^+ and for the relevant metal ions to ATP^{4-}, ADP^{3-}, and HPO_4^{2-}.

To avoid confusion between the two different types of equilibrium constants (K' and K) and to avoid ambiguity about whether specific species or sums of species are intended, the word "ammonia," for example, rather than NH_3 or NH_4^+, is used for total ammonia, and chemical formulas are used for specific chemical species. Other substances such as carbon dioxide (CO_2, HCO_3^-, and CO_3^{2-}), and phosphate ($H_2PO_4^-$, HPO_4^{2-}, and PO_4^{3-}) are treated in the same manner. Exceptions are made for water, which is always written as H_2O, and for gaseous hydrogen and oxygen, which are written as H_2(g) and O_2(g), respectively.

For symmetrical reactions, there is no concern about the units used to calculate the value of an equilibrium constant. However, care must be exercised for reactions that are not symmetrical. In such cases, the units "mol dm^{-3}" have been used for all concentrations. As stated above, a c° (1 mol dm^{-3}) is then used to make all equilibrium constants dimensionless.

All substances are assumed to be in aqueous solutions unless specified otherwise. Column definitions are as follows.

Column heading	Definition
Reaction	Enzyme-catalyzed reaction; see abbreviation list below for some reactants
K′	Apparent equilibrium constant of the reaction (see text above)
Enzyme Comm. No.	Enzyme Commission Number (Ref. 1) of the enzyme used to catalyze the reaction
T	Temperature at which K' is measured
pH	pH at which K' is measured
I	Ionic strength equilibrium mixture
pMg	pMg = $-\log_{10}[Mg^{2+}]$

Values of $\Delta_r G'^{\circ}$ and K' can also be calculated for many biochemical reactions by using the table "Standard Transformed Gibbs Energies of Formation for Biochemical Reactants" in this section of this *CRC Handbook*.

Abbreviations

ADP:	adenosine 5′-diphosphate
AMP:	adenosine 5′-monophosphate
ATP:	adenosine 5′-triphosphate
CoA:	coenzyme A
GDP:	guanosine 5′-diphosphate
GMP:	guanosine 5′-monophosphate
GTP:	guanosine 5′-triphosphate
IDP:	inosine 5′-diphosphate
IMP:	inosine 5′-monophosphate
ITP:	inosine 5′-triphosphate
NAD_{ox}:	β-nicotinamide-adenine dinucleotide, oxidized form
NAD_{red}:	β-nicotinamide-adenine dinucleotide, reduced form
$NADP_{ox}$:	β-nicotinamide-adenine dinucleotide phosphate, oxidized form
$NADP_{red}$:	β-nicotinamide-adenine dinucleotide phosphate, reduced form
UDP:	uridine 5′-diphosphate
UTP:	uridine 5′-triphosphate

References

1. Webb, E.C., *Enzyme Nomenclature 1992*, Academic Press, New York, 1992. See also <http://www.chem.qmul.ac.uk/iubmb/enzyme/>.
2. Akers, D.L., and Goldberg, R.N., *Mathematica J.*, 8, 86–113, 2001.
3. Alberty, R.A., *J. Biol. Chem.*, 243, 1337–1343, 1969.
4. Alberty, R.A., *Thermodynamics of Biochemical Reactions,* Wiley-Interscience, Hoboken, NJ, 2003. <https://doi.org/10.1002/0471332607>
5. Goldberg, R.N., Tewari, Y.B., Bell, D., Fazio, K., and Anderson, E., *J. Phys. Chem. Ref. Data*, 22, 515-582, 1993. <https://doi.org/10.1063/1.555939>

6. Goldberg, R.N., and Tewari, Y.B., *J. Phys. Chem. Ref. Data*, 23, 547-617, 1994. <https://doi.org/10.1063/1.555948>
7. Goldberg, R.N., and Tewari, Y.B., *J. Phys. Chem. Ref. Data*, 23, 1035-1103, 1994. <https://doi.org/10.1063/1.555957>
8. Goldberg, R.N., and Tewari, Y.B., *J. Phys. Chem. Ref. Data*, 24, 1669-1698, 1995. <https://doi.org/10.1063/1.555969>
9. Goldberg, R.N., and Tewari, Y.B., *J. Phys. Chem. Ref. Data*, 24, 1765-1801, 1995. <https://doi.org/10.1063/1.555970>
10. Goldberg, R.N., *J. Phys. Chem. Ref. Data*, 28, 931–965, 1999. <https://doi.org/10.1063/1.556041>
11. Goldberg, R.N., Tewari, Y.B., and Bhat, T.N., *Bioinformatics*, 20, 2874–2877, 2004; <http://xpdb.nist.gov/enzyme_thermodynamics/>. <https://doi.org/10.1093/bioinformatics/bth314>
12. Goldberg, R.N., Tewari, Y.B., and Bhat, T.N., *J. Phys. Chem. Ref. Data*, 36, 1347–1397, 2007. <https://doi.org/10.1063/1.2789450>

Apparent Equilibrium Constants for Enzyme-Catalyzed Reactions

Reaction	*K′*	Enzyme Comm. No.	*T*/K	pH	*I*/mol dm^{-3}	pMg
benzyl alcohol + NAD$_{ox}$ = benzaldehyde + NAD$_{red}$	$9.8\cdot10^{-4}$	1.1.1.1	298.15	7.5		
1-butanol + NAD$_{ox}$ = butanal + NAD$_{red}$	$1.8\cdot10^{-3}$	1.1.1.1	298.15	8.3		
cyclohexanol + NAD$_{ox}$ = cyclohexanone + NADH$_{red}$	0.09	1.1.1.1	298.15	7.2		
1-hexanol + NAD$_{ox}$ = hexanal + NAD$_{red}$	$2.87\cdot10^{-3}$	1.1.1.1	298.15	8.3		
1-octanol + NAD$_{ox}$ = octanal + NAD$_{red}$	1.1·10–3	1.1.1.1	298.15	8.3		
L-homoserine + NADP$_{ox}$ = L-aspartate 4-semialdehyde + NADP$_{red}$	$6.3\cdot10^{-4}$	1.1.1.3	298.15	7.9		
xylitol + NAD$_{ox}$ = L-xylulose + NAD$_{red}$	$2.97\cdot10^{-4}$	1.1.1.10	298.15	7.00		
D-sorbitol + NAD$_{ox}$ = D-fructose + NAD$_{red}$	0.032	1.1.1.14	298.15	7.0		
quinate + NAD$_{ox}$ = 5-dehydroquinate + NAD$_{red}$	$4.61\cdot10^{-3}$	1.1.1.24	305.15	7.2		
shikimate + NADP$_{ox}$ = 5-dehydroshikimate + NADP$_{red}$	0.036	1.1.1.25	303.15	7.0		
2-hydroxybutanoate + NAD$_{ox}$ = 2-oxobutanoate + NAD$_{red}$	$3.0\cdot10^{-3}$	1.1.1.27	298.65	8.0		
(*R*)-3-hydroxybutanoate + NAD$_{ox}$ = 3-oxobutanoate + NAD$_{red}$	$1.9\cdot10^{-3}$	1.1.1.30	298.15	7.0		
D-glucose 6-phosphate + NADP$_{ox}$ = D-glucono-1,5-lactone 6-phosphate + NADP$_{red}$	1.50	1.1.1.49	301.15	6.40		
5α-androstane-3α-ol-17-one + NAD$_{ox}$ = 5α-androstane-3,17-dione + NAD$_{red}$	0.058	1.1.1.50	298.15	7.0		
5α-pregnane-3α,17α,21-triol-20-one + NAD$_{ox}$ = 5α-pregnane-17α,21-diol-3,20-dione + NAD$_{red}$	0.0113	1.1.1.50	298.15	7.0		
5α-androstane-3β,17α-diol + NAD$_{ox}$ = 5α-androstane-17α-ol-3-one + NAD$_{red}$	0.0211	1.1.1.51	298.15	7.0		
4-androstene-17β-ol-3-one + NAD$_{ox}$ = 4-androstene-3,17-dione + NAD$_{red}$	0.378	1.1.1.51	298.15	7.0		
1,2-propanediol + NADP$_{ox}$ = L-lactaldehyde + NADP$_{red}$	$6.0\cdot10^{-5}$	1.1.1.55	298.15	8.4		
ribitol + NAD$_{ox}$ = D-ribulose + NAD$_{red}$	$3.1\cdot10^{-3}$	1.1.1.56	310.15	7.4		
3-hydroxypropanoate + NAD$_{ox}$ = 3-oxopropanoate + NAD$_{red}$	$9.0\cdot10^{-3}$	1.1.1.59	298.15	9.0		
estradiol-17β + NAD$_{ox}$ = estrone + NAD$_{red}$	0.18	1.1.1.62	298.15	7.00		
benzyl alcohol + NAD$_{ox}$ = benzaldehyde + NAD$_{red}$	0.097	1.1.1.90	300.15	9.5		
L-carnitine + NAD$_{ox}$ = 3-dehydrocarnitine + NAD$_{red}$	$1.3\cdot10^{-4}$	1.1.1.108	303.15	7.0		
L-threonate + NAD$_{ox}$ = 3-oxo-L-threonate + NAD$_{red}$	$3.42\cdot10^{-4}$	1.1.1.129	298.15	7.0		
prostaglandin E$_1$ + NAD$_{ox}$ = 15-oxo-prostaglandin E$_1$ + NAD$_{red}$	0.65	1.1.1.141	298.15	7.0		
7,8-dihydrobiopterin + NADP$_{ox}$ = sepiapterin + NADP$_{red}$	0.045	1.1.1.153	298.15	8.0		
glycine + acetaldehyde = L-threonine	56	2.1.2.1	310.15	7.6		
sedoheptulose 7-phosphate + D-glyceraldehyde 3-phosphate = D-ribose 5-phosphate + D-xylulose 5-phosphate	0.48	2.2.1.1	311.15	7.0	0.25	3.0
acetyl-CoA + choline = CoA + *O*-acetylcholine	1.60	2.3.1.7	298.15	7.0	0.25	
acetyl-CoA + acyl-carrier protein = CoA + acetyl-[acyl-carrier protein]	2.09	2.3.1.38	311.15	6.5		
UDPglucose + D-fructose = UDP + sucrose	6.7	2.4.1.13	298.15	7.5		
cellobiose + orthophosphate = D-glucose + α-D-glucose 1-phosphate	0.23	2.4.1.20	310.15	7.0		
laminaritriose + orthophosphate = laminaribiose + α-D-glucose 1-phosphate	0.26	2.4.1.31	310.15	6.5		
α,α-trehalose + orthophosphate = D-glucose + β-D-glucose 1-phosphate	0.24	2.4.1.64	310.15	7.0		
UDPglucose + sinapate = UDP + 1-sinapoyl-D-glucose	0.21	2.4.1.120	303.15	6.0		
inosine + orthophosphate = hypoxanthine + α-D-ribose 1-phosphate	0.0164	2.4.2.1	311.15	7.0	0.25	3.0
xanthosine + orthophosphate = xanthine + α-D-ribose 1-phosphate	0.0156	2.4.2.1	311.15	7.0	0.25	3.0
uridine + orthophosphate = uracil + α-D-ribose 1-phosphate	0.44	2.4.2.2	310.15	7.0		
adenine + 5-phospho-α-D-ribose 1-diphosphate = AMP + pyrophosphate	$2\cdot10^{3}$	2.4.2.7	311.15	7.4	0.25	3.0
GMP + hypoxanthine = IMP + guanine	0.38	2.4.2.8	310.15	7.4		
guanine + 5-phospho-α-D-ribose 1-diphosphate = GMP + pyrophosphate	$1\cdot10^{5}$	2.4.2.8	311.15	7.4	0.25	3.0
hypoxanthine + 5-phospho-α-D-ribose 1-diphosphate = IMP + pyrophosphate	$1\cdot10^{5}$	2.4.2.8	311.15	7.4	0.25	3.0
ATP + ammonium carbamate = ADP + carbamoyl phosphate	0.042	2.7.2.2	283.15	9.4		
ATP + creatine = ADP + phosphocreatine	$5.78\cdot10^{-3}$	2.7.3.2	310.15	7.11	0.25	2.47
ATP + L-arginine = ADP + N$^{\omega}$-phospho-L-arginine	0.10	2.7.3.3	285.15	7.25		
ATP + sulfate = adenosine 5′-phosphosulfate + pyrophosphate	$4\cdot10^{-8}$	2.7.7.4	303.15	7.5		
UTP + α-D-glucose 1-phosphate = pyrophosphate + UDPglucose	0.48	2.7.7.9	310.15	8.0		
succinyl-CoA + acetoacetate = succinate + acetoacetyl-CoA	$2.8\cdot10^{-3}$	2.8.3.5	303.15	7.0		
acetylcholine + H_2O = acetate + choline	$5.38\cdot10^{2}$	3.1.1.7	296.15	5.1		

Reaction	K'	Enzyme Comm. No.	T/K	pH	I/mol dm^{-3}	pMg
IMP + H_2O = inosine + orthophosphate	$1.58 \cdot 10^{2}$	3.1.3.1	298.15	8.55	1.53	4.44
phosphorylcholine + H_2O = choline + orthophosphate	49.9	3.1.3.1	311.15	6.90		
3-*O*-β-D-galactopyranosyl-D-arabinose + H_2O = D-galactose + D-arabinose	56	3.1.3.1	308.15	7.0		
cytidine 2′:3′-(cyclic)phosphate + H_2O = cytidine 3′-monophosphate	$1.06 \cdot 10^{3}$	3.1.27.5	298.15	6.0		
isomaltose + H_2O = 2 D-glucose	17.2	3.2.1.3	298.15	5.65		
β-gentiobiose + H_2O = 2 D-glucose	17.7	3.2.1.21	298.15	5.65		
3-*O*-β-D-galactopyranosyl-D-arabinose + H_2O = D-galactose + D-arabinose	$1.04 \cdot 10^{2}$	3.2.1.23	298.15	5.65		
lactulose + H_2O = D-galactose + D-fructose	$1.28 \cdot 10^{2}$	3.2.1.23	298.15	5.65		
4′,5′-anhydroadenosine + H_2O = adenosine	0.48	3.3.1.1	310.15	7.0		
pteroylglutamate + H_2O = pteroate + L-glutamate	15.6	3.4.19.9	310.15	7.3		
N-acetyl-L-phenylalanine methyl ester + H_2O = *N*-acetyl-L-phenylalanine + methanol	$5.88 \cdot 10^{2}$	3.4.21.1	293.15	5.5		
hippurylanilide + H_2O = hippuric acid + aniline	11	3.4.22.2	312.15	5.0		
ammonium carbamate + H_2O = 2 ammonia + carbon dioxide	$1.92 \cdot 10^{3}$	3.5.1.5	293.15	6.5		
ampicillin + H_2O = 6-aminopenicillanic acid + D(−)-α-aminophenylacetic acid	0.013	3.5.1.11	298.15	5.0		
cephalexin + H_2O = 7-aminodeacetoxycephalosporanic acid + cephem-4-carboxylic acid	0.044	3.5.1.11	298.15	5.8		
cephaloridine + H_2O = 2-thienylacetic acid + 7-amino-3-(1-pyridyl-methyl)-3-cephem-4-carboxylic acid	0.015	3.5.1.11	298.15	5.0		
penicillin G + H_2O = 6-aminopenicillanic acid + phenylacetic acid	0.445	3.5.1.11	298.15	6.71		
N-acetyl-L-alanine + H_2O = acetate + L-alanine	7	3.5.1.14	298.15	6.0		
ampicillin + H_2O = ampicillinoic acid	95	3.5.2.6	282.35	5.55		
penicillin G + H_2O = penicillinoic acid	2.9	3.5.2.6	298.15	6.01		
cytidine + H_2O = uridine + ammonia	$1.03 \cdot 10^{4}$	3.5.4.5	298.15	7.00		
N^4-methylcytidine + H_2O = uridine + methylamine	$4.88 \cdot 10^{2}$	3.5.4.5	298.15	7.50		
5,10-methenyltetrahydrofolate + H_2O = 10-formyltetrahydrofolate	50	3.5.4.9	298.15	7.0		
ITP + oxaloacetate + H_2O = IDP + phospho*enol*pyruvate + carbon dioxide	12	4.1.1.32	303.15	7.6		
2-deoxy-D-ribose 5-phosphate = D-glyceraldehyde 3-phosphate + acetaldehyde	$2.5 \cdot 10^{-4}$	4.1.2.4	295.15	7.5		
6-phospho-2-dehydro-3-deoxy-D-gluconate = pyruvate + D-glyceraldehyde 3-phosphate	$1.2 \cdot 10^{-3}$	4.1.2.14	298.15	8.0	0.37	
L-fuculose 1-phosphate = glycerone phosphate + (*S*)-lactaldehyde	$4.6 \cdot 10^{-4}$	4.1.2.17	310.15	7.2		
L-rhamnulose 1-phosphate = glycerone phosphate + (*S*)-lactaldehyde	0.083	4.1.2.19	310.15	7.5		
isocitrate = succinate + glyoxylate	$2.3 \cdot 10^{-3}$	4.1.3.1	303.15	7.7		
(*S*)-2-methylmalate = acetate + pyruvate	0.151	4.1.3.22	298.15	7.4	0.845	
isocitrate = citrate	14.7	4.2.1.3	298.15	7.4		
3-dehydroquinate = 3-dehydroshikimate + H_2O	15	4.2.1.10	302.15	7.4		
(3*R*)-3-hydroxybutanoyl-CoA = *cis*-but-2-enoyl-CoA + H_2O	0.18	4.2.1.17	298.15	7.5		
indole + D-glyceraldehyde 3-phosphate = 1-(indol-3-yl)glycerol 3-phosphate	$1.2 \cdot 10^{4}$	4.2.1.20	298.15	7.54		
(*R*)-malate = maleate + H_2O	$4.88 \cdot 10^{-4}$	4.2.1.31	298.15	7.00	0.10	
(*R*)-2-methylmalate = 2-methylmaleate + H_2O	0.0962	4.2.1.35	298.15	7.0	0.10	
D-glutamate = 5-oxo-D-proline + H_2O	24.3	4.2.1.48	293.4	7.9		
L-*threo*-3-methylaspartate = 2-methylfumarate + ammonia	0.238	4.3.1.2	298.15	7.9		
L-histidine = urocanate + ammonia	3.01	4.3.1.3	298.25	8.41	0.167	
L-phenylalanine = *trans*-cinnamate + ammonia	2.47	4.3.1.5	298.05	7.69		
ATP = adenosine 3′:5′-(cyclic)phosphate + diphosphate	0.065	4.6.1.1	298.2	7.0		
L,L-2,6-diaminoheptanedioate = *meso*-diaminoheptanedioate	1.9	5.1.1.7	310.15	7.0		
D-ribulose 5-phosphate = D-xylulose 5-phosphate	1.82	5.1.3.1	311.15	7.0	0.25	3.0
UDPglucose = UDPgalactose	0.33	5.1.3.2	298.15	8.7		
GDPmannose = GDP-L-galactose	0.52	5.1.3.18	310.15	8.0		
all-*trans*-retinal = 11-*cis*-retinal	0.05	5.2.1.3	309.15	7.0		
9-*cis*,12-*cis*-octadecadienoate = 9-*cis*,11-*trans*-octadecadienoate	61	5.2.1.5	308.15	7.0		
D-erythrose = D-erythrulose	2.3	5.3.1.2	308.15	5.8		
D-arabinose = D-ribulose	0.146	5.3.1.3	320.25	7.4		
L-fucose = L-fuculose	0.12	5.3.1.3	310.15	8.0		
L-arabinose = L-ribulose	0.11	5.3.1.4	298.15	7.0		
D-psicose = β-D-allose	2.15	5.3.1.4	317.25	7.4		
D-ribose 5-phosphate = D-ribulose 5-phosphate	0.83	5.3.1.6	311.15	7.0	0.25	3.0
D-rhamnose = D-rhamnulose	0.58	5.3.1.7	303.15	7.4		
D-mannose 6-phosphate = D-fructose 6-phosphate	0.99	5.3.1.8	298.15	8.50		
6-amino-D-glucose 6-phosphate = 6-amino-D-fructose 6-phosphate	0.202	5.3.1.9	278.85	8.7		
D-glucosamine 6-phosphate + H_2O = D-fructose 6-phosphate + ammonia	0.15	5.3.1.10	310.15	8.4		
D-lyxose = D-xylulose	0.23	5.3.1.15	298.15	7.0		
D-ribose = D-ribulose	0.391	5.3.1.20	313.15	7.4		

Reaction	K'	Enzyme Comm. No.	T/K	pH	I/mol dm^{-3}	pMg
keto-phenylpyruvate = *enol*-phenylpyruvate	0.1	5.3.2.1	298.15	7.8		
L-lysine = (3*S*)-3,6-diaminohexanoate	5.3	5.4.3.2	303.15	7.7		
(*R*)-methylmalonyl-CoA = succinyl-CoA	23.1	5.4.99.2	298.15	7.4		
(−)-4-carboxymethyl-Δ$^{\alpha}$-but-2-en-4-olide = *cis,trans*-hexadienedioate	4.0	5.5.1.1	303.15	8.0		
ATP + heptanoate + CoA = AMP + diphosphate + *n*-heptanoyl-CoA	1.11	6.2.1.2	311.15	8.0		
GTP + succinate + CoA = GDP + phosphate + succinyl-CoA	1.68	6.2.1.4	298.15	7.15	0.25	2.91
GTP + IMP + L-aspartate = GDP + phosphate + adenylosuccinate	2.9	6.3.4.4	310.15	8.0		
ATP + L-citrulline + L-aspartate = AMP + diphosphate + L-arginosuccinate	2.14	6.3.4.5	311.15	6.91		
ATP + propanoyl-CoA + carbon dioxide = ADP + phosphate + (*S*)-methylmalonyl-CoA	$8.1 \cdot 10^{-3}$	6.4.1.3	310.15	8.15		

APPARENT EQUILIBRIUM THERMODYNAMICS OF PROTEIN-LIGAND BINDING REACTIONS

Mark A. Williams

These tables contain values of the observed thermodynamics properties of the binding reactions of selected proteins with chemically related series of small molecule ligands at a specified temperature (T) and pH.

The standard Gibbs free energy change, $\Delta_b G'^\circ$, upon a ligand, L, binding a protein, P, determines the relative populations of protein, ligand and protein:ligand complex (P:L) at equilibrium. The apparent association constant (K_b') for the binding equilibrium is defined by

$$K_b' = [P{:}L]\, c^\circ / [P][L]$$

and the Gibbs energy determined from K_b' via

$$\Delta_b G'^\circ = -RT \ln(K_b')$$

where the molar gas constant R = 8.3144598 J K^{-1} mol^{-1}, and c° is the standard state concentration of unit molarity (1 mol dm^{-3}). The standard enthalpy change of the binding reaction ($\Delta_b H'^\circ$) is the heat released (negative) or taken up (positive) upon forming the complex. All reactions reported here are carried out at a constant 1 atmosphere pressure (0.1 MPa). Under these conditions, the standard entropy change, $\Delta_b S'^\circ$, associated with a reaction is determined by the Gibbs-Helmholtz relation,

$$\Delta_b G'^\circ = \Delta_b H'^\circ - T\,\Delta_b S'^\circ$$

All experimental data reported here are obtained by isothermal titration microcalorimetry (ITC); a method that takes advantage of the fact that almost all binding reactions are accompanied by a measurable exchange of heat with their environment. An ITC experiment determines both the Gibbs energy and enthalpy change of a reaction by analysis of a single titration. In the most common experimental arrangement, a titration of ligand against protein is performed in a series of small increments. At each increment, the ligand:protein ratio is increased, and the net heat change arising from the shift in the binding equilibrium is measured. As the binding site of the protein becomes saturated, the incremental heat change diminishes to zero. $\Delta_b H'^\circ$ is determined from the sum total heat change at saturation and the known concentrations of reactants. $\Delta_b G'^\circ$ is determined from the dependence of the incremental heat (which corresponds to the fractional saturation) on the ligand:protein concentration ratio according to the analysis described in Ref. 1. For very high affinity interactions, almost all added ligand binds at each increment and the protein will abruptly saturate at a 1:1 concentration ratio making the data difficult to analyze to obtain accurate values for $\Delta_b G'^\circ$. In such circumstances, the binding affinity of the ligand of interest is artefactually diminished in experiments by competition with a saturating excess of a lower affinity ligand that binds to the same site on the protein. The ligand of interest is titrated to displace the competitor, and the observed thermodynamics are then corrected for the known thermodynamics of the competitor binding according to procedures in Ref. 2. Those reactions carried out in this competitive manner are indicated in the tables. Column definitions for the table are as follows.

Column heading	Definition
Sequence modification	Protein system that binds to the ligand; includes species name as well as standard translated gene sequence code from the UNIPROT database (see text for further description of the protein systems)
Ligand	Common name of binding ligand
$\Delta_b G'^\circ$	Gibbs energy of protein:ligand binding, in units kJ mol^{-1}
$\Delta_b H'^\circ$	Enthalpy change of protein:ligand binding, in units kJ mol^{-1}
$-T\Delta_b S'^\circ$	Entropy change at specified temperature T (next column right) of protein:ligand binding, in units kJ mol^{-1}
T	Temperature of binding interaction, in K
Ref.	Reference
Buffer composition	Chemical composition of buffer in which binding interaction was measured
Alternative systematic ligand name	Systematic name for ligand, when given

Biochemical binding reactions may be complex events. An idealized scenario of two rigid, chemically distinct, molecular species simply coming into close proximity is rarely the case. The idealized binding event is frequently accompanied by changes in other conformational or chemical equilibria involving the protein or ligand, e.g., due to only one conformation being able to bind or due to altered interactions with other solution components between the bound and free states. The apparent binding thermodynamics reported is for the total of all processes that necessarily accompany the binding. Here we also report the buffer composition in which the reactions were carried out as it is possible that altered interactions with these components affect the observations. Within each selected protein-ligand series, protein and ligands have similar bound conformations and experiments have been carried out under identical or very similar conditions meaning that, in so far as possible, differences in the thermodynamic properties between compounds in a series reflect differences between the ligands and in their interactions with the protein only and are more amenable to structural interpretation (Ref. 20).

Proteins and many ligands contain a variety of titratable acidic and basic groups, and the fractional protonation state of those groups may change upon binding if the protonated state of each group is stabilized or destabilized by interactions between the ligand and protein. Such direct effects of binding are a necessary consequence of the interaction and are appropriately included in the apparent thermodynamic properties. However, any net change in protonation of protein and ligand is coupled to change in protonation of the buffer that introduces an artefactual buffer-dependent heat of ionization (see "Thermodynamic Quantities for the Ionization Reactions of Buffers in Water" in this section). Heats of reaction are consequently corrected for buffer ionization where appropriate.

The amino acid residue sequence of the protein is given by reference to the standard translated gene sequence found in the UNIPROT database, Ref. 3. Any differences from the standard sequence are noted.

References

1. Wiseman, T., Williston, S., Brandts, J. F., and Lin, L. N., *Anal. Biochem.*, 179, 131, 1989. <https://doi.org/10.1016/0003-2697(89)90213-3>
2. Sigurskjold, B.W., *Anal. Biochem.*, 277, 260, 2000. <https://doi.org/10.1006/abio.1999.4402>
3. The Uniprot Consortium, *Nucl. Acids Res.*, 43, D204, 2013. <https://doi.org/10.1093/nar/gku989>
4. Özen, C. and Serpersu, E. H., *Biochemistry*, 43, 14667, 2004. <https://doi.org/10.1021/bi0487286>
5. Martin, M. P., Zhu, J. Y., Lawrence, H. R., Pireddu, R., Luo, Y., Alam, R., Ozcan, S., Sebti, S. M., Lawrence, N. J., and Schonbrunn, E., *ACS Chem. Biol.*, 7, 698, 2012. <https://doi.org/10.1021/cb200508b>
6. Gloster, T. M., Meloncelli, P., Stick, R. V., Zechel, D., Vasella, A., and Davies, G. J., *J. Am. Chem. Soc.*, 129, 2345, 2007. <https://doi.org/10.1021/ja066961g>
7. Zechel, D. L., Boraston, A. B., Gloster, T., Boraston, C. M., Macdonald, J. M., Tilbrook, D. M. G., Stick, R. V., and Davies, G., *J. Am. Chem. Soc.*, 125, 14313, 2003. <https://doi.org/10.1021/ja036833h>
8. Reddy, S. G., Scapin, G., and Blanchard, J. S., *Biochemistry*, 35, 13294, 1996. <https://doi.org/10.1021/bi9615809>
9. Sharrow, S. D., Novotny, M. V., and Stone, M. J., *Biochemistry*, 42, 6302, 2003. <https://doi.org/10.1021/bi026423q>
10. Bingham, R. J., Findlay, J. B. C., Hsieh, S. -Y., Kalverda, A. P., Kjellberg, A., Perazzolo, C., Phillips, S. E. V., Seshadri, K., Trinh, C. H., Turnbull, W. B., Bodenhausen, G., and Homans, S.W., *J. Am. Chem. Soc.*, 126, 1675, 2004. <https://doi.org/10.1021/ja038461i>
11. Malham, R., Johnstone, S., Bingham, R. J., Barratt, E., Phillips, S. E. V., Laughton, C. A., and Homans, S. W., *J. Am. Chem. Soc.*, 127, 17061, 2005. <https://doi.org/10.1021/ja055454g>
12. Barratt, E., Bingham, R. J., Warner, D. J., Laughton, C. A., Phillips, S. E. V., and Homans, S. W., *J. Am. Chem. Soc.*, 127, 11827, 2005. <https://doi.org/10.1021/ja0527525>
13. Payne, J. W., Grail, B. M., Gupta, S., Ladbury, J. E., Marshall, N. J., O'Brien, R., and Payne, G. M., *Arch. Biochem. Biophys.*, 384, 9, 2000. <https://doi.org/10.1006/abbi.2000.2084>
14. Sleigh, S. H., Seavers, P. R., Wilkinson, A. J., Ladbury, J. E., and Tame, J. R. H., *J. Mol. Biol.*, 291, 393, 1999. <https://doi.org/10.1006/jmbi.1999.2929>
15. Baum, B., Mohamed, M., Zayed, M., Gerlach, C., Heine, A., Hangauer, D., and Klebe, G. J., *Mol. Biol.*, 390, 56, 2009. <https://doi.org/10.1016/j.jmb.2009.04.051>
16. Baum, B., Muley, L., Heine, A., Smolinski, M., Hangauer, D., and Klebe, G. J., *Mol. Biol.*, 391, 552, 2009. <https://doi.org/10.1016/j.jmb.2009.06.016>
17. Beila, A., Sielaff, F., Terwesten, F., Heine, A., Steinmetzer, T., and Klebe, G., *J. Med. Chem.*, 55, 6094, 2012. <https://doi.org/10.1021/jm300337q>
18. Talhout, R. and Engberts, J. B., *Eur. J. Biochem.*, 268, 1554, 2001. <https://doi.org/10.1046/j.1432-1327.2001.01991.x>
19. Talhout, R., Villa, A., Mark, A. E., and Engberts, J. B., *J. Am. Chem. Soc.*, 125, 10570, 2003. <https://doi.org/10.1021/ja034676g>
20. Olsson, T. S. G., Williams, M. A., Pitt, W. R., and Ladbury, J. E., *J. Mol. Biol.*, 384, 1002, 2008. <https://doi.org/10.1016/j.jmb.2008.09.073>

Description of Protein Systems

In the table following this section, thermodynamic data for a set of widely studied proteins binding to a variety of ligands are presented. Below are given more detailed descriptions of the proteins as well as structural and interaction details.

Enterococcus faecalis Aminoglycoside 3'-phosphotransferase

Uniprot ID: P0A3Y5

Aminoglycoside 3'-phosphotransferase is able to phosphorylate many aminoglycoside antibiotics rendering them inactive. The reaction proceeds by transferring a phosphate from ATP bound in a neighboring site to the antibiotic, e.g., ATP + kanamycin = ADP + kanamycin 3'-phosphate. The reaction is dependent on Mg^{2+}. Replacing magnesium with calcium ions leads to formation of the ternary ATP:aminoglycoside:enzyme complex, but no product is formed. The enzyme is monomeric, and crystallographic analysis shows that it undergoes only local structural change on ATP and/or aminoglycoside binding. The thermodynamics of formation of the binary complex by titration of aminoglycoside to enzyme and the ternary complex by titration of aminoglycoside into enzyme in the presence of Ca^{2+} and ATP have been measured. Data below are corrected for coupled protonation of the buffer according to procedures described in Ref. 4.

Homo sapiens Aurora A kinase

Uniprot ID: O14965

Aurora A kinase transfers phosphate from ATP to serine and threonine residues in other proteins in the context of specific amino acid sequences. Aurora A activity regulates processes in normal cell division. Increased concentrations, and thus activity, of Aurora A in cells leads to aberrant cell division and cancer. Consequently, inhibitors of its action may be useful as anticancer drugs. The series of (bis)anilinopyrimidine inhibitors has a common binding mode that prevents a conformational change required for enzyme activity.

Thermotoga maritima β-glucosidase A

Uniprot ID: Q08638

Glycoside hydrolases are a large family of related enzymes that act to degrade oligosaccharides in a variety of extracellular digestive and intracellular metabolic processes. Several members of the family are implicated in diseases. β-glucosidases attack the β-1-4 linkage of nonreducing (acetal) β-D-glucosyl residues releasing β-D-glucose. β-glucosidase A is an experimentally tractable and stable member of the family that is used to study the binding of a series of inhibitors that mimic the transition state of the reaction. At the pH of these experiments, the binding of 1-deoxynojirimycin and isofagomine are associated with the net release of a single proton upon binding (Ref. 6). This protonation event is thought to be associated with interaction with the glutamic acid, which acts as the general base in the catalytic mechanism. Given the high degree of structural similarity between the ligand complexes (Ref. 7), it is assumed all of the compounds in the series below undergo this same net protonation. Consequently, all of the enthalpies in this series are corrected here by -3.38 kJ mol^{-1} to account for the small heat of ionization of the citrate buffer.

Escherichia coli Dihydrodipicolinate reductase

Uniprot ID: P04036

Dihydrodipicolinate reductase catalyzes the cofactor-dependent reaction, NAD(P)H + dihydropicolinic acid = tetrahydropicolinic acid + $NAD(P)^+$, and is an essential component of the biosynthetic pathway of L-lysine and meso-diaminopimelic acid in bacteria. Given the requirement for these amino acids in, respectively, protein synthesis and the formation of peptidoglycan in the bacterial cell wall, inhibition of this enzyme is considered to be a possible therapeutic strategy. The native enzyme is a homotetramer. There is no evidence for cooperativity of binding of the cofactor and thermodynamic data are analyzed assuming that the four sites are equivalent. The thermodynamics of binding of the cofactor(s) and several analogs have been determined in the absence of the dihydropicolinic acid substrate.

Mus musculus Major urinary protein

Uniprot ID: P02762

Mouse major urinary proteins are a family of monomeric sequence-similar proteins that bind pheromones that are released from the urine of male mice. These pheromones affect the sexual behavior of females. The pheromone binding site is a deep pocket lined with nonpolar amino acids and has consequently been used as an archetypal hydrophobic binding site to investigate factors influencing hydrophobic small-molecule binding.

Escherichia coli Periplasmic dipeptide binding protein DppA

Uniprot ID: P23847

DppA is a component of the dipeptide ATP-binding cassette transporter, a multiprotein complex that is responsible for transport of dipeptides across the cytoplasmic inner membrane of enteric bacteria.

Salmonella typhimurium/enterica Oligopeptide binding protein OppA

Uniprot ID: P06202

OppA is a component of the oligopeptide permease peptide transport system. It is evolutionarily related to the dipeptide binding proteins but has a larger binding pocket capable of binding peptides up to five amino acids long with high affinity.

Homo sapiens Thrombin (in complex *Hirudo medicinalis* Hirudin)

Uniprot ID: P00734 (thrombin) & P01050 (hirudin)

Thrombin is a protease that is important in the coagulation of blood, in which process the protease specifically cleaves and activates fibrinogen, the clot-forming protein. Thrombin is a serine protease, in which the active site contains a serine-apartate-histidine catalytic triad, arranged such that the serine catalyzes hydrolysis of the peptide bond of a bound substrate protein or peptide. The structure of the active site is complementary to a limited number of peptide sequences. There is specificity for cleavage of peptides to the C-terminal side of a proline-arginine sequence. The series of ligands below contain or structurally mimic some of the features of this dipeptide sequence. The enzyme is inhibited by a peptide derived from the leech anticoagulant, hirudin and stabilized by Na^+ binding. Both these binding events are remote from the active site and there is no direct interaction with the titrated ligands.

Homo sapiens Thrombin (in complex *Hirudo medicinalis* Hirudin 2)

Uniprot ID: P00734 (thrombin) & P09945 (hirudin2)

For the ligands in this set of data, titrations in different buffers show a fractional net change in protonation state of the protein and ligand. Enthalpy changes have been corrected for buffer ionization effects as reported in Ref. 17. The thermodynamics of the very high affinity benzylsulfonyl-containing compounds have been determined via competition with the other two ligands.

Bos taurus Trypsin

Uniprot ID: P00760

Trypsin is a serine protease important in the digestion of proteins and also a useful model system due to it being readily purifiable from animal sources and its stability under laboratory conditions. Trypsin is evolutionarily related to thrombin and cleaves a similar but broader range of peptides, in this case preferentially to the C-terminal side of arginine or lysine residues. Benzamidine-based inhibitors mimic the arginine side-chain interactions. These inhibitors bind as the benzamidinium ion to the carboxylate side chain of an active-site aspartic acid. The enzyme is stabilized by Ca^{2+} ion binding, but this binding is remote from the active site with no direct interaction with substrates or inhibitors. Experiments with titrations in different buffers show no net change in protonation upon formation of the benzamadine complex at pH 8.0, although fractional protonation changes do occur at lower pH. Because of the chemical and structural similarity of this series of compounds, it is assumed that no protonation corrections are required for any of the compounds in the series.

Thermodynamic Data of Protein-Ligand Reactions

Sequence modification	Ligand	$\Delta_b G°$/ kJ mol^{-1}	$\Delta_b H°$/ kJ mol^{-1}	$-T\Delta_b S°$/ kJ mol^{-1}	T/K	Ref.	Buffer composition	Alternative systematic ligand name
***Enterococcus faecalis* Aminoglycoside 3'-phosphotransferase (P0A3Y5)**								
None	Amikacin	-23.8	-74.9	51.0	310.15	4	50 mM Tris, 100 mM KCl, pH 7.5	
	Kanamycin A	-31.0	-187.9	156.9	310.15	4	50 mM Tris, 100 mM KCl, pH 7.5	
	Kanamycin B	-38.5	-172.8	134.3	310.15	4	50 mM Tris, 100 mM KCl, pH 7.5	
	Lividomycin A	-36.0	-239.7	203.8	310.15	4	50 mM Tris, 100 mM KCl, pH 7.5	
	Neomycin B	-38.9	-233.0	194.1	310.15	4	50 mM Tris, 100 mM KCl, pH 7.5	
	Paramomycin I	-37.7	-183.7	146.0	310.15	4	50 mM Tris, 100 mM KCl, pH 7.5	
	Ribostamycin	-33.9	-107.1	73.2	310.15	4	50 mM Tris, 100 mM KCl, pH 7.5	
	Tobramycin	-36.4	-196.6	160.2	310.15	4	50 mM Tris, 100 mM KCl, pH 7.5	
	Amikacin	-26.4	-117.2	90.8	310.15	4	50 mM Tris, 100 mM KCl, 1.5 mM $CaCl_2$, 1 mM ATP, pH 7.5	
	Kanamycin A	-35.6	-129.7	94.1	310.15	4	50 mM Tris, 100 mM KCl, 1.5 mM $CaCl_2$, 1 mM ATP, pH 7.5	

Sequence modification	Ligand	$\Delta_b G°$/ kJ mol⁻¹	$\Delta_b H°$/ kJ mol⁻¹	$-T\Delta_b S°$/ kJ mol⁻¹	T/K	Ref.	Buffer composition	Alternative systematic ligand name
	Kanamycin B	-39.3	-121.8	82.4	310.15	4	50 mM Tris, 100 mM KCl, 1.5 mM $CaCl_2$, 1 mM ATP, pH 7.5	
	Lividomycin A	-36.4	-153.1	116.7	310.15	4	50 mM Tris, 100 mM KCl, 1.5 mM $CaCl_2$, 1 mM ATP, pH 7.5	
	Neomycin B	-41.0	-151.9	110.9	310.15	4	50 mM Tris, 100 mM KCl, 1.5 mM $CaCl_2$, 1 mM ATP, pH 7.5	
	Paramomycin I	-40.6	-116.7	76.1	310.15	4	50 mM Tris, 100 mM KCl, 1.5 mM $CaCl_2$, 1 mM ATP, pH 7.5	
	Ribostamycin	-38.9	-60.7	21.8	310.15	4	50 mM Tris, 100 mM KCl, 1.5 mM $CaCl_2$, 1 mM ATP, pH 7.5	
	Tobramycin	-39.3	-128.4	89.1	310.15	4	50 mM Tris, 100 mM KCl, 1.5 mM $CaCl_2$, 1 mM ATP, pH 7.5	
***Homo sapiens* Aurora A kinase (O14965)**								
Residues 123-390 of the full-length protein with a T287D mutation	4-{[4-(Phenylamino)pyrimidin-2-yl] amino}benzoic acid	-42.3	-58.6	16.3	298.15	5	100 mM Na/K phosphate, pH 7.4	
	2-({2-[(4-Carboxyphenyl)amino] pyrimidin-4- yl}amino)benzoic acid	-42.6	-56.9	14.3	298.15	5	100 mM Na/K phosphate, pH 7.4	
	4-{[4-(Biphenyl2-ylamino) pyrimidin-2-yl]amino}benzoic acid	-37.2	-13.7	-23.5	298.15	5	100 mM Na/K phosphate, pH 7.4	
	4-[(4-{[2-(Trifluoromethyl)phenyl]amino} pyrimidin- 2-yl)amino]benzoic acid	-41.7	-59.3	17.6	298.15	5	100 mM Na/K phosphate, pH 7.4	
	4-[(4-{[2-(Trifluoromethoxy)phenyl] amino}pyrimidin- 2-yl)amino]benzoic acid	-42.2	-46.0	3.8	298.15	5	100 mM Na/K phosphate, pH 7.4	
	4-({4-[(2-Fluorophenyl)amino] pyrimidin-2- yl}amino)benzoic acid	-44.5	-70.1	25.7	298.15	5	100 mM Na/K phosphate, pH 7.4	
	4-({4-[(2-Chlorophenyl)amino] pyrimidin-2- yl}amino)benzoic acid	-44.7	-71.1	26.5	298.15	5	100 mM Na/K phosphate, pH 7.4	
	4-({4-[(2-Bromophenyl)amino] pyrimidin-2-yl}amino)benzoic acid	-45.0	-62.3	17.3	298.15	5	100 mM Na/K phosphate, pH 7.4	
	4-({4-[(2-Cyanophenyl)amino] pyrimidin-2-yl}amino)benzoic acid	-43.4	-65.3	21.9	298.15	5		
***Thermotoga maritima* β-glucosidase A (Q08638)**								
Residues 2-446 of the full-length protein with a N-terminal hexa-His tag of sequence MGSSHHHHHHSSGLVPRGSHMAS	1-Deoxynojirimycin	-27.9	-22.6	-5.3	298.15	7	100 mM Na citrate, pH 5.8	(2*R*,3*R*,4*R*,5*S*)-2-(Hydroxymethyl) piperidine-3,4,5-triol
	Isofagomine	-41.6	-29.6	-12.0	298.15	7	100 mM Na citrate, pH 5.8	(3*R*,4*R*,5*R*)-5-(Hydroxymethyl) piperidine-3,4-diol
	Noeuromycin	-37.9	-41.4	3.5	298.15	7	100 mM Na citrate, pH 5.8	(2*R*,3*S*,4*R*,5*R*)-5-(Hydroxymethyl) piperidine-2,3,4-triol
	Tetrahydrooxazine	-35.8	-47.4	11.6	298.15	7	100 mM Na citrate, pH 5.8	(4*R*,5*S*,6*R*)-6-(Hydroxymethyl) oxazinane-4,5-diol
	Azafagomine	-41.0	-49.4	8.4	298.15	7	100 mM Na citrate, pH 5.8	(3*R*,4*R*,5*R*)-3-(Hydroxymethyl)-3,4,5,6-tetrahydropyridazine-4,5-diol
	Castanospermine	-32.4	-28.9	-3.5	298.15	7	100 mM Na citrate, pH 5.8	(1*S*,6*S*,7*R*,8*R*,8a*R*)-1,2,3,5,6,7,8,8a-Octahydroindolizine-1,6,7,8-tetrol
	Calystegine B2	-31.3	-15.7	-15.6	298.15	7	100 mM Na citrate, pH 5.8	(1*R*,2*S*,3*R*,4*S*,5*R*)-8-Azabicyclo[3.2.1] octane-2,3,4,5-tetrol
	Isofagomine lactam	-37.3	-60.1	22.8	298.15	7	100 mM Na citrate, pH 5.8	(3*S*,4*R*,5*R*)-3,4-Dihydroxy-5-(hydroxymethyl)piperidin-2-one
	Gluco-hydroximolactam	-36.6	-32.8	-3.9	298.15	7	100 mM Na citrate, pH 5.8	(2*Z*,3*S*,4*S*,5*R*,6*R*)-2-Hydroxyimino-6-(hydroxymethyl)piperidine-3,4,5-triol
	Galacto-hydroximolactam	-34.0	-27.4	-6.6	298.15	7	100 mM Na citrate, pH 5.8	(3*S*,4*S*,5*S*,6*R*)-2-Hydroxyimino-6-(hydroxymethyl)piperidine-3,4,5-triol
	Glucotetrazole	-37.8	-50.7	13.0	298.15	7	100 mM Na citrate, pH 5.8	(5*R*,6*R*,7*S*,8*S*)-5-(Hydroxymethyl)-5,6,7,8-tetrahydro-[1,2,3,4] tetrazolo[5,1-f]pyridine-6,7,8-triol
	Glucoimidazole	-41.4	-40.9	-0.5	298.15	7	100 mM Na citrate, pH 5.8	(5*R*,6*R*,7*S*,8*S*)-5-(Hydroxymethyl)-5,6,7,8-tetrahydro-1*H*-imidazo[2,1-f] pyridin-4-ium-6,7,8-triol
	2-Phenylethyl-glucoimidazole	-45.8	-22.8	-23.0	298.15	7	100 mM Na citrate, pH 5.8	(5*R*,6*R*,7*S*,8*S*)-5-(Hydroxymethyl)-2-phenethyl-5,6,7,8-tetrahydro-1*H*-imidazo[2,1-f]pyridin-4-ium-6,7,8-triol
	2-Phenylaminomethylglucoimidazole	-45.5	-38.7	-6.8	298.15	7	100 mM Na citrate, pH 5.8	(5*R*,6*R*,7*S*,8*S*)-5-(Hydroxymethyl)-2-(phenylazanylmethyl)-5,6,7,8-tetrahydroimidazo[2,1-f]pyridine-6,7,8-triol

Sequence modification	Ligand	$\Delta_b G°$/ kJ mol^{-1}	$\Delta_b H°$/ kJ mol^{-1}	$-T\Delta_b S°$/ kJ mol^{-1}	*T*/K	Ref.	Buffer composition	Alternative systematic ligand name
	2-Methoxycarbonylglucoimidazole	-40.7	-34.7	-6.0	298.15	7	100 mM Na citrate, pH 5.8	(5*R*,6*R*,7*S*,8*S*)-2-[(S)-Hydroxy-methoxy-methyl]-5-(hydroxymethyl)-5,6,7,8-tetrahydro-1*H*-imidazo[2,1-f] pyridin-4-ium-6,7,8-triol
	2-Methoxycarbonylmethylglucoimidazole	-41.8	-40.3	-1.5	298.15	7	100 mM Na citrate, pH 5.8	Methyl 2-[(5*R*,6*R*,7*S*,8*S*)-6,7,8-trihydroxy-5-(hydroxymethyl)-5,6,7,8-tetrahydro-1*H*-imidazo[1,2-a] pyridin-4-ium-2-yl]ethanoate
	2-Carboxylate-glucosimidazole	-36.3	-41.0	4.7	298.15	7	100 mM Na citrate, pH 5.8	(5*R*,6*R*,7*S*,8*S*)-6,7,8-Trihydroxy-5-(hydroxymethyl)-5,6,7,8-tetrahydro-1*H*-imidazo[1,2-a]pyridin-4-ium-2-carboxylic acid
	2-Carboxymethylglucosimidazole	-40.0	-48.2	8.2	298.15	7	100 mM Na citrate, pH 5.8	2-[(5*R*,6*R*,7*S*,8*S*)-6,7,8-Trihydroxy-5-(hydroxymethyl)-5,6,7,8-tetrahydro-1*H*-imidazo[1,2-a]pyridin-4-ium-2-yl] ethanoic acid
***Escherichia coli* Dihydrodipicolinate reductase (P04036)**								
None	β-Nicotinamide adenine dinucleotide 2'-phosphate, reduced	-32.7	-45.9	13.3	301.15	8	20 mM Hepes/KOH, pH 7.8	
	β-Nicotinamide hypoxanthine dinucleotide 2'-phosphate, reduced	-31.1	-35.7	4.6	301.15	8	20 mM Hepes/KOH, pH 7.8	
	β-Thionicotinamide adenine dinucleotide 2'-phosphate, reduced	-33.0	-53.1	20.1	301.15	8	20 mM Hepes/KOH, pH 7.8	
	β-Nicotinamide adenine dinucleotide 3'-phosphate, reduced	-30.5	-51.0	20.5	301.15	8	20 mM Hepes/KOH, pH 7.8	
	β-Nicotinamide adenine dinucleotide, reduced	-36.6	-37.4	0.8	301.15	8	20 mM Hepes/KOH, pH 7.8	
	β-Nicotinamide hypoxanthine dinucleotide, reduced	-33.6	-35.0	1.4	301.15	8	20 mM Hepes/KOH, pH 7.8	
	β-Thionicotinamide adenine dinucleotide, reduced	-37.7	-44.1	6.4	301.15	8	20 mM Hepes/KOH, pH 7.8	
	3-Acetylpyridine adenine dinucleotide, reduced	-37.7	-34.4	-3.2	301.15	8	20 mM Hepes/KOH, pH 7.8	
***Mus musculus* Major urinary protein (P02762)**								
Residues 19-180 of the full-length protein only	2-*sec*-Butyl-4,5-dihydrothiazole	-35.0	-46.9	11.9	298.75	9	10 mM Na/K phosphate, pH 6.3	2-(1-Methylpropyl)-4,5-dihydro-1,3-thiazole
	2-Isobutyl-4,5-dihydrothiazole	-37.8	-49.8	12.0	298.85	9	10 mM Na/K phosphate, pH 6.3	2-(2-Methylpropyl)-4,5-dihydro-1,3-thiazole
	2-Propyl-4,5-dihydrothiazole	-34.1	-44.8	10.7	298.65	9	10 mM Na/K phosphate, pH 6.3	2-Propyl-4,5-dihydro-1,3-thiazole
	2-Isopropyl-4,5-dihydrothiazole	-32.7	-45.2	12.5	298.55	9	10 mM Na/K phosphate, pH 6.3	2-(1-Methylethyl)-4,5-dihydro-1,3-thiazole
	2-Ethyl-4,5-dihydrothiazole	-28.9	-41.8	12.9	298.55	9	10 mM Na/K phosphate, pH 6.3	2-Ethyl-4,5-dihydro-1,3-thiazole
	2-Methyl-4,5-dihydrothiazole	-23.4	-40.6	17.1	298.45	9	10 mM Na/K phosphate, pH 6.3	2-Methyl-4,5-dihydro-1,3-thiazole
Residues 19-180 of the full-length protein with a N-terminal hexa-His tag (MRGSHHHHHHGS) and an N68K mutation	2-Methoxy-3-isopropylpyrazine	-33.9	-44.5	10.6	308.00	10	Phosphate buffered saline, pH 7.4	2-Methoxy-3-(1-methylethyl)-pyrazine
	2-Methoxy-3-isobutylpyrazine	-38.5	-47.9	9.4	308.00	10	Phosphate buffered saline, pH 7.4	2-Methoxy-3-(2-methylpropyl)-pyrazine
	Pentan-1-ol	-23.1	-41.0	17.9	300.00	11	Phosphate buffered saline, pH 7.4	
	Hexan-1-ol	-28.3	-47.6	19.3	300.00	11	Phosphate buffered saline, pH 7.4	
	Heptan-1-ol	-32.5	-53.4	20.9	300.00	11	Phosphate buffered saline, pH 7.4	
	Octan-1-ol	-35.6	-58.0	22.4	300.00	11	Phosphate buffered saline, pH 7.4	
	Nonan-1-ol	-38.6	-68.6	30.0	300.00	11	Phosphate buffered saline, pH 7.4	
Residues 19-180 of the full-length protein with a N-terminal hexa-His tag (MRGSHHHHHHGS) and N68K and Y138F mutations	2-Methoxy-3-isobutylpyrazine	-31.4	-35.3	3.9	298.00	12	Phosphate buffered saline, pH 7.4	2-Methoxy-3-(2-methylpropyl)-pyrazine
	2-Methoxy-3-isobutylpyrazine	-33.6	-35.9	2.3	308.00	12	Phosphate buffered saline, pH 7.4	2-Methoxy-3-(2-methylpropyl)-pyrazine
***Escherichia coli* Periplasmic dipeptide binding protein DppA (P23847)**								
None	L-Alanine-L-threonine	-44.4	-53.3	9.0	298.15	13	50 mM Na/K phosphate, pH 7.0	

Sequence modification	Ligand	$\Delta_b G^{\circ\prime}$/ kJ mol^{-1}	$\Delta_b H^{\circ}$/ kJ mol^{-1}	$-T\Delta_b S^{\circ}$/ kJ mol^{-1}	T/K	Ref.	Buffer composition	Alternative systematic ligand name
	L-Alanine-L-valine	-44.0	-41.3	-2.7	298.15	13	50 mM Na/K phosphate, pH 7.0	
	L-Lysine-L-alanine	-43.5	-49.1	5.6	298.15	13	50 mM Na/K phosphate, pH 7.0	
	L-Alanine-L-alanine	-43.4	-56.1	12.7	298.15	13	50 mM Na/K phosphate, pH 7.0	
	L-Alanine-L-serine	-43.3	-52.6	9.3	298.15	13	50 mM Na/K phosphate, pH 7.0	
	L-Aanine-L-lysine	-43.1	-45.1	2.0	298.15	13	50 mM Na/K phosphate, pH 7.0	
	L-Alanine-L-leucine	-42.8	-31.9	-10.8	298.15	13	50 mM Na/K phosphate, pH 7.0	
	L-Alanine-L-isoleucine	-41.3	-37.3	-4.0	298.15	13	50 mM Na/K phosphate, pH 7.0	
	L-Alanine-L-phenylalanine	-41.1	-51.1	10.0	298.15	13	50 mM Na/K phosphate, pH 7.0	
	L-Serine-L-alanine	-41.1	-53.9	12.8	298.15	13	50 mM Na/K phosphate, pH 7.0	
	L-Threonine-L-serine	-40.5	-75.6	35.1	298.15	13	50 mM Na/K phosphate, pH 7.0	
	L-Alanine-L-tyrosine	-38.6	-51.8	13.1	298.15	13	50 mM Na/K phosphate, pH 7.0	
	L-Alanine-L-proline	-38.6	-37.3	-1.3	298.15	13	50 mM Na/K phosphate, pH 7.0	
	L-Alanine-L-tryptophan	-37.7	-46.4	8.7	298.15	13	50 mM Na/K phosphate, pH 7.0	
	L-Alanine-L-aspartate	-33.9	-34.9	1.0	298.15	13	50 mM Na/K phosphate, pH 7.0	
	L-Glycine-L-leucine	-33.8	-43.6	9.7	298.15	13	50 mM Na/K phosphate, pH 7.0	
	L-Glycine-L-tyrosine	-31.7	-35.5	3.8	298.15	13	50 mM Na/K phosphate, pH 7.0	
	L-Leucine-L-tryptophan	-30.6	-53.1	22.5	298.15	13	50 mM Na/K phosphate, pH 7.0	
	L-Aspartate-L-alanine	-29.7	-43.2	13.4	298.15	13	50 mM Na/K phosphate, pH 7.0	
Salmonella typhimurium/enterica	**Oligopeptide binding protein OppA (P06202)**							
Residues 27-543 of the full-length protein only	L-Lysine-L-alanine-L-lysine	-41.4	20.1	-61.5	298.15	14	50 mM Sodium phosphate pH 7.0	
	L-Lysine-L-cysteine-L-lysine	-40.7	7.9	-48.6	298.15	14	50 mM Sodium phosphate pH 7.0	
	L-Lysine-L-aspartate-L-lysine	-29.8	8.1	-37.9	298.15	14	50 mM Sodium phosphate pH 7.0	
	L-Lysine-L-glutamate-L-lysine	-39.0	11.3	-50.3	298.15	14	50 mM Sodium phosphate pH 7.0	
	L-Lysine-L-phenylalanine-L-lysine	-41.5	22.0	-63.5	298.15	14	50 mM Sodium phosphate pH 7.0	
	L-Lysine-L-glycine-L-lysine	-33.6	14.1	-47.7	298.15	14	50 mM Sodium phosphate pH 7.0	
	L-Lysine-L-histidine-L-lysine	-39.3	20.6	-59.9	298.15	14	50 mM Sodium phosphate pH 7.0	
	L-Lysine-L-isoleucine-L-lysine	-38.2	20.5	-58.7	298.15	14	50 mM Sodium phosphate pH 7.0	
	L-Lysine-L-Lysine-L-lysine	-31.6	39.4	-71.0	298.15	14	50 mM Sodium phosphate pH 7.0	
	L-Lysine-L-leucine-L-lysine	-34.0	24.6	-58.6	298.15	14	50 mM Sodium phosphate pH 7.0	
	L-Lysine-L-methionine-L-lysine	-40.5	14.6	-55.1	298.15	14	50 mM Sodium phosphate pH 7.0	
	L-Lysine-L-asparagine-L-lysine	-40.2	7.7	-47.9	298.15	14	50 mM Sodium phosphate pH 7.0	
	L-Lysine-L-proline-L-lysine	-30.2	16.6	-46.8	298.15	14	50 mM Sodium phosphate pH 7.0	
	L-Lysine-L-glutamine-L-lysine	-42.4	11.4	-53.8	298.15	14	50 mM Sodium phosphate pH 7.0	
	L-Lysine-L-arginine-L-lysine	-33.8	36.0	-69.8	298.15	14	50 mM Sodium phosphate pH 7.0	
	L-Lysine-L-serine-L-lysine	-42.0	8.9	-50.9	298.15	14	50 mM Sodium phosphate pH 7.0	
	L-Lysine-L-threonine-L-lysine	-40.6	17.3	-57.9	298.15	14	50 mM Sodium phosphate pH 7.0	
	L-Lysine-L-valine-L-lysine	-41.9	22.4	-64.3	298.15	14	50 mM Sodium phosphate pH 7.0	

Sequence modification	Ligand	$\Delta_b G^{\circ}$/ kJ mol^{-1}	$\Delta_b H^{\circ}$/ kJ mol^{-1}	$-T\Delta_b S^{\circ}$/ kJ mol^{-1}	T/K	Ref.	Buffer composition	Alternative systematic ligand name
	L-Lysine-L-tryprophan-L-lysine	-39.3	29.3	-68.6	298.15	14	50 mM Sodium phosphate pH 7.0	
	L-Lysine-L-tyrosine-L-lysine	-37.6	20.7	-58.3	298.15	14	50 mM Sodium phosphate pH 7.0	
***Homo sapiens* Thrombin (P00734) in complex *Hirudo medicinalis* Hirudin (P01050)**								
Residues 328-363 and 364-622 of full-length thrombin. Residues 54-64 of hirudin where residue 63 is a non-standard amino acid O-sulpho-L-tyrosine.	Titrations of three of the ligands in the series below in different buffers show no difference in observed enthalpy changes, implying that there is no net change in protonation upon formation of the complex . Because of their equivalent complement of ligand titratable groups and crystallographic analyses showing a high degree of structural similarity of the complexes, it is assumed that no buffer ionization correction is required for any of these ligands.							
	D-Phenylalanyl-N-(3-fluorobenzyl)-L-prolinamide	-31.2	-13.1	-18.1	298.15	15	50 mM Tris,100 mM NaCl, 2.5% Dimethylsulfoxide, 0.1% Polyethylene glycol 8000, pH 7.8	
	D-Phenylalanyl-N-(3-chlorobenzyl)-L-prolinamide	-35.4	-37.1	1.7	298.15	15	50 mM Tris,100 mM NaCl, 2.5% Dimethylsulfoxide, 0.1% Polyethylene glycol 8000, pH 7.8	
	D-Phenylalanyl-N-(3-bromobenzyl)-L-prolinamide	-35.8	-34.5	-1.3	298.15	15	50 mM Tris,100 mM NaCl, 2.5% Dimethylsulfoxide, 0.1% Polyethylene glycol 8000, pH 7.8	
	D-Phenylalanyl-N-(3-iodobenzyl)-L-prolinamide	-34.5	-38.0	3.5	298.15	15	50 mM Tris,100 mM NaCl, 2.5% Dimethylsulfoxide, 0.1% Polyethylene glycol 8000, pH 7.8	
	D-Phenylalanyl-N-(3-methylbenzyl)-L-prolinamide	-34.8	-28.5	-6.3	298.15	15	50 mM Tris,100 mM NaCl, 2.5% Dimethylsulfoxide, 0.1% Polyethylene glycol 8000, pH 7.8	
	D-Phenylalanyl-N-(3-ethylbenzyl)-L-prolinamide	-32.9	-16.5	-16.4	298.15	15	50 mM Tris,100 mM NaCl, 2.5% Dimethylsulfoxide, 0.1% Polyethylene glycol 8000, pH 7.8	
	D-Phenylalanyl-N-(4-benzenecarboximidamide)-L-prolinamide	-46.1	-40.1	-6.0	298.15	15	50 mM Tris,100 mM NaCl, 2.5% Dimethylsulfoxide, 0.1% Polyethylene glycol 8000, pH 7.8	
	bis(3-Phenyl)-D-alanyl-N-(4-benzenecarboximidamide)-L-prolinamide	-48.5	-47.5	-1.0	298.15	16	50 mM Tris,100 mM NaCl, 2.5% Dimethylsulfoxide, 0.1% Polyethylene glycol 8000, pH 7.8	
	Methyl-D-alanyl-N-(4-benzenecarboximidamide)-L-prolinamide	-40.1	-38.7	-1.4	298.15	16	50 mM Tris,100 mM NaCl, 2.5% Dimethylsulfoxide, 0.1% Polyethylene glycol 8000, pH 7.8	
	D-Phenylalanyl-N-benzyl-L-prolinamide	-31.7	-13.6	-18.1	298.15	15	50 mM Tris,100 mM NaCl, 2.5% Dimethylsulfoxide, 0.1% Polyethylene glycol 8000, pH 7.8	
	D-Phenylalanyl-N-(3,6-dichloro-benzyl)-L-prolinamide	-38.4	-41.3	2.9	298.15	15	50 mM Tris,100 mM NaCl, 2.5% Dimethylsulfoxide, 0.1% Polyethylene glycol 8000, pH 7.8	
	D-Phenylalanyl-N-((3-chloro-6-methyl)-benzyl)-L-prolinamide	-37.2	-33.5	-3.7	298.15	15	50 mM Tris,100 mM NaCl, 2.5% Dimethylsulfoxide, 0.1% Polyethylene glycol 8000, pH 7.8	

Sequence modification	Ligand	$\Delta_b G°$/ kJ mol⁻¹	$\Delta_b H°$/ kJ mol⁻¹	$-T\Delta_b S°$/ kJ mol⁻¹	T/K	Ref.	Buffer composition	Alternative systematic ligand name
	D-Phenylalanyl-N-((3-chloro-6-fluoro)-benzyl)-L-prolinamide	-37.3	-41.0	3.7	298.15	15	50 mM Tris,100 mM NaCl, 2.5% Dimethylsulfoxide, 0.1% Polyethylene glycol 8000, pH 7.8	
	D-Phenylalanyl-N-(3,6-dimethylbenzyl)-L-prolinamide	-34.4	-31.9	-2.5	298.15	15	50 mM Tris,100 mM NaCl, 2.5% Dimethylsulfoxide, 0.1% Polyethylene glycol 8000, pH 7.8	
	D-Phenylalanyl-N-(3-chlorobenzyl)-L-prolinamide	-35.4	-37.1	1.7	298.15	16	50 mM Tris,100 mM NaCl, 2.5% Dimethylsulfoxide, 0.1% Polyethylene glycol 8000, pH 7.8	
	Bis(3-phenyl)-D-alanyl-N-(3-chlorobenzyl)-L-prolinamide	-39.6	-45.6	6.0	298.15	16	50 mM Tris,100 mM NaCl, 2.5% Dimethylsulfoxide, 0.1% Polyethylene glycol 8000, pH 7.8	
	3-Cyclohexyl-D-alanyl-N-(3-chlorobenzyl)-L-prolinamide	-38.9	-35.6	-3.3	298.15	16	50 mM Tris,100 mM NaCl, 2.5% Dimethylsulfoxide, 0.1% Polyethylene glycol 8000, pH 7.8	
	Methyl-D-alanyl-N-(3-chlorobenzyl)-L-prolinamide	-31.3	-33.5	2.2	298.15	16	50 mM Tris,100 mM NaCl, 2.5% Dimethylsulfoxide, 0.1% Polyethylene glycol 8000, pH 7.8	
***Homo sapiens* Thrombin (P00734) in complex *Hirudo medicinalis* Hirudin 2 (P09945)**								
Residues 328-363 and 364-622 of full-length thrombin where residue N416 is gycosylated by N-acetyl-D-glucosamine. Residues 53-65 of hirudin where residue 63 is a non-standard amino acid O-sulpho-L-tyrosine.	For the ligands below titrations in different buffers show a fractional net change in protonation state of the protein and ligand. Enthalpy changes have been corrected for buffer ionization effects as reported in [17]. The thermodynamics of the very high affinity benzylsulfonyl containing compounds have been determined via competition with the other two ligands.							
	N-(Benzylsulfonyl)-glycyl-N-[2-(aminomethyl)- 5-chlorobenzyl]-L-prolinamide	-47.8	-35.4	-12.4	298.15	17	50 mM Tris,100 mM NaCl, 3% Dimethylsulfoxide, 0.1% Polyethylene glycol 8000, pH 7.8	
	N-(Benzylsulfonyl)-D-alanyl-N-[2-(aminomethyl)- 5-chlorobenzyl]-L-prolinamide	-47.2	-32.1	-15.1	298.15	17	50 mM Tris,100 mM NaCl, 3% Dimethylsulfoxide, 0.1% Polyethylene glycol 8000, pH 7.8	
	N-(Benzylsulfonyl)-D-valyl-N-[2-(aminomethyl)- 5-chlorobenzyl]-L-prolinamide	-54.2	-36.2	-18.0	298.15	17	50 mM Tris,100 mM NaCl, 3% Dimethylsulfoxide, 0.1% Polyethylene glycol 8000, pH 7.8	
	N-(Benzylsulfonyl)-D-leucyl-N-[2-(aminomethyl)- 5-chlorobenzyl]-L-prolinamide	-54.3	-30.1	-24.2	298.15	17	50 mM Tris,100 mM NaCl, 3% Dimethylsulfoxide, 0.1% Polyethylene glycol 8000, pH 7.8	
	N-(Benzylsulfonyl)-3-cyclohexyl-D-alanyl-N-[2-(aminomethyl)-5-chlorobenzyl]-L-prolinamide	-54.5	-28.7	-25.8	298.15	17	50 mM Tris,100 mM NaCl, 3% Dimethylsulfoxide, 0.1% Polyethylene glycol 8000, pH 7.8	
	N-(Benzylsulfonyl)-glycyl-N-[4-carbamimidoylbenzyl]-L-prolinamide	-44.9	-14.3	-30.6	298.15	17	50 mM Tris,100 mM NaCl, 3% Dimethylsulfoxide, 0.1% Polyethylene glycol 8000, pH 7.8	
	N-(Benzylsulfonyl)-D-alanyl-N-[4-carbamimidoylbenzyl]-L-prolinamide	-42.6	-15.9	-26.7	298.15	17	50 mM Tris,100 mM NaCl, 3% Dimethylsulfoxide, 0.1% Polyethylene glycol 8000, pH 7.8	

Biochem

Sequence modification	Ligand	$\Delta_b G^{\circ}$/ kJ mol^{-1}	$\Delta_b H^{\circ}$/ kJ mol^{-1}	$-T\Delta_b S^{\circ}$/ kJ mol^{-1}	T/K	Ref.	Buffer composition	Alternative systematic ligand name
	N-(Benzylsulfonyl)-D-valyl-*N*-[4-carbamimidoylbenzyl]-L-prolinamide	-47.1	-13.8	-33.3	298.15	17	50 mM Tris,100 mM NaCl, 3% Dimethylsulfoxide, 0.1% Polyethylene glycol 8000, pH 7.8	
	N-(Benzylsulfonyl)-D-leucyl-*N*-[4-carbamimidoylbenzyl]-L-prolinamide	-51.9	-10.4	-41.5	298.15	17	50 mM Tris,100 mM NaCl, 3% Dimethylsulfoxide, 0.1% Polyethylene glycol 8000, pH 7.8	
	N-(Benzylsulfonyl)-3-cyclohexyl-D-alanyl-*N*-[4-carbamimidoylbenzyl]-L-prolinamide	-53.7	-11.4	-42.3	298.15	17	50 mM Tris,100 mM NaCl, 3% Dimethylsulfoxide, 0.1% Polyethylene glycol 8000, pH 7.8	
	D-Phenylalanyl-*N*-(3,6-dichloro-benzyl)-L-prolinamide	-37.6	-37.5	-0.1	298.15	17	50 mM Tris,100 mM NaCl, 3% Dimethylsulfoxide, 0.1% Polyethylene glycol 8000, pH 7.8	
	N-(Benzylethyl)-L-phenylanalyl-*N*-[4-carbamimidoylbenzyl]-L-prolinamide	-37.9	-17.1	-20.8	298.15	17	50 mM Tris,100 mM NaCl, 3% Dimethylsulfoxide, 0.1% Polyethylene glycol 8000, pH 7.8	
***Bos taurus* Trypsin (P00760)**								
Residues 24-246 of the full-length protein only	Benzamdine	-26.6	-18.9	-7.7	298.25	18	50 mM Tris, 10 mM $CaCl_2$, pH 8.0	Benzenecarboximidamide
	4-Aminobenzamidine	-29.1	-26.9	-2.2	298.25	18	50 mM Tris, 10 mM $CaCl_2$, pH 8.0	4-Aminobenzenecarboximidamide
	4-Methoxybenzamidine	-25.3	-15.7	-9.6	298.25	18	50 mM Tris, 10 mM $CaCl_2$, pH 8.0	4-Methoxybenzenecarboximidamide
	4-Carboxamidobenzamidine	-23.9	-12.3	-11.6	298.25	18	50 mM Tris, 10 mM $CaCl_2$, pH 8.0	4-Carboxyamidobenzenecarboximidamide
	4-Methylbenzamidine	-27.6	-18.5	-9.1	298.25	18	50 mM Tris, 10 mM $CaCl_2$, pH 8.0	4-Methylbenzenecarboximidamide
	4-Ethylbenzamidine	-25.4	-13.9	-11.5	298.15	19	50 mM Tris, 10 mM $CaCl_2$, pH 8.0	4-Ethylbenzenecarboximidamide
	4-Propylbenzamidine	-25.7	-12.7	-13.0	298.15	19	50 mM Tris, 10 mM $CaCl_2$, pH 8.0	4-Propylbenzenecarboximidamide
	4-Isopropylbenzamidine	-22.7	-7.0	-15.7	298.15	19	50 mM Tris, 10 mM $CaCl_2$, pH 8.0	4-(1-Methylethyl) benzenecarboximidamide
	4-Butylbenzamidine	-26.2	-9.9	-16.3	298.15	19	50 mM Tris, 10 mM $CaCl_2$, pH 8.0	4-Butylbenzenecarboximidamide
	4-Pentylbenzamidine	-27.2	-9.9	-17.3	298.15	19	50 mM Tris, 10 mM $CaCl_2$, pH 8.0	4-Pentylbenzenecarboximidamide
	4-Hexylbenzamidine	-29.2	-10.6	-18.6	298.15	19	50 mM Tris, 10 mM $CaCl_2$, pH 8.0	4-Hexylbenzenecarboximidamide

THERMODYNAMIC QUANTITIES FOR THE IONIZATION REACTIONS OF BUFFERS IN WATER

Robert N. Goldberg, Nand Kishore, and Rebecca M. Lennen

This table contains selected values for the pK, standard molar enthalpy of reaction $\Delta_r H°$, and standard molar heat-capacity change $\Delta_r C°_p$ for the ionization reactions of 64 buffers, many of which are relevant to biochemistry and to biology (Ref. 1). The values pertain to the temperature T = 298.15 K and the pressure p = 0.1 MPa. The standard state is the hypothetical ideal solution of unit molality. Column definitions for the table are as follows.

Column heading	Definition
Buffer	Name or common acronym of biological buffer, in bold above data line; acronyms used in this table are defined in the table "Biological Buffers" in this section
Reaction	Equation for ionization reaction
p*K*	Negative logarithm of equilibrium constant for ionization reaction
$\Delta_r H°$	Standard molar enthalpy of the ionization reaction, in kJ mol^{-1}
$\Delta_r C_p°$	Change to standard molar heat capacity at constant pressure, in J mol^{-1} K^{-1}

These data permit one to calculate values of the pK and of $\Delta_r H°$ at temperatures in the vicinity {$T \approx$ (274 K to 350 K)} of the reference temperature θ = 298.15 K by using the following equations (Ref. 2).

$$\Delta_r G_T^\circ = -RT \ln K_T = \ln(10) \cdot RT \cdot pK_T \quad (1)$$

$$R \ln K_T = -\left(\Delta_r G_\theta^\circ / \theta\right) + \Delta_r H_\theta^\circ \left\{(1/\theta) - (1/T)\right\} + \Delta_r C_{p\theta}^\circ \left\{(\theta/T) - 1 + \ln(T/\theta)\right\} \quad (2)$$

$$\Delta_r H_T^\circ = \Delta_r H_\theta^\circ + \Delta_r C_{p\theta}^\circ (T - \theta) \quad (3)$$

Here, $\Delta_r G°$ is the standard molar Gibbs energy change and K is the equilibrium constant for a reaction; R is the gas constant (8.314 462 J K^{-1} mol^{-1}). The subscripts T and θ denote the temperature to which a quantity pertains, the subscript p denotes constant pressure, and the subscript r denotes that the quantity refers to a reaction. Combination of equations (1) and (2) yields the following equation that gives pK as a function of temperature:

$$pK_T = -\{R \cdot \ln(10)\}^{-1}[-\{\ln(10) \cdot RT \cdot pK_\theta / \theta\} + \Delta_r H_\theta^\circ \{(1/\theta) - 1/T)\} \Delta_r C_{p\theta}^\circ \{(\theta/T) - 1 + \ln(T/\theta)\} \quad (4)$$

The above equations neglect higher order terms that involve temperature derivatives of $\Delta_r C°_p$. Also, it is important to recognize that the values of pK and $\Delta_r H°$ effectively pertain to ionic strength I = 0. However, the values of pK and $\Delta_r H°$ are almost always dependent on the ionic strength and the actual composition of the solution. These issues are discussed in Ref. 1, which also gives an approximate method for making appropriate corrections.

References

1. Goldberg, R. N., Kishore, N., and Lennen, R. M., Thermodynamic Quantities for the Ionization Reactions of Buffers, *J. Phys. Chem. Ref. Data*, 31, 231, 2002, <https://doi.org/10.1063/1.1416902>
2. Clarke, E. C. W., and Glew, D. N., *Trans. Faraday Soc.*, 62, 539-547, 1966, <https://doi.org/10.1039/tf9666200539>

Selected Values of Thermodynamic Quantities for the Ionization Reactions of Buffers in Water at T = 298.15 K and p = 0.1 MPa

Reaction	pK	$\Delta_r H°$/ kJ mol^{-1}	$\Delta_r C°_p$/ J mol^{-1} K^{-1}
ACES			
$HL^{\pm} = H^+ + L^-$, ($HL = C_4H_{10}N_2O_4S$)	6.847	30.43	-49
Acetate			
$HL = H^+ + L^-$, ($HL = C_2H_4O_2$)	4.756	-0.41	-142
ADA			
$H_3L^+ = H^+ + H_2L^{\pm}$, ($H_2L = C_6H_{10}N_2O_5$)	1.59		
$H_2L^{\pm} = H^+ + HL^-$	2.48	16.7	
$HL^- = H^+ + L^{2-}$	6.844	12.23	-144
2-Amino-2-methyl-1,3-propanediol			
$HL^+ = H^+ + L$, ($L = C_4H_{11}NO_2$)	8.801	49.85	-44
2-Amino-2-methyl-1-propanol			
$HL^+ = H^+ + L$, ($L = C_4H_{11}NO$)	9.694	54.05	-21[a]
3-Amino-1-propanesulfonic acid			
$HL = H^+ + L^-$, ($HL = C_3H_9NO_3S$)	10.2		
Ammonia			
$NH_4^+ = H^+ + NH_3$	9.245	51.95	8
AMPSO			
$HL^{\pm} = H^+ + L^-$, ($HL = C_7H_{17}NO_5S$)	9.138	43.19	-61
Arsenate			
$H_3AsO_4 = H^+ + H_2AsO_4^-$	2.31	-7.8	
$H_2AsO_4^- = H^+ + HAsO^{2-}$	7.05	1.7	
$HAsO_4^{2-} = H^+ + AsO_4^{3-}$	11.9	15.9	
Barbital			
$H_2L = H^+ + HL^-$, ($H_2L = C_8H_{12}N_2O_3$)	7.980	24.27	-135
$HL^- = H^+ + L^{2-}$	12.8		
BES			
$HL^{\pm} = H^+ + L^-$, ($HL = C_6H_{15}NO_5S$)	7.187	24.25	-2
Bicine			
$H_2L^+ = H^+ + HL^{\pm}$, ($HL = C_6H_{13}NO_4$)	2.0		
$HL^{\pm} = H^+ + L^-$	8.334	26.34	0
Bis-tris			
$H_3L^+ = H^+ + H_2L^{\pm}$, ($H_2L = C_8H_{19}NO_5$)	6.484	28.4	27

Reaction	pK	$\Delta_r H°$/ kJ mol^{-1}	$\Delta_r C°_p$/ J mol^{-1} K^{-1}
Bis-tris propane			
$H_2L^{2+} = H^+ + HL^+$, ($L = C_{11}H_{26}N_2O_6$)	6.65		
$HL^+ = H^+ + L$	9.10		
Borate			
$H_3BO_3 = H^+ + H_2BO_3^-$	9.237	13.8	-240[a]
Cacodylate			
$H_2L^+ = H^+ + HL$, ($HL = C_2H_6AsO_2$)	1.78	-3.5	
$HL = H^+ + L^-$	6.28	-3.0	-86
CAPS			
$HL^\pm = H^+ + L^-$, ($HL = C_9H_{19}NO_3S$)	10.499	48.1	57
CAPSO			
$HL^\pm = H^+ + L^-$, ($HL = C_9H_{19}NO_4S$)	9.825	46.67	21
Carbonate			
$H_2CO_3 = H^+ + HCO_3^-$	6.351	9.15	-371
$HCO_3^- = H^+ + CO^{2-}$	10.329	14.70	-249
CHES			
$HL^\pm = H^+ + L^-$, ($HL = C_8H_{17}NO_3S$)	9.394	39.55	9
Citrate			
$H_3L = H^+ + H_2L^-$, ($H_3L = C_6H_8O_7$)	3.128	4.07	-131
$H_2L^- = H^+ + HL^{2-}$	4.761	2.23	-178
$HL^{2-} = H^+ + L^{3-}$	6.396	-3.38	-254
L-Cysteine			
$H_3L^+ = H^+ + H_2L$, ($H_2L = C_3H_7NO_2S$)	1.71	-0.6[a]	
$H_2L = H^+ + HL^-$	8.36	36.1	-66[a]
$HL^- = H^+ + L^{2-}$	10.75	34.1	-204[a]
Diethanolamine			
$HL^+ = H^+ + L$, ($L = C_4H_{11}NO_2$)	8.883	42.08	36
Diglycolate			
$H_2L = H^+ + HL^-$, ($H_2L = C_4H_6O_5$)	3.05	-0.1	-142[a]
$HL^- = H^+ + L^{2-}$	4.37	-7.2	-138[a]
3,3-Dimethylglutarate			
$H_2L = H^+ + HL^-$, ($H_2L = C_7H_{12}O_4$)	3.70		
$HL^- = H^+ + L^{2-}$	6.34		
DIPSO			
$HL^\pm = H^+ + L^-$, ($HL = C_7H_{17}NO_6S$)	7.576	30.18	42
Ethanolamine			
$HL^+ = H^+ + L$, ($L = C_2H_7NO$)	9.498	50.52	26
***N*-Ethylmorpholine**			
$HL^+ = H^+ + L$, ($L = C_6H_{13}NO$)	7.77	27.4	
Glycerol 2-phosphate			
$H_2L = H^+ + HL^-$, ($H_2L = C_3H_9NO_6P$)	1.329	-12.2	-330
$HL^- = H^+ + L^{2-}$	6.650	-1.85	-212
Glycine			
$H_2L^+ = H^+ + HL^\pm$, ($HL = C_2H_5NO_2$)	2.351	4.00	-139
$HL^\pm = H^+ + L^-$	9.780	44.2	-57
Glycine amide			
$HL^+ = H^+ + L$, ($L = C_2H_6N_2O$)	8.04	42.9	

Reaction	pK	$\Delta_r H°$/ kJ mol^{-1}	$\Delta_r C°_p$/ J mol^{-1} K^{-1}
Glycylglycine			
$H_2L^+ = H^+ + HL^\pm$, ($HL = C_4H_8N_2O_3$)	3.140	0.11	-128
$HL^\pm = H^+ + L^-$	8.265	43.4	-16
Glycylglycylglycine			
$H_2L^+ = H^+ + HL^\pm$, ($HL = C_6H_{11}N_3O_4$)	3.224	0.84	
$HL^\pm = H^+ + L^-$	8.090	41.7	
HEPES			
$H_2L^+ = H^+ + HL^\pm$, ($HL = C_8H_{18}N_2O_4S$)	3.0[a]		
$HL^\pm = H^+ + L^-$	7.564	20.4	47
HEPPS			
$HL^\pm = H^+ + L^-$, ($HL = C_6H_{20}N_2O_4S$)	7.957	21.3	48
HEPPSO			
$HL^\pm = H^+ + L^-$, ($HL = C_9H_{20}N_2O_5S$)	8.042	23.70	47
L-Histidine			
$H_3L^{2+} = H^+ + H_2L^+$, ($HL = C_6H_9N_3O_2$)	1.54[b]	3.6	
$H_2L^+ = H^+ + HL$	6.07	29.5	176
$HL = H^+ + L^-$	9.34	43.8	-233
Hydrazine			
$H_2L^{2+} = H^+ + HL^+$, ($L = H_4N_2$)	-0.99	38.1	
$HL^+ = H^+ + L$	8.02	41.7	
Imidazole			
$HL^+ = H^+ + L$, ($L = C_3H_4N_2$)	6.993	36.64	-9
Maleate			
$H_2L = H^+ + HL^-$, ($H_2L = C_4H_4O_4$)	1.92	1.1	-21[a]
$HL^- = H^+ + L^{2-}$	6.27	-3.6	-31[a]
2-Mercaptoethanol			
$HL = H^+ + L^-$, ($HL = C_2H_6OS$)	9.75[b]	26.2	
MES			
$HL^\pm = H^+ + L^-$, ($HL = C_6H_{13}NO_4S$)	6.270	14.8	5
Methylamine			
$HL^+ = H^+ + L$, ($L = CH_5N$)	10.645	55.34	33
2-Methylimidazole			
$HL^+ = H^+ + L$, ($L = C_4H_6N_2$)	8.01[b]	36.8	
MOPS			
$HL^\pm = H^+ + L^-$, ($HL = C_7H_{15}NO_4S$)	7.184	21.1	25
MOPSO			
$H_2L^+ = H^+ + HL^\pm$, ($HL = C_7H_{15}NO_5S$)	0.060		
$HL^\pm = H^+ + L^-$	6.90	25.0	38[a]
Oxalate			
$H_2L = H^+ + HL^-$, ($H_2L = C_2H_2O_4$)	1.27	-3.9	-231[a]
$HL^- = H^+ + L^{2-}$	4.266	7.00	-231
Phosphate			
$H_3PO_4 = H^+ + H_2PO_4^-$	2.148	-8.0	-141
$H_2PO_4^- = H^+ + HPO_4^{2-}$	7.198	3.6	-230
$HPO_4^{2-} = H^+ + PO_4^{3-}$	12.35	16.0	-242
Phthalate			
$H_2L = H^+ + HL^-$, ($H_2L = C_8H_6O_4$)	2.950	-2.70	-91

Biochem

Reaction	pK	$\Delta_r H°$/ kJ mol^{-1}	$\Delta_r C°_p$/ J mol^{-1} K^{-1}
$HL^- = H^+ + L^{2-}$	5.408	-2.17	-295
Piperazine			
$H_2L^{2+} = H^+ + HL^+$, ($L = C_4H_{10}N_2$)	5.333	31.11	86
$HL^+ = H^+ + L$	9.731	42.89	75
PIPES			
$HL^{\pm} = H^+ + L^-$, ($HL = C_8H_{18}N_2O_6S_2$)	7.141	11.2	22
POPSO			
$HL^{\pm} = H^+ + L^-$, ($HL = C_{10}H_{22}N_2O_8S_2$)	8.0[a]		
Pyrophosphate			
$H_4P_2O_7 = H^+ + H_3P_2O_7^-$	0.83	-9.2	-90[a]
$H_3P_2O_7^- = H^+ + H_2P_2O_7^{2-}$	2.26	-5.0	-130[a]
$H_2P_2O_7^{2-} = H^+ + HP_2O_7^{3-}$	6.72	0.5	-136
$HP_2O_7^{3-} = H^+ + P_2O_7^{4-}$	9.46	1.4	-141
Succinate			
$H_2L = H^+ + HL^-$, ($H_2L = C_4H_6O_4$)	4.207	3.0	-121
$HL^- = H^+ + L^{2-}$	5.636	-0.5	-217
Sulfate			
$HSO_4^- = H^+ + SO_4^{2-}$	1.987	-22.4	-258
Sulfite			
$H_2SO_3 = H^+ + HSO_3^-$	1.857	-17.80	-272
$HSO_3^- = H^+ + SO_3^{2-}$	7.172	-3.65	-262
TAPS			
$HL^{\pm} = H^+ + L^-$, ($HL = C_7H_{17}NO_6S$)	8.44	40.4	15
TAPSO			
$HL^{\pm} = H^+ + L^-$, ($HL = C_7H_{17}NO_7S$)	7.635	39.09	-16
L(+)-Tartaric acid			
$H_2L = H^+ + HL^-$, ($H_2L = C_4H_6O_6$)	3.036	3.19	-147
$HL^- = H^+ + L^{2-}$	4.366	0.93	-218
TES			
$HL^{\pm} = H^+ + L^-$, ($HL = C_6H_{15}NO_6S$)	7.550	32.13	0
Tricine			
$H_2L^+ = H^+ + HL^{\pm}$, ($HL = C_6H_{13}NO_5$)	2.023	5.85	-196
$HL^{\pm} = H^+ + L^-$	8.135	31.37	-53
Triethanolamine			
$HL^+ = H^+ + L$, ($L = C_6H_{15}NO_3$)	7.762	33.6	50
Triethylamine			
$HL^+ = H^+ + L$, ($L = C_6H_{15}N$)	10.72	43.13	151
Tris			
$HL^+ = H^+ + L$, ($L = C_4H_{11}NO_3$)	8.072	47.45	-59

[a] Approximate value.
[b] Last digit uncertain.

BIOLOGICAL BUFFERS

This table of frequently used biological buffers gives the pK_a value at 20 °C, 25 °C, and 37 °C as well as the useful pH range of each buffer. The buffers are listed in order of increasing pH. Column definitions are as follows.

Column heading	Definition
Acronym	Common acronym of the biological buffer
Name	Systematic name of the biological buffer
Mol. wt.	Molecular weight of the biological buffer
pK_a(20 °C)	Negative logarithm of the acid dissociation constant at 20 °C
pK_a(25 °C)	Negative logarithm of the acid dissociation constant at 25 °C
pK_a(37 °C)	Negative logarithm of the acid dissociation constant at 37 °C
Useful pH range	Range of pHs buffer is useful

The table is reprinted with permission of Sigma Chemical Company, St. Louis, MO.

Biological Buffers

Acronym	Name	Mol. wt.	pK_a(20 °C)	pK_a(25 °C)	pK_a(37 °C)	Useful pH range
MES	2-(*N*-Morpholino)ethanesulfonic acid		6.16	6.10	5.97	5.5–6.7
BIS TRIS	Bis(2-hydroxyethyl)iminotris(hydroxymethyl)methane		n/a	6.50	6.36	5.8–7.2
ADA	*N*-(2-Acetamido)-2-iminodiacetic acid		6.65	6.59	6.46	6.0–7.2
ACES	2-[(2-Amino-2-oxoethyl)amino]ethanesulfonic acid	182.199	6.88	6.78	6.54	6.1–7.5
PIPES	Piperazine-*N,N′*-bis(2-ethanesulfonic acid)		6.80	6.76	6.66	6.1–7.5
MOPSO	3-(*N*-Morpholino)-2-hydroxypropanesulfonic acid		n/a	6.90	6.75	6.2–7.6
BIS TRISPROPANE	1,3-Bis[tris(hydroxymethyl)methylamino]propane		n/a	6.8	n/a	6.3–9.5
BES	*N,N*-Bis(2-hydroxyethyl)-2-aminoethanesulfonic acid	213.252	7.17	7.09	6.90	6.4–7.8
MOPS	3-(*N*-Morpholino)propanesulfonic acid		7.28	7.20	7.02	6.5–7.9
TES	*N*-Tris(hydroxymethyl)methyl-2-aminoethanesulfonic acid		7.50	7.40	7.16	6.8–8.2
HEPES	*N*-(2-Hydroxyethyl)piperazine-*N′*-(2-ethanesulfonic acid)		7.55	7.48	7.31	6.8–8.2
DIPSO	3-[*N,N*-Bis(2-hydroxyethyl)amino]-2-hydroxypropanesulfonic acid		n/a	7.60	7.35	7.0–8.2
MOBS	4-(*N*-Morpholino)butanesulfonic acid		n/a	7.60	n/a	6.9–8.3
TAPSO	3-[*N*-Tris(hydroxymethyl)methylamino]-2-hydroxypropanesulfonic acid		n/a	7.60	7.39	7.0–8.2
TRIZMA	Tris(hydroxymethyl)methylamine	121.135	8.20	8.06	7.72	7.0–9.1
HEPPSO	*N*-(2-hydroxyethyl)piperazine-*N′*-(2-hydroxypropanesulfonic acid)		n/a	7.80	6.66	7.1–8.5
POPSO	Piperazine-*N,N′*-bis(2-hydroxypropanesulfonic acid)		n/a	7.80	7.63	7.2–8.5
TEA	Tris(2-hydroxyethyl)amine	149.188	n/a	7.80	n/a	7.3–8.3
EPPS	4-(2-Hydroxyethyl)-1-piperazinepropanesulfonic acid			8.00	n/a	7.3–8.7
TRICINE	*N*-Tris(hydroxymethyl)methylglycine		n/a	8.05	7.80	7.4–8.8
Gly-Gly	*N*-Glycylglycine	132.118	n/a	8.20	n/a	7.5–8.9
BICINE	*N,N*-Bis(2-hydroxyethyl)glycine	163.172	8.35	8.26	8.04	7.6–9.0
HEPBS	*N*-(2-Hydroxyethyl)piperazine-*N′*-(4-butanesulfonic acid)		n/a	8.30	n/a	7.6–9.0
TAPS	*N*-Tris(hydroxymethyl)methyl-3-aminopropanesulfonic acid		8.49	8.40	8.18	7.7–9.1
AMPD	2-Amino-2-methyl-1,3-propanediol	105.136	n/a	8.80	n/a	7.8–9.7
TABS	*N*-Tris(Hydroxymethyl)methyl-4-aminobutanesulfonic acid		n/a	8.90	n/a	8.2–9.6
AMPSO	3-[(1,1-Dimethyl-2-hydroxyethyl)amino]-2-hydroxypropanesulfonic acid		n/a	9.00	9.10	8.3–9.7
CHES	2-(*N*-Cyclohexylamino)ethanesulfonic acid		9.55	9.49	9.36	8.6–10.0
CAPSO	3-(Cyclohexylamino)-2-hydroxy-1-propanesulfonic acid		n/a	9.60	9.43	8.9–10.3
AMP	2-Amino-2-methyl-1-propanol	89.136	n/a	9.70	n/a	9.0–10.5
CAPS	3-(Cyclohexylamino)-1-propanesulfonic acid		10.56	10.40	10.02	9.7–11.1
CABS	4-(Cyclohexylamino)-1-butanesulfonic acid		n/a	10.70	n/a	10.0–11.4

TYPICAL pH VALUES OF BIOLOGICAL MATERIALS AND FOODS

This table gives typical pH ranges for various biological fluids and common foods. It should be noted that actual values often fall outside these ranges due to differing physiological and other conditions. All values refer to 25 °C.

Typical pH Values of Biological Materials

Biological materials	pH
Blood, human	7.35–7.45
Blood, dog	6.9–7.2
Spinal fluid, human	7.3–7.5
Saliva, human	6.5–7.5
Gastric contents, human	1.0–3.0
Duodenal contents, human	4.8–8.2
Feces, human	4.6–8.4
Urine, human	4.8–8.4
Milk, human	6.6–7.6
Bile, human	6.8–7.0
Milk, cow	6.7–6.9
Saliva	6.7–7.3
Intracellular skeletal muscle, mammals	6.8–7.1
Intracellular fluid, human	7.0–7.4
Pancreatic juice	7.8–8.0

Typical pH Values of Common Foods

Foods	pH	Foods	pH
Apples	2.9–3.3	Milk, cows	6.3–6.6
Apricots	3.6–4.0	Olives	3.6–3.8
Asparagus	5.4–5.8	Oranges	3.0–4.0
Bananas	4.5–4.7	Oysters	6.1–6.6
Beans	5.0–6.0	Peaches	3.4–3.6
Beers	4.0–5.0	Pears	3.6–4.0
Beets	4.9–5.5	Peas	5.8–6.4
Blackberries	3.2–3.6	Pickles, dill	3.2–3.6
Bread, white	5.0–6.0	Pickles, sour	3.0–3.4
Butter	6.1–6.4	Pimento	4.6–5.2
Cabbage	5.2–5.4	Plums	2.8–3.0
Carrots	4.9–5.3	Potatoes	5.6–6.0
Cheese	4.8–6.4	Pumpkin	4.8–5.2
Cherries	3.2–4.0	Raspberries	3.2–3.6
Cider	2.9–3.3	Rhubarb	3.1–3.2
Corn	6.0–6.5	Salmon	6.1–6.3
Crackers	6.5–8.5	Sauerkraut	3.4–3.6
Dates	6.2–6.4	Shrimp	6.8–7.0
Eggs, fresh white	7.6–8.0	Soft drinks	2.0–4.0
Flour, wheat	5.5–6.5	Spinach	5.1–5.7
Gooseberries	2.8–3.0	Squash	5.0–5.4
Grapefruit	3.0–3.3	Strawberries	3.0–3.5
Grapes	3.5–4.5	Sweet potatoes	5.3–5.6
Hominy (lye)	6.8–8.0	Tomatoes	4.0–4.4
Jams, fruit	3.5–4.0	Tuna	5.9–6.1
Jellies, fruit	2.8–3.4	Turnips	5.2–5.6
Lemons	2.2–2.4	Vinegar	2.4–3.4
Limes	1.8–2.0	Water, drinking	6.5–8.0
Maple syrup	6.5–7.0	Wines	2.8–3.8

PROPERTIES AND FUNCTIONS OF COMMON DRUGS

Jessica L. Burger

These tables list the names, properties, categories, and therapeutic uses of selected single-component and multicomponent drugs. The drugs have been chosen to represent a variety of categories and are widely used throughout the world. Table 1 contains the generic (and in some cases chemical) name of single-component drugs with data on their melting point, boiling point, density, and solubility in various solvents, when available. Table 2 classifies the drugs by therapeutic categories and gives additional information about their uses. Table 3 provides an index that can be used to locate an individual single-component drug by its trade name. It should be noted that not all trade names are provided as many drugs are manufactured by more than one organization. Table 4 provides similar information for multicomponent drugs and links the components to the property data given in Table 1 through the ordering number. In all tables, the ordering number (indicated as the Drug no.) is unique to a specific single-component drug as listed in Table 1. It should be noted that most drug products in the marketplace have one or more inactive ingredients. These are not included in these tables.

References

1. *The Combined Chemical Dictionary on DVD, Version 18:1*, CRC Press, Boca Raton, FL, June 2014; also available on the Internet at <www.chemnetbase.com>.
2. Milne, G. W. A., *Drugs: Synonyms and Properties*, Ashgate Publishing, Aldershot, Hampshire, UK, 2000.
3. Corey, E. J., Czakó, B., and Kürti, L., *Molecules and Medicine*, John Wiley & Sons, Hoboken, NJ, 2007.
4. *Physicians' Desk Reference, 67th Edition*, Thomson PDR, Montvale, NJ, 2013.
5. O'Neil, M. J., ed., *The Merck Index, 15th Edition*, Royal Society of Chemistry, 2013.
6. Santa Cruz Biotechnology, Inc. <http://www.scbt.com>. Accessed August 2016.
7. Wishart, D. S. et al., *Nucleic Acids Res.* 34, D668, 2006. <https://doi.org/10.1093/nar/gkj067>
8. *ChemACX. Version 16.16.2;* CambridgeSoft Corporation: Cambridge, MA. Accessed August 2016.
9. *PubChem Database*, National Center for Biotechnology Information <http://pubchem.ncbi.nlm.nih.gov>. Accessed August 2016.
10. *EPI - Estimation Programs Interface Suite*™, v 4.11, 2017, United States Environmental Protection Agency, Washington, DC, USA, available at <https://www.epa.gov/tsca-screening-tools/epi-suitetm-estimation-program-interface>.

Physical Properties of Common Single-Component Drugs

Table 1 contains physical properties of 240 common single-component drugs. The data come from a variety of sources as listed in the references. The properties, units, and special notations are given in the table below. A few drugs have no properties listed except their physical state (in some) cases. These include monoclonal antibodies (indicated by an * after the name), drugs produced as recombinant DNA (indicated by a + after the name), and vaccines.

Table 1 Column heading	Definition
Drug. no.	Entry number; also used to link to information in other tables
Common name	Common name of the drug: generic or chemical
t_m	Melting point in °C; superscript p indicates predicted value; lit. indicates literature value
t_b	Boiling point in °C; superscript p indicates predicted value; all values are for 760 mmHg unless otherwise indicated; note that some of these values are high enough that decomposition will occur before boiling; nevertheless, such information can be used for correlations and separation design
ρ	Density in g mL^{-1}; all values are for 25 °C unless otherwise indicated by numerical superscript; superscript p indicates predicted value
Phys. state	Indication of whether the drug is a solid or liquid
Storage	Recommendations for normal storage; RT = room temperature
Solubility H_2O	Solubility in water in mg mL^{-1} unless otherwise indicated; temperature values, when available, are given in the superscript
Solubility CH_3OH	Solubility in methanol in mg mL^{-1} unless otherwise indicated; temperature values, when available, are given in the superscript
Solubility C_2H_5OH	Solubility in ethanol in mg mL^{-1} unless otherwise indicated; temperature values, when available, are given in the superscript
Solubility DMSO	Solubility in dimethylsulfoxide (DMSO) in mg mL^{-1} unless otherwise indicated; temperature values, when available, are given in the superscript
Solubility DMF	Solubility in dimethyl formamide (DMF) in mg mL^{-1} unless otherwise indicated; temperature values, when available, are given in the superscript

TABLE 1. Physical Properties of Common Single-Component Drugs

Drug no.	Common name	t_m/°C	t_b/°C	ρ/g mL^{-1}	Phys. state	Storage	Solubility H_2O/mg mL^{-1}	Solubility CH_3OH/mg mL^{-1}	Solubility C_2H_5OH/mg mL^{-1}	Solubility DMSO/mg mL^{-1}	Solubility DMF/mg mL^{-1}
1	Abatacept										
2	Abiraterone acetate	127-130	506.7^{p}	$\sim1.1^{p}$	Solid	-20 °C	$<1^{25}$	sparingly	28^{25}	~2	~16
3	Acetaminophen	168-172	387.83^{p}	1.29^{20}	Solid	RT	0.014^{20}	50	100 mmol L^{-1}	100 mmol L^{-1}	freely
4	Acetylsalicylic acid				Solid	RT			80	41	
5	Adalimumab: heavy chain*				Liquid						

Drug no.	Common name	t_m/°C	t_b/°C	ρ/g mL^{-1}	Phys. state	Storage	Solubility H_2O/mg mL^{-1}	Solubility CH_3OH/mg mL^{-1}	Solubility C_2H_5OH/mg mL^{-1}	Solubility DMSO/mg mL^{-1}	Solubility DMF/mg mL^{-1}
6	Adalimumab: light chain*				Liquid						
7	Aflibercept			1.081	Liquid	2-8 °C	>100				
8	Albuterol sulfate	153-155		1.4^{p}	Solid	-20 °C		freely	freely		
9	Alendronate sodium		617^{p}		Solid	RT					
10	Allopurinol				Solid	4 °C				freely	
11	Alprazolam	228-229.5			Solid		40 (pH 7)				
12	Amitriptyline hydrochloride	187-189.5			Solid		0.00971^{24}				
13	Amlodipine	133-135	527.2^{p}	1.2^{p}	Solid	RT	75.3	freely	25	25	25
14	Amlodipine besylate	133-135	527.2^{p}	1.2^{p}	Solid	RT	75.3	freely	25	25	25
15	Amoxicillin	329.9^{p}	702^{p}	1.6^{p}	Solid	4 °C					
16	Amphetamine aspartate monohydrate				Solid		slightly				
17	Amphetamine sulfate				Solid		slightly				
18	Apixaban	326.5^{p}	771^{p}	1.4^{p}	Solid	4 °C	$<1^{25}$		$<1^{25}$	18^{25}	~3
19	Aripiprazole	134-136			Solid	-20 °C			freely		
20	Atazanavir	207-209		1.2^{p}	Solid	4 °C		freely	~2	~16	~25
21	Atenolol	146-148			Solid	RT		freely		freely	
22	Atomoxetine				Solid	-20 °C		freely			
23	Atorvastatin calcium	176-178	722^{p}		Solid	RT	$<1^{25}$	freely	$<1^{25}$	100^{25}	
24	Azithromycin	114	822^{p}	1.2^{p}	Solid	-20 °C	$<1^{25}$		100^{25}	100^{25}	16
25	Bacitracin	221-225			Solid	4 °C	50				
26	Bendamustine hydrochloride	149-151	585^{p}		Solid	-20 °C	$\geq 15^{25}$	s	$\geq 17^{25}$	$\geq 79^{25}$	
27	Bevacizumab*				Liquid						
28	Bimatoprost	66-68	630^{p}	1.1^{p}	Solid	-20 °C		freely	~50	~25	
29	Bisacodyl	131.1-135	492.3^{p}	1.2^{p}	Solid	4 °C			slightly		
30	Bismuth subsalicylate				Solid		59.7				
31	Bortezomib	139-143	684^{p}	1.21^{p}	Solid	-20 °C	5-10 μmol L^{-1}(est. max)	freely	200	200	
32	Brimonidine tartrate	207-208	432.6		Solid	-20 °C	50 mmol L^{-1}	freely		freely	
33	Budesonide		600^{p}	1.28^{p}	Solid	RT	freely	freely	50 mmol L^{-1}	100 mmol L^{-1}	
34	Buprenorphine				Solid		0.0168				
35	Caffeine	238		1.23	Solid		666^{100}				
36	Canagliflozin	68-72	643^{p}	1.4^{p}	Solid	-20 °C	$<1^{25}$	freely	$<1^{25}$	88^{25}	
37	Capecitabine	115-120	517^{p}	1.49^{p}	Solid	RT	26	>40%	207	72^{25}	~14
38	Carisoprodol	89-91					0.300^{25}				
39	Carvedilol	117	655.2^{p}	1.25^{p}	Solid	RT	partly miscible	33.8	10	>25	
40	Cefdinir				Solid	-20 °C				freely	freely
41	Celecoxib	156-157	529.0^{p}	1.43^{p}	Solid	-20 °C	$<1^{5}$	freely	33^{25}	76^{25}	
42	Cephalexin				Solid	-20 °C		freely		freely	
43	Certolizumab pegol*										
44	Cetuximab*										
45	Chlorpheniramine maleate	130-135			Solid	RT	freely				
46	Cimetidine		488.0^{p}	1.27^{p}	Solid	4 °C	freely^{37}				
47	Cinacalcet		440.9		Solid	-20 °C	1^{25}	freely	33^{25}	79^{25}	
48	Ciprofloxacin	253-257	581^{p}	1.5^{p}	Solid	RT	30^{20}	freely	$<1^{25}$	$<1^{25}$	
49	Clavulanate potassium	117.5-118					300				
50	Clonazepam	238-240			Solid		0.100^{25}				
51	Clonidine hydrochloride	312	319.3^{p}		Solid	4 °C	50	freely	freely	75	
52	Clopidogrel bisulfate				Solid	RT				freely	
53	Cobicistat	87-105			Solid	4 °C					
54	Codeine	157.5					9^{20}		freely		
55	Colesevelam hydrochloride				Solid		freely				
56	Conjugated estrogens	173			Solid		0.0036				
57	Cyclobenzaprine hydrochloride	215-217			Solid	-20 °C	>200	freely	50		
58	Dabigatran etexilate mesylate				Solid	4 °C					
59	Daptomycin				Solid	-20 °C	100^{25}	5	50 mmol L^{-1}	100 mmol L^{-1}	freely
60	Darbepoetin alfa^{+}				Liquid						
61	Darunavir	74-76		1.3^{p}	Solid	-20 °C		freely			
62	Dasatinib	272-274			Solid	-20 °C	very poorly		very poorly	200	
62a	Delafloxacin	255.1^{p}	590.9^{p}		Solid	-20 °C	<1				
63	Denosumab*										
64	Desvenlafaxine				Solid	4 °C	>20			>20	
65	Dexlansoprazole				Solid		very slightly	freely	freely		
66	Dexmethylphenidate hydrochloride				Solid		0.82				
67	Dextroamphetamine saccharate				Solid		slightly				
68	Dextroamphetamine sulfate				Solid		slightly				
69	Dextromethorphan hydrobromide		427.5^{p}		Solid	RT	40 mmol L^{-1}		1:10		

Drug no.	Common name	t_m/°C	t_b/°C	ρ/g mL^{-1}	Phys. state	Storage	Solubility H_2O/mg mL^{-1}	Solubility CH_3OH/mg mL^{-1}	Solubility C_2H_5OH/mg mL^{-1}	Solubility DMSO/mg mL^{-1}	Solubility DMF/mg mL^{-1}
70	Diazepam	125-126			Solid						
71	Diclofenac sodium	275-277			Solid	RT	50	>24	~35	~35	~35
72	Digoxin	217-221					0.0648^{25}				
73	Dimethyl fumarate	102-106	192-193	1.37^{20}	Solid	RT	1.6^{25}	30-36	10^{25}	29^{25}	~12
74	Diphenhydramine hydrochloride	169-169.9		1.2	Solid	RT	1000	freely	500	58^{25}	~10
75	Dipyridamole	165-166	806^{p}		Solid	RT	slightly	freely	10 mmol L^{-1}	100 mmol L^{-1}	
76	Docusate sodium	153 to 157		1.120	Solid		15		freely		
77	Doxycycline	201			Solid	-20 °C	50	1:4	1:60	1	
78	Duloxetine	157^{p}	466.2	1.2	Liquid	-20 °C	slightly				
79	Dutasteride	247-248	620^{p}	1.3^{p}	Solid	4 °C	freely	64	44	62	
80	Eculizumab*										
81	Efavirenz	134-136	341^{p}	1.5^{p}	Solid	-20 °C		freely			
82	Elvitegravir	93-96			Solid	4 °C					
83	Emtricitabine	153-155	443.3^{p}	1.8^{p}	Solid	-20 °C	≥50		≥17	≥50	
84	Enalapril maleate	143-144			Solid	RT	25	20	8		
85	Enoxaparin sodium						>200				
86	Epoetin alfa[+]				Liquid						
87	Erlotinib hydrochloride	223-225			Solid	-20 °C	<1	sparingly	<1	3	
88	Erythromycin				Solid			freely	freely		
89	Escitalopram oxalate	155			Solid		very slightly				
90	Esomeprazole				Solid	-20 °C	$<1^{25}$	s	143^{25}	143^{25}	
91	Eszopiclone				Solid		Very slightly		slightly		
92	Etanercept[+]				Liquid/ solid						
93	Ethinyl estradiol	142-146			Solid				1 part in 6		
94	Etonogestrel	182-184			Solid	-20 °C					
95	Everolimus	110-115	998^{p}	1.18^{p}	Solid	-20 °C	$<1^{25}$	freely	100	100	
96	Ezetimibe	162-164	655^{p}	1.3^{p}	Solid	4 °C	$<1^{25}$	freely	83^{25}	82^{25}	
97	Febuxostat	208-210			Solid	4 °C				freely	
98	Fenofibrate	79-83	470^{p}	1.2^{p}	Solid	-20 °C	freely	slightly	1	15	30
99	Fexofenadine hydrochloride	193-196			Solid	-20 °C		freely			
100	Filgrastim[+]										
101	Fingolimod				Solid	-20	freely				
102	Fluconazole	138-142	79.8^{p}	1.5^{p}	Solid	RT	<1(25)		61^{25}	61^{25}	
103	Fluoxetine hydrochloride	158-159			Solid		4^{RT}	freely^{RT}		$>5^{RT}$	
104	Fluticasone propionate	275	568	1.3	Solid	4 °C				≥10	
105	Formoterol fumarate dihydrate	136-142			Solid	4 °C	freely			20	
106	Furosemide		582.1^{p}	1.6^{p}	Solid	RT	$<1^{25}$	50	$<1^{25}$	66^{25}	freely
107	Gabapentin	162-166	314.4^{p}	1.1^{p}	Solid	RT	10	freely	sparingly	sparingly	sparingly
108	Glatiramer acetate										
109	Glimepiride	212-215	750^{p}	1.3^{p}	Solid	RT	$<1^{25}$		$<1^{25}$	>10	
110	Glyburide	173-175		1.4^{p}	Solid	4 °C	freely	1:250	5	25	freely
111	Guaifenesin	79	215 (19 mmHg)	1.2^{p}	Solid	RT	50^{25}	freely	50	40^{25}	freely
112	Human papillomavirus quadrivalent										
113	Hydrochlorothiazide	273-277			Solid	RT		freely			
114	Hydrocodone	198			Solid		freely		freely		
115	Hydrocodone bitartrate	198			Solid		freely		freely		
116	Hydrocortisone	211-214			Solid	RT			15		
117	Ibuprofen	77-78	157 (4 mmHg)	1.0^{p}	Solid	RT	freely		25		
118	Imatinib mesylate	214-224			Solid	4 °C	200	freely	~0.2	200	~10
119	Infliximab*										
120	Influenza vaccine for intramuscular injection										
121	Insulin aspart (rDNA origin)				Solid		insoluble				
122	Insulin detemir (rDNA origin)				Solid		freely				
123	Insulin glargine	232-234			Solid		insoluble				
124	Insulin lispro				Solid		insoluble				
125	Interferon beta-1a[+]										
126	Ipilimumab*										
127	Ipratropium bromide	239-240			Solid	4 °C	83^{25}		83^{25}	83^{25}	
128	Isosorbide mononitrate	86-88	364.5^{p}	1.6^{p}	Solid	RT	freely	freely	freely		
129	Lansoprazole	173-175	555.8^{p}	1.5^{p}	Solid	4 °C	freely	freely	14^{25}	74^{25}	~30
130	Ledipasvir				Solid	-20 °C					
131	Lenalidomide	269-271	614^{p}	1.5^{p}	Solid	-20 °C	$<1^{25}$	freely	$<1^{25}$	$≥52^{25}$	

Biochem

Drug no.	Common name	t_m/°C	t_b/°C	ρ/g mL^{-1}	Phys. state	Storage	Solubility H_2O/mg mL^{-1}	Solubility CH_3OH/mg mL^{-1}	Solubility C_2H_5OH/mg mL^{-1}	Solubility DMSO/mg mL^{-1}	Solubility DMF/mg mL^{-1}
132	Levofloxacin	218-220			Solid	4 °C					
133	Levothyroxine sodium				Solid		0.015%				
134	Linagliptin	190-196	661^{p}	1.4^{p}	Solid	-20 °C	$<1^{25}$	freely	1^{25}	17^{25}	
135	Linezolid	171-175	586	1.3^{p}	Solid	4 °C	$<1^{25}$	freely	8^{25}	68^{25}	
136	Liraglutide				Solid		freely (pH>7)	68	1.1		
137	Lisdexamfetamine dimesylate				Solid		792				
138	Lisinopril	159^{p}	666^{p}	1.3^{p}	Solid	-20 °C			sparingly	sparingly	sparingly
139	Loperamide hydrochloride	223			Solid	RT	160	286	53.7	205	
140	Loratadine	134-136	531.3^{p}	1.3^{p}	Solid	RT	$<1^{25}$		77^{25}	50	~30
141	Lorazepam	192-194					80				
142	Losartan potassium	265	682^{p}		Solid	RT	>500		≥20	100 mmol L^{-1}	
143	Loteprednol etabonate	226-233			Solid	-20 °C					
144	Lovastatin	175.4	559^{p}	1.1^{p}	Solid	-20 °C	0.4 μg mL^{-1}	freely	10^{RT}	10^{RT}	freely
145	Lubiprostone				Solid		freely				
146	Lurasidone hydrochloride	198-205			Solid	4 °C	0.00789				
147	Medroxyprogesterone acetate	214.5			Solid		0.022				
148	Meloxicam	255	589	1.6	Solid	4 °C	22	slightly	slightly	25	freely
149	Memantine hydrochloride		239.8^{p}		Solid	RT	100 mmol L^{-1}		43^{25}	43^{25}	
150	Metformin hydrochloride	215-218			Solid	-20 °C		freely		freely	
151	Methylprednisolone				Solid	RT					
152	Metoprolol succinate	142.1-148.2	398.6		Solid	RT	miscible				
153	Miconazole nitrate	170-185			Solid	RT	$<1^{25}$	freely	10 mmol L^{-1}	95^{5}	
154	Mometasone furoate				Solid	-20 °C		freely			
155	Montelukast sodium	135.5	750^{p}		Solid	-20 °C	$>100^{25}$	freely	100 mmol L^{-1}	100 mmol L^{-1}	~30
156	Morphine sulfate				Solid	255	0.149^{20}				
157	Naloxone	184	532.8^{p}	1.4^{p}	Solid	RT					
158	Naproxen sodium	152-154 (lit.)	403.9	1.2	Solid	RT	$>3^{25}$	freely	freely	freely	freely
159	Nebivolol				Solid		0.0403				
160	Neomycin sulfate						500				
161	Norelgestromin						0.00605				
162	Norethindrone acetate	203.5					0.00704^{25}				
163	Norgestimate	230-232			Solid	-20 °C	freely	freely		5	
164	Octreotide acetate	153-156			Solid	-20 °C		freely		freely	
165	Olmesartan acid	189-192	738^{p}	1.3^{p}	Solid	-20 °C		sparingly			20 mmol L^{-1}
166	Olmesartan medoxomil	180-182			Solid	-20 °C			freely		
167	Olopatadine hydrochloride				Solid	4 °C	sparingly				
168	Omalizumab*										
169	Omeprazole magnesium	156			Solid		0.5	freely	freely		
170	Oseltamivir phosphate	197-212			Solid	-20 °C	>500		$<1^{25}$	2^{25}	
171	Oxycodone hydrochloride	219			Solid		slightly		slightly		
172	Oxymetazoline hydrochloride		431.9^{p}		Solid	RT	50	freely	59	59	
173	Paliperidone palmitate	114-116			Solid						
174	Palivizumab*										
175	Paroxetine hydrochloride	147-150			Solid					25	
176	Pegfilgrastim										
177	Pemetrexed disodium				Solid	-20 °C	freely	freely			
178	Pioglitazone	183-185	75^{p}	1.3^{p}	Solid	-20 °C	$<1^{25}$		$<1^{25}$	15^{25}	2.5
178a	Plecanatide										
179	Pneumococcal 13-valent conjugate vaccine										
180	Polyethylene glycol 3350						freely				
181	Polymyxin B				Solid		25				
182	Pramoxine hydrochloride				Solid						
183	Prasugrel	122	493.5^{p}	1.3^{p}	Solid	4 °C	30^{25}		7^{25}	$<1^{25}$	
184	Pravastatin sodium	171.2-173			Solid	4 °C	freely				
185	Prednisone	233-235 (d)			Solid		very slightly	slightly	freely		
186	Pregabalin						freely				
187	Promethazine hydrochloride	225-227			Solid	4 °C					
188	Quetiapine fumarate	173-175			Solid	RT	<1		<1	36	
189	Rabeprazole sodium	99-100	$538\text{-}669^{p}$	1.3^{p}	Solid	-20 °C	freely			freely	
190	Raloxifene hydrochloride	257.8	728^{p}		Solid	-20 °C	freely	freely	$<1^{25}$	50 mmol L^{-1}	
191	Raltegravir				Solid	-20 °C	$<1^{25}$		$<1^{25}$	88^{25}	
192	Ramipril	109	616^{p}	1.2^{p}	Solid	RT	$<1^{25}$	freely	83^{25}	83^{25}	
193	Ranibizumab										
194	Ranitidine	69-72	437.1^{p}	1.2^{p}	Solid		freely				

Drug no.	Common name	t_m/°C	t_b/°C	ρ/g mL^{-1}	Phys. state	Storage	Solubility H_2O/mg mL^{-1}	Solubility CH_3OH/mg mL^{-1}	Solubility C_2H_5OH/mg mL^{-1}	Solubility DMSO/mg mL^{-1}	Solubility DMF/mg mL^{-1}
195	Ranolazine	119-120	624	1.2	Solid			freely			
196	Rifaximin	218-228			Solid	4 °C	<1		157	47	
197	Rilpivirine	153-155	443.3^{p}	1.8^{p}	Solid	-20 °C	≥50		≥17	≥50	
198	Risedronate sodium	252-262	692^{p}		Solid	4 °C	5		<1	1	
199	Ritonavir	120-122	947^{p}	1.2^{p}	Solid	4 °C	partly	freely	~5	100^{25}	~15
200	Rituximab*										
201	Rivaroxaban	228-229	733^{p}	1.5^{p}	Solid	4 °C	<1^{25}		<1^{25}	87^{25}	
202	Rivastigmine tartrate	126					freely		s		
203	Rosuvastatin calcium	151-156			Solid	-20 °C	<1^{25}	freely	<1^{25}	100^{25}	
203a	Safinamide	452.6^{p}	190.4^{p}		Solid	-20 °C					
204	Salmeterol	243.5^{p}	603^{p}	1.1^{p}	Solid	4 °C			50 mmol L^{-1}	100 mmol L^{-1}	
205	Saxagliptin	96-102	548.7^{p}	1.4^{p}	Solid	-20 °C		freely		freely	
206	Sertraline hydrochloride	243-245			Solid	RT	3		10	>25	
207	Sevelamer hydrochloride				Solid		freely				
208	Sildenafil citrate	185-187			Solid	25 °C	0.46	freely		freely	
209	Simeprevir			1.4^{p}	Solid	4 °C				freely	
210	Simvastatin	127-132	565	1.1	Solid	-20 °C	0.03	freely	>25	>25	
211	Sitagliptin	110-120	529.9^{p}	1.6^{p}	Solid	RT					
212	Sofosbuvir	100-106			Solid	-20 °C					
213	Solifenacin succinate				Solid	-20 °C	96^{25}		5^{25}	3^{25}	
214	Spironolactone	207-208	597^{p}	1.2^{p}	Solid	RT	0.022^{25}	freely	50 mmol L^{-1}	100 mmol L^{-1}	
215	Sulfamethoxazole	168-170			Solid	4 °C	<1^{25}		23^{25}	51^{25}	
216	Tadalafil	276-279	679.1^{p}		Solid	-20 °C	<1^{25}	freely	<1^{25}	78^{25}	
217	Tamsulosin hydrochloride	226-228	596	1.2^{p}	Solid	RT	10 mmol L^{-1}			100 mmol L^{-1}	
218	Telmisartan	261-263	772^{p}	1.2^{p}	Solid	RT	freely		<1^{25}	13^{25}	~1.6
219	Tenofovir disoproxil fumarate	113-115			Solid	4 °C	<1^{25}	freely	44^{25}	128^{25}	
220	Testosterone	155			Solid		0.0234^{25}		1 part in 5		
221	Timolol	202-203	487.2^{p}		Solid		freely		freely	freely	
222	Tiotropium bromide	218-220			Solid	4 °C		freely		freely	
223	Tolterodine tartrate	205-207			Solid	4 °C	0.0212	freely	6	≥20	
224	Tramadol hydrochloride	180-181			Solid		freely				
225	Trastuzumab*										
226	Travoprost				Liquid		>16^{25}				
227	Trazodone hydrochloride	87			Solid		sparingly				
228	Triamcinolone acetonide	270			Solid		0.080^{25}				
229	Triamterene				Solid	RT	freely	freely			
230	Trimethoprim	234-236	405.2^{p}	1.3^{p}	Solid	4 °C	0.4	12.1	2.6^{20}		
231	Ustekinumab										
232	Valacyclovir hydrochloride	196-200			Solid	RT	170	freely	<1^{25}	14^{25}	
233	Valsartan	108-110	685^{p}	1.2^{p}	Solid			freely	freely		
234	Vardenafil hydrochloride	227-230			Solid	-20 °C					
235	Varenicline						51.5^{25}				
236	Venlafaxine hydrochloride	217	397.6^{p}		Solid	RT	>25			50 mmol L^{-1}	
237	Vilazodone hydrochloride	203-205			Solid	-20 °C					
238	Warfarin sodium	161			Solid	RT	freely		freely		
239	Zolpidem tartrate	196			Solid		23				
240	Zoster vaccine live										

Functions of Common Single-Component Drugs

Table 2 contains information on the therapeutic class, mechanism of action indications (use) of the common single-component drugs listed in Table 1. The information comes from a variety of sources as listed in the references. While every precaution has been taken to ensure this information is correct, it must not be used to make medical decisions of any type. Therapeutic class and Indication information are based on commonly occurring situations and must not be used in place of professional advice. This general information must not be interpreted as applicable to any specific individual health or medical situation.

Table 2 Column headings	Definition
Drug no.	Entry number; also used to link to information in other tables
Drug name	Common name of the drug: generic or chemical
Therapeutic class	Classification based on chemical characteristics, structure, and functionality; this information must not be used to make medical decisions of any type
Mechanism of action	General description of the mechanism(s) by which the drug commonly works; this information must not be used to make medical decisions of any type
Indications (Use)	General description of the result achieved by use of the drug; this information must not be used to make medical decisions of any type

TABLE 2. Functions of Common Single-Component Drugs

Drug no.	Drug name	Therapeutic class	Mechanism of action	Indications (Use)
Addiction				
34	Buprenorphine	Partial opioid agonist/opioid antagonist	Efficacy is due to its properties as a partial agonist at the mu-opioid receptor and an antagonist at the kappa-opioid receptor.	Used in the management of opioid dependence and should be used as part of a complete treatment plan that includes counseling and psychosocial support.
157	Naloxone		Efficacy is due to its properties as an antagonist at the mu-opioid receptor.	Used to reverse opioid overdoses.
235	Varenicline	Nicotinic acetylcholine receptor agonist	Works by binding to $\alpha4\beta2$ neuronal nicotinic acetylcholine receptors and providing agonist activity as well as limiting nicotine binding at these receptors.	Used to aid in the cessation of smoking.
Antiallergic and Asthma Control				
8	Albuterol sulfate	Short-acting $\beta2$-agonist	Use results in the relaxation of smooth muscles in the airways by activating $\beta2$-adrenergic receptors causing the activation of adenylcyclase and an increase in intracellular cAMP. Increased cAMP leads to activation of protein kinase A, which inhibits the phosphorylation of myosin and lowers intracellular ionic calcium concentrations.	Used in the treatment of acute bronchospasm caused by bronchial asthma, chronic bronchitis, and other bronchopulmonary disorders.
33	Budesonide	Corticosteroid/long-acting $\beta2$-agonist	Corticosteroid	Used in the management of asthma and airflow obstruction in patients with chronic obstructive pulmonary disease, including chronic bronchitis and emphysema.
45	Chlorpheniramine maleate	Antihistamine	Antihistamine	Used to provide temporary relief from allergy symptoms.
74	Diphenhydramine hydrochloride	Antihistamine	Antihistamine	Used to manage allergies.
99	Fexofenadine hydrochloride	Antihistamine	Antihistamine	
104	Fluticasone propionate	Corticosteroid	Corticosteroid	Used as a prophylactic therapy in the management of asthma.
105	Formoterol fumarate dihydrate		Use results in the relaxation of smooth muscles in the airways by activating $\beta2$-adrenergic receptors causing the activation of adenylcyclase and an increase in intracellular cAMP. Increased cAMP leads to activation of protein kinase A, which inhibits the phosphorylation of myosin and lowers intracellular ionic calcium concentrations.	
127	Ipratropium bromide	Anticholinergic/$\beta2$ agonist	Works by antagonizing the action of acetylcholine, which inhibits vagally mediated reflexes and stops the increase in intracellular concentration of Ca^{2+}, which is caused by interaction of acetylcholine with the muscarinic receptors on bronchial smooth muscle.	For use in the management of bronchial spasms associated with COPD and asthma.
140	Loratadine	Antihistamine	Antihistamine	Used to provide temporary relief of symptoms due upper respiratory allergies.
154	Mometasone furoate	Corticosteroid	Corticosteroid	
155	Montelukast sodium	Leukotriene receptor antagonist	Works by binding to cysteinyl leukotriene (CysLT) receptors found in the airway and on other proinflammatory cells and inhibiting the actions of leukotriene D4 at the CysLT type 1 receptor.	Used in the management of asthma, perennial allergic rhinitis, seasonal allergic rhinitis, and exercise-induced bronchoconstriction.
168	Omalizumab	Monoclonal antibody/IgE blocker	Works by reducing the binding of IgE to the high-affinity IgE receptor on the surface of mast cells and basophils, which reduces the release of mediators of allergic response.	Used in the management of moderate to severe persistent asthma associated with allergies. Also used in the management of chronic idiopathic urticaria in patients who remain symptomatic despite H1 antihistamine management.
172	Oxymetazoline hydrochloride	Decongestant	Vasoconstrictor	Used in the short-term reduction of nasal congestion and sinus pressure caused by the common cold, or upper respiratory allergies.
187	Promethazine hydrochloride	Phenothiazine derivative	Works as an H1 receptor-blocking agent with antihistaminic properties and delivers sedative and antiemetic effects.	Used in the treatment of allergic disorders, and nausea/vomiting.
204	Salmeterol		Works by increasing the activity of intracellular adenyl cyclase, which leads to an increase in the concentration of cAMP. Increased cAMP concentration causes the relaxation of bronchial smooth muscle and inhibition of release of mediators of immediate hypersensitivity from cells, particularly, mast cells.	
222	Tiotropium bromide	Anticholinergic	Works by inhibiting M3-receptors on smooth muscle, which leads to bronchodilation.	Used in the management of bronchospasm.
228	Triamcinolone acetonide	Corticosteroid	Corticosteroid	

Drug no.	Drug name	Therapeutic class	Mechanism of action	Indications (Use)
Antibiotics				
15	Amoxicillin	Semisynthetic ampicillin derivative	Bactericidal effect is due to the inhibition of the biosynthesis of cell wall.	Used to treat a range of infections caused by Gram-positive and a limited range of Gram-negative organisms, including those of the skin, ear, nose, throat, genitourinary tract, lower respiratory tract infections, acute uncomplicated gonorrhea, and in combination therapy for *Helicobacter pylori* eradication.
24	Azithromycin	Macrolide	The mechanism of action is due to the binding of 23S rRNA of the 50S ribosomal subunit, and interfering with bacterial protein synthesis by impeding the assembly of the 50S ribosomal subunit.	Used in the treatment of bacterial infections including acute bacterial sinusitis, chlamydia, gonorrhea, community-acquired pneumonia, pelvic inflammatory disease, pediatric otitis media, and pharyngitis.
25	Bacitracin	Antibiotic	Antibacterial action against a number of Gram-positive and a few Gram-negative organisms.	Polypeptide antibiotic derived from *Bacillus subtilis*.
40	Cefdinir	Cephalosporin (3rd generation)	Works by inhibiting bacterial cell-wall synthesis.	Used in the treatment of a variety of bacterial infections.
42	Cephalexin	Cephalosporin (1st generation)	Efficacy is mediated by the inhibition of bacterial cell-wall synthesis.	Used to treat a variety of bacterial infections including upper respiratory infections, ear infections, skin infections, and urinary tract infections. Mostly used against Gram-positive bacteria.
48	Ciprofloxacin	Fluoroquinolone	Works by inhibiting topoisomerase II, the bacterial enzymes required for DNA replication, transcription, repair, and recombination.	Used in the treatment of infections from a wide range of Gram-positive and Gram-negative microorganisms including: UTIs, chronic bacterial prostatitis, lower respiratory tract infections, acute exacerbations of chronic bronchitis, acute sinusitis, skin and skin structure infections, bone and joint infections, complicated intra-abdominal infections, infectious diarrhea, typhoid fever, and uncomplicated cervical and urethral gonorrhea. Also used to reduce the incidence or progression of postexposure inhalation of anthrax. Used in the treatment of plague, and as prophylaxis for plague.
49	Clavulanate potassium		Bactericidal effect is due to the inactivation of a variety of β-lactamase enzymes frequently found in microorganisms resistant to penicillin and cephalosporin.	
59	Daptomycin	Cyclic lipopeptide	Efficacy due to cell death caused by the loss of membrane potential as a result of daptomycin binding to bacterial cell membranes.	Used in the treatment of complicated skin and skin structure infections caused by susceptible Gram-positive microorganisms, and of *Staphylococcus aureus* bloodstream infections (bacteremia), including those with right-sided infective endocarditis, caused by methicillin-susceptible and methicillin-resistant isolates.
62a	Delafloxacin	Fluoroquinolone	Works by interfering with bacterial DNA replication by inhibiting the activity of DNA topoisomerase IV and DNA gyrase (topoisomerase II).	Used in the treatment of acute bacterial skin and skin structure infections caused by Gram-positive organisms.
77	Doxycycline	Tetracycline	Efficacy is thought to be due to a reduction in bacterial protein synthesis.	Used in the treatment of a variety of bacterial infections.
88	Erythromycin	Macrolide	Efficacy is due to the blocking of protein synthesis by binding to the 50S ribosomal subunit.	Used in the treatment of mild to moderate upper and lower respiratory tract infections, skin infections, listeriosis, pertussis, diphtheria infections, erythrasma, intestinal amebiasis, acute pelvic inflammatory disease, primary syphilis, and chlamydial infections.
132	Levofloxacin	Fluoroquinolone	Efficacy is due to inhibiting bacterial topoisomerase IV and DNA gyrase, enzymes necessary for DNA replication, transcription, repair, and recombination.	Used in the treatment of a variety of bacterial infections.
135	Linezolid	Oxazolidinone class antibacterial	Works by binding to a site on the bacterial 23S ribosomal RNA of the 50S subunit and inhibiting the formation of a functional 70S initiation complex, which is an essential component of the bacterial translation process.	Used in the treatment of nosocomial pneumonia; community-acquired pneumonia (including patients with concurrent bacteremia); complicated and uncomplicated skin and skin structure infections, and vancomycin-resistant *Enterococcus faecium* infections.
181	Polymyxin B	Antibacterial	Changes outer membrane of bacteria increasing permeability to water and cell death.	Used against Gram-negative infections.
196	Rifaximin	Semisynthetic rifampin analogue	Works by binding to β-subunit of bacterial DNA-dependent RNA polymerase and inhibiting transcription and thereby stemming protein synthesis and growth.	Used in the treatment of traveler's diarrhea, hepatic encephalopathy, and irritable bowel syndrome.
215	Sulfamethoxazole	Sulfonamide/tetrahydrofolic acid inhibitor	Works by reducing the bacterial synthesis of dihydrofolic acid by competing with para-aminobenzoic acid.	Used to treat a broad spectrum of bacterial infections.
230	Trimethoprim		Efficacy is mediated by reducing the production of tetrahydrofolic acid from dihydrofolic acid by inhibiting dihydrofolate reductase.	

Drug no.	Drug name	Therapeutic class	Mechanism of action	Indications (Use)
Anti-Diabetic				
36	Canagliflozin	Sodium-glucose cotransporter 2 (SGLT2) inhibitor	Efficacy is mediated by decreasing the reabsorption of filtered glucose and lowering the renal threshold for glucose leading to an increase in urinary glucose excretion.	Used to increase glycemic control in adults with type 2 diabetes mellitus.
109	Glimepiride	Sulfonylurea (2nd generation)	Efficacy is due to stimulating insulin release from pancreatic β cells which leads to a lowering of blood glucose.	Used in the treatment of type 2 diabetes mellitus.
110	Glyburide	Sulfonylurea (2nd generation)	Efficacy is due to stimulating insulin release from pancreatic β cells which leads to a lowering of blood glucose.	Used in the treatment of type 2 diabetes mellitus.
121	Insulin aspart (rDNA origin)	Insulin (rapid-acting)	Works by binding insulin receptors on muscle and fat cells increasing the cellular uptake of glucose while concurrently decreasing the output of glucose from the liver.	Used to increase glycemic control in patients with diabetes mellitus (DM).
122	Insulin detemir (rDNA origin)	Insulin (long-acting)	Long-acting insulin. Works by binding insulin receptors on muscle and fat cells increasing the cellular uptake of glucose while concurrently decreasing the output of glucose from the liver.	Used to increase glycemic control in patients with diabetes mellitus (DM).
123	Insulin glargine	Insulin (long-acting)	Long-acting insulin. Works by binding insulin receptors on muscle and fat cells increasing the cellular uptake of glucose while concurrently decreasing the output of glucose from the liver.	Used to increase glycemic control in patients with diabetes mellitus (DM).
124	Insulin lispro	Insulin (rapid-acting)	Works by binding insulin receptors on muscle and fat cells increasing the cellular uptake of glucose while concurrently decreasing the output of glucose from the liver.	Used to increase glycemic control in patients with diabetes mellitus (DM).
134	Linagliptin	Dipeptidyl peptidase-4 (DPP-4) inhibitor	Efficacy is due to an increase in the concentrations of active incretin hormones, stimulating the release of insulin in a glucose-dependent manner, and reducing glucagon levels.	Used to improve glycemic control in patients with type 2 diabetes mellitus.
136	Liraglutide	Glucagon-like peptide-1 (GLP-1) receptor agonist	Works by increasing intracellular cAMP, which causes insulin to be released in the presence of elevated glucose concentrations. Reduces glucagon secretion in a glucose-dependent manner and delays gastric emptying.	Used to improve glycemic control in adults with type 2 diabetes mellitus.
150	Metformin hydrochloride	Biguanide	Works by decreasing hepatic glucose production, intestinal absorption of glucose, and increasing peripheral glucose uptake and utilization.	Used in the management of type 2 diabetes mellitus.
178	Pioglitazone	Thiazolidinedione (glitazone)	Efficacy is due to a decrease in insulin resistance in the periphery and liver, leading to an increase in insulin-dependent glucose disposal and decreased hepatic glucose output.	Used in the management of type 2 diabetes mellitus.
205	Saxagliptin	Dipeptidyl peptidase-4 (DPP4) inhibitor	Increases the concentration of incretin in the blood by reducing the inactivation of incretin hormones. This reduces fasting and postprandial glucose concentrations in a glucose-dependent manner.	Used to increase glycemic control in patients with type 2 diabetes mellitus.
211	Sitagliptin	Dipeptidyl peptidase-4 (DPP4) inhibitor	Reduces the inactivation of incretin hormones, which promotes insulin release and decreases glucagon levels in the circulation in a glucose-dependent manner.	Used to increase glycemic control in adults with type 2 diabetes mellitus.
Anti-Inflammatory				
4	Acetylsalicylic acid	Nonsteroidal anti-inflammatory drug (NSAID)	Nonsteroidal anti-inflammatory drug (NSAID). The mechanism of action is not completely understood but is believed to involve the inhibition of COX-1 and COX-2.	Temporarily relieves minor aches and pains.
41	Celecoxib	Nonsteroidal anti-inflammatory Drug (NSAID)	Nonsteroidal anti-inflammatory drug (NSAID). The mechanism of action is not completely understood but is believed to involve the inhibition of COX-1 and COX-2.	Used in the management of pain associated with osteoarthritis, rheumatoid arthritis, and ankylosing spondylitis. Also used in the management of acute pain in adults, primary dysmenorrhea and juvenile rheumatoid arthritis.
71	Diclofenac sodium	Nonsteroidal anti-inflammatory drug (NSAID)	Nonsteroidal anti-inflammatory drug (NSAID). The mechanism of action is not completely understood but is believed to involve the inhibition of COX-1 and COX-2.	Used in the management of osteoarthritis and may be used topically.
117	Ibuprofen	Nonsteroidal anti-inflammatory drug (NSAID)	Nonsteroidal anti-inflammatory drug (NSAID). The mechanism of action is not completely understood but is believed to involve the inhibition of COX-1 and COX-2.	Used to reduce pains associated with muscular aches, headaches, backaches, menstrual cramps, toothaches, the common cold, and arthritis.
148	Meloxicam	Nonsteroidal anti-inflammatory drug (NSAID)	Nonsteroidal anti-inflammatory drug (NSAID). The mechanism of action is not completely understood but is believed to involve the inhibition of COX-1 and COX-2.	Used in the treatment of arthritis and osteoarthritis.
151	Methylprednisolone	Glucocorticoid	Works by decreasing the body's immune responses to various stimuli.	Used to treat inflammation.

Drug no.	Drug name	Therapeutic class	Mechanism of action	Indications (Use)
158	Naproxen sodium	Nonsteroidal anti-inflammatory drug (NSAID)	Nonsteroidal anti-inflammatory drug (NSAID). The mechanism of action is not completely understood but is believed to involve the inhibition of COX-1 and COX-2.	Used to reduce pains associated with muscular aches, headaches, backaches, menstrual cramps, toothaches, the common cold, and arthritis.
185	Prednisone	Glucocorticoid	Works by modifying the body's immune responses to various stimuli.	Used in the management of many undesirable inflammatory conditions.
Antiviral and Antifungal				
102	Fluconazole	Azole antifungal	Works by inhibiting fungal CYP450-dependent enzyme lanosterol 14-α-demethylase. This leads to an increase in concentration of 14-α-methylsterols in fungi and is believed to be responsible for the fungistatic activity.	Used in the treatment of *Candidiasis* and *Cryptococcal meningitis*.
130	Ledipasvir	HCV NS5A inhibitor/HCV nucleotide analogue NS5B polymerase inhibitor	Works by inhibiting HCV NS5A protein thereby preventing viral replication.	Used in the treatment of hepatitis C.
153	Miconazole nitrate	Azole antifungal	Azole antifungal	Used in the treatment of athlete's foot (*Tinea pedis*), jock itch (*Tinea cruris*), and ringworm (*Tinea corporis*).
170	Oseltamivir phosphate	Neuraminidase inhibitor	Efficacy is mediated by inhibiting influenza virus neuraminidase which limits budding from host cells.	Used in the treatment and prophylaxis of influenza.
174	Palivizumab	Monoclonal antibody/RSV F-protein blocker	Efficacy is mediated by the binding the RSV envelope fusion protein on the surface of the virus and blocking the membrane fusion process.	Used in the prevention of serious lower respiratory tract disease caused by respiratory syncytial virus (RSV) in children at high risk of RSV disease.
209	Simeprevir	HCV NS3/4A protease inhibitor	Acts as an antiviral agent against the virus that causes hepatitis C.	Used in the treatment of hepatitis C.
212	Sofosbuvir	HCV nucleotide analogue NS5B polymerase inhibitor	Works by inhibiting HCV NS5B RNA-dependent RNA polymerase thereby preventing viral replication.	Used in the treatment of hepatitis C.
232	Valacyclovir hydrochloride	Nucleoside analogue	Efficacy is due to valacyclovir's conversion to acyclovir which inhibits viral DNA polymerase by being incorporated into the DNA and causing the termination of the DNA growth.	Used in the treatment of cold sores (*Herpes labialis*), shingles (*Herpes zoster*), and genital herpes (*Herpes simplex*).
Autoimmune				
1	Abatacept	Selective costimulation modulator	Efficacy is mediated by inhibiting T-cell activation by binding to CD80 and CD86, thereby blocking interaction with CD28.	Used in the management of moderate to severe active rheumatoid arthritis and polyarticular juvenile idiopathic arthritis in order to limit the progression of structural damage, and to improve physical function.
5, 6	Adalimumab	Monoclonal antibody/TNF blocker	Works by binding to TNF-α and blocking its interaction with the p55 and p75 cell surface TNF receptors.	Used in the management of conditions where the suppression of immune response is desired. Reduces symptoms, inhibits progression of structural damage, and improves physical function in patients with rheumatoid arthritis, polyarticular juvenile idiopathic arthritis, ankylosing spondylitis, psoriatic arthritis, Crohn's disease, ulcerative colitis, and plaque psoriasis.
43	Certolizumab pegol	Tumor necrosis factor (TNF) blocker	Efficacy is mediated by the binding and neutralization of TNF-α, which is important in inflammatory processes.	Used in the management of Crohn's disease, moderate to severe rheumatoid arthritis, psoriatic arthritis, and ankylosing spondylitis.
80	Eculizumab	Monoclonal antibody/protein C5 blocker	Works by binding to the complement protein C5 with high affinity, which inhibits its cleavage to C5a and C5b and inhibits the production of the terminal complement complex C5b-9.	Used in the treatment of paroxysmal nocturnal hemoglobinuria and atypical hemolytic uremic syndrome.
92	Etanercept	Tumor necrosis factor (TNF) blocker	Works by reducing the binding of TNF-α and TNF-β to cell surface TNF-receptors, which results in TNF being biologically inactive.	Used in the management of rheumatoid arthritis, psoriatic arthritis, ankylosing spondylitis, and plaque psoriasis. Also may be used in the management of juvenile idiopathic arthritis.
119	Infliximab	Monoclonal antibody/TNF blocker	Efficacy is mediated by infliximab's binding to the soluble and transmembrane forms of TNF-α which inhibits the binding of TNF-α with its receptors.	Used in the management of autoimmune diseases. Decreases symptoms, inhibits progression of structural damage, and improves physical function with moderately to severely active Chrohn's disease, rheumatoid arthritis, psoriatic arthritis, ankylosing spondylitis, and plaque psoriasis.
231	Ustekinumab	Monoclonal antibody	Works by binding to the p40 protein subunit used by both the IL-12 and IL-23 cytokines.	Used in the management of moderate to severe plaque psoriasis.
Birth Control				
93	Ethinyl estradiol	Estrogen	Used with other birth control drugs to achieve efficacy primarily as mediated by suppressing ovulation.	Used in the prevention of pregnancy.
94	Etonogestrel	Estrogen/progestogen combination	Efficacy is primarily mediated by suppressing ovulation.	Used in the prevention of pregnancy.
161	Norelgestromin	Progestogen	Efficacy is primarily mediated by suppressing ovulation.	Used in combination with estrogen in the prevention of pregnancy.

Drug no.	Drug name	Therapeutic class	Mechanism of action	Indications (Use)
162	Norethindrone acetate	Progestogen	Works by binding to an intracellular protein, FKBP-12, creating an inhibitory complex formation with mTOR complex 1, which results in the reduction of mTOR kinase activity.	Used in combination with estrogen in the prevention of pregnancy.
163	Norgestimate	Estrogen/progestogen combination	Efficacy is primarily mediated by suppressing ovulation.	Used in the prevention of pregnancy.
Blood Treatment				
60	Darbepoetin alfa	Erythropoiesis-stimulating agent	Efficacy is due to increasing erythropoiesis.	Used in the management of anemia caused by chronic kidney disease and where anemia is caused by the effect of concomitant myelosuppressive chemotherapy.
86	Epoetin alfa	Erythropoiesis-stimulating agent (ESA)	Erythropoiesis-stimulating glycoprotein.	Used in the management of anemia associated with chronic kidney disease (including patients on dialysis), anemia caused by zidovudine in HIV-infected patients, and anemic patients with nonmyeloid malignancies.
100	Filgrastim	Granulocyte colony-stimulating factor (G-CSF)	Works by increasing proliferation, differentiation, commitment, and end cell activation by binding to specific cell surface receptors on hematopoietic cells.	Used to decrease the chance of infection by increasing the concentration of neutrophils, in patients receiving myelosuppressive anticancer treatments, and may also be used to increase survival in patients exposed to radiation.
176	Pegfilgrastim	Granulocyte colony-stimulating factor (G-CSF)	Works by stimulating proliferation, differentiation, commitment, and end cell activation by binding to specific cell surface receptors on hematopoietic cells.	Used to increase the number of neutrophils, thereby decreasing the chance of infection in patients with nonmyeloid malignancies receiving myelosuppressive anticancer treatments.
Cancer Therapy				
2	Abiraterone acetate	Antiandrogen	Efficacy is mediated by inhibiting 17 α-hydroxylase/C17,20-lyase (CYP17), an enzyme expressed in testicular, adrenal, and prostatic tumor tissues and required for androgen biosynthesis.	Used in combination with prednisone for the management of patients with metastatic castration-resistant prostate cancer.
26	Bendamustine hydrochloride	Alkylating agent	Efficacy is due to electrophilic alkyl groups that may form covalent bonds with electron-rich nucleophilic moieties. The result is DNA cross-links which leads to cell death. Active against both quiescent and dividing cells.	Used in the management of chronic lymphocytic leukemia. Also used in the management of indolent B-cell non-Hodgkin lymphoma that has progressed during or within 6 months of treatment with rituximab or a rituximab-containing regimen.
27	Bevacizumab	Vascular endothelial growth factor (VEGF) inhibitor	Efficacy is due to the binding of VEGF to the receptors Flt-1 and KDR, which leads to endothelial cell proliferation and new blood vessel formation.	1st- or 2nd-line treatment of some metastatic cancers.
31	Bortezomib	Proteasome inhibitor	Efficacy is mediated by reversibly inhibiting chymotrypsin-like activity of the 26S proteasome in cells.	Used in the management of multiple myeloma and mantle cell lymphoma.
37	Capecitabine	Fluoropyrimidine carbamate	Efficacy is mediated by binding to thymidylate synthase to form a covalently bound ternary complex that prevents the formation of thymidylate from 2'-deoxyuridylate, thereby preventing DNA synthesis, and also interferes with RNA processing and protein synthesis.	Used in the treatment of metastatic breast cancer and metastatic colorectal cancer.
44	Cetuximab	Monoclonal antibody/ epidermal growth factor receptor (EGFR) antagonist	Works by binding to EGFR on normal and tumor cells and decreases the binding of ligands such as epidermal growth factor and transforming growth factor-alpha.	Used in the treatment of squamous cell carcinoma of the head and neck and metastatic colorectal cancer.
62	Dasatinib	Kinase inhibitor	Efficacy is due to inhibiting BCR-ABL, SRC family, c-KIT, EPHA2, and PDGFRβ kinases.	Used in the management of newly diagnosed Philadelphia chromosome-positive chronic myeloid leukemia in chronic phase. Also used in the management of chronic, accelerated, or myeloid or lymphoid blast phase Ph+ CML with resistance or intolerance to prior therapy, including imatinib, and in the management of Ph+ acute lymphoblastic leukemia with resistance or intolerance to prior therapy.
87	Erlotinib hydrochloride	Kinase inhibitor	Efficacy is mediated by reversibly blocking the kinase activity of EGFR, this prevents autophosphorylation of tyrosine residues accompanying the receptor and reduces further downstream signaling.	Used in the treatment of metastatic non-small cell lung cancer where the tumors have epidermal growth factor receptor (EGFR) exon 19 deletions or exon 21 (L858R) substitution mutations.
95	Everolimus	Kinase inhibitor	Works by binding to an intracellular protein, FKBP-12, creating an inhibitory complex formation with mTOR complex 1, which results in the reduction of mTOR kinase activity.	Used in the treatment of advanced hormone receptor-positive, HER2-negative breast cancer, advanced neuroendocrine tumors, advanced renal cell carcinoma, renal angiomyolipoma with tuberous sclerosis complex, and subependymal giant cell astrocytoma with tuberous sclerosis complex.
118	Imatinib mesylate	Kinase inhibitor	Efficacy is mediated by inhibiting the BCR-ABL tyrosine kinase. This reduces the proliferation and induces apoptosis in BCR-ABL positive cell lines as well as leukemic cells. Inhibits the receptor tyrosine kinases for platelet-derived growth factor and stem cell factor, c-Kit, and inhibits their mediated cellular events.	Used in the treatment of some forms of cancer, including: Philadelphia chromosome positive chronic myeloid leukemia (Ph+ CML), Ph+ acute lymphoblastic leukaemia, myelodysplastic/myeloproliferative diseases, aggressive systemic mastocytosis, hypereosinophilic syndrome and/or chronic eosinophilic leukemia (CEL), dermatofibrosarcoma protuberans, and malignant gastrointestinal stromal tumors (GIST).

Drug no.	Drug name	Therapeutic class	Mechanism of action	Indications (Use)
126	Ipilimumab	Monoclonal antibody/CTLA-4 blocker	Efficacy is mediated by the binding of ipilimumab to CTLA-4 and blocking interactions with the ligands CD80/CD86. This causes an increase in T-cell activation and proliferation, including antitumor immune response.	Used in the treatment of unresectable or metastatic melanoma.
131	Lenalidomide	Thalidomide analogue	Works by inducing apoptosis and reducing the proliferation and of certain hematopoietic tumor cells. Immunomodulatory properties include activation and increased number of natural killer T cells, and inhibition of proinflammatory cytokines by monocytes.	Used in the treatment of myelodysplastic syndrome, multiple myeloma, and mantle cell lymphoma.
164	Octreotide acetate	Somatostatin analogue	Works by mimicking the hormone somatostatin. Octreotide suppresses luteinizing hormone response to gonadotropin-releasing hormone, decreases splanchnic blood flow, and reduces the release of serotonin, gastrin, vasoactive intestinal peptide, secretin, motilin, and pancreatic polypeptide.	Used in the management of acromegaly and in the reduction of diarrhea associated with metastatic carcinoid tumors, and vasoactive intestinal peptide-secreting tumors.
177	Pemetrexed disodium	Antifolate	Works by blocking thymidylate synthase, dihydrofolate reductase, and glycinamide ribonucleotide formyl transferase, which are required for cell replication.	Used in the treatment of metastatic nonsquamous non-small cell lung cancer and malignant pleural mesothelioma.
225	Trastuzumab	Monoclonal antibody/HER2 blocker	Works by inhibiting the proliferation of tumor cells characterized by HER2 overexpression.	Used in the treatment of HER2-overexpressing breast cancers. May be used in combination with cisplatin and capecitabine or 5-fluorouracil for the treatment of HER2-overexpressing metastatic gastric or gastroesophageal junction adenocarcinoma.
200	Rituximab	Monoclonal antibody/CD20 blocker	Works by binding to CD20 antigen expressed on the surface of B-lymphocytes and mediates B-cell lysis.	Used in treatment of diseases, which are characterized by excessive numbers of B cells, overactive B cells, or dysfunctional B cells such as CD20-positive non-Hodgkins lymphoma, chronic lymphocytic leukemia, and rheumatoid arthritis.
Cholesterol Treatment				
23	Atorvastatin calcium	HMG-CoA reductase inhibitor (statin)	Works by preventing the conversion of HMG-CoA to mevalonate (precursor of sterols).	Used in the management of high cholesterol and triglyceride levels.
55	Colesevelam hydrochloride	Bile acid sequestrant	Efficacy is due to LDL clearance from the blood due to the binding of bile acids in the intestine, which hinders their readsorption.	Used in order to decrease LDL levels either as monotherapy or in combination with a statin. Used in order to improve glycemic control in patients with type 2 diabetes mellitus.
96	Ezetimibe	Cholesterol absorption inhibitor	Works to reduce blood cholesterol by reducing the absorption of cholesterol by the sterol transporter, Niemann-Pick C1-Like 1 in the small intestine.	Used in the management of high cholesterol.
144	Lovastatin	HMG-CoA reductase inhibitor (statin)	Works by preventing the conversion of HMG-CoA to mevalonate (precursor of sterols).	Used in the management of high cholesterol and triglyceride levels.
184	Pravastatin sodium	HMG-CoA reductase inhibitor (statin)	Effect is mediated by inhibiting the conversion of HMG-CoA to mevalonate.	Used in the management of high cholesterol and triglyceride levels.
203	Rosuvastatin calcium	HMG-CoA reductase inhibitor (statin)	Works by increasing lipid-modifying effects by increasing the concentration of hepatic LDL receptors on the cell surface and enhancing the uptake and catabolism of LDL and by reducing hepatic synthesis of VLDL, which decreases the total number of VLDL and LDL particles.	Used in the management of high cholesterol and triglyceride levels.
210	Simvastatin	HMG-CoA reductase inhibitor (statin)	Efficacy is believed to be mediated by the inhibition of the conversion of HMG-CoA to mevalonate, which is a rate-limiting step in the biosynthetic pathway for cholesterol. Decreases concentrations of VLDL, TGs, and increases HDL.	Used in the management of high cholesterol and triglyceride levels.
Cold Therapy				
69	Dextromethorphan hydrobromide	Antitussive/expectorant	Cough suppressant	Used to treat cough caused by throat and bronchial irritation, and loosen phlegm and bronchial secretions.
111	Guaifenesin		Expectorant	
172	Oxymetazoline hydrochloride	Decongestant	Vasoconstrictor	Used in the short-term reduction of nasal congestion and sinus pressure caused by the common cold, or upper respiratory allergies.
Dementia Treatment				
149	Memantine hydrochloride	NMDA receptor antagonist	Efficacy is believed to be mediated by its action as a low to moderate affinity uncompetitive (open-channel) NMDA receptor antagonist.	Used in the management of moderate to severe dementia associated with Alzheimer's disease.
202	Rivastigmine tartrate	Acetylcholinesterase (AChE) inhibitor	Efficacy is believed to be mediated by an increase in cholinergic function by increasing concentration of acetylcholine through reversible inhibition of its hydrolysis by cholinesterase.	Used in the management of dementia accompanying Alzheimer's disease and Parkinson's disease.

Drug no.	Drug name	Therapeutic class	Mechanism of action	Indications (Use)
Depression and Anxiety Therapy				
11	Alprazolam	Benzodiazepine	Efficacy is believed to be mediated by binding at stereo specific receptors at multiple sites within the CNS.	Used in the management of anxiety symptoms and in anxiety disorders. Also used in the treatment of panic disorders, with or without agoraphobia.
12	Amitriptyline hydrochloride	Tricyclic antidepressant (TCA)	Efficacy is believed to be due to inhibiting the membrane pump mechanism responsible for uptake of norepinephrine and serotonin in adrenergic and serotonergic neurons.	Used mainly in the management of depression, but may also be used in the treatment of agitation and insomnia, chronic pain, diabetic neuropathy, irritable bowel syndrome, sleep disorders, and migraine prophylaxis.
16	Amphetamine aspartate monohydrate	CNS stimulant	Efficacy is believed to be mediated by reducing the reuptake of norepinephrine and dopamine into the presynaptic neuron and increasing the release of these monoamines into the extraneuronal space.	Used in the management of attention-deficit hyperactivity disorder and narcolepsy.
17	Amphetamine sulfate	CNS stimulant	Efficacy is believed to be mediated by reducing the reuptake of norepinephrine and dopamine into the presynaptic neuron and increasing the release of these monoamines into the extraneuronal space.	Used in the management of attention-deficit hyperactivity disorder and narcolepsy.
19	Aripiprazole	Atypical antipsychotic	Efficacy is mediated by its effect as a partial D2/5HT1A agonist and 5HT2A antagonist.	Used in the management of schizophrenia, irritability accompanying autistic disorder, and Tourette's disorder. Also used in the management of manic and mixed episodes accompanying bipolar I disorder. Used in adjunctive management of major depressive disorder (MDD).
22	Atomoxetine	Selective norepinephrine reuptake inhibitor	Efficacy is believed to be mediated by selectively inhibiting the presynaptic norepinephrine transporter.	Used in the management of attention deficit hyperactivity disorder.
50	Clonazepam	Benzodiazepine	Efficacy is believed to be mediated by its ability to enhance the activity of gamma-aminobutyric acid, the major inhibitory neurotransmitter in the CNS.	Used in the management of Lennox-Gastaut syndrome (petit mal variant), akinetic, and myoclonic seizures. Clonazepam has been used in patients with absence seizures (petit mal) who have failed to respond to succinimides and is also used in the treatment of panic disorder with or without agoraphobia.
64	Desvenlafaxine	Serotonin and norepinephrine reuptake inhibitor (SNRI)	Mechanism of efficacy has not been fully established. Efficacy may be associated with the potentiation of serotonin and norepinephrine in the CNS through inhibition of their reuptake.	Used in the management of major depressive disorder.
66	Dexmethylphenidate hydrochloride	CNS stimulant	Mechanism of efficacy has not been fully established. Efficacy is believed to be due to a reduction in the reuptake of norepinephrine and dopamine into the presynaptic neuron and an increase in the release of monoamines into the extraneuronal space.	Used in the management of attention deficit hyperactivity disorder (ADHD).
67	Dextroamphetamine saccharate	CNS stimulant	Efficacy is believed to be mediated by reducing the reuptake of norepinephrine and dopamine into the presynaptic neuron and increasing the release of these monoamines into the extraneuronal space.	Used in the management of attention deficit hyperactivity disorder and narcolepsy.
68	Dextroamphetamine sulfate	CNS stimulant	Efficacy is believed to be mediated by reducing the reuptake of norepinephrine and dopamine into the presynaptic neuron and increasing the release of these monoamines into the extraneuronal space.	Used in the management of attention deficit hyperactivity disorder and narcolepsy.
70	Diazepam	Benzodiazepine	Efficacy is believed to be mediated by facilitating the actions of gamma aminobutyric acid.	Used in the management of anxiety disorders or for temporary relief of the symptoms of anxiety. Used in the management of acute alcohol withdrawal. Used for relief of skeletal muscle spasm and used in the management of convulsive disorders.
78	Duloxetine	Serotonin and norepinephrine reuptake inhibitor (SNRI)	Efficacy is thought to be due to the potentiation of serotonergic and noradrenergic activity in the central nervous system.	Used in the management of major depressive disorder, generalized anxiety disorder, diabetic peripheral neuropathy, fibromyalgia, and chronic musculoskeletal pain.
89	Escitalopram oxalate	Selective serotonin reuptake inhibitor (SSRI)	Efficacy is believed to be due to a reduction of the CNS neuronal uptake of serotonin.	Used in the management of generalized anxiety disorder and major depressive disorder.
103	Fluoxetine hydrochloride	Selective serotonin reuptake inhibitor (SSRI)	Efficacy is believed to be mediated by inhibiting the CNS neuronal uptake of serotonin.	Used in the management of major depressive disorder, social anxiety disorder, panic disorders, with or without agoraphobia, obsessive-compulsive disorder, premenstrual dysphoric disorder, and post-traumatic stress disorder.
137	Lisdexamfetamine dimesylate	CNS stimulant	Works by reducing the reuptake of dopamine and norepinephrine into the presynaptic neuron and increasing their release into the extraneuronal space.	Used in the management of attention deficit hyperactivity disorder and moderate to severe binge eating disorder.
141	Lorazepam	Benzodiazepine	Efficacy is believed to be mediated by binding at stereo specific receptors at several sites within the CNS.	Used in the treatment of anxiety disorders.
146	Lurasidone hydrochloride	Atypical antipsychotic	Efficacy may be mediated by central dopamine type 2 and serotonin type 2-receptor antagonism.	Used in the management of schizophrenia, and major depressive episodes associated with bipolar I disorder.

Drug no.	Drug name	Therapeutic class	Mechanism of action	Indications (Use)
173	Paliperidone palmitate	Atypical antipsychotic	Efficacy is believed to be mediated by a combination of central dopamine type 2 and 5HT2A receptor antagonism.	Used in the management of schizophrenia.
175	Paroxetine hydrochloride	Selective serotonin reuptake inhibitor (SSRI)	Efficacy is believed to be mediated by inhibiting the CNS neuronal reuptake of serotonin.	Used in the management of major depressive disorder, panic disorder with or without agoraphobia, obsessive-compulsive disorder, social anxiety disorder, generalized anxiety disorder, post-traumatic stress disorder, and premenstrual dysphoric disorder.
188	Quetiapine fumarate	Atypical antipsychotic	Efficacy is believed to be mediated through a combination of dopamine type 2 and serotonin type 2A antagonism.	Used in the management of schizophrenia, acute manic, depressive, or mixed episodes accompanying bipolar I disorder, and as adjunctive therapy to antidepressants for the management of major depressive disorder in adults.
206	Sertraline hydrochloride	Selective serotonin reuptake inhibitor (SSRI)	Efficacy is believed to be mediated by reducing the CNS neuronal uptake of serotonin.	Used in the management of major depressive disorder, social anxiety disorder, panic disorders, premenstrual dysphoric disorder, post-traumatic stress disorder and obsessive-compulsive disorder.
227	Trazodone hydrochloride	Triazolopyridine derivative	Efficacy may be linked to the selective inhibition of neuronal reuptake of serotonin and its activity as an antagonist at 5-HT-2A/2C serotonin receptors.	Used in the treatment of major depressive disorder.
236	Venlafaxine hydrochloride	Serotonin and norepinephrine reuptake inhibitor (SNRI)	Mechanism of efficacy has not been fully established. Efficacy may be associated with the potentiation of serotonin and norepinephrine in the CNS through inhibition of their reuptake.	Used in the management of major depressive disorder, generalized anxiety disorder, social anxiety disorder, and panic disorder.
237	Vilazodone hydrochloride	Selective serotonin reuptake inhibitor (SSRI)/5-HT1A-receptor partial agonist	Efficacy may be associated with the potentiation of serotonin in the CNS through inhibition of its reuptake.	Used in the management of major depressive disorder.
Dermatology				
73	Dimethyl fumarate	Immunomodulatory agent	Efficacy is believed to be due to dimethyl fumarate and the metabolite (monomethyl fumarate [MMF]) activating the nuclear factor (erythroid-derived 2)-like 2 (Nrf2) pathway, which is important in the cellular response to oxidative stress.	
116	Hydrocortisone	Corticosteroid	Corticosteroid	Used in the management of skin conditions such as redness, itching, and swelling.
154	Mometasone furoate	Corticosteroid	Corticosteroid	Used in the management of asthma and the management of nasal symptoms of seasonal and perennial allergic rhinitis.
182	Pramoxine hydrochloride	Local anesthetic	Decreases permeability of membranes to Na influx.	Temporarily relieves itching and other minor skin irritations.
228	Triamcinolone acetonide	Corticosteroid	Corticosteroid	Used in the management of various disorders in which corticosteroids are indicated such as eczema, dermatitis, allergies, and rash.
Digestive Treatment				
29	Bisacodyl	Stimulant laxative	Stimulant laxative.	Used in the treatment of occasional constipation and irregularity.
76	Docusate sodium	Stool softener	Increases concentration of water in the stool.	Used to treat occasional constipation.
139	Loperamide hydrochloride	Antidiarrheal	Works by extending the transit time of intestinal contents.	Controls symptoms of diarrhea.
145	Lubiprostone	Chloride channel activator	Works by increasing chloride-rich intestinal fluid secretion, and motility in the intestine.	Used in the management of chronic idiopathic constipation, opioid-induced constipation, and irritable bowel syndrome with constipation.
178a	Plcanatide	Guanylate cyclase C (GC-C) agonist	Works by binding to GC-C and acting on the luminal surface of intestinal epithelial cells. This leads to an increase in intestinal fluid and accelerated transit.	Used in the management of chronic idiopathic constipation.
180	Polyethylene glycol 3350	Osmotic laxative	Osmotic laxative.	Used in the treatment of occasional constipation.
Heart Therapy				
13	Amlodipine		Works by reducing the transmembrane influx of Ca^{2+} ions into vascular smooth muscle and cardiac muscle. Reduces peripheral vascular resistance and BP by acting directly on vascular smooth muscle.	
14	Amlodipine besylate	Calcium channel blocker (CCB)	Works by reducing the transmembrane influx of Ca^{2+} ions into vascular smooth muscle and cardiac muscle. Reduces peripheral vascular resistance and BP by acting directly on vascular smooth muscle.	Used in the management of hypertension, angina, and coronary artery disease.
18	Apixaban	Selective factor Xa inhibitor	Efficacy is mediated by decreasing thrombin production and thrombus development by inhibiting FXa and does not require antithrombin III for antithrombotic activity.	Used to decrease the chance of stroke and systemic embolism associated with nonvalvular atrial fibrillation and to decrease the risk of developing venous thrombosis after some orthopedic surgical procedures.

Drug no.	Drug name	Therapeutic class	Mechanism of action	Indications (Use)
21	Atenolol	Selective β1-blocker	The exact mechanism of action is not fully understood. Believed to work by competitively antagonizing catecholamines at peripheral adrenergic neuron sites. This causes a decreased cardiac output, reduced sympathetic outflow to the periphery, and suppression of renin activity.	Used in the treatment of hypertension and angina pectoris.
39	Carvedilol	α1/β-adrenergic blocker	The mechanism of action has not been fully established. Carvedilol has α1-adrenoreceptor blocking activity and attenuates the pressor effects of phenylephrine, causes vasodilation, and reduces peripheral vascular resistance.	Used in the management of mild to moderate heart failure of ischemic or cardiomyopathic origin.
51	Clonidine hydrochloride	Alpha-adrenergic agonist	Efficacy is due to the stimulation of α-adrenoreceptors in the brain stem, decreasing sympathetic outflow from CNS and decreasing peripheral resistance, renal vascular resistance, HR, and BP.	Used in the treatment of hypertension and prophylaxis of vascular migraine headaches. Used in the management of vasomotor symptoms associated with menopause, rapid detoxification in the management of substance withdrawal, and in the treatment of attention deficit hyperactivity disorder.
52	Clopidogrel bisulfate	Antiplatelet agent	Efficacy is mediated by irreversibly and selectively inhibiting the binding of adenosine diphosphate to its platelet P2Y12 receptor and the consequent ADP-mediated activation of the glycoprotein GPIIb/IIIa complex.	Used in the treatment of myocardial infarction, recent stroke, peripheral arterial disease, and acute coronary syndrome.
58	Dabigatran etexilate mesylate	Direct thrombin inhibitor (DTI)	Works by inhibiting free and clot-bound thrombin and thrombin-induced platelet aggregation.	Used to decrease the risk of stroke and systemic embolism in patients with nonvalvular atrial fibrillation. Used in the management of deep venous thrombosis and pulmonary embolism in patients who have been treated with a parenteral anticoagulant for 5-10 days.
72	Digoxin	Cardiac glycoside	Works by inhibiting Na+-K+ ATPase.	Used in the management of congestive heart failure, arrhythmias, and heart failure.
75	Dipyridamole	Antiplatelet agent	Efficacy is due to a reduction in the uptake of adenosine into platelets, endothelial cells, and erythrocytes.	Used to decrease the chance of stroke in patients who have had transient ischemia of the brain or complete ischemic stroke due to thrombosis.
84	Enalapril maleate	ACE inhibitor	Efficacy is due to a decrease in plasma angiotensin II, which leads to reduced vasopressor activity and reduced aldosterone secretion.	Used in the treatment of hypertension and symptomatic congestive heart failure.
85	Enoxaparin sodium	Low molecular weight heparin	Enoxaparin sodium has antithrombotic properties.	Used in the treatment of myocardial infarction, unstable angina, and deep vein thrombosis.
98	Fenofibrate	Fibric acid derivative	Efficacy is due to the activation of the peroxisome proliferator-activated receptor α. This results in an increase in lipolysis and elimination of TG-rich particles from plasma by activating lipoprotein lipase and reducing production of apoprotein C-III.	Used in the management of high cholesterol and triglycerides.
128	Isosorbide mononitrate	Nitrate vasodilator	Works by relaxing vascular smooth muscle, producing dilatation of peripheral arteries and veins.	Used in the treatment of acute and chronic angina pectoris, hypertension, myocardial infarction, and in the prevention of angina pectoris due to coronary artery disease.
138	Lisinopril	ACE inhibitor	Mechanism of action is due to a reduction in plasma angiotensin II, which causes a decrease in vasopressor activity and decreased aldosterone secretion.	Used in the treatment of high blood pressure, heart failure, and acute myocardial infarction.
142	Losartan potassium	Angiotensin II receptor blocker (ARB)	Works by blocking the vasoconstrictor effects of angiotensin II by blocking the binding of angiotensin II to the AT1 receptor in vascular smooth muscle.	Used in the management of high blood pressure.
152	Metoprolol succinate	Selective β1-blocker	Efficacy is believed to be due to a decrease in cardiac output, reduced sympathetic outflow to the periphery, and suppressed renin activity.	Used in the management of hypertension. It is also used to prevent further heart problems after myocardial infarction.
159	Nebivolol	Selective β1-blocker.	Selective β1-blocker.	Used in the management of high blood pressure.
165	Olmesartan acid		Works by inhibiting the vasoconstrictor effects of angiotensin II by blocking the binding of angiotensin II to the AT1 receptor in vascular smooth muscle.	
166	Olmesartan medoxomil	Angiotensin II receptor blocker (ARB)	Works by blocking the vasoconstrictor effects of angiotensin II by selectively blocking the binding of angiotensin II to the AT1 receptor in vascular smooth muscle.	Used in the management of high blood pressure.
183	Prasugrel	Platelet aggregation inhibitor	Works by inhibiting platelet activation and aggregation through binding of its active metabolite to the P2Y12 class of ADP receptors on platelets.	Used to decrease the risk of atherothrombotic events in patients with acute coronary symptoms.
192	Ramipril	ACE inhibitor	Mechanism of action is due to a decrease in plasma angiotensin II, which leads to decreased vasopressor activity and decreased aldosterone secretion.	Used in the treatment of high blood pressure and heart failure.
195	Ranolazine	Miscellaneous antianginal	Antianginal	Used in the management of chronic angina.
201	Rivaroxaban	Selective factor Xa inhibitor	Efficacy is mediated by reducing factor Xa. Rivaroxaban decreases thrombin generation and indirectly inhibits platelet aggregation induced by thrombin.	Used to decrease the risk of stroke and systemic embolism in patients with nonvalvular atrial fibrillation. Also used to decrease the risk of the recurrence of deep vein thrombosis and pulmonary embolism, and used in the prophylaxis of deep vein thrombosis, which may lead to pulmonary embolism in patients undergoing knee or hip replacement surgery.

Drug no.	Drug name	Therapeutic class	Mechanism of action	Indications (Use)
218	Telmisartan	Angiotensin II receptor blocker (ARB)	Works by inhibiting angiotensin II's vasoconstrictor and aldosterone-secreting effects by selectively blocking the binding of angiotensin II to the AT1 receptor.	Used in the management of hypertension and to reduce the risk of MI, stroke, or death from cardiovascular causes.
233	Valsartan	Angiotensin II receptor blocker (ARB)	Works by inhibiting the vasoconstrictor and aldosterone-secreting effects of angiotensin II by blocking the binding of angiotensin II to the AT1 receptor.	Used in the management of high blood pressure.
238	Warfarin sodium	Vitamin K-dependent coagulation factor inhibitor	Efficacy is believed to be mediated by the inhibition of the C1 subunit of the vitamin K epoxide reductase enzyme complex, thus decreasing the regeneration of vitamin K1 epoxide.	Used in the management of venous thrombosis, pulmonary embolism, and thromboembolic complications associated with atrial fibrillation and/or cardiac valve replacement. Used to decrease the risk of death, recurrent myocardial infarction, and thromboembolic events, such as stroke or systemic embolization.
HIV Treatment				
20	Atazanavir	Protease inhibitor	Inhibits the development of mature virions. Inhibits virus-specific processing of viral Gag and Gag-Pol polyproteins in HIV-1 infected cells.	Used in the management of HIV-1 infection in combination with other antiretrovirals.
53	Cobicistat	CYP3A inhibitor/HIV integrase strand transfer inhibitor/ nucleoside reverse transcriptase inhibitor (NRTI)	Efficacy is due to an increase in systemic exposure to CYP3A substrates by inhibiting CYP3A-mediated metabolism (e.g., elvitegravir).	For use in the management of HIV-1 infection.
61	Darunavir	Protease inhibitor	Prevents the cleavage of HIV-1 encoded Gag-Pol polyproteins in infected cells, limiting the formation of mature virus particles.	Used in the management of HIV-1 infection in combination with ritonavir and other antiretroviral agents.
81	Efavirenz		Works by inhibiting HIV-1 transcriptase noncompetitively.	Used in the management of HIV-1 infection.
82	Elvitegravir		Limits HIV-1 infection by preventing the formation of HIV-1 provirus and therefore viral propagation by inhibiting the strand transfer activity of HIV-1 integrase.	Used in the management of HIV-1 infection.
83	Emtricitabine		Works by competing with deoxycytidine 5'-triphosphate and by being incorporated into nascent viral DNA, resulting in chain termination.	Used in the management of HIV-1 infection.
191	Raltegravir	HIV-integrase strand transfer inhibitor	Works by preventing the catalytic action of HIV-1 integrase, which inhibits the formation of HIV-1 provirus and limits the propagation of the viral infection.	Used in the management of HIV-1 infection.
197	Rilpivirine		Works by preventing HIV-1 replication by noncompetitive inhibition of HIV-1 RT.	Used in the management of HIV-1 infection.
199	Ritonavir	Protease inhibitor	Works by inhibiting the processing the Gag-Pol polyprotein precursor, which decreases the production of mature HIV-1 particles.	Used in the management of HIV-1 infection.
219	Tenofovir disoproxil fumarate	Nucleotide analogue reverse transcriptase inhibitor	Works by inhibiting HIV-1 reverse transcriptase and HBV reverse transcriptase by competing deoxyadenosine 5'-triphosphate and, following incorporation into DNA, by DNA chain termination.	Used in the management of HIV-1 infection and chronic hepatitis B.
Hormonal Treatment				
56	Conjugated estrogens	Estrogen	Works by increasing concentrations of estrogens, which bind to estrogen receptors in a number of tissues and regulate the pituitary secretions.	Used in the prevention of postmenopausal osteoporosis, and in the treatment of moderate to severe vasomotor symptoms and/or vulvar and vaginal atrophy caused by menopause. Used in the treatment of hypoestrogenism due to hypogonadism, castration, or primary ovarian failure.
133	Levothyroxine sodium	Thyroid replacement hormone	Efficacy is believed to be due to effects exerted through control of DNA transcription and protein synthesis.	Used in the management of congenital or acquired hypothyroidism.
147	Medroxyprogesterone acetate		Synthetic progestin (more potent than progesterone). Works by diffusing into cells in the female reproductive tract, mammary gland, hypothalamus and pituitary and binds progesterone receptor which slows the release of gonadotropin from the hypothalamus.	
Multiple Sclerosis Therapy				
73	Dimethyl fumarate	Immunomodulatory agent	Efficacy is believed to be due to dimethyl fumarate and the metabolite (monomethyl fumarate [MMF]) activating the nuclear factor (erythroid-derived 2)-like 2 (Nrf2) pathway, which is important in the cellular response to oxidative stress.	Used in the management of relapsing forms of multiple sclerosis.
101	Fingolimod	Sphingosine 1-phosphate receptor modulator	The exact mechanism is unknown but efficacy may be due to a reduction of lymphocyte migration into the CNS.	Used in the management of relapsing forms of multiple sclerosis.

Drug no.	Drug name	Therapeutic class	Mechanism of action	Indications (Use)
108	Glatiramer acetate	Immunomodulatory agent	Immunomodulatory agent.	Used in the management of multiple sclerosis to reduce the frequency of relapses.
125	Interferon beta-1a	Biological response modifier	Biological response modifier.	Used in the management of relapsing forms of multiple sclerosis.
Opthamalogic Treatment				
7	Aflibercept	Vascular endothelial growth factor inhibitor	Aflibercept binds VEGF-A and placental growth factor, and results in the decreased binding and activation of the cognate VEGF receptors.	Used in the treatment of macular edema, neovascular age-related macular degeneration, diabetic macular edema, and diabetic retinopathy.
28	Bimatoprost	Prostaglandin analogue	Efficacy is due to its ability to mimic prostamides.	Used in the treatment of elevated intraocular pressure in patients with open-angle glaucoma or ocular HTN.
32	Brimonidine tartrate	Selective α2 agonist	Works by decreasing aqueous humor production and increasing uveoscleral outflow.	Used in the management of increased pressure in the eye caused by open-angle glaucoma or a condition called hypertension of the eye (ocular hypertension).
143	Loteprednol etabonate	Corticosteroid	Corticosteroid.	Used in the management of inflammation and pain caused by several conditions, including allergies, allergic conjunctivitis, uveitis, acne rosacea, superficial punctate keratitis, herpes zoster keratitis, iritis, cyclitis, and selected infective conjunctivitides.
167	Olopatadine hydrochloride	H1 antagonist and mast cell stabilizer	Works by reducing the type 1 immediate hypersensitivity reaction.	Used in the management of ocular itching with accompanying allergic conjunctivitis.
193	Ranibizumab	Monoclonal antibody/vascular endothelial growth factor (VEGF)-A blocker	Works by inhibiting the interaction of VEGF-A with its receptors (VEGFR1 and VEGFR2) on the surface of endothelial cells, limiting endothelial cell proliferation, vascular leakage, and new blood vessel formation.	Used in the management of neovascular age-related macular degeneration, macular edema following retinal vein occlusion, diabetic macular edema, and diabetic retinopathy.
221	Timolol		Nonselective β-blocker.	
226	Travoprost	Prostaglandin analogue	Efficacy is believed to be mediated by travoprost being a selective FP prostanoid receptor agonist and is thought to decrease intraocular pressure by increasing uveoscleral outflow.	Used in the treatment of elevated intraocular pressure in patients with open-angle glaucoma or ocular HTN.
Osteoporosis				
9	Alendronate sodium	Bisphosphonate	Works by inhibiting osteoclast-mediated bone resorption by binding to hydroxyapatite found in bone.	Used in the treatment and prevention of osteoporosis and in the treatment of Paget's disease of bone.
63	Denosumab	IgG2 monoclonal antibody	Denosumab decreases bone resorption and increases bone mass and strength by binding to the receptor activator of nuclear factor kappa-B ligand (RANKL) and inhibiting RANKL from activating its receptor.	Used to increase bone mass in patients with a high risk of fracture.
190	Raloxifene hydrochloride	Selective estrogen receptor modulator	Works by binding to estrogen receptors, which results in activation of estrogenic pathways in some tissues and deactivation of estrogenic pathways in others. For example, it works as an estrogen agonist in bone; decreasing bone resorption and bone turnover, while increasing bone mineral density, and decreasing fracture incidence.	Used in management and prevention of osteoporosis in postmenopausal women. Also, reduces risk of invasive breast cancer in postmenopausal women.
198	Risedronate sodium	Bisphosphonate	Functions as an antiresorptive agent and inhibits osteoclasts.	Used in the management and prevention of osteoporosis, glucocorticoid-induced osteoporosis in patients who are either starting or continuing systemic glucocorticoid treatment for chronic diseases, and the treatment of Paget's disease.
Pain Relief				
3	Acetaminophen	Analgesic	Efficacy is believed to be due to effects on the CNS, increasing the pain threshold by inhibiting both isoforms of cyclooxygenase, COX-1, COX-2, and COX-3 enzymes involved in prostaglandin (PG) synthesis. In contrast to NSAIDs, acetaminophen does not inhibit cyclooxygenase in peripheral tissues and exhibits no peripheral anti-inflammatory actions.	Temporarily reduces fever and relieves minor aches and pains.
35	Caffeine		Stimulant; pain reliever aid.	
38	Carisoprodol	Skeletal muscle relaxant (centrally acting)	The mechanism of action is not completely understood. Efficacy may be due to altered interneuronal activity in the spinal cord and the descending reticular formation of the brain.	Used in the treatment of skeletal muscle spasms.
54	Codeine	Opioid analgesic	Opioid analgesic.	Used in the management of moderate to severe pain.
57	Cyclobenzaprine hydrochloride	Skeletal muscle relaxant (centrally acting)	Works by reducing tonic somatic motor activity, affecting both gamma and alpha motor systems.	Used in the treatment of skeletal muscle spasms.
107	Gabapentin	GABA analogue	While the mechanism of action is not fully known, efficacy is believed to be mediated by gabapentin's high-affinity binding to the α2-delta subunit of voltage-activated Ca^{2+} channels.	Used as an anticonvulsant and to treat peripheral neuropathic pain associated with diabetes and shingles.
114	Hydrocodone	Opioid analgesic and antitussive	The mechanism of action has not been fully established. Efficacy is believed to be related to the existence of opiate receptors in CNS.	Used in the management of severe pain and as an antitussive.

Drug no.	Drug name	Therapeutic class	Mechanism of action	Indications (Use)
115	Hydrocodone bitartrate	Opioid analgesic and antitussive	The mechanism of action has not been fully established. Efficacy is believed to be related to the existence of opiate receptors in CNS.	Used in the management of severe pain and as an antitussive.
156	Morphine sulfate	Opioid analgesic	Opioid analgesic.	Used in the management of severe pain.
171	Oxycodone hydrochloride	Opioid analgesic	Opioid agonist.	Used in the management of severe pain.
186	Pregabalin	GABA analogue	While the mechanism of action is not fully established, efficacy is believed to be due to its binding with high affinity to the α2-delta site in CNS tissues.	Used in the management of neuropathic pain accompanying diabetic peripheral neuropathy or spinal cord injury. Also used in the management of postherpetic neuralgia and fibromyalgia and as an adjunctive therapy for patients with partial onset seizures.
224	Tramadol hydrochloride	Centrally acting analgesic	Centrally acting analgesic.	Used in the treatment of moderate to severe pain.
Parkinson's Disease Treatment				
203a	Safinamide	MAO-B inhibitor	Works by selective and reversible inhibition of MAO-B with blockade of voltage-dependent Na^+ and Ca^{2+} channels and inhibition of glutamate release.	Used in the treatement of Parkinson's disease. It is frequently an add-on medication to levodopa.
Sleep Aid				
91	Eszopiclone	Nonbenzodiazepine hypnotic agent	Efficacy is believed to be mediated by its interaction with GABA-receptor complexes at binding domains located close to or allosterically coupled to benzodiazepine receptors.	Used in the management of insomnia.
239	Zolpidem tartrate	GABAA agonist	The mechanism of action is due to interactions with a gamma-aminobutyric acid-BZ receptor complex. Zolpidem binds the BZ1 receptor preferentially with a high-affinity ratio of the $\alpha1/\alpha5$ subunits.	Used in the management of insomnia.
Ulcer Treatment				
30	Bismuth subsalicylate	Salicylate	Salicylate.	Used in the treatment of diarrhea and upset stomach.
46	Cimetidine	H2 blocker	Works by competitively blocking the action of histamine at the histamine H2 receptors of the parietal cells.	Used in the management of heartburn, other symptoms accompanying GERD, and for the treatment of duodenal ulcers.
65	Dexlansoprazole	Proton pump inhibitor (PPI)	Works by reducing gastric acid secretion by specific inhibition of the (H^+/K^+)-ATPase in the gastric parietal cell, which is the last step in acid production.	Used in the management of gastroesophageal reflux disease (GERD), and in the management of heartburn. Used in the management of duodenal ulcers and used in combination with amoxicillin and clarithromycin for the treatment of *Helicobacter pylori* infection and DU disease for *H. pylori* eradication. Also used in long-term management of pathological hypersecretory conditions.
90	Esomeprazole	Proton pump inhibitor (PPI)	Works by reducing gastric acid secretion by specific inhibition of the (H^+/K^+)-ATPase in the gastric parietal cell.	Used in the treatment of erosive or ulcerative gastroesophageal reflux disease (GERD) and GERD symptoms. Also used in the treatment of duodenal ulcers (DUs) and in combination with amoxicillin and clarithromycin for the treatment of *Helicobacter pylori* infection.
129	Lansoprazole	Proton pump inhibitor (PPI)	Works to decrease gastric acid secretion by inhibition of the (H^+/K^+)-ATPase enzyme system at the secretory surface of the gastric parietal cell.	Used in the management of acid-reflux disorders, peptic ulcer disease, and *Helicobacter pylori* eradication.
169	Omeprazole magnesium	Proton pump inhibitor (PPI)	Limits the last step of gastric acid secretion. Works by reducing gastric acid secretion by specific inhibition of the (H^+/K^+)-ATPase in the gastric parietal cell.	Used in the treatment of erosive or ulcerative gastroesophageal reflux disease (GERD) and GERD symptoms. Also used in the treatment of duodenal ulcers (DUs) and in combination with amoxicillin and clarithromycin for the treatment of *Helicobacter pylori* infection.
189	Rabeprazole sodium	Proton pump inhibitor (PPI)	Impedes the last step of gastric acid secretion. Substituted benzimidazole limits acid secretion by reducing the gastric H^+/K^+-ATPase at the secretory surface of the gastric parietal cell.	Used in the treatment of erosive or ulcerative gastroesophageal reflux disease (GERD) and GERD symptoms. Also used in the treatment of duodenal ulcers (DUs) and in combination with amoxicillin and clarithromycin for the treatment of *Helicobacter pylori* infection.
194	Ranitidine	H2 blocker	Efficacy is mediated by blocking the action of histamine on parietal cells in the stomach, which limits the production of acid.	Used in the treatment of erosive or ulcerative gastroesophageal reflux disease (GERD) and GERD symptoms.
Urinary Tract and Reproductive Therapy				
10	Allopurinol	Xanthine oxidase inhibitor	Efficacy is mediated by decreasing the production of uric acid by acting on purine catabolism.	Used in the management of hyperuricemia connected to primary or secondary gout, or hyperuricemia and recurrent renal calculi associated with chemotherapy.
79	Dutasteride		Works by preventing the conversion of testosterone to dihydrotestosterone (the androgen mostly responsible for the enlargement of the prostate gland).	Used in the treatment of difficulties with urination caused by an enlarged prostate.
97	Febuxostat	Xanthine oxidase inhibitor	Efficacy is mediated by decreasing the production of uric acid.	Used in the management of hyperuricemia in patients with gout.

Drug no.	Drug name	Therapeutic class	Mechanism of action	Indications (Use)
106	Furosemide	Loop diuretic	Works by limiting the absorption of Na^+ and Cl^- not only in the proximal and distal tubules but also in the loop of Henle.	Used in the management of edema caused by congestive heart failure, kidney or liver disease, and other medical conditions.
113	Hydrochlorothiazide	Thiazide diuretic	Limits the reabsorption of Na^+ and Cl^- ions, which leads to an increase in the concentration of Na+ traversing the distal tubule and increases the volume of water excreted.	Used in the management of hypertension and edema.
207	Sevelamer hydrochloride	Phosphate binder	Lowers serum concentrations of phosphate by interacting with phosphate molecules through ionic and hydrogen bonding, and limiting absorption.	Used in the regulation of serum phosphorus in patients with chronic kidney disease.
208	Sildenafil citrate	Phosphodiesterase 5 (PDE5) inhibitor	Increases the effect of nitric oxide by inhibiting PDE-5, which leads to an increase in cGMP in the corpus cavernosum, resulting in smooth muscle relaxation and the inflow of blood.	Used in the management of erectile dysfunction.
213	Solifenacin succinate	Muscarinic antagonist	Works by inhibiting muscarinic receptors, which mediate contractions of the bladder smooth muscle.	Used in the management of overactive bladder with symptoms including urinary incontinence, urgency, and urinary frequency.
214	Spironolactone	Aldosterone blocker	Efficacy is due to competitively binding receptors at aldosterone-dependent Na^+ - K^+ exchange sites in distal convoluted renal tubule, this leads to an increase in Na^+ and water excretion and K^+ retention.	Used in the management of hypertension, edema, hypokalemia, severe heart failure, and primary hyperaldosteronism.
216	Tadalafil	Phosphodiesterase 5 (PDE5) inhibitor	Works by increasing the concentrations of cGMP that leads to smooth muscle relaxation and increased blood flow into the corpus cavernosum.	Used in the management of erectile dysfunction and benign prostatic hyperplasia.
217	Tamsulosin hydrochloride	α1 Antagonist	Efficacy is due to the selective blockade of α1 receptors in the prostate. This causes a relaxation of the smooth muscles of the bladder neck and prostate, which improves urine flow.	Used in the treatment of urinary obstruction with symptoms such as hesitancy, terminal dribbling of urine, interrupted, and weak stream.
220	Testosterone	Androgen	Steroid hormone which can activate the androgen receptor or be converted to estradiol and activation of certain estrogen receptors.	Used to treat conditions resulting from a deficiency or absence of endogenous testosterone.
223	Tolterodine tartrate	Muscarinic antagonist	Works by being a competitive antagonist of acetylcholine at postganglionic muscarinic receptors mediating urinary bladder contraction.	Used in the treatment of overactive bladder.
229	Triamterene	Potassium-sparing diuretic	Works by inhibiting the reabsorption of Na^+ ions in exchange for K^+ and H^+ ions at segment of distal tubule.	Used in the management of edema caused by congestive heart failure, kidney or liver disease, and other medical conditions.
234	Vardenafil hydrochloride	Phosphodiesterase-5 (PDE-5) inhibitor	Works by increasing the concentrations of cGMP that leads to smooth muscle relaxation and increased blood flow into the corpus cavernosum.	Used in the management of erectile dysfunction.
47	Cinacalcet	Calcimimetic agent	Decreases parathyroid hormone concentrations by increasing the sensitivity of the Ca^{2+}-sensing receptor to extracellular Ca^{2+}.	Used in the management of hyperparathyroidism (a consequence of chronic kidney disease on dialysis). Also used in the management of hypercalcemia in patients with parathyroid carcinoma or primary hyperparathyroidism.
Vaccine				
112	Human papillomavirus quadrivalent	Vaccine	Vaccine	Used in the vaccination of patients 9-26 yrs. of age for the prevention human papillomavirus (HPV). This may lead to the prevention of cervical, vulvar, vaginal, and anal cancer in women and in males may lead to the prevention of anal cancer.
120	Influenza vaccine for intramuscular injection	Vaccine	Vaccine	Used in the immunization of patients ages 6 months and older against infection by influenza virus subtypes A and type B present in the vaccine
179	Pneumococcal 13-valent conjugate vaccine	Vaccine	Vaccine	Streptococcus pneumoniae immunization.
240	Zoster vaccine live	Vaccine	Vaccine	Used in the prevention of herpes zoster (shingles) in individuals ≥50 yrs. of age.

Index of Trade Names of Common Single-Component Drugs

Table 3 gives the trade names of common single-component drugs and links them to the physical properties (Table 1) and function information (Table 2) through the drug number (Drug no.). Note that many trade name single-component drugs not only have a single active component, but also can contain one or more inactive ingredients. Solid drugs are often dissolved or suspended in water for ease of administration.

TABLE 3. Index of Trade Names of Common Drugs

Trade name	Drug no.	Drug name
Abilify	19	Aripiprazole
Aciphex	189	Rabeprazole sodium
Actonel	198	Risedronate sodium
Actos	178	Pioglitazone
Adcirca	216	Tadalafil
Adoxa	77	Doxycycline

Trade name	Drug no.	Drug name
Advil	117	Ibuprofen
Afinitor	95	Everolimus
Afluria	120	Influenza vaccine for intramuscular injection
Afrin	172	Oxymetazoline hydrochloride
Aggrenox	75	Dipyridamole
Aldactone	214	Spironolactone
Aleve	158	Naproxen sodium
Alimta	177	Pemetrexed disodium
Allegra	99	Fexofenadine hydrochloride
Alphagan P	32	Brimonidine tartrate
Altace	192	Ramipril
Amaryl	109	Glimepiride
Ambien	239	Zolpidem tartrate
Amitiza	145	Lubiprostone
Amitriptyline	12	Amitriptyline hydrochloride
Amoxicillin	15	Amoxicillin
Androderm	220	Testosterone
Antara	98	Fenofibrate
Aranesp	60	Darbepoetin alfa
Asmanex	154	Mometasone furoate
Aspirin	4	Acetylsalicylic acid
Atelvia	198	Risedronate sodium
Ativan	141	Lorazepam
Atripla	81	Efavirenz
Avastin	27	Bevacizumab
Avodart	79	Dutasteride
Avonex	125	Interferon beta-1a
Aygestin	162	Norethindrone acetate
Bacitracin	25	Bacitracin
Bactrim	215	Sulfamethoxazole
Baxdela	62a	Delafloxacin
Benadryl	74	Diphenhydramine hydrochloride
Benicar	166	Olmesartan medoxomil
Bystolic	159	Nebivolol
Caldolor	117	Ibuprofen
Catapres	51	Clonidine hydrochloride
Cefdinir	40	Cefdinir
Celebrex	41	Celecoxib
Chantix	235	Varenicline
Chlor-Trimeton	45	Chlorpheniramine maleate
Cialis	216	Tadalafil
Cimetidine	46	Cimetidine
Cimzia	43	Certolizumab pegol
Cipro	48	Ciprofloxacin
Claritin	140	Loratadine
Clonazepam	50	Clonazepam
Colace	76	Docusate sodium
Combivent Respimat	127	Ipratropium bromide
Copaxone	108	Glatiramer acetate
Coreg	39	Carvedilol
Cortaid	116	Hydrocortisone
Coumadin	238	Warfarin sodium
Cozaar	142	Losartan potassium
Crestor	203	Rosuvastatin calcium
Cubicin	59	Daptomycin
Cyclobenzaprine	57	Cyclobenzaprine hydrochloride
Cymbalta	78	Duloxetine
Depo-Provera	147	Medroxyprogesterone acetate

Trade name	Drug no.	Drug name
Detrol	223	Tolterodine tartrate
Dexilant	65	Dexlansoprazole
DiaBeta	110	Glyburide
Diflucan	102	Fluconazole
Diovan	233	Valsartan
Dulcolax	29	Bisacodyl
Duramorph	156	Morphine sulfate
Dyrenium	229	Triamterene
Edurant	197	Rilpivirine
Effient	183	Prasugrel
Eliquis	18	Apixaban
Elvitegravir	82	Elvitegravir
Embeda	156	Morphine sulfate
Emtriva	83	Emtricitabine
Enbrel	92	Etanercept
Epogen	86	Epoetin alfa
Erbitux	44	Cetuximab
Ery-Tab	88	Erythromycin
Estinyl	93	Ethinyl estradiol
Evista	190	Raloxifene hydrochloride
Exelon	202	Rivastigmine tartrate
Eylea	7	Aflibercept
Ezvio	157	Naloxone
Fenoglide	98	Fenofibrate
Flomax	217	Tamsulosin hydrochloride
Flovent HFA	104	Fluticasone propionate
Focalin	66	Dexmethylphenidate hydrochloride
Formoterol fumarate dihydrate	105	Formoterol fumarate dihydrate
Fortamet	150	Metformin hydrochloride
Fosamax	9	Alendronate sodium
Gardasil	112	Human papillomavirus quadrivalent
Gilenya	101	Fingolimod
Gleevec	118	Imatinib mesylate
Gralise	107	Gabapentin
Harvoni	130	Ledipasvir
Herceptin	225	Trastuzumab
Humalog	124	Insulin lispro
Humira	5, 6	Adalimumab: heavy chain
Hydrocodone	114	Hydrocodone
Hysingla	115	Hydrocodone bitartrate
Imdur	128	Isosorbide mononitrate
Imodium	139	Loperamide hydrochloride
Invega Sustenna	173	Paliperidone palmitate
Invokana	36	Canagliflozin
Isentress	191	Raltegravir
Januvia	211	Sitagliptin
Kadian	156	Morphine sulfate
Keflex	42	Cephalexin
Khedezla	64	Desvenlafaxine
Klonopin	50	Clonazepam
Lanoxin	72	Digoxin
Lantus	123	Insulin glargine
Lasix	106	Furosemide
Latuda	146	Lurasidone hydrochloride
Levaquin	132	Levofloxacin
Levemir	122	Insulin detemir (rDNA origin)
Levitra	234	Vardenafil hydrochloride
Lexapro	89	Escitalopram oxalate

Trade name	Drug no.	Drug name
Lipitor	23	Atorvastatin calcium
Lipofen	98	Fenofibrate
Lofibra	98	Fenofibrate
Lotemax	143	Loteprednol etabonate
Lotrimin AF	153	Miconazole nitrate
Lovastin	144	Lovastatin
Lovenox	85	Enoxaparin sodium
Lucentis	193	Ranibizumab
Lumigan	28	Bimatoprost
Lunesta	91	Eszopiclone
Lyrica	186	Pregabalin
Medrol	151	Methylprednisolone
Micardis	218	Telmisartan
Microzide	113	Hydrochlorothiazide
Midol	117	Ibuprofen
Miralax	180	Polyethylene glycol 3350
Mobic	148	Meloxicam
Morphine	156	Morphine sulfate
Motrin	117	Ibuprofen
Namenda	149	Memantine hydrochloride
Nasacort	228	Triamcinolone acetonide
Nasonex	154	Mometasone furoate
NeoProfen	117	Ibuprofen
Neulasta	176	Pegfilgrastim
Neupogen	100	Filgrastim
Neurotin	107	Gabapentin
Nexium	90	Esomeprazole
Norvasc	14	Amlodipine besylate
Norvir	199	Ritonavir
Novolog	121	Insulin aspart (rDNA origin)
Nuvaring	94	Etonogestrel
Oleptro	227	Trazodone hydrochloride
Olysio	209	Simeprevir
Olmesartan acid	165	Olmesartan acid
Onglyza	205	Saxagliptin
Oracea	77	Doxycycline
Orencia	1	Abatacept
Ortho Tri-Cyclen	163	Norgestimate
OxyContin	171	Oxycodone hydrochloride
Pataday	167	Olopatadine hydrochloride
Patanol	167	Olopatadine hydrochloride
Paxil	175	Paroxetine hydrochloride
Pepto-Bismol Caplets	30	Bismuth subsalicylate
Phenadoz	187	Promethazine hydrochloride
Plavix	52	Clopidogrel bisulfate
Polymyxin B	181	Polymyxin B
Pradaxa	58	Dabigatran etexilate mesylate
Pramocaine	182	Pramoxine hydrochloride
Pravachol	184	Pravastatin sodium
Premarin Tablets	56	Conjugated estrogens
Premarin Vaginal Cream	56	Conjugated estrogens
Prevacid	129	Lansoprazole
Prevnar 13	179	Pneumococcal 13-valent conjugate vaccine
Prezista	61	Darunavir
Prilosec	169	Omeprazole magnesium
Primolut-Nor	162	Norethindrone acetate
Primsol	230	Trimethoprim
Prinivil	138	lisinopril
Pristiq	64	Desvenlafaxine
Procrit	86	Epoetin alfa
Prolia	63	Denosumab
Promethazine	187	Promethazine hydrochloride
Promethegan	187	Promethazine hydrochloride
Proventil HFA	8	Albuterol sulfate
Provera	147	Medroxyprogesterone acetate
Prozac	103	Fluoxetine hydrochloride
Ranexa	195	Ranolazine
Rayos	185	Prednisone
Rebif	125	Interferon beta-1a
Remicade	119	Infliximab
Renagel	207	Sevelamer hydrochloride
Revlimid	131	Lenalidomide
Reyataz	20	Atazanavir
Rituxan	200	Rituximab
Robitussin	111	Guaifenesin
Sandostatin	164	Octreotide acetate
Sensipar	47	Cinacalcet
Serevent	204	Salmeterol
Seroquel	188	Quetiapine fumarate
Singulair	155	Montelukast sodium
Soliris	80	Eculizumab
Soma	38	Carisoprodol
Sovaldi	212	Sofosbuvir
Spiriva	222	Tiotropium bromide
Sprycel	62	Dasatinib
Stelara	231	Ustekinumab
Strattera	22	Atomoxetine
Stribild	53	Cobicistat
Suboxone	34	Buprenorphine
Sulfatrim	215	Sulfamethoxazole
Sustiva	81	Efavirenz
Symbicort	33	Budesonide
Synagis	174	Palivizumab (vh region medi-498)
Synthroid	133	Levothyroxine sodium
Tamiflu	170	Oseltamivir phosphate
Tarceva	87	Erlotinib hydrochloride
Tecfidera	73	Dimethyl fumarate
Tenormin	21	Atenolol
Timoptic	221	Timolol
Toprol-XL	152	Metoprolol succinate
Tradjenta	134	Linagliptin
Travatan Z	226	Travoprost
Treanda	26	Bendamustine hydrochloride
Triamcinolone Acetonide Cream	228	Triamcinolone acetonide
Tricor	98	Fenofibrate
Triglide	98	Fenofibrate
Trulance	178a	Plecanatide
Truvada	83	Emtricitabine
Tylenol	3	Acetaminophen
Uloric	97	Febuxostat
Ultram	224	Tramadol hydrochloride
Valium	70	Diazepam
Valtrex	232	Valacyclovir hydrochloride
Vasotec	84	Enalapril maleate
Velcade	31	Bortezomib
Venlafaxine Tablets	236	Venlafaxine hydrochloride
Ventolin HFA	8	Albuterol sulfate

Biochem

Trade name	Drug no.	Drug name
Vesicare	213	Solifenacin succinate
Viagra	208	Sildenafil citrate
Vibramycin	77	Doxycycline
Victoza	136	Liraglutide
Viibryd	237	Vilazodone hydrochloride
Viread	219	Tenofovir disoproxil fumarate
Voltaren Gel	71	Diclofenac sodium
Vyvanse	137	Lisdexamfetamine dimesylate
WelChol	55	Colesevelam hydrochloride
Xadago	203a	Safinamide
Xanax	11	Alprazolam
Xarelto	201	Rivaroxaban
Xeloda	37	Capecitabine
Xgeva	63	Denosumab
Xifaxan	196	Rifaximin
Xolair	168	Omalizumab
Yervoy	126	Ipilimumab
Zantac	194	Ranitidine
Zetia	96	Ezetimibe
Zmax	24	Azithromycin
Zocor	210	Simvastatin
Zohydro	115	Hydrocodone bitartrate
Zoloft	206	Sertraline hydrochloride
Zortress	95	Everolimus
Zostavax	240	Zoster vaccine live
Zyloprim	10	Allopurinol
Zytiga	2	Abiraterone acetate
Zyvox	135	Linezolid

Components of Common Trade Name Multicomponent Drugs

Many common drug products are multicomponent, designed to provide a combination of therapeutic effects. Table 4 lists the trade name of a number of multicomponent drugs, indexed to the individual components listed in Table 1 through the drug number. Note that multicomponent trade name drugs not only have two or more active components, but also can contain one or more inactive ingredients. Solid drugs are often dissolved or suspended in water for ease of administration.

TABLE 4. Components of Common Multicomponent Drugs

Trade name	Drug no.	Components
Advair Diskus	104	Fluticasone propionate
	204	Salmeterol
Aggrenox	75	Dipyridamole
	4	Acetylsalicylate acid
Asasantin	75	Dipyridamole
	4	Acetylsalicylate acid
Atripla	219	Tenofovir disoproxil fumarate
	81	Efavirenz
	83	Emtricitabine
Augmentin	15	Amoxicillin
	49	Clavulanate potassium
Azor	13	Amlodipine
	166	Olmesartan medoxomil
Bactrim	215	Sulfamethoxazole
	230	Trimethoprim
Benicar HCT	166	Olmesartan medoxomil
	113	Hydrochlorothiazide
Combigan	32	Brimonidine tartrate
	221	Timolol (as maleate)
Combivent	127	Ipratropium bromide
	8	Albuterol sulfate
Complera	83	Emtricitabine
	197	Rilpivirine
	219	Tenofovir disoproxil fumarate
DouNeb	127	Ipratropium bromide
	8	Albuterol sulfate
Dulera	105	Formoterol fumarate dihydrate
	154	Mometasone furoate
Excedrin	3	Acetaminophen
	4	Acetylsalicylic acid
	35	Caffeine
Exforge	233	Valsartan
	13	Amlodipine
Harvoni	130	Ledipasvir
	212	Sofosbuvir
Janumet	150	Metformin hydrochloride
	211	Sitagliptin
Lo Loestrin Fe	93	Ethinyl estradiol
	162	Norethindrone acetate
		Ferrous fumarate (no therapeutic purpose)
Minastrin 24 Fe	93	Ethinyl estradiol
	162	Norethindrone acetate
		Ferrous fumarate (no therapeutic purpose)
Mucinex DM	69	Dextromethorphan hydrobromide
	111	Guaifenesin
Neosporin +	160	Neomycin sulfate
	181	Polymyxin B
	182	Pramoxine hydrochloride
NuvaRing	93	Ethinyl estradiol
	94	Etonogestrel
Ortho Tri-Cyclin	93	Ethinyl estradiol
	163	Norgestimate
Percocet	3	Acetaminophen
	171	Oxycodone hydrochloride
Premphase/Prempro	56	Conjugated estrogens
	147	Medroxyprogesterone acetate
Robitussin DM	69	Dextromethorphan hydrobromide
	111	Guaifenesin
Septra	215	Sulfamethoxazole
	230	Trimethoprim
Seretide	104	Fluticasone propionate
	204	Salmeterol
Stribild	53	Cobicistat
	82	Elvitegravir
	83	Emtricitabine
	219	Tenofovir disoproxil fumarate
Suboxone	219	Tenofovir disoproxil fumarate
	157	Naloxone

Trade name	Drug no.	Components
Symbicort	33	Budesonide
	105	Formoterol fumarate dihydrate
Truvada	83	Emtricitabine
	219	Tenofovir disoproxil fumarate
Tylenol #3	3	Acetaminophen
	54	Codeine (as phosphate)
Tylenol with codeine	3	Acetaminophen
	54	Codeine (as phosphate)
Vicodin	3	Acetaminophen
	114	Hydrocodone
Vytorin	96	Ezetimibe
	210	Simvastatin
Xulane	93	Ethinyl estradiol
	161	Norelgestromin
ZenHale	105	Formoterol fumarate dihydrate
	154	Mometasone furoate
Zubsolv	34	Buprenorphine
	157	Naloxone

PROPERTIES OF CONTROLLED SUBSTANCES

Thomas J. Bruno

The following table provides some important information that is used in the design of laboratory procedures such as chemical analyses and separations on substances that are listed as controlled substances by the U.S. government (Ref. 1). In most cases these substances are addictive and cause harm to humans. The substances listed here are only a small subset of those listed in Ref. 1, and are from Schedules I, II/IIN, and III/IIIN. Substances included in these schedules are defined as follows (Ref. 1).

Schedule I Controlled Substances: Substances in this schedule have no currently accepted medical use in the United States, a lack of accepted safety for use under medical supervision, and a high potential for abuse.

Schedule II/IIN Controlled Substances: Substances in this schedule have a high potential for abuse which may lead to severe psychological or physical dependence.

Schedule III/IIIN Controlled Substances: Substances in this schedule have a potential for abuse less than substances in Schedules I or II and abuse may lead to moderate or low physical dependence or high psychological dependence.

Typically, very dilute samples of these substances are available as standards and laboratory samples for research, student experiments in chemistry and forensic science departments, and in forensic labs. The data in this table supports such analytical and educational work. Forensic labs, of course, also handle larger samples as evidence protected by a chain of custody.

Column definitions are as follows.

Column heading	Definition
Name	Common name of substance; substances are listed alphabetically by common name of base substance; derived salts are listed immediately below base substance
Mol. form.	Molecular formula of substance, in Hill order; formula is for the most common hydration state as appropriate
CAS Reg. No.	Chemical Abstracts Service Registry Number
Sched. No.	Number of Controlled Substance Schedule on which substance is listed (Ref. 1)
Rel. molar mass	Molar molecular mass
t_m	Melting point, in °C
pK_a	Acidity strength of substance; lower pK_a value indicates stronger acid
Log *P*	Logarithm base 10 of octanol-water partition coefficient
Solubility	Solvents (partial list) in which the substance is readily soluble
Insolubility or low solubility	Solvents (partial list) in which the substance is either insoluble or very sparingly soluble

There is some overlap with "Properties and Functions of Common Drugs" in this section, and those data can supplement what is presented here. Additional information on these substances can be found in Refs. 2-7. Reference 2 is actually a collection of monographs compiled by the U.S. Drug Enforcement Administration (DEA) but made available on the SWGDrug online database.

References

1. Title 21 Code of Federal Regulations, Part 1308 - Schedules of Controlled Substances, U.S. Department of Justice, Drug Enforcement Administration, Washington, DC, 2020. <www.deadiversion.usdoj.gov/21cfr/cfr/2108cfrt.htm>
2. Scientific Working Group for the Analysis of Seized Drugs (SWGDrug). <https://www.swgdrug.org/monographs.htm>
3. Bones, J., Thomas, K., and Paull, B., Using environmental analytical data to estimate levels of community consumption of illicit drugs and abused pharmaceuticals, *J. Environ. Monitoring*, 9, 701-707, 2007.
4. NIST Chemistry WebBook, NIST Standard Reference Database Number 69 (Last update to data: 2018).
5. Hill, A., Department of Chemistry, Metropolitan State University of Denver, private communication, 2020.
6. Kinnunen, M., Piirainen, P., Kokki, H., Lammi, P., and Kokki, M., Updated Clinical Pharmacokinetics and Pharmacodynamics of Oxycodone, *Clin. Pharmacokinetics* 58, 705–725, 2019.
7. Trivedi, M., Shaikh, S., Gwinnut, C, Pharmacology of Opioids, *Anesthesia Tutorial of the Week*, 118-124, 2007.

Properties of Controlled Substances

Name	Mol. form.	CAS Reg. No.	Sched. No.	Rel. molar mass	t_m/°C	pK_a	Log *P*	Solubility	Insolubility or low solubility
Amphetamine (base)	$C_9H_{13}N$	300-62-9	II	135.2	-98		1.76	acetone, methanol, ethers, hexanes	water
Amphetamine hydrochloride	$C_9H_{14}ClN$	405-41-4	II	171.6	decomposes	9.80		water, chloroform, methanol	acetone, ethers, hexanes
Amphetamine phosphate	$C_9H_{16}NOP_4$	139-10-6	II	233.2	decomposes after sintering at 150			water, methanol	ethers, hexanes
Amphetamine sulfate	$C_{18}H_{28}N_2O_4S$	60-13-9	II	368.5	decomposes			water, methanol	acetone, ethers, hexanes chloroform
Cocaine (base)	$C_{17}H_{21}NO_4$	50-36-2	II	303.4	98.0		2.31	acetone, methanol, chloroform, ethers	water, hexanes
Cocaine hydrochloride	$C_{17}H_{22}ClNO_4$	53-21-4	II	339.8	195.0	8.60		water, methanol, chloroform	ethers, hexanes
Fentanyl (base)	$C_{22}H_{28}N_2O$	437-38-7	II	336.5	83-84			methanol	water
Fentanyl citrate	$C_{28}H_{36}N_2O_8$	990-73-8	II	528.6	149-151			water, methanol	chloroform, ethers

Name	Mol. form.	CAS Reg. No.	Sched. No.	Rel. molar mass	t_m/°C	pK_a	Log *P*	Solubility	Insolubility or low solubility
Heroin (base)	$C_{21}H_{23}NO_5$	561-27-3	I	369.4	173		1.69	acetone, chloroform	water, ethers, hexanes, methanol
Heroin hydrochloride monohydrate	$C_{21}H_{26}ClNO_6$	1502-95-0 (anhydrous)	I	423.9	243-244	7.60		water, chloroform, methanol	acetone, ethers, hexanes
Ketamine (base)	$C_{13}H_{16}ClNO$	6740-88-1	III	237.7	92-93		2.88	methanol	
Ketamine hydrochloride	$C_{13}H_{17}Cl_2NO$	1867-66-9	III	274.2	258-261	7.50		water	
Lysergic acid diethylamide (LSD) base	$C_{20}H_{25}N_3O$	50-37-3	I	323.4	80-85/82.5	7.80	2.10		
Lysergic acid diethylamide (LSD) tartrate	$C_{46}H_{64}N_6O_{10}$	NA	I	861.0	198-200			acetone, chloroform, ethers	hexanes
3,4-Methylenedioxymethamphetamine (base) (MDMA, ecstasy)	$C_{11}H_{15}NO_2$	42542-10-9	I	193.2	oil at ambient temperature		-0.32	acetone, chloroform, ethers, methanol, hexanes	water
3,4-Methylenedioxymethamphetamine hydrochloride (MDMA, ecstasy)	$C_{11}H_{16}ClNO_2$	64057-70-1	I	229.7	147-153	9.90		water, chloroform, methanol	acetone, ethers, hexanes
3,4-Methylenedioxymethamphetamine phosphate (MDMA, ecstasy)		NA	I	291.24	184-185			water, methanol	acetone, chloroform, ethers, hexanes
Morphine (base)	$C_{17}H_{19}NO_3$	57-27-2	II	285.3	247-248 (decomposes)		0.96	methanol	acetone, chloroform, hexanes, ethers
Morphine hydrochloride	$C_{17}H_{22}ClNO_4$	52-26-6	II	375.8	200 (decomposes)	8.08		water	chloroform, ethers, hexanes
Oxycodone (base)	$C_{18}H_{21}NO_4$	76-42-6	II	351.9	270-272			methanol, chloroform	water, ethers
Oxycodone hydrochloride	$C_{18}H_{22}ClNO_4$	124-90-3	II	315.4	218-220	8.5		water	acetone, ethers, chloroform, hexanes
Δ9-Tetrahydrocannabinol (THC)	$C_{21}H_{30}O_2$	1972-08-3	I	314.5		10.60	6.48	acetone, ethers, alcohols, hexanes, chloroform	water

CHEMICAL CONSTITUENTS OF HUMAN BLOOD

This table lists typical concentrations of some of the chemical constituents of human blood. The table covers elements and compounds of relatively low molecular weight. Refs. 1 and 4 give extensive information on enzymes, hormones, vitamins, and other blood constituents.

The values given for the normal range refer to healthy adults who have not been exposed to unusual environmental agents. In keeping with IUPAC practice, all values refer to a volume of one liter, and thus are stated in units of g/L, mg/L, μg/L, or mmol/L. Many clinical test results, especially in the United States, are reported on a deciliter (dL) rather than on a liter basis; thus, the values in this table should be divided by 10 to place them on a dL basis. The symbols S (for serum), P (plasma), and WB (whole blood) in the second column indicate the nature of the blood sample to which the values apply. In some cases, only a single mean value has been reported, rather than a range; these are given in italics.

The total volume of blood in a 100 kg (220 lb) adult is 7.5 L for a male and 6.7 L for a female. The corresponding volume of plasma is 4.4 L and 4.3 L, respectively (Ref. 1).

Values from Ref. 1 are so-called "reference values" against which clinical tests of blood chemistry are compared. The Lower limit and Upper limit define the "normal range," which is understood to include about 95% of the healthy population. The remaining 5% may show values outside the normal range without necessarily implying a medical problem. Note that these reference values may vary slightly from one testing laboratory to another, depending on the detailed test procedure.

Accurate measurements on trace elements are very difficult to make, and wide variations can be found in the literature. Preferred measurement methods are discussed in Refs. 2 and 6. Values for the trace elements can also vary from one country to another, depending on dietary or environmental factors. Thus, cadmium levels tend to be higher in Japan because of the prevalence of seafood in the diet, and lead levels are higher in regions where lead additives are still used in gasoline. Variations with gender, age, geography, and occurrence of diseases are reviewed in Ref. 6.

The Critical values column gives levels that deviate far enough from the normal range to suggest a probable medical issue. Such values from Ref. 3 are the Biological Exposure Indexes (BEI) that are specified by the American Council of Government Industrial Hygienists (ACGIH) as danger signals for the levels of pollutants in the workplace.

References

1. Wallach, J., *Interpretation of Diagnostic Tests, Eight Edition,* Wolters Kluwer, Philadelphia, 2007.
2. IUPAC Commission on Toxicology, "Sample Collection Guidelines for Trace Elements in Blood and Urine," *Pure & Appl. Chem.*, 67, 1575, 1995. <https://doi.org/10.1351/pac199567081575>
3. *2008 TLV's and BEI's,* American Conference of Governmental Industrial Hygienists, 1330 Kemper Meadow Drive, Cincinnati, OH 45240–1634, 2008 <www.acgih.org>.
4. Altman, P. L., and Dittmer, D. S., Eds., *Biology Data Book, Second Edition, Vol. III,* Federation of American Societies for Experimental Biology, Bethesda, MD, 1974.
5. Bowen, H. J. M., *Trace Elements in Biochemistry,* Academic Press, New York, 1966.
6. Versieck, J., and Cornelis, R., *Trace Elements in Human Plasma or Serum,* CRC Press, Boca Raton, FL, 1989.

Constituents of Human Blood

Name	Source	Unit	Lower limit	Upper limit	Critical values	Ref.
Inorganic						
Aluminum	S	μg/L	1	10	>60	6,2
Ammonia	P	μg/L	190	600	>700	1
Antimony	S,P	μg/L		*1*		6
Arsenic	S	μg/L	0.5	5		6,2
Barium	S,P	μg/L		*79*		4,5
Beryllium	S,P	μg/L		*<4*		4,5
Bicarbonate (HCO_3^-)	WB	mmol/L	22	28	<10 or >40	1
Bromine	S,P	mg/L	2	11		6,4
Cadmium	S	μg/L	0.1	1	>5	6,2,3
Calcium, total	S	mg/L	90	105	<65 or >140	1
Calcium ion (Ca^{++})	WB	mg/L	30	45		1
Carbon dioxide	P	mmol/L	21	30	<11 or >40	1
Carbon monoxide*	WB	%CO-Hb	0	5%	30%	1
Cesium	S,P	μg/L	0.5	2.0		6
Chloride (Cl^-)	S	mmol/L	98	106	<80 or >115	1
Chromium	S	μg/L	0.1	0.4		6,2
Cobalt	S	μg/L	0.05	0.35	>1	6,2,3
Copper	S	mg/L	0.7	1.4		1,2,6
Fluorine	S,P	μg/L	33	236		6
Hydrogen ion (H^+)	WB	pH	7.38	7.44	<7.10 or >7.59	1
Iodine (total)	S,P	μg/L	59	76		4
Iron	S	mg/L	0.5	1.7		1
Lead	S	μg/L	5	100	>300	1,3,6
Lithium	S,P	μg/L		*8*		6
Magnesium	S	mg/L	18	30	<10 or >47	1

Biochem

Name	Source	Unit	Lower limit	Upper limit	Critical values	Ref.
Manganese	S	μg/L	0.3	1.0		6,2
Mercury	S	μg/L	0.5	3	>15	2,3
Molybdenum	S,P	μg/L	0.3	1.3		6
Nickel	S	μg/L	0.1	1.3		6,2
Oxygen (arterial)	WB	% saturation	96%	100%		1
Oxygen (venous)	WB	% saturation	60%	85%		1
Phosphorus (inorganic)	S	mg/L	30	45	<11	1
Potassium	S	mmol/L	3.5	5.0	<2.8 or >6.2	1
Potassium	S	mg/L	137	196		
Rubidium	S,P	μg/L	100	300		6
Selenium	S,P	μg/L	40	160		2,6
Silver	S,P	μg/L		*1*		6
Sodium	S	mmol/L	135	145	<120 or >160	1
Sodium	S	g/L	3.11	3.34		
Strontium	S,P	μg/L		*57*		4,5
Sulfur (total)	S,P	mg/L		*780*		4
Tellurium	S,P	μg/L		*30*		4,5
Titanium	S,P	μg/L		*33*		4,5
Tin	S,P	μg/L		*1*		4,5
Vanadium	S,P	μg/L	0.02	1.0		6
Zinc	S,P	mg/L	0.5	1.2		6,2,4
Zirconium	S,P	μg/L		*400*		4,5
Organic						
Acetoacetate ion	P	mg/L		*<10*		1
Acetone	S,P	mg/L	3	20		1
Alanine	S,P	mg/L	30	37		4
Arginine	S,P	mg/L	12	19		4
Asparagine	S,P	mg/L	5.4	6.5		4
Cholesterol, total	P	mg/L	1000	2000**	>2400	1,4
HDL Cholesterol	P	mg/L	400	600		1
LDL Cholesterol	P	mg/L	0	1000	>1900	1
Citrulline	S,P	mg/L	2.1	9.7		4
Creatine	S,P	mg/L	2.8	6.2		4
Creatinine	S	mg/L	5	15	>50	1
Fructose	WB	mg/L	5	50		4
Glucosamine	S,P	mg/L	760	1110		4
Glucose (fasting)	S	mg/L	600	1000	<450 or >1300	1
Glutamic acid	S,P	mg/L	4.3	11.5		4
Glutamine	S,P	mg/L	61	102		4
Glycine	S,P	mg/L	13.4	17.3		4
Histidine	S,P	mg/L	7.9	14.8		4
Homocysteine	P	mg/L	0.54	1.62		1
Isoleucine	S,P	mg/L	6.9	12.8		4
Lactate (venous)	P	mg/L	50	150		1
Leucine	S,P	mg/L	14	23		4
Lysine	S,P	mg/L	25	30		4
Methionine	S,P	mg/L	3.3	4.3		4
Ornithine	S,P	mg/L	6.2	8.0		4
Phenylalanine	S,P	mg/L	5.8	14.0		1
Proline	S,P	mg/L	20	33		4
Serine	S,P	mg/L	10.1	12.5		4
Taurine	S,P	mg/L	4.1	8.2		4
Threonine	S,P	mg/L	12	17		4
Triglyceride	S	mg/L	250	1750		1
Tyrosine	S,P	mg/L	8.1	14.5		4
Urea	S	mmol/L	3.5	7.0	<0.7 or >28	1
Urea nitrogen (BUN)	S	mg/L (of N)	100	200	<20 or >800	1
Uric acid (males)	S	mg/L	25	80		1
Uric acid (females)	S	mg/L	13	60		1
Valine	S,P	mg/L	24	37		4

* Measured as the percent of hemoglobin bound to CO. Typical value for heavy smokers is 5%–10%. Major symptoms begin around 30%, and respiratory failure sets in at >60%.

** This is the desirable upper limit. Values between 2000 and 2400 mg/L are considered borderline high.

CHEMICAL COMPOSITION OF THE HUMAN BODY

The Reference Man provides a compilation of many research reports and is periodically updated by the International Commission on Radiological Protection (Ref. 2). The current Reference Man is between 20-30 years of age, weighs 70 kg, is 170 cm in height, and lives in a climate with an average temperature of 10 to 20 °C. Reference Man is Caucasian and Western European or North American in habitat and custom. The elemental composition of the Reference Man ("standard man") is given below. The elemental composition of human beings with parameters, such as in gender, age, weight, height, habitat, race, etc., that are different from the Reference Man may vary significantly (Ref. 2).

Unfortunately, at present, a "Reference Person" for women, other races, ages, etc., does not exist. Ref. 3 describes difficulties in recent efforts to update these data.

References

1. Padikal, T. N., and Fivozinsky, S. P., *Medical Physics Data Book, National Bureau of Standards Handbook 138*, U.S. Government Printing Office, Washington, DC, 1981. <https://doi.org/10.6028/NBS.HB.138>
2. Snyder, W. S., et al., *Reference Man: Anatomical, Physiological, and Metabolic Characteristics*, Pergamon, New York, 1975.
3. Pierson, R. N., "Appendix: Reference Body Composition Tables" in *Human Body Composition, Second Edition*, Heymsfield, S. B.,, Lohman, T. G., Wang, Z. M., and Going, S., B, Eds., Human Kinetics, Champaign, IL, 2005.

Chemical Composition of the Human Body

Element	Amount in grams	Percent of total body mass
Oxygen	43,000	61
Carbon	16,000	23
Hydrogen	7000	10
Nitrogen	1800	2.6
Calcium	1000	1.4
Phosphorus	580	1.1
Sulfur	140	0.20
Potassium	140	0.20
Sodium	100	0.14
Chlorine	95	0.12
Magnesium	19	0.027
Silicon	18	0.026
Iron	4.2	0.006
Fluorine	2.6	0.0037
Zinc	2.3	0.0033
Rubidium	0.32	0.00046
Strontium	0.32	0.00046
Bromine	0.20	0.00029
Lead	0.12	0.00017
Copper	0.072	0.00010
Aluminum	0.061	0.00009
Cadmium	0.050	0.00007
Boron	<0.048	0.00007
Barium	0.022	0.00003
Tin	<0.017	0.00002
Manganese	0.012	0.00002
Iodine	0.013	0.00002
Nickel	0.010	0.00001
Gold	<0.010	0.00001
Molybdenum	<0.0093	0.00001
Chromium	<0.0018	0.000003
Cesium	0.0015	0.000002
Cobalt	0.0015	0.000002
Uranium	0.00009	0.0000001
Beryllium	0.000036	
Radium	$3.1 \cdot 10^{-11}$	

Biochem

Section 8
Analytical Chemistry

Abbreviations and Symbols Used in Analytical Chemistry 8-1
Basic Instrumental Techniques of Analytical Chemistry 8-6
Analytical Standardization and Calibration 8-10
Figures of Merit 8-17
Mass- and Volume-Based Concentration Units 8-18
Detection of Outliers in Measurements 8-19
Properties of Carrier Gases for Gas Chromatography 8-20
Common Symbols Used in Gas and Liquid Chromatographic Schematic Diagrams 8-21
Standard Fittings for Compressed Gas Cylinders 8-22
Stationary Phases for Porous-Layer Open Tubular Columns 8-23
Coolants for Cryotrapping 8-24
Properties of Common Cross-Linked Silicone Stationary Phases 8-25
Detectors for Gas Chromatography 8-27
Varieties of Hyphenated Gas Chromatography with Mass Spectrometry 8-29
Gas Chromatographic Retention Indices 8-31
Pressure Drop in Open Tubular Gas Chromatographic Columns 8-33
Phase Ratios for Capillary Columns 8-34
Minimum Recommended Injector Split Ratios for Capillary Columns 8-35
Eluotropic Values of Solvents on Octadecylsilane and Octylsilane 8-35
Instability of HPLC Solvents 8-36
Detectors for Liquid Chromatography 8-37
Solvents for Ultraviolet Spectrophotometry 8-38
Correlation Table for Ultraviolet Active Functionalities 8-39
Middle-Range Infrared Absorption Correlation Charts 8-42
Common Spurious Infrared Absorption Bands 8-48
Nuclear Spins, Moments, and Other Data Related to NMR Spectroscopy 8-49
Properties of Important NMR Nuclei 8-52
Proton NMR Absorption of Major Chemical Families 8-53
Proton NMR Correlation Chart for Major Organic Functional Groups 8-58
Proton NMR Shifts of Common Organic Liquids 8-59
Proton Chemical Shifts of Contaminants in Deuterated Solvents 8-65
^{13}C-NMR Absorptions of Major Functional Groups 8-66
^{13}C-NMR Chemical Shifts of Common Organic Solvents 8-68
^{15}N-NMR Chemical Shifts of Major Chemical Families 8-69
Natural Abundance of Important Isotopes 8-71
Common Mass Spectral Fragmentation Patterns of Organic Compound Families 8-72
Common Mass Spectral Fragments Lost 8-74
Major Reference Masses in the Spectrum of Heptacosafluorotributylamine (Perfluorotributylamine) 8-75
Mass Spectral Peaks of Common Organic Liquids 8-76
Common Spurious Signals Observed in Mass Spectrometers 8-83
Chlorine–Bromine Combination Isotope Intensities in Mass Spectral Patterns 8-84
Reduction of Weighings in Air to *In Vacuo* 8-85
Standards for Laboratory Weights 8-86
Indicators for Acids and Bases 8-88
Preparation of Special Analytical Reagents 8-90
Organic Analytical Reagents for the Determination of Inorganic Ions 8-95

ABBREVIATIONS AND SYMBOLS USED IN ANALYTICAL CHEMISTRY

Abbreviations Used in Analytical Chemistry

A: peak area; surface area of solid granular adsorbent; eddy diffusion term in the Van Deemter equation
AAA: absolute activation analysis
AAD: atomic absorption detector
AAS: atomic absorption spectroscopy
AC: alternating current, affinity chromatography
ACP: alternating current plasma
ADXPS: angular dependent x-ray photoelectron spectroscopy
AED: atomic emission detector
AEM: analytical electron microscope (microscopy)
AES: Auger electron spectroscopy, atomic emission spectroscopy
AFID: alkali flame ionization detector
AFM: atomic force microscopy
AFS: atomic force spectroscopy
AIS: average of individual samples
AL: action level
AM: amplitude modulation
AMS: accelerator mass spectrometry
AN: area normalization
ANRF: area normalization with response factors
AOTF: acousto-optical tunable filter
AP: analytical pyrolysis
APCI: atmospheric pressure chemical ionization
APD: azimuthal photoelectron diffraction
API: atmospheric pressure ionization
APSTM: analytical photon scanning tunneling microscope
APT: attached proton test
ARAES: angle resolved Auger electron spectroscopy
ARF: absolute response factor
ARM: atomic resolution microscopy
ARPES: angle resolved photoelectron spectroscopy
ARUPS: angle resolved ultraviolet photoelectron spectroscopy
ASE: accelerated solvent extraction
AsFIFFF (AF4): asymmetrical flow field flow fractionation
ATD: above-threshold dissociation
ATI: above-threshold ionization
ATR: attenuated total reflection
BB: band broadening
BE: magnetic sector – electric sector tandem mass spectrometer (note: also called a MIKE spectrometer)
BEE: magnetic sector – electric sector – electric sector mass spectrometer
BET: Brunauer-Emmett-Teller (adsorption isotherm)
BIFL: burst integrated fluorescence lifetime
BIS: Bremsstrahlung isochromat spectroscopy
BJH: Barrett-Joyner-Halenda (method)
BL: bioluminescence
BLRF: bispectral luminescence radiance factor
BQQ: magnetic sector-double quadrupole mass spectrometer BTOF
BTOF: magnetic sector – time-of-flight tandem mass spectrometer
CAD: collision-activated dissociation
CAR: continuous addition of reagent
CARS: coherent anti-Stokes Raman spectroscopy
CCC: counter-current chromatography
CCD: charge-coupled device
CCT: constant current topography
CD: circular dichroism
CE: capillary electrophoresis, counter electrode
CEC: capillary electrokinetic chromatography, capillary electrochromatography
CED: cohesive energy density
CFA: continuous flow analysis
CF-FAB: continuous flow-fast atom bombardment
CFM: chemical force microscopy
CGE: capillary gel electrophoresis
CHEMFET: chemical-sensing field effect transistor
CI: chemical ionization
CID: collision-induced dissociation
CIEF: capillary isoelectric focusing
CITP: capillary isotachophoresis
CL: chemiluminescence
CLLE: continuous liquid-liquid extraction
CMA: cylindrical mirror analyzer
COSY: correlation spectroscopy
CPAA: charged particle activation analysis
CP/MAS: cross polarization/magic angle spinning
CRDS: cavity ring-down spectroscopy
CRF: chromatographic response function
CRM: certified reference material
CS: carbon strip (adsorbent)
CT: cryogenic trapping
CTD: charge transfer device
CV: cyclic voltammetry
CV-ASS: cold vapor atomic absorption spectrometry
CVD: chemical vapor deposition
CW: continuous wave
CZE: capillary zone electrophoresis
DA: diode array
DAD: diode array detector (UV-Vis)
DADI: direct analysis of daughter ions
dB: de Boar t-plot
DBE: double bond equivalent
DC: direct current
DCI: desorption chemical ionization
DCP: direct-current plasma
DEP: differential electrolytic potentiometry
DEPT: distortionless enhancement by polarization transfer
DETA: dielectric thermal analysis
DIN: direct injection nebulizer
DLI: direct liquid introduction
DLS: dynamic light scattering
DMA: dynamic mechanical analysis
DME: dropping mercury electrode
DNMR: dynamic nuclear magnetic resonance
DPP: differential pulse polarography
DRIFT: diffuse-reflectance infrared Fourier transform
DSC: differential scanning calorimetry
DTA: differential thermal analysis
DTC: differential thermal calorimetry
EAES: electron-excited Auger electron spectroscopy
EB: electric sector – magnetic sector tandem mass spectrometer
EBE: electric sector – magnetic sector-electric sector tandem mass spectrometer

Analytical

EC: electrochemical
ECD: electron capture detector
ECMS: electron capture mass spectrometry
ECNIMS: electron capture negative ionization mass spectrometry
EDL: electrodeless discharge lamp
EDS: energy-dispersive spectrometer
EDXRF: energy-dispersive x-ray fluorescence
EELS: electron energy-loss spectroscopy
EFFF: electric field flow fractionation
EG: electrogravimetry
EGA: evolved gas analysis
EIA: enzyme-linked immunoassay
EI(I): electron impact (ionization)
EIMS: electron impact mass spectrometry
ELCD: electrolytic conductivity detector
ELISA: enzyme-linked immunosorbent assay
ELSD: evaporative light scattering detector
EM: electron microscopy
EMIRS: electrochemically modulated IR spectroscopy
EOF: electro-osmotic flow
EPL: enhanced photoactivated luminescence
EPMA: electron-probe microanalysis
EPR: electron paramagnetic resonance
EPXMA: electron-probe x-ray microanalysis
EQL: estimated quantitation limit
ERD: elastic recoil detection
ESA: electrostatic analyzer
ESCA: electron spectroscopy for chemical analysis
ESEM: environmental scanning electron microscope
ESI: electrospray ionization
ESP: electrospray
ESR: electron spin resonance
ET: electrometric titration
ETA: electrothermal analyzer, emanation thermal analysis
EXAFS: extended x-ray absorption fine structure
FAA: flame atomic absorption
FAAS: flame atomic absorption spectroscopy
FABMS: fast-atom bombardment mass spectrometry
FAES: flame atomic emission spectroscopy
FAFS: flame atomic fluorescence spectroscopy
FAM: field analytical method
FAS: flame absorption spectroscopy
FD: field desorption
FD/FI: field desorption/field ionization
FES: flame emission spectroscopy
FFEM: freeze-fracture electron microscopy
FFF: field-flow fractionation
FFFF: flow field-flow fractionation
FFM: friction force microscopy
FFS: flame fluorescence spectroscopy
FFT: fast Fourier transform
FGC: fast gas chromatography
FI: flow injection, field ionization
FIA: flow injection analysis
FIB: focused ion beam
FID: flame ionization detector, free-induction decay
FIM: field ion microscopy
FNAA: fast neutron activation analysis
FOCS: fiber optic chemical sensor
FPD: flame photometric detector
FSOT: fused silica open tubular (column)
FT: Fourier transform
FT-ICR: Fourier transform ion cyclotron resonance
FT-IR: Fourier transform infrared (often "FT/IR," "FTIR," "FT IR")
FT-IRRAS: FT-IR reflection-absorption spectroscopy
FT-MS: Fourier transform mass spectrometry
FWHM: full-width half-maximum
GC: gas chromatography
GC-IR: gas chromatography–infrared spectrometry
GCMS: gas chromatography mass spectrometry
GDL: glow discharge lamp
GDMS: glow discharge mass spectrometry
GE: gel electrophoresis, gradient elution
GEMBE: gradient elution moving boundary electrophoresis
GFAAS: graphite furnace atomic absorption spectroscopy
GLC: gas-liquid chromatography
GPC: gel permeation chromatography
GS: Gram-Schmidt (algorithm)
GSC: gas-solid chromatography
GSED: gaseous secondary electron detector
HCL: hollow cathode lamp
HCOT: helically coiled open tubular (column)
HDC: hydrodynamic chromatography
HETCOR: heteronuclear correlation
HETP: height equivalent of (a) theoretical plate(s)
HG: hydride generation
HIC: hydrophobic interaction chromatography
HMBC: heteronuclear multiple-bond correlation
HPAC: high-performance affinity chromatography
HPIAC: high-performance immunoaffinity chromatography
HPLC: high-performance liquid chromatography, high-pressure liquid chromatography
HPTLC: high-performance thin-layer chromatography
HRCGC: high-resolution capillary gas chromatography
HRGC: high-resolution gas(-liquid) chromatography
HS: headspace
HSA: hemispherical analyzer
HSC: heteronuclear shift correlation
HSQC: heteronuclear single quantum coherence
HTC: high-temperature combustion
IA: isocratic analysis
IAC: immunoaffinity chromatography
IAES: ion-excited Auger electron spectroscopy
IC: ion chromatography
ICMS: ion chromatography mass spectrometry
ICP: inductively coupled plasma
ICP-OES: ICP optical emission spectrometry
ICR: ion cyclotron resonance
IDMS: isotope dilution mass spectrometry
IEC: ion-exchange chromatography
IEF: isoelectric focusing
IF: intermediate frequency
IGC: inverse gas chromatography
IGF: inert gas fusion
ILDA: intensified linear diode array
IMAC: immobilized metal-ion affinity chromatography
INADEQUATE: incredible natural abundance double-quantum transfer experiment
INEPT: insensitive nuclei enhancement by polarization transfer
INAA: instrumental neutron activation analysis
IP: ion pairing
IPC: ion-pair chromatography
IPG: immobilized pH gradient
IPMA: ion probe microanalysis

IR: infrared (spectrophotometry)
IRN: indicator radionuclide(s)
IRS: internal reflection spectroscopy
ISCA: ionization spectroscopy for chemical analysis
ISE: ion selective electrode
ISP: ion spray
ISS: ion scattering spectrometry
LAMMS: laser micro mass spectrometry
LARIMS: laser atomization resonance ionization mass spectrometry
LARIS: laser atomization resonance ionization spectroscopy
LASER: light amplification by stimulated emission of radiation
LBB: Lambert-Beer-Bouguer law
LC: liquid chromatography
LC-LS: multidimensional liquid chromatography
LDMS: laser desorption mass spectrometry
LDR: linear dynamic range
LEAFS: laser-excited atomic fluorescence spectrometry
LED: light-emitting diode
LEED: low-energy electron diffraction
LEEM: low-energy electron microscopy
LEI: laser-enhanced ionization
LEISS: low-energy ion scattering spectrometry
LESS: laser-excited Shpol'skii spectroscopy
LFM: lateral force microscopy
LIDAR: light detection and ranging
LIFD: laser-induced fluorescence detection
LIMS: laboratory information management system
LLC: liquid-liquid chromatography
LLD: lower-limit detection
LLE: liquid-liquid extraction
LNRI: laser non-resonant ionization
LO: local oscillator
LOC: lab on a chip
LOD: limit of detection
LPDA: linear photodiode array
LPSIRS: linear potential-sweep IR reflectance spectroscopy
LRI: laser resonance ionization
LRMA: laser Raman microanalysis
LSC: liquid-solid chromatography
LSE: liquid-solid extraction
LTP: low-temperature phosphorescence
MAE: microwave assisted extraction
MALDI: matrix-assisted laser desorption/ionization
MAS: magic angle spinning
MCD: magnetic circular dichroism
MCP: microchannel plate
MDGC: multidimensional gas chromatography
MDL: method detection limit
MDM: minimum detectable mass
MDQ: minimum detectable quantity
MEIS: medium-energy ion scattering
MEKC: micellar electrokinetic chromatography
MFM: magnetic force microscopy
MID: multiple ion detection
MIKE: mass analyzed ion kinetic energy mass spectrometry
MIP: microwave-induced plasma, mercury intrusion porosimetry
MIRS: multiple internal reflection spectroscopy
MLC: micellar liquid chromatography
MLLSQ: multiple linear least squares
MMF: minimum mass fraction
MMLLE: microporous membrane liquid-liquid extraction
MPD: microwave plasma detector
MPI: multiphoton ionization
MRDL: maximum residual disinfectant level (in water analysis)
MRI: magnetic resonance imaging
MS: mass spectrometry
MS-MS: tandem mass spectrometry
MSPD: matric solid-phase dispersion
MSRTP: micelle-stabilized room-temperature phosphorescence
MWD: microwave (assisted) digestion
NAA: neutron activation analysis
NCIMS: negative chemical ionization mass spectrometry
NDP: neutron depth profiling
NEXAFS: near edge x-ray absorption fine structure
NHE: normal-hydrogen electrode
NICI: negative ion chemical ionization
NIR: near-infrared, near-IR
NIRA: near-infrared reflectance analysis
nm: nanometer
NMR: nuclear magnetic resonance
NOE (nOe): nuclear Overhauser effect
NOESY: nuclear Overhauser effect spectroscopy
NPD: nitrogen-phosphorus detector, normal photoelectron diffraction
NPLC: normal-phase liquid chromatography
ODMR: optically detected magnetic resonance
ODS: octadecylsilane
OES: optical emission spectrometry, optical emission spectroscopy
OID: optoelectronic imaging device
OMA: optical multichannel analyzer
OPO: optical parametric oscillator
OPTLC: over-pressured thin-layer chromatography
ORD: optical rotary dispersion
OTE: optically transparent electrodes
PA: proton affinity
PAA: photon activation analysis
PAGE: polyacrylamide gel electrophoresis
PAH: polycyclic aromatic hydrocarbon
PAS: photoacoustic spectroscopy
PB: particle beam
PC: paper chromatography
PCA: principal component analysis
PCR: polymerase chain reaction
PCS: photon correlation spectroscopy
PCSE: partially coherent solvent evaporation
PD: plasma desorption
PDA: photodiode array
PDHID: pulsed discharge helium ionization detector
PDMS: plasma desorption mass spectrometry, polydimethyl siloxane
PED: pulsed electrochemical detection, plasma emission detector, photoelectron diffraction
PES: photoelectron spectroscopy
PET: positron emission tomography
PFIA: process flow injection analysis
PGC: packed-column gas chromatography
pH: negative logarithm of hydrogen ion concentration
PICI: positive ion chemical ionization
PID: photoionization detector
PIXE: particle-induced x-ray emission
p*K*: negative logarithm of an equilibrium constant
PLE: pressurized liquid extraction
PLOT: porous-layer open tubular

PLOT-cryo: porous-layer open tubular (column) cryo-adsorption
PMT: photomultiplier tube
ppb: parts per billion
ppm: parts per million
ppt: parts per thousand, parts per trillion
PSD: position sensitive detector
PTFE: polytetrafluoroethylene
PTR: proton transfer reaction (in mass spectrometry)
PTV: programmable temperature vaporizer
PVD: pulsed voltammetric detection, physical vapor deposition
QCL: quantum cascade laser
QCM: quartz-crystal microbalance
QFAA: quartz furnace atomic adsorption
QIT: quadrupole ion trap
qNMR: quantitative nuclear magnetic resonance
QQQ, QqQ, Q_1qQ_3: triple quadrupole mass spectrometer
QTH: quartz tungsten halogen
QTOF: tandem quadrupole time-of-flight mass spectrometer
RAA: running annual average
RBS: Rutherford backscattering spectrometry
REELS: reflection electron energy loss spectrometry
RES: reflection electron spectrometry
RF: radio frequency
RHEED: reflection high-energy electron diffraction
RIC: reconstructed ion chromatogram
RI: refractive index, retention index
RID: refractive-index detector
RIMS: resonance ionization mass spectrometry
RIS: resonance ionization spectroscopy, range of individual samples
RM: reference material
RNAA: radiochemical neutron activation analysis
ROA: Raman optical activity
ROESY: rotating frame nuclear Overhauser effect spectroscopy
RPLC: reversed-phase liquid chromatography
RRDE: rotating ring-disk electrode
RS: Raman spectroscopy
RSF: relative sensitivity factor
RTIL: room-temperature ionic liquid
RTP: room-temperature phosphorescence
S/N: signal-to-noise ratio
SAE: sonication-assisted extraction
SAM: scanning Auger microscopy, self-assembly monolayers
SANS: small-angle neutron scattering
SAW: surface acoustic wave
SAXS: small-angle x-ray scattering
SBSE: stir bar sorptive extraction
SCE: standard calomel electrode, saturated calomel electrode
SCF: supercritical fluid
SCOT: support-coated open tubular
SDD: silicon drift detector
SdFFF: sedimentation field flow fractionation
SEC: size-exclusion chromatography
SEM: scanning electron microscope
SERS: surface-enhanced Raman spectroscopy
SFC: supercritical-fluid chromatography
SFE: supercritical-fluid extraction
SFFF: sedimentation field flow fractionation
SF-MS: sector field mass spectrometry
SFS: synchronous fluorescence spectroscopy
SHE: standard hydrogen electrode
SIA: sequential injection analysis
SIDA: stable isotope dilution assay
SIMS: secondary ion mass spectrometry
SIRIS: sputter-initiated resonance ionization spectroscopy
SMCL: secondary maximum contaminant level (in water analysis)
SMDE: static mercury drop electrode
SMSS: spark source mass spectrometry
SNIFTIRS: subtractively normalized interfacial FT-IR spectroscopy
SNMS: sputtered neutral mass spectrometry
SPAES: spin-polarized Auger electron spectroscopy
SPE: solid-phase extraction
SPME: solid-phase microextraction
SPR: surface plasmon resonance
SRE: stray radiant energy
SRM: standard reference material
SSMS: spark source mass spectrometry
SSRTF: solid-surface room-temperature fluorescence
SSRTP: solid-surface room-temperature phosphorescence
STEM: scanning transmission electron microscope
STM: scanning tunneling microscope
SVE: solvent vapor exit
SWE: supercritical-water extraction
TCA: thermochemical analysis
TCD: thermal-conductivity detector
TCT-GC-MS: thermal cold trap gas chromatography mass spectrometry thermodilatometry
TD: thermodilatometry
TDL: tunable diode laser
TEA: thermal energy analyzer
TED: thermionic emission detector
TEELS: transmission electron energy loss spectrometry
TEM: transmission electron microscope
TET: thermometric enthalpimetric titration
TFFF: thermal field flow fractionation
TGA: thermogravimetric analysis
TGA-IR: thermogravimetric analysis – infrared
THEED: transmission high-energy electron diffraction
ThFFF: thermal field flow fractionation
TIC: total ion current chromatogram, tentatively identified compound
TIMS: thermal ionization mass spectrometry
TLC: thin-layer chromatography
TLE: thin-layer electrode
TLM: thermal lens microscopy
TLV: threshold limit value
TMA: thermomechanical analysis
TMS: tetramethylsilane
TOCSY: total correlation spectroscopy
TOF: time-of-flight
TOF-MS: time-of-flight mass spectrometry
TSP: thermospray
UHV: ultrahigh vacuum
USE: ultrasonic extraction
UV: ultraviolet
UVPES, UPS: ultraviolet photoelectron spectroscopy
UV-VIS, UV-Vis: ultraviolet-visible
VAR: variable angle reflectance
Vis: visible (radiation)
VOC: volatile organic compound(s)
VOX: volatile organic halogens
VUV: vacuum ultraviolet
W: Wein filter (used in mass spectrometry)

WCOT: wall-coated open tubular
WDS: wavelength dispersive spectrometer
WWCOT: whisker-wall coated open tubular (column)
WWPLOT: whisker-wall coated porous layer open tubular (column)
WWSCOT: whisker-wall support-coated open tubular (column)
XAES: x-ray excited Auger electron spectroscopy, X-ray adsorption edge spectrometry
XANES: x-ray absorption near-edge spectroscopy
XPS: x-ray photoelectron spectroscopy
XRD: x-ray diffraction
XRF: x-ray fluorescence
XRFS: x-ray fluorescence spectroscopy
XRS: x-ray spectroscopy
ZAF: Z (element number) absorption fluorescence

Symbols Used in Analytical Chemistry

α: Auger yield, fine structure constant
a_0: Bohr radius
A: absorbance
A: peak asymmetry factor
B: magnetic field strength
[c]: concentration of component c
d_p: particle diameter (HPLC stationary phase)
D_{ab}: diffusion coefficient
e: electron elementary charge
ε: extinction coefficient
E: energy
E: electrode potential
E_b: binding energy
E_{ea}: electron affinity
E_i: ionization energy
ν: frequency
γ: gyromagnetic ratio
n: refractive index
h: Planck constant
H: enthalpy, plate height
I_0: incident intensity
J: coupling constant
k: coverage factor
k': capacity factor
λ: wavelength, thermal conductivity
m/z: mass-to-elementary-charge ratio (mass spectrometry)
Q_{crit}: Q value (outlier test)
q: quadrupole parameter (mass spectrometry)
r: correlation coefficient
R: resolution
R_∞: Rydberg constant
ρ: density
s: standard deviation
s^2: variance
δ: chemical shift
δ^*: solubility parameter
τ: true value of a measured quantity
τ_{crit}: Chauvanet's criterion (outlier test)
$t_{1/2}$: half-life
t_M: mobile-phase hold up
t_R: retention time
t_R^0: specific retention time
T: transmittance
$T1$: spin-lattice relaxation time
$T2$: spin-spin relaxation time
$\bar{u}$: carrier phase velocity
V_M: carrier hold-up volume
V_R: retention volume
V_R^0: specific retention volume

BASIC INSTRUMENTAL TECHNIQUES OF ANALYTICAL CHEMISTRY

Thomas J. Bruno

The following section provides a very brief description of the major instrumental methods of chemical analysis. Please note that these paragraphs are general descriptions and are not meant to convey a comprehensive knowledge on these topics. The reader is referred to one of many excellent texts on instrumental methods of chemical analysis for additional details.

Suggested Reading

1. Skoog, D. A., Holler, F. J., and Crouch, S. R., *Principles of Instrumental Analysis, Sixth Edition*, Thomson Brooks/Cole Publishing, Belmont, CA, 2007.
2. Robinson, J. W., Skelly Frame, E. M., and Frame II, G. M., *Undergraduate Instrumental Analysis, Sixth Edition*, CRC Press, Boca Raton, FL, 2004.
3. Pungor, E., *A Practical Guide to Instrumental Analysis*, CRC Press, Boca Raton, FL, 1994.

Separation Methods

Gas Chromatography (GC): A separation method in which the sample or solute is vaporized (usually in a solvent, but sometimes neat or free of solvent) and passed through a medium under the influence of a carrier gas. The medium is called the stationary phase, in contrast to the carrier gas, which is mobile. The most common modern stationary phases are based on open tubular or capillary columns, in which the separation medium coats the inside periphery of a tube (typically tenths of millimeters in inside diameter) that is between 25 and 60 meters long. Older media are packed columns, consisting of packed beds, which are still used for gas analysis. In these applications, a solid sorbent is very common, and this is called gas-solid chromatography. Some open tubular columns are available with solid sorbents as well. Interactions of the solute with the separation medium affect the separation of the components of the mixture. A wide variety of detectors is available for general or specific applications. One of the most useful combinations is gas chromatography coupled with mass spectrometry (GC-MS). Solutes amenable to analysis by GC are usually of moderate volatility and relative molecular mass, usually not exceeding a relative molecular mass of 400. The most common stationary phases are cross-linked polymers based on dimethyl polysiloxane, the backbone of which can be derivatized with ligands to provide specific interactions. It is also possible to incorporate stereogenic (chiral) stationary phases as well.

Liquid Chromatography (LC, HPLC): A separation method in which a sample or solute (usually in a solvent, but sometimes neat or free of solvent) is passed through a medium under the influence of a carrier liquid. The medium is called the stationary phase, in contrast to the carrier, which is a mobile phase. Unlike gas chromatography, where the carrier gas plays little role other than mass transfer, the mobile phase in liquid chromatography is a controllable variable whose polarity and other properties are varied, in addition to the interactions with the stationary phase, to affect separation. The stationary phase in liquid chromatography is usually a micrometer-size particle packed bed that requires a high-pressure solvent system to cause mass flow. Liquid chromatographic systems have therefore been called high-pressure liquid chromatography (HPLC), although the acronym is usually taken to mean high-performance liquid chromatography. Many variations of the method have been developed for specific analysis. For example, gel permeation chromatography is an adaptation used for the separation of polymers. Affinity chromatography is similar in concept to gel permeation chromatography but uses the specific interaction between an antibody and an antigen. The use of stereogenic (or chiral) stationary phases is also an important development, especially in the analysis of pharmaceuticals.

Thin-Layer Chromatography (TLC): A separation method in which a stationary phase (typically a polar adsorbent such as alumina or silica gel) coated on a sheet of plastic, aluminum, or glass is used with a mobile phase usually consisting of a solvent or mixture of solvents in a beaker. The solute is applied as a blotted spot just above the end of the adsorbent-coated plate, and then the end is immersed into the solvent (but not so far as to immerse the solute spot). Commercial plates are robust, plastic sheets that can be cut to the size desired. In earlier applications, filter paper has been used in TLC, giving rise to the term "paper chromatography," which is rarely used today. Solvent is then drawn up through the adsorbent coating by capillary action. Separation results from a combination of interactions with the adsorbent and solvent. Commonly, the separated components are rendered visible by spraying stains or reactants on the plate after separation has been completed. It is also common to view the "developed" plate under an ultraviolet lamp, to visualize spots that may be fluorescent.

Supercritical Fluid Chromatography and Extraction: A separation technique similar to other extraction and chromatographic methods, but in which the mobile phase is actually a fluid in its supercritical fluid state. A supercritical fluid is a fluid that is held above its critical temperature and pressure, and for which no application of additional pressure can result in the development of a liquid phase. Supercritical fluids are unique in that while they are chosen to have liquid-like densities, the mass transfer properties are very much similar to those of liquids. Supercritical fluid chromatography remains a niche method that is applicable to pharmaceuticals and other high relative molecular mass solutes. Supercritical extraction, on the other hand, is more widely used as a sample preparation method, especially in pharmaceutical analyses, polymers, and environmental analyses.

Electrophoresis: A family of separation methods based on the motion induced in particles by an applied, uniform electric field. The most common application of electrophoresis is gel electrophoresis, in which a sample (typically proteins, amino acids, nucleic acids, etc.) is applied to a channel that is formed in a cross-linked polymer, usually polyacrylamide or agarose (the gel). The speed at which the individual species move through the gel under the influence of the field is determined largely by the size of the species, as expressed by the mass-to-charge ratio. After separation, the individual species usually appear as discrete bands that may be better visualized by staining with ethidium bromide, silver, or Coomassie Brilliant Blue dye. Other related and more specific techniques include isoelectric focusing, pulsed-field gel electrophoresis, immunoelectrophoresis, and isotachophoresis.

Purge-and-Trap Sampling: Purge-and-trap sampling is a family of methods that are used to capture the headspace above a condensed phase for subsequent analysis, most often for complex mixtures, environmental samples, etc. The headspace is the vapor space that develops above any condensed (solid or liquid) phase.

Thermodynamics assures us that the concentration of a particular analyte found in the headspace will be different than that found in the condensed phase, but often the relationship is predictable. The value in the method comes from the simplicity; sample preparation is usually far simpler than the cleanup that is typically required for many complex mixtures. Purge-and-trap methods fall into two general categories: static and dynamic. The dynamic method typically uses a sweep gas to continuously purge vapor analytes into a cold (cryogenic) trap or an adsorbent. Modern dynamic purge-and-trap methods include porous layer open tubular (PLOT) column cryo-adsorption, which uses a combined adsorbent and cryo-trapping approach on a high-efficiency platform. Static methods typically employ a syringe to pressurize the headspace above a condensed phase, followed by uptake of the pressurized headspace into a trap. A modern static method (that usually is done without pressurization) is solid-phase microextraction (SPME). This method utilizes a fiber coated with a stationary phase (similar to stationary phases used for gas and liquid chromatography) at the end of a wire mounted in a syringe needle.

Spectroscopic Methods

Mass Spectrometry (MS): An analytical technique in which charged particles or radical ions are produced from a sample by either electron impact (bombardment with a stream of electrons) or chemical ionization (interaction with a small charged ion). Analysis on the basis of mass-to-charge ratio is performed on fragments of the molecule that develop after the initial ionization. The method is very useful for mass determination and structure determination on the basis of the induced fragmentation pattern. The charged fragments can be separated or analyzed by a magnetic sector, a quadrupole, an ion trap, by time of flight, or by cyclotroning. When coupled to separation techniques such as gas or liquid chromatography, a nearly universal qualitative detection capability is provided. Structure determination can be performed by comparison to well-known fragmentation patterns or characteristics. When a mass spectrometer is capable of high mass resolution, a nearly unequivocal identification of a compound formula is possible. While direct interfaces and gas chromatographs are the most common sample introduction techniques, many others are available for specific applications. Matrix-assisted laser desorption and ionization (MALDI) is often used in time-of-flight instruments and is especially useful for the analysis of biopolymers. Inductively coupled plasma ionization is capable of producing mass spectra at high sensitivity for many metals and some nonmetals. When coupled to liquid chromatography instrumentation, thermospray and electrospray methods have been used with HPLC for analytes that are not amenable to gas chromatography separation.

Ultraviolet Spectrophotometry (UV, UV-Vis): A spectroscopic technique that focuses on electronic transitions within the visible and ultraviolet regions of the electromagnetic spectrum for excitation and detection. The practical ultraviolet region extends from 190 nm to 400 nm in wavelength. The UV region can be divided into subranges: near-UV: 300 nm to 400 nm, mid-UV: 300 nm to 200 nm, far-UV: 200 nm to 122 nm, and vacuum-UV: 200 nm to 100 nm. Other divisions are possible, but these are less important for analytical chemistry. The visible region, so called because of the response of human vision, extends from about 390 nm to 750 nm in wavelength. Although it is possible to use ultraviolet visible spectrophotometry for structure determination, it is most often used as a quantitative tool. The wavelength-structure correlations are not as detailed, nor the spectra as sharp, as with other spectroscopic methods such as infrared spectrophotometry. The utility of UV-Vis absorptions for many organic compounds has led to this instrument being adapted as a detector for liquid chromatography. Related to ultraviolet spectrophotometry are fluorescence spectroscopic methods. In these methods, the energy emitted is at a different wavelength (usually longer, of lower energy), and this is typically detected perpendicularly to the incident excitation beam. This makes fluorescence spectroscopy more sensitive than UV-Vis spectroscopy. This type of instrument is also incorporated as a detector for liquid chromatography.

Infrared Spectrophotometry (IR, FTIR): A spectroscopic technique that focuses on molecular vibrations (with a concurrent change in dipole moment) within the infrared (IR) region of the electromagnetic spectrum for excitation and detection. This region is further divided into three separate but overlapping ranges. The near-infrared (high-energy IR, approximately 14 000 cm^{-1} to 4000 cm^{-1}, 0.8 µm to 2.5 µm wavelength) is used to study overtone or harmonic vibrations. The mid-infrared (mid-range energy IR, approximately 4000 cm^{-1} to 400 $^{-1}$, 2.5 µm to 25 µm wavelength) is used to study the fundamental vibrations and associated rotational-vibrational combinations. The far-infrared (low-energy IR, approximately 400 cm^{-1} to 10 cm^{-1}, 25 µm to 1000 µm wavelength) is close to the microwave region and is used to study rotational transitions. Most modern instruments use the Fourier transform (FT) technique to record the spectrum over all wavelengths, rather than by scanning through the wavelengths. The absorbances of the IR radiation are associated with specific chemical moieties, and a study of the spectra can be used to aid in structure determination. One often uses structure correlation charts to aid in the assignment of absorbance bands. An analysis of the intensity of the absorptions can also be used for quantitative analysis.

Nuclear Magnetic Resonance Spectrometry (NMR): A spectroscopic method that takes advantage of magnetic nuclei (nuclei with an odd number of protons and/or of neutrons, having an intrinsic magnetic moment and angular momentum), placed in a magnetic field will absorb pulses of electromagnetic radiation and then radiate this energy back out. For these nuclei, the energy and signal intensity are proportional to the applied magnetic field. The power of NMR results from the ability to probe the molecular environment around a particular nucleus, thus making it long a tool of the organic chemist. This is done by measurement of the chemical shift (or frequency) of an absorption, and by the analysis of splitting patterns, which are caused by the influence of adjacent nuclei. New high-field, high-sensitivity instruments have given this technique more applications in analytical chemistry. The most commonly studied nuclei are ^{1}H (proton, the most NMR-sensitive isotope after the radioactive ^{3}H) and ^{13}C. With high-field instruments, additional nuclei are accessible: ^{2}H, ^{10}B, ^{11}B, ^{14}N, ^{15}N, ^{17}O, ^{19}F, ^{23}Na, ^{29}Si, ^{31}P, ^{35}Cl, ^{113}Cd, ^{129}Xe, and ^{195}Pt.

Raman Spectroscopy: A vibrational spectroscopic method that arises from the inelastic scattering of monochromatic radiation by molecules that undergo a change in polarizability during the vibration. This is in contrast to infrared spectrophotometry, in which a change in the dipole moment occurs during the vibration. When radiation (typically light from a laser in the visible, near-infrared, or near-ultraviolet range) is scattered, a small fraction of the scattered radiation is observed to have a different frequency (the Raman effect). The variations of Raman spectroscopy are used to locate functional groups or chemical bonds in molecules. There are several variations in the approach to Raman spectroscopy. In resonance Raman spectroscopy, the excitation wavelength is matched to an electronic transition of the molecule, enhancing the vibrational modes. In coherent anti-Stokes Raman spectroscopy (CARS), two laser beams are used to generate a coherent anti-

Stokes frequency beam. In surface-enhanced Raman spectroscopy (SERS), surface plasmons (a quantum of plasma oscillation) on a silver or gold colloid on a surface (such as a mirror) are excited by the laser, resulting in an increase in the electric fields surrounding the metal.

Atomic Absorption and Emission Spectroscopy (AA, AES): Two related spectroscopic methods applied primarily to the analysis of inorganic compounds. Atomic absorption procedures use the absorption of optical radiation (light) by free atoms in the gaseous state. The light can be produced by a hollow cathode lamp, an electrodeless discharge lamp, or a deuterium lamp. The light is absorbed by the analyte during an electronic transition, the wavelength of which corresponds to only one element in the analyte, and the width of an absorption line is only of the order of a few picometers. This method can be used for the quantitative determination (on the basis of a calibration curve) of approximately 70 different elements in solution or directly in solid samples. Atomic emission spectroscopy (AES) uses the light emitted by a vaporized sample in a flame, plasma, arc, spark, or laser, at a particular wavelength, to determine the atomic spectrum (for determination of the elemental composition) of an element in a sample. The wavelength of the atomic spectral line gives the identity of the element while the intensity of the emitted light is proportional to the number of atoms of the element. No single source, as described above, is optimal for a given sample, and it is the choice of source that distinguishes the various techniques.

Miscellaneous Methods

Colorimetry, Spot Tests, Presumptive Tests: Rapid, simple tests based on color change are frequently used as the basis of preliminary conclusions or approximate concentration measurement. Colorimetric methods use simple comparative instruments to determine the concentration of colored compounds in solution. These devices, called colorimeters, can be manual or automatic, and use filtered light in the visible region (between 400 nm and 700 nm). In both cases, the operation depends upon having multiple solutions, including a blank, for comparison with a solution of unknown concentration. Manual colorimeters function by measuring the variable light path through the unknown solution as compared to a known solution until a match is achieved visually. The product of concentration and path length matches when the concentrations are the same, so an unknown concentration may be obtained by a proportion. Automated devices function similarly but with photocells. Colorimetric tests are usually done following four basic protocols: (1) an unknown can be treated/reacted with a reagent to form a new compound which is colored, (2) a chelate complex is formed having a different color than the starting compound, (3) a colored compound is oxidized or "bleached" by another compound, resulting in less color, or (4) an intermediate is formed that can be oxidized or reduced later to a colored compound. Related to colorimetric methods are spot tests, often called presumptive tests in forensic science (because they are used to establish probable cause for ordering and performing more sophisticated tests), which are the observation of color changes upon the addition of one or more reagents to an unknown. Spot tests are inherently simple and are done with a drop or two of reagent(s), on a filter paper or other suitable medium, and generate minimal waste. Usually instrumentation is not used, one merely notes a color change, although simple devices such as handheld color space analyzers (used in the paint industry) can be helpful. Other approaches include the recording of a digital photograph of the spot and the determination of the L*, a*, and b* axes in LAB color space. An example of spot testing for presumptive purposes includes the addition of cobalt thiocyanate ($CoC_2N_2S_2$) to cocaine, producing a blue color (the Scott test). These tests are presumptive rather than confirmatory because of the potential of false positives. Thus, the Scott test will also produce a blue coloration in the presence of lidocaine and diphenylhydramine.

Refractometry: Refractometry is the measurement of the degree to which the path of electromagnetic radiation, specifically light, is bent upon traversing from one medium to another. Indirectly, a refractometer measures the speed of light since the refractive index, the measurand provided by a refractometer, is defined as the speed of light in vacuo divided by the speed of light in a particular medium (which can be a gas, liquid, or solid). Refractive index typically varies between 1.3 and 1.7 for most compounds, and being a ratio, it is dimensionless. The refractive index of vacuum is by definition unity. Refraction is the result of differing densities of media; light passes through dense media slower than it passes through less dense media. Refractive index is dependent on temperature and the wavelength of light used for the measurement, and when both are known for a media, it can be used as an indication of composition or concentration. Thus, the refractive index is typically reported measured with white light from the sodium D-line (589 nm), and often at a temperature that is near ambient (20 °C). The measurand is thus indicated as n_D^{20}. As an analytical tool, refractometry is used in four primary ways. First, it is used to help identify compounds by comparison with known values. Second, it is used to assess purity by comparison with the refractive index of a pure material. Third, it is used as a measure of solute concentration of a solution, by reference to a calibration curve. Finally, it is used as the basis of universal detection in HPLC. Refractometers are commercially available as small handheld instruments, both manual and digital (that can be carried in a pocket), Abbe refractometers (the typical lab bench device usually equipped with a thermostat), and in-line refractometers such as those used for HPLC detectors.

Thermal Analysis (TA) Methods: A family of analytical techniques in which various properties of a sample are examined as a function of changing temperature at a particular rate of change. For chemical analysis, the most common thermal analysis techniques include differential thermal analysis (DTA) and thermogravimetric analysis (TGA). Many other thermal analysis methods are available, perhaps the most common of which is differential thermal analysis, used for determination of phase transitions. In chemical analysis, TGA determines a mass change as a function of temperature. This is used to determine decomposition or degradation temperatures, moisture content of materials (although Karl Fisher coulombic titrimetry is also used for this), the level of inorganic and organic components in materials, decomposition points of explosives, and solvent residues in materials. DTA, on the other hand, monitors temperature change rather than mass change. In this respect, it is useful as a complementary technique to probe the energetics of decomposition, moisture loss, etc. Other thermal analysis methods, such as differential scanning calorimetry (DSC), are not primarily analytical tools but rather thermophysical property measurement tools.

Electrochemical Methods: A family of techniques that analyze the effect and role of electricity in either creating or serving as an outcome of a chemical change. These techniques include electrolysis, electrogravimetry, cyclic voltammetry, linear sweep voltammetry, electrochemical titrations, and the newer area of nanoelectrochemistry. Electrolysis leads to the separation and isolation of metals originally in a molten or a solution (ionic) mixture on an electrode using a direct electric current (DC) and a voltage called the decomposition potential. Electrogravimetry includes electroplating, electrophoretic deposition, and underpotential deposition. These methods are closely related to coulometry, which quantitatively measures the amount of matter transformed during electrolysis by using Faraday's laws. Cyclic voltammetry is a potentiodynamic electrochemical measurement where the electrode potential is ramped linearly versus time. Once the set potential is reached, the working electrode's potential is ramped toward the opposite direction to return to the initial potential thus creating a cycle which can be repeated at will. Linear sweep voltammetry involves the measurement of the current at a working electrode while the potential between the working electrode and a reference electrode is plotted linearly against time. Electrometric titration refers to any technique that uses an electrometer, or an instrument that determines, or even detects, the magnitude of a potential difference or charge by the different electrostatic forces between charged species. Nano-electrochemistry is a recent branch of electrochemistry that studies the electrochemical properties of nanometer-size materials. It uses nano-electrodes whose size is in the order of 1 to100 nm and are made of metals or semiconducting materials. It has created a significant impact in the development of many sensors and the study of reactions that involve extremely low concentrations.

X-Ray Methods: A family of techniques that utilize radiation in the x-ray region, with wavelengths between 0.01 nm to 10 nm (corresponding to frequencies in the 3×10^{16} Hz to 3×10^{19} Hz) to analyze for the presence of elements in a sample. In x-ray fluorescence spectroscopy (XRS), short wavelength x-rays are used to excite secondary or fluorescent x-rays from a sample. The wavelength of x-rays used to excite the sample must be shorter than the expected fluorescence wavelength. The sample is typically presented as a powdered solid on a glass plate. Spectra are generated by changing the incident angle between the source and the sample with a device called a goniometer. The fluorescent signals obtained are very precise and specific and can rival wet chemical methods for the identification of elements. Auger electron spectroscopy (AES) is a surface analysis method that measures the electrons emitted from a surface by electron bombardment of the surface. This method is considered to be in the same family as x-ray methods because during electron bombardment, the surface can lose energy either by electron emission (the Auger effect, via Auger electron emission) or x-ray emission. These methods are in contrast to x-ray diffraction methods, in which a crystal structure measurement is desired.

Flow Injection Analysis (FIA): An analytical protocol that seeks to replace the manual "test tube and beaker" aspects of wet chemical analysis by injecting an analyte in a flowing stream of a carrier reactant. As the analyte flows with the reactant stream, it diffuses into the reactant and product forms. Ultimately, the product zone, under the influence of the moving reactant, is passed into a detector section. The detection devices can consist of the same wide variety as is used in HPLC. The major advantage of flow injection analysis is the automation and decreased uncertainty associated with sampling and reagent addition. Strict control of reagent concentration, flow rate, and analyte volume is possible. Modern applications of FIA include the sequential addition of analyte and reactant in a stream so that the two are "stacked" in an inert carrier. They then mix by the parabolic flow profile of a laminar flowing stream in a tube. This arrangement can be miniaturized within a sampling valve, forming the so-called lab-in-a-valve approach.

ANALYTICAL STANDARDIZATION AND CALIBRATION

Thomas J. Bruno

Overview

Most modern instrumental techniques used in analytical chemistry produce an output or signal that is not absolute; the signal or peak is not a direct quantitative measure of concentration or target analyte quantity. Thus, to perform quantitative analysis, one must convert the raw output from an instrument (information) into a quantity (knowledge). This is done by standardizing or calibrating the raw response from an instrument (Refs. 1-4). Here, we briefly summarize the most common methods applied in analytical chemistry, recognizing that this is a very large field. We note that the common use of the term "standardization" is not to be confused with the application of standard methods as specified by regulatory or consensus standard organizations.

Samples

In all of the discussion to follow, we assume that the sample has been properly drawn from the parent population material and properly prepared. Clearly, the most precise analytical methods and the most painstaking calibration methods are useless if applied to a sample that does not represent reality. Nevertheless, the term "sampling," which describes the process of obtaining the sample (from the population material), implies the existence of a sampling uncertainty (arising mainly from population material heterogeneity) (Refs. 5 and 6). Thus, the analytical result is an estimate of what would be obtained from the parent population material. The theory, concepts, and nomenclature regarding samples and sampling constitute a complex, statistically based, sub-specialty of analytical chemistry well beyond the scope presented here. We begin with some simplified definitions (Ref. 7):

Aliquot: An aliquot is a known fraction of a homogeneous mass or volume.

Amount of substance: The amount of substance is the fundamental quantity of material measured in the number of moles.

Analyte: The analyte is the target component or compound in the sample for which one desires a measurement.

Bias: As applied to sampling, bias refers to a systematic displacement, error, uncertainty, or mistake caused by a flaw in the sampling procedure.

Convenience sample: A sample chosen on the basis of accessibility, expediency, cost, efficiency, or other reason not directly concerned with sampling parameters.

Determination: The determination is the entire analytical procedure or method performed on a test portion.

Matrix: The matrix is the background or carrier material of the sample that includes all components except the analyte(s) of interest.

Phase: The phase describes the physical state of a substance: primarily solid, liquid, gas, but this term might include more detailed descriptions to include supercritical fluid, plasma, etc. Note that the term "vapor" typically refers to a gas phase above and in equilibrium with a condensed phase (often called the headspace in chemical analysis).

Population material: The population material is the entirety of the bulk material from which the sample is drawn. This might be a plot of earth, a warehouse full of sugar, or a tank of jet fuel.

Quantity: The quantity refers to the mass or volume of a substance.

Random sample: A sample selected so that any portion of the sample has an equal or known chance of being chosen for measurement.

Representative sample: A sample resulting from a sampling process that can be expected to adequately reflect the properties of interest of the parent population. A representative sample may be a random sample. The degree of representativeness of the sample is limited by cost or convenience.

Selective sample: A sample that is deliberately chosen by using a sampling plan that eliminates materials with certain characteristics or selects only a material with other relevant characteristics desired for an analysis.

Test portion: The test portion is the actual material removed from a sample for analysis.

Umpire sample: A sample taken, prepared, and stored in an agreed upon manner for the purpose of settling a dispute, arrived at by agreement that will include the test method and procedure, serving as the basis for acceptance, rejection, or economic adjustment. This is sometimes called a referee sample.

Unknown: The unknown is a term that describes the target measurement or unknown quantity that is desired for the analyte, or the analyte itself.

Sampling uncertainty is that part of the total uncertainty in an analytical procedure or determination that results from using only a fraction of the population material. In this respect, sampling by any method is an extrapolation process. Since the sampling uncertainty is usually ignored for an individual analysis on an individual test portion, the sampling uncertainty is considered as being due entirely to the variability of the test portion. It is therefore assessed, when necessary, by replication of the sampling from the parent population material, and statistically isolating the uncertainty thus introduced by analysis of the variance. Typically, the problems associated with liquid population material are less complex but must not be ignored. Sample stratification, concentration and thermal gradients, poor mixing, and gradients associated with flow are all real effects that must be considered. Sampling uncertainty is often minimized by field and laboratory processing, with procedures that can include mixing, reduction, coning and quartering, riffling, milling, and grinding.

Another aspect that must be considered subsequent to sampling is sample preservation and handling. The integrity of the sample must be preserved during the inevitable delay between sampling and analysis. Sample preservation may include the addition of preservatives or buffer solutions, pH adjustment, use of an inert gas "blanket," and cold storage or freezing.

Calibration and Standardization

External Standard Methods

The external standard method can be applied to nearly all instrumental techniques, within the general limits discussed

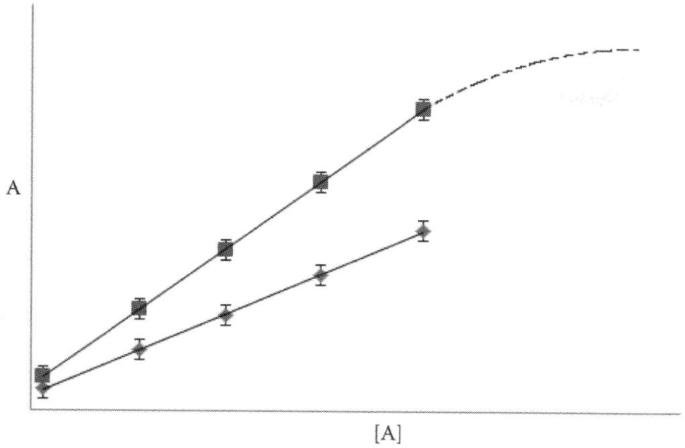

FIGURE 1a. An example calibration curve prepared by use of the external standard method. The instrument response is represented by *A*, and the concentration resulting in that response is [A]. While curves for two analytes are shown, in principle, one can plot as many analytes as desired. While five points per analyte have been shown, one can measure as many as required. Note that a region of nonlinearity is shown in the latter part of the curve for one of the components. One would require a larger number of points to adequately represent and fit any nonlinear areas.

here, and the specific limitations that may be applicable with individual techniques. This method is conceptually simple; the user constructs a calibration curve of instrumental response with prepared mixtures containing the analyte(s) over a range of concentrations, an example of which is shown in Figure 1a. Thus, the curve represents the raw instrumental response of the analyte as a function of analyte concentration or amount. Each point on this plot must be measured several times so that the repeatability can be assessed. Only random uncertainty should be observed in the replicates; trends of increasing or decreasing response (hysteresis) must be remedied by identifying the source and adjusting the method accordingly. The calibration solutions should be randomized (that is, measured in random order). Although called a calibration "curve," ideally the signal versus concentration plot is linear, or substantially linear (that is, areas of nonlinearity are unimportant, otherwise they are localized, minor, and properly treated by the measurement technique). In some cases, the response may be linearizable (for example, by calculating the logarithm of the raw response). If a curve shows nonlinearity in an area that is important for the analysis, one must measure more concentrations (data points) in the region of curvature.

In practice, the line that results from the calibration is fit with an appropriate model, and the desired value for the unknown concentration is calculated. The curve can be used graphically if approximation suffices. Mixtures prepared for external standard calibration can contain one or many analytes. Once a calibration curve is prepared, it can often be used for some time period, provided such a procedure has been previously validated (that is, the stability of the standards and the instrument over the time of use has been assessed). Otherwise, it is best to measure the unknown and the standards within a short period of time. Moreover, if any major change is made to the instrumentation (changing a detector or detector parameters, changing a chromatographic column, etc.), the standards must be remeasured.

To successfully use the method, the standard mixtures must be in a concentration range that is comparable to that of the unknown analyte, and ideally should bracket the unknown. Multiple measurements of each standard mixture should be made to establish repeatability of points on the curve. Many instrumental methods have operation ranges (frequency, temperature, etc.) in which the uncertainty is minimized, so components and concentrations for standard mixtures must respect this. The standard mixtures should be in the same matrix as the unknown, and the matrix must not interfere with the unknown or other standard mixture components. Any pretreatment of the unknown must also be reflected in the standard mixtures. As with any calibration method, components in the standard mixtures must be available at a high (or at least known) purity, they must be stable during preparation, and must be soluble in the required matrix. Unless the physical phenomenon of a measurement is well understood, extrapolation beyond the curve is not recommended (and indeed is usually strongly discouraged); nevertheless, extrapolation is occasionally done in practice. In those cases, one must be cautious, report exactly how the extrapolation was done, and assess any increase in uncertainty that may result. Note that the curve might not extrapolate through the origin. This is usually the result of adsorption (of components on container walls), carryover hysteresis, absorption (of components in seals or septa), or component degradation or evaporation.

A major consideration with external standardization is that typically, the sample size (for example, the injection volume in chromatography) must be maintained constant for standard mixtures and the unknowns. If the sample size varies slightly, it is often possible to apply a correction to the raw signal. One should not attempt to generate a calibration curve by varying the sample size (that is, for example, injecting increasing volumes into a gas chromatograph). This caution does not preclude serial dilution methods (see below), in which multiple solutions are generated for separate measurement. Other issues that can hinder successful application of the external standards method include instrumental aspects that might not be readily apparent. In chromatographic methods, for example, one can overload the column or detector. In older instruments, settings of signal attenuation were typically made manually, while in newer instruments, this may occur through software, sometimes without operator interaction or knowledge.

Note, *inter alia*, that in Figure 1a (and indeed all the examples presented here), the uncertainty is only indicated for the variable on the *y*-axis. In reality, we must recognize that there is uncertainty for the values plotted on the *x*-axis as well, but we often only treat the largest uncertainty, or the uncertainty that is most important for our application. Note, also, that it is critical to maintain the integrity of standards; decomposition, degradation, moisture uptake, etc., will adversely affect the validity of the calibration.

Abbreviated External Standard Methods

In many situations in chemical analysis, a full calibration curve is not prepared because of the complexity, time, or cost. In such situations, abbreviated external standard methods are often used. Under no circumstances can an abbreviated method be used if the raw signal response is nonlinear. Moreover, these methods are not generally appropriate for analyses in regulatory, forensic, or health care environments where the consequences can be far reaching.

Single Standard

This method uses a simple proportion approach to standardize an instrument response. It can be used only when the system has no constant, determinate error or bias,* and when the reagents

* Determinate error and bias are related terms that describe uncertainty that arises from a fixed cause, and that can, in principle, be eliminated if recognized. Determinate error (or systematic error) is most often associated with a measurement, while bias can be associated with either a measurement or with the sampling procedure.

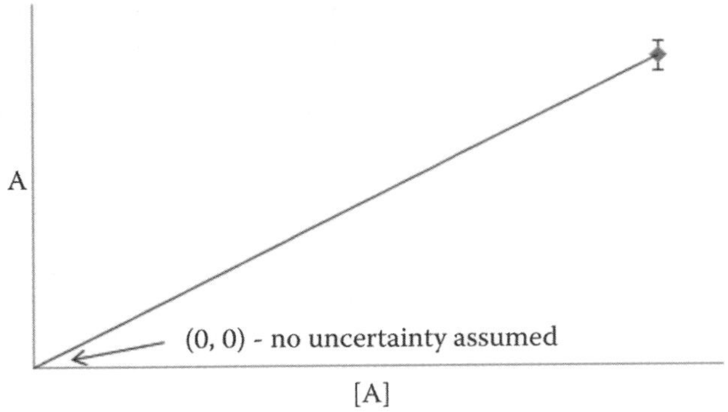

FIGURE 1b. An example of a single-point calibration curve. The instrument response is represented by *A*, and the concentration resulting in that response is [A]. The origin (0,0) is assigned as part of the curve and is assumed to have no uncertainty.

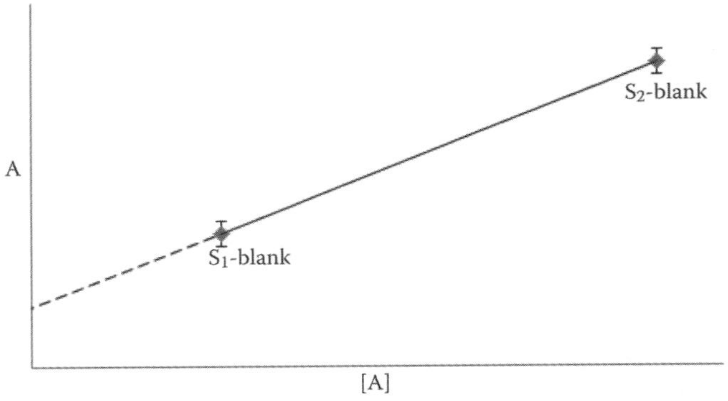

FIGURE 1c. An example of two standards plus a blank calibration curve. The blank is subtracted from each of the standards. The instrument response is represented by *A*, and the concentration resulting in that response is [A].

used give a zero blank response (that is, the instrument response from the matrix and measurement system only, without the analyte). A standard should be prepared such that the concentration is close to that of the unknown. One then calculates the concentration of the unknown, [X], as:

$$[X] = (A_x/A_s)\ [S] \quad (1)$$

where A_x is the instrument response of the unknown, A_s is the instrument response of the standard, and [S] is the concentration of the standard.

Single Standard Plus Assumed Zero

This method, illustrated schematically in Figure 1b, assumes that the blank reading will be zero. One uses a two-point calibration in which the origin is included as the first point. It is important to ensure, by experiment or experience, that such a method is adequate to the task.

Single Standard Plus Blank

If the analytical method has no determinate error or bias, but does produce a finite blank value, then one must also perform a blank measurement, which is subtracted from the instrument response of the standard and the unknown. Then the same procedure (Eq. 1) is used as for the single standard. If multiple samples are to be measured, it is important to measure the blank between each measurement.

Two Standards Plus Blank

When the analytical method has both a determinate error (or bias), and a finite blank value, at least three calibrations must be made: two standards and one blank. The standard concentrations are typically prepared widely spaced in concentration, and the higher concentration should be chosen to represent the limit of linearity of the instrument or method. If this is not practical, the higher concentration should simply be the highest expected concentration of the analyte (unknown). This method is illustrated schematically in Figure 1c. If multiple samples are to be measured, it is important to measure the blank between each measurement.

Internal Normalization Method

As mentioned above, the raw signal from an analytical instrument is typically not an absolute measure of concentration of the analyte(s), because the instrument may respond differently to each component. In some cases, such as with chromatographic methods, it is possible to apply response factors, determined from a standard mixture containing all constituents of the unknown sample, for standardization (Ref. 8). The standard mixture is gravimetrically prepared (with known mass percents for each component), and the instrument response is measured, for example as chromatographic areas. The total mass percent and the total area percent each sum to 100. One calculates the ratio of each mass percentage to each area, choosing one component as the reference, which is assigned a response factor of unity. To obtain the response factors of all the other components, one divides its (mass%-to-area ratio) with that of the reference. This is done for all components, producing a response factor for all components, except of course for the reference, defined as unity. When the unknown sample is measured, the response factor is multiplied by each raw area, and the resulting area percent provides the normalized mass percent of each component in the unknown.

This method corrects for minor variations in sample size (earlier defined as the test portion), although large differences in sample size must be avoided so that one is assured of consistent instrument performance. Although the method corrects for the different responses of samples, large differences must be avoided. This also means that the detector must respond linearly to the concentrations of each component, even if the concentrations are very different. This may require dilution or concentration of the sample in some situations. In chromatographic applications, all components of a mixture must be analyzed and standardized, since normalization must be performed on the entire sample.

Some techniques, such as gas chromatography with flame ionization detection and thermal conductivity detection, have well defined physical phenomena associated with output signals. With these techniques, there are some limited, published response factor data that can be used in an approximate way to standardize the response from these devices.

In Situ Standardization

While it is rare that an analytical method can be calibrated by use of a single solution, some instances of spectrophotometry and electroanalytical methods can qualify. To use this method one sequentially and incrementally adds known masses of standard analytes to a solution, with an instrument response being measured after each addition. This procedure can only be used if the analytical method itself does not change the analyte concentration (nondestructive) and does not lead to a loss of solution volume. A solid crystalline analyte is an example. One must also minimize changes in solution volume over the course of the standardization.

Standard Addition Methods

Samples presented for analysis often are contained in complex matrices with many impurities that may interact with the analyte, potentially enhancing or diminishing a signal from an instrumental technique. In such cases, the preparation of an external standard calibration curve will be impossible, because it might be very difficult to reproduce the matrix. In these cases, the standard addition method may be used. A standard solution containing the target analyte is prepared and added to the sample, thus maintaining the unknown impurities and their effects. While the quantity of target analyte in the target sample is unknown, the added quantity is known, and its incremental additive effect on the instrument signal can be measured. Then, the quantity of the unknown analyte is determined by what is effectively an extrapolation. In practice, the volume of the standard solution added is kept small to avoid dilution of the unknown impurities by no more than 1% of the total signal. This method can only be used if there is a verified linear relationship between the signal and quantity of analyte. If a determinate error is present, then the slope of the line must be known. Moreover, the sample cannot contain any components that can respond as the analyte (that is, masquerade).

Single Standard Addition

In the simplest case, one addition of analyte is made after first measuring the response of the analyte in the unknown sample. Thus, two measurements are required:

$$A_{xo} = m[X_0] \tag{2}$$

$$A_{xi} = m([X_0] + [S]) \tag{3}$$

where A_{xo} is the instrument response of the analyte in the unknown sample, $[X_0]$ is the concentration in the unknown sample, and A_{xi} is the instrument response upon the addition of the standard, [S] (additive in equation because X and S are the same compound). The assumed slope is the proportionality constant, m. The two equations are solved simultaneously for $[X_0]$. This technique is very rapid and economical, but there are serious drawbacks. There is no built in check for mistakes on the part of the analyst, there is no means to average random uncertainties, and there is no way to detect interference (mentioned above as masquerade).

Multiple Standard Addition

This standard addition method alleviates some of the problems inherent in single standard addition. Here, the unknown sample is first measured in the instrument. Then that sample is "spiked" with incrementally increasing concentrations of the analyte, generating a curve such as that shown in Figure 1d. The curve should extrapolate to zero signal at zero concentration. The concentration of the analyte in the unknown is read or calculated from the abscissa (x-axis).

Internal Standard Methods

An internal standard is a compound added to a sample at a known concentration, the purpose of which is to exhibit a similar signal when measured in an instrument but be distinguishable from the signal of the desired analyte. It provides the highest level of reliability in quantitation by chromatographic methods and is not affected by large differences in sample size (Ref. 8). Unlike the internal normalization method, it is not necessary to elute or measure all the components of the sample, one need focus only on the component(s) of interest. In atomic spectrometry, this method is not affected by changes in gas flow rates, sample aspiration rates, and flame suppression or enhancement. Another situation in which this method is valuable is when the sample matrix is either unknown or very complex, precluding the preparation of external standards.

Multiple Internal Standards

A set of calibration solutions is prepared by mass, containing the target analyte, X, and a standard that is not present in the unknown sample, A. The instrument response (for example, a chromatographic area) is measured for each calibration solution, and a plot is made to establish linearity as in Figure 1a. The ordinate axis is the ratio of the response of the unknown analyte component, A_x, to the response of the chosen standard, A_s. The abscissa is the ratio of mass of X to the mass of S for that standard mixture. Once the linearity is confirmed in the concentration range of interest, the unknown is spiked with a known mass of S, the instrument response is measured, and the area ratio A_x/A_s is calculated. Either the graph or a fit of the data on Figure 1e is then used to determine the corresponding mass fraction, from

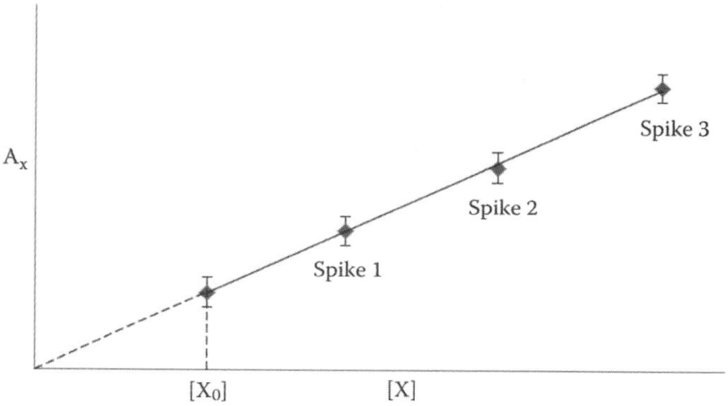

FIGURE 1d. An example of calibration by multiple standard addition. Three additions (spikes) of the analyte X are shown, as is the extrapolation to the unknown concentration, X_0.

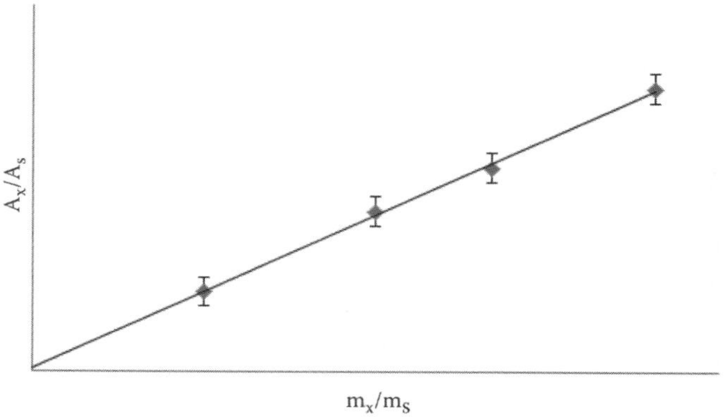

FIGURE 1e. An example of the multiple internal standard method. The ordinate (y) axis is the ratio of the response of the unknown analyte component, A_x, to the response of the chosen standard, A_s. The abscissa (x) axis is the ratio of mass of X to the mass of S for that standard mixture.

which the mass of X may be determined. Note that the calculations could be simplified if the same mass per volume of the internal standard is added to both the unknown samples and the calibration standards.

Single Internal Standard

In practice, once the linearity is established for a given mixture, it is no longer necessary to use multiple standards, although this is the most precise method. Subsequent to the verification of linearity, one standard solution can be used to fix the slope, provided it is close in concentration to that of the target analyte. In this case, the mass of the unknown can be found from:

$$X/S = (A_x/A_s)(1/R) \quad (4)$$

where X is the mass of the unknown analyte in the sample, S is the mass of the added internal standard in the sample, A_x and A_s are the instrument responses (areas) of the unknown and internal standard, respectively. R is a ratio determined from the standard solution prepared with both X and S: (mass, unknown analyte/mass, internal standard)/(signal, unknown analyte/signal, internal standard) = R.

$$\left(\frac{\left(\dfrac{\text{mass, unknown analyte}}{\text{mass, internal standard}}\right)}{\left(\dfrac{\text{signal, unknown analyte}}{\text{signal, internal standard}}\right)}\right) = R \quad (5)$$

Because R is the slope of the calibration curve discussed above, once linearity is established, one solution suffices. There are many conditions that must be fulfilled in order to use the internal standard method, and it is rare that all of them can actually be met. Indeed, in practice, one tries to meet as many as possible, but those that are mandatory are italicized. The compound chosen *must not be present* already in the unknown. The compound chosen *must be separable from the analyte* present in the unknown. An exception occurs when an isotopically labeled standard is used, in conjunction with mass discrimination or radioactive counting detection. In a chromatographic measurement, this is typically at least baseline resolution, although this would be a minimally acceptable degree of separation. On the other hand, the unknown analyte peak and the internal standard peak should be close to each other (temporally) on the chromatogram. The compound chosen *must be miscible* with the solvent at the temperature of reagent preparation and measurement. The compound chosen *must not react chemically* with the sample or solvent, or interfere in any way with the analysis. It is critical to maintain the integrity of standards; decomposition, degradation, moisture uptake, etc., will adversely affect the validity of the calibration. In the case of a chromatographic measurement, the same applies to interactions with the stationary phase. The compound chosen must be chemically similar (for example, in functionality, thermophysical properties) to the analyte. If such a compound is not available (for example, in a chromatographic measurement), an appropriate hydrocarbon should be chosen as a surrogate. The standard solution should be prepared at a similar concentration as in the unknown matrix; ratio correction of large differences is no substitute for an appropriate concentration. In a chromatographic measurement, the compound chosen must elute as closely as possible to the analyte and should not be the last peak to elute (the final peak often shows different geometry such as tailing). The compound chosen must be sufficiently nonvolatile to allow for storage as needed. When there is the potential for the unknown analyte to be lost by adsorption, absorption, or some other interaction with the matrix or container, a compound called a carrier is sometimes added in large excess. The carrier is similar, chemically and physically, to the unknown analyte, but easily separated from it. Its purpose is to saturate or season the matrix and prevent analyte loss.

Serial Dilution

Serial dilution is less a standardization method as it is a method of generating solutions to be used for standardizations. Nevertheless, its importance and utility, as well as the popularity of its application, warrants mention in this section. A serial dilution is the stepwise dilution of a substance, observant of a specified, constant progression, usually geometric (or logarithmic). One first prepares a known volume of stock solution of a known concentration, followed by withdrawing some small fraction of it to another container or vial. This subsequent container is then filled to the same volume as the stock solution with the same solvent or buffer. The process is then repeated for as many standard solutions as are desired. A ten-fold serial dilution could be 1 M, 0.1 M, 0.01 M, 0.001 M, etc. A ten-fold dilution for each step is called a logarithmic dilution or log-dilution, a 3.16 fold ($10^{0.5}$ fold) dilution is called a half-logarithmic dilution or half-log dilution, and a 1.78 fold ($10^{0.25}$ fold) dilution is called a quarter-logarithmic dilution or quarter-log dilution. In practice, the ten-fold dilution is the most common. The serial dilution procedure is not only used in chemical analysis but also in serological preparations in which cellular materials such as bacteria are diluted. A critical aspect of serial dilution is that the initial solution concentration must be prepared and determined with great care, since any mistake here will be propagated into all resulting solutions.

Traceability

Analytical measurements and certifications often contain a statement of traceability. Traceability describes the "result or measurement whereby it can be related to appropriate standards, generally international or national standards, through an unbroken chain of comparisons" (Ref. 9). Traceability typically includes the application of a reference material (RM) or a standard reference material (SRM) for instrument calibration before standardization for the analytes of interest. The true value of a measured quantity (τ) cannot typically be determined. The **true value** is defined as characterizing a quantity that is perfectly defined. It is an ideal value which could be arrived at only if all causes of measurement uncertainty were eliminated, and the entire population was sampled.

Uncertainty

As stated in Section 2, the result of a measurement is only an approximation or estimate of the true value of the measurand or quantity subject to measurement. In the determination of the combined standard uncertainty and ultimately the expanded uncertainty, it is critical to include the uncertainty of calibration in the process, as discussed above. The process of arriving at the uncertainty U_y of a quantity y that is based upon measured quantities $x_i,..,x_z$ is called the propagation of uncertainty. A full discussion of propagation of uncertainty is beyond the scope of this section; a simplified prescription, in the form of general and specific formulae, is provided here. In general, the propagated random uncertainty in y can be determined from:

$$U_y = \sqrt{\left(\frac{\partial y}{\partial x_i}\right)^2 + \ldots\left(\frac{\partial y}{\partial x_z} U_{x_z}\right)^2}$$

This approach can be used when the uncertainties are random (not systematic), are relatively small, and are independent or uncorrelated (that is, in the absence of covariance). Relatively large uncertainties (such as those approaching the magnitude of the measurand itself) cannot be treated with this approach, especially if the measurand is a nonlinear function of the measured quantity. Note that by convention the use of upper case U_y denotes the expanded uncertainty, which is the uncertainty multiplied by a coverage factor k in excess of unity (for the 95% confidence level, the coverage factor is 2). A coverage factor k = 1 represents the 68% confidence level. In scientific and technical reports and publications, the goal is to report measurements and the standard uncertainty (k = 1) or the expanded uncertainty (k > 1).

It is possible to reduce this general formulation to more specific formulae in the cases of common mathematical operations. These are provided in Table 1.

Compliance against Limits

It is often necessary or desirable to compare an analytical result against some established regulatory standard or limit, for example, to determine if the concentration of a toxic substance falls above or below a legal limit. It is imperative to consider the uncertainty when determining compliance against limits; indeed, most established limits are set with some consideration or allowance for uncertainty. A decision rule is often built into tests of compliance limits. A common decision rule is that a measured result indicates non-compliance if the measurand exceeds the limit by the expanded uncertainty. A similar approach is applied for measurands that fall below an established limit.

Uncertainty and Error Rates

A disconnect often occurs when scientific personnel attempt to convey their measured results in legal or regulatory venues. For example, the courts in the United States routinely must deal with scientific testimony based on measurements, and scientists are often asked to characterize their measurements in terms of error rates.

Statistically, the error rate is the frequency of type I and type II errors in null hypothesis significance testing. This has importance in forensic chemistry, e.g., when the blood alcohol content (BAC) of a sample is measured. Here, a null hypothesis might be the BAC of sample X is not below 0.08% (mass/mass). A measurement above that level, and therefore a failure to reject the null hypothesis, can result in a legal finding of intoxication. A type I error occurs when a rejected null hypothesis is correct (false positive); a type II error occurs when the accepted null hypothesis is false (false negative). Independent of the frequency of type I and type II errors (the statistical error rate), each measurement of BAC has an uncertainty. The uncertainty of each measurement is determined by the propagation of the contributions to uncertainty that is represented by the uncertainty budget, multiplied by the appropriate coverage factor. The error rate of a particular laboratory or technique is not so easily determined. In some large state forensic laboratories, error rates can be approximated by inserting known standard samples anonymously into the normal workflow, but even this approach has limitations.

It is important to understand that the concept of error rate is distinct from the frequency at which an analytical instrument "throws an error." For example, in the headspace gas chromatographic analysis of BAC, the instrument might report a sampling error, and the operator might notice a damaged needle, which is then replaced. The frequency of this type of error is different from the error rate mentioned above.

References

1. Chalmers, R. A., *Chapter 2: Standards and Standardization in Chemical Analysis*, Vol. 3, Elsevier, Amsterdam, 1975.
2. Danzer, K., and Currie, L. A., *Pure Appl. Chem.* 70, 993, 1998. <https://doi.org/10.1351/pac199870040993>
3. Danzer, K., Otto, M., and Currie, L. A., *Pure Appl. Chem.* 76, 1215, 2004. <https://doi.org/10.1351/pac200476061215>
4. Woodget, B. W., and Cooper, D., *Samples and Standards, Analytical Chemistry by Open Learning*, John Wiley & Sons, Chichester, 1987.
5. Gy, P., *Sampling for Analytical Purposes*, John Wiley & Sons, Chichester, 1998.
6. Vitt, J. E., and Engstrom, R. C., *J. Chem. Educ.* 76, 99, 1999. <https://doi.org/10.1021/ed076p99>
7. Horowitz, W., *Pure Appl. Chem.* 62, 1193, 1990. <https://doi.org/10.1351/pac199062061193>
8. Grob, R. L., *Modern Practice of Gas Chromatography*, Wiley Interscience, New York, 1995.
9. Inczedy, J., Lengyel, T., Ure, A. M., *Compendium of Analytical Nomenclature, Third Edition*, International Union of Pure and Applied Chemistry, 1997

TABLE 1. Specific Formulae for the Propagation of Random, Independent Uncertainty

Measurand argument	Arithmetic uncertainty formula
y (where y is a counted random event over a time interval)	$U_y = \sqrt{y}$
$y = A \times x$ (where A is a constant with no uncertainty)	$U_y = \lvert A \rvert \times U_x$
$y = x_1 + x_2$	$U_y = \sqrt{U_{x_1}^2 + U_{x_2}^2}$
$y = x_1/x_2 y = x_1 \times x_2$	$\frac{U_y}{\lvert y \rvert} = \sqrt{\left(\frac{U_{x_1}}{x_1}\right)^2 + \left(\frac{U_{x_2}}{x_2}\right)^2}$
$y = (x_1 \times x_2)/x_3$	$\frac{U_y}{\lvert y \rvert} = \sqrt{\left(\frac{U_{x_1}}{x_1}\right)^2 + \left(\frac{U_{x_2}}{x_2}\right)^2 + \left(\frac{U_{x_3}}{x_3}\right)^2}$
$y = \log(x)$	$U_y = \left(\frac{1}{2.303}\right)\left(\frac{U_x}{x}\right)$
$y = \ln(x)$	$U_y = \frac{U_x}{x}$
$y = e^x$	$U_y = y \times U_x$
$y = x^a$	$\frac{U_y}{\lvert y \rvert} = \lvert a \rvert \times \left(\frac{U_x}{x}\right)$
$y = 10^x$	$\frac{U_y}{\lvert y \rvert} = \lvert a \rvert \times \left(\frac{U_x}{x}\right)$

FIGURES OF MERIT

Thomas J. Bruno

In the interpretation of results from a particular technique, one often must evaluate the performance of the measurement on the basis of objective measures that are called figures of merit. The following table provides a description of the more common figures of merit used in analytical chemistry. Even more important than the evaluation of performance is the validation of metrology and the assessment of validation. The same terms used as figures of merit are critical in the validation process. Because many standardization bodies will issue their own definitions regarding these terms, it is not practical to be inclusive; the reader must be aware that other terms may sometimes be substituted for those used here.

References

1. General Requirements for the Competence of Testing and Calibration Laboratories, ISO/IEC 17025, International Organization for Standardization, 2005.
2. General Chapter 1058 Analytical Instrument Qualification, USP 32 – NF 27, U.S. Pharmacopeial Convention, 2009.

Figures of Merit Used in Analytical Chemistry

Parameter or metric	Definition	Comments
Accuracy	The deviation of a measured value from the true value (τ), which cannot typically be determined, but which is often approached by use of a standard method under standard conditions.	Can be expressed as: $\text{accuracy} = (x_{av} - \tau)/\tau$, where x_{av} is the mean of a series of measurements.
Precision	The reproducibility of a series of replicate measurements, typically under the same operating conditions in a relatively short period of time.	High precision does not imply high accuracy; reproducibility and repeatability often are used interchangeably, but the former typically refers to precision between two instruments or laboratories.
Sensitivity	Typically the ability to distinguish between small differences in concentration between samples at a desired confidence level.	Often specified by response to a *de facto* or consensus standard; a simple measure is the slope of a calibration curve adjusted for recovery in the sampling process.
Limit of Detection (LOD)	The lowest measureable concentration of an analyte by use of a particular metrology at a desired confidence level.	Typically defined as a response that is some multiplicative factor of the noise level; specified signal-to-noise ratios of 2 to 4 are typical; as with sensitivity, must adjust for recovery in sample preparation processes.
Linear Dynamic Range (LDR)	The linear range of a calibration curve obtainable with a particular metrology and analyte.	Within the LDR, the difference in response for two concentrations of a given compound is proportional to the difference in concentration of the two samples.
Selectivity (α)	The ability to distinguish the signal of an analyte from the signal of interferences, the matrix or degradants.	Typically expressed as a ratio, such as the ratio of capacity factors in chromatography; sometimes called specificity.
Analysis Speed	The time required for sample preparation and measurement.	Sometimes necessary to include data processing time.
Throughput	The number of samples that can be run in a given time.	Related to ease of automation, below.
Ease of Automation	The ability of a measurement to be performed without operator attention.	Related to throughput, above, since high throughput typically requires automation and minimal operator attention.
Ruggedness	The response of an instrument or technique to adverse conditions.	Adverse conditions include extremes in temperature, humidity, dust; includes rough handling; closely related to reproducibility, since it carries similar information content; distinguished from robustness, which is the ability of an instrument to remain unaffected by a small but intentional change in operating parameter, and still perform within specification.
Portability	The ability to use an instrument in other than a fixed location or installation.	Can include the ability to use a device in the field in addition to the laboratory.
Environmental Acceptability	The efficiency in terms of low waste generation, low power consumption.	Also described as sustainability, and often focused on the minimization of use of hazardous substances.
Economics	The sum of costs required to operate the sample preparation, analysis and data processing steps of a measurement.	Typically includes the costs of equipment, supplies, labor, utilities, and insurance; labor cost must consider the skill level required of personnel.

MASS- AND VOLUME-BASED CONCENTRATION UNITS

Thomas J. Bruno and Paris D. N. Svoronos

A variety of concentration units are used in analytical chemistry, and the most common are provided in this table (Ref. 1). The reference below provides additional details.

Reference

1. Bruno, T. J., and Svoronos, P. D. N., *CRC Handbook of Basic Tables for Chemical Analysis, Third Edition*, CRC Press, Boca Raton, FL, 2011. <https://doi.org/10.1201/b10385>

TABLE 1. Concentration Units Conversion – Parts per Million

Parts per Million		Percent
1 ppm	=	0.0001 %
10 ppm	=	0.001 %
100 ppm	=	0.01 %
1000 ppm	=	0.1 %
10 000 ppm	=	1.0 %
100 000 ppm	=	10.0 %
1 000 000 ppm	=	100.0 %

TABLE 2. Concentration Units Conversion – Parts per Billion

Parts per Billion		Percent
10	=	0.000 001 %
100	=	0.000 01 %
1000	=	0.0001 %
10 000	=	0.001 %
100 000	=	0.01 %
1 000 000	=	0.1 %

TABLE 3. Concentration Units Conversion – Parts per Trillion

Parts per Trillion		Percent
100	=	1×10^{-8} %
10 000	=	0.000 001 %
1 000 000	=	0.0001 %
100 000 000	=	0.01 %

Because the mass of one liter of water is approximately one kg, mg/L units of dilute aqueous solution are nearly equal to ppm units. The precise equivalence is obtained by dividing by the density, ρ:

$$\text{ppm} = (\text{mg/L})/\rho$$

where the solution density, ρ, is in g cm^{-3}. Some sources will substitute specific gravity for density in the above equation. The specific gravity is the ratio of the solution density to that of the density of pure water at 4 °C. Because the density of pure water at 4 °C is very nearly 1 g cm^{-3}, the specific gravity is numerically equal (within an uncertainty of 25 ppm) to the solution density when the latter is expressed in units of g cm^{-3}.

Concentration Units Nomenclature

The following table provides guidance in the use of base-10 concentration units (presented in the three preceding tables), since there are differences in usage worldwide.

TABLE 4. Notation for Large Numbers in Different Countries

Number	Number of Zeros	Name (Scientific Community)	Name (United Kingdom, France, Germany)
1000.	3	thousand	thousand
1 000 000.	6	million	million
1 000 000 000.	9	billion	milliard, or thousand million
1 000 000 000 000.	12	trillion	billion
1 000 000 000 000 000.	15	quadrillion	thousand billion

TABLE 5. Molar-Based Concentration Units

Molarity*, M:	(moles of solute)/(liters of solution)
Molality, *m*:	(moles of solute)/(kilograms of solvent)
Normality, N:	(equivalents* of solute)/(liters of solution)
Formality, F:	(moles of solute)/(kilograms of solution)

* This unit is temperature dependent.

To convert from ppm to formality units:

$$F = \text{ppm}/(1000\ RMM)$$

where *RMM* is the relative molecular mass of the solute *i*.

To convert from ppm to molality units:

$$m = [\text{ppm}/(1000\ RMM)]\ [1/(1 - tds/1\ 000\ 000)]$$

where *tds* is the total dissolved solids in ppm in the solution.

To convert from ppm to molarity units:

$$M = [\text{ppm}/(1000\ RMM)]\ \rho$$

where ρ is the solution density.

DETECTION OF OUTLIERS IN MEASUREMENTS

Thomas J. Bruno and Paris D. N. Svoronos

The field of outlier detection and treatment is considerable, and a rigorous mathematical discussion is well beyond any treatment that is possible here. Moreover, the practice in the treatment of analytical results is usually simplified, because the number of observations is often not very large. The two most common methods used by analysts to detect outliers in measured data are versions of the *Q*-test (Refs. 1–3, 6) and Chauvanet's criterion (Refs. 4–6), both of which assume that the data are sampled from a population that is normally distributed.

References

1. Dean, R. B., and Dixon, W. J., *Anal. Chem.* 23, 636, 1951. [https://doi.org/10.1021/ac60052a025]
2. Day, R. A., and Underwood, A.L., *Quantitative Analysis, Sixth Edition*, Prentice Hall, Englewood Cliffs, NJ, 1991.
3. Efstathiou, C. E., *Dixon's Q-test: Detection of a Single Outlier*, <http://www.chem.uoa.gr/applets/AppletQtest/Text_Qtest2.htm>, Laboratory of Analytical Chemistry, Department of Chemistry, National and Kapodistrian University of Athens, 2008.
4. Taylor, J. R., *An Introduction to Error Analysis, Second Edition*, University Science Books, Sausalito, CA, 1997.
5. Benziger, J. B., and Aksay, I.A., *Notes on Data Analysis*, <http://www.princeton.edu/~che346/Notes/Analysis.pdf>, Department of Chemical Engineering, Princeton University, 1999.
6. Bruno, T. J., and Svoronos, P. D. N., *CRC Handbook of Basic Tables for Chemical Analysis, Third Edition*, CRC Press, Boca Raton, FL, 2011. [https://doi.org/10.1201/b10385]

Q-Test

To perform the *Q*-test, one calculates the *Q* value given by:

$$Q = Q_{gap}/R$$

where Q_{gap} is the difference between the suspected outlier and the measured value closest to it, and *R* is the range of all the measured values in the data set. One then compares the calculated *Q* value with the critical *Q* values in Table 1.

TABLE 1. Critical *Q*–Test Values for Different Confidence Levels

Number of observations	Q_{crit}, 90% Confidence level	Q_{crit}, 95% Confidence level	Q_{crit}, 99% Confidence level
3	0.941	0.970	0.994
4	0.765	0.829	0.926
5	0.642	0.710	0.821
6	0.560	0.625	0.740
7	0.507	0.568	0.680
8	0.468	0.526	0.634
9	0.437	0.493	0.598
10	0.412	0.466	0.568

If the calculated value of *Q* is greater than the appropriate value of Q_{crit}, then the value is a suspected outlier.

Chauvenet's Criterion

To perform Chauvenet's test on a set of measurements, one first must calculate the mean and standard deviation of the data. Then one calculates:

$$\tau = (x_i - x_{ave})/\sigma$$

where x_i is the suspected outlier, x_{ave} is the mean of all the measurements, and σ is the standard deviation. One then compares the calculated value of τ with τ_{crit} in the following table.

TABLE 2. Critical Chauvenet's τ Values as a Function of the Number of Observations

Number of observations, *N*	τ_{crit}
5	1.65
6	1.73
7	1.81
8	1.86
9	1.91
10	1.96
15	2.12
20	2.24
25	2.33
50	2.57
100	2.81
150	2.93
200	3.02
500	3.29
1000	3.48

If the calculated value of τ is greater than the value of τ_{crit}, then the value is a suspected outlier.

For numbers of observations between those given in Table 2, especially for a large number of observations, one may use the following plot to estimate the value of Chauvenet's τ_{crit}.

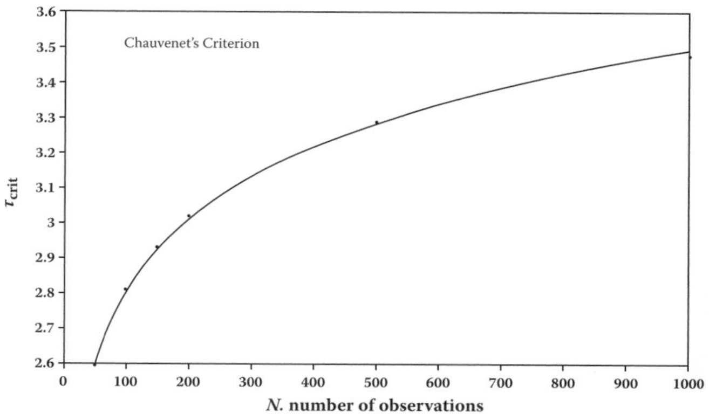

PROPERTIES OF CARRIER GASES FOR GAS CHROMATOGRAPHY

The following is a list of carrier gases sometimes used in gas chromatography, with properties relevant to the design of chromatographic systems. All data refer to normal atmospheric pressure (101.325 kPa). Additional properties related to the carrier gas, including split ratio, pressure drop in capillary columns, the Martin-James compressibility factor, and the Giddings plate height correction factor can be found in Ref. 2. Column definitions for the table are as follows.

Column heading	Definition
Name	Name of carrier gas
M_r	Molecular weight (relative molar mass)
ρ_{25}	Density at 25 °C, in g L^{-1}
c_p**(25 °C)**	Specific heat at 25 °C, in J g^{-1} °C^{-1}
λ**(25 °C)**	Thermal conductivity at 25 °C, in mW m^{-1} °C^{-1}
η**(25 °C)**	Viscosity at 25 °C, in μPa s (equal to 10^{-3} cP)
λ**(250 °C)**	Thermal conductivity at 250 °C, in mW m^{-1} °C^{-1}
η**(250 °C)**	Viscosity at 250 °C, in μPa s (equal to 10^{-3} cP)

References

1. Lide, D. R., and Kehiaian, H. V., *CRC Handbook of Thermophysical and Thermochemical Data*, CRC Press, Boca Raton, FL, 1994.
2. Bruno, T. J., and Svoronos, P. D. N., *CRC Handbook of Basic Tables for Chemical Analysis, Third Edition*, CRC Press, Boca Raton, FL, 2011. <https://doi.org/10.1201/b10385>

Properties of Carrier Gases for Gas Chromatography

Name	M_r	ρ_{25}/g L^{-1}	c_p(25 °C)/ J g^{-1} °C^{-1}	λ(25 °C)/ mW m^{-1} °C^{-1}	η(25 °C)/ μPa s	λ(250 °C)/ mW m^{-1} °C^{-1}	η(250 °C)/ μPa s
Hydrogen	2.016	0.0824	14.3	185.9	8.9	280	13.1
Helium	4.003	0.1636	5.20	154.6	19.9	230	29.5
Argon	39.948	1.6329	0.521	17.8	22.7	27.7	35.3
Nitrogen	28.014	1.1449	1.039	25.9	17.9	39.6	26.8
Oxygen	31.998	1.3080	0.919	26.2	20.7	42.6	31.8
Carbon monoxide	28.010	1.1449	1.039	24.8	17.8	40.7	26.5
Carbon dioxide	44.010	1.7989	0.843	16.7	14.9	35.5	24.9
Sulfur hexafluoride	146.055	5.9696	0.664	13.1	28.1	15.3	24.8
Methane	16.043	0.6556	2.23	34.5	11.1	75.0	17.6
Ethane	30.069	1.2291	1.75	20.9	9.4	57.7	15.5
Ethylene	28.053	1.1465	1.53	20.5	10.3	53.8	17.2
Propane	44.096	1.8025	1.67	17.9	8.3	49.2	14.0

COMMON SYMBOLS USED IN GAS AND LIQUID CHROMATOGRAPHIC SCHEMATIC DIAGRAMS

Thomas J. Bruno

The literature of gas and liquid chromatography frequently contains schematic diagrams that depict analytical apparatus and peripherals. The interpretation of such diagrams can be facilitated by the following graphics that show the most common symbols used to represent chromatographic components.

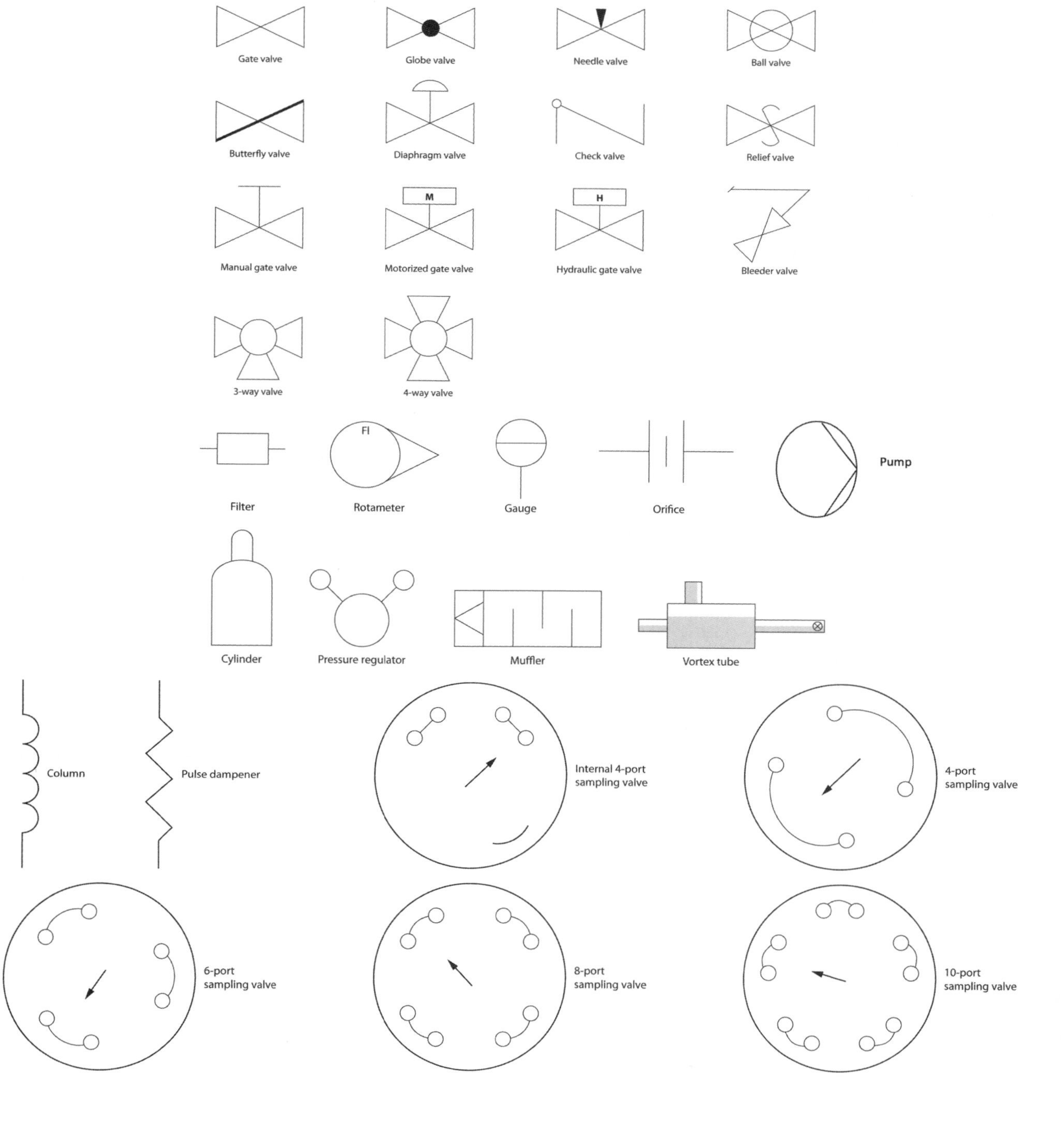

STANDARD FITTINGS FOR COMPRESSED GAS CYLINDERS

Thomas J. Bruno and Paris D. N. Svoronos

The following table presents a partial list of gases and the Compressed Gas Association (CGA) fittings that are required to use those gases when they are stored in, and dispensed from, compressed gas cylinders (Refs. 1 and 2). These fittings are matched to gas cylinder content for safe operation, and substitution and adaptation should be avoided.

References

1. CGA Pamphlet V-1-87, American Canadian and Compressed Gas Association Standard for Compressed Gas Cylinder Valve Outlet and Inlet Connections, ANSI-B57.1; CSA B96, 1987.
2. Bruno, T. J., and Svoronos, P. D. N., *CRC Handbook of Basic Tables for Chemical Analysis, Third. Edition,* CRC Press, Boca Raton, FL, 2011. <https://doi.org/10.1201/b10385>

Standard Fittings for Cylinders of Various Gases

Gas	Fitting
Acetylene	510
Air	346
Carbon dioxide	320
Carbon monoxide	350
Chlorine	660
Ethane	350
Ethylene	350
Ethylene oxide (Oxirane)	510
Helium	580
Hydrogen	350
Hydrogen chloride	330
Methane	350
Neon	580
Nitrogen	580
Nitrous oxide	326
Oxygen	540
Sulfur dioxide	660
Sulfur hexafluoride	590
Xenon	580

The following graphic shows the geometry and dimensions of common CGA fittings for compressed gas cylinders (Ref. 1).

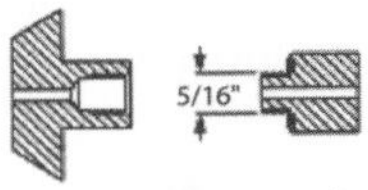

Connection 110 - Lecture bottle outlet for corrosive gases - 5/16" - 32 RH INT., with gasket

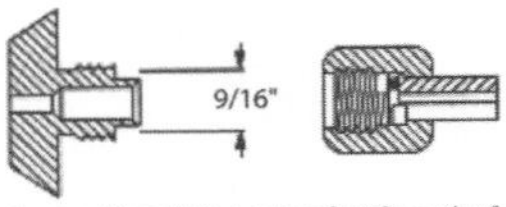

Connection 170 - Lecture bottle outlet for non-corrosive gases 9/16" - 18 RH EXT. and 5/16" - 32 RH INT., with gasket

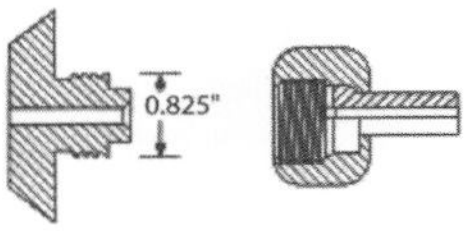

Connection 320 - 0.825" - 14 RH EXT., with gasket

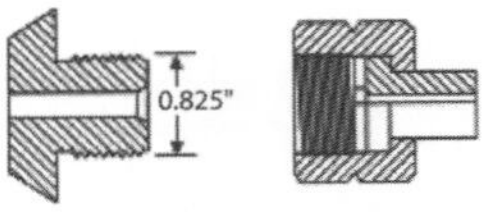

Connection 330 - 0.825" - 14 LH EXT., with gasket

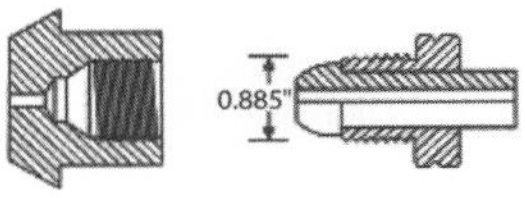

Connection 510 - 0.885" - 14 LH INT.

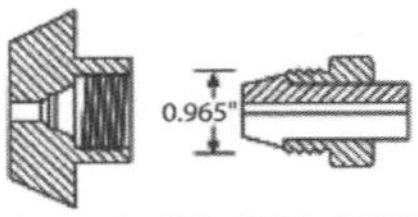

Connection 580 - 0.965" - 14 RH INT.

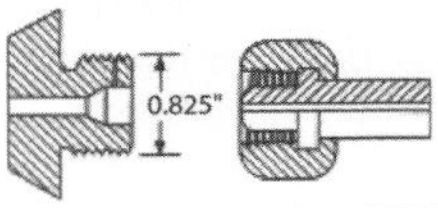

Connection 326 - 0.825" - 14 RH EXT.

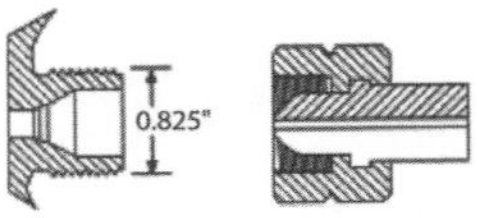

Connection 350 - 0.825" - 14 LH EXT.

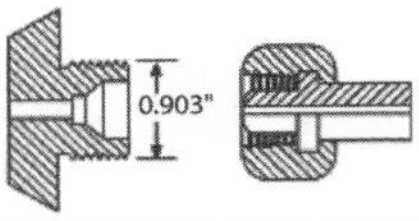

Connection 540 - 0.903" - 14 RH EXT.

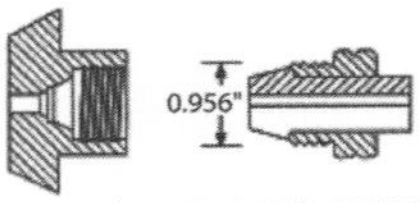

Connection 590 - 0.956" - 14 LH INT.

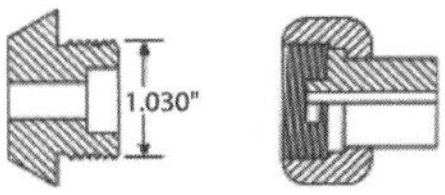

Connection 660 - 1.030" - 14 RH EXT., with gasket

Reproduced from the CGA Pamphlet V-1-87, American Canadian and Compressed Gas Association Standard for Compressed Gas Cylinder Valve Outlet and Inlet Connections, ANSI,B57.1; CSA B96, by permission of the Compressed Gas Association.

STATIONARY PHASES FOR POROUS-LAYER OPEN TUBULAR COLUMNS

Thomas J. Bruno and Paris D. N. Svoronos

The practical application of solid adsorbents in capillary chromatography typically makes use of porous-layer open tubular (PLOT) columns. Early capillary PLOT columns suffered from detector spiking due to particulates becoming dislodged and entering the detector. This was resolved by placing a short length of a polymeric column (usually polydimethyl siloxane) as a particle trap before the detector; however, recent advances in manufacturing methods have resulted in nearly complete binding of particles on the inside of the column. In this table, several common PLOT column stationary phases are listed along with their major separation applications and additional information. The maximum temperatures listed represent the point of severe resolution loss. The materials are often chemically stable to much higher temperatures. The user should also be aware that the adsorption of water during use will often change retention characteristics dramatically, sometimes resulting in a reversal of positions of adjacent peaks. This can result from traces of water in the carrier gas. Due to surface adsorption of solutes, some experimentation with temperature may be necessary to prevent tailing or to avoid statistical correlation (or a propagating error) among replicate analyses (Ref. 5). Note that PLOT columns have been used in HPLC for the separation of large molecules such as proteins; this table is limited to smaller molecules.

References

1. Jeffery, P. G., and Kipping, P. J., *Gas Analysis by Gas Chromatography*, Pergamon Press, Oxford, 1972.
2. Cowper, C. J., and DeRose, A. J., *The Analysis of Gases by Chromatography*, Pergamon Press, Oxford, 1983.
3. Breck, D. W., *Zeolite Molecular Sieves*, John Wiley & Sons, New York, 1973.
4. Poole, C. F., *The Essence of Chromatography*, Elsevier, Amsterdam, 2003.
5. Bruno, T. J., and Svoronos, P. D. N., *CRC Handbook of Basic Tables for Chemical Analysis, Third Edition*, CRC Press, Boca Raton, FL, 2011.

Characteristics of Stationary Phases

Phase	Max. temp./ °C	Application	Notes
Silica gel	250	H_2, Air, CO, C_1 to C_4 normal hydrocarbons, alkenes and alkynes, inorganic gases, volatile ethers	Very hydrophilic; requires activation; can be unpredictable; largely replaced by porous polymers; bonded versions are suitable for use with GC-MS because of the absence of particles
Alumina, deactivated with KCl	250	C_1 to C_8 hydrocarbons, especially useful for resolution of propadiene and butadiene from ethylene and propylene	Least polar of the alumina phases; lowest retention of olefins relative to the corresponding paraffin; specified in many standard methods; will adsorb water and CO_2
Alumina, deactivated with Na_2SO_4	250	C_1 to C_8 hydrocarbons, resolves acetylene from butane and propylene from isobutane	Medium and high polarity phases are available among the alumina phases; specified in many standard methods
Cyclodextrins	260	Light hydrocarbons (C_1 to C_{10}) and halocarbons	α- and β-cyclodextrins have been used; care should be taken with halocarbon analysis due to the potential of HF contamination of the sample; see the information on cyclodextrin phases in Ref. 5
Styrene – divinylbenzene	250	C_1 to C_3 hydrocarbons; paraffins up to C_{12}; CO from air, ethers, sulfur gases, water	See the information on porous polymers in Ref. 5
Divinylbenzene ethylene glycol dimethacrylate	280	C_1 to C_7 hydrocarbon isomers; CO_2, CH_4, amines, common solvents, alcohols, aldehydes, ketones	More polar than styrene – divinyl benzene phases
Molecular sieve, 5A	350	Air and light gas analysis; H_2, O_2, N_2, (CH_4, CO, NO, SF_6 co-elute); He, Ne, Xe, Kr (N_2 and Kr elute closely); thick coating phase can resolve Ar from O_2 at 35 °C; deuterated hydrocarbons	Synthetic calcium alumino-silicate (zeolite) having an effective pore diameter of 0.5 nm CO_2 is adsorbed strongly; 5A usually gives the best results of all synthetic zeolites; thick and thin film columns are available; easy to foul with hydrocarbons; will adsorb water and CO_2
Molecular sieve, 13X	350	Same separations as those performed on 5A, but with C_1 to C_4, alkanes, alkenes, and alkynes being separated as well	Sodium alumino-silicate (zeolite) having a larger pore size than 0.5 nm, thus producing lower retention times and less resolution; 28.6% (mass/mass) water capacity
Monolithic carbon	350	C_1 to C_5 hydrocarbon isomers; acetylene in ethylene; methane	Phase consists of a bonded carbon monolith; historically more suitable for use with GC-MS because of the absence of particles

COOLANTS FOR CRYOTRAPPING

Thomas J. Bruno and Paris D. N. Svoronos

The following table provides coolants, in some cases mixtures, that can be used to chill trapping sorbents (or any cold trap) for the purge-and-trap sampling of solutes (Refs. 1-4). In each case the ratio is mass/mass. Practical use of these methods can require optimization and proper insulation.

References

1. Gordon, A .J., and Ford, R. A., *The Chemists Companion – A Handbook of Practical Data, Techniques and References*, Wiley Interscience, New York, 1972.
2. Bruno, T. J., Chromatographic Cryofocusing and Cryotrapping with the Vortex Tube, *J. Chromatogr. Sci.* 32, 112, 1994. <https://doi.org/10.1093/chromsci/32.3.112>
3. Bruno, T. J., Simple, Quantitative Headspace Analysis by Cryoadsorption on a Short Alumina PLOT Column, *J. Chromatogr. Sci.* 47, 1, 2009. <https://doi.org/10.1093/chromsci/47.7.569>
4. Bruno, T. J., and Svoronos, P. D. N., *CRC Handbook of Basic Tables for Chemical Analysis, Third Edition*, CRC Press, Boca Raton, FL, 2011. <https://doi.org/10.1201/b10385>

Temperature Reached by Various Coolants Used for Cryotrapping

Name	Temperature/°C
Crushed ice + sodium chloride (3 : 1)	-21
Crushed ice + calcium chloride (1.2 : 2)	-39
Vortex tube	-40
Liquid nitrogen + butylamine (slush)	-50
Crushed ice + calcium chloride (1.4 : 2)	-55
Liquid nitrogen + chloroform (1 : 1)	-63
Liquid nitrogen + *tert*-butylamine (slush)	-68
Dry ice	-78
Liquid nitrogen + acetone (slush)	-95
Liquid nitrogen + ethanol (slush)	-120
Liquid nitrogen + methylcyclohexane (slush)	-126
Liquid nitrogen + pentane (slush)	-131
Liquid nitrogen + 1,5-hexadiene (slush)	-141
Liquid nitrogen + isopentane (slush)	-160
Liquid argon	-186
Liquid nitrogen	-196

PROPERTIES OF COMMON CROSS-LINKED SILICONE STATIONARY PHASES

Thomas J. Bruno and Paris D. N. Svoronos

Modern gas chromatography is most often performed with high-efficiency capillary (open tubular) columns coated with a cross-linked polymeric stationary phase. We provide chromatographic data on the two most common cross-linked phases (Ref. 1): 5% Phenyl dimethylpolysiloxane (Table 1) and Dimethylpolysiloxane (Table 2). These data are useful for the interpretation of chromatograms, and for quality control and assurance.

Tables 1 and 2 provide information for several probe compounds used to probe the characteristics of gas chromatography columns with two common stationary phases. Column definitions for these tables are as follows.

Column heading	Definition
Probe compound	Name of probe compound
McReynolds constant	Measure of the strength of attractive interaction between the probe compound and the specified stationary phase
McReynolds code	Code that indicates the type of molecular interaction between the probe compound and stationary phase (see code definitions below)
Kovats retention index	Number quantifying the retention of the probe compound in comparison to a pair of n-alkane adjacent peaks

The molecular interactions indicated by five McReynolds codes are given below. The other McReynolds codes (L, H, J, K) simply identify specific probe compounds.

McReynolds code	Type of molecular interaction
x'	pi-type
y'	electron-attracting effect
z'	dipole-dipole
u'	electron-donating effect
s'	non-bonding electron attraction, H-bonding

The retention indices presented in these tables were measured at 120 °C isothermally. Retention indices are temperature dependent; the temperature dependence of the Kovats indices have been studied for many compounds (Ref. 2). For more extensive information on other cross-linked phases, other silicone phases, mesogenic phases, and solid sorbents, the reader is advised to consult Ref. 1.

References

1. Bruno, T. J., and Svoronos, P. D. N., *CRC Handbook of Basic Tables for Chemical Analysis, Third Edition*, CRC Press, Boca Raton, FL, 2011. <https://doi.org/10.1201/b10385>
2. Bruno, T. J., Wertz, K. H., and Caciari, M., *Anal. Chem.* 68, 1347, 1996. <https://doi.org/10.1021/ac9510191>

Phase: 5% Phenyl Dimethylpolysiloxane

This phase is probably the most commonly used stationary phase in gas chromatography because it combines boiling point separation with a minor contribution of a specific interaction. The phase is typically used as the first phase in any method development; it is versatile for hydrocarbons and more polar compounds.

Temperature Range

−60 °C to 325 °C isothermally, −60 °C to 350 °C programmed for < 0.32 mm I.D. columns

−60 °C to 300 °C isothermally, −60 °C to 320 °C programmed for 0.53 mm I.D. columns

−60 °C to 260/280 °C for >2.0 μm films

Similar Phases

DB-5, Ultra-2, SPB-5, CP-Sil 8 CB, Rtx-5, BP-5, OV-5, 007-2 (MPS-5), SE-52, SE-54, XTI-5, PTE-5, HP-5MS, ZB-5, AT-5, MDN-5

TABLE 1. Chromatographic Properties of the 5% Phenyl Dimethylpolysiloxane Phase

Probe compound	McReynolds constant	McReynolds code	Kovats retention index
Hexane			600
1-Butanol	66	y'	656
Benzene	31	x'	684
2-Pentanone	61	z'	688
Heptane			700
1,4-Dioxane	64	L	718
2-Methyl-2-pentanol	41	H	731
1-Nitropropane	93	u'	745
Pyridine	62	s'	761
Octane			800
Iodobutane	22	J	840
2-Octyne	35	K	876
Nonane			900
Decane			1000

Phase: Dimethylpolysiloxane

This stationary phase is useful for the separation of hydrocarbons, pesticides, PCBs, phenols, sulfur compounds, flavors and fragrances, and some amines. The columns are typically stable and low bleed. This is a good all-purpose column used to begin method development protocols.

Temperature Ranges

−60 °C to 325 °C for normal operations, periodic operation to 350 °C can be used to facilitate column cleanup

−60 °C to 260/280 °C for >2.0 μm films

Similar Phases

DB-1, OV-1, HP-1, DB-1ms, HP-1ms, Rtx-1, Rtx-1ms, CP-Sil 5 CB Low Bleed/MS, MDN-1, AT-1

TABLE 2. Chromatographic Properties of the Dimethylpolysiloxane Phase

Probe compound	McReynolds constant	McReynolds code	Kovats retention index
Hexane			600
1-Butanol	54	y'	644
Benzene	16	x'	669
2-Pentanone	44	z'	671
Heptane			700
1,4-Dioxane	46	L	700
2-Methyl-2-pentanol	31	H	721
1-Nitropropane	62	u'	714
Pyridine	44	s'	743
Octane			800
Iodobutane	3	J	821
2-Octyne	23	K	864
Nonane			900
Decane			1000

DETECTORS FOR GAS CHROMATOGRAPHY

Thomas J. Bruno and Paris D. N. Svoronos

The following table provides some comparative data to aid in interpreting results from the more common detectors applied to capillary and packed-column gas chromatography (Refs. 1–8). For more detailed information regarding operation and interpretation of results, see Ref. 8.

References

1. Hill, H.H., and McMinn, D., eds., *Detectors for Capillary Chromatography*, John Wiley & Sons, New York, 1992.
2. Buffington, R., and Wilson, M. K., *Detectors for Gas Chromatography — A Practical Primer*, Hewlett Packard Corp., Avondale, PA, 1987.
3. Buffington, R., *GC-Atomic Emission Spectroscopy Using Microwave Plasmas*, Hewlett Packard Corp., Avondale, PA, 1988.
4. Liebrand, R. J., Ed., *Basics of GC/IRD and GC/IRD/MS*, Hewlett Packard Corp., Avondale, PA, 1993.
5. Bruno, T. J., *Sep. Purif. Method* 29, 63, 2000.
6. Bruno, T. J., *Sep. Purif. Method* 29, 27, 2000.
7. Sevcik, J., Detectors in Gas Chromatography, *Journal of Chromatography Library*, Vol. 4, Elsevier, Amsterdam, 1976.
8. Bruno, T. J., and Svoronos, P. D. N., *CRC Handbook of Basic Tables for Chemical Analysis, Third Edition*, CRC Press, Boca Raton, FL, 2011.

Characteristics of Detectors for Gas Chromatography

Detector	Limit of detection	Linearity	Selectivity	Comments
Thermal conductivity detector (TCD, katharometer)	1×10^{-10} g propane (in helium carrier gas)	1×10^{6}	Universal response, concentration detector	Ultimate sensitivity depends on analyte thermal conductivity difference with carrier gas Because thermal conductivity is temperature dependent, response depends on cell temperature Wire selection depends on chemical nature of analyte Helium is recommended as carrier and make-up gas; when analyzing mixtures containing hydrogen, one can use a mixture of 8.5% (mass/mass) hydrogen in helium
Gas density balance detector (GADE)	1×10^{-9} g: H_2 with SF_6 as carrier gas	1×10^{6}	Universal response, concentration detector	Response and sensitivity are based on difference in relative molecular mass of analyte with that of the carrier gas; approximate calibration can be done on the basis of relative density The sensing elements (hot wires) never touch sample, thus making GADE suitable for the analysis of corrosive analytes such as acid gases; gold-sheathed tungsten wires are most common Best used with SF_6 as a carrier gas, switched with nitrogen when analyses are required Detector can be sensitive to vibrations and should be isolated on a cushioned base
Flame ionization detector (FID)	1×10^{-11} g to 1×10^{-10} g	1×10^{7}	Organic compounds with C–H bonds	Ultimate sensitivity depends on the number of C–H bonds on analyte Nitrogen is recommended as carrier gas and make-up gas to enhance sensitivity Sensitivity depends on carrier make-up, and jet gas flow rates Column must be positioned 1 mm to 2 mm below the base of the flame tip Jet gases must be of high purity
Nitrogen-phosphorus detector (NPD, thermionic detector, alkali flame ionization detector)	4×10^{-13} g to 1×10^{-11} g of nitrogen compounds 1×10^{-13} g to 1×10^{-12} g of phosphorus compounds	1×10^{4}	10^5 to 10^6 by mass selectivity of N or P over carbon	Does not respond to inorganic nitrogen such as N_2 or NH_3 Jet gas flow rates are critical to optimization Response is temperature dependent Used for trace analysis only, and is very sensitive to contamination Avoid use of phosphate detergents or leak detectors Avoid tobacco use nearby Solvent-quenching is often a problem
Electron capture detector (ECD)	5×10^{-14} g to 1×10^{-12} g	1×10^{4}	Selective for compounds with high electron affinity, such as chlorinated organics; concentration detector	Sensitivity depends on number of halogen atoms on analyte Used with nitrogen or argon/methane (95/5, mass/mass) carrier and make-up gases Carrier and make-up gases must be pure and dry The radioactive ^{63}Ni source is subject to regulation and periodic inspection
Flame photometric detector (FPD)	2×10^{-11} g of sulfur compounds 9×10^{-13} g of phosphorus compounds	1×10^{3} for sulfur compounds 1×10^{4} for phosphorus compounds	10^5 to 1 by mass selectivity of S or P over carbon	Hydrocarbon quenching can result from high levels of CO_2 in the flame Self-quenching of S and P analytes can occur with large samples Gas flows are critical to optimization Response is temperature dependent Condensed water can be a source of window fogging and corrosion

Analytical

Detector	Limit of detection	Linearity	Selectivity	Comments
Photoionization detector (PID)	1×10^{-12} g to 1×10^{-11} g	1×10^{7}	Depends on ionization potentials of analytes	Used with lamps with energies of 10.0 eV to 10.2 eV Detector will have response to ionizable compounds such as aromatics and unsaturated organics, some carboxylic acids, aldehydes, esters, ketones, silanes, iodo- and bromoalkanes, alkylamines and amides, and some thiocyanates
Sulfur chemiluminescence detector (SCD)	1×10^{-12} g of sulfur in sulfur compounds	1×10^{4}	10^{7} by mass selectivity of S over carbon	Equimolar response to all sulfur compounds to within 10% Requires pure hydrogen and oxygen combustion gases Instrument generates ozone *in situ*, which must be catalytically destroyed at detector outlet Catalyst operates at 950 °C to 975 °C Detector operated at reduced pressure (10^{-3} Pa)
Electrolytic conductivity detector (ECD, Hall detector)	1×10^{-13} g to 1×10^{-12} g of chlorinated compounds 2×10^{-12} g of sulfur compounds 4×10^{-12} g of nitrogen compounds	1×10^{6} for chlorinated compounds, 10^{4} for sulfur and nitrogen compounds	10^{6} by mass selectivity of Cl over carbon, 10^{5} to 10^{6} by mass selectivity of S and N over carbon	Only high-purity solvents should be used Carbon particles in conductivity chamber can be problematic Frequent cleaning and maintenance is required Often used in conjunction with a photoionization detector For chlorine, use hydrogen as the reactant gas and 1-propanol as the electrolyte For nitrogen or sulfur, hydrogen or oxygen can be used as reactant gas, and water or methanol as the electrolyte Ultrahigh purity reactant gases are required
Ion mobility detector (IMD)	1×10^{-12} g	1×10^{3} to 1×10^{4}	10^{3}	Amenable to use in handheld instruments Linear dynamic range of 10^{3} for radioactive sources and 10^{5} for photoionization sources Selectivity depends on mobility differences of ions Has been used for a wide variety of compounds including amino acids, halogenated organics, explosives The radioactive ^{63}Ni source is subject to regulation and periodic inspection
Mass selective detector (MSD, mass spectrometer, MS)	1×10^{-11} g (single ion monitoring); 1×10^{-8} g (scan mode)	1×10^{5}	Universal	Single quadrupole, multiple quadrupole, ion trap, time-of-flight and magnetic sector instruments available (see separate table entitled: "Varieties of Hyphenated Gas Chromatography with Mass Spectrometry" in this section) Must operate under moderate vacuum (1×10^{-4} Pa) Requires a molecular jet separator to operate with packed columns Amenable to library searching for qualitative identification Requires tuning of electronic optics over the entire m/e range of interest
Infrared detector (IRD)	1×10^{-9} g of a strong infrared absorber	1×10^{3}	Universal for compounds with mid-infrared active functionality	A costly and temperamental instrument that requires high purity carrier gas, a nitrogen purge of optical components (purified air will, in general, not be adequate) Must be isolated from vibrations Presence of carbon dioxide is a typical impurity band at 2200 cm^{-1} to 2300 cm^{-1} Requires frequent cleaning and optics maintenance Amenable to library searching for qualitative identification
Atomic emission detector (AED)	1×10^{-13} g to 2×10^{-11} g of each element	1×10^{3} to 1×10^{4}	10^{3} to 10^{5}, element to element	Requires the use of ultrahigh-purity carrier and plasma gases Plasma produced in a microwave cavity operated at 2450 MHz Scavenger gases (H_2, O_2) are used as dopants Photodiode array is used to detect emitted radiation
Vacuum ultraviolet absorption detector (VUV)	1×10^{-11} to 1×10^{-9} g	1×10^{3} to 1×10^{4}	Universal except for He, H_2, Ar, N_2	Wavelength range of 120 to 430 nm, filter selectable Operable to 430 °C to prevent condensation of low volatility compounds Amenable to library search, though current libraries are limited Software can deconvolute multiple overlapping peaks Requires a ≈ 2 mL/min make-up gas of Ar, He, H_2, or N_2, which maintains constant pressure in flow cell

VARIETIES OF HYPHENATED GAS CHROMATOGRAPHY WITH MASS SPECTROMETRY

Thomas J. Bruno

Because of the very powerful and common application of mass spectrometry with gas chromatography, the various technologies warrant more detailed consideration (Refs. 1 to 4). The following table provides basic information on the capabilities and applicability of the most common approaches. Clearly, the use of the single quadrupole is by far the most prevalent and economical method, but the other techniques are important and advantageous. We exclude methods that are typically not interfaced with gas chromatographic separations and that are highly specific in research settings (such as ion cyclotron mass spectrometry, and combined quadrupoles beyond the triple quad, and the various hybrid sector-quad-time of flight-ion trap combinations) that are used in fundamental ion chemistry research. We also exclude magnetic sector instruments that are seldom used with chromatography.

References

1. March, R. E., and Todd, J. F., *Quadrupole Ion Trap Mass Spectrometry*, Wiley-Intersceince, New York, 2005.
2. Portoles, T., Sancho, J. V., Hernandez, A., Newton, A., and Hancock, P., Potential of Atmospheric Chemical Ionizationsource in GC-QTOF for Pesticide Residue Analysis, *J. Mass Spectrom.* 45, 926, 2010.
3. Busch, K. L., Glish, G. L., and McLuckey, S. A., *Mass Spectrometry Mass Spectrometry: Techniques and Applications of Tandem Mass Spectrometry*, VCH Publishers, New York, 1988.
4. Wong, P. S., and Cooks, R. G., Ion Trap Mass Spectrometry, *Curr. Sep.* 16, 85, 1997.

Method (with accepted acronyms and abbreviations)	Modes	Advantages	Limitations
Gas Chromatography Mass Spectrometry GC-MS (single quadrupole, SQ)	Scan, selected ion monitoring (SIM) and Scan/SIM	Relatively simple and relatively inexpensive; compound identification by library search; dynamic range = 10^5, but typically limited to $>10^4$; mass/charge range = 10^3 to 10^4, resolution (at m/z = 1000) 10^3 to 10^4 for most ions; well-developed hardware and software; used in standard protocols.	Coelution of compounds compromise library identifications; often user must interpret fragmentation patterns. SIM mode provides higher sensitivity but for target ions only, sensitivity decreases with increasing number of SIM ions. Scan/SIM methods must balance scan and SIM sensitivity requirements.
Gas Chromatography-Mass Spectrometry GC-MS (ion trap, IT)	Scan or selected ion monitoring (SIM); tandem MS-MS; gas or liquid chemical ionization	High sensitivity, compact design, tandem mass spectrometry is possible; dynamic range = 10^4, but typically limited to $< 10^4$; mass/charge range = 10^4 to 10^5, resolution (at m/z = 1000) 10^4; well-developed hardware and software; used in standard protocols.	Space charge effects can lead to relatively poor dynamic range, however, for "clean" samples, the dynamic range can be as high as single quadrupole units; specific libraries are limited; SIM is for target ions only with a sensitivity lower than that obtainable by a SIM analysis via single quadrupole, above; MS-MS analyses are limited to approximately 120 to 150 compounds per analysis; MS-MS analysis is typically slower than that done with a tandem MS-MS (process is done in time rather than space).
Gas Chromatography (time of flight) Mass Spectrometry, low resolution GC-TOF	Scan or selected ion monitoring	Identification possible with good chromatographic separations, dynamic range = 10^4, mass/charge range = 10^5, resolution (at m/z = 1000) 10^3 to 10^4.	Unit mass resolution limits identification capability; libraries are limited; if chromatographic separation is poor, comprehensive (GCXGC) might be needed; cannot distinguish neutral losses.
Comprehensive Gas Chromatography (time of flight) Mass Spectrometry, low resolution GCXGC-TOF	Scan or selected ion monitoring; normal column configuration (nonpolar – polar) or reversed column configuration (polar – nonpolar)	Identification possible with good chromatographic separations in two dimensions on a nonpolar (long) and a polar (short) column; identification of families with help of principal component analysis tools; dynamic range = 10^4, mass/charge range = 10^5, resolution (at m/z = 1000) 10^3 to 10^4, well-developed hardware and software; used in standard protocols.	Unit mass resolution limits identification capability; libraries are limited; GCXGC may not remedy all aspects of component coelution.

Method (with accepted acronyms and abbreviations)	Modes	Advantages	Limitations
Gas Chromatography Mass Spectrometry (triple quadrupole) GC-MS(QQQ, or QqQ)	Multiple and selected reaction monitoring	Provides product and precursor ion scans; very sensitive for target compounds or functional groups; developed to provide enhanced daughter ion resolution; relatively simple construction with straightforward scanning procedures; no high voltage arcing; dynamic range = 10^5, mass/charge range = 10^3 to 10^4, resolution (at m/z = 1000) 10^3 to 10^4.	Unit mass, making identification ambiguous due to multiple structures as source of breakdown mass; empirical formula determination can be difficult, spectra and fragmentations must often be interpreted manually; full scan data can be acquired (similar to the single quad procedure) by turning off the collision cell.
Gas Chromatography (time of flight) Mass Spectrometry, high resolution GC-TOF	Scan or selected ion monitoring, tandem MS-MS	Might obviate the need for GCXGC separations with accurate mass determinations; deconvolution software can aid in identification of multiple components under peaks; sensitivity intermediate between multiple reaction monitoring and product ion scan of a QQQ; dynamic range = 10^4, mass/charge range = 10^5, resolution (at m/z = 1000) 10^3 to 10^4.	Very large data files (currently approaching 2 Gb), software is currently developmental.
Gas Chromatography Tandem Mass Spectrometry GC-QTOF	Scan or selected ion monitoring, tandem MS-MS	Might obviate the need for GCXGC separations with accurate mass determinations; isolation of parent ions and subsequent fragmentation provides identification; will detect any daughter ion passed into the TOF; sensitivity intermediate between multiple reaction monitoring and product ion scan of a QQQ dynamic range = 10^4, mass/charge range = 10^5, resolution (at m/z = 1000) 10^3 to 10^4.	Large data files are produced; sophisticated software is needed for processing and deconvolution; requires accurate mass and high resolution.

GAS CHROMATOGRAPHIC RETENTION INDICES

Thomas J. Bruno

The interpretation of results from chromatographic measurements can often be augmented with an appropriate mathematical treatment of the solute retention that is observed. The goal of the treatment is to make the resulting metric as independent of the instrument as possible. A typical situation that arises from analysis by gas chromatography with mass spectrometry is that the library search routine produces "hits" that are ambiguous if not nonsensical (Ref. 1). The correct interpretation of the mass spectrum must then be done manually (Refs. 2, 3), and the mass spectral data should be augmented by additional analytical techniques. The specific techniques that should be used must be determined on a case-by-case basis by a qualified person. One additional datum that is typically already present in gas chromatography with mass spectrometry detection is chromatographic retention. The raw datum from a chromatographic measurement is a retention time, t_r, of each eluted peak, and a corresponding intensity. Here, we will not treat other aspects of the output, such as the width and shape of the chromatographic signal. The retention time (for a given stationary phase) is dependent on the column temperature, column pressure, column geometry (length and inside diameter; phase ratio), and ambient (atmospheric) pressure.

If the volumetric carrier gas flow rate (at the column exit) is measured and multiplied by the retention time, the retention volume, V_R, is obtained. The adjusted retention volume, V_R', is the retention volume corrected for the void volume (or mobile phase holdup) of the column. It is obtained by simply subtracting the retention volume of an unretained solute (V_M):

$$V_R' = V_R - V_M \tag{1}$$

While it is possible to calculate the corresponding adjusted retention time, t_r', by subtracting the retention time of the unretained peak, t_m, it is better to work with volumes since average flow-rate variations between individual analyses are then accounted for. Note that with each level of refinement beyond the raw retention time, a facet of instrument dependence from the resulting parameter is removed. V_R is independent of the flow rate; V_R' is, further, independent of column geometry. Continuing with this approach, the net retention volume, V_N, is defined, by applying a factor, j, to account for the pressure drop across the column:

$$V_N = jV_R' \tag{2}$$

where j is usually the Martin-James compressibility factor[3]:

$$j = \frac{3}{2}\left[\frac{\left(\frac{P_i}{P_o}\right)^2 - 1}{\left(\frac{P_i}{P_o}\right)^3 - 1}\right] \tag{3}$$

In Eq. 3, P_i is the inlet pressure (absolute) and P_o is the outlet pressure (usually atmospheric pressure). The net retention volume is important because it is independent of the inlet and outlet pressures, as well as being independent of the carrier gas flow rate and column geometry.

The specific retention volume, V_g, corrects the net retention volume for the amount of stationary phase actually on the column, and the column temperature is adjusted or corrected to 0 °C:

$$V_g = (273.15)\frac{V_N}{(W_s T_{col})} \tag{4}$$

where T_{col} is the column temperature (in K), and W_s is the mass of the stationary phase in the column. The V_g value is a characteristic for a particular solute on a particular stationary phase, and is instrument independent. This is a quantity that may be compared from instrument to instrument and laboratory to laboratory with a high level of confidence provided the stationary phase used is a single, pure compound or a well characterized mixture. If the mass of the stationary phase is not known, or is not meaningful, one may use the net retention volume directly, or one may correct the net retention volume to a column temperature of 0 °C (represented by V_N^0) by simply not including the term for W_s (that is, setting it equal to unity).

It is also extremely valuable to calculate a relative retention, $r_{a/b}$:

$$r_{a/b} = \left(\frac{V_g^b}{V_g^a}\right) = \left(\frac{V_N^b}{V_N^a}\right) \tag{5}$$

where the numerical superscripts refer to the retention volumes of solutes "a" and "b." In this case, solute "a" is a reference compound. The relative retention is dependent only on the column temperature and the type of stationary phase. For reasons of operational simplicity, this parameter is usually one of the best to use for qualitative analysis. It can account for small differences in the column temperature, stationary phase considerations, column history, and minor disturbances in the carrier gas flow rate. It is possible to account for the column temperature by plotting the logarithm of the retention parameters against $1/T$, where T is the thermodynamic temperature. The column pressure is accounted for by variations in the volume measurement; therefore, there is no pressure dependence to these parameters.

We can go beyond the simple retention parameters discussed earlier to incorporate a logarithmic interpolation on a uniform scale by use of the Kovats retention index (Ref. 4). The isothermal Kovats retention index is calculated by use of the following defining equation:

$$I_{sample}(T) = 100\left[\frac{\log X_S - \log X_L}{\log X_H - \log X_L} + n_L\right] \tag{6}$$

Here, I_{sample} is the dimensionless Kovats retention index that is a function of both temperature and the stationary phase employed. The terms represented by X are retention parameters of the sample and standards. Following this convention, X_S is the retention parameter of the sample under consideration. Any retention parameter, such as the adjusted retention time, t', the net retention volume, V_N, the adjusted net retention volume, V_N^0, and the relative retentions, $r_{a/b}$, can be used. X_S is the retention parameter of the sample under consideration, X_L is the retention parameter of a normal alkane (that is, straight chain or unbranched) of carbon number n_L that elutes earlier than the sample, and X_H is the retention parameter of a normal alkane having a carbon number greater than n_L+1 that elutes after the sample. The retention index of a sample is, therefore, 100 multiplied by the carbon number of a hypothetical normal alkane that shows the same retention parameter on the stationary phase at that temperature. Thus, a sample that has a retention index of 785, for example, would co-elute with a hypothetical normal alkane that has 7.85 carbon atoms. By definition, the retention indices of the normal alkanes (on any stationary phase) are equal to 100 multiplied by the carbon number. Thus, for *n*-hexane, I = 600, and similarly for the other *n*-alkanes in this homologous series. The zero point in the scale is defined for hydrogen, for which I = 0. Kovats retention indices have been determined for selected compounds on the most common stationary phases (Refs. 5-7).

The temperature dependence of I for a given sample is known to follow a hyperbolic form similar to the familiar Antoine equation used to represent vapor pressure:

$$I_{\text{sample}}(T) = A + \frac{B}{(T+C)} \quad (7)$$

In this equation, A, B, and C are empirically determined constants, and T is the temperature (in °C). A nonlinear fitting routine should be used to determine these constants. When retention indices are available for at least three temperatures, initial values for A, B, and C can be determined by use of the following equations:

$$C = \frac{(T_2 - T_1)(I_3T_3 - I_1T_1) + (T_3 - T_1)(I_1T_1 - I_2T_2)}{(T_3 - T_1)(I_2 - I_1) - (T_2 - T_1)(I_3 - I_1)} \quad (8)$$

$$A = \frac{I_2T_2 - I_1T_1 + C(I_2 - I_1)}{T_2 - T_1} \quad (9)$$

$$B = (I_2 - A)(T_2 + C) \quad (10)$$

In these equations, I_1, I_2, and I_3 are retention indices of the sample measured at temperatures T_1, T_2, and T_3. When additional retention indices are available at other temperatures, we advocate the use of minimum deviation estimates from these three equations to furnish the starting values for the nonlinear fit. When retention indices at four temperatures are available, the best starting values are obtained from the I_1, I_2, and I_3 triplet that minimizes the deviation with the experimental value with that produced by Eq. 7. This approach provides the fastest convergence, and also helps avoid converging to local minima. Predictions made by use of Eq. 7 can be used for retention indices within the measured temperature range as well as extrapolation somewhat beyond that range on a case-by-case basis.

It is also of value to report and use the temperature dependence as a slope coefficient, $\delta I_{\text{sample}}/10$ °C, the variation of I_{sample} for a particular stationary phase over a particular temperature range. While not as reliable as the Antoine-type fit, this coefficient is useful for predictions within the range of the measured results.

In all of the above discussion, one must understand that the column temperature is fixed. The Kovats retention indices can be made applicable to temperature programmed analysis by use of

$$I = 100 \times \left[n + (N - n)\frac{t_{r(\text{unknown})} - t_{r(n)}}{t_{r(N)} - t_{r(n)}} \right] \quad (11)$$

where I is the retention index, n is the number of carbon atoms in the smaller *n*-alkane, N is the number of carbon atoms in the larger *n*-alkane, and t_r is the retention time of the indicated peak.

A useful alternative to the Kovats system is the Lee retention index (isothermal and temperature dependent), based on the polynuclear aromatic hydrocarbon (PAH) standard compounds: naphthalene (I = 200), phenanthrene (I = 300), chrysene (I = 400), and picene or benzo(g,h,i)perylene (I = 500). Isothermal (and temperature-dependent) Kovats and Lee retention indices for many compounds are tabulated in the NIST Chemistry Web Book (Ref. 8).

References

1. NIST/EPA/NIH (NIST 11) Mass Spectral Database, NIST Standard Reference Database 1A, National Institute of Standards and Technology, Standard Reference Data Program, Gaithersburg, MD, 20899, 2011.
2. Bruno, T. J., and Svoronos, P. D. N., *CRC Handbook of Fundamental Spectroscopic Correlation Charts*, CRC Press, Boca Raton, FL, 2006 <https://doi.org/10.1201/9781420037685>.
3. Bruno, T. J., and Svoronos, P. D. N, *CRC Handbook of Basic Tables for Chemical Analysis, Third Edition*, CRC Press, Boca Raton, FL, 2011 <https://doi.org/10.1201/b10385>.
4. Kovats, E., *Helv. Chim. Acta* 41, 1915, 1958 <https://doi.org/10.1002/hlca.19580410703>.
5. Lubek, A. J., and Sutton, D. L., *Chromatogr. and Chromatogr. Commun.* 6, 328, 1983 <https://doi.org/10.1002/jhrc.1240060612>.
6. Bruno, T. J., Wertz, K. H., and Cacairi, M., *Anal. Chem.* 68, 1347, 1996 <https://doi.org/10.1021/ac9510191>.
7. Miller, K. E., and Bruno, T. J., *J. Chromatogr. A* 1007, 117, 2003 <https://doi.org/10.1016/S0021-9673(03)00958-0>.
8. NIST Chemistry Web Book, NIST Standard Reference Database Number 69, *Standard Reference Data Program, National Institute of Standards and Technology*, Gaithersburg, MD 20899, 2012.

PRESSURE DROP IN OPEN TUBULAR GAS CHROMATOGRAPHIC COLUMNS

The pressure drop across an open tubular or capillary column is often important for optimization of gas chromatographic analyses (Refs. 1,2). Column performance is typically assessed by the height equivalent to a theoretical plate (HETP), which is based on the average linear carrier gas velocity. As the average linear velocity increases, the head pressure and carrier gas flow rate increase as well. One may express the pressure drop across the column as:

$$\Delta p = p_i - p_o$$

where Δp is the pressure drop, p_i is the inlet or head pressure, and p_o is the outlet pressure. The head pressure is typically a gauge pressure measured electronically, while the outlet pressure is the barometric pressure, which can be measured (a) electronically, (b) with a mercury barometer, or (c) with an aneroid barometer. Concern for the spillage of mercury has caused an increase in the number of laboratories now employing an electronic measure for the outlet pressure. In relation to the average carrier gas velocity:

$$\Delta p = 8\eta Lu/r_c^2$$

where η is the carrier gas viscosity, L is the column length, u is the carrier gas velocity, and r_c is the column internal radius. For helium carrier gas at 100 °C, the following tables provide the pressure drop Δp as a function of carrier velocity u, in units of kPa, for 10 m columns (Table 1), 25 m columns (Table 2), and 50 m columns (Table 3) of various diameters $d = 2r_c$.

References

1. Hinshaw, J.V., Open tubular column pressures and flows, GC troubleshooting, *LC-GC — North America*, 7, 237–239, 1989.
2. Bruno, T.J., Svoronos, P.D.N., *CRC Handbook of Basic Tables for Chemical Analysis — Data Driven Methods and Interpretation, Fourth Edition*, CRC/Taylor & Francis, Boca Raton, FL, 2021.

TABLE 1. Pressure Drop Δp (in kPa, gauge) for Helium Carrier Gas in 10 m Open Tubular Gas Chromatographic Columns as a Function of Carrier Gas Velocity and Column Diameter

u/cm s^{-1}	d_c=0.750	d_c=0.530	d_c=0.320	d_c=0.200	d_c=0.100
10	1.3	2.6	6.9	18.6	69.0
20	2.6	5.2	14.5	36.5	146.2
30	3.9	7.6	21.4	54.5	219.3
40	5.2	10.3	28.3	73.1	291.7
60	7.6	15.9	42.7	109.6	437.8
80	10.3	20.7	57.2	146.9	584.0

TABLE 2. Pressure Drop Δp (in kPa, gauge) for Helium Carrier Gas in 25 m Open Tubular Gas Chromatographic Columns as a Function of Carrier Gas Velocity and Column Diameter

u/cm s^{-1}	d_c=0.750	d_c=0.530	d_c=0.320	d_c=0.200	d_c=0.100
10	3.2	6.5	17.9	45.5	182.7
20	6.5	13.1	35.9	91.7	364.7
30	9.7	19.3	53.8	136.5	547.5
40	13.1	26.2	71.0	182.7	
60	19.3	39.3	106.9	273.7	
80	26.2	51.7	142.7	364.7	

TABLE 3. Pressure Drop Δp (in kPa, gauge) for Helium Carrier Gas in 50 m Open Tubular Gas Chromatographic Columns as a Function of Carrier Gas Velocity and Column Diameter

u/cm s^{-1}	d_c=0.750	d_c=0.530	d_c=0.320	d_c=0.200	d_c=0.100
10	6.5	13.1	35.9	91.0	364.7
20	13.1	26.2	71.0	182.7	
30	19.3	39.3	106.9	273.7	
40	26.2	51.7	142.7	364.7	
60	38.6	77.9	213.7	547.5	
80	51.7	104.1	284.8	0.0	

PHASE RATIOS FOR CAPILLARY COLUMNS

The phase ratio is an important parameter used in the design of capillary (open tubular) column separations (Refs. 1-2). This quantity relates the partition coefficient (K) to the partition ratio (k):

$$K = k\beta \qquad (1)$$

where β is the phase ratio, defined as the ratio of the volume occupied by the gas or mobile phase (V_m) relative to that occupied by the liquid or stationary phase (V_s). For wall-coated open tubular columns, the phase ratio can be found from:

$$\beta = r/2d_f \qquad (2)$$

where r is the internal radius of the column, and d_f is the thickness of the stationary phase film. The following table provides the phase ratio for common combinations of column internal radii and stationary phase film thickness. These values are given to the nearest whole number, as only an approximate value is needed for most analytical applications. Column definitions are as follows.

Column heading	Definition
d_f	Thickness of the stationary phase film, in mm
r(0.025 mm), r(0.05 mm), etc.	Internal radius as specified in parentheses, in mm

References

1. Sandra, P., *High Resolution Gas Chromatography, Third Edition*, Hewlett Packard Corporation, Avondale, PA, USA, 1989.
2. Bruno, T. J., Svoronos, P. D. N., *CRC Handbook of Basic Tables for Chemical Analysis — Data Driven Methods and Interpretation*, CRC/Taylor & Francis, Boca Raton, FL, 2021.

Phase Ratios for Common Capillary Columns in Gas Chromatography

d_f	r(0.025 mm)	r(0.05 mm)	r(0.10 mm)	r(0.15 mm)	r(0.16 mm)	r(0.20 mm)	r(0.25 mm)	r(0.265 mm)
0.03	417	833	1667	2500	2667	3333	4167	4417
0.06	208	417	833	1250	1333	1667	2083	2208
0.1	125	250	500	750	800	1000	1250	1325
0.2	63	125	250	375	400	500	625	663
0.3	42	83	167	250	267	333	417	442
0.4	31	63	125	188	200	250	313	331
0.5	25	50	100	150	160	200	250	265
0.6	21	42	83	125	133	167	208	221
0.7	18	36	71	107	114	143	179	189
0.8	16	31	63	94	100	125	156	166
0.9	14	28	56	83	89	111	139	147
1.0	13	25	50	75	80	100	125	133
1.5	8	17	34	50	53	67	83	88
2.0	6.3	13	25	38	40	50	63	66
2.5	5	10	20	30	34	40	50	53
3.0	4	8	17	25	27	33	42	44
3.5	4	7	14	21	23	29	18	38
4.0	3	6	13	19	20	25	32	33
4.5	3	6	11	17	18	22	29	29
5.0	2.5	5	10	15	16	20	25	27
5.5	2	5	9	14	15	18	23	24
6.0	2	4	8	13	13	17	21	22
6.5	2	4	8	12	12	15	19	20
7.0	2	4	7	11	11	14	18	19
7.5	2	3	7	10	11	13	17	18
8.0	2	3	6	9	10	13	16	17
8.5	1	3	6	9	9	12	15	16
9.0	1	3	6	8	9	11	14	15

MINIMUM RECOMMENDED INJECTOR SPLIT RATIOS FOR CAPILLARY COLUMNS

Thomas J. Bruno

In order to avoid overloading high-efficiency open tubular or capillary columns (with theoretical plate counts between 400,000 and 600,000), it is necessary to split the flow in the injector. Split ratios that are too low will result in distorted peak shapes and poor analyses. As a first approximation, the lowest split ratio that can be used is dependent upon the column internal diameter. Secondary factors then include the solute properties (polarity, etc.), column temperature (or temperature program), linear volume, injector volume, and stationary phase properties. The following table provides the minimum split ratios that should be considered for typical capillary columns (Refs. 1-2).

Column diameter, mm	Minimum split ratio
0.18	1:25
0.20	1:20
0.25	1:15–1:20
0.32	1:10–1:12
0.53	1:3–1:5

References

1. Rood, D., *J. Chromatogr. Sci.*, 36(9), 476, 1998.
2. Bruno, T.J., Svoronos, P.D.N., *CRC Handbook of Basic Tables for Chemical Analysis — Data Driven Methods and Interpretation*, CRC/Taylor & Francis, Boca Raton, FL, 2021.

ELUOTROPIC VALUES OF SOLVENTS ON OCTADECYLSILANE AND OCTYLSILANE

Thomas J. Bruno and Paris D. N. Svoronos

The familiar eluotropic values of solvents in normal-phase liquid chromatography is a solvent series in order of increasing polarity; this series is used to explain or design solvent strength in liquid-solid or adsorption chromatography. In normal-phase chromatography, this series would be bounded by nonpolar solvents (such as n-pentane) at one end and the most polar solvent (usually water) at the other. Such series are important in reverse-phase high-performance liquid chromatography (HPLC) separations as well. The table below provides for comparative purposes eluotropic values on octadecylsilane (ODS) and octylsilane (OS) for common solvents (Refs. 1-4). These values were determined as the ratio of retention times of the indicated solvent relative to that of methanol (defined as unity) when water was used as the mobile phase. Note that in mixtures with water, these solvents will exhibit approximately the same ratios if the organic solvent concentration is the same.

For additional information on common, specific, and chiral stationary phases for HPLC, and for solvents, derivatizing reagents, and detectors, see Ref. 4.

References

1. Krieger, P. A., *High Purity Solvent Guide*, Burdick and Jackson Laboratories, McGaw Park, IL, 1984.
2. Ahuja, S., *Trace and Ultratrace Analysis by HPLC*, John Wiley and Sons, New York, 1992.
3. Karch, K., Sebastian, I., Halasz, I., Englehardt, H., *J. Chromatogr.*, 122, 71, 1976.
4. Bruno, T. J., and Svoronos, P. D. N., *CRC Handbook of Basic Tables for Chemical Analysis - Data-Driven Methods and Integration, Fourth Edition*, CRC Press, Boca Raton, FL, 2021.

Eluotropic Values of Solvents on Octadecylsilane and Octylsilane

Name	Eluotropic value, ODS	Eluotropic value, OS
Acetic acid		2.7
Acetone	8.8	9.3
Acetonitrile	3.1	3.3
1,4-Dioxane	11.7	13.5
N,N-Dimethylformamide	7.6	9.4
Methanol	1.0	1.0
Ethanol	3.1	3.2
1-Propanol	10.1	10.8
2-Propanol	8.3	8.4
Tetrahydrofuran	3.7	

INSTABILITY OF HPLC SOLVENTS

Thomas J. Bruno and Paris D. N. Svoronos

Solvents that are commonly used in high-performance liquid chromatography frequently have inherent chemical instabilities that must be considered when designing an analysis, or in the interpretation of results (Refs. 1-4). In many cases, such solvents are obtainable with stabilizers added to control the instability or to slow the reaction. Reactive solvents that do not have stabilizers must be used quickly or be given proper treatment. In either case, it is important to understand that the solvents (as they may be used in an analysis) are not necessarily pure materials.

Although not specifically considered below, the preparation of aqueous-organic solvents for HPLC can cause the introduction of impurities. Mixing acetonitrile and water is endothermic and can sometimes exacerbate air dissolution. Buffers are also a special case in that the pH can change due to exposure to air and the development of microbes (Ref. 4). Buffers should be used the same day of preparation and often incorporate the preservative sodium metabisulfite.

Column definitions for the table are as follows.

Column heading	Definition
Name	Solvent name
Mol. form.	Molecular formula of solvent
Contaminants, reaction products	Typical contaminants and reaction products found in the solvent
Stabilizers	Common stabilizers used with the solvent

References

1. Sadek, P. C., *The HPLC Solvent Guide, Second Edition*, Wiley Interscience, New York, 2002.
2. Bruno, T. J., and Straty, G. C., *J. Res. Natl. Bur. Stds. (U.S.)*, 91, 135, 1986. <https://doi.org/10.6028/jres.091.021>
3. Straty, G. C., Palavra, A. M. F., and Bruno, T. J., *Int. J. Thermophys.* 7, 1077, 1986. <https://doi.org/10.1007/BF00502379>
4. Bruno, T. J., and Svoronos, P. D. N., *CRC Handbook of Basic Tables for Chemical Analysis, Third Edition*, CRC Press, Boca Raton, FL, 2011. <https://doi.org/10.1201/b10385>

Characteristics of HPLC Solvents

Name	Mol. form.	Contaminants, reaction products	Stabilizers
Ethers			
Diethyl ether	$(C_2H_5)_2O$	Peroxides[1]	2–3% (vol/vol) ethanol[2] 1–10 ppm (mass/mass) BHT[3] (1.5–3.5% ethanol) + (0.2–0.5% water) + (5–10 ppm (mass/mass) BHT[3])
Diisopropyl ether	$C_6H_{14}O$	Peroxides[1]	0.01% (mass/mass) hydroquinone 5–100 ppm (mass/mass) BHT[3]
1,4-Dioxane	$C_4H_8O_2$	Peroxides[1]	25–1500 ppm (mass/mass) BHT[3]
Tetrahydrofuran	C_4H_8O	Peroxides[1]	25–250 ppm (mass/mass) BHT[3]
Chlorinated Alkanes			
Trichloromethane	$CHCl_3$	Hydrochloric acid, chlorine, phosgene (CCl_2O)	0.5–1% (vol/vol) ethanol 50–150 ppm (mass/mass) amylene[4] various ethanol amylene blends
Dichloromethane	CH_2Cl_2	Hydrochloric acid, chlorine, phosgene (CCl_2O)	25 ppm (mass/mass) amylene 25 ppm (mass/mass) cyclohexene 400–600 ppm (mass/mass) methanol various amylene methanol blends
Alcohols			
Ethanol	C_2H_5OH	Water, numerous denaturants are commonly added	
Methanol	CH_3OH	Water; formal dehydrate (at elevated temperature)	
Ketones			
Acetone	$(CH_3)_2CO$	Diacetone alcohol, and higher oligomers	

[1] The peroxide concentration that is usually considered hazardous is 250 ppm (mass/mass).
[2] Ethanol does not actually stabilize diethyl ether, nor is it a peroxide scavenger, although it was thought to be so in the past. It is still available in chromatographic solvents to preserve the utility of retention relationships and analytical methods.
[3] 2,6-Di-*tert*-butyl-*p*-cresol.
[4] Amylene is a generic name for 2-methyl-2-butene.

DETECTORS FOR LIQUID CHROMATOGRAPHY

Thomas J. Bruno and Paris D. N. Svoronos

The following table provides some comparative data for interpretation of results from the more common detectors applied to high-performance liquid chromatography (Refs. 1-6). In general, the operational parameters provided are for optimized systems and represent the maximum obtainable in terms of sensitivity and linearity. In this table, the molar extinction coefficient is represented by ε. Column definitions for the table are as follows.

Column heading	Definition
Detector	Type of detector
Sensitivity	Detector sensitivity, in terms of minimum mass detectable, in g
Linearity	Response linearity with respect to changes of concentration
Selectivity	Types of compounds for which the detector is most suitable
Comments	Additional information on the applicability and operation of the detector

References

1. Schwartz, M., *Journal of Liquid Chromatography & Related Technologies*, 33, 1130–1150, 2010. <https://doi.org/10.1080/10826076.2010.484356>
2. Dong, M.W., *Modern HPLC for Practicing Scientists*, Wiley Interscience, Hoboken, NJ, 2006. <https://doi.org/10.1002/0471973106>
3. Ahuja, S., *Trace and Ultratrace Analysis by HPLC*, Chemical Analysis Series, John Wiley & Sons, New York, 1991.
4. Snyder, L. R., Kirkland, J. J., and Glajch, J., *Practical HPLC Method Development*, John Wiley & Sons, New York, 1997. <https://doi.org/10.1002/9781118592014>
5. Bruno, T. J., *Sep. Purif. Meth.* 29, 63, 2000. <https://doi.org/10.1081/SPM-100100003>
6. Bruno, T. J., and Svoronos, P. D. N, *CRC Handbook of Basic Tables for Chemical Analysis, Third Edition*, CRC Press, Boca Raton, FL, 2011. <https://doi.org/10.1201/b10385>

Detector Properties for Liquid Chromatography

Detector	Sensitivity	Linearity	Selectivity	Comments
Ultraviolet spectrophotometer	1×10^{-9} g (for compounds of ε = 10 000 to 20 000)	1×10^{4}	For UV-active functionalities on the basis of absorptivity	Relatively insensitive to flow and temperature fluctuations; non-destructive; useful with gradient elution; use mercury lamp for 254 nm and quartz-iodine lamp for 350 nm to 700 nm; often a diode-array instrument is used to obtain entire UV-vis spectrum; solvents must be transparent in the UV region of interest
Refractive index detector (RID)	1×10^{-7} g	1×10^{4}	Universal, dependent on refractive-index difference with mobile phase	Relatively insensitive to flow fluctuations, but sensitive to temperature fluctuations; non-destructive; cannot be used with gradient elution; solvents must be degassed to avoid bubble formations; laser-based RI detectors offer higher sensitivity
Fluorometric detector	1×10^{-11} g	1×10^{5}	For fluorescent species with conjugated bonding and/or aromaticity	Relatively insensitive to temperature and flow fluctuations; non-destructive; can be used with gradient elution; often chemical derivatization is done on analytes to form fluorescent species; uses deuterium lamp for 190 nm to 400 nm, tungsten lamp for 350 nm to 600 nm, and the xenon flash lamp for 185 nm to 2000 nm
Amperometric detector	1×10^{-9} g	1×10^{4}	Responds to –OH functionalities	Used for aliphatic and aromatic –OH compounds, amines, and indoles; pulsed potential units are most sensitive, can be used with gradient elution and organic mobile phases; senses compounds in oxidative or reductive modes; mobile phases must be highly pure and purged of O_2
Conductivity detector	1×10^{-9} g	2×10^{4}	Specific to ionizable compounds	Most common detector for ion chromatography; requires suppression of solvent conductivity; uses post-column derivatization to produce ionic species; especially useful for certain halogen, sulfur, and nitrogen compounds
Radioactivity detector	1×10^{-11} g	10^{3} – 10^{4}	Highly selective to radiolabeled compounds	Based on liquid scintillation counting, either from an added post-column scintillator compound (homogeneous) or a bed of solid state scintillator (heterogeneous); relatively large volume flow cells can cause peak broadening
Light-scattering detector	1×10^{-9} g	10^{3}	Universal toward most nonvolatile solutes	Often called evaporative light-scattering detector (ELSD); universal response with higher sensitivity than the RI detector, response independent of solvent; requires use of volatile buffers to promote nebulization
Mass spectrometers	Interface dependent	Interface dependent	Universal, within limits imposed by interface	Complex, expensive devices highly dependent on an efficient interface; electrospray and thermospray interfaces are most common; linear response is difficult to achieve

SOLVENTS FOR ULTRAVIOLET SPECTROPHOTOMETRY

Thomas J. Bruno

This table lists some solvents commonly used for sample preparation for ultraviolet spectrophotometry as well as some relevant properties. Column definitions are as follows.

Column heading	Definition
Name	Solvent name
Mol. form.	Solvent molecular formula
λ_c	Cutoff wavelength, below which the solvent absorption becomes excessive, in nm
ε	Dielectric constant (relative permittivity); the temperature in °C is given as a superscript
t_b	Normal boiling point, in °C

References

1. Bruno, T. J., and Svoronos, P. D. N., *CRC Handbook of Basic Tables for Chemical Analysis*, CRC Press, Boca Raton, FL, 1989.
2. *Landolt-Börnstein, Numerical Data and Functional Relationships in Science and Technology, New Series*, IV/6, *Static Dielectric Constants of Pure Liquids and Binary Liquid Mixtures*, Springer–Verlag, Heidelberg, 1991.

Solvents for Ultraviolet Spectrophotometry

Name	Mol. form.	λ_c/nm	ε	t_b/°C
Acetic acid	CH_3COOH	260	6.20^{20}	117.9
Acetone	$(CH_3)_2CO$	330	21.01^{20}	56.08
Acetonitrile	CH_3CN	190	36.64^{20}	81.6
Benzene	C_6H_6	280	2.2825^{20}	80.08
2-Butanol	$C_4H_{10}O$	260	17.26^{20}	99.4
2-Butanone	C_4H_8O	330	18.56^{20}	79.6
Butyl acetate	$C_6H_{12}O_2$	254	5.07^{20}	126.0
Carbon disulfide	CS_2	380	2.6320^{20}	46.2
1-Chlorobutane	C_4H_9Cl	220	7.276^{20}	78.4
Cyclohexane	C_6H_{12}	210	2.0243^{20}	80.7
1,2-Dichloroethane	CH_2ClCH_2Cl	226	10.42^{20}	83.4
Dichloromethane	CH_2Cl_2	235	8.93^{25}	39.8
Diethyl ether	$(C_2H_5)_2O$	218	4.2666^{20}	34.4
1,2-Dimethoxyethane	$CH_3OC_2H_4OCH_3$	240	7.30^{24}	85.0
N,N-Dimethylacetamide	C_4H_9NO	268	38.85^{21}	165.9
N,N-Dimethylformamide	C_3H_7NO	270	38.25^{20}	152.8
Dimethyl sulfoxide	$(CH_3)_2S{=}O$	265	47.24^{20}	191.9
1,4-Dioxane	$C_4H_8O_2$	215	2.2189^{20}	101.2
Ethanol	C_2H_5OH	210	25.3^{20}	78.24
2-Ethoxyethanol	$C_4H_{10}O_2$	210	13.38^{25}	134.7
Ethyl acetate	$C_4H_8O_2$	255	6.0814^{20}	77.1
Glycerol	$CH_2OHCHOHCH_2OH$	207	46.53^{20}	289
Heptane	C_7H_{16}	197	1.9209^{20}	98.38
Hexadecane	$C_{16}H_{34}$	200	2.0460^{20}	286.9
Hexane	C_6H_{14}	210	1.8865^{20}	68.72
Methanol	CH_3OH	210	33.0^{20}	64.5
2-Methoxyethanol	$C_3H_8O_2$	210	17.2^{25}	124.3
Methylcyclohexane	C_7H_{14}	210	2.024^{20}	100.9
4-Methyl-2-pentanone	$C_6H_{12}O$	335	13.11^{20}	115.7
2-Methyl-1-propanol	$C_4H_{10}O$	230	17.93^{20}	107.84
N-Methyl-2-pyrrolidinone	C_5H_9NO	285	32.55^{20}	204.2
Nitromethane	CH_3NO_2	380	37.27^{20}	101.19
Pentane	C_5H_{12}	210	1.8371^{20}	36.06
Pentyl acetate	$C_7H_{14}O_2$	212	4.79^{20}	149.4
1-Propanol	$CH_3CH_2CH_2OH$	210	20.8^{20}	97.04
2-Propanol	$CH_3CHOHCH_3$	210	20.18^{20}	82.21
Pyridine	C_5H_5N	330	13.260^{20}	115.2
Tetrachloroethene	C_2Cl_4	290	2.268^{30}	121.2
Tetrachloromethane	CCl_4	265	2.2379^{20}	76.7
Tetrahydrofuran	C_4H_8O	220	7.52^{22}	66.0
Toluene	$C_6H_5CH_3$	286	2.379^{23}	110.60
Trichloromethane	$CHCl_3$	245	4.8069^{20}	61.2
1,1,2-Trichloro-1,2,2-trifluoroethane	$CFCl_2CF_2Cl$	231	2.41^{25}	47.6
2,2,4-Trimethylpentane	C_8H_{18}	215	1.943^{20}	99.2
Water	H_2O	191	80.1^{20}	99.974
o-Xylene	$C_6H_4(CH_3)_2$	290	2.562^{20}	144.4
m-Xylene	$C_6H_4(CH_3)_2$	290	2.359^{20}	139.1
p-Xylene	$C_6H_4(CH_3)_2$	290	2.2735^{20}	138.3

CORRELATION TABLE FOR ULTRAVIOLET ACTIVE FUNCTIONALITIES

Thomas J. Bruno and Paris D. N. Svoronos

The following tables present information about how organic compounds absorb ultraviolet (UV) radiation. While not as informative as infrared correlations, UV can often provide valuable qualitative information. Many common chromophores (that portion of a molecule that is responsible for its color) absorb UV light at characteristic wavelengths, as listed in Table 1. In some cases, extinction coefficients are given for multiple wavelengths. Column definitions for Table 1 are as follows.

Column heading	Definition
Chromophore	Name of chromophore
Functional group	Molecular formula of chromophore
λ_{max}	Wavelength at which the maximum absorption occurs, in nm
ε_{max}	Extinction coefficient at wavelength specified in column to the left

References

1. Willard, H. H., Merritt, Jr., L. L., Dean, J. A., and Settle, F. A., *Instrumental Methods of Analysis, Seventh Edition*, Wadsworth Publishing Co., Belmont, CA, 1988.
2. Silverstein, R. M., and Webster, F. X., *Spectrometric Identification of Organic Compounds, Sixth Edition*, Wiley, New York, 1998.
3. Lambert, J. B., Shurvell, H. F., Lightner D. A., Verbit, L., and Cooks, R. G., *Organic Structural Spectroscopy*, Prentice Hall, Upper Saddle River, NJ, 1998.
4. Bruno, T. J., and Svoronos, P. D. N., *CRC Handbook of Basic Tables for Chemical Analysis, Third Edition*, CRC Press, Boca Raton, FL, 2011. <https://doi.org/10.1201/b10385>
5. Woodward, R. B., *J. Am. Chem. Soc.* 63, 1123, 1941. <https://doi.org/10.1021/ja01849a066>
6. Woodward, R. B., *J. Am. Chem. Soc.* 64, 72, 1942. <https://doi.org/10.1021/ja01253a018>
7. Woodward, R. B., *J. Am. Chem. Soc.* 64, 76, 1942. <https://doi.org/10.1021/ja01253a019>
8. Fieser, L. F., and Fieser, M., *Natural Products Related to Phenanthrene, Third Edition*, Reinhold, New York, 1949.

TABLE 1. UV Wavelength Maxima and Extinction Coefficients for Different Chromophores

Chromophore	Functional group	λ_{max}/nm	ε_{max}	λ_{max}/nm	ε_{max}	λ_{max}/nm	ε_{max}
Ether	–O–	185	1000				
Thioether	–S–	194	4600	215	1600		
Amine	$-NH_2-$	195	2800				
Amide	$-CONH_2$	<210	—				
Thiol	–SH	195	1400				
Disulfide	–S–S–	194	5500	255	400		
Bromide	–Br	208	300				
Iodide	–I	260	400				
Nitrile	–C≡N	160	—				
Acetylide (alkyne)	–C≡C–	175–180	6000				
Sulfone	$-SO_2-$	180	—				
Oxime	–NOH	190	5000				
Azido	>C=N–	190	5000				
Alkene	–C=C–	190	8000				
Ketone	>C=O	195	1000	270–285	18–30		
Thioketone	>C=S	205	strong				
Esters	–COOR	205	50				
Aldehyde	–CHO	210	strong	280–300	11–18		
Carboxyl	–COOH	200–210	50–70				
Sulfoxide	>S→O	210	1500				
Nitro	$-NO_2$	210	strong				
Nitrite	–ONO	220–230	1000–2000	300–4000	10		
Azo	–N=N–	285–400	3–25				
Nitroso	–N=O	302	100				
Nitrate	$-ONO_2$	270 (shoulder)	12				
Conjugated hydrocarbon	$-(C=C)_2-$ (acyclic)	210–230	21 000				
Conjugated hydrocarbon	$-(C=C)_3-$	260	35 000				
Conjugated hydrocarbon	$-(C=C)_4-$	300	52 000				
Conjugated hydrocarbon	$-(C=C)_5-$	330	118 000				
Conjugated hydrocarbon	$-(C=C)_2-$ (alicyclic)	230–260	3000–8000				
Conjugated hydrocarbon	C=C–C≡C	219	6500				
Conjugated system	C=C–C=N	220	23 000				
Conjugated system	C=C–C=O	210–250	10 000–20 000			300–350	weak
Conjugated system	$C=C-NO_2$	229	9500				

Chromophore	Functional group	λ_{max}/nm	ε_{max}	λ_{max}/nm	ε_{max}	λ_{max}/nm	ε_{max}
Phenyl		184	46 700	202	6900	255	170
Diphenyl				246	20 000		
Naphthalene		220	112 000	275	5600	312	175
Anthracene		252	199 000	375	7900		
Pyridine		174	80 000	195	6000	251	1700
Quinoline		227	37 000	270	3600	314	2750
Isoquinoline		218	80 000	266	4000	317	3500

Wavenumber Adjustments for Bathochromic Shifts (Woodward's Rules)

Conjugated systems show bathochromic shifts (a change in the absorption spectra) in their $\pi \rightarrow \pi^*$ transition bands because a molecule has one or more structural features in addition to its base structure. Empirical methods for predicting these shifts were originally formulated by Woodward (Refs. 5–7) and Fieser and Fieser (Ref. 8). This section includes the most important conjugated system rules. The reader should consult Refs. 6 and 8 for more details on how to apply the wavelength increment data.

(a) Rules for Diene UV Absorption (Table 2)

The base UV absorption value for diene is 214 nm. Increments (in nm) for various structural components are given in Table 2.

TABLE 2. Woodward's Rules for UV Absorption by Dienes

Heteroannular diene	+ 0
Homoannular diene	+39
Extra double bond	+30
Alkyl substituent or ring residue	+5
Exocyclic double bond	+5
Polar groups:	
–OOCR	+0
–OR	+6
–S–R	+30
halogen	+5
–NR$_2$	+60
λ Calculated	= Total

(b) Rules for Enone UV Absorption (Table 3)

$$\underset{O}{\overset{\delta \quad \gamma \quad \beta \quad \alpha}{-C=C-C=C-C-}}$$

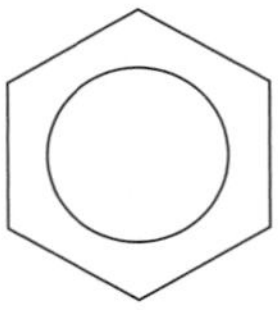

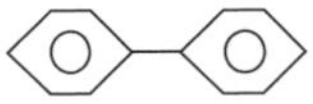

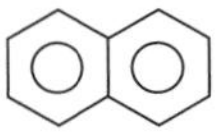

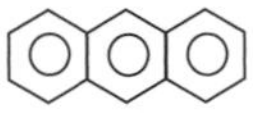

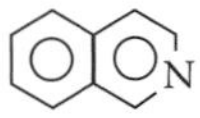

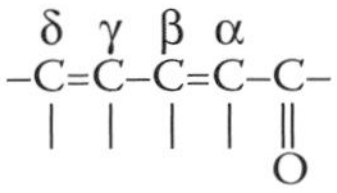

Base values for UV absorption of specific enone structures are as follows.

Structure	Base value
Acyclic (or six-membered) α,β-unsaturated ketone	215 nm
Five-membered α,β-unsaturated ketone	202 nm
α,β-Unsaturated aldehydes	210 nm
α,β-Unsaturated esters or carboxylic acids	195 nm

Solvent corrections (in nm) should be included as follows: water (-8), chloroform (+1), dioxane (+5), ether (+7), hexane (+11), and cyclohexane (+11). No correction is done for methanol or ethanol. Increments (in nm) for additional structural features in enones are given in Table 3.

TABLE 3. Woodward's Rules for UV Absorption by Enones

Heteroannular diene	+0
Homoannular diene	+39
Double bond	+30
Alkyl group:	
α–	+10
β–	+12
γ– and higher	+18
Polar groups:	
–OH	
α–	+35
β–	+30
δ–	+50
–OOCR	
$\alpha, \beta, \gamma, \delta$	+6
–OR	
α–	+35
β–	+30
γ–	+17
δ–	+31
–SR	
β–	+85
–Cl	
α–	+15

β–	+12
–Br	
α–	+25
β–	+30
$-NR_2$	
β–	+95
Exocyclic double bond	+5
λ Calculated	= Total

(c) Rules for UV Absorption by Monosubstituted Benzene Derivatives (Table 4)

The base UV absorption value for the parent chromophore (benzene) is 250 nm. Increments (in nm) for various monosubstituents on benzene are given in Table 4.

TABLE 4. Woodward's Rules for UV Absorption by Monosubstituted Benzene Derivatives

Substituent	Increment (in nm)
–R	-4
–COR	-4
–CHO	0
–OH	-16
–OR	-16
–COOR	-16

where R is an alkyl group, and the substitution is on C_6H_5-.

(d) Rules for UV Absorption by Disubstituted Benzene Derivatives (Table 5)

The base UV absorption value for the parent chromophore (benzene) is 250 nm. Increments (in nm) for various substituents in disubstituted bezene are given in Table 5.

TABLE 5. Woodward's Rules for UV Absorption by Disubstituted Benzene Derivatives

Substituent	*o*–	*m*–	*p*–
–R	+3	+3	+10
–COR	+3	+3	+10
–OH	+7	+7	+25
–OR	+7	+7	+25
$-O^-$	+11	+20	+78 (variable)
–Cl	+0	+0	+10
–Br	+2	+2	+15
$-NH_2$	+13	+13	+58
$-NHCOCH_3$	+20	+20	+45
$-NHCH_3$	—	—	+73
$-N(CH_3)_2$	+20	+20	+85

R indicates an alkyl group.

MIDDLE-RANGE INFRARED ABSORPTION CORRELATION CHARTS

Thomas J. Bruno and Paris D. N. Svoronos

The following charts provide characteristic middle-range infrared absorptions obtained from particular functional groups on molecules (Refs. 1 and 2). These include a general mid-range correlation chart, a chart for aromatic absorptions, and a chart for carbonyl moieties. Charts for near infrared absorptions and for inorganic moieties can be found in the cited references.

References

1. Bruno, T. J., and Svoronos, P. D. N., *CRC Handbook of Basic Tables for Chemical Analysis, Third Edition*, CRC Press, Boca Raton, FL, 2011.
2. Bruno, T. J., and Svoronos, P. D. N., *CRC Handbook of Fundamental Spectroscopic Correlation Charts*, CRC Press, Boca Raton, FL, 2006.

Notes:

AR = Aromatic
b = Broad
sd = Solid
sn = Solution
sp = Sharp
? = Unreliable

Strong
Medium
Weak

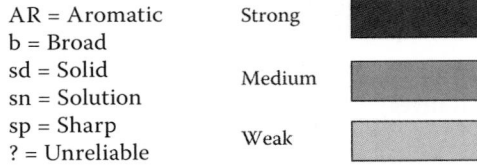

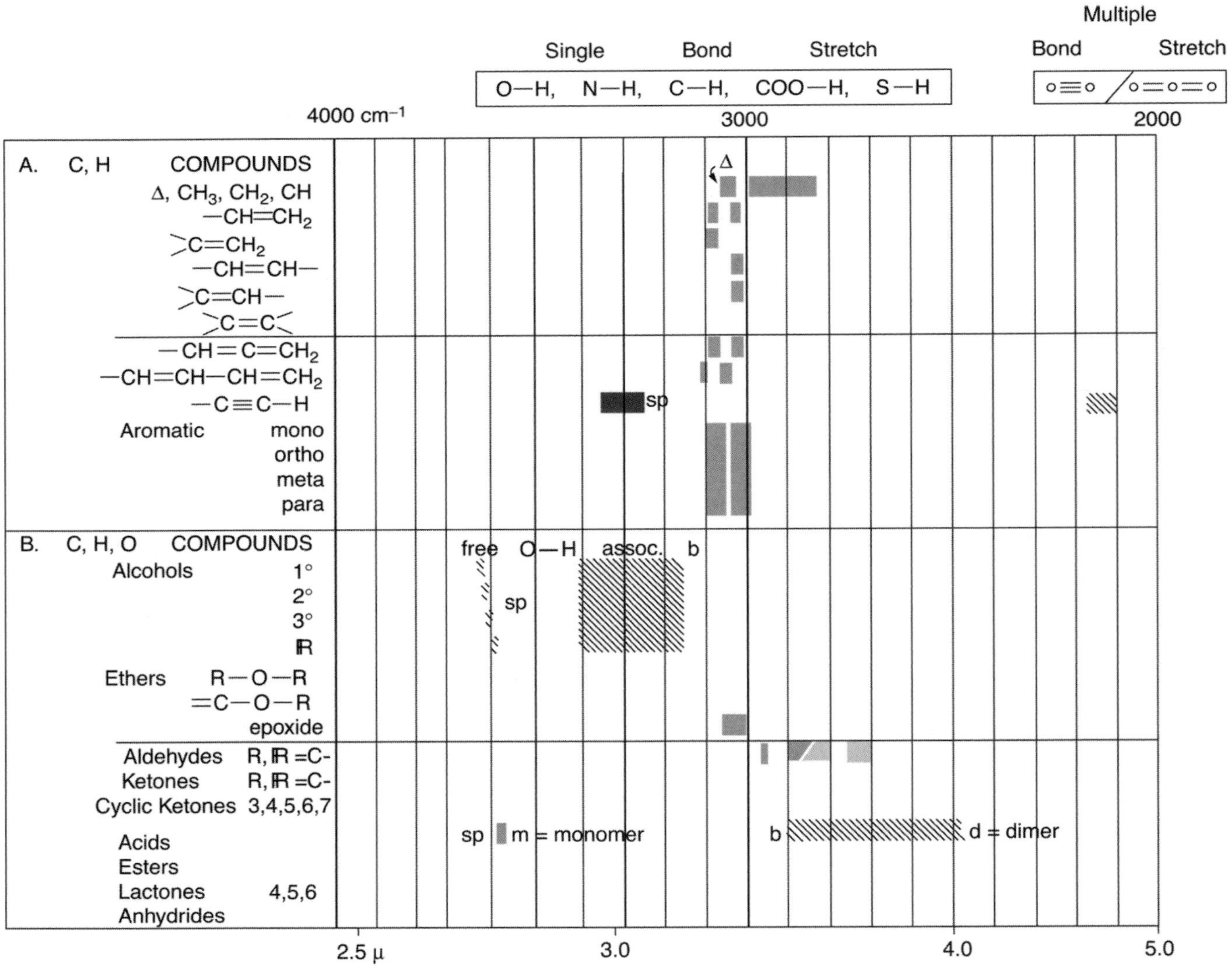

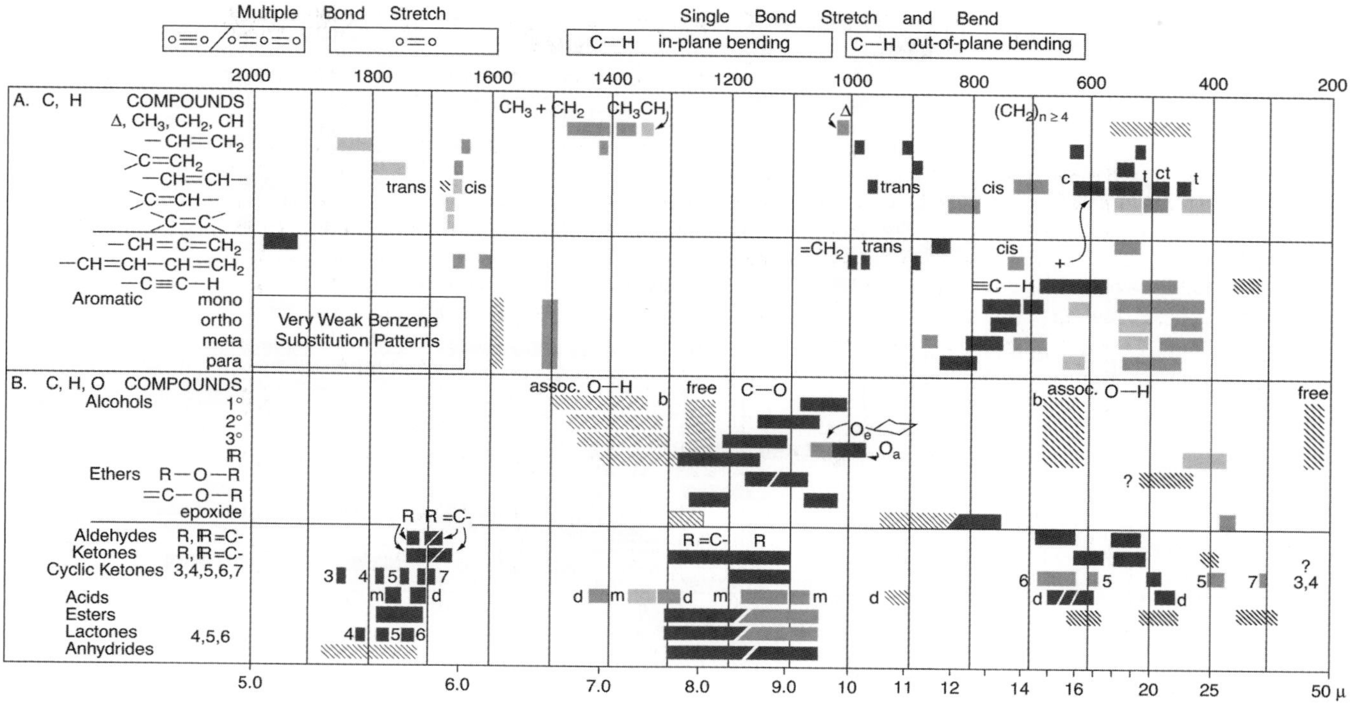
Multiple Bond Stretch
Single Bond Stretch and Bend
C—H in-plane bending
C—H out-of-plane bending
A. C, H COMPOUNDS
Δ, CH3, CH2, CH
Aromatic mono ortho meta para
Very Weak Benzene Substitution Patterns
B. C, H, O COMPOUNDS
Alcohols 1° 2° 3° R
Ethers R—O—R
epoxide
Aldehydes R, R=C-
Ketones R, R=C-
Cyclic Ketones 3,4,5,6,7
Acids
Esters
Lactones 4,5,6
Anhydrides
assoc. O—H
free
C—O
trans
cis
5.0 6.0 7.0 8.0 9.0 10 11 12 14 16 20 25 50 μ
2000 1800 1600 1400 1200 1000 800 600 400 200

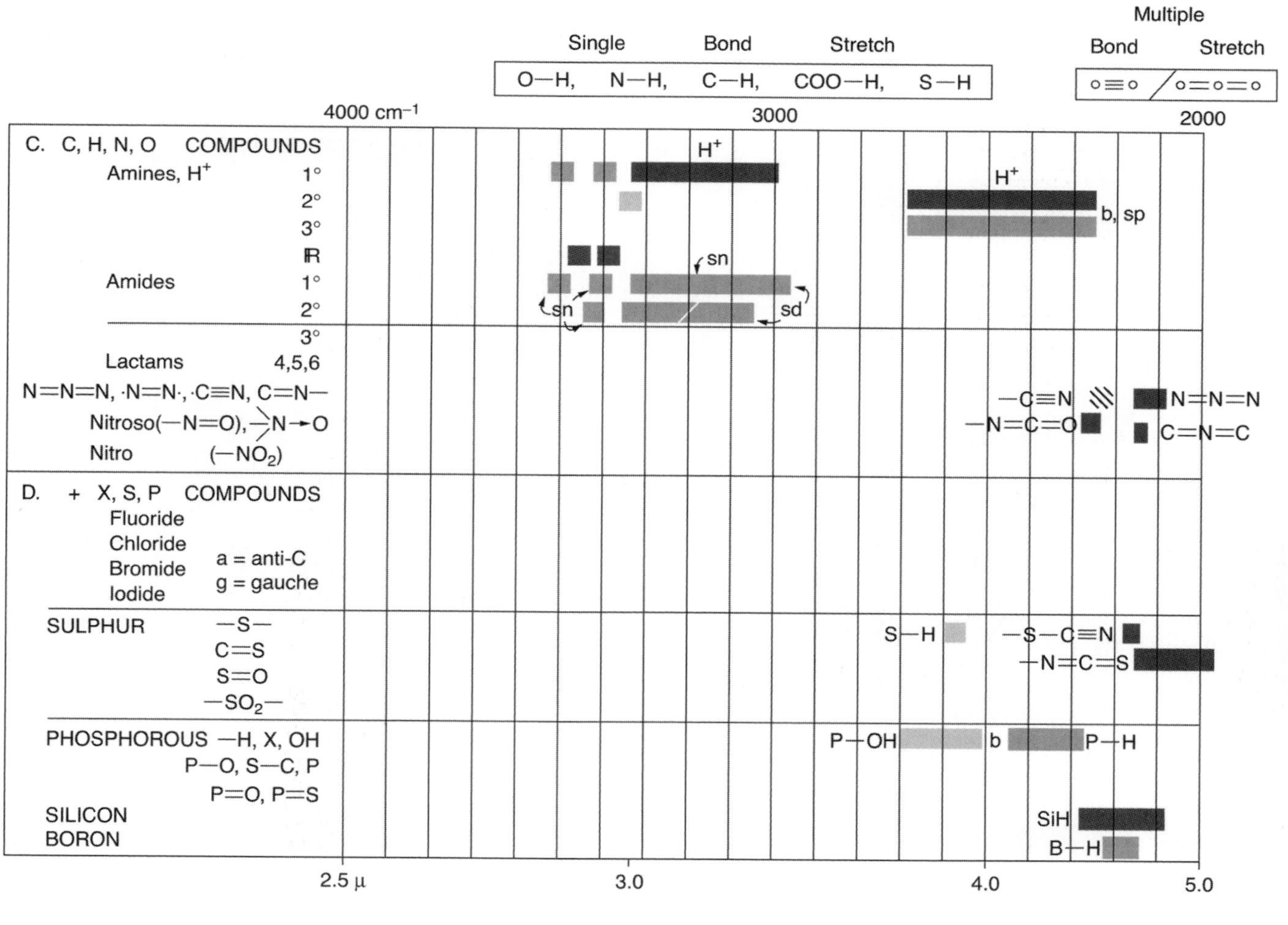
Single Bond Stretch
O—H, N—H, C—H, COO—H, S—H
Multiple Bond Stretch
4000 cm⁻¹ 3000 2000
C. C, H, N, O COMPOUNDS
Amines, H+ 1° 2° 3° R
Amides 1° 2° 3°
Lactams 4,5,6
N=N=N, ·N=N·, ·C≡N, C=N—
Nitroso(—N=O), N→O
Nitro (—NO2)
b, sp
sn
sd
—C≡N
—N=C=O
N=N=N
C=N=C
D. + X, S, P COMPOUNDS
Fluoride
Chloride
Bromide
Iodide
a = anti-C
g = gauche
SULPHUR —S— C=S S=O —SO2—
S—H
—S—C≡N
—N=C=S
PHOSPHOROUS —H, X, OH
P—O, S—C, P
P=O, P=S
P—OH
P—H
SILICON
BORON
SiH
B—H
2.5 μ 3.0 4.0 5.0

Analytical

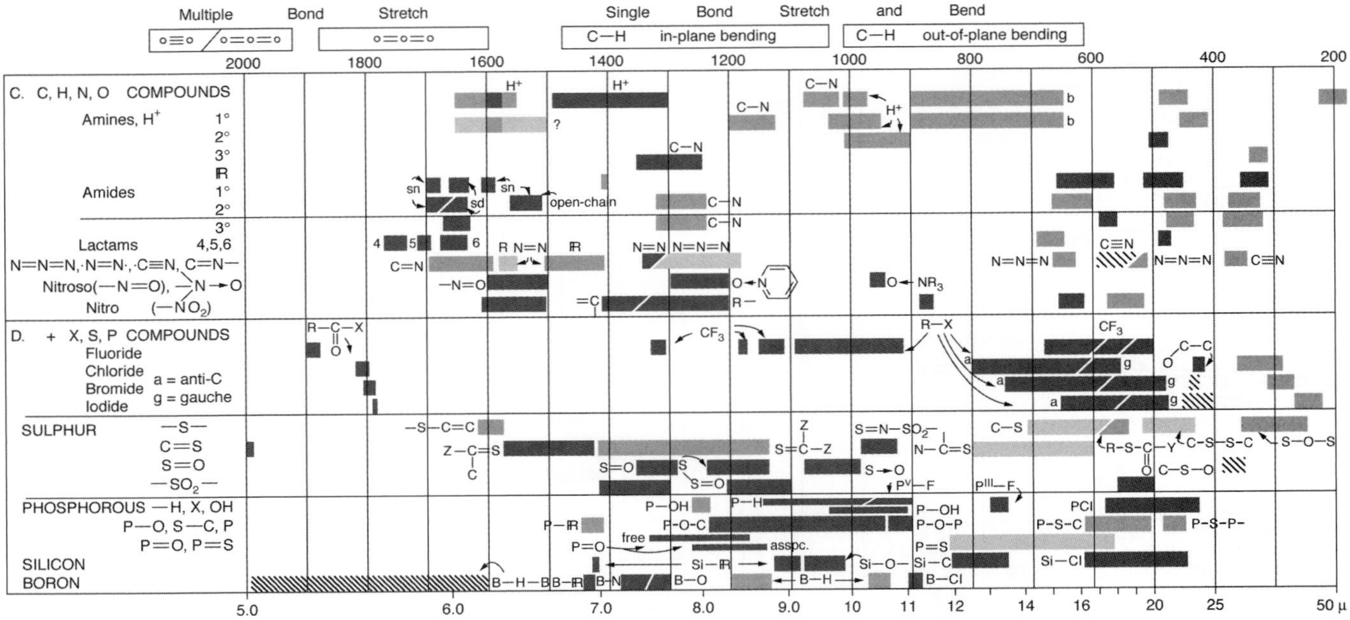

Aromatic Substitution Bands

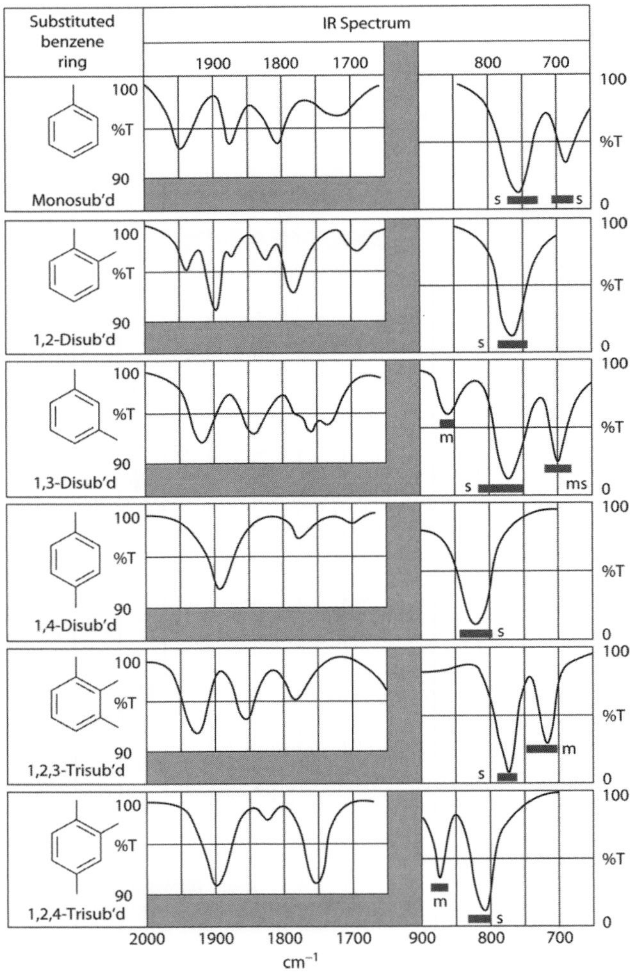

Aromatic Substitution Bands

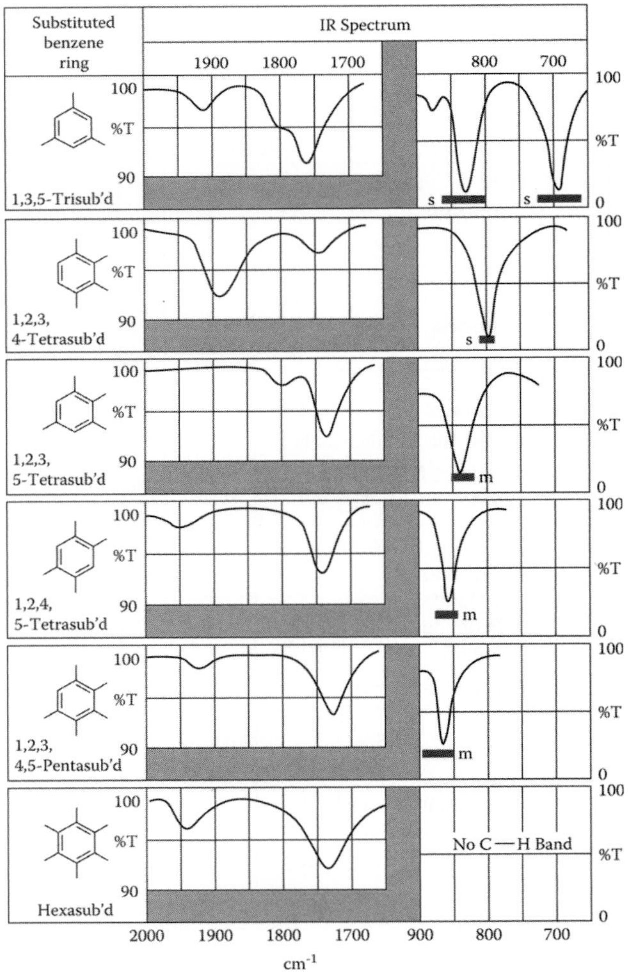

Carbonyl Group Absorptions

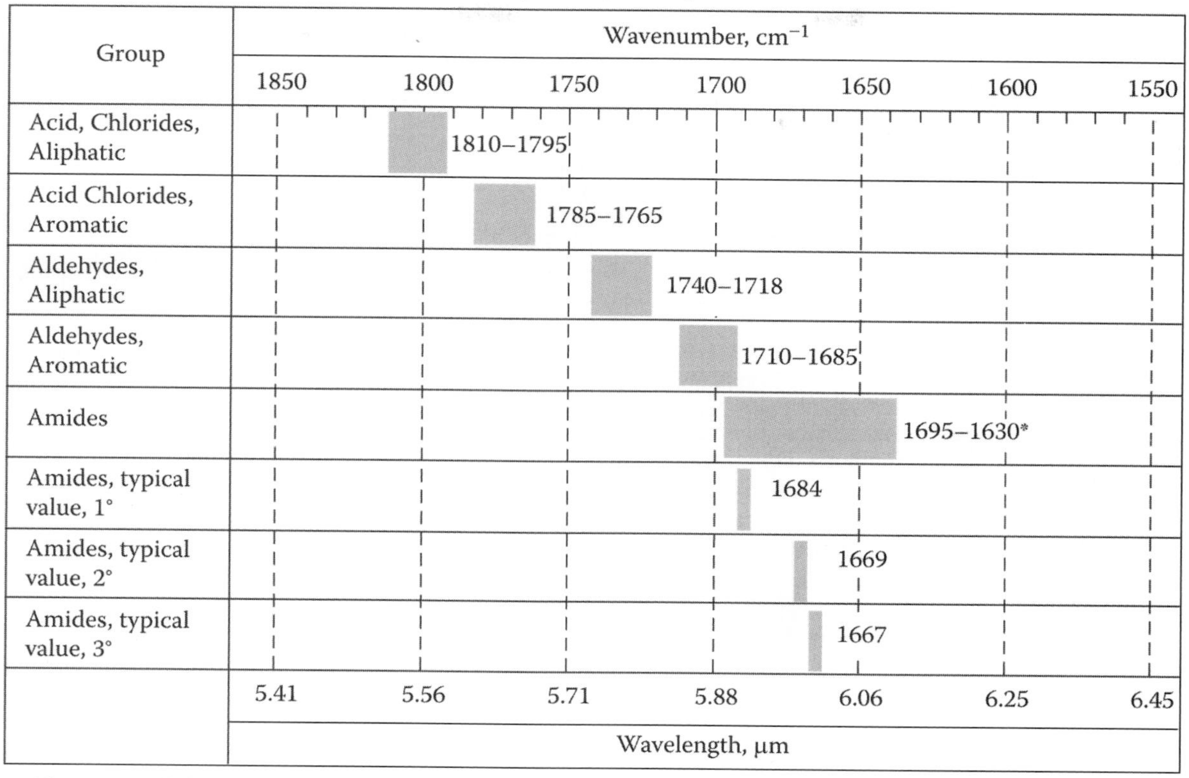
Group
Wavenumber, cm−1
1850
1800
1750
1700
1650
1600
1550
Acid, Chlorides, Aliphatic
1810–1795
Acid Chlorides, Aromatic
1785–1765
Aldehydes, Aliphatic
1740–1718
Aldehydes, Aromatic
1710–1685
Amides
1695–1630*
Amides, typical value, 1°
1684
Amides, typical value, 2°
1669
Amides, typical value, 3°
1667
5.41
5.56
5.71
5.88
6.06
6.25
6.45
Wavelength, μm

* Electron withdrawing groups at the α-position to the carbonyl will raise the wavenumber of the absorption.

Carbonyl Group Absorptions (continued)

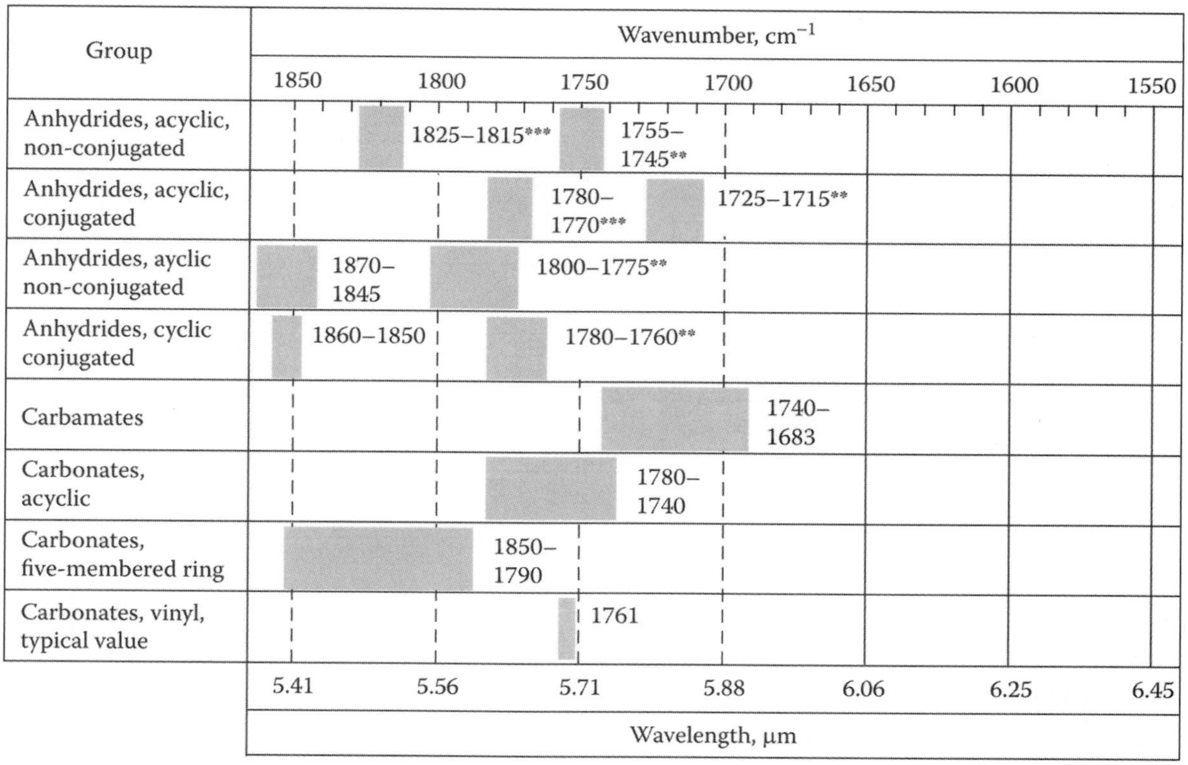
Group
Wavenumber, cm−1
1850
1800
1750
1700
1650
1600
1550
Anhydrides, acyclic, non-conjugated
1825–1815***
1755–1745**
Anhydrides, acyclic, conjugated
1780–1770***
1725–1715**
Anhydrides, ayclic non-conjugated
1870–1845
1800–1775**
Anhydrides, cyclic conjugated
1860–1850
1780–1760**
Carbamates
1740–1683
Carbonates, acyclic
1780–1740
Carbonates, five-membered ring
1850–1790
Carbonates, vinyl, typical value
1761
5.41
5.56
5.71
5.88
6.06
6.25
6.45
Wavelength, μm

** This band is the more intense of the two.
*** Intensity weakens as colinearity is approached.

Carbonyl Group Absorptions (continued)

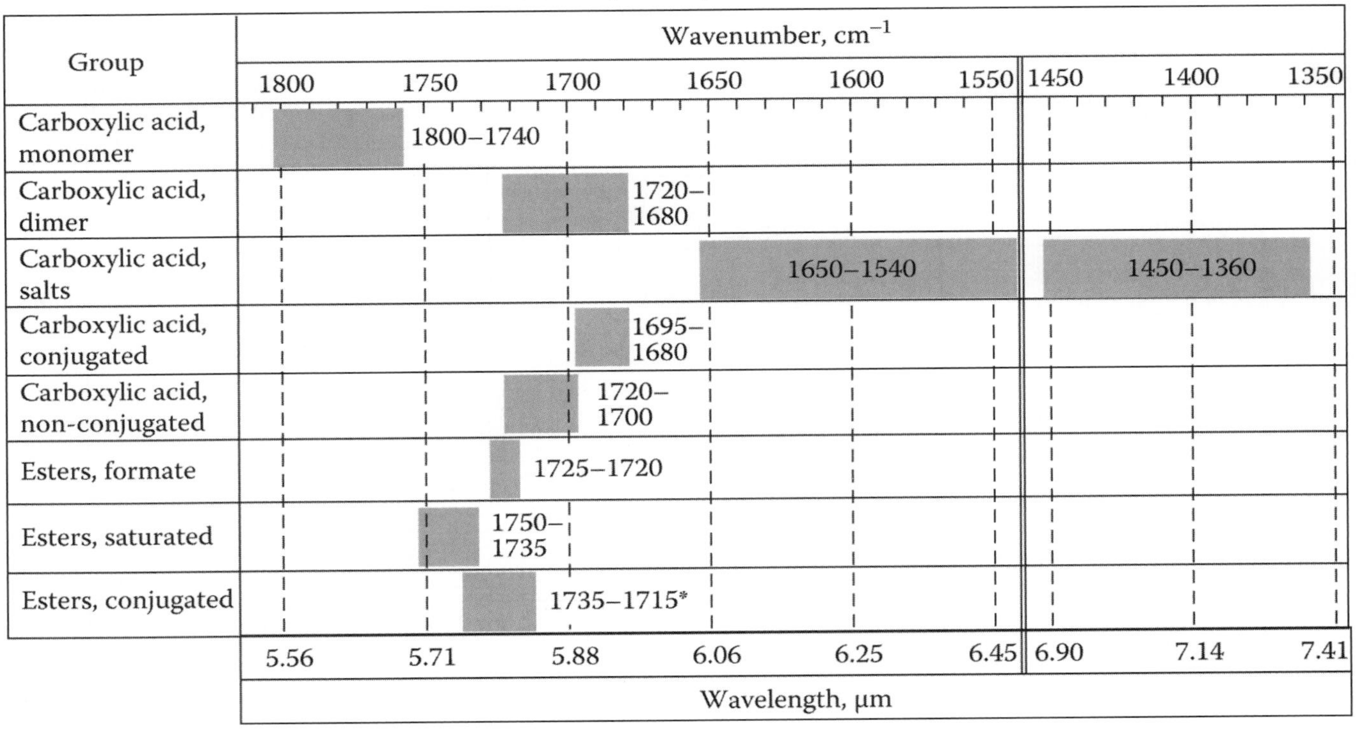
Group
Wavenumber, cm^{-1}
1800 1750 1700 1650 1600 1550 1450 1400 1350
Carboxylic acid, monomer
1800–1740
Carboxylic acid, dimer
1720–1680
Carboxylic acid, salts
1650–1540
1450–1360
Carboxylic acid, conjugated
1695–1680
Carboxylic acid, non-conjugated
1720–1700
Esters, formate
1725–1720
Esters, saturated
1750–1735
Esters, conjugated
1735–1715*
5.56 5.71 5.88 6.06 6.25 6.45 6.90 7.14 7.41
Wavelength, μm

* Electron withdrawing groups in the α-position to the carbonyl will raise the wavenumber adsorption.

Carbonyl Group Absorptions (continued)

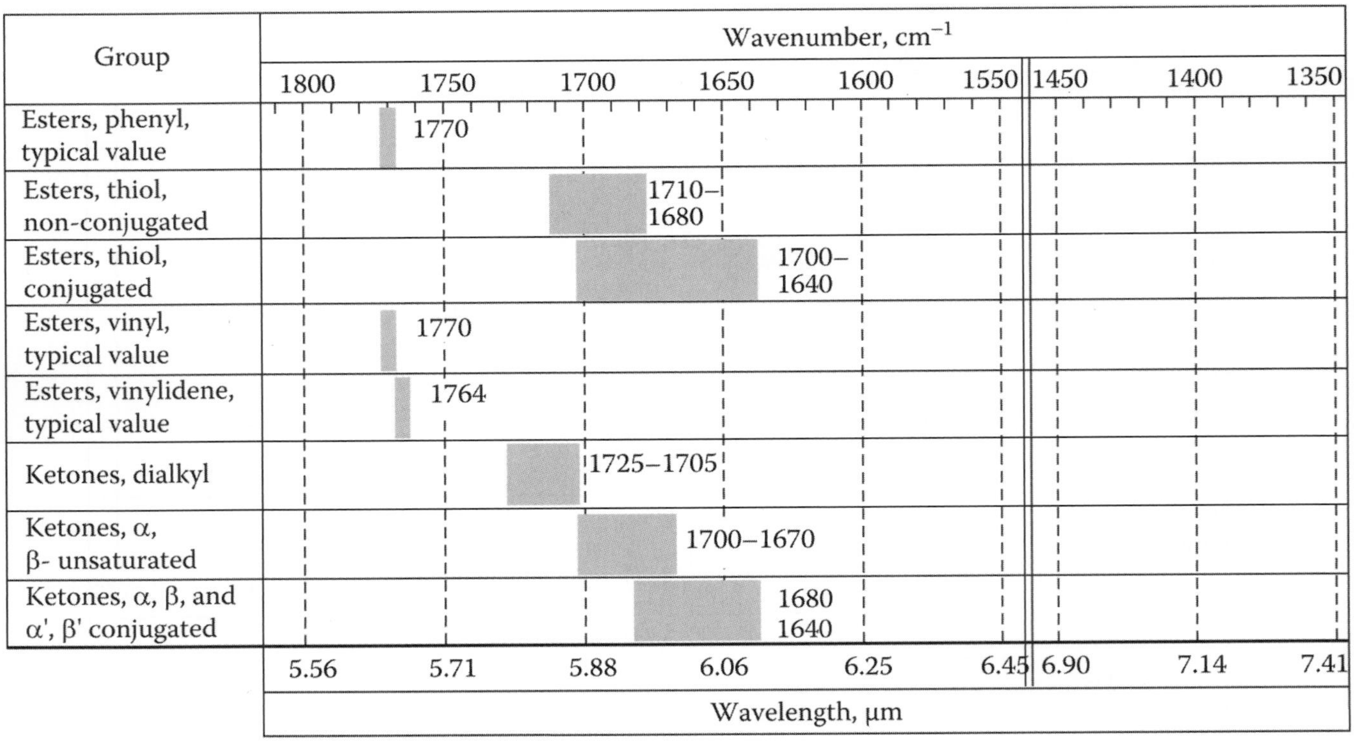
Group
Wavenumber, cm^{-1}
1800 1750 1700 1650 1600 1550 1450 1400 1350
Esters, phenyl, typical value
1770
Esters, thiol, non-conjugated
1710–1680
Esters, thiol, conjugated
1700–1640
Esters, vinyl, typical value
1770
Esters, vinylidene, typical value
1764
Ketones, dialkyl
1725–1705
Ketones, α, β- unsaturated
1700–1670
Ketones, α, β, and α', β' conjugated
1680 1640
5.56 5.71 5.88 6.06 6.25 6.45 6.90 7.14 7.41
Wavelength, μm

Analytical

Carbonyl Group Absorptions (continued)

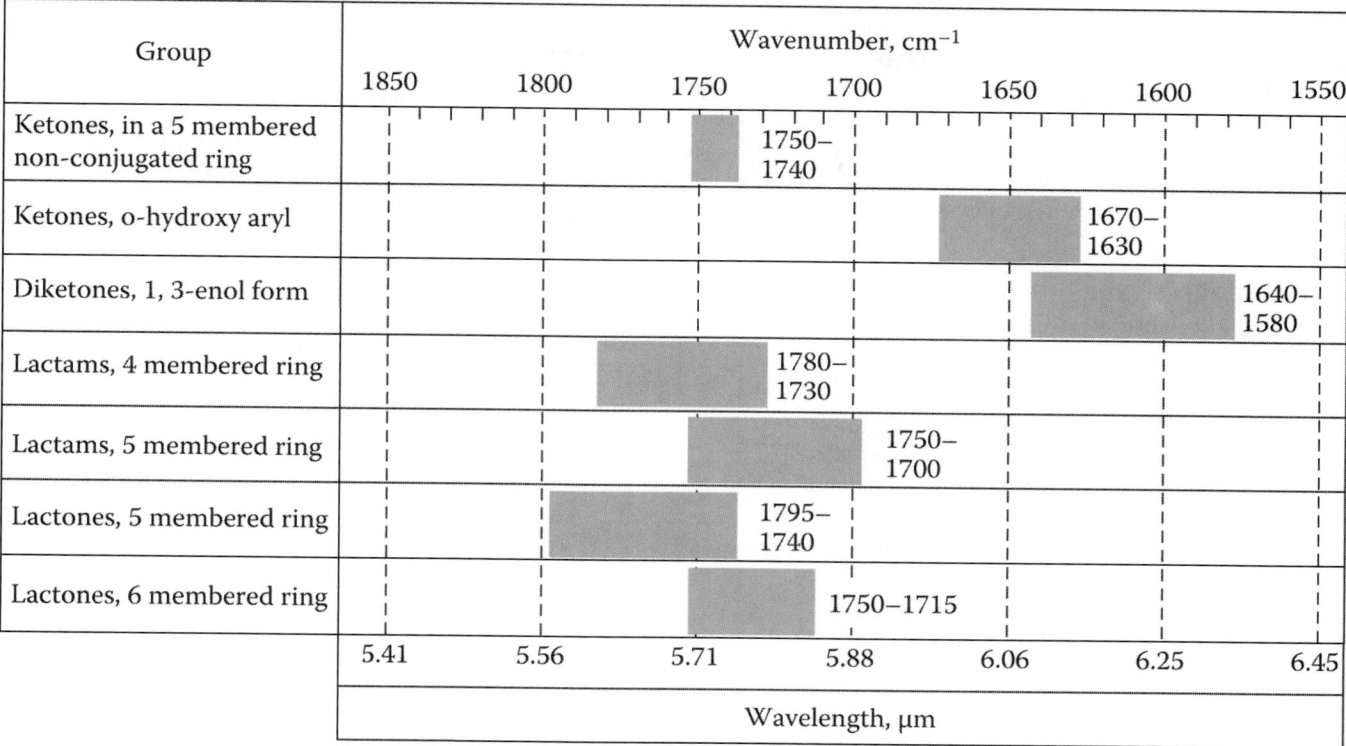

Group	Wavenumber, cm^{-1}
Ketones, in a 5 membered non-conjugated ring	1750–1740
Ketones, o-hydroxy aryl	1670–1630
Diketones, 1, 3-enol form	1640–1580
Lactams, 4 membered ring	1780–1730
Lactams, 5 membered ring	1750–1700
Lactones, 5 membered ring	1795–1740
Lactones, 6 membered ring	1750–1715

Wavenumber axis, cm^{-1}: 1850, 1800, 1750, 1700, 1650, 1600, 1550

Wavelength axis, μm: 5.41, 5.56, 5.71, 5.88, 6.06, 6.25, 6.45

COMMON SPURIOUS INFRARED ABSORPTION BANDS

Thomas J. Bruno and Paris D. N. Svoronos

The following table provides some of the common potential sources of spurious infrared absorptions that might appear on a spectrum (Refs. 1 and 2). Occasionally, the spectral lines of some impurities can be used as diagnostics; the reader is referred to the references for more details. Column definitions for the table are as follows.

Column heading	Definition
Approximate wavenumber	Approximate wavenumber of spurious infrared absorption, in cm^{-1}
Wavelength	Wavelength of spurious infrared absorption, in μm
Compound or group	Molecular formula of compound or group that is the source of spurious absorption
Origin	Possible origin of the compound or group

References

1. Bruno, T. J., and Svoronos, P. D. N., *CRC Handbook of Basic Tables for Chemical Analysis, Third Edition*, CRC Press, Boca Raton, FL, 2011. <https://doi.org/10.1201/b10385>
2. Bruno, T. J., and Svoronos, P. D. N., *CRC Handbook of Fundamental Spectroscopic Correlation Charts*, CRC Press, Boca Raton, FL, 2006. <https://doi.org/10.1201/9781420037685>

Common Spurious Infrared Absorption Bands

Approximate wavenumber in cm^{-1}	Wavelength in μm	Compound or group	Origin
3700	2.70	H_2O	Water in solvent (thick layers)
3650	2.74	H_2O	Water in some quartz windows
3450	2.9	H_2O	Hydrogen-bonded water, usually in KBr disks
2900	3.44	$-CH_3$, $=CH_2$	Paraffin oil, residual from previous mulls
2350	4.26	CO_2	Atmospheric absorption, or dissolved gas from a dry ice bath
2330	4.30	CO_2	
2300 and 2150	4.35 and 4.65	CS_2	Leaky cells, previous analysis of samples dissolved in carbon disulfide
1996	5.01	BO_2^-	Metaborate in the halide window
1400–2000	5–7	H_2O	Atmospheric absorption
1820	5.52	$COCl_2$	Phosgene, decomposition product in purified $CHCl_3$
1755	5.7	Phthalic anhydride	Decomposition product of phthalate esters or resins; paint off-gas product
1700–1760	5.7–5.9	=C=O	Bottle-cap liners leached by sample
1720	5.8	Phthalates	Phthalate polymer plastic tubing
1640	6.1	H_2O	Water of crystallization entrenched in sample
1520	6.6	CO_2	Leaky cells, previous analysis
1430	7.0	CO_3^{-2}	Contaminant in halide window
1360	7.38	NO_3^-	Contaminant in halide window
1270	7.9	=SiO–	Silicone oil or grease
1000–1110	9–10	≡Si–O–Si≡	Glass; silicones
980	10.2	SO_4^{-2}	From decomposition of sulfates in KBr pellets
935	10.7	$(CH_2O)_x$	Deposit from gaseous formaldehyde
907	11.02	≡C–Cl	Dissolved R-12 (Freon-12, CCl_2F_2)
837	11.95	NO_3^-	Contaminant in halide window
823	12.15	KNO_3	From decomposition of nitrates in KBr pellets
794	12.6	CCl_4 vapor	Leaky cells, from CCl_4 used as a solvent
788	12.7	CCl_4 liquid	Incomplete drying of cell or contamination, from CCl_4 used as a solvent
720 and 730	13.7 and 13.9	Polyethylene	Various experimental sources
728	13.75	≡Si–F	SiF_4, found in NaCl windows
667	14.98	CO_3^{-2}	Atmospheric carbon dioxide
Any	Any	Fringes	If refractive index of windows is too high, or if the cell is partially empty, or the solid sample is not fully pulverized

NUCLEAR SPINS, MOMENTS, AND OTHER DATA RELATED TO NMR SPECTROSCOPY

David R. Lide

This table presents the following data relevant to nuclear magnetic resonance spectroscopy. Column definitions for the table are as follows.

Column heading	Definition
Z	Atomic number
Isotope	Element symbol and mass number
Percent abundance	Natural abundance of the isotope in percent. An * indicates a radioactive nuclide; if no value is given, the nuclide is not present in nature or its abundance is highly variable
I	Nuclear spin
ν	Resonant frequency in megahertz for an applied field H_0 of 1 tesla (in cgs units, 10 kilogauss). The resonant frequency scales with H_0
Rel. sens.	Sensitivity relative to ^{1}H (=1) assuming an equal number of nuclei and constant temperature. Values were calculated from the expressions: For constant H_0: $0.0076508(\mu/\mu_N)^3(I+1)/I^2$ For constant ν: $0.23871(\mu/\mu_N)(I+1)$
μ	Nuclear magnetic moment, in units of the nuclear magneton μ_N
Q	Nuclear quadrupole moment in units of femtometers squared (1 fm^2 = 10^{-2} barn). Because the determination of quadrupole moments requires knowledge of the electron configuration near the nucleus, values of Q in the literature tend to scatter considerably. The values quoted here come mainly from the review of Pyykkö (Ref. 3), otherwise from Ref. 1

The table includes all stable nuclides of non-zero spin for which spin and magnetic moment values have been measured, as well as selected radioactive nuclides of current or potential interest. At least one isotope is included for each element through Z = 95 for which data are available. See Ref. 1 for a complete listing of spins and moments.

The assistance of P. Pyykkö in providing data on nuclear quadrupole moments is gratefully acknowledged.

References

1. Holden, N. E., "Table of the Isotopes," in *CRC Handbook of Chemistry and Physics, 98th Edition*, Rumble, J. R., Ed., CRC Press, Boca Raton, FL, 2017.
2. Raghavan, P., *At. Data Nucl. Data Tables* 42, 189, 1989. <https://doi.org/10.1016/0092-640X(89)90008-9>
3. Pyykkö, P., *Mol. Phys.* 106, 1965, 2008. <https://doi.org/10.1080/00268970802018367>
4. Stone, N. J., *At. Data Nucl. Data Tables* 90, 75, 2005. <https://doi.org/10.1016/j.adt.2005.04.001>
5. IUPAC Commission on Physiochemical Symbols, Terminology and Units, *Quantities, Units, and Symbols in Physical Chemistry, Third Edition*, Royal Society of Chemistry, Cambridge, 2007.

Nuclear Spins, Moments, and Other Useful NMR Data

Z	Isotope	Percent abundance	I	ν/MHz for H_0 = 1 T	Rel. sens. at const. H_0	Rel. sens. at const. ν	μ/μ_N	Q/fm^2
1	^{1}n		1/2	29.1647	0.32139	0.6850	-1.91304272	
1	^{1}H	99.885	1/2	42.5759	1.00000	1.0000	+2.792847337	
1	^{2}H	0.0115	1	6.53566	0.00965	0.4094	+0.857438228	+0.2860
1	^{3}H	*	1/2	45.4129	1.21354	1.0667	+2.9789625	
2	^{3}He	0.000134	1/2	32.4380	0.44220	0.7619	-2.127750	
3	^{6}Li	7.59	1	6.2661	0.00850	0.3925	+0.8220467	-0.0808
3	^{7}Li	92.41	3/2	16.5483	0.29356	1.9434	+3.25644	-4.01
4	^{9}Be	100	3/2	5.9842	0.01388	0.7028	-1.1776	+5.288
5	^{10}B	19.9	3	4.5752	0.01985	1.7193	+1.800645	+8.459
5	^{11}B	80.1	3/2	13.6630	0.16522	1.6045	+2.688649	+4.059
6	^{13}C	1.07	1/2	10.7054	0.01591	0.2515	+0.7024118	
7	^{14}N	99.636	1	3.0756	0.00101	0.1928	+0.4037610	+2.044
7	^{15}N	0.364	1/2	4.3142	0.00104	0.1014	-0.2831888	
8	^{17}O	0.038	5/2	5.772	0.02910	1.5822	-1.89379	-2.558
9	^{19}F	100	1/2	40.0541	0.83400	0.9413	+2.628868	
10	^{21}Ne	0.27	3/2	3.3631	0.00246	0.3949	-0.661797	+10.155
11	^{23}Na	100	3/2	11.2688	0.09270	1.3234	+2.217522	+10.4
12	^{25}Mg	10.00	5/2	2.6083	0.00268	0.7147	-0.85545	+19.94
13	^{27}Al	100	5/2	11.1031	0.20689	3.0424	+3.641507	+14.66
14	^{29}Si	4.685	1/2	8.4578	0.00786	0.1988	-0.55529	
15	^{31}P	100	1/2	17.235	0.06652	0.4052	+1.13160	
16	^{33}S	0.75	3/2	3.2654	0.00227	0.3842	+0.6438212	-6.78

Z	Isotope	Percent abundance	I	ν/MHz for H_0 = 1 T	Rel. sens. at const. H_0	Rel. sens. at const. ν	μ/μ_N	Q/fm^2
17	^{35}Cl	75.76	3/2	4.1717	0.00472	0.4905	+0.8218743	-8.165
17	^{37}Cl	24.24	3/2	3.4765	0.00272	0.4083	+0.6841236	-6.435
18	^{37}Ar	*	3/2	5.819	0.01276	0.6833	+1.145	+7.6
18	^{39}Ar	*	7/2	3.46	0.01130	1.7080	-1.59	-12
19	^{39}K	93.2581	3/2	1.9893	0.00051	0.2336	+0.3914662	+5.85
19	^{40}K	0.0117	4	2.4737	0.00523	1.5493	-1.298100	-7.3
19	^{41}K	6.7302	3/2	1.0919	0.00008	0.1282	+0.2148701	+7.11
20	^{43}Ca	0.135	7/2	2.8697	0.00643	1.4154	-1.317643	-4.08
21	^{45}Sc	100	7/2	10.3591	0.30244	5.1094	+4.756487	-22.0
22	^{47}Ti	7.44	5/2	2.4041	0.00210	0.6588	-0.78848	+30.2
22	^{49}Ti	5.41	7/2	2.4048	0.00378	1.1861	-1.10417	+24.7
23	^{50}V	0.250	6	4.2505	0.05571	5.5905	+3.345689	+21
23	^{51}V	99.750	7/2	11.2133	0.38360	5.5307	+5.1487057	-5.2
24	^{53}Cr	9.501	3/2	2.4115	0.00091	0.2832	-0.47454	-15
25	^{55}Mn	100	5/2	10.5763	0.17881	2.8981	+3.46872	+33
26	^{57}Fe	2.119	1/2	1.3816	0.00003	0.0324	+0.0906230	+16
27	^{59}Co	100	7/2	10.077	0.27841	4.9703	+4.627	+42
28	^{61}Ni	1.1399	3/2	3.8114	0.00359	0.4476	-0.75002	+16.2
29	^{63}Cu	69.15	3/2	11.3188	0.09393	1.3292	+2.2273456	-22.0
29	^{65}Cu	30.85	3/2	12.1027	0.11484	1.4213	+2.38161	-20.4
30	^{67}Zn	4.102	5/2	2.6685	0.00287	0.7312	+0.875205	+15.0
31	^{69}Ga	60.108	3/2	10.2478	0.06971	1.2035	+2.01659	+17.1
31	^{71}Ga	39.892	3/2	13.0208	0.14300	1.5291	+2.56227	+10.7
32	^{73}Ge	7.76	9/2	1.4897	0.00141	1.1547	-0.8794677	-19.6
33	^{75}As	100	3/2	7.3150	0.02536	0.8590	+1.439475	+31.4
34	^{77}Se	7.63	1/2	8.1568	0.00703	0.1916	+0.5350422	
35	^{79}Br	50.69	3/2	10.667	0.07945	1.2570	+2.106400	+31.3
35	^{81}Br	49.31	3/2	11.498	0.09951	1.3550	+2.270562	+26.2
36	^{83}Kr	11.500	9/2	1.6442	0.00190	1.2744	-0.970669	+25.9
37	^{85}Rb	72.17	5/2	4.1253	0.01061	1.1304	+1.35298	+27.6
37	^{87}Rb	27.83	3/2	13.9814	0.17704	1.6419	+2.75131	+13.35
38	^{87}Sr	7.00	9/2	1.8525	0.00272	1.4358	-1.093603	+30.5
39	^{89}Y	100	1/2	2.0949	0.00012	0.0492	-0.1374154	
40	^{91}Zr	11.22	5/2	3.9748	0.00949	1.0892	-1.30362	-17.6
41	^{93}Nb	100	9/2	10.4523	0.48821	8.1013	+6.1705	-32
42	^{95}Mo	15.90	5/2	2.7874	0.00327	0.7638	-0.9142	-2.2
42	^{97}Mo	9.56	5/2	2.8463	0.00349	0.7799	-0.9335	+25.5
43	^{99}Tc	*	9/2	9.6294	0.38174	7.4635	+5.6847	-12.9
44	^{99}Ru	12.76	5/2	1.9553	0.00113	0.5358	-0.6413	+7.9
44	^{101}Ru	17.06	5/2	2.1916	0.00159	0.6005	-0.7188	+45.7
45	^{103}Rh	100	1/2	1.3477	0.00003	0.0317	-0.08840	
46	^{105}Pd	22.33	5/2	1.957	0.00113	0.5364	-0.642	+66.0
47	^{107}Ag	51.839	1/2	1.7331	0.00007	0.0407	-0.1136796	
47	^{109}Ag	48.161	1/2	1.9924	0.00010	0.0468	-0.1306906	
48	^{111}Cd	12.80	1/2	9.0692	0.00966	0.2130	-0.5948861	
48	^{113}Cd	12.22	1/2	9.4871	0.01106	0.2228	-0.6223009	
49	^{113}In	4.29	9/2	9.3655	0.35121	7.2589	+5.5289	+75.9
49	^{115}In	95.71	9/2	9.3856	0.35348	7.2745	+5.5408	+77.0
50	^{115}Sn	0.34	1/2	14.0077	0.03561	0.3290	-0.91883	
50	^{117}Sn	7.68	1/2	15.2610	0.04605	0.3584	-1.00104	
50	^{119}Sn	8.59	1/2	15.9660	0.05273	0.3750	-1.04728	
51	^{121}Sb	57.21	5/2	10.2551	0.16302	2.8101	+3.3634	-54.3
51	^{123}Sb	42.79	7/2	5.5532	0.04659	2.7390	+2.5498	-69.2
52	^{123}Te	0.89	1/2	11.2349	0.01837	0.2639	-0.7369478	
52	^{125}Te	7.07	1/2	13.5446	0.03219	0.3181	-0.8884509	
53	^{127}I	100	5/2	8.5778	0.09540	2.3504	+2.813273	-69.6
54	^{129}Xe	26.4006	1/2	11.8604	0.02162	0.2786	-0.7779763	
54	^{131}Xe	21.2324	3/2	3.5159	0.00282	0.4129	+0.6918619	-11.4
55	^{133}Cs	100	7/2	5.6234	0.04838	2.7736	+2.582025	-0.343
56	^{135}Ba	6.592	3/2	4.2617	0.00501	0.5005	+0.838627	+16.0

Z	Isotope	Percent abundance	I	ν/MHz for H_0 = 1 T	Rel. sens. at const. H_0	Rel. sens. at const. ν	μ/μ_N	Q/fm²
56	^{137}Ba	11.232	3/2	4.7634	0.00700	0.5594	+0.937365	+24.5
57	^{138}La	0.090	5	5.6615	0.09404	5.3189	+3.713646	+45
57	^{139}La	99.910	7/2	6.0612	0.06058	2.9895	+2.7830455	+20.0
58	^{137}Ce	*	3/2	4.88	0.00752	0.5729	0.96	
58	^{139}Ce	*	3/2	5.39	0.01012	0.6326	1.06	
58	^{141}Ce	*	7/2	2.37	0.00364	1.1709	1.09	
59	^{141}Pr	100	5/2	13.0359	0.33483	3.5720	+4.2754	-5.9
60	^{143}Nd	12.2	7/2	2.319	0.00339	1.1440	-1.065	-63
60	^{145}Nd	8.3	7/2	1.429	0.00079	0.7047	-0.656	-33
61	^{143}Pm	*	5/2	11.59	0.23510	3.1748	+3.80	
61	^{147}Pm	*	7/2	5.62	0.04827	2.7714	+2.58	+74
62	^{147}Sm	14.99	7/2	1.7748	0.00152	0.8754	-0.8149	-26
62	^{149}Sm	13.82	7/2	1.4631	0.00085	0.7216	-0.6718	+7.4
63	^{151}Eu	47.81	5/2	10.5856	0.17929	2.9006	+3.4718	+90.3
63	^{153}Eu	52.19	5/2	4.6745	0.01544	1.2809	+1.5331	+241
64	^{155}Gd	14.80	3/2	1.312	0.00015	0.1541	-0.2582	+127
64	^{157}Gd	15.65	3/2	1.720	0.00033	0.2020	-0.3385	+135
65	^{159}Tb	100	3/2	10.23	0.06945	1.2019	+2.014	+143.2
66	^{161}Dy	18.889	5/2	1.4654	0.00048	0.4015	-0.4806	+250.7
66	^{163}Dy	24.896	5/2	2.0508	0.00130	0.5619	+0.6726	+265
67	^{165}Ho	100	7/2	9.0883	0.20423	4.4826	+4.173	+358
68	^{167}Er	22.869	7/2	1.2281	0.00050	0.6057	-0.5639	+356.5
69	^{169}Tm	100	1/2	3.531	0.00057	0.0829	-0.2316	-120
70	^{171}Yb	14.28	1/2	7.5261	0.00552	0.1768	+0.49367	
70	^{173}Yb	16.13	5/2	2.0730	0.00135	0.5680	-0.67989	+280
71	^{175}Lu	97.41	7/2	4.8626	0.03128	2.3984	+2.2327	+349
71	^{176}Lu	2.59	7	3.451	0.03975	6.0518	+3.169	+497
72	^{177}Hf	18.60	7/2	1.7282	0.00140	0.8524	+0.7935	+336.5
72	^{179}Hf	13.62	9/2	1.0856	0.00055	0.8414	-0.6409	+379.3
73	^{181}Ta	99.988	7/2	5.1627	0.03744	2.5464	+2.3705	+317
74	^{183}W	14.31	1/2	1.7716	0.00008	0.0422	+0.1177848	
75	^{185}Re	37.40	5/2	9.7176	0.13870	2.6628	+3.1871	+218
75	^{187}Re	62.60	5/2	9.8170	0.14300	2.6900	+3.2197	+207
76	^{187}Os	1.96	1/2	0.9856	0.00001	0.0231	+0.06465189	
76	^{189}Os	16.15	3/2	3.3536	0.00244	0.3938	+0.659933	+85.6
77	^{191}Ir	37.3	3/2	0.7658	0.00003	0.0899	+0.1507	+81.6
77	^{193}Ir	62.7	3/2	0.8319	0.00004	0.0977	+0.1637	+75.1
78	^{195}Pt	33.832	1/2	9.2922	0.01039	0.2182	+0.60952	
79	^{197}Au	100	3/2	0.7406	0.00003	0.0870	+0.145746	+54.7
80	^{199}Hg	16.87	1/2	7.7123	0.00594	0.1811	+0.5058855	
80	^{201}Hg	13.18	3/2	2.8469	0.00149	0.3343	-0.5602257	+38.7
81	^{203}Tl	29.52	1/2	24.7316	0.19598	0.5809	+1.6222579	
81	^{205}Tl	70.48	1/2	24.9749	0.20182	0.5866	+1.6382146	
82	^{207}Pb	22.1	1/2	9.0340	0.00955	0.2122	+0.59258	
83	^{209}Bi	100	9/2	6.9630	0.14433	5.3968	+4.1106	-51.6
84	^{209}Po	*	1/2	11.7	0.02096	0.2757	+0.77	
86	^{211}Rn	*	1/2	9.16	0.00997	0.2152	+0.601	
87	^{223}Fr	*	3/2	5.95	0.01362	0.6982	+1.17	+117
88	^{223}Ra	*	3/2	1.3746	0.00017	0.1614	+0.2705	+121
88	^{225}Ra	*	1/2	11.187	0.01814	0.2627	-0.7338	
89	^{227}Ac	*	3/2	5.6	0.01131	0.6565	+1.1	+170
90	^{229}Th	*	5/2	1.40	0.00042	0.3843	+0.46	+430
91	^{231}Pa	100	3/2	10.2	0.06903	1.1995	2.01	-172
92	^{235}U	0.7204	7/2	0.83	0.00015	0.4082	-0.38	+493.6
93	^{237}Np	*	5/2	9.57	0.13264	2.6234	+3.14	+388.6
94	^{239}Pu	*	1/2	3.09	0.00038	0.0727	+0.203	
95	^{243}Am	*	5/2	4.6	0.01446	1.2532	+1.5	+421

* Radioactive nuclide; natural abundance is undefined.

PROPERTIES OF IMPORTANT NMR NUCLEI

Thomas J. Bruno and Paris D. N. Svoronos

The following table lists the magnetic properties at higher field strengths required for choosing the nuclei to be used in NMR experiments (Refs. 1–15). The table shows NMR frequency at the indicated field strength for a variety of isotopes. The reader is referred to several excellent texts and the literature for guidelines in nucleus selection. For more detailed information on these and other less common nuclei at 10 kG, the reader should consult the table entitled "Nuclear Spins, Moments, and Other Data Related to NMR Spectroscopy" in this section. The column definitions for the table are as follows.

Column heading	**Definition**
Isotope	Atomic symbol for isotope
Abundance	Isotopic abundance (fractional) on Earth
Spin *I*	Nuclear spin
10.000 kG, 14.092 kG, etc.	NMR frequency at magnetic field specified, in MHz

References

1. Silverstein, R. M., Bassler, G. C., and Morrill, T. C., *Spectrometric Identification of Organic Compounds, Fifth Edition*, John Wiley and Sons, New York, 1991.
2. Yoder, C. H., and Shaeffer, C. D., *Introduction to Multinuclear NMR*, Benjamin/Cummings, Menlo Park, CA, 1987.
3. Gordon, A. J., and Ford, R. A., *The Chemist's Companion*, Wiley Interscience, New York, 1971.
4. Silverstein, R. M., and Webster F. X., *Spectrometric Identification of Organic Compounds, Sixth Edition*, John Wiley and Sons, New York, 1998.
5. Becker, E. D., *High Resolution NMR, Theory and Chemical Applications, Second Edition*, Academic Press, New York, 1980.
6. Gunther, H., *NMR Spectroscopy: Basic Principles, Concepts and Applications in Chemistry*, John Wiley and Sons, New York, 2003.
7. Rahman, A.-u., *Nuclear Magnetic Resonance*, Springer-Verlag, New York, 1986.
8. Harris, R. K., *Chem. Soc. Rev.* 5, 1, 1976. <https://doi.org/10.1039/cs97605fp001>
9. Kitamaru, R., *Nuclear Magnetic Resonance: Principles and Theory*, Elsevier Science, 1990.
10. Lambert, J. B., Holland, L. N., and Mazzola, E. P., *Nuclear Magnetic Resonance Spectroscopy: Introduction to Principles, Applications and Experimental Methods*, Prentice Hall, Englewood Cliffs, NJ, 2003.
11. Bovey, F. A., and Mirau, P. A., *Nuclear Magnetic Resonance Spectroscopy, Second Edition*, Academic Press, New York, 1988.
12. Harris, R. K., and Mann, B. E., *NMR and the Periodic Table*, Academic Press, London, 1978.
13. Hore, P. J., *Nuclear Magnetic Resonance*, Oxford University Press, Oxford, 1995.
14. Nelson, J. H., *Nuclear Magnetic Resonance Spectroscopy, Second Edition*, John Wiley and Sons, New York, 2003.
15. Bruno, T. J., and Svoronos, P. D. N., *CRC Handbook of Basic Tables for Chemical Analysis, Third Edition*, CRC Press, Boca Raton, FL, 2011. <https://doi.org/10.1201/b10385>

NMR Frequency of Important Nuclei (in MHz) at the Indicated Field Strength (in kG)**

Isotope	Abundance	Spin *I*	10.000 kG	14.092 kG	21.139 kG	23.487 kG	51.567 kG	93.950 kG	140.925 kG	223.131 kG
$^{1}_{1}H$	[0.999 72, 0.999 99]	1/2	42.5759	60.0000	90.0000	100.0000	220.0000	400.0000	600.0000	950.0000
$^{2}_{1}H$	[0.000 01, 0.000 28]	1	6.53566	9.21037	13.81555	15.35061	33.77134	61.40262	92.10380	145.9830
$^{3}_{1}H$*		1/2	45.4129	63.9980	95.9971	106.6634	234.6595	426.6542	639.9813	1013.3024
$^{13}_{6}C$	[0.0096, 0.0116]	1/2	10.7054	15.0866	22.6298	25.1443	55.3174	100.5735	150.8659	2388.5150
$^{14}_{7}N$	[0.995 78, 0.996 63]	1	3.0756	4.3343	6.5014	7.2238	15.924	28.9104	43.3615	68.6557
$^{15}_{7}N$	[0.003 37, 0.004 22]	1/2	4.3142	6.0798	9.1197	10.1330	22.2925	40.5306	60.7960	96.2601
$^{17}_{8}O$	[0.000 367, 0.000 400]	5/2	5.772	8.134	12.201	13.557	29.825	54.1811	81.3186	128.5801
$^{19}_{9}F$	1	1/2	40.0541	42.3537	63.5305	94.0769	206.9692	376.2515	564.3781	893.5963
$^{29}_{14}Si$	[0.046 45, 0.046 99]	1/2	8.4578	11.9191	17.8787	19.8652	43.7035	79.4638	119.1956	188.72
$^{31}_{15}P$	1	1/2	17.235	24.288	36.433	40.481	89.057	161.9828	242.9741	384.7086
$^{33}_{16}S$	[0.007 29, 0.007 97]	3/2	3.2654	4.6018	6.9026	7.6696	16.8731	30.6826	46.0238	72.8710
$^{35}_{16}S$*		3/2	5.08	7.16	10.74	11.932	26.250	47.7267	71.5875	113.3508
$^{35}_{17}Cl$	[0.755, 0.761]	3/2	4.1717	5.8790	8.8184	9.7983	21.5562	39.1948	58.7902	93.0876
$^{36}_{17}Cl$*		2	4.8931	6.8956	10.3434	11.4927	25.2838	45.9638	68.9432	109.1639
$^{76}_{35}Br$*		1	4.18	5.89	8.84	9.82	21.60	39.2768	58.9130	93.2822
$^{79}_{35}Br$	[0.505, 0.508]	3/2	10.667	15.032	22.549	25.054	55.119	100.2133	150.3202	238.0064
$^{81}_{35}Br$	[0.492, 0.495]	3/2	11.498	16.204	24.305	27.006	59.413	108.0258	162.0386	256.5608
$^{183}_{74}W$	0.1431(4)	1/2	1.7716	2.4966	3.7449	4.1610	9.1543	16.6430	24.9646	39.5272

* Nucleus is radioactive.
** 1 kG = 10^{-1} T, the corresponding SI unit.

PROTON NMR ABSORPTION OF MAJOR CHEMICAL FAMILIES

Thomas J. Bruno and Paris D. N. Svoronos

The following table gives the region of the expected nuclear magnetic resonance absorptions of major chemical families (Refs. 1–12). These absorptions are reported in the dimensionless units of parts per million (ppm) versus the standard compound tetramethylsilane (TMS), which is recorded as 0.0 ppm.

$$
\begin{array}{c}
\text{CH}_3 \\
| \\
\text{CH}_3-\text{Si}-\text{CH}_3 \\
| \\
\text{CH}_3
\end{array}
$$

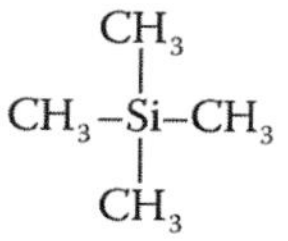

The use of this unit of measure makes the chemical shifts independent of the applied magnetic field strength or the radio frequency. For most proton NMR spectra, the protons in TMS are more shielded than almost all other protons. The chemical shift in this dimensionless unit system is then defined by:

$$\delta = \frac{\nu_s - \nu_r}{\nu_r} \times 10^6$$

where ν_s and ν_r are the absorption frequencies of the sample proton and the reference (TMS) protons (12, magnetically equivalent), respectively. In these tables, the proton(s) whose proton NMR shifts are cited are indicated by underscore. Ref. 1 provides additional details on the absorptions of other moieties, as well as correlation charts.

Major chemical families are included in the following tables.

Chemical family	Table
Hydrocarbons	1
Organic Oxygen Compounds	2
Organic Nitrogen Compounds	
Amides	3

Chemical family	Table
Amines	4
Cyano Compounds	5
Nitro Compounds	6
Organic Sulfur Compounds	
Benzothiopyrans, Disulfides, Isothiocyanates, Mercaptans, S-Methyl Salts, and Sulfates	7
Sulfides and Other Families	8

References

1. Bruno, T. J., and Svoronos, P. D. N., *CRC Handbook of Basic Tables for Chemical Analysis, Third Edition*, CRC Press, Boca Raton, FL, 2011. <https://doi.org/10.1201/b10385>
2. Silverstein, R. M., and Webster, F. X., *Spectrometric Identification of Organic Compounds, Sixth Edition*, Wiley, New York, 1998.
3. Rahman, A.-u., *Nuclear Magnetic Resonance*, Springer Verlag, New York, 1986.
4. Gordon, A. J., and Ford, R. A., *The Chemist's Companion*, Wiley Interscience, New York, 1971.
5. Becker, E. D., *High Resolution NMR, Theory and Chemical Applications, Second Edition*, Academic Press, New York, 1980.
6. Gunther, H., *NMR Spectroscopy: Basic Principles, Concepts and Applications in Chemistry*, Wiley, New York, 2003.
7. Kitamaru, R., *Nuclear Magnetic Resonance: Principles and Theory*, Elsevier Science, 1990.
8. Lambert, J. B., Holland, L. N., and Mazzola, E. P., *Nuclear Magnetic Resonance Spectroscopy: Introduction to Principles, Applications and Experimental Methods*, Prentice Hall, Englewood Cliffs, NJ, 2003.
9. Bovey, F. A., and Mirau, P. A., *Nuclear Magnetic Resonance Spectroscopy, Second Edition*, Academic Press, New York, 1988.
10. Hore, P. J., *Nuclear Magnetic Resonance*, Oxford University Press, Oxford, 1995.
11. Nelson, J. H., *Nuclear Magnetic Resonance Spectroscopy, Second Edition*, Wiley, New York, 2003.
12. Abraham, R. J., Fisher, J., and Loftus, P., *Introduction to NMR Spectroscopy*, Wiley, New York, 1988.

TABLE 1. Proton NMR Absorption of Hydrocarbons

Group	δ/ppm of underlined protons
Alkanes	
$\underline{CH_3}$-R	~0.8
-$\underline{CH_2}$-R	~1.1
>$\underline{CH}$-R	~1.4
Cyclopropane	0.2
Alkenes	
$\underline{CH_3}$-C=C<	~1.6
-$\underline{CH_2}$-C=C<	~2.1
>$\underline{CH}$- C=C<	~2.5
$\underline{CH_3}$-C-C=C<	~1.0
-$\underline{CH_2}$-C-C=C<	~1.4
>$\underline{CH}$-C-C=C<	~1.8
>C=C-$\underline{H}$	4.2 - 6.2

Group	δ/ppm of underlined protons
Alkynes	
$\underline{CH_3}$-C≡C-	~1.7
-$\underline{CH_2}$-C≡C-	~2.2
>$\underline{CH}$ -C≡C-	~2.7
$\underline{CH_3}$-C-C≡C-	~1.2
>$\underline{CH}$-C-C≡C-	~1.5
>$\underline{CH}$ -C-C≡C-	~1.8
R-C≡C-$\underline{H}$	~2.4
Aromatics	
$C_6\underline{H_5}$-G	8.5 - 6.9

TABLE 2. Proton NMR Absorption of Organic Oxygen Compounds

Group	Approximate δ/ppm of underlined protons
Alcohols	
CH_3-OH	3.2
RCH_2-OH	3.4
R_2CH-OH	3.6
CH_3-C-OH	1.2
RCH_2-C-OH	1,5
R_2CH-C-OH	1.8
R-O-H	1 to 5
Aldehydes	
CH_3-CHO	2.2
RCH_2-CHO	2.4
R_2CH-CHO	2.5
CH_3-C-CHO	1.1
RCH_2-C-CHO	1.6
Amides	
See organic nitrogen compounds	
Anhydrides, acylic	
CH_3-C(=O)O-	1.8
RCH_2-C(=O)O-	2.1
R_2CH-C(=O)O-	2.3
CH_3-C-C(=O)O-	1.2
RCH_2-C-C(=O)O-	1.8
R_2CH-C-C(=O)O-	2.0
Anhydrides, cyclic	
Succinic anhydride	3.0
Maleic anhydride	7.1
Carboxylic acids	
CH_3-COOH	2.1
RCH_2-COOH	2.3
R_2CH-COOH	2.5
CH_3-C-COOH	1.1
R-CH_2-C-COOH	1.6
R_2CH-C-COOH	2.0
R-COO-H	11 to 12
Cyclic ethers	
Oxirane	2.5
Oxetane	4.7 (adjacent to O), 2.7
Tetrahydrofuran	3.8 (adjacent to O), 1.9
Tetrahydropyran	3.6 (adjacent to O), 1.6, 1.6
1,4-Dioxane	3.6
1,3-Dioxane	4.7, 3.8 1.7
Furan	7.4 (adjacent to O), 6.3
Dihydropyran	6.2 (adjacent to O), 4.5, 1.9
Epoxides	
See cyclic ethers	
Esters	
CH_3-COOR (R=alkyl)	1.9
CH_3-COOR (R=aryl)	2.0
RCH_2-COOR (R=alkyl)	2.1
RCH_2-COOR (R=aryl)	2.2
R_2CH-COOR (R=alkyl)	2.3
R_2CH-COOR (R=aryl)	2.4
CH_3-C-COOR	1.1
RCH_2-C-COOR	1.7
R_2CH-C-COOR	1.9
CH_3-OOC-R	3.6
RCH_2-OOC-R	4.1
R_2CH-OOC-R	4.8
CH_3-C-OOC-R	1.3
RCH_2-C-OOC-R	1.6
R_2CH-C-OOC-R	1.8
Cyclic esters	
2H-Pyran-2-one	H_3=2.3, H_4=1.6, H_5=1.6, H_6=4.1
Ethers	
CH_3-O-R (R=alkyl)	3.2
CH_3-O-R (R=aryl)	3.9
RCH_2-O-R (R=alkyl)	3.4
RCH_2-O-R (R=aryl)	4.1
R_2CH-O-R (R=alkyl)	3.6
R_2CH-O-R (R=aryl)	4.5
CH_3-C-O-R (R=alkyl)	1.2
CH_3-C-O-R (R=aryl)	1.3
RCH_2-C-O-R (R=alkyl)	1.5
RCH_2-C-O-R (R=aryl)	1.6
R_2CH-C-O-R (R=alkyl)	1.8
R_2CH-C-O-R (R=aryl)	2.0
Isocyanates	
See nitrogen compounds	
Ketones	
CH_3-C(=O)-	1.9
CH_3-C(=O)-	2.4
RCH_2-C(=O)- (R=alkyl)	2.1
RCH_2-C(=O)- (R=aryl)	2.7
R_2CH-C(=O)- (R=alkyl)	2.3
R_2CH-C(=O)- (R=aryl)	3.4
CH_3-C(=O)-	1.1
CH_3-C(=O)-	1.2
RCH_2-C(=O)- (R=alkyl)	1.6
RCH_2-C(=O)- (R=aryl)	1.6
R_2CH-C(=O)- (R=alkyl)	2.0
R_2CH-C(=O)- (R=aryl)	2.1
Cyclic ketones (*n* = ring carbons):	
α-hydrogens, *n*=3	1.7
α-hydrogens, *n*=4	3.0
α-hydrogens, *n*>5	2.0 to 2.3
β-hydrogens	1.9 to 1.5
Lactones	
See cyclic esters	
Nitro compounds	
See organic nitrogen compounds	
Phenols	
Ar-O-H	9 to 10

Analytical

TABLE 3. Proton NMR Absorption of Organic Nitrogen Compounds: Amides

δ/ppm of underlined protons	Primary R–C(=O) NH_2δ/ppm	Secondary R–C(=O) NHR_1δ/ppm	Tertiary R–C(=O) NR_1R_2δ/ppm
(i) N-substitution			
R–C(=O)N–H	5–12	5–12	
(a) Alpha			
–C(=O)N–CH_3		~2.9	~2.9
–C(=O)N–CH_2–		~3.4	~3.4
–C(=O)N–CH–		~3.8	~3.8
(b) Beta			
–C(=O)N–C–CH_3	~1.1	~1.1	~1.1
–C(=O)N–C–CH_2–	~1.5	~1.5	~1.5
–C(=O)N–C–CH–	~1.9	~1.9	~1.9
(i) C-substitution			
(a) Alpha			
CH3–C(=O)N	~1.9	~2.0	~2.1
RCN_2–C(=O)N	~2.1	~2.1	~2.1
R2CH–C(=O)N	~2.2	~2.2	~2.2
(b) Beta			
CH_3–C–C(=O)N	~1.1	~1.1	~1.1
CH_2–C–C(=O)N	~1.5	~1.5	~1.5
–CH–C–C(=O)N	~1.8	~1.8	~1.8

TABLE 4. Proton NMR Absorption of Organic Nitrogen Compounds: Amines

δ of underlined protons	Primary R–NH_2δ/ppm	Secondary RN–HR δ/ppm	Tertiary RRRN δ/ppm
(i) Alpha protons			
>N–CH3	~2.5	2.3–3.0	~2.2
>N–CH2–	~2.7	2.6–3.4	~2.4
>N–CH<	~3.1	2.9–3.6	~2.8
(ii) Beta protons			
>N–C–CH3			~1.1
>N–C–CH_2–			~1.4
>N–C–CH<			~1.7

TABLE 5. Proton NMR Absorption of Organic Nitrogen Compounds: Cyano Compounds

α hydrogens (underlined)	δ/ppm	β hydrogens (underlined)	δ/ppm
Nitriles			
CH_3–C≡N	~2.1	CH_3–C–C≡N	~1.2
–CH_2–C≡N	~2.5	–CH_2–C–C≡N	~1.6
–CH–C≡N	~2.9	CH–C–C≡N	~2.0
Imides			
CH_3–C(=O)NHC(=O)–	~2.0	CH_3–C(=O)C–NH–C(=O)–	~1.2
CH_2–C(=O)NHC(=O)–	~2.1	CH_2–C(=O)C–NH–C(=O)–	~1.3
CH–C(=O)NHC(=O)–	~2.2	–CH–C(=O)C–NH–C(=O)–	~1.4
Isocyanates			
CH_3–N=C=O	~3.0		
–CH_2–N=C=O	~3.3		
–CH–N=C=O	~3.6		
Isocyanides (isonitriles)			
CH_3–N=C<	~2.9		
CH_2–N=C<	~3.3		
CH–N=C<	~4.9		
Isothiocyanates			
CH_3–N=C=S	~3.4		
CH_2–N=C=S	~3.7		
>CH–N=C=S	~4.0		
Nitroso			
–CH_2–O–N=O	~4.8		

TABLE 6. Proton NMR Absorption of Organic Nitrogen Compounds: Nitro Compounds

Nitro compounds (underlined)	δ/ppm
CH_3–NO_2	~ 4.1
CH_3–C–NO_2	~1.6
–CH_2–NO_2	~4.2
–CH_2–C–NO_2	~2.1
–CH–NO_2	~4.4
–CH–C–NO_2	~2.5

TABLE 7. Proton NMR Absorption of Organic Sulfur Compounds: Benzothiopyrans, Disulfides, Isothiocyanates, Mercaptans, S-Methyl Salts, and Sulfates

Family	δ/ppm of underlined protons
Benzothiopyrans	
2H–1– sp³ C–H	~3.3
4H–1–sp³ C–H	~3.2
2,3,4H–1–sp³ C–H	1.9–2.8
sp² C–H	5.8–6.4
sp² C–H	5.9–6.3
aromatic	~6.8
aromatic	~6.9
aromatic	~7.1
Disulfides	
$\underline{CH_3}$–S–S–R	~2.4
$\underline{CH_2}$–S–S–R	~2.7
$\underline{CH}$–S–S–R	~3.0
$\underline{CH_3}$–C–S–S–R	~1.2
$\underline{CH_2}$–C–S–S–R	~1.6
$\underline{CH}$–C–S–S–R	~2.0
Isothiocyanates	
$\underline{CH_3}$–N=C=S	~2.4
–$\underline{CH_2}$–N=C=S	~2.7
–$\underline{CH}$–N=C=S	~3.0
Mercaptans (thiols)	
$\underline{CH_3}$–S–H	~2.1
–$\underline{CH_2}$–S–H	~2.6
–$\underline{CH}$–S–H	~3.1
$\underline{CH_3}$–C–S–H	~1.3
–$\underline{CH_2}$–C–S–H	~1.6
–$\underline{CH}$–C–S–H	~1.7
S-methyl salts	
>S+–CH_3	~3.2
Sulfates	
($\underline{CH_3}$–O)$_2$S(=O)$_2$	~3.4

TABLE 8. Proton NMR Absorption of Organic Sulfur Compounds: Sulfides and Other Families

Family	δ/ppm of underlined protons
Sulfides	
$\underline{CH_3}$–S–	1.8–2.1
R–$\underline{CH_2}$–S–	1.9–2.4
R–$\underline{CHR}$–S–	2.8–3.4
Ar–$\underline{CH_2}$–S–	4.1–4.2
Ar–$\underline{CHR}$–S–	3.6–4.2
Ar_2–$\underline{CH}$–S–	5.1–5.2
$\underline{CH_3}$–CH_2–S–	1.1–1.2
$\underline{CH_3}$–CHR–S–	0.8–1.2
$\underline{CH_3}$–CHAr–S–	1.3–1.4
$\underline{CH_3}$–CR_2–S–	1.0
Ar–$\underline{CH_2}$–CHR–S–	3.0–3.2
>C=C–$\underline{CH_2}$–CHAr–S–	2.4–2.6
>C=C–$\underline{CH_2}$–CAr_2–S–	2.5
$R_2\underline{CH}$–CH_2–S–	2.6–3.0
$Ar_2\underline{CH}$–CH_2–S–	4.0–4.2
>C=C–$\underline{CHR}$–CHAr–S–	2.3–2.4
>C=C–$\underline{CHR}$–CAr_2–S–	2.8–3.2
Sulfilimines	
$\underline{CH_3}$(R)S=N–R^2	~2.5
Sulfonamides	
$\underline{CH_3}$–SO_2NH_2	~3.0
Sulfonates	
$\underline{CH_3}$–SO_2–OR	~3.0
Sulfones	
$\underline{CH_3}$–SO_2–R^2	~2.6
Sulfonic acids	
$\underline{CH_3}$–SO_3H	~3.0
Sulfoxides	
$\underline{CH_3}$–S(=O)R	~2.5
–$\underline{CH_2}$–S(=O)R	~3.1
Thiocyanates	
$\underline{CH_3}$–S–C≡N	~2.7
–$\underline{CH_2}$–S–C≡N	~3.0
–$\underline{CH}$–S–C≡N	~3.3
Thiols	
See mercaptans	

Note: Ar represents aryl.

Some Useful ¹H Coupling Constants

The following chart gives the values of some useful proton NMR coupling constants (in Hz). The single numbers indicate a typical or average value, while in some cases, the range is provided.

1. Freely rotating chains.

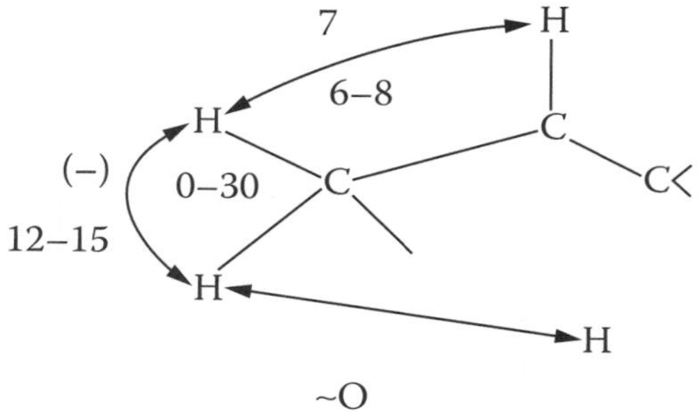

2. Alcohols with no exchange as in DMSO.

1° = triplet

2° = doublet (broad)

3° = singlet

Upon addition of TFA, a sharp singlet results.

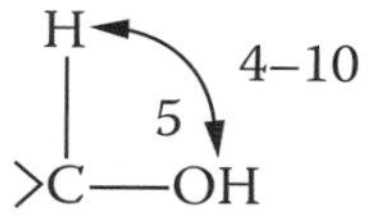

3. Alkenes

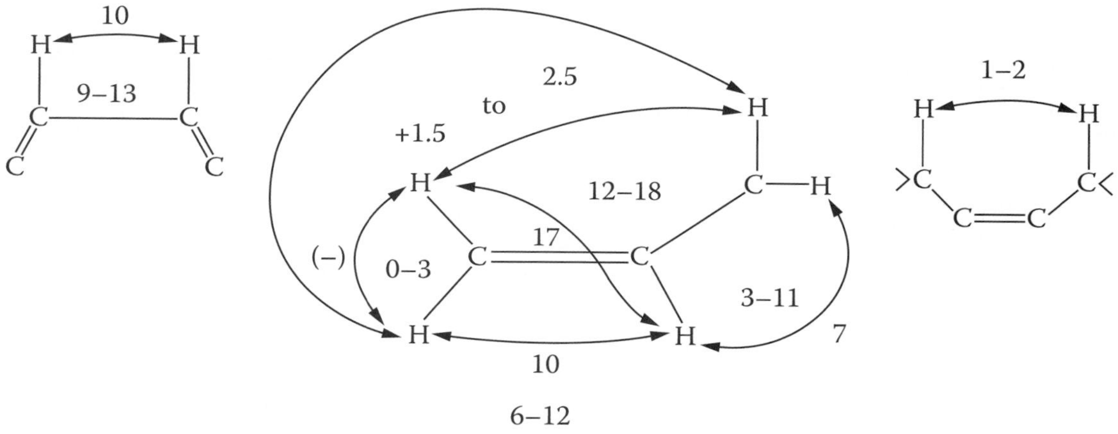

4. Alkynes

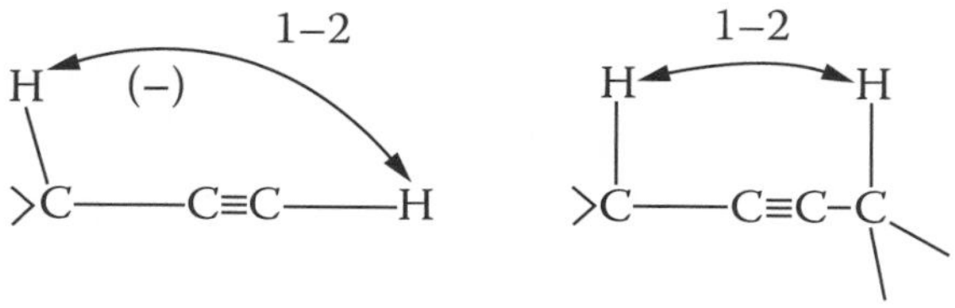

5. Aldehydes

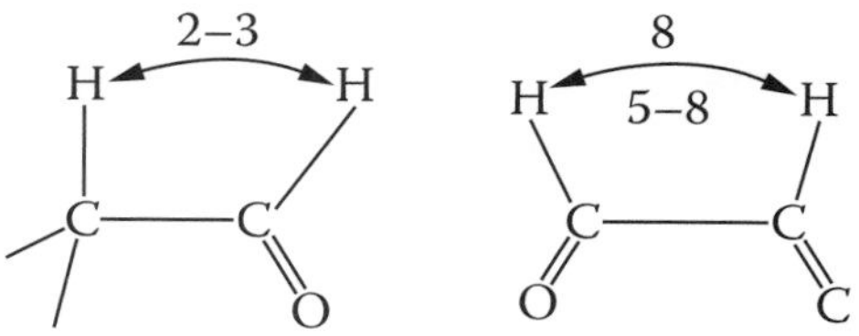

6. Aromatic

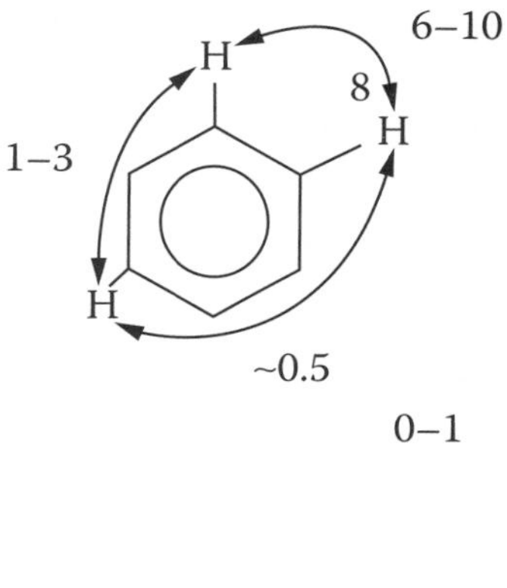

Analytical

PROTON NMR CORRELATION CHART FOR MAJOR ORGANIC FUNCTIONAL GROUPS

The chart below summarizes the range of chemical shifts for protons in several classes or organic compounds and substituent groups. The chemical shifts δ are given in parts per million relative to tetramethylsilane.

Reference

Mohacsi, E., *J. Chem. Edu.* 41, 38, 1964 (with permission).

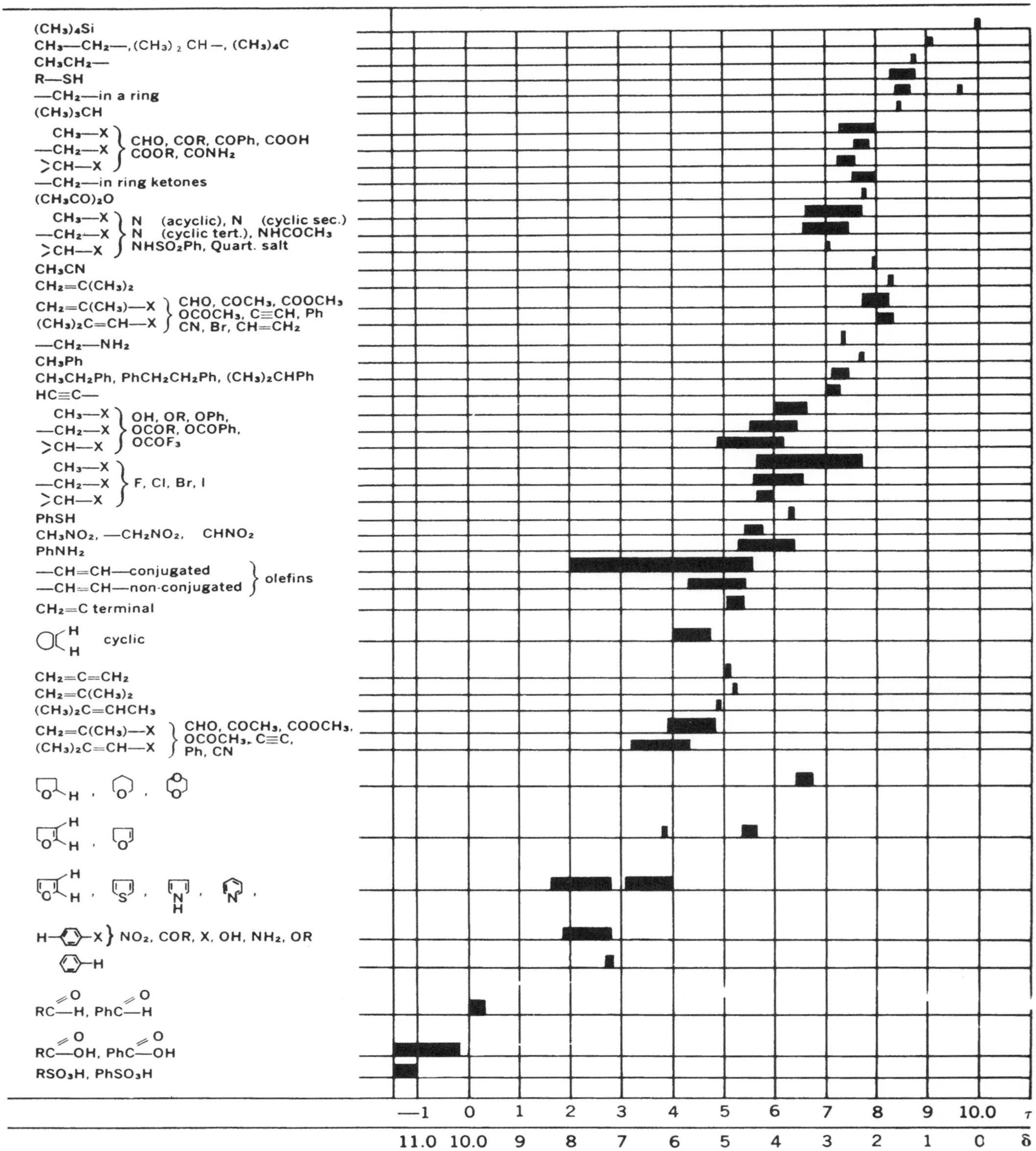

PROTON NMR SHIFTS OF COMMON ORGANIC LIQUIDS

The table below lists the ^{1}H chemical shifts for over 300 organic liquids. Column definitions for the table are as follows.

Column heading	Definition
Name	Name of organic liquid; compounds are listed alphabetically by name
Mol. form.	Molecular formula of organic liquid, written in Hill order
Solvent	Molecular formula of solvent, written in Hill order
Shift 1, Shift 2, etc.	Proton NMR shift, in parts per million relative to tetramethylsilane (TMS); listed in order of smallest to largest shift
Ref.	Reference

References

1. Lide, D. R., Editor, *Properties of Organic Compounds* <poc.chemnetbase.com>; Lide, D. R., and Milne, G. W. A., Editors, *Handbook of Data on Organic Compounds, Third Edition*, CRC Press, Boca Raton, FL, 1993.
2. Spectral Database for Organic Compounds, SDBS, National Institute of Advanced Industrial Science and Technology (AIST), Japan <sdbs.db.aist.go.jp/>.

^{1}H NMR Shifts of Common Organic Liquids in ppm Relative to TMS

Name	Mol. form.	Solvent	Shift 1	Shift 2	Shift 3	Shift 4	Shift 5	Shift 6	Ref.
Acetic acid	$C_2H_4O_2$	$CDCl_3$	2.10	11.4					2
Acetic anhydride	$C_4H_6O_3$	CCl_4	2.2						1
Acetone	C_3H_6O	$CDCl_3$	2.1						1
Acetonitrile	C_2H_3N	CCl_4	1.9						1
Acrolein	C_3H_4O	$CDCl_3$	6.4	9.5					1
Acrylonitrile	C_3H_3N	$CDCl_3$	6.3						1
Allyl alcohol	C_3H_6O	$CDCl_3$	3.6	4.1	5.1	5.3	6.0		1
Allylamine	C_3H_7N	$CDCl_3$	1.5	3.3	5.0	5.1	5.9		1
2-Amino-2-methyl-1-propanol	$C_4H_{11}NO$	CCl_4	1.1	2.8	3.2				1
Aniline	C_6H_7N	CCl_4	3.3	6.4	6.6	7.0			1
Anisole	C_7H_8O	$CDCl_3$	3.8	7.1					1
Benzaldehyde	C_7H_6O	$CDCl_3$	7.7	10.0					1
Benzene	C_6H_6	$CDCl_3$	7.34						2
Benzeneacetonitrile	C_8H_7N	CCl_4	0.8	0.9	1.6	2.3			1
Benzenethiol	C_6H_6S	CCl_4	3.2	6.9					1
Benzonitrile	C_7H_5N	CCl_4	7.5						1
Benzyl acetate	$C_9H_{10}O_2$	$CDCl_3$	2.1	5.1	7.3				1
Benzyl alcohol	C_7H_8O	$CDCl_3$	2.4	4.6	7.3				1
Bis(2-aminoethyl)amine	$C_4H_{13}N_3$	$CDCl_3$	1.23	2.69	2.79				2
Bis(2-chloroethyl) ether	$C_4H_8Cl_2O$	$CDCl_3$	3.66	3.77					2
Bis(2-ethylhexyl) phthalate	$C_{24}H_{38}O_4$	CCl_4	0.9	1.5	4.2	7.4	7.7		1
Bis(2-hydroxyethyl) sulfide	$C_4H_{10}O_2S$	$CDCl_3$	2.8	3.8	4.2				1
Bromobenzene	C_6H_5Br	CCl_4	7.1	7.4					1
1-Bromobutane	C_4H_9Br	CCl_4	1.0	1.4	1.8	3.4			1
2-Bromobutane, (±)-	C_4H_9Br	$CDCl_3$	1.1	1.7	1.8	4.1			1
1-Bromo-2-chloroethane	C_2H_4BrCl	$CDCl_3$	3.3	4.0					1
Bromochloromethane	CH_2BrCl	CCl_4	5.2						1
1-Bromodecane	$C_{10}H_{21}Br$	CCl_4	0.9	1.8	3.3				1
Bromoethane	C_2H_5Br	$CDCl_3$	1.7	3.4					1
2-Bromo-2-methylpropane	C_4H_9Br	CCl_4	1.8						1
1-Bromonaphthalene	$C_{10}H_7Br$	CCl_4	7.4	8.1					1
1-Bromopentane	$C_5H_{11}Br$	CCl_4	0.9	1.4	1.9	3.3			1
1-Bromopropane	C_3H_7Br	CCl_4	1.0	1.9	3.4				1
2-Bromopropane	C_3H_7Br	$CDCl_3$	1.7	4.3					1
2-Bromopropene	C_3H_5Br	$CDCl_3$	2.3	5.3	5.5				1
Butanal	C_4H_8O	$CDCl_3$	1.0	1.7	2.4	9.7			1
Butanenitrile	C_4H_7N	CCl_4	1.1	1.7	2.3				1
1-Butanethiol	$C_4H_{10}S$	$CDCl_3$	0.9	1.2	1.5	2.5			1
Butanoic acid	$C_4H_8O_2$	CCl_4	0.9	1.7	2.3	12.0			1
Butanoic anhydride	$C_8H_{14}O_3$	CCl_4	1.0	1.7	2.4				1
1-Butanol	$C_4H_{10}O$	$CDCl_3$	0.94	1.39	1.53				2
2-Butanol	$C_4H_{10}O$	$CDCl_3$	0.93	1.17	1.46				2

Name	Mol. form.	Solvent	Shift 1	Shift 2	Shift 3	Shift 4	Shift 5	Shift 6	Ref.
2-Butanone	C_4H_8O	$CDCl_3$	1.06	2.14	2.45				2
trans-2-Butenal	C_4H_6O	$CDCl_3$	2.0	6.1	6.9	9.5			1
2-Butoxyethanol	$C_6H_{14}O_2$	CCl_4	0.9	1.3	3.3	3.7			1
Butyl acetate	$C_6H_{12}O_2$	$CDCl_3$	0.9	1.4	2.0	4.1			1
Butylamine	$C_4H_{11}N$	$CDCl_3$	0.92	1.33	1.43				2
tert-Butylamine	$C_4H_{11}N$	$CDCl_3$	1.1	1.2					1
Butylbenzene	$C_{10}H_{14}$	CCl_4	0.9	1.4	2.6	7.1			1
sec-Butylbenzene, (±)-	$C_{10}H_{14}$	CCl_4	0.8	1.2	1.6	2.5	7.1		1
tert-Butylbenzene	$C_{10}H_{14}$	CCl_4	1.3	7.2					1
Butyl formate	$C_5H_{10}O_2$	$CDCl_3$	0.9	1.5	4.2	8.1			1
1-*tert*-Butyl-4-methylbenzene	$C_{11}H_{16}$	$CDCl_3$	1.30	2.31	7.11				2
Butyl vinyl ether	$C_6H_{12}O$	CCl_4	0.9	1.4	3.6	3.8	4.0	6.3	1
γ-Butyrolactone	$C_4H_6O_2$	$CDCl_3$	4.4						1
Caprolactam	$C_6H_{11}NO$	$CDCl_3$	1.7	2.4	3.2	7.8			1
2-Chloroaniline	C_6H_6ClN	CCl_4	3.8	6.8					1
Chlorobenzene	C_6H_5Cl	$CDCl_3$	7.3						2
2-Chlorobutane	C_4H_9Cl	CCl_4	1.1	1.5	1.7	3.9			1
Chloroethane	C_2H_5Cl	$CDCl_3$	1.5	3.6					1
2-Chloroethanol	C_2H_5ClO	$CDCl_3$	2.8	3.7	3.8				1
(Chloromethyl)benzene	C_7H_7Cl	CCl_4	4.5	7.3					1
1-Chloro-3-methylbutane	$C_5H_{11}Cl$	$CDCl_3$	0.9	1.7	3.6				1
1-Chloro-2-methylpropane	C_4H_9Cl	CCl_4	1.0	1.9	3.3				1
2-Chloro-2-methylpropane	C_4H_9Cl	CCl_4	1.6						1
1-Chloronaphthalene	$C_{10}H_7Cl$	CCl_4	7.1	7.5	8.2				1
1-Chlorooctane	$C_8H_{17}Cl$	CCl_4	0.9	1.3	1.8	3.5			1
1-Chloropentane	$C_5H_{11}Cl$	CCl_4	0.9	1.6	3.4				1
1-Chloropropane	C_3H_7Cl	CCl_4	1.0	1.8	3.4				1
3-Chloropropene	C_3H_5Cl	CCl_4	4.0	5.2	5.3	5.9			1
2-Chlorotoluene	C_7H_7Cl	$CDCl_3$	2.4	7.2					1
3-Chlorotoluene	C_7H_7Cl	CCl_4	2.3	7.1					1
Cyclohexane	C_6H_{12}	$CDCl_3$	1.43						2
Cyclohexanol	$C_6H_{12}O$	CCl_4	1.6	3.5	4.2				1
Cyclohexanone	$C_6H_{10}O$	CCl_4	1.8	2.3					1
Cyclohexene	C_6H_{10}	CCl_4	1.6	2.0	5.6				1
Cyclohexylamine	$C_6H_{13}N$	CCl_4	1.4	1.5	2.6				1
Cyclopentane	C_5H_{10}	CCl_4	1.5						1
Cyclopentanone	C_5H_8O	CCl_4	2.0						1
cis-Decahydronaphthalene	$C_{10}H_{18}$	$CDCl_3$	1.42	1.62					2
trans-Decahydronaphthalene	$C_{10}H_{18}$	$CDCl_3$	0.87	0.93	1.23				2
Decane	$C_{10}H_{22}$	CCl_4	0.9	1.3					1
Diacetone alcohol	$C_6H_{12}O_2$	$CDCl_3$	1.3	2.2	2.6	3.7			1
1,2-Dibromoethane	$C_2H_4Br_2$	$CDCl_3$	3.65						2
Dibromomethane	CH_2Br_2	CCl_4	4.9						1
1,2-Dibromopropane	$C_3H_6Br_2$	CCl_4	1.8	3.5	3.8	4.2			1
Dibutylamine	$C_8H_{19}N$	CCl_4	0.5	0.9	1.4	2.5			1
Dibutyl ether	$C_8H_{18}O$	CCl_4	0.9	1.4	3.3				1
Dibutyl sebacate	$C_{18}H_{34}O_4$	CCl_4	1.0	1.5	2.2	4.0			1
o-Dichlorobenzene	$C_6H_4Cl_2$	CCl_4	7.2						1
m-Dichlorobenzene	$C_6H_4Cl_2$	CCl_4	7.2	7.4					1
1,1-Dichloroethane	$C_2H_4Cl_2$	$CDCl_3$	2.06	5.90					2
1,2-Dichloroethane	$C_2H_4Cl_2$	CCl_4	3.7						1
1,1-Dichloroethene	$C_2H_2Cl_2$	CCl_4	5.5						1
cis-1,2-Dichloroethene	$C_2H_2Cl_2$	$CDCl_3$	6.28						2
trans-1,2-Dichloroethene	$C_2H_2Cl_2$	$(CH_3)_4Si$	6.24						2
Dichloromethane	CH_2Cl_2	CCl_4	5.3						1
(Dichloromethyl)benzene	$C_7H_6Cl_2$	CCl_4	6.6	7.4					1
1,2-Dichloropropane, (±)-	$C_3H_6Cl_2$	$CDCl_3$	1.61	3.59	3.74				2
2,4-Dichlorotoluene	$C_7H_6Cl_2$	CCl_4	2.3	7.0	7.3				1
3,4-Dichlorotoluene	$C_7H_6Cl_2$	CCl_4	2.3	7.0					1
Diethanolamine	$C_4H_{11}NO_2$	D_2O	2.7	3.7					1
1,1-Diethoxyethane	$C_6H_{14}O_2$	$CDCl_3$	1.2	1.3	3.5	3.7	4.7		2

Name	Mol. form.	Solvent	Shift 1	Shift 2	Shift 3	Shift 4	Shift 5	Shift 6	Ref.
1,2-Diethoxyethane	$C_6H_{14}O_2$	$CDCl_3$	1.22	3.54	3.58				1
Diethylamine	$C_4H_{11}N$	CCl_4	0.9	1.0	2.6				1
Diethyl carbonate	$C_5H_{10}O_3$	$CDCl_3$	1.3	4.2					1
Diethylene glycol	$C_4H_{10}O_3$	$CDCl_3$	3.7	4.2					1
Diethylene glycol dimethyl ether	$C_6H_{14}O_3$	$CDCl_3$	3.3	3.5					1
Diethylene glycol monoethyl ether	$C_6H_{14}O_3$	CCl_4	1.2	3.1	3.5	3.6			1
Diethylene glycol monoethyl ether acetate	$C_8H_{16}O_4$	$CDCl_3$	1.22	2.08	3.54				2
Diethylene glycol monomethyl ether	$C_5H_{12}O_3$	$CDCl_3$	3.3	3.4	3.6				1
Diethyl ether	$C_4H_{10}O$	$CDCl_3$	1.21	3.47					2
Diethyl sulfide	$C_4H_{10}S$	CCl_4	1.2	2.5					1
Diisopropylamine	$C_6H_{15}N$	CCl_4	0.7	1.0	2.9				1
Diisopropyl ether	$C_6H_{14}O$	CCl_4	1.0	3.5					1
1,2-Dimethoxybenzene	$C_8H_{10}O_2$	CCl_4	3.7	6.8					1
1,2-Dimethoxyethane	$C_4H_{10}O_2$	CCl_4	3.3	3.4					1
Dimethoxymethane	$C_3H_8O_2$	CCl_4	3.2	4.4					1
N,N-Dimethylacetamide	C_4H_9NO	$CDCl_3$	2.1	2.9	3.0				1
2,4-Dimethylaniline	$C_8H_{11}N$	CCl_4	2.0	2.2	3.4	6.4	6.7		1
2,2-Dimethylbutane	C_6H_{14}	CCl_4	0.9	1.1	1.3	1.1	1.3		1
2,3-Dimethylbutane	C_6H_{14}	CCl_4	0.9	1.5					1
N,N-Dimethylformamide	C_3H_7NO	$CDCl_3$	2.9	3.0	8.0				1
Dimethyl glutarate	$C_7H_{12}O_4$	$CDCl_3$	2.0	2.4	3.7				1
2,6-Dimethyl-4-heptanone	$C_9H_{18}O$	CCl_4	0.9	2.1					1
2,5-Dimethylhexane	C_8H_{18}	CCl_4	0.9	1.4					1
Dimethyl maleate	$C_6H_8O_4$	CCl_4	3.7	6.2					1
2,2-Dimethylpentane	C_7H_{16}	$CDCl_3$	0.9	0.9	1.2				1
2,4-Dimethylpentane	C_7H_{16}	CCl_4	0.9	1.1	1.6				1
2,4-Dimethyl-3-pentanone	$C_7H_{14}O$	CCl_4	1.0	2.6					1
2,4-Dimethylpyridine	C_7H_9N	$CDCl_3$	2.3	2.5	7.0	7.0	8.4		1
2,6-Dimethylpyridine	C_7H_9N	$CDCl_3$	2.51	6.93	7.42				2
Dimethyl sulfoxide	C_2H_6OS	$CDCl_3$	2.62						2
1,4-Dioxane	$C_4H_8O_2$	$CDCl_3$	3.69						2
1,3-Dioxolane	$C_3H_6O_2$	$CDCl_3$	3.88	4.90					2
Dipentyl ether	$C_{10}H_{22}O$	$CDCl_3$	0.9	1.4	3.4				1
Dipropylamine	$C_6H_{15}N$	$CDCl_3$	1.5	2.6					1
Dodecane	$C_{12}H_{26}$	CCl_4	0.9	1.3					1
1-Dodecene	$C_{12}H_{24}$	CCl_4	0.9	1.3	2.0	5.4			1
Epichlorohydrin	C_3H_5ClO	CCl_4	2.6	2.8	3.2	3.5	3.6		1
1,2-Epoxybutane	C_4H_8O	CCl_4	1.0	1.5	2.3	2.6	2.7		1
1,2-Ethanediamine	$C_2H_8N_2$	CCl_4	1.2	2.6					1
1,2-Ethanediol	$C_2H_6O_2$	D_2O	3.7						1
1,2-Ethanediol, diacetate	$C_6H_{10}O_4$	CCl_4	2.0	4.2					1
Ethanol	C_2H_6O	$CDCl_3$	1.23	2.61	3.69				2
Ethanolamine	C_2H_7NO	$CDCl_3$	2.7	2.8	3.5				1
Ethoxybenzene	$C_8H_{10}O$	CCl_4	1.3	3.9	6.9				1
2-Ethoxyethanol	$C_4H_{10}O_2$	$CDCl_3$	1.22	2.70	3.55				2
2-Ethoxyethyl acetate	$C_6H_{12}O_3$	CCl_4	1.2	2.0	3.4	3.5	4.1		1
Ethyl acetate	$C_4H_8O_2$	$CDCl_3$	1.26	2.04	4.12				2
Ethyl acetoacetate	$C_6H_{10}O_3$	$CDCl_3$	1.3	1.9	2.2	3.3	4.1	4.9	1
Ethyl acrylate	$C_5H_8O_2$	CCl_4	1.3	4.1	5.7	6.1	6.3		1
Ethylamine	C_2H_7N	D_2O	1.1	2.6					1
Ethylbenzene	C_8H_{10}	$CDCl_3$	1.3	2.7	7.2				1
Ethyl benzoate	$C_9H_{10}O_2$	CCl_4	1.3	4.3	7.4	8.0			1
Ethyl butanoate	$C_6H_{12}O_2$	CCl_4	0.9	1.2	1.7	2.2	4.1		1
Ethylcyanoacetate	$C_5H_7NO_2$	CCl_4	1.3	3.4	4.3				1
Ethylcyclohexane	C_8H_{16}	$CDCl_3$	0.9	1.9	1.4				1
Ethylene carbonate	$C_3H_4O_3$	$CDCl_3$	4.5						1
Ethyl formate	$C_3H_6O_2$	CCl_4	1.3	4.2	7.9				1
2-Ethyl-1,3-hexanediol	$C_8H_{18}O_2$	$CDCl_3$	1.0	1.4	3.8				1
2-Ethyl-1-hexanol	$C_8H_{18}O$	$CDCl_3$	0.9	1.3	1.8	3.5			1
Ethyl 3-methylbutanoate	$C_7H_{14}O_2$	$CDCl_3$	1.0	1.3	1.9	4.1			1
3-Ethyl-2-methylpentane	C_8H_{18}	CCl_4	0.9	0.9	1.5				1

Name	Mol. form.	Solvent	Shift 1	Shift 2	Shift 3	Shift 4	Shift 5	Shift 6	Ref.
Fluorobenzene	C_6H_5F	CCl_4	7.0						1
2-Fluorotoluene	C_7H_7F	CCl_4	2.2	6.9					1
3-Fluorotoluene	C_7H_7F	CCl_4	2.3	6.9					1
4-Fluorotoluene	C_7H_7F	CCl_4	2.2	6.8	7.0				1
Furan	C_4H_4O	$CDCl_3$	6.38	7.44					2
Furfural	$C_5H_4O_2$	$CDCl_3$	6.6	7.3	7.7	9.7			1
Furfuryl alcohol	$C_5H_6O_2$	$CDCl_3$	2.8	4.6	6.3	7.4			1
Glycerol	$C_3H_8O_3$	D_2O	3.6						1
Glycerol triacetate	$C_9H_{14}O_6$	$CDCl_3$	2.1	4.2	4.3	5.2			1
Heptane	C_7H_{16}	$CDCl_3$	0.88	1.27	1.30				2
1-Heptanol	$C_7H_{16}O$	CCl_4	0.9	1.4	3.4	3.5			1
3-Heptanol, (*S*)-	$C_7H_{16}O$	CCl_4	0.9	1.4	2.3	3.4			1
2-Heptanone	$C_7H_{14}O$	CCl_4	0.9	1.3	2.0	2.3			1
3-Heptanone	$C_7H_{14}O$	CCl_4	1.0	1.4	2.3				1
1-Heptene	C_7H_{14}	CCl_4	0.9	1.4	2.0	4.9	5.7		1
Hexane	C_6H_{14}	$CDCl_3$	0.89	1.27	1.29				2
Hexanedinitrile	$C_6H_8N_2$	$CDCl_3$	1.8	2.5					1
Hexanenitrile	$C_6H_{11}N$	$CDCl_3$	0.9	1.5	2.3				1
Hexanoic acid	$C_6H_{12}O_2$	$CDCl_3$	0.9	1.4	2.4	11.4			1
1-Hexanol	$C_6H_{14}O$	$CDCl_3$	0.90	1.32	1.56				2
Hexyl acetate	$C_8H_{16}O_2$	CCl_4	0.9	1.4	2.0	4.0			1
3-Hydroxypropanenitrile	C_3H_5NO	$CDCl_3$	2.6	3.4	3.9				1
Iodobenzene	C_6H_5I	CCl_4	6.8	7.5	7.7				1
1-Iodobutane	C_4H_9I	$CDCl_3$	1.0	1.7	1.9	4.2			1
2-Iodobutane, (±)-	C_4H_9I	$CDCl_3$	1.0	1.7	1.9	4.2			1
Iodoethane	C_2H_5I	$CDCl_3$	1.2	2.6	3.7				1
Iodomethane	CH_3I	$CDCl_3$	2.2						1
1-Iodopropane	C_3H_7I	CCl_4	1.0	1.8	3.2				1
2-Iodopropane	C_3H_7I	$CDCl_3$	1.9	4.3					1
Isobutanal	C_4H_8O	CCl_4	1.1	2.4	9.6				1
Isobutyl acetate	$C_6H_{12}O_2$	CCl_4	0.9	1.9	2.0	3.8			1
Isobutylbenzene	$C_{10}H_{14}$	CCl_4	0.9	1.9	2.4	7.1			1
Isobutyl formate	$C_5H_{10}O_2$	$CDCl_3$	1.0	2.0	3.9	8.0			1
Isobutyl isobutanoate	$C_8H_{16}O_2$	CCl_4	0.9	1.2	1.9	2.5	3.8		1
Isopentyl acetate	$C_7H_{14}O_2$	$CDCl_3$	0.9	1.5	2.0	4.0			1
Isophorone	$C_9H_{14}O$	$CDCl_3$	1.04	1.95	2.19				2
Isopropyl acetate	$C_5H_{10}O_2$	$CDCl_3$	0.9	1.4	1.6	2.4			1
Isopropylbenzene	C_9H_{12}	$CDCl_3$	1.3	2.4	2.9	7.3			1
1-Isopropyl-4-methylbenzene	$C_{10}H_{14}$	$CDCl_3$	1.2	2.3	2.9	7.1			1
Isoquinoline	C_9H_7N	$CDCl_3$	8.5	9.3					1
d-Limonene	$C_{10}H_{16}$	CCl_4	1.4	1.7	1.9	4.6	5.3		1
Mesityl oxide	$C_6H_{10}O$	$CDCl_3$	1.89	2.14	2.16				2
Methanol	CH_4O	$CDCl_3$	3.43	3.66					2
2-Methoxyethanol	$C_3H_8O_2$	$CDCl_3$	2.5	3.4	3.5	3.7	3.5	3.7	1
2-Methoxyethyl acetate	$C_5H_{10}O_3$	$CDCl_3$	2.09	3.39	3.59				2
Methyl acetate	$C_3H_6O_2$	CCl_4	2.0	3.7					1
2-Methylacrylonitrile	C_4H_5N	$CDCl_3$	2.0	5.7	5.8				1
2-Methylaniline	C_7H_9N	CCl_4	2.0	3.2	6.7				1
3-Methylaniline	C_7H_9N	CCl_4	2.2	3.3	6.4	6.9			1
N-Methylaniline	C_7H_9N	CCl_4	2.7	3.3	6.4	6.6	7.1		1
Methyl benzoate	$C_8H_8O_2$	CCl_4	3.8	7.4	8.0				1
3-Methylbutanoic acid	$C_5H_{10}O_2$	CCl_4	1.0	2.2	11.0				1
3-Methyl-1-butanol	$C_5H_{12}O$	CCl_4	0.9	1.5	3.5	4.1			1
Methyl cyanoacetate	$C_4H_5NO_2$	$CDCl_3$	3.5	3.8					1
Methylcyclohexane	C_7H_{14}	CCl_4	0.9	1.4					1
cis-4-Methylcyclohexanol	$C_7H_{14}O$	$CDCl_3$	0.9	1.5	2.9	3.9			1
N-Methylformamide	C_2H_5NO	$CDCl_3$	2.82	7.4	8.16				2
Methyl formate	$C_2H_4O_2$	$CDCl_3$	3.76	8.07					2
2-Methylheptane	C_8H_{18}	CCl_4	0.9	1.3					1
4-Methylheptane	C_8H_{18}	$CDCl_3$	0.8	0.9	1.4				1
2-Methylhexane	C_7H_{16}	$CDCl_3$	0.9	0.9	1.4				1

Name	Mol. form.	Solvent	Shift 1	Shift 2	Shift 3	Shift 4	Shift 5	Shift 6	Ref.
5-Methyl-2-hexanone	$C_7H_{14}O$	CCl_4	0.9	1.4	2.0	2.3			1
Methyl methacrylate	$C_5H_8O_2$	$CDCl_3$	2.0	3.8	5.6	6.1			1
2-Methyloctane	C_9H_{20}	CCl_4	0.9	1.0	1.3				1
2-Methylpentane	C_6H_{14}	CCl_4	0.8	0.9	1.5				1
3-Methylpentane	C_6H_{14}	CCl_4	0.8	1.5					1
2-Methyl-2,4-pentanediol	$C_6H_{14}O_2$	CCl_4	1.1	1.2	1.3	1.4	1.6	4.2	1
4-Methylpentanenitrile	$C_6H_{11}N$	CCl_4	1.0	1.6	2.3				1
4-Methyl-2-pentanol	$C_6H_{14}O$	CCl_4	0.9	1.1	1.3	1.8	3.5	3.7	1
3-Methyl-3-pentanol	$C_6H_{14}O$	CCl_4	0.9	1.1	1.4	1.8			1
4-Methyl-2-pentanone	$C_6H_{12}O$	$CDCl_3$	0.9	2.1	2.3				1
2-Methylpropanenitrile	C_4H_7N	$CDCl_3$	1.3	2.7					1
2-Methylpropanoic acid	$C_4H_8O_2$	CCl_4	1.2	2.6					1
2-Methyl-1-propanol	$C_4H_{10}O$	CCl_4	0.9	1.7	3.3	4.0			1
2-Methyl-2-propanol	$C_4H_{10}O$	$CDCl_3$	1.3	1.4					1
2-Methylpyridine	C_6H_7N	CCl_4	2.5	6.9	7.4	8.4			1
3-Methylpyridine	C_6H_7N	CCl_4	2.3	7.0	7.4	8.4			1
N-Methyl-2-pyrrolidinone	C_5H_9NO	$CDCl_3$	2.1	2.4	2.8	3.4			1
Methyl salicylate	$C_8H_8O_3$	CCl_4	3.9	6.7	6.9	7.3	7.7	10.6	1
2-Methylthiophene	C_5H_6S	$CDCl_3$	2.5	6.7	6.9	7.0			1
Morpholine	C_4H_9NO	$CDCl_3$	2.59	2.86	3.67				2
Nitrobenzene	$C_6H_5NO_2$	CCl_4	7.52	7.65	8.19				2
Nitroethane	$C_2H_5NO_2$	$CDCl_3$	1.6	4.3					1
Nitromethane	CH_3NO_2	CCl_4	4.2						1
1-Nitropropane	$C_3H_7NO_2$	$CDCl_3$	1.0	2.1	4.4				1
Nonane	C_9H_{20}	$CDCl_3$	0.9	1.3					1
Octane	C_8H_{18}	$CDCl_3$	0.88	1.26					1
1-Octanol	$C_8H_{18}O$	$CDCl_3$	0.88	1.29	1.5				2
2-Octanone	$C_8H_{16}O$	CCl_4	0.9	1.4	2.1	2.4			1
1-Octene	C_8H_{16}	CCl_4	0.9	1.3	2.1	4.8	4.9	5.7	1
Pentane	C_5H_{12}	$CDCl_3$	0.88	1.26	1.30				2
1,5-Pentanediol	$C_5H_{12}O_2$	D_2O	1.5	3.6					1
2,4-Pentanedione	$C_5H_8O_2$	CCl_4	2.0	2.2	3.5	5.4	14.7		1
Pentanenitrile	C_5H_9N	CCl_4	1.0	1.6	2.3				1
Pentanoic acid	$C_5H_{10}O_2$	CCl_4	0.9	1.5	2.3	11.7			1
1-Pentanol	$C_5H_{12}O$	CCl_4	0.9	1.4	3.5	4.4			1
2-Pentanol	$C_5H_{12}O$	$CDCl_3$	0.9	1.2	1.3	2.2	3.7		1
3-Pentanol	$C_5H_{12}O$	CCl_4	0.9	1.4	3.3	3.4			1
2-Pentanone	$C_5H_{10}O$	CCl_4	0.9	1.6	2.0	2.3			1
3-Pentanone	$C_5H_{10}O$	CCl_4	1.0	2.4					1
Pentyl acetate	$C_7H_{14}O_2$	CCl_4	0.9	1.4	2.0	4.0			1
Pentylamine	$C_5H_{13}N$	$CDCl_3$	0.9	1.4	2.7				1
α-Pinene	$C_{10}H_{16}$	$CDCl_3$	0.84	1.16	1.27				2
Piperidine	$C_5H_{11}N$	$CDCl_3$	1.53	2.18	2.79				2
Propanal	C_3H_6O	$CDCl_3$	1.1	2.4	9.8				1
1,2-Propanediol	$C_3H_8O_2$	$CDCl_3$	1.1	3.4	3.9	4.3			1
1,3-Propanediol	$C_3H_8O_2$	D_2O	1.8	3.7					1
Propanenitrile	C_3H_5N	CCl_4	1.3	2.3					1
Propanoic acid	$C_3H_6O_2$	$CDCl_3$	1.1	2.4	10.5				1
Propanoic anhydride	$C_6H_{10}O_3$	CCl_4	1.2	2.4					1
1-Propanol	C_3H_8O	$CDCl_3$	0.9	1.6	2.3	3.6			1
2-Propanol	C_3H_8O	$CDCl_3$	1.2	1.6	4.0				1
Propargyl alcohol	C_3H_4O	$CDCl_3$	2.5	2.8	4.3				1
Propyl acetate	$C_5H_{10}O_2$	CCl_4	0.9	1.6	2.0	4.0			1
Propylamine	C_3H_9N	CCl_4	0.9	1.5	2.1	2.7			1
Propylbenzene	C_9H_{12}	$CDCl_3$	0.9	1.6	2.6				1
Propyl formate	$C_4H_8O_2$	CCl_4	1.0	1.7	4.1	7.9			1
Pyridine	C_5H_5N	$CDCl_3$	7.23	7.62	8.59				2
Pyrrole	C_4H_5N	$CDCl_3$	6.2	6.7	8.0				1
Pyrrolidine	C_4H_9N	C_6H_6	1.5	2.4	2.7				1
2-Pyrrolidone	C_4H_7NO	$CDCl_3$	2.2	3.4	7.7				1
Quinoline	C_9H_7N	CCl_4	7.1	7.8	8.8				1

Name	Mol. form.	Solvent	Shift 1	Shift 2	Shift 3	Shift 4	Shift 5	Shift 6	Ref.
Salicylaldehyde	$C_7H_6O_2$	$CDCl_3$	7.0	7.4	9.8	11.0			1
Styrene	C_8H_8	CCl_4	5.1	5.6	6.6	7.2			1
Sulfolane	$C_4H_8O_2S$	$CDCl_3$	2.2	3.0					1
α-Terpinene	$C_{10}H_{16}$	$CDCl_3$	1.02	1.77	2.09				2
1,1,1,2-Tetrachloroethane	$C_2H_2Cl_4$	$CDCl_3$	4.29						2
1,1,2,2-Tetrachloroethane	$C_2H_2Cl_4$	$CDCl_3$	5.91						2
Tetraethylene glycol	$C_8H_{18}O_5$	CCl_4	3.5	3.6					1
Tetrahydrofuran	C_4H_8O	$CDCl_3$	1.84	3.73					2
1,2,3,4-Tetrahydronaphthalene	$C_{10}H_{12}$	$CDCl_3$	1.8	2.8	7.1				1
Tetrahydropyran	$C_5H_{10}O$	CCl_4	1.6	3.6					1
Tetrahydrothiophene	C_4H_8S	$CDCl_3$	1.9	2.8					1
Tetramethylsilane	$C_4H_{12}Si$	CCl_4	0.0						1
Tetramethylurea	$C_5H_{12}N_2O$	CCl_4	2.8						1
Thiophene	C_4H_4S	$CDCl_3$	7.1	7.3					1
Toluene	C_7H_8	$CDCl_3$	2.34	7.18					2
Tribromomethane	$CHBr_3$	CCl_4	6.8						1
Tributylamine	$C_{12}H_{27}N$	CCl_4	0.9	1.3	2.3				1
1,1,1-Trichloroethane	$C_2H_3Cl_3$	CCl_4	2.7						1
1,1,2-Trichloroethane	$C_2H_3Cl_3$	$CDCl_3$	4.0	5.8					1
Trichloroethene	C_2HCl_3	CCl_4	6.5						1
Trichloroethylsilane	$C_2H_5Cl_3Si$	CCl_4	1.3						1
Trichloromethane	$CHCl_3$	CCl_4	7.2						1
(Trichloromethyl)benzene	$C_7H_5Cl_3$	CCl_4	7.3	7.8					1
Tridecane	$C_{13}H_{28}$	CCl_4	0.9	1.3					1
Tris(2-hydroxyethyl)amine	$C_6H_{15}NO_3$	D_2O	2.7	3.6					1
Triethylamine	$C_6H_{15}N$	CCl_4	1.0	2.4					1
Triethylene glycol	$C_6H_{14}O_4$	$CDCl_3$	3.5	3.7					1
Triethyl phosphate	$C_6H_{15}O_4P$	$CDCl_3$	1.4	4.1					1
2,2,2-Trifluoroethanol	$C_2H_3F_3O$	$CDCl_3$	3.4	3.9					1
(Trifluoromethyl)benzene	$C_7H_5F_3$	CCl_4	7.5						1
Trimethylamine	C_3H_9N	CCl_4	2.12						2
1,2,3-Trimethylbenzene	C_9H_{12}	$CDCl_3$	2.2	2.3	7.0				1
1,2,4-Trimethylbenzene	C_9H_{12}	CCl_4	2.2	6.8					1
1,3,5-Trimethylbenzene	C_9H_{12}	$CDCl_3$	2.3	6.8					1
2,2,3-Trimethylbutane	C_7H_{16}	CCl_4	0.8	1.3					1
2,2,5-Trimethylhexane	C_9H_{20}	CCl_4	0.9	1.2					1
2,3,3-Trimethylpentane	C_8H_{18}	CCl_4	0.8	0.8	1.4				1
2,3,4-Trimethylpentane	C_8H_{18}	CCl_4	0.8	1.9					1
Trimethyl phosphate	$C_3H_9O_4P$	$CDCl_3$	3.78						2
2,4,6-Trimethylpyridine	$C_8H_{11}N$	CCl_4	2.2	2.4	6.6				1
1-Undecene	$C_{11}H_{22}$	CCl_4	0.9	1.3	2.0	4.8	4.9	5.6	1
Vinyl acetate	$C_4H_6O_2$	$CDCl_3$	2.1	4.6	4.9	7.3			1
o-Xylene	C_8H_{10}	$CDCl_3$	2.22	7.07					2
m-Xylene	C_8H_{10}	$CDCl_3$	2.28	6.95	7.11				2
p-Xylene	C_8H_{10}	$CDCl_3$	2.30	7.05					2

PROTON CHEMICAL SHIFTS OF CONTAMINANTS IN DEUTERATED SOLVENTS

Thomas J. Bruno

The following table provides the expected chemical shift of solutes that commonly occur as impurities in NMR spectra measured at 298 K. These chemical shifts can vary as a function of (deuterated) solvents. These data are adapted from the compilation of Gottlieb et al. (Ref. 1), as expanded by Fulmer et al. (Ref. 2). Only the more common solvents and impurities are considered here. The reader is referred to these sources for additional impurities and solvents. Note that the chemical shifts of the -OH protons will vary with concentration as a result of hydrogen bonding, so these have not been included. Column definitions for the table are as follows.

Column heading	Definition
Name	Contaminant name, listed alphabetically
Proton	Proton under consideration; when the same moiety contains multiple protons, boldface print indicates the protons under consideration; when protons differ by ring position, the appropriate protons are indicated parenthetically
Deuterated solvent name or formula (THF-d_8, CD_2Cl_2, etc.)	^1H chemical shift in specified solvent, in parts per million (ppm) relative to tetramethylsilane (TMS); THF is tetrahydrofuran

References

1. Gottlieb, H. E., Kotlyar, K., and Nudelman, A., *J. Org. Chem.* 62, 7512, 1997. <https://doi.org/10.1021/jo971176v>
2. Fulmer, G. R., Miller, A. J. M., Sherdan, N. H., Gottleib, H. E., Nudelman, A., Stoltz, B. M., Bercaw, J. E., and Goldberg, K. I., *Organometallics* 29, 2176, 2010. <https://doi.org/10.1021/om100106e>

^1H Chemical Shifts of Contaminants in Deuterated Solvents in ppm Relative to TMS

Name	Proton	THF-d_8	CD_2Cl_2	$CDCl_3$	Toluene-d_8	C_6D_6	CD_3OD	D_2O
Acetic acid	CH_3	1.89	2.06	2.1	1.57	1.52	4.87	
Acetone	CH_3	2.05	2.12	2.17	1.57	1.55	2.15	2.22
Acetonitrile	CH_3	1.95	1.97	2.1	0.69	0.58	2.03	2.06
Benzene	CH	7.31	7.35	7.36	7.12	7.15	7.33	
Cyclohexane	CH_2	1.44	1.44	1.43	1.4	1.4	1.45	
1,4-Dioxane	CH_2	3.56	3.65	3.71	3.33	3.35	3.66	3.75
Ethanol	CH_3	1.1	1.19	1.25	0.97	0.96	1.19	1.17
Ethanol	CH_2	3.51	3.66	3.72	3.36	3.39	3.6	3.65
Ethyl acetate	CH_3CO	1.94	2	2.05	1.69	1.65	2.01	2.07
Ethyl acetate	CH_2	4.04	4.08	4.12	3.87	3.89	4.09	4.14
Ethyl acetate	CH_3	1.19	1.23	1.26	0.94	0.92	1.24	1.24
Hexane	CH_3	0.89	0.89	0.88	0.88	0.89	0.9	
Hexane	CH_2	1.29	1.27	1.26	1.22	1.24	1.29	
2-Propanol	CH_3	1.08	1.17	1.22	0.95	0.95	1.5	1.17
2-Propanol	CH	3.82	3.97	4.04	3.65	3.67	3.92	4.02
Toluene	CH_3	2.31	2.34	2.36	2.11	2.11	2.32	
Toluene	CH (2,4,6)	7.1	7.15	7.17	6.99[(av)]	7.02	7.16	
Toluene	CH (3,5)	7.19	7.24	7.25	7.09	7.13	7.16	
Trichloromethane	CH	7.89	7.32	7.26	6.1	6.15	7.9	

^{13}C-NMR ABSORPTIONS OF MAJOR FUNCTIONAL GROUPS

Thomas J. Bruno and Paris D. N. Svoronos

The table below lists the ^{13}C chemical shift ranges (in ppm) with the corresponding functional groups in descending order. Some typical simple compounds for every family are given to illustrate the corresponding range. The shifts for the carbons of interest are given in parentheses, either for each carbon as it appears from left to right in the formula, or by the underscore (Refs. 1–17). In Ref. 1, the reader can find additional details, correlation charts, additivity rules, and the expected ^{13}C peaks attributed to common solvents. All shifts are relative to tetramethylsilane (TMS).

Column definitions for the table are as follows.

Column heading	Definition
δ range	^{13}C chemical shift range, in parts per million (ppm) relative to TMS
Group	Functional group exhibiting shifts within the specified range
Family	Families of compounds within the specified functional group
Example	Specific example compounds within the family
δ of underlined carbon	Actual ^{13}C chemical shift for the example compound, in ppm relative to TMS

References

1. Bruno, T. J., and Svoronos, P. D. N., *CRC Handbook of Basic Tables for Chemical Analysis, Third Edition*, CRC Press, Boca Raton, FL, 2011. <https://doi.org/10.1201/b10385>
2. Yoder, C. H., and Schaeffer, C. D., Jr., *Introduction to Multinuclear NMR: Theory and Application*, The Benjamin/Cummings Publishing Co., Menlo Park, CA, 1987.
3. Brown, D. W., *J. Chem. Educ.* 62, 209, 1985. <https://doi.org/10.1021/ed062p209>
4. Silverstein, R. M., and Webster F. X., *Spectrometric Identification of Organic Compounds, Sixth Edition*, John Wiley and Sons, New York, 1998.
5. Becker, E. D., *High Resolution NMR, Theory and Chemical Applications, Second Edition*, Academic Press, New York, 1980.
6. Gunther, H., *NMR Spectroscopy: Basic Principles, Concepts and Applications in Chemistry*, Wiley, New York, 2003.
7. Kitamaru, R., *Nuclear Magnetic Resonance: Principles and Theory*, Elsevier Science, 1990.
8. Lambert, J. B., Holland, L. N., and Mazzola, E. P., *Nuclear Magnetic Resonance Spectroscopy: Introduction to Principles, Applications and Experimental Methods*, Prentice Hall, Englewood Cliffs, NJ, 2003.
9. Bovey, F. A., and Mirau, P. A., *Nuclear Magnetic Resonance Spectroscopy, Second Edition*, Academic Press, New York, 1988.
10. Harris, R. K., and Mann, B. E., *NMR and the Periodic Table*, Academic Press, London, 1978.
11. Hore, P. J., *Nuclear Magnetic Resonance*, Oxford University Press, Oxford, 1995.
12. Nelson, J. H., *Nuclear Magnetic Resonance Spectroscopy, Second Edition*, Wiley, New York, 2003.
13. Levy, G. C., Lichter, R. L., and Nelson, G. L., *Carbon-13 Nuclear Magnetic Resonance Spectroscopy, Second Edition*, Wiley, New York, 1980.
14. Pihlaja, K., and Kleinpeter, E., *Carbon-13 NMR Chemical Shifts in Structural and Stereochemical Analysis*, VCH, New York, 1994.
15. *Aldrich Library of 1H and 13C FT-NMR Spectra*, Aldrich Chemical Company, Milwaukee, WI, 1996.
16. Balci, M., *Basic 1H- and 13C-NMR Spectroscopy*, Elsevier, London, 2005. <https://doi.org/10.1016/B978-044451811-8/50008-5>
17. <http://www.chem.wisc.edu/areas/reich/handouts/nmr-c13/cdata.htm>, 2009.

Chemical Shifts Relative to Tetramethylsilane for Major Functional Groups

δ range/ ppm	Group	Family	Example	δ of underlined carbon/ppm
220–165	>C=O	Ketones	$(CH_3)_2\underline{C}O$	(206.0)
			$(CH_3)_2CH\underline{C}OCH_3$	(212.1)
		Aldehydes	$CH_3\underline{C}HO$	(199.7)
		α,β-unsaturated carbonyls	$CH_3CH{=}CHCHO$	(192.4)
			$CH_2{=}CH\underline{C}OCH_3$	(169.9)
		Carboxylic acids	$H\underline{C}O_2H$	(166.0)
			$CH_3\underline{C}O_2H$	(178.1)
		Amides	$H\underline{C}ONH_2$	(165.0)
			$CH_3\underline{C}ONH_2$	(172.7)
		Esters	$CH_3\underline{C}O_2CH_2CH_3$	(170.3)
			$CH_2{=}CH\underline{C}O_2CH_3$	(165.5)
140–120	>C=C<	Aromatics, alkenes	C_6H_6	(128.5)
			$CH_2{=}CH_2$	(123.2)
			$CH_2{=}CHCH_3$	(115.9, 136.2)
			$CH_2{=}CHCH_2Cl$	(117.5, 133.7)
			$CH_3CH{=}\underline{C}HCH_2CH_3$	(132.7)
125–115	–C≡N	Nitriles	$CH_3{-}\underline{C}{\equiv}N$	(117.7)
80–70	–C≡C–	Alkynes	$H\underline{C}{\equiv}CH$	(71.9)
			$CH_3\underline{C}{\equiv}CCH_3$	(73.9)
70–45	>C–O–	Esters, alcohols	$\underline{C}H_3OOCCH_2CH_3$	(57.6, 67.9)

δ range/ppm	Group	Family	Example	δ of underlined carbon/ppm
	$>$C–OH		$HO\underline{C}H_3$	(49.0)
			$HO\underline{C}H_2CH_3$	(57.0)
40–20	$>$C–NH_2	Amines	$\underline{C}H_3NH_2$	(26.9)
			$CH_3\underline{C}H_2NH_2$	(35.9)
30–15	–S–CH_3	Sulfides (thioethers)	C_6H_5–S–$\underline{C}H_3$	15.6
30–(–2.3)	>CH–	Alkanes, cycloalkanes	$\underline{CH_4}$	(–2.3)
			$\underline{CH_3}CH_3$	(5.7)
			$\underline{CH_3CH_2}CH_3$	(15.8, 16.3)
			$CH_3CH_2CH_2\underline{CH_3}$	(13.4, 25.2)
			$\underline{CH_3CH_2CH_2}CH_2CH_3$	(13.9, 22.8, 34.7)
			Cyclohexane	(26.9)

^{13}C-NMR CHEMICAL SHIFTS OF COMMON ORGANIC SOLVENTS

The following table gives the expected carbon-13 (^{13}C) chemical shifts, in ppm relative to tetramethylsilane, for various useful NMR solvents. In some solvents, slight changes can occur with change of concentration (Refs. 2-3). Column definitions for the table are as follows.

Column heading	Definition
Name	Name of organic solvent, compounds are listed alphabetically by name
Mol. form.	Molecular formula of organic solvent
Chemical shift	^{13}C chemical shift, in parts per million (ppm) relative to tetramethylsilane (TMS)

References

1. Bruno, T. J., and Svoronos, P. D. N., *CRC Handbook of Basic Tables for Chemical Analysis, Third Edition*, CRC Press, Boca Raton, FL, 2011. <https://doi.org/10.1201/b10385>
2. Silverstein, R. M., Bassler, G. C., and Morrill, T. C., *Spectrometric Identification of Organic Compounds*, John Wiley & Sons, Now York, 1981.
3. Rahman, A.-U., *Nuclear Magnetic Resonance. Basic Principles*, Springer-Verlag, New York, 1986.
4. Pretsch, E., Clerc, T., Seibl, J., and Simon, W., *Spectral Data for Structure Determination of Organic Compounds, Second Edition*, Springer-Verlag, Heidelberg, 1989. <https://doi.org/10.1007/978-3-662-10207-7_1>

^{13}C-NMR Chemical Shifts of Common Organic Solvents (in ppm Relative to Tetramethylsilane)

Name	Mol. form.	Chemical shift (ppm)
Acetic acid-d_4	CD_3COOD	20.0 (CD_3) 205.8 (C=O)
Acetone	$(CH_3)_2C{=}O$	30.7 (CH_3) 206.7 (C=O)
Acetone-d_6	$(CD_3)_2C{=}O$	29.2 (CD_3) 204.1 (C=O)
Acetonitrile-d_3	$CD_3C{\equiv}N$	1.3 (CD_3) 117.1 (C≡N)
Benzene	C_6H_6	128.5
Benzene-d_6	C_6D_6	128.4
Carbon disulfide	CS_2	192.3
Carbon tetrachloride	CCl_4	96.0
Chloroform	$CHCl_3$	77.2
Chloroform-d_3	$CDC1_3$	77.05
Cyclohexane-d_{12}	C_6D_{12}	27.5
Dichloromethane-d_2	CD_2C1_2	53.6
Dimethylformamide-d_7	$(CD_3)_2NCDO$	31 (CD_3) 36 (CD_3) 162.4 (C=O)
Dimethylsulfoxide-d_6	$(CD_3)_2S{=}O$	39.6
Dioxane-d_8	$C_4D_3O_2$	67.4
Formic acid-d_2	DCOOD	165.5
Methanol-d_4	CD_3OD	49.3
Nitromethane-d_3	CD_3NO_2	57.3
Pyridine	C_5H_5N	123.6 (C_3) 135.7 (C_4) 149.8 (C_2)
Pyridine-d_5	C_5D_5N	123.9 (C_3) 135.9 (C_4) 150.2 (C_2)
1,1,2,2-Tetrachloroethane-d_2	$CDC1_2CDC1_2$	75.5
Tetrahydrofuran-d_8	C_4D_8O	25.8 (C_2) 67.9 (C_1)
Trichlorofluoromethane	$CFC1_3$	117.6

^{15}N-NMR CHEMICAL SHIFTS OF MAJOR CHEMICAL FAMILIES

Thomas J. Bruno and Paris D. N. Svoronos

The following table contains ^{15}N-NMR chemical shifts of various organic nitrogen compounds. Chemical shifts are expressed relative to different standards (NH_3, NH_4Cl, CH_3NO_2, NH_4NO_3, HNO_3, etc.) and are interconvertible. Chemical shifts are sensitive to hydrogen bonding and are solvent dependent as seen in the case of pyridine (see Footnote a below). Consequently, the reference as well as the solvent should always accompany chemical shift data. All shifts are relative to ammonia unless otherwise specified (Refs. 1–16). A section of "miscellaneous" data gives the chemical shift of special compounds relative to unusual standards. In Ref. 16, the reader can find additional details, correlation charts, and spin-spin coupling ranges.

References

1. Levy, G. C., and Lichter, R. L., *Nitrogen-15 Nuclear Magnetic Resonance Spectroscopy*, John Wiley and Sons, New York, 1979.
2. Yoder, C. H., and Schaeffer, C. D., Jr., *Introduction to Multinuclear NMR*, Benjamin/Cummings, Menlo Park, CA, 1987.
3. Duthaler, R. O., and Roberts, J. D., *J. Am. Chem. Soc.* 100, 4969, 1978. <https://doi.org/10.1021/ja00484a008>
4. Duthaler, R. O., and Roberts, J. D., *J. Am. Chem. Soc.* 100, 3889, 1978. <https://doi.org/10.1021/ja00480a038>
5. Kozerski, L., and von Philipsborn, W., *Org. Magn. Res.* 17, 306, 1981. <https://doi.org/10.1002/mrc.1270170418>
6. Duthaler, R. O., and Roberts, J. D., *J. Am. Chem. Soc.* 100, 3882, 1978. <https://doi.org/10.1021/ja00480a037>
7. Duthaler, R. O., and Roberts, J. D., *J. Magn. Res.* 34, 129, 1979. <https://doi.org/10.1016/0022-2364(79)90036-2>
8. Psota, L., Franzen-Sieveking, M., Turnier, J., and Lichter, R. L., *Org. Magn. Res.* 11, 401, 1978. <https://doi.org/10.1002/mrc.1270110808>
9. Subramanian, P. K., Chandra Sekara, N., and Ramalingam, K., *J. Org. Chem.* 47, 1933, 1982. <https://doi.org/10.1021/jo00349a020>
10. Schuster, I. I., and Roberts, J. D., *J. Org. Chem.* 45, 284, 1980. <https://doi.org/10.1021/jo01290a016>
11. Kupce, E., Liepins, E., Pudova, O., and Lukevics, E., *J. Chem. Soc., Chem. Commun.*, 581, 1984. <https://doi.org/10.1039/C39840000581>
12. Allen, M., and Roberts, J. D., *J. Org. Chem.* 45, 130, 1980. <https://doi.org/10.1021/jo01289a025>
13. Brownlee, R. T. C., and Sadek, M., *Magn. Res. Chem.* 24, 821, 1986. <https://doi.org/10.1002/mrc.1260240917>
14. Dega-Szafran, Z., Szafran, M., Stefaniak, L., Brevard, C., and Bourdonneau, M., *Magn. Res. Chem.* 24, 424, 1986. <https://doi.org/10.1002/mrc.1260240507>
15. Lambert, J. B., Shurvell, H. F., Verbit, L., Cooks, R. G., and Stout, G. H., *Organic Structural Analysis*, MacMillan, New York, 1976.
16. Bruno, T. J., and Svoronos, P. D. N., *CRC Handbook of Basic Tables for Chemical Analysis, Third Edition*, CRC Press, Boca Raton, FL, 2011. <https://doi.org/10.1201/b10385>

^{15}N-NMR Chemical Shifts of Major Chemical Families (Relative to NH_3 Unless Otherwise Indicated)

(δ) range/ppm	Family	Example	(δ) of N atom/ppm
<930	Nitroso compounds	C_6H_5–NO	(913, 930)
608	Sodium nitrite	$NaNO_2$	608
~500	Azo compounds	C_6H_5–N=N–C_6H_5	510
380–350	Nitro compounds	$C_6H_5NO_2$	370.3
		CH_3NO_2	380.2
		4-F–C_6H_4–NO_2	368.5
		1,3-$(NO_2)_2C_6H_4$	365.4
367	Nitric acid (8.57 M)	HNO_3	367
360–325	Nitramines	CH_3NHNO_2	355.6
		$CH_3O_2CNHNO_2$	334.9
350–300	Pyridines	C_5H_5N (gas)	317[a]
		4-CH_3–C_5H_4N	309.3
		4-NH_2–C_5H_4N	271.5
		4-NC–C_5H_4N	327.9
~310	Imines (aromatic)	$(C_6H_5)_2C$=NH	308
		C_6H_5CH=NCH_3	318
		C_6H_5CH=NC_6H_5	326
310.1	Nitrogen (gas)	N_2	310.1
250–200	Pyridinium salts	$C_5H_5NH^+$	215
260–175	Cyanides (nitriles)	CH_3CN	(239.5, 245)
		C_6H_5CN	258.7
		KCN	177.8
~160	Pyrroles, isonitriles	C_4H_4NH	158
		CH_3NC	162
~150	Thioamides	$CH_3C(=S)NH_2$	150.2
120–110	Lactams	$HN(CH_2)_3C$=O	114.7 (5-membered ring)
		$HN(CH_2)_6C$=O	117.7 (8-membered ring)
110–100	Amides	$C_6H_5CONH_2$	100
		CH_3CONH_2	103.4
		$CH_3CONHCH_3$	105.8

(δ) range/ppm	Family	Example	(δ) of N atom/ppm
		$CH_3CON(CH_3)_2$	103.8
		$HCONH_2$	108.5
125–90	Sulfonamides	$CH_3SO_2NH_2$	95
		$C_6H_5SO_2NH_2$	94.3
~100	Hydrazines	$C_6H_5NHNHC_6H_5$	96
110–60	Ureas	$[H_2N]_2CO$	(75, 82)
		$[(CH_3)_2N]_2CO$	63.5
		$[C_6H_5NH]_2CO$	107.7
100–70	Aminophospines, aminophosphine oxides	$C_6H_5NHP(CH_3)_2$	71.1
		$C_6H_5NHPO(CH_3)_2$	86.6
70–50	Aromatic amines	$C_6H_5NH_2$	(55, 59), (−322.3)[b]
		$C_6H_5NH_3^+$	48, −326.4[b], 26.1[c]
		p–O_2N–C_6H_4–NH_2	70
40–0	Aliphatic amines	CH_3NH_2	(1.3)[d], (−371)[b]
		$(CH_3)_2NH$	(−363.3)[b], (−364.9)[e], 6.7[d]
		$(CH_3)_3N$	(−356.9)[b], (−360.7)[e], 13.0[d]
50–10	Isonitriles	CH_3NCO	14.1
		C_6H_5NCO	46.5
65–20	Ammonium salts	NH_4Cl	26.1[d]
		CH_3NH_3Cl	24.5
		$(CH_3)_2NH_2Cl$	26.6
		$(CH_3)_3NHCl$	33.8
		$(CH_3)_4NCl$	44.7
~15	Isocyanates	CH_3NCO	14.1
Miscellaneous			
(−130)–(−110) and ~(−212)	Imidazoles	*N*-methylimdazole	-111.4 (pyridine N)[b]
			-215.7 (pyrrole N)[b]
(−345)–(−310)[b]	Piperidine, hydrochloride salts	piperidinium hydorchloride	-344.8
		2-methyl piperdinium hydrochloride	-322.1[e]
	Decahydroquinolines, hydrochloride salts	*trans*-decahydroquinolinium hydrochloride	-322.5
		cis-decahydroquinolinium hydrochloride	-328.5
(−293)–(−280)[f]	Enaminones	$CH_3C(=O)CH=CHNHCH_3$	
		(*E*)	−294.2
		(*Z*)	-285.9
35–15[g]	4-Aminotetrahydropyrans, 4-aminotetrahydro-thiopyrans	2,6-diphenyl 4-aminotetrahydropyran	34.5
		2,6-diphenyl 4-aminotetrahydrothiopyran	33.6
(−325)–(−310)[c]	1-Naphthylamines	8-nitro-1-naphthylamine	313.9
(−350)–(−300)[h]	Silylamines	$HN[Si(CH_3)_3]_2$	-354.2

[a] Varies with solvent. For instance: cyclohexane (315.5), benzene (312.1), chloroform (304.5), methanol (292.1), water (289), 2,2,2-trifluoroethanol (277.1). All chemical shifts are relative to ammonia.
[b] Upfield from external HNO_3 1M (CH_3OH).
[c] In ppm upfield from external 1M $D^{15}NO_3$ in D_2O (DMSO) (Ref. 10).
[d] Downfield from anhydrous liquid ammonia, ±0.2 ppm unless otherwise specified.
[e] Upfield from external HNO_3 1M (cyclohexane).
[f] Relative to external $CH_3{}^{15}NO_2$.
[g] With respect to an external standard of 5M $^{15}NH_4NO_3$ in 2M HNO_3 ($^{15}NH_4NO_3$ = 21.6 ppm relative to anhydrous ammonia) (Ref. 9).
[h] Relative to $N(SiH_3)_3$ (50% in $CDCl_3$) (Ref. 11).

NATURAL ABUNDANCE OF IMPORTANT ISOTOPES

Thomas J. Bruno and Paris D. N. Svoronos

The following table lists the atomic masses and uncertainty with their relative percent concentrations of naturally occurring isotopes of importance in mass spectroscopy (Refs. 1–6). The atomic masses, uncertainties, and relative percent concentrations are in accordance with the "Atomic Masses and Isotopic Abundances" in Section 1 of this *CRC Handbook* (Refs. 7-8). All masses are in atomic mass units μ, with μ of ^{12}C being exactly equal to 12.

References

1. de Hoffmann, E., and Stroobant, V., *Mass Spectrometry: Principles and Applications, Third Edition*, Wiley Interscience, Winchester, UK, 2007.
2. Johnstone, R. A. W., and Rose, M. E., *Mass Spectrometry for Chemists and Biochemists*, Cambridge University Press, 1996. <https://doi.org/10.1017/CBO9781139166522>
3. Lide, D. R., Ed., *CRC Handbook for Chemistry and Physics, 90th Edition*, CRC Press, Boca Raton, FL, 2010.
4. McLafferty, F. W., and Turecek, F., *Interpretation of Mass Spectra, Fourth Edition*, University Science Books, Mill Valley, CA, 1993.
5. Watson, J. T., *Introduction to Mass Spectrometry, Third Edition*, Lippincott-Raven, Philadelphia, PA, 1997.
6. Bruno, T. J., and Svoronos, P. D. N., *CRC Handbook of Basic Tables for Chemical Analysis, Third Edition*, CRC Press, Boca Raton, FL, 2011. <https://doi.org/10.1201/b10385>
7. Wang, M. et al., *Chinese Physics* C41, 030003, 2017.
8. See *CIAAW Web site*, <www.ciaaw.org/index.htm>.

Natural Abundance of More Prominent Isotopes (Mass in u, Percent Abundance)

Element	Primary isotope	Other isotopes	Other isotopes
Hydrogen	^{1}H (1.007 825 032 24(9), 99.972 – 99.999)	^{2}H (2.014 101 778 11(12), 0.001 – 0.028)	
Boron	^{10}B (10.012 936 862(16), 18.9 – 20.4)	^{11}B (11.009 305 167(13), 79.6 – 81.1)	
Carbon	^{12}C (12.000 00, 98.84 – 99.04)	^{13}C (13.003 354 835 21(23), 0.96 – 1.16)	
Nitrogen	^{14}N (14.003 074 004 46(21), 99.578 – 99.663)	^{15}N (15.0001088989(6), 0.337 – 0.422)	
Oxygen	^{16}O (15.994 914 619 60(17), 99.738 – 99.776)		^{18}O (17.999 159 6128(8), 0.187 – 0.222)
Fluorine	^{19}F (18.998 403 1629(9), ≈100.0)		
Silicon	^{28}Si (27.976 926 5350(5), 92.191 – 92.318)	^{29}Si (28.976 494 6653(6), 4.645 – 4.699)	^{30}Si (29.973 770 137(23), 3.037 – 3.110)
Phosphorus	^{31}P (30.973 761 9986(7), ≈100.0)		
Sulfur	^{32}S (31.972 071 1744(14), 94.41 – 95.29)	^{33}S (32.971 458 9099(15), 0.729 – 0.797)	^{34}S (33.967 867 01(5), 3.96 – 4.77)
Chlorine	^{35}Cl (34.968 852 69(4), 75.5 – 76.1)		^{37}Cl (36.965 902 58(6), 23.9 – 24.5)
Bromine	^{79}Br (78.918 3376(11), 50.5 – 50.8)		^{81}Br (80.916 2882(1), 49.2 – 49.5)
Iodine	^{127}I (126.904 472(4), ≈100.0)		

COMMON MASS SPECTRAL FRAGMENTATION PATTERNS OF ORGANIC COMPOUND FAMILIES

Thomas J. Bruno and Paris D. N. Svoronos

The following table provides a guide to the identification and interpretation of commonly observed mass spectral fragmentation patterns for common organic functional groups (Refs. 1–10). It is of course highly desirable to augment mass spectroscopic data with as much other structural information as possible. Especially useful in this regard will be the confirmatory information of infrared and ultraviolet spectrophotometry as well as nuclear magnetic resonance spectrometry.

The reader is referred to other tables in this book for additional information that can be used in assessing or predicting fragmentation patterns, particularly the tables of "Proton Affinities" and "Ionization Energies of Gas-Phase Molecules" in Section 10.

References

1. Bowie, J. H., Williams, D. H., Lawesson, S. O., Madsen, J. O., Nolde, C., and Schroll, G., *Tetrahedron* 22, 3515, 1966.
2. Johnstone, R. A. W., and Rose, M. E., *Mass Spectrometry for Chemical and Biochemists*, Cambridge University Press, New York, 1996.
3. Lee, T. A., *A Beginner's Guide to Mass Spectral Interpretation*, Wiley, New York, 1998
4. McLafferty, F. W., and Turecek, F., *Interpretation of Mass Spectra, Fourth Edition*, University Science Books, Mill Valley, CA, 1993.
5. Pasto, D. J., and Johnson, C. R., *Organic Structure Determination*, Prentice Hall, Englewood Cliffs, NJ, 1969.
6. Silverstein, R. M., Bassler, G. C., and Morrill, T. C., *Spectroscopic Identification of Organic Compounds, Sixth Edition*, John Wiley and Sons, New York, 1998.
7. Smakman, R., and deBoer, T. J., *Org. Mass Spec.* 3, 1561, 1970.
8. Smith, R. M., *Understanding Mass Spectra: A Basic Approach*, John Wiley and Sons, New York, 1999.
9. Watson, T. J., and Watson, J. T., *Introduction to Mass Spectrometry*, Lippincott, Williams and Wilkins, Philadelphia, PA, 1997.
10. Bruno, T. J., and Svoronos, P. D. N., *CRC Handbook of Basic Tables for Chemical Analysis, Third Edition*, CRC Press, Boca Raton, FL, 2011.

Mass Spectral Fragmentation Patterns of Organic Compound Families

Family	Molecular ion peak	Common fragments and characteristic peaks
Acetals		Cleavage of all C–O, C–H, and C–C bonds around the original aldehydic carbon
Alcohols	Weak for 1° and 2°; not detectable for 3°; strong for benzyl alcohols	Loss of 18 (H_2O — usually by cyclic mechanism); loss of H_2O and olefin simultaneously with four (or more) carbon-chain alcohols; prominent peak at $m/z = 31(CH_2\ddot{O}H)^+$ for 1° alcohols; prominent peak at $m/z = (RCH\ddot{O}H)^+$ for 2° alcohols; and $m/z = (R_2C\ddot{O}H)^+$ for 3° alcohols
Aldehydes	Low intensity	Loss of aldehydic hydrogen (strong M-1 peak, especially with aromatic aldehydes); strong peak at $m/z = 29(HC\equiv O^+)$; loss of chain attached to alpha carbon (beta cleavage); McLafferty rearrangement via beta cleavage if gamma hydrogen is present
Alkanes		
(a) Chain	Low intensity	Loss of 14 mass units (CH_2)
(b) Branched	Low intensity	Cleavage at the point of branch; low intensity ions from random rearrangements
(c) Alicyclic	Intense	Loss of 28 mass units (CH_2=CH_2) and side chains
Alkenes (olefins)	Rather high intensity (loss of π-electron) especially in case of cyclic olefins	Loss of units of general formula CnH_{2n-1}; formation of fragments of the composition CnH_{2n} (via McLafferty rearrangement); retro Diels-Alder fragmentation
Alkyl halides	Abundance of molecular ion F<Cl<Br<I; intensity decreases with increase in size and branching	Loss of fragments equal to the mass of the halogen until all halogens are cleaved off
(a) Fluorides	Very low intensity	Loss of 20 (HF); loss of 26 (C_2H_2) in case of fluorobenzenes
(b) Chlorides	Low intensity; characteristic isotope cluster	Loss of 35 (Cl) or 36 (HCl); loss of chain attached to the gamma carbon to the carbon carrying the Cl
(c) Bromides	Low intensity; characteristic isotope cluster	Loss of 79 (Br); loss of chain attached to the gamma carbon to the carbon carrying the Br
(d) Iodides	Higher than other halides	Loss of 127 (I)
Alkynes	Rather high intensity (loss of π-electron)	Fragmentation similar to that of alkenes
Amides	High intensity	Strong peak at $m/z = 44$ indicative of a 1° amide ($O=C=NH^+{}_2$); base peak at $m/z = 59$ ($CH_2=C(OH)N^+H_2$); possibility of McLafferty rearrangement; loss of 42 (C_2H_2O) for amides of the form $RNHCOCH_3$ when R is aromatic ring
Amines	Hardly detectable in case of acyclic aliphatic amines; high intensity for aromatic and cyclic amines	Beta cleavage yielding >C=N⁺<; base peak for all 1° amines at $m/z = 30$ ($CH_2=N^+H_2$); moderate M-1 peak for aromatic amines; loss of 27 (HCN) in aromatic amines; fragmentation at alpha carbons in cyclic amines
Aromatic hydrocarbons (arenes)	Intense	Loss of side chain; formation of RCH=CHR' (via McLafferty rearrangement); cleavage at the bonds beta to the aromatic ring; peaks at $m/z = 77$ (benzene ring; especially mono-substituted), 91 (tropyllium); the ring position of alkyl substitution has very little effect on the spectrum

Family	Molecular ion peak	Common fragments and characteristic peaks
Carboxylic acids	Weak for straight-chain monocarboxylic acids; large if aromatic acids	Base peak at m/z = 60 (CH_2=C(OH)$_2$) if alpha hydrogen is present; peak at m/z = 45 (COOH); loss of 17 (–OH) in case of aromatic acids or short-chain acids
Disulfides	Low intensity	Loss of olefins (m/z equal to R–S–S–$H^{+\cdot}$); strong peak at m/z = 66 ($HSSH^+$)
Phenols	Highly intense peak (base peak* generally)	Loss of 28 (C=O) and 29 (CHO); strong peak at m/z = 65 ($C_5H_5^+$)
Sulfides (thioethers)	Rather low intensity peak but higher than that of corresponding ether	Similar to those of ethers (–O– substituted by –S–); aromatic sulfides show strong peaks at m/z = 109 ($C_6H_5S^+$.); 65 ($C_5H_5^+$.); 91 (tropyllium ion)
Sulfonamides	Intense	Loss of m/z = 64 ($SONH_2$) and m/z = 27 (HCN) in case of benzenesulfonamide
Esters	Weak intensity	Base peak at m/z equal to the mass of R–C≡O^+; peaks at m/z equal to the mass of $^+$O≡C–OR', the mass of OR' and R'; McLafferty rearrangement possible in case of: (a) presence of a beta hydrogen in R' (peak at m/z equal to the mass of R–C($^+$OH)OH, and (b) presence of a gamma hydrogen in R (peak at m/z equal to the mass of (CH_2=C($^+$OH)OR); loss of 42 (CH_2=C=O) in case of benzyl esters; loss of ROH via the ortho effect in case of *o*–substituted benzoates
Ketones	High intensity	Loss of R–groups attached to the >C=O (alpha cleavage); peak at m/z = 43 for all methyl ketones (CH_3CO^+); McLafferty rearrangement via beta cleavage if gamma hydrogen is present; loss of m/z = 28 (C=O) for cyclic ketones after initial alpha cleavage and McLafferty rearrangement
Mercaptan (thiols)	Rather low intensity but higher than that of corresponding alcohol	Similar to those of alcohols (–OH substituted by –SH); loss of m/z = 45 (CHS) and m/z = 44 (CS) for aromatic thiols
Nitriles	Unlikely to be detected except in case of acetonitrile (CH_3CN) and propionitrile (C_2H_5CN)	M+1 ion may appear (especially at higher pressures); M-1 peak is weak but detectable (R–CH=C=N^+); base peak at m/z = 41 (CH_2=C=N^+H); McLafferty rearrangement possible; loss of HCN is case of cyanobenzenes
Nitrites	Absent (or very weak at best)	Base peak at m/z = 30 (NO^+); large peak at m/z = 60 (CH_2=O^+NO) in all unbranched nitrites at the alpha carbon; absence of m/z = 46 permits differentiation from nitro compounds
Nitro compounds	Seldom observed	Loss of 30 (NO); subsequent loss of CO (in case of aromatic nitro- compounds); loss of NO_2, m/z = 46, from molecular ion peak
Sulfones	High intensity	Similar to sulfoxides; loss of mass equal to RSO_2; aromatic heterocycles show peaks at M-32 (sulfur), M-48 (SO), M-64 (SO_2)
Sulfoxides	High intensity	Loss of 17 (OH); loss of alkene (m/z equal to $RSOH^+$.); peak at m/z = 63 (CH_2=SOH)$^+$; aromatic sulfoxides show peak at m/z = 125 ($^+$S–CH=CHCH=CHCHC=O), 97($C_5H_5S^+$), 93($C_6H_5O^\cdot$); aromatic heterocycles show peaks at M-16 (oxygen), M-29(COH); M-48(SO)

* The base peak is the most intense peak in the mass spectrum, and is often the molecular ion peak, M^+.

Analytical

COMMON MASS SPECTRAL FRAGMENTS LOST

Thomas J. Bruno and Paris D. N. Svoronos

The following table gives a list of neutral species that are most commonly lost when measuring the mass spectra of organic compounds. The list is suggestive rather than comprehensive and should be used in conjunction with other sources (Refs. 1–5). The listed fragments include only combinations of carbon, hydrogen, oxygen, nitrogen, sulfur, and the halogens.

References

1. Hamming, M., and Foster, N., *Interpretation of Mass Spectra of Organic Compounds*, Academic Press, New York, 1972.
2. McLafferty, F. W., and Turecek, F., *Interpretation of Mass Spectra, Fourth Edition*, University Science Books, Mill Valley, CA, 1993.
3. Silverstein, R. M., Bassler, G. C., and Morrill, T. C., *Spectroscopic Identification of Organic Compounds, Eighth Edition*, John Wiley & Sons, New York, 2014.
4. Bruno, T. J., *CRC Handbook for the Analysis and Identification of Alternative Refrigerants*, CRC Press, Boca Raton, FL, 1995.
5. Bruno, T. J., and Svoronos, P. D. N., *CRC Handbook of Basic Tables for Chemical Analysis, Third Edition*, CRC Press, Boca Raton, FL, 2011.

Common Mass Spectral Fragments Lost for Different Mass Losses

Mass lost	Fragment lost
1	$H^{\bullet}$
15	$CH_3^{\bullet}$
17	$OH^{\bullet}$
18	H_2O
19	$F^{\bullet}$
20	HF
26	$HC\equiv CH$; $^{\bullet}C\equiv N$
27	$CH_2{-}CH^{\bullet}$; $HC\equiv N$
28	$CH_2{=}CH_2$; $C{=}O$; (HCN and $H^{\bullet}$)
29	$CH_3CH_2^{\bullet}$; $H{-}C{=}O^{\bullet}$
30	$^{\bullet}CH_2NH_2$; $HCHO$; NO
31	$CH_3O^{\bullet}$; $^{\bullet}CH_2OH$; CH_3NH_2
32	CH_3OH; S
33	$HS^{\bullet}$
34	H_2S
35	$Cl^{\bullet}$
36	HCl
38	$C_3H_2^{\bullet}$; F_2
39	$^{\bullet}CH_2{-}C\equiv CH$
40	$CH_3C\equiv CH$
41	$CH_2{=}CHCH_2^{\bullet}$
42	$CH_2{=}CHCH_3$; $CH_2{=}C{=}O$; $(CH_2)_3$; NCO; $N\equiv CNH_2$
43	$C_3H_7^{\bullet}$ ($CH_3CH_2CH_2^{\bullet}$, $(CH_3)_2CH^{\bullet}$); $CH_3C{=}O^{\bullet}$; $CH_2{=}CH{-}O^{\bullet}$; HCNO
44	$CH_2{=}CHOH$; CO_2; N_2O; $^{\bullet}CONH_2$; $^{\bullet}NHCH_2CH_3$
45	$CH_3CH^{\bullet}OH$; $CH_3CH_2O^{\bullet}$; CO_2H; $CH_3CH_2NH_2$
46	CH_3CH_2OH; $^{\bullet}NO_2$
47	$CH_3S^{\bullet}$; CH_2SH
48	CH_3SH; SO; O_3
49	$^{\bullet}CH_2Cl$
51	$^{\bullet}CHF_2$
52	C_4H_4; C_2N_2
54	$CH_2{=}CHCH{=}CH_2$; $^{\bullet}CH_2CH_2CN$; $CH_3CH^{\bullet}{-}CN$
55	$CH_2{=}CH{-}CH^{\bullet}CH_3$; $^{\bullet}CH_2{-}CH{=}CHCH_3$
56	$CH_2{=}CH{-}CH_2CH_3$; $CH_3CH{=}CHCH_3$; CO (2 moles)
57	$C_4H_9^{\bullet}$ (various isomers); $C_2H_5{-}C\equiv O^{\bullet}$
58	$^{\bullet}NCS$; $(CH_3)_2C{=}O$; (NO and CO)
59	$CH_3OC{=}O^{\bullet}$; CH_3CONH_2; $C_2H_3S^{\bullet}$l $C_3H_7O^{\bullet}$
60	C_3H_7OH
61	$CH_3CH_2S^{\bullet}$; $CH_3SCH_2^{\bullet}$
62	[H_2S and $CH_2{=}CH_2$]
63	$^{\bullet}CH_2CH_2Cl$; $CH_3CH^{\bullet}Cl$
64	$S_2^{\bullet}$; SO_2; C_5H_4 (various isomers)
68	C_5H_8 (various isomers)
69	$CF_3^{\bullet}$; $C_5H_9^{\bullet}$ (various isomers)
71	$C_5H_{11}^{\bullet}$ (various isomers)l $C_3H_7{-}C\equiv O^{\bullet}$
73	$CH_3CH_2OC\equiv O^{\bullet}$; $C_4H_9O^{\bullet}$ (various isomers)
74	C_4H_9OH (various isomers)
77	$C_6H_5^{\bullet}$; $HCS_2^{\bullet}$
78	C_6H_6; $C_5H_4N^{\bullet}$ (various isomers)
79	$Br^{\bullet}$; C_5H_5N (various isomers)
80	HBr
85	$^{\bullet}CClF_2$
90	$HO{-}Si{-}(CH_3)_3$
91	$C_6H_5CH_2^{\bullet}$
97	$C_7H_{13}^{\bullet}$ (various isomers)
100	$CF_2{=}CF_2$
105	$C_6H_5C\equiv O^{\bullet}$; $C_6H_5CH_2CH_2^{\bullet}$; $C_6H_5CH^{\bullet}CH_3$
107	$C_6H_5{-}CH_2O^{\bullet}$
119	$CF_3CF_2^{\bullet}$
122	$C_6H_5CO_2H$
127	$I^{\bullet}$
128	HI

MAJOR REFERENCE MASSES IN THE SPECTRUM OF HEPTACOSAFLUOROTRIBUTYLAMINE (PERFLUOROTRIBUTYLAMINE)

Thomas J. Bruno and Paris D. N. Svoronos

The following list tabulates the major reference masses (with their relative intensities and formulas) of the mass spectrum of heptacosafluorotributylamine (CAS No. 311-89-7) (Ref. 1). This is one of the most widely used reference compounds in mass spectrometry. For guidance in the selection of additional reference compounds, the reader is advised to consult Ref. 1.

Reference

1. Bruno, T. J., and Svoronos, P. D. N., *CRC Handbook of Basic Tables for Chemical Analysis, Third Edition*, CRC Press, Boca Raton, FL, 2011. <https://doi.org/10.1201/b10385>

Reference Masses and Intensities in the Spectrum of Perfluorotributylamine

Mass	Relative intensity	Mol. form.
613.9647	2.6	$C_{12}F_{24}N$
575.9679	1.7	$C_{12}F_{22}N$
537.9711	0.4	$C_{12}F_{20}N$
501.9711	8.6	$C_9F_{20}N$
463.9743	3.8	$C_9F_{18}N$
425.9775	2.5	$C_9F_{16}N$
413.9775	5.1	$C_8F_{16}N$
375.9807	0.9	$C_8F_{14}N$
325.9839	0.4	$C_7F_{12}N$
313.9839	0.4	$C_6F_{12}N$
263.9871	10	$C_5F_{10}N$
230.9856	0.9	C_5F_9
225.9903	0.6	C_5F_8N
218.9856	62	C_4F_9
213.9903	0.6	C_4F_8N
180.9888	1.9	C_4F_7
175.9935	1.0	C_4F_6N
168.9888	3.6	C_3F_7
163.9935	0.7	C_3F_6N
161.9904	0.3	C_4F_6
149.9904	2.1	C_3F_6
130.9920	31	C_3F_5
118.9920	8.3	C_2F_5
113.9967	3.7	C_2F_4N
111.9936	0.7	C_3F_4
99.9936	12	C_2F_4
92.9952	1.1	C_3F_3
68.9952	100	CF_3
49.9968	1.0	CF_2
30.9984	2.3	CF

Analytical

MASS SPECTRAL PEAKS OF COMMON ORGANIC LIQUIDS

David R. Lide

The strongest peaks in the mass spectra of 374 important organic liquids are listed in this table. The *m/z* value for each peak is followed by the relative intensity in parentheses, with the strongest peak assigned an intensity of 100. The peaks for each compound are listed in order of decreasing intensity. Compounds are listed alphabetically by name.

Data on the physical properties of the same compounds may be found in Section 15 in the table "Laboratory Solvents and Other Liquid Reagents."

References

1. NIST 2011 Mass Spectral Library (NIST 11), National Institute of Standards and Technology, <www.nist.gov/srd/nist1a.cfm>.
2. Lide, D. R., and Milne, G. W. A., Editors, *Handbook of Data on Organic Compounds, Third Edition*, CRC Press, Boca Raton, FL, 1994.
3. Lide, D. R., Editor, *Properties of Organic Compounds*, <www.chemnetbase.com/tours/poc/intro.jsf>.

Mass Spectral Peaks of Common Organic Liquids (*m/z*; Relative Intensities in Parentheses)

Name										
Acetic acid	43(100)	45(87)	60(57)	15(42)	42(14)	29(13)	14(13)	28(7)	18(6)	16(6)
Acetic anhydride	43(100)	42(35)	45(29)	60(18)	29(9)	41(8)	40(2)	26(2)	87(1)	61(1)
Acetone	43(100)	15(34)	58(23)	27(9)	14(9)	42(8)	26(7)	29(5)	28(5)	39(4)
Acetonitrile	41(100)	40(46)	39(13)	14(9)	38(6)	28(4)	26(4)	25(3)	42(2)	27(2)
Acrolein	27(100)	56(74)	28(65)	26(54)	55(52)	29(37)	25(8)	53(5)	38(5)	57(4)
Acrylonitrile	53(100)	26(85)	52(79)	51(34)	27(13)	50(8)	25(7)	38(5)	54(3)	37(3)
Allyl alcohol	57(100)	31(34)	29(32)	28(31)	58(25)	39(22)	27(20)	30(16)	32(14)	26(11)
Allylamine	30(100)	56(80)	28(76)	57(33)	39(21)	29(20)	27(18)	26(13)	41(8)	18(8)
2-Amino-2-methyl-1-propanol	58(100)	41(18)	18(17)	42(13)	28(11)	56(10)	30(10)	29(8)	43(6)	59(5)
Aniline	93(100)	66(32)	65(16)	39(13)	92(10)	94(7)	41(5)	40(5)	67(4)	64(3)
Anisole	108(100)	65(76)	78(60)	39(44)	51(21)	77(20)	93(16)	79(14)	50(13)	63(12)
Benzaldehyde	51(100)	77(81)	50(55)	106(44)	105(43)	52(26)	78(16)	39(13)	27(10)	74(8)
Benzene	78(100)	77(20)	52(19)	51(17)	50(15)	39(12)	79(6)	76(5)	74(4)	38(4)
Benzeneacetonitrile	117(100)	90(43)	116(35)	89(22)	51(13)	39(11)	63(10)	118(9)	91(8)	50(8)
Benzenethiol	110(100)	66(28)	109(23)	39(15)	77(14)	51(14)	84(13)	69(11)	50(11)	45(11)
Benzonitrile	103(100)	76(34)	50(13)	104(9)	75(7)	51(7)	77(5)	52(4)	39(4)	74(3)
Benzyl acetate	108(100)	43(76)	91(60)	90(48)	79(27)	107(17)	77(17)	65(15)	51(15)	89(14)
Benzyl alcohol	79(100)	108(83)	77(74)	107(66)	51(35)	105(23)	106(22)	50(17)	78(16)	39(16)
Bis(2-aminoethyl)amine	44(100)	73(59)	30(35)	19(18)	56(16)	28(16)	27(16)	42(11)	99(8)	43(8)
Bis(2-chloroethyl) ether	93(100)	63(74)	27(38)	95(32)	65(24)	31(9)	49(4)	28(4)	94(3)	62(3)
Bis(2-ethylhexyl) phthalate	149(100)	57(32)	167(29)	71(21)	43(21)	70(18)	150(11)	113(10)	55(10)	41(9)
Bis(2-hydroxyethyl) sulfide	61(100)	45(68)	31(38)	104(36)	91(34)	47(26)	44(26)	27(24)	60(18)	43(17)
Bromobenzene	77(100)	158(64)	156(64)	51(39)	50(17)	78(8)	76(6)	75(6)	28(5)	159(4)
1-Bromobutane	57(100)	41(56)	29(45)	27(29)	56(13)	39(12)	28(12)	55(7)	43(7)	138(6)
2-Bromobutane, (±)-	57(100)	41(60)	29(57)	27(34)	39(20)	28(9)	26(9)	136(1)		
1-Bromo-2-chloroethane	63(100)	27(85)	65(31)	26(23)	144(8)	81(8)	79(8)	28(7)	142(6)	93(6)
Bromochloromethane	49(100)	130(67)	128(52)	51(31)	93(23)	81(20)	79(20)	95(17)	132(16)	47(8)
1-Bromodecane	43(100)	135(94)	137(91)	57(81)	41(58)	55(56)	71(38)	69(36)	85(33)	29(27)
Bromoethane	108(100)	110(97)	29(62)	27(51)	28(35)	26(14)	93(6)	32(6)	95(5)	81(5)
2-Bromo-2-methylpropane	57(100)	41(67)	29(45)	39(30)	27(18)	28(8)	40(5)	38(5)	58(4)	55(4)
1-Bromonaphthalene	44(100)	206(39)	127(39)	208(37)	36(31)	69(29)	131(13)	29(13)	126(12)	63(12)
1-Bromopentane	43(100)	71(80)	41(56)	27(51)	42(37)	29(34)	55(33)	39(30)	28(12)	26(9)
1-Bromopropane	43(100)	41(77)	28(69)	27(60)	39(49)	124(42)	122(41)	42(34)	32(20)	29(15)
2-Bromopropane	43(100)	27(47)	41(43)	39(22)	124(8)	122(8)	26(7)	81(6)	79(6)	38(6)
2-Bromopropene	41(100)	39(58)	122(37)	120(36)	38(12)	37(8)	40(6)	81(5)	79(5)	42(4)
Butanal	44(100)	43(74)	72(57)	41(56)	27(55)	29(48)	57(23)	39(22)	28(15)	42(11)
Butanenitrile	41(100)	29(62)	27(37)	28(10)	39(9)	26(7)	40(5)	42(4)	38(4)	15(4)
1-Butanethiol	56(100)	41(74)	90(66)	47(43)	27(43)	28(36)	29(33)	57(17)	39(16)	61(15)
Butanoic acid	60(100)	27(50)	73(27)	42(25)	41(24)	43(22)	29(21)	45(19)	39(15)	28(11)
Butanoic anhydride	71(100)	43(59)	27(26)	41(19)	42(10)	39(10)	73(9)	28(7)	72(5)	55(5)
1-Butanol	31(100)	56(81)	41(62)	43(60)	27(50)	42(31)	29(31)	28(17)	39(16)	55(12)
2-Butanol	45(100)	31(22)	27(22)	59(20)	29(18)	43(13)	41(12)	44(8)	18(8)	28(5)
2-Butanone	43(100)	72(24)	29(19)	27(12)	57(7)	42(5)	26(4)	28(3)	44(2)	39(2)
trans-2-Butenal	41(100)	39(97)	70(82)	69(65)	27(49)	29(39)	42(30)	38(29)	40(27)	37(18)

Name										
2-Butoxyethanol	57(100)	45(38)	29(35)	41(31)	87(16)	27(12)	56(11)	31(9)	75(7)	28(7)
2-Butoxyethyl acetate	57(100)	43(86)	56(50)	87(33)	41(26)	29(22)	85(18)	88(11)	44(11)	27(7)
Butyl acetate	43(100)	56(34)	41(17)	27(16)	29(15)	73(11)	61(10)	28(7)	55(6)	39(6)
sec-Butyl acetate	43(100)	56(21)	87(15)	41(14)	29(8)	57(6)	73(4)	61(4)	55(4)	27(4)
Butylamine	30(100)	73(10)	28(5)	41(3)	27(3)	18(3)	44(2)	42(2)	31(2)	29(2)
tert-Butylamine	58(100)	41(21)	42(15)	18(9)	30(8)	15(8)	39(7)	57(6)	28(6)	59(4)
Butylbenzene	91(100)	92(55)	134(20)	65(13)	27(10)	39(9)	105(8)	51(7)	78(6)	41(6)
sec-Butylbenzene, (±)-	105(100)	134(18)	91(14)	77(10)	27(9)	106(9)	51(7)	79(7)		
tert-Butylbenzene	119(100)	91(65)	41(40)	134(24)	39(15)	79(14)	77(13)	51(13)	120(11)	65(7)
Butyl butanoate	43(100)	71(90)	56(80)	89(69)	41(67)	27(52)	29(47)	57(29)	39(22)	60(19)
Butyl formate	56(100)	41(60)	31(58)	29(53)	27(45)	43(34)	28(21)	39(19)	55(11)	42(11)
1-*tert*-Butyl-4-methylbenzene	133(100)	105(38)	41(23)	148(18)	93(16)	91(14)	115(13)	134(11)	39(11)	116(10)
Butyl vinyl ether	29(100)	41(74)	56(56)	57(43)	27(42)	44(26)	15(16)	85(14)	39(14)	43(13)
γ-Butyrolactone	28(100)	42(74)	29(48)	27(33)	41(27)	56(25)	86(24)	26(18)	85(10)	39(10)
Caprolactam	55(100)	113(87)	30(81)	56(66)	84(60)	85(57)	42(51)	41(33)	28(26)	43(17)
Carbon disulfide	76(100)	32(22)	44(17)	78(9)	38(6)	28(5)	77(3)	64(1)	46(1)	39(1)
2-Chloroaniline	127(100)	129(32)	92(17)	65(16)	128(10)	91(9)	64(9)	39(8)	63(7)	99(6)
Chlorobenzene	112(100)	77(63)	114(33)	51(29)	50(14)	75(8)	113(7)	78(5)	76(5)	28(4)
1-Chlorobutane	56(100)	41(65)	27(50)	43(35)	29(24)	39(18)	28(16)	26(11)	55(8)	15(8)
2-Chlorobutane	56(100)	57(100)	41(90)	27(77)	29(57)	63(46)	39(34)	92(1)		
1-Chloro-1,1-difluoroethane	65(100)	45(31)	85(14)	31(10)	64(8)	44(7)	35(6)	26(6)	87(5)	81(4)
Chloroethane	64(100)	28(91)	29(84)	27(75)	66(32)	26(28)	49(25)	51(8)	63(6)	65(4)
2-Chloroethanol	31(100)	15(13)	29(10)	28(10)	27(9)	43(8)	44(7)	26(5)	18(5)	14(5)
(Chloromethyl)benzene	91(100)	126(20)	65(14)	92(9)	39(9)	63(8)	128(6)	45(6)	89(5)	125(3)
1-Chloro-3-methylbutane	43(100)	55(56)	41(55)	27(51)	70(49)	42(37)	29(34)	57(30)	39(30)	56(15)
1-Chloro-2-methylpropane	43(100)	41(67)	42(50)	27(33)	39(26)	15(11)	29(10)	56(7)	49(6)	38(5)
2-Chloro-2-methylpropane	57(100)	41(80)	77(44)	29(26)	39(24)	79(14)	27(14)	59(7)	56(7)	38(6)
1-Chloronaphthalene	162(100)	127(36)	164(30)	126(20)	163(10)	77(8)	101(6)	75(6)	128(5)	28(4)
1-Chlorooctane	91(100)	41(83)	43(76)	27(62)	29(56)	55(55)	39(34)	93(32)	57(32)	69(29)
Chloropentafluoroethane	85(100)	69(61)	31(38)	87(32)	50(17)	35(8)	119(6)	66(4)	100(3)	47(3)
1-Chloropentane	42(100)	41(90)	70(89)	55(87)	27(73)	29(55)	43(40)	39(40)	28(19)	57(18)
1-Chloropropane	42(100)	29(46)	27(37)	41(23)	28(15)	43(14)	39(12)	78(6)	63(6)	49(5)
3-Chloropropene	41(100)	39(73)	76(28)	38(16)	37(13)	40(12)	27(12)	26(11)	78(9)	49(5)
2-Chlorotoluene	91(100)	126(68)	89(23)	128(22)	90(18)	65(17)	125(15)	92(13)	127(10)	63(6)
3-Chlorotoluene	91(100)	126(27)	63(15)	65(12)	89(11)	39(11)	128(9)	125(8)	92(8)	62(6)
Cyclohexane	56(100)	84(71)	41(70)	27(37)	55(36)	39(35)	42(30)	69(23)	28(18)	43(14)
Cyclohexanol	57(100)	44(68)	41(68)	39(51)	32(40)	43(38)	31(32)	42(22)	67(18)	82(16)
Cyclohexanone	55(100)	42(85)	41(34)	27(33)	98(31)	39(27)	69(26)	70(20)	43(14)	28(14)
Cyclohexene	67(100)	54(72)	82(37)	41(35)	39(33)	27(15)	53(12)	81(9)	51(8)	79(6)
Cyclohexylamine	56(100)	43(23)	28(17)	99(10)	70(8)	57(6)	30(6)	93(5)	54(4)	41(4)
Cyclopentane	42(100)	70(30)	55(29)	41(29)	39(22)	27(15)	40(7)	29(5)	28(4)	43(3)
Cyclopentanone	55(100)	28(50)	84(42)	41(38)	56(29)	27(24)	39(19)	42(15)	26(9)	29(7)
cis-Decahydronaphthalene	67(100)	81(87)	41(81)	138(67)	96(62)	82(62)	39(50)	55(45)	27(44)	95(42)
trans-Decahydronaphthalene	41(100)	68(91)	67(88)	82(67)	27(65)	96(61)	95(55)	138(51)	81(51)	29(51)
Decane	43(100)	57(90)	41(41)	71(33)	29(30)	85(24)	27(20)	56(17)	55(14)	42(14)
Diacetone alcohol	43(100)	59(41)	58(17)	101(10)	41(9)	31(9)	83(6)	56(6)	55(6)	29(6)
1,2-Dibromoethane	27(100)	107(77)	109(72)	26(24)	28(10)	81(5)	79(5)	25(5)	95(4)	93(4)
Dibromofluoromethane	111(100)	113(98)	192(29)	43(16)	41(16)	190(15)	194(14)	81(9)	79(9)	122(7)
Dibromomethane	174(100)	93(96)	95(84)	172(53)	176(50)	91(11)	81(9)	79(9)	94(5)	65(5)
1,2-Dibromopropane	41(100)	121(66)	123(65)	39(48)	27(28)	107(11)	38(10)	26(10)	109(9)	42(9)
1,2-Dibromotetrafluoroethane	179(100)	181(97)	129(34)	131(33)	100(17)	31(13)	260(12)	50(8)	69(7)	262(6)
Dibutylamine	86(100)	72(52)	30(48)	44(40)	29(31)	57(24)	41(21)	73(15)	28(15)	43(13)
Dibutyl ether	57(100)	41(34)	29(30)	56(25)	87(21)	27(9)	58(8)	55(6)	39(5)	28(5)
Dibutyl oxalate	57(100)	41(61)	56(24)	44(9)	55(8)	43(7)	58(5)	42(4)	103(2)	73(2)
Dibutyl phthalate	149(100)	86(18)	57(18)	223(17)	205(17)	150(17)	104(17)	56(17)	41(17)	65(16)
Dibutyl sebacate	241(100)	185(71)	41(37)	56(35)	55(32)	57(31)	242(23)	143(21)	98(21)	125(20)
o-Dichlorobenzene	146(100)	148(64)	111(38)	75(23)	113(12)	74(12)	50(11)	150(10)	73(9)	147(7)
m-Dichlorobenzene	146(100)	148(65)	111(36)	75(25)	50(19)	74(16)	150(11)	113(11)	73(11)	147(8)
1,1-Dichloroethane	63(100)	27(71)	65(31)	26(19)	83(11)	85(7)	61(7)	35(6)	98(5)	62(5)
1,2-Dichloroethane	62(100)	27(91)	49(40)	64(32)	26(31)	63(19)	98(14)	51(13)	61(12)	100(9)
1,1-Dichloroethene	61(100)	96(61)	98(38)	63(32)	26(16)	60(15)	62(7)	25(7)	100(6)	35(6)

Name										
cis-1,2-Dichloroethene	61(100)	96(73)	98(47)	63(32)	26(30)	60(21)	25(13)	35(12)	62(9)	100(8)
trans-1,2-Dichloroethene	61(100)	96(67)	98(43)	26(34)	63(32)	60(24)	25(15)	62(10)	100(7)	47(7)
Dichlorofluoromethane	67(100)	69(32)	47(13)	35(13)	31(9)	32(8)	48(7)	83(5)	102(4)	49(4)
Dichloromethane	49(100)	84(64)	86(39)	51(31)	47(14)	48(8)	88(6)	50(3)	85(2)	83(2)
(Dichloromethyl)benzene	125(100)	127(32)	160(14)	89(13)	162(9)	63(9)	126(8)	62(7)	105(5)	39(5)
1,2-Dichloropropane, (±)-	63(100)	62(71)	27(57)	41(49)	39(32)	65(31)	76(27)	64(25)	49(13)	77(12)
1,2-Dichloro-1,1,2,2-tetrafluoroethane	85(100)	135(52)	87(33)	137(17)	101(9)	31(9)	103(6)	100(6)	50(5)	69(4)
2,4-Dichlorotoluene	125(100)	160(61)	162(40)	127(32)	89(23)	159(16)	161(14)	63(13)	62(11)	126(10)
3,4-Dichlorotoluene	125(100)	160(47)	127(32)	162(31)	89(15)	159(11)	161(10)	63(10)	126(8)	62(8)
Diethanolamine	30(100)	74(82)	28(77)	56(69)	18(50)	42(46)	29(36)	27(34)	45(30)	43(19)
1,1-Diethoxyethane	44(100)	43(92)	29(77)	31(76)	45(74)	27(52)	72(48)	73(23)	28(17)	46(15)
1,2-Diethoxyethane	31(100)	59(71)	29(58)	45(43)	27(33)	74(27)	43(15)	15(14)	28(12)	44(10)
Diethylamine	30(100)	58(81)	44(28)	73(18)	29(18)	28(17)	72(12)	42(11)	27(11)	59(4)
Diethyl carbonate	29(100)	45(70)	31(53)	27(39)	91(24)	28(15)	63(11)	26(10)	30(6)	43(5)
Diethylene glycol	45(100)	75(23)	31(20)	44(16)	27(14)	76(12)	29(12)	43(11)	42(9)	41(4)
Diethylene glycol dimethyl ether	59(100)	58(43)	31(34)	29(32)	45(28)	28(19)	89(15)	43(9)	27(5)	60(4)
Diethylene glycol monobutyl ether acetate	43(100)	87(93)	57(82)	41(28)	45(25)	56(18)	29(18)	72(14)	85(12)	101(10)
Diethylene glycol monoethyl ether	45(100)	59(56)	72(37)	73(22)	60(14)	31(13)	75(11)	44(9)	104(8)	103(7)
Diethylene glycol monoethyl ether acetate	43(100)	29(51)	31(42)	45(40)	59(24)	72(18)	44(10)	73(9)	42(9)	30(6)
Diethylene glycol monomethyl ether	45(100)	31(42)	59(41)	29(38)	28(32)	58(21)	43(14)	27(13)	44(11)	32(10)
Diethyl ether	31(100)	29(63)	59(40)	27(35)	45(33)	74(23)	15(17)	43(9)	28(9)	26(9)
Diethyl oxalate	29(100)	31(16)	45(14)	27(14)	74(11)	28(9)	43(4)	30(4)	73(3)	75(2)
Diethyl sulfide	75(100)	47(81)	90(73)	62(60)	29(55)	61(54)	27(48)	28(23)	46(17)	45(17)
Diisopropylamine	44(100)	86(30)	58(14)	42(13)	28(13)	41(12)	43(11)	27(11)	15(11)	39(6)
Diisopropyl ether	45(100)	43(39)	87(15)	41(12)	59(10)	27(8)	39(4)	69(3)	42(3)	31(3)
1,2-Dimethoxybenzene	138(100)	95(65)	77(48)	123(44)	52(42)	41(33)	65(30)	51(29)	39(19)	63(17)
1,2-Dimethoxyethane	45(100)	60(13)	29(13)	90(7)	58(6)	31(5)	28(5)	43(4)	59(3)	46(2)
Dimethoxymethane	45(100)	75(61)	29(59)	31(13)	30(6)	15(6)	47(5)	76(2)	46(2)	44(2)
N,N-Dimethylacetamide	44(100)	87(69)	43(46)	45(23)	42(19)	72(15)	15(11)	30(8)	28(5)	88(4)
Dimethylamine	44(100)	45(81)	18(32)	28(30)	43(19)	42(15)	15(9)	46(5)	41(5)	27(5)
2,4-Dimethylaniline	121(100)	120(78)	106(57)	77(15)	28(15)	91(12)	122(9)	18(8)	93(6)	118(5)
2,2-Dimethylbutane	43(100)	57(98)	71(73)	41(61)	29(51)	27(36)	56(29)	39(25)	55(15)	15(12)
2,3-Dimethylbutane	43(100)	42(97)	41(27)	71(25)	27(11)	86(10)	39(9)	29(6)	55(5)	44(4)
Dimethyl disulfide	94(100)	45(63)	79(59)	46(38)	47(26)	15(18)	48(14)	61(12)	64(11)	96(9)
N,N-Dimethylformamide	73(100)	44(86)	42(36)	30(22)	28(20)	29(8)	43(7)	72(6)	58(5)	74(4)
Dimethyl glutarate	59(100)	100(51)	55(49)	42(33)	129(32)	101(32)	41(26)	43(22)	128(19)	87(15)
2,6-Dimethyl-4-heptanone	57(100)	85(82)	41(46)	43(39)	58(33)	28(30)	26(30)	39(22)	42(12)	142(11)
2,5-Dimethylhexane	57(100)	43(93)	42(31)	41(26)	99(19)	71(19)	29(11)	70(10)	55(9)	27(9)
Dimethyl maleate	113(100)	59(83)	26(72)	29(40)	85(37)	54(31)	114(25)	53(24)	55(13)	82(9)
2,2-Dimethylpentane	57(100)	43(73)	41(46)	56(40)	85(34)	29(31)	27(23)	39(15)	15(7)	55(6)
2,4-Dimethylpentane	43(100)	57(73)	56(41)	41(35)	42(24)	85(17)	29(16)	27(12)	39(9)	58(3)
2,4-Dimethyl-3-pentanone	43(100)	71(31)	41(13)	27(9)	70(6)	39(6)	114(5)	42(5)	44(3)	72(2)
2,4-Dimethylpyridine	107(100)	106(63)	79(44)	92(22)	65(22)	51(19)	77(17)	80(12)	52(11)	50(11)
2,6-Dimethylpyridine	107(100)	39(39)	106(29)	66(22)	92(18)	65(18)	38(12)	27(11)	79(9)	63(9)
Dimethyl sulfoxide	63(100)	78(70)	15(40)	45(35)	29(16)	61(13)	46(12)	31(11)	48(10)	47(10)
1,4-Dioxane	28(100)	29(37)	88(31)	58(24)	31(17)	15(17)	27(15)	30(13)	43(11)	26(9)
1,3-Dioxolane	73(100)	29(56)	44(53)	45(28)	28(21)	43(20)	27(13)	31(7)	74(5)	42(3)
Dipentene	68(100)	93(50)	67(44)	94(22)	39(22)	107(18)	92(18)	53(18)	136(16)	79(16)
Dipentyl ether	71(100)	43(92)	29(43)	70(40)	41(36)	27(34)	42(23)	69(17)	55(16)	39(16)
Dipropylamine	30(100)	72(79)	44(40)	43(32)	27(25)	28(24)	41(22)	86(11)	58(10)	42(10)
Dodecane	57(100)	43(91)	71(53)	41(45)	85(31)	29(27)	55(19)	56(18)	28(16)	42(14)
1-Dodecene	43(100)	56(83)	55(83)	41(75)	69(62)	70(58)	57(57)	83(47)	29(38)	84(33)
Epichlorohydrin	57(100)	27(39)	29(32)	49(25)	31(22)	62(18)	28(16)	92(1)		
1,2-Epoxybutane	43(100)	44(58)	27(52)	29(50)	45(46)	26(27)	28(27)	72(5)		
1,2-Ethanediamine	30(100)	18(13)	42(6)	43(5)	27(5)	44(4)	29(4)	17(4)	15(4)	41(3)
1,2-Ethanediol	31(100)	33(35)	29(13)	32(11)	43(6)	27(5)	28(4)	62(3)	30(3)	44(2)
1,2-Ethanediol, diacetate	43(100)	86(11)	42(7)	15(7)	116(4)	73(4)	44(3)	29(3)	103(2)	45(2)
Ethanol	31(100)	45(44)	46(18)	27(18)	29(15)	43(14)	30(6)	42(3)	19(3)	14(3)
Ethanolamine	30(100)	18(30)	28(15)	42(7)	31(6)	17(6)	61(5)	15(5)	43(3)	29(3)
Ethoxybenzene	94(100)	122(39)	28(12)	66(11)	39(9)	77(8)	95(7)	65(7)	51(7)	29(6)
2-Ethoxyethanol	31(100)	29(52)	59(50)	27(27)	45(26)	72(14)	43(14)	15(14)	28(8)	26(6)

Name										
2-Ethoxyethyl acetate	43(100)	31(34)	59(31)	72(28)	44(25)	29(24)	45(12)	27(11)	15(11)	87(7)
Ethyl acetate	43(100)	29(46)	27(33)	45(32)	61(28)	28(25)	42(18)	73(11)	88(10)	70(10)
Ethyl acetoacetate	43(100)	29(24)	88(18)	28(16)	85(14)	27(12)	42(11)	60(9)	130(6)	45(6)
Ethyl acrylate	55(100)	27(32)	29(15)	56(12)	45(9)	73(8)	28(8)	26(6)	99(5)	85(5)
Ethylamine	30(100)	28(32)	44(20)	45(19)	27(13)	15(10)	42(9)	29(8)	41(5)	40(5)
Ethylbenzene	91(100)	106(31)	51(14)	39(10)	77(8)	65(8)	105(7)	92(7)	78(7)	27(6)
Ethyl benzoate	105(100)	77(65)	122(34)	51(34)	27(17)	150(16)	29(14)	50(13)	106(12)	78(6)
Ethyl butanoate	43(100)	71(88)	29(83)	27(43)	88(40)	41(28)	60(22)	45(20)	73(17)	42(17)
Ethylcyanoacetate	29(100)	68(59)	27(34)	40(21)	28(16)	41(14)	15(13)	45(10)	43(9)	26(9)
Ethylcyclohexane	83(100)	55(65)	82(42)	41(36)	112(23)	56(11)	67(10)	39(10)	42(9)	84(8)
Ethylene carbonate	29(100)	44(62)	43(54)	88(40)	30(16)	28(11)	45(7)	58(6)	42(6)	73(4)
Ethyl formate	31(100)	28(73)	27(51)	29(38)	45(34)	26(17)	74(11)	43(9)	47(8)	56(4)
2-Ethyl-1,3-hexanediol	56(100)	55(71)	41(60)	43(55)	29(46)	27(44)	31(34)	57(33)	73(32)	39(20)
2-Ethyl-1-hexanol	57(100)	43(41)	41(40)	29(29)	55(28)	83(27)	56(23)	70(20)	27(17)	31(13)
N-Ethyl-*N*-isopropyl-2-propanamine	72(100)	114(77)	44(41)	129(22)	42(21)	30(20)	43(18)	41(17)	27(15)	70(11)
Ethyl 3-methylbutanoate	29(100)	41(52)	27(51)	57(43)	43(42)	60(39)	88(38)	85(38)	61(26)	45(24)
3-Ethyl-2-methylpentane	43(100)	70(50)	41(27)	71(25)	29(19)	27(19)	85(18)	55(18)	57(15)	42(14)
Ethyl propanoate	57(100)	29(84)	102(17)	27(17)	75(15)	28(14)	45(13)	74(12)	73(7)	43(6)
Fluorobenzene	96(100)	70(17)	97(7)	95(6)	75(6)	50(6)	51(4)	39(4)	69(3)	57(3)
2-Fluorotoluene	109(100)	110(55)	83(18)	57(11)	63(10)	39(9)	51(6)	50(6)	107(5)	62(5)
3-Fluorotoluene	109(100)	110(54)	83(11)	57(5)	111(4)	107(4)	63(3)	39(3)	108(2)	89(2)
4-Fluorotoluene	109(100)	110(60)	83(13)	57(10)	108(8)	39(8)	63(7)	107(5)	51(5)	50(5)
Furan	68(100)	39(64)	40(9)	38(9)	42(6)	29(6)	37(5)	69(4)	34(2)	67(1)
Furfural	39(100)	96(55)	95(52)	38(38)	29(35)	37(29)	40(11)	97(9)	50(7)	42(7)
Furfuryl alcohol	98(100)	41(65)	39(59)	81(55)	53(53)	97(51)	42(49)	69(39)	70(36)	29(28)
Glycerol	61(100)	43(90)	31(57)	44(54)	29(38)	18(32)	27(12)	42(11)	60(10)	45(10)
Glycerol triacetate	43(100)	103(44)	145(34)	116(17)	115(13)	44(10)	86(9)	28(8)	73(7)	42(7)
Heptane	43(100)	41(56)	29(49)	57(47)	27(46)	71(45)	56(27)	42(26)	39(23)	70(18)
1-Heptanol	41(100)	70(87)	56(86)	31(78)	43(72)	29(70)	55(67)	27(65)	42(54)	69(41)
3-Heptanol, (*S*)-	59(100)	69(73)	87(37)	41(37)	29(33)	55(30)	31(24)	116(0)		
2-Heptanone	43(100)	58(60)	71(14)	41(11)	27(11)	59(9)	39(8)	29(8)	42(5)	114(4)
3-Heptanone	57(100)	29(76)	41(32)	85(29)	72(22)	43(15)	39(11)	114(10)	27(7)	55(5)
1-Heptene	41(100)	56(88)	29(70)	55(59)	42(52)	27(45)	39(41)	70(37)	69(27)	57(27)
Hexane	57(100)	43(78)	41(77)	29(61)	27(57)	56(45)	42(39)	39(27)	28(16)	86(14)
Hexanedinitrile	41(100)	68(50)	54(42)	40(21)	55(20)	27(17)	39(16)	28(13)	52(7)	42(6)
Hexanenitrile	41(100)	54(68)	27(59)	55(55)	29(44)	39(35)	57(30)	43(30)	68(24)	28(22)
Hexanoic acid	60(100)	73(42)	27(36)	41(33)	43(27)	29(26)	45(20)	39(16)	42(15)	55(14)
1-Hexanol	56(100)	43(78)	31(74)	41(71)	27(64)	29(59)	55(58)	42(53)	39(37)	69(27)
2-Hexanone	43(100)	58(60)	57(17)	100(16)	29(15)	41(13)	85(8)	27(8)	71(7)	59(5)
Hexyl acetate	43(100)	56(66)	41(38)	55(37)	61(35)	42(33)	84(31)	69(19)	73(17)	57(7)
3-Hydroxypropanenitrile	41(100)	31(97)	29(20)	42(17)	52(15)	40(13)	53(10)	51(9)	39(7)	26(7)
Iodobenzene	204(100)	77(82)	51(32)	50(20)	127(8)	205(6)	78(6)	74(6)	102(5)	76(5)
1-Iodobutane	57(100)	29(78)	41(56)	27(39)	184(36)	39(17)	28(16)	26(7)	127(6)	55(6)
2-Iodobutane, (±)-	57(100)	29(36)	184(36)	41(36)	27(10)	39(9)	127(6)	58(4)		
Iodoethane	156(100)	29(75)	27(63)	127(31)	26(14)	28(9)	128(8)	141(2)	25(2)	140(1)
Iodomethane	142(100)	127(38)	141(14)	15(13)	139(5)	140(4)	128(3)	14(1)	13(1)	71(0)
1-Iodopropane	43(100)	170(68)	41(35)	27(26)	127(14)	39(11)	44(4)	141(3)	128(3)	42(3)
2-Iodopropane	43(100)	170(46)	41(45)	27(44)	39(19)	127(17)	42(4)	38(4)	128(3)	44(3)
Isobutanal	43(100)	41(84)	27(47)	72(46)	39(30)	29(25)	42(10)	70(5)	38(5)	28(5)
Isobutyl acetate	43(100)	56(26)	73(15)	41(10)	29(5)	71(3)	57(3)	39(3)	27(3)	86(2)
Isobutylamine	30(100)	28(9)	41(6)	73(5)	27(5)	39(4)	29(3)	15(3)	58(2)	56(2)
Isobutylbenzene	91(100)	92(58)	43(22)	134(20)	65(12)	41(12)	39(11)	27(8)	51(7)	93(5)
Isobutyl formate	43(100)	56(82)	41(78)	31(65)	29(64)	27(53)	60(39)	39(36)	42(26)	15(14)
Isobutyl isobutanoate	43(100)	41(94)	56(76)	71(70)	57(68)	27(62)	29(59)	89(37)	39(32)	42(17)
Isopentyl acetate	43(100)	70(49)	55(38)	61(15)	42(15)	41(14)	27(12)	87(11)	29(10)	73(9)
Isophorone	82(100)	39(20)	138(17)	54(13)	27(12)	41(10)	53(8)	83(7)	29(7)	55(6)
Isopropyl acetate	43(100)	61(17)	41(14)	87(9)	59(8)	27(8)	42(7)	39(4)	45(3)	44(2)
Isopropylbenzene	105(100)	120(25)	77(13)	51(12)	79(10)	106(9)	39(9)	27(8)	103(6)	91(5)
1-Isopropyl-4-methylbenzene	119(100)	91(42)	134(33)	39(27)	41(20)	117(18)	65(18)	77(17)	27(16)	120(15)
Isoquinoline	129(100)	102(26)	51(20)	128(18)	50(11)	130(10)	75(10)	76(9)	103(8)	74(7)
d-Limonene	68(100)	93(50)	67(49)	41(22)	94(21)	79(21)	39(21)	136(20)	53(19)	121(16)

Name										
Mesityl oxide	55(100)	83(89)	43(73)	29(42)	98(36)	39(32)	27(28)	53(11)	41(10)	56(5)
Methanol	31(100)	29(72)	32(67)	15(42)	28(12)	14(10)	30(9)	13(6)	12(3)	16(2)
2-Methoxyethanol	45(100)	31(15)	29(14)	28(11)	47(9)	76(6)	43(6)	58(4)	46(4)	27(4)
2-Methoxyethyl acetate	43(100)	45(48)	58(42)	29(10)	42(4)	31(4)	73(3)	27(3)	59(2)	26(2)
Methyl acetate	43(100)	74(52)	28(38)	42(19)	59(17)	44(8)	32(8)	29(6)	31(4)	75(2)
2-Methylacrylonitrile	41(100)	39(54)	67(44)	40(26)	38(24)	52(23)	27(23)	37(20)	66(17)	51(12)
Methylamine	30(100)	31(87)	28(56)	29(19)	32(15)	15(12)	27(9)			
2-Methylaniline	106(100)	107(83)	77(17)	79(13)	39(12)	53(10)	52(10)	54(9)	51(9)	28(9)
3-Methylaniline	106(100)	107(84)	79(17)	77(17)	108(7)	78(6)	80(5)	89(4)	65(4)	53(4)
N-Methylaniline	106(100)	107(79)	77(23)	51(12)	79(11)	65(9)	39(9)	78(8)	108(7)	50(6)
Methyl benzoate	105(100)	77(81)	51(45)	136(24)	50(18)	106(8)	78(6)	28(6)	39(5)	27(5)
Methyl butanoate	43(100)	74(90)	71(66)	41(32)	27(31)	59(28)	87(19)	42(15)	28(15)	39(14)
3-Methylbutanoic acid	60(100)	43(61)	41(54)	27(33)	45(31)	29(27)	74(24)	39(24)	87(21)	57(20)
3-Methyl-1-butanol	55(100)	42(90)	43(82)	41(81)	70(71)	31(61)	29(59)	27(59)	39(44)	57(31)
Methyl *tert*-butyl ether	41(100)	73(78)	57(71)	29(60)	43(48)	39(34)	28(28)	56(23)	55(18)	45(16)
Methyl cyanoacetate	59(100)	15(65)	68(60)	40(38)	28(17)	29(16)	55(11)	54(10)	39(10)	67(8)
Methylcyclohexane	83(100)	55(82)	41(60)	98(44)	42(35)	56(30)	27(29)	39(27)	69(23)	70(22)
cis-4-Methylcyclohexanol	57(100)	58(54)	70(38)	81(36)	96(33)	55(25)	41(23)	114(17)	71(15)	56(12)
N-Methylformamide	59(100)	30(54)	28(34)	29(13)	58(8)	15(7)	60(3)	41(3)	27(3)	31(2)
Methyl formate	31(100)	29(63)	32(34)	60(28)	30(7)	28(7)	44(2)	18(2)	61(1)	59(1)
2-Methylheptane	43(100)	57(91)	42(39)	41(27)	29(16)	70(15)	27(13)	71(12)	99(10)	55(9)
4-Methylheptane	43(100)	71(53)	70(46)	41(27)	29(23)	27(23)	55(15)	57(14)	42(14)	39(10)
6-Methyl-1-heptanol	41(100)	55(94)	43(93)	69(84)	57(81)	56(73)	29(50)	84(48)	70(47)	27(46)
2-Methylhexane	43(100)	42(38)	41(35)	85(32)	57(26)	29(22)	27(22)	56(20)	39(11)	55(5)
5-Methyl-2-hexanone	43(100)	58(34)	27(14)	41(13)	15(13)	57(11)	39(9)	71(8)	59(8)	29(8)
Methyl methacrylate	41(100)	69(66)	39(40)	100(34)	15(20)	40(10)	59(8)	99(6)	38(6)	55(5)
2-Methyloctane	43(100)	57(51)	41(35)	71(28)	42(28)	29(23)	27(21)	85(17)	56(15)	84(13)
2-Methylpentane	43(100)	42(53)	41(35)	27(31)	71(29)	39(20)	29(18)	57(11)	15(10)	70(7)
3-Methylpentane	57(100)	56(76)	41(68)	29(60)	27(40)	43(29)	39(22)	55(9)	15(9)	28(8)
2-Methyl-2,4-pentanediol	59(100)	43(61)	56(25)	45(17)	41(16)	57(13)	42(13)	85(11)	61(10)	31(10)
4-Methylpentanenitrile	55(100)	41(52)	43(46)	27(39)	39(29)	57(27)	54(26)	29(22)	82(13)	28(12)
4-Methyl-2-pentanol	45(100)	43(47)	69(30)	41(27)	27(19)	39(13)	29(12)	87(11)	84(10)	57(10)
3-Methyl-3-pentanol	73(100)	55(38)	43(35)	45(28)	27(25)	29(21)	41(12)	87(11)	31(11)	15(9)
4-Methyl-2-pentanone	43(100)	58(84)	29(65)	41(56)	57(44)	27(42)	39(31)	85(19)	100(14)	42(14)
2-Methylpropanenitrile	42(100)	68(45)	28(45)	54(26)	41(26)	27(26)	29(25)	26(15)	39(13)	15(13)
Methyl propanoate	57(100)	29(72)	59(31)	88(26)	27(18)	28(9)	31(5)	44(4)	26(4)	58(3)
2-Methylpropanoic acid	43(100)	41(42)	27(40)	73(22)	39(15)	45(14)	42(11)	29(9)	88(7)	28(6)
2-Methyl-1-propanol	43(100)	33(73)	31(72)	41(66)	42(60)	27(43)	29(18)	39(17)	28(8)	74(6)
2-Methyl-2-propanol	59(100)	31(33)	41(22)	43(18)	29(13)	27(11)	57(10)	42(4)	60(3)	28(3)
2-Methylpyridine	93(100)	66(41)	39(31)	92(20)	78(19)	51(19)	65(16)	38(13)	50(12)	52(11)
3-Methylpyridine	93(100)	39(51)	66(46)	92(31)	65(29)	40(19)	38(18)	67(13)	63(11)	51(11)
N-Methyl-2-pyrrolidinone	99(100)	44(89)	98(80)	42(60)	41(38)	43(17)	28(17)	71(13)	39(11)	70(10)
Methyl salicylate	120(100)	92(59)	152(47)	121(32)	65(22)	39(22)	93(15)	64(14)	18(14)	63(13)
2-Methylthiophene	97(100)	98(57)	45(22)	39(14)	53(9)	99(8)	27(8)	69(6)	58(6)	59(5)
Morpholine	57(100)	29(100)	87(69)	28(69)	30(38)	56(33)	86(28)	31(28)	27(12)	15(7)
Nitrobenzene	77(100)	51(59)	123(42)	50(25)	30(15)	65(14)	39(10)	93(9)	74(7)	78(6)
Nitroethane	29(100)	30(12)	28(11)	26(9)	27(8)	43(5)	41(5)	14(5)	15(3)	46(2)
Nitromethane	30(100)	61(64)	46(39)	28(30)	45(8)	27(8)	44(7)	29(7)	60(5)	43(4)
1-Nitropropane	43(100)	27(93)	41(90)	39(34)	30(25)	44(20)	42(20)	26(20)	28(13)	54(12)
2-Nitropropane	43(100)	41(73)	27(71)	39(30)	30(18)	15(11)	42(9)	28(8)	26(8)	38(6)
Nonane	43(100)	57(75)	41(29)	84(26)	85(22)	29(22)	71(18)	56(16)	27(13)	42(12)
Octane	43(100)	57(30)	85(25)	41(25)	71(19)	29(17)	56(14)	70(10)	42(10)	27(10)
1-Octanol	41(100)	56(85)	43(82)	55(81)	31(69)	27(69)	29(68)	42(62)	70(53)	69(48)
2-Octanone	43(100)	58(79)	41(56)	59(52)	71(49)	27(46)	29(36)	39(27)	57(18)	55(17)
1-Octene	43(100)	41(82)	55(80)	56(67)	42(67)	70(54)	29(44)	27(31)	69(30)	39(29)
Pentachloroethane	167(100)	165(91)	117(90)	119(89)	83(58)	169(54)	130(43)	132(42)	60(40)	85(37)
Pentane	43(100)	42(55)	41(45)	27(42)	29(26)	39(19)	57(13)	28(9)	15(9)	72(8)
1,5-Pentanediol	31(100)	56(85)	41(67)	57(59)	55(51)	44(45)	29(37)	43(31)	68(29)	27(26)
2,4-Pentanedione	43(100)	85(31)	100(20)	27(12)	42(10)	29(10)	41(7)	39(7)	31(5)	26(5)
Pentanenitrile	41(100)	43(97)	54(54)	27(34)	55(21)	28(19)	29(16)	39(15)	42(5)	26(5)
Pentanoic acid	60(100)	73(34)	27(33)	29(28)	41(21)	43(17)	45(16)	28(14)	42(12)	39(12)

Name										
1-Pentanol	42(100)	70(72)	55(65)	41(56)	31(47)	29(41)	27(26)	57(22)	28(22)	43(21)
2-Pentanol	45(100)	43(20)	55(18)	27(17)	29(11)	41(9)	31(9)	15(9)	44(8)	39(8)
3-Pentanol	59(100)	31(83)	41(42)	27(35)	29(34)	58(15)	43(15)	57(14)	39(12)	15(9)
2-Pentanone	43(100)	41(17)	86(12)	42(12)	27(11)	39(8)	71(7)	58(7)	45(7)	44(3)
3-Pentanone	57(100)	29(50)	86(26)	27(13)	58(4)	56(4)	28(4)	26(3)	43(2)	42(2)
Pentyl acetate	43(100)	70(90)	42(52)	28(51)	61(50)	55(41)	73(21)	41(20)	29(14)	69(11)
Pentylamine	30(100)	87(8)	41(4)	28(4)	45(3)	42(3)	27(3)	56(2)	44(2)	43(2)
Pentyl lactate	43(100)	70(44)	71(32)	55(28)	41(22)	29(20)	27(15)	45(12)	42(9)	57(7)
α-Pinene	93(100)	92(30)	39(24)	41(23)	77(22)	91(21)	27(21)	79(18)	121(13)	53(10)
β-Pinene	93(100)	41(64)	69(47)	39(33)	27(31)	79(20)	77(18)	53(14)	94(13)	91(13)
Piperidine	84(100)	85(53)	56(46)	57(43)	28(41)	29(37)	44(34)	42(30)	30(30)	43(25)
Propanal	29(100)	58(59)	28(58)	27(39)	57(20)	31(4)	30(4)	42(3)	39(3)	59(2)
1,2-Propanediol	45(100)	18(46)	29(21)	43(19)	31(18)	27(17)	28(11)	19(8)	44(6)	61(5)
1,3-Propanediol	28(100)	58(93)	31(76)	57(70)	29(40)	27(26)	45(24)	43(23)	19(18)	30(17)
Propanenitrile	28(100)	54(63)	26(20)	27(17)	52(11)	55(10)	51(9)	15(9)	53(7)	25(7)
Propanoic acid	28(100)	29(84)	74(79)	27(62)	45(56)	73(48)	57(30)	26(21)	55(17)	56(16)
Propanoic anhydride	57(100)	29(55)	28(20)	27(20)	74(19)	73(12)	45(11)	26(5)	30(4)	58(3)
1-Propanol	31(100)	27(19)	29(18)	59(11)	42(9)	60(7)	41(7)	28(7)	43(3)	32(3)
2-Propanol	45(100)	43(19)	27(17)	29(12)	41(7)	31(6)	19(6)	42(5)	44(4)	59(3)
Propargyl alcohol	55(100)	39(25)	28(20)	27(19)	29(16)	38(14)	26(11)	37(8)	53(6)	56(4)
Propyl acetate	43(100)	61(19)	31(18)	27(15)	42(11)	73(9)	41(9)	29(9)	59(5)	39(5)
Propylamine	30(100)	28(13)	59(8)	27(7)	41(5)	42(3)	39(3)	29(3)	26(3)	18(3)
Propylbenzene	91(100)	120(21)	92(10)	38(10)	65(9)	78(6)	51(6)	27(5)	63(4)	105(3)
Propylene carbonate	28(100)	57(69)	43(66)	29(51)	27(45)	30(42)	26(19)	42(17)	31(16)	58(13)
Propyl formate	27(100)	29(92)	31(81)	42(60)	41(53)	43(29)	39(29)	26(28)	47(22)	30(16)
Pyridine	79(100)	52(62)	51(31)	50(19)	78(11)	53(7)	39(7)	80(6)	27(3)	77(2)
Pyrrole	67(100)	41(58)	39(58)	40(51)	28(42)	38(20)	37(12)	66(7)	68(5)	27(3)
Pyrrolidine	43(100)	28(52)	70(33)	71(26)	42(22)	41(20)	27(16)	39(15)	29(10)	30(9)
2-Pyrrolidone	85(100)	42(43)	41(36)	28(33)	30(29)	56(16)	84(14)	40(12)	27(12)	29(9)
Quinoline	129(100)	51(28)	76(25)	128(24)	44(24)	50(20)	32(19)	75(18)	74(12)	103(11)
Salicylaldehyde	28(100)	122(98)	121(92)	39(79)	65(52)	76(31)	32(31)	44(30)	93(26)	38(23)
Styrene	104(100)	103(41)	78(32)	51(28)	77(23)	105(12)	50(12)	52(11)	39(11)	102(10)
Sulfolane	41(100)	28(94)	56(82)	55(72)	120(37)	27(32)	39(19)	29(17)	26(11)	48(5)
α-Terpinene	121(100)	93(85)	136(43)	91(40)	77(34)	39(33)	27(33)	79(27)	41(26)	43(18)
1,1,1,2-Tetrachloro-2,2-difluoroethane	167(100)	169(96)	117(85)	119(82)	171(31)	85(29)	121(26)	82(14)	47(14)	101(13)
1,1,2,2-Tetrachloro-1,2-difluoroethane	101(100)	103(64)	167(54)	169(52)	117(19)	119(18)	171(17)	105(11)	31(11)	132(9)
1,1,1,2-Tetrachloroethane	131(100)	133(96)	117(76)	119(73)	95(34)	135(31)	121(23)	97(23)	61(19)	60(18)
1,1,2,2-Tetrachloroethane	83(100)	85(63)	95(11)	87(10)	168(8)	133(8)	131(8)	96(8)	61(8)	60(8)
Tetrachloroethene	166(100)	164(82)	131(71)	129(71)	168(45)	94(38)	47(31)	96(24)	133(20)	59(17)
Tetrachloromethane	117(100)	119(98)	121(31)	82(24)	47(23)	84(16)	35(14)	49(8)	28(8)	36(6)
Tetraethylene glycol	45(100)	89(10)	44(8)	43(6)	31(6)	29(6)	27(6)	101(5)	75(5)	28(5)
Tetrahydrofuran	42(100)	41(52)	27(33)	72(29)	71(27)	39(24)	43(22)	29(22)	40(13)	15(10)
1,2,3,4-Tetrahydronaphthalene	104(100)	132(53)	91(43)	51(17)	39(17)	131(15)	117(15)	115(14)	78(13)	77(13)
Tetrahydropyran	41(100)	28(64)	56(57)	45(57)	29(51)	27(49)	85(47)	86(42)	39(28)	55(23)
Tetrahydrothiophene	60(100)	88(54)	45(37)	46(32)	47(26)	27(24)	59(18)	87(16)	39(14)	54(13)
Tetramethylsilane	73(100)	43(14)	45(12)	74(8)	29(7)	15(5)	75(4)	44(4)	42(4)	31(4)
Tetramethylurea	72(100)	44(27)	116(24)	42(13)	15(13)	17(7)	28(5)	73(4)	56(4)	18(4)
1*H*-Tetrazole	42(100)	28(60)	27(25)	29(24)	41(13)	43(11)	70(8)	26(7)	40(2)	38(2)
Thiophene	84(100)	58(65)	45(58)	39(29)	57(13)	38(8)	69(7)	37(7)	83(6)	50(6)
Toluene	91(100)	92(73)	39(20)	65(14)	63(11)	51(11)	50(7)	27(6)	93(5)	90(5)
Tribromomethane	173(100)	171(50)	175(49)	93(22)	91(22)	79(18)	81(17)	94(13)	92(13)	254(11)
Tributylamine	142(100)	100(19)	143(11)	29(8)	185(7)	57(6)	44(6)	41(6)	30(5)	86(4)
1,2,4-Trichlorobenzene	180(100)	182(96)	184(30)	145(30)	109(26)	147(19)	75(11)	74(9)	111(8)	181(7)
1,1,1-Trichloroethane	97(100)	99(64)	61(58)	26(31)	27(24)	117(19)	63(19)	119(18)	35(17)	62(11)
1,1,2-Trichloroethane	97(100)	83(95)	99(62)	85(60)	61(58)	26(23)	96(21)	63(19)	27(17)	98(15)
Trichloroethene	95(100)	130(90)	132(85)	60(65)	97(64)	35(40)	134(27)	47(26)	62(21)	59(13)
Trichloroethylsilane	135(100)	133(100)	126(67)	128(46)	137(37)	98(22)	63(21)	35(16)	100(14)	127(11)
Trichlorofluoromethane	101(100)	103(66)	66(13)	105(11)	35(11)	47(9)	31(8)	82(4)	68(4)	37(4)
Trichloromethane	83(100)	85(64)	47(35)	35(19)	48(16)	49(12)	87(10)	37(6)	50(5)	84(4)
(Trichloromethyl)benzene	159(100)	161(64)	89(14)	163(11)	28(10)	63(9)	160(8)	123(8)	62(8)	124(6)
1,2,3-Trichloropropane	75(100)	39(58)	49(42)	110(38)	61(34)	77(33)	112(22)	27(16)	97(15)	38(15)

Name										
1,1,2-Trichloro-1,2,2-trifluoroethane	101(100)	151(68)	103(64)	85(45)	31(45)	153(44)	35(20)	66(19)	47(18)	87(14)
Tridecane	57(100)	43(91)	71(51)	41(34)	85(24)	29(23)	56(14)	55(12)	27(11)	42(10)
Tris(2-hydroxyethyl)amine	118(100)	56(69)	45(60)	42(56)	44(27)	43(25)	41(14)	116(8)	57(8)	86(7)
Triethylamine	86(100)	30(68)	58(37)	28(24)	29(23)	27(19)	44(18)	101(17)	42(16)	56(8)
Triethylene glycol	45(100)	58(11)	89(9)	31(8)	29(8)	75(7)	44(7)	43(7)	27(7)	28(5)
Triethyl phosphate	99(100)	81(71)	155(56)	82(45)	45(45)	109(44)	127(41)	43(24)	125(16)	111(14)
Trifluoroacetic acid	45(100)	69(70)	51(36)	28(28)	50(15)	44(11)	43(7)	97(5)	31(5)	29(5)
2,2,2-Trifluoroethanol	31(100)	33(24)	61(19)	29(19)	51(16)	69(9)	32(6)	49(5)	83(4)	81(4)
(Trifluoromethyl)benzene	146(100)	145(40)	127(34)	96(28)	77(10)	51(10)	147(8)	75(6)	50(6)	128(3)
Trimethylamine	58(100)	59(47)	30(29)	42(26)	44(17)	15(14)	28(10)	18(10)	43(8)	57(7)
1,2,3-Trimethylbenzene	105(100)	120(47)	39(22)	77(17)	91(14)	51(14)	27(14)	79(12)	119(11)	106(9)
1,2,4-Trimethylbenzene	105(100)	120(56)	119(17)	77(15)	39(15)	51(11)	91(10)	27(10)	106(9)	79(7)
1,3,5-Trimethylbenzene	105(100)	120(64)	119(15)	77(13)	39(11)	106(9)	91(9)	51(8)	27(7)	121(6)
2,2,3-Trimethylbutane	57(100)	43(71)	56(63)	41(53)	85(30)	29(26)	27(18)	39(14)	15(6)	55(5)
2,2,5-Trimethylhexane	57(100)	56(35)	71(18)	41(17)	43(14)	29(8)	70(4)	58(4)	113(3)	55(3)
2,2,4-Trimethylpentane	57(100)	41(31)	56(28)	43(24)	29(16)	27(9)	39(7)	58(4)	55(4)	99(2)
2,3,3-Trimethylpentane	43(100)	71(45)	70(36)	57(36)	41(29)	85(25)	27(18)	55(16)	29(16)	39(11)
2,3,4-Trimethylpentane	43(100)	71(62)	70(41)	41(25)	27(18)	55(17)	57(16)	29(14)	39(10)	42(7)
Trimethyl phosphate	110(100)	109(35)	79(34)	95(25)	80(23)	15(20)	140(18)	47(10)	31(7)	139(5)
2,4,6-Trimethylpyridine	121(100)	39(27)	79(26)	120(24)	106(17)	27(16)	77(13)	51(11)	42(11)	122(10)
1-Undecene	41(100)	43(87)	55(80)	70(67)	56(67)	69(55)	29(55)	83(51)	57(50)	27(46)
Vinyl acetate	43(100)	28(45)	42(26)	44(24)	86(20)	31(10)	32(7)	29(7)	45(2)	41(2)
o-Xylene	91(100)	106(40)	39(21)	105(17)	51(17)	77(15)	27(12)	65(10)	92(8)	79(8)
m-Xylene	91(100)	106(65)	105(29)	39(18)	51(15)	77(14)	27(10)	92(8)	79(8)	78(8)
p-Xylene	91(100)	106(62)	105(30)	51(16)	39(16)	77(13)	27(11)	92(7)	78(7)	65(7)

Analytical

COMMON SPURIOUS SIGNALS OBSERVED IN MASS SPECTROMETERS

Thomas J. Bruno and Paris D. N. Svoronos

The following table provides guidance in the recognition and interpretation of potentially spurious signals (*m/z* peaks) that are sometimes observed in measured mass spectra (Ref. 1). Often, the occurrence of these signals can be predicted by the recent history of the instrument or the method being used. This is especially true if the mass spectrometer is interfaced to a gas chromatograph.

Reference

1. Bruno, T. J., and Svoronos, P. D. N., *CRC Handbook of Basic Tables for Chemical Analysis, Third Edition*, CRC Press, Boca Raton, FL, 2011. <https://doi.org/10.1201/b10385>

Possible Sources for Spurious Signals Observed in Mass Spectrometers

Ions observed, *m/z*	Possible compound	Possible source
13, 14, 15, 16	Methane*	Chlorine reagent gas
18	Water*	Residual impurity, outgassing of ferrules, septa, and seals
14, 28	Nitrogen*	Residual impurity, outgassing of ferrules, septa, and seals; leaking seal
16, 32	Oxygen*	Residual impurity, outgassing of ferrules, septa, and seals; leaking seal
44	Carbon dioxide*	Residual impurity, outgassing of ferrules, septa, and seals; leaking seal; note it may be mistaken for propane in a sample
31, 51, 69, 100, 119, 131, 169, 181, 214, 219, 264, 376, 414, 426, 464, 502, 576, 614	Perfluorotributylamine (PFTBA), and related ions**	This is a common tuning compound; may indicate a leaking valve
31	Methanol	Solvent; can be used as a leak detector
43, 58	Acetone	Solvent; can be used as a leak detector
78	Benzene	Solvent; can be used as a leak detector
91, 92	Toluene	Solvent; can be used as a leak detector
105, 106	Xylenes	Solvent; can be used as a leak detector
151, 153	Trichloroethane	Solvent; can be used as a leak detector
69	Rough (fore) pump fluid, PFTBA	Back diffusion of fore pump fluid, possible leaking valve of tuning compound vial
73, 147, 207, 221, 281, 295, 355, 429	Dimethylpolysiloxane	Bleed from a column or septum, often during high-temperature program methods in GC-MS
77, 94, 115, 141, 168, 170, 262, 354, 446	Diffusion pump fluid	Back diffusion from diffusion pump, if present
149	Phthalates	Plasticizer in vacuum seals, gloves
X – 14 peaks	Hydrocarbons	Loss of a methylene group indicates a hydrocarbon sample

* It is possible to operate the analyzer to ignore these common background impurities. They will be present to contribute to poor vacuum if these impurities result from a significant leak.

** See the table "Major Reference Masses in the Spectrum of Heptacosafluorotributylamine (Perfluorotributylamine)" in this section for additional details.

CHLORINE–BROMINE COMBINATION ISOTOPE INTENSITIES IN MASS SPECTRAL PATTERNS

Thomas J. Bruno and Paris D. N. Svoronos

Because of the distinctive mass spectral patterns caused by the presence of chlorine and bromine in a molecule, interpretation of some mass spectra can be much easier because the relative isotopic concentrations are known. The following table provides ideal peak intensities, relative to the molecular ion (M^+) at an intensity normalized to 100%, for various combinations of chlorine and bromine atoms, assuming the absence of all other elements except carbon, hydrogen, and oxygen (Refs. 1-4). Note that in the table we refer to the base peak (the peak with the highest intensity) as P, as it might not always be the molecular ion. The mass abundance calculations were based upon the most recent atomic mass data (see the table "Atomic Masses and Isotopic Abundances" in Section 1 of this *CRC Handbook*).

Note that experimental uncertainty can change these ideal projections; many users will have an instrument capable of only unit mass resolution. In such cases, one must expect these patterns to be approximate. For example, if a molecule has no Cl and two Br, the P/(P+2)/(P+4) ratio is roughly 100/200/100 (P is assumed to be 100, and is not listed in the table). To confirm the presence of two bromine atoms, look for a peak at P-79, attributed to the loss of one bromine atom. Losing one bromine would lead to a "doublet" (roughly 1:1 ratio) which is indicative of a fragment containing only one bromine atom. Furthermore, at P-158 one would see that the 1:1 pattern has disappeared. These two observations will confirm the presence of two bromine atoms in the original P^+ base peak ion.

If a molecule has two Cl atoms and no Br atoms present, the P/(P+2)/(P+4) ratio is roughly 100/67/11 in a similar fashion. To confirm, one should consider the P-35 area of the spectrum (loss of one chlorine). Losing one chlorine would lead to a doublet (roughly 3:1 ratio). In general, no lost fragments of mass 35 or 79 are indicative of other atoms or fragments. Occasionally a halogen is lost as the hydrogen halide which will lead to the loss of mass 80 (HBr) or mass 36 (HCl), respectively, but the doublet pattern P/P+2 remains the same.

The spectrum becomes much more complicated when more than two bromine, or two chlorine atoms, or a combination of both atoms, is involved, as one can observe in the table. The rule of thumb, however, is that one ought to always look at the peaks (P-79) or (P-35) to alert the presence of a bromine or chlorine, respectively. We also note that the statistical likelihood of encountering molecules with high multiples of Cl and Br atoms is low outside of environmental analyses.

References

1. Bruno, T. J., and Svoronos, P. D. N., *CRC Handbook of Basic Tables for Chemical Analysis, Third Edition*, CRC Press, Boca Raton, FL, 2011. <https://doi.org/10.1201/b10385>
2. McLafferty, F. W., and Turecek, F., *Interpretation of Mass Spectra, Fourth Edition*, University Science Books, Mill Valley, CA, 1993.
3. Silverstein, R. H., Bassler, G. C., and Morrill, T. C., *Spectroscopic Identification of Organic Compounds, Sixth Edition*, John Wiley & Sons, New York, 1998.
4. Williams, D. H., and Fleming, I., *Spectroscopic Methods in Organic Chemistry, Fourth Edition*, McGraw-Hill, London, 1989.

Relative Intensities of Isotope Peaks for Combinations of Bromine and Chlorine in Mass Spectral Patterns (M^+ = 100%)

Cl_n	Br_0	Br_1	Br_2	Br_3	Br_4
Cl_0					
P + 2		98.0	196.0	294.0	390.8
P + 4			96.1	288.2	574.7
P + 6				94.1	375.3
P + 8					92.0
Cl_1					
P + 2	32.5	130.6	228.0	326.1	424.6
P + 4		31.9	159.0	383.1	704.2
P + 6			31.2	187.4	564.1
P + 8				30.7	214.8
P + 10					30.3
Cl_2					
P + 2	65.0	163.0	261.1	359.3	456.3
P + 4	10.6	74.4	234.2	490.2	840.3
P + 6		10.4	83.3	312.8	791.6
P + 8			10.2	91.7	397.5
P + 10				9.8	99.2
P + 12					10.1
Cl_3					
P + 2	97.5	195.3	294.0	393.3	489
P + 4	31.7	127.0	99.7	609.8	989
P + 6	3.4	34.4	159.4	473.8	1064
P + 8		3.3	37.1	193.9	654
P + 10			3.2	39.6	229
P + 12				3.0	42
P + 14					3.2
Cl_4					
P + 2	130.0	228.3	326.6	4.2	522
P + 4	63.3	190.9	414.9	735.3	1149
P + 6	13.7	75.8	263.1	670.0	1388
P + 8	1.2	14.4	88.8	347.1	1002
P + 10		1.1	15.4	102.2	443
P + 12			1.3	16.2	117
P + 14				0.7	17
Cl_5					
P + 2	162.6	260.7	358.9		
P + 4	105.7	265.3	520.8		
P + 6	34.3	137.9	397.9		
P + 8	5.5	39.3	174.5		
P + 10	0.3	5.8	44.3		
P + 12		0.3	5.7		
P + 14			0.5		
Cl_6					
P + 2	195.3				
P + 4	158.6				
P + 6	68.8				
P + 8	16.6				
P + 10	2.1				
P + 12	0.1				
Cl_7					
P + 2	227.8				
P + 4	222.1				
P + 6	120.3				
P + 8	39.0				
P + 10	7.5				
P + 12	0.8				
P + 14	0.05				

REDUCTION OF WEIGHINGS IN AIR TO *IN VACUO*

When the mass M of a body is determined in air, a correction is necessary for the buoyancy of the air. The corrected mass is given by $M + kM/1000$, where k is a function of the material used for the weights, given by

$$k = 1000\rho_{air}(1/\rho_{body} - 1/\rho_{weight})$$

and ρ is density. Table 1 has values of k for densities of three common masses used in laboratories: platinum-iridium (21.6 g cm^{-3}), brass (8.5 g cm^{-3}), and aluminum or quartz (2.65 g cm^{-3}) and is computed for an air density of 0.0012 g cm^{-3}.

TABLE 1. Value of k at Specified Density of Body for Buoyancy of Air Corrections for Masses of Different Materials

Density of body (g cm^{-3})	k (Pt-Ir)	k (Brass)	k (Quartz or Al)
0.5	2.34	2.26	1.95
0.6	1.94	1.86	1.55
0.7	1.66	1.57	1.26
0.8	1.44	1.36	1.05
0.9	1.28	1.19	0.88
1.0	1.14	1.06	0.75
1.1	1.04	0.95	0.64
1.2	0.94	0.86	0.55
1.3	0.87	0.78	0.47
1.4	0.80	0.72	0.40
1.5	0.74	0.66	0.35
1.6	0.69	0.61	0.30
1.7	0.65	0.56	0.25
1.8	0.61	0.53	0.21
1.9	0.58	0.49	0.18
2.0	0.54	0.46	0.15
2.5	0.42	0.34	0.03
3.0	0.34	0.26	-0.05
4.0	0.24	0.16	-0.15
6.0	0.14	0.06	-0.25
8.0	0.09	0.01	-0.30
10.0	0.06	-0.02	-0.33
15.0	0.02	-0.06	-0.37
20.0	0.00	-0.08	-0.39
22.0	0.00	-0.09	-0.40

For a more accurate calculation, Table 2 provides values of the density of air at the specified temperatures (assuming 50% relative humidity and 0.04% CO_2).

TABLE 2. Density of Air in g cm^{-3} at Indicated Pressure and Temperature

Name	P/kPa	10 °C	20 °C	30 °C
Air	85	0.001043	0.001005	0.000968
Air	90	0.001105	0.001065	0.001025
Air	95	0.001166	0.001124	0.001083
Air	100	0.001228	0.001184	0.001140
Air	105	0.001290	0.001243	0.001198

Formulas for calculating the density of air over more extended ranges of temperature, pressure, and humidity may be found in the references.

References

1. Kaye, G. W. C., and Laby, T. H., *Tables of Physical and Chemical Constants, 16th Edition*, pp. 25–28, Longman, London, 1995.
2. Giacomo, P., *Metrologia*, 18, 33, 1982. <https://doi.org/10.1088/0026-1394/18/1/006>
3. Davis, R. S., *Metrologia*, 29, 67, 1992. <https://doi.org/10.1088/0026-1394/29/1/008>

Analytical

STANDARDS FOR LABORATORY WEIGHTS

Thomas J. Bruno and Paris D. N. Svoronos

The following tables provide a summary of the requirements for metric weights and mass standards commonly used in chemical analysis (Refs. 1–3). The actual specifications are under the jurisdiction of ASTM Committee E-41 on General Laboratory Apparatus and are the direct responsibility of subcommittee E-41.06 that deals with weighing devices. These standards do not generally refer to instruments used in commerce. Weights are classified according to Type (Design), either Type I or Type II (Table 1); Grade (Physical Property) S, O, P, or Q (Table 2); and Tolerance (Deviation) 1, 2, 3, 4, 5, or 6 (Table 3). Details about each classification are included in the tables.

Information on the applications for various mass standards (Table 4) is presented to allow the user to make appropriate choices when using analytical weights for the calibration of electronic analytical balances, for making large-scale mass measurements (such as those involving gas cylinders), and in the use of dead-weight pressure balances. Some historical context can be found in Ref. 4.

References

1. *Annual Book of ASTM Standards, ANSI/ASTM E617-97 Standard Specification for Laboratory Weights and Precision Mass Standards*, Book of Standards Vol. 14.04, 2008.
2. Battino, R., and Williamson, A. G., *J. Chem. Educ.* 61, 51, 1984. <https://doi.org/10.1021/ed061p51>
3. Bruno, T. J., and Svoronos, P. D. N., *CRC Handbook of Basic Tables for Chemical Analysis, Third Edition*, CRC Press, Boca Raton, FL, 2011. <https://doi.org/10.1201/b10385>
4. NBS Circular 547, *Precision Laboratory Standards of Mass and Laboratory Weights*, Washington, DC, 1954 (reprinted as NIST-IR 78-1476, Washington, DC, 1978)

TABLE 1. Type — Classification by Design

Type I	One-piece construction; contains no added adjusting material; used for highest accuracy work
Type II	Can be of any appropriate and convenient design, incorporating plugs, knobs, rings, etc.; adjusting material can be added if it is contained so that it cannot become separated from the weight

TABLE 2. Grade — Classification by Physical Property

Grade S	Density:	7.7 g cm^{-3} to 8.1 g cm^{-3} (for 50 mg and larger)
	Surface area:	not to exceed that of a cylinder of equal height and diameter
	Surface finish:	highly polished
	Surface protection:	none permitted
	Magnetic properties:	no more magnetic than 300 series stainless steels
	Corrosion resistance:	same as 303 stainless steel
	Hardness:	at least as hard as brass
Grade O	Density:	7.7 g cm^{-3} to 9.1 g cm^{-3} (for 1 g and larger)
	Surface area:	same as grade S
	Surface finish:	same as grade S
	Surface protection:	may be plated with suitable material such as platinum or rhodium
	Magnetic properties:	same as grade S
	Corrosion resistance:	same as grade S
	Hardness:	at least as hard as brass when coated; smaller weights at least as hard as aluminum
Grade P	Density:	7.2 g cm^{-3} to 10 g cm^{-3} (for 1 g or larger)
	Surface area:	no restriction
	Surface finish:	smooth, no irregularities
	Surface protection:	may be plated or lacquered
	Magnetic properties:	same as grades S and O
	Corrosion resistance:	surface must resist corrosion and oxidation
	Hardness:	same as grade O
Grade Q	Density:	7.2 g cm^{-3} to 10 g cm^{-3} (for 1 g or larger)
	Surface area:	same as grade P
	Surface finish:	same as grade P
	Surface protection:	may be plated, lacquered, or painted
	Magnetic properties:	no more magnetic than unhardened unmagnetized steel
	Corrosion resistance:	same as grade P
	Hardness:	same as grades O and P

TABLE 3. Tolerance — Classification by Deviation[1]

CLASS 1

Grams	Individual tolerance/mg	Group tolerance/mg
500	1.2	
300	0.75	
200	0.50	
100	0.25	1.35
50	0.12	
30	0.074	
20	0.074	
10	0.050	0.16
5	0.034	
3	0.034	
2	0.034	
1	0.034	0.065

CLASS 2

Grams	Individual tolerance/mg	Group tolerance/mg
500	2.5	
300	1.5	
200	1.0	
100	0.5	2.7
50	0.25	
30	0.15	
20	0.10	
10	0.074	0.29
5	0.054	
3	0.054	
2	0.054	
1	0.054	0.105

CLASS 3

Grams	Tolerance/mg
500	5.0
300	3.0
200	2.0
100	1.0
50	0.6
30	0.45
20	0.35
10	0.25

CLASS 4

Grams	Tolerance/mg
500	10
300	6.0
200	4.0
100	2.0
50	1.2
30	0.9
20	0.7
10	0.5

CLASS 5

Grams	Tolerance/mg
500	30
300	20
200	15
100	9
50	5.6
30	4.0
20	3.0
10	2.0
5	1.3
3	0.95
2	0.75
1	0.50

CLASS 6

Grams	Tolerance/mg
500	50
300	30
200	20
100	10
50	7
30	5
20	3
10	2
5	2
3	2
2	2
1	2

[1] In simple terms, the permitted deviation between the assigned nominal mass value of the weight and the actual mass of the weight. Verification of tolerance should be possible on reasonably precise equipment, without using a buoyancy correction, within the political jurisdiction or organizational bounds of a given weight specification.

TABLE 4. Applications for Weights and Mass Standards[1]

Application	Type	Grade	Class
Reference standards used for calibrating other weights	I	S	1,2,3, or 4[1]
High-precision standards for calibration of weights and precision balances	I or II[2]	S or O[2]	1 or 2[3]
Working standards for calibration and precision analytical work, dead-weight pressure balances	I or II[2]	S or O	2
Laboratory weights for routine analytical work	II	O	2 or 3
Built-in weights, high-quality analytical balances	I or II	S	2
Moderate precision laboratory balances	II	P	3 or 4
Dial scales and trip balances	II	Q	4 or 5
Platform scales	II	Q	5 or 6

[1] Primary standards are for reference use only and should be calibrated. Since the actual values for each weight are stated, close tolerances are neither required nor desirable.

[2] Type I and Grade S will have a higher constancy but will probably be higher priced.

[3] Since working standards are used for the calibration of measuring instruments, the choice of tolerance depends upon the requirements of the instrument. The weights are usually used at the assumed nominal values and appropriate tolerances should be chosen.

Source: Reprinted (with modification) with permission of the ASTM International (formerly American Society for Testing and Materials), 100 Barr Harbor Drive, West Conshohocken, PA, USA.

INDICATORS FOR ACIDS AND BASES

Thomas J. Bruno and Paris D. N. Svoronos

The following table lists the most common indicators together with their pH range and colors in acidic and basic media. Because the color change is not instantaneous at the pK_a value, a pH range is given where a combination of colors is present. This pH range, which varies between indicators, generally falls between the pK_a with a spread or uncertainty of 1 pH unit. All solutions are either aqueous or ethanol/aqueous (% ethanol, vol/vol) (Refs. 1–4). Reference 4 provides additional solution properties that are related to buffer and solvent properties, and Ref. 5 lists the exact quantities needed for the indicator solutions. Column definitions for the table are as follows.

Column heading	Definition
Name	Name of acid-base indicator; names are listed alphabetically
Synonym	Common synonym of indicator
pH Range	Range of pHs at which indicators turn color (see discussion above)
Solvent	Name of solvent
Color change (acidic to basic)	Change of color as solution goes from more acidic (low end of pH range) to more basic (high end of pH range)

References

1. Lange, N. A., *Lange's Handbook of Chemistry, Eighth Edition*, Handbook Publishers, New York, 1952. <https://doi.org/10.1097/00010694-195301000-00013>
2. Kolthoff, I. M., and V. A. Stenger (translated in English by N. H. Furman), *Volumetric Analysis, Second Edition*, Interscience Publishers, New York, 1942.
3. Sabnis, R.W., *Handbook of Acid-Base Indicators*, CRC Press, Taylor & Francis, Boca Raton, FL, 2008. <https://doi.org/10.1201/9780849382192>
4. Bruno, T. J., and Svoronos, P. D. N., *CRC Handbook of Basic Tables for Chemical Analysis, Third Edition*, CRC Press, Boca Raton, FL, 2011. <https://doi.org/10.1201/b10385>
5. <http://www.csudh.edu/oliver/chemdata/ind-prep.htm>, accessed June 2011.

Indicators for Acids and Bases Listed in Order of pH Range

Name	Synonym	pH Range	Solvent	Color change (acidic to basic)
Crystal Violet	Gentian Violet	0.0-2.0	Aqueous	yellow to blue-violet
Pentamethoxy Red	2,4,2',4',2"-Pentamethoxytriphenylcarbinol	1.2-2.3	70% ethanol	red-violet to colorless
Thymol Blue		1.2-2.8	Aqueous	red to yellow
4'-Anilinoazobenzene-4-sulfonic acid, sodium salt	Tropaeolin OO	1.3-3.2	Aqueous	red to yellow
2,4-Dinitrophenol		2.4-4.0	50% ethanol	colorless to yellow
4-(Dimethylamino)azobenzene	Methyl Yellow	2.9-4.0	90% ethanol	red to yellow
Methyl Orange	Sodium *p*-dimethylaminoazobenzenesulfonate	3.1-4.4	Aqueous	red to orange
Bromophenol Blue	Bromphenol Blue	3.0-4.6	Aqueous	yellow to blue-violet
Tetrabromophenol Blue		3.0-4.6	Aqueous	yellow to blue
Congo Red		3.0-5.0	Aqueous	blue-violet to red
4-(Phenylazo)-1-naphthalenamine	α-Naphthyl Red	3.7-5.0	70% ethanol	red to yellow
Alizarin Red S	Sodium alizarinesulfonate	3.7-5.2	Aqueous	yellow to violet
Etoxazene	*p*-Ethoxychrysoidine	3.5-5.5	Aqueous	red to yellow
Bromocresol Green	3,3',5,5'-Tetrabromo-*m*-cresolsulfonephthalein	4.0-5.6	Aqueous	yellow to blue
Methyl Red	Benzoic acid, 2-[[4-(dimethylamino)phenyl]azo]-	4.4-6.2	Aqueous	red to yellow
Bromocresol Purple	Bromcresol Purple	5.2-6.8	Aqueous	yellow to purple
4-Nitrophenol		5.0-7.0	Aqueous	colorless to yellow
Chlorophenol Red		5.4-6.8	Aqueous	yellow to red
Azolitmin		5.0-8.0	Aqueous	red to blue
Bromothymol Blue	Bromthymol Blue	6.0-7.6	70% ethanol	yellow to blue
Bromophenol Blue	Bromphenol Blue	6.2-7.6	Aqueous	yellow to blue
Phenol Red	Phenolsulfonphthalein	6.4-8.0	Aqueous	yellow to red
Neutral Red		6.8-8.0	70% ethanol	red to yellow
Aurin	Rosolic acid	6.8-8.0	90% ethanol	yellow to red
Cresol Red	*o*-Cresolsulfonphthalein	7.2-8.8	Aqueous	yellow to red
p-Naphtholbenzein	α-Naphtholbenzein	7.3-8.7	70% ethanol	rose to green
Orange I	Tropaeolin 000	7.6-8.9	Aqueous	yellow to rose-red
Thymol Blue		8.0-9.6	Aqueous	yellow to blue
Phenolphthalein	3,3-Bis(4-hydroxyphenyl)-1(3*H*)-isobenzofuranone	8.0-10.0	70% ethanol	colorless to red

Name	Synonym	pH Range	Solvent	Color change (acidic to basic)
p-Naphtholbenzein	α-Naphtholbenzein	9.0-11.0	70% ethanol	yellow to blue
Thymolphthalein		9.4-10.6	90% ethanol	colorless to blue
Nile Blue	C.I. Basic Blue 12	10.1-11.1	Aqueous	blue to red
Alizarin Yellow R		10.0-12.0	Aqueous	yellow to lilac
4-Hydroxy-3'-nitroazobenzene-3-carboxylic acid, sodium salt	Salicylic Yellow	10.0-12.0	90% ethanol	yellow to orange-brown
4-[(4-Nitrophenyl)azo]-1,3-benzenediol	Diazo Violet	10.1-12.0	Aqueous	yellow to violet
4-[(2,4-Dihydroxyphenyl)azo] benzenesulfonic acid, sodium salt	Tropaeolin O	11.0-13.0	Aqueous	yellow to orange-brown
N-Methyl-*N*,2,4,6-tetranitroaniline	Nitramine	11.0-13.0	70% ethanol	colorless to orange-brown
Poirrier Blue C4B	C.I. Acid Blue 22	11.0-13.0	Aqueous	blue to violet-pink
2,4,6-Trinitrobenzoic acid		12.0-13.4	Aqueous	colorless to orange-red

PREPARATION OF SPECIAL ANALYTICAL REAGENTS

Paris D. N. Svoronos and Thomas J. Bruno

This listing of analytical reagents has been updated to include formulations based on more recent research, and to also include safety notes (Refs. 1–3). In the addition to specific cautions, the user must always observe sound laboratory practice and housekeeping to avoid exposure and environmental contamination. When less familiar reagents are described, we include a chemical formula and a CAS Registry Number to avoid ambiguity. When a reagent calls for 95% ethanol, the azeotrope with water is specified unless noted. Additional details can be found in the references listed below.

References

1. Lide, D. R., Ed., *CRC Handbook for Chemistry and Physics, 90th Edition*, CRC Press, Boca Raton, FL, 2010.
2. Bruno, T. J., and Svoronos, P. D. N., *CRC Handbook of Basic Tables for Chemical Analysis, Third Edition*, CRC Press, Boca Raton, FL, 2011. <https://doi.org/10.1201/b10385>
3. Svoronos, P. D. N., Sarlo, E., and Kulawiec, R., *Organic Chemistry Laboratory Manual*, WCB McGraw-Hill, New York, 1996.

Aluminon (qualitative test for aluminum). Aluminon is the name for the triammonium salt of aurintricarboxylic acid (5-[(3-carboxy-4-hydroxyphenyl)(3-carboxy-4-oxo-2,5-cyclohexadienylidene)methyl]-2-hydroxybenzoic acid triammonium salt, $C_{22}H_{23}N_3O_9$, CAS No. 569-58-4). Dissolve 1 g of the salt in 1 L of distilled water. Shake the solution well to ensure thorough mixing.

Bang's reagent (for glucose estimation). Dissolve (in the exact order) 100 g of potassium carbonate, 66 g of potassium chloride and 160 g of anhydrous potassium bicarbonate in approximately 700 mL of distilled water at 30 °C. Add (while stirring) 4.4 g of copper (II) sulfate and dilute to 1 L with distilled water after all CO_2 has been released. This solution should be shaken and saved in an air-tight flask. After 24 hours, 300 mL of it are diluted to 1 L with a saturated aqueous potassium chloride solution, shaken gently and used after 24 hours. 50 mL of this solution is equivalent to 10 mg glucose.

Barfoed's reagent (test for glucose). See Cupric acetate.

Baudisch's reagent. See Cupferron.

Benedict's solution (qualitative reagent for glucose). Dissolve 173 g of sodium citrate and 100 g of sodium carbonate in 800 mL of distilled water. Filter, if necessary, and dilute to 850 mL with distilled water. Dissolve 17.3 g of copper (II) sulfate pentahydrate in 100 mL of distilled water. Pour the latter solution, with constant stirring, into the carbonate-citrate solution and dilute to 1 L with distilled water.

Benzidine hydrochloride solution (for sulfate determination). Prepare a paste of 8 g of benzidine hydrochloride ($C_{12}H_8(NH_3)_2{\cdot}2HCl$) and 20 mL of distilled water, add 20 mL of 20% (mass/mass) HCl and dilute to 1 L with distilled water. Each mL of this solution is equivalent to 0.00357 g of H_2SO_4. Note that the reagent is often called benzidine dihydrochloride.

Bertrand's reagent (glucose estimation). Consists of the following solutions:

1. Dissolve 200 g of Rochelle salt (potassium sodium tartrate, $NaKC_4H_4O_6$) and 150 g of NaOH in sufficient distilled water to make 1.0 L of solution.
2. Dissolve 40 g of copper (II) sulfate pentahydrate in sufficient distilled water to a total of 1.0 L of solution.
3. Dissolve 50 g of iron (III) sulfate and 200 g of H_2SO_4 (sp. gr. 1.84) in sufficient distilled water to a total of 1.0 L of solution.
4. Dissolve 5 g of potassium permanganate in sufficient distilled water to a total of 1.0 L solution.

Bial's reagent (for pentose). Dissolve 1 g of orcinol (5-methyl-1,3-benzenediol, $C_7H_8O_2$, CAS No. 504-15-4) in 500 mL of 30% (mass/mass) HCl to which 30 drops of a 10% aqueous solution of iron (III) chloride have been added.

Boutron — Boudet soap solution. Consists of the following solutions:

1. Dissolve 100 g of pure castile soap (olive oil based) in about 2.5 L of 56% (vol/vol) aqueous ethanol.
2. Dissolve 0.59 g of barium nitrate in 1.0 L of water. Adjust the castile soap solution (1) so that 2.4 mL of it will produce a permanent lather with 40 mL of solution (2). When adjusted, 2.4 mL of this soap solution is equivalent to 220 parts per million of hardness (as calcium carbonate) for a 40 mL sample. See also Soap solution.

Brucke's reagent (protein precipitation). See Potassium iodide mercuric iodide.

Cobaltic cyanide paper (Rinnmann's test for zinc detection). Dissolve 4 g of potassium cobalt (III) hexacyanide ($K_3Co(CN)_6$) and 1 g of potassium chlorate in 100 mL of water. Soak filter paper in solution and dry at 100 °C. Apply a drop of the zinc solution and heat in an evaporating dish. A green disk on the filter paper is obtained if zinc is present.

Congo red. Dissolve 0.5 g of Congo red (3,3'-([1,1'-biphenyl]-4,4'-diyl)bis(4-aminonaphthalene-1-sulfonic acid, $C_{32}H_{22}N_6Na_2O_6S_2$, CAS No. 573-58-0) in 90 mL of distilled water and 10 mL of ethanol.

Cupferron (Baudisch's reagent for iron analysis). Dissolve 6 g of Cupferron, the ammonium salt of *N*-hydroxy-*N*-nitrosoaniline ($NH_4[C_6H_5N(O)NO]$) in 100 mL of distilled water. The reagent is good only for one week and must be kept in the dark.

Cupric acetate (Barfoed's reagent for reducing monosaccharides). Dissolve 66 g of copper (II) acetate and 10 mL of glacial acetic acid in water and dilute to 1.0 L.

Cupric oxide, ammoniacal; Schweitzer's reagent (dissolves cotton, linen, and silk, but not wool). Dissolve 5 g of cupric sulfate in 100 mL of boiling water and add sodium hydroxide until precipitation is complete. Wash the precipitate well and dissolve it in a minimal quantity of ammonium hydroxide. Bubble a slow stream of air through 300 mL of concentrated ammonium hydroxide solution containing 50 g of fine copper turnings. Continue stirring for 1 hour.

Cupric sulfate in glycerin-potassium hydroxide (reagent for silk). Dissolve 10 g of copper (II) sulfate pentahydrate in 100 mL of water and add 5 g of glycerol. Add 6N KOH solution slowly until a deep blue solution is obtained.

Cupron (precipitates copper). Dissolve 5 g of benzoinoxime in 100 mL 95% ethanol.

Cuprous chloride, acidic (reagent for CO in gas analysis).

1. Cover the bottom of a 2 L flask with a layer of copper (II) oxide about 1.5 cm deep, suspend a coil of copper wire so as to reach from the bottom to the top of the solution, and fill the flask with hydrochloric acid 20% (vol/vol) HCl(aq) with continuous stirring. When the solution becomes nearly colorless, transfer to reagent bottles, which should also contain copper wire. The stock bottle may be refilled with dilute (6N) HCl until either the cupric oxide or the copper wire is used up. Copper (II) sulfate may be substituted for copper oxide in the above procedure.
2. Dissolve 340 g of copper (II) chloride dihydrate in 600 mL of concentrated HCl and reduce the cupric chloride by adding 190 mL of a saturated solution of stannous chloride or until the solution is colorless. The stannous chloride is prepared by treating 300 g of metallic tin in a 500 mL flask with concentrated HCl until no more tin is dissolved.
3. (Winkler method). Add a mixture of 86 g of copper (II) chloride and 17 g of finely divided metallic copper, prepared by the reduction of 8 g to 12 g of cupper (II) oxide with hydrogen gas, to a solution of HCl, made by diluting 650 mL of concentrated HCl with 325 mL of distilled water. After the mixture has been added slowly and with frequent stirring, a spiral of copper wire is suspended in the bottle, reaching all the way to the bottom. The solution is ready to use when the solution becomes colorless.

Cuprous chloride, ammoniacal (reagent for CO in gas analysis).

1. The acid solution of copper (II) chloride as prepared above is neutralized with 6N ammonium hydroxide until the ammonia odor persists. An excess of metallic copper must be kept in the solution.
2. Pour 800 mL of acidic copper (II) chloride, prepared by the Winkler method, into approximately 4 L of water. Transfer the precipitate to a 250 mL graduated cylinder. After several hours, siphon off the liquid above the 50 mL mark and refill with 7.5% ammonium hydroxide solution which may be prepared by diluting 50 mL of concentrated ammonium hydroxide with 150 mL of distilled water. The solution, which should have a faint odor of ammonia should be well stirred and allowed to stand for several hours.*Note*: Proper precautions must be taken for the safe handling of gaseous samples that contain CO.

Dichlorofluorescein indicator. Dissolve 1 g dichlorofluorescein ($C_{20}H_{10}Cl_2O_5$, CAS No. 76-54-0) in 1 L 70% (vol/vol) ethanol or 1 g of the sodium salt in 1 L of distilled water.

Dimethyglyoxime, 0.01 N. Dissolve 0.6 g of dimethylglyoxime (2,3-butanedione oxime) in 500 mL of 95% ethanol. This is an especially sensitive test for nickel (II) and produces a very characteristic crimson color.

Diphenylamine (reagent for rayon). Dissolve 0.2 g diphenylamine in 100 mL of concentrated sulfuric acid.

Diphenylamine sulfonate (for titration of iron with $K_2Cr_2O_7$). Dissolve 0.32 g of the barium salt of diphenylamine sulfonic acid in 100 mL of distilled water, add 0.5 g of sodium sulfate and filter off the precipitated barium sulfate.

Diphenylcarbazide. Dissolve 0.2 g of diphenylcarbazide[(C_6H_5N HNH)$_2$CO] in 10 mL of glacial acetic acid and dilute to 100 mL with 95% ethanol.

2,4-Dinitrophenylhydrazine (2,4-DNP) reagent. Dissolve 3 g of 2,4-dinitrophenylhydrazine in 15 mL to 20 mL of concentrated sulfuric acid. Add this solution with stirring to a solution of 10 mL distilled water in 70 mL of 95% hydrous ethanol. Use the filtrate to test for the presence of aldehydes and ketones as follows: To 2 mL of the solution add 10 drops of the unknown. A precipitate (which may be recrystallized from 95% hydrous ethanol) is indicative of a positive test and its melting point can be matched to the corresponding melting points of various aldehydes and ketones.

Esbach's reagent (estimation of protein). Dissolve 10 g of picric acid and 20 g of citric acid in sufficient distilled water to make a 1.0 L of solution.

Eschka's compound. Two parts of calcinated ("light") magnesia are thoroughly mixed with 1 part of anhydrous sodium carbonate.

Fehling's solution (reagent for reducing sugars).

1. Copper (II) sulfate solution: Dissolve 34.66 g of copper (II) sulfate pentahydrate in distilled water and dilute to 500 mL.
2. Alkaline tartrate solution: Dissolve 173 g of potassium sodium tartrate (Rochelle salt, $KNaC_4H_4O_6{\cdot}4H_2O$) and 50 g of NaOH in distilled water, cool and dilute to 500 mL. Mix equal volumes of solutions (1) and (2) just before using.

Ferric-alum indicator. Dissolve 140 g of iron (III) ammonium sulfate crystals in 400 mL of hot distilled water. Cool, filter, and add enough 6N nitric acid until the 500 mL mark.

Folin's mixture (for uric acid). To 650 mL of distilled water add 500 g of ammonium sulfate, 5 g of uranium (VI) acetate, and 6 g of glacial acetic acid and then dilute to 1.0 L with distilled water.

Formaldehyde — sulfuric acid (Marquis' reagent for alkaloids). Add 100 mL of concentrated sulfuric acid to 5 mL of 37% (mass/mass) aqueous formaldehyde solution.

Froehde's reagent. See Sulfomolybdic acid.

Fuchsin (reagent for linen). Dissolve 1 g of fuchsin (an aniline dye) in 100 mL of ethanol.

Fuchsin — sulfurous acid (Schiff's reagent for aldehydes). Dissolve 0.5 g of fuchsin and 9 g of sodium bisulfite in 500 mL of distilled water,and add 10 mL of concentrated HCl. Keep in well-stoppered bottles and protect from light. The fuchsin used to make the Schiff reagent should incorporate a high content of pararosanilin.

Gunzberg's reagent (detection of HCl in gastric juice). Dissolve 4 g of phloroglucinol (1,3,5-benzenetriol, $C_6H_3(OH)_3$) and 2 g of vanillin ($C_8H_8O_3$) in 100 mL of absolute ethanol.

Hager's reagent. See Picric acid.

Hanus solution (for iodine number). Dissolve 13.2 g of resublimed iodine in 1 L of glacial acetic acid, which should pass the dichromate test for reducible matter. Add sufficient bromine to double the halogen content, as determined by titration (approximately 3 mL). The iodine may be dissolved by heating, but the solution should be cold when the bromine is added.

Hopkins-Cole reagent (test for presence of the indole ring in the tryptophan moiety of a protein or peptide). Add 5 mL cold water to 10 g of magnesium in a conical flask and add 250 mL of a cold saturated aqueous oxalic acid solution with vigorous stirring. Filter, add 25 mL of glacial acetic acid and dilute with distilled water to the 1.0 L mark. Place 10 drops of the saturated protein or peptide solution in a test tube, add 10 drops of concentrated nitric acid and heat in a water bath for 3 minutes. Allow the mixture to cool and add 4 mL 6N NaOH. A positive test is confirmed by the formation of a yellow color.

Iodine, tincture of. To 50 mL of distilled water add 70 g of iodine and 50 g of potassium iodide. Dilute to 1 L with absolute ethanol.

Iodo-potassium iodide (Wagner's reagent for alkaloids). Dissolve 2 g of iodine and 6 g of KI in 100 mL of distilled water.

Jones reagent (test is positive for 1° and 2° alcohols, aldehydes, mercaptans, sulfides, and sulfoxides). To 1 mL of 5% (mass/mass) sodium (or potassium) dichromate add 10 drops of the unknown compound. The formation of a green color (sometimes appearing as brown due to the mixture of the green chromium (III) and the unreacted orange chromium (VI) colors) indicates a positive test.

Litmus (indicator). Extract litmus powder three times with boiling alcohol (ethanol or isopropanol), each treatment lasting for an hour. Discard the alcoholic extract. Treat the residue with an equal mass of cold distilled water and filter; then add five times its weight of boiling distilled water. Cool, filter, and combine all aqueous extracts.

Magnesia mixture (reagent for phosphates and arsenates). Dissolve 50 g of magnesium chloride ($MgCl_2{\cdot}6H_2O$) and 100 g of ammonium chloride (NH_4Cl) in 500 mL of distilled water, add a slight excess of NH_4OH, and allow to stand overnight. Filter, make the solution just acidic with HCl, and dilute with distilled water to 1 L. Let stand for several days and separate the precipitated solid by decanting. The decantate should be used just prior to testing; otherwise, if stored for any period of time it becomes turbid.

Magnesium uranyl acetate. Dissolve 100 g of uranyl (VI) acetate dehydrate ($UO_2(C_2H_3O_2)_2{\cdot}2H_2O$) in 60 mL of glacial acetic acid and dilute to 500 mL with distilled water. Dissolve 330 g magnesium acetate tetrahydrate ($Mg(C_2H_3O_2)_2{\cdot}4H_2O$) in 60 mL of glacial acetic acid and dilute to 200 mL. Heat both solutions to boiling until clear; pour the magnesium acetate solution into the uranyl acetate solution, cool and dilute with distilled water to 1 L. Let stand overnight and filter if necessary.

Marme's reagent. See Potassium-cadmium iodide.

Marquis' reagent. See Formaldehyde-sulfuric acid.

Mayer's reagent (white precipitate with most alkaloids in slightly acid solutions). Dissolve 1.358 g of mercury (II) chloride in 60 mL of distilled water and pour into a solution of 5 g of potassium iodide in 10 mL of distilled water. Add sufficient distilled water to a final volume of 100 mL.

Methyl orange indicator. Dissolve 1 g of methyl orange in 1 L of distilled water. Filter, if necessary.

Methyl orange, modified. Dissolve 2 g of methyl orange and 2.8 g of xylene cyanole FF ($C_{25}H_{27}N_2NaO_6S_2$, CAS No: 2650-17-1) in 1 L of 50% (vol/vol) ethanol.

Methyl red indicator. Dissolve 1 g of methyl red in 600 mL of ethanol and dilute with 400 mL of distilled water.

Methyl red, modified. Dissolve 0.50 g of methyl red and 1.25 g of xylene cyanole FF ($C_{25}H_{27}N_2NaO_6S_2$) in 1 L of 90% (vol/vol) ethanol. Alternatively dissolve 1.25 g of methyl red and 0.825 g of methylene blue in 1 L of 90% (vol/vol) ethanol.

Millon's reagent (for albumins and phenols). Dissolve 1 part of mercury in 1 part of cold fuming nitric acid. Dilute with twice the volume of distilled water and decant the clear solution after several hours. All appropriate precautions must be observed when handling mercury.

Molisch's reagent. See 1-Naphthol.

1-Naphthol (Molisch's reagent for wool). Dissolve 15 g of 1-naphthol in 100 mL of 95% (vol/vol) ethanol or chloroform.

Nessler's reagent (for ammonia). Dissolve 50 g of potassium iodide in the smallest possible quantity of cold distilled water (50 mL). Add a saturated solution of mercury (II) chloride (about 22 g in 350 mL of distilled water will be needed) until an excess is indicated by the formation of a precipitate. Then add 200 mL of 5 N NaOH and dilute to 1 L with distilled water. Allow to settle and draw off the clear liquid.

Nickel oxide, ammoniacal (reagent for silk). Dissolve 5 g of nickel (II) sulfate in 100 mL of distilled water and add 6N NaOH solution until nickel hydroxide is completely precipitated. Wash the precipitate well and dissolve in 25 mL of concentrated ammonium hydroxide and 25 mL of distilled water.

Nitron (detection of the nitrate radical). Dissolve 10 g of nitron (1,4-diphenyl-3-(phenylamino)-1,2,4-triazolium hydroxide, $C_{20}H_{18}N_4O$) in 5 mL of glacial acetic acid and 95 mL of distilled water. The solution may be filtered with slight suction through an Alundum thimble crucible and kept in a dark bottle.

1-Nitroso-2-naphthol. Make a saturated solution 1-nitroso-2-naphthol in 50% (vol/vol) aqueous acetic acid. The reagent should be used as soon as it is prepared.

Nylander's solution (carbohydrates). Dissolve 20 g of bismuth subnitrate ($Bi_5O(OH)_9(NO_3)_4$) and 40 g of Rochelle salt (potassium sodium tartrate, $KNaC_4H_4O_6{\cdot}4H_2O$) in 1 L of 8% (mass/mass) aqueous NaOH solution. Cool and filter.

Obermayer's reagent (for indoxyl in urine). Dissolve 4 g of iron (III) chloride in 1 L of aqueous (vol/vol) 40% HCl.

Oxine. Dissolve 14 g of 8-hydroxyquinoline (C_9H_7NO) in 30 mL of glacial acetic acid. Warm slightly, if necessary, to dissolve. Dilute to 1 L with distilled water.

Oxygen absorbent. Dissolve 300 g of ammonium chloride in 1 L distilled water and add 1 L of concentrated ammonium hydroxide solution. Shake the solution thoroughly. For use as an oxygen absorbent, the gas to be tested is passed through a bottle that is half-full of copper turnings filled nearly to the top with this ammonium chloride — ammonium hydroxide solution.

Pasteur's salt solution. To 1 L of distilled water add 2.5 g of potassium phosphate, 0.25 g of calcium phosphate, 0.25 g of magnesium sulfate, and 12.00 g of ammonium tartrate.

Pavy's solution (glucose reagent). To 120 mL of Fehling's solution, add 300 mL of 6N ammonium hydroxide (sp. gr. 0.88) and dilute to 1 L with distilled water.

Phenanthroline ferrous ion indicator. Dissolve 1.485 g of 1,10-phenanthroline monohydrate ($C_{12}H_8N_2{\cdot}H_2O$) in 100 mL of 0.025 M aqueous iron (II) sulfate solution.

Phenolphthalein. Dissolve 1 g of phenolphthalein in 50 mL of ethanol and add 50 mL of distilled water.

Phenolsulfonic acid (determination of nitrogen as nitrate). Dissolve 25 g of phenol in 150 mL of concentrated H_2SO_4, add 75 mL of fuming sulfuric acid (approximately 15% SO_3), stir well and heat for 2 hours at 100 °C.

Phloroglucinol solution (pentosans). Make a 3% phloroglucinol (1,3,5-benzenetriol) solution in alcohol. Store in a dark bottle.

Phosphomolybdic acid (Sonnenschein's reagent for alkaloids).

1. Prepare ammonium phosphomolybdate and after washing with distilled water, boil with nitric acid to expel all ammonia. Evaporate to dryness and dissolve in 2 M HNO_3.
2. Dissolve ammonium molybdate in HNO_3 and treat with phosphoric acid. Filter, wash the precipitate, and boil with aqua regia (a mixture of concentrated nitric acid and concentrated hydrochloric acid, 1:3 vol/vol) until the ammonium salt is decomposed. Evaporate to dryness. The residue is dissolved in 10% (vol/vol) HNO_3. The solution constitutes Sonnenschein's reagent.

Phosphoric acid — sulfuric acid mixture. Dilute 150 mL of concentrated sulfuric acid and 100 mL of concentrated phosphoric acid with distilled water to a final volume of 1 L.

Phosphotungstic acid (Scheibler's reagent for alkaloids).

1. Dissolve 20 g of sodium tungstate and 15 g of sodium phosphate in 100 mL of distilled water slightly acidified with dilute nitric acid.
2. The reagent is a 10% (mass/mass) solution of phosphotungstic acid in distilled water. The phosphotungstic acid is prepared by evaporating a mixture of 10 g of sodium tungstate dissolved in 5 g of phosphoric acid (sp. gr. 1.13) and enough boiling water to a complete solution. Crystals of phosphotungstic acid will separate.

Picric acid (Hager's reagent for alkaloids, wool, and silk). Dissolve 1 g of picric acid in 100 mL of distilled water.

Potassium antimonate (reagent for sodium). Boil 22 g of potassium antimonate with 1 L of distilled water until nearly all of the salt has dissolved, cool quickly, and add 35 mL of 10% (mass/mass) potassium hydroxide. Filter after standing overnight.

Potassium-cadmium iodide (Marme's reagent for alkaloids). Add 2 g of cadmium (II) chloride to a boiling solution of 4 g of potassium iodide in 12 mL of distilled water, and then mix with 12 mL of saturated aqueous potassium iodide solution.

Potassium hydroxide (for CO_2 absorption). Dissolve 360 g of potassium hydroxide in distilled water and dilute to 1 L.

Potassium iodide — mercuric iodide (Brucke's reagent for proteins). Dissolve 50 g of potassium iodide in 500 mL of distilled water and saturate with mercury (II) iodide (about 120 g). Dilute to 1 L with distilled water.

Potassium pyrogallate (for oxygen absorption). For mixtures of gases containing less than 28% (mass/mass) oxygen, add 100 mL of potassium hydroxide solution (50 g of KOH to 100 mL of distilled water) to 5 g of pyrogallol. For mixtures containing more than 28% (mass/mass) oxygen, the KOH solution should contain instead 120 g of KOH in 100 mL of distilled water.

Pyrogallol, alkaline. The reagent is a mixture of two solutions

1. Dissolve 75 g of pyrogallic acid in 75 mL of distilled water.
2. Dissolve 500 g of KOH in 250 mL distilled water. When cooled, adjust the solution with distilled water until the concentration is 50% (vol/vol).

For use, add 270 mL of solution (2) to 30 mL of solution (1).

Rosolic acid (indicator). Dissolve 1 g of rosolic acid ($C_{19}H_{14}O_3$) in 10 mL of ethanol and add 100 mL of distilled water.

Sakaguchi reagent (for the presence of arginine in proteins or peptides). To 10 drops of the saturated protein or peptide solution, add 5 drops of 6 N NaOH followed by 5 drops of 0.05% (mass/mass) ethanolic 1-naphthol solution and 10 drops of 0.5% aqueous sodium hypochlorite. A positive test is indicated by the formation of a red color that fades away upon standing.

Scheibler's reagent. See Phosphotungstic acid.

Schiff's reagent. See Fuchsin-sulfurous acid.

Schweitzer's reagent. See Cupric oxide, ammoniacal.

Soap solution (reagent for hardness in water). Dissolve 100 g of dry castile soap in 1 L of 80% (vol/vol) ethanol (4 parts ethanol to 1 part distilled water). Allow to stand for several days and dilute with 70% ethanol until 6.4 mL of this solution produces a permanent lather with 20 mL of standardized calcium solution. The latter solution is made by dissolving 0.2 g of calcium carbonate in a small amount of dilute HCl, evaporating to dryness and then dissolving the precipitate with distilled water to a final volume of 1 L.

Sodium bismuthate (for the oxidation of manganese). Heat 20 parts of sodium hydroxide nearly to redness in an iron or nickel crucible and add slowly 10 parts of basic bismuth (III) nitrate which has been previously dried. Add 2 parts of sodium peroxide and pour the brownish-yellow fused mass onto an iron plate to cool. When cooled, break up in a mortar, extract with distilled water, and collect on an asbestos filter.

Sodium hydroxide (for CO_2 absorption). Dissolve 330 g of sodium hydroxide in distilled water and dilute to 1 L.

Sodium nitroprusside (reagent for hydrogen sulfide and wool). Use a freshly prepared solution of 1 g of sodium nitroferricyanide in 10 mL of distilled water.

Sodium oxalate (primary standard). Dissolve 30 g of the commercial salt of sodium oxalate in 1 L of distilled water, make slightly alkaline with sodium hydroxide, and let stand until clear. Filter and evaporate the filtrate to 100 mL. Cool and filter. Pulverize the residue and wash it several times with small volumes of distilled water. The procedure is repeated until the mother liquor is sulfate-free and is neutral to phenolphthalein.

Sodium plumbite (reagent for wool). Dissolve 5 g of sodium hydroxide in 100 mL distilled water. Add 5 g of litharge (lead (II) oxide) and boil until dissolved.

Sodium polysulfide. Dissolve 480 g of sodium sulfide nonahydrate in 500 mL of distilled water, add 40 g of NaOH and 18 g of sulfur. Stir thoroughly and dilute to 1 L with distilled water.

Sonnenschein's reagent. See Phosphomolybdic acid.

Starch solution.

1. Make a paste with 2 g of soluble starch and 0.01 g of mercury (II) iodide with a small amount of distilled water. Add the mixture slowly to 1 L of boiling distilled water and boil further for a few minutes. Keep in a glass stoppered bottle. If other than soluble starch is used, the solution will not be clear on boiling; it should then be allowed to stand, and the clear liquid decanted.
2. A solution of starch that keeps stable indefinitely is made as follows: Mix 500 mL of aqueous saturated NaCl solution (filtered), 80 mL of glacial acetic acid, 20 mL of distilled water and 3 g of starch. Bring slowly to a boil and further heat for 2 minutes.
3. Make a paste with 1 g of soluble starch and 5 mg of mercury (II) iodide using as little cold distilled water as possible. Then pour about 200 mL of boiling distilled water on the paste and stir immediately. This will give a clear solution if the paste is prepared correctly and the water is actually boiling. Cool and add 4 g of potassium iodide. Starch solution decomposes on standing due to bacterial action, but this solution will be stable if stored under a layer of toluene.

Stoke's reagent. Dissolve 30 g of iron (II) sulfate and 20 g of tartaric acid in distilled water and dilute to 1 L. Just before using, add concentrated ammonium hydroxide until the precipitate that is initially formed is redissolved.

Sulfanilic acid (reagent for nitrites). Dissolve 0.5 g of sulfanilic acid in a mixture of 15 mL of glacial acetic acid and 135 mL of recently boiled distilled water.

Sulfomolybdic acid (Froehde's reagent for alkaloids and glucosides). Dissolve 10 g of molybdic acid or sodium molybdate in 100 mL of concentrated sulfuric acid.

Tannic acid (reagent for albumin, alkaloids, and gelatin). Dissolve 10 g of tannic acid in 10 mL of ethanol and dilute with distilled water to 100 mL.

Titration mixture (residual chlorine in water analysis). Prepare 1 L of dilute HCl (100 mL of HCl (sp. gr. 1.19) in sufficient distilled water to make 1 L). Dissolve 1 g of *o*-tolidine (3,3'-dimethylbenzidine, CAS No. 119-93-7) in 100 mL dilute hydrochloride, stir well and dilute to 1 L using dilute HCl solution.

Tollen's reagent (confirming the presence of aldehydes). 5 drops of 5% NaOH is added to 2 mL 10% aqueous silver nitrate in a test tube. The insoluble silver (I) oxide dissolved by the drop wise addition of 10% aqueous ammonia (ammonium hydroxide) yields a clear solution. Excess ammonium hydroxide should be avoided as it may give false-positive result. Approximately 10 drops of the aldehyde will yield a silver mirror coating on the test tube inner wall, especially if the mixture is warmed up to 50 °C.

Trinitrophenol solution. See Picric acid.

Turmeric tincture (reagent for borates). Digest ground turmeric root with several quantities of distilled water which are discarded. Dry the residue and digest it several days with six times its weight of ethanol. Filter.

Uffelmann's reagent (turns yellow in presence of lactic acid). To a 2% solution of pure phenol in distilled water, add an aqueous solution of iron (III) chloride until the phenol solution becomes violet in color.

Wagner's reagent. See Iodo-potassium iodide.

Wagner's solution (used in phosphate rock analysis to prevent precipitation of iron and aluminum). Dissolve 25 g of citric acid and 1 g of salicylic acid in distilled water and dilute to 1 L. Use 50 mL of the reagent.

Wij's iodine monochloride solution (for iodine number). Dissolve 13 g of resublimed iodine in 1 L of glacial acetic acid that will pass the dichromate test for reducible matter. Set aside 25 mL of this solution. Bubble into the remainder of the solution dry chlorine gas (dried and washed by passing through concentrated sulfuric acid) until the characteristic color of free iodine has been dissipated. Add the 25 mL of the iodine solution that was set aside, until all free chlorine has been dissipated. A slight excess of iodine does little or no harm, but an excess of chlorine must be avoided. Preserve in well stoppered, amber-colored bottles. Avoid the use of solutions that have been stored for more than 30 days.

Wij's special solution (for iodine number). To 200 mL of glacial acetic acid that will pass the dichromate test for reducible matter, add 12 g of dichloramine T (*N,N*-dichloro-4-methylbenzenesulfonamide, CAS No. 473-34-7), and 16.6 g of dry potassium crystals (in small quantities with continual shaking until all the potassium iodide has dissolved). Dilute to 1 L with the same quality of acetic acid used above and preserve in a dark-colored bottle.

Zimmermann-Reinhardt reagent (determination of iron). Dissolve 70 g of manganese (II) sulfate tetrahydrate in 500 mL of water, add slowly 125 mL of concentrated sulfuric acid and 125 mL of 85% phosphoric acid, and dilute to 1 L with water.

Zinc chloride solution, basic (reagent for silk). Dissolve 1000 g of zinc chloride in 850 mL of distilled water and add 40 g of zinc oxide. Heat until complete dissolving is complete.

Zinc uranyl acetate (reagent for sodium). Dissolve 10 g of $UO_2(C_2H_3O_2)_2 \cdot 2H_2O$ in 6 g of 30% acetic acid, heat, if necessary, and dilute to 50 mL. Dissolve 30 g of zinc acetate dehydrate in 3 g of 30% acetic acid and dilute to 50 mL. Mix the two solutions, add 50 mg of sodium chloride, allow to stand overnight and filter. Use distilled water in all dilutions.

ORGANIC ANALYTICAL REAGENTS FOR THE DETERMINATION OF INORGANIC IONS

Paris D. N. Svoronos and Thomas J. Bruno

In this table of organic reagents used for determination of inorganic elements and compounds we present the major tests, reagents, and some guidance as to the expected result or method of observation. The abbreviations used here are defined in the abbreviations table in this section. In addition, for brevity, when a determination calls for a spectrophotometric measurement at a particular wavelength, for example at 500 nm, we denote this as: "spec λ = 500 nm." No wavelength is specified if it is variable, for example, with pH; here we indicate: "spec determination." When a determination calls for spectrofluorimetric determination, the excitation and emission wavelengths are provided, thus: specf λ_{ex} = 303.5 nm, λ_{em} = 353 nm. Note that a common surfactant used in many of these tests is cetyltrimethylammonium bromide, abbreviated (CTAB). Information and data on this and other surfactants commonly used in chemical analysis are provided in Section 6. When relevant, we provide an approximate limit of detection (LOD), and when the uncertainty can vary, this is expressed as a relative standard deviation (RSD). Some of the procedures listed here require the use of hazardous chemicals (carcinogens such as benzene, strong acids such as HF). Appropriate precautions must be observed.

While a great deal of the information presented here is from the recent literature, the reader is referred to several excellent reviews and monographs for additional information (Refs. 1–10).

References

1. Marczenko, Z., *Separation and Spectrophotometric Determination of Elements*, Ellis Horwood, Chichester, 1986.
2. Sandell, E. B., and Onishi, H., *Photometric Determination of Traces of Metals. General Aspects, Part I,Fourth Edition*, John Wiley & Sons, New York, 1986.
3. Onishi, H., *Photometric Determination of Traces of Metals. Part IIa: Individual Metals, Aluminium to Lithium,Fourth Edition*, John Wiley & Sons, New York, 1986.
4. Onishi, H., *Photometric Determination of Traces of Metals. Part IIb: Individual Metals, Magnesium to Zinc, Fourth Edition*, John Wiley & Sons, New York, 1986.
5. Townshend, A., Burns, D. T., Guilbault, G. G., Lobinski, R., Marczenko, Z., Newman, E., and Onishi, H., *Dictionary of Analytical Reagents*, Chapman & Hall, London, 1993.
6. West, T. S., and Nürnberg, H. W., *The Determination of Trace Metals in Natural Waters*, Blackwell, Oxford, 1988.
7. Savvin, S. B., Shtykov, S. N., and Mikhailova, A. V., *Russ. Chem. Rev.* 75, 341, 2006. <https://doi.org/10.1070/RC2006v075n04ABEH001189>
8. Ueno, K., Imamura, T., and Cheng, K. L., *Handbook of Organic Analytical Reagents*, CRC Press, Boca Raton, FL, 1992.
9. American Chemical Society, *Reagent Chemicals: Specifications and Procedures: American Chemical Society Specifications, Official from January 1, 2006*, American Chemical Society, Washington, DC, 2006.
10. Crompton, T. R., *Determination of Anions: A Guide for the Analytical Chemist*, Springer-Verlag, Berlin, Heidelberg, 1996.

TABLE 1. Organic Analytical Reagents for the Determination of Inorganic Cations

Reagents	Results
Aluminum	
Alizarin red S	Red color develops; spectrophotometric determination preferred
Aluminon	Lake pigment stabilized with CTAB
Chrome azurol S	Spec λ; = 500 nm, stabilized with CTAB
Chromazol KS	Spec λ = 625 nm, stabilized with cetylpyridinium bromide
Eriochrome cyanine R (also known as mordant blue 3) + CTAB	Red dye lake (pH = 6) stabilized with CTAB
Eriochrome cyanine R (also known as mordant blue 3) + *N,N*-dodecyltrimethylammonium bromide (DTAB)	Spec determination by use of cationic surfactants
8-Hydroxyquinoline	Produces tris(8-hydroxyquinolinato) aluminum (Alq3), found in organic light-emitting diodes (OLEDs)
Bromopyrogallol red + CTAB	Spec λ = 627 nm
Bromopyrogallol red + nonylphenol tetradecaethylene glycol ether	Spec λ = 612 nm
Pyrocatechol violet	Spec λ = 578 nm after separation of aluminum from the matrix materials by chloroform extraction of its acetylacetone complex (pH = 6.5), from an ammonium acetate-hydrogen peroxide medium
2,2′,3,4-Tetrahydroxy-3′,5′-disulfoazobenzene	Spec determination of the binary system (pH = 5)
Antimony	
Brilliant green	Isolated as the hexachloroantimonate (V) salt extracted by either toluene or benzene
Bromopyrogallol red	Used for the determination of antimony (III) with EDTA, cyanide, or fluoride ions as masking agents
Catechol violet	Ternary complex stabilized with CTAB
Malachite green (basic green 4)	Isolated as the hexachloantimonate (V) complex after benzene extraction
Phenyl fluorene	Sensitive color reaction with antimony (III) by use of cationic surfactants
Potassium iodide	Antimony (III)-iodide complex formation in the presence of ascorbic acid

Analytical

Reagents	Results
Pyrocatechol violet + tridodecylammonium bromide	Spec λ = 530 nm of the ternary complex with antimony (III)
Rhodamine B	Ion pair or ion association extraction by use of toluene or benzene as solvents
Silver diethyldithiocarbamate	Spec λ = 504 nm to avoid arsenic interference
Thiourea	Determination by hydride generation ICP-AES after reduction of antimony (V) to antimony (III) by thiourea
Arsenic	
Michler's ketone	Spec λ = 640 nm trophotometric determination of trace arsenic (V) in water
Silver diethyldithiocarbamate	Spec λ = 600 nm to avoid antimony interference
Thiourea	Determination by hydride generation ICP-AES after reduction of arsenic (V) to arsenic (III) by thiourea
Barium	
Dimethylsulfonazo-III (DMSA-III)	Spec λ = 662 nm of the chelate complex
Beryllium	
Aluminon	Lake pigment derivative
Ammonium bifluoride	Derivative detected by fluorescence
Beryllon II	A resin-phase spectrophotometric method that detects a change in absorbance of the resin phase immobilized with beryllon II
Beryllon III	Determination achieved via third-derivative spectrophotometry and decolorization of excess reagent
Chrome azurol S	Spec determination with chrome azurol S in the presence of EDTA or CTAB
Eriochrome cyanine R + (CTAB)	Spec λ = 590 nm of ternary complex
Sulfon black F	Derivative with a long color-development time
Bismuth	
Amberlite XAD-7	Determination system implemented with (HG-ICP-AES) associated with flow injection (FI)
Sodium azide	Azidodimethylbismuthine precipitate formed by the reaction of the corresponding bismuthine with sodium azide
Diethyldithiocarbamate	Heterometric micro-determination of lead with sodium diethyldithiocarbamate by use of a mixture of EDTA, cyanide, and ammonium hydroxide (λ = 400 nm)
Dithizone	Orange-red derivative that is extracted in carbon tetrachloride
Pyrocatechol violet	Spec determination of bismuth (III) with pyrocatechol violet in the presence of septonex CTAB
Pyrocatechol violet + tridodecylammonium bromide	Spec λ =564 of ternary complex
Quinolin-8-ol	Determination system implemented with HG-ICP-AES associated with flow injection (FI)
Thiourea	Determination by HG-ICP-AES
Xylenol orange	Derivative used for sol-gel thin films that serve as bismuth (III) sensors
Boron	
Azomethine H	Spec λ = 415 nm
Carminic acid	Spec λ = 615 nm
Curcumin	Spec λ = 555 nm determination of rosocyanine and rubrocurcumin formed by the reaction between borates and curcumin
Methylene blue	Spec determination of complex formed between fluoroborate ions and methylene blue after treatment with hydrofluoric and sulfuric acids and extraction with ethylene chloride
Cadmium	
2-(5-Bromo-2-pyridylazo)-5-diethylaminophenol (PAR)	Spec determination in the presence of cationic surfactant cetylpyridinium chloride
Cadion	Determination by β-correction spectrophotometry with cadion and surfactant Triton-X
Dithizone	Determination of dithizonate derivative by the extraction – spectrophotometric method
1-(2-Pyridylazo)-2-naphthol (PAN)	Two-dimensional absorption spec determination of complex in aqueous micellar solutions
4–(2-Pyridylazo)resorcinol	Preconcentration by cloud point extraction of the complex; determination by ICP optic emission spectrometry. Simultaneous spectrophotometric determination of cadmium and mercury
Calcium	
Alizarin S	Red dye used in staining bones for calcium determination
Chlorophosphonazo III	Spec λ = 667.5 nm (pH = 2.2)
Eriochrome black T + EDTA	Complexometric titration where the initially formed red calcium-eriochrome black T color is replaced with the blue calcium-EDTA color at the end point
Glyoxal-bis(2-hydroxyanil)	Spectrophotometric titration of the complex without preliminary extraction
Murexide	Spec λ = 506 nm (pH = 11.3)
Phthalein purple	High-performance chelation ion chromatography involving dye-coated resins
Cerium	
Butaperazine dimaleate propericiazine	Spec determination of the colored complex in a phosphoric acid medium
Persulfate oxidation to cerium (IV)	UV spec λ = 320 nm
Propericiazine	Spec determination of the colored complex in a phosphoric acid medium

Reagents	Results
Propionyl promazine phosphate (PPP)	Spec λ = 513 nm of the red-colored radical cation formed upon the reaction of PPP with cerium (IV) in a phosphoric acid medium
N-Benzoyl-*N*-phenylhydroxylamine	Spec titration of the cerium (IV) complex (pH = 8-10)
Sodium triphosphate	Specf λ_{ex} = 303.5 nm, λ_{em} = 353 nm of the cerium (III) complex
8-Hydroxyquinoline	Spec determination of the metal-ligand complex
Chromium	
Alizarin S	Lake pigment complex formation
1,5-Diphenylcarbazide	Spec determination during sonication in carbonated aqueous solutions saturated with CCl_4 that produces chlorine radicals
3-(2-Pyridyl)-5,6 bis(5-(2 furyl disulfonic acid))-1,2,4-triazine disodium salt (ferene-TM)	Indirect spec λ = 593 nm in aqueous samples with a chromogen ferene-TM
4-(2-Pyridylazo)resorcinol (PAR)	Spec determination of the ternary chromium-peroxo-PAR ternary complex
4-(2-Pyridylazo)resorcinol (PAR) + hydrogen peroxide	Spec determination of the ternary chromo-peroxo-PAR mixture after ethyl acetate extraction in 0.1 M sulfuric acid
4-(2-Pyridylazo)resorcinol (PAR) + xylometazolonium (XMH) chloride	Spec determination of the orange-red anionic complex formed in a heated acetate buffer medium (pH = 4.0-5.5) and extracted with the xylometazolonium (XMH) chloride
Sulfanilic acid	Spec λ = 360 nm on the catalytic effect of chromium (VI) in the oxidation of sulfanilic acid by hydrogen peroxide with *p*-aminobenzoic acid as an activator
Cobalt	
8-Hydroxyquinoline	Spec determination of the metal-ligand complex
p-Nitroso-*N,N*-dimethylaniline	Spec determination of the binary complex
Nitroso-R salt	AAS by use of a continuous on-line precipitation-dissolution procedure
1-Nitroso-2-naphthol	Spec determination by use of non-ionic surfactant Triton X-100
1-Nitroso-2-naphthol	Spec Tween 80 micellar determination
1-Nitroso-2-naphthol	AAS using a continuous on-line precipitation-dissolution procedure based on 1-nitroso-2-naphthol
2-Nitroso-1-naphthol	Spec λ = 530 nm after isoamyl acetate extraction
1-(2-Pyridylazo)-2-naphthol + surfactants	Spec λ = 620 nm of the cobalt complex in the presence of surfactants (Triton X-100 combined with sodium dodecylbenzene sulfonate [DBS]) and trace of ammonium persulfate (pH = 5.0)
2,4-Dinitroresorcinol (DNR)	Spec λ = 397 nm determination
4-(2-Pyridylazo)resorcinol (PAR)	Spec determination of complex at both pH = 7.2-7.9 and in 1 M H_2SO_4
4-(2-Pyridylazo)resorcinol (PAR) + triethanolamine	Ion-pair reversed-phase high-performance liquid chromatography of the complex
Copper	
Bathocuproine disulfonic acid	Spec λ = 470 nm - 550 nm of the bathocuproine-disulfonic acid complex after extraction with chloroform and methanol by use of partial least squares regression (PLS_2)
Dithizone	Spec determination of the dithizone complex at pH = 2.3
Neocuproine	Spec determination of the deep orange-red $Cu(neocuproine)_2^+$ complex color
Cuprizone	Spec λ = 595 nm determination of the highly chromogenic copper (III) cuprizone complex
p-Nitroso-*N,N*-dimethylaniline	Spec determination of the binary complex
1-Nitroso-2-naphthol	Spec determination using non-ionic surfactant Triton X-100
4-(2-Pyridylazo)resorcinol	Preconcentration of copper by cloud-point extraction of the complex and determination by ICP optic emission spectrometry
Europium	
1-Nitroso-2-naphthol	Spec Tween 80 micellar determination
ChromAsurol S	Spec λ = 510 nm and 655 nm determination of the binary complex
4-(2-Pyridylazo)resorcinol (PAR) + tetradecyldimethylbenzylammonium chloride (TDBA)	Spec λ = 510 nm of the ion-associate complex extracted with chloroform at pH = 9.7
Gallium	
Chrome azurol S + CTAB	Spec λ = 640 nm of the ternary complex after *n*-butyl acetate extraction from hydrobromic acid
Hematoxylin or its oxidized form + CTAB	Spec determination of the ternary complex of indium with hematoxylin or its oxidized form in the presence of cationic, anionic, and non-ionic surfactants such as CTAB
Pyrocatechol violet + diphenylguanidine	Spec λ = 345 nm determination of the ternary complex
8-hydroxyquinoline	Spec determination of the gallium complex
1-(2-Pyridylazo)-2-naphthol	Spectrofluorimetric determination of Ga(III) with 1-(2-pyridylazo)-2-naphthol in sodium dodecyl sulfate micellar medium
4-(2-Pyridylazo)resorcinol (PAR)	Extraction and spec λ = 510 nm of gallium with 4-(2-pyridylazo)resorcinol
Pyrocatechol violet + tridodecylammonium bromide	Spec λ = 595 nm determination of the ternary complex
Rhodamine B	Comparison of the determination of gallium by a rhodamine B spectrophotometric method and by an AA method based on preliminary solvent extraction
Xylenol orange + 8-hydroxyquinoline	Spectrophotometric determination of the ternary complex

Reagents	Results
Germanium	
Brilliant Green + Molybdate	Spec λ = 430 nm of the yellow germanomolybdic acid
Phenylfluorone	Spec λ = 597 nm determination of the complex previously extracted with carbon tetrachloride at pH = 3.1
Gold	
5-(4-Diethylaminobenzylidene) rhodanine	Immobilized 5-(4-dimethylamino-benzylidene) rhodanine serves as a stable solid sorbent for trace amounts of gold (III) ions (pH = 2-4)
Di(methylheptyl)methyl phosphonate (DMHMP)	Trace gold determination by on-line preconcentration with flow injection AAS using di(methylheptyl)methyl phosphonate (DMHMP) as the immobilized phase loaded onto a macroporous resin
2-Mercaptobenzothiazole	Separation and determination of gold complex matrices employing substoichiometric thermal neutron activation analysis
Molybdate + nile blue (NB)	A spectrophotometric method based on the reaction of gold(III) with molybdate and nile blue (NB) to form an ion-association complex in the presence of poly(vinyl alcohol)
Rhodamine B	Aqueous spec determination of gold with rhodamine B and surfactant
Hafnium	
Arsenazo III	Spec determination of the arsenazo III complex in 10 M HCl or H_2SO_4
Indium	
Alizarin red S	Spec λ = 415 nm determination after extraction in the presence of 1,3-diphenylguanidine
Bromopyrogallol red	Spec determination of indium after an ether extraction from hydrobromic acid and benzyl alcohol extraction of its complex with bromopyrogallol red (pH 9.0)
5-Bromine-salicylaldehyde salicyloylhydrazone (5-Br-SASH)	Spec λ_{ex} = 395 nm, λ_{em} = 461 nm of the indium: 5-Br-SASH chelate in a water–ethanol (63%) medium (pH = 4.6)
Chrome azurol S	Spec determination of the binary complex
Chrome azurol S + benzyldodecyldimethylammonium bromide (BDDMAB)	Spec determination of the mixed complex with chrome azurol S and BDDMAB
Chrome azurol S + CTAB	Spec λ = 630 nm of the ternary complex after *n*-butyl acetate extraction from hydrobromic acid
Chrome azurol S + cationic surfactants	Spec λ = 630 nm determination by use of chrome azurol S and surfactants such as CTA, CP, or zephiramine
4,4'-Dihydroxybenzophenone thiosemicarbazone dithizone	Spec λ = 415 nm determination of the complex after heating at 45 °C
Dithizone	Spec determination of the binary complex
Erichrome cyanine R and CTAB	Spec λ = 585 nm determination, interference from Be (II), Ga (III), Al (III), Fe (III), and V (IV)
Hematoxylin or its oxidized form + CTAB	Spec determination of the ternary complex of indium with hematoxylin or its oxidized form in the presence of cationic, anionic and non-ionic surfactants such as CTAB
2-Hydroxybenzaldehyde isonicotinoylhydrazone	Spec λ = 380 nm determination after extraction into 1-pentanol
8-Hydroxyquinoline	Spec determination of the binary complex
Methylthymol blue + zepheramine	Spec determination of the ternary complex
2-Oxoguanidine benzoic acid	Spec λ = 545 nm determination
1-(2-Pyridylazo)-2-naphthol (PAN)	Spec determination of the chelate complex after chloroform extraction (pH = 6)
4-(2-Pyridylazo)resorcinol	Spec λ = 520 nm after indium extraction from the aqueous phase (pH 5.0-5.5) into chloroform with *N-p*-chlorophenyl-2-furohydroxamic acid and formation of the 4-(2-pyridylazo)resorcinol red chelate; color develops after 10 minutes
Pyrocatechol violet	Spec determination of the binary complex
Pyrocatechol violet + tridodecylammonium bromide	Spec λ = 600 nm determination of the ternary complex
Quinalizarin (1,2,5,8-tetrahydroxyanthraquinone)	Spec λ = 565 nm of the binary 3:1 quinalizarin:In(III) colored complex in dimethylformamide-water solution
4,5,6,7-Tetrachlorogallein and cetyl pyridimium chloride	Spec λ = 620 nm determination, color reached after 30 minutes
3,5,7,4'-Tetrahydroxyflavone	Spec λ = 430 nm determination, interference by Al (III), Fe (III), Mo (VI), W (V), Sn (IV), Zr (IV), Ti (IV), and V (V)
2-(2-Thiazolylazo)-*p*-cresol	Spec λ = 580 nm determination of the complex, interference by Cu (II), Ni (II), Fe (II), Co (II), and Ti (IV)
1-(2-Thiazolylazo)-2-naphthol	Spec λ = 575 nm determination of the complex
2,2′,3,4-Tetrahydroxy-3′,5′-disulfoazobenzene	Spec determination of the binary system (pH = 5)
Xylenol orange	Spec λ = 560 nm determination of the ternary complex
Iridium	
1,5-Diphenylcarbazide	Spec determination of the complex (pH = 5.0)
1-(2-Pyridylazo)-2-naphthol	Spec λ = 550 nm of the red complex (pH = 5.1) after chloroform extraction
Iron	
Bathophenanthroline	Determination of iron(II) in the presence of thousand-to-one ratio of iron(III) using bathophenanthroline

Reagents	Results
Bathophenantroline-disulfonic acid	Spec λ = 470 nm - 550 nm of the bathophenantroline-disulfonic acid complex after extraction with chloroform and methanol
2,2′-Bipyridine	Spec determination of the iron (II)-2,2'-bipyridine dark red complex
FerroZine	Spec λ = 562 nm of iron (II)-ferrozine complex after all iron(III) has been reduced by ascorbic acid
Hematoxylin + (CTAB)	Spec of the ternary complex. Addition of CTAB shifts λ_{max} from 630 nm to 640 nm
1-Nitroso-2-naphthol	Spec determination by use of non-ionic surfactant Triton X-100
1-Nitroso-2-naphthol	Spec determination Tween 80 micellar determination
1,10-Phenanthroline (o-Phen)	Spec λ = 508 nm of the $Fe(o\text{-}Phen)_3^{+2}$ complex
1,10-Phenanthroline + bromothymol blue	Spec determination of the $Fe(o\text{-}Phen)_3^{+2}$ complex in the presence of bromothymol blue
Phenylfluorone	Spec λ = 530 nm of the binary complex (pH = 9.0)
Phenylfluorone + Triton X	Spec λ = 555 nm of the binary complex sensitized with Triton X-100 (pH = 9.0)
Xylenol	Spec λ = 560 nm of the binary complex
Lanthanum	
Ammonium purpurate	Spec determination of lanthanum with ammonium purpurate as a chromogenic reagent
Arsenazo III	Spec λ = 652 nm determination of the lanthanum complexation with reagents of the arsenazo III group on the solid phase of fibrous ion exchangers
Eriochrome cyanine R (ECR)	Spec λ = 540 nm of the binary complex
N-Phenylbenzohydroxamic acid + xylenol orange	Solvent extraction followed by the spec λ = 600 nm of the ternary complex (pH = 8.8-9.5)
Lead	
Dithizone	Orange-red derivative that is extracted in carbon tetrachloride
Sodium diethyldithiocarbamate	Electrothermal AAS determination of the diethyldithiocarbamate derivative extracted in carbon tetrachloride
Sodium diethyldithiocarbamate	Heterometric micro-determination of lead diethyldithiocarbamate derivative
4-(2-Pyridylazo)resorcinol	Spec λ = 520 nm of 4-(2-pyridylazo)-resorcinol:lead (1:1) complex in an ammonia-ammonium chloride medium at pH = 10 after extracting the lead in isobutyl methyl ketone
Lithium	
1-(*o*-Arsenophenylazo)-2-naphthol-3, 6-disulfonate (Thoron)	Spec λ = 480 nm of Thoron-lithium complex in an alkaline acetone medium is measured against the reagent as reference
Magnesium	
Chlorophosphonazo III	Spec λ = 669 nm (pH = 7.0)
Eriochrome black T + EDTA	Complexometric titration where the initially formed red magnesium-eriochrome black T color is replaced with the blue magnesium-EDTA color at the end point
8-Hydroxyquinoline	Volumetric, titrimetric, and colorimetric determination
8-Hydroxyquinoline + butylamine	Spec λ = 380 nm of the complex after chloroform extraction
Titan yellow	Spec determination of magnesium by titan yellow in biological fluids
Xylidyl blue	Spec determination of magnesium in biological fluids
Manganese	
Formaldoxime	Spec determination of the formaldoxime/ammonia complex (pH = 8.8-8.9) that is stable for 30 minutes
2-Hydroxy-4-methoxy acetophenone oxine	Spec λ = 410 nm determination of binary complex
1-(2-Pyridylazo)-2-naphthol (PAN)	Preconcentration determination by use of PAN anchored SiO_2 nanoparticles
Mercury	
Dithizone	Spec λ = 500 nm of dithizone complex after chloroform extraction at pH = 0.3
Michler's thioketone	Spec λ = 560 nm of the binary complex in acetate buffer
4-(2-Pyridylazo)resorcinol	Simultaneous spec determination of cadmium and mercury with 4-(2-pyridylazo) resorcinol
Rhodamine 6G	Photoelectrochemical determination of mercury (II) in aqueous solutions using a rhodamine 6G derivative (RS) and polyaniline (PANI) coated optical probe in a photoelectrochemical cell
Xylenol orange + amine buffer	Spec λ = 590 nm of Hg(II)/xylenol orange complex displays a sharp hyperchromic effect in the presence of amine buffers (pH= 7.5)
Xylenol orange + citric acid-phosphate buffer	Spec λ = 580 nm of Hg(II)/xylenol orange complex displays a sharp hypochromic effect upon substituting amine buffers with a citric acid-phosphate (pH = 7.5)
Molybdenum	
Bromopyrogallol red + cetylpyridium chloride	Sequential injection analysis to the determination of CPC-based on the sensitized molybdenum-bromopyrogallol red reaction
Phenylfluorone	Spec λ = 560 nm of molybdenum with phenylfluorone (pH of 1.5-3)
8-Hydroxyquinoline-5-sulfonic acid and phenylfluorone	Diffuse reflection spectrometry with 8-hydroxyquinoline-5-sulfonic acid and phenylfluorone after sorbing on a disk of an anion exchange fibrous material

Reagents	Results
Pyrocatechol violet	Preconcentration (using basic anion exchanger AV-17-10P) and determination via diffuse reflection spectroscopy. The colored surface compound to be determined involves Mo(VI) sorption on the resin and subsequent treatment of the concentrate obtained with pyrocatechol violet
Neodymium	
Semi-xylenol orange + cetylpyridinium chloride	Fourth-order derivative spectrophotometric determination of the ternary complex
Neptunium	
Arsenazo III	Spec λ after separation by use of thenoyltrifluoroacetone extraction method and determination in 5M HNO_3
Nickel	
2-(5-Bromo-2-pyridylazo)-5-diethylaminophenol	Spec λ = 520 nm and 560 nm of the red-violet complex in water-ethanol (pH = 5.5)
Dimethylglyoxime + ammonia	Spec λ = 543 nm of the complex
Dimethylglyoxime, voltammetry	Nickel voltammetric determination at a chemically modified electrode based on dimethylglyoxime-containing carbon paste
EDTA	Spec λ = 380 nm of the complex
2,2′-Furildioxime	Spec λ = 438 nm after separation by adsorption of its α-furildioxime complex on naphthalene
Hematoxylin	Spec λ = 595 nm of the binary system (pH = 7.8-8.3)
Hematoxylin + (CTAB)	Spec λ = 608 nm of the ternary system (pH = 7.4-8.1)
1-Nitroso-2-naphthol	Spec Tween 80 micellar determination
p-Nitroso-*N*,*N*-dimethylaniline	Spec determination of the binary complex
2-(2-Pyridylazo)-2-naphthol	Derivative spectrophotometry λ= 569 nm determination of nickel complex in Tween 80 micellar solutions
4-(2-Pyridylazo)resorcinol	Preconcentration of nickel by cloud point extraction of the complex and determination by ICP optic emission spectrometry
1-(2-Thiazolylazo)-2-naphthol (TAN)	Flow-injection solid-phase spectrophotometry by use of TAN immobilized on C_{18}-bonded silica (λ = 595 nm)
Xylenol orange	Spec determination by mean centering of ratio kinetic profile (pH = 5.3)
Niobium	
N-Benzoyl-*N*-phenylhydroxylamine	Separation and determination of the complex from a tartrate solution at pH > 2
Bromopyrogallol red	Spec λ = 610 nm of complex extracted into isopentyl acetate containing di-*n*-octylmethylamine
O-Hydroxyhydroquinonephthalein (Qnph) + hexadecyltrimethylammonium chloride (HTAC)	Spec λ = 520 nm of the complex in strong acidic media
4-(2-Pyridylazo)resorcinol + citrate	Determination of the ternary complex by ion-interaction reversed-phase HPLC on a C_{18} column with a 5 mM citrate buffer (pH = 6.5, λ = 540 nm)
Pyrocatechol violet	Extraction and spec determination of niobium complex in the presence of pyridine and trichloroacetic acid
Sulfochlorophenol S	Spec λ = 650 nm determination of the complex after extraction into amyl alcohol
Xylenol orange	Spec λ = 530 nm of chelate (pH = 5.0)
Osmium	
1,5-Diphenylcarbazide	Spec λ = 560 nm of the bluish-violet complex after extraction with isobutyl methyl ketone
Thiourea	Spec λ = 540 nm of the red-rose complex in acid medium
Palladium	
Ammonia + iodide	Thermogravimetric determination of palladium as $Pd(NH_3)_2I_2$
2,2- *bis*-[3-(2-Thiazolylazo)-4-4-hydroxyphenyl-propane], (TAPHP)	Application of TAPHP immobilized on silica beads to determine the palladium concentration in a trans-luminance configuration (pH = 2) by use of flow-through spectrophotometric sensing phase
2-(5-Bromo-2-pyridylazo)-5-diethylaminophenol	Spec determination of the complex in a sulfuric acid medium in the presence of ethanol
Dithizone	Volumetric determination of palladium with dithizone in acid medium after extraction of its dimethylglyoxime complex with chloroform; PAS determination of the dithizone extraction solution into a thermally thin solid film
Dithiazone + iodide	Spec determination of palladium with dithizone, by use of an iodide medium in the presence of sulfite at pH = 3-5, to separate platinum
Dithizone + stannous chloride	Extractive separation and spec determination of palladium in the presence of stannous chloride to separate platinum
Isonitrosobenzoylacetone	Radiochemical separation and determination of palladium in complex matrices employing substoichiometric thermal NAA
2-Nitroso-1-naphthol	Spec λ = 370 nm of the violet complex
4-(2-Pyridylazo)resorcinol	Spec determination of complex at both pH = 7.2-7.9 and in 1 M H_2SO_4
4-(2-Pyridylazo)resorcinol + diphenylguanidine	Spec determination after extraction of the red Pd(II) chelate with 4-(2-pyridylazo) resorcinol in the presence of *N*,*N*′-diphenylguanidine into *n*-butanol
Thioglycolic acid	Spec λ = 384 nm of the complex

Reagents	Results
Platinum	
N-Phenylbenzimidoylthiourea (PBITU)	Flow injection analysis spec λ = 345 nm of the Pd:PBITU complex in (0.2 M to 2.0 M) HCl in 10% (vol/vol) ethanol solution
Dithizone + iodide	Spec determination of platinum with dithizone, by use of an iodide medium in the presence of sulfite at pH = 3-5, to separate palladium
Dithizone + stannous chloride	Extractive separation and spectrophotometric determination of platinum in the presence of stannous chloride to separate palladium
2-Mercaptobenzothiazole	Radiochemical separation and determination of platinum complex matrices employing substoichiometric thermal NAA
Protactinium	
Arsenazo III	Spec λ = 680 nm of the complex after extraction with 7 N H_2SO_4 and isoamyl alcohol
EDTA + tannic acid	Gravimetric analysis of the tannic acid precipitate (pH = 5.0)
Rhenium	
2,2′-Furildioxime	Spec λ = 532 nm of the complex
N,N-Diethyl-*N*′-benzoylthiourea	Spec λ = 383 nm of the green complex in hydrochloric acid medium in the presence of tin(II) chloride
Tin(II)	Spectrophotometric titration for the determination of rhenium by use of tin(II) as the titrant
Rhodium	
1,5-Diphenylcarbazide	Spec determination of the complex after isobutyl alcohol extraction (pH = 5.0)
o-Methylphenyl thiourea	Spec λ = 320 nm of the binary complex
p-Nitrosodimethylaniline	Spec λ = 510 nm of the cherry-red binary complex (pH = 4.4)
1-(2-Pyridylazo)-2-naphthol	Spec λ = 598 nm of the green complex (pH = 5.1) after chloroform extraction
Ruthenium	
4-Benzylideneamino-3-mercapto-6-methyl-1,2,4-triazine (4H)-5-one	Spec λ = 620 nm
1,10-Phenanthroline	Chemiluminescence determination of chlorpheniramine by use of tris(1,10-phenanthroline)-ruthenium(II) peroxydisulfate system and sequential injection analysis; spec determination of the complex after reducing Ru(IV) to Ru(II)
Thiourea	Spec λ = 640 nm of ruthenium complex in carbon supported Pt-Ru-Ge catalyst in 5 M HCl
1,4-Diphenylthiosemicarbazide	Spec determination of the bright red complex
Scandium	
Alizarin red S	Cathodic adsorptive stripping of the scandium-alizarin red S complex onto a carbon paste electrode
Ammonium purpurate	Spec determination of scandium with ammonium purpurate as a chromogenic reagent
Chrome azurol S	Spec determination of the binary complex
Xylenol orange	Spec λ = 553 nm of the binary mixture
Selenium	
3,3′-Diaminobenzidine	Spec λ = 350 nm of the yellow complex (pH = 3.0)
2,3-Diaminonaphthaline	Fluorometric determination of the binary complex in water
Thiourea	Determination of selenium by hydride generation ICP-AES
Silver	
Dithizone	Graphite-furnace AAS determination of silver on suspended dithizone particles from acidic sample solutions with sonication to facilitate the separation
Eosin + 1,10-phenanthroline	Spec λ = 540 nm to 555 nm of the 1,10-phenanthroline (PHEN) and eosin (2,4,5,7-tetrabromofluorescein) association complexes
Strontium	
Phthalein purple	High-performance chelation ion chromatography involving dye-coated resins
Tantalum	
N-Benzoyl-*N*-phenylhydroxylamine	Separation and determination of the complex from a tartrate solution at pH < 1.5
Crystal violet (CV) + *N,N*′-diphenylbenzamidine (DPBA)	Spec λ = 600 nm of Ta(V)-F-CV+ cation complex with a benzene solution of DPBA from sulfuric acid solution
O-Hydroxyhydroquinonephthalein (Qnph) + hexadecyltrimethylammonium chloride (HTAC)	Spec λ = 510 nm of the complex in strong acidic media
Malachite green	Spec λ = 623 nm of the complex in HF after extraction with benzene or toluene
Methyl violet	Spec determination of the complex
4-(2-Pyridylazo)resorcinol + citrate	Determination of the ternary complex by ion-interaction reversed-phase HPLC on a C18 column in 5 mM citrate buffer (pH = 6.5, λ = 540 nm)
Phenylfluorone	Spectrophotometric determination from HF-HCl solution with methyl isobutyl ketone
Victoria blue	Spectrophotometric determination of the complex after benzene extraction
Tellurium	
Ammonium pyrrolidinedithiocarbamate	Differential determination of tellurium(IV) and tellurium(VI) by AAS with a carbon-tube atomizer

Reagents	Results
Bismuthiol II	Spec λ = 330 mm of the yellow complex in acidic medium (pH = 3.5) after chloroform extraction
Dithizone	Differential determination of tellurium(IV) and tellurium(VI) by AAS with a carbon-tube atomizer
Sodium diethyldithiocarbamate	Differential determination of tellurium(IV) and tellurium(VI) by AAS with a carbon-tube atomizer
Thiourea	Determination of selenium by hydride generation ICP-AES
Thallium	
Brilliant green	Spec determination of the binary complex after toluene extraction
Dithizone	Spec λ = 505 mm of the complex after chloroform extraction from a citrate-sulfite-cyanide medium at pH = 10.6
8-Hydroxyquinoline	Spectrophotometric determination of the metal-ligand complex
Rhodamine B	Fluorimetric determination of thallium (in silicate rocks) with rhodamine B after separation by adsorption on a crown ether polymer
Rhodamine B hydroxide	Spec λ = 565 nm via oxidation of rhodamine B hydrazide by thallium(I) in acidic medium to give a pinkish-violet radical cation
Thorium	
Arsenazo III	Spec determination of the arsenazo III complex in 10 M HCl or H_2SO_4
Eriochrome cyanine R (ECR)	Spec λ = 540 nm of the binary complex
Thoron	Spectrophotometric determination of binary complex in tartaric acid
Xylenol orange	Spec λ = 570 nm of the binary complex (pH = 4.0)
Xylenol orange + CTAB bromide	Spec λ = 600 nm of the complex sensitized by CTMAB (pH = 2.5)
Tin	
Catechol violet + CTAB bromide	Spectrophotometric determination of the green complex
Pyrocatechol violet (and + CTAB bromide)	Spec λ = 660 nm of the complex (pH = 2.0)
Gallein	Spectrophotometric determination of complex in acid medium
Phenylfluorone	Spec λ = 530 nm determination of the complex (pH = 1) in 36% aqueous ethanol after a preliminary solvent extraction of the tin as the iodide
Toluene-3,4-dithiol + dispersant	Spec determination of the complex
Titanium	
Chromotropic acid	Spec λ = 443 nm of the complex by use of a flow injection manifold
Diantipyrinylmethane	Spec λ = 390 nm of the yellow complex
3-Hydroxy-2-methyl-1-(4-tolyl)-4-pyridone (HY)	Spec λ = 355 nm of the ternary complex formed in perchloric acid and is extracted by chloroform
Methylene blue-ascorbic acid redox reaction	Spec λ = 665 nm of the titanium-methylene blue-ascorbic acid redox reaction
Tiron	Spec λ= 420 nm of the yellow Tiron derivative (pH = 5.2-5.6)
Tungsten	
Cyanate	Spectrophotometric determination of tungsten with thiocyanate after both tungsten(VI) and molybdenum(VI) are extracted into chloroform as benzoin α-oxime complexes
Pyrocatechol violet	Determination of the 2:1 green complex in acid medium that is fixed on a dextran-type anion-exchange resin (Sephadex QAEA-25) by first-derivative solid-phase spectrophotometry (λ = 674 nm)
Tetraphenylarsonium chloride + thiocyanate	Gravimetric determination of tungsten with tetraphenylarsonium chloride after its extraction as thiocyanate (pH = 2-4)
Toluene-3,5-dithiol	Spec λ = 630 nm of the toluene-3,4-dithiol derivative after extraction with isoamyl acetate
Uranium	
Arsenazo III (1,8-dihydroxynaphthalene-3,6-disulfonic acid-2,7-bis[(azo-2)-phenylarsonic acid])	Spec determination of the binary complex
2-(5-Bromo-2-pyridylazo)diethylaminophenol	Spec λ = 578 nm of the complex (pH = 7.6)
Chlorophosphonazo III	Spec λ = 673 nm of the complex after extraction into 3-methyl-1-butanol from (1.5 to 3.0) M HCl
8-Hydroxyquinoline	Spectrophotometric determination of the metal (UO_2 (II))-ligand complex
2-(2-Pyridylazo)-5-diethylaminophenol (PADAP)	Spec λ = 564 nm of the complex after extraction into methyl isobutyl ketone (pH = 8.2)
1-(2-Pyridylazo)-2-naphthol (PAN)	Spec λ = 560 nm of the deep red precipitate in ammoniacal solutions extracted with chloroform
4-(2-Pyridylazo)-resorsinol (PAR)	Spec λ = 530 nm of the intensely deep red complex after extraction into methyl isobutyl ketone (pH = 8.0)
2-(2-Thiazolylazo)-*p*-cresol + surfactants	Spec λ = 588 nm of the complex (pH = 6.5) with surfactants such as Triton X-100 or *N*-cetyl-*N,N,N*-trimethyl ammonium bromide (CTAB)
Vanadium	
Chrome azurol S	Spec λ = 598 - 603 nm determination of the binary complex
N-Benzoyl-*N*-phenylhydroxylamine	Spec λ = 416 nm determination of the violet complex in acidic medium

Reagents	Results
3,5-Dinitrocatechol (DNC) + brilliant green	Extraction-spectrophotometric determination of the system V(V)- 3,5-dinitrocatechol (DNC)-brilliant green chelate complex
8-Hydroxyquinoline (oxine)	Extraction with *n*-butanol and spec λ = 390 nm of a ternary complex (vanadium:oxine:*n*-butanol = 1:2:2)
8-Hydroxyquinoline-5-sulfonic acid and phenylfluorone	Diffuse reflection spectrometry with 8-hydroxyquinoline-5-sulfonic acid and phenylfluorone after sorbing on a disk of an anion exchange fibrous material
4-(2-Pyridylazo)resorcinol (PAR)	Spec determination of the binary complex based on the extraction tetraphenylphosphonium or tetraphenylarsonium chloride
Xylenol orange	Spec λ = 490 nm of chelate (pH = 5.0)
Yttrium	
Alizarin red S	Spec λ = 550 nm of the complex
Ammonium purpurate	Spec determination of yttrium with ammonium purpurate as a chromogenic reagent
Arsenazo III	Spec λ = 660 nm of the blue-colored complex after preliminary purification with hydroxide and subsequent acidification
Eriochrome cyanine R (ECR)	Spec λ = 540 nm of the binary complex
Pyrocatechol violet	Spec λ = 665 nm of the complex
Zinc	
2-(5-Bromo-2-pyridylazo)-5-diethylaminophenol	Spec determination by use of 2-(5-bromo-2-pyridylazo)-5-diethyl aminophenol in the presence of cationic surfactant cetylpyridinium chloride
Carbonic anhydrase	Enzymatic determination by use of carbonic anhydrase after removing zinc by dialysis against dipicolinic acid
Dithizone	Spec λ = 530 nm of binary complex after chloroform extraction
Eriochrome black T + EDTA	Complexometric titration where the initially formed red zinc-eriochrome black T color is replaced with the blue zinc-EDTA color at the end point (pH = 10)
7-(4-Nitrophenylazo)-8-hydroxyquinoline-5-sulfonic acid (*p*-NIAZOXS)	Spec λ = 520 nm of the zinc derivative (pH = 9.2, borax buffer)
Phenylglyoxal mono(2-pyridyl) hydrazone (PGMPH)	Spec λ = 464 nm - 470 nm of the yellow-orange complex (pH = 7.2-8.5) in 40% (v/v) ethanol
1-(2-Pyridylazo)-2-naphthol (PAN)	Two-dimensional absorption spec determination of complex in aqueous micellar solutions
1-(2-Pyridylazo)-2-naphthol (PAN)	Preconcentration determination by use of 1-(2-pyridylazo)-2-naphthol (PAN) anchored. SiO_2 nanoparticles
4-(2-Pyridylazo)resorcinol	Spec determination at pH = 7.0; preconcentration of zinc by cloud point extraction of the complex and determination by ICP optic emission spectrometry
1-(2-Thiazolylazo)-2-naphthol (TAN)	Flow-injection solid phase spec λ = 595 nm with TAN immobilized on C18-bonded silica
Xylenol orange	Spec determination of 1:1 zinc-xylenol orange red-violet complex (pH= 5.8-6.2)
Xylenol orange	Sequential injection analysis (SIA) based on the spec λ = 568 nm of zinc using xylenol orange as a color reagent
Xylenol orange	Spec determination by mean centering of ratio kinetic profile (pH = 5.3)
Xylenol orange + cetylpyridinium chloride	Spec λ = 580 nm of the 1:2:4 ratio for the metal:ligand:surfactant ternary complex (pH = 5.0-6.0)
Zirconium	
Alizarin red S	Spec λ = 520 nm determination of the binary complex after phosphate extraction (pH = 2.5)
Arsenazo III (1,8-dihydroxynaphthalene-3,6-disulfonic acid-2,7-bis[(azo-2)-phenylarsonic acid])	Spec determination of the binary complex
N-*p*-Chlorophenylbenzohydroxamic acid (*N*-*p*-Cl-BHA) + morin	Spec λ = 420 nm and fluorimetric determination of the *N*-*p*-Cl-BHA greenish-yellow complex after extraction with isoamyl alcohol from (0.2 to 0.5) N H_2SO_4 followed by morin addition; it is critical that the morin be very pure
Pyrocatechol violet	Spec λ = 650 nm of the blue complex in sulfuric acid solution
Pyrocatechol violet + tri-*n*-octylphosphine oxide (TOPO)	Spec λ = 655 nm of the blue complex after extraction with tri-*n*-octylphosphine oxide (TOPO) in cyclohexane
Morin	Fluorimetric determination of the complex after EDTA addition
Xylenol orange	Flow injection spectrophotometric determination in sulfuric acid medium

TABLE 2. Organic Analytical Reagents for the Determination of Inorganic Anions

Method	Results/Comments
Arsenate, AsO_4^{-3}	
X-ray absorption near edge structure (XANES)	Use of self-consistent multiple-scattering methods that measure the AsO_4^{-3} adsorption on TiO_2
Dual detection ion chromatography	Simultaneous determination of AsO_4^{-3} and AsO_3^{-3} (LOD = 0.044 mg/L)
Ion chromatography/inductively-coupled plasma/MS	Validation established after ultrasonic-assisted enzymatic extraction
Ion chromatography	Post-column generation and detection of the AsO_4^{-3} molybdate heteropoly ion
Surface-enhanced Raman scattering (SERS)	Use of Ag nanostructured multilayer films; LOD = 5 μg/L
Liquid chromatography-atomic fluorescence detection	Simultaneous determination of six As ions; LOD < 4 μg/L
LC-ICP/MS	Simultaneous determination of five As ions; LOD < 4 μg/L
HPLC/hydride generation atomic absorption	Simultaneous determination of six As ions; LOD = 3 ng to 9 ng
Ion exchange determination	Separation of AsO_4^{-3} and AsO_3^{-3} using the weakly basic anion exchanger A63-X4A by difference of respective pK_a values
Titration	As^{+5} is reduced by NH_2NH_2 and titration with Br^-/BrO_3^-
Bromide, Br^-	
Ion chromatography using electrochemical detection	Use of Ag working electrode; LOD = 10 ppb (mass/mass)
Carbonate, CO_3^{-2}	
GC	Carbonation of salt and use of a GC with methanation reformer and a flame ionization detector
GC-MS	Determination of pentafluorobenzyl derivative
FTIR	CO_3^{-2} measurement in organic matter (875 cm^{-1} and 712 cm^{-1})
Gas measurement	Conversion to CO_2 measured by a barometer or manometer
Ion chromatography	Measurement in the presence of formyl ions
Photoacoustic IR spectroscopy	Use of several multi-calibration methods typically in sedimentary rocks
Thermal optical analysis	Use of semi-continuous carbon elemental analyzer
Liquid chromatography	Use of a C18 HPLC column, MeOH/AcOH mobile phase and UV detection (λ = 290 nm)
Raman spectroscopy	Use of OH stretch band of H_2O as internal standard
Thermogravimetry	CO_3^{-2} measurement in organic matter after identification of evolved gaseous species
Chlorate, ClO_3^-	
Flow injection analysis	Reduction of methylene blue's leuco form to methylene blue followed by optical detection in the visible region
Spectrophotometry	Reduction to Cl_2 via Tl^{+3} which oxidizes the leuco form of methylene
FTIR	Absorption band at 973 cm^{-1} (λ = 10 277 nm)
Anodic electrochemical dissociation	Determination made during the anodic dissociation of Ti grade VT-1
Potentiometric titration	Reduction to Cl^- (pH = 7) using As^{+3} (OsO_4 as catalyst)
Titrimetric analysis	ClO_3^- converted to Cl^- which is titrated against $Hg(NO_3)_2$ (diphenylcarbazone as indicator)
Titrimetric analysis	Iodometric titration
Titrimetric analysis	Use of methyl orange (LOD = 10^{-2} μg, mass/mass)
Chromate, CrO_4^{-2}	
Potentiometric titration	Selective electrode based on the *N,N*-butylene bis(salicylideneiminato)copper(II); LOD = 3.0×10^{-6} mol/L
Spectrophotometry	Determination by oxidation of ascorbic acid (λ = 548 nm)
Ion exchange chromatography	Diaminetetraacetic acid (DCTA)/CH_3CN used as eluent
Ion pair chromatography	Tetrabutylammonium hydroxide (TBAH)/aq. CH_3CN used as eluent
Capillary electrophoresis	KH_2PO_4 buffer (pH = 7)
Titration	Iodometric titration
X-ray absorption near edge structure (XANES)	LOD = 10 ppm using a 150 μm synchrotron x-ray beam
Cyanide, CN^-	
Headspace GC/MS	Use of gas chromatography with a silica PLOT column coupled with mass selective detection
Headspace GC/MS	Use of gas chromatography with a nitrogen-phosphorus detector; LOD = 0.05 μg/mL to 5 μg/mL
Headspace GC/MS	Use of headspace solid-phase microextraction (SPME); LOD = 0.006 μg/mL
Headspace GC/MS	Determined as the pentafluorobenzylbromide derivative in the presence of TDMBC
Headspace GC	Use of GC with thermoionic detection (NPD)
Headspace GC/MS	Use of GC with the flame ionization detector (FID)
Headspace GC/MS	Determination after reaction with chloramine-T
	Derivatization by the pyridine-pyazolone method
Ion chromatography using electrochemical detection	Use of Ag working electrode; LOD = 1 μg/L
Cyclic voltammetry	Au electrode; LOD = 10^{-5} mol/L to 10^{-6} mol/L
Differential cathodic stripping voltammetry	LOD = 2×10^{-7} M in the presence of 5×10^{-5} M [S^{-2}] and 1×10^{-3} M [Cu^{+2}]

Method	Results/Comments
Differential electrolytic potentiometry (DEP)	DEP coupled with the sequential injection/flow analysis (SIA/FIA) of aqueous solutions; con. range 6 μg/mL to 39 μg/mL; LOD = 0.31 μg/mL; RSD = 2.55
Isotachophoresis	Determination as $[Au(CN)_2^-]$, $[Ag(CN)_2^-]$, or $[Cu(CN)_3^{-2}]$ complexes
Potentiometric determination	Use of PVC sensors responsive to $Ag(CN)_2^-$
Spectrofluorometry	Use of naphthoquinone imidazole boronic-based sensors (M-NQB and p-NQB) with CTAB (spec λ = 460 nm)
Fluorescence	Derivatization with naphthalene 2,3-dicarboxyaldehyde and taurine (specf λ_{ex} = 418 nm, λ_{em} = 460 nm); LOD = 0.002 μg/mL to 0.025 μg/mL
Argentometric method	Argentometric titration/quantification using KCNS ($FeNH_4SO_4$) as indicator
Capillary electrophoresis/UV spectrophotometry	Conversion of CN^- to SCN^- via rhodanese, separated by capillary electrophoresis and measured by UV spectrophotometry (spec λ = 200 nm); LOD = 3 μmol
Piezoelectric quartz crystal (PQC) sensors	Method makes use of silver coated sensors; LOD = 1.25×10^{-8} mol/L
Visible spectrophotometry colorimetric assay	Picrate/resorcinol derivative in bicarbonate, (spec λ = 488 nm)
Visible spectrophotometry colorimetric assay	Lithium/sodium picrate derivative (spec λ = 500 nm to 520 nm); LOD = 2.5 μg/100 mg sample
Visible spectrophotometry colorimetric assay	Isonicotinic acid-barbituric acid derivative after treatment with chloramine-T (spec λ = 600 nm)
Visible spectrophotometry colorimetric assay	4-hydroxy-3-(2-oxoindoline-3-ylideneamino)-2-thioxo-2*H*-1,3-thiazin-5(3*H*)-one) HOTT derivative (spec λ = 466 nm)
Visible spectrophotometry colorimetric assay	Ethanol extraction of the red complex formed on picric acid test strips (spec λ = 500 nm); LOD = 0.5 μg to 1.0 μg
Visible spectrophotometry colorimetric assay	Derivatization with *p*-nitrobenzaldehyde followed by reaction with *o*-dinitrobenzene to yield a purple color; LOD = 10 μg/L
Visible spectrophotometry colorimetric assay	Barbituric or N,N-dimethylbarbituric acid derivative
Visible spectrophotometry colorimetric assay	Spectrophotometric determination by use of a modified Konig reaction with NaOCl as the chlorinating agent
Visible spectrophotometry colorimetric assay	Complexation with pyridine or 4-pyridinecarboxylic acid
Visible spectrophotometry colorimetric assay	Reaction of CN^- with Pd-dimethylglyoximate complex that releases dimethylglyoxime, isolated as the Ni derivative
Visible spectrophotometry colorimetric assay	Determination (spec λ = 570 nm) of pyridine/barbituric acid derivative after chloramine-T derivative
Visible spectrophotometry colorimetric assay	Measurement of the pyridine-pyrazolone derivative
Visible spectrophotometry colorimetric assay	Measurement of the isocotinic acid/1-phenyl-3-methyl-5-pyrazolone derivative after reaction with chloramine-T (spec λ = 638 nm)
Visible spectrophotometry colorimetric assay	Measurement of the derivative with the Hg(II)-*p*-dimethylaminobenzylidenerhodanine complex (spec λ = 452 nm)
Visible spectrophotometry colorimetric assay	Measurement of the Hg(II)-diphenylcarbazone derivative (spec λ = 562 nm)
Visible spectrophotometry colorimetric assay	Measurement of the pyridine-benzidine complex; LOD = 0.02 μg
Visible spectrophotometry colorimetric assay	Measurement of the pentacyano(*N*-methylpyrazinium) ferrate (II) complex (spec λ = 658 nm)
Flow injection analysis	Use of Ag and Pt electrodes to determine CN^- as Cu complexes
Flow injection analysis	Analyzed as the teracyanonickelate (II) complex
Fluorescent sensors	Fluoresecein derivative emits a green light
Fluorescent sensors	Naphthoquinone-imidazole boronic-based sensors (spec λ = 460 nm); LOD = 1.4 μM
FTIR	Use of a horizontal attenuated total reflectance (HATR) accessory; detection range = 0.3 M to 1.0 M
Chemiluminescence	Measurement of intensity of the reaction with luminal, hemin, and peroxide
Amperometry/flow injection analysis	Use of a multimeter as a detection device; LOD = 200 μg/L
Photoacoustic IR detection	Applied to the determination of HCN and NCCl in air
X-ray diffraction	LOD = 0.1% in wastewater analyses
Coulometric determination	LOD = 10^{-9} mg/100 μL
Cavity ringdown spectroscopy	Use of external cavity tunable diode laser
CN^-/SCN^- (only)	
Chemical ionization GC/MS	Derivatization with pentafluorobenzyl bromide (PFBBr) via extraction with ethyl acetate and tetrabutylammonium sulfate (TBAS); LOD = 1 μM (CN^-) and LOD < 50 nM (SCN^-)
Dichromate, $Cr_2O_7^{-2}$	
FTIR	Absorption band at 950 cm^{-1}
Paper electrophoresis-photo density scanning	Photo density scanning determination after development with diphenyl carbazide
Cell voltage variation	Applied to the electrochemical synthesis of sodium dichromate from sodium chromate
Conductimetric determination	Performed in the presence of CrO_4^{-2}
Adsorptive stripping voltamemtry	Commonly applied to body fluids
Potentiometric titration	Determination of Pu via the $Fe^{+2}/Cr_2O_7^{-2}$ redox titration
Sequential injection analysis	$Cr_2O_7^{-2}$ determination based on oxidation of pindolol (spec λ = 640 nm)

Method	Results/Comments
Spectrophotometric analysis	Applied to the determination of pyridine derivatives
	Reduction to Cr^{+3} in organic carbon soil analysis
Titrimetric analysis	Reduction by Fe^{+3} titration
Thin-layer chromatography (TLC)	Applied to determination of unsaturated fats and further spectrophotometric analysis (spec λ = 350 nm)
X-ray absorption fine structure spectroscopy	Applied to the speciation of $CrO_4^{-2}/Cr_2O_7^{-2}$ equilibrium
Hypochlorite, ClO^-	
Titrimetric determination	Iodometric titration
Polarography	Determination by conversion of Br^- to BrO_3^-
Spectrophotometry	Oxidation of Ir^{+3} to Ir^{+4} (spec λ = 488 nm)
Spectrophotometry	Measurement of ClO^- absorbance at pH > 7
Spectrophotometry	Measurement of chloramine formed by addition of aq. NH_3
Voltammetry	Technique compared to iodometric titration
Potentiometric titration	ClO^- measured by subtraction of BrO^- in a ClO/BrO^-
Potentiometric titration	Reduction to Cl^- (pH < 7, OsO_4 as catalyst)
Iodide, I^-	
Ion chromatography using electrochemical detection	Use of Ag working electrode; LOD = 100 ppb
Isocyanates, NCO^-	
Mass spectrometry	Use of a proton transfer reaction (PTR), time-of-flight mass spectrometer, and high-sensitivity PTR quadruple mass spectrometer
FTIR	Use of sealed liquid cell for polymer analysis
HPLC	Analysis of ethanol derivative (spec λ = 245 nm)
Capillary electrophoresis	Use of tris(2,2'-bipyridine)ruthenium (II) by electrochemiluminescence
Potentiometric titration	Titration of carboxylic acid hydrazide derivative with $NaNO_2$
Spectrophotometric analysis	Conversion of CH_3NCO to CH_3NH_2 and reaction with 1 fluoro-2,4-dinitrobenzene to yield $CH_3NHC_6H_3(NO_2)_2$ (spec λ = 352 nm)
Manganate, MnO_4^{-2}	
Titrimetric analysis	Oxidation of As^{+3}, Cr^{+3}, Sb^{+3}, Te^{+4}, PO_3^{-3}, various carboxylic acids, trimethylammoniated compounds
Gravimetric analysis	Precipitation of $BaMnO_4$
Potentiometric titration	Use of a Pt electrode
Voltammetry	Use of glassy carbon working electrode
Spectrophotometry	LOD = 0.185 µg/mL; (spec λ = 635 nm)
Anodic oxidation	Rotating disk electrode
Gravimetric analysis	Precipitation of $BaMnO_4$
Nitrite, NO_2^-	
Ion chromatography	Use of high capacity, OH^--selective, anion exchange column with a guard column) in simultaneous determination of BrO_3^-; LOD = 0.05 µg/L; RSD = 3.61% to 4.77%
HPLC	Simultaneous determination of NO_3^-, SO_4^{-2}, and SCN^-; samples are ultrasonically extracted with $EtOH/CCl_3COOH$ and purified by C18 SPE column (highly retentive alkyl-bonded phase); LOD = 0.01 mg/kg to 0.02 mg/kg; RSD = 0.87 - 2.6 %
HPLC	KOH or NaOH aq. solution used as mobile phase for 10 µL to 25 µL sample size
Quantitative sequential injection analysis (SIA)	LOD = 0.002 mg/L
Diffuse UV-visible reflectance method	Reaction of nitrite with sulfadiazine and *o*-naphthol; LOD = 0.02 mg/L; RSD = 5% to 8%
Cyclic voltammetry (CV)/pulse voltammetry	Electropolymerization of Nile Blue at pre-polarized glassy carbon electrode (phosphate buffer, pH = 7.1) leading to a NO_2^- sensor
Cyclic voltammetry (CV)/pulse voltammetry	Use of a solid-state microelectrode formed through the electrochemical co-deposition of Pt-Fe nanoparticles on Au microelectrode
Cyclic voltammetry (CV)/pulse voltammetry	Use of a sol-gel electrode using 3-aminopropyltrimethoxy silane for covalent immobilization of toluidine blue
Capillary electrophoresis	Simultaneous determination of NO_3^- and $C_2O_4^{-2}$
Capillary zone electrophoresis	Simultaneous determination of NO_3^- and Cl^- by use of a photoiodie-array detector (PDA)
UV-visible spectrophotometry	Derivatization with $Zn(OH)_2$/basic fuchsin
UV-visible spectrophotometry	Derivatization with Griess reagent (spec λ = 537 nm); RSD ~1.5%
Microfluidic gradient elution moving boundary electrophoresis (GEMBE)	Simultaneous determination of NO_3^- and NH_4^+
Microfluidic gradient elution moving boundary electrophoresis (GEMBE)	Simultaneous determination of NO_3^-, Cl^-, $C_2O_4^{-2}$, NO_3^-, and SO_4^{-2}
Gasometric analysis	The NO_2^- concentration is quantified by the formation of N_2(g) produced *vis a vis* its reaction of with sulfamic acid
Seeping layer test paper	Derivatization with *p*-aminobenzenesulfonic acid/*N*-(1-naphthyl) ethylenediamine dihydrochloride

Method	Results/Comments
NO_2^- simultaneous NO_2^-/NO_3^- (only)	
HPLC	Pre-column derivatization with Griess reagents (spec λ = 540 nm)
UV-visible spectrophotometry	Following derivatization with HCl/naphthyl ethylenediamine, LOD = 0.5 mg to 18 mg nitrogen per L (NO_3^-) and 0.5 mg to 5 mg nitrogen per L (NO_2^-)
	LOD = 0.5 mg to 18 mg nitrogen per L (NO_3^-) and 0.5 mg to 5 mg nitrogen per L (NO_2^-)
HPLC-UV	LOD = 2.5 mg/kg (NO_3^-) and 5.0 mg/kg (NO_2^-)
GC-mass spectrometry	Quantification of ^{15}N-labeled nitrate and nitrite biological fluids
Chemiluminescence	LOD = 10 nmol/L; RSD = 3% (200 nmol/L) to 5% (400 nmol/L)
Oxalate, $C_2O_4^{-2}$	
Capillary electrophoresis	Use of acid run buffer to revert electroosmotic flow
Capillary gas chromatography	LOD = 500 μg by use of a flame ionization detector
Chromogenic strip	Test strip equipped with sorghum leaf oxalate oxidase cross-linked with glutaraldehyde and *o*-toluidine
FT-IR (attenuated)	LOD = 0.07% of the matrix
Capillary electrophoresis	Use of acid run buffer to revert electroosmotic flow
Capillary gas chromatography	LOD = 500 μg by use of a flame ionization detector
Flow injection analysis	Oxalate oxidase and peroxidase are immobilized on Au electrode
Gas chromatography	Analysis of methylated product
Ion exchange chromatography method	LOD = 6 μg/L to 9 μg/L
Isotachophoresis	Detection at λ = 254 nm; LOD = 20 nmol/L
LC-tandem MS	System attached to weak anion exchange chromatography
Raman spectroscopy	Intensity of peak at 1462 cm^{-1} is used
Spectrophotometric analysis	Analysis of catalytic effect of reaction products of methylene blue/$Cr_2O_7^{-2}$
Spectrophotometric analysis	Analysis of the NaDH → NAD^+ reaction with $C_2O_4^{-2}$
Titrimetric analysis	Titration against MnO_4^-
Titrimetric analysis	Titration against Ce^{+4}
Perchlorate, ClO_4^-	
Spectrophotometric analysis	Analysis of catalytic effect of reaction products of methylene blue/$Cr_2O_7^{-2}$
Spectrophotometric analysis	Analysis of the NaDH → NAD^+ reaction with $C_2O_4^{-2}$
Titrimetric analysis	Titration against MnO_4^-
Titrimetric analysis	Titration against Ce^{+4}
Spectrophotometry	Yb-pyrocatechol violet complex
Spectrophotometry	Determination of reddish-violet color produced by adding methylene blue/$ZnSO_4$ in presence of NO_3^-
HPLC/MS	LOD = 0.2 ppb
Capillary electrophoresis	Analysis typically performed on tobacco plants
MS	Determination of organic base-phosphate complex
Ion pair extraction-MS ESI	Use of cationic surfactants
Titrimetric analysis	Titration against *H*-tetramethylammonium sulfate with various indicators
Titrimetric analysis	Back titration of excess $TiCl_3$ against $Fe_2(SO_3)_2$
Phosphate, PO_4^{-3}	
FTIR	Use of 1037 cm^{-1} ($Ca_3(PO_4)_2$) and 1010 cm^{-1} ($MgNH_4PO_4$)
Phosphide, P^{-3}	
Gravimetric	Derivatization as Ag_3PO_3 after formation of Ag_3P and further oxidation with HNO_3
ESR	Derivatization to GaP
X-ray electron spectrometry	Derivatization to AlP
X-ray fluorescence	Derivatization to AlP
X-ray diffractometry	Derivatization to AlP
X-ray photoelectron spectroscopy	Derivatization to functionalized GaP
Fluorescence	Derivatization to Ca_3P_2
Phosphite, PO_3^{-3}	
Fluorescence	Enzymatic oxidation to phosphate with NAD^+ as co-substrate to produce fluorescence
^{31}P NMR	Estherification of OH^- in organic polymers
Laser Raman spectrometry	Ar-ion laser allows quantitative analysis of $H_2PO_3^-$ and HPO_3^{-2}
Ion exchange chromatography	Method makes use of conductivity detection
Visible spectrophotometry	Measurement of 3,5-$(O_2N)_2C_6H_3COOH$ derivative
Silcate, SiO_4^{-4}	
Raman spectroscopy	Use of *ab initio* quantum chemical simulation
Electron probe microanalysis (EPMA)	Analysis of Na, Mg, Al, Si, K, Ca, Ti, Mn, and Fe salts
Ion-exclusion chromatography	Use of ascorbate solution as both an eluent and a reducing agent following by molybdate derivatization

Method	Results/Comments
Laser-induced plasma spectroscopy	Use of a Q switched Nd:YAG laser and time-resolved spectroscopy
Time-of-flight secondary ion mass spectrometry	Use of C_{60} primary ions
Laser ablation-inductively coupled plasma mass spectrometry (LA-ICPMS)	No need for internal standard
Sulfide, S^{-2}	
Ion chromatography using electrochemical detection	Use of Ag working electrode; LOD = 30 ppb
Sulfite, SO_3^{-2}	
Amperometric titration	Use of iodine solution
Potentiometric titration	Pt wire coated with Hg is balanced against calomel electrode
Spectrophotometric analysis	Product of reaction with *p*-rosaniline acid bleached dye and HCHO (spec λ = 560 nm)
Ion chromatography	
Titration analysis	Titration against BrO_3^{-2}
Coulometric determination	LOD = 10^{-11} mg/100 μL
Thiocyanate, CNO^-	
Flow injection analysis	Formation of the 2-(5-bromo-2-pyridylazo)-5-diethylaminophenol derivative in $Cr_2O_7^{-2}/H^+$
Flow injection analysis	Analyzed as the teracyanonickelate (II) complex

Section 9
Molecular Structure and Spectroscopy

Bond Lengths in Crystalline Organic Compounds 9-1
Bond Lengths in Organometallic Compounds 9-17
Structure of Free Molecules in the Gas Phase 9-19
Characteristic Bond Lengths in Free Molecules 9-55
Atomic Radii of the Elements 9-56
Dipole Moments 9-58
Hindered Internal Rotation 9-68
Bond Dissociation Energies 9-73
Electronegativity 9-106
Force Constants for Bond Stretching 9-107
Fundamental Vibrational Frequencies of Small Molecules 9-108
Spectroscopic Constants of Diatomic Molecules 9-111

BOND LENGTHS IN CRYSTALLINE ORGANIC COMPOUNDS

Table 1 gives average interatomic distances for bonds between the elements H, B, C, N, O, F, Si, P, S, Cl, As, Se, Br, Te, and I as determined from x-ray and neutron diffraction measurements on organic crystals. The table has been derived from an analysis of high-precision structure data on about 10,000 crystals contained in the 1985 version of the Cambridge Structural Database, which is maintained by the Cambridge Crystallographic Data Center. Column definitions for the table are as follows.

Column heading	Definition
Bond	Specification of elements in the bond, with coordination number given in parentheses, and bond type (single, double, etc.). For carbon, the hybridization state is given
Substructure	Substructure in which the bond is found. The target bond is set in boldface. Where X is not specified, it denotes any element type. C# indicates any sp^3 carbon atom, and C* denotes any sp^3 carbon whose bonds, in addition to those specified in the linear formulation, are to C and H atoms only
d	Unweighted mean in Å units of all the values for that bond length found in the sample
m	Median in Å units of all values
σ	Standard deviation in the sample
q_l	Lower quartile for the sample (i.e., 25% of values are less than q1 and 75% exceed it)
q_u	Upper quartile for the sample
n	Number of observations in the sample
Note	Refers to the footnotes in Appendix 1

References to special cases are given in a shorthand form and listed in Appendix 2. Further information on the method of analysis of the data may be found in the reference cited below.

The table is reprinted with permission of the authors, the Royal Society of Chemistry, and the International Union of Crystallography.

Reference

Frank H. Allen, Olga Kennard, David G. Watson, Lee Brammer, A. Guy Orpen, and Robin Taylor, *J. Chem. Soc. Perkin Trans.* II, S1–S19, 1987.

TABLE 1. Bond Lengths in Crystalline Organic Compounds

Bond	Substructure	d	m	σ	q_l	q_u	n	Note
As(3)–As(3)	X_2–**As–As**–X_2	2.459	2.457	0.011	2.456	2.466	8	
As–B	see CUDLOC (2.065), CUDLUI (2.041)							
As–BR	see CODDEE, CODDII (2.346–3.203)							
As(4)–C	**X_3–As–CH_3**	1.903	1.907	0.016	1.893	1.916	12	
	$(X)_2$(C,O,S=)**As–C**sp^3	1.927	1.929	0.017	1.921	1.937	16	
	As–C*ar* in Ph_4As^+	1.905	1.909	0.012	1.897	1.912	108	
	$(X)_2$(C,O,S=)**As–C***ar*	1.922	1.927	0.016	1.908	1.934	36	
As(3)–C	X_2–**As–C**sp^3	1.963	1.965	0.017	1.948	1.978	6	
	X_2–**As–C***ar*	1.956	1.956	0.015	1.944	1.964	41	
As(3)–Cl	X_2–**As–Cl**	2.268	2.256	0.039	2.247	2.281	10	
As(6)–F	in **AsF_6^-**	1.678	1.676	0.020	1.659	1.695	36	
As(3)–I	see OPIMAS (2.579, 2.590)							
As(3)–N(3)	X_2–**As–N**–X_2	1.858	1.858	0.029	1.839	1.873	19	
As(4)=N(2)	see TPASSN (1.837)							
As(4)–O	$(X)_2$(O=)**As–OH**	1.710	1.712	0.017	1.695	1.726	6	
As(3)–O	see ASAZOC, PHASOC01 (1.787–1.845)							
As(4)=O	X_3–**As=O**	1.661	1.661	0.016	1.652	1.667	9	
As(3)=P(3)	see BELNIP (2.350, 2.362)							†
As(3)–P(3)	see BUTHAZ10 (2.124)							†
As(3)–S	X_2–**As–S**	2.275	2.266	0.032	2.247	2.298	14	
As(4)=S	X_3–**As=S**	2.083	2.082	0.004	2.080	2.086	9	
As(3)–Se(2)	see COSDIX, ESEARS (2.355–2.401)							†
As(3)–Si(4)	see BICGEZ, MESIAD (2.351–2.365)							†
As(3)–Te(2)	see ETEARS (2.571, 2.576)							†
B(n)–**B**(n)	n = 5–7 in boron cages	1.775	1.773	0.031	1.763	1.786	688	
B(4)–**B**(4)	see CETTAW (2.041)							
B(4)–**B**(3)	see COFVOI (1.698)							
B(3)–**B**(3)	X_2–**B–B**–X_2	1.701	1.700	0.014	1.691	1.712	8	
B(6)–BR		1.967	1.971	0.014	1.954	1.979	7	†
B(4)–BR		2.017	2.008	0.031	1.990	2.044	15	†
B(n)–C	n = 5–7: **B–C** in cages	1.716	1.717	0.020	1.707	1.728	96	
	n = 3–4: **B–C**sp^3 not cages	1.597	1.599	0.022	1.585	1.611	29	1
	n = 4: **B–C***ar*	1.606	1.607	0.012	1.596	1.615	41	
	n = 4: **B–C***ar* in Ph_4B^-	1.643	1.643	0.006	1.641	1.645	16	

Bond	Substructure	*d*	*m*	*σ*	q_l	q_u	*n*	Note
B(*n*)–C	*n* = 3: **B–C*ar***	1.556	1.552	0.015	1.546	1.566	24	
B(*n*)–Cl	**B**(5)**–Cl** and **B**(3)**–Cl**	1.751	1.751	0.011	1.743	1.761	14	
	B(4)**–Cl**	1.833	1.833	0.013	1.821	1.843	22	
B(4)–F	**B–F** (B neutral)	1.366	1.368	0.017	1.356	1.375	25	
	B⁻–F in BF_4^-	1.365	1.372	0.029	1.352	1.390	84	
B(4)–I	see TMPBTI (2.220, 2.253)							
B(4)–N(3)	X_3–**B–N**(=C)(X)	1.611	1.617	0.013	1.601	1.625	8	
	in pyrazaboles	1.549	1.552	0.015	1.536	1.560	10	
B(3)–N(3)	X_2–**B–N**–C_2: all coplanar	1.404	1.404	0.014	1.389	1.408	40	2
	for τ(BN) > 30° see BOGSUL, BUSHAY, CILRUK (1.434–1.530)							
	S_2–**B–N**–X_2	1.447	1.443	0.013	1.435	1.470	14	
B(4)–O	**B⁻–O** in BO_4^-	1.468	1.468	0.022	1.453	1.479	24	
	for neutral B–O see Note 3							3
B(3)–O(2)	X_2–**B–O**–X	1.367	1.367	0.024	1.349	1.382	35	
B(*n*)–P	*n* = 4: **B–P**	1.922	1.927	0.027	1.900	1.954	10	
	n = 3: see BUPSIB10 (1.892, 1.893)							
B(4)–S	**B**(4)**–S**(3)	1.930	1.927	0.009	1.925	1.934	10	
	B(4)**–S**(2)	1.896	1.896	0.004	1.893	1.899	6	
B(3)–S	N–**B–S**$_2$	1.806	1.806	0.010	1.799	1.816	28	
	(=X–)(N–)**B–S**	1.851	1.854	0.013	1.842	1.859	10	
Br–Br	see BEPZEB, TPASTB	2.542	2.548	0.015	2.526	2.551	4	
Br–C	**Br–C***	1.966	1.967	0.029	1.951	1.983	100	4
	Br–C*sp*^{3} (cyclopropane)	1.910	1.910	0.010	1.900	1.914	8	
	Br–C*sp*^{2}	1.883	1.881	0.015	1.874	1.894	31	4
	Br–C*ar* (mono-Br +*m.p*-Br$_2$)	1.899	1.899	0.012	1.892	1.906	119	4
	Br–C*ar* (o-Br$_2$)	1.875	1.872	0.011	1.864	1.884	8	4
⁻Br(2)–Cl	see TEACBR (2.362–2.402)							†
Br–I	see DTHIBR10 (2.646), TPHOSI (2.695)							
Br–N	see NBBZAM (1.843)							
Br–O	see CIYFOF	1.581	1.581	0.007	1.574	1.587	4	
Br–P	see CISTED (2.366)							
Br–S(2)	see BEMLIO (2.206)							†
Br–S(3)	see CIWYIQ (2.435, 2.453)							†
Br–S(3)$^+$	see THINBR (2.321)							†
Br–Se	see CIFZUM (2.508, 2.619)							
Br–Si	see BIZJAV (2.284)							
Br–Te	in **Br$_6$Te**$^{2-}$ see CUGBAH (2.692–2.716)							
	Br–Te(4) see BETUTE10 (3.079, 3.015)							
	Br–Te(3) see BTUPTE (2.835)							
C*sp*^{3}–C*sp*^{3}	C#–**CH$_2$–CH$_3$**	1.513	1.514	0.014	1.507	1.523	192	
	(C#)$_2$–**CH–CH$_3$**	1.524	1.526	0.015	1.518	1.534	226	
	(C#)$_3$–**C–CH$_3$**	1.534	1.534	0.011	1.527	1.541	825	
	C#–**CH$_2$–CH$_2$**–C#	1.524	1.524	0.014	1.516	1.532	2459	
	(C#)$_2$–**CH–CH$_2$**–C#	1.531	1.531	0.012	1.524	1.538	1217	
	(C#)$_3$–**C–CH$_2$**–C#	1.538	1.539	0.010	1.533	1.544	330	
	(C#)$_2$–**CH–CH**–(C#)$_2$	1.542	1.542	0.011	1.536	1.549	321	
	(C#)$_3$–**C–CH**–(C#)$_2$	1.556	1.556	0.011	1.549	1.562	215	
	(C#)$_3$–**C–C**–(C#)$_3$	1.588	1.580	0.025	1.566	1.610	21	
	C*–C* (overall)	1.530	1.530	0.015	1.521	1.539	5777	5,6
	in cyclopropane (any subst.)	1.510	1.509	0.026	1.497	1.523	888	7
	in cyclobutane (any subst.)	1.554	1.553	0.021	1.540	1.567	679	8
	in cyclopentane (C,H-subst.)	1.543	1.543	0.018	1.532	1.554	1641	
	in cyclohexane (C,H-subst.)	1.535	1.535	0.016	1.525	1.545	2814	
	cyclopropyl-C* (exocyclic)	1.518	1.518	0.019	1.505	1.531	366	7
	cyclobutyl-C* (exocyclic)	1.529	1.529	0.016	1.519	1.539	376	8
	cyclopentyl-C* (exocyclic)	1.540	1.541	0.017	1.527	1.549	956	
	cyclohexyl-C* (exocyclic)	1.539	1.538	0.016	1.529	1.549	2682	
	in cyclobutene (any subst.)	1.573	1.574	0.017	1.566	1.586	25	8
	in cyclopentene (C,H-subst.)	1.541	1.539	0.015	1.532	1.549	208	
	in cyclohexene (C,H-subst.)	1.541	1.541	0.020	1.528	1.554	586	
	in oxirane (epoxide)	1.466	1.466	0.015	1.458	1.474	249	9

Bond	Substructure	d	m	σ	q_l	q_u	n	Note
	in aziridine	1.480	1.481	0.021	1.465	1.496	67	9
	in oxetane	1.541	1.541	0.019	1.527	1.557	16	
	in azetidine	1.548	1.543	0.018	1.536	1.558	22	
	oxiranyl-C* (exocyclic)	1.509	1.507	0.018	1.497	1.519	333	9
	aziridinyl-C* (exocyclic)	1.512	1.512	0.018	1.496	1.526	13	9
Csp^3–Csp^2	**CH$_3$–C**=C	1.503	1.504	0.011	1.497	1.509	215	
	C#–**CH$_2$–C**=C	1.502	1.502	0.013	1.494	1.510	483	
	(C#)$_2$–**CH–C**=C	1.510	1.510	0.014	1.501	1.518	564	
	(C#)$_3$–**C–C**=C	1.522	1.522	0.016	1.511	1.533	193	
Csp^3–Csp^2	**C*–C**=C (overall)	1.507	1.507	0.015	1.499	1.517	1456	5
	C*–C=C (endocyclic)							
	in cyclopropene	1.509	1.508	0.016	1.500	1.516	20	10
	in cyclobutene	1.513	1.512	0.018	1.500	1.525	50	8
	in cyclopentene	1.512	1.512	0.014	1.502	1.521	208	
	in cyclohexene	1.506	1.505	0.016	1.495	1.516	391	
	in cyclopentadiene	1.502	1.503	0.019	1.490	1.515	18	
	in cyclohexa-1,3-diene	1.504	1.504	0.017	1.491	1.517	56	
	C*–C=C (exocyclic):							
	cyclopropenyl-C*	1.478	1.475	0.012	1.470	1.485	7	10
	cyclobutenyl-C*	1.489	1.483	0.015	1.479	1.496	11	8
	cyclopentenyl-C*	1.504	1.506	0.012	1.495	1.512	115	
	cyclohexenyl-C*	1.511	1.511	0.013	1.502	1.519	292	
	C*CH=O in aldehydes	1.510	1.510	0.008	1.501	1.518	7	
	(C*)$_2$–C=O							
	in ketones	1.511	1.511	0.015	1.501	1.521	952	11
	in cyclobutanone	1.529	1.530	0.016	1.514	1.545	18	
	in cyclopentanone	1.514	1.514	0.016	1.505	1.523	312	
	acyclic and 6 + rings	1.509	1.509	0.016	1.499	1.519	626	
	C*–COOH in carboxylic acids	1.502	1.502	0.014	1.495	1.510	176	
	C*–COO$^-$ in carboxylate anions	1.520	1.521	0.011	1.516	1.528	57	
	C*–C(=O)(–OC*)							
	in acyclic esters	1.497	1.496	0.018	1.484	1.509	553	12
	in β-lactones	1.519	1.519	0.020	1.500	1.538	4	13
	in γ-lactones	1.512	1.512	0.015	1.501	1.521	110	12
	in δ-lactones	1.504	1.502	0.013	1.495	1.517	27	12
	cyclopropyl (**C**)–**C**=O in ketones, acids, and esters	1.486	1.485	0.018	1.474	1.497	105	7
	C*–C(=O)(–NH$_2$) in acyclic amides	1.514	1.512	0.016	1.506	1.526	32	14
	C*–C(=O)(–NHC*) in acyclic amides	1.506	1.505	0.012	1.498	1.515	78	14
	C*–C(=O)[–N(C*)$_2$] in acyclic amides	1.505	1.505	0.011	1.496	1.517	15	14
Csp^3–C*ar*	**CH$_3$–C*ar***	1.506	1.507	0.011	1.501	1.513	454	
	C#–**CH$_2$–C*ar***	1.510	1.510	0.009	1.505	1.516	674	
	(C#)$_2$–**CH–C*ar***	1.515	1.515	0.011	1.508	1.522	363	
	(C#)$_3$–**C–C*ar***	1.527	1.530	0.016	1.517	1.539	308	
	C*–C*ar* (overall)	1.513	1.513	0.014	1.505	1.521	1813	
	cyclopropyl (**C**)–**C*ar***	1.490	1.490	0.015	1.479	1.503	90	7
Csp^3–Csp^1	**C*–C**≡C	1.466	1.465	0.010	1.460	1.469	21	15
	C#–C≡C	1.472	1.472	0.012	1.464	1.481	88	15
	C*–C≡N	1.470	1.469	0.013	1.463	1.479	106	7b
	cyclopropyl (**C**)–**C**≡N	1.444	1.447	0.010	1.436	1.451	38	7
Csp^2–Csp^2	C=**C–C**=C							
	(conjugated)	1.455	1.455	0.011	1.447	1.463	30	16,18
	(unconjugated)	1.478	1.476	0.012	1.470	1.479	8	17,18
	(overall)	1.460	1.460	0.015	1.450	1.470	38	
	C=**C–C**=**C–C**=C	1.443	1.445	0.013	1.431	1.454	29	18
	C=**C–C**=C (endocyclic in TCNQ)	1.432	1.433	0.012	1.424	1.441	280	19
	C=**C–C**(=O)(–C*)							
	(conjugated)	1.464	1.462	0.018	1.453	1.476	211	16,18
	(unconjugated)	1.484	1.486	0.017	1.475	1.497	14	17,18
	(overall)	1.465	1.462	0.018	1.453	1.478	226	
	C=**C–C**(=O)–**C**=C							
	in benzoquinone (C,H-subst. only)	1.478	1.476	0.011	1.469	1.488	28	

Bond	Substructure	*d*	*m*	*σ*	q_l	q_u	*n*	Note
	in benzoquinone (any subst.)	1.478	1.478	0.031	1.464	1.498	172	
	non-quinonoid	1.456	1.455	0.012	1.447	1.464	28	
	C=**C**–**C**OOH	1.475	1.476	0.015	1.461	1.488	22	
	C=**C**–**C**OOC*	1.488	1.489	0.014	1.478	1.497	113	
	C=**C**–**C**OO$^-$	1.502	1.499	0.017	1.488	1.510	11	
	HOO**C**–**C**OOH	1.538	1.537	0.007	1.535	1.541	9	
	HOO**C**–**C**OO$^-$	1.549	1.552	0.009	1.546	1.553	13	
	$^-$OO**C**–**C**OO$^-$	1.564	1.559	0.022	1.554	1.568	9	
	formal Csp^2–Csp^2 single bond in selected non-fused heterocycles:							
	in 1*H*-pyrrole (C3–C4)	1.412	1.410	0.016	1.401	1.427	29	
	in furan (C3–C4)	1.423	1.423	0.016	1.412	1.433	62	
	in thiophene (C3–C4)	1.424	1.425	0.015	1.415	1.433	40	
	in pyrazole (C3–C4)	1.410	1.412	0.016	1.400	1.418	20	
	in isoxazole (C3–C4)	1.425	1.425	0.016	1.413	1.438	9	
	in furazan (C3–C4)	1.428	1.427	0.007	1.422	1.435	6	
	in furoxan (C3–C4)	1.417	1.417	0.006	1.412	1.422	14	
Csp^2–*Car*	C=**C**–**C***ar*							
	(conjugated)	1.470	1.470	0.015	1.463	1.480	37	16,18
Csp^2–*Car*		1.488	1.490	0.012	1.480	1.496	87	17,18
	(overall)	1.483	1.483	0.015	1.472	1.494	124	
	cyclopropenyl (C=**C**)–**C***ar*	1.447	1.448	0.006	1.441	1.452	8	10
	C*ar*–**C**(=O)–C*	1.488	1.489	0.016	1.478	1.500	84	
	C*ar*–**C**(=O)–**C***ar*	1.480	1.481	0.017	1.468	1.494	58	
	C*ar*–**C**OOH	1.484	1.485	0.014	1.474	1.491	75	
	C*ar*–**C**(=O)(–OC*)	1.487	1.487	0.012	1.480	1.494	218	
	C*ar*–**C**OO$^-$	1.504	1.509	0.014	1.495	1.512	26	
	C*ar*–**C**(–O)–NH$_2$	1.500	1.503	0.020	1.498	1.510	19	
	C*ar*–**C**=N–C#							
	(conjugated)	1.476	1.478	0.014	1.466	1.486	27	16
	(unconjugated)	1.491	1.490	0.008	1.485	1.496	48	17
	(overall)	1.485	1.487	0.013	1.481	1.493	75	
	in indole (C3–C3a)	1.434	1.434	0.011	1.428	1.439	40	
Csp^2–Csp^1	C=**C**–**C**≡C	1.431	1.427	0.014	1.425	1.441	11	7b
	C=**C**–**C**≡N in TCNQ	1.427	1.427	0.010	1.420	1.433	280	19
Car–*Car*	in biphenyls (*ortho* subst. all **H**)	1.487	1.488	0.007	1.484	1.493	30	
	(≥1 non-**H** *ortho*-subst.)	1.490	1.491	0.010	1.486	1.495	212	
Car–Csp^1	**C***ar*–**C**≡C	1.434	1.436	0.006	1.430	1.437	37	
	C*ar*–**C**≡N	1.443	1.444	0.008	1.436	1.448	31	
Csp^1–Csp^1	C≡**C**–**C**=C	1.377	1.378	0.012	1.374	1.384	21	
Csp^2=Csp^2	C*–CH=CH$_2$	1.299	1.300	0.027	1.280	1.311	42	
	(**C***)$_2$–**C**=CH$_2$	1.321	1.321	0.013	1.313	1.328	77	
	C*–**CH**=**CH**–C*							
	(*cis*)	1.317	1.318	0.013	1.310	1.323	106	
	(*trans*)	1.312	1.311	0.011	1.304	1.320	19	
	(overall)	1.316	1.317	0.015	1.309	1.323	127	
	(C*)$_2$–**C**=**CH**–C*	1.326	1.328	0.011	1.319	1.334	168	
	(C*$_2$–**C**=**C**–(C*)$_2$	1.331	1.330	0.009	1.326	1.334	89	
	(C*,H)$_2$–**C**=**C**–(C*,H)$_2$ (overall)	1.322	1.323	0.014	1.315	1.331	493	5
	in cyclopropene (any subst.)	1.294	1.288	0.017	1.284	1.302	10	10
	in cyclobutene (any subst.)	1.335	1.335	0.019	1.324	1.347	25	8
	in cyclopentene (C,H-subst.)	1.323	1.324	0.013	1.314	1.331	104	
	in cyclohexene (C,H-subst.)	1.326	1.325	0.012	1.318	1.334	196	
	C=C=C (allenes, any subst.)	1.307	1.307	0.005	1.303	1.310	18	
	C=C–**C=C** (C,**H** subst., conjugated)	1.330	1.330	0.014	1.322	1.338	76	16
	C=C–**C=C**–**C=C** (C,**H** subst., conjugated)	1.345	1.345	0.012	1.337	1.350	58	16
	C=C–*Car* (C,H subst., conjugated)	1.339	1.340	0.011	1.334	1.346	124	16
	C=C in cyclopenta-1,3-diene (any subst.)	1.341	1.341	0.017	1.328	1.356	18	
	C=C in cyclohexa-1,3-diene (any subst.)	1.332	1.332	0.013	1.323	1.341	56	
	in **C=C**–C=O							
	(C,**H** subst., conjugated)	1.340	1.340	0.013	1.332	1.348	211	16,18
	(C,**H** subst., unconjugated)	1.331	1.330	0.008	1.326	1.339	14	17,18

Bond	Substructure	d	m	σ	q_l	q_u	n	Note
	(C,**H** subst., overall)	1.340	1.339	0.013	1.332	1.348	226	
	in cyclohexa-2,5-dien-1-ones	1.329	1.327	0.011	1.321	1.335	28	
	in *p*-benzoquinones							
	(C*,**H** subst.)	1.333	1.337	0.011	1.325	1.338	14	
	(any subst.)	1.349	1.339	0.030	1.330	1.364	86	
	in TCNQ							
	(endocyclic)	1.352	1.353	0.010	1.345	1.358	142	19
	(exocyclic)	1.392	1.391	0.017	1.379	1.405	139	19
	C=**C**–OH in enol tautomers	1.362	1.360	0.020	1.349	1.370	54	
	in heterocycles (any subst.):							
	1*H*-pyrrole (C2–C3, C4–C5)	1.375	1.377	0.018	1.361	1.388	58	
	furan (C2–C3, C4–C5)	1.341	1.342	0.021	1.329	1.351	125	
	thiophene (C2–C3, C4–C5)	1.362	1.359	0.025	1.346	1.377	60	
	pyrazole (C4–C5)	1.369	1.372	0.019	1.362	1.383	20	
	imidazole (C4–C5)	1.360	1.361	0.014	1.352	1.367	44	
	isoxazole (C4–C5)	1.341	1.336	0.012	1.331	1.355	9	
	indole (C2–C3)	1.364	1.363	0.012	1.355	1.371	40	
Car≃ *Car*	in phenyl rings with C*,**H** subst. only							
	H–**C**≃**C**–H	1.380	1.381	0.013	1.372	1.388	2191	
	C*–**C**≃**C**–H	1.387	1.388	0.010	1.382	1.393	891	
	C*–**C**≃**C**–C*	1.397	1.397	0.009	1.392	1.403	182	
	C≃**C** (overall)	1.384	1.384	0.013	1.375	1.391	3264	
	F–**C**≃**C**–F	1.372	1.374	0.011	1.366	1.380	84	4
	Cl–**C**≃**C**–Cl	1.388	1.389	0.014	1.380	1.398	152	4
	in naphthalene (D_{2h}, any subst.)							
	C1–C2	1.364	1.364	0.014	1.356	1.373	440	
	C2–C3	1.406	1.406	0.014	1.397	1.415	218	
	C1–C8a	1.420	1.419	0.012	1.412	1.426	440	
	C4a–C8a	1.422	1.424	0.011	1.417	1.429	109	
Car≃ *Car*	in anthracene ($D_{2h,}$ any subst.)							
	C1–C2	1.356	1.356	0.009	1.350	1.360	56	
	C2–C3	1.410	1.410	0.010	1.401	1.416	34	
	C1–C9a	1.430	1.430	0.006	1.426	1.434	56	
	C4a–C9a	1.435	1.436	0.007	1.429	1.440	34	
	C9–C9a	1.400	1.402	0.009	1.395	1.406	68	
	in pyridine (C,H subst.)	1.379	1.381	0.012	1.371	1.387	276	20
	(any subst.)	1.380	1.380	0.015	1.371	1.389	537	20
	in pyridinium cation							
	(N^+–H; C,H subst. on C)							
	C2–C3	1.373	1.375	0.012	1.368	1.380	30	
	C3–C4	1.379	1.380	0.011	1.371	1.388	30	
	(N^+–X; C,H subst. on C)							
	C2–C3	1.373	1.372	0.019	1.362	1.382	151	
	C3–C4	1.383	1.385	0.019	1.372	1.394	151	
	in pyrazine (H subst. on C)	1.379	1.377	0.010	1.370	1.388	10	
	(any subst. on C)	1.405	1.405	0.024	1.388	1.420	60	
	in pyrimidine (C,H subst. on C)	1.387	1.389	0.018	1.379	1.400	28	
Csp^1≡Csp^1	X–**C**≡**C**–X	1.183	1.183	0.014	1.174	1.193	119	15
	C,H–**C**≡**C**–C,H	1.181	1.181	0.014	1.173	1.192	104	15
	in **C**≡**C**–C(sp^2,*ar*)	1.189	1.193	0.010	1.181	1.195	38	15
	in **C**≡**C**–**C**≡**C**	1.192	1.192	0.010	1.187	1.197	42	15
	in **CH**≡**C**–C#	1.174	1.174	0.011	1.167	1.180	42	15
Csp^3–Cl	Omitting 1,2-dichlorides:							
	C–$\mathbf{CH_2}$–**Cl**	1.790	1.790	0.007	1.783	1.795	13	4
	C_2–**CH**–**Cl**	1.803	1.802	0.003	1.800	1.807	8	4
	C_3–**C**–**Cl**	1.849	1.856	0.011	1.837	1.858	5	4
	X–CH_2–Cl (X = C,H,N,O)	1.790	1.791	0.011	1.783	1.797	37	4
	X_2–CH–Cl (X = C,H,N,O)	1.805	1.803	0.014	1.800	1.812	26	4
	X_3–**C**–**Cl** (X = C,H,N,O)	1.843	1.838	0.014	1.835	1.858	7	4
	X_2–**C**–$\mathbf{Cl_2}$ (X = C,H,N,O)	1.779	1.776	0.015	1.769	1.790	18	4
	X–**C**–$\mathbf{Cl_3}$ (X = C,H,N,O)	1.768	1.765	0.011	1.761	1.776	33	4

Bond	Substructure	*d*	*m*	*σ*	q_1	q_u	*n*	Note
	Cl–CH(–C)–**CH**(–C)–**Cl**	1.793	1.793	0.013	1.786	1.800	66	4
	Cl–C(–C_2)–**C**(–C_2)–**Cl**	1.762	1.760	0.010	1.757	1.765	54	4
	cyclopropyl–Cl	1.755	1.756	0.011	1.749	1.763	64	
Csp^2–Cl	C=**C–Cl** (C,H,N,O subst. on C)	1.734	1.729	0.019	1.719	1.748	63	4
	C=**C–Cl_2** (C,H,N,O subst. on C)	1.720	1.716	0.013	1.708	1.729	20	4
	Cl–C=**C–Cl**	1.713	1.711	0.011	1.705	1.720	80	4
Car–Cl	***Car*–Cl** (mono–Cl + m,p-Cl_2)	1.739	1.741	0.010	1.734	1.745	340	4
	***Car*–Cl** (*o*–Cl_2)	1.720	1.720	0.010	1.713	1.717	364	4
Csp^1Cl	see HCLENE10 (1.634, 1.646)							
Csp^3–F	Omitting 1,2-difluorides							
	C–CH_2–**F** and C_2–CH–**F**	1.399	1.399	0.017	1.389	1.408	25	4
	C_3–C–F	1.428	1.431	0.009	1.421	1.435	11	4
	(C*,H)$_2$–**C–F_2**	1.349	1.347	0.012	1.342	1.356	58	4
	C*–**C–F_3**	1.336	1.334	0.007	1.330	1.344	12	4
	F–C*–C*–F	1.371	1.374	0.007	1.362	1.375	26	4
	X_3–**C–F** (X = C,H,N,O)	1.386	1.389	0.033	1.373	1.408	70	4
	X_2–**C–F_2** (X = C,H,N,O)	1.351	1.349	0.013	1.342	1.356	58	4
	X–**C–F_3** (X = C,H,N,O)	1.322	1.323	0.015	1.314	1.332	309	4
	F–C(–X)$_2$–**C**(–X)$_2$–**F** (X = C,H,N,O)	1.373	1.374	0.009	1.362	1.377	30	4
	F–C(–X)$_2$–NO_2 (X = any subst.)	1.320	1.319	0.009	1.312	1.327	18	
Csp^2–F	C=**C–F** (C,H,N,O subst. on C)	1.340	1.340	0.013	1.334	1.346	34	4
Car–F	***Car*–F** (mono-F +*m*,*p*-F_2)	1.363	1.362	0.008	1.357	1.368	38	4
	***Car*–F** (*o*-F_2)	1.340	1.340	0.009	1.336	1.344	167	4
Csp^3–H	C–**C–H_3** (methyl)	1.059	1.061	0.030	1.039	1.083	83	21
	C_2–**C–H_2** (primary)	1.092	1.095	0.013	1.088	1.099	100	21
	C_3–**C–H** (secondary)	1.099	1.097	0.004	1.095	1.103	14	21
	$C_{2,3}$–**C–H** (primary and secondary)	1.093	1.095	0.012	1.089	1.100	118	21
	X–**C–H_3** (methyl)	1.066	1.074	0.028	1.049	1.087	160	21
	X_2–**C–H_2** (primary)	1.092	1.095	0.012	1.088	1.099	230	21
	X_3–**C–H** (secondary)	1.099	1.099	0.007	1.095	1.103	117	21
	$X_{2,3}$–**C–H** (primary and secondary)	1.094	1.096	0.011	1.091	1.100	348	21
Csp^2–H	C–C=**C–H**	1.077	1.079	0.012	1.074	1.085	14	21
Car–H	***Car*–H**	1.083	1.083	0.011	1.080	1.087	218	21
Csp^3–I	**C*–I**	2.162	2.159	0.015	2.149	2.179	15	4
Car–I	***Car*–I**	2.095	2.095	0.015	2.089	2.104	51	4
Csp^3–N(4)	**C*–NH_3^+**	1.488	1.488	0.013	1.482	1.495	298	
	(C*)$_2$–NH_2^+	1.494	1.493	0.016	1.484	1.503	249	
	(C*)$_3$–NH^+	1.502	1.502	0.015	1.491	1.512	509	
	(C*)$_4$–N^+	1.510	1.509	0.020	1.496	1.523	319	
	C*–N^+ (overall)	1.499	1.498	0.018	1.488	1.510	1370	
Csp^3–N(3)	**C*–N^+** in N-subst. pyridinium	1.485	1.484	0.009	1.477	1.490	32	
	C*–NH_2 (Nsp^3: pyramidal)	1.469	1.470	0.010	1.462	1.474	19	22
	(C*)$_2$–NH (Nsp^3: pyramidal)	1.469	1.467	0.012	1.461	1.477	152	5,22
	(C*)$_3$–**N** (Nsp^3: pyramidal)	1.469	1.468	0.014	1.460	1.476	1042	5,22
	C*–Nsp^3 (overall)	1.469	1.468	0.014	1.460	1.476	1201	
	Csp^3–Nsp^3							
	in aziridine	1.472	1.471	0.016	1.464	1.482	134	
	in azetidine	1.484	1.481	0.018	1.472	1.495	21	
	in tetrahydropyrrole	1.475	1.473	0.016	1.464	1.483	66	
	in piperidine	1.473	1.473	0.013	1.460	1.479	240	
	Csp^3–Nsp^2 (N planar) in:							23
	acyclic amides **C*–NH**–C=O	1.454	1.451	0.011	1.446	1.461	78	14
	β-lactams **C*–N**(–X)–C=O (endo)	1.464	1.465	0.012	1.458	1.475	23	13
	γ-lactams							
	C*–NH–C=O (endo)	1.457	1.458	0.011	1.449	1.465	20	13
	C*–N(–C*)–C=O (endo)	1.462	1.461	0.010	1.453	1.466	15	13
	C*–N(–C*)–C=O (exo)	1.458	1.456	0.014	1.448	1.465	15	13
	δ-lactams							
	C*–NH–C=O (endo)	1.478	1.472	0.016	1.467	1.491	6	14
	C*–N(–C*)–C=O (endo)	1.479	1.476	0.007	1.475	1.482	15	14
	C*–N(–C*)–C=O (exo)	1.468	1.471	0.009	1.462	1.477	15	14

Bond	Substructure	*d*	*m*	*σ*	q_l	q_u	*n*	Note
	nitro compounds (1,2-dinitro omitted):							
	C–**CH$_2$–NO$_2$**	1.485	1.483	0.020	1.478	1.502	8	
	C$_2$–**CH–NO$_2$**	1.509	1.509	0.011	1.502	1.511	12	
	C$_3$–**C–NO$_2$**	1.533	1.533	0.013	1.530	1.539	17	
	C$_2$–**C**–(**NO$_2$**)$_2$	1.537	1.536	0.016	1.525	1.550	19	
	1,2-dinitro: **NO$_2$–C***–**C***–**NO$_2$**	1.552	1.550	0.023	1.536	1.572	32	
C*sp*^{3}–N(2)	C#–**N**=N	1.493	1.493	0.020	1.477	1.506	54	
	C*–N=C–C*ar*	1.465	1.468	0.011	1.461	1.472	75	
C*sp*^{2}–N(3)	C=**C–NH$_2$**N*sp*^{2} planar	1.336	1.344	0.017	1.317	1.348	10	23
	C=**C–NH–C**# N*sp*^{2} planar	1.339	1.340	0.016	1.327	1.351	17	23
	C=**C–N**–(C#)$_2$							
	N*sp*^{2} planar	1.355	1.358	0.014	1.341	1.363	22	23
	N*sp*^{3} pyramidal	1.416	1.418	0.018	1.397	1.432	18	22
	C*sp*^{2}–N*sp*^{2} (N planar) in:							23
	acyclic amides							
	NH$_2$–C=O	1.325	1.323	0.009	1.318	1.331	32	14
	C*–**NH–C**=O	1.334	1.333	0.011	1.326	1.343	78	14
	(C*)$_2$–**N–C**=O	1.346	1.342	0.011	1.339	1.356	5	14
	β-lactams C*–**NH–C**=O	1.385	1.388	0.019	1.374	1.396	23	13
	γ-lactams							
	C*–**NH–C**=O	1.331	1.331	0.011	1.326	1.337	20	13
	C*–N(–C*)–**C**=O	1.347	1.344	0.014	1.335	1.359	15	13
	δ-lactams							
	C*–**NH–C**=O	1.334	1.334	0.006	1.330	1.339	6	14
	(C*)–N(–C*)–**C**=O	1.352	1.353	0.010	1.344	1.356	15	14
	peptides C#–**N**(–X)–**C**(–C#)(=O)	1.333	1.334	0.013	1.326	1.340	380	24
	ureas							
	(**NH$_2$**)$_2$–**C**=O	1.334	1.334	0.008	1.329	1.339	48	25,26
	(C#–**NH**)$_2$–**C**=O	1.347	1.345	0.010	1.341	1.354	26	25
	[(C#)$_n$–**N**]$_2$–**C**=O	1.363	1.359	0.014	1.354	1.370	40	25,27
	thioureas	1.346	1.343	0.023	1.328	1.361	192	
	(X$_2$**N**)$_2$–**C**=S							
	imides							
	[C#–**C**(=O)]$_2$–**NH**	1.376	1.377	0.012	1.369	1.383	64	
	[C#–**C**(=O)]$_2$–**N**–C#	1.389	1.383	0.017	1.376	1.404	38	
	[C*sp*^{2}–**C**(=O)]$_2$–**N**–C#	1.396	1.396	0.010	1.389	1.403	46	
	[C*sp*^{2}–**C**(=O)]$_2$–**N**–C*sp*^{2}	1.409	1.406	0.020	1.391	1.419	28	
	guanidinium [**C**–(**NH$_2$**)$_3$]$^+$ (unsubst.)	1.321	1.320	0.008	1.314	1.327	39	
	(any subst.)	1.328	1.325	0.015	1.317	1.333	140	
	in heterocyclic systems (any subst.)							
	1*H*-pyrrole (N1–C2, N1–C5)	1.372	1.374	0.016	1.363	1.384	58	
	indole (N1–C2)	1.370	1.370	0.012	1.364	1.377	40	
	pyrazole (N1–C5)	1.357	1.359	0.012	1.347	1.365	20	
	imidazole (N1–C2)	1.349	1.349	0.018	1.338	1.358	44	
	imidazole (N1–C5)	1.370	1.370	0.010	1.365	1.377	44	
C*sp*^{2}–N(2)	in imidazole (N3–C4)	1.376	1.377	0.011	1.369	1.384	44	
C*ar*–N(4)	**C*ar*–N**$^+$–(C,H)$_3$	1.465	1.466	0.007	1.461	1.470	23	
C*ar*–N(3)	**C*ar*–NH$_2$**							
	(N*sp*^{2}: planar)	1.355	1.360	0.020	1.340	1.372	33	23
	(N*sp*^{3}: pyramidal)	1.394	1.396	0.011	1.385	1.403	25	22
	(overall)	1.375	1.377	0.025	1.363	1.394	98	28
C*ar*–N(3)	**C*ar*–NH**–C#							
	(N*sp*^{2}: planar)	1.353	1.353	0.007	1.347	1.359	16	23
	(N*sp*^{3}: pyramidal)	1.419	1.423	0.017	1.412	1.432	8	22
	(overall)	1.380	1.364	0.032	1.353	1.412	31	28
	C*ar*–N–(C#)$_2$							
	(N*sp*^{2}: planar)	1.371	1.370	0.016	1.363	1.382	41	23
	(N*sp*^{3}: pyramidal)	1.426	1.425	0.011	1.421	1.431	22	22
	(overall)	1.390	1.385	0.030	1.366	1.420	69	28
	in indole (N1–C7a)	1.372	1.372	0.007	1.367	1.376	40	
	C*ar*–NO$_2$	1.468	1.469	0.014	1.460	1.476	556	

Bond	Substructure	*d*	*m*	*σ*	q_l	q_u	*n*	Note
Car–N(2)	*Car*–**N**=N	1.431	1.435	0.020	1.422	1.442	26	
Csp^2=N(3)	in furoxan ($^+$N2=C3)	1.316	1.316	0.009	1.311	1.324	14	
Csp^2=N(2)	*Car*–**C=N**–C#	1.279	1.279	0.008	1.275	1.285	75	
	(C,H)$_2$–**C=N**–OH in oximes	1.281	1.280	0.013	1.273	1.288	67	
	S–**C=N**–X	1.302	1.302	0.021	1.285	1.319	36	
	in pyrazole (N2=C3)	1.329	1.331	0.014	1.315	1.339	20	
	in imidazole (C2=N3)	1.313	1.314	0.011	1.307	1.319	44	
	in isoxazole (N2=C3)	1.314	1.315	0.009	1.305	1.320	9	
	in furazan (N2=C3, C4=N5)	1.298	1.299	0.006	1.294	1.303	12	
	in furoxan (C4=N5)	1.304	1.306	0.008	1.300	1.308	14	
Car≃ N(3)	**C≃N**$^+$–H (pyrimidinium)	1.335	1.334	0.015	1.325	1.342	30	
	C≃N$^+$–C* (pyrimidinium)	1.346	1.346	0.010	1.340	1.352	64	
	C≃N$^+$–O$^-$ (pyrimidinium)	1.362	1.359	0.013	1.353	1.369	56	
Car≃ N(2)	**C≃N** (pyridine)	1.337	1.338	0.012	1.330	1.344	269	
	C≃N (pyrazine)	1.336	1.335	0.022	1.319	1.347	120	
	C≃N≃C (pyrimidine)	1.339	1.338	0.015	1.333	1.342	28	
	N≃C≃N (pyrimidine)	1.333	1.335	0.013	1.326	1.337	28	
	C≃N (pyrimidine) (overall) in any 6-membered N-containing aromatic ring:	1.336	1.337	0.014	1.331	1.339	56	
	H–C≃N≃C–H	1.334	1.334	0.014	1.327	1.341	146	
	H–C≃N≃C–C*	1.339	1.341	0.013	1.336	1.345	38	
	C*–C≃N≃C–C*	1.345	1.345	0.008	1.342	1.348	24	
	C≃N≃C (overall)	1.336	1.337	0.014	1.329	1.344	204	
Csp^1≡N(2)	X–S–N≡C$^-$ (isothiocyanide)	1.144	1.147	0.006	1.140	1.148	6	
Csp^1≡N(1)	C*–**C≡N**	1.136	1.137	0.010	1.131	1.142	140	
	C=C–**C≡N** in TCNQ	1.144	1.144	0.008	1.139	1.149	284	19
	Car–**C≡N**	1.138	1.138	0.007	1.133	1.143	31	
	X–**C≡N**	1.144	1.141	0.012	1.138	1.151	10	
	(S–**C≡N**)$^-$	1.155	1.156	0.012	1.147	1.165	14	
Csp^3–O(2)	in alcohols							
	CH$_3$–OH	1.413	1.414	0.018	1.395	1.425	17	
	C–**CH$_2$–OH**	1.426	1.426	0.011	1.420	1.431	75	
	C$_2$–**CH–OH**	1.432	1.431	0.011	1.425	1.439	266	
	C$_3$–**C–OH**	1.440	1.440	0.012	1.432	1.449	106	
	C*–OH (overall)	1.432	1.431	0.013	1.424	1.441	464	
	in dialkyl ethers							29
	CH$_3$–O–C*	1.416	1.418	0.016	1.405	1.426	110	
	C–**CH$_2$–O**–C*	1.426	1.424	0.011	1.418	1.435	34	
	C$_2$–**CH–O**–C*	1.429	1.430	0.010	1.420	1.437	53	
	C$_3$–**C–O**–C*	1.452	1.450	0.011	1.445	1.458	39	
	C*–O–C* (overall)	1.426	1.425	0.019	1.414	1.437	236	5
	in aryl alkyl ethers							29
	CH$_3$–O–*Car*	1.424	1.424	0.012	1.417	1.431	616	
	C–**CH$_2$–O**–*Car*	1.431	1.430	0.013	1.422	1.438	188	
	C$_2$–**CH–O**–*Car*	1.447	1.446	0.020	1.435	1.466	58	
	C$_3$–**C–O**–*Car*	1.470	1.469	0.018	1.456	1.483	55	
	C*–O–*Car* (overall)	1.429	1.427	0.018	1.419	1.436	917	
	in alkyl esters of carboxylic acids							12,29
	CH$_3$–O–C(=O)–C*	1.448	1.449	0.010	1.442	1.455	200	
	C–CH$_2$–O–C(=O)–C*	1.452	1.453	0.009	1.445	1.458	32	
	C$_2$–**CH–O**–C(=O)–C*	1.460	1.460	0.010	1.454	1.465	78	
	C$_3$–**C–O**–C(=O)–C*	1.477	1.475	0.008	1.472	1.484	6	
	C*–O–C(=O)–C* (overall)	1.450	1.451	0.014	1.442	1.459	314	
	in alkyl esters of α,β-unsaturated acids:							
	C*–O–C(=O)–C=C (overall)	1.453	1.452	0.013	1.444	1.459	112	
	in alkyl esters of benzoic acid							
	C*–O–C(=O)–C (phenyl) (overall)	1.454	1.454	0.012	1.446	1.463	219	
	in ring systems							
	oxirane (epoxides) (any subst.)	1.446	1.446	0.014	1.438	1.456	498	9
	oxetane (any subst.)	1.463	1.460	0.015	1.451	1.474	16	
	tetrahydrofuran (C,H subst.)	1.442	1.441	0.017	1.430	1.451	154	
Csp^3–O(2)	tetrahydropyran (C,H subst.)	1.441	1.442	0.015	1.431	1.451	22	

Bond	Substructure	*d*	*m*	σ	q_l	q_u	*n*	Note
	β-lactones: **C*–O**–C(=O)	1.492	1.494	0.010	1.481	1.501	4	16
	γ-lactones: **C*–O**–C(=O)	1.464	1.464	0.012	1.455	1.473	110	12
	δ-lactones: **C*–O**–C(=O)	1.461	1.464	0.017	1.452	1.473	27	12
	O–C–O system in *gem*-diols, and pyranose and furanose sugars:							30,31
	HO–C*–OH	1.397	1.401	0.012	1.388	1.405	18	
	C_5–O_5–C_1–O_1H in pyranoses							
	O_1 axial (α):							
	C_5–O_5	1.439	1.440	0.008	1.432	1.445	29	
	O_5–C_1	1.427	1.426	0.012	1.421	1.432	29	
	C_1–O_1	1.403	1.400	0.012	1.391	1.412	29	
	O_1 equatorial (β):							
	C_5–O_5	1.435	1.436	0.008	1.429	1.440	17	
	O_5–C_1	1.430	1.431	0.010	1.424	1.436	17	
	C_1–O_1	1.393	1.393	0.007	1.386	1.399	17	
	α + β (overall):							
	C_5–O_5	1.439	1.440	0.008	1.432	1.446	60	
	O_5–C_1	1.430	1.429	0.012	1.421	1.436	60	
	C_1–O_1	1.401	1.399	0.011	1.392	1.407	60	
	C_4–O_4–C_1–O_1H in furanoses (overall values)							
	C_4–O_4	1.442	1.446	0.012	1.436	1.449	18	
	O_4–C_1	1.432	1.432	0.012	1.421	1.443	18	
	C_1–O_1	1.404	1.405	0.013	1.397	1.409	18	
	C_5–O_5–C_1–O_1–C* in pyranoses							
	O_1 axial (α):							
	C_5–O_5	1.439	1.438	0.010	1.433	1.446	67	
	O_5–C_1	1.417	1.417	0.009	1.410	1.424	67	
	C_1–O_1	1.409	1.409	0.014	1.401	1.417	67	
	O_1–C*	1.435	1.435	0.013	1.427	1.443	67	
	O_1 equatorial (β):							
	C_5–O_5	1.434	1.435	0.006	1.429	1.439	39	
	O_5–C_1	1.424	1.424	0.008	1.418	1.431	39	
	C_1–O_1	1.390	1.390	0.011	1.381	1.400	39	
	O_1–C*	1.437	1.438	0.013	1.428	1.445	39	
	α + β (overall):							
	C_5–O_5	1.436	1.436	0.009	1.431	1.442	126	
	O_5–C_1	1.419	1.419	0.011	1.412	1.426	126	
	C_1–O_1	1.402	1.403	0.016	1.391	1.413	126	
	O_1–C*	1.436	1.436	0.013	1.428	1.445	126	
	C_4–O_4–C_1–O_1–C* in furanoses (overall values)							
	C_4–O_4	1.443	1.445	0.013	1.429	1.453	23	
	O_4–C_1	1.421	1.418	0.012	1.413	1.431	23	
	C_1–O_1	1.410	1.409	0.014	1.401	1.420	23	
	O_1–C*	1.439	1.437	0.014	1.429	1.449	23	
	Miscellaneous:							
	C#–O–SiX_3	1.416	1.416	0.017	1.405	1.428	29	
	C*–O–SO_2–C	1.465	1.461	0.014	1.454	1.475	33	
Csp^2–O(2)	in enols: C=**C–OH**	1.333	1.331	0.017	1.324	1.342	53	
	in enol esters: C=**C–O–C***	1.354	1.353	0.016	1.341	1.363	40	
	in acids:							
	C*–**C**(=O)–**OH**	1.308	1.311	0.019	1.298	1.320	174	
	C=C–**C**(=O)–**OH**	1.293	1.295	0.019	1.279	1.307	22	
	Car–**C**(=O)–**OH**	1.305	1.311	0.020	1.291	1.317	75	
	in esters:							
	C*–**C**(=O)–**O**–C*	1.336	1.337	0.014	1.328	1.346	551	12,29
	C=C–**C**(=O)–**O**–C*	1.332	1.331	0.011	1.324	1.339	112	
	Car–**C**(=O)–**O**–C*	1.337	1.335	0.013	1.329	1.344	219	12
	C*–**C**(=O)–**O**–C=C	1.362	1.359	0.018	1.351	1.374	26	
	C*–**C**(=O)–**O**–C=C	1.407	1.405	0.017	1.394	1.420	26	
	C*–**C**(=O)–**O**–*Car*	1.360	1.359	0.011	1.355	1.367	40	12
	in anhydrides: O=**C–O–C**=O	1.386	1.386	0.011	1.379	1.393	70	

Bond	Substructure	*d*	*m*	*σ*	q_l	q_u	*n*	Note
	in ring systems:							
	furan (O1–C2, O1–C5)	1.368	1.369	0.015	1.359	1.377	125	
	isoxazole (O1–C5)	1.354	1.354	0.010	1.345	1.360	9	
	β-lactones: C*–**C**(=O)–**O**–C*	1.359	1.359	0.013	1.348	1.371	4	13
	γ-lactones: C*–**C**(=O)–**O**–C*	1.350	1.349	0.012	1.342	1.359	110	12
	δ-lactones: C*–**C**(=O)–**O**–C*	1.339	1.339	0.016	1.332	1.347	27	12
Car–O(2)	in phenols: ***Car*–OH**	1.362	1.364	0.015	1.353	1.373	551	
	in aryl alkyl ethers: ***Car*–O**–C*	1.370	1.370	0.011	1.363	1.377	920	29,32
Car–O(2)	in diaryl ethers: ***Car*–O**–*Car*	1.384	1.381	0.014	1.375	1.391	132	
	in esters: ***Car*–O**–C(=O)–C*	1.401	1.401	0.010	1.394	1.408	40	12
Csp^2=O(1)	in aldehydes and ketones:							
	C*–**CH=O**	1.192	1.192	0.005	1.188	1.197	7	
	(C*)$_2$–**C=O**	1.210	1.210	0.008	1.206	1.215	474	5
	(C#)$_2$–**C=O**							
	in cyclobutanones	1.198	1.198	0.007	1.194	1.204	12	
	in cyclopentanones	1.208	1.208	0.007	1.203	1.212	155	
	in cyclohexanones	1.211	1.211	0.009	1.207	1.216	312	
	C=C–**C=O**	1.222	1.222	0.010	1.216	1.229	225	
	(C=C)$_2$–**C=O**	1.233	1.229	0.010	1.226	1.242	28	
	Car–**C=O**	1.221	1.218	0.014	1.212	1.229	85	
	(*Car*)$_2$–**C=O**	1.230	1.226	0.015	1.220	1.238	66	
	C=O in benzoquinones	1.222	1.220	0.013	1.211	1.231	86	
	delocalized double bonds in carboxylate anions:							
	H–**C≃O**$_2^-$ (formate)	1.242	1.243	0.012	1.234	1.252	24	
	C*–**C≃O**$_2^-$	1.254	1.253	0.010	1.247	1.261	114	
	C=C–**C≃O**$_2^-$	1.250	1.248	0.017	1.238	1.261	52	
	Car–**C≃O**$_2^-$	1.255	1.253	0.010	1.249	1.262	22	
	HOO**C–C≃O**$_2^-$ (hydrogen oxalate)	1.243	1.247	0.015	1.232	1.256	26	
	$^-$**O**$_2$**≃C–C≃O**$_2^-$ (oxalate)	1.251	1.251	0.007	1.248	1.254	18	
	in carboxylic acids (X–COOH)							
	C*–**C**(=**O**)–OH	1.214	1.214	0.019	1.203	1.224	175	
	C=C–**C**(=**O**)–OH	1.229	1.226	0.017	1.218	1.237	22	
	Car–**C**(=**O**)–OH	1.226	1.223	0.020	1.211	1.241	75	
	in esters:							
	C*–**C**(=**O**)–O–C*	1.196	1.196	0.010	1.190	1.202	551	12
	C=C–**C**(=**O**)–O–C*	1.199	1.198	0.009	1.193	1.203	113	
	Car–**C**(=**O**)–O–C*	1.202	1.201	0.009	1.196	1.207	218	12
	C*–**C**(=**O**)–O–C=C	1.190	1.190	0.014	1.184	1.198	26	
	C*–**C**(=**O**)–O–*Car*	1.187	1.188	0.011	1.181	1.195	40	12
	in anhydrides: **O=C**–O–**C=O**	1.187	1.187	0.010	1.184	1.193	70	
	in β-lactones: C*–**C**(=**O**)–O–C*	1.193	1.193	0.006	1.187	1.198	4	13
	γ-lactones: C*–**C**(=**O**)–O–C*	1.201	1.202	0.009	1.196	1.206	109	12
	δ-lactones: C*–**C**(=**O**)–O–C*	1.205	1.207	0.008	1.201	1.209	27	12
	in amides:							
	NH$_2$–**C**(–C*)=**O**	1.234	1.233	0.012	1.225	1.243	32	14
	(C*–)(C*,H–)N–**C**(–C*)=**O**	1.231	1.231	0.012	1.224	1.238	378	14
	β-lactams: C*–NH–**C=O**	1.198	1.200	0.012	1.193	1.204	23	13
	γ-lactams:							
	C*–NH–**C=O**	1.235	1.235	0.008	1.232	1.240	20	13
	C*–N(–C*)–C=O	1.225	1.226	0.011	1.217	1.233	15	13
	δ-lactams:							
	C*–NH–**C=O**	1.240	1.241	0.003	1.237	1.243	6	14
	C*–N(–C*)–**C=O**	1.233	1.233	0.007	1.229	1.239	15	14
	in ureas:							
	(NH)$_2$)$_2$–**C=O**	1.256	1.256	0.007	1.249	1.261	24	25,26
	(C#–NH)$_2$–**C=O**	1.241	1.237	0.011	1.235	1.245	13	25
	[(C#)$_n$–N]$_2$–**C=O**	1.230	1.230	0.007	1.224	1.234	20	25,27
Csp^3–P(4)	C$_3$–**P**$^+$–**C***	1.800	1.802	0.015	1.790	1.812	35	33
	C$_2$–**P**(=O)–**CH**$_3$	1.791	1.790	0.006	1.786	1.795	10	
	C$_2$–**P**(=O)–**CH**$_2$–C	1.806	1.806	0.009	1.801	1.813	45	
	C$_2$–**P**(=O)–**CH**–C$_2$	1.821	1.821	0.009	1.815	1.828	15	

Bond	Substructure	d	m	σ	q_1	q_u	n	Note
	C_2–**P**(=O)–**C**–C_3	1.841	1.842	0.008	1.835	1.847	14	
	C_2–**P**(=O)–**C*** (overall)	1.813	1.811	0.017	1.800	1.822	84	
*Csp*³–P(3)	C_2–**P–C***	1.855	1.857	0.019	1.840	1.870	23	
Car–P(4)	C_3–**P**⁺–***Car***	1.793	1.792	0.011	1.786	1.800	276	
	C_2–**P**(=O)–***Car***	1.801	1.802	0.011	1.796	1.807	98	
	Ph$_3$–**P**=N⁺=**P–Ph**$_3$	1.795	1.795	0.008	1.789	1.800	197	
Car–P(3)	C_2–**P–*Car***	1.836	1.837	0.010	1.830	1.844	102	
	(N≃)$_2$P–***Car*** (P ≃ N aromatic)	1.795	1.793	0.011	1.788	1.803	43	
*Csp*³–S(4)	**C*–SO**$_2$–C (C* = CH_3 excluded)	1.786	1.782	0.018	1.774	1.797	75	
	C*–SO$_2$–C (overall)	1.779	1.778	0.020	1.764	1.790	94	
	C*–SO$_2$–O–X	1.745	1.744	0.009	1.738	1.754	7	34
	C*–SO$_2$–N–X_2	1.758	1.756	0.018	1.746	1.773	17	34
*Csp*³–S(3)	**C*–S**(=O)–C (C* = CH_3 excluded)	1.818	1.814	0.024	1.802	1.829	69	
	C*–S(=O)–C (overall)	1.809	1.806	0.025	1.793	1.820	88	
	CH$_3$–**S**⁺–X_2	1.786	1.787	0.007	1.779	1.792	21	
	C*–S⁺–X_2 (C* = CH_3 excluded)	1.823	1.820	0.016	1.812	1.834	18	
	C*–S⁺–X_2 (overall)	1.804	1.794	0.025	1.788	1.820	41	
*Csp*³–S(2)	**C*–SH**	1.808	1.805	0.010	1.800	1.819	6	
	CH$_3$–**S**–C*	1.789	1.787	0.008	1.784	1.794	9	
*Csp*³–S(2)	C–**CH**$_2$–**S**–C*	1.817	1.816	0.013	1.808	1.824	92	
	C_2–**CH–S**–C*	1.819	1.819	0.011	1.811	1.825	32	
	C_3–**C–S**–C*	1.856	1.860	0.011	1.854	1.863	26	
	C*–S–C* (overall)	1.819	1.817	0.019	1.809	1.827	242	
	in thiirane	1.834	1.835	0.025	1.810	1.858	4	9
	in thiirane: see ZCMXSP (1.817, 1.844)							
	in tetrahydrothiophene	1.827	1.826	0.018	1.811	1.837	20	
	in tetrahydrothiopyran	1.823	1.821	0.014	1.812	1.832	24	
	C–**CH**$_2$–**S**–S–X	1.823	1.820	0.014	1.813	1.832	41	
	C_3–**C–S**–S–X	1.863	1.865	0.015	1.848	1.878	11	
	C*–**S–S**–X (overall)	1.833	1.828	0.022	1.818	1.848	59	
*Csp*²–S(2)	C=**C–S**–C*	1.751	1.755	0.017	1.740	1.764	61	
	C=**C–S**–C=C (in tetrathiafulvalene)	1.741	1.741	0.011	1.733	1.750	88	
	C=**C–S**–C=C (in thiophene)	1.712	1.712	0.013	1.703	1.722	60	
	O=**C–S**–C#	1.762	1.759	0.018	1.747	1.778	20	
Car–S(4)	***Car*–SO**$_2$–C	1.763	1.764	0.009	1.756	1.769	96	
	***Car*–SO**$_2$–O–X	1.752	1.750	0.008	1.749	1.756	27	
	***Car*–SO**$_2$–N–X_2	1.758	1.759	0.013	1.749	1.765	106	35
Car–S(3)	***Car*–S**(=O)–C	1.790	1.790	0.010	1.783	1.798	41	
	***Car*–S**⁺–X_2	1.778	1.779	0.010	1.771	1.787	10	
Car–S(2)	***Car*–S**–C*	1.773	1.774	0.009	1.765	1.779	44	
	***Car*–S**–*Car*	1.768	1.767	0.010	1.762	1.774	158	
	***Car*–S**–*Car* (in phenothiazine)	1.764	1.764	0.008	1.760	1.769	48	
	***Car*–S**–S–X	1.777	1.777	0.012	1.767	1.785	47	
*Csp*¹–S(2)	N≡**C–S**–X	1.679	1.683	0.026	1.645	1.698	10	
*Csp*¹–S(1)	(N≡**C–S**)⁻	1.630	1.630	0.014	1.619	1.641	14	
*Csp*²=S(1)	(C*)$_2$–**C=S**: see IPMUDS (1.599)							
	(*Car*)$_2$–**C=S**: see CELDOM (1.611)							
	(X)$_2$–**C=S** (X = C,N,O,S)	1.671	1.675	0.024	1.656	1.689	245	
	X_2N–**C**(=**S**)–S–X	1.660	1.660	0.016	1.648	1.674	38	
	(X_2N)$_2$–**C=S** (thioureas)	1.681	1.684	0.020	1.669	1.693	96	
	N–**C**(≃**S**)$_2$	1.720	1.721	0.012	1.709	1.731	20	
*Csp*³–Se	**C#–Se**	1.970	1.967	0.032	1.948	1.998	21	
*Csp*²–Se(2)	C=**C–Se–C**=C (in tetraselenafulvalene)	1.893	1.895	0.013	1.882	1.902	32	
Car–Se(3)	**Ph**$_3$–**Se**⁺	1.930	1.929	0.006	1.924	1.936	13	
*Csp*³–Si(5)	**C#–Si**⁻–X_4	1.874	1.876	0.015	1.859	1.884	9	
*Csp*³–Si(4)	**CH**$_3$–**Si**–X_3	1.857	1.857	0.018	1.848	1.869	552	
	C*–Si–X_3 (**C*** = CH_3 excluded)	1.888	1.887	0.023	1.872	1.905	124	
	C*–Si–X_3 (overall)	1.863	1.861	0.024	1.850	1.875	681	
Car–Si(4)	***Car*–Si**–X_3	1.868	1.868	0.014	1.857	1.878	178	
*Csp*¹–Si(4)	C≡**C–Si**–X_3	1.837	1.840	0.012	1.824	1.849	8	
*Csp*³–Te	**C#–Te**	2.158	2.159	0.030	2.128	2.177	13	

Bond	Substructure	*d*	*m*	*σ*	q_l	q_u	*n*	Note
Car–Te	*Car*–**Te**	2.116	2.115	0.020	2.104	2.130	72	
Csp^2=Te	see CEDCUJ (2.044)							
Cl–Cl	see PHASCL (2.306, 2.227)							
Cl–I	see CMBIDZ (2.563), HXPASC (2.541, 2.513), METAMM (2.552), BQUINI (2.416, 2.718)							
Cl–N	see BECTAE (1.743–1.757), BOGPOC (1.705)							
Cl–O(1)	in ClO^-_4	1.414	1.419	0.026	1.403	1.431	252	
Cl–P	$(N\simeq)_2$**P–Cl** (N ≃ P aromatic)	1.997	1.994	0.015	1.989	2.004	46	
	Cl–P (overall)	2.008	2.001	0.035	1.986	2.028	111	
Cl–S	**Cl–S** (overall)	2.072	2.079	0.023	2.047	2.091	6	
	see also longer bonds in CILSAR (2.283), BIHXIZ (2.357), CANLUY (2.749)							
Cl–Se	see BIRGUE10, BIRHAL10, CTCNSE(2.234–2.851)							
Cl–Si(4)	**Cl–Si**–X_3 (monochloro)	2.072	2.075	0.009	2.066	2.078	5	
	$\mathbf{Cl}_2$–**Si**–X_2 and $\mathbf{Cl}_3$–**Si**–X	2.020	2.012	0.015	2.007	2.036	5	
Cl–Te	Cl–Te in range 2.34–2.60	2.520	2.515	0.034	2.493	2.537	22	36
	see also longer bonds in BARRIV, BOJPUL, CETUTE, EPHTEA, OPNTEC10 (2.73–2.94)							
F–N(3)	**F–N**–C_2 and $\mathbf{F}_2$–**N**–C	1.406	1.404	0.016	1.395	1.416	9	
F–P(6)	in hexafluorophosphate, PF^-_6	1.579	1.587	0.025	1.563	1.598	72	
F–P(3)	$(N\simeq)_2$**P–F**(N ≃ P aromatic)	1.495	1.497	0.016	1.481	1.510	10	
F–S	43 observations in range 1.409–1.770 in a wide variety of environments; F–S(6) in F_2–SO_2–C_2 (see FPSULF10, BETJOZ)	1.640	1.646	0.011	1.626	1.649	6	
	F–S(4) in $\mathbf{F}_2$–**S**(=O)–N (see BUDTEZ)	1.527	1.528	0.004	1.524	1.530	24	37
F–Si(6)	in SiF^{2-}_6	1.694	1.701	0.013	1.677	1.703	6	
F–Si(5)	**F–Si**$^-$–X_4	1.636	1.639	0.035	1.602	1.657	10	
F–Si(4)	**F–Si**–X_3	1.588	1.587	0.014	1.581	1.599	24	
F–Te	see CUCPIZ (F–Te(6) = 1.942, 1.937), FPHTEL(F–Te(4) = 2.006)							
H–N(4)	X_3–**N**$^+$–**H**	1.033	1.036	0.022	1.026	1.045	87	21
H–N(3)	X_3–**N–H**	1.009	1.010	0.019	0.997	1.023	95	21
H–O(2)	in alcohols C*–**O–H**	0.967	0.969	0.010	0.959	0.974	63	21
	C#–**O–H**	0.967	0.970	0.010	0.959	0.974	73	21
	in acids O=C–**O–H**	1.015	1.017	0.017	1.001	1.031	16	21,38
I–I	in I^-_3	2.917	2.918	0.011	2.907	2.927	6	
I–N	see BZPRIB, CMBIDZ, HMTITI, HMTNTI, IFORAM, IODMAM (2.042–2.475)							
I–O	X–**I–O** (see BZPRIB, CAJMAB, IBZDAC11)	2.144	2.144	0.028	2.127	2.164	6	
	for IO^-_6 see BOVMEE (1.829–1.912)							
I–P(3)	see CEHKAB (2.490–2.493)							†
I–S	sec DTHIBR10 (2.687), ISUREA10 (2.629), BZTPPI (3.251)							
I–Te(4)	**I–Te**–X_3	2.926	2.928	0.026	2.902	2.944	8	
N(4)–N(3)	X_3–**N**$^+$–**N**^{0}–X_2 (N^0 planar)	1.414	1.414	0.005	1.412	1.418	13	
N(3)–N(3)	(C)(C,H)–$\mathbf{N}_a$–$\mathbf{N}_b$ (C)(C,H)							5,39
	$\mathbf{N}_a$, $\mathbf{N}_b$ pyramidal	1.454	1.452	0.021	1.444	1.457	44	40
	$\mathbf{N}_a$ pyramidal, N_b planar	1.420	1.420	0.015	1.407	1.433	68	40
	$\mathbf{N}_a$, N_b planar	1.401	1.401	0.018	1.384	1.418	40	40
	overall	1.425	1.425	0.027	1.407	1.443	139	
N(3)–N(2)	in pyrazole (N1–N2)	1.366	1.366	0.019	1.350	1.375	20	
	in pyridaznium (Nl$^+$≃N2)	1.350	1.349	0.010	1.345	1.361	7	
N(2)≃N(2)	**N≃N** (aromatic) in pyridazine							
	with C,H as *ortho* substituents	1.304	1.300	0.019	1.287	1.326	6	
	with N,Cl as *ortho* substituents	1.368	1.373	0.011	1.362	1.375	9	
N(2)=N(2)	C#–**N=N**–C#							
	cis	1.245	1.244	0.009	1.239	1.252	21	
	trans	1.222	1.222	0.006	1.218	1.227	6	
	(overall)	1.240	1.241	0.012	1.230	1.251	27	
	Car–**N=N**–*Car*	1.255	1.253	0.016	1.247	1.262	13	
	X–**N=N**=N (azides)	1.216	1.226	0.028	1.202	1.237	19	
N(2)=N(1)	X–N=**N=N** (azides)	1.124	1.128	0.015	1.114	1.137	19	
N(3)–O(2)	$(C,H)_2$–**N–OH** (Nsp^2: planar)	1.396	1.394	0.012	1.390	1.401	28	
	C_2–**N–O**–C							
	(Nsp^3: pyramidal)	1.463	1.465	0.012	1.457	1.468	22	
	(Nsp^2: planar)	1.397	1.394	0.011	1.388	1.409	12	
	in furoxan (N2–O1)	1.438	1.436	0.009	1.430	1.447	14	

Bond	Substructure	*d*	*m*	*σ*	q_l	q_u	*n*	Note
N(3)–O(1)	$(C\simeq)_2$**N**$^+$–**O**$^-$ in pyridine *N*-oxides	1.304	1.299	0.015	1.291	1.316	11	
	in furoxan ($^+$N2–O6$^-$)	1.234	1.234	0.008	1.228	1.240	14	
N(2)–O(2)	in oximes							
	$(C\#)_2$–C=**N–OH**	1.416	1.418	0.006	1.416	1.420	7	
	(H)(C*sp*^{2})–C=**N–OH**	1.390	1.390	0.011	1.380	1.401	20	
	(C#)(C*sp*^{2})–C=**N–OH**	1.402	1.403	0.010	1.393	1.410	18	
	(C*sp*^{2})$_2$–C=**N–OH**	1.378	1.377	0.017	1.365	1.393	16	
	(C,H)$_2$–C=**N–OH** (overall)	1.394	1.395	0.018	1.379	1.408	67	
	in furazan (O1–N2, O1–N5)	1.385	1.383	0.013	1.378	1.392	12	
	in furoxan (O1–N5)	1.380	1.380	0.011	1.370	1.388	14	
	in isoxazole (O1–N2)	1.425	1.425	0.010	1.417	1.434	9	
N(3)=O(1)	in nitrate ions NO_3^-	1.239	1.240	0.020	1.227	1.251	105	
	in nitro groups							
	C*–NO_2	1.212	1.214	0.012	1.206	1.221	84	
	C#–NO_2	1.210	1.210	0.011	1.203	1.218	251	
	Car–NO_2	1.217	1.218	0.011	1.211	1.215	1116	
	C–NO_2 (overall)	1.218	1.219	0.013	1.210	1.226	1733	
N(3)–P(4)	X_2–**P**(=X)–NX_2							
	N*sp*^{2}: planar	1.652	1.651	0.024	1.634	1.670	205	
	N*sp*^{3}: pyramidal	1.683	1.683	0.005	1.680	1.686	6	
	(overall)	1.662	1.662	0.029	1.639	1.682	358	
	subsets of this group are:							
	O_2–**P**(=S)–**NX_2**	1.628	1.624	0.015	1.615	1.634	9	
	C–**P**(=S)–(**NX_2**)$_2$	1.691	1.694	0.018	1.678	1.703	28	
	O–**P**(=S)–(**NX_2**)$_2$	1.652	1.654	0.014	1.642	1.664	28	
	P(=O)–(NX_2)$_3$	1.663	1.668	0.026	1.640	1.679	78	
N(3)–P(3)	–NX–**P**(–X)–**NX**–**P**(–X)–(P_2N_2ring)	1.730	1.721	0.017	1.716	1.748	20	
	–NX–**P**(=S)–**NX**–**P**(=S)–(P_2N_2ring)	1.697	1.697	0.015	1.690	1.703	44	
	in P-substituted phosphazenes:							
	(N≃)$_2$**P–N** (amino) (aziridinyl)	1.637	1.638	0.014	1.625	1.651	16	
		1.672	1.674	0.010	1.665	1.676	15	
N(2)=P(4)	Ph_3–**P=N**$^+$=P–Ph_3	1.571	1.573	0.013	1.563	1.580	66	
N(2)=P(3)	Ph_3–**P=N**–C,S	1.599	1.597	0.018	1.580	1.615	7	
N(2) ≃ P(3)	**N≃P**aromatic							
	in phosphazenes	1.582	1.582	0.019	1.571	1.594	126	
	in P ≃ N ≃ S	1.604	1.606	0.009	1.594	1.612	36	
N(3)–S(4)	C–SO_2–**NH_2**	1.600	1.601	0.012	1.591	1.610	14	35
	C–SO_2–**NH**–C#	1.633	1.633	0.019	1.615	1.652	47	35
	C–SO_2–**N**–C(#)$_2$	1.642	1.641	0.024	1.623	1.659	38	35
N(3)–S(2)	C–S–**NX_2**N*sp*^{2}: planar	1.710	1.707	0.019	1.698	1.722	22	23
	(for N*sp*^{3} pyramidal see MODIAZ: 1.765)							
	X–S–NX_2N*sp*^{2}: planar	1.707	1.705	0.012	1.699	1.715	30	23
N(2)–S(2)	C=**N–S**–X	1.656	1.663	0.027	1.632	1.677	36	
N(2) ≃ S(2)	**N≃S** aromatic in **P≃N≃S**	1.560	1.558	0.011	1.554	1.563	37	
N(2)=S(2)	**N=S** in **N=S=N** and **N=S=S**	1.541	1.546	0.022	1.521	1.558	37	
N(3)–Se	see COJCUZ (1.830), DSEMOR10 (1.846, 1.852), MORTRS10 (1.841)							
N(2)–Se	see SEBZQI (1.805), NAPSEZ10 (1.809, 1.820)							
N(2)=Se	see CISMUM (1.790, 1.791)							
N(3)–Si(5)	see DMESIP01, BOJLER, CASSAQ, CASYOK, CECXEN, CINTEY, CIPBUY, FMESIB, MNPSIL, PNPOSI (1.973–2.344)							
N(3)–Si(4)	X_3–**Si–NX_2** (overall)	1.748	1.746	0.022	1.735	1.757	170	
	subsets of this group are:							
	X_3–**Si–N**HX	1.714	1.719	0.014	1.702	1.727	16	
	X_3–**Si–N**X–Si–X_3 acyclic	1.743	1.744	0.016	1.731	1.755	45	
	N–Si–N in 4-membered rings	1.742	1.742	0.009	1.735	1.748	53	
	N–Si–N in 5-membered rings	1.741	1.742	0.019	1.726	1.749	33	
N(2)–Si(4)	X_3–**Si–N**$^-$–Si–X_3	1.711	1.712	0.019	1.693	1.729	15	
N–Te	see ACLTEP (2.402), BIBLAZ (1.980), CESSAU (2.023)							
O(2)–O(2)	C*–**O–O**–C*,H							
	τ(OO) = 70–85°	1.464	1.464	0.009	1.458	1.472	12	
	τ(OO)*ca.*180°	1.482	1.480	0.005	1.478	1.486	5	

Bond	Substructure	*d*	*m*	*σ*	q_l	q_u	*n*	Note
	overall	1.469	1.471	0.012	1.461	1.478	17	
	O=C–**O–O**–C=O see ACBZPO01 (1.446), CEYLUN (1.452), CIMHIP (1.454)							
	Si–**O–O**–Si	1.496	1.499	0.005	1.490	1.499	10	
O(2)–P(5)	X–**P**–(**O**X)$_4$							41
	trigonal bipyramidal:							
	axial	1.689	1.685	0.024	1.675	1.712	20	
	equatorial	1.619	1.622	0.024	1.604	1.628	20	
	square pyramidal	1.662	1.661	0.020	1.649	1.673	28	
O(2)–P(4)	C–**O–P**(≃ O)$_3^{2-}$	1.621	1.622	0.007	1.615	1.628	12	
	(H–**O**)$_2$–**P**(≃ O)$_2^-$	1.560	1.561	0.009	1.555	1.566	16	
	(C–**O**)$_2$–**P**(≃ O)$_2^-$	1.608	1.607	0.013	1.599	1.615	16	
	(C#–**O**)$_3$–**P**=O	1.558	1.554	0.011	1.550	1.564	30	
	(*Car*–**O**)$_3$–**P**=O	1.587	1.588	0.014	1.572	1.599	19	
	X–**O–P**(=O)–(C,N)$_2$	1.590	1.585	0.016	1.577	1.601	33	
	(X–**O**)$_2$–**P**(=O)–(C,N)	1.571	1.572	0.013	1.563	1.579	70	
O(2)–P(3)	(N≃)$_2$**P–O**–C (N ≃ P aromatic)	1.573	1.573	0.011	1.563	1.584	16	
O(1)=P(4)	C–O–**P**(≃**O**)$_3^{2-}$ (delocalized)	1.513	1.512	0.008	1.508	1.518	42	
	(H–O)$_2$–**P**(≃**O**)$_2^-$ (delocalized)	1.503	1.503	0.005	1.499	1.508	16	
	(C–O)$_2$–**P**(≃**O**)$_2^-$ (delocalized)	1.483	1.485	0.008	1.474	1.490	16	
	(C–O)$_3$–**P=O**	1.449	1.448	0.007	1.446	1.452	18	
	C$_3$–**P=O**	1.489	1.486	0.010	1.481	1.496	72	
	N$_3$–**P=O**	1.461	1.462	0.014	1.449	1.470	26	
	(C)$_2$(N)–**P=O**	1.487	1.489	0.007	1.479	1.493	5	
	(C,N)$_2$(O)–**P=O**	1.467	1.462	0.007	1.462	1.472	33	
	(C,N)(O)$_2$–**P=O**	1.457	1.458	0.009	1.454	1.462	35	
O(2)–S(4)	C–**O–SO**$_2$–C	1.577	1.576	0.015	1.566	1.584	41	
	C–**O–SO**$_2$–CH$_3$	1.569	1.569	0.013	1.556	1.582	7	
	C–**O–SO**$_2$–*Car*	1.580	1.578	0.015	1.571	1.588	27	
O(1)=S(4)	C–**SO**$_2$–C	1.436	1.437	0.010	1.431	1.442	316	42
	X–**SO**$_2$–NX$_2$	1.428	1.428	0.010	1.422	1.434	326	
	C–**SO**$_2$–N–(C,H)$_2$	1.430	1.430	0.009	1.425	1.435	206	
	C–**SO**$_2$–O–C	1.423	1.423	0.008	1.418	1.428	82	
	in **SO**$_4^{2-}$	1.472	1.473	0.013	1.463	1.481	104	
O(1)=S(3)	C–**S**(=**O**)–C	1.497	1.498	0.013	1.489	1.505	90	5
O–Se	see BAPPAJ, BIRGUE10, BIRHAL10, CXMSEO, DGLYSE, SPSEBU (1.597 for **O=Se** to 1.974 for **O–Se**)							
O(2)–Si(5)	(X–**O**)$_3$–**Si**–(N)(C)	1.663	1.658	0.023	1.650	1.665	21	
O(2)–Si(4)	X$_3$–**Si–O–X** (overall)	1.631	1.630	0.022	1.617	1.646	191	
O(2)–Si(4)	subsets of this group are:							
	X$_3$–**Si–O**–C#	1.645	1.647	0.012	1.634	1.652	29	
	X$_3$–**Si–O**–Si–X$_3$	1.622	1.625	0.014	1.614	1.631	70	
	X$_3$–**Si–O–O–Si**–X$_3$	1.680	1.676	0.008	1.673	1.688	10	
O(2)–Te(6)	(X–**O**)$_6$–**Te**	1.927	1.927	0.020	1.908	1.942	16	
O(2)–Te(4)	(X–**O**)$_2$–**Te**–X$_2$	2.133	2.136	0.054	2.078	2.177	12	
P(4)–P(4)	X$_3$–**P–P**–X$_3$	2.256	2.259	0.025	2.243	2.277	6	
P(4)–P(3)	see CECHEX (2.197), COZPIQ (2.249)							
P(3)–P(3)	X$_2$–**P–P**–X$_2$	2.214	2.210	0.022	2.200	2.224	41	
P(4)=P(4)	see BUTSUE (2.054)							
P(3)=P(3)	see BALXOB (2.034)							
P(4)=S(1)	C$_3$–**P=S**	1.954	1.952	0.005	1.950	1.957	13	
	(N,O)$_2$(C)–**P=S**	1.922	1.924	0.014	1.913	1.927	26	
	(N,O)$_3$–**P=S**	1.913	1.914	0.014	1.906	1.921	50	
P(4)=Se(1)	X$_3$–**P=Se**	2.093	2.099	0.019	2.075	2.108	12	
P(3)–Si(4)	X$_2$–**P–Si**–X$_3$: 3- and 4-rings excluded (see BOPFER, BOPFIV, CASTOF10, COZVIW: 2.201–2.317)	2.264	2.260	0.019	2.249	2.283	22	
P(4)=Te(1)	see MOPHTE (2.356), TTEBPZ (2.327)							
S(2)–S(2)	C–**S–S**–C							
	τ(SS) = 75–105°	2.031	2.029	0.015	2.021	2.038	46	
	τ(SS) = 0–20°	2.070	2.068	0.022	2.057	2.077	28	
	(overall)	2.048	2.045	0.026	2.028	2.068	99	
	in polysulfide chain **–S–S–S–**	2.051	2.050	0.022	2.037	2.065	126	
S(2)–S(1)	X–N=S–S	1.897	1.896	0.012	1.887	1.908	5	
S–Se(4)	see BUWZUO (2.264, 2.269)							

Bond	Substructure	d	m	σ	q_l	q_u	n	Note
S–Se(2)	X–**Se–S** (any)	2.193	2.195	0.015	2.174	2.207	9	
S(2)–Si(4)	X_3–**Si**–S–X	2.145	2.138	0.020	2.130	2.158	19	
S(2)–Te	X–**S–Te** (any)	2.405	2.406	0.022	2.383	2.424	10	
	X=**S–Te** (any)	2.682	2.686	0.035	2.673	2.694	28	
Se(2)–Se(2)	X–**Se–Se**–X	2.340	2.340	0.024	2.315	2.361	15	
Se(2)–Te(2)	see BAWFUA, BAWGAH (2.524–2.561)							†
Si(4)–Si(4)	X_3–**Si–Si**–X_3 3-membered rings excluded: see CIHRAM (2.511)	2.359	2.359	0.012	2.349	2.366	42	
Te–Te	see CAHJOK (2.751, 2.704)							

† The standard deviation in the sample for the bond type is greater than for the other entries.

Appendix 1. Footnotes to Table

1. Sample dominated by B–CH_3. For longer bonds in B^-–CH_3 see LITMEB10 [B(4)–CH_3 = 1.621–1.644Ŝ.
2. p(π)–p(π) Bonding with Bsp^2 and Nsp^2 coplanar (τBN = 0 ± 15°) predominates. See G. Schmidt, R. Boese, and D. Bläser, *Z. Naturforsch.*, 1982, **37b**, 1230.
3. 84 observations range from 1.38 to 1.61 Å and individual values depend on substituents on B and O. For a discussion of borinic acid adducts see S. J. Rettig and J. Trotter, *Can. J. Chem.*, 1982, **60**, 2957.
4. See M. Kaftory in *The Chemistry of Functional Groups. Supplement D: The Chemistry of Halides, Pseudohalides, and Azides*, S. Patai and Z. Rappoport, Eds., Wiley: New York, 1983, Part 2, ch. 24.
5. Bonds which are endocyclic or exocyclic to any 3- or 4-membered rings have been omitted from all averages in this section.
6. The overall average given here is for Csp^3–Csp^3 bonds which carry only C or H substituents. The value cited reflects the relative abundance of each 'substitution' group. The 'mean of means' for the 9 subgroups is 1.538 (σ = 0.022) Å.
7. See F. H. Allen, (a) *Acta Crystallogr.*, 1980, **B36**, 81; (b) 1981, **B37**, 890.
8. See F. H. Allen, *Acta Crystallogr.*, 1984, **B40**, 64.
9. See F. H. Allen, *Tetrahedron*, 1982, **38**, 2843.
10. See F. H. Allen, *Tetrahedron*, 1982, **38**, 645.
11. Cyclopropanones and cyclobutanones excluded.
12. See W. B. Schweizer and J. D. Dunitz, *Helv. Chim. Acta*, 1982, **65**, 1547.
13. See L. Norskov-Lauritsen, H.-B. Bürgi, P. Hoffmann, and H. R. Schmidt, *Helv. Chim. Acta*, 1985, **68**, 76.
14. See P. Chakrabarti and J. D. Dunitz, *Helv. Chim. Acta*, 1982, **65**, 1555.
15. See J. L. Hencher in *The Chemistry of the C≡C Triple Bond*, S. Patai, Ed., Wiley, New York, 1978, ch. 2.
16. Conjugated: torsion angle about central C–C single bond is 0 ± 20° (*cis*) or 180 ± 20° (*trans*).
17. Unconjugated: torsion angle about central C–C single bond is 20–160°.
18. Other conjugative substituents excluded.
19. TCNQ is tetracyanoquinodimethane.
20. No difference detected between C2 ≃ C3 and C3 ≃ C4 bonds.
21. Derived from neutron diffraction results only.
22. Nsp^3: pyramidal; mean valence angle at N is in range 108–114°.
23. Nsp^2: planar; mean valence angle at N is ≥ 117.5°.
24. Cyclic and acyclic peptides.
25. See R. H. Blessing, *J. Am. Chem. Soc.*, 1983, **105**, 2776.
26. See L. Lebioda, *Acta Crystallogr.*, 1980, **B36**, 271.
27. n = 3 or 4, i.e., tri- or tetra-substituted ureas.
28. Overall value also includes structures with mean valence angle at N in the range 115–118°.
29. See F. H. Allen and A. J. Kirby, *J. Am. Chem. Soc.*, 1984, **106**, 6197.
30. See A. J. Kirby, 'The Anomeric Effect and Related Stereoelectronic Effects at Oxygen,' Springer, Berlin, 1983.
31. See B. Fuchs, L. Schleifer, and E. Tartakovsky, *Nouv. J. Chim.*, 1984, **8**, 275.
32. See S. C. Nyburg and C. H. Faerman, *J. Mol. Struct.*, 1986, **140**, 347.
33. Sample dominated by P–CH_3 and P–CH_2–C.
34. Sample dominated by C* = methyl.
35. See A. Kalman, M. Czugler, and G. Argay, *Acta Crystallogr.*, 1981, **B37**, 868.
36. Bimodal distribution resolved into 22 'short' bonds and 5 longer outliers.
37. All 24 observations come from BUDTEZ.
38. 'Long' O–H bonds in centrosymmetric O---H---O H–bonded dimers are excluded.
39. N–N bond length also dependent on torsion angle about N–N bond and on nature of substituent C atoms; these effects are ignored here.
40. N pyramidal has average angle at N in range 100–113.5°; N planar has average angle of ≥ 117.5°.
41. See R. R. Holmes and J. A. Deiters, *J. Amer. Chem. Soc.*, 1977, **99**, 3318.
42. No detectable variation in S=O bond length with type of C-substituent.

Appendix 2. Reference Codes

The table below gives short-form codes to references cited in the main table. A full list of CSD bibliographic entries is given in SUP 56701.

Code	Reference
ACBZPO01	*J. Am. Chem. Soc.*, 1975, **97**, 6729.
ACLTEP	*J. Organomet. Chem.*, 1980, **184**, 417.
ASAZOC	*Dokl. Akad. Nauk SSSR*, 1979, **249**, 120.
BALXOB	*J. Am. Chem. Soc.*, 1981, **103**, 4587.
BAPPAJ	*Inorg. Chem.*, 1981, **20**, 3071.
BARRIV	*Acta Chem. Scand., Ser. A*, 1981, **35**, 443.
BAWFUA	*Cryst. Struct. Commun.*, 1981, **10**, 1345.
BAWGAH	*Cryst. Struct. Commun.*, 1981, **10**, 1353.
BECTAE	*J. Org. Chem.*, 1981, **46**, 5048, 1981.
BELNIP	*Z. Naturforsch., Teil B*, 1982, **37**, 299.
BEMLIO	*Chem. Ber.*, 1982, **115**, 1126.
BEPZEB	*Cryst. Struct. Commun.*, 1982, **11**, 175.
BETJOZ	*J. Am. Chem. Soc.*, 1982, **104**, 1683.
BETUTE10	*Acta Chem. Scand., Ser. A*, 1976, **30**, 719.
BIBLAZ	*Zh. Strukt. Khim.*, 1981, **22**, 118.
BICGEZ	*Z. Anorg. Allg. Chem.*, 1982, **486**, 90.
BIHXIZ	*J. Chem. Soc., Chem. Commun.*, 1982, 982.
BIRGUE10	*Z. Naturforsch., Teil B*, 1983, **38**, 20.
BIRHAL10	*Z. Naturforsch., Teil B*, 1982, **37**, 1410.
BIZJAV	*J. Organomet. Chem.*, 1982, **238**, C1.
BOGPOC	*Z. Naturforsch., Teil B*, 1982, **37**, 1402.
BOGSUL	*Z. Naturforsch., Teil B*, 1982, **37**, 1230.
BOJLER	*Z. Anorg. Allg. Chem.*, 1982, **493**, 53.
BOJPUL	*Acta Chem. Scand., Ser. A*, 1982, **36**, 829.
BOPFER	*Chem. Ber.*, 1983, **116**, 146.
BOPFIV	*Chem. Ber.*, 1983, **116**, 146.
BOVMEE	*Acta Crystallogr., Sect. B*, 1982, **38**, 1048.
BQUINI	*Acta Crystallogr., Sect. B*, 1979, **35**, 1930.
BTUPTE	*Acta Chem. Scand., Ser. A*, 1975, **29**, 738.
BUDTEZ	*Z. Naturforsch., Teil B*, 1983, **38**, 454.
BUPSIB10	*Z. Anorg. Allg. Chem.*, 1981, **474**, 31.
BUSHAY	*Z. Naturforsch., Teil. B*, 1983, **38**, 692.
BUTHAZ10	*Inorg. Chem.*, 1984, **23**, 2582.
BUTSUE	*J. Chem. Soc., Chem. Commun.*, 1983, 862.

Code	Reference
BUWZUO	*Acta Chem. Scand., Ser A,* 1983, **37,** 219.
BZPRIB	*Z. Naturforsch., Teil B,* 1981, **36,** 922.
BZTPPI	*Inorg. Chem.,* 1978, **17,** 894.
CAHJOK	*Inorg. Chem.,* 1983, **22,** 1809.
CAJMAB	*Chem. Z,* 1983, **107,** 169.
CANLUY	*Tetrahedron Lett.,* 1983, **24,** 4337.
CASSAQ	*J. Struct. Chem.,* 1983, **2,** 101.
CASTOF10	*Acta Crystallogr., Sect. C,* 1984, **40,** 1879.
CASYOK	*J. Struct. Chem.,* 1983, **2,** 107.
CECHEX	*Z. Anorg. Allg. Chem.,* 1984, **508,** 61.
CECXEN	*J. Struct. Chem.,* 1983, **2,** 207.
CEDCUJ	*J. Org. Chem.,* 1983, **48,** 5149.
CEHKAB	*Z. Naturforsch., Teil B,* 1984, **39,** 139.
CELDOM	*Acta Crystallogr., Sect. C,* 1984, **40,** 556.
CESSAU	*Acta Crystallogr., Sect. C,* 1984, **40,** 653.
CETTAW	*Chem. Ber.,* 1984, **117,** 1089.
CETUTE	*Acta Chem. Scand., Ser A,* 1975, **29,** 763.
CEYLUN	*Izv. Akad. Nauk SSSR, Ser. Khim.,* 1983, 2744.
CIFZUM	*Acta Chem. Scand., Ser A,* 1984, **38,** 289.
CIHRAM	*Angew. Chem., Int. Ed. Engl.,* 1984, **23,** 302.
CILRUK	*J. Chem. Soc., Chem. Commun.,* 1984, 1023.
CILSAR	*J. Chem. Soc., Chem. Commun.,* 1984, 1021.
CIMHIP	*Acta Crystallogr., C,* 1984, **40,** 1458.
CINTEY	*Dokl. Akad. Nauk SSSR,* 1984, **274,** 615.
CIPBUY	*J. Struct. Chem.,* 1983, **2,** 281.
CISMUM	*Z. Naturforsch., Teil B,* 1984, **39,** 485.
CISTED	*Z. Anorg. Allg. Chem.,* 1984, **511,** 95.
CIWYIQ	*Inorg. Chem.,* 1984, **23,** 1946.
CIYFOF	*Inorg. Chem.,* 1984, **23,** 1790.
CMBIDZ	*J. Org. Chem.,* 1979, **44,** 1447.
CODDEE	*Z. Naturforsch., Teil B,* 1984, **39,** 1257.
CODDII	*Z. Naturforsch., Teil B,* 1984, **39,** 1257.
COFVOI	*Z. Naturforsch., Teil B,* 1984, **39,** 1027.
COJCUZ	*Chem. Ber.,* 1984, **117,** 2686.
COSDIX	*Z. Naturforsch., Teil B,* 1984, **39,** 1344.
COZPIQ	*Chem. Ber.,* 1984, **117,** 2063.
COZVIW	*Z. Anorg. Allg. Chem.,* 1984, **515,** 7.
CTCNSE	*J. Am. Chem. Soc.,* 1980, **102,** 5430.
CUCPIZ	*J. Am. Chem. Soc.,* 1984, **106,** 7529.
CUDLOC	*J. Cryst. Spectrosc.,* 1985, **15,** 53.
CUDLUI	*J. Cryst. Spectrosc.,* 1985, **15,** 53.
CUGBAH	*Acta Crystallogr., Sect. C,* 1985, **41,** 476.
CXMSEO	*Acta Crystallogr., Sect. B,* 1973, **29,** 595.
DGLYSE	*Acta Crystallogr., Sect. B,* 1975, **31,** 1785.
DMESIP01	*Acta Crystallogr., Sect. C,* 1984, **40,** 895.
DSEMOR10	*J. Chem. Soc., Dalton Trans.,* 1980, 628.
DTHIBR10	*Inorg. Chem.,* 1971, **10,** 697.
EPHTEA	*Inorg. Chem.,* 1980, **19,** 2487.
ESEARS	*J. Chem. Soc. C,* 1971, 1511.
ETEARS	*J. Chem. Soc. C,* 1971, 1511.
FMESIB	*J. Organomet. Chem.,* 1980, **197,** 275.
FPHTEL	*J. Chem. Soc., Dalton Trans.,* 1980, 2306.
FPSULF10	*J. Am. Chem. Soc.,* 1982, **104,** 1683.
HCLENE10	*Acta Crystallogr., Sect. B,* 1982, **38,** 3139.
HMTITI	*Acta Crystallogr., Sect. B,* 1975, **31,** 1505.
HMTNTI	*Z. Anorg. Allg. Chem.,* 1974, **409,** 237.
HXPASC	*J. Chem. Soc., Dalton Trans.,* 1975, 1381.
IBZDAC11	*J. Chem. Soc., Dalton Trans.,* 1979, 854.
IFORAM	*Monatsh. Chem.,* 1974, **105,** 621.
IODMAM	*Acta Crystallogr., Sect. B,* 1977, **33,** 3209.
IPMUDS	*Acta Crystallogr., Sect. B,* 1973, **29,** 2128.
ISUREA10	*Acta Crystallogr., Sect. B,* 1972, **28,** 643.
LITMEB10	*J. Am. Chem. Soc.,* 1975, **97,** 6401.
MESIAD	*Z. Naturforsch., Teil B,* 1980, **35,** 789.
METAMM	*Acta Crystallogr.,* 1964, **17,** 1336.
MNPSIL	*J. Am. Chem. Soc.,* 1969, **91,** 4134.
MODIAZ	*J. Heterocycl. Chem.,* 1980, **17,** 1217.
MOPHTE	*Acta Chem. Scand., Ser. A,* 1980, **34,** 333.
MORTRS10	*J. Chem. Soc., Dalton Trans.,* 1980, 628.
NAPSEZ10	*J. Am. Chem. Soc.,* 1980, **102,** 5070.
NBBZAM	*Z. Naturforsch., Teil B,* 1977, **32,** 1416.
OPIMAS	*Aust. J. Chem.,* 1977, **30,** 2417.
OPNTEC10	*J. Chem. Soc., Dalton Trans.,* 1982, 251.
PHASCL	*Acta Crystallogr., Sect. B,* 1981, **37,** 1357.
PHASOC01	*Aust. J. Chem.,* 1975, **28,** 15.
PNPOSI	*J. Am. Chem. Soc.,* 1968, **90,** 5102.
SEBZQI	*J. Chem. Soc., Chem. Commun.,* 1977, 325.
SPSEBU	*Acta Chem. Scand., Ser. A,* 1979, **33,** 403.
TEACBR	*Cryst. Struct. Commun.,* 1974, **3,** 753.
THINBR	*J. Am. Chem. Soc.,* 1970, **92,** 4002.
TMPBTI	*Acta Crystallogr., Sect. B,* 1975, **31,** 1116.
TPASSN	*J. Chem. Soc., Dalton Trans.,* 1977, 514.
TPASTB	*Cryst. Struct. Commun.,* 1976, **5,** 39.
TPHOSI	*Z. Naturforsch., Teil B,* 1979, **34,** 1064.
TTEBPZ	*Z. Naturforsch., Teil B,* 1979, **34,** 256.
ZCMXSP	*Cryst. Struct. Commun.,* 1977, **6,** 93.

BOND LENGTHS IN ORGANOMETALLIC COMPOUNDS

These tables contain average values of interatomic distances of representative metal–ligand bonds. Sigma bonds between *d*- and *f*-block metals and carbon are given in Table 1. Sigma bond lengths between *d*- and *f*-block metals and the elements N, O, P, S, and As are given in Table 2. The values are extracted from a much larger list in Ref. 1. The tabulated values are the unweighted means of reported measurements on compounds in each category. If four or more measurements are available, the standard deviation is given in parentheses. All values are in angströms (10^{-10} m).

Column definitions for Table 1 are as follows.

Column heading	Definition
M	Atomic symbol for metal; metals are listed in atomic number order
M–CH$_3$	Metal–carbon bond length, in angströms (10^{-10} m)
M–CH$_2$R	Metal–carbon bond length, in angströms (10^{-10} m); R stands for any alkyl group
M–CR=CR$_2$	Metal–carbon bond length, in angströms (10^{-10} m); R stands for any alkyl group
M–C$_6$R$_5$	Metal–carbon bond length, in angströms (10^{-10} m); C$_6$R$_5$ indicates an aryl group
M–C(=O)R	Metal–carbon bond length, in angströms (10^{-10} m); C(=O)R is an acyl group

Column definitions for Table 2 are as follows.

Column heading	Definition
M	Atomic symbol for metal; metals are listed in atomic number order
M–NH$_3$	Metal–nitrogen bond length, in angströms (10^{-10} m)
M–OH$_2$	Metal–oxygen bond length, in angströms (10^{-10} m)
M–P(CH$_3$)$_3$	Metal–phosphorus bond length, in angströms (10^{-10} m)
M–SR	Metal–sulfur bond length, in angströms (10^{-10} m); R stands for any alkyl group
M–AsR$_3$	Metal–arsenic bond length, in angströms (10^{-10} m); R stands for any alkyl group

Reference

1. Orpen, A. G., Brammer, L., Allen, F. H., Kennard, O., Watson, D. G., and Taylor, R., *J. Chem. Soc. Dalton Trans.* 1989, S1-S83. <https://doi.org/10.1039/dt98900000s1>

TABLE 1. Metal-Carbon Bond Lengths in Å

M	M–CH$_3$	M–CH$_2$R	M–CR=CR$_2$	M–C$_6$R$_5$	M–C(=O)R
Ti		2.167	2.215(0.042)	2.148	
V				2.114(0.012)	
Cr	2.168		2.035(0.009)	2.075(0.019)	
Mn	2.095(0.030)	2.176(0.024)	2.007	2.064(0.021)	2.044
Fe	2.074	2.091(0.030)	1.991(0.039)	2.031(0.062)	1.997(0.033)
Co	2.014(0.023)	2.039(0.032)	1.934(0.019)	1.974	1.990
Ni	2.029	1.964	1.892(0.017)	1.917(0.038)	1.850(0.059)
Cu				2.020	
Zn		1.964			
Zr	2.292(0.049)		2.257		
Nb	2.336	1.319			
Mo	2.254(0.065)	2.250(0.061)	2.204(0.049)	2.193(0.054)	2.109
Ru	2.179(0.045)	2.036(0.010)	2.063	2.092(0.057)	2.091
Rh	2.092(0.027)	2.100	2.040(0.054)	2.011(0.026)	1.995(0.031)
Pd		2.028	2.000(0.024)	1.981(0.032)	1.982(0.029)
Hf	2.275(0.049)		2.205		
Ta	2.217(0.035)	2.225(0.056)		2.199(0.073)	
W	2.189(0.039)	2.175	2.224		
Re	2.173(0.051)	2.290		2.027	2.190(0.027)
Os		2.221	2.052	2.090(0.032)	2.161
Ir	2.175		2.071(0.044)	2.070(0.038)	2.019
Pt	2.083(0.045)	2.062(0.031)	2.024(0.037)	2.049(0.046)	1.991(0.025)
Au	2.066(0.045)		2.042	2.059(0.024)	
Hg	2.072(0.026)	2.125		2.086(0.040)	
Th	2.567				

TABLE 2. Bond Lengths of Metals to Elements Other Than Carbon

M	$M-NH_3$	$M-OH_2$	$M-P(CH_3)_3$	M-SR	$M-AsR_3$
Ti		2.066(0.052)		2.369	2.686
V		2.129(0.131)	2.510(0.010)	2.378(0.007)	
Cr	2.069(0.008)	1.997(0.070)	2.389(0.069)	2.362	2.460(0.040)
Mn		2.189(0.040)	2.455(0.164)	2.366(0.054)	2.400(0.013)
Fe		2.085(0.066)	2.246(0.042)	2.271(0.028)	2.352(0.043)
Co	1.965(0.021)	2.085(0.064)	2.217(0.043)	2.254(0.025)	2.323(0.021)
Ni	2.074(0.093)	2.079(0.038)	2.204(0.031)	2.187(0.007)	2.333(0.035)
Cu	1.987(0.017)	2.186(0.215)			2.367(0.016)
Zn	2.044	2.090(0.061)		2.295	
Y		2.398(0.068)			
Zr			2.692		
Nb		2.248(0.137)			2.741(0.008)
Mo	2.217	2.201(0.094)	2.462(0.046)	2.401(0.050)	2.582(0.036)
Ru	2.126(0.024)	2.074(0.051)	2.307(0.050)		2.446(0.031)
Rh	2.114(0.018)	2.190(0.096)	2.266(0.036)		2.416(0.039)
Pd	2.032	2.200	2.287(0.018)		2.386(0.052)
Ag		2.350			
Cd		2.318(0.065)		2.444	
La		2.556(0.062)			
Ce		2.565(0.063)			
Pr		2.518(0.038)			
Nd		2.533(0.058)			
Sm		2.459(0.050)			
Eu		2.441(0.055)			
Gd		2.443(0.074)			
Tb		2.455			
Dy		2.409(0.074)			
Ho		2.407(0.069)			
Er		2.404(0.083)			
Yb		2.353(0.066)			
Lu		2.404(0.116)			
Ta			2.589(0.044)		
W		2.115(0.065)	2.485(0.039)		
Re	2.253	2.199(0.091)	2.369(0.065)		2.575(0.006)
Os	2.136	2.166	2.328(0.029)		
Ir	2.050(0.021)		2.323(0.028)	2.461	
Pt			2.295(0.036)	2.320(0.015)	2.366(0.058)
Au		2.157		2.293	
Hg		2.690(0.083)		2.402(0.065)	
Th		2.483(0.032)			
U		2.455(0.047)			

STRUCTURE OF FREE MOLECULES IN THE GAS PHASE

David R. Lide

These tables give information on the geometric structure of selected molecules in the gas phase, including the overall geometry, interatomic distances, and bond angles. The molecules have been chosen to provide data on a wide variety of chemical bonds and to illustrate the influence of molecular environment on bond distances and angles. The tables are restricted to molecules with conventional covalent or ionic bonds, but it should be pointed out that structure data on many loosely bonded complexes of the van der Waals type have recently become available. The references below contain data on many molecules that are not included here and give additional information such as uncertainties and isotopic variations.

The two techniques for gas phase structure determination are spectroscopy and electron diffraction. The following codes are used to indicate the method used for each set of data:

ED – Gas-phase electron diffraction
MW – Microwave spectroscopy, including both measurements in bulk gases and molecular beams
IR – Infrared spectroscopy
R – Raman spectroscopy
UV – Electronic spectroscopy in the ultraviolet and visible regions, including fluorescence measurements
ESR – Electron spin resonance

In some cases, data from two sources have been combined to derive the structure; these are labeled by "ED, MW," for example.

Because of the internal vibrations that are present in all molecules, even in their lowest energy state, the definition of interatomic distance is not a simple matter. The ideal measure is the equilibrium distance in the hypothetical non-vibrating state, designated by r_e. This is the value of the separation of the atoms at the minimum of the potential function that describes the forces between the two atoms. All other measures represent some form of average, generally complex, over the vibrational motions. Since the potential function is asymmetric and less steep at distances beyond the potential minimum, the average distance is normally greater than r_e. Distances determined by electron diffraction (ED) represent an average over all vibrational states that are populated at the temperature of the measurement; the most common measure is designated r_g. Distances determined by spectroscopy (MW, IR, R, or UV) through measurements on the ground vibrational state of the molecule, designated by r_0, describe some form of average, not easily defined, over the zero-point vibrations. Another measure that is frequently used in microwave spectroscopy is the "substitution" distance r_s, which is operationally defined through a series of measurements on different isotopic species. In simple cases, r_s often lies between r_0 and r_e and is therefore a closer approximation to r_e. Several other types of averages have been used; good discussions can be found in Volumes II/25 and II/28 of the *Landolt-Börnstein* series (Ref. 1) and in Refs. 4 and 5.

Unless otherwise specified, distances and angles given in this table are r_0 values if the method is spectroscopic and r_g values if the method is electron diffraction. When given, equilibrium and substitution distances are designated by r_e and r_s, respectively.

Many interatomic distances and angles calculated by *ab initio* techniques have been reported in the recent literature. However, it should be emphasized that all data in this table are obtained from direct experimental measurements. In a few cases, *ab initio* calculations of vibration-rotation interaction constants have been combined with the primary experimental measurements to derive r_e values in the table.

The number of significant figures in the values is an indication of the precision of the measurement; thus, a distance quoted to three decimal places is probably reliable to about 0.005 Å or better. However, discrepancies between r_e, r_0, and r_g values for the same bond are often the order of 0.01 Å because of vibrational averaging considerations, so care must be taken in comparing bond distances in different molecules. Some distances in simple molecules are given here to four or five decimal places, but little chemical significance can be attached to differences beyond the third decimal place.

Table 1 covers molecules that do not contain carbon. Column definitions for Table 1 are as follows.

Column heading	Definition
Mol. form.	Molecular formula of compound; because many of the entries are free radicals or other transient species whose systematic chemical names are unfamiliar, the listing is in order of chemical formula
Name	Chemical name; shown in *Online Edition* only
Structure	Description of structure; either by words, point group of equilibrium geometry, or structural diagram
Bond/Angle 1	First bond distance or angle described
Value 1	Value of first bond distance (in angströms) or angle (in degrees)
Bond/Angle 2	Second bond distance or angle described
Value 2	Value of second bond distance (in angströms) or angle (in degrees)
Bond/Angle 3	Third bond distance or angle described
Value 3	Value of third bond distance (in angströms) or angle (in degrees)
Method	Method used for structure determination; see abbreviations and discussion above

Table 2 lists carbon-containing molecules. Column definitions for Table 2 are as follows.

Column heading	Definition
Name	Name of molecule; listed alphabetically; name given in **bold** above one or more structures
Structure	Description of structure; either by words, point group of equilibrium geometry, or structural diagram
Bond/Angle 1	First bond distance or angle described
Value 1	Value of first bond distance (in angströms) or angle (in degrees)
Bond/Angle 2	Second bond distance or angle described
Value 2	Value of second bond distance (in angströms) or angle (in degrees)
Bond/Angle 3	Third bond distance or angle described
Value 3	Value of third bond distance (in angströms) or angle (in degrees)
Method	Method used for structure determination; see abbreviations and discussion above

The contributions of Kozo Kuchitsu in preparing an earlier version of this table and in giving advice on the new version are gratefully acknowledged.

References

1. *Landolt-Börnstein Numerical Data and Functional Relationships in Science and Technology*, Springer-Verlag, Berlin. The following volumes are in the series *Structure Data of Free Polyatomic Molecules*: II/7, 1976 II/15, 1987 II/21, *Supplement to II/7 and II/15*, 1992 II/23, *Supplement to II/7, II/15, and II/21*, 1995 II/25A, *Inorganic Molecules*, 1998 II/25B, *Molecules Containing One or Two Carbon Atoms*, 1999 II/25C, *Molecules Containing Three or Four Carbon Atoms*, 2000 II/25D, *Molecules Containing Five or More Carbon Atoms*, 2003 II/28A, *Inorganic Molecules*, 2006 II/28B, *Molecules Containing One or Two Carbon Atoms*, 2006 II/28C, *Molecules Containing Three or Four Carbon Atoms*, 2007 II/28D, *Molecules Containing Five or More Carbon Atoms*, 2007.
2. Harmony, M. D., Laurie, V. W., Kuczkowski, R. L., Schwendeman, R. H., Ramsay, D. A., Lovas, F. J., Lafferty, W. J., and Maki, A. G., "Molecular Structure of Gas-Phase Polyatomic Molecules Determined by Spectroscopic Methods," *J. Phys. Chem. Ref. Data* 8, 619, 1979. <https://doi.org/10.1063/1.555605>
3. Huber, K. P., and Herzberg, G., *Molecular Spectra and Molecular Structure IV. Constants of Diatomic Molecules*, Van Nostrand Reinhold, London, 1979. <https://doi.org/10.1007/978-1-4757-0961-2_2>
4. Hargittai, M., "Molecular Structure of Metal Halides," *Chem. Rev.* 100, 2233-2301, 2000. <https://doi.org/10.1021/cr970115u>
5. Harmony, M. D., and Berry, R. J., *Struct. Chem.* 1, 49, 1989. <https://doi.org/10.1007/BF00675784>

TABLE 1. Structures of Non-Carbon Molecules: Bond Distances in Å and Angles in Degrees

Mol. form.	Structure	Bond/Angle 1	Value 1	Bond/Angle 2	Value 2	Bond/Angle 3	Value 3	Method
AgBr		Ag—Br (r_e)	2.3931					MW
AgCl		Ag—Cl (r_e)	2.2808					MW
AgF		Ag—F (r_e)	1.9832					MW
AgH		Ag—H (r_e)	1.617					UV
AgI		Ag—I (r_e)	2.5446					MW
AgLi		Ag—Li	2.41					UV
AgO		Ag—O (r_e)	2.0030					UV
AgOH	bent	Ag—O	2.016	O—H	0.952	∠HOAg	108.3[a]	MW
AlBr		Al—Br (r_e)	2.295					UV
$AlBr_3$	D_{3h}	Al—Br	2.221					ED
AlCa		Al—Ca	3.148					UV
AlCl		Al—Cl (r_e)	2.1301					MW
$AlCl_3$	D_{3h}	Al—Cl	2.063					ED
AlCo		Al—Co	2.283					UV
AlCu		Al—Cu	2.339					UV
AlF		Al—F (r_e)	1.6544					MW
AlF_3	D_{3h}	Al—F	1.633					ED
AlH		Al—H (r_e)	1.6482					UV
AlI		Al—I (r_e)	2.5371					MW
AlI_3	D_{3h}	Al—I	2.461					ED
AlK		Al—K	3.88					UV
AlMn		Al—Mn	2.638					UV
AlNi		Al—Ni	2.321					UV
AlO		Al—O (r_e)	1.6176					UV
AlS		Al—S (r_e)	2.029					UV
AlV		Al—V	2.620					UV
AlZn		Al—Zn	2.696					UV
Al_2		Al—Al (r_e)	2.701					UV
Al_2Br_6	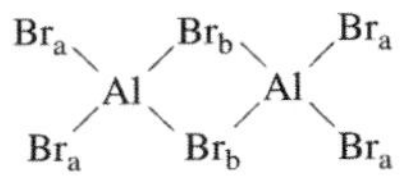	Al—Br_a	2.234	Al—Br_b	2.433			ED
	D_{2h}	∠Br_bAlBr_b	91.6	∠Br_aAlBr_a	122			
Al_2Cl_6	See Al_2Br_6	Al—Cl_a	2.061	Al—Cl_b	2.250			ED
	D_{2h}	∠Cl_bAlCl_b	90.0	∠Cl_aAlCl_a	122			
$AsBr_3$	C_{3v}	As—Br	2.324	∠BrAsBr	99.6			ED
$AsCl_3$	C_{3v}	As—Cl	2.165	∠ClAsCl	98.6			ED, MW
AsF_3	C_{3v}	As—F	1.710	∠FAsF	95.9			ED, MW
AsF_5	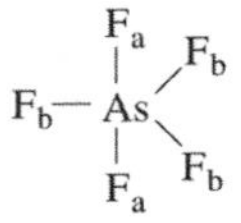 D_{3h}	As—F_a	1.711	As—F_b	1.656			ED
AsH		As—H (r_e)	1.5232					UV
AsH_3	C_{3v}	As—H (r_e)	1.511	∠HAsH (θ_e)	92.1			MW, IR
AsI_3	C_{3v}	As—I	2.557	∠IAsI	100.2			ED

Mol. form.	Structure	Bond/Angle 1	Value 1	Bond/Angle 2	Value 2	Bond/Angle 3	Value 3	Method
AsN		As—N (r_e)	1.6184					UV
AsO		As—O (r_e)	1.6236					UV
AsP		As—P (r_e)	1.99954					MW
As_2		As—As (r_e)	2.1026					UV
AuH		Au—H (r_e)	1.5237					UV
Au_2		Au—Au (r_e)	2.4719					UV
BBr		B—Br (r_e)	1.888					UV
BBr_3	D_{3h}	B—Br	1.893					ED
BCl		B—Cl (r_e)	1.7153					UV
$BClF_2$	C_{2v}	B—Cl (r_s)	1.728	B—F	1.315	∠FBF	118.1	MW
BCl_3	D_{3h}	B—Cl	1.742					ED
BF		B—F (r_e)	1.2626					UV
BF_2H		B—H	1.189	B—F	1.311	∠FBF	118.3	MW
BF_2OH	F_aF_bBOH	B—F_a (r_e)	1.3229	B—F_b (r_e)	1.3129	B—O (r_e)	1.3448	MW
	planar	∠FBF (θ_e)	118.36	∠F_aBO (θ_e)	122.25	∠BOH (θ_e)	113.14	
	F_a *cis* to OH	O—H (r_e)	0.9574					
BF_3	D_{3h}	B—F	1.313					ED, IR
BH		B—H (r_e)	1.2325					UV
BH_2NH_2	planar	B—N	1.391	B—H	1.195	N—H	1.004	MW
		∠HBH	122.2	∠HNH	114.2			
BH_3	planar	B—H	1.1900					IR
BH_3PH_3	staggered form	B—P	1.937	B—H	1.212	P—H	1.399	MW
		∠PBH	103.6	∠BPH	116.9	∠HBH	114.6	
		∠HPH	101.3					
BI_3	D_{3h}	B—I	2.118					ED
BN		B—N (r_e)	1.281					UV
BO		B—O (r_e)	1.2045					EPR
BO_2	linear	B—O	1.265					UV
BS		B—S	1.6091					UV
B_2		B—B (r_e)	1.590					UV
B_2H_6	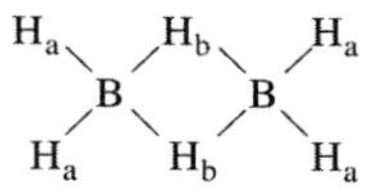	B—H_a	1.19	B—H_b	1.33	B···B	1.77	IR, ED
		∠H_aBH_a	122	∠H_bBH_b	97			
$B_3H_3O_3$		B—O	1.376	∠BOB	120	∠OBO	120	ED
$B_3H_6N_3$	C_2	B—N	1.435	B—H	1.26	N—H	1.05	ED
		∠BNB	121	∠NBN	118			
BaBr		Ba—Br (r_e)	2.8445					UV
$BaBr_2$		Ba—Br	2.912	∠BrBaBr	137.0			ED
BaCl		Ba—Cl (r_e)	2.6828					UV
BaF		Ba—F (r_e)	2.163					UV
BaH		Ba—H (r_e)	2.2318					UV
BaI		Ba—I (r_e)	3.0848					UV
BaI_2		Ba—I	3.150	∠IBaI	137.6			ED
BaO		Ba—O (r_e)	1.9397					MW
BaOH	linear	Ba—O	2.200	O—H	0.927			UV
BaS		Ba—S (r_e)	2.5074					MBE
$BeCl_2$	linear	Be—Cl (r_e)	1.791					ED,IR
BeF		Be—F (r_e)	1.3609					UV
BeF_2	linear	Be—F (r_e)	1.3730					IR
BeH		Be—H (r_e)	1.3431					UV
BeH_2	linear	Be—H (r_e)	1.3264					IR
BeO		Be—O (r_e)	1.3308					UV
BeS		Be—S (r_e)	1.7415					UV
BiBr		Bi—Br (r_e)	2.6095					MW
$BiBr_3$	C_{3v}	Bi—Br	2.577	∠BrBiBr	98.6			ED
BiCl		Bi—Cl (r_e)	2.4716					MW
$BiCl_3$	C_{3v}	Bi—Cl	2.424	∠ClBiCl	97.5			ED
BiF		Bi—F (r_e)	2.0516					MW
BiF_3	C_{3v}	Bi—F	1.987	∠FBiF	96.1			ED

Mol. form.	Structure	Bond/Angle 1	Value 1	Bond/Angle 2	Value 2	Bond/Angle 3	Value 3	Method
BiH		Bi—H (r_e)	1.805					UV
BiI		Bi—I (r_e)	2.8005					MW
BiI_3	C_{3v}	Bi—I	2.807	∠IBiI	99.5			ED
BiO		Bi—O (r_e)	1.934					UV
BiP		Bi—P (r_e)	2.29345					IR
Bi_2		Bi—Bi (r_e)	2.6596					UV
BrCl		Br—Cl (r_e)	2.1361					MW
BrF		Br—F (r_e)	1.7590					MW
BrF_3	F_a—Br—F_a | F_b	Br—F_a	1.810	$\angle F_{ax}BrF_{eq}$	85.1	$\angle F_aBrF_b$	86.2	MW
	C_{2v}	Br—F_b	1.721					
BrF_5	C_{4v}	Br—F (av.)	1.753	(Br—F_{eq}) − (Br—F_{ax})	0.069	$\angle F_{ax}BrF_{eq}$	85.1	ED, MW
BrN_3	$BrN_aN_bN_c$	N_a—N_b	1.113[a]	N_b—N_c	1.247	N_a—Br	1.899	ED
	planar	∠NNN	170.7	∠BrNN	109.7			
BrO		Br—O (r_e)	1.7172					MW
BrO_2	C_{2v}	Br—O (r_e)	1.644	∠OBrO (θ_e)	114.3			MW
Br_2		Br—Br (r_e)	2.2811					R
$CaBr_2$	linear	Ca—Br	2.62					ED
CaCl		Ca—Cl (r_e)	2.43676					UV
$CaCl_2$	linear	Ca—Cl	2.483					ED
CaF		Ca—F (r_e)	1.967					UV
CaH		Ca—H (r_e)	2.002					UV
CaI		Ca—I (r_e)	2.8286					UV
CaI_2	linear	Ca—I	2.840					ED
CaO		Ca—O (r_e)	1.8221					UV
CaOH	linear	Ca—O	1.985	O—H	0.921			UV
CaS		Ca—S (r_e)	2.3178					UV
CdH		Cd—H (r_e)	1.781					EPR
CdH_2	linear	Cd—H	1.6792					IR
$CdBr_2$	linear	Cd—Br	2.394					ED
$CdCl_2$	linear	Cd—Cl	2.284					ED
CdI_2	linear	Cd—I	2.582					ED
CeF_4	T_d	Ce—F	2.036					ED
CeI_3	quasiplanar	Ce—I	2.948					ED
ClBS	linear	B—Cl	1.681	B—S	1.606			MW
ClF		Cl—F (r_e)	1.6283					MW
ClF_3	F_a—Cl—F_a | F_b	Cl—F_a	1.698	Cl—F_b	1.598	$\angle F_aClF_b$	87.5	MW
ClN_3	$ClN_aN_bN_c$	N_a—N_b	1.253	N_b—N_c	1.113	N_a—Cl	1.746	MW
	planar	∠NNN	171.0	∠ClNN	108.7			
ClO		Cl—O (r_e)	1.5696					MW, UV
ClO_2	C_{2v}	Cl—O	1.470	∠OClO	117.38			MW
Cl_2		Cl—Cl (r_e)	1.9878					UV
Cl_2O	C_{2v}	Cl—O	1.6959	∠ClOCl	110.89			MW
$CoBr_2$	linear	Co—Br	2.241					ED
$CoCl_2$	linear	Co—Cl	2.113					ED
CoF_2	linear	Co—F	1.754	[Co—F (r_e)]	1.738			ED
CoF_3	D_{3h}	Co—F	1.732					ED
CoH		Co—H (r_e)	1.542					UV
CrF_2	linear	Cr—F	1.795					ED
CrF_3	D_{3h}	Cr—F	1.732					ED
CrF_4	T_d	Cr—F	1.706					ED
CrH		Cr—H (r_e)	1.656					UV
CrO		Cr—O (r_e)	1.615					UV
CsBr		Cs—Br (r_e)	3.0723					MW
CsCl		Cs—Cl (r_e)	2.9063					MW
CsF		Cs—F (r_e)	2.3454					MW
CsH		Cs—H (r_e)	2.4938					UV
CsI		Cs—I (r_e)	3.3152					MW

Mol. form.	Structure	Bond/Angle 1	Value 1	Bond/Angle 2	Value 2	Bond/Angle 3	Value 3	Method
CsO		Cs—O (r_e)	2.3007					MW
CsOH	linear; large amplitude bending mode	Cs—O (r_e)	2.395	O—H (r_e)	0.97			MW
Cs_2		Cs—Cs (r_e)	4.47					UV
CuBr		Cu—Br (r_e)	2.1734					MW
CuCl		Cu—Cl (r_e)	2.0512					MW
CuF		Cu—F (r_e)	1.7449					MW
CuF_2	linear	Cu—F	1.713					ED
CuH		Cu—H (r_e)	1.4626					UV
CuI		Cu—I (r_e)	2.3383					MW
CuLi		Cu—Li	2.26					UV
CuO		Cu—O (r_e)	1.7244					UV
CuOH	bent	Cu—O (r_s)	1.769	O—H	0.952	∠HOCu (θ_s)	110.24	MW
CuS		Cu—S	2.051					UV
Cu_2		Cu—Cu (r_e)	2.2197					UV
$DyBr_3$	quasiplanar	Dy—Br	2.609					ED
$DyCl_3$	quasiplanar	Dy—Cl	2.461					ED
FN_3	$FN_aN_bN_c$	N_a—N_b	1.253	N_b—N_c	1.132	N_a—F	1.439	MW
	planar	∠NNN	170.3	∠FNN	103.8			
F_2		F—F (r_e)	1.4119					R
$FeBr_2$	linear	Fe—Br	2.294					ED
$FeCl_2$	linear	Fe—Cl	2.132					UV,ED
FeF_2	linear	Fe—F	1.769	[Fe—F (r_e)]	1.755			ED
FeF_3	D_{3h}	Fe—F	1.763					ED
FeH		Fe—H	1.620					IR
FeO		Fe—O	1.444					UV
FeS		Fe—S	2.017					MW
GaBr		Ga—Br (r_e)	2.3525					MW
$GaBr_3$	D_{3h}	Ga—Br	2.249					ED
GaCl		Ga—Cl (r_e)	2.2017					MW
$GaCl_3$	D_{3h}	Ga—Cl	2.110					ED
GaF		Ga—F (r_e)	1.7744					MW
GaF_3	D_{3h}	Ga—F	1.725					ED
GaH		Ga—H (r_e)	1.663					UV
GaI		Ga—I (r_e)	2.5747					MW
GaI_3	D_{3h}	Ga—I	2.458					ED
GaO		Ga—O	1.744					UV
Ga_2Br_6	See Al_2Br_6	Ga—Br_a	2.250	Ga—Br_b	2.453			ED
	D_{2h}	∠Br_aGaBr_a	92.7	∠Br_bGaBr_b	123			
Ga_2Cl_6	See Al_2Br_6	Ga—Cl_a	2.116	Ga—Cl_b	2.305			ED
	D_{2h}	∠Cl_aGaCl_a	90	∠Cl_bGaCl_b	124.5			
$GdBr_3$	C_{3v}	Gd—Br	2.641					ED
$GdCl_3$	C_{3v}	Gd—Cl	2.488					ED
GdF_3	C_{3v}	Gd—F	2.053					ED
GdI_3	C_{3v}	Gd—I	2.840	∠IGdI	108			ED
$GeBrH_3$	C_{3v}	Ge—H	1.526	Ge—Br	2.299	∠HGeH	106.2	MW, IR
$GeBr_2$		Ge—Br (r_e)	2.359	∠BrGeBr	101.0			ED
$GeBr_4$	T_d	Ge—Br	2.272					ED
$GeClH_3$	C_{3v}	Ge—H	1.537	Ge—Cl	2.150	∠HGeH	111.0	IR, MW
$GeCl_2$		Ge—Cl (r_e)	2.186	∠ClGeCl	100.3			ED
$GeCl_4$	T_d	Ge—Cl	2.113					ED
$GeFH_3$	C_{3v}	Ge—H	1.522	Ge—F	1.732	∠HGeH	113.0	MW, IR
GeF_2		Ge—F (r_e)	1.7321	∠FGeF (θ_e)	97.15			MW
GeH		Ge—H (r_e)	1.5880					UV
GeHI		Ge—I	2.525	Ge—H	1.593	∠HGeI	93.5	UV
GeH_4	T_d	Ge—H	1.5251					IR, R
GeI_2		Ge—I	2.540	∠IGeI	102.1			ED
GeI_4	T_d	Ge—I	2.515					ED
GeO		Ge—O (r_e)	1.6246					MW
GeS		Ge—S (r_e)	2.0121					MW
GeSe		Ge—Se (r_e)	2.1346					MW

Mol. form.	Structure	Bond/Angle 1	Value 1	Bond/Angle 2	Value 2	Bond/Angle 3	Value 3	Method
GeTe		Ge—Te (r_e)	2.3402					MW
Ge_2H_6		Ge—Ge	2.403	Ge—H	1.541			ED
		∠HGeH	106.4	∠GeGeH	112.5			
HBr		H—Br (r_e)	1.4145					MW
HCl		H—Cl (r_e)	1.2746					MW
HClO	ClOH (bent)	Cl—O	1.690	O—H	0.975	∠HOCl	102.5	MW, IR
$HClO_4$	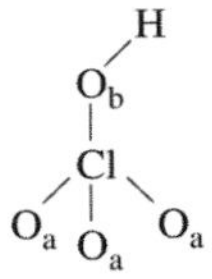	Cl—O_a	1.407	Cl—O_b	1.639			ED
		∠O_aClO_a	114.3	∠O_aClO_b	104.1			
HF		H—F (r_e)	0.9169					MW
HFO	FOH (bent)	F—O	1.442	O—H	0.96	∠HOF	97.2	MW
HI		H—I (r_e)	1.6090					MW
HIO	IOH (bent)	I—O	1.9941	O—H	0.967	∠HOI	103.9	MW
HNO	bent	N—O	1.212	N—H	1.063	∠HNO	108.6	UV
HNO_2	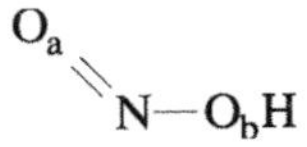	*s-trans* conformer		*s-cis* conformer				MW
		O_b—H	0.958	O_b—H	0.98			
		N—O_b	1.432	N—O_b	1.39			
		N—O_a	1.170	N—O_a	1.19			
		∠O_aNO_b	110.7	∠O_aNO_b	114			
		∠NO_bH	102.1	∠NO_bH	104			
HNO_3	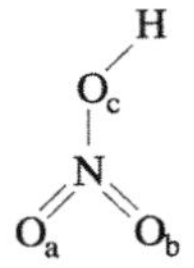	N—O_a	1.20	N—O_b	1.21	N—O_c	1.41	MW
		O_c—H	0.96	∠O_cNO_b	115.9	∠HO_cN	102.2	
	planar	∠O_cNO_a	113.9					
HNSO	planar *cis*	N—S	1.512	S—O	1.451	N—H	1.029	MW
		∠NSO	120.4	∠HNS	115.8			
HN_3	$HN_aNM_bN_c$	N_a—N_b	1.245	N_b—N_c	1.134	N_a—H	1.015	MW
	planar	∠NNN	171.8	∠HNN	109.2			
HPO		P—O	1.4843	P—H	1.473	∠HPO	104.57	MW
H_2		H—H (r_e)	0.74144					UV
H_2O	C_{2v}	O—H (r_e)	0.9575	∠HOH (θ_e)	104.51			MW, IR
H_2O_2	C_2	O—O	1.475	∠OOH	94.8	dihedral angle	119.8	IR
H_2S	C_{2v}	H—S (r_e)	1.3356	∠HSH (θ_e)	92.12			MW, IR
H_2SO_4	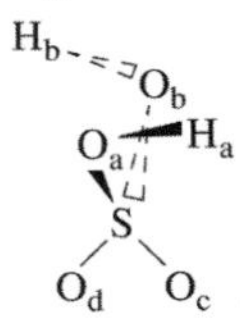 C_2	O—H	0.97	S—O_a	1.574	S—O_c	1.422	MW
		∠O_aSO_b	101.3	∠O_cSO_d	123.3	∠O_aSO_c	108.6	
		∠O_aSO_d	106.4	∠H_aO_aS	108.5	dihedral angle between the H_aO_aS and O_aSO_c planes	20.8	
	C_2	dihedral angle between the H_aO_aS and O_aSO_b planes	90.9	dihedral angle between the H_aSO_b and O_cSO_d planes	88.4			
H_2S_2	C_2	S—S	2.055	S—H	1.327	∠SSH	91.3	ED, MW
		dihedral angle	90.6					
$HfBr_4$	T_d	Hf—Br	2.450					ED
$HfCl_4$	T_d	Hf—Cl	2.316					ED
HfF		Hf—F	1.8596					UV

Mol. form.	Structure	Bond/Angle 1	Value 1	Bond/Angle 2	Value 2	Bond/Angle 3	Value 3	Method
HfF_4	T_d	Hf—F	1.909					ED
HfI_4	T_d	Hf—I	2.662					ED
$HgBr_2$	linear	Hg—Br	2.384					ED
$HgCl_2$	linear	Hg—Cl	2.252					ED
HgH		Hg—H (r_e)	1.7404					UV
HgI_2	linear	Hg—I	2.568					ED
$HoCl_3$		Ho—Cl	2.462					ED
HoF_3		Ho—F	2.007					ED
HoO		Ho—O	1.797					UV
IBr		I—Br (r_e)	2.4691					MW
ICl		I—Cl (r_e)	2.3210					MW
IF		I—F (r_e)	1.9098					UV
IF_5	C_{4v}	I—F (av.)	1.860	(I—F_{eq}) − (I—F_{ax})	0.03	$\angle F_{ax}IF_{eq}$	82.1	ED, MW
IO		I—O (r_e)	1.8676					MW
I_2		I—I (r_e)	2.6663					R
InBr		In—Br (r_e)	2.5432					MW
InCl		In—Cl (r_e)	2.4012					MW
$InCl_3$		In—Cl	2.291					ED
InF		In—F (r_e)	1.9854					MW
InH		In—H (r_e)	1.8376					UV
InI		In—I (r_e)	2.7537					MW
IrF_6	O_h	Ir—F	1.831					ED
KBH_4	$H_a(BH_3)K$ (C_{3v})	B—H (BH_3)	1.272	B—H_a	1.233	K—B	2.656	MW
KBr		K—Br (r_e)	2.8208					MW
KCl		K—Cl (r_e)	2.6667					MW
KF		K—F (r_e)	2.1716					MW
KH		K—H (r_e)	2.244					UV
KI		K—I (r_e)	3.0478					MW
KOH	linear; large amplitude bending mode	K—O	2.212	O—H	0.91			MW
K_2		K—K (r_e)	3.9051					UV
KrF_2	linear	Kr—F	1.89					ED
LaBr		La—Br (r_e)	2.65208					MW
$LaBr_3$	C_{3v}	La—Br	2.742					ED
LaCl		La—Cl (r_e)	2.49804					MW
$LaCl_3$	C_{3v}	La—Cl	2.589					ED
LaF		La—F (r_e)	2.02338					MW
LaI		La—I (r_e)	2.87885					MW
LaO		La—O (r_e)	1.82591					UV
$LiBH_4$	$H_a(BH_3)Li$ (C_{3v})	B—H (H_3)	1.257	B—H_a	1.218	Li—B	1.939	MW
LiBr		Li—Br (r_e)	2.1704					MW
LiCl		Li—Cl (r_e)	2.0207					MW
LiF		Li—F (r_e)	1.5639					MW
LiH		Li—H (r_e)	1.5949					MW
LiI		Li—I (r_e)	2.3919					MW
LiO		Li—O (r_e)	1.68822					UV
LiOH	linear	Li—O (r_e)	1.5776	O—H (r_e)	0.949			MW
Li_2		Li—Li (r_e)	2.6729					UV
Li_2Cl_2	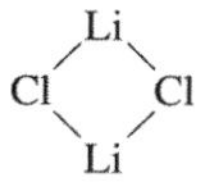	Li—Cl	2.23	Cl—Cl	3.61	∠ClLiCl	108	ED
Li_2O	linear	Li—O	1.606					UV
$LuBr_3$	C_{3v}	Lu—Br	2.557					ED
$LuCl_3$	C_{3v}	Lu—Cl	2.417	∠ClLuCl	112			ED
LuI_3	C_{3v}	Lu—I	2.768					ED
MgBr		Mg—Br (r_e)	2.34742					MW
MgCl		Mg—Cl (r_e)	2.1964					UV
$MgCl_2$	linear	Mg—Cl	2.179					ED
MgF		Mg—F (r_e)	1.7500					UV
MgF_2	linear	Mg—F	1.771					ED

Mol. form.	Structure	Bond/Angle 1	Value 1	Bond/Angle 2	Value 2	Bond/Angle 3	Value 3	Method
MgH		Mg—H (r_e)	1.7297					UV
MgO		Mg—O (r_e)	1.749					UV
MgOH	linear	Mg—O	1.770	O—H	0.912			UV
Mg_2		Mg—Mg (r_e)	3.891					UV
$MnBr_2$	linear	Mn—Br	2.344					ED
$MnCl_2$	linear	Mn—Cl	2.202					ED
MnF_2	linear	Mn—F	1.811	[Mn—F (r_e)]	1.797			ED
MnH		Mn—H (r_e)	1.7308					UV
MnI_2	linear	Mn—I	2.538					ED
$MoCl_4O$	C_{4v}	Mo—Cl	2.279	Mo—O	1.658			ED
		∠ClMoCl	87.2					
MoF_4		Mo—F	1.851					ED
MoF_6	O_h	Mo—F	1.821					ED
NBr		N—Br (r_e)	1.79					UV
NCl		N—Cl (r_e)	1.6107					UV
$NClH_2$		N—H	1.017	N—Cl	1.748			MW, IR
		∠HNCl	103.7	∠HNH	107			
NCl_3		N—Cl	1.759	∠ClNCl	107.1			ED
NF		N—F (r_e)	1.3170					UV
NF_2		N—F	1.3528	∠FNF	103.18			MW
NH_2		N—H	1.024	∠HNH	103.3			UV
NH_2NO_2		N—N	1.427	N—H	1.005			MW
		dihedral angle between NH_2 and NNO_2 planes	128.2	∠HNH	115.2	∠ONO	130.1	
NH_3	C_{3v}	N—H (r_e)	1.012	∠HNH (θ_e)	106.7			IR
NH_4Cl	H_3N····HCl (C_{3v})	N—Cl	3.136					MW
NH		N—H (r_e)	1.0362					LMR
NH_2OH	bisector of HNH angle is *trans* to OH bond	N—O	1.453	N—H	1.02	O—H	0.962	MW
		∠HNO	103.3	∠HNH	107	∠NOH	101.4	
NO		N—O (r_e)	1.1506					IR
NOCl		N—O	1.14	N—Cl	1.975	∠ONCl	113	MW
NOF		N—O	1.136	N—F	1.512	∠FNO	110.1	MW
NO_2		N—O	1.193	∠ONO	134.1			MW
NO_2Cl	C_{2v}	N—O	1.202	N—Cl	1.840	∠ONO	130.6	MW
NO_2F	C_{2v}	N—O	1.1798	N—F	1.467	∠ONO	136	MW
NS		N—S (r_e)	1.4940					IR
N_2		N—N (r_e)	1.0977					UV
N_2H_4	H_a atom is closer to the C_2 axis, H_b is farther from the C_2 axis	N—N	1.449	N—H	1.021	∠NNH_b	106	ED, MW
		∠HNH	106.6[a]	∠NNH_a	112			
		dihedral angle of internal rotation	91					
N_2O		N—N (r_e)	1.1284	N—O (r_e)	1.1841			MW, IR
N_2O_3	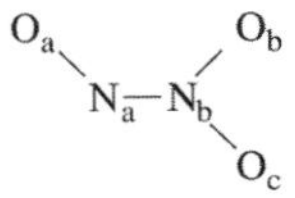	N_a—N_b	1.864	N_a—O_a	1.142			MW
		N_b—O_b	1.202	N_b—O_c	1.217			
		∠$O_aN_aN_b$	105.05	∠$N_aN_bO_b$	112.72	∠$N_aN_bO_c$	117.47	
N_2O_4	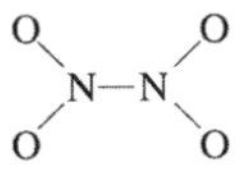 D_{2h}	N—N	1.782	N—O	1.190	∠ONO	135.4	ED
$NaBH_4$	$H_a(BH_3)Na$ (C_{3v})	B—H (BH_3)	1.278	B—H_a	1.238	Na—B	2.308	MW
NaBr		Na—Br (r_e)	2.5020					MW
NaCl		Na—Cl (r_e)	2.3609					MW
NaF		Na—F (r_e)	1.9260					MW

Mol. form.	Structure	Bond/Angle 1	Value 1	Bond/Angle 2	Value 2	Bond/Angle 3	Value 3	Method
NaH		Na—H (r_e)	1.8873					UV
NaI		Na—I (r_e)	2.7115					MW
NaO		Na—O (r_e)	2.05155					UV
Na_2		Na—Na (r_e)	3.0789					UV
$NbCl_4$	T_d	Nb—Cl	2.279					ED
$NbCl_5$	D_{3h}	Nb—Cl_{ax}	2.307	Nb—Cl_{eq}	2.276			ED
NbO		Nb—O (r_e)	1.691					UV
NdI_3	C_{3v}	Nd—I	2.879					ED
NiBr		Ni—Br	2.1963					UV
$NiBr_2$	linear	Ni—Br	2.201					ED
$NiCl_2$	linear	Ni—Cl	2.076					ED
NiF_2	linear	Ni—F	1.729	[Ni—F (r_e)]	1.715			ED
NiH		Ni—H (r_e)	1.476					UV
NiI		Ni—I	2.348					UV
NpF_6	O_h	Np—F	1.982					ED
OF		O—F (r_e)	1.3579					LMR
OF_2	C_{2v}	O—F (r_e)	1.4053	∠FOF (θ_e)	103.07			MW
OH		O—H (r_e)	0.96966					UV
$O(SiH_3)_2$		Si—H	1.486	Si—O	1.634	∠SiOSi	144.1	ED
O_2		O—O (r_e)	1.2074					MW
O_2F_2	C_2	O—O	1.217	F—O	1.575	∠OOF	109.5	MW
		dihedral angle of internal rotation	87.5					
O_3	C_{2v}	O—O (r_e)	1.2716	∠OOO (θ_e)	117.47			MW
OsF_6	O_h	Os—F	1.832					ED
OsO_4	T_d	Os—O	1.712					ED
PBr_3	C_{3v}	P—Br	2.220	∠BrPBr	101.0			ED
PCl		P—Cl (r_e)	2.01461					UV
PCl_3	C_{3v}	P—Cl	2.039	∠ClPCl	100.27			ED
PCl_5	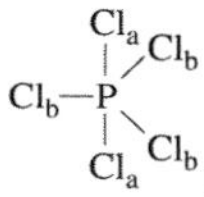 D_{3h}	P—Cl_a	2.124	P—Cl_b	2.020			ED
PF		P—F (r_e)	1.5896					UV
PF_3	C_{3v}	P—F	1.570	∠FPF	97.8			ED, MW
PF_5	D_{3h}	P—F_{eq}	1.534	P—F_{ax}	1.577			ED
PH		P—H (r_e)	1.4223					LMR
PH_2		P—H	1.418	∠HPH	91.70			UV
PH_3	C_{3v}	P—H	1.4200	∠HPH	93.345			MW
PN		N—P (r_e)	1.49087					MW
PO		O—P (r_e)	1.4759					UV
$POCl_3$	C_{3v}	P—O	1.449	P—Cl	1.993	∠ClPCl	103.3	ED
POF_3	C_{3v}	P—O	1.436	P—F	1.524	∠FPF	101.3	ED, MW
P_2		P—P (r_e)	1.8931					UV
P_2F_4	*trans* conformer	P—F	1.587	P—P	2.281	∠FPF	99.1	P_2F_4
		∠PPF	95.4					
P_4	T_d	P—P	2.21					ED
P_4O_6	T_d	P—O	1.638	∠POP	126.4			ED
$PbBr_2$	bent	Pb—Br (r_e)	2.598					ED
$PbCl_2$	bent	Pb—Cl (r_e)	2.444					ED
$PbCl_4$	T_d	Pb—Cl	2.369					ED
PbF		Pb—F (r_e)	2.0575					UV
PbF_2	bent	Pb—F (r_e)	2.041					ED
PbH		Pb—H (r_e)	1.839					UV
PbI_2	bent	Pb—I (r_e)	2.807					ED
PbO		Pb—O (r_e)	1.9218					MW
PbS		Pb—S (r_e)	2.2869					MW
PbSe		Pb—Se (r_e)	2.4022					MW
PbTe		Pb—Te (r_e)	2.5950					MW
$PrCl_3$	C_{3v}	Pr—Cl	2.554					ED
PrF_3	C_{3v}	Pr—F	2.091					ED

Mol. form.	Structure	Bond/Angle 1	Value 1	Bond/Angle 2	Value 2	Bond/Angle 3	Value 3	Method
PrI_3	C_{3v}	Pr—I	2.901	∠IPrI	113			ED
PtC		Pt—C (r_e)	1.6767					UV
PtH		Pt—H (r_e)	1.52852					UV
PtN		Pt—N (r_e)	1.682					MW
PtO		Pt—O (r_e)	1.7273					UV
PtS		Pt—S (r_e)	2.03983					MW
PtSi		Pt—Si (r_e)	2.0612					MW
PuF_6	O_h	Pu—F	1.972					ED
RbBr		Rb—Br (r_e)	2.9447					MW
RbCl		Rb—Cl (r_e)	2.7869					MW
RbF		Rb—F (r_e)	2.2703					MW
RbH		Rb—H (r_e)	2.367					UV
RbI		Rb—I (r_e)	3.1768					MW
RbO		Rb—O (r_e)	2.25420					UV
RbOH	linear; large amplitude bending mode	Rb—O	2.301	O—H	0.957			MW
$ReClO_3$	C_{3v}	Re—O	1.702	Re—Cl	2.229	∠ClReO	109.4	MW
$ReCl_5$	D_{3h}	Re—Cl_{eq}	2.238	Re—Cl_{ax}	2.263			ED
ReF_6	O_h	Re—F	1.832					ED
ReF_7	pseudorotation	Re—F	1.835					ED
RhB		Rh—B	1.691					UV
RhC		Rh—C	1.614					UV
RhS		Rh—S	2.059					UV
RuO_4	T_d	Ru—O	1.706					ED
SCl_2	C_{2v}	S—Cl	2.006	∠ClSCl	103.0			ED
SF		S—F (r_e)	1.6006					MW
SF_2		S—F	1.5921	∠FSF	98.20			MW
SF_6	O_h	S—F	1.561					ED
SH		S—H (r_e)	1.34066					UV
SO		S—O (r_e)	1.4811					MW
$SOCl_2$		S—O	1.44	S—Cl	2.072			MW
		∠ClSCl	97.2	∠OSCl	108.0			
SOF_2		S—O	1.420	S—F	1.583			ED
		∠FSF	92.2	∠OSF	106.2			
SOF_4	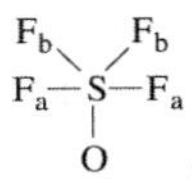 C_{2v}	S—O	1.403	S—F_a	1.575	S—F_b	1.552	ED
		∠OSF_a	90.7	∠OSF_b	124.9			
		∠F_aSF_b	89.6	∠F_bSF_b	110.2			
SO_2		S—O (r_e)	1.4308	∠OSO (θ_e)	119.329			MW
SO_2Cl_2	C_{2v}	S—Cl	2.011	S—O	1.404			ED
		∠ClSCl	100.0	∠OSO	123.5			
SO_2F_2	C_{2v}	S—F	1.530	S—O	1.397			ED
		∠FSF	97	∠OSO	123			
SO_3	D_{3h}	S—O	1.4198					IR
$S(SiH_3)_2$		Si—S	2.136	Si—H	1.494	∠SiSSi	97.4	ED
S_2		S—S (r_e)	1.8892					R
S_2Br_2	C_2	S—Br	2.24	S—S	1.98	∠SSBr	105	ED
		dihedral angle of internal rotation	83.5					
S_2Cl_2	C_2	S—Cl	2.057	S—S	1.931	∠SSCl	108.2	ED
		dihedral angle of internal rotation	84.1					
S_2O_2	planar *cis* form	S—S	2.025	S—O	1.458	∠OSS	112.8	MW
S_8	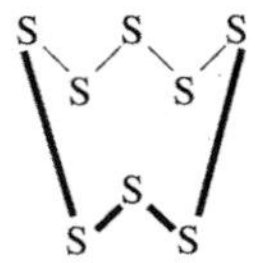	S—S	2.07	∠SSS (D_{4d})	105			ED
$SbBr_3$	C_{3v}	Sb—Br	2.490	∠BrSbBr	98.2			ED
$SbCl_3$	C_{3v}	Sb—Cl	2.334	∠ClSbCl	97.1			ED

Molecular

Mol. form.	Structure	Bond/Angle 1	Value 1	Bond/Angle 2	Value 2	Bond/Angle 3	Value 3	Method
$SbCl_5$	D_{3h}	Sb—Cl*eq*	2.277	Sb—Cl_{ax}	2.338			ED
SbF		Sb—F (r_e)	1.918					UV
SbF_3	C_{3v}	Sb—F	1.880	∠FSbF	94.9			ED
SbH		Sb—H	1.723					UV
SbH_3	C_{3v}	Sb—H	1.704	∠HSbH	91.6			MW
SbI_3	C_{3v}	Sb—I	2.721	∠ISbI	99.0			ED
SbO		Sb—O (r_e)	1.826					UV
SbP		Sb—P (r_e)	2.20544					MW
$ScCl_3$	D_{3h}	Sc—Cl	2.291					ED
ScF		Sc—F (r_e)	1.788					UV
ScF_3	D_{3h}	Sc—F	1.847					ED
SeF		Se—F	1.742					MW
SeF_6	O_h	Se—F	1.69					ED
SeH		Se—H (r_e)	1.48					UV
SeO		Se—O (r_e)	1.6393					MW
$SeOF_2$		Se—O	1.576	Se—F	1.730			MW
		∠OSeF	104.82	∠FSeF	92.22			
SeO_2		Se—O (r_e)	1.6076	∠OSeO (θ_e)	113.83			MW
SeO_3	D_{3h}	Se—O	1.69					ED
Se_2		Se—Se (r_e)	2.1660					UV
Se_6	six-membered ring with chair conformation	Se—Se	2.34	∠SeSeSe	102			ED
$SiBrF_3$	C_{3v}	Si—F	1.559	Si—Br	2.156	∠FSiBr	108.5	MW
$SiBrH_3$	C_{3v}	Si—Br	2.210	Si—H	1.486	∠HSiBr	107.8	MW
SiCl		Si—Cl (r_e)	2.058					UV
$SiClH_3$	C_{3v}	Si—Cl	2.049	Si—H	1.486	∠HSiCl	107.9	MW
$SiCl_4$	T_4	Si—Cl	2.019					ED
SiF		Si—F	1.6008					UV
$SiFH_3$	C_{3v}	Si—F	1.593	Si—H	1.486	∠HSiH	110.63	MW, IR
SiF_2		Si—F (r_e)	1.590	∠FSiF (θ_e)	100.8			MW
SiF_3H	C_{3v}	Si—H (r_e)	1.4468	Si—F (r_e)	1.5624	∠HSiF (θ_e)	110.64	MW
SiF_4	T_d	Si—F	1.553					ED
SiH		Si—H (r_e)	1.5201					UV
SiH_3I	C_{3v}	Si—I	2.437	Si—H	1.486	∠HSH	107.8	MW
SiH_4	T_d	Si—H	1.4798					IR
SiN		Si—N (r_e)	1.572					UV
SiO		Si—O (r_e)	1.5097					MW
SiS		Si—S (r_e)	1.9293					MW
SiSe		Si—Se (r_e)	2.0583					MW
Si_2		Si—Si (r_e)	2.246					UV
Si_2Cl_6		Si—Si	2.32	Si—Cl	2.009	∠ClSiCl	109.7	ED
Si_2F_6		Si—Si	2.317	Si—F	1.564	∠FSiF	108.6	ED
Si_2H_6		Si—Si	2.331	Si—H	1.492			ED
		∠SiSiH	110.3	∠HSiH	108.6			
$SnBr_2$		Sn—Br (r_e)	2.501	∠BrSnBr	100.0			ED
SnCl		Sn—Cl (r_e)	2.361					UV
$SnCl_2$		Sn—Cl (r_e)	2.335	∠ClSnCl	99.1			ED
$SnCl_4$	T_d	Sn—Cl	2.281					ED
SnF		Sn—F (r_e)	1.944					UV
SnH		Sn—H (r_e)	1.7815					UV
SnH_4	T_d	Sn—H	1.711					R, IR
SnI_2		Sn—I (r_e)	2.688					ED
SnO		Sn—O (r_e)	1.8325					MW,UV
SnS		Sn—S (r_e)	2.2090					MW
SnSe		Sn—Se (r_e)	2.3256					MW
SnTe		Sn—Te (r_e)	2.5228					MW
SrBr		Sr—Br (r_e)	2.7352					UV
$SrBr_2$	quasilinear	Sr—Br	2.783					
$SrCl_2$		Sr—Cl	2.630	∠ClSrCl	155			ED
SrF		Sr—F (r_e)	2.0754					UV

Mol. form.	Structure	Bond/Angle 1	Value 1	Bond/Angle 2	Value 2	Bond/Angle 3	Value 3	Method
SrH		Sr—H (r_e)	2.1456					UV
SrI		Sr—I (r_e)	2.9436					UV
SrI_2	linear	Sr—I	3.01					ED
SrO		Sr—O (r_e)	1.9198					MW
SrOH		Sr—O	2.111	O—H	0.922			UV
SrS		Sr—S (r_e)	2.4405					UV
$TaBr_5$	D_{3h}	Ta—Br_{eq}	2.412	Ta—Br_{ax}	2.473			ED
$TaCl_5$	D_{3h}	Ta—Cl_{eq}	2.268	Ta—Cl_{ax}	2.315			ED
TaO		Ta—O (r_e)	1.6875					UV
$TbCl_3$	C_{3v}	Tb—Cl	2.476					ED
TeF_6	O_h	Te—F	1.815					ED
TeH		Te—H	1.74					UV
TeO		Te—O (r_e)	1.825					UV
Te_2		Te—Te (r_e)	2.5574					UV
$ThCl_4$	T_d	Th—Cl	2.567					ED
ThF_4	T_d	Th—F	2.124					ED
ThO		Th—O (r_e)	1.84032					UV
$TiBr_4$	T_d	Ti—Br	2.339					ED
$TiCl_3$	D_{3h}	Ti—Cl	2.208					ED
$TiCl_4$	T_d	Ti—Cl	2.170					ED
TiF		Ti—F	1.8342					MW
TiF_4	T_d	Ti—F	1.756					ED
TiI_3	D_{3h}	Ti—I	2.568					ED
TiI_4	T_d	Ti—I	2.546					ED
TiO		Ti—O (r_e)	1.620					UV
TiS		Ti—S (r_e)	2.0825					UV
TlBr		Tl—Br (r_e)	2.6182					MW
TlCl		Tl—Cl (r_e)	2.4848					MW
TlF		Tl—F (r_e)	2.0844					MW
TlH		Tl—H (r_e)	1.870					UV
TlI		Tl—I (r_e)	2.8137					MW
UCl_4	T_d	U—Cl	2.506					ED
UCl_6	O_h	U—F	2.46					ED
UF_4	T_d	U—F	2.059					ED
UF_6	O_h	U—F	2.000					ED
UI_3	C_{3v}	U—I	2.88					ED
VCl_3O	C_{3v}	V—O	1.570	V—Cl	2.142	∠ClVCl	111.3	ED, MW
VBr_4	T_d (Jahn-Teller effect)	V—Br	2.276					ED
VCl_4	T_d (Jahn-Teller effect)	V—Cl	2.138					ED
VF_3	D_{3h}	V—F	1.751					ED
VF_5		V—F_{eq}	1.709	V—F_{ax}	1.736			ED
VMo		V—Mo	1.876					UV
VO		V—O (r_e)	1.5893					UV
$WClF_5$	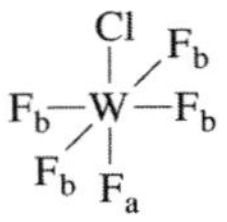	W—F (av.)	1.836	W—Cl	2.251	∠F_aWF_b	88.7	MW
WCl_5	D_{3h}	W—Cl_{eq}	2.243	W—Cl_{ax}	2.293			ED
WCl_6	O_h	W—Cl	2.290					ED
WF_4O	C_{4v}	W—O	1.666	W—F	1.847	∠FWF	86.2	ED
WF_6	O_h	W—F	1.833					ED
XeF_2	linear	Xe—F	1.977					IR
XeF_4	D_{4h}	Xe—F	1.94					ED
XeF_6	O_h	Xe—F	1.890					ED
XeO_4	T_d	Xe—O	1.736					ED
YCl		Y—Cl	2.385					UV
YCl_3		Y—Cl	2.437					ED
YF		Y—F (r_e)	1.9257					UV
YI_3		Y—I	2.817					ED
YO		Y—O (r_e)	1.790					UV
YbBr		Yb—Br (r_e)	2.6454					UV

Mol. form.	Structure	Bond/Angle 1	Value 1	Bond/Angle 2	Value 2	Bond/Angle 3	Value 3	Method
YbH		Yb—H (r_e)	2.0526					UV
$ZnBr_2$	linear	Zn—Br	2.204					ED
$ZnCl_2$	linear	Zn—Cl	2.072					ED
ZnF		Zn—F (r_e)	1.7677					MW
ZnF_2	linear	Zn—F	1.742	[Zn—F (r_e)]	1.729			ED
ZnH		Zn—H (r_e)	1.5949					UV
ZnI_2	linear	Zn—I	2.401					ED
$ZrBr_4$	T_d	Zr—Br	2.465					ED
$ZrCl_4$	T_d	Zr—Cl	2.328					ED
ZrF_4	T_d	Zr—F	1.902					ED
ZrI_4	T_d	Zr—I	2.660					ED
ZrO		Zr—O (r_e)	1.7116					UV

[a] Assumed value.

TABLE 2. Structures of Carbon-Containing Molecules: Bond Distances in Å and Angles in Degrees

Name/Structure	Bond/Angle 1	Value 1	Bond/Angle 2	Value 2	Bond/Angle 3	Value 3	Method
Acetaldehyde 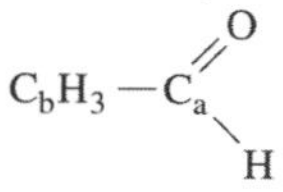	C_a—O	1.210	C_a—C_b	1.515			ED, MW
	C_a—H	1.128	C_b—H	1.107			
	$\angle C_bC_aO$	124.1	$\angle C_bC_aH$	115.3	$\angle HC_bH$	109.8	
Acetamide CH_3CONH_2	C—O	1.220	C—N	1.380			ED
	C—C	1.519	N—H	1.022	C—H	1.124	
	∠CCN	115.1	∠NCO	122.0			
Acetic acid 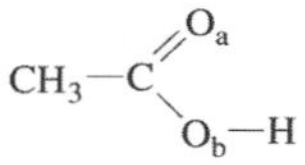	C—C	1.520	C—O_a	1.214	C—O_b	1.364	ED
	C—H	1.10	$\angle CCO_a$	126.6	$\angle CCO_b$	110.6	
Acetone $(CH_3)_2CO$	C—C	1.520	C—O	1.213	C—H	1.103	ED, MW
Symmetry axis of each CH_3 is tilted 2° from the C—C bond	∠CCC	116.0	∠HCH	108.5			
Acetonitrile CH_3CN (C_{3v})	C—N	1.159	C—C	1.468	C—H	1.107	ED, MW
	∠CCH	109.7					
Acetonitrile-*N*-oxide CH_3CNO (C_{3v})	C—C	1.442	C—N	1.169	N—O	1.217	MW
Acetyl chloride CH_3COCl	C—C	1.506	C—O	1.187	C—H	1.105	ED, MW
	C—Cl	1.798	∠HCH	108.6	∠OCCl	121.2	
	∠CCCl	111.6					
Acetylene HC≡CH	C—C (r_e)	1.203	C—H (r_e)	1.060			IR
Acrolein 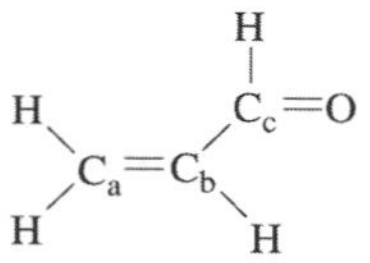 (planar *s-trans* form)	C_a—C_b	1.345	C_b—C_c	1.484	C_c—O	1.217	ED, MW
	C_a—H	1.10	C_c—H	1.13	$\angle HC_cC_b$	114	
	$\angle C_aC_bC_c$	120.3	$\angle C_bC_cO$	123.3	Other CCH (av.)	122	

Name/Structure	Bond/Angle 1	Value 1	Bond/Angle 2	Value 2	Bond/Angle 3	Value 3	Method
Acrylonitrile 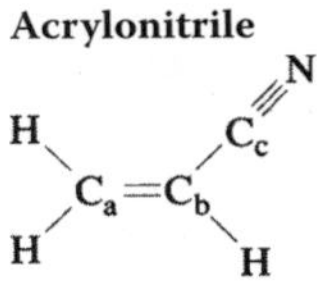	C_a—C_b	1.343	C_b—C_c	1.438	C_c—N	1.167	ED, MW
	C_a—H	1.114	∠C_bC_cN	178	∠$C_aC_bC_c$	121.7	
	∠HCC	120					
Allene							
CH_2=C=CH_2	C—C	1.3084	C—H	1.087	∠HCH	118.2	IR
Aniline							
$C_6H_5NH_2$	C—C	1.392	C—N	1.431	N—H	0.998	MW
	∠HNH	113.9	dihedral angle between NH_2 plane and N— C bond	140.6			
Azetidine CH_2—CH_2, CH_2—NH (ring)	C—N	1.482	C—C	1.553			ED
	C—H	1.107	N—H	1.03			
	∠CCC	86.9	∠CCN	85.8	∠CNC	92.2	
	dihedral angle between CCC and CNC planes	147					
Benzamide							
C_6H_5—C_aONH_2	C—C (ring)	1.401	C (ring)—C_a	1.511	C_a—O	1.225	ED
	C—H	1.112	C—N	1.380			
	∠CCN	117.8	∠CCC (ring) (ass.)	120	∠CCO	121.2	
Benzene							
C_6H_6	C—C	1.399	C—H	1.101			ED, IR
***p*-Benzoquinone** 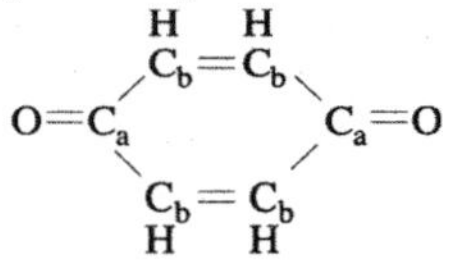	C_a—O	1.225	C_a—C_b	1.481	C_b—C_b	1.344	ED
	∠$C_bC_aC_b$	118.1					
Bicyclo[1.1.0]butane 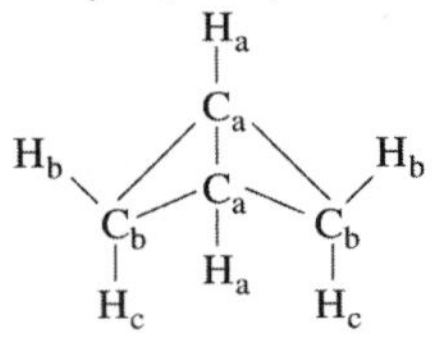	C_a—C_a	1.497	C_a—C_b	1.498	C_a—H_a	1.071	MW
	C_b—H_b	1.093	C_b—H_c	1.093	∠$H_bC_bH_c$	115.6	
	∠$C_bC_aH_a$	130.4	∠$C_aC_aH_a$	128.4	∠$C_aC_bC_a$	60.0	
	dihedral angle between the two $C_aC_aC_b$ planes	121.7					
Bicyclo[2.2.1]heptane							
See preceding structure C_7H_{12}	C_a—C_b	1.54	C_b—C_b	1.56	C_a—C_c	1.56	ED
	C—C (av.)	1.549	∠$C_aC_cC_a$	93.1			
	dihedral angle between the two $C_aC_bC_bC_a$ planes	113.1					
Bicyclo[2.2.0]hexa-2,5-diene 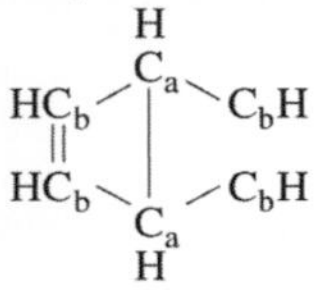	C_b—C_b	1.345	C_a—C_a	1.574	C_a—C_b	1.524	ED

Name/Structure	Bond/Angle 1	Value 1	Bond/Angle 2	Value 2	Bond/Angle 3	Value 3	Method
	dihedral angle between the two $C_aC_bC_bC_a$ planes	117.3					
Bicyclo[2.2.2]octane							
$HC_a(C_bH_2C_bH_2)_3C_aH$	C_a—C_b	1.54	C_b—C_b	1.55	C—C (av.)	1.542	ED
large-amplitude torsional motion about D_{3h} symmetry axis	$\angle C_aC_bC_b$	109.7					
Bicyclo[1.1.1]pentane							
C_5H_8	C—C	1.557	$\angle$CCC	74.2			ED
Bicyclo[2.1.0]pentane							
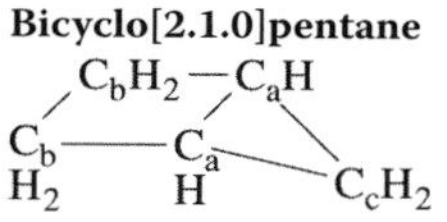	C_a—C_a	1.536	C_b—C_b	1.565	C_a—C_c	1.507	MW
	C_a—C_b	1.528	dihedral angle between the $C_aC_aC_bC_b$ and $C_aC_aC_c$ planes	112.7			
Biphenyl							
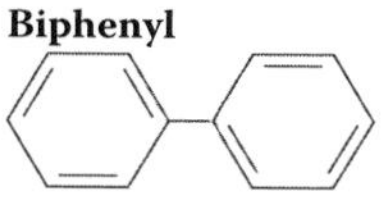	C—C (intraring)	1.396	C—C (interring)	1.49			ED
	torsional dihedral angle between the two rings	40[a]					
4,4′-Bipyridine							
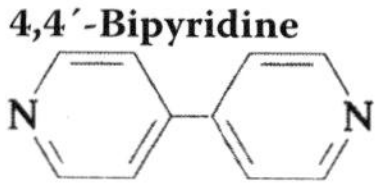	C—C (interring)	1.465	C—C (intraring)	1.375	C—N (intraring)	1.375	ED
	torsional dihedral angle between the two rings	37[a]					
Bis(cyclopentadienyl) beryllium							
$(C_5H_5)_2Be$ (C_{5v})	Be—(cyclopentadienyl plane)	[c]	C—C	1.423			ED
Bis(cyclopentadienyl) iron [Ferrocene]							
$(C_5H_5)_2Fe$ (D_{5h})	Fe—C	2.064	C—C	1.440	C—H	1.104	ED
Bis(cyclopentadienyl) lead							
$(C_5H_5)_2Pb$ (D_{5h})	Pb—C	2.79	C—C	1.430			ED
	dihedral angle between the two C_5H_5 planes	[d]					
Bis(cyclopentadienyl) manganese [Manganocene]							
$(C_5H_5)_2Mn$ (D_{5h})	Mn—C	2.383	C—C	1.429			ED
Bis(cyclopentadienyl) nickel [Nickelocene]							
$(C_5H_5)_2Ni$ (D_{5h})	Ni—C	2.196	C—C	1.430			ED
Bis(cyclopentadienyl) ruthenium [Ruthenocene]							
$(C_5H_5)_2Ru$ (D_{5h})	Ru—C	2.196	C—C	1.439			ED
Bis(cyclopentadienyl) tin							
$(C_5H_5)_2Sn$ (D_{5h})	Sn—C	2.71	C—C	1.431	C—H	1.14	ED
Borane carbonyl							
BH_3CO (C_{3v})	C—O	1.131	B—C	1.540	B—H	1.194	MW
	$\angle$BCO	180	$\angle$HBH	113.9			
Bromobenzene							
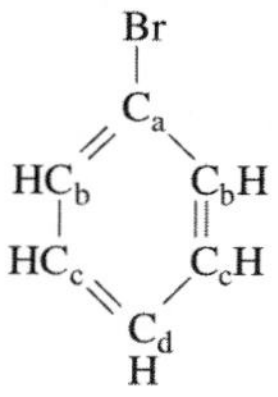	C_a—C_b	1.42	C_b—C_c	1.375	C_c—C_d	1.401	MW
	C—Br	1.85	C—H	1.072	$\angle C_bC_aC_b$	117.4	

Name/Structure	Bond/Angle 1	Value 1	Bond/Angle 2	Value 2	Bond/Angle 3	Value 3	Method
Bromochloroacetylene							
ClC≡CBr	C—Cl	1.636	C—Br	1.784	C—C	1.206	ED
Bromoiodoacetylene							
IC≡CBr	C—I	1.972	C—Br	1.795	C—C	1.206	ED
Bromomethane [Methyl bromide]							
CH_3Br	C—Br (r_e)	1.933	C—H (r_e)	1.086	∠HCH (θ_e)	111.2	MW, IR
Bromomethyl							
CH_2Br (planar)	C—Br	1.848	C—H	1.084	∠HCH (ass.)	124.5	MW
Bromomethylene							
CHBr (bent)	C—Br	1.857	C—H	1.110	∠HCH	101.0	UV
Bromomethylmercury							
CH_3HgBr (C_{3v})	C—Hg	2.07	Hg—Br	2.406			MW
1,3-Butadiene							
C_aH_2	C_a—C_b	1.349	C_b—C_b	1.467	C—H (av.)	1.108	ED
	∠CCC	124.4	∠C_bC_aH	120.9			
(C_{2h})							
1,3-Butadiyne							
$HC_a≡C_bC_b≡C_aH$ (linear)	C_a—C_b	1.218	C_b—C_b	1.384	C—H	1.09	ED
Butane							
$CH_3CH_2CH_2CH_3$	C—C	1.531	C—H	1.117	∠CCC	113.8	ED
	∠CCH	111.0	dihedral angle for the *gauche* conformer	65			
2,3-Butanedione							
$CH_3COCOCH_3$	C—O	1.215	C—C (av.)	1.524	C—H	1.108	ED
trans conformer	∠CCC	116.2	∠CCO	119.5			
2-Butanone							
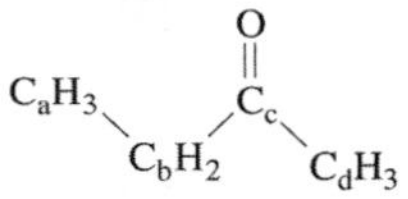	C—C (av.)	1.518	C_c—O	1.219	C—H (av.)	1.102	ED
trans conformer	∠$C_aC_bC_c$	113.5	∠C_bC_cO	121.9	∠C_dC_cO	121.9	
1,2,3-Butatriene							
H_2C_a=C_b=C_b=C_aH_2 (D_{2h})	C_a—C_b	1.32	C_b—C_b	1.28	C—H	1.08	ED
***cis*-2-Butene**							
$C_aH_3C_bH$=$C_bHC_aH_3$	C_a—C_b	1.506	C_b—C_b	1.346	∠$C_aC_bC_b$	125.4	ED
***trans*-2-Butene**							
$C_aH_3C_bH$=$C_bHC_aH_3$	C_a—C_b	1.508	C_b—C_b	1.347	∠$C_aC_bC_b$	123.8	ED
1-Buten-3-yne							
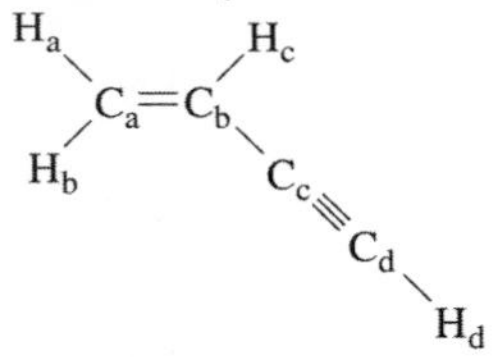	C_a—C_b	1.344	C_b—C_c	1.434	C_c—C_d	1.215	ED, MW
	C_a—H_a	1.11	C_d—H_d	1.09			
	∠$C_aC_bC_c$	123.1	∠$C_bC_cC_d$	178	∠$H_aC_aC_b$	119	
	∠$H_bC_aC_b$	122	∠$H_cC_bC_a$	122	∠$C_cC_dH_d$	182	
***tert*-Butyl chloride**							
$(CH_3)_3CCl$	C—C	1.528	C—Cl	1.828	C—H	1.102	ED, MW
	∠CCCl	107.3	∠CCH	110.8	∠CCC	111.6	
2-Butyne							
C_aH_3—C_b≡C_b—C_aH_3	C_b—C_b	1.214	C_a—C_b	1.468	C—H	1.116	ED
	∠C_bC_aH	110.7					
Carbon dimer							
C_2	C—C (r_e)	1.2425					UV

Name/Structure	Bond/Angle 1	Value 1	Bond/Angle 2	Value 2	Bond/Angle 3	Value 3	Method
Carbon trimer							
C_3 (linear)	C—C	1.277					UV
Carbon dioxide							
CO_2 (linear)	C—O (r_e)	1.1600					IR
Carbon disulfide							
CS_2 (linear)	C—S (r_e)	1.5526					IR
Carbon monobromide							
CBr	C—Br	1.8209					UV
Carbon monoselenide							
CSe	C—Se (r_e)	1.67609					UV
Carbon monosulfide							
CS	C—S (r_e)	1.5349					MW
Carbon monoxide							
CO	C—O (r_e)	1.1283					MW
Carbon oxyselenide							
OCSe (linear)	C—O	1.159	C—Se	1.709			MW
Carbon oxysulfide							
OCS (linear)	C—O (r_e)	1.1578	C—S (r_e)	1.5601			MW
Carbon phosphide							
CP	C—P (r_e)	1.562					UV
Carbon sulfide selenide							
SCSe (linear)	C—S	1.553	C—Se	1.693			MW
Carbon sulfide telluride							
SCTe (linear)	C—S	1.557	C—Te	1.904			MW
Carbon suboxide							
OCCCO (linear)	C—C	1.289	C—O	1.163			ED
Carbonyl bromide							
$COBr_2$	C—O	1.178	C—Br	1.923	∠BrCBr	112.3	ED, MW
Carbonyl chloride							
$COCl_2$	C—O	1.179	C—Cl	1.742	∠ClCCl	111.8	ED, MW
Carbonyl chloride fluoride							
COClF	C—O	1.173	C—F	1.334	C—Cl	1.725	ED, MW
	∠ClCO	127.5	∠FCCl	108.8			
Carbonyl dicyanide							
$CO(CN)_2$	C—O	1.209	C—C	1.466	C—N	1.153	ED, MW
	∠CCC	115	∠CCN	180			
Carbonyl fluoride							
COF_2	C—O	1.172	C—F	1.3157	∠FCF	107.71	ED, MW
Chloroacetylene							
HC≡CCl	C—Cl	1.6368	C—C	1.2033	C—H	1.0550	MW
Chlorobenzene							
C_6H_5Cl	C—C	1.400	C—Cl	1.737	C—H	1.083	ED
Chlorocyanoacetylene							
ClC≡C—CN	C—Cl	1.624	C—N	1.160	C—C	1.205	ED
	C—CN	1.362					
Chloroethane [Ethyl chloride]							
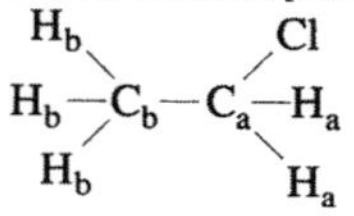	C—C	1.528	C—Cl	1.802	C—H	1.103	ED, MW
	∠CCCl	110.7	∠$H_bC_bH_b$	109.8	∠$H_aC_aH_a$	109.2	
	∠$C_bC_aH_a$	110.6	C_a—H_a= C_b—$_H$b (ass.)				

Name/Structure	Bond/Angle 1	Value 1	Bond/Angle 2	Value 2	Bond/Angle 3	Value 3	Method
2-Chloroethanol							
$ClCH_2CH_2OH$	C—O	1.413	C—C	1.519	C—Cl	1.801	ED
(*gauche*)	O—H	1.033	C—H	1.093			
	∠CCCl	110.7	∠CCO	113.8	dihedral angle of internal rotation	62.4	
Chloroiodoacetylene							
ClC≡CI	C—Cl	1.63	C—I	1.99	C—C (ass.)	1.209	MW
Chloromethane [Methyl chloride]							
CH_3Cl	C—Cl	1.785	C—H	1.090	∠HCH	110.8	MW, IR
Chloromethylidyne							
CCl	C—Cl	1.6512					UV
Chloromethylmercury							
CH_3HgCl (C_{3v})	C—Hg	2.06	Hg—Cl	2.282			MW
***trans*-1-Chloropropene**							
$CH_3CH{=}CHCl$	C—Cl	1.728	∠CCCl	121.9			MW
3-Chloropropene							
$CH_2ClCH{=}CH_2$ *cis* conformer	C—Cl	1.811	∠CCCl	115.2			MW
skew conformer	C—Cl	1.809	∠CCCl	109.6	dihedral angle of internal rotation	122.4	
Chlorotrifluoromethane							
$CClF_3$(C_{3v})	C—Cl	1.752	C—F	1.325	∠FCF	108.6	ED, MW
Chromium carbonyl							
Cr $(CO)_6$	Cr—C	1.92	C—O	1.16	∠CrCO	180	ED
Cobalt cyanide							
CoC≡N	Co—C	1.883	C—N	1.131			MW
Copper cyanide							
CuC≡N	Cu—C	1.832	C—N	1.158			MW
Cyanamide							
$H_2N_aCN_b$	N_a—C	1.346	C—N_b	1.160	N—H	1.00	MW
	∠HNH	114	dihedral angle between NH_2 plane and N— C bond	142			
Cyanide							
CN	C—N(r_e)	1.1718					MW
Cyanoacetylene							
$HC_a{\equiv}C_b{-}C_cN$	C_a—C_b	1.205	C_b—C_c	1.378	C—H	1.058	MW
	C_c—N	1.159					
Cyanocyclopropane [Cyclopropanecarbonitrile]							
$C_3H_5C_aN$	C—C (ring)	1.513	C—C_a	1.472	C_a—N	1.157	MW
	C—H	1.107	∠C_aCH	119.6	∠HCH	114.6	
Cyanogen							
N≡C—C≡N (linear)	C—N	1.163	C—C	1.393			ED
Cyanogen azide							
N≡C—N=N≡N	C—N	1.312	N=N	1.252	N≡N	1.133	MW
(planar)	C≡N	1.164	∠CNN	120.2	∠NCN	176.0	
Cyanogen bromide							
BrCN (linear)	C—N(r_e)	1.157	C—Br(r_e)	1.790			MW
Cyanogen chloride							
ClCN (linear)	C—Cl(r_e)	1.629	C—N(r_e)	1.160			MW
Cyanogen fluoride							
FCN (linear)	C—F	1.262	C—N	1.159			MW
Cyanogen iodide							
ICN (linear)	C—I	1.995	C—N	1.159			MW

Name/Structure	Bond/Angle 1	Value 1	Bond/Angle 2	Value 2	Bond/Angle 3	Value 3	Method
1-Cyano-2-propyne							
$HC_a{\equiv}C_bC_cH_2C_d{\equiv}N$	C_a—C_b (ass.)	1.207	C_b—C_c (ass.)	1.465	C_c—C_d	1.454	MW
	C_d—N (ass.)	1.159	C_a—H (ass.)	1.057	C_c—H (ass.)	1.090	
	∠$C_bC_cC_d$	113.4	∠HC_cH (ass.)	109.4	∠C_bC_cH	111.3	
Cyclobutane							
$(CH_2)_4$	C—C	1.555	C—H	1.113			ED
	dihedral angle between the two CCC planes	145					
Cyclobutanone							
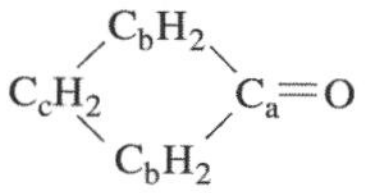	C_a—C_b	1.527	C_b—C_c	1.556			MW
	∠$C_bC_aC_b$	93.1	∠$C_aC_bC_c$	88.0			
Cyclobutene							
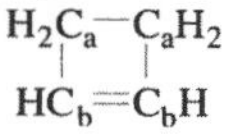	C_a—C_a	1.566	C_b—C_b	1.342	C_a—C_b	1.517	MW
	C_a—H	1.094	C_b—H	1.083			
	∠$C_aC_bC_b$	94.2	∠C_bC_bH	133.5	∠HC_aH	109.2	
	∠C_aC_aH	114.5	∠$C_aC_aC_b$	85.8	dihedral angle between CH_2 plane and C_a—C_a bond	135.8	
2,4,6-Cycloheptatrien-1-one							
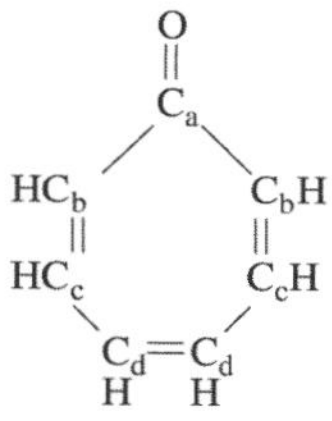 (C_{2v})	C_a—C_b	1.45	C_b—C_c	1.36	C_c—C_d	1.46	ED
	C_d—C_d	1.34	C_a—O	1.23	∠$C_bC_aC_b$	122	
	∠$C_aC_bC_c$	133	∠$C_bC_cC_d$	126	∠$C_cC_dC_d$	130	
Cyclohexane							
C_6H_{12} (chair form)	C—C	1.536	C—H	1.119	∠CCC	111.3	ED
Cyclohexene							
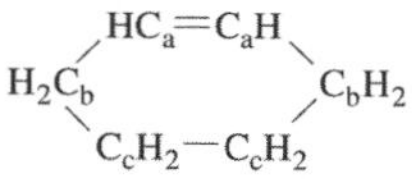 half-chair form (C_2)	C_a—C_a	1.334	C_a—C_b	1.50	C_b—C_c	1.52	ED
	C_c—C_c	1.54	∠$C_aC_aC_b$	123.4	∠C_aC_bCc	112.0	
	∠$C_bC_cC_c$	110.9					
Cyclooctatetraene							
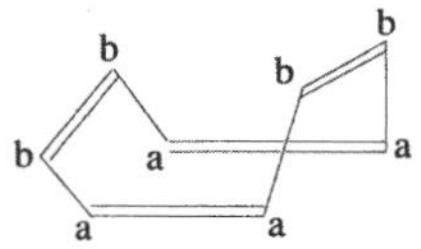 tub form (D_{2d})	C_a—C_b	1.476	C_a—C_a	1.340	C_b—C_b	1.340	ED
	C—H	1.100	∠$C_bC_aC_a$	126.1	∠$C_aC_bC_b$	126.1	
	dihedral angle between $C_aC_aC_aC_a$ and $C_aC_bC_bC_a$ planes	136.9					
1,3-Cyclopentadiene							
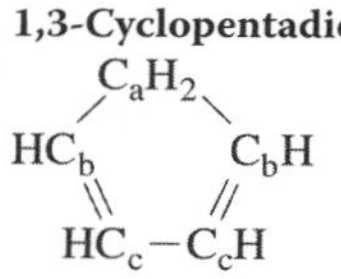	C_a—C_b	1.509	C_b—C_c	1.342	C_c—C_c	1.469	MW
	∠$C_aC_bC_c$	109.3	∠$C_bC_cC_c$	109.4	∠$C_bC_aC_b$	102.8	

Molecular

Name/Structure	Bond/Angle 1	Value 1	Bond/Angle 2	Value 2	Bond/Angle 3	Value 3	Method
Cyclopentadienylindium 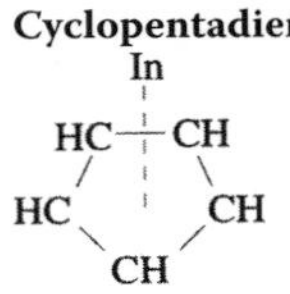	C—In	2.621	C—C	1.426	(C_{5v})		ED
Cyclopentane $(CH_2)_5$	C—C	1.546	C—H	1.114	∠CCH	111.7	ED
Cyclopentene 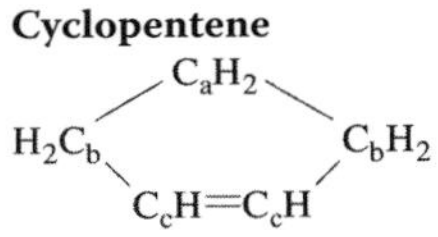	C_a—C_b	1.546	C_b—C_c	1.519	C_c—C_c	1.342	ED
	∠$C_aC_bC_c$	103.0	∠$C_bC_cC_c$	110.0	∠$C_bC_aC_b$	104.0	
	dihedral angle between $C_bC_aC_b$ and $C_bC_cC_cC_b$ planes	151.2					
Cyclopropane $(CH_2)_3$	C—C	1.512	C—H	1.083	∠HCH	114.0	R
Cyclopropanone 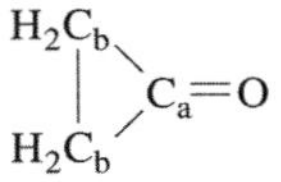	C_a—C_b	1.475	C_b—C_b	1.575	C_a—O	1.191	MW
	C—H	1.086	∠$C_aC_bC_b$	57.7	∠HC_bH	114	
	dihedral angle between CH_2 plane and C_b—C_b bond	151					
Cyclopropene 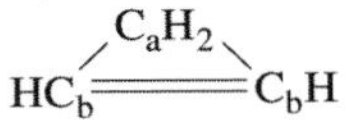	C_a—C_b	1.505	C_b—C_b	1.293	C_a—H	1.085	MW
	C_b—H	1.072	∠C_bC_bH	150	∠HC_aH	114.3	
Cyclopropenone 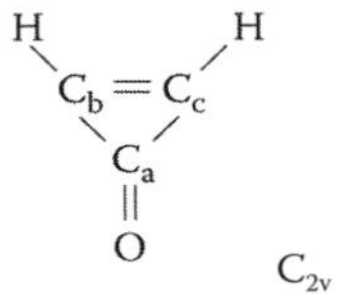	C_a—C_b (r_s)	1.423	C_b—C_c (r_s)	1.349	C_a—O (r_s)	1.212	MW
	C—H (r_s)	1.079	∠HC_bC_c (θ_s)	144.3	$C_bC_aC_c$ (θ_s)	56.6	
Decalin $C_{10}H_{18}$	C—C (av.)	1.530	C—H (av.)	1.113	∠CCC (av.)	111.4	ED
Diazirine 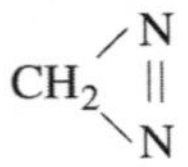	C—N	1.482	N—N	1.228	C—H	1.09	MW
	∠HCH	117					
Diazoacetonitrile 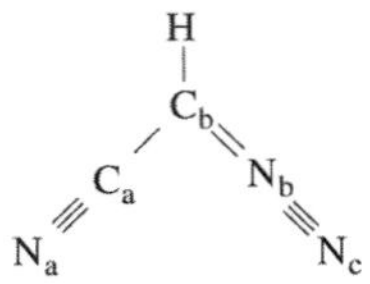	C_a—C_b	1.424	C_a—N_a	1.165	C_b—N_b	1.280	MW
	N_b—N_c	1.132	C—H	1.082			
	∠C_aC_bH	117	∠$C_aC_bN_b$	119.5			
Diazomethane CH_2N_2	C—N	1.32	N—N	1.12	C—H	1.075	MW, IR
	∠HCH	126.0					
1,2-Dibromoethane CH_2BrCH_2Br	C—C	1.506	C—Br	1.950	C—H	1.108	ED

Name/Structure	Bond/Angle 1	Value 1	Bond/Angle 2	Value 2	Bond/Angle 3	Value 3	Method
	∠CCBr	109.5	∠CCH	110			
Dibromomethane							
CH_2Br_2	C—Br	1.924	C—H	1.08	∠HCBr	109	ED
	∠BrCBr	113.2					
2,2'-Dichlorobiphenyl							
C_6H_4Cl—C_6H_4Cl	C—C (rings)	1.398	C—C (interring)	1.495	C—H	1.10	ED
	C—Cl	1.732	∠CCCl	121.4	∠CCH	126	
	dihedral angle between the two rings (defined as 0 for *cis* conformer)	74					
***trans*-1,4-Dichlorocyclohexane**							
$C_6H_{10}Cl_2$	C—C	1.530	C—Cl	1.810	C—H	1.102	ED
	∠CCC	111.5					
equatorial:	∠CCCl	108.6	∠HCCl	111.5			
axial:	∠CCCl	110.6	∠HCCl	107.6			
1,1-Dichloroethane							
$CHCl_2CH_3$	C—C	1.540	C—Cl	1.766			MW
	∠ClCCl	112.0	∠CCCl	111.0			
1,2-Dichloroethane							
CH_2ClCH_2Cl	C—C	1.531	C—Cl	1.790	C—H	1.11	ED
	∠CCCl	109.0	∠CCH	113			
1,1-Dichloroethene							
CH_2=CCl_2 (C_{2v})	C—C (ass.)	1.32	C—Cl	1.73			MW
	∠ClCC	123					
***cis*-1,2-Dichloroethene**							
CHCl=CHCl	C—C	1.354	C—Cl	1.718			ED
	∠ClCC	123.8					
Dichloromethane							
CH_2Cl_2	C—Cl (r_e)	1.765	C—H (r_e)	1.087			MW, IR
	∠ClCCl (θ_e)	112.0	∠HCH (θ_e)	111.5			
1,2-Dicyanocyclobutene							
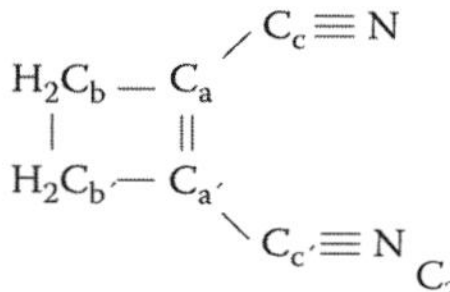 C_{2v}	C_a—$C_{a'}$	1.361	C_a—C_b	1.515	C_b—$C_{b'}$	1.567	MW
	C_a—C_c	1.420	C_c—N	1.157	C_b—H	1.088	
	$\angle C_{a'}C_aC_b$	93.9	$\angle C_aC_bC_{b'}$	86.1	$\angle C_aC_cN$	178.2	
	$\angle C_bC_aC_c$	133.3	$\angle C_aC_bH$	114.7	$\angle C_aC_aC_bH$	115.8	
Difluorocyanamide							
F_2N_b—C≡N_a	C—N_a	1.158	C—N_b	1.386	N_b—F	1.399	MW
	$\angle N_aCN_b$	174	$\angle CN_bF$	105.4	$\angle FN_bF$	102.8	
Difluorocyclopropenone							
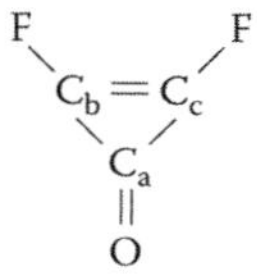 C_{2v}	C_a—C_b	1.453	C_b—C_c	1.324	C_a—O	1.192	MW
	C—F	1.314	$\angle FC_bC_c$	145.7			
Difluorodimethylsilane							
$(CH_3)_2SiF_2$	C—Si	1.844	Si—F	1.585	C—H (ass.)	1.093	MW
	∠CSiC	115.2	∠FSiF	106.1	∠SiCH (ass.)	110.8	
1,1-Difluoroethane							
CH_3CHF_2	C—C	1.498	C—F	1.364	C—H (av.)	1.081	ED
	∠CCF	110.7	∠CCH (av.)	111.0	dihedral angle between CCF planes	118.9	

Name/Structure	Bond/Angle 1	Value 1	Bond/Angle 2	Value 2	Bond/Angle 3	Value 3	Method
1,2-Difluoroethane							
CH_2FCH_2F	C—C	1.503	C—F	1.389	C—H	1.103	ED
	∠CCF	110.3	∠CCH	111	dihedral angle of internal rotation	109	
1,1-Difluoroethene							
$CH_2=CF_2$	C—C	1.340	C—F	1.315	C—H	1.091	ED, MW
	∠CCF	124.7	∠CCH	119.0			
***cis*-1,2-Difluoroethene**							
CHF=CHF	C—C	1.33	C—F	1.342	C—H	1.099	ED, MW
	∠CCF	122.0	∠CCH	124.1			
Difluoromethane							
CH_2F_2	C—F	1.357	C—H	1.093			MW
	∠FCF	108.3	∠HCH	113.7			
Dimethoxymethane							
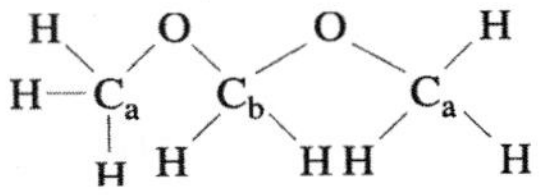	C_a—O	1.432	C_b—O	1.382	C—H (av.)	1.108	ED
	∠COC	114.6	∠OCO	114.3	∠OCH	110.3	
Dimethylamine							
$(CH)_2NH$	C—N	1.455	N—H	1.00	C—H	1.106	ED
	∠CNC	111.8	∠CNH	107	∠NCH	112	
	∠HCH	107					
Dimethylberyllium							
$(CH_3)_2Be$ (CBeC linear)	C—Be	1.698	C—H	1.127	∠BeCH	113.9	ED
Dimethyl cadmium							
$(CH_3)_2Cd$	C—Cd	2.112	∠HCH	108.4			R
Dimethyl carbonate							
$(C_aH_3O_a)_2C_b=O_b$	C_b—O_b	1.209	C_b—O_a	1.34	C_a—O_a	1.42	ED
	$\angle O_aC_bO_a$	107	$\angle C_bO_aC_a$	114.5			
Dimethylcyanamide							
$(C_aH_3)_2Na—C_b≡N_b$	C_b—N_b	1.161	C_a—N_a	1.463	C_b—N_a	1.338	ED
***trans*-Dimethyldiazene**							
$CH_3N=NCH_3$	C—N	1.482	N—N	1.247	∠CNN	112.3	ED
	$\angle C_aNC_a$	115.5	$\angle C_aNC_b$	116.0			
1,2-Dimethyldiborane							
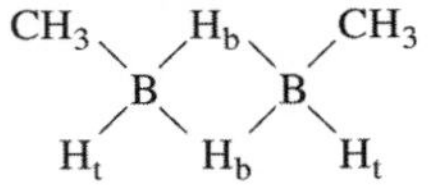	B—B	1.799	B—C	1.580			ED
	B—H_b (*cis*)	1.358	B—H_b (*trans*)	1.365	B—H_t	1.24	
	∠BBC (*cis*)	122.6	∠BBC (*trans*)	121.8			
Dimethyl diselenide							
$(CH_3)_2Se_2$	C—Se	1.95	Se—Se	2.326	C—H	1.13	ED
	∠CSeSe	98.9	∠HCSe	108	CSeSeC dihedral angle	88	
Dimethyl disulfide							
$(CH_3)_2S_2$	C—S	1.816	S—S	2.029	C—H	1.105	ED
	∠SSC	103.2	∠SCH	111.3	CSSC dihedral angle	85	
***S,S′*-Dimethyl dithiocarbonate**							
$C_aH_3SC_bSC_aH_3$ (with =O on C_b) *syn-syn* conformer	C_a—S	1.802	C_b—S	1.777	Cb—O	1.206	ED
	∠OCS	124.9	∠CSC	99.3			
Dimethyl ether							
$(CH_3)_2O$	C—O	1.416	C—H	1.121			ED
	∠COC	112	∠HCH	108			

Name/Structure	Bond/Angle 1	Value 1	Bond/Angle 2	Value 2	Bond/Angle 3	Value 3	Method
***N,N'*-Dimethylhydrazine**							
$CH_3NH—NHCH_3$	C—N	1.46	N—N	1.42	N—H	1.03	ED
	C—H	1.12	∠NNC	112	CNNC dihedral angle	90	
Dimethyl mercury							
$(CH_3)_2Hg$	C—Hg	2.083	C—H (ass.)	1.160	Hg···H	2.71	ED
Dimethylphosphine							
$(CH_3)_2PH$	C—P	1.848	P—H	1.419			MW
	∠CPC	99.7	∠CPH	97.0			
2,2-Dimethylpropanenitrile							
$(C_cH_3)_3C_b—C_a≡N$	$C_a—C_b$	1.495	$C_b—C_c$	1.536	$C_a—N$	1.159	MW
	$∠C_cC_bC_c$	110.5					
Dimethyl selenide							
$(CH_3)_2Se$	C—Se	1.943	C—H	1.093			MW
	∠CSeC	96.2	∠SeCH	108.7	∠HCH	110.3	
Dimethylsilane							
$(CH_3)_2SiH_2$	C—Si	1.868	C—H	1.089	Si—H	1.482	MW
	∠CSiC	110.9	∠CSiH	109.5	∠SiCH	110.9	
	∠HSiH	107.8					
Dimethyl sulfide							
$(CH_3)_2S$	C—S	1.802	C—H	1.090			ED, MW
	∠CSC	98.80	∠HCH	109.3			
Dimethyl sulfone							
$(CH_3)_2SO_2$	C—S	1.771	S—O	1.435	C—H	1.114	ED
	∠CSC	102	∠OSO	121			
Dimethyl sulfoxide							
$(CH_3)_2SO$	C—S	1.799	S—O	1.485	C—H	1.081	MW
	∠CSC	96.6	∠CSO	106.7	∠HCH	110.3	
	dihedral angle between SCC plane and S—O bond	115.5					
Dimethyl zinc							
$(CH_3)_2Zn$	C—Zn	1.929	∠HCH	107.7			R
1,4-Dioxane							
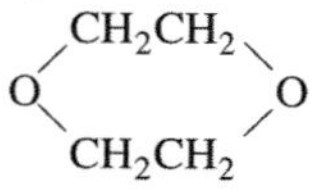	C—C	1.523	C—O	1.423	C—H	1.112	ED
	∠CCO	109.2	∠COC	112.45			
Ethane							
C_2H_6	C—C	1.5351	C—H	1.0940	∠CCH	111.17	MW
staggered conformation	C—C (r_e)	1.522					
1,2-Ethanediamine							
$H_2NCH_2CH_2NH_2$	C—C	1.545	C—N	1.469	C—H	1.11	ED
gauche conformer	∠CCN	110.2	dihedral angle between NCC and CCN planes	64			
Ethanethiol							
$C_bH_3—C_aH_2—SH$	$C_a—C_b$	1.530	$C_a—S$	1.829	S—H	1.350	MW
	$C_a—H$	1.090	$C_b—H$	1.093	$∠C_aSH$	96.4	
	$∠C_bC_aS$	108.3	$∠C_bC_aH$	109.6	$∠C_aC_bH$	109.7	
Ethanol							
$C_bH_3C_aH_2OH$	C—C	1.512	C—O	1.431	O—H	0.971	MW
staggered conformation	$C_a—H$	1.10	$C_b—H$	1.09	∠COH	105	
	∠CCO	107.8	$∠C_bC_aH$	111	$∠C_aC_bH$	110	
Ethylene							
$CH_2{=}CH_2$	C—C (r_s)	1.329	C—H (r_s)	1.082	∠HCH (θ_s)	117.2	MW, IR

Name/Structure	Bond/Angle 1	Value 1	Bond/Angle 2	Value 2	Bond/Angle 3	Value 3	Method
Ethyleneimine	C—C	1.481	N—C	1.475			MW
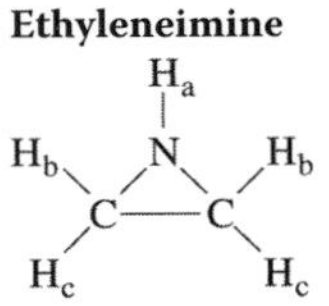							
	C—H	1.084	N—H	1.016			
	∠CNC	60.3	$\angle H_aNC$	109.3	$\angle H_bCH_c$	115.7	
	$\angle H_bCC$	117.8	$\angle H_bCN$	118.3	$\angle H_cCC$	119.3	
	$\angle H_cCN$	114.3					
Ethyl methyl ether							
$C_2H_5OCH_3$	C—C	1.520	C—O (av.)	1.418	C—H (av.)	1.118	ED
	∠COC	111.9	∠OCC	109.4	∠HCH	109.0	
Ethyl methyl sulfide							
$C_2H_5SCH_3$	C—C	1.536	C—S (av.)	1.813	C—H	1.111	ED
gauche conformer	∠CSC	97	∠SCC	114.0	∠HCH	110	
Fluoroketene							
HFC=C=O	C—C	1.317	C—O	1.167	C—F	1.360	MW
	C—H	1.102	∠CCO	178.0	∠CCF	119.5	
	∠CCH	122.3					
Fluoromethane [Methyl fluoride]							
CH_3F	C—F (r_e)	1.382	C—H (r_e)	1.095	∠HCH (θ_e)	110.45	MW, IR
Fluoromethylidyne							
CF	C—F (r_e)	1.2718					UV
(Fluoromethylidyne) phosphine							
FC≡P	C—F	1.285	C—P	1.541			MW
2-Fluoropropane							
CH_3CHFCH_3	C—C	1.522	C—F	1.398			MW
	∠CCC	113.4	∠CCF	108.2			
Formaldehyde							
H_2CO	C—O	1.208	C—H	1.116	∠HCH	116.5	MW
Formaldehyde azine							
H_2C=N—N=CH_2	C—N	1.277	N—N	1.418	C—H	1.094	ED
trans conformer	∠CNN	111.4	∠HCN	120.7			
Formaldehyde oxime							
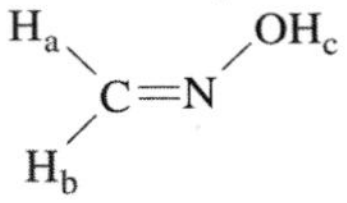	C—N	1.276	N—O	1.408	$O—H_c$	0.956	MW
	$C—H_a$	1.085	$C—H_b$	1.086	∠CNO	110.2	
	$\angle H_aCN$	121.8	$\angle H_bCN$	115.6	$\angle NOH_c$	102.7	
Formamide							
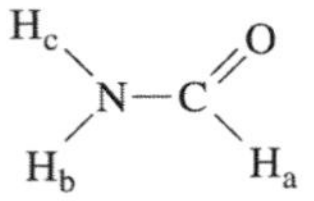	C—N	1.368	C—O	1.212	$C—H_a$	1.125	ED, MW
	N—H	1.027	∠CNH (av.)	119.2	∠NCO	125.0	
Formic acid							
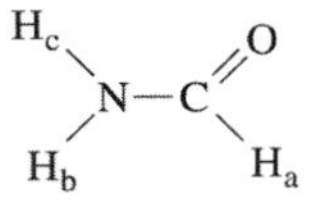	$C—O_a$	1.202	$C—O_b$	1.343	$O_b—H$	0.972	MW
	C—H	1.097					
	$\angle O_aCO_b$	124.9	$\angle HCO_a$	124.1	$\angle CO_bH$	106.3	

Molecular

Name/Structure	Bond/Angle 1	Value 1	Bond/Angle 2	Value 2	Bond/Angle 3	Value 3	Method
Formic acid dimer 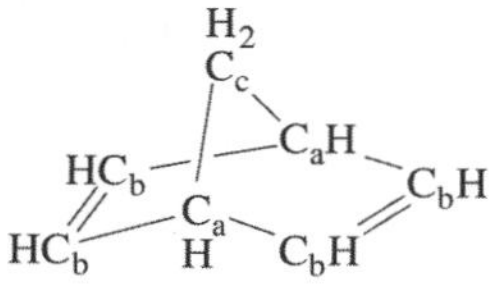	C—O_a	1.220	C—O_b	1.323	$O_a \cdots O_b$	2.703	ED
	$\angle O_aCO_b$	126.2	$\angle CO_aO_b$	108.5			
Formyl [Oxomethyl] HC=O	C—O	1.1712	C—H	1.110	∠HCO	127.43	MW
Fulvene 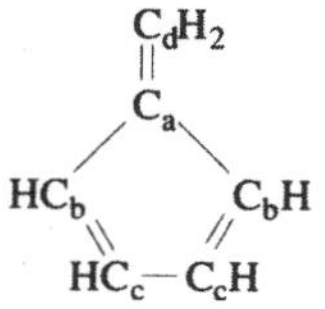	C_a—C_d	1.349	C_a—C_b	1.470	C_b—C_c	1.355	MW
	C_c—C_c	1.476	C_b—H	1.078	C_c—H	1.080	
	C_d—H	1.13	$\angle C_bC_aC_b$	106.6	$\angle C_bC_cC_c$	109	
	$\angle C_aC_bC_c$	107.7	$\angle C_aC_bH$	124.7	$\angle C_bC_cH$	126.4	
	$\angle HC_dH$	117					
Furan 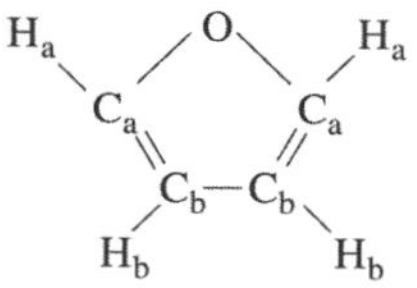	C_a—C_b	1.361	C_b—C_b	1.431	C_a—O	1.362	MW
	C_a—H_a	1.075	C_b—H_b	1.077			
	$\angle C_aC_bC_b$	106.1	$\angle C_bC_aO$	110.7	$\angle C_aOC_a$	106.6	
	$\angle C_bC_bH_b$	128.0	$\angle OC_aH_a$	115.9			
Furfural 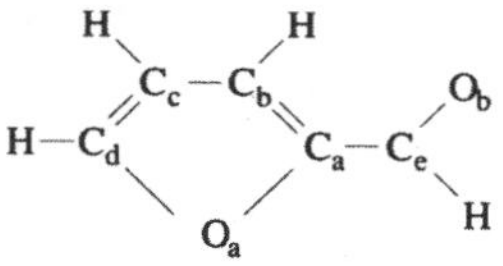	C_a—C_e	1.458	C_e—O_b	1.250	C_e—H	1.088	MW
	$\angle C_aC_eO$	121.6	$\angle C_eC_aC_b$	133.9	$\angle C_aC_eH$	116.9	
	trans conformer (with respect to O_a and O_b atoms)						
Glycolaldehyde 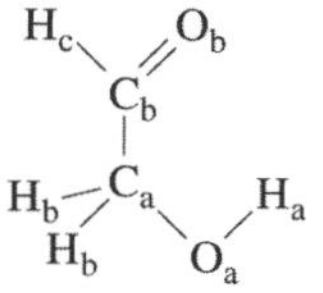	C_a—C_b	1.499	C_a—O_a	1.437	C_b—O_b	1.209	MW
	C_a—H_b	1.093	C_b—H_c	1.102	O_a—H_a	1.051	
	$\angle C_aC_bO_b$	122.7	$\angle C_bC_aO_a$	111.5			
	$\angle C_aC_bH_c$	115.3	$\angle C_bC_aH_b$	109.2	$\angle H_bC_aH_b$	107.6	
	$\angle C_aO_aH_a$	101.6	$\angle H_bC_aO_a$	109.7			
Glyoxal CHOCHO *trans* conformer	C—C	1.526	C—O	1.212	C—H	1.132	ED, UV
	∠CCO	121.2	∠HCO	112			
Hexachloroethane Cl_3CCCl_3	C—C	1.56	C—Cl	1.769	∠CCCl	110.0	ED
2,4-Hexadiyne $C_aH_3C_b{\equiv}C_cC_c{\equiv}C_bC_aH_3$	C_a—C_b	1.450	C_b—C_c	1.208	C_c—C_c	1.377	ED
	C_a—H	1.09					

Molecular

Name/Structure	Bond/Angle 1	Value 1	Bond/Angle 2	Value 2	Bond/Angle 3	Value 3	Method
Hexafluoroethane							
F_3CCF_3	C—C	1.545	CF	1.326	∠CCF	109.8	ED
Hexafluoropropene							
CF_2=$CFCF_3$	C—C	1.513	C=C (ass.)	1.329	C—F (ass.)	1.329	ED
	∠CCC	127.8	∠FCC (CF)	120	∠FCC(CF_2)	124	
	∠FCC(CF3)	110					
***trans*-1,3,5-Hexatriene**							
H_2C_a=C_bHC_cH=C_cHC_bH=C_aH_2	C_a—C_b	1.337	C_b—C_c	1.458	C_c—C_c	1.368	ED
	∠$C_aC_bC_c$	121.7	∠$C_bC_cC_c$	124.4			
Hydrogen cyanide							
HCN (linear)	C—H (r_e)	1.0655	C—N (r_e)	1.1532			MW, IR
Iminocyanide radical							
HNCN	N—H	1.034	N···N	2.470			UV
	∠HNC	116.5	∠NCN	180[a]			
Iodoacetylene							
IC≡CH	C—C	1.218	C—I	1.980	C—H	1.059	IR
Iodocyanoacetylene							
IC_a≡C_bC_c≡N	C_a—C_b	1.207	C_b—C_c	1.370	C_c—N	1.160	MW
(linear)	C_a—I	1.985					
Iodomethane [Methyl iodide]							
CH_3I	C—I (r_e)	2.132	C—H (r_e)	1.084	∠HCH(θ_e)	111.2	MW, IR
Iron pentacarbonyl							
$Fe(CO)_5$ (D_{3h})	F_e—C (av.)	1.821	$(F_e—C)_{eq}$ – $(Fe—C)_{ax}$	0.020	C—O (av.)	1.153	ED
Isobutane							
$(C_bH_3)_3C_aH$	C_a—C_b	1.535	C_a—H	1.122	C_b—H	1.113	ED, MW
	∠$C_bC_aC_b$	110.8	∠C_aC_bH	111.4			
Isobutene							
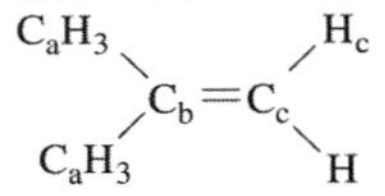	C_a—C_b	1.508	C_b—C_c	1.342	C_a—H	1.119	ED, MW
	C_c—H_c	1.10					
	∠$C_aC_bC_a$	115.6	∠$C_aC_bC_c$	122.2	∠C_bC_cH	121	
	∠HC_aC_b (av.)	111.4	∠HC_aH	107.9	∠$H_cC_cH_c$	118.5	
Isocyanic acid							
HNCO (bent)	N—C	1.209	C—O	1.166	N—H	0.986	MW
	∠NCO	180	∠HNC	128.0			
Isocyanomethane							
C_aH_3—N≡C_b	C_a—N	1.424	N—C_b	1.166	C_a—H	1.102	MW
	∠NC_aH	109.12					
	∠HCH	123.0					
Isofulminic acid							
HCNO (linear)	C—N	1.161	N—O	1.207	H—C	1.027	MW
Isothiocyanic acid							
HNCS	N—C	1.216	C—S	1.561	N—H	0.989	MW
	∠NCS	180	∠HNC	135.0			
Ketene							
H_2C=C=O	C—C	1.315	C—O	1.163			MW
	C—H	1.090	∠HCH	123.5			
Malononitrile							
$CH_2(CN)_2$	C—C	1.480	C—N	1.147	C—H	1.091	MW
	∠CCC	110.4	∠CCN	176.6	∠HCH	108.4	
Methane							
CH_4	C—H (r_e)	1.0870					IR

Name/Structure	Bond/Angle 1	Value 1	Bond/Angle 2	Value 2	Bond/Angle 3	Value 3	Method
Methanethioamide [Thioformamide]							
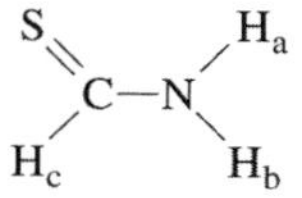	C—S	1.626	C—N	1.358	C—H_c	1.10	MW
	N—H_a	1.002	N—H_b	1.007			
	∠NCS	125.3	∠H_aN_C	117.9	∠H_bN_C	120.4	
	∠SCH_c	127	∠H_aNH_b	121.7	∠NCHc	108	
Methanethiol							
CH_3SH	C—S	1.819	S—H	1.34	C—H	1.09	MW
	∠HSC	96.5	∠HCH	109.8	angle between CH_3 symmetry axis and C—S bond	2.2	
Methanol							
CH_3OH	C—O	1.4246	C—H	1.0936	O—H	0.9451	MW
	∠COH	108.53	∠HCH	108.63	angle between CH_3 symmetry axis and C—O bond	3.27	
Methyl							
$\cdot CH_3$ planar (D_{3h})	C—H	1.076					R
***N*-Methylacetamide**							
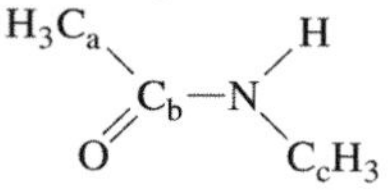	C_a—C_b	1.520	C_b—N	1.386	C_c—N	1.469	ED
	C_b—O	1.225	C—H	1.107			
	∠C_bNC_c	119.7	∠NC_bO	121.8	∠C_aC_bN	114.1	
Methylamine							
CH_3NH2	C—N	1.471	N—H	1.019	C—H	1.095	MW
	∠HNC	110.3	∠HNH	106.6	∠HCH	108.1	
	angle between CH_3 symmetry axis and C—N bond	4.3					
Methyl azide							
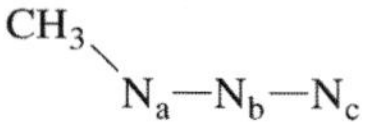	C—N_a	1.468	N_a—N_b	1.216	N_b—N_c	1.113	ED
	C—H	1.09	∠C_Na_Nb	116.8			
3-Methyl-3*H*-diazirine							
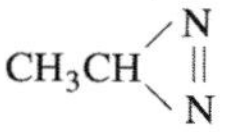	C—C	1.501	C—N	1.481	N—N	1.235	MW
	∠NCN	49.3	dihedral angle between CNN plane and C—C bond	122.3			
Methylene							
:CH2	C—H (r_e)	1.0748	∠HCH(θe)	133.84			IR,MW
Methylenecyclopropane							
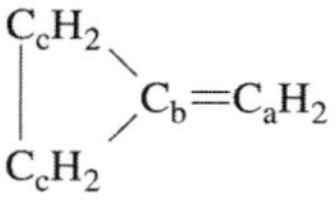	C_a—C_b	1.332	C_b—C_c	1.457	C_c—C_c	1.542	MW
	C_c—H	1.09	∠$C_cC_bC_c$	63.9	∠HC_aH	114.3	
	∠HC_cH	113.5	dihedral angle between C_cH_2 plane and C_c—C_c bond	150.8			
3-Methyleneoxetane							
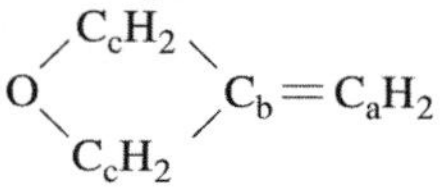	C_a—C_b	1.33	C_b—C_c	1.52	C_c—O	1.45	MW
	C—H (ass)	1.09	∠HC_cH (ass.)	114	∠HC_aH (ass.)	120	

Name/Structure	Bond/Angle 1	Value 1	Bond/Angle 2	Value 2	Bond/Angle 3	Value 3	Method
	∠$C_cC_bC_c$	87					
Methylenephosphine							
CH_cH_t=PH	C—P	1.673	C—H_c	1.09	C—H_t	1.09	MW
planar	P—H	1.420	∠CPH	97.4			
	∠HCH	117.2	∠PCH_c	124.4	∠PCH_t	118.4	
Methyl formate							
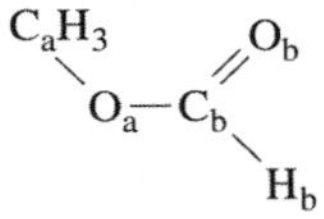	C_b—O_b	1.206	C—O (av.)	1.393	C_a—H	1.08	ED
	C_b—H (ass)	1.101					
	∠COC	114	∠$O_aC_bO_b$	127	∠O_aC_aH	110	
Methylgermane							
CH_3GeH_3	C—Ge	1.945	Ge—H	1.529	C—H	1.083	MW
	∠HGeH	109.3	∠HCH	108.4			
Methyl hypochlorite							
CH_3OCl	C—O	1.389	O—Cl	1.674	C—H	1.103	MW
	∠COC l	112.8	∠HCH	109.6			
Methylidyne							
:CH	C—H (r_e)	1.1198					UV
Methylidynephosphine							
HCP	C—P (r_e)	1.5398	C—H (r_e)	1.0692			MW
Methylketene [1-Propen-1-one]							
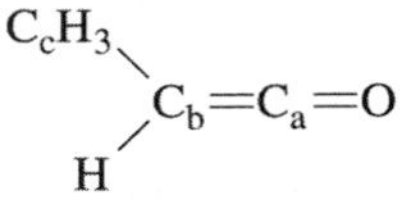	C_a—C_b	1.306	C_b—C_c	1.518	C_a—O	1.171	MW
	C_b—H	1.083	C_c—H	1.10			
	∠OC_aC_b	180.5	∠$C_aC_bC_c$	122.6	∠C_aC_bH	113.7	
	∠C_cC_bH	123.7	∠HCH	109.2			
Methyl nitrate							
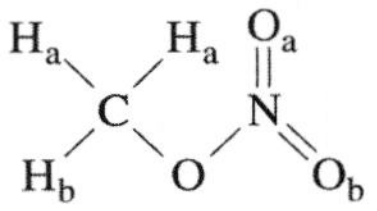	C—O	1.437	C—H_a	1.10	C—H_b	1.09	MW
	O—N	1.402	N—O_a	1.205	N—O_b	1.208	
	∠CON	112.7	∠ONO_a	118.1	∠ONO_b	112.4	
	∠OCH_a	110	∠OCH_b	103			
Methyloxirane							
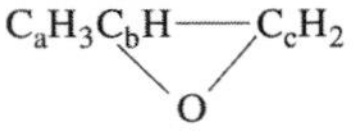	C_a—C_b	1.51	∠$C_aC_bC_c$	121.0	dihedral angle between C_bC_cO plane and $C_aC_b$bond	123.8	MW
Methylphosphine							
CH_3PH_2	C—P	1.858	C—H	1.094			ED
Methylphosphonic difluoride							
CH_3POF_2	C—P	1.770	P—O	1.444	P—F	1.545	ED,MW
	∠OPC	117.8	∠FPC	103.7	∠FPF	99.2	
Methylsilane							
CH_3SiH_3	C—Si	1.867	Si—H	1.485	C—H	1.093	MW
	∠HCH	107.7	∠HSiH	108.3			
Methylstannane							
CH_3SnH_3	C—Sn	2.143	Sn—H	1.700			MW
Methyl thiocyanate							
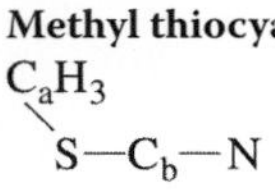	S—C_a	1.824	S—C_b	1.684	C_b—N	1.170	MW
	C—H	1.081					
	∠C_aSC_b	99.0	∠HCH	110.6	∠HCS	108.3	

Name/Structure	Bond/Angle 1	Value 1	Bond/Angle 2	Value 2	Bond/Angle 3	Value 3	Method
Methyltrioxorhenium							
CH_3ReO_3	Re—C	2.074	Re—O	1.703	C—H	1.088	MW
	∠ReCH	108.9	∠CReO	106.4			
Molybdenum carbide							
MoC	Mo—C	1.676					UV
Molybdenum carbonyl							
$Mo(CO)_6$ (O_h)	Mo—C	2.063	C—O	1.145			ED
Naphthalene 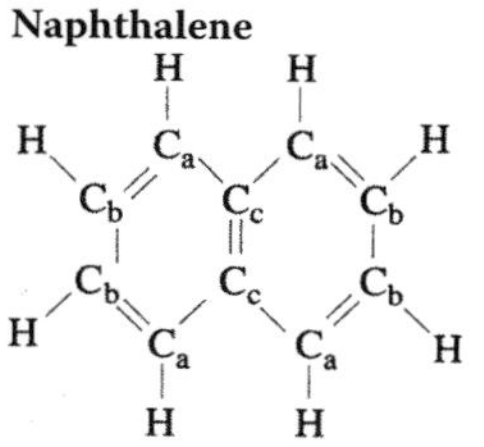	C_a—C_b	1.37	C_b—C_b	1.41	C_a—C_c	1.42	ED
	C_c—C_c	1.42	C—C (av.)	1.40	∠$C_aC_cC_c$	119.4	
Neopentane							
$C(CH_3)_4$	C—C	1.537	C—H	1.114	∠CCH	112	ED
Nickel carbonyl							
$Ni(CO)_4$ (T_d)	Ni—C	1.839	C—O	1.121			IR
Nickel monocarbonyl							
NiCO (linear)	Ni—C	1.64	C—O	1.19			IR
Nickel cyanide							
NiC≡N (linear)	Ni—C	1.828	C—N	1.158			MW
Nitromethane							
CH_3NO_2	C—N	1.489	N—O	1.224	C—H (ass.)	1.088	MW
	∠ONO	125.3	∠NCH	107			
***N*-Nitrosodimethylamine**							
$(CH_3)_2NNO$	C—N	1.461	N—O	1.235	N—N	1.344	ED
	∠CNC	123.2	∠CNN	116.4	∠ONN	113.6	
Nitrosomethane							
CH_3NO	C—N	1.49	N—O	1.22	C—H	1.084	MW
	∠CNO	112.6	∠NCH	109.0			
2,5-Norbornadiene 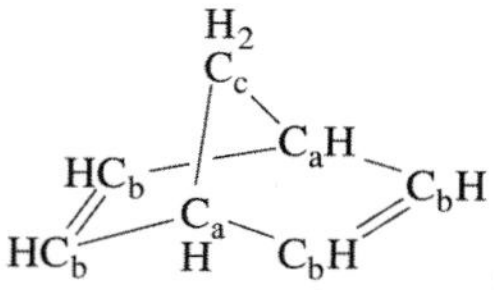 (C_{2v})	C_a—C_b	1.535	C_b—C_b	1.343	C_a—C_c	1.573	ED
	C—H	1.12	∠$C_aC_cC_a$	94			
	dihedral angle between the two $C_aC_bC_bC_a$ planes	115.6					
1,2,5-Oxadiazole 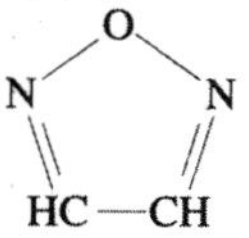	C—C	1.421	C—N	1.300	O—N	1.380	MW
	C—H	1.076	∠CCH	130.2	∠NCH	120.9	
	∠CCN	109.0	∠NON	110.4	∠ONC	105.8	
1,3,4-Oxadiazole 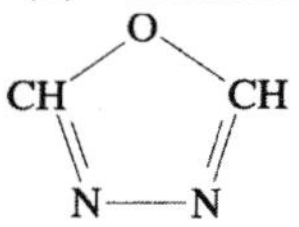	C—O	1.348	C—N	1.297	N—N	1.399	MW
	C—H	1.075	∠OCH	118.1	∠NCH	128.5	

Name/Structure	Bond/Angle 1	Value 1	Bond/Angle 2	Value 2	Bond/Angle 3	Value 3	Method
	∠CNN	105.6	∠COC	102.0	∠OCN	113.4	
Oxalic acid 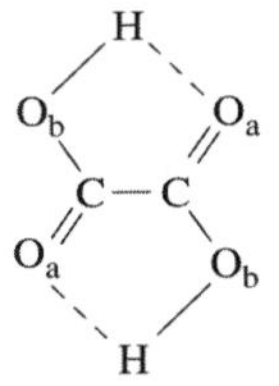	C—C	1.544	C—O_a	1.205	C—O_b	1.336	ED
	O_b—H	1.05					
	∠CCO_a	123.1	∠O_aCO_b	125.0	∠CO_bH	104	
Oxalyl chloride 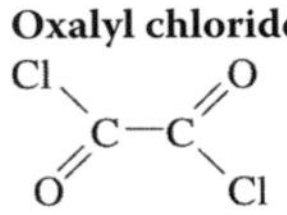	C—C	1.534	C—O	1.182	C—Cl	1.744	ED
	∠CCO	124.2	∠CCCl	111.7	68% *trans*, 32% *gauche* at 0 °C		
Oxetane 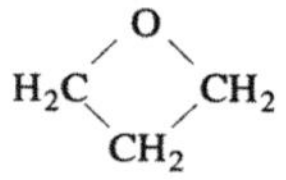	C—C	1.546	C—O	1.448	C—H (av.)	1.090	MW
	∠CCC	85	∠COC	92	∠OCC	92	
	∠HCH (av.)	109.9					
Oxirane 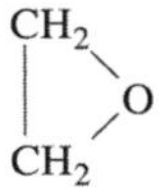	C—C	1.466	C—O	1.431	C—H	1.085	MW
	∠HCH	116.6	dihedral angle between NH_2 plane and N— C bond	158.0			
Phenol 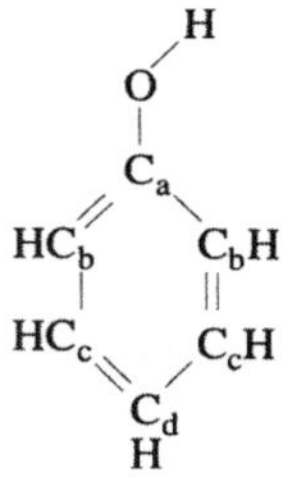	C—C (av.)	1.397	C_a—O	1.364	O—H	0.956	MW
	C_b—H	1.084	C_c—H	1.076	C_d—H	1.082	
	∠COH	109.0					
Phosphirane 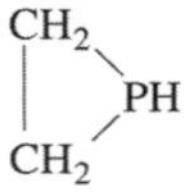	C—C	1.502	C—P	1.867	P—H	1.43	MW
	C—H	1.09	∠CPC	47.4	∠HPC	95.2	
	∠HCH	114.4	∠CCH	118	dihedral angle between PCC plane and PH bond	95.7	
Piperazine 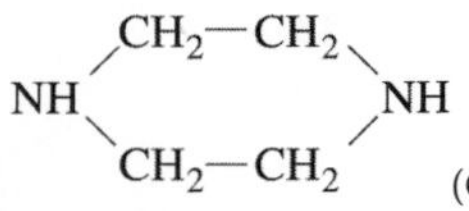 (C_{2h})	C—C	1.540	C—N	1.467	C—H	1.110	ED
	∠CNC	109.0	∠CCN	110.4			
Palladium carbide PdC	Pd—C	1.712					UV

Name/Structure	Bond/Angle 1	Value 1	Bond/Angle 2	Value 2	Bond/Angle 3	Value 3	Method
Platinum carbide							
PtC	Pt—C (*re*)	1.6767					UV
Potassium carbide							
KC	K—C	2.528					MW
Propane							
C_3H_8	C—C	1.532	C—H	1.107			ED
	∠CCC	112	∠HCH	107			
Propene							
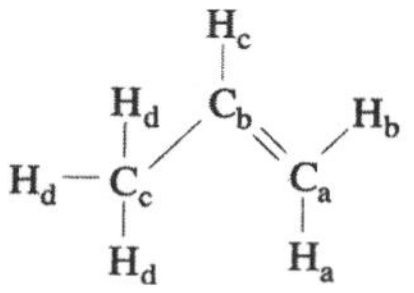	C_a—C_b	1.341	C_b—C_c	1.506			ED, MW
	C_a—H_a	1.104	C_c—H_d	1.117			
	$\angle C_aC_bC_c$	124.3	$\angle C_bC_aH_{a,b,c}$	121.3	$\angle C_bC_cH_d$	110.7	
2-Propenoyl chloride							
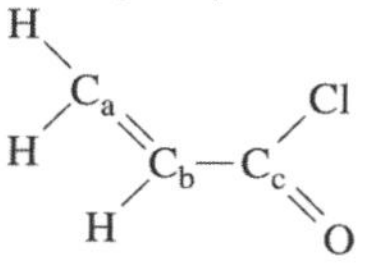	C_a—C_b	1.35	C_b—C_c	1.48	C_c—Cl	1.82	MW
	C_c—O	1.19	C—H (ass.)	1.086			
	$\angle C_aC_bC_c$	123	$\angle C_bC_cCl$	116	$\angle C_bC_cO$	127	
	$\angle C_aC_bH$ (ass.)	120	$\angle C_bC_aH$ (ass.)	121.5			
2-Propynal							
$H_aC_a{\equiv}C_b{-}C_cH_cO$ (planar)	C_a—C_b	1.211	C_b—C_c	1.453	C_c—O	1.214	ED, MW
	C_a—H_a	1.085	C_c—H_c	1.130			
	$\angle C_aC_bC_c$	178.6	$\angle C_bC_cO$	124.2	$\angle C_bC_cH_c$	113.7	
Propyne							
$H_3C_c{-}C_b{\equiv}C_aH$	C_c—C_b	1.459	C_b—C_a	1.206			MW
	C_a—H	1.056	C_c—H	1.105	$\angle HC_cC_b$	110.2	
Propynal isocyanide							
$H_3C_c{-}C_b{\equiv}C_a{-}N{\equiv}C$	C_c—C_b (r_s)	1.456	C_b—C_a (r_s)	1.206	C_a—N (r_s)	1.316	MW
	N—C (r_s)	1.175	C_c—H (r_s)	1.090	$\angle HC_cC_b$ (θ_s)	110.7	
Pyrazine							
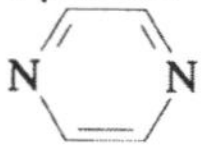	C—C	1.339	C—N	1.403	C—H	1.115	ED
	∠CCH	123.9	∠CCN	115.6			
Pyridazine							
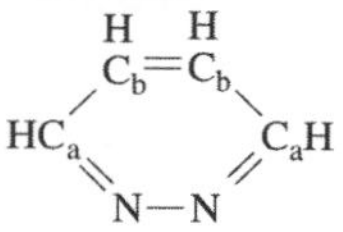	C_a—C_b	1.393	C_b—C_b	1.375	C_a—N	1.341	ED, MW
	N—N	1.330	∠NCC	123.7	∠NNC	119.3	
Pyridine							
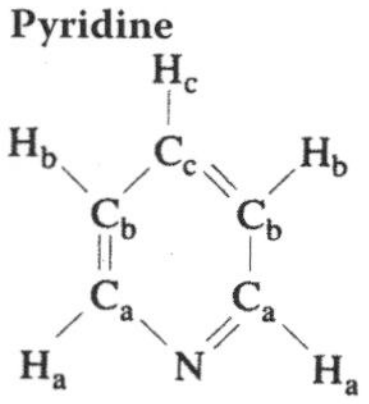	C_a—C_b	1.395	C_b—C_c	1.394	C_a—N	1.340	MW
	C_a—H_a	1.084	C_b—H_b	1.081	C_c—H_c	1.077	
	$\angle C_aC_bC_c$	118.5	$\angle C_bC_cC_b$	118.3	$\angle C_cC_bH_b$	121.3	
	$\angle C_aNC_a$	116.8	$\angle NC_aC_b$	123.9	$\angle NC_aH_a$	115.9	

Name/Structure	Bond/Angle 1	Value 1	Bond/Angle 2	Value 2	Bond/Angle 3	Value 3	Method
Pyrimidine 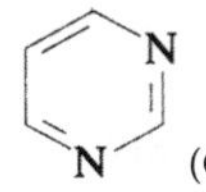 (C_{2v} assumed)	C—C	1.393	C—N	1.340			ED
	∠NCN	127.6	∠CNC	115.5			
Pyrrole 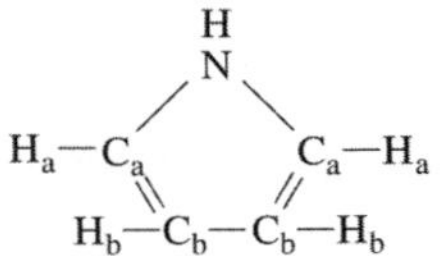	C_a—C_b	1.382	C_b—C_b	1.417	C_a—N	1.370	MW
	C_a—H_a	1.076	C_b—H_b	1.077	N—H	0.996	
	∠$C_aC_bC_b$	107.4	∠C_aNC_a	109.8	∠NC_aC_b	107.7	
	∠C_bC_bH	127.1	∠NC_aH_a	121.5			
Pyruvonitrile 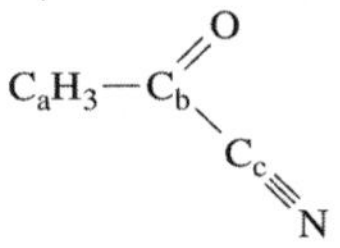	C_a—C_b	1.518	C_b—C_c	1.477	C—H	1.12	ED, MW
	C—N	1.17	C—O	1.208	∠HCH	109.2	
	∠$C_aC_bC_c$	114.2	∠C_aC_bO	124.5	∠CCN	179	
Ruthenium carbide RuC	Ru—C	1.607					UV
Silacyclobutane CH_2—CH_2 / CH_2—SiH_2	C—C	1.571	C—S_i	1.885	C—H	1.100	ED
	Si—H	1.47	∠CCC	99.8	∠CSiC	77.2	
	∠SiCC	84.8	dihedral angle between CCC and CSiC planes	146			
Silaethene $H_2Si{=}CH_2$	Si—C (r_e)	1.704	Si—H (r_e)	1.467	C—H (r_e)	1.082	MW
	∠HCSi	122.0	∠HSiC	122.4			
Silicon dicarbide CSiC (ring)	C—C(r_s)	1.269	Si—C (r_s)	1.832	∠CSiC (θ_s)	40.5	MW
Silylchloroacetylene $SiH_3C{\equiv}CCl$	C—C	1.234	Si—C	1.812	C—Cl	1.620	ED
	Si—H	1.488	∠HSiC	109.4			
Silyl cyanide $SiH_3C{\equiv}N$	Si—C	1.850	C—N	1.156	Si—H	1.487	ED,MW
	∠HSiC	107.25					
Sodium carbide NaC	Na—C	2.232					MW
Spiro[2.2]pentane 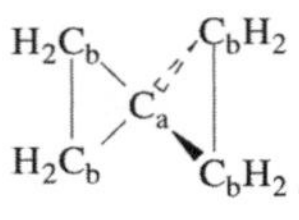 (D_{2d})	C_b—C_b	1.52	C_a—C_b	1.47	C—H	1.09	ED
	∠$C_bC_aC_b$	62	∠HCH	118			
Strontium methyl $SrCH_3$	Sr—C	2.487	C—H (ass.)	1.104	∠HCH	105.8	UV
Succinonitrile CH_2CN	C—C	1.561	C—C(N)	1.465	C—N	1.161	ED
	C—H	1.09	∠CCC	110.4	dihedral angle of CCCC for *gauche* conformer	75	

Name/Structure	Bond/Angle 1	Value 1	Bond/Angle 2	Value 2	Bond/Angle 3	Value 3	Method
Tetrabromomethane							
CBr_4 (T_d)	C—Br	1.935					ED
Tetrachloroethene							
CCl_2=CCl_2	C—C	1.354	C—C_l	1.718	∠ClCCl	115.7	ED
Tetrachloromethane							
CCl_4 (T_d)	C—Cl	1.767					ED
Tetracyanoethene							
$(CN)_2C$=$C(CN)_2$	C—C	1.435	C=C	1.357	C—N	1.162	ED
	∠CC =C	121.1					
2,2,4,4-Tetrafluoro-1,3-dithietane							
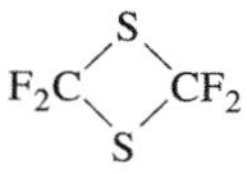 (D_{2h} assumed)	C—S	1.785	C—F	1.314	∠CSC	83.2	ED
	∠FCS	113.7					
Tetrafluoroethene							
CF_2=CF_2	C—C	1.31	C—F	1.319	∠CCF	123.8	ED
Tetrafluoromethane							
CF_4 (T_d)	C—F	1.323					ED
Tetrahydrofuran							
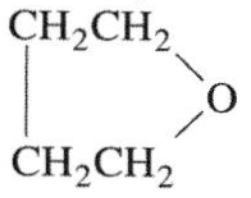	C—C	1.536	C—O	1.428	C—H	1.115	ED
Tetrahydropyran							
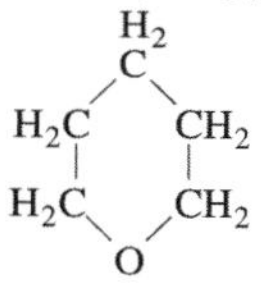	C—C	1.531	C—O	1.420	C—H	1.116	ED
	∠COC	111.5	∠OCC	111.8	∠CCC (C)	108	
	∠CCC (O)	111					
Tetrahydrothiophene							
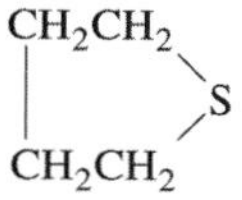	C—C	1.536	C—S	1.839	C—H	1.120	ED
	∠CCC	105.0	∠CSC	93.4	∠SCC	106.1	
Tetraiodomethane							
CI_4 (T_d)	C—I	2.15					ED
Tetramethylgermane							
$(CH_3)_4Ge$	C—G_e	1.945	C—H	1.12	∠GeCH	1.08	ED
Tetramethyl lead							
$(CH_3)_4Pb$	C—P_b	2.238					ED
Tetramethylsilane							
$(CH_3)_4Si$	C—Si	1.875	C—H	1.115	∠HCH	109.8	ED
Tetramethylstannane							
$(CH_3)_4Sn$	C—Sn	2.144	C—H	1.12			ED
1,2,5-Thiadiazole							
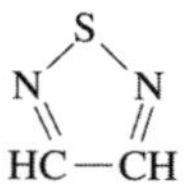	C—C	1.420	C—N	1.328	S—N	1.631	MW
	C—H	1.079					
	∠CCN	113.8	∠NSN	99.6	∠CCH	126.2	

Name/Structure	Bond/Angle 1	Value 1	Bond/Angle 2	Value 2	Bond/Angle 3	Value 3	Method
1,3,4-Thiadiazole 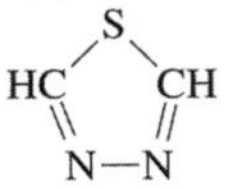	C—S	1.721	C—N	1.302	N—N	1.371	MW
	C—H	1.08	∠CSC	86.4	∠SCN	114.6	
	∠CCN	112.2	∠NCH	123.5	∠SCH	121.9	
Thietane 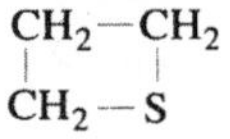	C—C	1.549	C—S	1.847	C—H (av.)	1.100	ED, MW
	∠CSC	76.8	∠HCH (av.)	112	dihedral angle between CCC and CSC planes	154	
Thiirane 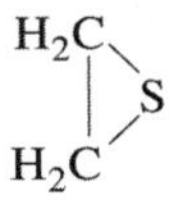	C—C	1.484	C—S	1.815	C—H	1.083	MW
	∠CSC	48.3	∠CCS	65.9	∠HCH	116	
	dihedral angle between CH_2 plane and C—C bond	152					
Thioacetaldehyde 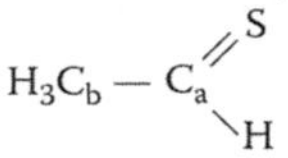	C_a—S (r_s)	1.610	C_a—C_b (r_s)	1.506			MW
	C_a—H (r_s)	1.089	C_b—H (r_s)	1.094[b]			
	∠C_bC_aS (θ_s)	125.3	∠C_bC_aH (θ_s)	119.4	∠HC_bC_a (θ_s)	110.6[b]	
Thiocarbonyl fluoride F_2CS	C—S	1.589	C—F	1.315	∠FCF	107.1	MW
Thioformaldehyde CH_2S	C—S	1.611	C—H	1.093	∠HCH	116.9	MW
Thioketene H_2C=C=S C_{2v}	C—C (r_s)	1.314	C—S (r_s)	1.554	C—H (r_s)	1.080	IR
	∠HCH (θ_s)	119.8					
Thiophene 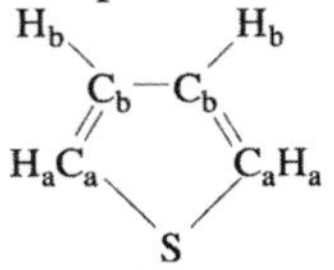	C_a—C_b	1.370	C_b—C_b	1.423	C_a—S	1.714	MW
	C_a—H_a	1.078	C_b—H_b	1.081			
	∠$C_aC_bC_b$	112.5	∠C_aSC_a	92.2	∠$SC_{aC}b$	115.5	
	∠S_Ca_Ha	119.9	∠$C_bC_bH_b$	124.3			
Toluene C_6H_5—CH_3	C—C (ring)	1.399	C—CH_3	1.524	C—H (av.)	1.11	ED
1,1,1-Tribromoethane CH_3CBr_3	C—C (ass.)	1.51	C—Br	1.93	C—H (ass.)	1.095	MW
	∠BrCBr	111	∠CC Br	1.8	∠CCH (ass.)	109.0	
Tribromomethane $CHBr_3$ (C_{3v})	C—Br	1.924	C—H	1.11	∠BrCBr	111.7	ED, MW
Tri-*tert*-butyl methane $HC_a[C_b(C_cH_3)_3]_3$	C_a—C_b	1.611	C_b—C_c	1.548	C—H	1.111	ED
	∠$C_aC_bC_c$	113.0					
Trichloroacetonitrile CCl_3CN	C—C	1.460	C—N	1.165	C—Cl	1.763	ED
	∠ClCCl	110.0					

Molecular

Name/Structure	Bond/Angle 1	Value 1	Bond/Angle 2	Value 2	Bond/Angle 3	Value 3	Method
1,1,1-Trichloroethane							
CH_3CCl_3	C—C	1.541	C—Cl	1.771	C—H	1.090	MW
	∠CCCl	109.6	∠ClCCl	109.4	∠HCH	110.0	
	∠CCH	108.9					
Trichlorofluoromethane							
CCl_3F	C—Cl	1.754	C—F	1.362	∠ClCCl	111	MW
Trichloromethane							
$CHCl_3$	C—Cl	1.758	C—H	1.100	∠ClCCl	111.3	MW
Trichloromethylgermane							
CH_3GeCl_3	C—Ge	1.89	Ge—Cl	2.132	C—H (ass.)	1.103	ED, MW
	∠ClGeCl	106.4	∠GeCH (ass.)	110.5			
Trichloromethylsilane							
CH_3SiCl_3	C—Si	1.876	Si—Cl	2.021			MW
Trichloromethylstannane							
CH_3SnCl_3	C—Sn	2.10	Sn—Cl	2.304	C—H	1.100	ED
1,1,1-Trichloro-2,2,2-trifluoroethane							
CF_3CCl_3	C—C	1.54	C—F	1.33	C—Cl	1.77	MW
(staggered configuration)	∠CCF	110	∠CCCl	109.6			
	∠CSnCl	113.9	∠ClSnCl	104.7	∠SnCH	108	
Triethylenediamine							
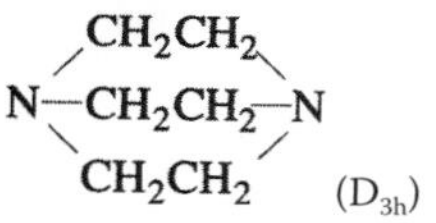 (D_{3h})	C—C	1.562	C—N	1.472	∠CNC	108.7	ED
	∠NCC	110.2					
Trifluoroacetic acid							
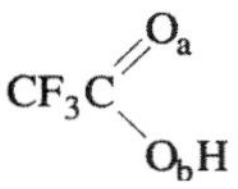	C—C	1.546	$C—O_a$	1.192	$C—O_b$	1.35	ED
	C—F	1.325	O—H (ass.)	0.96			
	$∠CCO_a$	126.8	$∠CCO_b$	111.1	∠CCF	109.5	
1,1,1-Trifluoroethane							
CH_3CF_3	C—C	1.494	C—F	1.340	C—H	1.081	ED
Trifluoroiodomethane							
CF_3I (C_{3v})	C—F	1.330	C—I	2.138	∠FCF	108.1	ED, MW
Trifluoromethane							
CHF_3 (C_{3v})	C—F	1.332	C—H	1.098	∠FCF	108.8	MW
Trifluoromethanesulfonyl fluoride							
$CF_3SO_2F_a$	C—S	1.835	C—F (av.)	1.325	S—O	1.410	ED
	$S—F_a$	1.543	$∠CSF_a$	95.4	∠CSO	108.5	
	∠OSO	124.1	∠FCF	109.8			
Trifluoromethyliminosulfur difluoride							
$CF_3N{=}SF_2$	C—N	1.409	S—N	1.477	S—F	1.594	ED,MW
	C—F	1.331	∠CNS	127.2	∠NSF	112.7	
	∠FSF	92.8	∠FCF	108.1			
Trifluoromethyl peroxide							
CF_3OOCF_3	O—O	1.42	C—O	1.399	C—F	1.320	ED
	∠COO	107	∠FCF	109.0	COOC dihedral angle of internal rotation	123	
	∠CCF	119.2	∠CCH	112			
Trimethyl aluminium							
$(CH_3)_3Al$	C—Al	1.957	C—H	1.113			ED
	∠CAlC	120	∠AlCH	111.7			

Name/Structure	Bond/Angle 1	Value 1	Bond/Angle 2	Value 2	Bond/Angle 3	Value 3	Method
Trimethylamine							
$(CH_3)_3N$	C—N	1.458	C—H	1.100			ED
	∠CNC	110.9	∠HCH	110			
Trimethylarsine							
$(CH_3)_3As$	C—As	1.979	∠CAsC	98.8	∠AsCH	111.4	ED
Trimethyl bismuth							
$(CH_3)_3Bi$	C—Bi	2.263	C—H	1.07	∠CBiC	97.1	ED
Trimethylborane							
$(CH_3)_3B$	C—B	1.578	C—H	1.114			ED
	∠CBC	120	∠BCH	112.5			
Trimethylphosphine							
$(CH_3)_3P$	C—P	1.847	C—H	1.091			ED
	∠CPC	98.6	∠PCH	110.7			
1,3,5-Trioxane							
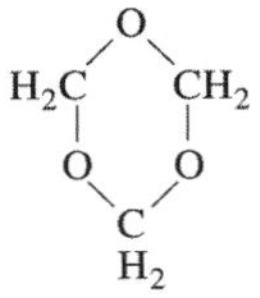	C—O	1.422	∠OCO	112.2	∠COC	110.3	MW
Triphenylamine							
$(C_6H_5)_3N$ (C_3)	C—C	1.392	C—N	1.42	∠CNC	116	ED
	torsional dihedral angle of phenyl rings	47					
Tungsten carbide							
WC	W—C	1.7135					UV
Tungsten carbonyl							
$W(CO)_6$ (O_h)	W—C	2.059	C—O	1.149			ED
Vanadium carbonyl							
$V(CO)_6$ (O_h, involving dynamic Jahn-Teller effect)	V—C	2.015	C—O	1.138			ED
Vinyl bromide [Bromoethene]							
See Vinyl chloride	C—C	1.3256	C—Br	1.8835	C—H_a	1.0780	MW
	C—H_b	1.0804	C—H_c	1.0794	∠CCBr	122.62	
	∠CCH_a	124.34	∠CCH_b	119.28	∠CCH_c	122.03	
Vinyl chloride [Chloroethene]							
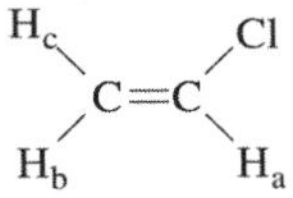	C—C	1.3262	C—Cl	1.7263	C—H_a	1.0783	MW
	C—H_b	1.0796	C—H_c	1.0796	∠CCCl	122.75	
	∠CCH_a	123.91	∠CCH_b	119.28	∠CCH_c	121.77	
Vinyl fluoride [Fluoroethene]							
See Vinyl chloride	C—C	1.3210	C—F	1.3428	C—H_a	1.0796	MW
	C—H_b	1.0774	C—H_c	1.0789	∠CCF	121.70	
	∠CCH_a	125.95	∠CCH_b	118.97	∠CCH_c	121.34	
Vinyl iodide [Iodoethene]							
See Vinyl chloride	C—C	1.3276	C—I	2.0830	C—H_a	1.0787	MW
	C—H_b	1.0823	C—H_c	1.0799	∠CCI	122.97	
	∠CCH_a	123.54	∠CCH_b	119.36	∠CCH_c	122.30	
Zinc cyanide							
ZnC≡N (linear)	Zn—C	1.955	C—N	1.146			MW

[a] Approximate.
[b] Average.
[c] Distances from Be to planes are 1.47 and 1.92 Å.
[d] 40°~50° (the two rings are not parallel).

CHARACTERISTIC BOND LENGTHS IN FREE MOLECULES

David R. Lide

These tables contain typical bond lengths in gas-phase molecules. The value given for each bond is near the mid-range of values found in simple molecules. Bond lengths usually vary by 1% or 2%, and often by more, depending on the nature of the other bonds attached to the two atoms in question. References 1 and 2 give measured bond lengths in individual gas-phase molecules, as determined by spectroscopic and electron diffraction methods. All bond lengths are given in Å (1 Å = 10^{-10} m).

Table 1 contains data for characteristic bond lengths for single bonds between select atoms. Table 2 has bond lengths for multiple bonds for non-ring molecules. Table 3 shows the effect of other bonds on the carbon-carbon single bond, and Table 4 has data for select metal-carbon bonds in gas-phase molecules.

References

1. "Structure of Free Molecules in the Gas Phase," *CRC Handbook of Chemistry and Physics, 102nd Edition*, CRC Press, Boca Raton, FL, 2021.
2. Harmony, M. D., Laurie, V. W., Kuczkowski, R. L., Schwendeman, R. H., Ramsay, D. A., Lovas, F. J., Lafferty, W. J., and Maki, A. G., *J. Phys. Chem. Ref. Data* 8, 619, 1979. <https://doi.org/10.1063/1.555605>
3. Lide, D. R., *Tetrahedron* 17, 125, 1962. <https://doi.org/10.1016/S0040-4020(01)99012-X>

TABLE 1. Characteristic Lengths (in Å) of Single Bonds

	As	B	Br	C	Cl	F	Ge	H	I	N	O	P	S	Se	Si
As	2.10														
Br	2.32	1.89	2.28												
C	1.98	1.58	1.93	1.53											
Cl	2.17	1.74	2.14	1.77	1.99										
F	1.71	1.31	1.76	1.35	1.63	1.41									
Ge			2.30	1.95	2.15	1.73	2.40								
H	1.51	1.19	1.41	1.09	1.28	0.92	1.53	0.74							
I		2.12	2.47	2.13	2.32	1.91	2.51	1.61	2.67						
N		1.39		1.46	1.90	1.37		1.02		1.45					
O		1.35		1.42	1.70	1.42		0.96		1.43	1.48				
P		1.94	2.22	1.85	2.04	1.57		1.42		1.65		2.25			
S			2.24	1.82	2.05	1.56		1.34					2.00		
Sb			2.49		2.33	1.88		1.70	2.27						
Se				1.95		1.71		1.47						2.33	
Si			2.21	1.87	2.05	1.58		1.48	2.44		1.63		2.14		2.33
Sn			2.50	2.14	2.30	1.94		1.71	2.67						
Te						1.82		1.66							

TABLE 2. Lengths of Multiple Bonds (Non-Ring Molecules)

Bond	Length/Å
C=C	1.34
C≡C	1.20
C=N	1.21
C≡N	1.16
C=O	1.21
C=S	1.61
N=N	1.24
N≡N	1.13
N=O	1.18
O=O	1.21
S=S	1.89
S=O	1.48
Se=Se	2.17
Se=O	1.65

TABLE 3. Effect of Environment on Carbon-Carbon Single-Bond Lengths (Other Single Bonds Not Shown) from Ref. 3

Configuration	C–C Length/Å	Examples of molecules
C–C	1.526	$H_3C–CH_3$
C–C=	1.501	$H_3C–CH=CH_2$
C–C≡	1.459	$H_3C–C≡CH$
=C–C=	1.467	$H_2C=CH–CH=CH_2$
≡C–C=	1.445	$HC≡C–CH=CH_2$
≡C–C≡	1.378	HC≡C–C≡CH

TABLE 4. Metal-Carbon Bond Lengths in Gas-Phase Molecules

Bond	Length/Å	Bond	Length/Å
Al–C	1.96	Hg–C	2.08
Be–C	1.70	Mo–C	2.06*
Bi–C	2.26	Ni–C	1.83
Cd–C	2.11	Pb–C	2.24
Co–C	1.88	Sn–C	2.14
Cr–C	1.92*	V–C	2.02*
Cu–C	1.83	W–C	2.06*
Fe–C	1.82*	Zn–C	1.93

* In carbonyl molecules.

ATOMIC RADII OF THE ELEMENTS

Manjeera Mantina, Rosendo Valero, Christopher J. Cramer, and Donald G. Truhlar

Atomic radii are not precisely defined but are nevertheless very widely used parameters in modeling and understanding molecular structure and interactions. Three main classes of radii may be identified: van der Waals radii (which include radii used to characterize steric interactions), covalent radii, and ionic radii. This table is concerned with the first two; ionic radii are covered in a table in Section 12 of the *CRC Handbook* called "Ionic Radii in Crystals."

There are many scales of van der Waals radii, but they are not fully consistent with one another. The van der Waals radii determined by Bondi (Ref. 1) from x-ray diffraction data, crystal densities, gas kinetic collision cross sections, critical densities, and liquid-state properties are the most widely used values, but Bondi recommended radius values for only 28 of the 44 main-group elements in the periodic table plus 9 transition metals and one actinide. Rowland and Taylor (Ref. 2) redetermined nine of the main-group radii from crystal structure data and recommended that Bondi's values be accepted except for H, for which they recommended a new value; we accepted their recommendation for H and adopted Bondi's 27 other values for main-group elements. Radii for the 16 remaining main-group elements were determined from electronic structure calculations on selected van der Waals molecules by analyzing the results in a way designed to yield radii compatible with Bondi's scale for main-group elements (Ref. 3). Bondi's values for the transition metals and actinide are smaller than expected based on ionization potentials and covalent radii, so we do not adopt them, but defer to later recommendations based on analysis of a more extensive set of crystal data. Van der Waals radii for the remaining elements through atomic number $Z = 93$ were taken from Hu et al. (Ref. 4), who determined them from bond valence parameters (this gives radii usually within 0.1 Å to 0.15 Å of the radii from average atomic volumes in crystals as obtained by statistical analysis of the Cambridge Structural Database (Ref. 5), but the results show a smoother variation with atomic number). Van der Waals radii for the elements with $Z = 94$ to 103 were based on the work of Guzei and Wendt (Ref. 6), who modeled the zero-potential distance for the interaction potentials involved in nonbonded steric interactions; we increased their values by 6%, which is the value they found is needed to make their steric radii scale most consistent with Bondi's van der Waals radii.

Covalent radii are more straightforward, especially for elements that tend to form only single bonds, although some researchers distinguish metallic bonds from covalent bonds. The covalent radii tabulated here are recommendations for single covalent bonds, and they are based on a comprehensive evaluation of experimental data by Cordero et al. (Ref. 7), who recommended covalent radii for all elements up to $Z = 96$, and on an analysis combining experimental data and theoretical calculations by Pyykkö and Atsumi (Ref. 8), who recommended single-bond covalent radii for all elements up to $Z = 118$. If one is interested in a specific coordination number, oxidation state, or type of ligand, one might find a more appropriate radius in the specialized literature or in the "Characteristic Bond Lengths in Free Molecules" table in this section since the values in the table below are generic average values. In particular, we give the Pyykkö-Atsumi values for carbon and for $Z = 97$ to 118 and an average of the values from the two sources for all other elements. For Mn, Fe, and Co, Cordero et al. give two values, and we used the lower of these on the average because it corresponds to a smoother periodic trend.

Please note that the van der Waals radii of bromine and lithium have typographical errors in Ref. 3; the correct values (1.85 for Br and 1.82 for Li) in the table below are from Ref. 1.

All values are rounded to the nearest 0.01 Å, but in most cases the uncertainty in the value is of the order of 0.1 Å. Column definitions are as follows.

Column heading	Definition
Name	Element name; elements are listed alphabetically by name
Symbol	Atomic symbol
R_{vdw}	van der Waals radius, in Å
R_{cov}	Covalent radius, in Å

References

1. Bondi, A., *J. Phys. Chem.* 68, 441, 1964. <https://doi.org/10.1021/j100785a001>
2. Rowland, R. S., and Taylor, R., *J. Phys. Chem.* 100, 7384, 1996. <https://doi.org/10.1021/jp953141+>
3. Mantina, M., Chamberlin, A. C., Valero, R., Cramer, C. J., and Truhlar, D. G., *J. Phys. Chem., A* 113, 5806, 2009. <https://doi.org/10.1021/jp8111556>
4. Hu, S.-Z., Zhou, Z.-H., and Robertson, B. E., *Z. Kristallogr.* 224, 375, 2009. <https://doi.org/10.1524/zkri.2009.1158>
5. Hu, S.-Z., Zhou, Z-.H., and Tsai, K.-R., *Acta Physico-Chimica Sinica* 19, 1073, 2003.
6. Guzei, I. A. and Wendt, M., *Dalton Trans.*, 2006, 3991, 2006. <https://doi.org/10.1039/b605102b>
7. Cordero, B., Gómez, V., Platero-Prats, A. E., Revés, M., Echeverría, J., Cremades, E., Barragán, F., and Alvarez, S., *Dalton Trans.* 2008, 2832, 2008. <https://doi.org/10.1039/b801115j>
8. Pyykkö, P., and Atsumi, M., *Chem. Eur. J.* 15, 186, 2009. <https://doi.org/10.1002/chem.200800987>

Van der Waals and Covalent Radii of the Elements

Name	Symbol	R_{vdw}/Å	R_{cov}/Å	Name	Symbol	R_{vdw}/Å	R_{cov}/Å	Name	Symbol	R_{vdw}/Å	R_{cov}/Å
Actinium	Ac	2.47	2.01	Berkelium	Bk	2.44	1.68	Californium	Cf	2.45	1.68
Aluminum	Al	1.84	1.24	Beryllium	Be	1.53	0.99	Carbon	C	1.70	0.75
Americium	Am	2.44	1.73	Bismuth	Bi	2.07	1.50	Cerium	Ce	2.42	1.84
Antimony	Sb	2.06	1.40	Bohrium	Bh		1.41	Cesium	Cs	3.43	2.38
Argon	Ar	1.88	1.01	Boron	B	1.92	0.84	Chlorine	Cl	1.75	1.00
Arsenic	As	1.85	1.20	Bromine	Br	1.85	1.17	Chromium	Cr	2.06	1.30
Astatine	At	2.02	1.48	Cadmium	Cd	2.18	1.40	Cobalt	Co	2.00	1.18
Barium	Ba	2.68	2.06	Calcium	Ca	2.31	1.74	Copernicium	Cn		1.22

Name	Symbol	R_{vdw}/Å	R_{cov}/Å
Copper	Cu	1.96	1.22
Curium	Cm	2.45	1.68
Darmstadtium	Ds		1.28
Dubnium	Db		1.49
Dysprosium	Dy	2.31	1.80
Einsteinium	Es	2.45	1.65
Erbium	Er	2.29	1.77
Europium	Eu	2.35	1.83
Fermium	Fm	2.45	1.67
Flerovium	Fl		1.43
Fluorine	F	1.47	0.60
Francium	Fr	3.48	2.42
Gadolinium	Gd	2.34	1.82
Gallium	Ga	1.87	1.23
Germanium	Ge	2.11	1.20
Gold	Au	2.14	1.30
Hafnium	Hf	2.23	1.64
Hassium	Hs		1.34
Helium	He	1.40	0.37
Holmium	Ho	2.30	1.79
Hydrogen	H	1.10	0.32
Indium	In	1.93	1.42
Iodine	I	1.98	1.36
Iridium	Ir	2.13	1.32
Iron	Fe	2.04	1.24
Krypton	Kr	2.02	1.16
Lanthanum	La	2.43	1.94
Lawrencium	Lr	2.46	1.61
Lead	Pb	2.02	1.45
Lithium	Li	1.82	1.30
Livermorium	Lv		1.75
Lutetium	Lu	2.24	1.74

Name	Symbol	R_{vdw}/Å	R_{cov}/Å
Magnesium	Mg	1.73	1.40
Manganese	Mn	2.05	1.29
Meitnerium	Mt		1.29
Mendelevium	Md	2.46	1.73
Mercury	Hg	2.23	1.32
Molybdenum	Mo	2.17	1.46
Moscovium	Mc		1.62
Neodymium	Nd	2.39	1.88
Neon	Ne	1.54	0.62
Neptunium	Np	2.39	1.80
Nickel	Ni	1.97	1.17
Nihonium	Nh		1.36
Niobium	Nb	2.18	1.56
Nitrogen	N	1.55	0.71
Nobelium	No	2.46	1.76
Oganesson	Og		1.57
Osmium	Os	2.16	1.36
Oxygen	O	1.52	0.64
Palladium	Pd	2.10	1.30
Phosphorus	P	1.80	1.09
Platinum	Pt	2.13	1.30
Plutonium	Pu	2.43	1.80
Polonium	Po	1.97	1.42
Potassium	K	2.75	2.00
Praseodymium	Pr	2.40	1.90
Promethium	Pm	2.38	1.86
Protactinium	Pa	2.43	1.84
Radium	Ra	2.83	2.11
Radon	Rn	2.20	1.46
Rhenium	Re	2.16	1.41
Rhodium	Rh	2.10	1.34
Roentgenium	Rg		1.21

Name	Symbol	R_{vdw}/Å	R_{cov}/Å
Rubidium	Rb	3.03	2.15
Ruthenium	Ru	2.13	1.36
Rutherfordium	Rf		1.57
Samarium	Sm	2.36	1.85
Scandium	Sc	2.15	1.59
Seaborgium	Sg		1.43
Selenium	Se	1.90	1.18
Silicon	Si	2.10	1.14
Silver	Ag	2.11	1.36
Sodium	Na	2.27	1.60
Strontium	Sr	2.49	1.90
Sulfur	S	1.80	1.04
Tantalum	Ta	2.22	1.58
Technetium	Tc	2.16	1.38
Tellurium	Te	2.06	1.37
Tennessine	Ts		1.65
Terbium	Tb	2.33	1.81
Thallium	Tl	1.96	1.44
Thorium	Th	2.45	1.90
Thulium	Tm	2.27	1.77
Tin	Sn	2.17	1.40
Titanium	Ti	2.11	1.48
Tungsten	W	2.18	1.50
Uranium	U	2.41	1.83
Vanadium	V	2.07	1.44
Xenon	Xe	2.16	1.36
Ytterbium	Yb	2.26	1.78
Yttrium	Y	2.32	1.76
Zinc	Zn	2.01	1.20
Zirconium	Zr	2.23	1.64

DIPOLE MOMENTS

David R. Lide

These tables give selected values of the electric dipole moment μ for over 900 molecules. When available, values determined by microwave spectroscopy, molecular beam electric resonance, and other high-resolution spectroscopic techniques were selected. Otherwise, the values come from measurements of the dielectric constant in the gas phase or, if these do not exist, in the liquid phase.

Entries are listed alphabetically by compound name. This is followed by a line formula for simple structures or, for more complex compounds, the molecular formula in Hill order.

The dipole moment is given in debye units (D). The conversion factor to SI units is 1 D = 3.33564×10^{-30} C m.

When the accuracy of a dipole moment value is explicitly stated, e.g., 1.234(12), where digit(s) in parentheses represent the uncertainty in the last digit(s) of the value, the stated uncertainty generally indicates two standard deviations. When no uncertainty is given, the value may be assumed to be precise to a few units in the last decimal place. However, if more than three decimal places are given, the exact interpretation of the final digits may require analysis of the vibrational averaging. Other information on molecules that have been studied by spectroscopy, such as the components of the dipole moment in the molecular framework and the variation with vibrational state and isotopic species, may be found in the references.

Values measured in the gas phase that are questionable because of undetermined error sources are preceded by ≈. Values obtained by liquid-phase measurements, which sometimes have large errors because of association effects, are followed by liq.

Table 1 gives dipole moments of individual molecules in the gas or liquid phase. Column definitions for Table 1 are as follows.

Column heading	Definition
Name	Name of compound; compounds are listed alphabetically by name
Mol. form.	Molecular formula of compound, in Hill order
μ	Dipole moment, in units D
Ref.	Reference number

Table 2 lists dipole moments of individual conformers of molecules that can exist in different structural configurations, related by internal rotation around single bonds. These conformers exist long enough to be measured by gas-phase microwave spectroscopic techniques but interconvert too rapidly to be isolated as bulk samples. The conformers are designated as *gauche*, *trans*, *axial*, etc. Because the notation of conformers is not well standardized, the references should be consulted for full details of the structure. Table 2 column definitions are as follows.

Column heading	Definition
Name	Name of compound; compounds are listed alphabetically by name
Mol. form.	Molecular formula of compound, in Hill order
Conformer	Conformer name; see text for discussion
μ	Dipole moment, in units D
Ref.	Reference number

References

1. Nelson, R. D., Lide, D. R., and Maryott, A. A., *Selected Values of Electric Dipole Moments for Molecules in the Gas Phase*, Natl. Stand. Ref. Data Ser. — Nat. Bur. Stnds. 10, 1967. <https://doi.org/10.6028/NBS.NSRDS.10>
2. *Landolt-Börnstein, Numerical Data and Functional Relationships in Science and Technology, New Series*, II/6 (1974), Springer-Verlag, Heidelberg.
3. *Landolt-Börnstein, Numerical Data and Functional Relationships in Science and Technology, New Series*, II/14a (1982), Springer-Verlag, Heidelberg.
4. *Landolt-Börnstein, Numerical Data and Functional Relationships in Science and Technology, New Series*, II/14b (1983), Springer-Verlag, Heidelberg.
5. *Landolt-Börnstein, Numerical Data and Functional Relationships in Science and Technology, New Series*, II/19c (1992), Springer-Verlag, Heidelberg.
6. *Landolt-Börnstein, Numerical Data and Functional Relationships in Science and Technology, New Series*, II/24c (2002), Springer-Verlag, Heidelberg.
7. Riddick, J. A., Bunger, W. B., and Sakano, T. K., *Organic Solvents, Fourth Edition*, John Wiley & Sons, New York, 1986.
8. Kasuya, T., Lafferty, W. J., and Lide, D. R., *J. Chem. Phys.* 48, 1, 1968. <https://doi.org/10.1063/1.1664452>
9. Kirchhoff, W. H., and Lide, D. R., *J. Chem. Phys.* 51, 467, 1969. <https://doi.org/10.1063/1.1671761>
10. Durig, J. R., Li, Y. S., and Rizzolo, J. J., *J. Chem. Phys.* 77, 5885, 1982. <https://doi.org/10.1063/1.443862>
11. Ogata, T., Mochizuki, A. and Yamashita, E., *J. Chem. Phys.* 87, 2531, 1987. <https://doi.org/10.1063/1.453092>
12. Rego, A., and Cox, A. P., *J. Chem. Phys.* 89, 124, 1988. <https://doi.org/10.1063/1.455514>
13. Tyblewski, M., et al., *J. Chem. Phys.* 97, 6168, 1992. <https://doi.org/10.1063/1.463725>
14. Kawashima, Y., et al., *J. Chem. Phys.* 99, 820, 1993. <https://doi.org/10.1063/1.465345>
15. Caminati, W., Melandri, S., and Favero, L., *J. Chem. Phys.* 100, 8569, 1994. <https://doi.org/10.1063/1.466761>
16. Cederberg, J., et al., *J. Chem. Phys.* 105, 3361, 1996. <https://doi.org/10.1063/1.472221>
17. Bauder, A., et al., *J. Chem. Phys.* 106, 7558, 1997. <https://doi.org/10.1063/1.473759>
18. Muller, H. S. P., Miller, C. E., and Cohen, E. A., *J. Chem. Phys.* 107, 8292, 1997. <https://doi.org/10.1063/1.475030>
19. Burgh, D. J., Suenram, R. D., and Stevens, W. J., *J. Chem. Phys.* 111, 3526, 1999. <https://doi.org/10.1063/1.479674>
20. Blake, T. A., et al., *J. Chem. Phys.* 98, 6031, 1993. <https://doi.org/10.1063/1.464842>
21. Ruoff, R. S., et al., *J. Chem. Phys.* 89, 138, 1988. <https://doi.org/10.1063/1.455515>
22. Muenter, J. S., *J. Chem. Phys.* 90, 4048, 1989. <https://doi.org/10.1063/1.456663>
23. Peterson, J. I., Suenram, R. D., and Lovas, F. J., *J. Chem. Phys.* 90, 5964, 1989. <https://doi.org/10.1063/1.456362>
24. Suenram, R. D., Lovas, F. J., and Matsumura, K., *Astrophys. J. Lett.* 342, 103, 1989. <https://doi.org/10.1086/185495>
25. Groner, P., et al., *J. Chem. Phys.* 91, 1434, 1989. <https://doi.org/10.1063/1.457103>
26. Suenram, R. D., Lovas, F. J., Fraser, G. T., and Matsumura, K., *J. Chem. Phys.* 92, 4724, 1990. <https://doi.org/10.1063/1.457690>
27. Andrews, A. M., et al., *J. Chem. Phys.* 93, 7030, 1990. <https://doi.org/10.1063/1.459425>
28. Peterson, K. I., Suenram, R. D., and Lovas, F. J., *J. Chem. Phys.* 94, 106, 1991. <https://doi.org/10.1063/1.460384>

29. Iida, M., Ohshima, Y., and Endo, Y., *J. Chem. Phys.* 95, 4772, 1991. <https://doi.org/10.1063/1.461719>
30. Andrews, A. M., Hillig, K. W., and Kuczkowski, R. L., *J. Chem. Phys.* 96, 1784, 1992. <https://doi.org/10.1063/1.462134>
31. Ruoff, R. S., et al., *J. Chem. Phys.* 96, 3441, 1992. <https://doi.org/10.1063/1.461947>
32. Germann, T. C., Tschopp, S. L., and Gutowsky, H. S., *J. Chem. Phys.* 97, 1619, 1992. <https://doi.org/10.1063/1.463203>
33. Taleb-Bendiab, A., Hillig, K. W., and Kuczkowski, R. L., *J. Chem. Phys.* 97, 2996, 1992. <https://doi.org/10.1063/1.463041>
34. Taleb-Bendiab, A., Hillig, K. W., and Kuczkowski, R. L., *J. Chem. Phys.* 98, 3627, 1993. <https://doi.org/10.1063/1.465069>
35. Xu, L-W., and Kuczkowski, R. L., *J. Chem. Phys.* 100, 15, 1994. <https://doi.org/10.1063/1.466987>
36. Peterson, K. I., Suenram, R. D., and Lovas, F. J., *J. Chem. Phys.* 102, 7807, 1995. <https://doi.org/10.1063/1.468981>
37. Tatamitani, Y., and Ogata, T., *J. Chem. Phys.* 121, 9885, 2004. <https://doi.org/10.1063/1.1809113>
38. Medvedev, I., et al., *Astrophys. J. Suppl.* 148, 593, 2003. <https://doi.org/10.1086/377259>
39. Lesarri, A., Suenram, R. D., and Brugh, D., *J. Chem. Phys.* 117, 9651, 2002. <https://doi.org/10.1063/1.1516797>
40. Arunan, E., et al., *J. Chem. Phys.* 117, 9766, 2002. <https://doi.org/10.1063/1.1518999>
41. Smith, T. C., Clouthier, D. J., and Steimle, T. C., *J. Chem. Phys.* 115, 817, 2001. <https://doi.org/10.1063/1.1378818>
42. Peebles, S. A., Sun, L., and Kuczkowski, R. L., *J. Chem. Phys.* 110, 6804, 1999. <https://doi.org/10.1063/1.478584>
43. Namiki, K. C., Robinson, J. S., and Steimle, T. C., *J. Chem. Phys.* 109, 5283, 1998. <https://doi.org/10.1063/1.477146>
44. Peebles, S. A., and Kuczkowski, R. L., *J. Chem. Phys.* 109, 5276, 1998. <https://doi.org/10.1063/1.477145>
45. Sauer, B. E., Wang, J., and Hinds, E. A., *J. Chem. Phys.* 105, 7412, 1996. <https://doi.org/10.1063/1.472569>
46. Fry, J. L., Drouin, B. J., and Miller, C. E., *J. Chem. Phys.* 124, 084304, 2006. <https://doi.org/10.1063/1.2163341>
47. Christiansen, J. J., *J. Mol. Spectrosc.* 231, 131, 2005. <https://doi.org/10.1016/j.jms.2005.01.004>
48. Kisiel, Z., et al., *Chem. Phys. Lett.* 325, 523, 2000. <https://doi.org/10.1016/S0009-2614(00)00729-6>
49. Lovas, F. J., et al., *Astrophys. J. Lett.* 455, 201, 1995. <https://doi.org/10.1086/309844>
50. Suenram, R. D., and Lovas, F. J., *Astrophys. J. Lett.* 429, 89, 1994. <https://doi.org/10.1086/187420>
51. Biermann, S., et al., *J. Chem. Phys.* 105, 9754, 1996. <https://doi.org/10.1063/1.472831>
52. McGlone, S., and Bauder, A., *J. Chem. Phys.* 109, 5383, 1998. <https://doi.org/10.1063/1.477157>
53. Peebles, S. A., and Kuczkowski, R. L., *J. Chem. Phys.* 111, 10511, 1999. <https://doi.org/10.1063/1.480404>
54. Plusquellic, D. F., et al., *J. Chem. Phys.* 115, 3057, 2001. <https://doi.org/10.1063/1.1385527>
55. Muller, H. S. P., and Cohen, E. A., *J. Chem. Phys.* 116, 2407, 2002. <https://doi.org/10.1063/1.1433002>
56. Andrews, A. M., and Kuczkowski, R. L., *J. Chem. Phys.* 98, 791, 1993. <https://doi.org/10.1063/1.464242>
57. Lovas, F. J., et al., *J. Chem. Phys.* 92, 891, 1990. <https://doi.org/10.1063/1.458123>
58. Klots, T. D., Emilsson, T., and Gutowsky, H. S., *J. Chem. Phys.* 97, 5335, 1992. <https://doi.org/10.1063/1.463793>
59. Careless, A. J., Kroto, H. W., and Landsberg, B. M., *Chem. Phys.* 1, 371, 1973. <https://doi.org/10.1016/0301-0104(73)80051-5>
60. Costain, C. C., and Kroto, H. W., *Can. J. Phys.* 50, 1453, 1972. <https://doi.org/10.1139/p72-199>
61. Kroto, H. W., Nixon, J. F., and Ohno, K., *J. Mol. Spectrosc.* 90, 367, 1981. <https://doi.org/10.1016/0022-2852(81)90134-X>
62. Kroto, H. W., Nixon, J. F., and Simmons, N. P. C., *J. Mol. Spectrosc.* 82, 185, 1980. <https://doi.org/10.1016/0022-2852(80)90108-3>
63. Kroto, H. W., Nixon, J. F., and Ohno, K., *J. Mol. Spectrosc.* 77, 270, 1979. <https://doi.org/10.1016/0022-2852(79)90108-5>
64. Cox, A. P., Ewart, I. C., and Gayton, T. R., *J. Mol. Spectrosc.* 125, 76, 1987. <https://doi.org/10.1016/0022-2852(87)90194-9>
65. Cohen, E. A., and Pickett, H. M., *J. Mol. Spectrosc.* 87, 582, 1981. <https://doi.org/10.1016/0022-2852(81)90430-6>
66. Peebles, S. A., and Peebles, R. A., *J. Mol. Struct.* 607, 19, 2002. <https://doi.org/10.1016/S0022-2860(01)00877-8>
67. Suenram, R. D., Lovas, F. J., and Pickett, H. M., *J. Mol. Spectrosc.* 116, 406, 1986. <https://doi.org/10.1016/0022-2852(86)90136-0>
68. Kroto, H. W., and Landsberg, B. M., *J. Mol. Spectrosc.* 62, 346, 1976. <https://doi.org/10.1016/0022-2852(76)90275-7>
69. Saito, S., and Makino, F., *Bull. Chem. Soc. Japan* 47, 1863, 1974. <https://doi.org/10.1246/bcsj.47.1863>
70. Sugie, M., Takeo, H., and Matsumura, C., *Chem. Phys. Lett.* 64, 573, 1979. <https://doi.org/10.1016/0009-2614(79)80247-X>
71. Borseth, D. G., Hillig, K. W., and Kuczkowski, R. L., *J. Am. Chem. Soc.* 106, 841, 1984. <https://doi.org/10.1021/ja00316a002>
72. Borseth, D. G., and Kuczkowski, R. L., *J. Phys. Chem.* 87, 5381, 1983. <https://doi.org/10.1021/j150644a015>
73. Hillig, K. W., Lattimer, R. P., and Kuczkowski, R. L., *J. Am. Chem. Soc.* 104, 988, 1982. <https://doi.org/10.1021/ja00368a012>
74. Gillies, C. W., and Kuczkowski, R. L., J. *Am. Chem. Soc.* 94, 6337, 1972. <https://doi.org/10.1021/ja00773a014>
75. Kisiel, Z., Pietrewicz, B. A., Fowler, P. W., Legon, A.C., and Steiner, E., *J. Phys. Chem. A* 104, 6970, 2000. <https://doi.org/10.1021/jp001156o>
76. Alonso, J. L., and Legon, A. C., *J. Chem, Soc. Faraday Trans.* 77, 2191, 1981. <https://doi.org/10.1039/F29817702191>
77. Kuczkowski, R. L., Suenram, R. D., and Lovas, F. J., *J. Am, Chem. Soc.* 103, 2561, 1981. <https://doi.org/10.1021/ja00400a013>
78. Kuczkowski, R. L., Lovas, F. J., Suenram, R. D., Lattimer, R. F., Hillig, K. W., and Ashe, R. J., *J. Mol. Struct.* 72, 143, 1981. <https://doi.org/10.1016/0022-2860(81)85014-4>
79. Lovas, F. J., McMahon, R. J., Grabow, J. U., Schnell, M., Mack, J., Scott, L. T., and Kuczkowski, R. L., *J. Am. Chem. Soc.* 127, 4345, 2005. <https://doi.org/10.1021/ja0426239>
80. Favero, L. B., Caminati, W., Arnason, I., and Kvaran, A., *J. Mol. Spec.* 229,188, 2005. <https://doi.org/10.1016/j.jms.2004.09.008>
81. Davies, A. P., Legon, A. C., Millen, D. J., and Roberts, A. J., *J. Mol. Struct.* 112, 9, 1984. <https://doi.org/10.1016/0022-2860(84)80238-0>
82. Maris, A., Giuliano, B. M., Melandri, S., Ottaviani, P., Caminati, W., Favero, L. B., and Velino, B., *Phys. Chem. Chem. Phys.* 7, 3317, 2005. <https://doi.org/10.1039/b507795h>
83. Zhang, R., Yu, Y., Steimle, T. C., and Cheng, L., *J. Chem. Phys.* 146, 064307, 2017. <https://doi.org/10.1063/1.4975816>
84. Steimle, T. C., Zhang, R., Qin, C., and Varberg, T. D., *J. Phys. Chem. A* 117, 11737, 2013. <https://doi.org/10.1021/jp402045k>
85. Zhang, R., Steimle, T. C., Cheng, L., and Stanton, J. F., *Mol. Phys.* 113, 2073, 2015. <https://doi.org/10.1080/00268976.2014.996619>
86. Collins, M. J., and Boggs, J. E., *J. Chem. Phys.* 57, 3811, 1972. <https://doi.org/10.1063/1.1678849>
87. Caminati, W., Corbelli, G., and Favero, L B., *J. Chem. Soc. Faraday Trans.* 89, 1631, 1993. <https://doi.org/10.1039/ft9938901631>
88. Foellmer, M. D., Murray, J. M., Serafin, M. M., Steber, A. L., Peebles, R. A., Peebles, S. A., Eichenberger, J. L., Guirgis, G. A., Wurrey, C. J., and Durig, J. R., *J. Phys. Chem. A* 113, 6077, 2009. <https://doi.org/10.1021/jp902033g>
89. Durig, J. R., Chang, M. S., Li, Y. S., Groner, P., and Stanley, A. E., *J. Phys. Chem.* 93, 3492, 1989. <https://doi.org/10.1021/j100346a027>
90. Lovas, F. J., Suenram, R. D., Ogata, T., and Yamamoto, S., *Astrophys J.* 399, 325, 1992. <https://doi.org/10.1086/171928>
91. Legon, A. C., and Willoughby, L. C., *Chem. Phys. Lett.* 111, 566, 1984. <https://doi.org/10.1016/0009-2614(84)80272-9>
92. Kisiel, Z., Legon, A. C., and Millen, D. J., *J. Chem. Phys.* 78, 2910, 1983. <https://doi.org/10.1063/1.445250>
93. Andrews, A. A., Hikkig, K. W., Kuczkowski, R. L., Legon, A. C., and Howard, N. W., *J. Chem. Phys*, 94, 6947, 1991. <https://doi.org/10.1063/1.460228>
94. Kawashima, Y., Suenram, R. D., and Hirota, E., *J. Chem. Phys.* 145, 114307, 2016. <https://doi.org/10.1063/1.4962363>

95. Groner, P., Winnewisser, M., Medvedev, I. R., De Lucia, F. C., Herbst, E., and Sastry, K. V. L. N., *Astrophys. J. Suppl. Ser.* 169, 28, 2007. <https://doi.org/10.1086/511133>
96. Mamleev, A. K., Galeev, R. V., Gunderova, L. N., Faizullin, M. G., and Shapkin, A. A., *J. Struct. Chem.* 48, 964, 2007. <https://doi.org/10.1007/s10947-007-0142-5>
97. Schnell, M., Hougen, J. T., and Grabow, J-U, *J. Mol. Spec.* 251, 38, 2008. <https://doi.org/10.1016/j.jms.2008.01.007>
98. Bizzocchi, L., Esposti, C. D., Dore, L., and Kisiel, Z., *J. Mol. Spec.* 251, 138, 2008. <https://doi.org/10.1016/j.jms.2008.02.009>
99. Wohlfart, K., Schnell, M., Grabow, J-U., and Küpper, J, *J. Mol. Spec.* 247, 119, 2008. <https://doi.org/10.1016/j.jms.2007.10.006>
100. Mokso, R., Møllendal, H., and Guillemin, J.-C., *J. Phys. Chem. A* 112, 4601, 2008. <https://doi.org/10.1021/jp801042p>
101. Samdal, S., Møllendal, H., and Guillemin, J.-C., *J. Phys. Chem. A* 117, 5073, 2013. <https://doi.org/10.1021/jp403374k>
102. Møllendal, H., Samdal, S., Matrane, A., and Guillemin, J.-C., *J. Phys. Chem. A* 115, 7978, 2011. <https://doi.org/10.1021/jp204296n>
103. Cazzoli, G., Puzzarini, C., Baldacci, A., and Baldan, A., *J. Mol. Spec.* 241, 112, 2007. <https://doi.org/10.1016/j.jms.2006.11.004>
104. Müller, H. S. P., Coutens, A., Walters, A., Grabow, J.-U., and Schlemmer, S., *J. Mol. Spec.* 267, 100, 2011. <https://doi.org/10.1016/j.jms.2011.02.011>
105. Kraśnicki, A., Pszczólkowski, L., and Kisiel, Z., *J. Mol. Spec.* 260, 57, 2010. <https://doi.org/10.1016/j.jms.2009.12.005>
106. Vogt, N., Rajappan Nair, K. P., Vogt, J., and Grabow, J.-U., *J. Mol. Spec.* 268, 16, 2011. <https://doi.org/10.1016/j.jms.2011.03.027>
107. Murray, J. M., Serafin, M. M., Steber, A. L., Peebles, S. A., Peebles, R. A., Wurrey, C. J., Durig, J. R., and Guirgis, G. A., *J. Mol. Struct.* 981, 54, 2010. <https://doi.org/10.1016/j.molstruc.2010.07.024>
108. Ka, S., Park, I., and Oh, J. J., *J. Mol. Spec.* 251, 374, 2008. <https://doi.org/10.1016/j.jms.2008.04.008>
109. Obenchain, D. A., Elliott, A. A., Steber, A. L., Peebles, R. A., Peebles, S. A., Wurrey, C. J., and Guirgis, G. A., *J. Mol. Spec.* 261, 35, 2010. <https://doi.org/10.1016/j.jms.2010.03.002>
110. Ilyushim, R. A., Motiyenko, R. A., Lovas, F. J., and Plusquellic, D. F., *J. Mol. Spec.* 251, 129, 2008. <https://doi.org/10.1016/j.jms.2008.02.005>
111. Møllendal, H., and Samdal, S., *J. Phys. Chem. A* 116, 12073, 2012. <https://doi.org/10.1021/jp309552m>
112. Filsinger, F., Wohlfart, K., Schnell, M., Grabow, J.-U., and Küpper, J., *Phys. Chem. Chem. Phys.* 10, 666, 2008. <https://doi.org/10.1039/B711888K>
113. Müller, H. S. P., Löblein, K., Hübner, H., Hüttner, W., and Brown, J. M., *J. Mol. Spec.* 251, 185, 2008. <https://doi.org/10.1016/j.jms.2008.02.014>
114. Huber, S., Grassi, G., and Bauder, A., *Mol. Phys.* 103, 1395, 2005. <https://doi.org/10.1080/00268970500038451>
115. Sedo, G., Schulz, J., and Leopold, K. R., *J. Mol. Spec.* 251, 4, 2008. <https://doi.org/10.1016/j.jms.2007.09.016>
116. Mamleev, A. K., Galeev, R. V., Gunderova, L. N., Faizullin, M. G., and Shapkin, A. A., *J. Struct. Chem.* 49, 639, 2008. <https://doi.org/10.1007/s10947-008-0088-2>
117. Margules, L., Huet, T. R., Demaison, J., Carvajal, M., Kleiner, I., Møllendal, H., Tercero, B., Marcelino, N., and Cernicharo, J., *Astrophys. J.* 714, 1120, 2010. <https://doi.org/10.1088/0004-637X/714/2/1120>
118. Carles, S., Møllendal, H., and Guillemin, J.-C., *Astron. Astrophys.* 558, A6, 2013. <https://doi.org/10.1051/0004-6361/201321427>
119. Ka, S., Kim, J., and Oh, J., *Bull. Korean Chem. Soc.* 32, 344, 2011. <https://doi.org/10.5012/bkcs.2011.32.1.344>
120. Mamleev, A. K., Galeev, R. V., and Faizullin, M. G., *J. Struct. Chem.* 52, 432, 2011. <https://doi.org/10.1134/S0022476611020272>
121. Mamleev, A. K., Galeev, R. V., and Faizullin, M. G., *J. Struct. Chem.* 53, 1056, 2012. <https://doi.org/10.1134/S0022476612060066>
122. Lovas, F. J., Suenram, R. D., Johnson, D. R., Clark, F. O., and Tiemann, E., *J. Chem. Phys.* 72, 4964, 1980. <https://doi.org/10.1063/1.439783>
123. Durig, J. R., Streusand, B. J., Li, Y. S., Richardson, L., and Laane, J., *J. Chem. Phys.* 73, 5564, 1980. <https://doi.org/10.1063/1.440075>
124. Vogelsanger, B., and Bauder, A., *J. Mol. Spec.* 130, 249, 1988. <https://doi.org/10.1016/0022-2852(88)90298-6>
125. Saito, S., and Matsumura, C., *J. Mol. Spec.* 80, 34, 1980. <https://doi.org/10.1016/0022-2852(80)90268-4>
126. Raw, T. T., and Gillies, C. W., *J. Mol. Spec.* 128, 195, 1988. <https://doi.org/10.1016/0022-2852(88)90217-2>
127. Sakaizumi, T., Mure, H., Ohashi, O., and Yamagauchi, I., *J. Mol. Spec.* 138, 375, 1989. <https://doi.org/10.1016/0022-2852(89)90005-2>
128. Fantoni, A. C., Filgueira, R. R., Boggia, L. M., and Caminati, W., *J. Mol. Spec.* 84, 493, 1980. <https://doi.org/10.1016/0022-2852(80)90039-9>
129. Langridge-Smith, P. R. R., Stevens, R., and Cox, A. P., *J. Chem. Soc. Faraday Trans.* 2, 76, 338, 1980. <https://doi.org/10.1039/f29807600330>
130. Hayashi, M., Nakagawa, J., and Kato, H., *J. Mol. Spec.* 84, 362, 1980. <https://doi.org/10.1016/0022-2852(80)90028-4>
131. Larsen, N. W., Pederson, T., and Soerensen, B. F., *J. Mol. Spec.* 128, 370, 1988. <https://doi.org/10.1016/0022-2852(88)90154-3>
132. Caminati, W., Fantoni, A.C., Meyer, R., Shaw, R. A., Smithson, T. L., and Wieser, H., *J. Mol. Spec.* 127, 450, 1988. <https://doi.org/10.1016/0022-2852(88)90133-6>
133. Wlodarczak, G., Martinache, L., Demaison, J., Marstokk, K. M., and Møllendal, H., *J. Mol. Spec.* 127, 178, 1988. <https://doi.org/10.1016/0022-2852(88)90017-3>
134. Caminati, W., Vogelsanger, B., and Bauder, A., *J. Mol. Spec.* 128, 384, 1988. <https://doi.org/10.1016/0022-2852(88)90155-5>
135. Kato, H., Nakagawa, J., and Hayashi, M., *J. Mol. Spec.* 80, 272, 1980. <https://doi.org/10.1016/0022-2852(80)90139-3>
136. Hayashi, M., Kato, H., and Oyamada, M., *J. Mol. Spec.* 83, 408, 1980. <https://doi.org/10.1016/0022-2852(80)90065-X>
137. Georgiou, K., and Kroto, H. W., *J. Mol. Spec.* 83, 94, 1980. <https://doi.org/10.1016/0022-2852(80)90313-6>
138. Mukhopadhyay, I., and Sastry, K. V. L. N., *J. Mol. Spec.* 312, 51, 2015. <https://doi.org/10.1016/j.jms.2015.03.009>
139. Kim, J., Jang, H., Ka, S., Obenchain, D. A., Peebles, R. A., and Peebles, S. A., *J. Mol. Spec.* 328, 50, 2016. <https://doi.org/10.1016/j.jms.2016.08.004>
140. Lovas, F. J., and Sprague, M. K., *J. Mol. Spec.* 316, 49, 2015. <https://doi.org/10.1016/j.jms.2015.07.005>
141. Thomas, A. J., Serafin, M. M., Ernst, A. A., Peebles, R. A., and Peebles, S. A., *J. Mol. Spec.* 289, 65, 2013. <https://doi.org/10.1016/j.jms.2013.03.007>
142. Dorosh, O., Bialkowska-Jaworska, E., Kisiel, Z., Pszczölkowski, L., Kanska, M., Krygowski, T. M., and Mäder, H., *J. Mol. Spec.* 335, 3, 2017. <https://doi.org/10.1016/j.jms.2017.01.009>
143. Kuze, N., Ohashi, O., and Sakaizumi, T., *J. Mol. Spec.* 337, 17, 2017. <https://doi.org/10.1016/j.jms.2017.04.003>
144. Jang, H., Shim, J.-S., and Oh, J. J., *J. Mol. Spec.* 337, 65, 2017. <https://doi.org/10.1016/j.jms.2017.04.021>
145. Jang, H., Ka, S., Dikkumbura, A. S., Peebles, R. A., Peebles, S. A., and Oh, J. J., *J. Mol. Struct.* 1133, 320, 2017. <https://doi.org/10.1016/j.molstruc.2016.11.096>
146. Munrow, M. R., *et al.*, *J. Mol. Spec.* 242, 129, 2007. <https://doi.org/10.1016/j.jms.2007.02.021>
147. Higgins, K. J., Freund, S. M., Klemperer, W., Apponi, A. J., and Ziurys, L. M., *J. Chem. Phys.* 121, 11715, 2004. <https://doi.org/10.1063/1.1814631>
148. Lovas, F. J., Plusquellic, D. F., Pate, B. H., Neill, J. L., Muckle, M. T., and Remijan, A. J., *J. Mol. Spec.* 257, 82, 2009. <https://doi.org/10.1016/j.jms.2009.06.013>
149. Christen, D., Ramme, K., Haas, B., and Oberhammer, H., *J. Chem. Phys.* 80, 4020, 1998. <https://doi.org/10.1063/1.447282>
150. Durig, J. R., Panikar, S. S., Groner, P., Hossein, N., Bürger, H., and Moritz, P., *J. Phys. Chem. A* 114, 4131, 2010. <https://doi.org/10.1021/jp911614d>
151. Ernst, A. A., Christenholz, C. L., Dhahir, Y. D., Peebles, S. A., and Peebles, R. A., *J. Phys. Chem. A* 119, 12999, 2015. <https://doi.org/10.1021/acs.jpca.5b09933>
152. Rajappan Nair, K. P., Herbers, S., Obenchain, D. A., Grabow, J.-U., and Lesarri, A., *J. Mol. Spec.* 344, 21, 2018. <https://doi.org/10.1016/j.jms.2017.10.003>
153. Penn, R. E., and Curl, R. F., *J. Chem. Phys.* 55, 651, 1971. <https://doi.org/10.1063/1.1676133>

TABLE 1. Dipole Moments of Gases and Liquids

Name	Mol. form.	μ/D	Ref.
Acenaphthene	$C_{12}H_{10}$	≈0.85	1
Acetaldehyde	CH_3CHO	2.750(6)	3
Acetamide	CH_3CONH_2	3.68(3)	5
Acetic acid	CH_3COOH	1.70(3)	2
Acetic anhydride	$C_4H_6O_3$	≈2.8	1
Acetone	$(CH_3)_2CO$	2.88(3)	1
Acetonitrile	CH_3CN	3.92519	5
Acetophenone	$C_6H_5C{=}OCH_3$	3.02(6)	1
Acetyl chloride	CH_3COCl	2.72(14)	1
Acetylene-carbon dioxide complex	$C_2H_2{\cdot}CO_2$	0.161(1)	22
Acetylene-carbon monoxide complex	$C_2H_2{\cdot}CO$	0.311(1)	32
Acetylene-carbon oxysulfide trimer complex	$C_2H_2{\cdot}(OCS)_3$	1.23(2)	53
Acetylene-hydrogen cyanide complex	$C_2H_2{\cdot}HCN$	3.29(3)	32
Acetylene-sulfur dioxide complex	$C_2H_2{\cdot}SO_2$	1.683(5)	93
Acetyl fluoride	CH_3COF	2.96(3)	1
Acrylonitrile	$CH_2{=}CHCN$	3.92(7)	5
Allenyl isocyanide	$CH_2{=}C{=}CHNC$	3.81(9)	102
Allyl alcohol	C_3H_6O	1.60(8)	1
Allylamine	$CH_2{=}CHCH_2NH_2$	≈1.2	1
Aluminum monofluoride	AlF	1.53(15)	1
Aminoborane	BH_2NH_2	1.844(15)	70
Ammonia	NH_3	1.4718(2)	5
Aniline	$C_6H_5NH_2$	1.13(2)	3
Anisole	C_7H_8O	1.38(7)	1
Arsenic(III) chloride	$AsCl_3$	1.59(8)	1
Arsenic(III) fluoride	AsF_3	2.59(5)	1
Arsine	AsH_3	0.217(3)	5
Azulene	$C_{10}H_8$	0.882(2)	114
Barium oxide	BaO	7.954(3)	5
Barium sulfide	BaS	10.86(2)	3
Benzaldehyde	C_6H_5CHO	3.0 liq	7
Benzeneacetonitrile	$C_6H_5CH_2CN$	3.5 liq	7
Benzene-hydrogen sulfide complex	$C_6H_6{\cdot}H_2S$	1.14(2)	40
Benzene-krypton complex	$C_6H_6{\cdot}Kr$	0.136(2)	58
Benzene-sulfur dioxide complex	$C_6H_6{\cdot}SO_2$	2.061(2)	33
Benzenethiol	C_6H_5SH	1.23 liq	7
1,3,2-Benzodioxaborole	$C_6H_5BO_2$	1.26(4)	87
1-Benzofuran	C_8H_6O	0.752(4)	82
Benzonitrile	C_6H_5CN	4.515(14)	99
Benzoyl fluoride	C_6H_5COF	3.57(7)	131
Benzyl acetate	$C_9H_{10}O_2$	1.22 liq	7
Benzyl alcohol	C_7H_8O	1.71(9)	1
Benzyl benzoate	$C_{14}H_{12}O_2$	2.06 liq	7
Bicyclo[3.1.0]hexan-2-one	C_6H_8O	3.74(9)	81
Bis(2-aminoethyl)amine	$C_4H_{13}N_3$	1.89 liq	7
Bis(2-chloroethyl) ether	$C_4H_8Cl_2O$	2.58 liq	7
Bis(2-ethylhexyl) phthalate	$C_{24}H_{38}O_4$	2.84 liq	7
Borane carbonyl	BH_3CO	1.698(20)	3
Bromine chloride	BrCl	0.519(4)	3
Bromine dioxide	BrO_2	2.8(1)	18
Bromine fluoride	BrF	1.422(16)	3
Bromine monoxide	BrO	1.76(4)	2
Bromine pentafluoride	BrF_5	1.51(15)	1
Bromoacetylene	HC≡CBr	0.22962	5
Bromobenzene	C_6H_5Br	1.70(3)	1
1-Bromobutane	C_4H_9Br	2.31(6)[a]	139
2-Bromobutane, (±)-	C_4H_9Br	2.23(11)	1
1-Bromo-2-chloroethane	CH_2BrCH_2Cl	1.2 liq	7
Bromochlorofluoromethane	CHBrClF	1.5(3)	17
Bromochloromethane	CH_2BrCl	1.66 liq	7
1-Bromodecane	$C_{10}H_{21}Br$	1.93 liq	7
Bromoethane	C_2H_5Br	2.04(2)	5
Bromoethene	$CH_2{=}CHBr$	1.42(3)	1
Bromofluoroacetylene	BrC≡CF	0.448(2)	5
Bromofluoromethane	CH_2BrF	1.739(18)	103
1-Bromoheptane	$C_7H_{15}Br$	2.16(11)	1
Bromomethane	CH_3Br	1.8203(4)	5
2-Bromo-2-methylpropane	$(CH_3)_3CBr$	2.17 liq	7
1-Bromonaphthalene	$C_{10}H_7Br$	1.55 liq	7
1-Bromopentane	$C_5H_{11}Br$	2.20(11)	1
1-Bromopropane	C_3H_7Br	2.18(11)	1
2-Bromopropane	C_3H_7Br	2.21(11)	1
2-Bromopropene	C_3H_5Br	1.51 liq	7
3-Bromopropene	C_3H_5Br	≈1.9	1
Bromosilane	SiH_3Br	1.319	3
Bromotrifluoromethane	CF_3Br	0.65(5)	1
Bromotrifluorosilane	$SiBrF_3$	0.835(7)	64
1,2-Butadiene	C_4H_6	0.403(2)	1
Butanal	C_4H_8O	2.72(5)	1
1,4-Butanediol	$C_4H_{10}O_2$	2.58 liq	7
1-Butanethiol	$C_4H_{10}S$	1.53 liq	7
Butanoic acid	$C_4H_8O_2$	1.65 liq	7
1-Butanol	C_4H_9OH	1.66(3)	1
2-Butanone	C_4H_8O	2.779(15)	2
trans-2-Butenal	C_4H_6O	3.67(7)	1
cis-2-Butene	C_4H_8	0.253(5)	2
cis-2-Butene-1,4-diol	$C_4H_8O_2$	2.48 liq	7
trans-2-Butene-1,4-diol	$C_4H_8O_2$	2.45 liq	7
trans-2-Butenoic acid	$C_4H_6O_2$	2.13 liq	7
cis-2-Buten-1-ol	C_4H_8O	1.96(3)	5
trans-2-Buten-1-ol	C_4H_8O	1.90(2)	5
1-Buten-3-yne	C_4H_4	0.22(2)	3
2-Butoxyethanol	$C_6H_{14}O_2$	2.08 liq	7
Butyl acetate	$C_6H_{12}O_2$	1.87 liq	7
sec-Butyl acetate	$C_6H_{12}O_2$	1.87 liq	7
Butylamine	$C_4H_9NH_2$	≈1.0	1
sec-Butylamine	$CH_3CH(NH_2)C_2H_5$	1.28 liq	7
tert-Butylamine	$(CH_3)_3CNH_2$	1.29 liq	7
tert-Butylbenzene	$(CH_3)_3CC_6H_5$	≈0.83	1
Butyl ethyl ether	$C_4H_9OC_2H_5$	1.24 liq	7
Butyl formate	C_4H_9OCHO	2.03 liq	7
Butyl stearate	$C_{22}H_{44}O_2$	1.88 liq	7
Butyl vinyl ether	$C_4H_9OC_2H_3$	1.25 liq	7
1-Butyne	C_4H_6	0.782(4)	5
γ-Butyrolactone	$C_4H_6O_2$	4.27(3)	3
Calcium methoxide	$CaOCH_3$	1.58(8)	43
Calcium monochloride	CaCl	≈3.6	4
Camphor, (+)	$C_{10}H_{16}O$	3.1 liq	7
Caprolactam	$C_6H_{11}NO$	3.9 liq	7
Carboimidic difluoride	CF_2NH	1.393(1)	5
Carbon dioxide-difluoromethane complex	$CO_2{\cdot}CH_2F_2$	1.968(14)	141
Carbon monoxide-sulfur dioxide complex	$CO{\cdot}SO_2$	1.599(3)	140
Carbon dioxide dimer-water complex	$(CO_2)_2{\cdot}H_2O$	1.989(2)	23
Carbon dioxide-mercury complex	$CO_2{\cdot}Hg$	0.107(3)	29
Carbon dioxide-water dimer complex	$CO_2{\cdot}(H_2O)_2$	1.746(10)	28
Carbon disulfide-sulfur dioxide complex	$CO_2{\cdot}SO_2$	1.096(1)	42
Carbon monoselenide	CSe	1.99(4)	3
Carbon monosulfide	CS	1.958(5)	2
Carbon monoxide	CO	0.10980	3
Carbon monoxide dimer-water complex	$(CO)_2{\cdot}H_2O$	1.57(5)	36
Carbon oxyselenide	OCSe	0.73(2)	1
Carbon oxysulfide	OCS	0.715189	5
Carbon oxysulfide-carbon dioxide dimer complex	$OCS{\cdot}(CO_2)_2$	0.69(5)	44
Carbon oxysulfide-water complex	$OCS{\cdot}H_2O$	2.668(3)	37
Carbonyl chloride	$COCl_2$	1.17(1)	1
Carbonyl fluoride	COF_2	0.95(1)	1
Cesium chloride	CsCl	10.387(4)	2
Cesium fluoride	CsF	7.884(1)	2
Cesium sodium	NaCs	4.75(20)	2
Chlorine fluoride	ClF	0.888061	5
Chlorine oxide (ClO)	ClO	1.297(1)	5

Name	Mol. form.	μ/D	Ref.
Chlorine trifluoride	ClF_3	0.6(1)	1
Chloroacetyl chloride	$C_2H_2Cl_2O$	2.23(11)	1
Chloroacetylene	$HC{\equiv}CCl$	0.44408	5
2-Chloroaniline	C_6H_6ClN	1.77 liq	7
Chlorobenzene	C_6H_5Cl	1.69(3)	1
Chloroborane	BH_2Cl	0.75(5)	14
1-Chlorobutane	C_4H_9Cl	2.05(4)	1
2-Chlorobutane	C_4H_9Cl	2.04(10)	1
1-Chloro-1,1-difluoroethane	CH_3CClF_2	2.14(4)	1
Chlorodifluoromethane	$CHClF_2$	1.42(3)	1
Chloroethane	C_2H_5Cl	2.05(2)	1
2-Chloroethanol	C_2H_5ClO	1.78(9)	1
Chloroethene	$CH_2{=}CHCl$	1.45(3)	1
1-Chloro-4-fluorobenzene	C_6H_4ClF	0.12(1)	66
1-Chloro-1-fluoroethane	CH_3CHClF	2.068(14)	3
Chlorofluoromethane	CH_2ClF	1.82(4)	1
Chlorogermane	GeH_3Cl	2.13(2)	1
Chloromethane	CH_3Cl	1.8963(2)	5
(Chloromethyl)benzene	C_7H_7Cl	1.82 liq	7
1-Chloro-3-methylbutane	$C_5H_{11}Cl$	1.92 liq	7
1-Chloro-2-methylpropane	C_4H_9Cl	2.00(10)	1
2-Chloro-2-methylpropane	$(CH_3)_3CCl$	2.13(4)	1
1-Chloronaphthalene	$C_{10}H_7Cl$	1.57 liq	7
1-Chloro-2-nitrobenzene	$C_6H_4ClNO_2$	4.64(9)	1
1-Chloro-3-nitrobenzene	$C_6H_4ClNO_2$	3.73(7)	1
1-Chloro-4-nitrobenzene	$C_6H_4ClNO_2$	2.83(6)	1
1-Chlorooctane	$C_8H_{17}Cl$	2.00 liq	7
Chloropentafluoroethane	CF_3CF_2Cl	0.52(5)	1
1-Chloropentane	$C_5H_{11}Cl$	2.16(11)	1
4-Chlorophenol	C_6H_5ClO	2.11(11)	1
1-Chloropropane	C_3H_7Cl	2.05(4)	1
2-Chloropropane	$CH_3CHClCH_3$	2.17(11)	1
cis-1-Chloropropene	C_3H_5Cl	1.67(8)	1
trans-1-Chloropropene	C_3H_5Cl	1.97(10)	1
2-Chloropropene	C_3H_5Cl	1.647(10)	3
3-Chloropropene	C_3H_5Cl	1.94(10)	1
4-Chloropyridine	C_5H_4ClN	0.756(5)	3
Chlorosilane	SiH_3Cl	1.31(1)	1
Chlorosyl fluoride	$FCl{=}O$	1.93(2)	55
2-Chlorotoluene	C_7H_7Cl	1.56(8)	1
3-Chlorotoluene	C_7H_7Cl	1.82 liq	7
4-Chlorotoluene	C_7H_7Cl	2.21(4)	1
Chlorotrifluoroethene	$CF_2{=}CFCl$	0.40(10)	1
Chlorotrifluoromethane	CF_3Cl	0.50(1)	1
Chlorotrifluorosilane	$SiClF_3$	0.636(4)	5
Chlorotrimethylstannane	$(CH_3)_3SnCl$	3.498(6)	97
Chromium monoxide	CrO	3.88(13)	5
Copper(I) fluoride	CuF	5.77(29)	2
Copper(II) oxide	CuO	4.5(5)	5
Corannulene	$C_{20}H_{10}$	2.07	79
o-Cresol	C_7H_8O	1.45 liq	7
m-Cresol	C_7H_8O	1.48 liq	7
p-Cresol	C_7H_8O	1.48 liq	7
Cyanamide	H_2NCN	4.28(10)	5
Cyanoacetylene	$HC{\equiv}CCN$	3.73172	5
Cyanoformamide	$C_2H_2N_2O$	4.10(12)	47
Cyanogen azide	N_3CN	2.96(7)	60
Cyanogen chloride	$ClCN$	2.8331(2)	3
Cyanogen fluoride	FCN	2.120(1)	3
Cyanogen iodide	ICN	3.67(2)	5
Cyanomethylmercury	CH_3HgCN	4.7(1)	12
Cyclobutanecarbonitrile	C_5H_7N	4.04(4)	5
Cyclobutanone	C_4H_6O	2.89(3)	2
Cyclobutene	C_4H_6	0.132(1)	1
1,3-Cycloheptadiene	C_7H_{10}	0.740	3
2,4,6-Cycloheptatrien-1-one	C_7H_6O	4.1(3)	3
3,5-Cyclohexadiene-1,2-dione	$C_6H_4O_2$	4.23(2)	3
Cyclohexanone	$C_6H_{10}O$	3.246(6)	5
Cyclohexylamine	$C_6H_{13}N$	1.26 liq	7
1,3-Cyclopentadiene	C_5H_6	0.419(4)	1
2,4-Cyclopentadien-1-one	C_5H_4O	3.132(7)	3
Cyclopentanone	C_5H_8O	≈3.3	1
Cyclopentene	C_5H_8	0.20(2)	1
Cyclopentene ozonide	$C_5H_8O_3$	2.48(1)	72
3-Cyclopenten-1-one	C_5H_6O	2.79(3)	3
Cyclopropanecarbonitrile	C_3H_5CN	4.220(6)	98
Cyclopropane-sulfur dioxide complex	$C_3H_6{\cdot}SO_2$	1.681(1)	30
Cyclopropanethiol	C_3H_5SH	1.52(10)	100
Cyclopropanone	C_3H_4O	2.67(13)	2
Cyclopropene	C_3H_4	0.454(10)	1
Cyclopropenylidene	C_3H_2	3.27(1)	90
Cyclopropenylidyne	C_3H	2.30(10)	90
Cyclopropylamine	$C_3H_5NH_2$	1.19(1)	2
Cyclopropylmethylgermane	$C_3H_5GeH_2CH_3$	0.680(5)[b]	107
Cyclopropyl methyl ketone	$C_3H_5C{=}OCH_3$	2.62(25)	2
Cyclopropylmethylsilane	$C_3H_5SiH_2CH_3$	0.790(13)	88
Diacetone alcohol	$C_6H_{12}O_2$	3.24 liq	7
Diazomethane	$CH_2{=}N{\equiv}N$	1.50(1)	1
Dibromodifluoromethane	CBr_2F_2	0.66(5)	1
1,2-Dibromoethane	CH_2BrCH_2Br	1.19 liq	7
Dibromomethane	CH_2Br_2	1.43(3)	1
1,2-Dibromopropane	$CH_2BrCHBrCH_3$	1.2 liq	7
Dibutylamine	$(C_4H_9)_2NH$	0.98 liq	7
Dibutyl ether	$(C_4H_9)_2O$	1.17(6)	1
Dibutyl phthalate	$C_{16}H_{22}O_4$	2.82 liq	7
Dibutyl sebacate	$C_{18}H_{34}O_4$	2.48 liq	7
Dibutyl sulfide	$(C_4H_9)_2S$	1.61 liq	7
o-Dichlorobenzene	$C_6H_4Cl_2$	2.50(5)	1
m-Dichlorobenzene	$C_6H_4Cl_2$	1.72(9)	1
1,4-Dichlorobutane	$C_4H_8Cl_2$	2.22(11)	1
1,1-Dichloro-2,2-difluoroethene	$F_2C{=}CCl_2$	0.50	7
Dichlorodifluoromethane	CF_2Cl_2	0.51(5)	1
1,1-Dichloroethane	CH_3CHCl_2	2.06(4)	1
1,2-Dichloroethane	CH_2ClCH_2Cl	1.83 liq	7
1,1-Dichloroethene	$CH_2{=}CCl_2$	1.34(1)	1
cis-1,2-Dichloroethene	$C_2H_2Cl_2$	1.90(4)	1
Dichlorofluoromethane	$CHCl_2F$	1.29(3)	1
Dichloromethane	CH_2Cl_2	1.60(3)	1
(Dichloromethyl)benzene	$C_6H_5CHCl_2$	2.07 liq	7
Dichloromethylborane	CH_3BCl_2	1.419(13)	5
1,2-Dichloropropane, (±)-	$C_3H_6Cl_2$	1.85 liq	7
1,3-Dichloropropane	$CH_2ClCH_2CH_2Cl$	2.08(4)	1
Dichlorosilane	SiH_2Cl_2	1.17(2)	1
1,2-Dichloro-1,1,2,2-tetrafluoroethane	$C_2Cl_2F_4$	≈0.5	1
2,4-Dichlorotoluene	$C_7H_6Cl_2$	1.70 liq	7
3,4-Dichlorotoluene	$C_7H_6Cl_2$	2.95 liq	7
Diethanolamine	$C_4H_{11}NO_2$	2.8 liq	7
1,1-Diethoxyethane	$(C_2H_5O)_2CHCH_3$	1.38 liq	7
Diethylamine	$(C_2H_5)_2NH$	0.92(5)	1
Diethyl carbonate	$(C_2H_5)_2CO_3$	1.10(6)	1
Diethylene glycol	$C_4H_{10}O_3$	2.31 liq	7
Diethylene glycol dimethyl ether	$C_6H_{14}O_3$	1.97 liq	7
Diethylene glycol monoethyl ether	$C_6H_{14}O_3$	1.6 liq	7
Diethylene glycol monoethyl ether acetate	$C_8H_{16}O_4$	1.8 liq	7
Diethylene glycol monomethyl ether	$C_5H_{12}O_3$	1.6 liq	7
Diethyl ether	$(C_2H_5)_2O$	1.098(1)	38
Diethyl malonate	$C_7H_{12}O_4$	2.54 liq	7
Diethyl oxalate	$(C_2H_5)_2C_2O_4$	2.49 liq	7
Difluoramine	NHF_2	1.92(2)	1
Difluorine dioxide	F_2O_2	1.44(7)	1
Difluoroamidogen	NF_2	0.136(10)	113
o-Difluorobenzene	$C_6H_4F_2$	2.46(5)	2
m-Difluorobenzene	$C_6H_4F_2$	1.51(2)	2
Difluoroborane	BHF_2	0.971(10)	8
1,1-Difluorocyclohexane	$C_6H_{10}F_2$	2.556(10)	3
3,3-Difluorocyclopropene	$C_3H_2F_2$	2.98(2)	3

Name	Mol. form.	μ/D	Ref.
cis-Difluorodiazine	N_2F_2	0.16(1)	1
1,1-Difluoroethane	CH_3CHF_2	2.27(5)	1
1,1-Difluoroethene	CH_2=CF_2	1.3893(2)	5
cis-1,2-Difluoroethene	CHF=CHF	2.42(2)	1
Difluoromethane	CH_2F_2	1.9785(21)	3
Difluoromethylborane	CH_3BF_2	1.668(3)	3
Difluoromethylene	CF_2	0.47(2)	3
1,1-Difluoro-1-propene	CH_3CH=CF_2	0.889(7)	2
Difluorosilane	SiH_2F_2	1.55(2)	1
Difluorosilylene	SiF_2	1.23(2)	2
2,4-Difluorotoluene	$C_7H_6F_2$	1.81(4)	152
3,6-Dihydro-1,2-dioxin	$C_4H_6O_2$	2.329(1)	3
2,3-Dihydro-1,4-dioxin	$C_4H_6O_2$	0.939(8)	3
2,3-Dihydrofuran	C_4H_6O	1.32(3)	2
2,5-Dihydrofuran	C_4H_6O	1.63(1)	5
Dihydro-3-methyl-2(3*H*)-furanone	$C_5H_8O_2$	4.56(2)	5
Dihydro-5-methyl-2(3*H*)-furanone, (±)-	$C_5H_8O_2$	4.71(5)	5
3,4-Dihydro-2*H*-pyran	C_5H_8O	1.400(8)	5
3,6-Dihydro-2*H*-pyran	C_5H_8O	1.283(5)	3
2,3-Dihydrothiophene	C_4H_6S	1.61(20)	5
2,5-Dihydrothiophene	C_4H_6S	1.75(1)	3
Diiodomethane	CH_2I_2	1.08 liq	7
Diisopentyl ether	$C_{10}H_{22}O$	1.23 liq	7
Diisopropylamine	$C_6H_{15}N$	1.15 liq	7
Diisopropyl ether	$C_6H_{14}O$	1.13(10)	1
Diketene	$C_4H_4O_2$	3.53(7)	1
1,2-Dimethoxybenzene	$C_8H_{10}O_2$	1.29 liq	7
Dimethoxymethane	$C_3H_8O_2$	0.74 liq	7
N,N-Dimethylacetamide	C_4H_9NO	3.7 liq	7
Dimethylamine	$(CH_3)_2NH$	1.01(2)	2
2,4-Dimethylaniline	$C_8H_{11}N$	1.40 liq	7
2,6-Dimethylaniline	$C_8H_{11}N$	1.63 liq	7
N,N-Dimethylaniline	$C_8H_{11}N$	1.68(17)	1
3,3-Dimethyl-1-butyne	$(CH_3)_3C$-C≡CH	0.661(4)	1
2,6-Dimethylcyclohexanone	$C_8H_{14}O$	2.862(2)	144
1,1-Dimethylcyclopropane	C_5H_{10}	0.142(1)	3
3,3-Dimethylcyclopropene	C_5H_8	0.287(3)	3
4,4-Dimethyl-1,3-dioxane	$C_6H_{12}O_2$	2.15(4)	120
Dimethyl disulfide	$C_2H_6S_2$	1.85 liq	7
Dimethyl ether	CH_3OCH_3	1.30(1)	1
N,N-Dimethylformamide	C_3H_7NO	3.82(8)	1
2,6-Dimethyl-4-heptanone	$C_9H_{18}O$	2.66 liq	7
Dimethyl maleate	$C_6H_8O_4$	2.48 liq	7
2,4-Dimethyl-3-pentanone	$C_7H_{14}O$	2.74 liq	7
2,2-Dimethylpropanal	$(CH_3)_3CCHO$	2.66(5)	2
2,2-Dimethylpropanenitrile	$(CH_3)_3CCN$	3.95(4)	1
2,4-Dimethylpyridine	C_7H_9N	2.30 liq	7
2,6-Dimethylpyridine	C_7H_9N	1.66 liq	7
Dimethyl sulfide	$(CH_3)_2S$	1.554(4)	3
Dimethyl sulfoxide	$(CH_3)_2S$=O	3.96(4)	1
1,3-Dioxane	$C_4H_8O_2$	2.06(4)	2
1,3-Dioxolane	$C_3H_6O_2$	1.19(6)	3
Dipentyl ether	$(C_5H_{11})_2O$	1.20 liq	7
Diphenyl ether	$(C_6H_5)_2O$	≈1.3	1
Dipropylamine	$(C_3H_7)_2NH$	1.03 liq	7
Dipropyl ether	$(C_3H_7)_2O$	1.21(6)	1
Disiloxane	$(SiH_3)_2O$	0.24(2)	1
1,3-Dithiane	$C_4H_8S_2$	2.14(4)	5
Divinyl ether	$(CH_2$=$CH)_2O$	0.78(5)	2
Epichlorohydrin	C_3H_5ClO	1.8 liq	7
1,2-Epoxybutane	C_4H_8O	1.891(11)	3
3,4-Epoxy-1-butyne	C_4H_4O	1.80(2)	86
1,2-Ethanediamine	$C_2H_8N_2$	1.99(10)	1
1,2-Ethanediol	$(CH_2OH)_2$	2.36(10)	5
1,2-Ethanediol, diacetate	$C_6H_{10}O_4$	2.34 liq	7
1,2-Ethanedithiol	$(CH_2SH)_2$	2.03(8)	5
Ethanol	C_2H_5OH	1.69(3)	1
Ethanolamine	$CH_2OHCH_2NH_2$	3.05(5)	153

Name	Mol. form.	μ/D	Ref.
Ethene ozonide	$C_2H_4O_3$	1.09(1)	74
Ethoxybenzene	$C_8H_{10}O$	1.45(15)	1
2-Ethoxyethanol	$C_4H_{10}O_2$	2.08 liq	7
2-Ethoxyethyl acetate	$C_6H_{12}O_3$	2.25 liq	7
Ethyl acetate	$C_4H_8O_2$	1.78(9)	1
Ethyl acrylate	$C_5H_8O_2$	1.96 liq	7
Ethylamine	$C_2H_5NH_2$	1.22(10)	1
Ethylbenzene	C_8H_{10}	0.59(5)	1
Ethyl benzoate	$C_9H_{10}O_2$	2.00(10)	1
Ethyl butanoate	$C_6H_{12}O_2$	1.74 liq	7
Ethyl *trans*-cinnamate	$C_{11}H_{12}O_2$	1.84 liq	7
Ethyl cyanate	C_2H_5CNO	4.72(20)	127
Ethylcyanoacetate	$C_5H_7NO_2$	2.17 liq	7
Ethylcyanoacetylene	C_5H_5N	5.75(6)	118
4-Ethylcyclohexanone	$C_8H_{14}O$	3.32(3)	108
Ethylene carbonate	$C_3H_4O_3$	4.9 liq	7
Ethyleneimine	C_2H_5N	1.90(1)	1
Ethylene-sulfur dioxide complex	$C_2H_4{\cdot}SO_2$	1.650(3)	27
Ethylene-water complex	$C_2H_4{\cdot}H_2O$	1.10(1)	56
Ethyl fluoromethyl ether	$C_2H_5OCH_2F$	1.806(12)	136
Ethyl formate	C_2H_5OCHO	1.93	1
2-Ethyl-1-hexanol	$C_8H_{18}O$	1.74 liq	7
2-Ethylhexyl acetate	$C_{10}H_{20}O_2$	1.8 liq	7
Ethyl lactate	$C_5H_{10}O_3$	2.4 liq	7
Ethyl propanoate	$C_5H_{10}O_2$	1.74 liq	7
Ethyl vinyl ether	C_2H_5OCH=CH_2	1.26 liq	7
Fluoramine	NH_2F	2.27(18)	5
Fluorine azide	FN_3	≈1.3	5
Fluorine monoxide	F_2O	0.308180	5
Fluorine oxide	FO	0.0043(4)	5
Fluoroacetylene	FC≡CH	0.7207(3)	3
Fluorobenzene	C_6H_5F	1.60(8)	1
4-Fluorobenzoyl fluoride	$C_7H_4F_2O$	2.16(8)	131
Fluoroborane(1)	BF	≈0.5	2
1-Fluorocyclohexene	C_6H_9F	1.942(10)	5
Fluoroethane	CH_3CH_2F	1.937(7)	5
Fluoroethene	CH_2=CHF	1.468(3)	5
Fluoroethene ozonide	$C_2H_3FO_3$	2.317(21)	73
Fluorogermane	GeH_3F	2.33(12)	2
Fluoromethane	CH_3F	1.858(2)	3
Fluoromethylidyne	CF	0.645(5)	3
(Fluoromethylidyne)phosphine	FC≡P	0.279(1)	62
Fluoromethylsilane	CH_2FSiH_3	1.674(8)	150
1-Fluoro-4-nitrobenzene	$C_6H_4FNO_2$	2.87(6)	1
2-Fluoropropane	CH_3CHFCH_3	1.958(1)	5
cis-1-Fluoropropene	CHF=$CHCH_3$	1.46(3)	1
trans-1-Fluoropropene	CHF=$CHCH_3$	≈1.9	1
2-Fluoropropene	CH_2=$CFCH_3$	1.61(3)	1
3-Fluoropropyne	HC≡CCH_2F	1.73(2)	5
3-Fluoropyridine	C_5H_4FN	2.09(26)	3
Fluorosilane	SiH_3F	1.2969(6)	5
2-Fluorotoluene	C_7H_7F	1.37(7)	1
3-Fluorotoluene	C_7H_7F	1.82(4)	2
4-Fluorotoluene	C_7H_7F	2.00(10)	1
Formaldehyde	HCHO	2.332(2)	3
Formaldehyde dimer	$(CH_2O)_2$	0.858(5)	57
Formamide	$HCONH_2$	3.73(7)	1
Formic acid	HCOOH	1.425(2)	5
Formyl fluoride	FCHO	2.081(1)	5
Fulminic acid	HONC	3.09934	5
Fulvene	C_6H_6	0.4236(13)	2
Fumaric acid	$C_4H_4O_4$	2.733(10)	106
4-Fluorophenylacetylene	C_8H_5F	0.8935(9)	145
Furan	C_4H_4O	0.66(1)	1
2-Furanone	$C_4H_4O_2$	4.905	76
Furfural	$C_5H_4O_2$	3.54 liq	7
Furfuryl alcohol	$C_5H_6O_2$	1.92 liq	7
Gallium monofluoride	GaF	2.45(5)	2

Name	Mol. form.	μ/D	Ref.
Germanium(II) fluoride	GeF_2	2.61(2)	2
Germanium(II) oxide	GeO	3.2823(1)	2
Germanium(II) selenide	GeSe	1.65(5)	2
Germanium(II) sulfide	GeS	2.00(6)	2
Germanium(II) telluride	GeTe	1.06(7)	2
Germylazide	GeH_3N_3	2.58(2)	25
Glycerol	$CH_2OHCHOHCH_2OH$	2.56 liq	7
Glycolaldehyde	CH_2OHCHO	2.73(5)	2
Gold(I) chloride	AuCl	3.69	84
Gold(I) fluoride	AuF	4.32	85
Gold monosulfide	AuS	2.22(5)	83
Gold monoxide	AuO	2.94(6)	83
Hafnium monoxide	HfO	3.431(5)	26
Hafnium(IV) oxide	HfO_2	7.92(1)	39
2-Heptanol, (±)-	$C_7H_{16}O$	1.71 liq	7
3-Heptanol, (*S*)-	$C_7H_{16}O$	1.71 liq	7
2-Heptanone	$C_7H_{14}O$	2.59 liq	7
3-Heptanone	$C_7H_{14}O$	2.78 liq	7
Hexaborane(10)	B_6H_{10}	2.50(5)	3
Hexamethylphosphoric triamide	$C_6H_{18}N_3OP$	5.5 liq	7
Hexanoic acid	$C_5H_{11}COOH$	1.13 liq	7
2-Hexanone	$C_6H_{12}O$	2.66 liq	7
sec-Hexyl acetate	$C_8H_{16}O_2$	1.9 liq	7
1-Hexyne	$C_4H_9C{\equiv}CH$	0.83(5)	1
Hydrazine	N_2H_4	1.75(9)	1
Hydrazoic acid	HN_3	1.70(9)	3
Hydrogen bromide	HBr	0.8272(3)	3
Hydrogen chloride	HCl	1.1086(3)	3
Hydrogen cyanide	HCN	2.985188	5
Hydrogen cyanide trimer	$(HCN)_3$	10.6	21
Hydrogen fluoride	HF	1.826178	2
Hydrogen iodide	HI	0.448(1)	2
Hydrogen isocyanide	HNC	3.05(15)	3
Hydrogen peroxide	H_2O_2	1.573(1)	65
Hydrogen sulfide	H_2S	0.97833	5
Hydroperoxy	HO_2	2.09(3)	125
p-Hydroquinone	$C_6H_6O_2$	2.38(5)	15
Hydroxyl	OH	1.655(1)	5
Hydroxylamine	NH_2OH	0.59(5)	2
Hypochlorous acid	HOCl	≈1.3	2
Hypofluorous acid	HFO	2.23(11)	3
Imidazole	$C_3H_4N_2$	3.8(4)	2
Imidogen	NH	1.39(7)	3
Indium(I) chloride	InCl	3.79(19)	2
Indium(I) fluoride	InF	3.40(7)	2
Iodine bromide	IBr	0.726(3)	5
Iodine chloride	ICl	1.24(2)	2
Iodine fluoride	IF	1.948(20)	3
Iodine monoxide	IO	2.45(5)	2
Iodine pentafluoride	IF_5	2.18(11)	1
Iodoacetylene	$HC{\equiv}CI$	0.02525	5
Iodobenzene	C_6H_5I	1.70(9)	1
1-Iodobutane	C_4H_9I	1.93 liq	7
2-Iodobutane, (±)-	C_4H_9I	2.12(11)	1
Iodoethane	C_2H_5I	1.976(2)	5
Iodoethene	$CH_2{=}CHI$	1.311(5)	5
Iodomethane	CH_3I	1.6406(4)	5
1-Iodo-2-methylpropane	C_4H_9I	1.87 liq	7
Iodomethylsilane	CH_2ISiH_3	1.862(5)	5
1-Iodopropane	C_3H_7I	2.04(10)	1
2-Iodopropane	CH_3CHICH_3	1.95 liq	7
Isobutane	$(CH_3)_3CH$	0.132(2)	1
Isobutene	$(CH_3)_2C{=}CH_2$	0.503(10)	1
Isobutyl acetate	$C_6H_{12}O_2$	1.86 liq	7
Isobutylamine	$(CH_3)_2CHCH_2NH_2$	1.27 liq	7
Isobutyl formate	$(CH_3)_2CHCH_2OCHO$	1.88 liq	7
Isobutyl isobutanoate	$C_8H_{16}O_2$	1.9 liq	7
Isobutyl nitrate	$(CH_3)_2CHCH_2NO_3$	3.74(4)	129

Name	Mol. form.	μ/D	Ref.
Isocyanic acid (HNCO)	HNCO	≈1.6	2
Isocyanobenzene	C_7H_5N	4.018(3)	5
Isocyanocyclopropane	C_4H_5N	4.03(10)	3
2-Isocyanopropane	$(CH_3)_2CHNC$	4.055(1)	5
Isopentane	C_5H_{12}	0.13(5)	1
Isopentyl acetate	$C_7H_{14}O_2$	1.86 liq	7
Isopropylamine	$(CH_3)_2CHNH_2$	1.19(6)	3
Isopropylbenzene	C_9H_{12}	≈0.79	1
Isopropyldifluorophosphine	$(CH_2)_2CHPF_2$	2.22(2)[g]	89
Isopropyl methyl ether	$(CH_3)_2CHOCH_3$	1.247(3)	5
Isoquinoline	C_9H_7N	2.73(14)	1
Isoxazole	C_3H_3NO	2.95(4)	3
Isoxazole-carbon monoxide complex	$C_3H_3NO{\cdot}CO$	2.873(4)	52
Ketene	$CH_2{=}C{=}O$	1.42215	3
Lanthanum monoxide	LaO	3.207(11)	26
Lead(II) oxide (litharge)	PbO	4.64(50)	2
Lead(II) sulfide	PbS	3.59(18)	2
Lithium bromide	LiBr	7.268(1)	2
Lithium chloride	LiCl	7.12887	2
Lithium fluoride	LiF	6.3274(2)	3
Lithium fluoride-sodium fluoride complex	LiF·NaF	2.62(2)	51
Lithium hydride	LiH	5.884(1)	2
Lithium hydroxide	LiOH	4.755(2)	147
Lithium iodide	LiI	7.428(1)	2
Lithium monoxide	LiO	6.84(3)	2
Lithium potassium	LiK	3.45(20)	2
Lithium rubidium	LiRb	4.0(1)	2
Lithium sodium	NaLi	0.463(2)	2
Magnesium oxide	MgO	6.2(6)	5
Mercapto	SH	0.7580(1)	3
Mesityl oxide	$C_6H_{10}O$	2.79 liq	7
Methacrylic acid	$C_4H_6O_2$	1.65 liq	7
Methanethiol	CH_3SH	1.52(8)	1
Methanol	CH_3OH	1.6792(9)	138
2-Methoxyethyl acetate	$C_5H_{10}O_3$	2.13 liq	7
1-Methoxy-1,2-propadiene	C_4H_6O	0.963(20)	5
3-Methoxy-1-propyne	C_4H_6O	1.170(18)	130
N-Methylacetamide	C_3H_7NO	4.3 liq	7
Methyl acetate	$C_3H_6O_2$	1.72(9)	1
Methyl acrylate	$C_4H_6O_2$	1.77 liq	7
2-Methylacrylonitrile	C_4H_5N	3.69(18)	1
Methylamine	CH_3NH_2	1.31(3)	1
2-Methylaniline	C_7H_9N	1.60 liq	7
3-Methylaniline	C_7H_9N	1.45 liq	7
4-Methylaniline	C_7H_9N	1.52 liq	7
Methyl azide	$CH_3N{=}N{\equiv}N$	2.17(4)	2
Methyl benzoate	$C_8H_8O_2$	1.94 liq	7
2-Methyl-1,3-butadiene	C_5H_8	0.25(1)	1
3-Methylbutanoic acid	$C_5H_{10}O_2$	0.63 liq	7
2-Methyl-1-butanol, (±)-	$C_5H_{12}O$	1.88 liq	7
2-Methyl-2-butanol	$C_5H_{12}O$	1.82 liq	7
3-Methyl-2-butenenitrile	C_5H_7N	4.61(13)	10
2-Methyl-1-buten-3-yne	C_5H_6	0.513(20)	2
Methyl carbamate	$CH_3O(C{=}O)NH_2$	2.30 (*syn conformer*)	95
Methyl cyanate	$CH_3OC{\equiv}N$	4.26(18)	5
cis-3-Methylcyclohexanol, (±)-	$C_7H_{14}O$	1.91 liq	7
trans-3-Methylcyclohexanol, (±)-	$C_7H_{14}O$	1.75 liq	7
2-Methylcyclohexanone, (±)-	$C_7H_{12}O$	3.07(3)	119
3-Methylcyclohexanone, (±)-	$C_7H_{12}O$	3.210(4)	119
4-Methylcyclohexanone	$C_7H_{12}O$	3.263(14)	119
3-Methylcyclopentanone, (±)-	$C_6H_{10}O$	3.14(3)	5
3-Methyl-2-cyclopenten-1-one	C_6H_8O	4.33(2)	5
Methylcyclopropane	C_4H_8	0.139(4)	2
Methyldiborane(6)	CH_8B_2	0.566(6)	3
Methyldifluorophosphine	CH_3PF_2	2.056(6)	3
4-Methyl-1,3-dioxane	$C_5H_{10}O_2$	2.03(1)	96
5-Methyl-1,3-dioxane	$C_5H_{10}O_2$	2.08(1)	116

Name	Mol. form.	μ/D	Ref.
Methylenecyclohexane	C_7H_{12}	0.62(1)	5
Methylenecyclopropene	C_4H_4	1.90(1)	5
Methylenephosphine	$CH_2{=}PH$	0.869(3)	61
N-Methylformamide	CH_3NHCHO	3.83(8)	1
Methyl formate	CH_3OCHO	1.793(22)	117
2-Methylfuran	C_5H_6O	0.65(5)	2
3-Methylfuran	C_5H_6O	1.03(2)	2
5-Methyl-2(3*H*)-furanone	$C_5H_6O_2$	4.08(2)	5
Methyl hydroperoxide	CH_3OOH	≈0.65	13
Methylidyne	CH	≈1.46	2
Methyl isocyanate	$CH_3N{=}C{=}O$	≈2.8	1
Methyl isothiocyanate	$CH_3N{=}C{=}S$	3.453(3)	5
4-Methylisoxazole	C_4H_5NO	3.583(5)	5
Methyl methacrylate	$C_5H_8O_2$	1.67 liq	7
2-Methyloxazole	C_4H_5NO	1.37(7)	5
4-Methyloxazole	C_4H_5NO	1.08(5)	5
5-Methyloxazole	C_4H_5NO	2.16(4)	5
Methyloxirane	C_3H_6O	2.01(2)	1
2-Methyl-2,4-pentanediol	$C_6H_{14}O_2$	2.9 liq	7
4-Methylpentanenitrile	$C_6H_{11}N$	3.5 liq	7
Methylphosphonic difluoride	$CH_3P{=}OF_2$	3.69(26)	3
N-Methylpropanamide	C_4H_9NO	3.61	7
2-Methylpropanenitrile	$(CH_3)_2CHCN$	4.07(5)	104
2-Methyl-2-propanethiol	$(CH_3)_3CSH$	1.66(3)	3
2-Methylpropanoic acid	$C_4H_8O_2$	1.08 liq	7
2-Methyl-1-propanol	$C_4H_{10}O$	1.64(8)	1
2-Methyl-2-propanol	$(CH_3)_3COH$	1.66 liq	7
2-Methylpropenal	C_4H_6O	2.68(13)	1
2-Methyl-2-propenal oxime, *trans*-(*E*)	C_4H_7NO	0.60(10)	143
Methyl propyl ether	$CH_3OC_3H_7$	1.107(13)	135
2-Methylpyridine	C_6H_7N	1.85(4)	2
3-Methylpyridine	C_6H_7N	2.40 liq	7
4-Methylpyridine	C_6H_7N	2.70(2)	2
2-Methylpyrimidine	$C_5H_6N_2$	1.676(10)	3
5-Methylpyrimidine	$C_5H_6N_2$	2.881(6)	3
N-Methylpyrrolidine	$C_5H_{11}N$	0.572(3)	5
N-Methyl-2-pyrrolidinone	C_5H_9NO	4.1 liq	7
Methyl salicylate	$C_8H_8O_3$	2.47 liq	7
Methylsilane	SiH_3CH_3	0.73456	5
Methyl silyl ether	$(CH_3)(SiH_3)O$	1.15(2)	2
Methylstannane	SnH_3CH_3	0.68(5)	1
2-Methylthiophene	C_5H_6S	0.674(5)	2
3-Methylthiophene	C_5H_6S	0.914(15)	3
Methyl vinyl ether	$(CH_3)(C_2H_3)O$	0.965(2)	5
Morpholine	C_4H_9NO	1.55(3)	3
Nitric acid	HNO_3	2.17(2)	1
Nitric oxide	NO	0.15872	2
2-Nitroanisole	$C_7H_7NO_3$	5.0 liq	7
Nitrobenzene	$C_6H_5NO_2$	4.22(8)	1
Nitroethane	$C_2H_5NO_2$	3.23(3)	2
Nitrogen dioxide	NO_2	0.316(10)	1
Nitrogen sulfide	NS	1.81(2)	2
Nitrogen trichloride	NCl_3	0.39(1)	3
Nitrogen trifluoride	NF_3	0.235(4)	1
Nitrogen trioxide	N_2O_3	2.122(10)	2
Nitromethane	CH_3NO_2	3.46(2)	1
1-Nitropropane	$C_3H_7NO_2$	3.66(7)	1
2-Nitropropane	$C_3H_7NO_2$	3.73(7)	1
Nitrosyl bromide	$NOBr$	≈1.8	1
Nitrosyl fluoride	NOF	1.730(3)	3
Nitrosyl hydride	HNO	1.62(3)	3
Nitrous oxide	N_2O	0.16083	3
Nitryl chloride	NO_2Cl	0.53	1
Nitryl fluoride	NO_2F	0.466(5)	2
Nonanoic acid	$C_8H_{17}COOH$	0.79 liq	7
2,5-Norbornadiene	C_7H_8	0.0587(2)	124
cis-9-Octadecenoic acid	$C_{18}H_{34}O_2$	1.18 liq	7
Octanoic acid	$C_7H_{15}COOH$	1.15 liq	7

Name	Mol. form.	μ/D	Ref.
1-Octanol	$C_7H_{15}CH_2OH$	1.76 liq	7
2-Octanol	$C_8H_{18}O$	1.71 liq	7
2-Octanone	$C_8H_{16}O$	2.70 liq	7
1,4-Oxathiane	C_4H_8OS	0.295(3)	3
Oxazole	C_3H_3NO	1.503(30)	3
Oxetane	C_3H_6O	1.94(1)	1
2-Oxetanone	$C_3H_4O_2$	4.18(3)	1
3-Oxetanone	$C_3H_4O_2$	0.887(5)	2
Oxirane	C_2H_4O	1.89(1)	1
Ozone	O_3	0.53373	3
Paraldehyde	$C_6H_{12}O_3$	1.43(7)	1
Pentaborane(9)	B_5H_9	2.13(4)	1
Pentachloroethane	CCl_3CHCl_2	0.92(5)	1
cis-1,3-Pentadiene	C_5H_8	0.500(15)	2
trans-1,3-Pentadiene	C_5H_8	0.585(10)	2
1,3-Pentadiyne	$CH_3C{\equiv}CC{\equiv}CH$	1.207(1)	5
1,4-Pentadiyne	$HC{\equiv}CCH_2C{\equiv}CH$	0.516(5)	78
1,5-Pentanediol	$CH_2OH(CH_2)_3CH_2OH$	2.5 liq	7
2,4-Pentanedione	$CH_3COCH_2COCH_3$	2.78 liq	7
Pentanenitrile	C_4H_9CN	4.12(8)	1
Pentanoic acid	C_4H_9COOH	1.61 liq	7
1-Pentanol	$C_5H_{11}OH$	1.7 liq	7
2-Pentanol	$CH_3CH(OH)C_3H_7$	1.66 liq	7
3-Pentanol	$C_2H_5CH(OH)C_2H_5$	1.64 liq	7
2-Pentanone	$C_5H_{10}O$	2.70 liq	7
3-Pentanone	$C_5H_{10}O$	2.82 liq	7
1,2,3-Pentatriene	C_5H_6	0.51(5)	11
1-Pentene	$C_3H_7CH{=}CH_2$	≈0.5	1
1-Penten-3-yne	$CH_2{=}CHC{\equiv}CCH_3$	0.66(2)	2
cis-3-Penten-1-yne	$HC{\equiv}CCH{=}CHCH_3$	0.78(2)	2
trans-3-Penten-1-yne	$HC{\equiv}CCH{=}CHCH_3$	1.06(5)	2
Pentyl acetate	$C_7H_{14}O_2$	1.75(10)	1
Pentyl formate	$C_5H_{11}OCHO$	1.90(10)	1
Perchloryl fluoride	ClO_3F	0.023(1)	3
Perfluoroisobutane	$(CF_3)_3CF$	0.0338(16)	146
Perfluoropyridine	C_5F_5N	0.98(8)	3
Peroxynitric acid	$HOONO_2$	1.99(2)	67
Peroxynitrous acid	$HOON{=}O$	1.07(2)	46
Phenol	C_6H_5OH	1.224(8)	3
Phenylacetylene	$C_6H_5C{\equiv}CH$	0.656(5)	3
Phenylsilane	C_6H_8Si	0.845(12)	3
1-Phosphapropyne	$CH_3C{\equiv}P$	1.499(1)	63
Phosphine	PH_3	0.5740(3)	3
Phosphine-hydrogen cyanide complex	$PH_3{\cdot}HCN$	4.05(5)[e]	91
3-Phospholene	C_4H_7P	1.11(2)	123
Phosphorothioc trifluoride	PSF_3	0.64(2)	1
Phosphorus(III) chloride	PCl_3	0.56(2)	2
Phosphorus(III) fluoride	PF_3	1.03(1)	1
Phosphorus monoxide	PO	1.88(7)	5
Phosphorus nitride	PN	2.7470(1)	2
Phosphoryl chloride	$POCl_3$	2.54(5)	2
Phosphoryl fluoride	POF_3	1.8685(1)	3
Piperidine	$C_5H_{11}N$	1.19 liq	3
Potassium bromide	KBr	10.628(1)	2
Potassium chloride	KCl	10.269(1)	2
Potassium fluoride	KF	8.585(3)	2
Potassium hydroxide	KOH	7.415(2)	16
Potassium iodide	KI	≈10.8	2
Potassium sodium	NaK	2.693(14)	3
Propanal	C_2H_5CHO	2.72	1
Propane	C_3H_8	0.084(1)	1
1,2-Propanediol	$C_3H_8O_2$	2.25 liq	7
1,3-Propanediol	$C_3H_8O_2$	2.55 liq	7
Propanenitrile	C_2H_5CN	4.05(3)	3
Propanoic acid	$C_3H_6O_2$	1.75(9)	1
Propargyl alcohol	C_3H_4O	1.13(6)	2
Propene	$CH_3CH{=}CH_2$	0.366(1)	1
Propene-sulfur dioxide complex	$C_3H_6{\cdot}SO_2$	1.34(3)	35

Name	Mol. form.	μ/D	Ref.
2-Propenethial	$CH_2{=}CH{-}CH{=}S$	2.61(2)	137
Propyl acetate	$C_5H_{10}O_2$	1.78 liq	7
Propylamine	$C_3H_7NH_2$	1.17(6)	1
Propylene carbonate	$C_4H_6O_3$	4.9 liq	7
Propyl formate	C_3H_7OCHO	1.89 liq	7
2-Propynal	$HC{\equiv}CCHO$	2.78(2)	5
Propyne	$CH_3C{\equiv}CH$	0.784(1)	3
Propyne-argon complex	$CH_3C{\equiv}CH{\cdot}Ar$	0.730(5)	20
Propyne--difluoromethane complex	$CH_3C{\equiv}CH{\cdot}CH_2F_2$	1.674(3)	151
4*H*-Pyran-4-one	$C_5H_4O_2$	3.79(2)	5
4*H*-Pyran-4-thione	C_5H_4OS	3.95(5)	5
1*H*-Pyrazole	$C_3H_4N_2$	2.20(1)	3
Pyridazine	$C_4H_4N_2$	4.22(2)	2
Pyridine	C_5H_5N	2.215(10)	3
2-Pyridinecarbonitrile	$C_6H_4N_2$	5.78(11)	3
3-Pyridinecarbonitrile	$C_6H_4N_2$	3.66(11)	3
4-Pyridinecarbonitrile	$C_6H_4N_2$	1.96(3)	3
2-Pyridinecarboxaldehyde	C_6H_5NO	3.56(7)	3
3-Pyridinecarboxaldehyde	C_6H_5NO	1.44	3
4-Pyridinecarboxaldehyde	C_6H_5NO	1.66	3
Pyrimidine	$C_4H_4N_2$	2.334(10)	2
Pyrrole	C_4H_5N	1.767(1)	5
Pyrrolidine	C_4H_9N	1.57 liq	7
2-Pyrrolidone	C_4H_7NO	3.5 liq	7
Pyruvonitrile	$CH_3(C{=}O)OCN$	3.468(4)	105
Quinoline	C_9H_7N	2.29(11)	1
Rubidium bromide	RbBr	≈10.9	2
Rubidium chloride	RbCl	10.510(5)	2
Rubidium fluoride	RbF	8.5465(5)	2
Rubidium iodide	RbI	≈11.5	2
Rubidium sodium	NaRb	3.1(3)	2
Salicylaldehyde	$C_7H_6O_2$	2.961(2)	142
Selenium dioxide	SeO_2	2.62(5)	2
Selenium tetrafluoride	SeF_4	1.78(9)	2
Selenoformaldehyde	CH_2Se	1.41(1)	5
Silacyclohexane	$C_5H_{12}Si$	0.80(2) (*chair*)	80
Silane-water complex	$SiH_4{\cdot}H_2O$	1.730(10)	94
Silicon dicarbide	SiC_2	2.393(6)	24
Silicon methylidyne	SiCH	0.066(2)	41
Silicon monosulfide	SiS	1.73(9)	2
Silicon monoxide	SiO	3.0982	2
Silver(I) bromide	AgBr	5.62(3)	5
Silver(I) chloride	AgCl	6.08(6)	5
Silver(I) fluoride	AgF	6.22(30)	2
Silver(I) iodide	AgI	4.55(5)	5
Sodium bromide	NaBr	9.1183(6)	2
Sodium chloride	NaCl	9.00117	2
Sodium fluoride	NaF	8.156(1)	2
Sodium iodide	NaI	9.236(3)	2
Stibine	SbH_3	0.12(5)	1
Strontium oxide	SrO	8.900(3)	2
Styrene	C_8H_8	0.125(5)	134
Succinonitrile	$C_4H_4N_2$	3.7 liq	7
Sulfolane	$C_4H_8O_2S$	4.8 liq	7
Sulfur dichloride	SCl_2	0.36(1)	3
Sulfur difluoride	SF_2	1.05(5)	2
Sulfur dioxide	SO_2	1.63305	3
Sulfur fluoride [SSF_2]	SSF_2	1.03(5)	1
Sulfur fluoride [FSSF]	FSSF	1.45(2)	1
Sulfuric acid	H_2SO_4	2.964(7)[f]	115
Sulfur monofluoride	SF	0.794(12)	3
Sulfur monoxide	SO	1.52(2)	90
Sulfur oxide (SSO)	S_2O	1.47(3)	1
Sulfur tetrafluoride	SF_4	0.632(3)	1
Sulfuryl chloride	SO_2Cl_2	1.81(4)	1
Sulfuryl fluoride	SO_2F_2	1.12(2)	1
Tetraborane(10)	B_4H_{10}	0.486(2)	3
1,1,2,2-Tetrabromoethane	$C_2H_2Br_4$	1.38 liq	7
1,1,2,2-Tetrachloroethane	$C_2H_2Cl_4$	1.32(7)	1
1,2,3,4-Tetrafluorobenzene	$C_6H_2F_4$	2.42(5)	3
1,2,3,5-Tetrafluorobenzene	$C_6H_2F_4$	1.46(6)	3
1,1,1,2-Tetrafluoroethane	CF_3CH_2F	1.80(22)	5
Tetrafluorosilane-ammonia complex	$SiF_4{\cdot}NH_3$	5.61(2)	31
Tetrahydrofuran	C_4H_8O	1.75(4)	2
Tetrahydrofurfuryl alcohol	$C_5H_{10}O_2$	2.1 liq	7
Tetrahydro-4*H*-pyran-4-one	$C_5H_8O_2$	1.720(3)	3
1,2,5,6-Tetrahydropyridine	C_5H_9N	1.007(3)	3
Tetrahydrothiophene	C_4H_8S	1.90 liq	7
Tetramethylurea	$C_5H_{12}N_2O$	3.5 liq	7
1*H*-Tetrazole	CH_2N_4	2.19(5)	3
Thallium(I) bromide	TlBr	4.49(5)	2
Thallium(I) chloride	TlCl	4.54299	2
Thallium(I) fluoride	TlF	4.2282(8)	2
Thallium(I) iodide	TlI	4.61(7)	2
Thiacyclohexane	$C_5H_{10}S$	1.781(10)	3
1,2,5-Thiadiazole	$C_2H_2N_2S$	1.579(7)	3
Thietane	C_3H_6S	1.85(9)	1
Thietane 1,1-dioxide	$C_3H_6O_2S$	4.8(1)	5
Thioacetaldehyde	CH_3CHS	2.33(2)	68
Thiocarbonyl fluoride	CF_2S	0.080	59
Thioformaldehyde	CH_2S	1.6491(4)	3
Thiomorpholine	C_4H_9NS	1.830(16)[h]	128
Thionitrosyl chloride	NSCl	1.87(2)	2
Thionitrosyl fluoride (NSF)	NSF	1.902(12)	2
Thionyl chloride	$SOCl_2$	1.45(3)	1
Thionyl fluoride	SOF_2	1.63(1)	1
Thiophene	C_4H_4S	0.55(1)	2
2-Thiophenecarbonitrile	C_5H_3NS	4.59(2)	3
3-Thiophenecarbonitrile	C_5H_3NS	4.13(2)	3
4*H*-Thiopyran-4-thione	$C_5H_4S_2$	3.9(2)	5
3-Thioxo-1,2-propadienylidene	C_3S	3.704(9)	50
Tin(II) oxide	SnO	4.32(22)	2
Tin(II) sulfide	SnS	3.18(16)	2
Titanium(II) oxide	TiO	2.96(5)	5
Toluene	$C_6H_5CH_3$	0.375(10)	3
Toluene-sulfur dioxide complex	$C_7H_8{\cdot}SO_2$	1.87(3)	34
1*H*-1,2,4-Triazole	$C_2H_3N_3$	2.7(1)	3
Tribromomethane	$CHBr_3$	0.99(2)	1
Tributylamine	$(C_4H_9)_3N$	0.78 liq	7
Tributyl borate	$(C_4H_9)_3BO_3$	0.77 liq	7
Tributyl phosphate	$C_{12}H_{27}O_4P$	3.07 liq	7
1,1,1-Trichloroethane	CH_3CCl_3	1.755(15)	2
1,1,2-Trichloroethane	$CHCl_2CH_2Cl$	1.4 liq	7
Trichloroethene	C_2HCl_3	0.8 liq	7
Trichloroethylsilane	$C_2H_5SiCl_3$	2.04 liq	7
Trichlorofluoromethane	CCl_3F	0.46(2)	2
Trichlorofluorosilane	$SiCl_3F$	0.49(1)	2
Trichloromethane	$CHCl_3$	1.04(2)	2
(Trichloromethyl)benzene	$C_6H_5CCl_3$	2.03 liq	7
Trichloromethylsilane	CH_3SiCl_3	1.91(1)	2
Trichlorosilane	$SiHCl_3$	0.86(1)	2
Tri-*o*-cresyl phosphate	$(C_7H_7)_3PO_4$	2.87 liq	7
Tri-*m*-cresyl phosphate	$(C_7H_7)_3PO_4$	3.05 liq	7
Tri-*p*-cresyl phosphate	$(C_7H_7)_3PO_4$	3.18 liq	7
Tris(2-hydroxyethyl)amine	$C_6H_{15}NO_3$	3.57 liq	7
Triethylamine	$(C_2H_5)_3N$	0.66(5)	1
Triethyl phosphate	$(C_2H_5)_3PO_4$	3.12 liq	7
Trifluoramine oxide	NOF_3	0.0390(4)	9
Trifluoroacetic acid	$C_2HF_3O_2$	2.28(25)	1
Trifluoroacetonitrile	CF_3CN	1.262(10)	3
1,2,4-Trifluorobenzene	$C_6H_3F_3$	1.402(9)	5
1,1,1-Trifluorodisilane	SiH_3SiF_3	2.03(10)	3
1,1,1-Trifluoroethane	CH_3CF_3	2.347(5)	3
Trifluoroethene	$CHF{=}CF_2$	1.32(3)	2
Trifluoroiodomethane	CF_3I	1.048(3)	3

Name	Mol. form.	μ/D	Ref.
Trifluoroiodosilane	SiF_3I	1.11(3)	5
Trifluoroisocyanomethane	CF_3NC	1.153(20)	149
Trifluoromethane	CHF_3	1.65150	3
(Trifluoromethyl)benzene	$C_6H_5CF_3$	2.86(6)	1
Trifluoromethylsilane	CH_3SiF_3	2.3394(2)	5
(Trifluoromethyl)silane	CF_3SiH_3	2.32(2)	5
Trifluorooxirane	C_2HF_3O	1.75(10)	126
3,3,3-Trifluoro-1-propene	$C_3H_3F_3$	2.43(2)	69
3,3,3-Trifluoro-1-propyne	$CF_3C{\equiv}CH$	2.317(13)	5
Trifluorosilane	$SiHF_3$	1.27(3)	1
Trimethylamine	$(CH_3)_3N$	0.612(3)	1
Trimethyl phosphate	$(CH_3)_3PO_4$	3.18 liq	7
2,4,6-Trimethylpyridine	$C_8H_{11}N$	2.05 liq	7
1,3,5-Trioxane	$(CH_2O)_3$	2.08(2)	1
Tris(trifluoromethyl)methane	$(CF_3)_3CH$	1.69(18)	146
1,2,4-Trithiacyclopentane	$C_2H_4S_3$	0.465(4)	71
Vinyl acetate	$C_4H_6O_2$	1.79 liq	7
Vinyl formate	$CH_2{=}CHOCHO$	1.49(1)	1
2-Vinylfuran	C_6H_6O	0.69(7)	5
Vinylsilane	$CH_2{=}CHSiH_3$	0.657(2)	5
Water	H_2O	1.8546(40)	3
Water dimer-hydrogen bromide complex	$(H_2O)_2{\cdot}HBr$	2.281(3)	48
Water dimer-hydrogen chloride complex	$(H_2O)_2{\cdot}HCl$	2.328(3)	48
Water-hydrogen chloride complex	$H_2O{\cdot}HCl$	3.437(4)	75
Water-hydrogen fluoride complex	$H_2O{\cdot}HF$	4.073(7)	92
o-Xylene	$C_6H_4(CH_3)_2$	0.640(5)	2
2,4-Xylenol	$C_8H_{10}O$	1.4 liq	7
2,5-Xylenol	$C_8H_{10}O$	1.45 liq	7
2,6-Xylenol	$C_8H_{10}O$	1.40 liq	7
3,4-Xylenol	$C_8H_{10}O$	1.56 liq	7
3,5-Xylenol	$C_8H_{10}O$	1.55 liq	7
Ytterbium monofluoride	YbF	3.91(4)	45
Yttrium monoxide	YO	4.524(7)	26
Zirconium(II) oxide	ZrO	2.551(1)	26
Zirconium(IV) oxide	ZrO_2	7.80(2)	19

[a] *anti-anti* conformer.
[b] *gauche* conformer.
[c] *cis* conformer.
[d] See Table 2 for individual conformers.
[e] For ^{15}N compound.
[f] C_2 conformer, believed to be dominant form.
[g] Average of *gauche* (2.205 D) and *trans* (2.231 D).
[h] *chair* conformer.

TABLE 2. Dipole Moments of Conformers

Name	Mol. form.	Conformer	μ/D	Ref.
Acrolein	C_3H_4O	*trans*	3.117(4)	5
Acrolein	C_3H_4O	*cis*	2.552(3)	5
Allyl alcohol	C_3H_6O	*gauche*	1.55(8)	3
3-Aminophenol	C_6H_7NO	*cis*	2.333(22)	112
3-Aminophenol	C_6H_7NO	*trans*	0.774(7)	112
Butanenitrile	C_4H_7N	*gauche*	3.91(8)	133
Butanenitrile	C_4H_7N	*anti*	3.73(12)	133
1-Butene	C_4H_8	*cis*	0.438(7)	2
1-Butene	C_4H_8	*skew*	0.359(11)	2
Chlorocyclohexane	$C_6H_{11}Cl$	*equitorial*	2.44(7)	5
Chlorocyclohexane	$C_6H_{11}Cl$	*axial*	1.91(2)	5
1-Chloropropane	C_3H_7Cl	*gauche*	2.02(3)	5
1-Chloropropane	C_3H_7Cl	*trans*	1.95(2)	5
Cyclohexene	C_6H_{10}	*half-chair*	0.332(12)	2
Cyclopropylmethyl isocyanide	C_5H_7N	*gauche (sc)*	4.05(4)	101
Diethyl sulfide	$C_4H_{10}S$	*trans-trans*	1.556(4)	54
Diethyl sulfide	$C_4H_{10}S$	*trans-gauche*	1.591(9)	54
Diethyl sulfide	$C_4H_{10}S$	*gauche-gauche*	1.645(1)	54
1,2-Difluoroethane	$C_2H_4F_2$	*gauche*	2.67(13)	2
3,3-Difluoropentane	$C_5H_{10}F_2$	*anti-gauche*	2.277(4)	109
3,3-Difluoropentane	$C_5H_{10}F_2$	*gauche-gauche*	2.419(8)	109
Ethanethiol	C_2H_6S	*gauche*	1.61(8)	3
Ethanethiol	C_2H_6S	*trans*	1.58(8)	3
Ethanimine	C_2H_5N	*cis*	2.059(13)	122
Ethanimine	C_2H_5N	*trans*	2.50(10)	122
Ethanol	C_2H_6O	*gauche*	1.68(3)	3
Ethanol	C_2H_6O	*trans*	1.44(3)	2
Ethylamine	C_2H_7N	*gauche*	1.210(15)	5
Ethylamine	C_2H_7N	*trans*	1.304(11)	5
Ethyl formate	$C_3H_6O_2$	*gauche*	1.81(2)	2
Ethyl formate	$C_3H_6O_2$	*trans*	1.98(2)	2
Ethyl methyl ether	C_3H_8O	*trans*	1.17(2)	3
Ethyl methyl sulfide	C_3H_8S	*gauche*	1.593(4)	5
Ethyl methyl sulfide	C_3H_8S	*trans*	1.56(3)	3
Fluorocyclohexane	$C_6H_{11}F$	*equitorial*	2.11(4)	2
Fluorocyclohexane	$C_6H_{11}F$	*axial*	1.81(4)	2
1-Fluoropropane	C_3H_7F	*gauche*	1.90(10)	1
1-Fluoropropane	C_3H_7F	*trans*	2.05(4)	1
3-Fluoropropene	C_3H_5F	*gauche*	1.939(15)	1
3-Fluoropropene	C_3H_5F	*cis*	1.765(14)	1
Glycerol	$C_3H_8O_3$	G'Gg'gg'	1.409(11)	110
Glycerol	$C_3H_8O_3$	GGt'g'g'	1.974(4)	110
Glycerol	$C_3H_8O_3$	GG't'g'g	3.192(16)	110
Glycerol	$C_3H_8O_3$	GGg'gt'	2.573(16)	110
Glycerol	$C_3H_8O_3$	GGgg'g'	3.54(5)	110
Glycine	$C_2H_5NO_2$	*Conformer I*	1.147(5)	49
Glycine	$C_2H_5NO_2$	*Conformer II*	5.45(5)	49
Glyoxal	$C_2H_2O_2$	*cis*	4.8(2)	2
3-Hydroxypropanenitrile	C_3H_5NO	*gauche*	3.17(2)	5
Isobutanal	C_4H_8O	*gauche*	2.69(1)	5
Isobutanal	C_4H_8O	*trans*	2.86(1)	5
Isopropyldifluorophosphine	$C_3H_7F_2P$	*gauche*	2.205(6)	89
Isopropyldifluorophosphine	$C_3H_7F_2P$	*trans*	2.231(4)	89
2-Methoxyethanol	$C_3H_8O_2$	*gauche*	2.36(5)	2
3-Methyl-1-butene	C_5H_{10}	*gauche*	0.398(4)	3
3-Methyl-1-butene	C_5H_{10}	*trans*	0.320(10)	3
2-Methyl-2-propenol	C_4H_8O	*skew*	1.295(22)	5
Methyl propyl ether	$C_4H_{10}O$	*trans-trans*	1.107(13)	3
3-Methylthietane	C_4H_8S	*equatorial*	2.046(9)	132
3-Methylthietane	C_4H_8S	*axial*	1.974(8)	132
Nitrous acid	HNO_2	*cis*	1.423(5)	2
Nitrous acid	HNO_2	*trans*	1.855(16)	2
1-Pentyne	C_5H_8	*gauche*	0.769(28)	2
1-Pentyne	C_5H_8	*trans*	0.842(10)	2
Piperidine	$C_5H_{11}N$	*equitorial*	0.82(2)	3
Piperidine	$C_5H_{11}N$	*axial*	1.19(2)	3
Propanal	C_3H_6O	*gauche*	2.86(1)	5
Propanal	C_3H_6O	*cis*	2.52(5)	1
1,2-Propanediol	$C_3H_8O_2$	*g'Gt*	2.287(6)	148
1,2-Propanediol	$C_3H_8O_2$	*gG't*	2.555(20)	148
1,2-Propanediol	$C_3H_8O_2$	*tG'g*	2.291(12)	148
1-Propanethiol	C_3H_8S	*gauche*	1.683(10)	3
1-Propanethiol	C_3H_8S	*trans*	1.60(8)	3
2-Propanethiol	C_3H_8S	*gauche*	1.53(3)	3
2-Propanethiol	C_3H_8S	*trans*	1.61(3)	3
Propanoic acid	$C_3H_6O_2$	*cis*	1.46(7)	2
1-Propanol	C_3H_8O	*gauche*	1.58(3)	2
1-Propanol	C_3H_8O	*trans*	1.55(3)	2
2-Propanol	C_3H_8O	*trans*	1.58(3)	2
Propyleneimine	C_3H_7N	*cis*	1.77(9)	2
Propyleneimine	C_3H_7N	*trans*	1.57(3)	2
Tetrafluorohydrazine	F_4N_2	*gauche*	0.257(2)	5
Tetrahydropyran	$C_5H_{10}O$	*chair*	1.58(3)	3
N-Vinylformamide	C_3H_5NO	*cis* (Form I)	3.06(4)	111
N-Vinylformamide	C_3H_5NO	*trans* (Form III)	3.59(10)	111

HINDERED INTERNAL ROTATION

I. Ozier and N. Moazzen-Ahmadi

In asymmetric rotors like methyl alcohol, CH_3OH, and symmetric rotors like CH_3SiH_3, the methyl group can undergo internal rotation relative to the rest of the molecule, traditionally called the frame (Refs. 1 and 2). Although various rotating groups are considered here, all have three-fold symmetry. In such cases, the potential V hindering the internal rotation can be written:

$$V(\alpha)= V_3(½)(1 - \cos 3\alpha) + V_6(½)(1 - \cos 6\alpha) + V_9(½)(1 - \cos 9\alpha) + ...,$$

where α is the deviation from equilibrium of the angle between the top and frame that measures the torsional motion. If only the first two terms are retained, then V_3 is the height of the hindering potential and V_6 is the shape parameter. For symmetric tops like CH_3CH_3 where the top and frame are identical, α is replaced by 2γ and the origin for γ is often taken as the eclipsed configuration. In the expansion, $-\cos 6n\gamma$ is then replaced by $(-1)^{n+1}\cos 6n\gamma$, where n = 1,2,... In cases where different forms of the expansion have been used in the original works, the values of the parameters published there have been converted to the conventions defined here.

In Tables 1 and 2, values are given for V_3 for a selection of asymmetric and symmetric rotors, respectively. Column definitions for Tables 1 and 2 are as follows.

Column heading	Definition
Mol. form.	Molecular formula; molecules are listed alphabetically in Hill order according to the molecular formula
Name	Name of molecule
Line formula	Formula written in terms of connectivity
Ref.	Reference number
V_3	Magnitude of hindering potential, in cm^{-1} (wavenumbers)
Comments	Additional information as appropriate including V_6 (shape potential), V_9 isomer, state, and/or isotopomer

For ethane, three symmetric top isotopomers are listed to illustrate the isotopic dependence of V_3 and V_6. In all other cases, only one isotopomer is listed, even if several have been studied. In all but one of these cases, the isotopomer reported is the one with the highest natural abundance. However, CH_3OCDO is listed because the results obtained are more precise than for CH_3OCHO.

The determinations listed for the potential parameters are effective values that incorporate to varying degrees effects from other molecular parameters. For example, the apparent value of V_3 can be changed significantly if the reduced rotational constant F is calculated from the structure, rather than being determined independently (Ref. 1). Other examples include such mechanisms as coupling to excited skeletal vibrations (Ref. 2) and redundancies connecting some of the torsional parameters (Refs. 3 and 4). The experimental uncertainties quoted are taken from the original works; no attempt has been made to standardize the definitions. All the potential parameters are given in cm^{-1}. Where the original work has reported these values in other units, the conversion to cm^{-1} has been carried out using standard factors: 1 calorie = 4.1868 joules and 1 calorie/mole = 0.34998915 cm^{-1} (Ref. 5).

A variety of different methods have been used to measure V_3, V_6, and V_9 (Refs. 1 and 2); only a few of the more important will be discussed here. For *asymmetric rotors*, both the pure rotational spectrum and its torsion-rotation counterpart are electric dipole allowed and are affected in lowest order by the leading terms in the torsional Hamiltonian. Both types of spectra have been used extensively to determine V_3 (Ref. 1). For *symmetric tops* with a single torsional degree of freedom, either the permanent electric dipole moment vanishes, as in CH_3CH_3, or the normal rotational spectrum is independent of V_3 in lowest order, as in CH_3SiH_3. In the latter case, the molecular beam avoided crossing method can often be used (Ref. 2). The torsion-rotation spectrum is forbidden in lowest order but becomes weakly allowed through interactions with the infrared active skeletal vibrations (Ref. 2). By employing long absorption path lengths, this spectrum has been used to determine V_3 in a number of molecules. For both asymmetric and symmetric tops, the most precise determinations of the molecular parameters have been made in cases where both rotational and torsion-rotation spectra have been investigated.

References

1. Lin, C. C., and Swalen, J. D., *Rev. Mod. Phys.* 31, 841 1959. <https://doi.org/10.1103/RevModPhys.31.841>
2. Ozier, I., and Moazzen-Ahmadi, N., Internal rotation in symmetric tops, in: Arimondo, E., Berman, P. R., and Lin, C. C., (Eds.), *Advances in Atomic, Molecular and Optical Physics*, vol. 54, p. 423, Elsevier, Amsterdam, 2007. <https://doi.org/10.1016/S1049-250X(06)54007-8>
3. Lees, R. M., and Baker, J. G., *J. Chem. Phys.* 48, 5299, 1968. <https://doi.org/10.1063/1.1668221>
4. Moazzen-Ahmadi, N., and Ozier, I., *J. Mol. Spectrosc.* 126, 99, 1987. <https://doi.org/10.1016/0022-2852(87)90080-4>
5. Demaison, J., and Wlodarczak, G., Hindered rotation-Asymmetric top molecules, in: Hüttner, W. (Ed), *Landolt-Börnstein Numerical Data and Functional Relationships in Science and Technology, New Series: Group II: Molecules and Radicals, volume 24, subvolume C, Molecular Constants mostly from Microwave, Molecular Beam, and Sub-Doppler Laser Spectroscopy,* Springer-Verlag, Heidelberg, 2002.
6. Suenram, R. D., Lovas, F. J., Plusquellic, D. F., Ellzy, M. W., Lochner, J. M., Jensen, J. O., and Samuels, A. C., *J. Mol. Spectrosc.* 235, 18, 2006. <https://doi.org/10.1016/j.jms.2005.10.001>
7. Sastry, K. V. L. N., Herbst, E., Booker, R. A., and De Lucia, F. C., *J. Mol. Spectrosc.* 116, 120, 1986. <https://doi.org/10.1016/0022-2852(86)90258-4>
8. Tyblewksi, M., Ha, T.-K., and Bauder, A., *J. Mol. Spectrosc.* 115, 353, 1986. <https://doi.org/10.1016/0022-2852(86)90052-4>
9. Durig, J. R., Guirgis, G. A., and Van Der Veken, B. J., *J. Raman Spectrosc.* 18, 549, 1987. <https://doi.org/10.1002/jrs.1250180804>
10. Eltayeb, S., Guirgis, G. A., Fanning, A. R., and Durig, J. R., *J. Raman Spectrosc.* 27, 111, 1996. <https://doi.org/10.1002/(SICI)1097-4555(199602)27:2<111::AID-JRS935>3.0.CO;2-Q>
11. Krisher, L. C., *J. Chem. Phys.* 33, 1237, 1960. <https://doi.org/10.1063/1.1731363>
12. Alonso, J. L., López, J. C., Blanco, S., and Guarnieri, A., *J. Mol. Spectrosc.* 182, 148, 1997. <https://doi.org/10.1006/jmsp.1996.7214>
13. Pierce, L., and Krisher, L. C., *J. Chem. Phys.* 31, 875, 1959. <https://doi.org/10.1063/1.1730542>
14. Moloney, M. J., and Krisher, L. C., *J. Chem. Phys.* 45, 3277, 1966. <https://doi.org/10.1063/1.1728102>

15. Kleiner, I., Hougen, J. T., Grabow, J.-U., Belov, S. P., Tretyakov, M. Yu., and Cosléou, J., *J. Mol. Spectrosc.* 179, 41, 1996. <https://doi.org/10.1006/jmsp.1996.0182>
16. Ilyushin, V. V., Alekseev, E. A., Dyubko, S. F., and Kleiner, I., *J. Mol. Spectrosc.* 220, 170, 2003. <https://doi.org/10.1016/S0022-2852(03)00073-0>
17. Fliege, E., Dreizler, H., Demaison, J., Boucher, D., Burie, J., and Dubrulle, A., *J. Chem. Phys.* 78, 3541, 1983 <https://doi.org/10.1063/1.445177>
18. Schnell, M., Grabow, J.-U., Hartwig, H., Heineking, N., Meyer, M., Stahl, W., and Caminati, W., *J. Mol. Spectrosc.* 229, 1, 2005. <https://doi.org/10.1016/j.jms.2004.08.005>
19. Niide, Y., and Hayashi, M., *J. Mol. Spectrosc.* 223, 152, 2004. <https://doi.org/10.1016/j.jms.2003.08.008>
20. Shiki, Y., Hasegawa, A., and Hayashi, M., *J. Mol. Structure* 78, 185, 1982. <https://doi.org/10.1016/0022-2860(82)80005-7>
21. Groner, P., Gillies, C. W., Gillies, J. Z., Zhang, Y., and Block, E., *J. Mol. Spectrosc.* 226, 169, 2004. <https://doi.org/10.1016/j.jms.2004.04.001>
22. Alonso, J. L., Lesarri, A., López, J. C., Blanco, S., Kleiner, I., and Demaison, *J., Molec. Phys.* 91, 731, 1997. <https://doi.org/10.1080/00268979709482763>
23. Demaison, J., Maes, H., van Eijck, B. P., Wlodarczak, G. and Lasne, M. C., *J. Mol. Spectrosc.* 125, 214, 1987. <https://doi.org/10.1016/0022-2852(87)90208-6>
24. Antolínez, S., López, J. C., and Alonso, J. L., *J. Chem. Soc., Faraday Trans.* 93, 1291, 1997. <https://doi.org/10.1039/a607041h>
25. Butcher, S. S., and Wilson, E. B., *J. Chem. Phys.* 40, 1671, 1964. <https://doi.org/10.1063/1.1725377>
26. Groner, P., *J. Mol. Structure* 550–551, 473, 2000. <https://doi.org/10.1016/S0022-2860(00)00507-X>
27. Marstokk, K.-M., Mollendal, H., Samdal, S., and Steinborn, D., *J. Mol. Structure* 567, 41, 2001. <https://doi.org/10.1016/S0022-2860(01)00532-4>
28. Stiefvater, O. L., *J. Chem. Phys.* 62, 233, 1975. <https://doi.org/10.1063/1.430268>
29. de Luis, A., Eugenia Sanz, M., Lorenzo, F. J., López, J. C., and Alonso, J. L., *J. Mol. Spectrosc.* 184, 60, 1997. <https://doi.org/10.1006/jmsp.1997.7298>
30. Kasten, W., and Dreizler, H., *Z. Naturforsch.* A 41, 944, 1986. <https://doi.org/10.1515/zna-1986-0913>
31. Vormann, K., and Dreizler, H., *Z. Naturforsch.* A 43, 338, 1988. <https://doi.org/10.1515/zna-1988-0315>
32. Marstokk, K.-M., Møllendal, H., and Samdal, S., *J. Mol. Structure* 376, 11, 1996. <https://doi.org/10.1016/0022-2860(95)09045-2>
33. Bestmann, G., Lalowski, W., and Dreizler, H., *Z. Naturforschung* A 40, 271, 1985. <https://doi.org/10.1515/zna-1985-0516>
34. Dreizler, H., and Scappini, F., *Z. Naturforsch.* A 36, 1187, 1981.
35. Typke, V., Botskor, I., and Wiedenmann, K.-H., *J. Mol. Spectrosc.* 120, 435, 1986. <https://doi.org/10.1016/0022-2852(86)90016-0>
36. Suenram, R. D., Lovas, F. J., Plusquellic, D. F., Lesarri, A., Kawashima, Y., Jensen, J. O., and Samuels, A. C., *J. Mol. Spectrosc.* 211, 110, 2002. <https://doi.org/10.1006/jmsp.2001.8486>
37. Ohashi, N., and Hougen, J. T., *J. Mol. Spectrosc.* 243, 162, 2007. <https://doi.org/10.1016/j.jms.2007.01.010>
38. Grabow, J.-U., Hartwig, H., Heineking, N., Jäger, W., Mäder, H., Nicolaisen, H. W., and Stahl, W., *J. Mol. Structure* 612, 349, 2002. <https://doi.org/10.1016/S0022-2860(02)00144-8>
39. Charro, M. E., and Alonso, J. L., *J. Mol. Spectrosc.* 176, 251, 1996. <https://doi.org/10.1006/jmsp.1996.0084>
40. Nair, K. P. R., Demaison, J., Wlodarczak, G., and Merke, I., *J. Mol. Spectrosc.* 237, 137, 2006. <https://doi.org/10.1016/j.jms.2006.03.011>
41. Suenram, R. D., DaBell, R. S., Hight Walker, A. R., Lavrich, R. J., Plusquellic, D. F., Ellzy, M. W., Lochner, J. M., Cash, L., Jensen, J. O., and Samuels, A. C., *J. Mol. Spectrosc.* 224, 176, 2004. <https://doi.org/10.1016/j.jms.2004.01.012>
42. Groner, P., Attia, G. M., Mohamad, A. B., Sullivan, J. F., Li, Y. S., and Durig, J. R., *J. Chem. Phys.* 91, 1434, 1989. <https://doi.org/10.1063/1.457103>
43. Varma, R., Ramaprasad, K. R., and Nelson, J. F., *J. Chem. Phys.* 63, 915, 1975. <https://doi.org/10.1063/1.431373>
44. Odom, J. D., Kalasinsky, V. F., and Durig, J. R., *Inorg. Chem.* 14, 2837, 1975. <https://doi.org/10.1021/ic50153a049>
45. Kuczkowski, R. L., and Lide, D. R., *J. Chem. Phys.* 46, 357, 1967. <https://doi.org/10.1063/1.1840394>
46. Durig, J. R., Li, Y. S., Carreira, L. A., and Odom, J. D., *J. Amer. Chem. Soc.* 95, 2491, 1973. <https://doi.org/10.1021/ja00789a013>
47. Lide, D. R., Johnson, D. R., Sharp, K. G., and Coyle, T. D., *J. Chem. Phys.* 57, 3699, 1972. <https://doi.org/10.1063/1.1678832>
48. Krisher, L. C., Watson, W. A., and Morrison, J. A., *J. Chem. Phys.* 61, 3429, 1974. <https://doi.org/10.1063/1.1682510>
49. Styger, C., Ozier, I., Wang, S.-X., and Bauder, A., *J. Mol. Spectrosc.* 239, 115, 2006. <https://doi.org/10.1016/j.jms.2006.06.005>
50. Laurie, V. W., *J. Chem. Phys.* 30, 1210, 1959. <https://doi.org/10.1063/1.1730158>
51. Cahill, P., and Butcher, S., *J. Chem. Phys.* 35, 2255, 1961. <https://doi.org/10.1063/1.1732258>
52. Wang, S.-X., Schroderus, J., Ozier, I., Moazzen-Ahmadi, N., Horneman, V.-M., Ilyushyn, V. V., Alekseev, E. A., Katrich, A. A., and Dyubko, S. F., *J. Mol. Spectrosc.* 214, 69, 2002. <https://doi.org/10.1006/jmsp.2002.8572>
53. Nakagawa, J., Yamada, K., Bester, M., and Winnewisser, G., *J. Mol. Spectrosc.* 110, 74, 1985. <https://doi.org/10.1016/0022-2852(85)90213-9>
54. Merke, I., Stahl, W., Kassi, S., Petitprez, D., and Wlodarczak, G., *J. Mol. Spectrosc.* 216, 437, 2002. <https://doi.org/10.1006/jmsp.2002.8632>
55. di Lauro, C., Bunker, P. R., Johns, J. W. C., and McKellar, A. R. W., *J. Mol. Spectrosc.* 184, 177, 1997. <https://doi.org/10.1006/jmsp.1997.7321>
56. Voges, K., Gripp, J., Hartwig, H., and Dreizler, H., *Z. Naturforsch.* A 51, 299, 1996. <https://doi.org/10.1515/zna-1996-0409>
57. Borvayeh, L., Moazzen-Ahmadi, N., and Horneman, V.-M., *J. Mol. Spectrosc.* 242, 77, 2007. <https://doi.org/10.1016/j.jms.2007.02.009>

TABLE 1. Asymmetric Rotor Potential Parameters

Mol. form.	Name	Line formula	Ref.	V_3/cm^{-1}	Comments
CHF_3S	Trifluoromethanethiol	CF_3SH	5	500.83 ± 0.03	
CH_3F_2OP	Methylphosphonic difluoride	$CH_3P(=O)F_2$	6	676 ± 25	
CH_4O	Methanol	CH_3OH	5	373.594 ± 0.007	$V_6 = -1.597 \pm 0.051$; $V_9 = 1.04 \pm 0.20$
CH_4S	Methanethiol	CH_3SH	7	443.029 ± 0.070	$V_6 = -1.6451 \pm 0.0144$
CH_4S_2	Methyldisulfane	CH_3SSH	8	609.0 ± 14.0	
C_2F_3NO	Trifluoromethyl isocyanate	$CF_3N{=}C{=}O$	5	47.8769 ± 0.0051	
C_2HF_3O	Trifluoroacetaldehyde	$CF_3C(H){=}O$	9	298 ± 10	
C_2HF_5	Pentafluoroethane	CF_3CHF_2	10	1190 ± 4	
C_2H_3BrO	Acetyl bromide	$CH_3C(Br){=}O$	11	456.7 ± 10.5	
$C_2H_3ClF_2$	1-Chloro-1,1-difluoroethane	CH_3CClF_2	12	1311.8 ± 1.4	
C_2H_3ClO	Acetyl chloride	$CH_3C(Cl){=}O$	5	442.74 ± 1.05	^{35}Cl

Mol. form.	Name	Line formula	Ref.	V_3/cm^{-1}	Comments
C_2H_3FO	Acetyl fluoride	$CH_3C(F)=O$	13	364.3 ± 2.1	
$C_2H_3FO_2$	Methyl fluoroformate	$CH_3OC(F)=O$	5	374.1 ± 0.2	
$C_2H_3F_3O$	Methyl trifluoromethyl ether	CH_3OCF_3	5	382 ± 10	CH_3
C_2H_3IO	Acetyl iodide	$CH_3C(=O)I$	14	455.3 ± 10.5	
C_2H_3NO	Methyl cyanate	$CH_3OC{\equiv}N$	5	399.0 ± 17.5	
C_2H_4ClF	1-Chloro-1-fluoroethane	CH_3CHClF	5	1334.9 ± 3.8	
$C_2H_4F_2$	1,1-Difluoroethane	CH_3CHF_2	5	1163.0 ± 2.5	
C_2H_4O	Acetaldehyde	$CH_3C(H)=O$	15	407.716 ± 0.010	$V_6 = -12.068 \pm 0.037$
C_2H_4OS	Thioacetaldehyde *S*-oxide	$CH_3C(H)=S=O$	5	285.6 ± 0.3	*Z* isomer
$C_2H_4O_2$	Acetic acid	CH_3COOH	16	170.1742 ± 0.0002	$V_6 = -6.4725 \pm 0.0001$
$C_2H_3DO_2$	Methyl formate	$CH_3OC(D)=O$	5	400.60 ± 0.03	deuterated
C_2H_5F	Fluoroethane	CH_3CH_2F	17	1172.1 ± 1.4	
C_2H_5NO	Nitrosoethane	$CH_3CH_2N=O$	5	903 ± 25	*gauche* conformer
C_2H_5NO		$CH_3CH_2N=O$	5	911 ± 25	*cis* conformer
C_2H_5NO	Acetamide	$CH_3C(NH_2)=O$	5	24.949 ± 0.008	
$C_2H_6F_2Si$	Difluorodimethylsilane	$(CH_3)_2SiF_2$	18	439.4 ± 2.5	
$C_2H_6N_2O$	*N*-Nitrosodimethylamine	$(CH_3)_2NN=O$	5	145.8 ± 0.25	*cis* CH_3
$C_2H_6N_2O$		$(CH_3)_2NN=O$	5	737.4 ± 13.3	*trans* CH_3
C_2H_6O	Ethanol	CH_3CH_2OH	5	1173.76 ± 2.20	*trans* isomer
C_2H_6O	Dimethyl ether	$(CH_3)_2O$	19	926.0 ± 3.5	
C_2H_6S	Dimethyl sulfide	$(CH_3)_2S$	19	751.1 ± 4.8	
C_2H_6Si	Vinylsilane	$SiH_3C(H)=CH_2$	20	520.1 ± 1.8	
$C_2H_6S_2$	Dimethyl disulfide	CH_3SSCH_3	5	535.1 ± 1.8	
$C_2H_6Se_2$	Dimethyl diselenide	$CH_3SeSeCH_3$	21	395 ± 2	
C_2H_8Si	Dimethylsilane	$(CH_3)_2SiH_2$	19	578.0 ± 3.5	
$C_3H_3F_3$	3,3,3-Trifluoropropene	$CF_3C(H)=CH_2$	22	653.06 ± 0.83	
$C_3H_3NO_2$	Methyl cyanoformate	$CH_3OC(C{\equiv}N)=O$	5	406.6 ± 1.1	*s-trans* conformer
C_3H_4S	(Methylthio)acetylene	$CH_3SC{\equiv}CH$	23	592.0 ± 3.3	
$C_3H_5F_3$	1,1,1-Trifluoropropane	$CH_3CH_2CF_3$	24	922.2 ± 1.4	
C_3H_5I	2-Iodopropene	$CH_3C(I)=CH_2$	5	905.8 ± 4.2	
C_3H_5N	Ethyl isocyanide	$CH_3CH_2N{\equiv}C$	5	1167.6 ± 18.2	
C_3H_6	Propene	$CH_3C(H)=CH_2$	5	697.499 ± 0.048	$V_6 = -13.0$ (fixed)
C_3H_6O	Propanal	$CH_3CH_2C(H)=O$	25	798 ± 39	*cis* conformer
C_3H_6O	Acetone	$(CH_3)_2C=O$	26	251.4 ± 2.6	$V_6 = -6.92 \pm 0.65$
C_3H_6S	(Methylthio)ethene	$CH_3SC(H)=CH_2$	27	1138 ± 13	
$C_3H_6O_2$	Propanoic acid	CH_3CH_2COOH	28	819.0 ± 10.5	*cis* conformer
$C_3H_6O_2S$	Methyl mercaptoacetate	$CH_3OC(=O)C(H_2)SH$	5	411 ± 8	state 0^+
$C_3H_6O_2S$		$CH_3OC(=O)C(H_2)SH$	5	412 ± 9	state 0^-
C_3H_7Br	2-Bromopropane	$(CH_3)_2CHBr$	5	1437.0 ± 2.5	^{79}Br
C_3H_7Cl	1-Chloropropane	$CH_3C(H_2)C(H_2)Cl$	29	1017.8 ± 1.4	*gauche* conformer
C_3H_7Cl		$CH_3C(H_2)C(H_2)Cl$	29	966.0 ± 7.0	*trans* conformer
C_3H_7Cl	2-Chloropropane	$(CH_3)_2CHCl$	5	1374.03 ± 1.00	^{35}Cl
C_3H_7F	1-Fluoropropane	$CH_3C(H_2)C(H_2)F$	30	965.3 ± 12.2	*gauche* conformer
C_3H_7F		$CH_3C(H_2)C(H_2)F$	30	948.5 ± 2.8	*trans* conformer
C_3H_7F	2-Fluoropropane	$(CH_3)_2CHF$	5	1162.79 ± 0.84	
C_4H_7N	Butanenitrile	$CH_3C(H_2)C(H_2)C{\equiv}N$	31	1087.4 ± 8.4	*gauche* conformer
C_4H_7N		$CH_3C(H_2)C(H_2)C{\equiv}N$	31	1088.5 ± 13.3	*trans* conformer
C_3H_7NO	Propanamide	$CH_3CH_2C(=O)NH_2$	32	761 ± 42	*syn* conformer
C_3H_7NO	*N,N*-Dimethylformamide	$(CH_3)_2NC(H)=O$	5	366.04 ± 0.26	*cis* CH_3
C_3H_7NO		$(CH_3)_2NC(H)=O$	5	772.4 ± 7.4	*trans* CH_3
C_3H_8	Propane	$(CH_3)_2CH_2$	33	1108.1 ± 9.5	
C_3H_8Ge	Cyclopropylgermane	$C(H_2)C(H_2)C(H)(GeH_3)$	5	466.6 ± 16.7	GeH_3
$C_3H_8N_2O$	*N*-Nitrosoethylmethylamine	$CH_3CH_2N(CH_3)N=O$	5	310 ± 30	*N*-methyl top, OGM conformer
C_3H_8O	1-Propanol	$CH_3C(H_2)C(H_2)OH$	34	956 ± 21	*trans* conformer
C_3H_8Si	Cyclopropylsilane	$C(H_2)C(H_2)C(H)(SiH_3)$	35	670.9 ± 1.5	
C_3H_8Si	Dimethyl(methylene)silane	$(CH_3)_2Si=CH_2$	5	351.4 ± 5.9	
$C_3H_9O_3P$	Dimethyl methylphosphonate	$(OCH_3)_2P(=O)CH_3$	36	662 ± 6	*P*-methyl top
$C_3H_9O_3P$		$(OCH_3)_2P(=O)CH_3$	37	278.82 ± 0.06	*O*-methyl top #1

Mol. form.	Name	Line formula	Ref.	V_3/cm^{-1}	Comments
$C_3H_9O_3P$		$(OCH_3)_2P(=O)CH_3$	37	181.82 ± 0.01	*O*-methyl top #2
C_4H_3FO	But-2-ynoyl fluoride	$CH_3C\equiv CC(F)=O$	5	2.20 ± 0.12	
C_4H_5N	*cis*-2-Butenenitrile	$CH_3C(H)=C(H)C\equiv N$	5	485.50 ± 0.25	
C_4H_5N	2-Methylacrylonitrile	$CH_2=C(CH_3)C\equiv N$	5	695.2 ± 2.1	
C_4H_5NO	2-Methyloxazole	$N=C(CH_3)OC(H)=C(H)$	5	251.70 ± 1.17	
C_4H_5NO	4-Methyloxazole	$N=C(H)OC(H)=C(CH_3)$	5	429.44 ± 0.33	
C_4H_5NO	5-Methyloxazole	$N=C(H)OC(CH_3)=C(H)$	5	477.90 ± 1.34	
C_4H_5NO	5-Methylisoxazole	$C(H)=NOC(CH_3)=C(H)$	5	272.05 ± 1.00	
C_4H_5NS	2-Methylthiazole	$N=C(CH_3)=SC(H)=C(H)$	38	34.938 ± 0.020	
C_4H_5NS	4-Methylisothiazole	$N=C(H)C(CH_3)=C(H)S$	5	105.767 ± 0.043	
$C_4H_6O_2$	4-Methyl-2-oxetanone	$OC(=O)C(H_2)C(H)(CH_3)$	5	1256.5 ± 10.5	
C_4H_7F	*trans*-1-Fluoro-2-butene	$CH_3C(H)=C(H)CH_2F$	5	596 ± 7	anticlinal conformer
C_4H_7N	1-Isocyanopropane	$CH_3C(H_2)C(H_2)N\equiv C$	5	1012.3 ± 8.4	*gauche* conformer
C_4H_7N		$CH_3C(H_2)C(H_2)N\equiv C$	5	1033.8 ± 7.7	*trans* conformer
C_4H_8	Isobutene	$(CH_3)_2C=CH_2$	5	761.58 ± 1.05	
C_4H_8	*cis*-2-Butene	$CH_3CH=CHCH_3$	5	259.89 ± 0.42	
C_4H_8O	3-Methoxy-1-propene	$CH_3OC(H_2)C(H)=CH_2$	5	728.0 ± 10.5	*skew-gauche* conformer
C_4H_8O		$CH_3OC(H_2)C(H)=CH_2$	5	829.5 ± 10.5	*syn-trans* conformer
C_4H_8O	2,2-Dimethyloxirane	$OC(CH_3)(CH_3)C(H_2)$	5	945.61 ± 0.75	
C_4H_8O	*cis*-2,3-Dimethyloxirane	$OC(H)(CH_3)C(H)(CH_3)$	5	577.80 ± 1.84	*cis* conformer
C_4H_8O		$OC(H)(CH_3)C(H)(CH_3)$	5	862.52 ± 1.84	*trans* conformer
C_4H_8O	2-Methyloxetane	$OC(H_2)C(H_2)C(H)(CH_3)$	5	1166.5 ± 4.9	
C_4H_8O	3-Methyloxetane	$OC(H_2)C(H)(CH_3)C(H_2)$	5	1149.4 ± 4.2	
C_4H_8OS	3-Methoxythietane	$SC(H_2)C(H)(OCH_3)C(H_2)$	5	1071.0 ± 10.5	
C_4H_8S	3-(Methylthio)-1-propene	$CH_3SC(H_2)C(H)=CH_2$	5	619 ± 28	
C_4H_8S	2,2-Dimethylthiirane	$SC(CH_3)(CH_3)C(H_2)$	5	1268.3 ± 3.0	
C_4H_{10}	Butane	$CH_3C(H_2)C(H_2)CH_3$	5	948 ± 24	
$C_4H_{10}N_2O$	*N*-Methyl-*N*-nitrosopropylamine	$CH_3C(H_2)C(H_2)N(CH_3)N=O$	5	320 ± 30	*N*-methyl top, conformer OMGA
$C_5H_8O_2$	Dihydro-3-methyl-2(3*H*)-furanone	$OC(=O)C(H)(CH_3)C(H_2)C(H_2)$	5	913.8 ± 2.5	
$C_5H_8O_2$	Dihydro-4-methyl-2(3*H*)-furanone	$OC(=O)C(H_2)C(H)(CH_3)C(H_2)$	39	1437.8 ± 8.4	
$C_5H_8O_2$	Dihydro-5-methyl-2(3*H*)-furanone	$OC(=O)C(H_2)C(H_2)C(H)(CH_3)$	39	1233.0 ± 4.2	
C_5H_9NO	*tert*-Butyl isocyanate	$(CH_3)_3C–N=C=O$	5	41.510 ± 0.015	$(CH_3)_3$ C group
$C_5H_{12}O$	Methyl *tert*-butyl ether	$(CH_3)_3COCH_3$	5	498.6 ± 1.5	*O*-methyl top
$C_6H_{10}O$	2-Methylcyclopentanone	$C(=O)C(H)(CH_3)C(H_2)C(H_2)C(H_2)$	5	844.2 ± 2.4	
$C_6H_{10}O$	3-Methylcyclopentanone	$C(=O)C(H_2)C(H)(CH_3)C(H_2)C(H_2)C(H_2)$	5	1233.8 ± 1.7	
$C_6H_{14}O$	*tert*-Butyl ethyl ether	$(CH_3)_3COC(H_2)CH_3$	5	1025 ± 3	ethyl CH_3
$C_7H_6F_2$	2,4-Difluorotoluene	$C(H)=C(CH_3)C(F)=C(H)C(F)=C(H)$	5	204.04 ± 0.23	
C_7H_7Cl	2-Chlorotoluene	$C(H)=C(H)C(Cl)=C(CH_3)C(H)=C(H)$	40	513.8 ± 2.7	^{35}Cl
C_7H_9N	2,6-Dimethylpyridine	$C(H)=C(H)C(CH_3)=NC(CH_3)=C(H)$	5	98.24 ± 0.27	
$C_7H_{16}FO_2P$	1,2,2-Trimethylpropyl methylphosphonofluoridate	$(CH_3)_3CC(H)(CH_3)OP(O)(F)CH_3$	41	821 ± 5	*P*-methyl top, conformer GD-I

Mol. form.	Name	Line formula	Ref.	V_3/cm^{-1}	Comments
$C_7H_{16}FO_2P$		$(CH_3)_3CC(H)(CH_3)OP(O)(F)CH_3$	41	738 ± 5	*P*-methyl top, conformer GD-II
GeH_3N_3	Germyl azide	GeH_3-N=N≡N	42	86.598 ± 0.062	
H_5PSi	Silylphospine	SiH_3PH_2	43	537.2 ± 14.0	

TABLE 2. Symmetric Rotor Potential Parameters

Mol. form.	Name	Line formula	Ref.	V_3/cm^{-1}	Comments
BF_3H_3P	Phosphine-trifluoroborane	H_3PBF_3	44	1169 ± 123	
BF_3H_3P	Trihydro(phosphorus trifluoride)boron	F_3PBH_3	45	1134 ± 53	
BH_6P	Trihydro(phosphine)boron	H_3PBH_3	46	864.5 ± 17.5	
CF_6Si	Trifluoro(trifluoromethyl)silane	CF_3SiF_3	47	489 ± 50	
CH_3F_3Ge	Trifluoromethylgermane	CF_3GeH_3	48	448 ± 53	
CH_3F_3Si	Trifluoromethylsilane	CH_3SiF_3	49	414.147 ± 0.030	
CH_6Ge	Methylgermane	CH_3GeH_3	50	433.6 ± 8.8	
CH_6Si	Methylsilane	CH_3SiH_3	2	603.3878 ± 0.0037	
CH_6Sn	Methylstannane	CH_3SnH_3	51	227 ± 10	
$C_2H_3F_3$	1,1,1-Trifluoroethane	CH_3CF_3	52	1112.24 ± 0.16	
C_2H_6	Ethane	CH_3CH_3	2	1013.28 ± 0.10	V_6 = 8.798 ± 0.041
$C_2H_3D_3$	Ethane-1,1,1-d_3	CH_3CD_3	2	1001.876 ± 0.023	V_6 = 9.328 ± 0.018
C_2D_6	Ethane-d_6	CD_3CD_3	2	989.946 ± 0.090	V_6 = 9.51 ± 0.10
C_3H_6Si	1-Silylpropyne	$CH_3C≡CSiH_3$	53	3.77 ± 0.70	
C_3H_9ClSi	Trimethylchlorosilane	$(CH_3)_3SiCl$	54	576.9 ± 0.9	
C_4H_6	2-Butyne	$CH_3C≡CCH_3$	55	6.067 ± 0.040	V_6 = 0.1240 ± 0.0144; V_9 = −0.0916 ± 0.0180
$C_5H_{10}Ge$	Ethynyltrimethylgermane	$(CH_3)_3GeC≡CH$	56	376.2 ± 16.7	
H_6Si_2	Disilane	SiH_3SiH_3	57	412.033 ± 0.010	

BOND DISSOCIATION ENERGIES

Jin-Pei Cheng

The bond dissociation energy (enthalpy) is also referred to as bond disruption energy, bond energy, bond strength, or binding energy (abbreviation: BDE, BE, or D). It is defined as the standard enthalpy change of the following fission: R–X → R + X. The BDE, denoted by $D°(R–X)$, is usually derived by the thermochemical equation, $D°(R–X) = \Delta_f H°(R) + \Delta_f H°(X) – \Delta_f H°(RX)$. The enthalpy of formation $\Delta_f H°$ of a large number of atoms, free radicals, ions, clusters, and compounds is available from the Web sites of NIST, NASA, CODATA, and IUPAC. Most authors prefer to use the BDE values at 298.15 K.

The following seven tables provide essential information of experimental BDE values of R–X and R^+–X bonds. The data in these tables have been revised through September 2020.

Table 1: Bond Dissociation Energies in Diatomic Molecules
Table 2: Enthalpy of Formation of Gaseous Atoms
Table 3: Bond Dissociation Energies in Polyatomic Molecules
Table 4: Enthalpies of Formation of Free Radicals and Other Transient Species
Table 5: Bond Dissociation Energies of Some Organic Molecules
Table 6: Bond Dissociation Energies in Diatomic Cations
Table 7: Bond Dissociation Energies in Polyatomic Cations

Bond Dissociation Energies in Diatomic Molecules

The BDEs in diatomic species have usually been measured by spectroscopy or mass spectrometry. In the absence of data on the enthalpy function, the values at 0 K, $D°(A–B)$, are converted to $D°_{298}$ by the approximate equation:

$$D°_{298}(A–B) \approx D°(A–B) + (3/2)RT = D°(A–B) + 3.7181 \text{ kJ mol}^{-1}$$

Column definitions for Table 1 are as follows.

Column heading	Definition
Mol. form.	Molecular formula for diatomic molecule A–B; table has been arranged in alphabetical order of atom A
$D°_{298}$	Bond dissociation energy, in units kJ mol^{-1}
Ref.	Reference number

TABLE 1. Bond Dissociation Energies in Diatomic Molecules

Mol. form.	$D°_{298}$/kJ mol^{-1}	Ref.
Ac–O	794	1
Ac–S	505 ± 68	22
Ag–Ag	162.9 ± 2.9	1
Ag–Al	183.7 ± 9.2	1
Ag–Au	202.5 ± 9.6	1
Ag–Bi	192 ± 42	1
Ag–Br	280.3 ± 1.3	1
Ag–Cl	279.1 ± 8.4	1
Ag–Cu	171.5 ± 9.6	1
Ag–D	226.8	1
Ag–Dy	130 ± 19	1
Ag–Eu	127 ± 13	1
Ag–F	356.9 ± 5.8	1
Ag–Ga	159 ± 17	1
Ag–Ge	174.5 ± 21	1
Ag–H	202.4 ± 9.1	1
Ag–Ho	124 ± 19	1
Ag–I	234 ± 29	1
Ag–In	166.5 ± 4.9	1
Ag–Li	186.1	1
Ag–Mn	99.2 ± 21	1
Ag–Na	133.1 ± 12.6	1
Ag–Nd	<213	1
Ag–O	221 ± 21	1
Ag–S	216.7 ± 14.6	1
Ag–Sb	156.3 ± 4.9	30
Ag–Se	210.0 ± 14.6	1
Ag–Si	185.1 ± 9.6	1
Ag–Sn	136 ± 21	1
Ag–Te	195.8 ± 14.6	1
Al–Al	140.5 ± 1.8	41
Al–Ar	5.69	1
Al–As	202.7 ± 7.1	1
Al–Au	325.9 ± 6.3	1
Al–Br	429.2 ± 5.8	1
Al–C	267.7	1
Al–Ca	52.7	1
Al–Cl	502	1
Al–Co	181.6 ± 0.2	1
Al–Cr	222.9 ± 0.9	1
Al–Cu	227.1 ± 1.2	1
Al–D	290.4	1
Al–F	675	1
Al–H	288 ± 13	1
Al–I	369.9 ± 2.1	1
Al–Kr	6.05	1
Al–Li	76.1	1
Al–N	≤368 ± 15	1
Al–Ne	3.9	1
Al–Ni	224.7 ± 4.8	1
Al–O	501.9 ± 10.6	1
Al–P	216.7 ± 12.6	1
Al–Pd	254.4 ± 12.1	1
Al–S	332 ± 10	1
Al–Sb	216.3 ± 6	1
Al–Se	318 ± 13	1
Al–Si	246.9 ± 12.6	1
Al–Te	268 ± 13	1
Al–Ti	263.4	1
Al–U	326 ± 29	1
Al–V	147.4 ± 1.0	1
Al–Xe	7.39	1
Am–O	582 ± 34	18
Am–S	375 ± 33	22
Ar–Ar	4.91	1
Ar–Au	5.50 ± 0.16	19
Ar–B	4.62	1
Ar–Br	~5.0	1
Ar–C	5.158	1
Ar–Ca	4.44 ± 0.60	1
Ar–Cd	5.57 ± 0.05	1
Ar–Ga	3.96	1
Ar–Ge	<5.4	1
Ar–He	3.96	1
Ar–Hg	5.32	1
Ar–I	~5.3	1
Ar–In	4.18	1
Ar–Kr	5.11	1
Ar–Li	~7.82	1
Ar–Mg	~3.7	1
Ar–Na	~4.2	1
Ar–Ne	4.27	1
Ar–Si	5.86	1
Ar–Sn	<5.1	1
Ar–Tl	4.09	1
Ar–Xe	5.28	1
Ar–Zn	5.0	1
As–As	385.8 ± 10.5	1
As–Cl	448	1
As–D	270.3	1
As–F	410	1
As–Ga	202.5 ± 4.8	1
As–H	274.0 ± 2.9	1
As–I	296.6 ± 24	1
As–In	201 ± 10	1
As–N	489 ± 2.1	1

Molecular

Mol. form.	$D°_{298}$/kJ mol^{-1}	Ref.
As–O	484 ± 8	1
As–P	433.5 ± 12.6	1
As–S	379.5 ± 6.3	1
As–Sb	330.5 ± 5.4	1
As–Se	96	1
As–Tl	198.3 ± 14.6	1
Au–Au	226.2 ± 0.5	1
Au–B	367.8 ± 10.5	1
Au–Ba	275.2 ± 6.3	25
Au–Be	237.7 ± 4.0	1
Au–Bi	293 ± 8.4	1
Au–Br	213 ± 21	1
Au–Ca	250.4 ± 4.0	1
Au–Ce	322 ± 18	1
Au–Cl	280 ± 13	1
Au–Co	218.0 ± 16.4	1
Au–Cr	223.7 ± 28.9	1
Au–Cs	253 ± 3.5	1
Au–Cu	227.1 ± 1.2	1
Au–D	322.2	1
Au–Dy	259 ± 24	1
Au–Eu	245 ± 12	1
Au–F	294.1	1
Au–Fe	187.0 ± 19.3	1
Au–Ga	290 ± 15	1
Au–Ge	273.2 ± 14.6	1
Au–H	300.5 ± 2.6	4
Au–Ho	267 ± 35	1
Au–I	276	1
Au–In	286.0 ± 5.7	1
Au–K	271.5 ± 19.3	28
Au–Kr	6.59 ± 0.23	19
Au–La	457 ± 28	1
Au–Li	284.5 ± 6.7	1
Au–Lu	332 ± 19	1
Au–Mg	179.1 ± 2.7	1
Au–Mn	197.7 ± 21	1
Au–Na	247.4 ± 7.8	27
Au–Nd	294 ± 29	1
Au–Ni	247 ± 16.4	1
Au–O	223 ± 21	1
Au–Pb	133 ± 42	1
Au–Pd	142.7 ± 21	1
Au–Pr	311 ± 25	1
Au–Rb	243 ± 3.5	1
Au–Rh	232.6 ± 29	1
Au–S	253.6 ± 14.6	1
Au–Sb	241.3 ± 5.8	30
Au–Sc	280 ± 40	1
Au–Se	251.0 ± 14.6	1
Au–Si	304.6 ± 6.0	1
Au–Sn	256.5 ± 7.2	1
Au–Sr	246.4 ± 4.8	25
Au–Tb	285 ± 33	1
Au–Te	237.2 ± 14.6	1
Au–U	318 ± 29	1
Au–V	246.0 ± 8.7	1
Au–Xe	11.33 ± 0.23	19
Au–Y	310 ± 12	1
B–B	290	1
B–Br	390.9 ± 0.5	1

Mol. form.	$D°_{298}$/kJ mol^{-1}	Ref.
B–C	448 ± 29	1
B–Cd	301.0	1
B–Ce	305 ± 21	1
B–Cl	427	1
B–Co	288.74 ± 0.29	44
B–D	341.0 ± 6.3	1
B–F	732	1
B–Fe	238.2 ± 1.9	44
B–H	345.2 ± 2.5	1
B–I	361	1
B–Ir	479.21 ± 4.7	44
B–La	335 ± 63	1
B–N	377.9 ± 8.7	1
B–Ni	334.76 ± 0.39	44
B–Ne	3.97	1
B–O	809	1
B–Os	426.14 ± 0.29	44
B–P	347 ± 16.7	1
B–Pt	508.83 ± 0.29	44
B–Rh	540.46 ± 0.29	44
B–Ru	468.30 ± 0.29	44
B–Pd	351.5 ± 16.7	1
B–Pt	477.8 ± 16.7	1
B–Rh	475.8 ± 21	1
B–Ru	446.9 ± 21	1
B–S	577 ± 9.2	1
B–Sc	272 ± 63	1
B–Se	462 ± 14.6	1
B–Si	317 ± 12	1
B–Te	354 ± 20	1
B–Th	297 ± 33	1
B–Ti	272 ± 63	1
B–U	322 ± 33	1
B–Y	289 ± 63	1
Ba–Br	359.9	23
Ba–Cl	439.3	23
Ba–D	≤193.7	1
Ba–F	580.0	23
Ba–H	192.0	1
Ba–I	321.0	23
Ba–O	562 ± 13.4	1
Ba–Pd	221.8 ± 5.0	1
Ba–Rh	259.4 ± 25	1
Ba–S	418 ± 21	1
Be–Be	59	1
Be–Br	316	1
Be–Cl	384	1
Be–D	203.1	1
Be–F	573	1
Be–H	221	1
Be–I	261	1
Be–O	437	1
Be–S	372 ± 59	1
Be–T	204.4	1
Bi–Bi	204.4	1
Bi–Br	240.2	1
Bi–Cl	300.4 ± 4.2	1
Bi–D	283.7	1
Bi–F	366.5 ± 12.5	1
Bi–Ga	158.6 ± 16.7	1
Bi–H	≤283.3	1

Mol. form.	$D°_{298}$/kJ mol^{-1}	Ref.
Bi–I	186.1 ± 5.8	1
Bi–In	153.6 ± 1.7	1
Bi–Li	149.4	1
Bi–O	337.2 ± 12.6	1
Bi–P	281.7 ± 13	1
Bi–Pb	142.4 ± 3.0	1
Bi–S	315.5 ± 4.6	1
Bi–Sb	252.7 ± 3.9	1
Bi–Se	280.3 ± 5.9	1
Bi–Sn	193 ± 13	1
Bi–Te	232.2 ± 11.3	1
Bi–Tl	120.9 ± 12.6	1
Bk–O	598	1
Br–Br	193.859 ± 0.120	1
Br–C	318.0 ± 8.4	1
Br–Ca	339	1
Br–Cd	159 ± 96	1
Br–Ce	375.2	23
Br–Cl	219.32 ± 0.05	1
Br–Co	326 ± 42	1
Br–Cr	328.0 ± 24.3	1
Br–Cs	389.1 ± 4.2	1
Br–Cu	331 ± 25	1
Br–D	370.74	1
Br–Dy	315.7	23
Br–Er	363.2	23
Br–Eu	328.8	23
Br–F	280 ± 12	1
Br–Fe	243 ± 84	1
Br–Ga	402 ± 13	1
Br–Gd	374.5	23
Br–Ge	347 ± 8	1
Br–H	366.16 ± 0.20	1
Br–Hg	74.9	1
Br–Ho	323.9	23
Br–I	179.1 ± 0.4	1
Br–In	409 ± 10	1
Br–K	379.1 ± 4.2	1
Br–La	448.6	23
Br–Li	418.8 ± 4.2	1
Br–Lu	303.3	23
Br–Mg	317.96	1
Br–Mn	314.2 ± 9.6	1
Br–Mo	313.4	1
Br–N	280.8 ± 21	1
Br–Na	363.1 ± 4.2	1
Br–Nd	341.8	23
Br–Ni	360 ± 13	1
Br–O	237.6 ± 0.4	1
Br–P	≤329	1
Br–Pb	248.5 ± 14.6	1
Br–Pm	337.6	23
Br–Pr	346.3	23
Br–Rb	380.7 ± 4.2	1
Br–S	218 ± 17	1
Br–Sb	314 ± 59	1
Br–Sc	444 ± 63	1
Br–Se	297 ± 84	1
Br–Si	358.2 ± 8.4	1
Br–Sm	334.3	23
Br–Sn	337 ± 13	1

Mol. form.	D°_{298}/kJ mol^{-1}	Ref.
Br–Sr	365	1
Br–T	372.77	1
Br–Tb	386.4	23
Br–Th	364	1
Br–Ti	373	1
Br–Tl	331 ± 21	1
Br–Tm	300.9	23
Br–U	377 ± 15	1
Br–V	439 ± 42	1
Br–W	329.3	1
Br–Xe	5.94 ± 0.02	1
Br–Y	481 ± 84	1
Br–Yb	297.7	23
Br–Zn	138 ± 29	1
Br–Zr	420	1
C–C	605.03 ± 0.28	33
C–Ce	443 ± 30	1
C–Cl	394.9 ± 13.4	1
C–Co	379.92 ± 1.25	45
C–D	341.4	1
C–F	513.8 ± 10.0	1
C–Fe	367.7 ± 4.2	20
C–Ge	455.7 ± 11	1
C–H	338.72 ± 0.11	33
C–Hf	430.74 ± 0.29	42
C–I	253.1 ± 35.6	1
C–Ir	631 ± 5	1
C–La	458.94 ± 0.39	45
C–Mo	482 ± 16	1
C–N	749.31 ± 0.14	33
C–Nb	545.95 ± 0.39	42
C–Ni	405.8 ± 0.4	40
C–O	1076.63 ± 0.06	33
C–Os	608 ± 25	1
C–P	507.5 ± 8.8	1
C–Pd	436 ± 20	1
C–Pt	577.8 ± 6.8	13
C–Rh	580 ± 4	1
C–Ru	648 ± 13	1
C–S	713.3 ± 1.2	1
C–Sc	297.23 ± 0.96	45
C–Se	590.4 ± 5.9	1
C–Si	447	1
C–Ta	483.72 ± 0.29	42
C–Tc	564 ± 29	1
C–Th	491.92 ± 0.29	42
C–Ti	423 ± 30	1
C–U	455 ± 15	1
C–V	396.42±0.24	36
C–W	514.03 ± 0.8	50
C–Y	333.7 ± 0.29	45
C–Zr	475.72 ± 0.17	42
Ca–Ca	16.52 ± 0.11	1
Ca–Cl	409 ± 8.7	1
Ca–D	≤169.9	1
Ca–F	529	1
Ca–H	223.8	1
Ca–I	284.7 ± 8.4	1
Ca–Kr	5.15 ± 0.72	1
Ca–Li	84.9 ± 8.4	1
Ca–O	383.3 ± 5.0	1
Ca–Pd	347 - 360	1
Ca–S	335 ± 21	1
Ca–Xe	7.31 ± 0.96	1
Cd–Cd	7.36	1
Cd–Cl	208.4	1
Cd–F	305 ± 21	1
Cd–H	69.0 ± 0.4	1
Cd–I	97.2 ± 2.1	1
Cd–In	134	1
Cd–K	7.3	1
Cd–Kr	5.17	1
Cd–Na	10.2	1
Cd–Ne	3.97	1
Cd–O	236 ± 84	1
Cd–S	208.5 ± 20.9	1
Cd–Se	127.6 ± 25.1	1
Cd–Te	100.0 ± 15.1	1
Cd–Xe	6.54	1
Ce–Ce	251.7	1
Ce–Cl	460.0	23
Ce–F	621.6	23
Ce–I	335.5	23
Ce–Ir	575 ± 9	1
Ce–N	519 ± 21	1
Ce–O	790	1
Ce–Os	524 ± 20	1
Ce–Pd	319 ± 21	1
Ce–Pt	550 ± 5	1
Ce–Rh	545 ± 7	1
Ce–Ru	494 ± 12	1
Ce–S	569	1
Ce–Se	494.5 ± 14.6	1
Ce–Te	189.4 ± 12.6	1
Cf–O	498	1
Cl–Cl	242.851 ± 0.096	8
Cl–Co	343.9	11
Cl–Cr	380.3	11
Cl–Cs	445.7 ± 7.7	1
Cl–Cu	377.8 ± 7.5	1
Cl–D	436.303 ± 0.011	1
Cl–Dy	395.1	23
Cl–Er	451.0	23
Cl–Eu	408.4	23
Cl–F	260.83	1
Cl–Fe	335.5	11
Cl–Ga	463 ± 13	1
Cl–Gd	453.9	23
Cl–Ge	390.8 ± 9.6	1
Cl–H	431.361 ± 0.013	1
Cl–Hg	92.0 ± 9.2	1
Cl–Ho	411.6	23
Cl–I	211.3 ± 0.4	1
Cl–In	436 ± 8	1
Cl–K	433.0 ± 8.4	1
Cl–La	524.4	23
Cl–Li	469 ± 13	1
Cl–Lu	383.3	23
Cl–Mg	312	1
Cl–Mn	337.6	11
Cl–N	333.9 ± 9.6	1
Cl–Na	412.1 ± 8.4	1
Cl–Nd	421.1	23
Cl–Ni	372.3	11
Cl–O	267.47 ± 0.08	1
Cl–P	≤376	1
Cl–Pb	301 ± 50	1
Cl–Pm	417.0	23
Cl–Pr	425.7	23
Cl–Ra	343 ± 75	1
Cl–Rb	427.6 ± 8.4	1
Cl–S	241.8	1
Cl–Sb	360 ± 50	1
Cl–Sc	331	1
Cl–Se	322	1
Cl–Si	416.7 ± 6.3	1
Cl–Sm	422.1	23
Cl–Sn	350 ± 8	1
Cl–Sr	409	1
Cl–T	438.64	1
Cl–Ta	544	1
Cl–Tb	474.2	23
Cl–Th	489	1
Cl–Ti	405.4 ± 10.5	1
Cl–Tl	372.8 ± 2.1	1
Cl–Tm	380.3	23
Cl–U	439	1
Cl–V	477 ± 63	1
Cl–W	372.10 ± 0.58	50
Cl–Xe	7.08	1
Cl–Y	523 ± 84	1
Cl–Yb	377.1	23
Cl–Zn	229 ± 8	1
Cl–Zr	530	1
Cm–O	709 ± 43	18
Cm–S	504 ± 25	22
Co–Co	<127	1
Co–Cu	161.1 ± 16.4	1
Co–D	270.2 ± 5.8	1
Co–F	431 ± 63	1
Co–Ge	230 ± 21	1
Co–H	244.9 ± 4.8	1
Co–I	280 ± 21	1
Co–Mn	50 ± 8	1
Co–Nb	267.02 ± 0.10	1
Co–O	397.4 ± 8.7	1
Co–S	331	1
Co–Sc	240.1	7
Co–Se	290.37±0.58	48
Co–Si	279.86 ± 0.29	43
Co–Ti	235.37 ± 0.10	1
Co–Y	253.71 ± 0.10	1
Co–Zr	306.39 ± 0.10	1
Cr–Cr	152.0 ± 6	1
Cr–Cu	154.4 ± 14.5	1
Cr–F	523 ± 19	1
Cr–Fe	~75	1
Cr–Ge	154 ± 7	1
Cr–H	207.7 ± 7	37
Cr–I	287.0 ± 24.3	1
Cr–N	377.8 ± 18.8	1
Cr–Nb	295.72 ± 0.06	1
Cr–O	452.28 ± 0.4	49

Mol. form.	$D°_{298}$/kJ mol^{-1}	Ref.
Cr–Pb	105 ± 2	1
Cr–S	331	1
Cr–Sn	141 ± 3	1
Cr–W	280.32±0.10	35
Cs–Cs	43.919 ± 0.010	1
Cs–F	517.1 ± 7.7	1
Cs–H	175.364	1
Cs–Hg	8	1
Cs–I	338.5 ± 2.1	1
Cs–Li	72.9 ± 1.2	5
Cs–Na	63.2 ± 1.3	1
Cs–O	293 ± 25	1
Cs–Rb	49.57 ± 0.01	1
Cu–Cu	182.0 ± 2.5	32
Cu–D	270.3	1
Cu–Dy	144 ± 18	1
Cu–F	414	1
Cu–Ga	215.9 ± 15	1
Cu–Ge	208.8 ± 21	1
Cu–H	254.8 ± 6	1
Cu–Ho	144 ± 19	1
Cu–I	289 ± 63	1
Cu–In	187.4 ± 7.9	1
Cu–Li	191.9	1
Cu–Na	176.1 ± 16.7	1
Cu–Ni	201.7 ± 9.6	1
Cu–O	293.4 ± 2.9	32
Cu–S	274.5 ± 14.6	1
Cu–Sb	186.7 ± 5.1	30
Cu–Se	255.2 ± 14.6	1
Cu–Si	221.3 ± 6.3	1
Cu–Sn	170 ± 10	1
Cu–Tb	191 ± 18	1
Cu–Te	230.5 ± 14.6	1
D–D	443.3197 ± 0.0003	1
D–F	576.236 ± 0.011	1
D–Ga	<276.5	1
D–Ge	≤322	1
D–H	439.2223 ± 0.0002	1
D–Hg	42.05	1
D–I	302.33	1
D–In	246	1
D–K	182.4	1
D–Li	240.24	1
D–Lu	302	1
D–Mg	161.33 ± 0.32	1
D–Mn	312 ± 6	1
D–N	341.6	1
D–Ni	≤302.9	1
D–O	429.64	1
D–P	299.0	1
D–Pt	≤350.2	1
D–S	350.62 ± 1.20	1
D–Si	302.5	1
D–Sr	167.7	1
D–T	444.91	1
D–Tl	193.0	1
D–Zn	88.7	1
Dy–Dy	70.3	1
Dy–F	531.1	23
Dy–I	277.2	23

Mol. form.	$D°_{298}$/kJ mol^{-1}	Ref.
Dy–O	615	1
Dy–S	414 ± 42	1
Dy–Se	322 ± 20	1
Dy–Te	234 ± 20	1
Er–Er	75 ± 29	1
Er–F	572.6	23
Er–I	317.6	23
Er–O	606	1
Er–S	418 ± 21	1
Er–Se	326 ± 20	1
Er–Te	238 ± 20	1
Es–O	460	1
Eu–Eu	45.2	1
Eu–F	543.0	23
Eu–I	290.4	23
Eu–Li	268.1 ± 12.6	1
Eu–O	473	1
Eu–Rh	238 ± 34	1
Eu–S	365.7 ± 13.4	1
Eu–Se	302.9 ± 14.6	1
Eu–Te	251.0 ± 14.6	1
F–F	158.670 ± 0.096	1
F–Fe	447	1
F–Ga	584 ± 13	1
F–Gd	594.6	23
F–Ge	523 ± 13	1
F–H	569.680 ± 0.011	1
F–Hf	650 ± 15	1
F–Hg	~180	1
F–Ho	517.7	23
F–I	≤271.5	1
F–In	516 ± 13	1
F–K	489.2	1
F–Kr	6.6	1
F–La	665.1	23
F–Li	577 ± 21	1
F–Lu	523.4	23
F–Mg	445.6	1
F–Mn	445.2 ± 7.5	1
F–Mo	464	1
F–N	≤349	1
F–Na	477.3	1
F–Nd	548.7	23
F–Ni	439.7 ± 5.9	2
F–Np	430 ± 50	1
F–O	220	1
F–P	≤405	1
F–Pb	355 ± 13	1
F–Pm	561.3	23
F–Pr	582.0	23
F–Pu	538 ± 29	1
F–Rb	494 ± 21	1
F–Ru	402	1
F–S	343.5 ± 6.7	1
F–Sb	439 ± 96	1
F–Sc	599.1 ± 13.4	1
F–Se	339 ± 42	1
F–Si	576.4 ± 17	1
F–Sm	565.2	23
F–Sn	476 ± 8	1
F–Sr	538	1

Mol. form.	$D°_{298}$/kJ mol^{-1}	Ref.
F–T	579.009 ± 0.108	1
F–Ta	573 ± 13	1
F–Tb	647.3	23
F–Th	652	1
F–Ti	569 ± 33	1
F–Tl	439 ± 21	1
F–Tm	509.1	23
F–U	648	1
F–V	590 ± 63	1
F–W	≤544	1
F–Xe	14.18	1
F–Y	685.3 ± 13.4	1
F–Yb	525.1	23
F–Zn	364 ± 63	1
F–Zr	627.2 ± 10.5	1
Fe–C	385.9 ± 1.8	40
Fe–Fe	118	1
Fe–Ge	210.9 ± 29	1
Fe–H	148 ± 3	1
Fe–I	123	1
Fe–O	407.0 ± 1.0	1
Fe–S	316.3 ± 0.4	40
Fe–Se	268.0 ± 0.4	40
Fe–Si	235.48 ± 0.29	43
Fm–O	443	1
Ga–Ga	<106.4	1
Ga–H	265.9 ± 5.9	4
Ga–I	334 ± 13	1
Ga–In	94.0 ± 3	1
Ga–Kr	4.08	1
Ga–Li	133.1 ± 14.6	1
Ga–O	374 ± 21	1
Ga–P	229.7 ± 12.6	1
Ga–Sb	192.0 ± 12.6	1
Ga–Te	265 ± 21	1
Ga–Xe	5.27	1
Gd–Gd	206.3 ± 67.5	1
Gd–I	336.0	23
Gd–O	715	1
Gd–S	526.8 ± 10.5	1
Gd–Se	430 ± 15	1
Gd–Te	341 ± 15	1
Ge–Ge	264.4 ± 6.8	1
Ge–H	263.2 ± 4.8	1
Ge–I	268 ± 25	1
Ge–Ni	290.3 ± 10.9	1
Ge–O	657.5 ± 4.6	4
Ge–Pb	142.2 ± 6.8	31
Ge–Pd	254.7 ± 10.5	1
Ge–S	534 ± 3	1
Ge–Sc	270 ± 11	1
Ge–Se	484.7 ± 1.7	1
Ge–Si	297	1
Ge–Sn	230.1 ± 13	1
Ge–Te	396.7 ± 3.3	1
Ge–Y	279 ± 11	1
H–H	435.7799 ± 0.0001	1
H–Hg	39.844	1
H–I	298.26 ± 0.10	1
H–In	243.1	1
H–K	174.576	1

Mol. form.	D°_{298}/kJ mol^{-1}	Ref.
H–Li	238.039 ± 0.006	1
H–Mg	127.18 ± 0.006	10
H–Mn	251 ± 5	1
H–Mo	202.5 ± 18.3	9
H–N	358.8 ± 0.2	26
H–Na	192.71 ± 0.01	34
H–Nb	>221.9 ± 9.6	1
H–Ni	240 ± 8	1
H–O	429.74 ± 0.03	33
H–P	297.0 ± 2.1	1
H–Pb	≤157	1
H–Pd	234 ± 25	1
H–Pt	330	1
H–Rb	172.6	1
H–Rh	241.0 ± 5.9	1
H–Ru	223 ± 15	1
H–S	353.57 ± 0.30	1
H–Sb	239.7 ± 4.2	1
H–Sc	205 ± 17	1
H–Se	312.5	1
H–Si	293.3 ± 1.9	1
H–Sn	264 ± 17	1
H–Sr	164 ± 8	1
H–T	440.49	1
H–Te	270.7 ± 1.7	1
H–Ti	204.6 ± 8.8	1
H–Tl	195.4 ± 4	1
H–V	226.7 ± 7	37
H–Yb	183.1 ± 2.0	1
H–Zn	85.8 ± 2	1
He–He	3.809	1
He–Hg	3.8	1
He–Xe	3.8	1
Hf–Hf	328 ± 58	1
Hf–N	535 ± 30	1
Hf–O	801 ± 13	1
Hf–S	561.4 ± 1.9	47
Hf–Se	501.0 ± 0.4	39
Hf–Si	289.73 ± 0.29	51
Hg–Hg	8.10 ± 0.18	1
Hg–I	34.69 ± 0.96	1
Hg–K	8.8	1
Hg–Kr	5.75	1
Hg–Li	13.16 ± 0.38	1
Hg–Na	10.8	1
Hg–Ne	4.14	1
Hg–O	269	1
Hg–Rb	8.4	1
Hg–S	217.3 ± 22.2	1
Hg–Se	144.3 ± 30.1	1
Hg–T	43.14	1
Hg–Te	<142	1
Hg–Tl	2.9	1
Hg–Xe	6.65	1
Hg–Zn	7.3	1
Ho–Ho	70.3	1
Ho–I	277.0	23
Ho–O	606	1
Ho–S	428.4 ± 14.6	1
Ho–Se	333 ± 15	1
Ho–Te	≤259 ± 15	1

Mol. form.	D°_{298}/kJ mol^{-1}	Ref.
I–I	152.25 ± 0.57	1
I–In	306.9 ± 1.1	1
I–K	322.5 ± 2.1	1
I–Kr	5.67	1
I–La	414.8	23
I–Li	345.2 ± 4.2	1
I–Lu	264.8	23
I–Mg	229	1
I–Mn	282.8 ± 9.6	1
I–Mo	266.9	1
I–N	159 ± 17	1
I–Na	304.2 ± 2.1	1
I–Nd	303.3	23
I–Ni	293 ± 21	1
I–O	233.4 ± 1.3	12
I–Pb	194 ± 38	1
I–Pm	299.1	23
I–Pr	307.8	23
I–Rb	318.8 ± 2.1	1
I–Si	243.1 ± 8.4	1
I–Sm	295.8	23
I–Sn	235 ± 3	1
I–Sr	301	1
I–Tb	339.6	23
I–Te	192 ± 42	1
I–Th	361 ± 25	1
I–Ti	306	1
I–Tl	285 ± 21	1
I–Tm	262.4	23
I–U	299 ± 27	1
I–Xe	~6.9	1
I–Y	422.6 ± 12.5	1
I–Yb	259.3	23
I–Zn	153.1 ± 6.3	1
I–Zr	127	1
In–In	82.0 ± 5.7	1
In–Kr	4.85	1
In–Li	92.5 ± 14.6	1
In–O	346 ± 30	1
In–P	197.9 ± 8.4	1
In–S	287.9 ± 14.6	1
In–Sb	151.9 ± 10.5	1
In–Se	245.2 ± 14.6	1
In–Te	215.5 ± 14.6	1
In–Xe	6.48	1
In–Zn	32.2	1
Ir–Ir	361 ± 68	1
Ir–La	577 ± 12	1
Ir–Nb	465 ± 25	1
Ir–O	410.1 ± 42.4	29
Ir–Se	350.20 ± 0.29	48
Ir–Si	481.52 ± 0.29	43
Ir–Th	574 ± 42	1
Ir–Ti	422 ± 13	1
Ir–Y	457 ± 15	1
K–K	56.96	1
K–Kr	4.6	1
K–Li	82.0 ± 4.2	1
K–Na	65.994 ± 0.008	1
K–Zn	6.5	1
K–O	271.5 ± 12.6	1

Mol. form.	D°_{298}/kJ mol^{-1}	Ref.
K–Rb	53.723 ± 0.005	1
K–Xe	5.0	1
Kr–Kr	5.39	1
Kr–Li	~12.1	1
Kr–Mg	6.71 ± 0.96	1
Kr–Na	~4.53	1
Kr–Ne	4.31	1
Kr–O	<8	1
Kr–Tl	4.14	1
Kr–Xe	5.66	1
Kr–Zn	5.0	1
La–La	244.9	1
La–N	519 ± 42	1
La–O	798	1
La–Pt	505 ± 12	1
La–Rh	550 ± 12	1
La–S	573.4 ± 1.7	1
La–Se	485.7 ± 14.6	1
La–Si	282.66 ± 0.48	45
La–Te	385.6 ± 15	1
La–Y	197 ± 21	1
Li–Li	105.0	1
Li–Mg	67.4 ± 6.3	1
Li–Na	87.181 ± 0.001	1
Li–O	340.5 ± 6.3	1
Li–Pb	78.7 ± 8	1
Li–S	312.5 ± 7.5	1
Li–Sb	169.0 ± 10.0	1
Li–Si	149	1
Li–Sm	193.3 ± 18.8	1
Li–Tm	276.1 ± 14.6	1
Li–Xe	~12.1	1
Li–Yb	143.5 ± 12.6	1
Lr–O	665	1
Lu–Lu	142 ± 33	1
Lu–O	669	1
Lu–Pt	402 ± 34	1
Lu–S	508.4 ± 14.4	1
Lu–Se	418 ± 15	1
Lu–Te	325 ± 15	1
Md–O	418	1
Mg–Mg	11.3	1
Mg–Ne	~4.1	1
Mg–O	358.2 ± 7.2	1
Mg–S	234	1
Mg–Xe	9.70 ± 1.79	1
Mn–Mn	61.6 ± 9.6	1
Mn–O	362 ± 25	1
Mn–S	301 ± 17	1
Mn–Se	239.3 ± 9.2	1
Mo–Mo	435.5 ± 1.0	1
Mo–Nb	452 ± 25	1
Mo–O	530.8 ± 1.6	38
N–N	944.87 ± 0.05	33
N–O	630.57 ± 0.06	33
N–P	617.1 ± 20.9	1
N–Pt	374.2 ± 9.6	1
N–Pu	469 ± 63	1
N–S	467 ± 24	1
N–Sb	460 ± 84	1
N–Sc	464 ± 84	1

Mol. form.	D°_{298}/kJ mol^{-1}	Ref.
N–Si	437.1 ± 9.9	1
N–Ta	607 ± 84	1
N–Th	577 ± 33	1
N–Ti	476 ± 33	1
N–U	531 ± 21	1
N–V	485.97 ± 36	36
N–Xe	26.9	1
N–Y	477 ± 63	1
N–Zr	565 ± 25	1
Na–Na	74.805 ± 0.586	1
Na–Ne	~3.8	1
Na–O	270 ± 4	1
Na–Rb	63.887 ± 0.024	1
Na–Xe	~5.12	1
Nb–Nb	513	1
Nb–Ni	271.9 ± 0.1	1
Nb–O	726.5 ± 10.6	1
Nb–S	541.33 ± 0.29	47
Nb–Se	470.1 ± 0.4	39
Nb–Si	300.90 ± 0.29	51
Nb–Ti	302.0 ± 0.1	1
Nb–V	369.3 ± 0.1	1
Nd–Nd	82.8	1
Nd–O	703	1
Nd–S	471.5 ± 14.6	1
Nd–Se	393.9	1
Nd–Te	305 ± 15	1
Ne–Ne	4.070	1
Ne–Xe	4.31	1
Ne–Zn	3.92	1
Ni–Ni	204	1
Ni–O	366 ± 30	1
Ni–Pd	140.9	1
Ni–Pt	273.7 ± 0.3	1
Ni–S	356.0 ± 0.4	40
Ni–Se	314.2 ± 0.4	40
Ni–Si	324.44 ± 0.29	43
Ni–V	206.3 ± 0.2	1
Ni–Y	283.92 ± 0.10	1
Ni–Zr	279.8 ± 0.1	1
No–O	268	1
Np–O	744 ± 21	18
Np–S	495 ± 55	22
O–O	498.458 ± 0.004	33
O–Os	575	1
O–P	589	1
O–Pa	801 ± 59	18
O–Pb	382.4 ± 3.3	4
O–Pd	238.1 ± 12.6	1
O–Pr	740	1
O–Pt	418.6 ± 11.6	13
O–Pu	656.1	1
O–Rb	276 ± 12.6	1
O–Re	627 ± 84	1
O–Rh	401.33 ± 0.29	49
O–Ru	472.93 ± 0.29	49
O–S	517.90 ± 0.05	1
O–Sb	434 ± 42	1
O–Sc	671.4 ± 1.0	1
O–Se	429.7 ± 6.3	1
O–Si	799.6 ± 13.4	1
O–Sm	573	1
O–Sn	528	1
O–Sr	426.3 ± 6.3	1
O–Ta	839	1
O–Tb	694	1
O–Tc	548	1
O–Te	377 ± 21	1
O–Th	871 ± 25	18
O–Ti	666.5 ± 5.6	1
O–Tl	213 ± 84	1
O–Tm	514	1
O–U	758 ± 13	18
O–V	635.21 ± 0.19	46
O–W	720 ± 71	1
O–Xe	36.4	1
O–Y	714.1 ± 10.2	1
O–Yb	387.7 ± 10	1
O–Zn	≤250	1
O–Zr	766.1 ± 10.6	1
Os–Os	415 ± 77	1
Os–Se	352.32 ± 0.29	48
Os–Si	439.45 ± 0.29	43
P–P	489.1	1
P–Pt	≤416.7 ± 16.7	1
P–Rh	353.1 ± 16.7	1
P–S	442 ± 10	1
P–Sb	356.9 ± 4.2	1
P–Se	363.7 ± 10.0	1
P–Si	363.6	1
P–Te	297.9 ± 10.0	1
P–Th	372 ± 29	1
P–Tl	209 ± 13	1
P–U	293 ± 21	1
P–W	305 ± 4	1
Pa–S	545 ± 91	22
Pb–Pb	86.6 ± 0.8	1
Pb–S	398	1
Pb–Sb	161.5 ± 10.5	1
Pb–Se	302.9 ± 4.2	1
Pb–Si	166.6 ± 10.4	31
Pb–Sn	124.8 ± 2.8	31
Pb–Te	249.8 ± 10.5	1
Pd–Pd	>136	1
Pd–Pt	191.0	1
Pd–Si	261 ± 12	1
Pd–Y	241 ± 15	1
Po–Po	187	1
Pr–Pr	129.1	1
Pr–S	492.5 ± 4.6	1
Pr–Se	446.4 ± 23.0	1
Pr–Te	326 ± 20	1
Pt–Pt	306.7 ± 1.9	1
Pt–Se	369.4 ± 6.7	48
Pt–Si	507.51 ± 0.87	43
Pt–Th	551 ± 42	1
Pt–Ti	397.5 ± 10.6	1
Pt–Y	474 ± 12	1
Pu–S	446 ± 30	22
Rb–Rb	48.898 ± 0.005	1
Re–Re	432 ± 30	1
Rh–Rh	235.85 ± 0.05	1
Rh–Sc	444 ± 11	1
Rh–Se	296.94 ± 0.87	48
Rh–Si	405.97 ± 0.29	43
Rh–Th	513 ± 21	1
Rh–Ti	390.8 ± 14.6	1
Rh–U	519 ± 17	1
Rh–V	364 ± 29	1
Rh–Y	446 ± 11	1
Ru–Ru	193.0 ± 19.3	1
Ru–Se	339.86 ± 0.29	48
Ru–Si	402.40 ± 0.29	43
Ru–Th	592 ± 42	1
Ru–V	414 ± 29	1
S–S	430.03 ± 0.03	17
S–Sb	378.7	1
S–Sc	471.9 ± 0.9	47
S–Se	371.1 ± 6.7	1
S–Si	617 ± 5	1
S-Sm	389	1
S–Sn	467	1
S–Sr	338.5 ± 16.7	1
S–Ta	538.44 ± 0.29	47
S–Tb	515 ± 42	1
S–Te	335 ± 42	1
S–Th	608 ± 77	22
S–Ti	456.23 ± 0.39	47
S–Tm	368 ± 21	1
S–U	510.4 ± 63	22
S–V	441.32 ± 0.24	36
S–W	479.87 ± 0.29	50
S–Y	523.87 ± 0.29	47
S–Yb	167	1
S–Zn	224.8 ± 12.6	1
S–Zr	549.83 ± 0.39	47
Sb–Sb	301.7 ± 6.3	1
Sb–Te	277.4 ± 3.8	1
Sb–Tl	126.7 ± 10.5	1
Sc–Sc	163 ± 21	1
Sc–Se	404.33 ± 0.29	48
Sc–Si	198.14 ± 0.29	45
Sc–Te	289 ± 17	1
Se–Se	330.5	1
Se–Si	538 ± 13	1
Se–Sm	331.0 ± 14.6	1
Se–Sn	401.2 ± 5.9	1
Se–Sr	251.0 ± 12.6	1
Se–Ta	457.7 ± 0.4	39
Se–Tb	423 ± 20	1
Se–Te	293.3	1
Se–Ti	389.4 ± 0.4	39
Se–Tm	274 ± 40	1
Se–V	378.4 ± 0.4	39
Se–W	421.79 ± 0.58	50
Se–Y	459.42 ± 0.29	48
Se–Zn	170.7 ± 25.9	1
Se–Zr	476.7 ± 0.4	39
Si–Si	310	1
Si–Sn	238.4 ± 3.6	31
Si–Ta	293.08 ± 0.29	51
Si–Te	429.2	3
Si–Ti	216.08 ± 0.29	51

Mol. form.	$D°_{298}$/kJ mol^{-1}	Ref.
Si–V	219.27 ± 0.29	51
Si–W	303.1 ± 1.0	50
Si–Y	240.11 ± 0.19	45
Si–Zr	288.35 ± 0.29	51
Sm–Sm	54 ± 21	1
Sm–Te	272.4 ± 14.6	1
Sn–Sn	187.1 ± 0.3	1
Sn–Te	338.1 ± 6.3	1
Sr–Sr	16.64 ± 1.12	1
T–T	446.67	1
Ta–Ta	390 ± 96	1
Tb–Tb	138.8	1
Tb–Te	339 ± 42	1
Tc–Tc	330	1
Te–Te	257.6 ± 4.1	1
Te–Ti	289 ± 17	1
Te–Tm	182 ± 40	1
Te–Y	339 ± 13	1
Te–Zn	117.6 ± 18.0	1
Th–Th	≤289 ± 33	1
Ti–Ti	117.6	1
Ti–V	203.2 ± 0.1	1
Ti–Zr	214.3 ± 0.1	1
Tl–Tl	59.4	1
Tl–Xe	4.18	1
Tm–Tm	54 ± 17	1
U–U	222 ± 21	1
V–V	269.3 ± 0.1	1
V–Zr	260.6 ± 0.3	1
W–W	666	1
Xe–Xe	6.023	1
Y–Y	~270 ± 39	1
Yb–Yb	16.3	1
Zn–Zn	22.2 ± 6.3	1
Zr–Zr	298.2 ± 0.1	1

References

1. Luo, Y. R., *Comprehensive Handbook of Chemical Bond Energies*, CRC Press, Boca Raton, FL, 2007. <https://doi.org/10.1201/9781420007282>
2. Hildenbrand, D. L., and Lau, K.H., *J. Phys. Chem. A* 110, 11886, 2006. <https://doi.org/10.1021/jp064171n>
3. Chattopadhyaya, S., Pramanik, A., Banerjee, A., and Das, K. K., *J. Phys. Chem. A* 110, 12303, 2006. <https://doi.org/10.1021/jp062610c>
4. Brutti, S., Balducci, G., and Gigli, G., *Rapid Commun. Mass Spectrom.* 21, 89, 2007. <https://doi.org/10.1002/rcm.2812>
5. Staanum, P., Pashov, A., Knöckel, H., and Tiemann, E., *Phys. Rev. A* 75, 042513, 2007. <https://doi.org/10.1103/PhysRevA.75.042513>
6. Ciccioli, A., Gigli, G., Meloni, G., and Testani, E., *J. Chem. Phys.* 127, 054303/1, 2007. <https://doi.org/10.1063/1.2752803>
7. Nagarajan, R., and Morse, M. D., *J. Chem. Phys.* 127, 074304/1, 2007. <https://doi.org/10.1063/1.2756533>
8. Li, J., Hao, Y., Yang, J., Zhou, C., and Mo, Y., *J. Chem. Phys.* 127, 104307/1, 2007. <https://doi.org/10.1063/1.2772273>
9. Armentrout, P. B., *Organometallics* 26, 5473, 2007. <https://doi.org/10.1021/om700579m>
10. Shayesteh, A., Henderson, R. D. E., Le Roy, R. J., and Bernath, P. F., *J. Phys. Chem. A* 111, 12495, 2007. <https://doi.org/10.1021/jp075704a>
11. Hildenbrand, D. H., *J. Phys. Chem. A* 112, 3813, 2008. <https://doi.org/10.1021/jp710621z>
12. Dooley, K. S., Geidosch, J. N., and North, S. W., *Chem. Phys. Lett.* 457, 303, 2008. <https://doi.org/10.1016/j.cplett.2008.04.009>
13. Citir, M., Metz, R. B., Belau, L., and Ahmed, M., *J. Phys. Chem. A* 112, 9584, 2008. <https://doi.org/10.1021/jp8024733>
14. Hildenbrand, D. L., Lau, K. H., Perez-Mariano, J., and Sanjurjo, A., *J. Phys. Chem. A* 112, 9978, 2008. <https://doi.org/10.1021/jp803711w>
15. Gibson, J. K., Haire, R. G., Santos, M., Pires de Matos, A., and Marçalo, J., *J. Phys. Chem. A* 112, 11373, 2008.
16. Ciccioli, A., Gigli, G., and Meloni, G., *Chem. Eur. J.* 15, 9543, 2009. <https://doi.org/10.1002/chem.200900804>
17. Frederix, P. W. J. M., Yang, C.-H., Groenenboom, G. C., Parker, D. H., Alnama, K., Western, C. M., and Orr-Ewing, A. J., *J. Phys. Chem. A* 113, 14995, 2009. <https://doi.org/10.1021/jp905104u>
18. Marçalo, J., and Gibson, J. K., *J. Phys. Chem. A* 113, 12599, 2009. <https://doi.org/10.1021/jp904862a>
19. Hopkins, W.S., Woodham, A. P., Plowright, R. J., Wright, T. G., and Mackenzie, S. R., *J. Chem. Phys.*, 132, 21403, 2010. <https://doi.org/10.1063/1.3432127>
20. Tzeli, D., and Mavridis, A., *J. Chem. Phys.*, 132, 194312, 2010. <https://doi.org/10.1063/1.3429612>
21. Huang, H.-Y., Lu, T.-L., Whang, T.-J., Chang, Y.-Y., and Tsai, C.-C., *J. Chem. Phys.* 133, 044301, 2010. <https://doi.org/10.1063/1.3458914>
22. Pereira, C. C. L., Marsden, C. J., Marçalo, J., and Gibson, J. K., *Phys. Chem. Chem. Phys.* 13, 12940, 2011. <https://doi.org/10.1039/c1cp20996e>
23. Mucklejohn, S. A., *J. Phys. D: Appl. Phys.* 44, 224010, 2011. <https://doi.org/10.1088/0022-3727/44/22/224010>
24. Guina, A. O., Lopatin, S. I., and Shugurov, S. M., *Inorg. Chem.*, 51, 4918–4924, 2012. <https://doi.org/10.1021/ic201644f>
25. Ciccioli, A., Gigli, G., and Lauricella, M., *J. Chem. Phys.*, 136, 184306, 2012. <https://doi.org/10.1063/1.4711085>
26. Ruscic, B., Active Thermochemical Table, <http://atct.anl.gov/index.html>, Dec. 16, 2010.
27. Ciccioli, A., and Gigli, G., *J. Phys. Chem. A* 117, 4956, 2013. <https://doi.org/10.1021/jp402374t>
28. Stangassinger. A., Knight, A. M., and Duncan, M. A., *J. Phys. Chem.A* 103, 1547, 1999. <https://doi.org/10.1021/jp984402t>
29. Armentrout, P. B., and Li, F.-X., *J. Phys. Chem. A* 117, 7754, 2013. <https://doi.org/10.1021/jp4063143>
30. Carta, V., Ciccioli, A., and Gigli, G., *J. Chem. Phys.* 140, 064305, 2014. <https://doi.org/10.1063/1.4864116>
31. Ciccioli, A., and Gigli, G., *ECS Trans.* 46, 99, 2013. <https://doi.org/10.1149/04601.0099ecst>
32. Parry, I. S., Hermes, A. C., Kartouzian, A., and Mackenzie, S. R., *Phys. Chem. Chem. Phys.* 16, 458, 2014. <https://doi.org/10.1039/C3CP53214C>
33. Ruscic, B., Feller, D., and Peterson, K. A., *Theor. Chem. Acc.*. 133, 1415, 2014. <https://doi.org/10.1007/s00214-013-1415-z>
34. Walji, S.-D., Sentjens, K. M., and Le Roy, R. J., *J. Chem. Phys.* 142, 044305, 2015. <https://doi.org/10.1063/1.4906086>
35. Matthew, D.J., et al., *J. Chem. Phys.*, 144, 214306, 2016. <https://doi.org/10.1063/1.4952453>
36. Johnson, E. L. et al., *J. Chem. Phys.* 144, 234306, 2016. <https://doi.org/10.1063/1.4953782>
37. Cheng, L., et al., *J. Chem. Theory Comput.* 13, 1044-1056, 2017. <https://doi.org/10.1021/acs.jctc.6b00970>
38. Cooper, G. A., et al., *J. Chem. Phys.* 147, 013921, 2017. <https://doi.org/10.1063/1.4979979>
39. Sorensen, J. J., et al., *J. Chem. Phys.* 145, 214308, 2016. <https://doi.org/10.1063/1.4968601>
40. Matthew, D. J., et al., *J. Chem. Phys.* 146, 144310, 2017. <https://doi.org/10.1063/1.4979679>
41. Johnston, M. D., et al., *J. Chem. Phys.* 148, 214308, 2018. <https://doi.org/10.1063/1.5034353>
42. Sevy, A., et al., *J. Chem. Phys.* 149, 044306, 2018. <https://doi.org/10.1063/1.5041422>
43. Sevy, A., et al., *J. Chem. Phys.* 149, 174307, 2018. <https://doi.org/10.1063/1.5050934>
44. Merriles, D. M., et al., *J. Chem. Phys.* 151, 044302, 2019. <https://doi.org/10.1063/1.5113511>
45. Sevy, A., et al., *J. Chem. Phys.* 151, 024302, 2019. <https://doi.org/10.1063/1.5098330>
46. Merriles, D. M., et al., *J. Chem. Phys.* 153, 024303, 2020.
47. Sorensen, J. J., et al., *J. Chem. Phys.* 152, 194307, 2020.
48. Sorensen, J. J., et al., *J. Chem. Phys.* 152, 124305, 2020.
49. Sorensen, J. J., et al., *J. Chem. Phys.* 153, 074303, 2020.
50. Sevy, A., et al., *J. Phys. Chem.* A 121, 9446-9457, 2017.
51. Sevy, A., et al., *J. Chem. Phys.* 147, 084301, 2017.

Enthalpy of Formation of Gaseous Atoms

Table 2 contains values of the enthalpy of formation of gaseous atoms. Column definitions for Table 2 are as follows.

Column heading	Definition
Atom	Atomic symbol; listed in alphabetical order
$\Delta_f H^\circ_{298}$	Enthalpy of formation at 298 °C, kJ mol^{-1}
Ref.	Reference number

TABLE 2. Enthalpy of Formation of Gaseous Atoms

Atom	$\Delta_f H^\circ_{298}$/kJ mol^{-1}	Ref.	Atom	$\Delta_f H^\circ_{298}$/kJ mol^{-1}	Ref.	Atom	$\Delta_f H^\circ_{298}$/kJ mol^{-1}	Ref.
Ac	406	5	Gd	397.5 ± 2.1	4	Pu	345	6
Ag	284.9 ± 0.8	2	Ge	372 ± 3	2	Ra	159	5
Al	330.9 ± 4.0	2	H	217.998 ± 0.006	2	Rb	80.9 ± 0.8	2
Am	284	6	Hf	618.4 ± 6.3	3	Re	774 ± 6.3	1
As	302.5 ± 13	1	Hg	61.38 ± 0.04	2	Rh	556 ± 4	1
Au	368.2 ± 2.1	1	Ho	300.6 ± 2.1	4	Ru	650.6 ± 6.3	1
B	565 ± 5	2	I	106.757 ± 0.002	9	S	277.17 ± 0.15	2
Ba	179.1 ± 5.0	3	In	243 ± 4	1	Sb	264.4 ± 2.5	1
Be	324 ± 5	2	Ir	669 ± 4	1	Sc	377.8 ± 4	1
Bi	209.6 ± 2.1	1	K	89.0 ± 0.8	2	Se	227.2 ± 4	1
Bk	310	6	La	431.0 ± 2.1	4	Si	450.0 ± 8	2
Br	111.85 ± 0.06	9	Li	159.3 ± 1.0	2	Sm	206.7 ± 2.1	4
C	716.87 ± 0.06	9	Lu	427.6 ± 2.1	4	Sn	301.2 ± 1.5	2
Ca	177.8 ± 0.8	2	Mg	147.1 ± 0.8	2	Sr	164.0 ± 1.7	3
Cd	111.80 ± 0.20	2	Mn	283.3 ± 4.2	3	Ta	782.0 ± 2.5	1
Ce	420.1 ± 2.1	4	Mo	658.98 ± 3.8	3	Tb	388.7 ± 2.1	4
Cf	196	6	N	472.44 ± 0.03	9	Tc	678	5
Cl	121.302 ± 0.002	9	Na	107.5 ± 0.7	3	Te	196.6 ± 2.1	1
Cm	386	6	Nb	733.0 ± 8	3	Th	602 ± 6	2
Co	426.7	3	Nd	326.9 ± 2.1	4	Ti	473 ± 3	2
Cr	397.48 ± 4.2	3	Ni	430.1 ± 8.4	3	Tl	182.2 ± 0.4	1
Cs	76.5 ± 1.0	2	Np	464.8	6	Tm	232.2 ± 2.1	4
Cu	337.4 ± 1.2	2	O	249.229 ± 0.002	7	U	533 ± 8	2
Dy	290.4 ± 2.1	4	Os	787 ± 6.3	1	V	515.5 ± 8	3
Er	316.4 ± 2.1	4	P	316.5 ± 1.0	2	W	851.0 ± 6.3	3
Es	133	6	Pa	563	5	Y	424.7 ± 2.1	4
Eu	177.4 ± 2.1	4	Pb	195.2 ± 0.8	2	Yb	155.6 ± 2.1	4
F	79.335 ± 0.068	8	Pd	376.6 ± 2.1	1	Zn	130.40 ± 0.40	2
Fe	415.5 ± 1.3	3	Pr	356.9 ± 2.1	4	Zr	610.0 ± 8.4	3
Ga	271.96 ± 2.1	3	Pt	565.7 ± 1.3	1			

References

1. Brewer, L., and Rosenblatt, G. M., *Adv. High Temp. Chem.* 2, 1, 1969. <https://doi.org/10.1016/S0065-2741(13)70007-1>
2. Cox, J. D., Wagman, D. D., and Medvedev, V. A., Eds., *CODATA Key Values for Thermodynamics*, Hemisphere Publishing Corporation, New York, 1989; updated e-version: <http://www.codata.org/codata>.
3. NIST Chemistry WebBook, <http://webbook.nist.gov>, *NIST-JANAF Thermochemical Table, Fourth Edition*, Chase, Jr., M. W., Ed., ACS, AIP, New York, 1998.
4. Chandrasekharaiah, M. S., and Gingerich, K. A., Thermodynamic properties of gaseous species, in *Handbook on the Chemistry and Physics of Rare Earths*, Gschneidner, Jr., K. A., and Eyring, L., Eds., Elsevier, Amsterdam, 1989, Vol. 12, Chap. 86, pp. 409–431. <https://doi.org/10.1016/S0168-1273(89)12010-8>
5. Lias, S. G., Bartmess, J. E., Liebman, J. F., Holmes, J. L., Levin, R. D., and Mallard, W. G., *J. Phys. Chem. Ref. Data* 17, Suppl. 1, 1988.
6. Kleinschmidt, P.D., Ward, J. W., Matlack, G. M., and Haire, R. G., *High Temp. Sci.* 19, 267, 1985.
7. Ruscic, B., Pinzon, R.E., Morton, M. E., Srinivasan, N. K., Su, M.-C., Sutherland, J. W., and Michael, J. V., *J. Phys. Chem. A* 110, 6592, 2006. <https://doi.org/10.1021/jp056311j>
8. Luo, Y. R., *University Chem.* 27, 80, 2012.
9. Ruscic, B., Active Thermochemical Table, <http://atct.anl.gov/index.html>, Dec. 16, 2010.

Bond Dissociation Energies in Polyatomic Molecules

The D°_{298} values in polyatomic molecules are notoriously difficult to measure accurately since the mechanism of the kinetic systems involved in many of the measurements are seldom straightforward. Thus, much lively controversy has taken place in the literature and is likely to continue for some time to come. We will continue updating and presenting our assessment of the most reliable BDE data every year.

The references relating to each of the D°_{298} values listed in Table 3 are contained in the *Comprehensive Handbook of*

Chemical Bond Energies, by Yu-Ran Luo, CRC Press, 2007. Many D°_{298} values in Table 3 are derived from the equation

$$D^\circ_{298}(R\text{-}X) = \Delta_f H^\circ(R) + \Delta_f H^\circ(X) - \Delta_f H^\circ(RX)$$

Here, the enthalpies of formation of the atoms and radicals are taken from Tables 2 and 4, respectively, and the enthalpies of formation of the molecules are from reference sources listed in the above *Comprehensive Handbook of Chemical Bond Energies*.

Table 3 is divided into several parts: (1) **H**-C, (2) **C**-C, (3) C-**halogen**, (4) **O**-X, (5) **N**-X, (6) **S**-X, (7) **Si**-, **Ge**-, **Sn**-, and **Pb**-X, (8) **P**-, **As**-, **Sb**-, and **Bi**-X, (9) **Se**- and **Te**-X, and (10) **metal**-X BDEs. The **boldface** in the species indicates the dissociated fragment. Column definitions for Table 3 are as follows.

Column heading	Definition
Bond	Molecular bond undergoing dissociation; **boldface** indicates the dissociated fragment; **metal**-X BDEs (subtable 10) are arranged on the basis of the Periodic Table with the new IUPAC notation for Groups 1 to 18, see the inside front cover of this *CRC Handbook*
D°_{298}	Bond dissociation energy for specified bond, in units kJ mol^{-1}
Ref.	Reference number

TABLE 3. Bond Dissociation Energies in Polyatomic Molecules

Bond	D°_{298}/kJ mol^{-1}	Ref.
(1) C–H BDEs		
$CH_3-\mathbf{H}$	439.3 ± 0.4	1
$CH_3CH_2-\mathbf{H}$	420.5 ± 1.3	1
$CH_3CH_2CH_2-\mathbf{H}$	422.2 ± 2.1	1
$CH_3\mathbf{CH_2}CH_3$	410.5 ± 2.9	1
$CH_3CH_2CH_2CH_2-\mathbf{H}$	421.3	1
$CH_3\mathbf{CH_2}CH_2CH_3$	411.1 ± 2.2	1
$(CH_3)_2CHCH_2-\mathbf{H}$	419.2 ± 4.2	1
$(CH_3)_3C-\mathbf{H}$	400.4 ± 2.9	1
$(CH_3)_3CCH_2-\mathbf{H}$	419.7 ± 4.2	1
$(CH_3CH_2)\mathbf{CH}(CH_3)_2$	400.8	1
$CH_3\mathbf{CH_2}(CH_2)_2CH_3$	415.1	1
$(C_3H_7)\mathbf{CH}(CH_3)_2$	396.2 ± 8.4	1
$CH_3CH(CH_3)\mathbf{CH}(CH_3)_2$	399.2 ± 13.0	1
$CH_3\mathbf{CH_2}(CH_2)_3CH_3$	410	1
$CH_3\mathbf{CH_2}(CH_2)_4CH_3$	410	1
$HCC-\mathbf{H}$	557.81 ± 0.30	1
$HCCCC-\mathbf{H}$	539 ± 12	1
$CHCCH_2-\mathbf{H}$	384.1 ± 4.2	1
$CH_3CCCH_2-\mathbf{H}$	379.5	1
$HCC\mathbf{CH_2}CH_3$	373.0	1
$CH_2{=}CHCCCH_2-\mathbf{H}$	363.3	1
$CH_3CC\mathbf{CH_2}CH_3$	365.3 ± 9.6	1
$HCC\mathbf{CH_2}CH_2CH_3$	349.8 ± 8.4	1
$HCC\mathbf{CH}(CH_3)_2$	345.2 ± 8.4	1
$CH_3CC\mathbf{CH}(CH_3)_2$	344.3 ± 11.3	1
$HCCCCCC-\mathbf{H}$	~543 ± 13	1
$H_2C{=}CH-\mathbf{H}$	464.2 ± 2.5	1
$CH_2{=}C{=}CH-\mathbf{H}$	371.1 ± 12.6	1
$CH_3CH{=}CH-\mathbf{H}$	464.8	1
$CH_2{=}CHCH_2-\mathbf{H}$	369 ± 3	1
$CH_2{=}CH\text{-}CH_2CH_2-\mathbf{H}$	410.5	1
$CH_2{=}CH\mathbf{CH_2}CH_3$	350.6	1
$CH_2{=}C(CH_3)CH_2-\mathbf{H}$	372.8	1
$CH_2{=}CHCH{=}CHCH_2\text{-}\mathbf{H}$	347.3 ± 12.6	1
$(CH_2{=}CH)_2CH-\mathbf{H}$	320.5 ± 4.2	1
$CH_2{=}CH\mathbf{CH_2}CH_2CH_3$	348.8	1
$CH_2{=}CH\mathbf{CH}(CH_3)_2$	332.6 ± 7.1	1
$CH_2{=}C(CH_3CH_2)CH_2-\mathbf{H}$	356.1 ± 8.4	1
$(CH_2{=}CH)_2C(CH_3)-\mathbf{H}$	322.2	1
H-*cyclo*-C_3H_5	444.8 ± 1.0	1
H-CH_2-*cyclo*-C_3H_5	407.5 ± 6.7	1
H-*cyclo*-C_4H_7	409.2 ± 1.3	1
H-*cyclo*-C_5H_9	400.0 ± 4.2	1
H-*cyclo*-C_6H_{11}	416.3	1
$\mathbf{H}-C_6H_5$	465.9 ± 0.6	12
$\mathbf{H}-CH_2C_6H_5$	375.5 ± 5.0	1
$\mathbf{H}-CH(CH_3)C_6H_5$	357.3 ± 6.3	1
$\mathbf{H}-CH(C_6H_5)_2$	353.5 ± 2.1	1
$\mathbf{H}\text{-}CH(C_6H_4\text{-}p\text{-}OH)_2$	375.8 ± 4.7	1
$\mathbf{H}-C(CH_3)_2C_6H_5$	348.1 ± 4.2	1
$\mathbf{H}-C(C_6H_5)_3$	338.9 ± 8.4	1
$1\text{-}\mathbf{H}\text{-}C_{10}H_7$	469.4 ± 5.4	1
$2\text{-}\mathbf{H}\text{-}C_{10}H_7$	468.2 ± 5.9	1
$\mathbf{H}-CF_3$	445.2 ± 2.9	1
$\mathbf{H}-CHF_2$	431.8 ± 4.2	1
$\mathbf{H}-CH_2F$	423.8 ± 4.2	1
$\mathbf{H}-CClF_2$	421.3 ± 8.4	1
$\mathbf{H}-CCl_2F$	410.9 ± 8.4	1
$\mathbf{H}-CBrF_2$	415.5 ± 12.6	1
$\mathbf{H}-CHClF$	421.7 ± 10.0	1
$\mathbf{H}-CCl_3$	392.5 ± 2.5	1
$\mathbf{H}-CHCl_2$	400.6 ± 2.0	1
$\mathbf{H}-CH_2Cl$	419.0 ± 2.3	1
$\mathbf{H}-CFClBr$	413 ± 21	1
$\mathbf{H}-CHClBr$	406.0 ± 2.4	1
$\mathbf{H}-CCl_2Br$	387 ± 21	1
$\mathbf{H}-CClBr_2$	371 ± 21	1
$\mathbf{H}-CBr_3$	399.2 ± 8.4	1
$\mathbf{H}-CHBr_2$	412.6 ± 2.7	3
$\mathbf{H}-CH_2Br$	427.2 ± 2.4	1
$\mathbf{H}-CI_3$	423 ± 29	1
$\mathbf{H}-CHI_2$	431.0 ± 8.4	1
$\mathbf{H}-CH_2I$	431.6 ± 2.8	1
$CF_3CF_2-\mathbf{H}$	429.7 ± 2.1	1
$CHF_2CF_2-\mathbf{H}$	431.0 ± 18.8	1
$CH_2FCF_2-\mathbf{H}$	433.0 ± 14.6	1
$CHF_2CFH-\mathbf{H}$	426.8 ± 14.6	1
$CF_3CH_2-\mathbf{H}$	446.4 ± 4.5	1
$CH_3CF_2-\mathbf{H}$	416.3 ± 4.2	1
$CH_2FCHF-\mathbf{H}$	413.4 ± 12.6	1
$CHF_2CH_2-\mathbf{H}$	433.0 ± 14.6	1
$CH_2FCH_2-\mathbf{H}$	433.5 ± 8.4	1
$CH_3CHF-\mathbf{H}$	410.9 ± 8.4	1
$CF_3CHCl-\mathbf{H}$	425.9 ± 6.3	1
$CF_3CClBr-\mathbf{H}$	404.2 ± 6.3	1
$CClF_2CHF-\mathbf{H}$	412.1 ± 2.1	1
$CCl_3CCl_2-\mathbf{H}$	397.5 ± 8.4	1
$CHCl_2CCl_2-\mathbf{H}$	393.3 ± 8.4	1
$CH_3CCl_2-\mathbf{H}$	397.9 ± 5.0	1
$CH_3CHCl-\mathbf{H}$	406.6 ± 1.5	1
$CH_2ClCH_2-\mathbf{H}$	423.1 ± 2.4	1
$CH_3CBr_2-\mathbf{H}$	397.1 ± 5.0	1
$CH_2BrCH_2-\mathbf{H}$	415.1 ± 8.4	1
$CH_3CHBr-\mathbf{H}$	415.0 ± 2.7	3
$CF_2{=}CF-\mathbf{H}$	464.4 ± 8.4	1
$CF_3CF_2CF_2-\mathbf{H}$	432.2	1
$CH_3CH_2CHCl-\mathbf{H}$	407.0 ± 3.5	1
$CH_2{=}CH\text{-}CHF-\mathbf{H}$	370.7 ± 4.6	1
$CH_2{=}CHCHCl-\mathbf{H}$	370.7 ± 4.6	1
$CH_2{=}CHCHBr-\mathbf{H}$	374.0 ± 4.6	1
$\mathbf{H}-C_6F_5$	487.4	1
$\mathbf{H}-CH_2OH$	401.92 ± 0.63	1
$\mathbf{CH_2}CHOH$	467 ± 11	1
$CH_3\mathbf{CH_2}OH$	401.2 ± 4.2	1
$\mathbf{CH_3}CH_2OH$	421.7 ± 8	1
$CH_3CH_2\mathbf{CH_2}OH$	392	1
$CH_3\mathbf{CH_2}CH_2OH$	394.6 ± 8.4	1
$\mathbf{CH_3}CH_2CH_2OH$	406.3 ± 8.4	1
$(CH_3)_2\mathbf{CH}OH$	383.7 ± 8.4	1
$(\mathbf{CH_3})_2CHOH$	394.6 ± 8.4	1
$CH_2{=}CH\mathbf{CH_2}OH$	341.4 ± 7.5	1
$(\mathbf{CH_3})_3COH$	418.4 ± 8.4	1
$(CH_2{=}CH)_2\mathbf{CH}OH$	288.7	1
$Ph_2\mathbf{CH}OH$	326	1
$CH_3\mathbf{CH}(OH)_2$	~385	1
$(\mathbf{CH_2}OH)_2$	385.3	1
$HOCH_2(CH_2)_2(OH)CH-\mathbf{H}$	399.2	1
$\mathbf{CH_3}OCH_3$	402.1	1
$\mathbf{CHF_2}OCF_3$	443.5 ± 4.2	1
$\mathbf{CHF_2}OCHF_2$	435.1 ± 4.2	1
$\mathbf{CH_3}OCF_3$	426.8 ± 4.2	1
$\mathbf{CH_3}OCH_2CH_3$	389.1	1
$(CH_3)_3COC(\mathbf{CH_3})_3$	402.1	1
$CH_3\mathbf{CH_2}OCH_2CH_3$	389.1	1
$CH_3\mathbf{CH_2}O$*t*-$C(CH_3)_3$	405.4	1
$\mathbf{CH_3}OPh$	385.0	1
H-2-oxiran-2-yl	420.5 ± 6.5	1
H-tetrahydrofuran-2-yl	385.3 ± 6.7	1
$HC(O)-\mathbf{H}$	368.40 ± 0.67	1
$FC(O)-\mathbf{H}$	423.0	1
$NH_2C(O)-H$	378.7 ± 5.4	27
$CH_3C(O)-\mathbf{H}$	374.0 ± 1.3	1
$CF_3C(O)-\mathbf{H}$	390.4	1
$C_2H_5C(O)-\mathbf{H}$	374.5	1
$CH_2{=}CHC(O)-\mathbf{H}$	372.8	1
$C_3H_7C(O)-\mathbf{H}$	371.2	1
iso-$C_3H_7C(O)-\mathbf{H}$	364.5	1
$C_4H_9C(O)-\mathbf{H}$	372.0	1
$(CH_3)_2CHCH_2C(O)-\mathbf{H}$	362.5	1
$C_2H_5CH(CH_3)C(O)-\mathbf{H}$	360.8	1
tert-BuC(O)–**H**	375.1	1
$Et_2CHC(O)-\mathbf{H}$	367.2	1
$CH_3(CH_2)_8C(O)-\mathbf{H}$	373.3	1
$C_6H_5C(O)-\mathbf{H}$	371.1 ± 10.9	1
$PhCH_2C(O)-\mathbf{H}$	362.0	1
$PhC(CH_3)_2C(O)-\mathbf{H}$	362.9	1
$\mathbf{H}-CH{=}C{=}O$	448.1	1
$\mathbf{CH_3}C(O)H$	394.5 ± 9.2	1

Bond	$D°_{298}$/kJ mol^{-1}	Ref.
CH$_3$C(O)Cl	≤423.4	1
CH$_3$**CH$_2$**C(O)H	383.7	1
CH$_3$COCH$_3$	401.2 ± 2.9	1
CF$_3$C(O)**CH$_3$**	465.6	1
CH$_3$COCH$_2$CH$_3$	403.8	1
MeCO**CH$_2$**Me	386.2 ± 7.1	1
EtCO**CH$_2$**Me	396.5 ± 2.8	1
CH$_3$CH$_2$COC$_6$H$_5$	402.8 ± 3.6	1
Me**CH$_2$**COPh	388.7	1
H–C(O)OH	404.2	1
CH$_3$C(O)OH	398.7 ± 12.1	1
Cl**CH$_2$**C(O)OH	398.9	1
H–C(O)OCH$_3$	399.2 ± 8.4	1
CH$_3$C(O)O**CH$_3$**	406.3 ± 10.5	1
CH$_3$C(O)OCH$_3$	404.6	1
CH$_3$C(O)OCH$_2$CH$_3$	401.7	1
CH$_3$C(O)OPh	419.2 ± 5.4	1
CH$_3$**CH$_2$**C(O)OEt	400	1
Ph**CH$_2$**C(O)OEt	370.7	1
Me$_2$**CH**C(O)OEt	387.4	1
Ph**CH**Me(C(O)OEt)	358.2	1
H-furaylmethyl	361.9 ± 8.4	1
CH$_3$NH$_2$	392.9 ± 8.4	1
CH$_3$N=CH$_2$	407.9 ± 14.6	1
CH$_3$**CH$_2$**NH$_2$	377.0 ± 8.4	1
C$_2$H$_5$**CH$_2$**NH$_2$	380.7 ± 8.4	1
C$_3$H$_7$**CH$_2$**NH$_2$	393.3 ± 8.4	1
C$_4$H$_9$**CH$_2$**NH$_2$	387.7 ± 8.4	1
HOCH$_2$**CH$_2$**NH$_2$	379.5 ± 8.4	1
(CH$_3$**CH$_2$**)$_2$NH	370.7 ± 8.4	1
(C$_3$H$_7$**CH$_2$**)$_2$NH	379.9 ± 8.4	1
(C$_4$H$_9$**CH$_2$**)$_2$NH	384.5 ± 8.4	1
(C$_2$H$_5$)$_2$N**CH$_2$**CH$_3$	379.5 ± 1.7	1
(C$_2$H$_5$**CH$_2$**)$_3$N	376.6 ± 8.4	1
((CH$_3$)$_2$C**CH$_2$**)$_3$N	388.3 ± 8.4	1
(Bu)$_2$N**CH$_2$**(*n*Pr)	381 ± 10.0	1
((CH$_3$)$_2$**CH**)$_3$N	387.0 ± 8.4	1
(CH$_3$)$_2$**CH**NH$_2$	372.0 ± 8.4	1
CH$_3$NHCH$_3$	364.0 ± 8.4	1
(**CH$_3$**)$_3$N	380.7 ± 8.4	1
tert-BuN(**CH$_3$**)$_2$	376.6 ± 8.4	1
((HOCH$_2$**CH$_2$**)$_2$(CH$_3$))N	364.4 ± 8.4	1
(HOCH$_2$**CH$_2$**)$_3$N	379.9 ± 8.4	1
((HOCH$_2$)**CH**(CH$_3$))$_3$N	379.9 ± 8.4	1
Ph**CH$_2$**NH$_2$	368.2	1
PhN(CH$_2$**CH$_3$**)$_2$	383.3 ± 4.2	1
Ph$_2$N**CH$_3$**	379.5 ± 1.7	1
PhN(**CH$_2$**Ph)$_2$	357.3 ± 8.8	1
N(CH$_2$Ph)$_3$	372.8 ± 2.5	1
PhN(**CH$_2$**CH=CH$_2$)$_2$	339.3 ± 2.9	1
N(CH$_2$CH=CH$_2$)$_3$	345.6 ± 3.3	1
H$_2$NNH(**CH$_3$**)	410	1
HNN(**CH$_3$**)$_2$	410	1
(**CH$_3$**)$_2$NC$_6$H$_5$	383.7 ± 5.4	1
H–CN	528.5 ± 0.8	1
CH$_3$CN	405.8 ± 4.2	1
CH$_3$**CH$_2$**CN	393.3 ± 12.6	1
Ph**CH$_2$**CN	338.5±9.6	18
C$_6$F$_5$**CH$_2$**CN	350.6	1
F**CH$_2$**CN	379.5 ± 11.7	19
CH$_2$(CN)$_2$	366.5	1
CH$_2$(CN)(NH$_2$)	355.2	1
(CH$_3$)$_2$**CH**CN	384.5	1
CH$_3$NC	389.1 ± 12.6	1
H–HCNN	405.8 ± 8.4	1
H–CNN	331 ± 17	1
CH$_3$NO$_2$	415.4	1
CH$_3$**CH$_2$**NO$_2$	410.5	1
C$_2$H$_5$**CH$_2$**NO$_2$	410.5	1
Me$_2$**CH**NO$_2$	394.9	1
C$_6$H$_5$C(NO$_2$)**CH**CH$_3$	357.3	1
H–C(S)H	399.6 ± 5.0	1
CH$_3$SH	392.9 ± 8.4	1
CH$_3$S**CH$_3$**	392.0 ± 5.9	1
PhS**CH$_3$**	389.1	1
Ph**CH$_2$**SPh	352.3	1
(PhS)$_2$**CH**Ph	341.0	1
PhS**CH**Ph$_2$	344.8	1
CH$_3$SOCH$_3$	393.3	1
CH$_3$SO$_2$CH$_3$	414.2	1
CH$_3$SO$_2$CF$_3$	431.0	1
CH$_3$SO$_2$Ph	414.2	1
Ph**CH$_2$**SO$_2$Me	380.7	1
Ph**CH$_2$**SO$_2$CF$_3$	372.4	1
Ph**CH$_2$**SO$_2$*t*Bu	376.6	1
Ph$_2$**CH**SO$_2$Ph	365.3	1
CH$_2$(SPh)$_2$	372.4	1
H–CH$_2$SiMe$_3$	418 ± 6.3	1
H–CH$_2$C(CH$_3$)$_2$SiMe$_3$	409 ± 5	1
H–CH$_2$SiMe$_2$Ph	410.1	1
H–CH((CH$_3$)$_3$Si)$_2$	397 ± 13	1
H–CH$_2$B(RO)$_2$	412.5	1
H–CH((CH$_3$)$_2$P)$_2$	385 ± 13	1
(η^5–C$_5$H$_5$)Fe(η^5–C$_5$H$_4$–**H**)	493.7 ± 5.4	30
(2) C–C BDEs		
CH$_3$-CH$_3$	377.4 ± 0.8	1
CH$_3$-C$_2$H$_5$	370.3 ± 2.1	1
CH$_3$-C$_3$H$_7$	372.0 ± 2.9	1
CH$_3$-*iso*-C$_3$H$_7$	369.0 ± 3.8	1
CH$_3$-C$_4$H$_9$	371.5 ± 2.9	1
CH$_3$-*iso*-C$_4$H$_9$	370.3 ± 4.6	1
CH$_3$-*sec*-C$_4$H$_9$	368.2 ± 2.9	1
CH$_3$-*tert*-C$_4$H$_9$	363.6 ± 2.9	1
CH$_3$-C$_5$H$_{11}$	368.4 ± 6.3	1
CH$_3$-CH(C$_2$H$_5$)$_2$	365.7 ± 4.2	1
CH$_3$-C(CH$_3$)$_2$(CH$_2$CH$_3$)	360.9 ± 6.3	1
CH$_3$-C$_6$H$_{13}$	368.2 ± 6.3	1
C$_2$H$_5$-C$_2$H$_5$	363.2 ± 2.5	1
C$_3$H$_7$-C$_3$H$_7$	366.1 ± 3.3	1
iso-**C$_3$H$_7$**-*iso*-C$_3$H$_7$	353.5 ± 4.6	1
C$_4$H$_9$-C$_4$H$_9$	364.0 ± 3.8	1
iso-**C$_4$H$_9$**-*iso*-C$_4$H$_9$	362.3 ± 6.3	1
sec-**C$_4$H$_9$**-*sec*-C$_4$H$_9$	348.5 ± 3.3	1
tert-**C$_4$H$_9$**-*tert*-C$_4$H$_9$	322.6 ± 4.2	1
CH$_3$-*cyclo*-C$_5$H$_9$	358.2 ± 5.0	1
CH$_3$-*cyclo*-C$_6$H$_{11}$	377.0 ± 7.5	1
cyclo-**C$_6$H$_{11}$**-*cyclo*-C$_6$H$_{11}$	369.0 ± 8.4	1
CH$_3$-CH$_2$C≡CH	320.5 ± 5.0	1
CH$_3$-CH$_2$C≡CCH$_3$	308.4 ± 6.3	1
CH$_3$-CH(CH$_3$)C≡CH	305.4 ± 8.4	1
CH$_3$-CH(CH$_3$)C≡CCH$_3$	320.9 ± 6.3	1
CH$_3$-C(CH$_3$)$_2$C≡CH	295.8 ± 6.3	1
CH$_3$-C(CH$_3$)$_2$C≡CCH$_3$	303.3 ± 6.3	1
CH$_3$-CHCH$_2$	426.3 ± 6.3	1
CH$_3$-CH=CCH$_2$	359.8 ± 5.9	1
CH$_3$-cyclopro-en-1-yl	340.6 ± 20.9	1
CH$_3$-CH$_2$CH=CH$_2$	317.6 ± 3.8	1
CH$_3$-CH$_2$C(CH$_3$)=CH$_2$	310.0 ± 4.2	1
CH$_3$-CH(CH$_3$)CH=CH$_2$	302.5 ± 6.3	1
CH$_3$-C(CH$_3$)$_2$CH=CH$_2$	282.4 ± 6.3	1
CH$_3$-*cyclo*-C$_5$H$_7$	299.2 ± 8.4	1
CH$_3$-C$_6$H$_5$	433.9 ± 4.2	20
HCC–C$_6$H$_5$	597.9 ± 5.9	20
C$_2$H$_3$-C$_6$H$_5$	489.1 ± 5.4	20
CH$_3$-CH$_2$C$_6$H$_5$	325.1 ± 4.2	1
CH$_3$-CH(CH$_3$)C$_6$H$_5$	318.8 ± 8.4	1
CH$_3$-C(CH$_3$)$_2$C$_6$H$_5$	303.3 ± 8.4	1
CH$_3$–CH$_2$CHCHPh	295.4	1
CH$_3$–CH(C$_6$H$_5$)$_2$	315.9 ± 6.3	1
CH$_3$–C(CH$_3$)(C$_6$H$_5$)$_2$	290.8 ± 8.4	1
C$_6$H$_5$-C$_6$H$_5$	492.9 ± 6.3	20
C$_6$H$_5$-CH$_2$C$_6$H$_5$	390.8 ± 8.4	20
C$_6$H$_5$CH$_2$-CH$_2$C$_6$H$_5$	272.8 ± 9.2	1
C$_6$H$_5$-CH(C$_6$H$_5$)$_2$	368.2 ± 8.4	20
C$_6$H$_5$-C(C$_6$H$_5$)$_3$	331.4 ± 12.6	20
Ph$_2$CH-CHPh$_2$	247.3 ± 8.4	1
PhCH$_2$-CPh$_3$	234.7 ± 14.6	1
R-R, π-dimer, R = phenalenyl	42	1
R-R, σ-dimer, R = phenalenyl	42.7	1
R-R, R = 9-phenylfluorenyl	63.6	1
CF$_3$-CF$_3$	413.0 ± 5.0	1
CF$_3$-CHF$_2$	399.6 ± 8.4	1
CF$_3$-CClF$_2$	373.6 ± 12.5	1
CF$_3$-CH$_2$F	397.5 ± 8.4	1
CF$_3$-CCl$_3$	332.2 ± 5.4	1
CF$_3$-CHBrCl	377.0 ± 10.5	1
CF$_3$-CH$_2$Br	399.6 ± 8.4	1
CF$_3$-CH$_2$I	408.4 ± 10.5	1
CF$_3$-CH$_3$	429.3 ± 5.0	1
CHF$_2$-CHF$_2$	382.4 ± 15.5	1
CClF$_2$-CClF$_2$	378.7 ± 12.6	1
CF$_2$Cl-CFCl$_2$	358.6 ± 12.6	1
CHF$_2$-CH$_2$F	394.1 ± 16.7	1
CH$_2$F-CH$_2$F	368.2 ± 8.4	1
CHF$_2$-CH$_3$	405.0 ± 8.4	1
CH$_2$F-CH$_3$	388.3 ± 8.4	1
CHClF–CH$_3$	399.6 ± 12.6	1
CF$_2$Br-CHClF	369.4	1
CF$_2$Br-CH$_3$	396.6 ± 15.1	1
CCl$_3$-CCl$_3$	285.8 ± 6.3	1
CCl$_3$-CClF$_2$	282.0 ± 12.6	1
CCl$_3$-CHCl$_2$	303.3 ± 6.3	1
CCl$_3$-CH$_2$Cl	323.8 ± 8.4	1
CCl$_3$-CH$_3$	362.3 ± 6.3	1
CHCl$_2$-CHCl$_2$	326.9 ± 4.1	1
CHCl$_2$-CH$_2$Cl	352.2 ± 5.9	1
CHCl$_2$-CH$_3$	361.3 ± 2.5	1
CHBrCl–CH$_3$	384.5	1
CHClBr–CHClBr	317.1 ± 12.6	1
CH$_2$Cl-CH$_2$Cl	360.7 ± 8.4	1
CH$_2$Cl-CH$_3$	375.7 ± 9.2	1
Br$_3$C-CH$_3$	356.9 ± 12.6	1
Br$_3$C-CBr$_3$	278.7 ± 16.7	1
CHBr$_2$-CH$_3$	372.8	1
CH$_2$Br-CH$_2$Cl	378.2	1
CH$_2$Br-CH$_2$Br	379.9 ± 8.4	1
CH$_2$I-CH$_2$I	387.0 ± 10.5	1
CH$_3$-CH$_2$Br	381.6 ± 8.4	1
CH$_3$-CH$_2$I	384.5 ± 8.4	1
CF$_3$-CF$_2$CF$_3$	424.3 ± 13.6	1
CF$_3$-CF=CF$_2$	420.5	1
CH$_3$-CH$_2$CH$_2$Cl	371.4 ± 2.8	1
CH$_3$-CHClCH$_3$	367.5 ± 2.0	1
CH$_2$Cl-CHClCH$_3$	356.5 ± 8.4	1
CH$_2$Cl-CH$_2$CClH$_2$	369.0 ± 8.4	1
CH$_3$-CCl$_2$CH$_3$	362.8 ± 8.4	1
CH$_2$Br-CHBrCH$_3$	369.4 ± 8.4	1
CH$_2$ClCH$_2$-CHClCH$_3$	364.4 ± 8.4	1
CH$_2$ClCH$_2$-CH$_2$CClH$_2$	369.0 ± 8.4	1
CH$_3$CHBr-CHBrCH$_3$	355.6 ± 8.4	1
CF$_3$-C$_6$H$_5$	470.3 ± 12.6	20
CCl$_3$-C$_6$H$_5$	395.8 ± 8.4	20
CH$_3$-C$_6$F$_5$	439.3	1

Bond	D°_{298}/kJ mol^{-1}	Ref.
CF$_3$-C$_6$F$_5$	435.1	1
CF$_3$-CH$_2$C$_6$H$_5$	365.7 ± 12.6	1
C$_6$F$_5$-C$_6$F$_5$	488.3	1
CF$_3$-CHPh$_2$	352.3 ± 16.7	1
CF$_3$-CPh$_3$	290.8 ± 16.7	1
CF$_2$CF-CFCF$_2$	558.1 ± 12.6	1
CH$_2$FCH$_2$-CPh$_3$	274.9 ± 16.7	1
CHF$_2$CH$_2$-CPh$_3$	264.0 ± 16.7	1
CH$_3$-CH$_2$OH	364.8 ± 4.2	1
CF$_3$-CH$_2$OH	405.4 ± 6.3	1
C$_2$H$_5$-CH$_2$OH	356.9 ± 5.0	1
C$_3$H$_7$-CH$_2$OH	357.3 ± 3.3	1
iso-**C$_3$H$_7$**-CH$_2$OH	354.8 ± 4.2	1
C$_4$H$_9$-CH$_2$OH	355.6 ± 4.2	1
sec-**C$_4$H$_9$**-CH$_2$OH	352.7 ± 4.2	1
iso-**C$_4$H$_9$**-CH$_2$OH	354.0 ± 5.4	1
C$_6$H$_5$-CH$_2$OH	409.0 ± 5.4	20
HOH$_2$C-CH$_2$OH	358.2 ± 6.3	1
NH$_2$CH$_2$-CH$_2$OH	335.6 ± 10.5	1
CH$_3$-CH$_2$OCH$_3$	363.2 ± 5.0	1
CH$_3$OCH$_2$-CH$_2$OCH$_3$	338.9 ± 10.5	1
CH$_3$-C(O)H	354.8 ± 1.7	1
CCl$_3$-C(O)H	309.2 ± 5.0	1
CH$_3$-C(O)F	417.6 ± 6.3	1
CH$_3$-C(O)Cl	367.8 ± 6.3	1
CCl$_3$-C(O)Cl	289.1 ± 6.3	1
CHCl$_2$-C(O)Cl	312.5 ± 8.4	1
CClH$_2$-C(O)Cl	340.2 ± 8.4	1
C$_2$H$_5$–C(O)H	339.8 ± 2.5	26
iso-C$_3$H$_7$–C(O)H	331.2 ± 2.5	26
C$_6$H$_5$-C(O)H	415.5 ± 4.2	20
C$_6$H$_5$-C(O)Cl	424.5 ± 6.3	20
CH$_3$-C(O)CH$_3$	351.9 ± 2.1	1
C$_2$H$_5$-C(O)CH$_3$	347.3 ± 2.9	1
C$_3$H$_7$-C(O)CH$_3$	348.5 ± 2.9	1
iso-**C$_3$H$_7$**-C(O)CH$_3$	340.2 ± 3.8	1
C$_4$H$_7$-C(O)CH$_3$	346.9 ± 5.4	1
tert-**C$_4$H$_9$**-C(O)CH$_3$	329.3 ± 4.2	1
C$_6$H$_5$-C(O)CH$_3$	413.8 ± 4.6	20
C$_6$H$_5$CH$_2$-C(O)CH$_3$	299.7 ± 8.4	1
HC(O)–C(O)H	295.8 ± 6.3	1
ClC(O)–C(O)Cl	292.5 ± 8.4	1
CH$_3$C(O)–C(O)H	302.5 ± 8.4	1
CH$_3$C(O)–C(O)CH$_3$	307.1 ± 4.2	1
C$_6$H$_5$C(O)–C(O)C$_6$H$_5$	288.3 ± 16.7	1
CH$_3$–C(O)OH	384.9 ± 8.4	1
CF$_3$–C(O)OH	370.7 ± 8.4	1
CCl$_3$–C(O)OH	310.5 ± 12.6	1
CClH$_2$–C(O)OH	357.7 ± 8.4	1
CH$_2$Br–C(O)OH	358.2 ± 8.4	1
NH$_2$CH$_2$–C(O)OH	349.4 ± 8.4	1
CH$_3$NHCH$_2$–C(O)OH	300.4 ± 8.4	1
C$_6$H$_5$–C(O)OH	436.8 ± 8.4	20
C$_6$F$_5$–C(O)OH	470.0 ± 10.5	1
HOCH$_2$–C(O)OH	371.5 ± 5.4	1
HOC(O)–C(O)OH	334.7 ± 6.3	1
CH$_3$NHCH$_2$–C(O)OH	301.2 ± 16.7	1
CH$_3$CH(NH$_2$)–C(O)OH	331.4 ± 16.7	1
NH$_2$CH$_2$–CH$_2$C(O)OH	325.5 ± 16.7	1
CN–**CN**	571.9 ± 6.7	1
HC(O)–**CN**	455.2 ± 8.4	1
HC(S)–**CN**	530.1 ± 8.4	1
CF$_3$–**CN**	469.0 ± 4.2	1
CH$_3$–**CN**	521.7 ± 9.2	1
NCC –**CN**	462.3	1
C$_2$H$_5$–**CN**	506.7 ± 7.5	1
CH$_3$–CH$_2$CN	348.1 ± 12.6	1
C$_6$H$_5$–CH$_2$CN	400.8 ± 8.4	20

Bond	D°_{298}/kJ mol^{-1}	Ref.
CH$_3$–CH(CH$_3$)CN	332.6 ± 8.4	1
CH$_3$–C(CH$_3$)$_2$CN	340.6 ± 16.7	1
CH$_3$–C(CH$_3$)(CN)C$_6$H$_5$	250.6	1
(Ph)$_2$(CN)C–C(CN)(Ph)$_2$	109.6	1
(NO$_2$)$_3$C–C(NO$_2$)$_3$	308.8	1
C$_{58}$-C$_2$	955.2 ± 14.5	1

(3) C–halogen BDEs

Bond	D°_{298}/kJ mol^{-1}	Ref.
F–CN	482.8	1
F–CF$_3$	546.8 ± 2.1	1
F–CHF$_2$	533.9 ± 5.9	1
F–CH$_2$F	496.2 ± 8.8	1
F–CF$_2$Cl	507.4 ± 9.0	17
F–CFCl$_2$	482.3 ± 9.0	17
F–CHFCl	462.3 ± 10.0	1
F–CCl$_3$	439.2 ± 4.2	17
F–CFClBr	501.6 ± 6.8	17
F–CH$_2$Cl	465.3 ± 9.6	1
F–CH$_3$	460.2 ± 8.4	1
F–C≡CH	521.3	1
F–C≡CF	519 ± 21	1
F–CF=CF$_2$	546.4 ± 12.6	1
F–CF$_2$CF$_3$	532.2 ± 6.3	1
F–CH$_2$CF$_3$	457.7	1
F–CF$_2$CH$_3$	522.2 ± 8.4	1
F–C$_2$H$_3$	517.6 ± 12.6	1
F–C$_2$H$_5$	467.4 ± 8.4	1
F–C$_3$H$_7$	474.9 ± 8.4	1
F–*iso*-C$_3$H$_7$	483.8 ± 8.4	1
F–*tert*-C$_4$H$_9$	495.8 ± 8.4	1
F–C$_6$H$_5$	532.0 ± 1.2	12
F–C$_6$F$_5$	485 ± 25	1
F–CH$_2$C$_6$H$_5$	412.8 ± 4.2	1
F–COH	497.9 ± 10.5	1
F–COF	510.3	1
F–COCl	484.5	1
F–C(O)CH$_3$	511.7 ± 12.6	1
Cl–CN	422.6 ± 8.4	1
Cl–CF$_3$	362.7 ± 3.9	17
Cl–CHF$_2$	364 ± 8	1
Cl–CH$_2$F	354.4 ± 11.7	1
Cl–CF$_2$Cl	334.3 ± 9.0	17
Cl–CF$_2$Br	354.4 ± 12.9	17
Cl–CFCl$_2$	320.9 ± 8.4	1
Cl–CHFCl	346.0 ± 13.4	1
Cl–CCl$_3$	288.7 ± 4.2	17
Cl–CHCl$_2$	311.1 ± 2.0	1
Cl–CH$_2$Cl	338.0 ± 3.3	1
Cl–CBrCl$_2$	287 ± 10.5	1
Cl–CH$_2$Br	332.8 ± 4.6	1
Cl–CH$_2$I	328.2 ± 6.9	1
Cl–CH$_3$	350.2 ± 1.7	1
Cl–C≡CCl	443 ± 50	1
Cl–C≡CH	435.6 ± 8.4	1
Cl–CH$_2$CN	267.4	1
Cl–CCl=CCl$_2$	383.7	1
Cl–CH=CH$_2$	394.1 ± 3.1	2
Cl–CF=CF$_2$	434.7 ± 8.4	1
Cl–CF$_2$CF$_3$	346.0 ± 7.1	1
Cl–CF$_2$CF$_2$Cl	331.4 ± 20.9	1
Cl–CCl$_2$CF$_3$	307.9	1
Cl–CCl$_2$CCl$_3$	303.8	1
Cl–CHClCCl$_3$	330.5 ± 4.2	1
Cl–CCl$_2$CHCl$_2$	311.7	1
Cl–CHClCH$_3$	327.9 ± 1.8	1
Cl–CH$_2$CH$_2$Cl	345.1 ± 5.0	1
Cl–CHBrCH$_3$	331.8 ± 8.4	1
Cl–CH$_2$CH$_3$	352.3 ± 3.3	1

Bond	D°_{298}/kJ mol^{-1}	Ref.
Cl–CH$_2$CH=CH$_2$	298.3 ± 5.0	1
Cl–C$_3$H$_7$	352.7 ± 4.2	1
Cl–CH$_2$CH$_2$CH$_2$Cl	348.9	1
Cl–*iso*-C$_3$H$_7$	354.8 ± 3.3	10
Cl–CH$_2$CHCH=CH$_2$	342.7	1
Cl–C$_4$H$_9$	350.6 ± 6.3	1
Cl–*sec*-C$_4$H$_9$	350.2 ± 6.3	1
Cl–*tert*-C$_4$H$_9$	351.9 ± 6.3	1
CH$_2$CHCH**Cl**(CH$_3$)	300.0 ± 6.3	1
Cl–C$_5$H$_{11}$	350.6 ± 6.3	1
Cl–C(CH$_3$)$_2$(C$_2$H$_5$)	352.7 ± 6.3	1
Cl–*cyclo*-C$_6$H$_{11}$	360.2 ± 6.5	1
Cl–C$_6$H$_5$	406.4 ± 0.8	12
Cl–C$_6$F$_5$	383.3 ± 8.4	1
Cl–CH$_2$C$_6$H$_5$	299.9 ± 4.3	1
Cl–C(O)Cl	318.8 ± 8.4	1
Cl–COF	376.6	1
Cl–C(O)CH$_3$	354.0 ± 8.4	1
Cl–C(O)CH$_2$CH$_3$	353.3 ± 6.3	1
Cl–C(O)C$_6$H$_5$	341.0 ± 8.4	1
Cl–CH$_2$C(O)C$_6$H$_5$	309	1
Cl–CH$_2$C(O)OH	310.9 ± 2.2	1
Cl–C(O)OC$_6$H$_5$	364	1
Cl–C(NO$_2$)$_3$	302.1	1
Br–CN	364.8 ± 4.2	1
Br–CF$_3$	296.2 ± 1.3	1
Br–CHF$_2$	288.7 ± 8.4	1
Br–CF$_2$Cl	290.5 ± 8.7	17
Br–CCl$_3$	231.4 ± 4.2	1
Br–CH$_2$Cl	277.3 ± 3.6	1
Br–CBr$_3$	242.3 ± 8.4	1
Br–CHBr$_2$	274.9 ± 13.0	1
Br–CH$_2$Br	276.1 ± 5.3	1
Br–CH$_2$I	274.5 ± 7.5	1
Br–CH$_3$	294.1 ± 2.1	1
Br–C≡CH	410.5	1
Br–CH=CH$_2$	338.3 ± 3.1	1
Br–CF$_2$CF$_3$	283.3 ± 6.3	1
Br–CClBrCF$_3$	251.0 ± 6.3	1
Br–CF$_2$CF$_2$Br	282.8 ± 6.7	1
Br–CHClCF$_3$	274.9 ± 6.3	1
Br–CF$_2$CH$_3$	287.0 ± 5.4	1
Br–CH$_2$CH$_2$Cl	292.5 ± 8.4	1
Br–CHClCH$_3$	272.0 ± 8.4	1
Br–C$_2$H$_5$	292.9 ± 4.2	1
Br–CH$_2$CH=CH$_2$	233.3 ± 5.3	17
Br–C$_3$H$_7$	298.3 ± 4.2	1
Br–*iso*-C$_3$H$_7$	295.1 ± 3.3	10
Br–CH$_2$CH$_2$CH$_2$Br	324.7	1
Br–CF$_2$CF$_2$CF$_3$	278.2 ± 10.5	1
CF$_3$CF**Br**CF$_3$	274.2 ± 4.6	1
Br–C$_4$H$_9$	296.6 ± 4.2	1
Br–*sec*-C$_4$H$_9$	300.0 ± 4.2	1
Br–*tert*-C$_4$H$_9$	292.9 ± 6.3	1
Br–C$_6$H$_5$	344.2 ± 1.3	12
Br–C$_6$F$_5$	~328	1
Br–CH$_2$C$_6$H$_5$	239.3 ± 6.3	1
Br–CH$_2$C$_6$F$_5$	225.1 ± 6.3	1
Br–1–C$_{10}$H$_7$	339.7	1
Br–2–C$_{10}$H$_7$	341.8	1
Br–anthracenyl	322.6	1
Br–C(O)CH$_3$	292.0 ± 8.4	1
Br–C(O)C$_6$H$_5$	276.6 ± 8.4	1
Br–CH$_2$C(O)CH$_3$	257.9 ± 10.5	1
Br–CH$_2$C(O)C$_6$H$_5$	271	1
Br–CH$_2$C(O)OH	257.4 ± 3.7	1
Br–C(NO$_2$)$_3$	218.4	1
I–CN	320.1	1

Bond	$D°_{298}$/kJ mol^{-1}	Ref.
I–CF_3	227.2 ± 1.3	1
I–CCl_3	168 ± 42	1
I–CH_2Cl	221.8 ± 4.2	1
I–CH_2Br	219.2 ± 5.4	1
I–CH_2I	216.9 ± 7.9	1
I–CH_3	232.4 ± 0.4	13
I–CH_2CN	187.0 ± 6.3	1
I–CF_2CF_3	219.2 ± 2.1	1
I–CF_2CF_2I	217.6 ± 6.7	1
I–CH_2CF_3	235.6 ± 4.2	1
I–$CHFCClF_2$	202 ± 2	1
I–CF_2CH_3	217.6 ± 4.2	1
I-$CFICH_3$	218.0 ± 4.2	1
CF_3CFICF_3	215.1	1
I–$CH{=}CH_2$	259.0 ± 4.2	1
I–C_2H_5	233.5 ± 6.3	1
I–$CH_2CH{=}CH_2$	185.8 ± 6.3	1
I–C_3H_7	236.8 ± 4.2	1
I–*iso*-C_3H_7	233.1 ± 3.3	10
I–C_4F_9	205.8	1
I–*tert*-C_4H_9	227.2 ± 6.3	1
I–C_6H_5	272.0 ± 4.2	1
I–C_6F_5	<301.7	1
I–$CH_2C_6H_5$	187.8 ± 4.8	1
I–1-naphthyl	274.5 ± 10.5	1
I–2-naphthyl	272.0 ± 10.5	1
I–CH_2CN	187.0 ± 8.4	1
I–CH_2OCH_3	229.4 ± 8.4	1
I–CH_2SCH_3	216.8 ± 6.3	1
I–$C(O)CH_3$	223.0 ± 8.4	1
I–$C(O)C_6H_5$	212.1 ± 8.4	1
I–$CH_2C(O)OH$	197.5 ± 2.7	1
I–$C(NO_2)_3$	144.8	1

(4) O–X BDEs

Bond	$D°_{298}$/kJ mol^{-1}	Ref.
HO–**H**	497.321 ± 0.02	16
FO–**H**	425.1	1
ClO–**H**	393.7	1
BrO–**H**	405	1
IO–**H**	403.3	1
CH_3O–**H**	436.8 ± 2.8	29
CF_3O–**H**	497.1	1
HC≡CO–**H**	443.1	1
C_2H_5O–**H**	434.5 ± 2.9	29
$CH_2{=}CHO$–**H**	355.6	1
CF_3CH_2O–**H**	447.7 ± 10.5	1
C_3H_7O–**H**	430.1 ± 4.4	29
iso-C_3H_7O–**H**	439.6 ± 2.8	29
C_4H_9O–**H**	432.3	1
sec-C_4H_9O–**H**	441.4 ± 4.2	1
tert-C_4H_9O–**H**	445.0 ± 2.9	29
tert-$BuCH_2O$–**H**	436.1	1
$C_6H_5CH_2O$–**H**	442.7 ± 8.8	1
$CH_2(OH)O$–**H**	435 ± 8	25
$CH_3CH(OH)O$–**H**	446.9 ± 6.3	1
$(CH_3)_2C(OH)O$–**H**	450.6 ± 6.3	1
$HC(O)O$–**H**	468.6 ± 12.6	1
$CH_3C(O)O$–**H**	468.6 ± 12.6	1
$C_2H_5C(O)O$–**H**	472.8	1
iso-$C_3H_7C(O)O$–**H**	472.8	1
$C_6H_5C(O)O$–**H**	464.4 ± 16.7	1
OO–**H**	205.73 ± 0.16	16
HOO–**H**	365.71 ± 0.17	16
CH_3OO–**H**	370.3 ± 2.1	1
CF_3OO–**H**	383	1
CH_2FOO–**H**	379	1
CCl_3OO–**H**	386	1
$CHCl_2OO$–**H**	383	1

Bond	$D°_{298}$/kJ mol^{-1}	Ref.
CH_2ClOO–**H**	379	1
CBr_3OO–**H**	383	1
CH_2BrOO–**H**	379	1
C_2H_5OO–**H**	354.8 ± 9.2	1
$CH_3CHClOO$–**H**	377	1
CH_3CCl_2OO–**H**	383	1
$CF_3CHClOO$–**H**	384	1
C_2Cl_5OO–**H**	383	1
iso-C_3H_7OO–**H**	356	1
$CH_2{=}CHCH_2OO$–**H**	372.4	1
tert-C_4H_9OO–**H**	352.3 ± 8.8	1
C_6H_5OO–**H**	384	1
$C_6H_5CH_2OO$–**H**	363	1
$(C_6H_5)_2CHOO$–**H**	370	1
trans-HC(O)OO–**H**	393 ± 13.8	6
cis-HC(O)OO–**H**	406.3 ± 13.8	6
trans-$CH_3C(O)OO$–**H**	381 ± 4	8
cis-$CH_3C(O)OO$–**H**	403 ± 14	8
$CCl_2(CN)OO$–**H**	384	1
$OHCH_2OO$–**H**	368	1
H–ONO	330.7	1
H–OONO	299.2	1
H–ONH_2	318	1
H–ONO_2	426.8	1
H–ONNOH	189	1
H–OPO_2	465.7 ± 12.6	1
H–OSO_2OH	441.4 ± 14.6	1
H–OSiMe3	495	1
$(CH_3)CHNO$-**H**	354.4	1
$(CH_3)_2CNO$-**H**	354.0	1
$(C_6H_5)CHNO$-**H**	368.6	1
PhO–**H**	362.8 ± 2.9	1
α-tocopherol RO-**H**	323.4	1
β-tocopherol RO-**H**	335.6	1
γ-tocopherol RO-**H**	335.1	1
δ-tocopherol RO-**H**	342.8	1
p-$C_6H_5CH_2$-C_6H_4O-**H**	356.2	1
HO–O	274.47 ± 1.6	16
O–O_2	107.483 ± 0.039	16
HO–OH	210.45 ± 0.04	16
HO–OF	199.7 ± 8.4	1
HO–OCl	~146	1
HO–OBr	138.5 ± 8.4	1
FO–OF	199.6	1
ClO–OCl	72.4 ± 2.8	1
IO–OI	74.9 ± 17	1
***trans-perp*-HO**-ONO	≤67.8 ± 0.4	1
***cis-cis*-HO**-ONO	83.3 ± 2.1	1
HO–ONO_2	163.2 ± 8.4	1
HO–OCH_3	189.1 ± 4.2	1
HO–OCF_3	201.3 ± 20.9	1
HO–OC_2H_5	178.7 ± 6.3	1
HO–O-*iso*-C_3H_7	185.8 ± 6.3	1
HO–O-*tert*-C_4H_9	186.2 ± 4.2	1
HO–$OC(O)CH_3$	169.9 ± 2.1	1
HO–$OC(O)C_2H_5$	169.9 ± 2.1	1
CH_3O–OCH_3	167.4 ± 6.3	1
CF_3O–OCF_3	198.7 ± 2.1	1
C_2H_5O–OC_2H_5	166.1	1
C_3H_7O–OC_3H_7	155.2 ± 4.2	1
iso-**C_3H_7O**-O-*iso*-C_3H_7	157.7	1
***sec*-C_4H_9O**-O-*sec*-C_4H_9	152.3 ± 4.2	1
***tert*-BuO**-O-*tert*-Bu	162.8 ± 2.1	1
***tert*-$BuCH_2O$**-OCH_2-*tert*-Bu	152.3	1
$EtC(Me)_2O$–$OC(Me)_2Et$	164.4 ± 4.2	1
$(CF_3)_3CO$–$OC(CF_3)_3$	148.5 ± 4.6	1
Ph_3CO–$OCPh_3$	131.4	1
SF_5O–OSF_5	155.6	1

Bond	$D°_{298}$/kJ mol^{-1}	Ref.
SF_5O–$OOSF_5$	126.8	1
$(CH_3)_3CO$–$OSi(CH_3)_3$	196.6	1
***tert*-BuO**-$OGeEt_3$	192.5	1
***tert*-BuO**-$OSnEt_3$	192.5	1
CF_3OO–OCF_3	126.8 ± 8.4	1
$HC(O)O$–**OH**	199.2 ± 8.4	1
FC(O)O–OC(O)F	96.2	1
$CH_3C(O)O$–ONO_2	131.4 ± 8.4	1
$CH_3C(O)O$–$OC(O)CH_3$	140.2 ± 21	1
$CF_3C(O)O$–$OC(O)CF_3$	125.5	1
$CF_3OC(O)O$–OC(O)F	121.3 ± 4.2	1
$CF_3OC(O)O$–OCF_3	142.3 ± 2.9	1
$CF_3OC(O)O$–$OC(O)OCF_3$	119.2	1
$C_2H_5C(O)O$–$OC(O)\ C_2H_5$	150.6	1
$C_3H_7C(O)O$–$OC(O)\ C_3H_7$	150.6	1
$FS(O)_2O$-$OS(O)_2F$	92–100	1
O–CO	532.18 ± 0.03	16
HO–CF_3	≤482.0 ± 1.3	1
FO–CF_3	408 ± 17	1
HO–CH_3	384.93 ± 0.71	1
HO–C_2H_5	391.2 ± 2.9	1
HO–CH_2CF_3	408.4 ± 8.4	1
HO–$CH_2CH{=}CH_2$	332.6 ± 4.2	1
HO–C_3H_7	392.0 ± 2.9	1
HO–*iso*-C_3H_7	397.9 ± 4.2	1
HO–C_4H_9	389.9 ± 4.2	1
HO–*sec*-C_4H_9	396.1 ± 4.2	1
HO–*iso*-C_4H_9	394.1 ± 4.2	1
HO–*tert*-C_4H_9	398.3 ± 4.2	1
HO-$CH(CH_3)(nC_3H_7)$	398.3 ± 4.2	1
HO–$CH(C_2H_5)_2$	399.2 ± 4.2	1
HO–$C(CH_3)_2(C_2H_5)$	395.8 ± 6.3	1
HO–C_6H_5	470.7 ± 4.2	20
HO–C_6F_5	446.9 ± 9.2	1
HO–$CH_2C_6H_5$	334.1 ± 2.6	1
HO–$C(CH_3)_2C_6H_5$	339.3 ± 6.3	1
cyclo-C_5H_9–**OH**	385.8 ± 6.3	1
1-$C_{10}H_7$–**OH**	468.6 ± 6.3	1
2-$C_{10}H_7$–**OH**	467.8 ± 6.3	1
$(CH_3)_2(NH_2)C$–**OH**	310.4 ± 6.3	1
$CH_3C(O)$-**OH**	459.4 ± 4.2	1
$HOCH_2$-**OH**	411.3	1
CH_3–OCH_3	351.9 ± 4.2	1
ICH_2–OCH_3	373.2 ± 12.6	1
CH_3O–**C_2H_5**	355.2 ± 5.4	1
CH_3O–**$CHClCH_3$**	370.3 ± 8.4	1
CH_3O–**C_3H_7**	358.6 ± 6.3	1
CH_3O-*iso*-**C_3H_7**	360.7 ± 4.2	1
CH_3O–**C_4H_9**	346.0 ± 6.3	1
CH_3O-*tert*-**C_4H_9**	353.1 ± 6.3	1
C_6H_5–OCH_3	425.9 ± 5.9	20
$C_6H_5CH(CH_3)$–OCH_3	313.4 ± 9.6	1
C_6H_5–OC_6H_5	333.9 ± 4.2	20
CH_3–OC(O)H	383.7 ± 12.6	1
HC(O)–OH	457.7 ± 2.1	1
$CH_3C(O)$–OH	459.4 ± 4.2	1
$C_6H_5C(O)$–OH	447.7 ± 10.5	1
HO–$CH_2C(O)OH$	368.2 ± 10.5	1
CH_3–$OC(O)CH_3$	380.3 ± 12.6	1
HC(O)–OCH_3	423.8 ± 4.2	1
$CH_3C(O)$–OCH_3	424.3 ± 6.3	1
$C_6H_5C(O)$–OCH_3	421.3 ± 12.6	1
$C_6H_5C(O)$–OC_6H_5	307.5 ± 8.4	1
CH_3OCH_2–OCH_3	367.5 ± 8.4	1
$CH_3C(O)$–$OC(O)CH_3$	382.4 ± 12.6	1
$C_6H_5C(O)$–$OC(O)C_6H_5$	384.9 ± 16.7	1
CH_3–OOH	300.4 ± 12.6	1
C_2H_5–OOH	332.2 ± 20.9	1

Bond	$D°_{298}$/kJ mol^{-1}	Ref.
C_3H_7–OOH	364.4	1
***iso*-C_3H_7**–OOH	298.3	1
***tert*-C_4H_9**–OOH	309.2 ± 4.2	1
CH_3–$OOCH_3$	292.5 ± 8.4	1
CF_3–$OOCF_3$	361.5 ± 8.4	1
CH_3–**OO**	137.0 ± 3.8	1
CF_3–**OO**	169.0	1
$CClF_2$–**OO**	127.6	1
CCl_2F–**OO**	124.7	1
CH_2Cl–**OO**	122.4 ± 10.5	1
$CHCl_2$–**OO**	108.2 ± 8.2	1
CCl_3–**OO**	92.0 ± 6.4	1
HC(O)–**OOH**	290.0	1
$CH_3C(O)$–$OOC(O)CH_3$	315.1	1
ClO–CF_3	≤369.9 ± 1.3	1
CH_3–**ONO**	245.2	1
C_2H_5–**ONO**	260.2	1
C_3H_7–**ONO**	249.4 ± 6.3	1
iso-C_3H_7–**ONO**	254.4 ± 6.3	1
C_4H_9–**ONO**	256.5 ± 6.3	1
iso-C_4H_9–**ONO**	254.0 ± 6.3	1
sec-C_4H_9–**ONO**	253.6 ± 6.3	1
tert-C_4H_9–**ONO**	252.7 ± 6.3	1
$(C_2H_5)(CH_3)_2C$–**ONO**	254.0 ± 8.4	1
CH_3–**ONO_2**	340.2	1
C_2H_5–ONO_2	344.8	1
CH_3O–CH_2CN	393.3	1
O–**N_2**	167.4 ± 0.4	1
O–**NO**	306.301	16
O–**NO_2**	206.3	1
NO–**NO**	40.6 ± 2.1	1
O_2N–**ONO**$_2$	95.4 ± 1.5	1
cis-HO–**NO**	207.0	1
trans-HO–**NO**	200.64 ± 0.19	1
FO–**NO**	132.5 ± 17	1
cis-ClO–**NO**	127.6 ± 8.4	1
trans-ClO–**NO**	116.6 ± 8.4	1
cis-BrO–**NO**	138.1 ± 8.4	1
trans-BrO–**NO**	121.6 ± 8.4	1
trans-perp-HOO–**NO**	114.2 ± 4	1
CH_3O–**NO**	176.6 ± 3.3	1
C_2H_5O–**NO**	185.4 ± 4.2	1
C_3H_7O–**NO**	179.1 ± 6.3	1
iso-C_3H_7O–**NO**	175.3 ± 4.2	1
C_4H_9O–**NO**	177.8 ± 6.5	1
iso-C_4H_9O–**NO**	175.7 ± 6.5	1
sec-C_4H_9O–**NO**	173.6 ± 3.3	1
tert-C_4H_9O–**NO**	176.1 ± 5.9	1
tert-AmO–**NO**	171.1 ± 0.4	1
C_6H_5O–**NO**	87.0	1
HO–**NO_2**	205.4	1
FO–**NO_2**	131.8 ± 12.6	1
ClO–**NO_2**	110.9	4
BrO–**NO_2**	118.0 ± 6.3	1
IO–**NO_2**	~100	1
CH_3O–**NO_2**	176.1 ± 4.2	1
C_2H_5O–**NO_2**	174.5 ± 4.2	1
C_3H_7O–**NO_2**	177.0 ± 4.2	1
iso-C_3H_7O-**NO_2**	175.7 ± 4.2	1
HOO–**NO_2**	99.2 ± 4.6	1
CH_3OO–**NO_2**	93.5 ± 0.3	21
C_4H_9OO–**NO_2**	93.5 ± 0.6	21
CF_3OO–**NO_2**	105	1
CF_2ClOO–**NO_2**	106.7	1
$CFCl_2OO$–**NO_2**	106.7	1
CCl_3OO–**NO_2**	95.8	1
$CH_3N(O)$-**O**	305.3 ± 4.4	1
$C_6H_5N(O)$-**O**	392 ± 8	1

Bond	$D°_{298}$/kJ mol^{-1}	Ref.
C_5H_5N-**O**	264.9 ± 2.0	1
$C_6H_5N{=}N(O)(C_6H_5)$-**O**	309.4 ± 3.5	1
$C_6H_5(O)N{=}N(O)(C_6H_5)$-**O**	309.4 ± 3.6	1
O–SO	551.1	1
O–SOF_2	513.3	1
O–$SOCl_2$	398.5	1
O–$S(OH)_2$	493.7 ± 25	1
HO–SH	293.3 ± 16.7	1
HO–SOH	313.4 ± 12.6	1
HO–$S(OH)O_2$	384.9 ± 8.4	1
HO–SCH_3	303.8 ± 12.6	1
HO–SO_2CH_3	360.2 ± 12.6	1
F–OH	215.1	1
F–OF	164.1	1
F–OCF_3	200.8 ± 4.2	1
F–OCH_3	>196.6	1
F–ONO_2	143.1	1
Cl–OH	233.5	1
Cl–OCl	142	1
Cl–OCF_3	≤220.9 ± 8.4	1
Cl–OCH_3	200.8	1
Cl–O-*tert*-C_4H_9	198.3	1
Cl–OOCl	91.2	1
Cl–ONO_2	172.0	1
Br–OH	209.6 ± 4.2	1
Br–OBr	125	1
Br–O-*tert*-C_4H_9	183.3	1
Br–ONO_2	143.1 ± 6.3	1
I–OH	213.4	1
I–OI	130.1	1
I-ONO_2	>140.6	1
(5) N–X BDEs		
H–NH	390.7 ± 0.2	16
H–NH_2	449.6 ± 0.2	16
H–NF_2	316.7 ± 10.5	1
H–NNH	254.4	1
H–N_3	≤389	1
H–$N{=}CH_2$	364 ± 25	1
H–NO	199.5	1
H–NHOH	341	1
H–NCO	460.7 ± 2.1	1
H–NCS	≤396.6 ± 4.6	1
H–NCN	347.3 ± 8.4	1
H–NSO	426.8 ± 2.1	24
CH_3N**H_2**	425.1 ± 8.4	1
tert-BuN**H_2**	397.5 ± 8.4	1
$C_6H_5CH_2N$**H_2**	418.4	1
$(CH_3)_2N$**H**	395.8 ± 8.4	1
H–$NHNH(CH_3)$	276 ± 21	1
H–$NHN(CH_3)_2$	356 ± 21	1
N**H_2**CN	414.2	1
(N**H_2**)$_2$C=O	464.4	1
(N**H_2**)$_2$C=S	389.1	1
CH_3CSN**H_2**	380.7	1
PhCSN**H_2**	380.7	1
(PhN**H**)$_2$C=S	364.0	1
(N**H_2**)$_2$C=NH	435.1	1
$Ph_2C{=}N$**H**	489.5	1
H–$N(SiMe_3)_2$	464	1
H–NHPh	375.3	1
C_6H_5N**H**OH	292	1
C_6H_5N**H**$(CONMe_2)$	387.9	1
H–NPh_2	364.8	1
HN–N_2	63	1
ON–N	480.7 ± 0.4	1
ON–NO	8.49 ± 0.12	1
ON–NO_2	42.5	1

Bond	$D°_{298}$/kJ mol^{-1}	Ref.
O_2N–NO_2	57.3 ± 1	1
H_2N–NH_2	277.0 ± 1.3	1
F_2N–NF_2	92.9 ± 12.6	1
H_2N–$NHCH_3$	275.8 ± 8.4	1
H_2N–$N(CH_3)_2$	259.8 ± 8.4	1
H_2N–NHC_6H_5	227.6 ± 8.4	1
H_2N–NO_2	230	1
$H_2NN(CH_3)$–**NO**	179.6	1
$(C_6H_5)_2N$-**NO**	94.6	1
N_3–CH_3	335.1 ± 20.5	1
N_3–C_6H_5	389.9 ± 20.9	20
N_3–$CH_2C_6H_5$	211.3 ± 14.2	1
CH_3–**NC**	413.0 ± 3.3	1
C_2H_5–**NC**	413.4 ± 8.4	1
iso-C_3H_7–**NC**	423.0 ± 8.4	1
tert-C_4H_9–**NC**	399.6 ± 5.4	1
NC–**NO**	204.4	1
CH_3–**NO**	172	1
CF_3–**NO**	167	1
CCl_3–**NO**	125	1
C_2H_5–**NO**	171.5	1
CH_2CHCH_2–**NO**	102.4 ± 3.2	15
iso-C_3H_7–**NO**	152.7 ± 12.6	1
tert-C_4H_9–**NO**	167	1
C_6H_5–**NO**	229.7 ± 1.4	12
C_6F_5–**NO**	211.3 ± 4.2	1
$C_6H_5CH_2$–**NO**	123	1
CH_3–**NO_2**	260.7 ± 2.1	1
C_2H_5–**NO_2**	254.4	1
C_3H_7–**NO_2**	256.5	1
iso-C_3H_7–**NO_2**	259.8	1
C_4H_9–**NO_2**	254.8	1
sec-C_4H_9–**NO_2**	263.2	1
tert-C_4H_9–**NO_2**	258.6	1
C_6H_5–**NO_2**	302.9 ± 4.2	20
$C_6H_5CH_2$–**NO_2**	210.3 ± 6.3	1
$(NO_2)CH_2$–**NO_2**	207.1	1
$(NO_2)_3C$–**NO_2**	176.1	1
CF_3–**NF_2**	280.7	1
$C_6H_5CH_2$–**NF_2**	237.2 ± 14.6	1
CH_3–**NH_2**	356.1 ± 2.1	1
C_2H_5–**NH_2**	352.3 ± 6.3	1
C_3H_7–**NH_2**	356.1 ± 2.9	1
iso-C_3H_7–**NH_2**	357.7 ± 3.8	1
C_4H_9–**NH_2**	356.1 ± 2.9	1
sec-C_4H_9–**NH_2**	359.0 ± 2.9	1
iso-C_4H_9–**NH_2**	254.8 ± 5.0	1
tert-C_4H_9–**NH_2**	355.6 ± 6.3	1
pyridin-2-yl–**NH_2**	431	1
C_6H_5–**NH_2**	436.4 ± 4.2	20
$C_6H_5CH_2$–**NH_2**	306.7 ± 6.3	1
$C_6H_5CH(CH)_3$–**NH_2**	307.5 ± 9.6	1
HC(O)–**NH_2**	421.7 ± 8.4	1
$CH_3C(O)$–**NH_2**	414.6 ± 8.4	1
HS–**NO**	138.9	1
CH_3S–**NO**	104.6 ± 4.2	1
tert-BuS–**NO**	115.1	1
$PhCH_2S$–**NO**	120.5	1
C_6H_5S-**NO**	81.2 ± 5.4	1
SCN–SCN	255.6	1
FSO_2–NF_2	163	1
F–NO	235.26	1
F–NO_2	221.3	1
F–NF_2	254.0	1
F–NH_2	286.6	1
Cl–NO	158.8 ± 0.8	1
Cl–NO_2	141.8 ± 1.3	1
Cl–NF_2	~134	1

Bond	$D°_{298}$/kJ mol^{-1}	Ref.
Cl–NH_2	253.1	1
Br–NO	120.1 ± 0.8	1
Br–NO_2	82.0 ± 7.1	1
Br–NF_2	<227.2	1
I–NO	75.6 ± 4	1
I–NO_2	79.6 ± 4	1

(6) S–X BDEs

Bond	$D°_{298}$/kJ mol^{-1}	Ref.
H–SH	381.18 ± 0.05	1
H–SCH_3	365.7 ± 2.1	1
H–$SCHCH_2$	351.5 ± 8.4	1
H–SC_2H_5	365.3	1
H–SC_3H_7	365.7	1
H–S-*iso*-C_3H_7	369.9 ± 8.4	1
H–S-*tert*-C_4H_9	362.3 ± 9.2	1
H–SOH	330.5 ± 14.6	1
H–$SCOCH_3$	370.7	1
H–SCO Ph	364	1
H–SO_2CH_3	≤397	1
H–$SSCH_3$	330.5 ± 14.6	1
H–SPh	349.4 ± 4.5	1
H–SSH	318.0 ± 14.6	1
H–SCS	73 ± 10	31
H–SSSH	292.9 ± 6.5	1
HS–SH	270.7 ± 8.4	1
FS–SF	362.3	1
ClS–SCl	329.7	1
HS–SCH_3	272.0	1
HS–SPh	255.2 ± 6.3	1
CH_3S–SCH_3	272.8 ± 3.8	1
C_2H_5S–SC_2H_5	276.6	1
MeS–SPh	272.0 ± 6.3	1
C_6H_5S–SC_6H_5	214.2 ± 12.6	1
F_5S–SF_5	305 ± 21	1
HS–CH_3	312.5 ± 4.2	1
HS–C_2H_5	307.9 ± 2.1	1
HS–C_3H_7	310.5 ± 2.9	1
HS–*iso*-C_3H_7	307.1 ± 3.8	1
HS–C_4H_9	309.2 ± 2.9	1
HS–*sec*-C_4H_9	307.5 ± 2.9	1
HS–*iso*-C_4H_9	310.0 ± 4.6	1
HS–*tert*-C_4H_9	301.2 ± 3.8	1
HS–C_6H_5	367.8 ± 6.3	20
HS–$CH_2C_6H_5$	258.2 ± 6.3	1
HS–C(O)H	309.6 ± 8.4	1
HS–C(O)CH_3	307.9 ± 6.3	1
CH_3S–**CH_3**	307.9 ± 3.3	1
HOS–**CH_3**	284.9 ± 12.6	1
CH_3SO–**CH_3**	221.8 ± 8.4	1
$HOSO_2$–**CH_3**	324.3 ± 12.6	1
CH_3SO_2–**CH_3**	279.5	1
F_5S–CF_3	392 ± 43	1
F–SF_5	391.6	1
F–SO_2(F)	379	1
Cl–SF_5	<272	1
Cl–SO_2CH_3	293	1
Cl-SO_2Ph	297	1
Br-SBr	259 ± 17	1
Br–SF_5	<230	1
I-SH	206.7 ± 8.4	1
I-SCH_3	206.3 ± 7.1	1

(7) Si-, Ge-, Sn-, and Pb–X BDEs

Bond	$D°_{298}$/kJ mol^{-1}	Ref.
SiH_3–**H**	383.7 ± 2.1	1
Me_3Si–**H**	396 ± 7	1
H_5Si_2–**H**	373 ± 8	1
$(C_2H_5)_3Si$–**H**	396 ± 4	1
$C_6H_5SiH_2$–**H**	382 ± 5	1
$(CH_3S)_3Si$–**H**	364.0	1
$(iPrS)_3Si$–**H**	376.6	1
$PhMe_2Si$–**H**	377 ± 7	1
Ph_2SiH–**H**	379 ± 7	1
Ph_2MeSi–**H**	361 ± 0	1
SiF_3–**H**	432 ± 5	1
$SiCl_3$–**H**	391	5
$SiBr_3$–**H**	334 ± 8	1
SiH_3–**SiH_3**	321 ± 4	1
SiH_3–Si_2H_5	313 ± 8	1
Ph_3Si–$SiPh_3$	368.2	1
F_3Si–SiF_3	453.1 ± 25	1
SiH_3–**CH_3**	375 ± 5	1
SiF_3–**CH_3**	355.6	1
H_3Si–**NO**	158.2 ± 5.7	1
H_3Si–**PH_2**	331.4	1
SiH_3–**F**	638 ± 5	1
SiH_3–**Cl**	458 ± 7	1
SiH_3–**Br**	376 ± 9	1
SiH_3–**I**	299 ± 8	1
GeH_3–**H**	348.9 ± 8.4	1
Me_3Ge–**H**	364.0	1
Ph_3Ge–**H**	359.8	1
$(CH_3)_3$Ge-Ge$(CH_3)_3$	294.7 ± 14.1	23
$(CH_3)_3$Ge–CH_3	340.5 ± 9.8	23
$(CH_3)_3$Ge–Cl	431.9 ± 10.6	23
$(CH_3)_3$Ge–Br	379.2 ± 9.9	23
Me_3Sn–H	309.5 ± 10.7	11
Ph_3Sn–**H**	318.0 ± 12.6	22
Ph_3Sn–$SnPh_3$	266.9 ± 15.5	22
$(CH_3)_3$Sn-Sn$(CH_3)_3$	252.6 ± 14.8	11
$(CH_3)_3$Sn-Cl	425 ± 7	1
$(CH_3)_3$Sn–CH_3	284.1 ± 9.9	11
$(CH_3)_3$Pb-Pb$(CH_3)_3$	228.4	1
Cl_3Pb-Cl	271 ± 4	1
$(CH_3)_3$Pb-CH_3	238 ± 21	1

(8) P-, As-, Sb-, Bi–X BDEs

Bond	$D°_{298}$/kJ mol^{-1}	Ref.
H_2P–**H**	351.0 ± 2.1	1
CH_3PH–**H**	322.2 ± 12.6	1
H_2P–PH_2	256.1	1
$(C_2H_5)_2$P–P$(C_2H_5)_2$	359.8	1
F_2P–F	549	1
Cl_2P–Cl	356 ± 8	1
Br_2P–Br	<259	1
I_2P–I	217	1
H_2P–SiH_3	331.4	1
H_2As–H	319.2 ± 0.8	1
H_2Sb–H	288.3 ± 2.1	1
F_2Bi–F	435 ± 19	1
Br_2Bi–Br	>297.1	1
$(CH_3)_2Bi$–CH_3	210 ± 7	32

(9) Se- and Te–X BDEs

Bond	$D°_{298}$/kJ mol^{-1}	Ref.
H–SeH	334.93 ± 0.75	1
H–SeC_6H_5	326.4 ± 16.7	1
PhSe–SePh	280 ± 9	1
H–TeH	277.0 ± 5.0	1
H–TeC_6H_5	≤264	1
PhTe–TePh	138.1 ± 12.6	1

(10) Metal-Centered BDEs

Arranged by the Periodic Table

(10.1) Group 1

Bond	$D°_{298}$/kJ mol^{-1}	Ref.
Li-OH	431.0	
Li–C_2H_5	214.6 ± 8.4	1
Li–nC_4H_9	197.9 ± 16.3	1
Na–OH	342.3	1
Na–O_2	<200	1
K–OH	359	1
Rb–OH	356.2 ± 4.2	1
Cs–OH	373	1

(10.2) Group 2

Bond	$D°_{298}$/kJ mol^{-1}	Ref.
BeO–H	469	1
Be(OH)–OH	476	1
MgO–H	441	1
Mg(OH)–OH	349	1
BrMg–CH_3	253	1
BrMg–CH_2CH_3	205	1
BrMg–i-C_3H_7	184	1
BrMg–t-C_4H_9	174	1
BrMg–C_6H_5	289	1
BrMg–$CH_2C_6H_5$	201	1
BrMg–C$(C_6H_5)_3$	180	1
Ca(OH)–OH	409	1
Sr(OH)–OH	407	1
Ba(OH)–OH	443	1

(10.3) Group 3

Bond	$D°_{298}$/kJ mol^{-1}	Ref.
Sc–CH_3	116 ± 29	1
Sc–C_6H_6	60.8	1
La(η^5-**C_5Me_5**)$_2$–CH$(SiMe_3)_2$	278.7 ± 10.5	1
Nd(η^5-**C_5Me_5**)$_2$–CH$(SiMe_3)_2$	236.8 ± 10.5	1
(η^5-**C_5Me_5)$_2$Sm**–H	226.8 ± 12.6	1
(η^5-**C_5Me_5)$_2$Sm**–OCH_3	343.1	1
(η^5-**C_5Me_5)$_2$Sm**–(η^3-C_3H_5)	188.3 ± 6.3	1
(η^5-**C_5Me_5)$_2$Sm**–S-nC_3H_7	295.4 ± 10.0	1
(η^5-**C_5Me_5)$_2$Sm**–N$(CH_3)_2$	201.7 ± 7.5	1
(η^5-**C_5Me_5)$_2$Sm**–SiH$(SiMe_3)_2$	179.9 ± 21	1
(η^5-**C5Me_5)$_2$Sm**–P(Et)$_2$	136.4 ± 8.4	1
(η^5-**C_5Me_5)$_2$Eu**–I	238.9 ± 8.4	1
(η^5-**C_5Me_5)$_2$Yb**–I	256.1 ± 6.3	1
Lu(η^5-**C_5Me_5**)$_2$–CH$(SiMe_3)_2$	279.1 ± 10.5	1
(η^5-**$C_5H_4SiMe_3$)$_3$Th**–H	277 ± 6	1
(η^5-**$C_5H_4SiMe_3$)$_3$Th**–O	371 ± 24	1
(η^5-**C_5H_5)$_3$Th**–CH_3	375 ± 9	1
(η^5-**C_5H_5)$_3$Th**–$CH_2Si(CH_3)_3$	369 ± 12	1
$(C_9H_7)_3$Th–$CH_2C_6H_5$	342 ± 9	1
(η^5-**C_5H_4tBu)$_3$U**–H	249.7 ± 5.7	1
(η^5-**$C_5H_4SiMe_3$)$_3$U**–H	253.7 ± 5.1	1
[HB(3,5-Me_2Pz)$_3$]U(Cl)$_2$–Cl	422.6	1
(η^5-**$C_5H_4SiMe_3$)$_3$U**–I	265.6 ± 4.3	1
(η^5-**C_5H_4tBu)$_3$U**–O	307 ± 9	1
(η^5-**$C_5H_4SiMe_3$)$_3$U**–CO	43.1 ± 0.8	1
$(C_9H_7)_3$U–CH_3	196.3 ± 6.6	1
(η^5-**C_5Me_5)$_2$U(Cl)**–C_6H_5	358 ± 11	1
(η^5-**$C_5H_4SiMe_3$)$_3$U**–THF	41.0 ± 0.8	1

(10.4) Group 4

Bond	$D°_{298}$/kJ mol^{-1}	Ref.
Ti(η^5-**C_5H_5**)$_2$–Cl	471	1
Ti(Cl)(η^5-**C_5H_5**)$_2$–Cl	390	1
Ti(η^5-**C_5Me_5**)$_2$–I	219	1
Ti(η^5-**C_5H_5**)$_2$–CO	174	1
Ti(CO)(η5-**C_5H_5**)$_2$–CO	170	1
Ti–CH_3	174 ± 29	1
Ti(Cl)(η^5-**C_5H_5**)$_2$–CH_3	276	1
Ti(Cl)((η^5-**C_5H_5**)$_2$–C_6H_5	292	1
Ti(C_6H_6)–C_6H_5	308.7	1
Zr(η^5-**C_5Me_5**)$_2$–H	351.0 ± 7.5	1
Zr(H)(η^5-**C_5Me_5**)$_2$–H	326.4 ± 4	1
Zr(η^5-**C_5Me_5**)$_2$–Cl	481.2	1
Zr(η^5-**C_5Me_5**)$_2$–Br	410.0	1
Zr(I)(η^5-**C_5Me_5**)$_2$–I	336.4 ± 2.1	1
Zr(η^5-**C_5Me_5**)$_2$**(Ph)**–OH	482.4 ± 6.3	1
Zr(η^5-**C_5Me_5**)$_2$**(Ph)(OH)**–OH	482.8 ± 10.5	1

Bond	$D°_{298}$/kJ mol^{-1}	Ref.
$Zr(\eta^5\text{-}C_5Me_5)_2(NH_2)H\text{-}NH_2$	421.3 ± 15.1	1
$Zr(\eta^5\text{-}C_5Me_5)_3\text{-}CH_3$	276 ± 10	1
$Zr(\eta^5\text{-}C_5H_5)_2(C_6H_5)\text{-}C_6H_5$	300 ± 10	1
$Zr(\eta^5\text{-}C_5H_5)2(Si(SiMe_3)_3)\text{-}SiMe_3$	188 ± 30	1
$Hf(H)(\eta^5\text{-}C_5Me_5)_2\text{-}H$	346.0 ± 7.9	1
$Hf(\eta^5\text{-}C_5Me_5)(C_4H_9)\text{-}C_4H_9$	274 ± 10	1

(10.5) Group 5

Bond	$D°_{298}$/kJ mol^{-1}	Ref.
$(\eta^5\text{-}C_5H_5)(CO)_3V\text{-}\eta^2H_2$	90 ± 20	1
$(\eta^5\text{-}C_5H_5)(CO)_3V\text{-}CO$	146 ± 21	1
$V\text{-}CH_3$	169 ± 18	1
$V\text{-}C_6H_6$	76.2	1
$V(C_6H_6)\text{-}C_6H_6$	307.8	1
$Nb(\eta^5\text{-}C_5H_5)_2H_3\text{-}TFE$	18.8 ± 1.3	1
$Ta(CH_3)_5\text{-}CH_3$	261 ± 5	1
$(Me_3SiCH_2)_4Ta\text{-}(CH_2SiMe_3)$	184.1 ± 8.4	1

(10.6) Group 6

Bond	$D°_{298}$/kJ mol^{-1}	Ref.
$[Cr(CO)_3(\eta^5\text{-}C_5Me_5)]_2\text{-}Hg$	61.5	1
$[Cr(CO)_3(\eta^5\text{-}C_5Me_5)]\text{-}Hg$	111.3	1
$Cr(CO)_5\text{-}Xe$	37.7 ± 3.8	1
$(CO)_2(PPh_3)(\eta^5\text{-}C_5H_5)Cr\text{-}H$	250.2 ± 4.2	1
$(\eta^5\text{-}C_5H_5)Cr(CO)_3\text{-}H$	257	1
$Cr(CO)_5\text{-}H_2$	78 ± 4	1
$(P(C_6H_{11})_3)_2(CO)_3Cr\text{-}H_2$	30.5 ± 0.4	1
$(\eta^6\text{-}C_6H_6)(CO)_3Cr\text{-}H_2$	251 ± 17	1
$Cr(CO)_5\text{-}N_2$	81 ± 4	1
$(P(C_6H_{11})_3)_2(CO)_3Cr\text{-}N_2$	38.9 ± 0.8	1
$(\eta^5\text{-}C_5Me_5)(CO)_3Cr\text{-}SH$	193	1
$Cr(CO)_5\text{-}CO$	154.0 ± 8.4	1
$Cr(CO)_5\text{-}CH_4$	~33.5 ± 8	1
$Cr\text{-}C_6H_6$	9.6 ± 5.8	1
$Cr(C_6H_6)\text{-}C_6H_6$	268.2 ± 15.4	1
$Cr(CO)_5\text{-}C_6H_6$	57.3 ± 3.3	1
$(P(C_6H_{11})_3)_2(CO)_3Cr\text{-}P(OMe_3)_3$	68.6 ± 2.5	1
$(\eta 5\text{-}C_5H_5)Mo(CO)_3\text{-}H$	290	1
$Mo(\eta^5\text{-}C_5H_5)_2\text{-}H$	246	1
$Mo(H)(\eta^5\text{-}C_5H_5)_2\text{-}H$	256.9 ± 8.4	1
$Mo(CO)_3(\eta^5\text{-}C_5H_5)\text{-}I$	216.7 ± 4.2	1
$(\eta^5\text{-}C_5Me_5)_2Mo\text{-}O$	272	1
$(P(C_6H_{11})_3)_2(CO)_3Mo\text{-}H_2$	27.2 ± 0.8	1
$(P(C_6H_{11})_3)_2(CO)_3Mo\text{-}N_2$	37.7 ± 2.5	1
$Mo(CO)_5\text{-}CO$	169.5 ± 8.4	1
$Mo(CO)_3(\eta^5\text{-}C_5H_5)\text{-}CH_3$	203 ± 8	1
$W(CO)_5\text{-}Xe$	35.1 ± 0.8	1
$W(CO)_3(\eta^5\text{-}C_5H_5)\text{-}H$	303	1
$W(H)(\eta^5\text{-}C_5H_5)_2\text{-}H$	310.9 ± 4.2	1
$W(I)(\eta^5\text{-}C_5H_5)_2\text{-}H$	273 ± 14	1
$(CO)_5W\text{-}H_2$	≥67	1
$(P(C_6H_{11})_3)(CO)_3W\text{-}(\eta^2\text{-}H_2)$	28.5 ± 2.1	1
$W(CO)_5\text{-}CO$	192.5 ± 8.4	1
$W(CH_3)(\eta^5\text{-}C_5H_5)_2\text{-}CH_3$	220.9 ± 4	1

(10.7) Group 7

Bond	$D°_{298}$/kJ mol^{-1}	Ref.
$F_3Mn\text{-}MnF_3$	210.9 ± 2.5	1
$(CO)_5Mn\text{-}Mn(CO)_5$	185 ± 8	1
$(CO)_5Mn\text{-}H$	284.5	1
$(PPh_3)Mn(CO)_4\text{-}H$	286.2	1
$MnBr(CO)_4\text{-}CO$	184	1
$(\eta^5\text{-}C_5H_5)(CO)_2Mn\text{-}CO$	195.8 ± 9.2	1
$Mn\text{-}CH_3$	>35 ± 12	1
$Mn(CO)_5\text{-}CH_3$	187.0 ± 3.8	1
$Mn(CO)_5\text{-}C_6H_5$	207 ± 11	1
$(CO)_5Mn\text{-}Re(CO)_5$	149 ± 11	1
$(\eta^5\text{-}C_5H_5)Mn(CO)_2\text{-}PhMe$	59.4 ± 3.3	1
$(CO)_5Tc\text{-}Tc(CO)_5$	177.5 ± 1.9	1
$(CO)_5Re\text{-}Re(CO)_5$	187 ± 4.8	1
$(CO)_5Re\text{-}H$	313	1
$(CO)_5Re\text{-}CH_3$	220 ± 8	1

(10.8) Group 8

Bond	$D°_{298}$/kJ mol^{-1}	Ref.
$(CO)_4Fe\text{-}Fe(CO)_5$	171.5	1
$(CO)_4Fe(H)_x\text{-}H$	259.4 ± 8.4	1
$(\eta^5\text{-}C_5H_5)(CO)_2Fe\text{-}H$	239	1
$Fe(CO)_3(N_2)\text{-}N_2$	37.7 ± 19.2	1
$Fe(C_2H_2)(CO)_4\text{-}CO$	88 ± 2.3	1
$Fe(CO)_2(PMe_3)\text{-}CO$	>125	1
$Fe(CO)_3(PPh_3)\text{-}CO$	<177.8 ± 5	1
$Fe\text{-}NH_3$	31.4 ± 4.2	1
$Fe\text{-}CH_2$	364 ± 29	1
$Fe\text{-}CH_3$	135 ± 29	1
$Fe(C_2H_4)(CO)_3\text{-}C_2H_4$	89.1 ± 8	1
$Fe\text{-}C_3H_5$	218	1
$Fe\text{-}C_3H_6$	79	1
$Fe(CO)_5\text{-}Ni(CO)_4$	37.7	1
$Fe(CO)_5\text{-}(\eta^3\text{-}C_3H_5)$	176	1
$Fe(C_3H_6)(CO)_3\text{-}C_3H_6$	~79.5	1
$(CO_2)(\eta^5\text{-}C_5H_5)Ru\text{-}H$	272	1
$(PMe_3)_2(\eta^5\text{-}C_5Me_5)Ru\text{-}H$	167.4	1
$(CO)_2(\eta^5\text{-}C_5Me_5)Ru\text{-}Cl$	337.6	1
$(\eta^5\text{-}C_5Me_5)(PMe_3)_2Ru\text{-}Cl$	<138	1
$(\eta^5\text{-}C_5Me_5)(PMe_3)_2Ru\text{-}OH$	204.6	1
$(CO)_4Ru\text{-}CO$	115 ± 1.7	1
$(\eta^5\text{-}C_5Me_5)(PMe_3)_2Ru\text{-}CH_3$	142.3	1
$Os(H)(CO)_4\text{-}H$	326.4	1
$(CO)_4Os\text{-}CO$	133 ± 2.6	1
$Os(C_2H_2)(CO)_4\text{-}CO$	99.5 ± 0.8	1

(10.9) Group 9

Bond	$D°_{298}$/kJ mol^{-1}	Ref.
$(CO)_4Co\text{-}Co(CO)_4$	83 ± 29	1
$(CO)_4Co\text{-}Mn(CO)_5$	96 ± 12	1
$(CO)_4Co\text{-}Re(CO)_5$	113 ± 15	1
$Co(CO)_4\text{-}H$	278	1
$Co(CO)_3(PPh_3)\text{-}H$	272	1
$(CO)_3HCo\text{-}CO$	~54	1
$(\eta^5\text{-}C_5H_5)Co(CO)\text{-}CO$	184.3 ± 4.8	1
$Co\text{-}CH_2$	331 ± 38	1
$Co\text{-}CH_3$	178 ± 8	1
cobalamin-CH_3	150.6	1
cobinamide-iC_4H_9	104	1
Co-C bonds in B_{12}	123.8 ± 6.3	1
$Cl(CO)_2Rh\text{-}Rh(CO)_2Cl$	94.6	1
HRh(m-xylyl)Rh-H	255.6 ± 1.7	1
$(PiPr_3)_2(Cl)Rh\text{-}H_2$	136.0	1
$(PiPr_3)_2(Cl)Rh\text{-}N_2$	69.0	1
$(PiPr_3)_2(Cl)Rh\text{-}CO$	201.7	1
HRh(m-xylyl)Rh-CH_2OH	195.4 ± 7.5	1
$Ir(Cl)(CO)(PMe_3)_2\text{-}H$	251	1
$Ir(H)(\eta^5\text{-}C_5Me_5)(PMe_3)\text{-}H$	310.5 ± 21	1
$Ir(Cl)(H)(CO)(PEt_3)_2\text{-}H$	243.1	1
$Ir(Cl)(H)(CO)(PPh_3)_2\text{-}H$	246.9	1
$(Cl)(CO)(PPh_3)_2Ir\text{-}H_2$	62.8	1
$(Cl)(CO)(PPh_3)_2Ir\text{-}CO$	45.2	1
$Ir(H)(\eta^5\text{-}C_5Me_5)(PMe_3)\text{-}C_6H_5$	321	1

(10.10) Group 10

Bond	$D°_{298}$/kJ mol^{-1}	Ref.
$Ni\text{-}H_2O$	~29	1
$Ni(CO)_3\text{-}N_2$	~42	1
$Ni(CO)_3\text{-}CO$	104.6 ± 8.4	1
$Ni\text{-}CH_3$	208 ± 8	1
$Ni\text{-}C_2H_2$	193 ± 25	1
$Ni\text{-}C_2H_4$	147.3 ± 17.6	1
Ni-propyne	155 ± 21	1
Ni-2-butyne	121 ± 21	1
Pd-OH	213	1
trans-$Pt(PPh_3)_2(Cl)\text{-}H$	307 ± 37	1
$[Ph_2PCH_2]_2MePt\text{-}H$	104.6	1
$[Ph_2PCH_2]_2MePt\text{-}OH$	167.4	1
$[Ph_2PCH_2]_2MePt\text{-}SH$	90.0	1
$Pt(\eta^5\text{-}C_5H_5)(CH_3)_2\text{-}CH_3$	163 ± 21	1
cis-$Pt(PEt_3)_2(CH_3)\text{-}CH_3$	269 ± 13	1

(10.11) Group 11

Bond	$D°_{298}$/kJ mol^{-1}	Ref.
Cu-OH	>406	1
Cu-CO	25 ± 5	1
$Cu\text{-}CH_3$	223 ± 5	1
$Cu\text{-}NH_3$	47 ± 15	1
$Cu(NH_3)\text{-}NH_3$	83.7 ± 4.2	1
$Cu\text{-}C_6H_6$	16.4 ± 12.5	1
$Cu(C_6H_6)\text{-}C_6H_6$	27.0 ± 19.3	1
$Ag\text{-}CH_3$	134.1 ± 6.8	1
$Ag\text{-}NH_3$	8 ± 13	1
$Ag(NH_3)\text{-}NH_3$	62.8 ± 4.2	1
Au-OH	>262	1
$Au\text{-}NH_3$	76 ± 6	1
$Au\text{-}CH_3$	≥191.6	1
$Au\text{-}C_6H_6$	8.4	1

(10.12) Group 12

Bond	$D°_{298}$/kJ mol^{-1}	Ref.
$Zn\text{-}CH_3$	70 ± 10	1
$Zn(CH_3)\text{-}CH_3$	266.5 ± 6.3	1
$Zn\text{-}C_2H_5$	92.0 ± 17.6	1
$Zn(C_2H_5)\text{-}C_2H_5$	219.2 ± 8.4	1
$Cd\text{-}CH_3$	63.6 ± 10.0	1
$Cd(CH_3)\text{-}CH_3$	234.3 ± 6.3	1
$Hg\text{-}CH_3$	22.6 ± 12.6	1
$Hg(CH_3)\text{-}CH_3$	239.3 ± 6.3	1
$ClHg\text{-}CH_3$	280.0 ± 12.6	1
$BrHg\text{-}CH_3$	270 ± 38	1
$IHg\text{-}CH_3$	258.6 ± 12.6	1

(10.13) Group 13

Bond	$D°_{298}$/kJ mol^{-1}	Ref.
$H_3B\text{-}BH_3$	172	1
$H_3B\text{-}NH_3$	130.1 ± 4.2	1
$(CH_3)_3B\text{-}NH_3$	57.7 ± 1.3	1
$F_3B\text{-}N(CH_3)_3$	130 ± 4.6	1
$Cl_3B\text{-}N(CH_3)_3$	127.6	1
$F_2B\text{-}CH_3$	397 - 418	1
Al-OH	547 ± 13	1
$Al\text{-}C_2H_2$	>54	1
$Cl_3Al\text{-}N(CH_3)_3$	198.7 ± 8.4	1
$(CH_3)_3Al\text{-}N(CH_3)_3$	130	1
$(CH_3)_3Al\text{-}O(CH_3)_2$	92	1
$CH_3Ga\text{-}GaH_3$	59 ± 16	1
$(CH_3)_3Ga\text{-}O(C_2H_5)_2$	50.6 ± 0.8	1
$Cl_3Ga\text{-}S(C_2H_5)_2$	235.1	1
$In\text{-}CH_3$	216.3	1
$In(CH_3)\text{-}CH_3$	318.8	1
$In(CH_3)_2\text{-}CH_3$	587.4	1
$In(CH_3)_3\text{-}N(CH_3)_3$	83.3 ± 2.1	1
Tl-OH	330 ± 30	1
Y-CH	399.51 ± 0.29	28

References

1. Luo, Y. R. *Comprehensive Handbook of Chemical Bond Energies*, CRC Press, Boca Raton, FL, 2007. <https://doi.org/10.1201/9781420007282>
2. Shuman, N. S., Ochieng, M. A., Sztáray, B., and Baer, T., *J. Phys. Chem. A* 112, 5647, 2008. <https://doi.org/10.1021/jp8007255>
3. Seetula, J. A., and Eskola, A. J., *Chem. Phys.* 351, 141, 2008. <https://doi.org/10.1016/j.chemphys.2008.04.008>
4. Golden, D. M., *Int. J. Chem. Kinet.* 41, 573, 2009. <https://doi.org/10.1002/kin.20432>
5. Shuman, N. S., Spencer, A. P., and Baer, T., *J. Phys. Chem. A* 113, 9458, 2009.
6. Villano, S. M., Eyet, N., Wren, S. W., Ellison, G. B., Bierbaum, V. M., and Lineberger, W. C., *J. Phys. Chem. A* 114, 191, 2010. <https://doi.org/10.1021/jp907569w>
7. Downs, A. J., Greene, T. M., Johnsen, E., Pulham, C. R., Robertson, H. E., and Wann, D. A., *Dalton Trans.* 39, 5637, 2010. <https://doi.org/10.1039/c000694g>
8. Villano, S. M., Eyet, N., Wren, S. W., Ellison, G. B., Bierbaum, V. M., and Lineberger, W. C., *Eur. J. Mass Spectrom.* 16, 255, 2010. <https://doi.org/10.1255/ejms.1055>
9. Bodi, A., Kercher, J. P., Bond, C., Meteesatien, P., Sztáray, B., and Baer, T., *J. Phys. Chem. A* 110, 13425, 2006. Derived from heats of formation of compounds.
10. Stevens, W. R., Bodi, A., and Baer, T., *J. Phys. Chem. A* 114, 11285, 2010. Derived from heats of formation of compounds.
11. Da'valos, J. Z., Herrero, R., Shuman, N. S., and Baer, T., *J. Phys. Chem. A* 115, 402, 2011. <https://doi.org/10.1021/jp111229d>
12. Stevens, W. R., Ruscic, B., and Baer, T., *J. Phys. Chem. A*, 114, 13134, 2010. <https://doi.org/10.1021/jp107561s>
13. Bodi, A., Shuman, N. S., and Baer, T., *Phys. Chem. Chem. Phys.*, 11, 11013, 2009. <https://doi.org/10.1039/b915400k>
14. Rissanen, M. P., Amedro, D., Krasnoperov, L., Marshall, P., and Timonen, R. S., *J. Phys. Chem. A*, 117, 793, 2013. <https://doi.org/10.1021/jp308621f>
15. Borkar, S., Sztáray, B., and Bodi, A., *Int. J. Mass Spectrom.*, 330-332, 100, 2012. Derived from heats of formation of compounds. <https://doi.org/10.1016/j.ijms.2012.08.014>
16. Ruscic, B., Active Thermochemical Table, <http://atct.anl.gov/index.html>, Dec. 16, 2010.
17. Harvey, J., Tuckett, R. P., and Bodi, A., *J. Phys. Chem. A*, 116, 9696, 2012. Derived from heats of formation of compounds. <https://doi.org/10.1021/jp307941k>
18. Dixon, A. R., Khuseynov, D., and Sanov, A., *Chem. Phys. Lett.* 614, 72-77, 2014. <https://doi.org/10.1016/j.cplett.2014.09.001>
19. Dixon, A. R., Khuseynov, D., and Sanov, A., *J. Phys. Chem. A* 118, 8533-8541, 2014. <https://doi.org/10.1021/jp5024229>
20. Re-derived based on updated $\Delta_f H°(C_6H_5)$ = 337.3±0.6 kJ/mol.
21. McKee, K., et al., *J. Phys. Chem. A* 120, 1408-1420, 2016.
22. Cai, X., et al., *Inorg. Chem.* 55, 10751-10766, 2016. <https://doi.org/10.1021/acs.inorgchem.6b01978>
23. Dávalos, J. Z., et al., *Chem. Phys. Lett.* 684, 298-303, 2017. <https://doi.org/10.1016/j.cplett.2017.06.035>
24. Lehman, J. H., et al., *J. Chem. Phys.* 147, 013943, 2017. <https://doi.org/10.1063/1.4984129>
25. Oliveira, A. M., et al., *J. Chem. Phys.* 145, 124317, 2016. <https://doi.org/10.1063/1.4963225>
26. Harrison, A. W., Kable, S. H., *J. Chem. Phys.* 148, 164308, 2018. <https://doi.org/10.1063/1.5019383>
27. Butkovskaya, N. I., Setser, D. W. *J. Phys. Chem.* A 122, 3735-3746, 2018. <https://doi.org/10.1021/acs.jpca.8b01512>
28. Sevy, A., et al., *J. Chem. Phys.* 151, 024302, 2019. <https://doi.org/10.1063/1.5098330>
29. Ghale, S. B., et al., *J. Phys* Chem. A 122, 7797, 2018. <https://doi.org/10.1021/acs.jpca.8b06851>
30. Speetzen, B., Kass, S. R., *J. Phys Chem. A* 123, 6016, 2019. <https://doi.org/10.1021/acs.jpca.9b04382>
31. Kerr, K. E., et al., *Proc. Combust. Instr.* 37, 373, 2019. <https://doi.org/10.1016/j.proci.2018.06.091>

Enthalpies of Formation of Free Radicals and Other Transient Species

Free radicals and other transient species are of increasing importance in modeling atmospheric, flame, and other chemical systems. Table 4 contains enthalpies of formation of these species. The table is divided into five sections: (1) carbon-centered species; (2) oxygen-centered species; (3) nitrogen-centered species; (4) sulfur-centered species; and (5) Si-, Ge-, Sn-, and Pb-centered species. Column definitions for Table 4 are as follows.

Column heading	Definition
Radical	Molecular formula, name, and other identifying information for free radical or transient species
$\Delta_f H°_{298}$	Enthalphy of formation of radical or transient species, in kJ mol^{-1}
Ref.	Reference number

TABLE 4. Enthalpies of Formation of Free Radicals and Other Transient Species

Radical	$\Delta_f H°_{298}$/kJ mol^{-1}	Ref.
(1) Carbon-Centered Species		
CH	595.8 ± 0.6	1
CH_2 (triplet)	391.2 ± 1.6	1
CH_2 (singlet)	428.8 ± 1.6	1
$^{\cdot}CH_3$, methyl	146.7 ± 0.3	1
$^{\cdot}C_2H$, acetenyl, $CH{\equiv}C^{\cdot}$	567.4 ± 2.1	1
$^{\cdot}C_2H_2$, vinylidene $CH_2{=}C^{\cdot\cdot}$	419.7 ± 16.7	1
$^{\cdot}C_2H_3$, vinyl, $CH_2{=}C^{\cdot}H$	299.6 ± 3.3	1
$^{\cdot}C_2H_5$, ethyl, $CH_3C^{\cdot}H_2$	118.8 ± 1.3	1
$^{\cdot}C_3H_3$, propargyl, $CH{\equiv}CC^{\cdot}H_2$	351.9	2
$^{\cdot}C_3H_3$, $CH_3C{\equiv}C^{\cdot}$	515 ± 13	1
$^{\cdot}C_3H_3$, $CH_2{=}C{=}CH^{\cdot} \leftrightarrow CH{\equiv}CC^{\cdot}H_2$	351.9	2
$^{\cdot}C_3H_3$, cyclopro-2-en-1-yl	497.5 ± 16.7	15
$^{\cdot}C_3H_5$, allyl, $CH_2{=}CHC^{\cdot}H_2$	171.0 ± 3.0	1
$^{\cdot}C_3H_5$, $CH_3CH{=}C^{\cdot}H$	267 ± 6	1
$^{\cdot}C_3H_5$, $CH_3C^{\cdot}{=}CH_2$	242.8±7.1	16
$^{\cdot}C_3H_5$, cyclopropyl	279.9 ± 10.5	1
n-C_3H_7•, *n*-propyl, CH_3CH_2C•H_2	100 ± 2	1
i-C_3H_7 •, *i*-propyl, CH_3C•HCH_3	88 ± 3	1
•*n*-C_4H_3, $CH{\equiv}CCH{=}C$•H	547.3	1
•*i*-C_4H_3, $CH_2{=}C$•$C{\equiv}CH$	499.2	1
$^{\cdot}C_4H_5$, $CH_3C{\equiv}CC^{\cdot}H_2$	304.5	1
$^{\cdot}C_4H_5$, $CH{\equiv}CC^{\cdot}HCH_3$	316.5	1
$^{\cdot}C_4H_5$, $^{\cdot}CH{=}CHCHCH_2$	364.4	1
$^{\cdot}C_4H_5$, $CH_2{=}CHC^{\cdot}CH_2$	313.3	1
$^{\cdot}C_4H_7$, $CH_3CH{=}CHC^{\cdot}H_2$	138.4±5.8	16
$^{\cdot}C_4H_7$, $CH_2{=}CHCH_2C^{\cdot}H_2$	209.2±4.2	16
$^{\cdot}C_4H_7$, $CH_2{=}C(CH_3)C^{\cdot}H_2$	137.9	1
$^{\cdot}C_4H_7$, $CH_2{=}CHC^{\cdot}HCH_3$	136.2	1
C_4H_7, $^{\cdot}CH{=}CHCH_2CH_3$	247.1±0.9	16
$^{\cdot}C_4H_7$, cyclopropylmethyl	213.8 ± 6.7	1

Radical	$\Delta_f H^\circ_{298}$/kJ mol^{-1}	Ref.
$\cdot C_4H_7$, cyclobutyl	219.2 ± 4.2	1
n-$C_4H_9\bullet$, n-butyl, $CH_3CH_2CH_2C\bullet H_2$	77.8 ± 2.1	1
i-$C_4H_9\bullet$, i-butyl, $(CH_3)_2CHC\bullet H_2$	70 ± 4	1
s-$C_4H_9\bullet$, s-butyl, $CH_3C\bullet HCH_2CH_3$	67.8 ± 2.1	1
t-$C_4H_9\bullet$, t-butyl, $(CH_3)_3C\bullet$	48 ± 3	1
$\cdot C_5H_3$, $CH\equiv C$-$C\equiv CC\cdot H_2$	579.1	1
$\cdot C_5H_3$, $(CH\equiv C)_2C\cdot H$	573.2	1
$\cdot C_5H_5$, $CH_2=CHC\equiv CC\cdot H_2$	351.5	1
$\cdot C_5H_5$, $CH_2=CH$-$C\cdot H$-$C\equiv CH$	372.4	1
$\cdot C_5H_5$, cyclopenta-1,3-dien-5-yl	274.1 ± 7.3	1
$\cdot C_5H_7$, $CH_3C\equiv CC\cdot HCH_3$	272.8 ± 9.2	1
$\cdot C_5H_7$, $CH\equiv CC\cdot HC_2H_5$	277.0 ± 8.4	1
$\cdot C_5H_7$, $CH\equiv CC\cdot(CH_3)_2$	257.3 ± 9.2	1
$\cdot C_5H_7$, $CH_2=CHCH=CHC\cdot H_2$	205.0 ± 12.6	1
$\cdot C_5H_7$, $(CH_2=CH)_2C\cdot H$	208.0 ± 4.2	1
$\cdot C_5H_7$, $CH_3CH=C=CHC\cdot H_2$	278.0	1
$\cdot C_5H_7$, spiropentyl	380.7 ± 4.2	1
$\cdot C_5H_7$, cyclopent-1-en-3-yl	160.7 ± 4.2	1
$\cdot C_5H_9$, cyclopentyl	105.9 ± 4.2	1
$\cdot C_5H_9$, $CH_2=CHC\cdot HCH_2CH_3$	109.6 ± 8.4	1
$\cdot C_5H_9$, $CH_3CH=CHC\cdot H(CH_3)$	92	1
$\cdot C_5H_9$, $CH_3CH=C(CH_3)C\cdot H_2$	92.0	1
$\cdot C_5H_9$, $CH_2=CHC\cdot(CH_3)_2$	87.0 ± 8.4	1
$\cdot C_5H_9$, $CH_2=C(CH_3)C\cdot H(CH_3)$	93.7	1
$\cdot C_5H_9$, $CH_2=C(C\cdot H_2)CH_2CH_3$	114.2	1
$\cdot C_5H_9$, $CH_2=CH(CH_2)_2C\cdot H_2$	179.5	1
$nC_5H_{11}\bullet$, $CH_3CH_2CH_2CH_2C\bullet H_2$	54.4	1
$\cdot C_5H_{11}$, $(C_2H_5)_2C\cdot H$	47.0	1
$\cdot C_5H_{11}$, $(nC_3H_7)(CH_3)C\cdot H$	50.2	1
$\cdot C_5H_{11}$, $(CH_3)_3C\cdot CH_2$	36.4 ± 8.4	1
$\cdot C_5H_{11}$, $(C_2H_5)(CH_3)_2C\cdot$	29	1
$\cdot C_6H_5$, phenyl	337.3 ± 30.6	11
$\cdot C_6H_7$, cyclohexa-1,3-dien-5-yl	199.2	1
$\cdot C_6H_7$, cyclohexa-1,4-dien-3-yl	208.0 ± 3.9	5
$\cdot C_6H_9$, $CH_3C\equiv CC\cdot(CH_3)_2$	221.8 ± 9.2	1
$\cdot C_6H_9$, $(CH_2=CH)_2C\cdot(CH_3)$	193.7	1
$\cdot C_6H_9$, cyclohexa-1-en-3-yl	119.7	1
$\cdot C_6H_{11}$, $CH_2=CH(CH_2)_3C\cdot H_2$	158.6	1
$\cdot C_6H_{11}$, $CH_2=CHC\cdot H(CH_2)_2CH_3$	89.0	1
$\cdot C_6H_{11}$, $CH_2=C(CH_3)C\cdot(CH_3)_2$	37.7 ± 6.3	1
$\cdot C_6H_{11}$, $(CH_3)_2C=C(CH_3)C\cdot H_2$	39.7 ± 6.3	1
$\cdot C_6H_{11}$, $(CH_3)_2C=CHC\cdot H(CH_3)$	47.3	1
$\bullet C_6H_{11}$, (Z)-$CH_3CH=CHC\bullet(CH_3)_2$	54.4	1
$\cdot C_6H_{11}$, cyclohexyl	75.3 ± 6.3	1
$nC_6H_{13}\bullet$, $CH_3CH_2CH_2CH_2CH_2C\bullet H_2$	33.5	1
$\bullet C_6H_{13}$, $(nC_4H_9)(CH_3)C\bullet H$	29.3	1
$\cdot C_6H_{13}$, 2-methyl-2-pentyl	3.3 ± 8.4	1
$\cdot C_6H_{13}$, 3-methyl-3-pentyl	14.2	1
$\cdot C_6H_{13}$, 2,3-dimethyl-2-butyl	3.1 ± 10	1
$\cdot C_7H_3$, $(CH\equiv C)_3C\cdot$	784.5	1
$\cdot C_7H_7$, benzyl, $C_6H_5C\cdot H_2$	208.0 ± 1.7	1
$\cdot C_7H_7$, quadricyclolan-5-yl	578.6 ± 5.4	1
$\cdot C_7H_7$, quadricyclolan-4-yl	587.4 ± 5.4	1
$\cdot C_7H_7$, norborna-2,5-dien-7-yl	511.7 ± 7.9	1
$\cdot C_7H_7$, cyclohepta-1,3,5-trien-7-yl	285.3 ± 12.6	1
$\cdot C_7H_9$, $CH_2=CH(CH=CH)_2CC\cdot H_2$	251.0	1
$\cdot C_7H_9$, $(CH_2=CH)_3C\cdot$	274.0	1
$\cdot C_7H_{11}$, norborn-1-yl	136.4 ± 10.5	1
$\cdot C_7H_{11}$, cycloheptenyl	119.2	1
$\cdot C_7H_{13}$, cycloheptyl	50.6 ± 4.2	1
$\cdot C_7H_{13}$, cyclo-$[C\cdot(CH_3)(CH_2)_5]$	22.6	1
$\cdot C_7H_{13}$, cyclo-$[C\cdot(CH_2CH_3)(CH_2)_4]$	47.0	1
$\bullet C_7H_{15}$, $(nC_5H_{11})(CH_3)CH\bullet$	8.4	1
$\cdot C_7H_{15}$, $(CH_3)_2CHCHC\cdot(CH_3)_2$	−21.8 ± 5.2	1
$\cdot C_8H_7$, cubyl	831.0 ± 16.7	1
$\cdot C_8H_7$, $C_6H_5C\cdot=CH_2$	309.6	1
$\cdot C_8H_7$, $C_6H_5CH=CH\cdot$	387.0	1
$\cdot C_8H_9$, $C_6H_5C\cdot H(CH_3)$	175.7 ± 7.5	1
$\cdot C_8H_9$, $C_6H_5CH_2C\cdot H_2$	236.0 ± 7.5	1
$\bullet C_8H_9$, p-$CH_3C_6H_4C\bullet H_2$	167.4	1
$\bullet C_8H_9$, m-$CH_3C_6H_4C\bullet H_2$	167.4	1
$\bullet C_8H_9$, o-$CH_3C_6H_4C\bullet H_2$	167.4	1
$\cdot C_8H_9$, 1-vinyl-cyclohexa-2,4-dienyl	247.7 ± 14.2	1
$\cdot C_8H_9$, 2-vinyl-cyclohexa-2,4-dienyl	249.8 ± 14.2	1
$\cdot C_8H_9$, 3-vinyl-cyclohexa-2,4-dienyl	269.4 ± 14.2	1
$\cdot C_8H_9$, 6-vinyl-cyclohexa-2,4-dienyl	284.5 ± 14.2	1
$\cdot C_8H_{13}$, $CH_2=CHCH=CHC\cdot H(CH_2)_2CH_3$	130.5	1
$\cdot C_8H_{13}$, $CH_2=CHC\cdot H(CH_2)_3CH=CH_2$	130.5	1
$\cdot C_8H_{13}$, bicyclooct-1-yl	92.0	1
$\cdot C_8H_{15}$, $CH_2=CHC\cdot H(CH_2)_4CH_3$	49.8	1
$\bullet C_8H_{15}$, (E)-$CH_3CH=C\bullet(CH_2)_4CH_3$	29.7	1
$\bullet C_8H_{15}$, (Z)-$(CH_3)_2C\bullet CH=CHCH(CH_3)_2$	9.2	1
$\cdot C_8H_{15}$, cyclooctanyl	59.4	1
$\cdot C_8H_{15}$, cyclo-$[C\cdot(CH_2CH_3)(CH_2)_5]$	10.0	1
$\cdot C_9H_7$, indenyl	297.1	1
$\cdot C_9H_9$, indanyl-1	204.2 ± 8.4	1
$\cdot C_9H_{11}$, 2,6-dimethylbenzyl	124.7	1
$\cdot C_9H_{11}$, 3,6-dimethylbenzyl	124.7	1
$\cdot C_9H_{11}$, 3,5-dimethylbenzyl	124.7	1
$\cdot C_9H_{11}$, $C_6H_5C\cdot(CH_3)_2$	133.9 ± 4.2	1
$\bullet C_9H_{11}$, o-$\bullet C_6H_4C_2H_5$	279.5 ± 7.5	1
$\cdot C_9H_{17}$, cyclononanyl	52.3	1
$\cdot C_{10}H_7$, naphth-1-yl	401.7 ± 5.4	1
$\cdot C_{10}H_7$, naphth-2-yl	400.4 ± 5.9	1
$\cdot C_{10}H_{11}$, tetralin-1-yl	154.8 ± 5.0	1
$\cdot C_{10}H_{13}$, 1-phenyl-but-4-yl	192.0	1
$\cdot C_{10}H_{13}$, $(C_6H_5CH_2)(C_2H_5)C\cdot H$	184.5	1
$\cdot C_{10}H_{13}$, $(C_6H_5CH_2CH_2)(CH_3)C\cdot H$	184.5	1
$\cdot C_{10}H_{13}$, $(C_6H_5C\cdot HCH_2CH_2CH_3$	134.7	1
$\cdot C_{10}H_{15}$, 1-adamantyl	51.5	1
$\cdot C_{10}H_{15}$, 2-adamantyl	61.9	1
$\cdot C_{10}H_{19}$, cyclodecanyl	32.2	1
$\cdot C_{11}H_9$, 1-naphthylmethyl	252.7	1
$\cdot C_{11}H_{21}$, cycloundecanyl	7.5	1
$\cdot C_{12}H_{23}$, cyclododecanyl	-38.5	1
$\cdot C_{13}H_9$, 9-fluorenyl	297.5	1
$\cdot C_{13}H_9$, 1-phenalenyl	264.3	17
$\cdot C_{13}H_{11}$, $(C_6H_5)_2C\cdot H$	302.1 ± 4.2	1
$\cdot C_{13}H_{11}$, 9-methyl-9-fluorenyl	268.2	1
$\cdot C_{14}H_{11}$, 9,10-dihydroanthracen-9-yl	261.0	1
$\cdot C_{15}H_{11}$, 9-anthracenylmethyl	337.6	1
$\cdot C_{15}H_{11}$, 9-phenanthrenylmethyl	311.3	1
$\cdot C_{16}H_{31}$, $CH_2=CHC\cdot H(CH_2)_{12}CH_3$	-118.8	1
$\cdot C_{19}H_{15}$, trityl, $(C_6H_5)_3C\cdot$	392.0 ± 8.4	1
$\cdot C_{35}H_{25}$, pentamethylcyclopentadienyl	67.4	1
CF	243.9 ± 3.9	13
CF_2	−199.2 ± 5.6	12
$FC\cdot(O)$	−161.2 ± 8.4	1
CHF	143.0 ± 12.6	1
CClF	31.0 ± 13.4	1
CCl	443.1 ± 13.0	1
CCl_2	226	1

Molecular

Radical	$\Delta_f H°_{298}$/kJ mol^{-1}	Ref.
$ClC\cdot(O)$	−21.8 ± 2.5	1
$CHCl$	326.4 ± 8.4	1
$CClBr$	267	1
CBr	510 ± 63	1
$CHBr$	373 ± 18	1
CBr_2	343.5	1
CI	570 ± 35	1
CI_2	468 ± 60	1
$\cdot CF_3$	−465.7 ± 2.1	1
$\cdot CHF_2$	−249.8 ± 6.6	13
$\cdot CH_2F$	−31.8 ± 4.2	1
$\cdot CClF_2$	−279.0 ± 8.4	1
$\cdot CCl_2F$	−89.0 ± 8.4	1
$\cdot CBrClF$	−35.5 ± 6.3	1
$\cdot CHClF$	−60.7 ± 10.0	1
$\cdot CBrF_2$	−224.7 ± 12.6	1
$\cdot CCl_3$	71.1 ± 2.5	1
$\cdot CHCl_2$	87.1 ± 1.6	1
$\cdot CH_2Cl$	117.2 ± 2.9	1
$\cdot CHBrCl$	140 ± 4	1
$\cdot CHBr_2$	199.1 ± 2.7	3
$\cdot CBr_2Cl$	163 ± 8	1
$\cdot CBrCl_2$	124 ± 8	1
$\cdot CBr_3$	214.8	1
$\cdot CH_2Br$	171.1 ± 2.7	1
$\cdot CI_3$	424.9 ± 2.8	1
$\cdot CHI_2$	314.4 ± 3.3	1
$\cdot CH_2I$	229.7 ± 8.4	1
$\cdot C_2F, FC\equiv C\cdot$	460.0 ± 21.0	1
$\cdot C_2Cl, ClC\equiv C\cdot$	568 ± 26	1
$\cdot C_2F_3, CF_2=C\cdot F$	−192.0 ± 8.4	1
$\cdot C_2F_2H, CF_2=C\cdot H$	−92.9 ± 8.4	1
$\cdot C_2F_2H, CHF=C\cdot F$	−50.6 ± 8.4	1
$\cdot CCl_2H, CHCl=C\cdot Cl$	234.7 ± 8.4	1
$\cdot CClH_2, CH_2=C\cdot Cl$	>251	1
$\cdot C_2F_5, CF_3C\cdot F_2$	−892.9 ± 4.2	1
$\cdot C_2HF_4, CF_3C\cdot HF$	−680.8 ± 9.6	1
$\cdot C_2HF_4, CHF_2C\cdot F_2$	-664.8	1
$\cdot C_2H_2F_3, CF_3C\cdot H_2$	−517.1 ± 8.4	1
$\cdot C_2H_2F_3, CHF_2C\cdot HF$	-456.0	1
$\cdot C_2H_2F_3, CH_2FC\cdot F_2$	-449.8	1
$\cdot C_2H_2F_2Cl, CF_2ClC\cdot H_2$	−310.9 ± 7.0	1
$\cdot C_2H_3F_2, CH_3C\cdot F_2$	−302.5 ± 8.4	1
$\cdot C_2H_3F_2, CHF_2C\cdot H_2$	-285.8	1
$\cdot C_2H_3F_2, CH_2FC\cdot HF$	-238.5	1
$\cdot C_2H_4F, CH_3C\cdot HF$	−70.3 ± 8.4	1
$\cdot C_2H_4F, CH_2FC\cdot H_2$	−59.4 ± 8.4	1
$\cdot C_2H_2F_2Cl, CF_2ClC\cdot H_2$	−315.2 ± 6	1
$\cdot C_2F_4Cl, CF_2ClC\cdot F_2$	-686.0	1
$\cdot C_2HF_3Cl, CClF_2C\cdot HF$	−450.6 ± 12.6	1
$\cdot C_2F_4Cl, CF_3C\cdot FCl$	-728.0	1
$\cdot C_2F_3Cl_2, CF_3C\cdot Cl_2$	-564.0	1
$\cdot C_2F_3ClBr, CF_3C\cdot ClBr$	−504.2 ± 8.4	1
$\cdot C_2Cl, ClC\equiv C\cdot$	534 ± 50	1
$\cdot C_2Cl_3, CCl_2=C\cdot Cl$	190 ± 50	1
$\cdot C_2Cl_5, CCl_3C\cdot Cl_2$	35.1 ± 5.4	1
$\cdot C_2HCl_4, CHCl_2C\cdot Cl_2$	23.4 ± 8.4	1
$\cdot C_2HCl_4, CCl_3C\cdot HCl$	51.0	1
$\cdot C_2H_2Cl_3, CH_2ClC\cdot Cl_2$	26.4	1
$\cdot C_2H_2Cl_3, CHCl_2C\cdot HCl$	46.4	1
$\cdot C_2H_2Cl_3, CCl_3C\cdot H_2$	71.5 ± 8	1
$\cdot C_2H_3Cl_2, CH_3C\cdot Cl_2$	42.5 ± 1.7	1
$\cdot C_2H_3Cl_2, CH_2ClC\cdot ClH$	65.3	1
$\cdot C_2H_3Cl_2, CHCl_2C\cdot H_2$	90.1 ± 0.8	1
$\cdot C_2H_4Cl, CH_3C\cdot HCl$	76.5 ± 1.6	1
$\cdot C_2H_4Cl, CH_2ClC\cdot H_2$	93.0 ± 2.4	1
$\cdot C_2H_3Br_2, CH_3C\cdot Br_2$	140.2 ± 5.4	1
$\cdot C_2H_4Br, BrCH_2C\cdot H_2$	135.1	1
$\cdot C_2H_4Br, CH_3C\cdot HBr$	133.4 ± 3.4	3
$\cdot C_2Br, CBrC\cdot$	623.8	1
$\cdot C_2Br_3, CBr_2C\cdot Br$	385.3	1
$\cdot C_2Br_5, CBr_3C\cdot Br_2$	283.3	1
$\cdot C_3H_6Cl, CH_3CH_2C\cdot HCl$	56.6	1
$\cdot C_3H_6Cl, CH_3C\cdot ClCH_3$	29.9 ± 0.6	1
$\cdot C_3H_6Br, C\cdot H_2CH_2CH_2Br$	120.1 ± 1.3	1
$\cdot C_3H_6Br, CH_3C\cdot HCH_2Br$	96.7 ± 5.9	1
$\cdot C_3H_6Br, CH_3CH_2C\cdot HBr$	107.5 ± 2.5	1
$\cdot C_6F_5$	−547.7 ± 8.4	1
$\cdot CH_3O, HOC\cdot H_2$	−17.0 ± 0.7	1
$\cdot CH_2ClO, HOC\cdot ClH$	−60.7 ± 7.5	1
$\cdot CHCl_2O, HOC\cdot Cl_2$	−94.1 ± 7.5	1
$\cdot CH_2ClO, ClOC\cdot H_2$	135.6 ± 9.2	1
$\cdot CH_2BrO, BrOC\cdot H_2$	151 ± 16	1
$\cdot C_2H_3O, C\cdot H=CHOH$	121 ± 11	1
$\cdot C_2H_3O, C\cdot H_2CHO$	13.0 ± 2	1
$\cdot C_2H_5O, CH_3C\cdot HOH$	-54.7±3.6	16
$\cdot C_2H_4ClO, CH_3C\cdot ClOH$	−108.4 ± 8.8	1
$\cdot C_2H_4ClO, C\cdot H_2CHClOH$	−73.2 ± 8.8	1
$\cdot C_2H_3Cl_2O, C\cdot H_2CCl_2OH$	−99.6 ± 8.8	1
$\cdot C_2H_5O, C\cdot H_2CH_2OH$	−28.9 ± 9.3	16
$\cdot C_2H_3O$, oxiran-2-yl	149.8 ± 6.3	1
$\cdot C_3H_5O, CH_2=CHC\cdot HOH$	0 ± 8.4	1
$\cdot C_3H_7O, CH_3CH_2C\cdot HOH$	−81 ± 4	1
$\cdot C_3H_7O, (CH_3)C\cdot HCH_2OH$	−78.7 ± 8.4	1
$\cdot C_3H_7O, HOCH_2CH_2C\cdot H_2$	−66.9 ± 8.4	1
$\cdot C_3H_7O, (CH_3)_2C\cdot OH$	-96.4	1
$\cdot C_3H_7O, \cdot CH_2CH(OH)CH_3$	−62.8 ± 11.7	1
$\cdot C_4H_9O, \cdot CH_2C(OH)(CH_3)_2$	−147.3 ± 8.4	1
$\cdot C_2H_5O_3, C\cdot H_2OCH_2OOH$	109.6 ± 4.2	1
$PhCH\cdot OH$	29.3 ± 8.4	1
$Ph_2C\cdot OH$	152.3 ± 6.3	1
$\cdot C_2H_5O, CH_3OC\cdot H_2$	0 ± 4.2	1
$\cdot C_3H_7O, CH_3OC\cdot HCH_3$	−57.7 ± 8.4	1
$\cdot C_3H_7O, CH_3CH_2OC\cdot H_2$	−45.2 ± 8.4	1
$\cdot C_3H_7O, C\cdot H_2CH_2OCH_3$	−7.1 ± 4.2	1
$\cdot C_4H_9O, (CH_3)_2CHOC\cdot H_2$	−70.3 ± 7.1	1
$\cdot C_4H_9O, CH_3CH_2OC\cdot HCH_3$	−81.2 ± 4.2	1
$\cdot C_4H_9O, C\cdot H_2CH(CH_3)OCH_3$	−42.3 ± 3.8	1
$\cdot C_4H_9O, (CH_3)_2C\cdot OCH_3$	−72.4 ± 10	1
$\cdot C_5H_{11}O, (CH_3)_3COC\cdot H_2$	−102.5 ± 8.4	1
$\cdot C_2H_5O_2, HOCH_2C\cdot HOH$	−220.1 ± 8.4	1
$C\cdot H=C=O$, ketenyl	177.5 ± 8.8	1
$HC\cdot(O)$	42.5 ± 0.5	1
$C\cdot CO$	381.2 ± 2.1	1
$CH_3C\cdot(O)$	−10.3 ± 1.8	1
$CF_3C\cdot(O)$	-608.7	1
$CH_2ClC\cdot(O)$	−21 ± 12.6	1
$CHCl_2C\cdot(O)$	−17.6 ± 23	1
$CCl_3C\cdot(O)$	-19.7	1
$CH_3CH_2C\cdot(O)$	−31.7 ± 3.4	1
$CH_2CHC\cdot(O)$	88.5	1
$CH_2C(CH_3)C\cdot(O)$	58.6 ± 16.7	1

Radical	$\Delta_f H^\circ_{298}$/kJ mol^{-1}	Ref.
$CH_3CH_2CH_2C^{\cdot}(O)$	54.4 ± 4.2	1
$(CH_3)_2CHC^{\cdot}(O)$	−64.0 ± 3.8	1
$(CH_3)_3CC^{\cdot}(O)$	−102.9 ± 6.3	1
$C_6H_5C^{\cdot}(O)$	116.3 ± 10.9	1
$HC(O)CH_2^{\cdot}$	10.5 ± 9.2	1
$ClC(O)CH_2^{\cdot}$	−52.7 ± 13	1
E-C•HClC(O)H	−27.2 ± 10.5	1
Z-C•HClC(O)H	−23.4 ± 10.5	1
$C^{\cdot}Cl_2C(O)H$	−55.6 ± 14.2	1
E-C•HClC(O)Cl	−88.7 ± 15.1	1
$C^{\cdot}H_2C(O)F$	−273.0 ± 5.8	1
Z-C•HClC(O)Cl	−84.9 ± 13.8	1
$C^{\cdot}Cl_2C(O)Cl$	−101.7 ± 15.5	1
$CH_3C(O)CH_2^{\cdot}$	−34 ± 3	1
$CH_3C(O)C^{\cdot}HCH_3$	−70.3 ± 7.1	1
$CH_3C(O)C^{\cdot}{=}CH_2$	113.4	1
$C_2H_5C(O)C^{\cdot}HCH_3$	−107.5 ± 20.9	1
*i*PrC(O)C•$(CH_3)_2$	−173.6 ± 20.9	1
t$C_4H_9C(O)C$•H_2	−115.5 ± 12.6	1
$PhC(O)C^{\cdot}H_2$	84.5 ± 12.6	1
$PhC(O)C^{\cdot}HCH_3$	41.4 ± 20.9	1
$PhC^{\cdot}HC(O)CH_2Ph$	134.3 ± 20.9	1
$PhC(O)OC^{\cdot}H_2$	-69.9	1
•C(O)OH-*trans*	≥−194.6 ± 2.9	1
•C(O)OH-*cis*	-219.7	1
$^{\cdot}C(O)OCH_3$	-161.5	1
$C^{\cdot}H_2C(O)OH$	−248.9 ± 12.0	1
$C^{\cdot}H(CH_3)C(O)OH$	−293 ± 3	1
$C^{\cdot}H_2C(O)OCH_3$	−236.8 ± 8.4	1
$C^{\cdot}H_2C(O)OCH_2CH_3$	−260.2 ± 12.6	1
$C^{\cdot}H_2C(O)OPh$	-28.0	1
$^{\cdot}C_4H_7O$, tetrahydrofuran-2-yl	−18.0 ± 6.3	1
$^{\cdot}C_4H_8O$, cyclopentanon-2-yl	−41.8 ± 12.6	1
$^{\cdot}C_4H_7O_2$, 1,4-dioxan-2-yl	−131.8 ± 12.6	1
$^{\cdot}C_7H_5O_2$, 2-C(O)OH-$^{\cdot}C_6H_4$	-33.0	1
$^{\cdot}C_7H_5O_2$, 3-C(O)OH-$^{\cdot}C_6H_4$	-35.0	1
$^{\cdot}C_7H_5O_2$, 4-C(O)OH-$^{\cdot}C_6H_4$	-36.0	1
$^{\cdot}CH_3O_2$, $C^{\cdot}H_2OOH$	66.1	1
$^{\cdot}C_2H_5O_2$, $C^{\cdot}H_2CH_2OOH$	46.0 ± 4.6	1
$^{\cdot}C_2H_5O_2$, $CH_3CH^{\cdot}OOH$	26.9	1
$^{\cdot}C_3H_7O_2$, $CH_3CH^{\cdot}CH_2OOH$	10.9 ± 5.4	1
$^{\cdot}C_3H_7O_2$, $C^{\cdot}H_2CH(OOH)CH_3$	2.9 ± 6.3	1
$^{\cdot}C_4H_9O_2$, $(CH_3)_2C^{\cdot}CH_2OOH$	−30.1 ± 5.4	1
$^{\cdot}C_4H_9O_2$, $C^{\cdot}H_2C(CH_3)_2OOH$	−26.8 ± 5.4	1
$^{\cdot}C_2H_3O_3$, $C^{\cdot}H_2C(O)OOH$	-137.9	1
$^{\cdot}CHN_2$	494.5	1
$^{\cdot}CH_2N{=}CH_2$	263.6 ± 12.6	1
$^{\cdot}CH_2NH_2$	151.9 ± 8.4	1
$CH_3C^{\cdot}HNH_2$	111.7 ± 8.4	1
$(CH_3)_2C^{\cdot}NH_2$	69.9 ± 8.4	1
$^{\cdot}CH_2NHCH_3$	156.6	1
$^{\cdot}CH_2N(CH_3)_2$	148.0	1
$(C_2H_5)_2NC^{\cdot}HCH_3$	68.6 ± 2.1	1
$^{\cdot}CH_2N(CH_3)Ph$	266.0 ± 12.6	1
$^{\cdot}CN$	439.3 ± 2.9	1
$^{\cdot}CH_2CN$	252.6 ± 4	1
$CH_3C^{\cdot}HCN$	226.7 ± 12.6	1
$^{\cdot}CH_2CH_2CN$	245.4 ± 12.6	1
$(CH_3)_2C^{\cdot}CN$	190.4 ± 12.6	1
$Ph(CH_3)C^{\cdot}CN$	248.5 ± 8.4	1
$NCC^{\cdot}HCH_2CN$	381.8 ± 12.6	1
$^{\cdot}CH_2NC$	334.7 ± 16.7	1
$^{\cdot}C(O)NC$	210.0 ± 10	1
$^{\cdot}C(O)NH_2$	−15.1 ± 4	1
$C^{\cdot}NN$	569 ± 21	1
$HC^{\cdot}NN$	460 ± 8	1
$H_2C^{\cdot}NN$	292.5 ± 2.1	1
$^{\cdot}CH_2NO$	157 ± 4	1
$^{\cdot}CH_2NO_2$	115.1 ± 12.6	1
$CH_3C^{\cdot}HNO_2$	61.9 ± 12.6	1
$(CH_3)_2C^{\cdot}NO_2$	6.3 ± 12.6	1
$PhC^{\cdot}HNO_2$	169.0 ± 12.6	1
$^{\cdot}C_6H_6N$, 3-NH_2-C_6H_4	320.1	1
$^{\cdot}C_6H_6N$, 4-NH_2-C_6H_4	327.8	1
$^{\cdot}C_6H_4NO_2$, 3-NO_2-C_6H_4	340.6 ± 10.0	1
$^{\cdot}C_6H_4NO_2$, 4-NO_2-C_6H_4	302.7	1
$^{\cdot}C_6H_4CH_3$, 2-Me-C_6H_4	315.1 ± 10.5	1
$^{\cdot}C_6H_4CH_3$, 4-Me-C_6H_4	296.6 ± 9.6	1
$^{\cdot}C_6H_3N_2O_4$, 3,5-$(NO_2)_2$-C_6H_3	305.4	1
$^{\cdot}C_7H_6NO_2$, 2-Me-4-NO_2-C_6H_3	295.4 ± 8.4	1
$^{\cdot}C_4H_3N$, pyrrol-2-yl	385.8	1
$^{\cdot}C_4H_3N$, pyrrol-3-yl	385.8	1
$^{\cdot}C_4H_8N$, pyrrolidin-2-yl	142.7 ± 12.6	1
$^{\cdot}C_5H_4N$, pyrid-2-yl	362.0	1
$^{\cdot}C_5H_4N$, pyrid-3-yl	391.0	1
$^{\cdot}C_5H_4N$, pyrid-4-yl	384.1 ± 8.4	12
$^{\cdot}C_4H_7N_2$, piperad-2-yl	119.7	1
$^{\cdot}C_4H_3N_2$, pyrazin-2-yl	428.0 ± 2.5	12
$^{\cdot}C_4H_3N_2$, pyrimid-2-yl	388.0 ± 12.6	1
•$C_4H_3N_2$, pyrimid-3-yl	529.5 ± 3.3	12
$^{\cdot}C_4H_3N_2$, pyrimid-4-yl	409.0 ± 12.6	1
$^{\cdot}C_4H_3N_2$, pyrimid-5-yl	452.3 ± 3.3	12
•$C_3H_2N_3$, pyrimid-5-yl	459.0 ± 3.3	12
$^{\cdot}CH(NO_2)_2$	139.1	1
$^{\cdot}C(NO_2)_3$	201.2	1
$^{\cdot}CH_2C(NO_2)_3$	150.6	1
$^{\cdot}CH_2CH(NO_2)_2$	103.3	1
$^{\cdot}CH_2CH_2C(NO_2)_3$	133.9	1
$^{\cdot}CH_2N(NO_2)CH_2C(NO_2)_3$	173.6	1
$^{\cdot}CH_2N(NO_2)CH_2CH(NO_2)_2$	126.4	1
$^{\cdot}CH_2CH_2N(NO_2)CH_2C(NO_2)_3$	168.6	1
$^{\cdot}CH_2CH_2ONO_2$	37.7	1
$^{\cdot}CH_2(ONO_2)CHCH_2ONO_2$	-25.5	1
$^{\cdot}CH(CH_2ONO_2)_2$	-57.3	1
$^{\cdot}CH_2C(CH_2ONO_2)_3$	-158.2	1
$^{\cdot}CH_2NHNO_2$	164.8	1
$^{\cdot}CH_2N(NO_2)CH_3$	149.4	1
$^{\cdot}CH_2N(NO_2)_2$	210.5	1
$^{\cdot}CH_2CH_2N(NO_2)CH_3$	144.3	1
$^{\cdot}CH_2N(NO_2)CH_2N(NO_2)CH_3$	202.1	1
$^{\cdot}CH_2N(NO_2)(CH_2)N(NO_2)CH_3$	173.2	1
$C^{\cdot}(S)H$	300.4 ± 8.4	1
$^{\cdot}CH_2SH$	151.9 ± 8.4	1
$^{\cdot}CH_2SCH_3$	136.8 ± 5.9	1
$^{\cdot}CH_2SPh$	268.6 ± 12.6	1
$^{\cdot}CH_2SOCH_3$	23.8 ± 12.6	1
$HOC^{\cdot}(S)S$	110.5	1
$^{\cdot}CH_2SO_2CH_3$	−177.0 ± 12.6	1
$^{\cdot}CH_2SO_2Ph$	−57.3 ± 12.6	1
$PhC^{\cdot}HSO_2CH_3$	−109.2 ± 12.6	1
$PhC^{\cdot}HSO_2Ph$	7 ± 12.6	1
$Ph_2C^{\cdot}SO_2Ph$	102 ± 12.6	1

Radical	$\Delta_f H^\circ_{298}$/kJ mol^{-1}	Ref.
$Ph_2C{\cdot}SPh$	435.6 ± 12.6	1
$NC{\cdot}(O)$	127.2	1
${\cdot}CNH$	207.9 ± 12.1	1
${\cdot}CNO$	323 ± 30	1
${\cdot}CH_2SiMe_3$	–32 ± 6	1
${\cdot}CH_2C(CH_3)_2SiMe_3$	-125	1
${\cdot}CP$	450 ± 9	1
(2) Oxygen-Centered Species		
$HO{\cdot}$	37.50 ± 0.03	14
$FO{\cdot}$	109 ± 10	1
$ClO{\cdot}$	101.63 ± 0.1	1
$BrO{\cdot}$	126.2 ± 1.7	1
$IO{\cdot}$	115.9 ± 5.0	1
$HOO{\cdot}$	12.27 ± 0.16	14
$FOO{\cdot}$	25.4 ± 2	1
$ClOO{\cdot}$	98.0 ± 4	1
$BrOO{\cdot}$	108 ± 40	1
$IOO{\cdot}$	96.6 ± 15	1
$OFO{\cdot}$	378.6 ± 20	1
$OClO{\cdot}$	95.4	1
$ClOOClO{\cdot}$	142 ± 12	1
$ClClO{\cdot}$	90 ± 30	1
$NCO{\cdot}$	184.1	1
$CNO{\cdot}$	386.6	1
$HONNO{\cdot}$	172	1
sym-ClO_3	217.2 ± 21	1
$HSO{\cdot}$	–21.8 ± 2.1	1
$HSOO{\cdot}$	112	1
$CH_3SOO{\cdot}$	76	1
$CF_3SO_2O{\cdot}$	-912	1
$NCO{\cdot}$	184.0	1
$O_2NO{\cdot}$	73.7 ± 1.4	1
$ONOO{\cdot}$	82.8	1
$HOS(O)_2O{\cdot}$	-511.7	1
$CH_3O{\cdot}$	21.0 ± 2.1	1
$CF_3O{\cdot}$	–635.1 ± 7.1	1
$CCl_3O{\cdot}$	–38.1 ± 9.2	1
$CH_2ClO{\cdot}$	–21.3 ± 9.2	1
$CHCl_2O{\cdot}$	–32.2 ± 9.2	1
$CH_2{=}CH\text{-}O{\cdot}$	18.4 ± 1.3	1
$CF_3CHFO{\cdot}$	-851.0	1
$C_2H_5O{\cdot}$	–13.6 ± 3.3	1
$CH_3CHClO{\cdot}$	–61.9 ± 12.1	1
$CH_3CCl_2O{\cdot}$	–91.6 ± 11.7	1
$nC_3H_7O{\bullet}$	–38.3 ± 9.6	16
$iC_3H_7O{\bullet}$	–48.5 ± 3.3	1
$(CH_3)_2CClO{\cdot}$	–108.4 ± 8.4	1
$nC_4H_9O{\bullet}$	-62.8	1
$sC_4H_9O{\bullet}$	-69.5	1
$tC_4H_9O{\bullet}$	–85.8 ± 3.8	1
$CH_2{=}CHCH_2O{\cdot}$	87.0	1
$C_6H_5O{\cdot}$	48.5 ± 2.9	1
o-Cl-$C_6H_4O{\bullet}$	30.6	1
$C_6Cl_5O{\cdot}$	~63	1
p-Cl-$C_6H_4O{\bullet}$	~9	1
o-OH-$C_6H_4O{\bullet}$	-186.3	1
p-OH-$C_6H_4O{\bullet}$	-143.6	1
o-CH_3O-$C_6H_4O{\bullet}$	-125.5	1
p-CH_3O-$C_6H_4O{\bullet}$	-81.1	1
$C_6H_5CH_2O{\cdot}$	136.0 ± 12.6	1

Radical	$\Delta_f H^\circ_{298}$/kJ mol^{-1}	Ref.
$C_{10}H_7O{\cdot}$, naphthoxy-1	165.3	1
$C_{10}H_7O{\cdot}$, naphthoxy-2	174.1	1
$HC(O)O{\cdot}$	–129.7 ± 12.6	1
$FC(O)O{\cdot}$	368.0	1
$CH_3C(O)O{\cdot}$	–179.9 ± 12.6	1
$CF_3C(O)O{\cdot}$	-797.0	1
$CF_3OC(O)O{\cdot}$	–958.1 ± 16.7	1
$C_6H_5C(O)O{\cdot}$	–50.2 ± 16.7	1
$CH_3OO{\cdot}$	20.1 ± 5.1	1
$C_2H_3OO{\cdot}$, $CH_2{=}CHOO{\cdot}$	101.7 ± 1.7	1
$C_2H_5OO{\cdot}$	–25.5 ± 5.6	16
$C_3H_5OO{\cdot}$, $CH_2{=}CHCH_2OO{\cdot}$	88.7	1
$nC_3H_7OO{\cdot}$	-43.6±2.1	16
$iC_3H_7OO{\cdot}$	–65.4 ± 11.3	1
$C_4H_7OO{\cdot}$, $CH_3CH{=}CHCH_2OO{\cdot}$	82.6 ± 5.3	1
$nC_4H_9OO{\cdot}$	-65.7±3.6	16
$tC_4H_9OO{\bullet}$	–109.7 ± 3.9	8
neo-$C_5H_{11}OO{\bullet}$	-115.5	1
$HOCH_2OO{\cdot}$	-162.1	1
$HOOCH_2CH_2OO{\cdot}$	100	1
$C_6H_5CH_2OO{\cdot}$	114.6 ± 4.2	1
c-$C_6H_{11}OO{\bullet}$	–25.0 ± 10.5	1
$(C_2H_5)N(CH3)CHOO{\cdot}$	–36.0 ± 12.6	1
$CF_3OO{\cdot}$	-635.0	1
$CF_2ClOO{\cdot}$	–406.7 ± 14.6	1
$CFCl_2OO{\cdot}$	-213.7	1
$CH_2ClOO{\cdot}$	–5.1 ± 13.6	1
$CHCl_2OO{\cdot}$	–19.2 ± 11.2	1
$CCl_3OO{\cdot}$	–20.9 ± 8.9	1
$CH_3CHClOO{\cdot}$	–54.7 ± 3.4	1
$CH_3CCl_2OO{\cdot}$	–63.8 ± 9.8	1
$CH_3OCH_2OO{\cdot}$	–142.2 ± 4.2	1
$CH_3C(O)CH_2OO{\cdot}$	–142.1 ± 4	1
cis-$HC(O)OO^*$	85.8 ± 14.6	7
trans-$HC(O)OO^*$	95.4 ± 14.6	7
$CH_3C(O)OO{\cdot}$	–154.4 ± 5.8	1
$HOOO{\cdot}$	19.3 ± 0.5	9
$CH_3OOO{\cdot}$	33.4 ± 12.6	1
$C_2H_5OOO{\cdot}$	5.4 ± 12.6	1
(3) Nitrogen-Centered Species		
ON	91.09 ± 0.06	14
NO_2	34.02 ± 0.07	14
N_2O	82.05 ± 0.4	1
NH	358.8 ± 0.2	14
${\cdot}NH_2$	186.1 ± 0.2	14
${\cdot}NNH$	249.5	1
${\cdot}NCO$	131.8	1
${\cdot}N_3$	414.2 ± 20.9	1
${\cdot}N_2H_3$	243.5	1
(*Z*)-N_2H_2	213.0 ± 10.9	1
NF	209.2	1
${\cdot}NF_2$	42.3 ± 8	1
${\cdot}NHF$	112 ± 15	1
NBr	301 ± 21	1
HNO	107.1 ± 2.5	1
FNO	–65.7 ± 1.7	1
ClNO	51.71 ± 0.42	1
BrNO	82.13 ± 0.8	1
INO	112.1 ± 20.9	1
NCO	120.9	1

Radical	$\Delta_f H^\circ_{298}$/kJ mol^{-1}	Ref.
NCN	464.8 ± 2.9	1
NSi	372 ± 63	1
$NH_2C(O)N{\cdot}H$	0.8 ± 12.6	1
$CH_3C(O)N{\cdot}H$	−6.7 ± 12.6	1
$NH_2C(S)N{\cdot}H$	194 ± 12.6	1
$CH_3C(S)N{\cdot}H$	173 ± 12.6	1
PhC(S)N·H	307 ± 12.6	1
HCON·H	49.8 ± 12.6	1
$NH_2C(NH)N{\cdot}H$	250.6 ± 12.6	1
·NHCN	319.2 ± 2.9	1
$CH_2N{\cdot}H$	104.6 ± 12.6	1
$CH_3N{\cdot}H$	184.1 ± 8.4	1
*t*BuN•H	95.4 ± 12.6	1
$C_6H_5CH_2N{\cdot}H$	288.3 ± 12.6	1
$C_6H_5N{\cdot}H$	244.3 ± 4.2	1
$(CH_3)_2N{\cdot}$	158.2 ± 4.2	1
$(C_6H_5)(CH_3)N{\cdot}$	241.0 ± 6.3	1
$(C_6H_5)_2N{\cdot}$	366.0 ± 6.3	1
1-pyrrolyl	269.2 ± 12.6	1
1-pyrazolyl	413.0 ± 2.1	1
carbazol-9-yl	383.3 ± 8.4	1
$CH_3N_2{\cdot}$	215.5 ± 7.5	1
$C_2H_5N_2{\cdot}$	187.4 ± 10.5	1
$iC_3H_7N_2{\bullet}$	146.0 ± 8.4	1
$nC_4H_9N_2{\bullet}$	140.6 ± 8.4	1
$tC_4H_9N_2{\bullet}$	97.5 ± 4.2	1
$(NO_2)HN{\cdot}$	162.3	1
$(CH_3)(NO_2)N{\cdot}$	139.0	1
$(NO_2)_2N{\cdot}$	200.0	1
$CH_3N{\cdot}CH_2N(NO_2)CH_3$	185.4	1
(4) Sulfur-Centered Species		
HOS·	−6.7 ± 2.1	1
HC(O)S·	56.5	1
$HS{\cdot}O_2$	-221.8	1
$HOS{\cdot}O_2$	-384.9	1
NCS·	300 ± 8	1
HS·	143.0 ± 0.8	1
$CH_3S{\cdot}$	124.7 ± 1.7	1
$C_2H_5S{\cdot}$	101	1
$nC_3H_7S{\bullet}$	80	1
$iC_3H_7S{\bullet}$	74.9 ± 8.4	1
$tC_4H_9S{\bullet}$	43.9 ± 8.4	1
$C_6H_5S{\cdot}$	242.7 ± 4.6	1
$C_6Cl_5S{\cdot}$	~184	1
$C_6H_5CH_2S{\cdot}$	246	1
$CH_3S{\cdot}O$	−67 ± 10	1
$CH_3S{\cdot}O_2$	-239.3	1
HSS·	115.5 ± 14.6	1
$CH_3SS{\cdot}$	68.6 ± 8.4	1
$C_2H_5SS{\cdot}$	43.5 ± 8.4	1
$iC_3H_7SS{\bullet}$	13.8 ± 8.4	1
$tC_4H_9SS{\bullet}$	−19.2 ± 8.4	1
HOC(S)S·	110.5 ± 4.6	1
HC(O)S·	56.5	1
SF	13.0 ± 6.3	1
SF_2	−296.7 ± 16.7	1
SF_3	−503.0 ± 33.5	1
SF_4	−763.2 ± 20.9	1
SF_5	−879.9 ± 15.1	1
ClS·	156.5 ± 16.7	1

Radical	$\Delta_f H^\circ_{298}$/kJ mol^{-1}	Ref.
SN	263.6 ± 105	1
SCl	156.5 ± 16.7	1
(5) Si-, Ge-, Sn-, Pb-Centered Species		
SiF	−20.1 ± 12.6	1
SiF_2	−638 ± 6	1
$\cdot SiF_3$	−987 ± 20	1
SiCl	198.3 ± 6.7	1
$SiCl_2$	−169 ± 3	1
$\cdot SiCl_3$	321 ± 8	6
SiBr	235 ± 46	1
$SiBr_2$	46 ± 8	1
$\cdot SiBr_3$	−201.7 ± 63	1
SiI	313.8 ± 42	1
SiI_2	92.5 ± 8.4	1
$\cdot SiI_3$	35.3 ± 63	1
SiH	376.6 ± 8.4	1
$SiH_2(^1A_1)$	273 ± 2	1
$SiH_2(^3B_1)$	360.7	1
$\cdot SiH_3$	200.4 ± 2.5	1
$MeSi{\cdot}H_2$	141 ± 6	1
$Me_2Si{\cdot}H$	78 ± 6	1
$Me_3Si{\cdot}$	15 ± 7	1
$\cdot Si_2H_3$	~402	1
$H_3SiSi{\cdot}H_2$	234 ± 6	1
$C_6H_5Si{\cdot}H_2$	274	1
$H_3SiSi{\cdot}H$	312 ± 8	1
MeSi·	302.2	1
MeSi·H	202 ± 6	1
$Me_2Si{\cdot\cdot}$	135 ± 8	1
SiN	313.8 ± 42	1
$\cdot GeH_3$	221.8 ± 8.4	1
GeF	−71 ± 10	1
GeF_2	−574 ± 20	1
$\cdot GeF_3$	−807 ± 50	1
GeCl	69 ± 18	1
$GeCl_2$	−171 ± 5	1
$\cdot GeCl_3$	−268 ± 50	1
GeBr	137 ± 5	1
$GeBr_2$	−61 ± 5	1
$\cdot GeBr_3$	−119 ± 50	1
GeI	211 ± 25	1
GeI_2	50.2 ± 4	1
$\cdot GeI_3$	42 ± 50	1
SnF	−95 ± 7.2	1
SnF_2	−511 ± 9.2	1
$\cdot SnF_3$	−647 ± 50	1
SnCl	35 ± 12	1
$SnCl_2$	−202.6 ± 7.1	1
$\cdot SnCl_3$	−292 ± 50	1
SnBr	76 ± 12	1
$SnBr_2$	−119 ± 2.8	1
$\cdot SnBr_3$	−159 ± 50	1
SnI	173 ± 12	1
SnI_2	−8.1 ± 4.2	1
$\cdot SnI_3$	−8 ± 50	1
$\cdot Sn(CH_3)_3$	116.6 ± 9.7	10
$\cdot Sn(C_6H_5)_3$	518.8 ± 21	1
PbH	236.2 ± 19.2	1
PbF	−80.3 ± 10.5	1
PbF_2	−435.1 ± 8.4	1

Radical	$\Delta_f H^\circ_{298}$/kJ mol^{-1}	Ref.
$\cdot PbF_3$	–490 ± 60	1
PbCl	15.1 ± 50	1
$PbCl_2$	–174.1 ± 1.3	1
$\cdot PbCl_3$	–178 ± 80	1
PbBr	70.9 ± 42	1
$PbBr_2$	–104.4 ± 6.3	1
$\cdot PbBr_3$	–104 ± 80	1
PbI	107.4 ± 37.7	1
PbI_2	–3.2 ± 4.2	1
$\cdot PbI_3$	22 ± 80	1

References

1. Luo, Y. R., *Comprehensive Handbook of Chemical Bond Energies*, CRC Press, Boca, Raton, FL, 2007. <https://doi.org/10.1201/9781420007282>
2. Wheeler, S. E., Robertson, K. A., Allen, W. D., Schaefer III, H. F., Bomble, Y. J., and Stanton, J. F., *J. Phys. Chem. A* 111, 3819, 2007. <https://doi.org/10.1021/jp0684630>
3. Seetula, J. A., and Eskola, A. J., *Chem. Phys.* 351, 141, 2008. <https://doi.org/10.1016/j.chemphys.2008.04.008>
4. Denis, P. A., and Ornellas, F. R., *J. Phys. Chem. A* 113, 499, 2009. <https://doi.org/10.1021/jp808795e>
5. Gao, Y., DeYonker, N. J., Garrett III, E. C., Wilson, A. K., Cundari, T. R., and Marshall, P., *J. Phys. Chem. A* 113, 6955, 2009. <https://doi.org/10.1021/jp901314y>
6. Shuman, N. S., Spencer, A. P., and Baer, T., *J. Phys. Chem. A* 113, 9458, 2009.
7. Villano, S. M., Eyet, N., Wren, S. W., Ellison, G. B., Bierbaum, V. M., and Lineberger, W. C., *J. Phys. Chem. A* 114, 191, 2010. <https://doi.org/10.1021/jp907569w>
8. Shuman, N. S., Bodi, A., and Baer, T., *J. Phys. Chem. A* 114, 232, 2010. <https://doi.org/10.1021/jp907767c>
9. Le Picard, S. D., Tizniti, M., Canosa, A., Sims, I. R., and Smith, I. W. M., *Science* 328, 1258, 2020. <https://doi.org/10.1126/science.1184459>
10. Da'valos, J. Z., Herrero, R., Shuman, N. S., and Baer, T., *J. Phys. Chem. A* 115, 402, 2011. <https://doi.org/10.1021/jp111229d>
11. Stevens, W. R., Ruscic, B., and Baer, T., *J. Phys. Chem. A*, 114, 13134, 2010. <https://doi.org/10.1021/jp107561s>
12. Wren, S. W., Vogelhuber, K. M., Garver, J. M., Kato, S., Sheps, L., Bierbaum, V. M., and Lineberger, W. C., *J. Am. Chem. Soc.*, 134, 6584, 2012. <https://doi.org/10.1021/ja209566q>
13. Harvey, J., Bodi, A., Tuckett, R. P., and Sztáray, B., *Phys. Chem. Chem. Phys.*, 14, 3935, 2012. <https://doi.org/10.1039/c2cp23878k>
14. Ruscic, B., Active Thermochemical Table, <http://atct.anl.gov/index.html>, Dec. 16, 2010.
15. Tian, Z., Lis, L., and Kass, S. R., *J. Org. Chem.* 78, 12650, 2013. <https://doi.org/10.1021/jo402263v>
16. Bukke, S. M., Simmie, J. M., and Curran, H. J., *J. Phys. Chem. Ref. Data* 44, 013101, 2015. <https://doi.org/10.1063/1.4902535>
17. Rossi, M. J., *Int. J. Chem. Kinet.* 40, 395–415, 2008.

Bond Dissociation Energies of Some Organic Molecules

D°_{298}(R-X)/kJ mol^{-1} of some organic compounds are listed below that allow comparison among similar bonds in different molecular environments. All data are from Tables 1 and 3.

TABLE 5. Bond Dissociation Energies (in kJ mol^{-1}) of Some Organic Molecules

	X=H	F	Cl	Br	I	OH	OCH_3	NH_2	NO	CH_3	$COCH_3$	CF_3	CCl_3
R=H	435.7799	569.658	431.361	366.16	298.26	497.10	440.2	450.08	199.5	439.3	374.0	445.2	392.5
CH_3	439.3	460.2	350.2	294.1	238.9	384.93	351.9	356.1	172.0	377.4	351.9	429.3	362.3
C_2H_5	420.5	447.4	352.3	292.9	233.5	391.2	355.2	355.2	171.5	370.3	347.3		
i-C_3H_7	410.5	483.8	354.8	295.1	233.1	397.9	360.7	357.7	152.7	369.0	340.2		
t-C_4H_9	400.4	495.8	351.9	292.9	227.2	398.3	353.1	355.6	167	363.6	329.3		
C_6H_5	433.9	525.5	399.6	336.4	272.0	470.7	425.9	436.4	226.8	433.9	413.8	470.3	395.8
$C_6H_5CH_2$	375.5	412.8	299.9	239.3	187.8	334.1		306.7	123	325.1	299.7	365.7	
CCl_3	392.5	439.2	288.7	231.4	168				125	362.3		332.2	285.8
CF_3	445.2	546.8	362.7	296.2	227.2	482.0[a]			167	429.3		413.0	332.2
C_2F_5	429.7	532.2	346.0	283.3	219.2							424.3	
CH_3CO	374.0	511.7	354.0	292.0	223.0	459.4	424.3	414.6		351.9	307.1		
CN	528.5	482.8	422.6	364.8	320.1				204.4	521.7		469.0	
C_6F_5	487.4	485	383.3	328[b]	301.7[a]	446.9			211.3	439.3		435.1	

[a] Upper limit.
[b] Approximate.

Bond Dissociation Energies in Diatomic Cations

Table 6 has bond dissociation energies for many diatomc cations. From thermochemistry, we have

$$D^\circ_{298}(A^+{-}B) = \Delta_f H^\circ(A^+) + \Delta_f H^\circ(B) - \Delta_f H^\circ(AB^+)$$
$$= D^\circ_{298}(A - B) + IP(A) - IP(AB)$$

Thus, $D^\circ_{298}(A^+{-}B)$ may be derived using Table 1 and the ionization potentials of species A and AB. Column definitions for Table 6 are as follows.

Column heading	Definition
$A^+{-}B$	Molecular formula of diatomic cation; cations are listed alphabetically by dissociating atom (A^+ in **boldface**)
D°_{298}	Dissociation energy, in kJ mol^{-1}
Ref.	Reference number

TABLE 6. Bond Dissociation Energies in Diatomic Cations

A$^+$–B	$D°_{298}$/kJ mol^{-1}	Ref.
Ac$^+$–S	465 ± 48	20
Ag$^+$–Ag	167.9 ± 8.7	1
Ag$^+$–Cl	32 ± 30	1
Ag$^+$–F	24 ± 27	1
Ag$^+$–H	43.5 ± 5.9	1
Ag$^+$–O	123 ± 5	1
Ag$^+$–S	138.8 ± 11.6	21
Al$^+$–Al	140.2 ± 1.8	38
Al$^+$–Ar	15.47	1
Al$^+$–Ca	148.5	1
Al$^+$–Cl	173 ± 42	1
Al$^+$–F	314 ± 21	1
Al$^+$–Kr	5.54	1
Al$^+$–O	166.7 ± 12.0	1
Al$^+$–Se	114 ± 49	1
Am$^+$–O	560 ± 28	14
Am$^+$–S	334 ± 27	20
Ar$^+$–Ar	130.323 ± 0.087	1
Ar$^+$–He	2.9 ± 0.8	1
Ar$^+$–Ne	7.5 ± 0.8	1
As$^+$–As	364 ± 22	1
As$^+$–H	290.8 ± 3.0	1
As$^+$–O	495	1
As$^+$–P	367 ± 59	1
As$^+$–S	433.2 ± 12.5	1
Au$^+$–Al	170 ± 30	1
Au$^+$–Au	215.99 ± 20.26	41
Au$^+$–B	329 ± 50	1
Au$^+$–Be	401 ± 29	1
Au$^+$–C	311.5 ± 7.7	4
Au$^+$–F	79	1
Au$^+$–Ge	292 ± 24	1
Au$^+$–H	209.2 ± 10.6	18
Au$^+$–I	230~280	1
Au$^+$–O	111.8 ± 7.7	19
Au$^+$–Xe	130 ± 13	1
B$^+$–Ar	32.7	1
B$^+$–B	187	1
B$^+$–Br	164 ± 21	1
B$^+$–C	284 ± 58	1
B$^+$–Cl	308 ± 21	1
B$^+$–F	460 ± 10	1
B$^+$–H	198 ± 5	1
B$^+$–O	326 ± 48	1
B$^+$–Pt	314 ± 98	1
B$^+$–Se	298 ± 98	1
B$^+$–Si	365 ± 15	1
Ba$^+$–Ar	11.85	1
Ba$^+$–Br	418 ± 10	1
Ba$^+$–Cl	468.2 ± 10	1
Ba$^+$–D	245.2 ± 9.6	1
Ba$^+$–F	640 ± 29	1
Ba$^+$–I	335 ± 10	1
Ba$^+$–O	441.4 ± 15	1
Be$^+$–Ar	49.0 ± 2.4	1
Be$^+$–Au	410 ± 29	1
Be$^+$–Be	196 ± 0.5	8
Be$^+$–Cl	417 ± 50	1
Be$^+$–F	575 ± 98	1
Be$^+$–H	307.3 ± 5.0	1
Be$^+$–O	362.0 ± 6.2	1
Bi$^+$–Bi	199 ± 10	1
Bi$^+$–O	174	1
Bi$^+$–S	179 ± 50	1
Bi$^+$–Se	184 ± 29	1
Bi$^+$–Te	125 ± 50	1
Bi$^+$–Tl	100 ± 42	1
Bk$^+$–O	610	1
Br$^+$–Br	318.858 ± 0.024	1
Br$^+$–C	451.5 ± 8.6	1
Br$^+$–Cl	303.000 ± 0.048	1
Br$^+$–F	251.5 ± 12.6	1
Br$^+$–H	379.26 ± 2.89	1
Br$^+$–O	365.7 ± 3.1	1
C$^+$–Ar	72.3	1
C$^+$–Br	398 ± 8.6	1
C$^+$–C	601.9 ± 19.3	1
C$^+$–Cl	614	1
C$^+$–F	721 ± 40	1
C$^+$–H	397.848 ± 0.013	1
C$^+$–N	524.5 ± 4.2	1
C$^+$–O	810.7 ± 0.8	1
C$^+$–P	587 ± 50	1
C$^+$–S	706.6 ± 2.1	1
C$^+$–Se	587 ± 50	1
Ca$^+$–Al	144.7	1
Ca$^+$–Ar	12.99 ± 0.60	1
Ca$^+$–Au	306 ± 29	1
Ca$^+$–Br	417.6 ± 10	1
Ca$^+$–Ca	104.1	1
Ca$^+$–Cl	433.4 ± 12	1
Ca$^+$–F	556.5 ± 8.4	1
Ca$^+$–H	284.2 ± 10	1
Ca$^+$–I	293.7 ± 10.8	1
Ca$^+$–Kr	18.60 ± 0.72	1
Ca$^+$–Ne	4.95 ± 0.06	1
Ca$^+$–O	348 ± 5	1
Ca$^+$–Xe	25.38 ± 0.96	1
Cd$^+$–Cd	122.5 ± 10	1
Cd$^+$–H	179.5	1
Ce$^+$–Au	278 ± 34	1
Ce$^+$–Br	341.0	1
Ce$^+$–C	254 ± 96	1
Ce$^+$–Ce	207 ± 42	1
Ce$^+$–Cl	429.5	1
Ce$^+$–F	586 ± 63	1
Ce$^+$–H	215.0 ± 8.7	52
Ce$^+$–I	295.5	1
Ce$^+$–Ir	530 ± 96	1
Ce$^+$–N	494 ± 63	1
Ce$^+$–O	852 ± 15	1
Ce$^+$–Pd	255 ± 53	1
Ce$^+$–Pt	467 ± 96	1
Ce$^+$–Rh	423 ± 96	1
Ce$^+$–S	524 ± 59	1
Cl$^+$–Ar	169	1
Cl$^+$–Cl	385.746 ± 0.096	6
Cl$^+$–D	457.284 ± 0.017	1
Cl$^+$–F	291 ± 10	1
Cl$^+$–H	452.714 ± 0.018	1
Cl$^+$–N	650 ± 10	1
Cl$^+$–O	468.0 ± 2.1	1
Cm$^+$–O	670 ± 38	14
Cm$^+$–S	455 ± 16	20
Co$^+$–Ar	52.89 ± 0.06	1
Co$^+$–Br	>289	1
Co$^+$–C	394.01 ± 1.25	44
Co$^+$–Cl	285 ± 12	1
Co$^+$–Co	269	1
Co$^+$–D	199.6 ± 5.8	1
Co$^+$–H	195 ± 6	1
Co$^+$–He	16.4 ± 0.4	1
Co$^+$–I	211.7 ± 8.4	1
Co$^+$–Kr	68.37 ± 0.18	1
Co$^+$–Ne	12.8 ± 0.4	1
Co$^+$–O	317.3 ± 4.8	1
Co$^+$–S	288.3 ± 8.7	1
Co$^+$–Si	317.1 ± 6.7	1
Co$^+$–Xe	85.7 ± 6.8	1
Cr$^+$–Ar	31.7 ± 3.9	1
Cr$^+$–C	277 ± 24	1
Cr$^+$–Cl	>211	1
Cr$^+$–Cr	129	1
Cr$^+$–D	135 ± 9	1
Cr$^+$–F	279 ± 42	1
Cr$^+$–H	136 ± 9	1
Cr$^+$–He	7.8 ± 0.4	1
Cr$^+$–Ne	9.5 ± 0.4	1
Cr$^+$–O	359	1
Cr$^+$–S	258.6 ± 16.4	1
Cr$^+$–Si	203 ± 15	1
Cr$^+$–Xe	71.9 ± 10.0	1
Cs$^+$–Ar	8.2	1
Cs$^+$–Br	60.5 ± 10	1
Cs$^+$–Cl	107.4 ± 10	1
Cs$^+$–Cs	62.6 ± 9.6	1
Cs$^+$–F	43.7 ± 10	1
Cs$^+$–He	5.1	1
Cs$^+$–I	29.3 ± 10	1
Cs$^+$–Kr	15.1	1
Cs$^+$–Na	48.1 ± 4.2	1
Cs$^+$–Ne	6.11	1
Cs$^+$–O	59	1
Cs$^+$–Rb	68.3 ± 10	1
Cs$^+$–Xe	14.7	1
Cu$^+$–Ar	51.9 ± 6.8	1
Cu$^+$–Cl	91 ± 10	1
Cu$^+$–Cu	155.2 ± 7.7	1
Cu$^+$–F	117 ± 21	1
Cu$^+$–Ge	231 ± 23	1
Cu$^+$–H	93 ± 13	1
Cu$^+$–Kr	24.3 ± 0.8	1
Cu$^+$–O	133.9 ± 11.6	1
Cu$^+$–S	203.3 ± 14.5	1
Cu$^+$–Si	260 ± 8	1
Cu$^+$–Xe	102.1 ± 5.8	1
D$^+$–D	263.4405 ± 0.0003	1
Dy$^+$–Br	324.2	1
Dy$^+$–Cl	407.9	1
Dy$^+$–Cu	196 ± 42	1

A^+–B	$D°_{298}$/kJ mol^{-1}	Ref.
Dy$^+$–F	535 ± 24	1
Dy$^+$–I	279.9	1
Dy$^+$–O	597 ± 15	1
Er$^+$–Br	315.8	1
Er$^+$–Cl	406.7	1
Er$^+$–F	546 ± 34	1
Er$^+$–I	271.6	1
Er$^+$–O	583 ± 15	1
Es$^+$–O	470 ± 60	1
Eu$^+$–Ag	85 ± 50	1
Eu$^+$–Au	252 ± 97	1
Eu$^+$–Br	333.8	1
Eu$^+$–Cl	430.7	1
Eu$^+$–F	543 ± 29	1
Eu$^+$–I	290.7	1
Eu$^+$–O	393 ± 15	1
Eu$^+$–S	257 ± 32	1
F$^+$–Ar	161.1	1
F$^+$–F	325.393 ± 0.096	1
F$^+$–He	181.62 ± 0.08	1
F$^+$–Kr	152.4	1
F$^+$–Xe	188	1
Fe$^+$–Ar	14.2 ± 7.7	1
Fe$^+$–Br	>293	1
Fe$^+$–C	415.7 ± 1.8	33
Fe$^+$–Cl	>343	1
Fe$^+$–Co	259 ± 21	1
Fe$^+$–Cr	209 ± 29	1
Fe$^+$–Cu	222 ± 29	1
Fe$^+$–D	227	1
Fe$^+$–F	360–423	1
Fe$^+$–Fe	272	1
Fe$^+$–H	211.2 ± 9.6	1
Fe$^+$–I	>239	1
Fe$^+$–Kr	33.5 ± 6.7	1
Fe$^+$–N	485	1
Fe$^+$–Nb	285 ± 21	1
Fe$^+$–Ni	268 ± 21	1
Fe$^+$–O	343.3 ± 1.9	13
Fe$^+$–S	295.2 ± 5.8	1
Fe$^+$–Sc	200 ± 21	1
Fe$^+$–Si	277 ± 9	1
Fe$^+$–Ta	301 ± 21	1
Fe$^+$–Ti	251 ± 25	1
Fe$^+$–V	314 ± 21	1
Fe$^+$–Xe	46.0 ± 5.8	1
Ga$^+$–Bi	62 ± 98	1
Ga$^+$–Br	56.5 ± 16	1
Ga$^+$–Cl	86 ± 21	1
Ga$^+$–F	136 ± 15	1
Ga$^+$–Ga	126.3	1
Ga$^+$–I	41.6 ± 15	1
Ga$^+$–O	46 ± 50	1
Ga$^+$–Sb	38 ± 96	1
Ga$^+$–Te	19 ± 29	1
Gd$^+$–C	310.5 ± 17.4	34
Gd$^+$–Cd	122.5 ± 10	1
Gd$^+$–H	214.1 ± 6.8	37
Gd$^+$–O	745.7 ± 9.7	34
Ge$^+$–Br	398 ± 42	1
Ge$^+$–C	223 ± 31	1
Ge$^+$–Cl	473 ± 50	1
Ge$^+$–F	565 ± 21	1
Ge$^+$–Ge	274 ± 10	1
Ge$^+$–H	377 ± 84	1
Ge$^+$–O	344 ± 21	1
Ge$^+$–S	283 ± 21	1
Ge$^+$–Se	234 ± 10	1
Ge$^+$–Si	268 ± 21	1
Ge$^+$–Te	233 ± 19	1
H$^+$–D	261.1021 ± 0.0002	1
H$^+$–H	259.4659 ± 0.0002	1
He$^+$–H	123.9	1
He$^+$–He	229.687 ± 0.019	1
Hf$^+$–C	311.5 ± 2.9	10
Hf$^+$–H	207.3 ± 7.7	17
Hf$^+$–O	670.4 ± 10.6	10
Hf$^+$–S	392.2 ± 1.9	49
Hg$^+$–Ar	22.2 ± 1.2	1
Hg$^+$–H	207	1
Hg$^+$–Hg	134	1
Hg$^+$–Kr	37.9 ± 1.3	1
Hg$^+$–Xe	72.2 ± 1.3	1
Ho$^+$–Ag	155 ± 61	1
Ho$^+$–Au	250 ± 60	1
Ho$^+$–Br	320.6	1
Ho$^+$–Cl	410.3	1
Ho$^+$–Cu	214 ± 35	1
Ho$^+$–F	542 ± 50	1
Ho$^+$–Ho	88 ± 96	1
Ho$^+$–I	270.4	1
Ho$^+$–O	551 ± 25	1
I$^+$–Br	184.90 ± 0.02	1
I$^+$–Cl	247.5 ± 0.4	1
I$^+$–F	262.9 ± 2.1	1
I$^+$–H	304.70 ± 0.10	1
I$^+$–I	262.90 ± 0.04	1
I$^+$–O	316.3 ± 10.5	1
In$^+$–Br	65.2 ± 12.6	1
In$^+$–Cl	193 ± 21	1
In$^+$–F	148 ± 50	1
In$^+$–I	51.5 ± 21	1
In$^+$–In	81 ± 30	1
In$^+$–S	171 ± 50	1
In$^+$–Sb	73 ± 50	1
In$^+$–Se	118 ± 50	1
In$^+$–Te	41 ± 50	1
Ir$^+$–C	635.8 ± 4.8	3
Ir$^+$–D	302.8 ± 5.8	1
Ir$^+$–H	305.7 ± 5.8	1
Ir$^+$–O	411.0 ± 8.7	24
K$^+$–Ar	14 ± 7	1
K$^+$–Br	35.7 ± 10.5	1
K$^+$–Cl	51 ± 19	1
K$^+$–He	6.00	1
K$^+$–I	18 ± 45	1
K$^+$–K	83.86 ± 0.15	1
K$^+$–Kr	15.8	1
K$^+$–Li	59.9 ± 5.9	1
K$^+$–Na	58.69 ± 0.08	1
K$^+$–Ne	7.79	1
K$^+$–O	13	1
K$^+$–Xe	19.5	1
Kr$^+$–Ar	55.31 ± 0.14	1
Kr$^+$–H	464	1
Kr$^+$–He	2.1 ± 0.8	1
Kr$^+$–Kr	110.967 ± 0.033	1
Kr$^+$–N	136.9 ± 13	1
Kr$^+$–Ne	3.8 ± 0.8	1
La$^+$–Au	436 ± 97	1
La$^+$–Br	425.9	1
La$^+$–C	427 ± 33	1
La$^+$–Cl	503.6	1
La$^+$–F	589 ± 34	1
La$^+$–H	243 ± 9	1
La$^+$–I	392.4	1
La$^+$–Ir	356 ± 97	1
La$^+$–O	875 ± 25	1
La$^+$–Pt	522 ± 78	1
La$^+$–Rh	345 ± 97	1
La$^+$–S	629 ± 96	1
La$^+$–Si	277.0 ± 9.6	1
Li$^+$–Ar	33 ± 14	1
Li$^+$–Bi	91 ± 50	1
Li$^+$–Br	41.8 ± 10.6	1
Li$^+$–Cl	66 ± 15	1
Li$^+$–F	7 ± 21	1
Li$^+$–He	10.66	1
Li$^+$–I	51.1 ± 6.3	1
Li$^+$–Kr	48.1	1
Li$^+$–Li	137.3 ± 6.3	1
Li$^+$–Ne	15.32	1
Li$^+$–O	38.9 ± 9.6	1
Li$^+$–Sb	129.6 ± 13.9	1
Li$^+$–Xe	56.4	1
Lu$^+$–Br	86.1	1
Lu$^+$–Cl	180.6	1
Lu$^+$–F	376.8	1
Lu$^+$–H	204 ± 15	1
Lu$^+$–I	40.7	1
Lu$^+$–O	524 ± 15	1
Lu$^+$–Si	107 ± 13	1
Mg$^+$–Ar	19.20	1
Mg$^+$–Au	267 ± 29	1
Mg$^+$–Cl	327 ± 6.5	1
Mg$^+$–D	203.6 ± 0.8	1
Mg$^+$–F	477 ± 50	1
Mg$^+$–H	190.8 ± 5.8	1
Mg$^+$–Kr	25.39	1
Mg$^+$–Mg	125	1
Mg$^+$–Ne	4.9 ± 0.6	1
Mg$^+$–O	245.2 ± 10	1
Mg$^+$–Xe	53.74	1
Mn$^+$–Cl	>211	1
Mn$^+$–F	321 ± 24	1
Mn$^+$–H	202.5 ± 5.9	1
Mn$^+$–I	>211	1
Mn$^+$–Mn	129	1
Mn$^+$–O	246 ± 5	45
Mn$^+$–S	247 ± 23	1
Mn$^+$–Se	165 ± 50	1
Mo$^+$–C	442.7 ± 13.5	1
Mo$^+$–F	376 ± 29	1

A^+–B	D°_{298}/kJ mol^{-1}	Ref.
Mo^+–H	170 ± 6	1
Mo^+–Mo	449.4 ± 1.0	1
Mo^+–O	488.2 ± 1.9	1
Mo^+–S	355.1 ± 5.8	1
Mo^+–Xe	>53.1 ± 6.8	1
N^+–Ar	208.4 ± 9.6	1
N^+–F	584 ± 42	1
N^+–H	≥435.67 ± 0.77	1
N^+–N	843.85 ± 0.10	1
N^+–O	115	1
Na^+–Ar	19 ± 8	1
Na^+–Br	58.2 ± 10.6	1
Na^+–Cl	20.3 ± 10	1
Na^+–He	7.55	1
Na^+–I	64.9 ± 3.0	1
Na^+–Kr	~24.9	1
Na^+–Li	95.8 ± 3.9	1
Na^+–Na	98.64 ± 0.29	1
Na^+–Na	6.4	1
Na^+–Ne	~9.04	1
Na^+–O	37 ± 19	1
Na^+–Xe	~28.6	1
Nb^+–Ar	40.87 ± 0.13	1
Nb^+–C	523.76 ± 0.29	40
Nb^+–Fe	>251	1
Nb^+–H	220 ± 7	1
Nb^+–Nb	576.8 ± 9.6	1
Nb^+–O	688 ± 11	1
Nb^+–S	501.7 ± 20.3	1
Nb^+–V	404.7 ± 0.2	1
Nb^+–Xe	73.28 ± 0.12	1
Nd^+–Au	267 ± 84	1
Nd^+–Br	352.9	1
Nd^+–C	255.55 ± 28.95	42
Nd^+–Cl	441.4	1
Nd^+–F	309.6	1
Nd^+–I	596 ± 32	1
Nd^+–O	706.14 ± 9.65	42
Ne^+–H	1239	1
Ne^+–He	13.0 ± 0.8	1
Ne^+–Ne	125.29 ± 1.93	1
Ni^+–Ar	53.9	1
Ni^+–Br	>289	1
Ni^+–C	335.1 ± 0.4	33
Ni^+–Cl	192 ± 4	1
Ni^+–D	166.0 ± 7.7	1
Ni^+–F	≥456	1
Ni^+–H	158.1 ± 7.7	1
Ni^+–He	12.4 ± 0.4	1
Ni^+–I	>297	1
Ni^+–Ne	9.9 ± 0.4	1
Ni^+–Ni	208	1
Ni^+–O	275.9 ± 7.7	1
Ni^+–S	241.0 ± 3.9	1
Ni^+–Si	326 ± 6.7	1
Np^+–F	730 ± 100	1
Np^+–O	760 ± 10	14
Np^+–S	491 ± 52	20
O^+–Ar	33.8	1
O^+–F	301.8 ± 8.4	1
O^+–H	485.26 ± 0.01	46
O^+–N	1050.64 ± 0.13	1
O^+–O	647.75 ± 0.17	1
Os^+–C	596.1 ± 13.5	35
Os^+–H	223.8 ± 6.6	28
Os^+–O	478 6 ± 1.9	26
Os^+–S	437.9 ± 16.4	50
P^+–C	512 ± 42	1
P^+–Cl	289	1
P^+–F	490.6 ± 8.4	1
P^+–H	329.6 ± 2.1	1
P^+–N	483 ± 21	1
P^+–O	791.3 ± 8.4	1
P^+–P	481 ± 50	1
P^+–S	606 ± 34	1
Pa^+–O	800 ± 50	14
Pa^+–S	525 ± 86	20
Pb^+–Br	260 ± 63	1
Pb^+–Cl	285 ± 63	1
Pb^+–F	347 ± 32	1
Pb^+–O	247 ± 8.4	1
Pb^+–Pb	214 ± 29	1
Pb^+–S	227.7 ± 10.6	12
Pb^+–Se	169.4 ± 6.3	1
Pb^+–Te	163 ± 63	1
Pd^+–C	528 ± 5	1
Pd^+–H	208.4 ± 8.7	1
Pd^+–O	145 ± 11	1
Pd^+–Pd	197 ± 29	1
Pd^+–S	197 ± 6	1
Pd^+–Si	289 ± 50	1
Pr^+–Au	317 ± 81	1
Pr^+–Br	357.7	1
Pr^+–Cl	445.0	1
Pr^+–F	557 ± 63	1
Pr^+–I	317.0	1
Pr^+–O	796 ± 15	1
Pt^+–Ar	36.4 ± 8.7	1
Pt^+–B	398 ± 105	1
Pt^+–C	530.5 ± 4.8	1
Pt^+–Cl	249.8 ± 14.5	1
Pt^+–H	275 ± 5	1
Pt^+–N	326.9 ± 9.6	1
Pt^+–O	318.4 ± 6.7	1
Pt^+–Pt	318 ± 23	1
Pt^+–Si	515 ± 50	1
Pt^+–Xe	86.6 ± 28.9	1
Pu^+–F	562 ± 50	1
Pu^+–O	651 ± 19	14
Pu^+–S	420 ± 23	20
Rb^+–Ar	12.0	1
Rb^+–Br	17.6v5.1	1
Rb^+–Cl	10.5 ± 10.5	1
Rb^+–I	27 ± 42	1
Rb^+–Kr	14.9	1
Rb^+–Na	50.1 ± 3.9	1
Rb^+–Ne	6.95	1
Rb^+–O	29	1
Rb^+–Rb	75.6 ± 9.6	1
Rb^+–Xe	21.5	1
Re^+–C	497.7 ± 3.9	1
Re^+–H	224.7 ± 6.7	1
Re^+–O	465.1 ± 4.8	25
Rh^+–C	414 ± 17	1
Rh^+–H	164.8 ± 3.8	1
Rh^+–O	295.0 ± 5.8	1
Rh^+–S	251.8 ± 11.6	11
Ru^+–C	527.6 ± 7.7	36
Ru^+–H	160.2 ± 5.0	1
Ru^+–O	372 ± 5	1
Ru^+–S	293.3 ± 9.6	15
S^+–C	620.8 ± 1.3	1
S^+–F	343.5 ± 4.8	1
S^+–H	348.2 ± 1.7	1
S^+–N	516 ± 34	1
S^+–O	524.3 ± 0.4	1
S^+–P	573 ± 21	1
S^+–S	522.4 ± 0.5	1
Sc^+–C	326 ± 6	1
Sc^+–Cl	410 ± 42	1
Sc^+–F	605 ± 32	1
Sc^+–Fe	201 ± 21	1
Sc^+–H	235 ± 8	1
Sc^+–O	689 ± 5	1
Sc^+–S	529.7 ± 17.4	1
Sc^+–Se	475.8 ± 8.4	1
Sc^+–Si	242.3 ± 10.5	1
Se^+–F	364 ± 42	1
Se^+–H	304	1
Se^+–P	514 ± 25	1
Se^+–S	392 ± 19	1
Se^+–Se	413 ± 19	1
Si^+–Au	175 ± 50	1
Si^+–B	351 ± 15	1
Si^+–Br	276 ± 96	1
Si^+–C	365 ± 50	1
Si^+–Cl	591.0 ± 0.6	1
Si^+–F	684.1 ± 5.4	1
Si^+–H	316.6 ± 2.1	1
Si^+–O	478 ± 13.4	1
Si^+–P	272 ± 50	1
Si^+–Pd	237 ± 50	1
Si^+–Pt	525 ± 50	1
Si^+–S	387.5 ± 6.0	1
Si^+–Si	334 ± 19	1
Si^+–Te	347 ± 50	1
Sm^+–Br	343.3	1
Sm^+–Cl	435.4	1
Sm^+–F	620.9	1
Sm^+–H	199.59 ± 5.78	43
Sm^+–I	299.1	1
Sm^+–O	556.1 ± 6.8	39
Sm^+–S	328.9 ± 8.8	39
Sn^+–Br	335 ± 50	1
Sn^+–Cu	184 ± 96	1
Sn^+–F	364 ± 29	1
Sn^+–O	281 ± 10	1
Sn^+–S	240 ± 19	1
Sn^+–Se	174 ± 6.3	1
Sn^+–Sn	193	1
Sn^+–Te	168.7 ± 8.4	1
Sr^+–Ar	13.32 ± 2.92	1
Sr^+–Br	378.1 ± 8.4	1

A^+–B	$D°_{298}$/kJ mol^{-1}	Ref.
Sr$^+$–Cl	427 ± 8.4	1
Sr$^+$–F	615 ± 50	1
Sr$^+$–H	209 ± 5	1
Sr$^+$–I	308.2	1
Sr$^+$–Kr	18.13 ± 6.94	1
Sr$^+$–Ne	4.52 ± 9.6	1
Sr$^+$–O	298.7	1
Sr$^+$–Sr	108.5 ± 1.6	1
Ta$^+$–C	369.4 ± 3.9	10
Ta$^+$–H	230 ± 6	1
Ta$^+$–O	688.7 ± 11.6	10
Ta$^+$–Ta	666	1
Tb$^+$–Cu	245 ± 34	1
Tb$^+$–O	722 ± 15	1
Tc$^+$–H	197.5	1
Tc$^+$–O	>167	1
Te$^+$–H	305 ± 12	1
Te$^+$–O	339 ± 50	1
Te$^+$–P	415 ± 97	1
Te$^+$–Se	342 ± 19	1
Te$^+$–Si	339.6	5
Te$^+$–Te	278 ± 29	1
Th$^+$–C	468.98±28.0	31
Th$^+$–Cl	499 ± 29	1
Th$^+$–F	682 ± 29	1
Th$^+$–H	240.1±6.8	29
Th$^+$–N	631.8 ± 7.7	47
Th$^+$–O	830.60 ± 13.5	31
Th$^+$–Pt	388 ± 193	1
Th$^+$–Rh	504 ± 67	1
Th$^+$–S	570 ± 75	20
Ti$^+$–C	398.24 ± 0.38	40
Ti$^+$–Cl	426.8	1
Ti$^+$–F	=456	1
Ti$^+$–H	226.6 ± 10.6	1
Ti$^+$–N	501 ± 13	1
Ti$^+$–O	667 ± 7	1
Ti$^+$–Pt	82 ± 96	1
Ti$^+$–S	461.1 ± 6.8	1
Ti$^+$–Si	249 ± 16	1
Ti$^+$–Ti	229	1
Tl$^+$–Br	52 ± 50	1
Tl$^+$–Cl	26 ± 4	1
Tl$^+$–F	13 ± 21	1
Tl$^+$–I	133 ± 21	1
Tl$^+$–Tl	22 ± 50	1
Tm$^+$–Br	312.2	1
Tm$^+$–Cl	407.9	1
Tm$^+$–F	537 ± 16	1
Tm$^+$–I	266.8	1
Tm$^+$–O	482 ± 15	1
U$^+$–Br	345 ± 29	1
U$^+$–C	300 ± 96	1
U$^+$–Cl	431 ± 34	1
U$^+$–D	283.4 ± 9.6	1
U$^+$–F	668 ± 29	1
U$^+$–H	284 ± 8	1
U$^+$–N	~485	1
U$^+$–O	774 ± 13	14
U$^+$–P	186	1
U$^+$–S	500 ± 60	20
V$^+$–Ar	39.39 ± 0.12	1
V$^+$–C	363.05 ± 0.24	32
V$^+$–D	206.3 ± 8.7	51
V$^+$–Fe	314 ± 21	1
V$^+$–H	203.4 ± 8.7	51
V$^+$–Kr	49.46 ± 0.18	1
V$^+$–N	468.98 ±0.19	32
V$^+$–Nb	403.5 ± 0.2	1
V$^+$–O	587.74 ± 0.19	48
V$^+$–S	358.9 ± 8.7	1
V$^+$–Si	229 ± 15	1
V$^+$–V	302	1
V$^+$–Xe	66.4 ± 0.6	1
W$^+$–C	464.9 ± 17.4	23
W$^+$–F	444 ± 96	1
W$^+$–H	222.5 ± 5	1
W$^+$–O	656.9 ± 6.8	10
Xe$^+$–Ar	13.4	1
Xe$^+$–F	188	1
Xe$^+$–H	355	1
Xe$^+$–Kr	41.65 ± 0.08	1
Xe$^+$–N	66.4 ± 9.6	1
Xe$^+$–Ne	2.1 ± 0.8	1
Xe$^+$–Xe	99.6	1
Y$^+$–C	281 ± 12	1
Y$^+$–F	677 ± 21	1
Y$^+$–H	260.5 ± 5.8	1
Y$^+$–O	718 ± 25	1
Y$^+$–Pt	466 ± 192	1
Y$^+$–S	533.9 ± 8	1
Y$^+$–Si	243 ± 13	1
Y$^+$–Te	360 ± 96	1
Y$^+$–Y	281 ± 21	1
Yb$^+$–Br	307.4	1
Yb$^+$–Cl	399.6	1
Yb$^+$–F	557.5 ± 14.4	1
Yb$^+$–I	271.9±17.4	30
Yb$^+$–O	376 ± 15	1
Yb$^+$–Yb	238 ± 96	1
Zn$^+$–Ar	28.7 ± 1.2	1
Zn$^+$–H	216 ± 15	1
Zn$^+$–O	161.1 ± 4.8	1
Zn$^+$–S	198 ± 12	1
Zn$^+$–Si	274.1 ± 9.6	1
Zn$^+$–Zn	60 ± 19	1
Zr$^+$–Ar	36.09 ± 0.24	1
Zr$^+$–C	445.8 ± 15.4	1
Zr$^+$–H	218.8 ± 9.6	1
Zr$^+$–N	443 ± 46	1
Zr$^+$–O	753 ± 11	1
Zr$^+$–S	549.0 ± 9.6	1
Zr$^+$–Zr	407.0 ± 9.6	1

References

1. Luo, Y. R., *Comprehensive Handbook of Chemical Bond Energies*, CRC Press, Boca Raton, FL, 2007. <https://doi.org/10.1201/9781420007282>
2. Parke, L. G., Hinton, C. S., and Armentrout, P. B., *Int. J. Mass Spectrom.* 254, 168, 2006. <https://doi.org/10.1016/j.ijms.2006.05.025>
3. Li, F.-X., Zhang, X.-G., and Armentrout, P. B., *Int. J. Mass Spectrom.* 255/256, 279, 2006. <https://doi.org/10.1016/j.ijms.2006.02.021>
4. Li, F.-X., and Armentrout, P. B., *J. Chem. Phys.* 125, 133114/1, 2006. <https://doi.org/10.1063/1.2220038>
5. Chattopadhyaya, S., Pramanik, A., Banerjee, A., and Das, K. K., *J. Phys. Chem. A* 110, 12303, 2006. <https://doi.org/10.1021/jp062610c>
6. Li, J., Hao, Y., Yang, J., Zhou, C., and Mo, Y., *J. Chem. Phys.* 127, 104307/1, 2007. <https://doi.org/10.1063/1.2772273>
7. Gibson, J. K., Haire, R. G., Santos, M., Pires de Matos, A., and Marçalo, J., *J. Phys. Chem. A* 112, 11373, 2008.
8. Merritt, J. M., Kaledin, A. L., Bondybey, V. E., and Heaven, M. C., *Phys. Chem. Chem. Phys.* 10, 4006, 2008. <https://doi.org/10.1039/b803975e>
9. Schröder, D., *J. Phys. Chem. A* 112, 13215, 2008. <https://doi.org/10.1021/jp8030804>
10. Hinton, C. S., Li, F.-X., and Armentrout, P. B., *Int. J. Mass Spectrom.* 280, 226, 2009. <https://doi.org/10.1016/j.ijms.2008.08.025>
11. Armentrout, P. B., and Kretzschmar, I., *J. Phys. Chem. A* 113, 10955, 2009. <https://doi.org/10.1021/jp907253r>
12. Armentrout, P. B., and Kretzschmar, I., *Inorg. Chem.* 48, 10371, 2009. <https://doi.org/10.1021/ic9015959>
13. Li, M., Liu, S.-R., and Armentrout, P. B., *J. Chem. Phys.* 131, 144310, 2009. <https://doi.org/10.1063/1.3246840>
14. Marçalo, J., and Gibson, J. K., *J. Phys. Chem. A* 113, 12599, 2009. <https://doi.org/10.1021/jp904862a>
15. Armentrout, P. B., and Kretzschmar, I., *Phys. Chem. Chem. Phys.* 12, 4078, 2010. <https://doi.org/10.1039/b926429a>
16. Tzeli, D., and Mavridis, A., *J. Chem. Phys.* 132, 194312, 2010. <https://doi.org/10.1063/1.3429612>
17. Hinton, C. S., and Armentrout, P. B., *J. Chem. Phys.* 133, 124307, 2010. <https://doi.org/10.1063/1.3482663>
18. Li, F.-X., Hinton, C. S., Citir, M., Liu, F., and Armentrout, P. B., *J. Chem. Phys.* 134, 024310, 2011. <https://doi.org/10.1063/1.3514899>
19. Li, F.-X., Gorham, K., and Armentrout, P. B., *J. Phys. Chem. A* 114, 11043, 2010.
20. Pereira, C. C. L., Marsden, C. J., Marçalo, J., and Gibson, J. K., *Phys. Chem. Chem. Phys.* 13, 12940, 2011. <https://doi.org/10.1039/c1cp20996e>
21. Armentrout, P. B., and Kretzschmar, I., *J. Chem. Phys.* 132, 024306, 2010. <https://doi.org/10.1063/1.3285837>
22. Hinton, C. S., Citir, M., and Armentrout, P. B., *J. Chem. Phys.* 135, 234302, 2011. <https://doi.org/10.1063/1.3669425>

23. Wnorowski, K., Stano, M., Barszczewska, W., Jówko, A., and Matejcik, Š., *Int. J. Mass Spectrom.* 314, 42, 2012. <https://doi.org/10.1016/j.ijms.2012.02.002>
24. Armentrout, P. B., and Li, F.-X., *J. Phys. Chem.* A 117, 7754, 2013. <https://doi.org/10.1021/jp4063143>
25. Armentrout, P. B., *J. Chem. Phys.* 139, 084305, 2013. <https://doi.org/10.1063/1.4818642>
26. Hinton, C. S., Citir, M., and Armentrout. P. B., *Int. J. Mass Spectrosc.* 354, 87, 2013. <https://doi.org/10.1016/j.ijms.2013.05.015>
27. Parry, I. S., Hermes, A. C., Kartouzian, A., and Mackenzie, S. R., *Phys. Chem. Chem. Phys.* 16, 458, 2014. <https://doi.org/10.1039/C3CP53214C>
28. Armentrout, P. B., Parke, L., Hinton, C., and Citir, M., *ChemPlusChem.* 78, 1157, 2013. <https://doi.org/10.1002/cplu.201300147>
29. Cox, R. M., et al., *J. Phys. Chem. B* 120, 1601-1614, 2016. <https://doi.org/10.1021/acs.jpcb.5b08008>
30. Sergeev, D. N., et al., *Int. J. Mass Spectrom.* 374, 1-3, 2014. <https://doi.org/10.1016/j.ijms.2014.09.014>
31. Cox, R. M., et al., *J. Chem. Phys.* 184309, 2016.
32. Johnson, E. L., et al., *J. Chem. Phys.* 144, 234306, 2016. <https://doi.org/10.1063/1.4953782>
33. Matthew, D. J., et al., *J. Chem. Phys.* 146, 144310, 2017. <https://doi.org/10.1063/1.4979679>
34. Demireva, M., et al., *J. Phys. Chem. A* 120, 8550-8563, 2016. <https://doi.org/10.1021/acs.jpca.6b09309>
35. Kim, J. S., et al., *J. Chem. Phys.* 145, 194305, 2016. <https://doi.org/10.1063/1.4967820>
36. Armentrout, P. B., et al., *Int. J. Mass Spectrom.* 413, 135-149, 2017. <https://doi.org/10.1016/j.ijms.2016.05.003>
37. Demireva, M., Armentrout, P. B., *J. Phys. Chem. A* 122, 750-761, 2018. <https://doi.org/10.1021/acs.jpca.7b11471>
38. Johnston, M. D., et al., *J. Chem. Phys.* 148, 214308, 2018. <https://doi.org/10.1063/1.5034353>
39. Armentrout, P. B., et al., *J. Phys. Chem. A* 122, 737-749, 2018. <https://doi.org/10.1021/acs.jpca.7b09905>
40. Sevy, A., et al., *J. Chem. Phys.* 149, 044306, 2018. <https://doi.org/10.1063/1.5041422>
41. Owen, C. J., et al., *J Chem. Phys.* 150, 174305, 2019. <https://doi.org/10.1063/1.5092957>
42. Ghiassee, M., et al., *J Chem. Phys.* 150, 144309, 2019. <https://doi.org/10.1063/1.5091679>
43. Demireva, M., and Armentrout, P. B., *J. Chem. Phys.* 149, 164304, 2018. <https://doi.org/10.1063/1.5053758>
44. Sevy, A., et al., *J. Chem. Phys.* 151, 024302, 2019. <https://doi.org/10.1063/1.5098330>
45. Johnston, M. D., et al., *J. Phys. Chem. A* 122, 8047, 2018. <https://doi.org/10.1021/acs.jpca.8b07849>
46. Hechtfischer, U., et al., *J. Chem. Phys.* 151, 044303, 2019. <https://doi.org/10.1063/1.5098321>
47. Cox, R. M., et al., *J. Chem. Phys.* 151, 034304, 2019. <https://doi.org/10.1063/1.5111534>
48. Merriles, D. M., et al., *J. Chem. Phys.* 153, 024303, 2020.
49. Sorensen, J. J., et al., *J. Chem. Phys.* 152, 194307, 2020.
50. Kim, J.; Armentrout, P. B., *J. Phys. Chem.* A 124, 6629-6644, 2020.
51. Armentrout, P. B., et al., *J. Phys. Chem.* A 124, 5306-5313, 2020.
52. Ghiassee, M.; Armentrout, P. B., *J. Phys. Chem.* A 124, 2560-2572, 2020.

Bond Dissociation Energies in Polyatomic Cations

Table 7 has bond dissociation energies for many polyatomic cations; the table has been arranged on the basis of the Periodic Table with the IUPAC notation for Groups 1 to 18, see the inside front cover of this *CRC Handbook*. Column definitions for Table 7 are as follows.

Column heading	Definition
Bond	Bond that is dissociating; **boldface** in the species indicates the dissociated fragment; bonds are grouped according to IUPAC group of dissociating fragment
$D°_{298}$	Dissociation energy, in kJ mol^{-1}
Ref.	Reference number

TABLE 7. Bond Dissociation Energies in Polyatomic Cations

Bond	$D°_{298}$/kJ mol^{-1}	Ref.
(1) Group 1		
Li$^+$–H_2	27.2	1
Li$^+$–CO	57 ± 13	1
Li$^+$–H_2O	139 ± 8	1
Li$^+$–NH_3	156 ± 8	1
Li$^+$–CH_4	130	1
Li$^+$–CH_3OH	156 ± 8	1
Li$^+$–CH_3OCH_3	167 ± 10	1
Li$^+$–pyridine	183.0 ± 14.5	1
Li$^+$–Gly (glycine)	220 ± 9	1
Na$^+$–H_2	10.4 ± 0.8	1
Na$^+$–N_2	33.5	1
Na$^+$–CO	31 ± 8	1
Na$^+$–CO_2	66.5	1
Na$^+$–SO_2	79.1	1
Na$^+$–O_3	52.3	1
Na$^+$–H_2O	91.2 ± 6.3	1
Na$^+$**(H$_2$O)**-H_2O	82.0 ± 5.8	1
Na$^+$**(H$_2$O)$_2$**–H_2O	66.1	1
Na$^+$**(H$_2$O)$_3$**–H_2O	52.7 ± 0.8	1
Na$^+$**(glycine)**–H_2O	75.1 ± 5.3	1
Na$^+$**(glutamine)**–H_2O	52 ± 1	1
Na$^+$–NH_3	106.2 ± 5.4	1
Na$^+$–HNO_3	86.2	1
Na$^+$–CH_4	30.1	1
Na$^+$–CH_3OH	98.8 ± 5.7	1
Na$^+$–CH_3CN	125.5 ± 9.6	1
Na$^+$–C_2H_4	44.6 ± 4.4	1
Na$^+$–CH_3OCH_3	101.4 ± 5.7	1
Na$^+$–$CH_3C(O)H$	114.4 ± 3.4	1
Na$^+$–MeCOMe	131.3 ± 4.1	1
Na$^+$–C_6H_6	97.0 ± 5.9	1
Na$^+$–pyrrole	103.7 ± 4.8	1
Na$^+$–Gly (glycine)	166.7 ± 5.1	1
Na$^+$–Ala (alanine)	167 ± 4	1
Na$^+$–GlyGly (glycylglycine)	203 ± 8	1
Na$^+$–(15-crown-5)	299.1 ± 15.5	12
Na$^+$–(18-crown-6)	300±19	1
K$^+$–H_2	6.1 ± 0.8	1
K$^+$–CO_2	35.6	1
K$^+$–H_2O	74.9	1
K$^+$**(H$_2$O)$_2$**-H_2O	67.4	1
K$^+$**(H$_2$O)$_3$**-H_2O	55.2	1
K$^+$**(H$_2$O)$_4$**-H_2O	11.8	1
K$^+$**(H$_2$O)$_5$**-H_2O	44.8	1

Bond	$D°_{298}$/kJ mol^{-1}	Ref.
$K^+(H_2O)_6$-H_2O	41.8	1
K^+–NH_3	79 ± 7	1
K^+–C_6H_6	80.3	1
K^+–adenine	95.1 ± 3.2	1
K^+–indole	104.6 ± 12.6	1
K^+–Phe (phenylalanine)	150.5 ± 5.8	1
K^+–Tyr (tyrosine)	165.0 ± 5.8	1
K^+–(15-crown-5)	217.9±10.6	12
K^+–(18-crown-6)	235±13	1
Rb^+–H_2O	66.9 ± 12.6	1
Rb^+–NH_3	78.2	1
Rb^+–CH_3CN	86.6 ± 1.3	1
Rb^+–C_6H_5OH	70.2 ± 3.7	1
Rb^+–(15-crown-5)	176.8±9.7	12
Rb+–(18-crown-6)	192±13	1
Cs^+–H_2O	57.3	1
Cs^+–$C_6H_5NH_2$	70.8 ± 4.5	1
Cs^+–(15-crown-5)	160.5±9.6	12
Cs^+–(18-crown-6)	170±9	1
(2) Group 2		
CH_3Be^+–CH_3	192.9 ± 13.4	1
tert-**$C(CH_3)_3Be^+$**–*tert*-$C(CH_3)_3$	121.8 ± 13.4	1
Mg^+–OH	314 ± 33	1
Mg^+–CO	43.1 ± 5.8	1
Mg^+–CO_2	58.4 ± 5.8	1
Mg^+–H_2O	122.5 ± 12.5	1
Mg^+–NH_3	158.9 ± 11.6	1
Mg^+–CH_4	29.8 ± 6.8	1
Mg^+–MeOH	147.6 ± 6.8	1
Mg^+–C_6H_6	155.2	1
Mg^+–pyridine	200.0 ± 6.4	1
Mg^+–imidazole	243.9 ± 10.4	1
$Mg^{2+}(H_2O)_5$-H_2O	101.3	1
$Mg^{2+}(Me_2CO)_5$-Me_2CO	93.3	1
Ca^+–OH	435.1 ± 14.5	1
Ca^+–H_2O	117.2	1
Ca^+–C_6H_6	134	1
Ca^+–imidazole	186.3 ± 3.9	1
$Ca^{2+}(H_2O)_4$-H_2O	110.0 ± 5.9	1
$Ca^{2+}(Me_2CO)_5$-Me_2CO	101.3	1
Sr^+–CO	20.3	1
Sr^+–CO_2	41.9	1
Sr^+–H_2O	144.3	1
Sr^+–C_6H_6	117	1
$Sr^{2+}(H_2O)_5$–H_2O	87.4	1
Ba^+–OH	530.7 ± 19.3	1
$Ba^{2+}(H_2O)_4$–H_2O	90.8	1
(3) Group 3		
Sc^+–H_2	23.0 ± 1.3	1
Sc^+–CH_2	412 ± 22	1
Sc^+–CH_3	233 ± 10	1
Sc^+–C_2H_2	240 ± 20	1
Sc^+–C_2H_4	≥131	1
Sc^+–C_6H_6	222 ± 21	1
Sc^+–H_2O	131	1
Sc^+–NH	483 ± 10	1
Sc^+–NH_2	347 ± 5	1
Sc^+–pyridine	231.5 ± 10.3	1
Sm^+–CO_2	40.5 ± 2.9	15
Sm^+(O)–S	133.1 ± 25.9	16
Sm^+(S)–O	359.8 ± 15.5	16
Y^+–CH_2	398 ± 13	1
Y^+–CH_3	249 ± 5.0	1
Y^+–C_2H_2	218 ± 13	1
Y^+–C_2H_4	>138	1
Y^+–CO	29.9 ± 10.6	1
Y^+–CS	137.0 ± 7.7	1
Y^+(O)–CO_2	86 ± 5	1
Gd^+–CO	62.7 ± 5.8	13
Gd^+(O)–CO	55.0 ± 5.0	17
Gd^+–OCO	36.7 ± 5.0	17
Gd^+–O_2	72.4 ± 10.6	14
La^+–CH	523 ± 33	1
La^+–CH_2	401 ± 7	1
La^+–CH_3	217 ± 15	1
La^+–C_2H_2	262 ± 30	1
La^+–C_2H_4	192.5	1
Nd^+(O)–O	204.55 ± 28.95	19
Nd^+(O)–CO	28.95 ± 20.26	19
Lu^+–CH_2	>230 ± 6	1
Lu^+–CH_3	176 ± 20	1
Th+(O)–D	224.8 ± 23.2	21
Th+(O)–O	476.6 ± 3.9	24
Th^+(O)–OD	578.9 ± 16.4	21
U^+(F)–F	552 ± 44	1
$U^+(F)_2$–F	523 ± 38	1
$U^+(F)_3$–F	381 ± 19	1
$U^+(F)_4$–F	243 ± 17	1
$U^+(F)_5$–F	26 ± 11	1
(4) Group 4		
Ti^+–CH	478 ± 5	1
Ti^+–CH_2	391 ± 15	1
Ti^+–CH_3	213.8 ± 3	1
Ti^+–CH_4	70.3 ± 2.5	1
Ti^+–C_2H_2	213 ± 13	1
Ti^+–C_2H_4	146 ± 11	1
Ti^+–C_6H_6	259 ± 9	1
Ti^+–CO	117.7 ± 5.8	1
Ti^+–H_2O	157.7 ± 5.9	1
Ti^+–NH	466 ± 12	1
Ti^+–NH_2	356 ± 13	1
Ti^+–NH_3	197 ± 7	1
Ti^+–pyridine	217.2 ± 9.3	1
Ti^+–imidazole	≤232.4 ± 8.2	1
Zr^+–CH	568 ± 13	1
Zr^+–CH_2	444.8 ± 5	1
Zr^+–CH_3	227.7 ± 9.6	1
Zr^+–C_2H_2	273 ± 14	1
Zr^+–CO	77 ± 10	1
Zr^+–CS	257.6 ± 10.6	1
Hf^+–CH	492.1 ± 14.5	2
Hf^+–CH_2	421.6 ± 6.8	2
Hf^+–CH_2	204.5 ± 25.1	2
Hf^+–C_2H_2	150.6	1
(5) Group 5		
$(CO)_6V^+$–H	220 ± 14	1
V^+–H_2	42.7 ± 2.1	1
V^+–CH	479.5 ± 8.8	10
V^+–CH_2	326 ± 6	1
V^+–CH_3	193 ± 7	1

Bond	D°_{298}/kJ mol^{-1}	Ref.
$V^+-C_2H_2$	172 ± 8	1
$V^+-C_2H_4$	124 ± 8	1
$V^+-(\eta^5-C_5H_5)$	530.7	1
$V^+-C_6H_6$	234 ± 10	1
V^+-CO	114.8 ± 2.9	1
V^+-CO_2	72.4 ± 3.8	1
V^+-H_2O	149.8 ± 5.0	1
V^+-NH	423 ± 29	1
V^+-NH_2	293 ± 6	1
V^+-NH_3	192 ± 11	1
V^+–pyridine	218.7 ± 13.5	1
V^+–imidazole	≤243.4 ± 8.0	1
Nb^+-H_2	61.9	1
Nb^+-CH	581 ± 19	1
Nb^+-CH_2	428.4 ± 8.7	1
Nb^+-CH_3	198.8 ± 10.6	1
$Nb^+-CH_3NH_2$	134	1
$Nb^+-C_3H_6$	117.7	1
$(NbFe)^+-C_3H_4$	>163	1
Nb^+-CO	95.5 ± 4.8	1
Nb^+-CS	242.2 ± 10.6	1
$Nb_7^+-N_2$	<215	1
Ta^+-CH	561.5 ± 15.4	6
Ta^+-CH_2	464.1 ± 2.9	6
Ta^+-CH_3	259.5 ± 13.5	6
$Ta^+-C_6H_6$	251~301	1

(6) Group 6

Bond	D°_{298}/kJ mol^{-1}	Ref.
$(CO)_6Cr^+-H$	230 ± 10	1
$(\eta^5-C_5H_5)(NO)(CO)_2Cr^+-H$	207.1 ± 14	1
Cr^+-H_2	31.8 ± 2.1	1
Cr^+-CH	294 ± 29	1
Cr^+-CH_2	216 ± 4	1
Cr^+-CH_3	110 ± 4	1
$Cr^+-C_6H_6$	170 ± 10	1
Cr^+–indole	196.6 ± 16.7	1
Cr^+-CO	89.7 ± 5.8	1
Cr^+-OH	298 ± 14	1
Cr^+-H_2O	132.6 ± 8.8	1
Cr^+-N_2	59 ± 4	1
Cr^+-NH_3	177.4 ± 1.2	20
$(CO)_6Mo^+-H$	260 ± 9	1
Mo^+-CH	513.3 ± 13.5	1
Mo^+-CH_2	344.4 ± 10	1
Mo^+-CH_3	151.5 ± 8.7	1
Mo^+-CO	193.9 ± 9.6	1
Mo^+-CO_2	49.2 ± 7	1
Mo^+-CS	162 ± 18	1
Mo^+-CS_2	67.5 ± 12.5	1
Mo^+-NH	<385	1
Mo^+–pyrrole	>289	1
$(CO)_6W^+-H$	257 ± 9	1
W^+-CH	580 ± 27	1
W^+-CH_2	456.4 ± 5.8	1
W^+-CH_3	~222.9 ± 9.6	1
$(PMe_3)_3(CO)_3W^+-H$	259.4	1
W^+–pyrrole	>209	1

(7) Group 7

Bond	D°_{298}/kJ mol^{-1}	Ref.
$(CO)_5Mn^+-H$	172 ± 10	1
Mn^+-H_2	7.9 ± 1.7	1
Mn^+-CH_2	295 ± 13	1
Mn^+-CH_3	215 ± 10	1
$Mn^+(CO)_5-CH_3$	132 ± 15	1
$Mn^+(CO)_5-CH_4$	>30	1
$Mn^+-(\eta^5-C_5H_5)$	326.1 ± 9.6	1
$Mn^+-C_6H_6$	145 ± 10	1
Mn^+-OH	332 ± 24	1
Mn^+-CO	25 ± 10	1
Mn^+-H_2O	121.8 ± 5.9	1
Mn^+-CH_3OH	134 ± 29	1
$Mn^+-OC(CH_3)_2$	159 ± 14	1
Mn^+-CS	80.0 ± 21	1
Mn^+-NH_2	254 ± 20	1
Mn^+-NH_3	147 ± 8	1
Tc^+-CH_2	<464	1
$Tc^+-C_2H_2$	<320	1
Re^+-O_2	544.8 ± 1.9	23
Re^+-SO	419.7 ± 13.5	23
ORe^+-O	583.7 ± 4.8	23
ORe^+-S	471.8 ± 18.3	23
$Re^+(CH_3)(CO)_5-H$	294 ± 13	1
$(PMe_3)(CO)_2Re^+-H$	300.4	1

(8) Group 8

Bond	D°_{298}/kJ mol^{-1}	Ref.
$Fe^+(O)-H$	444 ± 17	1
$Fe^+(CO)-H$	120 ± 23	1
$Fe^+(H_2O)-H$	215 ± 14	1
$Fe^+(\eta^5-C_5H_5)-H$	193 ± 21	1
$(CO_2)_5Fe^+-H$	299 ± 15	1
Fe^+-H_2	45.2 ± 2.5	1
Fe^+-CH	423 ± 29	1
Fe^+-CH_2	≤342 ± 2	1
Fe^+-CH_3	229 ± 5	1
Fe^+-CH_4	73.2	1
$Fe^+-C_2H_2$	159.0 ± 2.1	1
$Fe^+-C_2H_3$	238 ± 10	1
$Fe^+-C_2H_4$	145 ± 11	1
$Fe^+-C_2H_5$	233 ± 9	1
$Fe^+-C_2H_6$	64 ± 6	1
Fe^+-OH	322 ± 12	18
$Fe^+(H_2O) \pm OH$	420.7 ± 15.4	18
Fe^+-CO	129.3 ± 3.9	1
Fe^+D-CO	53 ± 13	1
Fe^+-CO_2	74.3 ± 7.7	1
Fe^+-H_2O	128.9 ± 0.8	1
Fe^+-N_2	53 ± 4	1
Fe^+-NH_3	184 ± 12	1
Fe^+-CS_2	166.1 ± 4.6	1
Fe^+–imidazole	246.1 ± 13.8	1
Fe^+-SiH	254 ± 13	1
Fe^+-SiH_2	181 ± 9	1
Fe^+-SiH_3	183 ± 9	1
$Ru^+(\eta^5-C_5H_5)_2-H$	292 ± 16	1
$(\eta^5-C_5Me_5)_2Ru^+-H$	284.5	1
Ru^+-CH	501.7 ± 11.6	1
Ru^+-CH_2	344.4 ± 4.8	1
Ru^+-CH_3	160.2 ± 5.8	1
Ru^+-CS	244.9 ± 17.4	9
Os^+-CH	653.2 ± 14.5	11
Os^+-CH_3	276.7 ± 14.5	11
Os^+–O2	496.9 ± 6.8	22
Os^+-SO	407.2 ± 10.6	22

Bond	$D°_{298}$/kJ mol^{-1}	Ref.
OsO_4^+–H	552 ± 13	1
(9) Group 9		
(η^5-**C_5H_5)(CO)$_2$Co$^+$**–H	245 ± 12	1
(CH_3OD)Co$^+$–H	147.6 ± 7.7	1
Co$^+$–H_2	76.1 ± 4.2	1
(η^5-**C_5H_5)Co$^+$**–H_2	67.8	1
Co$^+$–CH	420 ± 37	1
Co$^+$–CH_2	317 ± 5	1
Co$^+$–CH_3	203 ± 4	1
Co$^+$–CH_4	96.7	1
Co$^+$–C_{60}	243 ± 67	1
Co$^+$–CO	173.7 ± 6.7	1
Co$^+$–H_2O	164.4 ± 5.9	1
Co$^+$–CS	259 ± 33	1
Co$^+$–N_2	96.2 ± 7.1	1
Co$^+$–NH_2	247 ± 7	1
Co$^+$–NH_3	219 ± 16	1
Co$^+$–CH_3CN	>255 ± 17	1
Co$^+$–P(CH_3)$_3$	278 ± 11	1
Co$^+$–P(C_2H_5)$_3$	339 ± 16	1
(CH)Rh$^+$–H	372 ± 21	1
(η^5-**C_5H_5)(CO)$_2$Rh$^+$**–H	287 ± 12	1
Rh$^+$–CH	444 ± 12	1
Rh$^+$–CH_2	356 ± 8	1
Rh$^+$–CH_3	142 ± 6	1
Rh$^+$–NO	167 ± 21	1
Rh$^+$–CS	256.6 ± 18.3	8
(CO)(η^5-**C_5H_5)(PPh$_3$)Ir$^+$**–H	313.4	1
(CO)$_2$(η^5-**C_5Me_5)Ir$^+$**–H	298.3	1
Ir$^+$–CH	666.7 ± 22.2	3
Ir$^+$–CH_2	474.7 ± 2.9	3
Ir$^+$–CH_3	313.6 ± 17.4	3
Ir$^+$–C_2H_4	234.3	1
(10) Group 10		
(CO)$_4$Ni$^+$–H	248 ± 9	1
(η^5-**C_5H_5)(NO)Ni$^+$**–H	315 ± 14	1
(η^5-**C_5H_5)**(η^5-**C_5H_5)Ni$^+$**–H	215 ± 13	1
Ni$^+$–H_2	72.4 ± 1.3	1
Ni$^+$–CH	301.0 ± 11.6	1
Ni$^+$–CH_2	306 ± 4	1
Ni$^+$–CH_3	169.8 ± 6.8	1
Ni$^+$–CH_4	96.5 ± 4	1
Ni$^+$–OH	235 ± 19	1
Ni$^+$–CO	175 ± 11	1
Ni$^+$–CO_2	104 ± 1	1
Ni$^+$–H_2O	183.7 ± 3.3	1
Ni$^+$–CS	234.5 ± 9.6	1
Ni$^+$–N_2	110.9 ± 10.5	1
Ni$^+$–NO	227.6 ± 7.5	1
Ni$^+$–NH_2	232.5 ± 7.7	1
Ni$^+$–NH_3	238 ± 19	1
Pd$^+$–CH	536 ± 10	1
Pd$^+$–CH_2	463 ± 3	1
Pd$^+$–CH_3	258 ± 8	1
Pd$^+$–CH_4	170.8 ± 7.7	1
Pd$^+$–CS	200 ± 14	1
Pd$^+$–C_2H_2	>28.9 ± 4.8	1
Pt$^+$–H_2	146.7 ± 11.6	1
Pt$^+$–CH	536.4 ± 9.6	1
Pt$^+$–CH_2	471	1
Pt$^+$–CH_3	257.6 ± 7.7	1
Pt$^+$–CH_4	170.8 ± 7.7	1
Pt$^+$–O_2	64.6 ± 4.8	1
Pt$^+$–CO	218.1 ± 8.7	1
Pt$^+$–CO_2	59.8 ± 4.8	1
Pt$^+$–NH_3	274 ± 12	1
Pt$^+$–C_2H_4	229.7	1
(11) Group 11		
Cu$^+$–H_2	51.9 ± 0.4	1
Cu$^+$–CH_2	267.3 ± 6.8	1
Cu$^+$–CH_3	111 ± 7	1
Cu$^+$–C_2H_2	>21.2 ± 9.6	1
Cu$^+$–C_2H_4	176 ± 14	1
Cu$^+$–C_6H_6	218.0 ± 9.6	1
Cu$^+$–CO	149 ± 7	1
Cu$^+$–N_2	89 ± 30	1
Cu$^+$–NO	109.0 ± 4.8	1
Cu$^+$–H_2O	160.7 ± 7.5	1
Cu$^+$–NH_2	192 ± 13	1
Cu$^+$–NH_3	237 ± 15	1
Cu$^+$–CS	238.3 ± 11.6	1
Cu$^+$–SiH	246 ± 27	1
Cu$^+$–SiH_2	≥231 ± 7	1
Cu$^+$–SiH_3	97 ± 25	1
Ag$^+$–CH_2	≥107 ± 4	1
Ag$^+$–CH_3	66.6 ± 4.8	1
Ag$^+$–C_2H_5	65.7 ± 7.5	1
Ag$^+$–C_6H_6	167 ± 19	1
Ag$^+$–O_2	29.7 ± 0.8	1
Ag$^+$–CO	89 ± 5	1
Ag$^+$–H_2O	134 ± 8	1
Ag$^+$–CS	152 ± 20	1
Ag$^+$–NH_3	170 ± 13	1
Au$^+$–CH_2	357.0 ± 6.8	5
Au$^+$–CH_3	209.4 ± 23.2	5
Au$^+$–C_2H_4	344.5	1
Au$^+$–C_6H_6	289 ± 29	1
Au$^+$–CO	201 ± 8	1
Au$^+$–H_2O	164.0 ± 9.6	1
Au$^+$–H_2S	230 ± 25	1
Au$^+$–NH_3	297 ± 29	1
Au$^+$–PH_3	402 ± 33	1
(12) Group 12		
Zn$^+$–H_2	15.7 ± 1.7	1
Zn$^+$–CH_3	280 ± 7	1
Zn$^+$–OH	127.2	1
Zn$^+$–H_2O	163	1
Zn$^+$–NO	76.2 ± 9.6	1
Zn$^+$–pyrimidine	209.6 ± 7.7	1
Zn$^+$–CS	149 ± 23	1
Cd$^+$–CH_3	228 ± 3	1
Cd$^+$(CH_3)–CH_3	109 ± 3	1
Cd$^+$–C_6H_6	136 ± 19	1
Hg$^+$–CH_3	285 ± 3	1
Hg$^+$(CH_3)–CH_3	96 ± 3	1
(13) Group 13		
B$^+$–H_2	15.9 ± 0.8	1
HB$^+$–H_2	61.5 ± 2.1	1
(CH_3)$_2$B$^+$–CH_3	32.6 ± 4.2	1

Bond	$D°_{298}$/kJ mol^{-1}	Ref.
Al^+-H_2	5.6 ± 0.6	1
Al^+-N_2	5.6	1
Al^+-CO_2	≥29.3	1
Al^+-H_2O	104 ± 15	1
Al^+–MeOH	139.7	1
Al^+–EtC(O)Et	191.2	1
$Al^+-C_6H_6$	147.3 ± 8.4	1
Al^+–pyridine	190.3 ± 10.3	1
Al^+–phenol	154.8 ± 16.7	1
Al^+–imidazole	232.4 ± 8.2	1
Ga^+-NH_3	122.5	1
In^+-NH_3	111.0	1

(14) Group 14

Bond	$D°_{298}$/kJ mol^{-1}	Ref.
$C_{58}{}^+-C_2$	955 ± 15	1
$C_{60}{}^+-C_2$	822.0 ± 12.5	1
$C_{62}{}^+-C_2$	846.2 ± 10.6	1
$C_{78}{}^+-C_2$	938.8 ± 10.6	1
$HC_2{}^+-H$	574.749	1
$C_6H_5{}^+-H$	376.3 ± 4.8	1
$C_2H_3{}^+-Cl$	249 ± 1.0	7
$C_2H_5{}^+-Br$	206.3 ± 1.0	7
$C_6H_5{}^+-Br$	266.3	1
$C_2H_3{}^+-I$	196.2 ± 1.4	7
$CH_3{}^+-H_2$	186	1
$CH_5{}^+-H_2$	7.9 ± 0.4	1
$C_2H_5{}^+-H_2$	17	1
$CH_3{}^+-O_2$	80 ± 7	4
CO^+-N_2	67.5 ± 19.3	1
$H_2CH^+-N_2$	31.8	1
CO^+-CO	173.7 ± 14.6	1
$CO^+(CO)-CO$	52.3	1
$CO^+(CO)_2-CO$	30.2	1
$CO^+(CO)_3-CO$	18.4	1
$(CO_2)^+-CO_2$	70.3	1
$(CO_2)^+(CO_2)-CO_2$	34.7	1
$(CO_2)^+(CO_2)_2-CO_2$	21.3	1
$(CO_2)^+(CO_2)_3-CO_2$	20.1 ± 1.3	1
$CH_3{}^+-N_2O$	221.3	1
$CH_3{}^+-SO_2$	253.6	1
$CH_3{}^+-OCS$	239.3	1
$CH_3{}^+-CS_2$	251.9	1
$CH_3{}^+-H_2O$	279	1
$CH_3{}^+(H_2O)-H_2O$	106.3	1
$CH_3{}^+(H_2O)_2-H_2O$	87.9	1
$CH_3{}^+(H_2O)_3-H_2O$	61.9	1
$CH_3{}^+(H_2O)_4-H_2O$	48.5	1
$CH_3{}^+-H_2S$	344.8	1
$CH_2{}^+-CH_2O$	303.0 ± 2.9	1
$CH_3{}^+-NH_3$	431.4	1
$(CH_3)^+-CH_3$	209.2 ± 4.2	1
$CH_3{}^+-CH_4$	166.5	1
$CF_3{}^+-CH_4$	19.0	1
$(CH_5)^+-CH_4$	28.7 ± 1.3	1
$C_6H_6{}^+-CH_4$	12.0	1
$CH_3{}^+-CH_3F$	230	1
$CH_3{}^+-CF_3Cl$	221	1
$CH_3{}^+-CH_3Cl$	259	1
tert-$C_4H_9{}^+-CH_3OH$	63	1
tert-$C_4H_9{}^+-CH_3CN$	85	1
tert-$C_4H_9{}^+-SO_2F_2$	43.5	1
$CH_3{}^+-C_2H_3O$	338.7 ± 2.9	1
$CH_3{}^+-CF_3ClOCl$	252	1
tert-$C_4H_9{}^+-(CH_3)_2S$	185	1
tert-$C_4H_9{}^+-C_2H_5OH$	85	1
tert-$C_4H_9{}^+-C_3H_8$	27.6	1
tert-$C_4H_9{}^+$–t-C_4H_9Cl	339	1
tert-$C_4H_9{}^+-(CH_3)_3CH$	30.1	1
tert-$C_4H_9{}^+-C_6H_6$	92	1
$(C_6H_6)^+-C_6H_6$	73.6	1
$(C_6H_6)^+$–indole	54.8	1
$C_6F_6{}^+-C_6F_6$	30.1 ± 4	1
$C_{60}{}^+-C_{60}$	35.89 ± 7.72	1
$PhSiH_2{}^+-H$	159	1
$Si^+(CH_3)_3-Cl$	178.5 ± 1.9	1
$SiH_3{}^+-CO$	≥151	1
$SiF_3{}^+-CO$	174.1 ± 1.3	1
$(CH_3)_3Si^+-H_2O$	125.9 ± 7.9	1
$(CH_3)_3Si^+-NH_3$	194.6	1
$Si^+(CH_3)(Cl)_2-CH_3$	60.8 ± 2.9	1
$Si^+(CH_3)_2(Cl)-CH_3$	41.5 ± 1.9	1
Si^+-CH_3	413.9 ± 5.8	1
$Si^+(CH_3)-CH_3$	123 ± 48	1
$Si^+(CH_3)_2-CH_3$	513 ± 27	1
$Si^+(CH_3)_3-CH_3$	66.6 ± 5.8	1
$(CH_3)_3Si^+-CH_3OH$	164.0	1
$(CH_3)_3Si^+-(C_2H_5)_2O$	184.9	1
$(CH_3)_3Si^+-C_6H_6$	100.0	1
$(CH_3)_3Si^+-CH_3NH_2$	231.8	1
$(CH_3)_3Ge^+-H_2O$	119.7 ± 2.1	1
$(C_2H_5)_3Ge^+-H_2O$	104.2 ± 2.1	1
$(CH_3)_3Sn^+-NH_3$	154	1
$(CH_3)_3Sn^+-H_2O$	108	1
$(CH_3)_3Sn^+-(CH_3)_2CO$	157	1
$(CH_3)_3Sn^+-C_3H_7SH$	143	1
Pb^+-H_2O	93.7	1
Pb^+-NH_3	118.4 ± 0.8	1
Pb^+-CH_3OH	97.5 ± 0.8	1
$Pb^+-CH_3NH_2$	148.1 ± 1.3	1
$Pb^+-C_6H_6$	110 ± 2	1

(15) Group 15

Bond	$D°_{298}$/kJ mol^{-1}	Ref.
H_2N^+-H	544.43 ± 0.10	1
H_3N^+-H	515.1	1
Me_3N^+-H	376	1
Et_3N^+-H	362	1
(imidazole)$^+$ –Zn	216.1 ± 3.9	1
$N_2H^+-H_2$	24.7 ± 0.8	1
ON^+-O_2	14.2	1
N^+-N_2	303.8	1
ON^+-N_2	21.3	1
$N_2{}^+-N_2$	102.3 ± 14.6	1
$HN_2{}^+-N_2$	60.7	1
$N_3{}^+-N_2$	18.8 ± 1.3	1
$O_2N^+-N_2$	19.2 ± 1.3	1
$H_4N^+-N_2$	54 ± 21	1
ON^+-NO	59.4 ± 0.8	1
ON^+-CO	27.2 ± 1.3	1
ON^+-O_3	<58	1
ON^+-CO_2	32.2	1
$N_2O^+-ON_2$	72.8 ± 6.3	1
NO^+-ON_2	36.4 ± 0.8	1

Bond	$D°_{298}$/kJ mol^{-1}	Ref.
$(HON_2)^+$–ON_2	69.9 ± 4	1
ON^+–H_2O	95	1
$ON^+(H_2O)$–H_2O	67.4	1
$ON^+(H_2O)_2$–H_2O	56.5	1
H_4N^+–H_2O	86.2 ± 4.2	1
$H_4N^+(H_2O)$–H_2O	72.8 ± 4.2	1
$H_4N^+(H_2O)_2$–H_2O	57.3 ± 4.2	1
$H_4N^+(H_2O)_3$–H_2O	51.0	1
$H_4N^+(H_2O)_4$–H_2O	44.4	1
(glycine)H^+–H_2O	77.2 ± 11.0	1
(tryptophan)H^+–H_2O	31.2 ± 2.5	1
(tryptophanylglicine)H^+–H_2O	56.0 ± 5.3	1
H_4N^+–H_2S	47.7	1
$H^+(NH_3)$–NH_3	108.8	1
$H^+(NH_3)_2$–NH_3	69.5	1
$H^+(NH_3)_3$–NH_3	57.3	1
$H^+(NH_3)_4$–NH_3	49.0	1
$H^+(NH_3)_5$–NH_3	29.3	1
$H^+(NH_3)_6$–NH_3	27.2	1
NH_4^+–CH_4	15.0	1
ON^+–CH_3OH	97.6	1
O_2N^+–CH_3OH	80.3 ± 9.6	1
$(CH_3CNH)^+$–CH_3CN	130.1 ± 9.6	1
(pyridineH)$^+$–pyridine	105.4 ± 4	1
(valine H)$^+$–valine	86.6 ± 8.4	1
(betainH)$^+$–betaine	139.9 ± 4.8	1
H_4P^+–H_2O	54.4	1
$(H_4P)^+$–PH_3	48.1	1
AsH_2^+–H	257	1
I_2As^+–acetone	106 ± 17	1
I_2As^+–benzene	77 ± 17	1
Bi^+–H_2O	95.4	1
Bi^+–NH_3	149	1
Bi^+–C_6H_6	≤149	1
(16) Group 16		
$(H_3O)^+$–H_2	14.6 ± 2.1	1
O^+–O_2	179.5	1
$O^+(O_2)_1$–O_2	28.9	1
$O^+(O_2)_2$–O_2	3.9	1
O_2^+–O_2	38.3 ± 2.1	1
$O_2^+(O_2)$–O_2	24.6 ± 1.3	1
$O_2^+(O_2)_2$–O_2	10.4 ± 0.8	1
$O_2^+(O_2)_3$–O_2	9.0 ± 0.8	1
$O_2^+(O_2)_4$–O_2	8.0 ± 0.8	1
$O_2^+(O_2)_5$–O_2	7.9 ± 1.3	1
O^+–N_2	231.4	1
O_2^+–N_2	22.6	1
$(H_3O)^+$–N_2	22.2 ± 2.1	1
O_4^+–N_2	12.3	1
O_2^+–CO	31.8	1
O_2^+–CO_2	41.0 ± 2.1	1
CO_2^+–CO_2	65.3 ± 4	1
$(H_3O)^+$–CO_2	64.0	1
$(H_3O)^+(CO_2)$–CO_2	51.9	1
$(H_3O)^+(CO_2)_2$–CO_2	43.9	1
$(H_3O)^+(CO_2)_3$–CO_2	18.0	1
O_2^+–ON_2	56.1 ± 4	1
$(H_3O)^+$–ON_2	70.7 ± 6.5	1
$(H_3O)^+(H_2O)$–ON_2	50.6 ± 2.1	1
$(H_3O)^+(H_2O)_2$–ON_2	42.7 ± 2.1	1

Bond	$D°_{298}$/kJ mol^{-1}	Ref.
O_3^+–O_3	67.5 ± 39	1
$OClO^+$–OClO	246 ± 48	1
O_2^+–H_2O	>67	1
$(OH)^+(H_2O)_2$–H_2O	87.4	1
$(OH)^+(H_2SO_4)(H_2O)_4$–H_2O	56.9	1
$(OH)^+(H_2SO_4)(H_2O)_5$–H_2O	49.8	1
$(OH)^+(H_2SO_4)(H_2O)_6$–H_2O	44.8	1
$(H_2O)^+$–H_2O	164.0	1
$(H_3O)^+$–H_2O	140.2	1
$(H_3O)^+(H_2O)$–H_2O	93.3	1
$(H_3O)^+(H_2O)_2$–H_2O	71.1	1
$(H_3O)^+(H_2O)_3$–H_2O	64.0	1
$(H_3O)^+(H_2O)_4$–H_2O	54.4	1
$(H_3O)^+(H_2O)_5$–H_2O	49.0	1
$(H_3O)^+(H_2O)_6$–H_2O	43.1	1
(HCOOH)H^+–H_2O	100.8	1
$CH_3OH_2^+$–H_2O	115.6	1
CH_3CHOH^+–H_2O	104.6	1
$(CH_3)_2OH^+$–H_2O	100.4	1
(tetrahydrofuranH)$^+$–H_2O	82.8	1
(furanH)$^+$–H_2O	43.5	1
furane$^+$–H_2O	41.0	1
(phenol)$^+$–H_2O	78.0	1
(1-naphthol)$^+$–H_2O	66.4	1
H_3O^+–HC(O)H	137.7	1
H_3O^+–NH_3	229.3	1
$H_3O^+(NH_3)$–NH_3	77.0	1
$H_3O^+(NH_3)_2$–NH_3	71.5	1
$H_3O^+(NH_3)_3$–NH_3	62.8	1
H_3O^+–PH_3	144	1
H_3O^+–SO_3	74	1
$(HCOOH)^+$–HCOOH	96.5 ± 9.6	1
H_3O^+–CH_4	33.5	1
$(CH_3OH)^+$–CH_3OH	115.8 ± 19.3	1
$CH_3OH_2^+$–CH_3OH	136.4	1
H_3O^+–CH_3CN	195.4	1
furan$^+$–furan	94.1	1
BH^+–B, B = tetrahydofuran	125.1	1
S^+–CS_2	166	1
CS^+–CS_2	150.6	1
CS_2^+–CS_2	104.2	1
HCS_2^+–CS_2	46.4	1
OS^+–SO_2	57.7	1
O_2S^+–SO_2	63.6	1
OCS^+–OCS	100.0	1
OCS^+–CO_2	72.0	1
SO_2^+–CO_2	42.7	1
H_3S^+–H_2O	91.6	1
thiopheneH$^+$–H_2O	42.7	1
H_3S^+–H_2S	53.6 ± 6.3	1
H_3S^+–CH_4	16.3	1
$(CH_3)_2Se^{\cdot+}$–$Se(CH_3)_2$	~95 ± 3	1
$(CH_3)_2Te^{\cdot+}$–$Te(CH_3)_2$	97 ± 2	1
(17) Group 17		
HF^+–HF	≥138	1
$(H_2Cl)^+$ –Cl	39.6	1
HCl^+–HCl	83.9	1
Cl^+–CCl_3	446.7 ± 9.6	1
Cl^+–C_2H_3	685.0 ± 4.8	1
HBr^+–HBr	96	1

Bond	$D°_{298}$/kJ mol^{-1}	Ref.
I^+–CH_3	330.0	1
$I^+(CH_3I)$–CH_3	51.1	1
$I^+(CH_3I)_2$–CH_3	112.9	1
(18) Group 18		
$He^+(He)_1$–He	17.6	1
$He^+(He)_2$–He	2.7 ± 0.6	1
$Ne^+(Ne)$–Ne	10.3 ± 0.6	1
$Ne^+(Ne)_2$–Ne	3.3 ± 0.6	1
$Ar^+(Ar)$–Ar	20.4 ± 0.6	1
$Ar^+(Ar)_2$–Ar	7.0 ± 0.6	1
$Ar^+(N_2)$–Ar	25.1	1
$Ar^+(N_2)(Ar)$–Ar	7.1	1
$Ar^+(N_2)(Ar)_2$–Ar	7.1	1
$Kr^+(Kr)$–Kr	23.3 ± 0.6	1
$Kr^+(Kr)_2$–Kr	9.0 ± 0.6	1
$Xe^+(Xe)$–Xe	25.2 ± 0.6	1
$Xe^+(Xe)_2$–Xe	11.0 ± 0.6	1
Ar^+–H_2	93.7	1
Ar^+–N_2	127.6	1
$Ar^+(N_2)$–N_2	31.0	1
$Ar^+(N_2)_2$–N_2	10.9	1
Ar^+–CO	75 ± 17	1
$Ar^+(CO)$–CO	13	1
Kr^+–CO	103.3 ± 7.5	1
Kr^+–CO_2	79.1 ± 2.9	1
K^+–(18-crown-6)	235 ± 13	1

References

1. Luo, Y. R., *Comprehensive Handbook of Chemical Bond Energies*, CRC Press, Boca Raton, FL, 2007. <https://doi.org/10.1201/9781420007282>
2. Parke, L. G., Hinton, C. S., and Armentrout, P. B., *Int. J. Mass Spectrom.* 254, 168, 2006. <https://doi.org/10.1016/j.ijms.2006.05.025>
3. Li, F.-X., Zhang, X.-G., and Armentrout, P. B., *Int. J. Mass Spectrom.* 255/256, 279, 2006. <https://doi.org/10.1016/j.ijms.2006.02.021>
4. Meloni, G., Zou, P., Klippenstein, S. J., Ahmed, M., Leone, S. R., Taatjes, C. A., and Osborn, D. L., *J. Am. Chem. Soc.* 128, 13559, 2006. <https://doi.org/10.1021/ja064556j>
5. Li, F.-X., and Armentrout, P. B., *J. Chem. Phys.* 125, 133114/1, 2006. <https://doi.org/10.1063/1.2220038>
6. Parke, L. G., Hinton, C. S., and Armentrout, P. B., *J. Phys. Chem. C* 111, 17773, 2007. <https://doi.org/10.1021/jp070855z>
7. Shuman, N. S., Ochieng, M. A., Sztáray, B., and Baer, T., *J. Phys. Chem. A* 112, 5647, 2008. <https://doi.org/10.1021/jp8007255>
8. Armentrout, P. B., and Kretzschmar, I., *J. Phys. Chem. A* 113, 10955, 2009. <https://doi.org/10.1021/jp907253r>
9. Armentrout, P. B., and Kretzschmar, I., *Phys. Chem. Chem. Phys.* 12, 4078, 2010. <https://doi.org/10.1039/b926429a>
10. Luo, Z., Zhang, Z., Huang, H., Chang, Y.-C., and Ng, C. Y., *J. Chem. Phys.*, 140, 181101, 2014. <https://doi.org/10.1063/1.4876017>
11. Armentrout, P. B., Parke, L., Hinton, C., and Citir, M., *ChemPlusChem,* 78, 1157, 2013. <https://doi.org/10.1002/cplu.201300147>
12. Armentrout, P. B., Austin, C. A., and Rodgers, M. T., *J. Phys. Chem. A* 118, 8088-8097, 2014. <https://doi.org/10.1021/jp4116172>
13. Demireva, M., et al., *J. Phys. Chem. A* 120, 8550-8563, 2016. <https://doi.org/10.1021/acs.jpca.6b09309>
14. Demireva, M., et al., *J. Chem. Phys.* 146, 174302, 2017. <https://doi.org/10.1063/1.4982683>
15. Armentrout, P. B., et al., *Phys. Chem. Chem. Phys.* 19, 11075-11088, 2017. <https://doi.org/10.1039/C7CP00914C>
16. Armentrout, P. B., et al., *J. Chem. Phys.* 147, 214307, 2017. <https://doi.org/10.1063/1.5009916>
17. Demireva, M., Armentrout, P. B. *Top. Catal.* 61, 3-19, 2018. <https://doi.org/10.1007/s11244-017-0858-1>
18. Sander, O., and Armentrout, P. B., *J. Phys. Chem. A* 123, 1675-1688, 2019. <https://doi.org/10.1021/acs.jpca.8b12257>
19. Ghiassee, M., et al., *J. Chem. Phys.* 150, 144309, 2019. <https://doi.org/10.1063/1.5091679>
20. Ashraf, M. A., et al., *J. Chem. Phys.* 149, 174301, 2018. <https://doi.org/10.1063/1.5053691>
21. Katle, A., and Armentrout, P. B., *J. Phys. Chem. A*, 123, 5893-5905, 2019. <https://doi.org/10.1021/acs.jpca.9b03938>
22. Kim, J., Armentrout, P. B., *J. Phys. Chem.* A 124, 6629-6644, 2020.
23. Kim, J., et al., *Phys. Chem. Chem. Phys.* 22, 3191-3203, 2020.
24. Armentrout, P. B., Peterson, K. A., *Inorg. Chem.* 59, 3118-3131, 2020.

ELECTRONEGATIVITY

Electronegativity is a parameter originally introduced by Pauling which describes, on a relative basis, the tendency of an atom in a molecule to attract bonding electrons. While electronegativity is not a precisely defined molecular property, the electronegativity difference between two atoms provides a useful measure of the polarity and ionic character of the bond between them. This table gives the electronegativity χ, on the Pauling scale, for the most common oxidation state of most elements. Other scales are described in the references. Column definitions for the table are as follows.

Column heading	Definition
Z	Atomic number of element; elements are listed in atomic number order
Symbol	Element symbol
χ	Electronegativity (Pauling scale)

References

1. Pauling, L., *The Nature of the Chemical Bond, Third Edition*, Cornell University Press, Ithaca, NY, 1960.
2. Allen, L. C., *J. Am. Chem. Soc.* 111, 9003, 1989. <https://doi.org/10.1021/ja00207a003>
3. Allred, A. L., *J. Inorg. Nucl. Chem.* 17, 215, 1961. <https://doi.org/10.1016/0022-1902(61)80142-5>

Electronegativity Values on the Pauling Scale

Z	Symbol	χ	*Z*	Symbol	χ	*Z*	Symbol	χ	*Z*	Symbol	χ
1	H	2.20	25	Mn	1.55	49	In	1.78	73	Ta	1.5
2	He		26	Fe	1.83	50	Sn	1.96	74	W	1.7
3	Li	0.98	27	Co	1.88	51	Sb	2.05	75	Re	1.9
4	Be	1.57	28	Ni	1.91	52	Te	2.10	76	Os	2.2
5	B	2.04	29	Cu	1.90	53	I	2.66	77	Ir	2.2
6	C	2.55	30	Zn	1.65	54	Xe	2.60	78	Pt	2.2
7	N	3.04	31	Ga	1.81	55	Cs	0.79	79	Au	2.4
8	O	3.44	32	Ge	2.01	56	Ba	0.89	80	Hg	1.9
9	F	3.98	33	As	2.01	57	La	1.10	81	Tl	1.8
10	Ne		34	Se	2.55	58	Ce	1.12	82	Pb	1.8
11	Na	0.93	35	Br	2.96	59	Pr	1.13	83	Bi	1.9
12	Mg	1.31	36	Kr		60	Nd	1.14	84	Po	2.0
13	Al	1.61	37	Rb	0.82	61	Pm		85	At	2.2
14	Si	1.90	38	Sr	0.95	62	Sm	1.17	86	Rn	
15	P	2.19	39	Y	1.22	63	Eu		87	Fr	0.7
16	S	2.58	40	Zr	1.33	64	Gd	1.20	88	Ra	0.9
17	Cl	3.16	41	Nb	1.6	65	Tb		89	Ac	1.1
18	Ar		42	Mo	2.16	66	Dy	1.22	90	Th	1.3
19	K	0.82	43	Tc	2.10	67	Ho	1.23	91	Pa	1.5
20	Ca	1.00	44	Ru	2.2	68	Er	1.24	92	U	1.7
21	Sc	1.36	45	Rh	2.28	69	Tm	1.25	93	Np	1.3
22	Ti	1.54	46	Pd	2.20	70	Yb		94	Pu	1.3
23	V	1.63	47	Ag	1.93	71	Lu	1.0			
24	Cr	1.66	48	Cd	1.69	72	Hf	1.3			

FORCE CONSTANTS FOR BOND STRETCHING

David R. Lide

Representative force constants (f) for stretching of chemical bonds are listed in this table. Except where noted, all force constants are derived from values of the harmonic vibrational frequencies ω_e. Values derived from the observed vibrational fundamentals ν, which are noted by footnote [a], are lower than the harmonic force constants, typically by 2% to 3% in the case of heavy atoms (often by 5% to 10% if one of the atoms is hydrogen). Column definitions are as follows.

Column heading	Definition
Bond	Chemical bond undergoing stretching
Mol. form.	Molecular formula of representative compounds containing such a bond
f	Force constant, in units N cm^{-1}, which is identical to the commonly used cgs unit mdyn Å^{-1}

References

1. Huber, K. P., and Herzberg, G., *Molecular Spectra and Molecular Structure. IV. Constants of Diatomic Molecules*, Van Nostrand Reinhold, New York, 1979. <https://doi.org/10.1007/978-1-4757-0961-2_2>
2. Shimanouchi, T., The Molecular Force Field, in Eyring, H., Henderson, D., and Yost, W., Eds., *Physical Chemistry: An Advanced Treatise*, Vol. IV, Academic Press, New York, 1970.
3. Tasumi, M., and Nakata, M., *Pure and Appl. Chem.* 57, 121–147, 1985. <https://doi.org/10.1351/pac198557010121>

Force Constants for Bond Stretching

Bond	Mol. form.	f/N cm^{-1}
H-H	H_2	5.75
Be-H	BeH	2.27
B-H	BH	3.05
C-H	CH	4.48
C-H	CH_4	5.44[b]
C-H	C_2H_6	4.83[a, b, c]
C-H	CH_3CN	5.33[b]
C-H	CH_3Cl	5.02[a, b, c]
C-H	$CCl_2{=}CH_2$	5.57[b]
C-H	HCN	6.22
N-H	NH	5.97
O-H	OH	7.80
O-H	H_2O	8.45
P-H	PH	3.22
S-H	SH	4.23
S-H	H_2S	4.28
F-H	HF	9.66
Cl-H	HCl	5.16
Br-H	HBr	4.12
I-H	HI	3.14
Li-H	LiH	1.03
Na-H	NaH	0.78
K-H	KH	0.56
Rb-H	RbH	0.52
Cs-H	CsH	0.47
C-C	C_2	12.16
C-C	$CCl_2{=}CH_2$	8.43
C-C	C_2H_6	4.50[a, c]
C-C	CH_3CN	5.16
C-F	CF	7.42

Bond	Mol. form.	f/N cm^{-1}
C-F	CH_3F	5.71[a, c]
C-Cl	CCl	3.95
C-Cl	CH_3Cl	3.44[a, c]
C-Cl	$CCl_2{=}CH_2$	4.02[b]
C-Br	CH_3Br	2.89[a, c]
C-I	CH_3I	2.34[a, c]
C-O	CO	19.02
C-O	CO_2	16.00
C-O	OCS	16.14
C-O	CH_3OH	5.42[a, c]
C-S	CS	8.49
C-S	CS_2	7.88
C-S	OCS	7.44
C-N	CN	16.29
C-N	HCN	18.78
C-N	CH_3CN	18.33
C-N	CH_3NH_2	5.12[a, c]
C-P	CP	7.83
Si-Si	Si_2	2.15
Si-O	SiO	9.24
Si-F	SiF	4.90
Si-Cl	SiCl	2.63
N-N	N_2	22.95
N-N	N_2O	18.72
N-O	NO	15.95
N-O	N_2O	11.70
P-P	P_2	5.56
P-O	PO	9.45
O-O	O_2	11.77
O-O	O_3	5.74[a]

Bond	Mol. form.	f/N cm^{-1}
S-O	SO	8.30
S-O	SO_2	10.33[a]
S-S	S_2	4.96
F-F	F_2	4.70
Cl-F	ClF	4.48
Br-F	BrF	4.06
Cl-Cl	Cl_2	3.23
Br-Cl	BrCl	2.82
Br-Br	Br_2	2.46
I-I	I_2	1.72
Li-Li	Li_2	0.26
Li-Na	LiNa	0.21
Na-Na	Na_2	0.17
Li-F	LiF	2.50
Li-Cl	LiCl	1.43
Li-Br	LiBr	1.20
Li-I	LiI	0.97
Na-F	NaF	1.76
Na-Cl	NaCl	1.09
Na-Br	NaBr	0.94
Na-I	NaI	0.76
Be-O	BeO	7.51
Mg-O	MgO	3.48
Ca-O	CaO	3.61

[a] Derived from fundamental frequency, without anharmonicity correction.
[b] Average of symmetric and antisymmetric (or degenerate) modes.
[c] Calculated from Local Symmetry Force Field (see Ref. 2).

FUNDAMENTAL VIBRATIONAL FREQUENCIES OF SMALL MOLECULES

David R. Lide

These six tables list the fundamental vibrational frequencies of selected three-, four-, and five-atom molecules. Both stable molecules and transient free radicals are included. The data have been taken from evaluated sources. In general, the selected values are based on gas-phase infrared, Raman, or ultraviolet spectra; when these were not available, liquid-phase or matrix-isolation spectra were used.

Molecules are grouped by structural type. Within each group, related molecules appear together for convenient comparison.

The vibrational modes are described by their approximate character in terms of stretching, bending, deformation, etc. It should be emphasized, however, that most such descriptions are only approximate, and that the true normal mode usually involves a mixture of motions. Abbreviations for vibration types are as follows.

Abbreviation	Definition
sym.	Symmetric
antisym.	Antisymmetric
str.	Stretch
deform.	Deformation
scis.	Scissors
rock.	Rocking
deg.	Degenerate

In the case of free radicals, strong interactions may exist between the electronic and bending vibrational motions. Details can be found in Refs. 3 and 4. The references should be consulted for information on the accuracy of the data and for data on other molecules not listed here.

Column definitions for Tables 1-6 are as follows.

Column heading	Definition
Mol. form.	Molecular formula of molecule
Structure	Geometry of molecule (when appropriate)
Vibration type	Fundamental vibration frequency for the vibration type specified by column heading; in units of cm^{-1} (wavenumbers)

TABLE 1. Vibrational Frequencies (cm^{-1}) of XY_2 Molecules: Point Groups $D_{\infty h}$ (Linear) and C_{2v} (Bent)

Mol. form.	Structure	Sym. str.	Bend	Antisym. str.
CO_2	Linear	1333	667	2349
CS_2	Linear	658	397	1535
C_3	Linear	1224	63	2040
CNC	Linear		321	1453
NCN	Linear	1197	423	1476
BO_2	Linear	1056	447	1278
BS_2	Linear	510	120	1015
KrF_2	Linear	449	233	590
XeF_2	Linear	515	213	555
$XeCl_2$	Linear	316		481
H_2O	Bent	3657	1595	3756
D_2O	Bent	2671	1178	2788
F_2O	Bent	928	461	831
Cl_2O	Bent	639	296	686
O_3	Bent	1103	701	1042
H_2S	Bent	2615	1183	2626
D_2S	Bent	1896	855	1999
SF_2	Bent	838	357	813
SCl_2	Bent	525	208	535
SO_2	Bent	1151	518	1362
H_2Se	Bent	2345	1034	2358
D_2Se	Bent	1630	745	1696
NH_2	Bent	3219	1497	3301
NO_2	Bent	1318	750	1618
NF_2	Bent	1075	573	942
ClO_2	Bent	945	445	1111
CH_2	Bent		963	
CD_2	Bent		752	
CF_2	Bent	1225	667	1114
CCl_2	Bent	721	333	748
CBr_2	Bent	595	196	641
SiH_2	Bent	2032	990	2022
SiD_2	Bent	1472	729	1468
SiF_2	Bent	855	345	870
$SiCl_2$	Bent	515		505
$SiBr_2$	Bent	403		400
GeH_2	Bent	1887	920	1864
$GeCl_2$	Bent	399	159	374
SnF_2	Bent	593	197	571
$SnCl_2$	Bent	352	120	334
$SnBr_2$	Bent	244	80	231
PbF_2	Bent	531	165	507
$PbCl_2$	Bent	314	99	299
ClF_2	Bent	500		576

TABLE 2. Vibrational Frequencies (cm^{-1}) of XYZ Molecules: Point Groups $C_{\infty v}$ (Linear) and C_s (Bent)

Mol. form.	Structure	XY str.	Bend	YZ str.
HCN	Linear	3311	712	2097
DCN	Linear	2630	569	1925
FCN	Linear	1077	451	2323
ClCN	Linear	744	378	2216
BrCN	Linear	575	342	2198
ICN	Linear	486	305	2188
CCN	Linear	1060	230	1917
CCO	Linear	1063	379	1967
HCO	Bent	2485	1081	1868
HCC	Linear	3612		1848
OCS	Linear	2062	520	859
NCO	Linear	1270	535	1921
NNO	Linear	2224	589	1285
HNB	Linear	3675		2035
HNC	Linear	3653		2032
HNSi	Linear	3583	523	1198
HBO	Linear		754	1817
FBO	Linear		500	2075
ClBO	Linear	676	404	1958
BrBO	Linear	535	374	1937
FNO	Bent	766	520	1844
ClNO	Bent	596	332	1800
BrNO	Bent	542	266	1799
HNF	Bent		1419	1000
HNO	Bent	2684	1501	1565
HPO	Bent	2095	983	1179
HOF	Bent	3537	886	1393
HOCl	Bent	3609	1242	725
HOO	Bent	3436	1392	1098
FOO	Bent	579	376	1490
ClOO	Bent	407	373	1443
BrOO	Bent			1487
HSO	Bent		1063	1009
NSF	Bent	1372	366	640
NSCl	Bent	1325	273	414
HCF	Bent		1407	1181
HCCl	Bent		1201	815
HSiF	Bent	1913	860	834
HSiCl	Bent		808	522
HSiBr	Bent	1548	774	408

TABLE 3. Vibrational Frequencies (cm^{-1}) of Symmetric XY_3 Molecules: Point Groups D_{3h} (Planar) and C_{3v} (Pyramidal)

Mol. form.	Structure	Sym. str.	Sym. deform.	Deg. str.	Deg. deform.
NH_3	Pyramidal	3337	950	3444	1627
ND_3	Pyramidal	2420	748	2564	1191
PH_3	Pyramidal	2323	992	2328	1118
AsH_3	Pyramidal	2116	906	2123	1003
SbH_3	Pyramidal	1891	782	1894	831
NF_3	Pyramidal	1032	647	907	492
PF_3	Pyramidal	892	487	860	344
AsF_3	Pyramidal	741	337	702	262
PCl_3	Pyramidal	504	252	482	198
PI_3	Pyramidal	303	111	325	79
AsI_3	Pyramidal	219	94	224	71
$AlCl_3$	Pyramidal	375	183	595	150
SO_3	Planar	1065	498	1391	530
BF_3	Planar	888	691	1449	480

TABLE 4. Vibrational Frequencies (cm^{-1}) of Linear XYYX Molecules: Point Group $D_{\infty h}$

Mol. form.	Sym. XY str.	Antisym. XY str.	YY str.	Bend	Bend
C_2H_2	3374	3289	1974	612	730
C_2D_2	2701	2439	1762	505	537
C_2N_2	2330	2158	851	507	233

TABLE 5. Vibrational Frequencies (cm^{-1}) of Planar X_2YZ Molecules: Point Group C_{2v}

Mol. form.	Sym. XY str.	YZ str.	YX_2 scis.	Antisym. XY str.	YX_2 rock	YX_2 wag
H_2CO	2783	1746	1500	2843	1249	1167
D_2CO	2056	1700	1106	2160	990	938
F_2CO	965	1928	584	1249	626	774
Cl_2CO	567	1827	285	849	440	580
O_2NF	1310	822	568	1792	560	742
O_2NCl	1286	793	370	1685	408	652

TABLE 6. Vibrational Frequencies (cm^{-1}) of Tetrahedral XY_4 Molecules: Point Group T_d

Mol. form.	Sym. str.	Deg. deform. (e)	Deg. str. (f)	Deg. deform. (f)
CH_4	2917	1534	3019	1306
CD_4	2109	1092	2259	996
CF_4	909	435	1281	632
CCl_4	459	217	776	314
CBr_4	267	122	672	182
CI_4	178	90	555	125
SiH_4	2187	975	2191	914
SiD_4	1558	700	1597	681
SiF_4	800	268	1032	389
$SiCl_4$	424	150	621	221
GeH_4	2106	931	2114	819
GeD_4	1504	665	1522	596
$GeCl_4$	396	134	453	172
$SnCl_4$	366	104	403	134
$TiCl_4$	389	114	498	136
$ZrCl_4$	377	98	418	113
$HfCl_4$	382	102	390	112
RuO_4	885	322	921	336
OsO_4	965	333	960	329

References

1. T. Shimanouchi, Tables of Molecular Vibrational Frequencies, Consolidated Volume I, Natl. Stand. Ref. Data Ser., Natl. Bur. Stand. (U.S.), 39, 1972. <https://doi.org/10.6028/NBS.NSRDS.39>
2. T. Shimanouchi, Tables of Molecular Vibrational Frequencies, Consolidated Volume II, *J. Phys. Chem. Ref. Data* 6, 993, 1977. <https://doi.org/10.1063/1.555560>
3. G. Herzberg, *Electronic Spectra and Electronic Structure of Polyatomic Molecules*, D. Van Nostrand Co., Princeton, NJ, 1966.
4. M. E. Jacox, Ground state vibrational energy levels of polyatomic transient molecules, *J. Phys. Chem. Ref. Data* 13, 945, 1984. <https://doi.org/10.1063/1.555722>

Molecular

SPECTROSCOPIC CONSTANTS OF DIATOMIC MOLECULES

David R. Lide

This table lists the leading spectroscopic constants and equilibrium internuclear distance r_e in the ground electronic state for selected diatomic molecules. The constants are those describing the vibrational and rotational energy through the expressions:

$$E_{vib}/hc = \omega_e(v + ½) - \omega_e x_e(v + ½)^2 + \cdots$$

$$E_{rot}/hc = B_v J(J + 1) - D_v[J(J + 1)]^2 + \cdots$$

where

$$B_v = B_e - \alpha_e(v + ½) + \cdots$$

$$D_v = D_e + \cdots$$

Here v and J are the vibrational and rotational quantum numbers, respectively, h is Planck's constant, and c is the speed of light. In this customary formulation the constants ω_e, B_e, etc., have dimensions of inverse length; in this table they are given in units of cm^{-1}.

Users should note that higher order terms in the above energy expressions are required for very precise calculations. The references contain constants for many of these higher terms, as well as more precise values of the lower constants. Also, if the ground electronic state is not $^1\Sigma$, additional terms are needed to account for the interaction between electronic and pure rotational angular momentum. For some molecules in the table the data have been analyzed in terms of the Dunham series expansion:

$$E/hc = \Sigma_{lm} Y_{lm}(v + ½)^l J^m(J + 1)^m$$

In such cases it has been assumed that $Y_{10} = \omega_e$, $Y_{01} = B_e$, etc., although in the highest approximations these identities are not precisely correct. Some of the values of r_e in the table have been corrected for breakdown of the Born-Oppenheimer approximation, which can affect the last decimal place. Because of differences in the method of data analysis and limitations in the model, care should be taken in comparing r_e values for different molecules to a precision beyond 0.001 Å. Column definitions are as follows.

Column heading	Definition
Mol. form.	Molecular formula of diatomic molecule; molecules are listed in alphabetical order by formula as most commonly written. In most cases this form places the more electropositive element first, but there are exceptions such as OH, NH, CH, etc.
Molecule	Isotopic specification of the diatomic molecule; superscripts indicate isotope for each element of the diatomic molecule
State	Spectroscopic state symbol
ω_e	Vibrational constant—first term, in cm^{-1}; entries in the ω_e column that are marked by * give the interval between $v = 0$ and $v = 1$ states instead of a value of ω_e
$\omega_e x_e$	Vibrational constant—second term, in cm^{-1}
B_e	Rotational constant in equilibrium position, in cm^{-1}
α_e	Rotational constant—first term, in cm^{-1}
D_e	Centrifugal distortion constant, in cm^{-1}
r_e	Equilibrium internuclear distance, in Å
Ref.	Reference number

Refs. 1–5 are evaluated compilations covering many molecules and giving references to the original literature.

References

1. Huber, K. P., and Herzberg, G., *Molecular Spectra and Molecular Structure IV. Constants of Diatomic Molecules*, Van Nostrand Reinhold, New York, 1979. <https://doi.org/10.1007/978-1-4757-0961-2_2>
2. *Landolt-Börnstein, Numerical Data and Functional Relationships in Science and Technology, New Series*, II/6 (1974), II/14a (1982), II/14b (1983), II/19a (1992), II/19d-1 (1995), II/24a (1998), *Molecular Constants*, Springer-Verlag, Heidelberg.
3. Lovas, F. J., and Tiemann, E., *J. Phys. Chem. Ref. Data* 3, 609, 1974. <https://doi.org/10.1063/1.3253146>
4. Irikura, K. K., *J. Phys. Chem. Ref. Data* 35, 389, 2007. <https://doi.org/10.1063/1.2436891>
5. Lovas, F. J., Tiemann, E., Coursey, J. S., Kotochigova, S. A., Chang, J., Olsen, K., and Dragoset, R. A., *Diatomic Spectral Database* (version 2.0). Available: http://physics.nist.gov/Diatomic, National Institute of Standards and Technology, Gaithersburg, MD, November 2009.
6. Wormsbecher, R. F., Hessel, M. M., and Lovas, F. J., *J. Chem. Phys.* 74, 6893, 1981. <https://doi.org/10.1063/1.441067>
7. Hedderich, H. G., Dulick, M., and Bernath, P. F., *J. Chem. Phys.* 99, 8363, 1993. <https://doi.org/10.1063/1.465611>
8. Ram, R. S., and Bernath, P. F., *J. Mol. Spectrosc.* 176, 320, 1996. <https://doi.org/10.1006/jmsp.1996.0094>
9. Miller, C. E., and Drouin, B. J., *J. Mol. Spectrosc.*. 205, 312, 2001. <https://doi.org/10.1006/jmsp.2000.8257>
10. Tellinghuisen, P. C., Tellinghuisen, J., Coxon, J. A., Velazco, J. E., and Setser, D. W., *J. Chem. Phys.* 68, 5187, 1978. <https://doi.org/10.1063/1.435583>
11. Muntianu, A., Guo, B., and Bernath, P. F., *J. Mol. Spectrosc.* 176, 274, 1996. <https://doi.org/10.1006/jmsp.1996.0087>
12. Yamada, C., Chang, M. C., and Hirota, E., *J. Chem. Phys.* 86, 3804, 1987. <https://doi.org/10.1063/1.451938>
13. Wang, X., Magnes, J., Marjatta Lyyra, A., Ross, A. J., Martin, F., Dove, P. M., and Le Roy, R. J., *J. Chem. Phys.* 117, 9339, 2002. <https://doi.org/10.1063/1.1514670>
14. Yamada, C., and Hirota, E., *J. Chem. Phys.* 99, 8489, 1993. <https://doi.org/10.1063/1.465625>
15. James, A. M., Kowalczyk, P., Fournier, R., and Simard, B., *J. Chem. Phys.* 99, 8504, 1993. <https://doi.org/10.1063/1.465627>
16. Campbell, J. M., Dulick, M., Klapstein, D., White, J. B., and Bernath, P. F., *J. Chem. Phys.* 99, 8379, 1993. <https://doi.org/10.1063/1.465613>
17. White, J. B., Dulick, M., and Bernath, P. F., *J. Chem. Phys.* 99, 8371, 1993. <https://doi.org/10.1063/1.465612>
18. Shayesteh, A., Appadoo, D. R. T., Gordon, I., Le Roy, R. J., and Bernath, P. F., *J. Chem. Phys.* 120, 10002, 2004. <https://doi.org/10.1063/1.1724821>
19. Sanz, M. E., McCarthy, M. C., and Thaddeus, P., *J. Chem. Phys.* 119, 11715, 2003. <https://doi.org/10.1063/1.1612481>
20. Müller, W., and Meyer, W., *J. Chem. Phys.* 80, 3311, 1984. <https://doi.org/10.1063/1.447084>
21. Staanum, P., Pashov, A., Knöckel, H., and Tiemann, E., *Phys. Rev. A* 75, 042513, 2007. <https://doi.org/10.1103/PhysRevA.75.042513>
22. Lovas, F. J., Maki, A. G., and Olson, W. B., *J. Mol. Spectrosc.* 87, 449, 1981. <https://doi.org/10.1016/0022-2852(81)90416-1>
23. Bogey, M., Demuynck, C., and Destombes, J. L., *Chem. Phys.* 66, 99, 1982. <https://doi.org/10.1016/0301-0104(82)88010-5>
24. Babou, Y., Rivière, Ph., Perrin, M. Y., and Soufiani, A., *Int. J. Thermophys.* 30, 416, 2009. <https://doi.org/10.1007/s10765-007-0288-6>
25. Skatrud, D. D., DeLucia, F. C., Blake, G. A., and Sastry, K. V. L. N., *J. Mol. Spectrosc.* 99, 35, 1983. <https://doi.org/10.1016/0022-2852(83)90290-4>

26. Le Floch, A., *Mol. Phys.* 72, 133, 1991. <https://doi.org/10.1080/00268979100100081>
27. George, T., Urban, W., and Le Floch, A., *J. Mol. Spectrosc.* 165, 500, 1994. <https://doi.org/10.1006/jmsp.1994.1153>
28. Mürtz, P., Thümmel, H., Pfelzer, C., and Urban, W., *Mol. Phys.* 86, 1362, 1995.
29. Maki, A. G., Lovas, F. J., and Suenram, R. D., *J. Mol. Spectrosc.* 91, 424, 1982. <https://doi.org/10.1016/0022-2852(82)90155-2>
30. Engelke, F., Ennen, G., and Meiwes, K. H., *Chem. Phys.* 66, 391, 1982. <https://doi.org/10.1016/0301-0104(82)88039-7>
31. Tanaka, T., Tamura, M., and Tanaka, K., *J. Mol. Struct.* 413, 153, 1997. <https://doi.org/10.1016/S0022-2860(97)00183-X>
32. Ng, K. F., Zou, W., Liu, W., and Cheung, S.-C., *J. Chem. Phys.* 146, 094308, 2017. <https://doi.org/10.1063/1.4977215>

Spectroscopic Constants of Diatomic Molecules

Mol. form.	Molecule	State	ω_e/cm^{-1}	$\omega_e x_e$/cm^{-1}	B_e/cm^{-1}	α_e/cm^{-1}	D_e/10^{-6} cm^{-1}	r_e/Å	Ref.
AgBr	^{107}Ag^{79}Br	$^1\Sigma^+$	249.57	0.63	0.064833	0.0002361	0.0175	2.39311	1,2,3
AgCl	^{107}Ag^{35}Cl	$^1\Sigma^+$	343.49	1.17	0.12298388	0.00059541	0.06305	2.28079	1,2,3
AgD	^{107}Ag^{2}H	$^1\Sigma^+$	1250.70	17.17	3.2572	0.0722	85.9	1.6180	1,2
AgF	^{107}Ag^{19}F	$^1\Sigma^+$	513.45	2.59	0.2657020	0.0019206	0.284	1.98318	1,2,3
AgH	^{107}Ag^{1}H	$^1\Sigma^+$	1759.9	34.06	6.449	0.201	344	1.618	1,2
AgI	^{107}Ag^{127}I	$^1\Sigma^+$	206.50	0.46	0.04486821	0.0001414	0.00847	2.54463	1,2,3
AgO	^{107}Ag^{16}O	$^2\Pi_{1/2}$	490.2	3.1	0.3020	0.0025	0.45	2.003	1,2
AlBr	^{27}Al^{79}Br	$^1\Sigma^+$	378.0	1.28	0.15919713	0.00086045	0.11285	2.29481	1,2
AlCl	^{27}Al^{35}Cl	$^1\Sigma^+$	481.77	2.10	0.24393007	0.00161108	0.25017	2.13014	7
AlD	^{27}Al^{2}H	$^1\Sigma^+$	1211.77	15.06	3.3183929	0.0698773	99.42	1.64637	17
AlF	^{27}Al^{19}F	$^1\Sigma^+$	802.32	4.85	0.55248021	0.00498426	1.0464	1.65437	4,5
AlH	^{27}Al^{1}H	$^1\Sigma^+$	1682.37	29.05	6.3937842	0.1870527	368.53	1.64736	17
AlI	^{27}Al^{127}I	$^1\Sigma^+$	316.1	1.0	0.11769985	0.00055859		2.53710	1,2
AlO	^{27}Al^{16}O	$^2\Sigma^+$	979.49	7.01	0.6413856	0.0057796	1.08	1.61782	4
AlS	^{27}Al^{32}S	$^2\Sigma^+$	617.11	3.33	0.2800368	0.0017823	0.22	2.02828	4
Al$_2$	^{27}Al$_2$	$^3\Pi_u$	285.8	0.9	0.17127	0.0008		2.701	1,2
AsD	^{75}As^{2}H	$^3\Sigma^-$	1484*		3.6688		90	1.5306	1,2
AsH	^{75}As^{1}H	$^3\Sigma^-$	2130*		7.3067	0.2117	327	1.52315	1,2
AsN	^{75}As^{14}N	$^1\Sigma^+$	1068.54	5.41	0.54551	0.003366	0.53	1.6184	1,2
AsO	^{75}As^{16}O	$^2\Pi_{1/2}$	967.08	4.85	0.48482	0.003299	0.49	1.6236	1,2
As$_2$	^{75}As$_2$	$^1\Sigma_g^+$	429.55	1.12	0.10179	0.000333		2.1026	1,2
AuD	^{197}Au^{2}H	$^1\Sigma^+$	1634.98	21.65	3.6415	0.07614	70.9	1.5238	1,2
AuH	^{197}Au^{1}H	$^1\Sigma^+$	2305.01	43.12	7.2401	0.2136	279	1.5239	1,2
Au$_2$	^{197}Au$_2$	$^1\Sigma_g^+$	190.9	0.42	0.028013	0.0000723	0.00250	2.4719	1,2
BBr	^{11}B^{79}Br	$^1\Sigma^+$	684.31	3.52	0.4894	0.0035	1.00	1.888	1,2
BCl	^{11}B^{35}Cl	$^1\Sigma^+$	840.29	5.49	0.684282	0.006812	1.80	1.71528	1,29
BD	^{11}B^{2}H	$^1\Sigma^+$	1703.3	28	6.54	0.17	400	1.2324	1,2
BF	^{11}B^{19}F	$^1\Sigma^+$	1402.16	11.82	1.51674399	0.01904848	7.11	1.26267	4
BH	^{11}B^{1}H	$^1\Sigma^+$	2366.73	49.34	12.025755	0.421565	1242	1.23217	4
BN	^{11}B^{14}N	$^3\Pi$	1514.6	12.3	1.666	0.025	8.1	1.281	1,2
BO	^{11}B^{16}O	$^2\Sigma^+$	1885.29	11.69	1.781110	0.016516	6.32	1.20475	4
BS	^{11}B^{32}S	$^2\Sigma^+$	1179.91	6.25	0.79478	0.00578	1.4	1.60935	4
B$_2$	^{11}B$_2$	$^3\Sigma_g^-$	1051.3	9.35	1.212	0.014		1.590	1,2
BaBr	^{138}Ba^{79}Br	$^2\Sigma^+$	193.77	0.41	0.0415082	0.0001219	0.00762	2.84449	1,2
BaCl	^{138}Ba^{35}Cl	$^2\Sigma^+$	279.92	0.82	0.08396717	0.00033429	0.03022	2.68276	1,2
BaD	^{138}Ba^{2}H	$^2\Sigma^+$	829.77	7.32	1.7071	0.02363	28.77	2.2304	1,2
BaF	^{138}Ba^{19}F	$^2\Sigma^+$	468.9	1.79	0.2159	0.0012	0.175	2.163	1,2,3
BaH	^{138}Ba^{1}H	$^2\Sigma^+$	1168.31	14.50	3.38285	0.06599	112.67	2.23175	1,2
BaI	^{138}Ba^{127}I	$^2\Sigma^+$	152.14	0.27	0.02680587	0.00006634	0.00333	3.08476	1,2
BaO	^{138}Ba^{16}O	$^1\Sigma^+$	669.76	2.03	0.3126140	0.0013921	0.2724	1.93969	1,2,3
BaS	^{138}Ba^{32}S	$^1\Sigma^+$	379.42	0.88	0.10331	0.0003188	0.0306	2.5074	1,2
BeD	^{9}Be^{2}H	$^2\Sigma^+$	1530.32	20.71	5.6872	0.1225	313.8	1.3419	1,2
BeF	^{9}Be^{19}F	$^2\Sigma^+$	1247.36	9.12	1.4889	0.0176	8.28	1.3610	1,2
BeH	^{9}Be^{1}H	$^2\Sigma^+$	2061.24	37.33	10.31992	0.3084	1022	1.34241	4
BeO	^{9}Be^{16}O	$^1\Sigma^+$	1487.32	11.83	1.6510	0.0190	8.20	1.3309	1,2
BeS	^{9}Be^{32}S	$^1\Sigma^+$	997.94	6.14	0.79059	0.00664	2.00	1.7415	1
BiBr	^{209}Bi^{79}Br	O$^+$	209.62	0.52	0.04321526	0.00013269	0.007347	2.60950	5
BiCl	^{209}Bi^{35}Cl	O$^+$	308.18	1.09	0.9212553	0.0004020	0.0329	2.47152	5
BiD	^{209}Bi^{2}H	$^3\Sigma^-$	1173.32	16.1	2.592	0.054	50.6	1.804	1,2
BiF	^{209}Bi^{19}F	O$^+$	513.0	2.35	0.22998897	0.00150262	0.185	2.05154	5
BiH	^{209}Bi^{1}H	$^3\Sigma^-$	1635.73	31.6	5.137	0.148	183	1.805	1,2

Mol. form.	Molecule	State	ω_e/cm^{-1}	$\omega_e x_e/cm^{-1}$	B_e/cm^{-1}	α_e/cm^{-1}	$D_e/10^{-6}\ cm^{-1}$	r_e/Å	Ref.
BiI	$^{209}Bi^{127}I$	O^+	164.12	0.32	0.02722281	0.00006979	0.00300	2.80050	5
Bi_2	$^{209}Bi_2$	$^1\Sigma_g^+$	172.71	0.34	0.022781	0.000055	0.00150	2.6596	1,2
BrCl	$^{79}Br^{35}Cl$	$^1\Sigma^+$	444.28	1.84	0.152470	0.000770	0.07183	2.13607	1,2,3
BrF	$^{79}Br^{19}F$	$^1\Sigma^+$	670.75	4.05	0.35584	0.00261	0.401	1.75894	1,2,3
BrO	$^{79}Br^{16}O$	$^2\Pi_{3/2}$	779	6.8	0.429598	0.003639	0.523	1.717	1,2,3
Br_2	$^{79}Br_2$	$^1\Sigma_g^+$	325.32	1.08	0.082107	0.0003187	0.02092	2.2811	1,2
CCl	$^{12}C^{35}Cl$	$^2\Pi_{1/2}$	876.90	5.45	0.697137	0.006853	1.9	1.64518	4
CD	$^{12}C^2H$	$^2\Pi_{1/2}$	2101.05	34.73	7.8079823	0.212240	420	1.11887	4
CF	$^{12}C^{19}F$	$^2\Pi_{1/2}$	1307.93	11.08	1.41626	0.01844	6.6	1.27218	4
CH	$^{12}C^1H$	$^2\Pi_{1/2}$	2860.75	64.44	14.45988	0.53654	1450	1.1199	4
CN	$^{12}C^{14}N$	$^2\Sigma^+$	2068.65	13.10	1.8997830	0.0173717	6.4034	1.17181	24,25
CO	$^{12}C^{16}O$	$^1\Sigma^+$	2169.81	13.29	1.931280985	0.01750439	6.1216	1.12832	24,26,27
CP	$^{12}C^{31}P$	$^2\Sigma^+$	1239.79	6.83	0.79886775	0.00596933	1.33	1.56198	4
CS	$^{12}C^{32}S$	$^1\Sigma^+$	1285.15	6.50	0.82004356	0.00591835	1.336	1.53482	1,4
CSe	$^{12}C^{80}Se$	$^1\Sigma^+$	1035.36	4.86	0.5750	0.00379	0.71	1.67609	1,2,3
C_2	$^{12}C_2$	$^1\Sigma_g^+$	1855.01	13.56	1.82010	0.01801	6.96	1.24244	24
CaBr	$^{40}Ca^{79}Br$	$^2\Sigma^+$	285.3	0.86	0.09446622	0.00040360	0.0413	2.59358	1,5
CaCl	$^{40}Ca^{35}Cl$	$^2\Sigma^+$	367.53	1.31	0.1522302	0.0007990	0.1029	2.43676	1,2
CaD	$^{40}Ca^2H$	$^2\Sigma^+$	910*		2.1769	0.035	47.9	2.002	1,2
CaF	$^{40}Ca^{19}F$	$^2\Sigma^+$	581.1	2.74	0.339	0.0026	0.45	1.967	1,2
CaH	$^{40}Ca^1H$	$^2\Sigma^+$	1298.34	19.10	4.2766	0.0970	183.7	2.0025	1,2
CaI	$^{40}Ca^{127}I$	$^2\Sigma^+$	238.70	0.63	0.0693263	0.0002634	0.0234	2.82859	1,2
CaO	$^{40}Ca^{16}O$	$^1\Sigma^+$	732.03	4.83	0.444441	0.003282	0.6541	1.8221	1,2
CaS	$^{40}Ca^{32}S$	$^1\Sigma^+$	462.23	1.78	0.1766757	0.0008270	0.1032	2.31775	1,2
CdD	$^{114}Cd^2H$	$^2\Sigma^+$			2.704		76	1.775	1,2
CdH	$^{114}Cd^1H$	$^2\Sigma^+$	1337.1*		5.323		314	1.781	1,2
ClF	$^{35}Cl^{19}F$	$^1\Sigma^+$	783.45	4.95	0.5164805	0.0043585	0.88	1.62831	4
ClO	$^{35}Cl^{16}O$	$^2\Pi_{3/2}$	853.64	5.52	0.62345797	0.0059357	1.33	1.56962	4
Cl_2	$^{35}Cl_2$	$^1\Sigma_g^+$	559.75	2.69	0.24415	0.00152	0.186	1.9872	4
CrD	$^{52}Cr^2H$	$^6\Sigma^+$	1182*		3.14		88.8	1.664	1,2
CrH	$^{52}Cr^1H$	$^6\Sigma^+$	1581*		6.220	0.179	347	1.656	1,2
CrO	$^{52}Cr^{16}O$	$^5\Pi$	898.4	6.8	0.5231	0.0070		1.615	1,2
CsBr	$^{133}Cs^{79}Br$	$^1\Sigma^+$	149.66	0.37	0.03606925	0.00012401	0.00838	3.07225	1,2,3
CsCl	$^{133}Cs^{35}Cl$	$^1\Sigma^+$	214.17	0.73	0.07209149	0.00033756	0.03268	2.90627	1,2,3
CsD	$^{133}Cs^2H$	$^1\Sigma^+$	619.1*		1.354		20	2.505	1,2
CsF	$^{133}Cs^{19}F$	$^1\Sigma^+$	352.56	1.62	0.18436969	0.0011756	0.20168	2.34535	1,2,3
CsH	$^{133}Cs^1H$	$^1\Sigma^+$	891.0	12.9	2.7099	0.0579	113	2.4938	1,2
CsI	$^{133}Cs^{127}I$	$^1\Sigma^+$	119.18	0.25	0.02362736	0.00006826	0.00371	3.31519	1,2,3
CsO	$^{133}Cs^{16}O$	$^2\Sigma^+$	357.5*		0.223073	0.001303	0.348	2.3007	1,2
Cs_2	$^{133}Cs_2$	$^1\Sigma_g^+$	42.02	0.08	0.0127	0.0000264	0.00464	4.47	1,2
CuBr	$^{63}Cu^{79}Br$	$^1\Sigma^+$	314.8	0.96	0.10192625	0.00045214	0.04274	2.17344	1,2
CuCl	$^{65}Cu^{35}Cl$	$^1\Sigma^+$	415.29	1.58	0.17628802	0.00099647	0.12706	2.05118	1,2
CuD	$^{63}Cu^2H$	$^1\Sigma^+$	1384.14	18.97	4.0381	0.0917	136.2	1.4626	1,2
CuF	$^{63}Cu^{19}F$	$^1\Sigma^+$	622.7	3.95	0.3794029	0.0032298	0.563	1.74493	1,2,3
CuH	$^{63}Cu^1H$	$^1\Sigma^+$	1941.26	37.51	7.9441	0.2563	520	1.46263	1,2
CuI	$^{63}Cu^{127}I$	$^1\Sigma^+$	264.5	0.60	0.07328742	0.00028390	0.02244	2.33832	1,2
CuO	$^{63}Cu^{16}O$	$^2\Pi_{3/2}$	640.17	4.43	0.44454	0.00456	0.85	1.7244	1,2
CuS	$^{63}Cu^{32}S$	$^2\Pi_{3/2}$	415.0	1.75	0.1891		0.18	2.051	1,2
Cu_2	$^{63}Cu_2$	$^1\Sigma_g^+$	264.55	1.02	0.10874	0.000614	0.0716	2.2197	1,2
DBr	$^2H^{81}Br$	$^1\Sigma^+$	1884.75	22.72	4.245596	0.084	88.32	1.4145	1,2
DCl	$^2H^{35}Cl$	$^1\Sigma^+$	2145.16	27.18	5.448796	0.113292	140	1.27458	1,2
DF	$^2H^{19}F$	$^1\Sigma^+$	2998.19	45.76	11.0102	0.3017	594	0.91694	1,2
D_2	$^2H_2$	$^1\Sigma_g^+$	3115.50	61.82	30.444	1.0786	11410	0.74152	1,2
F_2	$^{19}F_2$	$^1\Sigma_g^+$	916.93	11.32	0.889294	0.0125952	3.3	1.41264	4
FeO	$^{56}Fe^{16}O$	$^5\Delta$	965*		0.650		0.72	1.444	1,2
GaBr	$^{69}Ga^{81}Br$	$^1\Sigma^+$	263.0	0.81	0.081839	0.0003207	0.032	2.35248	1,2,3
GaCl	$^{69}Ga^{35}Cl$	$^1\Sigma^+$	365.67	1.25	0.1499046	0.0007936	0.1008	2.20169	1,2,3
GaD	$^{69}Ga^2H$	$^1\Sigma^+$	1143.23	14.43	3.1218854	0.0689978	93.021	1.66113	16
GaF	$^{69}Ga^{19}F$	$^1\Sigma^+$	622.2	3.2	0.3595161	0.0028642	0.50	1.77437	1,2,3
GaH	$^{69}Ga^1H$	$^1\Sigma^+$	1603.94	28.41	6.1434095	0.1906376	359.70	1.66208	16
GaI	$^{69}Ga^{127}I$	$^1\Sigma^+$	216.38	0.47	0.0569359	0.0001897	0.015770	2.57464	1,2,3

Mol. form.	Molecule	State	ω_e/cm^{-1}	$\omega_e x_e$/cm^{-1}	B_e/cm^{-1}	α_e/cm^{-1}	D_e/10^{-6} cm^{-1}	r_e/Å	Ref.
GaO	$^{69}Ga^{16}O$	$^2\Sigma$	767.5	6.24	0.4271		0.37	1.744	1,2
GeBr	$^{74}Ge^{79}Br$	$^2\Pi_{1/2}$	295	0.7					1,2
GeCl	$^{74}Ge^{35}Cl$	$^2\Pi_{1/2}$	407.6	1.36					1,2
GeD	$^{72}Ge^2H$	$^2\Pi_{1/2}$	1320.09	19	3.415	0.070	83.2	1.5874	1,2
GeH	$^{72}Ge^1H$	$^2\Pi_{1/2}$	1833.77	37	6.726	0.192	326	1.5880	1,2
GeO	$^{74}Ge^{16}O$	$^1\Sigma^+$	986.49	4.47	0.4856981	0.0030787	0.4709	1.62464	1,2
GeS	$^{74}Ge^{32}S$	$^1\Sigma^+$	575.8	1.80	0.18656576	0.00074910	0.07883	2.01209	1,2
GeSe	$^{74}Ge^{80}Se$	$^1\Sigma^+$	408.7	1.36	0.09634051	0.00028904	0.02207	2.13463	1,2
GeTe	$^{74}Ge^{130}Te$	$^1\Sigma^+$	323.9	0.75	0.06533821	0.00017246	0.012	2.34017	1,2
HBr	$^1H^{81}Br$	$^1\Sigma^+$	2648.97	45.22	8.46488	0.23328	345.8	1.41444	1,2,3
HCl	$^1H^{35}Cl$	$^1\Sigma^+$	2990.92	52.80	10.5933002	0.3069985	531.94	1.27456	4
HF	$^1H^{19}F$	$^1\Sigma^+$	4138.39	89.94	20.953712	0.7933704	2150	0.91685	4
HI	$^1H^{127}I$	$^1\Sigma^+$	2309.01	39.64	6.4263650	0.1689	206.9	1.60916	1,2,3
H_2	$^1H_2$	$^1\Sigma_g^+$	4401.21	121.34	60.853	3.062	47100	0.74144	1,2
HgD	$^{202}Hg^2H$	$^2\Sigma^+$	896.12*		2.739		91	1.757	1,2
HgH	$^{202}Hg^1H$	$^2\Sigma^+$	1203.24*		5.3888		395.3	1.7662	1,2
IBr	$^{127}I^{79}Br$	$^1\Sigma^+$	268.64	0.81	0.0568325	0.0001969	0.0102	2.46899	1,2,3
ICl	$^{127}I^{35}Cl$	$^1\Sigma^+$	384.29	1.50	0.1141587	0.0005354	0.0403	2.32088	1,2,3
IF	$^{127}I^{19}F$	$^1\Sigma^+$	610.24	3.12	0.2797111	0.0018738	0.2356	1.90976	1,2,3
IO	$^{127}I^{16}O$	$^2\Pi_{3/2}$	681.5	4.3	0.34026	0.00270	0.36	1.8676	1,2
I_2	$^{127}I_2$	$^1\Sigma_g^+$	214.50	0.61	0.03737	0.000114	0.0043	2.666	1,2
InBr	$^{115}In^{81}Br$	$^1\Sigma^+$	221.0	0.65	0.05489468	0.00018672	0.01350	2.54315	1,2,3
InCl	$^{115}In^{35}Cl$	$^1\Sigma^+$	317.39	1.03	0.1090583	0.0005177	0.0515	2.40117	1,2,3
InD	$^{115}In^2H$	$^1\Sigma^+$	1048.2	12.4	2.523	0.051	58	1.837	1,2
InF	$^{115}In^{19}F$	$^1\Sigma^+$	535.4	2.6	0.2623241	0.0018798	0.252	1.98540	1,2,3
InH	$^{115}In^1H$	$^1\Sigma^+$	1476.0	25.61	4.995	0.143	223	1.8380	1,2
InI	$^{115}In^{127}I$	$^1\Sigma^+$	177.08	0.34	0.03686702	0.00010411	0.00639	2.75364	1,2,3
KBr	$^{39}K^{79}Br$	$^1\Sigma^+$	213	0.80	0.08122109	0.00040481	0.04462	2.82078	1,2,3
KCl	$^{39}K^{35}Cl$	$^1\Sigma^+$	281	1.30	0.1286348	0.0007899	0.1087	2.66665	1,2,3
KD	$^{39}K^2H$	$^1\Sigma^+$	707	7.7	1.754	0.0318	50	2.240	1,2
KF	$^{39}K^{19}F$	$^1\Sigma^+$	426.26	2.45	0.27993741	0.00233492	0.4829	2.17146	1,2,3
KH	$^{39}K^1H$	$^1\Sigma^+$	983.6	14.3	3.416400	0.085313	163.55	2.243	1,2
KI	$^{39}K^{127}I$	$^1\Sigma^+$	186.53	0.57	0.06087473	0.00026776	0.02593	3.04784	1,2,3
K_2	$^{39}K_2$	$^1\Sigma_g^+$	92.02	0.28	0.056743	0.000165	0.0863	3.9051	1
LaO	$^{139}La^{16}O$	$^2\Sigma^+$	812.8	2.22	0.35252001	0.00142365	0.2626	1.82591	1,2
LiBr	$^7Li^{79}Br$	$^1\Sigma^+$	563.2	3.5	0.555399	0.005644	2.159	2.17043	1,2,3
LiCl	$^7Li^{35}Cl$	$^1\Sigma^+$	642.95	4.47	0.7065225	0.0080102	3.409	2.02067	1,5
LiCs	$^7Li^{133}Cs$	$^1\Sigma^+$	184.70	1.00	0.188003	0.001248	0.7784	3.6681	21
LiD	$^7Li^2H$	$^1\Sigma^+$	1054.94	13.06	4.23308131	0.09149428	272	1.59526	4,5
LiF	$^7Li^{19}F$	$^1\Sigma^+$	910.57	8.21	1.34525715	0.02028749	11.75	1.56386	4,5
LiH	$^7Li^1H$	$^1\Sigma^+$	1405.50	21.17	7.5137315	0.2163911	859	1.59490	4,5
LiI	$^7Li^{127}I$	$^1\Sigma^+$	496.85	2.85	0.4431766	0.0040862	1.4104	2.39192	1,2,3
LiK	$^7Li^{39}K$	$^1\Sigma^+$	207		0.265			3.27	1,20
LiO	$^7Li^{16}O$	$^2\Pi$	814.62	7.78	1.212830	0.017899	0.1079	1.68822	3,14
Li_2	$^7Li_2$	$^1\Sigma_g^+$	351.41	2.58	0.672530	0.007046	9.79	2.6733	13
MgCl	$^{24}Mg^{35}Cl$	$^2\Sigma^+$	462.12*		0.2456154	0.0016204	0.2723	2.19639	1,2
MgD	$^{24}Mg^2H$	$^2\Sigma^+$	1077.30	15.52	3.034344	0.066607	96.25	1.72916	18
MgF	$^{24}Mg^{19}F$	$^2\Sigma^+$	711.69*		0.51922	0.00470	1.080	1.7500	1,2
MgH	$^{24}Mg^1H$	$^2\Sigma^+$	1492.78	29.85	5.825523	0.177298	354.56	1.72972	18
MgO	$^{24}Mg^{16}O$	$^1\Sigma^+$	785.21	5.13	0.5748414	0.0053223	1.233	1.74817	28,4
MgS	$^{24}Mg^{32}S$	$^1\Sigma^+$	528.74	2.70	0.26797	0.00176	0.276	2.1425	1
Mg_2	$^{24}Mg_2$	$^1\Sigma_g^+$	51.12	1.64	0.09287	0.00378	1.22	3.891	1
MnD	$^{55}Mn^2H$	$^7\Sigma$	1103	13.9	2.8957	0.051	79.5	1.7310	1,2
MnH	$^{55}Mn^1H$	$^7\Sigma$	1548.0	28.8	5.6841	0.1570	303.9	1.7311	1,2
NBr	$^{14}N^{79}Br$	$^3\Sigma^-$	691.75	4.72	0.444	0.0040		1.79	1,2
NCl	$^{14}N^{35}Cl$	$^3\Sigma^-$	827.96	5.30	0.64976739	0.00641432	1.596	1.61071	1,5
ND	$^{14}N^2H$	$^3\Sigma^-$	2399.13	42.11	8.9087	0.2546	491.7	1.03665	8
NF	$^{14}N^{19}F$	$^3\Sigma^-$	1141.37	8.99	1.205679	0.014889	5.4	1.31698	4
NH	$^{14}N^1H$	$^3\Sigma^-$	3282.72	79.04	16.66792	0.65038	1710	1.03719	4
NO	$^{14}N^{16}O$	$^2\Pi_{1/2}$	1904.20	14.07	1.67195	0.0171	0.5	1.15077	1,24
NS	$^{14}N^{32}S$	$^2\Pi_{1/2}$	1218.7	7.28	0.769602	0.0064	1.2	1.4940	1,2,3

Mol. form.	Molecule	State	ω_e/cm^{-1}	$\omega_e x_e$/cm^{-1}	B_e/cm^{-1}	α_e/cm^{-1}	D_e/10^{-6} cm^{-1}	r_e/Å	Ref.
N_2	$^{14}N_2$	$^1\Sigma_g^+$	2358.56	14.32	1.998236	0.017310	5.737	1.09769	24
NaBr	$^{23}Na^{79}Br$	$^1\Sigma^+$	302	1.5	0.1512533	0.0009410	0.1554	2.50204	1,2,3
NaCl	$^{23}Na^{35}Cl$	$^1\Sigma^+$	364.68	1.78	0.21806302	0.00162479	0.31202	2.36080	4,5
NaD	$^{23}Na^2H$	$^1\Sigma^+$	826.1*		2.557089	0.051600	93.46	1.88654	1,2
NaF	$^{23}Na^{19}F$	$^1\Sigma^+$	535.66	3.58	0.43690153	0.0045592	1.16296	1.92595	11
NaH	$^{23}Na^1H$	$^1\Sigma^+$	1171.97	19.70	4.90327	0.1370	343.8	1.8870	4,5
NaI	$^{23}Na^{127}I$	$^1\Sigma^+$	258	1.1	0.1178056	0.0006478	0.0973	2.71145	1,2,3
NaK	$^{39}K^{23}Na$	$^1\Sigma^+$	124.03	0.50	0.09519989	0.00044966	0.2206	3.49958	6
NaLi	$^7Li^{23}Na$	$^1\Sigma^+$	256.99	1.66	0.376833	0.003810	3.340	2.88851	4,30
NaO	$^{23}Na^{16}O$	$^2\Pi$	492.3		0.424630	0.004506	1.2638	2.05155	1,2
Na_2	$^{23}Na_2$	$^1\Sigma_g^+$	159.09	0.71	0.15473537	0.0086375	0.58	3.07858	1,4,20
NbO	$^{93}Nb^{16}O$	$^4\Sigma^-$	989.0	3.8	0.4321	0.0021	0.22	1.691	1,2
Nb_2	$^{93}Nb_2$	$^3\Sigma_g^-$	424.89	0.94	0.084054	0.000242	0.016	2.0778	15
NiD	$^{58}Ni^2H$	$^2\Delta_{5/2}$	1390.1	19	3.992	0.092	130	1.465	1,2
NiH	$^{58}Ni^1H$	$^2\Delta_{5/2}$	1926.6	38	7.700	0.23	481	1.476	1,2
OD	$^{16}O^2H$	$^2\Pi_{3/2}$	2720.24	44.05	10.021	0.276	537.4	0.9698	1,2
FO	$^{19}F^{16}O$	$^2\Pi_{3/2}$	1053.01	9.92	1.0587076	0.013295	4.2823	1.35411	9
OH	$^{16}O^1H$	$^2\Pi_{3/2}$	3737.76	84.88	18.911	0.7242	1938	0.96966	1,2,3
O_2	$^{16}O_2$	$^3\Sigma_g^-$	1580.19	11.98	1.445622	0.015933	4.839	1.20752	24
PCl	$^{31}P^{35}Cl$	$^3\Sigma^-$	551.38	2.23	0.2528748	0.0015119	0.2124	2.01461	1,2
PD	$^{31}P^2H$	$^3\Sigma^-$	1699.2	23.0	4.4081	0.0928	116	1.4220	1,2
PF	$^{31}P^{19}F$	$^3\Sigma^-$	846.73	4.49	0.5667427	0.004639	1.0156	1.58933	12
PH	$^{31}P^1H$	$^3\Sigma^-$	2363.77	43.91	8.53904	0.2534	4.462	1.42218	8
PN	$^{31}P^{14}N$	$^1\Sigma^+$	1336.95	6.90	0.7864844	0.0055337	1.091	1.49087	4
PO	$^{31}P^{16}O$	$^2\Pi_{1/2}$	1233.34	6.56	0.733223657	0.005466162	1.3	1.47637	1,4
P_2	$^{31}P_2$	$^1\Sigma_g^+$	780.77	2.84	0.30362	0.00149	0.188	1.8934	1
PbBr	$^{208}Pb^{79}Br$	$^2\Pi_{1/2}$	207.5	0.50					1,2
PbCl	$^{208}Pb^{35}Cl$	$^2\Pi_{1/2}$	303.9	0.88					1,2
PbF	$^{208}Pb^{19}F$	$^2\Pi_{1/2}$	502.73	2.28	0.22875	0.001473	0.183	2.0575	1,2
PbH	$^{208}Pb^1H$	$^2\Pi_{1/2}$	1564.1	29.75	4.971	0.144	201	1.839	1,2
PbO	$^{208}Pb^{16}O$	$^1\Sigma^+$	720.96	3.52	0.30730373	0.00190977	0.2138	1.92181	1,2,3
PbS	$^{208}Pb^{32}S$	$^1\Sigma^+$	429.17	1.26	0.11632307	0.00043510	0.03418	2.28678	1,2,3
PbSe	$^{208}Pb^{80}Se$	$^1\Sigma^+$	277.6	0.51	0.05059953	0.00012993	0.0070	2.40218	1,2,3
PbTe	$^{208}Pb^{130}Te$	$^1\Sigma^+$	212.0	0.43	0.03130774	0.00006743	0.0027	2.59492	1,2,3
Pb_2	$^{208}Pb_2$		110.5	0.35					1,2
PtC	$^{195}Pt^{12}C$	$^1\Sigma^+$	1051.13	4.86	0.53044	0.003273	0.546	1.6767	1,2
PtD	$^{195}Pt^2H$	$^2\Delta_{5/2}$	1644.3*		3.640	0.071	66	1.524	1,2
PtH	$^{195}Pt^1H$	$^2\Delta_{5/2}$	2294.68*		7.1963	0.1996	261	1.52852	1,2
RbBr	$^{85}Rb^{79}Br$	$^1\Sigma^+$	169.46	0.46	0.04752798	0.00018596	0.01496	2.94474	1,2,3
RbCl	$^{85}Rb^{35}Cl$	$^1\Sigma^+$	228	0.92	0.0876404	0.0004537	0.04947	2.78673	1,2,3
RbF	$^{85}Rb^{19}F$	$^1\Sigma^+$	376	1.9	0.2106640	0.0015228	0.2684	2.27033	1,2,3
RbH	$^{85}Rb^1H$	$^1\Sigma^+$	936.9	14.21	3.020	0.072	123	2.367	1,2
RbI	$^{85}Rb^{127}I$	$^1\Sigma^+$	138.51	0.33	0.03283293	0.00010946	0.00738	3.17688	1,2,3
RbO	$^{85}Rb^{16}O$	$^2\Sigma^+$	388.4*		0.246481	0.002174	0.397	2.25420	1,2
SD	$^{32}S^2H$	$^2\Pi_{3/2}$	1885	31	4.95130	0.10308	130	1.34049	1,2
SF	$^{32}S^{19}F$	$^2\Pi_{3/2}$	837.64	4.47	0.555173	0.004459	0.975	1.59624	4,5
SH	$^{32}S^1H$	$^2\Pi_{3/2}$	2696.25	48.74	9.60025	0.27990	480	1.34061	4
SO	$^{32}S^{16}O$	$^3\Sigma^-$	1149.2	5.6	0.7208171	0.005737	1.134	1.48109	1,19,23
S_2	$^{32}S_2$	$^3\Sigma_g^-$	725.71	2.86	0.29539516	0.00159754	0.19	1.88941	4
SbCl	$^{121}Sb^{35}Cl$	$^3\Sigma^-$	374.7	0.6					1,2
SbD	$^{121}Sb^2H$	$^3\Sigma^-$			2.8782		45	1.7194	1,2
SbF	$^{121}Sb^{19}F$	$^3\Sigma^-$	605.0	2.6	0.2792	0.0020	0.23	1.918	1,2
SbH	$^{121}Sb^1H$	$^3\Sigma^-$			5.684		240	1.723	1,2
SbN	$^{121}Sb^{14}N$	$^1\Sigma^+$	942.0	5.6					1,2
SbO	$^{121}Sb^{16}O$	$^2\Pi_{1/2}$	816	4.2	0.3580	0.0022	0.270	1.826	1,2
ScF	$^{45}Sc^{19}F$	$^1\Sigma^+$	735.6	3.8	0.3950	0.00266		1.788	1,2
SeD	$^{80}Se^2H$	$^2\Pi_{3/2}$	1708*		3.94			1.48	1,2
SeH	$^{80}Se^1H$	$^2\Pi_{3/2}$	2400*		8.02	0.23	330	1.48	1,2
SeO	$^{80}Se^{16}O$	$^3\Sigma^-$	914.69	4.52	0.4655	0.00323	0.5	1.648	1,2
Se_2	$^{80}Se_2$	$^3\Sigma_g^-$	385.30	0.96	0.08992	0.000288	0.024	2.166	1,2
SiCl	$^{28}Si^{35}Cl$	$^2\Pi_{1/2}$	535.59	2.18	0.256103	0.001582	0.25	2.05794	4

Mol. form.	Molecule	State	ω_e/cm^{-1}	$\omega_e x_e$/cm^{-1}	B_e/cm^{-1}	α_e/cm^{-1}	D_e/10^{-6} cm^{-1}	r_e/Å	Ref.
SiD	$^{28}Si^{2}H$	$^{2}\Pi_{1/2}$	1469.32	18.23	3.8840	0.0781	105.4	1.5199	1,2
SiF	$^{28}Si^{19}F$	$^{2}\Pi_{1/2}$	857.33	4.83	0.58125735	0.00503859	1.065	1.60100	31
SiH	$^{28}Si^{1}H$	$^{2}\Pi_{1/2}$	2042.52	36.06	7.503898	0.21814	400	1.51966	4
SiN	$^{28}Si^{14}N$	$^{2}\Sigma^{+}$	1151.28	6.46	0.730927	0.005685	1.2	1.57207	4
SiO	$^{28}Si^{16}O$	$^{1}\Sigma^{+}$	1241.54	5.97	0.7267521	0.0050379	0.9923	1.50975	1,19,22
SiS	$^{28}Si^{32}S$	$^{1}\Sigma^{+}$	749.64	2.58	0.30352788	0.00147308	0.201	1.92926	1,19
SiSe	$^{28}Si^{80}Se$	$^{1}\Sigma^{+}$	580.0	1.78	0.1920117	0.0007767	0.0842	2.05832	1,2,3
Si_2	$^{28}Si_2$	$^{3}\Sigma_g^{-}$	510.98	2.02	0.2390	0.0014	0.21	2.246	1
SnBr	$^{120}Sn^{79}Br$	$^{2}\Pi_{1/2}$	247.2	0.6					1,2
SnCl	$^{120}Sn^{35}Cl$	$^{2}\Pi_{1/2}$	351.1	1.06	0.1117	0.0004		2.361	1,2
SnD	$^{120}Sn^{2}H$	$^{2}\Pi_{1/2}$	1188.0*		2.6950	0.049	53.4	1.7770	1,2
SnF	$^{118}Sn^{19}F$	$^{2}\Pi_{1/2}$	577.6	2.69	0.2727	0.0014	0.26	1.944	1,2
SnH	$^{120}Sn^{1}H$	$^{2}\Pi_{1/2}$			5.31488		207.5	1.78146	1,2
SnI	$^{120}Sn^{127}I$	$^{2}\Pi_{1/2}$	199.0	0.6					1,2
SnO	$^{120}Sn^{16}O$	$^{1}\Sigma^{+}$	822.13	3.72	0.35571998	0.00214432	0.26638	1.83251	1,2,3
SnS	$^{120}Sn^{32}S$	$^{1}\Sigma^{+}$	487.26	1.36	0.13686139	0.00050563	0.0424	2.20898	1,2,3
SnSe	$^{120}Sn^{80}Se$	$^{1}\Sigma^{+}$	331.2	0.74	0.0649978	0.0001705	0.011	2.32557	1,2,3
SnTe	$^{120}Sn^{130}Te$	$^{1}\Sigma^{+}$	259.5	0.50	0.04247917	0.00009543	0.0055	2.52280	1,2,3
SrBr	$^{88}Sr^{79}Br$	$^{2}\Sigma^{+}$	216.60	0.52	0.0541847	0.0001827	0.01356	2.73522	1,2
SrCl	$^{88}Sr^{35}Cl$	$^{2}\Sigma^{+}$	302.3	0.95					1,2
SrD	$^{88}Sr^{2}H$	$^{2}\Sigma^{+}$	841	8.6	1.8609	0.0292	34.7	2.1449	1,2
SrF	$^{88}Sr^{19}F$	$^{2}\Sigma^{+}$	502.4	2.3	0.2505346	0.0015513	0.2498	2.07537	1,2
SrH	$^{88}Sr^{1}H$	$^{2}\Sigma^{+}$	1206.2	17.0	3.6751	0.0814	135	2.1456	1,2
SrI	$^{88}Sr^{127}I$	$^{2}\Sigma^{+}$	173.77	0.35	0.0367097	0.0001060	0.00655	2.94364	1,2
SrO	$^{88}Sr^{16}O$	$^{1}\Sigma^{+}$	653.5	3.96	0.33798	0.00219	0.36	1.91983	1,2
T_2	$^{3}H_2$	$^{1}\Sigma_g^{+}$	2546.5	41.23	20.335	0.5887		0.74142	1,2
TaF	$^{181}Ta^{19}F$	$^{3}\Sigma^{-}$	700.1		0.2965			1.8184	32
TaO	$^{181}Ta^{16}O$	$^{2}\Delta_{3/2}$	1028.69	3.51	0.40284	0.00182	0.2450	1.68746	1,2
TeH	$^{(130)}Te^{1}H$	$^{2}\Pi_{3/2}$			5.56			1.74	1,2
TeO	$^{130}Te^{16}O$	O^{+}	797.11	4.00	0.3554	0.00237	0.27	1.825	1,2
Te_2	$^{130}Te_2$	$^{3}\Sigma_g^{-}$	247.07	0.51	0.039681	0.000106	0.0044	2.5574	1,2
ThO	$^{232}Th^{16}O$	$^{1}\Sigma^{+}$	895.77	2.39	0.332644	0.001302	0.1833	1.84032	1,2
TiO	$^{48}Ti^{16}O$	$^{3}\Delta_1$	1009.02	4.50	0.53541	0.00301	0.603	1.6202	1,2
TlBr	$^{205}Tl^{81}Br$	$^{1}\Sigma^{+}$	192.10	0.39	0.0423899	0.0001276	0.0083	2.61817	1,2,3
TlCl	$^{205}Tl^{35}Cl$	$^{1}\Sigma^{+}$	284.71	0.86	0.09139702	0.00039784	0.0377	2.48483	1,2,3
TlD	$^{205}Tl^{2}H$	$^{1}\Sigma^{+}$	987.7	12.04	2.419	0.057	60	1.869	1,2
TlF	$^{205}Tl^{19}F$	$^{1}\Sigma^{+}$	476.86	2.24	0.22315014	0.00150380	0.1955	2.08439	1,2,3
TlH	$^{205}Tl^{1}H$	$^{1}\Sigma^{+}$	1390.7	22.7	4.806	0.154	254	1.870	1,2
TlI	$^{205}Tl^{127}I$	$^{1}\Sigma^{+}$	150*		0.0271676	0.0000664	0.0036	2.81361	1,2,3
VO	$^{51}V^{16}O$	$^{4}\Sigma^{-}$	1011.3	4.86	0.54825	0.00352	0.6	1.5893	1,2
XeF	$^{(132)}Xe^{19}F$	$^{2}\Sigma$	225.4	10.9	0.19326	0.00699	0.536	2.293	10
YCl	$^{89}Y^{35}Cl$	$^{1}\Sigma$	380.7	1.3	0.1160	0.0003	0.09	2.41	1,2
YF	$^{89}Y^{19}F$	$^{1}\Sigma^{+}$	631.29	2.50	0.29042	0.00163	0.237	1.9257	1,2
YO	$^{89}Y^{16}O$	$^{2}\Sigma^{+}$	861.0	2.9	0.3881	0.0018	0.32	1.790	1,2
YbD	$^{174}Yb^{2}H$	$^{2}\Sigma^{+}$	886.6	10.57	2.01162	0.03425	41.60	2.0516	1,2
YbH	$^{174}Yb^{1}H$	$^{2}\Sigma^{+}$	1249.54	21.06	3.9931	0.0957	161.8	2.0526	1,2
ZnCl	$^{64}Zn^{35}Cl$	$^{2}\Sigma$	390.5	1.6					1,2
ZnD	$^{64}Zn^{2}H$	$^{2}\Sigma^{+}$	1072	28	3.350		124	1.6054	1,2
ZnF	$^{64}Zn^{19}F$	$^{2}\Sigma$	628	3.5					1,2
ZnH	$^{64}Zn^{1}H$	$^{2}\Sigma^{+}$	1607.6	55.14	6.6794	0.2500	466	1.5949	1,2
ZnI	$^{64}Zn^{127}I$	$^{2}\Sigma$	223.4	0.6					1,2
ZrO	$^{90}Zr^{16}O$	$^{1}\Sigma^{+}$	969.8	4.9	0.42263	0.0023	0.319	1.7116	1,2

* Indicates a value for the interval between $v = 0$ and $v = 1$ states instead of a value of ω_e.

Section 10
Atomic, Molecular, and Optical Physics

Persistent Lines of the Neutral Atomic Elements 10-1
Atomic Transition Probabilities 10-51
Electron Affinities 10-54
Proton Affinities 10-76
Polarizabilities of Atoms and Molecules 10-95
Ionization Energies of Atoms and Atomic Ions 10-112
Ionization Energies of Gas-Phase Molecules 10-116
Attenuation Coefficients for High-Energy Electromagnetic Radiation 10-134
Classification of Electromagnetic Radiation 10-140
Sensitivity of the Human Eye to Light of Different Wavelengths 10-142
Blackbody Radiation 10-143
Characteristics of Infrared Detectors 10-145
Index of Refraction of Inorganic Crystals 10-146
Refractive Index and Transmittance of Representative Glasses 10-150
Index of Refraction of Water 10-151
Index of Refraction of Liquids for Calibration Purposes 10-152
Index of Refraction of Air 10-153
Index of Refraction of Gases 10-154

PERSISTENT LINES OF THE NEUTRAL ATOMIC ELEMENTS

Yuri Ralchenko and Alexander Kramida

The persistent lines in the spectra of atomic elements are the lines that are observed even at very low amounts of an element within a multielement sample. Such lines are useful for identifying the occurrence of an element even at quite small concentrations. Persistent lines are also called the "ultimate" lines for an element. The strongest persistent lines almost always contain the resonance lines, that is, the lines for transitions from a higher state to the ground state.

The data in these tables are from the NIST physical reference databases (Refs. 1-2) and include at least one of the resonance lines for each spectrum. The most sensitive or ultimate lines for many spectra lie in the vacuum-ultraviolet region (wavelength < 2000 Å). In such cases, NIST has included, when possible, some lines above 2000 Å in the persistent lines list. The data covers broader wavelength ranges than most tables of this sort. For all transitions with wavenumbers greater than 50,000 cm^{-1}, the wavelengths listed are vacuum wavelengths; for those less than 50,000 cm^{-1}, air wavelengths are given.

The list of persistent lines includes the energy levels involved in the transition, complete with configuration, term designations, and *J* values.

Persistent Line Spectra

These spectra cover the neutral atomic elements in order of atomic number Z from Hydrogen ($Z = 1$) to Einsteinium ($Z = 99$) and are grouped into eight tables, based on the common arrangement in the Periodic Table. Each spectrum is arranged in two rows. The upper row contains information about the line and transition probability as well as data on the lower state. The second row contains data on the upper state. The two-row set contains the following information.

Column heading	Definition
Intensity	Observed relative intensity on an arbitrary scale that is different for each element; the meaning of character symbols appearing after the numerical value of intensity is explained in the table below; see also NIST ASD Help pages <https://physics.nist.gov/PhysRefData/ASD/Html/lineshelp.html#OUTRELINT>
Wavelength	Observed wavelength in ångströms (Å); wavelengths between 2000 Å and 20000 Å are given in standard air (15 °C, 0.033% CO_2); outside of this range, the wavelengths are in air; conversion between vacuum and air was made by NIST using the 5-parameter formula in Ref. 3
Uncert. wavelength	Uncertainty of observed wavelength in ångströms (Å); blank if not available
A_{ki}	Radiative transition probability, in units of 10^8 s^{-1}
f_{ik}	Absorption oscillator strength (dimensionless) [Online Edition only]
S	Line strength (atomic units) [Online Edition only]
Accur.	Accuracy code for A_{ki}; see also <https://physics.nist.gov/PhysRefData/ASD/Html/lineshelp.html#OUTACC>
Energy levels	In the first row, energy of the lower level, in cm^{-1} (wavenumber); in the second row, the energy of the upper level, in cm^{-1} (wavenumber)
Configs.	Configuration and number of outermost electrons. In the first row, configuration of the lower level; in the second row, configuration of the upper level; if a single number, e.g., "5" appears as a configuration, the sub-shell is unknown
Terms	The spectroscopic Term Symbol of the element for the specified configuration; in the first row, the term symbol of the lower level; in the second row, the term symbol of the upper level (see discussion below)
J	The total angular momentum quantum number *J* for the specified configuration; in the first row, the *J* value for the lower level; in the second row, the value of *J* for the upper level
g	The statistical weight for the specified configuration; the first row is the statistical weight for the lower level; in the second row, the statistical weight of the upper level
DB Source	Reference to the primary source of the listed data (ASD = <https://physics.nist.gov/ASD>; HDBK = <https://physics.nist.gov/handbook>)

Intensities

The meanings of the character symbols appearing after the numerical value of intensity are as follows.

Symbol	Meaning
*	Intensity is shared by several lines (typically for multiply classified lines)
:	Observed value given is actually the rounded Ritz value, e.g., Ar IV, $\lambda = 443.40$ Å
-	Somewhat lower intensity than the value given
a	Observed in absorption
b	Band head
bl	Blended with another line that may affect the wavelength and intensity
B	Line or feature having large width due to autoionization broadening
c	Complex line
d	Diffuse line
D	Double line
E	Broad due to overexposure in the quoted reference
f	Forbidden line
g	Transition involving a level of the ground term
G	Line position roughly estimated
H	Very hazy line
h	Hazy line (same as "diffuse")
hfs	Line has hyperfine structure
i	Identification uncertain
j	Wavelength smoothed along isoelectronic sequence
l	Shaded to longer wavelengths; NB: This may look like a "one" at the end of the number
m	Masked by another line (no wavelength measurement)
p	Perturbed by a close line. Both wavelength and intensity may be affected
q	Asymmetric line
r	Easily reversed line
s	Shaded to shorter wavelengths
t	Tentatively classified line
u	Unresolved from a close line
w	Wide line
x	Extrapolated wavelength

Transition Probabilities

The values are listed as A_{ki} in units of 10^8 s^{-1}. These A_{ki} values can easily be converted to oscillator strengths, f_{ik}, $g_i f_{ik}$, or $\log(g_i f_{ik})$, or line strength, S, by using the following formula:

$$g_i f_{ik} = 1.499 \times 10^{-8} A_{ki} \lambda^2 g_k = 303.8 \lambda^{-1} S$$

where i refers to the lower energy level, k refers to the upper level, λ is the wavelength in angströms, and $g = 2J + 1$ for a given level.

The NIST transition-probability values include values with uncertainties ranging from 1% to larger than 50%, with an approximate uncertainty range being indicated for each line by an assigned letter. We have not included or assigned such letters here, but this information is given in the NIST publications and in the NIST online database for all transition-probability values taken from these sources. For data taken from the original literature or from non-NIST compilations, the user can consult the cited references for accuracy estimates. The reference list for the transition probabilities is given in the Online Edition of the *CRC Handbook.* See also the table "Atomic Transition Probabilities" in this section.

Accuracy

An estimated accuracy is listed for each transition strength, indicated by a code letter as given in the table below:

Accuracy code	Accuracy value
AAA	≤0.3%
AA	≤1%
A+	≤2%
A	≤3%
B+	≤7%
B	≤10%
C+	≤18%
C≤	25%
D+	≤40%
D	≤50%
E	>50%

The uncertainties are obtained from critical assessments, and in general, reflect estimates of predominantly systematic effects discussed in the NIST critical compilations, as cited in the references therein. If the accuracy is followed by a prime (′), then a multiplet in the original compilation has been separated into its component lines and the transition probability was derived from the compiled value assuming spin-orbit coupling. This may decrease the listed accuracy, especially for weaker transitions.

Energy-Level Classifications

Data pertaining to the two energy levels classifying each line are given with data for the lower level above that for the upper level. Included are the level values, configurations, term names, and J values for the levels classifying the line. The energy-level classifications for a few persistent lines are not known, as indicated by the absence of level values.

Terms

Terms of the lower and upper levels are displayed. A superscript "°" or an asterisk indicates odd parity. The J values represent the total electronic angular momentum of the lower and upper levels. g_i and g_k represent the statistical weights of the lower level ($g_i = 2J_i + 1$) and upper level ($g_k = 2J_k + 1$), respectively.

For atoms with more than two *d*-electrons, the normal $(2S + 1)L$ is not unique. Generally, a seniority number is used, but for some cases, the terms with the same $(2S + 1)L$ value are labeled by letters. The first even parity of a certain LS symmetry is labeled as "a ^{2S+1}L;" the second as "b ^{2S+1}L;" etc. The odd-parity terms are labeled in reverse alphabetical order starting with "z."

Integer numbers are often used instead of term symbols when the levels have not been theoretically interpreted or when the eigenvectors are so strongly mixed that assignment of proper term labels to each level is meaningless. For example, in the case of Ir I, there is no theoretical interpretation of highly excited odd-parity levels, so the labels are unknown. The integer labels for these levels of Ir I are the left four numerals of the level energy.

Such assignment (intended to provide a unique correspondence between the level and line lists) is not always done; in many cases the level labels are left blank; then the "°" symbol appearing in the Terms column represent the only information known about the nature of the level, i.e., parity is odd. It would be even if the label is completely blank.

References

1. Kramida, A., Ralchenko, Yu., Reader, J., and NIST ASD Team (2017). *NIST Atomic Spectra Database* (Version 5.5.1) [Online]. Available: <https://physics.nist.gov/asd> [Dec. 29, 2017]. National Institute of Standards and Technology, Gaithersburg, MD.
2. J. E. Sansonetti, W. C. Martin, and S. L. Young (2005), *Handbook of Basic Atomic Spectroscopic Data* (Version 1.1.2) [Online], Available: <http://physics.nist.gov/Handbook> [Dec. 29, 2017]. National Institute of Standards and Technology, Gaithersburg, MD.
3. E. R. Peck and K. Reeder, *J. Opt. Soc. Am.* 62, 958–962, 1972. <https://doi.org/10.1364/JOSA.62.000958>

TABLE 1. Persistent Lines of Neutral Atomic Elements — Hydrogen I (Z = 1) to Helium I (Z = 2)

Intensity	Wavelength/Å	Uncert. wavelength/Å	A_{ki}/(10^8 s^{-1})	Accur.	Energy levels/cm^{-1}	Configs.	Terms	J	g	DB Source
Hydrogen I (Z = 1)										
	949.742	0.004	0.34375	AAA	0.0000000000	1*s*	^{2}S	1/2	2	ASD
					105291.65209	5*p*	^{2}P°	3/2	4	
33000	949.742	0.004	0.041250	AAA	0.0000000000	1*s*	^{2}S	1/2	2	ASD
					105291.657	5			50	
	949.742	0.004	0.34375	AAA	0.0000000000	1*s*	^{2}S	1/2	2	ASD
					105291.62867	5*p*	^{2}P°	1/2	2	
83000	972.517	0.014	0.12785	AAA	0.0000000000	1*s*	^{2}S	1/2	2	ASD
					102823.904	4			32	
	972.541	0.019	0.68186	AAA	0.0000000000	1*s*	^{2}S	1/2	2	ASD
					102823.8943175	4*p*	^{2}P°	3/2	4	

Intensity	Wavelength/Å	Uncert. wavelength/Å	$A_{ki}/(10^8\ s^{-1})$	Accur.	Energy levels/cm^{-1}	Configs.	Terms	J	g	DB Source
	972.541	0.019	0.68186	AAA	0.0000000000	1s	^{2}S	1/2	2	ASD
					102823.8485825	4p	^{2}P^o	1/2	2	
	1025.728	0.003	1.6725	AAA	0.0000000000	1s	^{2}S	1/2	2	ASD
					97492.319611	3p	^{2}P^o	3/2	4	
250000	1025.728	0.003	0.55751	AAA	0.0000000000	1s	^{2}S	1/2	2	ASD
					97492.304	3			18	
	1025.728	0.003	1.6725	AAA	0.0000000000	1s	^{2}S	1/2	2	ASD
					97492.211200	3p	^{2}P^o	1/2	2	
	1215.6699	0.0020	6.2648	AAA	0.0000000000	1s	^{2}S	1/2	2	ASD
					82259.2850014	2p	^{2}P^o	3/2	4	
	1215.6699	0.0020	6.2649	AAA	0.0000000000	1s	^{2}S	1/2	2	ASD
					82258.9191133	2p	^{2}P^o	1/2	2	
840000	1215.6701	0.0021	4.6986	AAA	0.0000000000	1s	^{2}S	1/2	2	ASD
					82259.158	2			8	
			0.078548	AAA	82258.9191133	2p	^{2}P^o	1/2	2	ASD
					105291.651993	5d	^{2}D	3/2	4	
			0.0042955	AAA	82258.9191133	2p	^{2}P^o	1/2	2	ASD
					[105291.63094]	5s	^{2}S	1/2	2	
			0.049483	AAA	82258.9543992821	2s	^{2}S	1/2	2	ASD
					105291.65209	5p	^{2}P^o	3/2	4	
			0.049484	AAA	82258.9543992821	2s	^{2}S	1/2	2	ASD
					105291.62867	5p	^{2}P^o	1/2	2	
										ASD
90000	4340.472	0.006	0.025304	AAA	82259.158	2			8	ASD
					105291.657	5			50	
			0.094254	AAA	82259.2850014	2p	^{2}P^o	3/2	4	ASD
					[105291.659796]	5d	^{2}D	5/2	6	ASD
			0.015709	AAA	82259.2850014	2p	^{2}P^o	3/2	4	ASD
					105291.651993	5d	^{2}D	3/2	4	
			0.008592	AAA	82259.2850014	2p	^{2}P^o	3/2	4	ASD
					[105291.63094]	5s	^{2}S	1/2	2	
			0.17188	AAA	82258.9191133	2p	^{2}P^o	1/2	2	ASD
					102823.894250	4d	^{2}D	3/2	4	
	4861.28694917	0.00000008	0.096680	AAA	82258.9543992821	2s	^{2}S	1/2	2	ASD
					102823.8943175	4p	^{2}P^o	3/2	4	
	4861.29776054	0.00000012	0.096683	AAA	82258.9543992821	2s	^{2}S	1/2	2	ASD
					102823.8485825	4p	^{2}P^o	1/2	2	
180000	4861.35	0.05	0.084193	AAA	82259.158	2			8	ASD
					102823.904	4			32	
			0.20625	AAA	82259.2850014	2p	^{2}P^o	3/2	4	ASD
					102823.9094871	4d	^{2}D	5/2	6	
			0.034375	AAA	82259.2850014	2p	^{2}P^o	3/2	4	ASD
					102823.894250	4d	^{2}D	3/2	4	
	6562.709699	0.000024	0.53877	AAA	82258.9191133	2p	^{2}P^o	1/2	2	ASD
					97492.319433	3d	^{2}D	3/2	4	
	6562.724827	0.000004	0.22448	AAA	82258.9543992821	2s	^{2}S	1/2	2	ASD
					97492.319611	3p	^{2}P^o	3/2	4	
500000	6562.79	0.03	0.44101	AAA	82259.158	2			8	ASD
					97492.304	3			18	
	6562.85175	0.00006	0.64651	AAA	82259.2850014	2p	^{2}P^o	3/2	4	ASD
					97492.355566	3d	^{2}D	5/2	6	
			0.016377	AAA	97492.221701	3s	^{2}S	1/2	2	ASD
					105291.65209	5p	^{2}P^o	3/2	4	
32000	12818.072	0.008	0.022008	AAA	97492.304	3			18	ASD
					105291.657	5			50	
			0.042394	AAA	97492.319433	3d	^{2}D	3/2	4	ASD
					(105291.65983494)	5f	^{2}F^o	5/2	6	
			0.033915	AAA	97492.319611	3p	^{2}P^o	3/2	4	ASD
					[105291.659796]	5d	^{2}D	5/2	6	

Intensity	Wavelength/Å	Uncert. wavelength/Å	$A_{ki}/(10^8\ s^{-1})$	Accur.	Energy levels/cm^{-1}	Configs.	Terms	J	g	DB Source
			0.045421	AAA	97492.355566	$3d$	^2D	5/2	6	ASD
					105291.66370	$5f$	$^2F^o$	7/2	8	
			0.058647	AAA	97492.211200	$3p$	$^2P^o$	1/2	2	ASD
					102823.894250	$4d$	^2D	3/2	4	
			0.030650	AAA	97492.221701	$3s$	^2S	1/2	2	ASD
					102823.8943175	$4p$	$^2P^o$	3/2	4	
			0.12869	AAA	97492.319433	$3d$	^2D	3/2	4	ASD
					102823.90949	$4f$	$^2F^o$	5/2	6	
			0.070376	AAA	97492.319611	$3p$	$^2P^o$	3/2	4	ASD
					102823.9094871	$4d$	^2D	5/2	6	
			0.13788	AAA	97492.355566	$3d$	^2D	5/2	6	ASD
					102823.917091	$4f$	$^2F^o$	7/2	8	
(51000)	18751.3	1.0	0.089860	AAA	97492.304	3			18	ASD
					102823.904	4			32	
Helium I (Z = 2)										
100	522.186	0.020	2.4356	AAA	0.0000	$1s^2$	^1S	0	1	ASD
					[191492.711909]	$1s4p$	$^1P^o$	1	3	
400	537.0293	0.0005	5.6634	AAA	0.0000	$1s^2$	^1S	0	1	ASD
					[186209.364940]	$1s3p$	$^1P^o$	1	3	
1000	584.3339	0.0005	17.989	AAA	0.0000	$1s^2$	^1S	0	1	ASD
					[171134.896946]	$1s2p$	$^1P^o$	1	3	
		0.010	0.094746	AAA	[159855.9743297]	$1s2s$	^3S	1	3	ASD
					[185564.854540]	$1s3p$	$^3P^o$	0	1	
500*	3888.648	0.010	0.094746	AAA	[159855.9743297]	$1s2s$	^3S	1	3	ASD
					[185564.583895]	$1s3p$	$^3P^o$	1	3	
500*	3888.648	0.010	0.094746	AAA	[159855.9743297]	$1s2s$	^3S	1	3	ASD
					[185564.561920]	$1s3p$	$^3P^o$	2	5	
	5874.463	0.005	0.0001232	AA	[169086.8428979]	$1s2p$	$^3P^o$	1	3	ASD
					[186104.9666893]	$1s3d$	^1D	2	5	
500*	5875.621	0.010	0.70708	AAA	[169086.7664725]	$1s2p$	$^3P^o$	2	5	ASD
					[186101.5461767]	$1s3d$	^3D	3	7	
100	5875.966	0.010	0.39282	AAA	[169087.8308131]	$1s2p$	$^3P^o$	0	1	ASD
					[186101.5928903]	$1s3d$	^3D	1	3	
100	6678.151	0.010	0.63705	AAA	[171134.896946]	$1s2p$	$^1P^o$	1	3	ASD
					[186104.9666893]	$1s3d$	^1D	2	5	
200*	7065.190	0.010	0.15474	AAA	[169086.7664725]	$1s2p$	$^3P^o$	2	5	ASD
					[183236.79170]	$1s3s$	^3S	1	3	
200*	7065.190	0.010	0.092844	AAA	[169086.8428979]	$1s2p$	$^3P^o$	1	3	ASD
					[183236.79170]	$1s3s$	^3S	1	3	
30	7065.71	0.10	0.030948	AAA	[169087.8308131]	$1s2p$	$^3P^o$	0	1	ASD
					[183236.79170]	$1s3s$	^3S	1	3	
50	7281.349	0.010	0.18299	AAA	[171134.896946]	$1s2p$	$^1P^o$	1	3	ASD
					[184864.82932]	$1s3s$	^1S	0	1	
300	10829.088	0.010	0.10216	AAA	[159855.9743297]	$1s2s$	^3S	1	3	ASD
					[169087.8308131]	$1s2p$	$^3P^o$	0	1	
1000	10830.248	0.010	0.10216	AAA	[159855.9743297]	$1s2s$	^3S	1	3	ASD
					[169086.8428979]	$1s2p$	$^3P^o$	1	3	
2000	10830.337	0.010	0.10216	AAA	[159855.9743297]	$1s2s$	^3S	1	3	ASD
					[169086.7664725]	$1s2p$	$^3P^o$	2	5	
1000	20586.92	0.10	0.019746	AAA	[166277.440141]	$1s2s$	^1S	0	1	ASD
					[171134.896946]	$1s2p$	$^1P^o$	1	3	

TABLE 2. Persistent Lines of Neutral Atomic Elements — Lithium I (Z = 3) to Neon I (Z = 10)

Intensity	Wavelength/Å	Uncert. wavelength/Å	$A_{ki}/(10^8\ s^{-1})$	Accur.	Energy levels/cm^{-1}	Configs.	Terms	J	g	DB Source
Lithium I (Z = 3)										
13	4602.831		0.1935	AA	14903.622	$1s^22p$	$^2P^o$	1/2	2	NHBK,ASD
					36623.297	$1s^24d$	^2D	3/2	4	

Intensity	Wavelength/Å	Uncert. wavelength/Å	$A_{ki}/(10^8\ s^{-1})$	Accur.	Energy levels/cm^{-1}	Configs.	Terms	J	g	DB Source
30	4602.898		0.2322	AA	14903.957	$1s^22p$	$^2P^o$	3/2	4	NHBK,ASD
					36623.312	$1s^24d$	^2D	5/2	6	
320	6103.542		0.57138	AAA	14903.622	$1s^22p$	$^2P^o$	1/2	2	NHBK,ASD
					31283.018	$1s^23d$	^2D	3/2	4	
400	6103.654		0.68562	AAA	14903.957	$1s^22p$	$^2P^o$	3/2	4	NHBK,ASD
					31283.053	$1s^23d$	^2D	5/2	6	
1800	6707.775		0.36891	AAA	0.000	$1s^22s$	^2S	1/2	2	NHBK,ASD
					14903.957	$1s^22p$	$^2P^o$	3/2	4	
3600	6707.926		0.36890	AAA	0.000	$1s^22s$	^2S	1/2	2	NHBK,ASD
					14903.622	$1s^22p$	$^2P^o$	1/2	2	
48	8126.232		0.11156	AAA	14903.622	$1s^22p$	$^2P^o$	1/2	2	NHBK,ASD
					27206.066	$1s^23s$	^2S	1/2	2	
48	8126.453		0.22309	AAA	14903.957	$1s^22p$	$^2P^o$	3/2	4	NHBK,ASD
					27206.066	$1s^23s$	^2S	1/2	2	
10	26885.00		0.03738	AA	27206.066	$1s^23s$	^2S	1/2	2	NHBK,ASD
					30925.613	$1s^23p$	$^2P^o$	1/2	2	
5	26885.69		0.03737	AA	27206.066	$1s^23s$	^2S	1/2	2	NHBK,ASD
					30925.517	$1s^23p$	$^2P^o$	3/2	4	
Beryllium I (Z = 4)										
30	1661.478		0.0723	A	0.000	$1s^22s^2$	^1S	0	1	ASD
					60187.34	$1s^22s3p$	$^1P^o$	1	3	
40	2348.610		5.52	A	0.000	$1s^22s^2$	^1S	0	1	ASD
					42565.35	$1s^22s2p$	$^1P^o$	1	3	
20*	2494.728		1.92	A	21981.27	$1s^22s2p$	$^3P^o$	2	5	ASD
					62053.72	$1s^22s3d$	^3D	3	7	
20*	2494.728		0.482	A	21981.27	$1s^22s2p$	$^3P^o$	2	5	ASD
					62053.72	$1s^22s3d$	^3D	2	5	
20*	2494.728		0.0535	A	21981.27	$1s^22s2p$	$^3P^o$	2	5	ASD
					62053.72	$1s^22s3d$	^3D	1	3	
12	2650.454		1.06	A	21978.925	$1s^22s2p$	$^3P^o$	1	3	ASD
					59697.08	$1s^22p^2$	^3P	2	5	
11	2650.550		1.41	A	21978.28	$1s^22s2p$	$^3P^o$	0	1	ASD
					59695.07	$1s^22p^2$	^3P	1	3	
12	2650.613		1.06	A	21978.925	$1s^22s2p$	$^3P^o$	1	3	ASD
					59695.07	$1s^22p^2$	^3P	1	3	
15	2650.619		3.17	A	21981.27	$1s^22s2p$	$^3P^o$	2	5	ASD
					59697.08	$1s^22p^2$	^3P	2	5	
11	2650.694		4.23	A	21978.925	$1s^22s2p$	$^3P^o$	1	3	ASD
					59693.65	$1s^22p^2$	^3P	0	1	
13	2650.760		1.76	A	21981.27	$1s^22s2p$	$^3P^o$	2	5	ASD
					59695.07	$1s^22p^2$	^3P	1	3	
45	3321.340		0.851	B	21981.27	$1s^22s2p$	$^3P^o$	2	5	ASD
					52080.94	$1s^22s3s$	^3S	1	3	
30	4572.664		0.762	B	42565.35	$1s^22s2p$	$^1P^o$	1	3	ASD
					64428.31	$1s^22s3d$	^1D	2	5	
20	8254.070		0.338	B	42565.35	$1s^22s2p$	$^1P^o$	1	3	ASD
					54677.26	$1s^22s3s$	^1S	0	1	
Boron I (Z = 5)										
70	1825.89390	0.00017	1.70	A	0.0000	$2s^22p$	$^2P^o$	1/2	2	ASD
					54767.6944	$2s^23d$	^2D	3/2	4	
100	1826.39952	0.00017	2.04	A	15.287	$2s^22p$	$^2P^o$	3/2	4	ASD
					54767.8387	$2s^23d$	^2D	5/2	6	
90	2088.88851	0.00022	0.361	B+	0.0000	$2s^22p$	$^2P^o$	1/2	2	ASD
					47857.125	$2s2p^2$	^2D	3/2	4	
150	2089.57004	0.00022	0.433	B+	15.287	$2s^22p$	$^2P^o$	3/2	4	ASD
					47856.809	$2s2p^2$	^2D	5/2	6	
110	2496.7687	0.0003	0.840	A	0.0000	$2s^22p$	$^2P^o$	1/2	2	ASD
					40039.6907	$2s^23s$	^2S	1/2	2	
210	2497.7224	0.0003	1.68	A	15.287	$2s^22p$	$^2P^o$	3/2	4	ASD

Intensity	Wavelength/Å	Uncert. wavelength/Å	$A_{ki}/(10^8\ s^{-1})$	Accur.	Energy levels/cm^{-1}	Configs.	Terms	J	g	DB Source
					40039.6907	$2s^23s$	^{2}S	1/2	2	
200	11660.031	0.004	0.172	B	40039.6907	$2s^23s$	^{2}S	1/2	2	ASD
					48613.6486	$2s^23p$	^{2}P°	3/2	4	
100	11662.456	0.004	0.172	B	40039.6907	$2s^23s$	^{2}S	1/2	2	ASD
					48611.8663	$2s^23p$	^{2}P°	1/2	2	
Carbon I (Z = 6)										
15000000bl	1277.250	0.010	1.26	A	0.0000000	$2s^22p^2$	^{3}P	0	1	ASD
					78293.51263	$2s^22p3d$	^{3}D°	1	3	
24000000bl	1277.285	0.003	1.70	A	16.4167130	$2s^22p^2$	^{3}P	1	3	ASD
					78307.64625	$2s^22p3d$	^{3}D°	2	5	
18000000bl	1277.517	0.003	0.92	A	16.4167130	$2s^22p^2$	^{3}P	1	3	ASD
					78293.51263	$2s^22p3d$	^{3}D°	1	3	
140000000bl	1277.556	0.003	2.28	A	43.4134567	$2s^22p^2$	^{3}P	2	5	ASD
					78318.26602	$2s^22p3d$	^{3}D°	3	7	
68000000	1280.330	0.010	0.64	B+	43.4134567	$2s^22p^2$	^{3}P	2	5	ASD
					78148.10824	$2s^22p4s$	^{3}P°	2	5	
8100000000	1560.68148	0.00010	0.88	A	16.4167130	$2s^22p^2$	^{3}P	1	3	ASD
					64090.99351	$2s2p^3$	^{3}D°	2	5	
2400000000	1560.70782	0.00013	0.489	A	16.4167130	$2s^22p^2$	^{3}P	1	3	ASD
					64089.8990	$2s2p^3$	^{3}D°	1	3	
12000000000	1561.43727	0.00010	1.17	A	43.4134567	$2s^22p^2$	^{3}P	2	5	ASD
					64086.96961	$2s2p^3$	^{3}D°	3	7	
19000000000	1656.26674	0.00012	0.87	A	16.4167130	$2s^22p^2$	^{3}P	1	3	ASD
					60393.1693	$2s^22p3s$	^{3}P°	2	5	
15000000000	1656.92821	0.00013	1.16	A	0.0000000	$2s^22p^2$	^{3}P	0	1	ASD
					60352.6584	$2s^22p3s$	^{3}P°	1	3	
48000000000	1657.00751	0.00010	2.61	A	43.4134567	$2s^22p^2$	^{3}P	2	5	ASD
					60393.1693	$2s^22p3s$	^{3}P°	2	5	
11000000000	1657.37863	0.00014	0.87	A	16.4167130	$2s^22p^2$	^{3}P	1	3	ASD
					60352.6584	$2s^22p3s$	^{3}P°	1	3	
16000000000	1657.90661	0.00012	3.47	A	16.4167130	$2s^22p^2$	^{3}P	1	3	ASD
					60333.4476	$2s^22p3s$	^{3}P°	0	1	
21000000000	1658.12045	0.00012	1.44	A	43.4134567	$2s^22p^2$	^{3}P	2	5	ASD
					60352.6584	$2s^22p3s$	^{3}P°	1	3	
16000000	1751.8271	0.0008	0.84	B+	21648.030	$2s^22p^2$	^{1}S	0	1	ASD
					78731.2956	$2s^22p3d$	^{1}P°	1	3	
20000000000	1930.90540	0.00012	3.39	A	10192.657	$2s^22p^2$	^{1}D	2	5	ASD
					61981.83211	$2s^22p3s$	^{1}P°	1	3	
640000000	2478.5612	0.0010	0.28	C+	21648.030	$2s^22p^2$	^{1}S	0	1	ASD
					61981.83211	$2s^22p3s$	^{1}P°	1	3	
5900000	8335.1443	0.0006	0.351	B+	61981.83211	$2s^22p3s$	^{1}P°	1	3	ASD
					73975.92785	$2s^22p3p$	^{1}S	0	1	
30000000	9405.7281	0.0005	0.291	B+	61981.83211	$2s^22p3s$	^{1}P°	1	3	ASD
					72610.73530	$2s^22p3p$	^{1}D	2	5	
5900000	11753.30813	0.00014	0.26	B	69744.05213	$2s^22p3p$	^{3}D	3	7	ASD
					78249.96676	$2s^22p3d$	^{3}F°	4	9	
4400000	11754.77549	0.00014	0.240	B	69710.68728	$2s^22p3p$	^{3}D	2	5	ASD
					78215.54012	$2s^22p3d$	^{3}F°	3	7	
Nitrogen I (Z = 7)										
410	1199.550		4.07	A	0.000	$2s^22p^3$	^{4}S°	3/2	4	ASD
					83364.620	$2s^22p^2(^3P)3s$	^{4}P	5/2	6	
385	1200.223		4.03	A	0.000	$2s^22p^3$	^{4}S°	3/2	4	ASD
					83317.830	$2s^22p^2(^3P)3s$	^{4}P	3/2	4	
360	1200.710		4.00	A	0.000	$2s^22p^3$	^{4}S°	3/2	4	ASD
					83284.070	$2s^22p^2(^3P)3s$	^{4}P	1/2	2	
360	1243.179		3.22	A	19224.464	$2s^22p^3$	^{2}D°	5/2	6	ASD
					99663.427	$2s^22p^2(^1D)3s$	^{2}D	5/2	6	
700	1492.625		3.11	B+	19224.464	$2s^22p^3$	^{2}D°	5/2	6	ASD
					86220.510	$2s^22p^2(^3P)3s$	^{2}P	3/2	4	

Intensity	Wavelength/Å	Uncert. wavelength/Å	$A_{ki}/(10^8\ s^{-1})$	Accur.	Energy levels/cm^{-1}	Configs.	Terms	J	g	DB Source
775	1742.729		1.05	B	28839.306	$2s^22p^3$	$^2P^o$	3/2	4	ASD
					86220.510	$2s^22p^2(^3P)3s$	^2P	3/2	4	
785	7442.29		0.119	B+	83317.830	$2s^22p^2(^3P)3s$	^4P	3/2	4	ASD
					96750.840	$2s^22p^2(^3P)3p$	$^4S^o$	3/2	4	
900	7468.31		0.196	B+	83364.620	$2s^22p^2(^3P)3s$	^4P	5/2	6	ASD
					96750.840	$2s^22p^2(^3P)3p$	$^4S^o$	3/2	4	
700	8680.28		0.253	B+	83364.620	$2s^22p^2(^3P)3s$	^4P	5/2	6	ASD
					94881.820	$2s^22p^2(^3P)3p$	$^4D^o$	7/2	8	
650	8683.40		0.188	B+	83317.830	$2s^22p^2(^3P)3s$	^4P	3/2	4	ASD
					94830.890	$2s^22p^2(^3P)3p$	$^4D^o$	5/2	6	
920	12469.62		0.218	B	96864.050	$2s^22p^2(^3P)3p$	$^2D^o$	5/2	6	ASD
					104881.350	$2s^22p^2(^3P)3d$	^2F	7/2	8	
840	13581.33		0.0573	C+	86220.510	$2s^22p^2(^3P)3s$	^2P	3/2	4	ASD
					93581.550	$2s^22p^2(^3P)3p$	$^2S^o$	1/2	2	
Oxygen I (Z = 8)										
900	1302.168		3.41	A	0.000	$2s^22p^4$	^3P	2	5	ASD
					76794.978	$2s^22p^3(^4S^o)3s$	$^3S^o$	1	3	
600	1304.858		2.03	A	158.265	$2s^22p^4$	^3P	1	3	ASD
					76794.978	$2s^22p^3(^4S^o)3s$	$^3S^o$	1	3	
300	1306.029		0.676	A	226.977	$2s^22p^4$	^3P	0	1	ASD
					76794.978	$2s^22p^3(^4S^o)3s$	$^3S^o$	1	3	
870	7771.94		0.369	A	73768.200	$2s^22p^3(^4S^o)3s$	$^5S^o$	2	5	ASD
					86631.454	$2s^22p^3(^4S^o)3p$	^5P	3	7	
810	7774.17		0.369	A	73768.200	$2s^22p^3(^4S^o)3s$	$^5S^o$	2	5	ASD
					86627.778	$2s^22p^3(^4S^o)3p$	^5P	2	5	
750	7775.39		0.369	A	73768.200	$2s^22p^3(^4S^o)3s$	$^5S^o$	2	5	ASD
					86625.757	$2s^22p^3(^4S^o)3p$	^5P	1	3	
810	8446.25		0.322	B	76794.978	$2s^22p^3(^4S^o)3s$	$^3S^o$	1	3	ASD
					88631.303	$2s^22p^3(^4S^o)3p$	^3P	0	1	
1000	8446.36		0.322	B	76794.978	$2s^22p^3(^4S^o)3s$	$^3S^o$	1	3	ASD
					88631.146	$2s^22p^3(^4S^o)3p$	^3P	2	5	
935	8446.76		0.322	B	76794.978	$2s^22p^3(^4S^o)3s$	$^3S^o$	1	3	ASD
					88630.587	$2s^22p^3(^4S^o)3p$	^3P	1	3	
640	9266.01		0.445	A	86631.454	$2s^22p^3(^4S^o)3p$	^5P	3	7	ASD
					97420.630	$2s^22p^3(^4S^o)3d$	$^5D^o$	4	9	
700	13163.89		0.0714	B+	88630.587	$2s^22p^3(^4S^o)3p$	^3P	1	3	ASD
					96225.049	$2s^22p^3(^4S^o)4s$	$^3S^o$	1	3	
750	13164.85		0.119	B+	88631.146	$2s^22p^3(^4S^o)3p$	^3P	2	5	ASD
					96225.049	$2s^22p^3(^4S^o)4s$	$^3S^o$	1	3	
Fluorine I (Z = 9)										
500	951.87		2.42	B+	0.000	$2s^22p^5$	$^2P^o$	3/2	4	NHBK
					105056.283	$2s^22p^4(^3P)3s$	^2P	1/2	2	
1000	954.83		5.98	B+	0.000	$2s^22p^5$	$^2P^o$	3/2	4	NHBK
					104731.048	$2s^22p^4(^3P)3s$	^2P	3/2	4	
750	955.55		4.74	B+	404.141	$2s^22p^5$	$^2P^o$	1/2	2	NHBK
					105056.283	$2s^22p^4(^3P)3s$	^2P	1/2	2	
500	958.52		1.15	B+	404.141	$2s^22p^5$	$^2P^o$	1/2	2	NHBK
					104731.048	$2s^22p^4(^3P)3s$	^2P	3/2	4	
13000	6239.65		0.29	D	102405.714	$2s^22p^4(^3P)3s$	^4P	5/2	6	NHBK,ASD
					118427.814	$2s^22p^4(^3P)3p$	$^4S^o$	3/2	4	
10000	6348.51		0.18	D	102680.439	$2s^22p^4(^3P)3s$	^4P	3/2	4	NHBK,ASD
					118427.814	$2s^22p^4(^3P)3p$	$^4S^o$	3/2	4	
50000	6856.03		0.45	D	102405.714	$2s^22p^4(^3P)3s$	^4P	5/2	6	NHBK,ASD
					116987.391	$2s^22p^4(^3P)3p$	$^4D^o$	7/2	8	
15000	6902.48		0.31	D	102680.439	$2s^22p^4(^3P)3s$	^4P	3/2	4	NHBK,ASD
					117164.002	$2s^22p^4(^3P)3p$	$^4D^o$	5/2	6	
45000	7037.47		0.38	D	104731.048	$2s^22p^4(^3P)3s$	^2P	3/2	4	NHBK,ASD
					118936.791	$2s^22p^4(^3P)3p$	$^2P^o$	3/2	4	
30000	7127.89		0.30	D	105056.283	$2s^22p^4(^3P)3s$	^2P	1/2	2	NHBK,ASD

Intensity	Wavelength/Å	Uncert. wavelength/Å	$A_{ki}/(10^8\ s^{-1})$	Accur.	Energy levels/cm^{-1}	Configs.	Terms	J	g	DB Source
					119081.814	$2s^22p^4(^3P)3p$	$^2P^o$	1/2	2	
15000	7202.36		0.072	D	105056.283	$2s^22p^4(^3P)3s$	^2P	1/2	2	NHBK,ASD
					118936.791	$2s^22p^4(^3P)3p$	$^2P^o$	3/2	4	
15000	7311.02		0.27	D	104731.048	$2s^22p^4(^3P)3s$	^2P	3/2	4	NHBK,ASD
					118405.256	$2s^22p^4(^3P)3p$	$^2S^o$	1/2	2	
10000	7398.69		0.25	D	102405.714	$2s^22p^4(^3P)3s$	^4P	5/2	6	NHBK,ASD
					115917.904	$2s^22p^4(^3P)3p$	$^4P^o$	5/2	6	
18000	7754.70		0.35	D	104731.048	$2s^22p^4(^3P)3s$	^2P	3/2	4	NHBK,ASD
					117622.917	$2s^22p^4(^3P)3p$	$^2D^o$	5/2	6	
15000	7800.21		0.29	D	105056.283	$2s^22p^4(^3P)3s$	^2P	1/2	2	NHBK,ASD
					117872.917	$2s^22p^4(^3P)3p$	$^2D^o$	3/2	4	
Neon I (Z = 10)										
50	615.6283	0.0040	0.38	C	0.0000	$2s^22p^6$	^1S	0	1	ASD
					162435.6780	$2s^22p^5(^2P^o<1/2>)3d$	$^2[3/2]^o$	1	3	
50	618.6716	0.0040	0.93	C	0.0000	$2s^22p^6$	^1S	0	1	ASD
					161636.6175	$2s^22p^5(^2P^o<3/2>)3d$	$^2[3/2]^o$	1	3	
40	619.1023	0.0040	0.33	C	0.0000	$2s^22p^6$	^1S	0	1	ASD
					161524.1739	$2s^22p^5(^2P^o<3/2>)3d$	$^2[1/2]^o$	1	3	
60	626.8232	0.0040	0.74	C	0.0000	$2s^22p^6$	^1S	0	1	ASD
					159534.6196	$2s^22p^5(^2P^o<1/2>)4s$	$^2[1/2]^o$	1	3	
60	629.7388	0.0040	0.48	C	0.0000	$2s^22p^6$	^1S	0	1	ASD
					158795.9924	$2s^22p^5(^2P^o<3/2>)4s$	$^2[3/2]^o$	1	3	
300	735.8962	0.0040	6.11	C	0.0000	$2s^22p^6$	^1S	0	1	ASD
					135888.7173	$2s^22p^5(^2P^o<1/2>)3s$	$^2[1/2]^o$	1	3	
120	743.7195	0.0002	0.476	C	0.0000	$2s^22p^6$	^1S	0	1	ASD
					134459.2871	$2s^22p^5(^2P^o<3/2>)3s$	$^2[3/2]^o$	1	3	
10000	3520.4714	0.0005	0.093	C	135888.7173	$2s^22p^5(^2P^o<1/2>)3s$	$^2[1/2]^o$	1	3	ASD
					164285.8872	$2s^22p^5(^2P^o<1/2>)4p$	$^2[1/2]$	0	1	
20000	5400.5616	0.0004	0.0090	B	134459.2871	$2s^22p^5(^2P^o<3/2>)3s$	$^2[3/2]^o$	1	3	ASD
					152970.7328	$2s^22p^5(^2P^o<1/2>)3p$	$^2[1/2]$	0	1	
20000	5852.4878	0.0005	0.682	B	135888.7173	$2s^22p^5(^2P^o<1/2>)3s$	$^2[1/2]^o$	1	3	ASD
					152970.7328	$2s^22p^5(^2P^o<1/2>)3p$	$^2[1/2]$	0	1	
10000	6029.9968	0.0005	0.0561	B	134459.2871	$2s^22p^5(^2P^o<3/2>)3s$	$^2[3/2]^o$	1	3	ASD
					151038.4524	$2s^22p^5(^2P^o<1/2>)3p$	$^2[1/2]$	1	3	
10000	6074.3376	0.0005	0.603	C+	134459.2871	$2s^22p^5(^2P^o<3/2>)3s$	$^2[3/2]^o$	1	3	ASD
					150917.4307	$2s^22p^5(^2P^o<3/2>)3p$	$^2[1/2]$	0	1	
10000	6143.0627	0.0005	0.282	B	134041.8400	$2s^22p^5(^2P^o<3/2>)3s$	$^2[3/2]^o$	2	5	ASD
					150315.8612	$2s^22p^5(^2P^o<3/2>)3p$	$^2[3/2]$	2	5	
10000	6163.5937	0.0005	0.146	B	134818.6405	$2s^22p^5(^2P^o<1/2>)3s$	$^2[1/2]^o$	0	1	ASD
					151038.4524	$2s^22p^5(^2P^o<1/2>)3p$	$^2[1/2]$	1	3	
10000	6217.2812	0.0005	0.0637	B	134041.8400	$2s^22p^5(^2P^o<3/2>)3s$	$^2[3/2]^o$	2	5	ASD
					150121.5922	$2s^22p^5(^2P^o<3/2>)3p$	$^2[3/2]$	1	3	
10000	6266.4952	0.0005	0.249	B	134818.6405	$2s^22p^5(^2P^o<1/2>)3s$	$^2[1/2]^o$	0	1	ASD
					150772.1118	$2s^22p^5(^2P^o<1/2>)3p$	$^2[3/2]$	1	3	
10000	6382.9914	0.0005	0.321	B	134459.2871	$2s^22p^5(^2P^o<3/2>)3s$	$^2[3/2]^o$	1	3	ASD
					150121.5922	$2s^22p^5(^2P^o<3/2>)3p$	$^2[3/2]$	1	3	
20000	6402.2480	0.0010	0.514	B	134041.8400	$2s^22p^5(^2P^o<3/2>)3s$	$^2[3/2]^o$	2	5	ASD
					149657.0392	$2s^22p^5(^2P^o<3/2>)3p$	$^2[5/2]$	3	7	
15000	6506.5277	0.0005	0.30	B	134459.2871	$2s^22p^5(^2P^o<3/2>)3s$	$^2[3/2]^o$	1	3	ASD
					149824.2215	$2s^22p^5(^2P^o<3/2>)3p$	$^2[5/2]$	2	5	
10000	6598.9528	0.0005	0.232	B	135888.7173	$2s^22p^5(^2P^o<1/2>)3s$	$^2[1/2]^o$	1	3	ASD
					151038.4524	$2s^22p^5(^2P^o<1/2>)3p$	$^2[1/2]$	1	3	
100000	6929.4672	0.0004	0.174	B	135888.7173	$2s^22p^5(^2P^o<1/2>)3s$	$^2[1/2]^o$	1	3	ASD
					150315.8612	$2s^22p^5(^2P^o<3/2>)3p$	$^2[3/2]$	2	5	
85000	7032.4128	0.0004	0.253	B	134041.8400	$2s^22p^5(^2P^o<3/2>)3s$	$^2[3/2]^o$	2	5	ASD
					148257.7898	$2s^22p^5(^2P^o<3/2>)3p$	$^2[1/2]$	1	3	
77000	7173.9380	0.0004	0.0287	B	135888.7173	$2s^22p^5(^2P^o<1/2>)3s$	$^2[1/2]^o$	1	3	ASD
					149824.2215	$2s^22p^5(^2P^o<3/2>)3p$	$^2[5/2]$	2	5	
77000	7245.1665	0.0004	0.0935	B	134459.2871	$2s^22p^5(^2P^o<3/2>)3s$	$^2[3/2]^o$	1	3	ASD
					148257.7898	$2s^22p^5(^2P^o<3/2>)3p$	$^2[1/2]$	1	3	

Intensity	Wavelength/Å	Uncert. wavelength/Å	$A_{ki}/(10^8\ s^{-1})$	Accur.	Energy levels/cm^{-1}	Configs.	Terms	J	g	DB Source
76000	8377.6070	0.0010	0.51	C	149657.0392	$2s^22p^5(^2P^o<3/2>)3p$	$^2[5/2]$	3	7	ASD
					161590.3412	$2s^22p^5(^2P^o<3/2>)3d$	$^2[7/2]^o$	4	9	
64000	8654.3828	0.0009	0.445	C	150858.5079	$2s^22p^5(^2P^o<1/2>)3p$	$^2[3/2]$	2	5	ASD
					162410.1736	$2s^22p^5(^2P^o<1/2>)3d$	$^2[5/2]^o$	3	7	
57000	8780.6223	0.0004			150315.8612	$2s^22p^5(^2P^o<3/2>)3p$	$^2[3/2]$	2	5	ASD
					161701.4486	$2s^22p^5(^2P^o<3/2>)3d$	$^2[5/2]^o$	3	7	
43000	8783.7539	0.0004	0.313	C	151038.4524	$2s^22p^5(^2P^o<1/2>)3p$	$^2[1/2]$	1	3	ASD
					162419.9818	$2s^22p^5(^2P^o<1/2>)3d$	$^2[3/2]^o$	2	5	
26000	11143.0200	0.0005			149824.2215	$2s^22p^5(^2P^o<3/2>)3p$	$^2[5/2]$	2	5	ASD
					158795.9924	$2s^22p^5(^2P^o<3/2>)4s$	$^2[3/2]^o$	1	3	
49000	11177.5246	0.0005			149657.0392	$2s^22p^5(^2P^o<3/2>)3p$	$^2[5/2]$	3	7	ASD
					158601.1152	$2s^22p^5(^2P^o<3/2>)4s$	$^2[3/2]^o$	2	5	
33000	11522.7450	0.0014			150858.5079	$2s^22p^5(^2P^o<1/2>)3p$	$^2[3/2]$	2	5	ASD
					159534.6196	$2s^22p^5(^2P^o<1/2>)4s$	$^2[1/2]^o$	1	3	

TABLE 3. Persistent Lines of Neutral Atomic Elements — Sodium I (Z = 11) to Argon (Z = 18)

Intensity	Wavelength/Å	Uncert. wavelength/Å	$A_{ki}/(10^8\ s^{-1})$	Accur.	Energy levels/cm^{-1}	Configs.	Terms	J	g	DB Source
Sodium I (Z = 11)										
16	2852.811	0.004	0.00538	B+	0.00000	$2p^63s$	^2S	1/2	2	ASD
					35042.85	$2p^65p$	$^2P^o$	3/2	4	
15	2853.013	0.004	0.00531	B+	0.00000	$2p^63s$	^2S	1/2	2	ASD
					35040.38	$2p^65p$	$^2P^o$	1/2	2	
19	3302.369	0.005	0.0275	A	0.00000	$2p^63s$	^2S	1/2	2	ASD
					30272.58	$2p^64p$	$^2P^o$	3/2	4	
18	3302.979	0.005	0.0273	B+	0.00000	$2p^63s$	^2S	1/2	2	ASD
					30266.99	$2p^64p$	$^2P^o$	1/2	2	
80000	5889.950954	0.000015	0.616	AA	0.00000	$2p^63s$	^2S	1/2	2	ASD
					16973.36619	$2p^63p$	$^2P^o$	3/2	4	
40000	5895.924237	0.000015	0.614	AA	0.00000	$2p^63s$	^2S	1/2	2	ASD
					16956.17025	$2p^63p$	$^2P^o$	1/2	2	
5	8183.2556	0.0005	0.429	A+	16956.17025	$2p^63p$	$^2P^o$	1/2	2	ASD
					29172.887	$2p^63d$	^2D	3/2	4	
1	8194.7905	0.0005	0.0857	A	16973.36619	$2p^63p$	$^2P^o$	3/2	4	ASD
					29172.887	$2p^63d$	^2D	3/2	4	
9	8194.8237	0.0005	0.514	A+	16973.36619	$2p^63p$	$^2P^o$	3/2	4	ASD
					29172.837	$2p^63d$	^2D	5/2	6	
11	11381.45	0.03	0.0880	A	16956.17025	$2p^63p$	$^2P^o$	1/2	2	ASD
					25739.999	$2p^64s$	^2S	1/2	2	
12	11403.78	0.03	0.176	A	16973.36619	$2p^63p$	$^2P^o$	3/2	4	ASD
					25739.999	$2p^64s$	^2S	1/2	2	
(40)*	18465.39	0.03	0.140	A	29172.837	$2p^63d$	^2D	5/2	6	ASD
					34586.92	$2p^64f$	$^2F^o$	7/2	8	
(40)*	18465.39	0.03	0.00935	A	29172.837	$2p^63d$	^2D	5/2	6	ASD
					34586.92	$2p^64f$	$^2F^o$	5/2	6	
Magnesium I (Z = 12)										
9g	2025.824		0.612	B+	0.000	$2p^63s^2$	^1S	0	1	ASD
					49346.729	$3s4p$	$^1P^o$	1	3	
20*	2779.831		1.36	C+	21870.464	$3s3p$	$^3P^o$	1	3	ASD
					57833.40	$3p^2$	^3P	1	3	
20*	2779.831		4.09	B	21911.178	$3s3p$	$^3P^o$	2	5	ASD
					57873.94	$3p^2$	^3P	2	5	
50g	2852.127		4.91	A	0.000	$2p^63s^2$	^1S	0	1	ASD
					35051.264	$3s3p$	$^1P^o$	1	3	
36	3829.3549		0.899	B+	21850.405	$3s3p$	$^3P^o$	0	1	ASD
					47957.058	$3s3d$	^3D	1	3	
38*	3832.2996		0.674	B+	21870.464	$3s3p$	$^3P^o$	1	3	ASD
					47957.058	$3s3d$	^3D	1	3	

Intensity	Wavelength/Å	Uncert. wavelength/Å	$A_{ki}/(10^8\ s^{-1})$	Accur.	Energy levels/cm^{-1}	Configs.	Terms	J	g	DB Source
38*	3832.3037		1.21	B+	21870.464	$3s3p$	$^3P^o$	1	3	ASD
					47957.027	$3s3d$	^3D	2	5	
40	3838.2943		0.403	B+	21911.178	$3s3p$	$^3P^o$	2	5	ASD
					47957.027	$3s3d$	^3D	2	5	
42	5167.3216		0.113	B+	21850.405	$3s3p$	$^3P^o$	0	1	ASD
					41197.403	$3s4s$	^3S	1	3	
44	5172.6843		0.337	B+	21870.464	$3s3p$	$^3P^o$	1	3	ASD
					41197.403	$3s4s$	^3S	1	3	
45	5183.6042		0.561	A	21911.178	$3s3p$	$^3P^o$	2	5	ASD
					41197.403	$3s4s$	^3S	1	3	
2650	11828.171		0.222	A	35051.264	$3s3p$	$^1P^o$	1	3	ASD
					43503.333	$3s4s$	^1S	0	1	
38900	15024.997		0.135	A	41197.403	$3s4s$	^3S	1	3	ASD
					47851.162	$3s4p$	$^3P^o$	2	5	
Aluminum I (Z = 13)										
4g	2269.096		0.758	B	112.061	$3s^23p$	$^2P^o$	3/2	4	ASD
					44168.847	$3s^2$ 5d	^2D	5/2	6	
8g	2367.052		0.761	C+	0.000	$3s^23p$	$^2P^o$	1/2	2	ASD
					42233.742	$3s^24d$	^2D	3/2	4	
8g	2373.124		0.907	B	112.061	$3s^23p$	$^2P^o$	3/2	4	ASD
					42237.783	$3s^24d$	^2D	5/2	6	
10g	2575.094		0.360	C+	112.061	$3s^23p$	$^2P^o$	3/2	4	ASD
					38933.968	$3s^2$ nd	y ^2D	5/2	6	
24g	3082.1529		0.587	B+	0.000	$3s^23p$	$^2P^o$	1/2	2	ASD
					32435.453	$3s^23d$	^2D	3/2	4	
26g	3092.7099		0.729	B+	112.061	$3s^23p$	$^2P^o$	3/2	4	ASD
					32436.796	$3s^23d$	^2D	5/2	6	
20g	3092.8386		0.116	B	112.061	$3s^23p$	$^2P^o$	3/2	4	ASD
					32435.453	$3s^23d$	^2D	3/2	4	
24g	3944.0058		0.499	B+	0.000	$3s^23p$	$^2P^o$	1/2	2	ASD
					25347.756	$3s^24s$	^2S	1/2	2	
26g	3961.5200		0.985	B+	112.061	$3s^23p$	$^2P^o$	3/2	4	ASD
					25347.756	$3s^24s$	^2S	1/2	2	
Silicon I (Z = 14)										
400	1850.6717	0.0010	3.05	C+	223.157	$3s^23p^2$	^3P	2	5	HDBK,ASD
					54257.582	$3s^23p3d$	$^3D^o$	3	7	
400	1901.3370	0.0010	1.00	C	6298.850	$3s^23p^2$	^1D	2	5	HDBK,ASD
					58893.40	$3s^23p4d$	$^1F^o$	3	7	
400	1977.597	0.002	0.279	B	0.000	$3s^23p^2$	^3P	0	1	HDBK,ASD
					50566.397	$3s^23p3d$	$^3P^o$	1	3	
400	1979.206	0.002	0.869	B	77.115	$3s^23p^2$	^3P	1	3	HDBK,ASD
					50602.44	$3s^23p3d$	$^3P^o$	0	1	
300	1980.618	0.002	0.207	B	77.115	$3s^23p^2$	^3P	1	3	HDBK,ASD
					50566.397	$3s^23p3d$	$^3P^o$	1	3	
300	1983.232	0.002	0.218	B	77.115	$3s^23p^2$	^3P	1	3	HDBK,ASD
					50499.838	$3s^23p3d$	$^3P^o$	2	5	
500	1986.363	0.002	0.365	B	223.157	$3s^23p^2$	^3P	2	5	HDBK,ASD
					50566.397	$3s^23p3d$	$^3P^o$	1	3	
1000	1988.994	0.002	0.658	B	223.157	$3s^23p^2$	^3P	2	5	HDBK,ASD
					50499.838	$3s^23p3d$	$^3P^o$	2	5	
600	2058.1323	0.0010	0.708	C	6298.850	$3s^23p^2$	^1D	2	5	HDBK,ASD
					54871.031	$3s^23p5s$	$^1P^o$	1	3	
110	2207.9783	0.001463007	0.262	B	0.000	$3s^23p^2$	^3P	0	1	HDBK,ASD
					45276.188	$3s3p^3$	$^3D^o$	1	3	
115	2210.8940	0.001466871	0.346	B	77.115	$3s^23p^2$	^3P	1	3	HDBK,ASD
					45293.629	$3s3p^3$	$^3D^o$	2	5	
110	2211.7441	0.001468002	0.181	B	77.115	$3s^23p^2$	^3P	1	3	HDBK,ASD
					45276.188	$3s3p^3$	$^3D^o$	1	3	

Intensity	Wavelength/Å	Uncert. wavelength/Å	A_{ki}/(10^8 s^{-1})	Accur.	Energy levels/cm^{-1}	Configs.	Terms	J	g	DB Source
120	2216.6688	0.00098303	0.454	B	223.157	$3s^23p^2$	^{3}P	2	5	HDBK,ASD
					45321.848	$3s3p^3$	^{3}D^o	3	7	
120	2218.0569	0.000984262	0.109	B	223.157	$3s^23p^2$	^{3}P	2	5	HDBK,ASD
					45293.629	$3s3p^3$	^{3}D^o	2	5	
425	2506.8973	0.000502914	0.547	B	77.115	$3s^23p^2$	^{3}P	1	3	HDBK,ASD
					39955.053	$3s^23p4s$	^{3}P^o	2	5	
375	2514.3161	0.000695606	0.739	B	0.000	$3s^23p^2$	^{3}P	0	1	HDBK,ASD
					39760.285	$3s^23p4s$	^{3}P^o	1	3	
500	2516.1125	0.001899818	1.68	B	223.157	$3s^23p^2$	^{3}P	2	5	HDBK,ASD
					39955.053	$3s^23p4s$	^{3}P^o	2	5	
350	2519.2023	0.000507863	0.549	B	77.115	$3s^23p^2$	^{3}P	1	3	HDBK,ASD
					39760.285	$3s^23p4s$	^{3}P^o	1	3	
425	2524.1079	0.001147147	2.22	B	77.115	$3s^23p^2$	^{3}P	1	3	HDBK,ASD
					39683.163	$3s^23p4s$	^{3}P^o	0	1	
450	2528.5086	0.000447669	0.904	B	223.157	$3s^23p^2$	^{3}P	2	5	HDBK,ASD
					39760.285	$3s^23p4s$	^{3}P^o	1	3	
1000	2881.5792	0.000498356	2.17	B	6298.850	$3s^23p^2$	^{1}D	2	5	HDBK,ASD
					40991.884	$3s^23p4s$	^{1}P^o	1	3	
200	5948.545	0.004247396	0.0222	C	40991.884	$3s^23p4s$	^{1}P^o	1	3	HDBK,ASD
					57798.072	$3s^23p5p$	^{1}D	2	5	
400	7289.1730	0.0015944			45321.848	$3s3p^3$	^{3}D^o	3	7	HDBK,ASD
					59037.043	$3s^23p(^2P^o$<3/2>$)4f$	2[7/2]	4	9	
375	7405.7741	0.002194424	0.037	D	45276.188	$3s3p^3$	^{3}D^o	1	3	HDBK,ASD
					58775.451	$3s^23p(^2P^o$<1/2>$)4f$	2[5/2]	2	5	
425	7423.4969	0.00110247			45321.848	$3s3p^3$	^{3}D^o	3	7	HDBK,ASD
					58788.880	$3s^23p(^2P^o$<1/2>$)4f$	2[7/2]	4	9	
370	11984.177	0.022985609	0.140	B	39760.285	$3s^23p4s$	^{3}P^o	1	3	HDBK,ASD
					48102.323	$3s^23p4p$	^{3}D	2	5	
440	12031.481	0.01737557	0.173	B	39955.053	$3s^23p4s$	^{3}P^o	2	5	HDBK,ASD
					48264.292	$3s^23p4p$	^{3}D	3	7	
600	1674.591		0.40	D	0.00	$3s^23p^3$	^{4}S^o	3/2	4	ASD

Phosphorus I (Z = 15)

Intensity	Wavelength/Å	Uncert. wavelength/Å	A_{ki}/(10^8 s^{-1})	Accur.	Energy levels/cm^{-1}	Configs.	Terms	J	g	DB Source
					59715.921	$3s3p^4$	^{4}P	3/2	4	
600	1679.695		0.39	D	0.00	$3s^23p^3$	^{4}S^o	3/2	4	ASD
					59534.549	$3s3p^4$	^{4}P	5/2	6	
600	1774.951		2.17	C	0.00	$3s^23p^3$	^{4}S^o	3/2	4	ASD
					56339.656	$3s^23p^2(^3P)4s$	^{4}P	5/2	6	
500	1782.838		2.14	C	0.00	$3s^23p^3$	^{4}S^o	3/2	4	ASD
					56090.626	$3s^23p^2(^3P)4s$	^{4}P	3/2	4	
400	1787.656		2.13	C	0.00	$3s^23p^3$	^{4}S^o	3/2	4	ASD
					55939.421	$3s^23p^2(^3P)4s$	^{4}P	1/2	2	
400	2135.465		0.211	C	11361.02	$3s^23p^3$	^{2}D^o	3/2	4	ASD
					58174.366	$3s^23p^2(^3P)4s$	^{2}P	3/2	4	
400	2136.182		2.83	C	11376.63	$3s^23p^3$	^{2}D^o	5/2	6	ASD
					58174.366	$3s^23p^2(^3P)4s$	^{2}P	3/2	4	
400	2149.145		3.18	C	11361.02	$3s^23p^3$	^{2}D^o	3/2	4	ASD
					57876.574	$3s^23p^2(^3P)4s$	^{2}P	1/2	2	
280	2152.940		0.485	C	18722.71	$3s^23p^3$	^{2}P^o	1/2	2	ASD
					65156.242	$3s^23p^2(^1D)4s$	^{2}D	3/2	4	
500	2154.080		0.58	C	18748.01	$3s^23p^3$	^{2}P^o	3/2	4	ASD
					65157.126	$3s^23p^2(^1D)4s$	^{2}D	5/2	6	
750	2533.976		0.20	C	18722.71	$3s^23p^3$	^{2}P^o	1/2	2	ASD
					58174.366	$3s^23p^2(^3P)4s$	^{2}P	3/2	4	
950	2535.603		0.95	C	18748.01	$3s^23p^3$	^{2}P^o	3/2	4	ASD
					58174.366	$3s^23p^2(^3P)4s$	^{2}P	3/2	4	
750	2553.262		0.71	C	18722.71	$3s^23p^3$	^{2}P^o	1/2	2	ASD
					57876.574	$3s^23p^2(^3P)4s$	^{2}P	1/2	2	
1700	9525.73		0.14	D	56339.656	$3s^23p^2(^3P)4s$	^{4}P	5/2	6	ASD
					66834.648	$3s^23p^2(^3P)4p$	^{4}S^o	3/2	4	

Intensity	Wavelength/Å	Uncert. wavelength/Å	$A_{ki}/(10^8\,s^{-1})$	Accur.	Energy levels/cm^{-1}	Configs.	Terms	J	g	DB Source
1500	9545.18				56339.656	$3s^23p^2(^3P)4s$	^4P	5/2	6	HDBK,ASD
					66813.271	$3s^23p^2(^3P)4p$	$^2D^o$	3/2	4	
1700	9563.439		0.081	D	56090.626	$3s^23p^2(^3P)4s$	^4P	3/2	4	ASD
					66544.243	$3s^23p^2(^3P)4p$	$^4P^o$	5/2	6	
1500	9734.750		0.035	E	56090.626	$3s^23p^2(^3P)4s$	^4P	3/2	4	ASD
					66360.282	$3s^23p^2(^3P)4p$	$^4P^o$	3/2	4	
1500	9750.77		0.22	D	56090.626	$3s^23p^2(^3P)4s$	^4P	3/2	4	ASD
					66343.438	$3s^23p^2(^3P)4p$	$^4P^o$	1/2	2	
1700	9796.85		0.18	D	56339.656	$3s^23p^2(^3P)4s$	^4P	5/2	6	ASD
					66544.243	$3s^23p^2(^3P)4p$	$^4P^o$	5/2	6	
Sulfur I (Z = 16)										
8000g	1425.0301		3.79	C+	0.000	$3s^23p^4$	^3P	2	5	ASD
					70173.968	$3s^23p^3(^4S^o)3d$	$^3D^o$	3	7	
5000g	1433.2800		2.81	C+	396.055	$3s^23p^4$	^3P	1	3	ASD,HDBK
					70166.195	$3s^23p^3(^4S^o)3d$	$^3D^o$	2	5	
1500g	1433.3105		1.56	C	396.055	$3s^23p^4$	^3P	1	3	ASD
					70164.650	$3s^23p^3(^4S^o)3d$	$^3D^o$	1	3	
10000g	1473.9948		1.96	C+	0.000	$3s^23p^4$	^3P	2	5	ASD
					67842.867	$3s^23p^3(^2D^o)4s$	$^3D^o$	3	7	
17500	1666.6875		4.58	C+	9238.609	$3s^23p^4$	^1D	2	5	ASD
					69237.886	$3s^23p^3(^2D^o)4s$	$^1D^o$	2	5	
15000	1687.5305		0.690	D	22179.954	$3s^23p^4$	^1S	0	1	ASD
					81438.30	$3s^23p^3(^2D^o)3d$	$^1P^o$	1	3	
20000g	1807.3108		3.27	C+	0.000	$3s^23p^4$	^3P	2	5	ASD
					55330.811	$3s^23p^3(^4S^o)4s$	$^3S^o$	1	3	
17500g	1820.3426		1.71	C+	396.055	$3s^23p^4$	^3P	1	3	ASD
					55330.811	$3s^23p^3(^4S^o)4s$	$^3S^o$	1	3	
15000g	1826.2451		0.564	C	573.640	$3s^23p^4$	^3P	0	1	ASD
					55330.811	$3s^23p^3(^4S^o)4s$	$^3S^o$	1	3	
1500	9212.865		0.279	C	52623.640	$3s^23p^3(^4S^o)4s$	$^5S^o$	2	5	ASD
					63475.051	$3s^23p^3(^4S^o)4p$	^5P	3	7	
1050	9228.092		0.277	C	52623.640	$3s^23p^3(^4S^o)4s$	$^5S^o$	2	5	ASD
					63457.142	$3s^23p^3(^4S^o)4p$	^5P	2	5	
810	9237.538		0.277	C	52623.640	$3s^23p^3(^4S^o)4s$	$^5S^o$	2	5	ASD
					63446.065	$3s^23p^3(^4S^o)4p$	^5P	1	3	
1850	10455.451		0.217	B+	55330.811	$3s^23p^3(^4S^o)4s$	$^3S^o$	1	3	ASD
					64892.582	$3s^23p^3(^4S^o)4p$	^3P	2	5	
310	10456.757		0.218	B+	55330.811	$3s^23p^3(^4S^o)4s$	$^3S^o$	1	3	ASD
					64891.386	$3s^23p^3(^4S^o)4p$	^3P	0	1	
1300	10459.406		0.218	B+	55330.811	$3s^23p^3(^4S^o)4s$	$^3S^o$	1	3	ASD
					64888.964	$3s^23p^3(^4S^o)4p$	^3P	1	3	
Chlorine I (Z = 17)										
3000	1335.726		1.74	C	0.0000	$3s^23p^5$	$^2P^o$	3/2	4	ASD
					74865.667	$3s^23p^4(^3P)4s$	^2P	1/2	2	
10000	1347.240		4.19	C+	0.0000	$3s^23p^5$	$^2P^o$	3/2	4	ASD
					74225.846	$3s^23p^4(^3P)4s$	^2P	3/2	4	
5000	1351.657		3.23	C+	882.3515	$3s^23p^5$	$^2P^o$	1/2	2	ASD
					74865.667	$3s^23p^4(^3P)4s$	^2P	1/2	2	
12000	1363.447		0.75	C+	882.3515	$3s^23p^5$	$^2P^o$	1/2	2	ASD
					74225.846	$3s^23p^4(^3P)4s$	^2P	3/2	4	
20000	1379.528		0.11	D	0.0000	$3s^23p^5$	$^2P^o$	3/2	4	ASD
					72488.568	$3s^23p^4(^3P)4s$	^4P	3/2	4	
25000	1389.693		0.0023	D	0.0000	$3s^23p^5$	$^2P^o$	3/2	4	ASD
					71958.363	$3s^23p^4(^3P)4s$	^4P	5/2	6	
20000	1389.957		0.017	D	882.3515	$3s^23p^5$	$^2P^o$	1/2	2	ASD
					72827.038	$3s^23p^4(^3P)4s$	^4P	1/2	2	
12000	1396.527		0.015	D	882.3515	$3s^23p^5$	$^2P^o$	1/2	2	ASD
					72488.568	$3s^23p^4(^3P)4s$	^4P	3/2	4	

Intensity	Wavelength/Å	Uncert. wavelength/Å	A_{ki}/(10^8 s^{-1})	Accur.	Energy levels/cm^{-1}	Configs.	Terms	J	g	DB Source
11000	7547.072		0.12	D	72488.568	$3s^23p^4(^3P)4s$	^4P	3/2	4	ASD
					85735.091	$3s^23p^4(^3P)4p$	$^4S^o$	3/2	4	
10000	7744.97		0.063	C	72827.038	$3s^23p^4(^3P)4s$	^4P	1/2	2	ASD
					85735.091	$3s^23p^4(^3P)4p$	$^4S^o$	3/2	4	
18000	8212.04		0.079	D	71958.363	$3s^23p^4(^3P)4s$	^4P	5/2	6	ASD
					84132.262	$3s^23p^4(^3P)4p$	$^4D^o$	5/2	6	
20000	8221.74				72488.568	$3s^23p^4(^3P)4s$	^4P	3/2	4	ASD
					84648.100	$3s^23p^4(^3P)4p$	$^2D^o$	5/2	6	
18000	8333.31		0.16	D	72488.568	$3s^23p^4(^3P)4s$	^4P	3/2	4	ASD
					84485.309	$3s^23p^4(^3P)4p$	$^4D^o$	3/2	4	
99900	8375.94		0.28	D	71958.363	$3s^23p^4(^3P)4s$	^4P	5/2	6	ASD
					83894.037	$3s^23p^4(^3P)4p$	$^4D^o$	7/2	8	
15000	8428.25		0.24	D	72827.038	$3s^23p^4(^3P)4s$	^4P	1/2	2	ASD
					84688.637	$3s^23p^4(^3P)4p$	$^4D^o$	1/2	2	
20000	8575.24		0.12	D	72827.038	$3s^23p^4(^3P)4s$	^4P	1/2	2	ASD
					84485.309	$3s^23p^4(^3P)4p$	$^4D^o$	3/2	4	
75000	8585.97		0.19	D	72488.568	$3s^23p^4(^3P)4s$	^4P	3/2	4	ASD
					84132.262	$3s^23p^4(^3P)4p$	$^4D^o$	5/2	6	
Argon I (Z = 18)										
180	866.800		3.13	C+	0.0000	$3s^23p^6$	^1S	0	1	HDBK,ASD
					115366.871	$3s^23p^5(^2P^o<1/2>)3d$	$2[3/2]^o$	1	3	
150	869.754		0.350	C	0.0000	$3s^23p^6$	^1S	0	1	HDBK,ASD
					114975.024	$3s^23p^5(^2P^o<1/2>)5s$	$2[1/2]^o$	1	3	
180r	876.058		2.70	C+	0.0000	$3s^23p^6$	^1S	0	1	HDBK,ASD
					114147.737	$3s^23p^5(^2P^o<3/2>)3d$	$2[3/2]^o$	1	3	
180r	879.947		0.77	C	0.0000	$3s^23p^6$	^1S	0	1	HDBK,ASD
					113643.265	$3s^23p^5(^2P^o<3/2>)5s$	$2[3/2]^o$	1	3	
150	894.310				0.0000	$3s^23p^6$	^1S	0	1	HDBK,ASD
					111818.033	$3s^23p^5(^2P^o<3/2>)3d$	$2[1/2]^o$	1	3	
1000r	1048.220		5.1	C+	0.0000	$3s^23p^6$	^1S	0	1	HDBK,ASD
					95399.8329	$3s^23p^5(^2P^o<1/2>)4s$	$2[1/2]^o$	1	3	
500r	1066.660		1.19	C+	0.0000	$3s^23p^6$	^1S	0	1	HDBK,ASD
					93750.6031	$3s^23p^5(^2P^o<3/2>)4s$	$2[3/2]^o$	1	3	
10000	6965.431		0.0639	C	93143.7653	$3s^23p^5(^2P^o<3/2>)4s$	$2[3/2]^o$	2	5	HDBK,ASD
					107496.4219	$3s^23p^5(^2P^o<1/2>)4p$	$2[1/2]$	1	3	
10000	7067.218		0.0380	C	93143.7653	$3s^23p^5(^2P^o<3/2>)4s$	$2[3/2]^o$	2	5	HDBK,ASD
					107289.7054	$3s^23p^5(^2P^o<1/2>)4p$	$2[3/2]$	2	5	
20000	7503.869		0.445	C	95399.8329	$3s^23p^5(^2P^o<1/2>)4s$	$2[1/2]^o$	1	3	HDBK,ASD
					108722.6247	$3s^23p^5(^2P^o<1/2>)4p$	$2[1/2]$	0	1	
25000	7635.106		0.245	C	93143.7653	$3s^23p^5(^2P^o<3/2>)4s$	$2[3/2]^o$	2	5	HDBK,ASD
					106237.5571	$3s^23p^5(^2P^o<3/2>)4p$	$2[3/2]$	2	5	
20000	7948.176		0.186	C	94553.6705	$3s^23p^5(^2P^o<1/2>)4s$	$2[1/2]^o$	0	1	HDBK,ASD
					107131.7139	$3s^23p^5(^2P^o<1/2>)4p$	$2[3/2]$	1	3	
20000	8006.157		0.0490	C	93750.6031	$3s^23p^5(^2P^o<3/2>)4s$	$2[3/2]^o$	1	3	HDBK,ASD
					106237.5571	$3s^23p^5(^2P^o<3/2>)4p$	$2[3/2]$	2	5	
25000	8014.786		0.0928	C	93143.7653	$3s^23p^5(^2P^o<3/2>)4s$	$2[3/2]^o$	2	5	HDBK,ASD
					105617.2753	$3s^23p^5(^2P^o<3/2>)4p$	$2[5/2]$	2	5	
20000	8103.693		0.25	C	93750.6031	$3s^23p^5(^2P^o<3/2>)4s$	$2[3/2]^o$	1	3	HDBK,ASD
					106087.2651	$3s^23p^5(^2P^o<3/2>)4p$	$2[3/2]$	1	3	
35000	8115.311		0.331	C	93143.7653	$3s^23p^5(^2P^o<3/2>)4s$	$2[3/2]^o$	2	5	HDBK,ASD
					105462.7649	$3s^23p^5(^2P^o<3/2>)4p$	$2[5/2]$	3	7	
35000	9122.967		0.189	C	93143.7653	$3s^23p^5(^2P^o<3/2>)4s$	$2[3/2]^o$	2	5	HDBK,ASD
					104102.1043	$3s^23p^5(^2P^o<3/2>)4p$	$2[1/2]$	1	3	
25000	9657.786		0.0543	C	93750.6031	$3s^23p^5(^2P^o<3/2>)4s$	$2[3/2]^o$	1	3	HDBK,ASD
					104102.1043	$3s^23p^5(^2P^o<3/2>)4p$	$2[1/2]$	1	3	

TABLE 4. Persistent Lines of Neutral Atomic Elements — Potassium I (Z = 19) to Krypton (Z = 36)

Intensity	Wavelength/Å	Uncert. wavelength/Å	$A_{ki}/(10^8\ s^{-1})$	Accur.	Energy levels/cm^{-1}	Configs.	Terms	J	g	DB Source
Potassium I (Z = 19)										
18	4044.136		0.0116		0.0000	$3p^64s$	^2S	1/2	2	ASD
					24720.139	$3p^65p$	$^2P^o$	3/2	4	ASD
17	4047.208		0.0107		0.0000	$3p^64s$	^2S	1/2	2	ASD
					24701.382	$3p^65p$	$^2P^o$	1/2	2	ASD
25r	7664.899126		0.380		0.0000	$3p^64s$	^2S	1/2	2	ASD
					13042.896027	$3p^64p$	$^2P^o$	3/2	4	ASD
24r	7698.964562		0.375		0.0000	$3p^64s$	^2S	1/2	2	ASD
					12985.185724	$3p^64p$	$^2P^o$	1/2	2	ASD
17	11690.219		0.220	C	12985.185724	$3p^64p$	$^2P^o$	1/2	2	ASD
					21536.988	$3p^63d$	^2D	3/2	4	ASD
16	11769.637		0.0434	C	13042.896027	$3p^64p$	$^2P^o$	3/2	4	ASD
					21536.988	$3p^63d$	^2D	3/2	4	ASD
17	11772.838		0.259	C	13042.896027	$3p^64p$	$^2P^o$	3/2	4	ASD
					21534.680	$3p^63d$	^2D	5/2	6	ASD
Calcium I (Z = 20)										
7	2721.645		0.0027	D	0.000	$3p^64s^2$	^1S	0	1	HDBK,ASD
					36731.615	$3p^64s5p$	$^1P^o$	1	3	HDBK,ASD
50	4226.727		2.18	B+	0.000	$3p^64s^2$	^1S	0	1	HDBK,ASD
					23652.304	$3p^64s4p$	$^1P^o$	1	3	HDBK,ASD
25	4425.441		0.498	C	15157.901	$3p^64s4p$	$^3P^o$	0	1	HDBK,ASD
					37748.197	$3p^64s4d$	^3D	1	3	HDBK,ASD
26	4434.960		0.67	C	15210.063	$3p^64s4p$	$^3P^o$	1	3	HDBK,ASD
					37751.867	$3p^64s4d$	^3D	2	5	HDBK,ASD
25	4435.688		0.342	C	15210.063	$3p^64s4p$	$^3P^o$	1	3	HDBK,ASD
					37748.197	$3p^64s4d$	^3D	1	3	HDBK,ASD
30	4454.781		0.87	C	15315.943	$3p^64s4p$	$^3P^o$	2	5	HDBK,ASD
					37757.449	$3p^64s4d$	^3D	3	7	HDBK,ASD
28	4455.887		0.20	C	15315.943	$3p^64s4p$	$^3P^o$	2	5	HDBK,ASD
					37751.867	$3p^64s4d$	^3D	2	5	HDBK,ASD
20	4456.605		0.0245	C+	15315.943	$3p^64s4p$	$^3P^o$	2	5	HDBK,ASD
					37748.197	$3p^64s4d$	^3D	1	3	HDBK,ASD
27	5588.757		0.49	D	20371.000	$3p^63d4s$	^3D	3	7	HDBK,ASD
					38259.124	$3p^63d4p$	$^3D^o$	3	7	HDBK,ASD
27	6102.722		0.096	C	15157.901	$3p^64s4p$	$^3P^o$	0	1	HDBK,ASD
					31539.495	$3p^64s5s$	^3S	1	3	HDBK,ASD
29	6122.219		0.287	C	15210.063	$3p^64s4p$	$^3P^o$	1	3	HDBK,ASD
					31539.495	$3p^64s5s$	^3S	1	3	HDBK,ASD
30	6162.172		0.477	C	15315.943	$3p^64s4p$	$^3P^o$	2	5	HDBK,ASD
					31539.495	$3p^64s5s$	^3S	1	3	HDBK,ASD
35	6439.073		0.53	D	20371.000	$3p^63d4s$	^3D	3	7	HDBK,ASD
					35896.889	$3p^63d4p$	$^3F^o$	4	9	HDBK,ASD
34	6462.566		0.47	D	20349.260	$3p^63d4s$	^3D	2	5	HDBK,ASD
					35818.713	$3p^63d4p$	$^3F^o$	3	7	HDBK,ASD
32	6493.780		0.44	D	20335.360	$3p^63d4s$	^3D	1	3	HDBK,ASD
					35730.454	$3p^63d4p$	$^3F^o$	2	5	HDBK,ASD
23	6572.777		0.000026	D+	0.000	$3p^64s^2$	^1S	0	1	HDBK,ASD
					15210.063	$3p^64s4p$	$^3P^o$	1	3	HDBK,ASD
Scandium I (Z = 21)										
6g	3269.897		3.13	B+	0.00	$3d4s^2$	^2D	3/2	4	ASD
					30573.17	$3d4s(^3D)4p$	$^2P^o$	1/2	2	ASD
6g	3273.628		2.81	A'	168.34	$3d4s^2$	^2D	5/2	6	ASD
					30706.66	$3d4s(^3D)4p$	$^2P^o$	3/2	4	ASD
7g	3907.484		1.66	A'	0.00	$3d4s^2$	^2D	3/2	4	ASD
					25584.64	$3d4s(^3D)4p$	$^2F^o$	5/2	6	ASD
7g	3911.812		1.79	A'	168.34	$3d4s^2$	^2D	5/2	6	ASD
					25724.68	$3d4s(^3D)4p$	$^2F^o$	7/2	8	ASD

Intensity	Wavelength/Å	Uncert. wavelength/Å	$A_{ki}/(10^8\ s^{-1})$	Accur.	Energy levels/cm^{-1}	Configs.	Terms	J	g	DB Source
7g	3933.375		0.162	B+	168.34	$3d4s^2$	^2D	5/2	6	ASD
					25584.64	$3d4s(^3D)4p$	$^2F^o$	5/2	6	ASD
7g	3996.601		0.165	B+	0.00	$3d4s^2$	^2D	3/2	4	ASD
					25014.21	$3d4s(^3D)4p$	$^2D^o$	5/2	6	ASD
7g	4020.387		1.63	A'	0.00	$3d4s^2$	^2D	3/2	4	ASD
					24866.17	$3d4s(^3D)4p$	$^2D^o$	3/2	4	ASD
7g	4023.678		1.65	A'	168.34	$3d4s^2$	^2D	5/2	6	ASD
					25014.21	$3d4s(^3D)4p$	$^2D^o$	5/2	6	ASD
7g	4054.544				0.00	$3d4s^2$	^2D	3/2	4	ASD
					24656.72	$3d4s(^1D)4p$	$^2P^o$	1/2	2	ASD
7g	4082.387		0.273	B+	168.34	$3d4s^2$	^2D	5/2	6	ASD
					24656.88	$3d4s(^1D)4p$	$^2P^o$	3/2	4	ASD
Titanium I (Z = 22)										
15+	3635.4625	0.0016	0.909	A	0.000	$3d^24s^2$	a ^3F	2	5	ASD
					27498.982	$3d^2(^3F)4s4p(^1P^o)$	y $^3G^o$	3	7	ASD
15+	3642.6739	0.0016	0.895	A	170.1328	$3d^24s^2$	a ^3F	3	7	ASD
					27614.679	$3d^2(^3F)4s4p(^1P^o)$	y $^3G^o$	4	9	ASD
15+	3653.4949	0.0016	0.869	A	386.874	$3d^24s^2$	a ^3F	4	9	ASD
					27750.135	$3d^2(^3F)4s4p(^1P^o)$	y $^3G^o$	5	11	ASD
15+	3752.8588	0.0017	0.581	A	386.874	$3d^24s^2$	a ^3F	4	9	ASD
					27025.659	$3d^2(^1D)4s4p(^3P^o)$	x $^3F^o$	4	9	ASD
7000	3948.6708	0.0019	0.560	A	0.000	$3d^24s^2$	a ^3F	2	5	ASD
					25317.814	$3d^3(^4F)4p$	y $^3D^o$	1	3	ASD
8000	3956.3343	0.0019	0.346	A	170.1328	$3d^24s^2$	a ^3F	3	7	ASD
					25438.908	$3d^3(^4F)4p$	y $^3D^o$	2	5	ASD
8600	3958.2016	0.0019	0.488	A	386.874	$3d^24s^2$	a ^3F	4	9	ASD
					25643.701	$3d^3(^4F)4p$	y $^3D^o$	3	7	ASD
8800	3981.7616	0.0019	0.442	A	0.000	$3d^24s^2$	a ^3F	2	5	ASD
					25107.411	$3d^2(^3F)4s4p(^1P^o)$	y $^3F^o$	2	5	ASD
8800	3989.7586	0.0019	0.448	A	170.1328	$3d^24s^2$	a ^3F	3	7	ASD
					25227.222	$3d^2(^3F)4s4p(^1P^o)$	y $^3F^o$	3	7	ASD
10000	3998.6366	0.0019	0.481	A	386.874	$3d^24s^2$	a ^3F	4	9	ASD
					25388.331	$3d^2(^3F)4s4p(^1P^o)$	y $^3F^o$	4	9	ASD
6400	4305.9078	0.002			6842.962	$3d^3(^4F)4s$	a ^5F	5	11	ASD
					30060.340	$3d^3(^4F)4p$	x $^5D^o$	4	9	ASD
14000	4981.731	0.003	0.660	C+	6842.962	$3d^3(^4F)4s$	a ^5F	5	11	ASD
					26910.709	$3d^3(^4F)4p$	y $^5G^o$	6	13	ASD
13000	4991.067	0.003	0.584	C+	6742.756	$3d^3(^4F)4s$	a ^5F	4	9	ASD
					26772.968	$3d^3(^4F)4p$	y $^5G^o$	5	11	ASD
10000	5007.206	0.003	0.492	C+	6598.765	$3d^3(^4F)4s$	a ^5F	2	5	ASD
					26564.400	$3d^3(^4F)4p$	y $^5G^o$	3	7	ASD
11000	5014.186	0.003	0.0508	B	0.000	$3d^24s^2$	a ^3F	2	5	ASD
					19937.855	$3d^2(^3F)4s4p(^3P^o)$	z $^3D^o$	1	3	ASD
17000	5064.652	0.003	0.0437	A	386.874	$3d^24s^2$	a ^3F	4	9	ASD
					20126.062	$3d^2(^3F)4s4p(^3P^o)$	z $^3D^o$	3	7	ASD
21000	5210.384	0.003	0.0389	B	386.874	$3d^24s^2$	a ^3F	4	9	ASD
					19573.974	$3d^2(^3F)4s4p(^3P^o)$	z $^3F^o$	4	9	ASD
Vanadium I (Z = 23)										
16000000	3183.4093	0.0009	2.52	B+	137.383	$3d^34s^2$	a ^4F	5/2	6	ASD
					31541.167	$3d^3(^4F)4s4p(^1P^o)$	x $^4G^o$	7/2	8	ASD
27000000	3183.9603	0.0009	2.80	B+	323.432	$3d^34s^2$	a ^4F	7/2	8	ASD
					31721.780	$3d^3(^4F)4s4p(^1P^o)$	x $^4G^o$	9/2	10	ASD
19000000	3185.3979	0.0009	2.93	B+	552.955	$3d^34s^2$	a ^4F	9/2	10	ASD
					31937.131	$3d^3(^4F)4s4p(^1P^o)$	x $^4G^o$	11/2	12	ASD
21000000	3703.5676	0.0012	1.12	B+	2424.809	$3d^4(^5D)4s$	a ^6D	9/2	10	ASD
					29418.119	$3d^3(^4P)4s4p(^3P^o)$	y $^6P^o$	7/2	8	ASD
18000000	3855.8400	0.0015	0.57	B+	552.955	$3d^34s^2$	a ^4F	9/2	10	ASD
					26480.286†	$3d^4(^5D)4p$	y $^4D^o$	7/2	8	ASD
14000000	3902.2531	0.0015	0.261	B+	552.955	$3d^34s^2$	a ^4F	9/2	10	ASD

Intensity	Wavelength/Å	Uncert. wavelength/Å	$A_{ki}/(10^8\ s^{-1})$	Accur.	Energy levels/cm^{-1}	Configs.	Terms	J	g	DB Source
					26171.918	$3d^4(^5D)4p$	y $^4F^o$	9/2	10	ASD
53000000	4111.7787	0.0015	1.00	B+	2424.809	$3d^4(^5D)4s$	a ^6D	9/2	10	ASD
					26738.323	$3d^4(^5D)4p$	y $^6D^o$	9/2	10	ASD
25000000	4115.1778	0.0015	0.56	B+	2311.369	$3d^4(^5D)4s$	a ^6D	7/2	8	ASD
					26604.807†	$3d^4(^5D)4p$	y $^6D^o$	7/2	8	ASD
18000000	4128.0642	0.0015	0.72	B+	2220.156	$3d^4(^5D)4s$	a ^6D	5/2	6	ASD
					26437.754	$3d^4(^5D)4p$	y $^6D^o$	3/2	4	ASD
18000000	4131.9909	0.0015	0.52	B+	2311.369	$3d^4(^5D)4s$	a ^6D	7/2	8	ASD
					26505.953	$3d^4(^5D)4p$	y $^6D^o$	5/2	6	ASD
74000000	4379.2298	0.0018	1.15	B+	2424.809	$3d^4(^5D)4s$	a ^6D	9/2	10	ASD
					25253.457	$3d^4(^5D)4p$	y $^6F^o$	11/2	12	ASD
44000000	4384.7122	0.0018	0.92	B+	2311.369	$3d^4(^5D)4s$	a ^6D	7/2	8	ASD
					25111.473	$3d^4(^5D)4p$	y $^6F^o$	9/2	10	ASD
30000000	4389.9791	0.0018	0.70	B+	2220.156	$3d^4(^5D)4s$	a ^6D	5/2	6	ASD
					24992.909	$3d^4(^5D)4p$	y $^6F^o$	7/2	8	ASD
23000000	4395.2235	0.0018	0.52	B+	2153.221	$3d^4(^5D)4s$	a ^6D	3/2	4	ASD
					24898.804	$3d^4(^5D)4p$	y $^6F^o$	5/2	6	ASD
23000000	4408.1959	0.0012	0.51	B+	2220.156	$3d^4(^5D)4s$	a ^6D	5/2	6	ASD
					24898.804	$3d^4(^5D)4p$	y $^6F^o$	5/2	6	ASD
:	4408.493	0.0010	0.80	B+	2112.282	$3d^4(^5D)4s$	a ^6D	1/2	2	ASD
					24789.401	$3d^4(^5D)4p$	y $^6F^o$	1/2	2	ASD
29000000	4408.5139	0.0012	0.60	B+	2153.221	$3d^4(^5D)4s$	a ^6D	3/2	4	ASD
					24830.221	$3d^4(^5D)4p$	y $^6F^o$	3/2	4	ASD
8100000	6090.2090	0.002	0.257	B+	8715.747	$3d^4(^5D)4s$	a ^4D	7/2	8	ASD
					25131.002	$3d^4(^5D)4p$	z $^4P^o$	5/2	6	ASD
Chromium I (Z = 24)										
1380q	3578.704	0.018	1.48	B	0.0000	$3d^5(^6S)4s$	a ^7S	3	7	ASD
					27935.2412	$3d^4(^5D)4s4p(^3P^o)$	y $^7P^o$	4	9	ASD
1450q	3593.502	0.017	1.50	B	0.0000	$3d^5(^6S)4s$	a ^7S	3	7	ASD
					27820.1975	$3d^4(^5D)4s4p(^3P^o)$	y $^7P^o$	3	7	ASD
1240q	3605.345	0.016	1.62	B	0.0000	$3d^5(^6S)4s$	a ^7S	3	7	ASD
					27728.8110	$3d^4(^5D)4s4p(^3P^o)$	y $^7P^o$	2	5	ASD
2480q	4254.352	0.006	0.315	B	0.0000	$3d^5(^6S)4s$	a ^7S	3	7	ASD
					23498.8156	$3d^5(^6S)4p$	z $^7P^o$	4	9	ASD
2500q	4274.812	0.006	0.307	B	0.0000	$3d^5(^6S)4s$	a ^7S	3	7	ASD
					23386.3419	$3d^5(^6S)4p$	z $^7P^o$	3	7	ASD
2380q	4289.731	0.006	0.316	B	0.0000	$3d^5(^6S)4s$	a ^7S	3	7	ASD
					23305.0026	$3d^5(^6S)4p$	z $^7P^o$	2	5	ASD
5900q	5204.498	0.009	0.509	B	7593.1484	$3d^5(^6S)4s$	a ^5S	2	5	ASD
					26801.9009	$3d^5(^6S)4p$	z $^5P^o$	1	3	ASD
7100q	5206.023	0.009	0.514	B	7593.1484	$3d^5(^6S)4s$	a ^5S	2	5	ASD
					26796.2691	$3d^5(^6S)4p$	z $^5P^o$	2	5	ASD
7800q	5208.409	0.009	0.506	B	7593.1484	$3d^5(^6S)4s$	a ^5S	2	5	ASD
					26787.4640	$3d^5(^6S)4p$	z $^5P^o$	3	7	ASD
Manganese I (Z = 25)										
9700	1996.056		0.77		0.00	$3d^54s^2$	a ^6S	5/2	6	HDBK,ASD
					50099.03	$3d^5(^4P)4s4p(^3P^o)$	v $^6P^o$	3/2	4	HDBK,ASD
14000	1999.511		0.76		0.00	$3d^54s^2$	a ^6S	5/2	6	HDBK,ASD
					50012.50	$3d^5(^4P)4s4p(^3P^o)$	v $^6P^o$	5/2	6	HDBK,ASD
18000	2003.849		0.76		0.00	$3d^54s^2$	a ^6S	5/2	6	HDBK,ASD
					49888.01	$3d^5(^4P)4s4p(^3P^o)$	v $^6P^o$	7/2	8	HDBK,ASD
6200	2794.817		3.7	C	0.00	$3d^54s^2$	a ^6S	5/2	6	HDBK,ASD
					35769.97	$3d^5(^6S)4s4p(^1P^o)$	y $^6P^o$	7/2	8	HDBK,ASD
5100	2798.270		3.6	C	0.00	$3d^54s^2$	a ^6S	5/2	6	HDBK,ASD
					35725.85	$3d^5(^6S)4s4p(^1P^o)$	y $^6P^o$	5/2	6	HDBK,ASD
3700	2801.084		3.7	C	0.00	$3d^54s^2$	a ^6S	5/2	6	HDBK,ASD
					35689.98	$3d^5(^6S)4s4p(^1P^o)$	y $^6P^o$	3/2	4	HDBK,ASD
3200	3806.715		0.59	B	17052.29	$3d^6(^5D)4s$	a ^6D	9/2	10	HDBK,ASD
					43314.23	$3d^6(^5D)4p$	z $^6F^o$	11/2	12	HDBK,ASD

Intensity	Wavelength/Å	Uncert. wavelength/Å	$A_{ki}/(10^8\ s^{-1})$	Accur.	Energy levels/cm^{-1}	Configs.	Terms	J	g	DB Source
27000	4030.755		0.17	C+	0.00	$3d^54s^2$	a ^{6}S	5/2	6	HDBK,ASD
					24802.25	$3d^5(^6S)4s4p(^3P^o)$	z ^{6}P^o	7/2	8	HDBK,ASD
19000	4033.068		0.165	C+	0.00	$3d^54s^2$	a ^{6}S	5/2	6	HDBK,ASD
					24788.05	$3d^5(^6S)4s4p(^3P^o)$	z ^{6}P^o	5/2	6	HDBK,ASD
11000	4034.485		0.158	C+	0.00	$3d^54s^2$	a ^{6}S	5/2	6	HDBK,ASD
					24779.32	$3d^5(^6S)4s4p(^3P^o)$	z ^{6}P^o	3/2	4	HDBK,ASD
5600	4041.357		0.787	C+	17052.29	$3d^6(^5D)4s$	a ^{6}D	9/2	10	HDBK,ASD
					41789.48	$3d^6(^5D)4p$	z ^{6}D^o	9/2	10	HDBK,ASD
Iron I (Z = 26)										
38000	2483.2707	0.0005	4.80	B	0.000	$3d^64s^2$	a ^{5}D	4	9	ASD
					40257.314	$3d^6(^5D)4s4p(^1P^o)$	x ^{5}F^o	5	11	ASD
7200	2488.1423	0.0005	4.20	B	415.933	$3d^64s^2$	a ^{5}D	3	7	ASD
					40594.432	$3d^6(^5D)4s4p(^1P^o)$	x ^{5}F^o	4	9	ASD
18200	2490.64417	0.0002	3.44	B	704.007	$3d^64s^2$	a ^{5}D	2	5	ASD
					40842.154	$3d^6(^5D)4s4p(^1P^o)$	x ^{5}F^o	3	7	ASD
11700	2522.84927	0.0002	2.13	B	0.000	$3d^64s^2$	a ^{5}D	4	9	ASD
					39625.804	$3d^6(^5D)4s4p(^1P^o)$	x ^{5}D^o	4	9	ASD
10700	2719.02720	0.0002	1.42	B+	0.000	$3d^64s^2$	a ^{5}D	4	9	ASD
					36766.966	$3d^6(^5D)4s4p(^1P^o)$	y ^{5}P^o	3	7	ASD
4570	2788.10473	0.0002	0.592	B+	6928.268	$3d^7(^4F)4s$	a ^{5}F	5	11	ASD
					42784.352	$3d^6(^3H)4s4p(^3P^o)$	y ^{5}G^o	6	13	ASD
1350000	3440.6057	0.0004	0.171	B	0.000	$3d^64s^2$	a ^{5}D	4	9	ASD
					29056.324	$3d^6(^5D)4s4p(^3P^o)$	z ^{5}P^o	3	7	ASD
2500000r	3581.1928	0.0004	1.02	A	6928.268	$3d^7(^4F)4s$	a ^{5}F	5	11	ASD
					34843.957	$3d^7(^4F)4p$	z ^{5}G^o	6	13	ASD
2500000r	3719.9345	0.0004	0.162	A	0.000	$3d^64s^2$	a ^{5}D	4	9	ASD
					26874.550	$3d^6(^5D)4s4p(^3P^o)$	z ^{5}F^o	5	11	ASD
3000000r	3734.8636	0.0004	0.901	A	6928.268	$3d^7(^4F)4s$	a ^{5}F	5	11	ASD
					33695.397	$3d^7(^4F)4p$	y ^{5}F^o	5	11	ASD
2500000r	3737.1313	0.0004	0.141	A	415.933	$3d^64s^2$	a ^{5}D	3	7	ASD
					27166.820	$3d^6(^5D)4s4p(^3P^o)$	z ^{5}F^o	4	9	ASD
2510000	3745.5610	0.0004	0.115	A	704.007	$3d^64s^2$	a ^{5}D	2	5	ASD
					27394.691	$3d^6(^5D)4s4p(^3P^o)$	z ^{5}F^o	3	7	ASD
1910000	3748.2619	0.0004	0.0915	A	888.132	$3d^64s^2$	a ^{5}D	1	3	ASD
					27559.583	$3d^6(^5D)4s4p(^3P^o)$	z ^{5}F^o	2	5	ASD
1150000	3749.4851	0.0004	0.763	A	7376.764	$3d^7(^4F)4s$	a ^{5}F	4	9	ASD
					34039.516	$3d^7(^4F)4p$	y ^{5}F^o	4	9	ASD
740000	3758.2327	0.0004	0.634	A	7728.060	$3d^7(^4F)4s$	a ^{5}F	3	7	ASD
					34328.752	$3d^7(^4F)4p$	y ^{5}F^o	3	7	ASD
2000000r	3820.4249	0.0004	0.667	A	6928.268	$3d^7(^4F)4s$	a ^{5}F	5	11	ASD
					33095.941	$3d^7(^4F)4p$	y ^{5}D^o	4	9	ASD
2000000r	3859.9111	0.0004	0.0969	A	0.000	$3d^64s^2$	a ^{5}D	4	9	ASD
					25899.989	$3d^6(^5D)4s4p(^3P^o)$	z ^{5}D^o	4	9	ASD
550000	3886.2820	0.0005	0.0529	A	415.933	$3d^64s^2$	a ^{5}D	3	7	ASD
					26140.179	$3d^6(^5D)4s4p(^3P^o)$	z ^{5}D^o	3	7	ASD
1000000	4045.8122	0.0005	0.862	A	11976.239	$3d^7(^4F)4s$	a ^{3}F	4	9	ASD
					36686.176	$3d^7(^4F)4p$	y ^{3}F^o	4	9	ASD
141000	4383.5447	0.0006	0.500	A	11976.239	$3d^7(^4F)4s$	a ^{3}F	4	9	ASD
					34782.421	$3d^7(^4F)4p$	z ^{5}G^o	5	11	ASD
Cobalt I (Z = 27)										
5300	2407.256		3.6	C	0.000	$3p^63d^74s^2$	a ^{4}F	9/2	10	HDBK,ASD
					41528.455	$3p^63d^7(^4F)4s4p(^1P^o)$	x ^{4}G^o	11/2	12	HDBK,ASD
5300	2411.624				816.000	$3p^63d^74s^2$	a ^{4}F	7/2	8	HDBK,ASD
					42269.229	$3p^63d^7(^4F)4s4p(^1P^o)$	x ^{4}G^o	9/2	10	HDBK,ASD
4100	2424.935		3.2	C	0.000	$3p^63d^74s^2$	a ^{4}F	9/2	10	HDBK,ASD
					41225.710	$3p^63d^7(^4F)4s4p(^1P^o)$	x ^{4}F^o	9/2	10	HDBK,ASD
4300	2521.365		3.0	D	0.000	$3p^63d^74s^2$	a ^{4}F	9/2	10	HDBK,ASD
					39649.124	$3p^63d^7(^4F)4s4p(^1P^o)$	x ^{4}D^o	7/2	8	HDBK,ASD

Intensity	Wavelength/Å	Uncert. wavelength/Å	$A_{ki}/(10^8\ s^{-1})$	Accur.	Energy levels/cm^{-1}	Configs.	Terms	J	g	DB Source
11000	3405.118		1.0	C+	3482.780	$3p^63d^8(^3F)4s$	b ^4F	9/2	10	HDBK,ASD
					32841.916	$3p^63d^8(^3F)4p$	y $^4F^{\circ}$	9/2	10	HDBK,ASD
8800	3443.645		0.69	C+	4142.631	$3p^63d^8(^3F)4s$	b ^4F	7/2	8	HDBK,ASD
					33173.313	$3p^63d^8(^3F)4p$	$^{\circ}$	7/2	8	HDBK,ASD
21000	3453.510		1.1	C+	3482.780	$3p^63d^8(^3F)4s$	b ^4F	9/2	10	HDBK,ASD
					32430.535	$3p^63d^8(^3F)4p$	y $^4G^{\circ}$	11/2	12	HDBK,ASD
8000*	3473.974		0.034	C	0.000	$3p^63d^74s^2$	a ^4F	9/2	10	HDBK,ASD
					28777.236	$3p^63d^7(^4F)4s4p(^3P^{\circ})$	z $^4F^{\circ}$	7/2	8	HDBK,ASD
8000*	3474.042		0.56	B	4690.141	$3p^63d^8(^3F)4s$	b ^4F	5/2	6	HDBK,ASD
					33466.823	$3p^63d^8(^3F)4p$	$^{\circ}$	7/2	8	HDBK,ASD
9600	3502.280		0.80	C+	3482.780	$3p^63d^8(^3F)4s$	b ^4F	9/2	10	HDBK,ASD
					32027.440	$3p^63d^8(^3F)4p$	y $^4D^{\circ}$	7/2	8	HDBK,ASD
6400	3526.850		0.13	C	0.000	$3p^63d^74s^2$	a ^4F	9/2	10	HDBK,ASD
					28345.814	$3p^63d^7(^4F)4s4p(^3P^{\circ})$	z $^4F^{\circ}$	9/2	10	HDBK,ASD
7300	3529.808		0.46	C+	4142.631	$3p^63d^8(^3F)4s$	b ^4F	7/2	8	HDBK,ASD
					32464.688	$3p^63d^8(^3F)4p$	y $^4G^{\circ}$	9/2	10	HDBK,ASD
8800	3569.376		1.5	C	7442.399	$3p^63d^8(^3F)4s$	a ^2F	7/2	8	HDBK,ASD
					35450.505	$3p^63d^8(^3F)4p$	y $^2F^{\circ}$	7/2	8	HDBK,ASD
Nickel I (Z = 28)										
2000	2310.961				0.000	$3d^8(^3F)4s^2$	^3F	4	9	HDBK,ASD
					43258.726	$3d^8(^3F)4s4p(^1P^{\circ})$	$^3F^{\circ}$	4	9	HDBK,ASD
2600	2320.034		6.9	C	0.000	$3d^8(^3F)4s^2$	^3F	4	9	HDBK,ASD
					43089.578	$3d^8(^3F)4s4p(^1P^{\circ})$	$^3G^{\circ}$	5	11	HDBK,ASD
1200	2345.543		2.2	C	0.000	$3d^8(^3F)4s^2$	^3F	4	9	HDBK,ASD
					42620.994	$3d^8(^1D)4s4p(^3P^{\circ})$	$^3D^{\circ}$	3	7	HDBK,ASD
4000	3002.485		0.80	C+	204.787	$3d^9(^2D)4s$	^3D	3	7	HDBK,ASD
					33500.822	$3d^8(^3F)4s4p(^3P^{\circ})$	$^3D^{\circ}$	3	7	HDBK,ASD
3700	3012.001		1.3	C	3409.937	$3d^9(^2D)4s$	^1D	2	5	HDBK,ASD
					36600.791	$3d^8(^3F)4s4p(^3P^{\circ})$	$^1D^{\circ}$	2	5	HDBK,ASD
3500	3050.816		0.60	C+	204.787	$3d^9(^2D)4s$	^3D	3	7	HDBK,ASD
					32973.376	$3d^8(^3F)4s4p(^3P^{\circ})$	$^3F^{\circ}$	4	9	HDBK,ASD
2600	3101.557		0.63	C+	879.816	$3d^9(^2D)4s$	^3D	2	5	HDBK,ASD
					33112.334	$3d^8(^3F)4s4p(^3P^{\circ})$	$^3F^{\circ}$	3	7	HDBK,ASD
2900	3134.104		0.73	C+	1713.087	$3d^9(^2D)4s$	^3D	1	3	HDBK,ASD
					33610.890	$3d^8(^3F)4s4p(^3P^{\circ})$	$^3D^{\circ}$	2	5	HDBK,ASD
8200	3414.764		0.55	C	204.787	$3d^9(^2D)4s$	^3D	3	7	HDBK,ASD
					29480.989	$3d^9(^2D)4p$	$^3F^{\circ}$	4	9	HDBK,ASD
4800	3446.259		0.44	C+	879.816	$3d^9(^2D)4s$	^3D	2	5	HDBK,ASD
					29888.477	$3d^9(^2D)4p$	$^3D^{\circ}$	2	5	HDBK,ASD
5000	3458.460		0.61	C+	1713.087	$3d^9(^2D)4s$	^3D	1	3	HDBK,ASD
					30619.414	$3d^9(^2D)4p$	$^3F^{\circ}$	2	5	HDBK,ASD
5000	3461.652		0.27	C+	204.787	$3d^9(^2D)4s$	^3D	3	7	HDBK,ASD
					29084.456	$3d^8(^3F)4s4p(^3P^{\circ})$	$^5F^{\circ}$	4	9	HDBK,ASD
5500	3492.956		0.98	C+	879.816	$3d^9(^2D)4s$	^3D	2	5	HDBK,ASD
					29500.674	$3d^9(^2D)4p$	$^3P^{\circ}$	1	3	HDBK,ASD
6600	3515.052		0.42	C	879.816	$3d^9(^2D)4s$	^3D	2	5	HDBK,ASD
					29320.762	$3d^9(^2D)4p$	$^3F^{\circ}$	3	7	HDBK,ASD
8200	3524.536		1.0	C	204.787	$3d^9(^2D)4s$	^3D	3	7	HDBK,ASD
					28569.203	$3d^9(^2D)4p$	$^3P^{\circ}$	2	5	HDBK,ASD
5000	3566.372		0.56	C	3409.937	$3d^9(^2D)4s$	^1D	2	5	HDBK,ASD
					31441.635	$3d^9(^2D)4p$	$^1D^{\circ}$	2	5	HDBK,ASD
6600	3619.391		0.66	C	3409.937	$3d^9(^2D)4s$	^1D	2	5	HDBK,ASD
					31031.020	$3d^9(^2D)4p$	$^1F^{\circ}$	3	7	HDBK,ASD
Copper I (Z = 29)										
1300r	2165.093	0.021	0.55	C+	0.000	$3d^{10}4s$	^2S	1/2	2	ASD
					46172.89	$3d^9(^2D)4s4p(^3P^{\circ})$	$^{\circ}$	3/2	4	ASD
1600r	2178.944	0.021	0.91	C+	0.000	$3d^{10}4s$	^2S	1/2	2	ASD
					45879.32	$3d^9(^2D)4s4p(^3P^{\circ})$	$^2P^{\circ}$	3/2	4	ASD

Intensity	Wavelength/Å	Uncert. wavelength/Å	$A_{ki}/(10^8\ s^{-1})$	Accur.	Energy levels/cm^{-1}	Configs.	Terms	J	g	DB Source
1700r	2181.720	0.021	0.99	C+	0.000	$3d^{10}4s$	^2S	1/2	2	ASD
					45820.94	$3d^9(^2D)4s4p(^3P^o)$	$^2P^o$	1/2	2	ASD
2100r	2225.697	0.021	0.44	C+	0.000	$3d^{10}4s$	^2S	1/2	2	ASD
					44915.68	$3d^9(^2D)4s4p(^3P^o)$	$^4D^o$	1/2	2	ASD
2500r	2230.084	0.021			11202.618	$3d^94s^2$	^2D	5/2	6	ASD
					56029.89	$3d^9(^2D)4s4p(^1P^o)$	$^2F^o$	7/2	8	ASD
2300r	2244.265	0.021	0.0185	C+	0.000	$3d^{10}4s$	^2S	1/2	2	ASD
					44544.16	$3d^9(^2D)4s4p(^3P^o)$	$^4D^o$	3/2	4	ASD
1000r	2441.637	0.004	0.0201	B	0.000	$3d^{10}4s$	^2S	1/2	2	ASD
					40943.78	$3d^9(^2D)4s4p(^3P^o)$	$^4P^o$	1/2	2	ASD
2000r	2492.146	0.021	0.0279	B	0.000	$3d^{10}4s$	^2S	1/2	2	ASD
					40114.01	$3d^9(^2D)4s4p(^3P^o)$	$^4P^o$	3/2	4	ASD
10000r	3247.540	0.021	1.395	AA	0.000	$3d^{10}4s$	^2S	1/2	2	ASD
					30783.697	$3d^{10}4p$	$^2P^o$	3/2	4	ASD
10000r	3273.957	0.021	1.376	AA	0.000	$3d^{10}4s$	^2S	1/2	2	ASD
					30535.324	$3d^{10}4p$	$^2P^o$	1/2	2	ASD
1500	5105.541	0.009	0.020	C+	11202.618	$3d^94s^2$	^2D	5/2	6	ASD
					30783.697	$3d^{10}4p$	$^2P^o$	3/2	4	ASD
2000	5153.235	0.009	0.6	C+	30535.324	$3d^{10}4p$	$^2P^o$	1/2	2	ASD
					49935.195	$3d^{10}4d$	^2D	3/2	4	ASD
2500	5218.202	0.009	0.75	C+	30783.697	$3d^{10}4p$	$^2P^o$	3/2	4	ASD
					49942.051	$3d^{10}4d$	^2D	5/2	6	ASD
Zinc I (Z = 30)										
800r	2138.5735		7.09	B	0.000	$3d^{10}4s^2$	^1S	0	1	HDBK,ASD
					46745.404	$3d^{10}4s4p$	$^1P^o$	1	3	HDBK,ASD
400	2800.8635				32890.327	$3d^{10}4s4p$	$^3P^o$	2	5	HDBK,ASD
					68583.083	$3d^{10}4s5d$	^3D	3	7	HDBK,ASD
100	2801.0500				32890.327	$3d^{10}4s4p$	$^3P^o$	2	5	HDBK,ASD
					68580.705	$3d^{10}4s5d$	^3D	2	5	HDBK,ASD
150	3075.8971		0.000329	B	0.000	$3d^{10}4s^2$	^1S	0	1	HDBK,ASD
					32501.399	$3d^{10}4s4p$	$^3P^o$	1	3	HDBK,ASD
500r	3282.3256		0.90	B	32311.319	$3d^{10}4s4p$	$^3P^o$	0	1	HDBK,ASD
					62768.747	$3d^{10}4s4d$	^3D	1	3	HDBK,ASD
800	3302.5829		1.2	B	32501.399	$3d^{10}4s4p$	$^3P^o$	1	3	HDBK,ASD
					62772.014	$3d^{10}4s4d$	^3D	2	5	HDBK,ASD
700r	3302.9395		0.67	B	32501.399	$3d^{10}4s4p$	$^3P^o$	1	3	HDBK,ASD
					62768.747	$3d^{10}4s4d$	^3D	1	3	HDBK,ASD
800	3345.0134		1.7	B	32890.327	$3d^{10}4s4p$	$^3P^o$	2	5	HDBK,ASD
					62776.981	$3d^{10}4s4d$	^3D	3	7	HDBK,ASD
500	3345.5694		0.40	B	32890.327	$3d^{10}4s4p$	$^3P^o$	2	5	HDBK,ASD
					62772.014	$3d^{10}4s4d$	^3D	2	5	HDBK,ASD
150	3345.9353		0.045	B	32890.327	$3d^{10}4s4p$	$^3P^o$	2	5	HDBK,ASD
					62768.747	$3d^{10}4s4d$	^3D	1	3	HDBK,ASD
300	4680.1359				32311.319	$3d^{10}4s4p$	$^3P^o$	0	1	HDBK,ASD
					53672.240	$3d^{10}4s5s$	^3S	1	3	HDBK,ASD
400	4722.1569				32501.399	$3d^{10}4s4p$	$^3P^o$	1	3	HDBK,ASD
					53672.240	$3d^{10}4s5s$	^3S	1	3	HDBK,ASD
400	4810.5321				32890.327	$3d^{10}4s4p$	$^3P^o$	2	5	HDBK,ASD
					53672.240	$3d^{10}4s5s$	^3S	1	3	HDBK,ASD
1000w	6362.3458		0.474	C	46745.404	$3d^{10}4s4p$	$^1P^o$	1	3	HDBK,ASD
					62458.533	$3d^{10}4s4d$	^1D	2	5	HDBK,ASD
Gallium I (Z = 31)										
3	2874.235		1.17		0.000	$3d^{10}4s^24p$	$^2P^o$	1/2	2	ASD
					34781.66	$4s^24d$	^2D	3/2	4	ASD
3	2943.636		1.34		826.190	$3d^{10}4s^24p$	$^2P^o$	3/2	4	ASD
					34787.85	$4s^24d$	^2D	5/2	6	ASD
1	2944.173		0.261		826.190	$3d^{10}4s^24p$	$^2P^o$	3/2	4	ASD
					34781.66	$4s^24d$	^2D	3/2	4	ASD

Intensity	Wavelength/Å	Uncert. wavelength/Å	$A_{ki}/(10^8\,\text{s}^{-1})$	Accur.	Energy levels/cm⁻¹	Configs.	Terms	J	g	DB Source
4	4032.984		0.485		0.000	$3d^{10}4s^24p$	$^2P°$	1/2	2	ASD
					24788.530	$4s^25s$	^2S	1/2	2	ASD
5	4172.042		0.945		826.190	$3d^{10}4s^24p$	$^2P°$	3/2	4	ASD
					24788.530	$4s^25s$	^2S	1/2	2	ASD

Germanium I (Z = 32)

Intensity	Wavelength/Å	Uncert. wavelength/Å	$A_{ki}/(10^8\,\text{s}^{-1})$	Accur.	Energy levels/cm⁻¹	Configs.	Terms	J	g	DB Source
300	1988.268	0.005	0.412	B+	1409.9609	$4s^24P^2$	^3P	2	5	ASD,HDBK
					51705.020	$4s^24p4d$	$^3P°$	1	3	ASD,HDBK
500r	1998.887	0.005	1.24	B+	1409.9609	$4s^24P^2$	^3P	2	5	ASD,HDBK
					51437.802	$4s^24p4d$	$^3P°$	2	5	ASD,HDBK
1700	2019.0684	0.0010	0.492	B+	557.1341	$4s^24P^2$	^3P	1	3	ASD,HDBK
					50068.954	$4s^24p4d$	$^3F°$	2	5	ASD,HDBK
2400r	2041.7121	0.0010	1.51	B	0.0000	$4s^24P^2$	^3P	0	1	ASD,HDBK
					48962.783	$4s^24p4d$	$^3D°$	1	3	ASD,HDBK
1600r	2043.7695	0.0010	0.862	B+	1409.9609	$4s^24P^2$	^3P	2	5	ASD,HDBK
					50323.465	$4s^24p4d$	$^3F°$	3	7	ASD,HDBK
2600r	2068.6562	0.0010	1.37	B+	557.1341	$4s^24P^2$	^3P	1	3	ASD,HDBK
					48882.263	$4s^24p4d$	$^3D°$	2	5	ASD,HDBK
2000r	2094.2582	0.0010	1.70	B+	1409.9609	$4s^24P^2$	^3P	2	5	ASD,HDBK
					49144.397	$4s^24p4d$	$^3D°$	3	7	ASD,HDBK
340r	2198.7144	0.0010			7125.2989	$4s^24P^2$	^1D	2	5	ASD,HDBK
					52592.224	$4s^24p4d$	$^1F°$	3	7	ASD,HDBK
500	2592.5340	0.0010	0.604	B+	557.1341	$4s^24P^2$	^3P	1	3	ASD,HDBK
					39117.9021	$4s^24p5s$	$^3P°$	2	5	ASD,HDBK
1200	2651.1720	0.0010	1.60	B+	1409.9609	$4s^24P^2$	^3P	2	5	ASD,HDBK
					39117.9021	$4s^24p5s$	$^3P°$	2	5	ASD,HDBK
500	2651.5683	0.0010	0.694	B+	0.0000	$4s^24P^2$	^3P	0	1	ASD,HDBK
					37702.3054	$4s^24p5s$	$^3P°$	1	3	ASD,HDBK
500	2691.3411	0.0010	0.473	B+	557.1341	$4s^24P^2$	^3P	1	3	ASD,HDBK
					37702.3054	$4s^24p5s$	$^3P°$	1	3	ASD,HDBK
850	2709.6237	0.0010	2.08	A	557.1341	$4s^24P^2$	^3P	1	3	ASD,HDBK
					37451.6893	$4s^24p5s$	$^3P°$	0	1	ASD,HDBK
650	2754.5878	0.0010	0.792	B+	1409.9609	$4s^24P^2$	^3P	2	5	ASD,HDBK
					37702.3054	$4s^24p5s$	$^3P°$	1	3	ASD,HDBK
750	3039.0671	0.0010	2.04	A	7125.2989	$4s^24P^2$	^1D	2	5	ASD,HDBK
					40020.5604	$4s^24p5s$	$^1P°$	1	3	ASD,HDBK
110	3269.4889	0.0010	0.172	C+	7125.2989	$4s^24P^2$	^1D	2	5	ASD,HDBK
					37702.3054	$4s^24p5s$	$^3P°$	1	3	ASD,HDBK
1300	12069.201	0.005			37702.3054	$4s^24p5s$	$^3P°$	1	3	ASD,HDBK
					45985.592	$4s^24p5p$	^3D	1	3	ASD,HDBK
1050	12391.575	0.005			40020.5604	$4s^24p5s$	$^1P°$	1	3	ASD,HDBK
					48088.3504	$4s^24p5p$	^3P	1	3	ASD,HDBK

Arsenic I (Z = 33)

Intensity	Wavelength/Å	Uncert. wavelength/Å	$A_{ki}/(10^8\,\text{s}^{-1})$	Accur.	Energy levels/cm⁻¹	Configs.	Terms	J	g	DB Source
1000r	1890.43		2.66	C+	0.000	$4s^24p3$	$^4S°$	3/2	4	HDBK,ASD
					52898.056	$4s^24p^2(^3P)5s$	^4P	5/2	6	HDBK,ASD
800r	1937.59		2.19	C+	0.000	$4s^24p3$	$^4S°$	3/2	4	HDBK,ASD
					51610.393	$4s^24p^2(^3P)5s$	^4P	3/2	4	HDBK,ASD
585r	1972.62		2.02	C+	0.000	$4s^24p3$	$^4S°$	3/2	4	HDBK,ASD
					50693.897	$4s^24p^2(^3P)5s$	^4P	1/2	2	HDBK,ASD
230r	2003.35				10914.866	$4s^24p3$	$^2D°$	5/2	6	HDBK,ASD
					60815.218	$4s^24p^2(^1D)5s$	^2D	5/2	6	HDBK,ASD
350r	2288.12		2.8		10914.866	$4s^24p3$	$^2D°$	5/2	6	HDBK,ASD
					54605.491	$4s^24p^2(^3P)5s$	^2P	3/2	4	HDBK,ASD
350r	2349.84		3.1		10592.666	$4s^24p3$	$^2D°$	3/2	4	HDBK,ASD
					53135.750	$4s^24p^2(^3P)5s$	^2P	1/2	2	HDBK,ASD
100r	2370.77		0.42		18647.663	$4s^24p3$	$^2P°$	3/2	4	HDBK,ASD
					60815.218	$4s^24p^2(^1D)5s$	^2D	5/2	6	HDBK,ASD
170r	2456.53		0.072		10914.866	$4s^24p3$	$^2D°$	5/2	6	HDBK,ASD
					51610.393	$4s^24p^2(^3P)5s$	^4P	3/2	4	HDBK,ASD

Intensity	Wavelength/Å	Uncert. wavelength/Å	$A_{ki}/(10^8\ s^{-1})$	Accur.	Energy levels/cm^{-1}	Configs.	Terms	J	g	DB Source
170r	2780.22		0.78		18647.663	$4s^24p3$	$^2P^o$	3/2	4	HDBK,ASD
					54605.491	$4s^24p^2(^3P)5s$	^2P	3/2	4	HDBK,ASD
100r	2860.44		0.55		18186.328	$4s^24p3$	$^2P^o$	1/2	2	HDBK,ASD
					53135.750	$4s^24p^2(^3P)5s$	^2P	1/2	2	HDBK,ASD
Selenium I (Z = 34)										
500	1960.894		2.13	C	0.000	$4s^24p4$	^3P	2	5	HDBK,ASD
					50997.161	$4s^24p^3(^4S^o)5s$	$^3S^o$	1	3	HDBK,ASD
150	1995.111				22446.202	$4s^24p4$	^1S	0	1	HDBK,ASD
					72568.72	$4s^24p^3(^2P^o)5s$	$^1P^o$	1	3	HDBK,ASD
500	2039.842		0.979	C	1989.497	$4s^24p4$	^3P	1	3	HDBK,ASD
					50997.161	$4s^24p^3(^4S^o)5s$	$^3S^o$	1	3	HDBK,ASD
400	2062.779		0.330	C	2534.36	$4s^24p4$	^3P	0	1	HDBK,ASD
					50997.161	$4s^24p^3(^4S^o)5s$	$^3S^o$	1	3	HDBK,ASD
500	2074.784		0.0170	B+	0.000	$4s^24p4$	^3P	2	5	HDBK,ASD
					48182.420	$4s^24p^3(^4S^o)5s$	$^5S^o$	2	5	HDBK,ASD
500	2164.188		0.00324	B	1989.497	$4s^24p4$	^3P	1	3	HDBK,ASD
					48182.420	$4s^24p^3(^4S^o)5s$	$^5S^o$	2	5	HDBK,ASD
600	2413.535				9576.149	$4s^24p4$	^1D	2	5	HDBK,ASD
					50997.161	$4s^24p^3(^4S^o)5s$	$^3S^o$	1	3	HDBK,ASD
500	4730.78				48182.420	$4s^24p^3(^4S^o)5s$	$^5S^o$	2	5	HDBK,ASD
					69314.635	$4s^24p^3(^4S^o)6p$	^5P	3	7	HDBK,ASD
900	10327.26				50997.161	$4s^24p^3(^4S^o)5s$	$^3S^o$	1	3	HDBK,ASD
					60677.618	$4s^24p^3(^4S^o)5p$	^3P	2	5	HDBK,ASD
640	10386.36				50997.161	$4s^24p^3(^4S^o)5s$	$^3S^o$	1	3	HDBK,ASD
					60622.532	$4s^24p^3(^4S^o)5p$	^3P	1	3	HDBK,ASD
680	21448.41				60677.618	$4s^24p^3(^4S^o)5p$	^3P	2	5	HDBK,ASD
					65339.968	$4s^24p^3(^4S^o)4d$	$^3D^o$	3	7	HDBK,ASD
Bromine I (Z = 35)										
50000	1488.45		2.01		0.00	$4s^24p5$	$^2P^o$	3/2	4	HDBK,ASD
					67183.58	$4s^24p^4(^3P)5s$	^2P	3/2	4	HDBK,ASD
25000	1540.65		2.17		0.00	$4s^24p5$	$^2P^o$	3/2	4	HDBK,ASD
					64907.19	$4s^24p^4(^3P)5s$	^4P	3/2	4	HDBK,ASD
30000	1574.84		0.202		3685.24	$4s^24p5$	$^2P^o$	1/2	2	HDBK,ASD
					67183.58	$4s^24p^4(^3P)5s$	^2P	3/2	4	HDBK,ASD
75000	1633.40		0.226		3685.24	$4s^24p5$	$^2P^o$	1/2	2	HDBK,ASD
					64907.19	$4s^24p^4(^3P)5s$	^4P	3/2	4	HDBK,ASD
10000	7348.51		0.12		64907.19	$4s^24p^4(^3P)5s$	^4P	3/2	4	HDBK,ASD
					78511.60	$4s^24p^4(^3P)5p$	$^2D^o$	5/2	6	HDBK,ASD
40000	7512.96		0.12		63436.45	$4s^24p^4(^3P)5s$	^4P	5/2	6	HDBK,ASD
					76743.08	$4s^24p^4(^3P)5p$	$^4D^o$	3/2	4	HDBK,ASD
30000	7803.02		0.053		66883.87	$4s^24p^4(^3P)5s$	^4P	1/2	2	HDBK,ASD
					79695.89	$4s^24p^4(^3P)5p$	$^2P^o$	3/2	4	HDBK,ASD
30000c	7938.68		0.19		75890.33	$4s^24p^4(^1D)5s$	^2D	5/2	6	HDBK,ASD
					88483.42	$4s^24p^4(^1D)5p$	$^2D^o$	5/2	6	HDBK,ASD
75000c	8272.44				63436.45	$4s^24p^4(^3P)5s$	^4P	5/2	6	HDBK,ASD
					75521.50	$4s^24p^4(^3P)5p$	$^4D^o$	7/2	8	HDBK,ASD
10000	8343.70		0.22		66883.87	$4s^24p^4(^3P)5s$	^4P	1/2	2	HDBK,ASD
					78865.72	$4s^24p^4(^3P)5p$	$^4D^o$	1/2	2	HDBK,ASD
40000	8446.55		0.12		64907.19	$4s^24p^4(^3P)5s$	^4P	3/2	4	HDBK,ASD
					76743.08	$4s^24p^4(^3P)5p$	$^4D^o$	3/2	4	HDBK,ASD
Krypton I (Z = 36)										
71	1164.867		3.09	AA	0.0000	$4s^24p6$	^1S	0	1	ASD
					85846.7046	$4s^24p^5(^2P^o<1/2>)5s$	$2[1/2]^o$	1	3	ASD
100	1235.838		2.98	AA	0.0000	$4s^24p6$	^1S	0	1	ASD
					80916.7680	$4s^24p^5(^2P^o<3/2>)5s$	$2[3/2]^o$	1	3	ASD
2000	5570.28944		0.021	D	79971.7417	$4s^24p^5(^2P^o<3/2>)5s$	$2[3/2]^o$	2	5	ASD
					97919.1468	$4s^24p^5(^2P^o<1/2>)5p$	$2[1/2]$	1	3	ASD

Intensity	Wavelength/Å	Uncert. wavelength/Å	$A_{ki}/(10^8 s^{-1})$	Accur.	Energy levels/cm^{-1}	Configs.	Terms	J	g	DB Source
3000	5870.91599		0.018	D	80916.7680	$4s^24p^5(^2P^o<3/2>)5s$	$2[3/2]^o$	1	3	ASD
					97945.1664	$4s^24p^5(^2P^o<1/2>)5p$	2[3/2]	2	5	ASD
4000	8104.3660		0.13	D	79971.7417	$4s^24p^5(^2P^o<3/2>)5s$	$2[3/2]^o$	2	5	ASD
					92307.3786	$4s^24p^5(^2P^o<3/2>)5p$	2[5/2]	2	5	ASD
500	8112.9012		0.36	D	79971.7417	$4s^24p^5(^2P^o<3/2>)5s$	$2[3/2]^o$	2	5	ASD
					92294.4012	$4s^24p^5(^2P^o<3/2>)5p$	2[5/2]	3	7	ASD
300	8190.0566		0.11	D	80916.7680	$4s^24p^5(^2P^o<3/2>)5s$	$2[3/2]^o$	1	3	ASD
					93123.3409	$4s^24p^5(^2P^o<3/2>)5p$	2[3/2]	2	5	ASD
400	8263.2426		0.35	D	85846.7046	$4s^24p^5(^2P^o<1/2>)5s$	$2[1/2]^o$	1	3	ASD
					97945.1664	$4s^24p^5(^2P^o<1/2>)5p$	2[3/2]	2	5	ASD
500	8298.1099		0.32	D	80916.7680	$4s^24p^5(^2P^o<3/2>)5s$	$2[3/2]^o$	1	3	ASD
					92964.3943	$4s^24p^5(^2P^o<3/2>)5p$	2[3/2]	1	3	ASD
200	8508.8728		0.24	D	85846.7046	$4s^24p^5(^2P^o<1/2>)5s$	$2[1/2]^o$	1	3	ASD
					97595.9153	$4s^24p^5(^2P^o<1/2>)5p$	2[3/2]	1	3	ASD
300	8776.7505		0.27	D	80916.7680	$4s^24p^5(^2P^o<3/2>)5s$	$2[3/2]^o$	1	3	ASD
					92307.3786	$4s^24p^5(^2P^o<3/2>)5p$	2[5/2]	2	5	ASD
250	13634.2206				92294.4012	$4s^24p^5(^2P^o<3/2>)5p$	2[5/2]	3	7	ASD
					99626.882	$4s^24p^5(^2P^o<3/2>)6s$	$2[3/2]^o$	2	5	ASD
350	14426.7927				92964.3943	$4s^24p^5(^2P^o<3/2>)5p$	2[3/2]	1	3	ASD
					99894.0485	$4s^24p^5(^2P^o<3/2>)6s$	$2[3/2]^o$	1	3	ASD
340	16890.4538				92307.3786	$4s^24p^5(^2P^o<3/2>)5p$	2[5/2]	2	5	ASD
					98226.268	$4s^24p^5(^2P^o<3/2>)4d$	$2[7/2]^o$	3	7	ASD
940	18167.3273				92294.4012	$4s^24p^5(^2P^o<3/2>)5p$	2[5/2]	3	7	ASD
					97797.287	$4s^24p^5(^2P^o<3/2>)4d$	$2[7/2]^o$	4	9	ASD

TABLE 5. Persisent Lines of Neutrul Atomic Elements — Rubidium (Z = 37) to Xenon (Z = 54)

Intensity	Wavelength/Å	Uncert. wavelength/Å	$A_{ki}/(10^8 s^{-1})$	Accur.	Energy levels/cm^{-1}	Configs.	Terms	J	g	DB Source
Rubidium I (Z = 37)										
1000	4201.792	0.005	0.0177		0.000	$4p^65s$	^2S	1/2	2	ASD
					23792.591	$4p^66p$	$^2P^o$	3/2	4	ASD
500	4215.524	0.005	0.0150		0.000	$4p^65s$	^2S	1/2	2	ASD
					23715.081	$4p^66p$	$^2P^o$	1/2	2	ASD
90000c	7800.268	0.010	0.381		0.000	$4p^65s$	^2S	1/2	2	ASD
					12816.545	$4p^65p$	$^2P^o$	3/2	4	ASD
45000c	7947.603	0.010	0.361		0.000	$4p^65s$	^2S	1/2	2	ASD
					12578.950	$4p^65p$	$^2P^o$	1/2	2	ASD
1000	14752.410	0.010			12578.950	$4p^65p$	$^2P^o$	1/2	2	ASD
					19355.649	$4p^64d$	^2D	3/2	4	ASD
800	15288.430	0.010			12816.545	$4p^65p$	$^2P^o$	3/2	4	ASD
					19355.649	$4p^64d$	^2D	3/2	4	ASD
Strontium I (Z = 38)										
29000r	4607.3313	0.002	2.01	AA	0.000	$5s^2$	^1S	0	1	ASD
					21698.452	$5s5p$	$^1P^o$	1	3	ASD
2300	4811.8812	0.002	0.90	B+	14898.545	$5s5p$	$^3P^o$	2	5	ASD
					35674.637	$5p^2$	^3P	2	5	ASD
1500	4872.4931	0.002	0.48	A'	14504.334	$5s5p$	$^3P^o$	1	3	ASD
					35021.989	$5s5d$	^3D	2	5	ASD
2500	4962.2632	0.002	0.614	AA	14898.545	$5s5p$	$^3P^o$	2	5	ASD
					35045.019	$5s5d$	^3D	3	7	ASD
1200	4967.9437	0.002	0.128	B+	14898.545	$5s5p$	$^3P^o$	2	5	ASD
					35021.989	$5s5d$	^3D	2	5	ASD
3400	5256.8988	0.002	0.81	B+	18319.261	$5s4d$	^3D	3	7	ASD
					37336.591	$4d5p$	$^3P^o$	2	5	ASD
2700	5480.8651	0.002	0.79	A'	18319.261	$5s4d$	^3D	3	7	ASD
					36559.492	$4d5p$	$^3D^o$	3	7	ASD
3100	6408.463	0.003	0.24	C+	18319.261	$5s4d$	^3D	3	7	ASD
					33919.315	$4d5p$	$^3F^o$	4	9	ASD

Intensity	Wavelength/Å	Uncert. wavelength/Å	A_{ki}/(10^8 s^{-1})	Accur.	Energy levels/cm^{-1}	Configs.	Terms	J	g	DB Source
2100	6503.989	0.003	0.20	C+	18218.784	5s4d	^{3}D	2	5	ASD
					33589.709	4d5p	^{3}F°	3	7	ASD
7000	6791.022	0.003	0.089	B+'	14317.507	5s5p	^{3}P°	0	1	ASD
					29038.773	5s6s	^{3}S	1	3	ASD
12000	6878.313	0.003	0.27	B+	14504.334	5s5p	^{3}P°	1	3	ASD
					29038.773	5s6s	^{3}S	1	3	ASD
2300	6892.585	0.004	0.000469	A+	0.000	5s^2	^{1}S	0	1	ASD
					14504.334	5s5p	^{3}P°	1	3	ASD
14000	7070.071	0.004	0.42	B	14898.545	5s5p	^{3}P°	2	5	ASD
					29038.773	5s6s	^{3}S	1	3	ASD

Yttrium I (Z = 39)

Intensity	Wavelength/Å	Uncert. wavelength/Å	A_{ki}/(10^8 s^{-1})	Accur.	Energy levels/cm^{-1}	Configs.	Terms	J	g	DB Source
9400	4077.359		1.11	C+	0.000	4d5s^2	a ^{2}D	3/2	4	ASD
					24518.751	4d5s(^{1}D)5p	y ^{2}F°	5/2	6	ASD
9900	4102.364		1.27	C+	530.351	4d5s^2	a ^{2}D	5/2	6	ASD
					24899.632	4d5s(^{1}D)5p	y ^{2}F°	7/2	8	ASD
8900	4128.304		1.56	A	530.351	4d5s^2	a ^{2}D	5/2	6	ASD
					24746.573	4d5s(^{1}D)5p	y ^{2}D°	5/2	6	ASD
7500	4142.841		1.59	B+	0.000	4d5s^2	a ^{2}D	3/2	4	ASD
					24131.250	4d5s(^{1}D)5p	y ^{2}D°	3/2	4	ASD
	17422.838				11078.614	4d^2(^{3}F)5s	a ^{4}F	5/2	6	HDBK,ASD
					16816.641	4d5s(^{3}D)5p	z ^{4}D°	5/2	6	HDBK,ASD
	17903.209				11532.096	4d^2(^{3}F)5s	a ^{4}F	9/2	10	HDBK,ASD
					17116.156	4d5s(^{3}D)5p	z ^{4}D°	7/2	8	HDBK,ASD
	18049.810				11277.928	4d^2(^{3}F)5s	a ^{4}F	7/2	8	HDBK,ASD
					16816.641	4d5s(^{3}D)5p	z ^{4}D°	5/2	6	HDBK,ASD
	21266.247				11532.096	4d^2(^{3}F)5s	a ^{4}F	9/2	10	HDBK,ASD
					16234.382	4d5s(^{3}D)5p	z ^{4}F°	9/2	10	HDBK,ASD
	22549.980				11277.928	4d^2(^{3}F)5s	a ^{4}F	7/2	8	HDBK,ASD
					15712.522	4d5s(^{3}D)5p	z ^{4}F°	7/2	8	HDBK,ASD
	23996.996				11078.614	4d^2(^{3}F)5s	a ^{4}F	5/2	6	HDBK,ASD
					15245.803	4d5s(^{3}D)5p	z ^{4}F°	5/2	6	HDBK,ASD

Zirconium I (Z = 40)

Intensity	Wavelength/Å	Uncert. wavelength/Å	A_{ki}/(10^8 s^{-1})	Accur.	Energy levels/cm^{-1}	Configs.	Terms	J	g	DB Source
2000	3519.604				0.00	4d^{2}5s^2	a ^{3}F	2	5	HDBK,ASD
					28404.26	4d^{2}5s(a 3F)5p	x ^{3}G°	3	7	HDBK,ASD
1800	3547.683				570.41	4d^{2}5s^2	a ^{3}F	3	7	HDBK,ASD
					28749.80	4d^{2}5s(a 3F)5p	x ^{3}G°	4	9	HDBK,ASD
1100	3575.790		0.527	B	570.41	4d^{2}5s^2	a ^{3}F	3	7	HDBK,ASD
					28528.36	4d^{2}5s(a 3F)5p	w ^{3}F°	4	9	HDBK,ASD
3500	3601.191		1.33	B+	1240.84	4d^{2}5s^2	a ^{3}F	4	9	HDBK,ASD
					29001.65	4d^{2}5s(a 3F)5p	x ^{3}G°	5	11	HDBK,ASD
1100	3663.648		0.571	B	1240.84	4d^{2}5s^2	a ^{3}F	4	9	HDBK,ASD
					28528.36	4d^{2}5s(a 3F)5p	w ^{3}F°	4	9	HDBK,ASD
2200	3835.962				0.00	4d^{2}5s^2	a ^{3}F	2	5	HDBK,ASD
					26061.70	4d^{2}5s(b 3F)5p	x ^{3}F°	2	5	HDBK,ASD
2900	3863.872				570.41	4d^{2}5s^2	a ^{3}F	3	7	HDBK,ASD
					26443.88	4d^{2}5s(b 3F)5p	x ^{3}F°	3	7	HDBK,ASD
2900	3890.316				1240.84	4d^{2}5s^2	a ^{3}F	4	9	HDBK,ASD
					26938.42	4d^{2}5s(b 3F)5p	x ^{3}F°	4	9	HDBK,ASD
2000	3891.380		0.392	B+	1240.84	4d^{2}5s^2	a ^{3}F	4	9	HDBK,ASD
					26931.35	4d^{2}5s(a 3F)5p	z ^{5}G°	4	9	HDBK,ASD
2000	4072.698		1.19		5540.54	4d^3(^{4}F)5s	a ^{5}F	4	9	HDBK,ASD
					30087.33	4d^3(^{4}F)5p	x ^{5}D°	3	7	HDBK,ASD
2000	4081.209				5888.93	4d^3(^{4}F)5s	a ^{5}F	5	11	HDBK,ASD
					30384.50	4d^3(^{4}F)5p	x ^{5}D°	4	9	HDBK,ASD
2000	4227.750				5888.93	4d^3(^{4}F)5s	a ^{5}F	5	11	HDBK,ASD
					29535.48	4d^3(^{4}F)5p	y ^{5}F°	5	11	HDBK,ASD
2000	4239.309				5540.54	4d^3(^{4}F)5s	a ^{5}F	4	9	HDBK,ASD
					29122.71	4d^3(^{4}F)5p	y ^{5}F°	4	9	HDBK,ASD

Intensity	Wavelength/Å	Uncert. wavelength/Å	$A_{ki}/(10^8\ s^{-1})$	Accur.	Energy levels/cm^{-1}	Configs.	Terms	J	g	DB Source
2300	4687.799		0.85		5888.93	$4d^3(^4F)5s$	a ^5F	5	11	HDBK,ASD
					27214.89	$4d^3(^4F)5p$	y $^5G^o$	6	13	HDBK,ASD
1900	4710.074		0.64		5540.54	$4d^3(^4F)5s$	a ^5F	4	9	HDBK,ASD
					26765.66	$4d^3(^4F)5p$	y $^5G^o$	5	11	HDBK,ASD
680	6127.457		0.0170	B	1240.84	$4d^25s^2$	a ^3F	4	9	HDBK,ASD
					17556.26	$4d^25s(a\ ^4F)5p$	z $^3F^o$	4	9	HDBK,ASD
540	7097.727		0.051		5540.54	$4d^3(^4F)5s$	a ^5F	4	9	HDBK,ASD
					19625.58	$4d^25s(a\ ^4F)5p$	z $^5D^o$	3	7	HDBK,ASD
590	7169.092				5888.93	$4d^3(^4F)5s$	a ^5F	5	11	HDBK,ASD
					19833.78	$4d^25s(a\ ^4F)5p$	z $^5D^o$	4	9	HDBK,ASD
790	8070.099				5888.93	$4d^3(^4F)5s$	a ^5F	5	11	HDBK,ASD
					18276.92	$4d^25s(a\ ^4F)5p$	z $^5F^o$	5	11	HDBK,ASD
Niobium I (Z = 41)										
1300	3341.97				1142.79	$4d^35s^2$	a ^4F	3/2	4	HDBK
					31056.60	$4d^35s(^5F)5p$	x $^4G^o$	5/2	6	HDBK
1300	3343.71				1586.90	$4d^35s^2$	a ^4F	5/2	6	HDBK
					31485.20	$4d^35s(^5F)5p$	x $^4G^o$	7/2	8	HDBK
1700	3349.06				2154.11	$4d^35s^2$	a ^4F	7/2	8	HDBK
					32004.63	$4d^35s(^5F)5p$	x $^4G^o$	9/2	10	HDBK
1700	3358.42				2805.36	$4d^35s^2$	a ^4F	9/2	10	HDBK
					32572.72	$4d^35s(^5F)5p$	x $^4G^o$	11/2	12	HDBK
2000*	3535.30				695.25	$4d^4(^5D)5s$	a ^6D	7/2	8	HDBK
					28973.12	$4d^35s(^5P)5p$	y $^6P^o$	7/2	8	HDBK
2000*	3535.30				0.00	$4d^4(^5D)5s$	a ^6D	1/2	2	HDBK
					28278.25	$4d^35s(^5P)5p$	y $^6P^o$	3/2	4	HDBK
1500	3575.85				695.25	$4d^4(^5D)5s$	a ^6D	7/2	8	HDBK
					28652.66	$4d^35s(^5P)5p$	y $^6P^o$	5/2	6	HDBK
5000	3580.27				1050.26	$4d^4(^5D)5s$	a ^6D	9/2	10	HDBK
					28973.12	$4d^35s(^5P)5p$	y $^6P^o$	7/2	8	HDBK
2500	3739.80		0.59		695.25	$4d^4(^5D)5s$	a ^6D	7/2	8	HDBK
					27427.07	$4d^4(^5D)5p$	x $^6D^o$	7/2	8	HDBK
16000c*	4058.94		1.3		1050.26	$4d^4(^5D)5s$	a ^6D	9/2	10	HDBK
					25680.36	$4d^4(^5D)5p$	y $^6F^o$	11/2	12	HDBK
16000c*	4058.94				10237.51	$4d^35s^2$	a ^2D	5/2	6	HDBK
					34867.68	$4d^35s(^5P)5p$	x $^4P^o$	3/2	4	HDBK
12000	4079.73		0.99		695.25	$4d^4(^5D)5s$	a ^6D	7/2	8	HDBK
					25199.81	$4d^4(^5D)5p$	y $^6F^o$	9/2	10	HDBK
6700*	4100.92		0.76		391.99	$4d^4(^5D)5s$	a ^6D	5/2	6	HDBK
					24769.91	$4d^4(^5D)5p$	y $^6F^o$	7/2	8	HDBK
6700*	4100.92				13145.71	$4d^4$(b 3F)5s	b ^4F	9/2	10	HDBK
					37523.53	$4d^4$(c 3F)5p	u $^4G^o$	9/2	10	HDBK
5300	4123.81		0.58		154.19	$4d^4(^5D)5s$	a ^6D	3/2	4	HDBK
					24396.80	$4d^4(^5D)5p$	y $^6F^o$	5/2	6	HDBK
4400	4152.58		0.38		695.25	$4d^4(^5D)5s$	a ^6D	7/2	8	HDBK
					24769.91	$4d^4(^5D)5p$	y $^6F^o$	7/2	8	HDBK
4400	4163.66		0.65		154.19	$4d^4(^5D)5s$	a ^6D	3/2	4	HDBK
					24164.79	$4d^4(^5D)5p$	y $^6F^o$	3/2	4	HDBK
4000	4164.66		0.5		391.99	$4d^4(^5D)5s$	a ^6D	5/2	6	HDBK
					24396.80	$4d^4(^5D)5p$	y $^6F^o$	5/2	6	HDBK
3500	4168.13		0.91		0.00	$4d^4(^5D)5s$	a ^6D	1/2	2	HDBK
					23984.87	$4d^4(^5D)5p$	y $^6F^o$	1/2	2	HDBK
Molybdenum I (Z = 41)										
14000	3132.594	0.010	1.79	B+	0.000	$4d^5(^6S)5s$	a ^7S	3	7	ASD
					31913.171	$4d^45s(^6D)5p$	y $^7P^o$	4	9	ASD
6000	3158.166	0.010	0.463	B+	0.000	$4d^5(^6S)5s$	a ^7S	3	7	ASD
					31654.786	$4d^45s(^6D)5p$	z $^7D^o$	3	7	ASD
8700	3170.343	0.010	1.37	B+	0.000	$4d^5(^6S)5s$	a ^7S	3	7	ASD
					31533.206	$4d^45s(^6D)5p$	y $^7P^o$	3	7	ASD

Intensity	Wavelength/Å	Uncert. wavelength/Å	$A_{ki}/(10^8\ s^{-1})$	Accur.	Energy levels/cm^{-1}	Configs.	Terms	J	g	DB Source
7600	3193.978	0.010	1.53	B+	0.000	$4d^5(^6S)5s$	a ^7S	3	7	ASD
					31299.876	$4d^45s(^6D)5p$	y $^7P^o$	2	5	ASD
3000	3208.838	0.010	0.277	A	0.000	$4d^5(^6S)5s$	a ^7S	3	7	ASD
					31154.935	$4d^45s(^6D)5p$	z $^7D^o$	2	5	ASD
3200	3447.123	0.010	0.88	B+	12346.280	$4d^45s^2$	a ^5D	4	9	ASD
					41347.664	$4d^5(^4G)5p$	y $^5F^o$	5	11	ASD
29000	3798.252	0.010	0.69	B+	0.000	$4d^5(^6S)5s$	a ^7S	3	7	ASD
					26320.420	$4d^5(^6S)5p$	z $^7P^o$	4	9	ASD
29000	3864.103	0.010	0.624	A	0.000	$4d^5(^6S)5s$	a ^7S	3	7	ASD
					25871.887	$4d^5(^6S)5p$	z $^7P^o$	3	7	ASD
19000	3902.953	0.010	0.617	A	0.000	$4d^5(^6S)5s$	a ^7S	3	7	ASD
					25614.367	$4d^5(^6S)5p$	z $^7P^o$	2	5	ASD
2300	4069.881	0.010	0.325	B+	16783.856	$4d^5(^4G)5s$	a ^5G	6	13	ASD
					41347.664	$4d^5(^4G)5p$	y $^5F^o$	5	11	ASD
2500	4188.323	0.010	0.33	B	16784.522	$4d^5(^4G)5s$	a ^5G	5	11	ASD
					40653.701	$4d^5(^4G)5p$	z $^5H^o$	6	13	ASD
2500	4411.695	0.010	0.263	A	16784.522	$4d^5(^4G)5s$	a ^5G	5	11	HDBK
					39445.182	$4d^5(^4G)5p$	z $^5G^o$	5	11	HDBK
7800	5506.493	0.010	0.361	A+	10768.332	$4d^5(^6S)5s$	a ^5S	2	5	ASD
					28923.668	$4d^5(^6S)5p$	z $^5P^o$	3	7	ASD
5200	5533.031	0.010	0.372	A+	10768.332	$4d^5(^6S)5s$	a ^5S	2	5	ASD
					28836.592	$4d^5(^6S)5p$	z $^5P^o$	2	5	ASD
2500	5570.444	0.010	0.330	A	10768.332	$4d^5(^6S)5s$	a ^5S	2	5	ASD
					28715.242	$4d^5(^6S)5p$	z $^5P^o$	1	3	ASD
Technetium I (Z = 43)										
3000	3173.295		0.86		0.00	$4d^55s^2$	^6S	5/2	6	ASD
					31503.93	$4d^5(^6S)5s(^5S)5p$	$^6P^o$	3/2	4	ASD
2000	3182.367		0.39		0.00	$4d^55s^2$	^6S	5/2	6	ASD
					31414.05	$4d^6(^5D)5p$	$^6P^o$	7/2	8	ASD
2000	3183.108		0.56		0.00	$4d^55s^2$	^6S	5/2	6	ASD
					31406.74	$4d^5(^6S)5s(^5S)5p$	$^6P^o$	5/2	6	ASD
5000c,w	3466.278		1.22		2572.89	$4d^6(^5D)5s$	^6D	9/2	10	ASD
					31414.05	$4d^6(^5D)5p$	$^6P^o$	7/2	8	ASD
10000c*	3636.070		1.57		2572.89	$4d^6(^5D)5s$	^6D	9/2	10	ASD
					30067.29	$4d^6(^5D)5p$	$^6F^o$	11/2	12	ASD
10000	3718.861		1.05		3250.91	$4d^6(^5D)5s$	^6D	7/2	8	ASD
					30133.26	$4d^6(^5D)5p$	$^6F^o$	9/2	10	ASD
10000c	3984.967		0.42		2572.89	$4d^6(^5D)5s$	^6D	9/2	10	ASD
					27660.10	$4d^6(^5D)5p$	$^6D^o$	7/2	8	ASD
20000c,w	4031.626		1.01		2572.89	$4d^6(^5D)5s$	^6D	9/2	10	ASD
					27369.79	$4d^6(^5D)5p$	$^6D^o$	9/2	10	ASD
15000	4095.668		0.49		3250.91	$4d^6(^5D)5s$	^6D	7/2	8	ASD
					27660.10	$4d^6(^5D)5p$	$^6D^o$	7/2	8	ASD
5000c	4238.191		0.38		0.00	$4d^55s^2$	^6S	5/2	6	ASD
					23588.40	$4d^5(^6S)5s(^7S)5p$	$^6P^o$	3/2	4	ASD
10000	4262.270		0.40		0.00	$4d^55s^2$	^6S	5/2	6	ASD
					23455.21	$4d^5(^6S)5s(^7S)5p$	$^6P^o$	5/2	6	ASD
10000	4297.058		0.42		0.00	$4d^55s^2$	^6S	5/2	6	ASD
					23265.32	$4d^5(^6S)5s(^7S)5p$	$^6P^o$	7/2	8	ASD
20000	4853.588		0.49		10516.53	$4d^6(^5D)5s$	^4D	7/2	8	ASD
					31114.08	$4d^6(^5D)5p$	$^4F^o$	9/2	10	ASD
Ruthenium I (Z = 44)										
6400	3436.737		0.728		1190.64	$4d^7$(a ^4F)5s	a ^5F	4	9	HDBK,ASD
					30279.68	$4d^7$(a ^4F)5p	z $^5G^o$	5	11	HDBK,ASD
8300	3498.944		0.861		0.00	$4d^7$(a ^4F)5s	a ^5F	5	11	HDBK,ASD
					28571.89	$4d^7$(a ^4F)5p	z $^5G^o$	6	13	HDBK,ASD
6400	3589.220		0.911		3105.49	$4d^7$(a ^4F)5s	a ^5F	1	3	HDBK,ASD
					30958.80	$4d^7$(a ^4F)5p	z $^5G^o$	2	5	HDBK,ASD

Intensity	Wavelength/Å	Uncert. wavelength/Å	$A_{ki}/(10^8\ s^{-1})$	Accur.	Energy levels/cm^{-1}	Configs.	Terms	J	g	DB Source
6900	3593.029		0.817		2713.24	$4d^7$(a ^{4}F)5s	a ^{5}F	2	5	HDBK,ASD
					30537.06	$4d^7$(a ^{4}F)5p	z ^{5}G°	3	7	HDBK,ASD
6400	3596.185		0.42		2091.54	$4d^7$(a ^{4}F)5s	a ^{5}F	3	7	HDBK,ASD
					29890.91	$4d^7$(a ^{4}F)5p	z ^{3}G°	4	9	HDBK,ASD
8700	3726.926		0.753		1190.64	$4d^7$(a ^{4}F)5s	a ^{5}F	4	9	HDBK,ASD
					28014.79	$4d^7$(a ^{4}F)5p	z ^{5}F°	4	9	HDBK,ASD
11000	3728.026		0.820		0.00	$4d^7$(a ^{4}F)5s	a ^{5}F	5	11	HDBK,ASD
					26816.23	$4d^7$(a ^{4}F)5p	z ^{5}F°	5	11	HDBK,ASD
7100	3730.432		0.706		2091.54	$4d^7$(a ^{4}F)5s	a ^{5}F	3	7	HDBK,ASD
					28890.47	$4d^7$(a ^{4}F)5p	z ^{5}F°	3	7	HDBK,ASD
7600	3798.899		0.598		1190.64	$4d^7$(a ^{4}F)5s	a ^{5}F	4	9	HDBK,ASD
					27506.59	$4d^7$(a ^{4}F)5p	z ^{5}D°	3	7	HDBK,ASD
7600	3799.353		0.533		0.00	$4d^7$(a ^{4}F)5s	a ^{5}F	5	11	HDBK,ASD
					26312.83	$4d^7$(a ^{4}F)5p	z ^{5}D°	4	9	HDBK,ASD
7600	4199.892		0.399		6545.03	$4d^7$(a ^{4}F)5s	a ^{3}F	4	9	HDBK,ASD
					30348.45	$4d^7$(a ^{4}F)5p	z ^{3}F°	4	9	HDBK,ASD
Rhodium I (Z = 45)										
4200	3323.09		0.63	B	1529.97	$4d^8$(^{3}F)5s	a ^{4}F	7/2	8	ASD
					31613.78	$4d^8$(^{3}F)5p	z 2G°	9/2	10	ASD
5600	3396.82		0.65	B+	0.00	$4d^8$(^{3}F)5s	a ^{4}F	9/2	10	ASD
					29430.86	$4d^8$(^{3}F)5p	z ^{4}F°	9/2	10	ASD
8200	3434.89		1.32	B+	0.00	$4d^8$(^{3}F)5s	a ^{4}F	9/2	10	ASD
					29104.71	$4d^8$(^{3}F)5p	z ^{4}G°	11/2	12	ASD
5900	3502.52		0.43	B+	0.00	$4d^8$(^{3}F)5s	a ^{4}F	9/2	10	ASD
					28542.69	$4d^8$(^{3}F)5p	z ^{4}G°	9/2	10	ASD
2800	3507.32		0.336	B	2598.03	$4d^8$(^{3}F)5s	a ^{4}F	5/2	6	ASD
					31101.75	$4d^8$(^{3}F)5p	z ^{4}G°	7/2	8	ASD
8800	3528.02		0.85	B+	1529.97	$4d^8$(^{3}F)5s	a ^{4}F	7/2	8	ASD
					29866.34	$4d^8$(^{3}F)5p	z ^{4}F°	7/2	8	ASD
8200	3657.99		0.88	B+	1529.97	$4d^8$(^{3}F)5s	a ^{4}F	7/2	8	ASD
					28859.64	$4d^8$(^{3}F)5p	z ^{4}D°	5/2	6	ASD
9400	3692.36		0.91	B+	0.00	$4d^8$(^{3}F)5s	a ^{4}F	9/2	10	ASD
					27075.26	$4d^8$(^{3}F)5p	z ^{4}D°	7/2	8	ASD
7600	3700.91		0.39	B	1529.97	$4d^8$(^{3}F)5s	a ^{4}F	7/2	8	ASD
					28542.69	$4d^8$(^{3}F)5p	z ^{4}G°	9/2	10	ASD
4200	4374.80		0.164	B	5690.97	$4d^8$(^{3}F)5s	a ^{2}F	7/2	8	ASD
					28542.69	$4d^8$(^{3}F)5p	z ^{4}G°	9/2	10	ASD
Palladium I (Z = 46)										
1100	2447.9058		0.111		0.000	$4d^{10}$	^{1}S	0	1	ASD
					40838.874	$4d^9$(^{2}D<3/2>)5p	2[1/2]°	1	3	ASD
1700	2476.4127		0.118		0.000	$4d^{10}$	^{1}S	0	1	ASD
					40368.796	$4d^9$(^{2}D<3/2>)5p	2[3/2]°	1	3	ASD
1900	2763.0899		0.169		0.000	$4d^{10}$	^{1}S	0	1	ASD
					36180.677	$4d^9$(^{2}D<5/2>)5p	2[3/2]°	1	3	ASD
11000	3242.6983		0.77		6564.148	$4d^9$(^{2}D<5/2>)5s	2[5/2]	3	7	ASD
					37393.762	$4d^9$(^{2}D<5/2>)5p	2[5/2]°	3	7	ASD
24000	3404.5764		1.34		6564.148	$4d^9$(^{2}D<5/2>)5s	2[5/2]	3	7	ASD
					35927.948	$4d^9$(^{2}D<5/2>)5p	2[7/2]°	4	9	ASD
13000	3421.2215				7755.025	$4d^9$(^{2}D<5/2>)5s	2[5/2]	2	5	ASD
					36975.973	$4d^9$(^{2}D<5/2>)5p	2[5/2]°	2	5	ASD
7700	3460.7381		0.30		6564.148	$4d^9$(^{2}D<5/2>)5s	2[5/2]	3	7	ASD
					35451.443	$4d^9$(^{2}D<5/2>)5p	2[7/2]°	3	7	ASD
10000	3481.1516				10093.992	$4d^9$(^{2}D<3/2>)5s	2[3/2]	1	3	ASD
					38811.896	$4d^9$(^{2}D<3/2>)5p	2[5/2]°	2	5	ASD
12000	3516.9438		1.03		7755.025	$4d^9$(^{2}D<5/2>)5s	2[5/2]	2	5	ASD
					36180.677	$4d^9$(^{2}D<5/2>)5p	2[3/2]°	1	3	ASD
12000	3553.0803				11721.809	$4d^9$(^{2}D<3/2>)5s	2[3/2]	2	5	ASD
					39858.361	$4d^9$(^{2}D<3/2>)5p	2[5/2]°	3	7	ASD

Intensity	Wavelength/Å	Uncert. wavelength/Å	$A_{ki}/(10^8 s^{-1})$	Accur.	Energy levels/cm^{-1}	Configs.	Terms	J	g	DB Source
20000	3609.5457		0.82		7755.025	$4d^9(^2D<5/2>)5s$	2[5/2]	2	5	ASD
					35451.443	$4d^9(^2D<5/2>)5p$	2[7/2]°	3	7	ASD
20000	3634.6884				6564.148	$4d^9(^2D<5/2>)5s$	2[5/2]	3	7	ASD
					34068.977	$4d^9(^2D<5/2>)5p$	2[3/2]°	2	5	ASD
Silver I (Z = 47)										
55000r	3280.680		1.4		0.000000	$4d^{10}5s$	^2S	1/2	2	ASD
					30472.66516	$4d^{10}5p$	$^2P^o$	3/2	4	ASD
28000r	3382.887		1.3		0.000000	$4d^{10}5s$	^2S	1/2	2	ASD
					29552.05741	$4d^{10}5p$	$^2P^o$	1/2	2	ASD
1000	5209.078		0.75		29552.05741	$4d^{10}5p$	$^2P^o$	1/2	2	ASD
					48743.969	$4d^{10}5d$	^2D	3/2	4	ASD
1000	5465.497		0.86		30472.66516	$4d^{10}5p$	$^2P^o$	3/2	4	ASD
					48764.219	$4d^{10}5d$	^2D	5/2	6	ASD
500	7687.772				29552.05741	$4d^{10}5p$	$^2P^o$	1/2	2	ASD
					42556.147	$4d^{10}6s$	^2S	1/2	2	ASD
1000	8273.509				30472.66516	$4d^{10}5p$	$^2P^o$	3/2	4	ASD
					42556.147	$4d^{10}6s$	^2S	1/2	2	ASD
Cadmium I (Z = 48)										
1500r	2288.0225		5.3	C	0.000	$4d^{10}5s^2$	^1S	0	1	ASD
					43692.384	$4d^{10}5s5p$	$^1P^o$	1	3	ASD
800	3403.6521		0.77	C	30113.990	$4d^{10}5s5p$	$^3P^o$	0	1	ASD
					59485.768	$4d^{10}5s5d$	^3D	1	3	ASD
1000	3466.1996		1.2	D	30656.087	$4d^{10}5s5p$	$^3P^o$	1	3	ASD
					59497.868	$4d^{10}5s5d$	^3D	2	5	ASD
800	3467.6547		0.67	D	30656.087	$4d^{10}5s5p$	$^3P^o$	1	3	ASD
					59485.768	$4d^{10}5s5d$	^3D	1	3	ASD
1000	3610.5077		1.3	D	31826.952	$4d^{10}5s5p$	$^3P^o$	2	5	ASD
					59515.980	$4d^{10}5s5d$	^3D	3	7	ASD
800	3612.8729		0.35	D	31826.952	$4d^{10}5s5p$	$^3P^o$	2	5	ASD
					59497.868	$4d^{10}5s5d$	^3D	2	5	ASD
200	4678.1493		0.13	C	30113.990	$4d^{10}5s5p$	$^3P^o$	0	1	ASD
					51483.980	$4d^{10}5s6s$	^3S	1	3	ASD
300	4799.9123		0.41	C	30656.087	$4d^{10}5s5p$	$^3P^o$	1	3	ASD
					51483.980	$4d^{10}5s6s$	^3S	1	3	ASD
1000w	5085.8217		0.56	C	31826.952	$4d^{10}5s5p$	$^3P^o$	2	5	ASD
					51483.980	$4d^{10}5s6s$	^3S	1	3	ASD
2000	6438.4695		0.59	C	43692.384	$4d^{10}5s5p$	$^1P^o$	1	3	ASD
					59219.734	$4d^{10}5s5d$	^1D	2	5	ASD
Indium I (Z = 49)										
260	1711.1				2212.599	$5s^25p$	$^2P^o$	3/2	4	HDBK,ASD
					60652	$5s5p^2$	^2P	3/2	4	HDBK,ASD
260	1757.3				0.000	$5s^25p$	$^2P^o$	1/2	2	HDBK,ASD
					56906	$5s5p^2$	^2S	1/2	2	HDBK,ASD
1100	2560.150		0.20		0.000	$5s^25p$	$^2P^o$	1/2	2	ASD
					39048.53	$5s^26d$	^2D	3/2	4	ASD
1600	2710.265		0.27		2212.599	$5s^25p$	$^2P^o$	3/2	4	ASD
					39098.38	$5s^26d$	^2D	5/2	6	ASD
700	2753.878		0.13		0.000	$5s^25p$	$^2P^o$	1/2	2	ASD
					36301.864	$5s^27s$	^2S	1/2	2	ASD
1100	2932.630		0.23		2212.599	$5s^25p$	$^2P^o$	3/2	4	ASD
					36301.864	$5s^27s$	^2S	1/2	2	ASD
8000	3039.346		1.11		0.000	$5s^25p$	$^2P^o$	1/2	2	ASD
					32892.230	$5s^25d$	^2D	3/2	4	ASD
13000	3256.079		1.3		2212.599	$5s^25p$	$^2P^o$	3/2	4	ASD
					32915.539	$5s^25d$	^2D	5/2	6	ASD
3000	3258.551		0.30		2212.599	$5s^25p$	$^2P^o$	3/2	4	ASD
					32892.230	$5s^25d$	^2D	3/2	4	ASD

Intensity	Wavelength/Å	Uncert. wavelength/Å	$A_{ki}/(10^8\,s^{-1})$	Accur.	Energy levels/cm^{-1}	Configs.	Terms	J	g	DB Source
17000	4101.7504		0.50		0.000	$5s^25p$	$^2P^o$	1/2	2	HDBK,ASD
					24372.957	$5s^26s$	^2S	1/2	2	HDBK,ASD
18000	4511.2972		0.89		2212.599	$5s^25p$	$^2P^o$	3/2	4	HDBK,ASD
					24372.957	$5s^26s$	^2S	1/2	2	HDBK,ASD
10	12912.59				24372.957	$5s^26s$	^2S	1/2	2	HDBK,ASD
					32115.251	$5s^26p$	$^2P^o$	3/2	4	HDBK,ASD
9	13429.96				24372.957	$5s^26s$	^2S	1/2	2	HDBK,ASD
					31816.982	$5s^26p$	$^2P^o$	1/2	2	HDBK,ASD
Tin I (Z = 50)										
1700	2091.5900	0.0009			1691.806	$5s^25p^2$	^3P	1	3	ASD
					49487.127	$5s^25p5d$	$^3P^o$	0	1	ASD
6200r	2199.3455	0.0010	0.29	D	1691.806	$5s^25p^2$	^3P	1	3	ASD
					47145.684	$5s^25p5d$	$^1D^o$	2	5	ASD
7700r	2209.6597	0.0008	0.56	D	3427.673	$5s^25p^2$	^3P	2	5	ASD
					48669.409	$5s^25p5d$	$^3P^o$	2	5	ASD
7400r	2246.0573	0.0012	1.6	D	0.000	$5s^25p^2$	^3P	0	1	ASD
					44508.677	$5s^25p5d$	$^3D^o$	1	3	ASD
7800r	2268.9296	0.0010	1.2	D	3427.673	$5s^25p^2$	^3P	2	5	ASD
					47487.696	$5s^25p5d$	$^3D^o$	3	7	ASD
17000r	2354.8499	0.0012	1.7	D	1691.806	$5s^25p^2$	^3P	1	3	ASD
					44144.368	$5s^25p5d$	$^3D^o$	2	5	ASD
13000r	2421.6943	0.0005	2.5	D	8612.955	$5s^25p^2$	^1D	2	5	ASD
					49893.823	$5s^25p5d$	$^1F^o$	3	7	ASD
16000r	2429.4952	0.0007	1.5	C	3427.673	$5s^25p^2$	^3P	2	5	ASD
					44576.006	$5s^25p5d$	$^3F^o$	3	7	ASD
3700r	2661.2436	0.0004	0.11	D	1691.806	$5s^25p^2$	^3P	1	3	ASD
					39257.053	$5s^25p6s$	$^1P^o$	1	3	ASD
11000r	2706.5049	0.0009	0.66	D	1691.806	$5s^25p^2$	^3P	1	3	ASD
					38628.876	$5s^25p6s$	$^3P^o$	2	5	ASD
13000r	2839.9765	0.0008	1.7	D	3427.673	$5s^25p^2$	^3P	2	5	ASD
					38628.876	$5s^25p6s$	$^3P^o$	2	5	ASD
10000r	2863.3147	0.0012	0.54	D	0.000	$5s^25p^2$	^3P	0	1	ASD
					34914.282	$5s^25p6s$	$^3P^o$	1	3	ASD
12000r	3009.1333	0.0009	0.38	D	1691.806	$5s^25p^2$	^3P	1	3	ASD
					34914.282	$5s^25p6s$	$^3P^o$	1	3	ASD
10000r	3034.1150	0.0011	2,0	D	1691.806	$5s^25p^2$	^3P	1	3	ASD
					34640.758	$5s^25p6s$	$^3P^o$	0	1	ASD
15000r	3175.0354	0.0010	1.0	D	3427.673	$5s^25p^2$	^3P	2	5	ASD
					34914.282	$5s^25p6s$	$^3P^o$	1	3	ASD
15000r	3262.3310	0.0006	2.7	D	8612.955	$5s^25p^2$	^1D	2	5	ASD
					39257.053	$5s^25p6s$	$^1P^o$	1	3	ASD
6900r	3801.0108	0.0007	0.28	D	8612.955	$5s^25p^2$	^1D	2	5	ASD
					34914.282	$5s^25p6s$	$^3P^o$	1	3	ASD
2200	4524.7344	0.0008	0.26	D	17162.499	$5s^25p^2$	^1S	0	1	ASD
					39257.053	$5s^25p6s$	$^1P^o$	1	3	ASD
Antimony I (Z = 51)										
300r	1871.15				0.000	$5p3$	$^4S^o$	3/2	4	ASD
					53442.967	$5p^2(^3P)5d$	^4P	5/2	6	ASD
150r	2049.57				8512.125	$5p3$	$^2D^o$	3/2	4	ASD
					57287.052	$5p^2(^3P)5d$	^2F	5/2	6	ASD
1000r	2068.33		1.81		0.000	$5p3$	$^4S^o$	3/2	4	ASD
					48332.424	$5p^2(^3P)6s$	^4P	5/2	6	ASD
100r	2139.69		1.74		8512.125	$5p3$	$^2D^o$	3/2	4	ASD
					55232.963	$5p^2(^1D)6s$	^2D	3/2	4	ASD
100r	2144.86				8512.125	$5p3$	$^2D^o$	3/2	4	ASD
					55120.943	$5p^2(^3P)5d$	^4F	5/2	6	ASD
1500r	2175.81		1.75		0.000	$5p3$	$^4S^o$	3/2	4	ASD
					45945.340	$5p^2(^3P)6s$	^4P	3/2	4	ASD

Intensity	Wavelength/Å	Uncert. wavelength/Å	$A_{ki}/(10^8 s^{-1})$	Accur.	Energy levels/cm^{-1}	Configs.	Terms	J	g	DB Source
250r	2179.19		2.39		9854.018	$5p3$	$^2D^o$	5/2	6	ASD
					55728.264	$5p^2(^1D)6s$	^2D	5/2	6	ASD
300r	2208.45				9854.018	$5p3$	$^2D^o$	5/2	6	ASD
					55120.943	$5p^2(^3P)5d$	^4F	5/2	6	ASD
2500r	2311.47		1.69		0.000	$5p3$	$^4S^o$	3/2	4	ASD
					43249.337	$5p^2(^3P)6s$	^4P	1/2	2	ASD
2000r	2528.52		1.77		9854.018	$5p3$	$^2D^o$	5/2	6	ASD
					49391.133	$5p^2(^3P)6s$	^2P	3/2	4	ASD
1500r	2598.05		0.210		8512.125	$5p3$	$^2D^o$	3/2	4	ASD
					46991.058	$5p^2(^3P)6s$	^2P	1/2	2	ASD
1000r	2877.92		0.424		8512.125	$5p3$	$^2D^o$	3/2	4	ASD
					43249.337	$5p^2(^3P)6s$	^4P	1/2	2	ASD
600r	3232.52		0.307		18464.202	$5p3$	$^2P^o$	3/2	4	ASD
					49391.133	$5p^2(^3P)6s$	^2P	3/2	4	ASD
700r	3267.51		0.495		16395.359	$5p3$	$^2P^o$	1/2	2	ASD
					46991.058	$5p^2(^3P)6s$	^2P	1/2	2	ASD
Tellurium I (Z = 52)										
3400000	1822.151	0.025			0.000	$5p4$	^3P	2	5	ASD
					54880.070	$5p^3(^2D^o)6s$	$^3D^o$	2	5	ASD
500000	1857.279	0.025			4750.712	$5p4$	^3P	1	3	ASD
					58592.434	$5p^3(^4S^o)5d$	$^3D^o$	2	5	ASD
320000	1994.841	0.025			4750.712	$5p4$	^3P	1	3	ASD
					54880.070	$5p^3(^2D^o)6s$	$^3D^o$	2	5	ASD
530000	2002.019	0.025			4750.712	$5p4$	^3P	1	3	ASD
					54683.886	$5p^3(^2D^o)6s$	$^3D^o$	1	3	ASD
340000	2081.17	0.03			10557.877	$5p4$	^1D	2	5	ASD
					58592.434	$5p^3(^4S^o)5d$	$^3D^o$	2	5	ASD
7400000	2142.815	0.020	3.12		0.000	$5p4$	^3P	2	5	ASD
					46652.738	$5p^3(^4S^o)6s$	$^3S^o$	1	3	ASD
190000	2147.254	0.020			10557.877	$5p4$	^1D	2	5	ASD
					57114.206	$5p^3(^2D^o)6s$	$^1D^o$	2	5	ASD
970000	2259.031	0.020	0.128		0.000	$5p4$	^3P	2	5	ASD
					44253.000	$5p^3(^4S^o)6s$	$^5S^o$	2	5	ASD
930000	2383.28	0.03	0.404		4706.495	$5p4$	^3P	0	1	ASD
					46652.738	$5p^3(^4S^o)6s$	$^3S^o$	1	3	ASD
1200000	2385.79	0.03	0.808		4750.712	$5p4$	^3P	1	3	ASD
					46652.738	$5p^3(^4S^o)6s$	$^3S^o$	1	3	ASD
36000	9722.742	0.005			44253.000	$5p^3(^4S^o)6s$	$^5S^o$	2	5	ASD
					54535.345	$5p^3(^4S^o)6p$	^5P	3	7	ASD
25000	10051.414	0.005			44253.000	$5p^3(^4S^o)6s$	$^5S^o$	2	5	ASD
					54199.122	$5p^3(^4S^o)6p$	^5P	2	5	ASD
16000	11089.559	0.006			46652.738	$5p^3(^4S^o)6s$	$^3S^o$	1	3	ASD
					55667.758	$5p^3(^4S^o)6p$	^3P	2	5	ASD
6900	11487.230	0.007			46652.738	$5p^3(^4S^o)6s$	$^3S^o$	1	3	ASD
					55355.672	$5p^3(^4S^o)6p$	^3P	1	3	ASD
Iodine I (Z = 53)										
310000	1457.981	0.002			0.000	$5s^25p5$	$^2P^o$	3/2	4	ASD
					68587.859	$5s^25p^4(^1D<2>)6s$	2[2]	5/2	6	ASD
230000	1518.048	0.003			7602.970	$5s^25p5$	$^2P^o$	1/2	2	ASD
					73477.834	$5s^25p^4(^1D<2>)5d$	2[0]	1/2	2	ASD
460000	1702.071	0.003	2.05		7602.970	$5s^25p5$	$^2P^o$	1/2	2	ASD
					66355.093	$5s^25p^4(^3P<2>)5d$	2[1]	3/2	4	ASD
1300000	1782.757	0.003	2.71	C	0.000	$5s^25p5$	$^2P^o$	3/2	4	ASD
					56092.881	$5s^25p^4(^3P<2>)6s$	2[2]	3/2	4	ASD
9300000	1830.380	0.003	0.16	D	0.000	$5s^25p5$	$^2P^o$	3/2	4	ASD
					54633.460	$5s^25p^4(^3P<2>)6s$	2[2]	5/2	6	ASD
700000	1844.453	0.003	0.070		7602.970	$5s^25p5$	$^2P^o$	1/2	2	ASD
					61819.779	$5s^25p^4(^3P<1>)6s$	2[1]	3/2	4	ASD

Intensity	Wavelength/Å	Uncert. wavelength/Å	A_{ki}/(10^8 s^{-1})	Accur.	Energy levels/cm^{-1}	Configs.	Terms	J	g	DB Source
150000	2061.633	0.003	0.030		7602.970	$5s^25p5$	$^2P^o$	1/2	2	ASD
					56092.881	$5s^25p^4(^3P<2>)6s$	2[2]	3/2	4	ASD
2100	5119.287	0.003			56092.881	$5s^25p^4(^3P<2>)6s$	2[2]	3/2	4	ASD
					75621.427	$5s^25p^4(^3P<2>)7p$	$2[1]^o$	3/2	4	ASD
140000	8043.74	0.03			54633.460	$5s^25p^4(^3P<2>)6s$	2[2]	5/2	6	ASD
					67062.130	$5s^25p^4(^3P<2>)6p$	$2[1]^o$	3/2	4	ASD
33000	9058.3336	0.0020			54633.460	$5s^25p^4(^3P<2>)6s$	2[2]	5/2	6	ASD
					65669.988	$5s^25p^4(^3P<2>)6p$	$2[3]^o$	7/2	8	ASD
Xenon I (Z = 54)										
250a	1192.0376		5.93	B+	0.000	$5p6$	^1S	0	1	HDBK,ASD
					83889.971	$5p^5(^2P^o<3/2>)5d$	$2[3/2]^o$	1	3	HDBK,ASD
250a	1250.2091		0.149	B+	0.000	$5p6$	^1S	0	1	HDBK,ASD
					79986.618	$5p^5(^2P^o<3/2>)5d$	$2[1/2]^o$	1	3	HDBK,ASD
1000a	1295.5878		2.53	A+	0.000	$5p6$	^1S	0	1	HDBK,ASD
					77185.041	$5p^5(^2P^o<1/2>)6s$	$2[1/2]^o$	1	3	HDBK,ASD
600	1469.6123		2.73	A+	0.000	$5p6$	^1S	0	1	HDBK,ASD
					68045.156	$5p^5(^2P^o<3/2>)6s$	$2[3/2]^o$	1	3	HDBK,ASD
10000hfs	8231.6336	0.0005			67067.547	$5p^5(^2P^o<3/2>)6s$	$2[3/2]^o$	2	5	ASD
					79212.465	$5p^5(^2P^o<3/2>)6p$	2[3/2]	2	5	ASD
7000	8280.1162	0.0005			68045.156	$5p^5(^2P^o<3/2>)6s$	$2[3/2]^o$	1	3	ASD
					80118.962	$5p^5(^2P^o<3/2>)6p$	2[1/2]	0	1	ASD
5000	8819.4106	0.0005			67067.547	$5p^5(^2P^o<3/2>)6s$	$2[3/2]^o$	2	5	ASD
					78403.061	$5p^5(^2P^o<3/2>)6p$	2[5/2]	3	7	ASD
3000	9923.1983	0.0020			68045.156	$5p^5(^2P^o<3/2>)6s$	$2[3/2]^o$	1	3	ASD
					78119.798	$5p^5(^2P^o<3/2>)6p$	2[5/2]	2	5	ASD
2500	12623.399	0.005			77269.145	$5p^5(^2P^o<3/2>)6p$	2[1/2]	1	3	HDBK,ASD
					85188.777	$5p^5(^2P^o<3/2>)7s$	$2[3/2]^o$	2	5	HDBK,ASD
3000	14732.816	0.005			78403.061	$5p^5(^2P^o<3/2>)6p$	2[5/2]	3	7	HDBK,ASD
					85188.777	$5p^5(^2P^o<3/2>)7s$	$2[3/2]^o$	2	5	HDBK,ASD
6000	31077.70	0.02			79212.465	$5p^5(^2P^o<3/2>)6p$	2[3/2]	2	5	HDBK,ASD
					82430.204	$5p^5(^2P^o<3/2>)5d$	$2[5/2]^o$	3	7	HDBK,ASD
5000	35079.82	0.02			78119.798	$5p^5(^2P^o<3/2>)6p$	2[5/2]	2	5	HDBK,ASD
					80970.438	$5p^5(^2P^o<3/2>)5d$	$2[7/2]^o$	3	7	HDBK,ASD

TABLE 6. Persistent Lines of Neutral Atomic Elements — Cesium I (Z = 55), Barium I (Z = 56), and Hafnium I (Z = 72) to Radon I (Z = 86)

Intensity	Wavelength/Å	Uncert. wavelength/Å	A_{ki}/(10^8 s^{-1})	Accur.	Energy levels/cm^{-1}	Configs.	Terms	J	g	DB Source
Cesium I (Z = 55)										
3000	4555.2799	0.0005	0.01836	AA	0.0000	$5p^66s$	^2S	1/2	2	ASD,NHBK
					21946.397	$5p^67p$	$^2P^o$	3/2	4	ASD,NHBK
1500	4593.1721	0.0005	0.00794	A+	0.0000	$5p^66s$	^2S	1/2	2	ASD,NHBK
					21765.348	$5p^67p$	$^2P^o$	1/2	2	ASD,NHBK
190000	8521.13165305	0.0000002	0.3279	AAA	0.0000	$5p^66s$	^2S	1/2	2	ASD,NHBK
					11732.3071041	$5p^66p$	$^2P^o$	3/2	4	ASD,NHBK
160000	8761.4121	0.0008			11178.26815870	$5p^66p$	$^2P^o$	1/2	2	ASD,NHBK
					22588.8210	$5p^66d$	^2D	3/2	4	ASD,NHBK
600000	8943.47423876	0.00000006	0.2863	AAA	0.0000	$5p^66s$	^2S	1/2	2	ASD,NHBK
					11178.26815870	$5p^66p$	$^2P^o$	1/2	2	ASD,NHBK
190000	9172.3178	0.0008			11732.3071041	$5p^66p$	$^2P^o$	3/2	4	ASD,NHBK
					22631.6863	$5p^66d$	^2D	5/2	6	ASD,NHBK
380000	13588.293	0.003			11178.26815870	$5p^66p$	$^2P^o$	1/2	2	ASD,NHBK
					18535.5286	$5p^67s$	^2S	1/2	2	ASD,NHBK
550000	14694.9087	0.0021			11732.3071041	$5p^66p$	$^2P^o$	3/2	4	ASD,NHBK
					18535.5286	$5p^67s$	^2S	1/2	2	ASD,NHBK
28000	30111.48	0.03	0.00984	A+	11178.26815870	$5p^66p$	$^2P^o$	1/2	2	ASD
					14499.2568	$5p^65d$	^2D	3/2	4	ASD

Intensity	Wavelength/Å	Uncert. wavelength/Å	$A_{ki}/(10^8\ s^{-1})$	Accur.	Energy levels/ cm^{-1}	Configs.	Terms	J	g	DB Source
11000	34909.65	0.03	0.00781	AA	11732.3071041	$5p^6 6p$	^{2}P°	3/2	4	ASD
					14596.84232	$5p^6 5d$	^{2}D	5/2	6	ASD
Barium I (Z = 56)										
168	3071.5841	0.0010	0.42	C	0.000	$6s^2$	^{1}S	0	1	ASD
					32547.033	$6s7p$	^{1}P°	1	3	ASD
860	3501.1075	0.0010	0.35	B	0.000	$6s^2$	^{1}S	0	1	ASD
					28554.221	$5d6p$	^{1}P°	1	3	ASD
1830	5535.481	0.010	1.19	A+	0.000	$6s^2$	^{1}S	0	1	ASD
					18060.261	$6s6p$	^{1}P°	1	3	ASD
1060	6498.760	0.010	0.54	C+	9596.533	$6s5d$	^{3}D	3	7	ASD
					24979.834	$5d6p$	^{3}D°	3	7	ASD
890	6527.311	0.010	0.33	B	9215.501	$6s5d$	^{3}D	2	5	ASD
					24531.513	$5d6p$	^{3}D°	2	5	ASD
740	6595.325	0.010	0.38	B+	9033.966	$6s5d$	^{3}D	1	3	ASD
					24192.033	$5d6p$	^{3}D°	1	3	ASD
454	6693.842	0.010	0.146	B	9596.533	$6s5d$	^{3}D	3	7	ASD
					24531.513	$5d6p$	^{3}D°	2	5	ASD
770	7059.943	0.010	0.50	C	9596.533	$6s5d$	^{3}D	3	7	ASD
					23757.049	$5d6p$	^{3}F°	4	9	ASD
1280	7280.296	0.010	0.32	C+	9215.501	$6s5d$	^{3}D	2	5	ASD
					22947.423	$5d6p$	^{3}F°	3	7	ASD
560	7672.085	0.010	0.15	C	9033.966	$6s5d$	^{3}D	1	3	ASD
					22064.645	$5d6p$	^{3}F°	2	5	ASD
700	8559.998	0.010	0.20	C+	11395.350	$6s5d$	^{1}D	2	5	ASD
					23074.387	$5d6p$	^{1}D°	2	5	ASD
1300	14999.85	0.10	0.0025	B	11395.350	$6s5d$	^{1}D	2	5	ASD
					18060.261	$6s6p$	^{1}P°	1	3	ASD
9900	23259.91	0.10			9215.501	$6s5d$	^{3}D	2	5	ASD
					13514.745	$6s6p$	^{3}P°	2	5	ASD
10000	25521.84	0.10			9596.533	$6s5d$	^{3}D	3	7	ASD
					13514.745	$6s6p$	^{3}P°	2	5	ASD
Hafnium I (Z = 72)										
2100	2866.37				0.00	$5d^2 6s^2$	a ^{3}F	2	5	ASD
					34877.04	$5d^2(^3F)6s(^2F)6p$	^{1}F°	3	7	ASD
1800	2898.26				2356.68	$5d^2 6s^2$	a ^{3}F	3	7	ASD
					36850.04	$5d^2(^1G)6s(^2G)6p$	^{3}G°	4	9	ASD
1200	2904.41				4567.64	$5d^2 6s^2$	a ^{3}F	4	9	ASD
					38987.85	$5d^2(^3F)6s(^2F)6p$	^{1}G°	4	9	ASD
2000	2916.48				4567.64	$5d^2 6s^2$	a ^{3}F	4	9	ASD
					38845.45	$5d^2(^1G)6s(^2G)6p$	^{3}G°	5	11	ASD
2000	2940.77				0.00	$5d^2 6s^2$	a ^{3}F	2	5	ASD
					33994.86	$5d^2(^1G)6s(^2G)6p$	^{3}F°	2	5	ASD
2100	3072.88				0.00	$5d^2 6s^2$	a ^{3}F	2	5	ASD
					32533.30	$5d^2(^3F)6s(^2F)6p$	^{3}G°	3	7	ASD
2200	3682.24		0.26		0.00	$5d^2 6s^2$	a ^{3}F	2	5	ASD
					27149.64	$5d^2 6s(a\ ^4F)6p$	y ^{3}F°	2	5	ASD
1100	4174.34		0.046		2356.68	$5d^2 6s^2$	a ^{3}F	3	7	ASD
					26305.78	$5d^2 6s(a\ ^4F)6p$	z ^{5}D°	3	7	ASD
650	7237.10				4567.64	$5d^2 6s^2$	a ^{3}F	4	9	ASD
					18381.49	$5d^2 6s(b\ ^2D)6p$	z ^{3}D°	3	7	ASD
Tantalum I (Z = 73)										
1200	2559.43				0.00	$5d^3 6s^2$	a ^{4}F	3/2	4	ASD
					39059.52		390°	5/2	6	ASD
1400	2608.63				2010.10	$5d^3 6s^2$	a ^{4}F	5/2	6	ASD
					40333.03		403°	7/2	8	ASD
2400	2647.47				0.00	$5d^3 6s^2$	a ^{4}F	3/2	4	ASD
					37760.67		377°	3/2	4	ASD

Intensity	Wavelength/Å	Uncert. wavelength/Å	$A_{ki}/(10^8\ s^{-1})$	Accur.	Energy levels/ cm^{-1}	Configs.	Terms	J	g	DB Source
2600	2653.27				2010.10	$5d^36s^2$	a ^{4}F	5/2	6	ASD
					39688.20		396°	5/2	6	ASD
1900	2656.61				0.00	$5d^36s^2$	a ^{4}F	3/2	4	ASD
					37630.09		376°	5/2	6	ASD
1500	2661.34				5621.04	$5d^36s^2$	a ^{4}F	9/2	10	ASD
					43185.09		431°	11/2	12	ASD
1000	2698.30				2010.10	$5d^36s^2$	a ^{4}F	5/2	6	ASD
					39059.52		390°	5/2	6	ASD
2600	2714.67				0.00	$5d^36s^2$	a ^{4}F	3/2	4	ASD
					36825.97	$5d^36s(^3G)6p$	y $^4G^o$?	5/2	6	ASD
1000	2758.31				2010.10	$5d^36s^2$	a ^{4}F	5/2	6	ASD
					38253.39		382°	7/2	8	ASD
1500	2850.49				23512.34	$5d^4$(a ^{3}F)6s	b ^{4}F	5/2	6	ASD
					58583.88		585°	3/2	4	ASD
1900	2850.98				5621.04	$5d^36s^2$	a ^{4}F	9/2	10	ASD
					40686.42	$5d^36s(^3G)6p$	y $^4G^o$?	9/2	10	ASD
1700	2933.55				0.00	$5d^36s^2$	a ^{4}F	3/2	4	ASD
					34078.42		340°	3/2	4	ASD
1500	2963.32				2010.10	$5d^36s^2$	a ^{4}F	5/2	6	ASD
					35746.18		357°	7/2	8	ASD
1100	3311.162				5621.04	$5d^36s^2$	a ^{4}F	9/2	10	ASD
					35813.47		358°	11/2	12	ASD
450	4681.875		0.0152	B+	2010.10	$5d^36s^2$	a ^{4}F	5/2	6	ASD
					23363.09	$5d^36s(^5F)6p$	z $^4F^o$	5/2	6	ASD
330	5156.562				5621.04	$5d^36s^2$	a ^{4}F	9/2	10	ASD
					25008.83	$5d^36s(^5F)6p$	z $^6G^o$	11/2	12	ASD
Tungsten I (Z = 74)										
200r	2551.349		1.78	B	0.00	$5d^46s^2$	^{5}D	0	1	ASD
					39183.20		^{o}	1	3	ASD
400r	2681.422		0.74	B	2951.29	$5d^5(^6S)6s$	^{7}S	3	7	ASD
					40233.97		^{o}	4	9	ASD
300r	2724.352		1.05	B	2951.29	$5d^5(^6S)6s$	^{7}S	3	7	ASD
					39646.41		^{o}	3	7	ASD
250r	2831.379		0.49	B	2951.29	$5d^5(^6S)6s$	^{7}S	3	7	ASD
					38259.40		^{o}	4	9	ASD
400r	2896.442		1.3	D	2951.29	$5d^5(^6S)6s$	^{7}S	3	7	ASD
					37466.30		^{o}	2	5	ASD
300r	2944.398		1.08	B	2951.29	$5d^5(^6S)6s$	^{7}S	3	7	ASD
					36904.16		^{o}	2	5	ASD
300r	2946.989		0.82	B	2951.29	$5d^5(^6S)6s$	^{7}S	3	7	ASD
					36874.36		^{o}	3	7	ASD
200	3215.562		0.21	B	6219.33	$5d^46s^2$	^{5}D	4	9	ASD
					37309.16		^{o}	5	11	ASD
500	3617.515		0.11	B	2951.29	$5d^5(^6S)6s$	^{7}S	3	7	ASD
					30586.64	$5d^46s(^6D)6p$	$^5P^o$	3	7	ASD
1000	4008.7506		0.163	B	2951.29	$5d^5(^6S)6s$	^{7}S	3	7	ASD
					27889.68	$5d^5(^6S)6p$	$^7P^o$	4	9	ASD
600	4074.358		0.10	B	2951.29	$5d^5(^6S)6s$	^{7}S	3	7	ASD
					27488.11	$5d^5(^6S)6p$	$^7P^o$	3	7	ASD
800	4294.606		0.124	A	2951.29	$5d^5(^6S)6s$	^{7}S	3	7	ASD
					26229.77	$5d^5(^6S)6p$	$^7P^o$	2	5	ASD
200	4302.110		0.036	B	2951.29	$5d^5(^6S)6s$	^{7}S	3	7	ASD
					26189.20	$5d^46s(^6D)6p$	$^7D^o$	3	7	ASD
Rhenium I (Z = 75)										
25000	2003.53				0.00	$5d^56s^2$	^{6}S	5/2	6	NHBK
					49895.57		16°	3/2	4	NHBK
16000	2017.87				0.00	$5d^56s^2$	^{6}S	5/2	6	NHBK
					49540.96		15°	5/2	6	NHBK

Intensity	Wavelength/Å	Uncert. wavelength/Å	$A_{ki}/(10^8\ s^{-1})$	Accur.	Energy levels/ cm^{-1}	Configs.	Terms	J	g	DB Source
27000	2049.08				0.00	$5d^56s^2$	^{6}S	5/2	6	NHBK
					48786.35		14°	7/2	8	NHBK
4200	2074.70				0.00	$5d^56s^2$	^{6}S	5/2	6	NHBK
					48184.20		13°	7/2	8	NHBK
10000	2085.59				0.00	$5d^56s^2$	^{6}S	5/2	6	NHBK
					47932.55	$5d^6(^5D)6p$	^{6}F^o	7/2	8	NHBK
9800	2097.12				0.00	$5d^56s^2$	^{6}S	5/2	6	NHBK
					47669.01		12°	7/2	8	NHBK
4900	2167.94				0.00	$5d^56s^2$	^{6}S	5/2	6	NHBK
					46112.24		11°	5/2	6	NHBK
5500	2999.60				11754.52	$5d^6(^5D)6s$	6D	9/2	10	NHBK
					45082.63	$5d^6(^5D)6p$	^{6}F^o	11/2	12	NHBK
4000	3399.30				11754.52	$5d^6(^5D)6s$	6D	9/2	10	NHBK
					41163.91		10°	7/2	8	NHBK
8000	3424.62				11754.52	$5d^6(^5D)6s$	6D	9/2	10	NHBK
					40946.53	$5d^56s(^5P)6p$	6D^o	7/2	8	NHBK
16000c	3451.88		0.174	B	0.00	$5d^56s^2$	^{6}S	5/2	6	NHBK
					28961.55	$5d^56s(^7S)6p$	6P^o	3/2	4	NHBK
55000c	3460.46		0.313	B	0.00	$5d^56s^2$	^{6}S	5/2	6	NHBK
					28889.72	$5d^56s(^7S)6p$	6P^o	7/2	8	NHBK
40000c	3464.73		0.249	B	0.00	$5d^56s^2$	^{6}S	5/2	6	NHBK
					28854.18	$5d^56s(^7S)6p$	6P^o	5/2	6	NHBK
4000	3725.76				23631.82	$5d^56s(^7S)6p$	^{8}P^o	9/2	10	NHBK
					50464.34	$5d^56s(^7S)$6d	e ^{8}D	11/2	12	NHBK

Osmium I (Z = 76)

Intensity	Wavelength/Å	Uncert. wavelength/Å	$A_{ki}/(10^8\ s^{-1})$	Accur.	Energy levels/ cm^{-1}	Configs.	Terms	J	g	DB Source
29000	2018.14				0.00	$5d^66s^2$	^{5}D	4	9	NHBK
					49534.28		^{o}	5	11	NHBK
18000	2034.44				0.00	$5d^66s^2$	^{5}D	4	9	NHBK
					49138.11		^{o}	3	7	NHBK
26000	2045.36				0.00	$5d^66s^2$	^{5}D	4	9	NHBK
					48874.93		^{o}	3	7	NHBK
14000	2079.97				0.00	$5d^66s^2$	^{5}D	4	9	NHBK
					48062.22		^{o}	5	11	NHBK
3800	2637.13				0.00	$5d^66s^2$	^{5}D	4	9	NHBK
					37908.77	$5d^66s(^4D)6p$	^{5}D^o	4	9	NHBK
3000	2714.64				0.00	$5d^66s^2$	^{5}D	4	9	NHBK
					36826.39	$5d^66s(^6D)6p$	^{5}D^o	4	9	NHBK
2800	2806.91				0.00	$5d^66s^2$	^{5}D	4	9	NHBK
					35615.92	$5d^66s(^6D)6p$	^{5}P^o	3	7	NHBK
5100	2838.63				5143.92	$5d^7(^4F)6s$	^{5}F	5	11	NHBK
					40361.92	$5d^66s(^4D)6p$	^{3}F^o	4	9	NHBK
9600	2909.06				0.00	$5d^66s^2$	^{5}D	4	9	NHBK
					34365.33	$5d^66s(^6D)6p$	^{5}F^o	5	11	NHBK
4400	3018.04				0.00	$5d^66s^2$	^{5}D	4	9	NHBK
					33124.48	$5d^66s(^6D)6p$	^{7}P^o	3	7	NHBK
8600*	3058.66		0.290		0.00	$5d^66s^2$	^{5}D	4	9	NHBK
					32684.61	$5d^66s(^6D)6p$	7F^o	4	9	NHBK
8600*	3058.66				14091.37	$5d^7(^4F)6s$	^{3}F	3	7	NHBK
					46776.29	$5d^7(^4F)6p$	^{3}D^o	3	7	NHBK
3100	3262.29				4159.32	$5d^66s^2$	^{5}D	3	7	NHBK
					34803.82	$5d^66s(^6D)6p$	^{5}F^o	4	9	NHBK
3100	3267.94		0.0580		0.00	$5d^66s^2$	^{5}D	4	9	NHBK
					30591.45	$5d^66s(^6D)6p$	^{7}P^o	4	9	NHBK
7600	3301.56		0.0990		0.00	$5d^66s^2$	^{5}D	4	9	NHBK
					30279.95	$5d^66s(^6D)6p$	7F^o	5	11	NHBK
2100	3782.20				4159.32	$5d^66s^2$	^{5}D	3	7	NHBK
					30591.45	$5d^66s(^6D)6p$	^{7}P^o	4	9	NHBK
4900	4260.85				0.00	$5d^66s^2$	^{5}D	4	9	NHBK
					23462.90	$5d^66s(^6D)6p$	^{7}D^o	5	11	NHBK

Intensity	Wavelength/Å	Uncert. wavelength/Å	A_{ki}/(10^8 s^{-1})	Accur.	Energy levels/ cm^{-1}	Configs.	Terms	*J*	*g*	DB Source
4900	4420.47				0.00	$5d^66s^2$	^{5}D	4	9	NHBK
					22615.69	$5d^66s(^6D)6p$	^{7}D°	4	9	NHBK
Iridium I (*Z* = 77)										
15000	2033.57				0.00	$5d^76s^2$	a ^{4}F	9/2	10	ASD
					49158.61		4915°	9/2	10	ASD
17000	2088.82				0.00	$5d^76s^2$	a ^{4}F	9/2	10	ASD
					47858.47		4785°	11/2	12	ASD
7900	2158.05				2834.98	$5d^8(^3F)6s$	b ^{4}F	9/2	10	ASD
					49158.61		4915°	9/2	10	ASD
3500	2372.77				0.00	$5d^76s^2$	a ^{4}F	9/2	10	ASD
					42131.82		4213°	11/2	12	ASD
3300	2475.12		0.204	B+	0.00	$5d^76s^2$	a ^{4}F	9/2	10	ASD
					40389.83		4038°	9/2	10	ASD
4100	2502.98		0.38	B	0.00	$5d^76s^2$	a ^{4}F	9/2	10	ASD
					39940.27		3994°	11/2	12	ASD
7900	2543.97				2834.98	$5d^8(^3F)6s$	b ^{4}F	9/2	10	ASD
					42131.82		4213°	11/2	12	ASD
3500	2639.71		0.469	A	0.00	$5d^76s^2$	a ^{4}F	9/2	10	ASD
					37871.69	$5d^76s(^5F)6p$	z ^{4}F°	9/2	10	ASD
2700	2664.79		0.308	B+	0.00	$5d^76s^2$	a ^{4}F	9/2	10	ASD
					37515.31	$5d^76s(^5F)6p$	z 4D°	7/2	8	ASD
3000	2694.23		0.58	B	2834.98	$5d^8(^3F)6s$	b ^{4}F	9/2	10	ASD
					39940.27		3994°	11/2	12	ASD
3800	2849.72		0.226	B+	0.00	$5d^76s^2$	a ^{4}F	9/2	10	ASD
					35080.80	$5d^76s(^5F)6p$	z ^{6}G°	9/2	10	ASD
4400	2924.79		0.170	A+	0.00	$5d^76s^2$	a ^{4}F	9/2	10	ASD
					34180.48	$5d^76s(^5F)6p$	z ^{6}G°	11/2	12	ASD
5100	3220.78		0.197	B+	2834.98	$5d^8(^3F)6s$	b ^{4}F	9/2	10	ASD
					33874.43	$5d^76s(^5F)6p$	z ^{6}F°	7/2	8	ASD
3200	3513.64		0.031	B	0.00	$5d^76s^2$	a ^{4}F	9/2	10	ASD
					28452.32	$5d^76s(^5F)6p$	z ^{6}F°	11/2	12	ASD
Platinum I (*Z* = 78)										
1000*	2487.1685	0.0020			775.892	$5d^96s$	^{1}D	2	5	NHBK
					40970.165	$5d^8(^3F)6s6p(^3P°)$	^{5}G°	3	7	NHBK
1000*	2487.1685	0.0020			0.000	$5d^96s$	^{3}D	3	7	NHBK
					40194.228	$5d^8(^3F)6s6p(^3P°)$	^{5}F°	4	9	NHBK
1100	2628.0269	0.0020	0.482		775.892	$5d^96s$	^{1}D	2	5	NHBK
					38815.908	$5d^8(^1D)6s6p(^3P°)$	^{3}F°	2	5	NHBK
500	2650.8524	0.0020	0.0962		823.678	$5d^86s^2$	^{3}F	4	9	NHBK
					38536.160	$5d^8(^3F)6s6p(^3P°)$	^{5}G°	5	11	NHBK
2800	2659.4503	0.0020	0.890		0.000	$5d^96s$	^{3}D	3	7	NHBK
					37590.569	$5d^96p$	^{3}F°	4	9	NHBK
2000	2702.3995	0.0020	0.523		775.892	$5d^96s$	^{1}D	2	5	NHBK
					37769.073	$5d^96p$	^{3}D°	3	7	NHBK
1600	2705.8951	0.0020	0.380		823.678	$5d^86s^2$	^{3}F	4	9	NHBK
					37769.073	$5d^96p$	^{3}D°	3	7	NHBK
1800	2733.9567	0.0020	0.672		775.892	$5d^96s$	^{1}D	2	5	NHBK
					37342.101	$5d^96p$	^{3}P°	2	5	NHBK
1400	2830.2919	0.0020	0.168		0.000	$5d^96s$	^{3}D	3	7	NHBK
					35321.653	$5d^8(^3F)6s6p(^3P°)$	^{5}D°	3	7	NHBK
600	2893.8630	0.0020	0.0647		775.892	$5d^96s$	^{1}D	2	5	NHBK
					35321.653	$5d^8(^3F)6s6p(^3P°)$	^{5}D°	3	7	NHBK
1700	2929.7894	0.0020	0.185		0.000	$5d^96s$	^{3}D	3	7	NHBK
					34122.165	$5d^96p$	^{3}F°	3	7	NHBK
1800	2997.9622	0.0020	0.288		775.892	$5d^96s$	^{1}D	2	5	NHBK
					34122.165	$5d^96p$	^{3}F°	3	7	NHBK
800	3042.6318	0.0020	0.0769		823.678	$5d^86s^2$	^{3}F	4	9	NHBK
					33680.402	$5d^8(^3F)6s6p(^3P°)$	^{5}F°	5	11	NHBK

Intensity	Wavelength/Å	Uncert. wavelength/Å	$A_{ki}/(10^8\ s^{-1})$	Accur.	Energy levels/ cm^{-1}	Configs.	Terms	J	g	DB Source
3200	3064.7110	0.0020	0.644		0.000	$5d^9 6s$	^3D	3	7	NHBK
					32620.018	$5d^9 6p$	$^3P^o$	2	5	NHBK
500	3301.8596	0.0020	0.343		6567.461	$5d^9 6s$	^3D	2	5	NHBK
					36844.710	$5d^9 6p$	$^3P^o$	1	3	NHBK
60	3315.0419	0.0020	0.00176		0.000	$5d^9 6s$	^3D	3	7	NHBK
					30156.854	$5d^8(^3F)6s6p(^3P^o)$	$^5D^o$	4	9	NHBK
340	3408.1308	0.0020	0.0129		823.678	$5d^8 6s^2$	^3F	4	9	NHBK
					30156.854	$5d^8(^3F)6s6p(^3P^o)$	$^5D^o$	4	9	NHBK
80	3818.6874	0.0020			10116.729	$5d^8 6s^2$	^3F	3	7	NHBK
					36296.310	$5d^8(^3F)6s6p(^3P^o)$	$^5G^o$	4	9	NHBK
Gold I (Z = 79)										
150	1646.674				0.000	$5d^{10}6s$	^2S	1/2	2	ASD,NHBK
					60728.49	$5d^{10}7p$	$^2P^o$	3/2	4	ASD,NHBK
200	1699.339				0.000	$5d^{10}6s$	^2S	1/2	2	ASD,NHBK
					58845.414	$5d^9<3/2>6s<1/2>6p<1/2>$	$^2P^o$	3/2	4	ASD,NHBK
11000	2012.00				9161.177	$5d^9 6s^2$	^2D	5/2	6	ASD,NHBK
					58845.414	$5d^9<3/2>6s<1/2>6p<1/2>$	$^2P^o$	3/2	4	ASD,NHBK
2600	2021.38				9161.177	$5d^9 6s^2$	^2D	5/2	6	ASD,NHBK
					58616.764	$5d^9<3/2>6s<1/2>6p<1/2>$	$2F^o$	5/2	6	ASD,NHBK
2600	2427.95		1.99		0.000	$5d^{10}6s$	^2S	1/2	2	ASD,NHBK
					41174.613	$5d^{10}6p$	$^2P^o$	3/2	4	ASD,NHBK
3400	2675.954		1.64		0.000	$5d^{10}6s$	^2S	1/2	2	ASD,NHBK
					37358.991	$5d^{10}6p$	$^2P^o$	1/2	2	ASD,NHBK
1100	2748.253				9161.177	$5d^9 6s^2$	^2D	5/2	6	ASD,NHBK
					45537.195	$5d^9<5/2>6s<1/2>6p<1/2>$	$^4F^o$	7/2	8	ASD,NHBK
320	3029.204				9161.177	$5d^9 6s^2$	^2D	5/2	6	ASD,NHBK
					42163.530	$5d^9<5/2>6s<1/2>6p<1/2>$	$^4P^o$	5/2	6	ASD,NHBK
1600	3122.784		0.190		9161.177	$5d^9 6s^2$	^2D	5/2	6	ASD,NHBK
					41174.613	$5d^{10}6p$	$^2P^o$	3/2	4	ASD,NHBK
600	6278.170		0.034		21435.191	$5d^9 6s^2$	^2D	3/2	4	ASD,NHBK
					37358.991	$5d^{10}6p$	$^2P^o$	1/2	2	ASD,NHBK
Mercury I (Z = 80)										
5000r	1849.4994	0.0002	7.46	A	0.000	$5d^{10}6s^2$	^1S	0	1	ASD
					54068.6829	$5d^{10}6s6p$	$^1P^o$	1	3	ASD
900000	2536.5210	0.0010	0.0840	A+	0.000	$5d^{10}6s^2$	^1S	0	1	ASD
					39412.237	$5d^{10}6s6p$	$^3P^o$	1	3	ASD
3000	2967.2830	0.0010	0.46	D	37644.982	$5d^{10}6s6p$	$^3P^o$	0	1	ASD
					71336.005	$5d^{10}6s6d$	^3D	1	3	ASD
9000	3650.1580	0.0010	1.29	B+	44042.909	$5d^{10}6s6p$	$^3P^o$	2	5	ASD
					71431.180	$5d^{10}6s6d$	^3D	3	7	ASD
12000	4046.5650	0.0010	0.207	B	37644.982	$5d^{10}6s6p$	$^3P^o$	0	1	ASD
					62350.325	$5d^{10}6s7s$	^3S	1	3	ASD
12000	4358.3350	0.0010	0.56	B	39412.237	$5d^{10}6s6p$	$^3P^o$	1	3	ASD
					62350.325	$5d^{10}6s7s$	^3S	1	3	ASD
6000	5460.7500	0.0010	0.49	B	44042.909	$5d^{10}6s6p$	$^3P^o$	2	5	ASD
					62350.325	$5d^{10}6s7s$	^3S	1	3	ASD
1600	10139.75	0.05	0.271	B+	54068.6829	$5d^{10}6s6p$	$^1P^o$	1	3	ASD
					63928.120	$5d^{10}6s7s$	^1S	0	1	ASD
Thallium I (Z = 81)										
900w	2379.69		0.44	C	0.0	$6s^2 6p$	$^2P^o$	1/2	2	ASD
					42011.4	$6s^2 7d$	^2D	3/2	4	ASD
700	2580.14		0.18	D	0.0	$6s^2 6p$	$^2P^o$	1/2	2	ASD
					38745.9	$6s^2 8s$	^2S	1/2	2	ASD
4400bl	2767.87		1.26	C	0.0	$6s^2 6p$	$^2P^o$	1/2	2	ASD
					36117.9	$6s^2 6d$	^2D	3/2	4	ASD
2800	2918.32		0.42	C	7792.7	$6s^2 6p$	$^2P^o$	3/2	4	ASD
					42049.0	$6s^2 7d$	^2D	5/2	6	ASD

Intensity	Wavelength/Å	Uncert. wavelength/Å	$A_{ki}/(10^8\ s^{-1})$	Accur.	Energy levels/ cm^{-1}	Configs.	Terms	J	g	DB Source
1200	3229.75		0.173	C	7792.7	$6s^26p$	$^2P^o$	3/2	4	ASD
					38745.9	$6s^28s$	^2S	1/2	2	ASD
20000	3519.24		1.24	C	7792.7	$6s^26p$	$^2P^o$	3/2	4	ASD
					36199.9	$6s^26d$	^2D	5/2	6	ASD
5000	3529.43		0.220	C	7792.7	$6s^26p$	$^2P^o$	3/2	4	ASD
					36117.9	$6s^26d$	^2D	3/2	4	ASD
12000w	3775.72		0.625	B	0.0	$6s^26p$	$^2P^o$	1/2	2	ASD
					26477.5	$6s^27s$	^2S	1/2	2	ASD
18000	5350.46		0.705	B	7792.7	$6s^26p$	$^2P^o$	3/2	4	ASD
					26477.5	$6s^27s$	^2S	1/2	2	ASD
1000	11512.82				26477.5	$6s^27s$	^2S	1/2	2	ASD
					35161.1	$6s^27p$	$^2P^o$	3/2	4	ASD
700	13013.2				26477.5	$6s^27s$	^2S	1/2	2	ASD
					34159.9	$6s^27p$	$^2P^o$	1/2	2	ASD
Lead I (Z = 82)										
5r	2022.0162	0.0012	0.059	C+	0.000	$6s^26p^2$	(1/2,1/2)	0	1	ASD
					49439.6165	$6s^26p7s$	$(3/2,1/2)^o$	1	3	ASD
8r	2053.2836	0.0010	0.102	C	0.000	$6s^26p^2$	(1/2,1/2)	0	1	ASD
					48686.9340	$6s^26p8s$	$(1/2,1/2)^o$	1	3	ASD
500r	2170.0048	0.0011	1.47	C+	0.000	$6s^26p^2$	(1/2,1/2)	0	1	ASD
					46068.4385	$6s^26p(^2P^o<1/2>)6d$	$2[3/2]^o$	1	3	ASD
550r	2393.79228	0.00016			10650.3271	$6s^26p^2$	(3/2,1/2)	2	5	ASD
					52412.325	$6s^26p(^2P^o<1/2>)7d$	$2[5/2]^o$	3	7	ASD
900r	2614.17464	0.0002	1.98	C+	7819.2626	$6s^26p^2$	(3/2,1/2)	1	3	ASD
					46060.8364	$6s^26p(^2P^o<1/2>)6d$	$2[3/2]^o$	2	5	ASD
25000r	2801.99534	0.0002	1.61	C+	10650.3271	$6s^26p^2$	(3/2,1/2)	2	5	ASD
					46328.6668	$6s^26p(^2P^o<1/2>)6d$	$2[5/2]^o$	3	7	ASD
35000r	2833.05344	0.0002	0.49	C+	0.000	$6s^26p^2$	(1/2,1/2)	0	1	ASD
					35287.2244	$6s^26p7s$	$(1/2,1/2)^o$	1	3	ASD
50000r	3639.5677	0.0003	0.32	C+	7819.2626	$6s^26p^2$	(3/2,1/2)	1	3	ASD
					35287.2244	$6s^26p7s$	$(1/2,1/2)^o$	1	3	ASD
70000r	3683.4625	0.0003	1.37	B	7819.2626	$6s^26p^2$	(3/2,1/2)	1	3	ASD
					34959.9084	$6s^26p7s$	$(1/2,1/2)^o$	0	1	ASD
25000	3739.9353	0.0003	0.73	B	21457.7982	$6s^26p^2$	(3/2,3/2)	2	5	ASD
					48188.6296	$6s^26p7s$	$(3/2,1/2)^o$	2	5	ASD
95000	4057.8067	0.0003	0.9	C+	10650.3271	$6s^26p^2$	(3/2,1/2)	2	5	ASD
					35287.2244	$6s^26p7s$	$(1/2,1/2)^o$	1	3	ASD
14000	4062.13599	0.0002	0.92	B	21457.7982	$6s^26p^2$	(3/2,3/2)	2	5	ASD
					46068.4385	$6s^26p(^2P^o<1/2>)6d$	$2[3/2]^o$	1	3	ASD
Bismuth I (Z = 83)										
9000	1954.706		0.586		0.000	$6p^3$	$^4S^o$	3/2	4	ASD
					51158.494	$6p^2(^3P<0>)7d$	2[2]	5/2	6	ASD
7000	1960.049		0.154		0.000	$6p^3$	$^4S^o$	3/2	4	ASD
					51019.090	$6p^2(^3P<0>)7d$	2[2]	3/2	4	ASD
7000	2021.149		0.060		0.000	$6p^3$	$^4S^o$	3/2	4	ASD
					49460.910	$6p^2(^3P<2>)7s$	2[2]	3/2	4	ASD
9000	2061.634		0.96		0.000	$6p^3$	$^4S^o$	3/2	4	ASD
					48489.869	$6p^2(^3P<2>)7s$	2[2]	5/2	6	ASD
4600	2110.217		0.53		0.000	$6p^3$	$^4S^o$	3/2	4	ASD
					47373.477	$6p^2(^3P<0>)8s$	2[0]	1/2	2	ASD
360	2228.203		0.88		0.000	$6p^3$	$^4S^o$	3/2	4	ASD
					44865.076	$6p^2(^3P<1>)7s$	2[1]	3/2	4	ASD
1700	2230.602		2.34		0.000	$6p^3$	$^4S^o$	3/2	4	ASD
					44816.841	$6p^2(^3P<0>)6d$	2[2]	5/2	6	ASD
4000	2897.965		1.53		11419.039	$6p^3$	$^2D^o$	3/2	4	ASD
					45915.883	$6p^2(^3P<1>)7s$	2[1]	1/2	2	ASD
3200	2938.297		1.23		15437.501	$6p^3$	$^2D^o$	5/2	6	ASD
					49460.910	$6p^2(^3P<2>)7s$	2[2]	3/2	4	ASD

Intensity	Wavelength/Å	Uncert. wavelength/Å	$A_{ki}/(10^8\ s^{-1})$	Accur.	Energy levels/ cm^{-1}	Configs.	Terms	J	g	DB Source
2800	2989.019		0.54		11419.039	$6p^3$	$^2D^o$	3/2	4	ASD
					44865.076	$6p^2(^3P<1>)7s$	2[1]	3/2	4	ASD
700	2993.336		0.145		11419.039	$6p^3$	$^2D^o$	3/2	4	ASD
					44816.841	$6p^2(^3P<0>)6d$	2[2]	5/2	6	ASD
2400	3024.621		0.86		15437.501	$6p^3$	$^2D^o$	5/2	6	ASD
					48489.869	$6p^2(^3P<2>)7s$	2[2]	5/2	6	ASD
9000c	3067.700		1.67		0.000	$6p^3$	$^4S^o$	3/2	4	ASD
					32588.221	$6p^2(^3P<0>)7s$	2[0]	1/2	2	ASD
600c	4722.527		0.094		11419.039	$6p^3$	$^2D^o$	3/2	4	ASD
					32588.221	$6p^2(^3P<0>)7s$	2[0]	1/2	2	ASD
2000bl	9657.04				32588.221	$6p^2(^3P<0>)7s$	2[0]	1/2	2	ASD
					42940.519	$6p^2(^3P<0>)7p$	2[1]°	3/2	4	ASD
1500bl	11710.37				32588.221	$6p^2(^3P<0>)7s$	2[0]	1/2	2	ASD
					41124.985	$6p^2(^3P<0>)7p$	2[1]°	1/2?	2	ASD
Polonium I (Z = 84)										
1500w	2450.08				0.00	$6p^4$	^3P	2	5	ASD
					40802.70	$6p^3(^4S^o)7s$	$^3S^o$	1	3	ASD
1500w	2558.01				0.00	$6p^4$	^3P	2	5	ASD
					39081.19	$6p^3(^4S^o)7s$	5S°	2	5	ASD
2500w	3003.21				7514.69	$6p^4$	^3P	0	1	ASD
					40802.70	$6p^3(^4S^o)7s$	$^3S^o$	1	3	ASD
1200	4170.52				16831.61	$6p^4$	^3P	1	3	ASD
					40802.70	$6p^3(^4S^o)7s$	$^3S^o$	1	3	ASD
Astatine I (Z = 85)										
8	2162.25				0.0	$6p^5$	$^2P^o$	3/2	4	ASD
					46233.6	$6p^4(^3P)7s$	^4P	3/2	4	ASD
10	2244.01				0.0	$6p^5$	$^2P^o$	3/2	4	ASD
					44549.3	$6p^4(^3P)7s$	^4P	5/2	6	ASD
Radon I (Z = 86)										
	1451.56				0.0	$6p^6$	^1S	0	1	ASD
					68891.34	$6p^5(^2P^o<3/2>)6d$	2[1/2]°	1	3	ASD
	1786.07				0.0	$6p^6$	^1S	0	1	ASD
					55989.03	$6p^5(^2P^o<3/2>)7s$	2[3/2]°	1	3	ASD
100	4349.60				54620.35	$6p^5(^2P^o<3/2>)7s$	2[3/2]°	2	5	ASD
					77604.53	$6p^5(^2P^o<3/2>)8p$	2[5/2]	3	7	ASD
200	7055.42				54620.35	$6p^5(^2P^o<3/2>)7s$	2[3/2]°	2	5	ASD
					68789.93	$6p^5(^2P^o<3/2>)7p$	2[3/2]	2	5	ASD
100	7268.11				55989.03	$6p^5(^2P^o<3/2>)7s$	2[3/2]°	1	3	ASD
					69743.98	$6p^5(^2P^o<3/2>)7p$	2[1/2]	0	1	ASD
300	7450.00				54620.35	$6p^5(^2P^o<3/2>)7s$	2[3/2]°	2	5	ASD
					68039.48	$6p^5(^2P^o<3/2>)7p$	2[5/2]	3	7	ASD
100	7809.82				55989.03	$6p^5(^2P^o<3/2>)7s$	2[3/2]°	1	3	ASD
					68789.93	$6p^5(^2P^o<3/2>)7p$	2[3/2]	2	5	ASD
100	8099.51				55989.03	$6p^5(^2P^o<3/2>)7s$	2[3/2]°	1	3	ASD
					68332.10	$6p^5(^2P^o<3/2>)7p$	2[3/2]	1	3	ASD
100	8270.96				54620.35	$6p^5(^2P^o<3/2>)7s$	2[3/2]°	2	5	ASD
					66707.53	$6p^5(^2P^o<3/2>)7p$	2[5/2]	2	5	ASD
100	8600.07				54620.35	$6p^5(^2P^o<3/2>)7s$	2[3/2]°	2	5	ASD
					66244.97	$6p^5(^2P^o<3/2>)7p$	2[1/2]	1	3	ASD

TABLE 7. Persistent Lines of Neutral Atomic Elements — Lanthanum I (Z = 57) to Lutetium I (Z = 71)

Intensity	Wavelength/Å	Uncert. wavelength/Å	$A_{ki}/(10^8\ s^{-1})$	Accur.	Energy levels/cm^{-1}	Configs.	Terms	J	g	DB Source
Lanthanum I (Z = 57)										
200	3574.43		0.95	B+	0.000	$5d6s^2$	^2D	3/2	4	ASD
					27968.54	$4f5d(^3D^o)6s$	°	3/2	4	ASD

Intensity	Wavelength/Å	Uncert. wavelength/Å	$A_{ki}/(10^8\ s^{-1})$	Accur.	Energy levels/cm^{-1}	Configs.	Terms	J	g	DB Source
220	4060.33		0.56	B+	4121.572	$5d^2(^3F)6s$	^4F	9/2	10	ASD
					28743.24	$4f5d(^3G^o)6s$	$^4G^o$	11/2	12	ASD
280	4187.32		0.551	B+	0.000	$5d6s^2$	^2D	3/2	4	ASD
					23874.95	$5d6s(^3D)6p$	^{o}	5/2	6	ASD
300	4280.27		0.569	B+	1053.164	$5d6s^2$	^2D	5/2	6	ASD
					24409.68	$4f5d(^1G^o)6s$	^{o}	7/2	8	ASD
370	4949.77		0.90	B+	0.000	$5d6s^2$	^2D	3/2	4	ASD
					20197.34	$5d6s(^1D)6p$	^{o}	1/2	2	ASD
450	5145.42		0.82	B+	3010.002	$5d^2(^3F)6s$	^4F	5/2	6	ASD
					22439.36	$5d^2(^3F)6p$	$^4D^o$	3/2	4	ASD
580	5177.31		0.82	B+	3494.526	$5d^2(^3F)6s$	^4F	7/2	8	ASD
					22804.25	$5d^2(^3F)6p$	$^4D^o$	5/2	6	ASD
720	5211.86		0.497	B+	4121.572	$5d^2(^3F)6s$	^4F	9/2	10	ASD
					23303.26	$5d^2(^3F)6p$	$^4D^o$	7/2	8	ASD
520	5234.27		0.422	B+	4121.572	$5d^2(^3F)6s$	^4F	9/2	10	ASD
					23221.10	$5d6s(^3D)6p$	^{o}	7/2	8	ASD
500	5455.15		0.407	B+	1053.164	$5d6s^2$	^2D	5/2	6	ASD
					19379.40	$5d^2(^3F)6p$	$^2D^o$	5/2	6	ASD
470	5501.34		0.512	B+	0.000	$5d6s^2$	^2D	3/2	4	ASD
					18172.35	$5d^2(^3F)6p$	$^2D^o$	3/2	4	ASD
450	5791.34		0.393	B+	4121.572	$5d^2(^3F)6s$	^4F	9/2	10	ASD
					21384.00	$5d^2(^3F)6p$	$^4F^o$	9/2	10	ASD
720	6249.93		0.368	B+	4121.572	$5d^2(^3F)6s$	^4F	9/2	10	ASD
					20117.38	$5d^2(^3F)6p$	$^4G^o$	11/2	12	ASD
450	6394.23		0.301	B+	3494.526	$5d^2(^3F)6s$	^4F	7/2	8	ASD
					19129.31	$5d^2(^3F)6p$	$^4G^o$	9/2	10	ASD
Cerium I (Z = 58)										
170	4632.320		0.138	B+	0.000	$4f5d6s^2$	$^1G^o$	4	9	NHBK,ASD
					21581.408	$4f5d(^3H^o)6s6p(^3P^o)$		5	11	NHBK,ASD
210	5009.098		0.62	B+	3196.607	$4f(^2F^o)\ 5d^2(^3F)6s\ (^4F)$	$^5I^o$	4	9	NHBK,ASD
					23154.693			3	7	NHBK,ASD
280*	5159.686				5315.803	$4f(^2F^o)\ 5d^2(^3F)6s\ (^4F)$	$^5I^o$	7	15	NHBK,ASD
					24691.432			6	13	NHBK,ASD
280bl	5161.484				4766.323	$4f5d6s^2$	$^3D^o$	2	5	NHBK,ASD
					24135.490			3	7	NHBK,ASD
260	5223.461		0.65	B+	4762.718	$4f^26s^2$	^3H	4	9	NHBK,ASD
					23901.786	$4f^2(^3H<4>)6s6p(^1P^o<1>)$	$(4,1)^o$	4	9	NHBK,ASD
260	5245.916		0.491	B+	4762.718	$4f^26s^2$	^3H	4	9	NHBK,ASD
					23819.871	$4f^2(^3H<4>)6s6p(^1P^o<1>)$	$(4,1)^o$	5	11	NHBK,ASD
240	5601.280		0.65	B+	6809.128	$4f(^2F^o)\ 5d^2(^3F)6s\ (^4F)$	$^5I^o$	8	17	NHBK,ASD
					24657.245	$4f(^2F^o)5d^2(^3F)\ (^4I^o)6p$	^5I	8	17	NHBK,ASD
240	5669.959		0.347	B+	3210.583	$4f(^2F^o)\ 5d^2(^3F)6s\ (^4F)$	^{o}	5	11	NHBK,ASD
					20842.504	$4f(^2F^o)5d^2(^3F)\ (^4H^o)6p$	^5I	6	13	NHBK,ASD
300	5696.993		0.514	B+	5315.803	$4f(^2F^o)\ 5d^2(^3F)6s\ (^4F)$	$^5I^o$	7	15	NHBK,ASD
					22864.055	$4f(^2F^o)5d^2(^3F)\ (^4I^o)6p$	^5K	8	17	NHBK,ASD
370	5699.226				6809.128	$4f(^2F^o)\ 5d^2(^3F)6s\ (^4F)$	$^5I^o$	8	17	NHBK,ASD
					24350.503	$4f(^2F^o)5d^2(^3F)\ (^4I^o)6p$	^5K	9	19	NHBK,ASD
240	5719.031		0.507	B+	5802.108	$4f(^2F^o)\ 5d^2(^3F)6s\ (^4F)$	$^5H^o$	7	15	NHBK,ASD
					23282.722	$4f(^2F^o)5d^2(^3F)\ (^4H^o)6p$	^5I	8	17	NHBK,ASD
230	5940.857		0.277	B+	3764.008	$4f(^2F^o)\ 5d^2(^3F)6s\ (^4F)$	$^5I^o$	5	11	NHBK,ASD
					20591.937	$4f(^2F^o)5d^2(^3F)\ (^4I^o)6p$	^5K	6	13	NHBK,ASD
Praseodymium I (Z = 59)										
190	3959.44									NHBK
										NHBK
200	4639.55				1376.60	$4f^36s^2$	$^4I^o$	11/2	12	NHBK
					22924.39			9/2	10	NHBK
180	4687.80									NHBK
										NHBK
290	4695.77									NHBK
										NHBK

Intensity	Wavelength/Å	Uncert. wavelength/Å	A_{ki}/(10^8 s^{-1})	Accur.	Energy levels/cm^{-1}	Configs.	Terms	J	g	DB Source
180	4730.67		0.34		1376.60	$4f^36s^2$	$^4I^o$	11/2	12	NHBK
					22509.40			9/2	10	NHBK
250	4736.69		0.53		0.00	$4f^36s^2$	$^4I^o$	9/2	10	NHBK
					21105.88			7/2	8	NHBK
200	4924.60		0.59		1376.60	$4f^36s^2$	$^4I^o$	11/2	12	NHBK
					21677.15			11/2	12	NHBK
320	4939.74				2846.75	$4f^36s^2$	$^4I^o$	13/2	14	NHBK
					23085.08	$4f^36s6p$?	^4I?	13/2	14	NHBK
380	4951.37		0.83		0.00	$4f^36s^2$	$^4I^o$	9/2	10	NHBK
					20190.85	$4f^3(^4I^o)6s6p(^1P^o)$?	^4I?	9/2	10	NHBK
200	5019.76				2846.75	$4f^36s^2$	$^4I^o$	13/2	14	NHBK
					22762.55	$4f^36s6p$?	^4K?	15/2	16	NHBK
200	5026.96		0.40		1376.60	$4f^36s^2$	$^4I^o$	11/2	12	NHBK
					21263.71	$4f^3(^4I^o)6s6p$?	^4K?	13/2	14	NHBK
320	5045.52		0.71		4381.10	$4f^36s^2$	$^4I^o$	15/2	16	NHBK
					24195.13	$4f^36s6p$?	^4K?	17/2	18	NHBK
180	5087.12		0.44		2846.75	$4f^36s^2$	$^4I^o$	13/2	14	NHBK
					22498.81			15/2	16	NHBK
270	5133.44		0.38		0.00	$4f^36s^2$	$^4I^o$	9/2	10	NHBK
					19474.75	$4f^36s6p$?	^4I?	11/2	12	NHBK
Neodymium I (Z = 60)										
300	4621.94		0.56		2366.597	$4f^46s^2$	^5I	6	13	ASD
					23996.513		^{o}	6	13	ASD
510	4634.24		0.84		0.000	$4f^46s^2$	^5I	4	9	ASD
					21572.610	$4f^46s6p$	$^5H^o$	3	7	ASD
340	4641.10									ASD
										ASD
300	4649.67									ASD
										ASD
310	4683.45		0.52		0.000	$4f^46s^2$	^5I	4	9	ASD
					21345.572		^{o}	6	13	ASD
240	4719.02		0.33		0.000	$4f^46s^2$	^5I	4	9	ASD
					21184.881		^{o}	4	9	ASD
350	4883.81		0.88		5048.602	$4f^46s^2$	^5I	8	17	ASD
					25518.700	$4f^46s6p$	$^5K^o$	9	19	ASD
280	4896.93		0.59		1128.056	$4f^46s^2$	^5I	5	11	ASD
					21543.326	$4f^46s6p$	$^5K^o$	6	13	ASD
470	4924.53		0.90		0.000	$4f^46s^2$	^5I	4	9	ASD
					20300.875	$4f^46s6p$	$^5K^o$	5	11	ASD
260	4944.83		0.67		1128.056	$4f^46s^2$	^5I	5	11	ASD
					21345.572		^{o}	6	13	ASD
290	4954.78		0.29		0.000	$4f^46s^2$	^5I	4	9	ASD
					20176.912		^{o}	5	11	ASD
Promethium I (Z = 61)										
800r	4734.27				803.82	$4f^56s^2$	$^6H^o$	7/2	8	ASD
					21920.49			9/2	10	ASD
800r	4759.00				5089.79	$4f^56s^2$	$^6H^o$	15/2	16	ASD
					26096.75			17/2	18	ASD
700r	4762.57				5089.79	$4f^56s^2$	$^6H^o$	15/2	16	ASD
					26080.99			17/2	18	ASD
700r	4773.46				2797.10	$4f^56s^2$	$^6H^o$	11/2	12	ASD
					23740.42			13/2	14	ASD
700r	4798.98				2797.10	$4f^56s^2$	$^6H^o$	11/2	12	ASD
					23629.06			13/2	14	ASD
900r	4801.36				803.82	$4f^56s^2$	$^6H^o$	7/2	8	ASD
					21625.45			9/2	10	ASD
900r	4811.85				7497.99	$4f^56s^2$	$^6F^o$	9/2	10	NHBK,ASD
					28274.21			9/2	10	NHBK,ASD

Intensity	Wavelength/Å	Uncert. wavelength/Å	$A_{ki}/(10^8\ s^{-1})$	Accur.	Energy levels/cm^{-1}	Configs.	Terms	J	g	DB Source
800r	4837.66				1748.78	$4f^56s^2$	$^6H^o$	9/2	10	ASD
					22414.17			11/2	12	ASD
700r	4860.74									ASD
										ASD
700r	4892.52				803.82	$4f^56s^2$	$^6H^o$	7/2	8	ASD
					21237.49			9/2	10	ASD
600r	4932.99				0.00	$4f^56s^2$	$^6H^o$	5/2	6	ASD
					20265.98			5/2	6	ASD
700r	4959.46				0.00	$4f^56s^2$	$^6H^o$	5/2	6	ASD
					20157.85			7/2	8	ASD
500r	4997.10				0.00	$4f^56s^2$	$^6H^o$	5/2	6	ASD
					20006.04			3/2	4	ASD
300r	5058.31				803.82	$4f^56s^2$	$^6H^o$	7/2	8	ASD
					20567.76			5/2	6	ASD
400r	5127.34				2797.10	$4f^56s^2$	$^6H^o$	11/2	12	ASD
					22294.96			9/2	10	ASD
500r	5146.30				3919.03	$4f^56s^2$	$^6H^o$	13/2	14	ASD
					23345.07			11/2	12	ASD
900	6100.21									ASD
										ASD
1000bl	6520.45									ASD
										ASD
900	6598.15									ASD
										ASD
Samarium I (Z = 62)										
930	3745.465		0.27		1489.55	$4f^66s^2$	^7F	3	7	ASD,NHBK
					28180.95	$4f^5(^6F^o)5d6s^2$	$^7F^o$	4	9	ASD,NHBK
	3990.025		0.43		3125.46	$4f^66s^2$	^7F	5	11	ASD,NHBK
					28180.95	$4f^5(^6F^o)5d6s^2$	$^7F^o$	4	9	ASD,NHBK
1600	4296.743		1.00		4020.66	$4f^66s^2$	^7F	6	13	ASD,NHBK
					27287.58	$4f^6(^7F)6s6p(^1P^o)$	$^7G^o$	7	15	ASD,NHBK
880	4336.137				3125.46	$4f^66s^2$	^7F	5	11	ASD,NHBK
					26180.92		^{o}	6	13	ASD,NHBK
440	4362.912		0.49		0.00	$4f^66s^2$	^7F	0	1	ASD,NHBK
					22914.07	$4f^6(^7F)6s6p(^1P^o)$	$^7G^o$	1	3	ASD,NHBK
810	4470.886				2273.09	$4f^66s^2$	^7F	4	9	ASD,NHBK
					24633.75	$4f^5(^6F^o)5d6s^2$	^{o}	4	9	ASD,NHBK
730	4716.097				3125.46	$4f^66s^2$	^7F	5	11	ASD,NHBK
					24323.51	$4f^6(^7F)6s6p(^1P^o)$	$^7D^o$	5	11	ASD,NHBK
770	4728.423		0.49		1489.55	$4f^66s^2$	^7F	3	7	ASD,NHBK
					22632.30	$4f^6(^7F)6s6p(^1P^o)$	$^7D^o$	3	7	ASD,NHBK
730	4760.27		0.13		811.92	$4f^66s^2$	^7F	2	5	ASD,NHBK
					21813.22	$4f^6(^7F)6s6p(^1P^o)$	$^7D^o$	2	5	ASD,NHBK
970	4841.701		0.87		4020.66	$4f^66s^2$	^7F	6	13	ASD,NHBK
					24668.79	$4f^6(^7F)6s6p(^1P^o)$	^{o}	5	11	ASD,NHBK
730	4883.971				3125.46	$4f^66s^2$	^7F	5	11	ASD,NHBK
					23594.84	$4f^6(^7F)6s6p(^1P^o)$	$^7D^o$	4	9	ASD,NHBK
Europium I (Z = 63)										
950	3111.427		0.33	B	0.00	$4f^76s^2$	a $^8S^o$	7/2	8	ASD
					32130.25	$4f^6(^7F)5d6s^2$	^8P	9/2	10	ASD
1000	3212.804		0.341	B+	0.00	$4f^76s^2$	a $^8S^o$	7/2	8	ASD
					31116.38	$4f^6(^7F)5d6s^2$	^8P	7/2	8	ASD
950	3334.313		0.412	B+	0.00	$4f^76s^2$	a $^8S^o$	7/2	8	ASD
					29982.50	$4f^6(^7F)5d6s^2$	^8P	5/2	6	ASD
11000*	4594.03		1.61	B+	0.00	$4f^76s^2$	a $^8S^o$	7/2	8	ASD
					21761.26	$4f^7(^8S^o)6s6p(^1P^o)$	y ^8P	9/2	10	ASD
9800	4627.22		1.56	B+	0.00	$4f^76s^2$	a $^8S^o$	7/2	8	ASD
					21605.17	$4f^7(^8S^o)6s6p(^1P^o)$	y ^8P	7/2	8	ASD

Intensity	Wavelength/Å	Uncert. wavelength/Å	$A_{ki}/(10^8\ s^{-1})$	Accur.	Energy levels/cm^{-1}	Configs.	Terms	J	g	DB Source
8300	4661.88		1.52	B+	0.00	$4f^7 6s^2$	a $^8S^o$	7/2	8	ASD
					21444.58	$4f^7(^8S^o)6s6p(^1P^o)$	y ^8P	5/2	6	ASD
750	5215.10		0.396	B+	13778.68	$4f^7(^8S^o)5d\ (^9D^o)6s$	a $^{10}D^o$	13/2	14	ASD
					32948.41	$4f^7(^8S^o)5d\ (^9D^o)6p$	y ^{10}P	11/2	12	ASD
540*	5357.61		0.233	B+	13457.21	$4f^7(^8S^o)5d\ (^9D^o)6s$	a $^{10}D^o$	11/2	12	ASD
					32117.10	$4f^7(^8S^o)5d\ (^9D^o)6p$	z ^{10}D	13/2	14	ASD
600cw	5830.98		0.51	B+	13778.68	$4f^7(^8S^o)5d\ (^9D^o)6s$	a $^{10}D^o$	13/2	14	ASD
					30923.71	$4f^7(^8S^o)5d\ (^9D^o)6p$	z ^{10}F	15/2	16	ASD
Gadolinium I (Z = 64)										
2000	3684.13		0.91		0.000	$4f^7(^8S^o)5d6s^2$	$^9D^o$	2	5	ASD
					27135.695	$4f^7(^8S^o)5d\ (^9D^o)6s6p(^1P^o)$	^9P	3	7	ASD
2000	3713.57		0.92		215.124	$4f^7(^8S^o)5d6s^2$	$^9D^o$	3	7	ASD
					27135.695	$4f^7(^8S^o)5d\ (^9D^o)6s6p(^1P^o)$	^9P	3	7	ASD
2000	3717.48		0.68		532.977	$4f^7(^8S^o)5d6s^2$	$^9D^o$	4	9	ASD
					27425.245	$4f^7(^8S^o)5d\ (^9D^o)6s6p(^1P^o)$		4	9	ASD
2900	3783.05		0.94		999.121	$4f^7(^8S^o)5d6s^2$	$^9D^o$	5	11	ASD
					27425.245	$4f^7(^8S^o)5d\ (^9D^o)6s6p(^1P^o)$		4	9	ASD
2600	4053.64		0.62		999.121	$4f^7(^8S^o)5d6s^2$	$^9D^o$	5	11	ASD
					25661.340	$4f^7(^8S^o)5d\ (^7D^o)6s6p(^3P^o)$		6	13	ASD
2600	4058.22		0.57		215.124	$4f^7(^8S^o)5d6s^2$	$^9D^o$	3	7	ASD
					24849.514	$4f^7(^8S^o)5d\ (^7D^o)6s6p(^3P^o)$		4	9	ASD
2800	4078.70		0.59		532.977	$4f^7(^8S^o)5d6s^2$	$^9D^o$	4	9	ASD
					25043.649	$4f^7(^8S^o)5d\ (^9D^o)6s6p(^1P^o)$		5	11	ASD
2400	4175.54		0.42		1719.087	$4f^7(^8S^o)5d6s^2$	$^9D^o$	6	13	ASD
					25661.340	$4f^7(^8S^o)5d\ (^7D^o)6s6p(^3P^o)$		6	13	ASD
2200	4190.78		0.30		999.121	$4f^7(^8S^o)5d6s^2$	$^9D^o$	5	11	ASD
					24854.297	$4f^8(^7F)5d\ (^8G)6s$	^9G	6	13	ASD
4800	4225.85		0.89		1719.087	$4f^7(^8S^o)5d6s^2$	$^9D^o$	6	13	ASD
					25376.313	$4f^7(^8S^o)5d\ (^9D^o)6s6p(^1P^o)$		7	15	ASD
1800	4313.84		0.43		215.124	$4f^7(^8S^o)5d6s^2$	$^9D^o$	3	7	ASD
					23389.782	$4f^7(^8S^o)5d\ (^9D^o)6s6p(^1P^o)$		3	7	ASD
1900	4327.12		0.52		0.000	$4f^7(^8S^o)5d6s^2$	$^9D^o$	2	5	ASD
					23103.660	$4f^7(^8S^o)5d^2(^3F)\ (^{10}F^o)6p$		1	3	ASD
2200	4346.46				999.121	$4f^7(^8S^o)5d6s^2$	$^9D^o$	5	11	ASD
					23999.912	$4f^7(^8S^o)5d\ (^9D^o)6s6p(^1P^o)$		5	11	ASD
1400	4401.86		0.38		1719.087	$4f^7(^8S^o)5d6s^2$	$^9D^o$	6	13	ASD
					24430.425	$4f^7(^8S^o)5d\ (^9D^o)6s6p(^1P^o)$		6	13	ASD
1400	4422.41		0.28		215.124	$4f^7(^8S^o)5d6s^2$	$^9D^o$	3	7	ASD
					22820.895	$4f^7(^8S^o)5d\ (^7D^o)6s6p(^3P^o)$	^9D	4	9	ASD
1100	4430.63		0.30		0.000	$4f^7(^8S^o)5d6s^2$	$^9D^o$	2	5	ASD
					22563.824	$4f^7(^8S^o)5d\ (^7D^o)6s6p(^3P^o)$	^9D	3	7	ASD
1100	4519.66		0.38		215.124	$4f^7(^8S^o)5d6s^2$	$^9D^o$	3	7	ASD
					22334.508	$4f^7(^8S^o)5d\ (^9D^o)6s6p(^1P^o)$		2	5	ASD
910	5103.45				7947.294	$4f^7(^8S^o)5d^2(^3F)\ (^{10}F^o)6s$	$^{11}F^o$	7	15	ASD
					27536.397	$4f^7(^8S^o)5d^2(^3F)\ (^{10}F^o)6p$	^{11}G	8	17	ASD
860	5155.84				7480.348	$4f^7(^8S^o)5d^2(^3F)\ (^{10}F^o)6s$	$^{11}F^o$	6	13	ASD
					26870.393	$4f^7(^8S^o)5d^2(^3F)\ (^{10}F^o)6p$	^{11}G	7	15	ASD
Terbium I (Z = 65)										
1500	3830.261				285.500	$4f^8(^7F)5d6s^2$	^8G	13/2	14	ASD,NHBK
					26385.99		^{o}	13/2	14	ASD,NHBK
1600	3901.325				462.080	$4f^8(^7F)5d6s^2$	^8G	15/2	16	ASD,NHBK
					26087.13	$4f^85d6s6p$?	^{o}	17/2	18	ASD,NHBK
870	4032.284				2310.090	$4f^8(^7F)5d6s^2$	^8D	11/2	12	ASD,NHBK
					27102.94		^{o}	9/2	10	ASD,NHBK
1300	4061.558				462.080	$4f^8(^7F)5d6s^2$	^8G	15/2	16	ASD,NHBK
					25076.26		^{o}	15/2	16	ASD,NHBK
2200	4318.847				0.000	$4f^96s^2$	$^6H^o$	15/2	16	ASD,NHBK
					23147.92	$4f^9(^6H^o<15/2>)6s6p(^1P^o<1>)$	(15/2,1)	15/2	16	ASD,NHBK

Intensity	Wavelength/Å	Uncert. wavelength/Å	$A_{ki}/(10^8\,\text{s}^{-1})$	Accur.	Energy levels/cm^{-1}	Configs.	Terms	J	g	DB Source
3000	4326.472				0.000	$4f^96s^2$	$^6H°$	15/2	16	ASD,NHBK
					23107.25	$4f^9(^6H°<15/2>)6s6p(^1P°<1>)$	(15/2,1)	17/2	18	ASD,NHBK
870	4336.455				2771.675	$4f^96s^2$	$^6H°$	13/2	14	ASD,NHBK
					25825.53	$4f^9(^6H°<13/2>)6s6p(^1P°<1>)$	(13/2,1)	15/2	16	ASD,NHBK
1700	4338.435				0.000	$4f^96s^2$	$^6H°$	15/2	16	ASD,NHBK
					23043.43	$4f^9(^6H°<15/2>)6s6p(^1P°<1>)$	(15/2,1)	13/2	14	ASD,NHBK
870	4356.837				2771.675	$4f^96s^2$	$^6H°$	13/2	14	ASD,NHBK
					25717.68	$4f^9(^6H°<13/2>)6s6p(^1P°<1>)$	(13/2,1)	13/2	14	ASD,NHBK
160	5354.88				462.080	$4f^8(^7F)5d6s^2$	^8G	15/2	16	ASD
					19131.45	$4f^85d6s6p?$	°	17/2	18	ASD

Dysprosium I ($Z = 66$)

Intensity	Wavelength/Å	Uncert. wavelength/Å	$A_{ki}/(10^8\,\text{s}^{-1})$	Accur.	Energy levels/cm^{-1}	Configs.	Terms	J	g	DB Source
12000	4045.970		1.92	B+	0.000	$4f^{10}6s^2$	^5I	8	17	ASD,NHBK
					24708.971	$4f^{10}(^5I<8>)6s6p(^1P°<1>)$	(8,1)°	7	15	ASD,NHBK
5700	4167.974		1.92	B+	4134.222	$4f^{10}6s^2$	^5I	7	15	ASD,NHBK
					28119.931	$4f^{10}(^5I<7>)6s6p(^1P°<1>)$	(7,1)°	6	13	ASD,NHBK
12000	4186.821		1.26	B+	0.000	$4f^{10}6s^2$	^5I	8	17	ASD,NHBK
					23877.739	$4f^{10}(^5I<8>)6s6p(^1P°<1>)$	(8,1)°	8	17	ASD,NHBK
2200	4191.640		0.71	B+	4134.222	$4f^{10}6s^2$	^5I	7	15	ASD,NHBK
					27984.513	$4f^9(^6H°)5d^2(^3F)(^8L°)6s$	°	7	15	ASD,NHBK
6800	4194.846		0.88	B+	0.000	$4f^{10}6s^2$	^5I	8	17	ASD,NHBK
					23832.060	$4f^9(^6H°)5d^2(^3F)(^8K°)6s$	°	8	17	ASD,NHBK
16000	4211.714		2.08	B+	0.000	$4f^{10}6s^2$	^5I	8	17	ASD,NHBK
					23736.610	$4f^{10}(^5I<8>)6s6p(^1P°<1>)$	(8,1)°	9	19	ASD,NHBK
3700	4215.159		0.81	B+	4134.222	$4f^{10}6s^2$	^5I	7	15	ASD,NHBK
					27851.435	$4f^9(^6H°)5d^2(^3P)(^8I°)6s$	°	8	17	ASD,NHBK
4400	4218.092		1.20	B+	4134.222	$4f^{10}6s^2$	^5I	7	15	ASD,NHBK
					27834.934	$4f^{10}(^5I<7>)6s6p(^1P°<1>)$	(7,1)°	7	15	ASD,NHBK
4400	4221.110		1.28	B+	4134.222	$4f^{10}6s^2$	^5I	7	15	ASD,NHBK
					27818.000	$4f^{10}(^5I<7>)6s6p(^1P°<1>)$	(7,1)°	8	17	ASD,NHBK
2700	4225.154		1.95	B+	7050.603	$4f^{10}6s^2$	^5I	6	13	ASD,NHBK
					30711.717	$4f^{10}(^5I<6>)6s6p(^1P°<1>)$	(6,1)°	7	15	ASD,NHBK
2100	4589.364		0.137	B+	0.000	$4f^{10}6s^2$	^5I	8	17	ASD,NHBK
					21783.407	$4f^9(^6H°)5d6s^2$	$^5K°$	7	15	ASD,NHBK

Holmium I ($Z = 67$)

Intensity	Wavelength/Å	Uncert. wavelength/Å	$A_{ki}/(10^8\,\text{s}^{-1})$	Accur.	Energy levels/cm^{-1}	Configs.	Terms	J	g	DB Source
2700	4040.81		0.373		0.00	$4f^{11}6s^2$	$^4I°$	15/2	16	ASD,NHBK
					24740.52	$4f^{10}5d6s^2$		13/2	14	ASD,NHBK
8100	4053.93		1.62		0.00	$4f^{11}6s^2$	$^4I°$	15/2	16	ASD,NHBK
					24660.80	$4f^{11}(^4I°<15/2>)6s6p(^1P°<1>)$	(15/2,1)	15/2	16	ASD,NHBK
8900	4103.84		1.55		0.00	$4f^{11}6s^2$	$^4I°$	15/2	16	ASD,NHBK
					24360.81	$4f^{11}(^4I°<15/2>)6s6p(^1P°<1>)$	(15/2,1)	17/2	18	ASD,NHBK
2900	4108.62		2.00		5419.70	$4f^{11}6s^2$	$^4I°$	13/2	14	ASD,NHBK
					29751.91	$4f^{11}(^4I°<13/2>)6s6p(^1P°<1>)$	(13/2,1)	13/2	14	ASD,NHBK
4300	4127.16		2.1		5419.70	$4f^{11}6s^2$	$^4I°$	13/2	14	ASD,NHBK
					29642.60	$4f^{11}(^4I°<13/2>)6s6p(^1P°<1>)$	(13/2,1)	15/2	16	ASD,NHBK
8100	4163.03		0.897		0.00	$4f^{11}6s^2$	$^4I°$	15/2	16	ASD,NHBK
					24014.22	$4f^{11}(^4I°<15/2>)6s6p(^1P°<1>)$	(15/²,1)	13/2	14	ASD,NHBK
2500	4173.20		0.258		0.00	$4f^{11}6s^2$	$^4I°$	15/2	16	ASD,NHBK
					23955.69	$4f^{11}(^4I°<13/2>)6s6p(^3P°<2>)$	(13/²,2)	13/2	14	ASD,NHBK
2000	4227.13		1.1		5419.70	$4f^{11}6s^2$	$^4I°$	13/2	14	ASD,NHBK
					29069.78	$4f^{11}(^4I°<13/2>)6s6p(^1P°<1>)$	(13/²,1)	11/2	12	ASD,NHBK
1300c,w	4254.38		0.115		0.00	$4f^{11}6s^2$	$^4I°$	15/2	16	ASD,NHBK
					23498.57	$4f^{11}(^4I°<13/2>)6s6p(^3P°<2>)$	(13/²,2)	17/2	18	ASD,NHBK
1300	4350.73		0.089		0.00	$4f^{11}6s^2$	$^4I°$	15/2	16	ASD,NHBK
					22978.19	$4f^{10}(^5I<5>)5d<5/2>6s^2$	(5,5/²)	13/2	14	ASD,NHBK
290	4939.01		0.026		0.00	$4f^{11}6s^2$	$^4I°$	15/2	16	ASD,NHBK
					20241.31	$4f^{10}(^5I<6>)5d<5/2>6s^2$	(6,5/²)	13/2	14	ASD,NHBK
230c	5982.85		0.0093		0.00	$4f^{11}6s^2$	$^4I°$	15/2	16	ASD,NHBK
					16709.82	$4f^{11}(^4I°<15/2>)6s6p(^3P°<1>)$	(15/²,1)	17/2	18	ASD,NHBK

Intensity	Wavelength/Å	Uncert. wavelength/Å	$A_{ki}/(10^8\ s^{-1})$	Accur.	Energy levels/cm⁻¹	Configs.	Terms	*J*	*g*	DB Source
260	6604.91				0.00	$4f^{11}6s^2$	$^4I^o$	15/2	16	ASD,NHBK
					15136.06	$4f^{10}(^5I<7>)5d<3/2>6s^2$	$(^7,3/^2)$	15/2	16	ASD,NHBK
Erbium I (Z = 68)										
7500	3862.851		1.16		0.000	$4f^{12}6s^2$	^3H	6	13	ASD,NHBK
					25880.274	$4f^{12}(^3H)6s6p$	^{o}	6	13	ASD,NHBK
4200	3892.684		0.728		0.000	$4f^{12}6s^2$	^3H	6	13	ASD,NHBK
					25681.933	$4f^{11}(^4F^o)5d6s^2$	^{o}	5	11	ASD,NHBK
3200	3937.014		0.292		0.000	$4f^{12}6s^2$	^3H	6	13	ASD,NHBK
					25392.779	$4f^{11}(^4F^o)5d6s^2$	^{o}	6	13	ASD,NHBK
3200	3944.420				5035.193	$4f^{12}6s^2$	3F	4	9	ASD,NHBK
					30380.282	$4f^{11}(^4I^o)5d^2(^3F)\ (^4F^o)6s$	^{o}	5	11	ASD,NHBK
2700	3973.036		0.479		0.000	$4f^{12}6s^2$	^3H	6	13	ASD,NHBK
					25162.553	$4f^{11}(^4I^o)5d6s^2$	^{o}	5	11	ASD,NHBK
3200	3973.575		0.366		0.000	$4f^{12}6s^2$	^3H	6	13	ASD,NHBK
					25159.143		^{o}	7	15	ASD,NHBK
14000	4007.965		1.73		0.000	$4f^{12}6s^2$	^3H	6	13	ASD,NHBK
					24943.272	$4f^{12}(^3H)6s6p$	^{o}	7	15	ASD,NHBK
3000	4020.512				6958.329	$4f^{12}6s^2$	^3H	5	11	ASD,NHBK
					31823.748	$4f^{12}(^3H)6s6p$	^{o}	6	13	ASD,NHBK
3500	4087.632		0.303		0.000	$4f^{12}6s^2$	^3H	6	13	ASD,NHBK
					24457.139	$4f^{11}(^4I^o)5d6s^2$	^{o}	6	13	ASD,NHBK
6900	4151.108		0.960		0.000	$4f^{12}6s^2$	^3H	6	13	ASD,NHBK
					24083.166	$4f^{12}(^3H)6s6p$	^{o}	5	11	ASD,NHBK
1000	4606.606		0.0754		0.000	$4f^{12}6s^2$	^3H	6	13	ASD,NHBK
					21701.885	$4f^{11}(^4I^o<11/2>)5d<5/2>6s^2$	$(11/2,5/2)^o$	6	13	ASD,NHBK
430	5826.786		0.0104		0.000	$4f^{12}6s^2$	^3H	6	13	ASD,NHBK
					17157.307	$4f^{12}(^3H<6>)6s6p(^3P^o<1>)$	$(6,1)^o$	7	15	ASD,NHBK
360	6221.019		0.0084		0.000	$4f^{12}6s^2$	^3H	6	13	ASD,NHBK
					16070.095	$4f^{11}(^4I^o<13/2>)5d<3/2>6s^2$	$(13/2,3/2)^o$	6	13	ASD,NHBK
Thulium I (Z = 69)										
77000	3717.914	0.005	1.44	B+	0.000	$4f^{13}(^2F^o)6s^2$	$^2F^o$	7/2	8	ASD
					26889.125	$4f^{12}(^3H<5>)5d<5/2>6s^2$	(5,5/2)	9/2	10	ASD
50000	3744.064	0.005	0.99	B+	0.000	$4f^{13}(^2F^o)6s^2$	$^2F^o$	7/2	8	ASD
					26701.325	$4f^{13}(^2F^o<7/2>)6s6p(^1P^o<1>)$		7/2	8	ASD
68000	3883.132	0.005	1.06	B+	0.000	$4f^{13}(^2F^o)6s^2$	$^2F^o$	7/2	8	ASD
					25745.117	$4f^{13}(^2F^o<7/2>)6s6p(^1P^o<1>)$		5/2	6	ASD
54000	3887.348	0.005	0.372	B+	0.000	$4f^{13}(^2F^o)6s^2$	$^2F^o$	7/2	8	ASD
					25717.197	$4f^{12}(^3H<5>)5d<5/2>6s^2$	(5,5/2)	7/2	8	ASD
35000	3916.477	0.005	1.98	B+	8771.243	$4f^{13}(^2F^o)6s^2$	$^2F^o$	5/2	6	ASD
					34297.17	$4f^{13}(^2F^o<5/2>)6s6p(^1P^o<1>)$	(5/2,1)	7/2	8	ASD
100000	4094.187	0.005	0.98	B+	0.000	$4f^{13}(^2F^o)6s^2$	$^2F^o$	7/2	8	ASD
					24418.018	$4f^{13}(^2F^o<7/2>)6s6p(^1P^o<1>)$	(7/2,1)	5/2	6	ASD
95000	4105.841	0.005	0.64	B+	0.000	$4f^{13}(^2F^o)6s^2$	$^2F^o$	7/2	8	ASD
					24348.692	$4f^{12}(^3H<5>)5d<3/2>6s^2$	(5,3/2)	9/2	10	ASD
88000	4187.615	0.005	0.64	B+	0.000	$4f^{13}(^2F^o)6s^2$	$^2F^o$	7/2	8	ASD
					23873.207	$4f^{12}(^3F<4>)5d<5/2>6s^2$		7/2	8	ASD
60000	4203.727	0.005	0.243	B+	0.000	$4f^{13}(^2F^o)6s^2$	$^2F^o$	7/2	8	ASD
					23781.698	$4f^{12}(^3F<4>)5d<5/2>6s^2$	(4,5/2)	9/2	10	ASD
27000	4359.928	0.005	0.12	B+	0.000	$4f^{13}(^2F^o)6s^2$	$^2F^o$	7/2	8	ASD
					22929.717	$4f^{12}(^3H<5>)5d<5/2>6s^2$		5/2	6	ASD
14000	4386.434	0.005	0.0371	B+	0.000	$4f^{13}(^2F^o)6s^2$	$^2F^o$	7/2	8	ASD
					22791.176	$4f^{12}(^3H<5>)5d<3/2>6s^2$	(5,3/2)	7/2	8	ASD
6800	4733.335	0.005	0.020	B+	0.000	$4f^{13}(^2F^o)6s^2$	$^2F^o$	7/2	8	ASD
					21120.836	$4f^{12}(^3F<4>)5d<3/2>6s^2$	(4,3/2)	7/2	8	ASD
6500	5307.116	0.005	0.0217	B+	0.000	$4f^{13}(^2F^o)6s^2$	$^2F^o$	7/2	8	ASD
					18837.385	$4f^{12}(^3H<6>)5d<5/2>6s^2$	(6,5/2)	9/2	10	ASD
5200	5675.835	0.005	0.013	B+	0.000	$4f^{13}(^2F^o)6s^2$	$^2F^o$	7/2	8	ASD
					17613.659	$4f^{13}(^2F^o<7/2>)6s6p(^3P^o<1>)$	(7/2,1)	9/2	10	ASD

Intensity	Wavelength/Å	Uncert. wavelength/Å	$A_{ki}/(10^8\ s^{-1})$	Accur.	Energy levels/cm^{-1}	Configs.	Terms	J	g	DB Source
Ytterbium I (Z = 70)										
460	2464.50		1.00		0.000	$4f^{14}6s^2$	^1S	0	1	ASD,NHBK
					40563.97	$4f^{14}6s7p$	$^1P^{\circ}$	1	3	ASD,NHBK
390	2671.958		0.143		0.000	$4f^{14}6s^2$	^1S	0	1	ASD,NHBK
					37414.59	$4f^{13}(^2F^{\circ})5d6s^2$	$^{\circ}$	1	3	ASD,NHBK
2400	3464.37		0.683		0.000	$4f^{14}6s^2$	^1S	0	1	ASD,NHBK
					28857.014	$4f^{13}(^2F^{\circ}<7/2>)5d<5/2>6s^2$	$(7/2,5/2)^{\circ}$	1	3	ASD,NHBK
32000	3987.99		1.92		0.000	$4f^{14}6s^2$	^1S	0	1	ASD,NHBK
					25068.222	$4f^{14}6s6p$	$^1P^{\circ}$	1	3	ASD,NHBK
930	3990.885				19710.388	$4f^{14}6s6p$	$^3P^{\circ}$	2	5	ASD,NHBK
					44760.37	$4f^{14}6p^2$	^3P	2	5	ASD,NHBK
640	4576.209				17992.007	$4f^{14}6s6p$	$^3P^{\circ}$	1	3	ASD,NHBK
					39838.04	$4f^{14}6s6d$	^3D	2	5	ASD,NHBK
710	4935.500				19710.388	$4f^{14}6s6p$	$^3P^{\circ}$	2	5	ASD,NHBK
					39966.09	$4f^{14}6s6d$	^3D	3	7	ASD,NHBK
2400	5556.466		0.0115		0.000	$4f^{14}6s^2$	^1S	0	1	ASD,NHBK
					17992.007	$4f^{14}6s6p$	$^3P^{\circ}$	1	3	ASD,NHBK
690	6799.60				17992.007	$4f^{14}6s6p$	$^3P^{\circ}$	1	3	ASD,NHBK
					32694.692	$4f^{14}6s7s$	^3S	1	3	ASD,NHBK
750	7699.49				19710.388	$4f^{14}6s6p$	$^3P^{\circ}$	2	5	ASD,NHBK
					32694.692	$4f^{14}6s7s$	^3S	1	3	ASD,NHBK
5100h	3081.47		2.62	B+	1993.92	$5d6s^2$	^2D	5/2	6	ASD
					34436.49	$5d6s(^3D)6p$	$^2P^{\circ}$	3/2	4	ASD
3000	3118.43		2.03	B+	0.00	$5d6s^2$	^2D	3/2	4	ASD
					32058.10	$5d6s(^3D)6p$	$^2P^{\circ}$	1/2	2	ASD
3800	3278.97		1.00	B+	0.00	$5d6s^2$	^2D	3/2	4	ASD
					30488.62	$6s^27p$	$^2P^{\circ}$	3/2	4	ASD
7600	3281.74		3.05	B+	1993.92	$5d6s^2$	^2D	5/2	6	ASD
					32456.70	$5d6s(^1D)6p$	$^2D^{\circ}$	5/2	6	ASD
6200	3312.11		1.85	B+	0.00	$5d6s^2$	^2D	3/2	4	ASD
					30183.55	$5d6s(^3D)6p$	$^2F^{\circ}$	5/2	6	ASD
7600	3359.56		2.44	B+	1993.92	$5d6s^2$	^2D	5/2	6	ASD
					31751.17	$5d6s(^3D)6p$	$^2F^{\circ}$	7/2	8	ASD
6200	3376.50		1.68	B+	0.00	$5d6s^2$	^2D	3/2	4	ASD
					29608.01	$5d6s(^1D)6p$	$^2D^{\circ}$	3/2	4	ASD
4800	3567.84		0.69	B+	0.00	$5d6s^2$	^2D	3/2	4	ASD
					28020.11	$5d6s(^1D)6p$	$^2F^{\circ}$	5/2	6	ASD
2600	3647.77		0.92	B+	4136.13	$6s^26p$	$^2P^{\circ}$	1/2	2	ASD
					31542.24	$6s^26d$	^2D	3/2	4	ASD
2700	3841.18		0.265	B+	1993.92	$5d6s^2$	^2D	5/2	6	ASD
					28020.11	$5d6s(^1D)6p$	$^2F^{\circ}$	5/2	6	ASD
3100	4124.73		0.89	C+	7476.39	$6s^26p$	$^2P^{\circ}$	3/2	4	ASD
					31713.60	$6s^26d$	^2D	5/2	6	ASD
3300	4518.57		0.226	B+	0.00	$5d6s^2$	^2D	3/2	4	ASD
					22124.76	$5d6s(^3D)6p$	$^2D^{\circ}$	3/2	4	ASD
2700	5135.09		0.091	B+	1993.92	$5d6s^2$	^2D	5/2	6	ASD
					21462.38	$5d6s(^3D)6p$	$^2D^{\circ}$	5/2	6	ASD
1400	6004.52		0.49	B+	7476.39	$6s^26p$	$^2P^{\circ}$	3/2	4	ASD
					24125.99	$6s^27s$	^2S	1/2	2	ASD

TABLE 8. Persistent Lines of Neutral Atomic Elements — Francium I (Z = 87) to Einsteinium I (Z = 99)

Intensity	Wavelength/Å	Uncert. wavelength/Å	$A_{ki}/(10^8\ s^{-1})$	Accur.	Energy levels/cm^{-1}	Configs.	Terms	J	g	DB Source
Francium I (Z = 87)										
	4225.655	0.009	0.0282		0.000	$7s$	^2S	1/2	2	ASD
					23658.306	$8p$	$^2P^{\circ}$	3/2	4	
	4325.361	0.009	0.0264		0.000	$7s$	^2S	1/2	2	ASD
					23112.960	$8p$	$^2P^{\circ}$	1/2	2	

Intensity	Wavelength/Å	Uncert. wavelength/Å	$A_{ki}/(10^8\ s^{-1})$	Accur.	Energy levels/cm^{-1}	Configs.	Terms	*J*	*g*	DB Source
	7179.8660	0.0010	0.478		0.000	$7s$	^2S	1/2	2	ASD
					13923.998	$7p$	$^2P^o$	3/2	4	
	8169.4180	0.002	0.322		0.000	$7s$	^2S	1/2	2	ASD
					12237.409	$7p$	$^2P^o$	1/2	2	
Radium I (Z = 88)										
100	4825.92806	0.00012	1.770	AA	0.0000	$7s^2$	^1S	0	1	ASD
					20715.6142	$7s7p$	$^1P^o$	1	3	
50	5660.810	0.020	0.43	C+	14707.29	$7s6d$	^3D	3	7	ASD
					32367.70	$6d7p$	$^3F^o$	4	9	
30	6200.300	0.020	0.25	C+	13993.94	$7s6d$	^3D	2	5	ASD
					30117.72	$6d7p$	$^3F^o$	3	7	
50	7141.2167	0.0005	0.0237	B+	0.0000	$7s^2$	^1S	0	1	ASD
					13999.3569	$7s7p$	$^3P^o$	1	3	
Actinium I (Z = 89)										
400	3885.56				0.00	$6d7s^2$	^2D	3/2	4	NHBK
					25729.03	$6d7s(^3D)7p$	$^2P^o$	1/2	2	
1000	4179.98				0.00	$6d7s^2$	^2D	3/2	4	NHBK
					23916.84	$6d7s(^3D)7p$	$^2F^o$	5/2	6	
500	4183.12				0.00	$6d7s^2$	^2D	3/2	4	NHBK
					23898.86	$6d7s(^3D)7p$	$^4P^o$	5/2	6	
400	4194.40				2231.43	$6d7s^2$	^2D	5/2	6	NHBK
					26066.04	$6d7s(^1D)7p$	$^2D^o$	3/2	4	
300c	4384.53				0.00	$6d7s^2$	^2D	3/2	4	NHBK
					22801.10	$6d7s(^3D)7p$	$^4P^o$	3/2	4	
400	4396.71				2231.43	$6d7s^2$	^2D	5/2	6	NHBK
					24969.30	$6d7s(^3D)7p$	$^2F^o$	7/2	8	
400	4462.73				0.00	$6d7s^2$	^2D	3/2	4	NHBK
					22401.52	$6d7s(^3D)7p$	$^4P^o$	1/2	2	
300	4613.93				2231.43	$6d7s^2$	^2D	5/2	6	NHBK
					23898.86	$6d7s(^3D)7p$	$^4P^o$	5/2	6	
500	4716.58				0.00	$6d7s^2$	^2D	3/2	4	NHBK
					21195.87	$6d7s(^3D)7p$	$^4D^o$	5/2	6	
300c	5258.24				0.00	$6d7s^2$	^2D	3/2	4	NHBK
					19012.46	$6d7s(^3D)7p$	$^4D^o$	3/2	4	
300c	6359.86				2231.43	$6d7s^2$	^2D	5/2	6	NHBK
					17950.71	$6d7s(^3D)7p$	$^2D^o$	5/2	6	
200c	6691.27				0.00	$6d7s^2$	^2D	3/2	4	NHBK
					14940.72	$6d7s(^3D)7p$	$^4F^o$	5/2	6	
300c	7290.40				0.00	$6d7s^2$	^2D	3/2	4	NHBK
					13712.90	$6d7s(^3D)7p$	$^4F^o$	3/2	4	
Thorium I (Z = 90)										
14000	3304.23820	0.00003			0.000000	$6d^27s^2$	^3F	2	5	ASD
					30255.45106		^{o}	3	7	
9400	3348.76829	0.00007			0.000000	$6d^27s^2$	^3F	2	5	ASD
					29853.14415		^{o}	2	5	
11000	3421.20981	0.00007			4961.658832	$6d^27s^2$	^3F	4	9	ASD
					34182.7049		^{o}	5	11	
5600	3706.76712	0.00010			8111.004545	$6d^27s^2$	^1G	4	9	ASD
					35081.0176		^{o}	5	11	
13000	3719.43454	0.00004			0.000000	$6d^27s^2$	^3F	2	5	ASD
					26878.16179	$6d^27s7p$	$^3G^o$	3	7	
24000	3803.07488	0.00006			0.000000	$6d^27s^2$	^3F	2	5	ASD
					26287.04940	$6d^27s7p$	$^3D^o$	1	3	
14000	3828.38451	0.00004			0.000000	$6d^27s^2$	^3F	2	5	ASD
					26113.26894		^{o}	2	5	
7000	3895.41902	0.00011			7795.275291	$5f6d7s^2$	$^3H^o$	4	9	ASD
					33459.18355	$5f6d7s7p$	^3H	4	9	

Intensity	Wavelength/Å	Uncert. wavelength/Å	$A_{ki}/(10^8\ s^{-1})$	Accur.	Energy levels/cm^{-1}	Configs.	Terms	*J*	*g*	DB Source
6300	3967.39204	0.00003			4961.658832	$6d^{2}7s^{2}$	^{3}F	4	9	ASD
					30160.00425		°	4	9	
9500	4012.49498	0.00004			2869.259165	$6d^{2}7s^{2}$	^{3}F	3	7	ASD
					27784.36557		°	2	5	
18000	4112.754311	0.000014			0.000000	$6d^{2}7s^{2}$	^{3}F	2	5	ASD
					24307.74711		°	2	5	
3400	4115.75878	0.00005			5563.141519	$6d3(^{4}F)7s$	^{5}F	1	3	ASD
					29853.14415		°	2	5	
15000	5067.973811	0.000010			7795.275291	$5f^{6}d7s^{2}$	^{3}H^{o}	4	9	ASD
					27521.52753			4	9	
15000	5231.159556	0.000011			2558.056752	$6d^{2}7s^{2}$	^{3}P	0	1	ASD
					21668.957072	$6d^{2}7s7p$	^{5}P^{o}	1	3	
53000	5760.550581	0.000010			0.000000	$6d^{2}7s^{2}$	^{3}F	2	5	ASD
					17354.638583	$5f^{6}d7s^{2}$	^{3}D^{o}	1	3	
18000	6169.821970	0.000011			4961.658832	$6d^{2}7s^{2}$	^{3}F	4	9	ASD
					21165.098037	$5f^{6}d7s^{2}$	^{3}D^{o}	3	7	
44000	6457.282373	0.000013			7795.275291	$5f^{6}d7s^{2}$	^{3}H^{o}	4	9	ASD
					23277.38774	$5f^{6}d7s7p$	^{5}I	5	11	
31000	6989.655198	0.000015			7795.275291	$5f^{6}d7s^{2}$	^{3}H^{o}	4	9	ASD
					22098.18793	$5f^{6}d7s7p$	^{5}I	4	9	
19000	8967.639929	0.000016			8800.249723	$6d3(^{4}F)7s$	^{5}F	4	9	ASD
					19948.395175	$6d^{2}7s7p$	^{5}F^{o}	4	9	
Protactinium I (Z = 91)										
10000	3636.52				825.415	$5f^{2}(^{3}H)6d7s^{2}$	^{4}I	9/2	10	NHBK
					28316.365		°	9/2	10	
10000	3982.23				0.000	$5f^{2}(^{3}H)6d7s^{2}$	^{4}K	11/2	12	NHBK
					25104.455		°	11/2	12	
10000	6945.72				0.000	$5f^{2}(^{3}H)6d7s^{2}$	^{4}K	11/2	12	NHBK
					14393.410	$5f^{2}6d7s7p$	^{6}L^{o}	11/2	12	
10000s	7114.89				825.415	$5f^{2}(^{3}H)6d7s^{2}$	^{4}I	9/2	10	NHBK
					14876.545		°	9/2	10	
10000l	7368.25				825.415	$5f^{2}(^{3}H)6d7s^{2}$	^{4}I	9/2	10	NHBK
					14393.410	$5f^{2}6d7s7p$	^{6}L^{o}	11/2	12	
10000w	7493.15				3323.860	$5f^{2}(^{3}H)6d7s^{2}$		9/2	10	NHBK
					16665.690		°	7/2	8	
10000w	7608.20				3323.860	$5f^{2}(^{3}H)6d7s^{2}$		9/2	10	NHBK
					16463.970		°	9/2	10	
10000	7626.79				3711.625	$5f^{2}(^{3}H)6d7s^{2}$	^{4}K	13/2	14	NHBK
					16819.670		°	11/2	12	
10000s	7635.18				7512.695	$5f^{2}(^{3}H)6d7s^{2}$	^{4}K	15/2	16	NHBK
					20606.365		°	13/2	14	
10000	7669.34				4121.450	$5f^{2}(^{3}H)6d7s^{2}$	^{4}I	11/2	12	NHBK
					17156.785	$5f^{2}6d7s7p$	^{6}K^{o}	11/2	12	
10000w	7749.19				1618.325	$5f^{2}6d7s^{2}$	a ^{4}G	5/2	6	NHBK
					14519.355		°	5/2	6	
10000	8039.34				3711.625	$5f^{2}(^{3}H)6d7s^{2}$	^{4}K	13/2	14	NHBK
					16147.025	$5f^{3}7s^{2}$	^{4}I^{o}	11/2	12	
10000w	8099.84				4121.450	$5f^{2}(^{3}H)6d7s^{2}$	^{4}I	11/2	12	NHBK
					16463.970		°	9/2	10	
10000	8199.04				825.415	$5f^{2}(^{3}H)6d7s^{2}$	^{4}I	9/2	10	NHBK
					13018.610	$5f^{3}7s^{2}$	^{4}I^{o}	9/2	10	
10000	8271.87				825.415	$5f^{2}(^{3}H)6d7s^{2}$	^{4}I	9/2	10	NHBK
					12911.260		°	7/2	8	
10000w	8532.66				2966.530	$5f^{2}6d7s^{2}$	a ^{4}H	7/2	8	NHBK
					14683.005		°	5/2	6	
10000s	8572.96				7000.290	$5f^{2}(^{3}H)6d^{2}(^{3}F)(^{5}L)7s$	^{6}L	11/2	12	NHBK
					18661.700		°	9/2	10	
10000	8735.27				0.000	$5f^{2}(^{3}H)6d7s^{2}$	^{4}K	11/2	12	NHBK
					11444.705	$5f^{2}7s^{2}7p$<1/2>	(4,1/2)^{o}	9/2	10	

Intensity	Wavelength/Å	Uncert. wavelength/Å	$A_{ki}/(10^8\ s^{-1})$	Accur.	Energy levels/cm^{-1}	Configs.	Terms	J	g	DB Source
10000	11791.73				2966.530	$5f^2 6d7s^2$	a ^4H	7/2	8	NHBK
					11444.705	$5f^2 7s^2 7p$<1/2>	$(4,1/2)^o$	9/2	10	
10000	13522.40				7000.290	$5f^2(^3H)6d^2(^3F)(^5L)7s$	^6L	11/2	12	NHBK
					14393.410	$5f^2 6d7s7p$	$^6L^o$	11/2	12	
10000	14344.76				5149.465	$5f^2 6d7s^2$	b ^4H	7/2	8	NHBK
					12118.750	$5f^2 7s^2 7p$<1/2>	$(4,1/2)^o$	7/2	8	
Uranium I (Z = 92)										
1600	3489.3672		1.3e+07		0.000	$5f^3(^4I^o)6d7s^2$	$^5L^o$	6	13	NHBK
					28650.294			5	11	
1600	3514.6107		1.2e+07		0.000	$5f^3(^4I^o)6d7s^2$	$^5L^o$	6	13	NHBK
					28444.517			5	11	
1200	3561.8038		5.7e+06	C	0.000	$5f^3(^4I^o)6d7s^2$	$^5L^o$	6	13	NHBK,ASD
					28067.646			5	11	
2300	3566.5909		2.4e+07	B	620.323	$5f^3(^4I^o)6d7s^2$	$^5K^o$	5	11	NHBK,ASD
					28650.294			5	11	
3200	3584.8774		1.8e+07	B	0.000	$5f^3(^4I^o)6d7s^2$	$^5L^o$	6	13	NHBK,ASD
					27886.995	$5f^3 6d^2 7p$	^7N	7	15	
960	3659.1548				620.323	$5f^3(^4I^o)6d7s^2$	$^5K^o$	5	11	NHBK
					27941.253			6	13	
1900	3811.9911		1.6e+07		0.000	$5f^3(^4I^o)6d7s^2$	$^5L^o$	6	13	NHBK
					26225.569			6	13	
1200	3839.6255				3800.829	$5f^3(^4I^o)6d7s^2$	$^5L^o$	7	15	NHBK
					29837.643			7	15	
1500	3871.0353		1.9e+07		0.000	$5f^3(^4I^o)6d7s^2$	$^5L^o$	6	13	NHBK
					25825.565			6	13	
1200	3943.8161		2.1e+07		0.000	$5f^3(^4I^o)6d7s^2$	$^5L^o$	6	13	NHBK
					25348.977			6	13	
1000	4042.7496				620.323	$5f^3(^4I^o)6d7s^2$	$^5K^o$	5	11	NHBK
					25348.977			6	13	
880	4153.9710		1.2e+07		0.000	$5f^3(^4I^o)6d7s^2$	$^5L^o$	6	13	NHBK
					24066.566			7	15	
380	4156.6483				620.323	$5f^3(^4I^o)6d7s^2$	$^5K^o$	5	11	NHBK
					24671.388			6	13	
430w	4355.7400				620.323	$5f^3(^4I^o)6d7s^2$	$^5K^o$	5	11	NHBK
					23572.086			6	13	
430	4362.0510		1.1e+07		0.000	$5f^3(^4I^o)6d7s^2$	$^5L^o$	6	13	NHBK
					22918.553	$5f^3 6d7s7p$	^5L	7	15	
230	5915.385		4.5e+06		0.000	$5f^3(^4I^o)6d7s^2$	$^5L^o$	6	13	NHBK
					16900.386	$5f^3 6d7s7p$	^7M	7	15	
Neptunium I (Z = 93)										
300	3481.93				0.000	$5f^4(^5I)6d7s^2$	^6L	11/2	12	NHBK
					28711.465		^{o}	13/2	14	
300w	3501.50				0.000	$5f^4(^5I)6d7s^2$	^6L	11/2	12	NHBK
					28551.035		^{o}	13/2	14	
300l	3986.89				0.000	$5f^4(^5I)6d7s^2$	^6L	11/2	12	NHBK
					25075.145		^{o}	13/2	14	
300s	5044.66				2033.965	$5f^4(^5I)6d7s^2$	^6K	9/2	10	NHBK
					21851.300		^{o}	9/2	10	
300l	5878.04				0.000	$5f^4(^5I)6d7s^2$	^6L	11/2	12	NHBK
					17007.700	$5f^4 6d7s7p$	$^6M^o$	13/2	14	
1000s	6930.31				2033.965	$5f^4(^5I)6d7s^2$	^6K	9/2	10	NHBK
					16459.340	$5f^4(^5I)6d7s7p$	$^8L^o$	9/2	10	
3000s	6972.09				0.000	$5f^4(^5I)6d7s^2$	^6L	11/2	12	NHBK
					14338.880	$5f^4(^5I)6d7s7p$	$^8M^o$	11/2	12	
1000l	7735.14				6903.440	$5f^4(^5I)6d7s^2$	^6L	15/2	16	NHBK
					19827.885	$5f^4(^5I)6d7s7p$	$^8M^o$	15/2	16	
1000l	7765.75				7112.430	$5f^4(^5I)6d^2(^3F)7s$	^8M	11/2	12	NHBK
					19985.980		^{o}	9/2	10	

Intensity	Wavelength/Å	Uncert. wavelength/Å	$A_{ki}/(10^8 s^{-1})$	Accur.	Energy levels/cm^{-1}	Configs.	Terms	J	g	DB Source
1000l	7791.38				3502.855	$5f^4(^5I)6d7s^2$	^6L	13/2	14	NHBK
					16334.010	$5f^4(^5I)6d7s7p$	$^8M^o$	13/2	14	
1000l	8339.12				7112.430	$5f^4(^5I)6d^2(^3F)7s$	^8M	11/2	12	NHBK
					19100.810		°	9/2	10	
3000	8372.88				0.000	$5f^4(^5I)6d7s^2$	^6L	11/2	12	NHBK
					11940.075	$5f^4(^5I<4>)7s^27p<1/2>$	°	9/2	10	
3000	8529.96				6474.180	$5f^4(^5I)6d7s^2$	^6G	3/2	4	NHBK
					18194.400	$5f^4(^5I<4>)7s^27p<3/2>$	°	5/2	6	
10000l	9016.18				7112.430	$5f^4(^5I)6d^2(^3F)7s$	^8M	11/2	12	NHBK
					18200.530		°	9/2	10	
3000s	9379.33				2033.965	$5f^4(^5I)6d7s^2$	^6K	9/2	10	NHBK
					12692.765	$5f^4(^5I<4>)7s^27p<1/2>$	°	7/2	8	
3000l	9468.66				5185.015	$5f^4(^5I)6d7s^2$	^6K	11/2	12	NHBK
					15743.235	$5f^4(^5I<5>)7s^27p<1/2>$	°	11/2	12	
10000l	10091.99				2033.965	$5f^4(^5I)6d7s^2$	^6K	9/2	10	NHBK
					11940.075	$5f^4(^5I<4>)7s^27p<1/2>$	°	9/2	10	
10000s	10817.45				3450.995	$5f^4(^5I)6d7s^2$	^6I	7/2	8	NHBK
					12692.765	$5f^4(^5I<4>)7s^27p<1/2>$	°	7/2	8	
10000l	11776.64				3450.995	$5f^4(^5I)6d7s^2$	^6I	7/2	8	NHBK
					11940.075	$5f^4(^5I<4>)7s^27p<1/2>$	°	9/2	10	
10000s	12377.42				4615.670	$5f^4(^5I)6d7s^2$	^4G	5/2	6	NHBK
					12692.765	$5f^4(^5I<4>)7s^27p<1/2>$	°	7/2	8	
Plutonium I (Z = 94)										
5000	3753.628				0.000	$5f^67s^2$	^7F	0	1	NHBK
					26633.288	$5f^67s7p$		1	3	
5000	3755.940				2203.606	$5f^67s^2$	^7F	1	3	NHBK
					28820.548	$5f^67s7p$		0	1	
5000	3758.338				10238.473	$5f^67s^2$	^7F	6	13	NHBK
					36838.42			6	13	
5000	3774.384				2203.606	$5f^67s^2$	^7F	1	3	NHBK
					28690.480			1	3	
5000	3851.007				0.000	$5f^67s^2$	^7F	0	1	NHBK
					25959.849	$5f^56d^27s$		1	3	
5000	3878.540				2203.606	$5f^67s^2$	^7F	1	3	NHBK
					27979.161	$5f^67s7p$		2	5	
10000	4206.481				0.000	$5f^67s^2$	^7F	0	1	NHBK
					23766.139	$5f^57s7p$	$^7D^o$	1	3	
8000	6304.661				0.000	$5f^67s^2$	^7F	0	1	NHBK
					15856.888	$5f^5(^6F^o)6d7s^2$	$^7G^o$	1	3	
8000	6486.707				2203.606	$5f^67s^2$	^7F	1	3	NHBK
					17615.482	$5f^5(^6F^o)6d7s^2$	$^7F^o$	2	5	
10000	6488.853				0.000	$5f^67s^2$	^7F	0	1	NHBK
					15406.760	$5f^5(^6H^o)6d7s^2$	$^7G^o$	1	3	
8000	6887.710				6313.866	$5f^5(^6H^o)6d7s^2$	$^7K^o$	4	9	NHBK
					20828.477	$5f^56d7s(^8K^o)7p$	^9L	4	9	
10000	8630.189				6313.866	$5f^5(^6H^o)6d7s^2$	$^7K^o$	4	9	NHBK
					17897.919	$5f^5(^6H^o<5/2>)7s^27p<1/2>$	(5/2,1/2)	3	7	
Americium I (Z = 95)										
5000	3510.127				0.00	$5f^7(^8S^o)7s^2$	$^8S^o$	7/2	8	NHBK
					28480.87	$5f^7(^8S^o<7/2>)7s7p(^1P<1>)$	(7/2,1)	9/2	10	
5000	3569.163				0.00	$5f^7(^8S^o)7s^2$	$^8S^o$	7/2	8	NHBK
					28009.81	$5f^7(^8S^o<7/2>)7s7p(^1P<1>)$	(7/2,1)	7/2	8	
5000	3673.121				0.00	$5f^7(^8S^o)7s^2$	$^8S^o$	7/2	8	NHBK
					27217.14	$5f^7(^8S^o<7/2>)7s7p(^1P<1>)$	(7/2,1)	5/2	6	
5000	4289.258				0.00	$5f^7(^8S^o)7s^2$	$^8S^o$	7/2	8	NHBK
					23307.41			9/2	10	
5000l	4662.790				0.00	$5f^7(^8S^o)7s^2$	$^8S^o$	7/2	8	NHBK
					21440.37	$5f^7(^8S^o<7/2>)7s7p(^3P<2>)$	(7/2,2)	7/2	8	

Intensity	Wavelength/Å	Uncert. wavelength/Å	A_{ki}/(10^8 s^{-1})	Accur.	Energy levels/cm^{-1}	Configs.	Terms	*J*	*g*	DB Source
10000l	6054.64				0.00	$5f^7(^8S^o)7s^2$	$^8S^o$	7/2	8	NHBK
					16511.82	$5f^7(^8S^o<7/2>)7s7p(^3P<1>)$	(7/2,1)	9/2	10	
Curium I (Z = 96)										
10000	3109.690				1214.203	$5f^87s^2$	^7F	6	13	NHBK
					33362.424	$5f^76d7s^2$	^{o}	6	13	
10000	3116.411				0.000	$5f^76d7s^2$	$^9D^o$	2	5	NHBK
					32078.886	$5f^76d7s7p$	^5P	2	5	
10000	3224.226				1764.268	$5f^76d7s^2$	$^9D^o$	5	11	NHBK
					32770.516	$5f^76d7s7p$		4	9	
10000	3225.108				1764.268	$5f^76d7s^2$	$^9D^o$	5	11	NHBK
					32762.036	$5f^76d7s7p$		6	13	
10000	3304.849				1214.203	$5f^87s^2$	^7F	6	13	NHBK
					31464.053		^{o}	7	15	
10000	3452.922				0.000	$5f^76d7s^2$	$^9D^o$	6	13	NHBK
					22910.225	$5f^76d7s7p$		6	13	
10000	3458.338				0.000	$5f^87s^2$	^7F	6	13	NHBK
					22382.830		^{o}	5	11	
10000	3816.304				0.000	$5f^87s^2$	^7F	6	13	NHBK
					19179.210		^{o}	7	15	
10000	3936.666				6530.720	$5f^87s^2$	^7F	6	13	NHBK
					25493.793	$5f^8(^7F?<6>)7s7p(^1P^o<1>)$	$(6,1)^o$	7	15	
10000	3995.100				0.000	$5f^76d7s^2$	$^9D^o$	2	5	NHBK
					17665.980	$5f^76d7s7p$	^7D	3	7	
10000	4211.62				1214.203	$5f^87s^2$	^7F	6	13	NHBK
					24951.328	$5f^76d^27s$	$^9F^o$	7	15	
10000	4330.82				0.000	$5f^76d7s^2$	$^9D^o$	2	5	NHBK
					23083.822	$5f^76d7s7p$	^9D	3	7	
10000	4459.16				1214.203	$5f^87s^2$	^7F	6	13	NHBK
					23633.638	$5f^8(^7F?<6>)7s7p(^3P^o<2>)$	$(6,2)^o$	6	13	
10000	4608.40				1214.203	$5f^87s^2$	^7F	6	13	NHBK
					22907.611	$5f^8(^7F?<6>)7s7p(^3P^o<2>)$	$(6,2)^o$	7	15	
10000	5952.41				1214.203	$5f^87s^2$	^7F	6	13	NHBK
					18009.483	$5f^8(^7F<6>)7s7p(^3P^o<1>)$	$(6,1)^o$	7	15	
10000	6376.71				0.000	$5f^76d7s^2$	$^9D^o$	2	5	NHBK
					15677.750	$5f^76d7s7p$	^{11}F	3	7	
10000	6706.85				815.655	$5f^76d7s^2$	$^9D^o$	4	9	NHBK
					15721.679	$5f^7(^8S^o)7s^27p$	(7/2,3/2)	5	11	
10000	7162.69				1764.268	$5f^76d7s^2$	$^9D^o$	5	11	NHBK
					15721.679	$5f^7(^8S^o)7s^27p$	(7/2,3/2)	5	11	
10000	7720.47				9064.880	$5f^76d7s^2$	$^7D^o$	5	11	NHBK
					22013.913	$5f^76d7s7p$	^9F	5	11	
10000	8392.37				3809.355	$5f^76d7s^2$	$^9D^o$	6	13	NHBK
					15721.679	$5f^7(^8S^o)7s^27p$	(7/2,3/2)	5	11	
10000	10542.98				302.153	$5f^76d7s^2$	$^9D^o$	3	7	NHBK
					9784.543	$5f^7(^8S^o)7s^27p$	(7/2,1/2)	4	9	
10000	10792.25				0.000	$5f^76d7s^2$	$^9D^o$	2	5	NHBK
					9263.374	$5f^7(^8S^o)7s^27p$	(7/2,1/2)	3	7	
Berkelium I (Z = 97)										
10000l	3288.750				5416.690	$5f^97s^2$	$^6F^o$	11/2	12	NHBK
					35814.60			13/2	14	
10000	3289.347									NHBK
10000	3426.951				0.000	$5f^97s^2$	$^6H^o$	15/2	16	NHBK
					29172.103	$5f^9(^6H^o<13/2>)7s7p(^3P^o<2>)$	(13/2,2)	17/2	18	
10000	3442.664				0.000	$5f^97s^2$	$^6H^o$	15/2	16	NHBK
					29038.930	$5f^9(^6H^o<13/2>)7s7p(^3P^o<2>)$	(13/2,2)	15/2	16	
10000	3609.614				0.000	$5f^97s^2$	$^6H^o$	15/2	16	NHBK
					27695.870			15/2	16	

Intensity	Wavelength/Å	Uncert. wavelength/Å	$A_{ki}/(10^8\ s^{-1})$	Accur.	Energy levels/cm⁻¹	Configs.	Terms	J	g	DB Source
10000	4363.636				0.000	$5f^97s^2$	$^6H^o$	15/2	16	NHBK
					22910.225	$5f^9(^6H^o<15/2>)7s7p(^3P^o<2>)$	(15/2,2)	15/2	16	
10000	4466.457				0.000	$5f^97s^2$	$^6H^o$	15/2	16	NHBK
					22382.830	$5f^9(^6H^o<15/2>)7s7p(^3P^o<2>)$	(15/2,2)	17/2	18	
10000	5212.53				0.000	$5f^97s^2$	$^6H^o$	15/2	16	NHBK
					19179.210	$5f^9(^6H^o<15/2>)7s7p(^3P^o<1>)$	(15/2,1)	15/2	16	
10000	5271.95				6530.720	$5f^97s^2$	$^6H^o$	13/2	14	NHBK
					25493.793			13/2	14	
10000	5659.03				0.000	$5f^97s^2$	$^6H^o$	15/2	16	NHBK
					17665.980	$5f^9(^6H^o<15/2>)7s7p(^3P^o<1>)$	(15/2,1)	17/2	18	
10000	5910.71				0.000	$5f^97s^2$	$^6H^o$	15/2	16	NHBK
					16913.770	$5f^9(^6H^o<15/2>)7s7p(^3P^o<0>)$	(15/2,0)	15/2	16	
10000l	11293.14				9141.115	$5f^86d7s^2$	^8G	13/2	14	NHBK
					17993.628	$5f^8(^7F<6>)7s^27p<1/2>$	(6,1/2)°	13/2	14	
10000l	11500.30				9300.585	$5f^86d7s^2$	^8D	11/2	12	NHBK
					17993.628	$5f^8(^7F<6>)7s^27p<1/2>$	(6,1/2)°	13/2	14	
10000s	11793.09				9300.585	$5f^86d7s^2$	^8D	11/2	12	NHBK
					17777.808	$5f^8(^7F<6>)7s^27p<1/2>$	(6,1/2)°	11/2	12	
Californium I (Z = 98)										
10000	3298.14				0.000	$5f^{10}7s^2$	^5I	8	17	NHBK
					30311.400		°	7	15	
10000	3352.71				0.000	$5f^{10}7s^2$	^5I	8	17	NHBK
					29818.055	$5f^96d7s^2$	°	8	17	
10000	3598.77				0.000	$5f^{10}7s^2$	^5I	8	17	NHBK
					27779.345	$5f^{10}(^5I<8>)7s7p(^1P^o<1>)$	(8,1)°	9	19	
10000	4329.03				0.000	$5f^{10}7s^2$	^5I	8	17	NHBK
					23093.355	$5f^{10}(^5I<8>)7s7p(^3P^o<2>)$	(8,2)°	9	19	
10000	5408.88				0.000	$5f^{10}7s^2$	^5I	8	17	NHBK
					18483.060	$5f^{10}(^5I<8>)7s7p(^3P^o<1>)$	(8,1)°	9	19	
10000	8141.29				19907.460	$5f^9(^6H^o15/2)6d<3/2>7s^2$	(15/2,3/2)°	9	19	NHBK
					32187.145	$5f^9(^6H^o<15/2>)7s^27p<3/2>$	(15/2,3/2)	9	19	
10000	11941.33				16909.535	$5f^9(^6H^o15/2)6d<3/2>7s^2$	(15/2,3/2)°	8	17	NHBK
					25281.530	$5f^9(^6H^o<15/2>)7s^27p<1/2>$	(15/2,1/2)	8	17	
10000	12787.41				16909.535	$5f^9(^6H^o15/2)6d<3/2>7s^2$	(15/2,3/2)°	8	17	NHBK
					24727.600	$5f^9(^6H^o<15/2>)7s^27p<1/2>$	(15/2,1/2)	7	15	
Einsteinium I (Z = 99)										
10000s	3428.48				0.00	$5f^{11}7s^2$	$^4I^o$	15/2	16	NHBK
					29159.28			15/2,17/2	34	
10000s	3498.11				0.00	$5f^{11}7s^2$	$^4I^o$	15/2	16	NHBK
					28578.60	$5f^{10}6d7s^2$		15/2,17/2	34	
10000l	3514.33				0.00	$5f^{11}7s^2$	$^4I^o$	15/2	16	NHBK
					28447.02	$5f^{11}7s7p$		13/2	14	
10000	3521.38				10244.29	$5f^{11}7s^2$	°	13/2	14	NHBK
					38634.04			11/2	12	
10000s	3523.49				0.00	$5f^{11}7s^2$	$^4I^o$	15/2	16	NHBK
					28372.78	$5f^{10}6d7s^2$		13/2,15/2, 17/2	48	
10000l	3555.34				0.00	$5f^{11}7s^2$	$^4I^o$	15/2	16	NHBK
					28118.65	$5f^{10}6d7s^2$		13/2	14	
10000	3801.49				7894.54	$5f^{11}7s^2$	$^4F^o$	9/2	10	NHBK
					34192.6			9/2	10	
10000s	5161.74				0.00	$5f^{11}7s^2$	$^4I^o$	15/2	16	NHBK
					19367.85	$5f^{10}6d7s^2$	^6I	17/2	18	
10000l	5204.40				0.00	$5f^{11}7s^2$	$^4I^o$	15/2	16	NHBK
					19209.15	$5f^{11}7s7p$	^6I	17/2	18	
3000l	5615.51				0.00	$5f^{11}7s^2$	$^4I^o$	15/2	16	NHBK
					17803.09	$5f^{11}7s7p$	^6I	15/2	16	

Note: † for energy level indicates assignment is uncertain.

ATOMIC TRANSITION PROBABILITIES

J. R. Fuhr, W. L. Wiese, L. I. Podobedova, and D. E. Kelleher

This table contains critically evaluated atomic transition probabilities for 208 important lines of 50 selected elements for which reliable data are available on an absolute scale. The values are largely for neutral spectra, but also include some prominent lines of doubly and more highly charged ions of important elements.

Most of the data are obtained from comprehensive compilations of the Data Center on Atomic Transition Probabilities at the National Institute of Standards and Technology. Specifically, data have been taken from critical compilations on H, He, and Li (Ref. 1); on Be and B (Ref. 2); on neutral C and N (Ref. 3); and on Na (Ref. 4), Mg (Ref. 4), Al (Ref. 5), and Si (Ref. 6). Material from earlier compilations for the elements H through Ne (Refs. 7 and 8) and Na through Ca (Ref. 9) was supplemented by some more recent material taken directly from the original literature. Most of the original literature is cited in the above tables and in recent bibliographies (Refs. 10 and 11); for lack of space, individual literature references are not cited here.

The wavelength range for the neutral species is normally the visible spectrum or shorter wavelengths; only the very prominent near-infrared lines are included. The tabulation is limited to electric dipole lines, including intercombination, and comprises essentially the fairly strong transitions with estimated uncertainties in the 10% to 50% range. The column definitions are as follows.

Column heading	**Definition**
Spectrum	The atomic element symbol; all values in this table are for the neutral atom
λ	Wavelength in angströms; the energy of transition from an initial state i to a final state k
g_i	Statistical weight of the lower (i) state
g_k	Statistical weight of the upper (k) state
A_{ki}	The transition probability, in units of 10^8 s^{-1}; is listed with as many digits as is consistent with the indicated accuracy

Generally, the estimated uncertainties of the A values are in the range from 25% to 50% for two-digit numbers, 10% to 25% for three-digit numbers, and 1% or better for four- and five-digit numbers.

The product g_kA (or g_if) is needed for many applications. Whenever the wavelengths of individual lines within a multiplet are extremely close, only an average wavelength for the multiplet as well as the multiplet A value are given, and this is indicated by a footnote "c" on the wavelength. This also has been done when the transition probability for an entire multiplet has been taken from the literature and values for individual lines cannot be determined because of insufficient knowledge of the coupling of electrons. The wavelength data have been taken either from recent compilations or from the original literature cited in bibliographies published by the Atomic Energy Levels Data Center (Refs. 12 and 13) at the National Institute of Standards and Technology (NIST). Wavelength values are consistent with those given in the table "Line Spectra of the Elements," which appears in the Online Edition of the *CRC Handbook*.

In addition to the transition probability A, the atomic oscillator strength f and the line strength S are often used in the literature. The conversion factors between these quantities are (for electric-dipole transitions):

$$g_if = 1.499 \cdot 10^{-8}\lambda^2 g_k A = 303.8\ \lambda^{-1} S$$

where λ is in Å, A is in 10^8 s^{-1}, and S is in atomic units, which are $a_0^2e^2 = 7.188 \cdot 10^{-59}$ m^{2}C^2.

The table for hydrogen is presented first, followed by the tables for other elements in alphabetical sequence by element name (not symbol).

The transition probabilities for hydrogen are known precisely. Because of the hydrogen degeneracy, a "transition" is actually the sum of all fine-structure transitions between the principal quantum numbers; therefore, the hydrogen table gives weighted average A values. A more extensive compilation of over 12,000 lines is available in the Online Edition of the *CRC Handbook*, available at <hbcponline.com>. Additional transition probability data are available from NIST (Ref. 15).

References

1. Wiese, W. L., and Fuhr, J. R., *J. Phys. Chem. Ref. Data* 38, 565, 2009. <https://doi.org/10.1063/1.3077727>
2. Fuhr, J. R., and Wiese, W. L., *J. Phys. Chem. Ref. Data* 39, 013101, 2010. <https://doi.org/10.1063/1.3286088>
3. Wiese, W. L., and Fuhr, J. R., *J. Phys. Chem. Ref. Data* 36, 1287, 2007. <https://doi.org/10.1063/1.2740642>
4. Kelleher, D. E., and Podobedova, L. I., *J. Phys. Chem. Ref. Data* 37, 267, 2008. <https://doi.org/10.1063/1.2735328>
5. Kelleher, D. E., and Podobedova, L. I., *J. Phys. Chem. Ref. Data* 37, 709, 2008. <https://doi.org/10.1063/1.2734564>
6. Kelleher, D. E., and Podobedova, L. I., *J. Phys. Chem. Ref. Data* 37, 1285, 2008. <https://doi.org/10.1063/1.2734566>
7. Wiese, W. L., Smith, M. W., and Glennon, B. M., *Atomic Transition Probabilities (H through Ne—A Critical Data Compilation)*, National Standard Reference Data Series, National Bureau of Standards 4, Vol. I, U.S. Government Printing Office, Washington, DC, 1966. <https://doi.org/10.6028/NBS.NSRDS.4]
8. Wiese, W. L., Fuhr, J. R., and Deters, T. M., *Atomic Transition Probabilities of Carbon, Nitrogen, and Oxygen, J. Phys. Chem. Ref. Data, Monograph 7*, 1996.
9. Wiese, W. L., Smith, M. W., and Miles, B. M., *Atomic Transition Probabilities (Na through Ca—A Critical Data Compilation)*, National Standard Reference Data Series, National Bureau of Standards 22, Vol. II, U.S. Government Printing Office, Washington, DC, 1969.
10. Fuhr, J. R., Miller, B. J., and Martin, G. A., *Bibliography on Atomic Transition Probabilities (1914 through October 1997)*, National Bureau of Standards Special Publication 505, 1978; Miller, B. J., Fuhr, J. R., and Martin, G. A., *Bibliography on Atomic Transition Probabilities (November 1977 through February 1980)*, National Bureau of Standards Special Publication 505, Supplement 1, 1980.
11. Wiese, W. L., Reports on Astronomy, *Trans. Int. Astron. Union* 18A, 116, 1982; 19A, 122, 1985; 20A, 117, 1988, Reidel, D., Ed., Kluwer, Dordrecht, the Netherlands.
12. Moore, C. E., *Bibliography on the Analyses of Optical Atomic Spectra*, National Bureau of Standards Special Publication 306—Section 1, 1968; Sections 2-4, 1969.
13. Hagan, L., and Martin, W. C., *Bibliography on Atomic Energy Levels and Spectra (July 1968 through June 1971)*, National Bureau of Standards Special Publication 363, 1972; Hagan, L., *Bibliography on Atomic Energy Levels and Spectra (July 1971 through June 1975)*, National Bureau of Standards Special Publication 363, Supplement 1, 1977; Zalubas, R., and Albright, A., *Bibliography on Atomic Energy Levels and Spectra (July 1975 through June 1979)*, National Bureau of Standards Special Publication 363, Supplement 2, 1980; Musgrove, A., and Zalubas, R., *Bibliography on Atomic Energy Levels and Spectra (July 1979 through December 1983)*, National Bureau of Standards Special Publication 363, Supplement 3, 1985.

14. Kramida, A., Ralchenko, Yu., Reader, J., and NIST ASD Team (2013). NIST Atomic Spectra Database (Version 5.1), National Institute of Standards and Technology, Gaithersburg, MD, <www.nist.gov/pml/data/asd.cfm>.
15. <https://www.nist.gov/pml/basic-atomic-spectroscopic-data-handbook#II>.

Atomic Transition Probabilities

Spectrum	λ/Å[a,b]	g_i	g_k	$A_{ki}/10^8\ s^{-1}$
Hydrogen				
H I	1215.67g	2	8	4.6986
H I	1025.72g	2	18	$5.5751 \cdot 10^{-1}$
H I	972.537g	2	32	$1.2785 \cdot 10^{-1}$
H I	949.743g	2	50	$4.1250 \cdot 10^{-2}$
H I	937.803g	2	72	$1.6440 \cdot 10^{-2}$
H I	6562.83	8	18	$4.4101 \cdot 10^{-1}$
H I	4861.34	8	32	$8.4193 \cdot 10^{-2}$
H I	4340.47	8	50	$2.5304 \cdot 10^{-2}$
Aluminum				
Al I	3082.1529g	2	4	$5.87 \cdot 10^{-1}$
Al I	3092.7099g	4	6	$7.29 \cdot 10^{-1}$
Al I	3944.0058g	2	2	$4.99 \cdot 10^{-1}$
Al I	3961.52g	4	2	$9.85 \cdot 10^{-1}$
Argon				
Ar I	1048.22g	1	3	5.36
Ar I	4158.59	5	5	$1.40 \cdot 10^{-2}$
Ar I	4259.36	3	1	$3.98 \cdot 10^{-2}$
Ar I	7635.11	5	5	$2.45 \cdot 10^{-1}$
Ar I	7948.18	1	3	$1.86 \cdot 10^{-1}$
Ar I	8115.31	5	7	$3.31 \cdot 10^{-1}$
Arsenic				
As I	1890.4g	4	6	2.0
As I	1937.6g	4	4	2.0
As I	2288.1	6	4	2.8
As I	2349.8	4	2	3.1
Barium				
Ba I	5535.48g	1	3	1.19
Ba I	6498.76	7	7	$5.40 \cdot 10^{-1}$
Ba I	7059.94	7	9	$5.00 \cdot 10^{-1}$
Ba I	7280.3	5	7	$3.20 \cdot 10^{-1}$
Beryllium				
Be I	2348.61g	1	3	5.54
Be I[c]	2650.7	3	1	4.23
Bismuth				
Bi I	2228.3g	4	4	$8.9 \cdot 10^{-1}$
Bi I	2898	4	2	1.53
Bi I	2989	4	4	$5.5 \cdot 10^{-1}$
Bi I	3067.7g	4	2	2.07
Boron				
B I	1825.89g	2	4	1.7
B I	1826.4g	4	6	2.04
B I	2496.77g	2	2	$8.37 \cdot 10^{-1}$
B I	2497.72g	4	2	1.67
Bromine				
Br I	1488.5g	4	4	1.2
Br I	1540.7g	4	4	1.4
Br I	7348.5	4	6	$1.2 \cdot 10^{-1}$
Cadmium				
Cd I	2288gg	1	3	5.3
Cd I	3466.2	3	5	1.2
Cd I	3610.5	5	7	1.3
Cd I	5085.8	5	3	$5.6 \cdot 10^{-1}$
Calcium				
Ca I	4226.7g	1	3	2.18
Ca I	4302.5	5	5	1.36
Ca I	5588.8	7	7	$4.9 \cdot 10^{-1}$
Ca I	6162.2	5	3	$3.54 \cdot 10^{-1}$
Ca I	6439.1	7	9	$5.3 \cdot 10^{-1}$
Carbon				
C I	1561.44g	5	7	1.17
C I	1657.01g	5	5	2.61
C I	1930.9	5	3	3.39
C I	2478.56	1	3	$2.80 \cdot 10^{-1}$
Cesium				
Cs I	3876.1g	2	4	$3.8 \cdot 10^{-3}$
Cs I	4555.3g	2	4	$1.88 \cdot 10^{-2}$
Cs I	4593.2g	2	2	$8.0 \cdot 10^{-3}$
Cs I	8521.1g	2	4	$3.276 \cdot 10^{-1}$
Cs I	8943.5g	2	2	$2.87 \cdot 10^{-1}$
Chlorine				
Cl I	1347.2g	4	4	4.19
Cl I	1351.7g	2	2	3.23
Cl I	4526.2	4	4	$5.1 \cdot 10^{-2}$
Cl I	7256.6	6	4	$1.5 \cdot 10^{-1}$
Chromium				
Cr I	3578.68g	7	9	1.48
Cr I	3593.48g	7	7	1.5
Cr I	3605.32g	7	5	1.62
Cr I	4254.33g	7	9	$3.15 \cdot 10^{-1}$
Cr I	4274.81g	7	7	$3.07 \cdot 10^{-1}$
Cr I	5208.42	5	7	$5.06 \cdot 10^{-1}$
Cobalt				
Co I	3405.12	10	10	1.0
Co I	3453.51	10	12	1.1
Co I	3502.28	10	8	$8.0 \cdot 10^{-1}$
Co I	3569.37	8	8	1.6
Copper				
Cu I	2178.9g	2	4	$9.13 \cdot 10^{-1}$
Cu I	3247.5g	2	4	1.39
Cu I	3274g	2	2	1.37
Cu I	5218.2	4	6	$7.5 \cdot 10^{-1}$
Fluorine				
F I	954.83g	4	4	5.77
F I	6856	6	8	$4.94 \cdot 10^{-1}$
F I	7398.7	6	6	$2.85 \cdot 10^{-1}$
F I	7754.7	4	6	$3.82 \cdot 10^{-1}$
Gallium				
Ga I	2874.2g	2	4	1.2
Ga I	2943.6g	4	6	1.4
Ga I	4033g	2	2	$4.9 \cdot 10^{-1}$
Ga I	4172g	4	2	$9.2 \cdot 10^{-1}$
Germanium				
Ge I	2651.6g	1	3	$8.5 \cdot 10^{-1}$
Ge I	2709.6g	3	1	2.8
Ge I	2754.6g	5	3	1.1
Ge I	3039.1g	5	3	2.8
Gold				
Au I	2427.95g	2	4	1.99
Au I	2675.95g	2	2	1.64
Helium				
He I	537.03g	1	3	5.6634
He I	584.334g	1	3	$1.7989 \cdot 10^{1}$
He I	3888.64g[c]	3	9	$9.4746 \cdot 10^{-2}$
He I	4026.21[c]	9	15	$1.1600 \cdot 10^{-1}$
He I	4471.5[c]	9	15	$2.4578 \cdot 10^{-1}$
He I	5875.66[c]	9	15	$7.0703 \cdot 10^{-1}$
Indium				
In I	3039.4g	2	4	1.3
In I	3256.1g	4	6	1.3
In I	4101.8g	2	2	$5.6 \cdot 10^{-1}$
In I	4511.3g	4	2	1.02
Iodine				
I I	1782.8g	4	4	2.71
I I	1830.4g	4	6	$1.6 \cdot 10^{-1}$
Iron				
Fe I	3581.19g	11	13	1.02
Fe I	3719.93g	9	11	$1.62 \cdot 10^{-1}$
Fe I	3734.86	11	11	$9.01 \cdot 10^{-1}$
Fe I	3745.56g	5	7	$1.15 \cdot 10^{-1}$
Fe I	3859.91g	9	9	$9.69 \cdot 10^{-2}$
Fe I	4045.81	9	9	$8.62 \cdot 10^{-1}$
Krypton				
Kr I	5570.3	5	3	$2.1 \cdot 10^{-2}$
Kr I	5870.9	3	5	$1.8 \cdot 10^{-2}$
Kr I	7601.5	5	5	$3.1 \cdot 10^{-1}$
Kr I	8112.9	5	7	$3.6 \cdot 10^{-1}$
Lead				
Pb I	2802g	5	7	1.6
Pb I	2833.1g	1	3	$5.8 \cdot 10^{-1}$
Pb I	3683.5g	3	1	1.5
Pb I	4057.8g	5	3	$8.9 \cdot 10^{-1}$
Lithium				
Li I	3232.7g[c]	2	6	$1.002 \cdot 10^{-2}$

Spectrum	λ/Å[a,b]	g_i	g_k	$A_{ki}/10^8\ s^{-1}$
Li I	4602.9[c]	6	10	$2.322 \cdot 10^{-1}$
Li I	6103.6[c]	6	10	$6.8563 \cdot 10^{-1}$
Li I	6707.8g[c]	2	6	$3.6891 \cdot 10^{-1}$
Magnesium				
Mg I	2025.824g	1	3	$6.12 \cdot 10^{-1}$
Mg I	2852.127g	1	3	4.91
Mg I	4702.9909	3	5	$2.19 \cdot 10^{-1}$
Mg I	5183.6042	5	3	$5.61 \cdot 10^{-1}$
Manganese				
Mn I	2794.82g	6	8	3.7
Mn I	2798.27g	6	6	3.6
Mn I	2801.08g	6	4	3.7
Mn I	4030.76g	6	8	$1.7 \cdot 10^{-1}$
Mercury				
Hg I	2536.52g	1	3	$8.00 \cdot 10^{-2}$
Hg I	3125.66	3	5	$6.56 \cdot 10^{-1}$
Hg I	4358.34	3	3	$5.57 \cdot 10^{-1}$
Hg I	5460.75	5	3	$4.87 \cdot 10^{-1}$
Neon				
Ne I	735.9g	1	3	6.11
Ne I	743.72g	1	3	$4.86 \cdot 10^{-1}$
Ne I	5852.5	3	1	$6.82 \cdot 10^{-1}$
Ne I	6402.2	5	7	$5.14 \cdot 10^{-1}$
Nickel				
Ni I	3101.56	5	7	$6.3 \cdot 10^{-1}$
Ni I	3134.11	3	5	$7.3 \cdot 10^{-1}$
Ni I	3369.56g	9	7	$1.8 \cdot 10^{-1}$
Ni I	3414.76	7	9	$5.5 \cdot 10^{-1}$
Ni I	3524.54	7	5	1.0
Ni I	3619.39	5	7	$6.6 \cdot 10^{-1}$
Nitrogen				
N I	1199.55g	4	6	4.07
N I	1492.63	6	4	3.11
N I	7468.31	6	4	$1.96 \cdot 10^{-1}$
N I	8216.34	6	6	$2.26 \cdot 10^{-1}$
Oxygen				
O I	1302.17g	5	3	3.41
O I	4368.19[c]	3	1	$7.56 \cdot 10^{-3}$
O I	5436.86	7	5	$1.80 \cdot 10^{-2}$
O I	7771.94	5	7	$3.69 \cdot 10^{-1}$
Phosphorus				
P I	1775g	4	6	2.17
P I	1782.9g	4	4	2.14
P I	2136.2	6	4	2.83
P I	2535.6	4	4	$9.5 \cdot 10^{-1}$
Potassium				
K I	4044.1g	2	4	$1.24 \cdot 10^{-2}$
K I	4047.2g	2	2	$1.24 \cdot 10^{-2}$
K I	7664.9g	2	4	$3.87 \cdot 10^{-1}$
K I	7699g	2	2	$3.82 \cdot 10^{-1}$
Rubidium				
Rb I	4201.8g	2	4	$1.8 \cdot 10^{-2}$
Rb I	4215.5g	2	2	$1.5 \cdot 10^{-2}$
Rb I	7800.3g	2	4	$3.70 \cdot 10^{-1}$
Rb I	7947.6g	2	2	$3.40 \cdot 10^{-1}$
Scandium				
Sc I	3907.48g	4	6	1.66
Sc I	3911.81g	6	8	1.79
Sc I	4020.39g	4	4	1.63
Sc I	4023.68g	6	6	1.65
Silicon				
Si I	2506.9g	3	5	$5.47 \cdot 10^{-1}$
Si I	2516.113g	5	5	1.68
Si I	2881.579g	5	3	2.17
Si I	5006.061	3	5	$2.8 \cdot 10^{-2}$
Si I	5948.545	3	5	$2.2 \cdot 10^{-2}$
Silver				
Ag I	3280.7g	2	4	1.4
Ag I	3382.9	2	2	1.3
Ag I	5209.1	2	4	$7.5 \cdot 10^{-1}$
Ag I	5465.5	4	6	$8.6 \cdot 10^{-1}$
Strontium				
Sr I	2428.1g	1	3	$1.7 \cdot 10^{-1}$
Sr I	4607.3g	1	3	2.01
Sulfur				
S I	1474g	5	7	1.6
S I	1666.7	5	5	6.3
S I	1807.3g	5	3	3.8
S I	4694.1	5	7	$6.7 \cdot 10^{-3}$
Thallium				
Tl I	2767.9g	2	4	1.26
Tl I	3519.2g	4	6	1.24
Tl I	3775.7g	2	2	$6.25 \cdot 10^{-1}$
Tl I	5350.5g	4	2	$7.05 \cdot 10^{-1}$
Tin				
Sn I	2840g	5	5	1.7
Sn I	3034.1g	3	1	2.0
Sn I	3175.1g	5	3	1.0
Sn I	3262.3	5	3	2.7
Titanium				
Ti I	3642.68g	7	9	$7.74 \cdot 10^{-1}$
Ti I	3653.5g	9	11	$7.54 \cdot 10^{-1}$
Ti I	3998.64g	9	9	$4.08 \cdot 10^{-1}$
Ti I	4981.73	11	13	$6.60 \cdot 10^{-1}$
Uranium				
U I	3566.6g	11	11	$2.4 \cdot 10^{-1}$
U I	3571.6	17	15	$1.3 \cdot 10^{-1}$
U I	3584.9g	13	15	$1.8 \cdot 10^{-1}$
Vanadium				
V I	3183.41g	6	8	2.4
V I	4111.78	10	10	1.01
V I	4379.23	10	12	1.1
V I	4384.71	8	10	1.1
Xenon				
Xe I	1192g	1	3	6.2
Xe I	1295.6g	1	3	2.46
Xe I	1469.6g	1	3	2.81
Xe I	4671.2	5	7	$1.0 \cdot 10^{-2}$
Xe I	7119.6	7	9	$6.6 \cdot 10^{-2}$
Zinc				
Zn I	2138.6g	1	3	7.09
Zn I	3302.6	3	5	1.2
Zn I	3345	5	7	1.7
Zn I	6362.3	3	5	$4.74 \cdot 10^{-1}$

[a] A "g" after the wavelength means the line is a resonance line; that is, the lower level of the transition is the ground state.

[b] Wavelengths below 2000 Å are in vacuum; wavelengths above 2000 Å are in air.

[c] A superscript "c" indicates that the values for λ, E_k, g_i, g_k, and A_{ki} are multiplet values; i.e., wavelengths of the component lines almost overlap.

ELECTRON AFFINITIES

Thomas M. Miller

Electron affinity is defined as the energy difference between the lowest (ground) state of the neutral and the lowest state of the corresponding negative ion. The accuracy of electron affinity measurements has been greatly improved since the advent of laser photodetachment experiments with negative ions. Electron affinities can be determined with optical precision, though a detailed understanding of atomic and molecular states and splittings is required to specify the photodetachment threshold corresponding to the electron affinity.

Column definitions are as follows.

Column heading	Definition
Z	Atomic number; given only for atom entries
Mol. form.	Atomic or molecular formula; tables for molecular electron affinities are ordered by molecular formula
Name	Name of atom or molecule
Elec. aff. (eV)	Electron affinity; given in units of eV
Uncert. (eV)	Uncertainty of electron affinity value; given in units of eV
Method	Measurement technique; see table of techniques given below
Ref.	Source of electron affinity value as available in the Online Edition of the *CRC Handbook*, available at hbcponline.com; see discussion below

The references for this data table are in the Online Edition of the *CRC Handbook*, available at hbcponline.com. Atomic and molecular electron affinities are discussed in two excellent articles reviewing photodetachment studies which appear in *Gas Phase Ion Chemistry*, Vol. 3, Bowers, M. T., Ed., Academic Press, Orlando, 1984: Chapter 21 by Drzaic, P. S., Marks, J., and Brauman, J. I., "Electron Photodetachment from Gas Phase Negative Ions," p. 167, and Chapter 22 by Mead, R. D., Stevens, A. E., and Lineberger, W. C., "Photodetachment in Negative Ion Beams," p. 213. Persons interested in photodetachment details should consult these articles and the critical reviews of Andersen, T., Haugen, H. K., and Hotop, H., *J. Phys. Chem. Ref. Data* 28, 1511, 1999; Hotop, H., and Lineberger, W. C., *J. Phys. Chem. Ref. Data* 14, 731, 1985; and Andersen, T., Haugen, H. K., and Hotop, H. *J. Phys. Chem. Ref. Data* 28, 1511, 1999. For simplicity in the tables below, any electron affinity that was discussed in the articles by Drzaic et al. or Hotop and Lineberger is referenced to these sources, where original references are given.

The development of cluster-ion photodetachment apparatuses has brought an explosion of electron affinity estimates for atomic and molecular clusters. The policy in this tabulation is to list (in this order) the electron affinities for atoms, diatoms, triatoms, and selected polyatomic molecules and clusters. If the adiabatic electron affinity has been determined for a simple cluster, the value is given, but the reader is referred to original sources for higher-order clusters. Additional data on molecular electron affinities may be found in Lias, S. G., Bartmess, J. E., Liebman, J. F., Holmes, J. L., Levin, R. D., and Mallard, W. G., Gas Phase Ion and Neutral Thermochemistry, *J. Phys. Chem. Ref. Data* 17 (Supplement No. 1), 1988 and on the NIST WebBook at <http://webbook.nist.gov/>.

For the present tabulation, the 2014 CODATA value $e/(hc)$ = 8065.544005 ± 0.000050 cm^{-1} eV^{-1} (http://physics.nist.gov/constants/) has been used to convert electron affinities from the units used in spectroscopic work, cm^{-1}, into eV for this table. Experimental measurements have improved to the level that the 25 ppb uncertainty in $e/(hc)$ will make a difference in a few cases. For this reason, very accurate electron affinities will be given in cm^{-1} with the relevant references.

Abbreviations used for electron affinity measurement techniques are given below.

Abbreviation	Definition
calc	Calculated value
PT	Photodetachment threshold using a lamp as a light source
LPT	Laser photodetachment threshold
LPES	Laser photoelectron spectroscopy
DA	Dissociative attachment
attach	Electron attachment/detachment equilibrium
e-scat	Electron scattering
kinetic	Dissociation kinetics
Knud	Knudsen cell
CT	Charge transfer
CD	Collisional detachment
ZEKE	Zero electron kinetic energy spectroscopy

Electron Affinities of Atoms and Molecules

Z	Mol. form.	Name	Elec. aff. (eV)	Uncert. (eV)	Method	Ref.
Atoms (Listed by Atomic Number)						
1	H	Hydrogen (atomic)	0.75420817		calc	205
	H	Hydrogen (atomic)	0.754195	0.000019	LPT	89
	D	Deuterium (atomic)	0.754593	0.000074	LPT	89
	D	Deuterium (atomic)	0.75465629		calc	205
	T	Tritium (atomic)	0.75480545		calc	205
2	He	Helium	not stable		calc	1
3	Li	Lithium	0.618049	0.000020	LPT	185
4	Be	Beryllium	not stable		calc	1
5	B	Boron	0.279723	0.000025	LPES	191
6	C	Carbon	1.262119	0.000020	LPT	28
7	N	Nitrogen (atomic)	not stable		DA	1
8	O	Oxygen (atomic)	1.4611135	0.0000009	LPT	4
9	F	Fluorine (atomic)	3.4011897	0.0000024	LPT	227

Z	Mol. form.	Name	Elec. aff. (eV)	Uncert. (eV)	Method	Ref.
10	Ne	Neon	not stable		calc	1
11	Na	Sodium	0.547926	0.000025	LPT	1
12	Mg	Magnesium	not stable		e-scat	1
13	Al	Aluminum	0.43283	0.00005	LPES	208
14	Si	Silicon	1.3895211	0.0000007	LPES	4
15	P	Phosphorus	0.746607	0.000010	LPT	377
16	S	Sulfur	2.07710403 (^{32}S)	0.00000051	LPT	334
	S	Sulfur	2.0771043 (^{34}S)	0.0000011	LPT	334
17	Cl	Chlorine (atomic)	3.612725	0.000027	LPT	52
18	Ar	Argon	not stable		calc	1
19	K	Potassium	0.501459	0.000012	LPT	425
20	Ca	Calcium	0.02455	0.00010	LPT	44
21	Sc	Scandium	0.188	0.020	LPES	1
22	Ti	Titanium	0.07554	0.00004	LPES	504
23	V	Vanadium	0.525	0.012	LPES	1
24	Cr	Chromium	0.67584	0.00012	LPT	426
25	Mn	Manganese	not stable		calc	1
26	Fe	Iron	0.151	0.003	LPES	27
27	Co	Cobalt	0.662256	0.000046	LPES	417
28	Ni	Nickel	1.15716	0.00012	LPT	116
29	Cu	Copper	1.235775	0.000037	LPT	426
30	Zn	Zinc	not stable		e-scat	1
31	Ga	Gallium	0.43	0.03	LPES	183
32	Ge	Germanium	1.232712	0.000015	LPES	28
33	As	Arsenic	0.804	0.002	LPES	352
34	Se	Selenium	2.020670	0.000025	LPT	1
35	Br	Bromine (atomic)	3.3635882	0.0000019	LPT	74
36	Kr	Krypton	not stable		calc	1
37	Rb	Rubidium	0.48592	0.00002	LPT	1
38	Sr	Strontium	0.05206	0.00006	LPT	122
39	Y	Yttrium	0.307	0.012	LPES	1
40	Zr	Zirconium	0.426	0.014	LPES	1
41	Nb	Niobium	0.91740	0.00006	LPES	505
42	Mo	Molybdenum	0.74725	0.00025	LPT	426
43	Tc	Technetium	0.55	0.20	calc	1
44	Ru	Ruthenium	1.04638	0.00025	calc	427
45	Rh	Rhodium	1.14289	0.0002	LPT	116
46	Pd	Palladium	0.56214	0.00012	LPT	116
47	Ag	Silver	1.304475	0.000025	LPT	426
48	Cd	Cadmium	not stable		e-scat	1
49	In	Indium	0.404	0.009	LPT	428
50	Sn	Tin	1.112067	0.000015	LPES	28
51	Sb	Antimony	1.047401	0.000019	LPT	429
52	Te	Tellurium	1.970876	0.000007	LPT	261
53	I	Iodine (atomic)	3.0590368	0.0000010	LPES	92
54	Xe	Xenon	not stable		calc	1
55	Cs	Cesium	0.471626	0.000025	LPT	1
56	Ba	Barium	0.14462	0.00006	LPT	195
57	La	Lanthanum	0.55	0.02	LPT	184
58	Ce	Cerium	0.65	0.03	LPT	269
59	Pr	Praseodymium	0.962	0.024	LPES	225
60	Nd	Neodymium	>1.916		LPES	342
63	Eu	Europium	0.864	0.024	LPES	268
65	Tb	Terbium	>1.165		LPES	342
66	Dy	Dysprosium	>0		LPES	342
69	Tm	Thulium	1.029	0.022	LPES	264
70	Yb	Ytterbium	-0.020		calc	196
71	Lu	Lutetium	0.34	0.01	LPT	223
72	Hf	Hafnium	0.1780	0.0006	LPES	343
73	Ta	Tantalum	0.322	0.012	LPES	1
74	W	Tungsten	0.81626	0.00007	LPES	360

Z	Mol. form.	Name	Elec. aff. (eV)	Uncert. (eV)	Method	Ref.
75	Re	Rhenium	0.060396	0.000063	LPES	506
76	Os	Osmium	1.07780	0.00012	LPT	430
77	Ir	Iridium	1.5638	0.0005	LPT	141
78	Pt	Platinum	2.12510	0.00005	LPT	431
79	Au	Gold	2.30863	0.00003	LPT	1
80	Hg	Mercury	not stable		e-scat	1
81	Tl	Thallium	0.377	0.013	LPES	341
82	Pb	Lead	0.356743	0.000016	LPES	408
83	Bi	Bismuth	0.942362	0.000013	LPT	262
84	Po	Polonium	1.9	0.3	calc	1
85	At	Astatine	2.8	0.2	calc	1
86	Rn	Radon	not stable		calc	1
87	Fr	Francium	0.46		calc	82
89	Ac	Actinium	0.35		calc	207
90	Th	Thorium	0.60769	0.00006	LPES	503
114	Fl	Flerovium	<0		calc	354
118	Og	Oganesson	0.056	0.01	calc	140
121	Ubu	Unbiunium	0.57		calc	207
Atomic Families						
57-71		Lanthanides			calc	355
89-103		Actinides			calc	355
Diatomic Molecules (Listed by Formula)						
	Ag_2	Disilver	1.023	0.007	LPES	37
	AgF	Silver(I) fluoride	1.46	0.01	LPES	403
	AgO	Silver(II) oxide	1.654	0.002	LPES	297
	Al_2	Dialuminum	1.10	0.15	LPES	68
	AlO	Aluminum monoxide	2.60	0.02	LPES	143
	AlP	Aluminum phosphide	2.043	0.020	LPES	218
	AlS	Aluminum monosulfide	2.60	0.03	LPES	129
	As_2	Diarsenic	0.739	0.008	LPES	200
	AsH	Arsenic monohydride	1.0	0.1	PT	2
	AsO	Arsenic monoxide	1.286	0.008	LPES	198
	Au_2	Digold	1.9393	0.0003	LPES	37
	AuBi	Gold bismuth	1.38	0.04	LPES	432
	AuH	Gold hydride	0.758	0.020	LPES	276
	AuO	Gold monoxide	2.374	0.007	LPES	282
	AuPd	Gold palladium	1.88		LPES	220
	AuS	Gold monosulfide	2.4734	0.0005	LPES	435
	BN	Boron nitride	3.160	0.005	LPES	189
	BO	Boron monoxide	2.508	0.008	LPES	6
	BeH	Beryllium monohydride	0.7	0.1	PT	2
	Bi_2	Dibismuth	1.271	0.008	LPES	119
	Br_2	Bromine	2.55	0.10	CT	2
	BrO	Bromine monoxide	2.353	0.006	LPES	88
	C_2	Dicarbon	3.269	0.006	LPES	87
	CH	Methylidyne	1.238	0.008	LPES	2
	CN	Cyanide	3.862	0.004	LPES	111
	CP	Carbon phosphide	2.8508	0.0007	LPES	515
	CRh	Rhodium carbide (RhC)	1.46	0.02	LPES	206
	CS	Carbon monosulfide	0.205	0.021	LPES	2
	CaH	Calcium monohydride	0.93	0.05	PT	2
	CeO	Cerium monoxide	0.936	0.007	LPES	453
	Cl_2	Chlorine	2.38	0.10	CT	2
	ClO	Chlorine oxide (ClO)	2.2775	0.0013	LPES	339
	Co_2	Dicobalt	1.110	0.008	LPES	27
	CoH	Cobalt hydride (CoH)	0.671	0.010	LPES	29
	CoD	Cobalt hydride (CoD)	0.680	0.010	LPES	29
	Cr_2	Dichromium	0.505	0.005	LPES	114
	CrH	Chromium monohydride	0.563	0.010	LPES	29
	CrD	Chromium monohydride-*d*	0.568	0.010	LPES	29

Z	Mol. form.	Name	Elec. aff. (eV)	Uncert. (eV)	Method	Ref.
	CrO	Chromium monoxide	1.221	0.006	LPES	5
	Cs_2	Dicesium	0.469	0.015	LPES	104
	CsCl	Cesium chloride	0.455	0.010	LPES	30
	CsO	Cesium monoxide	0.273	0.012	LPES	133
	Cu_2	Dicopper	0.836	0.006	LPES	37
	CuH	Copper(I) hydride	0.444	0.006	LPES	308
	CuD	Copper(I) hydride-*d*	0.439	0.006	LPES	308
	CuO	Copper(II) oxide	1.777	0.006	LPES	118
	EuH	Europium monohydride	0.771	0.009	LPES	452
	EuO	Europium(II) oxide	0.69	0.12	LPES	452
	F_2	Fluorine	3.01	0.07	kinetic	331
	FO	Fluorine oxide	2.272	0.006	LPES	88
	Fe_2	Diiron	0.902	0.008	LPES	27
	FeH	Iron hydride (FeH)	0.934	0.011	LPES	9
	FeD	Iron hydride (FeD)	0.932	0.015	LPES	9
	FeO	Iron(II) oxide	1.493	0.005	LPES	45
	GaAs	Gallium arsenide	1.949	0.020	LPES	218
	GaO	Gallium monoxide	2.612	0.008	LPES	279
	GaP	Gallium phosphide	1.988	0.020	LPES	218
	Ge_2	Digermanium	2.035	0.001	LPES	123
	HfO	Hafnium monoxide	0.60	0.05	LPES	173
	I_2	Iodine	2.524	0.015	LPES	305
	IBr	Iodine bromide	2.512	0.003	LPES	350
	IO	Iodine monoxide	2.378	0.006	LPES	88
	InBi	Indium bismuthide	1.72		LPES	364
	InP	Indium phosphide	1.845	0.020	LPES	218
	K_2	Dipotassium	0.497	0.012	LPES	104
	KBr	Potassium bromide	0.642	0.010	LPES	30
	KCl	Potassium chloride	0.582	0.010	LPES	30
	KCs	Cesium potassium	0.471	0.020	LPES	104
	KI	Potassium iodide	0.728	0.010	LPES	30
	RbK	Potassium rubidium	0.486	0.020	LPES	104
	LiCl	Lithium chloride	0.593	0.010	LPES	30
	LiH	Lithium hydride	0.342	0.012	LPES	102
	LiD	Lithium hydride-*d*	0.337	0.012	LPES	102
	MgCl	Magnesium monochloride	1.589	0.011	LPES	31
	MgD	Magnesium monohydride-*d*	0.89	0.05	LPES	385
	MgH	Magnesium monohydride	0.90	0.05	LPES	385
	MgI	Magnesium monoiodide	1.899	0.018	LPES	31
	MgO	Magnesium oxide	1.630	0.025	LPES	178
	MnH	Manganese monohydride	0.869	0.010	LPES	9
	MnD	Manganese monohydride-*d*	0.866	0.010	LPES	9
	MnO	Manganese(II) oxide	1.375	0.010	LPES	158
	MoC	Molybdenum carbide (MoC)	1.360	0.003	LPES	386
	MoO	Molybdenum monoxide	1.290	0.006	LPES	127
	NH	Imidogen	0.370	0.004	LPT	32
	NO	Nitric oxide	0.0298	0.0012	LPES	73
	NRh	Rhodium nitride (RhN)	1.51	0.02	LPES	206
	NS	Nitrogen sulfide	1.194	0.011	LPES	2
	Na_2	Disodium	0.430	0.015	LPES	104
	NaBr	Sodium bromide	0.788	0.010	LPES	30
	NaCl	Sodium chloride	0.727	0.010	LPES	30
	NaF	Sodium fluoride	0.520	0.010	LPES	30
	NaI	Sodium iodide	0.865	0.010	LPES	30
	NaK	Potassium sodium	0.465	0.030	LPES	104
	NbN	Niobium nitride	1.450	0.003	LPES	386
	NbO	Niobium(II) oxide	1.29	0.02	LPES	174
	Ni_2	Dinickel	0.926	0.010	LPES	112
	NiCu	Copper nickel (CuNi)	0.889	0.010	LPES	128
	NiAg	Nickel silver (NiAg)	0.979	0.010	LPES	128
	NiH	Nickel monohydride	0.481	0.007	LPES	29

Z	Mol. form.	Name	Elec. aff. (eV)	Uncert. (eV)	Method	Ref.
	NiD	Nickel monohydride-*d*	0.477	0.007	LPES	29
	NiO	Nickel(II) oxide	1.455	0.005	LPES	146
	O_2	Oxygen	0.450	0.002	LPES	222
	OH	Hydroxyl	1.8276488	0.0000009	LPT	226
	OD	Hydroxyl-*d*	1.825533	0.000037	LPT	142
	P_2	Diphosphorus	0.589	0.025	LPES	42
	PH	Phosphorus monohydride	1.027	0.006	LPES	206
	PO	Phosphorus monoxide	1.092	0.010	LPES	2
	Pb_2	Dilead	1.366	0.010	LPES	117
	PbO	Lead(II) oxide (litharge)	0.722	0.006	LPES	105
	PbS	Lead(II) sulfide	1.049	0.010	LPES	228
	Pd_2	Dipalladium	1.685	0.008	LPES	112
	PdCo	Cobalt palladium (CoPd)	0.604	0.010	LPES	160
	PdO	Palladium(II) oxide	1.672	0.005	LPES	146
	PrO	Praseodymium monoxide	0.96	0.01	LPES	415
	Pt_2	Diplatinum	1.898	0.008	LPES	112
	PtN	Platinum nitride (PtN)	1.240	0.010	LPES	46
	PtO	Platinum(II) oxide	2.172	0.005	LPES	146
	Rb_2	Dirubidium	0.498	0.015	LPES	104
	RbCl	Rubidium chloride	0.544	0.010	LPES	30
	RbCs	Cesium rubidium	0.478	0.020	LPES	104
	Re_2	Dirhenium	1.571	0.008	LPES	33
	RhO	Rhodium monoxide	1.58	0.02	LPES	206
	S_2	Disulfur	1.670	0.015	LPES	53
	SAr	Sulfur argon	2.1129	0.0006	LPES	485
	SF	Sulfur monofluoride	2.285	0.006	LPES	93
	SH	Mercapto	2.3147282	0.0000015	LPT	47
	SD	Mercapto-*d*	2.315	0.002	LPES	10
	SKr	Sulfur krypton	2.1343	0.0006	LPES	485
	SNe	Sulfur neon	2.0828	0.0006	LPES	485
	SO	Sulfur monoxide	1.125	0.005	LPES	84
	Sb_2	Diantimony	1.282	0.008	LPES	108
	ScO	Scandium monoxide	1.35	0.02	LPES	171
	Se_2	Diselenium	1.94	0.07	LPES	38
	SeH	Selenium monohydride	2.212519	0.000025	LPT	48
	SeO	Selenium monoxide	1.456	0.020	LPES	41
	Si_2	Disilicon	2.201	0.010	LPES	100
	SiF	Fluorosilylidyne	0.81	0.02	LPES	278
	SiH	Silylidyne	1.277	0.009	LPES	2
	SiN	Silicon nitride	2.949	0.008	LPES	274
	SmO	Samarium monoxide	1.0581	0.0011	LPES	457
	Sn_2	Ditin	1.962	0.010	LPES	117
	SnO	Tin(II) oxide	0.598	0.006	LPES	168
	SnPb	Lead tin (PbSn)	1.569	0.008	LPES	117
	TaC	Tantalum carbide	1.928	0.056	LPES	416
	TaO	Tantalum monoxide	1.07	0.06	LPES	360
	Te_2	Ditellurium	1.92	0.07	LPES	38
	TeH	Tellurium monohydride	2.102	0.015	LPES	39
	TeO	Tellurium monoxide	1.697	0.022	LPES	40
	TiH	Titanium monohydride	0.881	0.015	LPES	504
	TiO	Titanium(II) oxide	1.30	0.03	LPES	172
	VO	Vanadium(II) oxide	1.229	0.008	LPES	170
	WC	Tungsten carbide	1.022	0.010	LPES	409
	YO	Yttrium monoxide	1.35	0.02	LPES	171
	ZnBr	Zinc monobromide	2.45	0.06	LPES	497
	ZnF	Zinc monofluoride	1.974	0.008	LPES	179
	ZnH	Zinc monohydride	<0.95		PT	2
	ZnO	Zinc oxide	2.087	0.008	LPES	179
	ZrO	Zirconium(II) oxide	1.26	0.05	LPES	173

Z	Mol. form.	Name	Elec. aff. (eV)	Uncert. (eV)	Method	Ref.
Triatomic Molecules (Listed by Formula)						
	Ag_3	Silver trimer	2.32	0.05	LPES	37
	AgCN	Silver(I) cyanide	1.588	0.010	LPES	163
	Al_3	Aluminum trimer	1.916	0.004	LPES	332
	$AlBi_2$	Dibismuth aluminum cluster	1.94	0.08	LPES	399
	AlO_2	Aluminum oxide [AlO_2]	4.23	0.02	LPES	143
	AlP_2	Aluminum phosphide [AlP_2]	1.933	0.007	LPES	217
	AlTiO	Aluminum titanium oxide [AlTiO]	1.70	0.08	LPES	359
	Al_2N	Aluminum nitride (Al_2N)	2.571	0.008	LPES	297
	Al_2P	Aluminum phosphide (Al_2P)	2.513	0.020	LPES	217
	Al_2S	Aluminum sulfide (Al_2S)	0.80	0.12	LPES	129
	As_3	Arsenic trimer	1.45	0.03	LPES	200
	AsH_2	Arsenic hydride [AsH_2]	1.27	0.03	PT	2
	Au_3	Gold trimer	3.7	0.3	LPES	37
	AuBO	Gold-boron oxide complex (AuBO)	1.46	0.02	LPES	336
	$AuBr_2$	Gold dibromide	4.46	0.07	LPES	294
	AuC_2	Gold dicarbide	3.2192	0.0007	LPES	406
	$AuCl_2$	Gold dichloride	4.60	0.07	LPES	294
	AuH_2	Gold dihydride	3.03	0.02	LPES	444
	AuI_2	Gold diiodide	4.226	0.010	LPES	372
	AuOH	Gold hydroxide (AuOH)	1.771	0.015	LPES	336
	AuO_2	Gold dioxide	3.40	0.03	LPES	373
	AuS_2	Gold disulfide (SAuS)	3.42	0.03	LPES	373
	AuS_2	Gold disulfide (AuSS)	2.24	0.03	LPES	373
	Au_2H	Gold hydride (Au_2H)	3.437	0.003	LPES	276
	Au_2Pd	Gold palladium (Au_2Pd)	3.80		LPES	220
	B_3	Boron trimer	2.82	0.02	LPES	221
	BiBO	Bismuth boron oxide cluster	1.84	0.03	LPES	432
	BO_2	Boron dioxide	4.46	0.03	LPES	338
	B_2N	Boron nitride (B_2N)	3.098	0.005	LPES	193
	Bi_3	Bismuth trimer	1.60	0.03	LPES	119
	$BiIn_2$	Bismuth-indium complex ($BiIn_2$)	2.13		LPES	364
	Bi_2In	Bismuth-indium complex (Bi_2In)	2.11		LPES	364
	C_3	Propadienediylidene	1.981	0.020	LPES	11
	CBr_2	Dibromomethylene	1.78	0.10	LPES	249
	CCl_2	Dichloromethylene	1.593	0.006	LPES	249
	CF_2	Difluoromethylene	0.180	0.020	LPES	235
	CH_2	Methylene	0.652	0.006	LPES	12
	CD_2	Methylene-d_2	0.645	0.006	LPES	12
	CI_2	Diiodomethylene	2.09	0.07	LPES	235
	CHBr	Bromomethylene	1.454	0.005	LPES	95
	CHCl	Chloromethylene	1.210	0.005	LPES	95
	CHF	Fluoromethylene	0.542	0.005	LPES	95
	CDF	Fluoromethylene-*d*	0.535	0.005	LPES	95
	CHI	Iodomethylene	1.42	0.17	LPES	95
	C_2H	Ethynyl	2.969	0.006	LPES	87
	CNC	Carbon nitride [CNC]	2.7489	0.0010	LPES	346
	C_2Nb	Niobium carbide [NbC_2]	1.380	0.025	LPES	243
	CCO	Dicarbon monoxide	2.3107	0.0006	LPES	323
	C_2P	Phosohaethenylidene	2.6328	0.0006	LPES	515
	CCS	Dicarbon monosulfide	2.7475	0.0006	LPES	323
	CeO_2	Cerium(IV) oxide	0.66	0.10	LPES	494
	Ce_2O	Dicerium monoxide	1.06	0.02	LPES	455
	Ce_3	Tricerium cluster	0.67	0.01	LPES	455
	CS_2	Carbon disulfide	0.5525	0.0013	LPES	278
	ClO_2	Chlorine dioxide	2.1451	0.0025	LPES	339
	ClO_2	Chlorine superoxide [ClOO]	1.6600	0.0002	LPES	339
	CoH_2	Cobalt hydride [CoH_2]	1.450	0.014	LPES	34
	CoD_2	Cobalt hydride-d_2 [CoD_2]	1.465	0.013	LPES	34

Z	Mol. form.	Name	Elec. aff. (eV)	Uncert. (eV)	Method	Ref.
	CoOH	Cobalt hydroxide	1.33	0.04	LPES	400
	CrH_2	Chromium hydride [CrH_2]	>2.5		LPES	34
	Cr_2H	Hydrodichromium	1.474	0.005	LPES	107
	Cr_2D	Hydro-*d*-dichromium	1.464	0.005	LPES	107
	CrO_2	Chromium oxide radical (OCrO)	2.413	0.008	LPES	144
	CrO_2	Chromium oxide radical [$Cr(O_2)$]	1.5	0.06	LPES	241
	Cr_2O	Chromium oxide (Cr_2O)	0.9	0.1	LPES	306
	Cs_3	Cesium trimer	0.864	0.030	LPES	18
	CsI_2	Cesium diiodide	4.52	0.02	LPES	372
	Cu_3	Copper trimer	2.11	0.05	LPES	37
	$CuBr_2$	Copper(II) bromide	4.35	0.05	LPES	177
	CuCN	Copper(I) cyanide	1.466	0.010	LPES	163
	$CuCl_2$	Copper(II) chloride	4.35	0.05	LPES	177
	CuH_2	Copper(II) hydride	2.60	0.05	LPES	308
	CuD_2	Copper(II) hydride-*d*	2.60	0.05	LPES	308
	CuI_2	Copper(II) iodide	4.256	0.010	LPES	372
	EuOH	Europium monohydroxide	0.700	0.011	LPES	452
	Fe_3	Iron trimer	1.43	0.06	LPES	149
	FeC_2	Iron carbide [FeC_2]	1.9782	0.0006	LPES	254
	FeCO	Iron carbonyl (FeCO)	1.157	0.005	LPES	103
	FeH_2	Iron hydride [FeH_2]	1.049	0.014	LPES	34
	FeD_2	Iron hydride-d_2 [FeD_2]	1.038	0.013	LPES	34
	FeO_2	Iron oxide [FeO_2]	2.358	0.030	LPES	130
	Fe_2H	Iron hydride (Fe_2H)	0.564	0.019	LPES	254
	Fe_2O	Iron oxide (Fe_2O)	1.60	0.02	LPES	152
	$GaAs_2$	Gallium arsenide [$GaAs_2$]	1.894	0.033	LPES	192
	GaP_2	Gallium phosphide [GaP_2]	1.666	0.041	LPES	192
	Ga_2As	Gallium arsenide (Ga_2As)	2.428	0.020	LPES	192
	Ga_2N	Gallium nitride (Ga_2N)	2.506	0.008	LPES	302
	Ga_2P	Gallium phosphide (Ga_2P)	2.481	0.015	LPES	192
	Ge_3	Germanium trimer	2.23	0.01	LPES	123
	GeH_2	Germylene	1.097	0.015	LPES	28
	HCO	Formyl	0.313	0.005	LPES	35
	DCO	Formyl-*d*	0.301	0.005	LPES	35
	HCl_2	Hydrogen chloride [HCl_2]	4.896	0.005	LPES	69
	HNO	Nitrosyl hydride	0.338	0.015	LPES	14
	DNO	Nitrosyl hydride-*d*	0.330	0.015	LPES	14
	HO_2	Hydroperoxy	1.078	0.006	LPES	15
	DO_2	Hydroperoxy-*d*	1.077	0.005	LPES	15
	HS_2	Thiosulfeno	1.916	0.015	LPES	347
	DS_2	Thiosulfeno-*d*	1.918	0.015	LPES	347
	HfOH	Hafnium hydroxide	1.70	0.05	LPES	173
	HfO_2	Hafnium(IV) oxide	2.1045	0.0005	LPES	319
	ICN	Cyanogen iodide	1.345	+0.040/-0.020	LPES	383
	I_3	Iodine trimer	4.226	0.013	LPES	162
	InP_2	Indium phosphide [InP_2]	1.61	0.05	LPES	137
	In_2P	Indium phosphide (In_2P)	2.36	0.05	LPES	137
	K_3	Potassium trimer	0.956	0.050	LPES	18
	MnH_2	Manganese hydride [MnH_2]	0.444	0.016	LPES	34
	MnD_2	Manganese hydride-d_2 [MnD_2]	0.465	0.014	LPES	34
	MnO_2	Manganese(IV) oxide	2.06	0.03	LPES	158
	N_3	Nitrogen trimer	2.70	0.12	PT	2
	N_3	Nitrogen trimer	2.68	0.01	LPT	255
	NCN	Cyanoimidogen	2.484	0.006	LPES	154
	NCO	Cyanate	3.609	0.005	LPES	111
	NCS	Thiocyanate	3.537	0.005	LPES	111
	NH_2	Amidogen	0.771	0.005	LPES	58
	N_2O	Nitrous oxide	-0.03	0.10	calc	59
	NOAr	Nitric oxide argon cluster	0.084	0.003	LPES	79
	NOKr	Nitric oxide krypton cluster	0.125	0.003	LPES	79

Z	Mol. form.	Name	Elec. aff. (eV)	Uncert. (eV)	Method	Ref.
	NO_2	Nitrogen dioxide	2.273	0.005	LPES	63
	NOO	Nitrogen dioxide peroxy isomer	0.42	0.04	LPES	73
	NOXe	Nitric oxide xenon cluster	0.193	0.003	LPES	79
	NSO	Thiazate radical	3.113	0.001	LPES	458
	Na_3	Trisodium	1.019	0.060	LPES	18
	Nb_3	Niobium trimer	1.032	0.010	LPES	175
	Ni_3	Nickel trimer	1.41	0.05	LPES	55
	NiCN	Nickel(I) cyanide	1.771	0.010	LPES	287
	NiCO	Nickel carbonyl (NiCO)	0.804	0.012	LPES	2
	NiH_2	Nickel hydride [NiH_2]	1.934	0.008	LPES	34
	NiD_2	Nickel hydride-d_2 [NiD_2]	1.926	0.007	LPES	34
	ONiO	Nickel oxide [NiO_2]	3.043	0.005	LPES	146
	NiO_2	Nickel peroxide	0.82	0.03	LPES	214
	O_3	Ozone	2.1028	0.0025	LPT	2
	O_2Ar	Argon oxide [ArO_2]	0.52	0.02	LPES	75
	OCS	Carbon oxysulfide	-0.04		LPES	272
	OIO	Iodine dioxide	2.577	0.008	LPES	88
	PH_2	Phosphino	1.263	0.006	LPES	281
	P_2H	Diphosphenyl	1.514	0.010	LPES	281
	PO_2	Phosphorus dioxide	3.42	0.01	LPES	124
	Pb_3	Lead trimer	1.70	0.09	LPES	345
	Pd_3	Palladium trimer	<1.5	0.1	LPES	55
	PdCN	Palladium(I) cyanide	2.543	0.007	LPES	287
	PdCO	Palladium carbonyl	0.606	0.010	LPES	293
	PdO_2	Palladium dioxide	3.086	0.005	LPES	146
	Pt_3	Platinum trimer	1.87	0.02	LPES	55
	PtCN	Platinum(I) cyanide	3.191	0.003	LPES	287
	PtCO	Platinum carbonyl	1.196	0.034	LPES	391
	PtO_2	Platinum dioxide	2.677	0.005	LPES	146
	Rb_3	Rubidium trimer	0.920	0.030	LPES	18
	ReO_2	Rhenium(IV) oxide	2.5	0.1	LPES	216
	S_3	Sulfur trimer	2.3630	0.0009	LPES	483
	SO_2	Sulfur dioxide	1.107	0.008	LPES	16
	S_2O	Sulfur oxide (SSO)	1.877	0.008	LPES	16
	Sb_3	Antimony trimer	1.85	0.03	LPES	108
	ScO_2	Scandium dioxide	2.32	0.02	LPES	171
	SeO_2	Selenium dioxide	1.823	0.050	LPES	38
	SiH_2	Silylene	1.124	0.020	LPES	2
	SiF_2	Difluorosilylene	0.10	0.10	LPES	278
	Si_2F	Fluorodisilynylium	1.99	0.28	LPES	17
	Si_2H	Disilynyl	2.31	0.01	LPES	182
	Si_3	Silicon trimer	2.29	0.02	LPES	110
	SmCeO	Samarium cerium monoxide	0.80	0.02	LPES	454
	Sm_2O	Disamarium oxide	0.71	0.01	LPES	454
	Sn_3	Tin trimer	2.24	0.01	LPES	289
	SnCN	Tin(I) cyanide	1.922	0.006	LPES	292
	SNO	Thionitrite radical	2.16	0.05	LPES	458
	Ta_3	Tantalum trimer	1.36	0.03	LPES	169
	TaO_2	Tantalum(IV) oxide	2.40	0.06	LPES	260
	TiC_2	Titanium carbide [TiC_2]	1.542	0.020	LPES	147
	TiO_2	Titanium(IV) oxide (rutile)	1.5892	0.0005	LPES	482
	UF_2	Uranium difluoride	1.16	0.03	LPES	412
	UFO	Uranium oxide fluoride	1.27	0.03	LPES	410
	UO_2	Uranium(IV) oxide	1.159	0.020	LPES	413
	V_3	Vanadium trimer	1.107	0.010	LPES	176
	VO_2	Vanadium(IV) oxide	1.8357	0.0005	LPES	481
	WO_2	Tungsten(IV) oxide	1.998	0.050	LPES	299
	YO_2	Yttrium dioxide	2.00	0.03	LPES	171
	ZrO_2	Zirconium(IV) oxide	1.6397	0.0005	LPES	482

Z	Mol. form.	Name	Elec. aff. (eV)	Uncert. (eV)	Method	Ref.
Larger Polyatomic Molecules (Listed by Formula)						
	Ag_n	Silver cluster	n=1-60		LPES	37
	Al_4	Tetraaluminum cluster	2.18	0.02	LPT	421
	Al_5	Aluminum pentamer	2.23	0.05	LPES	238
	Al_n	Aluminum cluster	n=3-32		LPES	68
	$Al_nB_mH_x$ n=1-7 m,x=1-4	Aluminum boride hydride cluster			LPES	382
	Al_nH_m n=3-8 m=1-4	Aluminum hydride cluster			LPES	382
	AlB_6	Aluminum hexaboride	2.49	0.03	LPES	511
	$AlTiO_2$	Aluminum titanium oxide [$AlTiO_2$]	1.70	0.08	LPES	358
	$AlTiO_3$	Aluminum titanium oxide [$AlTiO_3$]	2.47	0.08	LPES	359
	Al_2C_2	Aluminum carbide [Al_2C_2]	0.64	0.05	LPES	239
	Al_2O_2	Dialuminum dioxide	1.87904	0.00004	LPES	509
	Al_2TiO_2	Aluminum titanium oxide [Al_2TiO_2]	1.17	0.08	LPES	359
	Al_2TiO_3	Aluminum titanium oxide [Al_2TiO_3]	2.2	0.1	LPES	359
	Al_3C	Aluminum carbide (Al_3C)	2.56	0.06	LPES	161
	Al_3C_2	Aluminum carbide [Al_3C_2]	2.19	0.03	LPES	244
	Al_3Ge_2	Aluminum germanium [Al_3Ge_2]	2.43	0.03	LPES	244
	Al_3Si_2	Aluminum silicon [Al_3Si_2]	2.36	0.03	LPES	244
	Al_3O	Oxotrialuminum	1.00	0.15	LPES	68
	Al_5O_3	Pentaaluminum trioxide	2.0626	0.0004	LPES	509
	Al_5O_4	Pentaaluminum tetraoxide	3.50	0.05	LPES	283
	$Al_5O_4{\cdot}H_2O$	Aluminum oxide (Al_5O_4)-water complex	3.10	0.10	LPES	283
	Al_6N	Aluminum-nitrogen complex (Al_6N)	2.58	0.04	LPES	337
	Al_8N	Aluminum-nitrogen complex (Al_8N)	2.75	0.05	LPES	348
	Al_nO_m	Aluminum oxide cluster	n=1,2	m=1-5	LPES	143
	Al_nO_m	Aluminum oxide cluster	n=3-7	m=2-5	LPES	267
	Al_nP_m	Aluminum phosphide cluster	n=1-4	m=1-4	LPES	217
	Al_nS_m	Aluminum sulfide cluster	n=1-5	m=1-3	LPES	129
	$Ar(H_2O)_n$	Argon-water cluster	n=2,6,7		LPES	77
	Ar_nBr	Argon-bromine cluster	n=2-9		ZEKE	212
	Ar_nI	Argon-iodine cluster	n=2-19		ZEKE	212
	As_4	Arsenic tetramer	<0.8		LPES	200
	As_5	Arsenic pentamer	≈1.7		LPES	200
	As_5	Arsenic pentamer	≈3.5		LPES	253
	Au_n	Gold cluster	n=1-233		LPES	37
	$AuBO_2$	Gold-borate complex ($AuBO_2$)	2.8	0.1	LPES	371
	AuF_6	Gold fluoride [AuF_6]	7.5	estimate	CT	98
	$AuO(BO_2)$	Gold-borate complex ($AuO(BO_2)$)	4.0	0.1	LPES	371
	$Au(BO_2)_2$	Gold-borate complex ($Au(BO_2)_2$)	5.7	0.1	LPES	371
	AuB_3	Gold triboride cluster	2.29	0.02	LPES	440
	AuBrCN	Gold cyanide bromide complex	5.14	0.05	LPES	449
	AuCNCl	Gold cyanide chloride complex	5.38	0.05	LPES	449
	AuCNI	Gold cyanide iodide complex	4.75	0.05	LPES	449
	AuC_2H	Gold ethyne	1.475	0.001	LPES	405
	AuC_4	Gold tetracarbide	3.366	0.001	LPES	405
	AuC_4H	Gold buta-1,3-diyne	1.778	0.001	LPES	405
	AuC_6H	Gold 1,3,5-hexatriyne	1.962	0.001	LPES	405
	AuC_6	Gold hexacarbide	3.593	0.001	LPES	405
	$Au(SCH_3)_2$	Gold bi(methylthiol)	3.33	0.03	LPES	447
	Au_2Al_2	Gold aluminum cluster	1.4438	0.0008	LPES	436
	Au_2BO	Gold-boron oxide complex (Au_2BO)	4.32	0.02	LPES	336
	Au_2B_3	Bigold triboride cluster	3.17	0.03	LPES	440
	Au_2Br_7	Gold-bromine complex	3.52	0.02	LPES	301
	Au_2C_2	Gold carbon complex (Au_2C_2)	1.60	0.08	LPES	456
	Au_2C_2H	Digold dicarbon hydride	4.23	0.08	LPES	456
	$Au_2(SCH_3)_3$	Bigold trimethylthiol	4.06	0.03	LPES	447
	Au_3BO	Gold-boron oxide complex (Au_3BO)	3.08	0.02	LPES	336
	Au_3BO_2	Gold-borate complex (Au_3BO_2)	3.1	0.1	LPES	371
	$Au_3O(BO_2)$	Gold-borate complex ($Au_3O(BO_2)$)	4.9	0.1	LPES	371
	Au_3Pd	Gold palladium (Au_3Pd)	2.51		LPES	220
	Au_4	Gold cluster (Au_4)	2.7098	0.0006	LPES	441

Z	Mol. form.	Name	Elec. aff. (eV)	Uncert. (eV)	Method	Ref.
	Au_4Pd	Gold palladium (Au_4Pd)	2.69		LPES	220
	Au_4Ti	Gold titanium cluster (Au4Ti)	2.80	0.05	LPES	438
	Au_6	Gold hexamer	2.06	0.02	LPES	288
	$Au_6(CO)$	Gold hexamer carbonyl	2.04	0.05	LPES	288
	$Au_6(CO)_2$	Gold hexamer dicarbonyl	2.03	0.05	LPES	288
	$Au_6(CO)_3$	Gold hexamer tricarbonyl	1.95	0.05	LPES	288
	$Au_{12}Nb$	Gold-niobium complex	3.70	0.03	LPES	275
	$Au_{12}Ta$	Gold-tantalum complex	3.77	0.03	LPES	275
	$Au_{12}V$	Gold-vanadium complex	3.76	0.03	LPES	275
	B_5	Boron pentamer	2.33	0.02	LPES	245
	B_{11}	Boron cluster (B11)	3.401	0.002	LPES	492
	B_{12}	Boron cluster (B12)	2.221	0.002	LPES	492
	B_{37}	Boron cluster (B37)	3.91	0.05	LPES	490
	B_{38}	Boron cluster (B38)	3.76	0.05	LPES	490
	BH_3	Borane(3)	0.038	0.015	LPES	62
	BD_3	Borane(3)-d_3	0.027	0.014	LPES	62
	BO_2Fe_n	Boron dioxide-iron cluster	n=1-5		LPES	358
	B_3N	Boron nitride (B_3N)	2.098	0.035	LPES	193
	B_3O_3	Boron oxide complex (B_3O_3)	3.94	0.08	LPES	422
	B_3O_3H	Boron oxide hydride complex	1.50	0.08	LPES	422
	$B_3(BO)_2$	Boron oxide complex [$B_3(BO)_2$]	4.44	0.02	LPES	440
	B_3BO	Boron oxide complex (B_3BO)	2.71	0.02	LPES	440
	B_4O_4	Boron oxide complex (B_4O_4)	1.42	0.08	LPES	418
	$B_4(BO)$	Boron oxide complex ($B_4(BO)$)	3.45	0.01	LPES	446
	$B_4(BO)_n$,n=1-3	Boron oxide complex $B_4(BO)_n$,n-1,3			LPES	446
	Bi_n	Bismuth cluster	n=2-9		LPES	213
	$AlBi_2$	Dibismuth aluminum cluster	1.94	0.08	LPES	399
	Bi_2Al_2	Dibismuth dialuminum cluster	2.11	0.08	LPES	399
	Bi_2Al_3	Dibismuth trialuminum cluster	2.30	0.08	LPES	399
	Bi_2Al_4	Dibismuth tetraaluminum cluster	1.84	0.08	LPES	399
	Bi_2B_2	Bismuth-boron complex (Bi_2B_2)	2.2783	0.0054	LPES	513
	Bi_2B_3	Bismuth-boron complex (Bi_2B_3)	2.5805	0.0009	LPES	513
	Bi_2B_4	Bismuth-boron complex (Bi_2B_4)	2.0102	0.0004	LPES	513
	Bi_2In_2	Bismuth-indium complex (Bi_2In_2)	1.82		LPES	364
	Bi_2In_3	Bismuth-indium complex (Bi_2In_3)	2.36		LPES	364
	Bi_3Ga	Bismuth-gallium complex (Bi_3Ga)	1.87	0.06	LPES	363
	Bi_3Ga_2	Bismuth-gallium complex (Bi_3Ga_2)	2.39	0.05	LPES	363
	Bi_3Ga_3	Bismuth-gallium complex (Bi_3Ga_3)	1.84	0.06	LPES	363
	Bi_3Ga_4	Bismuth-gallium complex (Bi_3Ga_4)	2.29	0.08	LPES	363
	Bi_3In_2	Bismuth-indium complex (Bi_3In_2)	2.42		LPES	364
	Bi_4	Bismuth tetramer	1.05	0.010	LPES	119
	Bi_5	Bismuth pentamer	2.87	0.02	LPES	253
	$Br(CO_2)$	Bromine-carbon dioxide complex	3.582	0.017	LPES	131
	$Br(H_2O)_n$	Bromine-water cluster	n=1-4		LPES	250
	C_n	Carbon cluster	n=2-84		LPES	70
	C_nCr	Carbon-chromium cluster [C_nCr]	n=2-8		LPES	271
	C_nNb	Carbon-niobium cluster	n=2-7		LPES	243
	$(CO_2)_n$	Carbon dioxide cluster	n=1,2		LPES	75
	$(CS)_n$	Carbon monosulfide cluster	n=2		LPES	75
	$(CS_2)_n$	Carbon disulfide cluster	n=1,2		LPES	75
	CAl_3Ge	[μ_3-(Germanetetraylmethanetetrayl)]trialuminum	2.70	0.06	LPES	224
	CAl_3Si	[μ_3-(Methanetetraylsilanetetrayl)]trialuminum	2.77	0.06	LPES	224
	CB_9	Carbon boron complex (CB_9)	3.61	0.03	LPES	443
	CCl_4	Tetrachloromethane	≤1.14		CT	266
	CF_3	Trifluoromethyl	1.82	0.05	LPES	187
	CF_3Br	Bromotrifluoromethane	0.91	0.2	CD	2
	CF_3I	Trifluoroiodomethane	1.57	0.2	CD	2
	CFO_2	Fluorooxomethoxy	4.277	0.030	LPES	131
	$CHBr_2$	Dibromomethyl	1.9	0.2	LPES	367
	$CDBr_2$	Dibromomethyl-d	1.9	0.2	LPES	367
	$CHCl_2$	Dichloromethyl	1.3	0.2	LPES	367

Z	Mol. form.	Name	Elec. aff. (eV)	Uncert. (eV)	Method	Ref.
	$CDCl_2$	Dichloromethyl-*d*	1.3	0.2	LPES	367
	CHI_2	Diiodomethyl	1.9	0.2	LPES	367
	CHO_2	Formyloxyl	3.4961	0.0010	LPES	109
	CHO_2	*cis*-Hydroxyl formyl	1.511	0.005	LPES	395
	CHO_2	*trans*-Hydroxyl formyl	1.381	0.005	LPES	395
	CDO_2	Formyloxyl-*d*	3.5164	0.0010	LPES	109
	CHO_3	Formyldioxidanyl radical	2.493	0.006	LPES	466
	CH_2CN	Cyanomethyl radical	1.54583	0.00025	LPES	404
	CD_2CN	Cyanomethyl-d_2 radical	1.53765	0.00025	LPES	404
	CH_2S	Thioformaldehyde	0.465	0.023	LPES	53
	CD_3S	Methyl-d_3-sulfoniumylidene	1.856	0.006	LPT	2
	CH_3	Methyl	0.093	0.003	LPES	462
	CD_3	Methyl-d_3 radical	0.082	0.004	LPES	462
	$CHCl_3$	Trichloromethane	≤0.78		CT	266
	CH_3I	Iodomethane	0.11	0.02	LPES	277
	CH_3NO_2	Nitromethane	0.172	0.006	LPES	211
	CH_3O	Methoxy	1.5689	0.0007	LPES	475
	CD_3O	Methoxy-d_3	1.5548	0.0007	LPES	475
	$H_2C(OH)O$	Hydroxymethoxy radical	2.220	0.002	LPES	460
	CH_3O_2	Methyldioxy	1.161	0.005	LPES	188
	CD_3O_2	Methyl-d_3-dioxy	1.154	0.004	LPES	188
	CH_3S	Methylthio	1.867	0.004	LPES	166
	CH_3S_2	Methyldithio	1.757	0.022	LPES	53
	CD_3S_2	Methyldithio-d_3	1.748	0.022	LPES	53
	CH_3Si	Methylsilylidyne	0.852	0.010	LPES	97
	$CH_2{=}SiH$	Methylenesilyl	2.010	0.010	LPES	97
	CH_3NH	Methylamidogen	0.432	0.015	LPES	215
	$CH_2(NH_2)O$	Aminomethoxy radical	1.944	0.001	LPES	501
	CH_5Si	Methylsilyl	1.19	0.04	LPT	65
	CO_3	Carbon oxide [CO_3]	2.69	0.14	LPES	2
	$CO_3(H_2O)$	Carbon oxide (CO_3)-water complex	2.1	0.2	PT	2
	C_2B_8	Carbon boron complex (C_2B_8)	2.67	0.03	LPES	443
	C_2F_2	Difluorovinylidene	2.255	0.006	LPES	106
	C_2HF	Fluorovinylidene	1.718	0.006	LPES	106
	C_2HFe	Ethynyliron	1.4512	0.0025	LPES	254
	C_2HN	Cyanomethylene	2.001	0.015	LPES	465
	C_2DN	Cyanomethylene-*d*	1.998	0.015	LPES	465
	C_2HN	Isocyanomethylene	1.883	0.013	LPES	219
	C_2DN	Isocyanomethylene-*d*	1.877	0.010	LPES	219
	C_2HNi	Ethynylnickel	1.063	0.019	LPES	254
	PdC_2H	Palladium(I) acetylide	1.98	0.03	LPES	287
	PdC_2HN	Cyanomethylenepalladium	2.17	0.03	LPES	291
	PtC_2H	Platinum(I) acetylide	2.650	0.010	LPES	287
	C_2HO	Oxoethenyl	2.338	0.008	LPES	190
	C_2DO	Oxoethenyl-*d*	2.350	0.020	LPES	13
	C_2H_2	Vinylidene	0.4879	0.0009	LPES	499
	C_2HD	Vinylidene-d_1	0.489	0.006	LPES	83
	C_2D_2	Vinylidene-d_2	0.4886	0.0021	LPES	499
	$C_2H_2BrO_2$	Bromoacetate (CH_2BrCOO)	3.97	0.03	LPES	310
	$C_2H_2ClO_2$	Chloroacetate (CH_2ClCOO)	3.93	0.03	LPES	310
	C_2H_2FO	Acetyl fluoride enolate	2.22	0.09	PT	2
	$C_2H_2FO_2$	Fluoroacetate (CH_2FCOO)	3.80	0.03	LPES	310
	$HFeC_2H$	Ethynylhydroiron	1.328	0.019	LPES	254
	C_2H_2Mo	η^2-Acetylene-molybdenum complex (C_2H_2Mo)	0.718	0.010	LPES	495
	C_2H_2MoO	η^2-Acetylene-molybdenum complex (C_2H_2MoO)	1.680	0.07	LPES	495
	$C_2H_2MoO_2$	η^2-Acetylene-molybdenum complex ($C_2H_2MoO_2$)	2.51	0.07	LPES	495
	CH_2CN	Cyanomethyl	1.548	0.005	LPES	21
	CD_2CN	Cyanomethyl-d_2	1.538	0.012	LPES	21
	CH_2NC	Isocyanomethyl	1.059	0.024	LPES	21
	CD_2NC	Isocyanomethyl-d_2	1.070	0.024	LPES	21
	$C_2H_2N_3$	1,2,3-Triazolyl	3.447	0.004	LPES	316

Z	Mol. form.	Name	Elec. aff. (eV)	Uncert. (eV)	Method	Ref.
	$C_2H_2N_3$	2*H*-1,2,3-Triazol-4yl radical	1.865	0.004	LPES	316
	$C_2H_2N_3$	*EE*-Iminodiazomethyl radical	2.484	0.007	LPES	468
	$C_2H_2N_3$	*ZE*-Iminodiazomethyl radical	2.484	0.007	LPES	468
	$H_2C_2O_4$	Oxalic acid	0.72	0.05	LPES	393
	$HNiC_2H$	Ethynylhydronickel	2.531	0.005	LPES	287
	C_2H_3	Vinyl	0.667	0.024	LPES	90
	C_2H_3	1-Propynyl	2.7355	0.0010	LPES	311
	C_2H_3Fe	Ethynyldihydroiron	1.587	0.019	LPES	254
	C_2H_3Ni	Ethynyldihydronickel	1.103	0.019	LPES	254
	C_2H_3O	Acetaldehyde enolate	1.8249	0.0012	LPT	22
	C_2D_3O	Acetaldehyde-d_3 enolate	1.8191	0.0012	LPT	22
	CH_3COO	Acetyloxyl radical	3.2528	0.0010	LPES	402
	$CH_3C(O)OO$	Acetyldioxidanyl radical	2.381	0.007	LPES	467
	C_2H_5N	Ethylnitrene	0.56	0.01	PT	2
	$C_2H_5NO_2$	Nitroethane	0.3	0.2	LPES	322
	$(C_2H_5NO_2)_n$	Nitroethane cluster	n=0-4		LPES	322
	C_2H_5O	Ethoxy	1.712	0.004	LPES	194
	C_2D_5O	Ethoxy-d_5	1.699	0.004	LPES	194
	$C_2H_5O_2$	Ethyldioxy	1.186	0.004	LPES	188
	C_2H_5S	Ethylthio	1.953	0.006	LPT	2
	C_2H_5S	(Methylthio)methyl	0.868	0.051	LPES	53
	$C_2H_7O_2$	Methanol-methoxy complex	2.26	0.08	PT	50
	C_3Fe	1,2-Propadiene-1,3-diylideneferrate	1.69	0.08	LPES	132
	C_3H	Cyclopropenylidyne	1.999	0.003	LPES	315
	C_3D	Cyclopropenylidyne-*d*	1.997	0.005	LPES	315
	C_3HFe	2-Propynylidyneferrate	1.58	0.06	LPES	132
	CCHCN	Cyanovinyldyl radical	1.84	0.01	LPES	335
	CH_2CC	Propadienylidene	1.7957	0.0010	LPES	463
	HCCCH	Propargylene	1.156	+0.010, -0.095	LPES	396
	$C_3H_2F_3O$	1,1,1-Trifluoroacetone enolate	2.625	0.010	LPT	113
	$C_3H_2N_3$	1,3,5-Triazinyl	1.529	0.006	LPES	397
	CH_3CC	1-Propynyl	2.7355	0.0010	LPES	311
	CD_3CC	1-Propynyl-d_3	2.7300	0.0010	LPES	311
	CH_2CCH	2-Propynyl	0.918	0.008	LPES	153
	C_3H_2D	2-Propynyl-d_1	0.88	0.15	LPES	24
	C_3D_2H	2-Propynyl-d_2	0.907	0.023	LPES	24
	$C_3H_3N_2$	Vinyldiazomethyl radical	1.864	0.007	LPES	470
	$C_3H_3N_2$	1*H*-Pyrazolyl radical	2.938	0.005	LPES	472
	$C_3D_3N_2$	1*D*-Pyrazolyl-d_3 radical	2.935	0.006	LPES	473
	C_3H_4N	1-Cyanoethyl	1.247	0.012	LPES	21
	C_3H_5	Allyl	0.481	0.008	LPES	138
	C_3H_4D	Allyl-d_1	0.373	0.019	LPES	25
	$C_3H_4N_2$	Imidazolyl radical	2.613	0.006	LPES	290
	C_3H_4O	Oxyallyl radical	1.940	0.010	LPES	464
	C_3D_5	Allyl-d_5	0.464	0.006	LPES	138
	C_3H_5	Cyclopropyl	0.397	0.069	kinetic	155
	C_3H_5O	Acetone enolate	1.758	0.019	LPT	113
	C_3H_5O	Propanal enolate	1.621	0.006	LPT	113
	C_3H_5O	*i*-Methylvinoxy radical	1.747	0.002	LPES	486
	C_3H_5O	*trans-n*-Methylvinoxy radical	1.570	0.002	LPES	363
	C_3H_5O	*cis-n*-Methylvinoxy radical	1.6106	0.0008	LPES	363
	C_3D_5O	*trans-n*-Methylvinoxy-d_5 radical	1.561	0.001	LPES	363
	C_3D_5O	*cis-n*-Methylvinoxy-d_5 radical	1.603	0.002	LPES	363
	$C_3H_5O_2$	Methyl acetate enolate	1.80	0.06	PT	2
	$C_3H_7NO_2$	1-Nitropropane	0.223	0.006	LPES	392
	C_3H_7O	Propoxy	1.789	0.033	LPES	23
	C_3H_7O	Isopropoxy	1.847	0.004	LPES	194
	C_3H_7S	Propylthio	2.00	0.02	PT	2
	C_3H_7S	Isopropylthio	2.02	0.02	PT	2
	C≡C-C≡N	Cyanoethynyl	4.305	0.001	LPES	487
	C_3N_2	Dicyanocarbene	<3.25	0.05	LPES	237

Z	Mol. form.	Name	Elec. aff. (eV)	Uncert. (eV)	Method	Ref.
	O=C=C=C:	3-Oxo-1,3-propadienylidene	1.237	0.003	LPES	351
	C_3O_2	Carbon suboxide	0.85	0.15	LPES	11
	C_3S	3-Thioxo-1,2-propadienylidene	1.5957	0.0010	LPES	351
	$C_4F_4Cl_2$	1,2-Dichloro-3,3,4,4-tetrafluorocyclobutene	0.87	0.08	attach	258
	$C_4F_4O_3$	3,3,4,4-Tetrafluorodihydro-2,5-furandione	0.5	0.2	CD	2
	C-C_4F_8	Perfluorocyclobutane	0.63	0.05	attach	256
	C_4Fe	1,2,3-Butatriene-1,4-diylideneferrate	<2.2	0.2	LPES	132
	C_4H	1,3-Butadiynyl radical	3.5332	0.0010	LPES	314
	C_4D	1,4-Butadiynyl-*d*	3.5308	0.0012	LPES	314
	C_4HN	3-Cyano-1,2-propadienylidene	2.05	0.08	LPES	465
	HC_4Fe	1,3-Butadiynylferrate	1.67	0.06	LPES	132
	C_4H_2Fe	[(1,2-η)-1,3-Butadiyne]iron	1.633	0.019	LPES	254
	$C_4H_2O_3$	Maleic anhydride	1.44	0.10	CT	61
	C_4H_3Fe	Acetylene-iron complex	1.182	0.019	LPES	254
	$C_4H_3N_2$	1,2-Diazin-4-yl	1.850	0.006	LPES	397
	$C_4H_3N_2$	1,3-Diazin-5-yl	1.840	0.006	LPES	397
	$C_4H_3N_2$	1,4-Diazinyl	1.254	0.006	LPES	397
	$C_4H_3N_2O$	2-Hydroxypyrimidine (deprotonated)	3.2248	0.0006	LPES	491
	C_4H_3Ni	Acetylene-nickel complex	0.824	0.019	LPES	254
	C_4H_3O	α-Furanyl radical	1.8546	0.0004	LPES	477
	C_4H_3O	β-Furanyl radical	1.6566	0.0004	LPES	477
	C_4H_4	Vinylvinylidene	0.914	0.015	LPES	125
	C_4H_4Mo	η^2-Acetylene-molybdenum complex (C_4H_4Mo)	1.5	0.1	LPES	495
	C_4D_4	Vinylvinylidene-d_4	0.909	0.015	LPES	125
	C_4H_4N	1*H*-Pyrrol-1-yl	2.145	0.010	LPES	265
	$NO{\cdot}C_4H_4N_2$	Nitric oxide-pyrimidine complex	0.75		LPT	285
	$C_4H_4N_3O$	Cytosine, deprotonated	3.037	0.015	LPES	309
	$C_4H_5N_2$	*N*-Methyl-5-pyrazolyl radical	2.054	0.006	LPES	471
	$C_4H_5N_2$	*N*-Methyl-5-imidazolyl radical	1.987	0.008	LPES	471
	C_4H_5O	Cyclobutanone enolate	1.801	0.008	LPT	113
	C_4H_6	Trimethylenemethane	0.431	0.006	LPES	135
	$C_4H_6O_2$	2,3-Butanedione	0.69	0.10	CT	61
	C_4H_7	2-Methylallyl	0.505	0.006	LPES	138
	C_4H_6D	2-Methylallyl-d_7	0.493	0.008	LPES	138
	C_4H_7O	Butanal enolate	1.67	0.05	PT	2
	C_4H_5DO	2-Butanone-3-d_1 enolate	1.67	0.05	PT	2
	$C_4H_4D_2O$	2-Butanone-3,3-d_2 enolate	1.75	0.06	PT	2
	$C_4H_9NO_2$	1-Nitrobutane	0.240	0.006	LPES	392
	$C_4H_9O_2$	*tert*-Butyl peroxy radical	1.1962	0.0020	LPES	476
	C_4H_9O	*tert*-Butoxyl	1.909	0.004	LPES	194
	C_4H_9S	Butylthio	2.03	0.02	PT	2
	C_4H_9S	*tert*-Butylthio	2.07	0.02	PT	2
	C_4N	Carbon nitride (CCCCN)	3.1113	0.0010	LPES	346
	C_4O	4-Oxo-1,2,3-butatrienylidene	2.05	0.15	LPES	11
	C_4O_2	1,2,3-Butatriene-1,4-dione	2.0	0.2	LPES	11
	C_4Ti	Titanium carbide [TiC_4]	1.494	0.020	LPES	147
	C_5	Pentatetraenediylidene	2.8539	0.0004	LPES	99
	C_5F_5N	Perfluoropyridine	0.70	0.05	attach	259
	$C_5F_6O_3$	Hexafluoroglutaric anhydride	1.5	0.2	CD	2
	C_5H	2,4-Pentadiynylidyne	2.4225	0.0010	LPES	488
	C_5H	Pentacyclopentyl	2.857	0.028	LPES	317
	C_5HF_4N	Tetrafluoropyridine	0.40	0.08	attach	259
	C_5H_4N	4-Pyridyl	1.4797	0.0005	LPES	510
	C_5H_4N	3-Pyridyl	1.4473	0.005	LPES	510
	C_5H_4N	2-Pyridyl	0.8669	0.007	LPES	510
	C_5H_5	Cyclopentadienyl	1.808	0.006	LPES	469
	C_5D_5	Cyclopentadienyl-d_5	1.790	0.008	LPES	11
	$(C_5H_5N)_nCo_m$	Cobalt-pyridine cluster	n=1-4	m=1-3	LPES	327
	$O_2{\cdot}C_5H_5N$	Oxygen-pyridine complex	1.39		LPES	285
	$NO{\cdot}C_5H_5N$	Nitric oxide-pyridine complex	0.62		LPES	285
	$C_5H_5N_2O_2$	Thymine, deprotonated	3.2635	0.0006	LPES	434

Z	Mol. form.	Name	Elec. aff. (eV)	Uncert. (eV)	Method	Ref.
	C_5H_7	2,4-Pentadienyl	0.91	0.03	PT	2
	$O_2{\cdot}C_5H_5N{\cdot}H_2O$	Oxygen-pyridine-water complex	1.87		LPES	285
	C_5H_7O	Cyclopentanone enolate	1.598	0.007	LPT	113
	C_5H_9O	3-Pentanone enolate	1.69	0.05	PT	2
	$C_5H_{11}O$	Neopentoxyl	1.93	0.05	LPT	2
	$C_5H_{11}S$	Pentylthio	2.09	0.02	PT	2
	C_5N	4-Cyano-1,3-butadiyn-1-yl	4.45	0.03	LPES	487
	C_5O_2	1,2,3,4-Pentatetraene-1,5-dione	1.2	0.2	LPES	11
	C_5Ti	Titanium carbide (TiC_5)	1.748	0.050	LPES	147
	C_6	Carbon hexamer	4.185	0.006	LPT	8
	$C_6Br_4O_2$	Tetrabromobenzoquinone	2.44	0.20	CT	2
	$C_6Cl_4O_2$	2,3,5,6-Tetrachloro-2,5-cyclohexadiene-1,4-dione	2.78	0.10	CT	61
	$C_6F_4O_2$	Tetrafluorobenzoquinone	2.70	0.10	CT	61
	C_6F_5Br	Bromopentafluorobenzene	1.15	0.11	CT	67
	C_6F_5Cl	Chloropentafluorobenzene	0.75	0.05	attach	260
	C_6F_5I	Pentafluoroiodobenzene	1.41	0.11	CT	67
	$C_6F_5NO_2$	Perfluoronitrobenzene	1.52	0.11	CT	67
	C_6F_6	Hexafluorobenzene	0.53	0.05	attach	257
	C_6F_{10}	Perfluorocyclohexene	>1.4	0.3	CT	2
	C≡C-C≡C-C≡CH	1,3,5-Hexatriynyl	3.809	0.001	LPES	488
	$C_6H_2Cl_2O_2$	2,6-Dichloro-2,5-cyclohexadiene-1,4-dione	2.48	0.10	CT	61
	$C_6H_2O_2$	Benzynequinone	1.859	0.005	LPES	232
	$C_6H_3F_2NO_2$	2,4-Difluoro-1-nitrobenzene	1.17	0.10	CT	61
	$C_6H_3O_2$	Benzoquinonide	<2.18		LPES	232
	C_6H_4	Benzyne	0.560	0.010	LPES	36
	C_6D_4	Benzyne-d_4	0.551	0.010	LPES	36
	$C_6H_4BrNO_2$	1-Bromo-2-nitrobenzene	1.16	0.10	CT	61
	$C_6H_4BrNO_2$	1-Bromo-3-nitrobenzene	1.32	0.10	CT	61
	$C_6H_4BrNO_2$	1-Bromo-4-nitrobenzene	1.29	0.10	CT	61
	$C_6H_4ClNO_2$	1-Chloro-2-nitrobenzene	1.14	0.10	CT	61
	$C_6H_4ClNO_2$	1-Chloro-3-nitrobenzene	1.28	0.10	CT	61
	$C_6H_4ClNO_2$	1-Chloro-4-nitrobenzene	1.26	0.10	CT	61
	C_6H_4BrO	2-Bromophenoxyl	2.5480	0.0007	LPES	512
	C_6H_4ClO	2-Chlorophenoxyl	2.4917	0.0008	LPES	512
	C_6H_4FO	2-Fluorophenoxyl	2.2950	0.0006	LPES	512
	C_6H_4IO	2-Iodophenoxyl	2.2533	0.0004	LPES	512
	$C_6H_4FNO_2$	1-Fluoro-2-nitrobenzene	1.07	0.10	CT	61
	$C_6H_4FNO_2$	1-Fluoro-3-nitrobenzene	1.23	0.10	CT	61
	$C_6H_4FNO_2$	1-Fluoro-4-nitrobenzene	1.12	0.10	CT	61
	$C_6H_4N_2O_4$	1,2-Dinitrobenzene	1.65	0.10	CT	61
	$C_6H_4N_2O_4$	1,3-Dinitrobenzene	1.65	0.10	CT	61
	$C_6H_4N_2O_4$	1,4-Dinitrobenzene	2.00	0.10	CT	61
	$C_6H_4O_2$	*p*-Benzoquinone	1.860	0.005	LPES	284
	C_6H_5	Phenyl	1.096	0.006	LPES	26
	C_6D_5	Phenyl-d_5	1.092	0.020	LPES	26
	C_6H_5N	Phenylimidogen	1.429	0.011	LPT	115
	C_6D_5N	Phenylimidogen-d_5	1.44	0.02	LPES	96
	C_6H_5NH	Anilino	1.607	0.004	LPES	390
	$C_6H_5NO_2$	Nitrobenzene	1.00	0.01	LPES	164
	C_6H_5O	Phenoxy	2.2538	0.0008	LPES	484
	$HO(C_6H_4)O$	2-Hydroxyphenoxyl radical	2.3289	0.0006	LPES	419
	$HO(C_6H_4)O$	2-Hydroxyphenoxyl radical	2.3292	0.0004	LPES	479
	HOC_6H_4O	3-Hydroxyphenoxyl radical	2.330	0.010	LPES	375
	$C_6H_5O_2$	4-Hydroxyphenoxyl radical	1.990	0.010	LPES	375
	$HO(C_6H_4)O$	*anti*-3-Hydroxyphenoxyl radical	2.3454	0.0006	LPES	493
	$HO(C_6H_4)O$	*syn*-3-Hydroxyphenoxyl radical	2.3371	0.0006	LPES	493
	C_6H_5S	Phenylthio	2.3542	0.0006	LPES	484
	$(C_6H_6)_nCo_m$	Cobalt-benzene cluster	n=1-4	m=1-5	LPES	326
	$(C_6H_6)_nFe_m$	Iron-benzene cluster	n=1-4	m=1-7	LPES	329
	$NO{\cdot}C_6H_6$	Nitric oxide-benzene complex	0.44		LPES	285
	NbC_6H_6	Niobium benzene	0.893	0.006	LPES	311

Z	Mol. form.	Name	Elec. aff. (eV)	Uncert. (eV)	Method	Ref.
	$O_2{\cdot}C_6H_6$	Oxygen-benzene complex	1.06		LPES	285
	$(C_6H_6)_n$	Benzene cluster	*n*=53-124		LPES	248
	C_6H_7	Methylcyclopentadienyl	<1.67	0.04	PT	2
	$C_6H_7NO_2$	Ethyl 2-cyanoacrylate	0.9	0.2	LPES	388
	$(CH_2)_2CC(CH_2)_2$	Tetramethyleneethane	0.855	0.010	LPES	203
	C_6H_8Si	Phenylsilane	1.435	0.004	LPT	65
	$CH_2{=}C(CH_3)\text{-}C(CH_2)_2$	1,1,2-Trimethylene-2-methylethane	0.654	0.010	LPES	203
	C_6H_9O	Cyclohexanone enolate	1.526	0.010	LPT	113
	C_6H_{10}	*tert*-Butylvinylidene	0.645	0.015	LPES	126
	$C_6H_{11}O$	Pinacolone enolate	1.755	+0.05/-0.005	LPT	113
	$C_6H_{11}O$	3,3-Dimethylbutanal enolate	1.82	0.06	PT	2
	C_6N	Carbon nitride (CCCCCCN)	3.3715	0.0010	LPES	346
	C_6N_4	Tetracyanoethene	2.3	0.3	PT	2
	C_6O_6	Carbon monoxide trimer	2.54	0.05	LPES	333
		Carbon 7-cluster (linear)	3.3517	0.0004	LPES	508
	C_7F_5N	Pentafluorobenzonitrile	1.11	0.11	CT	67
	C_7F_8	Perfluorotoluene	0.86	0.11	CT	67
	C_7F_{14}	Perfluoromethylcyclohexane	1.06	0.10	CT	56
	C_7HF_5O	Pentafluorobenzaldehyde	1.10	0.11	CT	67
	$C_7H_3N_3O_4$	3,5-Dinitrobenzonitrile	2.16	0.10	CT	61
	$C_7H_4F_3NO_2$	1-Nitro-3-(trifluoromethyl)benzene	1.41	0.10	CT	61
	$C_7H_4N_2O_2$	2-Nitrobenzonitrile	1.61	0.10	CT	61
	$C_7H_4N_2O_2$	3-Nitrobenzonitrile	1.56	0.10	CT	61
	$C_7H_4N_2O_2$	4-Nitrobenzonitrile	1.72	0.10	CT	61
	C_7H_6Br	*o*-Bromobenzyl	1.308	0.008	LPES	167
	C_7H_6Br	*m*-Bromobenzyl	1.307	0.008	LPES	167
	C_7H_6Br	*p*-Bromobenzyl	1.229	0.008	LPES	167
	C_7H_6Cl	*o*-Chlorobenzyl	1.257	0.008	LPES	167
	C_7H_6Cl	*m*-Chlorobenzyl	1.272	0.008	LPES	167
	C_7H_6Cl	*p*-Chlorobenzyl	1.174	0.008	LPES	167
	C_6H_5CHCl	Chlorobenzyl	≤1.12		LPES	398
	C_7H	2,4,6-Heptatriynylidyne	2.8187	0.0010	LPES	488
	C_7H_6F	*o*-Fluorobenzyl	1.091	0.008	LPES	167
	C_7H_6F	*m*-Fluorobenzyl	1.173	0.008	LPES	167
	C_7H_6F	*p*-Fluorobenzyl	0.937	0.008	LPES	167
	C_7H_6FO	*m*-Fluoroacetophenone enolate	2.218	0.010	LPT	2
	C_7H_6FO	*p*-Fluoroacetophenone enolate	2.176	0.010	LPT	2
	$C_7H_6FeO_3$	η^4-1,3-Butadiene iron tricarbonyl	0.990	0.10	CT	120
	$C_6H_4(O)CH_2$	2-Methylenephenoxyl	1.217	0.12	LPES	502
	$C_6H_4(O)CH_2$	3-Methylenephenoxyl	2.227	0.008	LPES	502
	$C_6H_4(O)CH_2$	4-Methylenephenoxyl	1.096	0.007	LPES	502
	$C_7H_6N_2O_4$	4-Methyl-1,2-dinitrobenzene	1.77	0.05	PT	60
	$C_7H_6N_2O_4$	1-Methyl-2,3-dinitrobenzene	1.77	0.05	PT	60
	$C_7H_6N_2O_4$	1-Methyl-2,4-dinitrobenzene	1.60	0.05	PT	60
	$C_7H_6N_2O_4$	2-Methyl-1,3-dinitrobenzene	1.55	0.05	PT	60
	C_6H_5CHO	Benzaldehyde	0.35	0.05	LPES	384
	$C_7H_6O_2$	2-Methyl-2,5-cyclohexadiene-1,4-dione	1.85	0.10	CT	61
	C_7H_7	Benzyl	0.912	0.006	LPES	26
	C_7H_7	1-Quadricyclanide	0.868	0.006	LPES	136
	C_7H_7	2-Quadricyclanide	0.962	0.006	LPES	136
	C_7H_7	Norbornadienide	1.286	0.006	LPES	136
	C_7H_7	Cycloheptatrienide	0.39	0.04	LPES	136
	C_7H_7	1-(1,6-Heptadiynide)	3.046	0.006	LPES	136
	C_7H_7	3-(1,6-Heptadiynide)	>1.140	0.006	LPES	136
	$C_7H_7NO_2$	2-Nitrotoluene	0.92	0.10	CT	61
	$C_7H_7NO_2$	3-Nitrotoluene	0.99	0.10	CT	61
	$C_7H_7NO_2$	4-Nitrotoluene	0.95	0.10	CT	61
	$C_7H_7NO_3$	3-Nitroanisole	1.04	0.10	CT	61
	$C_7H_7NO_3$	4-Nitroanisole	0.91	0.10	CT	61
	C_7H_7O	2-Methylphenoxy	<2.36	0.06	PT	2
	$C_6H_4CH_2OH$	Methyl-(deprotonated)-2-methylphenol radical	1.024	0.008	LPES	459

Z	Mol. form.	Name	Elec. aff. (eV)	Uncert. (eV)	Method	Ref.
	$C_6H_4CH_3O$	(2-Methylphenyl)oxidanyl radical	2.1991	0.0014	LPES	459
	$C_6H_4CH_3O$	(3-Methylphenyl)oxidanyl radical	2.2177	0.0014	LPES	459
	$C_6H_4CH_3O$	(4-Methylphenyl)oxidanyl radical	2.1191	0.0014	LPES	459
	C_7H_7O	Benzyloxy	2.14	0.02	PT	50
	$(C_7H_8)_n$	Toluene cluster	n=33-139		LPES	248
	C_7H_8FO	Benzyloxy-hydrogen fluoride complex	<3.05	0.06	PT	50
	C_7H_9	2,4,6-Heptatrienyl	1.27	0.03	PT	2
	C_7H_9O	2-Norbornanone enolate	1.61	0.05	PT	2
	C_7H_9Si	Methylphenylsilyl	1.33	0.04	LPT	65
	$C_7H_{11}O$	Cycloheptanone enolate	1.598	0.007	LPT	113
	$C_7H_{11}O$	2,5-Dimethylcyclopentanone enolate	1.49	0.04	PT	2
	$C_7H_{13}O$	4-Heptanone enolate	1.72	0.06	PT	2
	$C_7H_{13}O$	Diisopropyl ketone enolate	1.46	0.04	PT	2
	$C_8F_{14}N_2$	2,3,5,6-Tetrafluoro-1,4-benzenedicarbonitrile	1.89	0.10	CT	51
	$C_6F_5COCH_3$	Pentafluoroacetophenone	0.88	0.11	CT	67
	C_8H	1,3,5,7-Octatetraynyl	3.9701	0.0010	LPES	488
	$C_8H_3F_6NO_2$	1-Nitro-3,5-bis(trifluoromethyl)benzene	1.79	0.10	CT	61
	$C_8H_4F_3N$	2-(Trifluoromethyl)benzonitrile	0.70	0.05	attach	263
	$C_8H_4F_3N$	3-(Trifluoromethyl)benzonitrile	0.67	0.05	attach	263
	$C_8H_4F_3N$	4-(Trifluoromethyl)benzonitrile	0.83	0.05	attach	263
	$C_8H_4O_3$	Phthalic anhydride	1.21	0.10	CT	61
	C_8H_6	1,3,5-Cyclooctatrien-7-yne	1.044	0.008	LPES	148
	C_8H_6N	Indoyl	2.4315	0.00	LPES	500
	C_6H_5CHCN	Cyanobenzyl	1,90	0.01	LPES	398
	C_8H_7	1,3,5,7-Cyclooctatetraen-1-yl	1.091	0.008	LPES	134
	C_8H_7O	Acetophenone enolate	2.057	0.010	PT	2
	C_8H_7O	Phenylacetaldehyde enolate	2.10	0.08	LPT	2
	C_8H_8	1,3,5,7-Cyclooctatetraene	0.55	0.02	CT	134
	C_8H_8	*m*-Xylylene	0.919	0.008	LPES	139
	$C_8H_9NO_2$	1,3-Dimethyl-2-nitrobenzene	2.61	0.05	PT	60
	$C_8H_9NO_2$	1,2-Dimethyl-3-nitrobenzene	0.86	0.10	CT	61
	$C_8H_9NO_2$	1,3-Dimethyl-5-nitrobenzene	1.21	0.05	PT	60
	$C_8H_{13}O$	Cyclooctanone enolate	1.63	0.06	PT	2
	C_8S_2	Carbon sulfide [C_8S_2]	0.049	0.005	LPES	230
	C_9	Carbon 9-cluster (linear)	3.6766	0.0014	LPES	508
	C_9H	2,4,6,8-Nonatriynylidyne	3.0971	0.0010	LPES	488
	C_9H_7	Indenyl	1.8019	0.0006	LPES	394
	$C_9H_8FeO_3$	η^4-1,3-Cyclohexadiene iron tricarbonyl	0.76	0.10	CT	120
	C_9H_9O	*m*-Methylacetophenone enolate	2.030	0.010	LPT	2
	$(CH_3)_3SiN$	Trimethylsilylimidogen	1.43	0.10	PT	2
	$C_9H_{11}NO_2$	1,3,5-Trimethyl-2-nitrobenzene	0.70	0.10	CT	61
	$C_9H_{15}O$	Cyclononanone enolate	1.69	0.06	PT	2
	$C_{10}H_4Cl_2O_2$	2,3-Dichloro-1,4-naphthalenedione	2.19	0.10	CT	61
	$C_{10}H_6N_2O_4$	1,3-Dinitronaphthalene	1.78	0.10	CT	61
	$C_{10}H_6N_2O_4$	1,5-Dinitronaphthalene	1.77	0.10	CT	61
	$C_{10}H_6O_2$	1,4-Naphthalenedione	1.81	0.10	CT	61
	$C_{10}H_7$	1-Naphthyl	1.403	0.015	LPES	197
	$C_{10}H_7$	α-Naphthyl radical	1.4095	0.0004	LPES	197
	$C_{10}H_7$	β-Naphthyl radical	1.3352	0.0002	LPES	197
	$C_{10}H_7NO_2$	1-Nitronaphthalene	1.23	0.10	CT	61
	$C_{10}H_7NO_2$	2-Nitronaphthalene	1.18	0.10	CT	61
	$C_{10}H_8$	Azulene	0.790	0.008	LPES	230
	$C_{10}H_8CrO_3$	η^4-1,3,5-Cycloheptatriene chromium tricarbonyl	0.93	0.10	CT	120
	$NO{\cdot}C_{10}H_8$	Nitric oxide-naphthalene complex	0.66		LPES	285
	$O_2{\cdot}C_{10}H_8$	Oxygen-naphthalene complex	1.41		LPES	285
	$O_2{\cdot}C_{10}H_8{\cdot}H_2O$	Oxygen-naphthalene-water complex	2.09		LPES	285
	$O_2{\cdot}C_{10}H_8{\cdot}(H_2O)_2$	Oxygen-naphthalene-water dimer complex	2.72		LPES	285
	$C_{10}H_8FeO_3$	η^4-1,3,5-Cycloheptatriene iron tricarbonyl	0.98	0.10	CT	120
	$C_5(CH_3)_5AlH$	Aluminum hydride pentamethylcyclopentadienyl	1.0	0.2	LPES	424
	$C_5(CH_3)_5Al_2H$	Dialuminum hydride pentamethylcyclopentadienyl	0.8	0.2	LPES	424
	$C_5(CH_3)_5Al_3H$	Trialuminum hydride pentamethylcyclopentadienyl	1.2	0.2	LPES	424

Z	Mol. form.	Name	Elec. aff. (eV)	Uncert. (eV)	Method	Ref.
	$C_{10}H_{17}O$	Cyclodecanone enolate	1.83	0.06	PT	2
	$C_{11}H_8FeO_3$	η^4-1,3,5,7-Cyclooctatetraene iron tricarbonyl	1.29	0.10	CT	120
	$C_{12}F_{10}$	2,2',3,3',4,4',5,5',6,6'-Decafluoro-1,1'-biphenyl	0.82	0.11	CT	67
	$C_{12}H_4N_4$	Tetracyanoquinodimethane	3.383	0.001	LPES	411
	$C_{12}H_{10}OP$	Diphenylphosphide oxide	2.2	0.1	LPES	401
	$C_{12}H_{10}P$	Diphenylphosphide	1.5	0.1	LPES	401
	$C_{12}H_{12}$	3-Bis(allyl)benzene	0.90	0.15	Kinetic	378
	$C_{12}H_{12}$	4-Bis(allyl)benzene	0.84	0.15	Kinetic	378
	$C_{13}H_9$	Perinaphthalenyl	1.07	0.10	PT	2
	$NO{\cdot}(C_6H_6)_2$	Nitric oxide-benzene dimer complex	0.79		LPES	285
	$C_{12}H_{15}O$	*tert*-Butylacetophenone enolate	2.032	0.010	LPT	2
	$C_{12}H_{21}O$	Cyclododecanone enolate	1.90	0.07	PT	2
	$C_{13}H_9$	Fluorenyl	1.8751	0.0003	LPES	394
	$C_{13}F_{10}O$	Decafluorobenzophenone	1.52	0.11	CT	67
	$C_{13}H_9FO$	4-Fluorobenzophenone	0.64	0.10	CT	61
	$(C_6H_5)_2CO$	Benzophenone	0.62	0.10	CT	61
	$C_{14}H_9$	1-Anthracenyl radical	1.5436	0.0002	LPES	478
	$C_{14}H_9$	2-Anthracenyl radical	1.4671	0.0002	LPES	478
	$C_{14}H_9$	9-Anthracenyl radical	1.7155	0.0002	LPES	478
	$C_{14}H_9NO_2$	9-Nitroanthracene	1.43	0.10	CT	61
	$C_{14}H_{10}$	Anthracene	0.530	0.005	LPES	286
	$C_{14}H_{10}{\cdot}H_2O$	Anthracene-water complex	0.770	0.005	LPES	286
	$(C_{14}H_{10})_n$	Anthracene cluster	n=1-16		LPES	231
	$C_{16}H_{10}$	Pyrene	0.406	0.010	LPES	270
	$(C_{16}H_{10})_nCo_m$	Cobalt-pyrene cluster	n=1,2	m=1,2	LPES	330
	$C_{18}H_{12}$	Chrysene	0.32	0.01	LPES	303
	$C_{18}H_{12}$	Naphthacene	1.058	0.005	CT	66
	$C_{20}H_{12}$	Benzo[*a*]pyrene	0.79	0.10	CT	66
	$C_{20}H_{12}$	Perylene	0.973	0.005	LPES	236
	$NO{\cdot}(C_{10}H_8)_2$	Nitric oxide-naphthalene dimer complex	1.06		LPES	285
	$C_{22}H_{14}$	Pentacene	1.35	0.10	CT	66
	$C_{24}H_{12}$	Coronene	0.47	0.09	LPES	328
	$C_{24}H_{12}Co$	Coronene-cobalt complex	1.15	0.15	LPES	324
	$C_{24}H_{12}Co_2$	Coronene-dicobalt complex	1.15	0.10	LPES	324
	$C_{24}H_{12}Fe$	Coronene-iron complex	1.06		LPES	324
	$C_{24}H_{12}Fe_2$	Coronene-diiron complex	1.59		LPES	324
	$C_{44}Cl_{28}FeN_4$	$FeTPPCl_{28}$	2.59	0.11	CT	186
	$C_{44}Cl_8F_{20}FeN_4$	$FeTPP\beta Cl_8$	3.21	0.03	CT	186
	$C_{44}Cl_9F_{20}FeN_4$	$FeTPPF_{20}\beta Cl_8Cl$	3.35	0.03	CT	186
	$C_{44}H_8F_{20}FeN_4$	$FeTPPF_{20}$	2.15	0.15	CT	186
	$C_{44}H_8ClF_{20}FeN_4$	$FeTPPF_{20}Cl$	3.14	0.03	CT	186
	$C_{44}H_8Cl_{21}FeN_4$	$FeTPPoCl_{20}Cl$	2.93	0.23	CT	186
	$C_{44}H_{12}Cl_{17}FeN_4$	$FeTPPoCl_8\beta Cl_8Cl$	3.14	0.03	CT	186
	$C_{44}H_{20}Cl_8FeN_4$	$FeTPPoCl_8$	1.86	0.03	CT	186
	$C_{44}H_{20}Cl_9FeN_4$	$FeTPPoCl_8Cl$	2.10	0.19	CT	186
	$C_{44}H_{28}FeN_4$	FeTPP	1.87	0.03	CT	186
	$C_{44}H_{28}NiN_4$	NiTPP	1.51	0.01	CT	186
	$C_{44}H_{28}ClFeN_4$	FeTPPCl	2.15	0.15	CT	186
	$C_{44}H_{30}N_4$	H_2TPP	1.69	0.01	CT	186
	$C_{45}H_{29}NiN_4O$	NiTPPCHO	1.74	0.01	CT	186
	$C_{48}H_{24}$	Coronene dimer	0.67	0.09	LPES	328
	$C_{48}H_{24}Fe$	Coronene dimer-iron complex	1.50		LPES	324
	$C_{48}H_{24}Fe_2$	Coronene dimer-diiron complex	1.50		LPES	324
	$C_{52}H_{39}FeN_7O$	FeTPP-val	1.97	0.03	CT	186
	C_{60}	Carbon [fullerene-C_{60}]	2.6835	0.0006	LPES	201
	$C_{60}F_2$	Difluorofullerene-C_{60}	2.74	0.07	Knud	202
	$C_{64}H_{64}FeN_8O_4$	FeTPP-piv	2.07	0.03	CT	186
	C_{70}	Carbon [fullerene-C_{70}]	2.676	0.001	LPT	201
	$C_{70}F_2$	Difluorofullerene-C_{70}	2.80	0.07	Knud	202
	$C_{60}(CF_3)_{10}$	Deca(trifluoromethyl)fullerene	2.81	0.14	Knud	376
	$C_{60}(CF_3)_{12}$	Dodeca(trifluoromethyl)fullerene	2.57	0.17	Knud	376

Z	Mol. form.	Name	Elec. aff. (eV)	Uncert. (eV)	Method	Ref.
	CeF_4	Cerium(IV) fluoride	3.8	0.4	CT	98
	CeO_3	Cerium trioxide	2.97	0.02	LPES	494
	CeO_3H_2	Cerium trioxide dihydride	1.19	0.02	LPES	494
	$Ce(OH)_2$	Cerium hydroxide	0.69	0.03	LPES	453
	$Ce(OH)_4$	Cerium(IV) tetrahydroxide	2.38	0.01	LPES	494
	CeO_2B_2	Cerium boron oxide complex (CeO_2B_2)	1.87	0.05	LPES	498
	CeO_2B_3	Cerium boron oxide complex (CeO_2B_3)	1.15	0.02	LPES	498
	$SmCeO_2$	Samarium cerium dioxide	0.95	0.10	LPES	496
	Ce_2O_2	Cerium oxide complex (Ce_2O_2)	0.83	0.02	LPES	454
	Ce_2O_2Sm	Cerium samarium oxide complex (Ce_2O_2Sm)	1.11	0.03	LPES	496
	Ce_2O_4	Dicerium tetraoxide complex (Ce_2O_4)	0.85	0.05	LPES	494
	Ce_3O	Cerium oxide cluster (Ce_3O)	1.3951	0.0004	LPES	480
	Ce_3O	Cerium oxide complex (Ce_3O)	0.89	0.05	LPES	455
	(Ce_3O_n) n=0-4	Cerium oxide complex (Ce_3O_n) n=0-4			LPES	455
	Ce_4	Tetracerium cluster	0.97	0.05	LPES	455
	Ce_4O	Cerium oxide complex (Ce_4O)	0.94	0.02	LPES	455
	Ce_4O_2	Cerium oxide complex (Ce_4O_2)	1.01	0.04	LPES	455
	Ce_5O	Cerium oxide complex (Ce_5O)	1.09	0.05	LPES	455
	Ce_5O_2	Cerium oxide complex (Ce_5O_2)	0.96	0.03	LPES	455
	$Cl(CO_2)$	Carbonochloridate	3.907	0.010	LPES	131
	$Cl(H_2O)_n$	Chlorine-water cluster	n=1-4		LPES	250
	ClO_3	Chlorine trioxide	4.25	0.10	LPES	340
	ClO_4	Chlorine tetroxide	5.25	0.10	LPES	340
	Co_n	Cobalt cluster	n=1-108		LPES	251
	CoB_{12}	Cobalt bororn complex (CoB_{12})	3.23	0.04	LPES	437
	$CoBr_3$	Cobalt bromide [$CoBr_3$]	4.6	0.1	LPES	249
	$CoCO_2NO$	Cobalt-carbon dioxide-nitric oxide complex	1.73	0.03	LPES	199
	$CoCl_3$	Cobalt chloride [$CoCl_3$]	4.7	0.1	LPES	249
	CoF_4	Cobalt fluoride [CoF_4]	6.4	0.3	CT	98
	CoOH	Cobalt hydroxide	1.33	0.04	LPES	400
	Co_2OH	Dicobalt hydroxide	1.14	0.08	LPES	400
	Co_3OH	Tricobalt hydroxide	1.56	0.08	LPES	400
	$Cr(CO)_3$	Chromium carbonyl [$Cr[CO]_3$]	1.349	0.006	LPES	94
	CrO_3	Chromium(VI) oxide	3.66	0.02	LPES	241
	CrO_4	Chromium oxide [CrO_4]	4.98	0.09	LPES	241
	CrO_5	Chromium oxide [CrO_5]	4.4	0.1	LPES	241
	Cr_2O_n	Chromium-oxygen cluster	n=1-7		LPES	306
	CsO_4	Cesium oxide [CsO_4]	2.5	0.2	LPES	252
	Cu_n	Copper cluster	n=1-411		LPES	37
	$CuBO_2$	Copper metaborate	1.90	0.08	LPES	362
	$Cu(BO_2)_2$	Copper bis(metaborate)	5.07	0.08	LPES	362
	Cu_2BO_2	Dicopper metaborate	3.53	0.08	LPES	362
	$Cu_2(BO_2)_2$	Dicopper bis(metaborate)	2.74	0.08	LPES	362
	$Cu_n(CN)_m$	Copper cyanide cluster	n=1-6	m=1-6	LPES	159
	$EuSi_n$	Europium-silicon cluster	n=3-17		LPES	321
	$F(H_2O)_n$	Fluorine-water cluster	n=1-4		LPES	242
	$F(H_2O)_n$	Fluorine-water cluster	n=1-4		LPES	250
	Fe_n	Iron cluster	n=3-34		LPES	149
	$Fe(CO)_2$	Iron carbonyl [$Fe(CO)_2$]	1.22	0.02	LPES	2
	$Fe(CO)_3$	Iron carbonyl [$Fe(CO)_3$]	1.8	0.2	LPES	2
	$Fe(CO)_4$	Iron carbonyl [$Fe(CO)_4$]	2.34	0.10	LPES	389
	$FeBr_3$	Iron(III) bromide	4.26	0.06	LPES	249
	$FeBr_4$	Iron bromide [$FeBr_4$]	5.50	0.08	LPES	249
	$FeCl_3$	Iron(III) chloride	4.22	0.06	LPES	249
	$FeCl_4$	Iron chloride [$FeCl_4$]	6.00	0.08	LPES	249
	FeF_3	Iron(III) fluoride	3.6	0.1	CT	98
	FeF_4	Iron fluoride [FeF_4]	6.0	estimate	CT	98
	Fe_2H_2	Iron hydride [Fe_2H_2]	0.942	0.019	LPES	254
	Fe_3O	Iron oxide cluster (Fe_3O)	1.4408	0.0003	LPES	480
	Fe_4O	Iron oxide cluster (Fe_4O)	1.6980	0.0003	LPES	407
	Fe_5O	Iron oxide cluster (Fe_5O)	1.8616	0.0003	LPES	407

Z	Mol. form.	Name	Elec. aff. (eV)	Uncert. (eV)	Method	Ref.
	Fe_nO_m	Iron oxide cluster	n=1-4	m=1-6	LPES	152
	Ga_2As_3	Gallium arsenide [Ga_3As_2]	2.783	0.024	LPES	192
	Ga_xAs_y	Gallium arsenide cluster	n=2-50	n=x+y	LPES	229
	Ga_2P_3	Gallium phosphide [Ga_2P_3]	2.991	0.026	LPES	192
	Ge_n	Germanium cluster	n=3-15		LPES	71
	Ge_xAs_y	Germanium-arsenic cluster	n=5-30	n=x+y	LPES	72
	GeH_3	Germyl	<1.74	0.04	PT	2
	$H(NH_3)_n$	Hydrogen-ammonia cluster	n=1,2		LPES	76
	HNO_2	*cis*-Nitrous acid	0.356	0.008	LPES	461
	HNO_3	Nitric acid	0.57	0.15	CD	2
	$(H_2O)_n$	Water cluster	n=2-19		LPES	77
	HfOOH	Hafnium oxide hydroxyl radical	1.73	0.05	LPES	173
	$I(CO_2)$	Iodine-carbon dioxide complex	3.225	0.001	LPES	131
	$I(CO_2)_n$	Iodine-carbon dioxide cluster	n=1-3		LPES	350
	$I(H_2O)_n$	Iodine-water cluster	n=1-4		LPES	250
	In_xP_y	Indium phosphide cluster	n=2-8	n=x+y	LPES	137
	IrB_3	Iridium boride (tetrahedral)	1.3147	0.0008	LPES	513
	IrF_4	Iridium fluoride [IrF_4]	4.7	0.3	CT	98
	IrF_6	Iridium(VI) fluoride	6.5	0.4	CT	98
	K_n	Potassium cluster	n=2-7		LPES	18
	KO_4	Potassium oxide [KO_4]	2.8	0.2	LPES	252
	$LaCl_4$	Lanthanum tetrachloride	7.03	0.01	LPES	145
	LiB_6	Lithium hexaboride	2.3	0.1	LPES	298
	LiO_4	Lithium oxide [LiO_4]	3.3	0.2	LPES	252
	$MnBr_3$	Manganese bromide [$MnBr_3$]	5.03	0.06	LPES	249
	$MnCl_3$	Manganese chloride [$MnCl_3$]	5.07	0.06	LPES	249
	MnF_4	Manganese fluoride [MnF_4]	5.5	0.2	CT	98
	MnO_3	Manganese oxide [MnO_3]	3.335	0.010	LPES	158
	$Mo(CO)_3$	Molybdenum carbonyl [$Mo[CO]_3$]	1.337	0.006	LPES	94
	MoF_5	Molybdenum(V) fluoride	3.5	0.2	CT	98
	MoF_6	Molybdenum(VI) fluoride	3.8	0.2	CT	98
	MoO_3	Molybdenum(VI) oxide	3.17	0.02	LPES	280
	MoO_4	Molybdenum oxide [MoO_4]	5.20	0.07	LPES	86
	MoO_5	Molybdenum oxide [MoO_5]	5.10	0.07	LPES	86
	MoW_2O_6	Molybdenum tungsten oxide [MoW_2O_6]	2.76	0.05	LPES	353
	Mo_2O_2	Molybdenum oxide [Mo_2O_2]	2.24	0.02	LPES	280
	Mo_2O_3	Molybdenum(III) oxide	2.33	0.07	LPES	280
	Mo_2O_4	Molybdenum oxide [Mo_2O_4]	2.13	0.04	LPES	280
	Mo_2WO_6	Molybdenum tungsten oxide [Mo_2WO_6]	2.85	0.05	LPES	353
	Mo_3O_6	Molybdenum oxide [Mo_3O_6]	2.75	0.02	LPES	353
	N_2CH	Cyanoamidogen	2.622	0.005	LPES	154
	N_2CD	Cyanoamidogen-*d*	2.622	0.005	LPES	154
	$(NH_3)_n$	Ammonia cluster	n=41-1100		LPES	77
	$NH_2(NH_3)_n$	Amidogen-ammonia cluster	n=1,2		LPES	78
	$NOAr_n n$=1,4	Nitric oxide argon cluster ($NOAr_n n$=1,4)			LPES	79
	$NO(C_2H_6O_2)_n n$=1-4	Nitric oxide ethylene glycol cluster [$NO(C_2H_6O_2)_n n$=1-4]			LPES	79
	$NO(H_2O)_n$	Nitric oxide-water cluster	n=1,2		LPES	75
	$NO(H_2S)$	Nitric oxide hydrogen sulfide cluster	0.268	0.020	LPES	79
	$NO(NH_3)$	Nitric oxide ammoniua cluster	0.48	0.02	LPES	79
	$NO(N_2O)_n$	Nitric oxide-nitrous oxide cluster	n=1,2		LPES	79
	NO_3	Nitrogen trioxide [NO_3]	3.937	0.014	LPES	85
	$NO_3(H_2O)_n$	Nitrogen trioxide-water cluster	n=0-6		LPES	240
	$(NO)_2$	Nitric oxide dimer	>2.1		LPES	75
	$(N_2O)_n$	Nitrous oxide cluster	n=1,2		LPES	81
	Na_n	Sodium cluster	n=2-5		LPES	18
	$NaCS_2$	Sodium-carbon disulfide complex [$NaCS_2$]	0.80	0.05	LPES	278
	Na_2CS_2	Sodium-carbon disulfide complex [Na_2CS_2]	0.25	0.05	LPES	278
	$(NaF)_n$	Sodium fluoride cluster	n=1-7,12		LPES	64
	$Na(NaF)_n$	Sodium-sodium fluoride cluster	n=5,7-12		LPES	64
	NaO_4	Sodium oxide [NaO_4]	3.1	0.2	LPES	252
	NaO_5	Sodium oxide [NaO_5]	3.2	0.2	LPES	252

Z	Mol. form.	Name	Elec. aff. (eV)	Uncert. (eV)	Method	Ref.
	$NaSO_3$	Sodium oxide sulfide (NaO_3S)	2.3	0.2	LPES	252
	$NaSn_n$, n=5-7	Sodium tin cluster ($NaSn_n$, n=5-7)	n=5-7		LPES	380
	Na_nSn (n=0,4)	Sodium tin cluster (Na_nSn, n=0,4)	n=0-4		LPES	380
	Nb_n	Niobium cluster	n=6-17		LPES	181
	Nb_8	Niobium octamer	1.513	0.008	LPES	157
	Nb_2O_2	Diniobium dioxide	0.97	0.01	LPES	370
	Nb_2O_3	Diniobium trioxide	1.61	0.01	LPES	370
	Nb_2O_4	Diniobium tetraoxide	1.62	0.01	LPES	370
	Nb_2O_5	Niobium(V) oxide	3.33	0.05	LPES	451
	Nb_2O_6	Diniobium hexaoxide	5.35	0.05	LPES	451
	Nb_2O_7	Diniobium heptaoxide	5.25	0.05	LPES	451
	Nb_3O	Niobium oxide (Nb_3O)	1.393	0.006	LPES	169
	Nb_3O_3	Triniobium trioxide	1.54	0.02	LPES	374
	Nb_3O_n (n=3,8)	Triniobium oxide cluster (Nb_3O_n, n=3,8)	n=3-8		LPES	374
	Ni_n	Nickel cluster	n=1-100		LPES	247
	$Ni_n(C_6H_6)_m$	Nickel-benzene complex	n=1-3	m=1,2	LPES	295
	$NiBr_3$	Nickel bromide [$NiBr_3$]	4.94	0.08	LPES	249
	$NiCl_3$	Nickel chloride [$NiCl_3$]	5.20	0.08	LPES	249
	Ni(CO)H	Nickel-carbon monoxide complex	1.126	0.010	LPES	293
	$Ni(CO)_2$	Nickel carbonyl [$Ni[CO]_2$]	0.643	0.014	LPES	2
	$Ni(CO)_3$	Nickel carbonyl [$Ni[CO]_3$]	1.077	0.013	LPES	2
	$NiAl(CO)_3$	Nickel aluminum tricarbonyl ($NiAl(CO)_3$)	1.064	0.063	LPES	423
	$NICa(CO)_3$	Nickel calcium tricarbonyl ($NiCa(CO)_3$)	1.050	0.064	LPES	423
	$NiMg(CO)_3$	Nickel magnesium tricarbonyl ($NiMg(CO)_3$)	1.541	0.040	LPES	423
	$NiC_8N_4S_4$	Nickel bis(dithiolene)	4.56	0.04	LPES	307
	O_2O_3	Oxygen ozonide	2.160	0.015	LPES	420
	$OH(H_2O)$	Hydroxyl-water complex	<2.95	0.15	PT	2
	$OH(NH_3)$	Hydroxyl-ammonia complex	2.35	0.07	LPES	234
	$OH(N_2O)$	Hydroxyl-nitrous oxide complex	2.14	0.02	LPES	209
	$OH(N_2O)_n$	Hydroxyl-nitrous oxide cluster	n=1-5		LPES	209
	OsF_4	Osmium(IV) fluoride	3.9	0.3	CT	98
	OsF_6	Osmium(VI) fluoride	6.0	0.3	CT	98
	P_5	Phosphorus pentamer	3.88	0.03	LPES	253
	PBr_3	Phosphorus(III) bromide	1.59	0.15	CD	2
	PBr_2Cl	Phosphorus(III) dibromide chloride	1.63	0.20	CD	2
	$PBrCl_2$	Phosphorus(III) bromide dichloride	1.52	0.20	CD	2
	PCl_3	Phosphorus(III) chloride	0.82	0.10	CD	2
	PF_5	Phosphorus(V) fluoride	0.75	0.15	CT	121
	P_2H_2	*cis*-Diphosphene	1.03	0.01	LPES	281
	P_2H_2	*trans*-Diphosphene	1.00	0.01	LPES	281
	PO_3	Phosphorus oxide [PO_3]	4.95	0.06	LPES	156
	$POCl_2$	Phosphorus oxychloride [$POCl_2$]	3.83	0.25	CD	2
	$POCl_3$	Phosphoryl chloride	1.41	0.20	CD	2
	Pb_4	Lead tetramer	1.55	0.09	LPES	345
	$PdC_8N_4S_4$	Palladium bis(dithiolene)	4.55	0.04	LPES	307
	PrB_7	Praseodymium heptaboron complex (PrB_7)	1.47	0.08	LPES	489
	$PtC_8N_4S_4$	Platinum bis(dithiolene)	4.45	0.04	LPES	307
	$Pt(CO)_2$	Platinum bicarbonyl	0.930	0.042	LPES	391
	$Pt(CO)_3$	Platinum tricarbonyl	1.253	0.032	LPES	391
	PtF_4	Platinum(IV) fluoride	5.5	0.3	CT	98
	PtF_6	Platinum(VI) fluoride	7.0	0.4	CT	98
	ReB_6	Rhenium hexaboride	2.3478	0.0008	LPES	511
	ReF_6	Rhenium(VI) fluoride	4.7	estimate	CT	98
	ReO_3	Rhenium(VI) oxide	3.53	0.05	LPES	450
	ReO_4	Rheium tetraoxide	5.58	0.03	LPES	450
	Re_3O_3	Trirhenium trioxide	2.54	0.02	LPES	448
	$RhBr_{12}$	Rhodium dodecabromide	3.45	0.02	LPES	437
	RhF_4	Rhenium fluoride [RhF_4]	5.4	0.3	CT	98
	RuF_4	Ruthenium(IV) fluoride	4.8	0.3	CT	98
	RuF_5	Ruthenium(V) fluoride	5.2	0.4	CT	98
	RuF_6	Ruthenium(VI) fluoride	7.5	0.3	CT	98

Z	Mol. form.	Name	Elec. aff. (eV)	Uncert. (eV)	Method	Ref.
	SF_4	Sulfur tetrafluoride	1.5	0.2	CT	91
	SF_5	Sulfur pentafluoride	4.23	0.12	e-scat	204
	SF_6	Sulfur hexafluoride	1.03	0.05	attach	318
	SO_3	Sulfur trioxide (α-form)	1.97	0.10	LPES	165
	$(SO_2)_2$	Sulfur dioxide dimer	0.6	0.2	LPES	80
	Sb_n	Antimony cluster	n=2-9		LPES	213
	Sb_5	Antimony pentamer	3.46	0.03	LPES	253
	$ScBr_4$	Scandium bromide [$ScBr_4$]	6.13	0.08	LPES	249
	$ScCl_4$	Scandium chloride [$ScCl_4$]	6.84	0.01	LPES	145
	Sc_2Sn_n	Scandium-tin cluster	n=2-6		LPES	356
	SeF_6	Selenium hexafluoride	2.9	0.2	CD	2
	SiC_3	Silicon tricarbon	2.827	0.007	LPES	474
	SiC_4	Silicon tetracarbide	2.543	0.006	LPES	474
	Si_4	Tetrasilicon	2.13	0.01	LPES	110
	Si_5	Silicon pentamer	2.59	0.02	LPES	110
	Si_7	Silicon heptamer	1.85	0.02	LPES	110
	Si_n	Silicon cluster	n=3-20		LPES	71
	SiF_3	Trifluorosilyl	2.41	0.22	LPES	17
	SiF_4	Tetrafluorosilane	≤0		LPES	17
	SiF_5	Pentafluorosilyl	≥4.66		LPES	17
	SiH_3	Silyl	1.406	0.014	LPES	43
	SiD_3	Silyl-d_3	1.386	0.022	LPES	43
	Si_2C_3	Disilicon tricarbide	1.766	0.012	LPES	296
	Si_3H	Silicon hydride (Si_3H)	2.53	0.01	LPES	182
	Si_4H	Silicon hydride (Si_4H)	2.68	0.01	LPES	182
	Si_nNa_m	Silicon-sodium cluster	n=4-11	m=1-3	LPES	210
	Si_nF	Silicon-fluorine cluster	n=2-11		LPES	17
	$SmCeO_2$	Samarium cerium dioxide	0.74	0.02	LPES	454
	Sm_2O_2	Disamarium dioxide	0.70	0.01	LPES	454
	Sm_3O_2	Trisamarium dioxide	0.87	0.03	LPES	496
	Sn_n	Tin cluster	n=1-12		LPES	289
	$SnCH_2CN$	Cyanomethylenetin	1.57	0.02	LPES	292
	$Sn(CN)_2$	Tin(II) cyanide	2.622	0.004	LPES	292
	$Sn(CN)(CH_2CN)$	Cyanomethylenetin cyanide	2.29	0.05	LPES	292
	Ta_2B_2	Ditantalum diboride	1.31	0.04	LPES	439
	Ta_2B_3	Ditantalum triboride	1.93	0.03	LPES	439
	Ta_2B_4	Ditantalum tetraboride	1.70	0.07	LPES	439
	Ta_2B_6	Ditantalum hexaboride	1.51	0.03	LPES	439
	Ta_2O_5	Tantalum(V) oxide	3.71	0.05	LPES	451
	Ta_2O_6	Ditantalum hexaoxide	5.28	0.05	LPES	451
	Ta_2O_7	Ditantalum heptaoxide	5.15	0.05	LPES	451
	Ta_3O	Tantalum oxide (Ta_3O)	1.583	0.010	LPES	169
	TeF_6	Tellurium hexafluoride	3.34	0.17	CD	2
	Ti_n	Titanium cluster	n=1-130		LPES	151
	C_3Ti	Titanium carbide [TiC_3]	1.561	0.015	LPES	147
	$TiO(OH)_2$	Titanium dihydroxide oxide	1.2529	0.0004	LPES	507
	$TiO_2\cdot(H_2O)_2$	Titanium dioxide dihydrate	0.77	0.08	LPES	381
	$TiO_2(H_2O)\cdot(H_2O)_n$,n=1-5	Titanium dioxide hydrate cluster			LPES	381
	TiO_3	Titanium oxide [TiO_3]	4.2		LPES	172
	UCl_6	Uranium(VI) chloride	5.3	0.2	LPES	414
	UF_3	Uranium(III) fluoride	1.09	0.03	LPES	412
	UF_4	Uranium(IV) fluoride	1.58	0.03	LPES	412
	UF_5	Uranium(V) fluoride	3.885	0.015	LPES	445
	UF_6	Uranium(VI) fluoride	5.1	0.2	CT	98
	UO_2Br_3	Uranyl tribromide	6.27	0.05	LPES	442
	UO_2Cl_3	Uranyl trichloride	6.64	0.05	LPES	442
	UO_2F_3	Uranyl trifluoride	6.25	0.05	LPES	442
	UO_2I_3	Uranyl triiodide	5.60	0.05	LPES	442
	UO_3	Uranium(VI) oxide	1.12	0.03	LPES	433
	UO_4	Uranium tetraoxide	3.60	0.03	LPES	433
	UO_5	Uranium pentaoxide	4.02	0.06	LPES	433

Z	Mol. form.	Name	Elec. aff. (eV)	Uncert. (eV)	Method	Ref.
	V_n	Vanadium cluster	n=3-65		LPES	150
	VF_4	Vanadium(IV) fluoride	3.5	0.2	CT	98
	VSi_n	Vanadium-silicon cluster [VSi_n]	n=3-6		LPES	357
	V_2O_n	Vanadium oxide cluster	n=3-7		LPES	246
	V_2Si_n	Vanadium-silicon cluster [V_2Si_n]	n=3-6		LPES	357
	V_3O	Vanadium oxide (V_3O)	1.218	0.008	LPES	169
	V_4O_{10}	Vanadium oxide [V_4O_{10}]	4.2	0.6	CT	101
	$WAlO_n$, n=2-4	Tungsten aluminum oxides			LPES	387
	$W(CO)_3$	Tungsten carbonyl [$W[CO]_3$]	1.859	0.006	LPES	94
	WF_5	Tungsten(V) fluoride	1.25	0.3	CD	18
	WF_6	Tungsten(VI) fluoride	3.5	0.1	CT	19
	WO_3	Tungsten(VI) oxide	3.62	0.05	LPES	86
	$(WO_3)_n$	Tungsten(IV) oxide cluster	n=7-10		LPES	300
	WO_4	Tungsten oxide [WO_4]	5.30	0.05	LPES	86
	WO_5	Tungsten oxide [WO_5]	5.1	0.1	LPES	86
	W_3O_4	Tritungsten tetraoxide	2.02	0.02	LPES	369
	W_3O_5	Tritungsten pentaoxide	1.9	0.1	LPES	369
	W_3O_6	Tungsten oxide [W_3O_6]	2.80	0.03	LPES	369
	W_3O_6	Tungsten oxide [W_3O_6]	2.85	0.02	LPES	353
	$YbBr_3$	Ytterbium(III) bromide	4.0	0.2	Knud	379
	YCl_4	Yttrium tetrachloride	7.02	0.01	LPES	145

PROTON AFFINITIES

Proton affinity is a useful parameter for describing gas-phase ion-molecule reactions in fields such as atmospheric chemistry, plasma chemistry, mass spectrometry, and astrophysics. The proton affinity E_{pa} (often designated in the literature as PA) of a molecular species M is defined as the negative of the enthalpy change for the gas-phase reaction

$$M + H^+ \rightarrow MH^+$$

A closely related quantity is the gas-phase basicity $\Delta_{base}G°$ (often designated as GB), which is the negative of the Gibbs energy change for the same reaction. Thus, the two are related by

$$\Delta_{base}G° = E_{pa} + T\Delta S$$

where T is the temperature and ΔS is the entropy change in the reaction (which can be calculated if the molecular structure of M and M^+ is known).

Direct measurement of the proton affinity is possible for only a few molecules, mainly olefins and carbonyl compounds. However, these measurements have been used to establish a scale of E_{pa} values that permits proton affinities to be determined for many other molecules, including unstable species and reaction intermediates. The basis for this scale is described by Hunter and Lias in Ref. 1.

The E_{pa} and $\Delta_{base}G°$ values at a temperature of 298 K are tabulated below for selected molecules. Many values are given to one decimal place, but the majority are not accurate to better than one or two kilojoules per mole. The methods of measurement are described in Ref. 1, which contains a much more extensive and detailed tabulation.

Column definitions for the table are as follows.

Column heading	Definition
Name	Name of compound; listed alphabetically by name
Mol. form.	Molecular formula of compound, in Hill order
E_{pa}	Proton affinity, in units kJ mol^{-1}
$\Delta_{base}G°$	Gas-phase basicity, in units kJ mol^{-1} (negative of the Gibbs energy change for the reaction)
Note	Reference or additional information of interest

References

1. Hunter, E. P. L., and Lias, S. G., *J. Phys. Chem. Ref. Data* 27, 413, 1998. <https://doi.org/10.1063/1.556018>
2. Hunter, E. P., and Lias, S. G., "Proton Affinity Evaluation", in *NIST Chemistry WebBook*, NIST Standard Reference Database No. 69, Linstrom, P. J., and Mallard, W. G., Eds., March 2003, National Institute of Standards and Technology, Gaithersburg, MD 20899, <http://webbook.nist.gov>.
3. Do, K., Klein, T. P., Pommerening, C. A., Bachrach, S. M., and Sunderlin, L. S., *J. Am. Chem. Soc.* 120, 6093, 1998. <https://doi.org/10.1021/ja970415t>
4. Kim, H.-T., Green, R. J., Qian, J., and Anderson, S. L., *J. Chem. Phys.* 112, 5717, 2000. <https://doi.org/10.1063/1.481146>
5. Park, S. T., Kim, S. K., and Kim, M. S., *J. Chem. Phys.* 114, 5568, 2001. <https://doi.org/10.1063/1.1353548>
6. Hiraoka, K., Mizuno, T., Eguchi, D., Takao, T., and Ino, S., *J. Chem. Phys.* 116, 7574, 2002. <https://doi.org/10.1063/1.1400787>
7. Oresmaa, L. O., Haukka, M., Vainiotalo, P., and Pakkanen, T. A., *J. Org. Chem.* 67, 8216, 2002. <https://doi.org/10.1021/jo026126r>
8. Wang, F., Ma, S., Zhang, D., and Cooks, R. G., *J. Phys. Chem. A* 102, 2988, 1998. <https://doi.org/10.1021/jp9804493>
9. Bouchoux, G., Gal, J.-F., Szulejko, J. E., McMahon, T. B., Tortajada, J., Luna, A., Yanez, M., and Mo, O., *J. Phys. Chem. A* 102, 9183, 1998. <https://doi.org/10.1021/jp982877e>
10. van Beelen, E., Koblenz, T. A., Ingemann, S. and Hammerum, S., *J. Phys. Chem. A* 108, 2728, 2004. <https://doi.org/10.1021/jp0375721>

Proton Affinities and Gas-Phase Basicities

Name	Mol. form.	E_{pa}/ kJ mol^{-1}	$\Delta_{base}G°$/ kJ mol^{-1}	Note
Acenaphthene	$C_{12}H_{10}$	851.7	821.0	
Acetaldehyde	C_2H_4O	768.5	736.5	
Acetamide	C_2H_5NO	863.6	832.6	
Acetic acid	$C_2H_4O_2$	783.7	752.8	
Acetic anhydride	$C_4H_6O_3$	844		Ref. 9
Acetohydroxamic acid	$C_2H_5NO_2$	854.0	823.0	
Acetone	C_3H_6O	812	782.1	
Acetonitrile	C_2H_3N	779.2	748	
Acetophenone	C_8H_8O	861.1	829.3	
4-Acetylanisole	$C_9H_{10}O_2$	895.6	863.7	
4-Acetylbenzonitrile	C_9H_7NO	826.8	795.0	
Acetylene	C_2H_2	641.4	616.7	
4-Acetylphenyl acetate	$C_{10}H_{10}O_3$	853.2	821.3	
4-Acetylthioanisole	$C_9H_{10}OS$	888.2	856.3	
Acridine	$C_{13}H_9N$	972.6	940.7	
Acrolein	C_3H_4O	797.0	765.1	
Acrylamide	C_3H_5NO	870.7	839.8	
Acrylonitrile	C_3H_3N	784.7	753.7	
Adenine	$C_5H_5N_5$	942.8	912.5	
Adenosine	$C_{10}H_{13}N_5O_4$	989.3	956.8	
L-Alanine	$C_3H_7NO_2$	901.6	867.7	

Name	Mol. form.	E_{pa}/ kJ mol⁻¹	$\Delta_{base}G°$/ kJ mol⁻¹	Note
N-L-Alanyl-L-alanine	$C_6H_{12}N_2O_3$		905.6	
N-(*N*-L-Alanyl-L-alanyl)-L-alanine	$C_9H_{17}N_3O_4$		924.1	
Allene	C_3H_4	775.3	745.8	
Allyl	C_3H_5	736	707.4	
Allylamine	C_3H_7N	909.5	875.5	
Allyldimethylamine	$C_5H_{11}N$	957.8	926.8	
Allyl ethyl ether	$C_5H_{10}O$	833.7	804.5	
N-Allyl-2-propen-1-amine	$C_6H_{11}N$	949.3	916.3	
2-Aminoacetamide	$C_2H_6N_2O$		882.3	
Aminoacetonitrile	$C_2H_4N_2$	824.9	791.0	
4-Aminobenzaldehyde	C_7H_7NO	910.4	878.6	
3-Aminobenzamide	$C_7H_8N_2O$	900.9	869.9	
4-Aminobenzamide	$C_7H_8N_2O$	927.9	896.9	
3-Aminobenzonitrile	$C_7H_6N_2$	842.3	810.4	
4-Amino-1-butanol	$C_4H_{11}NO$	984.5	932.1	
6-Amino-1-hexanol	$C_6H_{15}NO$	969.0	915.7	
2-Aminophenol	C_6H_7NO	898.8	866.9	
3-Aminophenol	C_6H_7NO	898.8	866.9	
1-(4-Aminophenyl)ethanone	C_8H_9NO	908.8	877.0	
3-Aminopropanenitrile	$C_3H_6N_2$	866.4	832.5	
3-Amino-1-propanol	C_3H_9NO	962.5	917.3	
Ammonia	H_3N	853.6	819.0	
Aniline	C_6H_7N	882.5	850.6	
Aniline-2-carboxylic acid	$C_7H_7NO_2$	901.5	869.0	
Aniline-3-carboxylic acid	$C_7H_7NO_2$	864.7	832.3	
Aniline-4-carboxylic acid	$C_7H_7NO_2$	864.7	832.3	
Anilino	C_6H_6N	949.8	917.4	
Anisole	C_7H_8O	839.6	807.2	
Anthracene	$C_{14}H_{10}$	877.3	846.6	
L-Arginine	$C_6H_{14}N_4O_2$	1051.0	1006.6	
Argon	Ar	369.2	346.3	
Arsenic(III) fluoride	AsF_3	636.7	604.2	
Arsine	AsH_3	747.9	712.0	
L-Asparagine	$C_4H_8N_2O_3$	929	891.5	
L-Aspartic acid	$C_4H_7NO_4$	908.9	875	
1-Azabicyclo[2.2.2]octane	$C_7H_{13}N$	983.3	952.5	
Azetidine	C_3H_7N	943.4	908.6	
2-Azetidinone	C_3H_5NO	852.6	821.7	
Azidobenzene	$C_6H_5N_3$	820	787.5	
Azulene	$C_{10}H_8$	925.2	896	
Barium oxide	BaO	1215.4	1187.6	
Benzaldehyde	C_7H_6O	834.0	802.1	
Benzamide	C_7H_7NO	892.1	861.2	
Benzene	C_6H_6	750.4	725.4	
Benzeneacetonitrile	C_8H_7N	805.5	774.8	
1,2-Benzenediamine	$C_6H_8N_2$	896.5	865.8	
1,3-Benzenediamine	$C_6H_8N_2$	929.9	899.2	
1,4-Benzenediamine	$C_6H_8N_2$	905.9	874.0	
Benzeneethanamine	$C_8H_{11}N$	936.2	902.3	
1*H*-Benzimidazole	$C_7H_6N_2$	953.8	920.5	
Benzoic acid	$C_7H_6O_2$	821.1	790.1	
Benzonitrile	C_7H_5N	811.5	780.9	
Benzo[*ghi*]perylene	$C_{22}H_{12}$	876.0	845.2	
Benzophenone	$C_{13}H_{10}O$	882.3	852.5	
p-Benzoquinone	$C_6H_4O_2$	799.1	769.3	
Benzoxazole	C_7H_5NO	891.6	859.8	
Benzyl	C_7H_7	831.4	800.7	
Benzyl alcohol	C_7H_8O	778.3	748.0	
Benzylamine	C_7H_9N	913.3	879.4	
Benzyl methyl ether	$C_8H_{10}O$	816.7	787.5	

Name	Mol. form.	E_{pa}/ kJ mol^{-1}	$\Delta_{base}G°$/ kJ mol^{-1}	Note
Benzyne	C_6H_4	841	808.5	
Bicyclo[2.2.1]heptan-2-one	$C_7H_{10}O$	847.4	815.5	
Bicyclo[2.2.1]hept-2-ene	C_7H_{10}	836.5	804.0	
Biphenyl	$C_{12}H_{10}$	813.6	782.9	
2,2'-Bipyridine	$C_{10}H_8N_2$	965		Ref. 7
Bis(2-cyanoethyl) ether	$C_6H_8N_2O$	813.8	786.4	
Bis(2-propynyl)amine	C_6H_7N	910.0	876.9	
Bis(2-propynyl) ether	C_6H_6O	783.9	756.5	
Bis(2,2,2-trifluoroethyl) ether	$C_4H_4F_6O$	702.3	674.9	
Borazine	$B_3H_6N_3$	802.5	772.8	
Boric acid	BH_3O_3	728.1	698.4	
Bromine (atomic)	Br	554.4	531.2	
3-Bromoaniline	C_6H_6BrN	873.2	841.4	
Bromobenzene	C_6H_5Br	754.1	725.8	
Bromoethane	C_2H_5Br	696.2	669.7	
2-Bromoethanol	C_2H_5BrO	766.1	735.7	
Bromomethane	CH_3Br	664.2	638.0	
2-Bromopyridine	C_5H_4BrN	904.8	873.0	
3-Bromopyridine	C_5H_4BrN	910.0	878.2	
4-Bromopyridine	C_5H_4BrN	917.8	886.0	
2-Bromotoluene	C_7H_7Br	775.3	745.8	
3-Bromotoluene	C_7H_7Br	782.0	752.5	
4-Bromotoluene	C_7H_7Br	775.3	745.8	
Bromotrifluoromethane	$CBrF_3$	580.0	550.3	
1-Bromo-3-vinylbenzene	C_8H_7Br	822.4	793.5	
1-Bromo-4-vinylbenzene	C_8H_7Br	838.7	809.8	
1,2-Butadiene	C_4H_6	778.9	749.8	
1,3-Butadiene	C_4H_6	783.4	757.6	
1,3-Butadiyne	C_4H_2	737.2	712.8	
Butanal	C_4H_8O	792.7	760.8	
1,4-Butanediamine	$C_4H_{12}N_2$	1005.6	954.3	
1,4-Butanediol	$C_4H_{10}O_2$	915.6	854.9	
2,3-Butanedione	$C_4H_6O_2$	801.9	770.1	
Butanenitrile	C_4H_7N	798.4	767.7	
1-Butanethiol	$C_4H_{10}S$	801.7	770.5	
1,2,4-Butanetriol	$C_4H_{10}O_3$	905.9	841	
1-Butanol	$C_4H_{10}O$	789.2	758.9	
2-Butanol	$C_4H_{10}O$	815.7	784.6	
2-Butanone	C_4H_8O	827.3	795.5	
2-Butenamide	C_4H_7NO	887.1	856.1	
trans-2-Butene	C_4H_8	747	719.9	
trans-2-Butenoic acid	$C_4H_6O_2$	824.0	793	
3-Buten-2-one	C_4H_6O	834.7	802.8	
Butylamine	$C_4H_{11}N$	921.5	886.6	
tert-Butylamine	$C_4H_{11}N$	934.1	899.9	
Butylbenzene	$C_{10}H_{14}$	791.9	764.2	
Butyldimethylamine	$C_6H_{15}N$	969.2	938.2	
tert-Butyl ethyl ether	$C_6H_{14}O$	856.0	826.9	
Butyl formate	$C_5H_{10}O_2$	806.0	775	
tert-Butyl isopropyl ether	$C_7H_{16}O$	870.7	841.5	
Butyl methyl ether	$C_5H_{12}O$	820.3	791.2	
tert-Butyl nitrite	$C_4H_9NO_2$	863.9	831.4	
1-(4-*tert*-Butylphenyl)ethanone	$C_{12}H_{16}O$	882.5	850.6	
2-*tert*-Butylpyridine	$C_9H_{13}N$	961.7	929.8	
4-*tert*-Butylpyridine	$C_9H_{13}N$	957.7	925.8	
Butyl trifluoroacetate	$C_6H_9F_3O_2$	764.8	733.8	
2-Butyne	C_4H_6	775.8	745.1	
γ-Butyrolactone	$C_4H_6O_2$	840.0	808.1	
Calcium oxide	CaO	1190.6	1162.3	
Carbon dioxide	CO_2	540.5	515.8	

Name	Mol. form.	E_{pa}/ kJ mol^{-1}	$\Delta_{base}G°$/ kJ mol^{-1}	Note
Carbon diselenide	CSe_2	725	700.9	
Carbon disulfide	CS_2	681.9	657.7	
Carbon [fullerene-C_{60}]	C_{60}		827.5	
Carbon [fullerene-C_{70}]	C_{70}		827.5	
Carbon monoselenide	CSe	831.8	800.2	Protonation at C
Carbon monosulfide	CS	791.5	760	
Carbon monoxide	CO	594	562.8	Protonation at C
Carbon monoxide	CO	426.3	402.2	Protonation at O
Carbonothioic dichloride	CCl_2S	752.5	721.8	
Carbon oxyselenide	COSe	670	644.1	Protonation at Se
Carbon oxysulfide	COS	628.5	602.6	Protonation at S
Carbonyl fluoride	CF_2O	666.7	637.0	
Cesium hydroxide	CsHO	1117.9	1092.2	
Cesium oxide	Cs_2O	1442.9	1412.2	
Chlorine (atomic)	Cl	513.6	490.1	
Chloroacetic acid	$C_2H_3ClO_2$	765.4	734.5	
Chloroacetonitrile	C_2H_2ClN	745.7	715.1	
3-Chloroaniline	C_6H_6ClN	868.1	836.3	
4-Chloroaniline	C_6H_6ClN	873.8	842.0	
3-Chlorobenzaldehyde	C_7H_5ClO	813.0	781.1	
4-Chlorobenzaldehyde	C_7H_5ClO	831.3	799.4	
3-Chlorobenzamide	C_7H_6ClNO	877.2	846.3	
4-Chlorobenzamide	C_7H_6ClNO	877.2	846.3	
Chlorobenzene	C_6H_5Cl	753.1	724.6	
4-Chloro-*N,N*-diethylaniline	$C_{10}H_{14}ClN$	931.0	899.2	
4-Chloro-*N,N*-dimethylaniline	$C_8H_{10}ClN$	922.9	896.4	
Chloroethane	C_2H_5Cl	693.4	666.9	
2-Chloroethanol	C_2H_5ClO	766.1	735.7	
1-Chloro-4-ethynylbenzene	C_8H_5Cl	832.4	801.7	
1-Chloro-4-isopropenylbenzene	C_9H_9Cl	854.3	825.4	
Chloromethane	CH_3Cl	647.3	621.1	
3-Chloro-4-methoxyacetophenone	$C_9H_9ClO_2$	883.7	851.9	
2-Chloro-6-methoxypyridine	C_6H_6ClNO	909.9	878.0	
3-(Chloromethyl)benzonitrile	C_8H_6ClN	811.2	780.6	
4-(Chloromethyl)benzonitrile	C_8H_6ClN	812.8	782.1	
Chloromethylene	CHCl	874.1	839.9	
2-Chloro-4-methylpyridine	C_6H_6ClN	921.2	889.4	
2-Chloro-6-methylpyridine	C_6H_6ClN	908.0	876.2	
1-(3-Chlorophenyl)ethanone	C_8H_7ClO	846.9	815.1	
1-(4-Chlorophenyl)ethanone	C_8H_7ClO	856.6	824.8	
3-Chloropropanenitrile	C_3H_4ClN	773.1	742.4	
6-Chloro-1*H*-purine	$C_5H_3ClN_4$	873.6	841.7	
2-Chloropyridine	C_5H_4ClN	900.9	869	
3-Chloropyridine	C_5H_4ClN	903.4	871.5	
4-Chloropyridine	C_5H_4ClN	916.1	884.2	
2-Chlorotoluene	C_7H_7Cl	790.5	761.1	
3-Chlorotoluene	C_7H_7Cl	783.9	754.5	
4-Chlorotoluene	C_7H_7Cl	762.9	735.2	
Chlorotrifluoromethane	$CClF_3$	571.3	541.5	
Chromium	Cr	791.3	768.4	
Chromium carbonyl	C_6CrO_6	739.2	714.6	
Chrysene	$C_{18}H_{12}$	840.9	810.1	
Cinnoline	$C_8H_6N_2$	936.3	904.4	
Cobalt	Co	742.7	719.8	
Copper	Cu	655.3	632.4	
Coronene	$C_{24}H_{12}$	861.3	835.0	
o-Cresol	C_7H_8O	832	800	Ref. 10
m-Cresol	C_7H_8O	841	809	Ref. 10
p-Cresol	C_7H_8O	814	782	Ref. 10
Cyanamide	CH_2N_2	805.6	774.9	

Atomic

Name	Mol. form.	E_{pa}/ kJ mol^{-1}	$\Delta_{base}G°$/ kJ mol^{-1}	Note
Cyanide	CN	>595	>564	Protonation at N
Cyanoacetylene	C_3HN	751.2	720.5	
Cyanogen	C_2N_2	674.7	645.8	
Cyanogen bromide	CBrN	749.8	719.2	
Cyanogen chloride	CClN	722.1	691.5	
Cyanogen fluoride	CFN	632	601.3	
Cyclobutanecarboxylic acid	$C_5H_8O_2$	817.4	786.4	
Cyclobutanone	C_4H_6O	802.5	772.7	
Cyclobutene	C_4H_6	784.4	753.6	
Cycloheptanone	$C_7H_{12}O$	845.6	815.9	
2,4,6-Cycloheptatrien-1-one	C_7H_6O	920.8	891.0	
1,3-Cyclohexadiene	C_6H_8	837	804.5	
1,4-Cyclohexadiene	C_6H_8	837	808.0	
Cyclohexane	C_6H_{12}	686.9	666.9	
Cyclohexanecarbonitrile	$C_7H_{11}N$	815.0	784.4	
Cyclohexanecarboxylic acid	$C_7H_{12}O_2$	823.8	792.8	
cis-1,3-Cyclohexanediol	$C_6H_{12}O_2$	882.2	849.7	
trans-1,3-Cyclohexanediol	$C_6H_{12}O_2$	828.6	797.9	
1,2-Cyclohexanedione	$C_6H_8O_2$	849.6	818.9	
1,3-Cyclohexanedione	$C_6H_8O_2$	881.2	849.4	
1,4-Cyclohexanedione	$C_6H_8O_2$	812.5	782.7	
Cyclohexanemethanamine	$C_7H_{15}N$	926.6	895.8	
Cyclohexanemethanol	$C_7H_{14}O$	802.1	771.7	
Cyclohexanone	$C_6H_{10}O$	841.0	811.2	
Cyclohexene	C_6H_{10}	784.5	752.0	
Cyclohexylamine	$C_6H_{13}N$	934.4	899.6	
Cyclohexyldimethylamine	$C_8H_{17}N$	983.6	952.6	
1-Cyclohexylethanone	$C_8H_{14}O$	841.4	809.5	
Cyclononanone	$C_9H_{16}O$	852.6	822.8	
Cyclooctanone	$C_8H_{14}O$	849.4	819.6	
1,3-Cyclopentadiene	C_5H_6	821.6	798.4	
Cyclopentadienyl	C_5H_5	831.5	799.1	
Cyclopentanecarboxylic acid	$C_6H_{10}O_2$	817.4	786.4	
cis-1,2-Cyclopentanediol	$C_5H_{10}O_2$	885.6	853.1	
Cyclopentanone	C_5H_8O	823.7	794.0	
Cyclopentene	C_5H_8	766.3	733.8	
N-(1-Cyclopenten-1-yl)pyrrolidine	$C_9H_{15}N$	1019.2	988.4	
Cyclopropane	C_3H_6	750.3	722.2	
Cyclopropanecarbonitrile	C_4H_5N	808.2	777.5	
Cyclopropanecarboxylic acid	$C_4H_6O_2$	821.4	790.4	
Cyclopropene	C_3H_4	818.5	787.8	
Cyclopropyl	C_3H_5	738.9	702.0	
Cyclopropylamine	C_3H_7N	904.7	869.9	
Cyclopropylbenzene	C_9H_{10}	834.9	802.4	
Cyclopropyl methyl ketone	C_5H_8O	854.9	823	
L-Cysteine	$C_3H_7NO_2S$	903.2	869.3	
Cytidine	$C_9H_{13}N_3O_5$	982.5	950.0	
Cytosine	$C_4H_5N_3O$	949.9	918	
Decylamine	$C_{10}H_{23}N$	930.4	896.5	
2'-Deoxyadenosine	$C_{10}H_{13}N_5O_3$	991.5	959.1	
Diacetone alcohol	$C_6H_{12}O_2$	822.9	791.1	
1,3-Diacetylbenzene	$C_{10}H_{10}O_2$	852.0	822.3	
1,4-Diacetylbenzene	$C_{10}H_{10}O_2$	850.8	821.0	
Diallyl ether	$C_6H_{10}O$	827.4	800.0	
N,N-Diallyl-2-propen-1-amine	$C_9H_{15}N$	972.3	941.3	
Diazomethane	CH_2N_2	858.9	826.7	
Diborane	B_2H_6	615	586.0	
Dibutylamine	$C_8H_{19}N$	968.5	935.3	
Di-*sec*-butylamine	$C_8H_{19}N$	980.7	947.5	
Dibutyl ether	$C_8H_{18}O$	845.7	818.3	

Name	Mol. form.	E_{pa}/ kJ mol⁻¹	$\Delta_{base}G°$/ kJ mol⁻¹	Note
Di-*sec*-butyl ether	$C_8H_{18}O$	865.9	838.5	
Di-*tert*-butyl ether	$C_8H_{18}O$	887.4	860.0	
2,5-Di-*tert*-butylfuran	$C_{12}H_{20}O$	894.7	863.9	
Di-*tert*-butyl ketone	$C_9H_{18}O$	861.3	831.5	
1,3-Di-*tert*-butyl-5-methylbenzene	$C_{15}H_{24}$	853.7	826.0	
2,4-Di-*tert*-butylpyridine	$C_{13}H_{21}N$	983.8	952.0	
2,6-Di-*tert*-butylpyridine	$C_{13}H_{21}N$	982.9	951	
Dibutyl sulfide	$C_8H_{18}S$	871.8	842.1	
Di-*tert*-butyl sulfide	$C_8H_{18}S$	893.8	864.0	
Dicarbon monoxide	C_2O	774.7	747.0	
Dichloromethylene	CCl_2	861	828.5	
m-Dicyanobenzene	$C_8H_4N_2$	779.3	750.4	
p-Dicyanobenzene	$C_8H_4N_2$	779.0	751.8	
Dicyclopropyl ketone	$C_7H_{10}O$	880.4	850.6	
Diethanolamine	$C_4H_{11}NO_2$	953	920	
N,N-Diethylacetamide	$C_6H_{13}NO$	925.4	894.4	
Diethylamine	$C_4H_{11}N$	952.4	919.4	
N,N-Diethylaniline	$C_{10}H_{15}N$	959.8	927.9	
Diethylene glycol dimethyl ether	$C_6H_{14}O_3$	918.8	870.9	
Diethyl ether	$C_4H_{10}O$	828.4	801	
Diethylmethylamine	$C_5H_{13}N$	971.0	940.0	
N,N-Diethyl-4-methylaniline	$C_{11}H_{17}N$	962.8	931.0	
Diethylpropylamine	$C_7H_{17}N$	978.8	947.9	
2,6-Diethylpyridine	$C_9H_{13}N$	972.3	940.4	
N,N-Diethyl-3-pyridinecarboxamide	$C_{10}H_{14}N_2O$	940.9	909.0	
Diethyl sulfide	$C_4H_{10}S$	856.7	827.0	
o-Difluorobenzene	$C_6H_4F_2$	731.2	703.5	
m-Difluorobenzene	$C_6H_4F_2$	749.7	722	
p-Difluorobenzene	$C_6H_4F_2$	718.7	692.8	
2,2-Difluoroethanol	$C_2H_4F_2O$	727.4	697.0	
1,1-Difluoroethene	$C_2H_2F_2$	734	705.1	
trans-1,2-Difluoroethene	$C_2H_2F_2$	688.6	657.9	
Difluoromethane	CH_2F_2	620.5	589.7	
Difluoromethylene	CF_2	765	732.5	
2,3-Dihydro-1,4-dioxin	$C_4H_6O_2$	823.5	792.8	
2,3-Dihydrofuran	C_4H_6O	866.9	834.4	
2,5-Dihydrofuran	C_4H_6O	823.4	796	
2,3-Dihydro-1*H*-indole	C_8H_9N	957.1	926.3	
3,4-Dihydro-2*H*-pyran	C_5H_8O	865.8	833.4	
Diisobutylamine	$C_8H_{19}N$	958.1	925.1	
Diisopropylamine	$C_6H_{15}N$	971.9	938.6	
Diisopropyl ether	$C_6H_{14}O$	855.5	828.1	
Diisopropyl sulfide	$C_6H_{14}S$	876.4	846.6	
Dilithium	Li_2	1162	1133.1	
Dimagnesium	Mg_2	919	886.5	
1,2-Dimethoxyethane	$C_4H_{10}O_2$	858.0	820.2	
N,N-Dimethylacetamide	C_4H_9NO	908.0	877.0	
Dimethylamine	C_2H_7N	929.5	896.5	
(Dimethylamino)acetonitrile	$C_4H_8N_2$	884.5	853.7	
4'-(Dimethylamino)acetophenone	$C_{10}H_{13}NO$	932.8	906.3	
4-(Dimethylamino)benzaldehyde	$C_9H_{11}NO$	924.8	898.3	
4-(Dimethylamino)benzonitrile	$C_9H_{10}N_2$	889.1	862.6	
1-[3-(Dimethylamino)phenyl]ethanone	$C_{10}H_{13}NO$	928.0	901.5	
3-(Dimethylamino)-1-propyne	C_5H_9N	940.3	909.5	
2,6-Dimethylaniline	$C_8H_{11}N$	901.7	869.8	
N,N-Dimethylaniline	$C_8H_{11}N$	941.1	909.2	
N,N-Dimethylbenzamide	$C_9H_{11}NO$	932.7	901.8	
N,N-Dimethyl-1,4-benzenediamine	$C_8H_{12}N_2$	955.0	928.4	
N,N-Dimethylbenzylamine	$C_9H_{13}N$	968.4	937.4	
2,3-Dimethyl-1,3-butadiene	C_6H_{10}	835.0	807.8	

Name	Mol. form.	E_{pa}/ kJ mol-1	$\Delta_{base}G°$/ kJ mol-1	Note
N,N-Dimethylbutanamide	$C_6H_{13}NO$	921.7	890.8	
3,3-Dimethyl-2-butanone	$C_6H_{12}O$	840.1	808.2	
2,3-Dimethyl-2-butene	C_6H_{12}	813.9	785.9	
Dimethyl carbonate	$C_3H_6O_3$	830.2	799.2	
Dimethylcyanamide	$C_3H_6N_2$	852.1	821.4	
1,2-Dimethylcyclopentene	C_7H_{12}	822.6	791.9	
3,3-Dimethylcyclopropene	C_5H_8	847.8	817.1	
trans-Dimethyldiazene	$C_2H_6N_2$	865.1	834.4	
Dimethyl disulfide	$C_2H_6S_2$	815.3	782.8	
N,N′-Dimethyl-1,2-ethanediamine	$C_4H_{12}N_2$	989.2	946.9	
Dimethyl ether	C_2H_6O	792	764.5	
N,N-Dimethylformamide	C_3H_7NO	887.5	856.6	
2,4-Dimethylfuran	C_6H_8O	894.7	862.3	
2,5-Dimethylfuran	C_6H_8O	865.9	835.2	
3,4-Dimethylfuran	C_6H_8O	869.0	838.3	
1,1-Dimethylhydrazine	$C_2H_8N_2$	927.1	894.7	
Dimethyl hydrogen phosphite	$C_2H_7O_3P$	894.8	862.4	
1,2-Dimethyl-1*H*-imidazole	$C_5H_8N_2$	984.7	952.6	
1,4-Dimethyl-1*H*-imidazole	$C_5H_8N_2$	976.7	944.9	
1,5-Dimethyl-1*H*-imidazole	$C_5H_8N_2$	977.6	945.8	
1,3-Dimethyl-2-imidazolidinone	$C_5H_{10}N_2O$	918.4	886.0	
Dimethyl isophthalate	$C_{10}H_{10}O_4$	843.5	814.3	
1,4-Dimethyl-7-isopropylazulene	$C_{15}H_{18}$	983.1	950.6	
Dimethyl mercury	C_2H_6Hg	771.6	740.8	
N,N-Dimethyl-3-nitroaniline	$C_8H_{10}N_2O_2$	894.1	867.6	
N,N-Dimethyl-4-nitroaniline	$C_8H_{10}N_2O_2$	896.7	870.2	
2,4-Dimethyl-1-nitrobenzene	$C_8H_9NO_2$	831.0	798.5	
2,4-Dimethyl-1,3-pentadiene	C_7H_{12}	886.5	857.6	
2,4-Dimethyl-3-pentanone	$C_7H_{14}O$	850.3	820.5	
2,4-Dimethyl-2-pentene	C_7H_{14}	812	783.1	
1-(2,4-Dimethylphenyl)ethanone	$C_{10}H_{12}O$	882.6	850.8	
1-(2,5-Dimethylphenyl)ethanone	$C_{10}H_{12}O$	873.5	841.6	
1-(3,4-Dimethylphenyl)ethanone	$C_{10}H_{12}O$	882.8	851.0	
Dimethylphosphine	C_2H_7P	912.0	877.9	
2,2-Dimethylpropanamide	$C_5H_{11}NO$	889.0	857.2	
N,N-Dimethyl-1-propanamine	$C_5H_{13}N$	962.8	931.9	
N,N-Dimethyl-1,3-propanediamine	$C_5H_{14}N_2$	1025.0	975.3	
2,2-Dimethylpropanenitrile	C_5H_9N	810.9	780.2	
2,2-Dimethyl-1-propanethiol	$C_5H_{12}S$	809.5	778.2	
2,2-Dimethyl-1-propanol	$C_5H_{12}O$	795.5	765.2	
N,N-Dimethyl-2-propenamide	C_5H_9NO	904.3	873.4	
2,2-Dimethylpropylamine	$C_5H_{13}N$	928.3	894.0	
2,6-Dimethyl-4*H*-pyran-4-one	$C_7H_8O_2$	941.5	907.3	
1,3-Dimethyl-1*H*-pyrazole	$C_5H_8N_2$	933.9	902.3	
1,5-Dimethyl-1*H*-pyrazole	$C_5H_8N_2$	934.3	902.8	
3,4-Dimethyl-1*H*-pyrazole	$C_5H_8N_2$	927.3	895.4	
3,5-Dimethyl-1*H*-pyrazole	$C_5H_8N_2$	933.5	900.1	
N,N-Dimethyl-2-pyridinamine	$C_7H_{10}N_2$	968.2	941.6	
N,N-Dimethyl-4-pyridinamine	$C_7H_{10}N_2$	997.6	971.1	
2,3-Dimethylpyridine	C_7H_9N	958.9	927.0	
2,4-Dimethylpyridine	C_7H_9N	962.9	930.8	
2,5-Dimethylpyridine	C_7H_9N	958.8	926.9	
2,6-Dimethylpyridine	C_7H_9N	963.0	931.1	
3,4-Dimethylpyridine	C_7H_9N	957.3	925.5	
3,5-Dimethylpyridine	C_7H_9N	955.4	923.5	
2,5-Dimethylpyrrole	C_6H_9N	918.7	887.1	
Dimethyl sulfide	C_2H_6S	830.9	801.2	
Dimethyl sulfoxide	C_2H_6OS	884.4	853.7	
Dimethyl terephthalate	$C_{10}H_{10}O_4$	843.2	812.3	
N,N-Dimethylthioacetamide	C_4H_9NS	925.3	894.4	

Name	Mol. form.	E_{pa}/ kJ mol^{-1}	$\Delta_{base}G°$/ kJ mol^{-1}	Note
N,N′-Dimethylthiourea	$C_3H_8N_2S$	926.0	895.1	
N,N-Dimethyltricyclo[3.3.1.1^{3,7}] decane-1-carboxamide	$C_{13}H_{21}NO$	949.4	917.6	
N,N′-Dimethylurea	$C_3H_8N_2O$	903.3	873.5	
1,3-Dioxane	$C_4H_8O_2$	825.4	796.2	
1,4-Dioxane	$C_4H_8O_2$	797.4	770.0	
Dipentyl ether	$C_{10}H_{22}O$	852.7	825.3	
1,4-Diphenylbutane	$C_{16}H_{18}$	822.0	779.8	
1,2-Diphenylethane	$C_{14}H_{14}$	801.8	774.1	
1,1-Diphenylethene	$C_{14}H_{12}$	885.7	856.9	
1,6-Diphenylhexane	$C_{18}H_{22}$	826.1	783.8	
Diphenylmethane	$C_{13}H_{12}$	802.0	769.5	
1,5-Diphenylpentane	$C_{17}H_{20}$	824.7	782.4	
1,3-Diphenylpropane	$C_{15}H_{16}$	820.1	787.6	
3,5-Diphenyl-1*H*-pyrazole	$C_{15}H_{12}N_2$	946.3	912.7	
Dipropylamine	$C_6H_{15}N$	962.3	929.3	
N,N-Dipropylaniline	$C_{12}H_{19}N$	963.0	931.1	
Dipropyl ether	$C_6H_{14}O$	837.9	810.5	
Dipropyl sulfide	$C_6H_{14}S$	864.7	834.9	
Disiloxane	H_6OSi_2	749	718.3	
Disodium	Na_2	1146.8	1118.2	
Ethane	C_2H_6	596.3	569.9	
1,2-Ethanediamine	$C_2H_8N_2$	951.6	912.5	
1,2-Ethanediol	$C_2H_6O_2$	815.9	773.6	
Ethanethiol	C_2H_6S	789.6	758.4	
Ethanimidamide	$C_2H_6N_2$	970.7	938.2	
Ethanol	C_2H_6O	776.4	746	
Ethanolamine	C_2H_7NO	930.3	896.8	
Ethenamine	C_2H_5N	898.9	866.5	
3-Ethoxypropanenitrile	C_5H_9NO	807.2	776.5	
N-Ethylacetamide	C_4H_9NO	898.0	867.0	
Ethyl acetate	$C_4H_8O_2$	835.7	804.7	
Ethylamine	C_2H_7N	912.0	878	
3-Ethylaniline	$C_8H_{11}N$	897.9	866.1	
N-Ethylaniline	$C_8H_{11}N$	924.8	892.9	
Ethylbenzene	C_8H_{10}	788.0	760.3	
Ethyl chloroformate	$C_3H_5ClO_2$	764.8	733.8	
Ethylcyanoformate	$C_4H_5NO_2$	745.7	714.7	
Ethyldimethylamine	$C_4H_{11}N$	960.1	929.1	
Ethyl 1,5-dimethyl-1*H*-pyrazole-3-carboxylate	$C_8H_{12}N_2O_2$	933.4	901.5	
Ethylene	C_2H_4	680.5	651.5	
Ethylene carbonate	$C_3H_4O_3$	814.2	784.4	
Ethyleneimine	C_2H_5N	905.5	872.5	
Ethyl formate	$C_3H_6O_2$	799.4	768.4	
N-Ethyl-*N*-hydroxyethanamine	$C_4H_{11}NO$	914.7	882.2	
Ethyl isocyanide	C_3H_5N	851.3	818.9	
Ethylisopropylamine	$C_5H_{13}N$	960.0	926.7	
Ethyl isopropyl ether	$C_5H_{12}O$	842.7	813.5	
N-Ethyl-*N*-isopropyl-2-propanamine	$C_8H_{19}N$	994.3	963.5	
Ethylmethylamine	C_3H_9N	942.2	909.2	
N-Ethyl-*N*-methylaniline	$C_9H_{13}N$	939.0	912.4	
Ethyl *N*-methylcarbamate	$C_4H_9NO_2$	888.8	857.8	
Ethyl methyl carbonate	$C_4H_8O_3$	842.7	810.8	
3-Ethyl-3-methyldiaziridine	$C_4H_{10}N_2$	903.8	871.3	
Ethyl methyl ether	C_3H_8O	808.6	781.2	
Ethyl methyl sulfide	C_3H_8S	846.5	815.3	
Ethyl nitrite	$C_2H_5NO_2$	818.9	786.4	
2-Ethylpyridine	C_7H_9N	952.4	920.6	
3-Ethylpyridine	C_7H_9N	947.4	915.5	
4-Ethylpyridine	C_7H_9N	951.1	919.2	

Name	Mol. form.	E_{pa}/ kJ mol^{-1}	$\Delta_{base}G°$/ kJ mol^{-1}	Note
Ethyl trichloroacetate	$C_4H_5Cl_3O_2$	790.4	759.4	
Ethyl trifluoroacetate	$C_4H_5F_3O_2$	758.8	727.9	
Ethyl vinyl ether	C_4H_8O	870.1	840.4	
Ethynyl	C_2H	753	720.8	
Ferrocene	$C_{10}H_{10}Fe$	863.6	841.3	
Fluoranthene	$C_{16}H_{10}$	828.6	800.9	
9*H*-Fluorene	$C_{13}H_{10}$	831.5	803.8	
Fluorine	F_2	332	305.5	
Fluorine (atomic)	F	340.1	315.1	
Fluorine oxide	FO	508.7	482.2	
Fluoroacetic acid	$C_2H_3FO_2$	765.4	734.5	
Fluoroacetylene	C_2HF	686	661.3	
3-Fluoroaniline	C_6H_6FN	867.3	835.5	
4-Fluoroaniline	C_6H_6FN	871.5	839.7	
3-Fluorobenzaldehyde	C_7H_5FO	814.3	782.5	
4-Fluorobenzaldehyde	C_7H_5FO	827.1	795.3	
3-Fluorobenzamide	C_7H_6FNO	877.2	846.3	
4-Fluorobenzamide	C_7H_6FNO	877.2	846.3	
Fluorobenzene	C_6H_5F	755.9	726.6	
m-Fluorobenzyl	C_7H_6F	836.5	804	
Fluoroethane	C_2H_5F	683.4	655.8	
2-Fluoroethanol	C_2H_5FO	715.6	685.2	
Fluoroethene	C_2H_3F	729	700.1	
Fluoromethane	CH_3F	598.9	571.5	
Fluoromethylene	CHF	797.9	763.8	
1-(4-Fluorophenyl)ethanone	C_8H_7FO	858.6	826.8	
1-Fluoro-2-propanone	C_3H_5FO	795.4	763.5	
2-Fluoropyridine	C_5H_4FN	884.6	852.7	
3-Fluoropyridine	C_5H_4FN	902.0	870.1	
2-Fluorotoluene	C_7H_7F	773.3	743.8	
3-Fluorotoluene	C_7H_7F	785.4	756.0	
4-Fluorotoluene	C_7H_7F	763.8	736.1	
Formaldehyde	CH_2O	712.9	683.3	
Formamide	CH_3NO	822.2	791.2	
Formic acid	CH_2O_2	742.0	710.3	
Formyl	CHO	636	601.8	
4-Formylbenzonitrile	C_8H_5NO	796.9	766.3	
Formyloxyl	CHO_2	623.4	590.9	
Fulminic acid	CHNO	758	725.5	
Furan	C_4H_4O	812	781	Ref. 10
Germane	GeH_4	713.4	687.1	
α-D-Glucose	$C_6H_{12}O_6$		778.9	
β-D-Glucose	$C_6H_{12}O_6$		778.9	
L-Glutamic acid	$C_5H_9NO_4$	913.0	879.1	
L-Glutamine	$C_5H_{10}N_2O_3$	937.8	900	
Glutaric anhydride	$C_5H_6O_3$	816		Ref. 9
Glycerol	$C_3H_8O_3$	874.8	820	
Glycine	$C_2H_5NO_2$	886.5	852.2	
N-Glycylglycine	$C_4H_8N_2O_3$		882	
N-(*N*-Glycylglycyl)glycine	$C_6H_{11}N_3O_4$	966.8	916.8	
Guanidine	CH_5N_3	986.3	949.4	
Guanine	$C_5H_5N_5O$	959.5	927.6	
Guanosine	$C_{10}H_{13}N_5O_5$	993.4	960.9	
Helium	He	177.8	148.5	
1,7-Heptanediamine	$C_7H_{18}N_2$	998.5	944.9	
4-Heptanone	$C_7H_{14}O$	845.0	815.3	
Heptylamine	$C_7H_{17}N$	923.2	889.3	
Hexafluorobenzene	C_6F_6	648.0	624.4	
1,1,1,3,3,3-Hexafluoro-2-propanol	$C_3H_2F_6O$	686.6	656.2	
Hexahydro-1*H*-azepine	$C_6H_{13}N$	956.7	923.5	

Name	Mol. form.	E_{pa}/ kJ mol^{-1}	$\Delta_{base}G°$/ kJ mol^{-1}	Note
2,3,4,6,7,8-Hexahydropyrrolo[1,2-*a*]pyrimidine	$C_7H_{12}N_2$	1038.3	1005.9	
Hexamethylbenzene	$C_{12}H_{18}$	860.6	836.0	
Hexamethyldisiloxane	$C_6H_{18}OSi_2$	846.4	816.2	
Hexamethylphosphoric triamide	$C_6H_{18}N_3OP$	958.6	928.7	
Hexamethylphosphorus triamide	$C_6H_{18}N_3P$	930.1	897.7	
1,6-Hexanediamine	$C_6H_{16}N_2$	999.5	946.2	
2,5-Hexanedione	$C_6H_{10}O_2$	892.0	851.8	
3-Hexanone	$C_6H_{12}O$	843.2	811.3	
1-Hexene	C_6H_{12}	805.2	776.3	
trans-3-Hexen-2-one	$C_6H_{10}O$	865.6	833.8	
Hexylamine	$C_6H_{15}N$	927.5	893.5	
2-Hexylpyridine	$C_{11}H_{17}N$	963.6	931.7	
1-Hexyne	C_6H_{10}	799.8	774.8	
2-Hexyne	C_6H_{10}	806.1	781.1	
Histamine	$C_5H_9N_3$	999.8	961.9	
L-Histidine	$C_6H_9N_3O_2$	988	950.2	
Hydrazine	H_4N_2	853.2	822.4	
Hydrazoic acid	HN_3	756.0	723.5	
Hydrogen	H_2	422.3	394.7	
Hydrogen bromide	BrH	584.2	557.7	
Hydrogen chloride	ClH	556.9	530.1	
Hydrogen cyanide	CHN	712.9	681.6	
Hydrogen fluoride	FH	484	456.7	
Hydrogen iodide	HI	627.5	601.3	
Hydrogen isocyanide	CHN	772.3	739.8	
Hydrogen peroxide	H_2O_2	674.5	643.8	
Hydrogen selenide	H_2Se	707.8	676.4	
Hydrogen sulfide	H_2S	705	673.8	
Hydrogen telluride	H_2Te	735.9	704.5	
Hydroperoxy	HO_2	660	627.5	
Hydroxyl	HO	593.2	564.0	
1-(3-Hydroxyphenyl)ethanone	$C_8H_8O_2$	863.6	831.8	
1-(4-Hydroxyphenyl)ethanone	$C_8H_8O_2$	883.7	851.9	
Hypoxanthine	$C_5H_4N_4O$	912.3	880.5	
Imidazole	$C_3H_4N_2$	942.8	909.2	
2-Imidazolidinethione	$C_3H_6N_2S$	921.9	891.2	
1*H*-Indazole	$C_7H_6N_2$	900.8	868.9	
Indene	C_9H_8	848.8	819.6	
1*H*-Indole	C_8H_7N	933.4	901.9	
Iodine (atomic)	I	608.2	583.5	
3-Iodoaniline	C_6H_6IN	878.7	846.8	
Iodoethane	C_2H_5I	724.8	698.3	
Iodomethane	CH_3I	691.7	665.5	
1-Iodo-2-methylbenzene	C_7H_7I	780.3	750.8	
Iron	Fe	754	731.1	
Iron(II) oxide	FeO	907	880.5	
Iron pentacarbonyl	C_5FeO_5	833.0	798.5	
Isobutanal	C_4H_8O	797.3	765.5	
Isobutane	C_4H_{10}	677.8	671.3	
Isobutene	C_4H_8	802.1	775.6	
Isobutylamine	$C_4H_{11}N$	924.8	890.8	
Isobutyldimethylamine	$C_6H_{15}N$	968.7	937.8	
1-Isobutylpiperidine	$C_9H_{19}N$	974.5	943.5	
Isocyanic acid (HNCO)	CHNO	753	718.8	
Isocyanobenzene	C_7H_5N	868.4	836.0	
Isocyanomethane	C_2H_3N	839.1	806.6	
2-Isocyano-2-methylpropane	C_5H_9N	870.7	838.3	
1-Isocyanopropane	C_4H_7N	856.8	824.3	
L-Isoleucine	$C_6H_{13}NO_2$	917.4	883.5	
Isophorone	$C_9H_{14}O$	893.5	861.6	

Name	Mol. form.	E_{pa}/ kJ mol⁻¹	$\Delta_{base}G°$/ kJ mol⁻¹	Note
Isopropenylbenzene	C_9H_{10}	864.2	835.3	
Isopropyl acetate	$C_5H_{10}O_2$	836.6	805.6	
Isopropylamine	C_3H_9N	923.8	889.0	
Isopropylbenzene	C_9H_{12}	791.6	763.9	
Isopropyl formate	$C_4H_8O_2$	811.3	780.3	
Isopropylmethylamine	$C_4H_{11}N$	952.4	919.4	
Isopropyl methyl ether	$C_4H_{10}O$	826.3	797.1	
Isopropyl nitrite	$C_3H_7NO_2$	845.5	813.0	
4-Isopropylpyridine	$C_8H_{11}N$	955.7	923.8	
Isoquinoline	C_9H_7N	951.7	919.9	
Isoxazole	C_3H_3NO	848.6	816.8	
Ketene	C_2H_2O	825.3	793.6	
Krypton	Kr	424.6	402.4	
Lanthanum	La	1013	991.9	
L-Leucine	$C_6H_{13}NO_2$	914.6	880.6	
Lithium bromide	BrLi	819	792.5	
Lithium chloride	ClLi	827	800.5	
Lithium hydride	HLi	1021.7	996.4	
Lithium hydroxide	HLiO	1000.1	972.1	
Lithium oxide	Li_2O	1206	1175.3	
Lutetium	Lu	992	970.6	
L-Lysine	$C_6H_{14}N_2O_2$	996	951.0	
Magnesium	Mg	819.6	797.3	
Magnesium oxide	MgO	988	959.4	
Malononitrile	$C_3H_2N_2$	723.0	694.1	
Manganese	Mn	797.3	774.4	
Manganese 2-methylcyclopentadienyl tricarbonyl	$C_9H_7MnO_3$	833.8	801.3	
Mesityl oxide	$C_6H_{10}O$	878.7	846.9	
Metaboric acid (α form)	BHO_2	763.0	730.5	
Methacrylic acid	$C_4H_6O_2$	816.7	785.7	
Methane	CH_4	543.5	520.6	
Methanesulfonic acid	CH_4O_3S	761.3	728.9	
Methanethiol	CH_4S	773.4	742	
Methanol	CH_4O	754.3	724.5	
L-Methionine	$C_5H_{11}NO_2S$	935.4	901.5	
Methoxyacetonitrile	C_3H_5NO	758.1	727.4	
2-Methoxyaniline	C_7H_9NO	905.2	873.3	
3-Methoxyaniline	C_7H_9NO	913.0	881.1	
4-Methoxyaniline	C_7H_9NO	900.3	868.5	
3-Methoxybenzaldehyde	$C_8H_8O_2$	844.1	812.2	
4-Methoxybenzaldehyde	$C_8H_8O_2$	881.1	849.3	
3-Methoxybenzamide	$C_8H_9NO_2$	900.9	869.9	
4-Methoxybenzamide	$C_8H_9NO_2$	900.3	869.4	
Methoxycyclohexane	$C_7H_{14}O$	840.5	811.3	
4-Methoxy-*N,N*-dimethylaniline	$C_9H_{13}NO$	949.1	922.4	
2-Methoxyethanol	$C_3H_8O_2$	768.8	729.8	
2-Methoxyethylamine	C_3H_9NO	928.6	894.6	
1-(3-Methoxyphenyl)ethanone	$C_9H_{10}O_2$	871.2	839.3	
2-Methoxy-1-propene	C_4H_8O	894.9	866.1	
2-Methoxypyridine	C_6H_7NO	934.7	902.8	
3-Methoxypyridine	C_6H_7NO	942.7	910.9	
4-Methoxypyridine	C_6H_7NO	961.7	929.8	
N-Methylacetamide	C_3H_7NO	888.5	857.6	
Methyl acetate	$C_3H_6O_2$	821.6	790.7	
4-Methylacetophenone	$C_9H_{10}O$	875.5	843.6	
Methyl acrylate	$C_4H_6O_2$	825.8	794.8	
2-Methylallyl	C_4H_7	778	747.3	
Methylamidogen	CH_4N	832.8	801.6	
Methylamine	CH_5N	899.0	864.5	
Methyl 4-aminobenzoate	$C_8H_9NO_2$	883.9	853.0	

Name	Mol. form.	E_{pa}/ kJ mol^{-1}	$\Delta_{base}G°$/ kJ mol^{-1}	Note
2-Methylaniline	C_7H_9N	890.9	859.1	
3-Methylaniline	C_7H_9N	895.8	864.0	
4-Methylaniline	C_7H_9N	896.7	864.8	
N-Methylaniline	C_7H_9N	916.6	890.1	
2-Methylanisole	$C_8H_{10}O$	850	818	Ref. 10
3-Methylanisole	$C_8H_{10}O$	860	828	Ref. 10
4-Methylanisole	$C_8H_{10}O$	841	809	Ref. 10
2-Methylanthracene	$C_{15}H_{12}$	887.5	855.1	
9-Methylanthracene	$C_{15}H_{12}$	896.5	865.8	
Methyl azide	CH_3N_3	833	800.5	
1-Methylaziridine	C_3H_7N	934.8	904.1	
3-Methylbenzaldehyde	C_8H_8O	840.0	808.1	
4-Methylbenzaldehyde	C_8H_8O	851.8	820.0	
3-Methylbenzamide	C_8H_9NO	900.9	869.9	
4-Methylbenzamide	C_8H_9NO	900.9	869.9	
1-Methyl-1*H*-benzimidazole	$C_8H_8N_2$	967.0	935.2	
Methyl benzoate	$C_8H_8O_2$	850.5	819.5	
2-Methylbenzofuran	C_9H_8O	859.6	827.2	
1-Methyl-1*H*-benzotriazole	$C_7H_7N_3$	931.2	898.7	
2-Methylbiphenyl	$C_{13}H_{12}$	815.9	783.4	
3-Methylbiphenyl	$C_{13}H_{12}$	828.0	795.5	
4-Methylbiphenyl	$C_{13}H_{12}$	817.9	785.4	
2-Methyl-1,3-butadiene	C_5H_8	826.4	797.6	
2-Methyl-2-butanamine	$C_5H_{13}N$	937.8	903.6	
Methyl butanoate	$C_5H_{10}O_2$	836.4	805.4	
3-Methyl-2-butanone	$C_5H_{10}O$	836.3	804.4	
3-Methyl-2-butenal	C_5H_8O	856.9	825.0	
2-Methyl-2-butene	C_5H_{10}	808.8	779.9	
Methyl *trans*-2-butenoate	$C_5H_8O_2$	851.3	820.4	
cis-2-Methyl-2-butenoic acid	$C_5H_8O_2$	822.5	791.5	
3-Methyl-2-butenoic acid	$C_5H_8O_2$	822.9	791.9	
3-Methyl-3-buten-2-one	C_5H_8O	843.1	811.3	
Methyl *tert*-butyl ether	$C_5H_{12}O$	841.6	812.4	
3-Methyl-1-butyne	C_5H_8	814.9	787.8	
Methyl 3-chlorobenzoate	$C_8H_7ClO_2$	835.4	804.4	
Methyl 4-chlorobenzoate	$C_8H_7ClO_2$	842.1	811.1	
1-Methylcyclobutene	C_5H_8	841.5	807.3	
Methyl cyclohexanecarboxylate	$C_8H_{14}O_2$	846.2	815.3	
4-Methylcyclohexanone	$C_7H_{12}O$	844.9	813.0	
1-Methylcyclohexene	C_7H_{12}	825.1	792.6	
1-Methylcyclopentene	C_6H_{10}	816.5	787.1	
Methyl cyclopropanecarboxylate	$C_5H_8O_2$	842.1	811.2	
2-Methyl-*N,N*-dimethylaniline	$C_9H_{13}N$	951.8	925.3	
3-Methyl-*N,N*-dimethylaniline	$C_9H_{13}N$	942.1	915.7	
4-Methyl-*N,N*-dimethylaniline	$C_9H_{13}N$	950.0	918.1	
Methyl 2,4-dimethylbenzoate	$C_{10}H_{12}O_2$	868.2	837.2	
Methyl 2,5-dimethylbenzoate	$C_{10}H_{12}O_2$	864.7	833.7	
Methyl 3,5-dimethylbenzoate	$C_{10}H_{12}O_2$	864.3	833.4	
Methyl 2,2-dimethylpropanoate	$C_6H_{12}O_2$	845.2	814.2	
Methyldiphenylphosphine	$C_{13}H_{13}P$	972.1	939.7	
Methyldipropylamine	$C_7H_{17}N$	983.5	950.9	
Methylenecyclopentane	C_6H_{10}	832.4	803.5	
N-Methylformamide	C_2H_5NO	851.3	820.3	
Methyl formate	$C_2H_4O_2$	782.5	751.5	
Methyl 4-formylbenzoate	$C_9H_8O_3$	832.9	801.9	
2-Methylfuran	C_5H_6O	865.9	833.5	
3-Methylfuran	C_5H_6O	854.0	821.5	
4-Methylglutaric anhydride	$C_6H_8O_3$	820		Ref. 9
Methylhydrazine	CH_6N_2	898.8	866.4	
Methyl 3-hydroxybenzoate	$C_8H_8O_3$	850.0	819.1	

Name	Mol. form.	E_{pa}/ kJ mol⁻¹	$\Delta_{base}G°$/ kJ mol⁻¹	Note
Methyl 4-hydroxybenzoate	$C_8H_8O_3$	863.4	832.5	
O-Methylhydroxylamine	CH_5NO	844.8	812.3	
1-Methylimidazol	$C_4H_6N_2$	959.6	927.7	
2-Methyl-1*H*-imidazole	$C_4H_6N_2$	963.4	929.6	
4-Methyl-1*H*-imidazole	$C_4H_6N_2$	952.8	920.9	
1-Methyl-1*H*-indazole	$C_8H_8N_2$	922.4	890.5	
2-Methyl-2*H*-indazole	$C_8H_8N_2$	941.4	909.6	
Methyl isobutanoate	$C_5H_{10}O_2$	836.6	805.7	
Methyl isocyanate	C_2H_3NO	764.4	732.0	
Methyl isothiocyanate	C_2H_3NS	799.2	766.7	
Methyl methacrylate	$C_5H_8O_2$	831.4	800.5	
Methyl 3-methoxybenzoate	$C_9H_{10}O_3$	856.7	825.8	
Methyl 4-methoxybenzoate	$C_9H_{10}O_3$	870.6	839.6	
Methyl 2-methylbenzoate	$C_9H_{10}O_2$	858.3	827.3	
Methyl 3-methylbenzoate	$C_9H_{10}O_2$	857.7	826.8	
Methyl 4-methylbenzoate	$C_9H_{10}O_2$	861.5	830.6	
1-Methyl-2-(1-methylvinyl)benzene	$C_{10}H_{12}$	857.8	828.9	
1-Methyl-3-(1-methylvinyl)benzene	$C_{10}H_{12}$	867.6	838.7	
1-Methyl-4-(1-methylvinyl)benzene	$C_{10}H_{12}$	881.8	852.9	
1-Methylnaphthalene	$C_{11}H_{10}$	834.8	805.3	
2-Methylnaphthalene	$C_{11}H_{10}$	831.9	802.4	
Methyl nitrate	CH_3NO_3	733.6	714.8	
Methyl nitrite	CH_3NO_2	798.9	766.4	
N-Methyl-4-nitroaniline	$C_7H_8N_2O_2$	891.6	865.1	
Methyl 3-nitrobenzoate	$C_8H_7NO_4$	815.7	784.7	
Methyl 4-nitrobenzoate	$C_8H_7NO_4$	813.2	782.3	
N-Methyl-*N*-nitromethanamine	$C_2H_6N_2O_2$	828.3	795.8	
2-Methyl-2-norbornene	C_8H_{12}	845	812.5	
2-Methyl-1,3-pentadiene	C_6H_{10}	864.9	836	
3-Methyl-1,3-pentadiene	C_6H_{10}	852.3	823.4	
2-Methyl-2-pentene	C_6H_{12}	812	783.1	
2-Methylphenoxy	C_7H_7O	874.5	842	
1-(3-Methylphenyl)ethanone	$C_9H_{10}O$	868.2	836.4	
1-Methyl-3-phenyl-1*H*-pyrazole	$C_{10}H_{10}N_2$	932.6	900.8	
1-Methyl-5-phenyl-1*H*-pyrazole	$C_{10}H_{10}N_2$	932.4	900.5	
Methyl phenyl sulfone	$C_7H_8O_2S$	812.7	780.3	
Methylphosphine	CH_5P	851.5	817.6	
1-Methylpiperidine	$C_6H_{13}N$	971.1	940.1	
1-Methyl-2-piperidinone	$C_6H_{11}NO$	924.4	892.6	
2-Methylpropanamide	C_4H_9NO	878.6	846.7	
N-Methylpropanamide	C_4H_9NO	920.4	889.4	
2-Methylpropanenitrile	C_4H_7N	803.6	772.8	
2-Methyl-1-propanethiol	$C_4H_{10}S$	802.6	771.4	
2-Methyl-2-propanethiol	$C_4H_{10}S$	816.4	785.1	
Methyl propanoate	$C_4H_8O_2$	830.2	799.2	
2-Methyl-1-propanol	$C_4H_{10}O$	793.7	762.2	
2-Methyl-2-propanol	$C_4H_{10}O$	802.6	772.2	
2-Methylpropenal	C_4H_6O	808.7	776.8	
2-Methyl-2-propenamide	C_4H_7NO	880.4	849.4	
Methyl propyl ether	$C_4H_{10}O$	814.9	785.7	
6-Methyl-1*H*-purine	$C_6H_6N_4$	939.2	907.3	
1-Methyl-1*H*-pyrazole	$C_4H_6N_2$	912.0	880.1	
3-Methyl-1*H*-pyrazole	$C_4H_6N_2$	906.0	874.2	
4-Methyl-1*H*-pyrazole	$C_4H_6N_2$	906.8	873.4	
2-Methylpyridine	C_6H_7N	949.1	917.3	
3-Methylpyridine	C_6H_7N	943.4	911.6	
4-Methylpyridine	C_6H_7N	947.2	915.3	
Methyl 3-pyridinecarboxylate	$C_7H_7NO_2$	925.6	893.8	
Methyl 4-pyridinecarboxylate	$C_7H_7NO_2$	926.6	894.7	
3-Methylpyridine-1-oxide	C_6H_7NO	935.2	902.8	

Name	Mol. form.	E_{pa}/ kJ mol⁻¹	$\Delta_{base}G°$/ kJ mol⁻¹	Note
1-Methyl-2(1*H*)-pyridinone	C_6H_7NO	925.8	894.8	
N-Methylpyrrolidine	$C_5H_{11}N$	965.6	934.8	
N-Methyl-2-pyrrolidinone	C_5H_9NO	923.5	891.6	
2-Methylstyrene	C_9H_{10}	855.2	826.3	
3-Methylstyrene	C_9H_{10}	849.4	820.5	
4-Methylstyrene	C_9H_{10}	861.7	832.8	
3-Methylsuccinic anhydride	$C_5H_6O_3$	807		Ref. 9
2-Methyltetrahydrofuran	$C_5H_{10}O$	840.8	811.6	
2-Methylthiazole	C_4H_5NS	930.6	898.7	
Methylthiirane	C_3H_6S	833.3	801.5	
(Methylthio)acetonitrile	C_3H_5NS	784.8	754.1	
(Methylthio)benzene	C_7H_8S	872.6	843.7	
Methyl thiocyanate	C_2H_3NS	796.7	766.1	
(Methylthio)ethene	C_3H_6S	858.2	829.3	
2-Methylthiophene	C_5H_6S	859.0	826.5	
Methyl trifluoroacetate	$C_3H_3F_3O_2$	740.5	709.6	
Methyl trifluoromethyl ether	$C_2H_3F_3O$	719.2	690.0	
Methyl 2,4,6-trimethylbenzoate	$C_{11}H_{14}O_2$	866.3	835.3	
(1-Methylvinyl)cyclopropane	C_6H_{10}	871.6	842.7	
Methyl vinyl ether	C_3H_6O	859.2	830.3	
Molybdenum carbonyl	C_6MoO_6	762.6	738.1	
Morpholine	C_4H_9NO	924.3	891.2	
Naphthacene	$C_{18}H_{12}$	905.5	876.5	
Naphthalene	$C_{10}H_8$	802.9	779.4	
1,8-Naphthalenediamine	$C_{10}H_{10}N_2$	944.5	912.1	
1-Naphthylamine	$C_{10}H_9N$	907.0	875.1	
Neon	Ne	198.8	174.4	
Nickel	Ni	737	714.1	
Nickel carbonyl [Ni(CO)$_4$]	C_4NiO_4	742.3	716.0	
Nickelocene	$C_{10}H_{10}Ni$	935.7	907.3	
L-Nicotine	$C_{10}H_{14}N_2$	963.4	932.6	
Nitramide	$H_2N_2O_2$	757.4	725.0	
Nitric acid	HNO_3	751.4	731.5	
Nitric oxide	NO	531.8	505.3	
4-Nitroaniline	$C_6H_6N_2O_2$	866.0	834.2	
4-Nitrobenzaldehyde	$C_7H_5NO_3$	795.1	763.2	
3-Nitrobenzamide	$C_7H_6N_2O_3$	854.2	823.2	
4-Nitrobenzamide	$C_7H_6N_2O_3$	845.3	814.4	
Nitrobenzene	$C_6H_5NO_2$	800.3	769.5	
4-Nitrobenzenemethanol	$C_7H_7NO_3$	810.5	778.0	
3-Nitrobenzonitrile	$C_7H_4N_2O_2$	781.4	750.7	
4-Nitrobenzonitrile	$C_7H_4N_2O_2$	775.7	745.1	
Nitroethane	$C_2H_5NO_2$	765.7	733.2	
Nitrogen	N_2	493.8	464.5	
Nitrogen (atomic)	N	342.2	318.7	
Nitrogen dioxide	NO_2	591.0	560.3	
Nitrogen trifluoride	F_3N	568.4	538.6	
Nitromethane	CH_3NO_2	754.6	721.6	
1-(3-Nitrophenyl)ethanone	$C_8H_7NO_3$	826.0	794.1	
1-(4-Nitrophenyl)ethanone	$C_8H_7NO_3$	824.3	792.5	
4-Nitropyridine	$C_5H_4N_2O_2$	874.3	842.5	
4-Nitropyridine 1-oxide	$C_5H_4N_2O_3$	868.0	837.3	
Nitrosobenzene	C_6H_5NO	854.3	823.6	
4-Nitrotoluene	$C_7H_7NO_2$	815.2	782.7	
Nitrous oxide	N_2O	549.8	523.3	Protonation at N
Nitrous oxide	N_2O	575.2	548.7	Protonation at O
5-Nonanone	$C_9H_{18}O$	853.7	821.9	
2,5-Norbornadiene	C_7H_8	849.3	820.3	
1,2,3,4,5,6,7,8-Octahydroanthracene	$C_{14}H_{18}$	845.4	814.7	
1,2,3,4,5,6,7,8-Octahydrophenanthrene	$C_{14}H_{18}$	846.2	815.5	

Name	Mol. form.	E_{pa}/ kJ mol⁻¹	$\Delta_{base}G°$/ kJ mol⁻¹	Note
Octylamine	$C_8H_{19}N$	928.9	895.0	
4-Octylaniline	$C_{14}H_{23}N$	894.5	862	
Osmium(VIII) oxide	O_4Os	676.9	650.6	
7-Oxabicyclo[2.2.1]heptane	$C_6H_{10}O$	844.2	816.8	
7-Oxabicyclo[4.1.0]heptane	$C_6H_{10}O$	848.1	815.6	
Oxazole	C_3H_3NO	876.4	844.5	
Oxepane	$C_6H_{12}O$	834.2	806.8	
Oxetane	C_3H_6O	801.3	773.9	
Oxirane	C_2H_4O	774.2	745.3	
Oxygen	O_2	421	396.3	
Oxygen (atomic)	O	485.2	459.6	
Ozone	O_3	625.5	595.9	
Palladium	Pd	696	673.4	
Pentaborane(9)	B_5H_9	699.4	666.9	
trans-1,3-Pentadiene	C_5H_8	834.1	804.4	
Pentafluorobenzene	C_6HF_5	690.4	662.7	
Pentamethylbenzene	$C_{11}H_{16}$	850.7	823.5	
Pentanal	$C_5H_{10}O$	796.6	764.8	
1,5-Pentanediamine	$C_5H_{14}N_2$	999.6	946.2	
2,4-Pentanedione	$C_5H_8O_2$	873.5	836.8	
Pentanenitrile	C_5H_9N	802.4	771.7	
2-Pentanone	$C_5H_{10}O$	832.7	800.9	
3-Pentanone	$C_5H_{10}O$	836.8	807	
trans-2-Pentenal	C_5H_8O	839.0	807.2	
Pentylamine	$C_5H_{13}N$	923.5	889.5	
2-Pentyne	C_5H_8	810.2	778.0	
Perfluoroacetone	C_3F_6O	670.4	639.7	
Perfluorocyclobutane	C_4F_8	>544		Ref. 6
Perfluoropyridine	C_5F_5N	764.9	733.0	
Perylene	$C_{20}H_{12}$	888.6	859.6	
Phenanthrene	$C_{14}H_{10}$	825.7	795.0	
Phenazine	$C_{12}H_8N_2$	938.4	908.3	
Phenol	C_6H_6O	817.3	786.3	
Phenoxy	C_6H_5O	873.2		Ref. 4
Phenyl	C_6H_5	884	851.5	
Phenylacetylene	C_8H_6	832.0	801.3	
L-Phenylalanine	$C_9H_{11}NO_2$	922.9	888.9	
1-Phenylpiperidine	$C_{11}H_{15}N$	952.9	926.4	
1-Phenyl-1-propanone	$C_9H_{10}O$	867.4	835.6	
1-Phenyl-2-propanone	$C_9H_{10}O$	842.6	810.8	
4-Phenylpyridine	$C_{11}H_9N$	939.7	907.8	
Phenyl-3-pyridinylmethanone	$C_{12}H_9NO$	934.1	902.3	
1-Phenylpyrrolidine	$C_{10}H_{13}N$	941.6	915.1	
Phosphine	H_3P	785	750.9	
Phosphino	H_2P	709.2	675.7	
Phosphorus	P	626.8	604.8	
Phosphorus(III) fluoride	F_3P	695.3	662.8	
Phosphorus monohydride	HP	670.3	639.6	
Phosphorus monoxide	OP	682	649.5	
Phosphorus nitride	NP	789.4	757.0	
Phosphoryl fluoride	F_3OP	694.0	664.2	
Picene	$C_{22}H_{14}$	851.3	820.6	
Piperazine	$C_4H_{10}N_2$	943.7	914.7	
Piperidine	$C_5H_{11}N$	954.0	921	
Potassium hydroxide	HKO	1101.8	1075.4	
Potassium oxide	K_2O	1342.5	1311.8	
L-Proline	$C_5H_9NO_2$	920.5	886.0	
Propadienediylidene	C_3	767.0	736.3	
Propanal	C_3H_6O	786.0	754.0	
Propanamide	C_3H_7NO	876.2	845.3	

Name	Mol. form.	E_{pa}/ kJ mol^{-1}	$\Delta_{base}G°$/ kJ mol^{-1}	Note
Propane	C_3H_8	625.7	607.8	
1,3-Propanediamine	$C_3H_{10}N_2$	987.0	940.0	
1,3-Propanediol	$C_3H_8O_2$	876.2	825.9	
Propanenitrile	C_3H_5N	794.1	763.0	
1-Propanethiol	C_3H_8S	794.9	763.6	
2-Propanethiol	C_3H_8S	803.6	772.3	
Propanoic acid	$C_3H_6O_2$	797.2	766.2	
1-Propanol	C_3H_8O	786.5	756.1	
2-Propanol	C_3H_8O	793.0	762.6	
Propene	C_3H_6	741.6		Ref. 5
1-Propen-1-one	C_3H_4O	834.1	803.4	
cis-1-Propenylbenzene	C_9H_{10}	836.4	807.5	
trans-1-Propenylbenzene	C_9H_{10}	834.2	805.3	
Propyl acetate	$C_5H_{10}O_2$	836.6	805.6	
Propylamine	C_3H_9N	917.8	883.9	
Propylbenzene	C_9H_{12}	790.1	762.4	
Propyleneimine	C_3H_7N	925.1	892.1	
Propyl formate	$C_4H_8O_2$	804.9	773.9	
2-Propylpyridine	$C_8H_{11}N$	955.7	923.8	
Propyl trifluoroacetate	$C_5H_7F_3O_2$	763.9	732.9	
2-Propyn-1-amine	C_3H_5N	887.4	853.5	
Propyne	C_3H_4	748.2	723.0	
2-Propynyl	C_3H_3	741	708.5	
1*H*-Purine	$C_5H_4N_4$	920.1	888.2	
Pyrazine	$C_4H_4N_2$	877.1	847.0	
1*H*-Pyrazol-3-amine	$C_3H_5N_3$	921.5	889.6	
1*H*-Pyrazol-4-amine	$C_3H_5N_3$	907.6	874.0	
1*H*-Pyrazole	$C_3H_4N_2$	894.1	860.5	
Pyrene	$C_{16}H_{10}$	869.2	840.1	
Pyridazine	$C_4H_4N_2$	907.2	877.1	
2-Pyridinamine	$C_5H_6N_2$	947.2	915.3	
3-Pyridinamine	$C_5H_6N_2$	954.4	922.6	
4-Pyridinamine	$C_5H_6N_2$	979.7	947.8	
Pyridine	C_5H_5N	930	898.1	
2-Pyridinecarbonitrile	$C_6H_4N_2$	872.9	841	
3-Pyridinecarbonitrile	$C_6H_4N_2$	877.0	845.1	
4-Pyridinecarbonitrile	$C_6H_4N_2$	880.6	848.8	
4-Pyridinecarboxaldehyde	C_6H_5NO	904.6	872.8	
3-Pyridinecarboxamide	$C_6H_6N_2O$	918.3	886.4	
Pyridine-1-oxide	C_5H_5NO	923.6	892.9	
3-Pyridinol	C_5H_5NO	929.5	897.7	
1-(3-Pyridinyl)ethanone	C_7H_7NO	916.2	884.3	
1-(4-Pyridinyl)ethanone	C_7H_7NO	914.7	882.9	
Pyrimidine	$C_4H_4N_2$	885.8	855.7	
2,4(1*H*,3*H*)-Pyrimidinedithione	$C_4H_4N_2S_2$	911.4	880.5	
Pyrrole	C_4H_5N	875.4	843.8	
Pyrrolidine	C_4H_9N	948.3	915.3	
3-(2-Pyrrolidinyl)pyridine, (*S*)-	$C_9H_{12}N_2$	964.0	931.0	
1*H*-Pyrrolo[2,3-*b*]pyridine	$C_7H_6N_2$	940.2	908.3	
Pyruvonitrile	C_3H_3NO	746.9	716.2	
Quinoline	C_9H_7N	953.2	921.4	
Quinoxaline	$C_8H_6N_2$	903.8	873.7	
Rhodium	Rh	768	745.4	
Ruthenium	Ru	774	751.4	
Ruthenocene	$C_{10}H_{10}Ru$	899.1	876.8	
Sarcosine	$C_3H_7NO_2$	921.2	888.7	
Scandium	Sc	914	892.0	
Selenoformaldehyde	CH_2Se	764.0	734.9	
L-Serine	$C_3H_7NO_3$	914.6	880.7	
Silane	H_4Si	639.7	613.4	

Name	Mol. form.	E_{pa}/ kJ mol⁻¹	$\Delta_{base}G°$/ kJ mol⁻¹	Note
Silicon	Si	837	814.1	
Silicon monosulfide	SSi	627	596.6	Protonation at Si
Silicon monosulfide	SSi	683	660.2	Protonation at S
Silicon monoxide	OSi	777.8	750.4	Protonation at O
Silicon monoxide	OSi	533	500.5	Protonation at Si
Silylene	H_2Si	839.2	804.1	
Sodium hydride	HNa	1095	1070.6	
Sodium hydroxide	HNaO	1071.8	1044.8	
Sodium oxide	Na_2O	1375.9	1345.2	
Strontium oxide	OSr	1209	1180.7	
Styrene	C_8H_8	839.5	809.2	
Succinic anhydride	$C_4H_4O_3$	797		Ref. 9
Sulfur	S	664.3	640.2	
Sulfur dioxide	O_2S	672.3	643.3	
Sulfur hexafluoride	F_6S	575.3	550.7	
Sulfuric acid	H_2O_4S	717	681	Ref. 3
Sulfur trioxide (α-form)	O_3S	588.3	560.3	
Sulfuryl fluoride	F_2O_2S	605.5	580.5	
Tetraborane(10)	B_4H_{10}	605	572.5	
Tetraethylene glycol	$C_8H_{18}O_5$		>910	
Tetraethylene glycol dimethyl ether	$C_{10}H_{22}O_5$	953.8	897.8	
Tetraethylhydrazine	$C_8H_{20}N_2$	964.3	935.3	
1,2,3,4-Tetrafluorobenzene	$C_6H_2F_4$	700.4	672.7	
1,2,3,5-Tetrafluorobenzene	$C_6H_2F_4$	747.3	719.6	
1,2,4,5-Tetrafluorobenzene	$C_6H_2F_4$	746.5	718.8	
Tetrafluoromethane	CF_4	529.3	503.7	
Tetrafluorosilane	F_4Si	502.9	476.6	
Tetrahydrofuran	C_4H_8O	822.1	794.7	
5,6,7,8-Tetrahydroisoquinoline	$C_9H_{11}N$	966.6	934.7	
1,2,3,4-Tetrahydronaphthalene	$C_{10}H_{12}$	809.7	782.1	
Tetrahydropyran	$C_5H_{10}O$	822.8	795.4	
5,6,7,8-Tetrahydroquinoline	$C_9H_{11}N$	966.0	934.1	
Tetrahydrothiophene	C_4H_8S	849.1	819.3	
N,N,2,6-Tetramethylaniline	$C_{10}H_{15}N$	954.1	923.2	
N,N,3,5-Tetramethylaniline	$C_{10}H_{15}N$	956.1	924.3	
1,2,3,5-Tetramethylbenzene	$C_{10}H_{14}$	845.6	816.5	
N,N,N',N'-Tetramethyl-1,2-benzenediamine	$C_{10}H_{16}N_2$	982.6	950.2	
1,4,7,7-Tetramethylbicyclo[2.2.1]heptan-2-one	$C_{11}H_{18}O$	863.3	831.4	
N,N,N',N'-Tetramethyl-1,4-butanediamine	$C_8H_{20}N_2$	1046.3	992.7	
1,1,3,3-Tetramethyldisiloxane	$C_4H_{14}OSi_2$	845.3	814.6	
N,N,N',N'-Tetramethyl-1,2-ethanediamine	$C_6H_{16}N_2$	1012.8	970.6	
1,1,3,3-Tetramethylguanidine	$C_5H_{13}N_3$	1031.6	997.4	
N,N,N',N'-Tetramethyl-1,6-hexanediamine	$C_{10}H_{24}N_2$	1035.8	982.2	
N,N,N',N'-Tetramethylmethanediamine	$C_5H_{14}N_2$	952.2	919.8	
2,2,6,6-Tetramethylpiperidine	$C_9H_{19}N$	987.0	953.9	
N,N,N',N'-Tetramethyl-1,3-propanediamine	$C_7H_{18}N_2$	1035.2	985.4	
Tetramethylstannane	$C_4H_{12}Sn$	823.7	797.4	
Tetramethylthiourea	$C_5H_{12}N_2S$	947.6	916.6	
Tetramethylurea	$C_5H_{12}N_2O$	930.6	899.6	
2,6,10,14-Tetraoxapentadecane	$C_{11}H_{24}O_4$		895.1	
Thiacyclohexane	$C_5H_{10}S$	855.8	826.0	
2-Thiazolamine	$C_3H_4N_2S$	930.6	898.7	
Thiazole	C_3H_3NS	904	872.1	
Thietane	C_3H_6S	834.8	805.0	
Thiirane	C_2H_4S	807.4	777.6	
Thioacetamide	C_2H_5NS	884.6	852.8	
Thiocyanate	CNS	751	718.5	
Thioformaldehyde	CH_2S	759.7	730.5	
Thiophene	C_4H_4S	815.0	784.3	
Thiourea	CH_4N_2S	893.7	863.9	

Name	Mol. form.	E_{pa}/ kJ mol^{-1}	$\Delta_{base}G°$/ kJ mol^{-1}	Note
L-Threonine	$C_4H_9NO_3$	922.5	888.5	
Thymidine	$C_{10}H_{14}N_2O_5$	948.3	915.9	
Thymine	$C_5H_6N_2O_2$	880.9	850.0	
Titanium	Ti	876	853.7	
Toluene	C_7H_8	784.0	756.3	
o-Toluic acid	$C_8H_8O_2$	838.8	807.8	
m-Toluic acid	$C_8H_8O_2$	829.8	798.8	
p-Toluic acid	$C_8H_8O_2$	836.7	805.7	
1,3,5-Triazine	$C_3H_3N_3$	848.8	819.6	
1*H*-1,2,3-Triazole	$C_2H_3N_3$	879.3	847.4	
1*H*-1,2,4-Triazole	$C_2H_3N_3$	886.0	855.9	
Tributylamine	$C_{12}H_{27}N$	998.5	967.6	
1,3,5-Tri-*tert*-butylbenzene	$C_{18}H_{30}$	848.8	822.3	
Trichloroacetaldehyde	C_2HCl_3O	722.3	690.5	
Trichloroacetic acid	$C_2HCl_3O_2$	770.0	739.1	
Trichloroacetonitrile	C_2Cl_3N	723.2	692.6	
2,2,2-Trichloroethanol	$C_2H_3Cl_3O$	729.3	698.9	
2,2,2-Trichloro-1-phenylethanone	$C_8H_5Cl_3O$	818.9	787.0	
1,1,1-Trichloro-2-propanone	$C_3H_3Cl_3O$	768.3	736.3	
Tricyclo[3.3.1.1^{3,7}]decan-1-amine	$C_{10}H_{17}N$	948.8	916.3	
Tricyclo[3.3.1.1^{3,7}]decane-1-carbonitrile	$C_{11}H_{15}N$	834.4	803.8	
1-Tricyclo[3.3.1.1^{3,7}]dec-1-ylethanone	$C_{12}H_{18}O$	864.9	833.1	
Triethylamine	$C_6H_{15}N$	981.8	951	
Triethylenediamine	$C_6H_{12}N_2$	963.4	934.6	
Triethylene glycol dimethyl ether	$C_8H_{18}O_4$	946.6	892.4	
Triethyl phosphate	$C_6H_{15}O_4P$	909.3	879.6	
Triethylphosphine	$C_6H_{15}P$	984.5	952.0	
Triethylphosphine oxide	$C_6H_{15}OP$	936.6	906.8	
Triethylsilanol	$C_6H_{16}OSi$	822.1	794.8	
Trifluoroacetic acid	$C_2HF_3O_2$	711.7	680.7	
1,1,1-Trifluoroacetone	$C_3H_3F_3O$	723.9	692.0	
Trifluoroacetonitrile	C_2F_3N	688.4	657.7	
Trifluoroacetyl chloride	C_2ClF_3O	681.6	649.8	
1,2,3-Trifluorobenzene	$C_6H_3F_3$	724.3	696.6	
1,2,4-Trifluorobenzene	$C_6H_3F_3$	729.5	699.4	
1,3,5-Trifluorobenzene	$C_6H_3F_3$	741.9	715.4	
2,2,2-Trifluoro-*N,N*-dimethylacetamide	$C_4H_6F_3NO$	849.0	818.0	
2,2,2-Trifluoroethanol	$C_2H_3F_3O$	700.2	669.9	
Trifluoroethene	C_2HF_3	699.4	666.9	
2,2,2-Trifluoroethylamine	$C_2H_4F_3N$	846.8	812.9	
2,2,2-Trifluoroethyl methyl ether	$C_3H_5F_3O$	747.6	718.4	
Trifluoroiodomethane	CF_3I	628.0	598.2	
Trifluoromethane	CHF_3	619.5	589.7	
Trifluoromethanesulfonic acid	CHF_3O_3S	699.4	666.9	
3-(Trifluoromethyl)aniline	$C_7H_6F_3N$	856.9	825.1	
4-(Trifluoromethyl)benzaldehyde	$C_8H_5F_3O$	805.6	773.8	
3-(Trifluoromethyl)benzonitrile	$C_8H_4F_3N$	791.4	760.8	
4-(Trifluoromethyl)benzonitrile	$C_8H_4F_3N$	787.2	758.3	
Trifluoronitrosomethane	CF_3NO	703.3	670.8	
2,2,2-Trifluoro-1-phenylethanone	$C_8H_5F_3O$	799.2	767.4	
1,3,5-Trimethoxybenzene	$C_9H_{12}O_3$	926.7	898.2	
Trimethylamine	C_3H_9N	948.9	918.1	
Trimethylamine oxide	C_3H_9NO	983.2	953.5	
Trimethylarsine	C_3H_9As	897.3	864.9	
N,N,3-Trimethylbenzamide	$C_{10}H_{13}NO$	927.0	896.0	
N,N,4-Trimethylbenzamide	$C_{10}H_{13}NO$	927.0	896.0	
1,3,5-Trimethylbenzene	C_9H_{12}	836.2	808.6	
Trimethyl borate	$C_3H_9BO_3$	815.8	783.4	
1,3,5-Trimethyl-2-nitrobenzene	$C_9H_{11}NO_2$	823.8	793.1	
2,2,4-Trimethyl-3-pentanone	$C_8H_{16}O$	856.9	825.0	

Name	Mol. form.	E_{pa}/ kJ mol^{-1}	$\Delta_{base}G°$/ kJ mol^{-1}	Note
Trimethyl phosphate	$C_3H_9O_4P$	890.6	860.8	
Trimethylphosphine	C_3H_9P	958.8	926.3	
Trimethyl phosphite	$C_3H_9O_3P$	929.7	899.9	
N,*N*,2-Trimethylpropenylamine	$C_6H_{13}N$	967.0	934.5	
1,3,5-Trimethyl-1*H*-pyrazole	$C_6H_{10}N_2$	949.3	917.4	
3,4,5-Trimethyl-1*H*-pyrazole	$C_6H_{10}N_2$	949.3	916.0	
Trimethylsilanol	$C_3H_{10}OSi$	814.0	781.5	
Triphenylamine	$C_{18}H_{15}N$	908.9	876.4	
Triphenylarsine	$C_{18}H_{15}As$	908.9	876.4	
Triphenylarsine oxide	$C_{18}H_{15}AsO$	906.2	876.4	
Triphenylene	$C_{18}H_{12}$	819.2	791.2	
Triphenylphosphine	$C_{18}H_{15}P$	972.8	940.4	
Triphenylphosphine oxide	$C_{18}H_{15}OP$	906.2	876.4	
Triphenylphosphine sulfide	$C_{18}H_{15}PS$	906.2	876.4	
Triphenylstibine	$C_{18}H_{15}Sb$	845.5	813.1	
Tripropylamine	$C_9H_{21}N$	991.0	960.1	
L-Tryptophan	$C_{11}H_{12}N_2O_2$	948.9	915	
Tungsten carbonyl	C_6O_6W	758.0	733.4	
L-Tyrosine	$C_9H_{11}NO_3$	926	892.1	
Uracil	$C_4H_4N_2O_2$	872.7	841.7	
Uranium	U	995.2	973.2	
Urea	CH_4N_2O	873.5	841.6	Protonation at O; Ref. 8
Uridine	$C_9H_{12}N_2O_6$	947.6	916.6	
L-Valine	$C_5H_{11}NO_2$	910.6	876.7	
Vanadium	V	859.4	836.8	
Vinyl acetate	$C_4H_6O_2$	813.9	782.9	
Vinylcyclopropane	C_5H_8	816.3	787.5	
4-Vinylpyridine	C_7H_7N	944.1	912.3	
Vinyltrimethylsilane	$C_5H_{12}Si$	833	804.1	
Water	H_2O	691	660.0	
Xenon	Xe	499.6	478.1	
o-Xylene	C_8H_{10}	796.0	768.3	
m-Xylene	C_8H_{10}	812.1	786.2	
p-Xylene	C_8H_{10}	794.4	766.8	
Yttrium	Y	967	945.9	
Zinc	Zn	608.6	586.0	

POLARIZABILITIES OF ATOMS AND MOLECULES

Thomas M. Miller

The *polarizability* of an atom or molecule describes the response of its electron cloud to an external field. The atomic or molecular energy shift ΔW due to an external electric field E is proportional to E^2 for external fields that are weak compared to the internal electric fields between the nucleus and electron cloud. The *electric dipole polarizability* α is the constant of proportionality defined by $\Delta W = -\alpha E^2/2$. The induced electric dipole moment is αE. *Hyperpolarizabilities*, coefficients of higher powers of E, are less often required. Technically, the polarizability is a tensor quantity but for spherically symmetric charge distributions reduces to a single number. In any case, an *average polarizability* is usually adequate in calculations. Frequency-dependent or *dynamic polarizabilities* are needed for electric fields that vary in time, except for frequencies that are much lower than electron orbital frequencies, for which *static polarizabilities* suffice.

Polarizabilities for atoms and molecules in excited states are found to be larger than for ground states and may be positive or negative. Molecular polarizabilities are very slightly temperature dependent because the size of the molecule depends on its rovibrational state. Only in the case of dihydrogen (H_2) has this effect been studied enough to warrant consideration.

Polarizabilities are normally expressed in cgs units of cm^3. Ground-state polarizabilities are in the range of 10^{-24} cm^3 = 1 Å^3 and hence are often given in Å^3 units. Theorists tend to use atomic units of a_0^3 where a_0 is the Bohr radius. The conversion is $\alpha(\text{cm}^3) = 0.148184 \times 10^{-24} \times \alpha(a_0^3)$. Polarizabilities are only recently encountered in SI units, C m^2/ V = J/(V/m)^2. The conversion from cgs units to SI units is $\alpha(\text{C m}^2/\text{ V}) = 4\pi\varepsilon_0 \times 10^{-6}\alpha(\text{cm}^3)$, where ε_0 is the permittivity of free space in SI units and the factor 10^{-6} simply converts cm^3 into m^3. Thus, $\alpha(\text{C m}^2/\text{ V}) = 1.11265 \times 10^{-16} \times \alpha(\text{cm}^3)$. Measurements of excited state polarizabilities by optical methods tend to be reported using units of MHz/(V/cm)^2, where the energy shift, ΔW, is expressed in frequency units with a factor of h understood. The polarizability is $-2\,\Delta W/E^2$. The conversion into cgs units is $\alpha(\text{cm}^3) = 5.955214 \times 10^{-16} \times \alpha[\text{MHz/(V/cm)}^2]$.

The polarizability appears in many formulas for low-energy processes involving the valence electrons of atoms or molecules. These formulas are given below in cgs units: the polarizability α is in cm^3; masses m or μ are in grams; energies are in ergs; and electric charges are in esu, where $e = 4.8032 \times 10^{-10}$ esu. The symbol $\alpha(\nu)$ denotes a frequency (ν) dependent polarizability, where $\alpha(\nu)$ reduces to the static polarizability α for $\nu = 0$. For further information, see Bonin, K. D., and Kresin, V. V., *Electric Dipole Polarizabilities of Atoms, Molecules, and Clusters*, World Scientific, Singapore, 1997; Bonin, K. D., and Kadar-Kallen, M. A., *Int. J. Mod. Phys.B*, 24, 3313, 1994; Miller, T. M., and Bederson, B., *Advances in Atomic and Molecular Physics*, 13, 1, 1977; and Gould, H., and Miller, T. M., *Advances in Atomic, Molecular, and Optical Physics*, 51, 243, 2005. A helpful listing of theoretical results for atomic polarizabilities is given by P. Schwerdtfeger at http://ctcp.massey.ac.nc/dipole-polarizabilities (accessed September 2016). Details on polarizability-related interactions, especially in regard to hyperpolarizabilities and nonlinear optical phenomena, are given by Bogaard, M. P., and Orr, B. J., in *Physical Chemistry, Series Two, Vol. 2, Molecular Structure and Properties*, Buckingham, A. D., Ed., Butterworths, London, 1975, pp. 149–194. A tabulation of tensor and hyperpolarizabilities is included.

An empirical additive formula for molecular polarizabilities at 589 nm frequency has been given in Bosque, R., and Sales, J., *J. Chem. Inf. Comput. Sci.* 42, 1154, 2002:

$$\alpha = 0.32 + 1.51\#C + 0.17\#H + 0.57\#O + 1.05\#N + 2.99\#S + 2.48\#P + 0.22\#F + 2.16\#Cl + 3.29\#Br + 5.45\#I$$

where #C denotes the number of carbon atoms in the molecule, etc. A helium-elimination additive method has been given by Kassimi, N. E.-B., and Thakkar, A. J., *Chem. Phys. Lett.* 472, 232, 2009.

All polarizabilities in this table are experimental values except those indicated as calculated in the Note column. The experimental polarizabilities are mostly determined by measurements of a dielectric constant or refractive index and are quite precise (0.5% or better). However, one should treat many of the results with several percent of caution because of the age of the data and because some of the results refer to optical frequencies rather than static. Comments given with the references are intended to allow one to judge the degree of caution required. Interested persons should consult these references. In many cases, the reference given is to a theoretical paper in which the experimental results are quoted. These papers, noted in the references, contain valuable information on polarizability calculations and experimental data, which often includes the tensor components of the polarizability.

The table first lists atoms in order of atomic number, followed by molecules in alphabetical order by name.

Formulas Involving Polarizability*

Description	Formula	Remarks
Lorentz-Lorenz relation	$\alpha(\nu) = \frac{3}{4\pi n}\left[\frac{\eta^2(\nu)-1}{\eta^2(\nu)+2}\right]$	For a gas of atoms or nonpolar molecules; the index of refraction is $\eta(\nu)$
Refraction by polar molecules	$\alpha(\nu) + \frac{d^2}{3kT} = \frac{3}{4\pi n}\left[\frac{n^2(\nu)-1}{n^2(\nu)+2}\right]$	The dipole moment is d, in esu·cm (= 10^{-18} D)
Dielectric constant (dimensionless)	$\kappa(\nu) = 1 + 4\pi n\,\alpha(\nu)$	From the Lorentz-Lorenz relation for the usual case of $\kappa(\nu) \approx 1$
Index of refraction (dimensionless)	$\mu(\nu) = 1 + 2\pi n\alpha(\nu)$	From $\eta^3(\nu) = \kappa(\nu)$

Description	Formula	Remarks
Diamagnetic susceptibility	$\chi_m = e^2(a_0 N\alpha)^{1/2} / 4m_e c^2$	From the approximation that the static polarizability is given by the variational formula $\alpha = (4/9a_0)\Sigma(N_i r_i^2)^2$; N is the number of electrons, m_e is the electron mass; a crude approximation is $\chi_m = (E_i/4m_e c^2)\alpha$, where E_i is the ionization energy
Long-range electron- or ion-molecule interaction energy	$V(r) = -e^2\alpha / 2r^4$	The target molecule polarizability is α
Ion mobility in a gas	$\kappa = 13.87 / (\alpha\mu)^{1/2} \mathrm{cm}^2 / \mathrm{V \cdot s}$	This one formula is not in cgs units. Enter α in Å^3 or 10^{-24} cm^3 units and the reduced mass μ of the ion-molecule pair in amu. Classical limit; pure polarization potential
Langevin capture cross section	$\sigma(v_0) = (2\pi e / v_0)(\alpha / \mu)^{1/2}$	The relative velocity of approach for an ion-molecule pair is v_0; the target molecular polarizability is α and the reduced mass of the ion-molecule pair is μ
Langevin reaction rate coefficient	$k = 2\pi e(\alpha / \mu)^{1/2}$	Collisional rate coefficient for an ion-molecule reaction
Rate coefficient for polar molecules	$kd = 2\pi e[(\alpha / \mu)^{1/2} + cd(2 / \mu\pi kT)^{1/2}]$	The dipole moment of the neutral is d in esu cm; the number c is a "locking factor" that depends on α and d, and is between 0 and 1
Modified effective range cross section for electron-neutral scattering	$\sigma(k) = 4\pi A^2 + 32\pi^4 \mu e^2 \alpha A k / 3h^2 + \ldots$	Here, k is the electron momentum divided by $h/2\pi$, where h is Planck's constant; A is called the "scattering length"; the reduced mass is μ
van der Waals constant between two systems A, B	$C_6 = \frac{3}{2}\left[\frac{\alpha^A\alpha^B E^A E^B}{E^A + E^B}\right]$	For the interaction potential term $V_6(r) = -C_6 r^6$; $E^{A,B}$ represents average dipole transition energies and $\alpha^{A,B}$ the respective polarizabilities of A, B
Dipole-quadrupole constant between two systems A, B	$C_8 = \frac{15}{4}\left[\frac{\alpha^A \alpha_q^B E^A E_q^B}{E^A + E_q^B}\right] + \frac{15}{4}\left[\frac{\alpha_q^A \alpha^B E_q^A E^B}{E_q^A + E^B}\right]$	For the interaction potential term $V_8(r) = -C_8 r^8$; $E_q^{A,B}$ represents average quadrupole transition energies and $\alpha_q^{A,B}$ are the respective quadrupole polarizabilities of A, B
van der Waals constant between an atom and a surface	$C_3 = \frac{\alpha g E^A E^S}{8(E^A + E^S)}$	For an interaction potential $V_3(r) = -C_3 r^3$; $E^{A,S}$ are characteristic energies of the atom and surface; $g = 1$ for a free-electron metal and $g = (\varepsilon_\infty - 1)/(\varepsilon_\infty + 1)$ for an ionic crystal
Relationship between $\alpha(\nu)$ and oscillator strengths	$\alpha(\nu) = \frac{e^2h^2}{4\pi^2 m_e}\Sigma\frac{f_k}{E_k^2 - (h\nu)^2}$	Here, f_k is the oscillator strength from the ground state to an excited state k, with excitation energy E_k. This formula is often used to estimate static polarizabilities ($\nu = 0$)
Dynamic polarizability	$\alpha(\nu) = \frac{\alpha E_r^2}{E_r^2 - (h\nu)^2}$	Approximate variation of the frequency-dependent polarizability $\alpha(\nu)$ from $\nu = 0$ up to the first dipole-allowed electronic transition, of energy E_r; the static dipole polarizability is $\alpha(0)$; infrared contributions ignored
Rayleigh scattering cross section	$\alpha(\nu) = \frac{8\pi}{9c^4}(2\pi\nu)^4 \times 3\alpha^2(\nu) + 2\gamma^2(\nu)/3$	The photon frequency is ν; the polarizability anisotropy (the difference between polarizabilities parallel and perpendicular to the molecular axis) is $\gamma(\nu)$
Verdet constant	$V(\nu) = \frac{\nu n}{2m_e c^2}\left[\frac{d\alpha(\nu)}{d\nu}\right]$	Defined from $\theta = V(\nu)B$, where θ is the angle of rotation of linearly polarized light through a medium of number density n, per unit length, for a longitudinal magnetic field strength B (Faraday effect)

* The gas number density, n, in the list of formulas is usually taken to be that of 1 atm at 0 °C in reporting experimental data.

Polarizability of Atoms and Molecules

Name	Mol. form.	$\alpha/10^{-24}$ cm^3	Ref.	Note	Z
Atoms					
Hydrogen (atomic)	H	0.666793	1	calculated ("exact")	1
Helium	He	0.2050522	2	calculated ("exact")	2
Helium	He	0.2050519	3	dielectric constant (±0.009%)	2
Lithium	Li	24.33	4	beam (±0.66%)	3
Beryllium	Be	5.60	42	calculated (±1.2%)	4
Boron	B	3.051	31	calculated (±0.07%)	5
Carbon	C	1.67	5	calculated (±2%)	6
Nitrogen (atomic)	N	1.13	2	index of refraction (±5.3%)	7
Nitrogen (atomic)	N	1.1	40	calculated (<2.5%)	
Oxygen (atomic)	O	0.77	2	index of refraction (±7.8%)	8
Fluorine (atomic)	F	0.557	1	calculated (±2%)	9
Neon	Ne	0.39432	6	dielectric constant (±0.003)	10

Name	Mol. form.	$\alpha/10^{-24}$ cm^3	Ref.	Note	Z
Sodium	Na	24.11	7	interferom (±0.50%)	11
Sodium	Na	24.11	8	interferom (±0.83%)	11
Magnesium	Mg	8.8	36	beam (±26%)	12
Magnesium	Mg	10.5	25	calculated (<2.9%)	12
Aluminum	Al	6.8	11	beam (±4.4%)	13
Aluminum	Al	8.22	31	calculated (±0.1%)	
Silicon	Si	5.53	5	calculated (±2%)	14
Phosphorus	P	3.69	40	calculated (<2.5%)	15
Sulfur	S	2.87	33	calculated (±0.5%)	16
Chlorine (atomic)	Cl	2.16	33	calculated (±0.5%)	17
Argon	Ar	1.6411	12,13	index/dielectric constant (±0.05%)	18
Potassium	K	42.93	32	interferom (±.16%)	19
Calcium	Ca	23.8	25	calculated (±2.9%)	20
Calcium	Ca	25.1	34	beam (±9%)	20
Scandium	Sc	14.4	36	beam (±9.7%)	21
Scandium	Sc	21.2	39	calculated (<12%)	21
Titanium	Ti	9.4	36	beam (±5.3%)	22
Titanium	Ti	16.9	39	calculated (<12%)	22
Vanadium	V	14.4	39	calculated (<12%)	23
Vanadium	V	10.1	36	beam (±7.96%)	23
Chromium	Cr	11.6	41	calculated (±1%)	24
Chromium	Cr	8.9	36	beam (±39%)	24
Manganese	Mn	9.9	41	calculated (±1%)	25
Iron	Fe	9.47	39	calculated (<12%)	26
Cobalt	Co	8.55	39	calculated (<12%)	27
Nickel	Ni	7.57	39	calculated (<12%)	28
Copper	Cu	6.03	41	calculated (±1%)	29
Copper	Cu	8.7	36	beam (±8%)	29
Zinc	Zn	5.75	15	index of refraction (±2%)	30
Zinc	Zn	5.73	44	calculated	30
Gallium	Ga	7.57	24	calculated (±2.9%)	31
Gallium	Ga	6.9	36	beam (±8.7%)	31
Germanium	Ge	5.84	5	calculated (±2%)	32
Arsenic	As	4.42	40	calculated (<2.5%)	33
Selenium	Se	3.89	35	calculated	34
Bromine (atomic)	Br	3.11	35	calculated	35
Krypton	Kr	2.4844	13	dielectric constant (±0.05%)	36
Rubidium	Rb	47.39	32	interferom (±.17%)	37
Strontium	Sr	27.6	1	beam (±7%)	38
Strontium	Sr	28.3	25	calculated (<2.9%)	38
Yttrium	Y	22.7	14	calculated (±25%)	39
Yttrium	Y	24.1	36	beam (±7.1%)	39
Zirconium	Zr	17.9	14	calculated (±25%)	40
Zirconium	Zr	16.6	36	beam (±11.4%)	40
Niobium	Nb	15.7	14	calculated (±25%)	41
Niobium	Nb	14.5	36	beam (±7.6%)	41
Molybdenum	Mo	10.7	41	calculated (±1%)	42
Molybdenum	Mo	12.9	36	beam (±7%)	42
Technetium	Tc	11.9	41	calculated (±1%)	43
Ruthenium	Ru	9.6	14	calculated (±25%)	44
Rhodium	Rh	8.6	14	calculated (±25%)	45
Rhodium	Rh	1.6	36	beam (±0.5%)	45
Palladium	Pd	3.874	46	calculated (±0.4%)	46
Silver	Ag	6.8	16	index (±11%)	47
Silver	Ag	6.8	36	beam (±16%)	47
Silver	Ag	5.44	41	calculated (±1%)	47
Cadmium	Cd	7.36	17	index of refraction (±3%)	48
Cadmium	Cd	6.82	47	calculated (±1%)	48
Indium	In	10.2	18	beam (±12%)	49
Indium	In	9.2	36	beam (±13%)	49
Indium	In	9.19	24	calculated (±3.1%)	49

Name	Mol. form.	$\alpha/10^{-24}$ cm³	Ref.	Note	Z
Tin	Sn	10.0	36	beam (±13%)	50
Tin	Sn	6.3	5	beam (±26%)	50
Tin	Sn	7.84	5	calculated	50
Antimony	Sb	6.25	40	calculated (<2.5%)	51
Tellurium	Te	5.5	14	calculated (±25%)	52
Iodine (atomic)	I	5.35	19	index of refraction (±25%)	53
Iodine (atomic)	I	5.11	21	calculated	53
Xenon	Xe	4.044	1	dielectric constant (±0.5%)	54
Cesium	Cs	59.42	20	beam (±0.13%)	55
Cesium	Cs	59.39	32	beam (±0.09%)	55
Barium	Ba	39.7	1	beam (±7%)	56
Barium	Ba	40.7	25	calculated (<2.9%)	56
Lanthanum	La	25.3	36	beam (±4.7	57
Lanthanum	La	31.7	30	calculated (<12%)	57
Cerium	Ce	30.3	30	calculated (<12%)	58
Cerium	Ce	28.4	36	beam (±10.6%)	58
Praseodymium	Pr	35.4	36	beam (±11.6%)	59
Praseodymium	Pr	32	30	calculated (<12%)	59
Neodymium	Nd	27.2	36	beam (±10.7%)	60
Neodymium	Nd	30.9	30	calculated (<12%)	60
Promethium	Pm	29.7	14	calculated	61
Samarium	Sm	23.2	36	beam (±10.3%)	62
Samarium	Sm	28.5	30	calculated (<12%)	62
Europium	Eu	21.6	36	beam (±17.1%)	63
Europium	Eu	27.3	30	calculated (<12%)	63
Gadolinium	Gd	26.1	36	beam (±14.9%)	64
Gadolinium	Gd	23.5	30	calculated (<12%)	64
Terbium	Tb	23.5	36	beam (±6.9%)	65
Terbium	Tb	25.1	30	calculated (<12%)	65
Dysprosium	Dy	23.3	51	radiative (±7%)	66
Dysprosium	Dy	24.1	30	calculated (<12%)	66
Holmium	Ho	21.5	36	beam (±7.9%)	67
Holmium	Ho	23.2	30	calculated (<12%)	67
Erbium	Er	23.0	50	radiative (±15%)	68
Erbium	Er	22.1	50	calculated (<15%)	68
Thulium	Tm	19.2	36	beam (±12.5%)	69
Thulium	Tm	21.4	30	calculated (<12%)	69
Ytterbium	Yb	20.6	48	radiative (±5%)	70
Ytterbium	Yb	20.9	49	calculated (<2%)	70
Lutetium	Lu	18.3	36	beam (±14.8%)	71
Lutetium	Lu	20.3	30	calculated (<12%)	71
Hafnium	Hf	12.4	36	beam (±22.6%)	72
Hafnium	Hf	16.2	14	calculated (±25%)	72
Tantalum	Ta	8.6	36	beam (±20.9%)	73
Tantalum	Ta	13.1	14	calculated (±25%)	73
Tungsten	W	11.1	14	calculated (±25%)	74
Rhenium	Re	9.05	41	calculated (±1%)	75
Osmium	Os	8.5	14	calculated (±25%)	76
Iridium	Ir	7.6	14	calculated (±25%)	77
Platinum	Pt	6.5	14	calculated (±25%)	78
Gold	Au	4.13	41	calculated (±1%)	79
Mercury	Hg	5.02	22	index of refraction (±1%)	80
Mercury	Hg	5.08	45	calculated (±0.5%)	80
Thallium	Tl	7.6	15	atomic beam deflection (±15%)	81
Thallium	Tl	7.3	52	calculated (±4%)	81
Lead	Pb	8.3	36	beam (±32.5%)	82
Lead	Pb	6.98	5	beam (±15%)	82
Lead	Pb	7.07	5	calculated	82
Bismuth	Bi	8.1	36	beam (±21%)	83
Bismuth	Bi	7.2	40	calculated (<2.5%)	83
Polonium	Po	6.8	14	calculated (±25%)	84

Name	Mol. form.	$\alpha/10^{-24}$ cm^3	Ref.	Note	Z
Astatine	At	6.39	46	calculated	85
Radon	Rn	4.83	40	calculated (<2.5%)	86
Francium	Fr	46.2	9	calculated (±2%)	87
Radium	Ra	36.48	26	calculated (±2%)	88
Actinium	Ac	32.1	14	calculated (±25%)	89
Thorium	Th	32.1	14	calculated (±25%)	90
Protactinium	Pa	25.4	14	calculated (±25%)	91
Uranium	U	24.9	27	atomic beam deflection (±6%)	92
Neptunium	Np	24.8	14	calculated (±25%)	93
Plutonium	Pu	24.5	14	calculated (±25%)	94
Americium	Am	23.3	14	calculated (±25%)	95
Curium	Cm	23.0	14	calculated (±25%)	96
Berkelium	Bk	22.7	14	calculated (±25%)	97
Californium	Cf	20.5	14	calculated (±25%)	98
Einsteinium	Es	19.7	14	calculated (±25%)	99
Fermium	Fm	23.8	14	calculated (±25%)	100
Mendelevium	Md	18.2	14	calculated (±25%)	101
Nobelium	No	15.6	30	calculated (<12%)	102
Dubnium	Db	6.22	29	calculated (±10%)	105
Seaborgium	Sg	5.93	29	calculated (±10%)	106
Bohrium	Bh	5.63	29	calculated (±11%)	107
Hassium	Hs	5.33	29	calculated (±11%)	108
Meitnerium	Mt	5.04	29	calculated (±9%)	109
Darmstadtium	Ds	4.74	29	calculated (±10%)	110
Roentgenium	Rg	4.45	29	calculated (±10%)	111
Copernicium	Cn	4.06	28	calculated (±2%)	112
Nihonium	Nh	4.43	38	calculated	113
Flerovium	Fl	4.72	5	calculated (±2%)	114
Flerovium	Fl	4.37	28	calculated (±2%)	114
Oganesson	Og		37	calculated	118
Ununennium	Uue	25.15	9	calculated (±2%)	119
Unbinilium	Ubn	24.09	9	calculated (±2%)	120
Molecules					
Acenaphthene	$C_{12}H_{10}$	20.61	27		
Acetaldehyde	C_2H_4O	4.6	2	molar refraction; see Ref.2	
Acetaldehyde	C_2H_4O	4.59	18		
Acetamide	C_2H_5NO	5.67	18		
Acetic acid	$C_2H_4O_2$	5.1	2	molar refraction; see Ref.2	
Acetic anhydride	$C_4H_6O_3$	8.9	2	molar refraction; see Ref.2	
Acetone	C_3H_6O	6.33	15		
Acetone	C_3H_6O	6.4	2	molar refraction; see Ref.2	
Acetone	C_3H_6O	6.39	18		
Acetone (1-methylethylidene)hydrazone	$C_6H_{12}N_2$	15.6	2		
Acetonitrile	C_2H_3N	4.40	2	molar refraction; see Ref.2	
Acetonitrile	C_2H_3N	4.48	18		
Acetophenone	C_8H_8O	15.0	2		
Acetyl chloride	C_2H_3ClO	6.62	2		
Acetylene	C_2H_2	3.33	3		
Acetylene	C_2H_2	3.93	2		
Acrolein	C_3H_4O	6.38	2	molar refraction; see Ref.2	
Acrylonitrile	C_3H_3N	8.05	2		
Allyl alcohol	C_3H_6O	7.65	2		
Aluminum 2,4-pentanedioate	$C_{15}H_{21}AlO_6$	51.9	2		
Ammonia	H_3N	2.81	20		
Ammonia	H_3N	2.10	2		
Ammonia	H_3N	2.26	3		
Ammonia	H_3N	2.22	33	calculated	
Ammonia-d_3	D_3N	1.70	2		
Aniline	C_6H_7N	12.1	2	molar refraction; see Ref.2	
Anisole	C_7H_8O	13.1	2	molar refraction; see Ref.2	

Name	Mol. form.	$\alpha/10^{-24}$ cm^3	Ref.	Note	Z
Anthracene	$C_{14}H_{10}$	25.4	17		
Anthracene	$C_{14}H_{10}$	25.93	27		
9,10-Anthracenedione	$C_{14}H_8O_2$	24.46	27		
Arsenic(III) chloride	$AsCl_3$	14.9	2		
Arsenic trinitride	AsN_3	5.75	2		
Benz[*a*]anthracene	$C_{18}H_{12}$	32.86	27		
Benzene	C_6H_6	10.0	25		
Benzene	C_6H_6	10.32	3		
Benzene	C_6H_6	10.74	2		
1,4-Benzenediamine	$C_6H_8N_2$	13.8	2	molar refraction; see Ref.2	
11*H*-Benzo[*b*]fluorene	$C_{17}H_{12}$	30.21	27		
Benzonitrile	C_7H_5N	12.5	2	molar refraction; see Ref.2	
p-Benzoquinone	$C_6H_4O_2$	14.5	2		
Beryllium hydride	BeH_2	4.34	14	calculated	
Beryllium 2,4-pentanedioate	$C_{10}H_{14}BeO_4$	34.1	2		
Bis(4-bromophenyl) ether	$C_{12}H_8Br_2O$	27.8	2	molar refraction; see Ref.2	
Bis(4-methylphenyl) ether	$C_{14}H_{14}O$	24.9	2	molar refraction; see Ref.2	
Borane(1)	BH	3.32	1	calculated	
Borazine	$B_3H_6N_3$	8.0	2	molar refraction; see Ref.2	
Boron trichloride	BCl_3	9.38	20		
Boron trifluoride	BF_3	3.31	2		
Bromine	Br_2	7.02	2		
Bromoacetylene	C_2HBr	7.39	2		
9-Bromoanthracene	$C_{14}H_9Br$	28.32	27		
Bromobenzene	C_6H_5Br	14.7	2		
Bromobenzene	C_6H_5Br	13.62	27		
1-Bromobutane	C_4H_9Br	13.9	2		
1-Bromobutane	C_4H_9Br	10.86	27		
1-Bromo-2-chloroethane	C_2H_4BrCl	9.5	2	molar refraction; see Ref.2	
1-Bromodecane	$C_{10}H_{21}Br$	21.60	27		
Bromodifluoromethane	$CHBrF_2$	5.7	2	molar refraction; see Ref.2	
1-Bromododecane	$C_{12}H_{25}Br$	25.18	27		
Bromoethane	C_2H_5Br	8.05	2		
Bromoethane	C_2H_5Br	7.28	27		
Bromoethene	C_2H_3Br	7.59	2		
1-Bromo-4-fluorobenzene	C_6H_4BrF	13.4	2	molar refraction; see Ref.2	
1-Bromoheptane	$C_7H_{15}Br$	16.8	2	molar refraction; see Ref.2	
1-Bromoheptane	$C_7H_{15}Br$	16.23	27		
1-Bromohexadecane	$C_{16}H_{33}Br$	32.34	27		
1-Bromohexane	$C_6H_{13}Br$	14.44	27		
Bromomethane	CH_3Br	5.87	20		
Bromomethane	CH_3Br	6.03	2		
Bromomethane	CH_3Br	5.55	15		
1-Bromo-4-(4-methylphenoxy)benzene	$C_{13}H_{11}BrO$	26.6	2	molar refraction; see Ref.2	
1-Bromonaphthalene	$C_{10}H_7Br$	20.34	27		
1-Bromononane	$C_9H_{19}Br$	19.81	27		
1-Bromooctadecane	$C_{18}H_{37}Br$	35.92	27		
1-Bromooctane	$C_8H_{17}Br$	18.02	27		
1-Bromopentane	$C_5H_{11}Br$	13.1	2	molar refraction; see Ref.2	
1-Bromo-4-phenoxybenzene	$C_{12}H_9BrO$	24.2	2	molar refraction; see Ref.2	
1-Bromopropane	C_3H_7Br	9.4	2	molar refraction; see Ref.2	
1-Bromopropane	C_3H_7Br	9.07	27		
2-Bromopropane	C_3H_7Br	9.6	2	molar refraction; see Ref.2	
4-Bromotoluene	C_7H_7Br	14.80	27		
1,3-Butadiene	C_4H_6	8.64	2		
Butanal	C_4H_8O	8.2	2	molar refraction; see Ref.2	
Butane	C_4H_{10}	8.20	2		
2,3-Butanedione	$C_4H_6O_2$	8.2	2	molar refraction; see Ref.2	
Butanenitrile	C_4H_7N	8.4	2	molar refraction; see Ref.2	
Butanoic acid	$C_4H_8O_2$	8.58	27		
1-Butanol	$C_4H_{10}O$	8.88	2		

Name	Mol. form.	$\alpha/10^{-24}$ cm³	Ref.	Note	Z
2-Butanone	C_4H_8O	8.13	15		
trans-2-Butenal	C_4H_6O	8.5	2	molar refraction; see Ref.2	
1-Butene	C_4H_8	7.97	2		
1-Butene	C_4H_8	8.52	2		
trans-2-Butene	C_4H_8	8.49	2		
trans-2-Butenedinitrile	$C_4H_2N_2$	11.8	2		
trans-2-Butenenitrile	C_4H_5N	8.2	2	molar refraction; see Ref.2	
Butylamine	$C_4H_{11}N$	13.5	2		
tert-Butylbenzene	$C_{10}H_{14}$	17.2	25		
tert-Butylbenzene	$C_{10}H_{14}$	17.8	2	molar refraction; see Ref.2	
tert-Butylcyclohexane	$C_{10}H_{20}$	19.8	2		
1-Butyne	C_4H_6	7.41	2	molar refraction; see Ref.2	
Carbon dioxide	CO_2	2.911	8		
Carbon disulfide	CS_2	8.74	3		
Carbon disulfide	CS_2	8.86	2		
Carbon [fullerene-C_{60}]	C_{60}	88.9	36		
Carbon [fullerene-C_{60}]	C_{60}	79	31		
Carbon [fullerene-C_{60}]	C_{60}	76.5	24		
Carbon [fullerene-C_{70}]	C_{70}	108.5	36		
Carbon [fullerene-C_{70}]	C_{70}	102	24		
Carbon monoxide	CO	1.95	3		
Carbonothioic dichloride	CCl_2S	10.2	2		
Carbon oxysulfide	COS	5.71	2		
Carbon oxysulfide	COS	5.2	15		
Carbonyl chloride	CCl_2O	7.29	2		
Carbonyl fluoride	CF_2O	1.88	17	calculated	
Cesium bromide dimer	Br_2Cs_2	54.5	16		
Cesium bromide cluster			38	*n*=3-32	
Cesium chloride dimer	Cl_2Cs_2	42.4	16		
Cesium fluoride dimer	Cs_2F_2	28.4	16		
Cesium iodide dimer	Cs_2I_2	51.8	16		
Cesium potassium	CsK	89	22		
Chlorine	Cl_2	4.61	3		
Chlorine trifluoride	ClF_3	6.32	2		
Chloroacetone	C_3H_5ClO	8.4	2	molar refraction; see Ref.2	
Chloroacetonitrile	C_2H_2ClN	6.10	18		
Chloroacetyl chloride	$C_2H_2Cl_2O$	8.92	2		
Chloroacetylene	C_2HCl	6.07	2		
9-Chloroanthracene	$C_{14}H_9Cl$	27.35	27		
Chlorobenzene	C_6H_5Cl	14.1	2		
Chlorobenzene	C_6H_5Cl	12.3	15		
4-Chloro-1,2-butadiene	C_4H_5Cl	10.0	2	molar refraction; see Ref.2	
1-Chlorobutane	C_4H_9Cl	11.3	2		
2-Chlorobutane	C_4H_9Cl	12.4	2		
2-Chlorobutanoic acid	$C_4H_7ClO_2$	10.87	27		
3-Chlorobutanoic acid	$C_4H_7ClO_2$	10.80	27		
4-Chlorobutanoic acid	$C_4H_7ClO_2$	10.69	27		
2-Chloro-1-butanol	C_4H_9ClO	10.70	27		
3-Chloro-1-butanol	C_4H_9ClO	10.38	27		
1-Chloro-1,1-difluoroethane	$C_2H_3ClF_2$	8.05	2		
Chlorodifluoromethane	$CHClF_2$	6.38	20		
Chlorodifluoromethane	$CHClF_2$	5.91	2		
Chloroethane	C_2H_5Cl	7.27	20		
Chloroethane	C_2H_5Cl	8.29	2		
Chloroethane	C_2H_5Cl	6.4	15		
2-Chloroethanol	C_2H_5ClO	7.1	2	molar refraction; see Ref.2	
2-Chloroethanol	C_2H_5ClO	6.88	27		
Chloroethene	C_2H_3Cl	6.41	2		
2-Chloroethyl ethyl ether	C_4H_9ClO	10.56	27		
1-Chloro-2-fluoroethane	C_2H_4ClF	6.5	2	molar refraction; see Ref.2	
Chlorogermane	$ClGeH_3$	6.7	2	molar refraction; see Ref.2	

Name	Mol. form.	$\alpha/10^{-24}$ cm^3	Ref.	Note	Z
Chloromethane	CH_3Cl	5.35	20		
Chloromethane	CH_3Cl	4.72	8		
1-Chloro-2-methoxyethane	C_3H_7ClO	8.71	27		
Chloromethyl methyl ether	C_2H_5ClO	7.1	2	molar refraction; see Ref.2	
1-Chloro-2-methylpropane	C_4H_9Cl	11.1	2		
2-Chloro-2-methylpropane	C_4H_9Cl	12.5	2	molar refraction; see Ref.2	
1-Chloro-2-methylpropene	C_4H_7Cl	10.8	2		
1-Chloronaphthalene	$C_{10}H_7Cl$	19.30	27		
2-Chloronaphthalene	$C_{10}H_7Cl$	19.58	27		
1-Chloro-2-nitrobenzene	$C_6H_4ClNO_2$	14.6	2	molar refraction; see Ref.2	
1-Chloro-3-nitrobenzene	$C_6H_4ClNO_2$	14.6	2	molar refraction; see Ref.2	
1-Chloro-4-nitrobenzene	$C_6H_4ClNO_2$	14.6	2	molar refraction; see Ref.2	
1-Chloro-1-nitroethane	$C_2H_4ClNO_2$	10.9	2		
Chloronitromethane	CH_2ClNO_2	6.9	2	molar refraction; see Ref.2	
1-Chloro-1-nitropropane	$C_3H_6ClNO_2$	10.4	2	molar refraction; see Ref.2	
Chloropentafluoroethane	C_2ClF_5	6.3	2	molar refraction; see Ref.2	
1-Chloropentane	$C_5H_{11}Cl$	12.0	2	molar refraction; see Ref.2	
2-Chloropentanoic acid, (±)-	$C_5H_9ClO_2$	12.69	27		
3-Chloropentanoic acid	$C_5H_9ClO_2$	12.57	27		
4-Chloropentanoic acid	$C_5H_9ClO_2$	12.53	27		
2-Chlorophenol	C_6H_5ClO	13.0	2	molar refraction; see Ref.2	
4-Chlorophenol	C_6H_5ClO	13.0	2	molar refraction; see Ref.2	
1-Chloropropane	C_3H_7Cl	10.0	2		
2-Chloro-1-propanol	C_3H_7ClO	8.89	27		
3-Chloro-1-propanol	C_3H_7ClO	8.84	27		
cis-1-Chloropropene	C_3H_5Cl	8.3	2		
trans-1-Chloropropene	C_3H_5Cl	8.3	2		
2-Chloropropene	C_3H_5Cl	8.3	2		
3-Chloropropene	C_3H_5Cl	8.3	2		
Chlorosilane	ClH_3Si	7.02	2		
4-Chlorotoluene	C_7H_7Cl	13.70	27		
Chlorotrifluoromethane	$CClF_3$	5.72	20		
Chlorotrifluoromethane	$CClF_3$	5.59	8		
Chromium(III) 2,4-pentanedioate	$C_{15}H_{21}CrO_6$	53.7	2		
Chrysene	$C_{18}H_{12}$	33.06	27		
Coronene	$C_{24}H_{12}$	42.50	27		
Cyanogen	C_2N_2	7.99	2		
Cyclohexane	C_6H_{12}	11.0	18		
Cyclohexane	C_6H_{12}	10.87	15		
Cyclohexanol	$C_6H_{12}O$	11.56	18		
Cyclohexene	C_6H_{10}	10.7	2	molar refraction; see Ref.2	
1,3-Cyclopentadiene	C_5H_6	8.64	2		
Cyclopentane	C_5H_{10}	9.15	18		
Cyclopropane	C_3H_6	5.66	2		
Decane	$C_{10}H_{22}$	19.10	27		
Dialuminum	Al_2	19	23		
Diboron hexanitride	B_2N_6	5.73	2		
Dibromodifluoromethane	CBr_2F_2	9.0	2	molar refraction; see Ref.2	
1,2-Dibromoethane	$C_2H_4Br_2$	10.7	2	molar refraction; see Ref.2	
Dibromomethane	CH_2Br_2	9.32	2		
Dibromomethane	CH_2Br_2	8.68	27		
1,4-Di-*tert*-butylbenzene	$C_{14}H_{22}$	24.5	25		
Dibutyl ether	$C_8H_{18}O$	17.2	2		
Dicesium	Cs_2	104	22		
o-Dichlorobenzene	$C_6H_4Cl_2$	14.17	27		
m-Dichlorobenzene	$C_6H_4Cl_2$	14.23	27		
p-Dichlorobenzene	$C_6H_4Cl_2$	14.20	27		
1,4-Dichlorobutane	$C_4H_8Cl_2$	12.0	2	molar refraction; see Ref.2	
2,5-Dichloro-2,5-cyclohexadiene-1,4-dione	$C_6H_2Cl_2O_2$	18.4	2		
1,1-Dichloro-2,2-difluoroethane	$C_2H_2Cl_2F_2$	8.4	2	molar refraction; see Ref.2	
Dichlorodifluoromethane	CCl_2F_2	7.93	20		

Name	Mol. form.	$\alpha/10^{-24}$ cm³	Ref.	Note	Z
Dichlorodifluoromethane	CCl_2F_2	7.81	2		
1,1-Dichloroethane	$C_2H_4Cl_2$	8.64	2		
1,2-Dichloroethane	$C_2H_4Cl_2$	8.0	2	molar refraction; see Ref.2	
1,1-Dichloroethene	$C_2H_2Cl_2$	7.83	27		
cis-1,2-Dichloroethene	$C_2H_2Cl_2$	8.03	27		
trans-1,2-Dichloroethene	$C_2H_2Cl_2$	8.15	27		
Dichlorofluoromethane	$CHCl_2F$	6.82	2		
Dichloromethane	CH_2Cl_2	6.48	3		
Dichloromethane	CH_2Cl_2	7.93	2		
1,2-Dichloropropane, (±)-	$C_3H_6Cl_2$	10.9	2	molar refraction; see Ref.2	
1,3-Dichloropropene (unspecified isomer)	$C_3H_4Cl_2$	10.1	2	molar refraction; see Ref.2	
Dichlorosilane	Cl_2H_2Si	8.92	2		
1,2-Dichloro-1,1,2,2-tetrafluoroethane	$C_2Cl_2F_4$	8.5	2	molar refraction; see Ref.2	
p-Dicyanobenzene	$C_8H_4N_2$	19.2	2		
1,1-Diethoxyethane	$C_6H_{14}O_2$	13.2	2	molar refraction; see Ref.2	
1,2-Diethoxyethane	$C_6H_{14}O_2$	11.3	2	molar refraction; see Ref.2	
Diethylamine	$C_4H_{11}N$	10.2	2		
Diethylamine	$C_4H_{11}N$	9.61	27		
Diethyl carbonate	$C_5H_{10}O_3$	11.3	2		
Diethyl ether	$C_4H_{10}O$	10.2	2		
Diethyl ether	$C_4H_{10}O$	8.73	15		
Diethyl succinate	$C_8H_{14}O_4$	16.8	2	molar refraction; see Ref.2	
Diethyl sulfide	$C_4H_{10}S$	10.8	2		
o-Difluorobenzene	$C_6H_4F_2$	9.80	27		
m-Difluorobenzene	$C_6H_4F_2$	10.3	2	molar refraction; see Ref.2	
p-Difluorobenzene	$C_6H_4F_2$	9.80	27		
1,1-Difluoroethene	$C_2H_2F_2$	5.01	20		
Diiodomethane	CH_2I_2	12.90	27		
Diketene	$C_4H_4O_2$	8.0	2	molar refraction; see Ref.2	
Dilithium	Li_2	32.8	29		
Dilithium	Li_2	34	22		
Dilithium disodium	Li_2Na_2	60.0	30		
Dilithium sodium	Li_2Na	35.4	30		
Dimethoxymethane	$C_3H_8O_2$	7.7	2	molar refraction; see Ref.2	
Dimethylamine	C_2H_7N	6.37	2		
Dimethylamine	C_2H_7N	5.90	33	calculated	
N,N-Dimethylaniline	$C_8H_{11}N$	16.2	2	molar refraction; see Ref.2	
2,3-Dimethyl-1,3-butadiene	C_6H_{10}	11.8	2	molar refraction; see Ref.2	
Dimethyl carbonate	$C_3H_6O_3$	7.7	2	molar refraction; see Ref.2	
2,5-Dimethyl-2,5-cyclohexadiene-1,4-dione	$C_8H_8O_2$	18.8	2		
Dimethyl ether	C_2H_6O	5.29	20		
Dimethyl ether	C_2H_6O	5.84	2		
Dimethyl ether	C_2H_6O	5.16	15		
N,N-Dimethylformamide	C_3H_7NO	7.81	18		
trans-2,3-Dimethyloxirane	C_4H_8O	8.22	17	calculated	
2,4-Dimethyl-3-pentanone	$C_7H_{14}O$	13.5	15		
2,2-Dimethylpropanenitrile	C_5H_9N	9.59	18		
(1,2-Dimethyl-1-propenyl)benzene	$C_{11}H_{14}$	19.64	27		
1,5-Dimethyl-1*H*-pyrazole	$C_5H_8N_2$	10.72	27		
2,3-Dimethylquinoxaline	$C_{10}H_{10}N_2$	18.70	27		
Dimethyl sulfone	$C_2H_6O_2S$	7.3	2	molar refraction; see Ref.2	
1,4-Dinitrobenzene	$C_6H_4N_2O_4$	18.4	2		
1,4-Dioxane	$C_4H_8O_2$	10.0	2		
1,4-Dioxane	$C_4H_8O_2$	8.60	18		
Dipotassium	K_2	77	22		
Dipotassium	K_2	72	21		
Dipropylamine	$C_6H_{15}N$	13.29	27		
Dipropyl ether	$C_6H_{14}O$	12.8	2		
Dipropyl ether	$C_6H_{14}O$	12.5	15		
Dirubidium	Rb_2	79	22		
Disilane	H_6Si_2	11.1	2		

Name	Mol. form.	$\alpha/10^{-24}$ cm^3	Ref.	Note	Z
Disodium	Na_2	40	22		
Disodium	Na_2	38	21		
Divinyl sulfide	C_4H_6S	10.9	2	molar refraction; see Ref.2	
Dodecane	$C_{12}H_{26}$	22.75	27		
Erbium(III) tris(cyclopentadienyl)	$C_{15}H_{15}Er$	28.44	37		
Ethane	C_2H_6	4.47	3		
Ethane	C_2H_6	4.43	2		
1,2-Ethanediamine	$C_2H_8N_2$	7.2	2	molar refraction; see Ref.2	
1,2-Ethanediol	$C_2H_6O_2$	5.7	2	molar refraction; see Ref.2	
1,2-Ethanediol	$C_2H_6O_2$	5.61	27		
Ethanethiol	C_2H_6S	7.41	2		
Ethanol	C_2H_6O	5.41	2		
Ethanol	C_2H_6O	5.11	18		
Ethoxybenzene	$C_8H_{10}O$	14.9	2		
2-Ethoxyethanol	$C_4H_{10}O_2$	9.28	27		
Ethyl acetate	$C_4H_8O_2$	9.7	2		
Ethyl acetate	$C_4H_8O_2$	8.62	27		
Ethyl acetoacetate	$C_6H_{10}O_3$	12.9	2	molar refraction; see Ref.2	
Ethylamine	C_2H_7N	7.10	2		
Ethylbenzene	C_8H_{10}	14.2	2		
Ethyl benzoate	$C_9H_{10}O_2$	16.9	2	molar refraction; see Ref.2	
Ethyl 2-chlorobutanoate	$C_6H_{11}ClO_2$	14.16	27		
Ethyl 3-chlorobutanoate	$C_6H_{11}ClO_2$	14.13	27		
Ethyl 4-chlorobutanoate	$C_6H_{11}ClO_2$	14.11	27		
Ethyl chloroformate	$C_3H_5ClO_2$	9.0	2	molar refraction; see Ref.2	
Ethylcyclohexane	C_8H_{16}	15.9	2		
Ethylene	C_2H_4	4.252	8		
Ethyl formate	$C_3H_6O_2$	8.01	2		
Ethyl formate	$C_3H_6O_2$	6.88	27		
Ethyl *trans,trans*-2,4-hexadienoate	$C_8H_{12}O_2$	17.2	2	molar refraction; see Ref.2	
Ethyl methyl ether	C_3H_8O	7.93	2		
1-Ethyl-5-methyl-1*H*-pyrazole	$C_6H_{10}N_2$	12.50	27		
1-Ethylnaphthalene	$C_{12}H_{12}$	21.19	27		
2-Ethylnaphthalene	$C_{12}H_{12}$	21.36	27		
Ethyl nitrite	$C_2H_5NO_2$	7.0	15		
1-Ethyl-1-phenylhydrazine	$C_8H_{12}N_2$	16.62	27		
Ethyl propanoate	$C_5H_{10}O_2$	10.41	27		
Ethyl propyl ether	$C_5H_{12}O$	10.68	27		
Ferrocene	$C_{10}H_{10}Fe$	17.1	26		
9*H*-Fluorene	$C_{13}H_{10}$	21.68	27		
Fluorine	F_2	1.38	7	calculated	
Fluoroanthracene (unspecified isomer)	$C_{14}H_9F$	28.34	27		
Fluorobenzene	C_6H_5F	10.3	2		
1-Fluorodecane	$C_{10}H_{21}F$	19.18	27		
1-Fluorododecane	$C_{12}H_{25}F$	22.83	27		
Fluoroethane	C_2H_5F	4.96	2		
1-Fluoroheptane	$C_7H_{15}F$	13.66	27		
2-Fluorohexane	$C_6H_{13}F$	11.80	27		
1-Fluoro-4-iodobenzene	C_6H_4FI	15.5	2	molar refraction; see Ref.2	
Fluoromethane	CH_3F	2.97	8		
1-Fluoro-4-nitrobenzene	$C_6H_4FNO_2$	12.8	2	molar refraction; see Ref.2	
1-Fluorononane	$C_9H_{19}F$	17.34	27		
1-Fluorooctane	$C_8H_{17}F$	15.46	27		
1-Fluoropentane	$C_5H_{11}F$	9.95	27		
1-Fluorotetradecane	$C_{14}H_{29}F$	26.57	27		
4-Fluorotoluene	C_7H_7F	11.70	27		
1-Fluoroundecane	$C_{11}H_{23}F$	21.00	27		
Formaldehyde	CH_2O	2.8	2	molar refraction; see Ref.2	
Formaldehyde	CH_2O	2.45	18		
Formamide	CH_3NO	4.2	2	molar refraction; see Ref.2	
Formamide	CH_3NO	4.08	18		

Name	Mol. form.	$\alpha/10^{-24}$ cm³	Ref.	Note	Z
Formic acid	CH_2O_2	3.4	2	molar refraction; see Ref.2	
Formic acid dimer	$C_2H_4O_4$	12.7	2		
Formyl fluoride	CHFO	1.76	17	calculated	
Gallium arsenide (variable composition)	GanAsm		28	*n*+*m*=4-30	
Germanium cluster (Ge_9)		90	35		
Germanium cluster (Ge_{10})		128	35		
Germanium cluster (Ge_{15})		167	35		
Germanium(IV) chloride	Cl_4Ge	15.1	2		
Heptane	C_7H_{16}	13.61	2		
1-Heptene	C_7H_{14}	13.51	27		
1-Heptyne	C_7H_{12}	12.8	2	molar refraction; see Ref.2	
Hexafluorobenzene	C_6F_6	9.58	27		
Hexafluoroethane	C_2F_6	6.82	2		
Hexamethylbenzene	$C_{12}H_{18}$	20.9	25		
Hexane	C_6H_{14}	11.9	2		
1-Hexene	C_6H_{12}	11.65	27		
1-Hexyne	C_6H_{10}	10.9	2	molar refraction; see Ref.2	
Hydrogen	H_2	0.8023	5	calc *v*=0,*J*=0	
Hydrogen	H_2	0.8045	5	calc 293 K	
Hydrogen	H_2	0.8042	6	293 K	
Hydrogen	H_2	0.8059	8	322 K	
Hydrogen-d_2	D_2	0.7921	5	calc *v*=0,*J*=0	
Hydrogen-d_2	D_2	0.7954	6	293 K	
Hydrogen-d_1	DH	0.7976	5	calc *v*=0,*J*=0	
Hydrogen bromide	BrH	3.61	3		
Hydrogen chloride	ClH	2.63	3		
Hydrogen chloride	ClH	2.77	2		
Hydrogen chloride-*d*	ClD	2.84	2		
Hydrogen cyanide	CHN	2.59	3		
Hydrogen cyanide	CHN	2.46	2		
Hydrogen fluoride	FH	0.80	27		
Hydrogen iodide	HI	5.44	3		
Hydrogen iodide	HI	5.35	2		
Hydrogen sulfide	H_2S	3.78	3		
Hydrogen sulfide	H_2S	3.95	2		
Iodine chloride	ClI	12.3	2		
Iodobenzene	C_6H_5I	15.5	2	molar refraction; see Ref.2	
1-Iodobutane	C_4H_9I	13.3	2	molar refraction; see Ref.2	
1-Iodobutane	C_4H_9I	12.65	27		
Iodoethane	C_2H_5I	10.0	2		
Iodoethene	C_2H_3I	9.3	2	molar refraction; see Ref.2	
Iodomethane	CH_3I	7.97	2		
1-Iodonaphthalene	$C_{10}H_7I$	22.41	27		
2-Iodonaphthalene	$C_{10}H_7I$	22.95	27		
1-Iodopropane	C_3H_7I	11.5	2	molar refraction; see Ref.2	
4-Iodotoluene	C_7H_7I	17.10	27		
Iron(III) 2,4-pentanedioate	$C_{15}H_{21}FeO_6$	58.1	2		
Isobutane	C_4H_{10}	8.14	27		
Isobutene	C_4H_8	8.29	2		
Isopropenylbenzene	C_9H_{10}	16.05	27		
Isopropylamine	C_3H_9N	7.77	27		
Isopropylbenzene	C_9H_{12}	16.0	2	molar refraction; see Ref.2	
Isopropylcyclohexane	C_9H_{18}	17.2	2		
Isoquinoline	C_9H_7N	16.43	27		
Ketene	C_2H_2O	4.4	2	molar refraction; see Ref.2	
Lithium bromide dimer	Br_2Li_2	18.9	16		
Lithium chloride	ClLi	3.46	10	calculated	
Lithium chloride dimer	Cl_2Li_2	13.1	16		
Lithium cluster	Li*n*		29	*n*=2-22	
Lithium disodium	$LiNa_2$	61.2	30		
Lithium fluoride	FLi	10.8	11	calculated	

Name	Mol. form.	$\alpha/10^{-24}$ cm^3	Ref.	Note	Z
Lithium fluoride dimer	F_2Li_2	6.9	16		
Lithium hydride	HLi	3.84	12	calculated	
Lithium hydride	HLi	3.68	13	calculated	
Lithium hydride	HLi	3.88	14	calculated	
Lithium iodide dimer	I_2Li_2	23.4	16		
Lithium sodium	LiNa	40	4		
Lithium trisodium	$LiNa_3$	75.6	30		
Malononitrile	$C_3H_2N_2$	5.79	18		
Mercury(II) bromide	Br_2Hg	14.5	2		
Mercury chloride	ClHg	7.4	9	calculated	
Mercury(I) chloride	Cl_2Hg_2	14.7	9		
Mercury(II) chloride	Cl_2Hg	11.6	2		
Mercury(II) iodide (red)	HgI_2	19.1	2		
Methane	CH_4	2.593	8		
Methanol	CH_4O	3.29	2		
Methanol	CH_4O	3.23	15		
Methanol	CH_4O	3.32	18		
2-Methoxyaniline	C_7H_9NO	14.2	2	molar refraction; see Ref.2	
Methoxycyclohexane	$C_7H_{14}O$	13.4	2	molar refraction; see Ref.2	
2-Methoxyethanol	$C_3H_8O_2$	7.44	27		
N-Methylacetamide	C_3H_7NO	7.82	18		
Methyl acetate	$C_3H_6O_2$	6.94	2		
Methyl acetate	$C_3H_6O_2$	6.81	27		
2-Methylacrylonitrile	C_4H_5N	8.0	2	molar refraction; see Ref.2	
Methylamine	CH_5N	4.7	2		
Methylamine	CH_5N	4.01	19		
Methylamine	CH_5N	4.01	33	calculated	
2-Methyl-1,3-butadiene	C_5H_8	9.99	2		
Methyl butanoate	$C_5H_{10}O_2$	10.41	27		
Methyl 2-chlorobutanoate	$C_5H_9ClO_2$	12.33	27		
Methyl 3-chlorobutanoate	$C_5H_9ClO_2$	12.31	27		
Methyl 4-chlorobutanoate	$C_5H_9ClO_2$	12.27	27		
Methyl chloroformate	$C_2H_3ClO_2$	7.1	2	molar refraction; see Ref.2	
Methylcyclohexane	C_7H_{14}	13.1	2		
2-Methyl-1,3-dioxolane	$C_4H_8O_2$	9.44	15		
3-Methylene-1-pentene	C_6H_{10}	11.8	2	molar refraction; see Ref.2	
N-Methylformamide	C_2H_5NO	5.91	18		
Methyl formate	$C_2H_4O_2$	5.05	27		
3-Methylheptane	C_8H_{18}	15.44	27		
1-Methylisoquinoline	$C_{10}H_9N$	18.28	27		
1-Methylnaphthalene	$C_{11}H_{10}$	19.35	27		
2-Methylnaphthalene	$C_{11}H_{10}$	19.52	27		
2-Methyl-2-nitropropane	$C_4H_9NO_2$	10.3	2	molar refraction; see Ref.2	
2-Methyl-1,3-pentadiene	C_6H_{10}	12.1	2	molar refraction; see Ref.2	
3-Methyl-1,3-pentadiene	C_6H_{10}	11.8	2	molar refraction; see Ref.2	
1-Methyl-1-phenylhydrazine	$C_7H_{10}N_2$	14.81	27		
2-Methylpropanenitrile	C_4H_7N	8.05	18		
2-Methylpropanenitrile	C_4H_7N	8.05	32	calculated	
Methyl propanoate	$C_4H_8O_2$	8.97	27		
2-Methyl-1-propanol	$C_4H_{10}O$	8.92	2		
2-Methylpropenal	C_4H_6O	8.3	2	molar refraction; see Ref.2	
Methyl propyl ether	$C_4H_{10}O$	8.86	27		
1-Methyl-1*H*-pyrazole	$C_4H_6N_2$	8.99	27		
2-Methylquinoline	$C_{10}H_9N$	18.65	27		
Naphthacene	$C_{18}H_{12}$	32.27	27		
Naphthalene	$C_{10}H_8$	16.5	17		
Naphthalene	$C_{10}H_8$	17.48	27		
1-Naphthalenecarboxaldehyde	$C_{11}H_8O$	19.75	27		
2-Naphthalenecarboxaldehyde	$C_{11}H_8O$	20.06	27		
1-Naphthylamine	$C_{10}H_9N$	19.50	27		
2-Naphthylamine	$C_{10}H_9N$	19.73	27		

Name	Mol. form.	$\alpha/10^{-24}$ cm^3	Ref.	Note	Z
Neodymium(III) tris(cyclopentadienyl)	$C_{15}H_{15}Nd$	33.05	37		
Neopentane	C_5H_{12}	10.20	18		
Nitric oxide	NO	1.70	2		
2-Nitroanisole	$C_7H_7NO_3$	15.7	2	molar refraction; see Ref.2	
3-Nitroanisole	$C_7H_7NO_3$	15.7	2	molar refraction; see Ref.2	
4-Nitroanisole	$C_7H_7NO_3$	15.7	2	molar refraction; see Ref.2	
Nitrobenzene	$C_6H_5NO_2$	14.7	2		
Nitrobenzene	$C_6H_5NO_2$	12.92	15		
4-Nitrobenzonitrile	$C_7H_4N_2O_2$	19.0	2		
1-Nitrobutane	$C_4H_9NO_2$	10.4	2	molar refraction; see Ref.2	
Nitroethane	$C_2H_5NO_2$	9.63	2		
Nitrogen	N_2	1.7403	6,8		
Nitrogen dioxide	NO_2	3.02	2	molar refraction; see Ref.2	
Nitrogen tetroxide	N_2O_4	6.69	2		
Nitrogen trifluoride	F_3N	3.62	2		
Nitromethane	CH_3NO_2	7.37	2		
1-Nitro-4-phenoxybenzene	$C_{12}H_9NO_3$	24.7	2	molar refraction; see Ref.2	
1-Nitropropane	$C_3H_7NO_2$	8.5	2	molar refraction; see Ref.2	
2-Nitropropane	$C_3H_7NO_2$	8.5	2	molar refraction; see Ref.2	
Nitrous oxide	N_2O	3.03	8		
Nonane	C_9H_{20}	17.36	27		
Octane	C_8H_{18}	15.9	2		
Osmium(VIII) oxide	O_4Os	8.17	2		
Oxirane	C_2H_4O	4.43	18		
Oxygen	O_2	1.5689	34		
Ozone	O_3	3.21	2		
Paraldehyde	$C_6H_{12}O_3$	17.9	2		
Pentachloroethane	C_2HCl_5	14.0	2		
trans-1,3-Pentadiene	C_5H_8	10	2		
2,4-Pentadienenitrile	C_5H_5N	10.5	2	molar refraction; see Ref.2	
Pentafluorobenzene	C_6HF_5	9.63	27		
Pentamethylbenzene	$C_{11}H_{16}$	19.1	25		
Pentane	C_5H_{12}	9.99	2		
2,4-Pentanedione	$C_5H_8O_2$	10.5	2	molar refraction; see Ref.2	
Pentanenitrile	C_5H_9N	10.4	2		
2-Pentanone	$C_5H_{10}O$	9.93	15		
3-Pentanone	$C_5H_{10}O$	9.93	15		
1-Pentene	C_5H_{10}	9.65	27		
2-Pentene (unspecified isomer)	C_5H_{10}	9.84	27		
Pentyl acetate	$C_7H_{14}O_2$	14.9	2		
Pentyl formate	$C_6H_{12}O_2$	14.2	2		
1-Pentyne	C_5H_8	9.12	2		
Perfluoronaphthalene	$C_{10}F_8$	17.64	27		
Phenanthrene	$C_{14}H_{10}$	36.8	17	calculated	
Phenanthrene	$C_{14}H_{10}$	24.70	27		
Phenazine	$C_{12}H_8N_2$	23.43	27		
Phenol	C_6H_6O	11.1	2	molar refraction; see Ref.2	
Phenol	C_6H_6O	9.94	17	calculated	
Phenylhydrazine	$C_6H_8N_2$	12.91	27		
Phosphine	H_3P	4.84	2		
Phosphorus(III) chloride	Cl_3P	12.8	2		
Phosphorus(V) fluoride	F_5P	6.10	2		
Potassium bromide dimer	Br_2K_2	42.0	16		
Potassium chloride dimer	Cl_2K_2	32.1	16		
Potassium cluster			21	n=2,5,7-9,11,	
Potassium fluoride dimer	F_2K_2	21.0	16		
Potassium iodide dimer	I_2K_2	36.3	16		
Potassium sodium	KNa	51	22		
Propanal	C_3H_6O	6.50	2		
Propane	C_3H_8	6.29	3		
Propane	C_3H_8	6.37	2		

Name	Mol. form.	$\alpha/10^{-24}$ cm^3	Ref.	Note	Z
Propanenitrile	C_3H_5N	6.70	2		
Propanenitrile	C_3H_5N	6.24	18		
Propanenitrile	C_3H_5N	6.27	32	calculated	
Propanoic acid	$C_3H_6O_2$	6.9	2	molar refraction; see Ref.2	
1-Propanol	C_3H_8O	6.74	2		
2-Propanol	C_3H_8O	7.61	2		
2-Propanol	C_3H_8O	6.97	18		
Propene	C_3H_6	6.26	2		
Propylamine	C_3H_9N	7.70	27		
Propylamine	C_3H_9N	9.20	2		
Propyne	C_3H_4	6.18	2		
1*H*-Pyrazole	$C_3H_4N_2$	7.23	27		
Pyrene	$C_{16}H_{10}$	28.22	27		
Pyridazine	$C_4H_4N_2$	9.27	17	calculated	
Pyridine	C_5H_5N	9.5	15		
Pyridine	C_5H_5N	9.18	27		
Pyrimidine	$C_4H_4N_2$	8.53	17	calculated	
Quinoline	C_9H_7N	15.70	27		
Quinoxaline	$C_8H_6N_2$	15.13	27		
Rubidium bromide dimer	Br_2Rb_2	48.2	16		
Rubidium chloride dimer	Cl_2Rb_2	43.2	16		
Rubidium fluoride dimer	F_2Rb_2	40.7	16		
Rubidium iodide dimer	I_2Rb_2	46.3	16		
Samarium(III) tris(cyclopentadienyl)	$C_{15}H_{15}Sm$	32.01	37		
Selenium hexafluoride	F_6Se	7.33	2		
Silane	H_4Si	5.44	2		
Sodium bromide dimer	Br_2Na_2	26.8	16		
Sodium chloride dimer	Cl_2Na_2	23.4	16		
Sodium cluster			39	n=7-93	
Sodium cluster			21	n=1-40	
Sodium fluoride dimer	F_2Na_2	20.7	16		
Sodium iodide dimer	I_2Na_2	26.9	16		
Styrene	C_8H_8	15.0	2		
Styrene	C_8H_8	14.41	27		
Succinonitrile	$C_4H_4N_2$	8.1	2	molar refraction; see Ref.2	
Sulfur decafluoride	$F_{10}S_2$	13.2	2		
Sulfur dioxide	O_2S	3.72	3		
Sulfur dioxide	O_2S	4.28	2		
Sulfur hexafluoride	F_6S	6.54	8		
Sulfur trioxide (α-form)	O_3S	4.84	2		
Sulfuryl chloride	Cl_2O_2S	10.5	2		
Tellurium hexafluoride	F_6Te	9.00	2		
1,1,2,2-Tetrachloroethane	$C_2H_2Cl_4$	12.1	2	molar refraction; see Ref.2	
Tetrachloromethane	CCl_4	11.2	2		
Tetrachloromethane	CCl_4	10.5	3		
1,2,3,4-Tetrafluorobenzene	$C_6H_2F_4$	9.69	27		
1,2,4,5-Tetrafluorobenzene	$C_6H_2F_4$	9.69	27		
Tetrafluoromethane	CF_4	3.838	8		
Tetrafluorosilane	F_4Si	5.45	2		
Tetramethoxymethane	$C_5H_{12}O_4$	13.0	2	molar refraction; see Ref.2	
1,2,4,5-Tetramethylbenzene	$C_{10}H_{14}$	17.3	25		
2,2,4,4-Tetramethyl-1,3-cyclobutanedione	$C_8H_{12}O_2$	18.6	2		
Tetranitromethane	CN_4O_8	15.3	2		
Tetraphosphorus	P_4	13.59	40		
Thiophene	C_4H_4S	9.67	2		
Thorium(IV) 2,4-pentanedioate	$C_{20}H_{28}O_8Th$	79.0	2		
Tin(IV) bromide	Br_4Sn	22.0	2		
Tin(IV) chloride	Cl_4Sn	18	2		
Tin(IV) chloride	Cl_4Sn	13.8	15		
Tin(IV) iodide	I_4Sn	32.3	2		
Titanium(IV) chloride	Cl_4Ti	16.4	2		

Name	Mol. form.	$\alpha/10^{-24}$ cm^3	Ref.	Note	Z
Toluene	C_7H_8	11.8	25		
Toluene	C_7H_8	12.26	15		
Toluene	C_7H_8	12.3	2		
Tribromomethane	$CHBr_3$	11.8	27		
1,3,5-Tri-*tert*-butylbenzene	$C_{18}H_{30}$	31.8	25		
Trichloroacetonitrile	C_2Cl_3N	10.42	18		
1,1,1-Trichloroethane	$C_2H_3Cl_3$	10.7	2		
Trichloroethene	C_2HCl_3	10.03	27		
1,1,2-Trichloro-2-fluoroethane	$C_2H_2Cl_3F$	10.2	2	molar refraction; see Ref.2	
Trichlorofluoromethane	CCl_3F	9.47	2		
Trichloromethane	$CHCl_3$	9.5	8		
Trichloromethane	$CHCl_3$	8.23	27		
Trichloronitromethane	CCl_3NO_2	10.8	2	molar refraction; see Ref.2	
Trichlorosilane	Cl_3HSi	10.7	2		
Triethylamine	$C_6H_{15}N$	13.1	2		
Triethylamine	$C_6H_{15}N$	13.38	27		
1,3,5-Trifluorobenzene	$C_6H_3F_3$	9.74	27		
1,1,1-Trifluoroethane	$C_2H_3F_3$	4.4	2	molar refraction; see Ref.2	
Trifluoromethane	CHF_3	3.52	20		
Trifluoromethane	CHF_3	3.57	8		
Triiodomethane	CHI_3	18.0	27		
Trilithium	Li_3	34.5	29		
Trilithium sodium	Li_3Na	54.8	30		
Trimethylamine	C_3H_9N	8.15	2		
Trimethylamine	C_3H_9N	7.78	33	calculated	
1,3,5-Trimethylbenzene	C_9H_{12}	15.5	25		
1,3,5-Trimethylbenzene	C_9H_{12}	16.14	27		
2,2,4-Trimethylpentane	C_8H_{18}	15.44	27		
Triphenylene	$C_{18}H_{12}$	31.07	27		
Tripropylamine	$C_9H_{21}N$	18.87	27		
Trisodium	Na_3	70	21		
Undecane	$C_{11}H_{24}$	21.03	27		
Uranium(VI) fluoride	F_6U	12.5	2		
Water	H_2O	1.45	2		
Water-d_2	D_2O	1.26	2		
o-Xylene	C_8H_{10}	14.9	2		
o-Xylene	C_8H_{10}	14.1	15		
m-Xylene	C_8H_{10}	14.2	15		
p-Xylene	C_8H_{10}	13.7	25		
p-Xylene	C_8H_{10}	14.2	15		
p-Xylene	C_8H_{10}	14.9	2		

References to Atoms

1. Miller, T. M., and Bederson, B., *Adv. At. Mol. Phys.* 13, 1, 1977. For simplicity, any value in Table 2 which has not changed since this 1977 review is referenced as 1. Persons interested in original references and further details should consult 1.
2. Alpher, R. A., and White, D. R., *Phys. Fluids* 2, 153, 1959. <https://doi.org/10.1063/1.1705906>
3. Schmidt, J. W., Glavioso, R. M., May, E. F., and Moldover, M. R., *Phys. Rev. Lett.* 98, 254504, 2007. <https://doi.org/10.1103/PhysRevLett.98.254504>
4. Miffre, A., Jacquey, M., Büchner, M., Trénec, G., and Vigué, J., *Phys. Rev. A* 73, 011603(R), 2006. <https://doi.org/10.1103/PhysRevA.73.011603>
5. Thierfelder, C., Assadollahzadeh, B., Schwerdtfeger, P., Schäfer, S., and Schäfer, R., *Phys. Rev. A* 78, 052506, 2008. Relativistic calculations for element 6 (C) through element 114 (Flerovium) and measurements for SN and Pb. The measurement uncertainties do not include that of the reference atom Ba. The polarizability of Ba has since been adjusted (see comment with reference 43). <https://doi.org/10.1103/PhysRevA.78.052506>
6. Gaiser, C., and Fellmuth, B., *Eur. Phys. Lett.* 90, 63002, 2010. <https://doi.org/10.1209/0295-5075/90/63002>
7. Ekstrom, C. R., Schmiedmayer, J., Chapman, M. S., Hammond, T. D., and Pritchard, D. E., *Phys. Rev. A* 51, 3883, 1995. See theoretical work by Thakkar, A. J., and Lupinetti, C., *Chem. Phys. Lett.* 402, 270, 2005. <https://doi.org/10.1016/j.cplett.2004.12.046>
8. Holmgren, W. F., Revelle, M. C., Lonij, V. P. A., and Cronin, A. D., *Phys. Rev. A* 81, 053607, 2010. <https://doi.org/10.1103/PhysRevA.81.053607>
9. Borschevsky, A., Pershina, V., Eliav, E., and Kaldor, U., *J. Chem. Phys.* 138, 124302, 2013; *Phys. Rev. A* 87, 022502, 2013. <https://doi.org/10.1063/1.4795433>
10. Bromley, M. W. J., and Mitroy, J., *Phys. Rev. A* 65, 062505, 2002; 062506, 2002. <https://doi.org/10.1103/PhysRevA.65.012505>
11. Milani, P., Moullet, I., and de Heer, W. A., *Phys. Rev. A* 42, 5150, 1990. See theoretical comments on this result, in Fuentealba, P., *Chem. Phys. Lett.* 397, 459, 2004, and in Lupinetti, C., and Thakkar, A. J., *J. Chem. Phys.* 122, 044301, 2005. <https://doi.org/10.1063/1.1834512>
12. Newell, A. C., and Baird, R. D., *J. Appl. Phys.* 36, 3751, 1965. <https://doi.org/10.1063/1.1713942>

13. Orcutt, R. H., and Cole, R. H., *J. Chem. Phys.* 46, 697, 1967; see also the later references from this group in Reference 8 following the tables for molecular polarizabilities. <https://doi.org/10.1063/1.1840728>
14. Doolen, G. D., Los Alamos National Laboratory, unpublished. A relativistic linear response method was used. The method is that described by Zangwill, A., and Soven, P., *Phys. Rev. A* 21, 1561, 1980. Adjustments of less than 10% across the periodic table have been made to these results to bring them into agreement with accurate experimental values when available, for the purpose of presenting "recommended" polarizabilities in Table 2.
15. Goebel, D., Holm, U., and Maroulis, G., *Phys. Rev. A* 54, 1973, 1996. <https://doi.org/10.1103/PhysRevA.54.1973>
16. Hu, M., and Kusse, B. R., *Phys. Rev. A* 66, 062506, 2002. Measured $11.4 \pm 1.2 \times 10^{-24}$ cm^3 at 532 nm and extrapolated to a static value.
17. Goebel, D., and Holm, U., *Phys. Rev. A* 52, 3691, 1995. <https://doi.org/10.1103/PhysRevA.52.3691>
18. Guella, T. P., Miller, T. M., Bederson, B., Stockdale, J. A. D., and Jaduszliwer, B., *Phys. Rev. A* 29, 2977, 1984. <https://doi.org/10.1103/PhysRevA.29.2977>
19. Atoji, M., *J. Chem. Phys.* 25, 174, 1956. <https://doi.org/10.1063/1.1742814>
20. Amini, J. M., and Gould, H., *Phys. Rev. Lett.* 91, 153001, 2003. <https://doi.org/10.1103/PhysRevLett.91.153001>
21. Fleig, T. and Sadlej, A. J., *Phys. Rev. A* 65, 032506, 2002.
22. Goebel, D. and Holm, U., *J. Chem. Phys.* 100, 7710, 1996. <https://doi.org/10.1021/jp960231l>
23. Preliminary value from the New York University group. See Guella, T. P., Miller, T. M., Bederson, B., Stockdale, J. A. D., and Jaduszliwer, B., *Phys. Rev. A* 29, 2977, 1984. <https://doi.org/10.1103/PhysRevA.29.2977>
24. Borschevsky, A., Zelovich, T., Eliav, E., and Kaldor, U., *Chem. Phys.* 395, 104, 2012. <https://doi.org/10.1016/j.chemphys.2011.05.011>
25. Chattopadhyay, S., Mani, B. K., and Angom, D., *Phys. Rev. A* 89, 022506, 2014. <https://doi.org/10.1103/PhysRevA.89.022506>
26. Łach, G., Jezionski, B., and Szalewicz, K., *Phys. Rev. Lett.* 92, 233001, 2004. <https://doi.org/10.1103/PhysRevLett.92.233001>
27. Kadar-Kallen, M. A., and Bonin, K. D., *Phys. Rev. Lett.* 72, 828, 1994. <https://doi.org/10.1103/PhysRevLett.72.828>
28. Pershina, V., Borschevsky, A., Eliav, E., and Kaldor, U., *J. Chem. Phys.* 128, 024707, 2008. <https://doi.org/10.1063/1.2814242>
29. Dzuba, V. A., *Phys. Rev. A* 93, 032519,2016. <https://doi.org/10.1103/PhysRevA.93.052517>
30. Dzuba, V. A., Koslov, A., and Flambaum, V. V., *Phys. Rev. A* 89, 042507, 2014. <https://doi.org/10.1103/PhysRevA.89.042507>
31. Fleig, T., *Phys. Rev. A* 72, 052506, 2005. <https://doi.org/10.1103/PhysRevA.72.052506>
32. Gregoire, M. D., Hromada, I., Holmgren, W. F., Trubko, R., and Cronin, A. D., *Phys. Rev. A* 92, 052513, 2015. <https://doi.org/10.1103/PhysRevA.92.052513>
33. Lupinetti, C., and Thakkar, A. J., *J. Chem. Phys.* 122, 044301, 2005. <https://doi.org/10.1063/1.1834512>
34. Miller, T. M., and Bederson, B., *Phys. Rev. A* 14, 1572, 1976. The polarizability ratio of α(Ca)/ α(Li) was measured to be 1.0316. Because α(Li) is now known more accurately, α(Ca) is reevaluated as $24.9 \pm 8\% \times 10^{-24}$ cm^3.
35. Medvedd', M., Fowler, P. W., and Hutson, J. M., *Mol. Phys.* 98, 453, 2000.
36. Ma, L., Indergaard, J., Zhang, B., Larkin, I., Moro, R., and deHeer, W. A., *Phys. Rev. A* 91, 010501, 2015. Polarizabilities were normalized to that measured for Al ($6.8 \pm 0.3 \times 10^{-24}$ cm^3). The uncertainties given in the table do not include that for Al. Note that all theoretical values tend to be 21% greater than the measured value for Al given above (see Reference 33).
37. Pershina, V., Borschevsky, A., Eliav, E., and Kaldor, U., *J. Chem. Phys.* 129, 144106, 2008. <https://doi.org/10.1063/1.2988318>
38. Pershina, V., Borschevsky, A., Eliav, E., and Kaldor, U., *J. Phys. Chem. A* 112, 13712, 2008. <https://doi.org/10.1021/jp8061306>
39. Pou-Amérigo, R., Merchán, M., Nebot-Gil, I., Widmark, P. -O., and Roos, B. O., *Theor. Chim Acta* (now *Theor. Chem Acc.*) 92, 149, 1995. <https://doi.org/10.1007/s002140050119>
40. Roos, B. O., Lindh, R., Malmqvist, P. -Å., Veryazov, V., and Widmark, P. -O., *J. Phys. Chem. A* 108, 2851, 2004. <https://doi.org/10.1021/jp031064+>
41. Roos, B. O., Lindh, R., Malmqvist, P. -Å., Veryazov, V., and Widmark, P. -O., *J. Phys. Chem. A* 109, 6575, 2005. <https://doi.org/10.1021/jp0581126>
42. Sahoo, B. K.,, and Das, B. P., *Phys. Rev. A* 77, 062516, 2008. <https://doi.org/10.1103/PhysRevA.77.062516>
43. Schwartz, H. L., Miller, T. M., and Bederson, B., *Phys. Rev. A* 10, 1924, 1974. The polarizability ratio α(Ba)/α(Li) was measured to be 1.6316 and that for α(Sr)/α(Li) = 1.1350. Because α(Li) is now known more accurately, α(Ba) is reevaluated as $39.3 \pm 6\%$ and α(Sr) = $39.3 \pm 6\%$.
44. Singh, Y., and Sahoo, B. K., *Phys. Rev. A* 90, 022511, 2014. <https://doi.org/10.1103/PhysRevA.90.022511>
45. Singh, Y., and Sahoo, B. K., *Phys. Rev. A* 91, 030501, 2015. <https://doi.org/10.1103/PhysRevA.91.030501>
46. Jerabek, P., Schwerdtfeger, P., and Nagle, J. K.,, *Phys. Rev. A* 98, 012508, 2018. <https://doi.org/10.1103/PhysRevA.98.012508>
47. Sahoo, B. K., and Yu, Y.-M., *Phys. Rev. A* 98, 012513, 2018. <https://doi.org/10.1103/PhysRevA.98.012513>
48. Beloy, K., *Phys. Rev. A* 86, 022521, 2012. <https://doi.org/10.1103/PhysRevA.86.051404>
49. Safronova, M. S., Porsev, S. G., and Clark, C. W., *Phys. Rev. Lett.* 109, 230802, 2012. <https://doi.org/10.1103/PhysRevLett.109.230802>
50. Becher, J. H., Baier, S., Aikawa, K., Lepers, M., Wyart, J.-F., Dulieu, O., and Ferlaino, F., *Phys. Rev. A* 97, 012509, 2018. The uncertainty in the static polarizability was estimated from those at the studied wavelengths. <https://doi.org/10.1103/PhysRevA.97.012509>
51. Li, H., Wyart, J.-F., Dulieu, O., Nascimbène, S., and Lepers, M., *J. Phys. B: At. Mol. Opt. Phys.*, 50, 014005, 2017. <https://doi.org/10.1088/1361-6455/50/1/014005>
52. Tang, Y.-B., Gao, N.-N., Lou, B.-Q., and Shi, T. Y., *Phys. Rev. A* 98, 062511, 2018. <https://doi.org/10.1103/PhysRevA.98.062511>

References to Molecules

1. McCullough, E. A., Jr., *J. Chem. Phys.*, 63, 5050, 1975. This calculation is for the parallel component, not the average polarizability. <https://doi.org/10.1063/1.431209>
2. Maryott, A. A., and Buckley, F., *U.S. National Bureau of Standards Circular No. 537*, 1953. A tabulation of dipole moments, dielectric constants, and molar refractions measured between 1910 and 1952, and used here to determine polarizabilities if no more recent result exists. The polarizability is $3/(4\pi N_A)$ times the molar polarization or molar refraction, where N_A is Avogadro's number. The value $3/(4\pi N_A) = 0.3964308 \times 10^{-24}$ cm^3 was used for this conversion. A dagger (†) following the reference number in the tables indicates that the polarizability was derived from the molar refraction and hence may not include some low-frequency contributions to the static polarizability; these "static" polarizabilities are therefore low by 1% to 30%.
3. Hirschfelder, J. O., Curtis, C. F., and Bird, R. B., *Molecular Theory of Gases and Liquids*, Wiley, New York, 1954, p. 950. Fundamental information on molecular polarizabilities.
4. Miller, T. M., and Bederson, B., *Adv. At. Mol. Phys.*, 13, 1, 1977. Review emphasizing atomic polarizabilities and measurement techniques. The data quoted in Table 3 are accurate to 8% to 12%.
5. Kolos, W., and Wolniewicz, L., *J. Chem. Phys.*, 46, 1426, 1967. Highly accurate molecular hydrogen calculations. See also recent work by Machado, A.M., and Masilli, M., *J. Chem. Phys.* 120, 7505, 2004. <https://doi.org/10.1063/1.1687677>
6. Newell, A. C., and Baird, R. C., *J. Appl. Phys.*, 36, 3751, 1965. Highly accurate refractive index measurements at 47.7 GHz (essentially static). <https://doi.org/10.1063/1.1713942>
7. Jao, T. C., Beebe, N. H. F., Person, W. B., and Sabin, J. R., *Chem. Phys. Lett.*, 26, 474, 1974. Tensor polarizabilities, derivatives, and other results are reported. <https://doi.org/10.1016/0009-2614(74)80394-5>

8. Orcutt, R. H., and Cole, R. H., *J. Chem. Phys.*, 46, 697, 1967 (He, Ne, Ar, Kr, H_2, N_2); Sutter, H., and Cole, R. H., *J. Chem. Phys.*, 52, 132, 1970 (CF_3H, CFH_3, $CClF_3$, $CClH_3$); Bose, T. K., and Cole, R. H., *J. Chem. Phys.*, 52, 140, 1970 (CO_2), and 54, 3829, 1971 (C_2H_4); Nelson, R. D., and Cole, R. H., *J. Chem. Phys.*, 54, 4033, 1971 (SF_6, $CClF_3$); Bose, T. K., Sochanski, J. S., and Cole, R. H., *J. Chem. Phys.*, 57, 3592, 1972 (CH_4, CF_4); Kirouac, S., and Bose, T. K., *J. Chem. Phys.*, 59, 3043, 1973 (N_2O), and 64, 1580, 1976 (He). Highly accurate dielectric constant measurements. These modern data give the most accurate polarizabilities available. A criticism of the interpretation of these data in the case of polar molecules is given in Ref. 20, p. 2905.
9. Huestis, D. L., Technical Report #MP 78-25, SRI International (project PYU 6158), Menlo Park, CA 94025. Molar refractions for mercury-chlorine compounds are analyzed.
10. Bounds, D. G., Clarke, J. H. R., and Hinchliffe, A., *Chem. Phys. Lett.*, 45, 367, 1977. Theoretical tensor polarizability for LiCl. <https://doi.org/10.1016/0009-2614(77)80291-1>
11. Kolker, H. J., and Karplus, M., *J. Chem. Phys.*, 39, 2011, 1963. Theoretical. <https://doi.org/10.1063/1.1734575>
12. Cutschick, V. P., and McKoy, V., *J. Chem. Phys.*, 58, 2397, 1973. Theoretical tensor polarizabilities. <https://doi.org/10.1063/1.1679518>
13. Gready, J. E., Bacskay, G. B., and Hush, N. S., *Chem. Phys.*, 22, 141, 1977, and 23, 9, 1977. Theoretical. <https://doi.org/10.1016/0301-0104(77)89039-3>
14. Amos, A. T., and Yoffe, J. A., *J. Chem. Phys.*, 63, 4723, 1975. Theoretical. <https://doi.org/10.1063/1.431258>
15. Stuart, H. A., *Landolt-Börnstein Zahlenwerte and Funktionen*, Vol. 1, Part 3, Eucken, A., and Hellwege, K. H., Eds., Springer-Verlag, Berlin, 1951, p. 511. Tabulation of molecular polarizabilities. Two misprints in the chemical symbols have been corrected.
16. Guella, T., Miller, T. M., Stockdale, J. A. D., Bederson, B., and Vuskovic, L., *J. Chem. Phys.*, 94, 6857, 1991. Beam measurements with accuracies between 12 and 24%. <https://doi.org/10.1063/1.460264>
17. Marchese, F. T., and Jaff, H. H., *Theoret. Chim. Acta* (Berlin), 45, 241, 1977. Theoretical and experimental tensor polarizabilities are tabulated in this paper. <https://doi.org/10.1007/BF00554533>
18. Applequist, J., Carl, J. R., and Fung, K.-K., *J. Am. Chem. Soc.*, 94, 2952, 1972. Excellent reference on the calculation of molecular polarizabilities, including extensive tables of tensor polarizabilities, both theoretical and experimental, at 589.3 nm wavelength. <https://doi.org/10.1021/ja00764a010>
19. Bridge, N. J., and Buckingham, A. D., *Proc. Roy. Soc.* (London), A295, 334, 1966. Measured tensor polarizabilities at 633 nm wavelength.
20. Barnes, A. N. M., Turner, D. J., and Sutton, L. E., *Trans. Faraday Soc.*, 67, 2902, 1971. Dielectric constants yielding polarizabilities accurate from 0.3%–8%. <https://doi.org/10.1039/tf9716702902>
21. Rayane, D., Allouche, A. R., Benichou, E., Antoine, R., Aubert-Frecon, M., Dugourd, Ph., Broyer, M., Ristori, C., Chandezon, F., Huber, B. A., and Guet, C., *Eur. Phys. J.* D 9, 243, 1999. See also Knight, W. D., Clemenger, K., de Heer, W. A., and Saunders, W. A., *Phys. Rev. B* 31, 2539, 1985. These data probably correspond to a very low internal temperature. <https://doi.org/10.1007/978-3-642-88188-6_46>
22. Tarnovsky, V., Bunimovicz, M., Vuskovic, L., Stumpf, B., and Bederson, B., *J. Chem. Phys.*, 98, 3894, 1993. These data correspond to internal temperatures 480-948 K. <https://doi.org/10.1063/1.464017>
23. Milani, P., Moullet, I., and de Heer, W. A., *Phys. Rev. A*, 42, 5150, 1990. Beam measurements accurate to 11%. <https://doi.org/10.1103/PhysRevA.42.5150>
24. Compagnon, I., Antoine, R., Broyer, M., Dugourd, P., Lermé, J., and Rayane, D., *Phys. Rev. A* 64, 025201, 2001. The uncertainties are ±8 Å^3 for C_{60} and ±14 Å^3 for C_{70}. <https://doi.org/10.1103/PhysRevA.64.025201>
25. Aroney, M. J., and Pratten, S. J., *J. Chem. Soc., Faraday Trans.*, 1, 80, 1201, 1984. Uncertainties in the range 1%–3%. <https://doi.org/10.1039/f19848001201>
26. Le Fevre, R. J. W., Murthy, D. S. N., and Saxby, J. D., *Aust. J. Chem.*, 24, 1057, 1971. Kerr effect. <https://doi.org/10.1071/CH9711177>
27. No, K. T., Cho, K. H., Jhon, M. S., and Scheraga, H. A., J. Am. Chem. Soc., 115, 2005, 1993. Theoretical; these results are quoted in numerous valuable papers on calculated polarizabilities, e.g., Miller, K. J., and Savchik, J. A., *J. Am. Chem. Soc.*, 101, 7206, 1979. <https://doi.org/10.1021/ja00518a014>
28. Schlecht, S., Schäfer, R., Woenckhaus, J., Becker, J. A., *Chem. Phys. Lett.*, 246, 315, 1995. <https://doi.org/10.1016/0009-2614(95)01095-Q>
29. Benichou, E., Antoine, R., Rayane, D., Vezin, B., Dalby, F. W., Dugourd, P., Ristori, C., Chandezon, F., Huber, B. A., Rocco, J. C., Blundell, S. A., and Guet, C., *Phys. Rev. A* 59, R1, 1999. See also Rayane, D., Allouche, A. R., Benichou, E., Antoine, R., Aubert-Frecon, M., Dugourd, Ph., Broyer, M., Ristori, C., Chandezon, F., Huber, B. A., and Guet, C., *Eur. Phys. J.* D 9, 243, 1999. <https://doi.org/10.1007/s100530050433>
30. Antoine, R., Rayane, D., Allouche, A. R., Aubert-Frécon, M., Benichou, E., Dalby, F. W., Dugourd, P., Broyer, M., and Guet, C., *J. Chem. Phys.*, 110, 5568, 1999. <https://doi.org/10.1063/1.478455>
31. Ballard, A., Bonin, K., and Louderback, J., *J. Chem. Phys.* 113, 5732, 2000. <https://doi.org/10.1063/1.1290472>
32. Ritchie, G.L.D., and Watson, J.N., *J. Phys. Chem. A* 108, 4515, 2004. These measurements are at 632.8 nm frequency, and are stated accurate to 0.4%.
33. Ritchie, G.L.D., and Blanch, E.W., *J. Phys. Chem. A* 107, 2093, 2003. These measurements are at 632.8 nm frequncy, and are stated accurate to better than 1%.
34. May, E. F., Moldover, M. R., and Schmidt, J. W., *Phys. Rev. A* 78, 032522 (2008). The uncertainty is 0.0003 Å^3. <https://doi.org/10.1103/PhysRevA.78.032522>
35. Heiles, S., Schäfer, S., and Schäfer, R., *J. Chem. Phys.* 135, 034303, 2011. The uncertainties are ±2.5 Å^3 for Ge_9 and Ge_{10}, and ±6.1 Å^3 for Ge_{15}. <https://doi.org/10.1063/1.3610390>
36. Berninger, M., Stefanov, A., Deachapunya, S., and Arndt, M., *Phys. Rev. A* 76, 013607, 2007. The uncertainties (statistical, systematic) are ±(0.9, 5.1) Å^3 for C_{60} and ±(2.0, 6.2) Å^3 for C_{70}.
37. Hohm, U., and Loose, A., *Chem. Phys. Lett.* 348, 375, 2001. The uncertainties are ±0.25 Å^3 for $Nd(C_5H_5)_3$, ±0.13 Å^3 for $Sm(C_5H_5)_3$, and ±0.45 Å^3 for $Er(C_5H_5)_3$. <https://doi.org/10.1016/S0009-2614(01)01167-8>
38. Rayane, D., Antoine, R., Dugourd, P., and Broyer, M., *J. Chem. Phys.* 113, 4501, 2000. A 10% precision in the measurements is stated. <https://doi.org/10.1063/1.1308562>
39. Tikhonov, G., Kasperovich, V., Wong, K., and Kresin, V. V., *Phys. Rev. A* 64, 063202, 2001. The uncertainties increase with cluster size, from ±2.8% (Na_7) to ±12.4% (Na_{93}).
40. Hohm, U., Loose, A., Maroulis, G., and Xenides, D., *Phys. Rev. A* 61, 053202, 2000. The uncertainty is ±0.14 Å^3. <https://doi.org/10.1103/PhysRevA.61.053202]

IONIZATION ENERGIES OF ATOMS AND ATOMIC IONS

The ionization energies (often called ionization potentials) of neutral and partially ionized atoms are listed in these tables. Data were obtained from the compilations cited below, supplemented by results from the recent research literature. Values for the first and second ionization energies come from Ref. 6. All values are given in electron volts (eV).

Following the traditional spectroscopic notation, columns are headed I, II, III, etc., up to XXX, where I indicates the neutral atom, II the singly ionized atom, III the doubly ionized atom, etc. The contents of the tables are as follows.

Table	Atomic Numbers, *Z*	Ionization States
1	1 – 104	I (Neutral) to VIII
2	9 – 42	IX to XVI
3	17 – 42	XVII to XXIV
4	25 – 42	XXV to XXX

References

1. Moore, C. E., *Ionization Potentials and Ionization Limits Derived from the Analysis of Optical Spectra,* Natl. Stand. Ref. Data Ser. — Natl. Bur. Stand. (U.S.) No. 34, 1970.
2. Martin, W. C., Zalubas, R., and Hagan, L., *Atomic Energy Levels — The Rare Earth Elements,* Natl. Stand. Ref. Data Ser. — Natl. Bur. Stand. (U.S.), No. 60, 1978.
3. Sugar, J. and Corliss, C., *Atomic Energy Levels of the Iron Period Elements: Potassium through Nickel, J. Phys. Chem. Ref. Data,* Vol. 14, Suppl. 2, 1985.
4. References to papers in *J. Phys. Chem. Ref. Data,* in the period 1973–91 covering other elements may be found in the cumulative index to that journal.
5. Martin, W. C., and Wiese, W. L., in *Atomic, Molecular, and Optical Physics Handbook,* Drake, G. W. F., Ed., AIP Press, New York, 1996.
6. Sansonetti, J. E., Martin, W. C., and Young, S. L., *Handbook of Basic Atomic Spectroscopic Data* (Version 1.1), NIST Physical Data Web site <http://physics.nist.gov/PhysRefData/Handbook> (October 2004); *J. Phys. Chem. Ref. Data,* 34, 1559, 2005.

TABLE 1. Ionization Energies in eV of Atoms *Z* = 1 to 104 in Ionization States I (Neutral) to VIII

Z	Formula	I	II	III	IV	V	VI	VII	VIII
1	H	13.598443							
2	He	24.587387	54.417760						
3	Li	5.391719	75.6400	122.45429					
4	Be	9.32270	18.21114	153.89661	217.71865				
5	B	8.29802	25.1548	37.93064	259.37521	340.22580			
6	C	11.26030	24.3833	47.8878	64.4939	392.087	489.99334		
7	N	14.5341	29.6013	47.44924	77.4735	97.8902	552.0718	667.046	
8	O	13.61805	35.1211	54.9355	77.41353	113.8990	138.1197	739.29	871.4101
9	F	17.4228	34.9708	62.7084	87.1398	114.2428	157.1651	185.186	953.9112
10	Ne	21.56454	40.96296	63.45	97.12	126.21	157.93	207.2759	239.0989
11	Na	5.139076	47.2864	71.6200	98.91	138.40	172.18	208.50	264.25
12	Mg	7.646235	15.03527	80.1437	109.2655	141.27	186.76	225.02	265.96
13	Al	5.985768	18.82855	28.44765	119.992	153.825	190.49	241.76	284.66
14	Si	8.15168	16.34584	33.49302	45.14181	166.767	205.27	246.5	303.54
15	P	10.48669	19.7695	30.2027	51.4439	65.0251	220.421	263.57	309.60
16	S	10.36001	23.33788	34.79	47.222	72.5945	88.0530	280.948	328.75
17	Cl	12.96763	23.8136	39.61	53.4652	67.8	97.03	114.1958	348.28
18	Ar	15.759610	27.62966	40.74	59.81	75.02	91.009	124.323	143.460
19	K	4.3406633	31.63	45.806	60.91	82.66	99.4	117.56	154.88
20	Ca	6.11316	11.87172	50.9131	67.27	84.50	108.78	127.2	147.24
21	Sc	6.56149	12.79977	24.75666	73.4894	91.65	110.68	138.0	158.1
22	Ti	6.82812	13.5755	27.4917	43.2672	99.30	119.53	140.8	170.4
23	V	6.74619	14.618	29.311	46.709	65.2817	128.13	150.6	173.4
24	Cr	6.76651	16.4857	30.96	49.16	69.46	90.6349	160.18	184.7
25	Mn	7.43402	15.6400	33.668	51.2	72.4	95.6	119.203	194.5
26	Fe	7.9024	16.1877	30.652	54.8	75.0	99.1	124.98	151.06
27	Co	7.88101	17.084	33.50	51.3	79.5	102.0	128.9	157.8
28	Ni	7.6398	18.16884	35.19	54.9	76.06	108	133	162
29	Cu	7.72638	20.2924	36.841	57.38	79.8	103	139	166
30	Zn	9.394199	17.96439	39.723	59.4	82.6	108	134	174
31	Ga	5.999301	20.51515	30.7258	63.241	86.01	112.7	140.9	169.9
32	Ge	7.89943	15.93461	34.2241	45.7131	93.5			
33	As	9.7886	18.5892	28.351	50.13	62.63	127.6		
34	Se	9.75239	21.19	30.8204	42.9450	68.3	81.7	155.4	
35	Br	11.8138	21.591	36	47.3	59.7	88.6	103.0	192.8
36	Kr	13.99961	24.35984	36.950	52.5	64.7	78.5	111.0	125.802
37	Rb	4.177128	27.2895	40	52.6	71.0	84.4	99.2	136

Z	Formula	I	II	III	IV	V	VI	VII	VIII
38	Sr	5.69485	11.0301	42.89	57	71.6	90.8	106	122.3
39	Y	6.2173	12.224	20.52	60.597	77.0	93.0	116	129
40	Zr	6.63390	13.1	22.99	34.34	80.348			
41	Nb	6.75885	14.0	25.04	38.3	50.55	102.057	125	
42	Mo	7.09243	16.16	27.13	46.4	54.49	68.8276	125.664	143.6
43	Tc	7.28	15.26	29.54					
44	Ru	7.36050	16.76	28.47					
45	Rh	7.45890	18.08	31.06					
46	Pd	8.3369	19.43	32.93					
47	Ag	7.57623	21.47746	34.83					
48	Cd	8.99382	16.90831	37.48					
49	In	5.78636	18.8703	28.03	54				
50	Sn	7.34392	14.6322	30.50260	40.73502	72.28			
51	Sb	8.60839	16.63	25.3	44.2	56	108		
52	Te	9.0096	18.6	27.96	37.41	58.75	70.7	137	
53	I	10.45126	19.1313	33					
54	Xe	12.12984	20.9750	32.1230					
55	Cs	3.893905	23.15744						
56	Ba	5.211664	10.00383						
57	La	5.5769	11.059	19.1773	49.95	61.6			
58	Ce	5.5387	10.85	20.198	36.758	65.55	77.6		
59	Pr	5.473	10.55	21.624	38.98	57.53			
60	Nd	5.5250	10.72	22.1	40.4				
61	Pm	5.582	10.90	22.3	41.1				
62	Sm	5.6437	11.07	23.4	41.4				
63	Eu	5.67038	11.25	24.92	42.7				
64	Gd	6.14980	12.09	20.63	44.0				
65	Tb	5.8638	11.52	21.91	39.79				
66	Dy	5.9389	11.67	22.8	41.47				
67	Ho	6.0215	11.80	22.84	42.5				
68	Er	6.1077	11.93	22.74	42.7				
69	Tm	6.18431	12.05	23.68	42.7				
70	Yb	6.25416	12.176	25.05	43.56				
71	Lu	5.42586	13.9	20.9594	45.25	66.8			
72	Hf	6.82507	15	23.3	33.33				
73	Ta	7.54957							
74	W	7.86403	16.1						
75	Re	7.83352							
76	Os	8.43823							
77	Ir	8.96702							
78	Pt	8.9588	18.563						
79	Au	9.22553	20.20						
80	Hg	10.4375	18.7568	34.2					
81	Tl	6.108194	20.4283	29.83					
82	Pb	7.41663	15.03248	31.9373	42.32	68.8			
83	Bi	7.2855	16.703	25.56	45.3	56.0	88.3		
84	Po	8.414							
85	At								
86	Rn	10.7485							
87	Fr	4.072741							
88	Ra	5.278423	10.14715						
89	Ac	5.17	11.75						
90	Th	6.3067	11.9	20.0	28.8				
91	Pa	5.89							
92	U	6.1941	10.6						
93	Np	6.2657							
94	Pu	6.0260	11.2						
95	Am	5.9738							
96	Cm	5.9914							
97	Bk	6.1979							

Z	Formula	I	II	III	IV	V	VI	VII	VIII
98	Cf	6.2817	11.8						
99	Es	6.42	12.0						
100	Fm	6.50							
101	Md	6.58							
102	No	6.65							
103	Lr	4.9							
104	Rf	6.0							

TABLE 2. Ionization Energies in eV for Atoms *Z* = 9 to 42 in Ionization States IX to XVI

Z	Formula	IX	X	XI	XII	XIII	XIV	XV	XVI
9	F	1103.1176							
10	Ne	1195.8286	1362.1995						
11	Na	299.864	1465.121	1648.702					
12	Mg	328.06	367.50	1761.805	1962.6650				
13	Al	330.13	398.75	442.00	2085.98	2304.1410			
14	Si	351.12	401.37	476.36	523.42	2437.63	2673.182		
15	P	372.13	424.4	479.46	560.8	611.74	2816.91	3069.842	
16	S	379.55	447.5	504.8	564.44	652.2	707.01	3223.78	3494.1892
17	Cl	400.06	455.63	529.28	591.99	656.71	749.76	809.40	3658.521
18	Ar	422.45	478.69	538.96	618.26	686.10	755.74	854.77	918.03
19	K	175.8174	503.8	564.7	629.4	714.6	786.6	861.1	968
20	Ca	188.54	211.275	591.9	657.2	726.6	817.6	894.5	974
21	Sc	180.03	225.18	249.798	687.36	756.7	830.8	927.5	1009
22	Ti	192.1	215.92	265.07	291.500	787.84	863.1	941.9	1044
23	V	205.8	230.5	255.7	308.1	336.277	896.0	976	1060
24	Cr	209.3	244.4	270.8	298.0	354.8	384.168	1010.6	1097
25	Mn	221.8	248.3	286.0	314.4	343.6	403.0	435.163	1134.7
26	Fe	233.6	262.1	290.2	330.8	361.0	392.2	457	489.256
27	Co	186.13	275.4	305	336	379	411	444	511.96
28	Ni	193	224.6	321.0	352	384	430	464	499
29	Cu	199	232	265.3	369	401	435	484	520
30	Zn	203	238	274	310.8	419.7	454	490	542
31	Ga	210.8	244.0	280.7	319.2	357.2	471.2	508.8	548.3
36	Kr	230.85	268.2	308	350	391	447	492	541
37	Rb	150	277.1						
38	Sr	162	177	324.1					
39	Y	146.2	191	206	374.0				
42	Mo	164.12	186.4	209.3	230.28	279.1	302.60	544.0	570

TABLE 3. Ionization Energies in eV for Atoms *Z* = 17 to 42 in Ionization States XVII to XXIV

Z	Formula	XVII	XVIII	XIX	XX	XXI	XXII	XXIII	XXIV
17	Cl	3946.2960							
18	Ar	4120.8857	4426.2296						
19	K	1033.4	4610.8	4934.046					
20	Ca	1087	1157.8	5128.8	5469.864				
21	Sc	1094	1213	1287.97	5674.8	6033.712			
22	Ti	1131	1221	1346	1425.4	6249.0	6625.82		
23	V	1168	1260	1355	1486	1569.6	6851.3	7246.12	
24	Cr	1185	1299	1396	1496	1634	1721.4	7481.7	7894.81
25	Mn	1224	1317	1437	1539	1644	1788	1879.9	8140.6
26	Fe	1266	1358	1456	1582	1689	1799	1950	2023
27	Co	546.58	1397.2	1504.6	1603	1735	1846	1962	2119
28	Ni	571.08	607.06	1541	1648	1756	1894	2011	2131
29	Cu	557	633	670.588	1697	1804	1916	2060	2182
30	Zn	579	619	698	738	1856			
36	Kr	592	641	786	833	884	937	998	1051
42	Mo	636	702	767	833	902	968	1020	1082

TABLE 4. Ionization Energies in eV for Atoms Z = 25 to 42 in Ionization States XXV to XXX

Z	Formula	XXV	XXVI	XXVII	XXVIII	XXIX	XXX
25	Mn	8571.94					
26	Fe	8828	9277.69				
27	Co	2219.0	9544.1	10012.12			
28	Ni	2295	2399.2	10288.8	10775.40		
29	Cu	2308	2478	2587.5	11062.38	11567.617	
36	Kr	1151	1205.3	2928	3070	3227	3381
42	Mo	1263	1323	1387	1449	1535	1601

IONIZATION ENERGIES OF GAS-PHASE MOLECULES

Sharon G. Lias

This table presents values for the first ionization energies (IP) of approximately 1000 molecules and atoms. Substances are listed by name. Values designated as "approx." are considered not to be well established. Data appearing in the 1988 reference were updated in 1996 for inclusion in the database of ionization energies available at the Internet site of the Standard Reference Data program of the National Institute of Standards and Technology (http://webbook.nist.gov). The list appearing here includes these updates. Column definitions for the table are as follows.

Column heading	Definition
Name	Name of gas-phase molecule
Mol. form.	Molecular formula of gas-phase molecule
IP	First ionization energy, in eV
Uncert.	Uncertainty in first ionization energy, in eV
$\Delta_f H_{ion}$	Enthalpy of formation of the ion at 298 K, in kJ mol^{-1}

The list also includes values for enthalpies of formation of the ions at 298 K, $\Delta_f H_{ion}$, given according to the ion convention used in mass spectrometry; to convert these values to the electron convention used in thermodynamics analysis, add 6 kJ mol^{-1}. Details on the calculation of $\Delta_f H_{ion}$, as well as data for a much larger number of molecules, may be found in the reference and on the NIST Internet site.

Reference

Lias, S. G., Bartmess, J. E., Liebman, J. F., Holmes, J. L., Levin, R. D., and Mallard, W. G., *Gas-Phase Ion and Neutral Thermochemistry, J. Phys. Chem. Ref. Data*, Vol. 17, Suppl. No. 1, 1988. Additional details at <webbook.nist.gov/chemistry/ion/>.

Ionization Energies of Gas-Phase Molecules

Name	Mol. form.	IP/eV	Uncert.	$\Delta_f H_{ion}$/kJ mol^{-1}
Acenaphthene	$C_{12}H_{10}$	7.75	± 0.07	903
Acenaphthylene	$C_{12}H_8$	8.22	approx.	1053
Acetaldehyde	C_2H_4O	10.229	± 0.0007	821
Acetamide	C_2H_5NO	9.65	± 0.03	693
Acetic acid	$C_2H_4O_2$	10.65	± 0.02	595
Acetic anhydride	$C_4H_6O_3$	10.0	approx.	398
Acetone	C_3H_6O	9.703	± 0.006	719
Acetonitrile	C_2H_3N	12.20	± 0.01	1253
Acetophenone	C_8H_8O	9.29	± 0.03	810
Acetyl chloride	C_2H_3ClO	10.82	± 0.04	801
Acetylene	C_2H_2	11.400	± 0.002	1328
Acetyl fluoride	C_2H_3FO	11.5	approx.	667
Acrolein	C_3H_4O	10.103	± 0.006	900
Acrylamide	C_3H_5NO	9.5	approx.	720
Acrylic acid	$C_3H_4O_2$	10.60		701
Acrylonitrile	C_3H_3N	10.91	± 0.01	1237
Actinium	Ac	5.17		905
Allene	C_3H_4	9.692	± 0.004	1126
Allyl alcohol	C_3H_6O	9.67	± 0.05	808
Allylamine	C_3H_7N	8.76	approx.	891
Aluminum	Al	5.98577		905
Aluminum bromide	$AlBr_3$	10.4	approx.	593
Aluminum chloride	$AlCl_3$	12.01	approx.	573
Aluminum fluoride	AlF_3	≤15.45		≤282
Aluminum iodide	AlI_3	9.1	approx.	673
Aluminum monobromide	AlBr	9.3	approx.	913
Aluminum monochloride	AlCl	9.4		855
Aluminum monofluoride	AlF	9.73	± 0.01	673
Aluminum monoiodide	AlI	9.3	± 0.3	965
Americium	Am	5.9738	± 0.0002	860
Amidogen	H_2N	11.14	± 0.01	1264
3-Amino-1-propanol	C_3H_9NO	9.0	approx.	651
Ammonia	H_3N	10.070	± 0.020	925
Aniline	C_6H_7N	7.720	± 0.002	832
Anisole	C_7H_8O	8.22	± 0.03	725
Anthracene	$C_{14}H_{10}$	7.439	± 0.006	948

Name	Mol. form.	IP/eV	Uncert.	$\Delta_f H_{ion}$/kJ mol^{-1}
Antimony	Sb	8.64		1096
Antimony(III) chloride	Cl_3Sb	≤10.7	approx.	≤719
Argon	Ar	15.75962		1521
Arsenic	As	9.8152		1250
Arsenic(III) chloride	$AsCl_3$	10.55	approx.	754
Arsenic(III) fluoride	AsF_3	12.84	approx.	452
Arsine	AsH_3	9.89	approx.	1021
trans-Azoxybenzene	$C_{12}H_{10}N_2O$	8.1	approx.	1123
Azulene	$C_{10}H_8$	7.38	± 0.05	1001
Barium	Ba	5.21170		683
Barium oxide	BaO	6.91	± 0.06	543
Benzaldehyde	C_7H_6O	9.49	± 0.02	878
Benzamide	C_7H_7NO	9.25	approx.	792
Benzene	C_6H_6	9.24378	± 0.00007	975
Benzeneacetic acid	$C_8H_8O_2$	8.26	approx.	479
1,2-Benzenediamine	$C_6H_8N_2$	7.2	approx.	787
1,3-Benzenediamine	$C_6H_8N_2$	7.14	approx.	777
1,4-Benzenediamine	$C_6H_8N_2$	6.87	approx.	759
Benzenethiol	C_6H_6S	8.32	approx.	915
Benzoic acid	$C_7H_6O_2$	9.3	approx.	604
Benzonitrile	C_7H_5N	9.70	± 0.01	1154
Benzophenone	$C_{13}H_{10}O$	9.08	± 0.05	926
p-Benzoquinone	$C_6H_4O_2$	10.01	± 0.06	844
Benzoyl chloride	C_7H_5ClO	9.53	approx.	815
Benzyl alcohol	C_7H_8O	8.3	approx.	701
Benzylamine	C_7H_9N	8.64	approx.	917
Berkelium	Bk	6.23		911
Beryllium	Be	9.32263		1224
Beryllium oxide	BeO	10.1	approx.	1111
Biphenyl	$C_{12}H_{10}$	8.23	± 0.10	977
Bismuth	Bi	7.2855		908
Bismuth trichloride	$BiCl_3$	10.4	approx.	736
Borane(1)	BH	9.77	approx.	1385
Borane(3)	BH_3	12.026	± 0.024	1261
Borane carbonyl	CH_3BO	11.14	± 0.02	962
Boron	B	8.29803		1363
Boron dioxide	BO_2	13.5	approx.	1001
Boron oxide	B_2O_3	13.5	± 0.15	460
Boron tribromide	BBr_3	10.51	approx.	809
Boron trichloride	BCl_3	11.60	± 0.02	718
Boron trifluoride	BF_3	15.7	± 0.3	365
Boron triiodide	BI_3	9.25	approx.	964
Bromine	Br_2	10.516	± 0.005	1046
Bromine (atomic)	Br	11.81381		1252
Bromine chloride	BrCl	11.01		1079
Bromine fluoride	BrF	11.86		1086
Bromine monoxide	BrO	10.46	± 0.02	1135
Bromine pentafluoride	BrF_5	13.172	± 0.002	840
Bromoacetylene	C_2HBr	10.31	± 0.02	1242
Bromobenzene	C_6H_5Br	9.00	± 0.02	971
1-Bromobutane	C_4H_9Br	10.12	approx.	869
2-Bromobutane, (±)-	C_4H_9Br	10.01	± 0.02	845
Bromochlorodifluoromethane	$CBrClF_2$	11.21	approx.	642
Bromochloromethane	CH_2BrCl	10.77	± 0.01	1085
2-Bromo-2-chloro-1,1,1-trifluoroethane	$C_2HBrClF_3$	11.0	approx.	363
Bromodichloromethane	$CHBrCl_2$	10.6		973
Bromoethane	C_2H_5Br	10.29	± 0.01	931
Bromoethene	C_2H_3Br	9.83	± 0.02	1028
Bromomethane	CH_3Br	10.541	± 0.003	979
1-Bromo-2-methylpropane	C_4H_9Br	10.09	± 0.02	861

Name	Mol. form.	IP/eV	Uncert.	$\Delta_f H_{ion}$/kJ mol^{-1}
2-Bromo-2-methylpropane	C_4H_9Br	9.92	± 0.03	823
1-Bromonaphthalene	$C_{10}H_7Br$	8.08	± 0.03	955
Bromopentafluorobenzene	C_6BrF_5	9.67	approx.	222
1-Bromopentane	$C_5H_{11}Br$	10.10	± 0.01	846
1-Bromopropane	C_3H_7Br	10.18	± 0.01	898
2-Bromopropane	C_3H_7Br	10.10	± 0.03	877
3-Bromopropene	C_3H_5Br	9.96	approx.	1008
Bromosilane	BrH_3Si	10.6		943
4-Bromotoluene	C_7H_7Br	8.67	± 0.02	908
Bromotrichloromethane	$CBrCl_3$	10.6	approx.	980
Bromotrifluoromethane	$CBrF_3$	11.40	approx.	451
1,2-Butadiene	C_4H_6	9.03	approx.	1034
1,3-Butadiene	C_4H_6	9.082	± 0.004	986
Butanal	C_4H_8O	9.84	± 0.02	742
Butane	C_4H_{10}	10.53	± 0.10	890
Butanenitrile	C_4H_7N	11.2	approx.	1110
1-Butanethiol	$C_4H_{10}S$	9.14	± 0.01	794
2-Butanethiol	$C_4H_{10}S$	9.10	approx.	781
Butanoic acid	$C_4H_8O_2$	10.17	± 0.05	509
1-Butanol	$C_4H_{10}O$	9.99	± 0.05	689
2-Butanol	$C_4H_{10}O$	9.88	± 0.03	658
2-Butanone	C_4H_8O	9.52	± 0.04	678
trans-2-Butenal	C_4H_6O	9.73	± 0.01	835
1-Butene	C_4H_8	9.55	± 0.06	921
cis-2-Butene	C_4H_8	9.11	± 0.01	871
trans-2-Butene	C_4H_8	9.10	± 0.01	866
cis-2-Butenoic acid	$C_4H_6O_2$	10.08	approx.	625
trans-2-Butenoic acid	$C_4H_6O_2$	9.9	approx.	604
1-Buten-3-yne	C_4H_4	9.58	± 0.02	1230
Butyl acetate	$C_6H_{12}O_2$	9.92	approx.	471
sec-Butyl acetate	$C_6H_{12}O_2$	9.90		453
Butylamine	$C_4H_{11}N$	8.7	± 0.1	748
sec-Butylamine	$C_4H_{11}N$	8.46	± 0.1	711
tert-Butylamine	$C_4H_{11}N$	8.46	± 0.1	695
Butylbenzene	$C_{10}H_{14}$	8.69	± 0.02	826
sec-Butylbenzene, (±)-	$C_{10}H_{14}$	8.68	± 0.02	820
tert-Butylbenzene	$C_{10}H_{14}$	8.68	± 0.05	816
Butylcyclohexane	$C_{10}H_{20}$	9.41	approx.	695
Butylcyclopentane	C_9H_{18}	9.95	approx.	793
Butyl ethyl ether	$C_6H_{14}O$	9.36	approx.	610
Butyl formate	$C_5H_{10}O_2$	10.52	± 0.02	584
1-*tert*-Butyl-4-methylbenzene	$C_{11}H_{16}$	8.12	approx.	730
Butyl methyl ether	$C_5H_{12}O$	9.4	approx.	648
tert-Butyl methyl sulfide	$C_5H_{12}S$	8.38	approx.	687
4-*tert*-Butylphenol	$C_{10}H_{14}O$	7.8	approx.	552
1-Butyne	C_4H_6	10.19	± 0.02	1148
2-Butyne	C_4H_6	9.59	± 0.03	1071
Cadmium	Cd	8.99367		980
Calcium	Ca	6.11316		768
Calcium monochloride	CaCl	5.86	± 0.07	462
Calcium oxide	CaO	6.66	± 0.18	668
Californium	Cf	6.30		805
Camphor, (+)	$C_{10}H_{16}O$	8.76	approx.	577
Caprolactam	$C_6H_{11}NO$	9.07	approx.	629
Carbazole	$C_{12}H_9N$	7.57	approx.	961
Carbon	C	11.26030		1803
Carbon dioxide	CO_2	13.773	± 0.002	935
Carbon disulfide	CS_2	10.0685	± 0.0020	1089
Carbon monosulfide	CS	11.33	± 0.01	1361
Carbon monoxide	CO	14.014	± 0.0003	1242

Name	Mol. form.	IP/eV	Uncert.	$\Delta_f H_{ion}$/kJ mol^{-1}
Carbon oxyselenide	$COSe$	10.36	± 0.01	929
Carbon oxysulfide	COS	11.18	± 0.01	936
Carbonyl chloride	CCl_2O	11.5	approx.	888
Carbonyl fluoride	CF_2O	13.035	± 0.030	617
Cerium	Ce	5.5387		957
Cesium	Cs	3.89390		452
Cesium chloride	$ClCs$	7.84	approx.	510
Cesium fluoride	CsF	8.80	approx.	489
Cesium sodium	$CsNa$	4.05	approx.	535
Chlorine	Cl_2	11.480	± 0.005	1108
Chlorine (atomic)	Cl	12.96764		1373
Chlorine difluoride	ClF_2	12.77	approx.	1128
Chlorine dioxide	ClO_2	10.33	± 0.02	1093
Chlorine fluoride	ClF	12.66	± 0.01	1171
Chlorine monoxide	Cl_2O	10.94		1135
Chlorine oxide (ClO)	ClO	10.95		1159
Chlorine trifluoride	ClF_3	12.65	approx.	1057
Chloroacetaldehyde	C_2H_3ClO	10.48	approx.	815
Chloroacetic acid	$C_2H_3ClO_2$	10.7	approx.	597
Chloroacetyl chloride	$C_2H_2Cl_2O$	≤10.3	approx.	815
Chloroacetylene	C_2HCl	10.58	± 0.02	1276
2-Chloroaniline	C_6H_6ClN	8.50	approx.	883
3-Chloroaniline	C_6H_6ClN	8.09	approx.	835
4-Chloroaniline	C_6H_6ClN	≤8.18	approx.	≤844
Chlorobenzene	C_6H_5Cl	9.07	± 0.02	930
1-Chlorobutane	C_4H_9Cl	10.67	± 0.03	875
2-Chlorobutane	C_4H_9Cl	10.53		857
Chlorodibromomethane	$CHBr_2Cl$	10.59	± 0.01	1030
1-Chloro-1,1-difluoroethane	$C_2H_3ClF_2$	11.98	approx.	626
1-Chloro-2,2-difluoroethene	C_2HClF_2	9.80	± 0.04	628
Chlorodifluoromethane	$CHClF_2$	12.2	approx.	693
Chloroethane	C_2H_5Cl	10.98	± 0.02	947
2-Chloroethanol	C_2H_5ClO	10.5	approx.	756
Chloroethene	C_2H_3Cl	9.99	± 0.02	985
Chlorofluoromethane	CH_2ClF	11.71	± 0.01	870
Chloromethane	CH_3Cl	11.22	± 0.01	1001
(Chloromethyl)benzene	C_7H_7Cl	9.10	± 0.02	897
Chloromethylene	$CHCl$	9.84		1247
Chloromethylidyne	CCl	8.9	approx.	1244
1-Chloro-2-methylpropane	C_4H_9Cl	10.73	± 0.07	877
2-Chloro-2-methylpropane	C_4H_9Cl	10.61	approx.	842
1-Chloronaphthalene	$C_{10}H_7Cl$	8.13	approx.	906
1-Chloro-3-nitrobenzene	$C_6H_4ClNO_2$	9.92	approx.	995
1-Chloro-4-nitrobenzene	$C_6H_4ClNO_2$	9.96	approx.	999
Chloropentafluorobenzene	C_6ClF_5	9.72	approx.	126
Chloropentafluoroethane	C_2ClF_5	12.6	approx.	99
3-Chlorophenol	C_6H_5ClO	8.655	± 0.001	680
4-Chlorophenol	C_6H_5ClO	≤8.69	approx.	≤692
1-Chloropropane	C_3H_7Cl	10.81	± 0.01	911
2-Chloropropane	C_3H_7Cl	10.79	± 0.02	896
3-Chloropropene	C_3H_5Cl	10.04	± 0.01	965
Chlorosilane	ClH_3Si	11.4		899
2-Chlorotoluene	C_7H_7Cl	8.7	approx.	856
3-Chlorotoluene	C_7H_7Cl	8.83	approx.	869
4-Chlorotoluene	C_7H_7Cl	8.69	approx.	855
Chlorotrifluoroethene	C_2ClF_3	9.81	± 0.03	373
Chlorotrifluoromethane	$CClF_3$	12.6	± 0.2	505
Chromium	Cr	6.76664		1050
Chromium(VI) dichloride dioxide	Cl_2CrO_2	11.6		580
Chrysene	$C_{18}H_{12}$	7.60	± 0.01	1017

Name	Mol. form.	IP/eV	Uncert.	$\Delta_f H_{ion}$/kJ mol^{-1}
Cobalt	Co	7.8810		1187
Copper	Cu	7.72638		1084
Copper(I) fluoride	CuF	10.15	± 0.02	984
Coronene	$C_{24}H_{12}$	7.29	± 0.01	1026
o-Cresol	C_7H_8O	8.24	approx.	670
m-Cresol	C_7H_8O	8.29	± 0.07	668
p-Cresol	C_7H_8O	8.3	approx.	675
Curium	Cm	6.02		966
Cyanamide	CH_2N_2	10.4	approx.	1137
Cyanate	CNO	11.76	± 0.01	1290
Cyanide	CN	13.5984		1748
Cyanoacetylene	C_3HN	11.64	± 0.01	1475
Cyanogen	C_2N_2	13.37	± 0.01	1597
Cyanogen chloride	CClN	12.34	± 0.01	1329
Cyanogen fluoride	CFN	13.34	± 0.02	1325
Cyclobutane	C_4H_8	9.82	approx.	976
Cyclobutanone	C_4H_6O	9.35	approx.	815
Cyclobutene	C_4H_6	9.43	± 0.02	1067
Cycloheptane	C_7H_{14}	9.97		844
Cyclohexane	C_6H_{12}	9.86	± 0.03	828
Cyclohexanol	$C_6H_{12}O$	9.75	approx.	651
Cyclohexanone	$C_6H_{10}O$	9.14	± 0.01	656
Cyclohexene	C_6H_{10}	8.945	± 0.01	859
Cyclohexylamine	$C_6H_{13}N$	8.86	approx.	750
Cyclohexylcyclohexane	$C_{12}H_{22}$	9.41	approx.	690
Cyclooctane	C_8H_{16}	9.75	± 0.05	816
1,3-Cyclopentadiene	C_5H_6	8.55	± 0.02	955
Cyclopentane	C_5H_{10}	10.33	approx.	918
Cyclopentanol	$C_5H_{10}O$	9.72	approx.	695
Cyclopentanone	C_5H_8O	9.26	± 0.01	701
Cyclopentene	C_5H_8	9.01	± 0.01	905
Cyclopropane	C_3H_6	9.86		1005
Cyclopropanecarbonitrile	C_4H_5N	10.25	approx.	1173
Cyclopropanone	C_3H_4O	9.1	approx.	895
Cyclopropene	C_3H_4	9.67	± 0.01	1209
Cyclopropylamine	C_3H_7N	8.8	approx.	926
Cyclopropylbenzene	C_9H_{10}	8.35	approx.	956
Cyclopropyl methyl ketone	C_5H_8O	≤9.46	approx.	796
cis-Decahydronaphthalene	$C_{10}H_{18}$	9.36	± 0.04	734
trans-Decahydronaphthalene	$C_{10}H_{18}$	9.34	± 0.04	720
Decane	$C_{10}H_{22}$	9.65	approx.	682
1-Decene	$C_{10}H_{20}$	9.42	± 0.05	786
Diazomethane	CH_2N_2	8.999	± 0.001	1098
Diborane	B_2H_6	11.38	± 0.05	1134
1,4-Dibromobutane	$C_4H_8Br_2$	10.15	approx.	879
Dibromodifluoromethane	CBr_2F_2	11.03	± 0.04	683
1,2-Dibromoethane	$C_2H_4Br_2$	10.35	± 0.04	961
Dibromomethane	CH_2Br_2	10.50	approx.	1013
1,2-Dibromopropane	$C_3H_6Br_2$	10.1	approx.	903
1,3-Dibromopropane	$C_3H_6Br_2$	≤10.2	approx.	≤919
1,2-Dibromotetrafluoroethane	$C_2Br_2F_4$	11.1	approx.	280
Dibutylamine	$C_8H_{19}N$	7.69	approx.	586
Dibutyl ether	$C_8H_{18}O$	9.28	approx.	≤560
Di-*sec*-butyl ether	$C_8H_{18}O$	9.11	approx.	511
Di-*tert*-butyl ether	$C_8H_{18}O$	8.88	± 0.07	493
Dibutyl sulfide	$C_8H_{18}S$	8.2	approx.	624
Di-*tert*-butyl sulfide	$C_8H_{18}S$	8.0	approx.	583
Dicarbon	C_2	11.4	approx.	2000
Dichloroacetyl chloride	C_2HCl_3O	10.9	approx.	809
Dichloroacetylene	C_2Cl_2	9.9		1165

Name	Mol. form.	IP/eV	Uncert.	$\Delta_f H_{ion}$/kJ mol^{-1}
o-Dichlorobenzene	$C_6H_4Cl_2$	9.06	± 0.02	907
m-Dichlorobenzene	$C_6H_4Cl_2$	9.10	± 0.02	906
p-Dichlorobenzene	$C_6H_4Cl_2$	8.92	± 0.02	885
Dichlorodifluoromethane	CCl_2F_2	12.05	± 0.24	685
Dichlorodimethylsilane	$C_2H_6Cl_2Si$	10.7	approx.	576
1,1-Dichloroethane	$C_2H_4Cl_2$	11.04	± 0.02	935
1,2-Dichloroethane	$C_2H_4Cl_2$	11.04	± 0.02	931
1,1-Dichloroethene	$C_2H_2Cl_2$	9.81	± 0.04	949
cis-1,2-Dichloroethene	$C_2H_2Cl_2$	9.66	± 0.01	936
trans-1,2-Dichloroethene	$C_2H_2Cl_2$	9.64	± 0.02	934
Dichlorofluoromethane	$CHCl_2F$	11.5	approx.	829
Dichloromethane	CH_2Cl_2	11.32	± 0.01	996
Dichloromethylene	CCl_2	9.27	approx.	1058
1,2-Dichloropropane, (±)-	$C_3H_6Cl_2$	10.8	± 0.1	886
1,3-Dichloropropane	$C_3H_6Cl_2$	10.89	± 0.04	892
Dichlorosilane	Cl_2H_2Si	11.4		765
Dichlorosilylene	Cl_2Si	10.93	approx.	887
1,2-Dichloro-1,1,2,2-tetrafluoroethane	$C_2Cl_2F_4$	12.2		252
Dicyclopropyl ketone	$C_7H_{10}O$	9.1	approx.	1041
1,1-Diethoxyethane	$C_6H_{14}O_2$	9.2	approx.	434
Diethylamine	$C_4H_{11}N$	7.85	± 0.1	684
o-Diethylbenzene	$C_{10}H_{14}$	≤8.51	approx.	≤804
m-Diethylbenzene	$C_{10}H_{14}$	8.49	approx.	798
p-Diethylbenzene	$C_{10}H_{14}$	8.40	approx.	790
Diethyl disulfide	$C_4H_{10}S_2$	8.27	approx.	724
Diethylene glycol dimethyl ether	$C_6H_{14}O_3$	≤9.8		≤448
Diethyl ether	$C_4H_{10}O$	9.51	± 0.03	666
Diethyl oxalate	$C_6H_{10}O_4$	9.8	approx.	205
Diethyl sulfide	$C_4H_{10}S$	8.43	approx.	730
Difluoramine	F_2HN	11.53	approx.	1046
Difluoroamidogen	F_2N	11.628	± 0.01	1155
o-Difluorobenzene	$C_6H_4F_2$	9.29	± 0.01	602
m-Difluorobenzene	$C_6H_4F_2$	9.33	± 0.01	591
p-Difluorobenzene	$C_6H_4F_2$	9.1589	± 0.0003	577
Difluoroborane(2)	BF_2	9.4	approx.	317
trans-Difluorodiazine	F_2N_2	12.8	approx.	1315
1,1-Difluoroethane	$C_2H_4F_2$	11.87	approx.	643
1,1-Difluoroethene	$C_2H_2F_2$	10.29	± 0.01	650
cis-1,2-Difluoroethene	$C_2H_2F_2$	10.23	± 0.02	690
Difluoromethane	CH_2F_2	12.71		774
Difluoromethylene	CF_2	11.44	± 0.03	899
Difluorosilane	F_2H_2Si	12.2	approx.	386
Difluorosilylene	F_2Si	10.78	± 0.05	450
3,4-Dihydro-2*H*-pyran	C_5H_8O	8.35	± 0.01	681
2,5-Dihydrothiophene	C_4H_6S	8.4	approx.	898
Diiodomethane	CH_2I_2	9.46	± 0.02	1030
Diisobutylamine	$C_8H_{19}N$	7.8	approx.	574
Diisobutyl sulfide	$C_8H_{18}S$	8.34	approx.	625
Diisopropylamine	$C_6H_{15}N$	7.73	approx.	602
Diisopropyl ether	$C_6H_{14}O$	9.20	± 0.05	569
Diisopropyl sulfide	$C_6H_{14}S$	8.2	approx.	649
Diketene	$C_4H_4O_2$	9.6	approx.	736
Dilithium	Li_2	5.1127	± 0.0003	709
1,2-Dimethoxyethane	$C_4H_{10}O_2$	9.3	approx.	558
Dimethoxymethane	$C_3H_8O_2$	9.7		588
N,N-Dimethylacetamide	C_4H_9NO	8.81	± 0.03	616
Dimethylamine	C_2H_7N	8.24	± 0.08	777
N,N-Dimethylaniline	$C_8H_{11}N$	7.12	± 0.02	787
2,2-Dimethylbutane	C_6H_{14}	10.06	approx.	787
2,3-Dimethylbutane	C_6H_{14}	10.02	approx.	791

Name	Mol. form.	IP/eV	Uncert.	$\Delta_f H_{ion}$/kJ mol^{-1}
3,3-Dimethyl-2-butanone	$C_6H_{12}O$	9.12	± 0.02	589
2,3-Dimethyl-1-butene	C_6H_{12}	9.07	approx.	812
2,3-Dimethyl-2-butene	C_6H_{12}	8.27	± 0.01	729
3,3-Dimethyl-1-butyne	C_6H_{10}	9.90	± 0.04	1060
1,1-Dimethylcyclohexane	C_8H_{16}	9.42	approx.	728
cis-1,2-Dimethylcyclohexane	C_8H_{16}	<9.78	approx.	772
trans-1,2-Dimethylcyclohexane	C_8H_{16}	9.41		728
cis-1,3-Dimethylcyclohexane	C_8H_{16}	<9.98	approx.	778
trans-1,3-Dimethylcyclohexane	C_8H_{16}	9.53		743
cis-1,4-Dimethylcyclohexane	C_8H_{16}	<9.93	approx.	782
trans-1,4-Dimethylcyclohexane	C_8H_{16}	9.56	approx.	738
cis-1,2-Dimethylcyclopentane	C_7H_{14}	9.92	approx.	828
trans-1,2-Dimethylcyclopentane	C_7H_{14}	9.7	± 0.2	799
Dimethyl disulfide	$C_2H_6S_2$	7.4	approx.	690
Dimethyl ether	C_2H_6O	10.025	± 0.025	783
N,N-Dimethylformamide	C_3H_7NO	9.12	approx.	688
2,6-Dimethyl-4-heptanone	$C_9H_{18}O$	9.01	± 0.06	512
1,1-Dimethylhydrazine	$C_2H_8N_2$	7.29	± 0.05	787
Dimethyl oxalate	$C_4H_6O_4$	10.0	approx.	287
2,4-Dimethyl-3-pentanone	$C_7H_{14}O$	8.95	± 0.01	552
Dimethyl phthalate	$C_{10}H_{10}O_4$	9.64	approx.	277
2,2-Dimethylpropanal	$C_5H_{10}O$	9.51	± 0.01	675
2,3-Dimethylpyridine	C_7H_9N	8.85	approx.	922
2,4-Dimethylpyridine	C_7H_9N	8.85	approx.	918
2,5-Dimethylpyridine	C_7H_9N	≤8.80	approx.	≤916
2,6-Dimethylpyridine	C_7H_9N	8.86	± 0.03	913
3,4-Dimethylpyridine	C_7H_9N	≤9.15	approx.	≤953
3,5-Dimethylpyridine	C_7H_9N	≤9.25	approx.	≤965
Dimethyl sulfide	C_2H_6S	8.69	± 0.02	801
Dimethyl sulfoxide	C_2H_6OS	9.10	± 0.03	727
1,3-Dioxane	$C_4H_8O_2$	9.8		607
1,4-Dioxane	$C_4H_8O_2$	9.19	± 0.01	571
1,3-Dioxolane	$C_3H_6O_2$	9.9	approx.	658
Diphenylacetylene	$C_{14}H_{10}$	7.94	± 0.03	1168
Diphenylamine	$C_{12}H_{11}N$	7.16	± 0.04	908
1,2-Diphenylethane	$C_{14}H_{14}$	8.9	± 0.1	1002
Diphenyl ether	$C_{12}H_{10}O$	8.09	approx.	766
Diphenylmethane	$C_{13}H_{12}$	8.55	approx.	963
Diphosphorus	P_2	10.53		1160
Dipotassium	K_2	4.0637	± 0.0002	519
Dipropylamine	$C_6H_{15}N$	7.84	approx.	641
Dipropyl ether	$C_6H_{14}O$	9.27	approx.	602
Dipropyl sulfide	$C_6H_{14}S$	8.30	± 0.02	676
Disilane	H_6Si_2	9.74	± 0.02	1019
Disodium	Na_2	4.894	± 0.003	614
Disulfur	S_2	9.356	± 0.002	1031
Divinyl ether	C_4H_6O	8.7	approx.	827
5,7-Dodecadiyne	$C_{12}H_{18}$	8.67	approx.	1079
Dysprosium	Dy	5.9389		862
Einsteinium	Es	6.42		753
Epichlorohydrin	C_3H_5ClO	10.64	approx.	919
1,2-Epoxybutane	C_4H_8O	≤10.15	approx.	862
Erbium	Er	6.1078		907
Ethane	C_2H_6	11.56	± 0.02	1031
1,2-Ethanediamine	$C_2H_8N_2$	8.6	approx.	812
1,2-Ethanediol	$C_2H_6O_2$	10.16		593
Ethanethiol	C_2H_6S	9.31	± 0.03	851
Ethanol	C_2H_6O	10.43	± 0.05	772
Ethanolamine	C_2H_7NO	8.96		664
Ethoxybenzene	$C_8H_{10}O$	8.13	approx.	683

Name	Mol. form.	IP/eV	Uncert.	$\Delta_f H_{ion}$/kJ mol^{-1}
2-Ethoxyethanol	$C_4H_{10}O_2$	9.6	approx.	529
Ethyl acetate	$C_4H_8O_2$	10.01	± 0.05	522
Ethyl acrylate	$C_5H_8O_2$	≤10.3	approx.	617
Ethylamine	C_2H_7N	8.86	± 0.02	808
N-Ethylaniline	$C_8H_{11}N$	≤7.67	approx.	≤794
Ethylbenzene	C_8H_{10}	8.77	± 0.01	876
Ethyl benzoate	$C_9H_{10}O_2$	8.9	approx.	537
2-Ethyl-1-butene	C_6H_{12}	9.06	approx.	818
Ethylcyclohexane	C_8H_{16}	9.54	approx.	748
Ethylene	C_2H_4	10.5138	± 0.0006	1067
Ethyleneimine	C_2H_5N	9.5	approx.	1044
Ethyl formate	$C_3H_6O_2$	10.61	± 0.01	639
Ethyl isopropyl sulfide	$C_5H_{12}S$	8.35	approx.	689
Ethyl methyl ether	C_3H_8O	9.72	± 0.07	722
Ethyl methyl sulfide	C_3H_8S	8.55	approx.	765
Ethyl pentyl ether	$C_7H_{16}O$	≤9.49	approx.	≤602
4-Ethylphenol	$C_8H_{10}O$	7.84	approx.	613
Ethyl propanoate	$C_5H_{10}O_2$	10.00	approx.	500
Ethyl propyl ether	$C_5H_{12}O$	9.45	approx.	640
Ethyl propyl sulfide	$C_5H_{12}S$	8.50	approx.	716
Ethyl vinyl ether	C_4H_8O	8.98	approx.	709
Ethynyl	C_2H	11.61	approx.	1685
Europium	Eu	5.6704		723
Fermium	Fm	6.50		627
Fluoranthene	$C_{16}H_{10}$	7.9	± 0.1	1052
9*H*-Fluorene	$C_{13}H_{10}$	7.91	± 0.02	952
Fluorine	F_2	15.697	± 0.003	1515
Fluorine (atomic)	F	17.42282		1761
Fluorine monoxide	F_2O	13.11	± 0.01	1290
Fluorine oxide	FO	12.78	± 0.03	1342
Fluorine superoxide [FOO]	FO_2	12.6	approx.	1228
Fluoroacetylene	C_2HF	11.26		1195
Fluorobenzene	C_6H_5F	9.20	± 0.01	772
Fluoroborane(1)	BF	11.12	± 0.01	957
Fluoroethane	C_2H_5F	11.78	approx.	873
Fluoroethene	C_2H_3F	10.36	± 0.01	861
Fluoromethane	CH_3F	12.47	± 0.02	956
Fluoromethylene	CHF	10.06	± 0.05	1121
Fluoromethylidyne	CF	9.11	± 0.01	1134
1-Fluoro-4-nitrobenzene	$C_6H_4FNO_2$	9.90	approx.	826
1-Fluoropropane	C_3H_7F	11.3	approx.	806
2-Fluoropropane	C_3H_7F	11.08	approx.	776
3-Fluoropropene	C_3H_5F	10.11	approx.	821
Fluorosilane	FH_3Si	11.7		752
2-Fluorotoluene	C_7H_7F	8.91	± 0.01	709
3-Fluorotoluene	C_7H_7F	8.91	± 0.01	709
4-Fluorotoluene	C_7H_7F	8.79	± 0.01	701
Formaldehyde	CH_2O	10.88	± 0.01	941
Formamide	CH_3NO	10.16	± 0.06	796
Formic acid	CH_2O_2	11.33	± 0.01	715
Formyl	CHO	8.55	approx.	826
Fulminic acid	CHNO	10.83	approx.	1263
Fulvene	C_6H_6	8.36	approx.	1031
Fumaric acid	$C_4H_4O_4$	10.7	approx.	355
Furan	C_4H_4O	8.883	± 0.003	822
Furfural	$C_5H_4O_2$	9.22	± 0.01	739
Gadolinium	Gd	6.1500		991
Gallium	Ga	5.99930		851
Gallium(III) bromide	Br_3Ga	10.40		711
Gallium(III) chloride	Cl_3Ga	11.52		664

Name	Mol. form.	IP/eV	Uncert.	$\Delta_f H_{ion}$/kJ mol^{-1}
Gallium(III) iodide	GaI_3	9.40		765
Gallium monofluoride	FGa	9.6	approx.	700
Germane	GeH_4	≤10.53		≤1108
Germanium	Ge	7.900		1139
Germanium(II) chloride	Cl_2Ge	10.20	approx.	813
Germanium(IV) chloride	Cl_4Ge	11.68	± 0.05	629
Germanium(II) fluoride	F_2Ge	≤11.65	approx.	551
Germanium(IV) fluoride	F_4Ge	15.5	approx.	307
Germanium(IV) iodide	GeI_4	9.42	approx.	850
Germanium(II) oxide	GeO	11.25	± 0.01	1044
Germanium(II) sulfide	GeS	9.98	approx.	1055
Glyoxal	$C_2H_2O_2$	10.2		773
Gold	Au	9.22567		1254
Hafnium	Hf	6.82507	± 0.00004	1278
Hafnium(IV) bromide	Br_4Hf	10.9	approx.	366
Hafnium(IV) chloride	Cl_4Hf	11.7	approx.	246
Helium	He	24.58741		2372
Heptanal	$C_7H_{14}O$	9.65	approx.	668
Heptane	C_7H_{16}	9.93	± 0.10	771
1-Heptanol	$C_7H_{16}O$	9.84	approx.	614
2-Heptanol, (±)-	$C_7H_{16}O$	9.70	approx.	580
3-Heptanol, (*S*)-	$C_7H_{16}O$	9.68	approx.	578
4-Heptanol	$C_7H_{16}O$	9.61	approx.	572
2-Heptanone	$C_7H_{14}O$	9.28	± 0.10	594
3-Heptanone	$C_7H_{14}O$	9.18	± 0.08	589
4-Heptanone	$C_7H_{14}O$	9.10	± 0.06	577
1-Heptene	C_7H_{14}	9.34	± 0.10	839
trans-3-Heptene	C_7H_{14}	8.92	approx.	790
Hexaborane(10)	B_6H_{10}	9.0	approx.	965
Hexachlorobenzene	C_6Cl_6	8.98	approx.	822
Hexachloroethane	C_2Cl_6	11.1	approx.	920
1,5-Hexadiene	C_6H_{10}	9.27	± 0.05	978
Hexafluorobenzene	C_6F_6	9.89	± 0.04	8
Hexafluoroethane	C_2F_6	13.6	approx.	-30
Hexamethylbenzene	$C_{12}H_{18}$	7.85	± 0.01	670
Hexanal	$C_6H_{12}O$	9.72	± 0.05	691
Hexane	C_6H_{14}	10.13		810
Hexanoic acid	$C_6H_{12}O_2$	≤10.12		≤463
1-Hexanol	$C_6H_{14}O$	9.89	approx.	639
2-Hexanol	$C_6H_{14}O$	9.80	approx.	611
3-Hexanol	$C_6H_{14}O$	9.63	approx.	599
2-Hexanone	$C_6H_{12}O$	9.3	± 0.1	626
3-Hexanone	$C_6H_{12}O$	9.12	± 0.02	600
1-Hexene	C_6H_{12}	9.44	± 0.04	869
cis-2-Hexene	C_6H_{12}	8.97	approx.	818
trans-2-Hexene	C_6H_{12}	8.97	approx.	814
Hexylamine	$C_6H_{15}N$	8.63	approx.	699
1-Hexyne	C_6H_{10}	10.03	± 0.05	1089
Holmium	Ho	6.0216		882
Hydrazine	H_4N_2	8.1	± 0.15	877
Hydrazoic acid	HN_3	10.72	± 0.025	1328
Hydrogen (atomic)	H	13.59844		1530
Hydrogen	H_2	15.42593	± 0.00005	1488
Hydrogen bromide	BrH	11.66	± 0.03	1087
Hydrogen chloride	ClH	12.749	± 0.009	1137
Hydrogen cyanide	CHN	13.60	± 0.01	1447
Hydrogen fluoride	FH	16.044	± 0.003	1276
Hydrogen iodide	HI	10.386	± 0.001	1028
Hydrogen isocyanide	CHN	12.5	approx.	1407
Hydrogen peroxide	H_2O_2	10.58	± 0.04	885

Name	Mol. form.	IP/eV	Uncert.	$\Delta_f H_{ion}$/kJ mol^{-1}
Hydrogen selenide	H_2Se	9.892	± 0.005	984
Hydrogen sulfide	H_2S	10.457	± 0.012	989
Hydroperoxy	HO_2	11.35	± 0.01	1106
p-Hydroquinone	$C_6H_6O_2$	7.94	± 0.01	503
Hydroxyl	HO	13.0170	± 0.0002	1294
Hydroxylamine	H_3NO	10.00	approx.	923
Hypochlorous acid	ClHO	11.12	approx.	993
Hypofluorous acid	FHO	12.71	± 0.01	1130
Imidazole	$C_3H_4N_2$	8.81	approx.	997
Imidogen	HN	≤13.49	± 0.01	1678
Indan	C_9H_{10}	8.3	approx.	864
Indene	C_9H_8	8.14	± 0.01	949
Indium	In	5.78636		802
Indium(I) chloride	ClIn	9.51	approx.	842
Indium(I) fluoride	FIn	9.6	approx.	740
1*H*-Indole	C_8H_7N	7.7602	± 0.0006	908
Iodine	I_2	9.3074	± 0.0002	960
Iodine (atomic)	I	10.45126		1115
Iodine bromide	BrI	9.790	± 0.004	986
Iodine chloride	ClI	10.088	± 0.01	991
Iodine fluoride	FI	10.54	± 0.01	922
Iodine pentafluoride	F_5I	12.943	± 0.005	408
Iodobenzene	C_6H_5I	8.685		1003
1-Iodobutane	C_4H_9I	9.23	± 0.01	840
2-Iodobutane, (±)-	C_4H_9I	9.10	± 0.02	815
Iodoethane	C_2H_5I	9.3492	± 0.0006	893
1-Iodohexane	$C_6H_{13}I$	9.179		794
Iodomethane	CH_3I	9.538		936
1-Iodo-2-methylpropane	C_4H_9I	9.19	± 0.01	824
2-Iodo-2-methylpropane	C_4H_9I	9.02	approx.	798
1-Iodopentane	$C_5H_{11}I$	9.20	± 0.01	817
1-Iodopropane	C_3H_7I	9.25	± 0.01	860
2-Iodopropane	C_3H_7I	9.19	± 0.02	845
Iridium	Ir	9.1		1543
Iron	Fe	7.9024		1177
Isobutanal	C_4H_8O	9.71	± 0.01	721
Isobutane	C_4H_{10}	10.57	approx.	886
Isobutene	C_4H_8	9.239	± 0.003	875
Isobutylamine	$C_4H_{11}N$	8.50	± 0.1	721
Isobutylbenzene	$C_{10}H_{14}$	8.69	± 0.02	817
Isocyanic acid (HNCO)	CHNO	11.595	± 0.005	1016
Isopentane	C_5H_{12}	10.32	± 0.05	843
Isophorone	$C_9H_{14}O$	≤9.07	approx.	≤670
Isophthalic acid	$C_8H_6O_4$	9.98	approx.	268
Isopropyl acetate	$C_5H_{10}O_2$	9.99	± 0.03	482
Isopropylamine	C_3H_9N	8.72	approx.	758
Isopropylbenzene	C_9H_{12}	8.73	± 0.01	847
Isopropylcyclohexane	C_9H_{18}	9.33	approx.	704
1-Isopropyl-4-methylbenzene	$C_{10}H_{14}$	8.29	approx.	771
Isopropyl methyl ether	$C_4H_{10}O$	9.45	± 0.04	661
Isopropyl methyl sulfide	$C_4H_{10}S$	8.7	approx.	749
Isoquinoline	C_9H_7N	8.53	± 0.03	1032
Isoxazole	C_3H_3NO	9.93	approx.	1038
Ketene	C_2H_2O	9.617	± 0.003	880
Krypton	Kr	13.99961		1351
Lanthanum	La	5.5770		969
Lead	Pb	7.41666		911
Lead(II) chloride	Cl_2Pb	10.2	approx.	791
Lead(II) fluoride	F_2Pb	11.5	approx.	679
Lead(II) oxide (massicot)	OPb	9.08	± 0.10	939

Name	Mol. form.	IP/eV	Uncert.	$\Delta_f H_{ion}$/kJ mol^{-1}
Lead(II) sulfide	PbS	8.5	approx.	954
Lithium	Li	5.39172		680
Lithium bromide	BrLi	8.7	approx.	685
Lithium chloride	ClLi	9.57		727
Lithium hydride	HLi	7.7		882
Lithium iodide	ILi	7.5	approx.	633
Lithium monoxide	LiO	8.44	approx.	894
Lithium potassium	KLi	4.57	± 0.04	512
Lithium rubidium	LiRb	4.3	± 0.1	486
Lithium sodium	LiNa	5.05	± 0.04	571
Lutetium	Lu	5.42585		950
Magnesium	Mg	7.64624		885
Magnesium fluoride	F_2Mg	13.6	approx.	588
Magnesium oxide	MgO	8.76	approx.	901
Maleic anhydride	$C_4H_2O_3$	10.8	approx.	645
Manganese	Mn	7.43402		998
Mendelevium	Md	6.58		635
Mercapto	HS	10.4219	± 0.0004	1145
Mercury	Hg	10.43750		1069
Mercury(II) bromide	Br_2Hg	10.560	± 0.003	935
Mercury(II) chloride	Cl_2Hg	11.380	± 0.003	952
Mercury(II) iodide (red)	HgI_2	9.5088	± 0.0022	900
Mesityl oxide	$C_6H_{10}O$	9.10	± 0.01	694
Methacrylic acid	$C_4H_6O_2$	10.15	approx.	611
Methane	CH_4	12.61	± 0.01	1143
Methanethiol	CH_4S	9.44	± 0.005	888
Methanol	CH_4O	10.85	± 0.01	845
Methoxy	CH_3O	10.72	approx.	1050
Methyl	CH_3	9.843	± 0.002	1095
Methyl acetate	$C_3H_6O_2$	10.25	± 0.02	579
Methyl acrylate	$C_4H_6O_2$	9.9	approx.	641
2-Methylacrylonitrile	C_4H_5N	10.34		1127
Methylamine	CH_5N	8.80	approx.	826
2-Methylaniline	C_7H_9N	7.44	approx.	772
3-Methylaniline	C_7H_9N	7.50	approx.	778
4-Methylaniline	C_7H_9N	7.24	approx.	753
N-Methylaniline	C_7H_9N	7.34	± 0.04	792
Methyl azide	CH_3N_3	9.81	± 0.02	1227
4-Methylbenzaldehyde	C_8H_8O	9.33	approx.	825
Methyl benzoate	$C_8H_8O_2$	9.32	± 0.03	611
2-Methylbenzonitrile	C_8H_7N	≤9.38	approx.	1085
3-Methylbenzonitrile	C_8H_7N	≤9.34	approx.	1085
4-Methylbenzonitrile	C_8H_7N	9.32	± 0.02	1083
2-Methyl-1,3-butadiene	C_5H_8	8.84	± 0.01	928
Methyl butanoate	$C_5H_{10}O_2$	10.07	approx.	520
3-Methylbutanoic acid	$C_5H_{10}O_2$	≤10.51	approx.	≤499
2-Methyl-1-butanol, (±)-	$C_5H_{12}O$	9.86	approx.	649
2-Methyl-2-butanol	$C_5H_{12}O$	9.8	approx.	615
3-Methyl-2-butanol, (±)-	$C_5H_{12}O$	9.88	approx.	637
3-Methyl-2-butanone	$C_5H_{10}O$	9.30	± 0.01	635
2-Methyl-1-butene	C_5H_{10}	9.12	± 0.01	844
3-Methyl-1-butene	C_5H_{10}	9.52	± 0.01	891
2-Methyl-2-butene	C_5H_{10}	8.69	± 0.01	796
2-Methyl-1-buten-3-yne	C_5H_6	9.25	± 0.02	1152
Methyl *tert*-butyl ether	$C_5H_{12}O$	9.24	approx.	608
Methyl chloroacetate	$C_3H_5ClO_2$	10.3	approx.	575
Methylcyclohexane	C_7H_{14}	9.64		775
1-Methylcyclohexanol	$C_7H_{14}O$	9.8	approx.	586
Methylcyclopentane	C_6H_{12}	9.85	approx.	845
Methylcyclopropane	C_4H_8	9.46	approx.	936

Name	Mol. form.	IP/eV	Uncert.	$\Delta_f H_{ion}$/kJ mol^{-1}
2-Methyldecane	$C_{11}H_{24}$	9.7	approx.	658
2-Methyl-*N,N*-dimethylaniline	$C_9H_{13}N$	7.40	± 0.02	814
Methyl 2,2-dimethylpropanoate	$C_6H_{12}O_2$	9.90	approx.	466
Methylene	CH_2	10.396	± 0.003	1392
N-Methylformamide	C_2H_5NO	9.83	± 0.04	760
Methyl formate	$C_2H_4O_2$	10.835	± 0.005	690
2-Methylfuran	C_5H_6O	8.38	± 0.02	729
3-Methylfuran	C_5H_6O	8.64	approx.	763
2-Methylheptane	C_8H_{18}	9.84	approx.	734
5-Methyl-2-hexanone	$C_7H_{14}O$	9.28	approx.	586
Methylhydrazine	CH_6N_2	7.7	± 0.15	835
Methylidyne	CH	10.64	± 0.01	1622
Methyl isocyanate	C_2H_3NO	10.67	approx.	900
Methyl methacrylate	$C_5H_8O_2$	9.7	approx.	589
1-Methylnaphthalene	$C_{11}H_{10}$	7.97	± 0.03	882
2-Methylnaphthalene	$C_{11}H_{10}$	7.91	± 0.08	877
Methyloxirane	C_3H_6O	10.22	approx.	892
2-Methylpentane	C_6H_{14}	10.12	approx.	802
3-Methylpentane	C_6H_{14}	10.08	approx.	801
3-Methyl-2-pentanone, (±)-	$C_6H_{12}O$	9.21	± 0.01	600
4-Methyl-2-pentanone	$C_6H_{12}O$	9.30	± 0.01	609
2-Methyl-3-pentanone	$C_6H_{12}O$	9.10	± 0.01	592
2-Methyl-1-pentene	C_6H_{12}	9.08	approx.	817
4-Methyl-1-pentene	C_6H_{12}	9.45	± 0.01	862
2-Methyl-2-pentene	C_6H_{12}	8.58	approx.	761
4-Methyl-*cis*-2-pentene	C_6H_{12}	8.98	± 0.01	809
4-Methyl-*trans*-2-pentene	C_6H_{12}	8.97	approx.	804
Methyl pentyl ether	$C_6H_{14}O$	≤9.67	approx.	≤657
2-Methylpropanenitrile	C_4H_7N	11.3	approx.	1115
2-Methyl-1-propanethiol	$C_4H_{10}S$	9.12	approx.	783
2-Methyl-2-propanethiol	$C_4H_{10}S$	9.03	approx.	762
Methyl propanoate	$C_4H_8O_2$	10.15	± 0.03	548
2-Methylpropanoic acid	$C_4H_8O_2$	10.33	± 0.03	516
2-Methyl-1-propanol	$C_4H_{10}O$	10.02	± 0.04	683
2-Methyl-2-propanol	$C_4H_{10}O$	9.90	± 0.02	642
2-Methylpropenal	C_4H_6O	9.92	approx.	834
Methyl propyl ether	$C_4H_{10}O$	9.41	± 0.07	670
Methyl propyl sulfide	$C_4H_{10}S$	8.8	approx.	767
2-Methylpyridine	C_6H_7N	9.02	approx.	970
3-Methylpyridine	C_6H_7N	9.04	approx.	979
4-Methylpyridine	C_6H_7N	9.04	approx.	976
N-Methylpyrrolidine	$C_5H_{11}N$	≤8.41	± 0.02	≤809
N-Methyl-2-pyrrolidinone	C_5H_9NO	≤9.17	approx.	≤676
Methylsilane	CH_6Si	10.7	approx.	1003
2-Methylstyrene	C_9H_{10}	8.20	approx.	908
3-Methylstyrene	C_9H_{10}	8.15	approx.	899
4-Methylstyrene	C_9H_{10}	8.1	approx.	895
2-Methylthiophene	C_5H_6S	8.14	approx.	867
3-Methylthiophene	C_5H_6S	8.40	approx.	893
Methyl vinyl ether	C_3H_6O	8.95	± 0.01	763
Molybdenum	Mo	7.09243		1343
Molybdenum(V) chloride	Cl_5Mo	8.7	approx.	392
Molybdenum(VI) fluoride	F_6Mo	14.5	approx.	-159
Morpholine	C_4H_9NO	8.2	approx.	841
Naphthalene	$C_{10}H_8$	8.1442	± 0.0009	936
1-Naphthol	$C_{10}H_8O$	7.76	± 0.03	719
2-Naphthol	$C_{10}H_8O$	7.87	± 0.06	729
Neodymium	Nd	5.5250		859
Neon	Ne	21.56454		2081
Neopentane	C_5H_{12}	≤10.2	approx.	≤818

Name	Mol. form.	IP/eV	Uncert.	$\Delta_f H_{ion}$/kJ mol^{-1}
Neptunium	Np	6.2657	± 0.0003	1069
Nickel	Ni	7.6398		1167
Nickel carbonyl [Ni(CO)$_4$]	C_4NiO_4	8.27	± 0.04	200
Niobium	Nb	6.75885		1384
Niobium(V) chloride	Cl_5Nb	10.97	approx.	356
Nitric acid	HNO_3	11.95	± 0.01	1019
Nitric oxide	NO	9.26438	± 0.00005	985
2-Nitroaniline	$C_6H_6N_2O_2$	8.27	approx.	861
3-Nitroaniline	$C_6H_6N_2O_2$	8.31	approx.	865
4-Nitroaniline	$C_6H_6N_2O_2$	8.34	approx.	859
Nitrobenzene	$C_6H_5NO_2$	9.86	± 0.02	1019
Nitroethane	$C_2H_5NO_2$	10.88	± 0.05	948
Nitrogen	N_2	15.5808		1503
Nitrogen (atomic)	N	14.53414		1875
Nitrogen dioxide	NO_2	9.586	± 0.002	958
Nitrogen pentoxide	N_2O_5	11.9	approx.	1161
Nitrogen sulfide	NS	8.87	± 0.01	1119
Nitrogen tetroxide	N_2O_4	10.8	approx.	1050
Nitrogen trichloride	Cl_3N	10.12	approx.	1244
Nitrogen trifluoride	F_3N	13.00	± 0.02	1125
Nitromethane	CH_3NO_2	11.08	± 0.07	994
2-Nitrophenol	$C_6H_5NO_3$	9.1	approx.	782
3-Nitrophenol	$C_6H_5NO_3$	9.0	approx.	755
4-Nitrophenol	$C_6H_5NO_3$	9.1	approx.	761
1-Nitropropane	$C_3H_7NO_2$	10.81	approx.	919
2-Nitropropane	$C_3H_7NO_2$	10.71	approx.	894
Nitrosyl bromide	BrNO	10.17	± 0.03	1065
Nitrosyl chloride	ClNO	10.87	± 0.01	1099
Nitrosyl fluoride	FNO	12.63	± 0.03	1152
Nitrosyl hydride	HNO	10.1	approx.	1075
2-Nitrotoluene	$C_7H_7NO_2$	9.24		946
3-Nitrotoluene	$C_7H_7NO_2$	9.45	± 0.1	941
4-Nitrotoluene	$C_7H_7NO_2$	9.46	± 0.05	942
Nitrous acid	HNO_2	≤11.3		≤1011
Nitrous oxide	N_2O	12.886		1325
Nitryl chloride	$ClNO_2$	11.84	approx.	1155
Nitryl fluoride	FNO_2	13.09	approx.	1154
Nobelium	No	6.65		642
Nonane	C_9H_{20}	9.71	± 0.10	709
2-Nonanone	$C_9H_{18}O$	9.16	approx.	545
5-Nonanone	$C_9H_{18}O$	9.07	approx.	530
Octane	C_8H_{18}	9.80	± 0.10	737
1-Octene	C_8H_{16}	9.43	± 0.01	829
1-Octyne	C_8H_{14}	9.95	approx.	1040
2-Octyne	C_8H_{14}	9.31	± 0.01	961
3-Octyne	C_8H_{14}	9.22	± 0.01	952
4-Octyne	C_8H_{14}	9.20	± 0.01	946
Osmium	Os	8.7		1630
Osmium(VIII) oxide	O_4Os	12.32	approx.	850
Oxazole	C_3H_3NO	9.9	approx.	940
Oxetane	C_3H_6O	9.65	± 0.01	851
2-Oxetanone	$C_3H_4O_2$	9.70	approx.	653
Oxirane	C_2H_4O	10.56	± 0.01	966
Oxygen	O_2	12.0697	± 0.0002	1165
Oxygen (atomic)	O	13.61806		1563
Ozone	O_3	12.43		1342
Palladium	Pd	8.3367		1181
Pentaborane(9)	B_5H_9	9.90	± 0.04	1028
Pentachloroethane	C_2HCl_5	11.0	approx.	919
cis-1,3-Pentadiene	C_5H_8	8.63	± 0.03	914

Name	Mol. form.	IP/eV	Uncert.	$\Delta_f H_{ion}$/kJ mol^{-1}
trans-1,3-Pentadiene	C_5H_8	8.59	± 0.02	905
1,4-Pentadiene	C_5H_8	9.60	± 0.02	1032
Pentafluorobenzene	C_6HF_5	9.63	approx.	122
Pentafluorophenol	C_6HF_5O	9.20	approx.	-71
2,3,4,5,6-Pentafluorotoluene	$C_7H_3F_5$	9.4	approx.	64
Pentanal	$C_5H_{10}O$	9.74	± 0.04	709
Pentane	C_5H_{12}	10.28	± 0.10	845
2,4-Pentanedione	$C_5H_8O_2$	8.85	± 0.01	469
Pentanoic acid	$C_5H_{10}O_2$	≤10.53	approx.	≤527
1-Pentanol	$C_5H_{12}O$	10.00	approx.	668
2-Pentanol	$C_5H_{12}O$	9.78	approx.	630
3-Pentanol	$C_5H_{12}O$	9.78		628
2-Pentanone	$C_5H_{10}O$	9.38	± 0.01	646
3-Pentanone	$C_5H_{10}O$	9.31	± 0.01	640
1-Pentene	C_5H_{10}	9.51	± 0.01	896
cis-2-Pentene	C_5H_{10}	9.01	± 0.03	843
trans-2-Pentene	C_5H_{10}	9.04	± 0.01	841
1-Penten-3-yne	C_5H_6	9.00	± 0.01	1119
cis-3-Penten-1-yne	C_5H_6	9.14	± 0.04	1137
trans-3-Penten-1-yne	C_5H_6	9.05	± 0.01	1128
1-Pentyne	C_5H_8	10.10	± 0.01	1119
Perchloryl fluoride	$ClFO_3$	12.945	approx.	1224
Perfluoroacetone	C_3F_6O	11.57	approx.	-282
Perfluorocyclohexane	C_6F_{12}	13.2	approx.	-1095
Perfluoronaphthalene	$C_{10}F_8$	8.85		-368
Perfluoropropane	C_3F_8	13.38	approx.	-491
Perfluoropropene	C_3F_6	10.60	± 0.03	-103
Perylene	$C_{20}H_{12}$	6.960	± 0.001	981
Phenanthrene	$C_{14}H_{10}$	7.8914	± 0.0006	966
Phenol	C_6H_6O	8.49	± 0.02	723
Phosphine	H_3P	9.869	± 0.002	958
Phosphorothioc trichloride	Cl_3PS	9.71	± 0.03	573
Phosphorothioc trifluoride	F_3PS	≤11.05	± 0.035	≤58
Phosphorus	P	10.48669		1328
Phosphorus(III) bromide	Br_3P	9.7		798
Phosphorus(III) chloride	Cl_3P	9.91		668
Phosphorus(V) chloride	Cl_5P	10.2	approx.	608
Phosphorus(III) fluoride	F_3P	11.60	± 0.05	161
Phosphorus(V) fluoride	F_5P	15.1	approx.	-137
Phosphorus nitride	NP	11.84	± 0.04	1247
Phosphoryl chloride	Cl_3OP	11.36	± 0.02	540
Phosphoryl fluoride	F_3OP	12.76	± 0.01	-24
Phthalic anhydride	$C_8H_4O_3$	10.1	approx.	603
α-Pinene	$C_{10}H_{16}$	8.07	approx.	808
Piperidine	$C_5H_{11}N$	8.03	± 0.11	726
Platinum	Pt	9.0		1433
Plutonium	Pu	6.025		926
Potassium	K	4.34066		508
Potassium bromide	BrK	7.85	± 0.1	578
Potassium chloride	ClK	8.0	approx.	557
Potassium iodide	IK	7.21	approx.	570
Potassium sodium	KNa	4.41636	± 0.00017	561
Praseodymium	Pr	5.464		883
Promethium	Pm	5.55		536
Propanal	C_3H_6O	9.96	± 0.01	772
Propane	C_3H_8	10.95	± 0.05	952
Propanenitrile	C_3H_5N	11.84	± 0.02	1194
1-Propanethiol	C_3H_8S	9.20	± 0.01	819
2-Propanethiol	C_3H_8S	9.145	± 0.005	806
Propanoic acid	$C_3H_6O_2$	10.525	± 0.003	568

Name	Mol. form.	IP/eV	Uncert.	$\Delta_f H_{ion}$/kJ mol^{-1}
1-Propanol	C_3H_8O	10.18	± 0.06	727
2-Propanol	C_3H_8O	10.17	± 0.02	709
Propargyl alcohol	C_3H_4O	10.49	± 0.02	1060
Propene	C_3H_6	9.73	± 0.02	959
Propyl acetate	$C_5H_{10}O_2$	≤9.92	approx.	501
Propylamine	C_3H_9N	8.78	approx.	777
Propylbenzene	C_9H_{12}	8.713	± 0.010	848
Propylcyclohexane	C_9H_{18}	9.46	approx.	720
Propylcyclopentane	C_8H_{16}	9.34	approx.	753
Propyleneimine	C_3H_7N	9.0	approx.	960
Propyl formate	$C_4H_8O_2$	10.52	± 0.02	555
2-Propynal	C_3H_2O	10.7	approx.	1145
Propyne	C_3H_4	10.37	± 0.01	1187
Protactinium	Pa	5.89		1133
Pyrene	$C_{16}H_{10}$	7.4256	± 0.0006	935
Pyridazine	$C_4H_4N_2$	8.67	± 0.03	1112
Pyridine	C_5H_5N	9.25		1031
Pyrimidine	$C_4H_4N_2$	9.23		1087
Pyrrole	C_4H_5N	8.207	± 0.005	900
Pyrrolidine	C_4H_9N	8.0	approx.	769
2-Pyrrolidone	C_4H_7NO	9.2	approx.	674
Quinoline	C_9H_7N	8.62	± 0.01	1041
Radium	Ra	5.27892		668
Radon	Rn	10.74850		1037
Rhenium	Re	7.88		1530
Rhenium(VII) oxide	O_7Re_2	12.7	approx.	125
Rhodium	Rh	7.45890		1276
Rubidium	Rb	4.17713		484
Rubidium bromide	BrRb	7.94	± 0.03	583
Rubidium chloride	ClRb	8.50	approx.	590
Rubidium sodium	NaRb	4.32	± 0.04	480
Ruthenium	Ru	7.36050		1355
Ruthenium(VIII) oxide	O_4Ru	12.15	± 0.03	988
Samarium	Sm	5.6437		751
Scandium	Sc	6.56144		1010
Selenium	Se	9.75238		1168
Silane	H_4Si	11.00	± 0.02	1095
Silicon	Si	8.15169		1238
Silicon monoxide	OSi	11.49	± 0.20	1008
Silver	Ag	7.57624		1016
Silver(I) chloride	AgCl	≤10.08	approx.	≤1065
Silver(I) fluoride	AgF	11.0	approx.	1071
Silylene	H_2Si	8.244	± 0.025	1084
Sodium	Na	5.13908		603
Sodium bromide	BrNa	8.31	± 0.1	660
Sodium chloride	ClNa	8.92	± 0.06	681
Sodium iodide	INa	7.64	± 0.02	659
Spiro[2.2]pentane	C_5H_8	9.26	approx.	1078
Stannane	H_4Sn	10.75	approx.	1200
Stibine	H_3Sb	9.54	± 0.03	1067
cis-Stilbene	$C_{14}H_{12}$	7.80	approx.	1005
trans-Stilbene	$C_{14}H_{12}$	7.656	± 0.001	973
Strontium	Sr	5.69484		713
Strontium oxide	OSr	6.6	± 0.2	623
Styrene	C_8H_8	8.464	± 0.001	964
Succinic anhydride	$C_4H_4O_3$	10.6	approx.	500
Succinonitrile	$C_4H_4N_2$	12.1	approx.	1377
Sulfolane	$C_4H_8O_2S$	9.8	approx.	577
Sulfur	S	10.36001		1277
Sulfur chloride pentafluoride	ClF_5S	12.335	approx.	144

Name	Mol. form.	IP/eV	Uncert.	$\Delta_f H_{ion}$/kJ mol^{-1}
Sulfur dichloride	Cl_2S	9.45	± 0.03	895
Sulfur difluoride	F_2S	10.08	approx.	676
Sulfur dioxide	O_2S	12.349	± 0.001	894
Sulfur hexafluoride	F_6S	15.32	± 0.02	258
Sulfur monofluoride	FS	10.09		986
Sulfur monoxide	OS	10.294	± 0.004	998
Sulfur oxide (SSO)	OS_2	10.584	± 0.005	971
Sulfur pentafluoride	F_5S	9.60	± 0.05	10
Sulfur tetrafluoride	F_4S	12.0	± 0.3	399
Sulfur trioxide (α-form)	O_3S	12.82	± 0.03	841
Sulfuryl chloride	Cl_2O_2S	12.05		807
Sulfuryl fluoride	F_2O_2S	13.04	± 0.01	501
Tantalum	Ta	7.89		1544
Tantalum(V) chloride	Cl_5Ta	11.08	approx.	303
Technetium	Tc	7.28		1380
Tellurium	Te	9.0096		1066
Terbium	Tb	5.8639		955
Terephthalic acid	$C_8H_6O_4$	9.86	approx.	232
o-Terphenyl	$C_{18}H_{14}$	7.99	approx.	1056
m-Terphenyl	$C_{18}H_{14}$	8.01	approx.	1057
p-Terphenyl	$C_{18}H_{14}$	7.80	± 0.03	1037
Tetraborane(10)	B_4H_{10}	10.76	± 0.04	1105
Tetrabromomethane	CBr_4	10.31	approx.	1079
1,1,2,2-Tetrachloro-1,2-difluoroethane	$C_2Cl_4F_2$	11.3	approx.	563
1,1,1,2-Tetrachloroethane	$C_2H_2Cl_4$	11.1	approx.	920
1,1,2,2-Tetrachloroethane	$C_2H_2Cl_4$	≤11.62	approx.	≤971
Tetrachloroethene	C_2Cl_4	9.326	± 0.001	887
Tetrachloromethane	CCl_4	11.47	± 0.01	1010
Tetrachlorosilane	Cl_4Si	11.79	± 0.01	527
Tetraethylsilane	$C_8H_{20}Si$	8.9	approx.	595
1,2,3,4-Tetrafluorobenzene	$C_6H_2F_4$	9.53	approx.	284
1,2,3,5-Tetrafluorobenzene	$C_6H_2F_4$	9.53	approx.	263
1,2,4,5-Tetrafluorobenzene	$C_6H_2F_4$	9.35	approx.	254
Tetrafluoroethene	C_2F_4	10.12	± 0.02	315
Tetrafluorohydrazine	F_4N_2	11.94	± 0.03	1119
Tetrafluorosilane	F_4Si	15.24	± 0.14	-144
Tetrahydrofuran	C_4H_8O	9.38	± 0.05	721
1,2,3,4-Tetrahydronaphthalene	$C_{10}H_{12}$	8.46	± 0.02	841
Tetrahydropyran	$C_5H_{10}O$	9.25	± 0.01	670
Tetrahydrothiophene	C_4H_8S	8.38		774
1,2,4,5-Tetramethylbenzene	$C_{10}H_{14}$	8.04	± 0.02	730
2,2,3,3-Tetramethylbutane	C_8H_{18}	9.8		720
Tetramethylsilane	$C_4H_{12}Si$	9.80	± 0.04	713
Tetramethylstannane	$C_4H_{12}Sn$	8.89	± 0.05	837
Thallium	Tl	6.10829		771
Thallium(I) bromide	BrTl	9.14	± 0.02	844
Thallium(I) chloride	ClTl	9.70	± 0.03	869
Thallium(I) fluoride	FTl	10.52		835
Thallium(I) iodide	ITl	8.47	± 0.02	826
Thiacyclohexane	$C_5H_{10}S$	8.2	approx.	728
Thietane	C_3H_6S	8.61		892
Thionitrosyl fluoride (NSF)	FNS	11.51	± 0.04	1090
Thionyl chloride	Cl_2OS	10.96		844
Thionyl fluoride	F_2OS	12.25		688
Thiophene	C_4H_4S	8.86	± 0.02	970
Thorium	Th	6.308	± 0.003	1207
Thorium(IV) oxide	O_2Th	8.7	approx.	342
Thulium	Tm	6.18431		827
Tin	Sn	7.34381		1011
Tin(II) bromide	Br_2Sn	9.0		839

Name	Mol. form.	IP/eV	Uncert.	$\Delta_f H_{ion}$/kJ mol^{-1}
Tin(IV) bromide	Br_4Sn	10.6		709
Tin(II) chloride	Cl_2Sn	10.0	approx.	760
Tin(IV) chloride	Cl_4Sn	11.7	± 0.2	656
Tin(II) fluoride	F_2Sn	11.1	approx.	586
Tin(II) oxide	OSn	9.60	± 0.02	944
Tin(II) sulfide	SSn	8.8	approx.	966
Titanium	Ti	6.8282		1127
Titanium(IV) bromide	Br_4Ti	10.3		375
Titanium(IV) chloride	Cl_4Ti	11.5	approx.	349
Titanium(IV) iodide	I_4Ti	9.1	approx.	602
Titanium(IV) oxide (rutile)	O_2Ti	9.54	approx.	623
Toluene	C_7H_8	8.8276	± 0.0006	901
o-Toluic acid	$C_8H_8O_2$	9.1	approx.	558
m-Toluic acid	$C_8H_8O_2$	9.43	approx.	579
p-Toluic acid	$C_8H_8O_2$	9.23	approx.	560
Tribromomethane	$CHBr_3$	10.48	± 0.02	1035
Tributylamine	$C_{12}H_{27}N$	7.4	approx.	492
Trichloroacetyl chloride	C_2Cl_4O	11.0	approx.	827
1,2,4-Trichlorobenzene	$C_6H_3Cl_3$	9.04	approx.	880
1,3,5-Trichlorobenzene	$C_6H_3Cl_3$	9.32	± 0.02	899
1,1,1-Trichloroethane	$C_2H_3Cl_3$	11.0	approx.	917
1,1,2-Trichloroethane	$C_2H_3Cl_3$	11.0	approx.	911
Trichloroethene	C_2HCl_3	9.46	± 0.02	894
Trichlorofluoromethane	CCl_3F	11.77	± 0.02	868
Trichloromethane	$CHCl_3$	11.37	± 0.02	992
(Trichloromethyl)benzene	$C_7H_5Cl_3$	≤9.60	approx.	≤914
Trichloromethylsilane	CH_3Cl_3Si	11.36	approx.	548
Trichlorosilane	Cl_3HSi	11.7	approx.	648
1,1,1-Trichloro-2,2,2-trifluoroethane	$C_2Cl_3F_3$	11.5		386
1,1,2-Trichloro-1,2,2-trifluoroethane	$C_2Cl_3F_3$	11.99	± 0.02	429
Tris(2-hydroxyethyl)amine	$C_6H_{15}NO_3$	7.9	approx.	206
Triethylamine	$C_6H_{15}N$	7.50	approx.	631
Trifluoramine oxide	F_3NO	13.31	± 0.06	1121
Trifluoroacetic acid	$C_2HF_3O_2$	11.46		75
Trifluoroacetonitrile	C_2F_3N	13.93	± 0.07	845
1,1,1-Trifluoroethane	$C_2H_3F_3$	13.3	± 0.5	536
Trifluoroethene	C_2HF_3	10.14		489
Trifluoroiodomethane	CF_3I	10.23		397
Trifluoromethane	CHF_3	13.86	approx.	643
Trifluoromethyl	CF_3	8.7	± 0.2	379
(Trifluoromethyl)benzene	$C_7H_5F_3$	9.685	± 0.005	335
3,3,3-Trifluoro-1-propene	$C_3H_3F_3$	10.9	approx.	437
Trifluorosilane	F_3HSi	14.0	approx.	150
Trifluorosilyl	F_3Si	9.99	approx.	-32
Triiodomethane	CHI_3	9.25	± 0.02	1010
Trimethylamine	C_3H_9N	7.82	± 0.06	731
1,2,3-Trimethylbenzene	C_9H_{12}	8.42	± 0.02	803
1,2,4-Trimethylbenzene	C_9H_{12}	8.27	± 0.01	784
1,3,5-Trimethylbenzene	C_9H_{12}	8.41	± 0.01	796
Trimethyl borate	$C_3H_9BO_3$	10.0	approx.	65
Trimethylchlorosilane	C_3H_9ClSi	10.15	approx.	624
2,2,4-Trimethylpentane	C_8H_{18}	9.86	approx.	713
2,2,4-Trimethyl-3-pentanone	$C_8H_{16}O$	8.80	approx.	511
2,4,6-Trimethylpyridine	$C_8H_{11}N$	≤8.9	approx.	≤880
1,3,5-Trioxane	$C_3H_6O_3$	10.3	approx.	528
Trisilane	H_8Si_3	9.2	approx.	1009
Tungsten	W	7.98		1621
Tungsten(VI) chloride	Cl_6W	9.5	approx.	348
Undecane	$C_{11}H_{24}$	9.56	approx.	650
Uranium	U	6.19405		1129

Name	Mol. form.	IP/eV	Uncert.	$\Delta_f H_{ion}$/kJ mol^{-1}
Uranium(VI) fluoride	F_6U	14.00	± 0.10	-796
Uranium(IV) oxide	O_2U	5.4	approx.	57
Uranium(VI) oxide	O_3U	10.5	approx.	214
Urea	CH_4N_2O	9.7		690
Vanadium	V	6.746	± 0.002	1166
Vanadium(IV) chloride	Cl_4V	9.2	approx.	361
Vanadyl trichloride	Cl_3OV	11.6	approx.	425
Vinyl acetate	$C_4H_6O_2$	9.19	± 0.05	572
Water	H_2O	12.6206	± 0.0020	976
Xenon	Xe	12.12987		1170
Xenon difluoride	F_2Xe	12.35	± 0.01	1083
Xenon tetrafluoride	F_4Xe	12.65	± 0.1	1016
o-Xylene	C_8H_{10}	8.56	± 0.01	844
m-Xylene	C_8H_{10}	8.56	± 0.01	843
p-Xylene	C_8H_{10}	8.44	± 0.01	832
2,3-Xylenol	$C_8H_{10}O$	8.26	approx.	640
2,4-Xylenol	$C_8H_{10}O$	8.0	approx.	609
2,6-Xylenol	$C_8H_{10}O$	8.05	approx.	615
3,4-Xylenol	$C_8H_{10}O$	8.09	approx.	624
Ytterbium	Yb	6.25416		754
Yttrium	Y	6.217		1022
Zinc	Zn	9.39405		1037
Zirconium	Zr	6.63390		1251
Zirconium(IV) bromide	Br_4Zr	10.7	approx.	388
Zirconium(IV) chloride	Cl_4Zr	11.2	approx.	210
Zirconium(IV) iodide	I_4Zr	9.3	approx.	500

ATTENUATION COEFFICIENTS FOR HIGH-ENERGY ELECTROMAGNETIC RADIATION

Martin J. Berger and John H. Hubbell

This table gives mass attenuation coefficients for photons for elements of Z = 1 to 100 at energies between 1 keV (soft x-rays) and 1 GeV (hard gamma rays). The mass attenuation coefficient μ describes the attenuation of electromagnetic radiation as it passes through matter by the relation

$$I(x)/I_o = e^{-\mu\rho x}$$

where I_o is the initial intensity, $I(x)$ the intensity after path length x, and ρ is the mass density of the element in question. To a high approximation the mass attenuation coefficient is additive for the elements present, independent of the way in which they are bound in chemical compounds. Therefore, these data can be used to calculate the shielding against x-rays and gamma rays provided by materials of known elemental composition and density.

Tables 1-4 give the attenuation coefficients for elements with atomic number 1–100. The power of ten is indicated beside each number in the table; i.e., 7.41 + 03 means 7.41 × 10^3. Plots of attenuation coefficient vs. energy (see Ref. 4) show a smooth decline with increasing energy until an absorption edge is reached, where the attenuation coefficient increases sharply and then continues to decline. Thus, interpolation is unreliable when the interval includes an absorption edge. Absorption edge energies are listed in Table 5.

The attenuation coefficients were calculated with the computer program XCOM (Refs. 1 and 3), which uses a cross-section database compiled at the Photon and Charged Particle Data Center at the National Institute of Standards and Technology. Their accuracy has been confirmed at all energies by extensive comparisons with experimental attenuation coefficients. Such comparisons for x-ray energies up to 100 keV can be found in Ref. 2.

References

1. Berger, M. J., and Hubbell, J. H., *National Bureau of Standards Report* NBSIR-87-3597, 1987.
2. Saloman, E. B., Hubbell, J. H., and Scofield, J. H., *Atomic Data and Nuclear Data Tables*, 38, 1, 1988. <https://doi.org/10.1016/0092-640X(88)90044-7>
3. XCOM: Photon Cross Sections Database, Version 3.1, NIST Standard Reference Database 8, November 2010 <www.nist.gov/pml/data/xcom/index.cfm>.
4. Tables of X-Ray Mass Attenuation Coefficients and Mass Energy-Absorption Coefficients from 1 keV to 20 MeV for Elements Z = 1 to 92 and 48 Additional Substances of Dosimetric Interest, NIST Standard Reference Database 26, July 2004 <www.nist.gov/pml/data/xraycoef/index.cfm>.

TABLE 1. Mass Attenuation Coefficients in cm^2 g^{-1} for H through Sn and Photon Energy from 0.001 to 0.5 MeV

	Atomic no.	0.001 MeV	0.002 MeV	0.005 MeV	0.01 MeV	0.02 MeV	0.05 MeV	0.1 MeV	0.2 MeV	0.5 MeV
H	1	7.21 + 00	1.06 + 00	4.19-01	3.85-01	3.69-01	3.36-01	2.94-01	2.43-01	1.73-01
He	2	6.08 + 01	6.86 + 00	5.77-01	2.48-01	1.96-01	1.70-01	1.49-01	1.22-01	8.71-02
Li	3	2.34 + 02	2.71 + 01	1.62 + 00	3.40-01	1.86-01	1.49-01	1.29-01	1.06-01	7.53-02
Be	4	6.04 + 02	7.47 + 01	4.37 + 00	6.47-01	2.25-01	1.55-01	1.33-01	1.09-01	7.74-02
B	5	1.23 + 03	1.60 + 02	9.68 + 00	1.25 + 00	3.01-01	1.66-01	1.39-01	1.14-01	8.07-02
C	6	2.21 + 03	3.03 + 02	1.91 + 01	2.37 + 00	4.42-01	1.87-01	1.51-01	1.23-01	8.72-02
N	7	3.31 + 03	4.77 + 02	3.14 + 01	3.88 + 00	6.18-01	1.98-01	1.53-01	1.23-01	8.72-02
O	8	4.59 + 03	6.95 + 02	4.79 + 01	5.95 + 00	8.65-01	2.13-01	1.55-01	1.24-01	8.73-02
F	9	5.65 + 03	9.05 + 02	6.51 + 01	8.21 + 00	1.13 + 00	2.21-01	1.50-01	1.18-01	8.27-02
Ne	10	7.41 + 03	1.24 + 03	9.34 + 01	1.20 + 01	1.61 + 00	2.58-01	1.60-01	1.24-01	8.66-02
Na	11	6.54 + 02	1.52 + 03	1.19 + 02	1.56 + 01	2.06 + 00	2.80-01	1.59-01	1.20-01	8.37-02
Mg	12	9.22 + 02	1.93 + 03	1.58 + 02	2.11 + 01	2.76 + 00	3.29-01	1.69-01	1.24-01	8.65-02
Al	13	1.19 + 03	2.26 + 03	1.93 + 02	2.62 + 01	3.44 + 00	3.68-01	1.70-01	1.22-01	8.44-02
Si	14	1.57 + 03	2.78 + 03	2.45 + 02	3.39 + 01	4.46 + 00	4.38-01	1.84-01	1.28-01	8.75-02
P	15	1.91 + 03	3.02 + 02	2.86 + 02	4.04 + 01	5.35 + 00	4.92-01	1.87-01	1.25-01	8.51-02
S	16	2.43 + 03	3.85 + 02	3.49 + 02	5.01 + 01	6.71 + 00	5.85-01	2.02-01	1.30-01	8.78-02
Cl	17	2.83 + 03	4.52 + 02	3.90 + 02	5.73 + 01	7.74 + 00	6.48-01	2.05-01	1.27-01	8.45-02
Ar	18	3.18 + 03	5.12 + 02	4.23 + 02	6.32 + 01	8.63 + 00	7.01-01	2.04-01	1.20-01	7.96-02
K	19	4.06 + 03	6.59 + 02	5.19 + 02	7.91 + 01	1.09 + 01	8.68-01	2.34-01	1.32-01	8.60-02
Ca	20	4.87 + 03	8.00 + 02	6.03 + 02	9.34 + 01	1.31 + 01	1.02 + 00	2.57-01	1.38-01	8.85-02
Sc	21	5.24 + 03	8.70 + 02	6.31 + 02	9.95 + 01	1.41 + 01	1.09 + 00	2.58-01	1.31-01	8.31-02
Ti	22	5.87 + 03	9.86 + 02	6.84 + 02	1.11 + 02	1.59 + 01	1.21 + 00	2.72-01	1.31-01	8.19-02
V	23	6.50 + 03	1.11 + 03	9.29 + 01	1.22 + 02	1.77 + 01	1.35 + 00	2.88-01	1.32-01	8.07-02
Cr	24	7.40 + 03	1.28 + 03	1.08 + 02	1.39 + 02	2.04 + 01	1.55 + 00	3.17-01	1.38-01	8.28-02
Mn	25	8.09 + 03	1.42 + 03	1.21 + 02	1.51 + 02	2.25 + 01	1.71 + 00	3.37-01	1.39-01	8.19-02
Fe	26	9.09 + 03	1.63 + 03	1.40 + 02	1.71 + 02	2.57 + 01	1.96 + 00	3.72-01	1.46-01	8.41-02
Co	27	9.80 + 03	1.78 + 03	1.54 + 02	1.84 + 02	2.80 + 01	2.14 + 00	3.95-01	1.48-01	8.32-02

	Atomic no.	0.001 MeV	0.002 MeV	0.005 MeV	0.01 MeV	0.02 MeV	0.05 MeV	0.1 MeV	0.2 MeV	0.5 MeV
Ni	28	9.86 + 03	2.05 + 03	1.79 + 02	2.09 + 02	3.22 + 01	2.47 + 00	4.44-01	1.58-01	8.70-02
Cu	29	1.06 + 04	2.15 + 03	1.90 + 02	2.16 + 02	3.38 + 01	2.61 + 00	4.58-01	1.56-01	8.36-02
Zn	30	1.55 + 03	2.37 + 03	2.12 + 02	2.33 + 02	3.72 + 01	2.89 + 00	4.97-01	1.62-01	8.45-02
Ga	31	1.70 + 03	2.52 + 03	2.27 + 02	3.42 + 01	3.93 + 01	3.08 + 00	5.20-01	1.62-01	8.24-02
Ge	32	1.89 + 03	2.71 + 03	2.47 + 02	3.74 + 01	4.22 + 01	3.34 + 00	5.55-01	1.66-01	8.21-02
As	33	2.12 + 03	2.93 + 03	2.71 + 02	4.12 + 01	4.56 + 01	3.63 + 00	5.97-01	1.72-01	8.26-02
Se	34	2.32 + 03	3.10 + 03	2.90 + 02	4.41 + 01	4.82 + 01	3.86 + 00	6.28-01	1.74-01	8.13-02
Br	35	2.62 + 03	3.41 + 03	3.21 + 02	4.91 + 01	5.27 + 01	4.26 + 00	6.86-01	1.84-01	8.33-02
Kr	36	2.85 + 03	3.60 + 03	3.43 + 02	5.26 + 01	5.55 + 01	4.52 + 00	7.22-01	1.87-01	8.23-02
Rb	37	3.17 + 03	3.41 + 03	3.74 + 02	5.77 + 01	5.98 + 01	4.92 + 00	7.80-01	1.96-01	8.36-02
Sr	38	3.49 + 03	2.59 + 03	4.06 + 02	6.27 + 01	6.39 + 01	5.31 + 00	8.37-01	2.04-01	8.44-02
Y	39	3.86 + 03	7.42 + 02	4.42 + 02	6.87 + 01	6.86 + 01	5.76 + 00	9.05-01	2.15-01	8.61-02
Zr	40	4.21 + 03	8.12 + 02	4.76 + 02	7.42 + 01	7.24 + 01	6.17 + 00	9.66-01	2.24-01	8.69-02
Nb	41	4.60 + 03	8.89 + 02	5.13 + 02	8.04 + 01	7.71 + 01	6.64 + 00	1.04 + 00	2.34-01	8.83-02
Mo	42	4.94 + 03	9.60 + 02	5.45 + 02	8.58 + 01	1.31 + 01	7.04 + 00	1.10 + 00	2.42-01	8.85-02
Tc	43	5.36 + 03	1.04 + 03	5.84 + 02	9.23 + 01	1.41 + 01	7.52 + 00	1.17 + 00	2.53-01	8.97-02
Ru	44	5.72 + 03	1.12 + 03	6.17 + 02	9.80 + 01	1.50 + 01	7.92 + 00	1.23 + 00	2.62-01	8.99-02
Rh	45	6.17 + 03	1.21 + 03	6.59 + 02	1.05 + 02	1.61 + 01	8.45 + 00	1.31 + 00	2.74-01	9.13-02
Pd	46	6.54 + 03	1.29 + 03	6.91 + 02	1.11 + 02	1.70 + 01	8.85 + 00	1.38 + 00	2.83-01	9.13-02
Ag	47	7.04 + 03	1.40 + 03	7.39 + 02	1.19 + 02	1.84 + 01	9.45 + 00	1.47 + 00	2.97-01	9.32-02
Cd	48	7.35 + 03	1.47 + 03	7.69 + 02	1.24 + 02	1.92 + 01	9.78 + 00	1.52 + 00	3.04-01	9.25-02
In	49	7.81 + 03	1.58 + 03	8.13 + 02	1.32 + 02	2.04 + 01	1.03 + 01	1.61 + 00	3.17-01	9.37-02
Sn	50	8.16 + 03	1.66 + 03	8.47 + 02	1.38 + 02	2.15 + 01	1.07 + 01	1.68 + 00	3.26-01	9.37-02

TABLE 2. Mass Attenuation Coefficients in $cm^2\,g^{-1}$ for H through Sn and Photon Energy from 1.0 to 1000.0 MeV

	Atomic no.	1.0 MeV	2.0 MeV	5.0 MeV	10.0 MeV	20.0 MeV	50.0 MeV	100.0 MeV	500.0 MeV	1000.0 MeV
H	1	1.26-01	8.77-02	5.05-02	3.25-02	2.15-02	1.42-02	1.19-02	1.14-02	1.16-02
He	2	6.36-02	4.42-02	2.58-02	1.70-02	1.18-02	8.61-03	7.78-03	7.79-03	7.95-03
Li	3	5.50-02	3.83-02	2.26-02	1.53-02	1.11-02	8.68-03	8.21-03	8.61-03	8.87-03
Be	4	5.65-02	3.94-02	2.35-02	1.63-02	1.23-02	1.02-02	9.94-03	1.08-02	1.12-02
B	5	5.89-02	4.11-02	2.48-02	1.76-02	1.37-02	1.19-02	1.19-02	1.32-02	1.37-02
C	6	6.36-02	4.44-02	2.71-02	1.96-02	1.58-02	1.43-02	1.46-02	1.64-02	1.70-02
N	7	6.36-02	4.45-02	2.74-02	2.02-02	1.67-02	1.57-02	1.63-02	1.85-02	1.92-02
O	8	6.37-02	4.46-02	2.78-02	2.09-02	1.77-02	1.71-02	1.79-02	2.06-02	2.13-02
F	9	6.04-02	4.23-02	2.66-02	2.04-02	1.77-02	1.75-02	1.86-02	2.14-02	2.21-02
Ne	10	6.32-02	4.43-02	2.82-02	2.20-02	1.95-02	1.96-02	2.11-02	2.43-02	2.51-02
Na	11	6.10-02	4.28-02	2.75-02	2.18-02	1.97-02	2.03-02	2.19-02	2.53-02	2.62-02
Mg	12	6.30-02	4.43-02	2.87-02	2.31-02	2.13-02	2.23-02	2.42-02	2.81-02	2.90-02
Al	13	6.15-02	4.32-02	2.84-02	2.32-02	2.17-02	2.31-02	2.52-02	2.93-02	3.03-02
Si	14	6.36-02	4.48-02	2.97-02	2.46-02	2.34-02	2.52-02	2.76-02	3.23-02	3.34-02
P	15	6.18-02	4.36-02	2.91-02	2.45-02	2.36-02	2.58-02	2.84-02	3.33-02	3.45-02
S	16	6.37-02	4.50-02	3.04-02	2.59-02	2.53-02	2.79-02	3.08-02	3.62-02	3.75-02
Cl	17	6.13-02	4.33-02	2.95-02	2.55-02	2.52-02	2.81-02	3.11-02	3.67-02	3.80-02
Ar	18	5.76-02	4.07-02	2.80-02	2.45-02	2.45-02	2.76-02	3.07-02	3.62-02	3.75-02
K	19	6.22-02	4.40-02	3.05-02	2.70-02	2.74-02	3.11-02	3.46-02	4.09-02	4.24-02
Ca	20	6.39-02	4.52-02	3.17-02	2.84-02	2.90-02	3.32-02	3.71-02	4.40-02	4.56-02
Sc	21	5.98-02	4.24-02	3.00-02	2.72-02	2.80-02	3.23-02	3.62-02	4.30-02	4.45-02
Ti	22	5.89-02	4.18-02	2.98-02	2.73-02	2.84-02	3.30-02	3.71-02	4.40-02	4.56-02
V	23	5.79-02	4.11-02	2.96-02	2.74-02	2.88-02	3.36-02	3.78-02	4.49-02	4.65-02
Cr	24	5.93-02	4.21-02	3.06-02	2.86-02	3.03-02	3.56-02	4.01-02	4.76-02	4.93-02
Mn	25	5.85-02	4.16-02	3.04-02	2.87-02	3.07-02	3.63-02	4.09-02	4.86-02	5.04-02
Fe	26	5.99-02	4.26-02	3.15-02	2.99-02	3.22-02	3.83-02	4.33-02	5.15-02	5.33-02
Co	27	5.91-02	4.20-02	3.13-02	3.00-02	3.26-02	3.88-02	4.40-02	5.23-02	5.41-02
Ni	28	6.16-02	4.39-02	3.29-02	3.18-02	3.48-02	4.17-02	4.73-02	5.61-02	5.81-02
Cu	29	5.90-02	4.20-02	3.18-02	3.10-02	3.41-02	4.10-02	4.66-02	5.53-02	5.72-02
Zn	30	5.94-02	4.24-02	3.22-02	3.18-02	3.51-02	4.24-02	4.82-02	5.72-02	5.91-02
Ga	31	5.77-02	4.11-02	3.16-02	3.13-02	3.48-02	4.22-02	4.80-02	5.70-02	5.89-02
Ge	32	5.73-02	4.09-02	3.16-02	3.16-02	3.53-02	4.30-02	4.89-02	5.80-02	6.00-02

	Atomic no.	1.0 MeV	2.0 MeV	5.0 MeV	10.0 MeV	20.0 MeV	50.0 MeV	100.0 MeV	500.0 MeV	1000.0 MeV
As	33	5.73-02	4.09-02	3.19-02	3.21-02	3.60-02	4.40-02	5.01-02	5.95-02	6.15-02
Se	34	5.62-02	4.01-02	3.14-02	3.19-02	3.60-02	4.41-02	5.03-02	5.97-02	6.17-02
Br	35	5.73-02	4.09-02	3.23-02	3.29-02	3.74-02	4.60-02	5.24-02	6.22-02	6.43-02
Kr	36	5.63-02	4.02-02	3.20-02	3.28-02	3.74-02	4.61-02	5.26-02	6.25-02	6.46-02
Rb	37	5.69-02	4.06-02	3.25-02	3.36-02	3.85-02	4.75-02	5.43-02	6.45-02	6.67-02
Sr	38	5.71-02	4.08-02	3.29-02	3.41-02	3.93-02	4.87-02	5.56-02	6.61-02	6.83-02
Y	39	5.80-02	4.14-02	3.35-02	3.50-02	4.05-02	5.03-02	5.75-02	6.83-02	7.06-02
Zr	40	5.81-02	4.15-02	3.38-02	3.55-02	4.12-02	5.13-02	5.87-02	6.98-02	7.22-02
Nb	41	5.87-02	4.18-02	3.44-02	3.63-02	4.22-02	5.27-02	6.03-02	7.17-02	7.42-02
Mo	42	5.84-02	4.16-02	3.44-02	3.65-02	4.26-02	5.33-02	6.10-02	7.26-02	7.51-02
Tc	43	5.88-02	4.19-02	3.48-02	3.71-02	4.35-02	5.45-02	6.24-02	7.43-02	7.68-02
Ru	44	5.85-02	4.16-02	3.48-02	3.73-02	4.39-02	5.50-02	6.30-02	7.51-02	7.77-02
Rh	45	5.89-02	4.20-02	3.53-02	3.80-02	4.48-02	5.63-02	6.45-02	7.69-02	7.94-02
Pd	46	5.85-02	4.16-02	3.52-02	3.80-02	4.50-02	5.66-02	6.49-02	7.73-02	8.00-02
Ag	47	5.92-02	4.21-02	3.58-02	3.88-02	4.61-02	5.81-02	6.67-02	7.93-02	8.20-02
Cd	48	5.83-02	4.14-02	3.54-02	3.85-02	4.59-02	5.79-02	6.64-02	7.91-02	8.18-02
In	49	5.85-02	4.15-02	3.56-02	3.90-02	4.65-02	5.88-02	6.75-02	8.04-02	8.32-02
Sn	50	5.80-02	4.11-02	3.55-02	3.90-02	4.66-02	5.90-02	6.78-02	8.07-02	8.35-02

TABLE 3. Mass Attenuation Coefficients in $cm^2 g^{-1}$ for Sb through Fm and Photon Energy from 0.001 to 0.5 MeV

	Atomic no.	0.001 MeV	0.002 MeV	0.005 MeV	0.01 MeV	0.02 MeV	0.05 MeV	0.1 MeV	0.2 MeV	0.5 MeV
Sb	51	8.58 + 03	1.77 + 03	8.85 + 02	1.46 + 02	2.27 + 01	1.12 + 01	1.76 + 00	3.38-01	9.45-02
Te	52	8.43 + 03	1.83 + 03	9.01 + 02	1.50 + 02	2.34 + 01	1.14 + 01	1.80 + 00	3.43-01	9.33-02
I	53	9.10 + 03	2.00 + 03	8.43 + 02	1.63 + 02	2.54 + 01	1.23 + 01	1.94 + 00	3.66-01	9.70-02
Xe	54	9.41 + 03	2.09 + 03	6.39 + 02	1.69 + 02	2.65 + 01	1.27 + 01	2.01 + 00	3.76-01	9.70-02
Cs	55	9.37 + 03	2.23 + 03	2.30 + 02	1.79 + 02	2.82 + 01	1.34 + 01	2.12 + 00	3.94-01	9.91-02
Ba	56	8.54 + 03	2.32 + 03	2.41 + 02	1.86 + 02	2.94 + 01	1.38 + 01	2.20 + 00	4.05-01	9.92-02
La	57	9.09 + 03	2.46 + 03	2.58 + 02	1.97 + 02	3.12 + 01	1.45 + 01	2.32 + 00	4.24-01	1.01-01
Ce	58	9.71 + 03	2.61 + 03	2.74 + 02	2.08 + 02	3.31 + 01	1.52 + 01	2.45 + 00	4.45-01	1.04-01
Pr	59	1.06 + 04	2.77 + 03	2.92 + 02	2.21 + 02	3.53 + 01	1.60 + 01	2.59 + 00	4.69-01	1.07-01
Nd	60	6.63 + 03	2.88 + 03	3.06 + 02	2.30 + 02	3.68 + 01	1.65 + 01	2.69 + 00	4.84-01	1.08-01
Pm	61	2.06 + 03	3.05 + 03	3.26 + 02	2.44 + 02	3.92 + 01	1.73 + 01	2.84 + 00	5.10-01	1.12-01
Sm	62	2.11 + 03	3.12 + 03	3.36 + 02	2.50 + 02	4.03 + 01	1.77 + 01	2.90 + 00	5.19-01	1.11-01
Eu	63	2.22 + 03	3.28 + 03	3.54 + 02	2.63 + 02	4.24 + 01	1.85 + 01	3.04 + 00	5.43-01	1.14-01
Gd	64	2.29 + 03	3.36 + 03	3.65 + 02	2.69 + 02	4.36 + 01	3.86 + 00	3.11 + 00	5.54-01	1.14-01
Tb	65	2.40 + 03	3.51 + 03	3.84 + 02	2.82 + 02	4.59 + 01	4.06 + 00	3.25 + 00	5.77-01	1.17-01
Dy	66	2.49 + 03	3.47 + 03	3.99 + 02	2.90 + 02	4.76 + 01	4.23 + 00	3.36 + 00	5.95-01	1.18-01
Ho	67	2.62 + 03	3.59 + 03	4.17 + 02	3.01 + 02	4.98 + 01	4.43 + 00	3.49 + 00	6.18-01	1.20-01
Er	68	2.75 + 03	3.52 + 03	4.36 + 02	3.13 + 02	5.20 + 01	4.63 + 00	3.63 + 00	6.41-01	1.23-01
Tm	69	2.90 + 03	3.69 + 03	4.57 + 02	2.83 + 02	5.45 + 01	4.87 + 00	3.78 + 00	6.68-01	1.26-01
Yb	70	3.02 + 03	3.80 + 03	4.72 + 02	2.94 + 02	5.63 + 01	5.04 + 00	3.88 + 00	6.86-01	1.27-01
Lu	71	3.19 + 03	3.45 + 03	4.94 + 02	2.21 + 02	5.88 + 01	5.28 + 00	4.03 + 00	7.13-01	1.30-01
Hf	72	3.34 + 03	3.60 + 03	5.11 + 02	2.30 + 02	6.09 + 01	5.48 + 00	4.15 + 00	7.34-01	1.32-01
Ta	73	3.51 + 03	3.77 + 03	5.33 + 02	2.38 + 02	6.33 + 01	5.72 + 00	4.30 + 00	7.60-01	1.35-01
W	74	3.68 + 03	3.92 + 03	5.53 + 02	9.69 + 01	6.57 + 01	5.95 + 00	4.44 + 00	7.84-01	1.38-01
Re	75	3.87 + 03	3.77 + 03	5.76 + 02	1.01 + 02	6.84 + 01	6.21 + 00	4.59 + 00	8.12-01	1.41-01
Os	76	4.03 + 03	2.22 + 03	5.93 + 02	1.04 + 02	7.04 + 01	6.41 + 00	4.70 + 00	8.33-01	1.43-01
Ir	77	4.24 + 03	1.03 + 03	6.18 + 02	1.09 + 02	7.32 + 01	6.69 + 00	4.86 + 00	8.63-01	1.46-01
Pt	78	4.43 + 03	1.08 + 03	6.40 + 02	1.13 + 02	7.57 + 01	6.95 + 00	4.99 + 00	8.90-01	1.49-01
Au	79	4.65 + 03	1.14 + 03	6.66 + 02	1.18 + 02	7.88 + 01	7.26 + 00	5.16 + 00	9.22-01	1.53-01
Hg	80	4.83 + 03	1.18 + 03	6.87 + 02	1.22 + 02	8.12 + 01	7.50 + 00	5.28 + 00	9.46-01	1.56-01
Tl	81	5.01 + 03	1.23 + 03	7.07 + 02	1.26 + 02	8.36 + 01	7.75 + 00	5.40 + 00	9.69-01	1.58-01
Pb	82	5.21 + 03	1.29 + 03	7.30 + 02	1.31 + 02	8.64 + 01	8.04 + 00	5.55 + 00	9.99-01	1.61-01
Bi	83	5.44 + 03	1.35 + 03	7.58 + 02	1.36 + 02	8.95 + 01	8.38 + 00	5.74 + 00	1.03 + 00	1.66-01
Po	84	5.72 + 03	1.42 + 03	7.93 + 02	1.43 + 02	9.35 + 01	8.80 + 00	5.99 + 00	1.08 + 00	1.71-01
At	85	5.87 + 03	1.49 + 03	8.25 + 02	1.49 + 02	9.70 + 01	9.19 + 00	6.17 + 00	1.12 + 00	1.77-01
Rn	86	5.83 + 03	1.49 + 03	8.16 + 02	1.48 + 02	9.56 + 01	9.12 + 00	6.09 + 00	1.10 + 00	1.73-01
Fr	87	6.08 + 03	1.56 + 03	8.49 + 02	1.54 + 02	9.93 + 01	9.52 + 00	1.66 + 00	1.14 + 00	1.78-01

	Atomic no.	0.001 MeV	0.002 MeV	0.005 MeV	0.01 MeV	0.02 MeV	0.05 MeV	0.1 MeV	0.2 MeV	0.5 MeV
Ra	88	6.20 + 03	1.62 + 03	8.74 + 02	1.59 + 02	1.02 + 02	9.85 + 00	1.71 + 00	1.17 + 00	1.82-01
Ac	89	6.47 + 03	1.70 + 03	8.69 + 02	1.65 + 02	1.06 + 02	1.03 + 01	1.79 + 00	1.21 + 00	1.87-01
Th	90	6.61 + 03	1.74 + 03	8.88 + 02	1.69 + 02	9.37 + 01	1.05 + 01	1.83 + 00	1.23 + 00	1.90-01
Pa	91	6.53 + 03	1.83 + 03	8.76 + 02	1.77 + 02	7.03 + 01	1.10 + 01	1.92 + 00	1.29 + 00	1.97-01
U	92	6.63 + 03	1.86 + 03	8.89 + 02	1.79 + 02	7.11 + 01	1.12 + 01	1.95 + 00	1.30 + 00	1.98-01
Np	93	6.95 + 03	1.96 + 03	9.32 + 02	1.87 + 02	7.45 + 01	1.18 + 01	2.05 + 00	1.35 + 00	2.05-01
Pu	94	7.19 + 03	2.04 + 03	9.65 + 02	1.94 + 02	7.71 + 01	1.22 + 01	2.13 + 00	1.39 + 00	2.10-01
Am	95	7.37 + 03	2.10 + 03	9.90 + 02	1.98 + 02	7.93 + 01	1.25 + 01	2.19 + 00	1.42 + 00	2.14-01
Cm	96	7.54 + 03	2.15 + 03	1.02 + 03	2.03 + 02	8.14 + 01	1.28 + 01	2.25 + 00	1.44 + 00	2.18-01
Bk	97	7.84 + 03	2.25 + 03	1.06 + 03	2.10 + 02	8.39 + 01	1.34 + 01	2.35 + 00	1.50 + 00	2.25-01
Cf	98	7.89 + 03	2.31 + 03	9.27 + 02	2.15 + 02	8.58 + 01	1.37 + 01	2.41 + 00	1.52 + 00	2.29-01
Es	99	7.79 + 03	2.40 + 03	9.59 + 02	2.22 + 02	4.01 + 01	1.42 + 01	2.51 + 00	1.57 + 00	2.36-01
Fm	100	7.13 + 03	2.46 + 03	9.77 + 02	2.26 + 02	4.09 + 01	1.45 + 01	2.57 + 00	1.59 + 00	2.39-01

TABLE 4. Mass Attenuation Coefficients in $cm^2 g^{-1}$ for Sb through Fm and Photon Energy from 1.0 to 1000.0 MeV

	Atomic no.	1.0 MeV	2.0 MeV	5.0 MeV	10.0 MeV	20.0 MeV	50.0 MeV	100.0 MeV	500.0 MeV	1000.0 MeV
Sb	51	5.80-02	4.10-02	3.56-02	3.92-02	4.70-02	5.96-02	6.85-02	8.16-02	8.44-02
Te	52	5.67-02	4.01-02	3.49-02	3.86-02	4.64-02	5.89-02	6.77-02	8.07-02	8.35-02
I	53	5.84-02	4.12-02	3.61-02	4.00-02	4.82-02	6.13-02	7.04-02	8.40-02	8.69-02
Xe	54	5.78-02	4.08-02	3.58-02	3.99-02	4.82-02	6.12-02	7.04-02	8.40-02	8.69-02
Cs	55	5.85-02	4.12-02	3.64-02	4.06-02	4.91-02	6.25-02	7.19-02	8.58-02	8.88-02
Ba	56	5.80-02	4.08-02	3.61-02	4.04-02	4.90-02	6.25-02	7.19-02	8.58-02	8.88-02
La	57	5.88-02	4.12-02	3.66-02	4.11-02	5.00-02	6.37-02	7.34-02	8.76-02	9.06-02
Ce	58	5.96-02	4.18-02	3.73-02	4.19-02	5.10-02	6.52-02	7.50-02	8.96-02	9.27-02
Pr	59	6.07-02	4.24-02	3.80-02	4.29-02	5.23-02	6.68-02	7.69-02	9.19-02	9.50-02
Nd	60	6.07-02	4.24-02	3.81-02	4.30-02	5.26-02	6.72-02	7.74-02	9.25-02	9.56-02
Pm	61	6.19-02	4.31-02	3.88-02	4.40-02	5.38-02	6.89-02	7.94-02	9.48-02	9.81-02
Sm	62	6.11-02	4.24-02	3.83-02	4.35-02	5.34-02	6.84-02	7.88-02	9.41-02	9.73-02
Eu	63	6.19-02	4.28-02	3.88-02	4.42-02	5.42-02	6.96-02	8.02-02	9.57-02	9.90-02
Gd	64	6.12-02	4.23-02	3.84-02	4.38-02	5.38-02	6.91-02	7.97-02	9.51-02	9.83-02
Tb	65	6.20-02	4.27-02	3.89-02	4.45-02	5.47-02	7.03-02	8.11-02	9.67-02	1.00-01
Dy	66	6.20-02	4.26-02	3.90-02	4.46-02	5.49-02	7.06-02	8.15-02	9.72-02	1.00-01
Ho	67	6.26-02	4.29-02	3.93-02	4.50-02	5.55-02	7.14-02	8.24-02	9.83-02	1.02-01
Er	68	6.32-02	4.32-02	3.96-02	4.55-02	5.61-02	7.23-02	8.34-02	9.95-02	1.03-01
Tm	69	6.40-02	4.36-02	4.01-02	4.61-02	5.70-02	7.35-02	8.48-02	1.01-01	1.04-01
Yb	70	6.40-02	4.35-02	4.00-02	4.61-02	5.70-02	7.35-02	8.49-02	1.01-01	1.04-01
Lu	71	6.48-02	4.39-02	4.05-02	4.66-02	5.77-02	7.45-02	8.60-02	1.02-01	1.06-01
Hf	72	6.50-02	4.39-02	4.05-02	4.68-02	5.80-02	7.48-02	8.64-02	1.03-01	1.06-01
Ta	73	6.57-02	4.41-02	4.08-02	4.72-02	5.85-02	7.56-02	8.73-02	1.04-01	1.07-01
W	74	6.62-02	4.43-02	4.10-02	4.75-02	5.89-02	7.62-02	8.80-02	1.05-01	1.08-01
Re	75	6.69-02	4.46-02	4.14-02	4.79-02	5.95-02	7.70-02	8.89-02	1.06-01	1.09-01
Os	76	6.71-02	4.46-02	4.13-02	4.79-02	5.96-02	7.71-02	8.90-02	1.06-01	1.10-01
Ir	77	6.79-02	4.50-02	4.17-02	4.84-02	6.02-02	7.80-02	9.01-02	1.07-01	1.11-01
Pt	78	6.86-02	4.52-02	4.20-02	4.87-02	6.06-02	7.86-02	9.08-02	1.08-01	1.12-01
Au	79	6.95-02	4.57-02	4.24-02	4.93-02	6.14-02	7.95-02	9.19-02	1.09-01	1.13-01
Hg	80	6.99-02	4.57-02	4.25-02	4.94-02	6.15-02	7.98-02	9.22-02	1.10-01	1.13-01
Tl	81	7.03-02	4.58-02	4.25-02	4.94-02	6.16-02	8.00-02	9.24-02	1.10-01	1.14-01
Pb	82	7.10-02	4.61-02	4.27-02	4.97-02	6.21-02	8.06-02	9.31-02	1.11-01	1.15-01
Bi	83	7.21-02	4.66-02	4.32-02	5.03-02	6.28-02	8.15-02	9.42-02	1.12-01	1.16-01
Po	84	7.39-02	4.75-02	4.40-02	5.12-02	6.40-02	8.32-02	9.61-02	1.15-01	1.18-01
At	85	7.54-02	4.82-02	4.46-02	5.20-02	6.49-02	8.44-02	9.76-02	1.16-01	1.20-01
Rn	86	7.30-02	4.65-02	4.30-02	5.01-02	6.26-02	8.14-02	9.42-02	1.12-01	1.16-01
Fr	87	7.45-02	4.72-02	4.36-02	5.08-02	6.35-02	8.26-02	9.56-02	1.14-01	1.18-01
Ra	88	7.53-02	4.75-02	4.38-02	5.10-02	6.38-02	8.31-02	9.61-02	1.15-01	1.19-01
Ac	89	7.69-02	4.82-02	4.44-02	5.17-02	6.47-02	8.43-02	9.75-02	1.16-01	1.20-01
Th	90	7.71-02	4.81-02	4.42-02	5.15-02	6.45-02	8.40-02	9.72-02	1.16-01	1.20-01
Pa	91	7.94-02	4.93-02	4.52-02	5.26-02	6.59-02	8.60-02	9.95-02	1.19-01	1.23-01
U	92	7.90-02	4.88-02	4.46-02	5.19-02	6.51-02	8.49-02	9.83-02	1.17-01	1.21-01

	Atomic no.	1.0 MeV	2.0 MeV	5.0 MeV	10.0 MeV	20.0 MeV	50.0 MeV	100.0 MeV	500.0 MeV	1000.0 MeV
Np	93	8.13-02	4.99-02	4.56-02	5.30-02	6.65-02	8.68-02	1.01-01	1.20-01	1.24-01
Pu	94	8.26-02	5.05-02	4.60-02	5.34-02	6.71-02	8.76-02	1.01-01	1.21-01	1.25-01
Am	95	8.33-02	5.06-02	4.60-02	5.34-02	6.70-02	8.77-02	1.02-01	1.21-01	1.25-01
Cm	96	8.41-02	5.08-02	4.60-02	5.34-02	6.70-02	8.77-02	1.02-01	1.21-01	1.26-01
Bk	97	8.62-02	5.18-02	4.68-02	5.42-02	6.81-02	8.92-02	1.03-01	1.24-01	1.28-01
Cf	98	8.70-02	5.20-02	4.68-02	5.42-02	6.81-02	8.92-02	1.04-01	1.24-01	1.28-01
Es	99	8.89-02	5.28-02	4.74-02	5.48-02	6.89-02	9.04-02	1.05-01	1.25-01	1.29-01
Fm	100	8.94-02	5.28-02	4.72-02	5.45-02	6.86-02	9.00-02	1.05-01	1.25-01	1.29-01

TABLE 5. Absorption Edges in keV

	Atomic no.	K	L1	L2	L3	M1	M2	M3	M4	M5
Na	11	1.07								
Mg	12	1.31								
Al	13	1.56								
Si	14	1.84								
P	15	2.15								
S	16	2.47								
Cl	17	2.82								
Ar	18	3.20								
K	19	3.61								
Ca	20	4.04								
Sc	21	4.49								
Ti	22	4.97								
V	23	5.47								
Cr	24	5.99								
Mn	25	6.54								
Fe	26	7.11								
Co	27	7.71								
Ni	28	8.33	1.01							
Cu	29	8.98	1.10							
Zn	30	9.66	1.19	1.04	1.02					
Ga	31	10.37	1.30	1.14	1.12					
Ge	32	11.10	1.41	1.25	1.22					
As	33	11.87	1.53	1.36	1.32					
Se	34	12.66	1.65	1.48	1.44					
Br	35	13.47	1.78	1.60	1.55					
Kr	36	14.33	1.92	1.73	1.68					
Rb	37	15.20	2.07	1.86	1.80					
Sr	38	16.10	2.22	2.01	1.94					
Y	39	17.04	2.37	2.16	2.08					
Zr	40	18.00	2.53	2.31	2.22					
Nb	41	18.99	2.70	2.47	2.37					
Mo	42	20.00	2.87	2.63	2.52					
Tc	43	21.04	3.04	2.79	2.68					
Ru	44	22.12	3.22	2.97	2.84					
Rh	45	23.22	3.41	3.15	3.00					
Pd	46	24.35	3.60	3.33	3.17					
Ag	47	25.51	3.81	3.52	3.35					
Cd	48	26.71	4.02	3.73	3.54					
In	49	27.94	4.24	3.94	3.73					
Sn	50	29.20	4.47	4.16	3.3					
Sb	51	30.49	4.70	4.38	4.13					
Te	52	31.81	4.94	4.61	4.34	1.01				
I	53	33.17	5.19	4.85	4.56	1.07				
Xe	54	34.56	5.45	5.10	4.78	1.15				
Cs	55	35.98	5.71	5.36	5.01	1.22	1.07			
Ba	56	37.44	5.99	5.62	5.25	1.29	1.14	1.06		
La	57	38.92	6.27	5.89	5.48	1.36	1.20	1.12		

	Atomic no.	K	L1	L2	L3	M1	M2	M3	M4	M5
Ce	58	40.44	6.55	6.16	5.72	1.44	1.27	1.19		
Pr	59	41.99	6.84	6.44	5.96	1.51	1.34	1.24		
Nd	60	43.57	7.13	6.72	6.21	1.58	1.40	1.30	1.01	
Pm	61	45.18	7.43	7.01	6.46	1.65	1.47	1.36	1.05	1.03
Sm	62	46.83	7.74	7.31	6.72	1.72	1.54	1.42	1.11	1.08
Eu	63	48.52	8.05	7.62	6.98	1.80	1.61	1.48	1.16	1.13
Gd	64	50.24	8.38	7.93	7.24	1.88	1.69	1.54	1.22	1.19
Tb	65	52.00	8.71	8.25	7.51	1.97	1.77	1.61	1.28	1.24
Dy	66	53.79	9.05	8.58	7.79	2.05	1.84	1.68	1.33	1.30
Ho	67	55.62	9.39	8.92	8.07	2.13	1.92	1.74	1.39	1.35
Er	68	57.49	9.75	9.26	8.36	2.21	2.01	1.81	1.45	1.41
Tm	69	59.39	10.12	9.62	8.65	2.31	2.09	1.88	1.52	1.47
Yb	70	61.33	10.49	9.98	8.94	2.40	2.17	1.95	1.58	1.53
Lu	71	63.31	10.87	10.35	9.24	2.49	2.26	2.02	1.64	1.59
Hf	72	65.35	11.27	10.74	9.56	2.60	2.37	2.11	1.72	1.66
Ta	73	67.42	11.68	11.14	9.88	2.71	2.47	2.19	1.79	1.74
W	74	69.53	12.10	11.54	10.21	2.82	2.58	2.28	1.87	1.81
Re	75	71.68	12.53	11.96	10.54	2.93	2.68	2.37	1.95	1.82
Os	76	73.87	12.97	12.39	10.87	3.05	2.79	2.46	2.03	1.96
Ir	77	76.11	13.42	12.82	11.22	3.17	2.91	2.55	2.12	2.04
Pt	78	78.39	13.88	13.27	11.56	3.30	3.03	2.65	2.20	2.12
Au	79	80.72	14.35	13.73	11.92	3.43	3.15	2.74	2.29	2.21
Hg	80	83.10	14.84	14.21	12.28	3.56	3.28	2.85	2.39	2.30
Tl	81	85.53	15.35	14.70	12.66	3.70	3.42	2.96	2.49	2.39
Pb	82	88.00	15.86	15.20	13.04	3.85	3.55	3.07	2.59	2.48
Bi	83	90.53	16.39	15.71	13.42	4.00	3.70	3.18	2.69	2.58
Po	84	93.10	16.94	16.24	13.81	4.15	3.85	3.30	2.80	2.68
At[a]	85	95.73	17.49	16.78	14.21	4.32	4.01	3.43	2.91	2.79
Rn[a]	86	98.40	18.05	17.34	14.62	4.48	4.16	3.54	3.02	2.89
Fr[a]	87	101.1	18.64	17.91	15.03	4.65	4.33	3.66	3.14	3.00
Ra[a]	88	103.9	19.24	18.48	15.44	4.82	4.49	3.79	3.25	3.10
Ac[a]	89	106.8	19.84	19.08	15.87	5.00	4.66	3.91	3.37	3.22
Th[a]	90	109.7	20.47	19.69	16.30	5.18	4.83	4.05	3.49	3.33
Pa[a]	91	112.6	21.10	20.31	16.73	5.37	5.00	4.17	3.61	3.44
U[a]	92	115.6	21.76	20.95	17.17	5.55	5.18	4.30	3.73	3.55
Np[a]	93	118.7	22.43	21.60	17.61	5.72	5.37	4.44	3.85	3.67
Pu[a]	94	121.8	23.10	22.27	18.06	5.93	5.54	4.56	3.97	3.78
Am[a]	95	125.0	23.77	22.94	18.50	6.12	5.71	4.67	4.09	3.89
Cm[a]	96	128.2	24.46	23.80	18.93	6.29	5.90	4.80	4.23	3.97
Bk[a]	97	131.6	25.27	24.39	19.45	6.56	6.15	4.98	4.37	4.13
Cf[a]	98	136.0	26.11	25.25	19.93	6.75	6.36	5.11	4.50	4.25
Es[a]	99	139.5	26.90	26.02	20.41	6.98	6.57	5.25	4.63	4.37
Fm[a]	100	143.1	27.70	26.81	20.90	7.21	6.79	5.40	4.77	4.50

[a] Additional absorption edges are present; see Refs. 3, 4.

CLASSIFICATION OF ELECTROMAGNETIC RADIATION

David R. Lide

Through historical usage and, in some cases, action of standardization bodies, the electromagnetic spectrum has been divided into frequency bands or regions where the character of the radiation is similar. Table 1 shows a division into decade frequency steps running from 3 Hz to 30 ZHz (30×10^{21} Hz), i.e., from low radio frequencies (RF) to γ-rays. The second column gives the approximate wavelength range corresponding to each frequency range. The wavenumber range and approximate photon energy range are also given. The last column gives the designation of each of the bands. The bands up to 3 THz frequency have been given letter designations by the International Telecommunication Union (ITU), but it should be noted that many other designations are in use. The higher frequency regions are labeled by historical terms like infrared (IR), visible, ultraviolet (UV), etc. There are no sharp boundaries between these regions, and usage varies.

The exact relations between these parameters are:

$$\nu = c/\lambda = ck$$
$$\lambda = c/\nu = 1/k$$
$$k = \nu/c = 1/\lambda$$
$$c = \text{speed of light} = 2.99\,792\,458 \times 10^8 \text{ m/s}$$

The major regions are illustrated graphically in Figure 1.

The microwave region, which is usually considered to extend from about 1 GHz to 300 GHz, has been subdivided into several bands that are given arbitrary letter designations. These bands, which are important in communications and radar technology, are listed in Table 2.

TABLE 1. Frequency, Wavelength, Wavenumber, and Energy Ranges of Radiation Bands in Decade Steps

Frequency (ν)	Wavelength (λ)	Wave number (k)	Approximate photon energy	Band name	Band designation
3 Hz—30 Hz	10^8—10^7 m	10^{-8}—10^{-7} m^{-1}		Extremely low frequency	ELF (ELF 1)
30 Hz—300 Hz	10^7—10^6 m	10^{-7}—10^{-6} m^{-1}		Super low frequency	SLF (ELF 2)
300 Hz—3 kHz	10^6—10^5 m	10^{-6}—10^{-5} m^{-1}		Ultralow frequency	ULF (ELF 3)
3 kHz—30 kHz	100 km—10 km	10^{-5}—10^{-4} m^{-1}		Very low frequency	VLF
30 kHz—300 kHz	10 km—1 km	10^{-4}—10^{-3} m^{-1}		Low frequnecy	LF
300 kHz—3 MHz	1 km—100 m	10^{-3}—10^{-2} m^{-1}		Medium frequency	MF
3 MHz—30 MHz	100 m—10 m	10^{-2}—10^{-1} m^{-1}		High frequency	HF
30 MHz—300 MHz	10 m—1 m	10^{-1}—10^0 m^{-1}		Very high frequency	VHF
300 MHz—3 GHz	1 m—100 mm	10^0—10^1 m^{-1}		Ultrahigh frequency	UHF
3 GHz—30 GHz	100 mm—10 mm	10^1—10^2 m^{-1}		Super high frequency	SHF, microwave
30 GHz—300 GHz	100 mm –10 mm	1 - 10 cm^{-1}		Extremely high frequency	EHF, microwave
300 GHz—3 THz	1 mm—100 μm	10—100 cm^{-1}		Far infrared	THF, far IR
3 THz—30 THz	100 μm—10 μm	100—1000 cm^{-1}		Mid infrared	Far IR, mid IR
30 THz—300 THz	10 μm—1 μm	1000—10,000 cm^{-1}	0.1 eV—1 eV	Near infrared	Mid IR, near IR
300 THz—3 PHz	1 μm—100 nm	1—10 μm^{-1}	1 eV—10 eV	Near ultraviolet, visible	Visible, near UV
3 PHz—30 PHz	100 nm—10 nm	10—100 μm^{-1}	10 eV—100 eV	Extreme ultraviolet	Far (vacuum) UV
30 PHz—300 PHz	10 nm—1 nm	100 μm^{-1}—1 nm^{-1}	100 eV—1 keV	Soft x-ray	Soft x-ray
300 PHz—3 EHz	1 nm—100 pm	1—10 nm^{-1}	1 keV—10 keV	Soft x-ray	Soft x-ray
3 EHz—30 EHz	100 pm—10 pm	10—100 nm^{-1}	10 keV—100 keV	Hard x-ray, soft γ-ray	Hard x-ray, soft γ-ray
30 EHz—300 EHz	10 pm—1 pm	100 nm^{-1}—1 pm^{-1}	100 keV—1 MeV	Soft γ-ray, hard γ-ray	Soft γ-ray, hard γ-ray
300 EHz—3 ZHz	1 pm—100 fm	1—10 pm^{-1}	1 MeV—10 MeV	Hard γ-ray, cosmic γ-ray	Hard γ-ray, cosmic γ-ray
3 ZHz—30 ZHz	100 pm—10 fm	10—100 pm^{-1}	10 Mev—100 MeV	Cosmic ray	Cosmic γ-ray

Abbreviations for radio frequency bands: EHF–extremely high frequency; ELF–extremely low frequency; HF–high frequency; ITU–International Telecommunications Union; LF–low frequency; MF–medium frequency; SHF–super high frequency; SLF–super low frequency; UHF–ultra high frequency; ULF–ultra low frequency; VHF–very high frequency; VLF–very low frequency.

TABLE 2. Designations of Microwave Bands (IEEE Standard 521-1984)

Band	Frequency (GHz)	Wavelength (cm)	Wavenumber (cm^{-1})
L	1—2	30—15	0.033—0.067
S	2—4	15—7.5	0.067—0.133
C	4—8	7.5—3.7	0.133—0.267
X	8—12	3.7—2.5	0.267—0.4
Ku	12—18	2.5—1.7	0.4—0.6
K	18—27	1.7—1.1	0.6—0.9
Ka	27—40	1.1—0.75	0.9—1.33
V	40—75	0.75—0.40	1.33—2.5
W	75—110	0.40—0.27	2.5—3.7

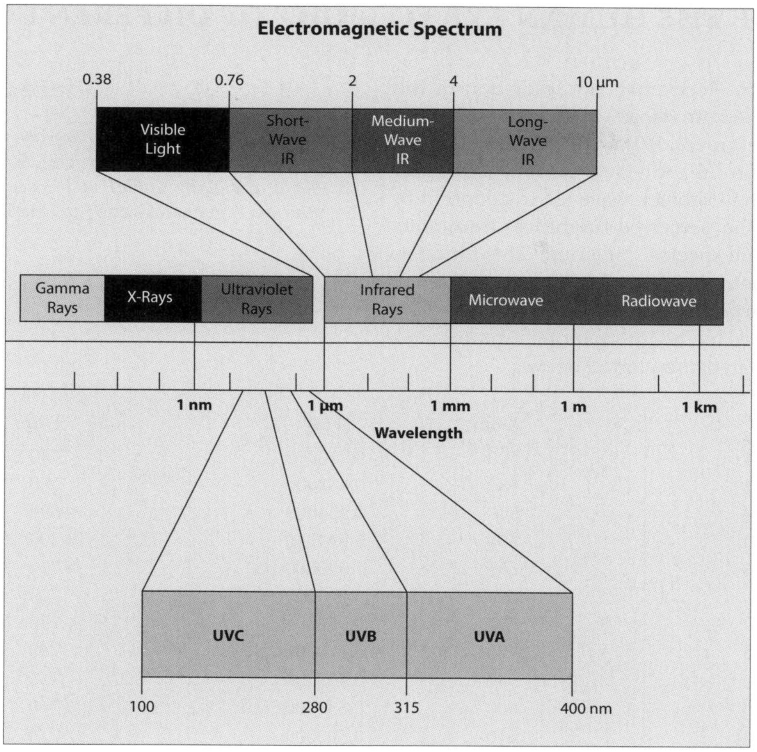

FIGURE 1. Subdivisions of the Electromagnetic Spectrum

SENSITIVITY OF THE HUMAN EYE TO LIGHT OF DIFFERENT WAVELENGTHS

The human eye responds to electromagnetic radiation in the wavelength range from about 360 nm (violet) to 820 nm (red), with a peak sensitivity near 555 nm (green). While the detailed shape of this response curve depends on the individual person, studies on representative samples of human subjects have led to adoption of a standard function relating the perceived brightness (luminous flux) to the actual power of the spectral radiation. This function is referred to as $V(\lambda)$, the photopic spectral luminous efficiency function, and it plays an important role in photometry.

The function $V(\lambda)$, as adopted by the International Commission on Illumination (CIE), is tabulated and plotted below.

References

1. *The Basis for Physical Photometry*, CIE Publication #18.2, 1983.
2. *CIE Standard Colorimetric Observers*, ISO/CIE #10527, 1991.
3. *Kaye and Laby Tables of Physical and Chemical Constants, 16th Edition*, Longman Group Ltd., Harlow, Essex, 1995.

λ/nm	$V(\lambda)$	λ/nm	$V(\lambda)$	λ/nm	$V(\lambda)$
360	0.000004	520	0.710000	670	0.032000
370	0.000012	530	0.862000	680	0.017000
380	0.000039	540	0.954000	690	0.008210
390	0.000120	550	0.994950	700	0.004102
400	0.000396	555	1.000000	710	0.002091
410	0.001210	560	0.995000	720	0.001047
420	0.004000	570	0.952000	730	0.000520
430	0.011600	580	0.870000	740	0.000249
440	0.023000	590	0.757000	750	0.000120
450	0.038000	600	0.631000	760	0.000060
460	0.060000	610	0.503000	770	0.000030
470	0.090980	620	0.381000	780	0.000015
480	0.139020	630	0.265000	790	0.000007
490	0.208020	640	0.175000	800	0.000004
500	0.323000	650	0.107000	810	0.000002
510	0.503000	660	0.061000	820	0.000001

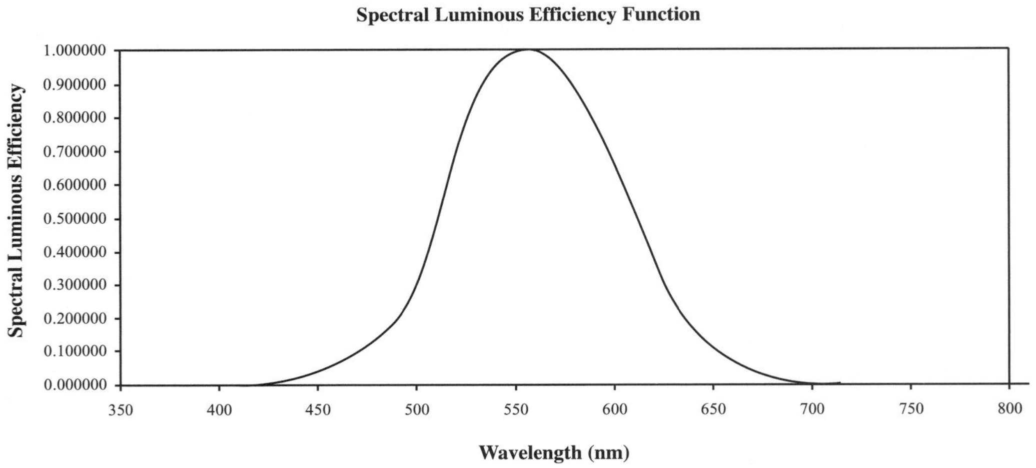

BLACKBODY RADIATION

The total power radiated from an ideal blackbody and the wavelength corresponding to maximum power are given here as a function of absolute temperature. The radiated power in a band $\Delta\lambda$ at λ_{max} may be calculated from:

$$P_{max} = 0.657548\ (\Delta\lambda/\lambda_{max})\ P_{tot}$$

Total Power and Wavelength of Maximum Power as a Function of Temperature

T/K	P_{tot}	Units	λ_{max}/μm
50	0.354398	W/m²	57.955
100	5.67037	W/m²	28.978
150	28.7063	W/m²	19.318
200	90.7260	W/m²	14.489
250	221.499	W/m²	11.591
273	314.965	W/m²	10.614
280	348.533	W/m²	10.349
290	401.055	W/m²	9.992
300	459.300	W/m²	9.659
310	523.671	W/m²	9.348
320	594.582	W/m²	9.055
330	672.461	W/m²	8.781
340	757.752	W/m²	8.523
350	850.910	W/m²	8.279
360	952.405	W/m²	8.049
370	1.06272	kW/m²	7.832
380	1.18235	kW/m²	7.626
390	1.31181	kW/m²	7.430
400	1.45162	kW/m²	7.244
410	1.60231	kW/m²	7.068
420	1.76445	kW/m²	6.899
430	1.93859	kW/m²	6.739
440	2.12531	kW/m²	6.586
450	2.32521	kW/m²	6.439
460	2.53888	kW/m²	6.299
470	2.76696	kW/m²	6.165
480	3.01007	kW/m²	6.037
490	3.26886	kW/m²	5.914
500	3.54398	kW/m²	5.796
510	3.83612	kW/m²	5.682
520	4.14596	kW/m²	5.573
530	4.47420	kW/m²	5.467
540	4.82155	kW/m²	5.366
550	5.18875	kW/m²	5.269
560	5.57653	kW/m²	5.175
570	5.98565	kW/m²	5.084
580	6.41688	kW/m²	4.996
590	6.87100	kW/m²	4.911
600	7.34880	kW/m²	4.830
610	7.85111	kW/m²	4.750
620	8.37873	kW/m²	4.674
630	8.93252	kW/m²	4.600
640	9.51331	kW/m²	4.528
650	10.1220	kW/m²	4.458
660	10.7594	kW/m²	4.391
670	11.4264	kW/m²	4.325
680	12.1240	kW/m²	4.261
690	12.8531	kW/m²	4.200
700	13.6146	kW/m²	4.140
710	14.4094	kW/m²	4.081
720	15.2385	kW/m²	4.025
730	16.1029	kW/m²	3.970
740	17.0035	kW/m²	3.916
750	17.9414	kW/m²	3.864
760	18.9176	kW/m²	3.813
770	19.9331	kW/m²	3.763
780	20.9889	kW/m²	3.715
790	22.0861	kW/m²	3.668
800	23.2258	kW/m²	3.622
810	24.4091	kW/m²	3.577
820	25.6370	kW/m²	3.534
830	26.9106	kW/m²	3.491
840	28.2312	kW/m²	3.450
850	29.5997	kW/m²	3.409
860	31.0174	kW/m²	3.369
870	32.4854	kW/m²	3.331
880	34.0050	kW/m²	3.293
890	35.5772	kW/m²	3.256
900	37.2033	kW/m²	3.220
910	38.8846	kW/m²	3.184
920	40.6222	kW/m²	3.150
930	42.4173	kW/m²	3.116
940	44.2714	kW/m²	3.083
950	46.1855	kW/m²	3.050
960	48.1611	kW/m²	3.018
970	50.1994	kW/m²	2.987
980	52.3017	kW/m²	2.957
990	54.4694	kW/m²	2.927
1000	56.7037	kW/m²	2.898
1020	61.3779	kW/m²	2.841
1040	66.3353	kW/m²	2.786
1060	71.5872	kW/m²	2.734
1080	77.1448	kW/m²	2.683
1100	83.0199	kW/m²	2.634
1120	89.2244	kW/m²	2.587
1140	95.7703	kW/m²	2.542
1160	102.670	kW/m²	2.498
1180	109.936	kW/m²	2.456
1200	117.581	kW/m²	2.415
1220	125.618	kW/m²	2.375
1240	134.060	kW/m²	2.337
1260	142.920	kW/m²	2.300
1280	152.213	kW/m²	2.264
1300	161.952	kW/m²	2.229
1320	172.150	kW/m²	2.195
1340	182.823	kW/m²	2.163
1360	193.985	kW/m²	2.131
1380	205.650	kW/m²	2.100
1400	217.833	kW/m²	2.070
1420	230.550	kW/m²	2.041
1440	243.816	kW/m²	2.012
1460	257.646	kW/m²	1.985
1480	272.056	kW/m²	1.958
1500	287.063	kW/m²	1.932
1520	302.682	kW/m²	1.906
1540	318.929	kW/m²	1.882
1560	335.823	kW/m²	1.858
1580	353.378	kW/m²	1.834
1600	371.614	kW/m²	1.811
1620	390.546	kW/m²	1.789
1640	410.192	kW/m²	1.767
1660	430.570	kW/m²	1.746
1680	451.699	kW/m²	1.725
1700	473.595	kW/m²	1.705
1720	496.278	kW/m²	1.685
1740	519.767	kW/m²	1.665
1760	544.079	kW/m²	1.646
1780	569.235	kW/m²	1.628
1800	595.253	kW/m²	1.610
1820	622.153	kW/m²	1.592
1840	649.954	kW/m²	1.575
1860	678.677	kW/m²	1.558
1880	708.342	kW/m²	1.541
1900	738.969	kW/m²	1.525
1920	770.578	kW/m²	1.509
1940	803.190	kW/m²	1.494
1960	836.827	kW/m²	1.478
1980	871.510	kW/m²	1.464
2000	907.260	kW/m²	1.449
2020	944.098	kW/m²	1.435
2040	982.047	kW/m²	1.420
2060	1.02113	MW/m²	1.407
2080	1.06137	MW/m²	1.393
2100	1.10278	MW/m²	1.380
2120	1.14539	MW/m²	1.367
2140	1.18923	MW/m²	1.354
2160	1.23432	MW/m²	1.342
2180	1.28067	MW/m²	1.329
2200	1.32832	MW/m²	1.317
2220	1.37728	MW/m²	1.305
2240	1.42759	MW/m²	1.294
2260	1.47926	MW/m²	1.282
2280	1.53233	MW/m²	1.271
2300	1.58680	MW/m²	1.260
2320	1.64272	MW/m²	1.249
2340	1.70010	MW/m²	1.238
2360	1.75897	MW/m²	1.228
2380	1.81936	MW/m²	1.218
2400	1.88129	MW/m²	1.207
2420	1.94479	MW/m²	1.197

T/K	P_{tot}	Units	λ_{max}/µm	*T*/K	P_{tot}	Units	λ_{max}/µm	*T*/K	P_{tot}	Units	λ_{max}/µm
2440	2.00988	MW/m^2	1.188	3400	7.57752	MW/m^2	0.852	4900	32.6886	MW/m^2	0.591
2460	2.07660	MW/m^2	1.178	3450	8.03319	MW/m^2	0.840	5000	35.4398	MW/m^2	0.580
2480	2.14496	MW/m^2	1.168	3500	8.50910	MW/m^2	0.828	5100	38.3612	MW/m^2	0.568
2500	2.21499	MW/m^2	1.159	3550	9.00586	MW/m^2	0.816	5200	41.4596	MW/m^2	0.557
2550	2.39758	MW/m^2	1.136	3600	9.52405	MW/m^2	0.805	5300	44.7420	MW/m^2	0.547
2600	2.59122	MW/m^2	1.115	3650	10.0643	MW/m^2	0.794	5400	48.2155	MW/m^2	0.537
2650	2.79637	MW/m^2	1.093	3700	10.6272	MW/m^2	0.783	5500	51.8875	MW/m^2	0.527
2700	3.01347	MW/m^2	1.073	3750	11.2134	MW/m^2	0.773	5600	55.7653	MW/m^2	0.517
2750	3.24297	MW/m^2	1.054	3800	11.8235	MW/m^2	0.763	5700	59.8565	MW/m^2	0.508
2800	3.48533	MW/m^2	1.035	3850	12.4582	MW/m^2	0.753	5800	64.1688	MW/m^2	0.500
2850	3.74103	MW/m^2	1.017	3900	13.1181	MW/m^2	0.743	5900	68.7100	MW/m^2	0.491
2900	4.01055	MW/m^2	0.999	3950	13.8038	MW/m^2	0.734	6000	73.4880	MW/m^2	0.483
2950	4.29437	MW/m^2	0.982	4000	14.5162	MW/m^2	0.724	6500	101.220	MW/m^2	0.446
3000	4.59300	MW/m^2	0.966	4100	16.0231	MW/m^2	0.707	7000	136.146	MW/m^2	0.414
3050	4.90694	MW/m^2	0.950	4200	17.6445	MW/m^2	0.690	7500	179.414	MW/m^2	0.386
3100	5.23671	MW/m^2	0.935	4300	19.3859	MW/m^2	0.674	8000	232.258	MW/m^2	0.362
3150	5.58282	MW/m^2	0.920	4400	21.2531	MW/m^2	0.659	8500	295.997	MW/m^2	0.341
3200	5.94582	MW/m^2	0.906	4500	23.2521	MW/m^2	0.644	9000	372.033	MW/m^2	0.322
3250	6.32623	MW/m^2	0.892	4600	25.3888	MW/m^2	0.630	9500	461.855	MW/m^2	0.305
3300	6.72461	MW/m^2	0.878	4700	27.6696	MW/m^2	0.617	10000	567.037	MW/m^2	0.290
3350	7.14152	MW/m^2	0.865	4800	30.1007	MW/m^2	0.604				

The curves below show, for various temperatures, the fraction of radiant power as a function of wavelength. The function plotted is $P_\lambda/\Delta\lambda P_{tot}$, where P_λ is the power at wavelength λ in a small interval $\Delta\lambda$ (in µm), and P_{tot} is the total power.

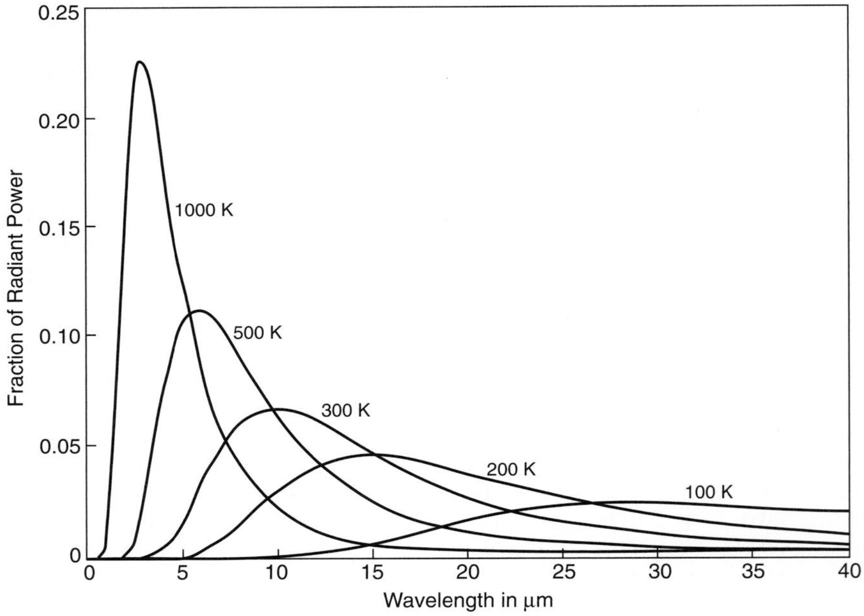

CHARACTERISTICS OF INFRARED DETECTORS

This graph summarizes the wavelength response of some semiconductors used as detectors for infrared radiation. The quantity $D^*(\lambda)$ is the signal-to-noise ratio for an incident radiant power density of 1 W/cm^2 and a bandwidth of 1 Hz (60° field of view). The Ge, InAs, and InSb detectors are photovoltaics, while the HgCdTe series are photoconductive devices. The cutoff wavelength of the latter can be varied by adjusting the relative amounts of Hg, Cd, and Te (three examples are shown at 77 K). The graph also shows the theoretical background limited sensitivity for ideal detectors which introduce no intrinsic noise.

Reference

Infrared Detectors 1995, EG&G Judson, Montgomeryville, PA.

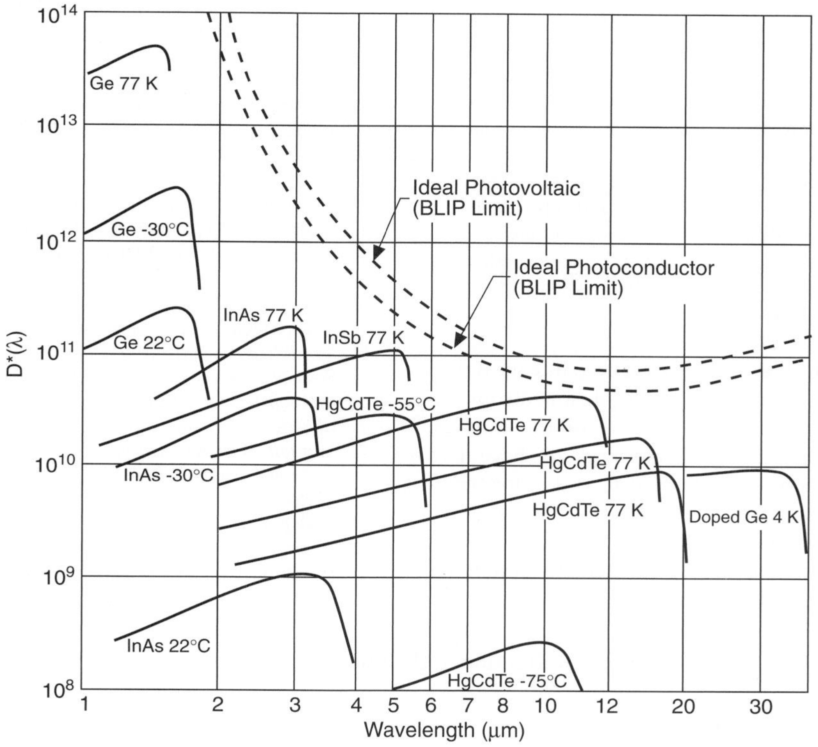

INDEX OF REFRACTION OF INORGANIC CRYSTALS

This table lists the index of refraction of selected crystalline inorganic compounds. When available, values are given as a function of wavelength in the range from the ultraviolet to the far infrared region. For most compounds a value at 589 nm, the wavelength of the principal sodium line, is given. The data have been taken from the references indicated; in many cases, data from a reference have been refitted to generate the index of refraction at the wavelengths used in this table. All values refer to ambient temperature. Entries marked by * are based on extrapolation beyond the range of available experimental data. Column definitions for the table are as follows.

Column heading	Definition
Mol. form.	Molecular formula of inorganic crystal, list ordered alphabetically
Cryst. syst.	Crystallographic system of the crystal
Ray	Notation for isotropic and anisotropic indexes; see below for definitions
***n* (200 nm), *n* (400 nm), etc.**	Index of refraction at the wavelength in nm specified within the parentheses
Ref.	Reference

Compounds belonging to the cubic crystal system have only a single refractive index value, but other systems are anisotropic, so that the crystal is characterized by two or three unique indexes. Hexagonal, rhombohedral, and tetragonal crystals have two unique indexes that are traditionally labeled n_o and n_e for "ordinary ray" and "extraordinary ray." Orthorhombic, monoclinic, and triclinic crystals are characterized by three indexes that here are called n_x, n_y, and n_z. The table indicates the crystal system for each entry in order to identify the material uniquely.

Compounds are listed in order of their common formula.

The refractive index and other optical properties for metals, semiconductors, and certain other compounds can be found in the tables "Optical Properties of Selected Elements" and "Optical Properties of Selected Inorganic and Organic Solids" in Section 12 of this *CRC Handbook.*

References

1. Li, H. H., Refractive Index of Alkali Halides and its Wavelength and Temperature Derivatives, *J. Phys. Chem. Ref. Data* 5, 329, 1976. <https://doi.org/10.1063/1.555536>
2. Li, H. H., Refractive Index of Alkaline Earth Halides and its Wavelength and Temperature Derivatives, *J. Phys. Chem. Ref. Data* 9, 161, 1980. <https://doi.org/10.1063/1.555616>
3. Li, H. H., Refractive Index of ZnS, ZnSe, and ZnTe and its Wavelength and Temperature Derivatives, *J. Phys. Chem. Ref. Data* 13, 103, 1984. <https://doi.org/10.1063/1.555705>
4. Shannon, R. D., Shannon, R. C., Medenbach, O., and Fischer, R. X., Refractive Index and Dispersion of Fluorides and Oxides, *J. Phys. Chem. Ref. Data* 31, 931, 2002. <https://doi.org/10.1063/1.1497384>
5. Gray, D. E., ed., *American Institute of Physics Handbook*, Sec. 6b, pp. 6–12, McGraw-Hill, New York, 1972.
6. *Landolt-Börnstein Numerical Data and Functional Relationships in Science and Technology, III/11, Elastic, Piezoelectric, Pyroelectric, Piezooptic, Electrooptic Constants, and Nonlinear Dielectric Susceptibilities of Crystals*, Springer-Verlag, Berlin, 1979.
7. *Landolt-Börnstein Numerical Data and Functional Relationships in Science and Technology, III/30A, High Frequency Properties of Dielectric Crystals. Piezooptic and Electrooptic Constants*, Springer-Verlag, Berlin, 1996.
8. Weber, M. J., *CRC Handbook of Laser Science and Technology*, Vol. IV. Optical Materials. Part 2: Properties, CRC Press, Boca Raton, FL, 1986.

Index of Refraction at the Indicated Wavelength

Mol. form	Cryst. syst.	Ray	*n* (200 nm)	*n* (300 nm)	*n* (589 nm)	*n* (750 nm)	*n* (1 μm)	*n* (2 μm)	*n* (5 μm)	*n* (10 μm)	*n* (20 μm)	Ref.
AgCl	cub	n			2.0668	2.0401	2.0224	2.0062	1.9975	1.9803	1.9069	5
$AlPO_4$	rhomb	n_o			1.5247	1.5203	1.5161	1.5034				6
$AlPO_4$	rhomb	n_e			1.5338	1.5290	1.5245	1.5116				6
Al_2O_3	hex	n_o			1.7673							4
Al_2O_3	hex	n_e			1.7598							4
As_2O_3[a]	cub	n			1.7537							4
BaF_2	cub	n	1.557	1.5010	1.4744	1.4712	1.4686	1.4647	1.4511	1.4014		2
BaO	cub	n			1.9841							4
$BaSO_4$	orth	n_x			1.6362							4
$BaSO_4$	orth	n_y			1.6374							4
$BaSO_4$	orth	n_z			1.6480							4
$BaTiO_3$	tetr	n_o			2.4405							4
$BaTiO_3$	tetr	n_e			2.3831							4
$BaWO_4$	tetr	n_o			1.8426							4
$BaWO_4$	tetr	n_e			1.8405							4
BeO	hex	n_o			1.7184							4
BeO	hex	n_e			1.7342							4
$BeSO_4 \cdot 4H_2O$	tetr	n_o			1.4713							4
$BeSO_4 \cdot 4H_2O$	tetr	n_e			1.4328							4
$CaCO_3$[b]	hex	n_o	1.9028	1.7216	1.6584	1.6503	1.6436	1.6249				5
$CaCO_3$[b]	hex	n_e	1.5765	1.5145	1.4864	1.4828	1.4801	1.4753				5
CaF_2	cub	n	1.495	1.4540	1.4338	1.4311	1.4289	1.4239	1.3990	1.299		2
CaO	cub	n			1.8396							4
$CaSO_4$	orth	n_x			1.5698							4

Mol. form	Cryst. syst.	Ray	n (200 nm)	n (300 nm)	n (589 nm)	n (750 nm)	n (1 μm)	n (2 μm)	n (5 μm)	n (10 μm)	n (20 μm)	Ref.
$CaSO_4$	orth	n_y			1.5755							4
$CaSO_4$	orth	n_z			1.6137							4
$CaSO_4·2H_2O$	monocl	n_x			1.5207							4
$CaSO_4·2H_2O$	monocl	n_y			1.5227							4
$CaSO_4·2H_2O$	monocl	n_z			1.5304							4
$CaWO_4$	tetr	n_o			1.9195							4
$CaWO_4$	tetr	n_e			1.9355							4
CdS	hex	n_o			2.507	2.390	2.334					5
CdS	hex	n_e			2.525	2.409	2.352					5
CdSe	hex	n_o				2.68*	2.5502	2.4682	2.4483	2.4331		7
CdSe	hex	n_e				2.69*	2.5696	2.4873	2.4676	2.4514		7
CdTe	cub	n								2.6724	2.6302	7
CeF_3	hex	n_o			1.6183							4
CeF_3	hex	n_e			1.6113							4
CsBr	cub	n		1.8047	1.6974	1.6861	1.6784	1.6711	1.6678	1.6630	1.6439	1
CsCl	cub	n	1.937	1.712	1.640	1.631	1.626	1.620	1.616	1.606	1.563	1
$CsClO_4$	orth	n_x			1.4752							4
$CsClO_4$	orth	n_y			1.4788							4
$CsClO_4$	orth	n_z			1.4804							4
CsF	cub	n	1.575*	1.506	1.477	1.474	1.472	1.469*	1.461*	1.436*	1.32*	1
CsI	cub	n		1.9790	1.7873	1.7694	1.7576	1.7465	1.7428	1.7396	1.7280	1
Cs_2SO_4	orth	n_x			1.5598							4
Cs_2SO_4	orth	n_y			1.5644							4
Cs_2SO_4	orth	n_z			1.5662							4
CuBr	cub	n			2.117							7
CuCl	cub	n			1.9727	1.9391				1.9245		7
$CuSO_4·5H_2O$	tricl	n_x			1.5140							4
$CuSO_4·5H_2O$	tricl	n_y			1.5367							4
$CuSO_4·5H_2O$	tricl	n_z			1.5436							4
Dy_2O_3	cub	n			1.9757							4
FeF_2	tetr	n_o			1.514							4
FeF_2	tetr	n_e			1.524							4
Gd_2O_3	cub	n			1.96							4
HgS	rhomb	n_o			2.9413	2.7770	2.7120	2.6305		2.6018		6
HgS	rhomb	n_e			3.3072	3.0896	3.0050	2.8776		2.8522		6
KBr	cub	n	2.099*	1.6482	1.5598	1.5498	1.5444	1.5383	1.5345	1.5264	1.4924	1
KCl	cub	n	1.717	1.5455	1.4902	1.4840	1.4798	1.4753	1.4704	1.4564	1.3946	1
$KClO_4$	orth	n_x			1.4730							4
$KClO_4$	orth	n_y			1.4736							4
$KClO_4$	orth	n_z			1.4768							4
KF	cub	n	1.426	1.380	1.362	1.360	1.358	1.355	1.344	1.304*	1.09*	1
KH_2AsO_4	tetr	n_o			1.5674							7
KH_2AsO_4	tetr	n_e			1.5179							7
KH_2PO_4	tetr	n_o	1.617*	1.5450	1.5093	1.5030	1.4957					5
KH_2PO_4	tetr	n_e	1.558*	1.4977	1.4682	1.4641	1.4606					5
KI	cub	n		1.834*	1.665	1.650	1.640	1.631	1.627	1.620	1.593	1
KIO_3	tricl	n_x			1.6959							7
KIO_3	tricl	n_y			1.8317							7
KIO_3	tricl	n_z			1.8343							7
KIO_4	tetr	n_o			1.6205							4
KIO_4	tetr	n_e			1.6476							4
$KNbO_3$	orth	n_x			2.2480	2.3395	2.2612					7
$KNbO_3$	orth	n_y			2.3464	2.2959	2.2622					7
$KNbO_3$	orth	n_x			2.1803	2.1457	2.1288					7
K_2SO_4	orth	n_x			1.4934							4
K_2SO_4	orth	n_y			1.4947							4
K_2SO_4	orth	n_z			1.4973							4
LaF_3	hex	n_o			1.605							4
LaF_3	hex	n_e			1.599							4
LiBr	cub	n	1.92*	1.810	1.783	1.781	1.778	1.774*	1.756*	1.68*	1.33	1
LiCl	cub	n	1.72	1.677	1.662	1.660	1.658	1.654*	1.62*	1.53*		1
$LiClO_4·3H_2O$	hex	n_o			1.4832							4
$LiClO_4·3H_2O$	hex	n_e			1.4384							4

Mol. form	Cryst. syst.	Ray	*n* (200 nm)	*n* (300 nm)	*n* (589 nm)	*n* (750 nm)	*n* (1 μm)	*n* (2 μm)	*n* (5 μm)	*n* (10 μm)	*n* (20 μm)	Ref.
LiF	cub	n	1.439	1.4087	1.3921	1.3895	1.3871	1.3786	1.3266	1.1005		1
LiI	cub	n		1.979	1.955	1.952	1.950	1.948*	1.940*	1.91*	1.77*	1
$LiIO_3$	hex	n_o			1.8875	1.8713	1.8589	1.8410				6
$LiIO_3$	hex	n_e			1.7400	1.7268	1.7179	1.7062				6
$LiNbO_3$	rhomb	n_o			2.3007	2.2632	2.2370					7
$LiNbO_3$	rhomb	n_e			2.2116	2.1804	2.1567					7
$LiTaO_3$	rhomb	n_o			2.1864	2.1590	2.1391	2.1066				7
$LiTaO_3$	rhomb	n_e			2.1908	2.1634	2.1432	2.1115				7
$Li_2SO_4 \cdot H_2O$	monocl	n_x			1.4615							4
$Li_2SO_4 \cdot H_2O$	monocl	n_y			1.4765							4
$Li_2SO_4 \cdot H_2O$	monocl	n_z			1.4863							4
Lu_2O_3	cub	n			1.9349							4
MgF_2	tetr	n_o	1.423	1.3930	1.3776	1.375	1.373	1.368	1.34	1.21		2
MgF_2	tetr	n_e	1.436	1.4055	1.3894	1.387	1.385	1.379	1.34	1.21		2
MgO	cub	n			1.7355	1.7283	1.7228	1.7084	1.6361			5
$MgSO_4 \cdot 7H_2O$	orth	n_x			1.4326							4
$MgSO_4 \cdot 7H_2O$	orth	n_y			1.4555							4
$MgSO_4 \cdot 7H_2O$	orth	n_z			1.4607							4
MnF_2	tetr	n_o			1.472							4
MnF_2	tetr	n_e			1.501							4
NaBr	cub	n		1.748	1.642	1.631	1.623	1.616	1.609	1.593*	1.520*	1
$NaBrO_3$	cub	n			1.6168							4
NaCl	cub	n	1.789*	1.6066	1.5441	1.5369	1.5320	1.5265	1.5188	1.4947	1.382*	1
$NaClO_3$	cub	n			1.5151							7
NaF	cub	n	1.383	1.3424	1.3252	1.3231	1.3214	1.3179	1.3017	1.2400		1
$NaH_2PO_4 \cdot 2H_2O$	orth	n_x			1.4400							7
$NaH_2PO_4 \cdot 2H_2O$	orth	n_y			1.4628							7
$NaH_2PO_4 \cdot 2H_2O$	orth	n_z			1.4814							7
NaI	cub	n		1.93*	1.774	1.758	1.74	1.73*	1.73*	1.71*	1.66*	1
$NaNO_2$	orth	n_x			1.6547							7
$NaNO_2$	orth	n_y			1.3455							7
$NaNO_2$	orth	n_z			1.4125							7
$NaNO_3$	rhomb	n_o			1.5840							5
$NaNO_3$	rhomb	n_e			1.3340							5
$Na_2HPO_4 \cdot 7H_2O$	monocl	n_x			1.4411							4
$Na_2HPO_4 \cdot 7H_2O$	monocl	n_y			1.4423							4
$Na_2HPO_4 \cdot 7H_2O$	monocl	n_z			1.4525							4
Na_2SO_4	orth	n_x			1.4669							4
Na_2SO_4	orth	n_y			1.4730							4
Na_2SO_4	orth	n_z			1.4809							4
Nd_2O_3	cub	n			1.92							4
NdF_3	hex	n_o			1.6191							4
NdF_3	hex	n_e			1.6132							4
$NH_4H_2AsO_4$	tetr	n_o		1.6401	1.5777	1.5704	1.5583					7
$NH_4H_2AsO_4$	tetr	n_e		1.5754	1.5232	1.5179	1.5101					7
$NH_4H_2PO_4$	tetr	n_o		1.5668	1.5247	1.5187	1.5084					7
$NH_4H_2PO_4$	tetr	n_e		1.5137	1.4797	1.4754	1.4694					7
NiF_2	tetr	n_o			1.526							4
NiF_2	tetr	n_e			1.561							4
$NiSO_4 \cdot 6H_2O$	tetr	n_o			1.5107							4
$NiSO_4 \cdot 6H_2O$	tetr	n_e			1.4870							4
PbF_2	cub	n		1.94*	1.767	1.754	1.745	1.73	1.70	1.66	1.32	5
$PbSO_4$	orth	n_x			1.8780							4
$PbSO_4$	orth	n_y			1.8834							4
$PbSO_4$	orth	n_z			1.8945							4
PrF_3	hex	n_o			1.6207							4
PrF_3	hex	n_e			1.6146							4
RbBr	cub	n		1.639	1.553	1.544	1.538	1.532	1.530	1.525	1.505*	1
RbCl	cub	n	1.738	1.549	1.493	1.487	1.483	1.479	1.475	1.465	1.424*	1
$RbClO_4$	orth	n_x			1.4691							4
$RbClO_4$	orth	n_y			1.4701							4
$RbClO_4$	orth	n_z			1.4732							4
RbF	cub	n	1.50*	1.428*	1.397	1.394	1.391	1.388	1.379	1.346	1.19*	1

Mol. form	Cryst. syst.	Ray	*n* (200 nm)	*n* (300 nm)	*n* (589 nm)	*n* (750 nm)	*n* (1 µm)	*n* (2 µm)	*n* (5 µm)	*n* (10 µm)	*n* (20 µm)	Ref.
RbH_2AsO_4	tetr	n_o		1.6183	1.5603	1.5538	1.5432					7
RbH_2AsO_4	tetr	n_e		1.5718	1.5232	1.5184	1.5121					7
RbH_2PO_4	tetr	n_o		1.5434	1.5078	1.5021	1.4941					7
RbH_2PO_4	tetr	n_e		1.5106	1.4791	1.4754	1.4704					7
RbI	cub	n		1.808	1.647	1.633	1.623	1.615	1.612	1.608	1.595	1
Rb_2SO_4	orth	n_x			1.5131							4
Rb_2SO_4	orth	n_y			1.5133							4
Rb_2SO_4	orth	n_z			1.5144							4
Sb_2O_3[c]	cub	n			2.8017							4
Sc_2O_3	cub	n			1.9943							4
SiO_2[d]	hex	n_o	1.6493	1.5733	1.5442	1.5394	1.5350	1.5209				5
SiO_2[d]	hex	n_e	1.6605	1.5882	1.5534	1.5484	1.5438	1.5291				5
SnO_2	tetr	n_o			1.993							4
SnO_2	tetr	n_e			2.088							4
SrF_2	cub	n	1.50	1.459	1.4380	1.435	1.433	1.429	1.412	1.35		2
SrO	cub	n			1.8710							4
$SrSO_4$	orth	n_x			1.6214							4
$SrSO_4$	orth	n_y			1.6231							4
$SrSO_4$	orth	n_z			1.6303							4
$SrTiO_3$	cub	n			2.4082	2.3525	2.3160	2.2676	2.1205			5
$SrWO_4$	tetr	n_o			1.8618							4
$SrWO_4$	tetr	n_e			1.8719							4
TbF_3	hex	n_o			1.6034							4
TbF_3	hex	n_e			1.5603							4
TeO_2	tetr	n_o			2.2738		2.2080					7
TeO_2	tetr	n_e			2.4295		2.3520					7
ThO_2	cub	n			2.1113							4
TiO_2[e]	tetr	n_o			2.614	2.533	2.485	2.399	2.290			5
TiO_2[e]	tetr	n_e			2.900	2.805	2.748					5
TiO_2[j]	tetr	n_o			2.562							4
TiO_2[j]	tetr	n_e			2.489							4
Tl_2SO_4	orth	n_x			1.8604							4
Tl_2SO_4	orth	n_y			1.8676							4
Tl_2SO_4	orth	n_z			1.8857							4
TlBr	cub	n			2.418	2.350	2.289	2.103	1.984	2.339	2.322	5
TlCl	cub	n			2.247	2.198	2.145	1.986	1.891	2.193		5
$TlClO_4$	orth	n_x			1.6427							4
$TlClO_4$	orth	n_y			1.6446							4
$TlClO_4$	orth	n_z			1.6542							4
Y_2O_3	cub	n			1.930							4
Yb_2O_3	cub	n			1.9468							4
ZnF_2	tetr	n_o			1.495							4
ZnF_2	tetr	n_e			1.525							4
ZnO	hex	n_o			2.0036	1.9662	1.9435	1.9197				7
ZnO	hex	n_e			2.0199	1.9821	1.9589	1.9330				7
ZnS[f]	cub	n			2.3691	2.3232	2.2932	2.2633				7
ZnS[g]	hex	n_o			2.372	2.331	2.303	2.26	2.25	2.20		3,5
ZnS[g]	hex	n_e			2.368	2.327	2.301					5
ZnSe	cub	n			2.6222	2.5384	2.4888	2.4462	2.4296	2.4065		3
ZnTe	cub	n			3.060	2.880	2.789	2.719	2.698	2.684		3
$ZrSiO_4$[h]	tetr	n_o			1.9255							4
$ZrSiO_4$[h]	tetr	n_e			1.9843							4

* Provisional value based on extrapolation beyond the range of experimental data.
[a] Arsenolite.
[b] Calcite.
[c] Senarmontite.
[d] α-Quartz.
[e] Rutile.
[f] Sphalerite.
[g] Wurtzite.
[h] Zircon.
[j] Anatase.

REFRACTIVE INDEX AND TRANSMITTANCE OF REPRESENTATIVE GLASSES

Typical values of the index of refraction and internal transmittance (fraction of light transmitted through a one centimeter thickness) are tabulated here for selected types of glasses, as well as for synthetic fused (vitreous) silica. Nominal compositions are given in Table 1. Table 2 contans the index of refraction, relative to air, and the internal transmittance for representative samples of each glass at wavelengths in the infrared, visible, and near-ultraviolet regions. It should be emphasized that a wide variation of these parameters may be found among subtypes of each glass. More detailed data may be found in Ref. 3.

Assuming that the Lambert-Beer Law is followed, the transmittance of a glass plate of thickness d (in centimeters) can be obtained by raising the transmittance value in the table to the power d.

References

1. Weber, M. J., *CRC Handbook of Laser Science and Technology*, Vol. IV, Part 2, CRC Press, Boca Raton, FL, 1988.
2. Gray, D. E., Ed., *American Institute of Physics Handbook, Third Edition*, McGraw-Hill, New York, 1972.
3. *Schott Optical Glass*, Schott Glass Technologies, Inc., 400 York Ave., Duryea, PA.
4. Kaye, G. W. C., and Laby, T. H., *Tables of Physical and Chemical Constants, 15th Edition*, Longman, London, 1986.

TABLE 1. Glass Composition in Percent by Mass

Type	Name	SiO_2	B_2O_3	Al_2O_3	Na_2O	K_2O	CaO	BaO	ZnO	PbO	P_2O_5
PK	Phosphate crown		3	10		12	5				70
PSK	Dense phosphate crown		3	5			4	28			60
BK	Borosilicate crown	70	10		8	8	1	3			
K	Crown	74			9	11	6				
ZK	Zinc crown	71			17				12		
BaK	Barium crown	60	3		3	10		19	5		
SK	Dense crown	39	15	5				41			
KF	Crown flint	67		2	16				3	12	
BaLF	Barium light flint	51			6	5		20	14	4	
SSK	Extra dense crown	35	10	5				42	8		
LLF	Extra light flint	63			5	8				24	
BaF	Barium flint	46				8		16	8	22	
LF	Light flint	53			5	8				34	
F	Flint	47			2	7				44	
BaSF	Dense barium flint	43			1	7		11	5	33	
SF	Dense flint	33				5				62	
KzFS	Short flint										
SiO_2	Fused silica	100									

TABLE 2. Refractive Index and Transmittance of a 1 cm Glass Plate at the Specified Wavelength

Type	Refractive index at 1.060 μm	Refractive index at 546.1 nm	Refractive index at 365.0 nm	Refractive index at 312.6 nm	Transmittance at 1.060 μm	Transmittance at 546.1 nm	Transmittance at 365.0 nm	Transmittance at 310 nm
PK	1.51519	1.52736	1.54503	1.5574	0.997	0.998	0.987	0.46
PSK	1.54154	1.55440	1.57342	1.5868	0.996	0.998	0.984	0.46
BK	1.50669	1.51872	1.53627	1.5486	0.999	0.998	0.987	0.35
K	1.50091	1.51314	1.53189	1.5454	0.998	0.998	0.988	0.40
ZK	1.52220	1.53534	1.55588	1.5708	0.996	0.998	0.976	0.27
BaK	1.55695	1.57124	1.59407	1.6108	0.998	0.997	0.986	0.28
SK	1.59490	1.60994	1.63398		0.998	0.998	0.959	0.28
KF	1.50586	1.51978	1.54251	1.5600	0.998	0.996	0.989	0.49
BaLF	1.57579	1.59166	1.61804		0.996	0.998	0.933	0.010
SSK	1.60402	1.61993	1.64595		0.999	0.998	0.915	0.010
LaK	1.69710	1.71616	1.74573		0.999	0.998	0.882	0.17
LLF	1.52775	1.54344	1.57038		0.998	0.997	0.990	0.32
BaF	1.56873	1.58565	1.61524		0.999	0.997	0.992	0.004
LF	1.56594	1.58482	1.61926		0.999	0.998	0.981	0.008
F	1.58636	1.60718	1.64606		0.997	0.998	0.959	
BaSF	1.60889	1.62987	1.66926		0.999	0.998	0.857	
SF	1.71350	1.74620	1.8145		0.998	0.997	0.650	
KzFS	1.59680	1.61639	1.64849	1.6739		0.998	0.672	0.012
SiO_2	1.44968	1.46008	1.47435[a]	1.53430[b]				

[a] At 366.3 nm.
[b] At 213.9 nm.

INDEX OF REFRACTION OF WATER

This table gives the index of refraction of liquid water at atmospheric pressure, relative to a vacuum, at several temperatures and wavelengths. It is generated from the formulation in Ref. 1, which covers a wide range of temperature, pressure, and wavelength. The wavelengths listed here correspond to prominent lines of cadmium (226.50 and 361.05 nm), potassium (404.41 nm), sodium (589.00 nm), Ne (632.80 nm, from a helium-neon laser), and mercury (1.01398 μm).

References

1. Schiebener, P., Straub, J., Levelt Sengers, J. M. H., and Gallagher, J. S., *J. Phys. Chem. Ref. Data*, 19, 677, 1990; 19, 1617, 1990.
2. Marsh, K. N., Editor, *Recommended Reference Materials for the Realization of Physicochemical Properties*, Blackwell Scientific Publications, Oxford, 1987.

Index of Refraction of Water as a Function of Wavelength

t/°C	226.50 nm	361.05 nm	404.41 nm	589.00 nm	632.80 nm	1.01398 μm
0	1.39450	1.34896	1.34415	1.33432	1.33306	1.32612
10	1.39422	1.34870	1.34389	1.33408	1.33282	1.32591
20	1.39336	1.34795	1.34315	1.33336	1.33211	1.32524
30	1.39208	1.34682	1.34205	1.33230	1.33105	1.32424
40	1.39046	1.34540	1.34065	1.33095	1.32972	1.32296
50	1.38854	1.34373	1.33901	1.32937	1.32814	1.32145
60	1.38636	1.34184	1.33714	1.32757	1.32636	1.31974
70	1.38395	1.33974	1.33508	1.32559	1.32438	1.31784
80	1.38132	1.33746	1.33284	1.32342	1.32223	1.31576
90	1.37849	1.33501	1.33042	1.32109	1.31991	1.31353
100	1.37547	1.33239	1.32784	1.31861	1.31744	1.31114

INDEX OF REFRACTION OF LIQUIDS FOR CALIBRATION PURPOSES

These tables give the index of refraction at three temperatures of six liquids that are available in highly pure form and whose index of refraction has been accurately measured as a function of wavelength and temperature. These values are therefore useful for calibration of refractometers. The estimated uncertainty in the values is:

trans-Decahydronaphthalene	±0.00008
Hexadecane	±0.00008
Methylcyclohexane	±0.00003
1-Methylnaphthalene	±0.00008
Toluene	±0.00003
2,2,4-Trimethylpentane	±0.00003

Full details are given in the references. This table is reprinted from Ref. 1 by permission of the International Union of Pure and Applied Chemistry.

References

1. Marsh, K. N., Editor, *Recommended Reference Materials for the Realization of Physicochemical Properties*, Blackwell Scientific Publications, Oxford, 1987.
2. Tilton, L. W., *J. Opt. Soc. Am.* 32, 71, 1941. <https://doi.org/10.1364/JOSA.32.000371>

TABLE 1. Index of Refraction for Calibration Purposes at 20 °C

Name	Mol. form.	667.81 nm	656.28 nm	589.26 nm	546.07 nm	501.57 nm	486.13 nm	435.83 nm
trans-Decahydronaphthalene	$C_{10}H_{18}$	1.46654	1.46688	1.46932	1.47141	1.47420	1.47535	1.48011
Hexadecane	$C_{16}H_{34}$	1.43204	1.43235	1.43453	1.43640	1.43888	1.43993	1.44419
Methylcyclohexane	C_7H_{14}	1.42064	1.42094	1.42312	1.42497	1.42744	1.42847	1.43269
1-Methylnaphthalene	$C_{11}H_{10}$	1.60828	1.60940	1.61755	1.62488	1.63513	1.63958	
Toluene	$C_6H_5CH_3$	1.4918	1.49243	1.49693	1.50086	1.50620	1.50847	1.51800
2,2,4-Trimethylpentane	C_8H_{18}	1.38916	1.38945	1.39145	1.39316	1.39544	1.39639	1.40029

TABLE 2. Index of Refraction for Calibration Purposes at 25 °C

Name	Mol. form.	667.81 nm	656.28 nm	589.26 nm	546.07 nm	501.57 nm	486.13 nm	435.83 nm
trans-Decahydronaphthalene	$C_{10}H_{18}$	1.46438	1.46472	1.46715	1.46923	1.47200	1.47315	1.47789
Hexadecane	$C_{16}H_{34}$	1.43001	1.43032	1.43250	1.43436	1.43684	1.43788	1.44213
Methylcyclohexane	C_7H_{14}	1.41812	1.41842	1.42058	1.42243	1.42488	1.42590	1.43010
1-Methylnaphthalene	$C_{11}H_{10}$	1.60592	1.60703	1.61512	1.62240	1.63259	1.63701	1.65627
Toluene	$C_6H_5CH_3$	1.48903	1.48966	1.49413	1.49803	1.50334	1.50559	1.51506
2,2,4-Trimethylpentane	C_8H_{18}	1.38670	1.38698	1.38898	1.39068	1.39294	1.39389	1.39776

TABLE 3. Index of Refraction for Calibration Purposes at 30 °C

Name	Mol. form.	667.81 nm	656.28 nm	589.26 nm	546.07 nm	501.57 nm	486.13 nm	435.83 nm
trans-Decahydronaphthalene	$C_{10}H_{18}$	1.46222	1.46256	1.46498	1.46705	1.46980	1.47095	1.47567
Hexadecane	$C_{16}H_{34}$	1.42798	1.42829	1.43047	1.43232	1.43480	1.43583	1.44007
Methylcyclohexane	C_7H_{14}	1.41560	1.41591	1.41806	1.41989	1.42233	1.42334	1.42752
1-Methylnaphthalene	$C_{11}H_{10}$	1.60360	1.60471	1.61278	1.62005	1.63022	1.63463	1.65386
Toluene	C_7H_8	1.48619	1.48682	1.49126	1.49514	1.50041	1.50265	1.51206
2,2,4-Trimethylpentane	C_8H_{18}	1.38424	1.38452	1.38650	1.38820	1.39044	1.39138	1.39523

INDEX OF REFRACTION OF AIR

This is a table of the index of refraction n of dry air at 15 °C and a pressure of 101.325 kPa and containing 0.045% by volume of carbon dioxide ("standard air"). The index of refraction is defined by $n = \lambda_{vac}/\lambda_{air}$ where λ is the wavelength of the radiation. The index is calculated from the expression

$$(n-1) \times 10^8 = 8342.54 + 2406147(130 - \sigma^2)^{-1} + 15998(38.9 - \sigma^2)^{-1}$$

where $\sigma = 1/\lambda_{vac}$ and λ_{vac} has units of μm. The equation is valid for λ_{vac} from 200 nm to 2 μm. The table also gives the correction $(n-1)\lambda_{air}$ that must be added to the wavelength in air to obtain λ_{vac}.

If the air is at a temperature t in °C (ITS-90) and a pressure p in Pa, a value of $(n - 1)$ from this table should be multiplied by

$$p[1 + p(60.1 - 0.972t) \times 10^{-10}]/96095.43(1 + 0.003661t)$$

References

1. Birch, K. P., and Downs, M. J., *Metrologia*, 31, 315, 1994.
2. Edlen, B., *Metrologia* 2, 71, 1966.

λ_{vac}	$(n-1) \times 10^8$	$\lambda_{vac} - \lambda_{air}$
200 nm	32409	0.06480 nm
210	31748	0.06665
220	31226	0.06868
230	30801	0.07082
240	30447	0.07305
250	30148	0.07535
260	29892	0.07769
270	29670	0.08009
280	29477	0.08251
290	29307	0.08497
300	29157	0.08745
310	29023	0.08995
320	28904	0.09247
330	28796	0.09500
340	28700	0.09755
350	28612	0.10011
360	28532	0.10269
370	28460	0.10527
380	28393	0.10786
390	28332	0.11046
400	28276	0.11307
410	28224	0.11569
420	28177	0.11831
430	28132	0.12094
440	28091	0.12357
450	28053	0.12620
460	28018	0.12885
470	27985	0.13149
480	27954	0.13414
490	27925	0.13679
500	27897	0.13945
510	27872	0.14211
520	27848	0.14477
530	27825	0.14743
540	27804	0.15010
550	27784	0.15277
560	27765	0.15544
570	27747	0.15811
580	27730	0.16079
590	27714	0.16347
600	27698	0.16614
610	27684	0.16882
620	27670	0.17151
630	27657	0.17419
640	27644	0.17688
650	27632	0.17956
660	27621	0.18225
670	27610	0.18494
680	27600	0.18763
690	27590	0.19032
700	27581	0.19301
710	27572	0.19570
720	27563	0.19840
730	27555	0.20109
740	27547	0.20379
750	27539	0.20649
760	27532	0.20918
770	27525	0.21188
780	27518	0.21458
790	27511	0.21728
800	27505	0.21998
810	27499	0.22268
820	27493	0.22538
830	27488	0.22808
840	27482	0.23079
850	27477	0.23349
860	27472	0.23619
870	27467	0.23890
880	27462	0.24160
890	27458	0.24431
900	27454	0.24701
910	27449	0.24972
920	27445	0.25243
930	27441	0.25513
940	27437	0.25784
950	27434	0.26055
960	27430	0.26326
970	27427	0.26597
980	27423	0.26868
990 nm	27420	0.27138 nm
1.00 μm	27417	0.0002741 μm
1.05	27402	0.0002876
1.10	27390	0.0003012
1.15	27379	0.0003148
1.20	27370	0.0003283
1.25	27361	0.0003419
1.30	27354	0.0003555
1.35	27347	0.0003691
1.40	27341	0.0003827
1.45	27336	0.0003963
1.50	27331	0.0004099
1.55	27327	0.0004234
1.60	27323	0.0004370
1.65	27319	0.0004506
1.70	27316	0.0004642
1.75	27313	0.0004778
1.80	27310	0.0004914
1.85	27307	0.0005050
1.90	27305	0.0005187
1.95	27303	0.0005323
2.00	27301	0.0005459

INDEX OF REFRACTION OF GASES

This table gives the index of refraction of several gases at selected wavelengths ranging from the blue to the red region of the spectrum. The entries at 0.5893 μm correspond to the prominent sodium D line in the yellow region. All values refer to gas at a pressure of one atmosphere (101.325 kPa) and at a temperature of 0 °C. Column definitions for the table are as follows.

Column heading	Definition
Name	Name of gas, listed alphabetically by name
Mol. form.	Molecular formula of gas
***n* (0.4360 μm), *n* (0.4861 μm), etc.**	Index of refraction at wavelength in μm specified within parentheses

References

1. Gray, D. E., ed., *American Institute of Physics Handbook, Third Edition*, p. 6-110, McGraw-Hill, New York, 1972.
2. Forsythe, W. E., *Smithsonian Physical Tables, Ninth Edition*, p. 533, Smithsonian Institution, Washington, DC, 1954.
3. *Kaye and Laby Tables of Physical and Chemical Constants, 16th Edition*, p. 131, Longman Group Ltd., Harlow, Essex, 1995.

Index of Refraction of Gases as a Function of Wavelength

Name	Mol. form.	*n* (0.4360 μm)	*n* (0.4861 μm)	*n* (0.5461 μm)	*n* (0.5790 μm)	*n* (0.5893 μm)	*n* (0.6563 μm)	*n* (0.6709 μm)
Air		1.0002966	1.0002947	1.0002932	1.0002926	1.0002924	1.0002915	1.0002913
Argon	Ar					1.000281		
Carbon dioxide	CO_2	1.0004563		1.0004506		1.0004493		1.0004471
Chlorine	Cl_2					1.000773		
Helium	He					1.000036		
Hydrogen	H_2	1.0001418	1.0001406	1.0001397	1.0001393	1.0001392	1.0001387	1.0001385
Methane	CH_4					1.000444		
Nitric oxide	NO					1.000297		
Nitrogen	N_2		1.0003012	1.0002998		1.0002990	1.0002982	
Nitrous oxide	N_2O					1.000516		
Oxygen	O_2	1.0002743	1.0002734	1.0002717	1.0002710	1.0002709	1.0002698	1.0002683
Sulfur dioxide	SO_2					1.000686		

Section 11
Nuclear and Particle Physics

Summary Tables of Particle Properties. 11-1
Table of the Isotopes 11-3
Neutron Scattering and Absorption Properties 11-43
Cosmic Radiation. 11-56

SUMMARY TABLES OF PARTICLE PROPERTIES

Fundamental Particles

Modern physical theory postulates the existence of several fundamental (elementary) particles, that is, physical particles that have no substructure and are not made of other particles. Over the last few decades, experimental evidence has identified a number of these particles, which have been classified using a framework developed in the Standard Model (Ref. 1). The highest level of classification is fermions and bosons. Fermions are particles that have a spin value of 1/2, 3/2, etc., and follow Fermi-Dirac statistics as well as obey the Pauli Exclusion Principle. Bosons are particles that have an integer value of spin, including zero (0), and follow Bose-Einstein statistics. Fermions are the particles that make up matter, and bosons are the particles that carry forces.

Fermions

Fermions can be divided into two primary categories: leptons and quarks. Leptons do not undergo strong interactions, while quarks do. Leptons are further divided into two classes: charged (electron, muon, and tau, which differ by mass) and uncharged (neutrinos, also of three types, electron neutrino, muon neutrino, and tau neutrino).

Quarks are characterized by three properties ("flavors"): up/down; charm/strange; and top/bottom and carry a fractional charge. To date, quarks are experimentally seen only as composite particles, such as protons and neutrons.

Each lepton and quark (matter particles) has an equivalent "antimatter" version. In addition, fermions are divided into families, usually called generations, in which particles in higher generations (second and third) have greater mass than in a lower generation: Mass (electron)<<Mass (muon)<<Mass (tau).

Bosons

Bosons can be elementary or composite particles. Composite bosons are made up of other bosons (spin zero or integer, such as a meson, which is made up of one quark and one antiquark) or fermions, such that the net spin is zero or an integer. For example, the carbon-12 nucleus has six protons and six neutrons with zero spin and is therefore a boson.

Elementary quarks are all gauge bosons, that is carriers of force, except the Higgs boson, which is a scalar boson. Photons are force carriers of the electromagnetic field. W and Z bosons are force carriers that mediate the weak force. Gluons are force carriers for the strong force. Some theories of quantum gravity postulate a force carrier for gravity, graviton, a boson of spin plus or minus two, but that has not been either proven or disproven experimentally yet.

Properties

The table summarizes data for the known fundamental particles. The references contain considerable additional data for particles as well as reviews of topics in particle physics and cosmology. In particular, the reviews of particle physics (Refs. 2-5) summarize much of particle physics and cosmology. It lists, evaluates, and averages measured properties of gauge bosons and the recently discovered Higgs boson, leptons, quarks, mesons, and baryons. Particle properties and search limits are listed in the Summary Tables (see Ref. 5). Complete information is published online on the Web site of the Particle Data Group <http://pdg.lbl.gov>.

References

1. Cottingham, W. N. and Greenwood, D. A., *An Introduction to the Standard Model of Particle Physics, Second Edition,* Cambridge University Press, 2008.
2. Zyla, P. A. et al. (Particle Physics Group), *Prog. Theor. Exp. Physics* 2020, 083C01, 2020.
3. *pdgLive*, 2020. <pdg.lbl.gov/Viewer.action>
4. Tanabashi, M. et al. (Particle Data Group), *Phys. Rev. D* 98, 030001, 2018.
5. Particle Physics Data Group, *2020 Particle Physics Booklet,* Physics Division, Lawrence Berkeley National Laboratory, U.S. Department of Energy, Berkeley, CA, 2020.

Properties of Fundamental Particles

Name	Symbol	F/B	Class	Gen.	Spin	Charge	Mass	Mass units	Mass uncert.	Mass uncert. units	Note
Electron	e	F	Matter	First	1/2	-1	$5.485\ 799\ 090\ 70 \times 10^{-4}$	u	$\pm 0.000\ 000\ 000\ 16 \times 10^{-4}$	u	
	e						0.5109 989 461	MeV	± 0.000 000 0031	MeV	a
	e						$9.109\ 383\ 7015 \times 10^{-31}$	kg	$\pm 0.000\ 000\ 0028 \times 10^{-31}$	kg	a
Muon	μ	F	Matter	Second	1/2	-1	0.113 428 925 7	u	± 0.000 000 0025	u	
	μ						105.6583 745	MeV	± 0.000 0024	MeV	a
	μ						$1.883\ 531\ 627 \times 10^{-28}$	kg	$\pm 0.000\ 000\ 042 \times 10^{-28}$	kg	a
Tau	τ	F	Matter	Third	1/2	-1	1.776.86	MeV	± 0.12	MeV	
	τ						$3.167\ 54 \times 10^{-27}$	kg	$\pm 0.000\ 21 \times 10^{-27}$	kg	a
Electron neutrino	ν_e	F	Matter	First		$< 4 \times 10^{-35}$	< 1.1	ev	90% confidence		
Muon neutrino	ν_μ	F	Matter	Second		$< 4 \times 10^{-35}$	< 0.19	MeV	90% confidence		
Tau neutrino	ν_τ	F	Matter	Third		$< 4 \times 10^{-37}$	< 18.2	MeV	90% confidence		
Up quark	u	F	Matter	First		2/3	2.16	MeV	+0.49;-0.26	MeV	
Down quark	d	F	Matter	First		-1/3	4.67	MeV	+0.48;-0.17	MeV	
Charm quark	c	F	Matter	Second		2/3	1.27	GeV	± 0.02	GeV	
Strange quark	s	F	Matter	Second		-1/3	93	MeV	+11;-5	MeV	
Top quark	t	F	Matter	Third		2/3	172.76	GeV	± 0.30	GeV	b
Bottom quark	b	F	Matter	Third		-1/3	4.18	MeV	+0.3;-0.2	MeV	
Photon	γ	B	Force			$< 1 \times 10^{-46}$	$< 1 \times 10^{-18}$	eV		eV	

Name	Symbol	F/B	Class	Gen.	Spin	Charge	Mass	Mass units	Mass uncert.	Mass uncert. units	Note
W boson	W	B	Force				80.379	GeV	± 0.012	GeV	
Z boson	Z	B	Force				91.1876	GeV	± 0.0021	GeV	
Gluon	g	B	Force				0				c
Graviton		B	Force				$< 6 \times 10^{-32}$	eV		eV	d
Higgs boson	H^0	B	Force				125.10	GeV	0.14	GeV	

[a] Conversion factor from u (atomic mass units) to MeV and kg is more uncertain than the mass of the electron in u (atomic mass units).
[b] Direct measurement cross-section value is 162.5 +2.1; -1.5.
[c] Theoretical value; a mass as large as a few MeV may not be precluded.

TABLE OF THE ISOTOPES

Norman E. Holden

This table presents an evaluated set of values for the experimental quantities that characterize the decay of radioactive nuclides. A list of the major references used in this evaluation is given below. When uncertainties are not listed, they are assumed to be five or less in the last digit quoted. If the uncertainty in the value exceeds five in the last digit, the value is preceded by an approximate sign.

For quasi-stable nuclides, the measured width, Γ, of the resonance is given. To estimate the approximate half-life, the Heisenberg relationship may be used: the half-life = 4.56×10^{-22} seconds/Γ (MeV). The effective literature cutoff date for data in this edition of the table is October 2019.

This table in the Print Edition of the *CRC Handbook* is condensed from the full table in the Online Edition available at <hbcponline.com>. The print table includes all stable isotopes and the radioactive isotopes with lifetimes of a few hours or longer, as well as selected shorter-lived isotopes of the transuranium elements.

Table Layout

Column number	Column heading	Description
1	**Elem. or Isot.**	For elements, the atomic number and chemical symbol are listed. For nuclides, the mass number and chemical symbol are listed. Isomers are indicated by the addition of m, m1, or m2.
2	**Natural abundance**	The abundance of an isotope in normal terrestrial samples of an element, listed in atom percent.
3	**Atomic mass or weight**	Atomic mass relative to ^{12}C = 12. Atomic weight of elements is given on the same scale.
4	**Half-life/Resonance width**	Half-life in decimal notation. μs = microseconds; ms = milliseconds; s = seconds; m = minutes; h = hours; d = days; and a = years. For quasi-stable nuclides, the measured width at half-maximum of the energy resonance is given.
5	**Decay mode/Energy**	Decay modes are α = alpha particle emission; β = negative beta emission; β+ = positron emission; ec = orbital electron capture; IT = isomeric transition from upper to lower isomeric state; n = neutron emission; p = proton emission; sf = spontaneous fission; ββ, ECEC, 2p, 2n, 3n = double beta, double EC, two proton and multiple neutron decay. Total disintegration energy in MeV units.
6	**Particle energy/ Intensity**	End point energies of beta transitions and discrete energies of alpha particles, neutrons, and protons are given in MeV units and their intensities in percent.
7	**Spin**	Nuclear spin or angular momentum of the nuclides in units of $h/2\pi$; parity is positive or negative.
8	**Nuclear magnetic mom.**	Magnetic dipole moments in nuclear magneton units. An absolute value is indicated in the absence of a positive or a negative sign.
9	**Elect. quadr. mom.**	Electric quadrupole moments in barn units (10^{-24} cm^2). An absolute value is indicated in the absence of a positive or a negative sign.
10	**γ-Energy/Intensity**	Gamma ray energies are given in MeV units and intensities in percent. Annihilation radiation (ann.rad.) refers to the 511.006 keV photons emitted in the annihilation of positrons in matter.

General Nuclear Data References

The following references represent the major sources of the nuclear data presented, along with subsequent published journal articles and reports.

1. G. Audi, F. G. Kondev, M. Wang, B. Pfeiffer, X. Sun, J. Blachot, M. MacCormick, The NUBASE2012 Evaluation of Nuclear Properties, *Chinese Physics C – High Energy and Nuclear Physics* 36, 1157 (2012). <https://doi.org/10.1088/1674-1137/36/12/001>
2. M. Wang, G. Audi, A. H. Wapstra, F. G. Kondev, M. MacCormick, X. Xu, B. Pfeiffer, The AME2012 Atomic Mass Evaluation (II., Tables, graphs and references), *Chinese Physics C – High Energy and Nuclear Physics* 36, 1603 (2012). <https://doi.org/10.1088/1674-1137/36/12/003>
3. J. Meija, T. B. Coplen, M. Berglund, W. A. Brand, P. De Bievre, M. Gröning, N. E. Holden, J. Irrgeher, R. D. Loss, T. Walczyk, T. Prohaska, Atomic Weights of the Elements — 2013, *Pure & Applied Chemistry* 88, 265 (2016). <https://doi.org/10.1515/pac-2015-0305>
4. International Union of Pure and Applied Chemistry (IUPAC) Standard Atomic Weight of Ytterbium Revised (press release, 24 August 2015) <https://iupac.org/standard-atomic-weight-of-ytterbium-revised/> (accessed December 2016).
5. J. Meija, T. B. Coplen, M. Berglund, W. A. Brand, P. De Bievre, M. Gröning, N. E. Holden, J. Irrgeher, R. D. Loss, T. Walczyk, T. Prohaska, Isotopic Composition of the Elements 2013 (IUPAC Technical Report), *Pure & Applied Chemistry* 88, 293 (2016). <https://doi.org/10.1515/pac-2015-0503>
6. E. M. Baum, H .D. Knox, T. R. Miller, *Chart of the Nuclides, 17th Edition*, Knolls Atomic Power Lab (2009).
7. N. E. Holden, Total and Spontaneous Fission Half-Lives for Uranium, Plutonium, Americium and Curium Nuclides, *Pure & Applied Chemistry* 61, 1483 (1989). <https://doi.org/10.1351/pac198961081483>
8. N. E. Holden, Half-lives of Selected Nuclides, *Pure & Applied Chemistry* 62, 941 (1990). <https://doi.org/10.1351/pac199062050941>
9. E. Brown, R. Firestone, *Radioactivity Handbook*, Wiley Interscience Press (1986).
10. J. K. Tuli, *Nuclear Wallet Cards*, Brookhaven National Laboratory (October 2011).
11. N. E. Holden, D.C. Hoffman, Spontaneous Fission Half-Lives for Ground State Nuclides, *Pure & Applied Chemistry* 72, 1525 (2000). <https://doi.org/10.1351/pac200072081525>
12. N. J. Stone, Table of Nuclear Magnetic Dipole and Electrical Quadrupole Moments, *Atomic Data Nuclear Data Tables* 90, 75 (2005). <https://doi.org/10.1016/j.adt.2005.04.001>
13. N. J. Stone, Table of Nuclear Electric Quadrupole Moments, *Atomic Data Nuclear Data Tables*, 111-112, 1 (2016). <https://doi.org/10.1016/j.adt.2015.12.002>
14. I. M. Villa, P. De Bièvre, N. E. Holden, P. R. Renne, Recommendation on the Half-Life of ^{87}Rb, *Geochimica Cosmochimica Acta*, 164, 382 (2015). <https://doi.org/10.1016/j.gca.2015.05.025>
15. I. M. Villa, M. L. Bonardi, P. De Bièvre, N. E. Holden, P. R Renne, IUPAC-IUGS Status Report on the Half-Lives of ^{238}U, ^{235}U and ^{234}U, *Geochimica Cosmochimica Acta* 172, 387 (2016). <https://doi.org/10.1016/j.gca.2015.10.011>

16. A. K. Jain, B. Maheshwari, S. Garg, M. Patial, B. Singh, Atlas of Nuclear Isomers, *Nuclear Data Sheets* 128, 1 (2015). <https://doi.org/10.1016/j.nds.2015.08.001>

This research was carried out under the auspices of the U.S. Department of Energy Contract No. DE-AC02-98CH10886. The author is at Brookhaven National Laboratory, Upton, NY, and can be contacted at holden@bnl.gov.

Table of the Isotopes

Elem. or Isot.	Natural abundance (Atom %)	Atomic mass or weight	Half-life/ Resonance width (MeV)	Decay mode/ Energy (MeV)	Particle energy/ Intensity (MeV/%)	Spin ($h/2\pi$)	Nuclear magnetic mom. (nm)	Elect. quadr. mom. (b)	γ-Energy/ Intensity (MeV/%)
^{1}H	99.972 - 99.999	1.007825032	>2.8 ×10^{23} a			1/2+	+2.7928473		
^{2}H	0.001-0.028	2.014101778				1+	+0.8574382	+2.86 mb	
^{3}H		3.016049282	12.31 a	β-/0.01859	0.01860/100.	1/2+	+2.9789624		
^{3}He	0.0002(2)	3.016029323				1/2+	-2.127750		
^{4}He	99.9998(2)	4.0026032541				0+			
^{6}Li	(1.9 - 7.8)	6.015122887				1+	+0.822047	-0.81 mb	
^{7}Li	(92.2 - 98.1)	7.01600344				3/2-	+3.25644	-0.0400	
^{7}Be		7.0169287	53.28 d	ec/0.8618		3/2-	-1.40		0.4776/10.4
^{9}Be	100.	9.0121831				3/2-	-1.1776	+0.0529	
^{10}Be		10.0135347	1.39 ×10^{6} a	β-/0.5559	0.555/100.	0+			
^{10}B	(18.9 - 20.4)	10.01293686				3+	+1.8006448	+0.0845	
^{11}B	(79.6 - 81.1)	11.00930517				3/2-	+2.688649	+0.0406	
^{12}C	(98.84-99.04)	12.000000000				0+			
^{13}C	(0.96-1.16)	13.0033548352				½-	+0.702412		
^{14}C		14.00324199	5.70 × 10^{3} a	β-/0.15648	0.1565/100.	0+			
^{14}N	(99.578 -99.663)	14.0030740045				1+	+0.403761	+0.02044	
^{15}N	(0.337-0.422)	15.000108899				½-	-0.283189		
^{16}O	(99.738 - 99.776)	15.9949146196				0+			
^{17}O	(0.0367 - 0.0400)	16.999131757				5/2+	-1.8938	-0.0256	
^{18}O	(0.187-0.222)	17.999159613				0+			
^{18}F		18.000937	1.8295 h	β+, ec/1.656	0.635/97.	1+			ann.rad.
^{19}F	100.	18.998403163				½+	+2.62887		
^{20}Ne	90.48(3)	19.992440176				0+			
^{21}Ne	0.27(1)	20.9938467				3/2+	-0.66180	+0.10	
^{22}Ne	9.25(3)	21.99138511				0+			
^{22}Na		21.9944374	2.601 a	β+ /90/2.842	0.545/90.	3+	+1.746	+0.18	ann.rad./
				ec/10/					1.2745/99.9
^{23}Na	100.	22.98976928				3/2+	+2.21752	+0.104	
^{24}Na		23.99096301	14.96 h	β- /5.5158	1.389/>99.	4+	+1.690		1.3686/100.
									2.754/100.
									(0.997–4.238)
^{24}Mg	(78.88-79.05)	23.98504170				0+			
^{25}Mg	(9.988 -10.034)	24.9858370				5/2+	- 0.85545	+0.199	
^{26}Mg	(10.96-11.09)	25.9825930				0+			
^{28}Mg		27.98377	20.9 h	β- /1.832	0.459/95.	0+			0.0306/95.
									0.4006/36.
									0.9418/36.
									1.342/54.
^{26}Al		25.9868919	7.1 ×10^{5} a	β+ /82/4.0042	1.16/	5+	+2.804	+0.26	ann.rad./
				ec/18					1.8087/99.8
^{27}Al	100.	26.9815384				5/2+	+3.641507	+0.147	
^{28}Si	(92.191 - 92.318)	27.976926535				0+			
^{29}Si	(4.645-4.699)	28.976494665				1/2+	-0.5553		
^{30}Si	(3.037-3.110)	29.97377014				0+			
^{31}Si		30.9753632	2.62 h	β- /1.4920	1.471/99.9	3/2+			1.2662(5)/0.05
^{32}Si		31.974152	1.5 ×10^{2} a	β- /0.224	0.213/100.	0+			
^{31}P	100.	30.973761999				1/2+	+1.13160		
^{32}P		31.9739076	14.27 d	β- /1.7106	1.710/100.	1+	-0.2524		
^{33}P		32.971726	25.3 d	β- /0.249	0.249/100.	1/2+			
^{32}S	(94.41-95.29)	31.972071174				0+			
^{33}S	(0.729-0.797)	32.971458910				3/2+	+0.64382	-0.068	
^{34}S	(3.96 -4.77)	33.9678670				0+			
^{35}S		34.9690323	87.2 d	β- /0.1672	0.1674/100.	3/2+	+1.00	+0.047	

Elem. or Isot.	Natural abundance (Atom %)	Atomic mass or weight	Half-life/ Resonance width (MeV)	Decay mode/ Energy (MeV)	Particle energy/ Intensity (MeV/%)	Spin ($h/2\pi$)	Nuclear magnetic mom. (nm)	Elect. quadr. mom. (b)	γ-Energy/ Intensity (MeV/%)
^{36}S	(0.0129 - 0.0187)	35.9670807				0+			
^{38}S		37.97116	2.84 h	β- /2.94	1.00/	0+			0.1962(4)/0.2
									1.9421(3)/84.
^{35}Cl	(75.5 - 76.1)	34.9688527				3/2+	+0.82187	-0.082	
^{36}Cl		35.9683068	3.01 ×10^5 a	β- /0.7086	0.7093/98.	2+	+1.28547	-0.18	
				β+, ec/1.1421	0.115/0.002				ann.rad./
^{37}Cl	(23.9 - 24.5)	36.9659026				3/2+	+0.68412	-0.064	
^{36}Ar	(0.00-2.08)	35.96754511				0+			
^{37}Ar		36.9667763	35.0 d	ec/.813		3/2+	+1.15	+0.08	
^{38}Ar	(0.00-4.33)	37.9627321				0+			
^{39}Ar		38.96431	268. a	β-/0.565	0.565/100.	7/2-	-1.59	-0.12	
^{40}Ar	(93.6-100.0)	39.962383124				0+			
^{42}Ar		41.96305	33. a	β-/0.60	0.60/100.	0+			
^{39}K	93.2581(44)	38.96370649				3/2+	+0.39146	+0.059	
^{40}K	0.0117(1)	39.9639982	1.248 ×10^9 a	β- /1.3111	1.312/89.	4-	-1.29810	-0.073	ann.rad./
				β+, ec/1.505	1.50/10.7				1.4608/10.5
^{41}K	6.7302(44)	40.96182526				3/2+	+0.21487	+0.071	
^{42}K		41.9624023	12.36 h	β- /3.525	1.97/19.	2-	-1.1425		0.31260(2)/0.3
					3.523/81.				1.5246(3)/18.1
^{43}K		42.960735	22.3 h	β- /1.82	0.465/8.	3/2+	+0.163		0.2211(2)/4.
					0.825/87.				0.3729(2)/88.
					1.24/3.5				0.3971(2)/11.
					1.814/1.3				0.6178(2)/81.
^{40}Ca	96.941(156)	39.96259087	> 9.9 ⊠10^{21} a	ec-ec		0+			
^{41}Ca		40.9622779	9.9 ×10^4 a	ec/0.4214		7/2-	-1.5948	-0.067	
^{42}Ca	0.647(23)	41.9586178				0+			
^{43}Ca	0.135(10)	42.9587664				7/2-	-1.31764	-0.041	
^{44}Ca	2.086(110)	43.9554815				0+			
^{45}Ca		44.956186	162.7 d	β- /0.257	0.257/100.	7/2-	-1.327	+0.04	
^{46}Ca	0.004(3)	45.953688	>0.4 ×10^{16} a	β-β-		0+			
^{47}Ca		46.954541	4.536 d	β-/1.992	0.684/84.	7/2-	-1.406	+0.08	1.297/75
					1.98/16.				(0.041–1.88)
^{48}Ca	0.187(21)	47.952529	4.4 ×10^{19} a	β-β-		0+			
			>2.5 ×10^{20} a	β-					
^{43}Sc		42.96115	3.89 h	β+, ec/2.221	0.82/22.	7/2-	+4.53	-0.27	ann.rad./
					1.22/78.				0.3729(1)/22.
^{44m}Sc			58.2 h	IT/0.27		6+	+3.83	0.21	0.27124(1)/87.
				ec/3.926					(1.00–1.16)
^{44}Sc		43.959403	4.04 h	β+, ec/3.653	1.47/	2+	+2.50	+0.10	ann.rad./
									1.157/100
^{45}Sc	100.	44.955908				7/2-	+4.75649	-0.220	
^{46}Sc		45.955168	83.81 d	β- /2.367	0.357/100.	4+	+3.04	+0.12	0.8893/100
									1.121/100
^{47}Sc		46.952403	3.349 d	β- /0.600	0.439/69.	7/2-	+5.34	-0.22	0.15938(1)/68.
					0.601/31.				
^{48}Sc		47.95222	43.7 h	β- /3.99	0.655/	6+	3.72		0.9835/100
									1.03750(1)/97.
									1.3121/100
^{44}Ti		43.959690	59. a	ec/0.268		0+			0.06787/91
									0.07832/97
^{45}Ti		44.958121	3.078 h	β+/86/2.062	1.04	7/2-	0.095	~ 0.015	ann.rad./
				ec/14/					(0.36–1.66)
^{46}Ti	8.25(3)	45.952629				0+			
^{47}Ti	7.44(2)	46.9517578				5/2-	-0.78848	+0.30	
^{48}Ti	73.72(3)	47.9479409				0+			
^{49}Ti	5.41(2)	48.9478646				7/2-	-1.10417	+0.25	
^{50}Ti	5.18(2)	49.9447858				0+			
^{48}V		47.952251	15.98 d	β+ /4.012	0.698/50.	4+	2.01		ann.rad./
									0.9835/100
									(1.3–2.4)

Elem. or Isot.	Natural abundance (Atom %)	Atomic mass or weight	Half-life/ Resonance width (MeV)	Decay mode/ Energy (MeV)	Particle energy/ Intensity (MeV/%)	Spin ($h/2\pi$)	Nuclear magnetic mom. (nm)	Elect. quadr. mom. (b)	γ-Energy/ Intensity (MeV/%)
^{49}V		48.948511	337. d	ec/0.602		7/2-	4.47		
^{50}V	0.250(10)	49.947159	2.7(2) ×10^{17} a	ec	/ > 93	6+	+3.34569	+0.21	
			> 1.9 ×10^{19} a	β-	/ < 7				
^{51}V	99.750(10)	50.943957				7/2-	+5.148706	-0.04	
^{48}Cr		47.95403	21.6 h	ec/1.66		0+			ann.rad./
									0.116(2)/95.
									0.305(10)/100.
^{50}Cr	4.345(13)	49.946041	>1.3 ×10^{18} a	β+ec		0+			
^{51}Cr		50.944765	27.70 d	ec/0.7527		7/2-	-0.934		0.3201/9.8
									0.00543/2.6
									0.00495/0.02
^{52}Cr	83.789(18)	51.940505				0+			
^{53}Cr	9.501(17)	52.940647				3/2-	-0.47454	~ -0.15	
^{54}Cr	2.365(7)	53.938878				0+			
^{52}Mn		51.945564	5.591 d	β+ /4.712	0.575/	6+	+3.063	+0.5	ann.rad./
				ec/					0.74421(1)/90.
									1.4341/100
^{53}Mn		52.941288	3.7 ×10^{6} a	ec/0.5970		7/2-	5.035	+0.17	
^{54}Mn		53.940356	312.2 d	ec/1.377		3+	+3.306	+0.37	0.8340/100
			6.7 ×10^{8} a	β+	//1.3 ×10^{-7}				
^{55}Mn	100.	54.9380432				5/2-	+3.4687	+0.33	
^{52}Fe		51.94811	8.28 h	β+ /57/2.37	0.804/	0+			ann.rad./
				ec/43/					0.16868(1)/99.
				IT/					0.377 (I.T.)/
^{54}Fe	5.845(105)	53.939608	>3.1 ×10^{22} a	ec-ec		0+			
^{55}Fe		54.938291	2.7563(4) a	ec/0.2314		3/2-			Mn x-ray
^{56}Fe	91.754(106)	55.9349356				0+			
^{57}Fe	2.119(29)	56.9353921				½-	+0.0906		
^{58}Fe	0.282(12)	57.933274				0+			
^{59}Fe		58.934874	44.50 d	β-/1.565	0.273/48.	3/2-	-0.336		1.099/57
					0.475/51.				1.292/43.
									(0.14–1.48)
^{60}Fe		59.934070	2.7 ×10^{6} a	β- /0.237	0.184/100.	0+			0.0586/100
^{55}Co		54.941997	17.53 h	β+ /3.4513	0.53/	7/2-	+4.822		ann.rad./
				ec/	1.03/				0.9312/75.
					1.50/				0.4772/20.
									(0.092–3.11)
^{56}Co		55.939838	77.3 d	β+/4.566	1.459/18.	4+	3.85	~+0.25	ann.rad./
				ec/					0.8468/99.9
									1.2383/68.
									(0.26–3.61)
^{57}Co		56.936290	271.8 d	ec/0.8361		7/2-	+4.72	+0.5	0.12206/86
									(0.014–0.706)
^{58m}Co			8.85(2) h	IT		5+			0.02489/0.035
^{58}Co		57.935751	70.88 d	β+ /2.307		2+	+4.04	+0.23	ann.rad./
				ec/					0.81076/99
^{59}Co	100.	58.933194				7/2-	+4.63	+0.42	
^{60}Co		59.933816	5.271 a	β- /2.824	0.315/99.7	5+	+3.799	+0.5	1.1732/100
									1.3325/100
^{56}Ni		55.942128	6.08 d	ec/2.14		0+			0.15838/99
				β+ /<10^{-6}					0.81185(3)/87.
									0.2695–0.7500
^{57}Ni		56.939792	35.6 h	β+ /3.264	0.712/10.	3/2-	-0.798		ann.rad./
				ec/	0.849/76.				1.3776/78.
									(0.127–3.177)
^{58}Ni	68.0769(190)	57.935342	>4 ×10^{19} a	ec-ec		0+			
^{59}Ni		58.934346	~ 7.6 ×10^{4} a	ec/		3/2-			
^{60}Ni	26.2231(150)	59.930785				0+			
^{61}Ni	1.1399(13)	60.931055				3/2-	-0.75002	+0.16	
^{62}Ni	3.6345(40)	61.928345				0+			

Elem. or Isot.	Natural abundance (Atom %)	Atomic mass or weight	Half-life/ Resonance width (MeV)	Decay mode/ Energy (MeV)	Particle energy/ Intensity (MeV/%)	Spin ($h/2\pi$)	Nuclear magnetic mom. (nm)	Elect. quadr. mom. (b)	γ-Energy/ Intensity (MeV/%)
^{63}Ni		62.929669	99. a	β- /0.066945	0.065/	½-	+0.05		
^{64}Ni	0.9256(19)	63.927966				0+			
^{66}Ni		65.92914	54.6 h	β- /0.23		0+			
^{61}Cu		60.933457	3.32 h	β+ /2.237	0.56/3.	3/2-	+2.109	- 0.22	ann.rad./
					0.94/5.				0.2830/13.
					1.15/2.				0.6560/11.
					1.220/51.				(0.067–2.123)
^{63}Cu	69.15(15)	62.929597				3/2-	+2.2273	-0.22	
^{64}Cu		63.929764	12.700 h	β- /0.579/38.5	0.578/	1+	-0.2164	+ 0.08	ann.rad./35.1
				β+ /1.6751/17.6	0.65/				1.3459(3)/0.475
				ec//43.5					
^{65}Cu	30.85(15)	64.927790				3/2-	+2.3817	-0.20	
^{67}Cu		66.92773	2.580 d	β- /0.58	0.395/56.	3/2-	+ 2.514	- 0.18	0.09125(1)/7.
					0.484/23.				0.09325(1)/17.
					0.577/20.				0.18453(1)/47.
^{62}Zn		61.934334	9.22 h	β+ /3/1.63	0.66/7.	0+			ann.rad./
				ec/93/					0.0408/25
									0.5967/26.
									(0.20–1.526)/
^{64}Zn	49.17(75)	63.929142	>7. ×10^{20} a	ec-β+		0+			
^{65}Zn		64.929241	244. d	β+ /98/1.3514	0.325/	5/2-	+0.7690	-0.023	ann.rad./
				ec/1.5/					1.1155/50.2
^{66}Zn	27.73(98)	65.926034				0+			
^{67}Zn	4.04(16)	66.927128				5/2-	+0.8753	+0.15	
^{68}Zn	18.45(63)	67.924844				0+			
^{69m}Zn			13.76 h	IT/99+/0.439		9/2+	1.157	-0.5	0.4390(2)/95.
^{70}Zn	0.61(10)	69.92532	>2.3 ×10^{17} a	β-β-		0+			
^{71m}Zn			4.13 h	β- /	1.45/	9/2+	1.05		0.3864/93.
^{72}Zn		71.92684	46.5 h	β- /0.46	0.25/14.	0+			0.0164(3)/8.
					0.30/86.				0.1447(1)/83.
									0.1915(2)/9.4
^{66}Ga		65.931590	9.31 h	β+ /56/5.175	0.74/1.	0+			ann.rad./
				ec/43/	1.84/54.				1.03935(8)/38.
					4.153/51.				2.7523(1)/23.
									(0.28–5.01)
^{67}Ga		66.928202	3.261 d	ec/1.001		3/2-	+1.8507	+ 0.197	0.09332/37.
									0.18459/20.
									0.30024/17.
									(0.091–0.89)
^{69}Ga	60.108(50)	68.925574				3/2-	+2.01659	+0.171	
^{71}Ga	39.892(50)	70.92470	>2.4 ×10^{26} a	β-		3/2-	+2.56227	+0.107	
^{72}Ga		71.92637	14.10 h	β- /4.001	0.64/40.	3-	-0.13224	+0.53	0.8340/95.53
					1.51/9.				2.202/26.9
					2.52/8.				0.630/26.2
					3.15/11.				(0.113–3.678)
^{73}Ga		72.92517	74.87 h	β- /1.59		1/2-		+ 0.209	0.05344(5)/10.
^{68}Ge		67.928095	270.8 d	ec/0.11		0+			Ga k x-ray/39.
^{69}Ge		68.927964	1.63 d	β+ /36/2.2273	0.70/	5/2-	0.735	+0.03	ann.rad./
				ec/64/	1.2/				0.574/13.
									1.1068/36.
									(0.2–2.04)
^{70}Ge	20.52(19)	69.924249				0+			
^{71}Ge		70.924952	11.2 d	ec/0.229		½-	+0.547		
^{72}Ge	27.45(15)	71.9220758				0+			
^{73}Ge	7.76(8)	72.9234590	>1.8 ×10^{23} a	β-		9/2+	-0.879468	-0.196	
^{74}Ge	36.52(12)	73.92117776				0+			
^{76}Ge	7.75(12)	75.92140273	1.5 ×10^{21} a	β-β-		0+			
^{77}Ge		76.9235498	11.25 h	β- /2.702	0.71/23.	7/2+			0.2110/29.
					1.38/35.				0.2155/27.

Elem. or Isot.	Natural abundance (Atom %)	Atomic mass or weight	Half-life/ Resonance width (MeV)	Decay mode/ Energy (MeV)	Particle energy/ Intensity (MeV/%)	Spin ($h/2\pi$)	Nuclear magnetic mom. (nm)	Elect. quadr. mom. (b)	γ-Energy/ Intensity (MeV/%)
					2.19/42.				0.2644/51.
									(0.15–2.35)
^{71}As		70.927114	2.72 d	β+ /32/2.013		5/2-	+1.6735	-0.02	ann.rad./
				ec/68/					0.1749(2)/84.
									1.0957(2)/4.2
^{72}As		71.926752	26.0 h	β+ /77/4.356	0.669/5.	2-	-2.1566	-0.08	ann.rad./
					1.884/12.				0.83395(5)/80.
					2.498/62.				1.0507(1)/9.6
					3.339/19.				(0.1–4.0)
^{73}As		72.923829	80.3 d	ec/0.341		3/2-			0.0133/0.1
									0.0534/10.5
									Se k x-ray/90.
^{74}As		73.923929	17.78 d	β+ /31/2.562	0.94/26.	2-	-1.597		ann.rad./
				ec/37/	1.53/3.				0.59588(1)/60.
				β- /1.353	0.71/16.				0.6084(1)/0.6
					1.35/16.				0.6348(1)/15.
^{75}As	100.	74.921595				3/2-	+1.43947	+0.31	
^{76}As		75.922392	26.3 h	β- /2.962	0.54/3.	2-	-0.903		0.5591(1)/45.
				Ec//<0.0002	1.785/8.				0.65703(5)/6.2
					2.410/36.				1.21602(1)/3.4
					2.97/51.				(0.3–2.67)
^{77}As		76.920648	38.8 h	β- /0.683	0.70/98.	3/2-	+1.295		0.2391(2)/1.6
^{72}Se		71.927141	8.5 d	ec/0.34		0+			0.0460(2)/57.
^{73}Se		72.92676	7.1 h	β+ /65/2.74	0.80/	9/2+	0.86		ann.rad
				ec/35/	1.32/95.				0.0670(1)/72.
					1.68/1.				0.3609(1)/97.
									(0.6–1.5)
^{74}Se	0.86(3)	73.92247594	>1.5×10^{19} a	ec-ec		0+			
^{75}Se		74.9225229	119.8 d	ec/0.864		5/2+	0.68	1.1	0.13600/55
									0.26465/58
									(0.024–0.821)
^{76}Se	9.23(7)	75.91921370				0+			
^{77}Se	7.60(7)	76.9199142				½-	+0.53506		
^{78}Se	23.69(22)	77.9173092				0+			
^{79}Se		78.9184993	2.8(2) ×10^{5} a	β- /0.151		7/2+	-1.02	+0.8	
^{80}Se	49.80(36)	79.916522				0+			
^{82}Se	8.82(15)	81.916700	9. ×10^{19} a	β-β-		0+			
^{76}Br		75.92454	16.0 h	β+ /57/4.96	1.9/	1-	0.54821	+0.255	ann.rad
					3.68/				0.55911
									1.85368
									(0.4–4.6)
^{77}Br		76.921379	2.376 d	ec/99/1.365		3/2-	0.973	+0.50	ann.rad.
									0.23898
									0.52069
									(0.08–1.2)
^{79}Br	(50.5-50.8)	78.918338				3/2-	+2.106400	+0.313	
^{80m}Br			4.421 h	IT	0.04879/100	5-	+1.3177	+0.71	Br k x-ray
									0.03705/39.1
									0.04885/0.3
^{81}Br	(49.2-49.5)	80.916288				3/2-	+2.270562	+0.262	
^{82}Br		81.916802	1.471 d	β- /3.093	0.444/	5-	+1.6270	+0.71	0.5544/71
									0.61905/43
									0.77649/84
									(0.013–1.96)
^{83}Br		82.915175	2.37 h	β- /0.972	0.395/1	3/2-			0.52964
					0.925/99				(0.12–0.68)
^{76}Kr		75.925911	14.8 h	ec/1.31		0+			Br k x-ray
									0.270/21
									0.3158/39
									(0.03–1.07)

Elem. or Isot.	Natural abundance (Atom %)	Atomic mass or weight	Half-life/ Resonance width (MeV)	Decay mode/ Energy (MeV)	Particle energy/ Intensity (MeV/%)	Spin (*h*/2π)	Nuclear magnetic mom. (nm)	Elect. quadr. mom. (b)	γ-Energy/ Intensity (MeV/%)
^{78}Kr	0.355(3)	77.9203663	9. ×10^{21} a	ec-ec		0+			
^{79}Kr		78.920083	1.455 d	β+ /7 /1.626		½-	+0.536		ann.rad.
				ec/93 /					0.2613/13
									0.39756/19
									0.6061/8
									(0.04–1.3)
^{80}Kr	2.286(10)	79.916378				0+			
^{81}Kr		80.916590	2.1 ×10^{5} a	ec/0.2807		7/2+	-0.908	+0.644	Br k x-ray
									0.2760
^{82}Kr	11.593(31)	81.91348116				0+			
^{83}Kr	11.500(19)	82.91412652				9/2+	-0.970669	+0.259	
^{84}Kr	56.987(15)	83.911497730				0+			
^{85m}Kr			4.48 h	β- /79 /	0.83/79	½-	+ 0.633		0.30487
				IT/21/	0.30487				0.15118
^{85}Kr		84.912527	10.68 a	β- /0.687	0.15/0.4	9/2+	1.005	+0.443	0.51399
^{86}Kr	17.279(41)	85.910610626				0+			
^{81}Rb		80.91899	4.57 h	β+ /27/2.24	1.05/	3/2-	+2.060	+0.5	ann.rad./
				ec/73					0.19030/64.
									(0.05–1.9)
^{82m}Rb			6.47 h	β+/26/	0.80/	5-	+1.51001	+1.2	ann.rad./
				ec/74/					0.5544/63.
									0.7765/85.
									(0.092–2.3)
^{83}Rb		82.915114	86.2 d	ec/0.91		5/2-	+1.425	+ ~ 0.24	Kr x-ray
									0.5205/46.
									(0.03–0.80)
^{84}Rb		83.914375	32.9 d	β+/22/2.681	0.780/11	2-	-1.32412	-0.02	ann.rad./
				ec/75 /	1.658/11				0.8817/68.
				β-/3/0.894	0.893/				(1.02–1.9)
^{85}Rb	72.17(2)	84.91178974				5/2-	+1.353	+0.276	
^{86}Rb		85.9111674	18.68 d	β-/1.775	1.774/8.8	2-	-1.6920	+0.2	1.0768/8.8
^{87}Rb	27.83(2)	86.90918053	4.97 ×10^{10} a	β-/0.283	0.273/100	3/2-	+2.7512	+0.134	
^{82}Sr		81.91840	25.36 d	ec/0.18		0+			Rb x-ray
^{83}Sr		82.91755	1.350 d	β+/24/2.28	0.465/	7/2+	-0.8298	+0.71	ann.rad./
				ec/76/	0.803/				0.3816/12.
					1.227/				0.3816
									0.7627/30.
									(0.094–2.15)
^{84}Sr	0.56(2)	83.913419				0+			
^{85}Sr		84.912932	64.85 d	ec/1.065		9/2+	-1.001	+0.26	0.51399/99.3
^{86}Sr	9.86(20)	85.90926073				0+			
^{87m}Sr			2.81 h	IT/0.3884		½-	+0.63		0.3884(IT)
^{87}Sr	7.00(20)	86.90887750				9/2+	-1.0936	+0.305	
^{88}Sr	82.58(35)	87.90561226				0+			
^{89}Sr		88.9074508	50.6 d	β-/1.497	1.492/100	5/2+	-1.148	-0.25	0.9092
^{90}Sr		89.907731	28.9 a	β-/0.546	0.546/100	0+			
^{91}Sr		90.91020	9.5 h	β-/2.70	0.61/7	5/2+	-0.885	+0.04	0.5556/61.
					1.09/33				0.7498/24.
					1.36/29				1.0243/33.
					2.66/26				(0.12–2.4)
^{85m}Y			4.9 h	β+/70/		9/2+	6.2		ann.rad./
				ec/30/					0.2317
									0.5356
									2.1238
									(0.1–3.1)
									0.7673
^{86}Y		85.91489	14.74 h	β+/5.24		4-	<0.6		ann.rad./
				ec/					0.3070
									0.6277
									1.0766

Elem. or Isot.	Natural abundance (Atom %)	Atomic mass or weight	Half-life/ Resonance width (MeV)	Decay mode/ Energy (MeV)	Particle energy/ Intensity (MeV/%)	Spin ($h/2\pi$)	Nuclear magnetic mom. (nm)	Elect. quadr. mom. (b)	γ-Energy/ Intensity (MeV/%)
									1.1531
									1.9207
									(0.1–3.8)
^{87m}Y			13.4 h	IT	0.3808/98	9/2+	+ 6.24	- 0.5	0.3807
				β+/0.7/	1.15/0.7				
				ec/					
^{87}Y		86.910876	3.35 d	ec/99+/1.862	0.78/	1/2-	- 0.19		0.3880
									0.4870
^{88}Y		87.909501	106.6 d	ec/99+/3.623	0.76/	4-	- 0.42	+ 0.16	ann.rad./
				β+/0.2/					0.89802
									1.83601
									2.73404
									3.2190
^{89}Y	100.	88.905841				1/2-	-0.13742		
^{90m}Y			3.20 h	IT/99+/	0.68204/100	7+	+ 5.28	~ - 0.65	0.2025
				β-/0.002/					0.4794
									0.6820
^{90}Y		89.907145	2.669 d	β-/2.282	2.28/	2-	-1.63	-0.13	(0.203-0.480)
^{91}Y		90.907298	58.5 d	β-/1.544	1.545/	1/2-	0.164		1.208
^{92}Y		91.90895	3.54 h	β-/3.63	3.64/	2-	- 0.67		0.4485
									0.5611
									0.9345
									1.4054
									(0.4–3.3)
^{93}Y		92.90958	10.2 h	β-/2.87	2.88/90	1/2-	- 0.12		0.2669
									0.9471
^{86}Zr		85.916297	16.5 h	ec/1.47		0+			0.0280
									0.243
									0.612
^{88}Zr		87.91022	83.4 d	ec/0.67		0+			0.3929
^{89}Zr		88.908882	3.27 d	β+ /23/2.832	0.9/	9/2+	-1.05	+ 0.3	ann.rad./
				ec/77/					0.9092
^{90}Zr	51.45(4)	89.9046988				0+			
^{91}Zr	11.22(5)	90.9056402				5/2+	-1.30362	-0.176	
^{92}Zr	17.15(3)	91.9050353				0+			
^{93}Zr		92.906471	1.5 ×10^{6} a	β- /0.091		5/2+			0.0304
^{94}Zr	17.38(4)	93.9063125	>10^{17} a	β-β-		0+			
^{95}Zr		94.908040	64.02 d	β- /1.125	0.366/55	5/2+	1.13	+0.22	0.7242
					0.400/44				0.7567
^{96}Zr	2.80(2)	95.9082776	2.3 ×10^{19} a	β-β-		0+			
			>2.4 ×10^{19} a	β-					
^{97}Zr		96.9109574	16.75 h	β- /2.658	1.91/	½-	- 0.937		0.7434
^{90}Nb		89.911259	14.6 h	β/53 /6.111	0.86/5	8+	4.961	~+ 0.01	ann.rad./
				ec/47 /	1.5/92				0.1412
									1.1292
									2.1862
									2.3189
									(0.1–3.3)
^{91m}Nb			60.9. d	IT	0.1046/97	1/2-			0.1045(IT)
				ec/3 /					1.2050
^{91}Nb		90.906990	7 ×10^{2} a	ec/1.253		9/2+		-0.25	Mo k x-ray
^{92m}Nb			10.15 d	ec/99+/		2+	6.14		0.9126
									0.9345
									1.8475
^{92}Nb		91.907189	3.7 ×10^{7} a	ec/2.006		7+		-0.35	0.5611
									0.9345
^{93m}Nb			16.1 a	IT	0.03077/100	1/2-			Nb x-ray
									0.0304
^{93}Nb	100.	92.906373				9/2+	+6.1705	-0.32	
^{94}Nb		93.907279	2.04 ×10^{4} a	β- /2.045	0.47/	6+			0.70263

Elem. or Isot.	Natural abundance (Atom %)	Atomic mass or weight	Half-life/ Resonance width (MeV)	Decay mode/ Energy (MeV)	Particle energy/ Intensity (MeV/%)	Spin ($h/2\pi$)	Nuclear magnetic mom. (nm)	Elect. quadr. mom. (b)	γ-Energy/ Intensity (MeV/%)
									0.87109
^{95m}Nb			3.61 d	IT	0.2357/94.4	1/2-			0.2040
				β- /2.5 /					0.2356
^{95}Nb		94.906831	34.97 d	β- /0.926	0.160/	9/2+	6.14		0.76578
^{96}Nb		95.9081016	23.4 h	β- /3.187	0.5/10	6+	4.976		0.7782
					0.75/90				0.2191–1.498
^{90}Mo		89.91391	5.7 h	β+ /25/2.489	1.085/	0+			ann.rad./
				ec/75 /					0.04274
									0.12237
									0.25734
^{92}Mo	14.649(106)	91.9068071	> 3 ×10^{17} a	β+-ec		0+			
^{93m}Mo			6.9 h	IT/99+ /2.425		21/2+	+9.9		0.26306(IT)
									0.68461
									1.47711
^{93}Mo		92.9068088	3.5 ×10^{3} a	ec/0.405		5/2+			0.0304
^{94}Mo	9.187(33)	93.9050836				0+			
^{95}Mo	15.873(30)	94.9058374				5/2+	-0.9142	-0.022	
^{96}Mo	16.673(8)	95.9046748				0+			
^{97}Mo	9.582(15)	96.9060169				5/2+	-0.9335	+0.26	
^{98}Mo	24.292(80)	97.9054036				0+			
^{99}Mo		98.9077073	2.747 d	β- /1.357	0.45/14	½+	0.375		0.144048
					0.84/2				0.18109
					1.21/84				0.36644
									0.73947
^{100}Mo	9.744(65)	99.9074680	7 ×10^{18} a	β-β-		0+			
^{93}Tc		92.910245	2.73 h	β+ /13/3.201	0.81	9/2+	6.3		ann.rad./
				ec/87/					1.3629
									1.4771
									1.5203
									(0.1–3.0)
^{94}Tc		93.909652	4.88 h	β+ /11/4.256		7+	5.12		ann.rad./
				ec/89/					0.4491
									0.7026
									0.8496
									0.8710
^{95m}Tc			61 d	IT/4/	0.0389	1/2-			ann.rad./
				β+ /0.3	0.5/				0.0389(IT)
				ec/96	0.7/				0.2041
									0.5821
									0.5821
									0.8351
^{95}Tc		94.90765	20.0 h	ec/100/1.691		9/2+	5.9		0.7657
									1.0738
^{96}Tc		95.90787	4.3 d	ec/2.973		7+	+5.1		Mo k x-ray
									0.7782
									0.8125
									0.8498
									1.12168
^{97m}Tc			91. d	IT	0.0956/96	1/2-			Tc k x-ray
				ec	/3.9				0.0965
^{97}Tc		96.906361	4.2 ×10^{6} a	ec/100/0.320		9/2+			Mo k x-ray
^{98}Tc		97.907211	~ 6.6 ×10^{6} a	β- /1.80	0.40/100	6+			0.65241/100
				ec	//<0.036				0.74535/100
^{99m}Tc			6.007 h	IT/100/0.142		1/2-			Tc k x-ray
									0.14049
									0.14261
^{99}Tc		98.906250	2.13 ×10^{5} a	β- /0.294	0.293/100	9/2+	+5.6847	~-0.129	
^{96}Ru	5.54(14)	95.9075889	>31.4 ×10^{20} a	β+β+		0+			
^{97}Ru		96.907546	2.84 d	ec/1.12		5/2+	-0.79		Tc k x-ray
									0.2157

Elem. or Isot.	Natural abundance (Atom %)	Atomic mass or weight	Half-life/ Resonance width (MeV)	Decay mode/ Energy (MeV)	Particle energy/ Intensity (MeV/%)	Spin ($h/2\pi$)	Nuclear magnetic mom. (nm)	Elect. quadr. mom. (b)	γ-Energy/ Intensity (MeV/%)
									0.3245
									0.4606
^{98}Ru	1.87(3)	97.90529				0+			
^{99}Ru	12.76(14)	98.905930				5/2+	-0.64	+0.079	
^{100}Ru	12.60(7)	99.904211				0+			
^{101}Ru	17.06(2)	100.9055731				5/2+	-0.72	+0.46	
^{102}Ru	31.55(14)	101.904340				0+			
^{103}Ru		102.906315	39.26 d	β- /0.763	0.223	3/2+	0.206	+0.62	0.05329
									0.29498
									0.4438
									0.49708
									0.55704
									0.61033
									(0.04–1.6)
^{104}Ru	18.62(27)	103.905425	> 6.6 ×10^{20} a			0+			
^{105}Ru		104.907746	4.44 h	β- /1.917	1.11/22	3/2+	-0.3		0.12968
					1.134/13				0.1491
					1.187/49				0.2629
									0.31664
									0.46943
									0.67634
									0.72420
									(0.1–1.8)
^{106}Ru		105.90733	1.020 a	β- /0.0394	0.0394/100	0+			
^{99m}Rh			4.7 h	β+ /8/	.74/	9/2+	5.67		ann.rad./
				ec/92/					0.2766/
									0.3408
									0.6178
									1.2612
^{99}Rh		98.90813	16. d	β+/4/2.10	0.54/	1/2-			ann.rad./
				ec/97/	0.68/				0.0894/
									0.3530
									0.5277
									(0.1–2.0)
^{100}Rh		99.90811	20.8 h	β+ /3.63	2.62/	1-			0.4462
				ec/	2.07/				0.5396
									0.5882
									0.8225
									1.5534
									2.3761
^{101m}Rh			4.34 d	ec/92.8/		9/2+	+5.47		Rh k x-ray
				IT/7.2/	0.1573				0.1272/
									0.3069
									0.5451
^{101}Rh		100.90616	3.3 a	ec/0.54		1/2-			Ru k x-ray
									0.1272
									0.1980
									0.3252
^{102m}Rh			3.74 a	ec/2.323		6(+)	4.04		0.4751
				IT/0.0419					0.6313
			> 1.2 ×10^{6} a	β^{+}	/<0.00025				0.6975
									0.7668
									1.0466
									1.1032
^{102}Rh		101.90682	207. d	ec/62		(2-)	~ 0.5		ann.rad./
				β- /19/					0.4686
				β+ /14/					0.4751
									0.5566
									0.6280
									1.1032

Elem. or Isot.	Natural abundance (Atom %)	Atomic mass or weight	Half-life/ Resonance width (MeV)	Decay mode/ Energy (MeV)	Particle energy/ Intensity (MeV/%)	Spin ($h/2\pi$)	Nuclear magnetic mom. (nm)	Elect. quadr. mom. (b)	γ-Energy/ Intensity (MeV/%)
									(0.4–1.6)
^{103}Rh	100.	102.905494				1/2-	-0.0884		
^{105}Rh		104.905688	35.3 h	β- /0.567	0.247/30	7/2+	+4.45		0.2801
					0.567/70				0.3061/4.8
									0.3189/17.0
^{106m}Rh			2.18 h	β- /	0.92/	(6)+			0.2217
									0.4510
									0.5119
									0.6162
									0.7173
									0.7484
									1.0458
									1.5277
^{100}Pd		99.90852	3.7 d	ec/0.36		0+			0.03271
									0.0748
									0.0840
^{101}Pd		100.90829	8.4 h	β+ /5/1.980	0.776/	5/2+	-0.66		ann.rad./
				ec/95/					0.0244
									0.2963
									0.5904
^{102}Pd	1.02(1)	101.905632	> 6.0 ×10^{18} a	Ec ec		0+			
^{103}Pd		102.906111	16.99 d	ec/0.543		5/2+			Rh k x-ray
									0.03975
									0.3575
									0.4971
^{104}Pd	11.14(8)	103.904030				0+			
^{105}Pd	22.33(8)	104.905080				5/2+	-0.642	+0.66	
^{106}Pd	27.33(3)	105.903480				0+			
^{107}Pd		106.905128	6.5 ×10^{6} a	β- /0.033	0.03/	5/2+			
^{108}Pd	26.46(9)	107.903892				0+			
^{109}Pd		108.905951	13.5 h	β- /1.116	1.028	5/2+			0.0880
									(0.08–1.0)
^{110}Pd	11.72(9)	109.905173	> 2.0 ×10^{20} a	β- β-		0+			
^{111m}Pd			5.57 h	IT/73/0.172		11/2-			0.0704
				β- /27/	0.35				0.1722
					0.77				0.3912
									(0.1–1.97)
^{112}Pd		111.90733	21.04 h	β- /0.29	0.28/	0+			0.018
^{105}Ag		104.90653	41.3 d	ec/1.35		1/2-	0.101		0.0640
									0.2804
									0.3445
									0.4434
^{106m}Ag			8.3 d	ec/		6+	3.71	+1.1	Pd k x-ray
									0.4510
									0.5118
									0.7173
									1.0458
^{107}Ag	51.839(8)	106.905092				1/2-	-0.113680		
^{108m}Ag			~440 a	ec/92/		6+	3.58	+1.3	Ag k x-ray
^{109}Ag	48.161(8)	108.904756				1/2-	-0.13069		
^{110m}Ag			249.8 d	β- /99/	0.087	6+	+3.61	+1.4	0.65774
				IT/1 /0.1164	0.530				0.76393
									0.88467
									0.93748
									1.38427
									(0.447–1.56)
^{111}Ag		110.905297	7.45 d	β- /1.037	1.035/	1/2-	-0.146		0.2454
									0.3421
^{112}Ag		111.907049	3.13 h	β- /3.96	3.94/	2(-)	0.0547		0.6067
					3.4				0.6174

Elem. or Isot.	Natural abundance (Atom %)	Atomic mass or weight	Half-life/ Resonance width (MeV)	Decay mode/ Energy (MeV)	Particle energy/ Intensity (MeV/%)	Spin ($h/2\pi$)	Nuclear magnetic mom. (nm)	Elect. quadr. mom. (b)	γ-Energy/ Intensity (MeV/%)
									1.3877
									(0.4–2.9)
^{113}Ag		112.90657	5.3 h	β- /2.02	2.01/	1/2-	0.159		0.2588
									0.2986
^{106}Cd	1.245(22)	105.906460	≥ 1.1x10^{21} a	ec, β+		0+			
^{107}Cd		106.906612	6.52 h	ec/99+/1.417		5/2+	-0.615055	+0.60	Ag k x-ray
				β+ /					0.0931
									0.8289
^{108}Cd	0.888(11)	107.904184	>4.1 ×10^{17} a	ec ec		0+			
^{109}Cd		108.904987	462.2 d	ec/0.214		5/2+	-0.827846	+0.60	Ag k x-ray
									0.08804/.0366
^{110}Cd	12.470(61)	109.9030075				0+			
^{111}Cd	12.795(38)	110.9041838				1/2+	-0.594886		
^{112}Cd	24.109(12)	111.9027639				0+			
^{113m}Cd			13.9 a	β- /99.9/0.59	0.59/99.9	11/2-	-1.087784	-0.61	0.2637
^{113}Cd	12.227(7)	112.9044081	8.04 ×10^{15} a	β-		1/2+	-0.622301		
^{114}Cd	28.754(81)	113.9033650	>1.3 ×10^{18} a	β-β-		0+			
^{115m}Cd			44.6 d	β- /1.629	0.68/1.6	11/2	-1.041034	-0.48	0.48450
					1.62/97				0.93381
									1.29064
^{115}Cd		114.9054387	2.228 d	β- /1.446	0.593/42	1/2+	-0.648426		0.23141
					1.11/58				0.26085
									0.33624
									0.49227
									0.52780
^{116}Cd	7.512(54)	115.9047632	3. ×10^{19} a	$\beta^-\beta^-$		0+			
^{117m}Cd			3.4 h	β- /2.66	0.72/	11/2-	- 0.998	- 0.32	0.1586
									0.5529
									0.37–2.42
^{117}Cd		116.907226	2.49 h	β- /2.52	0.67/51	1/2+	- 0.7436		0.2209
					2.2/10				0.2733
									0.3445
									1.3033
^{109}In		108.907150	4.17 h	β+ /8/2.02	0.79/	9/2+	+5.54	+0.80	ann.rad./
				ec/92/					Cd k x-ray
									0.2035
									0.6235
^{110m}In			4.9 h	ec/		7+	+4.71	+0.95	Cd k x-ray
									0.6577
									0.8847
									0.9375
									(0.1–1.98)
^{111}In		110.905107	2.8049 d	ec/0.866		9/2+	+5.50	+0.76	Cd k x-ray
									0.1712
									0.2453
^{113}In	4.281(52)	112.9040605				9/2+	+5.5289	+0.76	
^{114m}In			49.51 d	IT/97/0.190		5+	+4.65	+0.70	In k x-ray
				ec/3 /					0.19027
^{115m}In			4.486 h	IT/95/0.336		½-	-0.2440		In k x-ray
				β- /5 /0.83					0.3362
									0.4974
^{115}In	95.719(52)	114.90387877	4.4 ×10^{14} a	β- /0.495		9/2+	+5.5408	0.77	
^{110}Sn		109.90785	4.17 h	ec/0.64		0+			In k x-ray
									0.283
^{112}Sn	0.97(1)	111.9048249	> 9.7 ×10^{19} a	β+/ec		0+			
^{113}Sn		112.905176	115.1 d	ec/1.036		½+	-0.879		In k x-ray
									0.25511/3.2
									0.39169/100

Elem. or Isot.	Natural abundance (Atom %)	Atomic mass or weight	Half-life/ Resonance width (MeV)	Decay mode/ Energy (MeV)	Particle energy/ Intensity (MeV/%)	Spin ($h/2\pi$)	Nuclear magnetic mom. (nm)	Elect. quadr. mom. (b)	γ-Energy/ Intensity (MeV/%)
^{114}Sn	0.66(1)	113.9027801				0+			
^{115}Sn	0.34(1)	114.90334470				½+	-0.9188		
^{116}Sn	14.54(9)	115.9017428				0+			
^{117m}Sn			13.91 d	IT/0.3146		11/2-	-1.396	-0.4	Sn k x-ray
									0.15856
^{117}Sn	7.68(7)	116.902954				½+	-1.0010		
^{118}Sn	24.22(9)	117.901607				0+			
^{119m}Sn			293. d	IT/0.0896	0.0657/100	11/2-	-1.4	0.21	Sn k x-ray
									0.02387
^{119}Sn	8.59(4)	118.903311				½+	-1.0473		
^{120}Sn	32.58(9)	119.902202				0+			
^{121m}Sn			44. a	IT/78/0.006		11/2-	-1.388	-0.14	Sn k x-ray
				β- /22/	0.354/				0.03715
^{121}Sn		120.904243	1.128 d	β- /0.388	0.383/100	3/2+	0.698	~ -0.02	
^{122}Sn	4.63(3)	121.903444				0+			
									0.3814
^{123}Sn		122.905725	129.2 d	β- /1.404	1.42/99.4	11/2-	-1.370	~ +0.03	0.1603
									(1.030-1.089)
^{124}Sn	5.79(5)	123.905277	>2.2 ×10^{18} a	β$^-$β$^-$		0+			
^{125}Sn		124.907786	9.63 d	β- /2.364	2.35/82	11/2-	-1.348	~ +0.2	1.0671
									(0.2–2.3)
^{126}Sn	1 ×10^{-6}	125.90766	2.0 ×10^{5} a	β- /0.38	0.25/100	0+			0.0643
									0.0876
									(0.4148-0.6950)
^{118m2}Sb			5.00 h	ec/99/		8-	2.32		Sn k x-ray
									0.25368
									(1.051-1.230)
^{119}Sb		118.90395	38.1 h	ec/0.59		5/2+	+3.45	-0.4	Sn k x-ray
									0.0239
^{120m}Sb			5.76 d	ec/		8-	2.34		Sn k x-ray
									0.0898
									0.19730
									1.02301
									1.17121
^{121}Sb	57.21(5)	120.903810				5/2+	+3.3634	-0.54	
^{122}Sb		121.905168	2.72 d	β- /98/1.979	1.414/65	2-	-1.90	+1,3	0.56409
				β+ /2/1.620	1.980/26				0.69277
									(1.1495-1.257)
^{123}Sb	42.79(5)	122.904214				7/2+	+2.5498	-0.69	
^{124}Sb		123.905936	60.2 d	β- /2.905	0.61/52	3-	1.20	+2.80	0.60271/97.8
					2.301/23				1.69094/48.2
									(0.027–2.871)
^{125}Sb		124.905253	2.758 a	β- /0.767	0.13/30	7/2+	+2.63		0.0355
					0.302/45				0.17632
					0.62/13				0.38044
									0.42786
									0.46336
									0.60060
									0.63595
^{126}Sb		125.90725	12.4 d	β- /3.67	1.9	8-	1.3		0.2786
									0.4148
									0.6663
									(0.695-0.7205)
^{127}Sb		126.90692	3.84 d	β- /1.581	0.89/	7/2+	2.70		0.2524
					1.10/				0.2908
					1.50/				0.4121
									0.4370
									0.6857
									0.7837
^{128}Sb		127.90915	9.1 h	β- /4.38	2.3/	8-	1.3		0.2148

Elem. or Isot.	Natural abundance (Atom %)	Atomic mass or weight	Half-life/ Resonance width (MeV)	Decay mode/ Energy (MeV)	Particle energy/ Intensity (MeV/%)	Spin ($h/2\pi$)	Nuclear magnetic mom. (nm)	Elect. quadr. mom. (b)	γ-Energy/ Intensity (MeV/%)
									0.3141
									(0.5265-0.7540)
^{129}Sb		128.90915	4.37 h	β- /2.38	0.65/	7/2+	2.79		0.0278
									0.1808
									0.3594
									0.4596
									0.5447
									0.8128
									(0.9146-1.030)
^{118}Te		117.90585	6.00 d	ec/0.28		0+			Sb k x-ray
^{119m}Te			4.70 d	ec/		11/2-	0.89		Sb k x-ray
									0.15360
									0.2705
									1.21271
^{119}Te		118.90641	16.0 h	β+ /2/2.293	0.627/	½+	0.25		ann.rad.
				ec/98/					Sb k x-ray
									0.6440
									0.6998
^{120}Te	0.09(1)	119.904060	> 1.9×10^{17} a	β$^+$ ec		0+			
^{121m}Te			165.0(5) d	IT (89%)		11/2-	0.90		Te k x-ray
				ec (11%)					0.2122
^{121}Te		120.90494	16.8 d	ec/1.04		½+			Sb k x-ray
									0.5076
									0.5731
^{122}Te	2.55(12)	121.903043				0+			
^{123m}Te			119.7 d	IT/0.247		11/2-	-0.93		Te k x-ray
									0.1590/84.1
^{123}Te	0.89(3)	122.904270	>9.2 ×10^{16} a	ec/0.051		½+	-0.736948		
^{124}Te	4.74(14)	123.902817				0+			
^{125m}Te			57.4 d	IT/0.1448	0.1093/100	11/2-	-0.99	0.0	Te k x-ray
									0.0355
^{125}Te	7.07(15)	124.904430				½+	-0.8885		
^{126}Te	18.84(25)	125.903311				0+			
^{127m}Te			106. d	IT/98/0.088		11/2-	-1.04	~ 0.2	Te k x-ray
				β- /2/0.77/2.4					0.0883
^{127}Te		126.905226	9.30 h	β- /0.698	0.696/	3/2+	0.635		0.3603
^{128}Te	31.74(8)	127.904461	2.5 ×10^{24} a	β-β-		0+			
^{129m}Te			33.6 d	IT/63/0.105		11/2-	-1.09	0.40	Te k x-ray
				β- /37/	1.60/				0.45984
									0.6959
^{130}Te	34.08(62)	129.90622275	7. ×10^{20} a	β-β-		0+			
^{131m}Te			1.35 d	β- /78/2.4	0.42/	11/2-	-1.04	~ 0.25	0.0811
				IT/22/0.18					0.1021
									0.14973
									0.77369
									0.79375
									0.85225
^{132}Te		131.908547	3.26 d	β- /0.51	0.215	0+			0.049725
									0.11198
									0.22830
^{123}I		122.905589	13.22 h	ec/1.242		5/2+	2.82		Te k x-ray
									0.1590
^{124}I		123.906209	4.176 d	β+ /23/3.160	1.54/	2-	1.446		ann.rad./
				ec/77/	2.14/				Te k x-ray
					0.75/				0.6027/62.9
									0.7228/10.3
									1.6910/11.2
									(0.31–1.73)
^{125}I		124.904629	59.4 d	ec/0.1861		5/2+	2.82	-0.76	Te k x-ray
									0.0355

Elem. or Isot.	Natural abundance (Atom %)	Atomic mass or weight	Half-life/ Resonance width (MeV)	Decay mode/ Energy (MeV)	Particle energy/ Intensity (MeV/%)	Spin (h/2π)	Nuclear magnetic mom. (nm)	Elect. quadr. mom. (b)	γ-Energy/ Intensity (MeV/%)
^{126}I		125.905623	13.0 d	ec/		2-	1.438		ann.rad./
				β+ /2.155	1.13/				Te k x-ray
				β- /1.258/47	0.87/				0.3887
					1.25/				0.6622
^{127}I	100.	126.904472				5/2+	+2.8133	-0.70	
^{129}I		128.904984	1.7 ×10^{7} a	β- /0.194	0.15/	7/2+	+2.6210	-0.49	Xe k x-ray
									0.0396
^{130}I		129.906670	12.36 h	β- /2.949	1.04/	5+	3.35		0.4180
					0.62				0.5361
									(0.6685-0.7395)
^{131}I		130.906126	8.021 d	β- /0.971	0.606/	7/2+	+2.742	-0.34	0.08017
									0.28431
									0.36446
									0.63699
^{133}I		132.90783	20.8 h	β- /1.77	1.24/85	7/2+	+2.86	-0.23	0.51056
									0.52989
									0.87537
^{135}I		134.910059	6.57 h	β- /2.63	0.9/	7/2+	2.940		0.2884
					1.3/				0.41768
									0.52658
									1.13156
									1.26046
^{122}Xe		121.9037	20.1 h	ec/0.9		0+			I k x-ray
									0.3501
^{123}Xe		122.90848	2.00 h	β+ /23/2.68	1.51/	1/2+	-0.150		ann.rad./
				ec/77/					I k x-ray
									0.1489
									0.1781
									(0.1–2.1)
^{124}Xe	0.0952(3)	123.905892	> 10^{17} a	β-β-		0+			
			1.8(6) × 10^{22} a	ec-ec.					
^{125}Xe		124.906394	16.9(1) h	ec.	0.47/	½+	-0.269		I k x-ray
									0.1884
									0.2434
^{126}Xe	0.0890(2)	125.904297	> 4.3 × 10^{21} a	ec. ec		0+			
^{127}Xe		126.905183	36.34 d	ec/0.662		½+	-0.504		I k x-ray
									0.1721
									0.2029
									0.3750
^{128}Xe	1.9102(8)	127.903531				0+			
^{129m}Xe			8.89 d	IT/0.236	0.1966/100	11/2-	-0.89122	+0.63	1 k x-ray
									0.0396
									0.1966
^{129}Xe	26.4006(82)	128.90478086				½+	-0.77798		
^{130}Xe	4.0710(13)	129.90350935				0+			
^{131m}Xe			11.9 d	IT/0.164		11/2-	-0.99405	+0.72	1 k x-ray
									0.16398
^{131}Xe	21.2324(30)	130.90508414				3/2+	+0.69186	-0.114	
^{132}Xe	26.9086(33)	131.90415509				0+			
^{133m}Xe			2.20 d	IT/0.233		11/2-	-1.082	~+0.76	1 k x-ray
									0.23325
^{133}Xe		132.905911	5.243 d	β- /0.427	0.346/99	3/2+	+0.813	+0.14	Cs k x-ray
									(0.0810-0.1606)
^{134}Xe	10.4357(21)	133.9053903	>1.1 ×10^{16} a	β- β-		0+			
^{135}Xe		134.907232	9.10 h	β-/1.15	0.91/	3/2+	0.903	+0.21	0.24975/0.90
									0.60807
^{136}Xe	8.8573(44)	135.90721448	2.3 ×10^{21} a	β-β-		0+			
^{127}Cs		126.90742	6.2 h	β+ /96/2.08	0.65/	1/2+	+1.46		Xe k x-ray
				ec/4/	1.06				0.1247

Elem. or Isot.	Natural abundance (Atom %)	Atomic mass or weight	Half-life/ Resonance width (MeV)	Decay mode/ Energy (MeV)	Particle energy/ Intensity (MeV/%)	Spin ($h/2\pi$)	Nuclear magnetic mom. (nm)	Elect. quadr. mom. (b)	γ-Energy/ Intensity (MeV/%)
									0.4119
^{129}Cs		128.90607	1.336 d	ec/1.195		1/2+	+1.49		Xe k x-ray
									0.3719
									0.4115
^{131}Cs		130.90547	9.69 d	ec/0.352		5/2+	+3.543	-0.59	Xe k x-ray
^{132}Cs		131.906438	6.48 d	ec/98/		2 +	+2.22	+0.48	Xe k x-ray
				β+ /0.3/2.120					0.4646
				β- / /1.280					0.6302
									0.66769
^{133}Cs	100.	132.90545196				7/2+	+2.58291	-0.0034	
^{134}Cs		133.90671850	2.065 a	β- /2.059	0.089/27	4+	+2.994	+0.37	0.56327
					0.658/70				0.56935
				ec/1.22					0.60473
									0.79584
^{135}Cs		134.905977	~ 1.6 ×10^{6} a	β- /0.269	0.205/100	7/2+	+2.7324	+0.048	
^{136}Cs		135.907312	13.16 d	β- /2.548	0.341/	5+	+3.71	+0.21	0.06691
									0.34057
									0.81850
									1.04807
^{137}Cs		136.9070895	30.05 a	β- /1.176	0.514/95	7/2+	+2.851	+0.048	Ba k x-ray
									0.66164
^{128}Ba		127.90834	2.43 d	ec/0.52		0+			Cs k x-ray
									0.27344
^{130}Ba	0.106(1)	129.906321	2.2 ×10^{21} a	β+β+		0+			
^{131}Ba		130.906941	11.7 d	ec/1.37		1/2+	0.70811		Cs k x-ray
									0.12381/28.4
									0.21608/21.3
									0.49636/42.9
									(0.055–1.171)
^{132}Ba	0.101(1)	131.905061	1.3 ×10^{21} a	ec ec		0+			
^{133m}Ba			1.621 d	IT/0.288		11/2-	-0.91	+1.0	Ba k x-ray
									0.2761
^{133}Ba		132.906007	10.53 a	ec/0.517		1/2+	0.77167		Cs k x-ray
									(0.081-0.356)
^{134}Ba	2.417(18)	133.9045084				0+			
^{135m}Ba			1.20 d	IT/0.2682		11/2-	-1.00	+1.0	Ba k x-ray
									0.2682
^{135}Ba	6.592(12)	134.9056886				3/2+	+0.83863	+0.160	
^{136}Ba	7.854(24)	135.9045760				0+			
^{137}Ba	11.232(24)	136.9058274				3/2+	+0.93737	+0.245	
^{138}Ba	71.698(42)	137.9052472				0+			
^{140}Ba		139.91061	12.75 d	β- /1.05	0.48	0+			0.16268
					1.0/				0.30485
					1.02/				0.53727
^{132}La		131.91012	4.8 h	β+ /40/4.71	2.6/	2-			ann.rad./
				ec/60/	3.2				Ba k x-ray
					3.7/				0.4645
									0.5671
^{133}La		132.90822	3.91 h	β+ /4/2.2	1.2/	5/2+			Ba k x-ray
				ec/96/					0.2788
									(0.290-0.3024)
^{135}La		134.90699	19.5 h	ec/1.20		5/2+		~ -0.4	Ba k x-ray
									0.4805
^{137}La		136.906451	6 ×10^{4} a	ec/0.60		7/2+	+2.70	+0.21	0.2836
^{138}La	0.08881(71)	137.907118	1.06 ×10^{11} a			5+	+3.71365	+0.39	1.4358/65
									0.7887/35
^{139}La	99.91119(71)	138.906359				7/2+	+2.783046	+0.20	
^{140}La		139.909483	1.678 d	β- /3.762	1.35	3-	+0.73	+0.08	
					1.24/				
					1.67/				

Elem. or Isot.	Natural abundance (Atom %)	Atomic mass or weight	Half-life/ Resonance width (MeV)	Decay mode/ Energy (MeV)	Particle energy/ Intensity (MeV/%)	Spin ($h/2\pi$)	Nuclear magnetic mom. (nm)	Elect. quadr. mom. (b)	γ-Energy/ Intensity (MeV/%)
^{141}La		140.91097	3.90 h	β- /2.502	2.43/	7/2+			
^{132}Ce		131.91146	3.5 h	ec/1.3		0+			La k x-ray
									0.1554
									0.1821
^{133}Ce		132.91152	5.4 h	β+/8/2.9	1.3/	9/2-			ann.rad.
				ec/92/					0.0584
									0.1308
									0.4722
									0.5104
^{134}Ce		133.90893	3.16 d	ec/0.5		0+			La k x-ray
									0.1304
									0.1623
									0.6047
^{135}Ce		134.90916	17.7 h	β+/1 /2.026	0.8/	1/2(+)			La k x-ray
				ec/99 /					0.0345
									0.2656
									(0.300-0.6068)
^{136}Ce	0.185(2)	135.9071294	> 3. ×10^{16} a	ec ec		0+			
			> 4.2 ×10^{15} a	β$^-$ β$^-$					
^{137m}Ce			1.45 d	IT/99 /0.254		11/2-	1.0		Ce k x-ray
				ec/0.8 /					0.1693
									0.2543
^{137}Ce		136.907763	9.0 h	β+/1.222		3/2+	0.96		La k x-ray
									0.4472
^{138}Ce	0.251(2)	137.90599	>4.4 ×10^{16} a	ec ec		0+			
^{139}Ce		138.90666	137.6 d	ec/0.28		3/2+	1.06		La k x-ray
									1.320/72.1
									0.255/59.6
									(0.231-2.364)
^{140}Ce	88.450(51)	139.905446				0+			
^{141}Ce		140.908284	32.50 d	β-/0.581	0.436/69	7/2-	1.1		Pr k x-ray
					0.581/31				0.14544/48.0
^{142}Ce	11.114(51)	141.909250	>1.6 ×10^{17} a	β- β-		0+			
^{143}Ce		142.912392	1.38 d	β-/1.462	1.404/	3/2-	0.43		Pr k x-ray
					1.110/47				0.0574
									0.2933
^{144}Ce		143.913653	284.6 d	β-/0.319	0.185/20	0+			Pr k x-ray
					0.318/				0.0801
									0.1335
^{139}Pr		138.90894	4.41 h	β+ /8 /2.129	1.09/	5/2+			ann.rad./
				ec/92 /					Ce k x-ray
									0.2551
									(1.35-1.63)
^{141}Pr	100.	140.907658				5/2+	+4.275	-0.08	
^{142}Pr		141.910050	19.12 h	β- /2.162	0.58/4	2-	+0.234	+0.04	0.5088
				ec/0.744	2.16/96				1.57580
^{143}Pr		142.910823	13.57 d	β- /0.934	0.933/	7/2+	+2.70	+0.8	0.7420
^{145}Pr		144.91452	5.98 h	β- /1.81	1.80/97	7/2+			0.0725
									0.6758
									0.7483
^{138}Nd		137.91195	5.1 h	ec/1.1		0+			Pr k x-ray
									0.1995
									0.3258
^{139m}Nd			5.5 h	IT/12 /0.231	1.17/	11/2-			Nd k x-ray
				β+ /88 /					Pr k x-ray
									0.1139/34.
									0.7382/30.
^{140}Nd		139.909544	3.37 d	ec /0.22		0+			Pr k x-ray
^{142}Nd	27.152(40)	141.907729				0+			
^{143}Nd	12.174(26)	142.909820				7/2-	-1.07	-0.61	

Elem. or Isot.	Natural abundance (Atom %)	Atomic mass or weight	Half-life/ Resonance width (MeV)	Decay mode/ Energy (MeV)	Particle energy/ Intensity (MeV/%)	Spin $(h/2\pi)$	Nuclear magnetic mom. (nm)	Elect. quadr. mom. (b)	γ-Energy/ Intensity (MeV/%)
^{144}Nd	23.798(19)	143.910093	2.3(2) $\times 10^{15}$ a	α	1.83	0+			
^{145}Nd	8.293(12)	144.912579				7/2-	-0.66	-0.31	
^{146}Nd	17.189(32)	145.913123	> 1.6 $\times 10^{18}$ a	α		0+			
^{147}Nd		146.916106	10.98 d	β- /0.896	0.805/	5/2-	0.58	+0.9	Pr k x-ray
									0.53102
									0.09111–0.686
^{148}Nd	5.756(21)	147.916900	10^{20} a	$\beta^-\beta^-$		0+			
^{150}Nd	5.638(28)	149.920902	1.33 $\times 10^{20}$ a	β-β-		0+			
^{143}Pm		142.910938	265. d	ec/1.041		5/2+	3.8		Nd k x-ray
				β+ /< 6 $\times 10^{-6}$/					0.7420
^{144}Pm		143.912596	360. d	ec/2.332		5-	1.7		Nd k x-ray
				β+ /7 $\times 10^{-6}$/					0.6180
									0.6965
^{145}Pm		144.912756	17.7 a	ec/0.163		5/2+	+3.8	+0.2	Nd k x-ray
									0.0723
^{146}Pm		145.91470	5.53 a	ec/63 /1.472		3-			Nd k x-ray
				β- /37 /1.542	0.795/				0.4538
									0.7362
									0.7474
^{147}Pm		146.915145	2.623 a	β- /0.224	0.224/	7/2+	+2.6	+0.7	0.1213
									0.1974
^{148m}Pm			41.3 d	β- /95 /2.6	0.4/60	6-	1.8		0.5503/94.
				IT/5 /0.137	0.5/17				0.6300/89.
					0.7/21				0.7257/33
^{148}Pm		147.91748	5.37 d	β- /2.47	1.02/	1-	+2.0	~ +0.2	0.5503
					2.47/				(0.915-1.465)
^{149}Pm		148.918342	2.212 d	β- /1.071	0.78/9	7/2+	3.3		0.2859
					1.072/90				0.5909
									0.8594
^{151}Pm		150.92122	1.183 d	β- /1.187	0.84/	5/2+	+1.8	2.	0.1677/8
									0.2751/7
									0.3401/22
^{144}Sm	3.07(7)	143.912006	> 0.2 $\times 10^{20}$ a	ec-ec + ec-β+.		0+			
^{145}Sm		144.913417	340. d	ec/0.617		7/2-	-1.12	-0.6	Pm k x-ray
									0.0613
									0.4924
^{146}Sm		145.913047	~ 68. $\times 10^{8}$ a	α/	2.50/	0+			
^{147}Sm	14.99(18)	146.914904	1.06 $\times 10^{11}$ a	α/	2.23/	7/2-	-0.815	-0.26	
^{148}Sm	11.24(10)	147.914829	7 $\times 10^{15}$ a	α/	1.96/	0+			
^{149}Sm	13.82(7)	148.917191	10^{16} a	α/		7/2-	-0.672	+0.08	
^{150}Sm	7.38(1)	149.917282				0+			
^{151}Sm		150.919939	95 a	β- /0.0768	0.076/	5/2-	-0.363	+0.7	0.02154
^{152}Sm	26.75(16)	151.919739				0+			
^{153}Sm		152.922104	1.930 d	β- /0.808	0.64/	3/2+	-0.0216	+1.3	Eu k x-ray
					0.69/				0.0697/4.7
									0.10318/28
									0.075–0.714
^{154}Sm	22.75(29)	153.922216	> 6 $\times 10^{18}$ a	β-β-		0+			
^{156}Sm		155.92554	9.4 h	β- /0.72	0.43/	0+			0.0872
					0.71/				0.1657
									0.2038
^{145}Eu		144.916273	5.93 d	β+ /2 /2.660	0.79/	5/2+	+4.00	+0.29	ann.rad./
				ec/98 /1.71					Sm k x-ray
									0.6535
									(0.894-1.659)
^{146}Eu		145.91721	4.57 d	β+ /5 /3.88	1.47/	4-	+1.42	~ -0.18	ann.rad./
				ec/95 /					Sm k x-ray
									0.6336
									0.6341
									0.7470

Elem. or Isot.	Natural abundance (Atom %)	Atomic mass or weight	Half-life/ Resonance width (MeV)	Decay mode/ Energy (MeV)	Particle energy/ Intensity (MeV/%)	Spin ($\hbar$/2π)	Nuclear magnetic mom. (nm)	Elect. quadr. mom. (b)	γ-Energy/ Intensity (MeV/%)
									(0.27–2.64)
^{147}Eu		146.916752	24.4 d	ec/99. /1.722		5/2+	+3.73	+0.55	Sm k x-ray
				β+ /0.4 /					0.12113/20.6
									0.19725/24.0
									(0.601-1.077)
^{148}Eu		147.91809	54.5 d	ec/3.11	0.92	5-	+2.34	~ +0.35	Sm k x-ray
									0.5503/99.
									(0.067–2.17)
^{149}Eu		148.917937	93.1 d	ec/0.692		5/2+	+3.57	+0.75	Sm k x-ray
									0.2770/4.1
									0.3275/4.8
^{150}Eu		149.91971	36. a	ec/2.26		5-	+2.71	~ +1.13	Sm k x-ray
									0.3340
									0.4394
									0.5843
									(0.25–1.8)
^{150m}Eu			12.8 h	β- /92 /	1.013/	0-			Sm k x-ray
				β+ /0.4 /	1.24/				0.3339
				ec/8 /					0.4065
^{151}Eu	47.81(6)	150.919857	~ 5. × 10^{18} a	α		5/2+	+3.472	+0.90	
^{152m1}Eu			9.31 h	β- /72 /	1.85/	0-			Sm k x-ray
				ec/28 /	0.89/				0.12178
									0.84153
									0.96334
^{152}Eu		151.921751	13.5 a	ec/72 /1.874	0.69/	3-	-1.941	+2.72	Sm k x-ray
				β- /28 /1.818	1.47/				Gd k x-ray
									0.12178
									0.34427
									1.40802
									(0.252–1.528)
^{153}Eu	52.19(6)	152.921237	> 5.5 ×10^{17} a	α		5/2+	+1.533	+2.41	
^{154}Eu		153.922986	8.59 a	β- /99.9/1.969	0.27/29	3-	-2.01	+2.8	Gd k x-ray
				ec/0.02/0.717	0.58/38				0.12299/40.
					0.84/17				0.72331/20.
					0.98/4				1.2745/36
					1.87/11				(0.059-1.90)
^{155}Eu		154.922900	4.76 a	β- /0.252	0.15/	5/2+	+1.520	+2.50	Gd k x-ray
									0.0865/30
									0.1053/20
^{156}Eu		155.924763	15.2 d	β- /2.451	0.30/11	0+	≈1.1		0.08899/9.
					0.49/30				0.64623/7.
					1.2/12				0.723441/6.
					2.45/31				0.8118/10.
^{157}Eu		156.92543	15.13 h	β- /1.36	0.98/	5/2+	+1.50	+2.6	Gd k x-ray
					1.30/41				0.0639/100.
									0.3705/48.
									0.4107/76.
^{146}Gd		145.918319	48.3 d	ec/99.9 /1.03	0.35/	0+			Eu k x-ray
				β+ /0.2					0.1147
									0.1155
									0.1546
^{147}Gd		146.919101	1.588 d	ec/99.8 /2.188	0.93/	7/2-	1.0		Eu k x-ray
				ec/0.2 /					0.2293
									0.3699
									0.3960
									0.9289
									(0.1–1.8)
^{148}Gd		147.918122	71. a	α/3.27	3.1828/	0+			
^{149}Gd		148.919348	9.3 d	ec/1.32		7/2-	0.9		Eu k x-ray
									0.1496

Nuclear

Elem. or Isot.	Natural abundance (Atom %)	Atomic mass or weight	Half-life/ Resonance width (MeV)	Decay mode/ Energy (MeV)	Particle energy/ Intensity (MeV/%)	Spin ($h/2\pi$)	Nuclear magnetic mom. (nm)	Elect. quadr. mom. (b)	γ-Energy/ Intensity (MeV/%)
									0.2985
									0.3465
^{150}Gd		149.91866	1.8 ×10^{6} a	α/2.80	2.73/	0+			
^{151}Gd		150.920355	124. d	ec/0.464		7/2-	0.8		Eu k x-ray
									0.1536
									0.2432
^{152}Gd	0.20(1)	151.919799				0+			
^{153}Gd		152.921757	240. d	ec/0.485		3/2-	0.4		Eu k x-ray
									0.09743
									0.10318
^{154}Gd	2.18(3)	153.920873				0+			
^{155}Gd	14.80(12)	154.922630				3/2-	-0.258	+1.30	
^{156}Gd	20.47(9)	155.922131				0+			
^{157}Gd	15.65(2)	156.923968				3/2-	-0.340	+1.36	
^{158}Gd	24.84(7)	157.924112				0+			
^{159}Gd		158.926396	18.6 h	β- 0.971	0.971/58	3/2-	-0.44		Tb k x-ray
					0.913/29				0.36351
					0.607/12				0.058-0.855
^{160}Gd	21.86(19)	159.927062	>1.9 ×10^{19} a	β- β-		0+			
^{149}Tb		148.923254	4.13 h	β+ /4 /3.636	1.8/	½+	+1.35		Gd k x-ray
				α/16/	3.97/				0.1650
									0.3522
									0.3886
									(0.1–3.2)
^{150}Tb		149.9237	3.3 h	β+, ec/4.66		2-	-0.90		ann.rad./
									0.4963
									0.6380
									(0.3–4.29)
^{151}Tb		150.923109	17.61 h	β+/1 /2.565	0.70/	1/2(+)	+0.92		Gd k x-ray
				ec/99 /					0.1083
									0.2517
									0.2870
									(0.1–1.8)
^{152}Tb		151.92408	17.5 h	β+ /20 /3.99	2.5/	2-	-0.58	+0.3	ann.rad./
				ec/80 /	2.8/				Gd k x-ray
									0.3443
									(0.2–2.88)
^{153}Tb		152.923442	2.34 d	ec/1.570		5/2+	+3.44	+1.1	Gd k x-ray
									0.2119
									(0.05–1.1)
^{154m2}Tb			~ 22.7 h	ec/98 /		(7-)	0.9		Gd k x-ray
				IT/2 /					0.1231
									0.2479
									0.3467
									1.4199
^{154m1}Tb			10.0 h	β+ /78 /		(3-)	+1.7	+2	Gd k x-ray
				IT/22 /					0.1231
									0.2479
									0.5401
									(0.12–2.57)
^{154}Tb		153.92468	21.5 h	ec/99 /3.56	1.86/	0-			Gd k x-ray
				β+ /1 /	2.45				0.1231
									1.2744
									2.1872
									(0.12–3.14)
^{155}Tb		154.92351	5.3 d	ec/0.82		3/2+	+2.01	~ +1.41	Gd k x-ray
									0.08654
									0.10530
^{156m2}Tb			1.02 d	IT/		(7-)			Tb k x-ray
									0.0496

Elem. or Isot.	Natural abundance (Atom %)	Atomic mass or weight	Half-life/ Resonance width (MeV)	Decay mode/ Energy (MeV)	Particle energy/ Intensity (MeV/%)	Spin ($h/2\pi$)	Nuclear magnetic mom. (nm)	Elect. quadr. mom. (b)	γ-Energy/ Intensity (MeV/%)
^{156m1}Tb			5.3 h	IT/0.0884		(0+)			Tb k x-ray
									0.0884
^{156}Tb		155.924754	5.3 d	ec/2.444		3-	~ 1.7	+2.	Gd k x-ray
									0.08896
									0.19921
									0.53435
									1.22245
^{157}Tb		156.924032	1.1 ×10^{2} a	ec/0.0601		3/2+	+2.01	+1.4	Gd k x-ray
									0.0545
^{158}Tb		157.925420	1.8 ×10^{2} a	ec/80 /1.220		3-	+1.76	~ +2.7	Gd k x-ray
				β- /20 /0.937					0.0795
									0.9442
									0.9621
^{159}Tb	100.	158.925354				3/2+	+2.014	+1.43	
^{160}Tb		159.927175	72.3 d	β- /1.835	0.57/47	3-	+1.79	3.9	Dy k x-ray
					0.86/27				0.08678
									0.29857
									0.87936
									0.96615
^{161}Tb		160.927577	6.91 d	β- /0.593	0.46/23	3/2+	2.2	~ +1.3	Dy k x-ray
					0.52/66				0.02565
					0.6/10				0.04892
									0.07458
^{140m}Dy			7 μs	IT/2.166		(8-)			0.574
^{153}Dy		152.925772	6.3 h	β+ /1 /2.171	0.89/	7/2(-)	-0.78	~-0.15	Tb k x-ray
				ec/99 /					0.0807
				α /0.01 /	3.46/				0.0997
									0.2137
									(0.08–1.66)
^{154}Dy		153.92443	3. ×10^{6} a	α/2.95	2.87/	0+			
^{155}Dy		154.92576	9.9 h	β+ /2 /2.095	0.845/	3/2-	-0.385	+0.96	Tb k x-ray
				ec/98 /					0.0655
									0.2269
^{156}Dy	0.056(3)	155.924284	> 6.1 ×10^{14} a	ec-ec		0+			
^{157}Dy		156.92547	8.1 h	ec/1.34		3/2-	-0.301	+1.29	Tb k x-ray
									(0.061–1.319)
^{158}Dy	0.095(3)	157.924415	> 1. ×10^{15} a	ec-ec		0+			
^{159}Dy		158.925746	144. d	ec/0.366		3/2-	-0.354	+1.37	Tb k x-ray
									0.3262
^{160}Dy	2.329(18)	159.925203				0+			
^{161}Dy	18.889(42)	160.926939				5/2+	-0.480	+2.51	
^{162}Dy	25.475(36)	161.926804				0+			
^{163}Dy	24.896(42)	162.928737				5/2-	+0.673	+2.65	
^{164}Dy	28.260(54)	163.929181				0+			
^{166}Dy		165.932813	3.400 d	β- /0.486	0.40/	0+			Ho k x-ray
									0.0282
									0.0825
^{160m}Ho			5.0 h	IT/67/0.060		2-	+2.52	+1.8	0.0868
				ec/33/3.35					0.1970
									0.6464
									0.7281
									0.8791
									0.9619
									0.9658
^{163}Ho		162.928740	4.57 ×10^{3} a	ec/0.00258		7/2-	+4.23	~ +3.7	Dy M x-rays
^{165}Ho	100.	164.930328				7/2-	+4.17	+3.6	
			1.13 ×10^{3} a	β- /		7-	3.6	.	Er k x-ray
									0.18407
									0.71169
									0.81031

Nuclear

Elem. or Isot.	Natural abundance (Atom %)	Atomic mass or weight	Half-life/ Resonance width (MeV)	Decay mode/ Energy (MeV)	Particle energy/ Intensity (MeV/%)	Spin ($h/2\pi$)	Nuclear magnetic mom. (nm)	Elect. quadr. mom. (b)	γ-Energy/ Intensity (MeV/%)
^{166}Ho		165.932290	1.117 d	β- /1.855	1.776/48	0-			Er k x-ray
					1.855/51				(0.0806-1.379)
^{167}Ho		166.933140	3.1 h	β- /1.007	0.31/43	(7/2-)			Er k x-ray
					0.61/21				0.0793
					0.96/15				0.0835
					0.97/15				0.2379
									0.3213
									0.3465
^{160}Er		159.92908	1.191 d	ec/0.33		0+			Ho k x-ray
									(0.05–0.96)
^{161}Er		160.93000	3.21 h	ec/2.00		3/2-	-0.37	+1.36	Ho k x-ray
									0.8265
									(0.07–1.74)
^{162}Er	0.139(5)	161.928787				0+			
^{164}Er	1.601(3)	163.929207				0+			
^{165}Er		164.930733	10.36 h	ec/0.376		5/2-	+0.643	+2.71	Ho k x-ray
^{166}Er	33.503(36)	165.930299				0+			
^{167}Er	22.869(9)	166.932054				7/2+	-0.5639	+3.57	
^{168}Er	26.978(18)	167.932376				0+			
^{169}Er		168.934596	9.40 d	β- /0.351	0.35/~ 100	½-	+0.485		Tm k x-ray
									0.1098
									0.1182
^{170}Er	14.910(36)	169.935471				0+			
^{171}Er		170.938036	7.52 h	β- /1.491		5/2-	0.66	2.9	Tm k x-ray
									0.11160
									0.29591
									0.30832
									(0.08–1.4)
^{172}Er		171.939362	2.05 d	β-/0.891	0.28/48	0+			Tm k x-ray
					0.36/46				0.0597
									0.4073
									0.6101
^{165}Tm		164.932442	1.253 d	ec/1.593		½+	-0.139		Er k x-ray
									0.0472
									0.0544
									(0.297-0.8064)
^{166}Tm		165.93356	7.70 h	ec/98 /3.04		2+	+0.092	+2.14	Er k x-ray
				β+ /2 /					0.0806
									0.1844
									0.7789
									1.2734
									2.0524
^{167}Tm		166.932857	9.24 d	ec/0.748		½+	-0.197		Er k x-ray
									0.0571
									0.20778
^{168}Tm		167.934178	93.1 d	ec/1.679		3+	+0.23	+3.2	Er k x-ray
									0.19825
									0.4475
									0.81595
^{169}Tm	100	168.934218				½+	-0.231	-1.2	
^{170}Tm		169.935807	128.6 d	β- /99.8/0.968	0.883/24	1-	+0.246	+0.74	Yb k x-ray
^{171}Tm		170.936435	1.92 a	β- /0.096	0.03/2	½+	-0.228		0.06674
					0.096/98				
^{172}Tm		171.93841	2.65 d	β- /1.88	1.79/36	2-			Yb k x-ray
					1.88/29				0.07879
									1.38722
									1.46601
									(1.53-1.609)
^{173}Tm		172.93961	8.2 h	β- /1.298	0.80/21	½+			Yb k x-ray
					0.86/71				0.3988

Elem. or Isot.	Natural abundance (Atom %)	Atomic mass or weight	Half-life/ Resonance width (MeV)	Decay mode/ Energy (MeV)	Particle energy/ Intensity (MeV/%)	Spin ($h/2\pi$)	Nuclear magnetic mom. (nm)	Elect. quadr. mom. (b)	γ-Energy/ Intensity (MeV/%)
									0.4613
^{166}Yb		165.93387	2.363 d	ec/0.30		0+			Tm k x-ray
									0.0828
									0.1844
									0.7789
									1.2734
									2.0524
^{168}Yb	0.126(1)	167.933889	> 9 × 10^{13} a	β-β-		0+			
			> 2.8 × 10^{17} a	β-β-					
^{169}Yb		168.935182	32.02 d	ec/0.909		7/2+	-0.63	+3.5	0.1979/35.9
									0.3078/10.05
									0.0207–0.261
^{170}Yb	3.023(2)	169.93476725				0+			
^{171}Yb	14.216(7)	170.93633152				1/2-	+0.49367		
^{172}Yb	21.754(10)	171.93638666				0+			
^{173}Yb	16.098(9)	172.93821622				5/2-	-0.67989	+2.80	
^{174}Yb	31.896(26)	173.93886755				0+			
^{175}Yb		174.9412819	4.19 d	β- /0.470	0.466/73	7/2-	+0.77	+ 3.5	Lu k x-ray
					0.071/21				0.3963/13
					0.353/6.2				(0.114–0.28)
^{176}Yb	12.887(30)	175.94257471	> 4.5 × 10^{16} a	β-β-		0+	+ 0.15		
^{169}Lu		168.937644	1.419 d	ec/2.293	1.271/	7/2+	2.30	3.48	Yb k x-ray
									0.19121
									0.9606
									(0.08–2.1)
^{170}Lu		169.93848	2.01 d	ec/3.46	2.44/	0+			Yb k x-ray
									0.58711
									0.5908
									1.28029
									(0.1–3.38)
^{171}Lu		170.937919	8.24 d	ec/1.479	0.362/	7/2+	+2.293	+3.53	Yb k x-ray
									0.01939
									0.66744
									(0.02–1.3)
^{172}Lu		171.939091	6.64 d	ec/2.519		4-	2.90	+3.80	Yb k x-ray
									0.18156
									1.09367
									(0.07–2.2)
^{173}Lu		172.938936	1.37 a	ec/0.671		7/2+	+2.281	+3.53	Yb k x-ray
									0.07860
									0.27198
^{174m}Lu			142. d	IT/99.3/	0.17086	(6)-	+1.49	~ +4.80	Lu k x-ray
				ec/0.7 /					0.067055
^{174}Lu		173.940343	3.3 a	ec/1.374		(1)-	+1.988	~ +0.773	Yb k x-ray
									(0.0766-1.242)
^{175}Lu	97.401(13)	174.940777				7/2+	+2.232	+3.49	
^{176m}Lu			3.66 h	β- /1.315	1.229/	1-	+0.318	-1.45	Hf k x-ray
					1.317/				0.088372
^{176}Lu	2.599(13)	175.942692	3.73 ×10^{10} a	β- /1.192		7-	+3.169	+4.92	Hf k x-ray
				β+/ < 0.9					0.20187
				ec/<0.36					0.30691
^{177m2}Lu			160.4 d	IT/22/0.9702		23/2-	+2.32	+5.7	Lu k x-ray
				β- /78					Hf k x-ray
									0.11295/22.7
									0.20836/61.0
									0.37850/29.9
									(0.072-0.466)
^{177}Lu		176.943764	6.646 d	β- /0.498	0.497/78.4	7/2+	+2.238	+3.39	0.11295/0.062
					0.177/11.7				0.20836/0.104
					0.385/9.9				

Nuclear

Elem. or Isot.	Natural abundance (Atom %)	Atomic mass or weight	Half-life/ Resonance width (MeV)	Decay mode/ Energy (MeV)	Particle energy/ Intensity (MeV/%)	Spin (h/2π)	Nuclear magnetic mom. (nm)	Elect. quadr. mom. (b)	γ-Energy/ Intensity (MeV/%)
					0.249/0.002				
^{179}Lu		178.94733	4.6 h	β- /1.405	1.35/	7/2+	+2.38	+3.32	(0.214-0.3377)
^{170}Hf		169.93961	16.0 h	ec/1.1		0+			Lu k x-ray
									0.0985
									0.1202
									0.1647
									0.5729
									0.6207
^{171}Hf		170.94049	12.2 h	ec, β+ /2.4		7/2+	-0.67	+3.46	ann.rad./
									Lu k x-ray
									0.1221
									0.6620
									1.0714
^{172}Hf		171.93945	1.87 a	ec/0.35		0+			Lu k x-ray
									0.02399
									0.12582
									(0.082–0.123)
^{173}Hf		172.94051	23.6 h	ec/1.6		½-	+0.50		Lu k x-ray
									0.12367
									0.13963
									(0.1–2.1)
^{174}Hf	0.161(2)	173.940049	2.0 ×10^{15} a			0+			
^{175}Hf		174.941512	70.7 d	ec/0.686		5/2-	-0.68	+2.72	Lu k x-ray
									(0.0894-0.343)
^{176}Hf	5.24(14)	175.941410				0+			
^{177}Hf	18.58(9)	176.943230				7/2-	+0.794	+0.337	
^{178m2}Hf			31. a	IT/		16+	+8.16	+6.0	Hf k x-ray
									0.32555
									0.42635
									0.089–0.574
^{178}Hf	27.28(6)	177.943709				0+			
^{179m2}Hf			25.1 d	IT/1.1057		25/2-	7.4		Hf k x-ray
^{179}Hf	13.63(3)	178.945826				9/2+	-0.641	+3.79	
^{180m}Hf			5.47 h	IT/1.1416		8-	+9.	+4.6	Hf k x-ray
									0.2152
									0.3323
									0.4432
^{180}Hf	35.12(16)	179.946560				0+			
^{181}Hf		180.949111	42.4 d	β- /1.027	0.408/	1/2-			Ta k x-ray
									0.482
									(0.133-0.3459)
^{182}Hf		181.95056	8.9 ×10^{6} a	β- /0.37		0+			Ta k x-ray
									0.2704/79
									(0.098-0.270)
^{184}Hf		183.95545	4.1 h	β- /1.34	0.74/38	0+			Ta k x-ray
					0.85/16				0.0414
					1.10/46				0.1391
									0.3449
^{173}Ta		172.94375	3.6 h	β+ /24 /3.7		(5/2-)	1.70	-1.8	ann.rad./
				ec/76 /					Hf k x-ray
									0.06972
									0.17219
									(0.06 –2.7)
^{175}Ta		174.94374	10.5 h	ec/2.0		7/2+	2.27	+3.5	Hf k x-ray
									(0.208-0.339)
^{176}Ta		175.94486	8.1 h	ec/3.1		1-			Hf k x-ray
									0.08837
									1.15735
^{177}Ta		176.944482	2.356 d	ec/1.166		7/2+	2.25		Hf k x-ray
									0.11295

Elem. or Isot.	Natural abundance (Atom %)	Atomic mass or weight	Half-life/ Resonance width (MeV)	Decay mode/ Energy (MeV)	Particle energy/ Intensity (MeV/%)	Spin (h/2π)	Nuclear magnetic mom. (nm)	Elect. quadr. mom. (b)	γ-Energy/ Intensity (MeV/%)
									(0.07–1.06)
^{179}Ta		178.945939	1.8 a	ec/0.110		7/2+	+2.29	3.27	Hf k x-ray
^{180m}Ta	0.01201(32)		> 9.0 ×10^{16} a			(9-)	+4.82	+4.95	
			> 1.7 ×10^{17} a	β-					
			> 2.0 ×10^{17} a	ec					
^{180}Ta		179.947468	>4.5 × 10^{16} a	ec >2 × 10^{17} a		1+			Hf k x-ray
				β- >5.8 × 10^{16} a	0.61/3				W k x-ray
					0.71/10				(0.093-0.1034)
^{181}Ta	99.98799(32)	180.947999				7/2+	+2.370	+3.17	
^{182}Ta		181.950155	114.43 d	β- /1.814	0.25/30	3-	+3.02	+2.6	W k x-ray
					0.44/20				1.12127/100
					0.52/40				1.22138/79
									0.085–1.289
^{183}Ta		182.951376	5.1 d	β- /1.070	0.45/5	7/2+	+2.36		W k x-ray
					0.62/91				0.0847
									0.0991
									0.1079
									0.2461
									0.3540
^{184}Ta		183.95401	8.7 h	β- /2.87	1.11/15	(5-)			W k x-ray
					1.17/81				0.2528/44.
									0.4140/74.
									(0.09–1.4)
^{178}W		177.94589	21.6 d	ec/0.091		0+			Ta k x-ray
^{180}W	0.12(1)	179.946713	~ 1.5 ×10^{13} a	α/		0+			
^{181}W		180.948219	121.1 d	ec/0.188		9/2+			Ta k x-ray
									0.13617
									0.15221
^{182}W	26.50(16)	181.948206	>7.7 ×10^{21} a	α/		0+			
^{183}W	14.31(4)	182.950225	>4.1 ×10^{21} a	α/		½-	+0.1177848		
^{184}W	30.64(2)	183.950933	>8.9 ×10^{21} a	α/		0+			
^{185}W		184.953421	74.8 d	β- /0.433	0.433/99.9	3/2-	+ 0.54		0.12536
^{186}W	28.43(19)	185.954365	>8.2 ×10^{21} a	α/		0+			
^{187}W		186.957161	23.9 h	β- /1.311	0.624/66	3/2-	0.62		Re k x-ray
					1.315/16				0.68572/33
					0.081–1.18				0.134–0.773
^{188}W		187.958488	69.78 d	β- /0.349	0.349/99	0+			0.0636
									0.2271
									0.2907
^{181}Re		180.95006	20. h	ec /1.74		5/2+	3.19		W k x-ray
									0.3607
									0.3655
									0.6390
^{182m}Re			12.7 h	ec/	0.55/	2(+)	3.3	+1.8	W k x-ray
					1.74/				0.0677
									1.1214
									1.2215
									(0.06–2.2)
^{182}Re		181.9512	2.67 d	ec/2.8		7+	2.8	+4.1	W k x-ray
									0.0678
									0.2293
									(1.121-1.221)
^{183}Re		182.95082	70. d	ec/0.56		(5/2+)	+3.16	+2.3	W k x-ray
									0.16232
^{184m}Re			165. d	IT/75 /0.188		8+	+2.9		Re k x-ray
				ec/25 /					0.1047
									0.2165
									0.92093
									(0.10–1.1)
^{184}Re		183.95253	35. d	ec/1.48		3-	+2.53	+2.8	W k x-ray

Elem. or Isot.	Natural abundance (Atom %)	Atomic mass or weight	Half-life/ Resonance width (MeV)	Decay mode/ Energy (MeV)	Particle energy/ Intensity (MeV/%)	Spin ($h/2\pi$)	Nuclear magnetic mom. (nm)	Elect. quadr. mom. (b)	γ-Energy/ Intensity (MeV/%)
									0.79207
									08948
									0.90328
^{185}Re	37.40(2)	184.952958				5/2+	+3.1871	+2.18	
^{186m}Re			2.0 ×10^{5} a	IT/0.150		8+			Re k x-ray
									(0.059-0.148)
^{186}Re		185.954989	3.718 d	β- /92 /1.070	0.973/21	1-	+1.739	+0.62	W k x-ray
				ec/8 /0.582	1.07/71				0.1227/0.6
									0.1372/9.5
									(0.63–0.77)
^{187}Re	62.60(2)	186.955752	4.16 ×10^{10} a	β- /0.00266	0.0025/	5/2+	+3.2197	+2.07	
^{188}Re		187.958114	17.00 h	β- /2.120	1.962/20	1-	+1.788	+0.57	Os k x-ray
					2.118/79				0.15502
									0.309–2.022
^{189}Re		188.95923	24. h	β- /1.01	1.01/	(5/2+)			0.1471
									0.2167
									0.2194
									0.2451
^{190m}Re			3.2 h	β- /51 /		(6-)			Re k x-ray
				IT/49 /					0.1191
									0.2238
									0.6731
									(0.1–1.79)
^{182}Os		181.95211	21.5 h	ec/0.9		0+			Re k x-ray
									0.1802
									0.5100
^{183m}Os			9.9 h	ec/84 /		½-			Os k x-ray
				IT/16 /					Re k x-ray
									1.1020
									1.1080
^{183}Os		182.95312	13. h	ec/2.1		9/2+	-0.79	+3.1	Re k x-ray
									0.1144
									0.3818
^{184}Os	0.02(1)	183.952493	1.1 ×10^{13} a	α		0+			
			> 1.9 ×10^{14} a	ec,ec					
			> 2.5 ×10^{16} a	ec,β^{+}					
^{185}Os		184.954046	93.6 d	ec/1.013		½-			Re k x-ray
									0.6461
									0.8748
									0.8805
^{186}Os	1.59(3)	185.953838	2. ×10^{15} a	α/	~ 2.75/	0+			
^{187}Os	1.96(2)	186.955750				½-	+0.0646519		
^{188}Os	13.24(8)	187.955837				0+			
^{189m}Os			5.8 h	IT/0.0308		9/2-			Os L x-ray
^{189}Os	16.15(5)	188.958146				3/2-	+0.659933	+0.86	
^{190}Os	26.26(2)	189.958446				0+			
^{191m}Os			13.1 h	IT/0.0744		3/2-			Os k x-ray
									0.0744
^{191}Os		190.960928	15.4 d	β- /0.314	0.140/100	9/2-	+0.96	+2.5	Ir k x-ray
									0.1294
^{192}Os	40.78(19)	191.961479	> 5.3 × 10^{19} a	β^{+} β^{+}		0+			
^{193}Os		192.964150	30.0 h	β- /1.141	1.04/20	3/2-	+0.730	~ +0.48	Ir k x-ray
									0.1389
									0.4605
^{194}Os		193.965180	6.0 a	β- /0.097	0.054/33	0+			Ir L x-ray
					0.096/67				0.0429
^{184}Ir		183.95748	3.0 h	β+ /12 /4.6	2.3/	5-	0.70	+2.41	ann.rad./
				ec/88 /	2.9/				Os k x-ray
									0.11968
									0.2640

Elem. or Isot.	Natural abundance (Atom %)	Atomic mass or weight	Half-life/ Resonance width (MeV)	Decay mode/ Energy (MeV)	Particle energy/ Intensity (MeV/%)	Spin ($h/2\pi$)	Nuclear magnetic mom. (nm)	Elect. quadr. mom. (b)	γ-Energy/ Intensity (MeV/%)
									0.3904
^{185}Ir		184.95670	14. h	β+ /3 /2.4		(5/2-)	2.60	-1.8	ann.rad./
				ec/97 /					Os k x-ray
									0.2543
									1.8288
^{186}Ir		185.95795	15.7 h	ec/98 /3.83		5+	3.9	-2.55	Os k x-ray
				β+ /2 /					0.1372
									0.2968
									0.4348
									(0.13–3.0)
^{187}Ir		186.95754	10.5 h	ec/1.50		3/2+	+0.17	+0.94	Os k x-ray
									0.0743
									0.4009
									0.4271
									0.6109
									0.9128
^{188}Ir		187.95884	1.72 d	β+ /2.81	1.13/	1-	0.30	+0.48	Os k x-ray
				ec/99+ /	1.64/				0.1550
									0.4780
									0.6330
									2.2146
^{189}Ir		188.95872	13.2 d	ec/0.53		3/2+	+0.14	+0.8	Os k x-ray
									0.2449
^{190m2}Ir			3.09 h	β+, ec/95 /		(11-)			0.376
				IT/5 /					
^{190}Ir		189.960543	11.8 d	ec/2.0		(4+)	0.04	+2.9	Os k x-ray
									0.1867
									0.4072
									0.5186
									0.5580
									0.6051
									(0.2–1.4)
^{191}Ir	37.3(2)	190.960592				3/2+	+0.151	+0.82	
^{192m2}Ir			~ 241. a	IT/0.168	0.155	11-			Ir k x-ray
^{192}Ir		191.962603	73.83 d	β- /1.460		4+	+1.92	~ +2.15	Pt k x-ray
									0.31649/83.
									0.46806/48.
^{193m}Ir			10.53 d	IT/0.0802		11/2-			Ir L x-ray
									0.0803
^{193}Ir	62.7(2)	192.962924				3/2+	+0.164	+0.75	
^{194m}Ir			1.7×10^{2} d	β- /					Pt k x-ray
									0.3284
									0.4829
									0.5624
^{194}Ir		193.965076	19.2 h	β-/2.247	1.92/9	1-	+0.39	+0.34	0.2935
^{195m}Ir			3.7 h	β- /	0.41/	11/2-			Pt k x-ray
					0.97/				0.3199/9.6
									0.3649/9.5
									0.4329/9.6
									0.6849/9.6
^{188}Pt		187.95940	10.2 d	ec/0.51		0+			Ir k x-ray
									0.1876
									0.1951
^{189}Pt		188.96085	10.9 h	β+, ec/1.97		3/2-	-0.43	-0.95	Ir k x-ray
									0.0943
									0.6076
									0.7214
									(0.09–1.47)
^{190}Pt	0.012(2)	189.959950	$5.0(2)\times10^{11}$ a	α		0+			
			$>8.4\times10^{14}$ a	ec,ec					

Nuclear

Elem. or Isot.	Natural abundance (Atom %)	Atomic mass or weight	Half-life/ Resonance width (MeV)	Decay mode/ Energy (MeV)	Particle energy/ Intensity (MeV/%)	Spin ($h/2\pi$)	Nuclear magnetic mom. (nm)	Elect. quadr. mom. (b)	γ-Energy/ Intensity (MeV/%)
			> 9.2 ×10^{15} a	ec, β$^+$					
^{191}Pt		190.961676	2.86 d	ec/1.02		3/2-	-0.50	-0.87	Ir k x-ray
									0.3599
									0.4094
									0.5389
^{192}Pt	0.782(24)	191.961043				0+			
^{193m}Pt			4.33 d	IT/0.1498	0.136	13/2+	-0.75		Pt k x-ray
									0.1355
^{193}Pt		192.962985	60. a	ec/0.0566		(1/2-)	+0.60		Ir k x-rays
^{194}Pt	32.86(40)	193.962684				0+			
^{195m}Pt			4.01 d	IT/0.2952		13/2+	-0.61	~ +1.4	Pt k x-ray
									0.0989
^{195}Pt	33.78(24)	194.964794	7.3 ×10^{18} a	α		1/2-	+0.6095		
^{196}Pt	25.21(34)	195.964955				0+			
^{197}Pt		196.967343	19.9 h	β- /0.719		1/2-	0.51		Au k x-ray
^{198}Pt	7.356(130)	197.967897	> 4.7 ×10^{17} a	α		0+			
			> 3.5 ×10^{18} a	β$^-$β$^-$					
^{200}Pt		199.97145	12.5 h	β- /~ 0.66		0+			Au k x-ray
									0.13590
									0.22747
									0.24371
^{202}Pt		201.97564	1.8 d			0+			0.440
^{191}Au		190.9637	3.2 h	ec/1.83		3/2+	+0.137	+0.72	Pt k x-ray
									0.5864/16
									(0.088–1.30)
^{192}Au		191.96482	4.9 h	β+ /5 /3.52	2.19/	1-	-0.011	-0.23	ann.rad./
				ec/95 /	2.49/				Pt k x-ray
									0.2959
									0.3165
^{193}Au		192.96414	17.6 h	ec/1.07		3/2+	+0.140	+0.66	Pt k x-ray
									0.1862
									0.2556
^{194}Au		193.965419	1.64 d	β+ /3 /2.49	1.49/	1-	+0.076	-0.24	ann.rad./
				ec/97 /					Pt k x-ray
									0.3391
									(0.1015-2.413)
^{195}Au		194.965038	186.10 d	ec/0.227		3/2+	+0.149	+0.61	Pt k x-ray
^{196m2}Au			9.6 h	IT/0.5954	0.175	12-	5.7		Au k x-ray
									0.1478
									0.1883
^{196}Au		195.966571	6.17 d	ec/92 /1.506		2-	+0.591	~ +0.81	Pt k x-ray
^{197}Au	100.	196.966570				3/2+	+0.14575	+0.55	
^{198m}Au			2.27 d	IT/0.812	0.115/100	(12-)	+5.9		Au k x-ray
									0.0972
									0.1803
									0.2419
^{198}Au		197.968244	2.695 d	β- /1.372	0.290/1	2-	+0.5934	+0.64	Hg k x-ray
^{199}Au		198.968767	3.14 d	β- /0.453	0.25/22	3/2+	~ +0.2715	+0.51	Hg k x-ray
					0.292/72				0.15837
					0.462/6				0.20820
^{200m}Au			18.7 h	β- /84 /1.0	0.56/	12-	5.9		Au k x-ray
				IT/16 /					0.2559/71
									0.3680/77
									0.4978/73
									0.5793/72
									0.084–0.904)
^{192}Hg		191.96563	5.0 h	ec/~ 0.5		0+			Au k x-ray
									0.1572
									0.2748
									0.3065

Elem. or Isot.	Natural abundance (Atom %)	Atomic mass or weight	Half-life/ Resonance width (MeV)	Decay mode/ Energy (MeV)	Particle energy/ Intensity (MeV/%)	Spin ($h/2\pi$)	Nuclear magnetic mom. (nm)	Elect. quadr. mom. (b)	γ-Energy/ Intensity (MeV/%)
^{193m}Hg			11.8 h	β+, ec/91 /		13/2(+)	-1.058430	+0.92	Hg k x-ray
				IT/9 /0.2901					0.1866
									0.2580
									0.4076
									0.5733
									0.9324
									(0.1–1.96)
^{193}Hg		192.96665	3.8 h	ec, B+/2.34		3/2(-)	-0.6276	-0.7	0.1866
									0.2580
									0.8611
^{194}Hg		193.965449	520. a	ec/0.04		0+			Au L x-rays
^{195m}Hg			1.73 d	IT/(54)/0.3186		13/2+	-1.04465	+1.08	Hg k x-ray
				ec/(46)/					Au k x-ray
									0.2617
									0.5603
									0.7798
^{195}Hg		194.96671	10.5 h	ec/1.51		1/2-	+0.541475		Au k x-ray
									0.0614
									0.7798
^{196}Hg	0.155(12)	195.965833	>2.5 ×10^{18} a	α		0+			
^{197m}Hg			23.8 h	IT/(93)/0.2989		13/2+	-1.027684	+1.25	Hg k x-ray
									Au k x-ray
									0.13398
^{197}Hg		196.967214	2.69 d	ec/0.600		1/2-	+0.527374		Au k x-ray
									0.07735
^{198}Hg	10.038(16)	197.966769				0+			
^{199}Hg	16.938(39)	198.968281				1/2-	+0.505885		
^{200}Hg	23.138(65)	199.968327				0+			
^{201}Hg	13.170(66)	200.970303				3/2-	-0.560226	+0.39	
^{202}Hg	29.743(89)	201.970644				0+			
^{203}Hg		202.972872	46.61 d	β- /0.492	0.213/100	5/2-	+0.8489	+0.34	Tl k x-ray
									0.279188
^{204}Hg	6.818(35)	203.973494				0+			
^{198}Tl		197.97045	5.3 h	ec, β+ /(1)/3.5	1.4/	2-			Hg k x-ray
					2.1/				0.4118
					2.4/				0.6367
									0.6759
									(0.23–2.8)
^{199}Tl		198.96988	7.4 h	ec/1.4		1/2+	+1.60		Hg k x-ray
									0.2082
									0.2473
									0.4555
^{200}Tl		199.97096	1.087 d	ec/2.46	1.07/	2-	0.04		Hg k x-ray
					1.44/				0.36799
									1.2057
									(0.11–2.3)
^{201}Tl		200.97082	3.038 d	ec/0.48		1/2+	+1.605		Hg k x-ray
									0.13528
									0.16740/10.0
^{202}Tl		201.972109	12.47 d	ec/1.36		2-	0.06		Hg k x-ray
									0.43957
^{203}Tl	(29.44-29.59)	202.972344				1/2+	+1.622258		
^{204}Tl		203.973863	3.78 a	β- /97/0.7637	0.763/97	2-	0.09		Hg k x-ray
				ec/(3)/0.347					
^{205}Tl	(70.41-70.56)	204.974427				1/2+	+1.638215		
^{200}Pb		199.97182	21.5 h	ec/0.81		0+			Tl k x-ray
									0.14763
^{201}Pb		200.97287	9.33 h	ec/1.90		5/2-	+0.675	~ -0.01	Tl k x-ray
									0.33120

Elem. or Isot.	Natural abundance (Atom %)	Atomic mass or weight	Half-life/ Resonance width (MeV)	Decay mode/ Energy (MeV)	Particle energy/ Intensity (MeV/%)	Spin ($h/2\pi$)	Nuclear magnetic mom. (nm)	Elect. quadr. mom. (b)	γ-Energy/ Intensity (MeV/%)
									0.36131
									(0.11–1.8)
^{202m}Pb			3.53 h	IT/90/2.170		9-	-0.228	~ +0.58	Pb k x-ray
				β+ /10/					Tl k x-ray
									0.42219
									0.78700
									0.96271
^{202}Pb		201.972152	5.3 ×10^4 a	ec/0.05		0+			Tl L x-ray
^{203}Pb		202.97339	2.163 d	ec/0.98		5/2-	+0.686	~ +0.10	Tl k x-ray
									0.279188
^{204}Pb	1.4(1)	203.973043	> 1.4 ×10^{20} a	α		0+			
^{205}Pb		204.974482	1.51 ×10^7 a	ec/0.0512		5/2-	+0.712	+0.23	Tl L x-ray
^{206}Pb	24.1(1)	205.974465	> 2.5 ×10^{21} a	α		0+			
^{207}Pb	22.1(1)	206.975897	> 1.9 ×10^{21} a	α		1/2-	+0.59258		
^{208}Pb	52.4(1)	207.976652	> 2 ×10^{19} a	sf		0+			
			> 2.6 ×10^{21} a	α					
^{209}Pb		208.981090	3.23 h	β- /0.644	0.645/100	9/2+	-1.474	-0.3	
^{210}Pb		209.984188	22.6 a	β- /0.0635	0.017/81	0+			
					0.061/19				
				α	3.72				
^{212}Pb		211.991896	10.64 h	β- /0.574	0.28/83	0+			Bi k x-ray
					0.57/12				0.23858
^{203}Bi		202.97689	11.8 h	ec/99.8/3.25		9/2-	+4.02	-0.9	Pb k x-ray
				β+ /(0.2)/	1.35/				0.1865
									0.8203
									0.8969
									1.8475
									(0.1–2.9)
^{204}Bi		203.97784	11.2 h	ec/4.44		6+	+4.32	-0.7	Pb k x-ray
									0.37481
									0.89922
									0.98409
^{205}Bi		204.97739	14.9 d	ec/2.71		9/2-	+4.07	-0.81	Pb k x-ray
									0.70347
									1.76435
									(0.550-1.862)
^{206}Bi		205.97850	6.243 d	ec/3.76		6+	+4.36	-0.54	Pb k x-ray
									0.51619
									0.80313
									0.88100
^{207}Bi		206.978471	31.2 a	ec/2.399		9/2-	4.092	-0.76	Pb k x-ray
									0.56915
									1.06310
^{208}Bi		207.979742	3.68 ×10^5 a	ec/2.880		5+	4.6	-0.7	Pb k x-ray
									2.61435
^{209}Bi	100.	208.980399	2.0 ×10^{19} a	α	3.13/98.8	9/2-	+4.111	-0.52	0.204
^{210m}Bi			3.0 × 10^6 a	α/	4.420(3)/0.29	9-	+2.73	-0.7	Tl k x-ray
					4.59(3)/3.9				
					4.584(3)/1.4				0.3052
					4.908(4)/39				0.6502
					4.946(3)/55				
^{210}Bi		209.984120	5.01 d	β- /1.163	1.16/99	1-	-0.0445	+0.19	0.2661
									0.3.52
^{204}Po		203.98031	3.53 h	ec/2.34		0+			Bi k x-ray
				α	5.377/0.66				0.2702
									0.8844
									1.0162
									(0.11–1.9)
^{206}Po		205.980474	8.8 d	ec/(95)/1.85		0+			Bi k x-ray
				α/(5)/	5.223/5.5				0.28644

Elem. or Isot.	Natural abundance (Atom %)	Atomic mass or weight	Half-life/ Resonance width (MeV)	Decay mode/ Energy (MeV)	Particle energy/ Intensity (MeV/%)	Spin ($h/2\pi$)	Nuclear magnetic mom. (nm)	Elect. quadr. mom. (b)	γ-Energy/ Intensity (MeV/%)
									0.31156
									0.51134
									0.80737
									1.03228
									(0.11–1.5)
^{207}Po		206.98159	5.80 h	ec, β+/2.91		5/2-	~ +0.79	+0.28	Bi k x-ray
									0.74263
									0.91176
									0.99225
^{208}Po		207.981246	2.898 a	α/5.213	4.233/0.0002	0+			
					5.1158/100				
^{209}Po		208.982430	123. a	α/4.976	4.624/0.56	1/2-	+0.7		0.26049
					4.879/99.2				0.8964
^{210}Po		209.982874	138.4 d	α/5.407	4.516/0.001	0+			0.80313
					5.304/100				
^{209}At		208.98617	5.4 h	β+, ec/96/3.49		(6+)			Po k x-ray
				α/(4)/5.757	5.647/4.1				0.10422
									0.54503
									0.78189
									0.79020
									(0.1–2.6)
^{210}At		209.98715	8.1 h	ec/99.8/3.98		5+			Po k x-ray
				α/(0.2)/5.632	5.361/0.05				0.24535
					5.442/0.05				0.52758
									1.18143
									1.43678
									1.48335
									(0.04–2.4)
^{211}At		210.987496	7.21 h	ec/(58)/0.787		9/2+			Po k x-ray
				α/(42)/5.980	5.211/0.004				0.66956
					5.868/42				0.6870
									0.74263
^{211}Rn		210.99060	14.6 h	β+, ec/74/2.89		1/2-	+0.60		At k x-ray
				α/(26)/5.964	5.619(1)/0.7				0.16877
					5.784(1)/16.4				0.25022
					5.851(1)/8.8				0.37049
									0.67412
									0.67839
									1.36298
									(0.11–2.7)
^{218}Rn		218.005601	33.8 ms	α/7.267	6.534(1)/0.16	0+			0.6093
					7.133(1)/99.8				0.6653
^{219}Rn		219.0094879	3.96 s	α/6.946(1)	6.3130(5)/0.05	5/2+	-0.44	+1.2	Po k x-ray
					6.425(3)/7.5				0.13057
					6.5309(4)/0.12				0.27113
					6.5531(3)/12.2				0.40170
					6.8193(3)/81				(0.1–1.075)
^{220}Rn		220.011393	55.6 s	α/6.404	5.7486(5)/0.07	0+			
					6.2883(1)/99.9				
^{222}Rn		222.017576	3.823 d	α/5.590	4.987(1)/0.08	0+			0.510
					5.4897(3)/99.9				
			> 8.0 a	β-					
^{221}Fr		221.01425	4.81 m	α/6.457	5.9393(7)/0.17	5/2-	+1.58	-1.0	At k x-ray
					5.9797(7)/0.49				0.0995
					6.0751(7)/0.15				0.21798
					6.1270(7)/				0.4091
					6.2433(3)/1.3				
					6.3410(7)/83.4				
^{222}Fr		222.01758	14.3 m	β- /2.03	1.78/	2-	+0.63	+0.51	
				α/5.850					

Elem. or Isot.	Natural abundance (Atom %)	Atomic mass or weight	Half-life/ Resonance width (MeV)	Decay mode/ Energy (MeV)	Particle energy/ Intensity (MeV/%)	Spin ($\hbar/2\pi$)	Nuclear magnetic mom. (nm)	Elect. quadr. mom. (b)	γ-Energy/ Intensity (MeV/%)
^{223}Fr		223.019734	22.0 m	β- /1.149	α/5.291	3/2(-)	+1.17	+1.17	0.1509
				α//0.006	5.314				0.0589
					5.403				0.1453
^{223}Ra		223.018501	11.45 d	α/5.979	5.287(1)/0.15	3/2+	+0.270	+1.21	Rn k x-ray
					5.338(1)/0.13				0.12231
					5.365(1)/0.13				0.14418
					5.433(5)/2.3				0.15418
					5.502(1)/1.0				0.15859
					5.540(1)/9.2				0.26939
					5.607(3)/24				0.32388
					5.716(3)/52				0.33328
					5.747(1)/9				0.44494
					5.857(1)/0.32				(0.10–0.7)
					5.872(1)/0.85				
^{224}Ra		224.020211	3.66 d	α/5.789	5.034(10)/0.003	0+			Rn k x-ray
					5.047(1)/0.007				0.2407
					5.164(5)/0.007				0.4093
					5.449(2)/4.9				0.6501
					5.685(2)/95				
^{225}Ra		225.023611	14.9 d	β- /0.36	0.32/100	1/2+	-0.734		Ac k x-ray
				α	5.01 $\times 10^{-5}$				0.0434
					4.98 $\times 10^{-6}$				
^{226}Ra		226.025409	1599. a	α/4.870	4.194(1)/0.001	0+			Rn k x-ray
			> 4 $\times 10^{18}$ a	sf/4 $\times 10^{-14}$	4.343(1)/0.006				0.1861/3.64
					4.601(1)/6.16				0.2624
					4.784(1)/93.8				0.053–2.448
^{228}Ra		228.031069	5.76 a	β- /0.046	0.039/50	0+			0.0135
				βf//5$\times 10^{-12}$	0.014/30				(0.006–0.031)
					0.026/20				
^{225}Ac		225.02323	9.920 d	α/5.935	5.286(1)/0.2	3/2			Fr k x-ray
					5.444(3)/0.1				0.06296/0.48
					5.554(1)/0.1				0.09982/1.36
					5.608(1)/1.1				0.1084
					5.636(1)/4.5				0.1116
					5.681(1)/1.4				0.1451
					5.722(1)/2.9				0.150/0.691
					5.731(1)/10				0.15724
					5.791(1)/9				0.18795/0.54
					5.793(1)/18				0.0075–0.809
^{226}Ac		226.026097	1.224 d	ec/(17)/0.640		(1-)			Ra k x-ray
				β- /(83)/1.116					Th k x-ray
				α/(0.006)/5.51	5.399(5)/0.006				0.07218
									0.15816
									0.23034
^{227}Ac		227.027751	21.77 a	β- /98.6/0.045	0.045/54	3/2-	+1.1	+1.7	0.0838/23.
				α/(1.4)/5.043	4.869(1)/0.09				0.0811/14.
					4.938(1)/0.52				0.2696/13.
					4.951(1)/0.65				(0.044–1.27)
^{228}Ac		228.031020	6.15 h	β- /2.127	1.11/32	3+			Th L x-ray
					1.85/12				Th k x-ray
					2.18/11				0.12903
									0.33842
									0.91116
									0.96897
									(0.2–1.96)
^{227}Th		227.027703	18.68 d	α		(3/2+)			Ra L x-ray
									Ra k x-ray
									0.05014
									0.23597
									0.25624

Elem. or Isot.	Natural abundance (Atom %)	Atomic mass or weight	Half-life/ Resonance width (MeV)	Decay mode/ Energy (MeV)	Particle energy/ Intensity (MeV/%)	Spin ($h/2\pi$)	Nuclear magnetic mom. (nm)	Elect. quadr. mom. (b)	γ-Energy/ Intensity (MeV/%)
									(0.02–1.0)
^{228}Th		228.028740	1.913 a	α/5.520	5.1770(2)/0.18	0+			
					5.2114(1)/0.4				
					5.3405(1)/26.7				
					5.4233(1)/73				
^{229m}Th			13.9 h	α	4.83-5.08	(3/2+)			
^{229}Th		229.031761	7.9×10^{3} a	α/5.168	4.814/9.3	5/2+	+0.46	+4.	0.1935/4.3
					4.845(5)/56				0.21089/277
					4.9008(5)/10.2				0.13697/1.21
					4.689–5.077				0.011–0.6036
^{230}Th	0.02(02)	230.033132	7.56×10^{4} a	α/4.771	4.4383(6)/0.03	0+			0.0677/0.46
					4.4798(6)/0.12				0.1439/0.078
			> 2×10^{18} a	sf// < 4×10^{-12}	4.6211(6)/23.4				
					4.6876(6)/76.3				
^{231}Th		231.036303	1.063 d	β- /0.390	0.138/22	5/2+			Pa L x-ray
					0.218/20				Pa k x-ray
					0.305/52				0.02564
									0.084203/
									(0.02–0.3)
^{232}Th	99.98(02)	232.038054	1.40×10^{10} a	α/4.081	3.830(10)/0.2	0+			0.0590
			1.2×10^{21} a	sf/1.1×10^{-9}	3.952(5)/23				0.124
					4.010(5)/77				
^{234}Th		234.043600	24.10 d	β- /0.273	0.102/20	0+			Pa L x-ray
					0.198/72				0.06329/4.1
									0.09235/2.4
									0.09278/2.4
^{228}Pa		228.03105	22. h	ec/(98)/2.111		(3+)	+3.5		Th k x-ray
				α/(2)	5.779/0.23				0.409/100
					5.805/0.15				0.4631/222
					6.078/0.4				0.91116/242
					6.105/0.25				0.96464/120
					6.118/0.22				0.96897/149
									0.058–1.96
^{229}Pa		229.032096	1.5 d	ec/(99.8)/0.32		(5/2+)			0.04244
				α/(0.2)/5.836	5.536(2)/0.02				(0.024–0.18)
					5.579(2)/0.09				
					5.668(2)/0.05				
^{230}Pa		230.034540	17.4 d	ec/(90)/1.310	0.51/	(2-)	2.0		Th L x-ray
				β-/(10)/0.563					Th k x-ray
									0.4437
									0.45477
									0.89876
									0.91856
									0.95199
									(0.053–1.07)
^{231}Pa	100	231.035883	3.25×10^{4} a	α/5.148	4.6781(5)/1.5	3/2-	2.01	-1.7	Ac L x-ray
					4.7102(5)/1.0				Ac k x-ray
			> 2×10^{17} a	sf/< 1.6×10^{-15}	4.7343(5)/8.4				0.01899
					4.8513(5)/1.4				0.027396
					4.9339(5)/3				0.03823
					4.9505(5)/22.8				0.04639
					4.9858(5)/1.4				0.25586
					5.0131(5)/25.4				0.26029
					5.0292(5)/20				0.28367
					5.0318(5)/2.5				0.30007
					5.0587(5)/11				0.30264
									0.33007
									(0.02–0.61)
^{232}Pa		232.03859	1.31 d	β- /1.34		(2-)			U k x-ray
									0.10900

Elem. or Isot.	Natural abundance (Atom %)	Atomic mass or weight	Half-life/ Resonance width (MeV)	Decay mode/ Energy (MeV)	Particle energy/ Intensity (MeV/%)	Spin (h/2π)	Nuclear magnetic mom. (nm)	Elect. quadr. mom. (b)	γ-Energy/ Intensity (MeV/%)
									0.15009
									0.89439
									0.96934
									(0.10–1.17)
^{233}Pa		233.040247	26.97 d	β-/0.571	0.15/40	3/2-	+4.0	-3.0	U L x-ray
					0.256/60				U k x-ray
									0.30017
									0.31201/38.4
									(0.0286-0.456)
^{234}Pa		234.043306	6.69 h	β- /2.197	0.51/	(4+)			U L x-ray
									U k x-ray
									0.1312/0.03
									0.5695/0.02
									0.9256/0.02
									(0.02–1.99)
^{230}U		230.03394	20.23 d	α/5.992	5.5866(3)/0.01	0+			Th L x-ray
			> 4 ×10^{10} a	sf/< 10^{-10}	5.6624(3)/0.26				0.07218
					5.6663(3)/0.38				0.15421
					5.8178(3)/32				0.23034
					5.8887(3)/67				(0.081–0.8565)
^{231}U		231.036292	4.2 d	ec/0.36		(5/2-)			Pa L x-ray
				α/(10^{-3})	5.46/1.6 ×10^{-3}				Pa k x-ray
					5.47/1.4 ×10^{-3}				0.02564
					5.40/1. ×10^{-3}				0.08420
^{232}U		232.037155	70. a	α/5.414	4.9979(1)/0.003	0+			
			2.6 ×10^{15} a	sf/2.7 ×10^{-12}	5.1367(1)/0.3				
					5.2635(1)/31				
					5.3203(1)/69				
^{233}U		233.039634	1.590 ×10^{5} a	α/4.909	4.7830(8)/13.2	5/2+	+0.59	3.66	Th L x-ray
			>2.7 ×10^{17} a	sf/6 ×10^{-11}	4.8247(8)/84.4				0.04244
					4.510–4.804				0.09714
									(0.0252–1.119)
^{234}U	0.0054(5)	234.040950	2.453 ×10^{5} a	α/4.856	4.604(1)/0.24	0+			0.05323/0.156
			1.5 ×10^{16} a	sf/1.6 ×10^{-9}	4.7231(1)/27.5				0.12091
					4.776(1)/72.5				
^{235}U	0.7204(6)	235.043928	7.03 ×10^{8} a	α/4.6793	4.1525(9)/0.9	7/2-	-0.38	4.94	Th L x-ray
			1.0 ×10^{19} a	sf/7 ×10^{-9}	4.2157(9)/6.				Th k x-ray
					4.3237(9)/4.6				0.10917
					4.3641(9)/19.				0.14378/0.134
					4.370(4)/6				0.16338/0.067
					4.3952(9)/57.				0.18574/0.806
					4.4144(9)/2.1				0.1949/0.009
					4.5025(9)/1.7				0.20533/0.774
					4.5558(9)/4.2				0.2214/0.0014
					4.5970(9)/4.8				(0.03–0.79)
^{236}U		236.045566	2.342 ×10^{7} a	α/4.569	4.332(8)/0.26	0+			Th L x-ray
			2.5 ×10^{16} a	sf// 9 ×10^{-8}	4.445(5)/26				0.04946/100
					4.494(3)/74				0.11279/24.1
									0.17115/0.080
^{237}U		237.048728	6.75 d	β- /0.519	0.24/	1/2+			Np L x-ray
					0.25/				Np k x-ray
									0.05953
									0.20801
^{238}U	99.2742(10)	238.050787	4.47 ×10^{9} a	α	4.0395/0.23	0+			Th L x-ray
			8.2 ×10^{15} a	sf// 5 ×10^{-5}	4.147(5)/23				0.04955/.06
					4.196(5)/77				0.1135/.01
^{240}U		240.056592	14.1 h	β- /0.39	0.36/	0+			Np L x-ray
									0.04410
									0.05558
									0.06760

Elem. or Isot.	Natural abundance (Atom %)	Atomic mass or weight	Half-life/ Resonance width (MeV)	Decay mode/ Energy (MeV)	Particle energy/ Intensity (MeV/%)	Spin ($h/2\pi$)	Nuclear magnetic mom. (nm)	Elect. quadr. mom. (b)	γ-Energy/ Intensity (MeV/%)
^{234}Np		234.04289	4.4 d	β+, ec/1.81	0.79/	(0+)			U L x-ray
									U k x-ray
									1.5272
									1.5587
									1.6022
^{235}Np		235.044062	1.085 a	ec/99.9 /0.124		5/2+			U k x-ray
				α/0.001/5.191					
^{236m}Np			22.5 h	ec/52 /		(1-)			U L x-ray
				β- /48 /					Pu L x-ray
									U k x-ray
									0.64235
									0.68759
^{236}Np		236.04657	1.55×10^5 a	ec/91 /0.94		(6-)			U L x-ray
				β- /9 /0.49					U k x-ray
									0.10423
									0.16031
^{237}Np		237.048172	2.14×10^6 a	α/4.957	4.6395(5)/6.5	5/2+	+3.14	+3.87	Pa L x-ray
			1×10^{18} a	sf/2.1×10^{-10}	4.766(5)/9.7				Pa k x-ray
					4.7715(5)/22.7				0.029378/15
					4.7884(5)/47.8				0.08653/12
					4.558–4.873				(0.03–0.28)
^{238}Np		238.050945	2.117 d	β- /1.292	1.2/	2+			Pu L x-ray
									Pu k x-ray
									0.98447/25.2
									1.02855/18.3
									(.044–1.026)
^{239}Np		239.052938	2.355 d	β- /0.722	0.341/30	5/2+			Pu L x-ray
					0.438/48				Pu k x-ray
									0.10613
									0.228186/11
									0.27760/15
									(0.04–0.50)
^{234}Pu		234.04332	8.8 h	ec/94 /0.39		0+			
				α/6 /6.310	6.035(3)/0.024				
					6.149(3)/1.9				
					6.200(3)/4.				
^{236}Pu		236.046057	2.87 a	α/5.867	5.611/0.21	0+			0.0476/0.07
			1.5×10^9 a	sf/1.9×10^{-7}	5.7210/30.5				0.109/0.02
					5.7677(1)/69.3				(0.17–0.97)
^{237}Pu		237.048408	45.7 d	ec/99.9 /0.220		7/2-			Np L x-ray
				α/0.003 /5.747	5.334(4)/0.0015				Np k x-ray
					5.356(4)/0.0006				0.026344
					5.650(4)/0.0007				0.03319
									0.05954
									(0.03–0.5)
^{238}Pu		238.049558	87.7 a	α/5.593	5.3583(1)/0.10	0+			U k x-ray
			4.75×10^{10} a	sf/1.8×10^{-7}	5.465(1)/28.3				0.04347
					5.4992(1)/71.6				(0.04–1.1)
^{239}Pu		239.052162	2.410×10^4 a	α/5.244	5.055/0.047	1/2+	+0.203		U k x-ray
			$8. \times 10^{15}$ a	sf/3×10^{-10}	5.076/0.078				0.05162
					5.106/11.9				0.05682
					5.144/17.1				0.12928
					5.157/70.8				0.37502
					(4.74 –5.03)				0.41369
^{240}Pu		240.053812	6.56×10^3 a	α/5.255	5.0212(1)/0.07	0+			U L x-ray
			1.14×10^{11} a	sf/5.7×10^{-6}	5.1237(1)/26.4				0.04524
					5.1681(1)/73.5				0.10423
					(4.492-4.863)				(0.04–0.97)
^{241}Pu		241.056850	14.33 a	β-/99+/0.0208	4.853/3×10^{-4}	5/2+	-0.68	+6.	0.14854
				α/0.002 /5.139	4.897/0.002				0.1600

Elem. or Isot.	Natural abundance (Atom %)	Atomic mass or weight	Half-life/ Resonance width (MeV)	Decay mode/ Energy (MeV)	Particle energy/ Intensity (MeV/%)	Spin ($h/2\pi$)	Nuclear magnetic mom. (nm)	Elect. quadr. mom. (b)	γ-Energy/ Intensity (MeV/%)
			< 6 ×10^{16} a	sf/> 2.4 ×10^{-14}					
^{242}Pu		242.058741	3.75 ×10^{5} a	α/4.983	4.7546(7)/0.098	0+			U L x-ray
			6.77 ×10^{10} a	sf/5.5 ×10^{-4}	4.8564(7)/22.4				0.04491
					4.9006(7)/78				0.10350
^{243}Pu		243.062002	4.956 h	β- /0.582	0.49/21	7/2+			Am L x-ray
					0.58/60				0.0417
									0.0839
^{244}Pu		244.064204	8.12 ×10^{7} a	α/99.9/4.665	4.546(1)/19.4	0+			U L x-ray
			6.6 ×10^{10} a	sf/0.12	4.589(1)/80.5				0.0439
^{245}Pu		245.06783	10.5 h	β- /1.21	0.93/57	(9/2-)			Am L x-ray
					1.21/11				Am k x-ray
									0.2804 /
									0.30832
									0.32752
									0.56014
									(0.03–1.2)
^{246}Pu		246.07020	10.85 d	β- /0.40	0.150/85	0+			Am L x-ray
					0.35/10				Am k x-ray
									0.04379
									0.22371
^{247}Pu		247.0742	2.3 d						
^{239}Am		239.053023	11.9 h	ec/99.99/0.803		5/2-			Pu L x-ray
				α/0.01/5.924	5.734(2)/0.001				Pu k x-ray
					5.776(2)/0.008				0.18172
									0.22818
									0.27760
^{240}Am		240.05530	2.12 d	ec/1.38		(3-)			Pu L x-ray
				α/5.592	5.378/16 ×10^{-4}				Pu k x-ray
									0.88878
									0.98764
									(0.1–1.3)
^{241}Am		241.056827	432.7 a	α/5.637	5.2443(1)/0.002	5/2-	+1.58	+4.3	Np L x-ray
			1.2 ×10^{14} a	sf/3.6 ×10^{-10}	5.3221(1)/0.015				0.02634 /.024
					5.3884(1)/1.4				0.0332/.00126
					5.4431(1)/12.8				0.05954/0.359
					5.4857(1)/85.2				(0.03–1.128)
					5.5116(1)/0.20				
					5.5442(1)/0.34				
^{242m}Am			141. a	IT/99.5/0.048		5-	+1.0	+6.7	Am L x-ray
				α/0.5/5.62	5.141(4)/0.026				0.04863
			> 3 ×10^{12} a	sf/< 4.7 ×10^{-9}	5.2070(2)/0.4				0.08648
									0.10944
									0.16304
^{242}Am		242.059547	16.02 h	β- /83 /0.665	0.63/46	1-	+0.388	-2.44	Pu L x-ray
				ec/17 /0.750	0.67/37				Cm L x-ray
									Pu k x-ray
									0.0422
									0.04453
^{243}Am		243.061380	7.37 ×10^{3} a	α/5.438	5.1798(5)/1.1	5/2-	+1.50	+4.3	0.04354
			2. ×10^{14} a	sf/3.7 ×10^{-9}	5.2343(5)/11				0.07467
					5.2766(5)/88				0.08657
					5.394(5)/0.12				0.11770
					5.3500(5)/0.16				0.14197
^{244}Am		244.064283	10.1 h	β- /1.428					Am L x-ray
									Cm k x-ray
									0.7460
									0.9000
^{240}Cm		240.055528	27. d	α/6.397	5.989/0.014	0+			
					6.147/0.05				
			1.9 ×10^{6} a	sf/3.9 ×10^{-6}	6.2478(6) /28.8				

Elem. or Isot.	Natural abundance (Atom %)	Atomic mass or weight	Half-life/ Resonance width (MeV)	Decay mode/ Energy (MeV)	Particle energy/ Intensity (MeV/%)	Spin (h/2π)	Nuclear magnetic mom. (nm)	Elect. quadr. mom. (b)	γ-Energy/ Intensity (MeV/%)
					6.2906(6) /70.6				
^{241}Cm		241.057651	32.8 d	ec/99 /0.768		1/2+			Am k x-ray
				α/1 /6.184	5.8842(4)/0.12				0.13241
					5.9291(4)/0.18				0.16505
					5.9389(4)/0.69				0.18028
									0.43063
									0.47181
^{242}Cm		242.058834	162.8 d	α/6.216	5.9694(1)/0.035	0+			Pu L x-ray
					6.069(1)/25				0.04408
			7.0 ×10^{6} a	sf/6.4 ×10^{-6}	6.1129(1)/74				0.10189
									(0.04–1.2)
^{243}Cm		243.061387	29.1 a	α/6.167	5.6815(5) /0.2	5/2+	0.40		Pu L x-ray
					5.6856(5)/1.6				Pu k x-ray
			5.5 ×10^{11} a	sf/5.3 ×10^{-9}	5.7420(5)/10.6				0.10612
					5.7859(5)/73.5				0.20975
					5.9922(5)/6.5				0.22819
					6.0103(5)/1.0				0.27760
					6.0589(5)/5				0.28546
					6.0666(5)/1.5				0.33431
									(0.04–0.7)
^{244}Cm		244.062751	18.1 a	α/5.902	5.6656/0.02	0+			Pu L x-ray
					5.7528/23				0.04282
			1.32 ×10^{7} a	sf/1.4 ×10^{-4}	5.8050/77				0.09885
					5.515/0.004				0.15262
^{245}Cm		245.065491	8.48 ×10^{3} a	α/5.623	5.235(10)/0.3	7/2+	0.5		Pu L x-ray
					5.3038(10)/5.0				Pu k x-ray
			1.4 ×10^{12} a	sf/6.1 ×10^{-7}	5.3620(7)/93				0.04195
					5.4927(11)/0.8				0.13299
					5.5331(11)/0.6				0.13606
									0.17494
^{246}Cm		246.067222	4.75 ×10^{3} a	α/5.476	5.343(3)/21	0+			Pu L x-ray
			1.8 ×10^{7} a	sf/0.026	5.386(3)/79				0.04453
^{247}Cm		247.070353	1.56 ×10^{7} a	α/5.352	4.818(4)/4.7	9/2-	0.36		Pu k x-ray
					4.8690(20)/71				0.2792
					4.941(4)/1.6				0.2886
					4.9820(20)/2.0				0.3471
					5.1436(20)/1.2				0.4035
					5.2104(20)/5.7				
					5.2659(20)/13.8				
^{248}Cm		248.07235	3.48 ×10^{5} a	α/99.92 /5.162	4.931(5)/0.07	0+			
					5.0349(2)/16.5				
			4.15 ×10^{6} a	sf/8.38	5.0784(2)/ (75)/1				
^{250}Cm		250.0783	~ 9.7 ×10^{3} a	sf/85.8		0+			
				α/5.27					
^{243}Bk		243.06301	4.5 h	ec/99.8 /1.508	6.542(4)/0.03	(3/2-)			0.1466
				α/0.15 /6.871	6.5738(2)/0.04				0.1874
					6.7180(22)/0.02				0.755
					6.7581(20)/0.02				0.840
									0.946
^{244}Bk		244.06518	4.4 h	ec/99.99 /2.26		(4-)			0.1445
				α/0.01 /6.778	6.625(4)/0.003				0.1876
					6.667(4)/0.003				0.2176
									0.9815
									0.9215/
^{245}Bk		245.066360	4.94 d	ec/99.9 /0.810		3/2-			Cm L x-ray
				α/0.1 /6.453	5.8851(5)/0.03				Cm k x-ray
					6.1176(9)/0.01				0.25299
					6.1467(5)/0.02				0.3809
					6.3087(5)/0.014				0.3851

Elem. or Isot.	Natural abundance (Atom %)	Atomic mass or weight	Half-life/ Resonance width (MeV)	Decay mode/ Energy (MeV)	Particle energy/ Intensity (MeV/%)	Spin ($h/2\pi$)	Nuclear magnetic mom. (nm)	Elect. quadr. mom. (b)	γ-Energy/ Intensity (MeV/%)
					6.3492(5)/0.018				
^{246}Bk		246.0687	1.80 d	ec/1.35		(2-)			Cm L x-ray
									Cm k x-ray
									0.79881
									1.08142
^{247}Bk		247.07031	1.4×10^3 a	α/5.889	5.465(5)/1.5	(3/2-)			0.04175
					5.501(5)/7				0.0839
					5.532(5)/45				0.268
					5.6535(20)/5.5				
					5.678(2)/13				
					5.712(2)/17				
					5.753(2)/4.3				
					5.794(2)/5.5				
^{248}Bk		248.07309	23.7 h	β- /70 /0.87	0.86/	(1-)			Cm L x-ray
				ec/30 /0.72					Cf L x-ray
									Cm k x-ray
									Cf k x-ray
									0.5507
^{249}Bk		249.074983	320. d	β- /0.125	0.125/100	7/2+	2.0		0.327/10^{-5}
				α/0.001 /5.525	5.390(1)/0.0002				0.308/10^{-6}
			1.8×10^9 a	sf/4.9×10^{-8}	5.4174(6)/0.001				(0.028-0.280)
^{246}Cf		246.068804	1.49 d	α/6.869	6.6156(10)/0.18	0+			Cm L x-ray
					6.7086(7)/21.8				0.04221
			1.8×10^3 a	sf/2.3×10^{-4}	6.7501(7)/78.0				0.0945
									0.147
^{248}Cf		248.07218	334. d	α/6.369	6.220(5)/17	0+			
			3.2×10^4 a	sf/0.0029	6.262(5)/83				
^{249}Cf		249.074851	351. a	α/6.295	5.758/3.7	9/2-			Cm L x-ray
					5.812/85.7				Cm k x-ray
			$8. \times 10^{10}$ a	sf/4.4×10^{-7}	5.8488(2)/1.0				0.25299/2.5
					5.9029(2)/2.8				0.33351/13.6
					5.9451(2)/4.0				0.38832/63.6
					6.1401(2)/1.1				(0.0376–1.10)
					6.1940(2)/2.2				
^{250}Cf		250.076405	13.1 a	α/6.129	5.8913(4)/0.28	0+			Cm L x-ray
			1.7×10^4 a	sf/0.077	5.9889(4)/17.1				0.04285
^{251}Cf		251.079587	9.0×10^2 a	α/6.172	5.56448(7)/1.5	1/2+			0.109/19.8
					5.632(1)/4.5				0.1775/17.3
					5.648(1)/3.5				(0.0385-0.354)
					5.6773(6)/35				
					5.762(3)/3.8				
					5.7937(7)/2.0				
					5.8124(8)/4.2				
					5.8514(6)/27				
					6.0140(7)/11.6				
					6.0744(7)/2.7				
^{252}Cf		252.081627	2.65 a	α/96.9 /6.217	5.7977(1)/0.23	0+			Cm L x-ray
			86. a	sf/3.1/	6.0756(4)/15.2				0.04339
					6.1184(4)/81.6				0.1002
^{253}Cf		253.08513	17.8 d	β- /99.7 /0.29	0.27/100	(7/2+)			
				α/0.3 /6.126	5.921(5)/0.02				
^{254}Cf		254.08732	60.5 d	sf/99.7/		0+			
				α/0.3/5.930	5.792(5)/0.05				
					5.834(5)/0.26				
^{250}Es		250.0786	8.6 h	ec/2.1		(6+)			Cf L x-ray
									Cf k x-ray
									0.30339
									0.34948
									0.82883
^{251}Es		251.07999	1.38 d	ec/99.5 /0.38		(3/2-)			

Elem. or Isot.	Natural abundance (Atom %)	Atomic mass or weight	Half-life/ Resonance width (MeV)	Decay mode/ Energy (MeV)	Particle energy/ Intensity (MeV/%)	Spin ($h/2\pi$)	Nuclear magnetic mom. (nm)	Elect. quadr. mom. (b)	γ-Energy/ Intensity (MeV/%)
				α/0.5 /	6.462/0.05				
					6.492/0.4				
^{252}Es		252.08298	1.29 a	α/76 /	6.632/61.0	(5-)			
				ec/24 /1.26	6.562/10.3				
^{253}Es		253.084821	20.47 d	α/	6.633/89.8	7/2+	~ +4.10	7.	0.04180/5.6
			6.3 $\times10^{5}$ a	sf/8.9 $\times10^{-6}$	6.5916/6.6				0.3892/2.7
									(0.0309-1.106)
^{254m}Es			1.64 d	β- /99.6 /	0.475	2+	2.9	~ +3.7	Fm L x-ray
				α/0.3 /6.67	6.382	2+			Fm k x-ray
			> 10. a	sf/0.045					0.6488
									0.6938
^{254}Es		254.088021	276. d	α/	6.429	(7+)	4.35		0.064
			> 2.5 $\times10^{7}$ a	sf/< 3 $\times10^{-6}$					
^{255}Es		255.09027	40. d	β- /92 /0.29		(7/2+)			
				α/8 /	6.26				
			2.6 $\times10^{3}$ a	sf/0.0042	6.300				
^{256m}Es			7.6 h	β- /		(8+)			0.218
									0.232
									0.862
^{257}Es		257.0960	7.7 d	β-					
^{251}Fm		251.08154	5.3 h	ec/98 /1.47		(9/2-)			
				α/2 /	6.833				
^{252}Fm		252.08247	1.058 d	α/7.154	6.998/15	0+			
				sf/0.0023	7.039/85				
^{253}Fm		253.085181	3.0 d	ec/88/0.333	6.676/	½+			Es k x-ray
				α/12 /	6.943/				0.2719
^{254}Fm		254.086853	3.240 h	α/	7.150	0+			
				sf/0.059	7.192				
^{255}Fm		255.08996	20.1 h	α/	6.9635(5)/5.0	7/2+			0.08148/1.
			1.0 $\times10^{4}$ a	sf/2.3 $\times10^{-5}$	7.0225(5)/93.4				(0.041-0.900)
^{257}Fm		257.09511	100.5 d	α/99.79	6.519	(9/2+)			0.1794
				sf/0.21					0.2410
^{256}Md		256.0939	1.30 h	ec/89 /2.13	7.21/71				Fm k x-ray
				α/11 /	7.14/22				0.121/409
				sf// < 2.6	7.68/2.5				0.115/266
					7.25/2.5				0.136/143
					7.64/2.1				0.634/119
									0.141–1.37
^{257}Md		257.095538	5.5 h	ec/85 /0.41	7.074	(7/2-)			Fm k x-ray
				α/15, sf// < 1	7.014				(0.181–0.389)
^{258}Md		258.09843	51.5 d	α/7.40	6.718(2)/	(8-)			0.3678
				sf// < 0.003	6.763(4)/				0.057–0.448
^{259}Md		259.1005	1.64 h	sf// >98.7		7/2+			
				α// <1.3					
^{260}Md		260.103	~ 27.8 d	sf// 73–100					
^{253}No		253.09056	1.56 m	α/	8.00/96	(9/2-)			0.222/100
				ec/3.2	α/8.08/4				(0.058-0.670)
^{255}No		255.09319	3.1 m	α/62 /	8.095/58	½+			0.187
				ec/38/2.01	7.742/19				(0.163-0.358)
					7.903/11				
^{259}No		259.10100	58. m	α/78 /7.794	7.52	(9/2+)			
				ec/22/0.5	7.55				
				sf// <9.7					
^{251}Lr		251.0942	39 m	sf					
^{260}Lr		260.1055	3. m	α/	8.03				
^{261}Lr		261.1069	40. m	sf					
^{262}Lr		262.1096	3.6 h	ec/2.					
				sf// <10					
^{263}Rf		263.1124	11. m	sf, α					
^{265}Rf		265.1167	~ 2. m	α					

Elem. or Isot.	Natural abundance (Atom %)	Atomic mass or weight	Half-life/ Resonance width (MeV)	Decay mode/ Energy (MeV)	Particle energy/ Intensity (MeV/%)	Spin ($h/2\pi$)	Nuclear magnetic mom. (nm)	Elect. quadr. mom. (b)	γ-Energy/ Intensity (MeV/%)
^{267}Rf		267.1218	~ 1. h	sf					
^{266}Db		266.1210	~ 0.4 h	ec, sf					
^{267}Db		267.1225	1.2 h	sf					
^{268}Db		268.126	1.2 d	sf, ec					
^{270}Db		270.131	~ 0.7 d	sf					
^{267}Sg		267.1243	~ 1.3 m	α					
^{269}Sg		269.1286	~ 3 m	α	~ 8.50				
^{271}Sg		271.134	2. m	α/70	~ 8.54				
^{267}Bh		267.1275	~ 17 s	α	8.83				
^{270}Bh		270.1334	~ 1. m	α	9.0				
^{272}Bh		272.138	~ 11. s	α	8.9				
^{274}Bh		274.144	~ 0.9 m	α	8.76				
^{269}Hs		269.1337	~ 10. s	α	9.07				
					8.92				
^{270m}Hs			0.3 m	α	8.88				
^{270}Hs		270.1343	~ 7.6 s	α	9.02	0+			
^{271}Hs		271.1371	~ 4. s	α	9.13				
^{277m}Hs			~ 0.5 m	Sf					
^{270}Mt		270.1403	~ 0.8 s	α	10.0				
^{276m1}Mt			~6 s						
^{276}Mt		276.152	0.7 s	α	~ 9.71				
^{278}Mt		278.156	~ 5. s						
^{280}Ds		280.162	~ 7.6 s	sf		0+			
^{281}Ds		281.165	14. s	sf					
^{281}Rg		281.167	17. s	sf//90					
				α//10	9.28				
^{282}Rg		282.169	1. m	α	9.0-9.2				
^{283}Cn		283.173	4. s	sf// < 10					
				α// ~ 100	9.52				
^{285}Cn		285.177	~ 0.5 m	α	9.15				
^{285}Nh		285.180	4. s	α	9.74				
					9.48				
^{286}Nh		286.183	~ 13. s	α	9.6				
^{287}Fl		287.187	0.48 s	α	10.0				
^{288}Fl		288.188	~ 0.58 s	α	9.94				
^{289}Fl		289.191	~ 1.9 s	α	9.82				
^{288}Mc		288.193	0.17 s	α	10.5				
^{289}Mc		289.194	~ 0.33 s	α	10.3				
^{290}Mc		290.196	~ 0.2 s	α	9.95				
^{291}Lv		291.201	0.02 s	α	~ 10.74				
^{292}Lv		292.202	13. ms	α	~ 10.66				
^{293}Lv		293.205	0.06 s	α	10.5				
^{293}Ts		293.209	~ 22. ms	α	11.0				
^{294}Ts		294.211	~ 0.05 s	α	10.8				
^{294}Og		294.214	0.7 ms	α	11.7				

NEUTRON SCATTERING AND ABSORPTION PROPERTIES

Norman E. Holden

This table presents an evaluated set of values for experimental quantities that characterize the properties for scattering and absorption of neutrons. The neutron cross section is given for room temperature neutrons, 20.43 °C, corresponding to a thermal neutron energy of 0.0253 electron volts (eV) or a neutron velocity of 2200 meters/second. The neutron resonance integral is defined over the energy range from 0.5 eV to 0.1 × 10^6 eV, or 0.1 MeV. Bound neutron scattering lengths and neutron cross sections averaged over a Maxwellian spectrum at 30 keV for astrophysical applications are also presented. A list of the major references used is given below. The literature cutoff date is January 2003. Uncertainties are given in parentheses. Parentheses with two or more numbers indicate values to the excited state(s) and to the ground state of the product nucleus. Column definitions for the table are as follows.

Column heading	Definition
Elem. or Isot.	For elements, atomic number and chemical symbol are listed. For nuclides, mass number and chemical symbol are listed. Isomers are indicated by the notation m, m1, or m2
Natural abundance	Natural abundance, in atom percent
Half-life	Half-life in decimal notation. μs = microsecond; ms = millisecond; s = second; m = minute; h = hour; d = day; y = year
Thermal neut. cross section	Cross sections for neutron capture reactions in units of barns (10^{-24} cm^2) or millibarns (mb); proton, alpha production, and fission reactions are designated by σ_p, σ_α, σ_f, respectively; separate values are listed for isomeric production
Resonance integral	Resonance integrals for neutron capture reactions in barns (10^{-24} cm^2) or millibarns (mb); proton, alpha production, and fission reactions are designated by RI_p, RI_α, RI_f, respectively; separate values are listed for isomeric production
Coh. scat. length	Bound coherent scattering lengths for neutron scattering reactions in units of femtometers (fm), which is equal to fermis (10^{-13} cm)
σ(30 keV) Maxw. avg.	Astrophysical cross sections, averaged over a stellar neutron Maxwellian spectrum characterized by a thermal energy of 30 keV, expressed in barns (10^{-24} cm^2), millibarns (mb), or microbarns (μb)

General Nuclear Data References

The following references represent the major sources of the nuclear data.

1. Mughabghab, S.F., Divadeenam, M., Holden, N.E.; Neutron Cross Sections, Vol. 1 *Neutron Resonance Parameters and Thermal Cross Sections*, Part A, *Z* = 1–60. Academic Press, New York (1981); Mughabghab, S.F.; Part B, *Z* = 61–100. Academic Press, Orlando, FL (1984). <https://doi.org/10.1016/B978-0-12-509701-7.50006-7>
2. Holden, N. E., *Fifty Years with Nuclear Fission Conference*, Washington, DC, Gaithersburg, MD,. April 26–29, 1989, p. 946. American Nuclear Society, LaGrange Park, IL (1989).
3. Tuli, J. K., *Nuclear Wallet Cards*, Brookhaven National Laboratory (January 2000).
4. Holden, N. E., Half-Lives of Selected Nuclides, *Pure & Applied Chemistry* 62, 941 (1990). <https://doi.org/10.1351/pac199062050941>
5. Holden, N. E., Hoffman, D.C.; Spontaneous Fission Half-Lives for Ground State Nuclides, *Pure & Applied Chemistry* 72, 1525 (2000). <https://doi.org/10.1351/pac200072081525>
6. Koester, L., Rauch, H., Seymann, E., Neutron Scattering Lengths: A Survey of Experimental Data and Methods, *Atomic Data Nuclear Data Tables* 49, 65 (1991). <https://doi.org/10.1016/0092-640X(91)90012-S>
7. Sears, V. F., Neutron Scattering Lengths and Cross Sections, *Neutron News* 3, (3), 26 (1992). <https://doi.org/10.1080/10448639208218770>
8. Bao, Z. Y., Beer, H., Käpeler, F., Voss, F., Wisshak, K., Raucher, T., Neutron Cross Sections for Nucleo-Synthesis Studies, *Atomic Data Nuclear Data Tables* 76, 70 (2000). <https://doi.org/10.1006/adnd.2000.0838>

Neutron Scattering and Absorption Properties

Elem. or Isot.	Natural abundance (%)	Half-life	Thermal neut. cross section (barns)	Resonance integral (barns)	Coh. scat. length (fm)	σ (30 keV) Maxw. avg. (barns)
$_{1}$H			0.332(2)	0.149(1)	3.739(1)	
^{1}H	99.9885(70)	>2.8×10^{23} y	0.332(2)	0.149(1)	3.741(1)	0.25(2) mb*
^{2}H	0.0115(70)		0.51(1)mb	0.23(2) mb	6.671(4)	2.1(4) μb
^{3}H		12.33 y	< 6. μb		4.79(3)	
$_{2}$He			< 0.05		3.26(3)	
^{3}He	0.000134(3)		σ_p = 5.33(1)×10^3	RI_p = 2.39(1)×10^3	5.74(7)	
			0.05(1) mb			8.(1) μb*
^{4}He	99.999867(3)				3.26(3)	
$_{3}$Li			71.(2)	32.(1)	1.90(2)	
^{6}Li	7.59(4)		σ_t = 9.4(1)×10^2	RI_t = 422.(4)	2.0(1)	σ_t ≈ 1.
			39.(5) mb	17.(2) mb		0.06(1) mb*
^{7}Li	92.41(4)		45.(5) mb	20.(2) mb	2.22(2)	42.(3) μb
^{8}Li		0.84 s				< ≈ 5.5 μb

Elem. or Isot.	Natural abundance (%)	Half-life	Thermal neut. cross section (barns)	Resonance integral (barns)	Coh. scat. length (fm)	σ (30 keV) Maxw. avg. (barns)
$_4$Be			8.8(4) mb	3.9(2) mb	7.79(1)	
^{7}Be		53.28 d	$\sigma_p = 3.9(1)\times10^4$	$RI_p = 1.75(5)\times10^4$		$\sigma_p = 16(4)$*
			$\sigma_\alpha \approx 0.1$			
^{9}Be	100.		8.8(4) mb	3.9(2) mb	7.79(1)	
^{10}Be		1.52×10^6 y	<1. mb			
$_5$B			$7.6(1)\times10^2$	$3.4(1)\times10^2$	5.30(4)	
^{10}B	19.9(7)		$\sigma_\alpha = 38.4(1)\times10^2$	$RI_\alpha = 17.3(1)\times10^2$	0.1(3)	
			0.3(1)	0.13(4)		
			σ_p = 7.(1) mb			
			σ_t = 8.(2) mb			
^{11}B	80.1(7)		5.(3) mb	2.(1) mb	6.65(4)	
$_6$C			3.5(1) mb	1.6(1) mb	6.646(1)	
^{12}C	98.93(8)		3.5(1) mb	1.6(1) mb	6.651(2)	16.(1) μb*
^{13}C	1.07(8)		1.4(1) mb	1.7(2) mb	6.19(9)	0.021(4) mb
^{14}C		5715. y	<1.4 μb			3.(1) μb*
$_7$N			2.00(6)	0.90(3)	9.36(2)	
^{14}N	99.636(20)		σ_p = 1.93(5)	RI_p = 0.87(3)	9.37(2)	σ_p = 1.8(2) mb*
			0.080(1)	0.034(1)		0.04(1) mb
^{15}N	0.364(20)		0.04(1) mb	0.11(3) mb	6.44(3)	6.(1) μb*
$_8$O			0.29(1) mb	0.40(4) mb	5.805(4)	
^{16}O	99.757(16)		0.19(1) mb	0.36(4) mb	5.805(5)	34.(4) μb
^{17}O	0.038(1)		σ_α = 0.257(10)	0.11(1)	5.8(2)	σ_α = 3.9(5) mb*
			0.54(7) mb	0.39(5) mb		
^{18}O	0.205(14)		0.16(1) mb	0.81(4) mb	5.84(7)	9.(1) μb*
$_9$F			9.5(1) mb	21.(3) mb	5.65(1)	6.(1) mb
^{19}F	100.		9.5(1) mb	21.(3) mb	5.65(1)	6.(1) mb
$_{10}$Ne			42.(5) mb	19.(3) mb	4.566(6)	
^{20}Ne	90.48(3)		39.(5) mb	18.(3) mb	4.631(6)	0.12(1) mb
^{21}Ne	0.27(1)		0.7(1)	0.31(5)	6.7(2)	≈ 1.5 mb
			σ_α = 0.18(9) mb			
^{22}Ne	9.25(3)		51.(5) mb	23.(3) mb	3.87(1)	58.(4) μb*
$_{11}$Na			0.53(2)	0.32(2)	3.63(2)	2.1(2) mb
^{22}Na		2.605 y	$\sigma_p = 2.8(3)\times10^4$	$RI_p < 2.\times10^5$		
			$\sigma_\alpha = 2.6(4)\times10^2$	$RI_\alpha = 1.2(2)\times10^2$		
^{23}Na	100.		σ_m = 0.43(3)	RI_m = 0.30(6)	3.63(2)	2.1(2) mb
$_{12}$Mg			66.(6) mb	38.(5) mb	5.375(4)	
^{24}Mg	78.99(4)		0.053(6)	32.(4) mb	5.7(2)	3.3(4) mb
^{25}Mg	10.00(1)		0.20(1)	98.(15) mb	3.6(2)	6.4(4) mb
^{26}Mg	11.01(3)		0.038(1)	25.(2) mb	4.9(2)	0.13(1) mb*
^{27}Mg		9.45 m	0.07(2)	0.03(1)		
$_{13}$Al			0.230(2)	0.17(1)	3.45(1)	
^{26}Al		7.1×10^5 y	σ_p = 1.97(10)			0.14(2)
			σ_α = 0.34(1)			
^{27}Al	100.		0.230(2)	0.17(1)	3.45(1)	2.9(3) mb
$_{14}$Si			0.166(9)	0.12(2)	4.15(1)	
^{28}Si	92.223(19)		0.17(1)	0.11(2)	4.11(1)	2.9(3) mb
^{29}Si	4.685(8)		0.12(1)	0.08(2)	4.7(1)	7.9(9) mb
^{30}Si	3.092(11)		0.107(3)	0.62(6)	4.61(1)	3.2(3) mb*
^{31}Si		2.62 h	73.(6) mb	33.(3) mb		
^{32}Si		1.6×10^2 y	< 0.5			
$_{15}$P			0.17(1)	0.08(1)	5.13(1)	
^{31}P	100.		0.17(1)	0.08(1)	5.13(1)	1.7(1) mb
$_{16}$S			0.54(2)	0.24(2)	2.847(1)	
^{32}S	94.93(31)		0.55(5)	0.25(2)	2.804(2)	4.1(2) mb
			σ_α < 0.5 mb			
^{33}S	0.76(2)		0.46(3)	0.21(2)	4.7(2)	7.4(15) mb
			σ_α = 0.12(1)	RI_α = 0.05(1)		σ_α = 0.18(1)
			σ_p = 2. mb			

Elem. or Isot.	Natural abundance (%)	Half-life	Thermal neut. cross section (barns)	Resonance integral (barns)	Coh. scat. length (fm)	σ (30 keV) Maxw. avg. (barns)
^{34}S	4.29(28)		0.25(1)	0.13	3.48(3)	0.23(1) mb
^{36}S	0.02(1)		0.24(2)	0.26(3)		0.17(1) mb*
$_{17}$Cl			33.6(3)	15.(2)	9.58(1)	
^{35}Cl	75.78(4)		43.7(4)	20.(2)	11.7(1)	9.4(3) mb
			σ_p = 0.44(1)	RI_p = 0.2		σ_p = 1.7(2) mb*
			$\sigma_\alpha \approx$ 0.08 mb			
^{36}Cl		3.01×10^5 y	σ_p = 46.(2) mb	RI_p = 0.02		σ_p = 91.(8) mb
			<10.			
			σ_α = 0.59(7) mb			σ_α = 0.9(2) mb
^{37}Cl	24.22(4)		(0.05 + 0.38)	(0.04+0.26)	3.1(1)	2.0(2) mb
$_{18}$Ar			0.66(3)	0.42(5)	1.91(1)	
^{36}Ar	0.3365(30)		5.(1)	2.(1)	24.9(1)	
			σ_α = 5.4(3) mb			
			σ_p < 1.5 mb			
^{37}Ar		35.0 d	σ_α = 1.08(8)×10^3	RI_α = 900.		$\sigma_\alpha \approx$ 1.3
			σ_p = 37.(4)	RI_p = 31.		$\sigma_p \approx$ 0.04
^{38}Ar	0.0632(5)		0.8(2)	0.4(1)	3.5(35)	
^{39}Ar		268. y	6.(2)×10^2			
			σ_α <0.29			
^{40}Ar	99.6003(30)		0.64(3)	0.41(5)	1.83(1)	2.5(3) mb
^{41}Ar		1.82 h	0.5(1)	0.2(1)		
$_{19}$K			2.1(1)	1.0(1)	3.67(2)	
^{39}K	93.2581(44)		2.1(2)	0.9(1)	3.74(2)	11.8(4) mb
			σ_α = 4.3(5) mb			
			σ_p < 0.05 mb			
^{40}K	0.0117(1)	1.26×10^9 y	30.(8)	13.(4)		σ_p = 7.(1) mb
			σ_p = 4.4(4)	2.0(2)		σ_α = 40.(6) mb
			σ_α = 0.42(8)			
^{41}K	6.7302(44)		1.46(3)	1.4(2)	2.69(8)	22.(1) mb
$_{20}$Ca			0.43(2)	0.23(2)	4.70(2)	
^{40}Ca	96.941(156)		0.41(3)	0.22(4)	4.80(2)	6.7(7) mb
			σ_α = 0.13(4) mb			
^{41}Ca		1.02×10^5 y	≈ 4.			
			σ_α = 0.18(3)			
			σ_p = 7.(2) mb			
^{42}Ca	0.647(23)		0.65(10)	0.39(4)	3.4(1)	16.(2) mb
^{43}Ca	0.135(10)		6.(1)	3.9(2)	1.56(9)	51.(6) mb
^{44}Ca	2.086(110)		0.8(2)	0.56(1)	1.42(6)	9.(1) mb
^{45}Ca		162.7 d	≈ 15.			
^{46}Ca	0.004(3)	>4×10^{15} y	0.70(3)	0.9(1)	3.6(2)	5.3(5) mb*
^{48}Ca	0.187(21)	4.3×10^{19} y	1.0(1)	0.5(1)	0.39(9)	0.8(1) mb*
$_{21}$Sc			27.2(2)	12.(1)	12.3(1)	
^{45}Sc	100.		(10.+17.)	(5.6+6.4)	12.3(1)	69.(5) mb
^{46}Sc		83.81 d	8.(1)	3.6(5)		
$_{22}$Ti			6.1(1)	2.8(2)	3.438(2)	
^{44}Ti		60 y	1.1(2)			
			σ_p<0.2			
^{46}Ti	8.25(3)		0.6(2)	0.4(1)	4.93(6)	27.(3) mb
^{47}Ti	7.44(2)		1.6(2)	1.6(2)	3.63(1)	64.(8) mb
^{48}Ti	73.72(3)		7.9(9)	3.6(2)	6.09(2)	32.(5) mb
^{49}Ti	5.41(2)		1.9(5)	1.2(2)	1.04(5)	22.(2) mb
^{50}Ti	5.18(2)		0.179(3)	0.12(2)	6.18(8)	3.6(4) mb
$_{23}$V			5.0(2)	2.8(1)	0.382(1)	
^{50}V	0.250(4)	1.4×10^{17} y	21.(4)	50.(20)	7.6(6)	
			σ_p = 0.7(4) mb			
^{51}V	99.750(4)		4.9(1)	2.7(2)	0.402(2)	38.(4) mb
$_{24}$Cr			3.0(2)	1.7(1)	3.635(7)	
^{50}Cr	4.345(13)	>1.8×10^{17} y	15.(1)	8.(1)	4.5(1)	0.05(1)
^{51}Cr		27.70 d	< 10.			

Nuclear

Elem. or Isot.	Natural abundance (%)	Half-life	Thermal neut. cross section (barns)	Resonance integral (barns)	Coh. scat. length (fm)	σ(30 keV) Maxw. avg. (barns)
^{52}Cr	83.789(18)		0.8(1)	0.6(2)	4.91(2)	8.8(4) mb
^{53}Cr	9.501(17)		18.(2)	9.(1)	4.2(1)	0.06(1)
^{54}Cr	2.365(7)		0.36(4)	0.25(5)	4.6(1)	7.(2) mb
$_{25}$Mn			13.3(1)	14.0(3)	3.75(2)	
^{53}Mn		3.7×10^{6} y	70.(10)	32.(5)		
^{54}Mn		312.1 d	< 10.			
^{55}Mn	100.		13.3(1)	14.0(3)	3.75(2)	40.(3) mb
$_{26}$Fe			2.7(1)	1.4(2)	9.45(2)	
^{54}Fe	5.845(35)		2.3(2)	1.3(2)	4.2(1)	29.(2) mb
			σ_α = 10. μb	RI$_\alpha$ = 1.1(1) mb		
^{55}Fe		2.73 y	13.(2)	6.(1)		
			σ_α = 0.01			
^{56}Fe	91.754(36)		2.8(3)	1.4(2)	9.93(3)	11.7(5) mb
^{57}Fe	2.119(10)		1.4(2)	0.8(4)	2.3(1)	40.(4) mb
^{58}Fe	0.282(4)		1.3(1)	1.3(2)	15.(7)	12.(1) mb
^{59}Fe		44.51 d	13.(3)	6.(1)		
$_{27}$Co			37.19(8)	74.(2)	2.49(2)	
^{58m}Co		9.1 h	1.4(1)×10^{5}	2.5(10)×10^{5}		
^{58}Co		70.88 d	1.9(2)×10^{3}	7.(1)×10^{3}		
^{59}Co	100.		(20.7+16.5)	(39.+35.)	2.49(2)	38.(4) mb
^{60m}Co		10.47 m	58.(3)	230.(50)		
^{60}Co		5.271 y	2.0(2)	4.3(10)		
$_{28}$Ni			4.5(2)	2.3(2)	10.3(1)	
^{58}Ni	68.0769(89)	>4×10^{19} y	4.6(4)	2.3(2)	14.4(1)	41.(2) mb
			σ_α < 0.03 mb			
^{59}Ni		≈ 7.6×10^{4} y	σ_{abs} = 92.(4)	RI$_{abs}$ = 1.4(1)×10^{2}		
			σ_α = 14.(2)			
			σ_p = 2.(1)			
^{60}Ni	26.2231(77)		2.9(3)	1.5(2)	2.8(1)	25.(1) mb
^{61}Ni	1.1399(6)		2.5(5)	1.5(4)	7.60(6)	82.(8) mb
			σ_α = 0.03 mb			
^{62}Ni	3.6345(17)		15.(1)	6.8(3)	8.7(2)	13.(4) mb
^{63}Ni		100. y	20.(5)	9.(2)		
^{64}Ni	0.9256(9)		1.6(1)	1.2(2)	0.37(7)	9.(1) mb
^{65}Ni		2.517 h	22.(2)	10.(1)		
$_{29}$Cu			3.8(1)	4.1(4)	7.718(4)	
^{63}Cu	69.15(15)		4.5(2)	5.(1)	6.43(15)	0.09(1)
^{64}Cu		12.701 h	≈ 270.			
^{65}Cu	30.85(15)		2.17(3)	2.2(1)	10.61(19)	41.(5) mb
^{66}Cu		5.09 m	1.4(1)×10^{2}	60.(20)		
$_{30}$Zn			1.1(2)	2.8(4)	5.680(5)	
^{64}Zn	48.27(32)	>2.3×10^{18} y	0.74(5)	1.4(3)	5.23(4)	59.(5) mb
			σ_p < 12. μb			
			σ_α = 11.(3) μb			
^{65}Zn		243.8 d	66.(8)	30.(4)		
			σ_α = 2.0(2)			
^{66}Zn	27.977(77)		0.9(3)	1.8(2)	5.98(5)	35.(3) mb
			σ_α < 0.02 mb			
^{67}Zn	4.102(21)		6.9(1.4)	25.(5)	7.58(8)	0.15(2)
			σ_α = 0.4 mb			
^{68}Zn	19.02(12)		(0.072 + 0.8)	(0.2 + 2.9)	6.04(3)	19.(2) mb
			σ_α <0.02 mb			σ_m = 3.(1) mb
^{70}Zn	0.631(9)		(8.1+83.) mb	0.9(2)		0.02(1)
$_{31}$Ga			2.9(1)	22.(3)	7.288(2)	
^{69}Ga	60.108(9)		1.68(7)	16.(2)	7.88(4)	0.14(1)
^{71}Ga	39.892(9)	>2.4×10^{26} y	4.7(2)	31.(3)	6.40(3)	0.12(1)
			σ_m = 0.15(5)			
$_{32}$Ge			2.2(1)	6.(2)	8.19(2)	
^{68}Ge		270.8 d	1.0(5)			

Elem. or Isot.	Natural abundance (%)	Half-life	Thermal neut. cross section (barns)	Resonance integral (barns)	Coh. scat. length (fm)	σ (30 keV) Maxw. avg. (barns)
^{70}Ge	20.370(89)		(0.3 + 2.7)	2.3(1)	10.0(1)	88.(5) mb
^{72}Ge	27.380(60)		0.9(2)	0.8(3)	8.5(1)	0.07(2)
^{73}Ge	7.759(78)	>1.8×10^{23} y	15.(1)	66.(20)	5.02(4)	0.3(1)
^{74}Ge	36.656(80)		(0.14 + 0.28)	(0.4+0.5)	7.6(1)	53.(7) mb
^{76}Ge	7.835(81)	1.6×10^{21} y	(0.09 + 0.06)	(1.3+0.6)	8.2(15)	0.03(2)
$_{33}$As			4.0(4)	61.(5)	6.58(1)	
^{75}As	100.		4.0(4)	61.(5)	6.58(1)	0.57(4)
$_{34}$Se			12.(1)	14.(3)	7.970(9)	
^{74}Se	0.89(4)		50.(2)	520(50)	0.8(3)	0.2(1)
^{75}Se		119.78 d	3.3(10)×10^{2}			
^{76}Se	9.37(29)		(22. + 63.)	(9.+31.)	12.2(1)	0.16(1)
^{77}Se	7.63(16)		42.(4)	30.(5)	8.25(8)	0.3(1)
			σ_α = 0.97(3) μb			
^{78}Se	23.77(28)		σ_m = 0.38(2)	RI$_m$ = 4.3(4)	8.24(9)	0.1
^{80}Se	49.61(41)		(0.05+0.54)	(0.15+0.85)	7.48(3)	42.(3) mb
^{82}Se	8.73(22)	≈ 1×10^{20} y	(39.+ 5.2) mb	39.(4) mb	6.34(8)	0.04(2)
$_{35}$Br			6.8(2)	92.(8)	6.79(2)	
^{76}Br		16.0 h	224.(42)			
^{79}Br	50.69(7)		(2.5+8.3)	(36.+96.)	6.79(7)	0.63(4)
						σ_m = 0.08(1)
^{81}Br	49.31(7)		(2.4+0.24)	51.(5)	6.78(7)	0.31(2)
$_{36}$Kr			24.(1)	39.(6)	7.81(2)	
^{78}Kr	0.353(3)	>2.3×10^{20} y	(0.17+6.)	20.(1)		(0.11+0.19)
^{80}Kr	2.286(10)		(4.6+7.)	57.(6)		(0.09+0.18)
^{82}Kr	11.593(3)		(14.+7.)	130.(13)		90.(6) mb
^{83}Kr	11.500(19)		183.(30)	183.(20)		0.24(2)
^{84}Kr	56.987(15)		(σ_m+ σ_g) = 0.11	2.4(3)		(16.+33.) mb
			σ_m = 0.09			
^{85}Kr		10.73 y	1.7(2)	1.8(10)		0.07(2)
^{86}Kr	17.279(41)		3.(2) mb	≈ 1. mb	8.1(3)	3.2(4) mb
$_{37}$Rb			0.39(4)	6.(3)	7.08(2)	
^{84}Rb		32.9 d	σ_p = 12.(2)			
^{85}Rb	72.17(2)		(0.06+0.38)	(0.7+7.)	7.0(1)	0.24(1)
^{86}Rb		18.65 d	<20.			
^{87}Rb	27.83(2)	4.88×10^{10} y	0.10(1)	2.3(4)	7.3(1)	16.(1) mb
^{88}Rb		17.7 m	1.2(3)	0.5(1)		
$_{38}$Sr			1.2(1)	10.(1)	7.02(2)	
^{84}Sr	0.56(1)		(0.6+0.2)	(9.+1.)		0.4(1)
^{86}Sr	9.86(1)		σ_m = 0.81(4)	RI$_m$ = 4.(1)	5.68(5)	(48.+22.) mb
^{87}Sr	7.00(1)		16.(3)	118.(30)	7.41(7)	97.(5) mb
^{88}Sr	82.58(1)		5.8(4) mb	0.07(3)	7.16(6)	6.0(2) mb
^{89}Sr		50.52 d	0.42(4)	0.2		
^{90}Sr		29.1 y	10.(1) mb	0.10(2)		
$_{39}$Y			1.25(5)	1.0(1)	7.75(2)	
^{89}Y	100.		(0.001+1.25)	(0.006+1.0)	7.75(2)	19.(1) mb
^{90}Y		2.67 d	<6.5			
^{91}Y		58.5 d	1.4(3)	0.6(1)		
$_{40}$Zr			0.19(1)	0.95(9)	7.16(3)	
			σ_α <0.1 mb			
^{90}Zr	51.45(40)		≈ 0.014	0.2(1)	6.4(1)	21.(2) mb
^{91}Zr	11.22(5)		1.2(3)	5.(2)	8.8(1)	60.(8) mb
^{92}Zr	17.15(8)		0.2(1)	0.6(2)	7.5(2)	33.(4) mb
^{93}Zr		1.5×10^{6} y	<4.	16.(5)		0.10(1)
^{94}Zr	17.38(28)	>10^{17} y	0.049(6)	0.25(3)	8.3(2)	26.(1) mb
^{96}Zr	2.80(9)	>1.7×10^{18} y	0.020(3)	5.0(5)	5.5(1)	11.(1) mb
$_{41}$Nb			1.11(1)	8.5(6)	7.14(3)	
			σ_α <0.1 mb			
^{93}Nb	100.		1.1	(6.3+2.2)	7.14(3)	266.(5) mb
			σ_m = 0.86			

Elem. or Isot.	Natural abundance (%)	Half-life	Thermal neut. cross section (barns)	Resonance integral (barns)	Coh. scat. length (fm)	σ (30 keV) Maxw. avg. (barns)
^{94}Nb		2.4×10^{4} y	(σ_m+ σ_g) = 15.(1)	126.(13)		
			σ_m = 0.6(1)			
^{95}Nb		34.97 d	<7.	<200.		
$_{42}$Mo			2.5(1)	26.(5)	6.72(2)	
			σ_α <0.1 mb			
^{92}Mo	14.77(31)	>3×10^{17} y	0.06	≈ 0.8	6.93(8)	0.07(1)
			σ_m = 0.2 μb			
^{94}Mo	9.226(99)		0.02	≈ 0.8	6.82(7)	0.10(2)
^{95}Mo	15.900(85)		13.4(3)	109.(5)	6.93(6)	0.29(1)
			σ_α = 30.(4) μb			
^{96}Mo	16.674(12)		0.5	17.(3)	6.22(6)	0.11(1)
^{97}Mo	9.560(50)		2.5(2)	14.(3)	7.26(8)	0.34(1)
			σ_α = 0.4(2) μb			
^{98}Mo	24.20(25)		0.14(1)	7.2(7)	6.60(7)	0.10(1)
^{100}Mo	9.67(20)	≈ 1×10^{19} y	0.19(1)	3.6(3)	6.75(7)	0.11(1)
$_{43}$Tc						
^{98}Tc		≈ 6.6×10^{6} y	σ_m = 0.9(2)			
^{99}Tc		2.13×10^{5} y	23.(2)	4.0(4)×10^{2}	6.8(3)	0.93(5)
$_{44}$Ru			2.6 (1)	48.(5)	7.03(3)	
^{96}Ru	5.54(14)	>3.1×10^{16} y	0.23(4)	7.(2)		0.21(1)
^{98}Ru	1.87(3)		< 8.			0.3(1)
^{99}Ru	12.76(14)		4.(1)	195.(20)		1.2(3)
^{100}Ru	12.60(7)		5.8(6)	11.(2)		0.21(1)
^{101}Ru	17.06(2)		5.(1)	1.1(3)×10^{2}		1.00(4)
			σ_α <0.15 μb			
^{102}Ru	31.55(14)		1.2(1)	4.3(5)		0.15(1)
^{103}Ru		39.27 d	<20.	≈ 30.		
^{104}Ru	18.62(27)		0.49(2)	6.(2)		0.15(1)
^{105}Ru		4.44 h	0.29(3)	0.13(1)		
^{106}Ru		1.020 y	0.15(4)	2.0(6)		
$_{45}$Rh			145.(2)	1.2(1)×10^{3}	5.88(4)	
^{103}Rh	100.		(11.+ 134.)	(0.08+1.1)×10^{3}	5.88(4)	0.81(1)
^{104m}Rh		4.36 m	800.(100)			
^{104}Rh		42.3 s	40.(30)			
^{105}Rh		35.4 h	1.1(3)×10^{4}	1.7(4)×10^{4}		
$_{46}$Pd			7.(1)	82.(8)	5.91(6)	
^{102}Pd	1.02(1)		3.2(10)	10.(2)		0.3(1)
^{104}Pd	11.14(8)			16.(2)		0.29(3)
^{105}Pd	22.33(8)		22.(2)	60.(20)	5.5(3)	1.20(6)
			σ_α = 0.5(2) μb			
^{106}Pd	27.33(3)		(0.013+0.28)	(0.2+5.5)	6.4(4)	0.25(3)
^{107}Pd		6.5×10^{6} y	1.8(2)	108.(4)		1.34(6)
^{108}Pd	26.46(9)		(0.19+8.5)	(2.+240.)	4.1(3)	0.20(2)
^{110}Pd	11.72(9)		(0.033+0.7)	(0.7+8.)		0.15(2)
$_{47}$Ag			62.(1)	767.(60)	5.922(7)	
^{107}Ag	51.839(8)		(1.+35.)	(3.+105.)	7.56(1)	0.80(3)
^{109}Ag	48.161(8)		(4.1 + 87.)	(0.7+14.1)×10^{2}	4.17(1)	0.79(3)
^{110m}Ag		249.8 d	82.(11)	20.(4)		
^{111}Ag		7.47 d	3.(2)	105.(20)		
$_{48}$Cd			2.52(5)×10^{3}	73.(8)	4.87(5)	
^{106}Cd	1.25(6)	>2.6×10^{17} y	0.20(3)	4.(1)		0.30(2)
^{108}Cd	0.89(3)	>4.1×10^{17} y	1.	14.(3)	5.4(1)	0.20(1)
^{109}Cd		462.0 d	≈ 180.	6.7(12)×10^{3}		
			σ_α <0.05			
^{110}Cd	12.49(18)		(0.06+11.)	(6.+34.)	5.9(1)	(0.01+0.22)
^{111}Cd	12.80(12)		3.5(20)	51.(6)	6.5(1)	0.75(1)
^{112}Cd	24.13(21)		(0.012+2.2)	15.	6.4(1)	0.19(1)
^{113}Cd	12.22(12)	7.7×10^{15} y	2.06(4)×10^{4}	390.(40)	8.0(2)	0.67(1)
			σ_α <1. μb			

Elem. or Isot.	Natural abundance (%)	Half-life	Thermal neut. cross section (barns)	Resonance integral (barns)	Coh. scat. length (fm)	σ (30 keV) Maxw. avg. (barns)
^{114}Cd	28.73(42)		(0.04+0.29)	16.(7)	7.5(1)	(0.01+0.12)
^{116}Cd	7.49(18)	3.8×10^{19} y	(26.+52.) mb	1.2	6.3(1)	(12.+47.) mb
$_{49}$In			197.(4)	3.3(2)×10^{3}	4.07(2)	
^{113}In	4.29(5)		(3.1+5.0+3.9)	(220.+90.)	5.39(6)	(0.48+0.31)
^{115}In	95.71(5)	4.4×10^{14} y	(88.+73.+44.)	(1.5+1.2+0.7)×10^{3}	4.01(2)	(0.69+0.02)
$_{50}$Sn			0.61(3)	8.(2)	6.225(2)	
^{112}Sn	0.97(1)		(0.15+0.40)	(8.+19.)		0.21(1)
^{113}Sn		115.1 d	≈ 9.	210.(50)		
^{114}Sn	0.66(1)		≈ 0.12	5.(1)	6.2(3)	134.(3) mb
^{115}Sn	0.34(1)		σ_α = 0.06 mb	29.(6)		0.34(1)
^{116}Sn	14.54(9)		(0.006+0.14)	(0.5+11.)	5.93(5)	91.(2) mb
^{117}Sn	7.68(7)		1.1(1)	16.(5)	6.48(5)	319.(7) mb
^{118}Sn	24.22(9)		σ_m = 4. mb	4.7(5)	6.07(5)	62.(1) mb
^{119}Sn	8.59(4)		2.(1)	2.9(5)	6.12(5)	0.18(1)
^{120}Sn	32.58(9)		(0.001+0.13)	1.2(3)	6.49(5)	(0.5+36.) mb
^{122}Sn	4.63(3)		(0.15+0.001)	0.81(4)	5.74(5)	(18.+4.) mb
^{124}Sn	5.79(5)	>2.2×10^{18} y	(0.13+0.004)	(8.0+0.08)	5.97(5)	12.(2) mb
$_{51}$Sb			5.2(2)	169.(20)	5.57(3)	
^{121}Sb	57.21(5)		(0.4+5.8)	(13.+192.)	5.71(6)	0.53(2)
^{123}Sb	42.79(5)		(0.02+0.04+4.0)	(1.+119.)	5.38(7)	0.30(1)
^{124}Sb		60.20 d	17.(3)	≈ 8.		
$_{52}$Te			4.2(1)	47.(3)	5.80(3)	
^{120}Te	0.09(1)		(1.+5.)	≈ 1.	5.3(5)	0.4(1)
^{122}Te	2.55(12)		(0.4+3.)	(5.+75.)	3.8(2)	295.(3) mb
^{123}Te	0.89(3)	>5.3×10^{16} y	370.(40)	4.5(3)×10^{3}	0.05	0.83(1)
			σ_α = 0.05 mb			
^{124}Te	4.74(14)		(1.+6.)	(1.4+4.)	8.0(1)	155.(2) mb
^{125}Te	7.07(15)		1.1(2)	21.(4)	5.02(8)	431.(4) mb
^{126}Te	18.84(25)		(0.12+0.8)	(0.6+7.4)	5.56(7)	(28.+53.) mb
^{128}Te	31.74(8)	2.2×10^{24} y	(0.03+0.2)	(0.2+1.6)	5.89(7)	(3.+41.) mb
^{130}Te	34.08(62)	8.×10^{20} y	(0.01+0.19)	(0.03+0.3)	6.02(7)	(4.+11.) mb
$_{53}$I			6.2(1)	1.5(1)×10^{2}	5.28(2)	
^{125}I		59.4 d	900.(100)	1.4(2)×10^{4}		
^{127}I	100.		6.2(1)	1.5(1)×10^{2}	5.28(2)	0.64(3)
^{128}I		25.00 m	22.(4)	≈ 10.		
^{129}I		1.7×10^{7} y	(20.7+10.3)	36.(4)		0.44(2)
^{130}I		12.36 h	18.(3)	≈ 8.		
^{131}I		8.021 d	≈ 0.7	8.(4)		
$_{54}$Xe			25.(1)	263.(50)	4.92(3)	
^{124}Xe	0.0953(27)	>10^{17} y	(28.+137.)	(0.6+3.0)×10^{3}		(0.13+0.51)
^{125}Xe		17.1 h	σ_α < 0.03			
^{126}Xe	0.0890(14)		(0.45+3.)	(8.+52.)		(0.04+0.32)
^{127}Xe		36.34 d	σ_α ≤ 0.01			
^{128}Xe	1.910(22)		σ_m = 0.48	RI_m = 38.(10)		0.26(1)
^{129}Xe	26.40(18)		22.(5)	250.(50)		0.62(2)
^{130}Xe	4.071(53)		σ_m = 0.45	RI_m = 16.(4)		0.132(3)
^{131}Xe	21.233(62)		90.(10)	9.(1)×10^{2}		0.45(8)
^{132}Xe	26.9087(680)		(0.05+0.4)	(0.9+3.7)		(5.+60.) mb
^{133}Xe		5.243 d	190.(90)			
^{134}Xe	10.436(29)	>1.1×10^{16} y	(0.003 + 0.26)	0.40(4)		20.(2) mb
^{135}Xe		9.10 h	2.65(11)×10^{6}	7.6(5)×10^{3}		
^{136}Xe	8.858(33)	>8×10^{20} y	0.26(2)	0.7(2)		0.9(1) mb
$_{55}$Cs			30.4(8)	422.(50)	5.42(2)	
^{132}Cs		6.48 d	σ_α < 0.15			
^{133}Cs	100.		(2.7+27.3)	(32.+360.)	5.42(2)	(0.04+0.47)
^{134}Cs		2.065 y	140.(10)	54.(9)		
^{135}Cs		2.3×10^{6} y	8.3(3)	38.(3)		
^{137}Cs		30.2 y	(0.20+0.07)	0.36(7)		
$_{56}$Ba			1.3(2)	10.(2)	5.07(3)	

Elem. or Isot.	Natural abundance (%)	Half-life	Thermal neut. cross section (barns)	Resonance integral (barns)	Coh. scat. length (fm)	σ (30 keV) Maxw. avg. (barns)
^{130}Ba	0.106(1)	2.2×10^{21} y	(1.+8.)	(25.+200.)	3.6(6)	0.76(11)
^{132}Ba	0.101(1)	1.3×10^{21} y	(0.84+9.7)	(4.7+24.)	7.8(3)	0.6(1)
^{133}Ba		10.53 y	4.(1)	85.(30)		
^{134}Ba	2.417(18)		(0.1+1.3)	(5.6+18.)	5.7(1)	0.18(1)
^{135}Ba	6.592(12)		(0.014+5.8)	(0.47+131.)	4.7(1)	0.46(2)
^{136}Ba	7.854(24)		(0.010+0.44)	(0.1+1.5)	4.91(8)	61.(2) mb
^{137}Ba	11.232(24)		5.(1)	4.(1)	6.8(1)	76.(3) mb
^{138}Ba	71.698(42)		0.41(2)	0.4(1)	4.84(8)	4.0(2) mb
^{139}Ba		1.396 h	5.(1)	2.2(5)		
^{140}Ba		12.75 d	1.6(3)	14.(1)		
$_{57}$La			9.2(2)	12.(1)	8.24(4)	
^{138}La	0.090(1)	1.06×10^{11} y	57.(6)	4.1(9)×10^{2}		
^{139}La	99.910(1)		9.2(2)	12.(1)	8.24(4)	38.(3) mb
^{140}La		1.678 d	2.7(3)	69.(4)		
$_{58}$Ce			0.64(4)	0.71(6)	4.84(2)	
^{136}Ce	0.185(2)		(1.0+6.5)	58.(12)	5.80(9)	(0.028+0.3)
^{138}Ce	0.251(2)		(0.025+1.0)	(1.5+5.2)	6.70(9)	179.(5) mb
^{140}Ce	88.450(51)		0.58(4)	0.50(5)	4.84(9)	11.0(4) mb
^{141}Ce		32.50 d	29.(3)	13.(2)		
^{142}Ce	11.114(51)	>1.6×10^{17} y	0.97(3)	1.3(3)	4.75(9)	28.(1) mb
^{143}Ce		1.38 d	6.1(7)	2.7(3)		
^{144}Ce		284.6 d	1.0(1)	2.6(3)		
$_{59}$Pr			11.5(4)	14.(3)	4.58(5)	
^{141}Pr	100.		(4.+7.5)	14.(3)	4.58(5)	111.(2) mb
^{142}Pr		19.12 h	20.(3)	9.(1)		
^{143}Pr		13.57 d	90.(10)	190.(25)		
$_{60}$Nd			51.(2)	49.(5)	7.69(5)	
^{142}Nd	27.2(5)		19.(1)	34.(11)	7.7(3)	35.(1) mb
^{143}Nd	12.2(2)		330.(10) σ_α = 17. mb	128.(30)		0.24(1)
^{144}Nd	23.8(3)	2.1×10^{15} y	3.6(3)	3.9(5)	2.8(3)	81.(2) mb
^{145}Nd	8.3(1)		47.(6) σ_α = 12. μb	260.(40)		0.42(1)
^{146}Nd	17.2(3)		1.5(2)	3.0(4)	8.7(2)	91.(1) mb
^{147}Nd		10.98 d	440.(150)	200.		
^{148}Nd	5.7(1)		2.4(1)	13.(2)	5.7(3)	147.(2) mb
^{150}Nd	5.6(2)	≈ 1×10^{19} y	1.0(1)	14.(2)	5.3(2)	0.16(1)
$_{61}$Pm						
^{146}Pm		5.53 y	8.4(1.7)×10^{3}			
^{147}Pm		2.623 y	(84.+96.)	(1000.+1280.)	12.6(4)	2.(1)
^{148m}Pm		41.3 d	10600.(800)			
^{148}Pm		5.37 d	≈ 10^{3}	2.6(2.4)×10^{3}		
^{149}Pm		2.212 d	1400.(200)			
^{151}Pm		1.183 d	≈ 150.			
$_{62}$Sm			5.6(1)×10^{3}	1.4(2)×10^{3}		
^{144}Sm	3.07(7)		1.6(1)	2.4(3)		92.(6) mb
^{145}Sm		340. d	280.(20)	600.(90)		
^{147}Sm	14.99(18)	1.06×10^{11} y	56.(4), σ_α = 0.6 mb	710.(50)	14.(3)	0.97(1)
^{148}Sm	11.24(10)	7×10^{15} y	2.4(6)	27.(14)		241.(2) mb
^{149}Sm	13.82(7)	10^{16} y	4.01(6)×10^{4}, σ_α = 31. mb	3.1(5)×10^{3}		1.82(2)
^{150}Sm	7.38(1)		102.(5)	290.(30)	14.(3)	422.(4) mb
^{151}Sm		90. y	1.52(3)×10^{4}	3520.(60)		2.(1)
^{152}Sm	26.75(16)		206.(15)	3.0(3)×10^{3}	5.0(6)	473.(4) mb
^{153}Sm		1.929 d	420.(180)			
^{154}Sm	22.75(29)		7.5(3)	32.(6)	9.(1)	0.21(1)
$_{63}$Eu			4570.(100)	3.8(5)×10^{3}	5.3(3)	
^{151}Eu	47.81(6)		(4.+3150.+6000.) σ_α = 8.7(3) μb	(2.+4.)×10^{3}		(1.6+2.2)
^{152m1}Eu		9.30 h	6.8(15)×10^{4}	< 10^{5}		

Elem. or Isot.	Natural abundance (%)	Half-life	Thermal neut. cross section (barns)	Resonance integral (barns)	Coh. scat. length (fm)	σ (30 keV) Maxw. avg. (barns)
^{152}Eu		13.5 y	$1.1(2)\times10^{4}$	$1.6(2)\times10^{3}$		5.(2)
^{153}Eu	52.19(6)		300.(20), σ_α <1. μb	$1.8(4)\times10^{3}$	8.2(1)	2.8(1)
^{154}Eu		8.59 y	$1.5(3)\times10^{3}$	$1.6(2)\times10^{3}$		4.4(7)
^{155}Eu		4.76 y	$3.9(2)\times10^{3}$	$1.6(2)\times10^{4}$		1.3(1)
$_{64}$Gd			$48.8(6)\times10^{3}$	400.(10)	9.5(2)	
^{148}Gd		75. y	$1.40(14)\times10^{4}$			
^{152}Gd	0.20(1)	1.1×10^{14} y	700.(200), σ_α <7. mb	700.(200)		1.05(2)
^{153}Gd		240. d	$2.(1)\times10^{4}$, σ_α = 0.03			
^{154}Gd	2.18(3)		(0.035+60.)	230.(50)		1.03(1)
^{155}Gd	14.80(12)		$61.(1)\times10^{3}$, σ_α = .08 mb	1540.(100)		2.65(3)
^{156}Gd	20.47(9)		≈ 2.0	104.(15)	6.3(4)	615.(5) mb
^{157}Gd	15.65(2)		$2.54(3)\times10^{5}$, σ_α <0.05	800.(100)		1.37(2)
^{158}Gd	24.84(7)		2.3(3)	73.(7)	9.(2)	324.(3) mb
^{160}Gd	21.86(19)	>1.9×10^{19} y	1.5(7)	6.(1)	9.15(5)	0.15(2)
^{161}Gd		3.66 m	$2.0(6)\times10^{4}$			
$_{65}$Tb			23.2(5)	420.(50)	7.34(2)	
^{159}Tb	100.		23.2(5)	420.(50)	7.34(2)	1.6(2)
^{160}Tb		72.3 d	570.(110)			
$_{66}$Dy			$9.5(2)\times10^{2}$	$1.5(2)\times10^{3}$	16.9(3)	
^{156}Dy	0.056(3)		33.(3), σ_α < 9. mb	1000.(100)		1.6(2)
^{158}Dy	0.095(3)		43.(6), σ_α < 6. mb	120.(10)	6.1(5)	0.8(2)
^{159}Dy		144. d	$8.(2)\times10^{3}$			
^{160}Dy	2.39(18)		60.(10), σ_α < 0.3 mb	1100.(200)	6.7(4)	0.89(1)
^{161}Dy	18.889(42)		600.(50), σ_α < 1. μb	1100.(100)	10.3(4)	1.96(2)
^{162}Dy	25.475(36)		170.(20)	2755.(300)	- 1.4(5)	446.(4) mb
^{163}Dy	24.896(42)		120.(10), σ_α < 20. μb	1600.(400)	5.0(4)	1.11(1)
^{164}Dy	28.260(54)		$(1.7+1.0)\times10^{3}$	$(4.+2.)\times10^{2}$	49.4(2)	212.(3) mb
^{165m}Dy		1.26 m	$2.0(6)\times10^{3}$			
^{165}Dy		2.33 h	$3.5(3)\times10^{3}$	$2.2(3)\times10^{4}$		
$_{67}$Ho			61.(2)	670.(40)	8.01(8)	
^{163}Ho		4.57×10^{3} y				(0.4+1.7)
^{165}Ho	100.		(3.1+58.), σ_α < 20. μb	(?+670.)	8.01(8)	(0.8+0.5)
^{166m}Ho		1.2×10^{3} y	$3.1(8)\times10^{3}$	$10.(3)\times10^{3}$		
$_{68}$Er			$1.5(2)\times10^{2}$	730.(10)	7.79(2)	
^{162}Er	0.139(5)		19.(3), σ_α < 11. mb	480.(50)	8.8(2)	1.6(1)
^{164}Er	1.601(3)		13.(3), σ_α < 1.2 mb	105.(10)	8.2(2)	1.08(5)
^{166}Er	33.503(36)		(3.+14.), σ_α < 70. μb	96.(12)	10.6(2)	0.56(6)
^{167}Er	22.869(9)		$6.5(8)\times10^{2}$, σ_α = 3. μb	2970.(70)	3.0(3)	1.4(2)
^{168}Er	26.978(18)		2.3(3), σ_α = 0.09 mb	37.(5)	7.4(4)	0.34(4)
^{170}Er	14.910(36)		8.(2)	26.(4)	9.6(5)	0.17(1)
^{171}Er		7.52 h	370.(40)	170.(20)		
$_{69}$Tm			108.(4)	$1.5(2)\times10^{3}$	7.07(3)	
^{169}Tm	100		(8.+100.)	$1.5(2)\times10^{3}$	7.07(3)	1.13(6)
^{170}Tm		128.6 d	100.(20)	460.(50)		
^{171}Tm		1.92 y	≈ 160.	118.(6)		
$_{70}$Yb			52.(10)	$1.7(2)\times10^{2}$	12.43(3)	
^{168}Yb	0.13(1)		$2.4(2)\times10^{3}$, σ_α < 0.1 mb	$2.0(5)\times10^{4}$	-4.07(2)	0.7(4)
^{169}Yb		32.02 d	$3.6(3)\times10^{3}$	5200.(500)		
^{170}Yb	3.04(15)		12.(2), σ_α < 10. μb	320.(30)	6.8(1)	0.77(1)
^{171}Yb	14.28(57)		53.(5), σ_α < 1.5 μb	315.(30)	9.7(1)	1.21(1)
^{172}Yb	21.83(67)		≈ 1.3, σ_α < 1. μb	25.(3)	9.4(1)	0.34(1)
^{173}Yb	16.13(27)		16.(2), σ_α < 1. μb	380.(30)	9.56(7)	0.75(1)
^{174}Yb	31.83(92)		(46.+17.), σ_α < 0.02 mb	(13.+16.)	19.3(1)	151.(2) mb
^{176}Yb	12.76(41)		3.1(2), σ_α < 1. μb	8.(2)	8.7(1)	116.(2) mb
$_{71}$Lu			78.(7)	$8.3(7)\times10^{2}$	7.21(3)	
^{175}Lu	97.41(2)		(16.+8.)	(550.+270.)	7.24(3)	(1.04+0.11)
^{176}Lu	2.59(2)	3.73×10^{10} y	(2.+2100.)	(3.+930.)	6.1(2)	1.53(7)
^{177m}Lu		160.7 d	3.2(3)	1.4(2)		
^{177}Lu		6.65 d	1000.(300)			

Elem. or Isot.	Natural abundance (%)	Half-life	Thermal neut. cross section (barns)	Resonance integral (barns)	Coh. scat. length (fm)	σ (30 keV) Maxw. avg. (barns)
$_{72}$Hf			106.(3)	19.7(5)×10^2	7.8(1)	
^{174}Hf	0.16(1)	2.0×10^{15} y	600.(50)	400.(50)	11.(1)	0.8(2)
^{176}Hf	5.26(7)		23.(4)	700.(100)	6.6(2)	0.46(2)
^{177}Hf	18.60(9)		(1.+375.), σ_α < 20. μb	7170.(200)		1.5(1)
^{178m2}Hf		31. y	σ_{m2} = 45.(5)	RI$_{m2}$ = 8(1)×10^2		
^{178}Hf	27.28(7)		(54.+32.)	(0.9+1.0)×10^3	5.9(2)	0.31(1)
^{179}Hf	13.62(2)		(0.43+46.)	(6.8+620.)	7.5(2)	(0.01+0.95)
^{180}Hf	35.08(16)		13.0(5), σ_α < 13. μb	32.(1)	13.2(3)	179.(5) mb
^{181}Hf		42.4 d	30.(25)			
$_{73}$Ta			20.(1)	650(20.)	6.91(7)	
^{179}Ta		1.8 y	9.3(6)×10^2	1.22(7)×10^3		
^{180m}Ta	0.012(2)	> 1.2×10^{15} y	≈ 560.	1350.(100)		
^{181}Ta	99.988(2)		(0.012 + 20.), σ_α <1. μb	(0.4+650.)	6.91(7)	0.77(2)
^{182}Ta		114.43 d	8200.(600)	900.(90)		
$_{74}$W			18.(1)	3.6(3)×10^2	4.86(2)	
^{180}W	0.12(1)	7.4×10^{16} y	≈ 4.	210.(30)		0.54(6)
^{182}W	26.50(16)	8.3×10^{18} y	20.(1)	600.(90)	6.97(4)	274.(8) mb
^{183}W	14.31(4)	1.9×10^{18} y	10.5(3)	340.(50)	6.53(4)	0.52(2)
^{184}W	30.64(2)	4.0×10^{18} y	(0.002 + 2.0)	15.(2)	7.48(6)	0.22(1)
^{185}W		74.8 d	≈ 3.3	300.(50)		
^{186}W	28.43(19)	6.5×10^{18} y	37.(2)	510.(50)	0.72(4)	176.(5) mb
^{187}W		23.9 h	70.(10)	2760.(550)		
^{188}W		69.78 h	12.(1)			
$_{75}$Re			90.(4)	8.4(2)×10^2	9.2(3)	
^{185}Re	37.40(2)		(0.33+110.)	1700.(50)	9.0(3)	1.54(6)
^{187}Re	62.60(2)	4.2×10^{10} y	(2.+72.)	(9.+310.)	9.3(3)	1.16(6)
$_{76}$Os			17.(1)	1.5(1)×10^2	10.7(2)	
^{184}Os	0.02(1)	>5.6×10^{13} y	3.3(3)×10^3, σ_α <10. mb	1.4(1)×10^3		0.4(2)
^{186}Os	1.59(3)	2.×10^{15} y	≈ 80., σ_α < 0.1 mb	3.8(9)×10^2	12(2)	0.42(2)
^{187}Os	1.96(2)		2.(1)×10^2, σ_α < 0.1 mb	5.0(7)×10^2		0.90(3)
^{188}Os	13.24(8)		≈ 5., σ_α < 30. μb	1.5(2)×10^2	7.6(3)	0.40(2)
^{189}Os	16.15(5)		(0.00026+40.),σ_α<10. μb	(0.013+670.)	10.7(3)	1.17(5)
^{190}Os	26.26(2)		(9.+4.), σ_α < 20. μb	(22.+8.)	11.0(3)	0.30(5)
^{191}Os		15.4 d	3.8(6)×10^2	1.7(3)×10^2		
^{192}Os	40.78(19)		3.(1), σ_α < 10. μb	7.(1)	11.5(4)	0.31(5)
^{193}Os		30.5 h	2.5(5)×10^2	1.1(2)×10^2		
$_{77}$Ir			4.2(1)×10^2	2.8(4)×10^3	10.6(3)	
^{191}Ir	37.3(2)		(0.14+660.+260.)	(1.0+4.2)×10^3		1.35(4)
^{192}Ir		73.83 d	1.4(3)×10^3	4.8(7)×10^3		
^{193}Ir	62.7(2)		(0.04+6.+109.)	1.4(2)×10^3		0.99(7)
^{194}Ir		19.3 h	1.6(3)×10^3	7.(2)×10^2		
$_{78}$Pt			10.(1)	1.3(1)×10^2	9.60(1)	
^{190}Pt	0.014(1)	4.5×10^{11} y	1.5(1)×10^2, σ_α < 8. mb	70.(10)	9.(1)	0.7(2)
^{192}Pt	0.782(7)		(2.0+6.), σ_α < 0.2 mb	115.(20)	9.9(5)	0.6(1)
^{194}Pt	32.967(99)		(0.1+1.1), σ_α < 5. μb	(4.+?)	10.55(8)	(0.03+0.34)
^{195}Pt	33.832(10)		28.(1), σ_α < 5. μb	365.(50)	8.8(1)	0.9(2)
^{196}Pt	25.242(41)		(0.045+0.55)	7.(2)	9.89(8)	(0.01+0.19)
^{198}Pt	7.163(55)		(0.3+3.1)	(5.+53.)	7.8(1)	(3.+79.) mb
^{199}Pt		30.8 m	≈ 15.	≈ 7.		
$_{79}$Au			98.7(1)	1.55(3)×10^3	7.63(6)	
^{197}Au	100.		σ_{m+g} = 98.7(1) σ_m = 8.(2) mb	RI$_{m+g}$ = 1.55(3)×10^3 RI$_m$ = 0.06(2)	7.63(6)	582.(9) mb
^{198}Au		2.695 d	26.5(15)×10^3	≈ 4.×10^4		
^{199}Au		3.14 d	≈ 30.			
$_{80}$Hg			3.7(1)×10^2	87.(5)	12.69(2)	
^{196}Hg	0.15(1)	>2.5×10^{18} y	(105.+3000.)	(53.+410.)	30.(1)	0.4(2)
^{198}Hg	9.97(8)		(0.017+2.)	(1.7+70.)		0.17(2)
^{199}Hg	16.87(10)		2.1(2)×10^3	435(20)	16.9(4)	0.37(2)
^{200}Hg	23.10(16)		≈ 1.	2.1(5)		0.12(1)

Elem. or Isot.	Natural abundance (%)	Half-life	Thermal neut. cross section (barns)	Resonance integral (barns)	Coh. scat. length (fm)	σ (30 keV) Maxw. avg. (barns)
^{201}Hg	13.18(8)		≈ 8.	30.(3)		0.26(1)
^{202}Hg	29.86(20)		4.9(5)	4.5(2)	11.(1)	74.(6) mb
^{204}Hg	6.87(4)		0.4(1)	0.8(2)		42.(4) mb
$_{81}$Tl			3.3(1)	12.5(8)	8.776(5)	
^{203}Tl	29.524(14)		11.(1), σ_α < 0.3 mb	41.(2)	7.0(2)	124.(8) mb
^{204}Tl		3.78 y	22.(2)	90.(20)		0.14(5)
^{205}Tl	70.476(14)		0.11(2)	0.6(2)	9.52(7)	54.(4) mb
$_{82}$Pb			0.172(2)	0.14(4)	9.402(2)	
^{204}Pb	1.4(1)		0.68(7)	2.0(2)	10.9(1)	90.(6) mb
^{205}Pb		1.51×10^7 y	≈ 5.	≈ 2.		0.06(1)
^{206}Pb	24.1(1)		0.027(1)	0.10(1)	9.23(5)	16.(1) mb
^{207}Pb	22.1(1)		0.61(3)	0.38(1)	9.28(2)	10.(1) mb
^{208}Pb	52.4(1)	>2×10^{19} y	0.23(1) mb, σ_α < 8. μb	2.0(2) mb	9.50(3)	0.36(4) mb
^{210}Pb		22.6 y	< 0.5			
$_{83}$Bi			0.034(1)	0.19(2)	8.532(2)	
^{209}Bi	100.		(11.+23.) mb,σ_α<0.3 μb	0.19(2)	8.532(2)	2.7(5) mb
^{210m}Bi		3.0×10^6 y	54.(4) mb	0.20(3)		
$_{84}$Po						
^{210}Po		138.4 d	σ_m<0.5 mb, σ_α < 2. mb			
			σ_g<30. mb, σ_f< 0.1			
$_{85}$At						
$_{86}$Rn						
^{220}Rn		55.6 s	<0.2			
^{222}Rn		3.823 d	0.74(5)			
$_{88}$Ra						
^{223}Ra		11.43 d	1.3(2)×10^2, σ_f< 0.7			
^{224}Ra		3.66 d	12.0(5)			
^{226}Ra		1599. y	≈ 13., σ_f< 7. μb	280.(50)	10.(1)	
^{228}Ra		5.76 y	36.(5), σ_f< 2.			
$_{89}$Ac						
^{227}Ac		21.77 y	8.8(7)×10^2, σ_f< 0.35 mb	1.5(4)×10^3		
$_{90}$Th			7.4	85.(3)	10.31(3)	
^{227}Th		18.72 d	σ_f = 2.0(2)×10^2			
^{228}Th		1.913 y	1.2(2)×10^2, σ_f<0.3	1014.(400)		
^{229}Th		7.9×10^3 y	≈ 60.	1.0(2)×10^3		
			σ_f = 30.(3)	RI$_f$ = 466.(75)		
^{230}Th		7.54×10^4 y	23.4(5)	1.0(1)×10^3		
			σ_f< 0.5 mb			
^{232}Th	100.	1.40×10^{10} y	7.37(4)	85.(3)	10.31(3)	
			$\sigma_{f=}$ = 3.(1) μb			
			σ_α < 1. μb			
^{233}Th		22.3 m	1.5(1)×10^3	4.(1)×10^2		
			σ_f = 15.(2)			
^{234}Th		24.10 d	1.8(5)			
			σ_f< 0.01			
$_{91}$Pa						
^{230}Pa		17.4 d	1.5(3)×10^3			
^{231}Pa		3.25×10^4 y	2.0(1)×10^2	750.(80)	9.1(3)	
			σ_f = 20.(1) mb	RI$_f$ = 0.05(1)		
^{232}Pa		1.31 d	4.6(10)×10^2	300.(70)		
			σ_f = 1.5(5)×10^3	RI$_f$ = 1.0(1)×10^3		
^{233}Pa		27.0 d	39.(2)	(460.+440.)		
			σ_m = 20.(4)			
			σ_g = 19.(3)			
			σ_f< 0.1			
$_{92}$U			3.4(3); σ_f = 4.2(1)	280.(20),RI$_f$ = 2.0	8.417(5)	
^{230}U		20.8 d	σ_f ≈ 25.			
^{231}U		4.2 d	σ_f ≈ 250.			

Elem. or Isot.	Natural abundance (%)	Half-life	Thermal neut. cross section (barns)	Resonance integral (barns)	Coh. scat. length (fm)	σ (30 keV) Maxw. avg. (barns)
^{232}U		70. y	73.(2)	280.(15)		
			σ_f = 74.(8)	RI_f = 350.(30)		
^{233}U		1.592×10^5 y	47.(2)	137.(6)	10.1(2)	
			σ_f = $5.3(1)\times10^2$	RI_f = 760.(17)		
			σ_α < 0.2 mb			
^{234}U	0.0054(5)	2.455×10^5 y	96.(2)	660.(70)	12.(4)	
			σ_f = 0.07(2)	RI_f = 6.5		
^{235}U	0.7204(6)	7.04×10^8 y	95.(5)	144.(6)	10.47(4)	
			σ_f = 586.(2)	RI_f = 275(5)		
			σ_α < 0.1 mb			
^{236}U		2.342×10^7 y	5.1(3)	360.(15)		
			σ_f< 1.3 mb	RI_f = 4.38(50)		
^{237}U		6.75 d	≈ 10^2	1200.(200)		
			σ_f< 0.35			
^{238}U	99.2742(10)	4.47×10^9 y	2.7(1)	277.(3)	8.402(5)	
			σ_f≈ 3. μb	1.54(15) mb		
			σ_α = 1.4(5) μb			
^{239}U		23.5 m	22.(2)			
			σ_f= 15.(3)			
$_{93}$Np						
^{234}Np		4.4 d	σ_f= $9.(3)\times10^2$			
^{235}Np		1.085 y	$1.6(1)\times10^2$			
^{236m}Np		22.5 h	σ_f = $2.7(2)\times10^3$	$7.(4)\times10^2$		
^{236}Np		1.55×10^5 y	$\sigma_{f=}$ $3.0(2)\times10^3$	$1.35(30)\times10^3$		
^{237}Np		2.14×10^6 y	$1.7(1)\times10^2$	$6.5(3)\times10^2$	10.6(1)	
			σ_f= 20.(1) mb	RI_f = 4.7		
^{238}Np		2.117 d	σ_f = $2.6(3)\times10^3$	$1.4(3)\times10^3$		
^{239}Np		2.355 d	(32.+19.)			
			σ_f< 1.			
$_{94}$Pu						
^{236}Pu		2.87 y	σ_f = $1.6(3)\times10^2$	1000.(60)		
^{237}Pu		45.7 d	σ_f = $2.3(3)\times10^3$			
^{238}Pu		87.7 y	$5.1(2)\times10^2$	$1.6(2)\times10^2$	14.1(5)	
			σ_f = 17.(1)	RI_f = 26.(2)		
^{239}Pu		2.410×10^4 y	$2.7(1)\times10^2$	$2.0(2)\times10^2$	7.7(1)	
			σ_f = 752.(3)	$3.0(1)\times10^2$		
			σ_α ≤ 0.3 mb			
^{240}Pu		6.56×10^3 y	$2.9(1)\times10^2$	$8.4(3)\times10^3$	3.5(1)	
			σ_f≈ 59. mb	RI_f = 3.2		
^{241}Pu		14.4 y	$3.7(1)\times10^2$, σ_α <0.2 mb	$1.6(1)\times10^2$		
			σ_f = $1.01(1)\times10^3$	$5.7(4)\times10^2$		
^{242}Pu		3.75×10^5 y	19.(1)	$1.1(1)\times10^3$	8.1(1)	
			σ_f<0.2	RI_f = 0.23		
^{243}Pu		4.956 h	<100.			
			σ_f = $2.0(2)\times10^2$			
^{244}Pu		8.00×10^7 y	1.7(1)	41.(3)		
^{245}Pu		10.5 h	$1.5(3)\times10^2$	220.(40)		
$_{95}$Am						
^{241}Am		432.7 y	$(0.6+6.4)\times10^2$	$(1.+14.)\times10^2$		
			σ_f = 3.15(10)	14.(1)		
^{242m}Am		141. y	$1.7(4)\times10^3$	≈ 200.		
			σ_f = $5.9(3)\times10^3$	RI_f = $1.8(1)\times10^3$		
^{242}Am		16.02 h	σ_f = $2.1(2)\times10^3$	RI_f = < 300.		
			$3.3(5)\times10^2$	≈ 1.5×10^2		
^{243}Am		7.37×10^3 y	(75.+5.)	$(17.1+1.0)\times10^2$	8.3(2)	
			σ_f = 79.(2) mb	RI_f = 0.056		
^{244m}Am		≈ 26. m	σ_f = $1.6(3)\times10^3$			
^{244}Am		10.1 h	σ_f = $2.2(3)\times10^3$			
$_{96}$Cm						
^{242}Cm		162.8 d	≈ 20.	120.(50)		

Elem. or Isot.	Natural abundance (%)	Half-life	Thermal neut. cross section (barns)	Resonance integral (barns)	Coh. scat. length (fm)	σ (30 keV) Maxw. avg. (barns)
			$\sigma_f \approx 5.$			
^{243}Cm		29.1 y	1.3(1)×10^2	214.(20)		
			σ_f = 6.2(2)×10^2	RI_f = 1.6(1)×10^3		
^{244}Cm		18.1 y	15.(1)	640.(50)	9.5(3)	
			σ_f = 1.1(2)	RI_f = 10.8(8)		
^{245}Cm		8.48×10^3 y	3.5(2)×10^2	110.(10)		
			σ_f = 2.1(1)×10^3	RI_f = 8.(1)×10^2		
^{246}Cm		4.76×10^3 y	1.2(2)	120.(10)	9.3(2)	
			σ_f = 0.16(7)	13.(2)		
^{247}Cm		1.56×10^7 y	60.(30)	5.(1)×10^2		
			σ_f = 82.(5)	7.3(7)×10^2		
^{248}Cm		3.48×10^5 y	2.6(3)	270.(30)	7.7(2)	
			σ_f = 0.36(7)	13.(2)		
^{249}Cm		64.15 m	≈ 1.6			
^{250}Cm		≈ 9.7×10^3 y	≈ 80.			
$_{97}$Bk						
^{249}Bk		320. d	7.(1)×10^2	9.(1)×10^2		
			$\sigma_f \approx$ 0.1			
^{250}Bk		3.217 h	σ_f = 1.0(2)×10^3			
$_{98}$Cf						
^{249}Cf		351. y	5.0(3)×10^2	7.7(4)×10^2		
			σ_f = 1.7(1)×10^3	RI_f = 2.1(3)×10^3		
^{250}Cf		13.1 y	2.0(2)×10^3	12.(2)×10^3		
			σ_f = 110.(90)	RI_f = 160.(40)		
^{251}Cf		9.0×10^2 y	2.9(2)×10^3	1.6(1)×10^3		
			σ_f = 4.5(5)×10^3	RI_f = 5.5(3)×10^3		
^{252}Cf		2.65 y	20.(2)	43.(3)		
			σ_f = 32.(4)	RI_f = 1.1(3)×10^2		
^{253}Cf		17.8 d	18.(2)	8.(1)		
			σ_f = 1.3(2)×10^3			
^{254}Cf		60.5 d	4.5(10)	2.		
$_{99}$Es						
^{253}Es		20.47 d	(180.+5.8)	(37.5+1.1)×10^2		
^{254m}Es		1.64 d	σ_f = 1.8(1)×10^3			
^{254}Es		276. d	28.(3)	18.(2)		
			σ_f = 1.8(2)×10^3	RI_f = 1.2(3)×10^3		
^{255}Es		40. d	≈ 55.			
$_{100}$Fm						
^{255}Fm		20.1 h	26.(3)	14.(2)		
			σ_f = 3.3(2)×10^3			
^{257}Fm		100.5 d	σ_f = 3.0(2)×10^3			

* Extrapolated value.

COSMIC RADIATION

A. G. Gregory and R. W. Clay

The Nature of Cosmic Rays

Primary cosmic radiation, in the form of high-energy nuclear particles, electrons and photons from outside the solar system and from the Sun, continually bombards our atmosphere. Secondary radiation, resulting from the interaction of the primary cosmic rays with atmospheric gas, is present at sea level and throughout the atmosphere.

Secondary radiation is collimated by absorption and scattering in the atmosphere and consists of a number of components associated with different particle species. High-energy primary particles can produce large numbers of secondary particles forming an extensive air shower. Thus, a number of particles may then be detected simultaneously at sea level.

Primary particle energies accessible in the vicinity of the Earth range from $\sim10^8$ eV to $\sim10^{20}$ eV. At the lower energies, the limit is determined by the inability of charged particles to traverse the heliosphere to us through the outward-moving solar wind. The upper energy limit is set by the practicality of building detectors to record particles with the extremely low fluxes found at those energies (O. C. Allkofer, 1975a; J. G. Wilson, 1976).

Primary Cosmic Rays

Primary Particle Energy Spectrum

Figure 1 shows the spectrum of primary particle energies. This includes all particle species. In differential form it is roughly a power law of intensity versus energy with an index of ~ -3. There appears to be a knee (a steepening) at a little above 10^{15} eV and an ankle (a flattening) above $\sim10^{18}$ eV. Figure 2 emphasizes the features in the spectrum at the highest energies through multiplying the flux with a strongly rising power law of energy. This figure should be used with caution as errors for the two axes are not now independent.

Data on the high-energy cosmic ray spectrum are uncertain largely because of limited event statistics due to the very low flux which might best be measured in particles per square kilometer per century. The highest energy event recorded to 1995 had an energy of 3×10^{20} eV (D. J. Bird et al., 1993).

It is expected that the highest energy cosmic rays will interact with the 2.7 K cosmic microwave background through photoproduction or photodisintegration. These interactions will appreciably reduce the observed flux of cosmic rays with energies above 5×10^{19} eV if they travel further than ~150 million light-years. This process is known as the Greisen-Zatsepin-Kuz'min (GZK) cutoff (P. Sokolsky, 1989).

At energies below $\sim10^{13}$ eV, solar system magnetic fields and plasma can modulate the primary component and Figure 3 shows the extent of this modulation between solar maximum and minimum (E. Juliusson, 1975; J. Linsley, 1981).

Primary Particle Energy Density

If the above spectrum is corrected for solar effects, the energy density above a particle energy of 10^9 eV outside the solar system is found to be $\sim5 \times 10^5$ eV m^{-3}. As the threshold energy is increased, the energy density decreases rapidly, being 2×10^4 eV m^{-3} above 10^{12} eV and 10^2 eV m^{-3} above 10^{15} eV. The energy density at lower energies outside the heliosphere is unknown but may be substantially greater if the particle rest mass energy is included together with the kinetic energy (A. W. Wolfendale, 1979).

Primary Particle Isotropy

This is measured as an anisotropy $(I_{max} - I_{min})/(I_{max} + I_{min}) \times 100\%$, where I, the intensity (m^{-2}s^{-1}sr^{-1}), is usually measured with an angular resolution of a few degrees.

The measured anisotropy is small and energy dependent. It is roughly constant in amplitude at between 0.05 and 0.1% (with a phase of 0 to 6 hours in right ascension) for energies between 10^{11} eV and 10^{14} eV and appears to increase at higher energies roughly as $0.4 \times (\text{Energy(eV)}/10^{16})^{0.5}$ % up to $\sim10^{18}$ eV. The latter rise may well be an artifact of the progressively more limited statistics as the flux drops rapidly with energy. It appears possible that a real anisotropy has been observed at the highest energies (above a few times 10^{19} eV) with a directional preference for the supergalactic plane (this plane reflects the directions of galaxies within about 100 million light-years) (A. W. Wolfendale, 1979; R. W. Clay, 1987; T. Stanev et al., 1995).

Primary Particle Composition

The composition of low-energy cosmic rays is close to universal nuclear abundances except where propagation effects are present. For example, Li, Be, and B which are spallation products, are over-abundant by about six orders of magnitude.

Composition at 10^{11} eV per Nucleus

Charge	1	2	(3–5)	(6–8)	(10–14)	(16–24)	(26–28)	≥30
% Composition	50	25	1	12	7	4	4	0.1

Measurements at higher energies indicate that there is an increase in the relative abundances of nuclei with charges greater than 6 at energies above 50 TeV/nucleus (K. Asakimori et al., 1993) (1 TeV = 10^{12} eV).

Cosmic ray composition at low energies is often quoted at a fixed energy per nucleon. When presented in this way, protons constitute roughly 90% of the flux, helium nuclei about 10%, and the remainder sum to a total of about 1%.

Certain radioactive isotopic ratios show lifetime effects. The ratio of Be^{10}/B^9 abundances is used to measure an "age" of cosmic rays since Be^{10} is unstable with a half-life of about 1.6×10^6 years. A ratio of 0.6 is expected in the absence of Be^{10} decay and a ratio of about 0.2 is found experimentally (E. Juliusson, 1975; P. Meyer, 1981).

At higher energies, composition determinations are indirect and are rather contradictory and controversial. Experiments aim to differentiate between broad composition models. The measurement technique is based on studies of cosmic ray shower development. A rather direct technique for such studies is to use fluorescence observations of the shower development to determine the atmospheric depth of maximum development of the shower. Such observations suggest a heavy composition (large atomic number) at energies $\sim10^{17}$ eV which changes with increasing energy to a light composition (perhaps protonic) above $\sim10^{19}$ eV (T. K. Gaisser et al., 1993).

Primary Electrons

Primary electrons constitute about 1% of the cosmic ray beam. The positron to negative electron ratio is about 10% (J. M. Clem et al., 1995).

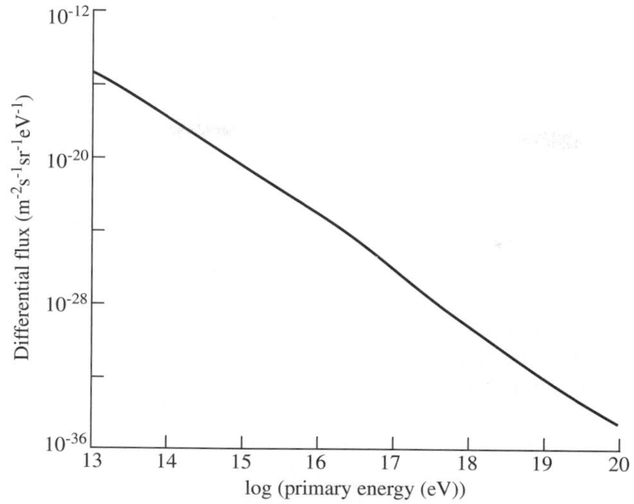

FIGURE 1. The energy spectrum of cosmic ray particles. This spectrum is of a differential form and can be converted to an integral spectrum by integration over all energies above a required threshold (E). Insofar as the spectrum approximates a power law of index −3, a simple conversion to the integral at an energy $E/1.8$ is obtained by multiplying the differential flux by the energy and dividing by 0.62.

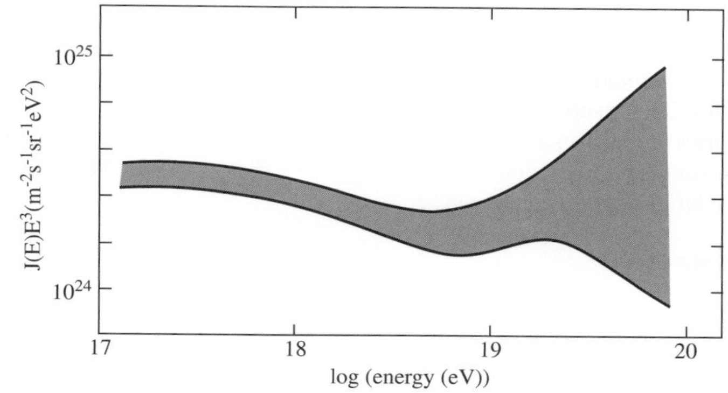

FIGURE 2. Energy spectrum at the highest energies. This spectrum (after Yoshida et al., 1995) has the differential spectrum multiplied by energy cubed. It is from a compilation of a number of measurements and indicates the good general agreement at the lower energies and a spread due to inadequate statistics at the highest energies.

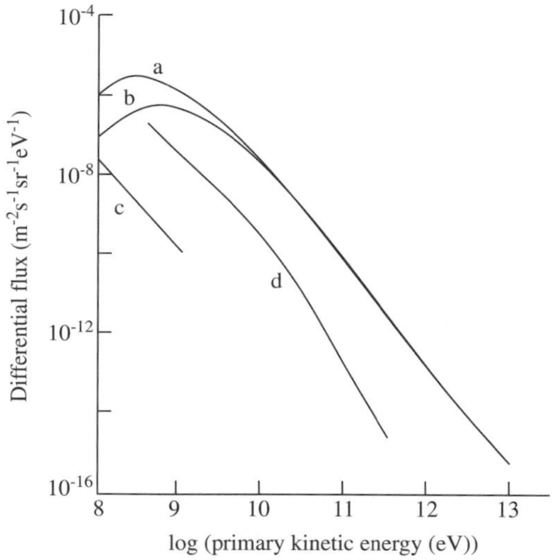

FIGURE 3. Energy spectrum of particles at lower energies. (a) Solar minimum proton energy spectrum. (b) Solar maximum proton energy spectrum. (c) Gamma ray energy spectrum. (d) Local interstellar electron spectrum.

Antimatter in the Primary Beam

The ratio of antiprotons to protons in the primary cosmic ray beam (at about 400 MeV) is about 10^{-5}. At about 10 GeV the ratio is about 10^{-3}. At the highest measured energies (10 TeV), the upper limit to the ratio is about 20% (M. Amenomori et al., 1995; S. Orito et al., 1995).

Primary Gamma Rays

The flux of primary gamma rays is low at high energies. At 1 GeV the ratio of gamma rays to protons is about 10^{-6}. The arrival directions of these gamma rays are strongly concentrated in the plane of the Milky Way although there is a diffuse, near isotropic background flux and some point sources have been detected.

Since the absorption cross section for gamma rays above 100 MeV is approximately 20 mbarn/electron, less than 10% of gamma rays reach mountain altitudes (A. W. Wolfendale, 1979; P. F. Michelson, 1994).

Sea-Level Cosmic Radiation

The sea-level cosmic ray dose is 300 millirad·yr^{-1} and the sea-level ionization is 2.2×10^6 ion pairs m^{-3}s^{-1}. The sea-level flux has a soft component, which can be absorbed in about 100 mm of lead (about 100 g·cm^{-2} of absorber) and a more penetrating (largely muon) hard component. The sea-level radiation is largely produced in the atmosphere and is a secondary component from interactions of the primary particles. The steep primary energy spectrum means that most secondaries at sea level are from rather low-energy primaries. Thus, the secondary flux is dependent on the solar cycle and the geomagnetic latitude of the observer.

Absolute Flux of the Hard Component

Vertical Integral Intensity $I(0)$ ~100 m^{-2} s^{-1} sr^{-1}
Angular Dependence $I(\theta) \sim I(0)\cos^2(\theta)$
Integrated Intensity ~200 m^{-2} s^{-1}
(O. C. Allkofer, 1975b)

Flux of the Soft Component

In free air, the soft component comprises about one-third of the total cosmic ray flux.

Latitude Effect

The geomagnetic field influences the trajectories of lower-energy cosmic rays approaching the Earth. As a result, the background flux is reduced by about 7% at the geomagnetic equator. The effect decreases toward the poles and is negligible at latitudes above about 40°.

Flux of Protons

The proton component is strongly attenuated by the atmosphere with an attenuation length (reduction by a factor of e) of about 120 g·cm^{-2}, constituting about 1% of the total vertical sea-level flux.

Absorption

The soft component is absorbed in about 100 g·cm^{-2} of matter. The hard component is absorbed much more slowly:

Absorption in lead, 6% per 100 g·cm^{-2}
Absorption in rock, 8.5% per 100 g·cm^{-2}
Absorption in water, 10% per 100 g·cm^{-2}
(Absorption for depths less than 100 g·pd cm^{-2} is given by K. Greisen, 1943)

Altitude Dependence

The cosmic ray background in the atmosphere has a maximum intensity of about 15 times that at sea level at a depth of about 150 g·cm^{-2} (15 km altitude). At maximum intensity, the soft and hard components contribute roughly equally but the hard component is then attenuated more slowly (S. Hayakawa, 1969).

Cosmic Ray Showers

High-energy cosmic rays produce particle cascades in the atmosphere which can be detected at sea level provided that their energy exceeds about 100 GeV (such low-energy cascades may be detected by using the most sensitive atmospheric Cerenkov detectors). The primary particle progressively loses energy which is transferred through the production of successive generations of secondary particles to a cascade of hadrons, an electromagnetic shower component (both positively and negatively charged electrons and gamma rays) and muons. The secondary particles are relativistic and all travel effectively at the speed of light. As a result, they reach sea level at approximately the same time but, due to Coulomb scattering (for the electrons) and production angles (for the pions producing the muons), are spread laterally into a disk-like shower front with a characteristic lateral width of several tens of meters and thickness (near the central shower core) of 2 to 3 m. The number of particles at sea level is roughly proportional to the primary particle energy:

$$\text{Number of particles at sea level} \sim 10^{-10} \times \text{energy (eV)}$$

At altitudes below a few kilometers, the number of particles in a shower attenuates with an *attenuation length* of about 200 g·cm^{-2}, i.e.,

$$\text{particle number} = \text{original number} \times \exp(-(\text{depth increase})/200)$$

The above applies to an individual shower. The rate of observation of showers of a given size (particle number at the detector) at different depths of absorber attenuates with an *absorption length* of about 100 g·cm^{-2} (J. G. Wilson, 1976).

Atmospheric Background Light from Cosmic Rays

Cosmic ray particles produce Cerenkov light in the atmosphere and produce fluorescent light through the excitation of atmospheric molecules.

Cerenkov Light

High-energy charged particles will cause the emission of Cerenkov light in air if their energies are above about 30 MeV (electrons). This threshold is pressure (and hence altitude) dependent. A typical Cerenkov light pulse (at sea level, 100 m from the central shower core) has a time spread of a few nanoseconds. Over this time, the photon flux between 430 and 530 nm would be ~10^{14} m^{-2}s^{-1} for a primary particle energy of 10^{16} eV. For comparison, the night sky background flux is ~6×10^{11} photons m^{-2}s^{-1}sr^{-1} in the same wavelength band (J. V. Jelley, 1967).

Fluorescence Light

Cosmic ray particles in the atmosphere excite atmospheric molecules which then emit fluorescence light. This is weak compared to the highly collimated Cerenkov component when viewed in the direction of the incident cosmic ray particle but is emitted isotropically. Typical pulse widths are longer than 50 ns and may be up to several microseconds for the total pulse from distant large showers (R. M. Baltrusaitis et al., 1985).

Effects of Cosmic Rays

Cerenkov Effects in Transparent Media

Background cosmic ray particles will produce Cerenkov light in transparent material with a photon yield between wavelengths λ_1 and λ_2

$$\sim (2\pi/137)\sin^2(\theta_c)\int_{\lambda_1}^{\lambda_2} d\lambda/\lambda^2 \text{ photons(unit length)}^{-1}$$

where θ_c (the Cerenkov angle) = $\cos^{-1}$ (1/refractive index).

This background light is known to affect light detectors, e.g., photomultipliers, and can be a major source of background noise (R. W. Clay and A. G. Gregory, 1977).

Effects on Electronic Components

If background cosmic ray particles pass through electronic components, they may deposit sufficient energy to affect the state of, e.g., a transistor flip-flop. This effect may be significant where reliability is of great importance or the background flux is high. For instance, it has been estimated that, in communication satellite operation, an error rate of about 2×10^{-3} per transistor per year may be found. Permanent damage may also result. A significant error rate may be found even at sea level in large electronic memories. This error rate is dependent on the sensitivity of the component devices to the deposition of electrons in their sensitive volumes (J. F. Ziegler, 1981).

Biophysical Significance

When cosmic rays interact with living tissue, they produce radiation damage. The amount of the damage depends on the total dose of radiation. At sea level, this dose is small compared with doses from other sources, but both the quantity and quality of the radiation change rapidly with altitude. Approximate dose rates under various conditions are:

Dose rates (mrem·yr^{-1})
Sea-level cosmic rays, 30
Cosmic rays at 10 km (subsonic jets), 2000
Cosmic rays at 18 km (supersonic transports), 10,000
(c.f., mean total sea-level dose, 300)

Astronauts would be subject to radiation from galactic (0.05 rads per day) and solar (a few hundred rads per solar flare) cosmic rays as well as large fluxes of low-energy radiation when passing through the Van Allen belts (about 0.3 rads per traverse).

Both astronauts and SST travelers would be subject to a small flux of low-energy heavy nuclei stopping in the body. Such particles are capable of destroying cell nuclei and could be particularly harmful in the early stages of the development of an embryo. The rates of heavy nuclei stopping in tissue in supersonic transports and spacecraft are approximately as follows:

Stopping nuclei ((cm^3 tissue)$^{-1}$ hr^{-1})
Supersonic transport (16 km), 0.0005
Supersonic transport (20 km), 0.005
Spacecraft, 0.15
(O. C. Allkofer, 1975a; O. C. Allkofer et al., 1974)

Carbon Dating

Radiocarbon is produced in the atmosphere due to the action of cosmic ray slow neutrons. Solar cycle modulation of the very low-energy cosmic rays causes an anticorrelation of the atmospheric ^{14}C activity with sunspot number with a mean amplitude of about 0.5%. In the long term, modulation of cosmic rays by a varying magnetic field may be important (A. A. Burchuladze et al., 1979).

Practical Uses of Cosmic Rays

There are few direct practical uses of cosmic rays. Their attenuation in water and snow have, however, enabled automatic monitors of water and snow depth to be constructed. A search for hidden cavities in pyramids has been carried out using a muon "telescope."

Other Effects

Stellar x-rays have been observed to affect the transmission times of radio signals between distant stations by altering the depth of the ionospheric reflecting layer. It has also been suggested that variations in ionization of the atmosphere due to solar modulation may have observable effects on climatic conditions.

References

1. O.C. Allkofer (1975a) *Introduction to Cosmic Radiation*, Verlag Karl Thiemig, Munchen, Germany.
2. O.O. Allkofer (1975b) *J. Phys. G: Nucl. Phys.*, 1, L51. <https://doi.org/10.1088/0305-4616/1/6/005>
3. O.C. Allkofer and W. Heinrich (1974) *Health Phys.*, 27, 543. <https://doi.org/10.1097/00004032-197412000-00001>
4. M. Amenomori et al. (1995) *Proc. 24th Int. Cosmic Ray Conf. Rome*, 3, 85. Universita La Sapienza, Roma.
5. K. Asakimori et al. (1993) *Proc. 23rd Int. Cosmic Ray Conf. Calgary*, 2, 25, University of Calgary, Canada.
6. R.M. Baltrusaitis et al. (1985) *Nucl. Inst. Meth.*, A420, 410.
7. D.J. Bird et al. (1993) *Phys. Rev. Lett.*, 71, 3401.
8. A.A. Burchuladze, S.V. Pagava, P. Povinec, G. I. Togonidze, S. Usacev (1979) *Proc. 16th Int. Cosmic Ray Conf. Kyoto*, 3, 201, Univ. of Tokyo, Japan.
9. R.W. Clay (1987) *Aust. J. Phys.*, 40, 423. <https://doi.org/10.1071/PH870423>
10. R.W. Clay and A.G. Gregory (1977) *J. Phys. A: Math. Gen.*, 10, 135. <https://doi.org/10.1088/0305-4470/10/1/026>
11. J.M. Clem et al. (1995) *Proc. 24th Int. Cosmic Ray Conf. Rome*, 3, 5, Universita La Sapienza, Roma.
12. T.K. Gaisser et al. (1993) *Phys. Rev. D*, 47, 1919.
13. K. Greisen (1943) *Phys. Rev.*, 63, 323. <https://doi.org/10.1103/PhysRev.63.323>
14. S. Hayakawa (1969) *Cosmic Ray Physics*, Wiley-Interscience, New York.
15. J.V. Jelley (1967) *Prog. in Elementary Particle and Cosmic Ray Physics*, 9, 41.
16. E. Juliusson (1975) *Proc. 14th Int. Cosmic Ray Conf. Munich*, 8, 2689, Max Planck Institute fur Extraterrestriche Physik, Munchen, Germany.
17. J. Linsley (1981) *Origin of Cosmic Rays*, I.A.U. Symposium 94, 53, D. Reidel Publishing Co. Dordrecht, Holland. <https://doi.org/10.1017/S0074180900074398>
18. P. Meyer (1981) *Origin of Cosmic Rays*, I.A.U. Symposium 94, 7, D. Reidel Publishing Co. Dordrecht, Holland. <https://doi.org/10.1007/978-94-009-8475-2_2>
19. P.F. Michelson (1994) in *Towards a Major Atmospheric Cerenkov Detector III*, 257, Ed. T. Kifune, Universal Academy Press Inc., Tokyo, Japan.
20. P. Sokolsky (1989) *Introduction to Ultrahigh Energy Cosmic Ray Physics*, Addison Wesley Publishing Company. <https://doi.org/10.1201/9780429499654-1>
21. T. Stanev et al. (1995) *Phys. Rev. Lett.*, 75, 3056. <https://doi.org/10.1103/PhysRevLett.75.3056>
22. S. Orito et al. (1995) *Proc. 24th Int. Cosmic Ray Conf. Rome*, 3, 76. Universita La Sapienza, Roma.
23. J.G. Wilson (1976) *Cosmic Rays*, Wykeham Pub. (London) Lt., U.K.
24. A.W. Wolfendale (1979) *Pramana*, 12, 631. <https://doi.org/10.1007/BF02846854>
25. S. Yoshida et al. (1995) *Astroparticle Phys.*, 3, 105.
26. J.F. Ziegler, (1981) *IEEE Trans. Electron Devices*, ED-28, 560. <https://doi.org/10.1109/T-ED.1981.20383>

Section 12
Properties of Solids

Techniques for Materials Characterization 12-1
Symmetry of Crystals 12-6
Ionic Radii in Crystals 12-11
Polarizabilities of Atoms and Ions in Solids 12-13
Crystal Structures and Lattice Parameters of Allotropes of the Elements 12-15
Phase Transitions in the Solid Elements at Atmospheric Pressure 12-19
The Madelung Constant and Crystal Lattice Energy 12-21
Elastic Constants of Single Crystals 12-22
Electrical Resistivity of Pure Metals 12-27
Electrical Resistivity of Selected Alloys 12-28
Electrical Resistivity of Graphite Materials 12-30
Permittivity (Dielectric Constant) of Inorganic Solids 12-31
Curie Temperature of Selected Ferroelectric Crystals 12-40
Properties of Antiferroelectric Crystals 12-41
Dielectric Constants of Glasses 12-41
Properties of Superconductors 12-42
High-Temperature Superconductors 12-58
Organic Superconductors 12-68
Properties of Semiconductors 12-70
Selected Properties of Semiconductor Solid Solutions 12-83
Properties of Organic Semiconductors 12-85
Diffusion Data for Semiconductors 12-89
Properties of Magnetic Materials 12-97
Organic Magnets 12-105
Electron Inelastic Mean Free Paths 12-106
Electron Stopping Powers 12-108
Electron Work Function of the Elements 12-110
Secondary Electron Emission 12-112
Optical Properties of Selected Elements 12-113
Optical Properties of Selected Inorganic and Organic Solids 12-117
Properties of Selected Materials at Cryogenic Temperatures 12-122
Heat Capacity of Selected Solids 12-128
Thermal and Physical Properties of Pure Metals 12-129
Thermal Conductivity of Metals and Semiconductors as a Function of Temperature 12-131
Thermal Conductivity of Alloys as a Function of Temperature 12-132
Thermal Conductivity of Crystalline Dielectrics 12-134
Thermal Conductivity of Ceramics and Other Insulating Materials 12-136
Thermal Conductivity of Glasses 12-138
Thermoelectric Properties of Metals and Semiconductors 12-141
Fermi Energy and Related Properties of Metals 12-143

TECHNIQUES FOR MATERIALS CHARACTERIZATION

H. P. R. Frederikse

The many experimental methods, originally designed to study the chemical and physical behavior of solids and liquids, have grown into a new field known as Materials Characterization (or Materials Analysis). During the past 30 years a host of techniques aimed at the study of surfaces and thin films has been added to the many tools for the analysis of bulk samples. The field has benefited particularly from the development of computers and microprocessors, which have vastly increased the speed and accuracy of the measuring devices and the recording of their output. Materials characterization was and is a very important tool in the search for new physical and chemical phenomena. It plays an essential role in new applications of solids and liquids in industry, communications, and medicine. Many of its techniques are used in quality control, in safety regulations, and in the fight against pollution.

In most Materials Characterization experiments the sample is subjected to some kind of radiation: electromagnetic, acoustic, thermal, or particles (electrons, ions, neutrons, etc.). The surface analysis techniques usually require a high vacuum. As a result of interactions between the solid (or liquid) and the incoming radiation a beam of a similar (or a different) nature will emerge from the sample. Measurement of the physical and/or chemical attributes of this emerging radiation will yield qualitative, and often quantitative, information about the composition and the properties of the material being probed.

The modern tendency of describing practically everything in this world by a combination of a few letters (acronyms) has also penetrated the field of Materials Characterization. The table below gives the meaning of the acronym for every technique listed, the form and size of the required sample (bulk, surface, film, liquid, powder, etc.), the nature of the incoming and of the emerging radiation, the depth and the lateral spatial resolution that can be probed, and the information obtained from the experiment. The last column lists one or two major references to the technique described.

Solids

Experimental Techniques Used to Determine the Composition, Structure, and Energy States of Solids and Liquids

Technique	Sample	In	Out	Depth	Lateral resolution	Information obtained	Ref.
Optical and Mass Spectroscopies for Chemical Analysis							
1. AAS Atomic Absorption Spectroscopy	Atomize (flame, electro, thermal, etc.)	Light, e.g., glow discharge	Absorption spectrum	–	–	Concentration of atomic species (quantitative, using standards)	1,2
2. ICP-AES Induct. Coupled Plasma – Atomic Emission Spectroscopy	Atomize (flame, electro, thermal, ICP, etc.)	–	Emission spectrum	–	–	Concentration of atomic species (quantitative, using standards)	3
3. Dynamic SIMS Dynamic Secondary Ion Mass Spectroscopy	Surface	Ion beam (1–20 keV)	Secondary ions; analysis with mass spectrometer	2 nm–1 μm (or deeper: ion milling)	0.50 nm	Elemental and isotopic analysis; depth profile (all elements); detection limits: ppb-ppm	4
4. Static SIMS Static Secondary Ion Mass Spectroscopy	Surface	Ion beam (0.5–20 keV)	Secondary ions, analysis with mass spectrometer	0.1–0.5 nm	10 μm	Elemental analysis of surface layers; molecular analysis; detection limits: ppb-ppm	4
5. SNMS Sputtered Neutral Mass Spectroscopy	Surface, bulk	Plasma discharge; noble gases: 0.5–20 keV	Sputtered atoms ionized by atoms or electrons; then mass analyzed	0.1–0.5 nm (or deeper: ion milling)	1 cm	Elemental analysis $Z \geq 3$; depth profile; detection limit: ppm	4,6
6. SALI Surface Analysis by Laser Ionization	Surface	e-beam, ion-beam, or laser for sputtering	Sputtered atoms ionized by laser; then mass analyzed	0.1–0.5 nm up to 3 μm in milling mode	60 nm	Surface analysis; depth profiling	7
7. LIMS Laser Ionization Mass Spectroscopy	Surface, bulk	u.v. laser (ns pulses)	Ionized species; analyzed with mass spectrometer	50–150 nm	5 μm–1 mm	Elemental (micro)analysis; detection limits: 1–100 ppm	8
8. SSMS Spark Source Mass Spectroscopy	Sample in the form of two electrodes	High voltage R.F. spark produces ions	Ions – analyzed in mass spectrometer	1–5 μm	–	Survey of trace elements; detection limit: 0.01–0.05 ppm	9
9. GDMS Glow Discharge Mass Spectroscopy	Sample forms the cathode for a D.C. glow discharge	Sputtered atoms ionized in plasma	Ions – analyzed in mass spectrometer	0.1–100 μm	3–4 mm	(Bulk) trace element analysis; detection limit: sub-ppb	9,10
10. ICPMS Induct. Coupled Plasma Mass Spectroscopy	Liquid-dissolved sample carried by gas stream into R.F. induction coil	Ions produced in argon plasma	Ions – analyzed in quadrupole mass spectrometer	–	–	High-sensitivity analysis of trace elements	11
Photons — Absorption, Reflection, and Electron Emission							
11. IRS Infrared Spectroscopy	Thin crystal, glass, liquid	I.R. light (W-filament, globar, Hg-arc)	I.R. spectrum	–	–	Electronic transitions (mainly in semiconductors and superconductors); vibrational modes (in crystals and molecules)	12,13,14

	Technique	Sample	In	Out	Depth	Lateral resolution	Information obtained	Ref.
12.	FTIR Fourier Transform I.R. Spectroscopy	Solid, liquid; transmission or reflection	White light (all frequencies)	Fourier transform of spectrum (interferometer)	–	–	Spectra obtained at higher speed and resolution	15
13.	ATR Attenuated Total Reflection	Surface or thin crystal	–	–	μm's	–	Atomic or molecular spectra of surfaces and films	16
14.	(μ)-RS (Micro-) Raman Spectroscopy	Solid, liquid (1 μm–1 cm)	Laser beam, e.g., Ar-line, YAG-line	Raman spectra	0.5 μm	0.5 μm	Molecular and crystal vibrations	12,14,17
15.	CARS Coherent Anti-Stokes Raman Spectroscopy	Solid, liquid (50 μm–3 cm)	Pump beam (ω_0)+ probe beam (ω_s)	Anti-Stokes spectrum	–	–	High-resolution Raman spectra	14
16.	Ellipsometry	Transparent films, crystals, adsorbed layers	Polarized light	Change in polarization	0.05 nm–5 μm	25 μm (or sample thickness)	Refractive index and absorption	18,19
17.	UPS Ultraviolet Photoelectron Spectroscopy	Surfaces, adsorbed layers	u.v. light, 10–100 eV; 200 eV (synchrotron)	Electrons	0.2–10 nm	0.1–10 nm	Energies of electronic states of surfaces and free molecules	20,21
18.	PSD Photon Stimulated Desorption	Surfaces with adsorbed species	Far u.v. light $E > 10$ eV	Ions – analyzed with mass spectrometer	0.1–2 nm	–	Structure and desorption kinetics of adsorbed atoms and molecules	22
	X-Rays							
19.	XRD X-Ray Diffraction	Single crystals, powders films	X-rays: $\lambda = 0.05$–0.2 nm (6–17 keV)	Diffracted x-ray beam	1–1000 μm	0.1–10 mm	Identification of crystallographic structures; all elements (low Z difficult)	23,24
20.	XRF/EDS X-Ray Fluorescence/Energy Dispersive Spectroscopy	Thin films, single layer	Prim. x-ray beam $\lambda =$ 0.02–0.1 nm 12–80 keV	Fluorescent x-rays	1–100 μm	10 mm	Elemental analysis; all elements except H, He, Li – (EDS also used in XRD, SEM, TEM, and EPMA)	25,26
21.	EXAFS Extended X-Ray Absorption Fine Structure	Films, foils	High intensity x-rays (synchrotron)	Spectrum near absorption edge	nm–μm	–	Local atomic structure: order/disorder in vicinity of absorbing atom	27
22.	XPS/ESCA X-Ray Photoelectron Spectroscopy/Electron Spect. for Chemical Analysis	Surfaces, thin films (≈20 atomic layers)	Soft x-rays (1–20 keV)	Core electrons; valence electrons	0.5–10 nm	5 nm–50 μm	(Quantitative) identification of all elements in surface layer or film	28,29
	Electrons							
23.	CL Cathode Luminescence	Insulators, semiconductors	Electrons 5–50 keV	Photons 0.1–5 eV	1 nm–2 μn	1 or 2 μm	Energy levels of impurities and point defects	30
24.	APS Appearance Potential Spectroscopy	Surface (≈20 atomic layers)	Electrons (energy scan) 50–2000 eV	X-rays to pinpoint electron energy threshold	–	–	Identification of surface species	21, see also C
25.	AES Auger Electron Spectroscopy	Thin films, surfaces	Electrons 3–10 keV	Auger electrons 20–2000 eV	0.3–3 nm	≈30 nm	Elemental composition of surface (except H, He); detection limit 0.1–1%	28,29
26.	EELS Electron Energy Loss Spectroscopy	Very thin samples (<200 nm)	Electrons (100–400 keV)	(Retarded) electrons; minus 1–1000 eV	<200 nm	1–100 nm	Local elemental concentration; electronic structure, chem. bonding; interatomic distances	31
27.	EXELFS Extended Electron Energy Loss Fine Structure	Thin films	Electrons (100–400 keV)	Electron energies 0–30 eV above edge	<200 nm	1–100 nm	Density of states of valence electrons (above Fermi level)	27,32
28.	ESD Electron Stimulated Desorption	Adsorbed species	Electrons $E > 10$ eV	Ions – analyzed with mass spectrometer	–	–	Structure and desorption properties of adsorbed atoms and molecules	22
29.	ESDIAD ESD-Ion Angular Distribution	(See ESD)	(See ESD)	Directional dependence of emitted ions	–	–	Geometries of adsorbed species (atoms or molecules)	22
30.	EPMA Electron Probe (X-Ray) Micro Analysis	Solid conductors and insulators <1 cm thick	Electrons 5–30 keV	Characteristic x-ray 0.1–15 keV	100 nm–5 μm	1 μm	Elemental analysis, $Z \leq 4$, major, minor, and trace amounts	33,34
31.	LEED Low-Energy Electron Diffraction	Surface	Mono-energetic electron beam 10–1000 eV	Diffracted electrons	0.4–2 nm	<5 μm	Crystallographic structure of surface; resolution: 0.01 nm	35
32.	RHEED Reflection High-Energy Electron Diffraction	Surface	Electron beam at grazing angle 5–50 keV	Reflected electrons	0.2–10 nm	<5 μm	Surface symmetry	36,37
33.	SEM Scanning Electron Microscopy	Bulk, films (conducting)	High-energy electrons usually ~30 keV	Secondary and backscattered electrons	1 nm–5 μm	1–20 nm	Surface image, defect structure; resolution 5–15 nm; magnification 300,000×	33,34

Technique	Sample	In	Out	Depth	Lateral resolution	Information obtained	Ref.
34. (S)TEM (Scanning) Transmission Electron Microscopy	Thin specimen – <200 nm	High-energy electrons typically 300 keV	Transmitted and diffracted electrons	(Sample thickness)	2–20 nm	(Defect) structure of cryst. solids; microchemistry; high resol.: 0.2 nm	33
35. FEM Field Emission Microscopy	Metals, alloys (sharp point)	–	Electron emission (with appl. electric field – 50 kV)	≈0.5 nm	10–100 nm	Surface image, crystallographic structure	34
36. STM Scanning Tunneling Microscopy	Polished or cleaved surface (conducting)	Tunneling current controls distance between sample and very sharp tip	Tunneling current controls distance between sample and very sharp tip	1–5 nm	2–10 nm	Atomic-scale relief map of surface; resolution: vert. 0.002 nm, hor. 0.2 nm	39
37. SPM Scanned Probe Microscopy	Very flat surface	Any field: e.g., mechan. vibration recorded with laser probe; same with magnetic, electric or thermal field	Any field: e.g., mechan. vibration recorded with laser probe; same with magnetic, electric or thermal field	1–100 nm	1–100 nm	Surface-magnetic field, surface-thermal conductivity, etc.	39a
38. AFM Atomic Force Microscopy	Very flat surface	Similar to STM; force measured with cantilever spring	Similar to STM; force measured with cantilever spring	0.5–5 nm	0.2–130 nm	Surface topography with atomic resolution; interatomic force	40
Ions and Neutrons							
39. ISS (or LEIS) Ion Scattering Spectroscopy (Low Energy Ion Scattering)	Surface	Ion beam He^+ or Ne^+ <3 keV	Sputtered ions (energy analysis)	0.1–0.5 nm	1–100 μm	Elemental analysis (better for low *Z*) detection limits: 0.01–1%	41
40. FIM Field Ion Microscopy	Surface: metals, alloys; very sharp tip	(He gas above sample)	He ions + high electric field produce image	≈0.1 nm	0.1–2 nm	Atomic structure of surface	34,42
41. RBS Rutherford Back Scattering	Solids, thin films	Mono-energetic ions (H^+ or He^{++}) 0.5–3 MeV	Backscattered ions	10 nm–1 μm	1 mm	Element identification (Li to U) detection limit: 0.01–1%	46
42. NRA Nuclear Reaction Analysis	Solids, thin films	Mono-energetic ions (Li, Be, B, etc.) 200 keV–6 MeV	Protons, deuterons ^3He, α-particles, γ-rays	0.1–5 μm	10 μm–10 mm	Element identification (all) detection limit: 10-12–10-2	47
43. PIXE Particle-Induced x-ray Emission	Thin films, surface layers	High-energy ions (H^+ or He^{++})	Characteristic x-rays	<10 μm	1 μm–2 mm	Trace impurities: $Z>3$ detection limit: 0.1–100 ppm (depending on sample thickness)	48
44. INS Ion Neutralization Spectroscopy	Surface	He-ions (≈5 eV)	Electrons	–	–	Energies of valence electrons	49
45. NAA Neutron Activation Analysis	Bulk, >0.5 g	Thermal neutrons	Characteristic γ-rays, (≈1 MeV)	Bulk	–	Trace concentrations (of isotopes) of elements: trans. metals, Pt-group; detection limit: 10^8–10^{14} atoms/cm^3	43
46. N(P)D Neutron (Powder) Diffraction	Crystalline solids	Thermal neutrons $E \approx 0.0025$ eV	Diffracted neutrons	Bulk	–	Crystallographic structure; porosity, particle size	44
47. SANS Small Angle Neutron Scattering	Inhomogeneous solids; powders; porous samples	Thermal neutrons $2\theta = 10^{-2}$–10^{-4}	Scattered neutrons	1–25 mm	–	Average size of inhomogeneities; range: 1 nm^{-1} mm	45
Acoustic							
48. SLAM Scanning Laser Acoustic Microscopy	Bulk, film	Acoustic wave produced by laser 1 MHz–1 GHz	Reflected acoustic wave	μm–cm	0.1–20 mm	Defect structure; thickness measurement	50
Thermal							
49. DTA Differential Thermal Analysis	Specimen and reference sample	Uniform heating	Temperature difference	Bulk	–	Phase transitions, crystallization	51
50. DSC Differential Scanning Calorimetry	Specimen and ref. sample	Controlled heating	Measure heat required for equal temperature	Bulk	–	Phase transitions, crystallization; activation energies	51
51. TGA Thermo Gravimetric Analysis	Bulk, 1–100 g	Controlled heating	Weight as function of temperature (and time)	Bulk	–	Decomposition, non-stoichiometry, kinetics of reaction	52
Resonance							
52. EPR (ESR) Electron Paramagnetic (Spin) Resonance	Paramagnetic solids or liquids	Microwave radiation in magnetic field 3–300 GHz; 1–100 kG	Microwave absorption (at resonance)	Bulk	–	Local environment of paramagnetic ion; concentration of paramagnetic, species; detection limit: 10^{11} spins/cm^3	53,54
53. ECR Electron Cyclotron Resonance	Semiconductors, metals; free electrons (low temperature)	Microwave radiation in magnetic field 10–30 GHz; 5–10 kG	Microwave absorption (at resonance)	Bulk	–	Electronic energy bands, effective masses	55

	Technique	Sample	In	Out	Depth	Lateral resolution	Information obtained	Ref.
54.	Mössbauer Effect	Source and absorber	Mono-energetic γ-rays: 5–100 keV	Mössbauer spectrum (Doppler shifted (lines)	50 m	1 cm	Interaction between nucleus and its environment (local electric, magnetic fields; bonds; valency; diffusion, etc.)	56
55.	NMR (MRI) Nuclear Magnetic Resonance (Magnetic Resonance Imaging)	Solids, liquids	R.F. radiation + magnetic field; e.g., for protons: 60 MHz, 14 kG	R.F. absorption	<1 cm	1 cm	Quant. analysis; local magnetic environment; diffusion; imaging	58
56.	ENDOR Electron Nuclear Double Resonance	Solids, liquids	R.F. + microwave radiation in magn. field.	Microwave absorption	–	–	Hyperfine interaction → local atomic structure	54
57.	NQR Nuclear Quadrupole Resonance	Solids	R.F. radiation 0.5–1000 MHz	R.F. absorption	–	–	Asymmetry of the charge distribution at the nucleus	55,59
Other								
58.	BET Brunauer-Emmett-Teller	(Large) surface area 1–20 m^2/g	Adsorbed gas (e.g., N_2 at low temp.) as function of pressure (monolayer coverage)	Adsorbed gas (e.g., N_2 at low temp.) as function of pressure (monolayer coverage)	–	–	Surface area measurement	60

Solids

General References

1. A. Wachtman, J. B., *Characterization of Materials*, Butterworth-Heinemann, Boston, 1993.
2. B. Brundle, C. R., Evans, C. A., and Wilson, S., Eds., *Encyclopedia of Materials*, Butterworth-Heinemann, Boston, 1992.
3. C. Woodruff, D. P. and Delchar, T. A., *Modern Techniques of Surface Science*, Cambridge University Press, Cambridge, 1986.
4. *D. Metals Handbook*, 9th Edition, Vol. 10, Materials Characterization, Whan, R. E., Coordinator, American Society for Metals, Metals Park, OH, 1986.

Specific References

1. Slavin, M., *Atomic Absorption Spectroscopy*, 2nd Edition, John Wiley & Sons, New York, 1978.
2. Schrenk, W. G., *Analytical Atomic Spectroscopy*, Plenum Press, New York, 1975.
3. Dean, J. A. and Rains, T. E., *Flame Emission and Atomic Absorption Spectroscopy*, Vols. 1–3, Marcel Dekker, New York, 1969.
4. Benninghoven, A., Rudenauer, F. G., and Werner, H. W., *Secondary Ion Mass Spectroscopy*, John Wiley & Sons, New York, 1987.
5. Bird, J. R. and Williams, J. S., Eds., in *Ion Beams for Materials Analysis*, Academic Press, New York, 1989, pp. 515–537.
6. Smith, G. C., *Quantitative Surface Analysis for Materials Science*, The Institute of Metals, London, 1991.
7. Becker, E. H., in *Ion Spectroscopies for Surface Analysis*, Czanderna, A. W. and Hercules, D. M., Eds., Plenum Press, New York, 1991, p. 273.
8. Simons, D. S., *Int. J. Mass Spectrometry and Ion Processes*, 55, 15, 1983.
9. White, F. A. and Wood, G. M., *Mass Spectrometry: Applications in Science and Engineering*, John Wiley & Sons, New York, 1986.
10. Harrison, W. W. and Bentz, B. L., *Prog. Anal. Spectrometry*, 11, 53, 1988.
11. Bowmans, P. W. J. M., *Inductively Coupled Plasma Emission Spectroscopy*, Parts I and II, John Wiley & Sons, New York, 1987.
12. Brame, Jr., E. G. and Grasselli, J., *Infrared and Raman Spectroscopy*, Practical Spectroscopy Series, Vol. I, Marcel Dekker, New York, 1976.
13. Hollas, J. M., *Modern Spectroscopy*, John Wiley & Sons, New York, 1987.
14. Turrell, G., *Infrared and Raman Spectroscopy of Crystals*, Academic Press, New York and London, 1972.
15. Griffith, P. R. and Haseth, J. A., *Fourier Transform Infrared Spectroscopy*, John Wiley & Sons, New York, 1986.
16. Barnowski, M. K., *Fundamentals of Optical Fiber Communications*, Academic Press, New York, 1976.
17. Long, D. A., *Raman Spectroscopy*, McGraw-Hill, New York, 1977.
18. Azzam, R. M. A., *Ellipsometry and Polarized Light*, Elsevier-North Holland, Amsterdam, 1977.
19. Hecht, E., *Optics*, 2nd Edition, Addison-Wesley, Reading MA, 1987.
20. Brundle, C. R., in *Molecular Spectroscopy*, West, A. R., Ed., Heyden, London, 1976.
21. Park, R. L., in *Experimental Methods in Catalytic Research*, Vol. III, Anderson, R. B. and Dawson, P. T., Academic Press, New York, 1976, pp. 1–39.
22. Madey, T. E. and Stockbauer, R., in *Solid State Physics: Surfaces*, Vol. 22 of Methods of Experimental Physics, Park, R.L. and Lagally, M. G., Eds., Academic Press, New York, 1985.
23. Cullity, B. D., *Elements of X-Ray Diffraction*, 2nd Edition, Addison-Wesley, Reading, MA, 1978.
24. Schwartz, L. H. and Cohen, J. B., *Diffraction from Materials*, Springer Verlag, Berlin, 1987.
25. deBoer, D. K. G., in *Advances in X-Ray Analysis*, Vol. 34, Barrett, C. S. et. al., Eds., Plenum Press, New York, 1991.
26. Birks, L. S., *X-Ray Spectrochemical Analysis*, 2nd Edition, John Wiley & Sons, New York, 1969.
27. Bonnelle, C. and Mande, C., *Advances inX-Ray Spectroscopy*, Pergamon Press, Oxford, 1982.
28. *Practical Surface Analysis by Auger and X-Ray Photo-Electric Spectroscopy*, Briggs, D. and Seah, M. P., Eds., John Wiley & Sons, New York, 1983.
29. Powell, C. J. and Seah, M. P., *J. Vac. Sci. Technol. A*, Vol. 8, 735, 1990.
30. Yacobi, G. G. and Holt, D. B., *Cathodeluminescence Microscopy of Inorganic Solids*, Plenum Press, New York, 1990.
31. Egerton, R. F., *Electron Energy Loss Spectroscopy in the Electron Microscope*, Plenum Press, New York, 1986.
32. Disko, M. M., Krivanek, O. L., and Rez, P., *Phys. Rev.*, B25, 4252, 1982.
33. Goldstein, J. I., et. al., *Scanning Electron Microscopy and X-Ray Microanalysis*, 2nd Edition, Plenum Press, New York, 1986.
34. Murr, L. E., *Electron and Ion Microscopy and Microanalysis*, Marcel Dekker, New York, 1982.
35. Armstrong, R. A., in *Experimental Methods in Catalytic Research*, Vol. III, Anderson, R. B., and Dawson, P. T., Eds., Academic Press, New York, 1976.
36. Dobson, P. J. et. al., *Vacuum*, 33, 593, 1983.
37. Rymer, T. B., *Electron Diffraction*, Methuen, London, 1970.
38. Reimer, L., *Transmission Electron Microscopy*, Springer-Verlag, Berlin, 1984.

39. *Scanning Tunneling Microscopy and Related Methods*, Behm, R. J., Garcia, N., and Rohrer, H., Eds., Kluwer Academic Publishers, Norwell, MA, 1990.
39a. Wikramasinghe, H.K., *Scientific American*, Vol. 261, No. 4, pp. 98–105, Oct. 1989.
40. Rugar, D. and Hansma, P., *Physics Today*, 43(10), pp. 23–30, 1990.
41. Feldman, C. C. and Mayer, J. W., *Fundamentals of Surface and Thin Film Analysis*, North-Holland, Amsterdam, 1986.
42. Muller, E. W. and Tsong, T. T., *Field Ion Microscopy*, Elsevier, Amsterdam, 1969.
43. Amiel, S., *Nondestructive Activation Analysis*, Elsevier, Amsterdam, 1981.
44. Bacon, G. E., *Neutron Diffraction*, 3rd Edition, Clarendon Press, Oxford, 1975.
45. Neutron Scattering, Part A., in *Methods of Experimental Physics*, Vol. 23, Skold, K. and Price, D. L., Eds., Academic Press, New York, 1986.
46. Chu, W. K., Mayer, J. W., and Nicolet, M. A., *Backscattering Spectroscopy*, Academic Press, New York, 1987.
47. Rickey, F. A., in *High Energy and Heavy Ion Beams in Materials Analysis*, Tesmer, J. R., et. al., Eds., MRS, 1990, pp. 3–26.
48. Johansson, S. A. E. and Campbell, J. L., *PIXE: A Novel Technique for Elemental Analysis*, John Wiley & Sons, New York, 1988.
49. Hagstrum, H. D., in *Inelastic Ion-Surface Collisions*, Tolk, N. H. et. al., Eds., Academic Press, New York, 1977, pp. 1–46.
50. Nikoonahad, M., in *Research Techniques in Nondestructive Testing*, Vol. VI, Sharpe, R.S., Ed., Academic Press, New York, 1984, pp. 217–257.
51. Gallagher, P. K., *Characterization of Materials by Thermoanalytical Techniques*, MRS - Bulletin, Vol. 13, No. 7, pp. 23–27, 1988.
52. Earnest, C. M., *Compositional Analysis by Thermogravimetry*, ASTM Special Technical Publication 997, 1988.
53. Poole, C. P., *Electron Spin Resonance – A Comprehensive Treatise on Experimental Techniques*, 2nd Edition, John Wiley & Sons, New York, 1983.
54. Atherton, N. M., *Principles of Electron Spin Resonance*, Ellis Horwood Ltd., Chichester, U.K., 1993.
55. Kittel, C., *Introduction to Solid State Physics*, 6th Edition, John Wiley & Sons, New York, 1986, p. 196.
56. Gibb, T. C., *Principles of Mössbauer Spectroscopy*, Chapman & Hall, London, 1976.
57. Slichter, C. P., *Principles of Magnetic Resonance*, 3rd Edition, Springer-Verlag, Berlin, 1990.
58. *NMR Spectroscopy Techniques*, Dybrowski, C. and Lichter, R. L., Eds., Marcel Dekker, New York, 1987.
59. Das, T. P. and Hahn, E. L., *Nuclear Quadrupole Resonance Spectroscopy*, Academic Press, New York, 1958.
60. Somorjai, G. A., *Principles of Surface Chemistry*, Prentice Hall, Englewood Cliffs, NJ, 1972, p. 216.

SYMMETRY OF CRYSTALS

L. I. Berger

The ability of a body to coincide with itself in its different positions regarding a coordinate system is called its symmetry. This property reveals itself in iteration of the parts of the body in space. The iteration may be done by reflection in mirror planes, rotation about certain axes, inversions and translations. These actions are called the symmetry operations. The planes, axes, points, etc., are known as symmetry elements. Essentially, mirror reflection is the only truly primitive symmetry operation. All other operations may be done by a sequence of reflections in certain mirror planes. Hence, the mirror plane is the only true basic symmetry element. But for clarity, it is convenient to use the other symmetry operations, and accordingly, the other aforementioned symmetry elements. The symmetry elements and operations are presented in Table 1.

The entire set of symmetry elements of a body is called its symmetry class. There are thirty-two symmetry classes that describe all crystals that have ever been noted in mineralogy or been synthesized (more than 150,000). The denominations and symbols of the symmetry classes are presented in Table 2.

There are several known approaches to classification of individual crystals in accordance with their symmetry and crystallochemistry. The particles that form a crystal are distributed in certain points in space. These points are separated by certain distances (translations) equal to each other in any chosen direction in the crystal. Crystal lattice is a diagram that describes the location of particles (individual or groups) in a crystal. The lattice parameters are three non-coplanar translations that form the crystal lattice. Three basic translations form the unit cell of a crystal. August Bravais (1848) has shown that all possible crystal lattice structures belong to one or another of fourteen lattice types (Bravais lattices). The Bravais lattices, both primitive and non-primitive, are the contents of Table 3.

Among the three-dimensional figures, there is a group of polyhedrons that are called regular, which have all faces of the same shape and all edges of the same size (regular polygons). It has been shown that there are only five regular polyhedrons. Because of their importance in crystallography and solid state physics, a brief description of these polyhedrons is included in Table 4.

The systematic description of crystal structures is presented primarily in the well-known *Strukturbericht*. The classification of crystals by the Strukturbericht does not reflect their crystal class, the Bravais lattice, but is based on the crystallochemical type. This makes it inconvenient to use the Strukturbericht categories for comparison of some individual crystals. Thus, there have been several attempts to provide a more convenient classification of crystals. Table 5 presents a compilation of different classifications which allows the reader to correlate the Strukturbericht type with the international and Schoenflies point and space groups and with Pearson's symbols, based on the Bravais lattice and chemical composition of the class prototype. The information included in Table 5 has been chosen as an introduction to a more detailed crystallophysical and crystallochemical description of solids.

TABLE 1. Symmetry Operations, Symmetry Elements, and Stereographic Projections

Symmetry operation	Name	International (Hermann-Mauguin)	Schoenflies	Parallel projection	Perpendicular projection
Reflection in a plane	Plane	m	C_s	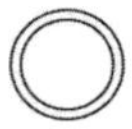	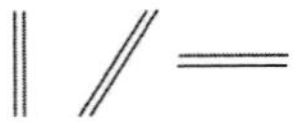
Rotation by angle $\alpha = 360°/n$ about an axis	Axis	n = 1, 2, 3, 4, or 6	C_n		
		n = 2	C_2		
		n = 3	C_3		
		n = 4	C_4		
		n = 6	C_6		
Rotation about an axis and inversion in a symmetry center lying on the axis	Inversion (improper) axis	$\bar{n} = \bar{3}, \bar{4}, \bar{6}$	C_{ni}		
		$\bar{n} = \bar{3}$	C_{3i}		
		$\bar{n} = \bar{4}$	C_{4i}		
		$\bar{n} = \bar{6}$	C_{6i}	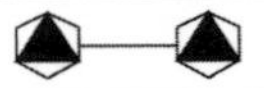	
Inversion in a point	Center	$\bar{1}$	C_i		
Parallel translation	Translation vector $\vec{a}, \vec{b}, \vec{c}$				
Reflection in a plane and translation parallel to the plane	Glide–plane	a, b, c, n, d			
Rotation about an axis and translation parallel to the axis	Screw axis	n_m (m = 1, 2, .., n − 1)			
Rotation about an axis and reflection in a plane perpendicular to the axis	Rotatory-reflection axis	$\tilde{n} = \tilde{1}, \tilde{2}, \tilde{3}, \tilde{4}, \tilde{6}$	S_n		

TABLE 2. The Thirty-Two Symmetry Classes

Triclinic	Int	Sch	Int	Sch	Int	Sch	Int	Sch	Int	Sch	Int	Sch	Int	Sch
Triclinic	1	C_1	1	C_i										
Monoclinic					m	C_s	2	C_2	2/m	C_{2h}				
Orthorhombic					mm2	C_{2v}	222	D_2	mmm	D_{2h}				
Trigonal	3	C_3	3	C_{3i}	3m	C_{3v}	32	D_3	$\bar{3}$m	C_{3d}				
Tetragonal	4	C_4	4/m	C_{4h}	4mm	C_{4v}	422	D_4	4/mmm	D_{4h}	$\bar{4}$	S_4	$\bar{4}$2m	D_{2d}
Hexagonal	6	C_6	6/m	C_{6h}	6mm	C_{6v}	622	D_6	6/mmm	D_{6h}	$\bar{6}$	C_{3h}	$\bar{6}$m2	D_{3h}
Cubic	23	T	m3	T_h	$\bar{4}$3m	T_h	432	O	m3m	O_h				

[a] Per Fedorov Institute of Crystallography, Russian Academy of Sciences, nomenclature.

TABLE 3. The Fourteen Possible Space Lattices (Bravais Lattices) (P, C, I, F, R Indicate Lattice Type; a, b, c, α, β, γ Indicate Characteristic Parameters)

Crystal system	Metric category of the system	No. of different lattices in the system	P	C	I	F	R	No. of identipoints per unit cell	a	b	c	α	β	γ	Description of characteristic parameters a⊂X, b⊂Y, c⊂Z α≡(b,c), β≡(a,c), γ≡(b,c)	Lattice symmetry (Int)	Lattice symmetry (Sch)
Triclinic	Trimetric	1	+					1	+	+	+	+	+	+	a ≠ b ≠ c, α ≠ β ≠ γ	1	C
Monoclinic	Trimetric	2	+	+				1 or 2	+	+	+		+		a ≠ b ≠ c, α = γ = 90° ≠ β	2/m	C_{2h}
Orthorhombic	Trimetric	4	+	+	+	+		1, 2, or 4	+	+	+				a ≠ b ≠ c, α = β = γ = 90°	mmm	D_{2h}
Trigonal (rhombohedral)	Dimetric	1					+	1	+			+			a = b = c, 120° > α = β = γ ≠ 90°	3m	D_{3d}
Tetragonal	Dimetric	2	+		+			1 or 2	+		+				a = b ≠ c, α = β = γ = 90°	4/mmm	D_{4h}
Hexagonal	Dimetric	1	+					1	+		+				a = b ≠ c, α = β = 90°, γ = 120°	6/mmm	D_{6h}
9Isometric (cubic)	Monometric	3	+		+	+		1, 2, or 4	+						a = b = c, α = β = γ = 90°	m3m	O_h

[a] Designations of the space-lattice types: P – primitive, C – side-centered (base-centered), I – body-centered, F – face-centered, R – rhombohedral.

TABLE 4. The Five Possible Regular Polyhedrons

Polyhedron	Symmetry class	Symmetry elements	Form of faces	No. of faces (F)	No. of edges (E)	No. of vertices (V)
Tetrahedron	T	$4C_33C_2$	Equilateral triangle	4	6	4
Cube (hexahedron)	O	$3C_44C_36C_2$	Square	6	12	8
Octahedron	O	$3C_44C_36C_2$	Equilateral triangle	8	12	6
Pentagonal dodecahedron	J	$6C_510C_315C_2$	Regular pentagon	12	30	20
Icosahedron	J	$6C_510C_315C_2$	Equilateral triangle	20	30	12

[a] Per formula by Leonhard Euler: F + V – E = 2.

TABLE 5. Classification of Crystals

Strukturbericht symbol	Structure name	Symmetry group (International)	Symmetry group (Schoenflies)	Pearson symbol[a]	Standard ASTM E157-82a symbol[b]
A1	Cu	Fm3m	$O^4{}_h$	cF4	F
A2	W	Im3m	$O^9{}_h$	cI2	B
A3	Mg	$P6_3$/mmc	$D^4{}_{6h}$	hP2	H
A4	C	Fd3m	$O^7{}_h$	cF8	F
A5	Sn	If_1/amd	$D^{19}{}_{4h}$	tI4	U
A6	In	I4/mmm	$D^{17}{}_{4h}$	tI2	U
A7	As	R$\bar{3}$m	$D^5{}_{3d}$	hR2	R
A8	Se	$P3_121$ or $P3_221$	$D^4{}_3$ ($D^6{}_3$)	hP3	H
A10	Hg	R$\bar{3}$m	$D^5{}_{3d}$	hR1	R
A11	Ga	Cmca	$D^{18}{}_{2h}$	oC8	Q
A12	α-Mn	I$\bar{4}$3m	$T^3{}_d$	cI58	B
A13	β-Mn	$P4_132$	O^7	cP20	C
A15	OW_3	Pm3n	$O^3{}_h$	cP8	C
A20	α-U	Cmcm	$D^{17}{}_{2h}$	oC4	Q
B1	ClNa	Fm3m	$O^5{}_h$	cF8	F
B2	ClCs	Pm3m	$O^1{}_h$	cP2	C
B3	SZn	F$\bar{4}$3m	$T^2{}_d$	cF8	F
B4	SZn	$P6_3$mc	$C^4{}_{6v}$	hP4	H
$B8_1$	AsNi	$P6_3$/mmc	$D^4{}_{6h}$	hP4	H
$B8_2$	$InNi_2$	$P6_3$/mmc	$D^4{}_{6h}$	hP6	H

Strukturbericht symbol	Structure name	Symmetry group (International)	Symmetry group (Schoenflies)	Pearson symbol[a]	Standard ASTM E157-82a symbol[b]
B9	HgS	$P3_121$ or $P3_221$	D^4_3 or D^6_3	hP6	H
B10	OPb	P4/nmm	D^7_{4h}	tP4	T
B11	γ-CuTi	P4/nmm	D^7_{4h}	tP4	T
B13	NiS	$R\bar{3}m$	D^5_{3d}	hR6	R
B16	GeS	Pnma	D^{16}_{2h}	oP8	O
B17	PtS	$P4_2/mmc$	D^9_{4h}	tP4	T
B18	CuS	$P6_3/mmc$	D^4_{6h}	hP12	H
B19	AuCd	Pmma	D^5_{2h}	oP4	O
B20	FeSi	$P2_13$	T^4	cP8	C
B27	BFe	Pnma	D^{16}_{2h}	oP8	O
B31	MnP	Pnma	D^{16}_{2h}	oP8	O
B32	NaTl	Fd3m	O^7_h	cF16	F
B34	Pds	$P4_2/m$	C^2_{4h}	tP16	T
B35	CoSn	P6/mmm	D^1_{6h}	hP6	H
B37	SeTl	I4/mcm	D^{18}_{4h}	tI16	U
B_e	CdSb	Pbca	D^{15}_{2h}	oP16	O
B_f(B33)	ξ-BCr	Cmcm	D^{17}_{2h}	oC8	Q
B_g	BMo	$I4_1/amd$	D^{19}_{4h}	tI4	U
B_h	CW	P6m2	D^1_{3h}	hP2	H
B_i	γ′CMo (AsTi)	$P6_3/mmc$	D^4_{6h}	hP8	H
C1	CaF_2	$Fm\bar{3}m$	O^5_h	cF12	F
$C1_b$	AgAsMg	$F\bar{4}3m$	T^2_d	cF12	F
C2	FeS_2	Pa3	T^6_h	cP12	C
C3	Cu_2O	Pn3m	O^4_h	cP6	C
C4	O_2Ti	$P4_2/mnm$	D^{14}_{4h}	tP6	T
C6	CdI_2	P3m1	D^3_{3d}	hP3	H
C7	MoS_2	$P6_3/mmc$	D^4_{6h}	hP6	H
$C11_a$	C_2Ca	I4/mmm	D^{17}_{4h}	tI6	U
$C11_b$	$MoSi_2$	I4/mmm	D^{17}_{4h}	tI6	U
C12	$CaSi_2$	$R\bar{3}m$	D^5_{3d}	hR6	R
C14	$MgZn_2$	$P6_3/mmc$	D^4_{6h}	hP12	H
C15	Cu_2Mg	Fd3m	O^7_h	cF24	F
$C15_b$	$AuBe_5$	$F\bar{4}3m$ or F23	T^2_d or T^2	cF24	F
C16	Al_2Cu	I4/mcm	D^{18}_{4h}	tI12	U
C18	FeS_2	Pnnm	D^{12}_{2h}	oP6	O
C19	$CdCl_2$	$R\bar{3}m$	D^5_{3d}	hR3	R
C22	Fe_2P	P2¯6m	D^1_{3h}	hP9	H
C23	Cl_2Pb	Pnma	D^{16}_{2h}	oP12	O
C32	AlB_2	P6/mmm	D^1_{6h}	hP3	H
C33	Bi_2STe_2	$R\bar{3}m$	D^5_{3d}	hR5	R
C34	$AuTe_2$	C2/m (P2/m)	C^3_{2h} (C^1_{2h})	mC6	N
C36	$MgNi_2$	$P6_3/mmc$	D^4_{6h}	hP24	H
C38	Cu_2Sb	P4/nmm	D^7_{4h}	tP6	T
C40	$CrSi_2$	$P6_222$	D^4_6	hP9	H
C42	SiS_2	Ibam	D^{26}_{2h}	oI12	P
C44	GeS_2	Fdd2	C^{19}_{2v}	oF72	S
C46	$AuTe_2$	Pma2	C^4_{2v}	oP24	O
C49	Si_2Zr	Cmcm	D^{17}_{2h}	oC12	Q
C54	Si_2Ti	Fddd	D^{24}_{2h}	oF24	S
C_c	Si_2Th	$I4_1/amd$	D^{19}_{4h}	tI12	U
C_e	$CoGe_2$	Aba2	C^{17}_{2v}	oC23	Q
DO_2	As_3Co	Im3	T^5_h	cI32	B
DO_3	BiF_3	Fm3m	O^5_h	cF16	F
DO_9	O_3Re	Pm3m	O^1_h	cP4	C
DO_{11}	CFe_3	Pnma	D^{16}_{2h}	oP16	O
DO_{18}	$AsNa_3$	$P6_3/mmc$	D^4_{6h}	hP8	H
DO_{19}	Ni_3Sn	$P6_3/mmc$	D^4_{6h}	hP8	H
DO_{20}	Al_3Ni	Pnma	D^{16}_{2h}	oP16	O
DO_{21}	Cu_3P	$P\bar{3}c1$	D^4_{3d}	hP24	H

Strukturbericht symbol	Structure name	Symmetry group (International)	Symmetry group (Schoenflies)	Pearson symbol[a]	Standard ASTM E157-82a symbol[b]
DO_{22}	Cu_3P	I4/mmm	D^{17}_{4h}	tI8	U
DO_{23}	Al_3Zr	I4/mmm	D^{17}_{4h}	tI16	U
DO_{24}	Ni_3Ti	$P6_3/mmc$	D^{4}_{6h}	hP16	H
DO_c	SiU_3	I4/mcm	D^{18}_{4h}	tI16	U
DO_e	Ni_3P	$I\bar{4}$	S^{2}_{4}	tI32	U
$D1_3$	Al_4Ba	I4/mmm	D^{17}_{4h}	tI10	U
$D1_a$	$MoNi_4$	I4/m	C^{5}_{4h}	tI10	U
$D1_b$	Al_4U	Imma	D^{28}_{2h}	oI20	P
$D1_c$	$PtSn_4$	Aba2	C^{17}_{2v}	oC20	Q
$D1_e$	B_4Th	P4/mbm	D^{5}_{4h}	tP20	T
$D1_f$	BMn_4	Fddd	D^{24}_{2h}	oF40	S
$D2_1$	B_6Ca	Pm3m	O^{1}_{h}	cP7	C
$D2_3$	$NaZn_{13}$	Fm3m	O^{5}_{h}	cF112	F
$D2_b$	$Mn_{12}Th$	I4/mmm	D^{17}_{4h}	tI26	U
$D2_c$	MnU_6	I4/mcm	D^{18}_{4h}	tI28	U
$D2_d$	$CaCu_5$	P6/mmm	D^{1}_{6h}	hP6	H
$D2_f$	$B_{12}U$	Fm3m	O^{5}_{h}	cF52	F
$D2_h$	Al_6Mn	Cmcm	D^{17}_{2h}	oC28	Q
$D5_1$	α-Al_2O_3	R3c	D^{6}_{3d}	hR10	R
$D5_2$	La_2O_3	$P\bar{3}m1$	D^{3}_{3d}	hP5	H
$D5_3$	Mn_2O_3	Ia3	T^{7}_{h}	cI80	B
$D5_8$	S_3Sb_2	Pnma	D^{16}_{2h}	oP20	O
$D5_9$	P_2Zn_3	$P4_2/mmc$	D^{9}_{4h}	tP40	T
$D5_{10}$	C_2C_3	Pnma	D^{16}_{2h}	oP20	O
$D5_{13}$	Al_3Ni_2	$P\bar{3}m1$	D^{3}_{3d}	hP5	H
$D5_a$	Si_2U_3	P4/mbm	D^{5}_{4h}	tP10	T
$D5_c$	C_3Pu_2	$I\bar{4}3d$	T^{6}_{d}	cI40	B
$D7_1$	Al_4C_3	$R\bar{3}m$	D^{5}_{3d}	hR7	R
$D7_3$	P_4Th_3	$I\bar{4}3d$	T^{6}_{d}	cI28	B
$D7_b$	B_4Ta_3	Immm	D^{25}_{2h}	oI14	P
$D8_1$	Fe_3Zn_{10}	Im3m	O^{9}_{h}	cI52	B
$D8_2$	Cu_5Zn_8	$I\bar{4}3m$	T^{3}_{d}	cI52	B
$D8_3$	Al_4Cu_9	P43m	T^{1}_{d}	cP52	C
$D8_4$	C_6Cr23	Fm3m	O^{5}_{h}	cF116	F
$D8_5$	Fe_7W_6	$R\bar{3}m$	D^{5}_{3d}	hR13	R
$D8_6$	$Cu_{15}Si_4$	$I\bar{4}3m$	T^{3}_{d}	cI76	B
$D8_8$	Mn_5Si_3	$P6_3/mcm$	D^{3}_{6h}	hP16	H
$D8_9$	Co_9S_8	Fm3m	O^{5}_{h}	cF68	F
$D8_{10}$	Al_8Cr_5	R3m	C^{5}_{3v}	hR26	R
$D8_{11}$	Al_5Co_2	$P6_3/mcm$	D^{3}_{6h}	hP28	H
$D8_a$	$Mn_{23}Th_6$	Fm3m	O^{5}_{h}	cF116	F
$D8_b$	σ-phase of Cr-Fe	$p\bar{4}_2/mnm$	D^{14}_{4h}	tP30	T
$D8_e$	$(Al,Zn)_{49}Mg_{32}$	Im3	T^{5}_{h}	cI162	B
$D8_f$	Ge_7Ir_3	Im3m	O^{9}_{h}	cI40	B
$D8_h$	B_5W_2	$P6_3/mmc$	D^{4}_{6h}	hP14	H
$D8_i$	B_5Mo_2	$R\bar{3}m$	D^{5}_{3d}	hR7	R
$D8_l$	B_3Cr_5	I4/mcm	D^{18}_{4h}	tI32	U
$D8_m$	Si_3W_5	I4/mcm	D^{18}_{4h}	tI32	U
$D10_1$	C_3Cr_7	P31c	C^{4}_{3v}	hP80	H
$D10_2$	Fe_3Th_7	$P6_3mc$	C^{4}_{6v}	hP20	H
$E0_1$	ClFPb	P4/nmm	D^{7}_{4h}	tP6	T
$E1_1$	$CuFeS_2$	$I\bar{4}2d$	D^{12}_{2d}	tI16	U
$E2_1$	CaO_3Ti	Pm3m	O^{1}_{h}	cP5	C
$E2_4$	S_3Sn_2	Pnma	D^{16}_{2h}	oP20	O
E3	Al_2CdS_4	$I\bar{4}$	S^{2}_{4}	tI14	U
$E9_3$	$SiFe_3W_3$	Fd3m	O^{7}_{h}	cF112	F

Strukturbericht symbol	Structure name	Symmetry group (International)	Symmetry group (Schoenflies)	Pearson symbol[a]	Standard ASTM E157-82a symbol[b]
$E9_a$	Al_7Cu_2Fe	P4/mnc	D^6_{4h}	tP40	T
$E9_b$	$AlLi_3N_2$	Ia3	T^7_h	cI96	B
$F0_1$	NiSSb	$P2_13$	T^4	cP12	C
$F5_1$	$CrNaS_2$	R3m or R32	D^5_{3d} or D^7_3	hR4	R
$F5_6$	CuS_2Sb	Pnma	D^{16}_{2h}	oP16	O
$H1_1$	Al_2MgO_4	Fd3m	O^7_h	cF56	F
$H2_4$	Cu_3S_4V	P43m	T^1_d	cP8	C
$H2_5$	$AsCu_3S_4$	$Pmn2_1$	C^7_{2v}	oP16	O
$L1_0$	AuCu	P4/mmm	D^1_{4h}	tP4	T
$L1_2$	$AlCu_3$	Pm3m	O^1_h	cP4	C
$L2_1$	$AlCu_2Mn$	Fm3m	O^5_h	cF16	F
$L2_2$	Sb_2Tl_7	Im3m	O^9_h	cI54	B
$L'2_b$	H_2Th	I4/mmm	D^{17}_{4h}	tI6	U
L'3	Fe_2N	$P6_3/mmc$	D^4_{6h}	hP3	H
$L6_0$	$CuTi_3$	P4/mmm	D^1_{4h}	tP4	T

[a] The first letter denotes the crystal system: triclinic (a), monoclinic (m), orthorhombic (o), tetragonal (t), hexagonal (h), and cubic (c). Trigonal (rhombohedral) system is denoted by combination hR. The second letter of Pearson's symbol denotes lattice type: primitive (P), edge-(base-) centered (C), body-centered (I), or face-centered (F). The following number denotes number of atoms in the crystal unit cell.

[b] Standard ASTM E157-82a has the Bravais lattice designations as follows: C – primitive cubic; B – body-centered cubic; F – face-centered cubic; T – primitive tetragonal; U – body-centered tetragonal; R – rhombohedral; H – hexagonal; O – primitive orthorhombic; P – body-centered orthorhombic; Q – base-centered orthorhombic; S – face-centered orthorhombic; M – primitive monoclinic; N – centered monoclinic; A – triclinic.

References

1. A. Schoenflies, *Kristallsysteme und Kristallstructur*, Leipzig, 1891.
2. E. S. Fedorow, Zusammenstellung der kristallographischen Resultate, *Zs. Krist.*, 20, 1892 <https://doi.org/10.1524/zkri.1892.20.1.25>.
3. P. Groth, *Elemente der physikalischen und chemischen Krystallographie*, R. Oldenbourg, München/Berlin, 1921.
4. N. V. Belov, *Class Method of Deriving Space Groups of Symmetry, Trudy Instituta Kristallodraffi imeni Fedorova (Transactions of the Fedorov Inst. of Crystallography)*, 5, 25, 1951, in Russian.
5. W. B. Pearson, *Handbook of Lattice Spacings and Structures of Metals and Alloys*, Vol. 1, Pergamon Press, 1958; Vol. 2, 1967.
6. Ch. Kittel, *Introduction to Solid State Physics*, John Wiley & Sons, 1956.
7. G. S. Zhdanov, *Fizika Tverdogo Tela (Solid State Physics)*, Moscow University Press, 1962, in Russian.
8. M. J. Buerger, *Elementary Crystallography*, John Wiley & Sons, 1963.
9. F. D. Bloss, *Crystallography & Crystal Chemistry*, Holt, Rinehart & Winston, 1971.
10. T. Janssen, *Crystallographic Groups*, North-Holland/American Elsevier, 1973.
11. M. P. Shaskolskaya, *Kristallografiya (Crystallography)*, Vysshaya Shkola, Moscow, 1976, in Russian.
12. T. Hahn, Ed., Internat. *Tables for Crystallography*, Vol. A, D. Reidel Publishing, Boston, 1983.
13. Crystal Data. Determinative Tables, Volumes 1–6, 1966–1983, JCPDS-Intern Centre for Diffraction Data and U.S. Dept. of Commerce.
14. R. W. G. Wyckoff, *Crystal Structures*, 2nd ed., Volumes 1–6, Interscience, New York, 1963.
15. C. J. Bradley and A. P. Cracknell, *The Mathematical Theory of Symmetry in Solids*, Clarendon Press, Oxford, 1972.
16. International Tables for Crystallography. Volume A, *Space–Group Symmetry*, T. Hahn, Ed., 1989; Volume B, *Reciprocal Space*, U. Schmueli, Ed.; Volume C, *Mathematical, Physical and Chemical Tables*, A. J. C. Wilson, Ed., Kluwer Academic Publishers, Dordrecht, 1989.
17. G. R. Desiraju, *Crystal Engineering: The Design of Organic Solids*, Elsevier, Amsterdam, 1989.
18. M. Senechal, *Crystalline Symmetries: An Informal Mathematical Introduction*, Adam Hilger Publ., Bristol, 1990.
19. C. Hammond, *Introduction to Crystallography*, Oxford University Press, 1990.
20. N.W. Alcock, *Bonding and Structure: Structural Principles in Inorganic and Organic Chemistry*, Ellis Norwood Publ., 1990.
21. T. C. W. Mak and G. D. Zhou. *Crystallography in Modern Chemistry: A Resource Book of Crystal Structures*, Wiley–Interscience, New York, 1992.
22. S. C. Abrahams, K. Mirsky, and R. M. Nielson, *Acta Cryst*, B52, 806 (1996); B52, 1057 (1996) <https://doi.org/10.1107/S0108768196004582>.
23. C. Marcos, A. Panalague, D. B. Morciras, S. Garcia-Granda and M. R. Dias. *Acta Cryst*, B52, 899 (1996) <https://doi.org/10.1107/S0108768196002996>.
24. A. C. Larson, *Crystallographic Computing*, Manksgaard, Copenhagen, 1970.
25. G. M. Sheldrick, SHELXS86. Crystallographic Computing 3, Clarendon Press, Oxford, 1986; SHELXL93. Program for the Refinement of Crystal Structures, University of Göttingen Press, 1993.
26. Inorganic Crystal Structure Database, CD–ROM. Sci. Inf. Service. E-mail: SISI@Delphi.com.

IONIC RADII IN CRYSTALS

Ionic radii are a useful tool for predicting and visualizing crystal structures. This table lists a set of ionic radii R_i in Å units for the most common coordination numbers *CN* of positive and negative ions. The values are based on experimental crystal structure determinations, supplemented by empirical relationships, and theoretical calculations. Anions are listed first, followed by cations.

The advice of Howard T. Evans and Marvin J. Weber in preparing this table is appreciated.

References

1. Shannon, R. D., *Acta Crystallogr.* A32, 751, 1976.
2. Jia, Y. Q., *J. Solid State Chem.* 95, 184, 1991.

Ionic Radii for Ions with Various Coordination Numbers

Ion	*CN*	R_i/Å
Anions		
F^{-1}	6	1.33
Cl^{-1}	6	1.81
Br^{-1}	6	1.96
I^{-1}	6	2.20
OH^{-1}	4	1.35
OH^{-1}	6	1.37
O^{-2}	2	1.21
O^{-2}	6	1.40
O^{-2}	8	1.42
S^{-2}	6	1.84
Se^{-2}	6	1.98
Te^{-2}	6	2.21
Cations		
Ac^{+3}	6	1.12
Ag^{+1}	4	1.00
Ag^{+1}	6	1.15
Ag^{+1}	8	1.28
Ag^{+2}	4 (square)	0.79
Ag^{+2}	6	0.94
Al^{+3}	4	0.39
Al^{+3}	5	0.48
Al^{+3}	6	0.54
Am^{+3}	6	0.98
Am^{+3}	8	1.09
Am^{+4}	6	0.85
Am^{+4}	8	0.95
As^{+3}	6	0.58
As^{+5}	4	0.34
As^{+5}	6	0.46
Au^{+1}	6	1.37
Au^{+3}	4 (square)	0.64
Au^{+3}	6	0.85
Ba^{+2}	6	1.35
Ba^{+2}	8	1.42
Ba^{+2}	12	1.61
Be^{+2}	4	0.27
Be^{+2}	6	0.45
Bi^{+3}	5	0.96
Bi^{+3}	6	1.03
Bi^{+3}	8	1.17
Bi^{+5}	6	0.76
Bk^{+3}	6	0.96
Bk^{+4}	6	0.83
Bk^{+4}	8	0.93
Br^{+5}	3 (pyramidal)	0.31
Br^{+7}	4	0.25
Br^{+7}	6	0.39
C^{+4}	4	0.15
C^{+4}	6	0.16
Ca^{+2}	6	1.00
Ca^{+2}	8	1.12
Ca^{+2}	10	1.23
Ca^{+2}	12	1.34
Cd^{+2}	4	0.78
Cd^{+2}	6	0.95
Cd^{+2}	8	1.10
Cd^{+2}	12	1.31
Ce^{+3}	6	1.01
Ce^{+3}	8	1.14
Ce^{+3}	10	1.25
Ce^{+3}	12	1.34
Ce^{+4}	6	0.87
Ce^{+4}	8	0.97
Ce^{+4}	10	1.07
Ce^{+4}	12	1.14
Cf^{+3}	6	0.95
Cf^{+4}	6	0.82
Cf^{+4}	8	0.92
Cl^{+5}	3 (pyramidal)	0.12
Cl^{+7}	4	0.08
Cm^{+3}	6	0.97
Cm^{+4}	6	0.85
Cm^{+4}	8	0.95
Co^{+2}	4	0.56
Co^{+2}	6	0.65
Co^{+2}	8	0.90
Co^{+3}	6	0.55
Cr^{+2}	6	0.73
Cr^{+3}	6	0.62
Cr^{+4}	4	0.41
Cr^{+4}	6	0.55
Cr^{+6}	4	0.26
Cr^{+6}	6	0.44
Cs^{+1}	6	1.67
Cs^{+1}	8	1.74
Cs^{+1}	10	1.81
Cs^{+1}	12	1.88
Cu^{+1}	2	0.46
Cu^{+1}	4	0.60
Cu^{+1}	6	0.77
Cu^{+2}	4 (square)	0.57
Cu^{+2}	6	0.73
Dy^{+2}	6	1.07
Dy^{+2}	8	1.19
Dy^{+3}	6	0.91
Dy^{+3}	8	1.03
Er^{+3}	6	0.89
Er^{+3}	8	1.00
Eu^{+2}	6	1.17
Eu^{+2}	8	1.25
Eu^{+2}	10	1.35
Eu^{+3}	6	0.95
Eu^{+3}	8	1.07
F^{+7}	6	0.08
Fe^{+2}	4	0.63
Fe^{+2}	6	0.61
Fe^{+2}	8	0.92
Fe^{+3}	4	0.49
Fe^{+3}	6	0.55
Fe^{+3}	8	0.78
Fr^{+1}	6	1.80
Ga^{+3}	4	0.47
Ga^{+3}	6	0.62
Gd^{+3}	6	0.94
Gd^{+3}	8	1.05
Ge^{+2}	6	0.73
Ge^{+4}	4	0.39
Ge^{+4}	6	0.53
Hf^{+4}	4	0.58
Hf^{+4}	6	0.71
Hf^{+4}	8	0.83
Hg^{+1}	6	1.19
Hg^{+2}	2	0.69
Hg^{+2}	4	0.96
Hg^{+2}	6	1.02
Hg^{+2}	8	1.14
I^{+5}	3 (pyramidal)	0.44
I^{+5}	6	0.95
I^{+7}	4	0.42
I^{+7}	6	0.53
In^{+3}	4	0.62
In^{+3}	6	0.80
Ir^{+3}	6	0.68
Ir^{+4}	6	0.63
Ir^{+5}	6	0.57
K^{+1}	4	1.37
K^{+1}	6	1.38
K^{+1}	8	1.51
K^{+1}	12	1.64
La^{+3}	6	1.03
La^{+3}	8	1.16
La^{+3}	10	1.27
La^{+3}	12	1.36
Li^{+1}	4	0.59
Li^{+1}	6	0.76
Li^{+1}	8	0.92
Lu^{+3}	6	0.86
Lu^{+3}	8	0.97
Mg^{+2}	4	0.57
Mg^{+2}	6	0.72
Mg^{+2}	8	0.89
Mn^{+2}	4	0.66
Mn^{+2}	6	0.83
Mn^{+2}	8	0.96

Ion	*CN*	R_i/Å
Mn^{+3}	6	0.58
Mn^{+4}	4	0.39
Mn^{+4}	6	0.53
Mn^{+5}	4	0.33
Mn^{+6}	4	0.26
Mn^{+7}	4	0.25
Mo^{+3}	6	0.69
Mo^{+4}	6	0.65
Mo^{+5}	4	0.46
Mo^{+5}	6	0.61
Mo^{+6}	4	0.41
Mo^{+6}	6	0.59
Mo^{+6}	7	0.73
N^{+3}	6	0.16
N^{+5}	6	0.13
Na^{+1}	4	0.99
Na^{+1}	6	1.02
Na^{+1}	8	1.18
Na^{+1}	9	1.24
Na^{+1}	12	1.39
Nb^{+3}	6	0.72
Nb^{+3}	8	0.79
Nb^{+4}	6	0.68
Nb^{+5}	4	0.48
Nb^{+5}	6	0.64
Nb^{+5}	8	0.74
Nd^{+3}	6	0.98
Nd^{+3}	8	1.12
Nd^{+3}	9	1.16
Nd^{+3}	12	1.27
Ni^{+2}	4 (square)	0.49
Ni^{+2}	6	0.69
Ni^{+3}	6	0.56
Np^{+3}	6	1.01
Np^{+4}	6	0.87
Np^{+5}	6	0.75
Np^{+6}	6	0.72
Os^{+4}	6	0.63
Os^{+5}	6	0.58
Os^{+6}	6	0.55
Os^{+8}	4	0.39
P^{+5}	4	0.17
P^{+5}	6	0.38
Pa^{+3}	6	1.04
Pa^{+4}	6	0.90
Pa^{+5}	6	0.78
Pb^{+2}	6	1.19
Pb^{+2}	8	1.29
Pb^{+2}	10	1.40
Pb^{+2}	12	1.49
Pb^{+4}	4	0.65
Pb^{+4}	6	0.78
Pb^{+4}	8	0.94
Pd^{+2}	4 (square)	0.64
Pd^{+2}	6	0.86
Pd^{+3}	6	0.76
Pd^{+4}	6	0.62
Pm^{+3}	6	0.97
Pm^{+3}	8	1.09
Po^{+4}	6	0.97
Pr^{+3}	6	0.99

Ion	*CN*	R_i/Å
Pr^{+3}	8	1.13
Pr^{+4}	6	0.85
Pr^{+4}	8	0.96
Pt^{+2}	4 (square)	0.60
Pt^{+2}	6	0.80
Pt^{+4}	6	0.63
Pu^{+3}	6	1.00
Pu^{+4}	6	0.86
Pu^{+5}	6	0.74
Pu^{+6}	6	0.71
Ra^{+2}	8	1.48
Ra^{+2}	12	1.70
Rb^{+1}	6	1.52
Rb^{+1}	8	1.61
Rb^{+1}	10	1.66
Rb^{+1}	12	1.72
Re^{+4}	6	0.63
Re^{+5}	6	0.58
Re^{+6}	6	0.55
Re^{+7}	4	0.38
Re^{+7}	6	0.53
Rh^{+3}	6	0.67
Rh^{+4}	6	0.60
Rh^{+5}	6	0.55
Ru^{+3}	6	0.68
Ru^{+4}	6	0.62
Ru^{+5}	6	0.57
Ru^{+7}	4	0.38
Ru^{+8}	4	0.36
S^{+4}	6	0.37
S^{+6}	4	0.12
S^{+6}	6	0.29
Sb^{+3}	4 (pyramidal)	0.76
Sb^{+3}	6	0.76
Sb^{+5}	6	0.60
Sc^{+3}	6	0.75
Sc^{+3}	8	0.87
Se^{+4}	6	0.50
Se^{+6}	4	0.28
Se^{+6}	6	0.42
Si^{+4}	4	0.26
Si^{+4}	6	0.40
Sm^{+2}	6	1.19
Sm^{+2}	8	1.27
Sm^{+3}	6	0.96
Sm^{+3}	8	1.08
Sm^{+3}	12	1.24
Sn^{+4}	4	0.55
Sn^{+4}	6	0.69
Sn^{+4}	8	0.81
Sr^{+2}	6	1.18
Sr^{+2}	8	1.26
Sr^{+2}	10	1.36
Sr^{+2}	12	1.44
Ta^{+3}	6	0.72
Ta^{+4}	6	0.68
Ta^{+5}	6	0.64
Tb^{+3}	6	0.92
Tb^{+3}	8	1.04
Tb^{+4}	6	0.76
Tb^{+4}	8	0.88

Ion	*CN*	R_i/Å
Tc^{+4}	6	0.65
Te^{+4}	4	0.66
Te^{+4}	6	0.97
Te^{+6}	4	0.43
Te^{+6}	6	0.56
Th^{+4}	6	0.94
Th^{+4}	8	1.05
Th^{+4}	10	1.13
Th^{+4}	12	1.21
Ti^{+2}	6	0.86
Ti^{+3}	6	0.67
Ti^{+4}	4	0.42
Ti^{+4}	6	0.61
Ti^{+4}	8	0.74
Tl^{+1}	6	1.50
Tl^{+1}	8	1.59
Tl^{+1}	12	1.70
Tl^{+3}	4	0.75
Tl^{+3}	6	0.89
Tl^{+3}	8	0.98
Tm^{+2}	6	1.01
Tm^{+2}	7	1.09
Tm^{+3}	6	0.88
Tm^{+3}	8	0.99
U^{+3}	6	1.03
U^{+4}	6	0.89
U^{+4}	8	1.00
U^{+4}	12	1.17
U^{+5}	6	0.76
U^{+6}	2	0.45
U^{+6}	4	0.52
U^{+6}	6	0.73
U^{+6}	8	0.86
V^{+2}	6	0.79
V^{+3}	6	0.64
V^{+4}	5	0.53
V^{+4}	6	0.58
V^{+4}	8	0.72
V^{+5}	4	0.36
V^{+5}	5	0.46
V^{+5}	6	0.54
W^{+4}	6	0.66
W^{+5}	6	0.62
W^{+6}	4	0.42
W^{+6}	5	0.51
W^{+6}	6	0.60
Y^{+3}	6	0.90
Y^{+3}	8	1.02
Y^{+3}	9	1.08
Yb^{+2}	6	1.02
Yb^{+2}	8	1.14
Yb^{+3}	8	0.99
Yb^{+3}	9	1.04
Zn^{+2}	4	0.60
Zn^{+2}	6	0.74
Zn^{+2}	8	0.90
Zr^{+4}	4	0.59
Zr^{+4}	6	0.72
Zr^{+4}	8	0.84
Zr^{+4}	9	0.89

POLARIZABILITIES OF ATOMS AND IONS IN SOLIDS

H. P. R. Frederikse

The polarization of a solid dielectric medium, $\boldsymbol{P}$, is defined as the dipole moment per unit volume averaged over the volume of a crystal cell. A component of $\boldsymbol{P}$ can be expanded as a function of the electric field $\boldsymbol{E}$:

$$P_i = \sum_j a_j E_j + \sum_{jk} b_{jk} E_j E_k$$

For relatively small electric fields in isotropic substances $\boldsymbol{P} = \chi_e \boldsymbol{E}$, where χ_e is the electric susceptibility. If the medium is made up of N atoms (or ions) per unit volume, the polarization is $\boldsymbol{P} = N\boldsymbol{p}_m$ where $\boldsymbol{p}_m$ is the average dipole moment per atom. The polarizability α can be defined as $\boldsymbol{p}_m = \alpha \boldsymbol{E}_0$, where $\boldsymbol{E}_0$ is the local field at the position of the atom. Using the Lorentz method to calculate the local field one finds:

$$\boldsymbol{P} = N\alpha(\boldsymbol{E} + 4\pi\boldsymbol{P}) = \chi_e \boldsymbol{E}$$

Together with the definition of the dielectric constant (relative permittivity), $\varepsilon = 1 + 4\pi\chi_e$, this leads to:

$$\alpha = (3/4\pi N)(\varepsilon - 1)/(\varepsilon + 2)$$

This expression is known as the Clausius-Mossotti equation.

The total polarization associated with atoms, ions, or molecules is due to three different sources:

Electronic polarization arises because the center of the local electronic charge cloud around the nucleus is displaced under the action of the field: $\boldsymbol{P}_e = N\alpha_e \boldsymbol{E}_0$ where α_e is the *electronic polarizability.*

Ionic polarization occurs in ionic materials because the electric field displaces cations and anions in opposite directions: $\boldsymbol{P}_i = N\alpha_i \boldsymbol{E}_0$, where α_i is the *ionic polarizability.*

Orientational polarization can occur in substances composed of molecules that have permanent electric dipoles. The alignment of these dipoles depends on temperature and leads to an *orientational polarizability* per molecule: $\alpha_{or} = p^2/3kT$, where p is the permanent dipole moment per molecule, k is the Boltzmann constant, and T is the temperature.

Because of the different nature of these three polarization processes the response of a dielectric solid to an applied electric field will strongly depend on the frequency of the field. The resonance of the electronic excitation in insulators (dielectrics) takes place in the ultraviolet part of the spectrum; the characteristic frequency of the lattice vibrations is located in the infrared, while the orientation of dipoles requires fields of much lower frequencies (below 10^{10} Hz). This response to electric fields of different frequencies is shown in Figure 1. Values of the electronic polarizabilities for selected atoms and ions are given in Table 1.

References

1. Kittel, C., *Introduction to Solid State Physics, Fourth Edition,* John Wiley & Sons, New York, 1971.
2. Lerner, R.G., and Trigg, G.L., Eds., *Encyclopedia of Physics, Second Edition,* VCH Publishers, New York, 1990.
3. Ralls, K.M., Courtney, T.H., and Wulff, J., *An Introduction to Materials Science and Engineering,* John Wiley & Sons, New York, 1976.

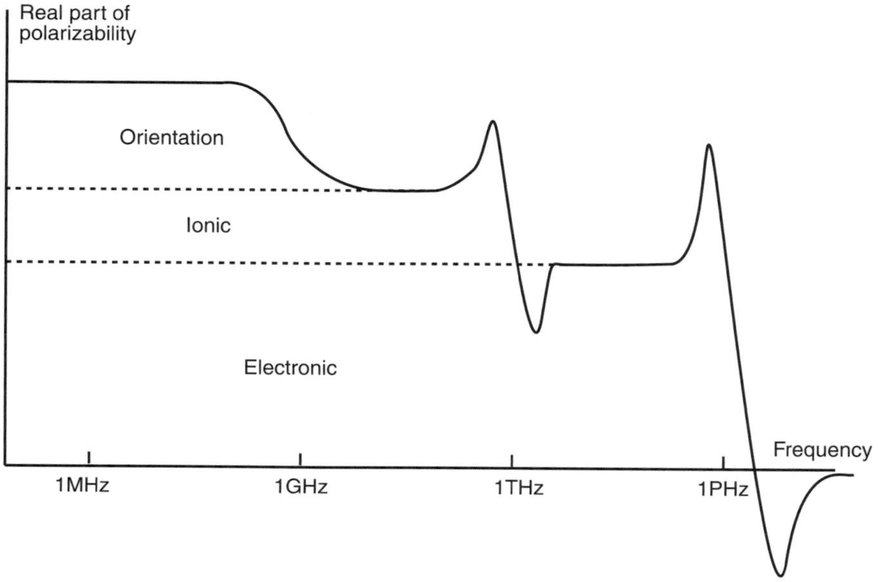

FIGURE 1. Schematic graph of the frequency dependence of the different contributions to polarizability.

TABLE 1. Electronic Polarizabilities in Units of 10^{-24} cm³

I	II	III	IV	VI	VII	VIII
						He
						0.205
Li^{+}	**Be^{2+}**	**B^{3+}**	**C^{4+}**	**O^{2-}**	**F^{-}**	**Ne**
0.029	0.008	0.003	0.0013	3.88	1.04	0.394
Na^{+}	**Mg^{2+}**	**Al^{3+}**	**Si^{4+}**	**S^{2-}**	**Cl^{-}**	**Ar**
0.179	0.094	0.052	0.0165	10.2	3.66	1.64
K^{+}	**Ca^{2+}**	**Sc^{3+}**	**Ti^{4+}**	**Se^{2-}**	**Br^{-}**	**Kr**
0.83	0.47	0.286	0.185	10.5	4.77	2.48
Rb^{+}	**Sr^{2+}**	**Y^{3+}**	**Zr^{4+}**	**Te^{2-}**	**I^{-}**	**Xe**
1.40	0.86	0.55	0.37	14.0	7.1	4.04
Cs^{+}	**Ba^{2+}**	**La^{3+}**	**Ce^{4+}**			
2.42	1.55	1.04	0.73			

Data from Pauling, L., *Proc. R. Soc. London*, A114, 181, 1927. See also Jaswal, S.S. and Sharma, T.P., *J. Phys. Chem. Solids*, 34, 509, 1973.

Values are appropriate for cgs units. To convert to SI, use the relation $\alpha(SI)/C\ m^2\ V^{-1} = 1.11265.10^{-16}\alpha(cgs)/cm^3$.

Solids

CRYSTAL STRUCTURES AND LATTICE PARAMETERS OF ALLOTROPES OF THE ELEMENTS

H. W. King

The crystal structures of the allotropic forms of the elements are presented in terms of the Pearson symbol, the Strukturbericht designation, and the prototype of the structure. The temperatures of the phase transformations are listed in degrees Celsius and the pressures are in GPa. A consistent nomenclature is used, whereby all allotropes are labeled by Greek letters. The lattice parameters of the units cells are given in nanometers (nm) and are considered to be accurate to ±2 in the last reported digit.

This compilation is restricted to changes in crystal structures that occur as a result of a change in temperature or pressure. Low-temperature structures are included for the diatomic and rare gases, which show many similarities with respect to the metallic elements.

Reprinted with the permission of ASM International from T.B. Massalski, Ed., *Binary Alloy Phase Diagrams*, ASM International, Metals Park, Ohio, 1986; certain data on rare-earth elements were provided by K.A. Gschneidner.

Element	*t*/°C	*p*/GPa	Pearson symbol	Space group	Strukturbericht designation	Prototype	*a*/nm	*b*/nm	*c*/nm
Ac	25	atm	*cF4*	*Fm3m*	*A1*	Cu	0.5311		
Ag	25	atm	*cF4*	*Fm3m*	*A1*	Cu	0.40857		
αAl	25	atm	*cF4*	*Fm3m*	*A1*	Cu	0.40496		
βAl	25	>20.5	*hP2*	$P6_3/mmc$	*A3*	Mg	0.2693		0.4398
α′Am	25	atm	*hP4*	$P6_3/mmc$	*A3′*	αLa	0.34681		1.1241
αAm	>769	atm	*cF4*	*Fm3m*	*A1*	Cu	0.4894		
βAm	>1074	atm	*cI2*	*Im3m*	*A2*	W	?		
γAm	25	>15	*oC4*	*Cmcm*	*A20*	αU	0.3063	0.5968	0.5169
αAr	<-189.35	atm	*cF4*	*Fm3m*	*A1*	Cu	0.5316		
(βAr)	<-189.40	atm	*hP2*	$P6_3/mmc$	*A3*	Mg	0.3760		0.6141
αAs	25	atm	*hR2*	*R3m*	*A7*	αAs	0.41319		
ϵAs	>448	atm	*oC8*	*Cmca*	...	P (black)	0.362	1.085	0.448
Au	25	atm	*cF4*	*Fm3m*	*A1*	Cu	0.40782		
βB	25	atm	*hR105*	*R3m*	...	βB	1.017		
αBa	25	atm	*cI2*	*Im3m*	*A2*	W	0.50227		
βBa	25	>5.33	*hP2*	$P6_3/mmc$	*A3*	Mg	0.3901		0.6154
γBa	25	>23	?	?	...	...			
αBe	25	atm	*hP2*	$P6_3/mmc$	*A3*	Mg	0.22859		0.35845
βBe	>1270	atm	*cI2*	*Im3m*	*A2*	W	0.25515		
γBe	25	>9.3	?	...	...	...			
αBi	25	atm	*hR2*	*r3m*	*A7*	αAs	0.47460		
βBi	25	>2.6	*mC4*	*C2/m*	...	βBi	0.6674	0.6117	0.3304
γBi	25	>3.0	*mP3*	?	...	...	0.605	0.42	0.465
σBi	25	>4.3	?	?	...	...			
ϵBi	25	>6.5	?	?	...	...			
ζBi	25	>9.0	*cI2*	*Im3m*	*A2*	W	0.3800		
αBk	25	atm	*hP4*	$P6_3/mmc$	*A3′*	αLa	0.3416		1.1069
βBk	>977	atm	*cF4*	*Fm3m*	*A1*	Cu	0.4997		
Br	<7.25	atm	*oC8*	*Cmca*	...	Cl	0.668	0.449	0.874
C (graphite)	25	atm	*hP4*	$P6_3/mmc$	*A9*	C (graphite)	0.24612		0.6709
C (diamond)	25	>60	*cF8*	*Fd3m*	*A4*	C (diamond)	0.35669		
C (hd)	25	HP	*hP4*	$P6_3/mmc$	...	C (hd)	0.2522		0.4119
αCa	25	atm	*cF4*	*Fm3m*	*A1*	Cu	0.55884		
βCa	>443	atm	*cI2*	*Im3m*	*A2*	W	0.4480		
γCa	25	>1.5	?	...	...	...			
Cd	25	atm	*hP2*	$P6_3/mmc$	*A3*	Mg	0.29793		0.56196
αCe	<-177	atm	*cF4*	*Fm3m*	*A1*	Cu	0.485		
βCe	25	atm	*hP4*	$P6_3/mmc$	*A3′*	αLa	0.36810		1.1857
γCe	25	atm	*cF4*	*Fm3m*	*A1*	Cu	0.51610		
δ-Ce	>726	atm	*cI2*	*Im3m*	*A2*	W	0.412		
α′Ce	25	>5.4	*oC4*	*Cmcm*	*A20*	αU	0.3049	0.5998	0.5215
αCf	25	atm	*hP4*	$P6_3/mmc$	*A3′*	αLa	0.339		1.1015

Element	t/°C	p/GPa	Pearson symbol	Space group	Strukturbericht designation	Prototype	a/nm	b/nm	c/nm
βCf	>590	atm	*cF4*	*Fm3m*	*A*1	Cu	?		
Cl	<-102	atm	*oC8*	*Cmca*	...	Cl	0.624	0.448	0.826
αCm	25	atm	*hP4*	$P6_3/mmc$	*A*3′	αLa	0.3496		1.1331
βCm	>1277	atm	*cF4*	*Fm3m*	*A*1	Cu	0.4382		
εCo	25	atm	*hP2*	$P6_3/mmc$	*A*3	Mg	0.25071		0.40686
αCo	>422	atm	*cF4*	*Fm3m*	*A*1	Cu	0.35447		
αCr	25	atm	*cI2*	*Im3m*	*A*2	W	0.28848		
α′Cr	25	HP	*tI2*	*I4/mmm*	...	α′Cr	0.2882		0.2887
aCs	25	atm	*cI2*	*Im3m*	*A*2	W	0.6141		
βCs	25	>2.37	*cF4*	*Fm3m*	*A*1	Cu	0.6465		
β′Cs	25	>4.22	*cF4*	*Fm3m*	*A*1	Cu	0.5800		
γCs	25	>4.27	?	...	...	...	?		
Cu	25	atm	*cF4*	*Fm3m*	*A*1	Cu	0.36146		
α′Dy	<-187	atm	*oC4*	*Cmcm*	...	α′Dy	0.3595	0.6184	0.5678
αDy	25	atm	*hP2*	$P6_3/mmc$	*A*3	Mg	0.35915		0.56501
βDy	>1381	atm	*cI2*	*Im3m*	*A*2	W	0.403		
γDy	25	>7.5	*hR3*	*R3m*	...	αSm	0.3436		2.483
Er	25	atm	*hP2*	$P6_3/mmc$	*A*3	Mg	0.35592		0.55850
αEs	25	atm	*hP4*	$P6_3/mmc$	*A*3′	αLa	?		
βEs	?	atm	*cF4*	*Fm3m*	*A*1	Cu	?		
Eu	25	atm	*cI2*	*Im3m*	*A*2	W	0.45827		
αF	<-227.6	atm	*mC8*	*C2/c*	...	αF	0.550	0.338	0.728
βF	<-219.67	atm	*cP16*	*Pm3n*	...	γO	0.667		
αFe	25	atm	*cI2*	*Im3m*	*A*2	W	0.28665		
γFe	>912	atm	*cF4*	*Fm3m*	*A*1	Cu	0.36467		
σFe	>1394	atm	*cI2*	*Im3m*	*A*2	W	0.29315		
εFe	25	>13	*hP2*	$P6_3/mmc$	*A*3	Mg	0.2468		0.396
αGa	25	atm	*oC8*	*Cmca*	*A*11	αGa	0.45186	0.76570	0.45258
βGa	25	>1.2	*tI2*	*I4/mmm*	*A*6	In	0.2808		0.4458
γGa	-53	>3.0	*oC40*	*Cmcm*	...	γGa	1.0593	1.3523	0.5203
αGd	25	atm	*hP2*	$P6_3/mmc$	*A*3	Mg	0.36336		0.57810
βGd	>1235	atm	*cI2*	*Im3m*	*A*2	W	0.406		
γGd	25	>3.0	*hR3*	*R3m*	...	αSm	0.361		2.603
αGe	25	atm	*cF8*	*Fd3m*	*A*4	C (diamond)	0.56574		
βGe	25	>12	*tI4*	$I4_1/amd$	*A*5	βSn	0.4884		0.2692
γGe	25	>12→atm	*tP12*	$P4_12_12$	...	σGe	0.593		0.698
σGe	LT	>12	*cI16*	*Im3m*	...	γSi	0.692		
αH	<-271.9	atm	*cF4*	*Fm3m*	*A*1	Cu	0.5338		
βH	<-259.34	atm	*hP2*	$P6_3/mmc$	*A*3	Mg	0.3776		0.6162
αHe	<-268.94	atm	*hP2*	$P6_3/mmc$	*A*3	Mg	0.3555		0.5798
βHe	>-258	0.125	*cF4*	*Fm3m*	*A*1	Cu	0.4240		
γHe	<-271.47	0.03	*cI2*	*Im3m*	*A*2	W	0.4110		
αHf	25	atm	*hP2*	$P6_3/mmc$	*A*3	Mg	0.31946		0.50510
βHf	>1995	atm	*cl2*	*Im3m*	*A*2	W	0.3610		
αHg	<-38.84	atm	*hR1*	*R3m*	*A*10	αHg	0.3005		
βHg	<-194	HP	*tI2*	*I4/mmm*	...	βHg	0.3995		0.2825
γHg	<-194	c.w.	*hR1*	?	...	...			
αHo	25	atm	*hP2*	$P6_3/mmc$	*A*3	Mg	0.35778		0.56178
βHo	25	>7.5	*hR3*	*R3m*	...	αSm	0.334		2.45
I	25	atm	*oC8*	*Cmca*	...	Cl	0.72697	0.47903	0.97942
In	25	atm	*tI2*	*I4/mmm*	*A*6	In	0.3253		0.49470
Ir	25	atm	*cF4*	*Fm3m*	*A*1	Cu	0.38392		
K	25	atm	*cI2*	*Im3m*	*A*2	W	0.5321		
Kr	<-157.39	atm	*cF4*	*Fm3m*	*A*1	Cu	0.5810		
αLa	25	atm	*hP4*	$P6_3/mmc$	*A*3′	αLa	0.37740		1.2171
βLa	>310	atm	*cF4*	*Fm3m*	*A*1	Cu	0.5303		
γLa	>865	atm	*cI2*	*Im3m*	*A*2	W	0.426		
β′La	25	>2.0	*cF4*	*Fm3m*	*A*1	Cu	0.517		
αLi	<-193	atm	*hP2*	$P6_3/mmc$	*A*3	Mg	0.3111		0.5093
βLi	25	atm	*cI2*	*Im3m*	*A*2	W	0.35093		

Element	t/°C	p/GPa	Pearson symbol	Space group	Strukturbericht designation	Prototype	a/nm	b/nm	c/nm
γLi	<-201	c.w.	*cF4*	$Fm3m$	*A1*	Cu	0.4388		
Lu	25	atm	*hP2*	$P6_3/mmc$	*A3*	Mg	0.35052		0.55494
Mg	25	atm	*hP2*	$P6_3/mmc$	*A3*	Mg	0.32094		0.52107
αMn	25	atm	cI58	$I43m$	*A12*	αMn	0.89126		
βMn	>710	atm	*cP20*	$P4_132$	*A13*	βMn	0.63152		
γMn	>1079	atm	*cF4*	$Fm3m$	*A1*	Cu	0.3860		
σMn	>1143	atm	*cI2*	$Im3m$	*A2*	W	0.3080		
Mo	25	atm	*cI2*	$Im3m$	*A2*	W	0.31470		
αN	<-237.6	atm	*cP8*	$Pa3$	...	αN	0.5661		
βN	<-210.00	atm	*hP4*	$P6_3/mmc$	...	βN	0.4050		0.6604
γN	<-253	>3.3	*tP4*	$P4_2/mnm$	...	γN	0.3957		0.5109
αNa	<-233	atm	*hP2*	$P6_3/mmc$	*A3*	Mg	0.3767		0.6154
βNa	25	atm	*cI2*	$Im3m$	*A2*	W	0.42906		
Nb	25	atm	*cI2*	$Im3m$	*A2*	W	0.33004		
αNd	25	atm	*hP4*	$P6_3/mmc$	*A3′*	αLa	0.36582		1.17966
βNd	>863	atm	*cI2*	$Im3m$	*A2*	W	0.413		
γNd	25	>5.0	*cF4*	$Fm3m$	*A1*	Cu	0.480		
Ne	<-243.59	atm	*cF4*	$Fm3m$	*A1*	Cu	0.4462		
Ni	25	atm	*cF4*	$Fm3m$	*A1*	Cu	0.35240		
αNp	25	atm	*oP8*	$Pnma$	A_c	αNp	0.6663	0.4723	0.4887
βNp	>280	atm	*tP4*	$P42_12$	A_d	βNp	0.4883		0.3389
γNp	>576	atm	*cI2*	$Im3m$	*A2*	W	0.352		
αO	<-243.3	atm	*mC4*	$C2m$	...	αO	0.5403	0.3429	0.5086
βO	<-229.6	atm	*hR2*	$R3m$	...	βO	0.4210		
γO	<-218.79	atm	*cP16*	$Pm3n$	...	γO	0.683		
Os	25	atm	*hP2*	$P6_3/mmc$	*A3*	Mg	0.27341		0.43918
P (black)	25	atm	*oC8*	$Cmca$	...	P (black)	0.33136	1.0478	0.43763
αPa	25	atm	*tI2*	$I4/mmm$	A_a	αPa	0.3921		0.3235
βPa	>1170	atm	*cI2*	$Im3m$	*A2*	W	0.381		
αPb	25	atm	*cF4*	$Fm3m$	*A1*	Cu	0.49502		
βPb	25	>10.3	*hP2*	$P6_3/mmc$	*A3*	Mg	0.3265		0.5387
Pd	25	atm	*cF4*	$Fm3m$	*A1*	Cu	0.38903		
αPm	25	atm	*hP4*	$P6_3/mmc$	*A3′*	αLa	0.365		1.165
βPm	>890	atm	*cI2*	$Im3m$	*A2*	W	0.410?		
αPo	25	atm	*cP1*	$Pm3m$	A_h	αPo	0.3366		
βPo	>54	atm	*hR1*	$R3m$	...	βPo	0.3373		
αPr	25	atm	*hP4*	$P6_3/mmc$	*A3′*	αLa	0.36721		1.18326
βPr	>795	atm	*cI2*	$Im3m$	*A2*	W	0.413		
γPr	25	>4.0	*cF4*	$Fm3m$	*A1*	Cu	0.488		
Pt	25	atm	*cF4*	$Fm3m$	*A1*	Cu	0.39236		
αPu	25	atm	*mP16*	$P2_1/m$	...	αPu	0.6183	0.4822	1.0963
βPu	>125	atm	*mI34*	$I2/m$	...	βPu	0.9284	1.0463	0.7859
γPu	>215	atm	*oF8*	$Fddd$	...	γPu	0.31587	0.57682	1.0162
σPu	>320	atm	*cF4*	$Fm3m$	*A1*	Cu	0.46371		
σ′Pu	>463	atm	*tI2*	$I4/mmm$	*A6*	In	0.33261		0.44630
ϵPu	>483	atm	*cI2*	$Im3m$	*A2*	W	0.36343		
Ra	25	atm	*cI2*	$Im3m$	*A2*	W	0.5148		
αRb	25	atm	*cI2*	$lm3m$	*A2*	W	0.5705		
βRb	25	>1.08	?	...	...	...			
γRb	25	>2.05	?	...	...	...			
Re	25	atm	*hP2*	$P6_3/mmc$	*A3*	Mg	0.27609		0.4458
Rh	25	atm	*cF4*	$Fm3m$	*A1*	Cu	0.38032		
Ru	25	atm	hP2	$P6_3/mmc$	*A3*	Mg	0.27058		0.42816
αS	25	atm	*oF128*	$Fddd$	*A16*	αS	1.0464	1.28660	2.44860
αSb	25	atm	*hR2*	$R3m$	*A7*	αAs	0.45067		
βSb	25	>5.0	*cP1*	$Pm3m$	A_h	αPo	0.2992		
γSb	25	>7.5	*hP2*	$P6_3/mmc$	*A3*	Mg	0.3376		0.5341
σSb	25	>14.0	*mP3*	?	...	...	0.556	0.404	0.422
αSc	25	atm	*hP2*	$P6_3/mm$	*A3*	Mg	0.33088		0.52680
βSc	>1337	atm	*cI2*	$Im3m$	*A2*	W	0.373?		

Element	t/°C	p/GPa	Pearson symbol	Space group	Strukturbericht designation	Prototype	a/nm	b/nm	c/nm
γSe	25	atm	*hP3*	$P3_121$	*A8*	γSe	0.43659		0.49537
αSi	25	atm	*cF8*	*Fd3m*	*A4*	C (diamond)	0.54306		
βSi	25	>9.5	*tI4*	$I4_1/amd$	*A5*	βSn	0.4686		0.2585
γSi	25	>16.0	*cI16*	*Im3m*	...	γSi	0.6636		
σSi	25	>16→atm	*hP4*	$P6_3/mmc$	*A3′*	αLa	0.380		0.628
αSm	25	atm	*hR3*	*R3m*	...	αSm	0.36290		2.6207
βSm	>734	atm	*hP2*	$P6_3/mmc$	*A3*	Mg	0.36630		0.58448
γ′Sm	>922	atm	*cI2*	*Im3m*	*A2*	W	0.410?		
σSm	25	>4.0	*hP4*	$P6_3/mmc$	*A3′*	αLa	0.3618		1.166
αSn	<13	atm	*cF8*	*Fd3m*	*A4*	C (diamond)	0.64892		
βSn	25	atm	*tI4*	$I4_1/amd$	*A5*	βSn	0.58318		0.31818
γSn	25	>9.0	*tI2*	?	...	γSn	0.370		0.337
αSr	25	atm	*cF4*	*Fm3m*	*A1*	Cu	0.6084		
βSr	>547	atm	*cI2*	*Im3m*	*A2*	W	0.487		
β′Sr	25	>3.5	*cI2*	*Im3m*	*A2*	W	0.4437		
Ta	25	atm	*cI2*	*Im3m*	*A2*	W	0.33030		
α′Tb	<-53	atm	*oC4*	*Cmcm*	...	α′Dy	0.3605	0.6244	0.5706
aTb	25	atm	*hP2*	$P6_3/mmc$	*A3*	Mg	0.36055		0.56966
βTb	>1289	atm	*cI2*	*Im3m*	*A2*	W	0.407?		
γTb	25	>6.0	*hR3*	*R3m*	...	αSm	0.341		2.45
Tc	25	atm	*hP2*	$P6_3/mmc$	*A3*	Mg	0.2738		0.4393
αTe	25	atm	*hP3*	$P3_121$	*A8*	γSe	0.44566		0.59264
βTe	25	>2.0	*hR2*	*R3m*	*A7*	αAs	0.469		
γTe	25	>7.0	*hR1*	*R3m*	...	βPo	0.3002		
αTh	25	atm	*cF4*	*Fm3m*	*A1*	Cu	0.50842		
βTh	>1360	atm	*cI2*	*Im3m*	*A2*	W	0.411		
αTi	25	atm	*hP2*	$P6_3/mmc$	*A3*	Mg	0.29506		0.46835
βTi	>882	atm	*cI2*	*Im3m*	*A2*	W	0.33065		
ωTi	25	HP→atm	*hP3*	*P6/mmm*	...	ωTi	0.4625		0.2813
αTl	25	atm	*hP2*	$P6_3/mmc$	*A3*	Mg	0.34566		0.55248
βTl	>230	atm	*cI2*	*Im3m*	*A2*	W	0.3879		
γTl	25	HP	*cF4*	*Fm3m*	*A1*	Cu	?		
Tm	25	atm	*hP2*	$P6_3/mmc$	*A3*	Mg	0.35375		0.55540
αU	25	atm	*oC4*	*Cmcm*	*A20*	αU	0.28537	0.58695	0.49548
βU	>668	atm	*tP30*	$P4_2/mnm$	A_b	βU	1.0759		0.5656
γU	>776	atm	*cI2*	*Im3m*	*A2*	W	0.3524		
V	25	atm	*cI2*	*Im3m*	*A2*	W	0.30240		
W	25	atm	*cI2*	*Im3m*	*A2*	W	0.31652		
Xe	<-111.76	atm	*cF4*	*Fm3m*	*A1*	Cu	0.6350		
αY	25	atm	*hP2*	$P6_3/mmc$	*A3*	Mg	0.36482		0.57318
βY	>1478	atm	*cI2*	*Im3m*	*A2*	W	0.410?		
αYb	<-3	atm	*hP2*	$P6_3/mmc$	*A3*	Mg	0.38799		0.63859
βYb	25	atm	*cF4*	*Fm3m*	*A1*	Cu	0.54848		
γYb	>795	atm	*cI2*	*Im3m*	*A2*	W	0.444		
Zn	25	atm	*hP2*	$P6_3/mmc$	*A3*	Mg	0.26650		0.49470
αZr	25	atm	*hP2*	$P6_3/mmc$	*A3*	Mg	0.32316		0.51475
βZr	>863	atm	*cI2*	*Im3m*	*A2*	W	0.36090		
ωZr	25	HP→atm	*hP2*	*P6/mmm*	...	ωTi	0.5036		0.3109

? Uncertain.

PHASE TRANSITIONS IN THE SOLID ELEMENTS AT ATMOSPHERIC PRESSURE

This table gives the phase transition temperatures for the elements that can exist in two or more crystalline forms (allotropes). The crystal phases are labeled by Greek letters in the most common conventions, although some variation is found. All data refer to normal atmospheric pressure.

References

1. Massalski, T. B., Ed., *Binary Alloy Phase Diagrams, Second Edition*, ASM International, Metals Park, OH, 1990.
2. Cordfunke, E. H. P., and Konings, R. J. M., Eds., *Thermochemical Data for Reactor Materials and Fission Products*, North-Holland, Amsterdam, 1990.
3. Greenwood, N. N., and Earnshaw, A., *Chemistry of the Elements, Second Edition*, Butterworth-Heinemann, Oxford, 1997.
4. Rhyne, J. J., Magnetic Phase Transitions of the Elements, *Bull. Alloy Phase Diag.* 3, 402, 1982.

Symbol	Transition	*t* /°C	Comments
Americium			
Am	α→β	769	
Am	β→γ	1077	
Am	γ→liq	1176	
Beryllium			
Be	α→β	1270	
Be	β→liq	1287	
Boron			
B	α→β	1100	
B	β→γ	1500	
B	γ→liq	2075	
Calcium			
Ca	α→β	443	
Ca	β→liq	842	
Californium			
Cf	α→β	590	
Cf	β→liq	900	
Cerium			
Ce	α→β	-177	
Ce	β→γ	61	β-Ce and γ-Ce are magnetic
Ce	γ→δ	726	
Ce	δ→liq	799	
Cobalt			
Co	ε→α	422	magnetic transition at 1115 °C
Co	α→liq	1495	
Curium			
Cm	α→β	1277	magnetic transition at −221 °C
Cm	β→liq	1345	
Dysprosium			
Dy	α'→α	-187	
Dy	α→β	1381	magnetic transitions in α-Dy at −184 °C and −94 °C
Dy	β→liq	1412	
Fluorine			
F_2	α→β	-227.60	
F_2	β→liq	-219.67	
Gadolinium			
Gd	α→β	1235	
Gd	β→liq	1313	
Hafnium			
Hf	α→β	1743	
Hf	β→liq	2233	
Iron			
Fe	α→γ	912	magnetic transition in α-Fe at 771 °C
Fe	γ→δ	1394	
Fe	δ→liq	1538	
Lanthanum			
La	α→β	277	
La	β→γ	860	
La	γ→liq	920	
Lithium			
Li	α→β	-193	
Li	β→liq	180.50	
Manganese			
Mn	α→β	727	magnetic transition in α-Mn at −100 °C
Mn	β→γ	1100	
Mn	γ→δ	1138	
Mn	δ→liq	1246	
Neodymium			
Nd	α→β	855	magnetic transition in α-Nd at −253 °C
Nd	β→liq	1016	
Neptunium			
Np	α→β	280	
Np	β→γ	576	
Np	γ→liq	644	
Nitrogen			
N_2	α→β	-237.54	
N_2	β→liq	-210.0	
Oxygen			
O_2	α→β	-249.29	
O_2	β→γ	-229.35	
O_2	γ→liq	-218.79	

Symbol	Transition	t/°C	Comments
Phosphorus			
P	brown→β-white	-190	several amorphous phases (red, black, gray) exist (Ref. 3)
P	β-white→α-white	-76.9	
P	α-white→liq	44.15	
Plutonium			
Pu	α→β	124.5	
Pu	β→γ	214.8	
Pu	γ→δ	320.0	
Pu	δ→δ'	462.9	
Pu	δ'→ε	482.6	
Pu	ε→liq	640	
Polonium			
Po	α→β	54	
Po	β→liq	254	
Praseodymium			
Pr	α→β	795	
Pr	β→liq	931	
Promethium			
Pm	α→β	890	magnetic transition in α-Pm at −175 °C
Pm	β→liq	1042	
Protactinium			
Pa	α→β	1170	
Pa	β→liq	1572	
Samarium			
Sm	α→β	734	magnetic transition in α-Sm at −167 °C
Sm	β→γ	922	
Sm	γ→liq	1072	
Scandium			
Sc	α→β	1337	
Sc	β→liq	1541	
Selenium			
Se	α-red→gray	180	many allotropes exist (Ref. 3)
Se	gray→liq	220.8	
Sodium			
Na	α→β	-233	
Na	β→liq	97.794	
Strontium			
Sr	α→β	547	
Sr	β→liq	777	
Sulfur			
S	α→β	95.3	many allotropes exist (Ref. 3)
S	β→liq	115.21	
Terbium			
Tb	α'→α	-53	
Tb	α→β	1289	magnetic transition in α-Tb at −230 °C
Tb	β→liq	1359	
Thallium			
Tl	α→β	230	
Tl	β→liq	304	
Thorium			
Th	α→β	1360	
Th	β→liq	1750	
Tin			
Sn	α (gray)→β (white)	13.2	
Sn	β (white)→liq	231.928	defining fixed point on ITS-90
Titanium			
Ti	α→β	882	
Ti	β→liq	1668	
Uranium			
U	α→β	669	
U	β→γ	776	
U	γ→liq	1135	
Ytterbium			
Yb	α→β	3	
Yb	β→γ	795	
Yb	γ→liq	824	
Yttrium			
Y	α→β	1478	
Y	β→liq	1522	
Zirconium			
Zr	α→β	866	
Zr	β→liq	1854.7	

THE MADELUNG CONSTANT AND CRYSTAL LATTICE ENERGY

If U is the crystal lattice energy and M is the Madelung constant, then[a]

$$U = \frac{NMz_i z_j e^2}{r}(1 - 1/n)$$

Substance	Ion type	Crystal form[b]	M
Sodium chloride, NaCl	M^+, X^-	FCC	1.74756
Cesium chloride, CsCl	M^+, X^-	BCC	1.76267
Calcium chloride, $CaCl_2$	M^{++}, $2X^-$	Cubic	2.365
Calcium fluoride (fluorite), CaF_2	M^{++}, $2X^-$	Cubic	2.51939
Cadmium chloride, $CdCl_2$	M^{++}, $2X^-$	Hexagonal	2.244[c]
Cadmium iodide (α), CdI_2	M^{++}, $2X^-$	Hexagonal	2.355[c]
Magnesium fluoride, MgF_2	M^{++}, $2X^-$	Tetragonal	2.381[c]
Cuprous oxide (cuprite), Cu_2O	$2M^+$, X^{--}	Cubic	2.22124
Zinc oxide, ZnO	M^{++}, X^{--}	Hexagonal	1.4985[c]
Sphalerite (zinc blende), ZnS	M^{++}, X^{--}	FCC	1.63806
Wurtzite, ZnS	M^{++}, X^{--}	Hexagonal	1.64132[c]
Titanium dioxide (anatase), TiO_2	M^{4+}, $2X^{--}$	Tetragonal	2.400[c]
Titanium dioxide (rutile), TiO_2	M^{4+}, $2X^{--}$	Tetragonal	2.408[c]
β-Quartz, SiO_2	M^{4+}, $2X^{--}$	Hexagonal	2.2197[c]
Corundum, Al_2O_3	$2M^{3+}$, $3X^{--}$	Rhombohedral	4.1719

[a] N is Avogadro's number, z_i and z_j are the integral charges on the ions (in units of e), and e is the charge on the electron in electrostatic units ($e = 4.803 \times 10^{-10}$ esu). r is the shortest distance between cation–anion pairs in centimeters. Then U is in ergs (1 erg = 10^{-7} J).

[b] FCC = face-centered cubic; BCC = body-centered cubic.

[c] For tetragonal and hexagonal crystals the value of M depends on the details of the lattice parameters.

The Born Exponent, n is:

Ion type	n
He, Li^+	5
Ne, Na^+, F^-	7
Ar, K^+, Cu^+, Cl^-	9
Kr, Rb^+, Ag^+, Br^-	10
Xe, Cs^+, Au^+, I^-	12

For a crystal with a mixed-ion type, an average of the values of n in this table is to be used (6 for LiF, for example).

ELASTIC CONSTANTS OF SINGLE CRYSTALS

H. P. R. Frederikse

This table gives selected values of elastic constants for single crystals. The values believed most reliable were selected from the original literature. The substances are arranged by crystal system and, within each system, alphabetically by name. A reference to the original literature is given for each value; a useful compilation of published values from many sources may be found in Ref. 1.

Data are given for the single-crystal density and for the elastic constants C_{ij}, in units of 10^{11} N/m^2, which is equivalent to 10^{12} dyn/cm^2.

General References

1. Simmons, G., and Wang, H., *Single Crystal Elastic Constants and Calculated Aggregate Properties: A Handbook, Second Edition*, MIT Press, Cambridge, MA, 1971.
2. Gray, D. E., Ed., *American Institute of Physics Handbook, Third Edition*, McGraw-Hill, New York, 1972.

Solids

Elastic Constants of Single Crystals: Cubic Crystals

Name	Formula	ρ/g cm^{-3}	T/K	C_{11}	C_{12}	C_{44}	Ref.
Aluminum	Al	2.6970	298	1.0675	0.6041	0.2834	1
Aluminum antimonide	AlSb	4.3600	300	0.8939	0.4427	0.4155	2
Ammonium bromide	NH_4Br	2.4314	300	0.3414	0.0782	0.0722	3
Ammonium chloride	NH_4Cl	1.5279	290	0.3814	0.0866	0.0903	4
Argon	Ar	1.7710	4.2	0.0529	0.0135	0.0159	5
Barium fluoride	BaF_2	4.8860	298	0.9199	0.4157	0.2568	6
Barium nitrate	$Ba(NO_3)_2$	3.2560	293	0.2925	0.2065	0.1277	7
Calcium fluoride	CaF_2	3.810	298	1.6420	0.4398	0.8406	8
Calcium telluride	CaTe	5.8544	298	0.5351	0.3681	0.1994	9
Cesium	Cs	1.9800	78	0.0247	0.0206	0.0148	10
Cesium bromide	CsBr	4.4560	298	0.3063	0.0807	0.0750	11
Cesium chloride	CsCl	3.9880	298	0.3644	0.0882	0.0804	11
Cesium iodide	CsI	4.5250	298	0.2446	0.0661	0.0629	11
Chromite	$FeCr_2O_4$	4.4500	[a]	3.2250	1.4370	1.1670	12
Chromium	Cr	7.20	298	3.398	0.586	0.990	13
Cobalt(II) oxide	CoO	6.44	298	2.6123	1.4699	0.8300	14
Cobalt zinc ferrite	$CoZnFeO_2$	5.43	303	2.660	1.530	0.780	12
Copper	Cu	8.932	298	1.683	1.221	0.757	15
Gallium antimonide	GaSb	5.6137	298	0.8839	0.4033	0.4316	16
Gallium arsenide	GaAs	5.3169	298	1.1877	0.5372	0.5944	17
Gallium phosphide	GaP	4.1297	300	1.4120	0.6253	0.7047	18
Germanium	Ge	5.313	298	1.2835	0.4823	0.6666	20
Gold	Au	19.283	296.5	1.9244	1.6298	0.4200	21
Indium antimonide	InSb	5.7890	298	0.6720	0.3670	0.3020	22
Indium arsenide	InAs	5.6720	293	0.8329	0.4526	0.3959	23
Indium phosphide	InP	4.78	[a]	1.0220	0.5760	0.4600	24
Iridium	Ir	22.52	300	5.80	2.42	2.56	25
Iron	Fe	7.8672	298	2.26	1.40	1.16	26
Lead	Pb	11.34	296	0.4966	0.4231	0.1498	27
Lead(II) fluoride	PbF_2	7.79	300	0.8880	0.4720	0.2454	28
Lead(II) nitrate	$Pb(NO_3)_2$	4.547	293	0.3729	0.2765	0.1347	29
Lead(II) telluride	PbTe	8.2379	303.2	1.0795	0.0764	0.1343	30
Lithium	Li	0.5326	298	0.1350	0.1144	0.0878	31
Lithium bromide	LiBr	3.47	[a]	0.3940	0.1880	0.1910	32
Lithium chloride	LiCl	2.068	295	0.4927	0.2310	0.2495	33
Lithium fluoride	LiF	2.638	[a]	1.1397	0.4767	0.6364	34
Lithium iodide	LiI	4.061	[a]	0.2850	0.1400	0.1350	32
Magnesium oxide	MgO	3.579	298	2.9708	0.9536	1.5613	20
Magnetite	Fe_3O_4	5.18	[a]	2.730	1.060	0.971	32
Manganese(II) oxide	MnO	5.39	298	2.23	1.20	0.79	35
Mercury(II) telluride	HgTe	8.079	290	0.548	0.381	0.204	36
Molybdenum	Mo	10.2284	273	4.637	1.578	1.092	37
Nickel	Ni	8.91	298	2.481	1.549	1.242	15
Niobium	Nb	8.578	300	2.4650	1.3450	0.2873	38
Palladium	Pd	12.038	300	2.2710	1.7604	0.7173	39
Platinum	Pt	21.50	300	3.4670	2.5070	0.7650	40

Name	Formula	ρ/g cm^{-3}	T/K	C_{11}	C_{12}	C_{44}	Ref.
Potassium	K	0.851	295	0.0370	0.0314	0.0188	41
Potassium bromide	KBr	2.740	298	0.3468	0.0580	0.0507	11
Potassium chloride	KCl	1.984	298	0.4069	0.0711	0.0631	11
Potassium cyanide	KCN	1.553	[a]	0.1940	0.1180	0.0150	32
Potassium fluoride	KF	2.480	295	0.6490	0.1520	0.1232	33
Potassium iodide	KI	3.128	300	0.2710	0.0450	0.0364	42
Pyrite	FeS_2	5.016	[a]	3.818	0.310	1.094	43
Rubidium	Rb	1.58	170	0.0296	0.0250	0.0171	44
Rubidium bromide	RbBr	3.350	300	0.3152	0.0500	0.0380	45
Rubidium chloride	RbCl	2.797	300	0.3624	0.0612	0.0468	45
Rubidium iodide	RbI	3.551	300	0.2556	0.0382	0.0278	45
Silicon	Si	2.331	298	1.6578	0.6394	0.7962	46
Silver	Ag	10.50	300	1.2399	0.9367	0.4612	47
Silver(I) bromide	AgBr	5.585	300	0.5920	0.3640	0.0616	48
Sodium	Na	0.971	299	0.0739	0.0622	0.0419	49
Sodium bromate	$NaBrO_3$	3.339	[a]	0.5450	0.1910	0.1500	32
Sodium bromide	NaBr	3.202	300	0.3970	0.1001	0.0998	33
Sodium chlorate	$NaClO_3$	2.485	[a]	0.4920	0.1420	0.1160	50
Sodium chloride	NaCl	2.163	298	0.4947	0.1288	0.1287	11
Sodium fluoride	NaF	2.804	300	0.9700	0.2380	0.2822	51
Sodium iodide	NaI	3.6689	300	0.3007	0.0912	0.0733	52
Spinel	$MgAl_2O_4$	3.6193	298	2.9857	1.5372	1.5758	53
Strontium fluoride	SrF_2	4.277	300	1.2350	0.4305	0.3128	54
Strontium nitrate	$Sr(NO_3)_2$	2.989	293	0.4255	0.2921	0.1590	29
Strontium oxide	SrO	4.99	300	1.601	0.435	0.590	55
Strontium titanate	$SrTiO_3$	5.123	[a]	3.4817	1.0064	4.5455	56
Tantalum	Ta	16.626	298	2.6023	1.5446	0.8255	57
Tantalum carbide	TaC	14.65	[a]	5.05	0.73	0.79	58
Thallium(I) bromide	TlBr	7.4529	298	0.3760	0.1458	0.0757	59
Thorium	Th	11.694	300	0.7530	0.4890	0.4780	60
Thorium(IV) oxide	ThO_2	9.991	298	3.670	1.060	0.797	61
Tin(II) telluride	SnTe	6.445	300	1.1250	0.0750	0.1172	62
Titanium carbide	TiC	4.940	[a]	5.00	1.13	1.75	107
Tungsten	W	19.257	297	5.2239	2.0437	1.6083	64
Uranium carbide	UC	13.63	300	3.200	0.850	0.647	65
Uranium(IV) oxide	UO_2	10.97	298	3.960	1.210	0.641	66
Vanadium	V	6.022	300	2.287	1.190	0.432	67
Yttrium iron oxide	$Y_3Fe_5O_{12}$	5.17	298	2.680	1.106	0.766	19
Zinc selenide	ZnSe	5.262	298	0.8096	0.4881	0.4405	68
Zinc sulfide (wurtzite)	ZnS	4.088	298	1.0462	0.6534	0.4613	68
Zinc telluride	ZnTe	5.636	298	0.7134	0.4078	0.3115	68
Zirconium carbide	ZrC	6.606	298	4.720	0.987	1.593	63

[a] Room temperature.

Elastic Constants of Single Crystals: Tetragonal Crystals

Name	Formula	ρ/g cm^{-3}	T/K	C_{11}	C_{12}	C_{13}	C_{16}	C_{33}	C_{44}	C_{66}	Ref.
Ammonium dihydrogen arsenate	$NH_4H_2AsO_4$	2.3110	298	0.6747	-0.106	0.1652		0.3022	0.0685	0.0639	69
Ammonium dihydrogen phosphate	$NH_4H_2PO_4$	1.8030	293	0.6200	-0.050	0.1400		0.3000	0.0910	0.0610	69
Barium titanate	$BaTiO_3$	5.9988	298	2.7512	1.7897	1.5156		1.6486	0.5435	1.1312	70
Calcium molybdate	$CaMoO_4$	4.255	298	1.447	0.664	0.466	0.134	1.265	0.369	0.451	79
Indium	In	7.300	[a]	0.4450	0.3950	0.4050		0.4440	0.0655	0.1220	71
Magnesium fluoride	MgF_2	3.177	[a]	1.237	0.732	0.536		1.770	0.552	0.978	72
Nickel(II) sulfate hexahydrate	$NiSO_4{\cdot}6H_2O$	2.070	[a]	0.3209	0.2315	0.0209		0.2931	0.1156	0.1779	73
Potassium dihydrogen arsenate	KH_2AsO_4	2.867	[a]	0.530	-0.060	-0.020		0.370	0.120	0.070	12
Potassium dihydrogen phosphate	KH_2PO_4	2.388	[a]	0.7140	-0.049	0.1290		0.5620	0.1270	0.0628	71
Rubidium dihydrogen phosphate	RbH_2PO_4	2.800	298	0.5562	-0.064	0.0279		0.4398	0.1142	0.0350	74
Tellurium dioxide	TeO_2	5.99	[a]	0.5320	0.4860	0.2120		1.0850	0.2440	0.5520	76
Tin (white)	Sn	7.29	288	0.7529	0.6156	0.4400		0.9552	0.2193	0.2336	77
Titanium(IV) oxide (rutile)	TiO_2	4.260	298	2.7143	1.7796	1.4957		4.8395	1.2443	1.9477	75
Zirconium(IV) orthosilicate	$ZrSiO_4$	4.70	[a]	2.585	1.791	1.542		3.805	0.733	1.113	78

[a] Room temperature.

Elastic Constants of Single Crystals: Orthorhombic Crystals

Name	Formula	ρ/g cm^{-3}	T/K	C_{11}	C_{12}	C_{13}	C_{22}	C_{23}	C_{33}	C_{44}	C_{55}	C_{66}	Ref.
Acenaphthene	$C_{12}H_{10}$	1.220	293	0.1380	0.0210	0.0410	0.1262	0.0460	0.1117	0.0265	0.0290	0.0185	80
Ammonium sulfate	$(NH_4)_2SO_4$	1.774	293	0.3607	0.1651	0.1580	0.2981	0.1456	0.3534	0.1025	0.0717	0.0974	81
Aragonite	$CaCO_3$	2.93	[a]	1.5958	0.3663	0.0197	0.8697	0.1597	0.8503	0.4132	0.2564	0.4274	82
Barite	$BaSO_4$	4.40	[a]	0.8941	0.4614	0.2691	0.7842	0.2676	1.0548	0.1190	0.2874	0.2778	82
Benzene	C_6H_6	1.061	250	0.0614	0.0352	0.0401	0.0656	0.0390	0.0583	0.0197	0.0378	0.0153	83
Benzophenone	$(C_6H_5)_2CO$	1.219	[a]	0.1070	0.0550	0.0169	0.1000	0.0321	0.0710	0.0203	0.0155	0.0353	32
Bronzite	$(Mg,Fe)SiO_4$	3.38	[a]	1.876	0.686	0.605	1.578	0.561	2.085	0.700	0.592	0.544	78
Calcium sulfate	$CaSO_4$	2.962	[a]	0.9382	0.1650	0.1520	1.845	0.3173	1.1180	0.3247	0.2653	0.0926	84
Celestite	$SrSO_4$	3.96	[a]	1.044	0.773	0.605	1.061	0.619	1.286	0.135	0.279	0.266	12
Cesium sulfate	Cs_2SO_4	4.243	293	0.4490	0.1958	0.1815	0.4283	0.1800	0.3785	0.1326	0.1319	0.1323	81
Forsterite	Mg_2SiO_4	3.224	298	3.2848	0.6390	0.6880	1.9980	0.7380	2.3530	0.6515	0.8120	0.8088	85
Iodic acid	HIO_3	4.630	[a]	0.3030	0.1194	0.1169	0.5448	0.0548	0.4359	0.1835	0.2193	0.1736	73
Lithium ammonium tartrate monohydrate	$LiNH_4C_4H_4O_6{\cdot}H_2O$	1.71	[a]	0.3864	0.1655	0.0875	0.5393	0.2007	0.3624	0.1190	0.0667	0.2326	12
Magnesium sulfate heptahydrate	$MgSO_4{\cdot}7H_2O$	1.68	[a]	0.325	0.174	0.182	0.288	0.182	0.315	0.078	0.156	0.090	86
Natrolite	$Na_2Al_2Si_3O_{10}{\cdot}2H_2O$	2.25	[a]	0.716	0.261	0.297	0.632	0.297	1.378	0.196	0.248	0.423	78
Nickel(II) sulfate heptahydrate	$NiSO_4{\cdot}7H_2O$	1.948	[a]	0.353	0.198	0.201	0.311	0.201	0.335	0.091	0.172	0.099	86
Olivine	$(Mg,Fe)SiO_4$	3.324	[a]	3.240	0.590	0.790	1.980	0.780	2.490	0.667	0.810	0.793	87
Potassium pentaborate tetrahydrate	$KB_5O_8{\cdot}4H_2O$	1.74	[a]	0.582	0.229	0.174	0.359	0.231	0.255	0.164	0.046	0.057	71
Potassium sulfate	K_2SO_4	2.665	293	0.5357	0.1999	0.2095	0.5653	0.1990	0.5523	0.195	0.1879	0.1424	81
Potassium sodium tartrate tetrahydrate	$KNaC_4H_4O_6{\cdot}4H_2O$	1.79	[a]	0.255	0.141	0.116	0.381	0.146	0.371	0.134	0.032	0.098	71
Rubidium sulfate	Rb_2SO_4	3.621	293	0.5029	0.1965	0.1999	0.5098	0.1925	0.4761	0.1626	0.1589	0.1407	81
Sodium ammonium tartrate tetrahydrate	$NaNH_4C_4H_4O_6{\cdot}4H_2O$	1.587	[a]	0.3685	0.2725	0.3083	0.5092	0.3472	0.5541	0.1058	0.0303	0.0870	12
Sodium tartrate	$Na_2C_4H_4O_6{\cdot}2H_2O$	1.794	[a]	0.461	0.286	0.320	0.547	0.352	0.665	0.124	0.031	0.098	12
Strontium formate dihydrate	$Sr(CHO_2)_2{\cdot}2H_2O$	2.25	[a]	0.4391	0.1037	-0.149	0.3484	-0.014	0.3746	0.1538	0.1075	0.1724	12
Sulfur (orthorhombic)	S_8	2.07	[a]	0.240	0.133	0.171	0.205	0.159	0.483	0.043	0.087	0.076	12
Thallium(I) sulfate	Tl_2SO_4	6.776	293	0.4106	0.2573	0.2288	0.3885	0.2174	0.4268	0.1125	0.1068	0.0751	81
Topaz	$Al_2(SiO_4)(F,OH)_2$	3.52	[a]	2.8136	1.2582	0.8464	3.8495	0.8815	2.9452	1.0811	1.3298	1.3089	82
Uranium	U	19.0453	293	2.1486	0.4622	0.2176	1.9983	1.0764	2.6763	1.2479	0.7379	0.7454	88
Zinc sulfate heptahydrate	$ZnSO_4{\cdot}7H_2O$	1.970	[a]	0.3320	0.1720	0.2000	0.2930	0.1980	0.3200	0.0780	0.1530	0.0830	86

[a] Room temperature.

Elastic Constants of Single Crystals: Monoclinic Crystals

Name	Formula	ρ/g cm^{-3}	T/K	C_{25}	C_{33}	C_{35}	C_{44}	C_{46}	C_{55}	C_{66}	Ref.
Acmite	$NaFe(SiO_3)_2$	3.50	RT	0.094	2.344	0.214	0.692	0.077	0.510	0.474	89
Anthracene	$C_{14}H_{10}$	1.28	RT	-0.0170	0.1522	-0.0187	0.0272	0.0138	0.0242	0.0399	90
Cobalt(II) sulfate heptahydrate	$CoSO_4{\cdot}7H_2O$	2.03	RT	-0.018	0.371	-0.047	0.060	0.016	0.058	0.101	86
Diopside	$CaMg(SiO_3)_2$	3.30	RT	-0.196	2.380	-0.336	0.675	-0.113	0.588	0.705	91
Feldspar	$K_2O{\cdot}Al_2O{\cdot}6SiO_2$	≈2.6	RT	-0.148	1.215	-0.131	0.143	-0.015	0.238	0.361	92
Iron(II) sulfate heptahydrate	$FeSO_4{\cdot}7H_2O$	1.895	RT	-0.019	0.360	-0.014	0.064	0.001	0.056	0.096	86
Lithium sulfate monohydrate	$Li_2SO_4{\cdot}H_2O$	2.06	RT	0.0571	0.5400	-0.0254	0.1400	-0.0054	0.1565	0.2770	32
Naphthalene	$C_{10}H_8$	1.0253	RT	-0.0270	0.1190	0.0290	0.0330	-0.0050	0.0210	0.0415	93
Potassium hydrogen tartrate	$KHC_4H_4O_6$	1.98	RT	0.0176	0.6816	0.0294	0.0961	-0.0044	0.1270	0.0841	12
Potassium tartrate	$K_2C_4H_4O_6$	1.99	RT	0.0182	0.5540	0.0710	0.0870	0.0072	0.1040	0.0826	32
Sodium thiosulfate	$Na_2S_2O_3$	1.69	RT	0.0983	0.4590	-0.0678	0.0569	-0.0268	0.1070	0.0598	12
trans-Stilbene	$(C_6H_5CH)_2$	0.9707	RT	-0.005	0.0790	-0.005	0.0325	0.0050	0.0640	0.0245	94
Triglycine sulfate	$(NH_2CH_2COOH)_3{\cdot}H_2SO_4$	1.68	RT	-0.0036	0.2630	-0.0500	0.0950	-0.0026	0.1110	0.0620	32

Elastic Constants of Single Crystals: Hexagonal Crystals

Name	Formula	ρ/g cm^{-3}	T/K	C_{11}	C_{12}	C_{13}	C_{33}	C_{55}	Ref.
Apatite	$Ca_5(PO_4)_3(OH,F,Cl)$	3.2	[a]	1.667	0.131	0.655	1.396	0.663	12
Beryl	$Be_3Al_2(SiO_3)_6$	2.64	[a]	2.800	0.990	0.670	2.480	0.658	12
Beryllium	Be	1.85	300	2.923	0.267	0.140	3.364	1.625	95
Beryllium oxide	BeO	3.01	[a]	4.70	1.68	1.19	4.94	1.53	96

Name	Formula	ρ/g cm^{-3}	T/K	C_{11}	C_{12}	C_{13}	C_{33}	C_{55}	Ref.
Cadmium	Cd	8.69	300	1.1450	0.3950	0.3990	0.5085	0.1985	97
Cadmium selenide	CdSe	5.81	298	0.7046	0.4516	0.3930	0.8355	0.1317	68
Cadmium sulfide	CdS	4.826	298	0.8431	0.5208	0.4567	0.9183	0.1458	98
Cobalt	Co	8.86	298	3.071	1.650	1.027	3.581	0.755	99
Dysprosium	Dy	8.55	298	0.7466	0.2616	0.2233	0.7871	0.2427	100
Erbium	Er	9.07	298	0.8634	0.3050	0.2270	0.8554	0.2809	100
Gadolinium	Gd	7.90	298	0.6667	0.2499	0.2132	0.7191	0.2089	101
Hafnium	Hf	13.3	298	1.881	0.772	0.661	1.969	0.557	102
Ice	H_2O	0.9167^0	250	0.1410	0.0660	0.0624	0.1515	0.0288	103
Indium	In	7.31	300	0.4535	0.4006	0.4151	0.4515	0.0651	104
Magnesium	Mg	1.74	298	0.5950	0.2612	0.2180	0.6155	0.1635	105
Rhenium	Re	20.8	298	6.1820	2.7530	2.0780	6.8350	1.6060	100
Ruthenium	Ru	12.1	298	5.6260	1.8780	1.6820	6.2420	1.8060	100
Thallium	Tl	11.8	300	0.4080	0.3540	0.2900	0.5280	0.0726	106
Titanium	Ti	4.506	298	1.6240	0.9200	0.6900	1.8070	0.4670	102
Titanium boride	TiB_2	4.38	[a]	6.90	4.10	3.20	4.40	2.50	107
Yttrium	Y	4.47	300	0.7790	0.2850	0.2100	0.7690	0.2431	108
Zinc	Zn	7.134	295	1.6368	0.3640	0.5300	0.6347	0.3879	109
Zinc oxide	ZnO	5.6	298	2.0970	1.2110	1.0510	2.1090	0.4247	110
Zinc sulfide (wurtzite)	ZnS	4.09	298	1.2420	0.6015	0.4554	1.4000	0.2864	96
Zirconium	Zr	6.52	298	1.434	0.728	0.653	1.648	0.320	102

[a] Room temperature.

Elastic Constants of Single Crystals: Trigonal Crystals

Name	Formula	Density	T/K	C_{11}	C_{12}	C_{13}	C_{14}	C_{33}	C_{44}	Ref.
Aluminum oxide (α)	Al_2O_3	3.99	300	4.9735	1.6397	1.1220	-0.2358	4.9911	1.4739	111
Aluminum phosphate	$AlPO_4$	2.56	[a]	1.0503	0.2934	0.6927	-0.1271	1.3353	0.2314	73
Antimony	Sb	6.70	295	1.0130	0.3450	0.2920	0.2090	0.4500	0.3930	112
Bismuth	Bi	9.79	295	0.6370	0.2490	0.2470	0.0717	0.3820	0.1123	112
Calcium carbonate (calcite)	$CaCO_3$	2.710	300	1.4806	0.5578	0.5464	-0.2058	0.8557	0.3269	113
Iron(III) oxide	Fe_2O_3	5.25	[a]	2.4243	0.5464	0.1542	-0.1247	2.2734	0.8569	82
Lithium niobate	$LiNbO_3$	4.30	[a]	2.030	0.530	0.750	0.090	2.450	0.600	114
Lithium tantalate	$LiTaO_3$	7.45	[a]	2.330	0.470	0.800	-0.110	2.750	0.940	114
Quartz (α)	SiO_2	2.65	298	0.8680	0.0704	0.1191	-0.1804	1.0575	0.5820	115
Selenium	Se	4.838	300	0.1870	0.0710	0.2620	0.0620	0.7410	0.1490	116
Sodium nitrate	$NaNO_3$	2.261		0.8670	0.1630	0.1600	0.0820	0.3740	0.2130	12
Tourmaline	$Na(Mg,Fe,Mn,Li,Al)_3Al_6Si_6O_{18}(BO_3)_3$	3.14	[a]	2.7066	0.6927	0.0872	-0.0774	1.6070	0.6682	82

[a] Room temperature.

References

3. Thomas, J. F., *Phys. Rev.*, 175, 955–962, 1968.
4. Bolef, D. I. and M. Menes, *J. Appl. Phys.*, 31, 1426–1427, 1960.
5. Garland, C. W. and C. F. Yarnell, *J. Chem. Phys.*, 44, 1112–1120, 1966.
6. Garland, C. W. and R. Renard, *J. Chem. Phys.*, 44, 1130–1139, 1966.
7. Gsänger, M., H. Egger and E. Lüscher, *Phys. Letters*, 27A, 695–696, 1968.
8. Wong, C. and D. E. Schuele, *J. Phys. Chem. Solids*, 29, 1309–1330, 1968.
9. Haussühl, S., *Phys. Stat. Sol.*, 3, 1072–1076, 1963.
10. Wong, C. and D. E. Schuele, *J. Phys. Chem. Solids*, 28, 1225–1231, 1967.
11. McSkimin, H. J. and D. G. Thomas, *J. Appl. Phys.*, 33, 56–59, 1962.
12. Kollarits, F. J. and J. Trivisonno, *J. Phys. Chem. Solids*, 29, 2133–2139, 1968.
13. Slagle, D. D. and H. A. McKinstry, *J. Appl. Phys.*, 38, 446–458, 1967.
14. Hearmon, R. F. S., *Adv. Phys.*, 5, 323–382, 1956.
15. Sumer, A. and J. F. Smith, *J. Appl. Phys.*, 34, 2691–2694, 1963.
16. Alexandrov, K. S. et al., *Sov. Phys. Sol. State*, 10, 1316–1321, 1968.
17. Epstein, S. G. and O. N. Carlson, *Acta Metal.*, 13, 487–491, 1965.
18. McSkimin, H. J., et al., *J. Appl. Phys.*, 39, 4127–4128, 1968.
19. McSkimin, H. J., et al., *J. Appl. Phys.*, 38, 2362–2364, 1967.
20. Weil, R. and W. O. Groves, *J. Appl. Phys.*, 39, 4049–4051, 1968.
21. Bateman, T. B., *J. Appl. Phys.*, 37, 2194–2195, 1966.
22. Bogardus, E. H., *J. Appl. Phys.*, 36, 2504–2513, 1965.
23. Golding, B., S. C. Moss and B. L. Averbach, *Phys. Rev.*, 158, 637–645, 1967.
24. Bateman, T. B., H. J. McSkimin and J. M. Whelan, *J. Appl. Phys.*, 30, 544–545, 1959.
25. Gerlich, D., *J. Appl. Phys.*, 35, 3062, 1964.
26. Hickernell, F. S. and W. R. Gayton, *J. Appl. Phys.*, 37, 462, 1966.
27. MacFarlane, R. E., et al., *Phys. Letters*, 20, 234–235, 1966.
28. Leese, J. and A. E. Lord Jr., *J. Appl. Phys.*, 39, 3986–3988, 1968.
29. Miller, R. A. and D. E. Schuele, *J. Phys. Chem. Solids*, 30, 589–600, 1969.
30. Wasilik, J. H. and M. L. Wheat, *J. Appl. Phys.*, 36, 791–793, 1965.
31. Haussühl, S., *Phys. Stat. Sol.*, 3, 1072–1076, 1963.
32. Houston, B., et al., *J. Appl. Phys.*, 39, 3913–3916, 1968.
33. Trivisonno, J. and C. S. Smith, *Acta Metal.*, 9, 1064–1071, 1961.
34. Alexandrov, K. S. and T. V. Ryzhova, *Sov. Phys. Cryst.*, 6, 228–252, 1961.
35. Lewis, J. T., A. Lehoczky and C. V. Briscoe, *Phys. Rev.*, 161, 877–887, 1967.
36. Drabble, J. R. and R. E. B. Strathen, *Proc. Phys. Soc.*, 92, 1090–1995, 1967.
37. Oliver, D. W., *J. Appl. Phys.*, 40, 893, 1969.
38. Alper, T., and G. A. Saunders, *J. Phys. Chem. Solids*, 28, 1637–1642, 1967.

39. Dickinson, J. M. and P. E. Armstrong, *J. Appl. Phys.*, 38, 602–606, 1967.
40. Bolef, D. I., *J. Appl. Phys.*, 32, 100–105, 1961.
41. Rayne, J. A., *Phys. Rev.*, 112, 1125–1130, 1958.
42. MacFarlane, R. E., et al., *Phys. Letters*, 18, 91–92, 1965.
43. Smith, P. A. and C. S. Smith, *J. Phys. Chem. Solids*, 26, 279–289, 1965.
44. Norwood, M. H. and C. V. Briscoe, *Phys. Rev.*, 112, 45–48, 1958.
45. Simmons, G. and F. Birch, *J. Appl. Phys.*, 34, 2736–2738, 1963.
46. Gutman, E. J. and J. Trivisonno, *J. Phys. Chem. Sol.*, 28, 805–809, 1967.
47. Ghafelehbashi, M., et al., *J. Appl. Phys.*, 41, 652–666, 1970.
48. McSkimin, H. J. and P. Andreatch, Jr., *J. Appl. Phys.*, 35, 2161–2165, 1964.
49. Neighbours, J. R. and G. A. Alers, *Phys. Rev.*, 111, 707–712, 1958.
50. Hidshaw, W., J. T. Lewis, and C. V. Briscoe, *Phys. Rev.*, 163, 876–881, 1967.
51. Daniels, W. B., *Phys. Rev.*, 119, 1246–1252, 1960.
52. Viswanathan, R., *J. Appl. Phys.*, 37, 884–886, 1966.
53. Miller, R. A. and C. S. Smith, *J. Phys. Chem. Sol.*, 25, 1279–1292, 1964.
54. Claytor, R. N. and B. J. Marshall, *Phys. Rev.*, 120, 332–334, 1960.
55. Schreiber, E., *J. Appl. Phys.*, 38, 2508–2511, 1967.
56. Gerlich, D., *Phys. Rev.*, 136, A1366–A1368, 1964.
57. Johnston, D. L., P. H. Thrasher and R. J. Kearney, *J. Appl. Phys.*, 41, 427–428, 1970.
58. Poindexter, E. and A. A. Giardini, *Phys. Rev.*, 110, 1069, 1958.
59. Soga, N., *J. Appl. Phys.*, 37, 3416–3420, 1966.
60. Bartlett, R. W. and C. W. Smith, *J. Appl. Phys.*, 38, 5428–5429, 1967.
61. Morse, G. E. and A. W. Lawson, *J. Phys. Chem. Sol.*, 28, 939–950, 1967.
62. Armstrong, P. E., O. N. Carlson and J. F. Smith, *J. Appl. Phys.*, 30, 36–41, 1959.
63. Macedo, P. M., W. Capps and J. B. Wachtman, *J. Am. Cer. Soc.*, 47, 651, 1964.
64. Beattie, A. G., *J. Appl. Phys.*, 40, 4818–4821, 1969.
65. Chang, R. and L. J. Graham, *J. Appl Phys.*, 37, 3778–3783, 1966.
66. Lowrie, R. and A. M. Gonas, *J. Appl. Phys.*, 38, 4505–4509, 1967.
67. Graham, L. J., H. Nadler and R. Chang, *J. Appl. Phys.*, 34, 1572–1573, 1963.
68. Wachtman, J. B., Jr., et al., *J. Nucl. Mat.*, 16, 39–41, 1965.
69. Bolef, D. I., *J. Appl. Phys.*, 32, 100–105, 1961.
70. Berlincourt, D., H. Jaffe and L. R. Shiozawa, *Phys. Rev.*, 129, 1009–1017, 1963.
71. Adhav. R. S. *J. Acoust. Soc. Am.*, 43, 835–838, 1968.
72. Berlincourt, D. and H. Jaffe, *Phys. Rev.*, 111, 143–148, 1958.
73. Huntington, H. B., in *Solid State Pysics, Vol. 7*, Seitz, F., and Turnbull, D., Ed., pp. 213–285, Academic Press, New York 1958.
74. Cutler, H. R., J. J. Gibson and K. A. McCarthy, *Sol. State Comm.*, 6, 431–433, 1968.
75. Mason, W. P., *Piezoelectric Crystals and Their Application to Ultrasonics*, D. Van Nostrand Co., Inc., New York, 1950.
76. Adhav, R. S., *J. Appl. Phys.*, 40, 2725–2727, 1969.
77. Manghnani, M. H., *J. Geophys. Res.*, 74, 4317–4328, 1969.
78. Uchida, N. and Y. Ohmachi, *J. Appl Phys.*, 40, 4692–4695, 1969.
79. House, D. G. and E. Y. Vernon, *Br. J. Appl. Phys.*, 11, 254–259, 1960.
80. Ryzhova, T. V., et al., *Bull. Acad. Sci. USSR, Earth Phys. Ser.*, English Transl., no. 2, 111–113, 1966.
81. Alton, W. J. and A. J. Barlow, *J. Appl. Phys.*, 38, 3817–3820, 1967.
82. Michard, F., et al., *C. R. Acad. Sci., Paris*, 265, 565–567, 1967.
83. Haussühl, S., *Acta Cryst.*, 18, 839–842, 1965.
84. Hearmon, R. F. S., *Rev. Mod. Phys.*, 18, 409–440, 1946.
85. Heseltine, J. C. W., D. W. Elliott and O. B. Wilson, *J. Chem. Phys.*, 40, 2584–2587, 1964.
86. Schwerdtner, W. M., et al., *Canad. J. Earth Sci.*, 2, 673–683, 1965.
87. Kumazawa, M. and O. L. Anderson, *J. Geophys. Res.*, 74, 5961–5972, 1969 .
88. Alexandrov, K. S., et al., *Sov. Phys. Cryst.*, 7, 753–755, 1963.
89. Verma, R. K., *J. Geophys. Soc.*, 65, 757–766, 1960.
90. McSkimin, H. J. and E. S. Fisher, *J. Appl. Phys.*, 31, 1627–1639, 1960.
91. Alexandrov, K. S. and T.V. Ryzhova, *Bull. Acad. Sci. USSR, Geophys. Ser.*, English Transl., no. 8, 871–875, 1961.
92. Afanaseva, G. K., et al, *Phys. Stat. Sol.*, 24, K61–K63, 1967.
93. Alexandrov, K. S., et al., *Sov. Phys. Cryst.*, 8, 589–591, 1964.
94. Alexandrov, K. S. and T. V Ryzhova, *Bull Acad. Sci. USSR, Geophys. Ser.*, English Transl., no. 2, 129–131, 1962.
95. Alexandrov, K. S., et al., *Sov. Phys. Cryst.*, 8, 164–166, 1963.
96. Teslenko, V. F., et al., *Sov. Phys. Cryst.*, 10, 744–747, 1966.
97. Smith, J. F. and C. L. Arbogast, *J. Appl. Phys.*, 31, 99–102, 1960.
98. Cline, C. F., H. L. Dunegan and G. M. Henderson, *J. Appl. Phys.*, 38, 1944–1948, 1967.
99. Chang, Y. A. and L. Himmel, *J. Appl. Phys.*, 37, 3787–3790, 1966.
100. Gerlich, D., *J. Phys. Chem. Solids*, 28, 2575–2579, 1967.
101. McSkimin, H. J., *J. Appl. Phys.*, 26, 406–409, 1955.
102. Fisher, E. S. and D. Dever, *Trans. Met. Soc. AIME*, 239, 48–57, 1967.
103. Fisher, E. S. and D. Dever, *Proc. Conf. Rare Earth Res.*, 6th, Gatlinburg, TN, 522–533, 1967.
104. Fisher, E. S. and C. J. Renken, *Phys. Rev.*, 135, A482–A494, 1964.
105. Proctor, T. M., Jr., *J. Acoust. Soc. Am.*, 39, 972–977, 1966.
106. Chandrasekhar, B. S. and J. A. Rayne, *Phys. Rev.*, 124, 1011–1041, 1961.
107. Wazzan, A. R. and L. B. Robinson, *Phys. Rev.*, 155, 586–594, 1967.
108. Ferris, R. W., et al., *J. Appl. Phys.*, 34, 768–770, 1963.
109. Gilman, J. J. and B. W. Roberts, *J. Appl. Phys.*, 32, 1405, 1961.
110. Smith, J. F. and J. A. Gjevre, *J. Appl. Phys.*, 31, 645–647, 1960.
111. Alers, G. A. and J. R. Neighbours, *J. Phys. Chem. Solids*, 7, 58–64, 1908.
112. Bateman, T. B., *J. Appl. Phys.*, 33, 3309–3312, 1962.
113. Tefft, W. E., *J. Res. Natl. Bur. Stand.*, 70A, 277–280, 1966.
114. DeBretteville, Jr., A. et al., *Phys. Rev.*, 148, 575–579, 1966.
115. Dandekar, D. P. and A. L. Ruoff, *J. Appl. Phys.*, 39, 6004–6009, 1968.
116. Warner, A. W., M. Onoe and G. A. Coquin, *J. Acoust. Soc. Am.*, 42, 1223–1231, 1967.
117. McSkimin, H. J., P. Andreatch and R. N. Thurston, *J. Appl. Phys.*, 36, 1624–1632, 1965.
118. Mort, J., *J. Appl. Phys.*, 38, 3414–3415, 1967.

ELECTRICAL RESISTIVITY OF PURE METALS

This table gives the electrical resistivity, in units of 10^{-8} Ω m, for 65 metallic elements at common room temperatures (293 K and 298 K). The data refer to polycrystalline samples. The number of significant figures indicates the accuracy of the values. The Online Edition of the *CRC Handbook* contains resistivity values at temperatures ranging from 1 K to 1000K. Note that at low temperatures (especially below 50 K) the electrical resistivity is extremely sensitive to sample purity. Thus, the low-temperature values refer to samples of specified purity and treatment. The references should be consulted for further information on this point, as well as for values at additional temperatures.

References

1. C. Y. Ho, et al., *J. Phys. Chem. Ref. Data*, 12, 183–322, 1983; 13, 1069–1096, 1984; 13, 1097–1130, 1984, 13, 1131–1172, 1984.
2. R. A. Matula, *J. Phys Chem. Ref. Data*, 8, 1147–1298, 1979 <https://doi.org/10.1063/1.555614>.
3. T. C. Chi, *J. Phys. Chem. Ref. Data*, 8, 339–438, 1979; 8, 439–498, 1979 <https://doi.org/10.1063/1.555598>.
4. K. H. Hellwege, Ed., *Landolt-Börnstein Numerical Data and Functional Relationships in Science and Technology*, Group III, Vol. 15, Subvolume a, Springer-Verlag, Heidelberg, 1982.
5. L. A. Hall, *Survey of Electrical Resistivity Measurements on 16 Pure Metals in the Temperature Range 0 to 273 K*, NBS Technical Note 365, U.S. Superintendent of Documents, 1968.

Electrical Resistivity of Pure Metals in 10^{-8} Ω m at the Indicated Temperature

Name	293 K	298 K
Aluminum	2.650	2.709
Antimony (gray)		
Barium	33.2	34.0
Beryllium	3.56	3.70
Bismuth		
Cadmium		
Calcium	3.36	3.42
Cerium (β form)	82.8	
Cerium (γ form)		74.4
Cesium	20.5	20.8
Chromium	12.5	12.6
Cobalt		
Copper	1.678	1.712
Dysprosium	92.6	
Erbium	86.0	
Europium	90.0	
Gadolinium	131	
Gallium		
Gold	2.214	2.255
Hafnium	33.1	33.7
Holmium	81.4	
Indium		
Iridium		
Iron	9.61	9.87
Lanthanum	61.5	
Lead	20.8	21.1
Lithium	9.28	9.47
Lutetium	58.2	
Magnesium	4.39	4.48
Manganese	144	144
Mercury		96.1
Molybdenum	5.34	5.47
Neodymium	64.3	
Nickel	6.93	7.12
Niobium		
Osmium		
Palladium	10.54	10.73
Platinum	10.5	10.7
Polonium		
Potassium	7.20	7.39
Praseodymium	70.0	
Promethium	75[e]	
Protactinium		
Rhenium		
Rhodium		
Rubidium	12.8	13.1
Ruthenium		
Samarium	94.0	
Scandium	56.2	
Silver	1.587	1.617
Sodium	4.77	4.88
Strontium	13.2	13.4
Tantalum	13.1	13.4
Terbium	115	
Thallium		
Thorium		
Thulium	67.6	
Tin (white)		
Titanium		
Tungsten	5.28	5.39
Uranium		
Vanadium	19.7	20.1
Ytterbium	25.0	
Yttrium	59.6	
Zinc	5.90	6.01
Zirconium	42.1	42.9

[e] Estimated.

ELECTRICAL RESISTIVITY OF SELECTED ALLOYS

These values were obtained by fitting all available measurements to a theoretical formulation describing the temperature and composition dependence of the electrical resistivity of metals. Some of the values listed here fall in regions of temperature and composition where no actual measurements exist. Details of the procedure may be found in the reference.

Values of the resistivity are given in units of 10^{-8} Ω m. General comments in the preceding table for pure metals also apply here.

Reference

Ho, C. Y., et al., *J. Phys. Chem. Ref. Data*, 12, 183–322, 1983.

Electrical Resistivity in Units of 10^{-8} Ω m at the Indicated Temperature

Wt%	100 K	273 K	293 K	300 K	350 K	400 K
Aluminum-Copper						
Wt% Al						
99[a]	0.531	2.51	2.74	2.82	3.38	3.95
95[a]	0.895	2.88	3.10	3.18	3.75	4.33
90[b]	1.38	3.36	3.59	3.67	4.25	4.86
85[b]	1.88	3.87	4.10	4.19	4.79	5.42
80[b]	2.34	4.33	4.58	4.67	5.31	5.99
70[b]	3.02	5.03	5.31	5.41	6.16	6.94
60[b]	3.49	5.56	5.88	5.99	6.77	7.63
50[b]	4.00	6.22	6.55	6.67	7.55	8.52
40[c]		7.57	7.96	8.10	9.12	10.2
30[c]		11.2	11.8	12.0	13.5	15.2
25[f]		16.3	17.2	17.6	19.8	22.2
15[h]			12.3			
10[g]	8.71	10.8	11.0	11.1	11.7	12.3
5[c]	7.92	9.43	9.61	9.68	10.2	10.7
1[b]	3.22	4.46	4.60	4.65	5.00	5.37
Aluminum-Magnesium						
Wt% Al						
99[c]	0.958	2.96	3.18	3.26	3.82	4.39
95[c]	3.01	5.05	5.28	5.36	5.93	6.51
90[c]	5.42	7.52	7.76	7.85	8.43	9.02
10[b]	14.0	17.1	17.4	17.6	18.4	19.2
5[b]	9.93	13.1	13.4	13.5	14.3	15.2
1[a]	2.78	5.92	6.25	6.37	7.20	8.03
Copper-Gold						
Wt% Cu						
99[c]	0.520	1.73	1.86	1.91	2.24	2.58
95[c]	1.21	2.41	2.54	2.59	2.92	3.26
90[c]	2.11	3.29	3.42	3.46	3.79	4.12
85[c]	3.01	4.20	4.33	4.38	4.71	5.05
80[c]	3.95	5.15	5.28	5.32	5.65	5.99
70[c]	5.91	7.12	7.25	7.30	7.64	7.99
60[c]	8.04	9.18	9.13	9.36	9.70	10.05
50[c]	9.88	11.07	11.20	11.25	11.60	11.94
40[c]	11.44	12.70	12.85	12.90	13.27	13.65
30[c]	12.43	13.77	13.93	13.99	14.38	14.78
25[c]	12.59	13.93	14.09	14.14	14.54	14.94
15[c]	11.38	12.75	12.91	12.96	13.36	13.77
10[c]	9.33	10.70	10.86	10.91	11.31	11.72
5[c]	5.91	7.25	7.41	7.46	7.87	8.28
1[c]	2.00	3.40	3.57	3.62	4.03	4.45
Copper-Nickel						
Wt% Cu						
99[c]	1.45	2.71	2.85	2.91	3.27	3.62
95[c]	6.19	7.60	7.71	7.82	8.22	8.62
90[c]	12.08	13.69	13.89	13.96	14.40	14.81
85[c]	18.01	19.63	19.83	19.90	20.32	20.70
80[c]	23.89	25.46	25.66	25.72	26.12	26.44
70[i]	35.73	36.67	36.72	36.76	36.85	36.89
60[i]	45.76	45.43	45.38	43.35	45.20	45.01
50[i]	50.22	50.19	50.05	50.01	49.73	49.50
40[c]	36.77	47.42	47.73	47.82	48.28	48.49
30[i]	26.73	40.19	41.79	42.34	44.51	45.40
25[c]	22.22	33.46	35.11	35.69	39.67	42.81
15[c]	13.49	22.00	23.35	23.85	27.60	31.38
10[c]	9.28	16.65	17.82	18.26	21.51	25.19
5[c]	5.20	11.49	12.50	12.90	15.69	18.78
1[c]	1.81	7.23	8.08	8.37	10.63	13.18
Copper-Palladium						
Wt% Cu						
99[c]	0.91	2.10	2.23	2.27	2.59	2.92
95[c]	2.99	4.21	4.35	4.40	4.74	5.08
90[c]	5.69	6.89	7.03	7.08	7.41	7.74
85[c]	8.30	9.48	9.61	9.66	10.01	10.36
80[c]	10.74	11.99	12.12	12.16	12.51	12.87
70[c]	15.67	16.87	17.01	17.06	17.41	17.78
60[c]	20.45	21.73	21.87	21.92	22.30	22.69
50[c]	26.07	27.62	27.79	27.86	28.25	28.64
40[c]	33.53	35.31	35.51	35.57	36.03	36.47
30[c]	45.03	46.50	46.66	46.71	47.11	47.47
25[c]	44.12	46.25	46.45	46.52	46.99	47.43
15[c]	31.79	36.52	36.99	37.16	38.28	39.35
10[c]	23.00	28.90	29.51	29.73	31.19	32.56
5[c]	13.09	20.00	20.75	21.02	22.84	24.54
1[c]	8.97	11.90	12.67	12.93	14.82	16.68
Copper-Zinc						
Wt% Cu						
99[b]	0.671	1.84	1.97	2.02	2.36	2.71
95[b]	1.54	2.78	2.92	2.97	3.33	3.69
90[b]	2.33	3.66	3.81	3.86	4.25	4.63
85[b]	2.93	4.37	4.54	4.60	5.02	5.44
80[b]	3.44	5.01	5.19	5.26	5.71	6.17
70[b]	4.08	5.87	6.08	6.15	6.67	7.19
Gold-Palladium						
Wt% Au						
99[c]	1.31	2.69	2.86	2.91	3.32	3.73
95[c]	3.88	5.21	5.35	5.41	5.79	6.17
90[i]	6.70	8.01	8.17	8.22	8.56	8.93
85[b]	9.14	10.50	10.66	10.72	11.10	11.48
80[b]	11.23	12.75	12.93	12.99	13.45	13.93
70[c]	16.44	18.23	18.46	18.54	19.10	19.67
60[b]	24.64	26.70	26.94	27.02	27.63	28.23
50[a]	23.09	27.23	27.63	27.76	28.64	29.42

Wt%	100 K	273 K	293 K	300 K	350 K	400 K
40[a]	19.40	24.65	25.23	25.42	26.74	27.95
30[b]	14.94	20.82	21.49	21.72	23.35	24.92
25[b]	12.72	18.86	19.53	19.77	21.51	23.19
15[a]	8.54	15.08	15.77	16.01	17.80	19.61
10[a]	6.54	13.25	13.95	14.20	16.00	17.81
5[a]	4.58	11.49	12.21	12.46	14.26	16.07
1[a]	3.01	10.07	10.85	11.12	12.99	14.80

Gold-Silver
Wt% Au

Wt%	100 K	273 K	293 K	300 K	350 K	400 K
99[b]	1.20	2.58	2.75	2.80	3.22	3.63
95[a]	3.16	4.58	4.74	4.79	5.19	5.59
90[j]	5.16	6.57	6.73	6.78	7.19	7.58
85[j]	6.75	8.14	8.30	8.36	8.75	9.15
80[j]	7.96	9.34	9.50	9.55	9.94	10.33
70[j]	9.36	10.70	10.86	10.91	11.29	11.68
60[j]	9.61	10.92	11.07	11.12	11.50	11.87
50[j]	8.96	10.23	10.37	10.42	10.78	11.14
40[j]	7.69	8.92	9.06	9.11	9.46	9.81
30[a]	6.15	7.34	7.47	7.52	7.85	8.19
25[a]	5.29	6.46	6.59	6.63	6.96	7.30
15[a]	3.42	4.55	4.67	4.72	5.03	5.34
10[a]	2.44	3.54	3.66	3.71	4.00	4.31
5[i]	1.44	2.52	2.64	2.68	2.96	3.25
1[b]	0.627	1.69	1.80	1.84	2.12	2.42

Iron-Nickel
Wt% Fe

Wt%	100 K	273 K	293 K	300 K	350 K	400 K
99[a]	3.32	10.9	12.0	12.4		18.7
95[c]	10.0	18.7	19.9	20.2		26.8
90[c]	14.5	24.2	25.5	25.9		33.2
85[c]	17.5	27.8	29.2	29.7		37.3
80[c]	19.3	30.1	31.6	32.2		40.0
70[b]	20.9	32.3	33.9	34.4		42.4
60[c]	28.6	53.8	57.1	58.2		73.9
50[d]	12.3	28.4	30.6	31.4		43.7
40[d]	7.73	19.6	21.6	22.5		34.0

Wt%	100 K	273 K	293 K	300 K	350 K	400 K
30[c]	5.97	15.3	17.1	17.7		27.4
25[b]	5.62	14.3	15.9	16.4		25.1
15[c]	4.97	12.6	13.8	14.2		21.1
10[c]	4.20	11.4	12.5	12.9		18.9
5[c]	3.34	9.66	10.6	10.9		16.1
1[b]	1.66	7.17	7.94	8.12		12.8

Silver-Palladium
Wt% Ag

Wt%	100 K	273 K	293 K	300 K	350 K	400 K
99[b]	0.839	1.891	2.007	2.049	2.35	2.66
95[b]	2.528	3.58	3.70	3.74	4.04	4.34
90[b]	4.72	5.82	5.94	5.98	6.28	6.59
85[k]	6.82	7.92	8.04	8.08	8.38	8.68
80[k]	8.91	10.01	10.13	10.17	10.47	10.78
70[k]	13.43	14.53	14.65	14.69	14.99	15.30
60[i]	19.4	20.9	21.1	21.2	21.6	22.0
50[k]	29.3	31.2	31.4	31.5	32.0	32.4
40[m]	40.8	42.2	42.2	42.2	42.3	42.3
30[b]	37.1	40.4	40.6	40.7	41.3	41.7
25[k]	32.4	36.67	37.06	37.19	38.1	38.8
15[i]	21.0	27.08	26.68	27.89	29.3	30.6
10[i]	14.95	21.69	22.39	22.63	24.3	25.9
5[b]	8.91	15.98	16.72	16.98	18.8	20.5
1[a]	3.97	11.06	11.82	12.08	13.92	15.70

[a] Uncertainty in resistivity is ± 2%.
[b] Uncertainty in resistivity is ± 3%.
[c] Uncertainty in resistivity is ± 5%.
[d] Uncertainty in resistivity is ± 7% below 300 K and ± 5% at 300 K and 400 K.
[e] Uncertainty in resistivity is ± 7%.
[f] Uncertainty in resistivity is ± 8%.
[g] Uncertainty in resistivity is ± 10%.
[h] Uncertainty in resistivity is ± 12%.
[i] Uncertainty in resistivity is ± 4%.
[j] Uncertainty in resistivity is ± 1%.
[k] Uncertainty in resistivity is ± 3% up to 300 K and ± 4% above 300 K.
[m] Uncertainty in resistivity is ± 2% up to 300 K and ± 4% above 300 K.

ELECTRICAL RESISTIVITY OF GRAPHITE MATERIALS

L. I. Berger

At normal conditions, the only stable crystallographic modification of carbon is graphite. The quasi-stable diamond turns into graphite starting from about 1000 °C in air. In industry, a graphitic material is commonly called either *carbon*, if it consists of small and low-oriented crystallites, or *graphite*, the material with a highly ordered structure. In the 1970s, the first carbon filaments of about 7 nm in diameter were grown by Morinobu Endo at the University of Orleans, France, by the vapor-growth technique. In 1985, Sir Harold Walter Kroto of Sussex University, UK, and Richard E. Smalley and coworkers at Rice University discovered spherical carbon molecules, C_{60} (or C-60), consisting of combinations of carbon atoms organized into hexagons and pentagons, named *buckminsterfullerenes* or *fullerenes* and possessing very promising mechanical and electrical properties. In 1991, Sumio Iijima, NEC Labs, Japan, and David S. Bethune, IBM Almaden Labs, observed the carbon atomic groups in the form of tubes capped by halves of the fullerene molecules and formed on the cathodes of carbon arc devices. The length of the tubes could be up to tens of micrometers and the diameter, naturally, is equal to that of the fullerene molecule. These tubes, called *nanotubes*, may be single wall (SWNT) or consist of several concentric tubes with a common axis (multi-walled nanotubes, MWNT). Two-dimensional *graphene* is another crystallographic modification of graphite (Saroj Nayak, Rensselaer U., 2004) that is a flat hexagonal network of carbon atoms with a thickness equal to the carbon atom size. The nanotube may be considered as formed by strips of graphene turned into a cylinder. The character of the electrical conductivity (metallic or semiconductive) of a SWNT depends on orientation of the carbon hexagons of the nanotube surface regarding its axis (the chiral angle [Ref. 1]). The following table contains some typical data on electrical and electronic properties of graphite materials.

Values in the table below refer to room-temperature measurements. Values of electrical resistivity in brackets [···] are in μΩ inch units.

References

1. M. S. Dresselhaus, G. Dresselhaus, and Ph. Avouris (Eds.), *Carbon Nanotubes. Synthesis, Structure, Properties, and Applications*, Springer-Verlag, 2001 <https://doi.org/10.1007/3-540-39947-X>.
2. ESPI Metals Catalog, 2007.
3. SPI Supplies Catalog, 2007.
4. F. L. Vogel, *J. Mater. Sci.* 12, 982–986, 1977 <https://doi.org/10.1007/BF00540981>.
5. K. S. Novoselov et al., *Nature* 438, 197–200, 2005 <https://doi.org/10.1038/nature04233>.
6. Y. Zhang et al., *Nature* 438, 201–204, 2005 <https://doi.org/10.1038/nature04235>.
7. N. Tombros et al., *Nature* 448, 571–574, 2007 <https://doi.org/10.1038/nature06037>.
8. H. Dai, in Ref. 1, pp. 29–53.
9. CTI Carbon Nanotube Cat., 2007.
10. L. Matija et al., *Sci. Forum* 413, 49–52, 2003 <https://doi.org/10.4028/www.scientific.net/MSF.413.49>.

Electrical Resistivity and Other Properties of Graphite Materials

Material	Electrical resistivity ρ/mΩ cm	Energy gap E/eV	Electron mobility μ/cm^2 V^{-1} s^{-1}	$(1/\rho)(d\rho/dt)$/ 10^{-4}°C^{-1}	Ref.
Bulk graphite					
Electromet graphite	1.90 [750] [e]			-5	2
Electro graphite	1.60 [630] [e]			-5	2
Aeromet graphite	1.47 [580] [e]			-5	2
ESPI Superconductive	1.75 [690] [e]			-5	2
Radioelectronics data	30 [11,800] [e]			-5.6	3
Highly ordered pyrolytic graphite	Parallel 0.04 [15.7] [e]; across 150 [59000] [e]				3
Single crystal graphite, normal to *c*-axis	$1\cdot10^{-6}$				4
Graphenes					
n-Graphene		≈5 (M); ≈10 (Γ) [c]	10^6		5,6
p-Graphene			10^4		7
Carbon nanotubes					
Metallic SWNT	12 kΩ [a]				1
Semiconducting SWNT		0.7 – 0.9 [b]		128 [d]	1
MWNT	10^2				9
Carbon fullerenes					
Fullerene (C_{60})	10^{12}	1.95			10

[a] Minimum resistance of individual nanotubes (Ref. 8).
[b] Est. from Ref. 1, p. 47.
[c] Est. from Ref. 1, p. 116.
[d] Est. from Ref. 1, p. 179.
[e] Value brackets are in μΩ inch units.

PERMITTIVITY (DIELECTRIC CONSTANT) OF INORGANIC SOLIDS

H. P. R. Frederikse

This table lists the permittivity ε, frequently called the dielectric constant, of a number of inorganic solids. When the material is not isotropic, the individual components of the permittivity are given. A superscript S indicates a measurement made under constant strain ("clamped" dielectric constant). If the constraint is removed, the measurement yields ε^T, the "unclamped" or free dielectric constant.

The temperature of the measurement is given when available; the symbol r.t. indicates a value at nominal room temperature. The frequency of the measurement is given in the last column (i.r. indicates a measurement in the infrared).

Substances are listed in alphabetical order by chemical formula.

Reference

Young, K. F., and Frederikse, H. P. R., *J. Phys. Chem. Ref. Data* 2, 313, 1973.

Permittivity of Inorganic Solids

Formula	Name	ε_{ijk}	T/K	ν/Hz
Ag_3AsS_3	Silver thioarsenate (Proustite)	ε_{11}^T = 16.5, ε_{11}^S = 14.5	r.t.	2 × 10^7
		ε_{33}^T = 20.0, ε_{11}^S = 18.0	r.t.	2 × 10^7
AgBr	Silver bromide	12.50	r.t.	
AgCN	Silver cyanide	5.6	r.t.	10^6
AgCl	Silver chloride	11.15	r.t.	
$AgNO_3$	Silver nitrate	9.0	293	5 × 10^5
$AgNa(NO_2)_2$	Silver sodium nitrite	4.5 ± 0.5	r.t.	9.4 ×10^9
Ag_2O	Silver oxide	8.8	r.t.	
$(AlF)_2SiO_4$	Aluminum fluorosilicate (Topaz)	ε_{11} = 6.62	297	7 × 10^3
		ε_{22} = 6.58	297	7 × 10^3
		ε_{33} = 6.95	297	7 × 10^3
Al_2O_3	Aluminum oxide (Alumina)	ε_{11} = ε_{22} = 9.34	298	10^2–8 × 10^9
		ε_{33} = 11.54	298	10^2–8 × 10^9
$AlPO_4$	Aluminum phosphate	ε_{11}^T = 6.05	r.t.	10^3
AlSb	Aluminum antimonide	11.21	300	i.r.
AsF_3	Arsenic trifluoride	5.7	r.t.	
BN	Boron nitride	7.1	r.t.	i.r.
$BaCO_3$	Barium carbonate	8.53	291	2 × 10^5
$Ba(COOH)_2$	Barium formate	ε_{11} = 7.9	r.t.	10^3
		ε_{22} = 5.9	r.t.	10^3
		ε_{33} = 7.5	r.t.	10^3
$BaCl_2$	Barium chloride	9.81	r.t.	
$BaCl_2 \cdot 2H_2O$	Barium chloride dihydrate	9.00	r.t.	10^3
BaF_2	Barium fluoride	7.32	292	5 × 10^2–10^11
$Ba(NO_3)_2$	Barium nitrate	4.95	292	2 × 10^5
$Ba_2NaNb_5O_{15}$	Barium sodium niobate ("Bananas")	ε_{11}^S = 222, ε_{11}^T = 235	296	10^4
		ε_{22}^S = 227, ε_{22}^T = 247	296	
		ε_{33}^S = 32, ε_{33}^T = 51	296	
BaO	Barium oxide (Baria)	34 ± 1	248, 333	60 × 10^7
BaO_2	Barium peroxide	10.7	r.t.	2 × 10^6
BaS	Barium sulfide	19.23	r.t.	7.25 × 10^6
$BaSO_4$	Barium sulfate	11.4	288	10^8
$BaSnO_3$	Barium stannate	18	298	25 × 10^5
$BaTiO_3$	Barium titanate	ε_{11}^T = 3600	298	10^5
		ε_{11}^T = 2300	298	2.5 × 10^8
		ε_{33}^T = 150	298	10^5
		ε_{33}^S = 80	298	2.5 × 10^8
$Ba_6Ti_2Nb_8O_{30}$	Barium titanium niobate	ε_{11} = ε_{22} ≈ 190	298	
		ε_{33}≈ 220	298	
$BaWO_4$	Barium tungstate	ε_{11} = ε_{22} = 35.5 ± 0.2	297.5	1.6 × 10^3
		ε_{33} = 37.2 ± 0.2	297.5	1.6 × 10^3
$BaZrO_3$	Barium zirconate	43	r.t.	
$Be_3Al_2Si_6O_{18}$	Beryllium aluminum silicate (Beryl)	ε_{33} = 5.95	297	7 × 10^3
		ε_{11} = ε_{22} = 6.86	297	7 × 10^3

Formula	Name	ε_{ijk}	T/K	ν/Hz
$BeCO_3$	Beryllium carbonate	9.7	291	2×10^5
BeO	Beryllium oxide (beryllia)	7.35 ± 0.2	293	2×10^6
$BiFeO_3$	Bismuth iron oxide	40 ± 3	300	9.4×10^9
$Bi_{12}GeO_{20}$	Bismuth germanite	$\varepsilon_{11}^{S} = 38$	r.t.	
$Bi_4(GeO_4)_3$	Bismuth germanate	16	293	
Bi_2O_3	Bismuth sesquioxide	18.2	r.t.	2×10^6
$Bi_4Ti_3O_{12}$	Bismuth titanate	112	r.t.	10^3
C	Diamond, Type I	5.87 ± 0.19	300	10^3
	Diamond, Type IIa	5.66 ± 0.04	300	10^3
$C_4H_4O_6$	Tartaric acid	$\varepsilon_{11} = \varepsilon_{22} = 4.3$	298	
		$\varepsilon_{33} = 4.5$	298	
		$\varepsilon_{13} = 0.55$	298	
$C_6H_{12}N_4$	Hexamethylene tetramine (HMTA)	2.6 ± 0.2	r.t.	10^9–10^{10}
$C_6H_{14}N_2O_6$	Ethylene diamine tartrate (EDT)	$\varepsilon_{11}^{T} = 5.0$	293	
		$\varepsilon_{22}^{T} = 8.3$	293	
		$\varepsilon_{33}^{T} = 6.0$	293	
		$\varepsilon_{13}^{T} = 0.7$	293	
$C_6H_{12}O_6 \cdot NaBr$	Dextrose sodium bromide	$\varepsilon_{11}^{T} = 4.0$	r.t.	10^3
$(CH_3NH_3)Al(SO_4)_2 \cdot 12H_2O$	Methyl ammonium alum (MASD)	19	197	
$Ca_2B_6O_{11} \cdot 5H_2O$	Calcium hexaborate pentahydrate	$\varepsilon_{11} = 20$	293	10^3
		$\varepsilon_{33} = 25$	293	10^3
$CaCO_3$	Calcium carbonate	$\varepsilon_{11} = 8.67$	r.t.	9.4×10^{10}
		$\varepsilon_{22} = 8.69$	r.t.	9.4×10^{10}
		$\varepsilon_{33} = 8.31$	r.t.	9.4×10^{10}
$CaCeO_3$	Calcium cerate	21	r.t.	
CaF_2	Calcium fluoride	6.81	300	5×10^2–10^{11}
$CaMoO_4$	Calcium molybdate	$\varepsilon_{11} = \varepsilon_{22} = 24.0 \pm 0.2$	297.5	<10
		$\varepsilon_{33} = 20.0 \pm 0.2$	297.5	<10
$Ca(NO_3)_2$	Calcium nitrate	6.54	292	2×10^5
$CaNb_2O_6$	Calcium niobate	$\varepsilon_{11} = 22.8 \pm 1.9$	r.t.	$(5–500) \times 10^3$
$Ca_2Nb_2O_7$	Calcium pyroniobate	~45	r.t.	5×10^7
CaO	Calcium oxide	11.8 ± 0.3	283	2×10^6
CaS	Calcium sulfide	6.699	r.t.	7.25×10^6
$CaSO_4 \cdot 2H_2O$	Calcium sulfate dihydrate	$\varepsilon_{11} = 5.10$	r.t.	
		$\varepsilon_{22} = 5.24$	r.t.	
		$\varepsilon_{33} = 10.30$	r.t.	
$CaTiO_3$	Calcium titanate	165	r.t.	
$CaWO_4$	Calcium tungstate	$\varepsilon_{11} = \varepsilon_{22} = 11.7 \pm 0.1$	297.5	1.59×10^3
		$\varepsilon_{33} = 9.5 \pm 0.2$	297.5	1.59×10^3
Cd_3As_2	Cadmium arsenide	$\varepsilon_{33} = 18.5$	4	
$CdBr_2$	Cadmium bromide	8.6	293	5×10^5
CdF_2	Cadmium fluoride	8.33 ± 0.08	300	10^5–10^7
CdS	Cadmium sulfide	$\varepsilon_{11} = \varepsilon_{22} = 8.7$	300	i.r.
		$\varepsilon_{33} = 9.25$	300	i.r.
		$\varepsilon_{11} = \varepsilon_{22} = 8.37$	8	i.r.
		$\varepsilon_{33} = 9.00$	8	i.r.
		$\varepsilon_{11}^{T} = 8.48$	77	10^4
		$\varepsilon_{33}^{T} = 9.48$	77	10^4
		$\varepsilon_{11}^{S} = 9.02, \varepsilon_{11}^{T} = 9.35$	298	10^4
		$\varepsilon_{33}^{S} = 9.53, \varepsilon_{33}^{T} = 10.33$	298	10^4
$CdSe$	Cadmium selenide	$\varepsilon_{11}^{S} = 9.53, \varepsilon_{11}^{T} = 9.70$	298	10^4
		$\varepsilon_{33}^{S} = 10.2, \varepsilon_{33}^{T} = 10.65$	298	10^4
$CdTe$	Cadmium telluride	$\varepsilon_{11} = \varepsilon_{22} = 10.60 \pm 0.15$	297	i.r.
		$\varepsilon_{33} = 7.05 \pm 0.05$	297	i.r.
$Cd_2Nb_2O_7$	Cadmium pyroniobate	500-580	293	10^3
CeO_2	Cerium oxide	7.0	r.t.	2×10^6
$CoNb_2O_6$	Cobalt niobate	$\varepsilon_{11} = 18.4 \pm 1.1$	r.t.	$(5–500) \times 10^3$
		$\varepsilon_{22} = 21.4 \pm 1.1$	r.t.	$(5–500) \times 10^3$
		$\varepsilon_{33} = 33.0 \pm 0.7$	r.t.	$(5–500) \times 10^3$
CoO	Cobalt oxide	12.9	298	10^2–10^{10}

Formula	Name	ε_{ijk}	T/K	ν/Hz
Cr_2O_3	Chromic sesquioxide	$\varepsilon_{11} = \varepsilon_{22} = 13.3$	298.5	10^3
		$\varepsilon_{33} = 11.9$	298.5	10^3
		8	315 (T_N)	6×10^{10}
$CsAl(SO_4)_2 \cdot 12H_2O$	Cesium alum	5.0	r.t.	$20–20 \times 10^3$
CsBr	Cesium bromide	6.38	298	1.6×10^3
Cs_2CO_3	Cesium carbonate	6.53	291	2×10^5
CsCl	Cesium chloride	7.2	298	
CsH_2AsO_4	Cesium dihydrogen arsenate (CDA)	4.8	273	9.5×10^9
CsH_2PO_4	Cesium dihydrogen phosphate (CDP)	6.15	285	9.5×10^9
$CsH_3(SeO_3)_2$	Cesium trihydrogen selenite	$\varepsilon_{11} = 80$	273	10^5
		$\varepsilon_{22} = 63$	273	10^5
		$\varepsilon_{33} = 12$	273	10^5
CsI	Cesium iodide	6.31	298	1.6×10^3
$CsNO_3$	Cesium nitrate	$\varepsilon_{11} = \varepsilon_{22} = 9.4$	r.t.	5×10^5
		$\varepsilon_{33} = 8.3$	r.t.	5×10^5
$CsPbCl_3$	Cesium lead chloride	14.37	300	$10^5–10^6$
CuBr	Cuprous bromide	8.0	293	5×10^5
CuCl	Cuprous chloride	9.8 ± 0.5	r.t.	10^3
CuO	Cupric oxide	18.1	r.t.	2×10^6
Cu_2O	Cuprous oxide (Cuprite)	7.60 ± 0.06	r.t.	10^5
$CuSO_4 \cdot 5H_2O$	Cupric sulfate pentahydrate	6.60	r.t.	
EuF_2	Europium fluoride	7.7 ± 0.2	298	$(1–300) \times 10^3$
$Eu_2(MoO_4)_3$	Europium molybdate	9.5	298	
EuS	Europium sulfide	13.10 ± 0.04	80	$5 \times 10^2–10^5$
FeO	Ferrous oxide	14.2	r.t.	2×10^6
Fe_2O_3	Ferric oxide	4.5	r.t.	$10^5–10^7$
Fe_2O_3-α	Ferric oxide (Hematite)	12		6×10^{10}
Fe_3O_4	Ferrosoferric oxide (Magnetite)	20	r.t.	$10^5–10^7$
GaAs	Gallium arsenide	13.13	300	
		12.90	4	i.r.
GaP	Gallium phosphide	11.1	r.t.	
		10.75 ± 0.1	1.6	i.r.
GaSb	Gallium antimonide	15.69	r.t.	
		15.7	4	i.r.
$Gd_2(MoO_4)_3$	Gadolinium molybdate	$\varepsilon^T = 10$	298	
		$\varepsilon^S = 9.5$	298	10^3
Ge	Germanium	16.0 ± 0.3	4	9.2×10^9
		15.8 ± 0.2	r.t.	$500–3 \times 10^{10}$
GeO_2	Germanium dioxide	$\varepsilon_{11} = \varepsilon_{22} = 7.44$	r.t.	i.r.
HIO_3	Iodic acid	$\varepsilon_{11} = 7.5$	r.t.	10^3
		$\varepsilon_{22} = 12.4$	r.t.	10^3
		$\varepsilon_{33} = 8.1$	r.t.	10^3
H_2O	Ice I (P = 0 kbar)	99	243	
	Ice III (P = 3 kbar)	117	243	
	Ice V (P = 5 kbar)	114	243	
	Ice VI (P = 8 kbar)	193	243	
HgCl	Mercury(I) chloride (Calomel)	$\varepsilon_{11} = \varepsilon_{22} = 14.0$	r.t.	10^{12}
$HgCl_2$	Mercuric chloride	6.5	r.t.	10^{12}
HgS	Mercurous sulfide (Cinnabar)	$\varepsilon_{11} = \varepsilon_{22} = 18.0$	r.t.	i.r.
		$\varepsilon_{33} = 32.5$	r.t.	i.r.
HgSe	Mercurous selenide	25.6	r.t.	$10^4–10^6$
I_2	Iodine	$\varepsilon_{11} = 6$	r.t.	$5 \times 10^4–10^7$
		$\varepsilon_{22} = 3$	r.t.	$5 \times 10^4–10^7$
		$\varepsilon_{33} = 40$	r.t.	$5 \times 10^4–10^7$
InAs	Indium arsenide	14.55 ± 0.3	r.t.	i.r.
		15.15	4	i.r.
InP	Indium phosphide	12.61	r.t.	i.r.
InSb	Indium antimonide	17.88	4	i.r.
$KAl(SO_4)_2 \cdot 12H_2O$	Potassium alum	6.5	r.t.	$20–20 \times 10^3$
KBr	Potassium bromide	4.88	300	
		4.53	4.2	

Formula	Name	ε_{ijk}	T/K	ν/Hz
$KBrO_3$	Potassium bromate	7.3	r.t.	2×10^6
KCN	Potassium cyanide	6.15	r.t.	2×10^6
K_2CO_3	Potassium carbonate	4.96	291	2×10^5
$K_2C_4H_4O_6 \cdot ½H_2O$	Dipotassium tartrate (DKT)	$\varepsilon_{11} = 6.44$	r.t.	
		$\varepsilon_{22} = 5.80$	r.t.	
		$\varepsilon_{33} = 6.49$	r.t.	
		$\varepsilon_{13} = 0.005$	r.t.	
KCl	Potassium chloride	4.86 ± 0.02	r.t.	5×10^3
		4.50	4.2	
$KClO_3$	Potassium chlorate	5.1	r.t.	2×10^6
$KClO_4$	Potassium perchlorate	5.9	r.t.	2×10^6
K_2CrO_4	Potassium chromate	7.3	r.t.	6×10^7
$KCr(SO_4)_2 \cdot 12H_2O$	Potassium chrome alum	6.5	100–240	175×10^3
KD_2AsO_4	Potassium dideuterium arsenate (KDDA)	$\varepsilon_{11} = 70$	298	
		$\varepsilon_{33} = 31$	298	
KD_2PO_4	Potassium dideuterium phosphate (KDDP)	50 ± 2	297	10^3
KF	Potassium fluoride	6.05		2×10^6
KH_2AsO_4	Potassium dihydrogen arsenate (KDA)	$\varepsilon_{11} = 60$	298	
		$\varepsilon_{33} = 24$	298	
KH_2PO_4	Potassium dihydrogen phosphate (KDP)	46	298	10^3
		$\varepsilon_{11} = 42$	r.t.	
		$\varepsilon_{33} = 21$	r.t.	
K_2HPO_4	Dipotassium monohydrogen orthophosphate	9.05	r.t.	2×10^6
KI	Potassium iodide	5.00	r.t.	9.4×10^{10}
KIO_3	Potassium iodate	170	255	10^5
		10	293	10^5
		$\varepsilon_{[101]} \approx 40, 70$	r.t.	10^5
		16.85	r.t.	2×10^6
$(K,H)Al_3(SiO_4)_3$	Mica (muscovite)	5.4	299	10^2–3×10^9
$(K,H)Mg_3Al(SiO_4)_3$	Mica (Canadian)	$\varepsilon_{11} = \varepsilon_{22} = 6.9$	298	10^2–10^4
		$\varepsilon_{33} = 7.3$	298	10^4
KNO_2	Potassium nitrite	25	305	
KNO_3	Potassium nitrate	4.37	293	2×10^5
$KNa(C_4H_4O_6) \cdot 4H_2O$	Sodium potassium tartrate tetrahydrate (Rochelle salt)	$\varepsilon_{11} = 170$	273	10^3
		$\varepsilon_{22} = 9.1$	273	10^3
$KNa(C_4H_2D_2O_6) \cdot 4D_2O$	Sodium potassium tartrate tetradeutrate (double deuterated Rochelle salt)	$\varepsilon_{11} = 70$	273	10^3
		$\varepsilon_{22} = 8.9$	273	10^3
$KNbO_3$	Potassium niobate	700	r.t.	
K_3PO_4	Potassium orthophosphate	7.75	r.t.	2×10^6
KSCN	Potassium thiocyanate	7.9	r.t.	2×10^6
K_2SO_4	Potassium sulfate	6.4	r.t.	2×10^6
$K_2S_3O_6$	Potassium trithionate	5.7	293	1.8×10^6
$K_2S_4O_6$	Potassium tetrathionate	5.5	293	1.8×10^6
$K_2S_5O_6 \cdot H_2O$	Potassium pentathionate	7.8	293	1.8×10^6
$K_2S_6O_6$	Potassium hexathionate	7.8	293	1.8×10^6
K_2SeO_4	Potassium selenate	$\varepsilon_{11} = 5.9$	r.t.	10^3
		$\varepsilon_{22} = 7.7$	r.t.	10^3
$KSr_2Nb_5O_{15}$	Potassium strontium niobate	$\varepsilon_{11} = \varepsilon_{11} \approx 1200$	298	
		$\varepsilon_{33} \approx 800$	298	
$KTaNbO_3$	Potassium tantalate niobate (KTN)	34,000	273	10^4
		6,000	293	10^4
$KTaO_3$	Potassium tantalate	242	298	2×10^5
$LaScO_3$	Lanthanum scandate	30	r.t.	
LiBr	Lithium bromide	12.1	r.t.	2×10^6
Li_2CO_3	Lithium carbonate	4.9	291	2×10^5
LiCl	Lithium chloride	11.05	r.t.	2×10^6
LiD	Lithium deuteride	14.0 ± 0.5	r.t.	i.r.
LiF	Lithium fluoride	9.00	298	10^2–10^7
		9.11	353	10^2–10^7
$LiGaO_2$	Lithium metagallate	$\varepsilon_{11}^T = 7.0, \varepsilon_{22}^T = 6.0$	r.t.	

Formula	Name	ε_{ijk}	T/K	ν/Hz
		ε_{33}^{T} = 9.5	r.t.	
		ε_{11}^{S} = 6.8, ε_{22}^{S} = 5.8	r.t.	
Li^6H	Lithium-6 hydride	13.2 ± 0.5	r.t.	
Li^7H	Lithium-7 hydride	12.9 ± 0.5	r.t.	
$LiH_3(SeO_3)_2$	Lithium trihydrogen selenite	29	298	10^4
		ε_{11} = 13.0	r.t.	
		ε_{22} = 12.9	r.t.	
		ε_{33} = 46	r.t.	
LiI	Lithium iodide	11.03	r.t.	2×10^6
$LiIO_3$	Lithium iodate	ε_{11} = ε_{22} = 65	294.5	10^3
		ε_{33} = 554	298	
$LiNH_4C_4O_6 \cdot H_2O$	Lithium ammonium tartrate (LAT)	ε_{11}^{T} = 7.2	298	
		ε_{22}^{T} = 8.0	298	
		ε_{33}^{T} = 6.9	298	
$LiNa_3CrO_4 \cdot 6H_2O$	Lithium trisodium chromate	8.0	r.t.	10^3
$LiNa_3MoO_4 \cdot 6H_2O$	Lithium trisodium molybdate	ε_{11} = 6.7	r.t.	10^3
		ε_{33} = 5.3	r.t.	10^3
$LiNbO_3$	Lithium niobate	ε_{11} = ε_{22} = 82	298	10^5
		ε_{33} = 30	298	10^5
$Li_2SO_4 \cdot H_2O$	Lithium sulfate monohydrate	ε_{11} = 5.6	298	
		ε_{22} = 10.3	298	
		ε_{33} = 6.5	298	
		ε_{13} = 0.07	298	
$LiTaO_3$	Lithium tantalate	ε_{11} = ε_{22} = 53	r.t.	10^5
		ε_{33} = 46	r.t.	10^5
		ε_{11}^{S} = ε_{22}^{S} = 41	r.t.	
		ε_{33}^{S} = 43	r.t.	
		ε_{11}^{T} = ε_{22}^{T} = 51	r.t.	
		ε_{33}^{T} = 45	r.t.	
$LiTIC_4H_4O_6 \cdot H_2O$	Lithium thallium tartrate (LTT)	ε_{11} ≈ 20	80	
$Mg_3B_7O_{13}Cl$	Magnesium borate monochloride (boracite)	ε_{11} = 14.1	r.t.	5×10^5
$MgCO_3$	Magnesium carbonate	8.1	291	2×10^5
$MgNb_2O_6$	Magnesium niobate	ε_{11} = 16.4 ± 0.5	r.t.	$(5–500) \times 10^3$
		ε_{22} = 20.9 ± 0.5	r.t.	$(5–500) \times 10^3$
		ε_{33} = 32.4 ± 0.5	r.t.	$(5–500) \times 10^3$
MgO	Magnesium oxide (Periclase)	9.65	298	$10^2–10^8$
$(MgO)_xAl_2O_3$	Spinel	8.6	r.t.	
$MgSO_4$	Magnesium sulfate	8.2	r.t.	
$MgSO_4 \cdot 7H_2O$	Magnesium sulfate heptahydrate	5.46	r.t.	
$MgTiO_3$	Magnesium titanate	13.5	r.t.	
$MgWO_4$	Magnesium tungstate	ε_{11} = 18.0 ± 1	r.t.	$(5–500) \times 10^3$
		ε_{22} = 18.0 ± 1	r.t.	$(5–500) \times 10^3$
$MnNb_2O_6$	Manganese niobate	ε_{11} = 17.4 ± 2	r.t.	$(5–500) \times 10^3$
		ε_{22} = 16.1 ± 0.5	r.t.	$(5–500) \times 10^3$
		ε_{33} = 30.7 ± 1	r.t.	$(5–500) \times 10^3$
MnO	Manganese oxide (Pyrolusite)	12.8	r.t.	6×10^{10}
MnO_2	Manganese dioxide	~10^4	298	10^4
Mn_2O_3	Manganese sesquioxide	8	r.t.	6×10^{10}
$MnWO_4$	Manganese tungstate	ε_{11} = 19.3 ± 1.3	r.t.	$(5–500) \times 10^3$
		ε_{22} = 14.3 ± 0.5	r.t.	$(5–500) \times 10^3$
		ε_{33} = 16.5 ± 1.1	r.t.	$(5–500) \times 10^3$
$N(CH_3)_4HgBr_3$	Tetramethylammonium tribromomercurate (TTM)	~10	233–373	
$N(CH_3)_4HgI_3$	Tetramethylammonium triiodo mercurate (TTM)	~10	233–373	
$(ND_4)_2BeF_4$	Deuteroammonium tetrafluoroberyllate	ε_{11} = 10	r.t.	
		ε_{22} = 9	r.t.	
		ε_{33} = 9	r.t.	
$(ND_4)_2SO_4$	Deuteroammonium sulfate	ε_{11} = 9	r.t.	
		ε_{22} = 10	r.t.	
		ε_{33} = 9	r.t.	

Formula	Name	ε_{ijk}	T/K	ν/Hz
$(NH_2\cdot CH_2COOH)_3\cdot H_2SO_4$	Triglycine sulfate (TGS)	$\varepsilon_{11} = 9$	273	10^4
		$\varepsilon_{22} = 30$	273	10^4
		$\varepsilon_{33} = 6.5$	273	10^4
$(NH_2\cdot CH_2COOH)_3\cdot H_2SeO_4$	Triglycine selenate (TGSe)	200	293	1.6×10^3
$(NH_2\cdot CH_2COOH)_3\cdot H_2BeF_4$	Triglycine tetrafluoroberyllate (TGFB)	$\varepsilon_{22} = 12$	273	10^4
$NH_4Al(SO_4)_2\cdot 12H_2O$	Ammonium alum	6	r.t.	10^{12}
$(NH_4)_2BeF_4$	Ammonium tetrafluoroberyllate	$\varepsilon_{11} = \varepsilon_{22} = 7.8$	123	10^5
		$\varepsilon_{33} = 7.1$	123	10^5
		$\varepsilon_{11} = \varepsilon_{22} = 8.8$	293	10^5
		$\varepsilon_{33} = 9.2$	293	10^5
NH_4Br	Ammonium bromide	7.1	r.t.	7×10^5
NH_4I	Ammonium iodide	9.8	r.t.	
$(NH_4)_2C_2H_6O_6$	Ammonium tartrate	$\varepsilon_{11} = 6.45$	r.t.	10^3
		$\varepsilon_{22} = 6.8$	r.t.	10^3
		$\varepsilon_{33} = 6.0$	r.t.	10^3
$(NH_4)_2Cd_2(SO_4)_3$	Ammonium cadmium sulfate	10.0	r.t.	10^4
NH_4Cl	Ammonium chloride	6.9	r.t.	7×10^5
$NH_4(ClCH_2COO)$	Ammonium monochloroacetate	5	r.t.	2×10^6
$NH_4H(ClCH_2COO)_2$	Hydrogen ammonium chloroacetate	$\varepsilon_{[102]} = 5.9$	r.t.	10^5
$NH_4Cr(SO_4)_2\cdot 12H_2O$	Ammonium chrome alum	6.5	r.t.	175×10^3
NH_4HSO_4	Ammonium hydrogen sulfate	165	273	5×10^4
$NH_4H_2AsO_4$	Ammonium dihydrogen arsenate (ADA)	5.1	265	9.5×10^9
		$\varepsilon_{11} = \varepsilon_{22} = 85$	298	10^3
		$\varepsilon_{33} = 22$	298	
$NH_4H_2PO_4$	Ammonium dihydrogen phosphate (ADP)	$\varepsilon_{11} = \varepsilon_{22} = 57.1 \pm 0.6$	294.5	10^5–35×10^9
		$\varepsilon_{33} = 14.0 \pm 0.3$	294	10^5–36×10^9
$ND_4D_2PO_4$	Ammonium dideuterium phosphate (ADDP)	$\varepsilon_{11} = \varepsilon_{22} = 74$, $\varepsilon_{33} = 24$	300	
NH_4NO_3	Ammonium nitrate	10.7	322	$(5–50) \times 10^3$
$(NH_4)_2SO_4$	Ammonium sulfate	$\varepsilon_{11} = \varepsilon_{22} = 8.0$	123	10^5
		$\varepsilon_{33} = 6.3$	123	10^5
		$\varepsilon_{11} = \varepsilon_{22} = 10.0$	293	10^5
		$\varepsilon_{33} = 9.3$	293	10^5
$(NH_4)_2UO_2(C_2O_4)_2$	Ammonium uranyl oxalate	8.03	r.t.	10^4–3.3×10^9
$(NH_4)_2UO_2(C_2O_4)_2\cdot 3H_2O$	Ammonium uranyl oxalate trihydrate	6.06	r.t.	10^4–3.3×10^9
NaBr	Sodium bromide	6.44	298	1.6×10^3
$NaBrO_3$	Sodium bromate	$\varepsilon_{11}^{T} = 5.70$	298	10^3
NaCN	Sodium cyanide	7.55	293	10^5
$NaCO_3$	Sodium carbonate	8.75	291	2×10^5
$NaCO_3\cdot 10H_2O$	Sodium carbonate decahydrate	5.3	r.t.	6×10^7
NaCl	Sodium chloride	5.9	298	10^2–10^7
		5.45	4.2	
$NaClO_3$	Sodium chlorate	$\varepsilon_{11}^{T} = 5.76$	301	10^3
		5.28	r.t.	10^3
$NaClO_4$	Sodium perchlorate	5.76	r.t.	10^3
NaF	Sodium fluoride	5.08 ± 0.02	r.t.	5×10^3
$NaH_3(SeO_3)_2$	Sodium trihydrogen selenite	$\varepsilon_{11} \approx 75$	273	2×10^5
$NaD_3(SeO_3)_2$	Sodium trideuterium selenite	$\varepsilon_{11} \approx 220$	273	2×10^5
NaI	Sodium iodide	7.28 ± 0.03	r.t.	
$NaNH_4(C_4H_4O_6)\cdot 4H_2O$	Sodium ammonium tartrate (Ammonium Rochelle salt)	$\varepsilon_{11} = 8.4$	298	
		$\varepsilon_{22} = 9.2$	298	
		$\varepsilon_{33} = 9.5$	298	
$NaNbO_3$	Sodium niobate	$\varepsilon_{33} = 670 \pm 13$	r.t.	
		$\varepsilon_{11} = \varepsilon_{22} = 76 \pm 2$	r.t.	
$NaNO_2$	Sodium nitrite	$\varepsilon_{11} = 7.4$	r.t.	5×10^5
		$\varepsilon_{22} = 5.5$	r.t.	5×10^5
		$\varepsilon_{33} = 5.0$	r.t.	5×10^5
$NaNO_3$	Sodium nitrate	6.85	292	2×10^5
$NaSO_4$	Sodium sulfate	7.90	r.t.	
$NaSO_4\cdot 10H_2O$	Sodium sulfate decahydrate	5.0	r.t.	

Formula	Name	ε_{ijk}	T/K	ν/Hz
$Na_2SO_4 \cdot 5H_2O$	Sodium thiosulfate pentahydrate	7	250–290	300–10^4
$Na_2UO_2(C_2O_4)_2$	Sodium uranyl oxalate	5.18	r.t.	
$NdAlO_3$	Neodymium aluminate	17.5	r.t.	
$NdScO_3$	Neodymium scandate	27	r.t.	
$Ni_3B_7O_{13}I$	Nickel iodine boracite	$\varepsilon_{11} = 14$	260	
$NiNb_2O_6$	Nickel niobate	$\varepsilon_{11} = 16.0 \pm 0.5$	r.t.	$(5–500) \times 10^3$
		$\varepsilon_{22} = 23.8 \pm 1.8$	r.t.	$(5–500) \times 10^3$
		$\varepsilon_{33} = 31.3 \pm 2.5$	r.t.	$(5–500) \times 10^3$
NiO	Nickel oxide	11.9	298	10^5
$NiSO_4 \cdot 6H_2O$	Nickel sulfate hexahydrate	$\varepsilon_{11} = 6.2$	r.t.	
		$\varepsilon_{33} = 6.8$	r.t.	
$NiWO_4$	Nickel tungstate	$\varepsilon_{11} = 17.4 \pm 2.4$	r.t.	$(5–500) \times 10^3$
		$\varepsilon_{22} = 13.6 \pm 1.0$	r.t.	$(5–500) \times 10^3$
		$\varepsilon_{33} = 19.7 \pm 0.6$	r.t.	$(5–500) \times 10^3$
P	Phosphorus (red)	4.1	r.t.	10^8
	Phosphorus (yellow)	3.6	r.t.	10^8
$[P(CH_3)_4]HgBr_3$	Tetramethylphosphonium tribromomercurate (TTM)	~10	233–373	
$PbBr_2$	Lead bromide	>30	293	$(0.5–3) \times 10^6$
$PbCO_3$	Lead carbonate	18.6	288	10.8
$Pb(C_2H_3O_2)_2$	Lead acetate	2.6	290–295	10^6
$PbCl_2$	Lead chloride	33.5	273	$(0.5–3) \times 10^6$
Pb_2CoWO_6	Lead cobalt tungstate	~250	r.t.	
PbF_2	Lead fluoride	26.3	r.t.	
$PbHfO_3$	Lead hafnate	390	300	10^5
		185	400	
PbI_2	Lead iodide	20.8	293	$(0.5–3) \times 10^6$
$Pb_3MgNb_2O_9$	Lead magnesium niobate	10,000	297	
$PbMoO_4$	Lead molybdate	$\varepsilon_{11} = 34.0 \pm 0.4$	297.5	1.6 ± 10^3
		$\varepsilon_{33} = 40.6 \pm 0.2$	297.5	1.6 ± 10^3
$Pb(NO_3)_2$	Lead nitrate	16.8	r.t.	$(0.5–3) \times 10^6$
$PbNb_2O_6$	Lead niobate	$\varepsilon^T_{33} = 180$	298	
PbO	Lead oxide	25.9	r.t.	2×10^6
PbS	Lead sulfide (Galena)	190	77	i.r.
		200 ± 35	r.t.	i.r.
$PbSO_4$	Lead sulfate	14.3	290—295	10^6
PbSe	Lead selenide	280	r.t.	i.r.
$PbTa_2O_6$	Lead tantalate	$\varepsilon_{11} = \varepsilon_{22} \approx 300$	r.t.	10^4
		$\varepsilon_{33} = 150$	r.t.	10^4
PbTe	Lead telluride	450	r.t.	i.r.
		40	77	10^4–15×10^4
		430	4.2	10^4–15×10^4
$PbTiO_3$	Lead titanate	~200	r.t.	10^3
$PbWO_4$	Lead tungstate	$\varepsilon_{11} = \varepsilon_{22} = 23.6 \pm 0.3$	297.5	1.59×10^3
		$\varepsilon_{33} = 31.0 \pm 0.4$	297.5	1.59×10^3
$Pb(Zn_{1/3}Nb_{2/3})O_3$	Lead zinc niobate	7	300	10^3, 300×10^3
$PbZrO_3$	Lead zirconate	200	400	
$RbAl(SO_4)_2 \cdot 12H_2O$	Rubidium alum	5.1	r.t.	10^{12}
RbBr	Rubidium bromide	4.83	300	
Rb_2CO_3	Rubidium carbonate	4.87 ± 0.02	r.t.	5×10^3
RbCl	Rubidium chloride	4.91 ± 0.02	r.t.	5×10^3
$RbCr(SO_4)_2 \cdot 12H_2O$	Rubidium chrome alum	5.0	r.t.	10^{12}
RbF	Rubidium fluoride	5.91	r.t.	2×10^6
$RbHSO_4$	Rubidium bisulfate	$\varepsilon_{11} = 7$	r.t.	10^5
		$\varepsilon_{22} = 8$	r.t.	10^5
		$\varepsilon_{33} = 10$	r.t.	10^5
RbH_2AsO_4	Rubidium dihydrogen arsenate (RDA)	3.90	273	9.5×10^9
RbH_2PO_4	Rubidium dihydrogen phosphate (RDP)	6.15	285	9.5×10^9
RbI	Rubidium iodide	4.94 ± 0.02	r.t.	5×10^3
$RbInSO_4$	Rubidium indium sulfate	6.85	r.t.	

Formula	Name	ε_{ijk}	T/K	ν/Hz
$RbNO_3$	Rubidium nitrate	20–380	433–488	10^6
		30	488–538	10^6
S	Sulfur	$\varepsilon_{11} = 3.75$	298	10^2–10^3
		$\varepsilon_{22} = 3.95$	298	10^2–10^3
		$\varepsilon_{33} = 4.44$	298	10^2–10^3
	(sublimed)	3.69	298	10^2–10^3
$SC(NH_2)_2$	Thiourea	$\varepsilon_{11} = \varepsilon_{22} \approx 3$	77–300	10^3
		$\varepsilon_{22} = 35$	300	10^3
Sb_2O_3	Antimonous sesquioxide	12.8	r.t.	$(1.5–2) \times 10^3$
Sb_2S_3	Antimonous sulfide (stibnite)	$\varepsilon_{11} = \varepsilon_{33} = 15$	r.t.	10^3
		$\varepsilon_{33} = 180$	r.t.	10^3
Sb_2Se_3	Antimonous selenide	~110	r.t.	$(10–16.5) \times 10^9$
SbSI	Antimonous sulfide iodide	2000	273	10^5
		$\varepsilon_{11} = \varepsilon_{22} \approx 25$	r.t.	10^3–10^5
		$\varepsilon_{33} \approx 5 \times 10^4$	295	10^3–10^5
Se	Selenium (monocrystal)	$\varepsilon_{11} = \varepsilon_{22} = 11$	300	24×10^9
		$\varepsilon_{33} = 21$	300	24×10^9
	(amorphous)	6.0	298	10^2—10^{10}
Si	Silicon	12.1	4.2	10^7—10^9
SiC	Silicon carbide, cubic	9.72	r.t.	i.r.
	Silicon carbide, 6H	$\varepsilon_{11} = \varepsilon_{22} = 9.66$	r.t.	i.r.
		$\varepsilon_{33} = 10.03$	r.t.	i.r.
		9.7 ± 0.1	1.8	i.r.
Si_3N_4	Silicon nitride	4.2 (film)	r.t.	10^3
SiO	Silicon monoxide	5.8	r.t.	10^3
SiO_2	Silicon dioxide	$\varepsilon_{11} = 4.42$	r.t.	9.4×10^{10}
		$\varepsilon_{22} = 4.41$	r.t.	9.4×10^{10}
		$\varepsilon_{33} = 4.60$	r.t.	9.4×10^{10}
$Sm_2(MoO_4)_3$	Samarium molybdate	12	298	
SnO_2	Stannic dioxide	$\varepsilon_{11} = \varepsilon_{22} = 14 \pm 2$	r.t.	10^4–10^{10}
		$\varepsilon_{33} = 9.0 \pm 0.5$	r.t.	10^4–10^{10}
SnSb	Tin antimonide	147	r.t.	10^4–10^6
SnTe	Tin telluride	1770 ± 300	r.t.	i.r.
$Sr(COOH)_2 \cdot 2H_2O$	Strontium formate dihydrate	6.1	r.t.	10^3
$SrCO_3$	Strontium carbonate	8.85	298	2×10^5
$SrCl_2$	Strontium chloride	9.19	r.t.	
$SrCl_2 \cdot 6H_2O$	Strontium chloride hexahydrate	8.52	r.t.	
SrF_2	Strontium fluoride	6.50	300	5×10^2–10^{11}
$SrMoO_4$	Strontium molybdate	$\varepsilon_{11} = \varepsilon_{22} = 31.7 \pm 0.2$	297.5	1.59×10^3
		$\varepsilon_{33} = 41.7 \pm 0.2$	297.5	1.59×10^3
$Sr(NO_3)_2$	Strontium nitrate	5.33	292	2×10^5
$Sr_2Nb_2O_7$	Strontium niobate	$\varepsilon_{11} = 75$	r.t.	10^3
		$\varepsilon_{22} = 46$	r.t.	10^3
		$\varepsilon_{33} = 43$	r.t.	10^3
SrO	Strontium oxide	13.3 ± 0.3	273	2×10^6
SrS	Strontium sulfide	11.3	r.t.	7.25×10^6
$SrSO_4$	Strontium sulfate	11.5	r.t.	
$SrTiO_3$	Strontium titanate	332	298	10^3
		2080	78	10^3
$SrWO_4$	Strontium tungstate	$\varepsilon_{11} = \varepsilon_{22} = 25.7 \pm 0.2$	297.5	1.6×10^3
		$\varepsilon_{33} = 34.1 \pm 0.2$	297.5	1.6×10^3
	Tantalum pentoxide, α phase	$\varepsilon_{11} = \varepsilon_{22} = 30$	77	10^3
		$\varepsilon_{33} = 65$	77	10^3
	Tantalum pentoxide, β phase	24	292	10^3
$Tb(MoO_4)_3$	Terbium molybdate	11	298	
		$\varepsilon_{11} = \varepsilon_{22} = 33$	100–200	9.4×10^9
		$\varepsilon_{33} = 53$	100–200	9.4×10^9

Formula	Name	ε_{ijk}	T/K	ν/Hz
Te	Tellurium	$\varepsilon_{11} = \varepsilon_{22} = 33$	r.t.	
		$\varepsilon_{33} = 54$	r.t.	
	polycrystalline	27.5	r.t.	i.r.
	monocrystalline	28.0	r.t.	i.r.
ThO_2	Thorium dioxide	18.9 ± 0.4	r.t.	3×10^5
TiO_2	Titanium dioxide (Rutile)	$\varepsilon_{11} = \varepsilon_{22} = 86$	300	10^4–10^6
		$\varepsilon_{33} = 170$	300	10^4–10^6
Ti_2O_3	Titanium sesquioxide	30	77	6×10^{10}
TlBr	Thallium bromide	30	293	10^3–10^7
TlCl	Thallous chloride	32.2 ± 0.2	293	10^3–10^5
TlI	Thallous iodide (orthorhombic)	20.7 ± 0.2	293	10^4
		37.3	193	10^7
$TlNO_3$	Thallous nitrate	16.5	293	5×10^5
$TlSO_4$	Thallous sulfate	25.5	293	5×10^5
UO_2	Uranium dioxide	24	r.t.	3×10^5
WO_3	Tungsten trioxide	300		
$YMnO_3$	Yttrium manganate	20	r.t.	2×10^7
Y_2O_3	Yttrium oxide	10	r.t.	10^6
$YbMnO_3$	Ytterbium manganate	20	r.t.	2×10^7
Yb_2O_3	Ytterbium oxide	5.0 (film)	r.t.	10^3
ZnO	Zinc oxide	$\varepsilon_{11}^{S} = 8.33$	r.t.	
		$\varepsilon_{33}^{S} = 8.84$	r.t.	
		$\varepsilon_{11}^{T} = 9.26$	r.t.	
		$\varepsilon_{33}^{T} = 11.0$	r.t.	
		$\varepsilon_{11} = 9.26$	r.t.	
		$\varepsilon_{33} = 8.2$	r.t.	
		8.15	r.t.	i.r.
ZnS	Zinc sulfide	$\varepsilon_{11}^{S} = 8.08 \pm 2\%$	77	10^4
		$\varepsilon_{11}^{S} = 8.32 \pm 2\%$	298	10^4
		$\varepsilon_{11}^{T} = 8.14 \pm 2\%$	77	10^4
		$\varepsilon_{11}^{T} = 8.37 \pm 2\%$	298	10^4
ZnSe	Zinc selenide	$\varepsilon_{11}^{T} = \varepsilon_{11}^{S} = 9.12 \pm 2\%$	298	10^4
ZnTe	Zinc telluride	$\varepsilon_{11}^{T} = \varepsilon_{11}^{S} = 10.10 \pm 2\%$	r.t.	
$ZnWO_4$	Zinc tungstate	$\varepsilon_{22} = 16.1 \pm 0.5$	r.t.	$(5–500) \times 10^3$
ZrO_2	Zirconium dioxide (zirconia)	12.5	r.t.	2×10^6

[a] Type I diamond.
[b] Type IIa diamond.
[c] γ phase.
[d] Ice I (P = 0 kbar).
[e] Ice III (P = 3 kbar).
[f] Ice V (P = 5 kbar).
[g] Ice VI (P = 8 kbar).
[h] Muscovite.
[i] Canadian mica.
[j] ^{6}LiH.
[k] ^{7}LiH.
[l] α phase.
[m] β phase.

CURIE TEMPERATURE OF SELECTED FERROELECTRIC CRYSTALS

H. P. R. Frederikse

The following table lists the major ferroelectric crystals and their Curie temperatures, T_C.

Reference

Young, K. F. and Frederikse, H. P. R., *J. Phys. Chem. Ref. Data*, 2, 313, 1973.

Curie Temperature of Ferroelectric Crystals

Systematic name	Acronym or common name	Formula	T_C/K
Potassium dihydrogen phosphate group			
Potassium dihydrogen phosphate	KDP	KH_2PO_4	123
Potassium dihydrogen arsenate	KDA	KH_2AsO_4	97
Potassium dideuterium phosphate	KDDP	KD_2PO_4	213
Potassium dideuterium arsenate	KDDA	KD_2AsO_4	162
Rubidium dihydrogen phosphate	RDP	RbH_2PO_4	146
Rubidium dihydrogen arsenate	RDA	RbH_2AsO_4	111
Rubidium dideuterium phosphate	RDDP	RbD_2PO_4	218
Rubidium dideuterium arsenate	RDDA	RbD_2AsO_4	178
Cesium dihydrogen phosphate	CDP	CsH_2PO_4	159
Cesium dihydrogen arsenate	CDA	CsH_2AsO_4	143
Cesium dideuterium arsenate	CDDA	CsD_2AsO_4	212
Rochelle salt group			
Potassium sodium tartrate tetrahydrate	Rochelle salt	$KNaC_4H_4O_6{\cdot}4H_2O$	255-297
Potassium sodium tartrate-d_2 tetrahydrate	Deuterated Rochelle salt	$KNaC_4H_2D_2O_6{\cdot}4H_2O$	251-308
Sodium ammonium tartrate tetrahydrate	Ammonium Rochelle salt	$NaNH_4C_4H_4O_6{\cdot}4H_2O$	109
Lithium ammonium tartrate monohydrate	LAT	$LiNH_4C_4H_4O_6{\cdot}H_2O$	106
Triglycine sulfate group			
Triglycine sulfate	TGS	$(NH_2CH_2COOH)_3{\cdot}H_2SO_4$	322
Triglycine selenate	TGSe	$(NH_2CH_2COOH)_3{\cdot}H_2SeO_4$	295
Triglycine tetrafluoroberyllate	TGFB	$(NH_2CH_2COOH)_3{\cdot}H_2BeF_4$	346
Ammonium tetrafluoroberyllate	AFB	$(NH_4)_2BeF_4$	176
Hydrogen ammonium chloroacetate		$NH_4H(CH_2ClCOO)_2$	128
Perovskites and related compounds			
Barium titanate	Barium metatitanate	$BaTiO_3$	406, 278, 193
Lead(II) titanate	Lead metatitanate	$PbTiO_3$	765
Potassium niobate		$KNbO_3$	712
Potassium tantalate niobate		$K_3Ta_2NbO_9$	241, 220, 170
Lithium niobate		$LiNbO_3$	1483
Lithium tantalate	Lithium tantalum trioxide	$LiTaO_3$	891
Barium titanium niobate		$Ba_6Ti_2Nb_8O_{30}$	521
Barium sodium niobate	BSN	$Ba_2Na(NbO_3)_5$	833
Potassium iodate		KIO_3	485, 343, 257–263, 83
Lithium iodate		$LiIO_3$	529
Potassium nitrate	Saltpeter	KNO_3	397
Sodium nitrate	Chile saltpeter	$NaNO_3$	548
Rubidium nitrate		$RbNO_3$	437-487
Miscellaneous compounds			
Cesium trihydrogen selenite		$CsH_3(SeO_3)_2$	143
Lithium trihydrogen selenite		$LiH_3(SeO_3)_2$	$T_C > T_{mp}$
Potassium selenate		K_2SeO_4	93
Methyl ammonium aluminum alum	MASD	$(CH_3NH_3)Al(SO_4)_2{\cdot}12H_2O$	177
Ammonium cadmium sulfate	Ammonium cadmium langbeinite	$(NH_4)_2Cd_2(SO_4)_3$	95
Ammonium hydrogen sulfate	Ammonium bisulfate	NH_4HSO_4	271
Ammonium sulfate	Mascagnite	$(NH_4)_2SO_4$	224
Ammonium nitrate		NH_4NO_3	398, 357, 305, 255
Calcium hexaborate pentahydrate	Colemanite	$2CaO{\cdot}3B_2O_3{\cdot}5H_2O$	266
Cadmium niobate		$Cd_2Nb_2O_7$	185
Gadolinium(III) molybdate		$Gd_2(MoO_4)_3$	432

PROPERTIES OF ANTIFERROELECTRIC CRYSTALS

H. P. R. Frederikse

Some important antiferroelectric crystals are listed here with their Curie temperatures T_C. The last column gives the constant T_0 which appears in the Curie–Weiss law describing the dielectric constant of these materials above the Curie temperature:

$$\varepsilon = \text{const.}/(T - T_0)$$

Name	Common name or acronym	Formula	T_C/K	T_0/K
Ammonium dihydrogen phosphate	ADP	$NH_4H_2PO_4$	148	
Ammonium dihydrogen arsenate	ADA	$NH_4H_2AsO_4$	216	
Ammonium dideuterium phosphate	ADDP	$NH_4D_2PO_4$	242[a]	
Ammonium dideuterium arsenate	ADDA	$NH_4D_2AsO_4$	299	
Ammonium-d_4 dideuterium phosphate	A_dDDP	$ND_4D_2PO_4$	243	
Ammonium-d_4 dideuterium arsenate	A_dDDA	$ND_4D_2AsO_4$	304	
Sodium niobate	Sodium niobate	$NaNbO_3$	911[b]	
Lead(II) hafnate	Lead hafnate	$PbHfO_3$	476	378
Lead(II) zirconate	Lead zirconate	$PbZrO_3$	503	475
Lead(II) niobate	Lead metaniobate	$Pb(NbO_3)_2$	843	530
Lead(II) tantalate	Lead metatantalate	$Pb(TaO_3)_2$	543	533
Tungsten(VI) oxide	Tungsten trioxide	WO_3	1010	
Potassium strontium niobate	Potassium strontium niobate	$KSr_2Nb_5O_{15}$	427	413
Sodium nitrite	Sodium nitrite	$NaNO_2$	437	437
Sodium trihydrogen selenite	Sodium trihydrogen selenite	$NaH_3(SeO_3)_2$	193	192
Sodium trideuterium selenite	Sodium trideuterium selenite	$NaD_3(SeO_3)_2$	271	245
Ammonium trihydrogen periodate	Ammonium trihydrogen periodate	$(NH_4)_2H_3IO_6$	245	

[a] Also reported as 245 K.
[b] Also reported as 793 K.

DIELECTRIC CONSTANTS OF GLASSES

Type	Dielectric constant at 100 MHz (20 °C)	Volume resistivity in MΩ cm (at 350 °C)	Loss factor[a]
Corning 0010	6.32	10	0.015
Corning 0080	6.75	0.13	0.058
Corning 0120	6.65	100	0.012
Pyrex 1710	6.00	2500	0.025
Pyrex 3320	4.71		0.019
Pyrex 7040	4.65	80	0.013
Pyrex 7050	4.77	16	0.017
Pyrex 7052	5.07	25	0.019
Pyrex 7060	4.70	13	0.018
Pyrex 7070	4.00	1300	0.0048
Vycor 7230	3.83		0.0061
Pyrex 7720	4.50	16	0.014
Pyrex 7740	5.00	4	0.040
Pyrex 7750	4.28	50	0.011
Pyrex 7760	4.50	50	0.0081
Vycor 7900	3.9	130	0.0023
Vycor 7910	3.8	1600	0.00091
Vycor 7911	3.8	4000	0.00072
Corning 8870	9.5	5000	0.0085
G. E. Clear (silica glass)	3.81	[d]	0.00038
Quartz (fused)	3.75[b]		0.0002[c]

[a] Power factor × dielectric constant equals loss factor.
[b] 4.1 at 1 MHz.
[c] At 1 MHz.
[d] Ranges from 4000 MΩ to 30000 MΩ.

PROPERTIES OF SUPERCONDUCTORS

L. I. Berger and B. W. Roberts

The following tables include superconductive properties of selected elements, compounds, and alloys. Individual tables are given for thin films, elements at high pressures, superconductors with high critical magnetic fields, and high critical temperature superconductors.

The historically first observed and most distinctive property of a superconductive body is the near total loss of resistance at a critical temperature (T_c) that is characteristic of each material. Figure 1(a) below illustrates schematically two types of possible transitions. The sharp vertical discontinuity in resistance is indicative of that found for a single crystal of a very pure element or one of a few well-annealed alloy compositions. The broad transition, illustrated by broken lines, suggests the transition shape seen for materials that are not homogeneous and contain unusual strain distributions. Careful testing of the resistivity limit for superconductors shows that it is less than 4×10^{-23} ohm cm, while the lowest resistivity observed in metals is of the order of 10^{-13} ohm cm. If one compares the resistivity of a superconductive body to that of copper at room temperature, the superconductive body is at least 10^{17} times less resistive.

The temperature interval ΔT_c, over which the transition between the normal and superconductive states takes place, may be of the order of as little as 2×10^{-5} K or several K in width, depending on the material state. The narrow transition width was attained in 99.9999% pure gallium single crystals.

A Type I superconductor below T_c, as exemplified by a pure metal, exhibits perfect diamagnetism and excludes a magnetic field up to some critical field H_c, whereupon it reverts to the normal state as shown in the *H-T* diagram of Figure 1(b).

The magnetization of a typical high-field superconductor is shown in Figure 1(c). The discovery of the large current-carrying capability of Nb_3Sn and other similar alloys has led to an extensive study of the physical properties of these alloys. In brief, a high-field superconductor, or Type II superconductor, passes from the perfect diamagnetic state at low-magnetic fields to a mixed state and finally to a sheathed state before attaining the normal resistive state of the metal. The magnetic-field values separating the four stages are given as H_{c1}, H_{c2}, and H_{c3}. The superconductive state below H_{c1} is perfectly diamagnetic, identical to the state of most pure metals of the Type I or "soft" superconductor. Between H_{c1} and H_{c2} a "mixed superconductive state" is found in which fluxons (a minimal unit of magnetic flux) create lines of normal flux in a superconductive matrix. The volume of the normal state is proportional to $-4\pi M$ in the "mixed state" region. Thus, at H_{c2} the fluxon density has become so great as to drive the interior volume of the superconductive body completely normal. Between H_{c2} and H_{c3} the superconductor has a sheath of current-carrying superconductive material at the body surface, and above H_{c3} the normal state exists. With several types of careful measurement, it is possible to determine H_{c1}, H_{c2}, and H_{c3}. Table 6 contains some of the available data on high-field superconductive materials.

High-field superconductive phenomena are also related to specimen dimension and configuration. For example, the Type I superconductor, Hg, has entirely different magnetization behavior in high-magnetic fields when contained in the very fine sets of filamentary tunnels found in an unprocessed Vycor glass. The great majority of superconductive materials are Type II. The elements in very pure form and a very few precisely stoichiometric and well annealed compounds are Type I with the possible exceptions of vanadium and niobium.

Metallurgical Aspects. The sensitivity of superconductive properties to the material state is most pronounced and has been used in a reverse sense to study and specify the detailed state of alloys. The mechanical state, the homogeneity, and the presence of impurity atoms and other electron-scattering centers are all capable of controlling the critical temperature and the current-carrying capabilities in high-magnetic fields. Well-annealed specimens tend to show sharper transitions than those that are strained or inhomogeneous. This sensitivity to mechanical state underlines a general problem in the tabulation of properties for superconductive materials. The occasional divergent values of the critical temperature and of the critical fields quoted for a Type II superconductor may lie in the variation in sample preparation. Critical temperatures of materials studied early in the history of superconductivity must be evaluated in light of the probable metallurgical state of the material, as well as the availability of less pure starting elements. It has been noted that recent work has given extended consideration to the metallurgical aspects of sample preparation.

Symbols in tables: T_c: Critical temperature; H_o: Critical magnetic field in the $T = 0$ limit; θ_D: Debye temperature; and γ: Electronic specific heat.

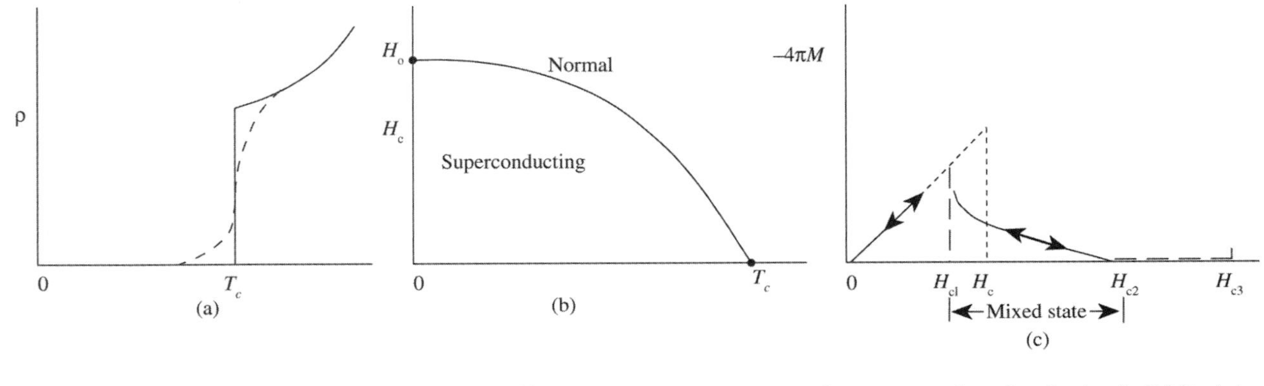

FIGURE 1. Physical properties of superconductors. (a) Resistivity vs. temperature for a pure and perfect lattice (solid line); impure and/or imperfect lattice (broken line). (b) Magnetic-field temperature dependence for Type I or "soft" superconductors. (c) Schematic magnetization curve for Type II or "hard" superconductors.

TABLE 1. Selective Properties of Superconductive Elements

Element	T_c/K	H_o/Oe	θ_D/K	γ/mJ mol^{-1}K^{-1}
Al	1.175 ± 0.002	104.9 ± 0.3	420	1.35
Am* (α,?)	0.6			
Am* (β,?)	1.0			
Be	0.026			0.21
Cd	0.517 ± 0.002	28 ± 1	209	0.69
Ga	1.083 ± 0.001	58.3 ± 0.2	325	0.60
Ga (β)	5.9, 6.2	560		
Ga (γ)	7	950, HF[a]		
Ga (Δ)	7.85	815, HF		
Hf	0.128	12.7		2.21
Hg (α)	4.154 ± 0.001	411 ± 2	87, 71.9	1.81
Hg (β)	3.949	339	93	1.37
In	3.408 ± 0.001	281.5 ± 2	109	1.672
Ir	0.1125 ± 0.001	16 ± 0.05	425	3.19
La (α)	4.88 ± 0.02	800 ± 10	151	9.8
La (β)	6.00 ± 0.1	1096, 1600	139	11.3
Lu	0.1 ± 0.03	350 ± 50		
Mo	0.915 ± 0.005	96 ± 3	460	1.83
Nb	9.25 ± 0.02	2060 ± 50, HF	276	7.80
Os	0.66 ± 0.03	70	500	2.35
Pa	1.4			
Pb	7.196 ± 0.006	803 ± 1	96	3.1
Re	1.697 ± 0.006	200 ± 5	4.5	2.35
Ru	0.49 ± 0.015	69 ± 2	580	2.8
Sn	3.722 ± 0.001	305 ± 2	195	1.78
Ta	4.47 ± 0.04	829 ± 6	258	6.15
Tc	7.8 ± 0.1	1410, HF	411	6.28
Th	1.38 ± 0.02	1.60 ± 3	165	4.32
Ti	0.40 ± 0.04	56	415	3.3
Tl	2.38 ± 0.02	178 ± 2	78.5	1.47
U	0.2			
V	5.40 ± 0.05	1408	383	9.82
W	0.0154 ± 0.0005	1.15 ± 0.03	383	0.90
Zn	0.85 ± 0.01	54 ± 0.3	310	0.66
Zr	0.61 ± 0.15	47	290	2.77
Zr (ω)	0.65, 0.95			

[a] HF denotes high-magnetic field superconductive properties.

TABLE 2. Range of Critical Temperatures Observed for Superconductive Elements in Thin Films Condensed Usually at Low Temperatures

Element	T_c/K range	Comments
Al	1.15–5.7	HF[a]
Be	5–9.75	HF
Bi	6.17–6.6	
Cd (Disordered)	0.79–0.91	
Cd (Ordered)	0.53–0.59	
Ga	2.5–8.5	HF
Hg	3.87–4.5	
In	2.2–5.6	HF
La	3.55–6.74	
Mo	3.3–8.0	
Nb	2.0–10.1	
Pb	1.8–7.5	
Re	1.7–7	
Sn	3.5–6	
Ta	<1.7–4.51	HF
Tc	4.6–7.7	
Ti	1.3 Max	
Tl	2.33–2.96	
V	1.8–6.02	
W	<1.0–4.1	
Zn	0.77–1.9	

[a] HF denotes high-magnetic field superconductive properties.

TABLE 3. Elements Exhibiting Superconductivity Under or After Application of High Pressure

Element	T_c/K Range	p/kbar
Al	1.98–0.075	0–62
As	0.31–0.5	220–140
As	0.2–0.25	140–100
Ba II	1–1.8	55–85
Ba III	1.8–5	85–144
Ba IV	4.5–5.4	144–190
Bi II	3.9	25–27
Bi III	6.55–7.25	28–38
Bi IV	7.0, 8.7–6.0	43, 43–62
Bi V	6.7, 8.3	48–80
Bi VI	8.55	90, 92–101
Bi VII(?)	8.2	30
Ce (α)	0.020–0.045	20–35
Ce (α′)	1.9–1.3	45–125
Cs V	1.5	>125
Ga II	6.38	≥35
Ga II′	7.5	≥35 then P removed
Ge	5.35	115
Lu	0.022–1.0	45–190
P	5.8	170
Pb II	3.55	160
Re II	2.3 Max.	"Plastic" compression
Sb (prepared 120 kbar, held below 77 K)	2.6–2.7	
Sb II	3.55–3.40	85–150
Se II	6.75, 6.95	130
Si	6.7–7.1	120–130
Sn II	5.2–4.85	125–160
Sn III	5.30	113
Te II	2.4–5.1	38–55
Te III	4.1–4.2	53–62
Te IV	4.72–4	63–80
Te IV	3.3–2.8	100–260
Tl (cubic form)	1.45	35
Tl (hexagonal form)	1.95	35
U	2.4–0.4	10–85
Y	1.7–2.5	110–160
Zr (omega form, metastable)	1–1.7	60–130

Properties of Superconductive Compounds and Alloys

In Tables 4 to 6 all compositions are denoted on an atomic basis, i.e., AB, AB_2, or AB_3 for compounds, unless noted. Solid solutions or odd compositions may be denoted as A_zB_{1-z} or A_zB. A series of three or more alloys is indicated as A_xB_{1-x} or by actual indication of the atomic fraction range, such as $A_{0-0.6}B_{1-0.4}$. The critical temperature of such a series of alloys is denoted by a range of values or possibly the maximum value.

The selection of the critical temperature from a transition in the effective permeability, or the change in resistance, or possibly the incremental changes in frequency observed by certain techniques is not often obvious from the literature. Most authors choose the mid-point of such curves as the probable critical temperature of the idealized material, while others will choose the highest temperature at which a deviation from the normal state property is observed. In view of the previous discussion concerning the variability of the superconductive properties as a function of purity and other metallurgical aspects, it is recommended that appropriate literature be checked to determine the most probable critical temperature or critical field of a given alloy.

A very limited amount of data on critical fields, H_o, is available for these compounds and alloys; these values are given in Table 5.

TABLE 4. Superconductive Compounds and Alloys

Substance	T_c/K	Crystal structure type
A. Superconductors with T_c < 10 K		
$Ag_{3.3}Al$	0.34	A12-cI58 (Mn)
$Ag_xAl_yZn_{1-x-y}$	0.15	Cubic
$AgBi_2$	2.87–3.0	
$Ag_7F_{0.25}N_{0.75}O_{10.25}$	0.85–0.90	
Ag_2F	0.0.066	
Ag_7FO_8	0.3	Cubic
$Ag_{0.8-0.3}Ga_{0.2-0.7}$	6.5–8	
Ag_4Ge	0.85	Hex., c.p.
$Ag_{0.438}Hg_{0.562}$	0.64	$D8_2$
$AgIn_2$	~2.4	C16
$Ag_{0.1}In_{0.9}Te$ ($n = 1.4 \times 10^{22}$)*	1.2–1.89	B1
$Ag_{0.2}In_{0.8}Te$ ($n = 1.07 \times 10^{22}$)	0.77–1.00	B1
AgLa	0.94	B2-cP2 (CsCl)
AgLa (9.5 kbar)	1.2	B2
AgLu	0.33	B2-cP2
$AgMo_4S_5$	9.1	hR15 (Mo_6PbS_8)
$Ag_{1.2}Mo_6Se_8$	5.9	Same
Ag_7NO_{11}	1.04	Cubic
Ag_xPb_{1-x}	7.2 max.	
Ag_4Sn	0.1	HCP+O
Ag_xSn_{1-x}	1.5–3.7	
Ag_xSn_{1-x} (film)	2.0–3.8	
$AgTe_3$	2.6	Cubic
AgTh	2.2	C16-tI12 (Al_2Cu)
$AgTh_2$	2.26	C16
$Ag_{0.03}Tl_{0.97}$	2.67	
$Ag_{0.94}Tl_{0.06}$	2.32	
AgY	0.33	B2-cP2 (CsCl)
Ag_xZn_{1-x}	0.5–0.845	
$AlAu_4$	0.4–0.7	Like A13
Al_2Au	0.1	C1-cF12 (CaF_2)
Al_2CMo_3	9.8–10.2	A13+trace 2nd. phase
Al_2CaSi	5.8	
$Al_{0.131}Cr_{0.088}V_{0.781}$	1.46	Cubic
$AlGe_2$	1.75	
Al_2Ge_2U	1.6	LI_2-cP4 (Cu_3Au)
$AlLa_3$	5.57	DO_{19}
Al_2La	3.23	C15
Al_2Lu	1.02	C15-cF24 (Cu_2Mg)
Al_3Mg_2	0.84	F.C.C.
$AlMo_3$	0.58	A15
$AlMo_6Pd$	2.1	
AlN	1.55	B4
Al_2NNb_3	1.3	A13
Al_3Nb	0.64	tI8 (Al_3Ti)
AlOs	0.39	B2
Al_3Os	5.90	
AlPb (film)	1.2–7	
Al_2Pt	0.48–0.55	C1
Al_5Re_{24}	3.35	A12
AlSb	2.8	B4-tI4 (Sn)
Al_2Sc	1.02	C15-cF24 (Cu_2Mg)
Al_2Si_2U	1.34	LI_2-cP4 (Cu_3Au)
$AlTh_2$	0.1	C16-tI12 (Al_2Cu)
Al_3Th	0.75	DO_{19}
$Al_xTi_yV_{1-x-y}$	2.05–3.62	Cubic
$Al_{0.108}V_{0.892}$	1.82	Cubic
Al_2Y	0.35	C15-cF24 (Cu_2Mg)
Al_3Yb	0.94	LI_2-cP4 (Cu_3Au)
Al_xZn_{1-x}	0.5–0.845	
$AlZr_3$	0.73	LI_2
AsBiPb	9.0	
AsBiPbSb	9.0	
AsHfOs	3.2	C22-hP9 (Fe_2P)
AsHfRu	4.9	same
$As_{0.33}InTe_{0.67}$ (n = 1.24×10^{22})	0.85–1.15	B1
$As_{0.5}InTe_{0.5}$ (n = 0.97×10^{22})	0.44–0.62	B1
As_4La_3	0.6	cI28 (Th_3P_4)
$AsNb_3$	0.3	$L1_2$-tP32
$As_{0.50}Ni_{0.06}Pd_{0.44}$	1.39	C2
$AsNi_{0.25}Pd_{0.75}$	1.6	$B8_1$-hP4 (NiAs)
AsOsZr	8.0	C22-hP9 (Fe_2P)
AsPb	8.4	
$AsPd_2$ (low-temp. phase)	0.60	Hexagonal
$AsPd_2$ (high-temp. phase)	1.70	C22
$AsPd_5$	0.46	Complex
As_3Pd_5	1.9	
AsRh	0.58	B31
$AsRh_{1.4-1.6}$	< 0.03–0.56	Hexagonal
AsSn	4.10	
AsSn (n = 2.14×10^{22})	3.41–3.65 3.5–3.6	B1
$As_{\sim 2}Sn_{\sim 3}$	1.21–1.17	
As_3Sn_4 (n = 0.56×10^{22})	1.16–1.19	Rhombohedral
AsV_3	0.20	A15-cP8 (Cr_3Si)
Au_5Ba	0.4–0.7	$D2_d$
AuBe	2.64	B20
Au_2Bi	1.80	C15
Au_5Ca	0.34–0.38	$C15_b$

Substance	T_c/K	Crystal structure type
$AuGa_2$	1.6	C1-cF12 (CaF_2)
AuGa	1.2	B31
$Au_{0.40–0.92}Ge_{0.60–0.08}$	<0.32–1.63	Complex
$AuIn_2$	0.2	C1-cF12
AuIn	0.4–0.6	Complex
AuLu	<0.35	B2
$AuNb_3$	1.2	A2
$AuPb_2$	3.15	
$AuPb_2$ (film)	4.3	
$AuPb_3$	4.40	
$AuPb_3$ (film)	4.25	
Au_2Pb	1.18; 6–7	C15
$AuSb_2$	0.58	C2
AuSn	1.25	$B8_1$
Au_xSn_{1-x} (film)	2.0–3.8	
Au_5Sn	0.7–1.1	A3
$AuTa_{4.3}$	0.55	A15-cP8 (Cr_3Si)
Au_3Te_5	1.62	Cubic
$AuTh_2$	3.08	C16
AuTl	1.92	
AuV_3	0.74	A15
Au_xZn_{1-x}	0.50–0.845	
$AuZn_3$	1.21	Cubic
Au_xZr_y	1.7–2.8	A3
$AuZr_3$	0.92	A15
$B_2Ba_{0.67}Pt_3$	5.60	hP12 (B_2BaPt_3)
$BCMo_2$	5.4	Orthorhombic
$BCMo_2$	5.3–7.0	Same
$B_2Ca_{0.67}Pt_3$	1.57	hP12
B_4ErIr_4	2.1	tP18 (B_4CeCo_4)
B_4ErRh_4	4.3	oC108 (B_4LuRh_4)
B_4ErRh_4	8.7	tP18 (B_4CeCo_4)
BHf	3.1	Cubic
B_4HoIr_4	2.0	tP18
B_4HoRh_4	1.4	oC108
B_2Ir_3La	1.65	hP6 ($CaCu_5$)
B_2Ir_3Th	2.09	Same
B_4Ir_4Tm	1.6	tP18
B_6La	5.7	
B_2LaRh_3	2.82	hP6
$B_{12}Lu$	0.48	
B_2LuOs	2.66	oP16 (B_2LuRu)
B_2LuOs_3	4.62	hP6
B_4LuRh_4	6.2	oC108
B_2LuRu	9.86	oP16
B_4LuRu_4	2.0	tI72 (B_4LuRu_4)
BMo	0.5[a]	
BMo_2	4.74	C16
BNb	8.25	B_f
B_4NdRh_4	5.3	tP18
B_2OsSc	1.34	oP16
B_2OsY	2.22	oP16
$B_2Pt_3Sr_{0.67}$	2.78	hP12 (B_2BaPt_3)
BRe_2	2.80; 4.6	
$B_4Rh_{3.4}Ru_{0.6}$	8.38	tI72
B_4Rh_4Sm	2.7	tP18
B_4Rh_4Th	4.3	Same
B_4Rh_4Tm	9.8	Same
B_4Rh_4Tm	5.4	oC108
$B_{0.3}Ru_{0.7}$	2.58	$D10_2$
B_4Ru_4Sc	7.2	tI72
B_2Ru_3Th	1.79	hP6
B_2Ru_3Y	2.85	Same
B_2Ru Y	7.80	oP16
B_4Ru_4Y	1.4	tI72
$B_{12}Sc$	0.39	
BTa	4.0	B_f
BTa_2	3.12	C16-tI12 (Al_2Cu)
B_6Th	0.74	
BW_2	3.1	C16
B_6Y	6.5–7.1	
$B_{12}Y$	4.7	
BZr	3.4	Cubic
$B_{12}Zr$	5.82	
$BaBi_3$	5.69	Tetragonal
$Ba_2Mo_{15}Se_{19}$	2.75	hP15 (Mo_6PbS_8)
$Ba_xO_3Sr_{1-x}Ti$ ($n = 4.2 \times 10^{19}$)	<0.1–0.55	
$Ba_{0.13}O_3W$	1.9	Tetragonal
$Ba_{0.14}O_3W$	<1.25–2.2	Hexagonal
$BaRh_2$	6.0	C15
$Be_{22}Mo$	2.51	Cubic ($Be_{22}Re$)
$Be_8Nb_5Zr_2$	5.2	
$Be_{0.98–0.92}Re_{0.02–0.08}$ (quenched)	9.5–9.75	Cubic
$Be_{0.957}Re_{0.043}$	9.62	Cubic ($Be_{22}Re$)
BeTc	5.21	Cubic
$Be_{22}W$	4.12	Cubic ($Be_{22}Re$)
$Be_{13}W$	4.1	Tetragonal
Bi_3Ca	2.0	
$Bi_{0.5}Cd_{0.13}Pb_{0.25}Sn_{0.12}$ (weight fractions)	8.2	
BiCo	0.42–0.49	
Bi_2Cs	4.75	C15
Bi_xCu_{1-x} (electrodeposited)	2.2	
BiCu	1.33–1.40	
Bi_3Fe	1.0	A7+A2
$Bi_{0.019}In_{0.981}$	3.86	
$Bi_{0.05}In_{0.95}$	4.65	α-phase
$Bi_{0.10}In_{0.90}$	5.05	Same
$Bi_{0.15–0.30}In_{0.85–0.70}$	5.3–5.4	α- and β-phases
$Bi_{0.34–0.48}In_{0.66–0.52}$	4.0–4.1	
Bi_3In_5	4.1	
$BiIn_2$	5.65	β-phase
Bi_2Ir	1.7–2.3	
Bi_2Ir (quenched)	3.0–3.96	
BiK	3.6	
Bi_2K	3.58	C15
BiLi	2.47	$L1_o$, α-phase
$Bi_{4–9}Mg$	0.7–~1.0	
Bi_3Mo	3–3.7	
BiNa	2.25	$L1_o$
$BiNb_3$	4.5	A15-cP8 (Cr_3Si)
$BiNb_3$ (high pressure and temperature)	3.05	A15
BiNi	4.25	$B8_1$
Bi_3Ni	4.06	Orthorhombic
$BiNi_{0.5}Rh_{0.5}$	3.0	$B8_1$-hP4 (AsNi)
$Bi_{0.5}NiSb_{0.5}$	2.0	Same
$Bi_{1-0}Pb_{0-1}$	7.26–9.14	
$Bi_{1-0}Pb_{0-1}$ (film)	7.25–8.67	
$Bi_{0.05–0.40}Pb_{0.95–0.60}$	7.35–8.4	H.C.P. to ε-phase
Bi_2Pb	4.25	HCP+CCP

Substance	T_c/K	Crystal structure type
BiPbSb	8.9	
$Bi_{0.5}Pb_{0.31}Sn_{0.19}$ (weight fractions)	8.5	
$Bi_{0.5}Pb_{0.25}Sn_{0.25}$	8.5	
$BiPd_2$	4.0	
$Bi_{0.4}Pd_{0.6}$	3.7–4	Hexagonal, ordered
BiPd	3.7	Orthorhombic
Bi_2Pd	1.70	Monoclinic, α-phase
Bi_2Pd	4.25	Tetragonal, β-phase
$BiPd_{0.45}Pt_{0.55}$	3.7	$B8_1$-hP4 (NiAs)
BiPdSe	1.0	C2
BiPdTe	1.2	C2
BiPt	1.21	$B8_1$
$Bi_{0.1}PtSb_{0.9}$	2.05; 1.5	$B8_1$-hP4 (NiAs)
BiPtSe	1.45	C2
BiPtTe	1.15	C2
Bi_2Pt	0.155	Hexagonal
Bi_2Rb	4.25	C15
$BiRe_2$	1.9–2.2	
BiRh	2.06	$B8_1$
Bi_3Rh	3.2	Orthorhombic (NiB_3)
Bi_4Rh	2.7	Hexagonal
BiRu	5.7	A7+a3
Bi_3Sn	3.6–3.8	
BiSn	3.8	
Bi_xSn_y	3.85–4.18	
Bi_3Sr	5.62	$L1_2$
Bi_3Te	0.75–1.0	
Bi_5Tl_3	6.4	
$Bi_{0.26}Tl_{0.74}$	4.4	Cubic, disordered
$Bi_{0.26}Tl_{0.74}$	4.15	$L1_2$, ordered (?)
Bi_2Y_3	2.25	
Bi_3Zn	0.8–0.9	
$Bi_{0.3}Zr_{0.7}$	1.51	
$BiZr_3$	2.4–2.8	
$BrMo_6Se_7$	7.1	hP15 (Mo_6PbS_8)
$Br_3Mo_6Se_5$	7.1	Same
CCs_x	0.020–0.135	Hexagonal
CFe_3	1.30	DO_{11}-oP16 (Fe_3C)
$CGaMo_2$	3.7–4.1	Hexagonal
$CHf_{0.5}Mo_{0.5}$	3.4	B1
$CHf_{0.3}Mo_{0.7}$	5.5	B1
$CHf_{0.25}Mo_{0.75}$	6.6	B1
$CHf_{0.7}Nb_{0.3}$	6.1	B1
$CHf_{0.6}Nb_{0.4}$	4.5	B1
$CHf_{0.5}Nb_{0.5}$	4.8	B1
$CHf_{0.4}Nb_{0.6}$	5.6	B1
$CHf_{0.25}Nb_{0.75}$	7.0	B1
$CHf_{0.2}Nb_{0.8}$	7.8	B1
$CHf_{0.9-0.1}Ta_{0.1-0.9}$	5.0–9.0	B1
CK (excess K)	0.55	Hexagonal
C_8K	0.39	Hexagonal
C_2La	1.66	tI6 (CaC_2)
C_2Lu	3.33	Same
$C_{0.40-0.44}Mo_{0.60-0.56}$	9–13	
C_3MoRe	3.8	B1-cF8
$C_{0.6}Mo_{4.8}Si_3$	7.6	$D8_8$
$CMo_{0.2}Ta_{0.8}$	7.5	B1
$CMo_{0.5}Ta_{0.5}$	7.7	B1
$CMo_{0.75}Ta_{0.25}$	8.5	B1
$CMo_{0.8}Ta_{0.2}$	8.7	B1
$CMo_{0.85}Ta_{0.15}$	8.9	B1
CMo_xV_{1-x}	2.9–9.3	B1
CMo_xZr_{1-x}	9.8	B1
$C_{0.984}Nb$	9.8	B1
CNb_2	9.1	
CNb_xTi_{1-x}	<4.2–8.8	B1
$CNb_{0.1-0.9}Zr_{0.9-0.1}$	4.2–8.4	B1
CRb_x (Au)	0.023–0.151	Hexagonal
$CRe_{0.06}W$	5.0	
CRu	2.00	hP2 (CW)
$C_{0.98}7Ta$	9.7	
$C_{0.848-0.987}$	2.04–9.7	
CTa (film)	5.09	B1
CTa_2	3.26	L'_3
$CTa_{0.4}Ti_{0.6}$	4.8	B1
$Cta_{1-0.4}W_{0-0.6}$	8.5–10.5	B1
$CTa_{0.2-0.9}Zr_{0.8-0.1}$	4.6–8.3	B1
CTc (excess C)	3.85	Cubic
$CTi_{0.5-0.7}W_{0.5-0.3}$	6.7–2.1	B1
CW	1.0	
CW_2	2.74	L'_3
CW_2	5.2	F.C.C.
C_2Y	3.88	tI6 (CaC_2)
$Ca_3Co_4Sn_{13}$	5.9	cP40 ($Pr_3Rh_2Sn_{13}$)
$Ca_3Ge_{13}Rh_4$	2.1	Same
CaHg	1.6	B2-cP2 (CsCl)
$CaHg_3$	1.6	hP8 (Ni_3Sn)
$CaIr_2$	6.15	C15
$Ca_3Ir_4Sn_{13}$	7.1	cP40
$Ca_xO_3Sr_{1-x}Ti$ (n = 3.7–11 × 10^{19})	< 0.1–0.55	
$Ca_{0.1}O_3W$	1.4–3.4	Hexagonal
CaPb	7.0	
$CaRh_2$	6.40	C15
$CaRh_{1.2}Sn_{4.5}$	8.7	cP40
$CaTl_3$	2.0	B2-cP2
$Cd_{0.3-0.5}Hg_{0.7-0.5}$	1.70–1.92	
CdHg	1.77; 2.15	Tetragonal
$Cd_{0.0075-0.05}In_{0.9925-0.95}$	3.24–3.36	Tetragonal
$Cd_{0.97}Pb_{0.03}$	4.2	
CdSn	3.65	
$Cd_{0.17}Tl_{0.83}$	2.3	
$Cd_{0.18}Tl_{0.82}$	2.54	
$CeCo_2$	0.84	C15
$CeCo_{1.67}Ni_{0.33}$	0.46	C15
$CeCo_{1.67}Rh_{0.33}$	0.47	C15
$Ce_xGd_{1-x}Ru_2$	3.2–5.2	C15
$CeIr_3$	3.34	
$CeIr_5$	1.82	
$Ce_{0.005}La_{0.995}$	4.6	
Ce_xLa_{1-x}	1.3–6.3	
$Ce_xPr_{1-x}Ru_2$	1.4–5.3	C15
Ce_xPt_{1-x}	0.7–1.55	
$CeRu_2$	6.0	C15
$Ce_3Mo_6Se_5$	5.7	hR15 (Mo_6PbS_8)
$Ce_2Mo_6Te_6$	1.7	Same
$Co_xFe_{1-x}Si_2$	1.4 (max.)	C1
$CoHf_2$	0.56	$E9_3$
$CoLa_3$	4.28	
$Co_4La_3Sn_{13}$	2.8	cP40
$CoLu_3$	~0.35	

Substance	T_c/K	Crystal structure type
Co_xLuSn_y	1.5	cP40
$Co_{0-0.01}Mo_{0.8}Re_{0.2}$	2–10	
$Co_{0.02-0.10}Nb_3Rh_{0.98-0.90}$	2.28–1.90	A15
$Co_xNi_{1-x}Si_2$	1.4 (max.)	C1
$Co_{0.5}Rh_{0.5}Si_2$	2.5	
$Co_xRh_{1-x}Si_2$	3.65 (max.)	
$Co_{\sim0.3}So_{\sim0.7}$	~0.35	
$Co_4Sc_5Si_{10}$	5.0	tP38 ($Co_4Sc_5Si_{10}$)
$CoSi_2$	1.40; 1.22	C1
Co_xSn_yYb	2.5	cP40
Co_3Th_7	1.83	$D10_2$
Co_xTi_{1-x}	2.8 (max.)	Co in α-Ti
Co_xTi_{1-x}	3.8 (max.)	Co in β-Ti
$CoTi_2$	3.44	$E9_3$
CoTi	0.71	A2
CoU	1.7	B2, distorted
CoU_6	2.29	$D2_c$
$Co_{0.28}Y_{0.72}$	0.34	
CoY_3	<0.34	
$CoZr_2$	6.3	C16
$Co_{0.1}Zr_{0.9}$	3.9	A3
$Cr_{0.6}Ir_{0.4}$	0.4	H.C.P.
$Cr_{0.65}Ir_{0.35}$	0.59	H.C.P.
$Cr_{0.7}Ir_{0.3}$	0.76	H.C.P.
$Cr_{0.72}Ir_{0.28}$	0.83	
Cr_3Ir	0.45	A15
$Cr_{0-0.1}Nb_{1-0.9}$	4.6–9.2	A2
$Cr_{0.80}Os_{0.20}$	2.5	Cubic
Cr_3Os	4.68	A15-cP8 (Cr_3Si)
Cr_xRe_{1-x}	1.2–5.2	
$Cr_{0.4}Re_{0.6}$	2.15	$D8_b$
$Cr_{0.8-0.6}Rh_{0.2-0.4}$	0.5–1.10	A3
Cr_3Rh	0.3	A15-cP8
Cr_3Ru (annealed)	3.3	A15
Cr_2Ru	2.02	$D8_b$
Cr_3Ru_2	2.10	$D8_b$-tP30 (CrFe)
$Cr_{0.1-0.5}Ru_{0.9-0.5}$	0.34–1.65	A3
Cr_xTi_{1-x}	3.6 (max.)	Cr in α-Ti
Cr_xTi_{1-x}	4.2 (max.)	Cr in β-Ti
$Cr_{0.1}Ti_{0.3}V_{0.6}$	5.6	
$Cr_{0.0175}U_{0.9825}$	0.75	β-phase
$Cs_{0.32}O_3W$	1.12	Hexagonal
$Cu_{0.15}In_{0.85}$ (film)	3.75	
$Cu_{0.04-0.08}In_{0.94-0.92}$	4.4	
CuLa	5.85	
$Cu_2Mo_6O_2S_6$	9	hR15 (Mo_6PbS_8)
$Cu_2Mo_6Se_8$	5.9	Same
Cu_xPb_{1-x}	5.7–7.7	
CuS	1.62	B18
CuS_2	1.48–1.53	C18
CuSSe	1.5–2.0	C18
$CuSe_2$	2.3–2.43	C18
CuSeTe	1.6–2.0	C18
Cu_xSn_{1-x}	3.2–3.7	
Cu_xSn_{1-x} (film, made at 10 K)	3.6–7	
Cu_xSn_{1-x} (film, made at 300 K)	2.8–3.7	
$CuTe_2$	<1.25–1.3	C18
$CuTh_2$	3.49	C16
$Cu_{0-0.027}V$	3.9–5.3	A2
CuY	0.33	B2-cP2 (CsCl)

Substance	T_c/K	Crystal structure type
Cu_xZn_{1-x}	0.5–0.845	
$DyMo_6S_8$	2.1	hR15
Er_xLa_{1-x}	1.4–6.3	
$ErMo_6S_8$	2.2	hR15
$ErMo_6Se_8$	6.2	hR15
$Fe_3Lu_2Si_5$	6.1	tP40 ($Fe_3Sc_2Si_5$)
$Fe_{0-0.04}Mo_{0.8}Re_{0.2}$	1–10	
$Fe_{0.05}Ni_{0.05}Zr_{0.90}$	~3.9	
Fe_3Re_2	6.55	$D8_b$-tP30 (FeCr)
$Fe_3Sc_2Si_5$	4.52	tP40
Fe_3Si_5Tm	1.3	Same
$Fe_3Si_5Y_2$	2.4	Same
Fe_3Th_7	1.86	D10
Fe_xTi_{1-x}	3.2 (max.)	Fe in α-Ti
Fe_xTi_{1-x}	3.7 (max.)	Fe in β-Ti
$Fe_xTi_{0.6}V_{1-x}$	6.8 (max.)	
FeU_6	3.86	$D2_c$
$Fe_{0.1}Zr_{0.9}$	1.0	A3
$Ga_{0.5}Ge_{0.5}Nb_3$	7.3	A15
Ga_2Ge_2U	0.87	B2-cP2
$GaHf_2$	0.21	C16-tI12 (Al_2Cu)
$GaLa_3$	5.84	
Ga_3Lu	2.3	B2-cP2
Ga_2Mo	9.5	
$GaMo_3$	0.76	A15
GaN (black)	5.85	B4
$Ga_{0.7}Pt_{0.3}$	2.9	C1
GaPt	1.74	B20
GaSb (120 kbar, 77 K, annealed)	4.24	A5
GaSb (unannealed)	~5.9	
$Ga_{0-1}Sn_{1-0}$ (quenched)	3.47–4.18	
$Ga_{0-1}Sn_{1-0}$ (annealed)	2.6–3.85	
GaTe	0.17	mC24 (GaTe)
Ga5V2	3.55	Tetragonal (Mn_2Hg_5)
$GaV_{4.5}$	9.15	
Ga_3Zr	1.38	
Ga_3Zr_5	3.8	D8b-hP16 (Mn_5Si_3)
Gd_xLa_{1-x}	< 1.0–5.5	
$GdMo_6S_8$	3.5	hR15
$GdMo_6Se_8$	5.6	hR15
$Gd_xOs_2Y_{1-x}$	1.4–4.7	
$Gd_xRu_2Th_{1-x}$	3.6 (max.)	C15
$Ge_{10}As_4Y_5$	9.06	tP38 ($C0_4Sc_5Si_{10}$)
GeIr	4.7	B31
GeIrLa	1.64	tI12 (LaPtSi)
$Ge_{10}Ir_4Lu_5$	2.60	tP38
$Ge_{10}Ir_4Y_5$	2.62	tP38
Ge_2La	1.49; 2.2	Orthorhombic, distorted (Mn_2Hg_5)
GeLaPt	3.53	tI12
$Ge_{13}Lu_3Os_4$	3.6	cP40 ($Pr_3Rh_2Sn_{13}$)
$Ge_{10}Lu_5Rh_4$	2.79	tP38
$Ge_{13}Lu_3Ru_4$	2.3	cP40
$GeMo_3$	1.43	A15
$GeNb_2$	1.9	
$Ge_{0.29}Nb_{0.71}$	6	A15
GePt	0.40	B31
Ge_3Rh_5	2.12	Orthorhombic, related to $InNi_2$
GeRh	0.96	B31-oP8 (MnP)
$Ge_{13}Rh_4Sc_3$	1.9	c P40

Substance	T_c/K	Crystal structure type
$Ge_{10}Rh_4Y_5$	1.35	tP38
$Ge_{13}Ru_4Y_3$	1.7	cP40
Ge_2So	1.3	
$GeTa_3$	8.0	A15-cP8 (Cr_3Si)
Ge_3Te_4 ($n = 1.06 \times 10^{22}$)	1.55–1.80	Rhombohedral
Ge_xTe_{1-x} ($n = 8.5–64 \times 10^{20}$)	0.07–0.41	R1
GeV_3	6.01	A15
Ge_2Y	3.80	C_c
$Ge_{1.62}Y$	2.4	
Ge_2Zr	0.30	oC12 ($ZrSi_2$)
$GeZr_3$	0.4	$L1_2$-tP32 (Ti_3P)
$H_{0.33}Nb_{0.67}$	7.28	B.C.C.
$H_{0.1}Nb_{0.9}$	7.38	Same
$H_{0.05}Nb_{0.95}$	7.83	Same
$H_{0.12}Ta_{0.88}$	2.81	B.C.C.
$H_{0.08}Ta_{0.92}$	3.26	Same
$H_{0.04}Ta_{0.96}$	3.62	Same
HfIrSi	3.50	C37-cP12 (Co_2Si)
$HfMo_2$	0.05	hP24 (Ni_2Mn)
$HfN_{0.989}$	6.6	B1
$Hf_{0-0.5}Nb_{1-0.5}$	8.3–9.5	A2
$Hf_{0.75}Nb_{0.25}$	> 4.2	
$HfOs_2$	2.69	C14
HfOsP	6.1	C22-hP9 (Fe_2P)
HfPRu	9.9	Same
$HfRe_2$	4.80	C14
$Hf_{0.14}Re_{0.86}$	5.86	A12
$Hf_{0.99-0.96}Rh_{0.01-0.04}$	0.85–1.51	
$Hf_{0-0.55}Ta_{1-0.45}$	4.4–6.5	A2
HfV_2	8.9–9.6	C15
Hg_xIn_{1-x}	3.14–4.55	
HgIn	3.81	
Hg_2K	1.20	Orthorhombic
Hg_3K	3.18	
Hg_4K	3.27	
Hg_8K	3.42	
Hg_3Li	1.7	Hexagonal
$HgMg_3$	0.17	hP8 (Na_3As)
Hg_2Mg	4.0	tI6 ($MoSi_2$)
Hg_3Mg_5	0.48	$D8_b$-hP16 (Mn_5Si_3)
Hg_2Na	1.62	Hexagonal
Hg_4Na	3.05	
Hg_xPb_{1-x}	4.14–7.26	
HgSn	4.2	
Hg_xTl_{1-x}	2.30–4.19	
Hg_5Tl_2	3.86	
Ho_xLa_{1-x}	1.3–6.3	
$Ho_{1.2}Mo_6Se_8$	6.1	$D10_2$-hR12 (Be_3Nb)
$In_{1-0.86}Mg_{0-0.14}$	3.395–3.363	
$In_2Mo_6Te_6$	2.6	hR15 (Mo_6PbS_8)
$InNb_3$ (high pressure and temp.)	4–8; 9.2	A15
$In_{0.5}Nb_3Zr_{0.5}$	6.4	
$In_{0.11}O_3W$	< 1.25–2.8	Hexagonal
$In_{0.95-0.85}Pb_{0.05-0.15}$	3.6–5.05	
$In_{0.98-0.91}Pb_{0.02-0.09}$	3.45–4.2	
InPb	6.65	
InPd	0.7	B2
InSb (quenched from 170 kbar into liquid N_2)	4.8	Like A5
InSb	2.1	B4
$(InSb)_{0.95-0.10}Sn_{0.05-0.90}$ (various heat treatments)	3.8–5.1	
$(InSb)_{0-0.07}Sn_{1-0.93}$	3.67–3.74	
In_3Sn	~5.5	
In_xSn_{1-x}	3.4–7.3	
$In_{0.82-1}Te$ ($n = 0.83–1.71 \times 10^{22}$)	1.02–3.45	B1
$In_{1.000}Te_{1.002}$	3.5–3.7	B1
In_3Te_4 ($n = 4.7 \times 10^{21}$)	1.15–1.25	Rhombohedral
In_xTl_{1-x}	2.7–3.374	
$In_{0.8}Tl_{0.2}$	3.223	
$In_{0.62}Tl_{0.38}$	2.760	
$In_{0.78-0.69}Tl_{0.22-0.31}$	3.18–3.32	Tetragonal
$In_{0.69-0.62}Tl_{0.31-0.38}$	2.98–3.3	F.C.C.
Ir_2La	0.48	C15
Ir_3La	2.32	$D10_2$
Ir_3La_7	2.24	$D10_2$
Ir_5La	2.13	
$IrLaSi_2$	2.03	oC16 ($CeNiSi_2$)
$IrLaSi_3$	2.7	tI10 ($BaNiSn_3$)
Ir_2Lu	2.47	C15
Ir_3Lu	2.89	C15
$Ir_4Lu_5Si_{10}$	3.9	tP38 ($Co_4Sc_5Si_{10}$)
IrMo	< 1.0	A3
$IrMo_3$	9.6	A15
$IrMo_3$	6.8	$D8_b$
$IrNb_3$	1.9	A15
$Ir_{0.4}Nb_{0.6}$	9.8	$D8_b$
$Ir_{0.37}Nb_{0.63}$	2.32	$D8_b$
IrNb	7.9	$D8_b$
$Ir_{1.15}Nb_{0.85}$	4.6	oP12 (IrTa)
$Ir_{0.02}Nb_3Rh_{0.98}$	2.43	A15
$Ir_{0.05}Nb_3Rh_{0.95}$	2.38	A15
$Ir_{0.287}O_{0.14}Ti_{0.573}$	5.5	$E9_3$
$Ir_{0.265}O_{0.035}Ti_{0.65}$	2.30	$E9_3$
Ir_xOs_{1-x}	0.3–0.98	
$Ir_{1.5}Os_{0.5}$	2.4	C14
IrOsY	2.6	C15
IrSiY	2.70	C37-oP12 (Co_2Si)
IrSiZr	2.04	Same
Ir_2Sc	2.07	C15
$Ir_{2.5}Sc$	2.46	C15
$Ir_4Sc_5Si_{10}$	8.46	tP38
Ir_2Si_2Th	2.14	tI10
$IrSi_3Th$	1.75	tI10
IrSiTh	6.50	tI12 (LaPtSi)
Ir_2Si_2Y	2.60	tI10 (Al4Ba)
$Ir_4Si_{10}Y_5$	3.10	tP38
$Ir_3Si_5Y_2$	2.83	oI40
$IrSn_2$	0.65–0.78	C1
Ir_2Sr	5.70	C15
Ir_7Ta_{13}	1.2	$D8_b$-tP30 (FeCr)
$Ir_{0.5}Te_{0.5}$	~3	
$IrTe_3$	1.18	C2
IrTh	< 0.37	B_f
Ir_2Th	6.50	C15
Ir_3Th	4.71	
Ir_3Th_7	1.52	$D10_2$
Ir_5Th	3.93	$D2_d$
$IrTi_3$	5.40	A15
IrV_2	1.39	A15
IrW_3	3.82	

Substance	T_c/K	Crystal structure type
$Ir_{0.28}W_{0.72}$	4.49	
Ir_2Y	2.18; 1.38	C15
$Ir_{0.69}Y_{0.31}$	1.98; 1.44	C15
$Ir_{0.70}Y_{0.30}$	2.16	C15
Ir_2Y_3	1.61	
Ir_3Y	3.50	$D10_2$-hR13 (Be_3Nb)
Ir_xY_{1-x}	0.3–3.7	
Ir_2Zr	4.10	C15
$Ir_{0.1}Zr_{0.9}$	5.5	A3
$K_2Mo_{15}S_{19}$	3.32	hR15
$K_{0.27-0.31}O_3W$	0.50	Hexagonal
$K_{0.40-0.57}O_3W$	1.5	Tetragonal
$La_{0.55}Lu_{0.45}$	2.2	Hexagonal, La type
$La_{0.8}Lu_{0.2}$	3.4	Same
$LaMg_2$	1.05	C15
$LaMo_6S_8$	7.1	hR15
LaN	1.35	
$LaOs_2$	6.5	C15
$LaPt_2$	0.46	C15
$La_{0.28}Pt_{0.72}$	0.54	C15
LaPtSi	3.48	tI12
$LaRh_3$	2.60	
$LaRh_5$	1.62	
La_7Rh_3	2.58	$D10_2$
$LaRhSi_2$	3.42	oC16 ($CeNiSi_2$)
$La_2Rh_3Si_5$	4.45	oI40 ($Co_3Si_5U_2$)
$LaRhSi_3$	2.7	tI10 ($BaNiSn_3$)
$LaRh_2Si_2$	3.90	tI10 (Al_4Ba)
$LaRu_2$	1.63	C15
La_3S_4	6.5	$D7_3$
La_3Se_4	8.6	$D7_3$
$LaSi_2$	2.3	C_c
La_xY_{1-x}	1.7–5.4	
LaZn	1.04	B2
$Li_2Mo_6S_8$	4.2	hR15
LiPb	7.2	
$LuOs_2$	3.49	C14
$Lu_{0.275}Rh_{0.725}$	1.27	C15
$LuRh_5$	0.49	
$Lu_5Rh_4Si_{10}$	3.95	tP38 ($Co_4So_5Si_{10}$)
$LuRu_2$	0.86	C14
$Mg_{1.14}Mo_{6.6}S_8$	3.5	hR15
Mg2Nb	5.6	
$Mg_{\sim 0.47}Tl_{\sim 0.53}$	2.75	B2
MgZn	0.9	A3-oP4 (AuCd)
Mn_xTi_{1-x}	2.3 (max.)	Mn in -Ti
Mn_xTi_{1-x}	1.1–3.0	Mn in -Ti
MnU_6	2.32	$D2_c$
Mo_2N	5.0	F.C.C.
$Mo_6Na_2S_8$	8.6	hR15
Mo_xNb_{1-x}	0.016–9.2	
$Mo_{5.25}Nb_{0.75}Se_8$	6.2	hR15
Mo_6NdSa_8	8.2	hR15
Mo_3Os	7.2	A15
$Mo_{0.62}Cs_{0.38}$	5.65	$D8_b$
Mo_3P	5.31	DO_e
$Mo_6Pb_{1.2}Se_8$	6.75	hR15
$Mo_{0.5}Pd_{0.5}$	3.52	A3
Mo_6PrSe_8	9.2	hR15
MoRe	7.8	$D8_b$-tP30
$MoRe_3$	9.25; 9.89	A12
Mo_xRe_{1-x}	1.2–12.2	
$Mo_{0.42}Re_{0.58}$	6.35	$D8_b$
MoRh	1.97	A3
Mo_xRh_{1-x}	1.5–8.2	B.C.C.
MoRu	9.5–10.5	A3
$Mo_{0.61}Ru_{0.39}$	7.18	$D8_b$
$Mo_{0.2}Ru_{0.8}$	1.66	A3
Mo_3Ru_2	7.0	$D8_b$-tP30
$Mo_4Ru_2Te_8$	1.7	hR15
Mo_6S_8	1.85	hR15
Mo_6S_8Sc	3.6	hR15
$Mo_6S_8Sm_{1.2}$	2.9	hR15
Mo_6S_8Tb	2.0	hR15
Mo_6S_8Tl	8.7	hR15
$Mo_6S_8Tm_{1.2}$	2.1	hR15
$Mo_6S_8Y_{1.2}$	3.0	hR15
Mo_6S_8Yb	9.2	hR15
$Mo_{6.6}S_8Zn_{11}$	3.6	hR15
Mo_3Sb_4	2.1	
Mo_6Se_8	6.3	hR15
$Mo_6Se_8Sm_{1.2}$	6.8	hR15
$Mo_6Se_8Sn_{1.2}$	6.8	hR15
Mo_6Se_8Tb	5.7	hR15
Mo_3Se_3Tl	4.0	hP14
$Mo_6Se_8Tm_{1.2}$	6.3	hR15
Mo_6Se_8Yb	6.2	hR15
Mo_3Si	1.30	A15
$MoSi_{0.7}$	1.34	
Mo_xSiV_{3-x}	4.54–16.0	A15
$Mo_{5.25}Ta_{0.75}Te_8$	1.7	hR15
Mo_6Te_8	1.7	hR15
$Mo_{0.16}Ti_{0.84}$	4.18; 4.25	
$Mo_{0.913}Ti_{0.087}$	2.95	
$Mo_{0.04}Ti_{0.96}$	2.0	Cubic
$Mo_{0.025}Ti_{0.975}$	1.8	
Mo_xU_{1-x}	0.7–2.1	
Mo_xV_{1-x}	0–~5.3	
Mo_2Zr	4.25–4.75	C15
NNb (film)	6–9	B1
$N_xO_yTi_z$	2.9–5.6	Cubic
$N_xO_yV_z$	5.8–8.2	Cubic
$N_{0.34}Re$	4–5	F.C.C.
NTa (film)	4.84	B1
$N_{0.6-0.987}Ti$	< 1.17–5.8	B1
$N_{0.82-0.99}V$	2.9–7.9	B1
NZr	9.8	B1
$N_{0.906-0.984}Zr$	3.0–9.5	B1
$Na_{0.28-0.35}O_3W$	0.56	Tetragonal
$Na_{0.28}Pb_{0.72}$	7.2	
NbO	1.25	
$NbOs_2$	2.52	A12
Nb_3Os	1.05	A15
$Nb_{0.6}Os_{0.4}$	1.89; 1.78	$D8_b$
$Nb_3Os_{0.02-0.10}Rh_{0.98-0.90}$	2.42–2.30	A15
Nb_3P	1.8	$L1_2$tP32 (Ti_3P)
NbPRh	4.08	C37-oP12 (Co_2Si)
$Nb_{0.6}Pd_{0.4}$	1.60	$D8_f$ plus cubic
$Nb_3Pd_{0.02-0.10}Rh_{0.92-0.90}$	2.49–2.55	A15
$Nb_{0.62}Pt_{0.38}$	4.21	$D8_b$

Substance	T_c/K	Crystal structure type
Nb_5Pt_3	3.73	$D8_b$
$Nb_3Pt_{0.02–0.98}Rh_{0.98–0.02}$	2.52–9.6	A15
$NbRe_3$	5.27	$D8_b$-tP30 (FeCr)
$Nb_{0.38–0.18}Re_{0.62–0.82}$	2.43–9.70	A15
NbRe	3.8	$D8_b$-tP30
NbReSi	5.1	oI36 (FeTiSi)
Nb_3Rh	2.64	A15
$Nb_{0.6}Rh_{0.40}$	4.21	$D8_b$ plus other
$Nb_{0.9}Rh_{1.1}$	3.07	A3-oP4 (AuCd)
$Nb_3Rh_{0.98–0.90}Ru_{0.02–0.10}$	2.42–2.44	A15
Nb_xRu_{1-x}	1.2–4.8	
NbRuSi	2.65	oI36
NbS_2	6.1–6.3	Hexagonal, $NbSe_2$ type
NbS_2	5.0–5.5	Hexagonal, three-layer type
Nb_3Sb	0.2	$L1_2$-tP32 (Ti_3P)
$Nb_3Sb_{0–0.7}Sn_{1–0.3}$	6.8–18	A15
$NbSe_2$	5.15–5.62	Hexagonal
$Nb_{1–1.05}Se_2$	2.2–7.0	Same
Nb_3Se_4	2.0	hP14
Nb_3Si	1.5	$L1_2$
Nb_3SiSnV_3	4.0	
$NbSn_2$	2.60	Orthorhombic
Nb_6Sn_5	2.8	oI44 (Sn_5Ti_6)
NbSnTaV	6.2	A15
NbSnV2	5.5	A15
Nb_2SnV	9.8	A15
Nb_xTa_{1-x}	4.4–9.2	A2
Nb_3Te_4	1.8	hP14
Nb_xTi_{1-x}	0.6–9.8	
$Nb_{0.6}Ti_{0.4}$	9.8	
Nb_xU_{1-x}	1.95 (max.)	
$Nb_{0.88}V_{0.12}$	5.7	A2
$Nb_{0.5}V_{1.5}Zr$	4.3	C15-hP12 ($MgZn_2$)
$Ni_{0.3}Th_{0.7}$	1.98	$D10_2$
$NiZr_2$	1.52	
$Ni_{0.1}Zr_{0.9}$	1.5	A3
$O_3Rb_{0.27–0.29}W$	1.98	Hexagonal
OSn	3.81	tP4 (PbO)
O_3SrTi ($n = 1.7–12.0 \times 10^{19}$)	0.12–0.37	
O_3SrTi ($n = 10^{18}–10^{21}$)	0.05–0.47	
O_3SrTi ($n = 10^{20}$)	0.47	
$O_3Sr_{0.08}W$	2–4	Hexagonal
OTi	0.58	
$O_3Tl_{0.30}W$	2.0–2.14	Hexagonal
OV_3Zr_3	7.5	$E9_3$
OW_3 (film)	3.35; 1.1	A15
OsPti	1.2	C22-hP9 (Fe_2P)
OsPZr	7.4	Same
OsReY	2.0	C14
Os2Sc	4.6	C14
OsTa	1.95	A12
Os_3Th_7	1.51	$D10_2$
Os_xW_{1-x}	0.9–4.1	
OsW_3	~3	
Os_2Y	4.7	C14
Os_2Zr	3.0	C14
Os_xZr_{1-x}	1.5–5.6	
PPb	7.8	
OsW_2	3.81	$D8_b$-tP30 (FeCr)
$PPd_{3.0–3.2}$	<0.35–0.7	DO_{11}

Substance	T_c/K	Crystal structure type
P_3Pd_7 (high temperature)	1.0	Rhombohedral
P_3Pd_7 (low temperature)	0.70	Complex
PRh	1.22	
PRh_2	1.3	C1
P_4Rh_5	1.22	oP28 ($CaFe_2O_4$)
PRhTa	4.41	C37-oP12 (Co_2Si)
PRhZr	1.55	Same
PRuTi	1.3	C22-hP9 (Fe_2P)
PRuZr	3.46	C37-oP12
PW_3	2.26	DO_e
Pb_2Pd	2.95	C16
Pb_4Pt	2.80	Related to C16
Pb_2Rh	2.66	C16
PbSb	6.6	
PbTe (plus 0.1 w/o Pb)	5.19	
PbTe (plus 0.1 w/o Te)	5.24–5.27	
$PbTl_{0.27}$	6.43	
$PbTl_{0.17}$	6.73	
$PbTl_{0.12}$	6.88	
$PbTl_{0.075}$	6.98	
$PbTl_{0.04}$	7.06	
$Pb_{1–0.26}Tl_{0–0.74}$	7.20–3.68	
$PbTl_2$	3.75–4.1	
Pb_3Zr_5	4.60	$D8_8$
$PbZr_3$	0.76	A15
$Pd_{0.9}Pt_{0.1}Te_2$	1.65	C6
$Pd_{0.05}Ru_{0.05}Zr_{0.9}$	~9	
$Pd_{2.2}S$ (quenched)	1.63	Cubic
$PdSb_2$	1.25	C2
PdSb	1.5	$B8_1$
PdSbSe	1.0	C2
PdSbTe	1.2	C2
Pd_4Se	0.42	Tetragonal
$Pd_{6–7}Se$	0.66	Like Pd_4Te
$Pd_{2.8}Se$	2.3	
Pd_xSe_{1-x}	2.5 (max.)	
PdSi	0.93	B31
PdSn	0.41	B31
$PdSn_2$	3.34	
Pd_2Sn	0.41	C37
Pd_3Sn	0.47–0.64	$B8_2$
Pd_2SnTm	1.77	DO_3-cF16 (BiF_3)
Pd_2SnY	4.92	Same
Pd_2SnYb	1.79	Same
PdTe	2.3; 3.85	$B8_1$
$PdTe_{1.02–1.08}$	2.56–1.88	$B8_1$
$PdTe_2$	1.69	C6
$PdTe_{2.1}$	1.89	C6
$PdTe_{2.3}$	1.85	C6
$Pd_{1.1}Te$	4.07	$B8_1$
Pd_3Te	0.76	cI2 (W)
$PdTh_2$	0.85	C16
$Pd_{0.1}Zr_{0.9}$	7.5	A3
PtSb	2.1	$B8_1$
PtSi	0.88	B31
PtSn	0.37	$B8_1$
$PtSn_4$	2.38	C16-oC20 ($PdSn_4$)
Pt_3Ta_7	1.5	$D8_b$-tP30
$PtTa_3$	0.4	A15-cP8 (Cr_3Si)
PtTe	0.59	Orthorhombic

Substance	T_c/K	Crystal structure type
PtTh	0.44	B_f
Pt_3Th_7	0.98	$D10_2$
Pt_5Th	3.13	
$PtTi_3$	0.58	A15
$Pt_{0.02}U_{0.98}$	0.87	β-phase
$PtV_{2.5}$	1.36	A15
PtV_3	2.87–3.20	A15
$PtV_{3.5}$	1.26	A15
$Pt_{0.5}W_{0.5}$	1.45	A1
Pt_xW_{1-x}	0.4–2.7	
Pt_2Y_3	0.90	
Pt_2Y	1.57; 1.70	C15
Pt_3Y_7	0.82	$D10_2$
PtZr	3.0	A3
Re_2Sc	4.2	C15-hP12 ($MgZn_2$)
$Re_{24}Sc_5$	2.2	A12-cI58 (Mg)
ReSiTa	4.4	oI36 (FeTiSi)
$Re_3Si_5Y_2$	1.76	tP40 ($Fe_3Sc_2Si_5$)
Re_3Ta_2	1.4	$D8_b$-tP30 (FeCr)
$Re_{0.64}Ta_{0.36}$	1.46	A12
Re_3Ta	6.78	A12-cI58 (Mn)
$Re_{24}Ti_5$	6.60	A12
Re_xTi_{1-x}	6.6 (max.)	
$Re_{0.76}V_{0.24}$	4.52	$D8_b$
Re_3V	6.26	$D8_b$-tP30
$Re_{0.92}V_{0.08}$	6.8	A3
$Re_{0.6}W_{0.4}$	6.0	
$Re_{0.5}W_{0.5}$	5.12	$D8_b$
$Re_{13}W_{12}$	5.2	$D8_b$-tP30
Re_3W	9.0	A12-cI58
Re_2Y	1.83	C14
Re_2Zr	5.9	C14
Re_3Zr	7.40	A12-cI58
Re_6Zr	7.40	Same
$Rh_{17}S_{15}$	5.8	Cubic
$Rh_{\sim0.24}Sc_{\sim0.76}$	0.88; 0.92	
$Rh_4Sc_5Si_{10}$	8.54	tP38
$Rh_4Sc_3Sn_{13}$	4.5	cP40
Rh_xSe_{1-x}	6.0 (max.)	
$RhSi_3Th$	1.76	tI10
$Rh_{0.86}Sc_{1.04}Th$	6.45	tI12
Rh_2Si_2Y	3.11	tI10
$Rh_3Si_5Y_2$	2.70	oI40
$Rh_4Sn_{13}Sr_3$	4.3	cP40
Rh_xSn_yTh	1.9	cI2 (W)
Rh_xSn_yTm	2.3	cP40
$Rh_4Sn_{13}Y_3$	3.2	cP40
Rh_2Sr	6.2	C15
$Rh_{0.4}Ta_{0.6}$	2.35	$D8_b$
$RhTe_2$	1.51	C2
$Rh_{0.67}Te_{0.33}$	0.49	
Rh_xTe_{1-x}	1.51 (max.)	
RhTh	0.36	B_f
Rh_3Th_7	2.15	$D10_2$
Rh_5Th	1.07	
Rh_xTi_{1-x}	2.25–3.95	
$Rh_{0.02}U_{0.98}$	0.96	
RhV_3	0.38	A15
RhW	~3.4	A3
RhY_3	0.65	

Substance	T_c/K	Crystal structure type
Rh_2Y_3	1.48	
Rh_3Y	1.07	C15
Rh_5Y	0.56	
Rh_3Y_7	0.32	hP20 (Fe_3Th_7)
$Rh_{0.005}Zr$ (annealed)	5.8	
$Rh_{0-0.45}Zr_{1-0.55}$	2.1–10.8	
$Rh_{0.1}Zr_{0.9}$	9.0	H.C.P.
Ru_2Sc	1.67	C14
RuSiTa	3.15	oI36
Ru_3Si_2Th	3.98	hP12
Ru_3Si_2Y	3.51	hP12
$Ru_{1.1}Sn_{3.1}Y$	1.3	cP40
Ru_2Th	3.56	C15
RuTi	1.07	B2
$Ru_{0.05}Ti_{0.95}$	2.5	
$Ru_{0.1}Ti_{0.9}$	3.5	
$Ru_xTi_{0.6}V_y$	6.6 (max.)	
Ru_3U	0.15	$L1_2$-cP4
$Ru_{0.45}V_{0.55}$	4.0	B2
RuW	7.5	A3
Ru_2Y	1.52	C14
Ru_2Zr	1.84	C14
$Ru_{0.1}Zr_{0.9}$	5.7	A3
STh	0.5	B1-cF8 (NaCl)
SbSn	1.30–1.42	B1 or distorted
$SbTa_3$	0.72	A15-cP8 (Cr_3Si)
$SbTi_3$	5.8	Same
Sb_2Ti_7	5.2	
$Sb_{0.01-0.03}V_{0.99-0.97}$	3.76–2.63	A2
SbV_3	0.80	A15
SeTh	1.7	B1-cF8
$SiMo_3$	1.4	A15-cP8
Si_2Th	3.2	C_c, α-phase
Si_2Th	2.4	C32, β-phase
$SiV_{2.7}Ru_{0.3}$	2.9	A15
Si_2W_3	2.8; 2.84	
$SiZr_3$	0.5	$L1_2$-tP32 (Ti_3P)
$Sn_{0.174-0.104}Ta_{0.826-0.896}$	6.5–< 4.2	A15
$SnTa_3$	8.35	A15, highly ordered
$SnTa_3$	6.2	A15, partially ordered
$SnTaV_2$	2.8	A15
$SnTa_2V$	3.7	A15
Sn_xTe_{1-x} ($n = 10.5–20 \times 10^{20}$)	0.07–0.22	B1
Sn_3Th	3.33	$L1_2$-cP4
$SnTi_3$	5.80	A15-cP8
Sn_xTl_{1-x}	2.37–5.2	
SnV_3	3.8	A15
$Sn_{0.02-0.057}V_{0.98-0.943}$	2.87–~1.6	A2
$SnZr_3$	0.92	A15-cP8
$Ta_{0.025}Ti_{0.975}$	1.3	Hexagonal
$Ta_{0.05}Ti_{0.95}$	2.9	Hexagonal
$Ta_{0.05-0.75}V_{0.95-0.25}$	4.30–2.65	A2
$Ta_{0.8-1}W_{0.2-0}$	1.2–4.4	A2
$Tc_{0.1-0.4}W_{0.9-0.6}$	1.25–7.18	Cubic
$Tc_{0.50}W_{0.50}$	7.52	α plus
$Tc_{0.60}W_{0.40}$	7.88	plus α
Tc_6Zr	9.7	A12
TeY	1.02	B1-cF8
$ThTl_3$	0.87	$L1_2$-cP4
$Th_{0-0.55}Y_{1-0.45}$	1.2–1.8	

Substance	T_c/K	Crystal structure type
$Ti_{0.70}V_{0.30}$	6.14	Cubic
Ti_xV_{1-x}	0.2–7.5	
$Ti_{0.5}Zr_{0.5}$ (annealed)	1.23	
$Ti_{0.5}Zr_{0.5}$ (quenched)	2.0	
Tl_3Y	1.52	$L1_2$-cP4
V_2Zr	8.80	C15
$V_{0.26}Zr_{0.74}$	5.9	
W_2Zr	2.16	C15
YZn	0.33	B2-cP2 (CsCl)
B. Superconductors with T_c > 10 K		
Al_2CMo_3	10.0	A13
$Al_{0.5}Ge_{0.5}Nb$	12.6	A15
$Al_{\sim0.8}Ge_{\sim0.2}Nb_3$	20.7	A15
$AlNb_3$	18.0	A15 (Cr_3Si)
$AlNb_3$	12.0	(FeCr)
$A1_xNb_{1-x}$	<4.2–13.5	$D8_b$
$A1_xNb_{1-x}$	12–17.5	A15
$A1_{0.27}Nb_{0.73-0.48}V_{0-0.25}$	14.5–17.5	A15
A1 Nb_xV_{1-x}	4.4–13.5	
$A1_{0.1}Si_{0.9}V_3$	14.05	
$A1V_3$	11.8	A15 (Cr_3Si)
$AuNb_3$	11.5	A15
$Au_{0-0.3}Nb_{1-0.7}$	1.1–11.0	
$Au_{0.02-0.98}Nb_3Rh_{0.98-0.02}$	2.53–10.9	A15
$AuNb_{3(1-x)}V_{3x}$	1.5–11.0	A15
$B_{0.03}C_{0.51}Mo_{0.47}$	12.5	
B_4LuRh_4	11.7	(B_4CeCo_4)
B_2LuRu	10	
B_4Rh_4Y	11.3	(B_4CeCo_4)
$B_{0.1}Si_{0.9}V_3$	15.8	A15
$BaBi_{0.2}O_3Pb_{0.8}$	13.2	
$Ba_2CaCu_2O_8T1_2$	120	
$Ba_2Cu_3LaO_6$	80	
$Ba_2Cu_3O_7Tm$	101	
$Ba_2Cu_3O_7Y$	90	
$(Ba,La)_2CuO_4$	36	A15 (K_2NiF_4)
$Bi_2CaCu_2O_8Sr_2$	110	
$Br_2Mo_6S_6$	13.8	(Mo_6PbS_8)
C_3La	11.0	(C_3Pu_2)
CMo	14.3	B1(NaCl)
CMo_2	12.2	cubic
$C_{0.5}Mo_xNb_{1-x}$	10.8–12.5	B1
CMo_xTi_{1-x}	10.2(max)	B1
$CMo_{0.83}Ti_{0.17}$	10.2	B1
$C_{0-0.38}N_{1-0.62}Ta$	10.0–11.3	
CNb (whiskers)	7.5–10.5	
CNb	11.5	B1
$C_{0.7-1.0}Nb_{0.3-0}$	6–11	B1
CNb_xTa_{1-x}	8.2–13.9	
$CNb_{0.6-0.9}W_{0.4-0.1}$	12.5–11.6	B1
$C_{0.1}Si_{0.9}V_3$	16.4	A15
CTa	10.3	B1
$CTa_{1-0.4}W_{0-0.6}$	8.5–10.5	B1
$C_{0.66}Th_{0.13}Y_{0.21}$	17	(C_3Pu_2)
C_3Y_2	11.5	(C_3Pu_2)
CW	10	B1
$(Ca,La)_2CuO_4$	18	(K_2NiF_4)
$Cu(La,Sr)_2O_4$	39	
$Cu_{1.8}Mo_6S_8$	10.8	(Mo_6PbS_8)
$Cr_{0.3}SiV_{2.7}$	11.3	A15

Substance	T_c/K	Crystal structure type
$GaNb_3$	14.5	A15 (Cr_3Si)
$Ga_xNb_3Sn_{1-x}$	14–18.37	A15
GaV_3	16.8	A15
$GaV_{2.1-3.5}$	6.3–14.45	A15
$GeNb_3$	23.2	A15
$GeNb_3$ (quenched)	6–17	A15
$Ge_xNb_3Sn_{1-x}$	17.6–18.0	A15
$Ge_{0.5}Nb_3Sn_{0.5}$	11.3	
$Ge_{0.1}Si_{0.9}V_3$	14.0	A15
GeV_3	11	A15
$InLa_3$	9.83; 10.4	LI_2 ($AuCu_3$)
$In_{0-0.3}Nb_3Sn_{1-0.7}$	18.0–18.19	A15
InV_3	13.9	A15
$Ir_{0.4}Nb_{0.6}$	10	(FeCr)
$LaMo_6Se_8$	11.4	(Mo_6PbS_8)
LiO_4Ti_2	13.7	($A1_2MgO_4$)
MgB_2	39.0±0.5	C32
MoN	12; 14.8	NaCl-B1
MoN (powder)	29.0	B1
Mo_3Os	12.7	A15
$Mo_6Pb_{0.9}S_{7.5}$	15.2	(Mo_6PbS_8)
Mo_3Re	10.0; 15	A15
Mo_xRe_{1-x}	1.2–12.2	
$Mo_{0.52}Re_{0.48}$	11.1	
$Mo_{0.57}Re_{0.43}$	14.0	
$Mo_{\sim0.60}Re_{0.395}$	10.6	
MoRu	9.5–10.5	A3
Mo_3Ru	10.6	A15
Mo_6Se_8T1	12.2	(Mo_6PbS_8)
$Mo_{0.3}SiV_{2.7}$	11.7	A15
Mn_3Si	12.5	A15
Mo_3Tc	15	A15
$Mo_{0.3}Tc_{0.7}$	12.0	A15
Mo_xTc_{1-x}	10.8–15.8	
$MoTc_3$	15.8	
NNb (whiskers)	10–14.5	
NNb (diffusion wires)	16.10	
$N_{0.988}Nb$	14.9; 17.3	B1
$N_{0.824-0.988}Nb$	14.4–15.3	B1
$N_{0.7-0.795}Nb$	11.3–12.9	
NNb_xO_y	13.5–17.0	B1
NNb_xO_y	6.0–11	
$N_{100-42w/o}Nb_{0-58w/o}Ti$	15–16.8	
$N_{100-75w/o}Nb_{0-25w/o}Zr$	12.5–16.35	
NNb_xZr_{1-x}	9.8–13.8	B1
$N_{0.93}Nb_{0.85}Zr_{0.15}$	13.8	B1
NTa	12–14	B1
NZr	10.7	B1
Nb_3Pt	10.9	A15
$Nb_{0.18}Re_{0.82}$	10	(Mn)
Nb_3Si	19	A15
$Nb_{0.3}SiV_{2.7}$	12.8	A15
Nb_3Sn	18.05	A15
$Nb_{0.8}Sn_{0.2}$	18.18; 18.5	A15
Nb_xSn_{1-x} (film)	2.6–18.5	BCC+I43m
Nb_3Sn_2	16.6	BCC+orthorh.
$NbSnTa_2$	10.8	A15
Nb_2SnTa	16.4	A15
$Nb_{2.5}SnTa_{0.5}$	17.6	A15
$Nb_{2.75}SnTa_{0.25}$	17.8	A15

Substance	T_c/K	Crystal structure type
$Nb_{3x}SnTa_{3(1-x)}$	6.0–18.0	
$Nb_2SnTa_{0.5}V_{0.5}$	12.2	A15
$NbTc_3$	10.5	A12
$Nb_{0.75}Zr_{0.25}$	10.8	
$Nb_{0.66}Zr_{0.33}$	10.8	
$PbTa_3$	17	A15
$RhTa_3$	10	A15
$RhZr_2$	10.8; 11.3	C16 (Al_2Cu)
$Rh_{0-0.45}Zr_{1-0.55}$	2.1–10.8	
$SiTi_{0.3}V_{2.7}$	10.9	A15
SiV_3	17.1	A15
$SiV_{2.7}Zr_{0.3}$	13.2	A15

[a] Extrapolated.

TABLE 5. Critical Field Data

Substance	H_o/Oe
Ag_2F	2.5
Ag_7NO_{11}	57
Al_2CMo_3	1700
$BaBi_3$	740
Bi_2Pt	10
Bi_3Sr	530
Bi_5Tl_3	>400
CdSn	>266
$CoSi_2$	105
$Cr_{0.1}Ti_{0.3}V_{0.6}$	1360
$In_{1-0.86}Mg_{0-0.14}$	272.4–259.2
InSb	1100
In_xTl_{1-x}	252–284
$In_{0.8}Tl_{0.2}$	252
$Mg_{0.47}Tl_{0.53}$	220
$Mo_{0.16}Ti_{0.84}$	<985
$NbSn_2$	620
$PbTl_{0.27}$	756
$PbTl_{0.17}$	796
$PbTl_{0.12}$	849
$PbTl_{0.075}$	880
$PbTl_{0.04}$	864

TABLE 6. High Critical Magnetic-Field Superconductive Compounds and Alloys

Substance	T_c/K	H_{c1}/kOe	H_{c2}/kOe	H_{c3}/kOe	T_{obs}/K[a]
Al_2CMo_3	9.8–10.2	0.091	156		1.2
$AlNb_3$		0.375			
$Ba_xO_3Sr_{1-x}Ti$	<0.1–0.55	0.0039 max.			
$Bi_{0.5}Cd_{0.1}Pb_{0.27}Sn_{0.13}$			>24		3.06
Bi_xPb_{1-x}	7.35–8.4	0.122 max.	30 max.		4.2
$Bi_{0.56}Pb_{0.44}$	8.8		15		4.2
$Bi_{7.5w/o}Pb_{92.5w/o}$[b]			2.32		
$Bi_{0.099}Pb_{0.901}$		0.29	2.8		
$Bi_{0.02}Pb_{0.98}$		0.46	0.73		
$Bi_{0.53}Pb_{0.32}Sn_{0.16}$			>25		3.06
$Bi_{1-0.93}Sn_{0-0.07}$			0–0.032		3.7
Bi_5Tl_3	6.4		>5.6		3.35
C_8K (excess K)	0.55		0.160 (H⊥c)		0.32
			0.730 (H|c)		0.32
C_8K	0.39		0.025 (H⊥c)		0.32
			0.250 (H⊥c)		0.32
$C_{0.44}Mo_{0.56}$	12.5–13.5	0.087	98.5		1.2
CNb	8–10	0.12	16.9		4.2
$CNb_{0.4}Ta_{0.6}$	10–13.6	0.19	14.1		1.2
CTa	9–11.4	0.22	4.6		1.2
$Ca_xO_3Sr_{1-x}Ti$	<0.1–0.55	0.002–0.004			
$Cd_{0.1}Hg_{0.9}$ (by weight)		0.23	0.34		2.04
$Cd_{0.05}Hg_{0.95}$		0.28	0.31		2.16

Solids

Substance	T_c/K	H_{c1}/kOe	H_{c2}/kOe	H_{c3}/kOe	T_{obs}/K[a]
$Cr_{0.10}Ti_{0.30}V_{0.60}$	5.6	0.071	84.4		0
GaN	5.85	0.725			4.2
Ga_xNb_{1-x}			>28		4.2
GaSb (annealed)	4.24		2.64		3.5
$GaV_{1.95}$	5.3		73[e]		
$GaV_{2.1\text{-}3.5}$	6.3–14.45		230–300[d]		0
GaV_3		0.4	350[c]		0
			500[d]		
$GaV_{4.5}$	9.15		121[c]		0
Hf^xNb^y			>52–>102		1.2
Hf^xTa^y			>28–>86		1.2
$Hg_{0.05}Pb_{0.95}$		0.235	2.3		
$Hg_{0.101}Pb_{0.899}$		0.23	4.3		4.2
$Hg_{0.15}Pb_{0.85}$	6.75		>13		2.93
$In_{0.98}Pb_{0.02}$	3.45	0.1		0.12	2.76
$In_{0.96}Pb_{0.04}$	3.68	0.1	0.12	0.25	2.94
$In_{0.94}Pb_{0.06}$	3.90	0.095	0.18	0.35	3.12
$In_{0.913}Pb_{0.087}$	4.2	~10.17	0.55	2.65	
$In_{0.316}Pb_{0.684}$		0.155	3.7		4.2
$In_{0.17}Pb_{0.83}$			2.8	5.5	4.2
$In_{1.000}Te_{1.002}$	3.5–3.7		1.2[c]		0
$In_{0.95}Tl_{0.05}$		0.263	0.263		3.3
$In_{0.90}Tl_{0.10}$		0.257	0.257		3.25
$In_{0.83}Tl_{0.17}$		0.242	0.39		3.21
$In_{0.75}Tl_{0.25}$		0.216	0.50		3.16
LaN	1.35	0.45			0.76
La_3S_4	6.5	≈0.15	>25		1.3
La_3Se_4	8.6	≈0.2	>25		1.25
$Mo_{0.52}Re_{0.48}$	11.1		14–21	22–33	4.2
			18–28	37–43	1.3
$Mo_{0.6}Re_{0.395}$	10.6		14–20	20–37	4.2
			19–26		
				26–37	1.3
$Mo_{0.5}Ti_{0.5}$			75[c]		0
$Mo_{0.16}Ti_{0.84}$	4.18	0.028	98.7[c]		0
$Mo_{0.16}Ti_{0.84}$			36–38		3.0
$Mo_{0.913}Ti_{0.087}$	2.95	0.060	15		4.2
$Mo_{0.1\text{-}0.3}U_{0.9\text{-}0.7}$	1.85–2.06		>25		
$Mo_{0.17}Zr_{0.83}$			30		
$N_{(12.8\ w/o)}Nb$	15.2		>9.5		13.2
NNb (wires)	16.1		153[c]		0
			132		4.2
			95		8
			53		12
NNb_xO_{1-x}	13.5–17.0		38		
NNb_xZr_{1-x}	9.8–13.8		4- >130		4.2
$N_{0.93}Nb_{0.85}Zr_{0.15}$	13.8		>130		4.2
$Na_{0.086}Pb_{0.914}$		0.19	6.0		
$Na_{0.016}Pb_{0.984}$		0.28	2.05		
Nb	9.15		2.020		1.4
			1.710		4.2
Nb		0.4–1.1	3–5.5		4.2
Nb (unstrained)		1.1–1.8	3.40	6–9.1	4.2
Nb (strained)		1.25–1.92	3.44	6.0–8.7	4.2
Nb (cold-drawn wire)		2.48	4.10	≈10	4.2
Nb (film)			>25		4.2
NbSc			>30		
Nb_3Sn		0.170	221		4.2
			70		14.15
			54		15
			34		16

Substance	T_c/K	H_{c1}/kOe	H_{c2}/kOe	H_{c3}/kOe	T_{obs}/K[a]
			17		17
$Nb_{0.1}Ta_{0.9}$		0.084	0.154		4.195
$Nb_{0.2}Ta_{0.8}$			10		4.2
$Nb_{0.65-0.73}Ta_{0.02-0.10}Zr_{0.25}$			>70–>90		4.2
Nb_xTi_{1-x}			148 max.		1.2
			120 max.		4.2
$Nb_{0.222}U_{0.778}$		1.98	23		1.2
Nb_xZr_{1-x}			127 max.		1.2
			94 max.		4.2
O_3SrTi	0.43	0.0049[c]	0.504[c]		0
O_3SrTi	0.33	0.00195[c]	0.420[c]		0
$PbSb_{1\ w/o}$(quenched)			>1.5		4.2
$PbSb_{1\ w/o}$(annealed)			>0.7		4.2
$PbSb_{2.8\ w/o}$(quenched)			>2.3		4.2
$PbSb_{2.8\ w/o}$(annealed)			>0.7		4.2
$Pb_{0.871}Sn_{0.129}$		0.45	1.1		
$Pb_{0.965}Sn_{0.035}$		0.53	0.56		
$Pb_{1-0.26}Tl_{0-0.74}$	7.20–3.68		2–6.9[c]		0
$PbTl_{0.17}$	6.73		4.5[c]		0
$Re_{0.26}W_{0.74}$			>30		
$Sb_{0.93}Sn_{0.07}$			0.12		3.7
SiV_3	17.0	0.55	156[e]		
Sn_xTe_{1-x}		0.00043–0.00236	0.005–0.0775		0.012–0.079
Ta (99.95%)		0.425	1.850		1.3
		0.325	1.425		2.27
		0.275	1.175		2.66
		0.090	0.375		3.72
$Ta_{0.5}Nb_{0.5}$			3.55		4.2
$Ta_{0.65-0}Ti_{0.35-1}$	4.4–7.8		>14–138		1.2
$Ta_{0.5}Ti_{0.5}$			138		1.2
Te	3.3	0.25[c]			0
Tc_xW_{1-x}	5.75–7.88		8–44		4.2
Ti				2.7	4.2
$Ti_{0.75}V_{0.25}$	5.3	0.029[c]	199[c]		0
$Ti_{0.775}V_{0.225}$	4.7	0.024[c]	172[c]		0
$Ti_{0.615}V_{0.385}$	7.07	0.050	34		4.2
$Ti_{0.516}V_{0.484}$	7.20	0.062	28		4.2
$Ti_{0.415}V_{0.585}$	7.49	0.078	25		4.2
$Ti_{0.12}V_{0.88}$			17.3	28.1	4.2
$Ti_{0.09}V_{0.91}$			14.3	16.4	4.2
$Ti_{0.06}V_{0.94}$			8.2	12.7	4.2
$Ti_{0.03}V_{0.97}$			3.8	6.8	4.2
Ti_xV_{1-x}			108 max.		1.2
V	5.31	0.8	3.4		1.79
		0.75	3.15		2
		0.45	2.2		3
		0.30	1.2		4
$V_{0.26}Zr_{0.74}$	≈5.9	0.238			1.05
		0.227			1.78
		0.185			3.04
		0.165			3.5
W (film)	1.7–4.1		>34		1

[a] Temperature of critical field measurement.
[b] w/o denotes weight percent.
[c] Extrapolated.
[d] Linear extrapolation.
[e] Parabolic extrapolation.

Solids

References

1. B. W. Roberts, in *Superconductive Materials and Some of Their Properties. Progress in Cryogenics*, Vol. IV, 1964, pp. 160–231.
2. B. W. Roberts, Superconductive Materials and Some of Their Properties, NBS Technical Notes 408 and 482, U.S. Government Printing Office, 1966 and 1969; B. W. Roberts, *J. Phys. Chem. Ref. Data*, 5, 581, 1976 <https://doi.org/10.1063/1.555540>.
3. B. W. Roberts, Properties of Selected Superconductive Materials, 1978 Supplement, NBS Technical Note 983, 1978 <https://doi.org/10.6028/NBS.TN.983>.
4. T. Claeson, *Phys. Rev.*, 147, 340, 1966 <https://doi.org/10.1103/PhysRev.147.340>.
5. C. J. Raub, W. H. Zachariasen, T. H. Geballe, and B. T. Matthias, *J. Phys. Chem. Solids*, 24, 1093, 1963 <https://doi.org/10.1016/0022-3697(63)90022-2>.
6. T. H. Geballe, B. T. Matthias, V. B. Compton, E. Corenzwit, G. W. Hull, Jr., and L. D. Longinotti, *Phys. Rev.*, 1A, 119, 1965 <https://doi.org/10.1103/PhysRev.137.A119>.
7. C. J. Raub, V. B. Compton, T. H. Geballe, B. T. Matthias, J. P. Maita, and G. W. Hull, Jr., *J. Phys. Chem. Solids*, 26, 2051, 1965 <https://doi.org/10.1016/0022-3697(65)90244-1>.
8. R. D. Blaugher, J. K. Hulm, and P. N. Yocom, *J. Phys. Chem. Solids*, 26, 2037, 1965 <https://doi.org/10.1016/0022-3697(65)90241-6>.
9. T. Claeson and H. L. Luo, *J. Phys. Chem. Solids*, 27, 1081, 1966 <https://doi.org/10.1016/0022-3697(66)90083-7>.
10. S. C. Ng and B. N. Brockhouse, *Solid State Comm.*, 5, 79, 1967 <https://doi.org/10.1016/0038-1098(67)90052-X>.
11. O. I. Shulishova and I. A. Shcherbak, *Izv. AN SSSR, Neorg. Materials*, 3, 1495, 1967.
12. T. F. Smith and H. L. Luo, *J. Phys. Chem. Solids*, 28, 569, 1967 <https://doi.org/10.1016/0022-3697(67)90087-X>.
13. A. C. Lawson, *J. Less-Common Metals*, 23, 103, 1971 <https://doi.org/10.1016/0022-5088(71)90016-6>.
14. R. Chevrel, M. Sergent, and J. Prigent, *J. Solid State Chem.*, 3, 515, 1971 <https://doi.org/10.1016/0022-4596(71)90095-8>.
15. M. Marezio, P. D. Dernier, J. P. Remeika, and B. T. Matthias, *Mat. Res. Bull.*, 8, 657, 1973 <https://doi.org/10.1016/0025-5408(73)90058-5>.
16. J. K. Hulm and R. D. Blaugher in *Superconductivity in d- and f-Band Metals*, D. H. Douglass, Ed., American Institute of Physics, 4, 1, 1972.
17. R. N. Shelton, A. C. Lawson, and D. C. Johnston, *Mat. Res. Bull.*, 10, 297, 1975 <https://doi.org/10.1016/0025-5408(75)90117-8>.
18. H. D. Wiesinger, *Phys. Status Sol.*, 41A, 465, 1977 <https://doi.org/10.1002/pssa.2210410216>.
19. O. Fisher, *Applied Phys.*, 16, 1, 1978 <https://doi.org/10.1007/BF00931416>.
20. D. C. Johnston, *Solid State Comm.*, 24, 699, 1977 <https://doi.org/10.1016/0038-1098(77)90078-3>.
21. H. C. Ku and R. H. Shelton, *Mat. Res. Bull.*, 15, 1441, 1980 <https://doi.org/10.1016/0025-5408(80)90099-9>.
22. H. Barz, *Mat. Res. Bull.*, 15, 1489, 1980 <https://doi.org/10.1016/0025-5408(80)90107-5>.
23. G. P. Espinosa, A. S. Cooper, H. Barz, and J. P. Remeika, *Mat. Res. Bull.*, 15, 1635, 1980 <https://doi.org/10.1016/0025-5408(80)90245-7>.
24. E. M. Savitskii, V. V. Baron, Yu. V. Efimov, M. I. Bychkova, and L. F. Myzenkova, in *Superconducting Materials*, Plenum Press, 1981, p. 107 <https://doi.org/10.1007/978-1-4615-8672-2_4>.
25. R. Fluckiger and R. Baillif, in Topics in *Current Physics*, O. Fischer and M. B. Maple, Eds., Springer Verlag, 34, 113, 1982.
26. R. N. Shelton, in *Superconductivity in d- and f-Band Metals*, W. Buckel and W. Weber, Eds., Kernforschungszentrum, Karlsruhe, 1982, p. 123.
27. D. C. Johnston and H. F. Braun, *Topics in Current Phys.*, 32, 11, 1982 <https://doi.org/10.1007/978-3-642-81894-3_2>.
28. R. Chevrel and M. Sergent, *Topics in Current Phys.*, 32, 25, 1982 <https://doi.org/10.1007/978-3-642-81868-4_2>.
29. G. P. Espinosa, A. S. Cooper, and H. Barz, *Mat. Res. Bull.*, 17, 963, 1982 <https://doi.org/10.1016/0025-5408(82)90121-0>.
30. R. Muller, R. N. Shelton, J. W. Richardson, Jr., and R. A. Jacobson, *J. Less-Comm. Met.*, 92, 177, 1983 <https://doi.org/10.1016/0022-5088(83)90240-0>.
31. You-Xian Zhao and Shou-An He, in *High Pressure in Science and Technology*, North Holland, 22, 51, 1983.
32. You-Xian Zhao and Shou-An He, *Solid State Comm.*, 24, 699, 1983.
33. G. P. Meisner and H. C. Ku, *Appl. Phys.*, A31, 201, 1983 <https://doi.org/10.1007/BF00614955>.
34. R. J. Cava, D. W. Murphy, and S. M. Zahurak, *J. Electrochem. Soc.*, 130, 2345, 1983 <https://doi.org/10.1149/1.2119583>.
35. R. N. Shelton, *J. Less-Comm. Met.*, 94, 69, 1983 <https://doi.org/10.1016/0022-5088(83)90141-8>.
36. B. Chevalier, P. Lejay, B. Lloret, Wang Xian–Zhong, J. Etourneau, and P. Hagenmuller, *Annales de Chemie*, 9, 191, 1984.
37. G. Venturini, M. Meot-Meyer, E. McRae, J. F. Mareche, and B. Rogues, *Mat. Res. Bull.*, 19, 1647, 1984 <https://doi.org/10.1016/0025-5408(84)90242-3>.
38. J. M. Tarascon, F. G. DiSalvo, D. W. Murphy, G. Hull, and J. V. Waszczak, *Phys. Rev.*, 29B, 172, 1984.
39. G. V. Subba and G. Balakrishnan, *Bull. Mat. Sci.*, 6, 283, 1984 <https://doi.org/10.1007/BF02743904>.
40. B. Batlog, *Physica*, 126B, 275, 1984 <https://doi.org/10.1016/0378-4363(84)90175-X>.
41. M. J. Johnson, Ames Lab (USA) Report IS-T-1140, 1984.
42. I. M. Chapnik, *J. Mat. Sci. Lett.*, 4, 370, 1985 <https://doi.org/10.1007/BF00719818>.
43. W. Rong-Yao, L. Q-Guang, and Z. Xiao, *Phys. Status Sol.*, 90A, 763, 1985 <https://doi.org/10.1002/pssa.2210900243>.
44. W. Xian-Zhong, B. Chevalier, J. Etourneau, and P. Hagenmuller, *Mat. Res. Bull.*, 20, 517, 1985 <https://doi.org/10.1016/0025-5408(85)90106-0>.
45. H. R. Ott, F. Hulliger, H. Rudigier, and Z. Fisk, *Phys. Rev.*, 31B, 1329, 1985 <https://doi.org/10.1103/PhysRevB.31.1329>.
46. P. Villars and L. D. Calver, *Pearson's Handbook of Crystallographic Data for Intermetallic Phases*, Vol. 1–3, ASM, 1985 <https://doi.org/10.1107/S010876738408689X>.
47. G. V. Subba Rao, K. Wagner, G. Balakhrishnan, J. Jakani, W. Paulus, and R. Scollhorn, *Bull. Mat. Sci.*, 7, 215, 1985 <https://doi.org/10.1007/BF02747575>.
48. J. G. Bednorz and K. A. Muller, *Zs. Physik*, B64, 189, 1986 <https://doi.org/10.1007/BF01303701>.
49. W. Rong-Yao, *Phys. Status Sol.*, 94A, 445, 1986 <https://doi.org/10.1002/pssa.2210940202>.
50. H. D. Yang, R. N. Shelton, and H. F. Braun, *Phys. Rev.*, 33B, 5062, 1986 <https://doi.org/10.1103/PhysRevB.33.5062>.
51. G. Venturini, M. Kanta, E. McRae, J. F. Mareche, B. Malaman, and B. Roques, *Mat. Res. Bull.*, 21, 1203, 1986 <https://doi.org/10.1016/0025-5408(86)90026-7>.
52. W. Rong-Yao, *J. Mat. Sci. Lett.*, 5, 87, 1986 <https://doi.org/10.1007/BF01671447>.
53. M. K. Wu, J. R. Ashburn, C. J. Torng, P. H. Hor, R. L. Meng, L. Gao, Z. J. Huang, Y. Q. Wang, and C. W. Chu, *Phys. Rev. Lett.*, 58, 908, 1987 <https://doi.org/10.1103/PhysRevLett.58.908>.
54. R. J. Cava, R. B. Van Dover, B. Batlog, and E. A. Rietman, *Phys. Rev. Lett.*, 58, 408, 1987 <https://doi.org/10.1103/PhysRevLett.58.408>.
55. L. C. Porter, T. J. Thorn, U. Geiser, A. Umezawa, H. H. Wang, W. K. Kwok, H-C. I. Kao, M. R. Monaghan, G. W. Crabtree, K. D. Carlson, and J. M. Williams, *Inorg. Chem.*, 26, 1645, 1987 <https://doi.org/10.1021/ic00258a004>.
56. A. M. Kini, U. Geiser, H-C. I. Kao, K. D. Carlson, H. H. Wang, M. R. Monaghan, and K. M. Williams, *Inorg. Chem.*, 26, 1834, 1987 <https://doi.org/10.1021/ic00259a004>.
57. T. Penney, S. von Molnar, D. Kaiser, F. Holtzberg, and A. W. Kleinsasser, *Phys. Rev.*, B38, 2918, 1988 <https://doi.org/10.1103/PhysRevB.38.2918>.
58. Y. K. Tao, J. S. Swinnea, A. Manthiram, J. S. Kim, J. B. Goodenoug, and H. Steinfink, *J. Mat. Res.*, 3, 248, 1988 <https://doi.org/10.1557/JMR.1988.0248>.
59. G. G. Peterson, B. R. Weinberger, L. Lynds, and H. A. Krasinski, *J. Mat. Res.*, 3, 605, 1988 <https://doi.org/10.1557/JMR.1988.0605>.

60. J. B. Torrance, Y. Tokura, A. Nazzai, and S. S. P. Parkin, *Phys. Rev. Lett.*, 60, 542, 1988 <https://doi.org/10.1103/PhysRevLett.60.542>.
61. K. Kourtakis, M. Robbins, P. K. Gallagher, and T. Teifel, *J. Mat. Res.*, 4, 1289, 1989 <https://doi.org/10.1557/JMR.1989.1289>.
62. J. C. Phillips, *Physics of High-T_c Superconductors*, Academic Press, 1989, p. 336.
63. Shui Wai Lin and L. I. Berger, *Rev. Sci. Instrum.*, 60, 507, 1989 <https://doi.org/10.1063/1.1140412>.
64. M. Tinkham, *Introduction to Superconductivity*, McGraw-Hill, New York, 1975.
65. O. Fischer and M. B. Maple, Eds., *Topics in Current Physics*, Volume 32: Superconductivity in Ternary Compounds I; Vol. 34: Superconductivity in Ternary Compounds II, Springer-Verlag, Berlin, 1982 <https://doi.org/10.1007/978-3-642-81868-4>.
66. K. J. Dunn and F. P. Bundy, *Phys. Rev.*, B25, 194, 1982 <https://doi.org/10.1103/PhysRevB.25.194>.
67. A. Barone and G. Paterno, *Physics and Applications of the Josephson Effect*, Wiley, New York, 1982 <https://doi.org/10.1002/352760278X>.
68. D. H. Douglass, Ed., *Superconductivity in d- and f-Band Metals*, Plenum Press, New York, 1976 <https://doi.org/10.1007/978-1-4615-8795-8>.
69. D. M. Ginsberg, Ed., *Physical Properties of High Temperature Superconductors*, (Volume II, 1990; Volume III, 1992; Volume V, 1996), World Scientific, Singapore <https://doi.org/10.1142/1577>.
70. T. Ishiguro and K. Yamji, *Organic Superconductors*, Springer-Verlag, Berlin, 1990 <https://doi.org/10.1007/978-3-642-97190-7>.
71. Sh. Okada, K. Shimizu, T. C. Kobayashi, K. Amaya, and Sh. Endo., *J. Phys. Soc. Jpn.*, 65, 1924, 1996 <https://doi.org/10.1143/JPSJ.65.1924>.
72. A. Bourdillon and N. X. Tan Bourdillon, *High Temperature Superconductors: Processing and Science*, Academic Press, 1994.
73. J. M. Williams, J. R. Ferraro, R. J. Thorn, K. Carlson, U. Geiser, H. H. Wang, A. M. Kini, and M.-H. Whangbo, *Organic Superconductors (Including Fullerenes): Synthesis, Structure, Properties, and Theory*, Prentice-Hall, 1992.
74. J. Nagamatsu, N. Nakagawa, T. Muranaka, Y. Zenitani, and J. Akimitsu, *Nature (London)*, 410, 63, 2001 <https://doi.org/10.1038/35065039>.
75. Y. Boguslavsky, G. K. Perkins, X. Qi, L. F. Cohen, and A. D. Caplin, *Nature (London)*, 410, 563, 2001 <https://doi.org/10.1038/35069029>.
76. B. Q. Fu, Y. Feng, G. Yan, Y. Zhao, A. K. Pradhan, C. H. Cheng, P. Ji, X. H. Liu, C. F. Liu, L. Zhou, and K. F. Yau, *J. Appl. Phys.*, 92, 7341, 2002 <https://doi.org/10.1063/1.1520725>.
77. M. G. T. Mentink *et al.*, *IEEE Trans. on Applied Superconductivity*, 23, (3), 2013.
78. R. Watanabe *et al.*, *J. Mater. Sci. Letters*, 5, 255, 1986 <https://doi.org/10.1007/BF01748069>.
79. S. Posen & D. L. Hall, *Supercond. Sci. & Techn.*, 30, 033004, 2017 <https://doi.org/10.1088/1361-6668/30/3/033004>.
80. N. S. Anatasou, *Modern Phys. Let.*, 11, 939, 1997 <https://doi.org/10.1142/S0217984997001158>.

HIGH-TEMPERATURE SUPERCONDUCTORS

J. Hänisch and S. C. Wimbush

The following tables list selected physical properties of most known high-temperature superconductors as of the date of compilation in 2020. The classification as a "high-temperature" superconductor is open to some degree of ambiguity. Naming was driven by excitement at the prospect of operation above the boiling point of liquid nitrogen (~77 K). However, the reality of superconductor applications today is that few can operate in this regime, while it is clearly illogical to exclude closely related materials of the same family that fall below such an arbitrary cut-off. Instead, we adopt here the classification that a "high-temperature" superconductor is a non-BCS type superconductor that is thereby able to operate beyond the BCS temperature limit of ~30 K. Until relatively recently, this limited the known materials to cuprate compounds; however, this has now been extended first by magnesium diboride and then by the extensive families of iron-based superconductors. We do not, however, include all unconventional superconductors, intentionally excluding those isolated examples and families where no related material is able to superconduct at an elevated temperature. Notably, this excludes all the so-called "heavy fermion" superconductors thus far discovered. For the same reason, we also exclude the organic superconductors, which are featured elsewhere in this volume.

The high-temperature superconductors are presented here in a series of tables, one for each distinctly identifiable family. Table 1 lists the detailed superconducting properties of a select number of materials drawn from families that have been studied in depth due to either importance or accessibility. Table 2 contains the "214" phases including $(La,Ba)_2CuO_4$, which was the original 1986 Bednorz and Müller discovery of high-temperature superconductivity. Here are also found the rare examples of electron-doped cuprates. Table 3 presents the rare-earth barium cuprates, including the most famous of the high-temperature superconductors, $YBa_2Cu_3O_7$. These materials are the core constituents of so-called second-generation commercial high-temperature superconducting wires. Table 4 supplements these materials with derivatives having missing rare-earth planes, while Tables 5 and 6 feature two distinct homologous series of cuprates based on Hg, Tl, Pb, Bi, and others, of which $Bi_2Ba_2Ca_2Cu_3O_{10}$ is the material used in the production of first-generation commercial high-temperature superconducting wires. Tables 7 and 8 move from the cuprates to the chalcogenide and pnictide families of iron-based superconductors, respectively, and Table 9 contains the few other non-cuprate high-temperature superconductors known, including MgB_2.

In compiling and modernizing these data tables, we have sought to be comprehensive in our coverage of the known materials, but selective in the properties to include. In guiding the latter selection, we aim to feature only those properties that are of genuine stand-alone use as figures of merit or in a comparative sense. Because no mere table of data can hope to adequately convey the nuances of materials as complex in structure and properties as the high-temperature superconductors, we avoid qualifications and list in each case the best value to which we have access. This matter of judgment must be supplemented by the reader referring in each case to a primary source to validate the entry. Lattice parameters have been rounded to four decimal places as a compromise between accuracy and generality. Critical temperatures are given to two significant figures for the same reason. The data given in Table 1 are ranges or averages of the most reliable values found in the literature.

These tables are current through 2020 and constitute a complete revision and update of previous data. The data listed here are compiled from a large number of sources spanning the primary literature. Individual references for each datum are too numerous to list. Secondary sources, as listed under the references, have been used to ensure the accuracy and completeness of the work.

References

1. Harshman, D. R. and Mills Jr., A. P., *Phys. Rev. B* 45, 10684, 1992.
2. Rao, C. N. R. et al., *Supercond. Sci. Technol.* 6, 1, 1993.
3. Martienssen, W. and Warlimont, H., Ed., *Springer Handbook of Condensed Matter and Materials Data*, Springer, Berlin and Heidelberg, Germany, 2005.
4. Narlikar, A. V., Ed., *Frontiers in Superconducting Materials*, Springer, Berlin and Heidelberg, Germany, 2005.
5. Chu, C. W. et al., *Physica C* 514, 290, 2015.
6. Hosono, H. et al., *Sci. Technol. Adv. Mater.* 16, 033503, 2015.
7. Hosono, H. and Kuroki, K., *Physica C* 514, 399, 2015.
8. Naito, M. et al., *Physica C* 523, 28, 2016.
9. Kleiner, R. and Buckel, W., *Superconductivity: An Introduction, Third Edition*, Wiley-VCH, Weinheim, Germany, 2016.

TABLE 1. Detailed Superconducting Properties of Selected High-Temperature Superconductors

		Coherence lengths		Penetration depths		Critical fields			
Mol. form.	T_c^{max}/K	ξ_{ab}/nm	ξ_c/nm	λ_{ab}/nm	λ_c/μm	B_{c1}^{ab}/mT	B_{c1}^{c}/mT	B_{c2}^{ab}/T	B_{c2}^{c}/T
$La_{1.85}Sr_{0.15}CuO_4$	39	2.2±0.1	0.06-0.3	219±10	<5	7	30	84±6	70±10
$Nd_{1.85}Ce_{0.15}CuO_{4-\delta}$	30	7.0±1.0	0.23	76±5				>100	7±1
$YBa_2Cu_3O_{7-\delta}$	93	1.5	0.14-0.3	90±10	0.64	3–5	30–50	240±25	150±30
$YBa_2Cu_4O_8$	81	1.9	0.2–1.1	130–200	0.16	34	17–32		90
$(Cu_{0.5}C_{0.5})Ba_2Ca_3Cu_4O_{11+\delta}$	117	1.6	1.0	120	0.22	26	63	195	121
$HgBa_2CuO_{4+\delta}$	97	2.1	1.2	120-200	0.45	8.2	12.9	125	70-100
$HgBa_2CaCu_2O_{6+\delta}$	127	1.5±0.1	0.4	190	0.83	21	50		110-170
$HgBa_2Ca_2Cu_3O_{8+\delta}$	135	1.3±0.1		180±30	3.5	5-10	35±10	350±50	100-200

Mol. form.	T_c^{max}/K	Coherence lengths ξ_{ab}/nm	ξ_c/nm	Penetration depths λ_{ab}/nm	λ_c/μm	Critical fields B_{c1}^{ab}/mT	B_{c1}^{c}/mT	B_{c2}^{ab}/T	B_{c2}^{c}/T
$Tl_2Ba_2CuO_{6+\delta}$	92	5	0.2	170±10	2.0		6	300	65
$Tl_2Ba_2CaCu_2O_{8+\delta}$	119	3	0.7	180±40	>25	60	28	>120	>100
$Tl_2Ba_2Ca_2Cu_3O_{10+\delta}$	128	1-3	<0.09	196±10	>20			200	>75
$Bi_2Sr_2CaCu_2O_{8+\delta}$	96	0.8-0.9	≤0.05	150±20	10-40	0.25-0.5	6-10	>250	220±30
$Bi_2Sr_2Ca_2Cu_3O_{10+\delta}$	110	0.6-0.9	0.02-0.09	120±10	1.0	0.94	54	>250	184
$Fe_{1+\delta}Se$	8.5	4.7	2.8	400			2.5	25	15
$FeSe_{0.5}Te_{0.5}$	15	2.8	3.0	460±100	1.1±0.3	2.2	4.5	42	45
(Li,Fe)OHFeSe	42	2.7	0.24	200	1.8		4.5	98	67
LiFeAs	18	4.4±0.4	1.8±0.2	200	0.25	16	19	30	24
$LaFeAsO_{1-x}F_x$	26	2.8	1.0±0.1	310			6	63	42
$NdFeAsO_{1-x}F_x$	47	2.2±0.2	0.4±0.1	200		18	25	130	70
$Ba_{0.6}K_{0.4}Fe_2As_2$	38	1.6	0.75	200				90	75
$Ba(Fe_{0.92}Co_{0.08})_2As_2$	22	2.45	1.4±0.1	200			20	65	55
$BaFe_2(As_{0.7}P_{0.3})_2$	31	2.3	1.3±0.2	300			63	90	60
$CaKFe_4As_4$	35	1.83±0.03	1.87±0.06	100		120	25	95	102
MgB_2	39	10±1	2.0±0.2	85±6	0.09±0.01	35±5		17±2	3.0±0.5

TABLE 2. Crystal Structures and Critical Temperatures of the Infinite Layer and "214" High-Temperature Superconductors

Mol. form.	Structure (lattice parameters in nm)	T_c^{max}/K	Comments
0011, i.e., infinite layer structures	**P4/mmm ($BaCuO_2$ structure)**		
$Sr_{1-x}CuO_2$	a = 0.3927, c = 0.3435	60	
$Sr_{1-x}Ca_xCuO_2$	a = 0.3902, c = 0.3350	110	x_{max} = 0.3
$Sr_{1-x}Ba_xCuO_2$	a = 0.3922, c = 0.343	90	x_{max} = 0.1
$Sr_{1-x}La_xCuO_2$	a = 0.3951, c = 0.3409	42	x_{max} = 0.1
$Sr_{1-x}Pr_xCuO_2$	a = 0.3942, c = 0.3391	43	x_{max} = 0.12
$Sr_{1-x}Nd_xCuO_2$	a = 0.3935, c = 0.3413	45	x_{max} = 0.12
$Sr_{1-x}Sm_xCuO_2$	a = 0.3942, c = 0.3391	42	x_{max} = 0.1
0201, i.e., "214" T structures	**I4/mmm (K_2NiF_4 structure)**		
$La_2CuO_{4+\delta}$	a = 0.379, c = 1.319	0	
$La_2CuO_{4+\delta}$	Bmab; a = 0.5345, b = 0.5433, c = 1.3252	42	
$La_2CuO_4F_x$	Phase uncertain	35	
$(Ca_{1-x}Na_x)_2CuO_2Cl_2$	a = 0.3855, c = 1.510	26	x_{max} = 0.04
$(La_{1-x}Na_x)_2CuO_4$	a = 0.3775, c = 1.3170	30	x_{max} = 0.3-0.5
$(La_{1-x}K_x)_2CuO_4$	a = 3.7683, c = 1.3259	41	
$(La_{1-x}Rb_x)_2CuO_4$	a = 0.3878, c = 1.3276	22	
$(La_{1-x}Ca_x)_2CuO_4$	Cmca; a = 0.5341, b = 0.5359, c = 1.3170	24	x_{max} = 0.05
$(La_{1-x}Sr_x)_2CuO_4$	a = 0.3779, c = 1.3226	39	x_{max} = 0.075
$(La_{1-x}Ba_x)_2CuO_4$	a = 0.3779, c = 1.323	28	
0021, i.e., "214" T' structures	**I4/mmm (Nd_2CuO_4 structure)**		
$(La_{1-x}RE_x)_2CuO_{4+\delta}$		up to 21	Thin films; RE = Y, Lu, Sm, Eu, Gd, Tb
$(La_{1-x}Ce_x)_2CuO_{4-\delta}$	a = 0.4007, c = 1.244	30	Electron-doped, x_{max} = 0.065
$(Pr_{1-x}Ce_x)_2CuO_{4-\delta}$		17	Electron-doped
$Pr_{1-x}Ce_{0.15}Sr_xCuO_{3.94}$	a = 0.396, c = 1.216	21	Electron-doped, x_{max} = 0.06
$(Nd_{1-x}Ce_x)_2CuO_{4-\delta}$	a = 0.3947, c = 1.2078	30	Electron-doped, x_{max} = 0.075
$Nd_2CuO_{4-\delta}F_y$	a = 0.3951, c = 1.2115	25	Electron-doped, x_{max} = 0.2
$(Sm_{1-x}Ce_x)_2CuO_{4-\delta}$		15	Electron-doped
$(Eu_{1-x}Ce_x)_2CuO_{4-\delta}$		13	Electron-doped, x_{max} = 0.075-0.085
0222, i.e., "214" T* structures	**P4/nmm (alternating T and T')**		
$La_{1-x/2}Eu_{1-x/2}Sr_xCuO_4$	a = 0.3871, c = 1.2597	25	x_{max} = 0.14
$(Nd,Sr,Ce)_2CuO_4$	a = 0.3856, c = 1.2484	25	
$SmLa_{1-x}Sr_xCuO_{4-\delta}$		33	x_{max} = 0.15

Solids

TABLE 3. Crystal Structures and Critical Temperatures of the $RE_2Ba_4Cu_{5+n}O_{13+n}$ Series of High-Temperature Superconductors

Mol. form.	Structure (lattice parameters in nm)	T_c^{max}/K	Comments
123 (n = 1)	**Pmmm**		
$YBa_2Cu_3O_{7-\delta}$	a = 0.3820, b = 0.3885, c = 1.1676	93	
$YSr_2Cu_3O_{7-\delta}$	P4/mmm, a = 0. 3786, c = 1.1386	63	
$YCa_2Cu_3O_{7-\delta}$	a = 0.3643, b = 0.3838, c = 1.1759	84	
$La_{1+x}Ba_{2-x}Cu_3O_{7-\delta}$	a = 0.3885, b = 0.3938, c = 1.1817	93	$x > 0.25$ for phase stability
$CeBa_2Cu_3O_{7-\delta}$	Forms $BaCeO_3$ due to tetravalent Ce	—	
$Pr_{1+x}Ba_{2-x}Cu_3O_{7-\delta}$	a = 0.3866, b = 0.3933, c = 1.1724	0	$x > 0.15$ for phase stability
$Nd_{1+x}Ba_{2-x}Cu_3O_{7-\delta}$	a = 0.3878, b = 0.3913, c = 1.1753	94	$x > 0.1$ for phase stability
$PmBa_2Cu_3O_{7-\delta}$	Not synthesized due to radioactive Pm	—	
$SmBa_2Cu_3O_{7-\delta}$	a = 0.3902, b = 0.3844, c = 1.1725	95	
$EuBa_2Cu_3O_{7-\delta}$	a = 0.3897, b = 0.3838, c = 1.1707	95	
$GdBa_2Cu_3O_{7-\delta}$	a = 0.3895, b = 0.3835, c = 1.1699	95	
$TbBa_2Cu_3O_{7-\delta}$	Forms $BaTbO_3$ by Ba-Tb intermixing	—	
$TbSr_2Cu_{2.7}Mo_{0.3}O_{7-\delta}$	P4/mmm, a = 0.3871, c = 1.15784	80	Mo necessary for phase stability
$DyBa_2Cu_3O_{7-\delta}$	a = 0.3887, b = 0.3825, c = 1.1686	93	
$HoBa_2Cu_3O_{7-\delta}$	a = 0.3885, b = 0.3819, c = 1.1677	93	
$ErBa_2Cu_3O_{7-\delta}$	a = 0.3878, b = 0.3813, c = 1.1664	93	
$TmBa_2Cu_3O_{7-\delta}$	a = 0.3875, b = 0.3809, c = 1.1666	90	
$YbBa_2Cu_3O_{7-\delta}$	a = 0.3871, b = 0.3802, c = 1.1658	90	Multiphase due to small Yb ion
$LuBa_2Cu_3O_{7-\delta}$	a = 0.387, b = 0.380, c = 1.169	93	Mostly multiphase due to small Lu ion
247 (n = 2)	**Ammm**		
$Y_2Ba_4Cu_7O_{15-\delta}$	a = 0.3854, b = 0.3874, c = 5.040	55	
$Pr_2Ba_4Cu_7O_{15-\delta}$	a = 0.3892, b = 0.3902, c = 5.081	10	
$N_d2Ba_4Cu_7O_{15-\delta}$	a = 0.3894, b = 0.3901, c = 5.079	40	
$Eu_2Ba_4Cu_7O_{15-\delta}$	a = 0.3879, b = 0.3886, c = 5.039	45	
$Gd_2Ba_4Cu_7O_{15-\delta}$	a = 0.3872, b = 0.3879, c = 5.036	45	
$Dy_2Ba_4Cu_7O_{15-\delta}$	a = 0.3864, b = 0.3879, c = 5.039	50	
$Ho_2Ba_4Cu_7O_{15-\delta}$	a = 0.3857, b = 0.3879, c = 5.040	55	
$Er_2Ba_4Cu_7O_{15-\delta}$	a = 0.3847, b = 0.3873, c = 5.044	55	
$Yb_2Ba_4Cu_7O_{15-\delta}$	a = 0.381, b = 0.386, c = 5.045	86	
124 (n = 3)	**Ammm**		
$YBa_2Cu_4O_8$	a = 0.3840, b = 0.3870, c = 2.7231	81	
$LaBa_2Cu_4O_8$	a = 0.3854, b = 0.3874, c = 2.710	?	Superconductivity not reported
$CeBa_2Cu_4O_8$	No reports on phase formation	—	
$PrBa_2Cu_4O_8$	a = 0.3884, b = 0.3903, c = 2.7293	0	
$NdBa_2Cu_4O_8$	Small volume fraction, multiphase	~57	
$PmBa_2Cu_4O_8$	Not synthesized due to radioactive Pm	?	
$SmBa_2Cu_4O_8$	a = 0.3872, b = 0.3886, c = 2.7308	69	
$EuBa_2Cu_4O_8$	a = 0.3865, b = 0.3884, c = 2.7279	78	
$GdBa_2Cu_4O_8$	a = 0.3863, b = 0.3881, c = 2.7259	74	
$TbBa_2Cu_4O_8$	Single synthesis attempt	?	Superconductivity not reported
$DyBa_2Cu_4O_8$	a = 0.3846, b = 0.3873, c = 2.7237	77	
$HoBa_2Cu_4O_8$	a = 0.3840, b = 0.3870, c = 2.7221	80	
$ErBa_2Cu_4O_8$	a = 0.3837, b = 0.3869, c = 2.7230	78	
$TmBa_2Cu_4O_8$	a = 0.3827, b = 0.3864, c = 2.718	79	
$YbBa_2Cu_4O_8$	a = 0.3846, b = 0.3871, c = 2.7231	80	
$Yb(Ba_{0.8}Sr_{0.2})_2Cu_4O_8$	a = 0.3805, b = 0.3855, c = 2.7072	83	Sr improves phase stability
$(Yb_{0.95}Ca_{0.05})Ba_2Cu_4O_8$	a = 0.3822, b = 0.3853, c = 2.7175	85	Ca improves phase stability
$LuBa_2Cu_4O_8$	a = 0.3844, b = 0.3871, c = 2.7225	82	

TABLE 4. Crystal Structures and Critical Temperatures of the $RE_nBa_mCu_{n+m}O_{2(n+m+1)}$ Series of High-Temperature Superconductors

Mol. form.	Structure (lattice parameters in nm)	T_c^{max}/K	Comments
***n* = 2**	**Pmm2**		
$YBa_2Cu_3O_{7-\delta}$	*a* = 0.3820, *b* = 0.3885, *c* = 1.1676	93	cf. Table 3 (123 compounds)
	Pmmm		
$YBa_3Cu_4O_{9-\delta}$	*a* = 0.3802, *b* = 0.3885, *c* = 1.5256	88	
$YBa_4Cu_5O_{11-\delta}$	*a* = 0.3802, *b* = 0.3865, *c* = 1.9382	91	
$YBa_5Cu_6O_{13-\delta}$	*a* = 0.3801, *b* = 0.3891, *c* = 2.2944	85	
***n* = 3**	**Pmm2**		
$Y_3Ba_5Cu_8O_{18}$	*a* = 0.3888, *b* = 0.3823, *c* = 3.1013	102	
$Nd_3Ba_5Cu_8O_{18}$	*a* = 0.3922, *b* = 0.3862, *c* = 3.5211	95	
$Sm_3Ba_5Cu_8O_{18}$	*a* = 0.3899, *b* = 0.3852, *c* = 3.5146	97	
$Eu_3Ba_5Cu_8O_{18}$	*a* = 0.3864, *b* = 0.3888, *c* = 3.113	66	
$Gd_3Ba_5Cu_8O_{18}$	*a* = 0.3851, *b* = 0.3877, *c* = 3.112	97	
$Dy_3Ba_5Cu_8O_{18}$	*a* = 0.3833, *b* = 0.3867, *c* = 3.1025	81	
$Ho_3Ba_5Cu_8O_{18}$	*a* = 0.3832, *b* = 0.3875, *c* = 3.101	84	
$(Yb_{0.6}Sm_{0.4})_3Ba_5Cu_8O_{18}$	*a* = 0.3891, *b* = 0.3829, *c* = 3.1209	88	Sm required for phase stability
$Y_3Ba_8Cu_{11}O_{24}$	*a* = 0.3814, *b* = 0.3885, *c* = 4.2699	91	
***n* = 4**	**Pmm2**		
$Y_2Ba_3Cu_5O_{11-\delta}$	*a* = 0.3820, *b* = 0.3868, *c* = 1.8930	92	
$Gd_2Ba_3Cu_5O_{11-\delta}$	*a* = 0.3882, *b* = 0.3874, *c* = 1.9355	16	
$Y_2Ba_5Cu_7O_{15-\delta}$	*a* = 0.3832, *b* = 0.3851, *c* = 2.8683	98	
$Y_2Ba_5Cu_8O_{17-\delta}$	*a* = 0.3871, *b* = 0.3848, *c* = 2.7160	105	Extra CuO_2 layer
$Y_2Ba_5Cu_9O_{19-\delta}$	*a* = 0.3821, *b* = 0.3898, *c* = 2.3320	94	Extra CuO_2 layers
***n* > 4**	**Pmm2**		
$Gd_5Ba_7Cu_{12}O_{26-\delta}$	*a* = 0.3868, *b* = 0.3876, *c* = 4.6507	22	
$Y_5Ba_8Cu_{13}O_{28-\delta}$	*a* = 0.3819, *b* = 0.3897, *c* = 5.0461	91	
$Y_7Ba_{11}Cu_{18}O_{38-\delta}$	*a* = 0.3824, *b* = 0.3880, *c* = 6.9870	91	
$Y_{13}Ba_{20}Cu_{33}O_{68-\delta}$	*a* = 0.3815, *b* = 0.3878, *c* = 12.8116	89	

TABLE 5. Crystal Structures and Critical Temperatures of the $M_mAE_2Ca_{n-1}Cu_nO_{2+m+2n}$ (*M* = B, Cu, Au, Hg, Tl, Pb, Bi; *AE* = Ca, Sr, Ba, La) Series of High-Temperature Superconductors

Mol. form.	Structure (lattice parameters in nm)	T_c^{max}/K	Comments
02(*n*-1)*n* (*m* = 0)	**I4/mmm**		
Ba_2CuO_4	*a* = 0.385, *c* = 1.46	85	Thin film; cf. Table 2 (T structures)
$Ba_2CaCu_2O_6$	*a* = 0.385, *c* = 2.20	103	
$Ba_2Ca_2Cu_3O_8$	*a* = 0.385, *c* = 2.83	126	
$Ba_2Ca_3Cu_4O_{10}$	*a* = 0.385, *c* = 4.04	117	
$Ba_2CaCu_2O_4(O_{1-x}F_x)_2$	*a* = 0.3879, *c* = 1.471	90	
$Ba_2Ca_2Cu_3O_6(O_{1-x}F_x)_2$	*a* = 0.3861, *c* = 2.108	120	
$Ba_2Ca_3Cu_4O_8(O_{1-x}F_x)_2$	*a* = 0.3856, *c* = 2.745	105	
$Ba_2Ca_4Cu_5O_{10}(O_{1-x}F_x)_2$	*a* = 0.3860, *c* = 3.382	90	
$Sr_2CuO_{3+\delta}$	*a* = 0.3764, *c* = 1.2548	70	
$Sr_2CuO_{4-\delta}$	*a* = 0.3795, *c* = 1.2507	95	cf. Table 2 (T structures)
$Sr_2CuO_2F_{2+\delta}$	Fmmm; *a* = 0.5394, *b* = 0.5513, *c* = 1.3468	46	
$Sr_2CaCu_2O_6$	*a* = 0.386, *c* = 2.04	70	
$Sr_3Cu_2O_{5+\delta}$ (i.e. $Sr_2SrCu_2O_{5+\delta}$)	*a* = 0.3902, *c* = 2.1085	100	
$(Sr,Ca)_2CaCu_2O_4Cl_2$	*a* = 0.387, *c* = 2.216	80	
$Sr_2Ca_2Cu_3O_8$	*a* = 0.386, *c* =2.72	90	
$Sr_2Ca_3Cu_4O_{10}$	*a* = 0.386, *c* = 3.40	70	
$(La_{1-x}Ca_x)_2CaCu_2O_{6+\delta}$	*a* = 0.3821, *c* = 1.953	60	x_{max} = 0.075
$(La_{1-x}Sr_x)_2CaCu_2O_{6-\delta}$	*a* = 0.3821, *c* = 1.9599	60	x_{max} = 0.2
$(La_{1-x}Sr_x)_2CaCu_2O_4Cl_2$	*a* = 0.3827, *c* = 1.942	45	x_{max} = 0.2

Mol. form.	Structure (lattice parameters in nm)	T_c^{max}/K	Comments
12(*n*-1)*n* (*m* = 1)	**P4/mmm**		
$BSr_2Ca_2Cu_3O_9$	a = 0.3821, c = 1.3854	75	
$BSr_2Ca_3Cu_4O_{11}$	a = 0.3836, c = 1.7082	110	
$BSr_2Ca_4Cu_5O_{13}$	a = 0.3837, c = 2.022	85	
$CuBa_2Ca_2Cu_3O_9$	Not synthesized in pure form	–	
$(Cu_{0.5}C_{0.5})Ba_2Ca_2Cu_3O_9$	a = 0.3859, c = 1.4766	118	
$CuBa_2Ca_3Cu_4O_{11+\delta}$	Not synthesized in pure form	–	
$(Cu_{0.5}C_{0.5})Ba_2Ca_3Cu_4O_{11+\delta}$	a = 0.3855, c = 1.7930	117	
$Au(Ba,La)CuO_{5+\delta}$	Pmmm; a = 0.3798, b = 0.3851, c = 0.8575	19	
$AuBa_2(Y_{1-x}Ca_x)Cu_2O_7$	Pmmm; a = 0.3826, b = 0.3850, c = 1.2075	84	x_{max} = 0.4
$AuBa_2Ca_2Cu_3O_9$	Pmmm; a = 0.3812, b = 0.3856, c = 1.5443	30	
$AuBa_2Ca_3Cu_4O_{11}$	Pmmm; a = 0.3827, b = 0.3851, c = 1.8494	99	
$HgSr_2CuO_{4+\delta}$	Not synthesized in pure form	–	
$(Hg_{1-x}Mo_x)Sr_2CuO_x$	a = 0.3787, c = 0.8844	78	x_{max} = 0.15
$(Hg_{0.9}Re_{0.1})Sr_2CuO_x$	a = 0.378, c = 0.884	66	x_{max} = 0.15
$HgBa_2CuO_{4+\delta}$	a = 0.3883, c = 0.9513	97	
$HgBa_2CaCu_2O_{6+\delta}$	a = 0.3853, c = 1.2637	127	
$HgBa_2Ca_2Cu_3O_{8+\delta}$	a = 0.3850, c = 1.5784	135	
$HgBa_2Ca_3Cu_4O_{10+\delta}$	a = 0.3854, c = 1.9006	127	
$HgBa_2Ca_4Cu_5O_{12+\delta}$	a = 0.3851, c = 2.2136	110	
$HgBa_2Ca_5Cu_6O_{14+\delta}$	a = 0.3851, c = 2.5251	107	
$HgBa_2Ca_6Cu_7O_{16+\delta}$	a = 0.3851, c = 2.8406	89	
$HgBa_2Ca_7Cu_8O_{18+\delta}$	a = 0.3847, c = 3.1583	<90	
$TlSr_2CuO_{5-\delta}$	Pmmm; a = 0.3661, b = 0.3793, c = 0.899	0	
$(Tl_{0.5}Pb_{0.5})Sr_2CuO_{5\pm\delta}$	a = 0.3736, c = 0.9022	60	
$Tl(Sr,Ba)_2CuO_{5-\delta}$	a = 0.3805, c = 0.9120	43	
$Tl(Sr,La)_2CuO_5$	a = 0.37, c = 0.9	40	
$TlSr_2CaCu_2O_7$	a = 0.3797, c = 1.2092	55	
$(Tl_{0.5}Pb_{0.5})Sr_2CaCu_2O_7$	a = 0.3802, c = 1.2107	85	
$TlSr_2(Ca_{0.5}Y_{0.5})Cu_2O_7$	a = 0.380, c = 1.210	90	
$(Tl_{0.5}Pb_{0.5})Sr_2(Ca_{0.8}Y_{0.2})Cu_2O_7$	a = 0.3808, c = 1.2014	107	
$TlBa_2CuO_{5-\delta}$	a = 0.3859, c = 0.9261	9.5	
$Tl(Ba_{0.6}La_{0.4})_2CuO_5$	a = 0.3848, c = 0.9091	52	
$TlBa_2CaCu_2O_7$	a = 0.3857, c = 1.2754	103	
$TlBa_2Ca_{1-x}Y_xCu_2O_7$	a = 0.3850, c = 1.265	86	x_{max} = 0.25
$TlBa_2Ca_2Cu_3O_9$	a = 0.385, c = 1.59	130	
$TlBa_2Ca_3Cu_4O_{11}$	a = 0.385, c = 1.91	122	
$TlBa_2Ca_4Cu_5O_{13}$	a = 0.385, c = 2.23	117	
$PbSr_2CuO_{5-\delta}$	Tetragonal; a = 0.381, c = 0.881	40	Thin film
$PbSr_2CaCu_2O_7$	Orthorhombic; a = 0.381, b = 0.383, c = 1.21	70	
$PbSr_2Ca_2Cu_3O_9$	Pseudotetragonal; a = 0.3834, c = 1.529	122	
$PbSr_2Ca_3Cu_4O_{11}$	Not synthesized in pure form	–	
$(Pb_{0.6}Sr_{0.3}Cu_{0.1})Sr_2(Ca,Sr)_3Cu_4O_{11}$	a = 0.3833, c = 1.8442	107	
22(*n*-1)*n* (*m* = 2)	**I4/mmm**		
$Cu_2Ba_2Ca_2Cu_3O_{11}$	Not synthesized in pure form	–	
$(Cu_{0.5}C_{0.5})_2Ba_3Ca_3Cu_4O_{13}$	P4/mmm; a = 0.3855, c = 2.187	113	Extra BaO layer
$(Cu_{0.5}C_{0.5})_2Ba_3Ca_4Cu_5O_{15}$	P4/mmm; a = 0.3857, c = 2.5067	~110	Extra BaO layer
$Hg_2Ba_2CaCu_2O_8$	Unstable	—	
$Hg_2Ba_2(Y_{0.6}Ca_{0.4})Cu_2O_{8-\delta}$	a = 0.3855, c = 2.894	45	
$(Hg_{0.7}Tl_{0.3})_2Ba_2(Y_{1-x}Ca_x)Cu_2O_{8-\delta}$	a = 0.386, c = 2.90	84	x_{max} = 0.4; T_c = 12 K for x = 0
$(Hg_{0.6}Tl_{0.4})_2Ba_2Ca_2Cu_3O_{10}$	a = 0.3840, c = 3.569	45	
$(Hg_{0.6}Tl_{0.4})_2Ba_2Ca_3Cu_4O_{12}$	a = 0.3845, c = 4.206	114	
$Tl_2Ba_2CuO_{6+\delta}$	a = 0.3866, c = 2.3239	92	
$Tl_2Ba_2CaCu_2O_{8+\delta}$	a = 0.3855, c = 2.9318	119	
$Tl_2Ba_2Ca_2Cu_3O_{10+\delta}$	a = 0.3850, c = 3.588	128	
$Tl_2Ba_2Ca_3Cu_4O_{12+\delta}$	a = 0.3581, c = 4.199	119	

Mol. form.	Structure (lattice parameters in nm)	T_c^{max}/K	Comments
$Pb_2(Sr,La)_2Cu_2O_6$	P222$_1$; $a = 0.5333$, $b = 0.5421$, $c = 1.2609$	32	
$Pb_2Sr_2(Y_{1-x}Ca_x)Cu_3O_8$	Cmmm; $a = 0.5393$, $b = 0.5431$, $c = 1.5733$	80	
$Pb_2Sr_2La_{0.5}Ca_{0.5}Cu_3O_8$	Cmmm; $a = 0.5435$, $b = 0.5463$, $c = 1.5817$	70	
$Bi_2Sr_2CuO_{6+\delta}$	Cmmm; $a = 0.5361$, $b = 0.5370$, $c = 2.4369$	9	
$Bi_2(Sr_{1-x}La_x)_2CuO_{6+\delta}$	Cmmm; $a = 0.5417$, $b = 0.5381$, $c = 2.439$	33	$x_{max} \approx 0.2$
$Bi_{1.6}Pb_{0.4}Sr_{1.6}La_{0.4}CuO_{6+\delta}$		39	
$Bi_2Sr_2CaCu_2O_{8+\delta}$	Fmmm; $a = 0.5413$, $b = 0.5411$, $c = 3.091$	96	
$Bi_2Sr_2Ca_2Cu_3O_{10+\delta}$	Fmmm; $a = 0.539$, $b = 0.539$, $c = 3.71$	110	
$(Bi_{0.8}Pb_{0.2})_2Sr_2Ca_2Cu_3O_{10+\delta}$	Fmmm; $a = 0.5413$, $b = 0.5413$, $c = 3.710$	110	Pb stabilizes phase formation
$Bi_2Sr_2Ca_3Cu_4O_{12+\delta}$		47	Thin film
$Bi_2Sr_2(Ln_{1-x}Ce_x)_2Cu_2O_{10+y}$	Cmma; $a = 0.55$, $b = 0.55$, $c = 1.79$	~25	Ln = Sm, Eu, Gd; $x_{max} = 0.15$

TABLE 6. Crystal Structures and Critical Temperatures of the $M_mAE_2RE_sCu_2O_{4+m+2s}$ (M = Fe, Co, Cu, Ga, Nb, Ru, Tl, Pb, Bi, AE = Sr, Ba, RE = Rare-Earth or Ca, Sr) Series of High-Temperature Superconductors

Mol. form.	Structure (lattice parameters in nm)	T_c^{max}/K	Comments
12s2			
$FeSr_2YCu_2O_{6+\delta}$	Ortho; $a = 0.5409$, $b = 0.5458$, $c = 2.292$	60	
$FeSr_2NdCu_2O_{6+\delta}$	P4/mmm; $a = 0.3843$, $c = 1.1458$	50	
$CoSr_2(Y_{1-x}Ca_x)Cu_2O_{7-\delta}$	Ima2; $a = 0.540$, $b = 0.541$, $c = 2.253$	40	
$CuSr_2YCu_2O_{7-\delta}$	P4/mmm, $a = 0.3786$, $c = 1.1386$	63	$YSr_2Cu_3O_{7-\delta}$, cf. Table 3 (123 compounds)
$(Cu_{1-x}Au_x)Sr_2YCu_2O_{7-\delta}$	P4/mmm; $a = 0.394$, $c = 1.201$	80	
$(Cu_{0.75}Mo_{0.25})Sr_2YCu_2O_{7+\delta}$	P4/mmm; $a = 0.3815$, $c = 1.15$	88	
$(Cu_{0.75}Mo_{0.25})Sr_2(Ce_{0.50}Y_{0.50})_2Cu_2O_{9+\delta}$	I4/mmm; $a = 0.3825$, $c = 1.40$	58	
$(Cu_{0.75}Mo_{0.25})Sr_2(Ce_{0.67}Y_{0.33})_3Cu_2O_{11+\delta}$	P4/mmm; $a = 0.3829$, $c = 1.68$	55	
$(Cu_{0.75}Mo_{0.25})Sr_2(Ce_{0.75}Y_{0.25})_4Cu_2O_{13+\delta}$	I4/mmm; $a = 0.3828$, $c = 1.98$	55	
$Cu(Ba,Eu)_2(Ce,Eu)_2Cu_2O_9$	I4/mmm; $a = 0.386$, $c = 2.848$	62	
$GaSr_2(Y_{1-x}Ca_x)Cu_2O_{7-\delta}$	Ima2; $a = 0.539$, $b = 0.548$, $c = 2.275$	70	
$(Ga_{1-y}Cu_y)Sr_2(Y_{1-x}Ca_x)Cu_2O_7$	Pmmm; $a = 0.539$, $b = 0.548$, $c = 2.281$	50	
$(Nb_{1-x}Cd_x)Sr_2EuCu_2O_{8-\delta}$	P4/mmm, $a = 0.3874$, $c = 1.1629$	43	$x_{max} = 0.2$
$RuSr_2YCu_2O_{8-\delta}$	P4/mbm; $a = 0.3820$, $c = 1.1518$	39	Ferromagnetic superconductor
$RuSr_2SmCu_2O_{8-\delta}$	P4/mbm; $a = 0.3852$, $c = 1.156$	12	Ferromagnetic superconductor
$RuSr_2EuCu_2O_{8-\delta}$	P4/mbm; $a = 0.3843$, $c = 1.155$	36	Ferromagnetic superconductor
$RuSr_2GdCu_2O_{8-\delta}$	P4/mbm; $a = 0.3838$, $c = 1.153$	45	Ferromagnetic superconductor
$RuSr_2[(Eu,Gd)_{0.7}Ce_{0.3}]_2Cu_2O_{10-\delta}$	I4/mmm; $a = 0.3844$, $c = 2.8615$	35	Ferromagnetic superconductor
$TaSr_2(Gd_{1+x}Ce_{1-x})_2Cu_2O_9$	I4/mmm, $a = 0.3858$, $c = 2.881$	30	$x_{max} = 0.6$
$TlBa_2(Eu_{1-x}Ce_x)_2Cu_2O_9$	I4/mmm, $c = 3.05$	40	
$(Pb,Cu)(Sr,Eu)_2(Eu,Ce)_2Cu_2O_9$	I4/mmm; $a = 0.3837$, $c = 2.901$	25	
$(Bi_{0.4}Pb_{0.35}Cu_{0.05})Sr_2(Y_{0.5}Ca_{0.5})Cu_2O_{7+\delta}$	P4/mmm; $a = 0.3819$, $c = 1.181$	102	
22s2	**P4/nmm**		cf. Table 5 (2212 compounds)
$Bi_2Sr_2(Y_{1-x}Ce_x)_2Cu_2O_{10}$	$a = 0.3836$, $c = 1.785$	20	$x = 0.18$
$Bi_2Sr_2(Nd_{1-x}Ce_x)_2Cu_2O_{10}$	$a = 0.3881$, $c = 1.793$	14	$x = 0.18$
$Bi_2Sr_2(Sm_{1-x}Ce_x)_2Cu_2O_{10}$	$a = 0.3863$, $c = 1.790$	16	$x = 0.18$
$Bi_2Sr_2(Eu_{1-x}Ce_x)_2Cu_2O_{10}$	$a = 0.3854$, $c = 1.788$	27	$x = 0.18$
$Bi_2Sr_2(Gd_{1-x}Ce_x)_2Cu_2O_{10}$	$a = 0.3851$, $c = 1.788$	34	$x = 0.18$
$Bi_2Sr_2(Dy_{1-x}Ce_x)_2Cu_2O_{10}$	$a = 0.3844$, $c = 1.787$	27	$x = 0.18$
$Bi_2Sr_2(Ho_{1-x}Ce_x)_2Cu_2O_{10}$	$a = 0.3840$, $c = 1.786$	24	$x = 0.18$

TABLE 7. Crystal Structures and Critical Temperatures of the Chalcogenide Fe-Based Superconductors

Mol. form.	Structure (lattice parameters in nm)	T_c^{max}/K	Comments
11	**P4/nmm (PbO structure)**		
FeS	$a = 0.3680$, $c = 0.5031$	4.5	
$FeSe_{1-x}S_x$	$a = 0.377$, $c = 0.552$	11	
$Fe_{1+\delta}Se$	$a = 0.3762$, $c = 0.5502$	8.5	β-FeSe
$Fe_{1-x}Cr_xSe$	$a = 0.3773$, $c = 0.5524$	12	$x_{max} = 0.02$

Solids

Mol. form.	Structure (lattice parameters in nm)	T_c^{max}/K	Comments
$Fe_{1-x}Nb_xSe$		14	x_{max} = 0.04
$FeSe_{1-x}Te_x$	*a* = 0.3798, *c* = 0.6038	15	x_{max} = 0.4-0.6
FeTe	*a* = 0.3825, *c* = 0.6291	0	Antiferromagnetic
$FeTe_{1-x}S_x$	*a* = 0.3812, *c* = 0.6244	9.4	x_{max} = 0.06-0.12
122	**I4/mmm ($ThCr_2Si_2$ structure)**		
$Li_xFe_2Se_2$	*a* = 0.3775, *c* = 1.704	44	
$Na_xFe2Se_{2-\delta}$	*a* = 0.3785, *c* = 1.7432	46	
$K_xFe_2Se_2$	Multiphase	40	
$(Tl_{1-x}K_x)Fe_{2-y}Se_2$	*a* = 0.388, *c* = 1.405	31	*x* = 0.25-0.46, y_{max} = 0.12-0.22
$Rb_xFe_{2-y}Se_2$		33	
$Cs_xFe_{2-y}Se_2$	*a* = 0.3850, *c* = 1.5647	30	
$Ca_xFe_2Se_2$	Multiphase	~40	
$Sr_xFe_2Se_2$	Multiphase	38	
$Ba_xFe_2Se_{2-\delta}$	*a* = 0.3778, *c* = 1.6843	40	
$Eu_xFe_2Se_2$	Multiphase	40	
$Yb_xFe_2Se_2$	Multiphase	42	
Intercalated FeSe			
$(C_2H_8N_2)_xFeSe$	Amma; *a* = 0.3865, *b* = 0.3897, *c* = 2.1700	30	
$(C_4H_4N_2O_2S)_{0.3}FeSe$	*c* = 1.55	48	
$(C_{40}H_{54}O_{27})_{0.3}FeSe$	*c* = 1.45	45	
(Li,Fe)OHFeSe	P4/nmm; *a* = 0.378, *c* = 0.930	42	
$A_x(NH_3)_yFe_2Se_2$	I4/mmm	up to 44	*A* = **Li**, Na, **K**, Cs
$Li_x(NH_2)_y(NH_3)_{1-y}Fe_2Se_2$	I4/mmm; *a* = 0.3825, *c* = 1.6527	43	x_{max} = 0.6, y_{max} = 0.2
$Li_x[(CH_2)_n(NH_2)_2]_yFe_2Se_2$	I4/mmm	up to 41	*n* = 1,**2**,3
$Na_x[(CH_2)_n(NH_2)_2]_yFe_2Se_2$	I4/mmm	up to 47	*n* = 1,**2**,3
$Sr_x[(CH_2)_n(NH_2)_2]_yFe_2Se_2$	I4/mmm	up to 37	*n* = 1,**2**,3

TABLE 8. Crystal Structures and Critical Temperatures of the Pnictide Fe-Based Superconductors

Mol. form.	Structure (lattice parameters in nm)	T_c^{max}/K	Comments
111	**P4/nmm (anti-PbFCl structure)**		
LiFeP	*a* = 0.3692, *c* = 0.6031	~6	
LiFeAs	*a* = 0.3772, *c* = 0.6357	18	
$Na_{1-\delta}FeAs$	*a* = 0.3949, *c* = 0.7040	15	
$Na(Fe_{1-x}Co_x)As$	*c* = 0.704	20	x_{max} = 0.3
112	**P21/m**		
$CaFeAs_2$		0	
$(Ca_{1-x}RE_x)FeAs_2$		up to 43	*RE* = **La**, Pr, Nd, Sm, Eu, Gd
$(Ca_{1-x}RE_x)Fe(As_{1-y}Sb_y)_2$		up to 47	*RE* = **La**, Ce, Pr, Nd
$EuFeAs_2$		0	
$Eu(Fe_{1-x}Ni_x)As_2$	*a* = 0.3987, *b* = 0.3908, *c* = 1.0645	18	x_{max} = 0.4
$(Eu_{1-x}La_x)FeAs_2$	*a* = 0.3980, *b* = 0.3900, *c* = 1.0643	11	x_{max} = 0.15
1111	**P4/nmm (ZrCuSiAs structure)**		
LaOFeP	*a* = 0.3964, *c* = 0.8512	~4	
$LaOFeAs_{1-x}P_x$		11	x_{max} = 0.25 and 0.7
$La_{1-x}OFeP$	*a* = 0.3961, *c* = 0.8506	~7	x_{max} = 0.1
*Ln*OFeP		~4	*Ln* = Pr, Nd
$(Sr_{1-x}RE_x)FeAsF$		up to 56	*RE* = La, Nd, **Sm**
$Sr(Fe_{1-x}Co_x)AsF$	*a* = 0.4002, *c* = 0.8943	~4	x_{max} = 0.125
$(Ca_{1-x}RE_x)FeAsH$		up to 47	*RE* = **La**, Sm
$Ca(Fe_{1-x}Co_x)AsH$	*a* = 0.382, *c* = 0.822	24	x_{max} = 0.15
$Ca(Fe_{1-x}Co_x)AsF$	*a* = 0.3880, *c* = 0.8552	22	x_{max} = 0.1
$Ca(Fe_{1-x}Ni_x)AsF$	*a* = 0.3879, *c* = 0.8578	15	x_{max} = 0.05
$CaFeAs(F_{1-x}H_y)$	*a* = 0.3896, *c* = 0.8669	29	x_{max} ≈ 0.3

Solids

Mol. form.	Structure (lattice parameters in nm)	T_c^{max}/K	Comments
$YFeAsO_{1-\delta}$	a = 0.3842, c = 0.8303	47	
$LaFeAsO_{1-\delta}$	a = 0.4026, c = 0.8719	28	
$CeFeAsO_{1-\delta}$	a = 0.3995, c = 0.8631	39	
$PrFeAsO_{1-\delta}$	a = 0.3963, c = 0.8572	48	
$NdFeAsO_{1-\delta}$	a = 0.3943, c = 0.8529	53	
$PmFeAsO_{1-\delta}$	Not synthesized due to radioactive Pm	–	
$SmFeAsO_{1-\delta}$	a = 0.3922, c = 0.8452	53	
$EuFeAsO_{1-\delta}$	Not reported	–	
$GdFeAsO_{1-\delta}$	a = 0.3891, c = 0.8393	54	
$TbFeAsO_{1-\delta}$	a = 0.3878, c = 0.8353	52	
$DyFeAsO_{1-\delta}$	a = 0.3863, c = 0.8322	51	
$HoFeAsO_{1-\delta}$	a = 0.3846, c = 0.8295	50	
$YFeAsO_{1-x}F_x$		10	x_{max} = 0.1
$LaFeAsO_{1-x}F_x$	a = 0.4036, c = 0.8739	26	
$CeFeAsO_{1-x}F_x$	a = 0.3989, c = 0.8631	41	x_{max} = 0.16
$PrFeAsO_{1-x}F_x$	a = 0.3967, c = 0.8561	47	x_{max} = 0.15
$NdFeAsO_{1-x}F_x$	a = 0.3954, c = 0.8540	47	x_{max} = 0.15
$PmFeAsO_{1-x}F_x$	Not synthesized due to radioactive Pm	–	
$SmFeAsO_{1-x}F_x$	a = 0.3934, c = 0.8468	58	x_{max} = 0.2
SmFeAsF	a = 0.3940, c = 0.8503	56	
$EuFeAsO_{1-x}F_x$		11	x_{max} = 0.15
$GdFeAsO_{1-x}F_x$	a = 0.3915, c = 0.8457	40	x_{max} = 0.25
$TbFeAsO_{1-x}F_x$	a = 0.3860, c = 0.8332	51	x_{max} ≈ 0.1
$DyFeAsO_{1-x}F_x$	a = 0.3843, c = 0.8284	46	x_{max} ≈ 0.1
$HoFeAsO_{1-x}F_x$	a = 0.3830, c = 0.8270	36	x_{max} ≈ 0.1
$Gd_{1-x}Th_xFeAsO$	a = 0.3916, c = 0.8439	56	x_{max} = 0.2
$Tb_{1-x}Th_xFeAsO$	a = 0.3903, c = 0.8413	52	x_{max} = 0.2
$La(Fe_{1-x}Co_x)AsO$	a = 0.4035, c = 0.8724	13	x_{max} = 0.06
$Sm(Fe_{1-x}Co_x)AsO$	a = 0.3939, c = 0.8467	17	x_{max} = 0.1
$LaFeAsO_{1-x}H_x$	a = 0.399, c = 0.865	33	x_{max} = 0.36
$CeFeAsO_{1-x}H_x$	a = 0.397, c = 0.861	48	x_{max} = 0.25
$NdFeAsO_{1-x}H_x$		54	
$SmFeAsO_{1-x}H_x$	a = 0.391, c = 0.845	56	x_{max} = 0.22
$GdFeAsO_{1-x}H_x$	a = 0.389, c = 0.840	54	x_{max} = 0.1
$DyFeAsO_{1-x}H_x$		52	x_{max} = 0.17
$ErFeAsO_{1-x}H_x$	a = 0.3822, c = 0.8281	41	x_{max} = 0.05
ThFeAsN	a = 0.4037, c = 0.8526	30	Undoped superconductor
122	**I4/mmm ($ThCr_2Si_2$ structure)**		
$NaFe_2As_2$	Metastable a = 0.3809, c = 1.2441	25	
$(Ce_{0.5-x}Na_{0.5+x})Fe_2As_2$	a = 0.3841, c = 1.2239	26	x_{max} = 0.3
$(Pr_{0.5-x}Na_{0.5+x})Fe_2As_2$	a = 0.3839, c = 1.2193	25	x_{max} = 0.25
AFe_2As_2		up to 3.8	A = **K**, Rb, Cs
$(Ca_{1-x}Na_x)Fe_2As_2$	a = 0.3841, c = 1.22	33	x_{max} = 0.66
$(Ca_{1-x}RE_x)Fe_2As_2$		up to 49	RE = La, Ce, **Pr**
$Ca(Fe_{1-x}M_x)_2As_2$		up to 20	M = **Co**, Ni, Rh, Pd, Ir
$CaFe_2(As_{1-x}P_x)_2$	a = 0.390, c = 1.165	15	x_{max} = 0.05
$(Sr_{1-x}A_x)Fe_2As_2$		up to 37	A = Na, K, **Cs**
$(Sr_{1-x}La_x)Fe_2As_2$	a = 0.395, c = 1.21	22	x_{max} = 0.3. Thin film; metastable
$Sr(Fe_{1-x}M_x)_2As_2$		up to 24	M = Co, Ni, Ru, Rh, Pd, **Ir**, Pt
$SrFe_2(As_{1-x}P_x)_2$	a = 0.390, c = 1.21	33	x_{max} = 0.35
$(Ba_{1-x}A_x)Fe_2As_2$		up to 38	A = Na, **K**, Rb
$(Ba_{1-x}RE_x)Fe_2As_2$		up to 22	RE = **La**, Ce, Pr, Nd
$Ba(Fe_{1-x}M_x)_2As_2$		up to 24	M = Co, Ni, Ru, Rh, Pd, **Ir**, Pt
$BaFe_2(As_{1-x}P_x)_2$		31	x_{max} = 0.32
$Ba_2Ti_2Fe_2As_4O$	a = 0.4028, c = 2.7344	21	Intergrowth of $BaFe_2As_2$ and $BaTi_2As_2O$
$LaFe_2As_2$	a = 0.3938, c = 1.1714	12	
$(La_{0.5-x}Na_x)Fe_2As_2$	a = 0.3841, c = 1.2325	27	x_{max} = 0.3
$(La_{0.5-x}Na_xK_{0.5})Fe_2As_2$	a = 0.3850, c = 1.321	23	x_{max} = 0.25

Mol. form.	Structure (lattice parameters in nm)	T_c^{max}/K	Comments
$(Eu_{1-x}A_x)Fe_2As_2$		up to 35	A = Na, **K**, x_{max} = 0.5, antiferromagnetic
$(Eu_{0.78}La_{0.27})Fe_2As_2$		13	Antiferromagnetic
$Eu(Fe_{1-x}M_x)_2As_2$			M = Co, **Ir**, Ru, Rh, antiferromagnetic
$EuFe_2(As_{1-x}P_x)_2$	a = 0.389, c = 1.1835	30	x_{max} = 0.3, antiferromagnetic
1144 (2×122)	**P4/mmm**		
$(La,Na)RbFe_4As_4$	a = 0.3861, c = 1.326	26	
$(La,Na)CsFe_4As_4$	a = 0.3880, c = 1.360	24	
$CaKFe_4As_4$	a = 0.3866, c = 1.2817	35	
$CaRbFe_4As_4$	a = 0.3876, c = 1.3104	35	
$CaCsFe_4As_4$	a = 0.3891, c = 1.3414	32	
$SrRbFe_4As_4$	a = 0.3897, c = 1.3417	35	
$SrCsFe_4As_4$	a = 0.3910, c = 1.3729	37	
$BaCsFe_4As_4$	Possibly 122 structure	26	
$EuRbFe_4As_4$	a = 0.3889, c = 1.330	36	
$EuCsFe_4As_4$	a = 0.3901, c = 1.361	35	
12442 (122+2×1111)	**I4/mmm**		
$KGd_2Fe_4As_4O_2$	a = 0.3897, c = 3.0670	37	
$KTb_2Fe_4As_4O_2$	a = 0.3886, c = 3.0621	37	
$KDy_2Fe_4As_4O_2$	a = 0.3874, c = 3.0598	37	
$KHo_2Fe_4As_4O_2$	a = 0.3866, c = 3.0597	36	
$RbSm_2Fe_4As_4O_2$	a = 0.3921, c = 3.1381	36	
$RbGd_2Fe_4As_4O_2$	a = 0.3901, c = 3.1343	35	
$RbTb_2Fe_4As_4O_2$	a = 0.3890, c = 3.1277	35	
$RbDy_2Fe_4As_4O_2$	a = 0.3879, c = 3.1265	34	
$RbHo_2Fe_4As_4O_2$	a = 0.3869, c = 3.1242	34	
$CsNd_2Fe_4As_4O_2$	a = 0.3949, c = 3.223	35	
$CsSm_2Fe_4As_4O_2$	a = 0.3926, c = 3.2124	35	
$CsGd_2Fe_4As_4O_2$	a = 0.3907, c = 3.2051	33	
$CsTb_2Fe_4As_4O_2$	a = 0.3895, c = 3.1982	33	
$CsDy_2Fe_4As_4O_2$	a = 0.3888, c = 3.1961	33	
$CsHo_2Fe_4As_4O_2$	a = 0.3876, c = 3.1949	33	
$KCa_2Fe_4As_4F_2$	a = 0.3868, c = 3.1007	33	
$RbCa_2Fe_4As_4F_2$	a = 0.3872, c = 3.1667	30	
$CsCa_2Fe_4As_4F_2$	a = 0.3881, c = 3.2363	28	
$BaTh_2Fe_4As_4(N_{0.7}O_{0.3})_2$	a = 0.3989, c = 2.9853	30	
42214 ("221"+2×1111)	**I4/mmm**		
$Pr_4Fe_2As_2Te_{0.88}O_4$	a = 0.4016, c = 2.9857	25	
$Sm_4Fe_2As_2Te_{0.92}O_4$	a = 0.3964, c = 2.9509	25	
$Sm_4Fe_2As_2Te_{0.72}O_{4-y}F_y$	a = 0.3960, c = 2.9268	40	
$Gd_4Fe_2As_2Te_{0.90}O_4$	a = 0.3935, c = 2.9369	25	
$Gd_4Fe_2As_2Te_{0.92}O_{4-y}F_y$	a = 0.3936, c = 2.9350	45	
10-3-8 and 10-4-8	**$P\bar{1}$**		
$(Ca_{1-x}RE_xFeAs)_{10}Pt_3As_8$	a = 0.8749, b = 0.8753, c = 1.0714	up to 33	x_{max} = 0.14
$(CaFe_{1-x}M_xAs)_{10}Pt_3As_8$		up to 15	M = **Co**, Ni, Pd, **Pt**
$(CaFeAs)_{10}Pt_4As_8$	a = 0.8755, b = 0.8764, c = 1.0690	35	
$(CaFe_{1-x}Ir_xAs)_{10}Ir_4As_8$	P4/n; a = 0.8732, c = 1.0391	16	
32522	**I4/mmm**		
$Ca_3Al_2O_5Fe_2P_2$	a = 0.3715, c = 2.5236	16	
$Ca_3Al_2O_5Fe_2As_2$	a = 0.3742, c = 2.6078	30	
$Ca_{n+1}(Mg,Ti)nO_yFe_2As_2$		up to 47	n = 3, **4**; $y \approx 3n$-1
$Ca_{n+1}(Sc,Ti)nO_yFe_2As_2$		up to 42	n = 3, 4, **5**; $y \approx 3n$-1
$Sr_3Sc_2O_5Fe_2As_2$	a = 0.4069, c = 2.6876	0	
$Sr_4(Sc,Ti)_3O_8Fe_2As_2$		28	
$Ba_3Sc_2O_5Fe_2As_2$	a = 0.4133, c = 2.8355	0	
$Ba_4Sc_3O_{7.5}Fe_2As_2$	a = 0.4123, c = 3.7565	11	

Mol. form.	Structure (lattice parameters in nm)	T_c^{max}/K	Comments
42622	**P4/mmm**		
$Ca_4Al_2O_6Fe_2P_2$	a = 0.3693, c = 1.4927	17	
$Ca_4Al_2O_6Fe_2As_2$	a = 0.3713, c = 1.5404	28	
$Ca_5(Al,Ti)_3O_9Fe_2As_2$		39	
$Ca_6(Al,Ti)_4O_{12}Fe_2As_2$		36	
$Ca_8(Mg,Ti)_6O_{18}Fe_2As_2$		40	
$Sr_4Sc_2O_6Fe_2P_2$	a = 0.4016, c = 1.5543	17	
$Sr_4V_2O_6Fe_2As_2$	a = 0.3930, c = 1.5673	37	
$Sr_4(Mg,Ti)_2O_6Fe_2As_2$	a = 0.3935, c = 1.5952	39	

TABLE 9. Crystal Structures and Critical Temperatures of Other Non-Cuprate High-Temperature Superconductors

Mol. form.	Structure (lattice parameters in nm)	T_c^{max}/K
MgB_2	P6/mmm; a = 0.3074, c = 0.3534	39
$Ba_{0.6}K_{0.4}BiO_3$	Pm$\bar{3}$m; a = 0.4287	32

ORGANIC SUPERCONDUCTORS

H. P. R. Frederikse

Although the vast majority of organic compounds are insulators, a small number of organic solids show considerable electrical conductivity. Some of these materials appear to be superconductors. The superconducting organics fall primarily into two groups: those containing fulvalenes (pentagonal rings containing sulfur or selenium) and those based on fullerenes, involving the nearly spherical cluster C_{60}.

The transition temperatures T_c of the fulvalene derivatives are shown in Table 1. The abbreviations of the various molecular groups are listed in Table 2 and their chemical structures are depicted in Figure 1. Most of the T_c's are between 1 and 12 K. Several of the compounds only show superconductivity under pressure.

The fullerenes are A_3C_{60} compounds, where A represents a single or a combination of alkali atoms. The C_{60} cluster is shown in Figure 2a, while Figure 2b illustrates how the alkali atoms fit into the A_3C_{60} molecule to form the A15 crystallographic structure. Their superconducting transition temperatures range from 8 to 31.3 K (see Table 3).

References

1. Ishigura, T. and Yamaji, K., *Organic Superconductors*, Springer-Verlag, Berlin, 1990.
2. Williams, J. M. et al., *Organic Superconductors (Including Fullerenes)*, Prentice Hall, Englewood Cliffs, NJ, 1992.
3. *The Fullerenes*, Ed.: Krato, H. W., Fisher, J. E., and Cox, D. E., Pergammon Press, Oxford, 1993.
4. Schluter, M. et al., in *The Fullerenes* (Ref. 3), p. 303.

TABLE 1. Critical Pressure and Maximum Critical Temperature of Organic Superconductors

Material	P_c/kbar	T_c/K
$(TMTSF)_2PF_6$	6.5	1.2
$(TMTSF)_2AsF_6$	9	1.3
$(TMTSF)_2SbF_6$	11	0.4
$(TMTSF)_2TaF_6$	12	1.4
$(TMTSF)_2ClO_4$	0	1.4
$(TMTSF)_2ReO_4$	9.5	1.3
$(TMTSF)_2FSO_3$	5	3
$(ET)_4(ReO_4)_2$	4.5	2
β_L-$(ET)_2I_3$	0	1.4
β_H-$(ET)_2I_3$	0	8.1
γ-$(ET)_3I_{2.5}$	0	2.5
ε-$(ET)_2I_3(I_8)_{0.5}$	0	2.5
α-$(ET)_2I_3I_2$-doped	0	3.3
α_t-$(ET)_2I_3$	0	8
$\varepsilon\rightarrow\beta$-$(ET)_2I_3$[a]	0	6
θ-$(ET)_2I_3$	0	3.6
κ-$(ET)_2I_3$	0	3.6
β-$(ET)_2IBr_2$	0	2.8
β-$(ET)_2AuI_2$	0	4.8
$(ET)_4Hg_{2.89}Cl_8$	0	4.2
$(ET)_4Hg_{2.89}Br_8$	12	1.8
$(ET)_3Cl_2(H_2O)_2$	16	2
κ-$(ET)_2Cu(NCS)_2$	0	10.4
κ-$(d\text{-}ET)_2Cu(NCS)_2$	0	11.4
$(DMET)_2Au(CN)_2$	1.5	0.9
$(DMET)_2AuI_2$	5	0.6
$(DMET)_2AuBr_2$	0	1.9
$(DMET)_2AuCl_2$	0	0.9
$(DMET)_2I_3$	0	0.6
$(DMET)_2IBr_2$	0	0.7
$(MDT\text{-}TTF)_2AuI_2$	0	3.5
$TTF[Ni(dmit)_2]_2$	2	1.6[b]
$TTF[Pd(dmit)_2]_2$	20	6.5
$(CH_3)_4N[Ni(dmit)_2]_2$	7	5

[a] Converted from ε-type to β-type by thermal treatment.
[b] For 7 kbar.
From Ishigura, T. and Yamaji, K., *Organic Superconductors*, Springer-Verlag, Berlin, 1990. With permission.

TABLE 2. List of Symbols and Abbreviations

TTF	tetrathiafulvalene
TMTSF	tetramethyltetraselenafulvalene
BEDT-TTF or "ET"	bis(ethylenedithio)tetrathiafulvalene
MDT-TTF	methylenedithiotetrathiafulvalene
DMET	[dimethyl(ethylenedithio)diselenadithiafulvalene]
dmit	4,5-dimercapto-1,3-dithiole-2-thione
T_c	transition temperature to superconducting state
P_c	minimum pressure required for superconducting transition

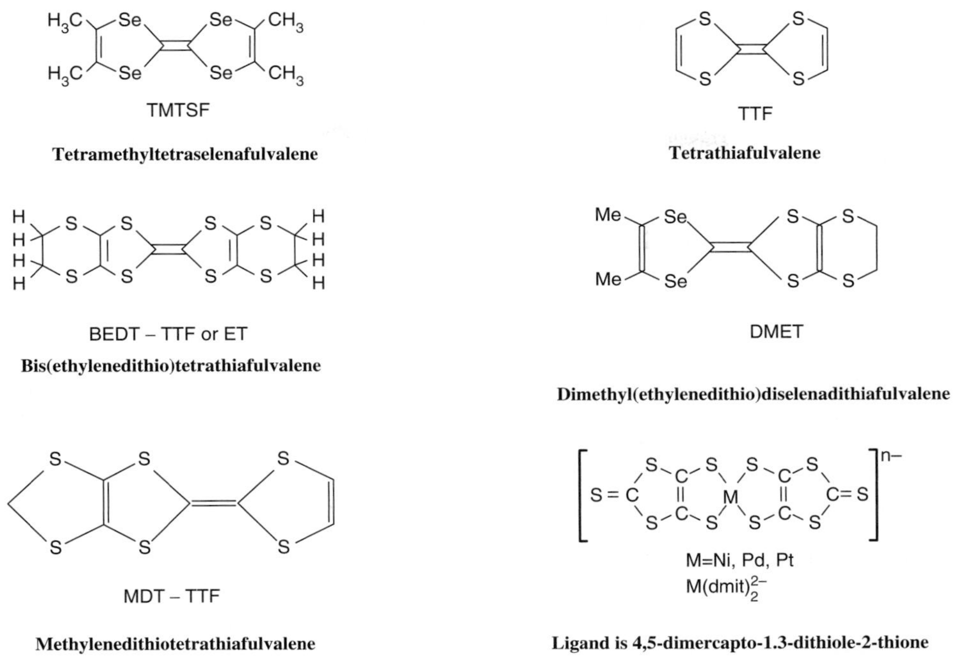

FIGURE 1. Structures of various donor molecules and acceptor species.

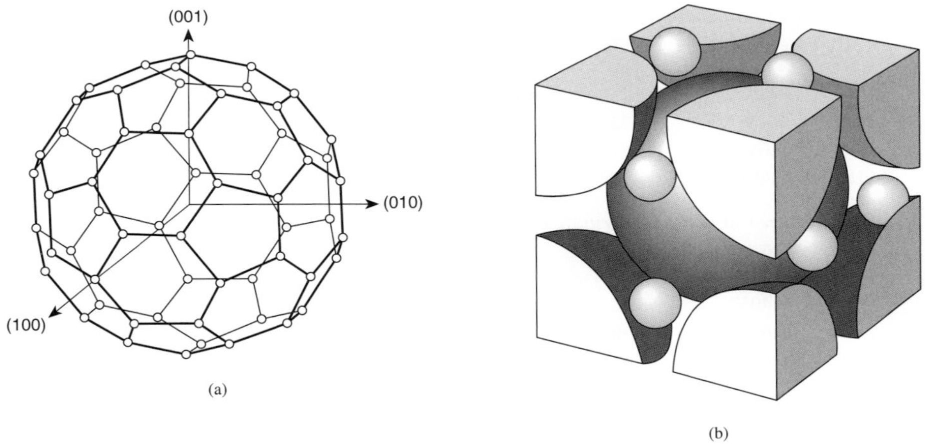

FIGURE 2. (a) C_{60} cluster placed in a fcc lattice. Each crystal axis crosses a double bond shared by two hexagons. (b) A hypothetical A_3C_{60} with the A15 structure. The structure can be seen to be an ordered defect structure of A_6C_{60}.

TABLE 3. Unit Cell and T_c for FCC A_3C_{60} Crystals

	Lattice parameter(s) (Å)	T_c/K
$Na_2Rb_{0.5}Cs_{0.5}C_{60}$	14.148(3)	8.0
Na_2CsC_{60} No. 1[a]	14.132(2)	10.5
Na_2CsC_{60} No. 2[a]	14.176(9)	14.0
K_3C_{60}	14.253(3)	19.3
K_2RbC_{60}	14.299(2)	21.8
Rb_2KC_{60} No. 1[a]	14.336(1)	24.4
Rb_2KC_{60} No. 2[a]	14.364(5)	26.4
Rb_3C_{60}	14.436(2)	29.4
Rb_2CsC_{60}	14.493(2)	31.3

[a] Samples labeled No. 1 and No. 2 have the same nominal composition.
From Schluter, M. et al., *The Fullerenes*, Krato, H. W., Fisher, J. E., and Cox, D. E., Eds., Pergamon Press, Oxford, 1993. With permission.

PROPERTIES OF SEMICONDUCTORS

L. I. Berger

The term *semiconductor* is applied to a material in which electric current is carried by electrons or holes and whose electrical conductivity, when extremely pure, rises exponentially with temperature and may be increased from its low "intrinsic" value by many orders of magnitude by "doping" with electrically active impurities.

Semiconductors are characterized by an energy gap in the allowed energies of electrons in the material that separates the normally filled energy levels of the *valence band* (where "missing" electrons behave like positively charged current carriers "holes") and the *conduction band* (where electrons behave rather like a gas of free negatively charged carriers with an effective mass dependent on the material and the direction of the electrons' motion). This energy gap depends on the nature of the material and varies with direction in anisotropic crystals. It is slightly dependent on temperature and pressure, and this dependence is usually almost linear at normal temperatures and pressures.

Data are presented in five tables. Table 1 lists the main crystallographic and semiconducting properties of a large number of semiconducting materials in three main categories: "Tetrahedral Semiconductors" in which every atom is tetrahedrally coordinated to four nearest neighbor atoms (or atomic sites) as for example in the diamond structure; "Octahedral Semiconductors" in which every atom is octahedrally coordinated to six nearest neighbor atoms—as for example the halite structure; and "Other Semiconductors."

Table 2 gives electrical, magnetic, and optical properties, while Tables 3 and 4 give more details on the semiconducting properties and band structures of the most common semiconductors. Table 5 lists semiconducting minerals with typical resistivity ranges.

TABLE 1. Physico-Chemical Properties of Semiconductors (Listed by Crystal Structure)

Substance	Molecular weight	Average atomic weight	Lattice parameters (Å, room temp.)	Density (g/cm³)	Melting point (K)	Microhardness, N/mm² (M-Mohs scale)	Specific heat, J/kg·K (300 K)	Debye temp. (K)	Coefficient of thermal linear expansion [10^{-6} K^{-1} (300K)]	Thermal conductivity [mW cm^{-1}·K^{-1} (300K)]
1.1. Tetrahedral (Adamantine) Semiconductors										
1.1.1. Diamond Structure Elements (Strukturbericht Symbol A4, Space Group $Fd3m$-O_h^7)										
C (Diamond)		12.01	3.56683	3.513	≈4713 (12.4 GPa) Transition to graphite > 980	10 (M)	471.5	2340	1.18	9900(I) 23200(IIA) 13600(IIB)
Si		28.09	5.43072	2.329	1687	11270	702	645	2.6	1240
Ge		72.64	5.65754	5.323	1211.35	7644	321.9	374	5.8	640
α-Sn		118.71	6.4912	5.769	505.1 (Tr. 286.4)		213	230	5.4 (220 K)	
1.1.2. Sphalerite (Zinc Blende) Structure Compounds (Strukturbericht Symbol B3 Space Group $F\bar{4}3m$-T_d^2)										
I-VII Compounds										
CuF	82.54	41.27	4.255		1181					
CuCl	98.99	49.49	5.4057	3.53	695	2.3 (M)	490	240	12.1	8.4
CuBr	143.45	71.73	5.6905	4.98	770	2.5 (M)	381	207	15.4	12.5
Cul	190.45	95.23	6.60427	5.63	878	192	276	181	19.2	16.8
AgBr	187.77	93.89		6.473	>1570 (Tr. 410)	2.5 (M)	270			
Agl	234.77	117.39	6.502	5.67	831	2.5 (M)	232	134	-2.5	4.2
II-VI Compounds										
BeS	41.08	20.54	4.865	2.36	dec.	3120				
BeSe	87.97	43.99	5.139	4.315		2540				
BeTe	136.61	68.31	5.626	5.090		1500				
BePo	(2318)	(109)	5.838	7.3						
ZnO	81.39	40.69	4.63	5.675	2248	5.0 (M)	494	416	2.9	234
ZnS	97.46	48.72	5.4093	4.079	2100 (Tr. 1295)	1780	472	530	6.36	251
ZnSe	144.34	72.17	5.6676	5.42	1790	1350	339	400	7.2	140
ZnTe	192.99	96.5	6.101	6.34	1568	900	264	223	8.19	108
ZnPo	(274)	(137)	6.309							
CdS	144.48	72.24	5.832	4.826	1750	1250	330	219	4.7	200
CdSe	191.37	95.68	6.05	5.674	1512	1300	255	181	3.8	90
CdTe	240.01	120.00	6.477	5.86	1365	600	205	200	4.9	58.5
CdPo	(321)	(161)	6.665							
HgS	232.66	116.33	5.8517	7.73	1820	3 (M)	210			
HgSe	279.55	139.78	6.084	8.25	1070	2.5 (M)	178	151	5.46	10
HgTe	328.19	164.10	6.4623	8.17	943	300	164	242	4.6	20

Substance	Molecular weight	Average atomic weight	Lattice parameters (Å, room temp.)	Density (g/cm³)	Melting point (K)	Microhardness, N/mm² (M-Mohs scale)	Specific heat, J/kg·K (300 K)	Debye temp. (K)	Coefficient of thermal linear expansion [10^{-6} K^{-1} (300K)]	Thermal conductivity [mW cm^{-1}·K^{-1} (300K)]
III-V Compounds										
BN	24.82	12.41	3.615	3.49	3239	10 (M)	793	≈1900		200
BP(L.T.)	41.78	20.87	4.538	2.9	1398 (dec)	37000		≈980		
BAs	85.73	42.87	4.777		≈2300	19000		≈625		
AlP	57.95	28.98	5.451	2.42	≈2100	5.5 (M)		588		920
AlAs	101.90	50.95	5.6622	3.81	2013	5000		417	3.5	840
AlSb	148.74	74.37	6.1355	4.218	1330	4000		292	4.2	600
GaP	100.70	50.35	5.4495	4.13	1750	9450		446	5.3	752
GaAs	144.64	72.32	5.65315	5.316	1510	7500		344	5.4	560
GaSb	191.48	95.74	6.0954	5.619	980	4480	320	265	6.1	270
InP	145.79	72.90	5.86875	4.787	1330	4100		321	4.6	800
InAs	189.74	94.87	6.05838	5.66	1215	3300	268	249	4.7	290
InSb	236.58	118.29	6.47877	5.775	798	2200	144	202	4.7	160
Other Sphalerite Structure Compounds										
MnS	87.00	43.5	5.011							
MnSe	133.90	66.95	5.82							
β-SiC (3-C SiC)	40.10	20.1	4.348	3.21	3070				2.9	4.9
Ga_2Se_3	376.32	75.26	5.429	4.92	1020	3160			8.9	50
Ga_2Te_3	522.24	104.45	5.899	5.75	1063	2370				47
In_2Te_3(H.T.)	608.44	121.7	6.173	5.8	940	1660				69
$MgGeP_2$	158.84	39.71	5.652							
$ZnSnP_2$	246.00	61.5	5.65		1200					
$ZnSnAs_2$(H.T.)	333.90	82.38	5.851	5.53	1050					76
$ZnSnSb_2$	427.56	106.89	6.281	5.67	870	2500				76

1.1.3. Wurtzite (Zincite) Structure Compounds (Strukturbericht Symbol B4, Space Group $P\,6_3mc\text{-}C_{6v}^4$)

Substance	Molecular weight	Average atomic weight	Lattice parameters (Å, room temp.)	Density (g/cm³)	Melting point (K)	Microhardness, N/mm² (M-Mohs scale)	Specific heat, J/kg·K (300 K)	Debye temp. (K)	Coefficient of thermal linear expansion [10^{-6} K^{-1} (300K)]	Thermal conductivity [mW cm^{-1}·K^{-1} (300K)]
I-VII Compounds										
CuCl	99.0	49.5	3.91; 6.42		703					
CuBr	143.45	71.73	4.06; 6.66		770					
Cul	190.45	95.23	4.31; 7.09							
AgI	234.77	117.40	4.580; 7.494							
II-VI Compounds										
BeO	25.01	12.51	2.698; 4.380		2800					
MgTe	151.9	76.0	4.54; 7.39	3.85	≈2800					
ZnO	81.37	40.69	3.24950; 5.2069	5.66	2250					600
ZnS	97.43	48.72	3.8140; 6.2576	4.1	2100					460
ZnTe	192.99	46.50	4.27; 6.99		1568					
CdS	144.48	72.23	4.1348; 6.7490	4.82	1748					401
CdSe	191.37	95.68	4.299; 7.010	5.66	1512					316
CdTe	240.01	120.00	4.57; 7.47							
III-V Compounds										
BP(H.T.)	41.79	20.90	3.562; 5.900						3.2	
AlN	40.99	20.50	3.111; 4.978	3.26	≈2500					823
GaN	83.73	41.87	3.190; 5.189	6.10	1500					656
InN	128.83	64.42	3.533; 5.693	6.88	1200					556
Other Wurtzite Structure Compounds										
MnS	87.00	43.5	3.985; 6.45	3.248						
MnSe	133.90	66.95	4.12; 6.72							
SiC	40.10	20.1	3.076; 5.048							
MnTe	182.54	91.27	4.078; 6.701							
Al_2S_3	150.14	30.03	3.579; 5.829	2.55	1400					
Al_2Se_3	290.84	58.17	3.890; 6.30	3.91	1250					

1.1.4. Chalcopyrite Structure Compounds (Strukturbericht Symbol $E1_1$, Space Group $I\bar{4}d$-D^{12}_{14})

I-III-VI$_2$ Compounds

Substance	Molecular weight	Average atomic weight	Lattice parameters (Å, room temp.)	Density (g/cm³)	Melting point (K)	Microhardness, N/mm² (M-Mohs scale)	Specific heat, J/kg·K (300 K)	Debye temp. (K)	Coefficient of thermal linear expansion [10^{-6} K^{-1} (300K)]	Thermal conductivity [mW cm^{-1}·K^{-1} (300K)]
$CuAlS_2$	154.65	38.66	5.323; 10.44	3.47	2500					
$CuAlSe_2$	248.45	62.11	5.617; 10.92	4.70	2260					
$CuAlTe_2$	345.73	86.43	5.976; 11.80	5.50	2550					
$CuGaS_2$	197.39	49.53	5.360; 10.49	4.35	2300					
$CuGaSe_2$	291.19	72.80	5.618; 11.01	5.56	1970	4200		275	5.4	42
$CuGaTe_2$	388.47	97.12	6.013; 11.93	5.99	2400	3500			6.9	27
$CuInS_2$	242.49	60.62	5.528; 11.08	4.75	1400	2550				
$CuInSe_2$	336.29	84.07	5.785; 11.56	5.77	1600	2050			6.6	37
$CuInTe_2$	433.57	108.39	6.179; 12.365	6.10	1660	400		195	7.1	49
$CuTlS_2$	322.05	83.01	5.580; 11.17	6.32						
$CuTlSe_2$(L.T.)	425.85	106.46	5.844; 11.65	7.11	900					
$CuFeS_2$	183.51	45.88	5.29; 10.32	4.088	1135					
$CuFeSe_2$	277.31	69.33			850					
$CuLaS_2$	266.58	66.65	5.65; 10.86							
$AgAlS_2$	198.97	49.74	5.707; 10.28	3.94						
$AgAlSe_2$	292.77	73.19	5.968; 10.77	5.07	1220					
$AgAlTe_2$	390.05	97.51	6.309; 11.85	6.18	1000					
$AgGaS_2$	241.71	60.43	5.755; 10.28	4.72						
$AgGaSe_2$	335.51	83.88	5.985; 10.90	5.84	1120	4400				
$AgGaTe_2$	432.79	108.2	6.301; 11.96	6.05	990	1800		212		10
$AgInS_2$(L.T.)	286.87	71.70	5.828; 11.19	5.00		2250				
$AgInSe_2$	380.61	95.15	6.102; 11.69	5.81	1053	1850				30
$AgInTe_2$	477.89	119.47	6.42; 12.59	6.12	965				9.49, 0.69	
$AgFeS_2$	227.83	56.96	5.66; 10.30	4.53						

II-IV-V$_2$ Compounds

Substance	Molecular weight	Average atomic weight	Lattice parameters (Å, room temp.)	Density (g/cm³)	Melting point (K)	Microhardness, N/mm² (M-Mohs scale)	Specific heat, J/kg·K (300 K)	Debye temp. (K)	Coefficient of thermal linear expansion [10^{-6} K^{-1} (300K)]	Thermal conductivity [mW cm^{-1}·K^{-1} (300K)]
$ZnSiP_2$	155.40	38.85	5.400; 10.441	3.39	1640	1100				
$ZnGeP_2$	199.90	49.98	5.465; 10.771	4.17	1295	8100				180
$ZnSnP_2$	246.00	61.5				6500				
$CdSiP_2$	202.43	50.61	5.678; 10.431	4.00	≈1470	10500		282		
$CdGeP_2$	246.94	61.74	5.741; 10.775	4.48	1049	5650				110
$CdSnP_2$	243.03	73.26	5.900; 11.518		840	5000		195		140
$ZnSiAs_2$	242.20	60.55	5.61; 10.88	4.70	1311	9200				
$ZnGeAs_2$	287.80	71.95	5.672; 11.153	5.32	1150	6800		263		110
$ZnSnAs_2$	333.90	83.48	5.8515; 11.704	5.53	1048	4550		271		150
$CdSiAs_2$	290.34	72.58	5.884; 10.882		>1120	6850				
$CdGeAs_2$	334.83	83.71	5.9427; 11.2172	5.60	938	4700				48
$CdSnAs_2$	380.93	95.23	6.0944; 11.9182	5.72	880	3450				40

1.1.5. Other Ternary Semiconductors with Tetrahedral Coordination

I$_2$-IV-VI$_3$ Compounds

Substance	Molecular weight	Average atomic weight	Lattice parameters (Å, room temp.)	Density (g/cm³)	Melting point (K)	Microhardness, N/mm² (M-Mohs scale)	Specific heat, J/kg·K (300 K)	Debye temp. (K)	Coefficient of thermal linear expansion [10^{-6} K^{-1} (300K)]	Thermal conductivity [mW cm^{-1}·K^{-1} (300K)]
Cu_2SiS_3(H.T.)	251.36	41.89	3.684; 6.004	3.81	1200					23
Cu_2SiS_3(L.T.)			5.290; 10.156	3.63						
Cu_2SiTe_3	537.98	89.66	5.93	5.47						
Cu_2GeS_3(H.T.)	295.88	49.31	5.317	4.45	1210	4550	510	254	7.2	12
Cu_2GeS_3(L.T.)			5.327; 5.215	4.46						
Cu_2GeSe_3	436.56	72.76	5.589; 5.485	5.57	1030	3840	340	168	8.4	24
Cu_2GeTe_3	582.51	97.09	5.958; 5.935	5.92		2890				130
Cu_2SnS_3	341.98	57.00	5.436	5.02	1110	2770	440	214	7.8	28
$CuSnSe_3$	482.66	80.44	5.687	5.94	960	2510	310	148	8.9	35
Cu_2SnTe_3	628.61	104.77	6.048	6.51	680	1970				144
Ag_2GeSe_3	525.21	87.54			810					
Ag_2SnSe_3	571.31	95.22								

Substance	Molecular weight	Average atomic weight	Lattice parameters (Å, room temp.)	Density (g/cm³)	Melting point (K)	Microhardness, N/mm² (M-Mohs scale)	Specific heat, J/kg·K (300 K)	Debye temp. (K)	Coefficient of thermal linear expansion [10^{-6} K^{-1} (300K)]	Thermal conductivity [mW cm^{-1}·K^{-1} (300K)]
Ag_2GeTe_3	671.13	111.86			600					
Ag_2SnTe_3	717.23	119.54								
I_3-V-VI_4-Compounds										
Cu_3PS_4	349.85	40.73	7.44; 6.19							
Cu_3AsS_4	393.79	49.22	6.43; 6.14	4.37	931				3.2	30.2
Cu_3AsSe_4	581.37	72.67	5.570; 10.957	5.61	733			169	9.5	19
Cu_3SbS_4	440.64	55.08	5.38; 16.76	4.90	830					
Cu_3SbSe_4	628.22	78.53	5.654; 11.256	6.0	700			131	12.4	14.6
I-IV_2-V_3 Compounds										
$CuSi_2P_3$	212.64	35.44	5.25							
$CuGe_2P_3$	301.65	50.28	5.375	4.318	1113	8500	429	8.21	37.6	
$AgGe_2P_3$	345.97	57.66			1015	6150				
1.1.6. "Defect Chalcopyrite" Structure Compounds (Strukturbericht Symbol E3, Space Group $I\bar{4}$-S_4^2)										
$ZnAl_2Se_4$	435.18	62.17	5.503; 10.90	4.37						
$ZnAl_2Te_4$(?)	629.74	84.96	5.904; 12.05	4.95						
$ZnGa_2S_4$(?)	333.06	47.58	5.274; 10.44	3.80						
$ZnGa_2Se_4$(?)	520.66	74.38	5.496; 10.99	5.21						
$ZnGa_2Te_4$(?)	715.22	102.17	5.937; 11.87	5.67						
$ZnIn_2Se_4$	610.86	87.27	5.711; 11.42	5.44	1250					
$ZnIn_2Te_4$	805.42	115.06	6.122; 12.24	5.83	1075					
$CdAl_2S_4$	294.61	42.09	5.564; 10.32	3.06						
$CdAl_2Se_4$	482.21	68.89	5.747; 10.68	4.54						
$CdAl_2Te_4$(?)	676.77	97.68	6.011; 12.21	5.10						
$CdGa_2S_4$	380.09	54.30	5.577; 10.08	4.03						
$CdGa_2Se_4$	567.69	81.10	5.743; 10.73	5.32						
$CdGa_2Te_4$	762.25	108.89	6.093; 11.81	5.77						
$CdIn_2Te_4$	852.45	121.78	6.205; 12.41	5.9	1060					
$HgAl_2S_4$	382.79	54.68	5.488; 10.26	4.11						
$HgAl_2Se_4$	570.39	82.48	5.708; 10.74	5.05						
$HgAl_2Te_4$(?)	764.48	109.28	6.004; 12.11	5.81						
$HgGa_2S_4$	468.27	66.90	5.507; 10.23	5.00						
$HgGa_2Se_4$	655.87	93.70	5.715; 10.78	6.18						
$HgIn_2Se_4$	746.07	106.58	5.764; 11.80	6.3	1100					
$HgIn_2Te_4$(?)	940.63	134.38	6.186; 12.37	6.3	980					
1.1.7. Other Adamantine Compounds										
α–SiC	40.10	20.10	3.0817; 15.12	3.21	3070					
$Hg_5Ga_2Te_8$	2163.19	144.21	6.235							
$Hg_5In_2Te_8$	2253.39	150.23	6.328							
$CdIn_2Se_4$	657.89	93.98	a = c = 5.823							
1.2. Octahedral Semiconductors										
1.2.1. Halite Structure Semiconductors (Strukturbericht Symbol B1, Space Group Fm3m-O_h^5)										
GeTe	200.21	100.10	5.98	6.14						
SnSe	197.67	98.83	6.020		1133					
SnTe	246.31	123.15	6.313	6.45	1080 (max)					91
PbS	239.3	119.63	5.9362	7.61	1390					23
PbSe	286.2	143.08	6.1243	8.15	1340					17
PbTe	334.8	167.4	6.454	8.16	1180					23
1.2.2. Selected Other Binary Halites										
BiSe	287.94	143.97	5.99	7.98	880					
BiTe	336.58	168.29	6.47							
EuSe	230.92	115.46	6.191		2300					2.4
GdSe	236.21	118.11	5.771		2400					
NiO	74.69	37.35	4.1684	6.6	2260					
CdO	128.41	64.21	4.6953		1700					7

Substance	Molecular weight	Average atomic weight	Lattice parameters (Å, room temp.)	Density (g/cm³)	Melting point (K)	Microhardness, N/mm² (M-Mohs scale)	Specific heat, J/kg·K (300 K)	Debye temp. (K)	Coefficient of thermal linear expansion [10^{-6} K^{-1} (300K)]	Thermal conductivity [mW cm^{-1}·K^{-1} (300K)]
SrS	119.69	59.84	6.0199	3.643	3000					
1.3. Other Semiconductors										
1.3.1. Antifluorite Structure Compounds ($Fm3m$-O_h^5))										
Mg_2Si	76.70	25.57	6.338	1.88	1375				11.5	
Mg_2Ge	121.22	40.4	6.380	3.08	1388				15.0	
Mg_2Sn	167.32	55.77	6.765	3.53	1051				9.9	92
Mg_2Pb	225.81	85.27	6.836	5.1	823				10.0	
1.3.2. Tetradymite Structure Compounds ($R3m$-D_{3d}^5)										
Sb_2Te_3	626.3	125.26	4.25; 30.3	6.44	895					
Bi_2Se_3	654.84	130.97	4.14; 28.7	7.51	979	167				24
Bi_2Te_3	800.76	160.15	4.38; 30.45	7.73	858	155	16			30
1.3.3. Skutterudite Structure Compounds ($R3m$-D_{3d}^5)										
CoP_3	151.85	37.96	7.7073		>1270					
$CoAs_3$	286.70	71.65	8.2060	6.73	1230					
$CoSb_3$	424.18	106.05	9.0385		1123			307		50
$NiAs_3$	283.45	70.86	8.330	6.43						
RhP_3	195.83	48.96	7.9951		>1470					
$RhAs_3$	327.67	81.92	8.4427		>1270					100
$RhSb_3$	468.16	117.04	9.2322		1170					
IrP_3	285.14	71.29	8.0151	7.36	>1470					
$IrAs_3$	416.98	104.25	8.4673	9.12	>1470					90
$IrSb_3$	557.47	139.37	9.2533	9.35	1170			303		
1.3.4. Selected Multinary Compounds										
$AgSbSe_2$	387.54	96.88	5.786	6.60	910					10.5
$AgSbTe_2$ (or $Ag_{19}Sb_{29}Te_{52}$)	484.82	121.2	6.078	7.12	830					86
$AgBiS_2$(H.T.)	380.97	95.24	5.648							
$AgBiSe_2$(H.T.)	474.77	118.69	5.82							
$AgBiTe_2$(H.T.)	572.05	143.01	6.155							
Cu_2CdSnS_4	486.43	60.80	5.586	10.83						
1.3.5. Some Elemental Semiconductors										
B		10.81	4.91; 12.6	2.34	2348	9.5 (M)	1277	1370	8.3	600
Se(gray)		78.96	4.36; 4.95	4.81	493	350	292.6		(‖C) 17.89 (⊥C) 74.09	(‖C) 45.2 (⊥C) 13.1
Te		127.60	4.45; 5.91	6.23	723		196.5		16.8	(‖C) 33.8 (⊥C) 19.7
1.3.6 III_2VI_3 Oxides										
Ga_2O_3	187.44	37.49	a=4.9825; c=13.433	6.44	2173					
In_2O_3	277.64	55.53	10.117	7.179	2180					
Ta_2O_3	456.77	91.35		10.19	990					

Solids

TABLE 2. Basic Thermodynamic, Electrical, and Magnetic Properties of Semiconductors (Listed by Crystal Structure)

Substance	Heat of formation at 300 K (kJ/mol)	Volume compressibility (10^{-10} m²/N)	Static dielectric constant	Atomic magnetic susceptibility (10^{-6} cgs)	Index of refraction	Minimum room temperature energy gap (eV)	Electron mobility at 300 K (cm²/V·s)	Hole mobility at 300 K (cm²/V·s)	Optical transition[a]	Breakdown voltage (kV/mm)	Remarks
2.1. Adamantine Semiconductors											
2.1.1. Diamond Structure Elements (Strukturbericht Symbol A4, Space Group Fd3m-O_h^7)											
C	714.4	18	5.7	-5.88	2.419 (589 nm)	5.4	1800	1400	i	500	
Si	324	0.306	11.9	-3.9	3.49 (589 nm)	1.12	1900	500	i	30	
Ge	291	0.768	16	-0.12	3.99 (589 nm)	0.67	3800	1820	i		
α-Sn	267.5		24		2.75 (589 nm)	0.0; 0.8	2500	2400			
2.1.2. Sphalerite (Zinc Blende) Structure Compounds (Strukturbericht Symbol B3 Space Group F$\bar{4}$3m-T_d^2)											
I-VII Compounds											
CuF											
CuCl	481	0.26	7.9		1.93	3.17			d		Nantokite
CuBr	481	0.26	7.9		2.12	2.91			d		
CuI	439	0.27	6.5		2.346	2.95			d		Marshite
AgBr	486		12.4		2.253	2.50	4000		i		Bromirite
AgI	389	0.41	10		2.22	2.22	30		d		Miersite
II-VI Compounds											
BeS					4.17				i		
BeSe					3.61				i		
BeTe					1.45		20		d		
BePo											
ZnO											See 2.1.3.
ZnS	477		8.9	-9.9	2.356	3.54	180	5(400°C)	d		See also 2.1.3.
ZnSe	422		9.2		2.89	2.58	540	28	d		
ZnTe	376		10.4		3.56	2.26	340	100	d		
ZnP											
CdS											See 2.1.3.
CdSe											See 2.1.3.
CdTe	339		7.2		2.50	1.44	1200	50	d		
CdPo											
HgS					2.85		250		d		Metacinnabarite
HgSe	247					2.10 (α)	20000	≈1.5	s		Tiemannite
HgTe	242					-0.06	25000	350	s		Coloradoite
III-V Compounds											
BN	815					4.6					Borazone
BP(L.T.)						≈2.1	500	70			Ignites 470K
BAs						≈1.5					
AlP						2.45	80		i		
AlAs	627		10.9			2.16	1200	420	i		
AlSb	585	0.571	11		3.2	1.60	200-400	550	i		
GaP	635	0.110	11.1	-13.8	3.2	2.24	300	150	i		
GaAs	535	0.771	13.2	-16.2	3.30	1.35	8800	400	d		
GaSb	493	0.457	15.7	-14.2	3.8	0.67	4000	1400	d		
InP	560	0.735	12.4	-22.8	3.1	1.27	4600	150	d		
InAs	477	0.549	14.6	-27.7	3.5	0.36	33000	460	d		
InSb	447	0.442	17.7	-32.9	3.96	0.163	78000	750	d		
Other Sphalerite Structure Compounds											
MnS											See also 2.1.3.
MnSe											See also 2.1.3.
β-SiC					2.697	2.3	4000				
Ga_2Te_3	271			-13.5		1.35	50				
In_2Te_3 (H.T.)	198			-13.6		1.04	50				
$MgGeP_2$											El–T^{d12}
$ZnSnP_2$						2.1					Same
$ZnSnAs_2$ (H.T.)						≈0.7					Same
$ZnSnSb_2$						0.4					Same
2.1.3. Wurtzite (Zincite) Structure Compounds (Strukturbericht Symbol B4, Space Group P6$_3$mc-C_{6v}^4)											
I-VII Compounds											
CuCl											
CuBr											
CuI											

Solids

Substance	Heat of formation at 300 K (kJ/mol)	Volume compressibility (10^{-10} m^2/N)	Static dielectric constant	Atomic magnetic susceptibility (10^{-6} cgs)	Index of refraction	Minimum room temperature energy gap (eV)	Electron mobility at 300 K ($cm^2/V{\cdot}s$)	Hole mobility at 300 K ($cm^2/V{\cdot}s$)	Optical transition[a]	Breakdown voltage (kV/mm)	Remarks
AgI						2.63					Iodargirite
II-VI Compounds											
BeO											
MgTe											
ZnO	-350					3.2	180				
ZnS	-206					3.67					
ZnTe	-163										
CdS				8.45; 9.12	2.32	2.42	350	40	d		Greenockide
CdSe						1.74	900	50	d		Cadmoselite
CdTe						1.50	650				
III-V Compounds											
BP(H.T.)											
AlN						6.02					
GaN						3.34					
InN						2.0					
Other Wurtzite Structure Compounds											
MnS											
MnSe											
SiC					2.654						
MnTe						≈1.0					
Al_2S_3	426					4.1					
Al_2Se_3	367					3.1					
2.1.4. Chalcopyrite Structure Compounds (Strukturbericht Symbol E11, Space Group $I\bar{4}2d$-D_{2d}^{12})											
I-III-VI_2 Compounds											
$CuAlS_2$		0.106				2.5					
$CuAlSe_2$						2.67					
$CuAlTe_2$						0.88					
$CuCaS_2$		0.106				2.38					
$CuGaSe_2$		0.141				0.96, 1.63					
$CuGaTe_2$		0.227				0.82, 1.0					
$CuInS_2$		0.141				1.2					
$CuInSe_2$		0.187				0.86, 0.92					
$CuInTe_2$		0.278				0.95					
$CuTlS_2$											
$CuTlSe_2$(L.T.)						1.07					
$CuFeS_2$						0.53					Chalcopyrite
$CuFeSe_2$						0.16					
$CuLaS_2$											
$AgAlS_2$											
$AgAlSe_2$						0.7					
$AgAlTe_2$						0.56					
$AgGaS_2$		0.150				1.66					
$AgGaSe_2$		0.182				1.1					
$AgGaTe_2$		0.280				1.9					
$AgInS_2$(L.T.)		0.185				1.18					
$AgInSe_2$		0.238				0.96, 0.52					
$AgInTe_2$		0.338									
$AgFeS_2$											
II-IV-V_2 Compounds											
$ZnSiP_2$	312					2.3	1000				
$ZnGeP_2$	293					2.2					
$ZnSnP_2$	275					1.45					
$CdSiP_2$		0.103				2.2	1000				
$CdGeP_2$	289					1.8					
$CdSnP_2$	270					1.5					
$ZnSiAs_2$	290					1.7		50			
$ZnGeAs_2$	271			-14.4		0.85					
$ZnSnAs_2$	252			-18.4		0.65		300			Disorders at 910 K
$CdSiAs_2$		0.143				1.6					
$CdGeAs_2$	266			-23.4		0.53	70	25			Disorders at 903 K
$CdSnAs_2$	247		13.7	-21.5		0.26	22000	250			

Substance	Heat of formation at 300 K (kJ/mol)	Volume compressibility (10^{-10} m²/N)	Static dielectric constant	Atomic magnetic susceptibility (10^{-6} cgs)	Index of refraction	Minimum room temperature energy gap (eV)	Electron mobility at 300 K (cm²/V·s)	Hole mobility at 300 K (cm²/V·s)	Optical transition[a]	Breakdown voltage (kV/mm)	Remarks
2.1.5. Other Ternary Semiconductors with Tetrahedral Coordination											
II_2-IV-VI_3 Compounds											
Cu_2SiS_3 (H.T.)											Wurtzite
Cu_2SiS_3 (L.T.)											Tetragonal
Cu_2SiTe_3											Cubic
Cu_2GeS_3 (H.T.)				-18.7							Cubic
Cu_2GeS_3 (L.T.)							360				Tetragonal
Cu_2GeSe_3	211.5			-21.3		0.94	238				Same
Cu_2GeTe_3	190.2			-23.4							Same
Cu_2SnS_3				-18.2		0.91	405				Cubic
$CuSnSe_3$				-21.0		0.66	870				Cubic
Cu_2SnTe_3				-28.4							Cubic
Ag_2GeSe_3				-29.6		0.91 (77K)					
Ag_2SnSe_3				-29.5		0.81					
Ag_2GeTe_3				-31.4		0.25					
Ag_2SnTe_3				-31.0		0.08					
II_3-V-VI_4 Compounds											
Cu_3PS_4											Enargite
Cu_3AsS_4	269.6			-15.8		1.24					
Cu_3AsSe_4	161.3			-13.1		0.88					Famatinite
Cu_3SbS_4				-8.3		0.74					Famatinite
Cu_3SbSe_4	127.1			-20.5		0.31					
II-IV_2-V_3 Compounds											
$CuSi_2P_3$											El
$CuGe_2P_3$		0.12				0.90					El
$AgGe_2P_3$											
2.1.6. "Defect Chalcopyrite" Structure Compounds (Strukturbericht Symbol E3, Space Group $I\bar{4} - S_4^2$)											
$ZnAl_2Se_4$											
$ZnAl_2Te_4$ (?)											
$ZnGa_2S_4$ (?)						≈3.4					
$ZnGa_2Se_4$ (?)						≈2.2					
$ZnGa_2Te_4$ (?)						1.35					
$ZnIn_2Se_4$	206					1.82	35				
$ZnIn_2Te_4$	198					1.2					
$CdAl_2S_4$											
$CdAl_2Se_4$											
$CdAl_2Te_4$ (?)											
$CdGa_2S_4$	256					3.44	60				
$CdGa_2Se_4$	216					2.43	33				
$CdGa_2Te_4$											
$CdIn_2Te_4$	195					(1.26 or 0.9)	4000				
$HgAl_2S_4$											
$HgAl_2Se_4$											
$HgAl_2Te_4$ (?)											
$HgGa_2S_4$	249					2.84					
$HgGa_2Se_4$	204					1.95	400				
$HgIn_2Se_4$	196					0.6	290				
$HgIn_2Te_4$ (?)	188					0.86	200				
2.1.7. Other Adamantine Compounds											
α–SiC			10.2	-6.4	2.67	2.86	400				6H structure
$Hg_5Ga_2Te_8$											B_3 with superlattice
$Hg_5In_2Te_8$						0.7	2000				B_3 with superlattice
$CdIn_2Se_4$						1.55					
2.2. Octahedral Semiconductors											
2.2.1. Halite Structure Semiconductors (Strukturbericht Symbol B1, Space Group Fm3m – O_h^5)											
GeTe											
SnSe											
SnTe											
PbS	435					0.5	600	600			

Solids

Substance	Heat of formation at 300 K (kJ/mol)	Volume compressibility (10^{-10} m²/N)	Static dielectric constant	Atomic magnetic susceptibility (10^{-6} cgs)	Index of refraction	Minimum room temperature energy gap (eV)	Electron mobility at 300 K (cm²/V·s)	Hole mobility at 300 K (cm²/V·s)	Optical transition[a]	Breakdown voltage (kV/mm)	Remarks
PbSe	393		161			0.37	1000	900			
PbTe	393		280			0.26	1600	600			Altaite
			360			0.25					

2.2.2. Selected Other Binary Halites

BiSe											
BiTe						0.4					
EuSe											
GdSe						1.8	4				
NiO						2.0 or 3.7	100				
CdO	531					2.5					
SrS						4.1					

2.3. Other Semiconductors

2.3.1. Antifluorite Structure Compounds (Fm3m – O_h^5)

Mg_2Si	79.08					0.77	405	70			
Mg_2Ge						0.74	520	110			
Mg_2Sn	76.57					0.36	320	260			
Mg_2Pb	52.72					0.1					

2.3.2. Tetradymite Structure Compounds ($R\bar{3}$ – D_{3d}^5)

Sb_2Te_3						0.3		360			
Bi_2Se_3						0.35	600				
Bi_2Te_3						0.21	1140	680			R3m (166)

2.3.3. Skutterudite Structure Compounds (Im3 – T_h^5)

CoP_3						0.43					
$CoAs_3$						0.69		~4000			
$CoSb_3$						0.63	70	~3000			
RhP_3								700			
$RhAs_3$						0.85		~3000			
$RhSb_3$						0.80		~7000			
$IrSb_3$						1.18		1500			

2.3.4. Selected Multinary Compounds

$AgSbSe_2$						0.58					
$AgSbTe_2$ (or $Ag_{19}Sb_{29}Te_{52}$)						0.7, 0.27					
$AgBiS_2$ (H.T.)											
$AgBiSe_2$ (H.T.)											
$AgBiTe_2$ (H.T.)											
Cu_2CdSnS_4						1.16	<2				

2.3.5. Some Elemental Semiconductors

B	397.1			-6.7	3.4	1.55	10				
Se (gray)			6.6	-22.1	2.5	1.5		5			$P3_121$(152)
			(0.1 GHz)								
Te				-39.5	3.3	0.33	1700	1200			Same

[a] i = indirect, d = direct, s = semimetal.

TABLE 3. Semiconducting Properties of Selected Materials

Substance	Min. energy gap at 300 K (eV)	Min. energy gap at 0 K (eV)	10^4 dE/dT (eV K^{-1})	dE/dT (meV kbar^{-1})	Electron effective mass m_{da} (m_o)	Electron mobility μ_a (cm^2 V^{-1} s^{-1})	Electron temp. dependence $-x$	Hole effective mass m_{dp} (m_a)	Hole mobility μ_p (cm^2 V^{-1} s^{-1})	Hole temp. dependence $-x$
Elements										
Si	1.110	1.169	-2.8	-1.41	1.026	1900	2.6	0.056	500	2.3
Ge	0.664	0.744	-3.7	5.1	0.0823	3800	1.66	0.0438	1820	2.33
α-Sn	0.08	0.094	-0.5		0.0236	2500	1.65	0.195	z	2.0
Se	2.11	2.48								
Te	0.335				0.08	1100		0.19	560	
III-V Compounds										
AlAs	2.2	2.3				1200			420	
AlSb	1.6	1.7	-3.5	-1.6	0.09	200	1.5	0.4	500	1.8
GaP	2.272	2.350	-3.7	10.5	0.35	300	1.5	0.5	150	1.5
GaAs	1.441	1.579	-3.9	11.3	0.068	9000	1.0	0.5	500	2.1
GaSb	0.70	0.812	-3.7	14.5	0.050	5000	2.0	0.23	1400	0.9
InP	1.34	1.4236	-2.9	9.1	0.067	5000	2.0		200	2.4
InAs	0.356	0.418	-3.4	10.0	0.022	33 000	1.2	0.41	460	2.3
InSb	0.180	0.235	-2.8	15.7	0.014	78 000	1.6	0.4	750	2.1
II-VI Compounds										
ZnO	3.2	3.4376	-9.5	0.6	0.38	180	1.5			
ZnS	3.80	3.91	-4.7	-5.8		180			5(400 °C)	
ZnSe	2.713	2.820	-4.5	0.7		540			28	
ZnTe	2.26	2.391	-5.2	8.3		340			100	
CdO	1.20		-6		0.1	120				
CdS	2.485	2.585	-4.1	4.5	0.165	400		0.8		
CdSe	1.751	1.841	-3.6	5.0	0.13	650	1.0	0.6		
CdTe	1.43	1.606	-5.4	8	0.14	1200		0.35	50	
HgSe	-0.061				0.030	20 000	2.0			
HgTe	-0.141	-0.3025			0.017	25 000		0.5	350	
Halite Structure Compounds										
PbS	0.41	0.286	4		0.16	800		0.1	1000	2.2
PbSe	0.278	0.145	4		0.3	1500		0.34	1500	2.2
PbTe	0.310	0.187	4	-7	0.21	1600		0.14	750	2.2
Others										
ZnSb	0.50	0.56			0.15	10				1.5
CdSb	0.459	0.57	-5.4		0.15	300			2000	1.5
Bi_2S_3	1.3	1.45				200			1100	
Bi_2Se_3	0.160					600			675	
Bi_2Te_3	0.13		-0.95		0.58	1200	1.68	1.07	510	1.95
Mg_2Si		0.77	-6.4		0.46	400	2.5		70	
Mg_2Ge	0.54	0.74	-9			280	2		110	
Mg_2Sn	0.18	0.36	-3.5		0.37	320			260	
Mg_2Sb_2		0.32				20			82	
Zn_3As_2	0.93					10	1.1		10	
Cd_3As_2	0.55				0.046	100 000	0.88			
GaSe	2.021	2.1275	-3.8						20	
GaTe	1.694	1.799	-3.6			14				
InSe	1.172	1.32				900				
TlSe	0.745		-3.9		0.3	30		0.6	20	1.5
$CdSnAs_2$	0.26				0.05	25 000	1.7			
Ga_2Te_2	1.22	1.55	-4.8							
α-In_2Te_2	0.92	1.2			0.7				50	1.1
β-In_2Te_2	1.0								5	
$Hg_5In_2Te_8$	0.5								11 000	
SnO_2	3.47	3.596							78	

Solids

TABLE 4.1. Band Properties of Semiconductors: Valence Bands

Substance	Heavy holes[a]	Light holes[a]	"Split-off" band holes[a]	Energy separation of "split-off" band (eV)	Measured (light) hole mobility ($cm^2/V \cdot s$)
Semiconductors with Valence Band Maximum at the Center of the Brillouin Zone ("Γ")					
Si	0.52	0.16	0.25	0.044	500
Ge	0.34	0.043	0.08	0.3	1820
Sn	0.3				2400
AlAs					
AlSb	0.4			0.7	550
GaP				0.13	100
GaAs	0.8	0.12	0.20	0.34	400
GaSb	0.23	0.06		0.7	1400
InP				0.21	150
InAs	0.41	0.025	0.083	0.43	460
InSb	0.4	0.015		0.85	750
CdTe	0.35				50
HgTe	0.5				350

[a] Band curvature effective mass expressed as fraction of free electron mass.

TABLE 4.2. Band Properties of Semiconductors: Multiple Band Maxima

Substance	Number of equivalent valleys and direction	Longitudinal[a] m_L	Transverse[a] m_T	Anistrophy $K = m_L/m_T$	Measured (light) hole mobility ($cm^2/V \cdot s$)
Semiconductors with Multiple Band Maxima					
PbSe	4 "L" [111]	0.095	0.047	2.0	1500
PbTe	4 "L" [111]	0.27	0.02	10	750
Bi_2Te_3	6	0.207	~0.045	4.5	515

[a] Band curvature effective mass expressed as fraction of free electron mass.

TABLE 4.3. Band Properties of Semiconductors: Conduction Bands

Substance	Energy gap (eV)	Effective mass (m_o)	Mobility ($cm^2/V \cdot s$)	Comments
Single Valley Semiconductors				
GaAs	1.35	0.067	8500	3(or 6?) equivalent [100] valleys 0.36 eV above this maximum with a mobility of ~50.
InP	1.27	0.067	5000	3(or 6?) equivalent [100] valleys 0.4 eV above this minimum.
InAs	0.36	0.022	33,000	Equivalent valleys ~1.0 eV above this minimum.
InSb	0.165	0.014	78,000	
CdTe	1.44	0.11	1000	4(or 8?) equivalent [111] valleys 0.51 eV above this minimum.

TABLE 4.4. Band Properties of Semiconductors: Conduction Bands

Substance	Energy gap (eV)	Number of equivalent valleys and direction	Longitudinal[a] m_L	Transverse[a] m_T	Anisotropy $K = m_L/m_T$
Multivalley Semiconductors					
Si	1.107	6 in [100] "Δ"	0.00	0.192	4.7
Ge	0.67	4 in [111] at "L"	1.588	0.0815	19.5
GaSb	0.67	as Ge (?)	~1.0	~0.2	~5
PbSe	0.26	4 in [111] at "L"	0.085	0.05	1.7
PbTe	0.25	4 in [111] at "L"	0.21	0.029	5.5
Bi_2Te_3	0.13	6			~0.05

[a] Band curvature effective mass expressed as fraction of free electron mass.

TABLE 5. Resistivity of Semiconducting Minerals

Mineral	$\rho/\Omega\cdot m$
Diamond (C)	2.7
Sulfides	
Argentite, Ag_2S	1.5 to 2.0 × 10^{-3}
Arsenopyrite, FeAsS	20 to 300 × 10^{-6}
Bismuthinite, Bi_2S_3	3 to 570
Bornite, $Fe_2S_3{\cdot}nCu_2S$	1.6 to 6000 × 10^{-6}
Chalcocite, Cu_2S	80 to 100 × 10^{-6}
Chalcopyrite, Fe_2S_3 Cu_2S	150 to 9000 × 10^{-6}
Cobaltite, CoAsS	6.5 to 130 × 10^{-3}
Covellite, CuS	0.30 to 83 × 10^{-6}
Enargite, Cu_3AsS_4	0.2 to 40 × 10^{-3}
Galena, PbS	6.8 × 10^{-6} to 9.0 × 10^{-2}
Gersdorfite, NiAsS	1 to 160 × 10^{-6}
Glaucodote, (Co, Fe)AsS	5 to 100 × 10^{-6}
Haverite, MnS_2	10 to 20
Marcasite, FeS_2	1 to 150 × 10^{-3}
Metacinnabarite, HgS	2 × 10^{-6} to 1 × 10^{-3}
Millerite, NiS	2 to 4 × 10^{-7}
Molybdenite, MoS_2	0.12 to 7.5
Pentlandite, $(Fe, Ni)_4S_4$	1 to 11 × 10^{-6}
Pyrite, FeS_2	1.2 to 600 × 10^{-3}
Pyrrhotite, Fe_7S_4	2 to 160 × 10^{-6}
Sphalerite, ZnS	2.7 × 10^{-3} to 1.2 × 10^{4}
Antimonide	
Dyscrasite, Ag_3Sb	0.12 to 1.2 × 10^{-6}
Arsenides	
Allemonite, $SbAs_3$	70 to 60,000
Lollingite, $FeAs_2$	2 to 270 × 10^{-6}
Nicollite, NiAs	0.1 to 2 × 10^{-6}
Skutterudite, $CoAs_3$	1 to 400 × 10^{-6}
Smaltite, $CoAs_2$	1 to 12 × 10^{-6}
Tellurides	
Altaite, PbTe	20 to 200 × 10^{-6}
Calavarite, $AuTe_2$	6 to 12 × 10^{-6}
Coloradoite, HgTe	4 to 100 × 10^{-6}
Hessite, Ag_2Te	4 to 100 × 10^{-6}
Nagyagite, $Pb_6Au(S,Te)_{14}$	20 to 80 × 10^{-6}
Sylvanite, $AgAuTe_4$	4 to 20 × 10^{-6}
Antimony-sulfur compounds	
Berthierite, $FeSb_2S_4$	0.0083 to 2.0
Boulangerite, $Pb_5Sb_3S_{11}$	2 × 10^{3} to 4 × 10^{4}
Cylindrite, $Pb_3Sn_4Sb_2S_{14}$	2.5 to 60
Franckeite, $Pb_5Sn_3Sb_2S_{14}$	1.2 to 4
Hauchecornite, $Ni_4(Bi, Sb)_2S_{14}$	1 to 83 × 10^{-6}
Jamesonite, $Pb_4FeSb_6S_{14}$	0.020 to 0.15
Tetrahedrite, Cu_3SbS_3	0.30 to 30,000
Oxides	
Braunite, Mn_2O_3	0.16 to 1.0
Cassiterite, SnO_2	4.5 × 10^{-4} to 10,000
Cuprite, Cu_2O	10 to 50
Hollandite, (Ba, Na, K) Mn_8O_{16}	2 to 100 × 10^{-3}
Ilmenite, $FeTiO_3$	0.001 to 4
Magnetite, Fe_3O_4	52 × 10^{-6}
Manganite, MnO OH	0.018 to 0.5
Melaconite, CuO	6000
Psilomelane, $BaMn_9O_{18}{\cdot}2H_2O$	0.04 to 6000
Pyrolusite, MnO_2	0.007 to 30
Rutile, TiO_2	29 to 910
Uraninite, UO_2	1.5 to 200

References

1. Beer, A. C., *Galvanomagnetic Effects in Semiconductors*, Academic Press, New York, 1963.
2. Goryunova, N. A., *The Chemistry of Diamond-Like Semiconductors*, MIT Press, Cambridge, MA, 1965 <https://doi.org/10.1007/978-1-4899-4944-8_1>.
3. Abrikosov, N. Kh., Bankina, V. F., Poretskaya, L. E., Shelimova, L. E., and Skudnova, E.V., *Semiconducting II-VI, IV-VI, and V-VI Compounds*, Plenum Press, New York, 1969 <https://doi.org/10.1007/978-1-4899-6373-4_1>.
4. Berger, L. I. and Prochukhan, V. D., *Ternary Diamond-Like Semiconductors*, Cons. Bureau/Plenum Press, New York, 1969 <https://doi.org/10.1007/978-1-4757-0040-4_1>.
5. Shay, J. L. and Wernick, J. H., *Ternary Chalcopyrite Semiconductors: Growth, Electronic Properties, and Applications*, Pergammon Press, 1975 <https://doi.org/10.1016/B978-0-08-017883-7.50012-3>.
6. Bergman, R., *Thermal Conductivity in Solids*, Clarendon, Oxford, 1976.
7. *Handbook of Semiconductors*, Vol. 1, Moss, T. S. and Paul, W., Eds., *Band Theory and Transport Properties*; Vol. 2, Moss, T. S. and Balkanski, M., Eds., *Optical Properties of Solids*; Vol. 3, Moss, T. S. and Keller, S. P., Eds., *Materials Properties and Preparation*, North-Holland, Amsterdam, 1980.
8. Böer, K. W., *Survey of Semiconductor Physics*, Van Nostrand Reinhold, 1990 <https://doi.org/10.1007/978-1-4615-9744-5_1>.
9. Rowe, D. M., Ed., *CRC Handbook of Thermoelectrics*, CRC Press, Boca Raton, FL, 1995 <https://doi.org/10.1201/9781420049718>.
10. Berger, L. I., *Semiconductor Materials*, CRC Press, Boca Raton, FL, 1997.
11. Glazov, V. M., Chizhevskaya, S. N., and Glagoleva, N. N., *Liquid Semiconductors*, Plenum Press, New York, 1969 <https://doi.org/10.1007/978-1-4899-6451-9_1>.
12. Phillips, J. C., *Bonds and Bands in Semiconductors*, Academic Press, New York, 1973 <https://doi.org/10.1016/B978-0-12-553350-8.50005-1>.
13. Harrison, W. A., *Electronic Structure and the Properties of Solids*, Freeman, San Francisco, 1980.
14. Balkanski, M., Ed., *Optical Properties of Solids*, North-Holland, Amsterdam, 1980.
15. *Landolt-Börnstein. Numerical Data and Functional Relationships in Science and Technology, New Series, Group III: Crystal and Solid State Physics*, Hellwege, K.-H. and Madelung, O., Eds., Volumes 17 and 22, Springer Verlag, Berlin, 1984 (and further).
16. Shklovskii, B. L. and Efros, A. L., *Electronic Processes in Doped Semiconductors*, Springer Verlag, Berlin, 1984 <https://doi.org/10.1007/978-3-662-02403-4>.
17. Cohen, M. L. and Chelikowsky, J. R., *Electronic Structure and Optical Properties of Semiconductors*, Springer Verlag, New York, 1988 <https://doi.org/10.1007/978-3-642-97080-1>.
18. Glass, J. T., Messier, R. F., and Fujimori, N., Eds., *Diamond, Silicon Carbide, and Related Wide Bandgap Semiconductors*, MRS Symposia Proc. 1652, Mater. Res. Soc., Pittsburgh, 1990.
19. Palik, E., Ed., *Handbook of Optical Constants of Solids II*, Academic Press, New York, 1991.
20. Reed, M., Ed., *Semiconductors and Semimetals*, Volume 35, Academic Press, Boston, 1992.

21. Haug, H. and Koch, S. W., *Quantum Theory of the Optical and Electronic Properties of Semiconductors*, 2nd Edition, World Scientific, Singapore, 1993 <https://doi.org/10.1142/1977>.
22. Lockwood, D. J., Ed., *Proc. 22nd Intl. Conf. on the Physics of Semiconductors, Vancouver, 1994*, World Scientific, Singapore, 1994 <https://doi.org/10.1142/9789814533638>.
23. Morelli, D. T., Caillat, T., Fleurial, J.-P., Borschchevsky, A., Vandersande, J., Chen, B., and Uher, C., *Phys. Rev.*, B51, 9622, 1995 <https://doi.org/10.1103/PhysRevB.51.9622>.
24. Caillat, T., Borshchevsky, A., and Fleurial, J.-P., *J. Appl. Phys.*, 80, 4442, 1996 <https://doi.org/10.1063/1.363405>.
25. Fleurial, J.-P.,Caillat, T., and Borshchevsky, A., *Proc. XVI Intl. Conf. Thermoelectrics*, Dresden, Germany, August 26–29, 1997 (in print).
26. Borshchevsky, A. et al., U.S. Patents 5,610,366 (March 1977) and 5,831,286 (March 1998).
27. Jarrendahl, K. and Davis, R. F., *Semiconductors and Semimetals*, Vol. 52, Y. S. Park, Ed., 1998, pp. 1–20 <https://doi.org/10.1016/S0080-8784(08)62843-4>.
28. Bettini, M., *Solid State Comm.*, 13, 599, 1973 <https://doi.org/10.1016/S0038-1098(73)80021-3>.
29. Chen, A. and Sher, A., *Semiconductor Alloys, Physics and Material Engineering*, Plenum Press, New York, 1995.
30. Holloway, P. H. and McGuire, G. E., Eds., *Handbook of Compound Semiconductors*, Noyes Publ., Park Ridge, NJ, 1995.
31. Madelung, O., Ed., *Semiconductors: Group IV Elements and III-V Compounds (Data in Science and Technology)*, Springer-Verlag, Berlin, Heidelberg, 1991.
32. Madelung, O., Ed., *Semiconductors: Other Than Group IV Elements and III-V Compounds (Data in Science and Technology)*, Springer-Verlag, Berlin, Heidelberg, 1992.
33. Levinshtein, M., Rumyantsev, S., and Shur, M., Eds., Handbook Series on Semiconductor Parameters. Vol. 1, *Si, Ge, C (Diamond), GaAs, GaP, GaSb, InAs, InP, InSb*, World Scientific, Singapore, 1996 <https://doi.org/10.1142/9789812832078_0001>.
34. Levinshtein, M., Rumyantsev, S., and Shur, M., Eds., Handbook Series on Semiconductor Parameters. Vol. 2, *Ternary and Quaternary III-V Compounds*, World Scientific, Singapore, 1996 <https://doi.org/10.1142/2046-vol2>.
35. *Physical Properties of Semiconductors*, NSM Archive: <http://www.ioffe.ru/SVA/NSM/Semicond/>.
36. Kumar, V., et al, *Indian J. Appl. Phys.*, 53, 429, 2015.
37. Stepanov, S. I. et al, *Rev. Adv. Mater. Sci.*, 44, 63, 2016.
38. Li, S. et al, *Appl. Phys. Lett.*, 115, 69, 2019.
39. Goudarzi, M., *J. Mol. Liquids*, 219, 120, 2016.

SELECTED PROPERTIES OF SEMICONDUCTOR SOLID SOLUTIONS

L. I. Berger

Alloy system	Limits of solubility	Energy gap in eV (300 K)	Remarks, references
	Adamantine Semiconductors IV-IV		
Si_xGe_{1-x}	$0\le x\le 1$	$0.8941+0.0421x+0.1691x^2$ $0.7596+1.0860x+0.3306x^2$	Transition Γ - X [Ref.1] Trans. Γ - L [Ref. 1]
	Adamantine Semiconductors III-V/III-V		
Common Anion			
$Al_xGa_{1-x}N$	$0\le x\le 1$		Wurtzite Structure [Refs. 2 and 3]
$Al_xGa_{1-x}P$	$0\le x\le 0.5$	$2.28+0.16x$	[Ref. 2]
$Al_xIn_{1-x}P$	$0\le x\le 0.44$	at Γ: $134+2.23x$; at X: $2.24+0.18x$	[Ref. 2]
$Al_xGa_{1-x}As$	$0\le x\le 0.5$	$1.42=0.75x$ [Ref.3]; $1.424+1.429x-0.14x^2$ [Ref.4]	
$Al_xIn_{1-x}As$	$0\le x\le 1$	at Γ: $0.37+1.91x+0.74x^2$; at X: $1.8+0.4x$	[Refs. 2 and 6]
$Al_xGa_{1-x}Sb$	$0\le x<\le 1$	$0.73+1.10x+0.47x^2$	Trans. Γ_{8v} - Γ_{6c} [Ref. 2]
$Al_xIn_{1-x}Sb$	$0\le x\le 1$		[Ref. 6]
$Ga_xIn_{1-x}N$	$0\le x\le 1$	$1.950+1.487x-1.000x(1-x)$	Wurtzite [Refs. 8 and 10]
$Ga_xIn_{1-x}P$	$0\le x\le 1$		[Ref. 2]
$Ga_xIn_{1-x}As$	$0\le x\le 1$	$0.360+0.629x+0.436x^2$	[Ref. 5]
$Ga_xIn_{1-x}Sb$	$0<x<1$	$0.235+0.1653x+0.413x^2$	[Ref. 2, see also Ref. 9]
Common Cation			
GaN_xAs_{1-x}	$0\le x\le 0.05$	$1.42-9.9x$	[Ref. 2]
GaP_xAs_{1-x}	$0<x<1$	$2.270-0.846x$	[Ref. 2]
GaP_xAs_{1-x}	$0\le x\le 0.05$	$1.515+1.172x+0.186x^2$ $1.9715+0.144x+0.211x^2$	(at 2 K, Γ–Γ) [Ref. 7] [Ref. 2]
$GaAs_xSb_{1-x}$	$0\le x\le 0.45$, $0.6\le x\le 1$	$1.43-1.9x+1.2x^2$	[Ref. 5]
InP_xAs_{1-x}	$0<x<1$	$0.356+0.675x+0.32x^2$	[Ref. 2]
	Adamantine Binary Semiconductors II-VI/II-VI [Refs. 3 and 6]		
Common Anion			
$Zn_xCd_{1-x}S$	$0\le x\le 1$		Wurtzite Structure
$Zn_xHg_{1-x}S$	$0\le x\le 1$		
$Cd_xHg_{1-x}S$	$0\le x\le 1$		Wurtzite Structure at $x<0.6$
$Zn_xCd_{1-x}Se$	$0.7\le x\le 1$		
$Zn_xHg_{1-x}Se$	$0\le x\le 1$		
$Cd_xHg_{1-x}Se$	$0\le x\le 0.7$ and $0.75\le x^*\le 1$		x^*- Wurtzite Structure
$Zn_xCd_{1-x}Te$	$0\le x\le 1$		
$Zn_xHg_{1-x}Te$	$0\le x\le 1$		
$Cd_xHg_{1-x}Te$	$0\le x\le 1$		
Common Cation			
ZnS_xSe_{1-x}	$0\le x\le 1$		
ZnS_xTe_{1-x}	$0\le x\le 0.1$ and $0.9\le x^*\le 1$		x^*- Wurtzite Structure
$ZnSe_xTe_{1-x}$	$0\le x\le 1$		
CdS_xSe_{1-x}	$0\le x\le 1$		Wurtzite Structure
CdS_xTe_{1-x}	$0\le x\le 0.25$ and $0.8\le x^*\le 1$		x^*- Wurtzite Structure
$CdSe_xTe_{1-x}$	$0\le x\le 0.4$ and $0.6\le x^*\le 1$		x^*- Wurtzite Structure
HgS_xSe_{1-x}	$0\le x\le 1$		
HgS_xTe_{1-x}	$0\le x\le 1$		
$HgSe_xTe_{1-x}$	$0\le x\le 1$		
	Quaternary Adamantine Semiconductors II-VI/III-V [Ref. 6]		
$(ZnS)_x(AlP)_{1-x}$	$0.99\le x\le 1$		
$(ZnSe)_x(GaAs)_{1-x}$	$0\le x<1$		
$(CdTe)_x(InAs)_{1-x}$	$0<x\le 0.2$ and $0.7\le x\le 1$		
$(CdTe)_x(AlSb)_{1-x}$	$0\le x\le 1$		
$(HgTe)_x(InAs)_{1-x}$	$0\le x\le 1$		

Alloy system	Limits of solubility	Energy gap in eV (300 K)	Remarks, references
	Quaternary Adamantine Semiconductors III_x-III_{1-x}-V_y - V_{1-y}		
$Ga_xIn_{1-x}As_yP_{1-y}$	$0 \le x \le 1$, $0 \le x \le 1$	$1.35+0.668x-1.068y+0.758x^2+0.078y^2-0.069xy-0.322x^2y+0.03xy^2$	[Refs. 2 and 6]
	Quaternary Adamantine Semiconductors III_{1-x-y}-III_x-III_y-V		
$Al_xGa_yIn_{1-x-y}Sb$	$0 \le x \le 1$, $0 \le y \le 1$	$0.095+1.76x+0.28y+0.345(x^2+y^2)+0.085(1-x-y)^2+xy\ (1-x-y)(23-28y)$	[Refs. 2 and 6]

References

1. Krishnamurti, S., Sher, A., and Chen, A. *Appl. Phys. Lett.* 47, 160, 1985.
2. Madelung, O., Ed., *Semiconductors Group IV Elements and III-V Compounds*, Springer, 1991; *Semiconductors Other Than Group IV Elements and III-V Compounds*, Springer, 1992.
3. Goryunova, N. A., Kesamanly, F. P., and Nasledov, D. N., *Semiconductors and Semimetals,* Vol. 4, 1968, 413.
4. El Allali, M., Sorensen, C. B., Veje, E., and Tideman-Peterson, P., *Phys. Rev. B* 48, 4398, 1993.
5. Nahorny, R. E., Pollack, M. A., Johnson, W. D., and Barns, R. L. *Appl. Phys. Lett.* 33, 695, 1978.
6. Goryunova, N. A., *Multicomponent Diamond-Like Semiconductors*, Sov. Radio, Moscow, 1968 (in Russian).
7. Capizzi, M., Modesti, S., Martelli, F., and Frova, A., *Solid State Comm.* 39, 333, 1981.
8. Nakamura, S., Pearton, S., and Fasol, G., *The Blue Laser Diode,* 2nd Ed., Springer, 2000.
9. Roth, A. P., Keeler, W. J., and Fortin, E. *Canad. J. Phys.* 58, 560, 1980.
10. Wu, J., Walukiewicz, Yu, K. M., Ager, J. W., Haller, E. E., Lu, H., and Schaff, W. J., *Appl. Phys. Lett.* 80, 4741, 2002.

PROPERTIES OF ORGANIC SEMICONDUCTORS

L. I. Berger

Substance	Energy gap E/eV	Electrical resistivity[a] ρ/Ω cm	Mobility μ/cm^2 V^{-1} s^{-1}	Sign of majority carriers	Temperature range °C	Ref.
Metal-Free Molecular Crystals						
3-Acetylamino-*N*-methylphthalimide	3.46				67 to 100	1
3-Acetylamino-*N*-phenylphthalimide	3.50				54 to 124	1
4-Amino-*N*-cyclomethylphthalimide	2.90				73 to 100	1
4-Aminophthalimide	2.78				123 to 151	1
Acridine	3.90					1
Anthanthrene	1.67	$1.5 \cdot 10^{19}$			40 to 105	1
Anthanthrene	0.84	$1.5 \cdot 10^{19}$ (15 °C)				2
Anthanthrone	1.70	$7.7 \cdot 10^{18}$			20 to 150	1
Anthracene	0.83	$1.3 \cdot 10^{14}$ (15 °C)				2
Anthracene	2.50	$1.5 \cdot 10^{11}$	2.3	+	20 to 130	1
Anthracene	3.88 to 4.1	$>10^{15}$	1.74(n), 2.07(p)	+ & −		4
1,2-Benzanthracene	1.04	10^{16} (30 °C)				2
Benzanthrone	3.12	$1.6 \cdot 10^{16}$				1
Benzene (liquid)	0.41					2
Benzene (amorphous)	0.84	10^{15}			−14 to 5	1
Benzene (cryst.)	7		2	−	−23	4
Benzimidazole	3.0 to 4.0	$5 \cdot 10^{13}$			84 to 144	1
Benzophenone	3.34				−23 to 14	1
Benzo[*f*]quinoline	2.77				30 to 50	1
Benzo[*h*]quinoline	2.72					1
3-Benzoylamino-*N*-methylphthalimide	3.28				84 to 112	1
Benzpentacene	1.72				0 to 150	1
Biphenyl	1.46	$1.7 \cdot 10^{15}$ (50 °C)				2
Biphenyl	1.45				20 to 70	1
o-Chloranil	3.0	10^{15}				1
p-Chloranil	0.61					1
Chlorpromazine	2.1	10^{12} (32 °C)		+ & −	32 to 80	1
Chrysene	1.1	$4 \cdot 10^{19}$ (15 °C)				2
Chrysene	2.20	$4 \cdot 10^{19}$			25 to 90	1
Circumanthracene	1.8	$6 \cdot 10^{12}$				1
Coronene	1.7	$1.7 \cdot 10^{17}$			60 to 80	1
Coronene	0.85	$1.7 \cdot 10^{17}$ (15 °C)				2
Cyananthrone	0.20	$1.2 \cdot 10^{7}$			30 to 125	1
1,6-Diaminopyrene	0.6	10^{8}				
Dibenzpentacene	1.50				0 to 150	1
Dinaphthopyrene	1.60				25 to 90	1
1,8-Diphenyl-1,3,5,7-octatetraene	1.7				72 to 191	1
Diphenylpentacene	1.60				0 to 150	1
4,4′-Diphenylstilbene	1.56				160 to 280	1
4,4′-Diphenylstilbene	0.80					2
Ferrocene	1.22	10^{14}		+		1
Flavanthrone	0.70	$1.4 \cdot 10^{11}$				1
Fluorene	2.7					2
Fluorene	1.4					2
Fluoridine	1.6	$6 \cdot 10^{13}$		+	20 to 140	1
Hexacene	0.57	$3.8 \cdot 10^{10}$ (50 °C)				2
Hexacene	1.3					1
Hexamethylbenzene	1.86			+	20 to 140	1
3-Hydroxy-*N*-methylphthalimide	3.80				60 to 91	1
Imidazole	2.6	10^{11}			28 to 68	1

Substance	Energy gap E/eV	Electrical resistivity[a] ρ/Ω cm	Mobility μ/cm^2 V^{-1} s^{-1}	Sign of majority carriers	Temperature range °C	Ref.
Indanthrazine	0.66	$1.4 \cdot 10^{15}$			30 to 125	1
Indanthrone	0.64	$7.5 \cdot 10^{14}$			30 to 125	1
Indanthrone (black)	0.56	$2.5 \cdot 10^{8}$			30 to 125	1
Mesitylene (liquid)	0.19					2
Mesonaphthodianthracene	0.6	$4.0 \cdot 10^{18}$ (15 °C)				2
Mesonaphthodianthrene	1.48				45 to 250	1
Mesonaphthodianthrone	0.86				5 to 110	1
3-Methoxy-*N*-methyl-phthalimide	3.18				54 to 78	1
Naphthacene	1.7	$1 \cdot 10^{15}$				1
Naphthalene	3.5	10^{14}			27 to 47	1
Naphthalene	1.15	$2.8 \cdot 10^{14}$ (50 °C)				2
Naphthalene	4.9 to 5.1		0.64(n), 1.50(p)	+ & –		4
m-Naphthodianthrene	1.20	$4 \cdot 10^{18}$			40 to 150	1
m-Naphthodianthrone	1.30	$1.5 \cdot 10^{18}$			40 to 150	1
β-Naphthol	2.36	$2 \cdot 10^{5}$			60 to 110	1
β-Naphthoquinoline	2.77					1
1-Naphthylamine	2.2				25 to 42	1
1-Naphthylamine picrate	2.7				28 to 98	1
2-Naphthylphenyl sulphone	3.5				67 to 102	1
1-Nitronaphthalene	2.5				25 to 44	1
Ovalene	1.13	$2.3 \cdot 10^{15}$				1
Pentacene	0.58	$2.4 \cdot 10^{9}$ (50 °C)				2
Pentacene	1.5	$6 \cdot 10^{13}$			20 to 140	1
Perylene	2.1	$4.1 \cdot 10^{13}$			40 to 100	1
Perylene	3.10		5.53(n), 87.4(p)	+ & –	–213	4
Phenanthrene	1.15	$1.3 \cdot 10^{14}$			12 to 72	1
Phenanthrene	0.65					2
1,10-Phenanthroline	2.73				50 to 90	1
Phenazine	2.1	$7 \cdot 10^{14}$ (100 °C)			98 to 143	1
Phenazine	1.1			–		4
Phenothiazine	1.6	10^{11}			50 to 150	1
Phenothiazine			2.45(n), 0.02(p)	+ & –		4
Phenylanthranilic acid	3.30				87 to 119	1
4-Phenylstilbene	1.74				140 to 220	1
4-Phenylstilbene	0.86					2
Phosphonitrilic chloride trimer	1.68	10^{15}				1
Phthalocyanine, PcH_2	1.66	10^{13}	0.1 to 0.4	+	26 to 350	1
Phthalocyanine, PcH_2	2	10^{7}	1.2(n), 1.1(p)	+ & –	100	4
Pyranthrene	1.11	$1 \cdot 10^{15}$				1
Pyranthrene	0.51	$4.5 \cdot 10^{16}$ (15 °C)				2
Pyranthrone	1.06	$3.9 \cdot 10^{15}$			40 to 150	1
Pyrene	2.02	$5 \cdot 10^{17}$		–		1
Pyrene			0.50	+		4
5,6-*N*-Pyridine-1,9-benzanthrone	1.60					2
p-Quaterphenyl	0.89	$1.0 \cdot 10^{15}$ (50 °C)				2
Quaterrylene	0.6	10^{5}		–		1
p-Quinquiphenyl	0.91	$2.0 \cdot 10^{15}$ (50 °C)				2
α-Resorcin	2.10	$2 \cdot 10^{16}$			30 to 94	1
β-Resorcin	3.27	$2 \cdot 10^{18}$			30 to 94	1
Salanil	4.1	10^{4}			20 to 40	1
p-Sexiphenyl	0.91	$7.0 \cdot 10^{14}$ (50 °C)				2
cis-Stilbene	2.4			+	at 20	1
trans-Stilbene	1.80		2.4	+	70 to 120	1
trans-Stilbene	0.91					2
trans-Stilbene	1.4			+		4
o-Terphenyl		$3 \cdot 10^{-5}$		+		1
m-Terphenyl			10^{-5}	+		1
p-Terphenyl	0.6	10^{14} (25 °C)				2
p-Terphenyl	1.2		0.025	+		1

Substance	Energy gap E/eV	Electrical resistivity[a] ρ/Ω cm	Mobility μ/cm^2 V^{-1} s^{-1}	Sign of majority carriers	Temperature range °C	Ref.
p-Terphenyl			1.2(n), 0.80(p)	+ & −		4
Tetracene	0.66	$3.2 \cdot 10^{12}$ (50 °C)				2
Tetracene	1.7					1
Tetracene	3.4		0.85	+		4
1,1,10,10-Tetracyanodecapentaene	2.24	10^{13}			>68	1
1,1,6,6-Tetracyanohexatriene	1.54	10^{14}		−		1
Tetracyanoethylene			0.26(max)	+		4
7,7,8,8-Tetracyanoquinodimethane			0.65	−		4
1,1,8,8-Tetracyanooctatetraene	1.42	10^{12}		−		1
Tetraphenylpentacene	1.62				0 to 150	1
Tetrathiotetracene	0.46	10^4				1
Triphenodioxazine	1.65	$5 \cdot 10^{14}$		−	20 to 140	1
Triphenyldiamine			$2 \cdot 10^{-2}$		at 20	1
Violanthrene	0.85	$2.1 \cdot 10^{14}$			40 to 105	1
Isoviolanthrene	0.82	$8.4 \cdot 10^{13}$			40 to 150	1
Violanthrone	0.78	$2.3 \cdot 10^{10}$			40 to 150	1
Isoviolanthrone	0.76	$5.7 \cdot 10^9$			40 to 150	1
o-Xylene (liquid)	0.45					2
m-Xylene (liquid)	0.41					2
p-Xylene (liquid)	0.41					2
Long-Chain Compounds and Polymers						
Acrylic acid-amylproparylaniline copolymers		10^9–10^{10}				3
Acrylic acid-methylproparylaniline copolymers		10^9–10^{10}				3
Acrylic acid-octylproparylaniline copolymers		10^9–10^{10}				3
Anthrone polymers		0.28 ≥100 at 1.8 kbar ≥2 at 33 kbar				3
$[CH(AsF_5)_{0.1}]_x$		0.0005		+		3
$[CH \cdot I_{0.22}]_x$	1.9	*trans* 10^5, *cis* 10^9				3
1,6-Diacetylenes (cyclopolymerized)		10^{10}–10^{14}				3
Ionene elastomers		$2.7 \cdot 10^7$ to $2.2 \cdot 10^8$			−80 to 60	3
1,3,4-Oxydiazole polymers	0.81	$3 \cdot 10^{12}$			20 to 140	3
Oxypyrrole polymer films		0.125				3
Phenylformaldehyde polymeric pyrolysates						3
a) Pyrolysis temperature 600 °C		27	0.0014	−		
b) 1200 °C		0.0044	7.84	+		
Phenylthiocyanate polymers	0.5 to 0.8	10^5–10^8				3
Polyacetylene (undoped)		10^{10}				3
Polyacetylene (I_2 doped)		0.04				3
Polyacetylene (*cis*-rich, undoped)		10^7				3
trans-Polyacetylene (I_2 doped, 0.22 mole %)				+		3
Polyacrylonitrile (heat treated 700 °C)			0.01	−	−100 to 100	3
Poly-5,5′-biisatyl thiophene-indophene	air 0.84 vacuum 1.0	air $2.6 \cdot 10^9$ vacuum $3.1 \cdot 10^9$		+	20 to 140	3
Poly bis(amino)-phosphazenes	1.75	$1.8 \cdot 10^{11}$			20 to 180	3
Poly-5,5′-diisatylmetane-thiophene-indophenine	0.45	$7.3 \cdot 10^4$		+	20 to 140	3
Polyethylene	2.74				20 to 70	3
Polyethylene (low density)	0.17	$4 \cdot 10^9$			above T_g	3
Polyimide	2.84					3
Polymalonitrile	1.72			−		3
Poly(metalphthalocyanines) :Cu	0.12	$7 \cdot 10^6$				3
:Fe	0.15	$1.1 \cdot 10^6$				3
:Ni	0.46	100				3
:Sb		$3.1 \cdot 10^6$				3
:Zn	0.12	$5.3 \cdot 10^3$				3
Poly-*N*-methylpyrrole		$2 \cdot 10^6$				3
Polyoxypyrrole (black)	0.044	1790			−173 to 27	3
Polyphthalocyanines	0.01	7 to 58				3
Polypyrrole	0.01				−193 to 250	3

Substance	Energy gap E/eV	Electrical resistivity[a] ρ/Ω cm	Mobility μ/cm^2 V^{-1} s^{-1}	Sign of majority carriers	Temperature range °C	Ref.
Polypyrroline II	1.74					3
Polyselenomethylene	0.7 to 2.62	$>10^{13}$			20 to 120	3
Poly(2-vinylpyridine):I_2 (1:2)	0.12	1000			−73 to 27	3
PVC (commercial)	2.84–3.04				$T<T_g$	3
PVC (commercial)	1.24–1.96				$T>T_g$	3
PVC (pure)	1.0±0.1				0 to 30	3
Salicylal-*N*-alkyliminate-Cu	1.62	$1.7 \cdot 10^{14}$				4
TTF-acetylacetonate polymers		$1.6 \cdot 10^{4}$				3
TTF-metal polymers		$1.6 \cdot 10^{4}$				3

Solids

References

1. F. Gutman and L. E. Lyons, *Organic Semiconductors*, John Wiley & Sons, New York, 1967.
2. Y. Okamoto and W. Brenner, *Organic Semiconductors*, Reinhold Publ. Corp., New York, 1964.
3. F. Gutman, H. Keyzer, L. E. Lyons, and R. B. Somoano, *Organic Semiconductors, Part B*, R. E. Krieger Publ. Co., Melbourne, FL, 1983.
4.. L. I. Berger, *Semiconductor Materials*, CRC Press, Boca Raton, FL, 1997.

DIFFUSION DATA FOR SEMICONDUCTORS

B. L. Sharma

The diffusion coefficient D in many semiconductors may be expressed by an Arrhenius-type relation

$$D = D_o \exp(-Q/kT)$$

where D_o is a frequency factor, Q is the activation energy for diffusion, k is the Boltzmann constant, and T is the absolute temperature. This table lists D_o and Q for various diffusants in common semiconductors.

Abbreviations used in the table are

AES – Auger Electron Spectroscopy
DLTS – Deep Level Transient Spectroscopy
SEM – Scanning Electron Microscopy
SIMS – Secondary Ion Mass Spectrometry
$D(c)$ – Concentration Dependent Diffusion Coefficient
D_{max} – Maximum Diffusion Coefficient
(f) – Fast Diffusion Component
(i) – Interstitial Diffusion Component
(s) – Slow Diffusion Component
(‖) – Parallel to c Direction
(⊥) – Perpendicular to c Direction

Solids

Semiconductor	Diffusant	Frequency factor, D_o (cm²/s)	Activation energy, Q(eV)	Temperature range (°C)	Method of measurement	Ref.
Si	H	6×10^{-1}	1.03	120–1207	Electrical and SIMS	1
	Li	2.5×10^{-3}	0.65	25–1350	Electrical	2
	Na	1.65×10^{-3}	0.72	530–800	Electrical and flame photometry	3
	K	1.1×10^{-3}	0.76	740–800	Electrical and flame photometry	3
	Cu	4×10^{-2}	1.0	800–1100	Radioactive	4
		4.7×10^{-3}	0.43 (i)	300–700	Radioactive	5
	Ag	2×10^{-3}	1.6	1100–1350	Radioactive	6
	Au	2.4×10^{-4}	0.39 (i)	700–1300	Radioactive	7
		2.75×10^{-3}	2.05 (s)			
	Be	$(D \sim 10^{-7})$	–	1050	Electrical	8
	Ca	$(D \sim 6 \times 10^{-14})$	–	1100	Electrical and SIMS	1
	Zn	1×10^{-1}	1.4	980–1270	Electrical	9
	B	2.46	3.59	1100–1250	Electrical	10
		2.4×10^{1}	3.87	840–1250	Electrical	11
	Al	1.38	3.41	1119–1390	Electrical	12
		1.8	3.2	1025–1175	Electrical	13
	Ga	3.74×10^{-1}	3.39	1143–1393	Electrical	12
		6×10^{1}	3.89	900–1050	Radioactive	14
	In	7.85×10^{-1}	3.63	1180–1389	Electrical	12
		1.94×10^{1}	3.86	1150–1242	Radioactive	15
	Tl	1.37	3.7	1244–1338	Electrical	12
		1.65×10^{1}	3.9	1105–1360	Electrical	16
	Sc	8×10^{-2}	3.2	1100–1250	Radioactive	1
	Ce	$(D \sim 3.9 \times 10^{-13})$	–	1050	SIMS	1
	Pr	2.5×10^{-7}	1.74	1100–1280	Electrical	1
	Pm	7.5×10^{-9}	1.2 (s)	730–1270	Radioactive	1
		4.2×10^{-12}	0.13 (f)			
	Er	2×10^{-3}	2.9	1100–1250	Radioactive	1
	Tm	8×10^{-3}	3.0	1100–1280	Radioactive	1
	Yb	2.8×10^{-5}	0.95	947–1097	Neutron activation	1
	Ti	1.45×10^{-2}	1.79	950–1200	DLTS	17
	C	3.3×10^{-1}	2.92	1070–1400	Radioactive	18
	Si (self)	1.54×10^{2}	4.65	855–1175	SIMS	19
		1.6×10^{3}	4.77	1200–1400	Radioactive	20
	Ge	3.5×10^{-1}	3.92	855–1000	Radioactive	21
		2.5×10^{3}	4.97	1030–1302	Radioactive	21
		7.55×10^{3}	5.08	1100–1300	SIMS	22
	Sn	3.2×10^{1}	4.25	1050–1294	Neutron activation	23
	N	2.7×10^{-3}	2.8	800–1200	Out diffusion; SIMS	1

Semiconductor	Diffusant	Frequency factor, D_o (cm²/s)	Activation energy, Q(eV)	Temperature range (°C)	Method of measurement	Ref.
	P	2.02×10^{1}	3.87	1100–1250	Electrical	10
		1.1	3.4	900–1200	Radioactive	24
		7.4×10^{-2}	3.3	1130–1405	Electrical	25
	As	6.0×10^{1}	4.2	950–1350	Radioactive	26
		6.55×10^{-2}	3.44	1167–1394	Electrical	27
		2.29×10^{1}	4.1	900–1250	Electrical	28
	Sb	1.29×10^{1}	3.98	1190–1398	Radioactive	29
		2.14×10^{-1}	3.65	1190–1405	Electrical	27
	Bi	1.03×10^{3}	4.64	1220–1380	Electrical	16
		1.08	3.85	1190–1394	Electrical	27
	Cr	1×10^{-2}	1	1100–1250	Radioactive	30
	Mo	($D \sim 2 \times 10^{-10}$)	–	1000	DLTS	1
	W	($D \sim 10^{-12}$)	–	1100	DLTS	1
	O	7×10^{-2}	2.44	700–1250	SIMS	31
		1.4×10^{-1}	2.53	700–1160	SIMS	32
	S	5.95×10^{-3}	1.83	975–1200	Radioactive	33
	Se	9.5×10^{-1}	2.6	1050–1250	Electrical	34
	Te	5×10^{-1}	3.34	900–1250	SIMS	1
	Mn	6.9×10^{-4}	0.63	900–1200	Radioactive	35
	Fe	1.3×10^{-3}	0.68	30–1250	Radioactive	36
	Co	2×10^{-3}	0.69	700–1300	Radioactive	37
	Ni	2×10^{-3}	0.47	800–1300	Radioactive	38
	Ru	($D \sim 5 \times 10^{-7}$ – 5×10^{-6})	–	1000–1280	Electrical	1
	Rh	($D \sim 10^{-6}$–10^{-4})	–	1000–1200	Electrical	39
	Pd	2.95×10^{-4}	0.22 (i)	702–1320	Nuclear activation	1
	Pt	1.5×10^{2}	2.22	800–1000	Electrical	1
	Os	($D \sim 2 \times 10^{-6}$)	–	1280	Electrical	40
	Ir	4.2×10^{-2}	1.3	950–1250	Electrical	41
Ge	Li	1.3×10^{-3}	0.46	350–800	Electrical	42
		9.1×10^{-3}	0.57	800–500	Electrical	43
	Na	3.95×10^{-1}	2.03	700–850	Radioactive	44
	Cu	1.9×10^{-4}	0.18 (i)	750–900	Radioactive	45
		4×10^{-2}	0.99 (s)	600–700		
		4×10^{-3}	0.33 (i)	350–750	Radioactive	5
	Ag	4.4×10^{-2}	1.0 (i)	700–900	Radioactive	46, 47
		4×10^{-2}	2.23 (s)	800–900	Radioactive	48
	Au	2.25×10^{2}	2.5	600–900	Radioactive	49
	Be	5×10^{-1}	2.5	720–900	Electrical	50
	Mg	($D \sim 8 \times 10^{-9}$)	–	900	Electrical	1
	Zn	5	2.7	600–900	Radioactive and electrical	51
	Cd	1.75×10^{9}	4.4	760–915	Radioactive	52
	B	1.8×10^{9}	4.55	600–900	Electrical	51
	Al	1.0×10^{3}	3.45	554–905	SIMS	53
		$\sim 1.6 \times 10^{2}$	~3.24	750–850	Electrical	54
	Ga	1.4×10^{2}	3.35	554–916	SIMS	55
		3.4×10^{1}	3.1	600–900	Electrical	51
	In	1.8×10^{4}	3.67	554–919	SIMS	56
		3.3×10^{1}	3.02	700–855	Radioactive	57
	Tl	1.7×10^{3}	3.4	800–930	Radioactive	58
	Si	2.4×10^{-1}	2.9	650–900	(γ) Resonance	59
	Ge (self)	2.48×10^{1}	3.14	549–891	Radioactive	60
		7.8	2.95	766–928	Radioactive	61
	Sn	1.7×10^{-2}	1.9	–	Radioactive	45
	P	3.3	2.5	600–900	Electrical	51
	As	2.1	2.39	700–900	Electrical	62
	Sb	3.2	2.41	700–855	Radioactive	57
		1.0×10^{1}	2.5	600–900	Radioactive and electrical	51

Semiconductor	Diffusant	Frequency factor, D_o (cm²/s)	Activation energy, Q(eV)	Temperature range (°C)	Method of measurement	Ref.
	Bi	3.3	2.57	650–850	–	63
	O	4×10^{-1}	2.08	–	Optical	64
	S	$(D \sim 10^{-9})$	–	920	–	65
	Se	$(D \sim 10^{-10})$	–	920	–	65
	Te	5.6	2.43	750–900	Radioactive	66
	Fe	1.3×10^{-1}	1.08	750–900	Radioactive	67
	Co	1.6×10^{-1}	1.12	750–850	Radioactive	47
	Ni	8×10^{-1}	0.9	670–900	Electrical	68
GaAs	Li	5.3×10^{-1}	1.0	250–500	Electrical and chemical	69
	Cu	3×10^{-2}	0.53	100–500	Radioactive	69
		6×10^{-2}	0.98	450–750	Ultrasonic	69
		1.5×10^{-3}	0.6	800–1000	Radioactive	69
	Ag	4×10^{-4}	0.8	500–1150	Radioactive	69
	Au	1×10^{-3}	1.0	740–1025	Radioactive	69
	Be	7.3×10^{-6}	1.2	800–990	Electrical	69
	Mg	4×10^{-5}	1.22	800–1200	Electrical	69
	Zn	1.5×10^{1}	2.49	600–980	Radioactive	69
		2.5×10^{-1}	3.0	750–1000	Radioactive	69
	Cd	1.3×10^{-3}	2.2	800–1100	Radioactive	69
		5×10^{-2}	2.43	868–1149	Radioactive	69
	Hg	$(D \sim 5 \times 10^{-14})$	–	1100	Radioactive	69
	Al	$(D \sim 4 \times 10^{-18}–10^{-14})$	4.3	850–1100	AES	70
	Ga (self)	4×10^{-5}	2.6	1025–1100	Radioactive	69
		1×10^{7}	5.6	1125–1230	Radioactive	69
	In	$(D \sim 7 \times 10^{-11})$	–	1000	Radioactive	69
	C	$(D \sim 1.04 \times 10^{-16})$	–	825	SIMS	69
	Si	1.1×10^{-1}	2.5	850–1050	SIMS	69
	Ge	1.6×10^{-5}	2.06	650–850	SIMS	69
	Sn	6×10^{-4}	2.5	1060–1200	Radioactive	69
		1×10^{-5}	2	800–1000	Radioactive	69
	P	$(D \sim 10^{-12}–10^{-10})$	2.9	800–1150	Reflectance measurements	69
	As (self)	7×10^{-1}	3.2	–	Radioactive	69
	Cr	2.04×10^{-6}	0.83 (f)	750–1000	SIMS	69
			1.7 (s)	700–900		
		7.9×10^{-3}	2.2	800–1100	Chemical analysis	69
	O	2×10^{-3}	1.1	700–900	Mass spectroscopy	69
	S	1.85×10^{-2}	2.6	1000–1300	Radioactive	69
		1.1×10^{1}	2.95	750–900	Electrical	69
	Se	3×10^{3}	4.16	1025–1200	Radioactive	69
	Te	1.5×10^{-1}	3.5	1000–1150	Radioactive	69
	Mn	6.5×10^{-1}	2.49	850–1100	Radioactive	69
	Fe	4.2×10^{-2}	1.8	850–1150	Radioactive	69
		2.2×10^{-3}	2.32	750–1050	Radioactive	69
	Co	5×10^{2}	2.5	800–1000	Radioactive	69
		1.2×10^{-1}	2.64	750–1050	Radioactive	69
	Tm	2.3×10^{-16}	1.0	800–1000	Radioactive	69
GaSb	Li	2.3×10^{-4}	1.9 (s)	527–657	Electrical and flame photometry	69
		1.2×10^{-1}	0.7 (f)	277–657		
	Cu	4.7×10^{-3}	0.9	470–650	Radioactive	69
	Zn	$(D \sim 2 \times 10^{-13} – 1 \times 10^{-11})$	2	510–600	Radioactive	69
	Cd	1.5×10^{-6}	0.72	640–800	Electrical	69
	Ga (self)	3.2×10^{3}	3.15	658–700	Radioactive	69
	In	1.2×10^{-7}	0.53	320–650	Radioactive	69
	Sn	2.4×10^{-5}	0.8	320–650	Radioactive	69
		1.3×10^{-5}	1.1	500–650	Radioactive	69
	Sb (self)	3.4×10^{4}	3.45	658–700	Radioactive	69
	Se	$(D \sim 2.4 \times 10^{-13} – 1.37 \times 10^{-11})$	–	400–500	Radioactive	69

Semiconductor	Diffusant	Frequency factor, D_o (cm²/s)	Activation energy, Q(eV)	Temperature range (°C)	Method of measurement	Ref.
	Te	3.8×10^{-4}	1.20	320–650	Radioactive	69
	Fe	5×10^{-2}	1.9 (I)	500–650	Radioactive	69
		5×10^{2}	2.3 (II)	500–650		
GaP	Ag	–	–	1000–1300	Radioactive	69
	Au	8	2.5 (I)	1050–1250	Radioactive	69
		20	2.4 (II)	1100–1250	Diffusion (I) A face and (II) B face	
	Be	($D_{max} \sim 2.4 \times 10^{-9} – 8.5 \times 10^{-8}$)	–	900–1000	Atomic absorption analysis	69
	Mg	5×10^{-5}	1.4	700–1050	Electrical	69
	Zn	1.0	2.1	700–1300	Radioactive	69
	Ge	–	–	900–1000	Radioactive	69
	Cr	6.2×10^{-4}	1.2	900–1130	Radioactive; ESR	69
	S	3.2×10^{3}	4.7	1120–1305	Radioactive	69
	Mn	2.1×10^{9}	4.7	T < 950	Radioactive; ESR	69
		1.1×10^{-6}	0.9	950–1130		
	Fe	1.6×10^{-1}	2.3	980–1180	Radioactive	69
	Co	2.8×10^{-3}	2.9	850–1100	Radioactive	69
InP	Cu	3.8×10^{-3}	0.69	600–900	Radioactive	69
	Ag	3.6×10^{-4}	0.59	500–900	Radioactive	69
	Au	1.32×10^{-5}	0.48	600–820	Radioactive	69
		1.37×10^{-4}	0.73	600–900	Radioactive	69
	Zn	1.6×10^{-8}	0.3	750–900	Electrical	69
		($D \sim 2 \times 10^{-9} – 4 \times 10^{-8}$)	–	700–900	Radioactive	69
	Cd	1.8	1.9	700–900	Radioactive	69
		1.1×10^{-7}	0.72	700–900	Electrical	69
		($D \sim 7 \times 10^{-13} – 2 \times 10^{-10}$)	–	450-650	Electrical	69
	In (self)	1×10^{5}	3.85	830–990	Radioactive	69
	Sn	($D \sim 3 \times 10^{-8}$)	–	550	Etching and cathodo-luminescence	69
	P (self)	7×10^{10}	5.65	900–1000	Radioactive	69
	Cr	–	–	600–900	Radioactive	69
	S	3.6×10^{-4}	1.94	585–708	Electrical	69
	Se	($D \sim 2 \times 10^{-8}$)	–	550	Cathodoluminescence	69
	Mn	–	2.9	650–750	SIMS	69
	Fe	3	2	600–950	Radioactive	69
		6.8×10^{5}	3.4	600–700	SIMS	69
	Co	9×10^{-1}	1.8	600–950	Radioactive	69
InAs	Cu	3.6×10^{-3}	0.52	342–875	Radioactive	69
		2.2×10^{-2}	0.54	525–890	Radioactive	69
	Ag	7.3×10^{-4}	0.26	450–900	Radioactive	69
	Au	5.8×10^{-3}	0.65	600–900	Radioactive	69
	Mg	1.98×10^{-6}	1.17	600–900	Electrical	69
	Zn	4.2×10^{-3}	0.96	600–900	Radioactive	69
		3.11×10^{-3}	1.17	600–900	Electrical	69
	Cd	7.4×10^{-4}	1.15	650–900	Radioactive	69
	Hg	1.45×10^{-5}	1.32	650–850	Radioactive	69
	In (self)	6×10^{5}	4.0	740–900	Radioactive	69
	Ge	3.74×10^{-6}	1.17	600–900	Electrical	69
	Sn	1.49×10^{-6}	1.17	600–900	Electrical	69
	As (self)	3×10^{7}	4.45	740–900	Radioactive	69
	S	6.78	2.2	600–900	Electrical	69
	Se	12.6	2.2	600–900	Electrical	69
	Te	3.43×10^{-5}	1.28	600–900	Electrical	69
InSb	Li	7×10^{-4}	0.28	0–210	Electrical	69
	Cu	9×10^{-4}	1.08	200–500	Radioactive	69
		3×10^{-5}	0.37	230–490	Radioactive	69
	Ag	1×10^{-7}	0.25	440–510	Radioactive	69
	Au	7×10^{-4}	0.32	140–510	Radioactive	69
	Zn	5×10^{-1}	1.35	362–508	Radioactive	69

Semiconductor	Diffusant	Frequency factor, D_o (cm²/s)	Activation energy, Q(eV)	Temperature range (°C)	Method of measurement	Ref.
		–	1.5	355–455	SIMS	69
	Cd	1×10^{-5}	1.1	250–500	Radioactive	69
		1.3×10^{-4}	1.2	360–500	Electrical	69
	Hg	4×10^{-6}	1.17	425–500	Radioactive	69
	In (self)	6×10^{-7}	1.45	400–500	Radioactive	69
		1.8×10^{13}	4.3	475–517	Radioactive	69
	Sn	5.5×10^{-8}	0.75	390–512	Radioactive	69
	Pb	($D \sim 2.7 \times 10^{-15}$)	–	500	Radioactive	71
	Sb (self)	5.35×10^{-4}	1.91	400–500	Radioactive	69
		3.1×10^{13}	4.3	475–517	Radioactive	69
	S	9×10^{-2}	1.4	360–500	Electrical	69
	Se	1.6	1.87	380–500	Electrical	69
	Te	1.7×10^{-7}	0.57	300–500	Radioactive	69
	Fe	1×10^{-7}	0.25	440–510	Radioactive	69
	Co	2.7×10^{-11}	0.39	420–500	Radioactive	69
AlAs	Ga	($D \sim 2 \times 10^{-18}$– 10^{-15})	3.6	850–1100	AES	70
	Zn	($D \sim 9 \times 10^{-11}$)	–	557	SEM	69
AlSb	Cu	3.5×10^{-3}	0.36	150–500	Radioactive	69
	Zn	3.3×10^{-1}	1.93	660–860	Radioactive	69
	Cd	D(c) ~ 4×10^{-12}– 3×10^{-10}	–	900	Radioactive	69
	Al (self)	2	1.88	570–620	X-ray	69
	Sb (self)	1	1.7	570–620	X-ray	69
ZnS	Cu	2.6×10^{-3}	0.79	470–750	Radioactive	69
		4.3×10^{-4}	0.64	250–1200	Electroluminescence	69
		9.75×10^{-3}	1.04	400–800	Luminescence	69
	Au	1.75×10^{-4}	1.16	500–800	Radioactive	69
	Zn (self)	3×10^{-4}	1.5	$925<T<940$	Radioactive	69
		1.5×10^{4}	3.26	$940<T<1030$		
		1×10^{16}	6.5	$1030<T<1075$		
	Cd	($D \sim 10^{-10}$)	–	1100	Luminescence	72
	Al	5.69×10^{-4}	1.28	800–1000	Luminescence	69
	In	3×10^{1}	2.2	750–1000	Radioactive	69
	S (self)	2.16×10^{4}	3.15	600–800	Radioactive	69
		8×10^{-5}	2.2	740–1100	Radioactive	69
	Se	($D \sim 5 \times 10^{-13}$)	–	1070	X-ray microprobe	69
	Mn	2.3×10^{3}	2.46	500–800	Radioactive	69
ZnSe	Li	2.66×10^{-6}	0.49	950–980	Electrical	69
	Cu	1×10^{-4}	0.66	400–800	Luminescence	69
		1.7×10^{-5}	0.56	200–570	Radioactive	69
	Ag	2.2×10^{-2}	1.18	400–800	Luminescence	69
	Zn (self)	9.8	3.0	760–1150	Radioactive	69
	Cd	6.39×10^{-4}	1.87	700–950	Photoluminescence	69
	Al	2.3×10^{-2}	1.8	800–1100	Luminescence	69
	Ga	1.81×10^{2}	3.0	900–1100	Luminescence	69
		–	1.3	700–850	Electron probe	69
	In	($D \sim 2 \times 10^{-12}$)	–	940	–	69
	S	($D \sim 8 \times 10^{-12}$)	–	1060	X-ray microprobe	69
	Se (self)	1.3×10^{1}	2.5	860–1020	Radioactive	69
		2.3×10^{-1}	2.7	1000–1050	Radioactive	69
	Ni	($D \sim 1.5 \times 10^{-8}$– 1.7×10^{-7})	–	740–910	Luminescence	69
ZnTe	Li	2.9×10^{-2}	1.22 (s)	400–700	Nuclear and chemical analysis	69
		1.7×10^{-4}	0.78 (f)			
	Zn (self)	2.34	2.56	760–860	Radioactive	69
		1.4×10^{1}	2.69	667–1077	Radioactive	69
	Al	–	2.0	700–1000	Electrical and optical	69
	In	4	1.96	1100–1300	Radioactive	69
	Te (self)	2×10^{4}	3.8	727–977	Radioactive	69
CdS	Li	3×10^{-6}	0.68	610–960	Microhardness	69
	Na	($D \sim 3 \times 10^{-7}$)	–	800	Radioactive	69

Semiconductor	Diffusant	Frequency factor, D_o (cm^2/s)	Activation energy, Q(eV)	Temperature range (°C)	Method of measurement	Ref.
	Cu	1.5×10^{-3}	0.76	400–700	Radioactive	69
		1.2×10^{-2}	1.05	300–700	Ultrasonic	69
		8×10^{-5}	0.72	20–200	Electrical	69
	Ag	2.5×10^{1}	1.2 (s)	300–500	Radioactive	69
		2.4×10^{-1}	0.8 (f)			
	Au	2×10^{2}	1.8	500–800	Radioactive	69
	Zn	1.27×10^{-9}	0.86 (s)	720–1000	Radioactive	69
		1.22×10^{-8}	0.66 (f)			
	Cd (self)	3.4	2.0	700–1100	Radioactive	69
	Ga	–	–	667–967	Optical and microprobe	69
	In	6×10^{1}	2.3 (‖)	650–930	Radioactive, optical, and microprobe	69
		1×10^{1}	2.03 (⊥)			
	P	6.5×10^{-4}	1.6	800–1100	Radioactive	69
	S (self)	1.6×10^{-2}	2.05	800–900	Radioactive	69
		–	2.4	750–1050	Radioactive	69
	Se	($D \sim 1.2 \times 10^{-9}$)	–	900	Radioactive	69
	Te	1.3×10^{-7}	10.4	700–1000	Radioactive	69
	Cl	($D \sim 3 \times 10^{-10}$)	–	800	Electrical	69
	I	($D \sim 5 \times 10^{-12}$)	–	1000	Radioactive	69
	Ni	6.75×10^{-3}	10.9	570–900	Luminescence	69
	Yb	($D \sim 1.3 \times 10^{-9}$)	–	960	Photoluminescence	69
CdSe	Ag	2×10^{-4}	0.53	22–400	Ultrasonic	69
	Cd (self)	1.6×10^{-3}	1.5	700–1000	Radioactive	69
		6.3×10^{-2}	1.25 (I)	600–900	Radioactive;	69
		4.12×10^{-2}	2.18 (II)	600–900	(I) saturated Cd and (II) saturated Se pressure	
	P	($D \sim 5.3 \times 10^{-12}$– 6×10^{-11})	–	900–1000	Radioactive	69
	Se (self)	2.6×10^{3}	1.55	700–1000	Radioactive; saturated Se pressure	69
CdTe	Li	($D \sim 1.5 \times 10^{-10}$)	–	300	Ion microprobe	69
	Cu	3.7×10^{-4}	0.67	97–300	Radioactive	69
		8.2×10^{-8}	0.64	290–350	Ion backscattering	69
	Ag	–	–	700–800	Electrical and photo-luminescence	69
	Au	6.7×10^{1}	2.0	600–1000	Radioactive	69
	Cd (self)	1.26	2.07	700–1000	Radioactive	69
		3.26×10^{2}	2.67 (I)	650–900	Radioactive;	69
		1.58×10^{1}	2.44 (II)		(I) saturated Cd and (II) saturated Te pressure	
	In	8×10^{-2}	1.61	650–1000	Radioactive	69
		1.17×10^{2}	2.21 (I)	500–850	Radioactive; (I) saturated Cd and (II) saturated Te pressure	69
		6.48×10^{-4}	1.15 (II)			
	Sn	8.3×10^{-2}	2.2	700–925	Radioactive	69
	P	($D \sim 1.2 \times 10^{-10}$)	–	900	Radioactive	69
	As	–	–	850	–	69
	O	5.6×10^{-9}	1.22	200–650	Mass spectrometry	69
		6.0×10^{-10}	0.29	650–900		
	Se	1.7×10^{-4}	1.35	700–1000	Radioactive	69
	Te (self)	8.54×10^{-7}	1.42 (I)	600–900	Radioactive; (I) saturated Cd and (II) saturated Te pressure	69
		1.66×10^{-4}	1.38 (II)	500–800		
	Cl	7.1×10^{-2}	1.6	520–800	Radioactive	69
	Fe	($D \sim 4 \times 10^{-8}$)	0.77	900	Radioactive	69

Semiconductor	Diffusant	Frequency factor, D_o(cm²/s)	Activation energy, Q(eV)	Temperature range (°C)	Method of measurement	Ref.
HgSe	Sb	6.3×10^{-5}	0.85	540–630	Radioactive	69
	Se (self)	–	–	200–400	Radioactive	69
HgTe	Ag	6×10^{-4}	0.8	250–350	Radioactive	69
	Zn	5×10^{-8}	0.6	250–350	Radioactive	69
	Cd	3.1×10^{-4}	0.66	250–350	Radioactive	69
	Hg (self)	2×10^{-8}	0.6	200–350	Radioactive	69
	In	6×10^{-6}	0.9	200–300	Radioactive	69
	Sn	1.72×10^{-6}	0.66 (s)	200–300	Radioactive	69
		1.8×10^{-3}	0.80 (f)			
	Te (self)	10^{-6}	1.4	200–400	Radioactive	69
	Mn	1.5×10^{-4}	1.3	250–350	Radioactive	69
PbS	Cu	4.6×10^{-4}	0.36	150–450	Electrical	69
		5×10^{-3}	0.31	100–400	Electrical	69
	Pb (self)	8.6×10^{-5}	1.52	500–800	Radioactive	69
	S (self)	6.8×10^{-5}	1.38	500–750	Radioactive	69
	Ni	1.78×10^{1}	0.95	200–500	Electrical	69
PbSe	Na	1.5×10^{1}	1.74 (s)	400–850	Radioactive	69
		5.6×10^{-6}	0.4 (f)			
	Cu	2×10^{-5}	0.31	93–520	Radioactive	69
	Ag	7.4×10^{-4}	0.35	400–850	Radioactive	69
	Pb (self)	4.98×10^{-6}	0.83	400–800	Radioactive	69
	Sb	3.4×10^{-1}	2.0	650–850	Radioactive	69
	Se (self)	2.1×10^{-5}	1.2	650–850	Radioactive	69
	Cl	1.6×10^{-8}	0.45	400–850	Radioactive	69
	Ni	$(D \sim 1 \times 10^{-10})$	–	700	Radioactive	69
PbTe	Na	1.7×10^{-1}	1.91	600–850	Radioactive	69
	Sn	3.1×10^{-2}	1.56	500–800	Radioactive	69
	Pb (self)	2.9×10^{-5}	0.6	250–500	Radioactive	69
	Sb	4.9×10^{-2}	1.54	500–800	Radioactive	69
	Te	2.7×10^{-6}	0.75	500–800	Radioactive	69
	Cl	$(D > 2.3 \times 10^{-10})$	–	700	Radioactive	69
	Ni	$(D > 1 \times 10^{-6})$	–	700	Radioactive	69

References

1. N. A. Stolwijk and H. Bracht, in *Diffusion in Semiconductors and Non-Metallic Solids*, D. L. Beke, Ed., Springer-Verlag, Berlin, 1998, 2-1.
2. E. M. Pell, *Phys. Rev.*, 119, 1960; 119, 1014, 1960.
3. L. Svob, *Solid State Electron*, 10, 991, 1967.
4. B. I. Boltaks and I. I. Sosinov, *Zh. Tekh. Fiz.*, 28, 3, 1958.
5. R. N. Hall and J. N. Racette, *J. Appl. Phys.*, 35, 379, 1964.
6. B. I. Boltaks and Hsueh Shih-Yin, *Sov. Phys. Solid State*, 2, 2383, 1961.
7. W. R. Wilcox and T. J. LaChapelle, *J. Appl. Phys.*, 35, 240, 1964.
8. E. A. Taft and R. O. Carlson, *J. Electrochem. Soc.*, 117, 711, 1970.
9. R. Sh. Malkovich and N. A. Alimbarashvili, *Sov. Phys. Solid State*, 4, 1725, 1963.
10. R. N. Ghoshtagore, *Solid State Electron*, 15, 1113, 1972.
11. C. Hill, *Semiconductor Silicon 1981*, H. R. Huff, R. J. Kreiger, and Y. Takeishi, Eds., p. 988, *Electrochem. Soc.*, 1981.
12. R. N. Ghoshtagore, *Phys. Rev. B*, 3, 2507, 1971.
13. W. Rosnowski, *J. Electrochem. Soc.*, 125, 957, 1978.
14. J. S. Makris and B. J. Masters, *J. Appl. Phys.*, 42, 3750, 1971.
15. M. F. Millea, *J. Phys. Chem. Solids*, 27, 315, 1965 (refer Reference 2).
16. C. S. Fuller and J. A. Ditzenberger, *J. Appl. Phys.*, 27, 544, 1956.
17. S. Hocine and D. Mathiot, *Appl. Phys. Lett.*, 53, 1269, 1988.
18. R. C. Newman and J. Wakefield, *J. Phys. Chem. Solids*, 19, 230, 1961.
19. L. Kalinowski and R. Seguin, *Appl. Phys. Lett.*, 35, 211, 1979; *Appl. Phys. Lett.*, 36, 171, 1980.
20. R. F. Peart, *Phys. Stat. Sol.*, 15, K 119, 1966.
21. G. Hettich, H. Mehrer and K. Maler, *Inst. Phys. Conf. Ser.*, 46, 500, 1979.
22. M. Ogina, Y. Oana and M. Watanabe, *Phys. Stat. Sol. (a)*, 72, 535, 1982.
23. T. H. Yeh, S. M. Hu, and R. H. Kastl, *Appl. Phys.*, 39, 4266, 1968.
24. I. Franz and W. Langheinrich, *Solid State Electron*, 14, 835, 1971.
25. R. N. Ghoshtagore, *Hys. Rev. B*, 3, 389, 1971.
26. B. J. Masters and J. M. Fairfield, *J. Appl. Phys.*, 40, 2390, 1969.
27. R. N. Goshtagore, *Phys. Rev. B*, 3, 397, 1971.
28. R. S. Fair and J. C. C. Tsai, *J. Electrochem. Soc.*, 122, 1689, 1975.
29. J. J. Rohan, N. E. Pickering, and J. Kennedy, *J. Electrochem. Soc.*, 106, 705, 1969.
30. W. Wuerker, K. Roy, and J. Hesse, *Matsr. Res. Bull.*, 9, 971, 1974.
31. J. C. Mikkelsen, Jr., *Appl. Phys. Lett.*, 40, 336, 1982.
32. S. Tang Lee and D. Nicols, *Appl. Phys. Lett.*, 47, 1001, 1985.
33. P. L. Gruzin, S. V. Zemskii, A. D. Bullkin, and N. M. Makarov, *Sov. Phys. Sem.*, 7, 1241, 1974.
34. N. S. Zhdanovich and Yu. I. Kozlov, *Svoistva Legir, Poluprovodn.*, V. S. Zemskov, Ed., Nauka, Moscow, 1977, 115–120; *Fiz Tekh. Poluprovod.*, 9, 1594, 1975.
35. D. Gilles, W. Bergholze, and W. Schroeter, *J. Appl. Phys.*, 59, 3590, 1986.
36. E. R. Weber, *Appl. Phys.* A, 30, 1, 1983.
37. E. R. Weber, Properties of Silicon, EMIS Datareviews Ser. No. 4, INSPEC Publications, 1988, 409–451.
38. M. K. Bakhadyrkhanov, S. Zainabidinov, and A. Khamidov, *Sov. Phys. Sem.*, 14, 243, 1980.
39. S. A. Azimov, M. S. Yunosov, F. K. Khatamkulov, and G. Nasyrov, *Poluprovod.*, N. Kh. Abrikosov and V. S. Zemskov, Eds., Nauka, Moscow, 1975, 21–23.
40. S. A. Azimov, M. S. Yunosov, G. Nurkuziev, and F. R. Karimov, *Sov. Phys. Sem.*, 12, 981, 1978.
41. S. A. Azimov, B. V. Umarov, and M. S. Yunusov, *Sov. Phys. Sem.*, 10, 842, 1976.

42. C. S. Fuller and J. A. Ditzenberger, *Phys. Rev.*, 91, 193, 1953.
43. B. Pratt and F. Friedman, *J. Appl. Phys.*, 37, 1893, 1966.
44. M. Stojic, V. Spiric, and D. Kostoski, *Inst. Phys. Conf. Ser.*, 31, 304, 1976.
45. B. I. Boltaks, *Diffusion in Semiconductors*, Inforsearch, London, 1963, 162.
46. A. A. Bugai, V. E. Kosenko, and E. G. Miselyuk, *Zh. Tekh. Fiz.*, 27, 67, 1957.
47. L. Y. Wei, *J. Phys. Chem. Solids*, 18, 162, 1961.
48. V. E. Kosenko, *Sov. Phys. Solid State*, 4, 42, 1962.
49. W. C. Dunlap, Jr., *Phys. Rev.*, 97, 614, 1955
50. Yu. I. Belyaev and V. A. Zhidkov, *Sov. Phys. Solid State*, 3, 133, 1961.
51. W. C. Dunlap, Jr. *Phys. Rev.*, 94, 1531, 1954.
52. V. E. Kosenko, *Sov. Phys. Solid State*, 1, 1481, 1960.
53. P. Dorner, W. Gust, A. Lodding, H. Odelius, B. Predel, and U. Roll, *Acta Metall.*, 30, 941, 1982.
54. W. Meer and D. Pommerrening, *Z. Agnew. Phys.*, 23, 369, 1967.
55. U. Sodervall, H. Odelius, A. Lodding, U. Roll, B. Predel, W. Gust, and P. Dorner, *Phil. Mag. A*, 54, 539, 1986.
56. P. Dorner, W. Gust, A. Lodding, H. Odelius, B. Predel, and U. Roll, *Z. Metalkd.*, 73, 325, 1982.
57. P. V. Pavlov, *Sov. Phys. Solid State*, 8, 2377, 1967.
58. V. I. Tagirov and A. A. Kuliev, *Sov. Phys. Solid State*, 4, 196, 1962.
59. J. Raisanen, J. Hirvonen, and A. Anttila, *Solid State Electron.*, 24, 333, 1981.
60. C. Vogel, G. Hettich, and H. Mehrer, *J. Phys. C.*, 16, 6197, 1983.
61. H. Letaw, Jr., W. M. Portnoy, and L. Slifkin, *Phys. Rev.*, 102, 363, 1956.
62. W. Bosenberg, *Z. Naturforsch.*, 10a, 285, 1955.
63. V. M. Glazov and V. S. Zemskov, Physicochemical Principles of Semiconductor Doping, Israel Program for Scientific Translation, Jerusalem, 1968.
64. J. W. Corbett, R. S. McDonald, and G. D. Watkins, *J. Phys. Chem. Solids*, 25, 873, 1964.
65. W. W. Tyler, *J. Phys. Chem. Solids*, 8, 59, 1959.
66. V. D. Ignatkov and V. E. Kosenko, *Sov. Phys. Solid State*, 4, 1193, 1962.
67. A. A. Bugal, V. E. Kosenko, and E. G. Miseluk, *Zh. Tekh. Fiz.*, 27, 210, 1957.
68. F. van der Maesen and J. A. Brenkman, *Phillips Res. Rep.*, 9, 255, 1954.
69. M. B. Dutt and B. L. Sharma, in *Diffusion in Semiconductors and Non-Metallic Solids*, D. L. Beke, Ed., Springer-Verlag, Berlin, 1998, 3-1.
70. L. L. Chang and A. Koma, *Appl. Physics Lett.*, 29, 138, 1976.
71. D. L. Kendall, *Semiconductors and Semimetals*, Vol. 4, R. K. Willardson and A. C. Beer, Eds., Academic, 1968, 255.
72. H. J. Biter and F. Williams, *J. Luminescence*, 3, 395, 1971.

PROPERTIES OF MAGNETIC MATERIALS

H. P. R. Frederikse

Glossary of Symbols

Quantity	Symbol	SI Units	emu Units
Magnetic field	H	A m^{-1}	Oe (oersted)
Magnetic induction	B	T (tesla)	G (gauss)
Magnetization	M	A m^{-1}	emu cm^{-3}
Spontaneous magnetization	M_s	A m^{-1}	emu cm^{-3}
Saturation magnetization	M_0	A m^{-1}	emu cm^{-3}
Magnetic flux	Φ	Wb (weber)	maxwell
Magnetic moment	m, μ	A m^2	erg/G
Coercive field	H_c	A m^{-1}	Oe
Remanence	B_r	T	G
Saturation magnetic polarization	J_s	T	G
Magnetic susceptibility	χ		
Magnetic permeability	μ	H m^{-1} (henry/meter)	
Magnetic permeability of free space	μ_0	H m^{-1}	
Saturation magnetostriction	λ ($\Delta l/l$)		
Curie temperature	T_C	K	K
Néel temperature	T_N	K	K

Magnetic moment $\mu = \gamma\hbar J = g\mu_B J$

where

γ = gyromagnetic ratio
J = angular momentum
g = spectroscopic splitting factor (~2)
μ_B = Bohr magneton = $9.27401 \cdot 10^{-24}$
J/T = $9.2741 \cdot 10^{-21}$ erg/G
Earth's magnetic field H = 56 A m^{-1} = 0.7 Oe
For iron: $M_0 = 1.7 \cdot 10^6$ A m^{-1}; $B_r = 0.8 \cdot 10^6$ A m^{-1}
1 Oe = $(1000/4\pi)$ A m^{-1}; 1 G = 10^{-4} T; 1 emu cm^{-3} = 10^3 A m^{-1}
1 Maxwell = 10^{-8} Wb
$\mu_0 = 4\pi \cdot 10^{-7}$ H m^{-1}

Relation between Magnetic Induction and Magnetic Field

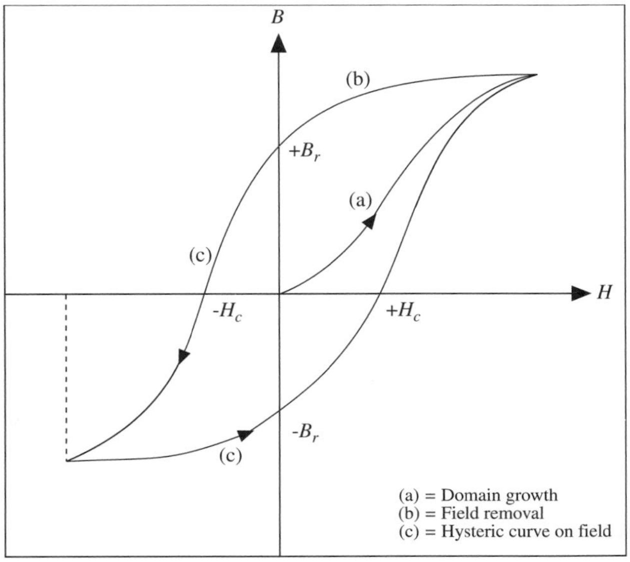

FIGURE 1. Typical curve representing the dependence of magnetic induction B on magnetic field H for a ferromagnetic material. When H is first applied, B follows curve (a) as the favorably oriented magnetic domains grow. This curve flattens as saturation is approached. When H is then reduced, B follows curve (b), but retains a finite value (the remanence B_r) at $H = 0$. In order to demagnetize the material, a negative field $-H_c$ (where H_c is called the coercive field or coercivity) must be applied. As H is further decreased and then increased to complete the cycle (curve c), a hysteresis loop is obtained. The area within this loop is a measure of the energy loss per cycle for a unit volume of the material.

Solids

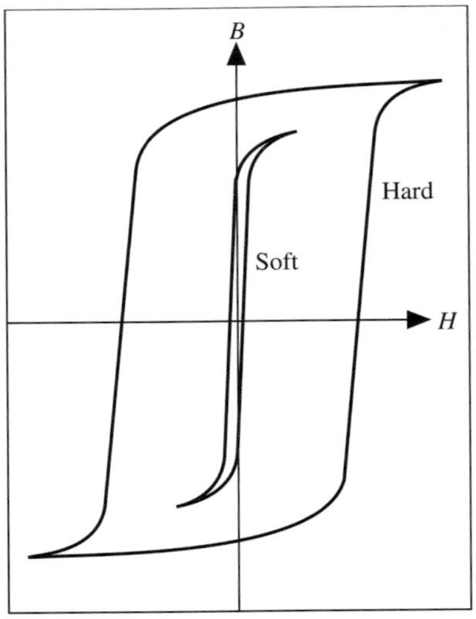

FIGURE 2. Schematic curve illustrating the *B* vs. *H* dependence for hard and soft magnetic materials. Hard materials have a larger remanence and coercive field, and a correspondingly large hysteresis loss.

Reference

Ralls, K. M., Courtney, T. H., and Wulff, J., *Introduction to Materials Science and Engineering*, J. Wiley & Sons, New York, 1976, p. 577, 582. With permission.

Magnetic Susceptibility of the Elements

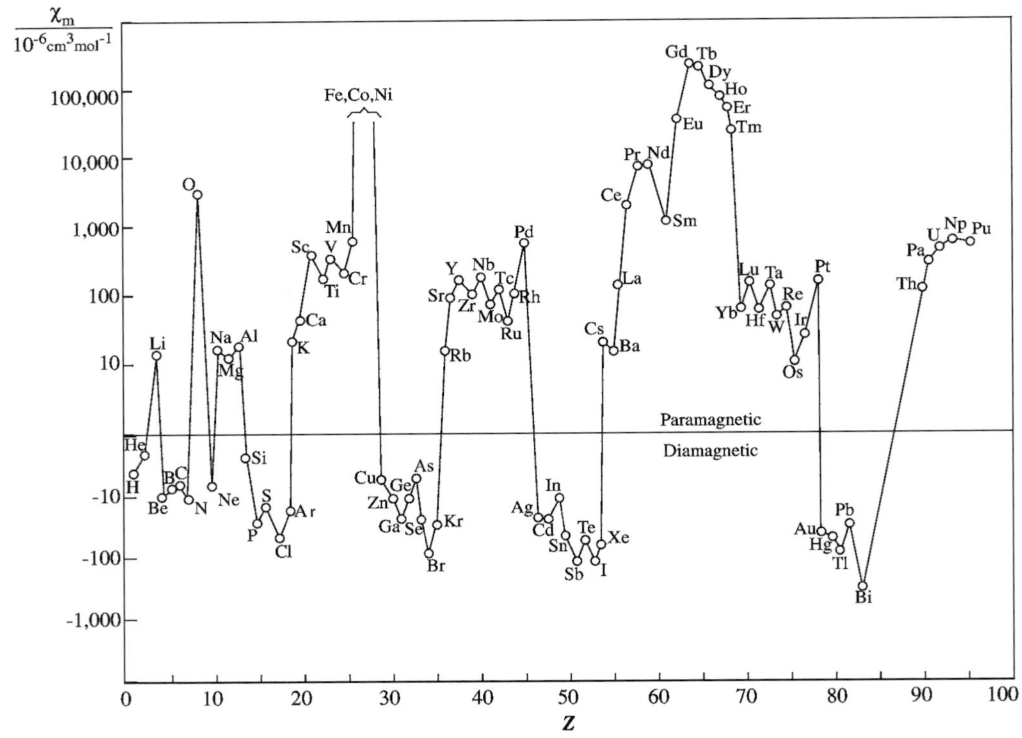

FIGURE 3. Molar susceptibility of the elements at room temperature (cgs units of 10^{-6} cm^3/mol). Values are not available for Z = 9, 61, and 84–89; Fe, Co, and Ni (Z = 26–28) are ferromagnetic. Data taken from the table "Magnetic Susceptibility of the Elements and Inorganic Compounds" in Section 4.

Reference

Gray, D. E., Ed., *American Institute of Physics Handbook, Third Edition*, McGraw-Hill, New York, 1972, p. 5-224. With permission.

Ground State of Ions with Partly Filled *d* or *f* Shells

Z	Element	n	S	L	J	Gr. state	p_{calc}	p_{calc}[b]	p_{meas}
22	Ti^{+3}	1	1/2	2	3/2	$^2D_{3/2}$	1.73[a]	1.55	1.8
23	V^{+4}	1	1/2	2	3/2	$^2D_{3/2}$	1.73[a]	1.55	1.8
23	V^{+3}	2	1	3	2	$^3F_2$	2.83[a]	1.63	2.8
23	V^{+2}	3	3/2	3	3/2	$^4F_{3/2}$	3.87[a]	0.77	3.8
24	Cr^{+3}	3	3/2	3	3/2	$^4F_{3/2}$	3.87[a]	0.77	3.7
25	Mn^{+4}	3	3/2	3	3/2	$^4F_{3/2}$	3.87[a]	0.77	4.0
24	Cr^{+2}	4	2	2	0	$^5D_0$	4.90[a]	0	4.9
25	Mn^{+3}	4	2	2	0	$^5D_0$	4.90[a]	0	5.0
25	Mn^{+2}	5	5/2	0	5/2	$^6S_{5/2}$	5.92[a]	5.92	5.9
26	Fe^{+3}	5	5/2	0	5/2	$^6S_{5/2}$	5.92[a]	5.92	5.9
26	Fe^{+2}	6	2	2	4	$^5D_4$	4.90[a]	6.70	5.4
27	Co^{+2}	7	3/2	3	9/2	$^4F_{9/2}$	3.87[a]	6.54	4.8
28	Ni^{+2}	8	1	3	4	$^3F_4$	2.83[a]	5.59	3.2
29	Cu^{+2}	9	1/2	2	5/2	$^2D_{5/2}$	1.73[a]	3.55	1.9
58	Ce^{+3}	1	1/2	3	5/2	$^2F_{5/2}$	2.54[c]		2.4
59	Pr^{+3}	2	1	5	4	$^3H_4$	3.58[c]		3.5
60	Nd^{+3}	3	3/2	6	9/2	$^4I_{9/2}$	3.62[c]		3.5
61	Pm^{+3}	4	2	6	4	$^5I_4$	2.68[c]		
62	Sm^{+3}	5	5/2	5	5/2	$^6H_{5/2}$	0.84[c]		1.5
63	Eu^{+3}	6	3	3	0	$^7F_0$	0.0[c]		3.4
64	Gd^{+3}	7	7/2	0	7/2	$^8S_{7/2}$	7.94[c]		8.0
65	Tb^{+3}	8	3	3	6	$^7F_6$	9.72[c]		9.5
66	Dy^{+3}	9	5/2	5	15/2	$^6H_{15/2}$	10.63[c]		10.6
67	Ho^{+3}	10	2	6	8	$^5I_8$	10.60[c]		10.4
68	Er^{+3}	11	3/2	6	15/2	$^4I_{15/2}$	9.59[c]		9.5
69	Tm^{+3}	12	1	5	6	$^3H_6$	7.57[c]		7.3
70	Yb^{+3}	13	1/2	3	7/2	$^2F_{7/2}$	4.54[c]		4.5

[a] $p_{calc} = 2[S(S+1)]^{1/2}$.
[b] $p_{calc} = 2[J(J+1)]^{1/2}$.
[c] $p_{calc} = g[S(S+1)]^{1/2}$.

References

1. Jiles, D., *Magnetism and Magnetic Materials*, Chapman & Hall, London, 1991, p. 243.
2. Kittel, C., *Introduction to Solid State Physics, Sixth Edition*, J. Wiley & Sons, New York, 1986, pp. 405–406.
3. Ashcroft, N. W. and Mermin, N. D., *Solid State Physics*, Holt, Rinehart, and Winston, New York, 1976, p. 652.

Solids

Ferro- and Antiferromagnetic Elements

M_0 is the saturation magnetization at $T = 0$ K
n_B is the number of Bohr magnetons per atom
T_C is the Curie temperature
T_N is the Néel temperature

Curie Temperature, Néel Temperature, and Other Properties of Ferro- and Antiferromagnetic Elements

Element	M_0/gauss	n_B	T_C/K	T_N/K	Comments
Fe	22020	2.22	1043		
Co	18170	1.72	1388		
Ni	6410	0.62	627		
Cr				311	
Mn				100	
Ce				12.5	*c*-Axis antiferromagnetic
Nd				19.2	Basal plane modulation on hexagonal sites
				7.8	Cubic sites order (periodicity different from high-*T* phase)
Sm				106	Ordering on hexagonal sites
				13.8	Cubic site order
Eu				90.5	Spiral along cube axis
Gd	24880	7	293		
Tb		9	220		Basal plane ferromagnet
				230.2	Basal plane spiral
Dy		10	87		Basal plane ferromagnet
				176	Basal plane spiral
Ho		10	20		Bunched cone structure
				133	Basal plane spiral
Er		9	32		*c*-Axis ferrimagnetic cone structure
				80	*c*-Axis modulated structure
Tm		7	32		*c*-Axis ferrimagnetic cone structure
				56	*c*-Axis modulated structure

References

1. Ashcroft, N. W., and Mermin, N. D., *Solid State Physics*, Holt, Rinehart, and Winston, New York, 1976, p. 652.
2. Gschneidner, K. A., and Eyring, L., *Handbook on the Physics and Chemistry of Rare Earths*, North-Holland, Amsterdam, 1978.

Selected Ferromagnetic Compounds

M_0 is the saturation magnetization at $T = 293$ K
T_C is the Curie temperature

Saturation Magnetization and Curie Temperature of Selected Ferromagnetic Compounds

Compound	Formula	M_0/gauss	T_C/K	Crystal system
Manganese boride (MnB)	MnB	152	578	orthorh(FeB)
Manganese arsenide [MnAs]	MnAs	670	318	hex(FeB)
Manganese bismuthide [MnBi]	MnBi	620	630	hex(FeB)
Manganese antimonide (MnSb)	MnSb	710	587	hex(FeB)
Manganese nitride [Mn_4N]	Mn_4N	183	743	
Manganese silicide	MnSi		34	cub(FeSi)
Chromium monotelluride	CrTe	247	339	hex(NiAs)
Chromium(III) bromide	$CrBr_3$	270	37	hex(BiI_3)
Chromium(III) iodide	CrI_3		68	hex(BiI_3)
Chromium(IV) oxide	CrO_2	515	386	tetr(TiO_2)
Europium(II) oxide	EuO	1910*	77	cub
Europium(II) sulfide	EuS	1184*	16.5	cub
Gadolinium(III) chloride	$GdCl_3$	550*	2.2	orthorh
Iron boride (FeB)	FeB		598	orthorh
Iron boride (Fe_2B)	Fe_2B		1043	tetr ($CuAl_2$)
Iron beryllide [$FeBe_5$]	$FeBe_5$		75	cub($MgCu_2$)
Iron carbide	Fe_3C		483	orthorh
Iron phosphide (FeP)	FeP		215	orthorh (MnP)

* At $T = 0$ K.

References

1. Kittel, C., *Introduction to Solid State Physics, Sixth Edition*, J. Wiley & Sons, New York, 1986.
2. Ashcroft, N. W., and Mermin, N. D., *Solid State Physics*, Holt, Rinehart, and Winston, New York, 1976.

Magnetic Properties of High-Permeability Metals and Alloys (Soft)

μ_i is the initial permeability
μ_m is the maximum permeability
H_c is the coercive force
J_s is the saturation polarization
W_H is the hysteresis loss per cycle
T_C is the Curie temperature

Properties of High-Permeability Metals and Alloys (Soft)

Material	Composition (mass %)	μ_i/μ_0	μ_m/μ_0	H_c/A m^{-1}	J_s/T	W_H/J m^{-3}	T_C/K
Iron	Commercial 99Fe	200	6000	70	2.16	500	1043
Iron	Pure 99.9Fe	25000	350000	0.8	2.16	60	1043
Silicon-iron	96Fe-4Si	500	7000	40	1.95	50–150	1008
Silicon-iron (110) [001]	97Fe-3Si	9000	40000	12	2.01	35–140	1015
Silicon-iron {100} <100>	97Fe-3Si		100000	6	2.01		1015
Mild steel	Fe-0.1C-0.1Si-0.4Mn	800	1100	200			
Hypernik	50Fe-50Ni	4000	70000	4	1.60	22	753
Deltamax {100} <100>	50Fe-50Ni	500	200000	16	1.55		773
Isoperm {100} <100>	50Fe-50Ni	90	100	480	1.60		
78 Permalloy	78Ni-22Fe	4000	100000	4	1.05	50	651
Supermalloy	79Ni-16Fe-5Mo	100000	1000000	0.15	0.79	2	673
Mumetal	77Ni-16Fe-5Cu-2Cr	20000	100000	4	0.75	20	673
Hyperco	64Fe-35Co-0.5Cr	650	10000	80	2.42	300	1243
Permendur	50Fe-50Co	500	6000	160	2.46	1200	1253
2V-Permendur	49Fe-49Co-2V	800	4000	160	2.45	600	1253

Material	Composition (mass %)	μ_i/μ_0	μ_m/μ_0	H_c/A m^{-1}	J_s/T	W_H/J m^{-3}	T_C/K
Supermendur	49Fe-49Co-2V		60000	16	2.40	1150	1253
25Perminvar	45Ni-30Fe-25Co	400	2000	100	1.55		
7Perminvar	70Ni-23Fe-7Co	850	4000	50	1.25		
Perminvar (magnet. annealed)	43Ni-34Fe-23Co		400000	2.4	1.50		
Alfenol (or Alperm)	84Fe-16Al	3000	55000	3.2	0.8		723
Alfer	87Fe-13Al	700	3700	53	1.20		673
Aluminum-Iron	96.5Fe-3.5Al	500	19000	24	1.90		
Sendust	85Fe-10Si-5Al	36000	120000	1.6	0.89		753

References

1. McCurrie, R. A., *Structure and Properties of Ferromagnetic Materials*, Academic Press, London, 1994, p. 42.
2. Gray, D. E., Ed., *American Institute of Physics Handbook, Third Edition*, McGraw-Hill, New York, 1972, p. 5-224.

Applications of High-Permeability Materials

Applications	Requirements
Power applications	
Distribution and power transformers	Low core losses, high permeability, high saturation magnetic polarization
High-quality motors and generators, stators and armatures, switched-mode power supplies	
Instrument transformers	
Audiofrequency transformers	Low core losses, high permeability, high magnetic polarization
Pulse transformers	High permeability
Cores for inductor coils	
Audiofrequency	Low hysteresis, high permeability
Carrier frequency	Very low hysteresis and eddy current loss
Radiofrequency	High permeability at low fields
Miscellaneous	
Relays, switches, earth leakage circuit	High permeability, low remanence, low coercivity
Magnetic shielding	Low core loss for AC applications
Magnetic recording heads	High initial permeability, low or zero remanence
Magnetic amplifiers, saturable reactors, saturable transformers, transformer cores	Rectangular hysteresis loops, low hysteresis loss
Magnetic shunts for temperature compensation in magnetic circuits	Low Curie temperature, appropriate decrease in permeability with increase in temperature
Electromagnets in indicating instruments, fire detection, quartz watches, electromechanical devices	High permeability, high saturation magnetic polarization
Magnetic yokes in permanent magnet devices, such as lifting and holding magnets, loudspeakers	High permeability, high saturation magnetic polarization

Reference

McCurrie, R. A., *Structure and Properties of Ferromagnetic Materials*, Academic Press, London, 1994. With permission.

Saturation Magnetostriction of Selected Materials

The tabulated parameter λ_s is related to the fractional change in length $\Delta l/l$ by $\Delta l/l = (3/2)\lambda_s(\cos^2\theta - 1/3)$, where θ is the angle of rotation.

Material	$\lambda_s \times 10^6$
Iron	-7
Fe - 3.2% Si	9
Nickel	-33
Cobalt	-62
45 Permalloy, 45% Ni - 55% Fe	27
Permalloy, 82% Ni - 18% Fe	0
Permendur, 49% Co - 49% Fe - 2% V	70
Alfer, 87% Fe - 13% Al	30
Magnetite, Fe_3O_4	40
Cobalt ferrite, $CoFe_2O_4$	-110
$SmFe_2$	-1560
$TbFe_2$	1753
$Tb_{0.3}Dy_{0.7}Fe_{1.93}$ (Terfenol D)	2000
$Fe_{66}Co_{18}B_{15}Si$ (amorphous)	35
$Co_{72}Fe_3B_6A_{13}$ (amorphous)	0

Reference

McCurrie, R.A., *Structure and Properties of Ferromagnetic Materials*, Academic Press, London, 1994, p. 91; additional data provided by A.E. Clark, Adelphi, MD.

Properties of Various Permanent Magnetic Materials (Hard)

B_r is the remanence
${}_BH_c$ is the flux coercivity
${}_iH_c$ is the intrinsic coercivity
$(BH)_{max}$ is the maximum energy product
T_C is the Curie temperature
T_{max} is the maximum operating temperature

Composition	B_r/T	${}_BH_c/10^3$A m^{-1}	${}_iH_c/10^3$A m^{-1}	$(BH)_{max}$/kJ m^{-3}	T_C/°C	T_{max}/°C
Alnico1 20Ni;12Al;5Co	0.72		35	25		
Alnico2 17Ni;10Al;12.5Co;6Cu	0.72		40-50	13-14		
Alnico3 24-30Ni;12-14Al;0-3Cu	0.5–0.6		40-54	10		
Alnico4 21-28Ni;11-13Al;3-5Co;2-4Cu	0.55–0.75		36-56	11-12		
Alnico5 14Ni;8Al;24Co;3Cu	1.25	53	54	40	850	520
Alnico6 16Ni;8Al;24Co;3Cu;2Ti	1.05		75	52		
Alnico8 15Ni;7Al;35Co;4Cu;5Ti	0.83	1.6	160	45		
Alnico9 15Ni;7Al;35Co;4Cu;5Ti	1.10	1.45	1.45	75	850	520
Alnico12 13.5Ni;8Al;24.5Co;2Nb	1.20		64	76.8		
$BaFe_{12}O_{19}$ (Ferroxdur)	0.4	1.6	192	29	450	400
$SrFe_{12}O_{19}$	0.4	2.95	3.3	30	450	400
$LaCo_5$	0.91			164	567	
$CeCo_5$	0.77			117	380	
$PrCo_5$	1.20			286	620	
$NdCo_5$	1.22			295	637	
$SmCo_5$	1.00	7.9	696	196	700	250
$Sm(Co_{0.76}Fe_{0.10}Cu_{0.14})_{6.8}$	1.04	4.8	5	212	800	300
$Sm(Co_{0.65}Fe_{0.28}Cu_{0.05}Zr_{0.02})_{7.7}$	1.2	10	16	264	800	300
$Nd_2Fe_{14}B$ sintered	1.22	8.4	1120	280	300	100
Fe;52Co;14V (Vicalloy II)	1.0	42		28	700	500
Fe;24Cr;15Co;3Mo (anisotropic)	1.54	67		76	630	500
Fe;28Cr;10.5Co (Chromindur II)	0.98	32		16	630	500
Fe;23Cr;15Co;3V;2Ti	1.35	4		44	630	500
Cu;20Ni;20Fe (Cunife)	0.55	4		12	410	350
Cu;21Ni;29Fe (Cunico)	0.34	0.5		8		
Pt;23Co	0.64	4		76	480	350
Mn;29.5Al;0.5C (anisotropic)	0.61	2.16	2.4	56	300	120

References

1. McCurrie, R. A., *Structure and Properties of Ferromagnetic Materials*, Academic Press, London, 1994, p. 204.
2. Gray, D. E., Ed., *American Institute of Physics Handbook, Third Edition*, McGraw-Hill, New York, 1972, p. 5-165.
3. Jiles, D., *Magnetism and Magnetic Materials*, Chapman & Hall, London, 1991.

Selected Ferrites

J_s is the saturation magnetic polarization
T_C is the Curie temperature
ΔH is the line width

Properties and Applications of Ferrites

Material	J_s/T	T_C/°C	ΔH/kA m^{-1}	Applications
Spinels				
γ-Fe_2O_3	0.52	575		
Fe_3O_4	0.60	585		
$NiFe_2O_4$	0.34	575	350	Microwave devices
$MgFe_2O_4$	0.14	440	70	
$NiZnFe_2O_4$	0.50	375	120	Transformer cores
$MnFe_2O_4$	0.50	300	50	Microwave devices
$NiCoFe_2O_4$	0.31	590	140	Microwave devices
$NiCoAlFe_2O_4$	0.15	450	330	Microwave devices

Material	J_s/T	T_C/°C	ΔH/kA m^{-1}	Applications
$NiAl_{0.35}Fe_{1.65}O_4$	0.12	430	67	Microwave devices
$NiAlFe_2O_4$	0.05	1860	32	Microwave devices
$Mg_{0.9}Mn_{0.1}Fe_2O_4$	0.25	290	56	Microwave devices
$Ni_{0.5}Zn_{0.5}Al_{0.8}Fe_{1.2}O_4$	0.14		17	Microwave devices
$CuFe_2O_4$	0.17	455		Electromechanical transducers
$CoFe_2O_4$	0.53	520		
$LiFe_5O_8$	0.39	670		Microwave devices
Garnets				
$Y_3Fe_5O_{12}$	0.178	280	55	Microwave devices
$Y_3Fe_5O_{12}$ (single crys.)	0.178	292	0.5	Microwave devices
$(Y,Al)_3Fe_5O_{12}$	0.12	250	80	Microwave devices
$(Y,Gd)_3Fe_5O_{12}$	0.06	250	150	Microwave devices
$Sm_3Fe_5O_{12}$	0.170	305		Microwave devices
$Eu_3Fe_5O_{12}$	0.116	293		Microwave devices
$GdFe_5O_{12}$	0.017	291		Microwave devices
Hexagonal crystals				
$BaFe_{12}O_{19}$	0.45	430	1.5	Permanent magnets
$Ba_3Co_2Fe_{24}O_{41}$	0.34	470	12	Microwave devices
$Ba_2Zn_2Fe_{12}O_{22}$	0.28	130	25	Microwave devices
$Ba_3Co_{1.35}Zn_{0.65}Fe_{24}O_{41}$		390	16	Microwave devices
$Ba_2Ni_2Fe_{12}O_{22}$	0.16	500	8	Microwave devices
$SrFe_{12}O_{19}$	0.4	450		Permanent magnets

Reference

McCurrie, R. A., *Structure and Properties of Ferromagnetic Materials*, Academic Press, London, 1994.

Spinel Structure (AB_2O_4)

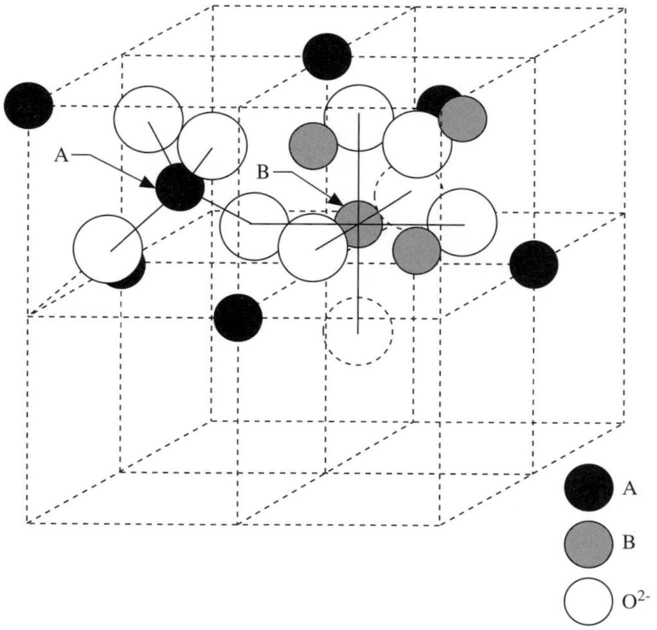

FIGURE 4. Arrangement of metal ions in the two octants A and B, showing tetrahedrally (A) and octahedrally (B) coordinated sites. (Reprinted from McCurrie, R.A., *Ferromagnetic Materials*, Academic Press, London, 1994. With permission.)

Crystal Structure and Néel Temperature of Selected Antiferromagnetic Solids

Material	Structure	T_N/K
Binary oxides		
MnO	cub(fcc)	122
FeO	cub(fcc)	198
CoO	cub(fcc)	291
NiO	cub(fcc)	525
α-Mn_2O_3	cub	90
CuO	monocl	230
UO_2	cub	30.8
Er_2O_3	cub	3.4
Gd_2O_3	cub	1.6
Perovskites		
$LaCrO_3$	orth	282
$LaMnO_3$	orth	100
$LaFeO_3$	orth	750
$NdCrO_3$	orth	224
$NdFeO_3$	orth	760
$YbCrO_3$	orth	118
$CaMnO_3$	cub	110
$EuTiO_3$	cub	5.3
$YCrO_3$	orth	141
$BiFeO_3$	cub*	673
$KCoF_3$	cub	125
$KMnF_3$	cub*	88.3
$KFeF_3$	cub	115
$KNiF_3$	cub	275
$NaMnF_3$	cub*	60
$NaNiF_3$	orth	149
$RbMnF_3$	cub	82
Spinels		
Co_3O_4	cub	40
$NiCr_2O_4$	tetr	65
$ZnCr_2O_4$	cub	15
$ZnFe_2O_4$	cub	9
$GeFe_2O_4$	cub	10
MgV_2O_4	cub	45
$MnGa_2O_4$	cub	33
NiAs and related structures		
Cr_2As	orth	300
CrSb	hex	714[a]
CrSe	hex	300
MnTe	hex	321[b]
NiS	hex	263
CrS	monocl	460
Rutile and related structures		
CoF_2	tetr	38
CrF_2	monocl	53
FeF_2	tetr	79
MnF_2	tetr	67
NiF_2	tetr	83
$CrCl_2$	orth	20
MnO_2	tetr	84
FeOF	tetr	315
Corundum and related structures		
Cr_2O_3	rhomb	318
α-Fe_2O_3	rhomb	948
$FeTiO_3$	rhomb	68
$MnTiO_3$	rhomb	41
$CoTiO_3$	rhomb	38
VF_3 and related structures		
CoF_3	rhomb	460
CrF_3	rhomb	80
FeF_3	rhomb	394
MnF_3	monocl	43
MoF_3	rhomb	185
Miscellaneous		
K_2NiF_4	tetr	97
MnI_2	hex	3.4
$CoUO_4$	orth	12
$CaMn_2O_4$	orth	225
CrN	cub*	273
CeC_2	tetr	33
$FeSn_2$	hex	373
Mn_2P	hex	103

* Distorted.
[a] Range of 705-723 K.
[b] Range of 320-323 K.

References

1. Gray, D. E., Ed., *American Institute of Physics Handbook, Third Edition*, McGraw-Hill, New York, 1972, pp. 5-168 to 5-183.
2. Kittel, C., *Introduction to Solid State Physics, Sixth Edition*, J. Wiley & Sons, New York, 1986.
3. Ashcroft, N. W., and Mermin, N. D., *Solid State Physics*, Holt, Rinehart, and Winston, New York, 1976, p. 697.

ORGANIC MAGNETS

J. S. Miller

Magnetic ordering, e.g., ferromagnetism, like superconductivity, is a property of a solid, not of an individual molecule or ion, and very rarely occurs for organic compounds. In contrast to superconductivity, where all electron spins pair to form a perfect diamagnetic material, magnetic ordering requires unpaired electron spins; hence, superconductivity and ferromagnetism are mutually exclusive.

The vast majority of organic compounds are diamagnetic (i.e., all electron spins are paired). Relatively few, possess unpaired electrons and are paramagnetic (PM), i.e., they are oriented in random directions. Some organic solids, however, exhibit strong magnetic behavior and magnetically order as ferromagnets (FO), with all spins aligned in the same direction. In some cases, the spins align in the opposite direction and compensate to form an antiferromagnet (AF). In other cases, these spins are not opposed to each other and do not compensate and lead to a canted antiferromagnet or weak ferromagnet (WF). If the number of spins that align in one direction differs from the number of spins that align in the opposite direction, the spins cannot compensate and a ferrimagnet (FI) results. Metamagnets (MM) are antiferromagnets in which all the spins become aligned like a ferromagnet in an applied magnetic field. Above the ordering or critical temperature, T_c, all magnets are paramagnets (PM). Organic magnets all possess electron spins in *p*-orbitals, but these may be in conjunction with metal ion-based spins.

Summary of the Critical Temperature, T_c, Saturation Magnetization, M_s, Coercive Field, H_{cr}, and Remanent Magnetization, M_r, for Selected Organic-Based Magnets

Magnet	Type	T_c/K	M_s/A m^{-1}	H_{cr}/T	M_r/A m^{-1}
α-1,3,5,7-Tetramethyl-2,6-diazaadamantane-*N*,*N*'-doxyl	FO	1.48	48,300	<0.00001	
β-2-(4'-Nitrophenyl)-4,4,5,5-tetramethyl-4,5-dihydro-1H-imidazol-1-oxyl-3-*N*-oxide	FO	0.6	22,300	0.00008	<200
$\{Fe^{III}[C_5(CH_3)_5]_2\}$[TCNE]	FO	4.8	37,600	0.10	2,300
$\{Mn^{III}[C_5(CH_3)_5]_2\}$[TCNE]	FO	8.8	58,200	0.12	3,700
$\{Cr^{III}[C_5(CH_3)_5]_2\}$[TCNE]	FO	3.65	46,300		
α-$\{Fe^{III}[C_5(CH_3)_5]_2\}$[TCNQ]	MM	2.55	34,200		
β-$\{Fe^{III}[C_5(CH_3)_5]_2\}$[TCNQ]	FO	3.0	21,600		
Tanol subarate	MM	0.38	20,700		
$NCC_6F_4CN_2S_2$	WF	35.5	45	0.00009	
$Mn^{II}(hfac)_2NITC_2H_5$	FI	7.8	39,400	0.03	27,600
$Mn^{II}(hfac)_2NIT(i\text{-}C_3H_8)$	FI	7.6	42,400	<0.0005	<420
$[Mn(hfac)_2]_3[\{ON[C_6H_3(t\text{-}C(CH_3)_3]_2NO]_2\}$	FI	46	24,400		
$[MnTPP][TCNE]\cdot 2C_6H_5CH_3$	FI	13	18,400	2.4	10,300
$V[TCNE]_x\cdot yCH_2Cl_2(x \sim 2; y \sim 0.5)$	FI	~400	28,200	0.0015 - 0.006	1,650
$Mn[TCNE]_x\cdot yCH_2Cl_2(x \sim 2; y \sim 0.5)$	FI	75	52,000	0.002	270
$Fe[TCNE]_x\cdot yCH_2Cl_2(x \sim 2; y \sim 0.5)$	FI	97	46,300	0.23	3
$Co[TCNE]_x\cdot yCH_2Cl_2(x \sim 2; y \sim 0.5)$	FI	44	22,000	0.65	

List of Symbols and Abbreviations

M_s: Saturation magnetization at 2 K
H_{cr}: Coercive field
T_c: Critical temperature
M_r: Remanent magnetization at 2 K
TCNE: Tetracyanoethylene
TCNQ: 7,7,8,8-Tetracyano-*p*-quinodimethane
hfac: Hexafluoroacetonate
NIT: Nitronyl nitroxide
FO: Ferromagnet
FI: Ferrimagnet
MM: Metamagnet
WF: Weak ferromagnet

References

1. Miller, J. S. and Epstein, A. J., *Angew. Chem. Internat. Ed.*, 33, 385, 1994 <https://doi.org/10.1002/anie.199403851>.
2. Chiarelli, R., Rassat, A., Dromzee, Y., Jeannin, Y., Novak, M. A., and Tholence, J. L., *Phys. Scrip.*, T49, 706, 1993 <https://doi.org/10.1088/0031-8949/1993/T49B/055>.
3. Kinoshita, M., *Jap. J. Appl. Phys.*, 33, 5718, 1994 <https://doi.org/10.1143/JJAP.33.5718>.
4. Gatteschi, D., *Adv. Mat.*, 6, 635, 1994 <https://doi.org/10.1002/adma.19940060903>.
5. Miller, J. S. and Epstein, A. J., *J. Chem. Soc., Chem. Commun.*, 1319, 1998 <https://doi.org/10.1039/a800922h>.
6. Broderick, W. E., Eichorn, D. M., Lu, X., Toscano, P. J., Owens, S. M. and Hoffman, B. M., *J. Am. Chem. Soc.*, 117, 3641, 1995 <https://doi.org/10.1021/ja00117a045>.
7. Banister, A. J., Bricklebank, N., Lavander, I., Rawson, J., Gregory, C. I., Tanner, B. K., Clegg, W. J., Elsegood, M. R., and Palacio, F., *Angew. Chem. Internat. Ed.*, 35, 2533, 1996 <https://doi.org/10.1002/anie.199625331>.

ELECTRON INELASTIC MEAN FREE PATHS

Cedric J. Powell

The inelastic mean free path (IMFP) of electrons impinging on a solid surface is defined as the average of distances, measured along the trajectories, that electrons with a given energy travel between inelastic collisions in the substance. It is an important parameter in analyzing results from surface characterization techniques such as Auger electron spectroscopy, x-ray photoelectron spectroscopy, low-energy electron diffraction, and others. IMFPs can be measured by the elastic-peak electron spectroscopy technique and other methods, and they can be calculated from optical data. A detailed analysis of the experimental and theoretical considerations in obtaining reliable IMPF values can be found in Refs. 4 and 5.

Table 1 below gives recommended IMFP values for 41 elemental solids in the energy range 50 eV to 30 000 eV. All values in Table 1 are taken from Ref. 1. Table 2 gives IMFP values for several inorganic compounds and organic materials in the range 50 eV to 2000 eV. The entries in Table 1 are listed by atomic number, while substances in Table 2 are listed in alphabetical order by name, with inorganic compounds preceding the organic materials. IMFP values are given in Ångström units (1 Å = 10^{-10} m).

References

1. Tanuma, S., Powell, C. J., and Penn, D. R., *Surf. Interface Anal.* (in press); doi 10.1002/sia.3522.
2. Tanuma, S., Powell, C. J., and Penn, D. R., *Surf. Interface Anal.* 17, 927, 1991.
3. Tanuma, S., Powell, C. J., and Penn, D. R., *Surf. Interface Anal.* 21, 165, 1994.
4. Powell, C. J., and Jablonski, A., *J. Phys. Chem. Ref. Data* 28, 19, 1999.
5. Powell, C. J., and Jablonski, A., *Nucl. Instr. Meth. Phys. Res. A* 601, 54, 2009.

TABLE 1. Electron Inelastic Mean Free Paths of Elemental Solids in Å (10^{-10} m)

	Electron Energy in eV														
Element	**50**	**100**	**250**	**500**	**750**	**1000**	**1250**	**1500**	**1750**	**2000**	**3000**	**5000**	**10000**	**15000**	**20000**	**30000**
Li	4.6	7.0	13.3	22.7	31.4	39.7	47.8	55.7	63.4	71.0	100.4	156.2	286.6	410.1	529.6	760.6
Be	3.5	4.3	7.2	11.7	15.9	19.9	23.8	27.6	31.2	34.9	48.8	75.3	136.7	194.7	250.6	358.4
C-graphite	4.8	4.4	6.8	10.8	14.6	18.2	21.7	25.0	28.3	31.6	44.0	67.6	122.2	173.7	223.3	318.9
C-diamond	6.9	4.7	6.4	9.9	13.2	16.4	19.4	22.4	25.3	28.1	39.0	59.6	107.3	152.1	195.3	278.4
C-glassy	5.8	6.3	10.0	16.0	21.7	27.0	32.1	37.1	42.0	46.8	65.4	100.4	181.9	258.6	332.7	475.4
Na	5.0	7.6	13.8	23.1	31.8	40.0	48.0	55.8	63.5	71.0	100.1	155.1	283.4	404.7	522.0	748.5
Mg	4.0	5.4	9.4	15.3	20.7	25.9	30.9	35.8	40.6	45.3	63.5	97.9	177.7	253.0	325.7	465.9
Al	3.5	4.6	7.9	12.7	17.1	21.3	25.4	29.4	33.3	37.2	52.0	79.9	144.8	206.0	265.1	378.8
Si	4.1	5.3	9.0	14.6	19.7	24.5	29.2	33.7	38.2	42.6	59.5	91.5	165.7	235.7	303.1	433.2
K	7.1	10.0	18.3	31.4	43.5	54.8	65.7	76.4	86.8	97.1	136.8	212.0	387.5	553.5	714.0	1024.0
Sc	4.6	4.9	7.4	11.9	16.2	20.2	24.1	27.9	31.5	35.1	49.0	75.2	135.9	193.1	248.2	354.3
Ti	4.2	4.4	6.6	10.4	14.1	17.5	20.9	24.1	27.3	30.4	42.3	64.7	116.8	165.8	213.0	303.8
V	4.6	4.7	6.7	10.3	13.8	17.1	20.3	23.4	26.4	29.4	40.7	62.1	111.7	158.3	203.1	289.3
Cr	4.5	4.4	6.4	9.9	13.2	16.3	19.4	22.3	25.2	28.0	38.8	59.2	106.4	150.7	193.4	275.6
Fe	4.3	4.6	6.7	10.2	13.5	16.7	19.7	22.7	25.7	28.5	39.5	60.1	107.8	152.7	195.8	278.8
Co	5.1	4.6	6.1	9.0	11.8	14.5	17.2	19.8	22.3	24.8	34.2	51.9	92.9	131.4	168.3	239.5
Ni	4.9	4.6	6.2	9.2	12.1	14.9	17.6	20.2	22.8	25.3	35.0	53.1	95.0	134.3	172.0	244.7
Cu	5.0	5.0	7.0	10.4	13.6	16.7	19.6	22.6	25.4	28.2	39.0	59.2	105.7	149.5	191.5	272.4
Ge	4.0	5.0	8.2	12.9	17.2	21.2	25.2	29.0	32.8	36.5	50.7	77.5	139.5	197.9	254.2	362.5
Y	5.1	5.5	8.8	14.4	19.4	24.1	28.6	33.0	37.3	41.5	57.8	88.6	159.9	227.0	291.7	416.2
Nb	5.7	5.3	7.3	11.4	15.2	18.8	22.2	25.5	28.8	31.9	44.1	67.1	120.4	170.3	218.3	310.7
Mo	5.0	4.4	6.1	9.5	12.7	15.6	18.4	21.2	23.8	26.5	36.6	55.6	99.6	141.0	180.7	257.1
Ru	4.9	4.1	5.4	8.3	11.1	13.7	16.1	18.5	20.8	23.1	31.9	48.5	86.7	122.6	157.1	223.5
Rh	4.8	4.0	5.2	7.9	10.6	13.0	15.4	17.7	19.9	22.1	30.5	46.2	82.7	117.0	149.9	213.2
Pd	4.8	4.0	5.3	8.1	10.8	13.4	15.8	18.2	20.5	22.7	31.3	47.6	85.2	120.4	154.3	219.5
Ag	6.1	4.9	5.9	8.7	11.6	14.2	16.8	19.2	21.6	24.0	33.0	50.0	89.3	126.2	161.6	229.7
In	4.7	5.4	8.0	12.3	16.5	20.6	24.4	28.1	31.8	35.3	48.9	74.5	133.8	189.5	243.1	346.3
Sn	5.8	6.4	9.2	13.8	18.4	22.8	27.0	31.1	35.1	38.9	53.8	81.8	146.5	207.3	265.8	378.2
Cs	6.3	9.3	17.4	29.3	40.3	50.8	61.0	71.0	80.7	90.2	126.8	196.1	357.6	510.4	657.9	942.8
Gd	4.1	4.6	7.4	11.8	15.7	19.5	23.2	26.8	30.3	33.7	46.9	71.6	128.8	182.6	234.4	334.1
Tb	4.1	4.3	6.5	10.2	13.7	16.9	20.1	23.2	26.2	29.2	40.6	61.9	111.3	157.8	202.5	288.6
Dy	4.5	4.6	6.8	10.6	14.1	17.5	20.8	23.9	27.0	30.0	41.7	63.5	114.0	161.5	207.2	295.1
Hf	5.5	5.5	7.5	11.3	14.9	18.2	21.5	24.6	27.7	30.7	42.3	64.2	114.4	161.5	206.7	293.8
Ta	4.9	4.5	6.1	9.2	12.1	14.8	17.4	19.9	22.4	24.8	34.2	51.7	92.2	130.0	166.5	236.4

Element	Electron Energy in eV															
	50	100	250	500	750	1000	1250	1500	1750	2000	3000	5000	10000	15000	20000	30000
W	5.4	4.4	5.8	8.7	11.3	13.9	16.3	18.6	20.9	23.2	31.8	48.1	85.5	120.5	154.1	218.8
Re	5.4	4.2	5.4	8.0	10.5	12.8	15.0	17.2	19.3	21.3	29.3	44.3	78.7	111.0	142.0	201.5
Os	5.6	4.4	5.6	8.4	10.9	13.3	15.6	17.8	20.0	22.1	30.3	45.7	81.1	114.3	146.2	207.3
Ir	5.3	4.2	5.4	8.0	10.4	12.7	15.0	17.1	19.2	21.2	29.1	43.9	78.1	110.1	140.8	199.7
Pt	5.0	4.2	5.5	8.2	10.7	13.1	15.4	17.6	19.8	21.9	30.0	45.4	80.7	113.8	145.5	206.6
Au	5.1	4.3	5.6	8.4	11.0	13.5	15.9	18.2	20.5	22.7	31.1	47.1	83.9	118.4	151.5	215.1
Bi	4.9	5.5	8.0	12.3	16.4	20.2	23.9	27.5	31.0	34.4	47.5	72.1	129.3	182.8	234.4	333.6

TABLE 2. Electron Inelastic Mean Free Paths of Other Materials in Å (10^{-10} m)

Substance	Formula	Electron Energy in eV														Ref.
		50	100	150	200	300	400	600	800	1000	1200	1400	1600	1800	2000	
Gallium phosphide	GaP	5.6	6.5	7.8	9.0	11.4	13.7	18.1	22.3	26.3	30.2	34.1	37.8	41.5	45.2	2
Indium phosphide	InP	4.8	4.9	5.6	6.4	8.1	9.7	12.8	15.7	18.7	21.4	24.2	26.8	29.4	32.0	2
Lead(II) sulfide	PbS	4.8	5.6	6.7	7.8	10.0	12.1	16.1	19.8	23.6	27.1	30.6	33.9	37.2	40.5	2
Lead(II) telluride	PbTe	4.3	5.5	6.6	7.7	9.8	11.9	15.8	19.6	23.4	26.9	30.3	33.7	37.0	40.2	2
Potassium chloride	KCl	7.5	7.8	9.3	10.9	14.2	17.3	23.2	28.7	34.0	39.2	44.2	49.1	54.0	58.8	2
Silicon carbide	SiC	4.7	4.9	5.8	6.8	8.7	10.5	13.9	17.1	20.3	23.3	26.3	29.2	32.1	35.0	2
Silicon dioxide (vitreous)	SiO_2	8.0	7.7	8.8	10.0	12.6	15.2	20.0	24.7	29.3	33.7	38.0	42.2	46.4	50.5	2
Zinc sulfide	ZnS	5.8	6.5	7.7	8.9	11.3	13.6	17.9	22.0	26.0	29.8	33.6	37.3	40.9	44.5	2
Adenine	$C_5H_5N_5$	6.4	6.6	7.8	9.2	11.8	14.4	19.2	24.1	28.6	33.1	37.4	41.6	45.8	49.9	3
Bovine plasma albumin		7.3	7.2	8.5	9.9	12.7	15.4	20.7	25.8	30.8	35.6	40.2	44.8	49.4	53.8	3
β-Carotene	$C_{40}H_{56}$	6.4	7.0	8.5	10.0	13.0	15.9	21.4	26.9	32.0	37.0	41.9	46.6	51.3	56.0	3
Deoxyribonucleic acid (DNA)		7.3	7.3	8.5	9.8	12.6	15.4	20.7	25.9	30.8	35.6	40.3	44.9	49.4	53.8	3
1,6-Diphenyl-1,3,5-hexatriene	$C_{18}H_{16}$	6.4	7.0	8.4	9.9	12.9	15.8	21.3	26.7	31.8	36.8	41.7	46.4	51.1	55.7	3
Guanine	$C_5H_5N_5O$	6.2	6.2	7.2	8.4	10.8	13.1	17.5	21.8	25.9	29.9	33.8	37.6	41.4	45.1	3
Hexacosane	$C_{26}H_{54}$	7.0	7.6	9.2	10.9	14.1	17.2	23.2	29.2	34.7	40.1	45.4	50.6	55.7	60.7	3
Kapton		7.0	6.8	7.9	9.2	11.7	14.2	19.0	23.7	28.2	32.5	36.7	40.9	44.9	49.0	3
Polyacetylene		5.3	5.7	6.8	7.9	10.2	12.5	16.9	21.1	25.1	29.0	32.8	36.5	40.2	43.8	3
Poly(butene-1-sulfone)		7.1	7.2	8.5	9.9	12.7	15.4	20.7	25.8	30.6	35.3	39.9	44.4	48.8	53.2	3
Polyethylene		6.9	7.2	8.6	10.1	13.0	15.9	21.4	26.8	31.8	36.8	41.6	46.3	51.0	55.6	3
Poly(methyl methacrylate)		7.8	7.9	9.3	10.8	13.9	16.9	22.7	28.3	33.7	38.8	43.9	48.9	53.8	58.6	3
Polystyrene		6.9	7.3	8.7	10.2	13.2	16.1	21.6	27.1	32.2	37.2	42.1	46.9	51.6	56.2	3
Poly(2-vinylpyridine)		6.9	7.3	8.7	10.3	13.3	16.2	21.8	27.3	32.5	37.5	42.4	47.3	52.0	56.7	3

ELECTRON STOPPING POWERS

Cedric J. Powell

Numerical data are given for mass collision electron stopping powers in 41 elemental solids for energies between 100 eV and 30 keV. These stopping powers were determined with an algorithm that utilizes experimental optical data for each solid.

The stopping power for electrons and other charged particles in matter is often needed in calculations of electron transport in a medium, particularly in radiation physics and in descriptions of signal generation in analytical techniques such as electron-probe microanalysis and Auger electron spectroscopy. The stopping power is defined as the average rate at which the charged particles lose energy at any point along their trajectories. For electrons, it is customary to separate the total stopping power into two components, the collision stopping power due to inelastic-scattering events of the electrons in a medium and the radiative stopping power due to the emission of bremsstrahlung in the electric field of the atomic nucleus and atomic electrons (Ref. 1). For electron energies less than 30 keV, the radiative stopping power is less than 1% of the collision stopping power (Ref. 1) and is neglected in the numerical data given here.

Numerical data for collision and radiative stopping powers at electron energies between 10 keV and 1 GeV have been published for materials of interest in radiation physics and dosimetry (Ref. 1). Similar data can also be obtained from a Web site of the National Institute of Standards and Technology (Ref. 2). The collision stopping powers were calculated from the theory of Bethe (Refs. 3,4) and recommended values of the one material-dependent parameter, the mean excitation energy (Ref. 1). While the Bethe theory is expected to be valid for electron energies much larger than the largest K-shell binding energy of atoms in the particular material, the Bethe stopping-power equation is frequently utilized to calculate stopping powers for energies of 10 keV and above (Refs. 1,2). Detailed analyses of the Bethe stopping-power theory have been published (Refs. 1,4).

The table below gives mass collision stopping powers for 41 elemental solids at energies between 100 eV and 30 keV (Ref. 5). The mass collision stopping power is the collision stopping power divided by the mass density of the solid. The mass collision stopping powers in the table were determined by interpolation with a clamped cubic spline from the published data (Ref. 5), which had been calculated with an algorithm that utilizes experimental optical data for each solid. Comparisons with collision stopping powers from the Bethe stopping-power equation showed root-mean-square differences of 9.1% and 8.7% for energies of 9.897 keV and 29.733 keV, respectively (Ref. 5). This level of agreement was considered satisfactory on account of uncertainties of the algorithm and optical data used for the calculations as well as uncertainties of the mean excitation energies used with the Bethe equation. The mass collision stopping powers in the table are given in units of MeV cm^2/g for a range of relativistic kinetic energies. The elemental solids are listed in order of element symbol.

References

1. *Stopping Powers for Electrons and Positrons*, ICRU Report 37 (International Commission on Radiation Units and Measurements, Bethesda, MD, 1984).
2. Berger, M. J., Coursey, J. S., Zucker, M. A., and Chang, J., *Stopping-Power and Range Tables for Electrons, Positrons, and Helium Ions*, Version 1.2.3, 2005 <http://www.nist.gov/pml/data/star/index.cfm>.
3. Bethe, H., *Ann. Physik* 5, 325, 1930 <https://doi.org/10.1002/andp.19303970303>.
4. Inokuti, M., *Rev. Mod. Phys.* 43, 297, 1971 <https://doi.org/10.1103/RevModPhys.43.297>.
5. Shinotsuka, H., Tanuma, S., Powell, C. J., and Penn, D. R., *Nucl. Instr. Methods Phys. Res. B* 270, 75, 2012 <https://doi.org/10.1016/j.nimb.2011.09.016>.

Values of Mass Collision Stopping Powers in MeV cm^2/g for the Indicated Relativistic Electron Kinetic Energies (Ref. 5)

Element	100 eV	200 eV	300 eV	400 eV	500 eV	1000 eV	1500 eV	2000 eV	3000 eV	4000 eV	5000 eV	10000 eV	15000 eV	20000 eV	30000 eV
Ag	66.9	94.9	95.5	87.0	78.7	54.7	44.3	38.5	31.1	26.2	22.8	14.3	10.7	8.71	6.54
Al	187.6	142.8	127.1	118.1	110.7	83.8	67.5	56.7	43.4	35.5	30.2	18.2	13.5	10.9	8.08
Au	41.7	52.1	49.7	45.5	41.7	31.1	26.0	22.8	18.7	16.0	14.1	9.21	7.12	5.94	4.58
Be	342.2	257.8	210.0	183.5	165.7	113.5	87.4	71.6	53.3	42.9	36.2	21.0	15.2	12.1	8.82
Bi	47.8	56.8	56.1	51.9	47.5	34.2	28.2	24.6	20.2	17.4	15.3	10.0	7.65	6.34	4.87
C[a]	251.3	218.8	183.9	158.4	139.8	92.5	73.9	62.4	48.1	39.4	33.6	20.0	14.7	11.8	8.60
C[b]	343.5	296.1	246.2	210.6	184.6	118.9	92.1	76.2	57.5	46.7	39.5	23.2	16.9	13.5	9.84
C[c]	262.2	249.6	212.4	183.6	162.0	106.1	83.0	69.0	52.4	42.7	36.2	21.4	15.6	12.5	9.12
Co	82.0	97.7	96.4	90.8	84.7	62.3	49.5	41.5	32.1	26.8	23.3	14.7	11.0	8.96	6.67
Cr	97.4	104.1	96.9	88.5	81.0	57.4	45.3	38.1	30.1	25.5	22.3	14.0	10.5	8.51	6.34
Cs	80.3	63.5	59.6	58.1	54.1	37.8	29.8	25.1	19.8	16.9	14.9	9.90	7.60	6.26	4.74
Cu	68.1	76.4	76.5	73.9	70.7	55.5	45.3	38.5	30.0	25.0	21.8	13.9	10.5	8.60	6.43
Dy	75.7	78.4	70.3	63.6	58.6	43.9	35.6	30.3	23.8	19.8	17.2	11.1	8.52	7.02	5.32
Fe	80.4	86.2	83.7	78.7	73.7	54.8	43.9	37.0	28.9	24.4	21.3	13.6	10.2	8.31	6.20
Gd	72.2	66.5	58.9	53.4	49.8	37.8	30.7	26.2	20.6	17.3	15.0	10.0	7.74	6.42	4.89
Ge	90.4	79.3	73.7	69.8	66.5	53.6	44.7	38.4	30.2	25.1	21.7	13.8	10.5	8.55	6.41
Hf	42.6	51.0	48.0	44.7	41.9	32.9	27.7	24.2	19.6	16.6	14.5	9.48	7.39	6.15	4.72
In	62.6	74.7	75.4	69.7	63.4	43.7	34.7	29.5	23.9	20.5	18.1	11.7	8.88	7.27	5.50
Ir	39.3	47.0	44.6	40.9	37.9	29.4	25.1	22.2	18.3	15.7	13.9	9.07	7.04	5.88	4.53
K	189.9	156.2	127.5	108.8	95.8	64.1	52.4	45.2	35.8	29.8	25.7	15.7	11.7	9.42	7.01
Li	336.6	263.5	233.4	207.0	185.3	122.0	92.2	74.7	55.0	44.1	37.0	21.3	15.4	12.2	8.86

Element	100 eV	200 eV	300 eV	400 eV	500 eV	1000 eV	1500 eV	2000 eV	3000 eV	4000 eV	5000 eV	10000 eV	15000 eV	20000 eV	30000 eV
Mg	203.8	175.9	165.9	154.5	143.2	103.5	81.6	67.8	51.3	41.7	35.4	21.3	15.7	12.6	9.32
Mo	74.1	87.1	76.2	66.5	59.2	41.5	34.8	30.6	25.0	21.3	18.6	11.8	8.95	7.36	5.57
Na	161.5	160.4	153.7	143.3	132.6	94.4	73.5	60.7	45.7	37.1	31.5	19.0	14.0	11.3	8.30
Nb	70.8	86.3	75.1	65.5	58.5	41.5	34.8	30.5	24.8	21.1	18.4	11.7	8.84	7.27	5.51
Ni	77.4	90.3	91.5	87.6	82.7	62.2	49.9	42.0	32.5	27.0	23.5	14.9	11.2	9.13	6.82
Os	36.9	44.9	43.0	39.8	37.1	29.1	24.8	21.9	18.0	15.4	13.6	8.88	6.91	5.77	4.44
Pd	69.8	88.2	82.8	73.3	65.5	44.6	35.7	30.9	25.3	21.6	19.0	12.2	9.19	7.53	5.70
Pt	40.4	47.1	44.8	41.2	38.1	29.4	25.0	22.1	18.3	15.7	13.9	9.08	7.04	5.88	4.53
Re	42.3	50.5	47.7	44.0	40.9	31.7	26.8	23.5	19.2	16.4	14.4	9.35	7.27	6.06	4.65
Rh	70.1	87.0	82.0	72.9	65.3	44.9	36.3	31.7	25.9	22.1	19.4	12.4	9.35	7.66	5.80
Ru	68.9	83.9	76.6	67.6	60.4	41.6	34.1	29.9	24.5	21.0	18.4	11.7	8.88	7.28	5.52
Sc	204.5	204.0	171.7	148.1	130.7	85.4	66.0	55.5	43.2	35.8	30.7	18.7	13.8	11.1	8.20
Si	199.4	148.9	125.7	115.9	109.2	84.8	68.6	57.8	44.3	36.3	30.9	18.6	13.8	11.1	8.23
Sn	56.3	68.5	71.5	67.5	62.0	43.3	34.4	29.2	23.5	20.2	17.8	11.6	8.79	7.20	5.44
Ta	43.3	51.8	49.0	45.4	42.5	33.2	28.0	24.4	19.8	16.9	14.8	9.58	7.44	6.20	4.75
Tb	85.4	85.3	75.1	67.2	61.5	45.1	36.3	30.7	24.0	20.0	17.3	11.2	8.58	7.07	5.35
Ti	147.7	153.7	132.6	116.2	103.6	69.3	53.8	45.2	35.5	29.7	25.6	15.8	11.7	9.46	7.01
V	101.1	119.0	112.1	102.0	92.8	64.5	50.6	42.5	33.5	28.2	24.4	15.2	11.3	9.15	6.79
W	41.5	49.3	46.7	43.3	40.5	31.8	26.9	23.7	19.3	16.5	14.4	9.40	7.31	6.09	4.68
Y	117.2	106.8	88.0	75.6	67.1	48.4	40.5	35.2	28.2	23.7	20.5	12.8	9.66	7.92	5.97

[a] Glassy carbon.
[b] Graphite.
[c] Diamond.

ELECTRON WORK FUNCTION OF THE ELEMENTS

The electron work function Φ is a measure of the minimum energy required to extract an electron from the surface of a solid. It is defined more precisely as the energy difference between the state in which an electron has been removed to a distance from the surface of a single crystal face that is large enough that the image force is negligible but small compared to the distance to any other face (typically about 10^{-4} cm) and the state in which the electron is in the bulk solid. In general, Φ differs for each face of a monocrystalline sample.

Because Φ is dependent on the cleanliness of the surface, measurements reported in the literature often cover a considerable range (Refs. 1-4). This table contains selected values for the electron work function of the elements which may be regarded as typical values for a reasonably clean surface. The table contains recommended values from a recent review that includes estimated uncertainties (see Ref. 1 for more details). Those values, marked R in the table, come from analysis of results from different measurement methods. The method of measurement is indicated for each value. The following abbreviations appear.

R – Recommended value from Ref. 1
TE – Thermionic emission
PE – Photoelectric effect
FE – Field emission
CPD – Contact potential difference
polycr – Polycrystalline sample
amorp – Amorphous sample

Column definitions for the table are as follows.

Column heading	Definition
Mol. form.	Element symbol
Plane	Crystallographic plane; phase is indicated within square brackets, when applicable
Φ	Electron work function, in eV
Uncert.	Uncertainty, in eV, when available; values in parentheses are approximate
Method	Measurement method or source; see list above

References

1. Derry, G. N., Kern, M. E., and Worth, E. H., Recommended Value of Clean Metal Surface Work Functions, *J. Vac. Sci. Technol.*, A33, 060801, 2015. <doi.org/10.1116/1.4934685>.
2. Hölzl, J., and Schulte, F. K., Work Functions of Metals, in *Solid Surface Physics*, Höhler, G., Ed., Springer-Verlag, Berlin, 1979. <https://doi.org/10.1007/BFb0048919>
3. Riviere, J. C., Work Function: Measurements and Results, in *Solid State Surface Science, Vol. 1*, Green, M., Ed., Decker, New York, 1969.
4. Michaelson, H. B., *J. Appl. Phys.*, 48, 4729, 1977. <https://doi.org/10.1063/1.323539>

Electron Work Functions of Crystalline Elements

Mol. form.	Plane	Φ/eV	Uncert./eV	Method
Ag	100	4.36	0.05	R
	110	4.10	0.15	R
	111	4.53	0.07	R
Al	100	4.31	0.18	R
	110	4.23	0.13	R
	111	4.32	0.06	R
As	polycr	(3.75)		PE
Au	100	5.22	0.31	R
	110	5.16	0.22	R
	111	5.33	0.06	R
B	polycr	(4.45)		TH
Ba	polycr	2.52		TH
Be	polycr	4.98		PE
Bi	polycr	4.34		PE
C	polycr	5.00		CPD
Ca	polycr	2.87		PE
Cd	polycr	4.08		CPD
Ce	polycr	2.90		PE
Co	polycr	5.00		PE
Cr	polycr	4.50		PE
Cs	polycr	1.95		PE
Cu	100	4.73	0.10	R
	110	4.56	0.10	R
	111	4.90	0.02	R
	112	4.53		PE
Eu	polycr	2.50		PE
Fe	100	4.60	0.33	R
	110	5.07	0.04	R
	111	4.81	0.29	R
Ga	polycr	4.32		PE
Gd	polycr	2.90		CPD
Ge	polycr	5.00		CPD
Hf	polycr	3.90		PE
Hg	liquid	4.475		PE
In	polycr	4.09		PE
Ir	100[1X1]	5.97	0.23	R
	100[5X1]	5.95	0.25	R
	110	5.42	0.32	R
	111	5.78	0.04	R
	210	5.00		PE
K	polycr	2.29		PE
La	polycr	3.50		PE
Li	polycr	2.93		FE
Lu	polycr	(3.30)		CPD
Mg	polycr	3.66		PE
Mn	polycr	4.10		PE
Mo	100	4.46	0.11	R
	110	4.92	0.05	R
	111	4.37	0.24	R
	112	4.36		PE
	114	4.50		PE
	332	4.55		PE
Na	polycr	2.36		PE
Nb	100	4.08	0.17	R
	110	4.63	0.17	R
	111	4.37	0.19	R
	112	4.63		TH
	113	4.29		TH
	116	3.95		TH
	310	4.18		TH

Mol. form.	Plane	Φ/eV	Uncert./eV	Method
Nd	polycr	3.20		PE
Ni	100	5.17	0.11	R
	110	4.72	0.13	R
	111	5.24	0.07	R
Os	polycr	5.93		PE
Pb	polycr	4.25		PE
Pd	polycr	5.22		PE
	100	5.48	0.23	R
	110	5.07	0.20	R
	111	5.67	0.12	R
Pt	polycr	5.64		PE
	100[1X1]	5.75	0.13	R
	100[5X1]	5.67	0.14	R
	110	5.53	0.13	R
	111	5.91	0.08	R
	320	5.22		FE
	331	5.12		FE
Rb	polycr	2.261		PE
Re	polycr	4.72		TE
Rh	polycr	4.98		PE
	100	5.30	0.15	R
	110	4.86	0.21	R
	111	5.46	0.09	R
Ru	polycr	4.71		PE
	100	5.40	0.11	R
	1010	4.60	0.28	R
Sb	amorp	4.55		
	100	4.70		
Sc	polycr	3.50		PE
Se	polycr	5.90		PE
Si	n	4.85		CPD
	p 100	(4.91)		CPD
	p 111	4.60		PE
Sm	polycr	2.70		PE
Sn	polycr	4.42		CPD
Sr	polycr	(2.59)		TH
Ta	polycr	4.25		TH
	100	4.10	0.25	R
	110	4.74	0.09	R
	111	3.50	0.21	R
Tb	polycr	3.00		PE
Te	polycr	4.95		PE
Th	polycr	3.40		TH
Ti	polycr	4.33		PE
Tl	polycr	(3.84)		CPD
U	polycr	3.63		PE
	100	3.73		PE
	110	3.90		PE
	113	3.67		PE
V	polycr	4.30		PE
W	polycr	4.55		CPD
	100	4.70	0.06	R
	110	5.44	0.14	R
	111	4.44	0.03	R
	113	4.46		FE
	116	4.32		TH
	211	4.84	0.07	R
Y	polycr	3.10		PE
Zn	polycr	3.63		PE
	polycr	(4.90)		CPD
Zr	polycr	4.05		PE

SECONDARY ELECTRON EMISSION

The secondary emission yield, or secondary emission ratio, δ, is the average number of secondary electrons emitted from a bombarded material for every incident primary electron. It is a function of the primary electron energy E_p. The maximum yield δ_{max} corresponds to a primary electron energy E_{pmax} (see figure). The two primary electron energies corresponding to a yield of unity are denoted by the first and second crossovers (E_I and E_{II}). An insulating target, or a conducting target that is electrically floating, will charge positively or negatively depending on the primary electron energy. For $E_I < E_p < E_{II}$, $\delta > 1$ and the surface charges positively provided there is a collector present that is positive with respect to the target. For $E_p < E_I$ or $E_p > E_{II}$, $\delta < 1$, and the surface charges negatively with respect to the potential of the source of primary electrons.

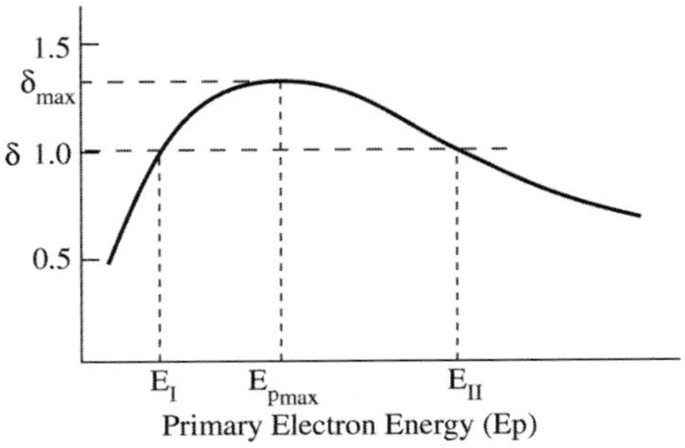

Secondary Electron Emission of the Elements

Element	δ_{max}	E_{pmax}/eV	E_I/eV	E_{II}/eV
Ag	1.5	800	200	>2000
Al	1.0	300	300	300
Au	1.4	800	150	>2000
B	1.2	150	50	600
Ba	0.8	400	None	None
Bi	1.2	550	None	None
Be	0.5	200	None	None
C (diamond)	2.8	750	None	>5000
C (graphite)	1.0	300	300	300
C (carbon black)	0.45	500	None	None
Cd	1.1	450	300	700
Co	1.2	600	200	None
Cs	0.7	400	None	None
Cu	1.3	600	200	1500
Fe	1.3	400	120	1400
Ga	1.55	500	75	None
Ge	1.15	500	150	900
Hg	1.3	600	350	>1200
K	0.7	200	None	None
Li	0.5	85	None	None
Mg	0.95	300	None	None
Mo	1.25	375	150	1200
Na	0.82	300	None	None
Nb	1.2	375	150	1050
Ni	1.3	550	150	>1500
Pb	1.1	500	250	1000
Pd	1.3[a]	250[b]	120	None
Pt	1.8	700	350	3000
Rb	0.9	350	None	None
Sb	1.3	600	250	2000
Si	1.1	250	125	500
Sn	1.35	500	None	None
Ta	1.3	600	250	>2000
Th	1.1	800	None	None
Ti	0.9	280	None	None
Tl	1.7	650	70	>1500
W	1.4	650	250	>1500
Zr	1.1	350	None	None

[a] ≤1.3.
[b] ≥250.

Secondary Electron Emission of Inorganic Compounds

Material	δ_{max}	E_{pmax}/eV
Alkali halides		
CsCl	6.5	
KBr (crystal)	14	1800
KCl (crystal)	12	1600
KCl (layer)	7.5	1200
KI (crystal)	10	1600
KI (layer)	5.6	
LiF (crystal)	8.5	
LiF (layer)	5.6	700
NaBr (crystal)	24	1800
NaBr (layer)	6.3	
NaCl (crystal)	14	1200
NaCl (layer)	6.8	600
NaF (crystal)	14	1200
NaF (layer)	5.7	
NaI (crystal)	19	1300
NaI (layer)	5.5	
RbCl (layer)	5.8	
Oxides		
Ag_2O	1.0	
Al_2O_3 (layer)	2–9	
BaO (layer)	2.3–4.8	400
BeO	3.4	2000
CaO	2.2	500
Cu_2O	1.2	400
MgO (crystal)	20–25	1500
MgO (layer)	3–15	400–1500
MoO_2	1.2	
SiO_2 (quartz)	2.1–4	400
SnO_2	3.2	640
Sulfides		
MoS_2	1.1	
PbS	1.2	500
WS_2	1.0	
ZnS	1.8	350
Others		
BaF_2 (layer)	4.5	
CaF_2 (layer)	3.2	
$BiCs_3$	6	1000
BiCs	1.9	1000
GeCs	7	700
Rb_3Sb	7.1	450
$SbCs_3$	6	700
Mica	2.4	350
Glasses	2–3	300–450

OPTICAL PROPERTIES OF SELECTED ELEMENTS

J. H. Weaver and H. P. R. Frederikse

These tables list the index of refraction *n*, the extinction coefficient *k*, and the normal incidence reflection R ($\phi = 0$) as a function of photon energy *E*, which is expressed in electron volts (eV). To convert the energy in eV to the wavelength in μm, use $\lambda = 1.2398/E$. To compute the dielectric function $\tilde{\varepsilon} = \varepsilon_1 + i\varepsilon_2$ from the complex index of refraction $\tilde{N} = n + ik$, use $\varepsilon_1 = n^2 - k^2$ and $\varepsilon_2 = 2nk$.

This table in the Print Edition of the *CRC Handbook* is condensed from the longer table in the Online Edition, which gives values of the optical properties at finer energy intervals.

The symbol ***E*** indicates the electric field vector of the incident light, which may be either parallel or perpendicular to the unit vector perpendicular to the crystal face.

The optical constants in these tables are abridged from three more extensive tabulations:

- *Optical Properties of Metals* (OPM), Volumes I and II, *Physics Data, Nr.* 18-1 and 18-2, J. H. Weaver, C. Krafka, D. W. Lynch, and E. E. Koch, Fachinformationzentrum, Karlsruhe, Germany.
- *Handbook of Optical Constants* (HOC), Vol. I, 1985, and Vol. II, 1991. E. D. Palik, Ed., Academic Press, London.
- *American Institute of Physics Handbook (AIPH), Third Edition*, D. E. Gray, Ed., McGraw-Hill, New York, 1972.

The first two of these major sources provide detailed comparisons of all optical data available in the literature at the time of the compilation. For critical applications the reader should refer to the original work. References for individual metals and semiconductors are listed at the end of the tables. Generally, tabulated values for the optical properties are accurate to better than 10%. Extrapolated or interpolated values are marked with a footnote [a]. For most elements the spectral range covered is from the far infrared (0.010 or 0.10 eV) to the far ultraviolet (10, 30, or 300 eV). The intervals between successive energies in the tables are chosen in such a way that the major spectral features are preserved.

Very small values of *k* are expressed in exponential notation, e.g., 1.23E-5 means 1.23×10^{-5}.

The following table is convenient for associating the energy entries in these tables with the corresponding wavelengths.

Relation between Energy and Wavelength

λ	*E*/eV
1 mm	0.00124
500 μm	0.00248
100 μm	0.01240
50 μm	0.02480
10 μm	0.12398
5 μm	0.24797
1 μm	1.240
700 nm	1.771
600 nm	2.066
500 nm	2.480
400 nm	3.100
300 nm	4.133
200 nm	6.199
100 nm	12.398
50 nm	24.797

Refractive Index *n*, Extinction Coefficient *k*, and Normal Incidence Reflection *R* as a Function of Energy

Energy (eV)	*n*	*k*	$R(\phi = 0)$
Aluminum[1]			
0.050	74.997	172.199	0.9915
0.100	34.464	105.600	0.9889
0.500	3.072	25.581	0.9817
1.000	1.212	12.464	0.9697
2.200	1.018	6.846	0.9200
2.400	0.826	6.283	0.9228
10.000	0.044	1.178	0.9286
50.000	0.969	0.006	0.0003
100.000	0.991	0.030	0.0002
Carbon (diamond)[2]			
0.06199	2.3741		0.166
0.4959	2.3801		0.167
2.271	2.4210		0.173
2.480	2.4299		0.174
10.00	3.453	1.258	0.355
Cesium (evaporated)[3]			
2.271	0.278	0.950	0.561
3.064	0.540	0.320	0.127
4.889	0.916	0.143	0.007
Chromium[4]			
0.06	21.19	42.00	0.962
0.10	11.81	29.76	0.955
0.54	3.92	7.06	0.788
1.00	4.47	4.43	0.639
2.20	3.18	4.41	0.656
2.40	2.75	4.46	0.677
2.60	2.22	4.36	0.698
10.00	0.98	0.73	0.120
Cobalt, single crystal, $\vec{E} \parallel \hat{c}$[5]			
0.10	6.71	37.87	0.982
0.50	4.41	7.19	0.782
1.00	4.46	5.86	0.722
2.20	2.07	3.70	0.642
2.30	2.01	3.59	0.634
2.40	1.95	3.49	0.627
2.50	1.88	3.40	0.622
2.60	1.81	3.32	0.618
Cobalt, single crystal, $\vec{E} \perp \hat{c}$[5]			
0.10	5.83	32.36	0.979
0.50	5.17	5.75	0.709
1.00	4.83	5.94	0.721
2.20	2.04	3.72	0.646
2.30	1.99	3.56	0.632
2.40	1.95	3.44	0.620
2.50	1.90	3.34	0.611
2.60	1.86	3.26	0.605
Copper[6]			
0.10	29.69	71.57	0.980
0.50	1.71	17.63	0.979
1.00	0.44	8.48	0.976
2.20	0.83	2.60	0.673
2.30	1.04	2.59	0.618
2.40	1.12	2.60	0.602
2.60	1.15	2.50	0.577
10.00	1.04	0.82	0.139
50.00	0.95	0.13	0.005
Gallium (liquid)[7]			
1.425	2.40	9.20	0.900
2.066	1.25	6.60	0.897
2.480	0.89	5.60	0.898
3.100	0.59	4.50	0.896

Energy (eV)	n	k	R(φ = 0)
Germanium, single crystal[8]			
0.05083		1.40E-03	
0.1	4.0063	3.70E-05	0.361
0.5	4.074		0.367
1.0	4.325	8.09E-02	0.390
2.2	5.283	2.049	0.516
2.3	5.062	2.318	0.519
2.4	4.610	2.455	0.508
2.5	4.340	2.384	0.492
2.6	4.180	2.309	0.480
10.0	0.93	0.86	0.167
Gold, electropolished, Au (110)[9]			
0.10	8.17	82.83	0.995
0.50	0.39	16.61	0.994
1.00	0.13	8.03	0.992
2.20	0.24	2.54	0.880
2.40	0.50	1.86	0.647
2.50	0.82	1.59	0.438
2.60	1.24	1.54	0.331
10.00	1.37	0.80	0.126
Hafnium, single crystal, $\vec{E} \parallel \hat{c}$[10]			
0.52	1.48	4.11	0.747
1.00	4.45	3.00	0.545
2.20	3.53	2.99	0.526
2.30	3.34	3.09	0.534
2.40	3.15	3.11	0.537
2.50	2.99	3.13	0.540
2.60	2.83	3.12	0.542
10.00	1.36	1.22	0.235
Hafnium, single crystal, $\vec{E} \perp \hat{c}$[10]			
0.52	2.25	4.65	0.723
1.00	4.76	3.76	0.602
2.20	3.18	3.36	0.563
2.30	2.99	3.39	0.568
2.40	2.78	3.35	0.569
2.50	2.65	3.26	0.562
2.60	2.54	3.22	0.560
10.00	1.32	1.21	0.230
Iridium[11]			
0.10	28.49	60.62	0.975
0.50	2.98	14.06	0.944
1.00	3.15	7.31	0.822
2.20	2.29	4.38	0.695
2.30	2.18	4.26	0.692
2.40	2.07	4.14	0.689
2.50	1.98	4.00	0.682
2.60	1.91	3.86	0.673
10.00	1.45	1.01	0.175
Iron[5]			
0.10	6.41	33.07	0.978
0.50	4.14	8.02	0.817
1.00	3.43	4.79	0.678
2.20	2.74	3.33	0.563
2.30	2.65	3.34	0.567
2.40	2.56	3.31	0.567
2.50	2.46	3.31	0.570
2.60	2.34	3.30	0.576
10.00	0.88	0.97	0.213

Energy (eV)	n	k	R(φ = 0)
10.17	0.87	0.94	0.203
Lithium[12]			
0.54	0.661	12.6	0.984
1.05	0.448	5.58	0.946
2.25	0.206	2.48	0.892
2.55	0.217	2.11	0.854
10.1	0.726	0.108	0.029
Magnesium (evaporated)[13]			
2.270	0.57	3.47	0.843
2.522	0.53	2.92	0.805
Manganese[14]			
0.64	3.89	5.95	0.738
1.02	3.48	4.74	0.673
2.26	2.39	3.33	0.577
2.38	2.32	3.23	0.567
2.50	2.25	3.14	0.559
6.60	1.48	1.47	0.288
Mercury (liquid)[15]			
0.5	8.528	9.805	0.818
1.0	4.962	7.643	0.789
2.2	1.620	4.751	0.780
2.4	1.384	4.407	0.779
2.6	1.186	4.090	0.779
10.0	1.062	0.567	0.071
10.2	1.054	0.569	0.072
Molybdenum[16]			
0.10	18.53	68.51	0.985
0.50	1.37	13.55	0.971
1.00	1.94	5.58	0.805
2.20	3.76	3.41	0.562
2.30	3.79	3.61	0.578
2.40	3.59	3.78	0.594
2.50	3.36	3.73	0.591
2.60	3.22	3.61	0.582
10.00	0.77	0.99	0.250
Nickel[17]			
0.10	9.54	45.82	0.983
0.50	4.03	9.64	0.864
1.00	3.06	5.74	0.753
2.20	1.80	3.33	0.620
2.30	1.75	3.19	0.605
2.40	1.71	3.06	0.590
2.50	1.67	2.93	0.575
2.60	1.65	2.81	0.557
10.00	0.95	0.87	0.166
50.00	0.92	0.10	0.004
Niobium[18]			
1.05	1.44	5.86	0.857
2.25	2.93	2.87	0.505
2.35	2.92	2.88	0.506
2.45	2.89	2.90	0.509
2.55	2.83	2.92	0.512
10.00	1.18	1.05	0.194
Osmium (polycrystalline)[9]			
0.10	4.08	50.23	0.994
0.50	2.41	9.97	0.913
1.00	2.09	4.41	0.712

Energy (eV)	n	k	R(φ = 0)
2.20	4.58	1.62	0.457
2.30	4.84	1.76	0.479
2.40	5.10	2.01	0.506
2.50	5.28	2.38	0.532
2.60	5.36	2.82	0.557
10.00	1.16	1.10	0.209
Palladium[19]			
0.10	4.13	54.15	0.994
0.50	4.10	11.44	0.896
1.00	2.99	6.89	0.811
2.20	1.60	3.88	0.707
2.30	1.53	3.75	0.700
2.40	1.47	3.61	0.693
2.50	1.41	3.48	0.685
2.60	1.37	3.36	0.676
10.00	1.14	0.65	0.088
Platinum[20]			
0.10	13.21	44.72	0.976
0.50	3.91	7.71	0.813
1.00	4.25	6.62	0.762
2.20	2.17	3.77	0.642
2.30	2.10	3.67	0.636
2.40	2.03	3.54	0.626
2.50	1.96	3.42	0.616
2.60	1.91	3.30	0.605
10.00	1.46	1.15	0.209
Potassium[21]			
0.55	0.139	7.10	0.989
1.05	0.044	3.58	0.987
2.27	0.049	1.43	0.938
2.45	0.046	1.28	0.933
Rhenium, single crystal, $\vec{E} \parallel \hat{c}$[9]			
0.10	6.06	51.03	0.991
0.50	4.53	10.88	0.878
1.00	2.45	6.36	0.813
2.20	3.06	2.84	0.501
2.30	3.07	2.82	0.499
2.40	3.06	2.81	0.498
2.50	3.02	2.80	0.497
2.60	2.96	2.77	0.493
10.00	1.23	1.26	0.252
50.00	0.80	0.30	0.038
Rhenium, single crystal, $\vec{E} \perp \hat{c}$[9]			
0.10	4.25	42.83	0.991
0.50	4.34	8.26	0.821
1.00	3.09	4.96	0.701
2.20	3.83	2.38	0.472
2.30	3.93	2.44	0.481
2.40	4.00	2.55	0.492
2.50	4.01	2.70	0.505
2.60	3.90	2.84	0.514
10.00	1.24	1.24	0.244
50.00	0.80	0.30	0.039
Rhodium[11]			
0.10	18.48	69.43	0.986
0.50	4.20	16.07	0.941
1.00	3.71	8.67	0.848
2.20	2.00	5.11	0.772

Energy (eV)	n	k	$R(\varphi = 0)$
2.30	1.94	4.94	0.765
2.40	1.90	4.78	0.756
2.50	1.88	4.65	0.748
2.60	1.85	4.55	0.743
10.00	1.17	0.69	0.098
Ruthenium, single crystal, $\vec{E} \parallel \hat{c}$ [9]			
0.10	11.50	51.38	0.984
0.50	3.18	11.04	0.909
1.00	3.39	5.33	0.715
2.20	3.35	4.82	0.683
2.30	3.09	4.70	0.681
2.40	2.89	4.55	0.677
2.50	2.74	4.40	0.671
2.60	2.64	4.25	0.663
10.00	0.88	0.76	0.144
Ruthenium, single crystal, $\vec{E} \perp \hat{c}$ [5]			
0.10	11.85	50.81	0.983
0.50	3.27	11.63	0.915
1.00	3.17	4.59	0.670
2.20	3.94	5.00	0.681
2.30	3.69	4.97	0.684
2.40	3.44	4.88	0.684
2.50	3.27	4.77	0.681
2.60	3.14	4.66	0.677
10.00	0.86	0.81	0.163
Selenium, single crystal, $\vec{E} \parallel \hat{c}$ [22]			
0.04959	3.403	2.79E-03	0.298
0.09919	3.409	3.23E-04	0.299
0.4959	3.442	1.41E-04	0.302
2.2	4.49	1.19	0.431
2.4	4.28	1.21	0.417
2.6	4.40	1.32	0.430
10.0	0.92	1.07	0.238
Selenium, single crystal, $\vec{E} \perp \hat{c}$ [22]			
0.04959	2.627	3.58E-03	0.201
0.09919	2.637	8.95E-05	0.203
0.4959	2.654	4.58E-05	0.205
2.20	3.07	0.73	0.282
2.40	2.93	0.61	0.259
2.60	3.00	0.53	0.263
10.00	1.72	0.95	0.171
Silicon, single crystal [23]			
0.04959	3.4201	9.15E-05	0.300
0.09919		1.77E-04	
0.4568	3.4393	2.50E-09	0.302
1.1	3.5341	3.30E-05	0.312
2.2	4.042	0.032	0.364
2.3	4.123	0.048	0.372
2.4	4.215	0.060	0.380

Energy (eV)	n	k	$R(\varphi = 0)$
2.5	4.320	0.073	0.390
2.6	4.442	0.090	0.400
10.0	0.306	1.38	0.661
Silver [6]			
0.10	9.91	90.27	0.995
0.50	0.67	18.32	0.992
1.00	0.28	9.03	0.987
2.50	0.24	3.09	0.914
10.00	1.46	0.56	0.082
50.00	0.88	0.29	0.027
100.00	0.87	0.04	0.005
Sodium [24]			
1.05	0.078	5.11	0.989
2.27	0.059	2.23	0.961
2.45	0.063	2.07	0.953
Tantalum [16]			
0.10	10.14	66.39	0.984
0.50	1.37	14.26	0.974
1.00	0.89	6.47	0.992
2.20	2.36	1.81	0.351
2.30	2.56	1.86	0.365
2.40	2.68	1.92	0.378
2.50	2.75	1.98	0.388
2.60	2.80	2.02	0.395
10.00	1.20	1.37	0.286
Tellurium, $\vec{E} \parallel \hat{c}$ [25]			
0.05083		7.38E-04	
0.09919		3.68E-04	
0.5	6.53	2.30E-02	0.539
1.0	7.70	1.56	0.606
2.2	4.94	5.16	0.681
2.4	3.94	5.08	0.686
2.6	3.25	4.77	0.681
10.0	0.86	0.86	0.181
Tellurium, $\vec{E} \perp \hat{c}$ [25]			
0.05083		6.79E-04	
0.09919		4.28E-05	
0.5	4.90		0.437
1.0	5.35	0.45	0.472
2.2	5.10	3.61	0.594
2.4	4.24	3.77	0.593
2.6	3.57	3.75	0.591
10.0	0.99	1.04	0.215
Titanium (polycrystalline) [14]			
0.10	5.03	23.38	0.965
0.50	4.43	3.22	0.555
1.00	3.62	3.52	0.570
2.20	1.92	2.67	0.509

Energy (eV)	n	k	$R(\varphi = 0)$
2.30	1.86	2.56	0.495
2.40	1.81	2.47	0.483
2.50	1.78	2.39	0.471
2.60	1.75	2.34	0.462
10.00	0.89	0.88	0.180
Tungsten [27]			
0.10	14.06	54.71	0.983
0.50	1.40	10.52	0.952
1.00	3.14	4.32	0.649
2.20	3.49	2.76	0.497
2.30	3.49	2.72	0.494
2.40	3.45	2.72	0.493
2.50	3.38	2.68	0.487
2.60	3.34	2.62	0.480
10.00	1.13	1.34	0.287
Vanadium [9]			
0.10	12.83	45.89	0.978
0.52	1.16	8.93	0.945
1.00	1.34	3.80	0.730
2.20	2.28	2.80	0.510
2.30	2.23	2.83	0.516
2.40	2.15	2.88	0.528
2.50	2.02	2.91	0.540
2.60	1.89	2.92	0.552
10.00	0.91	0.74	0.133
Zinc, $\vec{E} \parallel \hat{c}$ [28]			
0.7514	1.9241	7.5619	0.883
1.033	1.5407	5.3192	0.823
2.275	0.7737	3.9129	0.832
2.445	0.6395	3.4013	0.821
4.678	0.2354	1.6357	0.776
Zinc, $\vec{E} \perp \hat{c}$ [28]			
0.751	1.4469	7.4158	0.905
1.033	1.2889	5.4001	0.850
2.275	0.9725	4.2879	0.825
2.455	0.7568	3.7627	0.824
4.678	0.2806	1.7997	0.770
Zirconium (polycrystalline) [28]			
0.10	6.18	1.76	0.300
0.50	4.13	1.44	0.175
1.00	3.66	1.35	0.143
2.20	1.87	0.97	0.034
2.30	1.78	0.94	0.030
2.40	1.71	0.92	0.027
2.50	1.62	0.90	0.024
2.60	1.54	0.88	0.022
10.00	1.47	0.86	0.020

[a] Interpolated or extrapolated.

References

1. Shiles, E., Sasaki, T., Inokuti, M., and Smith, D. Y., *Phys. Rev. Sect. B*, 22, 1612, 1980.
2. Edwards, D. F., and Philipp, H. R., in *HOC-I*, p. 665.
3. Ives, H. E., and Briggs, N. B., *J. Opt. Soc. Am.*, 27, 395, 1937.
4. Bos, L. W., and Lynch, D. W., *Phys. Rev. Sect. B*, 2, 4567, 1970.
5. Weaver, J. H., Colavita, E., Lynch, D. W., and Rosei, R., *Phys. Rev. Sect. B*, 19, 3850, 1979.
6. Hagemann, H. J., Gudat, W., and Kunz, C., *J. Opt. Soc. Am.*, 65, 742, 1975.
7. Schulz, L. G., *J. Opt. Soc. Am.*, 47, 64, 1957.
8. Potter, R. F., in *HOC-I*, p. 465.
9. Olson, C. G., Lynch, D. W., and Weaver, J. H., unpublished.
10. Lynch, D. W., Olson, C. G., and Weaver, J. H., unpublished.
11. Weaver, J. H., Olson, C. G., and Lynch, D. W., *Phys. Rev. Sect. B*, 15, 4115, 1977.
12. Lynch, D. W., and Hunter, W. R., in *HOC-II*, p. 345.
13. Priol, M. A., Daudé, A., and Robin, S., *Compt. Rend.*, 264, 935, 1967.
14. Johnson, P. B., and Christy, R. W., *Phys. Rev. Sect. B*, 9, 5056, 1974.
15. Arakawn, E. T., and Inagaki, T., in *HOC-II*, p. 461.
16. Weaver, J. H., Lynch, D. W., and Olson, D. G., *Phys. Rev. Sect. B*, 10, 501, 1973.
17. Lynch, D. W., Rosei, R., and Weaver, J. H., *Solid State Commun.*, 9, 2195, 1971.
18. Weaver, J. H., Lynch, D. W., and Olson, C. G., *Phys. Rev. Sect. B*, 7, 4311, 1973.
19. Weaver, J. H., and Benbow, R. L., *Phys. Rev. Sect. B*, 12, 3509, 1975.
20. Weaver, J. H., *Phys. Rev., Sect. B*, 11, 1416, 1975.
21. Lynch, D. W., and Hunter, W. R., in *HOC-II*, p. 364.
22. Palik, E. D., in *HOC-II*, p. 691.
23. Edwards, D. F., in *HOC-I*, p. 547.
24. Lynch, D. W., and Hunter, W. R., in *HOC-II*, p. 354.
25. Palik, E. D., in *HOC-II*, p. 709.
26. Lynch, D. W., Olson, C. G., and Weaver, J. H., *Phys. Rev. Sect. B*, 11, 3671, 1975.
27. Weaver, J. H., Lynch, D. W., and Olson, C. G., *Phys. Rev. Sect. B*, 12, 1293, 1975.
28. Lanham, A. P., and Terherne, D. M., *Proc. Phys. Soc.*, 83, 1059, 1964.

OPTICAL PROPERTIES OF SELECTED INORGANIC AND ORGANIC SOLIDS

L. I. Berger

Optical properties of materials are closely related to their dielectric properties. The complex dielectric function (relative permittivity) of a material is equal to

$$\varepsilon(\omega) = \varepsilon'(\omega) - j\varepsilon''(\omega)$$

where $\varepsilon'(\omega)$ and $\varepsilon''(\omega)$ are its real and imaginary parts, respectively, and ω is the angular frequency of the applied electric field. For a non-absorbing medium, the index of refraction is $n = (\varepsilon\mu)^{1/2}$, where μ is the relative magnetic permeability of the medium (material); in the majority of dielectrics, $\mu \cong 1$.

For many applications, the most important optical properties of materials are the index of refraction, the extinction coefficient, k, and the reflectivity, R. The common index of refraction of a material is equal to the ratio of the phase velocity of propagation of an electromagnetic wave of a given frequency in vacuum to that in the material. Hence, $n \geq 1$. The optical properties of highly conductive materials like metals and semiconductors (at photon energy range above the energy gap) differ from those of optically transparent media. Free electrons absorb the incident electromagnetic wave in a thin surface layer (a few hundred nanometers thick) and then release the absorbed energy in the form of secondary waves reflected from the surface. Thus, the light reflection becomes very strong; for example, highly conductive sodium reflects 99.8% of the incident wave (at 589 nm). Introduction of the effective index of refraction, $n_{eff} = (\varepsilon')^{1/2} = n - jk$, where $\varepsilon' = \varepsilon - j\delta/\omega\, \varepsilon_o$, δ is the electrical conductivity of the material in S/m, and $\varepsilon_o = 8.8542{\cdot}10^{-12}$ F/m is the permittivity of vacuum, allows one to apply the expressions of the optics of transparent media to the conductive materials. It is clear that the effective index of refraction may be smaller than 1. For example, $n = 0.05$ for pure sodium and $n = 0.18$ for pure silver (at 589.3 nm). At very high photon energies, the quantum effects, such as the internal photoeffect, start playing a greater role, and the optical properties of these materials become similar to those of insulators (low reflectance, existence of Brewster's angle, etc.).

The extinction coefficient characterizes absorption of the electromagnetic wave energy in the process of propagation of a wave through a material. The wave intensity, I, after it passes a distance x in an isotropic medium is equal to

$$I = I_0\exp(-\alpha x)$$

where I_0 is the intensity at $x = 0$ and α is called the absorption coefficient. For many applications, the extinction coefficient, k, which is equal to

$$k = \alpha\frac{\lambda}{4\pi}$$

where λ is the wavelength of the wave in the medium, is more commonly used for characterization of the electromagnetic losses in materials.

Reflection of an electromagnetic wave from the interface between two media depends on the media indices of refraction and on the angle of incidence. It is characterized by the reflectivity, which is equal to the ratio of the intensity of the wave reflected back into the first medium to the intensity of the wave approaching the interface. For polarized light and two non-absorbing media,

$$R = \frac{(N_1 - N_2)^2}{(N_1 + N_2)^2}$$

where $N_1 = n_1/\cos\theta_1$ and $N_2 = n_2/\cos\theta_2$ for the wave polarized in the plane of incidence, and $N_1 = n_1\cos\theta_1$ and $N_2 = n_2\cos\theta_2$ for the wave polarized normal to the plane of incidence; θ_1 and θ_2 are the angles between the normal to the interface in the point of incidence and the directions of the beams in the first and second medium, respectively. The reflectivity at normal incidence in this case is

$$R = [(n_1 - n_2)/(n_1 + n_2)]^2$$

For any two opaque (absorbing) media, the normal incidence reflectivity is

$$R = \frac{(n_1 - n_2)^2 + k_2^2}{(n_1 + n_2)^2 + k_2^2}$$

In the majority of experiments, the first medium is air ($n \approx 1$), and hence,

$$R = \frac{(1 - n)^2 + k^2}{(1 + n)^2 + k^2}$$

The data on n and k in the following table are abridged from the sources listed in the references. The reflectivity at normal incidence, R, has been calculated from the last equation. For convenience, the energy E, wavenumber $\bar{\nu}$, and wavelength λ are given for the incidence radiation.

This table in the Print Edition of the *CRC Handbook* is condensed from the longer table in the Online Edition, which gives values of the optical properties at finer energy intervals. The full table should be consulted for cases where there are resonances or regions of sharp changes of optical properties with wavelength. The full table also contains complete references to data sources.

Refractive Index n, Extinction Coefficient k, and Reflectivity at Normal Incidence R.*
The Incident Light Is Characterized by Energy E, Wavenumber $\bar{\nu}$, and Wavelength λ

E/eV	$\bar{\nu}$/cm $^{-1}$	λ/µm	n	n_a	n_c	k	k_a	k_c	R	R_a	R_c
Crystalline Arsenic Selenide (As_2Se_3) - [Ref. 1]											
2.000	16130	0.620						0.082			
0.06199	500.0	20.0		3.2	2.9		$1.7{\cdot}10^{-3}$	$1.8{\cdot}10^{-3}$		0.27	0.24
0.01240	100.0	100.0		4.4	2.7		0.22	0.81		0.40	0.25

E/eV	$\bar{\nu}$/cm^{-1}	λ/μm	n	n_a	n_c	k	k_a	k_c	R	R_a	R_c
0.006199	50.0	200.0		3.8	3.6		0.0091	0.019		0.34	0.32
Vitreous Arsenic Selenide (As_2Se_3) - [Ref. 1]											
2.006	16180	0.618				0.099					
1.033	8333	1.20	2.88								
0.2455	1980	5.05				$1.6 \cdot 10^{-7}$					
0.05021	405	24.69	2.7			$9.4 \cdot 10^{-4}$			0.21		
0.007606	61.35	163.0	3.3			0.12			0.29		
Vitreous Arsenic Sulfide (As_2S_3) - [Ref. 2]											
4.959	40000	0.2500	2.48			1.21			0.27		
3.100	25000	0.40	3.09			0.34			0.27		
0.2480	2000	5.0	2.41						0.17		
0.04959	400.0	25.0	1.79			0.2			0.085		
Cadmium Telluride (CdTe) - [Ref. 3]											
4.9	39520	0.2530	2.48			2.04			0.39		
3.0	24200	0.4133	3.37			0.861			0.32		
1.00	8065	1.240	2.7793						0.22		
0.30	2420	4.133	2.6800						0.21		
0.05	403.3	24.80	2.5801						0.19		
Gallium Arsenide (GaAs) - [Ref. 4]											
5.00	40330	0.2480	2.273			4.084			0.67		
3.00	24200	0.4133	4.509			1.948			0.47		
2.00	16130	0.6199	3.878			0.211			0.35		
1.00	8065	1.240	3.4232						0.30		
0.25	2016	4.959	3.2978						0.29		
0.0495	399.2	25.05	3.058			$2.07 \cdot 10^{-3}$			0.26		
Gallium Phosphide (GaP) - [Ref. 5]											
15.0		0.0826	0.748			0.628					
5.5	44360	0.2254	1.543			3.556			0.68		
3.00	24200	0.4133	4.081			0.224			0.37		
2.000	16130	0.62	3.3254						0.29		
0.1907	1538	6.5	2.995			$4.29 \cdot 10^{-4}$			0.25		
0.06199	500	20	2.615			$7.16 \cdot 10^{-3}$			0.20		
Indium Antimonide (InSb) - [Ref. 6]											
10		0.1240	0.74			0.88					
5.00	40330	0.2480	1.307			2.441			0.53		
3.34	26940	0.3712	3.528			2.280			0.45		
1.50	12100	0.8266	4.418			0.643			0.41		
0.2480	2000	5.0	4.14			$9.1 \cdot 10^{-2}$			0.37		
0.06199	500	20.00	3.869			$2.0 \cdot 10^{-3}$			0.35		
Indium Arsenide (InAs) - [Ref. 7]											
10		0.1240				0.835			1.071		
5.0	40330	0.2480	1.524			2.871			0.58		
3.0	24200	0.4133	3.197			2.034			0.41		
1.0	8065	1.240	3.548						0.31		
0.04959	400	25.00	3.264						0.28		
Indium Phosphide (InP) - [Ref. 8]											
10		0.1240	0.806			1.154					
5.0	40330	0.2480	2.131			3.495			0.61		
3.0	24200	0.4133	4.395			1.247			0.43		
2.0	16130	0.6199	3.549			0.317			0.32		
1.00	8068	1.239	3.220						0.28		
0.30	2420	4.131	3.089						0.26		
0.050	403.4	24.79	2.429			$3.35 \cdot 10^{-2}$			0.17		
Lead Selenide (PbSe) - [Ref. 9]											
10		0.1240	0.68			0.50					
5	40330	0.2480	0.54			1.2					

E/eV	$\bar{\nu}$/cm^{-1}	λ/μm	n	n_a	n_c	k	k_a	k_c	R	R_a	R_c
2.0	16130	0.6199	3.65			2.9			0.51		
1.0	8065	1.240	4.65			1.1			0.44		
0.04959	400	25.00	4.49			$1.77 \cdot 10^{-2}$			0.40		
Lead Sulfide (PbS) - [Ref. 10]											
10.0		0.1240	0.879			1.050					
4.95	39920	0.2505	1.52			2.10			0.43		
3.00	24200	0.4133	3.88			3.00			0.53		
2.00	16130	0.6199	4.29			1.48			0.43		
1.03	8333	1.2	4.30			0.458			0.39		
0.2480	2000	5	4.115			$9.25 \cdot 10^{-4}$			0.37		
0.04959	400	25.00	3.53						0.31		
Lead Telluride (PbTe) - [Ref. 11]											
10		0.1240	0.66			0.60					
5.0	40330	0.2480	0.72			1.0					
3.0	24200	0.4133	1.0			2.2					
1.0	8065	1.240	4.55			2.2			0.49		
0.30	2420	4.133	5.95			$3.55 \cdot 10^{-2}$			0.51		
Lithium Fluoride (LiF) - [Ref. 12]											
10.00		0.12398	1.606			$7.70 \cdot 10^{-7}$			0.05		
4.959	40000	0.250	1.4189						0.03		
2.952	23810	0.42	1.3978						0.03		
2.000	16130	0.62	1.3915						0.03		
0.9919	8000	1.25	1.3851								
0.2480	2000	5.0	1.3266			$1.8 \cdot 10^{-6}$			0.02		
0.04959	400	25.00	0.208			2.71			0.91		
Potassium Chloride (KCl) - [Ref. 13]											
10.0		0.1240	1.16			0.38			0.035		
4.959	40000	0.25	1.58972								
2.952	23810	0.42	1.50701								
2.066	16670	0.60	1.48969						0.039		
1.033	8333	1.2	1.47813						0.037		
0.2480	2000	5.0	1.47048						0.036		
0.04959	400.0	25.0	1.34059			$6.57 \cdot 10^{-4}$			0.021		
Silicon Dioxide (Glass) - [Ref. 14]											
10.00		0.1240	2.330			0.323			0.17		
4.9939	40278.4	0.248272	1.50841						0.041		
3.0640	24712.3	0.404656	1.46961						0.036		
1.9257	15531.6	0.643847	1.45671						0.035		
1.0985	8860.06	1.12866	1.44888						0.034		
0.2480	2000	5.000	1.342			$3.98 \cdot 10^{-3}$			0.021		
0.04959	400	25.0	2.739			0.397			0.23		
Silicon Monoxide (Noncrystalline) - [Ref. 15]											
10		0.1240	1.378			0.6843			0.10		
5	40330	0.2480	2.001			0.6052			0.15		
3	24200	0.4133	2.116			0.1211			0.13		
2	16130	0.6199	1.969			0.01175			0.11		
1.240	10000	1.000	1.87						0.092		
0.2480	2000	5.000	1.75						0.074		
0.08856	714.3	14.00	2.01			0.30			0.12		
Noncrystalline Silicon Nitride (Si_3N_4) - [Ref. 16]											
10	80650	0.1240	2.000			1.49			0.29		
5	40330	0.2480	2.278			$4.9 \cdot 10^{-3}$			0.15		
3	24200	0.4133	2.066						0.12		
2	16130	0.6199	2.022						0.11		
1	8065	1.240	1.998						0.11		
Sodium Chloride (NaCl) - [Ref. 17]											
10.0		0.1240	1.55			0.71			0.12		

E/eV	$\bar{\nu}$/cm^{-1}	λ/μm	n	n_a	n_c	k	k_a	k_c	R	R_a	R_c
5.00	40330	0.2480	1.65						0.060		
2.952	23810	0.42	1.56324						0.048		
2.000	16130	0.62	1.54228						0.045		
1.033	8333	1.2	1.53000						0.044		
0.2480	2000	5.0	1.51883						0.042		
0.04959	400	25.0	1.27			$3.5 \cdot 10^{-3}$			0.014		
Cubic Zinc Sulfide (ZnS) - [Ref. 18]											
9.919		0.125	1.02			1.36			0.31		
3.00	24200	0.4133	2.54			$4 \cdot 10^{-2}$			0.19		
2.00	16130	0.6199	2.3576						0.16		
1.00	8065	1.240	2.2795						0.15		
0.30	2420	4.133	2.2529						0.15		
0.05	403.3	24.80	1.6866						0.065		
Polytetrafluoroethylene (Teflon) - [Ref. 19] (Total diffuse hemispherical reflectance)											
4.960	40000	0.250							0.970		
3.351	27027	0.370							0.993		
2.067	16667	0.600							0.992		
0.9920	8000	1.250							0.990		
0.4960	4000	2.500							0.945		

* Indices a and c relate to the radiation electric field parallel to the a and c axes of the crystal, respectively.

References

1. Arsenic Selenide

D. J. Treacy in *Handbook of Optical Constants of Solids*, E. D. Palik, Editor, Academic Press, 1985, p. 623. (Hereafter abbreviated as *HOCS*.)

R. Zallen, R. E. Drews, R. L. Emerald, and M. L. Slade, *Phys. Rev. Lett.* 26, 1564 (1971).

R. Zallen, M. L. Slade , and A. T. Ward, *Phys. Rev.* B 3, 4257 (1971).

U. Strom and P. C. Taylor, *Phys. Rev.* B 16, 5512 (1977).

G. Lucovsky, *Phys. Rev.* B 6, 1480 (1972).

C. T. Moynihan, P. B. Macedo, M. S. Maklad, R. K. Mohr, and R. E. Howard, *J. Non-Cryst. Solids*, 17, 369 (1975).

Y. Ohmachi, *J. Opt. Soc. Am.* 63, 630 (1973).

2. Arsenic Sulfide

D. J. Treacy in *HOCS*, 1985, p. 641.

P. A. Young, *J. Phys.* C 4, 93 (1971).

W. S. Rodny, I. H. Malitson, and T. A. King, *J. Opt. Soc. Am.* 48, 633 (1958).

R. Zallen, R.E. Drew, R. L. Emerald, and M.L. Slade, *Phys. Rev. Lett.* 26, 1564 (1971).

M. S. Maklad, R. K. Mohr, R. E. Howard, P. B. Macedo, and C. T. Moynihan, *Solid State Commun.* 15, 855 (1974).

P. B. Klein, P. C. Taylor, and D. J. Treacy, *Phys. Rev.* B16, 4511 (1977).

G. Lucovsky, *Phys. Rev.* B 6, 1480 (1972).

3. Cadmium Telluride

E. D. Palik in *HOCS*, 1985, p. 409.

D. T. F. Marple and H. Ehrenreich, *Phys. Lett.* 8, 87 (1962).

T. H. Myers, S. W. Edwards, and J. F. Schetzina, *J. Appl. Phys.* 52, 4231 (1981).

D. T. F. Marple, *Phys. Rev.* 150, 728 (1966).

A. N. Pikhtin and A. D. Yas'kov, *Sov. Phys. Semicond.* 12, 622 (1978).

L. S. Ladd, *Infrared Phys.* 6, 145 (1966).

J. E. Harvey and W. L. Wolfe, *J. Opt. Soc. Am.* 65, 1267 (1975).

A. Manabe, A. Mitsuishi, and H. Yoshinaga, *Jpn. J. Appl. Phys.* 6, 593 (1967).

A. Manabe, A. Mitsuishi, H. Oshinaga, Y. Ueda, and H. Sei, *Technol. Rep. Osaka Univ. Jpn.* 17, 263 (1967).

J. R. Birch and D. K. Murrey, *Infrared Phys.* 18, 283 (1978).

4. Gallium Arsenide

E. D. Palik in *HOCS*, 1985, p. 429.

M. Cardona, W. Gudat, B. Sonntag, and P. Y. Yu, in *Proc. Intl. Conf. Phys. Semicond.*, 10th. Cambridge, 1970, p. 208. US Atom. Energy Commission, Oak Ridge, TN, 1970.

H. R. Philipp and H. Ehrenreich, *Phys. Rev.* 129, 1550 (1963).

J. B. Theeten, D. E. Aspnes, and R. P. H. Chang, *J. Appl. Phys.* 49, 6097 (1978).

H. C. Casey, D. D. Sell, and K. W. Wecht, *J. Appl. Phys.* 46, 250 (1975).

A. H. Kachare, W. G. Spitzer, F. K. Euler, and A. Kahan, *J. Appl. Phys.* 45, 2938 (1974).

R. T. Holm, J. W. Gibson, and E. D. Palik, *J. Appl. Phys.* 48, 212 (1977).

W. Cochran, S. J. Fray, F. A. Johnson, J. E. Quarrington, and N. Williams, *J. Appl. Phys. Suppl.* 32, 2102 (1961).

C. P. Christensen, R. Joiner, S. K. T. Nieh, and W. H. Steier, *J. Appl. Phys.* 45, 4957 (1974).

R. H. Stolen, *Phys. Rev.* B 11, 767 (1975); *Appl. Phys. Lett.* 15, 74 (1969).

5. Gallium Phosphide

A. Borghesi and G. Guizzetti in *HOCS*, 1985, p. 445.

M. Cardona, W. Gudat, B. Sonntag, and P. Y. Yu, *Proc. Intl. Conf. Phys. Semicond.* Cambridge, 1970, p. 208. US Atom. Energy Commission, Oak Ridge, TN, 1970.

M. Cardona, W. Gudat, E. E. Koch, M. Skibowski, B. Sonntag, and P. Yu. *Phys. Rev. Lett.* 25, 659 (1970).

S. E. Stokowski and D. D. Sell, *Phys. Rev.* B 5, 1636 (1972).

S. A. Abagyan, G. A. Ivanov, Y. E. Shanurin, and V. I. Amosov, *Sov. Phys. Semicond.* 5, 889 (1971).

P. G. Dean, G. Kaminsky, and R. B. Zetterstorm, *J. Appl. Phys.* 38, 3551 (1967).

D. E. Aspnes and A. A. Studna, *Phys. Rev.* B 27, 985 (1983).

6. Indium Antimonide

R. T. Holm in *HOCS*, 1985, p. 491.

M. Cardona, W. Gudat, B. Sonntag, and P. Y. Yu, *Proc. Int. Conf. Phys. Semicond.*, 10th. Cambridge, 1970, p. 208. US Atom. Comm., Oak Ridge, TN, 1970.

H. R. Philipp and H. Ehrenreich, *Phys. Rev.* 129, 1550 (1963).

D. E. Aspnes and A. A. Studna, *Phys. Rev.* B 27, 985 (1983).

T. S. Moss, S. D. Smith, and T. D. F. Hawkins, *Proc. Phys. Soc. London* 70B, 776 (1957).

H. Yoshinaga and R. A. Oetjen, *Phys. Rev.* 101, 526 (1956).

R. B. Sanderson, *J. Phys. Chem. Solids* 26, 803 (1965).

7. Indium Arsenide

E. D. Palick and R. T. Holm in *HOCS*, 1985, p. 479.

H. R. Philipp and H. Ehrenreich, *Phys. Rev.* 129, 1550 (1963).

B. O. Seraphin and H. E. Bennett in *Semiconductors and Semimetals* (R. K. Willardson and A. C. Beer, Eds.), vol. 3, Academic, 1967, p. 499.
D. E. Aspnes and A. A. Studna, *Phys. Rev.* B 27, 985 (1983).
J. R. Dixon and J. M. Ellis, *Phys. Rev.* 123, 1560 (1961).
A. Memon, T. J. Parker, and J. R. Birch, *Proc. SPIE*, 289, 20 (1981).

8. Indium Phosphide
O. J. Glembocki and H. Piller in *HOCS*, 1985, p. 503.
M. Cardona, *J. Appl. Phys.* 32, 958 (1961); 36, 2181 (1965).
D. E. Aspnes and A. A. Studna, *Phys. Rev.* B 27, 985 (1983).
G. D. Pettit and W. J. Turner, *J. Appl. Phys.* 36, 2081 (1965).
R. Newman, *Phys. Rev.* 111, 1518 (1958).
W. N. Reynolds, M. T. Lilburne, and R. M. Dell, *Proc. Phys. Soc. London* 71, 416 (1958).
H. Jamshidi and T. J. Parker, Int. Meet. Infrared Mm. Waves, 7th., Marseilles, 1983.

9. Lead Selenide
G. Bauer and H. Krenn in *HOCS*, 1985, p. 517.
M. Cardona and D. L. Greenaway, *Phys. Rev.* A 133, 1685 (1964).
T. S. Moss, *Optical Properties of Semiconductors*, Butterworth, 1959, p. 189.
J. N. Zemel, J. D. Jensen, and R. B. Schoolar, *Phys. Rev.* A 140, 330 (1965).
W. W. Scanlon, *J. Phys. Chem. Solids*, 8, 423 (1959).
K. V. Vyatkin and A. P. Shotov, *Sov. Phys. Semicond.* 14, 785 (1980); Fiz. Tekh. Poluprovodn. 14, 1331 (1980).

10. Lead Sulfide
G. Guizzetti and A. Borghesi in *HOCS*, 1985, p. 525.
M. Cardona and R. Haensel, *Phys. Rev.* B 1, 2605 (1970).
M. Cardona and D. L. Greenaway, *Phys. Rev.* A 133, 1685 (1964).
M. Cardona, C. M. Penchina, E. E. Koch, and P. Y. Yu, *Phys. Status Solidi* B 53, 327 (1972).
P. R. Wessel, *Phys. Rev.* 153, 836 (1967).
C. E. Rossi and W. Paul, *J. Appl. Phys.* 38, 1803 (1967).
J. N. Zemel, J. D. Jensen, and R. B. Schoolar, *Phys. Rev.* A 140, 330 (1965).

11. Lead Telluride
G. Bauer and H. Krenn in *HOCS*, 1985, p. 535.
M. Cardona and R. Haensel, *Phys. Rev.* B 1, 2605 (1970).
M. Cardona and D. L. Greenaway, *Phys. Rev.* 133, A1685 (1964).
D. M. Korn and R. Braunstein, *Phys. Rev.* B 5, 4837 (1972).
W. W. Scanlon, *J. Phys. Chem. Solids* 8, 423 (1959).
J. N. Zemel, J. D. Jensen, and R. B. Schoolar, *Phys. Rev.* 140, A330 (1965).

12. Lithium Fluoride
E. D. Palik and W. R. Hunter in *HOCS*, 1985, p. 675.
B. L. Henke, P. Lee, T. J. Tanaka, R. L. Shimabukuro, and B. K. Fujikawa, *Low Energy X-ray Diagnostics-1981* (D. T. Attwood and B. L. Henke, Eds.), AIP Conf. Proc. No. 75, 1981.
A. P. Lukirskii, E. P. Savinov, O. A. Ershov, and Y. F. Shepelev, *Opt. Spektrosk.* 16, 168 (1964); 16, 310 (1964).
F. C. Brown, C. Gahwiller, A. B. Kunz, and N. O. Lipari, *Phys. Rev.* Lett. 25, 927 (1970).
A. Milgram and M. P. Givens, *Phys. Rev.* 125, 1506 (1962).
T. Tomiki and T. Miyata, *J. Phys. Soc. Jpn.* 27, 658 (1969).
A. Kachare, G. Andermann, and L. R. Brantley, *J. Phys. Chem. Solids* 33, 467 (1972).

13. Potassium Chloride
E. D. Palik in *HOCS*, 1985, p. 703.
O. Aita, I. Nagakura, and T. Sagawa, *J. Phys. Soc. Jpn.* 30, 1414 (1971).
A. P. Lukirskii, E. P. Savinov, O. A. Ershov, and Y. F. Shepelev, *Opt. Spectrosc.* 16, 168 (1964); *Opt. Spektrosk.* 16, 310 (1964).
T. Tomika, *J. Phys. Soc. Jpn.* 22, 463 (1967).
M. Antinori, A. Balzarotti, and M. Piacentini, *Phys. Rev.* B 7, 1541 (1973).
H. H. Li, *J. Phys. Chem. Ref. Data* 5, 329 (1976).
S. D. Allen and J. A. Harrington, *Appl. Opt.* 17, 1679 (1978).
K. W. Johnson and E. E. Bell, *Phys. Rev.* 139A, 1295 (1965).

14. Silicon Dioxide
H. R. Philipp in *HOCS*, 1985, p. 749.
J. Rife and J. Osantowski, *J. Opt. Soc. Am.* 70, 1513 (1980).
B. L. Henke, P. Lee, T. J. Tanaka, R. L. Shimabukuro, and B. K. Fujikawa, *Low Energy X-ray Diagnostics-1981* (D. T. Attwood and B. L. Henke, Eds.), AIP Conf. Proc. No. 75, 1981.
H. R. Philipp, *Solid State Commun.* 4, 73 (1966); *J. Phys. Chem. Solids*, 32, 1935 (1971).
P. L. Lamy, *Appl. Opt.* 16, 2212 (1977).
H. R. Philipp, *J. Appl. Phys.* 50 1053 (1979).
D. G. Drummond, *Proc. Roy. Soc. London*, 153, 328 (1935).

15. Silicon Monoxide
H. R. Philipp in *HOCS*, 1985, p. 765.
H. R. Philipp, *J. Phys. Chem. Solids*, 32, 1935 (1971).
G. Hass and C. D. Salzberg, *J. Opt. Soc. Am.* 44, 181 (1954).
E. Cremer, T. Kraus, and E. Ritter, *Zs. Electrochem.* 62, 939 (1958).
A. P. Bradford, G. Hass, M. McFarland, and E. Ritter, *Appl. Opt.* 4, 971 (1965).

16. Silicon Nitride
H. R. Philipp in *HOCS*, 1985, p. 771.
H. R. Philipp, *J. Electrochem. Soc.* 120, 295 (1973).
J. B. Theeten, D. E. Aspnes, F. Simondet, M. Errman, and P. C. Mürau, *J. Appl. Phys.* 52, 6788 (1981).
J. Bauer, *Phys. Status Solidi*, A 39, 411 (1977).

17. Sodium Chloride
J. E. Eldridge and E. D. Palik in *HOCS*, p. 775.
J. A. Harrington, C. J. Duthler, F. W. Patten, and M. Hass, *Solid State Commun.* 18, 1043 (1976).
T. Miyata and T. Tomiki, *J. Phys. Soc. Jpn.* 24, 1286 (1968); ibid., 22, 209 (1967).
D. M. Roessler and W. C. Walker, *J. Opt. Soc. Am.* 58, 279 (1968).
D. M. Roessler and W. C. Walker, *Phys. Rev.* 166, 599 (1968).
S. Allen and J. A. Harrington, *Appl. Opt.* 17, 1679 (1978).
O. Aita, I. Nagakura, and T. Sagawa, *J. Phys. Soc. Jpn.* 30, 1414 (1971).

18. Zinc Sulfide
E. D. Palik and A. Addamiano in *HOCS*, 1985, p. 597.
B. L. Henke, P. L. Lee, T. J. Tanaka, R. L. Shimabukuro, and B. F. Fujikawa, *Low Energy X-ray Diagnostics-1981* (D. T. Attwood and B. L. Henke, Eds.), AIP Conf. Proc. No. 75, 1981.
M. Cardona and G. Harbeke, *Phys. Rev.* 137, A1467 (1965).
Eastman Kodak, Publ. No. U-72, Rochester, New York (1981).
C. A. Klein and R. N. Donadio, *J. Appl. Phys.* 51, 797 (1980).
T. Deutsch, *Proc. Int. Conf. Phys. Semicond.*, 6th Exeter 1962, p. 505. The Inst. of Physics and the Physical Soc., London, 1962.
A. Manabe, A. Mitsuishi, and H. Yoshinaga, *Jpn. J. Appl. Phys.* 6, 593 (1967).
W. W. Piper, D. T. F. Marple, and P. D. Johnson, *Phys. Rev.* 110, 323 (1958).

19. Polytetrafluoroethylene
J. W. L. Thomas (NIST), Private communication.
NIST Certificate, STM 2044.
P. Y. Barnes, E. A. Early, and A. C. Parr, *NIST Special Publ. 250-48*, NIST Measurement Services: Spectral Reflectance.
Diffuse Reflectance Coatings and Materials Sections, Labsphere Catalog, 1996.
A. Arecchi and C. Ryder (Labsphere, North Sutten, NJ), private communication.

PROPERTIES OF SELECTED MATERIALS AT CRYOGENIC TEMPERATURES

Peter E. Bradley and Ray Radebaugh

The design of systems for operation at cryogenic temperatures requires the use of material properties at these low temperatures. The properties at cryogenic temperatures can be much different than the room-temperature values. In addition, some properties can be strong functions of temperature. Property data at cryogenic temperatures are not easy to find. Many measurements were made at the National Institute of Standards and Technology (NIST) and other laboratories about 50 years ago. Some of the results were published in reports that are now out of print, which makes the results unavailable to most researchers. To correct that problem, NIST initiated a program to critically evaluate cryogenic material properties and to fit the available data for temperatures in the range of about 4 K to 300 K. The parameters for the fit, as well as a graph of the curve, are available on the NIST cryogenics Web site <www.cryogenics.nist.gov>. The properties available include thermal conductivity, specific heat, linear thermal expansion relative to 293 K, thermal expansion coefficient, and Young's modulus. Not all properties are available for all materials. The materials currently in the database are ones commonly used in the construction of cryogenic hardware.

Five tables are given for important properties for selected materials at cryogenic temperatures, as listed below. The Online Edition of the *CRC Handbook* has additional temperatures.

Table	Contents
1	Thermal Conductivity λ, in units W m^{-1} K^{-1}
2	Specific Heat at Constant Pressure c_p, in units J kg^{-1} K^{-1}
3	Linear Thermal Expansion (Fractional Change in Length Relative to 293 K)
4	Coefficient of Thermal Expansion α, in units of 10^{-6} K^{-1} or 10^{-8} K^{-1} as specified
5	Young's Modulus E in GPa

The tables presented here as a function of temperature are the calculated values using the equations given on the NIST cryogenics Web site. In general, the equations fit a single set of data to within about 1% to 2%, but often several sets of data are used in determining the best fit, in which case deviations can be significantly higher, as much as 5%. The NIST cryogenics Web site specifies the deviation of the fit relative to the experimental data for each property and each material. Uncertainties in the experimental data usually are in the range of 2% to 5%, and variations from sample to sample can also lead to similar uncertainties, especially in thermal conductivity. Some well-characterized materials, such as silicon, are used for standard reference materials. Thus, uncertainties in the experimental data for the thermal expansion coefficient of silicon are usually less than 0.2%, and the standard deviation of the fit to the data is less than about 0.2% over most of the temperature range. All values refer to ambient pressure, i.e., pressure in the neighborhood of 101.3 kPa.

Copper referred to here is of very high purity 99.99% (4N or better) and may be considered oxygen-free (sometimes referred to as OFHC – oxygen-free high conductivity). Values are given with respect to the RRR (Residual Resistivity Ratio) which correlates the thermal resistivity and electrical resistivity as the impurity effect and is primarily additive in resistivity. Higher RRR values indicate higher purity and lower electrical and thermal resistance leading to higher thermal conductivity. Standard high-purity copper such as grade 101 or 102 has an RRR value of approximately 100. Higher values may be obtained with considerable effort at minimizing trace impurities by special annealing techniques that can achieve an RRR of about 1000 or greater in some special instances. Specially obtained high RRR value copper is often used only when very low temperatures (<40 K) and necessarily high thermal conduction at low temperatures are required.

Ti 15-3-3-3 has a nominal composition of 15% V, 3% Cr, 3% Sn, 3% Al, balance Ti. For the specific measurements documented by Canavan and Tuttle (Ref. 29), the exact composition is 14.88% V, 3.13% Cr, 2.88% Sn, 3.01% Al, balance Ti. The composition for brass is 65% Cu, 32% Zn, 3% Pb which is free machining. The composition for BeCu is 2% Be, 0.3% Co, balance Cu.

References

1. Mann, D., Ed., *LNG Materials and Fluids, First Edition*, Cryogenics Division, National Bureau of Standards, Boulder, CO, 1977. [Al, Invar, FeNi, Polystyrene, Polyurethane, PVC, Stainless steel, Be, G-10]
2. Veres, H. M., Ed., *Thermal Properties Database for Materials at Cryogenic Temperatures*, Vol. 1. [Al, G-10, Nylon, Teflon]
3. Touloukian, Y. S., *Recommended Values of the Thermophysical Properties of Eight Alloys, Major Constituents and Their Oxides*, Purdue University, West Lafayette, IN, 1965. [Al, Stainless steel]
4. Touloukian, Y. S., Powell, R. W., Ho, C. Y., and Klemens, P. G., *The TPRC Data Series: Vol. 1, Thermal Conductivity-Metallic Elements and Alloys*, Shackelford, J., and Alexander, W., Eds., New York, Washington, 1970. [Be, Mo]
5. Johnson, V. J., Ed., A Compendium of the Properties of Materials at Low Temperature (Phase l), *Part II: Properties of Solids, Wadd Technical Report 60-56*, National Bureau of Standards, Boulder, CO, 1960. [BeCu, Pb, Pt, Nylon, Be, In, Al, Cu]
6. Berman, R., Foster, E. L., and Rosenberg, H. M., *Brit. J. Appl. Phys.* 6, 181, 1955. [BeCu]
7. Powell, R.L., Rogers, W.M., and Roder, H.M., Thermal Conductivities of Copper and Copper Alloys, *Adv. Cryog. Eng.* 2, 166, 1956. [Brass, Cu]
8. Simon, N. J., Drexler, E. S., and Reed, R. P., Properties of Copper and Copper Alloys at Cryogenic Temperature, *NIST Monograph 177*, 1992. [Brass, Cu, Phosphor-bronze, BeCu]
9. He, G. H., Wang, B. Q., Guo, X. N., Yang, F., Guo, J. D., and Zhou, B. L., Investigation of Thermal Expansion Measurement of Brass Strip H62 after High Current Density Electropulsing by the CCD Technique, *Mater. Sci. Eng., A* 292, 183, 2000. [Brass]
10. Hust, J. G., Thermal Conductivity of Glass Fiber/Epoxy Composite Support Bands for Cryogenic Dewars, Phase II, *NBSIR 84-3003*, National Bureau of Standards, Boulder, CO, 1984. [G-10]
11. Child, G., Erics, L. J., and Powell, R. L., Thermal Conductivity of Solids at Room Temperature and Below, *NBS Monograph 131*, 1973. [G-10]
12. Mechanical, Thermal, Electrical, and Magnetic Properties of Structural Materials, *Wadd Technical Report: Handbook on Materials for Superconducting Machinery Metals and Ceramics*, Information Center, Battelle, Columbus Laboratories, 1974 (with 1975 and 1977 Supplements). [Inconel]
13. Hust, J. G., Low-Temperature Thermal Conductivity of Two Fibre-Epoxy Composites, *Cryogenics* 15, 126, 1975. [Kevlar]
14. Foster, W. G., Naes, L. G., and Barnes, C.B., Thermal Conductivity Measurements of Fiberglass/Epoxy Structural Tubes from 4 K to 320 K, *AlAAS Paper 75-711*, American Institute of Aeronautics and Astronautics (A1AA), 10th Thermophysics Conference, Denver, CO, 1975. [Kevlar]

15. Harris, J. P., Yates, B., Batchelor, J., and Garrington, P. J., *J. Mater. Sci.* 17, 2925, 1982. [Kevlar]
16. Ventura, G., and Martelli, V., Thermal Conductivity of Kevlar 49 between 7 and 290 K, *Cryogenics* 49, 735, 2009. [Kevlar]
17. Ventura, G., and Martelli, V., Very Low Temperature Thermal Conductivity of Kevlar 49, *Cryogenics* 49, 376, 2008. [Kevlar]
18. Poulaert, B., Chieliens, J. C., Vandehande, C., and Legras, R., Temperature Variation of the Thermal Conductivity of Kevlar, *Polym. Commun.*, vol. 26, 1985 (digitized data). [Kevlar]
19. Hartwig, G., and Knaak, S., Fibre-Epoxy Composites at Low Temperatures, *Cryogenics* 24, 11, 1984. [Kevlar]
20. Shackelford, J. F., and Alexander, W., eds., *CRC Materials Science and Engineering Handbook, Third Edition*, CRC Press, Boca Raton, FL, 2001. [Mo]
21. Choy, C. L., and Grieg, D., The Low Temperature Thermal Conductivity of a Semi-crystalline Polymer, Polyethylene Terephthalate, *J. Phys. C: Solid State Phys.* 8, 3121, 1975. [Mylar]
22. Rule, D.L., Smith, D.R., and Sparks, L.L., Thermal Conductivity of a Polyimide Film between 4.2 and 300 K, with and without Alumina Particles as Filler, *NISTIR 3948*, 1990. [Kapton]
23. Touloukian, Y. S, and Ho, C. Y., Eds., *Thermophysical Properties of Selected Aerospace Materials, Part II: Thermophysical Properties of Seven Materials*, Plenum Press, New York, 1976. [Stainless steel]
24. Hust, J. G., and Sparks, L. L., Thermal Conductivity of Austentic Stainless Steel, SRM 735, from 5 to 280 K, *NBS Special Publication 260-35*, 1972. [Stainless steel 735]
25. Hust, J. G., and Giarratano, P. J., Thermal Conductivity and Electrical Resistivity Standard Reference Materials: Austentic Stainless Steel, SRM 735 and 798, From 4 to 1200 K, *NBS Special Publication 260-46*, 1975. [Stainless steel 735]
26. Ventura, G. Bianchini, G., Gottardi, E., Peroni, I., and Peruzzi, A., Thermal Expansion and Thermal Conductivity of Torlon at Low Temperatures, *Cryogenics* 39, 481, 1999. [Torlon]
27. Barucci, M., Olivieri, E., Pasca, E., Risegari, L., and Ventura, G., Thermal Conductivity of Torlon between 4.2 and 300 K, *Cryogenics* 45, 295, 2005. [Torlon]
28. Ziegler, W. T., Mullins, J. C., and Hwa, S. C. P., Specific Heat and Thermal Conductivity of Four Commercial Titanium Alloys from 20 to 300 K. *Adv. Cryog. Eng.* 8, 268, 1963. [Ti-6Al-4V]
29. Canavan, E. R., and Tuttle, J. G., Thermal Conductivity and Specific Heat Measurements of Candidate Structural Materials for the JWST Optical Bench, *Adv. Cryog. Eng.* 52, 233, 2006. [Ti 15-3-3-3]
30. Bunting, J. G., Ashworth, T., and Steeple, H., The Specific Heat of Apiezon N Grease, *Cryogenics* 9, 385, 1969. [Apiezon N Grease]
31. Bevolo, A. J., Heat Capacity of Apiezon N Grease from 1 to 50 K, *Cryogenics* 14, 661, 1974. [Apiezon N Grease]
32. Wun, M., and Phillips, N. E., Low Temperature Specific Heat of Apiezon N Grease, *Cryogenics* 15, 36, 1975. [Apiezon N Grease]
33. Kreitman, M. M., Ashworth, T., and Rechowicz, M., A Correlation between Thermal Conductance and Specific Heat Anomalies and the Glass Temperature of Apiezon N and T Greases, *Cryogenics* 12, 32, 1972. [Apiezon N Grease]
34. Schnelle, W., Engelhardt, J., and Gemlin, E., Specific Heat Capacity of Apiezon N High Vacuum Grease and of Duran Borosilicate Glass, *Cryogenics* 39, 271, 1999. [Apiezon N Grease]
35. Touloukian, Y.S., and Buyco, E.H., *The TPRC Data Series: Vol. 4, Specific Heat-Metallic Elements and Alloys*, New York, Washington, 1970. [Be]
36. Touloukian, Y. S., Kirby, R. K, Taylor, R. E., and Desai, P. D, *The TPRC Data Series: Vol. 12, Thermal Expansion-Metallic Elements and Alloys*, Plenum Press, New York, 1970. [Be, Mo, Ti-6Al-4V]
37. Corruccini, R. J., and Gniewek, J. J., Thermal Expansion of Technical Solids at Low Temperatures, *NBS Monograph 29*, National Bureau of Standards, Boulder, CO, 1961. [BeCu, Nylon, Stainless steel, Teflon]
38. Reed, R. P., and Clark, A. F., *Materials at Low Temperatures*, American Society of Metals, Metals Park, OH, 1983. [G-10, NiTi, Nylon]
39. Arp, V., Wilson, J. H., Winrich, L., and Sikora, P., Thermal Expansion of Some Engineering Materials from 20 to 293 K, *Cryogenics* 2, 230, 1962. [Sapphire, Al]
40. Apostolescu, D. E., Gaal, P. S., and Chapman, A. S., A Proposed High Temperature Thermal Expansion Reference Material, Standard Reference Material, pp. 637-646. [Sapphire]
41. Swenson, C. A., Linear Thermal Expansivity (1.5-300 K) and Heat Capacity (1.2-90 K) of Stycast 2850FT, *Rev. Sci. Instrum.* 68, 1312, 1997. [Stycast epoxy]
42. Taylor, C. T., Notcutt, M., Wong, E. K., Mann, A. G., and Blair, D. G., Measurement of the Thermal Expansion Coefficient of an All-Sapphire Optical Cavity, *IEEE Trans. Instrum. Meas.* 46, 183, 1977. [Sapphire]
43. Taylor, C. T., Notcutt, M., Wong, E. K., Mann, A. G., and Blair, D. G, Measurement of the Thermal Expansion Coefficient of a Cryogenic All-Sapphire Optical Cavity, *Opt. Commun.* 131, 311, 1996. [Sapphire]
44. Lyon, K. G., Salinger, G. L., Swenson, C. A., and White, G. K., Linear Thermal Expansion Measurements on Silicon from 6 to 340 K, *J. Appl. Phys.* 48, 865, 1977. [Silicon]
45. Karlmann, P. B., Klein, K. J., Halverson, P. G., Peters, R. D., Levine, M. B., van Buren, D., and Dudik, M. J., Linear Thermal Expansion Measurements of Single Crystal Silicon for Validation of Interferometer Based Cryogenic Dilatometer, *Adv. Cryog. Eng.* 52, 35, 2006. [Silicon]
46. Roberts, R. B., Thermal Expansion Reference Data: Silicon 300-850 K, *J. Phys. D: Appl. Phys.* 14, L163, 1981. [Silicon]
47. White, G. K., and Minges, M. L., *Int. J. Thermophys.* 18, 1269, 1997. [Silicon]
48. Swenson, C. A., Recommended Values for the Thermal Expansivity of Silicon from 0 to 1000 K, *J. Phys. Chem. Ref. Data* 12, 179, 1983. [Silicon]

Solids

TABLE 1. Thermal Conductivity λ in W m^{-1} K^{-1} for Selected Materials at Cryogenic Temperatures

Material	1 K	2 K	4 K	6 K	8 K	10 K	12 K	16 K
Al 1100			54.11	83.26	113.5	141.8	170.1	228.0
Al 3003			10.81	16.77	22.81	28.94	35.15	47.49
Al 5083			3.295	4.982	6.685	8.427	10.19	13.73
Al 6061			5.347	8.268	11.23	14.20	17.15	22.91
Al 6063			34.36	51.64	69.70	86.51	103.5	139.2
Be	181.4	355.0	741.4	1082	1410	1754	2120	2887
Be-Cu	0.3161	0.8999	1.879	2.850	3.877	4.955	6.069	8.357
Brass				3.143	4.400	5.702	7.028	9.696
Cu[a] RRR=50			320.4	466.8	622.3	778.1	927.3	1185
Cu[a] RRR=100			642.3	931.7	1239	1540	1814	2226
Cu[a] RRR=300				2810	3636	4320	4829	5276
G-10[b] (norm-dir)			0.07229	0.09112	0.1023	0.1122	0.1217	0.1398
G-10[b] (warp-dir)			0.07322	0.09693	0.1179	0.1361	0.1518	0.1775
Inconel-718			0.4624	0.8644	1.199	1.519	1.832	2.426
Invar (Fe-36Ni)			0.2419	0.4001	0.5630	0.7339	0.9114	1.279

Material	1 K	2 K	4 K	6 K	8 K	10 K	12 K	16 K
Kevlar-49 (composite)	0.002234	0.008663	0.03089	0.06021	0.09263	0.1261	0.1595	0.2234
Kevlar-49 (fiber)	0.003775	0.01457	0.05042	0.09448	0.1405	0.1865	0.2320	0.3213
Pb			2682	700.1	313.3	177.1	118.3	72.82
Mo			56.74	90.04	118.4	146.3	175.4	234.7
Fe-2.25 Ni			1.119	1.787	2.460	3.137	3.809	5.121
Fe-3.5 Ni			0.9289	1.455	1.999	2.535	3.064	4.107
Fe-5.0 Ni			0.7754	1.223	1.679	2.138	2.599	3.523
Fe-9.0 Ni			0.6256	0.9814	1.349	1.716	2.080	2.804
Pt			921.7	1210	1329	1240	1053	687.0
Nylon			0.01245	0.01983	0.02880	0.03902	0.05014	0.07388
Kapton			0.01079	0.01309	0.01802	0.02345	0.02883	0.03884
Mylar[c]		0.03092	0.03764	0.03978	0.04334	0.04791	0.05295	0.06337
PS[d] 12.66 kg m^{-3}								
PS[d] 32.04 kg m^{-3}								
PS[d] 49.98 kg m^{-3}				0.004852	0.004842	0.004862	0.004902	0.005018
PS[d] 99.96 kg m^{-3}			0.009431	0.008950	0.009014	0.009055	0.009098	0.009226
PU[e] 31.88 kg m^{-3}								
PU[e] 32.04 kg m^{-3}								
PU[e] 49.02 kg m^{-3}								
PU[e] 64.08 kg m^{-3}								
PVC[f] 20.02 kg m^{-3}								
PVC[f] 56.06 kg m^{-3}								
Stainless Steel 304			0.2724	0.4653	0.6770	0.9039	1.143	1.647
Stainless Steel 304L			0.2724	0.4653	0.6770	0.9039	1.143	1.647
Stainless Steel 310			0.2409	0.3960	0.5654	0.7457	0.9348	1.331
Stainless Steel 316			0.2724	0.4653	0.6770	0.9039	1.143	1.647
Stainless Steel 735	0.4028	0.5637	1.052	1.315	1.583	1.854	2.126	3.729
Teflon			0.04599	0.06624	0.08244	0.09545	0.1066	0.1258
Torlon 4203	0.004330	0.008974	0.01280	0.01410	0.01467	0.01477	0.01871	0.04760
Ti-6Al-4V								
Ti-15V-3Cr-3Al-3Sn		0.05608	0.1010	0.1450	0.1904	0.2354	0.2796	0.3673

Material	20 K	40 K	60 K	80 K	100 K	160 K	200 K	300 K
Al 1100	282.6	389.5	338.0	283.3	249.7	218.2	215.5	211.8
Al 3003	59.37	104.7	128.6	140.6	147.4	158.9	165.5	177.8
Al 5083	17.21	32.89	45.85	56.81	66.26	88.26	99.24	119.3
Al 6061	28.43	52.23	70.76	85.56	97.70	123.9	136.0	155.3
Al 6063	175.4	276.6	265.6	235.4	213.8	195.5	198.6	200.5
Be	3611	4571	2968	1693	1014	390.7	298.0	194.9
Be-Cu	10.66	21.48	30.42	37.15	41.41			
Brass	12.33	24.12	33.36	40.84	47.45			
Cu[a] RRR=50	1368	1163	670.0	500.3	443.9	406.0	400.1	392.4
Cu[a] RRR=100	2423	1485	741.2	529.3	461.5	415.0	407.0	396.3
Cu[a] RRR=300	5052	1833	801.8	551.0	471.1	421.2	414.6	397.9
G-10[b] (norm-dir)	0.1565	0.2176	0.2553	0.2837	0.3089	0.3846	0.4399	0.5951
G-10[b] (warp-dir)	0.1982	0.2740	0.3366	0.3942	0.4477	0.5899	0.6741	0.8636
Inconel-718	2.956	4.744	5.784	6.532	7.119	8.264	8.720	9.793
Invar (Fe-36Ni)	1.652	3.437	5.005	6.386	7.611	10.524	11.915	13.835
Kevlar-49 (composite)	0.2819	0.5080	0.6794	0.8269	0.9579	1.274	1.436	1.735
Kevlar-49 (fiber)	0.4077	0.8031	1.157	1.482	1.779	2.523	2.917	3.664
Pb	56.95	41.97	38.73	36.98	36.23	35.96	35.46	
Mo	286.9	345.0	267.3	207.9	174.5	143.7	141.4	138.6
Fe-2.25 Ni	6.390	12.28	17.42	21.71	25.23	32.62	35.84	39.17
Fe-3.5 Ni	5.139	10.11	14.52	18.25	21.38	28.10	31.01	34.43
Fe-5.0 Ni	4.442	8.825	12.68	15.98	18.79	24.98	27.72	31.32
Fe-9.0 Ni	3.521	7.022	10.28	13.21	15.81	21.77	24.46	27.83
Pt	451.5	137.0	94.57	82.32	76.69	70.43	70.65	68.42
Nylon	0.09811	0.1999	0.2619	0.2974	0.3179	0.3405	0.3431	0.3368
Kapton	0.04782	0.08319	0.1090	0.1280	0.1419	0.1659	0.1749	0.1919
Mylar[c]	0.07349	0.1158	0.1443	0.1567				

Material	20 K	40 K	60 K	80 K	100 K	160 K	200 K	300 K
PS[d] 12.66 kg m^{-3}					0.01362	0.02032	0.02576	0.04323
PS[d] 32.04 kg m^{-3}		0.005073	0.007232	0.009382	0.01151	0.01829	0.02320	0.03615
PS[d] 49.98 kg m^{-3}	0.005158	0.005905	0.006644	0.007437	0.008328	0.01180	0.01486	0.02509
PS[d] 99.96 kg m^{-3}	0.009393	0.01026	0.01121	0.01239	0.01378	0.01890	0.02309	0.04112
PU[e] 31.88 kg m^{-3}				0.01018	0.01212	0.01734	0.02013	0.02181
PU[e] 32.04 kg m^{-3}					0.01206	0.02005	0.02550	0.04027
PU[e] 49.02 kg m^{-3}		0.01238	0.01465	0.01684	0.01900	0.02554	0.02964	0.03540
PU[e] 64.08 kg m^{-3}			0.007880	0.009551	0.01179	0.01827	0.02184	0.02608
PVC[f] 20.02 kg m^{-3}					0.01680	0.02072	0.02416	0.04093
PVC[f] 56.06 kg m^{-3}					0.02844	0.03503	0.04015	0.05756
Stainless Steel 304	2.169	4.670	6.647	8.114	9.224	11.479	12.633	15.309
Stainless Steel 304L	2.169	4.670	6.647	8.114	9.224	11.48	12.63	15.31
Stainless Steel 310	1.739	3.648	5.093	6.135	6.917	8.568	9.473	11.63
Stainless Steel 316	2.169	4.670	6.647	8.114	9.224	11.48	12.63	15.31
Stainless Steel 735	5.985	7.618	8.783	9.650	10.33	11.89	12.77	15.01
Teflon	0.1422	0.1953	0.2203	0.2341	0.2434	0.2608	0.2672	0.2728
Torlon 4203	0.06377	0.08344	0.09491	0.1079	0.1143	0.1477	0.1684	0.2480
Ti-6Al-4V	0.8426	1.905	2.917	3.531	3.804	4.846	5.750	7.555
Ti-15V-3Cr-3Al-3Sn	0.4559	0.9301	1.420	1.881	2.300	3.412	4.133	6.041

[a] Oxygen-free high conductivity (OFHC) copper.
[b] G-10 is a glass-epoxy composite laminate material.
[c] Polyethylene terephthalate.
[d] Polystyrene.
[e] Polyurethane.
[f] Polyvinylchloride.

TABLE 2. Specific Heat at Constant Pressure c_p in J kg^{-1} K^{-1} of Selected Materials at Cryogenic Temperatures

Material	1 K	2 K	4 K	6 K	8 K	10 K	12 K	16 K
Al 3003			0.2920	0.6047	1.048	1.573	2.268	4.581
Al 5083			0.292	0.605	1.048	1.573	2.268	4.581
Al 6061			0.2920	0.6047	1.048	1.573	2.268	4.581
Apiezon-N Grease	0.02400	0.2118	2.031	6.761	14.38	24.32	35.98	62.76
Be							0.5348	0.9528
Cu (OFHC)			0.09942	0.2303	0.4639	0.8558	1.470	3.640
G-10[a]			2.016	5.463	10.03	15.36	21.20	33.85
In			0.9463	3.505	8.387	15.28	23.56	42.06
Invar (Fe-36Ni)			0.9749	1.475	2.179	3.130	4.334	7.621
Pb (normal)						13.69	21.43	37.50
Pb (superconducting)	2.570	11.78	75.27	297.0	764.6	1539		
Fe-2.25Ni								
Fe-3.5Ni								
Fe-5.0Ni								
Fe-9.0Ni								
Pt	0.02435	0.05529	0.1524	0.3300	0.6349	1.115	1.814	4.009
Nylon			1.560	4.986	11.01	20.00	31.89	62.87
Kapton			0.7916	2.708	6.306	11.68	18.70	36.61
PS[b] (99.96 kg m^{-3})								
PS[b] (9.93 kg m^{-3})								
PS[b] (6.07 kg m^{-3})								
PU[c] (49.02 kg m^{-3})								
PU[c] (389.25 kg m^{-3})								
PVC[d]						9.335	13.41	21.52
Stainless Steel 304			1.849	2.945	4.041	5.163	6.394	9.456
Stainless Steel 304L			1.674	2.557	3.525	4.571	5.740	8.399
Stainless Steel 310			2.164	3.166	4.160	5.226	6.350	8.780
Stainless Steel 316			1.977	2.963	4.047	5.203	6.494	9.615
Teflon			2.232	3.690	9.104	18.02	29.27	53.77

Material	20 K	40 K	60 K	80 K	100 K	160 K	200 K	300 K
Al 3003	8.854	81.96	223.6	368.7	492.2	744.5	835.2	953.9
Al 5083	8.854	81.96	223.6	368.7	492.2	744.5	835.2	953.9
Al 6061	8.854	81.96	223.6	368.7	492.2	744.5	835.2	953.9
Apiezon-N Grease	92.37	252.8	405.6	539.1	654.9	960.6	1194	2276
Be	1.607	9.544	33.66	90.79	192.7	700.1	1097	1883
Cu (OFHC)	7.491	57.63	135.2	203.8	253.5	329.4	355.0	384.0
G-10[a]	47.17	115.4	182.5	249.4	316.9	524.3	664.4	998.7
In	61.08	138.6	179.9	196.9	203.1	213.5	225.0	
Invar (Fe-36Ni)	12.34							
Pb (normal)	52.25	94.45	109.0	115.3	118.7	123.9	126.3	130.1
Pb (superconducting)								
Fe-2.25Ni			90.95	162.6	225.6	346.9	397.6	457.7
Fe-3.5Ni			90.95	162.6	225.6	346.9	397.6	457.7
Fe-5.0Ni			90.95	162.6	225.6	346.9	397.6	457.7
Fe-9.0Ni			90.95	162.6	225.6	346.9	397.6	457.7
Pt	7.376	36.78	66.88	86.73	98.57	115.7	122.9	
Nylon	100.2	297.2	461.8	602.2	729.0	1058	1247	1684
Kapton	57.92	172.9	270.4	348.9	413.7	557.1	626.8	754.5
PS[b] (99.96 kg m^{-3})					441.8	719.3	898.5	1337
PS[b] (9.93 kg m^{-3})					521.4	705.6	834.6	1189
PS[b] (6.07 kg m^{-3})					472.7	670.8	811.4	1201
PU[c] (49.02 kg m^{-3})	42.26	111.9	207.2	310.0	419.6	783.7	1035	1537
PU[c] (389.25 kg m^{-3})					555.6	796.3	959.0	1351
PVC[d]	28.80	64.72	105.4	148.9	195.3	365.2	514.3	1167
Stainless Steel 304	13.65	58.22	135.7	216.1	279.0	373.6	410.0	491.3
Stainless Steel 304L	11.88							
Stainless Steel 310	11.78	55.98	134.2	197.7	247.5	348.8	395.0	472.8
Stainless Steel 316	13.61	57.55	141.8	214.7	273.0	380.3	424.8	490.2
Teflon	76.79	164.4	239.0	312.0	384.6	599.6	738.9	1032

[a] G-10 is a glass-epoxy composite laminate material.
[b] Polystyrene.
[c] Polyurethane.
[d] Polyvinylchloride.

TABLE 3. Linear Thermal Expansion (10^5 × Fractional Change in Length Relative to 293 K) of Selected Materials at Cryogenic Temperatures

Material	4 K	6 K	8 K	10 K	12 K	16 K	20 K	40 K	60 K	80 K	100 K	140 K	200 K	240 K	300 K
Al 3003	-413.8	-414.3	-414.6	-414.9	-415.2	-415.4	-415.4	-411.5	-401.6	-386.1	-365.4	-310.8	-202.6	-118.6	15.81
Al 5083							-415.4	-411.5	-401.6	-386.1	-365.4	-310.8	-202.6	-118.6	15.81
Al 6061	-413.8	-414.3	-414.6	-414.9	-415.2	-415.4	-415.4	-411.5	-401.6	-386.1	-365.4	-310.8	-202.6	-118.6	15.81
Be (a-axis)										-0.1480	-0.1450	-0.1316	-0.0942	-0.0588	0.0079
Be (c-axis)										-0.0970	-0.0960	-0.0892	-0.0660	-0.0419	0.0063
Be (polycrystalline)										-0.1313	-0.1287	-0.1172	-0.0844	-0.0529	0.0073
Be-Cu								-314.6	-306.7	-293.8	-276.8	-232.9	-150.9	-89.16	12.43
G-10[a] (norm-dir)				-714.6	-713.3	-710.7	-707.9	-690.1	-666.5	-637.2	-602.4	-516.4	-348.2	-211.2	29.44
G-10[a] (warp-dir)					-244.0	-242.8	-241.6	-233.9	-224.2	-212.4	-198.8	-166.6	-108.5	-65.01	4.398
Inconel-718							-238.9	-236.9	-231.2	-222.1	-209.9	-177.6	-114.4	-66.17	8.536
Invar (Fe-36Ni)	-40	-40	-40	-40	-40	-40	-40	-40	-40	-40.21	-36.13	-27.45	-14.53	-7.062	0.6163
Mo							-92.4	-92.04	-89.98	-86.45	-81.64	-68.91	-44.34	-25.80	3.390
Nb-Ti							-188.2	-184.9	-177.4	-166.9	-154.1	-139.8	-117.0	-101.3	-77.77
Fe-2.25Ni	-210.5	-210.6	-210.5	-210.5	-210.4	-210.1	-209.6	-205.4	-198.3	-188.8	-177.3	-149.6	-99.05	-59.90	9.770
Fe-3.5Ni	-210.5	-210.6	-210.5	-210.5	-210.4	-210.1	-209.6	-205.4	-198.3	-188.8	-177.3	-149.6	-99.05	-59.90	9.770
Fe-5.0Ni	-210.5	-210.6	-210.5	-210.5	-210.4	-210.1	-209.6	-205.4	-198.3	-188.8	-177.3	-149.6	-99.05	-59.90	9.770
Fe-9.0Ni	-210.5	-210.6	-210.5	-210.5	-210.4	-210.1	-209.6	-205.4	-198.3	-188.8	-177.3	-149.6	-99.05	-59.90	9.770
Nylon	-1389	-1389	-1388	-1388	-1387	-1384	-1381	-1352	-1306	-1246	-1173	-997.9	-671.9	-412.3	64.02
PS[b] (51.42 kg m^{-3})	-1697	-1698	-1698	-1697	-1696	-1691	-1685	-1624	-1525	-1398	-1257	-979.1	-571.7	-307.4	39.90
PS[b] (102.2 kg m^{-3})	-1742	-1738	-1735	-1732	-1730	-1724	-1718	-1679	-1605	-1489	-1346	-1059	-640.6	-372.2	53.03
PU[c] (32.04 kg m^{-3})							-808.3	-795.4	-769.9	-733.5	-688.1	-575.9	-371.2	-218.9	28.96
PU[c] (64.07 kg m^{-3})							-1050	-1019	-975.6	-921.0	-857.3	-709.3	-453.1	-266.8	36.16
PVC[d]	-1025	-1020	-1016	-1011	-1006	-996.6	-986.7	-934.0	-876.8	-816.2	-752.7	-619.1	-402.2	-241.7	43.09
Sapphire						-79.16	-79.21	-79.25	-78.73	-77.49	-75.36	-68.01	-48.44	-29.84	4.523
Stainless Steel 304	-297.0	-297.6	-298.1	-298.6	-299.0	-299.6	-300.0	-297.9	-290.2	-277.7	-261.2	-218.6	-139.1	-80.50	10.86
Stainless Steel 304L	-297.0	-297.6	-298.1	-298.6	-299.0	-299.6	-300.0	-297.9	-290.2	-277.7	-261.2	-218.6	-139.1	-80.50	10.86
Stainless Steel 310	-297.0	-297.6	-298.1	-298.6	-299.0	-299.6	-300.0	-297.9	-290.2	-277.7	-261.2	-218.6	-139.1	-80.50	10.86

Material	4 K	6 K	8 K	10 K	12 K	16 K	20 K	40 K	60 K	80 K	100 K	140 K	200 K	240 K	300 K
Stainless Steel 316	-297.0	-297.6	-298.1	-298.6	-299.0	-299.6	-300.0	-297.9	-290.2	-277.7	-261.2	-218.6	-139.1	-80.50	10.86
Stycast 2850 FT[e]	-44.57	-44.53	-44.48	-44.43	-44.38	-44.25	-44.11	-43.15	-41.80	-40.08	-37.99	-32.76	-22.43	-13.87	1.579
Teflon	-2127	-2127	-2127	-2127	-2126	-2123	-2119	-2077	-2012	-1930	-1839	-1639	-1265	-883.3	162.2
Ti-6Al-4V	-171.9	-172.3	-172.5	-172.8	-173.0	-173.4	-173.6	-172.5	-168.2	-161.1	-151.6	-126.4	-78.55	-43.92	5.285

[a] G-10 is a glass-epoxy composite laminate material.
[b] Polystyrene.
[c] Polyurethane.
[d] Polyvinylchloride.
[e] Epoxy.

TABLE 4. Coefficient of Thermal Expansion α for Selected Materials at Cryogenic Temperatures

T/K	Cu (OFHC) $\alpha/10^{-6}$ K^{-1}	Sapphire $\alpha/10^{-6}$ K^{-1}	Si[a] $\alpha/10^{-8}$ K^{-1}	Phosphor-bronze $\alpha/10^{-6}$ K^{-1}
0.1			0.00000004800	
0.2			0.0000003840	
0.4			0.000003072	
0.6			0.00001037	
0.8			0.00002458	
1			0.00000600	
2			0.00004800	
3			0.0003840	
4	0.002281	0.0001060	0.001296	
6	0.009244	0.0003781	0.003072	
8	0.01893	0.0008953	0.01037	
10	0.03338	0.0016535	0.02458	0.06000
12	0.05522	0.0027127	0.04800	0.1048
16	0.1321	0.0061990	0.1331	0.2444
20	0.2715	0.01249	-0.04541	0.4599
40	2.229	0.1152	-5.330	2.667
60	5.426	0.3323	-29.23	5.807
80	8.335	0.8437	-46.61	8.653
100	10.49		-42.69	10.80
120	12.02		-21.48	12.32
160	13.97		49.33	14.17
200	15.21		123.6	15.29
240	16.05		185.1	16.12
300	16.58		252.6	17.04

[a] Note unit differs from other columns.

TABLE 5. Young's Modulus E in GPa for Selected Materials at Cryogenic Temperatures

Material	2 K	4 K	6 K	8 K	10 K	12 K	16 K	20 K	40 K	60 K	80 K	100 K	140 K	200 K	300 K
Al 5083	80.85	80.87	80.89	80.90	80.91	80.92	80.93	80.93	80.82	80.54	80.11	79.56	78.21	75.82	71.62
Al 6061	77.73	77.75	77.76	77.78	77.79	77.80	77.81	77.81	77.71	77.46	77.08	76.61	75.48	73.56	70.10
Invar	141.6	141.7	141.7	141.7	141.7	141.7	141.7	141.7	141.4	140.8	140.1	139.7	139.7	142.8	152.4
Fe-2.25Ni	217.7	217.7	217.7	217.7	217.7	217.7	217.7	217.7	217.6	217.2	216.7	216.0	214.3	211.2	205.8
Fe-3.5Ni	215.4	215.4	215.4	215.4	215.4	215.4	215.4	215.4	215.3	214.9	214.3	213.6	211.9	208.9	203.6
Fe-5.0Ni	209.2	209.2	209.3	209.3	209.3	209.4	209.4	209.4	209.5	209.4	209.1	208.7	207.3	204.3	198.2
Fe-9.0Ni	205.4	205.4	205.4	205.5	205.5	205.5	205.5	205.5	205.5	205.2	204.8	204.2	202.7	199.9	194.9
Stainless Steel 304		210.1	210.2	210.2	210.2	210.2	210.1	210.1	211.7	214.6	214.4	213.7	211.5	207.2	198.8
Stainless Steel 310		206.7	206.7	206.7	206.7	206.7	206.7	206.6	206.3	205.7	204.9	204.0	201.7	197.7	190.3
Stainless Steel 316				207.9	207.8	207.8	207.9	208.1	209.3	209.5	209.0	208.2	206.0	201.8	194.1

Solids

HEAT CAPACITY OF SELECTED SOLIDS

This table gives the molar heat capacity at constant pressure of representative metals, semiconductors, and other crystalline solids as a function of temperature in the range 200 to 600 K.

References

1. Chase, M. W., et al., *JANAF Thermochemical Tables, 3rd ed., J. Phys. Chem. Ref. Data,* 14, Suppl. 1, 1985.
2. Garvin, D., Parker, V. B., and White, H. J., *CODATA Thermodynamic Tables,* Hemisphere Press, New York, 1987.
3. DIPPR Database of Pure Compound Properties, Design Institute for Physical Properties Data, American Institute of Chemical Engineers, New York, 1987.

Name	Formula	C_p(200 K)/ J mol^{-1} K^{-1}	C_p(250 K)/ J mol^{-1} K^{-1}	C_p(300 K)/ J mol^{-1} K^{-1}	C_p(350 K)/ J mol^{-1} K^{-1}	C_p(400 K)/ J mol^{-1} K^{-1}	C_p(500 K)/ J mol^{-1} K^{-1}	C_p(600 K)/ J mol^{-1} K^{-1}
Aluminum	Al	21.33	23.08	24.25	25.11	25.78	26.84	27.89
Aluminum oxide (α)	Al_2O_3	51.12	67.05	79.45	88.91	96.14	106.17	112.55
Anthracene	$C_{14}H_{10}$	138.6	173.9	210.7	248.8	288.4		
Benzoic acid	C_6H_5COOH	102.7	123.5	147.4	172.0			
Beryllium	Be	9.98	13.58	16.46	18.53	19.95	21.94	23.34
Biphenyl	$C_{12}H_{10}$	131.0	162.5	197.2				
Boron	B	5.99	8.82	11.40	13.65	15.69	18.72	20.78
Calcium	Ca	24.54	25.41	25.94	26.32	26.87	28.49	30.38
Calcium carbonate (calcite)	$CaCO_3$	66.50	75.66	83.82	91.51	96.97	104.52	109.86
Calcium oxide	CaO	33.64	38.59	42.18	45.07	46.98	49.33	50.72
Carbon (graphite)	C	5.01	6.82	8.58	10.24	11.81	14.62	16.84
Cesium chloride	CsCl	50.13	51.34	52.48	53.58	54.68	56.90	59.10
Chromium	Cr	19.86	22.30	23.47	24.39	25.23	26.63	27.72
Cobalt	Co	22.23	23.98	24.83	25.68	26.53	28.20	29.66
Copper	Cu	22.63	23.77	24.48	24.95	25.33	25.91	26.48
Copper(II) oxide	CuO	34.80		42.41	44.95	46.78	49.19	50.83
Copper(II) sulfate	$CuSO_4$	77.01	89.25	99.25	107.65	114.93	127.19	136.31
Germanium	Ge			23.25	23.85	24.31	24.96	25.45
Gold	Au			25.41	25.37	25.51	26.06	26.65
Hexachlorobenzene	C_6Cl_6	162.7	183.6	202.4				
Iodine	I	51.57	53.24	54.51	58.60			
Iron	Fe	21.59	23.74	25.15	26.28	27.39	29.70	32.05
Lead	Pb	25.87	26.36	26.85	27.30	27.72	28.55	29.40
Lithium	Li	21.57	23.42	24.64	25.96	27.60	29.28	
Lithium chloride	LiCl	43.35	46.08	48.10	49.66	50.97	53.34	55.59
Magnesium	Mg	22.72	24.02	24.90	25.57	26.14	27.17	28.18
Magnesium oxide	MgO			37.38	40.59	42.77	45.56	47.30
Manganese	Mn	23.05	24.95	26.35	27.52	28.53	30.29	31.90
Naphthalene	$C_{10}H_8$	105.8	134.1	167.8	204.1			
Potassium	K	27.00	28.01	29.60				
Potassium chloride	KCl	48.44	50.10	51.37	52.31	53.08	54.71	56.35
Silicon	Si	15.64	18.22	20.04	21.28	22.14	23.33	24.15
Silicon dioxide (α-quartz)	SiO_2	32.64	39.21	44.77	49.47	53.43	59.64	64.42
Silver	Ag			25.36	25.55	25.79	26.36	26.99
Sodium	Na	22.45	27.01	28.20	30.14			
Sodium chloride	NaCl	46.89	48.85	50.21	51.25	52.14	53.96	55.81
Tantalum	Ta	24.08	24.86	25.31	25.60	25.84	26.35	26.84
Titanium	Ti	22.37	24.07	25.28	26.17	26.86	27.88	28.60
Tungsten	W	22.49	23.69	24.30	24.65	24.92	25.36	25.79
Vanadium	V	21.88	23.70	24.93	25.68	26.23	26.94	27.49
Zinc	Zn	24.05	25.02	25.45	25.88	26.35	27.39	28.59
Zirconium	Zr	23.87	24.69	25.22	25.61	25.93	26.56	27.28

THERMAL AND PHYSICAL PROPERTIES OF PURE METALS

This table gives the following properties for the metallic elements under ambient conditions. Column definitions for the table are as follows.

Column heading	Definition
Name	Metal name; metals are listed alphabetically by name
Mol. form.	Atomic symbol
Mol. wt.	Molecular weight
t_m	Melting point, in °C
t_b	Normal boiling point, in °C, at a pressure of 101.325 kPa (760 Torr)
$\Delta_{fus}H$	Enthalpy of fusion at the melting point, in J g^{-1}
ρ	Density at 25 °C, in g cm^{-3}
α	Coefficient of linear expansion at 25 °C in K^{-1} (the quantity listed is $10^6 \times \alpha$)
c_p	Specific heat capacity at constant pressure at 25 °C, in units J g^{-1} K^{-1}
λ	Thermal conductivity at 27 °C, in units W m^{-1} K^{-1}
ρ_{elec}	Electrical conductivity, in μΩ cm

References

1. Dinsdale, A. T., *CALPHAD* 15, 317, 1991 [melting point, enthalpy of fusion]. <https://doi.org/10.1016/0364-5916(91)90030-N>
2. Touloukian, Y. S., *Thermophysical Properties of Matter*, Vol. 12, Thermal Expansion, IFI/Plenum, New York, 1975 [coefficient of expansion, density].
3. Ho, C. Y., Powell, R. W., and Liley, P. E., *J. Phys. Chem. Ref. Data* 3, Suppl. 1, 1974 [thermal conductivity].
4. Cox, J. D., Wagman, D. D., and Medvedev, V. A., *CODATA Key Values for Thermodynamics*, Hemisphere Publishing Corp., New York, 1989 [heat capacity].
5. Glushko, V. P., Ed., *Thermal Constants of Substances*, VINITI, Moscow, 1965-1981 [enthalpy of fusion, heat capacity].
6. Wagman, D. D., Evans, W. H., Parker, V. B., Schumm, R. H., Halow, I., Bailey, S. M., Churney, K. L., and Nuttall, R. L., *The NBS Tables of Chemical Thermodynamic Properties, J. Phys. Chem. Ref. Data* 11, Suppl. 2, 1982 [heat capacity].
7. Gschneidner, K. A., *Bull. Alloy Phase Diagrams* 11, 216–224, 1990 [various properties of the rare earth metals]. <https://doi.org/10.1007/BF03029283>
8. Hellwege, K. H., Ed., *Landolt Börnstein, Numerical Values and Functions in Physics, Chemistry, Astronomy, Geophysics, and Technology*, Vol. 2, Part 1, Mechanical-Thermal Properties of State, 1971 [density].
9. *Physical Encyclopedic Dictionary*, Vol. 1–5, Encyclopedy Publishing House, Moscow, 1960–66.

Thermal and Physical Properties of Pure Metals

Name	Mol. form.	Mol. wt.	t_m/°C	t_b/°C	$\Delta_{fus}H$/J g^{-1}	ρ/g cm^{-3}	$10^6 \times \alpha$/K^{-1}	c_p/J g^{-1} K^{-1}	λ/W m^{-1} K^{-1}	ρ_{elec}/μΩ cm
Actinium	Ac	227	1050	≈3200	52.9	10		0.120		
Aluminum	Al	26.982	660.323	2519	396.9	2.70	23.1	0.897	237 sol	2.709
Antimony (gray)	Sb	121.760	630.628	1587	162.5	6.68	11.0	0.207	24.3 sol	39
Barium	Ba	137.327	727	≈1845	51.85	3.62	20.6	0.204	18.4 sol	34.0
Beryllium	Be	9.012	1287	2468	876.1	1.85	11.3	1.825	200	3.70
Bismuth	Bi	208.980	271.402	1564	53.144	9.79	13.4	0.122	7.87	107
Cadmium	Cd	112.411	321.069	767	55.2	8.69	30.8	0.231	96.8	6.8
Calcium	Ca	40.078	842	1484	213.1	1.54	22.3	0.647	200	3.42
Cerium	Ce	140.116	799	3443	38.97	6.770	6.3	0.192	11.3	82.8
Cesium	Cs	132.905	28.5	671	15.7	1.873	97	0.242	35.9	20.8
Chromium	Cr	51.996	1907	2671	403.9	7.15	4.9	0.449	93.7	12.6
Cobalt	Co	58.933	1495	2927	274.9	8.86	13.0	0.421	100	5.6
Copper	Cu	63.546	1084.62	2560	208.7	8.96	16.5	0.385	401	1.712
Dysprosium	Dy	162.500	1412	2567	69.85	8.55	9.9	0.173	10.7	92.6
Erbium	Er	167.259	1529	2868	118.98	9.07	12.2	0.168	14.5	86.0
Europium	Eu	151.964	822	1529	60.61	5.24	35.0	0.182	13.9	90.0
Gadolinium	Gd	157.25	1313	3273	61.49	7.90	9.4	0.235	10.5	131
Gallium	Ga	69.723	29.7646	2229	80.10	5.91	18	0.373	40.6	13.6
Gold	Au	196.967	1064.18	2836	63.72	19.3	14.2	0.129	317	2.255
Hafnium	Hf	178.49	2233	4600	152.39	13.3	5.9	0.144	23.0	33.7
Holmium	Ho	164.930	1472	2700	71.30	8.80	11.2	0.165	16.2	81.4
Indium	In	114.818	156.5985	2027	28.66	7.31	32.1	0.233	81.6	8.0
Iridium	Ir	192.217	2446	4428	213.93	22.562[a]	6.4	0.131	147	4.7
Iron	Fe	55.845	1538	2861	247.3	7.87	11.8	0.449	80.2 sol	9.87
Lanthanum	La	138.905	920	3464	44.64	6.15	12.1	0.195	13.4	61.5
Lead	Pb	207.2	327.462	1749	23.04	11.3	28.9	0.130	35.3	21.1
Lithium	Li	6.94	180.50	1342	432	0.534	46	3.582	84.7	9.47
Lutetium	Lu	174.967	1663	3402	106.59	9.84	9.9	0.154	16.4	58.2
Magnesium	Mg	24.305	650	1090	348.9	1.74	24.8	1.023	156 sol	4.48
Manganese	Mn	54.938	1246	2061	235.0	7.3	21.7	0.479	7.82	144
Mercury	Hg	200.59	-38.8290	356.619	11.44	13.53359	60.4	0.140	8.514 liq	96.1

Name	Mol. form.	Mol. wt.	t_m/°C	t_b/°C	$\Delta_{fus}H$/J g^{-1}	ρ/g cm^{-3}	$10^6 \times \alpha$/K^{-1}	c_p/J g^{-1} K^{-1}	λ/W m^{-1} K^{-1}	ρ_{elec}/μΩ cm
Molybdenum	Mo	95.96	2622	4639	390.6	10.2	4.8	0.251	138	5.47
Neodymium	Nd	144.242	1016	3074	49.50	7.01	9.6	0.190	16.5	64.3
Neptunium	Np	237	644	≈3902	13.50	20.2			6.3	
Nickel	Ni	58.693	1455	2913	297.8	8.90	13.4	0.444	90.7	7.12
Niobium	Nb	92.906	2477	4741	323	8.57	7.3	0.265	53.7	15.2
Osmium	Os	190.23	3033	5008	304.11	22.587[a]	5.1	0.130	87.6	8.1
Palladium	Pd	106.42	1554.8	2963	157.3	12.0	11.8	0.244	71.8	10.73
Platinum	Pt	195.084	1768.2	3825	113.669	21.5	8.8	0.133	71.6	10.7
Plutonium	Pu	244	640	3228	11.57	19.7	46.7		6.74	
Polonium	Po	209	254	962	47.8	9.20	23.5		20	40
Potassium	K	39.098	63.5	759	59.72	0.89	83.3	0.757	102.4	7.39
Praseodymium	Pr	140.908	931	3520	48.90	6.77	6.7	0.193	12.5	70.0
Promethium	Pm	145	1042	≈3000		7.26	11	0.19	15	75
Protactinium	Pa	231.036	1572		53.41	15.4				17.7
Radium	Ra	226	696		34.1	5				
Rhenium	Re	186.207	3185	5590	183.02	20.8	6.2	0.137	47.9	17.2
Rhodium	Rh	102.906	1963	3695	258.4	12.4	8.2	0.243	150	4.3
Rubidium	Rb	85.468	39.30	688	25.6	1.53		0.363	58.2	13.1
Ruthenium	Ru	101.07	2333	4147	381.8	12.1	6.4	0.238	117	7.1
Samarium	Sm	150.36	1072	1794	57.33	7.52	12.7	0.196	13.3	94.0
Scandium	Sc	44.956	1541	2836	313.6	2.99	10.2	0.568	15.8	56.2
Silver	Ag	107.868	961.78	2162	104.8	10.5	18.9	0.235	429	1.617
Sodium	Na	22.990	97.794	882.940	113.1	0.97	71	1.228	141	4.88
Strontium	Sr	87.62	777	1377	84.8	2.64	22.5	0.306	35.3	13.4
Tantalum	Ta	180.948	3017	5455	202.10	16.4	6.3	0.140	57.5	13.4
Technetium	Tc	98	2157	4262	339.7	11			50.6	
Terbium	Tb	158.925	1359	3230	63.87	8.23	10.3	0.182	11.1	115
Thallium	Tl	204.38	304	1473	20.27	11.8	29.9	0.129	46.1	15
Thorium	Th	232.038	1750	4785	59.52	11.7	11.0	0.118	54.0	14.7
Thulium	Tm	168.934	1545	1950	99.68	9.32	13.3	0.160	16.9	67.6
Tin (white)	Sn	118.710	231.928	2586	60.2	7.287	22.0	0.227	66.6	11.5
Titanium	Ti	47.867	1670	3287	295.6	4.506	8.6	0.524	21.9	39
Tungsten	W	183.84	3414	5555	284.54	19.3	4.5	0.132	174	5.39
Uranium	U	238.029	1135	4131	38.40	19.1	13.9	0.116	27.6	28
Vanadium	V	50.942	1910	3407	422	6.0	8.4	0.489	30.7	20.1
Ytterbium	Yb	173.054	824	1196	44.26	6.90	26.3	0.155	38.5	25.0
Yttrium	Y	88.906	1522	3345	128.1	4.47	10.6	0.298	17.2	59.6
Zinc	Zn	65.38	419.527	907	108.11	7.134	30.2	0.388	116	6.01
Zirconium	Zr	91.224	1854	4406	230.2	6.52	5.7	0.278	22.7	42.9

[a] At 20 °C.

THERMAL CONDUCTIVITY OF METALS AND SEMICONDUCTORS AS A FUNCTION OF TEMPERATURE

These tables give the temperature dependence of the thermal conductivity of several metals and of carbon, germanium, and silicon. For graphite, separate entries are given for the thermal conductivity parallel (||) and perpendicular (⊥) to the layer planes. The thermal conductivity of all these materials is very sensitive to impurities at low temperatures, especially below 100 K. Therefore, the values given here should be regarded as typical values for a highly purified specimen; the thermal conductivity of different specimens can vary by more than an order of magnitude in the low-temperature range. See Ref. 2 for details.

Table 1 contains thermal conductivity values for temperatures 1 K to 90 K. Table 2 has thermal conductivity values for temperatures 100 K to 2000 K, for germanium and silicon, values below 300 K are typical values.

References

1. Ho, C. Y., Powell, R. W., and Liley, P. E., *J. Phys. Chem. Ref. Data*, 1, 279, 1972.
2. White, G. K., and Minges, M. L., *Thermophysical Properties of Some Key Solids*, CODATA Bulletin No. 59, 1985.

TABLE 1. Thermal Conductivity in W cm^{-1} K^{-1} of Metals and Semiconductors as a Function of Temperature (1 K to 90 K)

Material	1 K	2 K	3 K	4 K	5 K	6 K	7 K	8 K	9 K	10 K	15 K	20 K	30 K	40 K	50 K	60 K	70 K	80 K	90 K
Ag	39.4	78.3	115	147	172	187	193	190	181	168	96.0	51.0	19.3	10.5	7.0	5.5	4.97	4.71	4.60
Al	41.1	81.8	121	157	188	213	229	237	239	235	176	117	49.5	24.0	13.5	8.5	5.85	4.32	3.42
Au	5.46	10.9	16.1	20.9	25.2	28.5	30.9	32.3	32.7	32.4	24.6	15.8	7.55	5.15	4.21	3.74	3.48	3.32	3.28
Diamond (Type I)		0.0138	0.0461	0.108	0.206	0.344	0.523	0.762	1.05	1.40	3.96	7.87	18.8	29.4	35.3	37.4	36.9	35.1	32.7
Diamond (Type IIa)		0.033	0.111	0.261	0.494	0.820	1.24	1.77	2.41	3.17	8.65	16.8	38.9	65.9	92.1	112	119	117	109
Diamond (Type IIb)		0.0200	0.0676	0.160	0.307	0.510	0.778	1.12	1.53	2.03	5.66	11.2	26.5	44.0	59.1	67.5	69.1	65.7	60.0
Pyrolytic Graphite ||										0.811		4.20	9.86	16.4	23.1	29.8	36.6	42.8	47.5
Pyrolytic Graphite ⊥										0.0116		0.0397	0.0786	0.120	0.152	0.173	0.181	0.181	0.176
Cr	0.402	0.803	1.20	1.60	2.00	2.39	2.27	3.14	3.50	3.85	5.24	5.93	5.49	4.25	3.17	2.48	2.07	1.84	1.69
Cu	42.2	84.0	125	162	195	222	239	248	249	243	171	108	44.5	21.7	12.5	8.29	6.47	5.57	5.08
Fe	1.71	3.42	5.11	6.77	8.39	9.93	11.4	12.7	13.9	14.8	17.0	15.4	10.0	6.23	4.05	2.85	2.16	1.75	1.50
Ge[a]	0.274	2.06	5.35	8.77	11.6	13.9	15.5	16.6	17.3	17.7	17.3	14.9	10.8	7.98	6.15	4.87	3.93	3.25	2.70
Mg	9.86	19.6	29.0	37.6	45.0	50.8	54.7	56.7	57.0	55.8	41.1	27.2	12.9	7.19	4.65	3.27	2.49	2.02	1.78
Ni	2.17	4.34	6.49	8.59	10.6	12.5	14.2	15.8	17.1	18.1	19.5	16.5	9.56	5.82	4.00	3.08	2.50	2.10	1.83
Pb	27.9	44.6	35.8	22.2	13.8	8.10	4.86	3.20	2.30	1.78	0.845	0.591	0.477	0.451	0.436	0.425	0.416	0.409	0.403
Pt	2.31	4.60	6.79	8.8	10.5	11.8	12.6	12.9	12.8	12.3	8.41	4.95	2.15	1.39	1.09	0.947	0.862	0.815	0.789
Si[a]	0.0693	0.454	1.38	2.97	5.27	8.23	11.7	15.5	19.5	23.3	41.6	49.8	48.1	35.3	26.8	21.1	16.8	13.4	10.8
Sn	183	323	297	181	117	76	52	36	26	19.3	6.3	3.2	1.79	1.33	1.15	1.04	0.96	0.915	0.880
Ti	0.0144	0.0288	0.0432	0.0575	0.0719	0.0863	0.101	0.115	0.129	0.143	0.212	0.275	0.365	0.390	0.374	0.355	0.340	0.326	0.315
W	14.4	28.7	42.8	56.3	68.7	79.5	88.0	93.8	96.8	97.1	72.0	40.5	14.4	6.92	4.27	3.14	2.58	2.29	2.17

[a] Values below 300 K are typical values.

TABLE 2. Thermal Conductivity in W cm^{-1} K^{-1} of Metals and Seminconductors as a Function of Temperature (100 K to 2000 K)

Material	100 K	150 K	200 K	250 K	300 K	350 K	400 K	500 K	600 K	800 K	1000 K	1200 K	1400 K	1600 K	1800 K	2000 K
Ag	4.50	4.32	4.30	4.29	4.29	4.27	4.25	4.19	4.12	3.96	3.79	3.61				
Al	3.02	2.48	2.37	2.35	2.37	2.40	2.40	2.36	2.31	2.18						
Au	3.27	3.25	3.23	3.21	3.17	3.14	3.11	3.04	2.98	2.84	2.70	2.55				
Diamond (Type I)	30.0	19.5	14.1	11.0	8.95	7.55	6.5									
Diamond (Type IIa)	100	60.2	40.3	29.7	23.0	18.5	15.4									
Diamond (Type IIb)	54.2	32.5	22.6	17.0	13.5	11.1	9.32									
Pyrolytic Graphite ||	49.7	45.1	32.3	24.4	19.5	16.2	13.9	10.8	8.92	6.67	5.34	4.48	3.84	3.33	2.93	2.62
Pyrolytic Graphite ⊥	0.168	0.125	0.0923	0.0711	0.0570	0.0477	0.0409	0.0322	0.0268	0.0201	0.0160	0.0134	0.0116	0.0100	0.00895	0.00807
Cr	1.59	1.29	1.11	1.00	0.937	0.929	0.909	0.860	0.807	0.713	0.654	0.619	0.588	0.556	0.526	0.494
Cu	4.82	4.29	4.13	4.06	4.01	3.96	3.93	3.86	3.79	3.66	3.52	3.39				
Fe	1.34	1.04	0.94	0.865	0.802	0.744	0.695	0.613	0.547	0.433	0.323	0.283	0.312	0.330	0.345	
Ge[a]	2.32	1.32	0.968	0.749	0.599	0.495	0.432	0.338	0.273	0.198	0.174	0.174				
Mg	1.69	1.61	1.59	1.57	1.56	1.55	1.53	1.51	1.49	1.46						
Ni	1.64	1.22	1.07	0.975	0.907	0.850	0.802	0.722	0.656	0.676	0.718	0.762	0.804			
Pb	0.397	0.379	0.367	0.360	0.353	0.347	0.340	0.328	0.314							
Pt	0.775	0.740	0.726	0.718	0.716	0.717	0.718	0.723	0.732	0.756	0.787	0.826	0.871	0.919	0.961	0.994
Si[a]	8.84	4.09	2.64	1.91	1.48	1.19	0.989	0.762	0.619	0.422	0.312	0.257	0.235	0.221		
Sn	0.853	0.779	0.733	0.696	0.666	0.642	0.622	0.596								
Ti	0.305	0.270	0.245	0.229	0.219	0.210	0.204	0.197	0.194	0.197	0.207	0.220	0.236	0.253	0.270	
W	2.08	1.92	1.85	1.80	1.74	1.67	1.59	1.46	1.37	1.25	1.18	1.12	1.08	1.04	1.01	0.98

[a] Values below 300 K are typical values.
* Extrapolated.

THERMAL CONDUCTIVITY OF ALLOYS AS A FUNCTION OF TEMPERATURE

This table lists the thermal conductivity λ of selected alloys at various temperatures. The indicated compositions refer to weight percent. Because the thermal conductivity is sensitive to exact composition and processing history, especially at low temperatures, these values should be considered approximate.

References

1. Powell, R. L., and Childs, G. E., in *American Institute of Physics Handbook, Third Edition*, Gray, D. E., Ed., McGraw-Hill, New York, 1972.
2. Ho, C. Y., et al., *J. Phys. Chem. Ref. Data*, 7, 959, 1978. <https://doi.org/10.1063/1.555583>

Thermal Conductivity in W m^{-1} K^{-1}

Alloy	λ(4 K)/ W m^{-1} K^{-1}	λ(20 K)/ W m^{-1} K^{-1}	λ(77 K)/ W m^{-1} K^{-1}	λ(194 K)/ W m^{-1} K^{-1}	λ(273 K)/ W m^{-1} K^{-1}	λ(373 K)/ W m^{-1} K^{-1}	λ(573 K)/ W m^{-1} K^{-1}	λ(973 K)/ W m^{-1} K^{-1}
Aluminum								
1100	50	240	270	220	220			
2024	3.2	17	56	95	130			
3003	11	58	140	150	160			
5052	4.8	25	77	120	140			
5083, 5086	3	17	55	95	120			
Duralumin	5.5	30	91	140	160	180		
Bismuth								
Rose metal		5.5	8.3	14	16			
Wood's metal	4	17	23					
Copper								
electrolytic tough pitch	330	1300	550	400	390	380	370	350
free cutting, leaded	200	800	460	380	380			
phosphorus, deoxidized	7.5	42	120	190	220			
brass, leaded	2.3	12	39	70	120			
bronze, 68% Cu; 32% Zn	2.3	16	48	92	110			
beryllium	2	17	36	70	90	113	172	
German silver	0.75	7.5	17	20	23	25	30	40
silicon bronze A		3.4	11	23	30			
manganin	0.48	3.2	14	17	22			
constantan	0.9	8.6	17	19	22			
Gold								
colbalt thermocouple	1.2	8.6	20					
65% Au; 35% Ag		12	24		61	89		
Indium								
85.5% In; 14.5% Pb	1.9	7.8	24	41				
Iron								
commercial pure iron	15	72	106	82	76	66	54	34
plain carbon steel (AISI 1020)	13	20	58	65	65			
plain carbon steel (AISI 1095)		8.5	31	41	45			
3% Ni; 0.7% Cr; 0.6% Mo		6	22		33	35	36	30
4% Si					20	24	28	26
stainless steel	0.3	2	8	13	14	16	19	25
27% Ni; 15% Cr		1.7	55		11	12	16	21
Lead								
60% Pb; 40% Sn (soft solder)		28	44					
64.35% Pb; 35.65% In	0.8	3.26	9.1		20.2			
Nickel								
80% Ni; 20% Cr					12	14	17	23
contracid	0.2	2	7.3	9.5	13			
inconel	0.5	4.2	12.5	13	15	16	19	26
monel	0.9	7.1	15	20	21	24	30	43

Alloy	λ(4 K)/ W m⁻¹ K⁻¹	λ(20 K)/ W m⁻¹ K⁻¹	λ(77 K)/ W m⁻¹ K⁻¹	λ(194 K)/ W m⁻¹ K⁻¹	λ(273 K)/ W m⁻¹ K⁻¹	λ(373 K)/ W m⁻¹ K⁻¹	λ(573 K)/ W m⁻¹ K⁻¹	λ(973 K)/ W m⁻¹ K⁻¹
Platinum								
90% Pt; 10% Ir					31	31.4		
90% Pt; 10% Rh					30.1	30.5		
Silver								
silver solder		12	34	58				
normal Ag thermocouple	48	230	310					
Tin								
60% Sn; 40% Pb	16	55	51					
Titanium								
5.5% Al; 2.5% Sn; 0.2% Fe		1.8	4.3	6.4	7.8	8.4	10.8	
4.7% Mn; 3.99% Al; 0.14% C		1.7	4.5	6.5	8.5			

THERMAL CONDUCTIVITY OF CRYSTALLINE DIELECTRICS

This table lists the thermal conductivity, *k*, of a number of crystalline dielectrics, including some that find use as optical materials. Values are given at temperatures for which data are available.

Reference

Powell, R. L., and Childs, G. E., in *American Institute of Physics Handbook, Third Edition*, Gray, D. E., Ed., McGraw-Hill, New York, 1972.

Thermal Conductivity *k* at Various Temperatures

Material	Formula or other information	T/K	k/W m^{-1} K^{-1}
Aluminum oxide (α)	Al_2O_3 (sintered)	4.2	0.5
		20	23
		77	150
		194	48
		273	35
		373	26
		973	8
Aluminum sulfide	As_2S_3 (glass)	283	0.16
		323	0.21
		373	0.27
Ammonium chloride	NH_4Cl	77	17
		194	23
		230	38
		273	27
Ammonium dihydrogen phosphate	‖ to optic axis	315	0.71
		339	0.71
	⊥ to optic axis	313	1.26
		342	1.34
Argon	Ar	8	6.0
		10	3.7
		20	1.4
		77	0.31
Barium fluoride	BaF_2	225	20
		260	13.4
		305	10.9
		370	10.5
Barium titanate	$BaTiO_3$	5	4.2
		30	24.0
		40	25.0
		100	12.0
		250	4.8
		300	6.2
Beryl	Al, Be silicate	315	6.4
Beryllium oxide	BeO	4.2	0.3
		20	16
		77	270
		373	210
		573	120
		1273	29
Bismuth telluride	Bi_2Te_3	80	6.4
		204	2.8
		303	3.6
		370	4.6
Boron nitride	BN	1047	36.2
		1475	22.7
		1928	21.9
		2111	18.5
Cadmium telluride	CdTe	160	7.0
		297	3.6
		422	2.9
Calcium carbonate (calcite)	‖ to *c* axis	83	25
		273	5.5
	⊥ to *c* axis	83	17
		194	6.5
		273	4.6
		373	3.6
Calcium fluoride	CaF_2 (fluorite)	83	39
		223	18
		273	10
		323	9.2
		373	9
Calcium tungstate	$CaWO_4$ (scheelite)	422	11.3
Carbon (diamond)	Type 1	4.2	13
		20	800
		77	3550
		194	1450
		273	1000
Cesium bromide	CsBr	223	1.2
		273	0.94
		323	0.81
		373	0.77
Cesium iodide	CsI	223	1.4
		273	1.2
		323	1
		373	0.95
Copper(I) oxide	Cu_2O (cuprite)	102	3.74
		163	7.76
		299	5.58
		360	4.86
Garnet	Al,Fe silicate	315	35.8
		358	35.4
		377	35.6
Glass	phoenix	4.2	0.095
		20	0.13
		77	0.37
	plastic perspex	4.2	0.058
		20	0.074
	pyrex	77	0.44
		194	0.88
		273	1
Helium	^3He (high pressure)	0.6	25
		1	2
		1.5	0.57
		2	0.21
	^4He (high pressure)	0.5	42
		0.8	120
		1	24
		2	0.18
Hydrogen	H_2 (para + 0.5% ortho)	2.5	100

Material	Formula or other information	T/K	k/W m⁻¹ K⁻¹
		3	150
		4	200
		6	30
		10	3
Ice	H_2O	173	3.5
		223	2.8
		273	2.2
Iodine	I_2	300	0.45
		325	0.42
		350	0.4
Iron(II,III) oxide	Fe_3O_4 (magnetite)	4.5	27.4
		20.5	293.0
		126.5	7.4
		304	7.0
Krypton	Kr	4.2	0.48
		10	1.7
		20	1.2
		77	0.36
Lanthanum fluoride	LaF_3	78	7.8
		197	5.0
		274	5.4
Lithium fluoride	LiF	4.2	620
		20	1800
		77	150
Manganese(II) oxide	MnO	4.2	0.25
		40	55
		120	8
		573	3.5
Neon	Ne	2	3.0
		3	4.6
		4.2	4.2
		10	0.8
		20	0.3
Nickel(II) oxide	NiO	4.2	5.9
		40	400
		194	82
Potassium bromide	KBr	2	150
		4.2	360
		100	12
		273	5
		323	4.8
		373	4.8
Potassium chloride	KCl	4.2	500
		25	140
		80	35
		194	10
		273	7.0
		323	6.5
		373	6.3
Potassium iodide	KI	4.2	700
		80	13
		194	4.6
		273	3.1
Sapphire	36° to *c* axis	4.2	110

Material	Formula or other information	T/K	k/W m⁻¹ K⁻¹
		20	3500
		35	6000
		77	1100
	⊥ to *c* axis	373	2.6
		523	3.9
		773	5.8
Silicon dioxide (α-quartz)	‖ to *c* axis	20	720
		194	20
		273	12
	⊥ to *c* axis	20	370
		194	10
		273	6.8
Silicon dioxide (vitreous)	SiO_2 (fused silica)	4.2	0.25
		20	0.7
		77	0.8
		194	1.2
		273	1.4
		373	1.6
		673	1.8
Silver(I) chloride	AgCl	223	1.3
		273	1.2
		323	1.1
		373	1.1
Sodium chloride	NaCl	4.2	440
		20	300
		77	30
		273	6.4
		323	5.6
		373	5.4
Sodium fluoride	NaF	5	1100
		50	250
		100	90
Spinel	$MgO{\cdot}Al_2O_3$	373	13
		773	8.5
Strontium titanate	$SrTiO_3$	5	2.4
		30	21.0
		40	19.2
		100	18.5
		250	12.5
		300	11.2
Thallium(I) bromide	TlBr	316	0.59
Thallium(I) chloride	TlCl	311	0.75
	‖ to optic axis	4.2	200
		20	1000
		273	13
	⊥ to optic axis	4.2	160
		20	690
		273	9
Topaz	‖ to *c* axis	315	17.7
		358	15.6
		417	13.3
Tourmaline	‖ to *c* axis	398	2.9
		540	3.2
		723	3.5

THERMAL CONDUCTIVITY OF CERAMICS AND OTHER INSULATING MATERIALS

Thermal conductivity values λ for ceramics, refractory oxides, and miscellaneous insulating materials are given here. The thermal conductivity refers to samples with the density indicated in the second column. Because most of these materials have highly variable composition and structure, the values should only be considered as a rough guide. Column definitions for the table are as follows.

Column heading	Definition
Name	Name or approximate molecular formula of material, materials are listed alphabetically by name
ρ	Density, in g cm^{-3}, for which the thermal conductivity value is applicable
t	Temperature, in °C, for which the thermal conductivity value is applicable
λ	Thermal conductivity, in units W m^{-1} K^{-1}, at specified density and temperature

References

1. Powell, R. L., and Childs, G. E., in *American Institute of Physics Handbook, Third Edition*, Gray, D. E., Ed., McGraw-Hill, New York, 1972.
2. Perry, R. H., and Green, D., *Perry's Chemical Engineers' Handbook, Sixth Edition*, McGraw-Hill, New York, 1984.

Thermal Conductivity at Specified Values of Density and Temperature

Name	ρ/g cm^{-3}	t/°C	λ/W m^{-1} K^{-1}
Alumina (Al_2O_3)	3.8	100	30
	3.8	400	13
	3.8	1300	6
	3.8	1800	7.4
	3.5	100	17
	3.5	800	7.6
Al_2O_3 + MgO		100	15
		400	10
		1000	5.6
Asbestos	0.4	-100	0.07
	0.4	0	0.09
	0.4	100	0.10
Asbestos + 85% MgO	0.3	30	0.08
Asphalt	2.1	20	0.06
Beryllia (BeO)	2.8	100	210
	2.8	400	90
	2.8	1000	20
	2.8	1800	15
	1.85	50	64
	1.85	200	40
	1.85	600	23
Brick, dry	1.54	0	0.04
Brick, refractory, alosite		1000	1.3
Brick, refractory, aluminous	1.99	400	1.2
	1.99	1000	1.3
Brick, refractory, diatomaceous	0.77	100	0.2
	0.77	500	0.24
	0.4	100	0.08
	0.4	500	0.1
Brick, refractory, fireclay	2	400	1
	2	1000	1.2
Brick, refractory, silicon carbide	2	200	2
	2	600	2.4
Brick, refractory, vermiculite	0.77	200	0.26
	0.77	600	0.31
Calcium oxide		100	16
		400	9
		1000	7.5
Cement mortar	2	90	0.55
Charcoal	0.2	20	0.055
Coal	1.35	20	0.26
Concrete	1.6	0	0.8
Cork	0.05	0	0.03
	0.05	100	0.04
	0.35	0	0.06
	0.35	100	0.08
Cotton wool	0.08	30	0.04
Diatomite	0.2	0	0.05
	0.2	400	0.09
	0.5	0	0.09
	0.5	400	0.16
Ebonite	1.2	0	0.16
Felt, flax	0.2	30	0.05
	0.3	30	0.04
Fuller's earth	0.53	30	0.1
Glass wool	0.2	20[a]	0.005
	0.2	50	0.04
	0.2	100	0.05
	0.2	300	0.08
Graphite, 100 mesh	0.48	40	0.18
Graphite, 20-40 mesh	0.7	40	1.29
Linoleum cork	0.54	20	0.08
Magnesia (MgO)		100	36
		400	18
		1200	5.8
		1700	9.2
MgO + SiO_2		100	5.3
		400	3.5
		1500	2.3
Mica, muscovite		100	0.72
		300	0.65
		600	0.69
Mica, phlogopite		100	0.66
Mica, Canadian		300	0.19
		600	0.2
Micanite		30	0.3

Name	ρ/g cm^{-3}	t/°C	λ/W m^{-1} K^{-1}
Mineral wool	0.15	30	0.04
Perlite, expanded	0.1	20[a]	0.002
Plastics, bakelite	1.3	20	1.4
Plastics, celluloid	1.4	30	0.02
Plastics, polystyrene foam	0.05	20[a]	0.033
Plastics, mylar foil	0.05	20[a]	0.0001
Plastics, nylon		-253	0.10
		-193	0.23
		25	0.30
Plastics, polytetrafluoroethylene		-253	0.13
		-193	0.16
		25	0.26
		230	2.5
Plastics, urethane foam	0.07	20	0.06
Porcelain		90	1
Rock, basalt		20	2
Rock, chalk		20	0.92
Rock, granite	2.8	20	2.2
Rock, limestone	2	20	1
Rock, sandstone	2.2	20	1.3
Rock, slate, ⊥		95	1.4
Rock, slate, ||		95	2.5
Rubber, sponge	0.2	20	0.05
Rubber, 92%		25	0.16
Sand, dry	1.5	20	0.33
Sawdust	0.2	30	0.06
Shellac		20	0.23
Silica aerogel	0.1	20[a]	0.003
Snow	0.25	0	0.16
Steel wool	0.1	55	0.09
Thoria (ThO_2)		100	10
		400	5.8
		1500	2.4
Titanium dioxide		100	6.5
		400	3.8
		1200	3.3
Uranium dioxide		100	9.8
		400	5.5
		1000	3.4
Wood, balsa, ⊥	0.11	30	0.04
Wood, fir, ⊥	0.54	20	0.14
Wood, fir, ||	0.54	20	0.35
Wood, oak		20	0.16
Wood, plywood		20	0.11
Wood, pine, ⊥	0.45	60	0.11
Wood, pine, ||	0.45	60	0.26
Wood, walnut, ⊥	0.65	20	0.14
Wool	0.09	30	0.04
Zinc oxide		200	17
		800	5.3
Zirconia (ZrO_2)		100	2
		400	2
		1500	2.5
Zirconia + silica		200	5.6
		600	4.6
		1500	3.7

[a] -200 to 20 °C.

THERMAL CONDUCTIVITY OF GLASSES

This table gives the composition of various types of glasses and the thermal conductivity k as a function of temperature. Because of the variability of glasses, the data should be regarded as only approximate.

Composition and Thermal Conductivity of Glasses

Type of glass	wt% SiO_2	Other oxides	wt% Other oxides	t/°C	k/W m^{-1} K^{-1}
Vitreous silica	100			-150	0.85
				-100	1.05
				-50	1.20
				0	1.30
				50	1.40
				100	1.50
Vycor glass	96	B_2O_3	3	-100	1.00
				0	1.25
				100	1.40
Pyrex type chemically resistant borosilicate glasses	80–81	B_2O_3	12–13	-100	0.90
		Na_2O	4	0	1.10
		Al_2O_3	2	100	1.25
Crown glasses					
Borosilicate crown glasses	60–65	B_2O_3	15–20	-100	0.65–0.75
				0	0.90–0.95
				100	1.00–1.05
	65–70	B_2O_3	10–15	-100	0.75–0.80
				0	0.95–1.00
				100	1.05–1.15
	70–75	B_3O_3	5–10	-100	0.80–0.85
				0	1.05–1.10
				100	1.15–1.20
Zinc crown glasses (i)	55–65	ZnO	5–15	-100	0.88–0.92
		Remainder:		0	1.10–1.15
		B_2O_3, Al_2O_3		100	1.15–1.25
		ZnO	5–15	-100	0.60–0.70
		Remainder:		0	0.70–0.90
		Na_2O, K_2O		100	0.85–0.95
		ZnO	15–25	-100	0.88–0.92
		Remainder:		0	1.10–1.15
		B_2O_3, Al_2O_3		100	1.15–1.20
		ZnO	15–25	-100	0.65–0.80
		Remainder:		0	0.85–0.95
		Na_2O, K_2O		100	0.90–1.05
Zinc crown glasses (ii)	65–75	ZnO	5–15	-100	0.88–0.92
		Remainder:		0	1.15–1.15
		B_2O_3, Al_2O_3		100	1.20–1.30
		ZnO	5–15	-100	0.70–0.85
		Remainder:		0	0.90–1.05
		Na_2O, K_2O		100	1.00–1.15
		ZnO	15–25	-100	0.90–0.95
		Remainder:		0	1.15–1.15
		B_2O_3, Al_2O_3		100	1.20–1.25
		ZnO	15–25	-100	0.65–0.85
		Remainder:		0	0.85–1.00
		Na_2O, K_2O		100	1.05–1.20
Barium crown glasses	31	B_2O_3	12	-100	0.55
		Al_2O_3	8	0	0.70
		BaO	48	100	0.80
	41	B_2O_3	6	-100	0.60
		Al_2O_3	2	0	0.75
		ZnO	8	100	0.85
		BaO	43		
	47	B_2O_3	4	-100	0.65

Type of glass	wt% SiO_2	Other oxides	wt% Other oxides	t/°C	k/W m^{-1} K^{-1}
		Na_2O	1	0	0.75
		K_2O	7	100	0.90
		ZnO	8		
		BaO	32		
	65	B_2O_3	2	-100	0.70
		Na_2O	5	0	0.90
		K_2O	15	100	1.00
		ZnO	2		
		BaO	10		
Borate glasses					
Borate flint glass	9	B_2O_3	36	-100	0.55
		Na_2O	1	0	0.65
		K_2O	2	100	0.80
		PbO	36		
		Al_2O_3	10		
		ZnO	6		
Borate flint glass	0	B_2O_3	56	-100	0.50
		Al_2O_3	12	0	0.65
		PbO	32	100	0.85
Borate flint glass	0	B_2O_3	43	-100	0.40
		Al_2O_3	5	0	0.55
		PbO	52	100	0.70
Borate glass	4	B_2O_3	55	-100	0.65
		Al_2O_3	14	0	0.80
		PbO	11	100	0.90
		K_2O	4		
		ZnO	12		
Borate crown glass	0	B_2O_3	64	-100	0.50
		Na_2O	8	0	0.65
		K_2O	3	100	0.85
		BaO	4		
		PbO	3		
		Al_2O_3	18		
Light borate crown glass	0	B_2O_3	69	-100	0.55
		Na_2O	8	0	0.70
		BaO	5	100	0.90
		Al_2O_3	18		
Zinc borate glass	0	B_2O_3	40	-100	0.65
		ZnO	60	0	0.75
				100	0.85
Phosphate crown glasses					
Potash phosphate glass	0	P_2O_5	70	0	0.75
		B_2O_3	3	100	0.85
		K_2O	12		
		Al_2O_3	10		
		MgO	4		
Baryta phosphate glass	0	P_2O_5	60	45	0.75
		B_2O_3	3		
		Al_2O_3	8		
		BaO	28		
Soda-lime glasses	75	Na_2O	17	-100	0.75
		CaO	8	0	0.95
				100	1.10
	75	Na_2O	12	-100	0.90
		CaO	13	0	1.10
				100	1.15
	72	Na_2O	15	-100	0.80
		CaO	11	0	1.00
		Al_2O_3	2	100	1.15
	65	Na_2O	25	-100	0.65
		CaO	10	0	0.85
				100	0.95

Type of glass	wt% SiO_2	Other oxides	wt% Other oxides	t/°C	k/W m^{-1} K^{-1}
	65	Na_2O	15	-100	0.85
		CaO	20	0	1.00
				100	1.10
	60	Na_2O	20	-100	0.75
		CaO	20	0	0.90
				100	1.00
Other crown glasses					
Crown glass	75	Na_2O	9	-100	0.80
		K_2O	11	0	1.00
		CaO	5	100	1.10
High dispersion crown glass	68	Na_2O	16	-100	0.65
		ZnO	3	0	0.85
		PbO	13	100	1.00
Miscellaneous flint glasses					
(i) Silicate flint glasses	65	PbO	25	-100	0.65–0.70
Light flint glass		Others	10	0	0.88–0.92
				100	1.00–1.05
Light flint glass	55	PbO	35	-100	0.60–0.65
		Others	10	0	0.75–0.85
				100	0.88–0.92
Ordinary flint glass	45	PbO	45	-100	0.50–0.60
		Others	10	0	0.65–0.75
				100	0.80–0.85
Heavy flint glass	35	PbO	60	-100	0.45–0.50
		Others	5	0	0.60–0.65
				100	0.70–0.75
Very heavy flint glass	25	PbO	73	-100	0.40–0.45
		Others	2	0	0.55–0.60
				100	0.63–0.67
Very heavy flint glass	20	PbO	80	-100	0.40
				0	0.50
				100	0.60
(ii) Borosilicate flint glass	33	B_2O_3	31	-100	0.65
		PbO	25	0	0.85
		Al_2O_3	7	100	0.95
		K_2O	3		
		Na_2O	1		
(iii) Barium flint glass	50	BaO	24	-100	0.60
		PbO	6	0	0.70
		K_2O	8	100	0.85
		Na_2O	3		
		ZnO	8		
		Sb_2O_3	1		
Other glasses					
Potassium glass	59	K_2O	33	50	0.88–0.92
		CaO	8		
Iron glasses	63	Fe_2O_3	10	-100	0.80
		Na_2O	17	0	0.95
		MgO	4	100	1.05
		CaO	3		
		Al_2O_3	2		
	67	Fe_2O_3	15	0	0.88–0.92
		Na_2O_3	18	100	1.00–1.05
	62	Fe_2O_3	20	0	0.85–0.90
		Na_2O	18	100	0.95–1.00
Rock glasses					
Obsidian				0	1.35
Artificial diabase				100	1.25

THERMOELECTRIC PROPERTIES OF METALS AND SEMICONDUCTORS

Lev I. Berger

There are three thermoelectric phenomena that result from correlation between propagation of heat through a conductor and displacement of the current carriers in the conductor. The Seebeck effect (Ref. 1) consists of formation of an electric current in an electrical circuit formed by two dissimilar conductors if the contacts between the conductors are held at different temperatures. A reverse phenomenon, the Peltier effect (Ref. 2), consists of formation of a temperature difference between the contacts in a circuit of this type if an electric current is created in the circuit by an external current source to which the circuit is connected. W. Thomson (Lord Kelvin), who explained both effects (Refs. 3,4), predicted and experimentally confirmed the existence of another thermoelectric phenomenon, named the Thomson effect, which consists of absorption or release of heat in a uniform conductor with a current passing through it when a temperature gradient (positive or negative) is present along the current direction.

The electromotive force, ΔU, which creates the Seebeck current in the circuit, is the algebraic sum of the emf's created in each of the conductors, and is proportional to the temperature difference, ΔT, between the electrical contact points: $\Delta U = \Delta U_1 + \Delta U_2 = \alpha_1\Delta T + \alpha_2\Delta T$. The coefficient of proportionality, α, called the Seebeck coefficient or thermoelectric power or thermal electromotive force (thermal emf), of each of the two materials depends on the electrical properties and temperature of the material. The Peltier effect is measured by the amount of heat, ΔQ, released or absorbed in a unit of time (in addition to the Joule heat) at a contact of two dissimilar conductors with electric current ΔI passing through the contact: $\Delta Q = \Pi{\cdot}\Delta I$. Thomson showed that $\Pi = \alpha T$. The Thomson effect's heat, dQ, released or absorbed in a unit of time along a part of a conductor of length dx is proportional to the current magnitude I, the temperature gradient along the conductor $\partial T/\partial x$, and the increment dx: $dQ = \tau I(\partial T/\partial x)dx$. Thomson showed that the magnitude of the coefficient of proportionality, τ, later named the Thomson coefficient, depends on only the properties of the conductor and the ambient temperature and correlates with the other thermoelectric parameters of a material through the equation $\tau = T(\partial\alpha/\partial T)$.

Another thermoelectric phenomenon, called the Bridgman effect or the internal Peltier effect (Ref. 5), occurs when an electric current passes through an anisotropic crystal, resulting in absorption or liberation of heat because of non-uniformity in current distribution.

In view of the correlations between α, Π, and τ, we need only to present data for one of these parameters, namely, thermal emf α and its dependence on temperature. These values are presented below, first for metals and then for semiconductors. In accordance with modern theory of solids, thermal emf in semiconductors is up to three or even four orders of magnitude higher than that in metals (Ref. 9).

The accuracy of the data presented in this table is dependent on a number of factors. The thermal emf of a material is sensitive to negligibly small amounts of impurities in the material, which may be below the limits of sensitivity of the chemical analysis; to orientation of crystal grains in a sample of the material, and to thermal processing of the material.

References

1. Seebeck, T. J., *Abhand. Deut. Akad. Wiss. Berlin*, 265–373, 1822.
2. Peltier, J. C. A, *Ann. Chem.*, LVI, 371–387, 1834.
3. Thomson, W., *Proc. Roy. Soc. Edinburgh*, 91–98, 1851.
4. Thomson, W., *Math. and Phys. Papers, Cambridge*, 1, 558, 1882; 2, 306, 1882.
5. Bridgman, P. W., *Proc. Natl. Acad. Sci. USA*, 13(2), 46–50, 1927; *Phys. Rev.*30, 911–921 (1927).
6. Blatt, F. J., *Thermoelectric Power of Metals*. Plenum Press, NY 1976.
7. Foiles, C. L., Thermopower of Pure Metals and Dilute Alloys, in *Landolt–Bornstein. Numerical Data and Functional Relationships in Science and Technology. New Series.* Group III, v. 15, Metals. Springer-Verlag, NY, 1985.
8. Burkov, A. T., and Vedernikov, M. V., in *CRC Handbook of Thermoelectrics*, D. M. Rowe, Ed., CRC Press, Boca Raton, FL, 1995, pp. 387–399.
9. Ioffe, A. F., *Semiconductor Thermoelements and Thermoelectric Cooling*, Infosearch Ltd., 1957.
10. Berger, L. I., and Prochuchan, V. D., *Ternary Diamond-Like Semiconductors*, Cons. Bureau, Plenum Press, New York, 1969.
11. Rowe, D. M., Ed., *Thermoelectrics Handbook Macro to Nano*, Taylor & Francis, Boca Raton, 2006.
12. Berger, L. I., *Semiconductor Materials*, CRC Press, Boca Raton, FL, 1996.
13. Glazov, V. M., Tshizhevskaya, S. N., and Glagoleva, N. N., *Liquid Semiconductors*, Nauka Publ. House, Moscow, 1967.
14. Shay, J. L., and Wernick, J. H., *Ternary Chalcopyrite Semiconductors: Growth, Electronic Properties and Applications*, Pergamon Press, New York, 1975.
15. Heikes, R. R., and Ure, R. W., *Thermoelectricity: Science and Engineering*, Interscience Publ., New York, 1961.
16. Goland, A. N., and Ewald, A. W., *Phys. Rev.* 104, 948 (1956).
17. Tauc J., *Photo and Thermoelectric Effects in Semiconductors*, Pergamon, New York, 1962.
18. Dugdale, J. S., *The Electrical Properties of Metals and Alloys*, Edward Arnold, London, 1977.
19. Rowe, D. M., Ed., *CRC Handbook of Thermoelectrics*, CRC Press, Boca Raton, FL, 1994.

Thermoelectric Properties of Elemental Metals

	Thermal emf α(T) in μV/K at Temperature T				
	100 K	300 K	500 K	1000 K	1500 K
Ag	0.73	1.51	2.82	7.95	
Al	−2.2	−1.66	1.96		
Au	0.82	1.94	2.86	3.85	
Ba	−4	12.1	28.5		
Be	−2.5	1.7	2.7	7.9	
Ca	1.05	10.3	17.1		
Cd	−0.05	2.55			
Ce	13.6	6.2	5.2	−4.8	
Co	−8.43	−30.8	−44.8	−35.9	−7.8
Cr	5	21.8	16.6	17.9	5.7
Cs		−0.9			
Cu	1.19	1.83	2.83	5.36	
Dy	−4.1	−1.8	0.9	2.3	
Er	−3.8	−0.1	1.9	4.2	
Eu	5.3	24.5	46		
Fe	11.6	15	3	0.4	
Ga	0.5				
Gd	−4.6	−1.6	−0.5	−0.8	
Hf	0	5.5	5.7	−0.5	
Ho	−6.7	−1.6	1.4	2.8	

	Thermal emf α(T) in μV/K at Temperature T				
	100 K	300 K	500 K	1000 K	1500 K
In	0.56	1.68			
Ir	1.42	0.86	−0.1	−2.7	−5.7
K	−5.2	−13.7			
La	0.1	1.7	2	−1.7	
Li	4.3				
Lu	−6.9	−4.3	−2.6	0	
Mg	−2.1	−1.46			
Mn	−2.5	−9.8	−8.4	−1.5	
Mo	0.1	5.6	11.4	17.4	13.7
Na	−2.6	−6.3			
Nb	1.05	−0.44	−1.1	0.45	3.2
Nd	−4	−2.3	0	−1.2	
Ni	−8.5	−19.5	−25.8	−29.9	
Np	8.9	−3.1			
Os	−3.2	−4.4	−4.7	−6.3	−8.5
Pb	−0.58	−1.05	−1.5		
Pd	1.1	−10.7	−16.3	−32.3	−46.4
Pt	−4.1	−5.3	−7.9	−8.2	
Pu	12				
Rb	−3.6	−10			
Re	−1.4	−5.9	−5.9	−1.9	1.8
Rh	0.8	0.6	0.5	−1.5	
Ru	0.3	−1.4	−1.8	−4.2	−7.5
Sc	−14.3	−19	−17.5	−5.4	10.2
Sm	0.7	1.2	0.6	−3	
Sn	−0.04	−1			
Sr	−3	1.1	4.2		
Ta	0.7	−1.9	−2.3	1.6	7.2
Tb	−1.6	−1	0.3	0.6	
Th	0.6	−3.2	−9.2	−14.3	−10.4
Ti	−2	9.1	5.3	−3.1	−0.5
Tl	0.6	0.3	−1.5		
Tm	−1.3	1.9	2.7	2.2	
U	3	7.1	11	16.7	
V	2.9	0.23	1.1	4.6	
W	−4.4	0.9	9	19.8	21.3
Y	−5.1	−0.7	0.3	2.9	6.6
Yb	5.1	30	20.3	12.3	
Zn	0.7	2.4			
Zr	4.4	8.9	4.6	−3	1.1

Thermoelectric Properties of Selected Semiconductors; Values near Room Temperature Unless Otherwise Indicated

Material	$\alpha/\mu V\ K^{-1}$	Material	$\alpha/\mu V\ K^{-1}$	Material	$\alpha/\mu V\ K^{-1}$
		Elemental Semiconductors			
B	600 (500 K)	n-Si	300	p-Si	−500
n-Ge	600	p-Ge	−830	α-Sn	−40 (250 K)
		I-VI Compounds			
Cu_2S	327	Cu_2Se	135	Cu_2Te	40
Ag_2Te	120				
		II-VI Compounds			
ZnO	300	CdS	700	ZnSe	55
CdSe	200				
		III-V Compounds			
GaN	70	GaP	1200	InP	−400
AlAs	70	n-GaAs	380	p-GaAs	−310
InAs	200	AlSb	500	n-GaSb	250
p-GaSb	−55	n-InSb	240	p-InSb	200
		V-VI Compounds			
Sb_2Te_3	110	n-Bi_2Te_3	224	p-Bi_2Te_3	−227
		I-III-VI Compounds			
$CuAlS_2$	50	$AgInSe_2$	−370	$CuTlTe_2$	80
CuGaSe	40	$AgTlSe_2$	800	$AgAlTe_2$	321
$CuInSe_2$	340	$CuGaTe_2$	340	$AgGaTe_2$	950
$CuTlSe_2$	−5	$CuInTe_2$	260	$AgInTe_2$	298
$AgGaSe_2$	90	$CuTlTe_2$	80		
		I-IV-VI Compounds			
Cu_2GeS_3	300	Cu_2GeSe_3	100	Cu_2GeTe_3	10
Cu_2SnS_3	600	Cu_2SnSe_3	250	Cu_2SnTe_3	30
		I-V-VI Compounds			
Cu_3AsS_4	130	Cu_3AsSe_4	120	Cu_3SbSe_4	200
		II-IV-V Compounds			
$ZnGeP_2$	1200	$ZnSiAs_2$	1100	$CdGeAs_2$	190
$CdSnAs_2$	600				

FERMI ENERGY AND RELATED PROPERTIES OF METALS

Lev I. Berger

In the classical Drude theory of metals, the Maxwell-Boltzmann velocity distribution of electrons is used. It states that the number of electrons per unit volume with velocities in the range of $d\vec{v}$ about any magnitude $\vec{v}$ at temperature T is

$$f_B(\vec{v})d\vec{v} = n\left(\frac{m}{2\pi k_B T}\right)\exp\left(-\frac{mv^2}{2k_B T}\right)d\vec{v}$$

where n is the total number of conduction electrons in a unit volume of a metal, m is the free electron mass, and k_B is the Boltzmann constant. In an attempt to explain a substantial discrepancy between the experimental data on the specific heat of metals and the values calculated on the basis of the Drude model, Sommerfeld suggested a model of the metal in which the Pauli exclusion principle is applied to free electrons. In this case, the Maxwell-Boltzmann distribution is replaced by the Fermi-Dirac distribution:

$$f(\vec{v})d\vec{v} = 2\left(\frac{m}{h}\right)^3 d\vec{v}\left\{\exp\left[\left(\frac{mv^2}{2} - k_B T_0\right)\Big/ k_B T\right] + 1\right\}^{-1}$$

Here h is the Planck constant and T_0 is a characteristic temperature which is determined by the normalization condition

$$n = \int d\vec{v}\cdot f(\vec{v})$$

The magnitude of T_0 is quite high; usually, $T_0 > 10^4$ K. So, at common temperatures ($T < 10^3$ K), the free electron density of a metal is much smaller than in the case of the Maxwell-Boltzmann distribution. This allows us to explain why the experimental data on specific heat for metals are close to those for insulators.

The maximum kinetic energy the electrons of a metal may possess at $T = 0$ K is called the Fermi energy, e.g.,

$$E_F = \frac{\hbar^2 k_F^2}{2m} = \left(\frac{e^2}{2k_B}\right)(k_F r_B)^2$$

where k_F is the Fermi momentum or the Fermi wave vector

$$k_F = (3\pi^2 n)^{1/3}$$

e is the electron charge, and r_B is the Bohr radius

$$r_B = \hbar^2/me^2 = 0.529\cdot 10^{-10}\ \text{m}$$

Another, more common expression for the Fermi energy is

$$E_F = \tfrac{1}{2} m v_F^2$$

where $v_F = \hbar k_F/m$ is the Fermi velocity which can be expressed using the concept of the electron radius, r_s. It is equal to radius of a sphere occupied by one free electron. If the total volume of a metal sample is V and the number of conduction electrons in this volume is N, then the volume per electron is equal to

$$\frac{V}{N} = \frac{1}{n} = \frac{4}{3}\pi r_S^3$$

and

$$r_S = \left(\frac{3}{4\pi n}\right)^{1/3}$$

The following table contains information pertinent to the Sommerfeld model for some metals. The magnitudes of T_0 are calculated using the expression

$$T_0 = \frac{E_F}{k_B} = \frac{58.2\cdot 10^4}{(r_S/r_B)^2}\ \text{K}$$

Fermi Energy and Other Ground-State Properties of the Electron Gas in Metals[d]

Metal	Name	Valency	$n/10^{28}$ m^{-3}	r_S/pm	r_S/r_B	E_F/eV	$T_0/10^4$ K	$k_F/10^{10}$ m^{-1}	$v_F/10^6$ m s^{-1}
Li[a]	Lithium	1	4.70	172	3.25	4.74	5.51	1.12	1.29
Na[b]	Sodium	1	2.65	208	3.93	3.24	3.77	0.92	1.07
K[b]	Potassium	1	1.40	257	4.86	2.12	2.46	0.75	0.86
Rb[b]	Rubidium	1	1.15	275	5.20	1.85	2.15	0.70	0.81
Cs[b]	Cesium	1	0.91	298	5.62	1.59	1.84	0.65	0.75
Cu	Copper	1	8.47	141	2.67	7.00	8.16	1.36	1.57
Ag	Silver	1	5.86	160	3.02	5.49	6.38	1.20	1.39
Au	Gold	1	5.90	159	3.01	5.53	6.42	1.21	1.40
Be	Beryllium	2	24.7	99	1.87	14.3	16.6	1.94	2.25
Mg	Magnesium	2	8.61	141	2.66	7.08	8.23	1.36	1.58
Ca	Calcium	2	4.61	173	3.27	4.69	5.44	1.11	1.28
Sr	Strontium	2	3.55	189	3.57	3.93	4.57	1.02	1.18
Ba	Barium	2	3.15	196	3.71	3.84	4.23	0.98	1.13
Nb	Niobium	1	5.56	163	3.07	5.32	6.18	1.18	1.37
Fe	Iron	2	17.0	112	2.12	11.1	13.0	1.71	1.98
Mn[c]	Manganese	2	16.5	113	2.14	10.9	12.7	1.70	1.96
Zn	Zinc	2	13.2	122	2.30	9.47	11.0	1.58	1.83
Cd	Cadmium	2	9.27	137	2.59	7.47	8.88	1.40	1.62

Metal	Name	Valency	$n/10^{28}$ m^{-3}	r_S/pm	r_S/r_B	E_F/eV	$T_0/10^4$ K	$k_F/10^{10}$ m^{-1}	$v_F/10^6$ m s^{-1}
Hg[a]	Mercury	2	8.65	140	2.65	7.13	8.29	1.37	1.58
Al	Aluminum	3	18.1	110	2.07	11.7	13.6	1.75	2.03
Ga	Gallium	3	15.4	118	2.19	10.4	12.1	1.66	1.92
In	Indium	3	11.5	127	2.41	8.63	10.0	1.51	1.74
Tl	Thallium	3	10.5	131	2.48	8.15	9.46	1.46	1.69
Sn	Tin (white)	4	14.8	117	2.22	10.2	11.8	1.64	1.90
Pb	Lead	4	13.2	122	2.30	9.47	11.0	1.58	1.83
Bi	Bismuth	5	14.1	119	2.25	9.90	11.5	1.61	1.87
Sb	Antimony	5	16.5	113	2.14	10.9	12.7	1.70	1.96

[a] At 78 K.
[b] At 5 K.
[c] α-phase.
[d] The data in the table are for atmospheric pressure and room temperature unless otherwise noted.

References

1. Drude, P., *Ann. Physik*, 1, 566, 1900; ibid., 3, 369, 1900.
2. Sommerfeld, A. and Bethe, H., *Handbuch der Physik*, Chapter 3, Springer, 1933.
3. Wyckoff, R. W. G., *Crystal Structures*, 2nd. ed., Interscience, 1963.
4. Ashcroft, N. W. and Mermin, N. D., *Solid State Physics*, Holt, Rinehart and Winston, 1976.

Section 13
Polymer Properties

Abbreviations Used in Polymer Science and Technology 13-1
Physical Properties of Selected Polymers 13-3
Nomenclature for Organic Polymers 13-5
Solvents for Common Polymers 13-9
Glass Transition Temperature for Selected Polymers 13-10
Dielectric Constant of Selected Polymers 13-17
Pressure-Volume-Temperature Relationships for Polymer Melts 13-18
Vapor Pressures (Solvent Activities) for Binary Polymer Solutions 13-22
Solubility Parameters of Selected Polymers 13-27

ABBREVIATIONS USED IN POLYMER SCIENCE AND TECHNOLOGY

ABA	triblock copolymers; acrylonitrile-butadiene acrylate
ABS	copolymer of acrylonitrile, butadiene, and styrene
ACS	acrylonitrile-chlorinated polyethylene styrene terpolymer
AIBN	2,2'-azobisisobutyronitrile
AMA	acrylate-maleic anhydride terpolymer
AMMA	acrylate-methyl methacrylate copolymer
AN	acrylonitrile
AP	ethylene-propylene copolymers
APO	amorphous polyolefin
AS	acrylonitrile-styrene copolymer
ASA	acrylonitrile-styrene-acrylonitrile block
ATR	attenuated total reflectance spectroscopy
AU	polyurethane
BMC	bulk molding compound
BMI	bis-maleimide
BPO	benzoyl peroxide
CA	cellulose acetate
CAB	cellulose acetate butyrate
CAP	cellulose acetate proprionate
CAR	carbon fiber
CED	cohesive energy density
CFRP	carbon-reinforced plastics
CMC	carboxymethylcellulose
CN	cellulose nitrate
COC	cycloolefin copolymer
COP	copolyester thermoplastic elastomer
CPE	chloronated polyethylene
CPVC	chlorinated poly(vinyl chloride)
CR	neoprene
CTA	cellulose triacetate
CTFE	chlorotrifluoroethylene
C_s	chain transfer constant
DAIP	diallyl isophthalate plasticizer
DAP	diallyl phthalate plasticizer
DNA	deoxyribonucleic acid
DP	degree of polymerization
DRS	dynamic reflectance spectroscopy
DS	degree of substitution
EAA	ethylene-acrylic acid copolymer
EC	ethyl cellulose
ECTFE	ethylene-chlorotrifluoroethylene copolymer
EEA	ethylene-ethyl acetate copolymer
EGG	Einstein-Guth-Gold equation
EMAC	ethylene-methyl acrylate copolymer
EnBA	ethylene butyl acetate
EP	epoxy resin
EPDM	poly(ethylene-*co*-propylene) cross-linked
EPM	ethylene-propylene copolymer
EPR	ethylene-propylene rubber
EPS	expanded polystyrene
ET	thiokol
ETFE	ethylene-tetrafluoroethylene polymer
EU	polyether polyurethane
EVA	ethylene-vinyl acetate copolymer
EVOH	ethylene-vinyl alcohol copolymer
FEP	fluorinated ethylene propylene
FRP	fibrous glass-reinforced polyester; fiber-reinforced plastic
G	molar attraction constant
GF	glass reinforced
GRS	poly(butadiene-*co*-styrene)
HDPE	high-density polyethylene
HIPS	high-impact polystyrene
HMC	high-strength molding compound
HMWHDPE	high-molecular-weight high-density polyethylene
I	ionomer
IIR	butyl rubber
IPN	interpenetrating polymer network
K	constant in Mark-Houwink equation
LC	liquid crystal
LCP	liquid crystal polymer
LDPE	low-density polyethylene
LLDPE	linear low-density polyethylene
LPE	linear polyethylene
MA	maleic anhydride
MABS	methyl methacrylate ABS copolymer
MBS	methyl methacrylate butadiene styrene terpolymer
MDPE	medium-density polyethylene
MDI	methylene diphenylisocyanate
MF	melamine-formaldehyde resin
MP	melamine phenolic
MWD	molecular-weight distribution
M_n	number-average molecular weight
M_v	viscosity-average molecular weight
M_w	weight-average molecular weight
M_z	*Z*-average molecular weight
NBR	poly(butadiene-co-acrylonitrile); nitrile butadiene rubber
NR	natural rubber
OSA	olefin-modified styrene acrylonitrile
P	phenolic
PA	polyamide; nylon
PA6	polyamide 6, nylon 6
PA11	polyamide 11, nylon 11
PA12	polyamide 12, nylon 12
PA46	polyamide 46, nylon 46
PA66	polyamide 66, nylon 6,6
PA66/6T	polyamide 66/6T
PA610	polyamide 610, nylon 6,10
PA612	polyamide 612, nylon 6,12
PA666	polyamide 666
PAA	poly(acrylic acid)
PAEK	polyaryletherketone
PAI	polyamide-imide
PAK	polyester alkyd
PAL	polyanaline
PAN	polyacrylonitrile
PARA	polyaryl amide
PAS	polyarylsulfone
PB	polybutylene
PBAN	polybutylene-acrylonitrile copolymer
PBD	polybutadine
PBI	polybenzimidazole
PBN	poly(butylene napthalate)
PBS	polybutadiene-styrene copolymer
PBT	poly(butylene terephthalate)
PC	polycarbonate

PC/ABS	polycarbonate/acrylonitrile butadiene styrene blend
PCB	polychlorinated biphenyl
PCCE	poly(cyclohexylene dimethylene cyclohexanedicarboxylate), glycol and acid comonomer
PCL	polycaprolactone
PCT	poly(cyclohexylene terephthalate)
PCTA	poly(cyclohexylene dimethylene terephthalate) copolyester
PCTFE	polychlorotrifluoroethylene
PCTG	poly(cyclohexylene dimethylene terephthalate) copolyester
PCT-G	glycol-modified polycyclohexyl terephthallate
PE	polyethylene
PEBA	polyether block amide or polyester block amide
PEEK	poly(ether ether ketone)
PEG	poly(ethylene glycol)
PEI	polyetherimide
PEK	polyetherketone
PEKEKK	polyetherketone etherketone ketone
PEKK	polyetherketoneketone
PEN	poly(ethylene napthalene)
PEO	poly(ethylene oxide)
PES	polyethersulfone
PET	poly(ethylene terephthalate)
PET-G	glycol-modified poly(ethylene terephthalate)
PEX	cross-linked polyethylene
PF	phenol-formaldehyde resin
PFA	perfluoroalkoxy
PI	polyimide, polyisoprene
PIB	polyisobutylene
PIR	polyisocyanurate
PK	polyketone
PLGA	poly(lactic-*co*-glycolic acid)
PMAN	polymethactylonitrile
PMMA	poly(methyl methacrylate)
PMP	polymethylpentene
PMS	polymethylstyrene
PNF	poly(phosphonitrilic fluorides)
PO	polyolefin
POM	polyoxymethylene, polyformaldehyde, acetals
PP	polypropylene
PPA	polyphthalamide
PPC	chlorinated polypropylene, polyphthalate carbonate
PPE	poly(phenylene ether)
PPI	polymeric polyisocyanate
PPO	poly(phenylene oxide)
PPOX	poly(propylene oxide)
PPS	poly(phenylene sulfide)
PPSU	poly(phenylene sulfone)
PPT	poly(propylene terephthalate)
PS	polystyrene
PS-b-PI	polystyrene/polyisoprene block copolymer
PSO, PSU	polysulfone
PTFE	polytetrafluoroethylene, Teflon
PTME	poly(tetramethylene terephthalate)
PTMT	poly(tetramethylene terephthalate)
PU	polyurethane
PUR	polyurethane rubber
PVA	poly(vinyl alcohol); sometimes poly(vinyl acetate)
PVAc	poly(vinyl acetate)
PVB	poly(vinyl butyral)
PVC	poly(vinyl chloride)
PVCA	copolymer of vinyl chloride and vinyl acetate
PVDA	polyvinylidene acetate
PVDC	poly(vinylidene chloride)
PVDF	poly(vinylidene fluoride)
PVF	poly(vinyl fluoride)
PVK	poly(vinyl carbazole)
PVOH	poly(vinyl alcohol)
PVP	poly(vinyl pyrrolidone)
RIM	reaction injection molding
RNA	ribonucleic acid
ROMP	ring-opening metathesis polymerization
ROP	ring-opening polymerization
S	radius of gyration
SAN	poly(styrene-*co*-acrylonitrile)
SB	styrene butadiene copolymer
SBR	poly(butadiene-*co*-styrene) elastomer
SBS	styrene butadiene styrene block copolymer
SEBS	styrene ethylene butylene styrene block copolymer
SI	silicon
SIS	styrene isoprene styrene block copolymer
SMA	poly(styrene-*co*-maleic anhydride)
SMC	sheet-molding compound
SMMA	styrene methyl methacrylate copolymer
SMS	styrene/α-methyl styrene
SN	sulfur nitride
SR	synthetic rubber
SRP	styrene-rubber plastics
SVA	styrene vinyl acrylonitrile
TDI	toluenediisocyanate
TEO	thermoplastic elastic olefin
TGA, TG	thermal gravimetric analysis
TMC	thick molding compound
TMMV	threshold molecular-weight value
TPA	polyamide thermoplastic elastomer
TPC	copolyester thermoplastic elastomer
TPE	thermoplastic elastomer
TPE-O, TPO	thermoplastic elastomer - olefinic
TPE-S, TPS	thermoplastic elastomer - styrenic
TPU	thermoplastic urethane
TPX	poly-4-methylpentene
TVO	thermoplastic vulcanites
T_c	ceiling temperature; cloud-point temperature
T_g	glass transition temperature
T_m	melting-point temperature
UF	urea-formaldehyde resin
UHMWPE	ultrahigh molecular-weight polyethylene
ULDPE	ultralow-density polyethylene
ULPE	ultra-linear polyethylene
UP, UPE	unsaturated polyester (thermoset)
VA	vinyl acetate
VAE	vinyl acetate ethylene
VLDPE	very low-density polyethylene
WLF	Williams-Landel-Ferry equation
WS	polyurethane
XLPE	cross-linked polyethylene
XPS	expandable polystyrene

PHYSICAL PROPERTIES OF SELECTED POLYMERS

The physical properties of polymers are important parameters in determining their behavior and performance in a wide range of applications. This table lists some examples of general representative physical properties (including mechanical properties) of representative polymeric compounds. For glass transition temperatures, see the table "Glass Transition Temperature for Selected Polymers" in this section. Some of the properties in this table are defined as follows.

Property	Definition
Heat deflection temperature (HDT) or heat distortion temperature	Temperature at which a polymer or plastic sample deforms under a specified load (normally either 0.455 MPa or 1.82 MPa)
Crystalline melting point	Temperature (or temperature range) at which a crystalline solid changes its state from solid to liquid. Although the phrase would suggest a specific temperature, most crystalline compounds actually melt over a range of a few °C or less
Coefficient of linear thermal expansion	Fractional change in length per °C change in temperature at constant pressure
Compressive strength	Maximum uniaxial compressive stress (compressive force per unit area) reached when a material fails completely on being subjected to a load that pushes it together
Tensile strength	Measure of the ability of a material to withstand pulling stresses; defined as the stress (stretching force per unit area) required to break a specimen; polymers are approximately 20% stronger in compression than in tension
Flexural strength or cross-breaking strength	Measure of the bending strength or stiffness of a material specimen expressed as the stress required to break a specimen by exerting a torque on it
Impact strength (toughness)	Measure of the energy needed to break a sample; the notched Izod impact test is a single point test that measures the resistance of a material to impact from a swinging pendulum; Izod impact is defined as the kinetic energy needed to initiate fracture and continue the fracture until the specimen is broken; Izod specimens are notched to prevent deformation of the specimen upon impact; this test can be used as a quick and easy quality control check to determine if a material meets specific impact properties or to compare materials for general toughness
Ultimate elongation	Measure of how far a material will stretch before breaking, expressed as a percentage of its original length

The properties of the following polymers are presented in these tables.

Table 1	
PET	poly(ethylene terephthalate)
PBT	poly(butylene terephthalate)
PC	polycarbonate
Nylon 6,6	poly(iminoadipoyliminohexamethylene)
Nylon 6	poly[imino(1-oxohexamethylene)]
PPO	poly(phenylene ether)
Table 2	
POM	polyoxymethylene
LDPE	low-density polyethylene
HDPE	high-density polyethylene
UHMWPE	ultrahigh molecular weight polyethylene
iPP	isotactic polypropylene
ABS	copolymer of acrylonitrile, butadiene, and styrene (extrusion grade)
Table 3	
PTFE	polytetrafluoroethylene, Teflon
PCTFE	polymonochlorotrifluoroethylene
PVDF	poly(vinylidene fluoride)
PVF	poly(vinyl fluoride)
PVC (rigid)	poly(vinyl chloride)
PVC (plasticized)	poly(vinyl chloride)
PMMA	poly(methyl methacrylate)

The assistance of Charles E. Carraher, Jr. in providing these data is gratefully acknowledged.

Reference

Carraher, Jr., C.E., *Seymour/Carraher's Polymer Chemistry, Seventh Edition*, CRC Press, Taylor & Francis Group, Boca Raton, FL, 2008. <https://doi.org/10.1201/9781420051032>

TABLE 1. Physical Properties of Polymers

Property	PET	PBT	PC	Nylon 6,6	Nylon 6	PPO
Heat deflection temperature at 1820 kPa °C	100	65	130	75	80	100
Maximum resistance to continuous heat °C	100	60	115	120	125	80
Crystalline melting point °C	—	—	225	265	225	215
Coefficient of linear expansion (10^{-5}/°C)	6.5	7.0	6.8	8.0	8.0	5.0
Compressive strength (kPa)	8.6×10^4	7.5×10^4	8.6×10^4	1×10^5	9.7×10^4	9.6×10^4
Flexural strength (kPa)	1.1×10^5	9.6×10^4	9.3×10^4	1×10^5	9.7×10^4	8.9×10^4
Impact strength (Izod: cm N/cm of notch)	26	53	530	80	160	270
Tensile strength (kPa)	6.2×10^4	5.5×10^4	7.2×10^4	8.3×10^4	6.2×10^4	5.5×10^4
Ultimate elongation (%)	100	100	110	30	—	50
Density (g cm^{-3})	1.35	1.35	1.2	1.2	1.15	1.1

TABLE 2. Physical Properties of Polymers

Property	POM	LDPE	HDPE	UHMWPE	iPP	ABS
Heat deflection temperature at 1820 kPa (°C)	125	40	50	85	55	90
Maximum resistance to continuous heat (°C)	100	40	80	80	100	90
Crystalline melting point (°C)	180	—	—	—	—	—
Coefficient of linear expansion (10^{-5}/°C)	10.0	10	12	12	9	9.5
Compressive strength (kPa)	1.1×10^5	—	3×10^4	—	—	4.8×10^4
Flexural strength (kPa)	9.7×10^4	—	—	—	5×10^4	6.2×10^4
Impact strength (Izod: cm N/cm of notch)	80	No break	30	No break	27	320
Tensile strength (kPa)	6.9×10^4	5×10^3	2×10^4	6×10^4	3.5×10^4	3.4×10^4
Ultimate elongation (%)	30	—	—	—	100	60
Density (g cm^{-3})	1.4	0.91	0.96	0.93	0.90	1.0

TABLE 3. Physical Properties of Polymers

Property	PTFE	PCTFE	PVDF	PVF	Rigid PVC	Plasticized PVC	PMMA
Heat deflection temperature at 1820 kPa (°C)	100	100	80	90	75	—	95
Maximum resistance to continuous heat (°C)	250	200	150	125	60	35	75
Crystalline melting point (°C)	—	—	—	—	170	—	—
Coefficient of linear expansion (10^{-5}/°C)	10	14	8.5	10	6	12	7.0
Compressive strength (kPa)	2.7×10^4	3.8×10^4	—	—	6.8×10^4	6×10^3	1×10^5
Flexural strength (kPa)	—	6×10^4	—	—	9×10^4	—	9.6×10^4
Impact strength (Izod: cm N/cm of notch)	160	130	—	—	27	—	21
Tensile strength (kPa)	2.4×10^4	3.4×10^4	5.5×10^4	—	4.4×10^4	1×10^4	6.5×10^4
Ultimate elongation (%)	200	100	200	—	50	200	4
Density (g cm^{-3})	2.16	2.1	1.76	1.4	1.4	1.3	1.2

NOMENCLATURE FOR ORGANIC POLYMERS

Robert B. Fox and Edward S. Wilks

Organic polymers have traditionally been named on the basis of the monomer used, a hypothetical monomer or a semi-systematic structure. Alternatively, they may be named in the same way as organic compounds, i.e., on the basis of a structure as drawn. The former method, often called "source-based nomenclature" or "monomer-based nomenclature," sometimes results in ambiguity and multiple names for a single material. The latter method, termed "structure-based nomenclature," generates a sometimes cumbersome unique name for a given polymer, independent of its source. Within their limitations, both types of names are acceptable and well-documented.[1] The use of stereochemical descriptors with both types of polymer nomenclature has been published.[2]

Traditional Polymer Names

Monomer-Based Names

"Polystyrene" is the name of a homopolymer made from the single monomer styrene. When the name of a monomer comprises two or more words, the name should be enclosed in parentheses, as in "poly(methyl methacrylate)" or "poly(4-bromostyrene)" to identify the monomer more clearly. This method can result in several names for a given polymer: thus, "poly(ethylene glycol)," "poly(ethylene oxide)," and "poly(oxirane)" describe the same polymer. Sometimes, the name of a hypothetical monomer is used, as in "poly(vinyl alcohol)." Even though a name like "polyethylene" covers a multitude of materials, the system does provide understandable names when a single monomer is involved in the synthesis of a single polymer. When one monomer can yield more than one polymer, e.g., 1,3-butadiene or acrolein, some sort of structural notation must be used to identify the product, and one is not far from a formal structure-based name.

Copolymers, Block Polymers, and Graft Polymers. When more than one monomer is involved, monomer-based names are more complex. Some common polymers have been given names based on an apparent structure, as with "poly(ethylene terephthalate)." A better system has been approved by the IUPAC.[1] With this method, the arrangement of the monomeric units is introduced through use of an italicized connective placed between the names of the monomers. For monomer names represented by A, B, and C, the various types of arrangements are shown in Table 1.

Table 2 contains examples of common or semi-systematic names of copolymers. The systematic names of comonomers may also be used; thus, the polyacrylonitrile-*block*-polybutadiene-*block*-polystyrene polymer in Table 2 may also be named poly(prop-2-enenitrile)-*block*-polybuta-1,3-diene-*block*-poly(ethenylbenzene). IUPAC does not require alphabetized names of comonomers within a polymer name; many names are thus possible for some copolymers.

These connectives may be used in combination and with small, non-repeating (i.e., non-polymeric) junction units; see, for example, Table 2, line 8. A long dash may be used in place of the connective -*block*-; thus, in Table 2, the polymers of lines 7 and 8 may also be written as shown on lines 9 and 10.

IUPAC also recommends an alternative scheme for naming copolymers that comprises use of "copoly" as a prefix followed by the names of the comonomers, a solidus (an oblique stroke) to separate comonomer names, and the addition before "copoly" of any applicable connectives listed in Table 2 except -*co*-.

Table 3 gives the same examples shown in Table 2 but with the alternative format. Comonomer names need not be parenthesized.

TABLE 1. IUPAC Source-Based Copolymer Classification

No.	Copolymer type	Connective	Example
1	Unspecified or unknown	-*co*-	poly(A-*co*-B)
2	Random (obeys Bernoullian distribution)	-*ran*-	poly(A-*ran*-B)
3	Statistical (obeys known statistical laws)	-*stat*-	poly(A-*stat*-B)
4	Alternating (for two monomeric units)	-*alt*-	poly(A-*alt*-B)
5	Periodic (ordered sequence for 2 or more monomeric units)	-*per*-	poly(A-*per*-B-per-C)
6	Block (linear block arrangement)	-*block*-	polyA-*block*-polyB
7	Graft (side chains connected to main chains)	-*graft*-	polyA-*graft*-polyB

TABLE 2. Examples of Source-Based Copolymer Nomenclature

No.	Copolymer name
1	poly(propene-*co*-methacrylonitrile)
2	poly[(acrylic acid)-*ran*-(ethyl acrylate)]
3	poly(butene-*stat*-ethylene-*stat*-styrene)
4	poly[(sebacic acid)-*alt*-butanediol]
5	poly[(ethylene oxide)-*per*-(ethylene oxide)-*per*-tetrahydrofuran]
6	polyisoprene-*graft*-poly(methacrylic acid)
7	polyacrylonitrile-*block*-polybutadiene-*block*-polystyrene
8	polystyrene-*block*-dimethylsilylene-*block*-polybutadiene
9	polyacrylonitrile—polybutadiene—polystyrene
10	polystyrene—dimethylsilylene—polybutadiene

TABLE 3. Examples of Source-Based Copolymer Nomenclature (Alternative Format)

No.	Copolymer name
1	copoly(propene/methacrylonitrile)
2	*ran*-copoly(acrylic acid/ethyl acrylate)
3	*stat*-copoly(butene/ethylene/styrene)
4	*alt*-copoly(sebacic acid/butanediol)
5	*block*-copoly(acrylonitrile/butadiene/styrene)
6	*per*-copoly(ethylene oxide/ethylene oxide/tetrahydrofuran)
7	*graft*-copoly(isoprene/methacrylic acid)

Source-based nomenclature for nonlinear macromolecules and macromolecular assemblies is covered by a 1997 IUPAC document.[11] The types of polymers in these classes, together with their connectives, are given in Table 4; the terms shown may be used as connectives, prefixes, or both to designate the features present.

TABLE 4. Connectives for Nonlinear Macromolecules and Macromolecular Assemblies

No.	Type	Connective
1	Branched (type unspecified)	branch
2	Branched with branch point of functionality f	f-branch
3	Comb	comb
4	Cross-link	ι (Greek iota)
5	Cyclic	cyclo
6	Interpenetrating polymer network	ipn
7	Long-chain branched	l-branch
8	Network	net
9	Polymer blend	blend
10	Polymer-polymer complex	compl
11	Semi-interpenetrating polymer network	sipn
12	Short-chain branched	sh-branch
13	Star	star
14	Star with f-arms	f-star

Nonlinear polymers are named by using the italicized connective as a *prefix* to the source-based name of the polymer component or components to which the prefix applies; some examples are listed in Table 5.

TABLE 5. Nonlinear Macromolecules

No.	Polymer name	Polymer structural features
1	poly(methacrylic acid)-*comb*-polyacrylonitrile	Comb polymer with a poly(methacrylic acid) backbone and polyacrylonitrile side chains
2	*comb*-poly[ethylene-*stat*-(vinyl chloride)]	Comb polymer with unspecified backbone composition and statistical ethylene/vinyl chloride copolymer side chains
3	polybutadiene-*comb*-(polyethylene; polypropene)	Comb polymer with butadiene backbone and side chains of polyethylene and polypropene
4	*star*-(polyA; polyB; polyC; polyD; polyE)	Star polymer with arms derived from monomers A, B, C, D, and E, respectively
5	*star*-(polyA-*block*-polyB-*block*-polyC)	Star polymer with every arm comprising a triblock segment derived from comonomers A, B, and C
6	*star*-poly(propylene oxide)	A star polymer prepared from propylene oxide
7	5-*star*-poly(propylene oxide)	A 5-arm star polymer prepared from propylene oxide
8	*star*-(polyacrylonitrile; polypropylene) (M_r 10000: 25000)	A star polymer containing polyacrylonitrile arms of MW 10000 and polypropylene arms of MW 25000

Macromolecular assemblies held together by forces other than covalent bonds are named by inserting the appropriate italicized connective between names of individual components; Table 6 gives examples.

TABLE 6. Examples of Polymer Blends and Nets

No.	Polymer name
1	polyethylene-*blend*-polypropene
2	poly(methacrylic acid)-*blend*-poly(ethyl acrylate)
3	*net*-poly(4-methylstyrene-ι-divinylbenzene)
4	*net*-poly[styrene-*alt*-(maleic anhydride)]-ι-(polyethylene glycol; polypropylene glycol)
5	*net*-poly(ethyl methacrylate)-*sipn*-polyethylene
6	[*net*-poly(butadiene-*stat*-styrene)]-*ipn*-[*net*-poly(4-methylstyrene-ι-divinylbenzene)]

Structure-Based Polymer Nomenclature

Regular Single-Strand Polymers

Structure-based nomenclature has been approved by the IUPAC[4] and is currently being updated; it is used by *Chemical Abstracts*.[5] Monomer names are not used. To the extent that a polymer chain can be described by a repeating unit in the chain, it can be named "poly(repeating unit)." For regular single-strand polymers, "repeating unit" is a bivalent group; for regular double-strand (ladder and spiro) polymers, "repeating unit" is usually a tetravalent group.[9]

Since there are usually many possible repeating units in a given chain, it is necessary to select one, called the "constitutional repeating unit" (CRU) to provide a unique and unambiguous name, "poly(CRU)," where "CRU" is a recitation of the names of successive units as one proceeds through the CRU from left to right. For this purpose, a portion of the main chain structure that includes at least two repeating sequences is written out. These sequences will typically be composed of bivalent subunits such as -CH_2-, -O-, and groups from ring systems, each of which can be named by the usual nomenclature rules.[6,7]

Where a chain is simply one long sequence comprising repetition of a single subunit, that subunit is itself the CRU, as in "poly(methylene)" or "poly(1,4-phenylene)." In chains having more than one kind of subunit, a seniority system is used to determine the beginning of the CRU and the direction in which to move along the main chain atoms (following the shortest path in rings) to complete the CRU. Determination of the first, most senior, subunit is based on a descending order of seniority: (1) heterocyclic rings, (2) hetero atoms, (3) carbocyclic rings, and lowest, (4) acyclic carbon chains.

Within each of these classes, there is a further order of seniority that follows the usual rules of nomenclature.

Heterocycles: A nitrogen-containing ring system is senior to a ring system not containing nitrogen.[4,9] Further descending order of seniority is determined by:

(i) the highest number of rings in the ring system
(ii) the largest individual ring in the ring system
(iii) the largest number of hetero atoms
(iv) the greatest variety of hetero atoms

Hetero atoms: The senior bivalent subunit is the one nearest the top right-hand corner of the Periodic Table; the order of seniority is: O, S, Se, Te, N, P, As, Sb, Bi, Si, Ge, Sn, Pb, B, Hg.

Carbocycles: Seniority[4] is determined by:

(i) the highest number of rings in the ring system
(ii) the largest individual ring in the ring system
(iii) degree of ring saturation; an unsaturated ring is senior to a saturated ring of the same size

Carbon chains: Descending order of seniority is determined by:

(i) chain length (longer is senior to shorter)
(ii) highest degree of unsaturation
(iii) number of substituents (higher number is senior to lower number)
(iv) ascending order of locants
(v) alphabetical order of names of substituent groups

Among equivalent ring systems, preference is given to the one having lowest locants for the free valences in the subunit, and among otherwise identical ring systems, the one having least hydrogenation is senior. Lowest locants in unsaturated chains are also given preference. Lowest locants for substituents are the final determinant of seniority.

Direction within the repeating unit depends upon the shortest path, which is determined by counting main chain atoms, both cyclic and acyclic, from the most senior subunit to another subunit of the same kind or to a subunit next lower in seniority. When identification and orientation of the CRU have been accomplished, the CRU is named by writing, in sequence, the names of the largest possible subunits within the CRU from left to right. For example, the main chain of the polymer traditionally named "poly(ethylene terephthalate)" has the structure shown in Figure 1.

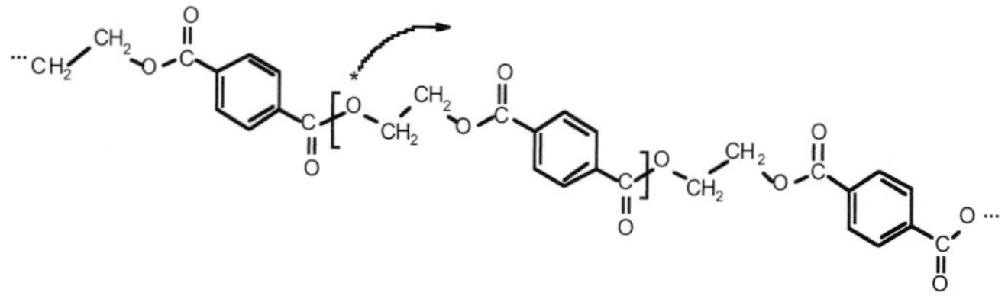

FIGURE 1. Structure-based name: poly(oxyethyleneoxyterephthaloyl); traditional name: poly(ethylene terephthalate).

The CRU in Figure 1 is enclosed in brackets and read from left to right. It is selected because (1) either backbone oxygen atom qualifies as the "most senior subunit," (2) the shortest path length from either -O- to the other -O- is via the ethylene subunit. Orientation of the CRU is thus defined by (1) beginning at the -O- marked with an asterisk, and (2) reading in the direction of the arrow. The structure-based name of this polymer is therefore "poly(oxyethyl eneoxyterephthaloyl)," not much longer than the traditional name and much more adaptable to the complexities of substitution. As organic nomenclature evolves, more systematic names may be used for subunits, e.g., "ethane-1,2-diyl" instead of "ethylene." IUPAC still prefers "ethylene" for the -CH_2-CH_2- unit, however, but also accepts "ethane-1,2-diyl."

Structure-based nomenclature can also be used when the CRU backbone has no carbon atoms. An example is the polymer traditionally named "poly(dimethylsiloxane)," which on the basis of structure would be named "poly(oxydimethylsilylene)" or "poly(oxydimethylsilanediyl)." This nomenclature method has also been applied to inorganic and coordination polymers[8] and to double-strand (ladder and spiro) organic polymers.[9]

Irregular Single-Strand Polymers

Polymers that cannot be described by the repetition of a single CRU or comprise units not all connected identically in a directional sense can also be named on a structure basis.[10] These include copolymers, block and graft polymers, and star polymers. They are given names of the type "poly(A/B/C...)," where A, B, C, etc., are the names of the component constitutional units, the number of which are minimized. The constitutional units may include regular or irregular blocks as well as atoms or atomic groupings, and each is named by the method described above or by the rules of organic nomenclature.

The solidus denotes an unspecified arrangement of the units within the main chain.[10] For example, a statistical copolymer derived from styrene and vinyl chloride with the monomeric units joined head-to-tail is named "poly(l-chloroethylene/l-phenylethylene)." A polymer obtained by 1,4-polymerization and both head-to-head and head-to-tail 1,2-polymerization of 1,3-butadiene would be named "poly(but-1-ene-l,4-diyl/l-vinylethylene/2-vinylethylene)."[12] In graphic representations of these polymers, shown in Figure 2, the hyphens or dashes at each end of each CRU depiction are shown *completely within* the enclosing parentheses; this indicates that they are not necessarily the terminal bonds of the macromolecule.

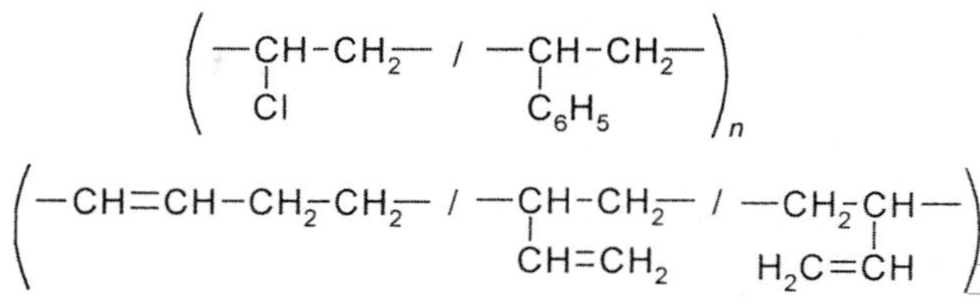

FIGURE 2. Graphic representations of copolymers.

A long hyphen is used to separate components in names of block polymers, as in "poly(A)—poly(B)—poly(C)," or "poly(A)—X—poly(B)" in which X is a non-polymeric junction unit, e.g., dimethylsilylene.

In graphic representations of these polymers, the blocks are shown connected when the bonding is known (Figure 3, for example); when the bonding between the blocks is unknown, the blocks are separated by solidi and are shown *completely within* the outer set of enclosing parentheses (Figure 4, for example).[10,13]

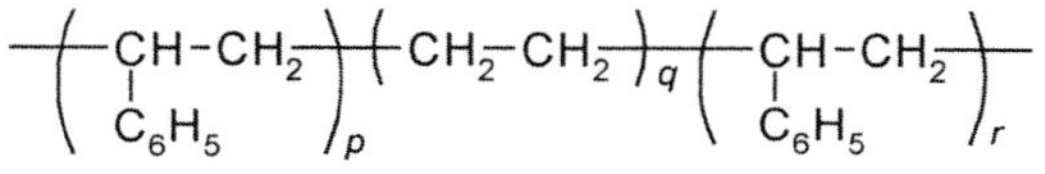

FIGURE 3. polystyrene—polyethylene—polystyrene.

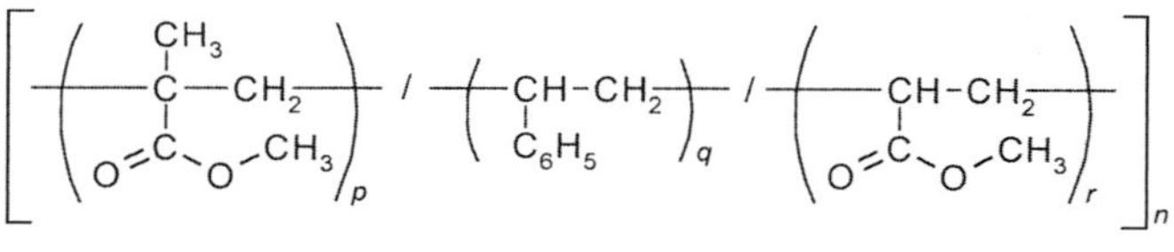

FIGURE 4. poly[poly(methyl methacrylate)—polystyrene—poly(methyl acrylate)].

Graft polymers are named in the same way as a substituted polymer but without the ending "yl" for the grafted chain; the name of a regular polymer, comprising Z units in which some have grafts of "poly(A)," is "poly[Z/poly(A)Z]." Star polymers are treated as a central unit with substituent blocks, as in "tetrakis(polymethylene) silane."[10,13]

Other Nomenclature Articles and Publications

In addition to the *Chemical Abstracts* and IUPAC documents cited above and listed below, other articles on polymer nomenclature are available. A 1999 article lists significant documents on polymer nomenclature published during the last 50 years in books, encyclopedias, and journals by *Chemical Abstracts*, IUPAC, and individual authors.[14] A comprehensive review of source-based and structure-based nomenclature for all of the major classes of polymers,[15] and a short tutorial on the correct identification, orientation, and naming of most commonly encountered constitutional repeating units were both published in 2000.[16]

References and Notes

1. International Union of Pure and Applied Chemistry, *Compendium of Macromolecular Nomenclature*, Blackwell Scientific Publications, Oxford, 1991.
2. International Union of Pure and Applied Chemistry, Stereochemical Definitions and Notations Relating to Polymers (Recommendations 1980), *Pure Appl. Chem.*, 53, 733–752 (1981).
3. International Union of Pure and Applied Chemistry, Source-Based Nomenclature for Copolymers (Recommendations 1985), *Pure Appl. Chem.*, 57, 1427–1440 (1985).
4. International Union of Pure and Applied Chemistry, Nomenclature of Regular Single-Strand Organic Polymers (Recommendations 1975, *Pure Appl. Chem.*, 48, 373–385 (1976).
5. Chemical Abstracts Service, Naming and Indexing of Chemical Substances for Chemical Abstracts, Appendix IV, *Chemical Abstracts 1999 Index Guide.*
6. International Union of Pure and Applied Chemistry, *A Guide to IUPAC Nomenclature of Organic Compounds* (1993), Blackwell Scientific Publications, Oxford, 1993.
7. International Union of Pure and Applied Chemistry, *Nomenclature of Organic Chemistry, Sections A, B, C, D, E, F, and H*, Pergamon Press, Oxford, 1979.
8. International Union of Pure and Applied Chemistry, Nomenclature of Regular Double-Strand and Quasi-Single-Strand Inorganic and Coordination Polymers (Recommendations 1984), *Pure Appl. Chem.*, 57, 149–168 (1985).
9. International Union of Pure and Applied Chemistry, Nomenclature of Regular Double-Strand (Ladder and Spiro) Organic Polymers (Recommendations 1993), *Pure Appl. Chem.*, 65, 1561–1580 (1993).
10. International Union of Pure and Applied Chemistry, Structure-Based Nomenclature for Irregular Single-Strand Organic Polymers (Recommendations 1994), *Pure Appl. Chem.*, 66, 873–889 (1994).
11. International Union of Pure and Applied Chemistry, Source-Based Nomenclature for Non-Linear Macromolecules and Macromolecular Assemblies (Recommendations 1997). *Pure Appl. Chem.*, 69, 2511–2521 (1997).
12. Poly(1,3-butadiene) obtained by polymerization of 1,3-butadiene in the so-called 1,4- mode is frequently drawn incorrectly in publications as $\text{-(CH}_2\text{-CH=CH-CH}_2)_n\text{-}$; the double bond should be assigned the lowest locant possible, i.e., the structure should be drawn as $\text{-(CH=CH-CH}_2\text{-CH}_2)_n\text{-}$.
13. International Union of Pure and Applied Chemistry, Graphic Representations (Chemical Formulae) of Macromolecules (Recommendations 1994). *Pure Appl. Chem.*, 66, 2469–2482 (1994).
14. Wilks, E. S. Macromolecular Nomenclature Note No. 17: "Whither Nomenclature?" *Polym. Prepr.* 40(2), 6–11 (1999).
15. Wilks, E. S. Polymer Nomenclature: The Controversy between Source-Based and Structure-Based Representations (A Personal Perspective). *Prog. Polym. Sci.* 25, 9–100 (2000).
16. Wilks, E. S. Macromolecular Nomenclature Note No. 18: "SRUs: Using the Rules." *Polym. Prepr.* 41(1), 6a–11a (2000).

SOLVENTS FOR COMMON POLYMERS

The following table lists one or more solvents for common polymers. Polymer solubility is complex, though a few general rules apply, such as polar solvents work best with polar polymers; nonpolar solvents work better with nonpolar polymers. Factors such as molecular weight, cross-linking, crystallinity, and branch length also play important roles. This table contains qualitative information on solvents used with common polymers. See "Solubility Parameters of Selected Polymers" in this section for more quantitative information. Solvent abbreviations used in the table are as follows.

Abbreviation	Full name
HC	Hydrocarbons
MEK	Methyl ethyl ketone
THF	Tetrahydrofuran
DMF	Dimethylformamide
DMSO	Dimethylsulfoxide

Solvents for Common Polymers

Polymer	Solvent
Polyethylene (HDPE)	HC and halogenated HC
Polypropylene (atactic)	HC and halogenated HC
Polybutadiene	HC, THF, ketones
Polystyrene	ethylbenzene, $CHCl_3$, CCl_4, THF, MEK
Polyacrylates	aromatic HC, chlorinated HC, THF, esters, ketones
Polymethacrylates	aromatic HC, chlorinated HC, THF, esters, MEK
Polyacrylamide	water
Poly(vinyl ethers)	halogenated HC, MEK, butanol
Poly(vinyl alcohol)	glycols (hot), DMF
Poly(vinyl acetate)	aromatic HC, chlorinated HC, THF, esters, DMF
Poly(vinyl chloride)	THF, DMF, DMSO
Poly(vinylidene chloride)	THF (hot), dioxane, DMF
Poly(vinyl fluoride)	DMF, DMSO (hot)
Polyacrylonitrile	DMF, DMSO
Poly(oxyethylene)	aromatic HC, $CHCl_3$, alcohols, esters, DMF
Poly(2,6-dimethylphenylene oxide)	aromatic HC, halogenated HC
Poly(ethylene terephthalate)	phenol, DMSO (hot)
Polyurethanes (linear)	aromatic HC, THF, DMF
Polyureas	phenol, formic acid
Polysiloxanes	HC, THF, DMF
Poly[bis(2,2,2-trifluoroethoxy)-phosphazene]	THF, ketones, ethyl acetate

Polymers

GLASS TRANSITION TEMPERATURE FOR SELECTED POLYMERS

Robert B. Fox

Polymer names are based on the IUPAC structure-based nomenclature system described in the table "Nomenclature for Organic Polymers" in this section. Within each category, names are listed in alphabetical order. Source-based and trivial names are also given (in italics) for the most common polymers. The table does not include polymers for which T_g is not clearly defined because of variability of structure or because of reactions taking place near the glass transition.

All values of T_g cited in this table have been determined by differential scanning calorimetry (DSC) except those values indicated by:

(D)	Dynamic method
(Dil)	Dilatometry
(M)	Mechanical method

Polymer name	**Glass transition temperature (T_g/K)**
ACYCLIC CARBON CHAINS	
Polyalkadienes	
Poly(alkenylene) *Polyalkadiene* –[CH=CHCH$_2$CH$_2$]–	
Poly(*cis*-1-butenylene)	171
cis-1,3-polybutadiene [PBD]	
Poly(*trans*-1-butenylene)	215
trans-1,3-polybutadiene [PBD]	
Poly(1-chloro-*cis*-1-butenylene)	253
cis-1,3-polychloroprene	
Poly(1-chloro-*trans*-1-butenylene)	233
trans-1,3-polychloroprene	
Poly(1-methyl-*cis*-1-butenylene)	200
cis-1,3-polyisoprene	
Poly(1-methyl-*trans*-1-butenylene)	207
trans-1,3-polyisoprene	
Poly(1,4,4-trifluoro-1-butenylene)	238
Polyalkenes	
Poly(alkylethylene) *Poly(alkylethylene)* -[RCHCH$_2$]-	
Poly(1-benzylethylene)	333
Poly(1-butylethylene)	223
Poly(1-cyclohexylethylene) (atactic)	393
Poly(1-cyclohexylethylene) (isotactic)	406 (D)
Poly(1,1-dimethylethylene)	200
Polyisobutylene [PIB]	
Poly(ethylene)	148
Poly(methylene)	155
Poly(1-phenethylethylene)	283
Poly(propylene) (isotactic)	272
Poly(propylene) (syndiotactic)	ca. 265
Poly[1-(2-pyridyl)ethylene]	377
Poly[1-(4-pyridyl)ethylene]	415
Poly(1-vinylethylene)	273
Polyacrylics	
Poly[1-(alkoxycarbonyl)ethylene] *Poly(alkyl acrylate)* –[(ROCO)CHCH$_2$]–	
Poly[1-(benzyloxycarbonyl)ethylene]	279
Poly[1-(butoxycarbonyl)ethylene]	219 (M)
Poly(butyl acrylate) [PBA]	
Poly[1-(*sec*-butoxycarbonyl)ethylene]	251
Poly[1-(butoxycarbonyl)-1-cyanoethylene]	358
Poly[1-(butylcarbamoyl)ethylene]	319 (M)
Poly(1-carbamoylethylene)	438
Polyacrylamide [PAM]	
Poly(1-carboxyethylene)	379
Poly(acrylic acid) [PAA]	
Poly[1-(2-chlorophenoxycarbonyl)ethylene]	326

Polymers

Polymer name	Glass transition temperature (T_g/K)
Poly[1-(4-chlorophenoxycarbonyl)ethylene]	331
Poly[1-(4-cyanobenzyloxycarbonyl)ethylene]	317
Poly[1-(2-cyanoethoxycarbonyl)ethylene]	277
Poly[1-(cyanomethoxycarbonyl)ethylene)]	433 Dil
Poly[1-(4-cyanophenoxycarbonyl)ethylene]	363
Poly[1-(cyclohexyloxycarbonyl)ethylene]	292
Poly[1-(2,4-dichlorophenoxycarbonyl)ethylene]	333
Poly[1-(dimethylcarbamoyl)ethylene]	362
Poly[1-(ethoxycarbonyl)ethylene]	249
Poly(ethyl acrylate) [PEA]	
Poly[1-(ethoxycarbonyl)-1-fluoroethylene]	316
Poly[1-(2-ethoxycarbonylphenoxycarbonyl)ethylene]	303
Poly[1-(3-ethoxycarbonylphenoxycarbonyl)ethylene]	297
Poly[1-(4-ethoxycarbonylphenoxycarbonyl)ethylene]	310
Poly[1-(2-ethoxyethoxycarbonyl)ethylene]	223
Poly[1-(3-ethoxypropoxycarbonyl)ethylene]	218
Poly[1-(isopropoxycarbonyl)ethylene]	267–270
Poly[1-(methoxycarbonyl)ethylene]	283
Poly(methyl acrylate) [PMA]	
Poly[1-(2-methoxycarbonylphenoxycarbonyl)ethylene]	319
Poly[1-(3-methoxycarbonylphenoxycarbonyl)ethylene]	311
Poly[1-(4-methoxycarbonylphenoxycarbonyl)ethylene]	340
Poly[1-(2-methoxyethoxycarbonyl)ethylene]	223
Poly[1-(4-methoxyphenoxycarbonyl)ethylene]	324
Poly[1-(3-methoxypropoxycarbonyl)ethylene]	198
Poly[1-(2-naphthyloxycarbonyl)ethylene]	358
Poly[1-(pentachlorophenoxycarbonyl)ethylene]	420
Poly[1-(phenethoxycarbonyl)ethylene]	270
Poly[1-(phenoxycarbonyl)ethylene]	330
Poly[1-(*m*-tolyloxycarbonyl)ethylene]	298
Poly[1-(*o*-tolyloxycarbonyl)ethylene]	325
Poly[1-(*p*-tolyloxycarbonyl)ethylene]	316
Poly[1-(2,2,2-trifluorethoxycarbonyl)ethylene]	263
Polymethacrylics	
Poly[1-(alkoxycarbonyl)-1-methylethylene] *Poly(alkyl methacrylate)* –[(ROCO)(Me)CCH$_2$]–	
Poly[1-(benzyloxycarbonyl)-1-methylethylene]	327
Poly[1-(2-bromoethoxycarbonyl)-1-methylethylene]	325
Poly[(1-(butoxycarbonyl)-1-methylethylene]	293
Poly(butyl methacrylate) [PBMA]	
Poly[1-(*sec*-butoxycarbonyl)-1-methylethylene]	333
Poly[1-(*tert*-butoxycarbonyl)-1-methylethylene)]	391
Poly[1-(2-chloroethoxycarbonyl)-1-methylethylene]	ca 315
Poly[1-(2-cyanoethoxycarbonyl)-1-methylethylene]	364
Poly[1-(4-cyanophenoxycarbonyl)-1-methylethylene]	428
Poly[1-(cyclohexyloxycarbonyl)-1-methylethylene] (atactic)	356
Poly[1-(cyclohexyloxycarbonyl)-1-methylethylene)] (isotactic)	324
Poly[1-(dimethylaminoethoxycarbonyl)-1-methylethylene]	292
Poly[1-(ethoxycarbonyl)-1-ethylethylene]	300
Poly[1-(ethoxycarbonyl)-1-methylethylene] (atactic) *Poly(ethyl methacrylate)* [PEMA]	338
Poly[1-(ethoxycarbonyl)-1-methylethylene] (isotactic)	285
Poly[1-(ethoxycarbonyl)-1-methylethylene)] (syndiotactic)	339
Poly[1-(hexyloxycarbonyl)-1-methylethylene]	268
Poly[1-(isobutoxycarbonyl)-1-methylethylene]	326
Poly[1-(isopropoxycarbonyl)-1-methylethylene]	354
Poly[1-(methoxycarbonyl)-1-methylethylene] (atactic) *Poly(methyl methacrylate)* [PMMA]	378
Poly[1-(methoxycarbonyl)-1-methylethylene)] (isotactic)	311
Poly[1-(methoxycarbonyl)-1-methylethylene)] (syndiotactic)	378
Poly[1-(4-methoxycarbonylphenoxy)-1-methylethylene]	379
Poly[1-(methoxycarbonyl)-1-phenylethylene)] (atactic)	391
Poly[1-(methoxycarbonyl)-1-phenylethylene)] (isotactic)	397

Polymer name	Glass transition temperature (T_g/K)
Poly[1-methyl-1-(phenethoxycarbonyl)ethylene]	299
Poly[1-methyl-1-(phenoxycarbonyl)ethylene]	383
Polyvinyl ethers, alcohols, and ketones	
Poly(1-alkoxyethylene) *Poly(alkyl vinyl ether)* –[$ROCHCH_2$]–	
Poly(1-hydroxyethylene) *Poly(vinyl alcohol)* –[$HOCHCH_2$]–	
Poly(1-alkanoylethylene) *Poly(alkyl vinyl ketone)* –[$RCOCHCH_2$]–	
Poly(1-butoxyethylene)	218
Poly(1-*sec*-butoxyethylene)	253
Poly(1-*tert*-butoxyethylene)	361
Poly[1-(butylthio)ethylene]	253
Poly(1-ethoxyethylene)	230
Poly[1-(4-ethylbenzoyl)ethylene]	325
Poly(1-hydroxyethylene) *Poly(vinyl alcohol)* [PVA]	358 (D)
Poly(hydroxymethylene)	407
Poly(1-isopropoxyethylene)	270
Poly[1-(4-methoxybenzoyl)ethylene]	319 (M)
Poly(1-methoxyethylene) *Poly(methyl vinyl ether)* [PMVE]	242
Poly[1-(methylthio)ethylene]	272
Poly(1-propoxyethylene)	224
Poly[1-(trifluoromethoxy)trifluoroethylene]	268
Polyvinyl halides and nitriles	
Poly(1-haloethylene) *Poly(vinyl halide)* –[$XCHCH_2$]–	
Poly(1-cyanoethylene) *Poly(acrylonitrile)* –[$NCCHCH_2$]–	
Poly(1-chloroethylene) *Poly(vinyl chloride)* [PVC]	354
Poly(chlorotrifluoroethylene)	373
Poly(1-cyanoethylene) *Polyacrylonitrile* [PAN]	370
Poly(1-cyano-1-methylethylene) *Polymethacrylonitrile*	393
Poly(1,1-dichloroethylene) *Poly(vinylidene chloride)*	255
Poly(1,1-difluoroethylene) *Poly(vinylidene fluoride)*	ca 233
Poly(1-fluoroethylene) *Poly(vinyl fluoride)*	314 (M)
Poly(1-hexafluoropropylene)	425
Poly[1-(2-iodoethyl)ethylene]	343
Poly(tetrafluoroethylene)	(160)
Poly[1-(trifluoromethyl)ethylene]	300
Polyvinyl esters	
Poly[1-(alkanoyloxy)ethylene] *Poly(vinyl alkanoate)* –[$RCOOCHCH_2$]–	
Poly(1-acetoxyethylene) *Poly(vinyl acetate)* [PVAc]	305
Poly[1-(benzoyloxy)ethylene]	344
Poly[1-(4-bromobenzoyloxy)ethylene]	365
Poly[1-(2-chlorobenzoyloxy)ethylene]	335
Poly[1-(3-chlorobenzoyloxy)ethylene]	338
Poly[1-(4-chlorobenzoyloxy)ethylene]	357
Poly[1-(cyclohexanoyloxy)ethylene]	349 (M)
Poly[1-(4-ethoxybenzoyloxy)ethylene]	343
Poly[1-(4-ethylbenzoyloxy)ethylene]	326
Poly[1-(4-isopropylbenzoyloxy)ethylene]	342
Poly[1-(2-methoxybenzoyloxy)ethylene]	338

Polymer name	Glass transition temperature (T_g/K)
Poly[1-(3-methoxybenzoyloxy)ethylene]	ca 317
Poly[1-(4-methoxybenzoyloxy)ethylene]	360
Poly[1-(4-methylbenzoyloxy)ethylene]	343
Poly[1-(4-nitrobenzoyloxy)ethylene]	395
Poly[1-(propionoyloxy)ethylene]	283 (M)
Polystyrenes	
Poly(1-phenylethylene) *Polystyrene* $–[C_6H_5CHCH_2]–$	
Poly[1-(4-acetylphenyl)ethylene]	389 (M)
Poly[1-(4-benzoylphenyl)ethylene]	371 (M)
Poly[1-(4-bromophenyl)ethylene]	391
Poly[1-(4-butoxyphenyl)ethylene]	ca 320 (M)
Poly[1-(4-butoxycarbonylphenyl)ethylene]	349 (M)
Pol[(1-(4-butylphenyl)ethylene]	279
Poly[1-(4-carboxyphenyl)ethylene]	386 (M)
Poly[1-(2-chlorophenyl)ethylene]	392
Poly[1-(3-chlorophenyl)ethylene]	363
Poly[1-(4-chlorophenyl)ethylene]	383
Poly[1-(2,4-dichlorophenyl)ethylene]	406
Poly[1-(2,5-dichlorophenyl)ethylene]	379
Poly[1-(2,6-dichlorophenyl)ethylene]	440
Poly[1-(3,4-dichlorophenyl)ethylene]	401
Poly[1-(2,4-dimethylphenyl)ethylene]	385
Poly[1-(4-(dimethylamino)phenyl)ethylene]	398 (M)
Poly[1-(4-ethoxyphenyl)ethylene]	ca 359 (M)
Poly[1-(4-ethoxycarbonylphenyl)ethylene]	367 (M)
Poly[1-(4-fluorophenyl)ethylene]	368
Poly[1-(4-iodophenyl)ethylene]	429
Poly[1-(4-methoxyphenyl)ethylene]	386
Poly[1-(4-methoxycarbonylphenyl)ethylene]	386 (M)
Poly(1-methyl-1-phenylethylene) *Poly(α-methylstyrene)*	373
Poly[1-(2-(methylamino)phenyl)ethylene]	462 (M)
Poly(1-phenylethylene) *Polystyrene* [PS]	373
Poly[1-(4-propoxyphenyl)ethylene]	343 (M)
Poly[1-(4-propoxycarbonylphenyl)ethylene]	365 (M)
Poly(1-*o*-tolylethylene)	409
CHAINS WITH CARBOCYCLIC UNITS	
Poly(arylenealkylene) $–[–Ar–(CH_2)_n]–$	
Poly[1-(2-bromo-1,4-phenylene)ethylene]	353 (M)
Poly[1-(2-chloro-1,4-phenylene)ethylene]	343 (M)
Poly[1-(2-cyano-1,4-phenylene)ethylene]	363 (M)
Poly[1-(2,5-dimethyl-1,4-phenylene)ethylene]	373 (M)
Poly[1-(2-ethyl-1,4-phenylene)ethylene]	298 (M)
Poly[1-(1,4-naphthylene)ethylene]	433 (M)
Poly[1-(1,4-phenylene)ethylene]	ca 353 (M)
CHAINS WITH HETEROATOM UNITS	
Main chain oxide units	
Poly(oxyalkylene) *Poly(alkylene oxide)* $–[O(CH_2)_n]–$	
Poly[oxy(1,1-bis(chloromethyl)trimethylene)]	265
Poly[oxy(1-(bromomethyl)ethylene)]	259
Poly[oxy(1-(butoxymethyl)ethylene)]	194
Poly[oxy(1-butylethylene)]	203
Poly[oxy(1-*tert*-butylethylene)]	308
Poly[oxy(1-(chloromethyl)ethylene)] *Poly(epichlorohydrin)*	251
Poly[oxy(2,6-dimethoxy-1,4-phenylene)]	440

Polymer name	Glass transition temperature (T_g/K)
Poly[oxy(1,1-dimethylethylene)]	264
Poly[oxy(2,6-dimethyl-1,4-phenylene)]	482
Poly[oxy(2,6-diphenyl-1,4-phenylene)]	493
Poly[oxy(1-ethylethylene)]	203
Poly(oxyethylidene)	243
Polyacetaldehyde	
Poly[oxy(1-(methoxymethyl)ethylene)]	211
Poly[oxy(2-methyl-6-phenyl-1,4-phenylene)]	428
Poly[oxy(1-methyltrimethylene)]	223 (D)
Poly[oxy(2-methyltrimethylene)]	218
Poly(oxy-1,4-phenylene)	358
Poly(phenylene oxide) [PPO]	
Poly[oxy(1-phenylethylene)]	313
Poly(oxytetramethylene)	189
Poly(tetrahydrofuran) [PTMO]	
Poly(oxytrimethylene)	195
Main chain ester or anhydride units	
Poly(oxyalkyleneoxyalkanedioyl) *Poly(alkylene alkanedioate)*–[O(CH_2)$_m$OCO(CH_2)$_n$CO]–	
Poly(oxyadipoyloxydecamethylene)	217
Poly(oxyadipoyloxy-1,4-phenyleneisopropylidene-1,4-phenylene)	341
Poly(oxycarbonyloxy-1,4-phenylene-isopropylidene-1,4-phenylene)	422
Bisphenol A polycarbonate	
Poly(oxycarbonylpentamethylene)	213
Poly(oxycarbonyl-1,4-phenylenemethylene-1,4-phenylene)	395
Poly(oxycarbonyl-1,4-phenyleneisopropylidene-1,4-phenylene)	333
Poly[oxy(2,6-dimethyl-1,4-phenyleneisopropylidene-3,5-dimethyl-1,4-phenylene)oxysebacoyl]	318
Poly(oxyethylenecarbonyl-1,4-cyclohexylenecarbonyl) (trans)	291
Poly(oxyethyleneoxycarbonyl-1,4-naphthylenecarbonyl)	337
Poly(oxyethyleneoxycarbonyl-1,5-naphthylenecarbonyl)	344
Poly(oxyethyleneoxycarbonyl-2,6-naphthylenecarbonyl)	386
Poly(oxyethyleneoxycarbonyl-2,7-naphthylenecarbonyl)	392
Poly(oxyethyleneoxyterephthaloyl)	342
Poly(ethylene terephthalate) [PET]	
Poly(oxyisophthaloyl)	403 (D)
Poly(oxy(1-oxo-2,2-dimethyltrimethylene))	263
Poly(pivalolactone)	
Poly(oxy-1,4-phenyleneisopropylidene-1,4-phenyleneoxysebacoyl)	280
Poly(oxy-1,4-phenyleneoxy-1,4-phenyleneoxy-carbonyl-1,4-phenylene) [PEEK]	416
Poly(oxypropyleneoxyterephthaloyl)	341
Poly[oxyterephthaloyloxy(2,6-dimethyl-1,4-phenyleneisopropylidene-3,5-dimethyl-1,4-(D)phenylene)]	498
Poly(oxyterephthaloyloxyoctamethylene)	318 (D)
Poly(oxyterephthaloyloxy-1,4-phenyleneisopropylidene-1,4-phenylene)	478
Poly(bisphenol A terephthalate)	
Poly(oxytetramethyleneoxyterephthaloyl)	323
Poly(butylene terephthalate) [PBT]	
Main chain amide units	
Poly(iminoalkyleneiminoalkanedioyl) *Poly(alkylene alkanediamide)*–[NH(CH_2)$_m$NHCO(CH_2)$_n$CO]–	
Poly(iminoadipoyliminodecamethylene)	313
Nylon 10,6	
Poly(iminoadipoyliminohexamethylene)	ca 323
Nylon 6,6	
Poly(iminoadipoyliminooctamethylene)	318
Nylon 8,6	
Poly[iminoadipoyliminotrimethylene(methylimino)trimethylene]	278
Poly(iminocarbonyl-1,4-cyclohexylenemethylene)	466
Poly[iminocarbonyl-1,4-phenylene(2-oxoethylene)iminohexamethylene]	377
Poly(iminoethylene-1,4-phenyleneethyleneiminosebacoyl)	378 (D)
Poly(iminohexamethyleneiminoazelaoyl)	331
Nylon 6,9	

Polymer name	Glass transition temperature (T_g/K)
Poly(iminohexamethyleneiminododecanedioyl) *Nylon 6,12*	319
Poly(iminohexamethyleneiminopimeloyl) *Nylon 6,7*	331
Poly(iminohexamethyleneiminosebacoyl) *Nylon 6,10*	323
Poly(iminohexamethyleneiminosuberoyl) *Nylon 6,8*	330
Poly(iminoisophthaloylimino-4,4′-biphenylylene)	558
Poly(iminoisophthaloyliminohexamethylene)	390
Poly(iminoisophthaloyliminomethylene-1,4-cyclohexylenemethylene)	481
Poly(iminoisophthaloyliminomethylene-1,3-phenylenemethylene)	438 (M)
Poly[iminomethylene(2,5-dimethyl-1,4-phenylene)methyleneiminosuberoyl]	351
Poly(imino-1,5-naphthyleneiminoisophthaloyl)	598
Poly(imino-1,5-naphthyleneiminoterephthaloyl)	578
Poly(iminooctamethyleneiminodecanedioyl) *Nylon 8,10*	333
Poly(iminooxalyliminohexamethylene) *Nylon 6,2*	430
Poly[imino(1-oxohexamethylene)] *Nylon 6*	326
Poly[imino(1-oxodecamethylene)] *Nylon 10*	315
Poly[imino(1-oxoheptamethylene)] *Nylon 7*	325
Poly[imino(1-oxo-3-methyltrimethylene]	369
Poly[imino(1-oxononamethylene)] *Nylon 9*	319
Poly[imino(1-oxooctamethylene)] *Nylon 8*	323
Poly[imino(1-oxotrimethylene)] *Nylon 3*	384
Poly(iminopentamethyleneiminoadipoyl) *Nylon 5,6*	318
Poly[iminopentamethyleneiminocarbonyl-1,4-phenylene(2-oxoethylene)]	376
Poly(imino-1,3-phenyleneiminoisophthaloyl)	553 (M)
Poly(imino-1,4-phenyleneiminoterephthaloyl)	618
Poly(iminopimeloyliminoheptamethylene) *Nylon 7,7*	328
Poly(iminoterephthaloylimino-4,4′-biphenylylene)	613
Poly(iminotetramethyleneiminoadipoyl) *Nylon 4,6*	316
Poly[iminotetramethyleneiminocarbonyl-1,4-phenylene(2-oxoethylene)]	357
Poly(iminotrimethyleneiminoadipoyliminotrimethylene)	307
Poly[iminotrimethyleneiminocarbonyl-1,4-phenylene(2-oxoethylene)]	382
Poly(oxy-1,4-phenyleneiminoterephthaloyl-imino-1,4-phenylene)	613
Poly(sulfonylimino-1,4-phenyleneiminoadipoylimino-1,4-phenylene)	467
Main chain urethane units	
Poly(oxyalkyleneoxycarbonyliminoalkyleneiminocarbonyl)–[O(CH$_2$)$_m$OCONH(CH$_2$)$_n$NHCO]–	
Poly(oxyethyleneoxycarbonyliminohexamethyleneiminocarbonyl)	329
Poly[oxyethyleneoxycarbonylimino(6-methyl-1,3-phenylene)iminocarbonyl]	325
Poly(oxyethyleneoxycarbonylimino-1,4-phenylenemethylene-1,4-phenyleneiminocarbonyl)	412
Poly(oxyhexamethyleneoxycarbonyliminohexamethyleneiminocarbonyl)	332
Poly[oxyhexamethyleneoxycarbonylimino(6-methyl-1,3-phenylene)iminocarbonyl]	305
Poly(oxyhexamethyleneoxycarbonylimino-1,4-phenylenemethylene-1,4-phenyleneiminocarbonyl)	364
Poly(oxyoctamethyleneoxycarbonyliminohexamethyleneiminocarbonyl)	331
Poly[oxyoctamethyleneoxycarbonylimino(6-methyl-1,3-phenylene)iminocarbonyl]	337
Poly(oxyoctamethyleneoxycarbonylimino-1,4-phenylenemethylene-1,4-phenyleneiminocarbonyl)	352
Poly(oxytetramethyleneoxycarbonyliminohexamethyleneiminocarbonyl)	332
Poly[oxytetramethyleneoxycarbonylimino(6-methyl-1,3-phenylene)iminocarbonyl]	315

Polymer name	Glass transition temperature (T_g/K)
Poly(oxytetramethyleneoxycarbonylimino-1,4-phenylenemethylene-1,4-phenyleneiminocarbonyl)	382
Main chain siloxanes	
Poly[oxy(dialkylsilylene)] *Poly(dialkylsiloxane)* –[O(R_2Si)]–	
Poly[oxy(dimethylsilylene)]	148
Poly(dimethylsiloxane) [PDMS]	
Poly[oxy(dimethylsilylene)oxy-1,4-phenylene]	363 (M)
Poly[oxy(dimethylsilylene)oxy-1,4-phenyleneisopropylidene-1,4-phenylene]	318 (M)
Poly[oxy(diphenylsilylene)]	238
Poly(diphenylsiloxane)	
Poly[oxy(diphenylsilylene)-1,3-phenylene]	ca 331
Poly[oxy((methyl)phenylsilylene)]	187
Poly[oxy((methyl)-3,3,3-trifluoropropylsilylene]	<193
Main chain sulfur-containing units	
Poly(dithioethylene)	223
Poly(dithiomethylene-1,4-phenylenemethylene)	296
Poly(oxy-4,4′-biphenylylene-1,4-phenylenesulfonyl-1,4-phenylene)	503 (M)
Poly(oxycarbonyloxy-1,4-phenylenethio-1,4-phenylene)	ca 383
Poly(oxyethylenedithioethylene)	220 (M)
Poly[oxy(2-hydroxytrimethylene)oxy-1,4-phenylenesulfonyl-1,4-phenylene]	428
Poly(oxymethyleneoxyethylenedithioethylene)	214
Poly(oxy-1,4-phenylenesulfinyl-1,4-phenyleneoxy-1,4-phenylenecarbonyl-1,4-phenylene)	478 (M)
Poly(oxy-1,4-phenylenesulfinyl-1,4-phenyleneoxy-1,4-phenyleneisopropylidene-1,4-phenylene)	438 (M)
Poly(oxy-1,4-phenylenesulfonyl-1,4-phenylene)	487
Poly(oxy-1,4-phenylenesulfonyl-4,4′-biphenylylenesulfonyl-1,4-phenylene)	533
Poly[oxy-1,4-phenylenesulfonyl-1,4-phenyleneoxy(2,6-dimethyl-1,4-phenylene)isopropylidene (3,5-dimethyl-1,4-phenylene)]	508 (M)
Poly(oxy-1,4-phenylenesulfonyl-1,4-phenyleneoxy-1,4-phenylenecarbonyl-1,4-phenylene)	478 (M)
Poly[oxy-1,4-phenylenesulfonyl-1,4-phenyleneoxy-1,4-phenylene(hexafluoroisopropylidene)1,4-phenylene]	478 (M)
Poly(oxy-1,4-phenylenesulfonyl-1,4-phenyleneoxy-1,4-phenyleneisopropylidene-1,4-phenylene)	449
Poly(oxy-1,4-phenylenesulfonyl-1,4-phenyleneoxy-1.4-phenylenemethylene-1,4-phenylene)	453 (M)
Poly(oxy-1,4-phenylenesulfonyl-1,4-phenyleneoxy-1.4-phenylenethio-1,4-phenylene)	448 (M)
Poly(oxy-1,4-phenylenesulfonyl-1,4-phenyleneoxyterephthaloyl)	522
Poly(oxytetramethylenedithiotetramethylene)	197
Poly(sulfonyl-1,2-cyclohexylene)	401
Poly(sulfonyl-1,3-cyclohexylene)	381
Poly(sulfonyl-1,4-phenylenemethylene-1,4-phenylene)	497
Poly(thio-1,3-cyclohexylene)	221
Poly[thio(difluoromethylene)]	155
Poly(thioethylene)	223
Poly[thio(1-ethylethylene]	218
Poly[thio(1-methyl-3-oxotrimethylene)]	285
Poly[thio(1-methyltrimethylene)]	214
Pol[(thio(1-oxohexamethylene)]	292
Poly(thio-1,4-phenylene)	370
Poly(thiopropylene)	226
Main chain heterocyclic units	
Poly(1,3-dioxa-4,6-cyclohexylenemethylene)	378
Poly(vinyl formal)	
Poly[(2,6-dioxopiperidine-1,4-diyl)trimethylene]	363
Poly[(2-methyl-1,3-dioxa-4,6-cyclohexylene)methylene]	355
Poly(vinyl acetal)	
Poly(1,4-piperazinediylcarbonyloxyethyleneoxycarbonyl)	333
Poly(1,4-piperazinediylisophthaloyl)	465 (M)
Poly[(2-propyl-1,3-dioxa-4,6-cyclohexylene)methylene]	322
Poly(vinyl butyral)	
Poly(3,6-pyridazinediyloxy-1,4-phenyleneisopropylidene-1,4-phenyleneoxy)	453 (M)
Poly(2,5-pyridinediylcarbonyliminohexamethyleneiminocarbonyl)	322

Polymers

DIELECTRIC CONSTANT OF SELECTED POLYMERS

This table lists typical values of the dielectric constant (more properly called relative permittivity) of some important polymers. Values are given for frequencies of 1 kHz, 1 MHz, and 1 GHz; in most cases the dielectric constant at frequencies below 1 kHz does not differ significantly from the value at 1 kHz. Since the dielectric constant of a polymeric material can vary with density, degree of crystallinity, and other details of a particular sample, the values given here should be regarded as only typical or average values.

References

1. Gray, D. E., Ed., *American Institute of Physics Handbook, Third Edition*, p. 5-132, McGraw-Hill, New York, 1972.
2. Anderson, H. L., Ed., *A Physicist's Desk Reference*, American Institute of Physics, New York, 1989.
3. Brandrup, J., and Immergut, E. H., *Polymer Handbook, Third Edition*, John Wiley & Sons, New York, 1989.

Name	*t*/°C	1 kHz	1 MHz	1 GHz
Polyacrylonitrile	25	5.5	4.2	
Polyamides (nylons)	25	3.50	3.14	2.8
	84	11	4.4	2.8
Polybutadiene	25	2.5		
Polycarbonate	23	2.92	2.8	
Polychloroprene (neoprene)	25	6.6	6.3	4.2
Polychlorotrifluoroethylene	23	2.65	2.46	2.39
Polyethylene	23	2.3		
Poly(ethylene terephthalate) (Mylar)	23	3.25	3.0	2.8
Polyisoprene (natural rubber)	27	2.6	2.5	2.4
Poly(methyl methacrylate)	27	3.12	2.76	2.6
	80	3.80	2.7	2.6
Polyoxymethylene (polyformaldehyde)	25	3.8		
Poly(phenylene oxide)	23	2.59	2.59	
Polypropylene	25	2.3	2.3	2.3
Polystyrene	25	2.6	2.6	2.6
Polysulfones	25	3.13	2.10	
Polytetrafluoroethylene (teflon)	25	2.1	2.1	2.1
Poly(vinyl acetate)	50		3.5	
	150		8.3	
Poly(vinyl chloride)	25	3.39	2.9	2.8
	100	5.3	3.3	2.7
Poly(vinylidene chloride)	23	4.6	3.2	2.7
Poly(vinylidene fluoride)	23	12.2	8.9	4.7

PRESSURE-VOLUME-TEMPERATURE RELATIONSHIPS FOR POLYMER MELTS

Christian Wohlfarth

Numerous theoretical equations of state for polymer liquids have been developed. These, at the minimum, have to provide accurate fitting functions to experimental data. However, for the purpose of this table, the empirical Tait equation along with a polynomial expression for the zero pressure isobar is used. This equation is able to represent the experimental data for the melt state within the limits of experimental errors, i.e., the maximum deviations between measured and calculated specific volumes are about 0.001-0.002 cm^3/g.

The general form of the Tait equation is:

$$V(P,T) = V(0,T)\{1 - C\ln[1 + P/B(T)]\} \quad (1)$$

where the coefficient C is usually taken to be a universal constant equal to 0.0894. T is the absolute temperature in K and P the pressure in MPa. The volume V is the specific volume in cm^3/g. The Tait parameter $B(T)$ has the very simple meaning that it is inversely proportional to the compressibility κ at constant temperature and zero pressure:

$$\kappa(0,T) = -[1/V(0,T)](dV/dP) = C/B(T) \quad (2)$$

The $B(T)$ function is usually given by:

$$B(T) = B_0 \exp[-B_1(T\text{-}273.15)] \quad (3)$$

though sometimes a polynomial expression is used:

$$B(T) = b_0 + b_1(T\text{-}273.15) + b_2(T\text{-}273.15)^2 \quad (4)$$

The zero-pressure isobar $V(0,T)$ is usually given by:

$$V(0,T) = A_0 + A_1(T\text{-}273.15) + A_2(T\text{-}273.15)^2 \quad (5)$$

where A_0, A_1, A_2 are specific constants for a given polymer (the expression T-273.15 is used because fitting to the zero-pressure isobar is usually done in terms of Celsius temperature). Other forms for $V(0,T)$ are also found in the literature, such as

$$V(0,T) = A_3 \exp[A_4(T\text{-}273.15)] \quad (6)$$

or

$$V(0,T) = A_5 \exp(A_6 T^{1.5}) \quad (7)$$

where A_3 and A_4 or A_5 and A_6 are again specific constants for a given polymer.

The Tait equation is particularly useful to calculate derivative quantities, such as the isothermal compressibility and the thermal expansivity coefficients. The isothermal compressibility $\kappa(P,T)$ is derived from equation (1) as:

$$\kappa(P,T) = -(1/V)(dV/dP) = 1/\{[P + B(T)][1/C - \ln(1 + P/B(T))]\} \quad (8)$$

and the thermal expansivity $\alpha(P,T)$ as:

$$\alpha(P,T) = (1/V)(dV/dT) = \alpha(0,T) - PB_1\kappa(P,T) \quad (9)$$

where $\alpha(0,T)$ represents the thermal expansivity at zero (atmospheric) pressure and is calculated from any suitable fit for the zero-pressure volume, such as equations (5) through (7) above.

Because polymer melt PVT-behavior depends only slightly on polymer molar mass above the oligomeric region, usually no information is given in the original literature for the average molar mass of the polymers.

Table 1 summarizes the polymers or copolymers considered here and the experimental ranges of pressure and temperature over which data are available. In Table 2 the Tait equation functions, with parameters obtained from the fit, are given for 90 polymer or copolymer melts.

References

1. Zoller, P., *J. Appl. Polym. Sci.*, 23, 1051–1056, 1979 <https://doi.org/10.1002/app.1979.070230410>.
2. Starkweather, H. W., Jones, G. A., and Zoller, P., *J. Polym. Sci., Pt. B Polym. Phys.*, 26, 257–266, 1988 <https://doi.org/10.1002/polb.1988.090260204>.
3. Fakhreddine, Y. A., and Zoller, P., *J. Polym. Sci., Pt. B Polym. Phys.*, 29, 1141–1146, 1991 <https://doi.org/10.1002/polb.1991.090290913>.
4. Rodgers, P. A., *J. Appl. Polym. Sci.*, 48, 1061–1080, 1993 <https://doi.org/10.1002/app.1993.070480613>.
5. Rodgers, P. A., *J. Appl. Polym. Sci.*, 48, 2075–2083, 1993 <https://doi.org/10.1002/app.1993.070501205>.
6. Yi, Y. X., and Zoller, P., *J. Polym. Sci., Pt. B Polym. Phys.*, 31, 779–788, 1993 <https://doi.org/10.1002/polb.1993.090310705>.
7. Callaghan, T. A., and Paul, D. R., *Macromolecules*, 26, 2439–2450, 1993 <https://doi.org/10.1021/ma00062a008>.
8. Wang, Y. Z., Hsieh, K. H., Chen, L. W.,and Tseng, H. C., *J. Appl. Polym. Sci.*, 53, 1191–1201, 1994 <https://doi.org/10.1002/app.1994.070530906>.
9. Privalko, V. P., Arbuzova, A. P., Korskanov, V. V., and Zagdanskaya, N. E., *Polym. Intern.*, 35, 161–169, 1994 <https://doi.org/10.1002/pi.1994.210350206>.
10. Sachdev, V. K., Yashi, U., and Jain, R. K., *J. Polym. Sci., Pt. B Polym. Phys.*, 36, 841–850, 1998 <https://doi.org/10.1002/(SICI)1099-0488(19980415)36:5<841::AID-POLB11>3.0.CO;2-9>.

TABLE 1. Names of the Polymers, Abbreviation Used, and Range of Experimental Data Applied in the Determination of the Equation Constants

Polymer	Symbol	T/K	P/MPa	Ref.
Ethylene/propylene copolymer [50 wt%]	EP50	413-523	0.1-63	4
Ethylene/vinyl acetate copolymer [18 wt% vinyl acetate]	EVA18	385-491	0.1-177	4
Ethylene/vinyl acetate copolymer [25 wt% vinyl acetate]	EVA25	367-506	0.1-177	4
Ethylene/vinyl acetate copolymer [28 wt% vinyl acetate]	EVA28	367-508	0.1-177	4
Ethylene/vinyl acetate copolymer [40 wt% vinyl acetate]	EVA40	348-508	0.1-177	4
Polyamide-6	PA6	509-569	0.1-196	4
Polyamide-11	PA11	478-542	0.1-200	5
Polyamide-66	PA66	519-571	0.1-196	4
cis-1,4-Polybutadiene	cPBD	277-328	0.1-284	4
Polybutadiene, 8% 1,2-content	PBD-8	298-473	0.1-200	6
Polybutadiene, 24% 1,2-content	PBD-24	298-473	0.1-200	6
Polybutadiene, 40% 1,2-content	PBD-40	298-473	0.1-200	6
Polybutadiene, 50% 1,2-content	PBD-50	298-473	0.1-200	6
Polybutadiene, 87% 1,2-content	PBD-87	298-473	0.1-200	6
Poly(1-butene), isotactic	iPB	406-519	0.1-196	4
Poly(butyl methacrylate)	PnBMA	307-473	0.1-200	4
Poly(butylene terephthalate)	PBT	508-576	0.1-200	3
Poly(ε-caprolactone)	PCL	373-421	0.1-200	4
Polycarbonate-bisphenol-A	PC	424-613	0.1-177	4
Polycarbonate-bisphenol-chloral	BCPC	428-557	0.1-200	4
Polycarbonate-hexafluorobisphenol-A	HFPC	432-553	0.1-200	4
Polycarbonate-tetramethylbisphenol-A	TMPC	491-563	0.1-160	4
Poly(cyclohexyl methacrylate)	PcHMA	396-471	0.1-200	4
Poly(2,5-dimethylphenylene oxide)	PPO	473-593	0.1-177	4
Poly(dimethyl siloxane)	PDMS	298-343	0.1-100	4
Poly(dimethyl siloxane) M_n = 1000	PDMS-10	304-420	0.1-250	10
Poly(dimethyl siloxane) M_n = 4000	PDMS-40	298-418	0.1-250	10
Poly(dimethyl siloxane) M_n = 6000	PDMS-60	291-423	0.1-250	10
Poly(epichlorohydrin)	PECH	333-413	0.1-200	4
Poly(ether ether ketone)	PEEK	619-671	0.1-200	4
Poly(ethyl acrylate)	PEA	310-490	0.1-196	4
Poly(ethyl methacrylate)	PEMA	386-434	0.1-196	4
Polyethylene, high density	HDPE	413-476	0.1-196	4
Polyethylene, linear	LPE	415-473	0.1-200	4
Polyethylene, linear, high MW	HMLPE	410-473	0.1-200	4
Polyethylene, branched	BPE	398-471	0.1-200	4
Polyethylene, low density	LDPE	394-448	0.1-196	4
Polyethylene, low density, type A	LDPE-A	385-498	0.1-196	1
Polyethylene, low density, type B	LDPE-B	385-498	0.1-196	1
Polyethylene, low density, type C	LDPE-C	385-498	0.1-196	1
Poly(ethylene oxide)	PEO	361-497	0.1- 68	4
Poly(ethylene terephthalate)	PET	547-615	0.1-196	4
Poly(4-hexylstyrene)	P4HS	303-403	30-100	4
Polyisobutylene	PIB	326-383	0.1-100	4
Polyisoprene, 8% 3,4-content	PI-8	298-473	0.1-200	6
Polyisoprene, 14% 3,4-content	PI-14	298-473	0.1-200	6
Polyisoprene, 41% 3,4-content	PI-41	298-473	0.1-200	6
Polyisoprene, 56% 3,4-content	PI-56	298-473	0.1-200	6
Poly(methyl acrylate)	PMA	310-493	0.1-196	4
Poly(methyl methacrylate)	PMMA	387-432	0.1-200	4
Poly(4-methyl-1-pentene)	P4MP	514-592	0.1-196	4
Poly(α-methylstyrene)	PαMS	473-533	0.1-170	7
Poly(*o*-methylstyrene)	PoMS	412-471	0.1-180	4
Polyoxymethylene	POM	463-493	0.1-196	2
Phenoxy[a]	PH	341-573	0.1-177	4
Polysulfone[b]	PSF	475-644	0.1-196	4
Polyarylate[c]	PAr	450-583	0.1-177	4
Polypropylene, atactic	aPP	353-393	0.1-100	4
Polypropylene, isotactic	iPP	443-570	0.1-196	4

Polymer	Symbol	T/K	P/MPa	Ref.
Polystyrene	PS	388-469	0.1-200	4
Poly(tetrafluoroethylene)	PTFE	603-645	0.1-39	4
Poly(tetrahydrofuran)	PTHF	335-439	0.1-78	4
Poly(vinyl acetate)	PVAc	308-373	0.1-80	4
Poly(vinyl chloride)	PVC	373-423	0.1-200	4
Poly(vinyl methyl ether)	PVME	303-471	0.1-200	4
Poly(vinylidene fluoride)	PVdF	451-521	0.1-200	5
Styrene/acrylonitrile copolymer [2.7 wt% acrylonitrile]	SAN3	378-539	0.1-200	4
Styrene/acrylonitrile copolymer [5.7 wt% acrylonitrile]	SAN6	370-540	0.1-200	4
Styrene/acrylonitrile copolymer [15.3 wt% acrylonitrile]	SAN15	405-531	0.1-200	4
Styrene/acrylonitrile copolymer [18.0 wt% acrylonitrile]	SAN18	377-528	0.1-200	4
Styrene/acrylonitrile copolymer [40 wt% acrylonitrile]	SAN40	373-543	0.1-200	4
Styrene/acrylonitrile copolymer [70 wt% acrylonitrile]	SAN70	373-544	0.1-200	4
Styrene/butadiene copolymer [10 wt% styrene]	SBR10	393-533	0.1-196	8
Styrene/butadiene copolymer [23.5 wt% styrene]	SBR23	393-533	0.1-196	8
Styrene/butadiene copolymer [60 wt% styrene]	SBR60	393-533	0.1-196	8
Styrene/butadiene copolymer [85 wt% styrene]	SBR85	393-533	0.1-196	8
Styrene/methyl methacrylate copolymer [20 wt% methyl methacrylate]	SMMA20	383-543	0.1-200	4
Styrene/methyl methacrylate copolymer [60 wt% methyl methacrylate]	SMMA60	383-543	0.1-200	4
N-Vinylcarbazole/4-ethylstyrene copolymer [50 mol% ethylstyrene]	VCES50	393-443	30-100	9
N-Vinylcarbazole/4-hexylstyrene copolymer [80 mol% hexylstyrene]	VCHS80	313-423	30-100	9
N-Vinylcarbazole/4-hexylstyrene copolymer [67 mol% hexylstyrene]	VCHS67	333-423	30-100	9
N-Vinylcarbazole/4-hexylstyrene copolymer [60 mol% hexylstyrene]	VCHS60	383-453	30-100	9
N-Vinylcarbazole/4-hexylstyrene copolymer [50 mol% hexylstyrene]	VCHS50	373-443	30-100	9
N-Vinylcarbazole/4-hexylstyrene copolymer [40 mol% hexylstyrene]	VCHS40	423-493	30-100	9
N-Vinylcarbazole/4-hexylstyrene copolymer [33 mol% hexylstyrene]	VCHS33	463-523	30-100	9
N-Vinylcarbazole/4-hexylstyrene copolymer [20 mol% hexylstyrene]	VCHS20	473-523	30-100	9
N-Vinylcarbazole/4-octylstyrene copolymer [50 mol% octylstyrene]	VCOS50	403-453	30-100	9
N-Vinylcarbazole/4-pentylstyrene copolymer [50 mol% pentylstyrene]	VCPS50	383-443	30-100	9

[a] Phenoxy = Poly(oxy-2-hydroxytrimethyleneoxy-1,4-phenyleneisopropylidene-1,4-phenylene).
[b] Polysulfone = Poly(oxy-1,4-phenylenesulfonyl-1,4-phenyleneoxy-1,4-phenyleneisopropylidene-1,4-phenylene).
[c] Polyarylate = Poly(oxyterephthaloyl/isophthaloyl T/I=50/50)oxy-1,4-phenyleneisopropylidene-1,4-phenylene.

TABLE 2. Tait Equation Parameter Functions for Polymer Melts

Polymer	$V(0,T)$/cm^{3}g^{-1}	$B(T)$/MPa
EP50	$1.2291 + 5.799{\cdot}10^{-5}(T-273.15) + 1.964{\cdot}10^{-6}(T-273.15)^2$	$487.0\ \exp[-8.103{\cdot}10^{-3}(T-273.15)]$
EVA18	$1.02391\ \exp(2.173{\cdot}10^{-5}T^{1.5})$	$188.2\ \exp[-4.537{\cdot}10^{-3}(T-273.15)]$
EVA25	$1.00416\ \exp(2.244{\cdot}10^{-5}T^{1.5})$	$184.4\ \exp[-4.734{\cdot}10^{-3}(T-273.15)]$
EVA28	$1.00832\ \exp(2.241{\cdot}10^{-5}T^{1.5})$	$183.5\ \exp[-4.457{\cdot}10^{-3}(T-273.15)]$
EVA40	$1.06332\ \exp(2.288{\cdot}10^{-5}T^{1.5})$	$205.1\ \exp[-4.989{\cdot}10^{-3}(T-273.15)]$
PA6	$0.7597\ \exp[4.701{\cdot}10^{-4}(T-273.15)]$	$376.7\ \exp[-4.660{\cdot}10^{-3}(T-273.15)]$
PA11	$0.9581\ \exp[6.664{\cdot}10^{-4}(T-273.15)]$	$254.7\ \exp[-4.178{\cdot}10^{-3}(T-273.15)]$
PA66	$0.7657\ \exp[6.600{\cdot}10^{-4}(T-273.15)]$	$316.4\ \exp[-5.040{\cdot}10^{-3}(T-273.15)]$
cPBD	$1.0970\ \exp[6.600{\cdot}10^{-4}(T-273.15)]$	$177.7\ \exp[-3.593{\cdot}10^{-3}(T-273.15)]$
PBD-8	$1.1004 + 6.718{\cdot}10^{-4}(T-273.15) + 6.584{\cdot}10^{-7}(T-273.15)^2$	$200.0\ \exp[-4.606{\cdot}10^{-3}(T-273.15)]$
PBD-24	$1.1049 + 6.489{\cdot}10^{-4}(T-273.15) + 7.099{\cdot}10^{-7}(T-273.15)^2$	$193.0\ \exp[-4.519{\cdot}10^{-3}(T-273.15)]$
PBD-40	$1.1013 + 6.593{\cdot}10^{-4}(T-273.15) + 5.776{\cdot}10^{-7}(T-273.15)^2$	$188.0\ \exp[-4.437{\cdot}10^{-3}(T-273.15)]$
PBD-50	$1.1037 + 5.955{\cdot}10^{-4}(T-273.15) + 7.789{\cdot}10^{-7}(T-273.15)^2$	$183.0\ \exp[-4.425{\cdot}10^{-3}(T-273.15)]$
PBD-87	$1.1094 + 6.729{\cdot}10^{-4}(T-273.15) + 4.470{\cdot}10^{-7}(T-273.15)^2$	$175.0\ \exp[-4.538{\cdot}10^{-3}(T-273.15)]$
iPB	$1.1417\ \exp[6.751{\cdot}10^{-4}(T-273.15)]$	$167.5\ \exp[-4.533{\cdot}10^{-3}(T-273.15)]$
PnBMA	$0.9341 + 5.5254{\cdot}10^{-4}(T-273.15) + 6.5803{\cdot}10^{-6}(T-273.15)^2 + 1.5691{\cdot}10^{-10}(T-273.15)^3$	$226.7\ \exp[-5.344{\cdot}10^{-3}(T-273.15)]$
PBT	$0.9640 - 1.017{\cdot}10^{-3}(T-273.15) + 3.065{\cdot}10^{-6}(T-273.15)^2$	$263.0\ \exp[-3.444{\cdot}10^{-3}(T-273.15)]$
PCL	$0.9049\ \exp[6.392{\cdot}10^{-4}(T-273.15)]$	$189.0\ \exp[-3.931{\cdot}10^{-3}(T-273.15)]$
PC	$0.73565\ \exp(1.859{\cdot}10^{-5}T^{1.5})$	$310.0\ \exp[-4.078{\cdot}10^{-3}(T-273.15)]$
BCPC	$0.6737 + 3.634{\cdot}10^{-4}(T-273.15) + 2.370{\cdot}10^{-7}(T-273.15)^2$	$363.4\ \exp[-4.921{\cdot}10^{-3}(T-273.15)]$
HFPC	$0.6111 + 4.898{\cdot}10^{-4}(T-273.15) + 1.730{\cdot}10^{-7}(T-273.15)^2$	$236.6\ \exp[-5.156{\cdot}10^{-3}(T-273.15)]$
TMPC	$0.8497 + 5.073{\cdot}10^{-4}(T-273.15) + 3.832{\cdot}10^{-7}(T-273.15)^2$	$231.4\ \exp[-4.242{\cdot}10^{-3}(T-273.15)]$
PcHMA	$0.8793 + 4.0504{\cdot}10^{-4}(T-273.15) + 7.774{\cdot}10^{-7}(T-273.15)^2 - 7.7534{\cdot}10^{-10}(T-273.15)^3$	$295.2\ \exp[-5.220{\cdot}10^{-3}(T-273.15)]$
PPO	$0.78075\ \exp(2.151{\cdot}10^{-5}T^{1.5})$	$227.8\ \exp[-4.290{\cdot}10^{-3}(T-273.15)]$
PDMS	$1.0079\ \exp[9.121{\cdot}10^{-4}(T-273.15)]$	$89.4\ \exp[-5.701{\cdot}10^{-3}(T-273.15)]$
PDMS-10	$0.8343 + 5.991{\cdot}10^{-4}(T-273.15) + 5.734{\cdot}10^{-7}(T-273.15)^2$	$542.63\ \exp[-6.69{\cdot}10^{-3}(T-273.15)]$

Polymer	$V(0,T)$/cm³g⁻¹	$B(T)$/MPa
PDMS-40	$0.8018 + 7.072\cdot10^{-4}(T-273.15) + 3.635\cdot10^{-7}(T-273.15)^2$	$482.73 \exp[-6.09\cdot10^{-3}(T-273.15)]$
PDMS-60	$0.8146 + 5.578\cdot10^{-4}(T-273.15) + 5.774\cdot10^{-7}(T-273.15)^2$	$482.73 \exp[-6.09\cdot10^{-3}(T-273.15)]$
PECH	$0.7216 \exp[5.825\cdot10^{-4}(T-273.15)]$	$238.3 \exp[-4.171\cdot10^{-3}(T-273.15)]$
PEEK	$0.7158 \exp[6.690\cdot10^{-4}(T-273.15)]$	$388.0 \exp[-4.124\cdot10^{-3}(T-273.15)]$
PEA	$0.8756 \exp[7.241\cdot10^{-4}(T-273.15)]$	$193.2 \exp[-4.839\cdot10^{-3}(T-273.15)]$
PEMA	$0.8614 \exp[7.468\cdot10^{-4}(T-273.15)]$	$260.9 \exp[-5.356\cdot10^{-3}(T-273.15)]$
HDPE	$1.1595 + 8.0394\cdot10^{-4}(T-273.15)$	$179.9 \exp[-4.739\cdot10^{-3}(T-273.15)]$
LPE	$0.9172 \exp[7.806\cdot10^{-4}(T-273.15)]$	$176.7 \exp[-4.661\cdot10^{-3}(T-273.15)]$
HMLPE	$0.8992 \exp[8.502\cdot10^{-4}(T-273.15)]$	$168.3 \exp[-4.292\cdot10^{-3}(T-273.15)]$
BPE	$0.9399 \exp[7.341\cdot10^{-4}(T-273.15)]$	$177.1 \exp[-4.699\cdot10^{-3}(T-273.15)]$
LDPE	$1.1944 + 2.841\cdot10^{-4}(T-273.15) + 1.872\cdot10^{-6}(T-273.15)^2$	$202.2 \exp[-5.243\cdot10^{-3}(T-273.15)]$
LDPE-A	$1.1484 \exp[6.950\cdot10^{-4}(T-273.15)]$	$192.9 \exp[-4.701\cdot10^{-3}(T-273.15)]$
LDPE-B	$1.1524 \exp[6.700\cdot10^{-4}(T-273.15)]$	$196.6 \exp[-4.601\cdot10^{-3}(T-273.15)]$
LDPE-C	$1.1516 \exp[6.730\cdot10^{-4}(T-273.15)]$	$186.7 \exp[-4.391\cdot10^{-3}(T-273.15)]$
PEO	$0.8766 \exp[7.087\cdot10^{-4}(T-273.15)]$	$207.7 \exp[-3.947\cdot10^{-3}(T-273.15)]$
PET	$0.6883 + 5.90\cdot10^{-4}(T-273.15)$	$369.7 \exp[-4.150\cdot10^{-3}(T-273.15)]$
P4HS	$0.8251 + 6.77\cdot10^{-4}T$	$103.1 \exp[-2.417\cdot10^{-3}(T-273.15)]$
PIB	$1.0750 \exp[5.651\cdot10^{-4}(T-273.15)]$	$200.3 \exp[-4.329\cdot10^{-3}(T-273.15)]$
PI-8	$1.1030 + 6.488\cdot10^{-4}(T-273.15) + 5.125\cdot10^{-7}(T-273.15)^2$	$188.0 \exp[-4.541\cdot10^{-3}(T-273.15)]$
PI-14	$1.0943 + 6.293\cdot10^{-4}(T-273.15) + 6.231\cdot10^{-7}(T-273.15)^2$	$202.0 \exp[-4.653\cdot10^{-3}(T-273.15)]$
PI-41	$1.0951 + 6.188\cdot10^{-4}(T-273.15) + 6.629\cdot10^{-7}(T-273.15)^2$	$199.0 \exp[-4.622\cdot10^{-3}(T-273.15)]$
PI-56	$1.0957 + 6.655\cdot10^{-4}(T-273.15) + 5.661\cdot10^{-7}(T-273.15)^2$	$200.0 \exp[-4.644\cdot10^{-3}(T-273.15)]$
PMA	$0.8365 \exp[6.795\cdot10^{-4}(T-273.15)]$	$235.8 \exp[-4.493\cdot10^{-3}(T-273.15)]$
PMMA	$0.8254 + 2.8383\cdot10^{-4}(T-273.15) + 7.792\cdot10^{-7}(T-273.15)^2$	$287.5 \exp[-4.146\cdot10^{-3}(T-273.15)]$
P4MP	$1.4075 - 9.095\cdot10^{-4}(T-273.15) + 3.497\cdot10^{-6}(T-273.15)^2$	$37.67 + 0.2134(T-273.15)] - 7.0445\cdot10^{-4}(T-273.15)^2$
PαMS	$0.89365 + 3.4864\cdot10^{-4}(T-273.15) + 5.0184\cdot10^{-7}(T-273.15)^2$	$297.7 \exp[-4.074\cdot10^{-3}(T-273.15)]$
PoMS	$0.9396 \exp[5.306\cdot10^{-4}(T-273.15)]$	$261.9 \exp[-4.114\cdot10^{-3}(T-273.15)]$
POM	$0.7484 \exp[6.770\cdot10^{-4}(T-273.15)]$	$305.6 \exp[-4.326\cdot10^{-3}(T-273.15)]$
PH	$0.76644 \exp(1.921\cdot10^{-5}T^{1.5})$	$359.9 \exp[-4.378\cdot10^{-3}(T-273.15)]$
PSF	$0.7644 + 3.419\cdot10^{-4}(T-273.15) + 3.126\cdot10^{-7}(T-273.15)^2$	$365.9 \exp[-3.757\cdot10^{-3}(T-273.15)]$
PAr	$0.73381 \exp(1.626\cdot10^{-5}T^{1.5})$	$296.9 \exp[-3.375\cdot10^{-3}(T-273.15)]$
aPP	$1.1841 - 1.091\cdot10^{-4}(T-273.15) + 5.286\cdot10^{-6}(T-273.15)^2$	$162.1 \exp[-6.604\cdot10^{-3}(T-273.15)]$
iPP	$1.1606 \exp[6.700\cdot10^{-4}(T-273.15)]$	$149.1 \exp[-4.177\cdot10^{-3}(T-273.15)]$
PS	$0.9287 \exp[5.131\cdot10^{-4}(T-273.15)]$	$216.9 \exp[-3.319\cdot10^{-3}(T-273.15)]$
PTFE	$0.3200 + 9.5862\cdot10^{-4}(T-273.15)$	$425.2 \exp[-9.380\cdot10^{-3}(T-273.15)]$
PTHF	$1.0043 \exp[6.691\cdot10^{-4}(T-273.15)]$	$178.6 \exp[-4.223\cdot10^{-3}(T-273.15)]$
PVAc	$0.82496 + 5.820\cdot10^{-4}(T-273.15) + 2.940\cdot10^{-7}(T-273.15)^2$	$204.9 \exp[-4.346\cdot10^{-3}(T-273.15)]$
PVC	$0.7196 + 5.581\cdot10^{-5}(T-273.15) + 1.468\cdot10^{-6}(T-273.15)^2$	$294.2 \exp[-5.321\cdot10^{-3}(T-273.15)]$
PVME	$0.9585 \exp[6.653\cdot10^{-4}(T-273.15)]$	$215.8 \exp[-4.588\cdot10^{-3}(T-273.15)]$
PVdF	$0.5790 \exp[8.051\cdot10^{-4}(T-273.15)]$	$244.0 \exp[-5.210\cdot10^{-3}(T-273.15)]$
SAN3	$0.9233 + 3.936\cdot10^{-4}(T-273.15) + 5.685\cdot10^{-7}(T-273.15)^2$	$239.8 \exp[-4.376\cdot10^{-3}(T-273.15)]$
SAN6	$0.9211 + 4.370\cdot10^{-4}(T-273.15) + 5.846\cdot10^{-7}(T-273.15)^2$	$226.9 \exp[-4.286\cdot10^{-3}(T-273.15)]$
SAN15	$0.9044 + 4.207\cdot10^{-4}(T-273.15) + 4.077\cdot10^{-7}(T-273.15)^2$	$238.4 \exp[-3.943\cdot10^{-3}(T-273.15)]$
SAN18	$0.9016 + 4.036\cdot10^{-4}(T-273.15) + 4.206\cdot10^{-7}(T-273.15)^2$	$240.4 \exp[-3.858\cdot10^{-3}(T-273.15)]$
SAN40	$0.8871 + 3.406\cdot10^{-4}(T-273.15) + 4.938\cdot10^{-7}(T-273.15)^2$	$289.3 \exp[-4.431\cdot10^{-3}(T-273.15)]$
SAN70	$0.8528 + 3.616\cdot10^{-4}(T-273.15) + 2.634\cdot10^{-7}(T-273.15)^2$	$335.4 \exp[-3.923\cdot10^{-3}(T-273.15)]$
SBR10	$0.9053 \exp(2.437\cdot10^{-5}T^{1.5})$	$530.3 \exp[-3.99\cdot10^{-3}(T-273.15)]$
SBR23	$0.8986 \exp(2.317\cdot10^{-5}T^{1.5})$	$551.6 \exp[-4.17\cdot10^{-3}(T-273.15)]$
SBR60	$0.8812 \exp(2.031\cdot10^{-5}T^{1.5})$	$486.0 \exp[-4.34\cdot10^{-3}(T-273.15)]$
SBR85	$0.8704 \exp(1.846\cdot10^{-5}T^{1.5})$	$356.7 \exp[-4.24\cdot10^{-3}(T-273.15)]$
SMMA20	$0.9063 + 3.570\cdot10^{-4}(T-273.15) + 6.532\cdot10^{-7}(T-273.15)^2$	$232.0 \exp[-4.143\cdot10^{-3}(T-273.15)]$
SMMA60	$0.8610 + 3.350\cdot10^{-4}(T-273.15) + 6.980\cdot10^{-7}(T-273.15)^2$	$261.0 \exp[-4.611\cdot10^{-3}(T-273.15)]$
VCES50	$0.6676 + 6.63\cdot10^{-4}T$	$5281.7 \exp[-9.264\cdot10^{-3}(T-273.15)]$
VCHS80	$0.7753 + 6.17\cdot10^{-4}T$	$247.6 \exp[-2.604\cdot10^{-3}(T-273.15)]$
VCHS67	$0.8028 + 6.50\cdot10^{-4}T$	$581.7 \exp[-4.553\cdot10^{-3}(T-273.15)]$
VCHS60	$0.8213 + 6.23\cdot10^{-4}T$	$229.1 \exp[-2.133\cdot10^{-3}(T-273.15)]$
VCHS50	$0.7827 + 5.05\cdot10^{-4}T$	$136.0 \exp[-1.083\cdot10^{-3}(T-273.15)]$
VCHS40	$0.7805 + 4.92\cdot10^{-4}T$	$155.0 \exp[-1.605\cdot10^{-3}(T-273.15)]$
VCHS33	$0.7710 + 4.86\cdot10^{-4}T$	$460.4 \exp[-3.453\cdot10^{-3}(T-273.15)]$
VCHS20	$0.6416 + 5.42\cdot10^{-4}T$	$489.8 \exp[-3.193\cdot10^{-3}(T-273.15)]$
VCOS50	$0.7081 + 7.40\cdot10^{-4}T$	$666.5 \exp[-4.503\cdot10^{-3}(T-273.15)]$
VCPS50	$0.7814 + 4.36\cdot10^{-4}T$	$880.1 \exp[-4.393\cdot10^{-3}(T-273.15)]$

VAPOR PRESSURES (SOLVENT ACTIVITIES) FOR BINARY POLYMER SOLUTIONS

Christian Wohlfarth

The vapor pressure of a binary polymer solution is given by the activity of the solvent A, a_A. Solvent activities in polymer solutions are measured either by the isopiestic method applying a reference system whose solvent activity is precisely known or by determining the solvent partial pressure, P_A, and calculating the activity of the solvent by equation (1):

$$a_A = \left(P_A / P_A^s\right) \exp \frac{\left(B_{AA} - V_A^L\right)\left(P - P_A^s\right)}{RT} \qquad (1)$$

where B_{AA} is the second virial coefficient, P_A^s is the saturation vapor pressure, and V_A^L is the molar volume of the pure solvent A at the measuring temperature T. The exponential term is neglected in quite a lot of original papers, however, and only the reduced vapor pressures are given (such data are indicated by an asterisk in the table below). Vapor pressures of polymer solutions have been measured since the 1940s, but the amount of experimental data for polymer solutions is still relatively small in comparison to low-molecular mixtures and solutions. The data scatter with respect to temperature, concentration, molar mass, and other polymer characterization variables. Furthermore, the concentration range for measuring vapor pressures in good thermodynamic quality is often limited to the polymer mass fraction range between 0.4 and 0.85. A recent review on methods for the measurement of vapor pressures/solvent activities of polymer solutions and on related problems is given in Ref. 1. Experimental data have been collected in several books (Refs. 2-6).

The table in this *CRC Handbook* provides data for a number of polymer solutions as smoothed values over the complete range of solvent activities between 0 (polymer mass fraction = 1) and 1 (polymer mass fraction = 0). For this purpose, the data were selected from data books (Refs. 4-6) as well as from a number of original sources (Refs. 7-22) which are not included in these books. The appropriate data were smoothed. The final table provides then the polymer mass fractions at given fixed solvent activities between 0.1 and 0.9. Of course, the user must keep in mind that the activity vs. concentration range of the experimental data is sometimes smaller than the below given complete range, thus the smoothed data should be used with sufficient care.

Generally, vapor pressures or solvent activities of binary polymer solutions depend on molar mass. However, for high molecular weight polymers (well above the oligomer region), this molar-mass dependence can be neglected in many cases. Therefore, the table below presents only data for polymer solutions where the number average molar mass, M_n, is in the order of 10^5 g/mol or even higher, therefore, the molar mass is not specified. The temperature is stated, even though the temperature dependence of a_A is relatively small for the temperature ranges where most of the experimental data exist.

References

1. Wohlfarth, C., Methods for the measurement of solvent activity of polymer solutions, in *Handbook of Solvents*, Wypych, G., Ed., ChemTec Publishing, Toronto, 2000, 146.
2. Wen, H., Elbro, H. S., and Alessi, P., *Polymer Solution Data Collection.* I. Vapor-liquid equilibrium; II. Solvent activity coefficients at infinite dilution; III. Liquid-liquid equilibrium, Chemistry Data Series, Vol. 15, DECHEMA, Frankfurt am Main, 1992.
3. Danner, R. P. and High, M. S., *Handbook of Polymer Solution Thermodynamics*, American Institute of Chemical Engineers, New York, 1993.
4. Wohlfarth, C., *Vapour-Liquid Equilibrium Data of Binary Polymer Solutions: Physical Science Data*, 44, Elsevier, Amsterdam, 1994.
5. Wohlfarth, C., *CRC Handbook of Thermodynamic Data of Copolymer Solutions*, CRC Press, Boca Raton, FL, 2001.
6. Wohlfarth, C., *CRC Handbook of Thermodynamic Data of Aqueous Polymer Solutions*, CRC Press, Boca Raton, FL, 2003.
7. Wang, K., Hu, Y., and Wu, D. T., *J. Chem. Eng. Data*, 39, 916, 1994.
8. Choi, J. S., Tochigi, K., and Kojima, K., *Fluid Phase Equil.*, 111, 143, 1995.
9. Tochigi, K., Kurita, S., Ohashi, M., and Kojima, K., *Kagaku Kogaku Ronbunshu*, 23, 720, 1997.
10. Wong, H. C., Campbell, S. W., and Bhethanabotla, V. R., *Fluid Phase Equil.*, 139, 371, 1997.
11. Kim, J., Joung, K. C., Hwang, S., Huh, W., Lee, C. S., and Yoo, K.-P., *Korean J. Chem. Eng.*, 15, 199, 1998.
12. Kim, N. H., Kim, S.J., Won, Y. S., and Choi, J. S., *Korean J. Chem. Eng.*, 15, 141, 1998.
13. Feng, W., Wang, W., and Feng, Z., *J. Chem. Ind. Eng. (China)*, 49, 271, 1998.
14. French, R. N. and Koplos, G. J., *Fluid Phase Equil.*, 160, 879, 1999.
15. Striolo, A. and Praunsitz, J. M., *Polymer*, 41, 1109, 2000.
16. Fornasiero, F., Halim, M., and Prausnitz, J. M., *Macromolecules*, 33, 8435, 2000.
17. Wong, H. C., Campbell, S. W., and Bhethanabotla, V. R., *Fluid Phase Equil.*, 179, 181, 2001.
18. Wibawa, G., Takahashi, M., Sato, Y., Takishima, S., and Masuoka, H., *J. Chem. Eng. Data*, 47, 518, 2002.
19. Wibawa, G., Hatano, R., Sato, Y., Takishima, S., and Masuoka, H., *J. Chem. Eng. Data*, 47, 1022, 2002.
20. Pfohl, O., Riebesell, C., and Dohrn, R., *Fluid Phase Equil.*, 202, 289, 2002.
21. Jung, J. K., Joung, S. N., Shin, H.Y., Kim, S. Y., Yoo, K.-P., Huh, W., Lee, C. S., *Korean J. Chem. Eng.*, 19, 296, 2002.
22. Kang, S., Huang, Y., Fu, J., Liu, H., and Hu, Y., *J. Chem. Eng. Data*, 47, 788, 2002.

Solvent Activity a_A as a Function of Temperature and Mass Fraction

Polymer/Solvent	T/K	Mass Fraction of the Polymer 0.1	0.2	0.3	0.4	0.5	0.6	0.7	0.8	0.9
Acrylonitrile/Styrene Copolymer (28 wt% Acrylonitrile)										
Benzene*	343.15	0.982	0.962	0.940	0.915	0.886	0.851	0.809	0.753	0.670
1,2-Dimethylbenzene*	398.15	0.983	0.964	0.942	0.918	0.890	0.857	0.817	0.764	0.685
1,3-Dimethylbenzene*	398.15	0.983	0.965	0.944	0.921	0.893	0.861	0.821	0.769	0.690
1,4-Dimethylbenzene*	398.15	0.983	0.964	0.942	0.918	0.890	0.857	0.817	0.763	0.684

Polymer/Solvent	T/K	Mass Fraction of the Polymer								
		0.1	0.2	0.3	0.4	0.5	0.6	0.7	0.8	0.9
Propylbenzene*	398.15	0.987	0.972	0.955	0.935	0.913	0.885	0.851	0.804	0.732
Toluene*	343.15	0.982	0.962	0.940	0.915	0.886	0.851	0.809	0.753	0.669
Butadiene/Styrene Copolymer (41 wt% Styrene)										
Benzene*	343.15	0.968	0.934	0.896	0.853	0.805	0.748	0.680	0.591	0.461
Cyclohexane*	343.15	0.978	0.953	0.925	0.893	0.856	0.811	0.754	0.678	0.556
Ethylbenzene*	398.15	0.974	0.945	0.912	0.875	0.831	0.779	0.713	0.625	0.491
Mesitylene*	398.15	0.977	0.950	0.921	0.887	0.847	0.799	0.738	0.656	0.526
Toluene*	343.15	0.970	0.936	0.899	0.857	0.808	0.751	0.682	0.591	0.456
Cellulose Triacetate										
Dichloromethane	298.15	0.979	0.956	0.930	0.899	0.863	0.819	0.762	0.683	0.554
Trichloromethane	298.15	0.978	0.953	0.924	0.892	0.853	0.806	0.747	0.665	0.533
Dextran										
Water	313.15	0.988	0.975	0.960	0.942	0.921	0.894	0.860	0.810	0.725
Hydroxyethylcellulose										
Water	368.15	0.988	0.974	0.958	0.939	0.915	0.884	0.841	0.775	0.650
Hydroxypropylstarch										
Water	293.15	0.989	0.977	0.963	0.947	0.927	0.903	0.872	0.827	0.749
Nitrocellulose										
Ethyl acetate	293.15	0.938	0.885	0.835	0.786	0.737	0.685	0.627	0.560	0.471
Ethyl formate	293.15	0.958	0.916	0.873	0.828	0.780	0.728	0.668	0.595	0.494
Ethyl propionate	293.15	0.941	0.889	0.839	0.789	0.739	0.685	0.625	0.555	0.460
Methyl acetate	293.15	0.890	0.820	0.763	0.711	0.660	0.609	0.554	0.490	0.406
2-Propanone	293.15	0.922	0.861	0.807	0.756	0.706	0.653	0.596	0.530	0.443
Propyl acetate	293.15	0.937	0.881	0.827	0.775	0.722	0.665	0.602	0.528	0.426
Polybutadiene (random cis-trans-vinyl)										
Benzene	298.15	0.964	0.925	0.884	0.839	0.788	0.731	0.663	0.578	0.455
Cyclohexane	298.15	0.974	0.945	0.913	0.876	0.833	0.782	0.719	0.635	0.507
Dichloromethane	298.15	0.951	0.902	0.852	0.800	0.745	0.684	0.616	0.532	0.415
Hexane	298.15	0.984	0.965	0.943	0.916	0.881	0.837	0.775	0.683	0.534
Tetrachloromethane	298.15	0.932	0.865	0.799	0.731	0.660	0.585	0.503	0.409	0.288
Toluene	298.15	0.969	0.935	0.898	0.856	0.809	0.754	0.688	0.603	0.476
Trichloromethane	298.15	0.925	0.855	0.788	0.720	0.650	0.578	0.498	0.406	0.289
1,4-cis-Polybutadiene										
Benzene	298.15	0.966	0.930	0.890	0.846	0.796	0.738	0.668	0.580	0.450
Cyclohexane	298.15	0.977	0.951	0.922	0.888	0.849	0.803	0.747	0.677	0.581
Dichloromethane	298.15	0.948	0.898	0.848	0.796	0.742	0.683	0.616	0.536	0.424
Hexane	298.15	0.983	0.963	0.941	0.916	0.886	0.850	0.804	0.741	0.639
Tetrachloromethane	298.15	0.936	0.871	0.805	0.736	0.665	0.588	0.505	0.409	0.287
Toluene	298.15	0.969	0.936	0.900	0.860	0.815	0.763	0.701	0.622	0.506
Trichloromethane	298.15	0.915	0.840	0.770	0.702	0.634	0.562	0.485	0.396	0.283
Poly(butyl acrylate)										
Benzene	298.15	0.964	0.926	0.887	0.845	0.799	0.749	0.691	0.619	0.519
Dichloromethane	298.15	0.868	0.801	0.744	0.690	0.636	0.577	0.511	0.430	0.318
Tetrachloromethane	298.15	0.932	0.868	0.805	0.742	0.677	0.607	0.529	0.438	0.317
Toluene	298.15	0.967	0.932	0.893	0.849	0.801	0.744	0.676	0.590	0.463
Trichloromethane	298.15	0.901	0.811	0.733	0.662	0.595	0.529	0.459	0.381	0.282
Poly(butyl methacrylate)										
Benzene	313.15	0.971	0.939	0.902	0.861	0.813	0.756	0.685	0.592	0.453
1-Butanol	313.15	0.991	0.980	0.968	0.953	0.936	0.914	0.885	0.842	0.762
2-Butanol	313.15	0.992	0.982	0.969	0.953	0.933	0.906	0.869	0.815	0.719
2-Butanone	313.15	0.982	0.963	0.940	0.914	0.884	0.846	0.799	0.732	0.623
Butyl acetate*	308.15	0.982	0.961	0.936	0.908	0.875	0.836	0.789	0.730	0.652
Cyclohexane	313.15	0.985	0.968	0.948	0.925	0.899	0.866	0.823	0.764	0.666
Cyclopentane	313.15	0.984	0.965	0.944	0.918	0.886	0.846	0.792	0.714	0.579
Diethyl ether*	298.15	0.987	0.973	0.956	0.937	0.914	0.885	0.848	0.795	0.703
1,4-Dimethylbenzene	333.15	0.971	0.940	0.905	0.866	0.822	0.770	0.706	0.622	0.497

Polymer/Solvent	T/K	Mass Fraction of the Polymer 0.1	0.2	0.3	0.4	0.5	0.6	0.7	0.8	0.9
Ethylbenzene	333.15	0.969	0.935	0.899	0.859	0.815	0.764	0.704	0.627	0.517
Methyl acetate	313.15	0.984	0.965	0.944	0.920	0.891	0.856	0.811	0.748	0.645
2-Methyl-1-propanol	333.15	0.988	0.974	0.958	0.940	0.919	0.893	0.860	0.815	0.744
Octane	313.15	0.988	0.974	0.959	0.942	0.921	0.896	0.865	0.823	0.758
1-Propanol	333.15	0.990	0.980	0.967	0.952	0.934	0.911	0.881	0.834	0.746
2-Propanol	313.15	0.991	0.981	0.970	0.956	0.939	0.918	0.889	0.845	0.755
2-Propanone	313.15	0.989	0.976	0.961	0.944	0.921	0.892	0.850	0.783	0.647
Propyl acetate	313.15	0.980	0.957	0.932	0.903	0.870	0.830	0.780	0.714	0.612
Toluene	313.15	0.971	0.939	0.903	0.863	0.818	0.764	0.698	0.613	0.485
Poly(ε-caprolacton)										
Tetrachloromethane*	338.15	0.956	0.910	0.864	0.815	0.762	0.704	0.637	0.554	0.438
Poly(dimethylsiloxane)										
Chlorodifluoromethane	298.15	0.976	0.950	0.921	0.888	0.850	0.805	0.750	0.677	0.565
Cyclohexane	303.15	0.979	0.955	0.928	0.898	0.863	0.822	0.770	0.702	0.596
Hexane	303.15	0.982	0.962	0.939	0.912	0.880	0.842	0.793	0.724	0.611
Pentane	308.15	0.982	0.962	0.940	0.913	0.881	0.842	0.791	0.720	0.600
Pentane	423.15	0.984	0.966	0.946	0.922	0.893	0.858	0.813	0.749	0.641
Poly(ethyl acrylate)										
Benzene	298.15	0.970	0.939	0.904	0.866	0.823	0.774	0.716	0.641	0.533
Dichloromethane	298.15	0.900	0.830	0.768	0.709	0.648	0.584	0.512	0.427	0.313
Tetrachloromethane	298.15	0.950	0.900	0.848	0.794	0.736	0.672	0.598	0.509	0.385
Toluene	298.15	0.972	0.942	0.910	0.874	0.833	0.786	0.730	0.659	0.555
Trichloromethane	298.15	0.866	0.776	0.701	0.632	0.566	0.499	0.428	0.349	0.248
Poly(ethylene oxide)										
Benzene	323.15	0.972	0.942	0.908	0.869	0.824	0.771	0.706	0.620	0.490
2-Butanone	353.15	0.981	0.959	0.934	0.902	0.863	0.813	0.746	0.651	0.503
Cyclohexane	353.15	0.989	0.976	0.960	0.943	0.921	0.893	0.855	0.798	0.688
Methanol	303.15	0.964	0.927	0.887	0.844	0.797	0.744	0.682	0.604	0.494
2-Propanone	353.15	0.979	0.947	0.896	0.815	0.719	0.625	0.532	0.434	0.315
Water	293.15	0.977	0.951	0.923	0.890	0.852	0.806	0.748	0.671	0.550
Poly(ethylenimine)										
Water	353.15	0.975	0.947	0.917	0.883	0.845	0.801	0.748	0.680	0.581
Poly(ethyl methacrylate)										
Benzene	298.15	0.970	0.938	0.903	0.864	0.821	0.771	0.712	0.637	0.529
Dichloromethane	298.15	0.912	0.838	0.769	0.703	0.636	0.567	0.491	0.404	0.292
Tetrachloromethane	298.15	0.935	0.873	0.812	0.750	0.686	0.616	0.540	0.449	0.328
Toluene	298.15	0.974	0.945	0.913	0.877	0.836	0.787	0.727	0.647	0.527
Trichloromethane	298.15	0.859	0.760	0.678	0.604	0.533	0.464	0.392	0.313	0.217
Polyisobutylene										
Benzene	313.15	0.984	0.965	0.945	0.921	0.892	0.858	0.813	0.751	0.645
Cyclohexane	313.15	0.976	0.950	0.921	0.888	0.850	0.805	0.749	0.676	0.563
Cyclopentane	313.15	0.977	0.952	0.924	0.892	0.855	0.812	0.758	0.687	0.579
1,4-Dimethylbenzene	313.15	0.979	0.955	0.929	0.899	0.863	0.821	0.767	0.694	0.579
2,2-Dimethylbutane	298.15	0.983	0.964	0.942	0.917	0.887	0.852	0.806	0.743	0.640
Ethylbenzene	313.15	0.979	0.955	0.927	0.895	0.857	0.810	0.750	0.668	0.535
Heptane	298.15	0.983	0.964	0.942	0.917	0.887	0.851	0.804	0.741	0.637
Hexane	298.15	0.980	0.959	0.934	0.906	0.873	0.834	0.784	0.715	0.606
Octane	298.15	0.983	0.963	0.940	0.914	0.883	0.845	0.797	0.729	0.617
Tetrachloromethane	298.15	0.962	0.921	0.877	0.829	0.776	0.715	0.643	0.552	0.423
Toluene	313.15	0.984	0.966	0.944	0.918	0.884	0.840	0.779	0.688	0.537
Trichloromethane	298.15	0.969	0.935	0.899	0.858	0.813	0.761	0.698	0.619	0.503
2,4,4-Trimethylpentane	298.15	0.981	0.961	0.937	0.911	0.879	0.842	0.794	0.730	0.628
1,4-cis-Polyisoprene										
Benzene	313.15	0.982	0.962	0.937	0.908	0.873	0.827	0.766	0.679	0.537
2-Butanone	353.15	0.986	0.970	0.953	0.933	0.910	0.883	0.850	0.808	0.746
Cyclohexane	313.15	0.978	0.954	0.928	0.899	0.865	0.825	0.778	0.716	0.625

Polymer/Solvent	T/K	Mass Fraction of the Polymer 0.1	0.2	0.3	0.4	0.5	0.6	0.7	0.8	0.9
Dichloromethane	298.15	0.969	0.935	0.898	0.857	0.811	0.757	0.693	0.610	0.488
1,4-Dimethylbenzene	313.15	0.977	0.951	0.923	0.892	0.857	0.816	0.767	0.704	0.613
Ethylbenzene	313.15	0.978	0.954	0.928	0.898	0.864	0.823	0.774	0.709	0.612
Methyl acetate	313.15	0.968	0.935	0.900	0.862	0.820	0.773	0.717	0.649	0.554
Octane	313.15	0.984	0.967	0.948	0.926	0.901	0.871	0.834	0.785	0.711
Propyl acetate	333.15	0.983	0.964	0.942	0.916	0.886	0.850	0.803	0.738	0.633
Tetrachloromethane	298.15	0.929	0.864	0.800	0.737	0.672	0.602	0.526	0.435	0.316
Toluene	313.15	0.978	0.954	0.927	0.898	0.865	0.827	0.782	0.725	0.645
Trichloromethane	298.15	0.930	0.867	0.807	0.747	0.685	0.620	0.547	0.462	0.346
Poly(methyl acrylate)										
Benzene	298.15	0.979	0.956	0.930	0.901	0.867	0.826	0.776	0.710	0.608
Dichloromethane	298.15	0.917	0.851	0.791	0.732	0.671	0.605	0.532	0.444	0.326
Tetrachloromethane	298.15	0.963	0.924	0.882	0.838	0.788	0.733	0.668	0.586	0.470
Toluene	298.15	0.981	0.960	0.936	0.909	0.878	0.840	0.792	0.727	0.626
Trichloromethane	298.15	0.912	0.830	0.753	0.678	0.603	0.527	0.446	0.357	0.248
Poly(methyl methacrylate)										
Benzene	298.15	0.982	0.961	0.938	0.912	0.881	0.843	0.795	0.729	0.622
2-Butanone*	308.15	0.989	0.976	0.961	0.945	0.925	0.900	0.869	0.825	0.751
Cyclohexanone*	323.15	0.978	0.954	0.928	0.899	0.866	0.827	0.781	0.723	0.640
Dichloromethane	298.15	0.939	0.882	0.825	0.766	0.704	0.637	0.560	0.468	0.343
Ethyl acetate*	308.15	0.986	0.969	0.950	0.928	0.902	0.869	0.826	0.763	0.649
Toluene	298.15	0.981	0.959	0.935	0.908	0.877	0.841	0.795	0.736	0.646
Trichloromethane	298.15	0.924	0.848	0.771	0.694	0.616	0.536	0.451	0.358	0.246
Poly(α-methylstyrene)										
Cumene	338.15	0.984	0.965	0.944	0.918	0.887	0.848	0.796	0.721	0.593
α-Methylstyrene	338.15	0.978	0.954	0.927	0.896	0.859	0.816	0.761	0.687	0.570
Poly(propylene oxide)										
Benzene	333.15	0.967	0.932	0.893	0.850	0.801	0.744	0.675	0.588	0.460
Methanol	298.15	0.992	0.982	0.970	0.955	0.936	0.910	0.872	0.812	0.689
Polystyrene										
Benzene	333.15	0.978	0.953	0.924	0.891	0.852	0.804	0.742	0.657	0.521
2-Butanone*	298.15	0.986	0.971	0.954	0.935	0.912	0.885	0.851	0.804	0.724
Cyclohexane	313.15	0.990	0.978	0.965	0.949	0.931	0.908	0.877	0.833	0.754
Cyclohexanone*	313.15	0.970	0.937	0.900	0.858	0.810	0.753	0.684	0.593	0.459
Dichloromethane	298.15	0.949	0.899	0.849	0.797	0.743	0.684	0.617	0.536	0.423
1,3-Dimethylbenzene*	323.15	0.980	0.956	0.926	0.891	0.846	0.791	0.723	0.638	0.524
1,4-Dimethylbenzene	423.15	0.974	0.944	0.911	0.872	0.826	0.770	0.698	0.601	0.452
Ethyl acetate*	313.15	0.976	0.948	0.918	0.882	0.841	0.791	0.728	0.642	0.507
Hexane	423.15	0.980	0.958	0.933	0.904	0.869	0.827	0.772	0.697	0.574
2-Propanone	323.15	0.991	0.980	0.969	0.955	0.938	0.918	0.892	0.854	0.788
Propyl acetate	343.15	0.983	0.965	0.943	0.919	0.891	0.858	0.815	0.758	0.667
Tetrachloromethane	298.15	0.961	0.917	0.869	0.814	0.751	0.678	0.592	0.486	0.344
Toluene	313.15	0.981	0.959	0.933	0.901	0.861	0.809	0.738	0.638	0.481
Trichloromethane	298.15	0.949	0.898	0.847	0.793	0.736	0.675	0.604	0.519	0.400
Poly(tetramethylene glycol)										
Methanol	303.15	0.981	0.961	0.938	0.913	0.883	0.849	0.806	0.751	0.671
Poly(vinyl acetate)										
Benzene	313.15	0.985	0.967	0.945	0.919	0.886	0.844	0.784	0.696	0.548
1-Butanol	313.15	0.992	0.982	0.971	0.958	0.942	0.923	0.896	0.856	0.779
2-Butanol	313.15	0.987	0.972	0.956	0.937	0.915	0.889	0.856	0.813	0.747
2-Butanone	313.15	0.980	0.958	0.934	0.906	0.873	0.835	0.787	0.724	0.626
1,2-Dichloroethane*	300.15	0.955	0.906	0.851	0.790	0.722	0.644	0.556	0.450	0.315
1,4-Dimethylbenzene	313.15	0.990	0.978	0.964	0.948	0.928	0.903	0.868	0.814	0.705
Ethylbenzene	313.15	0.990	0.979	0.966	0.950	0.932	0.910	0.880	0.836	0.759
Methanol	333.15	0.990	0.978	0.965	0.949	0.931	0.908	0.877	0.834	0.757
Methyl acetate	313.15	0.976	0.949	0.919	0.886	0.849	0.805	0.752	0.684	0.583
2-Methyl-1-propanol	353.15	0.984	0.966	0.946	0.924	0.899	0.868	0.832	0.784	0.715

Polymer/Solvent	*T*/K	Mass Fraction of the Polymer								
		0.1	0.2	0.3	0.4	0.5	0.6	0.7	0.8	0.9
1-Propanol	353.15	0.987	0.972	0.955	0.936	0.914	0.888	0.856	0.815	0.753
2-Propanol	353.15	0.988	0.974	0.958	0.940	0.919	0.894	0.863	0.820	0.754
2-Propanone	333.15	0.983	0.963	0.940	0.913	0.880	0.838	0.784	0.707	0.578
Propyl acetate	333.15	0.979	0.955	0.930	0.901	0.869	0.831	0.786	0.728	0.645
Tetrahydrofuran	323.15	0.973	0.943	0.911	0.874	0.831	0.781	0.720	0.640	0.519
Toluene	333.15	0.983	0.965	0.944	0.920	0.891	0.857	0.815	0.756	0.664
Poly(vinyl chloride)										
2-Butanone*	313.15	0.976	0.949	0.920	0.887	0.849	0.804	0.749	0.676	0.566
Cyclohexanone*	333.15	0.971	0.934	0.889	0.839	0.781	0.714	0.635	0.536	0.397
Poly(vinyl methyl ether)										
Benzene*	298.15	0.969	0.935	0.897	0.855	0.807	0.751	0.683	0.596	0.466
Chlorobenzene*	343.15	0.972	0.941	0.906	0.867	0.822	0.769	0.705	0.620	0.494
1,2-Dimethylbenzene*	363.15	0.973	0.943	0.910	0.871	0.826	0.772	0.705	0.616	0.478
Ethylbenzene*	343.15	0.978	0.954	0.927	0.895	0.857	0.811	0.753	0.672	0.542
Propylbenzene*	373.15	0.977	0.951	0.923	0.890	0.852	0.808	0.752	0.678	0.563
Poly(4-vinylpyridine)										
Methanol	343.15	0.986	0.971	0.953	0.931	0.905	0.871	0.825	0.756	0.627
2-Propanol	343.15	0.989	0.977	0.964	0.948	0.928	0.904	0.872	0.826	0.743
Poly(1-vinyl-2-pyrrolidinone)										
Water	368.15	0.984	0.966	0.946	0.924	0.899	0.870	0.835	0.790	0.727
Starch (amorphous)										
Water	383.15	0.991	0.981	0.970	0.956	0.939	0.918	0.889	0.845	0.754
Styrene/Methyl methacrylate Copolymer (41.45 wt% Styrene)										
Benzene*	308.15	0.982	0.963	0.940	0.913	0.881	0.841	0.789	0.716	0.590
Vinyl acetate/Vinyl chloride Copolymer (12 wt% Vinyl acetate)										
Benzene	398.15	0.976	0.949	0.918	0.883	0.841	0.791	0.728	0.643	0.509
Chlorobenzene	398.15	0.984	0.965	0.944	0.920	0.891	0.856	0.810	0.746	0.638
1,4-Dimethylbenzene	398.15	0.989	0.977	0.963	0.946	0.926	0.899	0.863	0.807	0.692
Ethylbenzene	398.15	0.989	0.976	0.961	0.944	0.924	0.899	0.866	0.818	0.735
Octane	398.15	0.992	0.982	0.971	0.958	0.942	0.922	0.893	0.847	0.739

* $a_A = P_A / P_A^S$.

SOLUBILITY PARAMETERS OF SELECTED POLYMERS

Charles E. Carraher, Jr.

Physical properties of polymers, including solubility, are related to the strength of covalent bonds, the stiffness of the segments in the polymer backbone, the amount of crystallinity or amorphous nature of the polymer, and the intermolecular forces between the polymer chains. The strength of the intermolecular forces is directly related to the cohesive energy density (CED), which is the molar energy of vaporization per unit volume. Because intermolecular attractions of solvent and solute must be overcome when a solute (here a polymer) dissolves, CED values can be used to predict solubility.

This has led to the concept of a **solubility parameter** δ, introduced by Hildebrand, which is the square root of CED. The solubility parameter δ for nonpolar solvents is equal to the square root of the heat of vaporization per unit volume.

$$\delta = (\Delta E/V)^{1/2} = (\mathrm{CED})^{1/2}, \text{ or } \delta^2 = \mathrm{CED}$$

V is the molar volume, and ΔE is the molar energy of vaporization. Units for the solubility parameter are $(\mathrm{MPa})^{1/2} = (\mathrm{J\ cm^{-3}})^{1/2} = 0.4887(\mathrm{cal\ cm^{-3}})^{1/2}$. The polymer community often uses the unit Hildebrand (H) = $(\mathrm{cal\ cm^{-3}})^{1/2}$.

The heat of mixing a solute and a solvent, ΔH_m, is proportional to the square of the difference in solubility parameters, as shown below, where ϕ is the partial volume of each component, namely solvent ϕ_1 and solute (polymer) ϕ_2. Because typically the entropy term favors solution and the enthalpy term acts counter to solution, the objective is to match solvent and solute so that the difference between their δ values is small, resulting in a small enthalpy acting against solubility occurring.

$$\Delta H_m = \phi_1\phi_2 (\delta_1 - \delta_2)^2$$

The energy of vaporization is not accessible for polymers, but cohesive energy density of polymers can be determined from *PVT* data. However, common ways for determining polymer solubility parameters use thermodynamic properties of polymer solutions and their relations to excess enthalpy or excess Gibbs energy per unit volume.

The solubility parameter concept predicts the heat of mixing for liquids and amorphous polymers. It has been experimentally found that generally any nonpolar amorphous polymer dissolves in a liquid or mixture of liquids having a solubility parameter δ that does not differ by more than $\pm 1.8\ (\mathrm{cal\ cm^{-3}})^{1/2}$.

Sometimes the Flory-Huggins solvent-polymer interaction parameter is applied. References 1–3 give details for such procedures as well as extensive tables of polymer solubility parameters. Methods for calculating solubility parameters can be found in References 4–7.

Table 1 gives solubility parameters for a variety of typical solvents, including those that are poorly hydrogen bonding, moderately hydrogen bonding, and strongly hydrogen bonding. Table 2 provides solubility parameter values for selected common polymers covering all three hydrogen bonding scenarios. Table 3 gives solubility characteristics of a number of common polymers with various solvents.

References

1. Barton, A.F.M., *CRC Handbook of Polymer-Liquid Interaction Parameters and Solubility Parameters*, CRC Press, Boca Raton, FL, 1991.
2. Brandrup, J., Immergut, E.H., Grulke, E.A. (eds.), *Polymer Handbook, Fourth Edition*, J. Wiley & Sons, New York, 1999.
3. Wohlfarth, C., *Thermodynamic Properties of Polymer Solutions*, in Landolt-Börnstein, New Series, Group VIII, Vol. 6D, Lechner, M.D., (ed.), Springer Verlag, Berlin, Heidelberg, 2010.
4. Hildebrand, J.H., Prausnitz, J.M., Scott, R.L., *Regular and Related Solutions*, Van Nostrand Reinhold Co., New York, 1970.
5. [Van] Krevelen, D.W., *Properties of Polymers, Third Edition*, Elsevier, Amsterdam, 1990.
6. Bicerano, J., *Prediction of Polymer Properties, Third Edition*, CRC Press, Boca Raton, FL, 2002 <https://doi.org/10.1201/9780203910115>.
7. Hansen, C.M., *Hansen Solubility Parameters: A User's Handbook, Second Edition*, CRC Press, Boca Raton, FL, 2007 <https://doi.org/10.1201/9781420006834>.

TABLE 1. Solubility Parameter (δ) for Typical Solvents

Solvent	δ (cal cm^{-3})$^{1/2}$
Poorly Hydrogen Bonding	
Dimethylsiloxane monomer	5.5
Dichlorodifluoromethane	5.5
Neopentane	6.3
1-Nitrooctane	7.0
Pentane	7.0
Octane	7.6
Turpentine	8.1
Cyclohexane	8.2
1-Isopropyl-4-methylbenzene	8.2
Tetrachloromethane	8.6
Propylbenzene	8.6
4-Chlorotoluene	8.8
Decahydronaphthalene (unspecified isomer)	8.8
Xylene (unspecified isomer)	8.8
Benzene	9.2
Styrene	9.3
1,2,3,4-Tetrahydronaphthalene	9.4
Chlorobenzene	9.5
1,2-Dichloroethane	9.8
p-Dichlorobenzene	10.0
Nitroethane	11.1
Acetonitrile	11.9
Nitroethane	12.7
Moderately Hydrogen Bonding	
Diisopropyl ether	6.9
Diethyl ether	7.4
Isopentyl acetate	7.8
2,6-Dimethyl-4-heptanone	7.8
Dipropyl ether	7.8
sec-Butyl acetate	8.2
Isopropyl acetate	8.4
2-Heptanone	8.5
Ethyl acetate	9.0

Solvent	δ (cal cm^{-3})$^{1/2}$
2-Butanone	9.3
2-Butoxyethanol	9.5
Methyl acetate	9.6
Bis(2-chloroethyl) ether	9.8
Acetone	9.9
1,4-Dioxane	10.0
Cyclopentanone	10.4
2-Ethoxyethanol	10.5
N,N-Dimethylacetamide	10.8
Propylene carbonate	13.3
Ethylene carbonate	14.7
Strongly Hydrogen Bonding	
Diethylamine	8.0
Pentylamine	8.7
2-Ethyl-1-hexanol	9.5
3-Methyl-1-butanol	10.0
Acetic acid	10.1
m-Cresol	10.2
Aniline	10.3
1-Octanol	10.3
2-Methyl-2-propanol	10.9
1-Pentanol	10.9
1-Butanol	11.4
2-Propanol	11.5
Diethylene glycol	12.1
Furfuryl alcohol	12.5
Ethanol	12.7
N-Ethylformamide	13.9
Methanol	14.5
1,2-Ethanediol	14.6
Glycerol	16.5
Water	23.4

TABLE 2. Approximate Solubility Parameter (δ) Values of Selected Polymers

Polymer	Poorly H bonding δ (cal cm^{-3})$^{1/2}$	Moderately H bonding δ (cal cm^{-3})$^{1/2}$	Strongly H bonding δ (cal cm^{-3})$^{1/2}$
Polytetrafluoroethylene	5.8-6.4		
Poly(vinyl ethyl ether)	7.0-11.0	7.4-10.8	9.5-14.0
Poly(butyl acrylate	7.0-12.5	7.4-11.5	
Poly(butyl methacrylate)	7.4-11.0	7.4-10.0	9.5-11.2
Polyisobutylene	7.5-8.0		
Polyethylene	7.7-8.2		
Poly(vinyl butyl ether)	7.8-10.6	7.5-10.0	9.5-11.2
Natural rubber	8.1-8.5		
Polystyrene	8.5-10.6	9.1-9.4	
Poly(vinyl acetate)	8.5-9.5		
Poly(vinyl chloride)	8.5-11.0	7.8-10.5	
Buna N	8.7-9.3		
Poly(methyl methacrylate)	8.9-12.7	8.5-13.3	
Poly(ethylene oxide)	8.9-12.7	8.5-14.5	9.5-14.5
Poly(ethylene sulfide)	9.0-10.0		
Polycarbonate	9.5-10.6	9.5-10.0	
Poly(ethylene terephthate)	9.5-10.8	9.3-9.9	
Polyurethane	9.8-10.3		
Polymethacrylonitrile		10.6-11.0	
Cellulose acetate	11.1-12.5	10.0-14.5	
Nitrocellulose	11.1-12.5	8.0-14.5	12.5-14.5
Polyacrylonitrile		12.0-14.0	
Poly(vinyl alcohol)			12.0-13.0
Nylon-66			13.5-15.0
Cellulose			14.5-16.5

Table 3 gives solubility characteristcs of a number of common polymers with various solvents. Abbreviations used in Table 3 are as follows.

Abbreviation	Solvent
CCl_4	Carbon tetrachloride
DMA	Dimethylacetamide
DMF	Dimethylformamide
DMSO	Dimethyl sulfoxide
MEK	Methyl ethyl ketone (butanone)
THF	Tetrahydrofuran

TABLE 3. Representative Solubility Characteristics of Common Polymers

Polymer	Soluble	Insoluble
Amylose	Nitromethane, DMSO, water (hot), ethylenediamine	Aliphatic alcohols, diethylether
Amylopectin	Water	Ethanol, diethylether, acetone
Cellulose	Trifluoroacetic acid, sodium xanthate, tetramethylammonium hydroxide, cupri-ammonium hydroxide, sulfuric acid	Hydrocarbons, water
Gutta percha	Benzene, chloroform	Water, alcohol
Natural rubber (non-cross-linked)	Aromatic and aliphatic hydrocarbons	Acetone, organic acids, alcohols
Nylon 6 & 6,6	Aromatic alcohols, organic acids, sulfuric acid, phosphoric acid	Hydrocarbons, chloroform, alcohols, ethers, ketones, esters
Nylon-poly(ethylene terephthalamide)	Sulfuric acid	Chloroform, aromatic alcohols, DMF, DMA, trifluoroacetic acid
Polyacrylamide	Morpholine, water	Hydrocarbons, alcohols, THF, DMF, diethylether
Poly(acrylic acid)	Low alcohols, ethylene glycol, diethyl ether	Aliphatic and aromatic hydrocarbons, ketones
Polyacrylonitrile	1-Methyl-2-pyridone, DMF, many organic nitriles, strong acids	Hydrocarbons, chlorinated hydrocarbons, many amides, DMSO, ketones
Polybutadiene	Most aliphatics, halogenated hydrocarbons, benzene	Water, lower alcohols, nitromethane
Poly(dimethoxy-phosphazene)	Chloroform, THF, DMF, acetonitrile, pyridine, dioxane	Water, acetone, benzene, ethanol, diethylether
Poly(ethylene oxide)	Water, chloroform, CCl_4, esters, DMF, acetonitrile, alcohols	Aliphatic hydrocarbons, dioxane, ethers
Poly(ethylene terephthalate)	Phenol, DMSO, nitrobenzene, halogenated aliphatic carboxylic acids	Hydrocarbons, esters, ethers, aliphatic alcohols
Poly(methyl acrylate)	Aromatic and chlorinated hydrocarbons, THF, esters, ketones	Aliphatic hydrocarbons, diethylether, CCl_4, lower alcohols
Poly(methyl methacrylate)	Aromatic hydrocarbons, chlorinated lower esters, nitroethane	CCl_4, higher esters, diethylether
Poly(methylene oxide)	Phenol (with heat), aniline, DMF, DMA, diphenol ether, chlorophenols	Aliphatic hydrocarbons, lower alcohols, diethylether, lower esters
i-Polypropylene	Hydrocarbons, chlorinated hydrocarbons, diethylether	More polar organic solvents
Polystyrene	Aromatic hydrocarbons, lower chlorinated aliphatic hydrocarbons, THF, MEK, dioxane, many phthalates	Saturated hydrocarbons, alcohols, diols, acetone, acetic acid, diethylether
Polyurethanes - general	Aromatic alcohols, formic acid, sulfuric acid	Saturated hydrocarbons, diethylether, alcohols
Poly(vinyl acetate)	Aromatic hydrocarbons, chlorinated hydrocarbons, dioxane, acetone, DMF, DMSO, chloroform, nitromethane, acetonitrile	Saturated hydrocarbons, ethanol, CCl_4, ethylene glycol, diethylether, water, higher esters, carbon disulfide
Poly(vinyl alcohol)	Water, DMF, hot DMSO, piperazine	Hydrocarbons, chlorinated hydrocarbons, THF, dioxane
Poly(vinyl chloride)	THF, DMF, DMSO, MEK	Aliphatic and aromatic hydrocarbons, alcohols, glycols, acetone

Section 14
Geophysics, Astronomy, and Acoustics

Astronomical Constants 14-1
Properties of the Solar System 14-2
Satellites of the Planets........ 14-6
Interstellar Molecules........ 14-9
Mass, Dimensions, and Other Parameters of the Earth 14-14
Geological Time Scale........ 14-16
Acceleration Due to Gravity 14-17
Density, Pressure, and Gravity as a Function of Depth within the Earth 14-17
Ocean Pressure as a Function of Depth and Latitude 14-18
Properties of Seawater 14-19
Abundance of Elements in the Earth's Crust and in the Sea 14-21
Solar Irradiance at the Earth........ 14-22
U.S. Standard Atmosphere (1976) 14-23
Geographical and Seasonal Variations in Solar Radiation 14-29
Major World Earthquakes 14-30
Infrared Absorption by the Earth's Atmosphere 14-34
Atmospheric Concentration of Carbon Dioxide, 1959–2020........ 14-35
Global Temperature Trend, 1880–2020 14-36
Global Warming Potential of Greenhouse Gases........ 14-37
Speed of Sound in Various Media........ 14-39
Attenuation and Speed of Sound in Air as a Function of Humidity and Frequency........ 14-41
Speed of Sound in Dry Air........ 14-42
Allocation of Frequencies in the Radio Spectrum 14-43

ASTRONOMICAL CONSTANTS

The constants in this table are based originally on the set of constants adopted by the International Astronomical Union (IAU) in 1976. Updates have been made when new data were available (Refs. 1-4), with the latest updates being 2020. The left-most column of the table gives the most current reference for the cited value. All values are given in SI units; thus, masses are expressed in kilograms and distances in meters or kilometers as appropriate.

The astronomical unit of time is a time interval of one day (1 d) equal to 86400 s. An interval of 36525 d is one Julian century (1 cy). See Refs. 5-7 for an explanation of the way in which the various distance units used in astronomy are related.

Data on the sun, the Earth, and other planets in the solar system are found in "Properties of the Solar System" in this section.

References

1. See "International System of Units" in Section 1.
2. Astronomical Constants, Chapter K6 in *The Astronomical Almanac 2019*, Department of the Navy, Nautical Almanac Office, Washington, DC, 2018, retrieved January 31, 2021. <asa.hmnao.com/static/files/2018/Astronomical_Constants_2018.pdf>
3. *Astronomical Constants, Solar System Dynamics*, Jet Propulsion Laboratory, California Institute of Technology, Pasadena, CA, retrieved January 31, 2021. <ssd.jpl.nasa.gov/constants>
4. Pitjeva, E.V., *J. Phys. Chem. Ref. Data* 44, 031210, 2015.
5. Seidelmann, P. K., *Explanatory Supplement to the Astronomical Almanac*, University Science Books, Mill Valley, CA, 1990.
6. Lang, K. R., *Astrophysical Data: Planets and Stars*, Springer-Verlag, New York, 1992. <https://doi.org/10.1007/978-1-4684-0640-5_3>
7. International Astronomical Union, <www.iau.org/static/resolutions/IAU2012_English.pdf>, 2012.
8. Measuring the Universe—The IAU and Astronomical Units, <www.iau.org/public/themes/measuring/>.

Astronomical Constants

Quantity	Value	Ref.
Defining constants		
Astronomical unit	au = 149 597 870 700 m (exact)	1
Speed of light	c = 299 792 458 m s^{-1} (exact)	1
Primary constants		
Light-time for unit distance (1 au)	τ_A = 499.004 783 836 s	2
Geocentric gravitational constant	GE = 3.986 004 418 × 10^{14} m^3 s^{-2} ± 8 × 10^5	2
Constant of gravitation	G = 6.674 30 × 10^{-11} m^3 kg^{-1} s^{-2} ± 0.000 15 × 10^{-11}	1
Obliquity of the ecliptic at standard epoch J2000	ε = 23°26′21″.448	2
Derived constants		
Constant of nutation at standard epoch J2000	N = 9″.2052 331	2
Solar parallax (π_0 = arcsin(a_e/ua))	π_0 = 8″.794 143	2
Constant of aberration for standard epoch J2000	κ = 20″.495 51	2
Heliocentric gravitational constant ($GS = A^3k^2/D^2$)	GS = 1.327 124 400 18 × 10^{20} ± 8 × 10^9 m^3 s^{-2}	3
	GS = 1.327 124 400 42 × 10^{20} ± 10 × 10^9 m^3 s^{-2}	4

PROPERTIES OF THE SOLAR SYSTEM

Our solar system consists of eight major planets and numerous other objects (Ref. 1) that have been observed orbiting the sun. These include the following.

- Dwarf planets: five (Ref. 2)
- Minor planets: 1 026 572 through December 14, 2020 (Ref. 1)
- Asteroids: more than 1 047 527 through December 14, 2020 (Ref. 3)
- Asteroid belts: between Mars and Jupiter
- Satellites (moons): associated with specific planets; see "Satellites of the Planets" in this section
- Comets: 4535 through December 14, 2020 (Ref. 1)

Improved optical and non-optical telescopes (Ref. 4) constantly add new objects to the list.

Dwarf planets are defined by the International Astronomical Union (Ref. 2) as bodies in orbit around the sun massive enough to adopt a near-spherical shape because of their self-gravity, but are appreciably smaller than the eight major planets. Plutoids form a subset of the dwarf planets, with orbits larger than that of Neptune (Ref. 5). As of 2020, the IAU has recognized the names of four plutoids: Pluto, Eris, Haumea, and Makemake. Because Ceres has an orbit much smaller than Neptune, it is classified as a dwarf planet but not as a plutoid. Data in these tables are drawn from many references as indicated. See Refs. 17–19 for additional information.

The following tables contain various properties about our solar system.

Table 1. Properties of the Planetary System
Table 2. Properties of the Sun
Table 3. Orbital Properties of the Major and Important Dwarf Planets
Table 4. Physical Properties of the Major and Important Dwarf Planets and the Moon
Table 5. Mean Surface Temperature and Pressure and Atmospheric Composition of the Major Planets and Pluto

References

1. IAU Minor Planet Center, retrieved 31 January 2021. <www.minorplanetcenter.net>
2. *Dwarf Planets,* IAU Minor Planet Center, retrieved January 3, 2021. <www.minorplanetcenter.net/dwarf_planets>
3. *Asteroids,* NASA Science Solar System Exploration, retrieved January 31, 2021. <solarsystem.nasa.gov/asteroids-comets-and-meteors/asteroids/overview/?page=0&per_page=40&order=name+asc&search=&condition_1=101%3Aparent_id&condition_2=asteroid%3Abody_type%3Ailike>
4. *The Multiwavelength Milky Way: Telescopes,* NASA, Goddard Space Flight Center, retrieved January 31, 2021. <asd.gsfc.nasa.gov/archive/mvmw/mmw_telescope.html>
5. IAU Press Release, <www.iau.org/public_press/news/release/iau0804>, June 2008.
6. Lang, K. R., *Astrophysical Data: Planets and Stars,* Springer-Verlag, New York, 1992. <https://doi.org/10.1007/978-1-4684-0640-5_3>
7. Cox, A. N., *Allen's Astrophysical Quantities, Fourth Edition,* Springer-Verlag, New York, 2000; this is a revision of Allen, C. W., *Astrophysical Quantities, Third Edition,* 1983.
8. *Planetary Fact Sheet—Metric,* NASA Goddard Space Flight Center, retrieved January 31, 2021. <nssdc.gsfc.nasa.gov/planetary/factsheet>
9. *Sun Fact Sheet,* NASA Goddard Space Flight Center, retrieved January 31, 2021. <nssdc.gsfc.nasa.gov/planetary/factsheet/sunfact.html>
10. See individual fact sheets for each planet at *Planetary Facts Sheets,* National Space Science Data Center, NASA Goddard Space Flight Center, retrieved January 31, 2021. <nssdc.gsfc.nasa.gov/planetary/planetfact.html>
11. Data available at *JPL Small-Body Database Browser* search engine using name of dwarf planet, Jet Propulsion Laboratory, California Institute of Technology, Pasadena, CA, retreived January 31, 2021. <ssd.jpl.nasa.gov/sbdb.cgi>
12. The Planetary Society, <www.planetary.org/explore/topics/groups/our_solar_system/>.
13. *The Astronomical Almanac for the Year 2019,* U.S. Government Printing Office, Washington, DC, 2005; available online at <asa.usno.navy.mil/>.
14. Arnet, B., The Nine Planets, <www.nineplanets.org>.
15. Onasch, B., Our Solar System, <www.onasch.de/astro/>.
16. Killen, R., Cremonese, G., Lammer, H. et al, *Space Science Reviews* 132, 433–509, 2007. <doi: 10.1007/s11214-007-9232-0>
17. K. A. Olive et al. (Particle Data Group), *Chin. Phys. C* 38(9), 090001, 2014.; section on astrophysical constants available at <pdg.lbl.gov/2014/reviews/astrorpp.pdf>, <https://doi.org/10.1088/1674-1137/38/9/090001>
18. Seidelmann, P. K., Editor, *Explanatory Supplement to the Astronomical Almanac,* University Science Books, Mill Valley, CA, 1992.
19. Kopp, G., and Lean, J. L., *Geophys. Res. Lett.* 38 (1), 2011. <https://doi.org/10.1029/2010GL045777>

Properties of the Planetary System

Our planetary system, which includes all major and dwarf planets as well as other objects that orbit the sun, has been the object of research and wonder since ancient times. Ancient observers clearly recognized that several objects in the night sky had different motion from more fixed objects (i.e., stars) and over the centuries, considerable effort was expended to explain this motion. Nicolas Copernicus' work first published in 1542 provided the foundation for our present-day helio-centric view of the solar system. Many refinements to the theory have been made since Copernicus' first publication.

Table 1 presents data for several properties of the planetary system as a whole (Refs. 6–8). The mass of the earth is included as it is often used as a reference point for masses of objects in the solar system.

TABLE 1. Properties of the Planetary System

Property	Value
Mass of the earth (M_e)	5.9723×10^{24} kg
Total mass of planetary system	2.669×10^{27} kg
	447 M_e
Total angular momentum of planetary system	3.148×10^{48} kg m^2 s^{-1}
Total kinetic energy of planets	1.99×10^{35} J
Total rotational energy of planets	0.7×10^{35} J

Properties of the Sun

Our sun is a star categorized as a G-type, yellow-dwarf, main sequence star. Table 2 contains important properties of the sun (Ref. 9) except as noted.

TABLE 2. Properties of the Sun

Property	Value
Mass	1.9885×10^{30} kg
	3329483 M_e
Radius	6.95700×10^5 km
Surface area	6.079×10^{12} km^2
Volume	1.412×10^{18} km^3
Mean density	1.408 g cm^{-3}
Surface gravity	274.0 m s^{-2}
Surface escape velocity	617.6×10^5 km s^{-1}
Effective temperature	5772 K
Luminosity (Total radiant power emitted)	3.828×10^{26} W
Flux of radiant energy at the earth (Solar constant) (Ref. 11)	1360.8 W m^{-2}
Surface flux of radiant energy*	6.293×10^7 W m^{-2}
Mass conversion rate	4260×10^6 kg s^{-1}
Mean energy production	0.195×10^{-3} J kg^{-1} s^{-1}
Sidereal rotation period (at 16 deg. latitude)	609.12 h

* Calculated with Stefan-Boltzmann law with T = 5772 K.

Orbital Properties of the Major and Important Dwarf Planets

Table 3 has orbital and related properties for the major (Ref. 10) and dwarf planets (Ref. 11). Additional property information is available in Refs. 12–15. These properties for the major planets are well established, though given that the planets are dynamic, i.e., bodies that evolve over time with consequent small changes in orbits and rotation. Column definitions for Table 3 are as follows. Note: 1 astronomical unit (au) = 149 597 870.700 km.

Column heading	Definition
Planet	Name of planet; the IAU number for plutoids is given in parentheses; Ceres is designated as minor planet 1
Year of discovery	Year of first observation; note Galileo made the first telescopic observations of Venus, Jupiter, and Saturn in 1609–1610
Dist. to sun	Distance to sun, in au
Sidereal orbit period	Time, in years, it takes the planet to make revolution around the sun relative to the fixed stars; for Pluto, it is the time from the last zero longitude crossing to the next (July 24, 1820 to July 2, 2068); all planet orbits are prograde, that is, in the direction of the sun's rotation
Sidereal rotational period (days)	Time, in days, for one rotation of the planet on its axis relative to the fixed stars; a minus sign indicates retrograde rotation, that is, the rotation is opposite to the orbital motion
Sidereal rotational period (hours)	Sidereal rotational period, in hours
Perihelion	Point in the planet's orbit closest to the sun, in au
Aphelion	Point in the planet's orbit farthest from the sun, in au
Semi-major axis	Mean distance from the sun, in au, from center to center
Orbit eccentricity	Measure of the circularity of the orbit, equal to the difference between the aphelion and perihelion distance divided by twice the semi-major axis; also expresses the shape of the ellipse
Orbit inclination	Inclination, in degrees, of the orbit to the ecliptic; for planets, the ecliptic is the earth's orbit; sometimes the orbit inclination is simply called the *ecliptic*
Inclination of equator	Angle, in degrees, between the equator and orbital plane with north defined as pole axis above (north by the right-hand rule) the plane of the solar system; also known as the axial tilt

TABLE 3. Orbital Properties of the Major and Important Dwarf Planets

Planet	Year of discovery	Dist. to sun/ au	Sidereal orbit period/ y	Sidereal rotational period (days)/ d	Sidereal rotational period (hours)/ hr	Perihelion/ au	Aphelion/ au	Semi-major axis/ au	Orbit eccentricity (ε)	Orbit Inclination/ deg.	Inclination of equator/ deg.
Major Planets											
Mercury	Ancient	0.3871	0.2408467	58.6462	1407.51	0.307720094	0.467025212	0.387369308	0.20563	7.0049°	0.01°
Venus	Ancient	0.723332	0.61519726	-243.018	-5832.6	0.718936673	0.728723084	0.723839912	0.006773	3.3947°	177.36°
Earth		1	1.0000174	0.99726968	23.9345	0.983992562	1.017438944	1.000715753	0.01671	0°	23.45°
Mars	Ancient	1.52366	1.8808476	1.02595676	24.6229	1.382118226	1.667161673	1.52463995	0.09341	1.8506°	25.19°
Jupiter	700-800 BCE	5.20336	11.862615	0.41354	9.925	4.953556354	5.462583532	5.208069943	0.048392	1.3053°	3.12°
Saturn	700 BCE	9.53707	29.447498	0.44401	10.656	9.04761427	10.1309359	9.589271738	0.05415	2.484°	26.73°
Uranus	1781	19.19126	84.016846	-0.71833	-17.24	18.33732683	20.09207789	19.21469902	0.04717	0.7699°	97.86°
Neptune	1846	30.06896	164.79132	0.67125	16.11	29.7301479	30.40724984	30.06869887	0.008586	1.769°	28.32°

Planet	Year of discovery	Dist. to sun/ au	Sidereal orbit period/ y	Sidereal rotational period (days)/ d	Sidereal rotational period (hours)/ hr	Perihelion/ au	Aphelion/ au	Semi-major axis/ au	Orbit eccentricity (ε)	Orbit Inclination/ deg.	Inclination of equator/ deg.
Dwarf Planets											
Pluto (134340)	1930	39.48168	247.74	-6.3872	-153.2928	29.666	48.860	39.263	0.2444	17.14°	57.47°
Eris (136199)	2005	67.7	559.07	1.0792		38.272	97.456	67.864	0.4360	44.0°	
Haumea (136108)	2004	43.13	283.77	0.163139	3.8	34.767	51.598	43.182	0.1949	28.2°	
Makemake (136472)	2005	45.79	306.21	0.951108		38.105	52.756	45.431	0.1613	29.0°	
Ceres (1)	1801	2.77	4.61	0.3780904	9.074	2.5587	2.9796	2.7692	0.0760	10.587°	3°

Physical Properties of the Major and Important Dwarf Planets and the Moon

Table 4 has physical and related properties for the major and important dwarf planets. The moon is included for ease of comparison. These properties, for the most part, are well established for the major planets. For the dwarf planets, the properties are not well established or not measured at the present time. Refs. 10-11 contain further information.

Column definitions for Table 4 are as follows.

Column heading	Definition
Planet	Name of planet; the IAU number for plutoids is given in parentheses; Ceres is designated minor planet 1
Mass	Mass of the planet, in units 10^{24} kg
Equit. radius	Radius of the planet at the equator, in km; * indicates mean radius
Polar radius	Radius of the planet at the poles, in km
Volume	Volume of planet, in units 10^{10} km^3
Density	Mean density of the planet (mass/volume), in g cm^{-3}
Flattening	Ratio of (equatorial radius – polar radius)/(equatorial radius)
Surface gravity	Equatorial gravitational acceleration at the surface of the planet or the 1 bar level, not including the effect of rotation, in units m s^{-2}
Escape velocity	Initial velocity required to escape the planet's gravitational pull, in km s^{-1}
Geometric albedo	Ratio of the planet's brightness at a phase angle of zero to the brightness of a perfectly diffusing disk with the same position and apparent size; earth is highly variable
No. of satellites	Number of detected satellites orbiting the planet

TABLE 4. Physical Properties of the Major and Important Dwarf Planets and the Moon

Planet	Mass/ 10^{24} kg	Equit. radius/ km	Polar radius/ km	Volume/ 10^{10} km^3	Density/ g cm^{-3}	Flattening	Surface gravity/ m s^{-2}	Escape velocity/ km s^{-1}	Geometric albedo	No. of satellites
Mercury	0.33011	2440.53	2439.7	6.083	5.427	0	3.70	4.25	0.142	0
Venus	4.8675	6051.8	6051.8	92.843	5.243	0	8.87	10.36	0.689	0
Earth	5.9724	6378.1366	6356.8	108.321	5.514	0.003353	9.80	11.186	0.434	1
(Moon)	0.073483	1738.1	1736.0	2.1968	3.344	0.0012	1.62	2.38	0.12	
Mars	0.64171	3396.2	3376.2	16.318	3.934	0.00589	3.71	5.03	0.170	2
Jupiter	1898.19	71492	66854	143128	1.326	0.06487	24.79	60.20	0.538	79
Saturn	568.34	60268	54364	82713	0.687	0.09796	10.44	36.09	0.499	82
Uranus	86.813	25559	24973	6833	1.270	0.02293	8.87	21.38	0.51	27
Neptune	102.413	24764	24341	6254	1.638	0.01708	11.15	23.56	0.442	14
Pluto (134340)	0.01303	1188	1188	0.702	1.89	0	0.62	1.21	0.52	5
Eris (136199)	0.01466	1163*		0.659	2.43		0.82	1.38	0.96	1
Haumea (136108)	0.004006	780-798*		0.198	1.75-2.02		0.401	0.809	<0.51	
Makemake (136472)	0.0031	715-739*		0.153	1.7-2.1		<0.57	<0.91	0.81	
Ceres (1)	0.000943	469*		0.434	2.162		27	0.51	0.09	

Other Characteristics of the Major Planets and Pluto

Table 5 contains data on the mean surface temperature and pressure as well as the atmospheric composition for the major planets and Pluto (Ref. 10). Data for Mercury are from Ref. 16.

TABLE 5. Mean Surface Temperature and Pressure and Atmospheric Composition of the Major Planets and Pluto

Planet	T_{sur}/K	P_{sur}/bar	CO_2	N_2	O_2	H_2O	H_2	He	Ar	Ne	CO	CH_4
Mercury	440	$< \sim 5 \times 10^{-15}$	Trace	Trace	$<9 \times 10^{14}$ cm^{-2}*	$< 1 \times 10^{12}$ cm^{-2}*	Trace	$< 3 \times 10^{11}$ cm^{-2}*	$\sim 1.3 \times 10^{9}$ cm^{-2}*	Trace		
Venus	737	92	96.5%	3.5%	69 ppm	20 ppm		12 ppm	70 ppm	7 ppm	17 ppm	
Earth	288	1.014	410 ppm	78.084%	20.946%	0 to 3%	0.55 ppm	5.24 ppm	9340 ppm	18.18 ppm	1 ppm	1.7 ppm
Mars	210	0.004 to 0.0087[a]	95.17%	2.59%	0.16%	210 ppm			1.94%	2.5 ppm	0.06%	
Jupiter	165	>>1000				4 ppm	89.8%	10.2%				3000 ppm
Saturn	134	>>1000					96.3%	3.25%				4500 ppm
Uranus	76	>>1000					82.5%	15.2%				2.3%
Neptune	72						80.0%	19.0%				1.5%
Pluto	31[b]	~0.000013		99.0%							0.05%	0.5%

* These values are for column density, that is, the number of atoms or molecules in a vertical column of 1 cm^2.
[a] The surface temperature on Mars varies with its seasons.
[b] Surface temperature values for Pluto range from 24 to 38 K.

SATELLITES OF THE PLANETS

This table gives characteristics of the known satellites of the planets. The column definitions are as follows.

Column heading	Definition
Planet	Name of planet about which satellites orbit
No.	Satellite number
Satellite	Satellite name
Orb. period	Orbital period, in units of Earth days, an R following the value indicates a retrograde motion
Distance	Distance from the planet, in 10^3 km, as measured by the semimajor axis of the orbit
Eccentricity	Eccentricity of the orbit
Inclination	Inclination of the satellite orbit with respect to the equator of the planet
Mass	Mass of the satellite, in kg
Radius	Radius of the satellite, in km
Albedo	Geometric albedo, which is a measure of the fraction of incident sunlight reflected by the satellite

It should be noted that at least 79 satellites of Jupiter have been discovered, but not all have been verified. Included here are data on the first 49 of Jupiter's satellites. Because this is a very active field of research, the Internet sites listed below should be consulted for the most recent data.

References

1. *Solar System Dynamics*, Jet Propulsion Laboratory, California Institute of Technology, <ssd.jpl.nasa.gov/?phys_data>, June 2008.
2. The Planetary Society, <www.planetary.org/explore/topics/groups/our_solar_system/>.
3. Arnet, B., *The Nine Planets*, <www.nineplanets.org>.
4. Onasch, B., *Our Solar System*, <www.onasch.de/astro/>.
5. Sheppard, S. S., *The Giant Satellite and Moon Page*, <www.dtm.ciw.edu/sheppard/satellites/>.
6. *Gazetteer of Planetary Nomenclature*, U.S. Geological Survey, <planetarynames.wr.usgs.gov/append7.html>.
7. Seidelmann, P. K., Editor, *Explanatory Supplement to the Astronomical Almanac*, University Science Books, Mill Valley, CA, 1992.
8. Lang, K. R., *Astrophysical Data: Planets and Stars*, Springer-Verlag, New York, 1992. <https://doi.org/10.1007/978-1-4684-0640-5_3>
9. Allen, C. W., *Astrophysical Quantities, Second Edition*, Athlone Press, London, 1955.

Satellites of the Planets

Planet	No.	Satellite	Orb. period/ Earth days	Distance/ 10^3 km	Eccentricity	Inclination	Mass/kg	Radius/km	Albedo
Earth		Moon	27.321661	384.400	0.054900489	18.28–28.58°	$7.3483 \cdot 10^{22}$	1737.5	0.12
Mars	I	Phobos	0.31891023	9.378	0.0151	1.0°	$1.06 \cdot 10^{16}$	13.5×10.8×9.4	0.07
	II	Deimos	1.2624407	23.460	0.0005	0.9–2.7°	$2.4 \cdot 10^{15}$	7.5×6.1×5.5	0.07
Jupiter	I	Io	1.769137786	421.8	0.0041	0.04°	$8.932 \cdot 10^{22}$	1821.6	0.63
	II	Europa	3.551181041	671.1	0.0101	0.47°	$4.8 \cdot 10^{22}$	1560.8	0.67
	III	Ganymede	7.15455296	1070.4	0.0015	0.21°	$1.4819 \cdot 10^{23}$	2631.2	0.43
	IV	Callisto	16.6890184	1882.7	0.007	0.51°	$1.0759 \cdot 10^{23}$	2410.3	0.17
	V	Amalthea	0.49817905	181.4	0.003	0.40°	$7.17 \cdot 10^{18}$	131×73×67	0.09
	VI	Himalia	250.5662	11460	0.162	27.63°	$9.56 \cdot 10^{18}$	85	0.04
	VII	Elara	259.6528	11737	0.217	24.77°	$7.77 \cdot 10^{17}$	40	0.04
	VIII	Pasiphae	743.63 R	23620	0.409	145°	$1.91 \cdot 10^{17}$	18	0.04
	IX	Sinope	758.90 R	23940	0.250	153°	$7.77 \cdot 10^{16}$	14	0.04
	X	Lysithea	259.20	11720	0.112	29.02°	$7.77 \cdot 10^{16}$	12	0.04
	XI	Carme	734.17 R	23400	0.253	164°	$9.56 \cdot 10^{16}$	15	0.04
	XII	Ananke	629.77 R	21280	0.244	147°	$3.82 \cdot 10^{16}$	10	0.04
	XIII	Leda	240.92	11170	0.164	26.07°	$5.68 \cdot 10^{15}$	5	0.04
	XIV	Thebe	0.6745	221.9	0.018	0.8°	$7.77 \cdot 10^{17}$	55×45	0.05
	XV	Adrastea	0.29826	129	0.0015		$1.91 \cdot 10^{16}$	13×10×8	0.10
	XVI	Metis	0.29478	128	0.0002		$9.56 \cdot 10^{16}$	20	0.06
	XVII	Callirrhoe	758.77	24100	0.283		$8.7 \cdot 10^{14}$	4	0.04
	XVIII	Themisto	130.02	7507	0.242		$6.9 \cdot 10^{14}$	4	0.04
	XIX	Megaclite	752.86	23810	0.425		$2.1 \cdot 10^{14}$	2.7	0.04
	XX	Taygete	732.41	23360	0.251		$1.6 \cdot 10^{14}$	2.5	0.04
	XXI	Chaldene	723.72	23180	0.238		$7.5 \cdot 10^{13}$	1.9	0.04
	XXII	Harpalyke	623.32	21110	0.227		$1.2 \cdot 10^{14}$	2.2	0.04
	XXIII	Kalyke	742.06	23580	0.243		$1.9 \cdot 10^{14}$	2.6	0.04
	XXIV	Iocaste	631.60	21270	0.218		$1.9 \cdot 10^{14}$	2.6	0.04
	XXV	Erinome	728.46	23280	0.270		$4.5 \cdot 10^{13}$	1.6	0.04
	XXVI	Isonoe	726.63	23220	0.261		$7.5 \cdot 10^{13}$	1.9	0.04
	XVII	Praxidike	625.39	21150	0.220		$4.3 \cdot 10^{14}$	3.4	0.04
	XXVIII	Autonoe	760.95	23039	0.334		$9.0 \cdot 10^{13}$	2.0	0.04
	XXIX	Thyone	627.21	20940	0.229		$9.0 \cdot 10^{13}$	2.0	0.04
	XXX	Hermippe	633.90	21131	0.210		$9.0 \cdot 10^{13}$	2.0	0.04
	XXXI	Aitne	730.18	23231	0.264		$4.5 \cdot 10^{13}$	1.5	0.04

Planet	No.	Satellite	Orb. period/ Earth days	Distance/ 10^3 km	Eccentricity	Inclination	Mass/kg	Radius/km	Albedo
	XXXII	Eurydome	717.33	22685	0.276		$4.5 \cdot 10^{13}$	1.5	0.04
	XXXIII	Euanthe	620.49	20721	0.232		$4.5 \cdot 10^{13}$	1.5	0.04
	XXXIV	Euporie	550.74	19302	0.144		$1.5 \cdot 10^{13}$	1	0.04
	XXXV	Orthosie	622.56	20721	0.281		$1.5 \cdot 10^{13}$	1	0.04
	XXXVI	Sponde	748.34	23487	0.312		$1.5 \cdot 10^{13}$	1	0.04
	XXXVII	Kale	729.47	23217	0.260		$1.5 \cdot 10^{13}$	1	0.04
	XXXVIII	Pasithee	719.44	23096	0.267		$1.5 \cdot 10^{13}$	1	0.04
	XXXIX	Hegemone	739.6	23947	0.328		$4.5 \cdot 10^{13}$	1.5	
	XL	Mneme	620.0	21069	0.227		$1.5 \cdot 10^{13}$	1	
	XLI	Aoede	761.5	23981	0.432		$9.0 \cdot 10^{13}$	2.0	
	XLII	Thelxinoe	628.1	21162	0.221		$1.5 \cdot 10^{13}$	1	
	XLIII	Arche	723.9	22931	0.259		$4.5 \cdot 10^{13}$	1.5	
	XLIV	Kallichore	764.7	24043	0.264		$1.5 \cdot 10^{13}$	1	
	XLV	Helike	634.8	21263	0.156		$9.0 \cdot 10^{13}$	2.0	
	XLVI	Carpo	456.1	16989	0.430		$4.5 \cdot 10^{13}$	1.5	
	XLVII	Eukelade	746.4	23661	0.272		$9.0 \cdot 10^{13}$	2.0	
	XLVIII	Cyllene	737.8	24349	0.319		$1.5 \cdot 10^{13}$	1	
	XLIX	Kore	779.2	24543	0.325			1	
Saturn	I	Mimas	0.942421813	185.52	0.0202	1.53°	$3.75 \cdot 10^{19}$	196	0.5
	II	Enceladus	1.370217855	238.02	0.00452	1.86°	$6.50 \cdot 10^{19}$	250	1.0
	III	Tethys	1.887802160	294.66	0.00000	1.86°	$6.27 \cdot 10^{20}$	530	0.9
	IV	Dione	2.736914742	377.40	0.002230	0.02°	$1.10 \cdot 10^{21}$	560	0.7
	V	Rhea	4.517500436	527.04	0.00100	0.35°	$2.31 \cdot 10^{21}$	765	0.7
	VI	Titan	15.94542068	1221.83	0.029192	0.33°	$1.3455 \cdot 10^{23}$	2575	0.21
	VII	Hyperion	21.2766088	1481.1	0.104	0.43°	$1.59 \cdot 10^{19}$	205×130×110	0.3
	VIII	Iapetus	79.3301825	3561.3	0.02828	14.72°	$1.59 \cdot 10^{21}$	730	0.6
	IX	Phoebe	550.31 R	12952	0.16326	177°	$7.2 \cdot 10^{18}$	110	0.08
	X	Janus	0.6945	151.472	0.007	0.14°	$1.92 \cdot 10^{18}$	110×100×80	0.6
	XI	Epimetheus	0.6942	151.422	0.009	0.34°	$5.4 \cdot 10^{17}$	70×60×50	0.5
	XII	Helene	2.7369	377.40	0.005	0.0°	$2.5 \cdot 10^{16}$	18×16×15	0.6
	XIII	Telesto	1.8878	294.66			$7.2 \cdot 10^{15}$	17×14×13	1.0
	XIV	Calypso	1.8878	294.66			$3.6 \cdot 10^{15}$	17×11×11	0.7
	XV	Atlas	0.6019	137.670		0.3°	$1.1 \cdot 10^{16}$	20×10	0.4
	XVI	Prometheus	0.6130	139.353	0.0024	0.0°	$3.3 \cdot 10^{17}$	70×50×40	0.6
	XVII	Pandora	0.6285	141.70	0.0042	0.0°	$1.9 \cdot 10^{17}$	55×45×35	0.5
	XVIII	Pan	0.5750	133.583			$2.7 \cdot 10^{15}$	10	0.5
	XIX	Ymir	1315.14	23096	0.470		$4.9 \cdot 10^{15}$	8	0.06
	XX	Paaliaq	686.95	15199	0.364		$8.2 \cdot 10^{15}$	9.5	0.06
	XXI	Tarvos	926.23	18247	0.536		$2.7 \cdot 10^{15}$	6.5	0.06
	XXII	Ijiraq	451.42	11440	0.322		$1.2 \cdot 10^{15}$	5	0.06
	XXIII	Suttungr	1016.67	19463	0.114		$2.1 \cdot 10^{14}$	2.8	0.06
	XXIV	Kiviuq	449.22	11365	0.334		$3.3 \cdot 10^{15}$	7	0.06
	XXV	Mundilfari	952.77	18709	0.208		$2.1 \cdot 10^{14}$	2.8	0.06
	XXVI	Albiorix	783.45	16404	0.478		$2.1 \cdot 10^{16}$	13	0.06
	XXVII	Skathi	728.20	15647	0.270		$3.1 \cdot 10^{14}$	3.2	0.06
	XXVIII	Erriapus	871.19	17616	0.474		$7.6 \cdot 10^{14}$	4.3	0.06
	XXIX	Siarnaq	895.53	18160	0.295		$3.9 \cdot 10^{16}$	16	0.06
	XXX	Thrymr	1094.11	20382	0.470		$2.1 \cdot 10^{14}$	2.8	0.06
	XXXI	Narvi	1003.86	19007	0.431		$4.9 \cdot 10^{15}$	3.3	0.04
	XXXII	Methone	1.010	194			$1.65 \cdot 10^{13}$	1.5	
	XXXIII	Pallene	1.154	211			$3.92 \cdot 10^{13}$	2	
	XXXIV	Polydeuces	2.737	377.4				4	
	XXXV	Daphnis	0.594	136.5				3.5	
	XXXVI	Aegir	1117.52	20735				3.5	
	XXXVII	Bebhionn	834.84	17119				3	
	XXXVIII	Bergelmir	1005.74	19338				3	
	XXXIX	Bestla	1088.72	20129				3.5	
	XL	Farbauti	1085.55	20390				2.5	
	XLI	Fenrir	1260.35	22453				2	
	XLII	Fornjot	1494.20	25108				3	

Planet	No.	Satellite	Orb. period/ Earth days	Distance/ 10^3 km	Eccentricity	Inclination	Mass/kg	Radius/km	Albedo
	XLIII	Hati	1038.61	19856				3	
	XLIV	Hyrrokkin	931.86	18437				4	
	XLV	Kari	1230.97	22118				3.5	
	XLVI	Loge	1311.36	23065				3	
	XLVII	Skoll	878.29	17665				3	
	XLVIII	Surtur	1297.36	22707				3	
	XLIX	Anthe		197.7				1	
	L	Jarnsaxa		18600				3	
	LI	Greip		18105				3	
	LII	Tarqeq		19720				3.5	
Uranus	I	Ariel	2.52037935	191.02	0.0034	0.3°	$1.35 \cdot 10^{21}$	579	0.39
	II	Umbriel	4.1441772	266.30	0.0050	0.36°	$1.17 \cdot 10^{21}$	584.7	0.21
	III	Titania	8.7058717	435.91	0.0022	0.14°	$3.52 \cdot 10^{21}$	788.9	0.27
	IV	Oberon	13.4632389	583.52	0.0008	0.10°	$3.01 \cdot 10^{21}$	761.4	0.23
	V	Miranda	1.41347925	129.39	0.0027	4.2°	$6.59 \cdot 10^{19}$	236	0.32
	VI	Cordelia	0.335033	49.77	0.0003	0.1°	$5.4 \cdot 10^{16}$	20.1	0.07
	VII	Ophelia	0.376409	53.79	0.0099	0.1°	$5.4 \cdot 10^{16}$	21.4	0.07
	VIII	Bianca	0.434577	59.17	0.0009	0.2°	$9.3 \cdot 10^{16}$	25.7	0.07
	IX	Cressida	0.463570	61.78	0.0004	0.0°	$3.4 \cdot 10^{17}$	39.8	0.07
	X	Desdemona	0.473651	62.68	0.0001	0.2°	$1.8 \cdot 10^{17}$	32.0	0.07
	XI	Juliet	0.493066	64.35	0.0007	0.1°	$5.6 \cdot 10^{17}$	46.8	0.07
	XII	Portia	0.513196	66.09	0.0001	0.1°	$1.7 \cdot 10^{18}$	67.6	0.07
	XIII	Rosalind	0.558459	69.94	0.0001	0.3°	$2.6 \cdot 10^{17}$	36.0	0.07
	XIV	Belinda	0.623525	75.26	0.0001	0.0°	$3.6 \cdot 10^{17}$	40.3	0.07
	XV	Puck	0.761832	86.01	0.0001	0.31°	$2.9 \cdot 10^{18}$	81.0	0.07
	XVI	Caliban	579.73	7231	0.1587		$7.4 \cdot 10^{17}$	49	0.07
	XVII	Sycorax	1288.30	12179	0.5224		$5.4 \cdot 10^{18}$	95	0.07
	XVIII	Prospero	1978.29	16256	0.4448		$2.1 \cdot 10^{16}$	15	0.07
	XIX	Setebos	2225.21	17418	0.5914		$2.1 \cdot 10^{16}$	15	0.07
	XX	Stephano	677.36	8004	0.2292		$6.0 \cdot 10^{15}$	10	0.07
	XXI	Trinculo	749.24	8504	0.2200		$7.5 \cdot 10^{14}$	5	0.04
	XXII	Francisco	266.56	4276	0.146		$1.3 \cdot 10^{15}$	11	
	XXIII	Margaret	1687.01	14345	0.661		$1.0 \cdot 10^{15}$	5.5	
	XXIV	Ferdinand	2887.21	20901	0.368		$1.3 \cdot 10^{15}$	6	
	XXV	Perdita	0.638	76.42	0.0		$4.0 \cdot 10^{17}$	40	
	XXVI	Mab	0.923	97.73	0.0		$4.0 \cdot 10^{15}$	8	
	XXVII	Cupid	0.613	74.8	0.0		$1.2 \cdot 10^{15}$	6	
Neptune	I	Triton	5.8768541 R	354.76	0.000016	157.345°	$2.147 \cdot 10^{22}$	1353.4	0.76
	II	Nereid	360.13619	5513.4	0.7512	27.6°	$3.1 \cdot 10^{19}$	170	0.15
	III	Naiad	0.294396	48.227	0.0003	4.74°	$1.3 \cdot 10^{17}$	33	0.07
	IV	Thalassa	0.311485	50.075	0.0002	0.21°	$3.5 \cdot 10^{17}$	41	0.09
	V	Despina	0.334655	52.526	0.0001	0.07°	$2.3 \cdot 10^{18}$	75	0.09
	VI	Galatea	0.428745	61.953	0.0001	0.05°	$2.7 \cdot 10^{18}$	88	0.08
	VII	Larissa	0.554654	73.548	0.0014	0.20°	$4.8 \cdot 10^{18}$	104×89	0.09
	VIII	Proteus	1.122315	117.647	0.0004	0.55°	$4.9 \cdot 10^{19}$	218×208×201	0.10
	IX	Halimede	1879.08	16611	0.2646				
	X	Psamathe	9074.30	48096	0.3809		$1.5 \cdot 10^{16}$	14	
	XI	Sao	2912.72	22228	0.1365				
	XII	Laomedeia	3171.33	23567	0.3969				
	XIII	Neso	9740.73	49285	0.5714				
Pluto*	I	Charon	6.387	17.536	0.0022	99°	$1.6 \cdot 10^{21}$	593	0.37
	II	Nix	24.86	48.708	0.0030		$5 \cdot 10^{16}$	22–65	
	III	Hydra	38.20	64.749	0.0051		$5 \cdot 10^{16}$	22–65	
Eris*	I	Dysnomia		30				100–200	

* In June 2008 the International Astronomical Union decided on the name *plutoid* for the category of transneptunian dwarf planets. Plutoids are celestial bodies in orbit around the sun at a semimajor axis greater than that of Neptune and sufficiently massive to adopt a near-spherical shape. See <www.iau.org/public_press/news/release/iau0804/>.

INTERSTELLAR MOLECULES

Frank J. Lovas and Holger S. P. Müller

A number of molecules have been detected in the interstellar medium, in circumstellar envelopes around evolved stars, and comae and tails of comets through observation of their microwave, infrared, or optical spectra. The following list gives the molecules and the particular isotopic species that have been reported so far. Molecules are listed by molecular formula in the Hill order.* All species not footnoted otherwise are observed in interstellar clouds, while some are also found in comets and circumstellar clouds. The list was last updated through November 2020 and lists 241 molecules (473 isotopic forms). Information on structural bond lengths, angles, and configuration for many of the molecules listed here can be found in "Structure of Free Molecules in the Gas Phase" in Section 9. See Ref. 1 for transition frequencies reported for a typical molecular cloud and Ref. 2 for a discussion of comet observations. Laboratory measurements and predicted frequencies may be found in Refs. 3 and 4.

* The molecular formula gives the number of atoms of each element present in the molecule, without regard to the way in which the atoms are bonded. The Hill order specifies that C (if present) comes first; H (if present) is second; and the other element symbols follow in alphabetical order.

References

1. F. J. Lovas, Recommended Rest Frequencies for Observed Interstellar Molecule Microwave Transitions—2002 Revision, *J. Phys. Chem. Ref. Data* 33, 177-355 (2004); and last update appearing at <www.nist.gov/pml/data/micro/index.cfm>. <https://doi.org/10.1063/1.1633275>
2. Anita L. Cochran, Anny-Chantal Levasseur-Regourd, Martin Cordiner, Edith Hadamcik, Jérémie Lasue, Adeline Gicquel, David G. Schleicher, Steven B. Charnley, Michael J. Mumma, Lucas Paganini, Dominique Bockelée-Morvan, Nicolas Biver, and Yi-Jehng Kuan, The Composition of Comets, *Space Science Reviews*, pp. 1-38, First online: July 2015, <link.springer.com/article/10.1007%2Fs11214-015-0183-6>.
3. Splatalogue, <www.splatalogue.net>.
4. Cologne Database for Molecular Spectroscopy, <https://cdms.astro.uni-koeln.de/classic>.

Interstellar Molecules Observed to Date (November 2020)

Mol. form.	Name	Isotopic species*
AlCl	Aluminum monochloride	$Al^{35}Cl$[a]
		$Al^{37}Cl$[a]
AlF	Aluminum monofluoride	AlF[a]
		^{26}AlF[a]
AlHO	Aluminum hydroxide	AlOH[a]
AlO	Aluminum monoxide	AlO[a]
ArH^+	Argonium	$^{36}ArH^+$
		$^{38}ArH^+$
CAlN	Aluminum isocyanide	AlNC[a]
CCaN	Calcium isocyanide	CaNC[a]
CF^+	Fluoromethyliumylidene	CF^+
CFeN	Iron cyanide	FeCN
CH	Methylidyne	CH
	Methylidyne	^{13}CH
CH^+	Methyliumylidene	CH^+
		$^{13}CH^+$
CHMgN	Magnesium hydride isocyanide	HMgNC[a]
CHN	Hydrogen cyanide	HCN
		$H^{13}CN$
		$HC^{15}N$
		$H^{13}C^{15}N$
		DCN
CHN	Hydrogen isocyanide	HNC
		$H^{15}NC$
		$HN^{13}C$
		DNC
CHNO	Cyanic acid	HOCN
CHNO	Isocyanic acid	HNCO
		$HN^{13}CO$
		DNCO
CHNO	Fulminic acid	HCNO
CHNS	Thiocyanic acid	HSCN
CHNS	Isothiocyanic acid	HNCS
CHO	Formyl	HCO
CHO^+	Oxomethylium	HCO^+
		$H^{13}CO^+$
		$HC^{17}O^+$
		$HC^{18}O^+$
		DCO^+
		$D^{13}CO^+$
CHO^+	Hydroxymethyliumylidene	HOC^+
CHO_2^+	Hydroxyoxomethylium	$HOCO^+$
		$DOCO^+$
CHP	Phosphaethyne	HCP
CHS	Methanethione	HCS
CHS	Mercaptomethylidyne	HSC
CHS^+	Thiooxomethylium	HCS^+
		DCS^+
CH_2	Methylene	CH_2
CH_2N^+	Iminomethylium	$HCNH^+$
CH_2N	Methylene amidogen	CH_2N
CH_2N_2	Cyanamide	NH_2CN
		$NH_2{}^{13}CN$
		NHDCN
CH_2N_2	Methane diimine	HNCNH
CH_2NO^+	Aminooxomethylium	H_2NCO^+
CH_2O	Formaldehyde	H_2CO
		$H_2{}^{13}CO$
		$H_2C^{17}O$
		$H_2C^{18}O$
		HDCO
		$HDC^{18}O$
		D_2CO
		$D_2{}^{13}CO$

Geo-Astro

Mol. form.	Name	Isotopic species*
CH_2O_2	*trans*-Formic acid	*t*-HCOOH
		t-$H^{13}COOH$
		t-HCOOD
		t-DCOOH
CH_2O_2	*cis*-Formic acid	*c*-HCOOH
CH_2S	Thioformaldehyde	H_2CS
		$H_2{}^{13}CS$
		$H_2C^{34}S$
		HDCS
		D_2CS
CH_3	Methyl	CH_3
CH_3Cl	Chloromethane	$CH_3{}^{35}Cl$
		$CH_3{}^{37}Cl$
CH_3N	Methanimine	CH_2NH
		$^{13}CH_2NH$
CH_3NO	Formamide	NH_2CHO
		$NH_2{}^{13}CHO$
		$NH_2CH^{18}O$
		$^{15}NH_2CHO$
		NH_2CDO
		s-NHDCHO
		a-NHDCHO
CH_3NSi	Silyl cyanide	SiH_3CN
CH_3O	Methoxy	CH_3O
CH_3O^+	Hydroxymethylium	H_2COH^+
CH_4	Methane	CH_4
CH_4N_2O	Urea	$(NH_2)_2CO$
CH_4O	Methanol	CH_3OH
		$^{13}CH_3OH$
		$CH_3{}^{18}OH$
		CH_2DOH
		CH_3OD
		CHD_2OH
		CD_3OH
CH_4S	Methanethiol	CH_3SH
CH_5N	Methylamine	CH_3NH_2
CH_6Si	Methylsilane	CH_3SiH_3
CKN	Potassium cyanide	KCN
CMgN	Magnesium cyanide	MgCN[a]
CMgN	Magnesium isocyanide	MgNC[a]
		$^{25}MgNC$[a]
		$^{26}MgNC$[a]
CN	Cyanide radical	CN
		^{13}CN
		$C^{15}N$
CN^+	Cyanide cation	CN^+
CN^-	Cyanide anion	CN^-[a]
CNNa	Sodium cyanide	NaCN[a]
CNO	Cyanato	NCO
CNSi	Silicon cyanide	SiCN[a]
CNSi	Silicon isocyanide	SiNC[a]
CN_2	Cyanoimidogen	NCN[b]
CO	Carbon monoxide	CO
		^{13}CO
		$C^{17}O$
		$C^{18}O$
		$^{13}C^{17}O$
		$^{13}C^{18}O$
CO^+	Carbon monoxide ion	CO^+
COS	Carbon oxysulfide	OCS
		$OC^{33}S$
		$OC^{34}S$
		$O^{13}CS$
		$O^{13}C^{34}S$
		^{18}OCS
CO_2	Carbon dioxide	CO_2
CO_2^+	Carbon dioxide ion	CO_2^+[b]
CP	Carbon phosphide	CP[a]
CS	Carbon monosulfide	CS
		$C^{33}S$
		$C^{34}S$
		$C^{36}S$
		^{13}CS
		$^{13}C^{34}S$
CS_2	Carbon disulfide	CS_2
CSi	Silicon carbide	SiC
CSi_2	Disilylidynemethylene	
C_2	Dicarbon	C_2
C_2H	Ethynyl	C_2H
		^{13}CCH
		$C^{13}CH$
		C_2D
C_2HMg	Magnesium acetylide	MgCCH[a]
C_2HN	Cyanomethylene	HCCN
C_2HNO	Cyanoformaldehyde	CNCHO
$C_2HN_2^+$	Protonated cyanogen	$NCCNH^+$
C_2HO	Ketenyl radical	HCCO
C_2H_2	Acetylene	HCCH[a]
		$HC^{13}CH$[a]
C_2H_2N	Cyanomethyl	CH_2CN
$C_2H_2N_2$	*E*-Cyanomethanimine	*E*-HNCHCN
	Z-Cyanomethanimine	*Z*-HNCHCN
C_2H_2O	Ketene	H_2CCO
		$H_2{}^{13}CCO$
		$H_2C^{13}CO$
		HDCCO
C_2H_3N	Acetonitrile	CH_3CN
		$^{13}CH_3CN$
		$CH_3{}^{13}CN$
		$CH_3C^{15}N$
		$^{13}CH_3{}^{13}CN$
		CH_2DCN
		CHD_2CN
C_2H_3N	Isocyanomethane	CH_3NC
C_2H_3N	Ketenimine	CH_2CNH
C_2H_3NO	Methyl isocyanate	CH_3NCO
C_2H_3NO	2-Hydroxyacetonitrile	$HOCH_2CN$
C_2H_4	Ethylene	H_2CCH_2
$C_2H_4N_2$	Aminoacetonitrile	NH_2CH_2CN
C_2H_4O	Acetaldehyde	CH_3CHO
		$^{13}CH_3CHO$
		$CH_3{}^{13}CHO$
		CH_3CDO
		CH_2DCHO
C_2H_4O	Ethylene oxide	c-C_2H_4O
C_2H_4O	*anti*-Ethenol	*a*-CH_2CHOH
C_2H_4O	*syn*-Ethenol	*s*-CH_2CHOH
$C_2H_4O_2$	Methyl formate	CH_3OCHO
		$^{13}CH_3OCHO$
		$CH_3O^{13}CHO$

Mol. form.	Name	Isotopic species*
		CH_2DOCHO
		CHD_2OCHO
		CH_3OCDO
		$CH_3{}^{18}OCHO$
		$CH_3OCH^{18}O$
$C_2H_4O_2$	Acetic acid	CH_3COOH
$C_2H_4O_2$	Glycolaldehyde	CH_2OHCHO
		$^{13}CH_2OHCHO$
		$CH_2OH^{13}CHO$
		$CHDOHCHO$
		CH_2ODCHO
		CH_2OHCDO
C_2H_5N	Ethanimine	CH_3CHNH
C_2H_5NO	Acetamide	CH_3CONH_2
C_2H_5NO	*N*-Methyl formamide	CH_3NHCHO
C_2H_6	Ethane	CH_3CH_3[b]
C_2H_6O	*trans*-Ethanol	*t*-CH_3CH_2OH
		t-$^{13}CH_3CH_2OH$
		t-$CH_3{}^{13}CH_2OH$
		t-a-CH_2DCH_2OH
		t-s-CH_2DCH_2OH
		t-CH_3CHDOH
C_2H_6O	*gauche*-Ethanol	*g*-CH_3CH_2OH
C_2H_6O	Dimethyl ether	CH_3OCH_3
		$CH_3O^{13}CH_3$
		anti-CH_2DOCH_3
		sym-CH_2DOCH_3
$C_2H_6O_2$	*aGg'*-Ethylene glycol	*aGg'*-$HOCH_2CH_2OH$
$C_2H_6O_2$	*gGg'*-Ethylene glycol	*gGg'*-$HOCH_2CH_2OH$
$C_2H_6O_2$	Methoxymethanol	CH_3OCH_2OH
C_2N	Cyanomethylidyne	CCN[a]
C_2NP	Cyanophosphaethyne	$NCCP$[a]
C_2N_2	Cyanogen cyanide	$CNCN$
C_2O	Oxoethenylidene	CCO
C_2P	Phosphaethenylidene	CCP[a]
C_2S	Thioxoethenylidene	CCS
		$CC^{34}S$
		^{13}CCS
		$C^{13}CS$
C_2Si	Silacyclopropynylidene	c-SiC_2
		c-$^{29}SiC_2$
		c-$^{30}SiC_2$
		c-$Si^{13}CC$
C_3	Propadienediylidene	C_3
		^{13}CCC
		$C^{13}CC$
C_3H	Cyclopropenylidyne	c-C_3H
		c-$CC^{13}CH$
C_3H	Propenylidyne	l-C_3H
C_3H		l-C_3D
C_3H^+	2-Propyn-1-ylium-1-ylidene	l-C_3H^+
C_3HN	Cyanoacetylene	$HCCCN$
		$H^{13}CCCN$
		$HC^{13}CCN$
		$HCC^{13}CN$
		$HCCC^{15}N$
		$H^{13}C^{13}CCN$
		$H^{13}CC^{13}CN$
		$HC^{13}C^{13}CN$
		$DCCCN$

Mol. form.	Name	Isotopic species*
C_3HN	Isocyanoacetylene	$HCCNC$
		$H^{13}CCNC$
		$HC^{13}CNC$
		$HCCN^{13}C$
		$DCCNC$
C_3HO	Iminopropadienylidene	$HNCCC$
		$DNCCC$
C_3HO^+	Ethynyloxomethylium	$HCCCO^+$
C_3H_2	Cyclopropenylidene	c-C_3H_2
		c-$H^{13}CCCH$
		c-$HC^{13}CCH$
		c-$HCC^{13}CD$
		c-$H^{13}CCCD$
		c-C_3HD
		c-C_3D_2
C_3H_2	Propadienylidene	l-H_2CCC
		l-$HDCCC$
$C_3H_2N^+$	Protonated cyanoacetylene	$HCCCNH^+$
C_3H_2O	2-Propynal	$HCCCHO$
C_3H_2O	Cyclopropenone	c-C_3H_2O
C_3H_3N	Acrylonitrile (vinyl cyanide)	CH_2CHCN
		$^{13}CH_2CHCN$
		$CH_2{}^{13}CHCN$
		$CH_2CH^{13}CN$
C_3H_3N	Propargylimine	$HCCC(H)NH$
C_3H_4	Propyne	CH_3CCH
		$CH_3C^{13}CH$
		$^{13}CH_3CCH$
		$CH_3{}^{13}CCH$
		CH_2DCCH
		CH_3CCD
C_3H_4O	Propenal	CH_2CHCHO
C_3H_5N	Propanenitrile (ethyl cyanide)	CH_3CH_2CN
		$^{13}CH_3CH_2CN$
		$CH_3{}^{13}CH_2CN$
		$CH_3CH_2{}^{13}CN$
		$^{13}CH_3{}^{13}CH_2CN$
		$^{13}CH_3CH_2{}^{13}CN$
		$CH_3{}^{13}CH_2{}^{13}CN$
		$CH_3CH_2C^{15}N$
		CH_2DCH_2CN
		CH_3CHDCN
C_3H_6	Propylene	CH_2CHCH_3
C_3H_6O	Acetone	$(CH_3)_2CO$
C_3H_6O	Propanal	CH_3CH_2CHO
C_3H_6O	Methyloxirane (propylene oxide)	$CH_3CH\_CH_2O\_$
$C_3H_6O_2$	Methyl acetate	CH_3COOCH_3
$C_3H_6O_2$	*trans*-Ethyl formate	*t*-CH_3CH_2OCHO
	gauche-Ethyl formate	*g*-CH_3CH_2OCHO
$C_3H_6O_2$	Hydroxyacetone	$CH_3C(O)CH_2OH$
C_3H_8O	*trans*-Methoxyethane	*t*-$C_2H_5OCH_3$
C_3MgN	Cyanoethynyl magnesium	$MgCCCN$
C_3N	Cyanoethynyl	$CCCN$
		$^{13}CCCN$
		$C^{13}CCN$
		$CC^{13}CN$
C_3N^-	Cyanoethynyl anion	$CCCN^-$
C_3O	3-Oxo-1,2-propadienylidene	$CCCO$
C_3S	3-Thioxo-1,2-propadienylidene	$CCCS$
		$CCC^{34}S$

Mol. form.	Name	Isotopic species*
		$C^{13}CCS$
C_3Si	Silicon tricarbon	c-SiC_3[a]
C_4H	1,3-Butadiynyl radical	HCCCC
		$H^{13}CCCC$
		$HC^{13}CCC$
		$HCC^{13}CC$
		$HCCC^{13}C$
		DCCCC
C_4H^-	1,3-Butadiynyl anion	$HCCCC^-$
C_4HMg	1,3-Butadiyn-1-yl magnesium	MgC_4H
C_4HN	3-Cyano-1,2-propadienylidene	HCCCCN
C_4H_2	Butatrienylidene	H_2CCCC
		HDCCCC
C_4H_2	1,3-Butadiyne (diacetylene)	HCCCCH[a]
C_4H_3N	2-Butynenitrile	CH_3CCCN
C_4H_3N	Cyanoallene	CH_2CCHCN
C_4H_3N	Propargyl cyanide	$HCCCH_2CN$
C_4H_7N	Isopropylcyanide	*i*-$CH_3CH(CN)CH_3$
C_4H_7N	Propyl cyanide	C_3H_7CN
C_4Si	Silicon tetracarbide	SiC_4[a]
C_5	Pentatetraenediylidene	C_5[a]
C_5H	2,4-Pentadiynylidyne	HCCCCC
C_5HN	2,4-Pentadiynenitrile	HCCCCCN
		$H^{13}CCCCCN$
		$HC^{13}CCCCN$
		$HCC^{13}CCCN$
		$HCCC^{13}CCN$
		$HCCCC^{13}CN$
		DCCCCCN
		$HCCCCC^{15}N$
C_5HN	Isocyanodiacetylene	HCCCCNC
C_5HO	1-Oxo-pentadiynyl	HCCCCCO
$C_5H_2N^+$	Butadiynyliminomethylium	$HCCCCCNH^+$
C_5H_4	1,3-Pentadiyne	CH_3C_4H
C_5N	4-Cyano-1,3-butadiyn-1-yl	C_5N[a]
C_5N^-	2,4-Pentadiynenitrile anion	C_5N^-
C_5S	5-Thioxo-1,2,3,4-pentatetraen-1-ylidene	CCCCCS[a]
C_6H	1,3,5-Hexatriynyl	HCCCCCC
C_6H^-	1,3,5-Hexatriynyl anion	$HCCCCCC^-$
C_6H_2	1,3,5-Hexatriyne	HCCCCCCH
C_6H_2	1,2,3,4,5-hexapentaenylidene	$H_2CCCCCC$
C_6H_3N	Methylcyanodiacetylene	CH_3C_4CN
C_6H_5N	1-Cyano-1,3-cyclopentadiene	c-C_5H_5CN
C_6H_6	Benzene	C_6H_6
C_7H	2,4,6-Heptatriynylidyne	HCCCCCCC
C_7HN	2,4,6-Heptatriynenitrile	HC_7N
		$HC_6{}^{13}CN$
		$HC_5{}^{13}CCN$
		$HC_4{}^{13}CC_2N$
		$HC_3{}^{13}CC_3N$
		$HC_2{}^{13}CC_4N$
		$HC^{13}CC_5N$
		DC_7N
C_7HO	1-Oxo-hepta-2,4,6-triynyl radical	HC_7O
C_7H_4	Methyltriacetylene	CH_3C_6H
C_7H_5N	Benzonitrile	C_6H_5CN
C_8H	1,3,5,7-Octatetraynyl	HC_8
C_8H^-	1,3,5,7-Octatetraynyl anion	HC_8^-
C_9HN	2,4,6,8-Nonatetraynenitrile	HC_9N

Mol. form.	Name	Isotopic species*
$C_{11}HN$	Unidecimpentaynenitrile	$HC_{11}N$
C_{60}	Fullerene-C_{60}	C_{60}[a]
C_{60}^+	Fullerene-C_{60} cation	C_{60}^+
C_{70}	Fullerene-C_{70}	C_{70}[a]
ClH	Hydrogen chloride	$H^{35}Cl$
		$H^{37}Cl$
ClH^+	Hydrogen chloride ion	$H^{35}Cl^+$
		$H^{37}Cl^+$
ClH_2^+	Chloronium	$H_2{}^{35}Cl^+$
		$H_2{}^{37}Cl^+$
ClK	Potassium chloride	$^{39}K^{35}Cl$[a]
		$^{41}K^{35}Cl$[a]
		$^{39}K^{37}Cl$
ClNa	Sodium chloride	$Na^{35}Cl$[a]
		$Na^{37}Cl$[a]
CrO	Chromium monoxide	CrO[a]
FH	Hydrogen fluoride	HF
FeO	Iron monoxide	FeO
HHe^+	Helium hydride	HeH^+
HN	Imidogen	HN
		ND
HNO	Nitrosyl hydride	HNO
HNO_2	Nitrous acid	HONO
HN_2^+	Hydrodinotrogen(1+)	N_2H^+
		$^{15}NNH^+$
		$N^{15}NH^+$
		N_2D^+
HO	Hydroxyl	OH
		^{17}OH
		^{18}OH
		OD
HO^+	Oxoniumylidene	OH^+
HO_2	Hydroperoxo	HO_2
HS	Mercapto	SH
HS^+	Sulfanylium	HS^+
		$H^{34}S^+$
HS_2	Thiosulfeno	HSS
H_2	Hydrogen	H_2
		HD
H_2N	Amidogen	NH_2
		$^{15}NH_2$[b]
		NHD
		ND_2
H_2O	Water	H_2O
		$H_2{}^{17}O$
		$H_2{}^{18}O$
		HDO
		$HD^{18}O$
		D_2O
H_2O^+	Oxoniumyl	H_2O^+
H_2O_2	Hydrogen peroxide	HOOH
H_2S	Hydrogen sulfide	H_2S
		$H_2{}^{34}S$
		HDS
		$HD^{34}S$
		D_2S
H_3^+	Trihydrogen ion	H_3^+
		H_2D^+
		D_2H^+
H_3N	Ammonia	NH_3

Mol. form.	Name	Isotopic species*
		$^{15}NH_3$
		NH_2D
		$^{15}NH_2D$
		NHD_2
		ND_3
H_3NO	Hydroxylamine	NH_2OH
H_3O^+	Oxonium	H_3O^+
H_3P	Phosphine	PH_3[a]
H_4N^+	Ammonium-*d*	NH_3D^+
H_4Si	Silane	SiH_4[a]
NO	Nitric oxide	NO
NO^+	Nitric oxide cation	NO^+
NP	Phosphorus nitride	PN
NS	Nitrogen sulfide	NS
		$N^{34}S$
NS^+	Thionitrosyl ion	NS^+
NSi	Silicon nitride	SiN
N_2	Nitrogen	N_2
N_2^+	Nitrogen ion	N_2^+[b]
N_2O	Nitrous oxide	N_2O
OP	Phosphorus monoxide	PO
OS	Sulfur monoxide	SO
		^{34}SO
		^{33}SO
		$S^{17}O$
		$S^{18}O$
OS^+	Sulfur monoxide ion	SO^+
OSi	Silicon monoxide	SiO
		$Si^{17}O$
		$Si^{18}O$
		^{29}SiO
		$^{29}Si^{18}O$
		^{30}SiO
		$^{30}Si^{18}O$
OTi	Titanium monoxide	TiO[a]
OV	Vanadium monoxide	VO[a]
O_2	Oxygen	O_2
O_2S	Sulfur dioxide	SO_2
		$^{33}SO_2$
		$^{34}SO_2$
		$OS^{17}O$
		$OS^{18}O$
O_2Ti	Titanium dioxide	TiO_2[a]
SSi	Silicon monosulfide	SiS
		$Si^{33}S$
		$Si^{34}S$
		^{29}SiS
		$^{29}Si^{34}S$
		^{30}SiS
		$^{30}Si^{34}S$
S_2	Sulfur	S_2[b]

* l- Before the isotopic species indicates a linear configuration, while c- indicates a cyclic molecule.

\- Underscore in formula indicates a ring compound.

[a] Reported only in circumstellar clouds.

[b] Reported only in comets.

[c] *cis* isomer.

[t] *trans* isomer.

MASS, DIMENSIONS, AND OTHER PARAMETERS OF THE EARTH

This table is a collection of data on various properties of the Earth. Most of the values are given in SI units. Note that 1 au (astronomical unit) = 149 597 870 km.

References

1. Seidelmann, P. K., Ed., *Explanatory Supplement to the Astronomical Almanac*, University Science Books, Mill Valley, CA, 1992.
2. Lang, K. R., *Astrophysical Data: Planets and Stars*, Springer-Verlag, New York, 1992.
3. *The Astronomical Almanac 2019*, Department of the Navy, Nautical Almanac Office, Washington, DC, April 2018.

Properties of the Earth and Its Orbit

Quantity	Symbol	Value	Unit
Mass	M	$5.972\ 2 \cdot 10^{27}$	g
Major orbital semi-axis	a_{orb}	1.000000	au
		$1.4959787 \cdot 10^{8}$	km
Distance from Sun at perihelion	r_π	0.9833	au
Distance from Sun at aphelion	r_α	1.0167	au
Moment of perihelion passage	T_π	Jan. 2, 4 h 52 min	
Moment of aphelion passage	T_α	July 4, 5 h 05 min	
Siderial rotation period around Sun	P_{orb}	$31.5581 \cdot 10^{6}$	s
		365.25636	d
Mean rotational velocity	U_{orb}	29.78	km/s
Mean equatorial radius	$\bar{a}$	6 378.136 6	km
Mean polar compression (flattening factor)	α	1/298.257	
Difference in equatorial and polar semi-axes	$a - c$	21.385	km
Compression of meridian of major equatorial axis	α_a	1/295.2	
Compression of meridian of minor equatorial axis	α_b	1/298.0	
Equatorial compression	ε	1/30 000	
Difference in equatorial semi-axes	$a - b$	213	m
Difference in polar semi-axes	$c_N - c_S$	~70	m
Polar asymmetry	η	$\sim 1.10^{-5}$	
Mean acceleration of gravity at equator	g_e	9.78036	m/s²
Mean acceleration of gravity at poles	g_p	9.83208	m/s²
Difference in acceleration of gravity at pole and at equator	$g_p - g_e$	5.172	cm/s²
Mean acceleration of gravity for entire surface of terrestrial ellipsoid	g	9.7978	m/s²
Mean radius	R	6371.0	km
Area of surface	S	$5.10 \cdot 10^{8}$	km²
Volume	V	$1.0832 \cdot 10^{12}$	km³
Mean density	ρ	5.515	g/cm³
Siderial rotational period	P	86,164.09	s
Rotational angular velocity	ω	$7.292116 \cdot 10^{-5}$	rad/s
Mean equatorial rotational velocity	v	0.46512	km/s
Rotational angular momentum	L	$5.861 \cdot 10^{33}$	J s
Rotational energy	E	$2.137 \cdot 10^{29}$	J
Ratio of centrifugal force to force of gravity at equator	q_c	0.0034677 = 1/288	
Moment of inertia	I	$8.070 \cdot 10^{37}$	kg m²

Quantity	Symbol	Value	Unit
Relative braking of Earth's rotation due to tidal friction	$\Delta\omega_e/\omega$	$-4.2 \cdot 10^{-8}$	century^{-1}
Relative secular acceleration of Earth's rotation	$\Delta\omega_i/\omega$	$+1.4 \cdot 10^{-8}$	century^{-1}
Non-secular braking of Earth's rotation	$\Delta\omega/\omega$	$-2.8 \cdot 10^{-8}$	century^{-1}
Probable value of total energy of tectonic deformation of Earth	E_t	$\sim 1 \cdot 10^{23}$	J/century
Secular loss of heat of Earth through radiation into space	$\Delta' E_k$	$1 \cdot 10^{23}$	J/century
Portion of Earth's kinetic energy transformed into heat as a result of lunar and solar tides in the hydrosphere	$\Delta'' E_k$	$1.3 \cdot 10^{23}$	J/century
Differences in duration of days in March and August	ΔP	0.0025 (March-Aug.)	s
Corresponding relative annual variation in Earth's rotational velocity	$\Delta^*\omega/\omega$	$2.9 \cdot 10^{-8}$ (Aug.-March)	
Presumed variation in Earth's radius between August and March	$\Delta^* R$	−9.2 (Aug.-March)	cm
Annual variation in level of world ocean	Δh_o	~10 (Sept.-March)	cm
Area of continents	S_C	$1.49 \cdot 10^{8}$	km²
		29.2	% of surface
Area of world ocean	S_O	$3.61 \cdot 10^{8}$	km²
		70.8	% of surface
Mean height of continents above sea level	h_C	875	m
Mean depth of world ocean	h_O	3794	m
Mean thickness of lithosphere within the limits of the continents	$h_{C.l.}$	35	km
Mean thickness of lithosphere within the limits of the ocean	$h_{O.l.}$	4.7	km
Mean rate of thickening of continental lithosphere	$\Delta h/\Delta t$	10–40	m/10⁶ y
Mean rate of horizontal extension of continental lithosphere	$\Delta l/\Delta t$	0.75–20	km/10⁶ y
Mass of crust	m_l	$2.36 \cdot 10^{22}$	kg
Mass of mantle		$4.05 \cdot 10^{24}$	kg
Amount of water released from the mantle and core in the course of geological time		$3.40 \cdot 10^{21}$	kg

Quantity	Symbol	Value	Unit
Total reserve of water in the mantle		$2 \cdot 10^{23}$	kg
Present content of free and bound water in the Earth's lithosphere		$2.4 \cdot 10^{21}$	kg
Mass of hydrosphere	m_h	$1.664 \cdot 10^{21}$	kg
Amount of oxygen bound in the Earth's crust		$1.300 \cdot 10^{21}$	kg
Amount of free oxygen		$1.5 \cdot 10^{18}$	kg
Mass of atmosphere	m_a	$5.136 \cdot 10^{18}$	kg
Mass of biosphere	m_b	$1.148 \cdot 10^{16}$	kg
Mass of living matter in the biosphere		$3.6 \cdot 10^{14}$	kg
Density of living matter on dry land		0.1	g/cm²
Density of living matter in ocean		$15 \cdot 10^{-8}$	g/cm³
Age of the Earth		$4.55 \cdot 10^{9}$	y
Age of oldest rocks		$4.0 \cdot 10^{9}$	y
Age of most ancient fossils		$3.4 \cdot 10^{9}$	y

GEOLOGICAL TIME SCALE

Geological Time Scale for the Earth

Period or epoch	Beginning and end, in 10^6 years	Key events
Cenozoic era		
Quaternary		
Holocene - Meghalayan	present – 0.0042	
Holocene - Northgrippian	0.0042 – 0.0082	
Holocene - Greenlandian	0.0082 – 0.0117	
Pleistocene*	0.0117 – 2.58	Homo erectus breakout
Tertiary		
Pliocene*	2.58 – 5.333	Ape man fossils
Miocene	5.333 – 23.03	Origin of grass
Oligocene	23.03 – 33.9	Rise of cats, dogs, pigs
Eocene	33.9 – 56.0	Debut of hoofed mammals
Paleocene	56.0 – 66.0	Earliest primates
Mesozoic era		
Cretaceous	66.0 – 145.0	Demise of dinosaurs
Jurassic	145.0 – 201.3 ± 0.2	First birds
Triassic	201.3 ± 0.2 – 251.902 ± 0.024	Appearance of dinosaurs
Paleozoic era		
Permian	251.902 ± 0.024 – 298.9 ± 0.15	Flowers, insect pollination
Carboniferous	298.9 ± 0.15 – 358.9 ± 0.4	First conifers
Devonian	358.9 ± 0.4 – 419.2 ± 3.2	First vertebrates ashore
Silurian	419.2 ± 3.2 – 443.8 ± 1.5	Spore-bearing plants
Ordovician	443.8 ± 1.5 – 485.4 ± 1.9	First animals ashore
Cambrian	485.4 ± 1.9 – 541.0 ± 1.0	Vertebrates appear
Pre-Cambrian		
Pre-Cambrian III (Proterozoic)	541.0 ± 1.0 – 2500	First plants, jellyfish
Pre-Cambrian II (Archean)	2500 – 4000	Photosynthetic bacteria
Pre-Cambrian I (Hadean)	4000 – 4600	Earth formed 4600 million years ago

* Some authorities place the boundary between the Pleistocene and Pliocene at $1.81 \cdot 10^6$ years.

References

1. U.S. Geological Survey Geologic Names Committee, 2007, Divisions of Geologic Time—Major Chronostratigraphic and Geochronologic Units: U.S. Geological Survey Fact Sheet 2010-3059. Available on the Internet at <https://pubs.usgs.gov/fs/2010/3059/pdf/FS10-3059.pdf>.
2. Walker, J. D., and Geissman, J. W., Compilers, 2009, Geologic Time Scale: Geological Society of America, <www.geosociety.org/science/timescale/timescl.pdf>.
3. Calder, N., *Timescale — An Atlas of the Fourth Dimension,* Viking Press, New York, 1983.
4. International Chronostratigraphic Chart, v2018/08, available on the Internet at <http://www.stratigraphy.org/index.php/ics-chart-timescale>.

ACCELERATION DUE TO GRAVITY

The acceleration due to gravity is tabulated here as a function of latitude and height above the Earth's surface. Values were calculated from the expression

$$g/(\text{m/s}^2) = 9.780356 \, (1 + 0.0052885 \sin^2 \phi - 0.0000059 \sin^2 2\phi) - 0.003086 \, H$$

where ϕ is the latitude and H is the height in kilometers.

Reference

Jursa, A. S., Ed., *Handbook of Geophysics and the Space Environment, Fourth Edition*, Air Force Geophysics Laboratory, 1985, pp. 14–17.

ϕ	$H = 0$	$H = 1$ km	$H = 5$ km	$H = 10$ km
0	9.78036	9.77727	9.76493	9.74950
5	9.78075	9.77766	9.76532	9.74989
10	9.78191	9.77882	9.76648	9.75105
15	9.78381	9.78072	9.76838	9.75295
20	9.78638	9.78330	9.77095	9.75552
25	9.78956	9.78647	9.77413	9.75870
30	9.79324	9.79016	9.77781	9.76238
35	9.79732	9.79424	9.78189	9.76646
40	9.80167	9.79858	9.78624	9.77081
45	9.80616	9.80307	9.79073	9.77530

ϕ	$H = 0$	$H = 1$ km	$H = 5$ km	$H = 10$ km
50	9.81065	9.80757	9.79522	9.77979
55	9.81501	9.81193	9.79958	9.78415
60	9.81911	9.81602	9.80368	9.78825
65	9.82281	9.81972	9.80738	9.79195
70	9.82601	9.82292	9.81058	9.79515
75	9.82860	9.82551	9.81317	9.79774
80	9.83051	9.82743	9.81508	9.79965
85	9.83168	9.82860	9.81625	9.80082
90	9.83208	9.82899	9.81665	9.80122

DENSITY, PRESSURE, AND GRAVITY AS A FUNCTION OF DEPTH WITHIN THE EARTH

This table gives the density ρ, pressure p, and acceleration due to gravity g as a function of depth below the Earth's surface, as calculated from the model of the structure of the Earth in Ref. 1. The model assumes a radius of 6371 km for the Earth. The boundary between the crust and mantle (the Mohorovicic discontinuity) is taken as 21 km, while in reality it varies considerably with location.

References

1. Anderson, D. L., and Hart, R. S., *J. Geophys. Res.* 81, 1461, 1976.
2. Carmichael, R. S., *CRC Practical Handbook of Physical Properties of Rocks and Minerals*, CRC Press, Boca Raton, FL, 1989, p. 467.

Depth km	ρ/g cm³	p/kbar	g/cm s²
Crust			
0	1.02	0	981
3	1.02	3	982
3	2.80	3	982
21	2.80	5	983
Mantle (solid)			
21	3.49	5	983
41	3.51	12	983
61	3.52	19	984
81	3.48	26	984
101	3.44	33	984
121	3.40	39	985
171	3.37	56	987
221	3.34	73	989
271	3.37	89	991
321	3.47	106	993
371	3.59	124	994
571	3.95	199	999
871	4.54	328	997
1171	4.67	466	992
1471	4.81	607	991

Depth km	ρ/g cm³	p/kbar	g/cm s²
1771	4.96	752	994
2071	5.12	903	1002
2371	5.31	1061	1017
2671	5.45	1227	1042
2886	5.53	1352	1069
Outer core (liquid)			
2886	9.96	1352	1069
2971	10.09	1442	1050
3371	10.63	1858	953
3671	11.00	2154	874
4071	11.36	2520	760
4471	11.69	2844	641
4871	11.99	3116	517
5156	12.12	3281	427
Inner core (solid)			
5156	12.30	3281	427
5371	12.48	3385	355
5771	12.52	3529	218
6071	12.53	3592	122
6371	12.58	3617	0

OCEAN PRESSURE AS A FUNCTION OF DEPTH AND LATITUDE

The following table is based upon an ocean model which takes into account the equation of state of standard seawater and the dependence on latitude of the acceleration of gravity. The tabulated pressure value is the excess pressure over the ambient atmospheric pressure at the surface.

References

1. *International Oceanographic Tables, Volume 4,* Unesco Technical Papers in Marine Science No. 40, Unesco, Paris, 1987.
2. Saunders, P. M., and Fofonoff, N. P., *Deep-Sea Res.* 23, 109–111, 1976.

Pressure in MPa at the Specified Latitude

Depth (meters)	0°	15°	30°	45°	60°	75°	90°
0	0.0000	0.0000	0.0000	0.0000	0.0000	0.0000	0.0000
500	5.0338	5.0355	5.0404	5.0471	5.0537	5.0586	5.0605
1000	10.0796	10.0832	10.0930	10.1064	10.1198	10.1296	10.1333
1500	15.1376	15.1431	15.1577	15.1778	15.1980	15.2127	15.2182
2000	20.2076	20.2148	20.2344	20.2613	20.2882	20.3080	20.3153
2500	25.2895	25.2985	25.3231	25.3568	25.3905	25.4153	25.4244
3000	30.3831	30.3940	30.4236	30.4641	30.5047	30.5345	30.5453
3500	35.4886	35.5012	35.5358	35.5832	35.6307	35.6654	35.6782
4000	40.6056	40.6201	40.6598	40.7140	40.7683	40.8082	40.8229
4500	45.7342	45.7505	45.7952	45.8564	45.9176	45.9626	45.9791
5000	50.8742	50.8924	50.9421	51.0102	51.0785	51.1285	51.1469
5500	56.0255	56.0456	56.1004	56.1755	56.2508	56.3059	56.3262
6000	61.1882	61.2100	61.2700	61.3521	61.4344	61.4947	61.5168
6500	66.3619	66.3857	66.4508	66.5399	66.6292	66.6947	66.7187
7000	71.5467	71.5724	71.6427	71.7388	71.8352	71.9059	71.9318
7500	76.7426	76.7701	76.8456	76.9488	77.0523	77.1282	77.1560
8000	81.9493	81.9788	82.0594	82.1697	82.2804	82.3614	82.3911
8500	87.1669	87.1983	87.2841	87.4016	87.5193	87.6057	87.6373
9000	92.3950	92.4284	92.5194	92.6440	92.7689	92.8606	92.8941
9500	97.6346	97.6698	97.7661	97.8978	98.0300	98.1269	98.1624
10000	102.8800	102.9170	103.0185	103.1572	103.2961	103.3981	103.4355

PROPERTIES OF SEAWATER

In addition to the dependence on temperature and pressure, the physical properties of seawater vary with the concentration of the dissolved constituents. A convenient parameter for describing the composition is the salinity, *S*, which is defined in terms of the electrical conductivity of the seawater sample. The defining equation for the practical salinity is:

$$S = a_0 + a_1K^{1/2} + a_2K + a_3K^{3/2} + a_4K^2 + a_5K^{5/2}$$

where *K* is the ratio of the conductivity of the seawater sample at 15 °C and atmospheric pressure to the conductivity of a potassium chloride solution in which the mass fraction of KCl is 0.0324356, at the same temperature and pressure. The values of the coefficients are:

$$a_0 = 0.0080$$
$$a_1 = -0.1692$$
$$a_2 = 25.3851$$
$$a_3 = 14.0941$$
$$a_4 = -7.0261$$
$$a_5 = 2.7081$$
$$\Sigma a_i = 35.0000$$

Thus, when $K = 1$, $S = 35$ exactly (*S* is normally quoted in units of ‰, i.e., parts per thousand). The value of *S* can be roughly equated with the mass of dissolved material in grams per kilogram of seawater. Salinity values in the open oceans at midlatitudes typically fall between 34 and 36.

It is customary in oceanography to define the pressure at a given point as the pressure due to the column of water between that point and the surface. Thus, by convention $P = 0$ at the sea surface. To a good approximation the pressure in decibars (dbar) can be equated to the depth in meters. Thus at 45° latitude the pressure is 5000 dbar at 4902 m, 10000 dbar at 9700 m.

The first table below gives several properties of seawater as a function of temperature for a salinity of 35. The second and third give density and electrical conductivity as a function of salinity at several temperatures, and the fourth lists typical concentrations of the main constituents of seawater as a function of salinity. The final table gives the freezing point as a function of salinity and pressure.

References

1. *The Practical Salinity Scale 1978 and the International Equation of State of Seawater 1980*, Unesco Technical Papers in Marine Science No. 36, Unesco, Paris, 1981; sections No. 37, 38, 39, and 40 in this series give background papers and detailed tables.
2. Kennish, M. J., *CRC Practical Handbook of Marine Science*, CRC Press, Boca Raton, FL, 1989.
3. Poisson, A. *IEEE J. Ocean. Eng.* OE-5, 50, 1981.
4. Webster, F., in *AIP Physics Desk Reference*, E. R. Cohen, D. R. Lide, and G. L. Trigg, Eds., Springer-Verlag, New York, 2003.

Properties of Seawater

Column heading	Definition
t	Temperature in °C
ρ	Density in g cm^{-3}
β	$(1/\rho)(d\rho/dS)$ = Fractional change in density per unit change in salinity (S)
α	$-(1/\rho)(d\rho/dt)$ = Fractional change in density per unit change in temperature (°C^{-1})
κ	Electrical conductivity in S cm^{-1} (siemens per cm)
η	Viscosity in mPa s (equal to cP)
c_p	Specific heat in J kg °C^{-1}
v	Speed of sound in m s^{-1}

TABLE 1. Properties of Seawater as a Function of Temperature at Salinity S = 35 and Normal Atmospheric Pressure

t/°C	ρ/g cm^{-3}	$10^7\cdot\beta$	$10^7\cdot\alpha$/°C^{-1}	κ/S cm^{-1}	η/mPa s	c_p/J kg^{-1} °C^{-1}	v/m s^{-1}
0	1.028106	7854	526	0.029048	1.892	3986.5	1449.1
5	1.027675	7717	1136	0.033468	1.610		
10	1.026952	7606	1668	0.038103	1.388	3986.3	1489.8
15	1.025973	7516	2141	0.042933	1.221		
20	1.024763	7444	2572	0.047934	1.085	3993.9	1521.5
25	1.023343	7385	2970	0.053088	0.966		
30	1.021729	7338	3341	0.058373	0.871	4000.7	1545.6
35	1.019934	7300	3687				
40		7270	4004			4003.5	1563.2

TABLE 2. Density of Surface Seawater in g cm^{-3} as a Function of Temperature (t) and Salinity (S)

t/°C	S = 0	S = 5	S = 10	S = 15	S = 20	S = 25	S = 30	S = 35	S = 40
0	0.999843	1.003913	1.007955	1.011986	1.016014	1.020041	1.024072	1.028106	1.032147
5	0.999967	1.003949	1.007907	1.011858	1.015807	1.019758	1.023714	1.027675	1.031645
10	0.999702	1.003612	1.007501	1.011385	1.015269	1.019157	1.023051	1.026952	1.030862
15	0.999102	1.002952	1.006784	1.010613	1.014443	1.018279	1.022122	1.025973	1.029834
20	0.998206	1.002008	1.005793	1.009576	1.013362	1.017154	1.020954	1.024763	1.028583
25	0.997048	1.000809	1.004556	1.008301	1.012050	1.015806	1.019569	1.023343	1.027128
30	0.995651	0.999380	1.003095	1.006809	1.010527	1.014252	1.017985	1.021729	1.025483
35	0.994036	0.997740	1.001429	1.005118	1.008810	1.012509	1.016217	1.019934	1.023662
40	0.992220	0.995906	0.999575	1.003244	1.006915	1.010593	1.014278	1.017973	1.021679

TABLE 3. Electrical Conductivity of Seawater in S cm^{-1} as a Function of Temperature (t) and Salinity (S)

t/°C	S = 5	S = 10	S = 15	S = 20	S = 25	S = 30	S = 35	S = 40
0	0.004808	0.009171	0.013357	0.017421	0.021385	0.025257	0.029048	0.032775
5	0.005570	0.010616	0.015441	0.020118	0.024674	0.029120	0.033468	0.037734
10	0.006370	0.012131	0.017627	0.022947	0.028123	0.033171	0.038103	0.042935
15	0.007204	0.013709	0.019905	0.025894	0.031716	0.037391	0.042933	0.048355
20	0.008068	0.015346	0.022267	0.028948	0.035438	0.041762	0.047934	0.053968
25	0.008960	0.017035	0.024703	0.032097	0.039276	0.046267	0.053088	0.059751
30	0.009877	0.018771	0.027204	0.035330	0.043213	0.050888	0.058373	0.065683

TABLE 4. Composition of Seawater and Ionic Strength at Various Salinities (S) (Ref. 2)

Constituent	Molality S = 30	Molality S = 35	Molality S = 40	g/kg of Seawater S = 30	g/kg of Seawater S = 35	g/kg of Seawater S = 40
Cl^-	0.482	0.562	0.650	16.58	19.33	22.36
Br^-	0.00074	0.00087	0.00100	0.057	0.067	0.078
F^-		0.00007			0.001	
SO_4^{2-}	0.0104	0.0114	0.0122	0.97	1.06	1.14
HCO_3^-	0.00131	0.00143	0.00100	0.078	0.085	0.059
$NaSO_4^-$	0.0085	0.0108	0.0139	0.98	1.25	1.60
KSO_4^-	0.00010	0.00012	0.00015	0.013	0.016	0.020
Na^+	0.405	0.472	0.544	9.03	10.53	12.13
K^+	0.00892	0.01039	0.01200	0.338	0.394	0.455
Mg^{2+}	0.0413	0.0483	0.0561	0.974	1.139	1.323
Ca^{2+}	0.00131	0.00143	0.00154	0.051	0.056	0.060
Sr^{2+}	0.00008	0.00009	0.00011	0.007	0.008	0.009
$MgHCO_3^+$	0.00028	0.00036	0.00045	0.023	0.030	0.037
$MgSO_4$	0.00498	0.00561	0.00614	0.582	0.655	0.717
$CaSO_4$	0.00102	0.00115	0.00126	0.135	0.152	0.166
$NaHCO_3$	0.00015	0.00020	0.00024	0.012	0.016	0.020
H_3BO_3	0.00032	0.00037	0.00042	0.019	0.022	0.025
Ionic strength	0.5736	0.6675	0.7701			

TABLE 5. Freezing Point of Seawater in °C as a Function of Salinity (S) and Pressure (P)

P/dbar [a]	S = 0	S = 5	S = 10	S = 15	S = 20	S = 25	S = 30	S = 35	S = 40
0	0.000	−0.274	−0.542	−0.812	−1.083	−1.358	−1.638	−1.922	−2.212
50	−0.038	−0.311	−0.580	−0.849	−1.121	−1.396	−1.676	−1.960	−2.250
100	−0.075	−0.349	−0.618	−0.887	−1.159	−1.434	−1.713	−1.998	−2.287
500	−0.377	−0.650	−0.919	−1.188	−1.460	−1.735	−2.014	−2.299	−2.589

[a] Pressure above atmospheric.

ABUNDANCE OF ELEMENTS IN THE EARTH'S CRUST AND IN THE SEA

This table gives the estimated abundance of the elements in the continental crust (in mg/kg, equivalent to parts per million by mass) and in seawater near the surface (in mg/L). Values represent the median of reported measurements. The concentrations of the less abundant elements may vary with location by several orders of magnitude.

References

1. Carmichael, R. S., Ed., *CRC Practical Handbook of Physical Properties of Rocks and Minerals,* CRC Press, Boca Raton, FL, 1989.
2. Bodek, I., et al., *Environmental Inorganic Chemistry*, Pergamon Press, New York, 1988.
3. Ronov, A. B., and Yaroshevsky, A. A., "Earth's Crust Geochemistry," in *Encyclopedia of Geochemistry and Environmental Sciences,* Fairbridge, R. W., Ed., Van Nostrand, New York, 1969.

	Abundance	
	Crust	Sea
Element	mg/kg	mg/L
Ac	5.5×10^{-10}	
Ag	7.5×10^{-2}	4×10^{-5}
Al	8.23×10^{4}	2×10^{-3}
Ar	3.5	4.5×10^{-1}
As	1.8	3.7×10^{-3}
Au	4×10^{-3}	4×10^{-6}
B	1.0×10^{1}	4.44
Ba	4.25×10^{2}	1.3×10^{-2}
Be	2.8	5.6×10^{-6}
Bi	8.5×10^{-3}	2×10^{-5}
Br	2.4	6.73×10^{1}
C	2.00×10^{2}	2.8×10^{1}
Ca	4.15×10^{4}	4.12×10^{2}
Cd	1.5×10^{-1}	1.1×10^{-4}
Ce	6.65×10^{1}	1.2×10^{-6}
Cl	1.45×10^{2}	1.94×10^{4}
Co	2.5×10^{1}	2×10^{-5}
Cr	1.02×10^{2}	3×10^{-4}
Cs	3	3×10^{-4}
Cu	6.0×10^{1}	2.5×10^{-4}
Dy	5.2	9.1×10^{-7}
Er	3.5	8.7×10^{-7}
Eu	2.0	1.3×10^{-7}
F	5.85×10^{2}	1.3
Fe	5.63×10^{4}	2×10^{-3}
Ga	1.9×10^{1}	3×10^{-5}
Gd	6.2	7×10^{-7}
Ge	1.5	5×10^{-5}
H	1.40×10^{3}	1.08×10^{5}
He	8×10^{-3}	7×10^{-6}
Hf	3.0	7×10^{-6}
Hg	8.5×10^{-2}	3×10^{-5}
Ho	1.3	2.2×10^{-7}
I	4.5×10^{-1}	6×10^{-2}
In	2.5×10^{-1}	2×10^{-2}
Ir	1×10^{-3}	
K	2.09×10^{4}	3.99×10^{2}
Kr	1×10^{-4}	2.1×10^{-4}
La	3.9×10^{1}	3.4×10^{-6}
Li	2.0×10^{1}	1.8×10^{-1}
Lu	8×10^{-1}	1.5×10^{-7}
Mg	2.33×10^{4}	1.29×10^{3}
Mn	9.50×10^{2}	2×10^{-4}
Mo	1.2	1×10^{-2}

	Abundance	
	Crust	Sea
Element	mg/kg	mg/L
N	1.9×10^{1}	5×10^{-1}
Na	2.36×10^{4}	1.08×10^{4}
Nb	2.0×10^{1}	1×10^{-5}
Nd	4.15×10^{1}	2.8×10^{-6}
Ne	5×10^{-3}	1.2×10^{-4}
Ni	8.4×10^{1}	5.6×10^{-4}
O	4.61×10^{5}	8.57×10^{5}
Os	1.5×10^{-3}	
P	1.05×10^{3}	6×10^{-2}
Pa	1.4×10^{-6}	5×10^{-11}
Pb	1.4×10^{1}	3×10^{-5}
Pd	1.5×10^{-2}	
Po	2×10^{-10}	1.5×10^{-14}
Pr	9.2	6.4×10^{-7}
Pt	5×10^{-3}	
Ra	9×10^{-7}	8.9×10^{-11}
Rb	9.0×10^{1}	1.2×10^{-1}
Re	7×10^{-4}	4×10^{-6}
Rh	1×10^{-3}	
Rn	4×10^{-13}	6×10^{-16}
Ru	1×10^{-3}	7×10^{-7}
S	3.50×10^{2}	9.05×10^{2}
Sb	2×10^{-1}	2.4×10^{-4}
Sc	2.2×10^{1}	6×10^{-7}
Se	5×10^{-2}	2×10^{-4}
Si	2.82×10^{5}	2.2
Sm	7.05	4.5×10^{-7}
Sn	2.3	4×10^{-6}
Sr	3.70×10^{2}	7.9
Ta	2.0	2×10^{-6}
Tb	1.2	1.4×10^{-7}
Te	1×10^{-3}	
Th	9.6	1×10^{-6}
Ti	5.65×10^{3}	1×10^{-3}
Tl	8.5×10^{-1}	1.9×10^{-5}
Tm	5.2×10^{-1}	1.7×10^{-7}
U	2.7	3.2×10^{-3}
V	1.20×10^{2}	2.5×10^{-3}
W	1.25	1×10^{-4}
Xe	3×10^{-5}	5×10^{-5}
Y	3.3×10^{1}	1.3×10^{-5}
Yb	3.2	8.2×10^{-7}
Zn	7.0×10^{1}	4.9×10^{-3}
Zr	1.65×10^{2}	3×10^{-5}

SOLAR IRRADIANCE AT THE EARTH

The solar luminosity (total radiant power emitted by the Sun) is $3.828 \cdot 10^{26}$ W, of which about 1361 W m^{-2} (the solar irradiance or "solar constant") reaches the top of the Earth's atmosphere (Ref. 6). To a zeroth approximation, the sun can be considered a black body with an effective temperature of 5780 K, which implies a peak in the radiation at around 0.520 μm (5200 Å). The actual solar spectral emission is more complex, especially at ultraviolet and shorter wavelengths. The graph in Figure 1, which was taken from Ref. 1, summarizes the solar irradiance at the top of the atmosphere in the range 0.3 μm to 10 μm.

The solar irradiance undergoes both long-term and short-term variations. Figure 2, which is taken from Ref. 4, shows the short-term variation over the 1976-2010 period. At the time of those measurements the mean value of the solar constant was believed to be about 1366 W/m^2. The more recent measurements in Ref. 6 show that the mean value (which is more difficult to measure than the short-term variations) is closer to 1361 W/m^2. Thus, the curve in Figure 1 should be shifted down by about 5 W/m^2.

References

1. Jursa, A. S., Ed., *Handbook of Geophysics and the Space Environment*, Air Force Geophysics Laboratory, 1985.
2. Pierce, A. K., and Allen, R. G., "The Solar Spectrum between 0.3 and 10 μm", in *The Solar Output and Its Variation*, White, O. R., Ed., Colorado Associated University Press, Boulder, CO, 1977.
3. Lang, K. R., *Astrophysical Data. Planets and Stars*, Springer-Verlag, New York, 1992 <https://doi.org/10.1007/978-1-4684-0640-5_3>.
4. Hansen, J. E., and Sato, M., <www.columbia.edu/~mhs119/Solar>, 2011.
5. Frohlich, C., and Lean, J., *Astron. Astrophys. Rev.* 12, 273, 2004 <https://doi.org/10.1007/s00159-004-0024-1>.
6. Kopp, G., and Lean, J. L., *Geophys. Res. Lett.* 38 (1), 2011 <https://doi.org/10.1029/2010GL045777>.

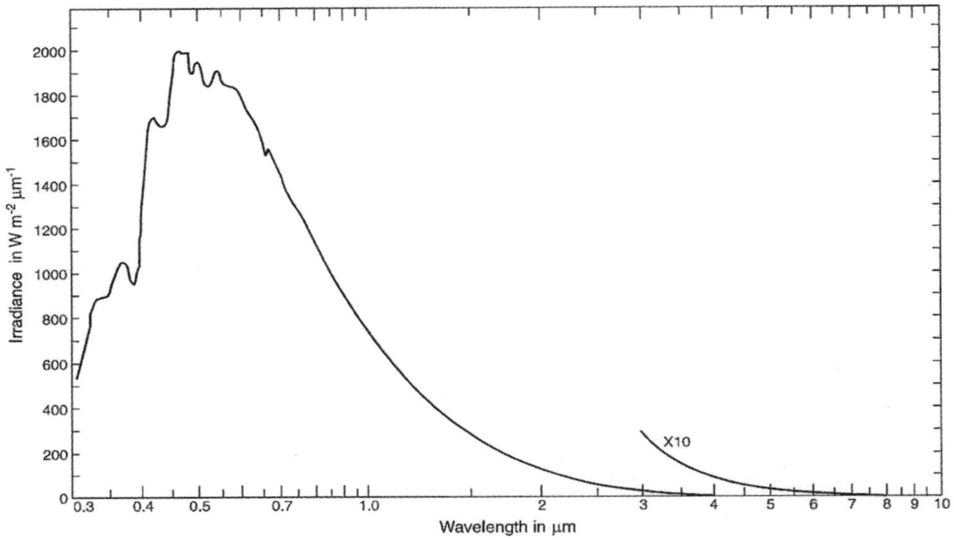

FIGURE 1. Wavelength dependence of solar irradiance.

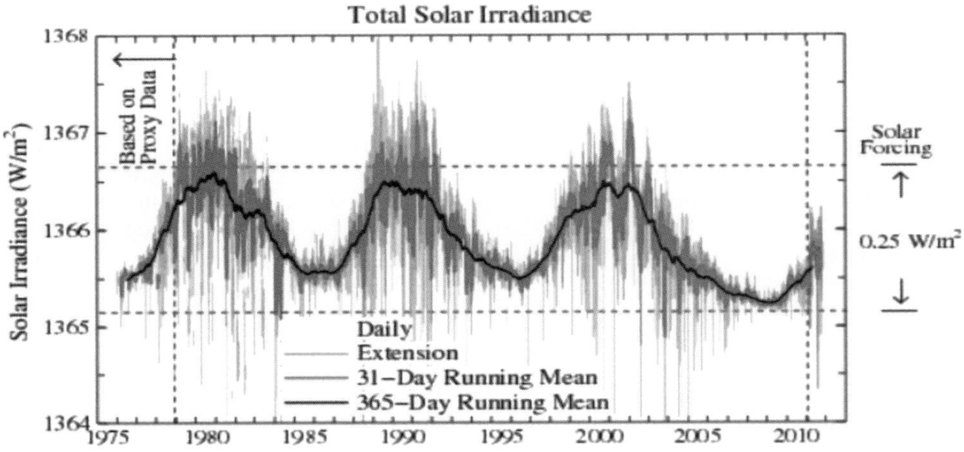

FIGURE 2. Variation of solar irradiance in the period 1976–2010. More recent measurements suggest that the curve should be shifted down by about 5 W/m^2.

Geo-Astro

U.S. STANDARD ATMOSPHERE (1976)

A Standard Atmosphere is a hypothetical vertical distribution of atmospheric temperature, pressure, and density that is roughly representative of year-round, midlatitude conditions. Typical uses are to serve as a basis for pressure altimeter calibrations, aircraft performance calculations, aircraft and rocket design, ballistic tables, meteorological diagrams, and various types of atmospheric modeling. The air is assumed to be dry and to obey the perfect gas law and the hydrostatic equation which, taken together, relate temperature, pressure, and density with vertical position. The atmosphere is considered to rotate with the Earth and to be an average over the diurnal cycle, the semiannual variation, and the range from active to quiet geomagnetic and sunspot conditions.

The U.S. Standard Atmosphere (1976) is an idealized, steady-state representation of mean annual conditions of the Earth's atmosphere from the surface to 1000 km at latitude 45° N, as it is assumed to exist during a period with moderate solar activity. The defining meteorological elements are sea-level temperature and pressure and a temperature-height profile to 1000 km. The 1976 Standard Atmosphere uses the following sea-level values that have been standard for many decades:

Temperature — 288.15 K (15 °C)
Pressure — 101325 Pa (1013.25 mbar, 760 mm of Hg, or 29.92 in. of Hg)
Density — 1225 g/m³ (1.225 g/L)
Mean molar mass — 28.964 g/mol

The parameters included in this condensed version of the U.S. Standard Atmosphere are:

Z — Height (geometric) above mean sea level in meters
T — Temperature in kelvins
P — Pressure in pascals (1 Pa = 0.01 millibars)
ρ — Density in kilograms per cubic meter (1 kg/m³ = 1 g/L)
n — Number density in molecules per cubic meter
ν — Mean collision frequency in collisions per second
l — Mean free path in meters
η — Absolute viscosity in pascal seconds (1 Pa s = 1000 cP)
k — Thermal conductivity in joules per meter second kelvin (W/m K)
v_s — Speed of sound in meters per second
g — Acceleration of gravity in meters per second square

The sea-level composition (percent by volume) is taken to be:

N_2 — 78.084%	He — 0.000524
O_2 — 20.9476	Kr — 0.000114
Ar — 0.934	Xe — 0.0000087
CO_2 — 0.0314	CH_4 — 0.0002
Ne — 0.001818	H_2 — 0.00005

The T and P columns for the troposphere and lower stratosphere were generated from the following formulas:

	T/K	P/Pa
$H \le 11000$ m	$288.15 - 0.0065\,H$	$101325(288.15/T)^{-5.25577}$
11000 m $< H \le$ 20000 m	216.65	$22632\, e^{-0.00015768832(H-11000)}$
20000 m $< H \le$ 32000 m	$216.65 + 0.0010(H-20000)$	$5474.87(216.65/T)^{34.16319}$

where $H = rZ/(r + Z)$ is the geopotential height in meters and r is the mean Earth radius at 45° N latitude, taken as 6356766 m. For altitudes up to 32 km, $\rho = 0.003483677(P/T)$ in the units used here. Formulas for the other quantities may be found in the references.

References

1. COESA, *U.S. Standard Atmosphere*, 1976, U.S. Government Printing Office, Washington, DC, 1976.
2. Jursa, A. S., Ed., *Handbook of Geophysics and the Space Environment*, Air Force Geophysics Laboratory, 1985.

Geo-Astro

Z/m	T/K	P/Pa	ρ/kg m^{-3}	n/m^{-3}	ν/s^{-1}	l/m	η/Pa s	k/W m^{-1} K^{-1}	v_s/m s^{-1}	g/m s^{-2}
−5000	320.68	1.778E+05	1.931	4.015E+25	1.151E+10	4.208E−08	1.942E−05	0.02788	359.0	9.822
−4500	317.42	1.685E+05	1.849	3.845E+25	1.096E+10	4.395E−08	1.927E−05	0.02763	357.2	9.830
−4000	314.17	1.596E+05	1.770	3.680E+25	1.044E+10	4.592E−08	1.912E−05	0.02738	355.3	9.819
−3500	310.91	1.511E+05	1.693	3.520E+25	9.933E+09	4.800E−08	1.897E−05	0.02713	353.5	9.818
−3000	307.66	1.430E+05	1.619	3.366E+25	9.448E+09	5.019E−08	1.882E−05	0.02688	351.6	9.816
−2500	304.41	1.352E+05	1.547	3.217E+25	8.982E+09	5.252E−08	1.867E−05	0.02663	349.8	9.814
−2000	301.15	1.278E+05	1.478	3.102E+25	8.623E+09	5.447E−08	1.852E−05	0.02638	347.9	9.813
−1500	297.90	1.207E+05	1.411	2.935E+25	8.106E+09	5.757E−08	1.836E−05	0.02613	346.0	9.811
−1000	294.65	1.139E+05	1.347	2.801E+25	7.693E+09	6.032E−08	1.821E−05	0.02587	344.1	9.810
−500	291.40	1.075E+05	1.285	2.672E+25	7.298E+09	6.324E−08	1.805E−05	0.02562	342.2	9.808
0	288.15	1.013E+05	1.225	2.547E+25	6.919E+09	6.633E−08	1.789E−05	0.02533	340.3	9.807
500	284.90	9.546E+04	1.167	2.427E+25	6.556E+09	6.961E−08	1.774E−05	0.02511	338.4	9.805
1000	281.65	8.988E+04	1.112	2.311E+25	6.208E+09	7.310E−08	1.758E−05	0.02485	336.4	9.804
1500	278.40	8.456E+04	1.058	2.200E+25	5.874E+09	7.680E−08	1.742E−05	0.02459	334.5	9.802
2000	275.15	7.950E+04	1.007	2.093E+25	5.555E+09	8.073E−08	1.726E−05	0.02433	332.5	9.801
2500	271.91	7.469E+04	0.957	1.990E+25	5.250E+09	8.491E−08	1.710E−05	0.02407	330.6	9.799
3000	268.66	7.012E+04	0.909	1.891E+25	4.959E+09	8.937E−08	1.694E−05	0.02381	328.6	9.797
3500	265.41	6.579E+04	0.863	1.795E+25	4.680E+09	9.411E−08	1.678E−05	0.02355	326.6	9.796
4000	262.17	6.166E+04	0.819	1.704E+25	4.414E+09	9.917E−08	1.661E−05	0.02329	324.6	9.794
4500	258.92	5.775E+04	0.777	1.616E+25	4.160E+09	1.046E−07	1.645E−05	0.02303	322.6	9.793
5000	255.68	5.405E+04	0.736	1.531E+25	3.918E+09	1.103E−07	1.628E−05	0.02277	320.6	9.791
5500	252.43	5.054E+04	0.697	1.450E+25	3.687E+09	1.165E−07	1.612E−05	0.02250	318.5	9.790
6000	249.19	4.722E+04	0.660	1.373E+25	3.467E+09	1.231E−07	1.595E−05	0.02224	316.5	9.788
6500	245.94	4.408E+04	0.664	1.299E+25	3.258E+09	1.302E−07	1.578E−05	0.02197	314.4	9.787
7000	242.70	4.111E+04	0.590	1.227E+25	3.058E+09	1.377E−07	1.561E−05	0.02170	312.3	9.785
7500	239.46	3.830E+04	0.557	1.159E+25	2.869E+09	1.458E−07	1.544E−05	0.02144	310.2	9.784
8000	236.22	3.565E+04	0.526	1.093E+25	2.689E+09	1.545E−07	1.527E−05	0.02117	308.1	9.782
8500	232.97	3.315E+04	0.496	1.031E+25	2.518E+09	1.639E−07	1.510E−05	0.02090	306.0	9.781
9000	229.73	3.080E+04	0.467	9.711E+24	2.356E+09	1.740E−07	1.493E−05	0.02063	303.9	9.779
9500	226.49	2.858E+04	0.440	9.141E+24	2.202E+09	1.848E−07	1.475E−05	0.02036	301.7	9.777
10000	223.25	2.650E+04	0.414	8.598E+24	2.056E+09	1.965E−07	1.458E−05	0.02009	299.5	9.776
10500	220.01	2.454E+04	0.389	8.079E+24	1.918E+09	2.091E−07	1.440E−05	0.01982	297.4	9.774
11000	216.77	2.270E+04	0.365	7.585E+24	1.787E+09	2.227E−07	1.422E−05	0.01954	295.2	9.773
11500	216.65	2.098E+04	0.337	7.016E+24	1.653E+09	2.408E−07	1.422E−05	0.01953	295.1	9.771
12000	216.65	1.940E+04	0.312	6.486E+24	1.528E+09	2.605E−07	1.422E−05	0.01953	295.1	9.770
12500	216.65	1.793E+04	0.288	5.996E+24	1.412E+09	2.818E−07	1.422E−05	0.01953	295.1	9.768
13000	216.65	1.658E+04	0.267	5.543E+24	1.306E+09	3.048E−07	1.422E−05	0.01953	295.1	9.767
13500	216.65	1.533E+04	0.246	5.124E+24	1.207E+09	3.297E−07	1.422E−05	0.01953	295.1	9.765
14000	216.65	1.417E+04	0.228	4.738E+24	1.116E+09	3.566E−07	1.422E−05	0.01953	295.1	9.764
14500	216.65	1.310E+04	0.211	4.380E+24	1.032E+09	3.857E−07	1.422E−05	0.01953	295.1	9.762
15000	216.65	1.211E+04	0.195	4.049E+24	9.538E+08	4.172E−07	1.422E−05	0.01953	295.1	9.761
16000	216.65	1.035E+04	0.166	3.461E+24	8.153E+08	4.881E−07	1.422E−05	0.01953	295.1	9.758
17000	216.65	8.850E+03	0.142	2.959E+24	6.969E+08	5.710E−07	1.422E−05	0.01953	295.1	9.754
18000	216.65	7.565E+03	0.122	2.529E+24	5.958E+08	6.680E−07	1.422E−05	0.01953	295.1	9.751
19000	216.65	6.467E+03	0.104	2.162E+24	5.093E+08	7.814E−07	1.422E−05	0.01953	295.1	9.748
20000	216.65	5.529E+03	8.891E−02	1.849E+24	4.354E+08	9.139E−07	1.422E−05	0.01953	295.1	9.745
21000	217.58	4.729E+03	7.572E−02	1.574E+24	3.716E+08	1.073E−06	1.427E−05	0.01961	295.1	9.742
22000	218.57	4.048E+03	6.451E−02	1.341E+24	3.173E+08	1.260E−06	1.432E−05	0.01970	296.4	9.739
23000	219.57	3.467E+03	5.501E−02	1.144E+24	2.712E+08	1.477E−06	1.438E−05	0.01978	297.1	9.736
24000	220.56	2.972E+03	4.694E−02	9.759E+23	2.319E+08	1.731E−06	1.443E−05	0.01986	297.7	9.733
25000	221.55	2.549E+03	4.008E−02	8.334E+23	1.985E+08	2.027E−06	1.448E−05	0.01995	298.4	9.730
26000	222.54	2.188E+03	3.426E−02	7.123E+23	1.700E+08	2.372E−06	1.454E−05	0.02003	299.1	9.727
27000	223.54	1.880E+03	2.930E−02	6.092E+23	1.458E+08	2.773E−06	1.459E−05	0.02011	299.7	9.724
28000	224.53	1.610E+03	2.508E−02	5.214E+23	1.250E+08	3.240E−06	1.465E−05	0.02020	300.4	9.721
29000	225.52	1.390E+03	2.148E−02	4.466E+23	1.073E+08	3.783E−06	1.470E−05	0.02028	301.1	9.718
30000	226.51	1.197E+03	1.841E−02	3.828E+23	9.219E+07	4.414E−06	1.475E−05	0.02036	301.7	9.715
31000	227.50	1.031E+03	1.579E−02	3.283E+23	7.925E+07	5.146E−06	1.481E−05	0.02044	302.4	9.712
32000	228.49	8.891E+02	1.356E−02	2.813E+23	6.818E+07	5.995E−06	1.486E−05	0.02053	303.0	9.709
33000	230.97	7.673E+02	1.157E−02	2.406E+23	5.852E+07	7.021E−06	1.499E−05	0.02073	304.7	9.706
34000	233.74	6.634E+02	9.887E−03	2.056E+23	5.030E+07	8.218E−06	1.514E−05	0.02096	306.5	9.703
35000	236.51	5.746E+02	8.463E−03	1.760E+23	4.331E+07	9.601E−06	1.529E−05	0.02119	308.3	9.700

Z/m	T/K	P/Pa	ρ/kg m^{-3}	n/m^{-3}	ν/s^{-1}	l/m	η/Pa s	k/W m^{-1} K^{-1}	v_s/m s^{-1}	g/m s^{-2}
36000	239.28	4.985E+02	7.258E−03	1.509E+23	3.736E+07	1.120E−05	1.543E−05	0.02142	310.1	9.697
38000	244.82	3.771E+02	5.367E−03	1.116E+23	2.794E+07	1.514E−05	1.572E−05	0.02188	313.7	9.690
40000	250.35	2.871E+02	3.996E−03	8.308E+22	2.104E+07	2.034E−05	1.601E−05	0.02233	317.2	9.684
42000	255.88	2.200E+02	2.995E−03	6.227E+22	1.594E+07	2.713E−05	1.629E−05	0.02278	320.7	9.678
44000	261.40	1.695E+02	2.259E−03	4.697E+22	1.215E+07	3.597E−05	1.657E−05	0.02323	324.1	9.672
46000	266.93	1.313E+02	1.714E−03	3.564E+22	9.318E+06	4.740E−05	1.685E−05	0.02376	327.5	9.666
48000	270.65	1.023E+02	1.317E−03	2.738E+22	7.208E+06	6.171E−05	1.704E−05	0.02397	329.8	9.660
50000	270.65	7.978E+01	1.027E−03	2.135E+22	5.620E+06	7.913E−05	1.703E−05	0.02397	329.8	9.654
52000	269.03	6.221E+01	8.056E−04	1.675E+22	4.397E+06	1.009E−04	1.696E−05	0.02384	328.8	9.648
54000	263.52	4.834E+01	6.390E−04	1.329E+22	3.452E+06	1.272E−04	1.660E−05	0.02340	325.4	9.642
56000	258.02	3.736E+01	5.045E−04	1.049E+22	2.696E+06	1.611E−04	1.640E−05	0.02296	322.0	9.636
58000	252.52	2.872E+01	3.963E−04	8.239E+21	2.095E+06	2.051E−04	1.612E−05	0.02251	318.6	9.632
60000	247.02	2.196E+01	3.097E−04	6.439E+21	1.620E+06	2.624E−04	1.584E−05	0.02206	315.1	9.624
65000	233.29	1.093E+01	1.632E−04	3.393E+21	8.294E+05	4.979E−04	1.512E−05	0.02093	306.2	9.609
70000	219.59	5.221	8.283E−05	1.722E+21	4.084E+05	9.810E−04	1.438E−05	0.01978	297.1	9.594
75000	208.40	2.388	3.992E−05	8.300E+20	1.918E+05	2.035E−03	1.376E−05	0.01883	289.4	9.579
80000	198.64	1.052	1.846E−05	3.838E+20	8.656E+04	4.402E−03	1.321E−05	0.01800	282.5	9.564
85000	188.89	4.457E−01	8.220E−06	1.709E+20	3.766E+04	9.886E−03	1.265E−05	0.01716	275.5	9.550
90000	186.87	1.836E−01	3.416E−06	7.116E+19	1.560E+04	2.370E−02				9.535
95000	188.42	7.597E−02	1.393E−06	2.920E+19	6.440E+03	5.790E−02				9.520
100000	195.08	3.201E−02	5.604E−07	1.189E+19	2.680E+03	1.420E−01				9.505
110000	240.00	7.104E−03	9.708E−08	2.144E+18	5.480E+02	7.880E−01				9.476
120000	360.00	2.538E−03	2.222E−08	5.107E+17	1.630E+02	3.310				9.447
130000	469.27	1.251E−03	8.152E−09	1.930E+17	7.100E+01	8.800				9.418
140000	559.63	7.203E−04	3.831E−09	9.322E+16	3.800E+01	1.800E+01				9.389
150000	634.39	4.542E−04	2.076E−09	5.186E+16	2.300E+01	3.300E+01				9.360
160000	696.29	3.040E−04	1.233E−09	3.162E+16	1.500E+01	5.300E+01				9.331
170000	747.57	2.121E−04	7.815E−10	2.055E+16	1.000E+01	8.200E+01				9.302
180000	790.07	1.527E−04	5.194E−10	1.400E+16	7.200	1.200E+02				9.274
190000	825.16	1.127E−04	3.581E−10	9.887E+15	5.200	1.700E+02				9.246
200000	854.56	8.474E−05	2.541E−10	7.182E+15	3.900	2.400E+02				9.218
220000	899.01	5.015E−05	1.367E−10	4.040E+15	2.300	4.200E+02				9.162
240000	929.73	3.106E−05	7.858E−11	2.420E+15	1.400	7.000E+02				9.106
260000	950.99	1.989E−05	4.742E−11	1.515E+15	9.300E−01	1.100E+03				9.051
280000	965.75	1.308E−05	2.971E−11	9.807E+14	6.100E−01	1.700E+03				8.997
300000	976.01	8.770E−06	1.916E−11	6.509E+14	4.200E−01	2.600E+03				8.943
320000	983.16	5.980E−06	1.264E−11	4.405E+14	2.900E−01	3.800E+03				8.889
340000	988.15	4.132E−06	8.503E−12	3.029E+14	2.000E−01	5.600E+03				8.836
360000	991.65	2.888E−06	5.805E−12	2.109E+14	1.400E−01	8.000E+03				8.784
380000	994.10	2.038E−06	4.013E−12	1.485E+14	1.000E−01	1.100E+04				8.732
400000	995.83	1.452E−06	2.803E−12	1.056E+14	7.200E−02	1.600E+04				8.680
450000	998.22	6.447E−07	1.184E−12	4.678E+13	3.300E−02	3.600E+04				8.553
500000	999.24	3.024E−07	5.215E−13	2.192E+13	1.600E−02	7.700E+04				8.429
550000	999.67	1.514E−07	2.384E−13	1.097E+13	8.400E−03	1.500E+05				8.307
600000	999.85	8.213E−08	1.137E−13	5.950E+12	4.800E−03	2.800E+05				8.188
650000	999.93	4.887E−08	5.712E−14	3.540E+12	3.100E−03	4.800E+05				8.072
700000	999.97	3.191E−08	3.070E−14	2.311E+12	2.200E−03	7.300E+05				7.958
750000	999.98	2.260E−08	1.788E−14	1.637E+12	1.700E−03	1.000E+06				7.846
800000	999.99	1.704E−08	1.136E−14	1.234E+12	1.400E−03	1.400E+06				7.737
850000	1000.00	1.342E−08	7.824E−15	9.717E+11	1.200E−03	1.700E+06				7.630
900000	1000.00	1.087E−08	5.759E−15	7.876E+11	1.000E−03	2.100E+06				7.525
950000	1000.00	8.982E−09	4.453E−15	6.505E+11	8.700E−04	2.600E+06				7.422
1000000	1000.00	7.514E−09	3.561E−15	5.442E+11	7.500E−04	3.100E+06				7.322

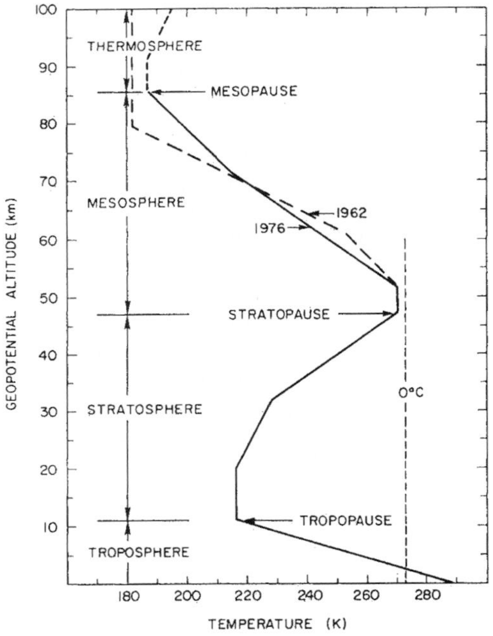

FIGURE 1. Temperature-height profile for U.S. Standard Atmosphere.

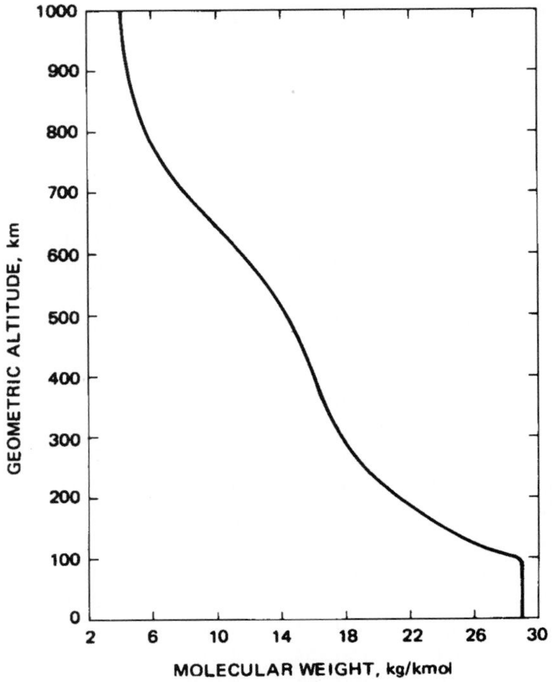

FIGURE 3. Mean molecular weight as a function of geometric altitude.

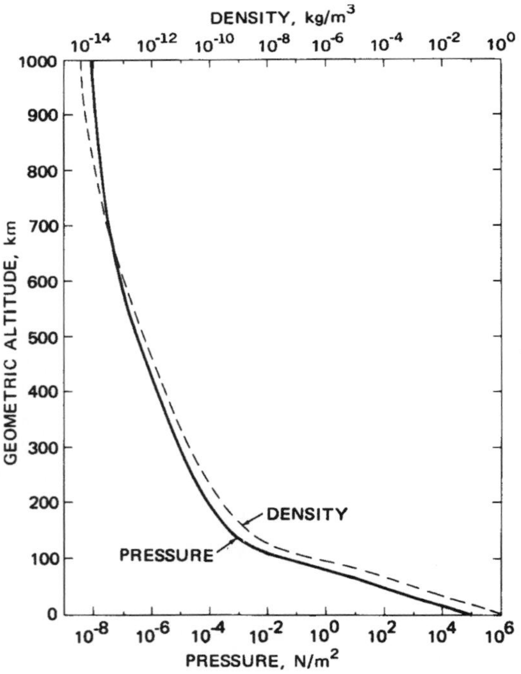

FIGURE 2. Total pressure and mass density as a function of geometric altitude.

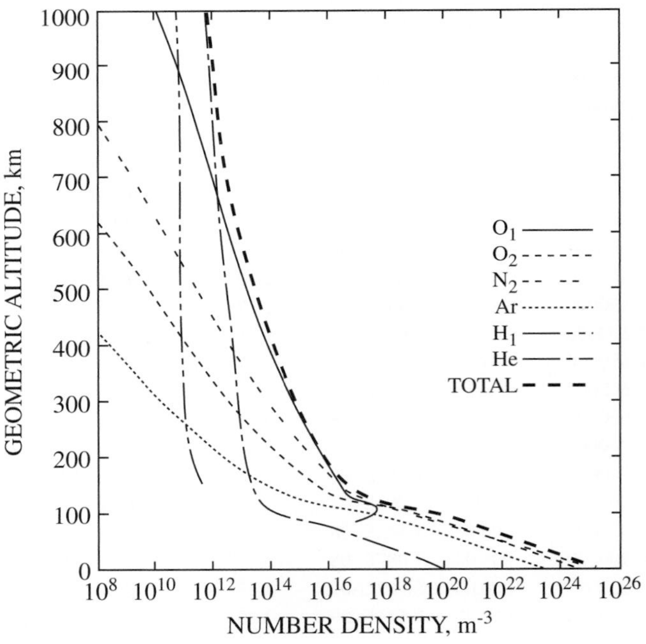

FIGURE 4. Number density of individual species and total number density as a function of geometric altitude.

Geo-Astro

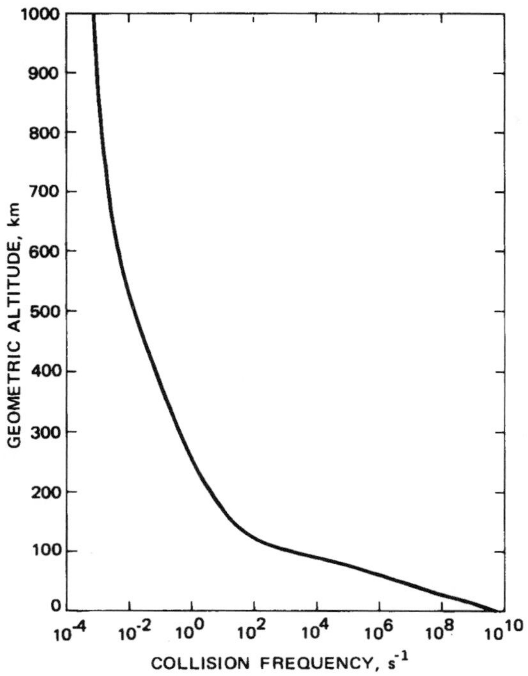

FIGURE 5. Collision frequency as a function of geometric altitude.

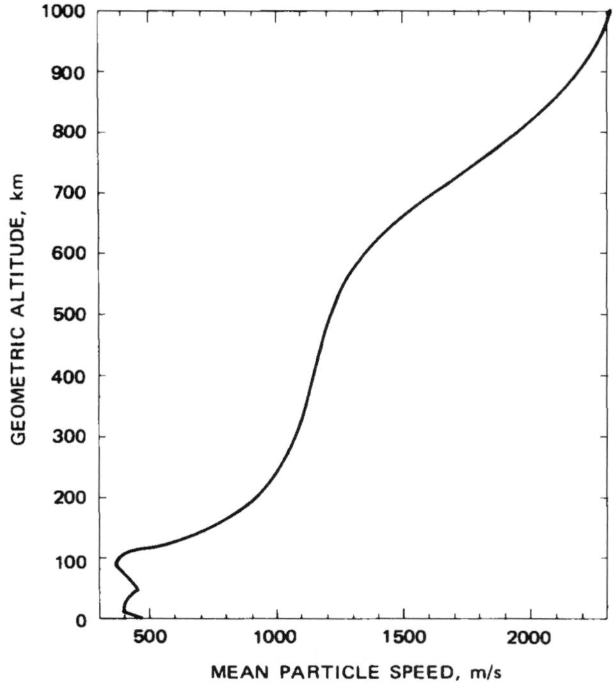

FIGURE 7. Mean air-particle speed as a function of geometric altitude.

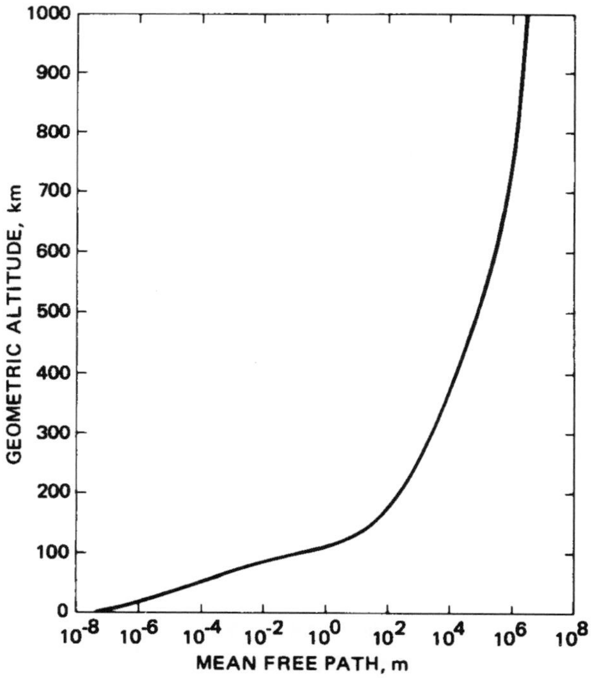

FIGURE 6. Mean free path as a function of geometric altitude.

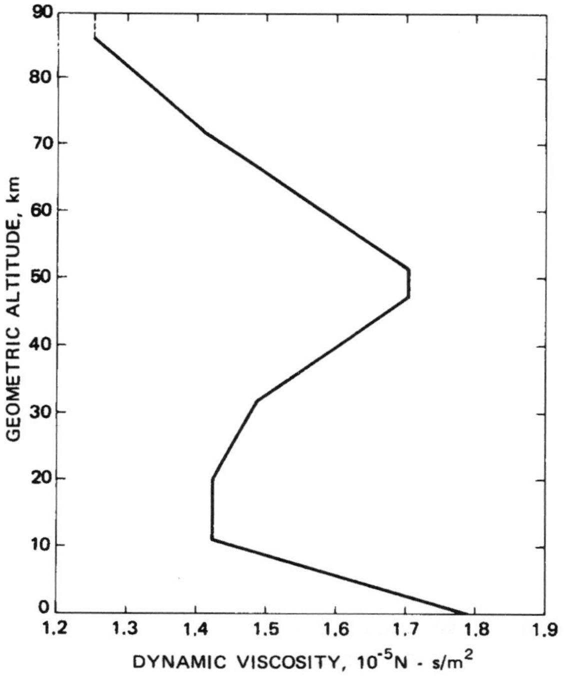

FIGURE 8. Dynamic viscosity as a function of geometric altitude.

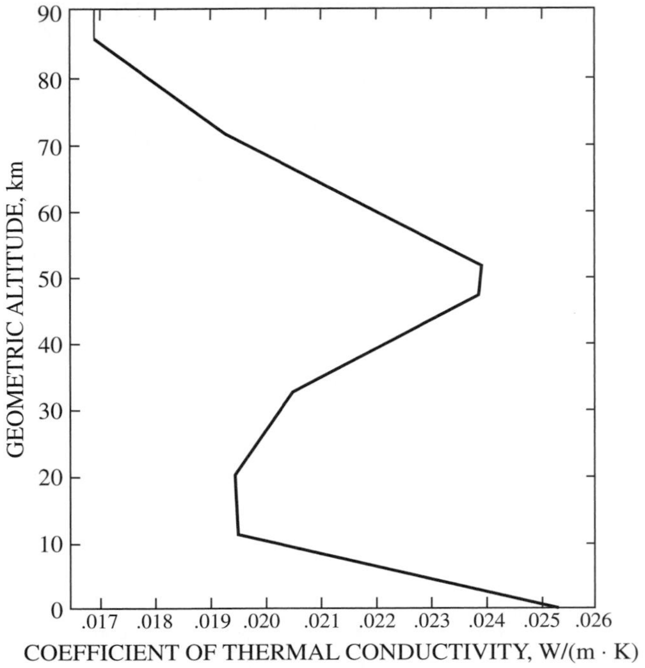

FIGURE 9. Coefficient of thermal conductivity as a function of geometric altitude.

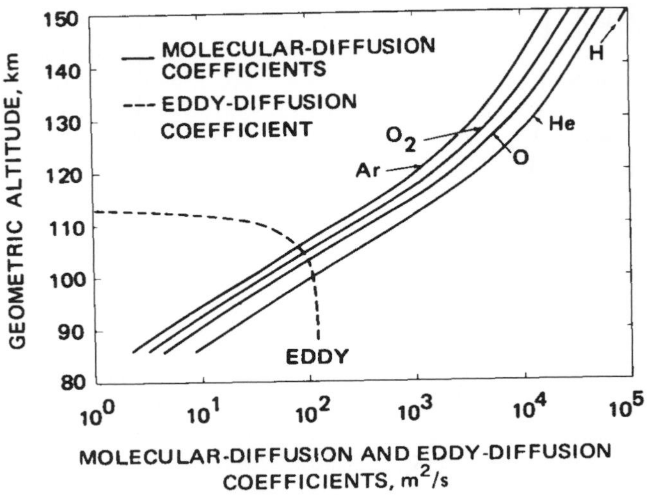

FIGURE 11. Molecular-diffusion and eddy-diffusion coefficients as a function of geometric altitude.

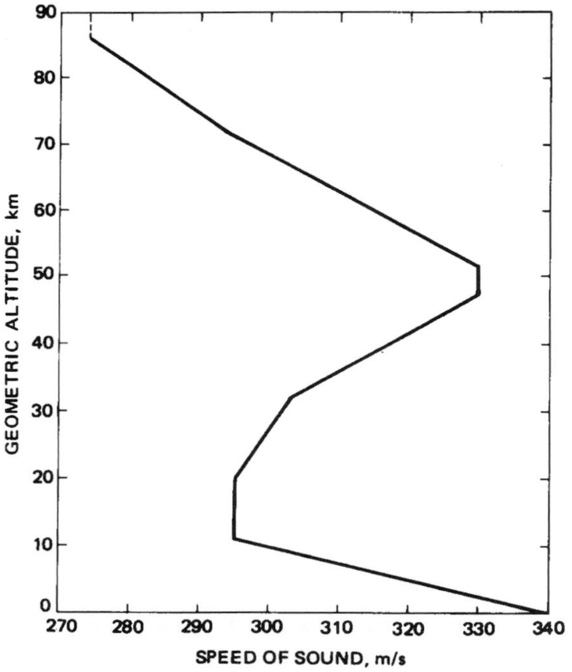

FIGURE 10. Speed of sound as a function of geometric altitude.

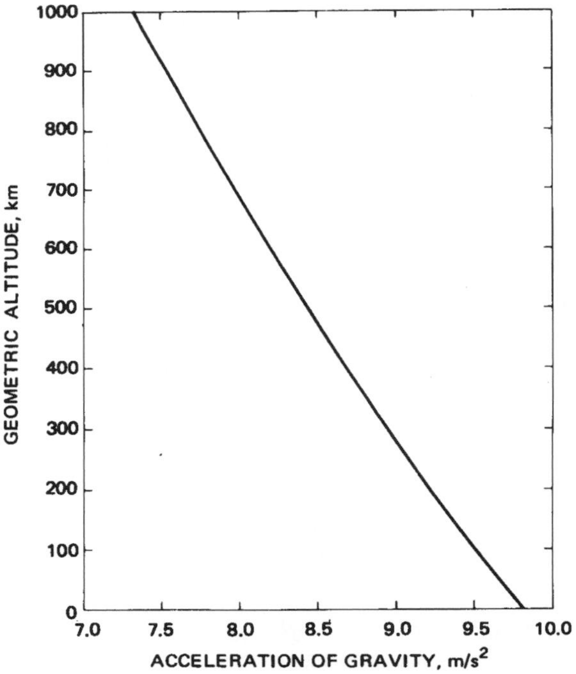

FIGURE 12. Acceleration of gravity as a function of geometric altitude.

Geo-Astro

GEOGRAPHICAL AND SEASONAL VARIATIONS IN SOLAR RADIATION

This table gives the amount of solar radiation reaching a unit area at the top of the Earth's atmosphere per day as a function of latitude and approximate date. It is based upon a solar constant (total energy per unit area at the Earth's average orbital distance) of 1373 W/m^2. Absorption of radiation by the atmosphere is not taken into consideration.

More recent measurements indicate the value of the solar constant is closer to 1361 W/m^2. Thus, the values in this table should be reduced by about 1%.

Reference

List, R. J., *Smithsonian Meteorological Tables, Seventh Edition*, Smithsonian Institution Press, Washington, DC, 1962.

Daily Solar Radiation in MJ/m^2

Lat.	Mar. 21	Apr. 13	May 6	May 29	Jun. 2	Jul. 15	Aug. 8	Aug. 31	Sep. 23	Oct. 16	Nov. 8	Nov. 30	Dec. 22	Jan. 13	Feb. 4	Feb. 26
90°		18.0	32.8	42.4	45.7	42.2	32.5	17.7								
80	6.6	18.0	32.3	41.8	45.0	41.6	32.0	17.7	6.5	0.3						0.3
70	13.0	22.3	31.8	39.9	43.0	39.7	31.5	22.0	12.9	5.5	1.0				1.0	5.6
60	19.0	27.0	34.4	39.7	41.6	39.4	34.0	26.7	18.8	11.6	6.2	3.1	2.1	3.1	6.2	11.7
50	24.4	31.1	36.8	40.7	42.0	40.5	36.5	30.8	24.1	17.6	12.1	8.7	7.5	8.7	12.3	17.8
40	29.1	34.3	38.6	41.3	42.1	41.1	38.3	33.9	28.7	23.1	18.2	14.8	13.5	14.9	18.4	23.5
30	32.9	36.7	39.4	41.1	41.4	40.8	39.1	36.3	32.5	28.2	23.9	20.9	19.8	21.0	24.1	28.4
20	35.7	38.0	39.2	39.7	39.7	39.5	38.9	37.5	35.3	32.3	29.1	26.6	25.7	26.7	29.3	32.7
10	37.4	38.1	37.9	37.4	37.1	37.2	37.6	37.7	37.0	35.5	33.5	31.8	31.1	31.9	33.8	35.9
0	38.0	37.1	35.5	34.1	33.5	34.0	35.2	36.6	37.6	37.6	36.9	36.1	35.8	36.3	37.3	38.0
-10	37.4	35.0	32.3	30.0	29.2	29.9	32.0	34.6	37.0	38.6	39.4	39.5	39.6	39.7	39.7	39.1
-20	35.7	31.8	28.0	25.2	24.1	25.1	27.8	31.5	35.3	38.5	40.7	42.0	42.4	42.2	41.1	39.0
-30	32.9	27.8	23.1	19.7	18.5	19.7	22.8	27.4	32.5	37.2	40.9	43.3	44.2	43.5	41.3	37.7
-40	29.1	22.8	17.5	14.0	12.6	13.9	17.4	22.6	28.7	34.8	40.1	43.6	45.0	43.8	40.5	35.2
-50	24.4	17.3	11.7	8.2	7.0	8.2	11.6	17.2	24.1	31.5	38.3	43.1	44.8	43.2	38.6	31.9
-60	19.0	11.4	5.9	2.9	2.0	2.9	5.9	11.3	18.8	27.3	35.7	41.9	44.4	42.1	36.0	27.7
-70	13.0	5.4	1.0				1.0	5.3	12.9	22.6	33.0	42.2	45.9	42.4	33.3	22.9
-80	6.6	0.3						0.3	6.5	18.2	33.5	44.2	48.1	44.4	33.8	18.4
-90										18.2	34.0	44.8	48.8	45.1	34.4	18.4

Geo-Astro

MAJOR WORLD EARTHQUAKES

The United States Geological Survey maintains a database of historic earthquakes throughout the world (Ref. 1). The table below is extracted from that database; it includes more than 408 major earthquakes, based upon the magnitude and the degree of destruction. All recorded earthquakes of magnitude 7.5 or greater are listed, even if the fatalities are unknown or small. For recent years, smaller earthquakes of magnitude 6.0 or larger have been included if the death toll is significant. The death toll is often a rough estimate; in many cases the true toll could be much greater. More details on the exact location and degree of destruction can be found in Refs. 1 and 2.

The magnitude is given on the Richter scale, which was developed in 1935 by Charles F. Richter of the California Institute of Technology as a mathematical device to compare the size of earthquakes. The magnitude of an earthquake is measured by the logarithm of the amplitude of waves recorded by seismographs. Adjustments are included for the variation in the distance between the various seismographs and the epicenter of the earthquake. On the Richter scale, magnitude is expressed in whole numbers and decimal fractions, e.g., 6.3. Because of the logarithmic basis of the scale, each whole number increase in magnitude represents a tenfold increase in measured amplitude; as an estimate of energy, each whole number step in the magnitude scale corresponds to the release of about 31 times more energy than the amount associated with the preceding whole number value.

References

1. Earthquakes, <earthquake.usgs.gov/earthquakes/>.
2. <en.wikipedia.org/wiki/List_of_earthquakes_in_2019>.

Major World Earthquakes, 856–2020

Date	Location	Magnitude	Fatalities
856/12/22	Damghan, Iran		200,000
893/03/23	Ardabil, Iran		150,000
1138/08/09	Aleppo, Syria		230,000
1268	Silicia, Asia Minor		60,000
1290/09/27	Chihli, China		100,000
1556/01/23	Shensi, China	8.0	830,000
1619/02/14	Trujillo, Peru	7.7	350
1667/11	Shemakha, Caucasia		80,000
1668/08/17	Anatolia, Turkey	8.0	
1687/10/20	Lima, Peru	8.5	
1693/01/11	Sicily, Italy	7.5	60,000
1700/01/26	Cascadia Subduction Zone (Oregon to British Columbia)	9.0	
1727/11/18	Tabriz, Iran		77,000
1730/07/08	Valparasio, Chile	8.7	
1755/11/01	Lisbon, Portugal	8.7	70,000
1783/02/04	Calabria, Italy		50,000
1787/05/02	Puerto Rico	8.0	
1811/12/16	New Madrid region, Missouri	8.1	
1812/02/07	New Madrid region, Missouri	8.0	
1812/03/26	Caracas, Venezuela	7.7	
1812/12/08	Southwest of San Bernardino County, California	6.9	40
1812/12/21	West of Ventura, California	7.1	1
1821/07/10	Camana, Peru	8.2	162
1835/02/20	Concepcion, Chile	8.2	500
1843/02/08	Leeward Islands	8.3	
1855/01/23	Wellington, New Zealand	8.0	4
1857/01/09	Fort Tejon, California	7.9	1
1868/04/03	Ka'u District, Island of Hawaii	7.9	77
1868/08/13	Arica, Peru (now Chile)	9.0	400
1872/03/26	Owens Valley, California	7.4	27
1877/05/10	Offshore Tarapaca, Chile	8.3	34
1886/09/01	Charleston, South Carolina	7.3	60
1887/05/03	Northern Sonora, Mexico	7.4	51
1891/10/27	Mino-Owari, Japan	8.0	
1892/02/24	Imperial Valley, California	7.8	
1896/06/15	Sanriku, Japan	8.5	28,000
1897/06/12	Assam, India	8.3	1,500
1899/09/04	Cape Yakataga, Alaska	7.9	
1899/09/10	Yakutat Bay, Alaska	8.0	
1900/10/09	Kodiak Island, Alaska	7.7	
1902/04/19	Quezaltenango and San Marcos, Guatemala	7.5	2,000
1902/12/16	Eastern Uzbekistan (Turkestan)	6.4	4,700
1903/04/28	Malazgirt, Turkey	7.0	3,500
1903/05/28	Gole, Turkey (Ottoman Empire)	5.8	1,000
1903/08/11	Southern Greece	8.3	
1905/04/04	Kangra, India	7.5	19,000
1905/07/09	Mongolia	8.4	
1905/09/08	Calabria, Italy	7.9	557
1906/01/31	Off the coast of Esmeraldas, Ecuador	8.8	1,000
1906/03/16	Chia-i, Taiwan	6.8	1,250
1906/04/18	San Francisco, California	7.8	3,000
1906/08/17	Valparaiso, Chile	8.2	20,000
1907/01/14	Kingston, Jamaica	6.5	800-1,000
1907/04/15	Guerrero, Mexico	7.7	
1907/10/21	Qaratog, Tajikistan, Russia	8.0	12,000
1908/12/12	Off the coast of Central Peru	8.2	
1908/12/28	Messina, Italy	7.2	72,000
1909/01/23	Silakhor, Iran (Persia)	7.3	5,000-6,000
1910/04/12	Taiwan region	7.6	
1911/01/03	Chong-Kemin, Kyrgyzstan	7.8	450
1911/02/18	Sarez, Tajikistan	7.4	90
1911/06/07	Off Guerrero, Mexico	7.7	45
1912/08/09	Murefte, Turkey (Ottoman Empire)	7.8	2,800
1914/10/03	Burdur, Turkey (Ottoman Empire)	7.0	4,000
1915/01/13	Avezzano, Italy	7.0	32,610
1917/01/20	Bali, Indonesia		1,500
1917/07/30	Daguan, Yunnan, China	7.5	1,800
1918/02/13	Nan'ao, Guangdong, (Kwangtung), China	7.3	1,000
1918/10/11	Mona Passage	7.5	116
1920/06/05	Taiwan region	8.0	
1920/12/16	Haiyuan, Ningxia, China	7.8	200,000
1922/11/11	Chile-Argentina border	8.5	100
1923/02/03	Kamchatka Peninsula	8.5	
1923/03/24	Near Luhuo, Sichuan, China	7.3	3,500
1923/05/25	Torbat-e Heydariyeh, Iran	5.7	2,200
1923/09/01	Kanto (Kwanto), Japan	7.9	142,800
1925/03/16	Yunnan, China	7.1	5,800

Date	Location	Magnitude	Fatalities
1927/03/07	Tango, Japan	7.6	3,020
1927/05/22	Tsinghai (Kansu), China	7.6	40,900
1928/12/01	Talca, Chile	7.6	225
1929/03/07	Fox Islands, Aleutian Islands, Alaska	7.8	
1929/05/01	Koppeh Dagh, Iran (Persia)	7.4	3,800
1930/05/06	Salmas, Iran (Persia)	7.2	2,500
1930/07/23	Irpinia, Italy	6.5	1,400
1931/01/15	Oaxaca, Mexico	7.8	114
1931/02/02	Hawke's Bay, New Zealand	7.9	256
1931/03/31	Managua, Nicaragua	6.0	2,500
1931/04/27	Zangezur Mountains, Armenia — Azerbaijan border	5.7	2,800
1931/08/10	Xinjiang, China	8.0	10,000
1932/06/03	Jalisco, Mexico	8.1	45
1932/06/18	Colima, Mexico	7.8	
1932/12/25	Gansu, China	7.6	275
1933/03/02	Sanriku, Japan	8.4	3,000
1933/03/11	Long Beach, California	6.4	115
1933/08/25	Sichuan, China	7.4	9,300
1934/01/15	Bihar, India — Nepal	8.1	10,700
1935/04/20	Taiwan (Formosa)	7.1	3,270
1935/05/30	Quetta, Pakistan	7.5	30,000
1935/07/16	Taiwan (Formosa)	6.5	2,740
1938/02/01	Banda Sea, Indonesia	8.5	
1938/11/10	Shumagin Islands, Alaska	8.2	
1939/01/25	Chillan, Chile	7.8	28,000
1939/12/26	Erzincan, Turkey	7.8	32,700
1940/05/19	Imperial Valley, California	7.1	9
1940/05/24	Callao, Peru	8.2	249
1940/11/10	Vrancea, Romania	7.3	1,000
1942/08/06	Guatemala	7.9	38
1942/08/24	Off the coast of central Peru	8.2	30
1942/11/26	Turkey	7.6	
1942/12/20	Erbaa, Turkey	7.3	1,100
1943/04/06	Illapel — Salamanca, Chile	8.2	25
1943/09/10	Tottori, Japan	7.4	1,190
1943/11/26	Ladik, Turkey	7.6	4,000
1944/01/15	San Juan, Argentina	7.4	8,000
1944/02/01	Gerede, Turkey	7.4	2,790
1944/12/07	Tonankai, Japan	8.1	998
1945/01/12	Mikawa, Japan	7.1	1,961
1945/11/27	Makran coast, Pakistan	8.0	4,000
1946/04/01	Unimak Island, Alaska	8.1	165
1946/05/31	Ustukran, Turkey	5.9	840-1,300
1946/08/04	Samana, Dominican Republic	8.0	100
1946/11/10	Ancash, Peru	7.3	1,400
1946/12/20	Nankaido, Japan	8.1	1,362
1947/11/01	Satipo, Peru	7.3	233
1948/05/11	Moquegua, Peru	7.4	70
1948/05/25	Sichuan, China	7.3	800
1948/06/28	Fukui, Japan	7.3	3,769
1948/10/05	Ashgabat, Turkmenistan	7.3	110,000
1949/04/13	Puget Sound, Washington	7.1	8
1949/07/10	Khait, Tajikistan	7.5	12,000
1949/08/05	Ambato, Ecuador	6.8	5,050
1949/08/22	Queen Charlotte Islands, British Columbia, Canada	8.1	
1950/08/15	Assam — Tibet	8.6	1,526
1951/08/02	Cosiguina, Nicaragua	5.8	1,000
1952/07/21	Kern County, California	7.3	12
1952/11/04	Kamchatka Peninsula	9.0	
1953/02/12	Torud, Iran	6.5	970
1953/03/18	Yenice-Gonen, Turkey	7.3	1,070
1953/08/12	Kefallinia, Greece	7.1	455
1953/12/12	Tumbes, Peru	7.4	7
1954/03/29	Spain	7.9	
1954/04/30	Greece	7.1	31
1954/09/09	Orleansville, Algeria	6.8	1,250
1957/03/09	Andreanof Islands, Alaska	8.6	
1957/04/25	Fethiye, Turkey	7.1	15
1957/05/26	Bolu Province, Turkey	7.1	66
1957/06/27	Stanovoy Mountains, Russia (USSR)	7.6	1,200
1957/07/02	Mazandaran, Iran	7.1	1,200
1957/07/28	Guerrero, Mexico	7.9	68
1957/12/04	Gobi-Altay, Mongolia	8.1	30
1957/12/13	Sahneh, Iran	7.1	1,130
1958/01/15	Arequipa, Peru	7.3	26
1958/07/10	Lituya Bay, Alaska	7.7	5
1958/11/06	Kuril Islands	8.3	
1959/04/26	Taiwan region	7.5	2
1959/08/18	Hebgen Lake, Montana	7.3	28
1960/01/13	Arequipa, Peru	7.5	57
1960/02/29	Agadir, Morocco	5.7	12,000-15,000
1960/05/21	Arauco Peninsula, Chile	7.9	
1960/05/22	Chile (off coast)	9.5	1,655
1962/05/11	Guerrero, Mexico	7.0	4
1962/05/19	Guerrero, Mexico	7.1	3
1962/09/01	Qazvin, Iran	7.1	12,225
1963/07/26	Skopje, Macedonia	6.0	1,100
1963/10/13	Kuril Islands	8.5	
1964/03/28	Prince William Sound, Alaska	9.2	128
1964/06/16	Niigata, Japan	7.5	26
1964/10/06	Western Turkey	7.0	36
1965/01/24	Sanana, Indonesia (Ceram Sea)	7.6	71
1965/02/04	Rat Islands, Alaska	8.7	
1965/03/14	Hindu Kush, Afghanistan	7.8	
1965/03/28	La Ligua, Chile	7.4	400
1965/03/31	Central Greece	7.1	6
1965/08/23	Oaxaca, Mexico	7.3	6
1966/03/07	Hebei, China	7.0	1,000
1966/03/22	Hebei, China	6.9	1,000
1966/08/19	Varto, Turkey	6.8	2,529
1966/10/17	Near the coast of Peru	8.1	125
1967/07/22	Mudurnu Valley, Turkey	7.3	173
1968/05/23	Inangahua, New Zealand	7.1	2
1968/08/02	Oaxaca, Mexico	7.1	18
1968/08/31	Dasht-e Bayaz, Iran	7.3	12,000
1969/02/28	Portugal-Morocco area	7.8	13
1969/07/25	Guangdong, China	5.9	3,000
1970/01/04	Yunnan Province, China	7.5	10,000
1970/03/28	Gediz, Turkey	6.9	1,086
1970/05/31	Chimbote, Peru	7.9	70,000
1970/07/31	Colombia	8.0	1
1971/02/09	San Fernando, California	6.6	65
1971/05/22	Eastern Turkey	6.9	1,000
1971/07/09	Valparaiso region, Chile	7.5	90
1972/01/25	Taiwan region	7.5	
1972/04/10	Southern Iran	7.1	5,054
1972/04/24	Taiwan region	7.2	4
1972/07/30	Sitka, Alaska	7.6	
1972/12/23	Nicaragua	6.2	5,000
1974/05/10	Near Zhaotong, China	6.8	20,000
1974/07/13	Panama-Colombia border region	7.3	11
1974/10/03	Near the coast of central Peru	8.1	78
1974/10/08	Leeward Islands	7.5	

Date	Location	Magnitude	Fatalities
1974/12/28	Northern Pakistan	6.2	5,300
1975/02/02	Near Islands, Alaska	7.6	
1975/02/04	Haicheng, China	7.0	2,000
1975/09/06	Diyarbakir Province, Turkey	6.7	2,300
1975/11/29	Kalapana, Hawaii	7.2	2
1976/02/04	Guatemala	7.5	23,000
1976/05/06	Northeastern Italy	6.5	1,000
1976/06/25	Papua, Indonesia	7.1	422
1976/07/27	Tangshan, China	7.5	255,000
1976/08/16	Mindanao, Philippines	7.9	8,000
1976/11/24	Turkey-Iran border region	7.3	5,000
1977/03/04	Romania	7.2	1,500
1978/09/16	Iran	7.8	15,000
1979/02/28	Mt. St. Elias, Alaska	7.5	
1980/10/10	El Asnam (formerly Orleansville), Algeria	7.7	5,000
1981/02/24	Greece	6.8	3,000
1981/06/11	Southern Iran	6.9	3,000
1981/07/28	Southern Iran	7.3	1,500
1982/12/13	Yemen	6.0	2,800
1983/10/30	Erzurum Province, Turkey	6.9	1,342
1985/03/03	Offshore Valparaiso, Chile	7.8	177
1985/09/19	Michoacan, Mexico	8.0	9,500
1986/05/07	Andreanof Islands, Alaska	7.9	
1986/10/10	El Salvador	5.5	1,000
1987/03/06	Colombia-Ecuador	7.0	1,000
1987/11/30	Gulf of Alaska	7.8	
1988/03/06	Gulf of Alaska	7.7	
1988/08/20	Nepal-India border region	6.8	1,000
1988/12/07	Spitak, Armenia	6.8	25,000
1989/10/18	Loma Prieta, California	6.9	63
1990/06/20	Western Iran	7.4	50,000
1990/07/16	Luzon, Philippine Islands	7.7	1,621
1991/04/22	Costa Rica	7.6	47
1991/10/19	Northern India	6.8	2,000
1992/09/02	Nicaragua	7.6	116
1992/12/12	Flores region, Indonesia	7.8	2,500
1993/08/08	South of the Mariana Islands	7.8	
1993/09/29	Latur-Killari, India	6.2	9,748
1994/01/17	Northridge, California	6.7	60
1994/06/09	Bolivia	8.2	10
1995/01/16	Kobe, Japan	6.9	5,502
1995/05/27	Sakhalin Island	7.1	1,989
1996/06/10	Andreanof Islands, Alaska	7.9	
1997/05/10	Northern Iran	7.3	1,567
1997/10/14	South of Fiji Islands	7.8	
1997/12/05	Near east coast of Kamchatka	7.8	
1998/01/04	Loyalty Islands region	7.5	
1998/02/04	Afghanistan-Tajikistan border region	5.9	2,323
1998/03/25	Balleny Islands region (off Antarctica)	8.1	
1998/05/03	Southeast of Taiwan	7.5	
1998/05/30	Afghanistan-Tajikistan border region	6.6	4,000
1998/07/17	Near north coast of New Guinea, Papua New Guinea	7.0	2,183
1999/01/25	Colombia	6.1	1,185
1999/08/17	Izmit, Turkey	7.6	17,118
1999/09/20	Taiwan	7.6	2,400
1999/09/30	Oaxaca, Mexico	7.5	
1999/11/12	Duzce, Turkey	7.2	894
2000/06/04	Southern Sumatera, Indonesia	7.9	103
2000/06/18	South Indian Ocean	7.9	
2000/11/16	New Ireland Region, Papua New Guinea	8.0	2
2001/01/01	Mindanao, Philippines	7.5	
2001/01/13	El Salvador	7.7	852
2001/01/26	Gujarat, India	7.6	20,085
2001/02/13	El Salvador	6.6	315
2001/06/23	Near the coast of Peru	8.4	138
2002/03/03	Hindu Kush region, Afghanistan	7.4	166
2002/03/05	Mindanao, Philippines	7.5	15
2002/03/25	Hindu Kush region, Afghanistan	6.1	1,000
2002/03/31	Taiwan region	7.1	5
2002/08/19	Fiji Islands	7.7	
2002/09/08	New Guinea, Papua New Guinea	7.6	
2002/10/10	Irian Jaya, Indonesia	7.6	8
2002/11/02	Northern Sumatera, Indonesia	7.4	3
2002/11/03	Denali Fault, Alaska	7.9	
2003/01/22	Offshore Colima, Mexico	7.6	29
2003/05/21	Northern Algeria	6.8	2,226
2003/05/26	Halmahera, Indonesia	7.0	1
2003/07/15	Carlsberg Ridge	7.6	
2003/08/04	Scotia Sea	7.6	
2003/09/25	Hokkaido, Japan region	8.3	
2003/09/27	Southwestern Siberia, Russia	7.3	3
2003/11/17	Rat Islands, Aleutian Islands, Alaska	7.8	
2003/12/26	Southeastern Iran	6.6	31,000
2004/02/05	Irian Jaya, Indonesia	7.0	37
2004/11/11	Kepulauan Alor, Indonesia	7.5	34
2004/11/26	Papua, Indonesia	7.1	32
2004/12/23	North of Macquarie Island, New Zealand	8.1	
2004/12/26	Sumatra-Andaman Islands	9.1	227,898
2005/03/28	Northern Sumatra, Indonesia	8.6	1,313
2005/06/13	Tarapaca, Chile	7.8	11
2005/09/09	New Ireland region, Papua New Guinea	7.6	
2005/09/26	Northern Peru	7.5	5
2005/10/08	Pakistan	7.6	86,000
2006/01/27	Banda Sea	7.6	
2006/02/22	Mozambique	7.0	4
2006/04/20	Koryakia, Russia	7.6	
2006/05/03	Tonga	8.0	
2006/05/26	Java, Indonesia	6.3	5,749
2006/07/17	South of Java, Indonesia	7.7	730
2006/11/15	Kuril Islands	8.3	
2006/12/26	Taiwan region	7.1	2
2007/01/13	East of the Kuril Islands	8.1	
2007/01/21	Molucca Sea	7.5	
2007/04/01	Solomon Islands	8.1	40
2007/08/08	Java, Indonesia	7.5	
2007/08/15	Near the coast of central Peru	8.0	514
2007/09/12	Southern Sumatra, Indonesia	8.5	25
2007/09/12	Kepulauan Mentawai region, Indonesia	7.9	
2007/09/28	Mariana Islands region	7.5	
2007/11/14	Antofagasta, Chile	7.7	2
2007/12/09	South of the Fiji Islands	7.8	
2008/05/12	Eastern Sichuan, China	7.9	87,587
2008/07/05	Sea of Okhotsk	7.7	
2009/01/03	Near the north coast of Papua, Indonesia	7.7	5
2009/03/19	Tonga region	7.6	
2009/04/06	Central Italy	6.3	295
2009/07/15	Off west coast of the South Island, New Zealand	7.8	

Date	Location	Magnitude	Fatalities
2009/08/10	Andaman Islands, India region	7.5	
2009/09/29	Samoa Islands region	8.1	192
2009/09/30	Southern Sumatra, Indonesia	7.5	1117
2009/10/07	Santa Cruz Islands	7.8	
2009/10/07	Vanuatu, Coral Sea	7.7	
2010/01/12	Haiti region	7.0	316,000
2010/02/27	Offshore Maule, Chile	8.8	547
2010/03/08	Eastern Turkey	6.1	42
2010/04/04	Baja California, Mexico	7.2	2
2010/04/13	Southern Qinghai, China	6.9	2,968
2010/07/23	Moro Gulf, Mindanao, Philippines	7.6	
2011/02/21	South Island of New Zealand	6.1	181
2011/03/11	Near the east coast of Honshu, Japan	9.0	20,896
2012/02/06	Negros-Cebu region, Philippines	6.7	113
2012/04/11	Off west coast of northern Sumatra	8.6[a]	10
2012/08/11	Northwestern Iran	6.4	306
2012/11/07	Offshore Guatamala	7.4	139
2013/01/05	Southeastern Alaska	7.5	
2013/02/06	Santa Cruz Islands	8.0	18
2013/04/06	Khash, Iran	7.7	35
2013/04/20	Linqiong, China	6.6	193
2013/05/24	Sea of Okhotsk	8.3	
2013/09/24	Awaran, Pakistan	7.7	825
2013/10/15	Sagbayan, Philippines	7.1	222
2013/11/17	Scotia Sea, Antarctica	7.8	
2014/04/03	Off west coast of northern Chile	7.7	6
2014/04/12	Solomon Islands	7.6	
2014/04/19	Panguna, Papua New Guinea	7.5	1
2014/06/23	Little Sitkin Island, Alaska	7.9	
2014/08/03	Wewnping, China	6.2	617
2015/04/25	E of Khudi, Nepal	7.8	9.000
2015/05/12	SE of Kodan, Nepal	7.3	218
2015/05/30	WNW of Chici-shima, Japan	7.8	0
2015/09/16	W of Illapel, Chile	8.3	19
2015/10/25	N of Alaqahdari-ye Kiran wa Munyan, Afghanistan	7.5	398
2016/02/06	SE of Yujing, Taiwan	6.4	117
2016/03/02	Southwest of Sumatra, Indonesia	7.8	
2016/04/16	SSE of Muisne, Ecuador	7.8	673
2016/07/29	SW of Agrihan, Northern Mariana Islands	7.7	
2016/08/19	South Georgia Island region	7.4	
2016/08/24	SE of Norcia, Italy	6.2	299
2016/11/13	NNE of Amberley, New Zealand	7.8	2
2016/12/08	Kirakira, Solomon Islands	7.8	
2016/12/17	E of Taron, Papua New Guinea	7.9	
2016/12/25	SW of Puerto Quellon, Chile	7.6	
2017/01/10	SSE of Tabiauan, Philippines	7.3	
2017/01/22	WNW of Panguna, Papua New Guinea	7.9	
2017/07/17	ESE of Nikol'skoye, Russia	7.7	
2017/09/08	SSW of Tres Picos, Mexico	8.2	98
2017/09/19	E of Ayutla, Mexico	7.1	370
2017/11/12	S of Halabjah, Iraq	7.3	630
2017/11/19	ENE of Tadine, New Caledonia	7.0	
2018/01/23	SE of Kodial, Alaska, United States	7.9	
2018/02/06	NNE of Hualin, Taiwan	6.4	17
2018/02/16	S of San Pedro Jicayan, Mexico	7.2	14
2018/02/25	SW of Porgera, Papua, New Guinea	7.5	160
2018/03/04	SW of Porgera, Papua, New Guinea	6.0	11
2018/03/06	SW of Porgera, Papua, New Guinea	6.7	25
2018/07/28	WNW of Obelobel, Indonesia	6.4	20
2018/05/04	SSW of Leilani Estates, Hawaii	6.9	
2018/08/05	SW of Loloan, Indonesia	6.9	513
2018/08/19	S of Belanting, Indonesia	6.9	14
2018/09/05	ENE of Tomakomai, Japan	6.6	41
2018/09/28	N of Paulu, Indonesia	7.5	2256
2018/10/07	WNW of Ti Port-de-Paix, Haiti	5.9	18
2018/11/30	N of Anchorage, Alaska, United States	7.0	
2018/12/05	ESE of New Caledonia, Loyalty Islands	7.5	
2018/12/11	South Georgia and the South Sandwich Islands, 48 km north of Bristol Island	7.1	
2018/12/20	Russia, Komandorski Islands offshore, 83 km west of Nikolskoye	7.3	
2018/12/29	Philippines offshore, 83 km east southeast of Pundaguitan	7.0	
2019/02/22	ESE Palora, Ecuador	7.5	1
2019/03/01	NNE Azangaro, Peru	7.0	1
2019/03/17	Sembalunbumbung, Indonesia	5.6	6
2019/04/22	Gutad, Central Luzon, Philippines	6.1	18
2019/05/14	SSE Namatanai, Papua, New Guinea	7.6	
2019/05/26	SE Lagunas, Peru	8.0	2
2019/06/17	S Changing, Sichuan, China	5.8	13
2019/07/06	Ridgecrest, California	7.1	
2019/07/14	NNE Laiwui, Indonesia	7.2	14
2019/07/26	Itbayat, Batanes, Philippines	5.9	9
2019/08/02	Batan, Indonesia	6.9	8
2019/09/24	Mirpur, Azad Kasmir, Pakistan	5.6	40
2019/09/25	Kairatu, Maluku, Indonesia	6.5	41
2019/10/16	Columbo, Soccsksargen, Philippines	6.3	7
2019/10/29	Bual, Soccsksargen, Philippines	6.6	14
2019/10/31	Kisante, Soccsksargen, Philippines	6.5	10
2019/11/07	Hashtrud, East Azerbaijan, Iran	5.9	7
2019/11/14	E Bitung, Indonesia	7.1	1
2019/11/26	Durres, Albania	6.4	51
2019/12/15	Davao, Philippines	6.8	13
2020/01/28	Luca, Jamaica	7.7	
2020/06/18	Kermadec Islands, New Zealand	7.4	
2020/06/23	Santa María Xadani, Mexico	7.4	10
2020/07/17	Popondetta, Papua New Guinea	7.0	1
202/07/22	Perryville, Alaska	7.8	
2020/10/19	Sand Point, Alaska	7.6	
2020/10/30	Néon Karlovásion, Greece	7.0	119
2020/12/29	Petrinja, Sisak-Moslavina, Croatia	6.4	7

[a] Initial shock of magnitude 8.2.

INFRARED ABSORPTION BY THE EARTH'S ATMOSPHERE

Several constituents of the Earth's atmosphere absorb infrared radiation. At ground level the strongest absorbers are H_2O and CO_2, but 30 to 40 other compounds can make significant contributions. The centers of the most important absorption bands are listed below:

Molecule	Vibrational mode	Band center in cm^{-1}
H_2O	Bend	1595
H_2O	Symmetric O-H stretch	3657
H_2O	Antisymmetric O-H stretch	3756
CO_2	Bend	667
CO_2	Antisymmetric C-O stretch	2349
O_3	Bend	701
O_3	Antisymmetric O-O stretch	1042
O_3	Symmetric O-O stretch	1103
N_2O	Bend	589
N_2O	N-O stretch	1285
N_2O	N-N stretch	2224
CO	C-O stretch	2143
CH_4	Degenerate deformation	1306
CH_4	Degenerate stretch	3019

The HITRAN Molecular Spectroscopy Database (Refs. 1 and 2) is a compilation of wavenumbers and intensities of more than 1.7 million spectral lines of atmospheric constituents. It is a valuable resource for calculating transmission of the atmosphere, radiative energy transfer, and other phenomena. The graph below, which was supplied by Walter J. Lafferty (Ref. 3), gives the transmittance of the atmosphere for one set of conditions.

References

1. Rothman, L. S., et al., *J. Quant. Spectros. Radiat. Trans*fer 82, 5, 2003; *ibid.*, to be published, 2005.
2. HITRAN Molecular Spectroscopy Database, <http://cfa-www.Harvard.edu/HITRAN/hitrandata04/>.
3. Lafferty, W. J., Some Aspects of High Resolution Molecular Spectroscopy, in *Lectures on Molecular Physics*, Institute for the Structure of Matter, Centro de Fisica Miguel A. Catalan, Madrid, 1997.

Transmittance of U.S. Standard Atmosphere at Ground Level for a Path of 1 km at 296 K

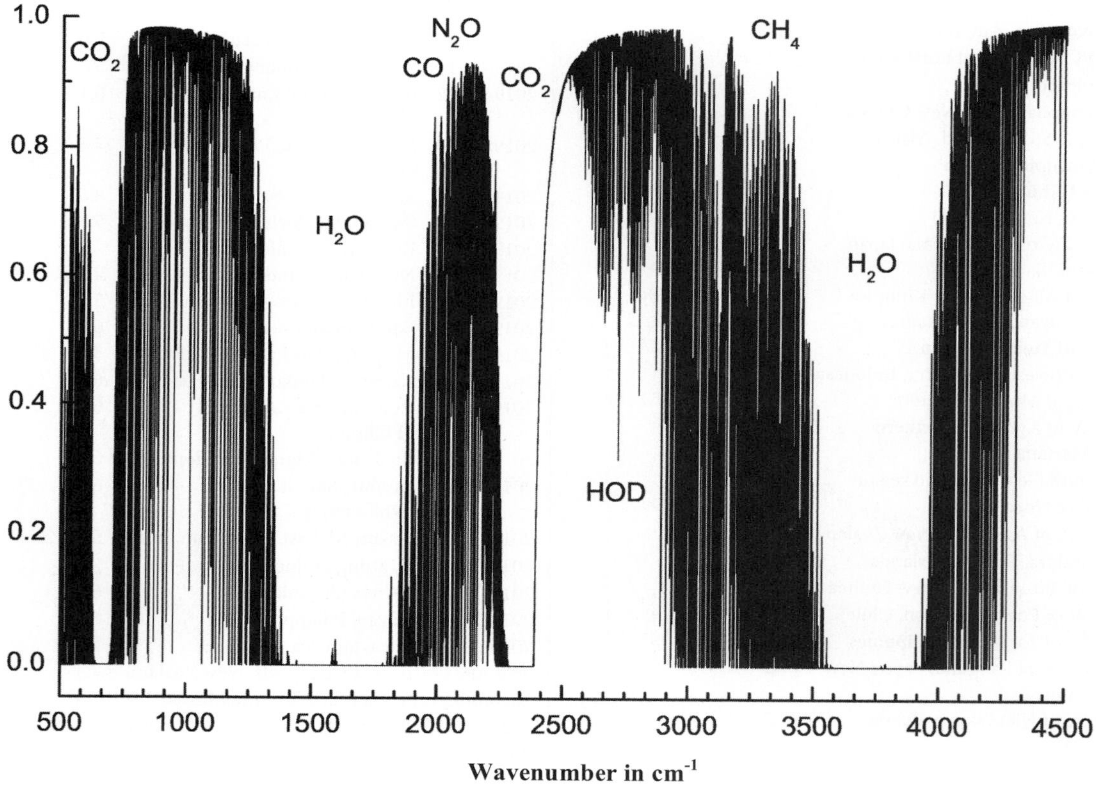

ATMOSPHERIC CONCENTRATION OF CARBON DIOXIDE, 1959–2020

The data summarized in this table were taken at the Mauna Loa Observatory in Hawaii and represent averages over each year. The concentration of CO_2 is given in parts per million (ppm) by volume (i.e., μmol/mol of dry air). The 2020 value from Ref. 1 is from January 18, 2021, and may change very slightly with further analysis. The standard deviation of the concentration values is estimated to be 0.12 ppm. Data from other measurement sites, as well as monthly data from Mauna Loa, may be found in Ref. 2. Additional data on atmospheric CO_2 is available in Refs. 3-4.

The first graph (Ref. 1) illustrates the seasonal variation of CO_2 concentration and the steady increase over the last 62 years. The second graph (Ref. 4) summarizes the growth in global emissions of CO_2 into the atmosphere as a result of burning of fossil fuels.

References

1. NOAA-ESRL Data, <esrl.noaa.gov/gmd/webdata/ccgg/trends/co2/co2_annmean_mlo.txt>, January 18, 2021.
2. Trends in Atmospheric Carbon Dioxide, <http://www.esrl.noaa.gov/gmd/ccgg/trends/>, 2020.
3. ESS-DIVE, <ess-dive.lbl.gov>.
4. CO_2 Earth, <www.co2.earth>.

Mean Annual CO_2 Concentration at Mauna Loa 1959–2020

Year	Mean CO_2 conc. in ppm	Year	Mean CO_2 conc. in ppm	Year	Mean CO_2 conc. in ppm	Year	Mean CO_2 conc. in ppm
1959	315.97	1975	331.11	1991	355.61	2007	383.79
1960	316.91	1976	332.04	1992	356.45	2008	385.60
1961	317.64	1977	333.83	1993	357.10	2009	387.43
1962	318.45	1978	335.40	1994	358.83	2010	389.90
1963	318.99	1979	336.84	1995	360.82	2011	391.65
1964	319.62	1980	338.75	1996	362.61	2012	393.85
1965	320.04	1981	340.11	1997	363.73	2013	396.52
1966	321.38	1982	341.45	1998	366.70	2014	398.65
1967	322.16	1983	343.05	1999	368.38	2015	400.83
1968	323.04	1984	344.65	2000	369.55	2016	404.24
1969	324.62	1985	346.12	2001	371.14	2017	406.55
1970	325.68	1986	347.42	2002	373.28	2018	408.52
1971	326.32	1987	349.18	2003	375.80	2019	411.43
1972	327.45	1988	351.57	2004	377.52	2020	414.01
1973	329.68	1989	353.12	2005	379.80		
1974	330.18	1990	354.39	2006	381.90		

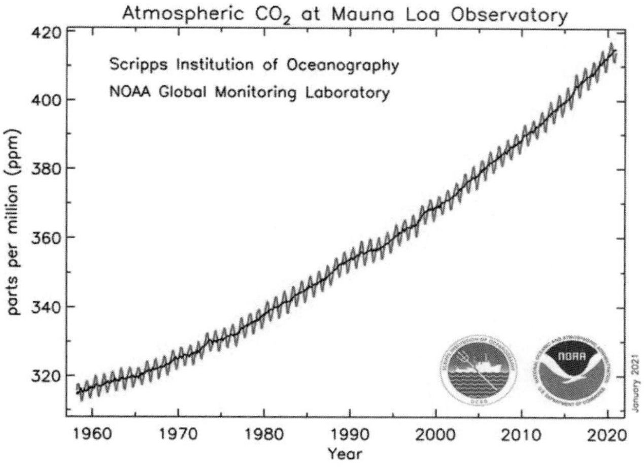

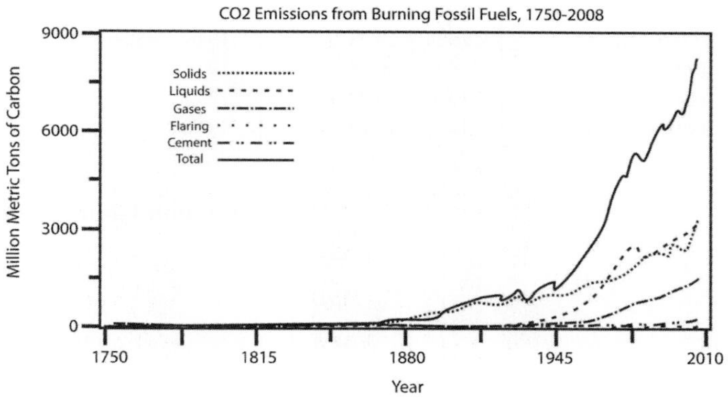

GLOBAL TEMPERATURE TREND, 1880–2020

This table summarizes the trend in annual mean global surface temperature from 1880 to 2020. The values were calculated from the global mean temperature anomalies in Ref. 1 by assuming an absolute global mean of 14.00 °C, which is the best estimate for the 1951–1980 period. The 95% confidence interval for comparing the annual mean temperature values for recent years is 0.05 °C. The values given here are for the annual mean land-ocean temperature.

Ref. 2 gives links to more extensive data sets that cover individual months, seasons, and zonal regions. Ref. 3 describes aspects of modeling global temperature.

References

1. Annual Mean Land-Ocean Temperature Index in 0.01 Degrees Celsius: Selected Zonal Means, <data.giss.nasa.gov/gistemp/tabledata_v4/GLB.Ts+dSST.txt>, January 18, 2021.
2. NASA Goddard Institute for Space Studies, <data.giss.nasa.gov/gistemp>.
3. Lenssen, N., Schmidt, G., Hansen, J., Menne, M., Persin, A., Ruedy,R., and Zyss, D., Improvements in the GISTEMP Uncertainty Model, *J. Geophys. Res. Atmos.*, 124, 6307-6326, 2019. <doi:10.1029/2018JD029522>.

Mean Global Temperature 1880–2020

Year	Mean t/°C	Year	Mean t/°C	Year	Mean t/°C	Year	Mean t/°C	Year	Mean t/°C	Year	Mean t/°C
1880	13.84	1904	13.54	1928	13.80	1952	14.01	1976	13.90	2000	14.41
1881	13.93	1905	13.74	1929	13.64	1953	14.08	1977	14.18	2001	14.54
1882	13.90	1906	13.78	1930	13.84	1954	13.87	1978	14.07	2002	14.63
1883	13.83	1907	13.62	1931	13.91	1955	13.86	1979	14.16	2003	14.62
1884	13.72	1908	13.58	1932	13.84	1956	13.81	1980	14.26	2004	14.54
1885	13.68	1909	13.52	1933	13.72	1957	14.05	1981	14.32	2005	14.68
1886	13.70	1910	13.57	1934	13.87	1958	14.06	1982	14.14	2006	14.64
1887	13.65	1911	13.56	1935	13.80	1959	14.03	1983	14.31	2007	14.67
1888	13.84	1912	13.64	1936	13.85	1960	13.98	1984	14.15	2008	14.55
1889	13.91	1913	13.66	1937	13.97	1961	14.06	1985	14.11	2009	14.66
1890	13.66	1914	13.85	1938	14.00	1962	14.04	1986	14.18	2010	14.72
1891	13.78	1915	13.86	1939	13.98	1963	14.05	1987	14.32	2011	14.61
1892	13.74	1916	13.64	1940	14.13	1964	13.80	1988	14.38	2012	14.65
1893	13.69	1917	13.54	1941	14.19	1965	13.89	1989	14.27	2013	14.68
1894	13.71	1918	13.71	1942	14.07	1966	13.94	1990	14.45	2014	14.75
1895	13.79	1919	13.73	1943	14.09	1967	13.98	1991	14.40	2015	14.90
1896	13.90	1920	13.73	1944	14.20	1968	13.92	1992	14.22	2016	15.02
1897	13.90	1921	13.81	1945	14.09	1969	14.05	1993	14.23	2017	14.93
1898	13.74	1922	13.72	1946	13.93	1970	14.02	1994	14.31	2018	14.85
1899	13.83	1923	13.74	1947	13.97	1971	13.92	1995	14.44	2019	14.99
1900	13.93	1924	13.73	1948	13.89	1972	14.01	1996	14.32	2020	15.02
1901	13.85	1925	13.78	1949	13.89	1973	14.16	1997	14.46		
1902	13.73	1926	13.89	1950	13.83	1974	13.93	1998	14.61		
1903	13.63	1927	13.78	1951	13.93	1975	13.99	1999	14.41		

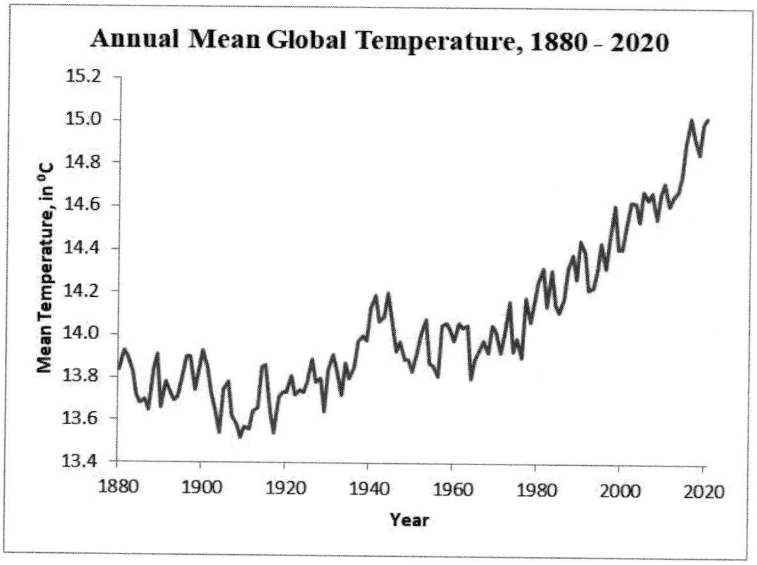

GLOBAL WARMING POTENTIAL OF GREENHOUSE GASES

The Global Warming Potential (GWP) of a gas is a measure of the degree, relative to carbon dioxide, to which the presence of that gas in the atmosphere will contribute to a long-term increase in global temperature. The calculation of the GWP for a given gas takes into account the efficiency of the gas in absorbing solar radiation (primarily determined by the infrared spectrum of the compound) and the time the compound will remain in the atmosphere before it is removed by natural processes. Thus, if a pulse of 1 kg of the gas is emitted to the atmosphere at the same time as a pulse of 1 kg of CO_2, the GWP compares the warming effect of the gas relative to CO_2 over various time horizons.

This table, which is taken from the 2014 and 2007 reports of the Intergovernmental Panel of Climate Change (IPCC) (Refs. 1 and 2), gives the lifetime in years and the radiative efficiency for the major compounds identified in the Kyoto Protocol as contributing to global climate change. The data columns are defined as follows.

Column heading	Definition
Name	Compound chemical name
Synonym/Code	Common synynom or industrial designation
Mol. form.	Molecular formula
Lifetime	Lifetime of the compound, in years
Radiative efficiency	A measure of the radiative forcing that influences the energy balance in the Earth–atmosphere system; in units of watts per square meter for a concentration of one part per billion ($W\ m^{-2}\ ppb^{-1}$) (see Refs. 1 and 2)
GWP 100-yr est. (2014)	Global Warming Potential for a 100-year horizon; as estimated by the IPCC in 2014 (Ref. 2)
GWP 100-yr est. (2007)	Global Warming Potential for a 100-year horizon; as estimated by the IPCC in 2007 (Ref. 1)
GWP 100-yr est. (1995)	Global Warming Potential for a 100-year horizon; as estimated by the IPCC in 1995
GWP 20-yr est. (2007)	Global Warming Potential for a 20-year horizon; as estimated by the IPCC in 2007 (Ref. 1)
GWP 500-yr est. (2007)	Global Warming Potential for a 500-year horizon; as estimated by the IPCC in 2007 (Ref. 1)

The calculation of a GWP involves a number of assumptions, and other measures have been proposed (see Refs 1 and 2).

The list of compounds includes those identified in the Montreal Protocol as contributing to ozone depletion, since these compounds also contribute to global warming. It also includes compounds used or proposed as replacements for the ozone-depleting compounds that still have global warming potential.

References

1. Forster, P., V. Ramaswamy, P. Artaxo, T. Berntsen, R. Betts, D. W. Fahey, J. Haywood, J. Lean, D.C. Lowe, G. Myhre, J. Nganga, R. Prinn, G. Raga, M. Schulz, and R. Van Dorland, 2007: Changes in Atmospheric Constituents and in Radiative Forcing. In: *Climate Change 2007: The Physical Science Basis. Contribution of* Working *Group I to the Fourth Assessment Report of the Intergovernmental Panel on Climate Change* [Solomon, S., D. Sin, M. Manning, Z. Chen, M. Marquis, K. B. Averyt, M. Tignor and H.L. Miller (eds.)], Cambridge University Press, Cambridge, United Kingdom, and New York, NY, <http://ipcc-wg1.ucar.edu/wg1/wg1-report.html>.
2. Myhre, G., D. Shindell, F.-M. Bréon, W. Collins, J. Fuglestvedt, J. Huang, D. Koch, J.-F. Lamarque, D. Lee, B. Mendoza, T. Nakajima, A. Robock, G. Stephens, T. Takemura and H. Zhang, 2013: Anthropogenic and Natural Radiative Forcing. In: *Climate Change 2013: The Physical Science Basis. Contribution of Working Group I to the Fifth Assessment Report of the Intergovernmental Panel on Climate Change* [Stocker, T.F., D. Qin, G.-K. Plattner, M. Tignor, S.K. Allen, J. Boschung, A. Nauels, Y. Xia, V. Bex and P.M. Midgley (eds.)], Cambridge University Press, Cambridge, United Kingdom, and New York, NY, <https://www.ipcc.ch/pdf/assessment-report/ar5/wg1/WG1AR5_Chapter08_FINAL.pdf>.

Lifetime, Radiative Efficiency, and Global Warming Potential (GWP) of Greenhouse Gases

Name	Synonym/Code	Mol. form.	Lifetime (years)	Radiative efficiency/ $W\ m^{-2}\ ppb^{-1}$	GWP 100-yr est. (2014)	GWP 100-yr est. (2007)	GWP 100-yr est. (1995)	GWP 20-yr est. (2007)	GWP 500-yr est. (2007)
Natural atmospheric constituents									
Carbon dioxide		CO_2		0.000014	1	1	1	1	1
Methane		CH_4	12	0.00037	28	25	21	72	7.6
Nitrous oxide		N_2O	114	0.00303		298	310	289	153
Substances controlled by the Montreal Protocol									
Trichlorofluoromethane	CFC-11	CCl_3F	45	0.25	4600	4750	3800	6730	1620
Dichlorodifluoromethane	CFC-12	CF_2Cl_2	100	0.32	10200	10900	8100	11000	5200
Chlorotrifluoromethane	CFC-13	CF_3Cl	640	0.25	13900	14400		10800	16400
1,1,2-Trichloro-1,2,2-trifluoroethane	CFC-113	$CFCl_2CF_2Cl$	85	0.3	5820	6130	4800	6540	2700
1,2-Dichloro-1,1,2,2-tetrafluoroethane	CFC-114	$C_2Cl_2F_4$	300	0.31	8590	10000		8040	8730
Chloropentafluoroethane	CFC-115	CF_3CF_2Cl	1700	0.18	7670	7370		5310	9990
Bromotrifluoromethane	Halon-1301	CF_3Br	65	0.32	6290	7140	5400	8480	2760
Bromochlorodifluoromethane	Halon-1211	CF_2BrCl	16	0.3	1750	1890		4750	575
1,2-Dibromotetrafluoroethane	Halon-2402	CF_2BrCF_2Br	20	0.33	1470	1640		3680	503
Tetrachloromethane	Carbon tetrachloride	CCl_4	26	0.13	1730	1400	1400	2700	435
Bromomethane	Methyl bromide	CH_3Br	0.7	0.01	2	5		17	1
1,1,1-Trichloroethane	Methyl chloroform	CH_3CCl_3	5	0.06	160	146		506	45
Chlorodifluoromethane	HCFC-22	$CHClF_2$	12	0.2	1760	1810	1500	5160	549
2,2-Dichloro-1,1,1-trifluoroethane	HCFC-123	$CHCl_2CF_3$	1.3	0.14	79	77	90	273	24

Name	Synonym/Code	Mol. form.	Lifetime (years)	Radiative efficiency/ W m^{-2} ppb^{-1}	GWP 100-yr est. (2014)	GWP 100-yr est. (2007)	GWP 100-yr est. (1995)	GWP 20-yr est. (2007)	GWP 500-yr est. (2007)
1-Chloro-1,2,2,2-tetrafluoroethane	HCFC-124	$CHClFCF_3$	5.8	0.22	527	609	470	2070	185
1,1-Dichloro-1-fluoroethane	HCFC-141b	CH_3CCl_2F	9.3	0.14	782	725		2250	220
1-Chloro-1,1-difluoroethane	HCFC-142b	CH_3CClF_2	17.9	0.2	1980	2310	1800	5490	705
3,3-Dichloro-1,1,1,2,2-pentafluoropropane	HCFC-225ca	$CHCl_2CF_2CF_3$	1.9	0.2	127	122		429	37
1,3-Dichloro-1,1,2,2,3-pentafluoropropane	HCFC-225cb	$CHClFCF_2CClF_2$	5.8	0.32	525	595		2030	181
Hydrofluorocarbons									
Trifluoromethane	HFC-23	CHF_3	270	0.19	12400	14800	11700	12000	12200
Difluoromethane	HFC-32	CH_2F_2	4.9	0.11	677	675	650	2330	205
Pentafluoroethane	HFC-125	CF_3CF_2H	29	0.23	3170	3500	2800	6350	1100
1,1,2,2-Tetrafluoroethane	HFC-134	CHF_2CHF_2			1120		1000		
1,1,1,2-Tetrafluoroethane	HFC-134a	CF_3CH_2F	14	0.16	1300	1430	1300	3830	435
1,1,2-Trifluoroethane	HFC-134	CHF_2CH_2F			328		300		
1,1,1-Trifluoroethane	HFC-143a	CH_3CF_3	52	0.13	4800	4470	3800	5890	1590
1,1-Difluoroethane	HFC-152a	CH_3CHF_2	1.4	0.09	138	124	140	437	38
Fluoroethane	HFC-161	CH_3CH_2F			4				
1,1,1,2,3,3,3-Heptafluoropropane	HFC-227ea	CF_3CHFCF_3	34.2	0.26	3350	3220	2900	5310	1040
1,1,1,2,2,3-Hexafluoropropane	HFC-236cb	$CH_2FCF_2CF_3$			1210				
1,1,1,2,3,3-Hexafluoropropane	HFC-236ea				1330				
1,1,1,3,3,3-Hexafluoropropane	HFC-236fa	$CF_3CH_2CF_3$	240	0.28	8060	9810	6300	8100	7660
1,1,2,2,3-Pentafluoropropane	HFC-245ca	CH2FCF2CHF2			716				
1,1,1,3,3-Pentafluoropropane	HFC-245fa	$CHF_2CH_2CF_3$	7.6	0.28	858	1030		3380	314
1,1,1,3,3-Pentafluorobutane	HFC-365mfc	$CH_3CF_2CH_2CF_3$	8.6	0.21	804	794		2520	241
1,1,1,2,3,4,4,5,5,5-Decafluoropentane	HFC-43-10mee	$CF_3CHFCHFCF_2CF_3$	15.9	0.4	1650	1640	1300	4140	500
Perfluorinated compounds									
Sulfur hexafluoride		SF_6	3200	0.52	23500	22800	23900	16300	32600
Nitrogen trifluoride		NF_3	740	0.21	16100	17200		12300	20700
Tetrafluoromethane	PFC-14	CF_4	50000	0.10	6630	7390	6500	5210	11200
Hexafluoroethane	PFC-116	CF_3CF_3	10000	0.26	11100	12200	9200	8630	18200
Perfluoropropane	PFC-218	$CF_3CF_2CF_3$	2600	0.26	8900	8830	7000	6310	12500
Perfluorocyclobutane	PFC-318	$c-C_4F_8$	3200	0.32	9540	10300	8700	7310	14700
Perfluorobutane	PFC-3-1-10	C_4F_{10}	2600	0.33	9200	8860	7000	6330	12500
Perfluoropentane	PFC-4-1-12	C_5F_{12}	4100	0.41	8550	9160		6510	13300
Perfluorohexane	PFC-5-1-14	C_6F_{14}	3200	0.49	7910	9300	7400	6600	13300
Perfluorodecalin	PFC-9-1-18	$C_{10}F_{18}$	>1,000	0.56	7190	>7500		>5500	>9500
(Trifluoromethyl)sulfur pentafluoride		SF_5CF_3	800	0.57	17400	17700		13200	21200
Fluorinated ethers									
Trifluoromethyl difluoromethyl ether	HFE-125	CF_3OCF_2H	136	0.44	12400	14900		13800	8490
Bis(difluoromethyl) ether	HFE-134	$(CF_2H)_2O$	26	0.45	5560	6320		12200	1960
Methyl trifluoromethyl ether	HFE-143a	CH_3OCF_3	4.3	0.27	523	756		2630	230
2-Chloro-2-(difluoromethoxy)-1,1,1-trifluoroethane	HCFE-235da2	$CHF_2OCHClCF_3$	2.6	0.38	491	350		1230	106
Methyl 1,1,2,2-tetrafluoroethyl ether	HFE-245cb2	$CH_3OCF_2CHF_2$	5.1	0.32	654	708		2440	215
2-(Difluoromethoxy)-1,1,1-trifluoroethane	HFE-245fa2	$CHF_2OCH_2CF_3$	4.9	0.31	812	659		2280	200
Methyl pentafluoroethyl ether	HFE-254cb2	$CH_3OCF_2CHF_2$	2.6	0.28		359		1260	109
Perfluoropropyl methyl ether	HFE-347mcc3	$CH_3OCF_2CF_2CF_3$	5.2	0.34	530	575		1980	175
1,1,2,2-Tetrafluoroethyl 1,1,1-trifluoroethyl ether	HFE-347pcf2	$CF_2HCF_2OCH_2CF_3$	7.1	0.25	889	580		1900	175
1-Methoxy-1,1,2,2,3,3-hexafluoropropane	HFE-356pcc3	$CH_3OCF_2CF_2CHF_2$	0.33	0.93	413	110		386	33
Methyl nonafluorobutyl ether	HFE-449sl (HFE-7100)	$C_4F_9OCH_3$	3.8	0.31	421	297		1040	90
1-Ethoxy-1,1,2,2,3,3,4,4,4-nonafluorobutane	HFE-569sf2 (HFE-7200)	$C_4F_9OC_2H_5$	0.77	0.3	57	59		207	18
1-(Difluoromethoxy)-2-[(difluoromethoxy) difluoromethoxy]-1,1,2,2-tetrafluoroethane	HFE-43-10pccc124 (H-Galden 1040x)	$CHF_2OCF_2OC_2F_4OCHF_2$	6.3	1.37	2820	1870		6320	569
Bis(difluoromethoxy)difluoromethane	HFE-236ca12 (HG-10)	$CHF_2OCF_2OCHF_2$	12.1	0.66	5350	2800		8000	860
1,2-Bis(difluoromethoxy)-1,1,2,2-tetrafluoroethane	HFE-338pcc13 (HG-01)	$CHF_2OCF_2CF_2OCHF_2$	6.2	0.87	2910	1500		5100	460
Perfluoropolymethylisopropyl ether	PFPMIE	$CF_3OCF(CF_3)CF_2OCF_2OCF_3$	800	0.65	9710	10300		7620	12400
Other compounds - Direct effects									
Dimethyl ether	Methyl ether	CH_3OCH_3	0.015	0.02		1		1	<<1
Dichloromethane	Methylene chloride	CH_2Cl_2	0.38	0.03	9	8.7		31	2.7
Chloromethane	Methyl chloride	CH_3Cl	1	0.01	12	13		45	4

Geo-Astro

SPEED OF SOUND IN VARIOUS MEDIA

The speed of sound in various solids, liquids, and gases is given in these tables. While only a single parameter v is needed for liquids and gases, sound propagation in isotropic solids is characterized by three velocity parameters. For a solid of infinite extent (or of finite extent if all dimensions are much larger than a wavelength), there are two relevant quantities.

v_l: Velocity of longitudinal waves
v_s: Velocity of shear waves

For a cylindrical rod with diameter much smaller than a wavelength,

v_{ext}: Velocity of extensional waves along the rod. (Torsional waves in the rod are propagated at the same speed as sheer waves in an infinite solid.)

Table 1 lists values for a variety of solid materials. Table 2 covers liquids and gases; values for cryogenic liquids are given at the normal boiling point. Table 3 gives the speed of sound in pure water and in seawater of salinity S = 3.5% as a function of temperature. All values are in meters per second and are given for normal atmospheric pressure.

References

1. Gray, D. E., Ed., *American Institute of Physics Handbook, Third Edition*, McGraw-Hill, New York, 1972.
2. Anderson, H.L., Ed., *A Physicist's Desk Reference*, American Institute of Physics, New York, 1989.
3. Mason, W. P., *Physical Acoustics and the Properties of Solids*, D. Van Nostrand Co., Princeton, NJ, 1958.
4. Landolt-Börnstein, *Numerical Data and Functional Relationships in Science and Technology, New Series, II/5, Molecular Acoustics*, Springer-Verlag, Heidelberg, 1967.
5. Lemmon, E.W., Huber, M.L., and McLinden, M.O., *NIST Standard Reference Database 23: Reference Fluid Thermodynamic and Transport Properties-REFPROP*, Version 9.0, National Institute of Standards and Technology, Standard Reference Data Program, Gaithersburg, MD, 2010.
6. International Association for the Properties of Water and Steam (IAPWS), *Release on the IAPWS Formulation 2008 for the Thermodynamic Properties of Seawater* (2008), available from <http://www.iapws.org>.
7. Feistel, R., A Gibbs Function for Seawater Thermodynamics for −6 to 80 °C and Salinity up to 120 g kg−1, *Deep-Sea Res. I* 55, 1639-1671 (2008).
8. Millero, F.J., Feistel, R., Wright, D.G., and McDougall, T.J., The Composition of Standard Seawater and the Definition of the Reference-Composition Salinity Scale, *Deep-Sea Res. I* 55, 50-72 (2008).

TABLE 1. Speed of Sound in Solids at Room Temperature

Name	v_l/m s^{-1}	v_s/m s^{-1}	v_{ext}/m s^{-1}
Metals			
Aluminum, rolled	6420	3040	5000
Beryllium	12890	8880	12870
Brass (70 Cu, 30 Zn)	4700	2110	3480
Constantan	5177	2625	4270
Copper, annealed	4760	2325	3810
Copper, rolled	5010	2270	3750
Duralumin 17S	6320	3130	5150
Gold, hard-drawn	3240	1200	2030
Iron, cast	4994	2809	4480
Iron, electrolytic	5950	3240	5120
Iron, Armco	5960	3240	5200
Lead, annealed	2160	700	1190
Lead, rolled	1960	690	1210
Magnesium, annealed	5770	3050	4940
Molybdenum	6250	3350	5400
Monel metal	5350	2720	4400
Nickel	6040	3000	4900
Platinum	3260	1730	2800
Silver	3650	1610	2680
Steel (1% C)	5940	3220	5180
Steel, 347 Stainless	5790	3100	5000
Steel, K9	5940	3250	5250
Tin, rolled	3320	1670	2730
Titanium	6070	3125	5090
Tungsten, annealed	5220	2890	4620
Tungsten, drawn	5410	2640	4320
Zinc, rolled	4210	2440	3850
Other materials			
Fused silica	5968	3764	5760
Glass, heavy silicate flint	3980	2380	3720
Glass, light borate crown	5100	2840	4540
Glass, pyrex	5640	3280	5170
Lucite	2680	1100	1840
Nylon 6-6	2620	1070	1800
Polyethylene	1950	540	920
Polystyrene	2350	1120	1840
Rubber, butyl	1830		
Rubber, gum	1550		
Rubber, neoprene	1600		
Tungsten carbide	6655	3980	6220

TABLE 2. Speed of Sound in Liquids and Gases

Name	Synonym	t/°C	v/m s^{-1}
Liquids			
Acetone	2-Propanone	20	1187
Argon		-185.8	838
Benzene	[6]Annulene	25	1302
Bromobenzene	Phenyl bromide	20	1169
Butane		-0.5	1038
1-Butanol	Butyl alcohol	20	1262
Carbon disulfide	Carbon bisulfide	25	1140
Chlorobenzene	Phenyl chloride	20	1311
Cyclohexane	Hexahydrobenzene	19	1299
1-Decene		20	1250
Diethyl ether	Ethyl ether	25	951
Ethane		-88.6	1316
Ethanol	Ethyl alcohol	20	1159
Ethylene	Ethene	-103.8	1307
Ethylene glycol	Ethylene glycol	25	1639
Fluorobenzene	Phenyl fluoride	20	1183
Glycerol	1,2,3-Propanetriol	25	1904
Helium		-268.9	177
Heptane		20	1149
1-Heptene		20	1128
Hexane		20	1078
Hydrogen		-252.8	1119
Iodobenzene	Phenyl iodide	20	1114
Mercury	Quicksilver	25	1450
Methane		-161.5	1338
Methanol	Methyl alcohol	20	1116
Nitrobenzene		25	1463
Nitrogen		-195.8	851
1-Nonene	1-Nonylene	20	1218
Octane		20	1166
1-Octene	Caprylene	20	1184
Oxygen		-183.0	904
1-Pentadecene		20	1351
Pentane		20	1036
Propane	LPG	-42.1	1161
1-Propanol	Propyl alcohol	20	1223
Tetrachloromethane	Carbon tetrachloride	25	930
Trichloromethane	Chloroform	25	987
1-Undecene		20	1275
Water		25	1497
Water (sea, S = 3.5%)		25	1534
Gases at 1 atm			
Air, dry		25	346
Ammonia	R-717	0	414
Argon		27	323
Carbon monoxide	Carbon oxide	0	337
Carbon dioxide	Carbonic anhydride	0	258
Chlorine		0	205
Deuterium	Deuterium	0	888
Ethane		27	312
Ethylene	Ethene	27	331
Helium		0	973
Hydrogen		27	1320
Hydrogen bromide	Hydrobromic acid	0	200
Hydrogen chloride	Hydrochloric acid	0	294
Hydrogen iodide	Hydroiodic acid	0	157
Hydrogen sulfide		0	295
Methane		27	450
Neon		0	433
Nitric oxide		10	330
Nitrogen		27	353
Nitrous oxide		0	256
Oxygen		27	330
Sulfur dioxide		0	209
Water (steam)		100	472

TABLE 3. Speed of Sound in m/s in Water and Reference Seawater (Practical Salinity S = 35; Absolute Salinity S_A = 35.16504 g kg^{-1}) at Different Temperatures

t/°C	Water	Seawater
0	1402.4	1449.0
10	1447.3	1489.8
20	1482.3	1521.5
25	1496.7	1534.3
30	1509.2	1545.5
40	1528.9	1563.4
50	1542.6	
60	1551.0	
70	1554.7	
80	1554.4	
90	1550.4	
100	1543.5	

ATTENUATION AND SPEED OF SOUND IN AIR AS A FUNCTION OF HUMIDITY AND FREQUENCY

This table gives the attenuation and speed of sound as a function of frequency at various values of relative humidity. All values refer to still air at 20 °C.

References

1. Tables of Absorption and Velocity of Sound in Still Air at 68 °F (20 °C), AD-738576, National Technical Information Service, Springfield, VA.
2. Evans, L. B., Bass, H. E., and Sutherland, L. C., *J. Acoust. Soc. Am.*, 51, 1565, 1972.

Frequency (Hz)	Attenuation (dB/km)	Speed (m/s)
Relative humidity 0%		
20	0.51	343.477
40	1.07	343.514
50	1.26	343.525
63	1.43	343.536
100	1.67	343.550
200	1.84	343.559
400	1.96	343.561
630	2.11	343.562
800	2.27	343.562
1250	2.82	343.562
2000	4.14	343.562
4000	8.84	343.564
6300	14.89	343.565
10000	26.28	343.566
12500	35.81	343.566
16000	52.15	343.567
20000	75.37	343.567
40000	267.01	343.567
63000	644.66	343.567
80000	1032.14	343.567
Relative humidity 30%		
20	0.03	343.807
40	0.11	343.808
50	0.17	343.810
63	0.25	343.810
100	0.50	343.814
200	1.01	343.821
400	1.59	343.826
630	2.24	343.827
800	2.85	343.828
1250	5.09	343.828
2000	10.93	343.829
4000	38.89	343.831
6300	90.61	343.836
10000	204.98	343.846
12500	294.08	343.854
16000	422.51	343.865
20000	563.66	343.877
40000	1110.97	343.911
63000	1639.47	343.924
80000	2083.08	343.929
Relative humidity 60%		
20	0.02	344.182
40	0.06	344.183
50	0.09	344.183
63	0.15	344.184
100	0.34	344.185
200	0.99	344.190
400	1.94	344.197
630	2.57	344.200
800	2.94	344.201
1250	4.01	344.202
2000	6.55	344.203
4000	18.73	344.204
6300	42.51	344.204
10000	101.84	344.206
12500	155.67	344.208
16000	247.78	344.211
20000	373.78	344.215
40000	1195.37	344.238
63000	2220.64	344.262
80000	2951.71	344.274
Relative humidity 100%		
20	0.01	344.685
40	0.04	344.685
50	0.06	344.685
63	0.09	344.685
100	0.22	344.686
200	0.77	344.689
400	2.02	344.695
630	3.05	344.699
800	3.57	344.701
1250	4.59	344.704
2000	6.29	344.705
4000	13.58	344.706
6300	27.72	344.706
10000	63.49	344.706
12500	96.63	344.707
16000	154.90	344.708
20000	237.93	344.709
40000	884.28	344.718
63000	1973.62	344.731
80000	2913.01	344.742

SPEED OF SOUND IN DRY AIR

Eric W. Lemmon

These values were calculated from the equation of state for dry air (average molecular weight 28.96) treated as a real gas. Values refer to standard atmospheric pressure. The speed of sound varies only slightly with pressure; at two atmospheres and −100 °C the value decreases by 0.16%, while at two atmospheres and 80 °C the speed increases by 0.05%. For additional values, see the table in Section 6 labeled "Thermophysical Properties of Air."

Reference

Lemmon, E.W., Jacobsen, R.T, Penoncello, S.G., and Friend, D.G., Thermodynamic Properties of Air and Mixtures of Nitrogen, Argon, and Oxygen from 60 to 2000 K at Pressures to 2000 MPa, *J. Phys. Chem. Ref. Data* 29, 331, 2000.

t/°C	v_s/m s^{-1}	t/°C	v_s/m s^{-1}	t/°C	v_s/m s^{-1}
−100	263.5	−30	312.7	40	354.9
−95	267.3	−25	315.9	45	357.7
−90	271.1	−20	319.1	50	360.4
−85	274.8	−15	322.2	55	363.2
−80	278.5	−10	325.4	60	365.9
−75	282.1	−5	328.4	65	368.7
−70	285.7	0	331.5	70	371.3
−65	289.2	5	334.5	75	374.0
−60	292.7	10	337.5	80	376.7
−55	296.1	15	340.5	85	379.3
−50	299.5	20	343.4	90	381.9
−45	302.9	25	346.3	95	384.5
−40	306.2	30	349.2	100	387.0
−35	309.5	35	352.0		

ALLOCATION OF FREQUENCIES IN THE RADIO SPECTRUM

In the United States, the National Telecommunications and Information Administration (NTIA) has responsibility for assigning each portion of the radio spectrum (9 kHz to 300 GHz) for different uses. These assignments must be compatible with the rules of the International Telecommunications Union (ITU), to which the United States is bound by treaty. The current assignments are given in a wall chart (Ref. 1) and may also be found on the NTIA Web site (Ref. 2). The list below summarizes the broad features of the spectrum allocation, with particular attention to those sections of scientific interest. The references should be consulted for details of the allocations in the frequency bands listed here, which in some cases are quite complex.

References

1. *United States Frequency Allocations*, 1996 Spectrum Wall Chart, Stock No. 003-000-00652-2, U.S. Government Printing Office, P.O. Box 371954, Pittsburgh, PA 15250-7954.
2. http://www.ntia.doc.gov/files/ntia/publications/2003-allochrt.pdf.

Frequency Allocations for Commercial and Scientific Applications

Frequency range	Allocation
9–19.95 kHz	Maritime communication, navigation
19.95–20.05 kHz	Standard frequency and time signal (also at 60 kHz and 2.5, 5, 10, 15, 20, 25 MHz)
20.05–535 kHz	Maritime and aeronautical communication, navigation
535–1605 kHz	AM radio broadcasting
1605–3500 kHz	Mobile communication and navigation, amateur radio (1800–1900 kHz)
3.5–4.0 MHz	Amateur radio
4.0–5.95 MHz	Mobile communication
5.95–13.36 MHz	Mobile communication, amateur, short-wave broadcasting
13.36–13.41 MHz	Radioastronomy
13.41–25.55 MHz	Mobile communication, amateur, short-wave broadcasting
25.55–25.67 MHz	Radioastronomy
25.67–37.5 MHz	Mobile communication, amateur, short-wave broadcasting
37.5–38.25 MHz	Radioastronomy
38.25–50.0 MHz	Mobile communication
50.0–54.0 MHz	Amateur
54.0–72.0 MHz	TV channels 2–4
72.0–73.0 MHz	Mobile communication
73.0–74.6 MHz	Radioastronomy
74.6–76.0 MHz	Mobile communication
76.0–88.0 MHz	TV channels 5–6
88.0–108.0 MHz	FM radio broadcasting
108.0–118.0 MHz	Aeronautical navigation
118.0–174.0 MHz	Mobile communication, space research, meteorological satellites
174.0–216.0 MHz	TV channels 7–13
216.0–400.05 MHz	Mobile communication
400.05–400.15 MHz	Standard frequency and time satellite (also 20 and 25 GHz)
400.15–406.1 MHz	Meteorological aids (radiosonde)
406.1–410.0 MHz	Radioastronomy
410.0–470.0 MHz	Mobile communication, amateur
470.0–512.0 MHz	TV channels 14–20
512.0–608.0 MHz	TV channels 21–36
608.0–614.0 MHz	Radioastronomy
614.0–806.0 MHz	TV channels 38–69
806–1400 MHz	Mobile communication, navigation
1400–1427 MHz	Radioastronomy, space research
1427–1660 MHz	Various navigation and satellite applications
1660–1710 MHz	Radioastronomy, space research, meteorology
1710–2655 MHz	Various navigation and satellite applications
2655–2700 MHz	Radioastronomy, space research
2.7–4.99 GHz	Various navigation and satellite applications
4.99–5.0 GHz	Radioastronomy, space research
5.0–10.6 GHz	Various navigation and satellite applications
10.6–10.7 GHz	Radioastronomy, space research
10.7–15.35 GHz	Various navigation and satellite applications
15.35–15.4 GHz	Radioastronomy, space research
15.4–22.21 GHz	Various navigation and satellite applications
22.21–22.5 GHz	Radioastronomy, space research
22.25–23.6 GHz	Various navigation and satellite applications
23.6–24.0 GHz	Radioastronomy, space research
24.0–31.3 GHz	Various navigation and satellite applications
31.3–31.8 GHz	Radioastronomy, space research
31.8–42.5 GHz	Various navigation and satellite applications
42.5–43.5 GHz	Radioastronomy
43.5–51.4 GHz	Various navigation and satellite applications
51.4–54.25 GHz	Radioastronomy, space research
54.25–58.2 GHz	Space research
58.2–59.0 GHz	Radioastronomy, space research
59.0–64.0 GHz	Satellite applications
64.0–65.0 GHz	Radioastronomy, space research
65.0–72.77 GHz	Various navigation and satellite applications
72.77–72.91 GHz	Radioastronomy, space research
72.91–86.0 GHz	Various navigation and satellite applications
86.0–92.0 GHz	Radioastronomy, space research
92.0–105.0 GHz	Various navigation and satellite applications
105.0–116.0 GHz	Radioastronomy, space research
116.0–164.0 GHz	Various navigation and satellite applications
164.0–168.0 GHz	Radioastronomy, space research
168.0–182.0 GHz	Various navigation and satellite applications
182.0–185.0 GHz	Radioastronomy, space research
185.0–217.0 GHz	Various navigation and satellite applications
217.0–231.0 GHz	Radioastronomy, space research
231.0–265.0 GHz	Various navigation and satellite applications
265.0–275.0 GHz	Radioastronomy
275.0–300.0 GHz	Mobile communications

Section 15
Practical Laboratory Data

Standard ITS-90 Thermocouple Tables.......15-1
Reference Points on the ITS-90 Temperature Scale.......15-10
Relative Sensitivity of Bayard-Alpert Ionization Gauges to Various Gases.......15-12
Laboratory Solvents and Other Liquid Reagents.......15-13
Miscibility of Organic Solvents.......15-23
Density of Solvents as a Function of Temperature.......15-27
Dependence of Boiling Point of Organic Liquids on Pressure.......15-28
Ebullioscopic Constants for Calculation of Boiling Point Elevation.......15-30
Cryoscopic Constants for Calculation of Freezing Point Depression.......15-31
Freezing Point Lowering by Electrolytes in Aqueous Solution.......15-32
Correction of Barometer Readings to 0 °C Temperature.......15-33
Determination of Relative Humidity from Dew Point.......15-34
Determination of Relative Humidity from Wet and Dry Bulb Temperatures.......15-35
Constant Humidity Solutions.......15-36
Standard Salt Solutions for Humidity Calibration.......15-37
Low-Temperature Baths for Maintaining Constant Temperature.......15-37
Metals and Alloys with Low-Melting Temperature.......15-38
Characteristics of Particles and Particle Dispersoids.......15-40
Density of Various Solids.......15-41
Density of Sulfuric Acid.......15-42
Density of Ethanol–Water Mixtures.......15-44
Dielectric Strength of Insulating Materials.......15-45
Coefficient of Friction.......15-50

STANDARD ITS-90 THERMOCOUPLE TABLES

The Instrument Society of America (ISA) has assigned standard letter designations to a number of thermocouple types having specified emf-temperature relations. These designations and the approximate metal compositions that meet the required relations, as well as the useful temperature ranges, are given below.

TABLE 1. Designation and Composition of Various Metals Used in Thermocouples

Type	Composition	*t*
Type B	(Pt + 30% Rh) vs. (Pt + 6% Rh)	0 to 1820 °C
Type E	(Ni + 10% Cr) vs. (Cu + 43% Ni)	−270 to 1000 °C
Type J	Fe vs. (Cu + 43% Ni)	−210 to 1200 °C
Type K	(Ni + 10% Cr) vs. (Ni + 2% Al + 2% Mn + 1% Si)	−270 to 1372 °C
Type N	(Ni + 14% Cr + 1.5% Si) vs. (Ni + 4.5% Si + 0.1% Mg)	−270 to 1300 °C
Type R	(Pt + 13% Rh) vs. Pt	−50 to 1768 °C
Type S	(Pt + 10% Rh) vs. Pt	−50 to 1768 °C
Type T	Cu vs. (Cu + 43% Ni)	−270 to 400 °C

The compositions are given in weight percent, and the positive leg is listed first. It should be emphasized that the standard letter designations do not imply a precise composition but rather that the specified emf-temperature relation is satisfied.

Tables 2, 4, 6, 8, 10, 12, 14, and 16 contain, for each thermocouple type, the emf as a function of temperature on the International Temperature Scale of 1990 (ITS-90). The left-most column has temperature *t* at 100 °C intervals. Columns to the right have temperatures at *t* ± n °C intervals where n = 10, 20, 30, etc., as indicated in the column heading. The values in these tables were calculated using the following equation.

$$E = c_0 + c_1 t + c_2 t^2 + c_3 t^3 + \ldots c_n t^n$$

where E is the emf in millivolts, t is the temperature in degrees Celsius (ITS-90), and c_0, c_1, c_2, c_3, etc., are the coefficients, which have been extracted from Ref. 1. For Type K thermocouples, an additional term in the equation is required as discussed in the Note. The coefficients in the equation used to generate the table are also given in Tables 3, 5, 7, 9, 11, 13, 15, and 17, respectively.

Tables 18–25 give the inverse relationships, i.e., the coefficients in the polynomial equation that expresses the temperature as a function of thermocouple emf. The accuracy of these equations is also stated.

Further details and tables at closer intervals may be found in Ref. 1.

References

1. Burns, G. W., Seroger, M. G., Strouse, G. F., Croarkin, M. C., and Guthrie, W. F., *Temperature-Electromotive Force Reference Functions and Tables for the Letter-Designated Thermocouple Types Based on the ITS-90*, Monograph 175, Natl. Inst. Stand. Tech. (U.S.), Washington, DC, 1993.
2. Schooley, J. F., *Thermometry*, CRC Press, Boca Raton, FL, 1986. <https://doi.org/10.6028/NBS.IR.85-3133>

Type B Thermocouples: emf-Temperature (°C) Reference Table and Equations

TABLE 2. Type B Thermocouple emf in mV as a Function of Temperature (*t*) in °C (ITS-90); Reference Junctions at 0 °C

t/°C	*t*+0/°C	*t*+10/ °C	*t*+20/ °C	*t*+30/ °C	*t*+40/ °C	*t*+50/ °C	*t*+60/ °C	*t*+70/ °C	*t*+80/ °C	*t*+90/ °C	*t*+100/ °C
0	0.000	-0.002	-0.003	-0.002	-0.000	0.002	0.006	0.011	0.017	0.025	0.033
100	0.033	0.043	0.053	0.065	0.078	0.092	0.107	0.123	0.141	0.159	0.178
200	0.178	0.199	0.220	0.243	0.267	0.291	0.317	0.344	0.372	0.401	0.431
300	0.431	0.462	0.494	0.527	0.561	0.596	0.632	0.669	0.707	0.746	0.787
400	0.787	0.828	0.870	0.913	0.957	1.002	1.048	1.095	1.143	1.192	1.242
500	1.242	1.293	1.344	1.397	1.451	1.505	1.561	1.617	1.675	1.733	1.792
600	1.792	1.852	1.913	1.975	2.037	2.101	2.165	2.230	2.296	2.363	2.431
700	2.431	2.499	2.569	2.639	2.710	2.782	2.854	2.928	3.002	3.078	3.154
800	3.154	3.230	3.308	3.386	3.466	3.546	3.626	3.708	3.790	3.873	3.957
900	3.957	4.041	4.127	4.213	4.299	4.387	4.475	4.564	4.653	4.743	4.834
1000	4.834	4.926	5.018	5.111	5.205	5.299	5.394	5.489	5.585	5.682	5.780
1100	5.780	5.878	5.976	6.075	6.175	6.276	6.377	6.478	6.580	6.683	6.786
1200	6.786	6.890	6.995	7.100	7.205	7.311	7.417	7.524	7.632	7.740	7.848
1300	7.848	7.957	8.066	8.176	8.286	8.397	8.508	8.620	8.731	8.844	8.956
1400	8.956	9.069	9.182	9.296	9.410	9.524	9.639	9.753	9.868	9.984	10.099
1500	10.099	10.215	10.331	10.447	10.563	10.679	10.796	10.913	11.029	11.146	11.263
1600	11.263	11.380	11.497	11.614	11.731	11.848	11.965	12.082	12.199	12.316	12.433
1700	12.433	12.549	12.666	12.782	12.898	13.014	13.130	13.246	13.361	13.476	13.591
1800	13.591	13.706	13.820								

TABLE 3. Coefficients in the Equation for Type B Thermocouples

Coeff.	0 °C to 630.615 °C	630.615 °C to 1820 °C
c_0	0.000 000 000 0	−3.893 816 862 1 ...
c_1	$-2.465\,081\,834\,6 \times 10^{-4}$	$2.857\,174\,747\,0 \times 10^{-2}$
c_2	$5.904\,042\,117\,1 \times 10^{-6}$	$-8.488\,510\,478\,5 \times 10^{-5}$
c_3	$-1.325\,793\,163\,6 \times 10^{-9}$	$1.578\,528\,016\,4 \times 10^{-7}$
c_4	$1.566\,829\,190\,1 \times 10^{-12}$	$-1.683\,534\,486\,4 \times 10^{-10}$
c_5	$-1.694\,452\,924\,0 \times 10^{-15}$	$1.110\,979\,401\,3 \times 10^{-13}$
c_6	$6.299\,034\,709\,4 \times 10^{-19}$	$-4.451\,543\,103\,3 \times 10^{-17}$
c_7	....	$9.897\,564\,082\,1 \times 10^{-21}$
c_8	....	$-9.379\,133\,028\,9 \times 10^{-25}$

Type E Thermocouples: emf-Temperature (°C) Reference Table and Equations

TABLE 4. Type E Thermocouple emf in mV as a Function of Temperature (*t*) in °C (ITS-90); Reference Junctions at 0 °C

t/°C	*t*-0/°C	*t*-10/°C	*t*-20/°C	*t*-30/°C	*t*-40/°C	*t*-50/°C	*t*-60/°C	*t*-70/°C	*t*-80/°C	*t*-90/°C	*t*-100/°C
-200	-8.825	-9.063	-9.274	-9.455	-9.604	-9.718	-9.797	-9.835			
-100	-5.237	-5.681	-6.107	-6.516	-6.907	-7.279	-7.632	-7.963	-8.273	-8.561	-8.825
0	0.000	-0.582	-1.152	-1.709	-2.255	-2.787	-3.306	-3.811	-4.302	-4.777	-5.237

t/°C	*t*+0/°C	*t*+10/°C	*t*+20/°C	*t*+30/°C	*t*+40/°C	*t*+50/°C	*t*+60/°C	*t*+70/°C	*t*+80/°C	*t*+90/°C	*t*+100/°C
0	0.000	0.591	1.192	1.801	2.420	3.048	3.685	4.330	4.985	5.648	6.319
100	6.319	6.998	7.685	8.379	9.081	9.789	10.503	11.224	11.951	12.684	13.421
200	13.421	14.164	14.912	15.664	16.420	17.181	17.945	18.713	19.484	20.259	21.036
300	21.036	21.817	22.600	23.386	24.174	24.964	25.757	26.552	27.348	28.146	28.946
400	28.946	29.747	30.550	31.354	32.159	32.965	33.772	34.579	35.387	36.196	37.005
500	37.005	37.815	38.624	39.434	40.243	41.053	41.862	42.671	43.479	44.286	45.093
600	45.093	45.900	46.705	47.509	48.313	49.116	49.917	50.718	51.517	52.315	53.112
700	53.112	53.908	54.703	55.497	56.289	57.080	57.870	58.659	59.446	60.232	61.017
800	61.017	61.801	62.583	63.364	64.144	64.922	65.698	66.473	67.246	68.017	68.787
900	68.787	69.554	70.319	71.082	71.844	72.603	73.360	74.115	74.869	75.621	76.373
1000	76.373										

TABLE 5. Coefficients in the Equation for Type E Thermocouples

Coeff.	−270 to 0 °C	0 °C to 1000 °C
c_0	0.000 000 000 0 ...	0.000 000 000 0 ...
c_1	$5.866\,550\,870\,8 \times 10^{-2}$	$5.866\,550\,871\,0 \times 10^{-2}$
c_2	$4.541\,097\,712\,4 \times 10^{-5}$	$4.503\,227\,558\,2 \times 10^{-5}$
c_3	$-7.799\,804\,868\,6 \times 10^{-7}$	$2.890\,840\,721\,2 \times 10^{-8}$
c_4	$-2.580\,016\,084\,3 \times 10^{-8}$	$-3.305\,689\,665\,2 \times 10^{-10}$
c_5	$-5.945\,258\,305\,7 \times 10^{-10}$	$6.502\,440\,327\,0 \times 10^{-13}$
c_6	$-9.321\,405\,866\,7 \times 10^{-12}$	$-1.919\,749\,550\,4 \times 10^{-16}$
c_7	$-1.028\,760\,553\,4 \times 10^{-13}$	$-1.253\,660\,049\,7 \times 10^{-18}$
c_8	$-8.037\,012\,362\,1 \times 10^{-16}$	$2.148\,921\,756\,9 \times 10^{-21}$
c_9	$-4.397\,949\,739\,1 \times 10^{-18}$	$-1.438\,804\,178\,2 \times 10^{-24}$
c_{10}	$-1.641\,477\,635\,5 \times 10^{-20}$	$3.596\,089\,948\,1 \times 10^{-28}$
c_{11}	$-3.967\,361\,951\,6 \times 10^{-23}$	....
c_{12}	$-5.582\,732\,872\,1 \times 10^{-26}$	....
c_{13}	$-3.465\,784\,201\,3 \times 10^{-29}$	....

Type J Thermocouples: emf-Temperature (°C) Reference Table and Equations

TABLE 6. Type J Thermocouple emf in mV as a Function of Temperature (*t*) in °C (ITS-90); Reference Junctions at 0 °C

t/°C	*t*-0/°C	*t*-10/°C	*t*-20/°C	*t*-30/°C	*t*-40/°C	*t*-50/°C	*t*-60/°C	*t*-70/°C	*t*-80/°C	*t*-90/°C	*t*-100/°C
-200	-7.890	-8.095									
-100	-4.633	-5.037	-5.426	-5.801	-6.159	-6.500	-6.821	-7.123	-7.403	-7.659	-7.890
0	0.000	-0.501	-0.995	-1.482	-1.961	-2.431	-2.893	-3.344	-3.786	-4.215	-4.633

t/°C	*t*+0/°C	*t*+10/°C	*t*+20/°C	*t*+30/°C	*t*+40/°C	*t*+50/°C	*t*+60/°C	*t*+70/°C	*t*+80/°C	*t*+90/°C	*t*+100/°C
0	0.000	0.507	1.019	1.537	2.059	2.585	3.116	3.650	4.187	4.726	5.269
100	5.269	5.814	6.360	6.909	7.459	8.010	8.562	9.115	9.669	10.224	10.779
200	10.779	11.334	11.889	12.445	13.000	13.555	14.110	14.665	15.219	15.773	16.327
300	16.327	16.881	17.434	17.986	18.538	19.090	19.642	20.194	20.745	21.297	21.848
400	21.848	22.400	22.952	23.504	24.057	24.610	25.164	25.720	26.276	26.834	27.393
500	27.393	27.953	28.516	29.080	29.647	30.216	30.788	31.362	31.939	32.519	33.102
600	33.102	33.689	34.279	34.873	35.470	36.071	36.675	37.284	37.896	38.512	39.132
700	39.132	39.755	40.382	41.012	41.645	42.281	42.919	43.559	44.203	44.848	45.494
800	45.494	46.141	46.786	47.431	48.074	48.715	49.353	49.989	50.622	51.251	51.877
900	51.877	52.500	53.119	53.735	54.347	54.956	55.561	56.164	56.763	57.360	57.953
1000	57.953	58.545	59.134	59.721	60.307	60.890	61.473	62.054	62.634	63.214	63.792
1100	63.792	64.370	64.948	65.525	66.102	66.679	67.255	67.831	68.406	68.980	69.553
1200	69.553										

TABLE 7. Coefficients in the Equation for Type J Thermocouples

Coeff.	−260 °C to 760 °C	760 °C to 1200 °C
c_0	0.000 000 000 0 ...	$2.964\,562\,568\,1 \times 10^{2}$
c_1	$5.038\,118\,781\,5 \times 10^{-2}$	−1.497 612 778 6 ...
c_2	$3.047\,583\,693\,0 \times 10^{-5}$	$3.178\,710\,392\,4 \times 10^{-3}$
c_3	$1.322\,819\,529\,5 \times 10^{-10}$	$1.572\,081\,900\,4 \times 10^{-9}$
c_5	$-1.705\,295\,833\,7 \times 10^{-13}$	$-3.069\,136\,905\,6 \times 10^{-13}$
c_6	$2.094\,809\,069\,7 \times 10^{-16}$	....
c_7	$-1.253\,839\,533\,6 \times 10^{-19}$	....
c_8	$1.563\,172\,569\,7 \times 10^{-23}$	....

Type K Thermocouples: emf-Temperature (°C) Reference Table and Equations

TABLE 8. Type K Thermocouple emf in mV as a Function of Temperature (*t*) in °C (ITS-90); Reference Junctions at 0 °C

t/°C	*t*-0/°C	*t*-10/°C	*t*-20/°C	*t*-30/°C	*t*-40/°C	*t*-50/°C	*t*-60/°C	*t*-70/°C	*t*-80/°C	*t*-90/°C	*t*-100/°C
-200	-5.891	-6.035	-6.158	-6.262	-6.344	-6.404	-6.441	-6.458			
-100	-3.554	-3.852	-4.138	-4.411	-4.669	-4.913	-5.141	-5.354	-5.550	-5.730	-5.891
0	0.000	-0.392	-0.778	-1.156	-1.527	-1.889	-2.243	-2.587	-2.920	-3.243	-3.554

t/°C	*t*+0/°C	*t*+10/°C	*t*+20/°C	*t*+30/°C	*t*+40/°C	*t*+50/°C	*t*+60/°C	*t*+70/°C	*t*+80/°C	*t*+90/°C	*t*+100/°C
0	0.000	0.397	0.798	1.203	1.612	2.023	2.436	2.851	3.267	3.682	4.096
100	4.096	4.509	4.920	5.328	5.735	6.138	6.540	6.941	7.340	7.739	8.138
200	8.138	8.539	8.940	9.343	9.747	10.153	10.561	10.971	11.382	11.795	12.209
300	12.209	12.624	13.040	13.457	13.874	14.293	14.713	15.133	15.554	15.975	16.397
400	16.397	16.820	17.243	17.667	18.091	18.516	18.941	19.366	19.792	20.218	20.644
500	20.644	21.071	21.497	21.924	22.350	22.776	23.203	23.629	24.055	24.480	24.905
600	24.905	25.330	25.755	26.179	26.602	27.025	27.447	27.869	28.289	28.710	29.129
700	29.129	29.548	29.965	30.382	30.798	31.213	31.628	32.041	32.453	32.865	33.275
800	33.275	33.685	34.093	34.501	34.908	35.313	35.718	36.121	36.524	36.925	37.326
900	37.326	37.725	38.124	38.522	38.918	39.314	39.708	40.101	40.494	40.885	41.276
1000	41.276	41.665	42.053	42.440	42.826	43.211	43.595	43.978	44.359	44.740	45.119
1100	45.119	45.497	45.873	46.249	46.623	46.995	47.367	47.737	48.105	48.473	48.838
1200	48.838	49.202	49.565	49.926	50.286	50.644	51.000	51.355	51.708	52.060	52.410
1300	52.410	52.759	53.106	53.451	53.795	54.138	54.479	54.819			

Note: In the 0 °C to 1372 °C range there is also an exponential term that must be evaluated and added to the equation. The exponential term is of the form $c_0 e^{c_1 (t-126.9686)^2}$, where t is the temperature in °C, e is the natural logarithm base, and c_0 and c_1 are the coefficients. These coefficients are extracted from Ref. 1.

TABLE 9. Coefficients in the Equation for Type K Thermocouples

Coeff.	−270 °C to 0 °C	0 °C to 1372 °C	0 °C to 1372 °C (Exponential term)
c_0	0.000 000 000 0	−1.760 041 368 6 × 10^{-2}	1.185 976 × 10^{-1}
c_1	3.945 012 802 5 × 10^{-2}	3.892 120 497 5 × 10^{-2}	−1.183 432 × 10^{-4}
c_2	2.362 237 359 8 × 10^{-5}	1.855 877 003 2 × 10^{-5}	....
c_3	−3.285 890 678 4 × 10^{-7}	−9.945 759 287 4 × 10^{-8}	....
c_4	−4.990 482 877 7 × 10^{-9}	3.184 094 571 9 × 10^{-10}	....
c_5	−6.750 905 917 3 × 10^{-11}	−5.607 284 488 9 × 10^{-13}	....
c_6	−5.741 032 742 8 × 10^{-13}	5.607 505 905 9 × 10^{-16}	....
c_7	−3.108 887 289 4 × 10^{-15}	−3.202 072 000 3 × 10^{-19}	....
c_8	−1.045 160 936 5 × 10^{-17}	9.715 114 715 2 × 10^{-23}	....
c_9	−1.988 926 687 8 × 10^{-20}	−1.210 472 127 5 × 10^{-26}	....
c_{10}	−1.632 269 748 6 × 10^{-23}	....	....

Type N Thermocouples: emf-Temperature (°C) Reference Table and Equations

TABLE 10. Type N Thermocouple emf in mV as a Function of Temperature (*t*) in °C (ITS-90); Reference Junctions at 0 °C

t/°C	*t*-0/°C	*t*-10/°C	*t*-20/°C	*t*-30/°C	*t*-40/°C	*t*-50/°C	*t*-60/°C	*t*-70/°C	*t*-80/°C	*t*-90/°C	*t*-100/°C
-200	3.990	-4.083	-4.162	-4.226	-4.277	-4.313	-4.336	-4.345			
-100	-2.407	-2.612	2.808	2.994	3.171	3.336	-3.491	3.634	-3.766	-3.884	-3.990
0	0.000	-0.260	-0.518	-0.772	-1.023	-1.269	-1.509	1.744	-1.972	-2.193	-2.407

t/°C	*t*+0/°C	*t*+10/°C	*t*+20/°C	*t*+30/°C	*t*+40/°C	*t*+50/°C	*t*+60/°C	*t*+70/°C	*t*+80/°C	*t*+90/°C	*t*+100/°C
0	0.000	0.261	0.525	0.793	1.065	1.340	1.619	1.902	2.189	2.480	2.774
100	2.774	3.072	3.374	3.680	3.989	4.302	4.618	4.937	5.259	5.585	5.913
200	5.913	6.245	6.579	6.916	7.255	7.597	7.941	8.288	8.637	8.988	9.341
300	9.341	9.696	10.054	10.413	10.774	11.136	11.501	11.867	12.234	12.603	12.974
400	12.974	13.346	13.719	14.094	14.469	14.846	15.225	15.604	15.984	16.366	16.748
500	16.748	17.131	17.515	17.900	18.286	18.672	19.059	19.447	19.835	20.224	20.613
600	20.613	21.003	21.393	21.784	22.175	22.566	22.958	23.350	23.742	24.134	24.527
700	24.527	24.919	25.312	25.705	26.098	26.491	26.883	27.276	27.669	28.062	28.455
800	28.455	28.847	29.239	29.632	30.024	30.416	30.807	31.199	31.590	31.981	32.371
900	32.371	32.761	33.151	33.541	33.930	34.319	34.707	35.095	35.482	35.869	36.256
1000	36.256	36.641	37.027	37.411	37.795	38.179	38.562	38.944	39.326	39.706	40.087
1100	40.087	40.466	40.845	41.223	41.600	41.976	42.352	42.727	43.101	43.474	43.846
1200	43.846	44.218	44.588	44.958	45.326	45.694	46.060	46.425	46.789	47.152	47.513
1300	47.513										

TABLE 11. Coefficients in the Equation for Type N Thermocouples

Coeff.	−270 °C to 0 °C	0 °C to 1300 °C
c_0	0.000 000 000 0 ...	0.000 000 000 0 ...
c_1	2.615 910 596 2 × 10^{-2}	2.592 939 460 1 × 10^{-2}
c_2	1.095 748 422 8 × 10^{-5}	1.571 014 188 0 × 10^{-5}
c_3	−9.384 111 155 4 × 10^{-8}	4.382 562 723 7 × 10^{-8}
c_4	−4.641 203 975 9 × 10^{-11}	−2.526 116 979 4 × 10^{-10}
c_5	−2.630 335 771 6 × 10^{-12}	6.431 181 933 9 × 10^{-13}
c_6	−2.265 343 800 3 × 10^{-14}	−1.006 347 151 9 × 10^{-15}
c_7	−7.608 930 079 1 × 10^{-17}	9.974 533 899 2 × 10^{-19}
c_8	−9.341 966 783 5 × 10^{-20}	−6.086 324 560 7 × 10^{-22}
c_9	....	2.084 922 933 9 × 10^{-25}
c_{10}	....	−3.068 219 615 1 × 10^{-29}

Type R Thermocouples: emf-Temperature (°C) Reference Table and Equations

TABLE 12. Type R Thermocouple emf in mV as a Function of Temperature (*t*) in °C (ITS-90); Reference Junctions at 0 °C

t/°C	*t*-0/°C	*t*-10/°C	*t*-20/°C	*t*-30/°C	*t*-40/°C	*t*-50/°C	*t*-60/°C	*t*-70/°C	*t*-80/°C	*t*-90/°C	*t*-100/°C
0	0.000	-0.051	-0.100	-0.145	-0.188	-0.226					

t/°C	*t*+0/°C	*t*+10/°C	*t*+20/°C	*t*+30/°C	*t*+40/°C	*t*+50/°C	*t*+60/°C	*t*+70/°C	*t*+80/°C	*t*+90/°C	*t*+100/°C
0	0.000	0.054	0.111	0.171	0.232	0.296	0.363	0.431	0.501	0.573	0.647
100	0.647	0.723	0.800	0.879	0.959	1.041	1.124	1.208	1.294	1.381	1.469
200	1.469	1.558	1.648	1.739	1.831	1.923	2.017	2.112	2.207	2.304	2.401
300	2.401	2.498	2.597	2.696	2.796	2.896	2.997	3.099	3.201	3.304	3.408
400	3.408	3.512	3.616	3.721	3.827	3.933	4.040	4.147	4.255	4.363	4.471
500	4.471	4.580	4.690	4.800	4.910	5.021	5.133	5.245	5.357	5.470	5.583
600	5.583	5.697	5.812	5.926	6.041	6.157	6.273	6.390	6.507	6.625	6.743
700	6.743	6.861	6.980	7.100	7.220	7.340	7.461	7.583	7.705	7.827	7.950
800	7.950	8.073	8.197	8.321	8.446	8.571	8.697	8.823	8.950	9.077	9.205
900	9.205	9.333	9.461	9.590	9.720	9.850	9.980	10.111	10.242	10.374	10.506
1000	10.506	10.638	10.771	10.905	11.039	11.173	11.307	11.442	11.578	11.714	11.850
1100	11.850	11.986	12.123	12.260	12.397	12.535	12.673	12.812	12.950	13.089	13.228
1200	13.228	13.367	13.507	13.646	13.786	13.926	14.066	14.207	14.347	14.488	14.629
1300	14.629	14.770	14.911	15.052	15.193	15.334	15.475	15.616	15.758	15.899	16.040
1400	16.040	16.181	16.323	16.464	16.605	16.746	16.887	17.028	17.169	17.310	17.451
1500	17.451	17.591	17.732	17.872	18.012	18.152	18.292	18.431	18.571	18.710	18.849
1600	18.849	18.988	19.126	19.264	19.402	19.540	19.677	19.814	19.951	20.087	20.222
1700	20.222	20.356	20.488	20.620	20.749	20.877	21.003				

TABLE 13. Coefficients in the Equation for Type R Thermocouples

Coeff.	−50 °C to 1064.18 °C	1064.18 °C to 1664.5 °C	1664.5 °C to 1768.1 °C
c_0	0.000 000 000 00 ...	2.951 579 253 16 ...	1.522 321 182 09 × 10^{2}
c_1	5.289 617 297 65 × 10^{-3}	−2.520 612 513 32 × 10^{-3}	−2.688 198 885 45 × 10^{-1}
c_2	1.391 665 897 82 × 10^{-5}	1.595 645 018 65 × 10^{-5}	1.712 802 804 71 × 10^{-4}
c_3	−2.388 556 930 17 × 10^{-8}	−7.640 859 475 76 × 10^{-9}	−3.458 957 064 53 × 10^{-8}
c_4	3.569 160 010 63 × 10^{-11}	2.053 052 910 24 × 10^{-12}	−9.346 339 710 46 × 10^{-15}
c_5	−4.623 476 662 98 × 10^{-14}	−2.933 596 681 73 × 10^{-16}	....
c_6	5.007 774 410 34 × 10^{-17}	....	....
c_7	−3.731 058 861 91 × 10^{-20}	....	....
c_8	1.577 164 823 67 × 10^{-23}	....	....
c_9	−2.810 386 252 51 × 10^{-27}	....	....

Type S Thermocouples: emf-Temperature (°C) Reference Table and Equations

TABLE 14. Type S Thermocouple emf in mV as a Function of Temperature (*t*) in °C (ITS-90); Reference Junctions at 0 °C

t/°C	*t*-0/°C	*t*-10/°C	*t*-20/°C	*t*-30/°C	*t*-40/°C	*t*-50/°C	*t*-60/°C	*t*-70/°C	*t*-80/°C	*t*-90/°C	*t*-100/°C
0	0.000	-0.053	-0.103	-0.150	-0.194	-0.236					

t/°C	*t*+0/°C	*t*+10/°C	*t*+20/°C	*t*+30/°C	*t*+40/°C	*t*+50/°C	*t*+60/°C	*t*+70/°C	*t*+80/°C	*t*+90/°C	*t*+100/°C
0	0.000	0.055	0.113	0.173	0.235	0.299	0.365	0.433	0.502	0.573	0.646
100	0.646	0.720	0.795	0.872	0.950	1.029	1.110	1.191	1.273	1.357	1.441
200	1.441	1.526	1.612	1.698	1.786	1.874	1.962	2.052	2.141	2.232	2.323
300	2.323	2.415	2.507	2.599	2.692	2.786	2.880	2.974	3.069	3.164	3.259
400	3.259	3.355	3.451	3.548	3.645	3.742	3.840	3.938	4.036	4.134	4.233
500	4.233	4.332	4.432	4.532	4.632	4.732	4.833	4.934	5.035	5.137	5.239
600	5.239	5.341	5.443	5.546	5.649	5.753	5.857	5.961	6.065	6.170	6.275
700	6.275	6.381	6.486	6.593	6.699	6.806	6.913	7.020	7.128	7.236	7.345
800	7.345	7.454	7.563	7.673	7.783	7.893	8.003	8.114	8.226	8.337	8.449
900	8.449	8.562	8.674	8.787	8.900	9.014	9.128	9.242	9.357	9.472	9.587
1000	9.587	9.703	9.819	9.935	10.051	10.168	10.285	10.403	10.520	10.638	10.757
1100	10.757	10.875	10.994	11.113	11.232	11.351	11.471	11.590	11.710	11.830	11.951

t/°C	*t*-0/°C	*t*-10/°C	*t*-20/°C	*t*-30/°C	*t*-40/°C	*t*-50/°C	*t*-60/°C	*t*-70/°C	*t*-80/°C	*t*-90/°C	*t*-100/°C
1200	11.951	12.071	12.191	12.312	12.433	12.554	12.675	12.796	12.917	13.038	13.159
1300	13.159	13.280	13.402	13.523	13.644	13.766	13.887	14.009	14.130	14.251	14.373
1400	14.373	14.494	14.615	14.736	14.857	14.978	15.099	15.220	15.341	15.461	15.582
1500	15.582	15.702	15.822	15.942	16.062	16.182	16.301	16.420	16.539	16.658	16.777
1600	16.777	16.895	17.013	17.131	17.249	17.366	17.483	17.600	17.717	17.832	17.947
1700	17.947	18.061	18.174	18.285	18.395	18.503	18.609				

TABLE 15. Coefficients in the Equation for Type S Thermocouples

Coeff.	−50 °C to 1064.18 °C	1064.18 °C to 1664.5 °C	1664.5 °C to 1768.1 °C
c_0	0.000 000 000 00 ...	1.329 004 440 85 ...	$1.466\ 282\ 326\ 36 \times 10^{2}$
c_1	$5.403\ 133\ 086\ 31 \times 10^{-3}$	$3.345\ 093\ 113\ 44 \times 10^{-3}$	$-2.584\ 305\ 167\ 52 \times 10^{-1}$
c_2	$1.259\ 342\ 897\ 40 \times 10^{-5}$	$6.548\ 051\ 928\ 18 \times 10^{-6}$	$1.636\ 935\ 746\ 41 \times 10^{-4}$
c_3	$-2.324\ 779\ 686\ 89 \times 10^{-8}$	$-1.648\ 562\ 592\ 09 \times 10^{-9}$	$-3.304\ 390\ 469\ 87 \times 10^{-8}$
c_4	$3.220\ 288\ 230\ 36 \times 10^{-11}$	$1.299\ 896\ 051\ 74 \times 10^{-14}$	$-9.432\ 236\ 906\ 12 \times 10^{-15}$
c_5	$-3.314\ 651\ 963\ 89 \times 10^{-14}$	....	....
c_6	$2.557\ 442\ 517\ 86 \times 10^{-17}$	....	....
c_7	$-1.250\ 688\ 713\ 93 \times 10^{-20}$	....	....
c_8	$2.714\ 431\ 761\ 45 \times 10^{-24}$	....	....

Type T Thermocouples: emf-Temperature (°C) Reference Table and Equations

TABLE 16. Type T Thermocouple emf in mV as a Function of Temperature (*t*) in °C (ITS-90); Reference Junctions at 0 °C

t/°C	*t*-0/°C	*t*-10/°C	*t*-20/°C	*t*-30/°C	*t*-40/°C	*t*-50/°C	*t*-60/°C	*t*-70/°C	*t*-80/°C	*t*-90/°C	*t*-100/°C
-200	-5.603	-5.753	-5.888	-6.007	-6.105	-6.180	-6.232	-6.258			
-100	-3.379	-3.657	-3.923	-4.177	-4.419	-4.648	-4.865	-5.070	-5.261	-5.439	-5.603
0	0.000	-0.383	-0.757	-1.121	-1.475	-1.819	-2.153	-2.476	-2.788	-3.089	-3.379

t/°C	*t*+0/°C	*t*+10/°C	*t*+20/°C	*t*+30/°C	*t*+40/°C	*t*+50/°C	*t*+60/°C	*t*+70/°C	*t*+80/°C	*t*+90/°C	*t*+100/°C
0	0.000	0.391	0.790	1.196	1.612	2.036	2.468	2.909	3.358	3.814	4.279
100	4.279	4.750	5.228	5.714	6.206	6.704	7.209	7.720	8.237	8.759	9.288
200	9.288	9.822	10.362	10.907	11.458	12.013	12.574	13.139	13.709	14.283	14.862
300	14.862	15.445	16.032	16.624	17.219	17.819	18.422	19.030	19.641	20.255	20.872
400	20.872										

TABLE 17. Coefficients in the Equation for Type T Thermocouples

Coeff.	−270 °C to 0 °C	0 °C to 400 °C
c_0	0.000 000 000 0 ...	0.000 000 000 0 ...
c_1	$3.874\ 810\ 636\ 4 \times 10^{-2}$	$3.874\ 810\ 636\ 4 \times 10^{-2}$
c_2	$4.419\ 443\ 434\ 7 \times 10^{-5}$	$3.329\ 222\ 788\ 0 \times 10^{-5}$
c_3	$1.184\ 432\ 310\ 5 \times 10^{-7}$	$2.061\ 824\ 340\ 4 \times 10^{-7}$
c_4	$2.003\ 297\ 355\ 4 \times 10^{-8}$	$-2.188\ 225\ 684\ 6 \times 10^{-9}$
c_5	$9.013\ 801\ 955\ 9 \times 10^{-10}$	$1.099\ 688\ 092\ 8 \times 10^{-11}$
c_6	$2.265\ 115\ 659\ 3 \times 10^{-11}$	$-3.081\ 575\ 877\ 2 \times 10^{-14}$
c_7	$3.607\ 115\ 420\ 5 \times 10^{-13}$	$4.547\ 913\ 529\ 0 \times 10^{-17}$
c_8	$3.849\ 393\ 988\ 3 \times 10^{-15}$	$-2.751\ 290\ 167\ 3 \times 10^{-20}$
c_9	$2.821\ 352\ 192\ 5 \times 10^{-17}$	....
c_{10}	$1.425\ 159\ 477\ 9 \times 10^{-19}$	....
c_{11}	$4.876\ 866\ 228\ 6 \times 10^{-22}$	....
c_{12}	$1.079\ 553\ 927\ 0 \times 10^{-24}$	....
c_{13}	$1.394\ 502\ 706\ 2 \times 10^{-27}$	....
c_{14}	$7.979\ 515\ 392\ 7 \times 10^{-31}$	....

Coefficients in the Equations Expressing Temperature as a Function of Thermocouple emf *E*/mV

Often it is useful to calculate the temperature as the function of emf. Tables 19 to 25 contain, for each thermocouple type, the temperature for a given emf for various emf ranges using the following equation.

$$t/°C = c_0 + c_1E + c_2E^2 + c_3E^3 + c_4E^4 + \cdots$$

where t is the calculated temperature in degrees Celsius (ITS-90), E is a given emf in millivolts, and c_0, c_1, c_2, c_3, c_4, etc., are the coefficients, which have been extracted from Ref. 1. Each table is divided into 2 to 4 emf ranges to improve the fit within the specified emf range. Uncertainty estimates are given in notes for each table.

TABLE 18. Type B Thermocouples: Coefficients (c_i) of Polynomials for the Computation of Temperatures in °C as a Function of the Thermocouple emf in Various Temperature and emf Ranges

Coeff.	*t*: 250 °C to 700 °C emf: 0.291 mV to 2.431 mV	700 °C to 1820 °C 2.431 mV to 13.820 mV
c_0 =	$9.842\ 332\ 1 \times 10^{1}$	$2.131\ 507\ 1 \times 10^{2}$
c_1 =	$6.997\ 150\ 0 \times 10^{2}$	$2.851\ 050\ 4 \times 10^{2}$
c_2 =	$-8.476\ 530\ 4 \times 10^{2}$	$-5.274\ 288\ 7 \times 10^{1}$
c_3 =	$1.005\ 264\ 4 \times 10^{3}$	9.916 080 4 ...
c_4 =	$-8.334\ 595\ 2 \times 10^{2}$	−1.296 530 3 ...
c_5 =	$4.550\ 854\ 2 \times 10^{2}$	$1.119\ 587\ 0 \times 10^{-1}$
c_6 =	$-1.552\ 303\ 7 \times 10^{2}$	$-6.062\ 519\ 9 \times 10^{-3}$
c_7 =	$2.988\ 675\ 0 \times 10^{1}$	$1.866\ 169\ 6 \times 10^{-4}$
c_8 =	−2.474 286 0 ...	$-2.487\ 858\ 5 \times 10^{-6}$

Note: The above coefficients are extracted from Ref. 1 and are for an expression of the form shown in Section 10.3.2. They yield approximate values of temperature that agree within ±0.03 °C with the values given in Table 10.2.

TABLE 19. Type E Thermocouples: Coefficients (c_i) of Polynomials for the Computation of Temperatures in °C as a Function of the Thermocouple emf in Various Temperature and emf Ranges

Coeff.	*t*: −200 °C to 0 °C emf: −8.825 mV to 0.0 mV	0 °C to 1000 °C 0.0 mV to 76.373 mV
c_0 =	0.000 000 0 ...	0.000 000 0 ...
c_1 =	$1.697\ 728\ 8 \times 10^{1}$	$1.705\ 703\ 5 \times 10^{1}$
c_2 =	$-4.351\ 497\ 0 \times 10^{-1}$	$-2.330\ 175\ 9 \times 10^{-1}$
c_3 =	$-1.585\ 969\ 7 \times 10^{-1}$	$6.543\ 558\ 5 \times 10^{-3}$
c_4 =	$-9.250\ 287\ 1 \times 10^{-2}$	$-7.356\ 274\ 9 \times 10^{-5}$
c_5 =	$-2.608\ 431\ 4 \times 10^{-2}$	$-1.789\ 600 \times 10^{-6}$
c_6 =	$-4.136\ 019\ 9 \times 10^{-3}$	$8.403\ 616\ 5 \times 10^{-8}$
c_7 =	$-3.403\ 403\ 0 \times 10^{-4}$	$-1.373\ 587\ 9 \times 10^{-9}$
c_8 =	$-1.156\ 489\ 0 \times 10^{-5}$	$1.062\ 982\ 3 \times 10^{-11}$
c_9 =	....	$-3.244\ 708\ 7 \times 10^{-14}$

Note: The above coefficients are extracted from Ref. 1 and yield approximate values of temperature that agree within ±0.02 °C with the values given in Table 2.

TABLE 20. Type J Thermocouples: Coefficients (c_i) of Polynomials for the Computation of Temperatures in °C as a Function of the Thermocouple emf in Various Temperature and emf Ranges

Coeff.	*t*: −210 °C to 0 °C emf: −8.095 mV to 0.0 mV	0 °C to 760 °C 0.0 mV to 42.919 mV	760 °C to 2100 °C 42.919 mV to 69.553 mV
c_0 =	0.000 000 0...	0.000 000 ...	$-3.113\ 581\ 87 \times 10^{3}$
c_1 =	$1.952\ 826\ 8 \times 10^{1}$	$1.978\ 425 \times 10^{1}$	$3.005\ 436\ 84 \times 10^{2}$
c_2 =	-1.228 618 5 ...	$-2.001\ 204 \times 10^{-1}$	-9.947 732 30 ...
c_3 =	-1.075 217 8 ...	$1.036\ 969 \times 10^{-2}$	$1.702\ 766\ 30 \times 10^{-1}$
c_4 =	$-5.908\ 693\ 3 \times 10^{-1}$	$-2.549\ 687 \times 10^{-4}$	$1.430\ 334\ 68 \times 10^{-3}$
c_5 =	$-1.725\ 671\ 3 \times 10^{-1}$	$3.585\ 153 \times 10^{-6}$	$4.438\ 860\ 84 \times 10^{-6}$
c_6 =	$-2.813\ 151\ 3 \times 10^{-2}$	$-5.344\ 285 \times 10^{-8}$	....
c_7 =	$-2.396\ 337\ 0 \times 10^{-3}$	$5.099\ 890 \times 10^{-10}$	....
c_8 =	$-8.382\ 332\ 1 \times 10^{-5}$	....	....

Note: The above coefficients are extracted from Ref. 1 and yield approximate values of temperature that agree within ±0.5 °C with the values given in Table 4.

TABLE 21. Type K Thermocouples: Coefficients (c_i) of Polynomials for the Computation of Temperatures in °C as a Function of the Thermocouple emf in Various Temperature and emf Ranges

Coeff.	*t*: −200 °C to 0 °C emf: −5.891 mV to 0.0 mV	0 °C to 500 °C 0.0 mV to 20.644 mV	500 °C to 1372 °C 20.644 mV to 54.886 mV
c_0 =	0.000 000 0 ...	0.000 000 ...	$-1.318\ 058 \times 10^{2}$
c_1 =	$2.517\ 346\ 2 \times 10^{1}$	$2.508\ 355 \times 10^{1}$	$4.830\ 222 \times 10^{1}$
c_2 =	−1.166 287 8 ...	$7.860\ 106 \times 10^{-2}$	−1.646 031 ...
c_3 =	−1.083 363 8 ...	$-2.503\ 131 \times 10^{-1}$	$5.464\ 731 \times 10^{-2}$
c_4 =	$-8.977\ 354\ 0 \times 10^{-1}$	$8.315\ 270 \times 10^{-2}$	$-9.650\ 715 \times 10^{-4}$
c_5 =	$-3.734\ 237\ 7 \times 10^{-1}$	$-1.228\ 034 \times 10^{-2}$	$8.802\ 193 \times 10^{-6}$
c_6 =	$-8.663\ 264\ 3 \times 10^{-2}$	$9.804\ 036 \times 10^{-4}$	$3.110\ 810 \times 10^{-8}$
c_7 =	$-1.045\ 059\ 8 \times 10^{-2}$	$-4.413\ 030 \times 10^{-5}$	....
c_8 =	$-5.192\ 057\ 7 \times 10^{-4}$	$1.057\ 734 \times 10^{-6}$	....
c_9 =	....	$-1.052\ 755 \times 10^{-8}$	....

Note: The above coefficients are extracted from Ref. 1 and yield approximate values of temperature that agree within ±0.05 °C with the values given in Table 6.

TABLE 22. Type N Thermocouples: Coefficients (c_i) of Polynomials for the Computation of Temperatures in °C as a Function of the Thermocouple emf in Various Temperature and emf Ranges

Coeff.	t: −200 °C to 0 °C emf: −3.990 mV to 0.0 mV	0 °C to 600 °C 0.0 mV to 20.613 mV	600 °C to 1300 °C 20.613 mV to 47.513 mV
c_0 =	0.000 000 0 ...	0.000 00 ...	1.972 485 × 10^{1}
c_1 =	3.843 684 7 × 10^{1}	3.868 96 × 10^{1}	3.300 943 × 10^{1}
c_2 =	1.101 048 5 ...	-1.082 67 ...	-3.915 159 × 10^{-1}
c_3 =	5.222 931 2 ...	4.702 05 × 10^{-2}	9.855 391× 10^{-3}
c_4 =	7.206 052 5 ...	-2.121 69 × 10^{-6}	-1.274 371 × 10^{-4}
c_5 =	5.848 858 6 ...	-1.172 72 × 10^{-4}	7.767 022 × 10^{-7}
c_6 =	2.775 491 6 ...	5.392 80 × 10^{-6}	....
c_7 =	7.707 516 6 × 10^{-1}	-7.981 56 × 10^{-8}	....
c_8 =	1.158 266 5 × 10^{-1}	....	....
c_9 =	7.313 886 8 × 10^{-3}	....	....

Note: The above coefficients are extracted from Ref. 1 and yield approximate values of temperature that agree within ±0.04 °C with the values given in Table 8.

TABLE 23. Type R Thermocouples: Coefficients (c_i) of Polynomials for the Computation of Temperatures in °C as a Function of the Thermocouple emf in Various Temperature and emf Ranges

Coeff.	t: −50 °C to 250 °C emf: −0.226 mV to 1.923 mV	250 °C to 1200 °C 1.923 mV to 13.228 mV	1064 °C to 1664.5 °C 11.361 mV to 19.739 mV	1664.5 °C to 1768.1 °C 19.739 mV to 21.103 mV
c_0 =	0.000 000 0 ...	1.334 584 505 × 10^{1}	-8.199 599 416 × 10^{1}	3.406 177 836 × 10^{4}
c_1 =	1.889 138 0 × 10^{2}	1.472 644 573 × 10^{2}	1.553 962 042 × 10^{2}	-7.023 729 171 × 10^{3}
c_2 =	-9.383 529 0 × 10^{1}	-1.844 024 844 × 10^{1}	-8.342 197 663	5.582 903 813 × 10^{2}
c_3 =	1.306 861 9 × 10^{2}	4.031 129 726 ...	4.279 433 549 × 10^{-1}	-1.952 394 635 × 10^{1}
c_4 =	-2.270 358 0 × 10^{2}	-6.249 428 360 × 10^{-1}	-1.191 577 910 × 10^{-2}	2.560 740 231 × 10^{-1}
c_5 =	3.514 565 9 × 10^{2}	6.468 412 046 × 10^{-2}	1.492 290 091 × 10^{-4}	....
c_6 =	-3.895 390 0 × 10^{2}	-4.458 750 426 × 10^{-3}	....	....
c_7 =	2.823 947 1 × 10^{2}	1.994 710 149 × 10^{-4}	....	....
c_8 =	-1.260 728 1 × 10^{2}	-5.313 401 790 × 10^{-6}	....	....
c_9 =	3.135 361 1 × 10^{1}	6.481 976 217 × 10^{-8}	....	....
c_{10} =	-3.318 776 9 ...	....	....	....

Note: The above coefficients are extracted from Ref. 1 and yield approximate values of temperature that agree within ±0.02 °C with the values given in Table 10.

TABLE 24. Type S Thermocouples: Coefficients (c_i) of Polynomials for the Computation of Temperatures in °C as a Function of the Thermocouple emf in Various Temperature and emf Ranges

Coeff.	t: −50 °C to 250 °C emf: −0.235 mV to 1.874 mV	250 °C to 1200 °C 1.874 mV to 11.950 mV	1064 °C to 1664.5 °C 10.332 mV to 17.536 mV	1664.5 °C to 1768.1 °C 17.536 mV to 18.693 mV
c_0 =	0.000 000 00 . . .	1.291 507 177 × 10^{1}	-8.087 801 117 ×10^{1}	5.333 875 126 × 10^{4}
c_1 =	1.849 494 60 × 10^{2}	1.466 298 863 × 10^{2}	1.621 573 104 × 10^{2}	-1.235 892 298 × 10^{4}
c_2 =	-8.005 040 62 × 10^{1}	-1.534 713 402 ×10^{1}	-8.536 869 453 . . .	1.092 657 613 × 10^{3}
c_3 =	1.022 374 30 × 10^{2}	3.145 945 973 . . .	4.719 686 976 × 10^{-1}	-4.265 693 686 × 10^{1}
c_4 =	-1.522 485 92 × 10^{2}	-4.163 257 839 × 10^{-1}	-1.441 693 666 × 10^{-2}	6.247 205 420 × 10^{-1}
c_5 =	1.888 213 43 × 10^{2}	3.187 963 771 × 10^{-2}	2.081 618 890 × 10^{-4}	....
c_6 =	-1.590 859 41 × 10^{2}	-1.291 637 500 × 10^{-3}	....	....
c_7 =	8.230 278 80 × 10^{1}	2.183 475 087 × 10^{-5}	....	....
c_8 =	-2.341 819 44 × 10^{1}	-1.447 379 511 × 10^{-7}	....	....
c_9 =	2.797 862 60 . . .	8.211 272 125 × 10^{-9}	....	....

Note: The above coefficients are extracted from Ref. 1 and yield approximate values of temperature that agree within ±0.02 °C with the values given in Table 12.

TABLE 25. Type T Thermocouples: Coefficients (c_i) of Polynomials for the Computation of Temperatures in °C as a Function of the Thermocouple emf in Various Temperature and emf Ranges

Coeff.	*t*: −200 °C to 0 °C emf: −5.603 mV to 0.0 mV	0 °C to 400 °C 0.0 mV to 20.872 mV
c_0 =	0.000 000 0 . . .	0.000 000 . . .
c_1 =	$2.594\,919\,2 \times 10^{1}$	$2.592\,800 \times 10^{1}$
c_2 =	$-2.131\,696\,7 \times 10^{-1}$	$-7.602\,961 \times 10^{-1}$
c_3 =	$7.901\,869\,2 \times 10^{-1}$	$4.637\,791 \times 10^{-2}$
c_4 =	$4.252\,777\,7 \times 10^{-1}$	$-2.165\,394 \times 10^{-3}$
c_5 =	$1.330\,447\,3 \times 10^{-1}$	$6.048\,144 \times 10^{-5}$
c_6 =	$2.024\,144\,6 \times 10^{-2}$	$-7.293\,422 \times 10^{-7}$
c_7 =	$1.266\,817\,1 \times 10^{-3}$	....

Note: The above coefficients are extracted from Ref. 1 and yield approximate values of temperature that agree within ±0.04 °C with the values given in Table 14.

Lab Data

REFERENCE POINTS ON THE ITS-90 TEMPERATURE SCALE

The International Temperature Scale of 1990 is described in Section 1 of this *CRC Handbook*, where the defining fixed points are listed. The Consultative Committee on Thermometry (CCT) of the International Committee on Weights and Measures (CIPM), which oversees the temperature scale, has recommended a number of secondary reference points whose values have been accurately determined with respect to the primary fixed points. The most accurate of these, referred to as "first quality points," satisfy several criteria involving purity of the material, reproducibility, and documentation of the measurements. The CCT also lists "second quality points" that do not yet satisfy all the criteria but are still useful. Taken together, these secondary reference points, help fill in the gaps between the primary fixed points.

The table below describes these secondary reference points, along with the primary fixed points. Column definitions for the table are as follows.

Column heading	Definition
Name	Name of substance
Type of transition	Transition used as reference point
T	Transition temperature, in K; entries are listed in order of increasing temperature
t	Transition temperature, in °C
Uncert.	Estimate of the uncertainty
Note	Information on the quality of the reference point, see text above for discussion

Full details are given in the reference.

Reference

Bedford, R. E., Bonnier, G., Maas, H., and Pavese, F., *Metrologia* 33, 133, 1996. <https://doi.org/10.1088/0026-1394/33/2/3>

Reference Points on the ITS-90 Temperature Scale

Name	Type of transition	*T*/K	*t*/°C	Uncert.	Note
Zinc	Superconductive transition	0.8500	-272.3000	0.0030	1st Quality
Aluminum	Superconductive transition	1.1810	-271.9690	0.0025	1st Quality
Helium	Superfluid transition (^{4}He)	2.1768	-270.9732	0.0001	1st Quality
Indium	Superconductive transition	3.4145	-269.7355	0.0025	1st Quality
Lead	Superconductive transition	7.1997	-265.9503	0.0025	1st Quality
Niobium	Superconductive transition	9.2880	-263.8620	0.0025	1st Quality
Hydrogen	Triple point (equilibrium H_2)	13.8033	-259.3467		Primary
Hydrogen	Triple point (normal H_2)	13.952	-259.198	0.002	2nd Quality
Hydrogen-d_2	Triple point (equilibrium D_2)	18.689	-254.461	0.001	1st Quality
Hydrogen-d_2	Triple point (normal D_2)	18.724	-254.426	0.001	1st Quality
Hydrogen	Boiling point (normal H_2)	20.388	-252.762	0.002	2nd Quality
Oxygen	α-β transition	23.868	-249.282	0.005	2nd Quality
Neon	Triple point (^{20}Ne)	24.541	-248.609	0.001	1st Quality
Neon	Triple point (normal Ne)	24.5561	-248.5939		Primary
Neon	Boiling point	27.097	-246.053	0.001	1st Quality
Nitrogen	α-β transition	35.614	-237.536	0.006	2nd Quality
Oxygen	β-γ transition	43.796	-229.354	0.001	2nd Quality
Oxygen	Triple point	54.3584	-218.7916		Primary
Nitrogen	Triple point	63.151	-209.999	0.001	1st Quality
Nitrogen	Boiling point	77.352	-195.798	0.002	1st Quality
Argon	Triple point	83.8058	-189.3442		Primary
Argon	Boiling point	87.303	-185.847	0.001	1st Quality
Oxygen	Condensation point	90.197	-182.953	0.001	1st Quality
Methane	Triple point	90.694	-182.456	0.001	1st Quality
Krypton	Triple point	115.775	-157.375	0.001	2nd Quality
Xenon	Triple point	161.405	-111.745	0.001	1st Quality
Carbon dioxide	Sublimation point	194.686	-78.464	0.003	2nd Quality
Carbon dioxide	Triple point	216.592	-56.558	0.001	1st Quality
Sulfur hexafluoride	Triple point	223.554	-49.596	0.005	2nd Quality
Mercury	Triple point	234.3156	-38.8344		Primary
Mercury	Freezing point	234.3210	-38.8290	0.0005	1st Quality
Water	Ice point	273.15	0		1st Quality
Water	Triple point	273.16	0.01		Primary
Gallium/20% indium	Eutectic melting point	288.800	15.650	0.001	2nd Quality
Gallium/8% tin	Eutectic melting point	293.626	20.476	0.002	2nd Quality
Diphenyl ether	Triple point	300.014	26.864	0.001	2nd Quality
Gallium	Melting point	302.9146	29.7646		Primary
Gallium	Triple point	302.9166	29.7666	0.0001	1st Quality

Name	Type of transition	T/K	t/°C	Uncert.	Note
Ethylene carbonate	Triple point	309.465	36.315	0.001	2nd Quality
Succinonitrile	Triple point	331.215	58.065	0.002	2nd Quality
Sodium	Freezing point	370.944	97.794	0.005	2nd Quality
Water	Boiling point	373.124	99.974	0.001	1st Quality
Benzoic acid	Triple point	395.486	122.336	0.002	2nd Quality
Benzoic acid	Freezing point	395.502	122.352	0.007	2nd Quality
Indium	Triple point	429.7436	156.5936	0.0002	1st Quality
Indium	Freezing point	429.7485	156.5985		Primary
Tin (white)	Freezing point	505.078	231.928		Primary
Bismuth	Freezing point	544.552	271.402	0.001	1st Quality
Cadmium	Freezing point	594.219	321.069	0.001	1st Quality
Lead	Freezing point	600.612	327.462	0.001	1st Quality
Mercury	Boiling point	629.811	356.619	0.004	2nd Quality
Zinc	Freezing point	692.677	419.527		Primary
Sulfur	Boiling point	717.674	444.614	0.002	2nd Quality
Copper/66.9% aluminum	Eutectic melting point	821.308	548.158	0.010	2nd Quality
Silver/30% aluminum	Eutectic melting point	840.957	567.807	0.002	2nd Quality
Antimony	Freezing point	903.778	630.628	0.001	1st Quality
Aluminum	Freezing point	933.473	660.323		Primary
Copper/71.9% silver	Eutectic melting point	1052.78	779.63	0.05	1st Quality
Sodium chloride	Freezing point	1075.168	802.018	0.011	2nd Quality
Sodium	Boiling point	1156.090	882.940	0.005	2nd Quality
Silver	Freezing point	1234.93	961.78		Primary
Gold	Freezing point	1337.33	1064.18		Primary
Copper	Freezing point	1357.77	1084.62		Primary
Nickel	Freezing point	1728	1455	1	2nd Quality
Cobalt	Freezing point	1768	1495	3	2nd Quality
Iron	Freezing point	1811	1538	3	2nd Quality
Palladium	Freezing point	1828.0	1554.8	0.1	1st Quality
Titanium	Melting point	1943	1670	2	2nd Quality
Platinum	Freezing point	2041.3	1768.2	0.4	1st Quality
Zirconium	Melting point	2127	1854	8	2nd Quality
Rhodium	Freezing point	2236	1963	3	1st Quality
Aluminum oxide (α)	Melting point	2326	2053	2	2nd Quality
Ruthenium	Melting point	2606	2333	10	2nd Quality
Iridium	Freezing point	2719	2446	6	1st Quality
Molybdenum	Melting point	2895	2622	4	1st Quality
Tungsten	Melting point	3687	3414	7	1st Quality

RELATIVE SENSITIVITY OF BAYARD-ALPERT IONIZATION GAUGES TO VARIOUS GASES

Paul Redhead

Bayard-Alpert hot-cathode ionization gauges are widely used for pressure measurements in the 10^{-3} Torr to 10^{-10} Torr (≈1.33 × 10^{-7} kPa to ≈1.33 × 10^{-14} kPa) pressure range. The ion current I_+ in a hot-cathode ionization gauge is given by $I_+ = KI_eP$. The gauge constant is $K = (I_+/I_e)(1/P)$, where I_e is the electron current, and P the pressure. The sensitivity is given by $S = KI_e = I_+/P$. The constant K is independent of pressure below about 10^{-3} Torr (≈1.33 × 10^{-7} kPa), but it depends on the gas that is present. This table gives the sensitivity relative to nitrogen for several different gases.

Relative sensitivities for different Bayard-Alpert ionization gauges may differ by as much as ±15% as a result of differences in applied voltages, electron current, and electrode structure. The table presents the average of the measurements of 12 experimenters on Bayard-Alpert ionization gauges in various gases.

References

1. Hollanda, R., *J. Vac. Sci. Technol.*, 10, 1133, 1973. <https://doi.org/10.1116/1.1318508>
2. Nakayama, K., and Hojo, H., *Jap. J. Appl. Phys.*, Suppl. 2, part 1, p. 113, 1974. <https://doi.org/10.7567/JJAPS.2S1.113>
3. Tilford, C. R., *J. Vac. Sci. Technol. A*, 1, 152, 1983. <https://doi.org/10.1116/1.572063>
4. Tilford, C. R., in *Physical Methods of Chemistry, Vol.6, Determination of Thermodynamic Properties*, B. W. Rossiter and R. C. Baetzoid, Eds., p. 101-173, John Wiley, New York, 1992.

Sensitivity of Bayard-Alpert Ionization Gauges to Various Gases Relative to Nitrogen

Name	Mol. form.	Relative sensitivity $S/S(N_2)$
Nitrogen	N_2	**1.00**
Acetylene	HC≡CH	0.61
Allene	$CH_2=C=CH_2$	1.3
Argon	Ar	1.4
Benzene	C_6H_6	3.8
Butane	C_4H_{10}	4.3
Carbon dioxide	CO_2	1.4
Carbon monoxide	CO	1.0
Ethane	C_2H_6	2.6
Ethylene	$CH_2=CH_2$	1.3
Helium	He	0.18
Hydrogen	H_2	0.43
Krypton	Kr	1.9
Mercury	Hg	3.5
Methane	CH_4	1.6
Neon	Ne	0.31
Oxygen	O_2	0.96
Propane	C_3H_8	3.5
Propene	$CH_3CH=CH_2$	1.8
Propyne	$CH_3C≡CH$	1.4
Sulfur hexafluoride	SF_6	2.3
Water	H_2O	0.93
Xenon	Xe	2.7

LABORATORY SOLVENTS AND OTHER LIQUID REAGENTS

This table summarizes the properties of 574 liquids that are commonly used in the laboratory as solvents or chemical reagents. Column definitions for the table are as follows.

Column heading	Definition
Name	Name of solvent or reagent
Mol. form.	Molecular formula of solvent or reagent
M_r	Molecular weight
mp	Melting point, in °C
bp	Normal boiling point, in °C
ρ	Density, in g cm^{-3}, at the temperature in °C indicated by the superscript
ε	Dielectric constant (permittivity) at the temperature in °C indicated by the superscript (otherwise at ambient temperature)
μ	Dipole moment, in Debye units
Vap. pres. (25 °C)	Vapor pressure at 25 °C, in kPa (1 kPa = 7.5006 mmHg)
FP	Flash point, in °C
Flam. limits	Flammable (explosive) limits in air, in percent by volume
IT	Autoignition temperature, in °C
TWA and STEL	Threshold limits for allowable airborne concentration, in parts per million, by volume at 25 °C and atmospheric pressure; TWA refers to an 8-hour time-weighted average and STEL to a short-term limit; more details on threshold limits can be found in Section 16

Data on three additional properties, refractive index, specific heat capacity, and viscosity, for these liquids are included in the Online Edition of the *CRC Handbook*. Data on the temperature dependence of viscosity, dielectric constant, and vapor pressure can be found in the pertinent tables in Section 6.

References

1. Lide, D. R., *Handbook of Organic Solvents,* CRC Press, Boca Raton, FL, 1994.
2. Lide, D. R., and Kehiaian, H. V., *Handbook of Thermophysical and Thermochemical Data,* CRC Press, Boca Raton, FL, 1994.
3. Riddick, J. A., Bunger, W. B., and Sakano, T. K., *Organic Solvents, Fourth Edition,* John Wiley & Sons, New York, 1986.
4. *Fire Protection Guide to Hazardous Materials, 11th Edition,* National Fire Protection Association, Quincy, MA, 1994.
5. Urben, P. G., Ed., *Bretherick's Handbook of Reactive Chemical Hazards, Fifth Edition,* Butterworth-Heinemann, Oxford, 1995.
6. *2015 TLV's and BEI's,* American Conference of Governmental Industrial Hygienists, Cincinnati, OH, 2015.

Physical Properties of Common Solvents and Liquid Reagents

Name	Mol. form.	M_r	mp/°C	bp/°C	ρ/g cm^{-3}	ε	μ/D	Vap. pres. (25 °C)/ kPa	FP/°C	Flam. limits (%)	IT/°C	TWA/ ppm	STEL/ ppm
Acetaldehyde	C_2H_4O	44.052	-123.4	20.8	0.7834^{18}	21.0^{18}	2.750	120	-39	4.0-60%	175		25
Acetic acid	$C_2H_4O_2$	60.052	17	117.9	1.0510^{20}	6.20^{20}	1.70	2.07	39	4.0-19.9%	463	10	15
Acetic anhydride	$C_4H_6O_3$	102.089	-73.4	139.5	1.082^{20}	22.45^{20}	≈2.8	0.680	49	2.7-10.3%	316	1	3
Acetone	C_3H_6O	58.079	-94.9	56.08	0.7902^{20}	21.01^{20}	2.88	30.8	-20	2.5-12.8%	465	250	500
Acetone cyanohydrin	C_4H_7NO	85.105	-19	180	0.932^{19}				74	2.2-12.0%	688		
Acetonitrile	C_2H_3N	41.052	-44	81.6	0.7825^{20}	36.64^{20}	3.92519	11.9	6	3.0-16.0%	524	20	
Acetophenone	C_8H_8O	120.149	19.4	202.1	1.0281^{20}	17.44^{25}	3.02	0.049	77		570	10	
Acetyl bromide	C_2H_3BrO	122.948	-96.5	74	1.6625^{16}			16.2					
Acetyl chloride	C_2H_3ClO	78.497	-112.7	51	1.1051^{20}	15.8^{22}	2.72	38.4	4		390		
Acrolein	C_3H_4O	56.063	-87.8	52.3	0.840^{20}		[a]	36.2	-26	2.8-31%	220		0.1
Acrylic acid	$C_3H_4O_2$	72.063	13.56	142	1.0511^{20}			0.53	50	2.4-8.0%	438	2	
Acrylonitrile	C_3H_3N	53.063	-83.51	77.2	0.8007^{25}	33.0^{20}	3.92	14.1	0	3.0-17.0%	481	2	
Allyl alcohol	C_3H_6O	58.079	-129	96.9	0.8540^{20}	19.7^{20}	1.60	3.14	21	2.5-18.0%	378	0.5	
Allylamine	C_3H_7N	57.095	-88.2	54	0.758^{20}		≈1.2	33.1	-29	2.2-22%	374		
2-Amino-2-methyl-1-propanol	$C_4H_{11}NO$	89.136	25.5	163.8	0.934^{20}				67				
3-Amino-1-propanol	C_3H_9NO	75.109	12.1	185	0.9824^{26}				80				
Aniline	C_6H_7N	93.127	-6.0	184.1	1.0250^{20}	7.06^{20}	1.13	0.090	70	1.3-11%	615	2	
Anisole	C_7H_8O	108.138	-37.3	153.6	0.9940^{20}	4.30^{21}	1.38	0.472	52		475		
Antimony(V) chloride	Cl_5Sb	299.025	4	140 dec	2.34	3.222^{20}							
Antimony(V) fluoride	F_5Sb	216.752	8.3	141	3.10								
Arsenic(III) chloride	$AsCl_3$	181.281	-16	130	2.150		1.59	5.38					
Benzaldehyde	C_7H_6O	106.122	-57.12	178.7	1.0401^{25}	17.85^{20}	3.0	0.169	63		192		
Benzene	C_6H_6	78.112	5.538	80.08	0.8788^{20}	2.2825^{20}	0	12.7	-11	1.2-7.8%	498	0.5	2.5
Benzeneacetonitrile	C_8H_7N	117.149	-22.1	232	1.0205^{15}	17.87^{26}	3.5	0.012	113				
Benzeneethanamine	$C_8H_{11}N$	121.180	<0	204	0.9640^{25}								
Benzeneethanol	$C_8H_{10}O$	122.164	-19	220	1.0202^{20}	12.31^{20}		0.01	96				
Benzenemethanethiol	C_7H_8S	124.204	-30	199	1.058^{20}	4.705^{25}							
Benzenesulfonyl chloride	$C_6H_5ClO_2S$	176.621	14.5	252	1.3470^{15}	28.90^{50}		0.008					
Benzenethiol	C_6H_6S	110.177	-14.87	169.1	1.0775^{20}	4.26^{30}	1.23	0.26				0.1	
Benzonitrile	C_7H_5N	103.122	-12.82	191	1.0093^{15}	25.9^{20}	4.515	0.11					

Name	Mol. form.	M_r	mp/°C	bp/°C	ρ/g cm^{-3}	ε	μ/D	Vap. pres. (25 °C)/ kPa	FP/°C	Flam. limits (%)	IT/°C	TWA/ ppm	STEL/ ppm
Benzoyl chloride	C_7H_5ClO	140.567	-0.5	201	1.2120^{20}	23.0^{20}		0.084	72				0.5
Benzyl acetate	$C_9H_{10}O_2$	150.174	-51.5	215	1.0550^{20}	5.34^{30}	1.22	0.022	90		460	10	
Benzyl alcohol	C_7H_8O	108.138	-15.5	205.3	1.0419^{24}	11.916^{30}	1.71	0.015	93		436		
Benzylamine	C_7H_9N	107.153		185	0.9813^{20}	5.18^{20}		0.096					
2,2'-Bioxirane	$C_4H_6O_2$	86.090	2.0	144	1.113^{20}								
Bis(2-aminoethyl)amine	$C_4H_{13}N_3$	103.166	-39	206.5	0.9569^{20}	12.62^{20}	1.89	0.03	98	2-6.7%	358	1	
N,N'-Bis(2-aminoethyl)-1,2-ethanediamine	$C_6H_{18}N_4$	146.234	12	266.5		10.76^{20}			135				
Bis(2-chloroethyl) ether	$C_4H_8Cl_2O$	143.012	-46.9	178	1.22^{20}	21.20^{20}	2.58	0.143	55	>2.7%	369	5	10
Bis(chloromethyl) ether	$C_2H_4Cl_2O$	114.958	-41.5	104	1.323^{15}	3.51^{20}						0.001	
Bis(2-ethylhexyl) phthalate	$C_{24}H_{38}O_4$	390.557	-55	384	0.981^{25}		2.84	0.00000005	218				
Bis(2-hydroxyethyl) sulfide	$C_4H_{10}O_2S$	122.186	-10.2	282	1.1793^{25}	28.61^{20}		0.08	160		298		
Boron tribromide	BBr_3	250.523	-46	91.3	2.6	2.58^{0}	0						0.7
Boron trichloride	BCl_3	117.170	-107.3	12.5	0.004789^{25} gas		0	156					0.7
Bromine	Br_2	159.808	-7.2	58.8	3.1028	3.1484^{25}		28.2				0.1	0.2
Bromobenzene	C_6H_5Br	157.008	-30.74	155.9	1.4950^{20}	5.45^{20}	1.70	0.556	51		565		
1-Bromobutane	C_4H_9Br	137.018	-112.5	101.4	1.2758^{20}	7.315^{10}	2.31	5.26	18	2.6-6.6%	265		
2-Bromobutane, (±)-	C_4H_9Br	137.018	-112.6	91	1.2585^{20}	8.64^{25}	2.23	9.32	21				
Bromochloromethane	CH_2BrCl	129.384	-87.9	67.9	1.9344^{20}		1.66	19.5				200	
Bromodichloromethane	$CHBrCl_2$	163.829	-56.0	90	1.980^{20}								
Bromoethane	C_2H_5Br	108.965	-118.4	38.2	1.4604^{20}	9.01^{25}	2.04	62.5		6.8-8.0%	511	5	
Bromoethene	C_2H_3Br	106.949	-139.5	16	1.4933^{20}		1.42	141		9-15%	530	0.5	
2-Bromo-2-methylpropane	C_4H_9Br	137.018	-16.8	73.3	1.4278^{20}	10.98^{20}	2.17	17.7					
1-Bromopentane	$C_5H_{11}Br$	151.045	-88.0	126	1.2182^{20}	6.31^{26}	2.20	1.68	32				
1-Bromopropane	C_3H_7Br	122.992	-110.1	70.8	1.3537^{20}	8.09^{20}	2.18	18.6			490	0.1	
2-Bromopropane	C_3H_7Br	122.992	-88.9	59.34	1.3140^{20}	9.46^{20}	2.21	28.9					
3-Bromopropene	C_3H_5Br	120.976	-119.3	70.1	1.398^{20}	7.0^{20}	≈1.9	18.6	-1	4.4-7.3%	295	0.1	0.2
2-Bromotoluene	C_7H_7Br	171.035	-27.5	182	1.4232^{20}	4.641^{20}		0.17	79				
Bromotrichloromethane	$CBrCl_3$	198.274	-5.6	103	2.012^{25}	2.405^{20}		5.35					
Butanal	C_4H_8O	72.106	-96.86	74.8	0.8016^{20}	13.45^{25}	2.72	15.7	-22	1.9-12.5%	218		
1,3-Butanediol	$C_4H_{10}O_2$	90.121	-77	208.2	1.0053^{20}	28.8^{25}		0.008	121		395		
1,4-Butanediol	$C_4H_{10}O_2$	90.121	20.43	229.5	1.0171^{20}	31.9^{25}	2.58	0.002	121				
2,3-Butanediol	$C_4H_{10}O_2$	90.121	7	178	1.0033^{20}			0.02			402		
2,3-Butanedione	$C_4H_6O_2$	86.090	-1.2	87.5	0.9808^{18}	4.04^{25}		7.45	27			0.01	0.02
Butanenitrile	C_4H_7N	69.106	-111.76	117.6	0.7936^{20}	24.83^{20}	≈3.8	2.55	24	>1.6%	501		
1-Butanethiol	$C_4H_{10}S$	90.187	-115.66	98.4	0.8416^{20}	5.204^{15}	1.53	6.07	2			0.5	
2-Butanethiol	$C_4H_{10}S$	90.187	-165	85.0	0.8295^{20}	5.645^{15}		10.8	-23				
Butanoic acid	$C_4H_8O_2$	88.106	-5.12	163.7	0.9528^{25}	2.98^{14}	1.65	0.221	72	2.0-10.0%	443		
Butanoic anhydride	$C_8H_{14}O_3$	158.195	-75.0	195	0.9668^{20}	12.8^{20}		0.07	54	0.9-5.8%	279		
1-Butanol	$C_4H_{10}O$	74.121	-88.60	117.6	0.8148^{20}	17.84^{20}	1.66	0.86	37	1.4-11.2%	343	20	
2-Butanol	$C_4H_{10}O$	74.121	-88.44	99.4	0.8063^{20}	17.26^{20}		2.32	24	1.7-9.8%	405	100	
2-Butanone	C_4H_8O	72.106	-86.67	79.6	0.7999^{25}	18.56^{20}	2.779	12.6	-9	1.4-11.4%	404	200	300
trans-2-Butenal	C_4H_6O	70.090	-76.6	102.2	0.8516^{20}		3.67	4.92	13	2.1-15.5%	232		0.3
cis-2-Butenoic acid	$C_4H_6O_2$	86.090	15	169	1.0267^{20}			0.06					
2-Butoxyethanol	$C_6H_{14}O_2$	118.174	-74.8	171	0.9015^{20}		2.08	0.15	69	4-13%	238	20	
Butyl acetate	$C_6H_{12}O_2$	116.158	-77.0	126.0	0.8825^{20}	5.07^{20}	1.87	1.66	22	1.7-7.6%	425	50	150
sec-Butyl acetate	$C_6H_{12}O_2$	116.158	-98.9	108	0.8748^{20}	5.135^{20}	1.87		31	1.7-9.8%		50	150
Butyl acrylate	$C_7H_{12}O_2$	128.169	-63.6	146.6	0.8898^{20}	5.25^{28}		0.731	29	1.7-9.9%	292	2	
Butylamine	$C_4H_{11}N$	73.137	-49	77.0	0.7417^{20}	4.71^{20}	≈1.0	12.2	-12	1.7-9.8%	312		5
sec-Butylamine	$C_4H_{11}N$	73.137	-104.5	62.71	0.7246^{20}		1.28		-9				
tert-Butylamine	$C_4H_{11}N$	73.137	-66.92	44.02	0.6958^{20}		1.29	48.4	-9	1.7-8.9%	380		
Butylbenzene	$C_{10}H_{14}$	134.218	-87.81	183.3	0.8601^{20}	2.359^{20}		0.150	71	0.8-5.8%	410		
tert-Butylbenzene	$C_{10}H_{14}$	134.218	-57.84	169.1	0.8665^{20}	2.359^{20}	≈0.83	0.280	60	0.7-5.7%	450		
Butyl benzoate	$C_{11}H_{14}O_2$	178.228	-22.4	249	1.000^{20}	5.52^{30}		0.005	107				
tert-Butyl ethyl ether	$C_6H_{14}O$	102.174	-94.0	72.7	0.736^{25}			16.5				25	
tert-Butyl hydroperoxide	$C_4H_{10}O_2$	90.121	6	89 dec	0.8960^{20}				27				
1-*tert*-Butyl-4-methylbenzene	$C_{11}H_{16}$	148.245	-52.49	193	0.8612^{20}			0.09	68			1	
Butyl vinyl ether	$C_6H_{12}O$	100.158	-92	94	0.7888^{20}		1.25	6.65	-9		255		
γ-Butyrolactone	$C_4H_6O_2$	86.090	-43.36	204.6	1.1296^{20}	39.0^{20}	4.27	0.43	98				
Carbon disulfide	CS_2	76.141	-111.7	46.2	1.2632^{20}	2.6320^{20}	0	48.2	-30	1.3-50.0%	90	1	
Chloroacetaldehyde	C_2H_3ClO	78.497	-16.3	87	1.19								1
Chloroacetone	C_3H_5ClO	92.524	-44.5	116	1.15^{20}			2					1
Chloroacetyl chloride	$C_2H_2Cl_2O$	112.942	-21.7	106.0	1.4202^{20}		2.23	3.33				0.05	0.15
2-Chloroaniline	C_6H_6ClN	127.572	-2.3	209	1.2114^{22}	13.40^{20}	1.77	0.034					
3-Chloroaniline	C_6H_6ClN	127.572	-10.3	230	1.2161^{20}	13.3^{20}		0.0156			705		

Name	Mol. form.	M_r	mp/°C	bp/°C	ρ/g cm^{-3}	ε	μ/D	Vap. pres. (25 °C)/ kPa	FP/°C	Flam. limits (%)	IT/°C	TWA/ ppm	STEL/ ppm
Chlorobenzene	C_6H_5Cl	112.557	-45.2	131.6	1.1058^{20}	5.6895^{20}	1.69	1.6	28	1.3-9.6%	593	10	
2-Chloro-1,3-butadiene	C_4H_5Cl	88.536	-130	59	0.956^{20}	4.914^{20}		29.5	-20	4.0-20.0%		1	
1-Chlorobutane	C_4H_9Cl	92.567	-123.1	78.4	0.8857^{20}	7.276^{20}	2.05	13.7	-12	1.9-10.1%	240		
2-Chlorobutane	C_4H_9Cl	92.567	-131.3	71	0.8732^{20}	8.564^{20}	2.04	21.0	-10				
Chlorocyclohexane	$C_6H_{11}Cl$	118.604	-45	142.6	1.000^{20}	7.9505^{30}	a	1.0	32				
Chlorodibromomethane	$CHBr_2Cl$	208.280	-20	120	2.451^{20}								
Chloroethane	C_2H_5Cl	64.514	-138.4	12.3	0.9239^{0}	9.45^{20}	2.05	160	-50	3.8-15.4%	519	100	
2-Chloroethanol	C_2H_5ClO	80.513	-68	126	1.2019^{20}	25.80^{20}	1.78	1.2	60	4.9-15.9%	425		1
2-Chloroethyl vinyl ether	C_4H_7ClO	106.551	-70	108	1.0495^{20}				27				
(Chloromethyl)benzene	C_7H_7Cl	126.584	-39.4	174	1.1004^{20}	6.854^{20}	1.82	0.164	67	>1.1%	585	1	
Chloromethyl methyl ether	C_2H_5ClO	80.513	-103.5	59	1.063^{10}			24.9					
1-Chloro-2-methylpropane	C_4H_9Cl	92.567	-130.3	69	0.8773^{20}	7.027^{20}	2.00	19.9	-6	2.0-8.7%			
2-Chloro-2-methylpropane	C_4H_9Cl	92.567	-25.60	50.9	0.8420^{20}	9.663^{20}	2.13	42.7	0				
1-Chloronaphthalene	$C_{10}H_7Cl$	162.616	-6.0	259	1.1880^{25}	5.04^{25}	1.57	0.003	121		>558		
1-Chlorooctane	$C_8H_{17}Cl$	148.674	-57.8	183	0.8734^{20}	5.05^{25}	2.00	0.11	70				
1-Chloropentane	$C_5H_{11}Cl$	106.594	-99.0	107.9	0.8820^{20}	6.654^{20}	2.16	4.36	13	1.6-8.6%	260		
2-Chlorophenol	C_6H_5ClO	128.556	8	173.4	1.2634^{20}	7.40^{23}		0.308	64				
1-Chloropropane	C_3H_7Cl	78.541	-122.9	46.2	0.8899^{20}	8.588^{20}	2.05	45.8	<-18	2.6-11.1%	520		
2-Chloropropane	C_3H_7Cl	78.541	-117.1	35.0	0.8617^{20}		2.17	68.9	-32	2.8-10.7%	593		
3-Chloro-1,2-propanediol	$C_3H_7ClO_2$	110.540		221	1.325^{18}	31.0^{20}							
3-Chloropropanenitrile	C_3H_4ClN	89.524	-51.4	175.5	1.1573^{20}				76				
2-Chloropropene	C_3H_5Cl	76.525	-137.4	23	0.9017^{20}	8.92^{26}	1.647	110	-37	4.5-16%			
3-Chloropropene	C_3H_5Cl	76.525	-136	44.8	0.9376^{20}	8.2^{20}	1.94	48.9	-32	2.9-11.1%	485	1	2
Chlorosulfonic acid	$ClHO_3S$	116.524	-80	152	1.75			0.42					
2-Chlorotoluene	C_7H_7Cl	126.584	-35.9	158.8	1.0825^{20}	4.721^{20}	1.56	0.482				50	
4-Chlorotoluene	C_7H_7Cl	126.584	7.4	161.8	1.0697^{20}	6.25^{20}	2.21	0.4					
Chromium(VI) dichloride dioxide	Cl_2CrO_2	154.901	-96.5	117	1.91							0.025	
trans-Cinnamaldehyde	C_9H_8O	132.159	-7.5	246	1.0497^{20}	17.72^{33}		0.005					
o-Cresol	C_7H_8O	108.138	31.0	191.0	1.0327^{35}	6.76^{25}	1.45	0.041	81	>1.4%	599		
m-Cresol	C_7H_8O	108.138	12.2	202.2	1.0339^{20}	12.44^{25}	1.48	0.019	86	>1.1%	558		
p-Cresol	C_7H_8O	108.138	34.77	201.9	1.0185^{40}	13.05^{25}	1.48	0.017	86	>1.1%	558		
Cyanogen chloride	CClN	61.471	-6.55	13	1.186^{20}		2.8331						0.3
Cyclobutane	C_4H_8	56.107	-90.7	12.5	0.7038^{0}		0	157	<10	>1.8%			
Cyclohexane	C_6H_{12}	84.159	6.7	80.7	0.7786^{20}	2.0243^{20}		13.0	-20	1.3-8%	245	100	
Cyclohexanol	$C_6H_{12}O$	100.158	26	160.9	0.9624^{20}	16.40^{20}		0.10	68	1-9%	300	50	
Cyclohexanone	$C_6H_{10}O$	98.142	-27.93	155.4	0.9478^{20}	16.1^{20}	3.246	0.53	44	1.1-9.4%	420	20	50
Cyclohexene	C_6H_{10}	82.143	-103.5	82.9	0.8110^{20}	2.2176^{20}	a	11.8	-12	>1.2%	310	300	
Cyclohexylamine	$C_6H_{13}N$	99.174	-17.7	133.6	0.8191^{20}	4.547^{20}	1.26	1.20	31	1.9-9.4%	293	10	
1,3-Cyclopentadiene	C_5H_6	66.102	-96.54	41	0.8021^{20}		0.419	58.5				75	
Cyclopentane	C_5H_{10}	70.133	-93.4	49.2	0.7457^{20}	1.9687^{20}		42.3	-25	>1.5%	361	600	
Cyclopentanol	$C_5H_{10}O$	86.132	-17	140.4	0.9488^{20}	18.5^{15}		0.294	51				
Cyclopentanone	C_5H_8O	84.117	-51.70	130.5	0.9487^{20}	13.58^{25}	≈3.3	1.55	26				
cis-Decahydronaphthalene	$C_{10}H_{18}$	138.250	-42.9	195.8	0.8965^{20}	2.219^{20}		0.10					
trans-Decahydronaphthalene	$C_{10}H_{18}$	138.250	-30.35	187.3	0.8659^{25}	2.184^{20}		0.164	54	0.7-5.4%	255		
Decamethylcyclopentasiloxane	$C_{10}H_{30}O_5Si_5$	370.770	-37.0	213	0.9593^{20}	2.50^{20}		0.02					
Decanal	$C_{10}H_{20}O$	156.265	-3.9	212	0.830^{15}			0.02					
Decane	$C_{10}H_{22}$	142.282	-29.61	174.1	0.7303^{20}	1.9853^{20}		0.170	51	0.8-5.4%	210		
Decanoic acid	$C_{10}H_{20}O_2$	172.265	31.39	270	0.8858^{40}								
1-Decanol	$C_{10}H_{22}O$	158.281	7	229	0.8294^{20}	7.93^{20}		0.009	82		288		
1-Decene	$C_{10}H_{20}$	140.266	-66.21	171	0.7408^{20}	2.136^{20}		0.210	<55		235		
Diacetone alcohol	$C_6H_{12}O_2$	116.158	-47	167.9	0.9387^{20}	18.2^{25}	3.24	0.224	58	1.8-6.9%	643	50	
Dibenzyl ether	$C_{14}H_{14}O$	198.260	1.8	298	1.0428^{20}	3.821^{20}			135				
Dibromodifluoromethane	CBr_2F_2	209.816	-110.1	22.79		2.939^{0}	0.66	110				100	
1,2-Dibromoethane	$C_2H_4Br_2$	187.861	9.8	131.3	2.1683^{25}	4.9612^{20}	1.19	1.55					
Dibromomethane	CH_2Br_2	173.835	-52.1	97.0	2.4969^{20}	7.77^{10}	1.43	6.12					
1,2-Dibromotetrafluoroethane	$C_2Br_2F_4$	259.823	-110.1	47.1	2.149^{25}	2.34^{25}		43.4					
Dibutylamine	$C_8H_{19}N$	129.244	-61.8	162	0.7670^{20}	2.765^{20}	0.98	0.34	47	1.1-6%			
Dibutyl ether	$C_8H_{18}O$	130.228	-96	141.6	0.7684^{20}	3.0830^{20}	1.17	0.898	25	1.5-7.6%	194		
Di-*tert*-butyl peroxide	$C_8H_{18}O_2$	146.228	-40	110.0	0.704^{20}			3.43	18				
Dibutyl phthalate	$C_{16}H_{22}O_4$	278.344	-35	338	1.0465^{20}	6.58^{20}	2.82		157	>0.5%	402		
Dibutyl sebacate	$C_{18}H_{34}O_4$	314.461	-9.2	356	0.9405^{15}	4.54^{20}	2.48		178	>0.4%	365		
Dibutyl sulfide	$C_8H_{18}S$	146.294	-74.97	168	0.8386^{20}	4.29^{25}	1.61	0.09	76				
Dichloroacetic acid	$C_2H_2Cl_2O_2$	128.942	12	193	1.5634^{20}	8.33^{20}		0.03				0.5	
o-Dichlorobenzene	$C_6H_4Cl_2$	147.002	-17.0	180.2	1.3059^{20}	10.12^{20}	2.50	0.18	66	2.2-9.2%	648	25	50
m-Dichlorobenzene	$C_6H_4Cl_2$	147.002	-24.8	172	1.2884^{20}	5.02^{20}	1.72	0.252	72				
trans-1,4-Dichloro-2-butene	$C_4H_6Cl_2$	124.997	3	155.4	1.183^{25}							0.005	

Lab Data

Name	Mol. form.	M_r	mp/°C	bp/°C	ρ/g cm^{-3}	ε	μ/D	Vap. pres. (25 °C)/ kPa	FP/°C	Flam. limits (%)	IT/°C	TWA/ ppm	STEL/ ppm
Dichlorodimethylsilane	$C_2H_6Cl_2Si$	129.061	-16	70.5	1.064^{25}			18.9	<21	3.4-9.5%			
1,1-Dichloroethane	$C_2H_4Cl_2$	98.959	-96.93	56.3	1.1757^{20}	10.10^{25}	2.06	30.5	-17	5.4-11.4%	458	100	
1,2-Dichloroethane	$C_2H_4Cl_2$	98.959	-35.6	83.4	1.2454^{25}	10.42^{20}	1.83	10.6	13	6.2-16%	413	10	
1,1-Dichloroethene	$C_2H_2Cl_2$	96.943	-122.5	31.6	1.213^{20}	4.60^{20}	1.34	80.0	-28	6.5-15.5%	570	5	
cis-1,2-Dichloroethene	$C_2H_2Cl_2$	96.943	-80.0	60	1.2837^{20}	9.20^{25}	1.90	26.8	6	3-15%	460	200	
trans-1,2-Dichloroethene	$C_2H_2Cl_2$	96.943	-49.8	47.64	1.2565^{20}	2.14^{20}	0	44.2	2	6-13%	460	200	
Dichloromethane	CH_2Cl_2	84.933	-94.9	39.8	1.3232^{20}	8.93^{25}	1.60	58.2		13-23%	556	50	
(Dichloromethyl)benzene	$C_7H_6Cl_2$	161.029	-17.0	205	1.26^{25}	6.9^{20}	2.07	0.06					
1,1-Dichloropropane	$C_3H_6Cl_2$	112.986		88.4	1.1321^{20}			9.09					
1,2-Dichloropropane, (±)-	$C_3H_6Cl_2$	112.986	-100.53	96.4	1.1560^{20}	8.37^{20}	1.85	6.62	21	3.4-14.5%	557	10	
1,3-Dichloropropane	$C_3H_6Cl_2$	112.986	-99.5	120.8	1.1785^{25}	10.27^{30}	2.08	2.44					
2,3-Dichloropropene	$C_3H_4Cl_2$	110.970	10	93.0	1.211^{20}				15	2.6-7.8%			
2,4-Dichlorotoluene	$C_7H_6Cl_2$	161.029	-13.5	200	1.2476^{20}	5.68^{28}	1.70	0.055					
Dicyclohexylamine	$C_{12}H_{23}N$	181.318	-0.1	251	0.9123^{20}			0.003	>99				
Diethanolamine	$C_4H_{11}NO_2$	105.136	27.9	271.2	1.0966^{20}	25.75^{20}	2.8	<0.01	172	2-13%	662		
1,1-Diethoxyethane	$C_6H_{14}O_2$	118.174	-106.1	102	0.8254^{20}		1.38	3.68	-21	1.6-10.4%	230		
1,2-Diethoxyethane	$C_6H_{14}O_2$	118.174	-74.0	120.6	0.8351^{25}	3.90^{20}		4.33	27		205		
Diethylamine	$C_4H_{11}N$	73.137	-50	55.4	0.7056^{20}	3.680^{20}	0.92	30.1	-23	1.8-10.1%	312	5	15
N,N-Diethylaniline	$C_{10}H_{15}N$	149.233	-21.3	216	0.9307^{20}	5.15^{30}		0.025	85		630		
o-Diethylbenzene	$C_{10}H_{14}$	134.218	-31.4	183.4	0.8800^{20}	2.594^{20}		0.13	57		395		
m-Diethylbenzene	$C_{10}H_{14}$	134.218	-83.9	181.1	0.8602^{20}	2.369^{20}		0.14	56		450		
p-Diethylbenzene	$C_{10}H_{14}$	134.218	-43.3	184	0.8620^{20}	2.259^{20}		0.13	55	0.7-6.0%	430		
Diethyl carbonate	$C_5H_{10}O_3$	118.131	-43	125.9	0.9692^{25}	2.820^{24}	1.10	1.63	25				
Diethylene glycol	$C_4H_{10}O_3$	106.120	-10.3	245.5	1.1197^{15}	31.82^{20}	2.31	0.001	124	2-17%	224		
Diethylene glycol diethyl ether	$C_8H_{18}O_3$	162.227	-44.3	185	0.9063^{20}			0.10	82				
Diethylene glycol dimethyl ether	$C_6H_{14}O_3$	134.173	-64.0	162	0.9434^{20}	7.23^{25}	1.97	0.315	67				
Diethylene glycol monobutyl ether	$C_8H_{18}O_3$	162.227	-68	232	0.9553^{20}			0.0032				10	
Diethylene glycol monoethyl ether	$C_6H_{14}O_3$	134.173		202	0.9885^{20}		1.6	0.017	96				
Diethylene glycol monoethyl ether acetate	$C_8H_{16}O_4$	176.211	-25	218	1.0096^{20}		1.8	0.029	110		425		
Diethylene glycol monomethyl ether	$C_5H_{12}O_3$	120.147		194	1.035^{20}		1.6	0.024	96	1.38-22.7%	240		
Diethyl ether	$C_4H_{10}O$	74.121	-116.22	34.4	0.7135^{20}	4.2666^{20}	1.098	71.7	-45	1.9-36.0%	180	400	500
Diethyl maleate	$C_8H_{12}O_4$	172.179	-8.8	222	1.0662^{20}	7.560^{25}		0.015	121		350		
Diethyl malonate	$C_7H_{12}O_4$	160.168	-50	200	1.0551^{20}	7.550^{31}	2.54	0.048	93				
Diethyl oxalate	$C_6H_{10}O_4$	146.141	-40.6	186	1.0785^{20}	8.266^{20}	2.49	0.030	76				
Diethyl phthalate	$C_{12}H_{14}O_4$	222.237	-40.5	298	1.232^{14}	7.86^{20}		0.002	161	>0.7%	457		
Diethyl succinate	$C_8H_{14}O_4$	174.195	-21.6	217	1.0402^{20}	6.098^{20}		0.15	90				
Diethyl sulfate	$C_4H_{10}O_4S$	154.185	-26.0	208	1.172^{25}	29.2^{20}		0.05	104		436		
Diethyl sulfide	$C_4H_{10}S$	90.187	-103.9	92.1	0.8362^{20}	5.723^{25}	[a]	7.78					
Diiodomethane	CH_2I_2	267.836	6.0	182	3.3211^{20}	5.32^{25}	1.08	0.172					
Diiodosilane	H_2I_2Si	283.911	-1	150									
Diisobutylamine	$C_8H_{19}N$	129.244	-73.5	139.6				0.972	29				
Diisopentyl ether	$C_{10}H_{22}O$	158.281		172	0.7777^{20}	2.817^{20}	1.23	0.210					
Diisopropylamine	$C_6H_{15}N$	101.190	-61	84	0.7153^{20}		1.15	10.7	-1	1.1-7.1%	316	5	
Diisopropyl ether	$C_6H_{14}O$	102.174	-85.37	68.4	0.7192^{25}	3.805^{30}	1.13	19.9	-28	1.4-7.9%	443	250	310
1,2-Dimethoxyethane	$C_4H_{10}O_2$	90.121	-69.0	85.0	0.8637^{25}	7.30^{24}		9.93	-2		202		
Dimethoxymethane	$C_3H_8O_2$	76.095	-105.11	42.3	0.8593^{20}	2.644^{20}	0.74	53.1	-32	2.2-13.8%	237	1000	
Dimethylacetal	$C_4H_{10}O_2$	90.121	-113.2	63	0.8501^{20}			22.9					
N,N-Dimethylacetamide	C_4H_9NO	87.120	-19	165.9	0.9372^{25}	38.85^{21}	3.7	0.075	70	1.8-11.5%	490	10	
2,3-Dimethylaniline	$C_8H_{11}N$	121.180	3	223	0.9931^{20}				97	>1.0%		0.5	
2,6-Dimethylaniline	$C_8H_{11}N$	121.180	11.0	211	0.9842^{20}		1.63	0.45	96			0.5	
N,N-Dimethylaniline	$C_8H_{11}N$	121.180	2.1	193	0.9562^{20}	4.90^{25}	1.68	0.107	63		371	5	10
2,2-Dimethylbutane	C_6H_{14}	86.175	-99.0	49.7	0.6444^{25}	1.869^{20}		42.5	-48	1.2-7.0%	405	500	1000
2,3-Dimethylbutane	C_6H_{14}	86.175	-128.1	58.0	0.6616^{20}	1.889^{20}		31.3	-29	1.2-7.0%	405	500	1000
3,3-Dimethyl-2-butanone	$C_6H_{12}O$	100.158	-51.40	106.1	0.7229^{25}	12.73^{20}		4.27					
Dimethylcarbamic chloride	C_3H_6ClNO	107.539	-33	154	1.168^{25}								
Dimethyl disulfide	$C_2H_6S_2$	94.199	-84.67	109.72	1.0625^{20}	9.6^{25}	1.85	3.82	24				
N,N-Dimethylethanolamine	$C_4H_{11}NO$	89.136	-65	130.7	0.8866^{20}			0.9					
N,N-Dimethylformamide	C_3H_7NO	73.094	-60.3	152.8	0.9445^{25}	38.25^{20}	3.82	0.439	58	2.2-15.2%	445	10	
2,6-Dimethyl-4-heptanone	$C_9H_{18}O$	142.238	-46.0	157	0.8062^{20}	9.91^{20}	2.66	0.23	49	0.8-7.1%	396	25	
1,1-Dimethylhydrazine	$C_2H_8N_2$	60.098	-57.15	62.4	0.791^{22}			20.9	-15	2-95%	249	0.01	
Dimethyl phthalate	$C_{10}H_{10}O_4$	194.184	1.03	282.7	1.1905^{20}	8.66^{20}		0.001	146	>0.9%	490		
2,6-Dimethylpyridine	C_7H_9N	107.153	-6.12	144.0	0.9226^{20}	7.33^{20}	1.66	0.746					
Dimethyl sulfate	$C_2H_6O_4S$	126.132	-31.8	186	1.3322^{20}	55.0^{25}		0.13	83		188	0.1	
Dimethyl sulfide	C_2H_6S	62.134	-98.26	37.32	0.8483^{20}	6.70^{21}	1.554	64.4	-37	2.2-19.7%	206	10	
Dimethyl sulfoxide	C_2H_6OS	78.133	18.52	191.9	1.1010^{25}	47.24^{20}	3.96	0.084	95	2.6-42%	215		

Name	Mol. form.	M_r	mp/°C	bp/°C	ρ/g cm^{-3}	ε	μ/D	Vap. pres. (25 °C)/ kPa	FP/°C	Flam. limits (%)	IT/°C	TWA/ ppm	STEL/ ppm
1,4-Dioxane	$C_4H_8O_2$	88.106	11.75	101.2	1.0337^{20}	2.2189^{20}	0	4.95	12	2.0-22%	180	20	
1,3-Dioxolane	$C_3H_6O_2$	74.079	-97.21	75.3	1.060^{20}		1.19	14.6	2			20	
Dipentyl ether	$C_{10}H_{22}O$	158.281	-69.2	187	0.7833^{20}	2.798^{25}	1.20	0.13	57		170		
Dipropylamine	$C_6H_{15}N$	101.190	-63	107.5	0.7400^{20}	2.923^{20}	1.03	3.21	17		299		
Dipropylene glycol monomethyl ether	$C_7H_{16}O_3$	148.200	-80	203	0.95							100	150
Dipropyl ether	$C_6H_{14}O$	102.174	-114.8	90.1	0.7466^{20}	3.38^{24}	1.21	8.35	21	1.3-7.0%	188		
Dodecane	$C_{12}H_{26}$	170.334	-9.55	216.3	0.7495^{20}	2.0120^{20}		0.016	74	>0.6%	203		
1-Dodecanol	$C_{12}H_{26}O$	186.333	24.2	264.1	0.8309^{24}	5.82^{30}		0.000016	127		275		
1-Dodecene	$C_{12}H_{24}$	168.319	-35.19	213.4	0.7584^{20}	2.152^{20}		0.019	79				
Epichlorohydrin	C_3H_5ClO	92.524	-26	111.99	1.1812^{20}	22.6^{20}	1.8	2.2	31	3.8-21.0%	411	0.5	
1,2-Epoxybutane	C_4H_8O	72.106	-150	63.4	0.8297^{20}		1.891	31.7	-22	1.7-19%	439		
1,2-Epoxy-4-(epoxyethyl)cyclohexane	$C_8H_{12}O_2$	140.180	<-55	227	1.0966^{20}							0.1	
1,2-Ethanediamine	$C_2H_8N_2$	60.098	11.14	116.9	0.8979^{20}	13.82^{20}	1.99	1.62	40	2.5-12.0%	385	10	
1,2-Ethanediol	$C_2H_6O_2$	62.068	-13	197.5	1.1135^{20}	41.4^{20}	2.36	0.01	111	3.2-22%	398	25	50
1,2-Ethanediol, diacetate	$C_6H_{10}O_4$	146.141	-31	184	1.1043^{20}	7.7^{17}	2.34	0.030	88	1.6-8.4%	482		
1,2-Ethanediol, dinitrate	$C_2H_4N_2O_6$	152.062	-22.5	199	1.4918^{20}	28.26^{20}		0.009				0.05	
1,2-Ethanedithiol	$C_2H_6S_2$	94.199	-41.2	144	1.234^{20}	7.26^{20}	2.03						
Ethanethiol	C_2H_6S	62.134	-147.89	35.0	0.8315^{25}	6.667^{25}	[a]	70.3	-17	2.8-18.0%	300	0.5	
Ethanol	C_2H_6O	46.068	-114.14	78.24	0.7893^{20}	25.3^{20}	1.69	7.87	13	3.3-19%	363		1000
Ethanolamine	C_2H_7NO	61.083	10.4	170.3	1.0180^{20}	31.94^{20}	3.05	0.05	86	3.0-23.5%	410	3	6
4-Ethoxyaniline	$C_8H_{11}NO$	137.179	4.7	254	1.0652^{16}	7.43^{25}		0.0007	116				
Ethoxybenzene	$C_8H_{10}O$	122.164	-29.6	169.8	0.9651^{20}	4.216^{20}	1.45	0.204	63				
2-Ethoxyethanol	$C_4H_{10}O_2$	90.121	-70	134.7	0.9253^{25}	13.38^{25}	2.08	0.71	43	3-18%	235	5	
2-Ethoxyethyl acetate	$C_6H_{12}O_3$	132.157	-61.7	156.6	0.9740^{20}	7.567^{30}	2.25	0.24	56	2-8%	379	5	
Ethyl acetate	$C_4H_8O_2$	88.106	-83.8	77.1	0.9006^{20}	6.0814^{20}	1.78	12.6	-4	2.0-11.5%	426	400	
Ethyl acetoacetate	$C_6H_{10}O_3$	130.141	-45	180	1.0368^{10}	14.0^{20}		0.095	57	1.4-9.5%	295		
Ethyl acrylate	$C_5H_8O_2$	100.117	-71	98.9	0.9234^{20}	6.05^{30}	1.96	5.14	10	1.4-14%	372	5	15
Ethylamine	C_2H_7N	45.084	-81	16.6	0.689^{15}	8.7^{0}	1.22	141	-16	3.5-14%	385	5	15
N-Ethylaniline	$C_8H_{11}N$	121.180	-63.4	204	0.9625^{20}	5.87^{20}		0.039	85				
Ethylbenzene	C_8H_{10}	106.165	-94.95	136.2	0.8668^{20}	2.4463^{20}	0.59	1.28	21	0.8-6.7%	432	20	
Ethyl benzoate	$C_9H_{10}O_2$	150.174	-34.5	212.5	1.0415^{25}	6.20^{20}	2.00	0.04	88		490		
Ethyl butanoate	$C_6H_{12}O_2$	116.158	-97	121.1	0.8735^{25}	5.18^{28}	1.74	2.01	24		463		
2-Ethyl-1-butanol	$C_6H_{14}O$	102.174	<-15	155	0.8326^{20}	6.19^{89}		0.206	57				
Ethyl chloroacetate	$C_4H_7ClO_2$	122.551	-21	144	1.1585^{20}			0.640	64				
Ethyl chloroformate	$C_3H_5ClO_2$	108.524	-80.6	91	1.1352^{20}	9.736^{36}			16		500		
Ethylcyanoacetate	$C_5H_7NO_2$	113.116	-26.1	216	1.0654^{20}	31.62^{-10}	2.17	0.003	110				
Ethyleneimine	C_2H_5N	43.068	-78.0	54	0.832^{25}	18.3^{25}	1.90	28.9	-11	3.3-54.8%	320	0.05	0.1
Ethyl formate	$C_3H_6O_2$	74.079	-79.6	54.09	0.9218^{20}	8.57^{15}	1.93	32.3	-20	2.8-16.0%	455		100
2-Ethylhexanal	$C_8H_{16}O$	128.212	<-100	161	0.8540^{20}				44	0.85-7.2%	190		
2-Ethyl-1,3-hexanediol	$C_8H_{18}O_2$	146.228	-40	243	0.9325^{22}	18.73^{20}			127		360		
2-Ethyl-1-hexanol	$C_8H_{18}O$	130.228	-70	186.2	0.8319^{25}	7.58^{25}	1.74	0.019	73	0.88-9.7%	231		
2-Ethylhexyl acetate	$C_{10}H_{20}O_2$	172.265	-80	200	0.8718^{20}		1.8	0.09	71	0.76-8.14%	268		
Ethyl lactate	$C_5H_{10}O_3$	118.131	-26	151	1.0328^{20}	15.4^{30}	2.4		46	>1.5%	400		
Ethyl 3-methylbutanoate	$C_7H_{14}O_2$	130.185	-99.3	135	0.8656^{20}	4.71^{20}		1.07					
Ethyl 2-methylpropanoate	$C_6H_{12}O_2$	116.158	-97.8	111	0.868^{20}			3.25	13				
Ethyl nitrite	$C_2H_5NO_2$	75.067		17.5	0.899^{15}			135	-35	4.0-50%	90		
Ethyl propanoate	$C_5H_{10}O_2$	102.132	-73.6	98.9	0.8895^{20}	5.76^{20}	1.74	4.97	12	1.9-11%	440		
Ethyl silicate	$C_8H_{20}O_4Si$	208.329	-82.2	168	0.9320^{20}	2.50^{20}		1.17	52			10	
Eucalyptol	$C_{10}H_{18}O$	154.249	1.4	176	0.9267^{20}	4.57^{25}		0.260	48				
Fluorobenzene	C_6H_5F	96.102	-42.18	84.7	1.0225^{20}	5.465^{20}	1.60	10.4	-15				
Fluorosulfonic acid	FHO_3S	100.069	-89	163	1.726			0.08					
Formamide	CH_3NO	45.041	2.57	217	1.1334^{20}	111.0^{20}	3.73	0.01	154			10	
Formic acid	CH_2O_2	46.026	8.3	101	1.220^{20}	51.1^{25}	1.425	5.75	50	18-57%	434	5	10
Furan	C_4H_4O	68.074	-85.58	31.3	0.9514^{20}	2.94^{25}	0.66	80.0	-36	2.3-14.3%			
Furfural	$C_5H_4O_2$	96.085	-38.3	161.5	1.1594^{20}	42.1^{20}	3.54	0.29	60	2.1-19.3%	316	0.2	
Furfuryl alcohol	$C_5H_6O_2$	98.101	-14.5	168	1.1296^{20}	16.85^{25}	1.92	0.097	75	1.8-16.3%	491	0.2	
Germanium(IV) chloride	Cl_4Ge	214.45	-51.50	86.55	1.88	2.463^{0}	0						
Glycerol	$C_3H_8O_3$	92.094	18.2	289	1.2613^{20}	46.53^{20}	2.56	<0.01	199	3-19%	370		
Glycerol triacetate	$C_9H_{14}O_6$	218.203	-78	259	1.1583^{20}	7.11^{20}		<0.01	138	>1.0%	433		
Glycerol trioleate	$C_{57}H_{104}O_6$	885.432	5.3		0.915^{15}	3.109^{20}							
Heptanal	$C_7H_{14}O$	114.185	-43.94	153	0.8132^{25}	9.07^{22}		0.46					
Heptane	C_7H_{16}	100.202	-90.549	98.38	0.6837^{20}	1.9209^{20}		6.09	-4	1.05-6.7%	204	400	500
Heptanoic acid	$C_7H_{14}O_2$	130.185	-7.17	222	0.9181^{20}	3.04^{15}		0.001			275		
1-Heptanol	$C_7H_{16}O$	116.201	-33.2	178	0.8219^{20}	11.75^{20}		0.0044					
2-Heptanone	$C_7H_{14}O$	114.185	-34.7	151.0	0.8111^{20}	11.95^{20}	2.59	0.49	39	1.1-7.9%	393	50	
3-Heptanone	$C_7H_{14}O$	114.185	-37.2	146	0.8183^{20}	12.7^{20}	2.78	0.5	46			50	75

Name	Mol. form.	M_r	mp/°C	bp/°C	ρ/g cm^{-3}	ε	μ/D	Vap. pres. (25 °C)/ kPa	FP/°C	Flam. limits (%)	IT/°C	TWA/ ppm	STEL/ ppm
4-Heptanone	$C_7H_{14}O$	114.185	-32.1	144	0.8174^{20}	12.60^{20}		0.164	49			50	
1-Heptene	C_7H_{14}	98.186	-118.83	94	0.6970^{20}	2.092^{20}		7.52	-1		260		
Hexachloro-1,3-butadiene	C_4Cl_6	260.761	-21	216	1.556^{25}	2.55^{20}		0.13			610	0.02	
Hexachloro-1,3-cyclopentadiene	C_5Cl_6	272.772	-9	239	1.7019^{25}							0.01	
Hexafluorobenzene	C_6F_6	186.054	5.10	80.2	1.6175^{20}	2.029^{25}	0	11.3					
Hexamethyldisiloxane	$C_6H_{18}OSi_2$	162.377	-68.2	100.5	0.7638^{20}	2.179^{20}		5.57					
Hexamethylphosphoric triamide	$C_6H_{18}N_3OP$	179.200	7.2	235	1.03^{20}	31.3^{20}	5.5						
Hexanal	$C_6H_{12}O$	100.158	-58.2	129.6	0.8335^{20}			1.48	32				
Hexane	C_6H_{14}	86.175	-95.27	68.72	0.6593^{20}	1.8865^{20}		20.2	-22	1.1-7.5%	225	50	
Hexanedinitrile	$C_6H_8N_2$	108.141	2.2	295	0.9676^{20}			<0.01	93	>1.0%	550	2	
Hexanoic acid	$C_6H_{12}O_2$	116.158	-4.1	204.9	0.9274^{20}	2.600^{25}	1.13	0.005	102		380		
1-Hexanol	$C_6H_{14}O$	102.174	-46.4	156.9	0.8195^{20}	13.03^{20}		0.11	63		290		
2-Hexanone	$C_6H_{12}O$	100.158	-55.45	127.6	0.8113^{20}	14.56^{20}	2.66	1.54	25	1-8%	423	5	10
1-Hexene	C_6H_{12}	84.159	-139.76	63.4	0.6685^{25}	2.077^{21}		24.8	-26	1.2-6.9%	253	50	
Hexyl acetate	$C_8H_{16}O_2$	144.212	-61.0	171.1	0.8779^{15}	4.42^{20}		0.185	45				
Hydrazine	H_4N_2	32.045	1.54	113.55	1.0036	51.7^{25}	1.75	1.91	38	5-100%		0.01	
Hydrazoic acid	HN_3	43.028	-80	35.7			1.70	68.2					0.11
Hydrogen cyanide	CHN	27.026	-13.28	25.63	0.6876^{20}	114.9^{20}	2.985188	98.8	-18	6-40%	538		4.7
Hydrogen peroxide	H_2O_2	34.015	-0.43	150.2	1.44	74.6^{17}	1.573	0.26				1	
3-Hydroxypropanenitrile	C_3H_5NO	71.078	-46	218	1.0404^{25}		[a]	0.010	129				
Indan	C_9H_{10}	118.175	-51.34	177.8	0.9639^{20}			0.2					
Indene	C_9H_8	116.160	-1.45	182.5	0.9960^{25}			0.220				5	
Iodine bromide	BrI	206.808	40	116 dec	4.3		0.726						
Iodine chloride	ClI	162.357	27.38	97.0 dec	3.24		1.24	3.59					
Iodobenzene	C_6H_5I	204.008	-30.7	188.5	1.8308^{20}	4.59^{20}	1.70	0.133					
1-Iodobutane	C_4H_9I	184.018	-103.5	130	1.6154^{20}	6.27^{20}	1.93	1.85					
Iodoethane	C_2H_5I	155.965	-111.0	72	1.9357^{20}	7.82^{20}	1.976	18.2					
Iodomethane	CH_3I	141.939	-66	42.4	2.2789^{20}	6.97^{20}	1.6406	53.9				2	
1-Iodopropane	C_3H_7I	169.992	-101.4	102	1.7489^{20}	7.07^{20}	2.04	5.75					
2-Iodopropane	C_3H_7I	169.992	-90.4	89	1.7042^{20}	8.19^{25}	1.95	9.36					
Iron pentacarbonyl	C_5FeO_5	195.896	-20	103	1.5^{20}	2.602^{20}		4				0.1	0.2
Isobutanal	C_4H_8O	72.106	-72.1	64.1	0.7891^{20}		[a]	23.0	-18	1.6-10.6%	196		
Isobutyl acetate	$C_6H_{12}O_2$	116.158	-97.1	116.9	0.8712^{20}	5.068^{20}	1.86	2.39	18	1.3-10.5%	421	150	
Isobutyl acrylate	$C_7H_{12}O_2$	128.169	-61	137.0	0.8896^{20}				30		427		
Isobutylamine	$C_4H_{11}N$	73.137	-86	68.8	0.724^{25}		1.27	19.0	-9	2-12%	378		
Isobutylbenzene	$C_{10}H_{14}$	134.218	-51.6	172.7	0.8532^{20}	2.318^{20}		0.257	55	0.8-6.0%	427		
Isobutyl formate	$C_5H_{10}O_2$	102.132	-95.5	98.4	0.8776^{20}	6.41^{20}	1.88	5.34	5	2-9%	320		
Isobutyl isobutanoate	$C_8H_{16}O_2$	144.212	-80.6	148	0.8542^{20}		1.9	0.552	38	0.96-7.59%	432		
Isopentane	C_5H_{12}	72.149	-159.8	27.83	0.6201^{20}	1.845^{20}	0.13	91.7	-51	1.4-7.6%	420	600	
Isopentyl acetate	$C_7H_{14}O_2$	130.185	-78.5	141.6	0.876^{15}	4.72^{20}	1.86	0.728	25	1.0-7.5%	360	50	100
Isophorone	$C_9H_{14}O$	138.206	-8.1	214.8	0.9255^{20}			0.06	84	0.8-3.8%	460		5
Isopropenyl acetate	$C_5H_8O_2$	100.117	-92.9	97.4	0.9090^{20}			6.02	26		432		
Isopropenylbenzene	C_9H_{10}	118.175	-22.36	165.4	0.9106^{20}	2.28^{20}		0.40	54	1.9-6.1%	574	10	
Isopropyl acetate	$C_5H_{10}O_2$	102.132	-73.4	88.6	0.8662^{25}			7.88	2	1.8-8%	460	100	200
Isopropylamine	C_3H_9N	59.110	-95.119	31.8	0.6891^{20}	5.6268^{20}	1.19	78.0	-37		402	5	10
Isopropylbenzene	C_9H_{12}	120.191	-96.01	152.4	0.8615^{20}	2.381^{20}	≈0.79	0.61	36	0.9-6.5%	424	50	
Isopropylbenzene hydroperoxide	$C_9H_{12}O_2$	152.190		153 expl	1.03^{20}			0.004	79				
1-Isopropyl-2-methylbenzene	$C_{10}H_{14}$	134.218	-71.5	178	0.8766^{20}			0.2					
1-Isopropyl-3-methylbenzene	$C_{10}H_{14}$	134.218	-63.8	175	0.8610^{20}			0.22					
1-Isopropyl-4-methylbenzene	$C_{10}H_{14}$	134.218	-68.1	177	0.8573^{20}	2.2322^{25}		0.19	47	0.7-5.6%	436		
Isoquinoline	C_9H_7N	129.159	26.46	243.2	1.0910^{30}	11.0^{25}	2.73	0.007					
d-Limonene	$C_{10}H_{16}$	136.234	-74.0	177.6	0.8411^{20}	2.3746^{25}		0.277	45	0.7-6.1%	237		
l-Limonene	$C_{10}H_{16}$	136.234		178	0.843^{20}	2.3738^{25}		0.254					
Mesityl oxide	$C_6H_{10}O$	98.142	-52.8	129.7	0.8653^{20}	15.6^{0}	2.79	1.47	31	1.4-7.2%	344	15	25
Methacrylic acid	$C_4H_6O_2$	86.090	14.6	160	1.0153^{20}		1.65	0.12	77	1.6-8.8%	68	20	
Methanol	CH_4O	32.042	-97.5	64.5	0.7909^{20}	33.0^{20}	1.6792	16.9	11	6.0-36%	464	200	250
2-Methoxyaniline	C_7H_9NO	123.152	6.2	221	1.0923^{20}	5.230^{30}		0.013	118				
4-Methoxybenzaldehyde	$C_8H_8O_2$	136.149	0	255	1.119^{15}	22.0^{30}		0.004					
2-Methoxyethanol	$C_3H_8O_2$	76.095	-85.1	124.3	0.9647^{20}	17.2^{25}	[a]	1.31	39	1.8-14%	285	0.1	
2-Methoxyethyl acetate	$C_5H_{10}O_3$	118.131	-70	142	1.0074^{19}		2.13	0.67	49	1.5-12.3%	392	0.1	
Methyl acetate	$C_3H_6O_2$	74.079	-98.2	56.7	0.9346^{20}	7.07^{15}	1.72	28.8	-10	3.1-16%	454	200	250
Methyl acrylate	$C_4H_6O_2$	86.090	-75.6	80.1	0.9535^{20}	7.03^{30}	1.77	11.0	-3	2.8-25%	468	2	
2-Methylacrylonitrile	C_4H_5N	67.090	-35.8	90	0.8001^{20}		3.69	8.26	1	2-6.8%		1	
2-Methylaniline	C_7H_9N	107.153	-14.41	200.0	0.9984^{20}	6.138^{25}	1.60	0.043	85		482	2	
3-Methylaniline	C_7H_9N	107.153	-30.8	203.3	0.9889^{20}	5.816^{25}	1.45	0.036				2	
N-Methylaniline	C_7H_9N	107.153	-57	197	0.9859^{20}	5.96^{20}		0.05			533	0.5	

Lab Data

Name	Mol. form.	M_r	mp/°C	bp/°C	ρ/g cm^{-3}	ε	μ/D	Vap. pres. (25 °C)/ kPa	FP/°C	Flam. limits (%)	IT/°C	TWA/ ppm	STEL/ ppm
Methyl benzoate	$C_8H_8O_2$	136.149	-12.35	199	1.0837^{25}	6.642^{30}	1.94	0.052	83				
2-Methyl-1,3-butadiene	C_5H_8	68.118	-146.1	34.0	0.679^{20}	2.098^{20}	0.25	73.4	-54	1.5-8.9%	395		
Methyl butanoate	$C_5H_{10}O_2$	102.132	-85.8	101.9	0.8984^{20}	5.48^{28}		4.30	14				
3-Methylbutanoic acid	$C_5H_{10}O_2$	102.132	-29.6	176.5	0.931^{20}		0.63	0.067			416		
3-Methyl-1-butanol	$C_5H_{12}O$	88.148	-117.2	130.8	0.8104^{20}	15.63^{20}		0.315	43	1.2-9.0%	350	100	125
2-Methyl-2-butanol	$C_5H_{12}O$	88.148	-8.7	102.4	0.8096^{20}	5.78^{25}	1.82	2.19	19	1.2-9.0%	437		
3-Methyl-2-butanol, (±)-	$C_5H_{12}O$	88.148		113.7	0.8180^{20}	12.1^{25}		1.20	38				
3-Methyl-2-butanone	$C_5H_{10}O$	86.132	-93.13	94.2	0.8051^{20}	10.37^{20}		6.99				20	
2-Methyl-1-butene	C_5H_{10}	70.133	-137.53	31.1	0.6504^{20}	2.180^{20}		81.4	-20				
2-Methyl-2-butene	C_5H_{10}	70.133	-133.72	38.5	0.6623^{20}	1.979^{23}		62.1	-20				
Methyl *tert*-butyl ether	$C_5H_{12}O$	88.148	-108.6	55.1	0.7353^{25}			33.6				50	
Methyl chloroacetate	$C_3H_5ClO_2$	108.524	-32.3	130	1.236^{20}	12.0^{20}		1.0	57	7.5-18.5%			
Methylcyclohexane	C_7H_{14}	98.186	-126.6	100.9	0.7694^{20}	2.024^{20}		6.18	-4	1.2-6.7%	250	400	
Methylcyclopentane	C_6H_{12}	84.159	-142.419	71.8	0.7486^{20}	1.9853^{20}		18.3	-29	1.0-8.35%	258		
N-Methylformamide	C_2H_5NO	59.067	-2.5	186	1.011^{19}	189.0^{20}	3.83	0.03					
Methyl formate	$C_2H_4O_2$	60.052	-99.7	31.6	0.9739^{20}	9.20^{15}	1.793	78.1	-19	4.5-23%	449	50	100
5-Methyl-2-hexanone	$C_7H_{14}O$	114.185		139	0.888^{20}	13.53^{20}		0.691	36	1.0-8.2%	191	20	50
Methylhydrazine	CH_6N_2	46.072	-52.3	83				6.61	-8	2.5-92%	194	0.01	
Methyl isocyanate	C_2H_3NO	57.051	-45	38.3	0.9588^{20}	21.75^{16}	≈2.8	57.7	-7	5.3-26%	534	0.02	0.06
Methyl lactate, (±)-	$C_4H_8O_3$	104.105		144.8	1.0928^{20}			0.62	49	>2.2%	385		
Methyl methacrylate	$C_5H_8O_2$	100.117	-47.55	100.6	0.9377^{25}	6.32^{30}	1.67	5.10	10	1.7-8.2%		50	100
1-Methylnaphthalene	$C_{11}H_{10}$	142.197	-30.43	244.4	1.0202^{20}	2.915^{20}		0.009			529	0.5	
Methyloxirane	C_3H_6O	58.079	-111.9	35	0.859^{0}		2.01	71.7	-37	3.1-27.5%	449	2	
2-Methylpentane	C_6H_{14}	86.175	-153.60	60.21	0.650^{25}	1.886^{20}		28.2	<-29	1.0-7.0%	264	500	1000
3-Methylpentane	C_6H_{14}	86.175	-162.89	63.3	0.6598^{25}	1.886^{20}		25.3	-7	1.2-7.0%	278	500	1000
2-Methyl-2,4-pentanediol	$C_6H_{14}O_2$	118.174	-50	197.9	0.923^{15}	25.86^{20}	2.9	<0.01	102	1-9%	306		25
2-Methyl-1-pentanol	$C_6H_{14}O$	102.174		157	0.8263^{20}			0.236	54	1.1-9.65%	310		
4-Methyl-2-pentanol	$C_6H_{14}O$	102.174	-90	132.0	0.8075^{20}			0.698	41	1.0-5.5%		25	40
4-Methyl-2-pentanone	$C_6H_{12}O$	100.158	-85	115.7	0.7965^{25}	13.11^{20}		2.64	18	1.2-8.0%	448	20	75
2-Methylpropanenitrile	C_4H_7N	69.106	-71.5	102	0.7704^{20}	24.42^{20}	4.07		8		482		
2-Methyl-2-propanethiol	$C_4H_{10}S$	90.187	1.27	64.2	0.7943^{25}	5.475^{20}	1.66	24.2	<-29				
Methyl propanoate	$C_4H_8O_2$	88.106	-87.5	78.6	0.9150^{20}	6.200^{20}		11.5	-2	2.5-13%	469		
2-Methylpropanoic acid	$C_4H_8O_2$	88.106	-46	154.4	0.9681^{20}	2.58^{20}	1.08	0.17	56	2.0-9.2%	481		
2-Methyl-1-propanol	$C_4H_{10}O$	74.121	-101.96	107.84	0.8018^{20}	17.93^{20}	1.64	1.39	28	1.7-10.6%	415	50	
2-Methyl-2-propanol	$C_4H_{10}O$	74.121	25.81	82.3	0.7887^{20}	12.47^{25}	1.66	5.52	11	2.4-8.0%	478	100	
2-Methylpyridine	C_6H_7N	93.127	-66.65	129.4	0.9443^{20}	10.18^{20}	1.85	1.5	39		538		
3-Methylpyridine	C_6H_7N	93.127	-18.1	144.1	0.9566^{20}	11.10^{30}	2.40	0.795					
4-Methylpyridine	C_6H_7N	93.127	3.68	145.3	0.9548^{20}	12.2^{20}	2.70	0.759	57				
N-Methyl-2-pyrrolidinone	C_5H_9NO	99.131	-24.0	204.2	1.0230^{25}	32.55^{20}	4.1	0.04	96	1-10%	346		
Methyl salicylate	$C_8H_8O_3$	152.148	-8.5	222.6	1.181^{25}	8.80^{41}	2.47	0.015	96		454		
4-Methylstyrene	C_9H_{10}	118.175	-37.8	172	0.9173^{25}			0.245	53	0.8-11.0%	538	50	100
Morpholine	C_4H_9NO	87.120	-4.8	128.2	1.0005^{20}	7.42^{25}	1.55	1.34	37	1.4-11.2%	290	20	
β-Myrcene	$C_{10}H_{16}$	136.234		171	0.8013^{15}	2.3^{25}		0.280					
Nickel carbonyl [Ni(CO)$_4$]	C_4NiO_4	170.734	-19.3	43 (exp 60)	1.31^{25}								0.05
L-Nicotine	$C_{10}H_{14}N_2$	162.231	-79	246	1.0097^{20}	8.937^{20}							
Nitric acid	HNO_3	63.013	-41.6	83	1.5129^{20}		2.17	8.34				2	4
2-Nitroanisole	$C_7H_7NO_3$	153.136	9.4	272	1.2540^{20}	45.75^{20}	5.0	0.002					
Nitrobenzene	$C_6H_5NO_2$	123.110	5.65	210.7	1.2037^{20}	35.6^{20}	4.22	0.03	88	>1.8%	482	1	
Nitroethane	$C_2H_5NO_2$	75.067	-89.42	114.1	1.0448^{25}	29.11^{15}	3.23	2.79	28	3.4-17%	414	100	
Nitromethane	CH_3NO_2	61.041	-28.7	101.19	1.1371^{20}	37.27^{20}	3.46	4.79	35	>7.3%	418	20	
1-Nitropropane	$C_3H_7NO_2$	89.094	-104.3	131.2	0.9961^{25}	24.70^{15}	3.66	1.36	36	>2.2%	421	25	
2-Nitropropane	$C_3H_7NO_2$	89.094	-91.3	120.2	0.9821^{25}	26.74^{15}	3.73	2.3	24	2.6-11.0%	428	10	
N-Nitrosodiethylamine	$C_4H_{10}N_2O$	102.134		172	0.9422^{20}								
N-Nitrosodimethylamine	$C_2H_6N_2O$	74.081		146	1.0048^{20}			0.73					
2-Nitrotoluene	$C_7H_7NO_2$	137.137	-3.6	220.9	1.1611^{19}	26.26^{20}		0.0014	106			2	
3-Nitrotoluene	$C_7H_7NO_2$	137.137	15.9	232.1	1.1581^{20}	24.95^{30}		0.03	106			2	
Nonane	C_9H_{20}	128.255	-53.47	150.8	0.7179^{20}	1.9722^{20}		0.570	31	0.8-2.9%	205	200	
Nonanoic acid	$C_9H_{18}O_2$	158.238	12.38	256	0.9052^{20}	2.475^{22}	0.79	0.00005					
1-Nonanol	$C_9H_{20}O$	144.254	-5.0	213.7	0.8280^{20}	8.83^{20}		0.00050			260		
1-Nonene	C_9H_{18}	126.239	-81.24	146.9	0.7253^{25}	2.180^{20}		0.714	26				
4-Nonylphenol	$C_{15}H_{24}O$	220.351	42	317	0.950^{20}								
cis,cis-9,12-Octadecadienoic acid	$C_{18}H_{32}O_2$	280.446	-6.9		0.9022^{20}	2.754^{20}							
cis-9-Octadecenoic acid	$C_{18}H_{34}O_2$	282.462	14	360	0.8935^{20}	2.336^{20}	1.18	0.000001	189		363		
Octane	C_8H_{18}	114.229	-56.73	125.62	0.7022^{20}	1.948^{20}		1.86	13	1.0-6.5%	206	300	375
Octanoic acid	$C_8H_{16}O_2$	144.212	16.51	240	0.9106^{20}	2.85^{15}	1.15	0.0002					

Name	Mol. form.	M_r	mp/°C	bp/°C	ρ/g cm^{-3}	ε	μ/D	Vap. pres. (25 °C)/ kPa	FP/°C	Flam. limits (%)	IT/°C	TWA/ ppm	STEL/ ppm
1-Octanol	$C_8H_{18}O$	130.228	-14.7	194.7	0.8262^{25}	10.30^{20}	1.76	0.01	81		270		
2-Octanol	$C_8H_{18}O$	130.228	-31.6	179	0.8193^{20}	8.13^{20}	1.71		88		265		
2-Octanone	$C_8H_{16}O$	128.212	-20.31	173	0.820^{20}	9.51^{20}	2.70	0.12	52				
1-Octene	C_8H_{16}	112.213	-101.66	121.3	0.7149^{20}	2.113^{20}		2.30	21		230		
Oxetane	C_3H_6O	58.079	-97	47.6	0.8930^{25}		1.94						
2-Oxetanone	$C_3H_4O_2$	72.063	-33.283	161	1.1460^{20}		4.18	0.3	74	>2.9%		0.5	
Oxirane	C_2H_4O	44.052	-112.46	10.4	0.8821^{10}	12.42^{20}	1.89	175	-20	3.0-100%	429	1	
Oxiranemethanol, (±)-	$C_3H_6O_2$	74.079	-45	156	1.1143^{25}							2	
Paraldehyde	$C_6H_{12}O_3$	132.157	12	124	0.9943^{20}		1.43	1.6	36	>1.3%	238		
Parathion	$C_{10}H_{14}NO_5PS$	291.261	6.1	375	1.2681^{20}								
Pentachloroethane	C_2HCl_5	202.294	-29.0	161	1.6796^{20}	3.716^{25}	0.92	0.478					
cis-1,3-Pentadiene	C_5H_8	68.118	-140.81	44.0	0.6910^{20}	2.319^{25}	0.500	50.6					
trans-1,3-Pentadiene	C_5H_8	68.118	-87.5	42.0	0.6710^{25}		0.585	54.7					
Pentanal	$C_5H_{10}O$	86.132	-81.5	103	0.8095^{20}	10.00^{20}		4.58	12		222	50	
Pentane	C_5H_{12}	72.149	-129.67	36.06	0.6262^{20}	1.8371^{20}		68.3	-40	1.4-8.0%	260	1000	
Pentanedial	$C_5H_8O_2$	100.117	-14	176									0.05
1,5-Pentanediol	$C_5H_{12}O_2$	104.148	-20	241	0.9914^{20}	26.2^{20}	2.5	0.001	129		335		
2,4-Pentanedione	$C_5H_8O_2$	100.117	-18.3	140.7	0.9721^{25}	26.524^{30}	2.78	1.02	34		340		
1-Pentanethiol	$C_5H_{12}S$	104.214	-75.69	126.6	0.850^{20}	4.847^{20}		1.83	18				
Pentanoic acid	$C_5H_{10}O_2$	102.132	-33.63	186.1	0.9389^{20}	2.661^{21}	1.61	0.024	96		400		
1-Pentanol	$C_5H_{12}O$	88.148	-77.58	137.6	0.8144^{20}	15.13^{25}	1.7	0.259	33	1.2-10.0%	300		
2-Pentanol	$C_5H_{12}O$	88.148	-73	119.1	0.8094^{20}	13.71^{25}	1.66	0.804	34	1.2-9.0%	343		
3-Pentanol	$C_5H_{12}O$	88.148	-69.9	123	0.8203^{20}	13.35^{25}	1.64	1.10	41	1.2-9.0%	435		
2-Pentanone	$C_5H_{10}O$	86.132	-76.83	102.2	0.809^{20}	15.45^{20}	2.70	4.97	7	1.5-8.2%	452		150
3-Pentanone	$C_5H_{10}O$	86.132	-38.98	101.9	0.8098^{25}	17.00^{20}	2.82	4.72	13	>1.6%	450	200	300
1-Pentene	C_5H_{10}	70.133	-165.13	30.0	0.6405^{20}	2.011^{20}	≈0.5	85.0	-18	1.5-8.7%	275		
cis-2-Pentene	C_5H_{10}	70.133	-151.35	36.9	0.6556^{20}			66.0	<-20				
trans-2-Pentene	C_5H_{10}	70.133	-140.20	36.3	0.6431^{25}			67.4	<-20				
Pentyl acetate	$C_7H_{14}O_2$	130.185	-70.9	149.4	0.8756^{20}	4.79^{20}	1.75	0.60	16	1.1-7.5%	360	50	100
Pentylamine	$C_5H_{13}N$	87.164	-51	104.7	0.7544^{20}	4.27^{20}		4.00	-1	2.2-22%			
Perchloric acid	$ClHO_4$	100.459	-112	≈90 dec	1.77								
Peroxyacetic acid	$C_2H_4O_3$	76.051	-0.2	110	1.226^{15}			1.93	41				0.4
Phenol	C_6H_6O	94.111	40.89	181.8	1.0545^{45}	12.40^{30}	1.224	0.055	79	1.8-8.6%	715	5	
2-Phenoxyethanol	$C_8H_{10}O_2$	138.164	12	246	1.102^{22}			0.001	121				
Phenylhydrazine	$C_6H_8N_2$	108.141	20	244	1.0986^{20}	7.15^{20}		0.003	88			0.1	
1-Phenyl-2-propylamine, (±)-	$C_9H_{13}N$	135.206		198	0.9306^{25}			0.06	<100				
Phosphinic acid	H_3O_2P	65.997	26.5	130	1.49								
Phosphoric acid	H_3O_4P	97.995	42.4	407									
Phosphorothioc trichloride	Cl_3PS	169.398	-36.2	125	1.635	4.94^{25}							
Phosphorus(III) bromide	Br_3P	270.686	-41.5	173.2	2.8			0.38					
Phosphorus(III) chloride	Cl_3P	137.333	-93	76	1.574	3.498^{17}	0.56	16.1				0.2	0.5
Phosphoryl chloride	Cl_3OP	153.332	1.18	105.5	1.645	14.1^{20}	2.54	4.97				0.1	
α-Pinene	$C_{10}H_{16}$	136.234	-74	156.3	0.8539^{25}	2.1787^{25}		0.64	33		255	20	
β-Pinene	$C_{10}H_{16}$	136.234	-50.0	165.8	0.860^{25}	2.4970^{25}		0.61	38		275	20	
Piperidine	$C_5H_{11}N$	85.148	-11.05	106.19	0.8606^{20}	4.33^{20}	1.19	4.28	16	1-10%			
Propanal	C_3H_6O	58.079	-80	48.0	0.8657^{25}	18.5^{17}	2.72	42.2	-30	2.6-17%	207	20	
1,2-Propanediol	$C_3H_8O_2$	76.095	-60	187.3	1.0361^{20}	27.5^{30}	2.25	0.02	99	2.6-12.5%	371		
1,3-Propanediol	$C_3H_8O_2$	76.095	-27.6	214.7	1.0538^{20}	35.1^{20}	2.55	0.007			400		
Propanenitrile	C_3H_5N	55.079	-93	97.3	0.7818^{20}	29.7^{20}	4.05	6.14	2	3.1-14%	512		
Propanoic acid	$C_3H_6O_2$	74.079	-20.5	141.5	0.9882^{25}	3.44^{25}	1.75	0.553	52	2.9-12.1%	465	10	
Propanoic anhydride	$C_6H_{10}O_3$	130.141	-45.0	168	1.0110^{20}	18.30^{20}		0.45	63	1.3-9.5%	285		
1-Propanol	C_3H_8O	60.095	-124.39	97.04	0.8048^{20}	20.8^{20}	[a]	2.76	23	2.2-13.7%	412	100	400
2-Propanol	C_3H_8O	60.095	-87.91	82.21	0.7855^{20}	20.18^{20}	[a]	6.02	12	2.0-12.7%	399	200	400
Propargyl alcohol	C_3H_4O	56.063	-51.8	113	0.9478^{20}	20.8^{20}	1.13		36			1	
Propyl acetate	$C_5H_{10}O_2$	102.132	-93	101.0	0.8885^{20}	5.62^{20}	1.78	4.49	13	1.7-8%	450	200	250
Propylamine	C_3H_9N	59.110	-84.78	47.21	0.7173^{20}	5.08^{23}	1.17	42.1	-37	2.0-10.4%	318		
Propylbenzene	C_9H_{12}	120.191	-99.52	159.2	0.8619^{20}	2.370^{20}		0.45	30	0.8-6.0%	450		
Propyl butanoate	$C_7H_{14}O_2$	130.185	-95.2	144	0.8730^{20}	4.3^{20}		0.618	37				
Propylene carbonate	$C_4H_6O_3$	102.089	-48.8	241.6	1.2047^{20}	66.14^{20}	4.9	0.05	135				
Propyl formate	$C_4H_8O_2$	88.106	-92.9	80.6	0.9053^{20}	6.92^{30}	1.89	10.9	-3		455		
Propyl propanoate	$C_6H_{12}O_2$	116.158	-75.9	122.2	0.8755^{25}	5.249^{20}		1.88	79				
Pyridine	C_5H_5N	79.101	-41.63	115.2	0.9819^{20}	13.260^{20}	2.215	2.76	20	1.8-12.4%	482	1	
Pyrrole	C_4H_5N	67.090	-23.39	129.74	0.9698^{20}	8.00^{20}	1.767	1.10	39				
Pyrrolidine	C_4H_9N	71.121	-57.79	86.6	0.8586^{20}	8.30^{20}	1.57	8.40	3				
2-Pyrrolidone	C_4H_7NO	85.105	25.92	251.2	1.120^{20}	28.18^{25}	3.5		129				
Quinoline	C_9H_7N	129.159	-14.78	237.1	1.0977^{15}	9.16^{20}	2.29	0.011			480		

Name	Mol. form.	M_r	mp/°C	bp/°C	ρ/g cm^{-3}	ε	μ/D	Vap. pres. (25 °C)/ kPa	FP/°C	Flam. limits (%)	IT/°C	TWA/ ppm	STEL/ ppm
Safrole	$C_{10}H_{10}O_2$	162.185	11.2	235	1.1000^{20}			0.01	100				
Salicylaldehyde	$C_7H_6O_2$	122.122	-7	208	1.1674^{20}	18.35^{20}	2.961	0.075	78				
Selenium chloride	Cl_2Se_2	228.83	-85	127 dec	2.774								
Selenium oxychloride	Cl_2OSe	165.87	8.5	177	2.44	46.2^{20}		0.02					
Selenium oxyfluoride	F_2OSe	132.96	15	125	2.8			0.56					
Styrene	C_8H_8	104.150	-30.65	145.3	0.9016^{25}	2.4737^{20}	0.125	0.81	31	0.9-6.8%	490	20	40
Sulfolane	$C_4H_8O_2S$	120.171	28.45	286	1.2723^{18}		4.8	<0.01	177				
Sulfur chloride [ClSSCl]	Cl_2S_2	135.036	-77	137	1.69	4.79^{15}		1.27					1
Sulfur dichloride	Cl_2S	102.971	-122	59.6	1.62	2.915^{25}	0.36	17.9					
Sulfuric acid	H_2O_4S	98.079	10.31	337	1.8305^{20}		2.964						
Sulfuryl chloride	Cl_2O_2S	134.970	-51	69.4	1.680	9.1^{20}	1.81	18.7					
α-Terpinene	$C_{10}H_{16}$	136.234		174	0.8375^{19}	2.4526^{25}							
1,1,2,2-Tetrabromoethane	$C_2H_2Br_4$	345.653	0	248	2.9655^{20}	6.72^{30}	1.38	0.003			335	0.1	
Tetrabromosilane	Br_4Si	347.702	5.39	154	2.8		0						
1,1,2,2-Tetrachloro-1,2-difluoroethane	$C_2Cl_4F_2$	203.830	26.54	92.83	1.5951^{50}	2.52^{35}		7.51				50	
1,1,1,2-Tetrachloroethane	$C_2H_2Cl_4$	167.849	-70.2	130.2	1.5406^{20}	9.22^{-66}		1.6	47	5-12%			
1,1,2,2-Tetrachloroethane	$C_2H_2Cl_4$	167.849	-42.4	146.0	1.5953^{20}	8.50^{20}	1.32	0.622	62	20-54%		1	
Tetrachloroethene	C_2Cl_4	165.833	-22.2	121.2	1.6230^{20}	2.268^{30}		2.42	45			25	100
Tetrachloromethane	CCl_4	153.823	-22.8	76.7	1.5940^{20}	2.2379^{20}		15.2				5	10
Tetrachlorosilane	Cl_4Si	169.898	-68.74	57.65	1.5	2.248^{0}	0	31.3					
Tetradecane	$C_{14}H_{30}$	198.388	5.87	253.5	0.7596^{20}	2.0343^{20}		0.002	112	>0.5%	200		
Tetraethylene glycol	$C_8H_{18}O_5$	194.226	-9.4	315	1.1285^{15}	20.44^{20}		0.000001	182				
Tetrafluoroboric acid	BF_4H	87.813		130 dec	≈1.8								
Tetrahydrofuran	C_4H_8O	72.106	-108.38	66.0	0.8833^{25}	7.52^{22}	1.75	21.6	-14	2-11.8%	321	50	100
Tetrahydrofurfuryl alcohol	$C_5H_{10}O_2$	102.132	<-80	176.3	1.0524^{20}	13.48^{30}	2.1	0.100	75	1.5-9.7%	282		
1,2,3,4-Tetrahydronaphthalene	$C_{10}H_{12}$	132.202	-35.76	207.2	0.9645^{25}	2.771^{25}		0.05	71	0.8-5.0%	385		
Tetrahydropyran	$C_5H_{10}O$	86.132	-49.1	88.0	0.8814^{20}	5.66^{20}	[a]	9.54	-20				
Tetrahydrothiophene	C_4H_8S	88.172	-96.13	121.1	0.9987^{20}		1.90	2.45					
Tetramethylsilane	$C_4H_{12}Si$	88.224	-99.063	26.7	0.648^{19}	1.921^{20}	0	94.2					
Tetramethylurea	$C_5H_{12}N_2O$	116.161	-2.67	177.1	0.9687^{20}	23.10^{20}	3.5	0.138	77				
Tetranitromethane	CN_4O_8	196.033	13.9	125.6	1.6380^{20}	2.317^{20}	0	1.13				0.005	
Thionyl bromide	Br_2OS	207.872	-50	140		9.06^{20}		0.84					
Thionyl chloride	Cl_2OS	118.970	-101	75.6	1.631	8.675^{25}	1.45	16.0					0.2
Thiophene	C_4H_4S	84.140	-38.12	84.1	1.0649^{20}	2.739^{20}	0.55	10.6	-1				
Tin(IV) chloride	Cl_4Sn	260.522	-34.07	114.15	2.234	3.014^{0}	0						
Titanium(IV) chloride	Cl_4Ti	189.679	-24.12	136.45	1.73	2.843^{-16}							
Toluene	C_7H_8	92.139	-95.0	110.60	0.8668^{20}	2.379^{23}	0.375	3.79	4	1.1-7.1%	480	20	
Toluene-2,4-diisocyanate	$C_9H_6N_2O_2$	174.156	20.5	251	1.2244^{20}	8.433^{20}		0.003	127	0.9-9.5%		0.001	0.005
Tribromomethane	$CHBr_3$	252.731	8.69	149.2	2.8788^{25}	4.404^{10}	0.99	0.726	83			0.5	
Tributylamine	$C_{12}H_{27}N$	185.349	-70	207	0.7770^{20}	2.340^{20}	0.78	0.01	63	1-5%			
Tributyl borate	$C_{12}H_{27}BO_3$	230.151	<-70	233.8	0.8567^{20}	2.23^{20}	0.77		93				
Tributyrin	$C_{15}H_{26}O_6$	302.363	-75	307.5	1.0350^{20}	5.72^{10}			180	>0.5%	407		
Trichloroacetaldehyde	C_2HCl_3O	147.387	-57.5	98	1.512^{20}	6.8^{25}		6.66					
1,2,4-Trichlorobenzene	$C_6H_3Cl_3$	181.447	17.0	213.5	1.459^{25}			0.057	105	2.5-6.6%	571		5
1,1,1-Trichloroethane	$C_2H_3Cl_3$	133.404	-30	74.02	1.3390^{20}	7.243^{20}	1.755	16.5	-1	8-10.5%	500	350	450
1,1,2-Trichloroethane	$C_2H_3Cl_3$	133.404	-36.3	113	1.4397^{20}	7.1937^{25}	1.4	3.1	32	6-28%	460	10	
Trichloroethene	C_2HCl_3	131.388	-84.7	86.8	1.4642^{20}	3.390^{28}	0.8	9.91	32	8-10.5%	420	10	25
Trichloroethylsilane	$C_2H_5Cl_3Si$	163.506	-105.6	98.7	1.2373^{20}		2.04	6.29	22				
Trichlorofluoromethane	CCl_3F	137.368	-110.44	23.7	1.4879^{20}	3.00^{20}	0.46	106					1000
Trichloromethane	$CHCl_3$	119.378	-63.3	61.2	1.4890^{20}	4.8069^{20}	1.04	26.2				10	
(Trichloromethyl)benzene	$C_7H_5Cl_3$	195.474	-17.0	221	1.3723^{20}		2.03	0.35	127		211		0.1
Trichloromethylsilane	CH_3Cl_3Si	149.480	-75.77	66	1.273^{20}		1.91	22.5	-9	7.6->20%	>404		
Trichloronitromethane	CCl_3NO_2	164.376	-69.4	112	1.6558^{20}	7.319^{20}		3.18				0.1	
1,2,3-Trichloropropane	$C_3H_5Cl_3$	147.431	-13.8	158	1.3889^{20}	7.5^{20}		0.492	71	3.2-12.6%		0.005	
Trichlorosilane	Cl_3HSi	135.453	-128.2	33	1.331		0.86		-50		104		
1,1,2-Trichloro-1,2,2-trifluoroethane	$C_2Cl_3F_3$	187.375		47.6	1.5635^{25}	2.41^{25}		44.8				1000	1250
Tri-*o*-cresyl phosphate	$C_{21}H_{21}O_4P$	368.363	11	410	1.1955^{20}	6.7^{25}	2.87	0.0000002	225		385		
Tridecane	$C_{13}H_{28}$	184.361	-5.35	235.4	0.7564^{20}	2.0213^{20}		0.005	79				
1-Tridecene	$C_{13}H_{26}$	182.345	-23.07	232.8	0.7658^{20}	2.139^{20}		0.0047	79				
Tris(2-hydroxyethyl)amine	$C_6H_{15}NO_3$	149.188	21.5	350	1.1242^{20}		3.57	<0.01	179	1-10%			
Triethylamine	$C_6H_{15}N$	101.190	-114.7	88.8	0.7275^{20}	2.418^{20}	0.66	7.70	-7	1.2-8.0%	249	0.5	1
Triethylene glycol	$C_6H_{14}O_4$	150.173	-9.4	288.6	1.1274^{15}	23.69^{20}		0.0002	177	0.9-9.2%	371		
Triethylene glycol dimethyl ether	$C_8H_{18}O_4$	178.227	-43.8	218	0.986^{20}	7.62^{25}			111				
Triethyl phosphate	$C_6H_{15}O_4P$	182.154	-56.4	216	1.0695^{20}	13.20^{25}	3.12		115		454		
Trifluoroacetic acid	$C_2HF_3O_2$	114.023	-15.2	72	1.5351^{25}	8.42^{20}	2.28	15.1					
(Trifluoromethyl)benzene	$C_7H_5F_3$	146.110	-28.99	102.0	1.1884^{20}	9.22^{25}	2.86	5.14	12				

Name	Mol. form.	M_r	mp/°C	bp/°C	ρ/g cm^{-3}	ε	μ/D	Vap. pres. (25 °C)/ kPa	FP/°C	Flam. limits (%)	IT/°C	TWA/ ppm	STEL/ ppm
1,2,3-Trimethylbenzene	C_9H_{12}	120.191	-25.32	176.0	0.8944^{20}	2.656^{20}		0.20	44	0.8-6.6%	470	25	
1,2,4-Trimethylbenzene	C_9H_{12}	120.191	-43.8	169.4	0.8758^{20}	2.377^{20}		0.30	44	0.9-6.4%	500	25	
1,3,5-Trimethylbenzene	C_9H_{12}	120.191	-44.69	164.7	0.8615^{25}	2.279^{20}	0	0.33	50	1-5%	559	25	
Trimethyl borate	$C_3H_9BO_3$	103.912	-29.3	67.4	0.915^{25}	2.2762^{20}		17.2	-8				
Trimethylchlorosilane	C_3H_9ClSi	108.642	-55.17	57.6	0.856^{25}	10.21^{0}		30.7	-28		395		
2,2,4-Trimethylpentane	C_8H_{18}	114.229	-107.36	99.2	0.6919^{20}	1.943^{20}		6.50	-12		418	300	
2,3,3-Trimethylpentane	C_8H_{18}	114.229	-101.2	114.7	0.7262^{20}	1.9780^{20}		3.60	<21		425	300	
Trimethyl phosphate	$C_3H_9O_4P$	140.074	-46	197.2	1.2144^{20}	20.6^{20}	3.18	0.11	107				
2,4,6-Trimethylpyridine	$C_8H_{11}N$	121.180	-44.3	170	0.9166^{22}	7.807^{25}	2.05	4.1					
Trinitroglycerol	$C_3H_5N_3O_9$	227.087	12.8	218 exp	1.5931^{20}	19.25^{20}		0.00005			270	0.05	
Undecane	$C_{11}H_{24}$	156.309	-25.54	195.9	0.7402^{20}	1.9972^{20}		0.05	69				
Vanadium(IV) chloride	Cl_4V	192.754	-28	151	1.816	3.05^{25}							
Vanadyl trichloride	Cl_3OV	173.300	-79	127	1.829	3.4^{25}							
Vinyl acetate	$C_4H_6O_2$	86.090	-100.2	72.6	0.9256^{25}		1.79	15.4	-8	2.6-13.4%	402	10	15
4-Vinylcyclohexene	C_8H_{12}	108.181	-108.9	130	0.8299^{20}			1.87	16		269	0.1	
Water	H_2O	18.015	0.00	99.974	0.9970^{25}	80.1^{20}	1.8546	3.17					
o-Xylene	C_8H_{10}	106.165	-25.16	144.4	0.8802^{20}	2.562^{20}	0.640	0.88	32	0.9-6.7%	463	100	150
m-Xylene	C_8H_{10}	106.165	-47.85	139.1	0.8641^{20}	2.359^{20}		1.13	27	1.1-7.0%	527	100	150
p-Xylene	C_8H_{10}	106.165	13.3	138.3	0.8610^{20}	2.2735^{20}	0	1.19	27	1.1-7.0%	528	100	150
2,4-Xylenol	$C_8H_{10}O$	122.164	25	210.94	0.9650^{20}	5.060^{30}	1.4	0.022					

[a] See the table "Dipole Moments" in Section 9 for values for individual conformers.

MISCIBILITY OF ORGANIC SOLVENTS

The chart below gives qualitative information on the miscibility of pairs of organic liquids. Two liquids are considered *miscible* if mixing equal volumes produces a single liquid phase. If two phases separate, they are considered *immiscible*. An entry of *partially misc.* indicates two phases whose volumes differ appreciably, suggesting a partial miscibility of the components. The entry *reaction* indicates a chemical reaction between the components. All data refer to room temperature.

References

1. Drury, J. S., *Ind. Eng. Chem.* 44, 2744, 1959.
2. Jackson, W. M., and Drury, J. S., *Ind. Eng. Chem.* 51, 1491, 1959.

Miscibility of Organic Solvents

Name	Acetone	Benzaldehyde	Benzene	Butyl acetate	Butyl alcohol	Carbon tetrachloride	2-Chloroethanol	Chloroform	*o*-Cresol
Acetone	-	miscible	miscible	miscible	miscible	miscible	miscible	miscible	
2-Amino-2-methyl-1-propanol	miscible	miscible	miscible	miscible	miscible	miscible			
Benzaldehyde	miscible	-	miscible	miscible	miscible	miscible			
Benzene	miscible	miscible	-	miscible	miscible	miscible	miscible	miscible	
Benzenemethanethiol	miscible		miscible		miscible	miscible			
Benzonitrile	miscible		miscible		miscible	miscible			
Benzothiazole	miscible		miscible		miscible	miscible			
Benzyl alcohol	miscible	miscible	miscible	miscible	miscible	miscible			
Bis(2-aminoethyl)amine	miscible		miscible				reaction	miscible	
N,N'-Bis(2-aminoethyl)-1,2-ethanediamine	miscible		miscible				miscible	miscible	
2-Bromoethyl acetate	miscible		partially misc.					miscible	
1,3-Butanediol	miscible		immiscible					miscible	miscible
2,3-Butanediol	miscible		partially misc.					miscible	miscible
1-Butanol	miscible	miscible	miscible	miscible	-	miscible			
2-Butoxyethanol	miscible		miscible					miscible	miscible
Butyl acetate	miscible	miscible	miscible	-	miscible	miscible			
2-Chloroethanol	miscible		miscible				-		miscible
3-Chloro-1,2-propanediol	miscible		immiscible					miscible	miscible
trans-Cinnamaldehyde	miscible		miscible				miscible	miscible	
o-Cresol							miscible		-
Diacetone alcohol	miscible	miscible	partially misc.	miscible	miscible	partially misc.			
Dibenzyl ether	miscible		miscible				miscible	miscible	
Dibutylamine							reaction		
Dibutyl carbonate	miscible		miscible					miscible	
Dibutyl ether	miscible	miscible	miscible	miscible	miscible	miscible			
Diethanolamine	miscible	immiscible	immiscible	immiscible	miscible	immiscible			
Diethylene glycol dibutyl ether	miscible		miscible					miscible	miscible
Diethylene glycol diethyl ether	miscible		miscible					miscible	miscible
Diethylene glycol monobutyl ether	miscible		miscible					miscible	miscible
Diethylene glycol monoethyl ether	miscible		miscible					miscible	miscible
Diethylene glycol monomethyl ether	miscible		miscible					miscible	
Diethyl ether	miscible	miscible	miscible	miscible	miscible	miscible	miscible	miscible	
N,N-Diethylformamide	miscible		miscible				miscible	miscible	
Dihexyl ether	miscible		miscible				miscible	miscible	
Diisopentyl sulfide	miscible		miscible		miscible	miscible			
Diisopropylamine	miscible		miscible				reaction	miscible	
N,N-Dimethylaniline	miscible	miscible	miscible	miscible	miscible	miscible			
Dimethyl disulfide	miscible		miscible		miscible	miscible			
2,6-Dimethyl-4-heptanone	miscible		miscible				miscible	miscible	
Dipentylamine	miscible		miscible					miscible	
N,N-Dipropylaniline	miscible		miscible		miscible	miscible	miscible	miscible	
Dipropylene glycol	miscible		miscible				miscible	miscible	miscible
1,2-Ethanediol	miscible	partially misc.	immiscible	partially misc.	miscible	immiscible		partially misc.	miscible
Ethanol	miscible	miscible	miscible	miscible	miscible	miscible		miscible	
2-Ethoxyaniline	miscible		miscible				miscible	miscible	
2-Ethoxyethanol	miscible		miscible					miscible	miscible
Ethyl benzoate	miscible	miscible	miscible	miscible	miscible	miscible		miscible	
2-Ethylbutanoic acid	miscible		miscible				miscible	miscible	
Ethyl chloroacetate	miscible		miscible				miscible	miscible	
Ethyl *trans*-cinnamate	miscible		miscible					miscible	
2-Ethyl-1-hexanol	miscible	miscible	miscible	miscible	miscible	miscible			

Name	Acetone	Benzaldehyde	Benzene	Butyl acetate	Butyl alcohol	Carbon tetrachloride	2-Chloroethanol	Chloroform	*o*-Cresol
Ethyl phenylacetate	miscible		miscible				miscible	miscible	
Ethyl thiocyanate	miscible		miscible		miscible	miscible			
Formamide	miscible	miscible	immiscible	immiscible	miscible	immiscible			
Furfuryl alcohol	miscible	miscible	miscible	miscible	miscible	miscible			
Glycerol	immiscible	partially misc.	immiscible	immiscible	miscible	immiscible		immiscible	miscible
1-Heptadecanol	miscible		miscible					miscible	
3-Heptanol, (*S*)-	miscible		miscible				miscible	miscible	
Heptyl acetate	miscible		miscible				miscible	miscible	
Hexanedinitrile	miscible		miscible		miscible	immiscible			
Hexanenitrile	miscible		miscible				miscible	miscible	
Isopentyl acetate	miscible		miscible				miscible	miscible	
4-Methoxybenzaldehyde							miscible		
2-Methoxyethanol	miscible		miscible						miscible
3-Methyl-1-butanol	miscible	miscible	miscible	miscible	miscible	miscible			
3-Methyl-2-butanone	miscible		miscible				miscible	miscible	
4-Methylpentanoic acid	miscible		miscible					miscible	
4-Methyl-2-pentanone	miscible	miscible	miscible	miscible	miscible	miscible			
2-Methyl-1-propanethiol	miscible		miscible		miscible	miscible			
Nitromethane	miscible	miscible	immiscible	miscible	miscible	miscible			
1-Octanol	miscible	miscible	miscible	miscible	miscible	miscible			
1,2-Propanediol	miscible		immiscible					miscible	miscible
1,3-Propanediol	miscible	miscible	immiscible	partially misc.	miscible	immiscible		miscible	miscible
Pyridine	miscible	miscible	miscible	miscible	miscible	miscible	miscible		
Tetrachloromethane	miscible	miscible	miscible	miscible	miscible	-			
1-Tetradecanol	miscible		miscible				miscible	miscible	
Tributyl phosphate	miscible		miscible				miscible	miscible	
Trichloromethane	miscible		miscible				miscible	-	
Triethylene glycol	miscible		partially misc.					miscible	miscible
Triethyl phosphate	miscible		miscible					miscible	miscible
2,6,8-Trimethyl-4-nonanone	miscible		miscible				miscible	miscible	

Name	Diethyl ether	*N,N*-Dimethylaniline	Dipentylamine	Ethyl alcohol	Ethylene glycol	Ethylene glycol monoethyl ether	Formamide	Furfuryl alcohol
Acetone	miscible	miscible	miscible	miscible	miscible	miscible	miscible	miscible
2-Amino-2-methyl-1-propanol	miscible	miscible		miscible	miscible		miscible	miscible
Benzaldehyde	miscible	miscible		miscible	partially misc.		miscible	miscible
Benzene	miscible	miscible	miscible	miscible	immiscible	miscible	immiscible	miscible
Benzenemethanethiol	miscible	miscible		miscible	immiscible		immiscible	miscible
Benzonitrile	miscible	miscible		miscible	immiscible		immiscible	miscible
Benzothiazole	miscible	miscible		miscible	miscible		immiscible	miscible
Benzyl alcohol	miscible	miscible		miscible	miscible		miscible	miscible
Bis(2-aminoethyl)amine			immiscible	miscible	miscible	miscible		
N,N'-Bis(2-aminoethyl)-1,2-ethanediamine	miscible		immiscible	miscible	miscible	miscible		
2-Bromoethyl acetate			reaction	miscible				
1,3-Butanediol	partially misc.		miscible	miscible				
2,3-Butanediol	miscible		miscible	miscible				
1-Butanol	miscible	miscible		miscible	miscible		miscible	miscible
2-Butoxyethanol	miscible		miscible	miscible				
Butyl acetate	miscible	miscible		miscible	partially misc.		immiscible	miscible
2-Chloroethanol	miscible		miscible					
3-Chloro-1,2-propanediol	miscible		reaction	miscible				
trans-Cinnamaldehyde			miscible	miscible	immiscible	miscible		
o-Cresol					miscible	miscible		
Diacetone alcohol	miscible	miscible		miscible	miscible		miscible	miscible
Dibenzyl ether			miscible		immiscible	miscible		
Dibutylamine					miscible	miscible		
Dibutyl carbonate			miscible	miscible				
Dibutyl ether	miscible	miscible		miscible	immiscible		immiscible	miscible
Diethanolamine	immiscible	partially misc.		miscible	miscible		miscible	miscible
Diethylene glycol dibutyl ether	miscible		reaction	miscible				
Diethylene glycol diethyl ether	miscible		miscible	miscible				
Diethylene glycol monobutyl ether	miscible		miscible	miscible				
Diethylene glycol monoethyl ether	miscible		miscible	miscible				
Diethylene glycol monomethyl ether	miscible		miscible	miscible				
Diethyl ether	-	miscible	miscible	miscible	immiscible	miscible	immiscible	miscible
N,N-Diethylformamide			reaction	miscible	miscible	miscible		

Name	Diethyl ether	*N,N*-Dimethylaniline	Dipentylamine	Ethyl alcohol	Ethylene glycol	Ethylene glycol monoethyl ether	Formamide	Furfuryl alcohol
Dihexyl ether			miscible	miscible	immiscible	miscible		
Diisopentyl sulfide	miscible	miscible		miscible	immiscible		immiscible	immiscible
Diisopropylamine			miscible	miscible	miscible	miscible		
N,N-Dimethylaniline	miscible	-		miscible	immiscible		immiscible	miscible
Dimethyl disulfide	miscible	miscible		miscible	immiscible		immiscible	miscible
2,6-Dimethyl-4-heptanone			miscible	miscible	immiscible	miscible		
Dipentylamine	miscible		-	miscible	partially misc.	miscible		
N,N-Dipropylaniline	miscible	miscible	miscible	miscible	immiscible	miscible	immiscible	miscible
Dipropylene glycol	miscible			miscible				
1,2-Ethanediol	immiscible	immiscible	partially misc.	miscible	-		miscible	miscible
Ethanol	miscible	miscible	miscible	-	miscible	miscible	miscible	miscible
2-Ethoxyaniline			miscible	miscible	miscible	miscible		
2-Ethoxyethanol	miscible		miscible	miscible		-		
Ethyl benzoate	miscible	miscible	miscible	miscible	immiscible		immiscible	miscible
2-Ethylbutanoic acid			reaction	miscible	miscible	miscible		
Ethyl chloroacetate			miscible	miscible	immiscible	miscible		
Ethyl *trans*-cinnamate			miscible	miscible	immiscible	miscible		
2-Ethyl-1-hexanol	miscible	miscible		miscible	miscible		immiscible	miscible
Ethyl phenylacetate			miscible		immiscible	miscible		
Ethyl thiocyanate	miscible	miscible		miscible	immiscible		immiscible	miscible
Formamide	immiscible	immiscible		miscible	miscible		-	miscible
Furfuryl alcohol	miscible	miscible		miscible	miscible		miscible	-
Glycerol	immiscible	immiscible	partially misc.	miscible	miscible		miscible	miscible
1-Heptadecanol			miscible	miscible				
3-Heptanol, (*S*)-			miscible	miscible	miscible	miscible		
Heptyl acetate			miscible		immiscible	miscible		
Hexanedinitrile	immiscible	miscible		miscible	immiscible		miscible	miscible
Hexanenitrile			miscible	miscible	immiscible	miscible		
Isopentyl acetate			miscible	miscible	immiscible	miscible		
4-Methoxybenzaldehyde					immiscible	miscible		
2-Methoxyethanol	miscible		miscible	miscible				
3-Methyl-1-butanol	miscible	miscible		miscible	miscible		miscible	miscible
3-Methyl-2-butanone	miscible		miscible	miscible	immiscible	miscible		
4-Methylpentanoic acid			miscible	miscible	miscible	miscible		
4-Methyl-2-pentanone	miscible	miscible		miscible	immiscible		partially misc.	miscible
2-Methyl-1-propanethiol	miscible	miscible		miscible	immiscible		immiscible	miscible
Nitromethane	miscible	miscible		miscible	immiscible		miscible	miscible
1-Octanol	miscible	miscible		miscible	miscible		immiscible	miscible
1,2-Propanediol	partially misc.		miscible	miscible				
1,3-Propanediol	immiscible	immiscible	miscible	miscible	miscible		miscible	miscible
Pyridine	miscible	miscible		miscible	miscible	miscible	miscible	miscible
Tetrachloromethane	miscible	miscible		miscible	immiscible		immiscible	miscible
1-Tetradecanol			miscible	miscible	immiscible	miscible		
Tributyl phosphate			miscible	miscible	partially misc.	miscible		
Trichloromethane	miscible		miscible	miscible	partially misc.	miscible		
Triethylene glycol	immiscible		partially misc.	miscible				
Triethyl phosphate	miscible		miscible	miscible				
2,6,8-Trimethyl-4-nonanone			miscible	miscible	immiscible	miscible		

Name	Glycerol	Methyl isopropyl ketone	Nitromethane	1-Octanol	1,3-Propanediol	Pyridine	Triethylenetetramine	Triethyl phosphate
Acetone	immiscible	miscible	miscible	miscible	miscible	miscible	miscible	miscible
2-Amino-2-methyl-1-propanol	miscible		miscible	miscible	miscible	miscible		
Benzaldehyde	partially misc.		miscible	miscible	miscible	miscible		
Benzene	immiscible	miscible	immiscible	miscible	immiscible	miscible	miscible	miscible
Benzenemethanethiol	immiscible				immiscible	miscible		
Benzonitrile	immiscible				immiscible	miscible		
Benzothiazole	immiscible				miscible	miscible		
Benzyl alcohol	miscible		miscible	miscible	miscible	miscible		
Bis(2-aminoethyl)amine	miscible	reaction					miscible	miscible
N,N'-Bis(2-aminoethyl)-1,2-ethanediamine	miscible	reaction			miscible		-	miscible
2-Bromoethyl acetate		miscible					reaction	
1,3-Butanediol		miscible				miscible	miscible	
2,3-Butanediol		miscible				miscible	miscible	
1-Butanol	miscible		miscible	miscible	miscible	miscible		

Name	Glycerol	Methyl isopropyl ketone	Nitromethane	1-Octanol	1,3-Propanediol	Pyridine	Triethylenetetramine	Triethyl phosphate
2-Butoxyethanol		miscible				miscible	miscible	
Butyl acetate	immiscible		miscible	miscible	partially misc.	miscible		
2-Chloroethanol		miscible				miscible	miscible	miscible
3-Chloro-1,2-propanediol		miscible				miscible	reaction	
trans-Cinnamaldehyde	immiscible	miscible			immiscible		reaction	miscible
o-Cresol	miscible				miscible			miscible
Diacetone alcohol	immiscible		miscible	miscible	miscible	miscible		
Dibenzyl ether	immiscible	miscible					miscible	miscible
Dibutylamine	partially misc.							miscible
Dibutyl carbonate		miscible					immiscible	
Dibutyl ether	immiscible		immiscible	miscible	immiscible	miscible		
Diethanolamine	miscible		immiscible	miscible	miscible	miscible		
Diethylene glycol dibutyl ether		miscible				miscible		
Diethylene glycol diethyl ether		miscible				miscible	miscible	
Diethylene glycol monobutyl ether		miscible				miscible	miscible	
Diethylene glycol monoethyl ether		miscible				miscible	miscible	
Diethylene glycol monomethyl ether		miscible				miscible	miscible	
Diethyl ether	immiscible	miscible	miscible	miscible	immiscible	miscible	miscible	miscible
N,N-Diethylformamide	miscible	miscible			miscible		reaction	miscible
Dihexyl ether	immiscible	miscible			immiscible		immiscible	
Diisopentyl sulfide	immiscible				immiscible	miscible		
Diisopropylamine	miscible	miscible			miscible		miscible	miscible
N,N-Dimethylaniline	immiscible		miscible	miscible	immiscible	miscible		
Dimethyl disulfide	immiscible				reaction	miscible		
2,6-Dimethyl-4-heptanone	immiscible	miscible			immiscible		miscible	miscible
Dipentylamine	partially misc.	miscible			miscible		immiscible	miscible
N,N-Dipropylaniline	immiscible	miscible			immiscible	miscible	miscible	miscible
Dipropylene glycol		miscible				miscible	miscible	
1,2-Ethanediol	miscible	immiscible	immiscible	miscible	miscible	miscible	miscible	
Ethanol	miscible	miscible	miscible	miscible	miscible	miscible	miscible	miscible
2-Ethoxyaniline		miscible			miscible		miscible	
2-Ethoxyethanol		miscible				miscible	miscible	
Ethyl benzoate	immiscible	miscible	miscible	miscible	partially misc.	miscible	miscible	
2-Ethylbutanoic acid	immiscible	miscible			miscible		reaction	miscible
Ethyl chloroacetate	immiscible	miscible			immiscible		reaction	miscible
Ethyl *trans*-cinnamate	immiscible	miscible			immiscible		miscible	miscible
2-Ethyl-1-hexanol	immiscible		immiscible	miscible	miscible	miscible		
Ethyl phenylacetate	immiscible	miscible			immiscible		miscible	miscible
Ethyl thiocyanate	immiscible				immiscible	miscible		
Formamide	miscible		miscible	immiscible	miscible	miscible		
Furfuryl alcohol	miscible		miscible	miscible	miscible	miscible		
Glycerol	-	immiscible	immiscible	immiscible	miscible	miscible	miscible	
1-Heptadecanol		miscible					miscible	
3-Heptanol, (*S*)-	immiscible				miscible		miscible	miscible
Heptyl acetate	immiscible	miscible			immiscible		reaction	miscible
Hexanedinitrile	immiscible				miscible	miscible		
Hexanenitrile	immiscible	miscible			immiscible		miscible	miscible
Isopentyl acetate	immiscible	miscible			immiscible		miscible	miscible
4-Methoxybenzaldehyde	immiscible				immiscible			miscible
2-Methoxyethanol						miscible	miscible	
3-Methyl-1-butanol	immiscible		miscible	miscible	miscible	miscible		
3-Methyl-2-butanone	immiscible	-			miscible		reaction	miscible
4-Methylpentanoic acid	immiscible	miscible			miscible		reaction	miscible
4-Methyl-2-pentanone	immiscible		miscible	miscible	immiscible	miscible		
2-Methyl-1-propanethiol	immiscible				reaction	miscible		
Nitromethane	immiscible		-	partially misc.	immiscible	miscible		
1-Octanol	immiscible		partially misc.	-	miscible	miscible		
1,2-Propanediol		miscible				miscible	miscible	
1,3-Propanediol	miscible	miscible	immiscible	miscible	-	miscible	miscible	
Pyridine	miscible		miscible	miscible	miscible	-		miscible
Tetrachloromethane	immiscible		miscible	miscible	immiscible	miscible		
1-Tetradecanol	immiscible	miscible			partially misc.		miscible	miscible
Tributyl phosphate	immiscible	miscible			miscible		miscible	miscible
Trichloromethane	immiscible	miscible			miscible		miscible	miscible
Triethylene glycol		miscible				miscible	miscible	
Triethyl phosphate		miscible				miscible	miscible	-
2,6,8-Trimethyl-4-nonanone	immiscible	miscible			immiscible		immiscible	miscible

DENSITY OF SOLVENTS AS A FUNCTION OF TEMPERATURE

The table below lists the density of 46 common solvents in the temperature range from 0 °C to 100 °C. The values have been calculated from the Rackett equation, using parameters from Ref. 1, or from the REFPROP program (Ref. 2). Density values refer to the liquid at its saturation vapor pressure; thus, entries for temperatures above the normal boiling point are for pressures greater than atmospheric pressure.

References

1. Lide, D. R., and Kehiaian, H. V., *Handbook of Thermophysical and Thermochemical Data*, CRC Press, Boca Raton, FL, 1994.
2. Lemmon, E.W., Huber, M.L., McLinden, M.O., *NIST Standard Reference Database 23: Reference Fluid Thermodynamic and Transport Properties-REFPROP*, Version 9.0, National Institute of Standards and Technology, Standard Reference Data Program, Gaithersburg, MD, 2010.

Density ρ of Solvents at the Temperature in °C Indicated by Superscript

Name	Mol. form.	ρ^{0}/ g cm^{-3}	ρ^{10}/ g cm^{-3}	ρ^{20}/ g cm^{-3}	ρ^{30}/ g cm^{-3}	ρ^{40}/ g cm^{-3}	ρ^{50}/ g cm^{-3}	ρ^{60}/ g cm^{-3}	ρ^{70}/ g cm^{-3}	ρ^{80}/ g cm^{-3}	ρ^{90}/ g cm^{-3}	ρ^{100}/ g cm^{-3}	Ref.
Acetic acid	$C_2H_4O_2$			1.0510	1.0380	1.0250	1.0120	0.9993	0.9861	0.9728	0.9592	0.9454	1
Acetone	C_3H_6O	0.8121	0.8012	0.7902	0.7790	0.7677	0.7561	0.7443	0.7322	0.7198	0.7070	0.6938	2
Acetonitrile	C_2H_3N			0.7825	0.7707	0.7591	0.7473	0.7353	0.7231	0.7106	0.6980	0.6851	1
Aniline	C_6H_7N	1.0410	1.0330	1.0250	1.0160	1.0080	1.0000	0.9909	0.9823	0.9735	0.9646	0.9557	1
Benzene	C_6H_6		0.8894	0.8788	0.8681	0.8574	0.8466	0.8357	0.8247	0.8135	0.8022	0.7907	2
1-Butanol	$C_4H_{10}O$	0.8347	0.8246	0.8148	0.8051	0.7955	0.7860	0.7766	0.7671	0.7577	0.7481	0.7385	2
Butylamine	$C_4H_{11}N$	0.7606	0.7512	0.7417	0.7320	0.7221	0.7120	0.7017	0.6911	0.6803	0.6693	0.6579	1
Carbon disulfide	CS_2	1.2900	1.2770	1.2630	1.2480	1.2340							1
Chlorobenzene	C_6H_5Cl	1.1270	1.1160	1.1060	1.0960	1.0850	1.0740	1.0640	1.0530	1.0420	1.0300	1.0190	1
Cyclohexane	C_6H_{12}		0.7880	0.7786	0.7691	0.7595	0.7499	0.7401	0.7303	0.7203	0.7101	0.6997	2
Decane	$C_{10}H_{22}$	0.7459	0.7381	0.7303	0.7226	0.7148	0.7070	0.6992	0.6913	0.6834	0.6754	0.6674	2
1-Decanol	$C_{10}H_{22}O$			0.8294	0.8229	0.8162	0.8093	0.8024	0.7955	0.7884	0.7813	0.7740	1
Dichloromethane	CH_2Cl_2	1.3632	1.3433	1.3232	1.3031	1.2828	1.2624	1.2417	1.2207	1.1995	1.1777	1.1555	2
Diethyl ether	$C_4H_{10}O$	0.7358	0.7248	0.7135	0.7020	0.6901	0.6778	0.6652	0.6521	0.6384	0.6242	0.6093	2
N,N-Dimethylaniline	$C_8H_{11}N$		0.9638	0.9562	0.9483	0.9401	0.9318	0.9234	0.9150	0.9064	0.8978	0.8890	1
Ethanol	C_2H_6O	0.8063	0.7979	0.7893	0.7807	0.7720	0.7631	0.7540	0.7446	0.7348	0.7247	0.7140	2
Ethyl acetate	$C_4H_8O_2$	0.9245	0.9126	0.9006	0.8884	0.8759	0.8632	0.8503	0.8370	0.8234	0.8095	0.7952	1
Ethylbenzene	C_8H_{10}	0.8842	0.8755	0.8668	0.8581	0.8493	0.8405	0.8316	0.8226	0.8135	0.8044	0.7951	2
Ethyl formate	$C_3H_6O_2$	0.9472	0.9346	0.9218	0.9087	0.8954	0.8818	0.8678	0.8535	0.8389	0.8238	0.8082	1
Ethyl propanoate	$C_5H_{10}O_2$	0.9113	0.9005	0.8895	0.8784	0.8671	0.8556	0.8439	0.8319	0.8197	0.8072	0.7944	1
Heptane	C_7H_{16}	0.7004	0.6921	0.6837	0.6753	0.6667	0.6581	0.6493	0.6404	0.6314	0.6221	0.6127	2
Hexane	C_6H_{14}	0.6773	0.6684	0.6593	0.6501	0.6407	0.6313	0.6216	0.6117	0.6016	0.5913	0.5806	2
1-Hexanol	$C_6H_{14}O$	0.8359	0.8278	0.8195	0.8111	0.8027	0.7941	0.7854	0.7766	0.7676	0.7585	0.7492	1
Isopropylbenzene	C_9H_{12}	0.8769	0.8696	0.8615	0.8533	0.8450	0.8366	0.8280	0.8194	0.8106	0.8017	0.7927	1
Methanol	CH_4O	0.8096	0.8003	0.7909	0.7815	0.7721	0.7625	0.7528	0.7428	0.7326	0.7220	0.7109	2
Methyl acetate	$C_3H_6O_2$	0.9606	0.9478	0.9346	0.9211	0.9074	0.8933	0.8790	0.8643	0.8491	0.8336	0.8176	1
N-Methylaniline	C_7H_9N	1.0010	0.9933	0.9859	0.9785	0.9709	0.9633	0.9556	0.9478	0.9399	0.9319	0.9239	1
Methylcyclohexane	C_7H_{14}	0.7865	0.7780	0.7694	0.7607	0.7520	0.7433	0.7344	0.7255	0.7164	0.7072	0.6979	2
Methyl formate	$C_2H_4O_2$	1.003	0.9887	0.9739	0.9588	0.9433	0.9275	0.9112	0.8945	0.8772	0.8594	0.8409	1
Methyl propanoate	$C_4H_8O_2$	0.9383	0.9268	0.9150	0.9030	0.8907	0.8783	0.8656	0.8526	0.8393	0.8257	0.8117	1
Nitromethane	CH_3NO_2			1.1390	1.1250	1.1110	1.0970	1.0830	1.0690	1.0550	1.0400	1.0260	1
Nonane	C_9H_{20}	0.7337	0.7258	0.7179	0.7101	0.7021	0.6942	0.6862	0.6781	0.6700	0.6617	0.6534	2
Octane	C_8H_{18}	0.7182	0.7102	0.7022	0.6941	0.6860	0.6779	0.6696	0.6631	0.6528	0.6430	0.6442	2
Pentanoic acid	$C_5H_{10}O_2$	0.9563	0.9476	0.9389	0.9301	0.9211	0.9121	0.9029	0.8937	0.8843	0.8748	0.8652	1
1-Propanol	C_3H_8O	0.8252	0.8151	0.8048	0.7943	0.7837	0.7729	0.7619	0.7506	0.7391	0.7273	0.7152	1
2-Propanol	C_3H_8O	0.8003	0.7931	0.7885	0.7774	0.7689	0.7599	0.7504	0.7402	0.7295	0.7182	0.7062	2
Propyl acetate	$C_5H_{10}O_2$	0.9101	0.8994	0.8885	0.8775	0.8662	0.8548	0.8432	0.8313	0.8192	0.8069	0.7942	1
Propylbenzene	C_9H_{12}	0.8779	0.8700	0.8619	0.8538	0.8456	0.8373	0.8289	0.8204	0.8117	0.8030	0.7943	1
Propyl formate	$C_4H_8O_2$	0.9275	0.9166	0.9053	0.8938	0.8821	0.8702	0.8581	0.8457	0.8330	0.8201	0.8068	1
Tetrachloromethane	CCl_4	1.6290	1.6110	1.5930	1.5750	1.5570	1.5380	1.5180	1.4990	1.4790	1.4580	1.4370	1
Toluene	C_7H_8	0.8854	0.8761	0.8668	0.8575	0.8481	0.8387	0.8291	0.8195	0.8098	0.7999	0.7899	2
Trichloromethane	$CHCl_3$	1.5240	1.5070	1.4890	1.4710	1.4520	1.4330	1.4140	1.3940				1
2,2,4-Trimethylpentane	C_8H_{18}	0.7082	0.7001	0.6919	0.6837	0.6753	0.6668	0.6583	0.6496	0.6407	0.6316	0.6224	2
o-Xylene	C_8H_{10}	0.8969	0.8886	0.8802	0.8717	0.8633	0.8547	0.8461	0.8375	0.8287	0.8199	0.8109	2
m-Xylene	C_8H_{10}	0.8811	0.8727	0.8641	0.8555	0.8469	0.8381	0.8293	0.8204	0.8113	0.8022	0.7929	2
p-Xylene	C_8H_{10}			0.8610	0.8523	0.8436	0.8348	0.8260	0.8171	0.8081	0.7990	0.7898	2

DEPENDENCE OF BOILING POINT OF ORGANIC LIQUIDS ON PRESSURE

The normal boiling point of a liquid is defined as the temperature at which the vapor pressure reaches standard atmospheric pressure, 101.325 kPa. Calculated values of the change in boiling point of organic liquids with small changes in pressure, $\Delta t/\Delta p$, are given in this table in units of °C kPa^{-1} and °C $mmHg^{-1}$. The values from Ref. 1 were calculated from the representation of the vapor pressure by the Antoine equation,

$$\ln p/p_0 = A_1 - A_2/(T + A_3)$$

where p is the vapor pressure in kPa, p_0 is 101.325 kPa, T is the absolute temperature, and A_1, A_2, and A_3 are constants, taken from Ref. 1. The correction to the boiling point is generally accurate to 0.1 °C to 0.2 °C as long as the pressure is within 10% of standard atmospheric pressure. Values from Ref. 2 were calculated from the REFPROP program, which employs a more sophisticated equation.

A slightly less accurate estimate of $\Delta t/\Delta p$ may be obtained from the Clausius-Clapeyron equation, with the assumption that the change in volume upon vaporization equals the ideal-gas volume of the vapor. This leads to the equation,

$$\Delta t/\Delta p = RT_b^2/[p_0\Delta_{vap}H(T_b)]$$

where R is the molar gas constant, T_b is the normal boiling point temperature (absolute), and $\Delta_{vap}H(T_b)$ is the molar enthalpy of vaporization at the normal boiling point. Values of the last quantity may be obtained from the table "Enthalpy of Vaporization" in Section 6. Column definitions for the table are as follows.

Column heading	Definition
Name	Name of liquid
t_b	Normal boiling point, in °C
$\Delta t/\Delta p$	Change of boiling point with small change of pressure (in kPa), in units °C kPa^{-1}
$\Delta t/\Delta p$	Change of boiling point with small change of pressure (in mmHg), in units °C $mmHg^{-1}$
Ref.	Reference

References

1. Lide, D. R., and Kehiaian, H. V., *CRC Handbook of Thermophysical and Thermochemical Data*, CRC Press, Boca Raton, FL, 1994, pp. 49–59.
2. Lemmon, E.W., Huber, M.L., and McLinden, M.O., *NIST Standard Reference Database 23: Reference Fluid Thermodynamic and Transport Properties-REFPROP*, Version 9.1, National Institute of Standards and Technology, Standard Reference Data Program, Gaithersburg, MD, 2013.

Dependence of Boiling Point on Pressure

Name	t_b/°C	$\Delta t/\Delta p$/ °C kPa^{-1}	$\Delta t/\Delta p$/ °C $mmHg^{-1}$	Ref.
Acetaldehyde	20.8	0.261	0.0348	1
Acetic acid	117.9	0.324	0.0432	1
Acetic anhydride	139.5	0.343	0.0457	1
Acetone	56.08	0.289	0.0385	2
Acetonitrile	81.6	0.316	0.0421	1
Ammonia	-33.33	0.197	0.0262	2
Aniline	184.1	0.378	0.0504	1
Anisole	153.6	0.367	0.0489	1
Benzaldehyde	178.7	0.392	0.0523	1
Benzene	80.08	0.321	0.0427	2
Bromine	58.8	0.300	0.0400	1
1,3-Butadiene	-4.6	0.254	0.0339	1
Butane	-0.5	0.260	0.0346	2
1-Butanol	117.6	0.286	0.0381	2
2-Butanone	79.6	0.309	0.0412	1
Carbon disulfide	46.2	0.304	0.0405	1
Chlorine	-34.04	0.224	0.0299	2
Chlorobenzene	131.6	0.365	0.0487	1
1-Chlorobutane	78.4	0.321	0.0428	1
Chloroethane	12.3	0.262	0.0349	1
Chloroethene	-13.8	0.241	0.0321	1
o-Cresol	191.0	0.381	0.0508	1
m-Cresol	202.2	0.372	0.0496	1
p-Cresol	201.9	0.372	0.0496	1
Cyclohexane	80.7	0.329	0.0438	2
Cyclohexanol	160.9	0.344	0.0459	1
Cyclohexanone	155.4	0.382	0.0509	1
Decane	174.1	0.388	0.0518	2
Dibutyl ether	141.6	0.363	0.0484	1
Dichloromethane	39.8	0.282	0.0377	2
Diethyl ether	34.4	0.278	0.0370	2
Dimethyl sulfoxide	191.9	0.379	0.0505	1
1,4-Dioxane	101.2	0.321	0.0428	1
Dipropyl ether	90.1	0.326	0.0435	1
1,2-Ethanediol	197.5	0.333	0.0444	2
Ethanol	78.24	0.250	0.0334	2
Ethyl acetate	77.1	0.300	0.0400	1
Heptane	98.38	0.336	0.0448	2
Hexafluorobenzene	80.2	0.305	0.0407	1
Hexane	68.72	0.314	0.0419	2
1-Hexanol	156.9	0.318	0.0424	1
Hydrogen fluoride	20	0.276	0.0368	1
Iodomethane	42.4	0.291	0.0388	1
Isobutane	-11.7	0.252	0.0336	2
Isopropylbenzene	152.4	0.381	0.0508	1
Methanol	64.5	0.251	0.0334	2
Methyl acetate	56.7	0.282	0.0376	1
N-Methylaniline	197	0.396	0.0528	1
N-Methylformamide	186	0.371	0.0495	1
Methyl formate	31.6	0.582	0.0776	1
Nitrobenzene	210.7	0.418	0.0557	1
Nitromethane	101.19	0.320	0.0427	1
1-Nonanol	213.7	0.346	0.0461	1
Pentane	36.06	0.289	0.0386	2
1-Pentanol	137.6	0.296	0.0395	1
Phenol	181.8	0.349	0.0465	1
Propane	-42.11	0.224	0.0298	2
1-Propanol	97.04	0.261	0.0348	1

Name	t_b/°C	$\Delta t/\Delta p$/ °C kPa^{-1}	$\Delta t/\Delta p$/ °C mmHg^{-1}	Ref.
2-Propanol	82.21	0.247	0.0329	2
Pyridine	115.2	0.340	0.0453	1
Pyrrole	129.74	0.330	0.0440	1
Pyrrolidine	86.6	0.309	0.0412	1
Styrene	145.3	0.369	0.0492	1
Sulfur dioxide	-10.02	0.221	0.0294	2
Tetrachloroethene	121.2	0.354	0.0472	1
Tetrachloromethane	76.7	0.325	0.0433	1
Toluene	110.60	0.348	0.0463	2
Trichloroethene	86.8	0.330	0.0440	1
Trichloromethane	61.2	0.302	0.0403	1
Trimethylamine	2.8	0.248	0.0331	1
Water	99.974	0.277	0.0369	2
o-Xylene	144.4	0.373	0.0497	2
m-Xylene	139.1	0.368	0.0490	2
p-Xylene	138.3	0.369	0.0492	2

Lab Data

EBULLIOSCOPIC CONSTANTS FOR CALCULATION OF BOILING POINT ELEVATION

The boiling point T_b of a dilute solution of a non-volatile, non-dissociating solute is elevated relative to that of the pure solvent. If the solution is ideal (i.e., follows Raoult's Law), the amount of elevation depends only on the number of particles of solute present. Hence the change in boiling point ΔT_b can be expressed as

$$\Delta T_b = E_b m_2$$

where m_2 is the molality (moles of solute per kilogram of solvent) and E_b is the Ebullioscopic Constant, a characteristic property of the solvent. The Ebullioscopic Constant may be calculated from the relation

$$E_b = RT_b^2 M/\Delta_{vap}H$$

where R is the molar gas constant, T_b is the normal boiling point temperature (absolute) of the solvent, M the molar mass of the solvent, and $\Delta_{vap}H$ the molar enthalpy (heat) of vaporization of the solvent at its normal boiling point.

This table lists E_b values for some common solvents, as calculated from data in the table "Enthalpy of Vaporization" in Section 6. Column definitions for the table are as follows.

Column heading	Definition
Name	Name of liquid
Mol. form.	Molecular formula of liquid
t_b	Normal boiling point, in °C
E_b	Ebullioscopic constant, in units K kg mol^{-1}

Ebullioscopic Constants for Calculation of Boiling Point Elevation

Name	Mol. form.	t_b/°C	E_b/K kg mol^{-1}
Acetic acid	$C_2H_4O_2$	117.9	3.22
Acetone	C_3H_6O	56.08	1.80
Acetonitrile	C_2H_3N	81.6	1.44
Aniline	C_6H_7N	184.1	3.82
Anisole	C_7H_8O	153.6	4.20
Benzaldehyde	C_7H_6O	178.7	4.24
Benzene	C_6H_6	80.08	2.64
1-Butanol	$C_4H_{10}O$	117.6	2.17
Carbon disulfide	CS_2	46.2	2.42
Chlorobenzene	C_6H_5Cl	131.6	4.36
1-Chlorobutane	C_4H_9Cl	78.4	3.13
Cyclohexane	C_6H_{12}	80.7	2.92
Cyclohexanol	$C_6H_{12}O$	160.9	3.5
Decane	$C_{10}H_{22}$	174.1	6.10
Dichloromethane	CH_2Cl_2	39.8	2.42
Diethyl ether	$C_4H_{10}O$	34.4	2.20
Dimethyl sulfoxide	C_2H_6OS	191.9	3.22
1,4-Dioxane	$C_4H_8O_2$	101.2	3.01
1,2-Ethanediol	$C_2H_6O_2$	197.5	2.26
Ethanol	C_2H_6O	78.24	1.23
Ethyl acetate	$C_4H_8O_2$	77.1	2.82
Heptane	C_7H_{16}	98.38	3.62
Hexane	C_6H_{14}	68.72	2.90
Iodomethane	CH_3I	42.4	4.31
Methanol	CH_4O	64.5	0.86
Methyl acetate	$C_3H_6O_2$	56.7	2.21
N-Methylaniline	C_7H_9N	197	4.3
N-Methylformamide	C_2H_5NO	186	2.2
Nitrobenzene	$C_6H_5NO_2$	210.7	5.2
Nitromethane	CH_3NO_2	101.19	2.09
1-Octanol	$C_8H_{18}O$	194.7	5.06
Phenol	C_6H_6O	181.8	3.54
1-Propanol	C_3H_8O	97.04	1.66
2-Propanol	C_3H_8O	82.21	1.58
Pyridine	C_5H_5N	115.2	2.83
Pyrrole	C_4H_5N	129.74	2.33
Pyrrolidine	C_4H_9N	86.6	2.32
Tetrachloroethene	C_2Cl_4	121.2	6.18
Tetrachloromethane	CCl_4	76.7	5.26
Toluene	C_7H_8	110.60	3.40
Trichloroethene	C_2HCl_3	86.8	4.52
Trichloromethane	$CHCl_3$	61.2	3.80
Water	H_2O	99.974	0.513
o-Xylene	C_8H_{10}	144.4	4.25

CRYOSCOPIC CONSTANTS FOR CALCULATION OF FREEZING POINT DEPRESSION

The freezing point T_f of a dilute solution of a non-volatile, non-dissociating solute is depressed relative to that of the pure solvent. If the solution is ideal (i.e., follows Raoult's Law), this lowering is a function only of the number of particles of solute present. Thus, the absolute value of the lowering of freezing point ΔT_f can be expressed as

$$\Delta T_f = E_f m_2$$

where m_2 is the molality (moles of solute per kilogram of solvent) and E_f is the Cryoscopic Constant, a characteristic property of the solvent. The Cryoscopic Constant may be calculated from the relation

$$E_f = RT_f^2 M / \Delta_{fus}H$$

where R is the molar gas constant, T_f is the freezing point temperature (absolute) of the solvent, M the molar mass of the solvent, and $\Delta_{fus}H$ the molar enthalpy (heat) of fusion of the solvent.

This table lists cryoscopic constants for selected substances, as calculated from data in the table "Enthalpy of Fusion" in Section 6. Column definitions for the table are as follows.

Column heading	Definition
Name	Name of solute
Mol. form.	Molecular formula of solute
t_f	Normal freezing point of solute, in °C
E_f	Cyoscopic constant, in units K kg mol^{-1}

Cryoscopic Constants for Common Reagents

Name	Mol. form.	t_f/°C	E_f/K kg mol^{-1}
Acetamide	C_2H_5NO	80.16	3.92
Acetic acid	$C_2H_4O_2$	17	3.63
Acetophenone	C_8H_8O	19.4	5.16
Aniline	C_6H_7N	-6.0	5.23
Benzene	C_6H_6	5.538	5.07
Benzonitrile	C_7H_5N	-12.82	5.35
Benzophenone	$C_{13}H_{10}O$	48.0	8.58
Camphor, (+)	$C_{10}H_{16}O$	178.7	37.8
1-Chloronaphthalene	$C_{10}H_7Cl$	-6.0	7.68
o-Cresol	C_7H_8O	31.0	5.92
m-Cresol	C_7H_8O	12.2	7.76
p-Cresol	C_7H_8O	34.77	7.20
Cyclohexane	C_6H_{12}	6.7	20.8
Cyclohexanol	$C_6H_{12}O$	26	42.2
cis-Decahydronaphthalene	$C_{10}H_{18}$	-42.9	6.42
trans-Decahydronaphthalene	$C_{10}H_{18}$	-30.35	4.70
Dibenzyl ether	$C_{14}H_{14}O$	1.8	6.17
p-Dichlorobenzene	$C_6H_4Cl_2$	53.1	7.57
Diethanolamine	$C_4H_{11}NO_2$	27.9	3.16
Dimethyl sulfoxide	C_2H_6OS	18.52	3.85
1,4-Dioxane	$C_4H_8O_2$	11.75	4.63
Diphenylamine	$C_{12}H_{11}N$	53.2	8.38
1,2-Ethanediol	$C_2H_6O_2$	-13	3.11
Formamide	CH_3NO	2.57	4.25
Formic acid	CH_2O_2	8.3	2.38
Glycerol	$C_3H_8O_3$	18.2	3.56
4-Methylaniline	C_7H_9N	43.3	4.91
Methylcyclohexane	C_7H_{14}	-126.6	2.60
Naphthalene	$C_{10}H_8$	80.22	7.45
Nitrobenzene	$C_6H_5NO_2$	5.65	6.87
Phenol	C_6H_6O	40.89	6.84
Pyridine	C_5H_5N	-41.63	4.26
Quinoline	C_9H_7N	-14.78	6.73
Succinonitrile	$C_4H_4N_2$	57.985	19.3
1,1,2,2-Tetrabromoethane	$C_2H_2Br_4$	0	21.4
1,1,2,2-Tetrachloro-1,2-difluoroethane	$C_2Cl_4F_2$	26.54	41.0
Toluene	C_7H_8	-95.0	3.55
Tribromomethane	$CHBr_3$	8.69	15.0
Water	H_2O	0.00	1.86
p-Xylene	C_8H_{10}	13.3	4.31

FREEZING POINT LOWERING BY ELECTROLYTES IN AQUEOUS SOLUTION

The freezing point of a solution of water and a non-volatile solute, such as an inorganic salt, is depressed relative to that of pure water. While for ideal solutions, the freezing point depression is linear with molality, for non-ideal solutions, deviations from linearity are observed. This table contains the freezing point lowering by selected electrolytes in water as a function of molality. The lowered freezing point is in °C.

Reference

Forsythe, W. E., *Smithsonian Physical Tables, Ninth Edition*, Smithsonian Institution, Washington, DC, 1956.

Lowering of Freezing Point (in °C) by Electrolytes in Water as a Function of Molality *m* (in mol kg^{-1})

Mol. form.	Name	0.05 m	0.10 m	0.25 m	0.50 m	0.75 m	1.00 m	1.50 m	2.00 m	2.50 m	3.00 m
$CaCl_2$	Calcium chloride	0.25	0.49	1.27	2.66	4.28	6.35	10.78	15.27	20.42	28.08
$CuSO_4$	Copper(II) sulfate	0.13	0.23	0.47	0.96						
HCl	Hydrogen chloride	0.18	0.36	0.90	1.86	2.90	4.02	6.63	9.94		
HNO_3	Nitric acid	0.18	0.35	0.88	1.80	2.78	3.80	5.98	8.34	10.95	13.92
H_2SO_4	Sulfuric acid	0.20	0.39	0.96	1.95	3.04	4.28	7.35	11.35	16.32	
KBr	Potassium bromide	0.18	0.36	0.92	1.78						
KCl	Potassium chloride	0.17	0.35	0.86	1.68	2.49	3.29	4.88	6.50	8.14	9.77
KNO_3	Potassium nitrate	0.17	0.33	0.78	1.47	2.11	2.66				
K_2SO_4	Potassium sulfate	0.23	0.43	1.01	1.87						
LiCl	Lithium chloride	0.18	0.35	0.88	1.80	2.78					
$MgSO_4$	Magnesium sulfate	0.13	0.24	0.55	1.01	1.50	2.08	3.41			
NH_4Cl	Ammonium chloride	0.17	0.34	0.85	1.70	2.55					
NaCl	Sodium chloride	0.18	0.35	0.85	1.68	2.60					
$NaNO_3$	Sodium nitrate	0.18	0.36	0.80	1.62	2.63	3.10				

CORRECTION OF BAROMETER READINGS TO 0 °C TEMPERATURE

The following corrections are used to reduce the reading of a mercury barometer with a brass scale to 0 °C. The number in the table should be subtracted from the observed height of the mercury column to give the true pressure in mmHg (1 mmHg = 133.322 Pa). The table is calculated from the formula

$$\Delta h = -0.0001634\, ht/(1+0.0001818\, t)$$

where h is the observed column height in mm and t the Celsius temperature. This relation is based on thermal expansion coefficients of $181.8 \cdot 10^{-6}$ °C^{-1} for mercury and $18.4 \cdot 10^{-6}$ °C^{-1} for brass.

Correction in mm to Be Applied to the Indicated Observed Height in mm

t/°C	620	630	640	650	660	670	680	690	700	710	720	730	740	750	760	770	780	790	800
0	0.00	0.00	0.00	0.00	0.00	0.00	0.00	0.00	0.00	0.00	0.00	0.00	0.00	0.00	0.00	0.00	0.00	0.00	0.00
1	0.10	0.10	0.10	0.11	0.11	0.11	0.11	0.11	0.11	0.12	0.12	0.12	0.12	0.12	0.12	0.13	0.13	0.13	0.13
2	0.20	0.21	0.21	0.21	0.22	0.22	0.22	0.23	0.23	0.23	0.24	0.24	0.24	0.25	0.25	0.25	0.25	0.26	0.26
3	0.30	0.31	0.31	0.32	0.32	0.33	0.33	0.34	0.34	0.35	0.35	0.36	0.36	0.37	0.37	0.38	0.38	0.39	0.39
4	0.40	0.41	0.42	0.42	0.43	0.44	0.44	0.45	0.46	0.46	0.47	0.48	0.48	0.49	0.50	0.50	0.51	0.52	0.52
5	0.51	0.51	0.52	0.53	0.54	0.55	0.56	0.56	0.57	0.58	0.59	0.60	0.60	0.61	0.62	0.63	0.64	0.64	0.65
6	0.61	0.62	0.63	0.64	0.65	0.66	0.67	0.68	0.69	0.70	0.71	0.71	0.72	0.73	0.74	0.75	0.76	0.77	0.78
7	0.71	0.72	0.73	0.74	0.75	0.77	0.78	0.79	0.80	0.81	0.82	0.83	0.85	0.86	0.87	0.88	0.89	0.90	0.91
8	0.81	0.82	0.84	0.85	0.86	0.87	0.89	0.90	0.91	0.93	0.94	0.95	0.97	0.98	0.99	1.01	1.02	1.03	1.04
9	0.91	0.92	0.94	0.95	0.97	0.98	1.00	1.01	1.03	1.04	1.06	1.07	1.09	1.10	1.12	1.13	1.15	1.16	1.17
10	1.01	1.03	1.04	1.06	1.08	1.09	1.11	1.13	1.14	1.16	1.17	1.19	1.21	1.22	1.24	1.26	1.27	1.29	1.30
11	1.11	1.13	1.15	1.17	1.18	1.20	1.22	1.24	1.26	1.27	1.29	1.31	1.33	1.35	1.36	1.38	1.40	1.42	1.44
12	1.21	1.23	1.25	1.27	1.29	1.31	1.33	1.35	1.37	1.39	1.41	1.43	1.45	1.47	1.49	1.51	1.53	1.55	1.57
13	1.31	1.34	1.36	1.38	1.40	1.42	1.44	1.46	1.48	1.50	1.53	1.55	1.57	1.59	1.61	1.63	1.65	1.67	1.70
14	1.41	1.44	1.46	1.48	1.51	1.53	1.55	1.57	1.60	1.62	1.64	1.67	1.69	1.71	1.73	1.76	1.78	1.80	1.83
15	1.52	1.54	1.56	1.59	1.61	1.64	1.66	1.69	1.71	1.74	1.76	1.78	1.81	1.83	1.86	1.88	1.91	1.93	1.96
16	1.62	1.64	1.67	1.69	1.72	1.75	1.77	1.80	1.82	1.85	1.88	1.90	1.93	1.96	1.98	2.01	2.03	2.06	2.09
17	1.72	1.74	1.77	1.80	1.83	1.86	1.88	1.91	1.94	1.97	1.99	2.02	2.05	2.08	2.10	2.13	2.16	2.19	2.22
18	1.82	1.85	1.88	1.91	1.93	1.96	1.99	2.02	2.05	2.08	2.11	2.14	2.17	2.20	2.23	2.26	2.29	2.32	2.35
19	1.92	1.95	1.98	2.01	2.04	2.07	2.10	2.13	2.17	2.20	2.23	2.26	2.29	2.32	2.35	2.38	2.41	2.44	2.48
20	2.02	2.05	2.08	2.12	2.15	2.18	2.21	2.25	2.28	2.31	2.34	2.38	2.41	2.44	2.47	2.51	2.54	2.57	2.60
21	2.12	2.15	2.19	2.22	2.26	2.29	2.32	2.36	2.39	2.43	2.46	2.50	2.53	2.56	2.60	2.63	2.67	2.70	2.73
22	2.22	2.26	2.29	2.33	2.36	2.40	2.43	2.47	2.51	2.54	2.58	2.61	2.65	2.69	2.72	2.76	2.79	2.83	2.86
23	2.32	2.36	2.40	2.43	2.47	2.51	2.54	2.58	2.62	2.66	2.69	2.73	2.77	2.81	2.84	2.88	2.92	2.96	2.99
24	2.42	2.46	2.50	2.54	2.58	2.62	2.66	2.69	2.73	2.77	2.81	2.85	2.89	2.93	2.97	3.01	3.05	3.08	3.12
25	2.52	2.56	2.60	2.64	2.68	2.72	2.77	2.81	2.85	2.89	2.93	2.97	3.01	3.05	3.09	3.13	3.17	3.21	3.25
26	2.62	2.66	2.71	2.75	2.79	2.83	2.88	2.92	2.96	3.00	3.04	3.09	3.13	3.17	3.21	3.26	3.30	3.34	3.38
27	2.72	2.77	2.81	2.85	2.90	2.94	2.99	3.03	3.07	3.12	3.16	3.20	3.25	3.29	3.34	3.38	3.42	3.47	3.51
28	2.82	2.87	2.91	2.96	3.00	3.05	3.10	3.14	3.19	3.23	3.28	3.32	3.37	3.41	3.46	3.51	3.55	3.60	3.64
29	2.92	2.97	3.02	3.06	3.11	3.16	3.21	3.25	3.30	3.35	3.39	3.44	3.49	3.54	3.58	3.63	3.68	3.72	3.77
30	3.02	3.07	3.12	3.17	3.22	3.27	3.32	3.36	3.41	3.46	3.51	3.56	3.61	3.66	3.71	3.75	3.80	3.85	3.90
31	3.12	3.17	3.22	3.27	3.32	3.37	3.43	3.48	3.53	3.58	3.63	3.68	3.73	3.78	3.83	3.88	3.93	3.98	4.03
32	3.22	3.28	3.33	3.38	3.43	3.48	3.54	3.59	3.64	3.69	3.74	3.79	3.85	3.90	3.95	4.00	4.05	4.11	4.16
33	3.32	3.38	3.43	3.48	3.54	3.59	3.64	3.70	3.75	3.81	3.86	3.91	3.97	4.02	4.07	4.13	4.18	4.23	4.29
34	3.42	3.48	3.53	3.59	3.64	3.70	3.75	3.81	3.87	3.92	3.98	4.03	4.09	4.14	4.20	4.25	4.31	4.36	4.42
35	3.52	3.58	3.64	3.69	3.75	3.81	3.86	3.92	3.98	4.03	4.09	4.15	4.21	4.26	4.32	4.38	4.43	4.49	4.55
36	3.62	3.68	3.74	3.80	3.86	3.92	3.97	4.03	4.09	4.15	4.21	4.27	4.32	4.38	4.44	4.50	4.56	4.62	4.68
37	3.72	3.78	3.84	3.90	3.96	4.02	4.08	4.14	4.20	4.26	4.32	4.38	4.44	4.50	4.56	4.62	4.68	4.74	4.80
38	3.82	3.88	3.95	4.01	4.07	4.13	4.19	4.25	4.32	4.38	4.44	4.50	4.56	4.62	4.69	4.75	4.81	4.87	4.93
39	3.92	3.99	4.05	4.11	4.18	4.24	4.30	4.37	4.43	4.49	4.56	4.62	4.68	4.75	4.81	4.87	4.94	5.00	5.06
40	4.02	4.09	4.15	4.22	4.28	4.35	4.41	4.48	4.54	4.61	4.67	4.74	4.80	4.87	4.93	5.00	5.06	5.13	5.19

DETERMINATION OF RELATIVE HUMIDITY FROM DEW POINT

The relative humidity of a water vapor–air mixture is defined as 100 times the partial pressure of water divided by the saturation vapor pressure of water at the same temperature. The relative humidity may be determined from the dew point t_{dew}, which is the temperature at which liquid water first condenses when the mixture is cooled from an initial temperature t. This table gives relative humidity as a function of the dew point depression $t - t_{dew}$ for several values of the dew point. Values are calculated from the table "Vapor Pressure" in Section 6.

Relative Humidity as a Function of Ambient Temperature t and Dew Point t_{dew}

$(t-t_{dew})$/°C	t_{dew} = −10 °C	t_{dew} = 0 °C	t_{dew} = 10 °C	t_{dew} = 20 °C	t_{dew} = 30 °C
0.0	100	100	100	100	100
0.2	99	99	99	99	99
0.4	97	97	97	98	98
0.6	95	96	96	96	97
0.9	94	94	95	95	96
1.0	92	93	94	94	94
1.2	91	92	92	93	93
1.4	90	90	91	92	92
1.6	88	89	90	91	91
1.8	87	88	89	90	90
2.0	86	87	88	88	89
2.2	84	85	86	87	89
2.4	83	84	85	86	87
2.6	82	83	84	85	86
2.8	80	82	83	84	85
3.0	79	81	82	83	84
3.2	78	80	81	82	83
3.4	77	79	80	81	82
3.6	76	77	79	80	82
3.8	75	76	78	79	81
4.0	73	75	77	78	80
4.2	72	74	76	77	79
4.4	71	73	75	77	78
4.6	70	72	74	76	77
4.8	69	71	73	75	76
5.0	69	70	72	74	75
5.2	67	69	71	73	75
5.4	66	68	70	72	74
5.6	65	67	69	71	73
5.9	64	66	69	70	72
6.0	63	66	68	70	71
6.2	62	65	67	69	71
6.4	61	64	66	68	70
6.6	60	63	65	67	69
6.8	60	62	64	66	68
7.0	59	61	63	66	68
7.2	58	60	63	65	67
7.4	57	60	62	64	66
7.6	56	59	61	63	65
7.8	55	58	60	63	65
8.0	54	57	60	62	64
8.2	54	56	59	61	63
8.4	53	56	58	60	63
8.6	52	55	57	60	62
8.8	51	54	57	59	61
9.0	51	53	56	58	61
9.2	50	53	55	58	60
9.4	49	52	55	57	59
9.6	48	51	54	56	59
9.8	48	51	53	56	58
10.0	47	50	53	55	57
10.5	45	48	51	54	56
11.0	44	47	49	52	55
11.5	42	45	48	51	53
12.0	41	44	47	49	52
12.5	39	42	45	48	50
13.0	38	41	44	46	49
13.5	37	40	43	45	48
14.0	35	38	41	44	47
14.5	34	37	40	43	45
15.0	33	36	39	42	44
15.5	32	35	38	40	
16.0	31	34	37	39	
16.5	30	33	36	38	
17.0	29	32	35	37	
17.5	28	31	34	36	
18.0	27	30	33	35	
18.5	26	29	32	34	
19.0	25	28	31	33	
19.5	24	27	30	33	
20.0	24	26	29	32	
21.0	22	25	27	30	
22.0	21	23	26	29	
23.0	19	22	24	27	
24.0	18	21	23	26	
25.0	17	19	22	24	
26.0	16	18	21	23	
27.0	15	17	20	22	
28.0	14	16	19	21	
29.0	13	15	18	20	
30.0	12	14	17	19	

DETERMINATION OF RELATIVE HUMIDITY FROM WET AND DRY BULB TEMPERATURES

Relative humidity may be determined by comparing temperature readings of wet and dry bulb thermometers. The following table, extracted from more extensive U.S. National Weather Service tables, gives the relative humidity as a function of air temperature t_d (dry bulb) and the difference $t_d - t_w$ between dry and wet bulb temperatures. The data assume a pressure near normal atmospheric pressure and an instrumental configuration with forced ventilation.

	$(t_d - t_w)$/°C											
t_d/°C	**0.5**	**1.0**	**1.5**	**2.0**	**2.5**	**3.0**	**3.5**	**4.0**	**4.5**	**5.0**	**5.5**	**6.0**
−10	83	67	51	35	19							
−8	86	71	57	43	29	15						
−6	88	74	61	49	37	25	8					
−4	89	77	66	55	44	33	23	12				
−2	90	79	69	60	50	40	31	22	12			
0	91	81	72	64	55	46	38	29	21	13	5	
2	91	84	76	68	60	52	44	37	29	22	14	7
4	92	85	78	71	63	57	49	43	36	29	22	16
6	93	86	79	73	66	60	54	48	41	35	29	24
8	93	87	81	75	69	63	57	51	46	40	35	29
10	94	88	82	77	71	66	60	55	50	44	39	34
12	94	89	83	78	73	68	63	58	53	48	43	39
14	95	90	85	79	75	70	65	60	56	51	47	42
16	95	90	85	81	76	71	67	63	58	54	50	46
18	95	91	86	82	77	73	69	65	61	57	53	49
20	96	91	87	83	78	74	70	66	63	59	55	51
22	96	92	87	83	80	76	72	68	64	61	57	54
24	96	92	88	84	80	77	73	69	66	62	59	56
26	96	92	88	85	81	78	74	71	67	64	61	58
28	96	93	89	85	82	78	75	72	69	65	62	59
30	96	93	89	86	83	79	76	73	70	67	64	61
35	97	94	90	87	84	81	78	75	72	69	67	64
40	97	94	91	88	85	82	80	77	74	72	69	67

	$(t_d - t_w)$/°C											
t_d/°C	**6.5**	**7.0**	**7.5**	**8.0**	**8.5**	**9.0**	**10.0**	**11.0**	**12.0**	**13.0**	**14.0**	**15.0**
4	9											
6	17	11	5									
8	24	19	14	8								
10	29	24	20	15	10	6						
12	34	29	25	21	16	12	5					
14	38	34	30	26	22	18	10					
16	42	38	34	30	26	23	15	8				
18	45	41	38	34	30	27	20	14	7			
20	48	44	41	37	34	31	24	18	12	6		
22	50	47	44	40	37	34	28	22	17	11	6	
24	53	49	46	43	40	37	31	26	20	15	10	5
26	54	51	49	46	43	40	34	29	24	19	14	10
28	56	53	51	48	45	42	37	32	27	22	18	13
30	58	55	52	50	47	44	39	35	30	25	21	17
32	60	57	54	51	49	46	41	37	32	28	24	20
34	61	58	56	53	51	48	43	39	35	30	26	23
36	62	59	57	54	52	50	45	41	37	33	29	25
38	63	61	58	56	54	51	47	43	39	35	31	27
40	64	62	59	57	54	53	48	44	40	36	33	29

CONSTANT HUMIDITY SOLUTIONS

Anthony Wexler

An excess of a water soluble salt in contact with its saturated solution and contained within an enclosed space produces a constant relative humidity and water vapor pressure according to

$$RH = A \exp(B/T)$$

where RH is the percent relative humidity (generally accurate to ±2%), T is the temperature in kelvin, and the constants A and B and the range of valid temperatures are given in the table below. The vapor pressure, p, can be calculated from

$$p = (RH/100) \times p_0$$

where p_0 is the vapor pressure of pure water at temperature T as given in the table in Section 6 titled "Vapor Pressure and Other Saturation Properties of Water."

References

1. Wexler, A. S. and Seinfeld, J. H., *Atmospheric Environment*, 25A, 2731, 1991. <https://doi.org/10.1016/0960-1686(91)90203-J>
2. Greenspan, L., *J. Res. National Bureau of Standards*, 81A, 89, 1977. <https://doi.org/10.6028/jres.081A.011>
3. Broul, et al., *Solubility of Inorganic Two-Component Systems*, Elsevier, New York, 1981.
4. Wagman, D. D. et al., *J. Phys. Chem. Ref. Data,* Vol. 11, Suppl. 2, 1982.

Coefficients in the Equation for Relative Humidity above Aqueous Salt Solutions

Mol. form.	Name	Temperature range (°C)	A	B	RH (25 °C)
$NaOH \cdot H_2O$	Sodium hydroxide monohydrate	15–60	5.48	27	6
$LiBr \cdot 2H_2O$	Lithium bromide dihydrate	10–30	0.23	996	6
$ZnBr_2 \cdot 2H_2O$	Zinc bromide dihydrate	5–30	1.69	455	8
$KOH \cdot 2H_2O$	Potassium hydroxide dihydrate	5–30	0.014	1924	9
$LiCl \cdot H_2O$	Lithium chloride monohydrate	20–65	14.53	-75	11
$CaBr_2 \cdot 6H_2O$	Calcium bromide hexahydrate	11–22	0.17	1360	16
$LiI \cdot 3H_2O$	Lithium iodide trihydrate	15–65	0.15	1424	18
$CaCl_2 \cdot 6H_2O$	Calcium chloride hexahydrate	15–25	0.11	1653	29
$MgCl_2 \cdot 6H_2O$	Magnesium chloride hexahydrate	5–45	29.26	34	33
$NaI \cdot 2H_2O$	Sodium iodide dihydrate	5–45	3.62	702	38
$Ca(NO_3)_2 \cdot 4H_2O$	Calcium nitrate tetrahydrate	10–30	1.89	981	51
$Mg(NO_3)_2 \cdot 6H_2O$	Magnesium nitrate hexahydrate	5–35	25.28	220	53
$NaBr \cdot 2H_2O$	Sodium bromide dihydrate	0–35	20.49	308	58
NH_4NO_3	Ammonium nitrate	10–40	3.54	853	62
KI	Potassium iodide	5–30	29.35	254	69
$SrCl_2 \cdot 6H_2O$	Strontium chloride hexahydrate	5–30	31.58	241	71
$NaNO_3$	Sodium nitrate	10–40	26.94	302	74
NaCl	Sodium chloride	10–40	69.20	25	75
NH_4Cl	Ammonium chloride	10–40	35.67	235	79
KBr	Potassium bromide	5–25	40.98	203	81
$(NH_4)_2SO_4$	Ammonium sulfate	10–40	62.06	79	81
KCl	Potassium chloride	5–25	49.38	159	84
$Sr(NO_3)_2 \cdot 4H_2O$	Strontium nitrate tetrahydrate	5–25	28.34	328	85
$BaCl_2 \cdot 2H_2O$	Barium chloride dihydrate	5–25	69.99	75	90
CsI	Cesium iodide	5–25	70.77	75	91
KNO_3	Potassium nitrate	0–50	43.22	225	92
K_2SO_4	Potassium sulfate	10–50	86.75	34	97

STANDARD SALT SOLUTIONS FOR HUMIDITY CALIBRATION

Saturated aqueous solutions of inorganic salts are convenient secondary standards for calibration of instruments for measurement of relative humidity. The International Union of Pure and Applied Chemistry has recommended salt solutions for calibrations in the range of 10% to 90% relative humidity, and the American Society for Testing and Materials has published similar standards. The data in this table are taken from the IUPAC recommendations, except for K_2CO_3 and K_2SO_4, which are ASTM recommendations.

Details on the preparation and use of these standards may be found in Refs. 1 and 2. Data for other salts are given in Ref. 3.

References

1. Marsh, K. N., Editor, *Recommended Reference Materials for the Realization of Physicochemical Properties*, Blackwell Scientific Publications, Oxford, 1987, pp. 157–162.
2. *Standard Practice for Maintaining Constant Relative Humidity by Means of Aqueous Solutions*, ASTM Standard E 104-85, Reapproved 1991.
3. Greenspan, L., *J. Res. Nat. Bur. Stand.*, 81A, 89, 1977. <https://doi.org/10.6028/jres.081A.011>

Relative Humidity in % for Humidity Calibration with Standard Salt Solutions as a Function of Temperature

Mol. form.	0 °C	10 °C	20 °C	30 °C	40 °C	50 °C	60 °C	70 °C	80 °C
LiCl			11.31 ± 0.31	11.28 ± 0.24	11.21 ± 0.21	11.10 ± 0.22	10.95 ± 0.26	10.75 ± 0.33	10.51 ± 0.44
$MgCl_2$	33.66 ± 0.33	33.47 ± 0.24	33.07 ± 0.18	32.44 ± 0.14	31.60 ± 0.13	30.54 ± 0.14	29.26 ± 0.18	27.77 ± 0.25	26.05 ± 0.34
K_2CO_3	43.1 ± 0.7	43.1 ± 0.4	43.2 ± 0.3	43.2 ± 0.5					
$Mg(NO_3)_2$	60.35 ± 0.55	57.36 ± 0.33	54.38 ± 0.23	51.40 ± 0.24	48.42 ± 0.37	45.44 ± 0.60			
NaCl	75.51 ± 0.34	75.67 ± 0.22	75.47 ± 0.14	75.09 ± 0.11					
KCl	88.61 ± 0.53	86.77 ± 0.39	85.11 ± 0.29	83.62 ± 0.25	82.32 ± 0.25	81.20 ± 0.31	80.25 ± 0.41	79.49 ± 0.57	78.90 ± 0.77
K_2SO_4	98.8 ± 2.1	98.2 ± 0.8	97.6 ± 0.5	97.0 ± 0.4	96.4 ± 0.4	95.8 ± 0.5			

LOW-TEMPERATURE BATHS FOR MAINTAINING CONSTANT TEMPERATURE

A liquid–solid slurry is a convenient means of maintaining a constant temperature environment below room temperature. The following is a list of readily available organic liquids suitable for this purpose, arranged in order of their melting (freezing) points t_m. The normal boiling points t_b are also given.

Useful Baths for Maintaining Constant Temperature

Name	Comment	t_m/°C	t_b/°C
Isopentane	2-Methylbutane	-159.8	27.83
Methylcyclopentane		-142.4	71.8
3-Chloropropene	Allyl chloride	-136	44.8
Pentane		-129.7	36.06
Allyl alcohol	2-Propen-1-ol	-129	96.9
Ethanol		-114.1	78.24
Carbon disulfide		-111.7	46.2
2-Methyl-1-propanol	Isobutyl alcohol	-102.0	107.84
Toluene		-95.0	110.60
Acetone		-94.9	56.08
Ethyl acetate		-83.8	77.1
Carbon dioxide	Dry ice + acetone	-78.5	-78.464
1-Isopropyl-4-methylbenzene	*p*-Cymene	-68.1	177
Trichloromethane	Chloroform	-63.3	61.2
N-Methylaniline		-57	197
Chlorobenzene		-45.2	131.6
Anisole	Methoxybenzene	-37.3	153.6
Bromobenzene		-30.7	155.9
Tetrachloromethane	Carbon tetrachloride	-22.8	76.7
Benzonitrile		-12.8	191

METALS AND ALLOYS WITH LOW-MELTING TEMPERATURE

L. I. Berger

Low-melting metals and alloys are often used to join and cast parts near or below room temperature. These metals can liquify and resolidify at lower temperatures, making it easier to avoid parts being exposed to high temperatures during processing. The table has the melting points for a number of these metals and alloys, including tin-, gallium-, bismuth-, and indium-based alloys, as well as a few others. Composition information is also given.

References

1. Zintle, E. and Hauke, W., *Z. Electrochem.*, 44, 104, 1938.
2. Rinck, E., *Compt. Rend.*, 199, 1217, 1934.
3. Krier, C. A., Craign, R. S., and Wallace, W. E., *J. Phys. Chem.*, 61, 522, 1957. <https://doi.org/10.1021/j150551a004>
4. Goria, C., *Gazz. Chim. Ital.*, 65, 865, 1935.
5. Baker, H., Ed., *ASM Handbook, Volume 3: Alloy Phase Diagrams*, ASM Intl., Materials Park, OH, 1992.
6. Sedlacek, V., *Non-Ferrous Metals and Alloys*, Elsevier, 1986.
7. Villars, P., Prince, A., Okamoto, H., Eds., *Handbook of Ternary Alloy Phase Diagrams*, ASM Intl., 1994.
8. Palatnik, L. S., Kosevich, V. M., and Tyrina, L. V., *Phys. Metals Metallog. (USSR)*, 11, 75, 1961.
9. Neumann, T. and Alpout, O., *J. Less-Common Metals*, 6, 108, 1964. <https://doi.org/10.1016/0022-5088(64)90115-8>
10. Neumann, T. and Predel, B., *Z. Metallk.*, 50, 309, 1959.
11. Roy, P., Orr, R. L., and Hultgren, R., *J. Phys. Chem.*, 64, 1034, 1960. <https://doi.org/10.1021/j100837a017>
12. Dobovicek, B. and Smajic, N., *Rudarsko-Met. Zbornik*, 4, 353, 1962.
13. Massalski, T. B., Okamoto, H., Subramanian, P. R., and Kacprzak, L., Eds., *Binary Alloy Phase Diagrams, Second Edition*, ASM Intl., 1990.
14. Dobovicek, B. and Straus, B., *Rudarsko-Met. Zbornik*, 3, 273, 1960.
15. Schurmann, E. and Gilhaus, F. J., *Arch. Eisenhuettenw.*, 32, 867, 1961. <https://doi.org/10.1002/srin.196103281>
16. Rosenblatt, G. M. and Birchenall, C. E., *Trans. AIME*, 224, 481, 1962.
17. Evans, D. S. and Prince, A., in *Alloy Phase Diagrams*, MRS Simposia Proc., Vol. 19, North-Holland, 1983, p. 383. <https://doi.org/10.1557/PROC-19-383>
18. Umanskiy, M. M., *Zh. Fiz. Khim.*, 14, 846, 1940.
19. Homer, C. E. and Plummer, H., *J. Inst. Met.*, 64, 169, 1939.

Melting Temperature for Pure Metals and Alloys

Metal or alloy system	Composition in weight %	Composition in atom %	Melting temperature (°C)	Comments	Ref.
Hg	100	100	-38.829		
Cs–K	77.0–23.0	50.0–50.0	-37.5	Eutectic (?)	1
Cs–Na	94.5–5.5	75.0–25.0	-30.0	Eutectic	2
K–Na	76.7–23.3	65.9–34.1	-12.65	Eutectic	3
Na–Rb	8.0–92.0	24.4–75.6	-5	Eutectic	4
Ga–In–Sn	62.5–21.5–16.0	73.6–15.3–11.1	11	Eutectic	5
Ga–Sn–Zn	82.0–12.0–6.0	86.0–7.3–6.7	17	Eutectic	5
Cs	100	100	28.44		
Ga	100	100	29.7646		
K–Rb	32.0–68.0	50–50	33	Eutectic	4
Bi–Cd–In–Pb–Sn	44.7–5.3–19.1–22.6–8.3	35.1–8.2–27.3–17.9–11.5	46.7	Eutectic	6
Bi–In–Pb–Sn	49.5–21.3–17.6–11.6	39.2–30.7–14.0–16.2	58.2	Eutectic	6
Bi–In–Sn	32.5–51.0–16.5	21.1–60.1–18.8	60.5	Eutectic	7
K	100	100	63.38		
Bi–Cd–Pb–Sn	50.0–12.5–25.0–12.5	41.5–19.3–21.0–18.2	70	Wood's alloy	6
Bi–In	33.0–67.0	21.3–78.7	72	Eutectic	8
Bi–Cd–Pb	51.6–8.2–40.2	48.1–14.2–37.7	91.5	Eutectic	6
Bi–Pb–Sn	52.5–32.0–15.5	46.8–28.7–24.5	95	Eutectic	6
Na	100	100	97.8		
Bi–Cd–Sn	54.0–20.0–26.0	39.4–27.2–33.4	102.5	Eutectic	6
In–Sn	51.8–48.2	52.6–47.4	119	Eutectic	9
Cd–In	25.3–74.7	25.7–74.3	120	Eutectic	10
Bi–Pb	55.5–44.5	55.3–44.7	124	Eutectic	11
Bi–Sn–Zn	56.0–40.0–4.0	40.2–50.6–9.2	130	Eutectic	6, 7
Bi–Sn	70-30	57.0–43.0	138.5	Eutectic	6, 12
Bi–Cd	60.3–39.7	45.0–55.0	145.5	Eutectic	13, 14
In	100	100	156.6		
Li	100	100	180.5		
Pb–Sn	38.1–61.9	26.1–73.9	183	Eutectic	6,15
Bi–Tl	48.0–52.0	47.5–52.5	185	Eutectic	13
Sn–Zn	91.0–9.0	85.0–15.0	198	Eutectic	14

Lab Data

Metal or alloy system	Composition in weight %	Composition in atom %	Melting temperature (°C)	Comments	Ref.
Sb–Sn	8.0–92.0	7.8–92.2	199	White Metal	16
Au–Pb	14.6–85.4	15.2–84.8	212	Eutectic	17
Ag–Sn	3.5–96.5	3.8–96.2	221	Eutectic	13,18
Bi–Pb–Sb–Sn	48.0–28.5–9.0–14.5	40.8–24.5–13.1–21.6	226	Matrix Alloy	6
Cu–Sn	0.75–99.25	1.3–98.7	227	Eutectic	13, 19
Sn	100	100	231.928		

Useful expressions for correlations between the atomic and weight concentrations of alloy components are:

$$f(\mathrm{a}, A_k) = \frac{f(\mathrm{w}, A_k)}{M_k \sum_{i=1}^{N} \frac{f(\mathrm{w}, A_i)}{M_i}} \quad \text{and} \quad f(\mathrm{w}, A_k) = \frac{M_k \cdot f(\mathrm{a}, A_k)}{\sum_{i=1}^{N} M_i \cdot f(\mathrm{a}, A_i)} \quad (i = 1, \ldots, k, \ldots, N)$$

where $f(\mathrm{a}, A_i)$ and $f(\mathrm{w}, A_i)$ are the atomic and weight concentrations of component A_i, respectively, and M_i is the atomic weight of this component.

CHARACTERISTICS OF PARTICLES AND PARTICLE DISPERSOIDS

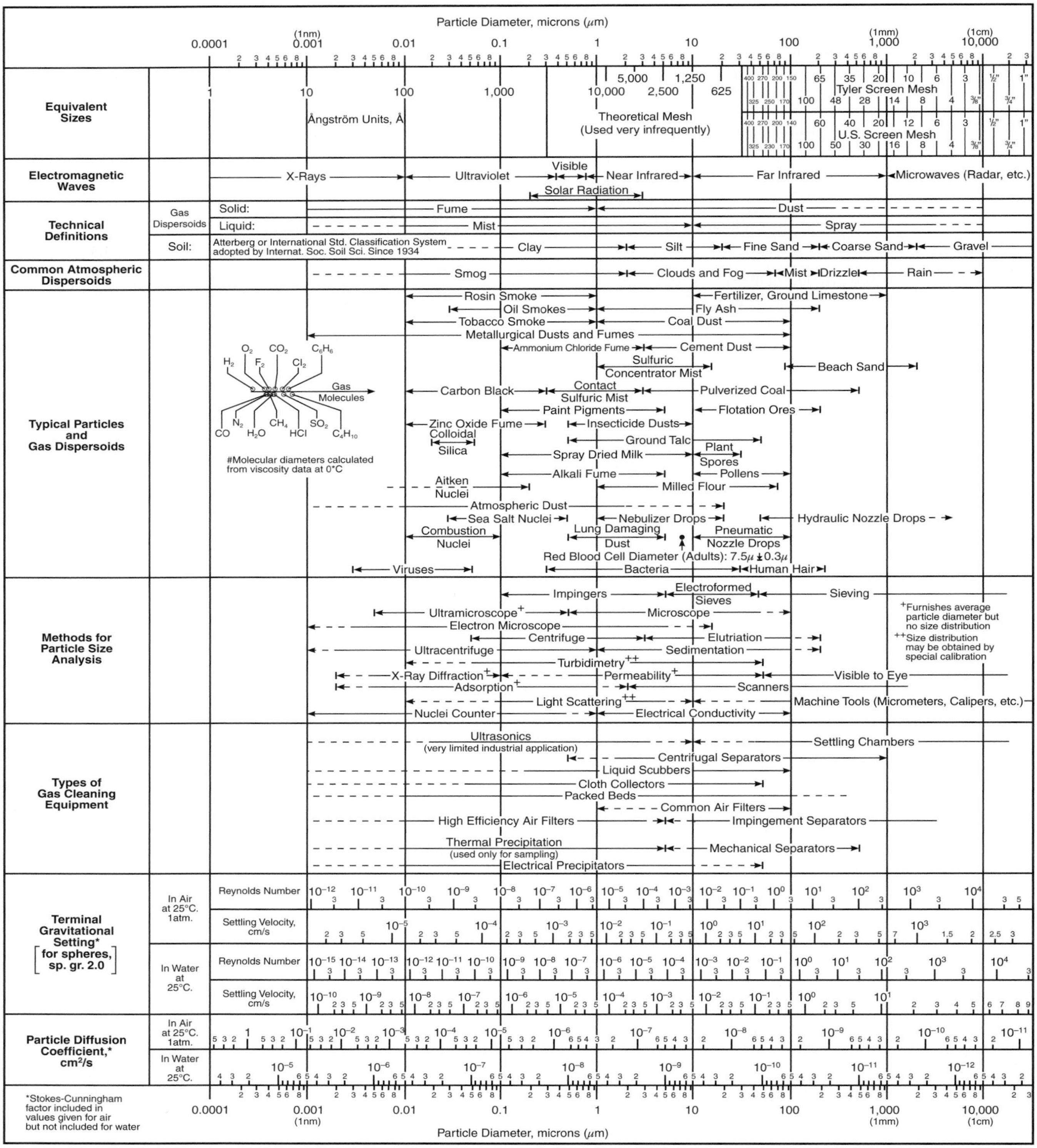

From C.E. Lapple, *Stanford Research Institute Journal*, Vol. 5, p. 95 (Third Quarter) 1961.

Lab Data

DENSITY OF VARIOUS SOLIDS

This table gives the range of density for miscellaneous solid materials whose characteristics depend on the source or method of preparation. The entries for different types of wood refer to seasoned wood.

References

1. Forsythe, W. E., *Smithsonian Physical Tables, Ninth Edition*, Smithsonian Institution, Washington, DC, 1956.
2. Kaye, G. W. C., and Laby, T. H., *Tables of Physical and Chemical Constants, 16th Edition*, Longman, London, 1995.
3. Brandrup, J., and Immergut, E. H., *Polymer Handbook, Third Edition*, John Wiley & Sons, New York, 1989.

Density Ranges for Solid Materials

Material	ρ/g cm^{-3}
Agate	2.5 - 2.7
Alabaster, carbonate	2.69 - 2.78
Alabaster, sulfate	2.26 - 2.32
Albite	2.62 - 2.65
Amber	1.06 - 1.11
Amphiboles	2.9 - 3.2
Anorthite	2.74 - 2.76
Asbestos	2.0 - 2.8
Asbestos slate	1.8
Asphalt	1.1 - 1.5
Basalt	2.4 - 3.1
Beeswax	0.96 - 0.97
Beryl	2.69 - 2.70
Biotite	2.7 - 3.1
Bone	1.7 - 2.0
Brasses	8.44 - 8.75
Brick	1.4 - 2.2
Bronzes	8.74 - 8.89
Butter	0.86 - 0.87
Calamine	4.1 - 4.5
Calcspar	2.6 - 2.8
Camphor	0.99
Cardboard	0.69
Celluloid	1.4
Cement, set	2.7 - 3.0
Chalk	1.9 - 2.8
Charcoal, oak	0.57
Charcoal, pine	0.28 - 0.44
Cinnabar	8.12
Clay	1.8 - 2.6
Coal, anthracite	1.4 - 1.8
Coal, bituminous	1.2 - 1.5
Coke	1.0 - 1.7
Copal	1.04 - 1.14
Cork	0.22 - 0.26
Corundum	3.9 - 4.0
Diamond	3.51
Dolomite	2.84
Ebonite	1.15
Emery	4.0
Epidote	3.25 - 3.50
Feldspar	2.55 - 2.75
Flint	2.63
Fluorite	3.18
Galena	7.3 - 7.6
Garnet	3.15 - 4.30
Gelatin	1.27
Glass, common	2.4 - 2.8
Glass, lead	3 - 4
Glass, Pyrex	2.23
Granite	2.64 - 2.76
Graphite	2.30 - 2.72
Gum arabic	1.3 - 1.4
Gypsum	2.31 - 2.33
Halite	2.18
Hematite	4.9 - 5.3
Hornblende	3.0
Ice	0.917
Iron, cast	7.0 - 7.4
Ivory	1.83 - 1.92
Kaolin	2.6
Leather, dry	0.86
Lime, slaked	1.3 - 1.4
Limestone	2.68 - 2.76
Linoleum	1.18
Magnetite	4.9 - 5.2
Malachite	3.7 - 4.1
Marble	2.60 - 2.84
Meerschaum	0.99 - 1.28
Mica	2.6 - 3.2
Muscovite	2.76 - 3.00
Ochre	3.5
Opal	2.2
Paper	0.7 - 1.15
Paraffin	0.87 - 0.91
Peat blocks	0.84
Pitch	1.07
Polyamides	1.15 - 1.25
Polyethylene	0.91 - 0.96
Poly(methyl methacrylate)	1.19
Polypropylene	0.90 - 0.94
Polystyrene	1.06 - 1.12
Polytetrafluoroethylene	2.16 - 2.30
Poly(vinyl acetate)	1.19
Poly(vinyl chloride)	1.39 - 1.42
Porcelain	2.3 - 2.5
Porphyry	2.6 - 2.9
Pyrite	4.95 - 5.10
Quartz	2.65
Resin	1.07
Rubber, hard	1.19
Rubber, soft	1.1
Rubber, pure gum	0.91 - 0.93
Rubber, Neoprene	1.23 - 1.25
Sandstone	2.14 - 2.36
Serpentine	2.50 - 2.65
Silica, fused	2.21
Silicon carbide	3.16
Slag	2.0 - 3.9
Slate	2.6 - 3.3
Soapstone	2.6 - 2.8
Solder	8.7 - 9.4
Starch	1.53
Steel, stainless	7.8
Sugar	1.59
Talc	2.7 - 2.8
Tallow, beef	0.94
Tar	1.02
Topaz	3.5 - 3.6
Tourmaline	3.0 - 3.2
Tungsten carbide	14.0 - 15.0
Wax, sealing	1.8
Wood, alder	0.42 - 0.68
Wood, apple	0.66 - 0.84
Wood, ash	0.65 - 0.85
Wood, balsa	0.11 - 0.14
Wood, bamboo	0.31 - 0.40
Wood, basswood	0.32 - 0.59
Wood, beech	0.70 - 0.90
Wood, birch	0.51 - 0.77
Wood, blue gum	1.00
Wood, box	0.95 - 1.16
Wood, butternut	0.38
Wood, cedar	0.49 - 0.57
Wood, cherry	0.70 - 0.90
Wood, dogwood	0.76
Wood, ebony	1.11 - 1.33
Wood, elm	0.54 - 0.60
Wood, hickory	0.60 - 0.93
Wood, holly	0.76
Wood, juniper	0.56
Wood, larch	0.50 - 0.56
Wood, locust	0.67 - 0.71
Wood, logwood	0.91
Wood, mahogany	0.66 - 0.85
Wood, maple	0.62 - 0.75
Wood, oak	0.60 - 0.90
Wood, pear	0.61 - 0.73
Wood, pine, pitch	0.83 - 0.85
Wood, pine, white	0.35 - 0.50
Wood, pine, yellow	0.37 - 0.60
Wood, plum	0.66 - 0.78
Wood, poplar	0.35 - 0.50
Wood, satinwood	0.95
Wood, spruce	0.48 - 0.70
Wood, sycamore	0.40 - 0.60
Wood, teak, Indian	0.66 - 0.98
Wood, walnut	0.64 - 0.70
Wood, water gum	1.00
Wood, willow	0.40 - 0.60
Woods metal	9.70

DENSITY OF SULFURIC ACID

This table gives the density of aqueous sulfuric acid solutions as a function of concentration (in mass percent of H_2SO_4) and temperature.

Reference

Washburn, E. W., Ed., *International Critical Tables of Numerical Data of Physics, Chemistry, and Technology,* Vol. 3, p. 56, McGraw-Hill, New York, 1926–1932.

Density ρ of Sulfuric Acid at the Temperature in °C Indicated by the Superscript

Mass %	ρ^{0}/g cm^{-3}	ρ^{10}/g cm^{-3}	ρ^{15}/g cm^{-3}	ρ^{20}/g cm^{-3}	ρ^{25}/g cm^{-3}	ρ^{30}/g cm^{-3}	ρ^{40}/g cm^{-3}	ρ^{50}/g cm^{-3}	ρ^{60}/g cm^{-3}	ρ^{80}/g cm^{-3}	ρ^{100}/g cm^{-3}
1	1.0074	1.0068	1.0060	1.0051	1.0038	1.0022	0.9986	0.9944	0.9895	0.9779	0.9645
2	1.0147	1.0138	1.0129	1.0118	1.0104	1.0087	1.0050	1.0006	0.9956	0.9839	0.9705
3	1.0219	1.0206	1.0197	1.0184	1.0169	1.0152	1.0113	1.0067	1.0017	0.9900	0.9766
4	1.0291	1.0275	1.0264	1.0250	1.0234	1.0216	1.0176	1.0129	1.0078	0.9961	0.9827
5	1.0364	1.0344	1.0332	1.0317	1.0300	1.0281	1.0240	1.0192	1.0140	1.0022	0.9888
6	1.0437	1.0414	1.0400	1.0385	1.0367	1.0347	1.0305	1.0256	1.0203	1.0084	0.9950
7	1.0511	1.0485	1.0469	1.0453	1.0434	1.0414	1.0371	1.0321	1.0266	1.0146	1.0013
8	1.0585	1.0556	1.0539	1.0522	1.0502	1.0481	1.0437	1.0386	1.0330	1.0209	1.0076
9	1.0660	1.0628	1.0610	1.0591	1.0571	1.0549	1.0503	1.0451	1.0395	1.0273	1.0140
10	1.0735	1.0700	1.0681	1.0661	1.0640	1.0617	1.0570	1.0517	1.0460	1.0338	1.0204
12	1.0886	1.0846	1.0825	1.0802	1.0780	1.0756	1.0705	1.0651	1.0593	1.0469	1.0335
14	1.1039	1.0994	1.0971	1.0947	1.0922	1.0897	1.0844	1.0788	1.0729	1.0603	1.0469
16	1.1194	1.1145	1.1120	1.1094	1.1067	1.1040	1.0985	1.0927	1.0868	1.0740	1.0605
18	1.1351	1.1298	1.1271	1.1243	1.1215	1.1187	1.1129	1.1070	1.1009	1.0879	1.0744
20	1.1510	1.1453	1.1424	1.1394	1.1365	1.1335	1.1275	1.1215	1.1153	1.1021	1.0885
22	1.1670	1.1609	1.1579	1.1548	1.1517	1.1486	1.1424	1.1362	1.1299	1.1166	1.1029
24	1.1832	1.1768	1.1736	1.1704	1.1672	1.1640	1.1576	1.1512	1.1448	1.1313	1.1176
26	1.1996	1.1929	1.1896	1.1862	1.1829	1.1796	1.1730	1.1665	1.1599	1.1463	1.1325
28	1.2160	1.2091	1.2057	1.2023	1.1989	1.1955	1.1887	1.1820	1.1753	1.1616	1.1476
30	1.2326	1.2255	1.2220	1.2185	1.2150	1.2115	1.2046	1.1977	1.1909	1.1771	1.1630
32	1.2493	1.2421	1.2385	1.2349	1.2314	1.2278	1.2207	1.2137	1.2068	1.1928	1.1787
34	1.2661	1.2588	1.2552	1.2515	1.2479	1.2443	1.2371	1.2300	1.2229	1.2088	1.1946
36	1.2831	1.2757	1.2720	1.2684	1.2647	1.2610	1.2538	1.2466	1.2394	1.2251	1.2109
38	1.3004	1.2929	1.2891	1.2855	1.2818	1.2780	1.2707	1.2635	1.2561	1.2418	1.2276
40	1.3179	1.3103	1.3065	1.3028	1.2991	1.2953	1.2880	1.2806	1.2732	1.2589	1.2446
42	1.3357	1.3280	1.3242	1.3205	1.3167	1.3129	1.3055	1.2981	1.2907	1.2762	1.2619
44	1.3538	1.3461	1.3423	1.3384	1.3346	1.3308	1.3234	1.3160	1.3086	1.2939	1.2796
46	1.3724	1.3646	1.3608	1.3569	1.3530	1.3492	1.3417	1.3343	1.3269	1.3120	1.2976
48	1.3915	1.3835	1.3797	1.3758	1.3719	1.3680	1.3604	1.3528	1.3455	1.3305	1.3159
50	1.4110	1.4029	1.3990	1.3951	1.3911	1.3872	1.3795	1.3719	1.3644	1.3494	1.3348
52	1.4310	1.4228	1.4188	1.4148	1.4109	1.4069	1.3991	1.3914	1.3837	1.3687	1.3540
54	1.4515	1.4431	1.4391	1.4350	1.4310	1.4270	1.4191	1.4113	1.4036	1.3884	1.3735
56	1.4724	1.4640	1.4598	1.4557	1.4516	1.4475	1.4396	1.4317	1.4239	1.4085	1.3934
58	1.4937	1.4852	1.4809	1.4768	1.4726	1.4685	1.4604	1.4524	1.4446	1.4290	1.4137
60	1.5154	1.5067	1.5024	1.4983	1.4940	1.4898	1.4816	1.4735	1.4656	1.4497	1.4344
62	1.5375	1.5287	1.5243	1.5200	1.5157	1.5115	1.5031	1.4950	1.4869	1.4708	1.4554
64	1.5600	1.5510	1.5465	1.5421	1.5378	1.5335	1.5250	1.5167	1.5086	1.4923	1.4766
66	1.5828	1.5736	1.5691	1.5646	1.5602	1.5558	1.5472	1.5388	1.5305	1.5140	1.4981
68	1.6059	1.5965	1.5920	1.5874	1.5829	1.5785	1.5697	1.5611	1.5528	1.5359	1.5198
70	1.6293	1.6198	1.6151	1.6105	1.6059	1.6014	1.5925	1.5838	1.5753	1.5582	1.5417
72	1.6529	1.6433	1.6385	1.6338	1.6292	1.6246	1.6155	1.6067	1.5981	1.5806	1.5637
74	1.6768	1.6670	1.6622	1.6574	1.6526	1.6480	1.6387	1.6297	1.6209	1.6031	1.5857
76	1.7008	1.6908	1.6858	1.6810	1.6761	1.6713	1.6619	1.6526	1.6435	1.6252	1.6074
78	1.7247	1.7144	1.7093	1.7043	1.6994	1.6944	1.6847	1.6751	1.6657	1.6469	1.6286
80	1.7482	1.7376	1.7323	1.7272	1.7221	1.7170	1.7069	1.6971	1.6873	1.6680	1.6493
82	1.7709	1.7599	1.7544	1.7491	1.7437	1.7385	1.7281	1.7180	1.7080	1.6882	1.6692
84	1.7916	1.7804	1.7748	1.7693	1.7639	1.7585	1.7479	1.7375	1.7274	1.7072	1.6878
86	1.8095	1.7983	1.7927	1.7872	1.7818	1.7763	1.7657	1.7552	1.7449	1.7245	1.7050
88	1.8243	1.8132	1.8077	1.8022	1.7968	1.7914	1.7809	1.7705	1.7602	1.7397	1.7202
90	1.8361	1.8252	1.8198	1.8144	1.8091	1.8038	1.7933	1.7829	1.7729	1.7525	1.7331

Mass %	ρ^{0}/g cm^{-3}	ρ^{10}/g cm^{-3}	ρ^{15}/g cm^{-3}	ρ^{20}/g cm^{-3}	ρ^{25}/g cm^{-3}	ρ^{30}/g cm^{-3}	ρ^{40}/g cm^{-3}	ρ^{50}/g cm^{-3}	ρ^{60}/g cm^{-3}	ρ^{80}/g cm^{-3}	ρ^{100}/g cm^{-3}
91	1.8410	1.8302	1.8248	1.8195	1.8142	1.8090	1.7986	1.7883	1.7783	1.7581	1.7388
92	1.8453	1.8346	1.8293	1.8240	1.8188	1.8136	1.8033	1.7932	1.7832	1.7633	1.7439
93	1.8490	1.8384	1.8331	1.8279	1.8227	1.8176	1.8074	1.7974	1.7876	1.7681	1.7485
94	1.8520	1.8415	1.8363	1.8312	1.8260	1.8210	1.8109	1.8011	1.7914		
95	1.8544	1.8439	1.8388	1.8337	1.8286	1.8236	1.8137	1.8040	1.7944		
96	1.8560	1.8457	1.8406	1.8355	1.8305	1.8255	1.8157	1.8060	1.7965		
97	1.8569	1.8466	1.8414	1.8364	1.8314	1.8264	1.8166	1.8071	1.7977		
98	1.8567	1.8463	1.8411	1.8361	1.8310	1.8261	1.8163	1.8068	1.7976		
99	1.8551	1.8445	1.8393	1.8342	1.8292	1.8242	1.8145	1.8050	1.7958		
100	1.8517	1.8409	1.8357	1.8305	1.8255	1.8205	1.8107	1.8013	1.7922		

DENSITY OF ETHANOL–WATER MIXTURES

This table gives the density ρ of mixtures of ethanol and water as a function of composition and temperature. The composition is specified in weight percent of ethanol, i.e., mass of ethanol per 100 g of solution. Values from the reference have been converted to true densities.

Reference

Washburn, E. W., Ed., *International Critical Tables of Numerical Data of Physics, Chemistry, and Technology*, Vol. 3, McGraw-Hill, New York, 1926–1932.

Density ρ of Ethanol–Water Mixtures at the Temperature in °C Indicated by Superscript

Mass % ethanol	ρ^{10}/g cm^{-3}	ρ^{15}/g cm^{-3}	ρ^{20}/g cm^{-3}	ρ^{25}/g cm^{-3}	ρ^{30}/g cm^{-3}	ρ^{35}/g cm^{-3}	ρ^{40}/g cm^{-3}
0	0.99970	0.99910	0.99820	0.99705	0.99565	0.99403	0.99222
5	0.99095	0.99029	0.98935	0.98814	0.98667	0.98498	0.98308
10	0.98390	0.98301	0.98184	0.98040	0.97872	0.97682	0.97472
15	0.97797	0.97666	0.97511	0.97331	0.9713	0.96908	0.96667
20	0.97249	0.97065	0.96861	0.96636	0.96392	0.96131	0.95853
25	0.96662	0.96421	0.96165	0.95892	0.95604	0.95303	0.94988
30	0.95974	0.95683	0.95379	0.95064	0.94738	0.94400	0.94052
35	0.95159	0.94829	0.94491	0.94143	0.93787	0.93422	0.93048
40	0.94235	0.93879	0.93515	0.93145	0.92767	0.92382	0.91989
45	0.93223	0.92849	0.92469	0.92082	0.91689	0.91288	0.90881
50	0.92159	0.91773	0.91381	0.90982	0.90577	0.90165	0.89747
55	0.91052	0.90656	0.90255	0.89847	0.89434	0.89013	0.88586
60	0.89924	0.89520	0.89110	0.88696	0.88275	0.87848	0.87414
65	0.88771	0.88361	0.87945	0.87524	0.87097	0.86664	0.86224
70	0.87599	0.87184	0.86763	0.86337	0.85905	0.85467	0.85022
75	0.86405	0.85985	0.85561	0.85131	0.84695	0.84254	0.83806
80	0.85194	0.84769	0.84341	0.83908	0.83470	0.83027	0.82576
85	0.83948	0.83522	0.83093	0.82658	0.82218	0.81772	0.81320
90	0.82652	0.82225	0.81795	0.81360	0.80920	0.80476	0.80026
95	0.81276	0.80850	0.80422	0.79989	0.79553	0.79112	0.78668
100	0.79782	0.79358	0.78932	0.78504	0.78073	0.77639	0.77201

DIELECTRIC STRENGTH OF INSULATING MATERIALS

L. I. Berger

The loss of the dielectric properties by a sample of a gaseous, liquid, or solid insulator as a result of application to the sample of an electric field* greater than a certain critical magnitude is called *dielectric breakdown*. The critical magnitude of electric field at which the breakdown of a material takes place is called the *dielectric strength* of the material (or *breakdown voltage*). The dielectric strength of a material depends on the specimen thickness (as a rule, thin films have greater dielectric strength than that of thicker samples of a material), the electrode shape,† the rate of the applied voltage increase, the shape of the voltage vs. time curve, and the medium surrounding the sample, e.g., air or other gas (or a liquid — for solid materials only).

Below are given the dielectric strength of gases (Table 1), liquids (Table 2), and solids (Table 3) commonly used in insulating situations.

Breakdown in Gases

The current carriers in gases are free electrons and ions generated by external radiation. The equilibrium concentration of these particles at normal pressure is about 10^3 cm^{-3}, and hence the electrical conductivity is very small, of the order of 10^{-16} – 10^{-15} S/cm. But in a strong electric field, these particles acquire kinetic energy along their free path, large enough to ionize the gas molecules. The new charged particles ionize more molecules; this avalanche-like process leads to formation between the electrodes of channels of conducting plasma (streamers), and the electrical resistance of the space between the electrodes decreases virtually to zero.

Because the dielectric strength (breakdown voltage) of gases strongly depends on the electrode geometry and surface condition and the gas pressure, it is generally accepted to present the data for a particular gas as a fraction of the dielectric strength of either nitrogen or sulfur hexafluoride measured at the same conditions. In Table 1, the data are presented in comparison with the dielectric strength of nitrogen, which is considered equal to 1.00. For convenience to the reader, a few average magnitudes of the dielectric strength of some gases are expressed in kilovolts per millimeter. The data in the table relate to the standard conditions, unless indicated otherwise.

Breakdown in Liquids

If a liquid is pure, the breakdown mechanism in it is similar to that in gases. If a liquid contains liquid impurities in the form of small drops with greater dielectric constant than that of the main liquid, the breakdown is the result of formation of ellipsoids from these drops by the electric field. In a strong enough electric field, these ellipsoids merge and form a high-conductivity channel between the electrodes. The current increases the temperature in the channel, liquid boils, and the current along the steam canal leads to breakdown. Formation of a conductive channel (bridge) between the electrodes is observed also in liquids with solid impurities. If a liquid contains gas impurities in the form of small bubbles, breakdown is the result of heating of the liquid in strong electric fields. In the locations with the highest current density, the liquid boils, the size of the gas bubbles increases, they merge and form gaseous channels between the electrodes, and the breakdown medium is again the gas plasma.

Breakdown in Solids

It is known that the current in solid insulators does not obey Ohm's law in strong electric fields. The current density increases almost exponentially with the electric field, and at a certain field magnitude it jumps to very high magnitudes at which a specimen of a material is destroyed. The two known kinds of electric breakdown are thermal and electrical breakdowns. The former is the result of material heating by the electric current. Destruction of a sample of a material happens when, at a certain voltage, the amount of heat produced by the current exceeds the heat release through the sample surface; the breakdown voltage in this case is proportional to the square root of the ratio of the thermal conductivity and electrical conductivity of the material. A semi-empirical expression for dependence of the breakdown voltage, V_B, on the physical properties and geometry of a sample of a solid material for the one-dimensional case is

$$V_B = [A\rho\kappa/a\phi(d)]^{1/2}$$

where A is a numerical constant related to the system of units used, ρ and κ are the volume resistivity and thermal conductivity of the sample material, a is a constant related to the chemical bond nature and crystal structure of the sample material, and $\phi(d)$ is a function of the sample geometry, first of all, thickness, d (see, e.g., Ref. R6). In the majority of materials, $\phi(d)$ increases with d, hence, the magnitude of V_B is greater in the thinner samples of a particular material.

The electrical breakdown results from the tunneling of the charge carriers from electrodes or from the valence band or from the impurity levels into the conduction band, or by the impact ionization. The tunnel effect breakdown happens mainly in thin layers, e.g., in thin p-n junctions. Otherwise, the impact ionization mechanism dominates. For this mechanism, the dielectric strength of an insulator can be estimated using Boltzmann's kinetic equation for electrons in a crystal.

In the following tables, the dielectric strength values are for room temperature and normal atmospheric pressure, unless indicated otherwise.

* The unit of electric field in the SI system is newton per coulomb or volt per meter.

† For example, the U.S. standard ASTM D149 is based on use of symmetrical electrodes, while per U.K. standard BS2918 one electrode is a plane and the other is a rod with the axis normal to the plane.

TABLE 1. Dielectric Strength of Gases

Material	Dielectric strength*	Ref.
Nitrogen, N_2	**1.00**	
Hydrogen, H_2	0.50	1,2
Helium, He	0.15	1
Oxygen, O_2	0.92	2
Air	0.97	6
Air[b]	3.0[a]	3
Air	0.4-0.7[a]	4
Air	1.40[a]	5
Neon, Ne	0.25	1
	0.16	2
Argon, Ar	0.18	2
Chlorine, Cl_2	1.55	1
Carbon monoxide, CO	1.02	1
	1.05	2
Carbon dioxide, CO_2	0.88	1
	0.82	2
	0.84	6
Nitrous oxide, N_2O	1.24	2
Sulfur dioxide, SO_2	2.63	2
	2.68	6
Sulfur monochloride, S_2Cl_2[c]	1.02	1
Thionyl fluoride, SOF_2	2.50	1
Sulfur hexafluoride, SF_6	2.50	1
	2.63	2
Sulfur hexafluoride, SF_6	8.50[a]	7
	9.8	8
Perchloryl fluoride, ClO_3F	2.73	1
Tetrachloromethane, CCl_4	6.33	1
	6.21	2
Tetrafluoromethane, CF_4	1.01	1
Methane, CH_4	1.00	1
	1.13	2
Bromotrifluoromethane, CF_3Br	1.35	1
	1.97	2
Bromomethane, CH_3Br	0.71	2
Chloromethane, CH_3Cl	1.29	2
Iodomethane, CH_3I	3.02	2
Iodomethane, CH_3I[d]	2.20	7
Dichloromethane, CH_2Cl_2	1.92	2
Dichlorodifluoromethane, CCl_2F_2	2.42	1
	2.63	2,6
Chlorotrifluoromethane, $CClF_3$	1.43	1
	1.53	2
Trichlorofluoromethane, CCl_3F	3.50	1
	4.53	2
Trichloromethane, $CHCl_3$	4.2	1
	4.39	2
Methylamine, CH_3NH_2	0.81	1
Difluoromethane, CH_2F_2	0.79	2
Trifluoromethane, CHF_3	0.71	2
Bromochlorodifluoromethane, CF_2ClBr	3.84	2
Chlorodifluoromethane, $CHClF_2$	1.40	1
	1.11	2
Dichlorofluoromethane, $CHCl_2F$	1.33	1
	2.61	2
Chlorofluoromethane, CH_2ClF	1.03	1
Hexafluoroethane, C_2F_6	1.82	1
	2.55	2
Ethyne (Acetylene), C_2H_2	1.10	1
	1.11	2
Chloropentafluoroethane, C_2ClF_5	2.3	1
	3.0	6
Dichlorotetrafluoroethane, $C_2Cl_2F_4$	2.52	1
Chlorotrifluoroethylene, C_2ClF_3	1.82	2
1,1,1-Trichloro-2,2,2-trifluoroethane	6.55	2
1,1,2-Trichloro-1,2,2-trifluoroethane	6.05	2
Chloroethane, C_2H_5Cl	1.00	1
1,1-Dichloroethane	2.66	2
Trifluoroacetonitrile, CF_3CN	3.5	1
Acetonitrile, CH_3CN	2.11	2
Dimethylamine, $(CH_3)_2NH$	1.04	1
Ethylamine, $C_2H_5NH_2$	1.01	1
Ethylene oxide (oxirane), CH_3CHO	1.01	1
Perfluoropropene, C_3F_6	2.55	2
Octafluoropropane, C_3F_8	2.19	1
	2.47	2
3,3,3-Trifluoro-1-propene, CH_2CHCF_3	2.11	2
Pentafluoroisocyanoethane, C_2F_5NC	4.5	1
1,1,1,4,4,4-Hexafluoro-2-butyne, CF_3CCCF_3	5.84	2
Octafluorocyclobutane, C_4F_8	3.34	2
1,1,1,2,3,4,4,4-Octafluoro-2-butene	2.8	1
Decafluorobutane, C_4F_{10}	3.08	1
Perfluorobutanenitrile, C_3F_7CN	5.5	1
Perfluoro-2-methyl-1,3-butadiene, C_5F_8	5.5	1
Hexafluorobenzene, C_6F_6	2.11	2
Perfluorocyclohexane, C_6F_{12}, (saturated vapor)	6.18	2

* Relative to nitrogen, unless units of kV/mm are indicated.
[a] In kV/mm.
[b] Flat electrodes.
[c] At 12.5 torr.
[d] At 370 torr.

TABLE 2. Dielectric Strength of Liquids

Material	Dielectric strength kV mm^{-1}	Ref.
Helium, He, liquid, 4.2 K	10	9
Static	10	11
Dynamic	5	11
	23	12
Nitrogen, N_2, liquid, 77 K		
Coaxial cylinder electrodes	20	10
Sphere to plane electrodes	60	10
Water, H_2O, distilled	65-70	13
Carbon tetrachloride, CCl_4	5.5	14
	16.0	15
Hexane, C_6H_{14}	42.0	16
Two 2.54 cm diameter spherical electrodes, 50.8 μm space	156	17,18
Cyclohexane, C_6H_{12}	42-48	16
2-Methylpentane, C_6H_{14}	149	17,18
2,2-Dimethylbutane, C_6H_{14}	133	17,18

Material	Dielectric strength kV mm^{-1}	Ref.
2,3-Dimethylbutane, C_6H_{14}	138	17,18
Benzene, C_6H_6	163	17,18
Chlorobenzene, C_6H_5Cl	7.1	14
	18.8	15
2,2,4-Trimethylpentane, C_8H_{18}	140	17,18
Phenylxylylethane	23.6	19
Heptane, C_7H_{16}	166	17,18
2,4-Dimethylpentane, C_7H_{16}	133	17,18
Toluene, $C_6H_5CH_3$	199	17,18
	46	16
	12.0	14
	20.4	15
Octane, C_8H_{18}	16.6	14
	20.4	15
	179	17,18
Ethylbenzene, C_8H_{10}	226	17,18
Propylbenzene, C_9H_{12}	250	17,18
Isopropylbenzene, C_9H_{12}	238	17,18
Decane, $C_{10}H_{22}$	192	17,18
Synthetic Paraffin Mixture Synfluid 2cSt PAO	29.5	37
Butylbenzene, $C_{10}H_{14}$	275	17,18
Isobutylbenzene, $C_{10}H_{14}$	222	17,18
Silicone oils—polydimethylsiloxanes, $(CH_3)_3Si\text{-}O\text{-}[Si(CH_3)_2]_x\text{-}O\text{-}Si(CH_3)_3$		
Polydimethylsiloxane silicone fluid	15.4	20
Dimethyl silicone	24.0	21,22
Phenylmethyl silicone	23.2	22
Silicone oil, Basilone M50	10-15	23
Mineral insulating oils	11.8	6
Polybutene oil for capacitors	13.8	6
Transformer dielectric liquid	28-30	6
Isopropylbiphenyl capacitor oil	23.6	6
Transformer oil	110.7	24
Transformer oil Agip ITE 360	9-12.6	23
Perfluorinated hydrocarbons		
Fluorinert FC 6001	8.0	23
Fluorinert FC 77	10.7	23
Perfluorinated polyethers		
Galden XAD (Mol. wt. 800)	10.5	23
Galden D40 (Mol. wt. 2000)	10.2	23
Castor oil	65	25

TABLE 3. Dielectric Strength of Solids

Material	Dielectric strength kV mm^{-1}	Ref
Sodium chloride, NaCl, crystalline	150	26
Potassium bromide, KBr, crystalline	80	26
Ceramics		
Alumina (99.9% Al_2O_3)	13.4	6,27a
Aluminum silicate, Al_2SiO_5	5.9	6
Berillia (99% BeO)	13.8	6,27b
Boron nitride, BN	37.4	6
Cordierite, $Mg_2Al_4Si_5O_{18}$	7.9	6,27c
Forsterite, Mg_2SiO_4	9.8	28
Porcelain	35-160	26
Steatite, $Mg_3Si_4O_{11}\cdot H_2O$	9.1-15.4	6
Titanates of Mg, Ca, Sr, Ba, and Pb	20-120	3
Barium titanate, glass bonded	>30	36
Zirconia, ZrO_2	11.4	29
Glasses		
Fused silica, SiO_2	470-670	26
Alkali-silicate glass	200	26
Standard window glass	9.8-13.8	28
Micas		
Muscovite, ruby, natural	118	6
Phlogopite, amber, natural	118	6
Fluorophlogopite, synthetic	118	6
Glass-bonded mica	14.0-15.7	6
Thermoplastic Polymers		
Polypropylene	23.6	6
Amide polymer nylon 6/6, dry	23.6	6
Polyamide-imide copolymer	22.8	6
Modified polyphenylene oxide	21.7	6
Polystyrene	19.7	6
Polymethyl methacrylate	19.7	6
Polyetherimide	18.9	6
Amide polymer nylon 11(dry)	16.7	6
Polysulfone	16.7	6
Styrene-acrylonitrile copolymer	16.7	6
Acrylonitrile-butadiene-styrene	16.7	6
Polyethersulfone	15.7	6
Polybutylene terephthalate	15.7	6
Polystyrene-butadiene copolymer	15.7	6
Acetal homopolymer	15.0	6
Acetal copolymer	15.0	6
Polyphenylene sulfide	15.0	6
'Polycarbonate	15.0	6
Acetal homopolymer resin (molding resin)	15.0	6
Acetal copolymer resin	15.0	6
Thermosetting Molding Compounds		
Glass-filled allyl (Type GDI-30 per MIL-M-14G)	15.7	6
Glass-filled epoxy, electrical grade	15.4	6
Glass-filled phenolic (Type GPI-100 per MIL-M-14G)	15.0	6
Glass-filled alkyd/polyester (Type MAI-60 per MIL-M-14G)	14.8	6
Glass-filled melamine (Type MMI-30 per MIL-M-14G)	13.4	6
Extrusion Compounds for High-Temperature Insulation		
Polytetrafluoroethylene	19.7	6
Perfluoroalkoxy polymer	21.7	6
Fluorinated ethylene-propylene copolymer	19.7	6
Ethylene-tetrafluoroethylene copolymer	15.7	6
Polyvinylidene fluoride	10.2	6
Ethylene-chlorotrifluoroethylene copolymer	19.3	6
Polychlorotrifluoroethylene	19.7	6
Extrusion Compounds for Low-Temperature Insulation		
Polyvinyl chloride		
Flexible	11.8-15.7	30

Material	Dielectric strength kV mm^{-1}	Ref
Rigid	13.8-19.7	30
Polyethylene	18.9	28
Polyethylene, low-density	21.7	6
	300	31
Polyethylene, high-density	19.7	6
Polypropylene/polyethylene copolymer	23.6	6
Embedding Compounds		
Basic epoxy resin: bisphenol A/epichlorohydrin polycondensate	19.7	6
Cycloaliphatic epoxy: alicyclic diepoxy carboxylate	19.7	6
Polyetherketone	18.9	30
Polyurethanes		
Two-component, polyol-cured	25.4	6
Two-part solventless, polybutylene-based	24.0	6
Silicones		
Clear two-part heat curing electrical grade silicone embedding resin	21.7	6
Red insulating enamel (MIL-E-22118)		
Dry	47.2	6
Wet	11.8	6
Enamels		
Red enamel, fast cure		
Standard conditions	78.7	6
Immersion conditions	47.2	6
Black enamel		
Standard conditions	70.9	6
Immersion conditions	47.2	6
Varnishes		
Vacuum-pressure impregnated baking type solventless polyester varnish		
Rigid, two-part	70.9	6
Semiflexible high-bond thixotropic	78.7	6
Rigid high-bond high-flash freon-resistant	68.9	6
Baking type epoxy varnish		
Solventless, rigid, low viscosity, one-part	90.6	6
Solventless, semiflexible, one-part	82.7	6
Solventless, semirigid, chemical resistant, low dielectric constant	106.3	6
Solvable, for hermetic electric motors	181.1	6
Polyurethane coating		
Clear conformal, fast cure		
Standard conditions	78.7	6
Immersion conditions	47.2	6
Insulating Films and Tapes		
Low-density polyethylene film (40 μm thick)	300	31
Poly-*p*-xylylene film	410-590	32
Aromatic polymer films		

Material	Dielectric strength kV mm^{-1}	Ref
Kapton H (Du Pont)	389-430	33
Ultem (GE Plastic and Roem AG)	437-565	33
Hostaphan (Hoechst AG)	338-447	33
Amorphous Stabar K2000 (ICI film)	404-422	33
Stabar S100 (ICI film)	353-452	33
Polyetherimide film (26 μm)	486	34
Parylene N/D (poly-*p*-xylylene/polydichloro-*p*-xylylene) 25 μm film	275	6
Cellulose acetate film	157	6
Cellulose triacetate film	157	6
Polytetrafluoroethylene film	87-173	6
Perfluoroalkoxy film	157-197	6
Fluorinated ethylene-propylene copolymer film	197	6
Ethylene-tetrafluoroethylene film	197	6
Ethylene-chlorotrifluoroethylene copolymer film	197	6
Polychlorotrifluoroethylene film	118-153.5	6
High-voltage rubber insulating tape	28	6
Composites		
Isophthalic polyester (vinyl toluene monomer) filled with		
Calcium carbonate, $CaCO_3$	15.0	38
Gypsum, $CaSO_4$	14.4	38
Alumina trihydrate	15.4	38
Clay	14.4	38
BPA fumarate polyester (vinyl toluene monomer) filled with		
Calcium carbonate	6.1	38
Gypsum	5.9	38
Alumina trihydrate	11.8	38
Clay	12.6	38
Polysulfone resin—30% glass fiber	16.5-18.7	38
Polyamid resin (Nylon 66)—30% carbon fiber		
Polyimide thermoset resin, glass reinforced	12.0	39
Polyester resin (thermoplastic)—40% glass fiber	20.0	38
Epoxy resin (diglycidyl ether of bisphenol A), glass reinforced	16.0	40
Various Insulators		
Rubber, natural	100-215	26
Butyl rubber	23.6	6
Neoprene	15.7-27.6	6
Silicone rubber	26-36	6
Room-temperature vulcanized silicone rubber	9.2-10.9	35
Ureas (from carbamide to tetraphenylurea)	11.8-15.7	28
Dielectric papers		
Aramid paper, calendered	28.7	6
Aramid paper, uncalendered	12.2	6
Aramid with Mica	39.4	6

References

1. Vijh, A. K., *IEEE Trans.*, EI-12, 313, 1997. <https://doi.org/10.1109/TEI.1977.297984>
2. Brand, K. P., *IEEE Trans.*, EI-17, 451, 1982. <https://doi.org/10.1109/TEI.1982.298489>
3. *Encyclopedic Dictionary in Physics*, Vedensky, B. A. and Vul, B. M., Eds., Vol. 4, Soviet Encyclopedia Publishing House, Moscow, 1965.
4. Kubuki, M., Yoshimoto, R., Yoshizumi, K., Tsuru, S., and Hara, M., *IEEE Trans.*, DEI-4, 92, 1997. <https://doi.org/10.1109/94.590874>
5. Al-Arainy, A. A., Malik, N. H., and Qureshi, M. I., *IEEE Trans.*, DEI-1, 305, 1994. <https://doi.org/10.1109/94.300263>
6. Shugg, W. T., *Handbook of Electrical and Electronic Insulating Materials*, Van Nostrand Reinhold, New York, 1986.
7. Devins, J. C., *IEEE Trans.*, EI-15, 81, 1980. <https://doi.org/10.1109/TEI.1980.298243>
8. Xu, X., Jayaram, S., and Boggs, S. A., *IEEE Trans.*, DEI-3, 836, 1996. <https://doi.org/10.1109/94.556569>
9. Okubo, H., Wakita, M., Chigusa, S., Nayakawa, N., and Hikita, M., *IEEE Trans.*, DEI-4, 120, 1997. <https://doi.org/10.1109/94.590880>
10. Hayakawa, H., Sakakibara, H., Goshima, H., Hikita, M., and Okubo, H., *IEEE Trans.*, DEI-4, 127, 1997. <https://doi.org/10.1109/94.590883>

11. Okubo, H., Wakita, M., Chigusa, S., Hayakawa, N., and Hikita, M., *IEEE Trans.*, DEI-4, 220, 1997. <https://doi.org/10.1109/94.590880>
12. Von Hippel, A. R., *Dielectric Materials and Applications*, MIT Press, Cambridge, MA, 1954.
13. Jones, H. M. and Kunhards, E. E., *IEEE Trans.*, DEI-1, 1016, 1994. <https://doi.org/10.1109/94.368641>
14. Nitta, Y. and Ayhara, Y., *IEEE Trans.*, EI-11, 91, 1976. <https://doi.org/10.1109/TEI.1976.297905>
15. Gallagher, T. J., *IEEE Trans.*, EI-12, 249, 1977. <https://doi.org/10.1109/TEI.1977.298029>
16. Wong, P. P. and Forster, E. O., in *Dielectric Materials. Measurements and Applications*, IEE Conf. Publ. 177, 1, 1979.
17. Kao, K. C. *IEEE Trans.*, EI-11, 121, 1976. <https://doi.org/10.1109/TEI.1976.297919>
18. Sharbaugh, A. H., Crowe, R. W., and Cox, E. B., *J. Appl. Phys.*, 27, 806, 1956. <https://doi.org/10.1063/1.1722486>
19. Miller, R. L., Mandelcorn, L., and Mercier, G. E., in *Proc. Intl. Conf. on Properties and Applications of Dielectric Materials*, Xian, China, June 24-28, 1985; cited in Ref. 6, p. 492.
20. Hakim, R. M., Oliver, R. G., and St-Onge, H., *IEEE Trans.*, EI-12, 360, 1977. <https://doi.org/10.1109/TEI.1977.298043>
21. Hosticka, C., *IEEE Trans.*, 389, 1977. <https://doi.org/10.1109/TEI.1977.297989>
22. Yasufuku, S., Umemura, T., and Ishioka, Y., *IEEE Trans.*, EI-12, 402, 1977. <https://doi.org/10.1109/TEI.1977.297991>
23. Forster, E. O., Yamashita, H., Mazzetti, C., Pompini, M., Caroli, L., and Patrissi, S., *IEEE Trans.*, DEI-1, 440, 1994. <https://doi.org/10.1109/94.300287>
24. Bell, W. R., *IEEE Trans.*, 281, 1977. <https://doi.org/10.1109/TEI.1977.297980>
25. Ramu, T. C. and Narayana Rao, Y., in *Dielectric Materials. Measurements and Applications*, IEE Conf. Publ. 177, 37.
26. Skanavi, G. I., *Fizika Dielektrikov; Oblast Silnykh Polei* (Physics of Dielectrics; Strong Fields). Gos. Izd. Fiz. Mat. Nauk (State Publ. House for Phys. and Math. Scis.), Moscow, 1958.
27. Kleiner, R. N., in *Practical Handbook of Materials Science*, Lynch, C. T., Ed., CRC Press, 1989; 27a: p. 304; 27b: p. 300; 27c: p. 316.
28. *Materials Selector Guide. Materials and Methods*, Reinhold Publ., New York, 1973.
29. Flinn, R. A. and Trojan, P. K., *Engineering Materials and Their Applications, Second Edition*, Houghton Mifflin, 1981, p. 614.
30. Lynch, C. T., Ed., *Practical Handbook of Materials Science*, CRC Press, Boca Raton, FL, 1989.
31. Suzuki, H., Mukai, S., Ohki, Y., Nakamichi, Y., and Ajiki, K., *IEEE Trans.*, DEI-4, 238, 1997. <https://doi.org/10.1109/94.595252>
32. Mori, T., Matsuoka, T., and Muzitani, T., *IEEE Trans.*, DEI-1, 71, 1994. <https://doi.org/10.1109/94.300233>
33. Bjellheim, P. and Helgee, B., *IEEE Trans.*, DEI-1, 89, 1994. <https://doi.org/10.1109/94.300236>
34. Zheng, J. P., Cygan, P. J., and Jow, T. R., *IEEE Trans.*, DEI-3, 144, 1996. <https://doi.org/10.1109/94.485527>
35. Danukas, M. G., *IEEE Trans.*, DEI-1, 1196, 1994. <https://doi.org/10.1109/94.368643>
36. Burn, I. and Smithe, D. H., *J. Mater. Sci.*, 7, 339, 1972. <https://doi.org/10.1007/BF00555636>
37. Hope, K.D., Chevron Chemical, Private Communication.
38. *Engineering Materials Handbook*, Vol. 1, Composites, C.A. Dostal, Ed., ASM Intl., 1987.
39. 1985 Materials Selector, *Mater. Eng.*, (12) 1984.
40. *Modern Plastics Encyclopedia*, McGraw-Hill, v. 62 (No. 10A) 1985–1986.
41. *Review Literature on the Subject*
42. R1. Kuffel, E. and Zaengl, W. S., *HV Engineering Fundamentals*, Pergamon, 1989.
43. R2. Kok, J. A., *Electrical Breakdown of Insulating Liquids*, Phillips Tech. Library, Cleaver-Hum, London, 1961.
44. R3. Gallagher, T. J., *Simple Dielectric Liquids*, Clarendon, Oxford, 1975.
45. R4. Meek, J. M. and Craggs, J. D., Eds., *Electric Breakdown in Gases*, John Wiley & Sons, 1976.
46. R5. Von Hippel, A. R., *Dielectric Materials and Applications*, MIT Press, Cambridge, MA, 1954.
47. R6. O'Dwyer, J. J. *The Theory of Dielectric Breakdown of Solids*, Clarendon Press, 1964.

COEFFICIENT OF FRICTION

The coefficient of friction between two surfaces is the ratio of the force required to move one over the other to the force pressing the two together. Thus, if F is the minimum force needed to move one surface over the other, and W is the force pressing the surfaces together, the coefficient of friction μ is given by $\mu = F/W$. A greater force is generally needed to initiate movement from rest than to continue the motion once sliding has started. Thus, the static coefficient of friction μ(static) is usually larger than the sliding or kinetic coefficient μ(sliding).

This table gives characteristic values of both the static and sliding coefficients of friction for a number of material combinations. In each case Material 1 is moving over the surface of Material 2. The type of lubrication or any other special condition is indicated in the third column. All values refer to room temperature unless otherwise indicated. It should be emphasized that the coefficient of friction is very sensitive to the condition of the surface, so that these values represent only a rough guide.

References

1. Minshall, H., in *CRC Handbook of Chemistry and Physics, 73rd Edition,* Lide, D. R., Ed., CRC Press, Boca Raton, FL, 1992.
2. Fuller, D. D., in *American Institute of Physics Handbook, Third Edition,* Gray, D. E., Ed., McGraw-Hill, New York, 1972.

Coefficient of Friction between Selected Materials

Material 1	Material 2	Conditions	μ(static)	μ(sliding)
Metals				
Hard steel	Hard steel	Dry	0.78	0.42
Hard steel	Hard steel	Castor oil	0.15	0.081
Hard steel	Hard steel	Steric acid	0.005	0.029
Hard steel	Hard steel	Lard	0.11	0.084
Hard steel	Hard steel	Light mineral oil	0.23	
Hard steel	Hard steel	Graphite		0.058
Hard steel	Graphite	Dry	0.21	
Mild steel	Mild steel	Dry	0.74	0.57
Mild steel	Mild steel	Oleic acid		0.09
Mild steel	Phosphor bronze	Dry		0.34
Mild steel	Cast iron	Dry		0.23
Mild steel	Lead	Dry	0.95	0.95
Mild steel	Lead	Mineral oil	0.5	0.3
Mild steel	Brass	Dry	0.35	
Cast iron	Cast iron	Dry	1.10	0.15
Aluminum	Aluminum	Dry	1.05	1.4
Aluminum	Mild steel	Dry	0.61	0.47
Brass	Mild steel	Dry	0.51	0.44
Brass	Mild steel	Castor oil	0.11	
Brass	Cast iron	Dry		0.30
Bronze	Cast iron	Dry		0.22
Cadmium	Mild steel	Dry		0.46
Copper	Copper	Dry	1.6	
Copper	Mild steel	Dry	0.53	0.36
Copper	Mild steel	Oleic acid		0.18
Copper	Cast iron	Dry	1.05	0.29
Copper	Glass	Dry	0.68	0.53
Lead	Cast iron	Dry		0.43
Magnesium	Magnesium	Dry	0.6	
Magnesium	Mild steel	Dry		0.42
Magnesium	Cast iron	Dry		0.25
Nickel	Nickel	Dry	1.10	0.53
Nickel	Mild steel	Dry		0.64
Tin	Cast iron	Dry		0.32
Zinc	Cast iron	Dry	0.85	0.21
Nonmetals				
Diamond	Diamond	Dry	0.1	
Diamond	Metals	Dry	0.12	
Garnet	Mild steel	Dry		0.39
Glass	Glass	Dry	0.94	0.4
Glass	Nickel	Dry	0.78	0.56
Graphite	Graphite	Dry	0.1	
Mica	Mica	Freshly cleaved	1.0	
Nylon	Nylon	Dry	0.2	
Nylon	Steel	Dry	0.40	
Polyethylene	Polyethylene	Dry	0.2	
Polyethylene	Steel	Dry	0.2	
Polystyrene	Polystyrene	Dry	0.5	
Polystyrene	Steel	Dry	0.3	
Sapphire	Sapphire	Dry	0.2	
Teflon	Teflon	Dry	0.04	0.04
Teflon	Steel	Dry	0.04	0.04
Tungsten carbide	Tungsten carbide	Dry, room temp.	0.17	
Tungsten carbide	Tungsten carbide	Dry, 1000 °C	0.45	
Tungsten carbide	Tungsten carbide	Dry, 1600 °C	1.8	
Tungsten carbide	Tungsten carbide	Oleic acid	0.12	
Tungsten carbide	Graphite	Dry	0.15	
Tungsten carbide	Steel	Dry	0.5	
Tungsten carbide	Steel	Oleic acid	0.08	
Miscellaneous Materials				
Cotton	Cotton	Threads	0.3	
Leather	Cast iron	Dry	0.6	0.56
Leather	Oak	Parallel	0.61	0.52
Oak	Oak	Parallel	0.62	0.48
Oak	Oak	Perpendicular	0.54	0.32
Silk	Silk	Clean	0.25	
Wood	Wood	Dry	0.35	
Wood	Wood	Wet	0.2	
Wood	Brick	Dry	0.6	
Wood	Leather	Dry	0.35	
Various Materials on Ice and Snow				
Ice	Ice	Clean, 0 °C	0.1	0.02
Ice	Ice	Clean, -12 °C	0.3	0.035
Ice	Ice	Clean, -80 °C	0.5	0.09
Aluminum	Snow	Wet, 0 °C	0.4	
Aluminum	Snow	Dry, 0 °C	0.35	
Brass	Ice	Clean, 0 °C		0.02
Brass	Ice	Clean, -80 °C		0.15
Nylon	Snow	Wet, 0 °C	0.4	
Nylon	Snow	Dry, -10 °C	0.3	
Teflon	Snow	Wet, 0 °C	0.05	
Teflon	Snow	Dry, 0 °C	0.02	
Wax, ski	Snow	Wet, 0 °C	0.1	
Wax, ski	Snow	Dry, 0 °C	0.04	
Wax, ski	Snow	Dry, -10 °C	0.2	

Section 16
Health and Safety Information

Chemical Safety 16-1
Abbreviations Used in the Assessment and Presentation of Laboratory Hazards 16-3
Incompatible Chemicals 16-5
Explosion (Shock) Hazards 16-7
Water-Reactive Chemicals 16-8
Testing Requirements for Peroxidizable Compounds 16-8
Tests for the Presence of Peroxides 16-9
Pyrophoric Compounds – Compounds That Are Reactive with Air 16-9
Flammability Hazards of Common Solvents 16-10
Flammability of Chemical Substances 16-12
Materials Compatible with and Resistant to 72% Perchloric Acid 16-30
Selection of Laboratory Gloves 16-31
Selection of Protective Laboratory Garments 16-33
Selection of Respirators and Respirator Cartridges and Filters 16-34
Protective Clothing Levels 16-36
Selection of Hearing Protection Devices 16-37
Chemical Fume Hoods and Biological Safety Cabinets 16-39
Gas Cylinder Safety and Stamped Markings 16-41
Compressed Air Safety 16-42
Safety in the Use of Cryogens 16-43
Nanomaterial Safety Guidelines 16-46
Threshold Limits for Airborne Contaminants 16-48
Chemical Carcinogens 16-76
Laser Hazards in the Laboratory 16-86
General Characteristics of Ionizing Radiation for the Purpose of Practical Application of Radiation Protection 16-88
Radiation Safety Units 16-89
Relative Dose Ranges from Ionizing Radiation 16-91
Annual Limits on Intakes of Radionuclides 16-93

CHEMICAL SAFETY

Maria J. Doa

Information on chemical hazards and methods to reduce exposure are important elements to safely working with hazardous chemicals in laboratories as well as in other workplaces. This document focuses on hazard communication and tools and requirements to minimize exposures to chemical hazards.

Hazard Communication

The OSHA Hazard Communication Standard (Ref. 1) requires evaluating the potential hazards of chemicals and communicating information concerning those hazards and appropriate protective measures to employees. The standard includes provisions for the following:

- Developing and maintaining a written hazard communication program for the workplace, including lists of hazardous chemicals present
- Labeling of containers of chemicals in the workplace
- Preparation and distribution of safety data sheets to workers
- Development and implementation of worker training programs regarding hazards of chemicals
- The use of appropriate protective measures when handling the chemicals

Safety Data Sheets

The Safety Data Sheets (SDS) for each hazardous chemical must include general information about the chemical for each of the following topics:

- Chemical identification
- Hazard(s) identification
- Composition and mixture information
- Safe handling and storage practices
- First aid measures
- Emergency control measures (e.g., firefighting)
- Physical and chemical properties
- Stability and reactivity information
- Toxicological information
- Exposure control and personal protection information

The SDS may also include information on proper disposal, environmental hazards, and regulatory information. For each topic, the SDS must also indicate if no relevant information was found.

Hierarchy of Controls

Occupational safety and health professionals use a framework called "hierarchy of controls" to select ways of dealing with workplace hazards, such as chemical hazards (Ref. 2). In this hierarchy, the types of measures that may be used to protect laboratory workers, prioritized from the most effective to the least effective, are:

1. Engineering controls
2. Administrative controls
3. Work practices
4. Personal protective equipment

Often a combination of control methods is necessary to minimize exposures to chemical hazards. A description of each type of control follows.

1. Engineering Controls

Engineering controls involve making changes to the work environment to reduce work-related hazards. These types of controls are preferred over all others because they make permanent changes that reduce exposure to hazards and do not rely on worker behavior. Engineering controls include protective equipment such as chemical fume hoods and glove bags. Specific measures should be taken to ensure proper and adequate performance of protective equipment.

2. Administrative Controls

Administrative controls are those that modify workers' work schedules and tasks in ways that minimize their exposure to workplace hazards. Examples of administrative controls are:

- Developing a Chemical Hygiene Plan, which provides guidelines for prudent practices and procedures for the use of chemicals, including the use of equipment, personal protective equipment, work practices capable of protecting workers, and provisions for medical consultation and examination when exposure to a hazardous chemical has or may have taken place
- Developing Standard Operating Procedures for chemical handling

3. Work Practices

Work practices are procedures for safe and proper work that are used to reduce the duration, frequency, or intensity of exposure to a hazard. An example of a work practice is chemical substitution where feasible (e.g., selecting a less hazardous chemical for a specific procedure).

4. Personal Protective Equipment

Personal protective equipment (PPE) is protective gear needed to keep workers safe while working with a chemical. Examples of PPE include respirators, face shields, goggles, protective clothing, and disposable gloves. While engineering and administrative controls and proper work practices are considered to be more effective in minimizing exposure to many chemical hazards, the use of PPE may also be very important. It is critical that the appropriate PPE be used for the type of chemical being used and for the level of exposure to the chemical. PPE should be:

- Selected based upon the hazard to the worker
- Properly fitted and in some cases periodically refitted (e.g., respirators)
- Conscientiously and properly worn
- Regularly maintained and replaced in accord with the manufacturer's specifications
- Properly removed and disposed of to avoid contamination of self, others, or the environment
- If reusable, properly removed, cleaned, disinfected, and stored

Laboratory Standard

Many research laboratories are subject to regulation for the safety of workers in the laboratory (Ref. 3). The OSHA Occupational Exposure to Hazardous Chemicals in Laboratories standard, commonly referred to as the Laboratory Standard, applies to laboratories in the United States that conduct research and development and related analytical work. It does not apply to most quality control laboratories. The purpose of the Laboratory Standard is to ensure that workers in non-production laboratories are informed about the hazards of chemicals in the workplace and are protected from chemical exposures.

The Laboratory Standard consists of five major elements:

- Chemical hazard identification
- Chemical Hygiene Plan
- Information and training
- Exposure monitoring
- Medical consultation and examinations

The Laboratory Standard requires that the employer designate a Chemical Hygiene Officer and have a written Chemical Hygiene Plan that is actively verified for continuing effectiveness. The Chemical Hygiene Plan must include provisions for worker training, chemical exposure monitoring where appropriate, medical consultation when exposure occurs, criteria for the use of PPE and engineering controls, and special precautions for particularly hazardous substances. The Chemical Hygiene Officer should be required to be responsible for implementation of the Chemical Hygiene Plan.

The Chemical Hygiene Plan must be tailored to reflect the specific chemical hazards present in the laboratory where it is to be used. Laboratory personnel must receive training regarding the Laboratory Standard, the Chemical Hygiene Plan, and other laboratory safety practices, including exposure detection, physical and health hazards associated with chemicals, and protective measures.

References

1. OSHA Hazard Communication Standard, 29 CFR 1910.1200.
2. OSHA, OSHA Brief: Hazard Communication Standard: Safety Data Sheets.
3. OSHA, Laboratory Safety Guidance, OSHA 3404-11R, 2011.

ABBREVIATIONS USED IN THE ASSESSMENT AND PRESENTATION OF LABORATORY HAZARDS

Thomas J. Bruno and Paris D. N. Svoronos

The following abbreviations are commonly encountered in presentations of laboratory and industrial hazards. The reader is urged to consult the references (Refs. 1 and 2) for additional information.

References

1. Furr, A.K., ed., *CRC Handbook of Laboratory Safety, Fifth Edition*, CRC Press, Boca Raton, FL, 2000.
2. Bruno, T. J., and Svoronos, P. D. N., *CRC Handbook of Basic Tables for Chemical Analysis, Third Edition*, CRC Press, Boca Raton, FL, 2011.

ABBREVIATION: DEFINITION

ALARA: As low as reasonably achievable

ALARP: As low as reasonably practicable

CC: Closed Cup; method for the measurement of the flash point. With this method, sample vapors are not allowed to escape as they can with the open cup method. Because of this, flash points measured with the CC method are usually a few degrees lower than those measured with the OC. The choice between CC and OC is dependent on the (usually ASTM) standard method chosen for the test.

COC: Cleveland Open Cup, see OC: Open Cup

IDLH: Immediately Dangerous to Life and Health; the maximum concentration of chemical contaminants, normally expressed as parts per million (ppm, mass/mass), from which one could escape within 30 minutes without a respirator, and without experiencing any escape impairment (severe eye irritation) or irreversible health effects. Set by NIOSH. Note that this term is also used to describe electrical hazards.

LEL: Lower Explosion Limit; the minimum concentration of a chemical in air at which detonation can occur.

LFL: Lower Flammability Limit; the minimum concentration of a chemical in air at which flame propagation occurs.

MSDS: Material Safety Data Sheet; a (legal) document that must accompany any supplied chemical that provides information on chemical content, physical properties, hazards, and treatment of hazards. The MSDS should be considered only a minimal source of information, and cannot replace additional information available in other, more comprehensive sources. (See also SDS [safety data sheets].)

NOEL: No Observed Effect Level; the maximum dose of a chemical at which no signs of harm are observed. This term can also be used to describe hazards other than chemical hazards, such as electrical hazards.

OC: Open Cup; also called Cleveland Open Cup. This refers to the test method for determining the flash point of common compounds. It consists of a brass, aluminum or stainless steel cup, a heater base to heat the cup, a thermometer in a fixture, and a test flame applicator. The flash point is the lowest temperature at which a material will form a flammable mixture with air above its surface. The lower the flash point, the easier it is to ignite.

PEL: Permissible Exposure Level; an exposure limit that is published and enforced by OSHA as a legal standard. The PEL may be expressed as a time-weighted average (TWA) exposure limit (for an 8-hour workday), a 15-minute short-term exposure limit (STEL), or a ceiling (C, or CEIL, or TLV-C).

REL: Recommended Exposure Level; average concentration limit recommended for up to a 10-hour workday during a 40-hour workweek, by NIOSH.

RTECS: Registry of Toxic Effects of Chemical Substances; a database maintained by the National Institute of Occupational Safety and Health (NIOSH). The goal of the database is to include data on all known toxic substances, along with the concentration at which toxicity is known to occur. There are approximately 140,000 compounds listed.

SDS: Safety Data Sheet; a (legal) document that must accompany any supplied chemical that provides information on chemical content, physical properties, hazards, and treatment of hazards. The SDS should be considered only a minimal source of information, and cannot replace additional information available in other, more comprehensive sources. This document was formerly referred to as the material safety data sheet (MSDS).

STEL: Short-Term Exposure Level; an exposure limit for a short-term – 15-minute – exposure that cannot be exceeded during the workday, enforced by OSHA as a legal standard. Short-term exposures below the STEL level generally will not cause irritation, chronic or reversible tissue damage, or narcosis.

TLV: Threshold Limit Value; guidelines suggested by the American Conference of Governmental Industrial Hygienists to assist industrial hygienists in limiting hazards of chemical exposures in the workplace.

TLV-C: Threshold Limit Ceiling Value; an exposure limit which should not be exceeded under any circumstances.

TWA: Time-weighted average concentration for a conventional 8-hour workday and a 40-hour workweek. It is the concentration to which it is believed possible that nearly all workers can be exposed without adverse health effects.

UEL: Upper Explosion Limit; the maximum concentration of a chemical in air at which detonation can occur.

UFL: Upper Flammability Limit; the maximum concentration of a chemical in air at which flame propagation can occur.

WEEL: Workplace Environmental Exposure Limit; set by the American Industrial Hygiene Association (AIHA).

Some abbreviations that are sometimes used on material safety data sheets, safety data sheets, and in other sources, are ambiguous. The most common meanings of some of these vague abbreviations are provided below, but the reader is cautioned that these are only suggestions:

EST: Established; estimated

MST: Mist

N/A, NA: Not applicable

ND: None determined; not determined

NE: None established; not established
NEGL: Negligible
NF: None found; not found
N/K, NK: Not known
N/P, NP: Not provided
SKN: Skin
TS: Trade secret
UKN: Unknown

INCOMPATIBLE CHEMICALS

The term "incompatible chemicals" refers to chemicals that can react with each other

- Violently
- With evolution of substantial heat
- To produce flammable products
- To produce toxic products

Good laboratory safety practice requires that incompatible chemicals be stored, transported, and disposed of in ways that will prevent their coming together in the event of an accident. Tables 1 and 2 give some basic guidelines for the safe handling of acids, bases, reactive metals, and other chemicals. Neither of these tables is exhaustive, and additional information on incompatible chemicals can be found in the following references.

References

1. Urben, P. G., Ed., *Bretherick's Handbook of Reactive Chemical Hazards, Fifth Edition*, Butterworth-Heinemann, Oxford, 1995.
2. Luxon, S. G., Ed., *Hazards in the Chemical Laboratory, Fifth Edition*, Royal Society of Chemistry, Cambridge, 1992.
3. *Fire Protection Guide to Hazardous Materials, 11th Edition*, National Fire Protection Association, Quincy, MA, 1994.

TABLE 1. General Classes of Incompatible Chemicals

A	B
Acids	Bases, reactive metals
Oxidizing agents[a]	Reducing agents[a]
Chlorates	Ammonia, anhydrous and aqueous
Chromates	Carbon
Chromium trioxide	Metals
Dichromates	Metal hydrides
Halogens	Nitrites
Halogenating agents	Organic compounds
Hydrogen peroxide	Phosphorus
Nitric acid	Silicon
Nitrates	Sulfur
Perchlorates	
Peroxides	
Permanganates	
Persulfates	

[a] The examples of oxidizing and reducing agents are illustrative of common laboratory chemicals; they are not intended to be exhaustive.

TABLE 2. Examples of Incompatible Chemicals

Chemical	Is incompatible with
Acetic acid	Chromic acid, nitric acid, hydroxyl compounds, ethylene glycol, perchloric acid, peroxides, permanaganates
Acetylene	Chlorine, bromine, copper, fluorine, silver, mercury
Acetone	Concentrated nitric and sulfuric acid mixtures
Alkali and alkaline earth metals (such as powdered aluminum or magnesium, calcium, lithium, sodium, potassium)	Water, carbon tetrachloride or other chlorinated hydrocarbons, carbon dioxide, halogens
Ammonia (anhydrous)	Mercury (in manometers, for example), chlorine, calcium hypochlorite, iodine, bromine, hydrofluoric acid (anhydrous)
Ammonium nitrate	Acids, powdered metals, flammable liquids, chlorates, nitrites, sulfur, finely divided organic or combustible materials
Aniline	Nitric acid, hydrogen peroxide
Arsenical materials	Any reducing agent
Azides	Acids
Bromine	Ammonia, acetylene, butadiene, butane, methane, propane (or other petroleum gases), hydrogen, sodium carbide, benzene, finely divided metals, turpentine
Calcium oxide	Water
Carbon (activated)	Calcium hypochlorite, all oxidizing agents
Carbon tetrachloride	Sodium
Chlorates	Ammonium salts, acids, powdered metals, sulfur, finely divided organic or combustible materials
Chromic acid and chromium troixide	Acetic acid, naphthalene, camphor, glycerol, alcohol, flammable liquids in general

Safety

Chemical	Is incompatible with
Chlorine	Ammonia, acetylene, butadiene, butane, methane, propane (or other petroleum gases), hydrogen, sodium carbide, benzene, finely divided metals, turpentine
Chlorine dioxide	Ammonia, methane, phosphine, hydrogen sulfide
Copper	Acetylene, hydrogen peroxide
Cumene hydroperoxide	Acids (organic or inorganic)
Cyanides	Acids
Flammable liquids	Ammonium nitrate, chromic acid, hydrogen peroxide, nitric acid, sodium peroxide, halogens
Fluorine	Everything
Hydrocarbons (such as butane, propane, benzene)	Fluorine, chlorine, bromine, chromic acid, sodium peroxide
Hydrocyanic acid	Nitric acid, alkali
Hydrofluoric acid (anhydrous)	Ammonia (aqueous or anhydrous)
Hydrogen peroxide	Copper, chromium, iron, most metals or their salts, alcohols, acetone, organic materials, aniline, nitro-methane, combustible materials
Hydrogen sulfide	Fuming nitric acid, oxidizing gases
Hypochlorites	Acids, activated carbon
Iodine	Acetylene, ammonia (aqueous or anhydrous), hydrogen
Mercury	Acetylene, fulminic acid, ammonia
Nitrates	Sulfuric acid
Nitric acid (concentrated)	Acetic acid, aniline, chromic acid, hydrocyanic acid, hydrogen sulfide, flammable liquids, flammable gases, copper, brass, any heavy metals
Nitrites	Acids
Nitroparaffins	Inorganic bases, amines
Oxalic acid	Silver, mercury
Oxygen	Oils, grease, hydrogen, flammable liquids, solids, or gases
Perchloric acid	Acetic anhydride, bismuth and its alloys, alcohol, paper, wood, grease, oils
Peroxides, organic	Acids (organic or mineral), avoid friction, store cold
Phosphorus (white)	Air, oxygen, alkalis, reducing agents
Potassium	Carbon tetrachloride, carbon dioxide, water
Potassium chlorate	Sulfuric and other acids
Potassium perchlorate (see also chlorates)	Sulfuric and other acids
Potassium permanganate	Glycerol, ethylene glycol, benzaldehyde, surfuric acid
Selenides	Reducing agents
Silver	Acetylene, oxalic acid, tartartic acid, ammonium compounds, fulminic acid
Sodium	Carbon tetrachloride, carbon dioxide, water
Sodium nitrite	Ammonium nitrate and other ammonium salts
Sodium peroxide	Ethyl or methyl alcohol, glacial acetic acid, acetic anhydride, benzaldehyde, carbon disulfide, glycerin, ethylene glycol, ethyl acetate, methyl acetate, furfural
Sulfides	Acids
Sulfuric acid	Potassium chlorate, potassium perchlorate, potassium permanganate (similar compounds of light metals, such as sodium, lithium)
Tellurides	Reducing agents

EXPLOSION (SHOCK) HAZARDS

Table 1 lists some common classes of laboratory chemicals that have the potential for producing a violent explosion when subjected to shock or friction. These chemicals should never be disposed of as such, but should be handled by procedures given in Ref. 1, especially Chapter 8. Additional information on these, as well as on some less common classes of explosives, can be found in Ref. 2.

Table 2 lists some illustrative combinations of common laboratory reagents that can produce explosions when they are brought together or that form reaction products that can explode without any apparent external initiating action. This list is not exhaustive, and additional information on potentially explosive reagent combinations can be found in Ref. 3.

References

1. National Research Council, *Prudent Practices in the Laboratory: Handling and Management of Chemical Hazards, Updated Version*, National Academy Press, Washington, DC, 2011. <doi.org.10.17226/12654>
2. Bretherick, L., *Handbook of Reactive Chemical Hazards, Eighth Edition*, Elsevier, 2017.
3. *Manual of Hazardous Chemical Reactions: A Compilation of Chemical Reactions Reported to Be Potentially Hazardous*, NFPA 491M, National Fire Protection Association, Quincy, MA, 1991.

TABLE 1. Shock-Sensitive Compounds

Acetylenic compounds, especially polyacetylenes, haloacetylenes, and heavy metal salts of acetylenes (copper, silver, and mercury salts are particularly sensitive)
Acyl nitrates
Alkyl nitrates, particularly polyol nitrates such as nitrocellulose and nitroglycerine
Alkyl and acyl nitrites
Alkyl perchlorates
Amminemetal oxosalts: metal compounds with coordinated ammonia, hydrazine, or similar nitrogenous donors and ionic perchlorate, nitrate, permanganate, or other oxidizing group
Azides, including metal, nonmetal, and organic azides
Chlorite salts of metals, such as $AgClO_2$ and $Hg(ClO_2)_2$
Diazo compounds such as CH_2N_2
Diazonium slats, when dry
Fulminates (silver fulminate, AgCNO, can form in the reaction mixture from the Tollens' test for aldehydes if it is allowed to stand for some time; this can be prevented by adding dilute nitric acid to the test mixture as soon as the test has been completed)
Hydrogen peroxide becomes increasingly treacherous as the concentration rises above 30%, forming explosive mixtures with organic materials and decomposing violently in the presence of traces of transition metals
N–Halogen compounds such as difluoroamino compounds and halogen azides
N–Nitro compounds such as *N*–nitromethylamine, nitrourea, nitroguanidine, and nitric amide
Oxo salts of nitrogenous bases: perchlorates, dichromates, nitrates, iodates, chlorites, chlorates, and permanganates of ammonia, amines, hydroxylamine, guanidine, etc.
Perchlorate salts. Most metal, nonmetal, and amine perchlorates can be detonated and may undergo violent reaction in contact with combustible materials
Peroxides and hydroperoxides, organic (see Chapter 6, Section II.P)
Peroxides (solid) that crystallize from or are left from evaporation of peroxidizable solvents (see Chapter 6 and Appendix I)
Peroxides, transition–metal salts
Picrates, especially salts of transition and heavy metals, such as Ni, Pb, Hg, Cu, and Zn; picric acid is explosive but is less sensitive to shock or friction than its metal salts and is relatively safe as a water–wet paste (see Chapter 7)
Polynitroalkyl compounds such as tetranitromethane and dinitroacetonitrile
Polynitroaromatic compounds, especially polynitro hydrocarbons, phenols, and amines

Safety

TABLE 2. Potentially Explosive Combinations of Some Common Reagents

Acetone + chloroform in the presence of base
Acetylene + copper, silver, mercury, or their salts
Ammonia (including aqueous solutions) + Cl_2, Br_2, or I_2
Carbon disulfide + sodium azide
Chlorine + an alcohol
Chloroform or carbon tetrachloride + powdered Al or Mg
Decolorizing carbon + an oxidizing agent
Diethyl ether + chlorine (including a chlorine atmosphere)
Dimethyl sulfoxide + an acyl halide, $SOCl_2$ or $POCl_3$
Dimethyl sulfoxide + CrO_3
Ethanol + calcium hypochlorite
Ethanol + silver nitrate
Nitric acid + acetic anhydride or acetic acid
Picric acid + a heavy–metal salt, such as of Pb, Hg, or Ag
Silver oxide + ammonia + ethanol
Sodium + a chlorinated hydrocarbon
Sodium hypochlorite + an amine

WATER-REACTIVE CHEMICALS

This table lists some common laboratory chemicals that react violently with water and that should always be stored and handled so that they do not come into contact with liquid water or water vapor. Procedures for decomposing laboratory quantities are given in Ref. 1, Chapter 6; the pertinent section of that chapter is given in parentheses. Additional information is available in Ref. 2.

References

1. Joyce, R., and McKusick, B. C., *Prudent Practices for Disposal of Chemicals from Laboratories*, National Academy Press, Washington, DC, 1983.
2. National Research Council, *Prudent Practices in the Laboratory: Handling and Management of Chemical Hazards, Updated Version*, National Academy Press, Washington, DC, 2011. <doi.org.10.17226/12654>.

Examples of Water-Reactive Chemicals

Alkali metals (III.D)
Alkali metal hydrides (III.C.2)
Alkali metal amides (III.C.7)
Metal alkyls, such as lithium alkyls and aluminum alkyls (IV.A)
Grignard reagents (IV.A)
Halides of nonmetals, such as BCl_3, BF_3, PCl_3, PC_5, SiC_4, S_2Cl_2 (III.F)
Inorganic acid halides, such as $POCl_3$, $SOCl_2$, SO_2Cl_2 (III.F)
Anhydrous metal halides, such as $AlCl_3$, $TiCl_4$, $ZrCl_4$, $SnCl_4$ (III.E)
Phosphorus pentoxide (III.I)
Calcium carbide (IV.E)
Organic acid halides and anhydrides of low-molecular weight (II.J)

TESTING REQUIREMENTS FOR PEROXIDIZABLE COMPOUNDS

Thomas J. Bruno and Paris D. N. Svoronos

Because some compounds form peroxides more easily or faster than others, prudent practices require testing the supply on hand in the laboratory on a periodic basis. The following list provides guidelines on test scheduling (Refs. 1 and 2). The peroxide hazard of the compounds listed in Group 1 is on the basis of time in storage. The compounds in Group 2 present a peroxide hazard primarily due to concentration, mainly by evaporation of the liquid. The compounds listed in Group 3 are hazardous because of the potential of peroxide-initiated polymerization. When stored as liquids, the peroxide formation may increase, and therefore these compounds should be treated as Group 1 peroxidizable compounds. See Ref. 3 for additional information.

References

1. Ringen, S., Environmental Health and Safety Manual - Chemical Safety, Sec. 4-50, University of Wyoming, June 2000.
2. Bruno, T. J., and Svoronos, P. D. N., *CRC Handbook of Basic Tables for Chemical Analysis, Third Edition*, CRC Press, Boca Raton, FL, 2011. <https://doi.org/10.1201/b10385>
3. National Research Council, *Prudent Practices in the Laboratory: Handling and Management of Chemical Hazards, Updated Version*, National Academy Press, Washington, DC, 2011. <doi.org.10.17226/12654>.

Group 1 Test every 3 months
- Divinylacetylene
- Isopropyl ether
- Potassium
- Sodium amide
- Vinylidene chloride

Group 2 Test every 6 months
- Acetal
- Cumene
- Cyclohexene
- Diacetylene
- Dicyclopentadiene
- Diethyl ether
- Dimethyl ether
- 1,4-Dioxane
- Ethylene glycol dimethyl ether (glyme)
- Methylacetylene
- Methyl isobutyl ketone
- Methylcyclopentane
- Tetrahydrofuran
- Tetrahydronaphthalene (tetralin)
- Vinyl ethers

Group 3 Test every 12 months
- Acrylic acid
- Acrylonitrile
- Butadiene
- Chloroprene
- Chlorotrifluoroethene
- Methyl methacrylate
- Styrene
- Tetrafluoroethylene
- Vinyl acetate
- Vinyl acetylene
- Vinyl chloride
- Vinyl pyridine

TESTS FOR THE PRESENCE OF PEROXIDES

Thomas J. Bruno and Paris D. N. Svoronos

Peroxides may be detected qualitatively with one of the following test procedures (Refs. 1 and 2).

References

1. Gordon, A. J., and Ford, R. A., *The Chemist's Companion*, John Wiley and Sons, New York, 1972.
2. Bruno, T. J., and Svoronos, P. D. N., *CRC Handbook of Basic Tables for Chemical Analysis, Third Edition*, CRC Press, Boca Raton, FL, 2011 <https://doi.org/10.1201/b10385>.

Ferrithiocyanate Test

Reagent Preparation, in sequence:

Add 9 g $FeSO_4 \cdot 7H_2O$ to 50 mL 18% (vol/vol) $HCl_{(aq)}$
Add 1 mg to 3 mg granular Zn
Add 5 g NaSCN

After the red color fades, add an additional 12 g NaSCN, decant leaving unreacted Zn.

Upon mixing this reagent with a peroxide-containing liquid, the colorless solution will produce a red color, the result of the conversion of ferrothiocyanide to ferrithiocyanide. This test is very sensitive and can be used to detect peroxides at a concentration of 0.001% (mass/mass).

Potassium Iodide Test

Reagent Preparation:
Make a 10% (mass/mass) solution of KI in water.
Upon mixing this reagent with a peroxide-containing liquid, a yellow color will appear within one minute.

Acidic Iodide Test

Reagent Preparation:
To 1 mL of glacial acetic acid, add 100 mg KI or NaI.
Upon mixing this reagent with an equal volume of a peroxide-containing liquid, a yellow coloration will appear. The color will appear dark or even brown if the peroxide concentration is very high.

Perchromate Test

Reagent Preparation:
Dissolve 1 mg of $Na_2Cr_2O_7$ in 1 mL of water, add a drop of dilute H_2SO_4 (aq).
Upon mixing this reagent with a peroxide-containing liquid, a blue color will develop in the organic layer indicating the formation of the perchromate ion.

PYROPHORIC COMPOUNDS – COMPOUNDS THAT ARE REACTIVE WITH AIR

Thomas J. Bruno and Paris D. N. Svoronos

The following listing provides the classes of compounds, with some examples, that can undergo spontaneous reaction upon exposure to air (Refs. 1 to 4). In some cases, the reaction is vigorous, while in others, the reaction is more subdued, or will only occur if other conditions (such as temperature, humidity, or a reactive surface) are present. The reader is advised to check the literature for more specific information.

References

1. Bretherick, L., *Handbook of Reactive Chemical Hazards, Eighth Edition*, Elsevier, 2017.
2. Pyrophoric Materials, <www.safety.science.tamu.edu/pyrophorics.html>, Texas A&M University, 2003.
3. Virginia Commonwealth University Office of Environmental Health & Safety, Chemical/Biological Safety Section, Pyrophoric Chemicals: Safe Work Practices Information Page, <http://www.vcu.edu/oehs/ chemical/biosafe/pyrophorics.pdf>.
4. Bruno, T. J., and Svoronos, P. D. N., *CRC Handbook of Basic Tables for Chemical Analysis, Third Edition*, CRC Press, Boca Raton, FL, 2011. <https://doi.org/10.1201/b10385>

Alkali metals* (sodium, potassium, potassium/sodium alloy, lithium/tin alloys)
Alkylaluminum derivatives (diethylaluminum hydride)
Alkylated metal alkoxides (diethylethoxyaluminum)
Alkylboranes
Alkylhaloboranes (bromodimethyl borane)
Alkylhalophosphines
Alkylhalosilanes
Alkyl metals
Alkyl nonmetal hydrides
Boranes (diborane)
Carbonyl metals (pentacarbonyl iron, octacarbonyl dicobalt, nickel carbonyl)
Complex acetylides
Complex hydrides (diethylaluminum hydride)
Finely divided metals* (calcium, zirconium)
Grignard reagents
Haloacetylene derivatives
Hexamethylnitrato dialluminum salts
Metal hydrides (germane, sodium hydride, lithium aluminum hydride)
Nonmetal hydrides
Some nonmetal (organic) halides (dichloro(methyl)silane)
Spent hydrogenation catalysts (can be especially hazardous because of adsorbed hydrogen; for example, Raney nickel)
White phosphorus

* Note that the reactivity depends on the particle size and the ease at which oxides are formed on the metal surface.

FLAMMABILITY HAZARDS OF COMMON SOLVENTS

Thomas J. Bruno and Paris D. N. Svoronos

The following table lists relevant data regarding the flammability of common organic solvents. The codes for entries in the last column are given in the footnotes. Refs. 1-2 contain additional information.

See "Flammability of Chemical Substances" in this section for a more extensive list of compounds.

References

1. Turner, C. F., and McCreery, J. W., *The Chemistry of Fire and Hazardous Materials*, Allyn & Bacon, Boston, 1981.
2. *Fire Protection Guide to Hazardous Materials, 11th Edition*, National Fire Protection Association, Quincy, MA, 2010.

Flammability Hazards of Common Solvents

Name	Mol. form.	Density/g cm^{-3}	Boiling point/°C	Flash point/°C	Autoignition point/°C	How to extinguish fires
Acetaldehyde	CH_3CHO	0.7834^{18}	20.8	-39	175	a,b,c
Acetone	$(CH_3)_2CO$	0.7902^{20}	56.08	-20	465	a,b
Acetonitrile	CH_3CN	0.7825^{20}	81.6	6	524	a,c,d
Acrolein	$CH_2{=}CHCHO$	0.840^{20}	52.3	-26	220	a,b,c
Acrylonitrile	$CH_2{=}CHCN$	0.8007^{25}	77.2	0	481	a,c,d
Allylamine	$CH_2{=}CHCH_2NH_2$	0.758^{20}	54	-29	374	a,b
Aniline	$C_6H_5NH_2$	1.0250^{20}	184.1	70	615	a,b,c; use masks
Anisole	C_7H_8O	0.9940^{20}	153.6	52	475	a,b,c
Benzaldehyde	C_6H_5CHO	1.0401^{25}	178.7	63	192	a,b,c
Benzene	C_6H_6	0.8788^{20}	80.08	-11	498	a,b,c
Butanal	C_4H_8O	0.8016^{20}	74.8	-22	218	a,b,c
2,3-Butanedione	$C_4H_6O_2$	0.9808^{18}	87.5	27		a,b,c
1-Butanol	C_4H_9OH	0.8148^{20}	117.6	37	343	a,b,c
2-Butanone	C_4H_8O	0.7999^{25}	79.6	-9	404	a,b,c
2-Butanone peroxide	$C_8H_{16}O_4$		63 exp			a,b
trans-2-Butenal	C_4H_6O	0.8516^{20}	102.2	13	232	a,b,c
tert-Butyl peroxyacetate	$C_6H_{12}O_3$	>1.0	expl	<27		a,b,c
tert-Butyl peroxybenzoate	$C_{11}H_{14}O_3$	1.021^{25}	118 expl	>88	8	b,c
Carbon disulfide	CS_2	1.2632^{20}	46.2	-30	90	b,d; use masks
Cyclohexanone	$C_6H_{10}O$	0.9478^{20}	155.4	44	420	a,b,c
Diethanolamine	$C_4H_{11}NO_2$	1.0966^{20}	271.2	172	662	b,c
Diethylene glycol diethyl ether	$C_8H_{18}O_3$	0.9063^{20}	185	82		a,e
Diethyl ether	$(C_2H_5)_2O$	0.7135^{20}	34.4	-45	180	a,b,e
Diisopropyl ether	$C_6H_{14}O$	0.7192^{25}	68.4	-28	443	a,b
Dimethyl sulfate	$C_2H_6O_4S$	1.3322^{20}	186	83	188	a,b,c,d
Dimethyl sulfide	$(CH_3)_2S$	0.8483^{20}	37.32	-37	206	b,c
1,4-Dioxane	$C_4H_8O_2$	1.0337^{20}	101.2	12	180	a,b,c
1,2-Ethanediamine	$C_2H_8N_2$	0.8979^{20}	116.9	40	385	a,b,c
1,2-Ethanediol	$(CH_2OH)_2$	1.1135^{20}	197.5	111	398	a,b,c,d
Ethanol	C_2H_5OH	0.7893^{20}	78.24	13	363	a,b,c
Ethylamine	$C_2H_5NH_2$	0.689^{15}	16.6	-16	385	a,b,c
Formaldehyde	HCHO	0.815^{-20}	-19.1	85	424	a,b,c
Furfural	$C_5H_4O_2$	1.1594^{20}	161.5	60	316	a,b,c,d
Furfuryl alcohol	$C_5H_6O_2$	1.1296^{20}	168	75	491	a,b,c
Gasoline[f]		≈0.8	38-204	-43	280	a,b,c
Hexylamine	$C_6H_{13}NH_2$	0.7660^{20}	132	29		a,b
Isopropylbenzene hydroperoxide	$C_9H_{12}O_2$	1.03^{20}	153 expl	79		a,b,c
Kerosene[f]		<1.0	150-300	38-72	210	a,b,c
Methanol	CH_3OH	0.7909^{20}	64.5	11	464	a,b
Methylamine	CH_3NH_2	0.656^{25} (p>1 atm)	-6.4	0	430	a,b,c
N-Methylaniline	C_7H_9N	0.9859^{20}	197		533	a,b
Naphtha[f]		<1.0	100-160[h]	0-30[h]	230[h]	a,b,c
Paraldehyde	$C_6H_{12}O_3$	0.9943^{20}	124	36	238	a,b,c
2,4-Pentanedione	$CH_3COCH_2COCH_3$	0.9721^{25}	140.7	34	340	a,b,c
1-Pentanethiol	$C_5H_{11}SH$	0.850^{20}	126.6	18		a,b
2-Pentanone	$C_5H_{10}O$	0.809^{20}	102.2	7	452	a,b,c

Name	Mol. form.	Density/g cm^{-3}	Boiling point/°C	Flash point/°C	Autoignition point/°C	How to extinguish fires
2-Pentanone	$C_5H_{10}O$	0.809^{20}	102.2	7	452	
3-Pentanone	$C_5H_{10}O$	0.8098^{25}	101.9	13	450	a,b,c
3-Pentenenitrile	C_5H_7N	0.83^{i}	14.5	40		b,c,d
Peroxyacetic acid	$C_2H_4O_3$	1.226^{15}	110	41		a,b,c
Petroleum ether[f]		≈0.6	35-60	<-18	288	a,b,c
Propanal	C_2H_5CHO	0.8657^{25}	48.0	-30	207	a,b,c
1,3-Propanediol	$C_3H_8O_2$	1.0538^{20}	214.7		400	a,b
2-Propanol	$CH_3CHOHCH_3$	0.7855^{20}	82.21	12	399	b,c
Propylamine	$C_3H_7NH_2$	0.7173^{20}	47.21	-37	318	a,b,c
Sulfur dichloride	SCl_2	1.62	59.6			b,c
Sulfuryl chloride	SO_2Cl_2	1.680	69.4			a,b,c
Tetrahydrofuran	C_4H_8O	0.8833^{25}	66.0	-14	321	a,b,c
Thionyl chloride	$SOCl_2$	1.631	75.6			
Toluene	$C_6H_5CH_3$	0.8668^{20}	110.60	4	480	a,b,c
Tris(2-hydroxyethyl)amine	$C_6H_{15}NO_3$	1.1242^{20}	350	179		b,c,d
Triethylamine	$(C_2H_5)_3N$	0.7275^{20}	88.8	-7	249	a,b,c
o-Xylene	$C_6H_4(CH_3)_2$	0.8802^{20}	144.4	32	463	a,b,c
m-Xylene	$C_6H_4(CH_3)_2$	0.8641^{20}	139.1	27	527	a,b,c
p-Xylene	$C_6H_4(CH_3)_2$	0.8610^{20}	138.3	27	528	a,b,c

[a] Extinguish using foam.
[b] Extinguish using carbon dioxide.
[c] Extinguish using dry chemical (ABC).
[d] Extinguish using water.
[e] Extinguish using halons.
[f] Mixture of C_5 to C_{10} hydrocarbons, mostly aliphatic (C_5 to C_6 for petroleum ether).
[g] Aqueous solution.
[h] Varies widely with manufacturer.

FLAMMABILITY OF CHEMICAL SUBSTANCES

These tables give properties related to the flammability for chemical compounds and mixtures. The column definitions are as follows. Not all columns are in each table.

Column heading	Definition
Name	Name of chemical substance; substances are listed alphabetically by name
Mol. form.	Molecular formula of chemical substance
t_b	Normal boiling point, in °C, at 101.325 kPa pressure ("sp" indicates sublimation point)
FP	Flash point, in °C, which is the minimum temperature at which the vapor pressure of a liquid is sufficient to form an ignitable mixture with air near the surface of the liquid. Flash point is not an intrinsic physical property but depends on the conditions of measurement (see Ref. 1)
Flam. limits	Flammable limits (often called explosive limits), which specify the range of concentration of the vapor in air (in percent by volume) for which a flame can propagate; below the lower flammable limit, the gas mixture is too lean to burn; above the upper flammable limit, the mixture is too rich; values refer to ambient temperature and pressure and are dependent on the precise test conditions; a ? indicates that one of the limits is not known
IT	Ignition temperature (sometimes called autoignition temperature), in °C, which is the minimum temperature required for self-sustained combustion in air in the absence of an external ignition source; as in the case of flash point, the value depends on specified test conditions
VP	Vapor pressure at 25 °C, in kPa

Even in cases where very careful measurements of flash point have been replicated in several laboratories, observed values can differ by 3 to 6 °C (Ref. 4). For more typical measurements, larger uncertainties should be assumed in both flash points and autoignition temperatures. The absence of a flash point entry in this table does not mean that the substance is nonflammable, but only that no reliable value is available.

Table 1 covers selected fuels and other mixtures. Table 2 gives data for pure compounds.

References

1. *Fire Protection Guide to Hazardous Materials, 11th Edition*, National Fire Protection Association, Quincy, MA, 1994.
2. Urben, P. G., Ed., *Bretherick's Handbook of Reactive Chemical Hazards, Fifth Edition*, Butterworth-Heinemann, Oxford, 1995.
3. Daubert, T. E., Danner, R. P., Sibul, H. M., and Stebbins, C. C., *Physical and Thermodynamic Properties of Pure Compounds: Data Compilation*, extant 1994 (core with 4 supplements), Taylor & Francis, Bristol, PA.
4. *Report of Investigation: Flash Point Reference Materials*, National Institute of Standards and Technology, Standard Reference Materials Program, Gaithersburg, MD, 1995.
5. Alternative Fuels Data Center-Fuel Properties Comparison, <www.afdc.energy.gov/fuels/fuel_comparison_chart.pdf>, October 2014.

TABLE 1. Flammability of Common Mixtures

Name	Type	FP/°C	IT/°C	Flam. limits
Gasoline	ethanol ≤10%	-43	257	1.4-7.6%
Diesel fuel	petroleum	74	≈320	
Diesel fuel	bio	100-170	≈150	
Fuel oil	light	69-169		
Fuel oil	heavy	71-121		
Natural gas			≈540	5-15%
Kerosene		38-72	210	0.7-5%
Naphtha		<-18	288	1.1-5.9%
Ethanol	96% with H_2O	17		
Ethanol	45% (90 proof)	25		
Ethanol	12%	45		

TABLE 2. Flash Point, Flammability Limits, and Ignition Temperature of Pure Substances

Name	Mol. form.	t_b/°C	FP/°C	Flam. limits	IT/°C	VP (25 °C)/kPa
Acetaldehyde	C_2H_4O	20.8	-39	4.0-60%	175	120
Acetanilide	C_8H_9NO	292	169		530	
Acetic acid	$C_2H_4O_2$	117.9	39	4.0-19.9%	463	2.07
Acetic anhydride	$C_4H_6O_3$	139.5	49	2.7-10.3%	316	0.680
Acetoacetanilide	$C_{10}H_{11}NO_2$		185			
Acetone	C_3H_6O	56.08	-20	2.5-12.8%	465	30.8
Acetone cyanohydrin	C_4H_7NO	180	74	2.2-12.0%	688	
Acetonitrile	C_2H_3N	81.6	6	3.0-16.0%	524	11.9

Name	Mol. form.	t_b/°C	FP/°C	Flam. limits	IT/°C	VP (25 °C)/ kPa
Acetophenone	C_8H_8O	202.1	77		570	0.049
Acetyl chloride	C_2H_3ClO	51	4		390	38.4
Acetylene	C_2H_2	-84.7 sp		2.5-100%	305	
N-Acetylethanolamine	$C_4H_9NO_2$		179		460	
3-Acetyl-6-methyl-2*H*-pyran-2,4(3*H*)-dione	$C_8H_8O_4$	270	157			
4-Acetylmorpholine	$C_6H_{11}NO_2$		113			
Acrolein	C_3H_4O	52.3	-26	2.8-31%	220	36.2
Acrylic acid	$C_3H_4O_2$	142	50	2.4-8.0%	438	0.53
Acrylonitrile	C_3H_3N	77.2	0	3.0-17.0%	481	14.1
Allyl acetate	$C_5H_8O_2$	104	22		374	
Allyl alcohol	C_3H_6O	96.9	21	2.5-18.0%	378	3.14
Allylamine	C_3H_7N	54	-29	2.2-22%	374	33.1
4-Allyl-1,2-dimethoxybenzene	$C_{11}H_{14}O_2$	254.7	99			
Allyl hexanoate	$C_9H_{16}O_2$	186	66			
Allyltrichlorosilane	$C_3H_5Cl_3Si$	118	35			
Allyl vinyl ether	C_5H_8O	66	<20			
2-Aminobiphenyl	$C_{12}H_{11}N$	298.3			450	
2-Amino-1-butanol, (±)-	$C_4H_{11}NO$	178	74			
N-(2-Aminoethyl)ethanolamine	$C_4H_{12}N_2O$	242	132			
2-Amino-2-methyl-1-propanol	$C_4H_{11}NO$	163.8	67			
3-Amino-1-propanol	C_3H_9NO	185	80			
1-Amino-2-propanol	C_3H_9NO	141	77		374	0.165
Ammonia	H_3N	-33.33		16-25%		
Aniline	C_6H_7N	184.1	70	1.3-11%	615	0.090
Aniline hydrochloride	C_6H_8ClN		193			
Anisole	C_7H_8O	153.6	52		475	0.472
Anthracene	$C_{14}H_{10}$	341.3	121	>0.6%	540	
9,10-Anthracenedione	$C_{14}H_8O_2$	377	185			
Benzaldehyde	C_7H_6O	178.7	63		192	0.169
Benzene	C_6H_6	80.08	-11	1.2-7.8%	498	12.7
Benzeneacetaldehyde	C_8H_8O	202	71			0.052
Benzeneacetic acid	$C_8H_8O_2$	268	>100			
Benzeneacetonitrile	C_8H_7N	232	113			0.012
1,2-Benzenediamine	$C_6H_8N_2$	257	156	>1.5%		
1,3-Benzenedicarbonyl dichloride	$C_8H_4Cl_2O_2$	276	180			
1,4-Benzenedicarbonyl dichloride	$C_8H_4Cl_2O_2$	258	180			
Benzeneethanol	$C_8H_{10}O$	220	96			0.01
Benzoic acid	$C_7H_6O_2$	250.2	121		570	
Benzoyl chloride	C_7H_5ClO	201	72			0.084
Benzyl acetate	$C_9H_{10}O_2$	215	90		460	0.022
Benzyl alcohol	C_7H_8O	205.3	93		436	0.015
Benzyl benzoate	$C_{14}H_{12}O_2$	321.3	148		480	
Benzyl methyl ether	$C_8H_{10}O$	175	135			
2-(Benzyloxy)ethanol	$C_9H_{12}O_2$	256	129		352	
Benzyl salicylate	$C_{14}H_{12}O_3$	320	>100			
[1,1'-Bicyclohexyl]-1-ol	$C_{12}H_{22}O$		132			
Biphenyl	$C_{12}H_{10}$	255.2	113	0.6-5.8%	540	
Bis(2-aminoethyl)amine	$C_4H_{13}N_3$	206.5	98	2-6.7%	358	0.03
N,N'-Bis(2-aminoethyl)-1,2-ethanediamine	$C_6H_{18}N_4$	266.5	135			
1,2-Bis(2-chloroethoxy)ethane	$C_6H_{12}Cl_2O_2$	214	121			
Bis(2-chloroethyl) ether	$C_4H_8Cl_2O$	178	55	>2.7%	369	0.143
Bis(2-ethylhexyl)amine	$C_{16}H_{35}N$		132			
Bis(2-ethylhexyl) azelate	$C_{25}H_{48}O_4$		227	>0.3%	374	
Bis(2-ethylhexyl) hexanedioate	$C_{22}H_{42}O_4$		206	>0.4%	377	
Bis(2-ethylhexyl) phthalate	$C_{24}H_{38}O_4$	384	218			0.00000005
Bis(2-hydroxyethyl)methylamine	$C_5H_{13}NO_2$	245	127			
Bis(2-hydroxyethyl) sulfide	$C_4H_{10}O_2S$	282	160		298	0.08
Bis(2-methoxyethyl) phthalate	$C_{14}H_{18}O_6$		187			
Borneol, (±)-	$C_{10}H_{18}O$	213	66			
Bromobenzene	C_6H_5Br	155.9	51		565	0.556

Name	Mol. form.	t_b/°C	FP/°C	Flam. limits	IT/°C	VP (25 °C)/ kPa
4-Bromobiphenyl	$C_{12}H_9Br$	309	144			
1-Bromobutane	C_4H_9Br	101.4	18	2.6-6.6%	265	5.26
2-Bromobutane, (±)-	C_4H_9Br	91	21			9.32
1-Bromo-2-butene	C_4H_7Br	98		4.6-12.0%		
1-Bromododecane	$C_{12}H_{25}Br$	275	144			
Bromoethane	C_2H_5Br	38.2		6.8-8.0%	511	62.5
Bromoethene	C_2H_3Br	16		9-15%	530	141
Bromomethane	CH_3Br	3.4		10-16%	537	217
1-Bromopentane	$C_5H_{11}Br$	126	32			1.68
1-Bromopropane	C_3H_7Br	70.8			490	18.6
3-Bromopropene	C_3H_5Br	70.1	-1	4.4-7.3%	295	18.6
3-Bromo-1-propyne	C_3H_3Br	73	10	>3.0%	324	
Bromosilane	BrH_3Si	1.9	<0		≈20	
2-Bromotoluene	C_7H_7Br	182	79			0.17
4-Bromotoluene	C_7H_7Br	184	85			
1,3-Butadiene	C_4H_6	-4.6		2.0-12.0%	420	
Butanal	C_4H_8O	74.8	-22	1.9-12.5%	218	15.7
Butanal oxime	C_4H_9NO	152	58			
Butane	C_4H_{10}	-0.5	-60	1.9-8.5%	287	242
1,2-Butanediol, (±)-	$C_4H_{10}O_2$	196.42	40			
1,3-Butanediol	$C_4H_{10}O_2$	208.2	121		395	0.008
1,4-Butanediol	$C_4H_{10}O_2$	229.5	121			0.002
2,3-Butanediol	$C_4H_{10}O_2$	178			402	0.02
2,3-Butanedione	$C_4H_6O_2$	87.5	27			7.45
Butanenitrile	C_4H_7N	117.6	24	>1.6%	501	2.55
1-Butanethiol	$C_4H_{10}S$	98.4	2			6.07
2-Butanethiol	$C_4H_{10}S$	85.0	-23			10.8
Butanoic acid	$C_4H_8O_2$	163.7	72	2.0-10.0%	443	0.221
Butanoic anhydride	$C_8H_{14}O_3$	195	54	0.9-5.8%	279	0.07
1-Butanol	$C_4H_{10}O$	117.6	37	1.4-11.2%	343	0.86
2-Butanol	$C_4H_{10}O$	99.4	24	1.7-9.8%	405	2.32
2-Butanone	C_4H_8O	79.6	-9	1.4-11.4%	404	12.6
2-Butanone oxime	C_4H_9NO	151.5	≈70			0.396
trans-2-Butenal	C_4H_6O	102.2	13	2.1-15.5%	232	4.92
1-Butene	C_4H_8	-6.3		1.6-10.0%	385	296
cis-2-Butene	C_4H_8	3.72		1.7-9.0%	325	214
trans-2-Butene	C_4H_8	0.88		1.8-9.7%	324	234
trans-2-Butenenitrile	C_4H_5N	120	16			
2-Buten-1-ol (unspecified isomer)	C_4H_8O	122	27	4.2-35.3%	349	
3-Buten-2-one	C_4H_6O	81	-7	2.1-15.6%	491	12.0
2-Butenylbenzene	$C_{10}H_{12}$	182	71			
1-Buten-3-yne	C_4H_4	6.0		21-100%		
2-Butoxyethanol	$C_6H_{14}O_2$	171	69	4-13%	238	0.15
N-Butylacetamide	$C_6H_{13}NO$	229	116			
Butyl acetate	$C_6H_{12}O_2$	126.0	22	1.7-7.6%	425	1.66
sec-Butyl acetate	$C_6H_{12}O_2$	108	31	1.7-9.8%		
Butyl acetoacetate	$C_8H_{14}O_3$		85			
Butyl acrylate	$C_7H_{12}O_2$	146.6	29	1.7-9.9%	292	0.731
Butylamine	$C_4H_{11}N$	77.0	-12	1.7-9.8%	312	12.2
sec-Butylamine	$C_4H_{11}N$	62.71	-9			
tert-Butylamine	$C_4H_{11}N$	44.02	-9	1.7-8.9%	380	48.4
N-tert-Butylaminoethyl methacrylate	$C_{10}H_{19}NO_2$		96			
N-Butylaniline	$C_{10}H_{15}N$	254	107			
Butylbenzene	$C_{10}H_{14}$	183.3	71	0.8-5.8%	410	0.150
sec-Butylbenzene, (±)-	$C_{10}H_{14}$	173.3	52	0.8-6.9%	418	
tert-Butylbenzene	$C_{10}H_{14}$	169.1	60	0.7-5.7%	450	0.280
4-*tert*-Butyl-1,2-benzenediol	$C_{10}H_{14}O_2$	286	130			
Butyl benzoate	$C_{11}H_{14}O_2$	249	107			0.005
2-Butyl-1,1'-biphenyl	$C_{16}H_{18}$	292	>100		430	
Butyl butanoate	$C_8H_{16}O_2$	164.95	53			

Name	Mol. form.	t_b/°C	FP/°C	Flam. limits	IT/°C	VP (25 °C)/ kPa
Butyl citrate	$C_{18}H_{32}O_7$		157		368	
Butylcyclohexane	$C_{10}H_{20}$	180.9			246	
tert-Butylcyclohexane	$C_{10}H_{20}$	171.6			342	
Butylcyclohexylamine	$C_{10}H_{21}N$		93			
Butylcyclopentane	C_9H_{18}	156			250	
Butylethylamine	$C_6H_{15}N$	104.8	18			3.19
Butyl ethyl ether	$C_6H_{14}O$	89	4			7.46
2-Butyl-2-ethyl-1,3-propanediol	$C_9H_{20}O_2$	269.0	138			
Butyl formate	$C_5H_{10}O_2$	106.1	18	1.7-8.2%	322	3.53
tert-Butyl hydroperoxide	$C_4H_{10}O_2$	89 dec	27			
Butyl *cis*-12-hydroxy-9-octadecenoate, (*R*)-	$C_{22}H_{42}O_3$		110			
Butyl methacrylate	$C_8H_{14}O_2$	163.7	52			
Butylmethylamine	$C_5H_{13}N$	91	13			7.28
1-*tert*-Butyl-4-methylbenzene	$C_{11}H_{16}$	193	68			0.09
Butyl 3-methylbutanoate	$C_9H_{18}O_2$		53			
4-*tert*-Butyl-2-methylphenol	$C_{11}H_{16}O$	256	118			
1-Butylnaphthalene	$C_{14}H_{16}$	288	360			
Butyl nitrate	$C_4H_9NO_3$	133	36			
2-Butyl-1-octanol	$C_{12}H_{26}O$	248	110			
Butyl oleate	$C_{22}H_{42}O_2$		180			
tert-Butyl peroxybenzoate	$C_{11}H_{14}O_3$	118 expl	>88		8	
N-Butyl-*N*-phenylacetamide	$C_{12}H_{17}NO$	281	141			
Butyl phenyl ether	$C_{10}H_{14}O$	210	82			
Butyl propanoate	$C_7H_{14}O_2$	145.1	32		426	
Butyl stearate	$C_{22}H_{44}O_2$	343	160		355	
Butyltrichlorosilane	$C_4H_9Cl_3Si$	148.5	54			
1-*tert*-Butyl-4-vinylbenzene	$C_{12}H_{16}$		81			
Butyl vinyl ether	$C_6H_{12}O$	94	-9		255	6.65
2-Butyne	C_4H_6	27.1	-31	>1.4%		94.3
γ-Butyrolactone	$C_4H_6O_2$	204.6	98			0.43
Camphor, (+)	$C_{10}H_{16}O$	209	66	0.6-3.5%	466	0.032
Caprolactam	$C_6H_{11}NO$	270.8	125			0.0003
Carbon disulfide	CS_2	46.2	-30	1.3-50.0%	90	48.2
Carbon monoxide	CO	-191.51		12.5-74%	609	
Carbon oxysulfide	COS	-50.2		12-29%		
α-Chloroacetophenone	C_8H_7ClO	247	118			0.001
4-Chlorobenzaldehyde	C_7H_5ClO	213.5	88			
Chlorobenzene	C_6H_5Cl	131.6	28	1.3-9.6%	593	1.6
2-Chloro-1,3-butadiene	C_4H_5Cl	59	-20	4.0-20.0%		29.5
1-Chlorobutane	C_4H_9Cl	78.4	-12	1.9-10.1%	240	13.7
2-Chlorobutane	C_4H_9Cl	71	-10			21.0
2-Chloro-1-butene	C_4H_7Cl	58	-19	2.3-9.3%		
2-Chloro-4-*tert*-butylphenol	$C_{10}H_{13}ClO$		107			
Chlorocyclohexane	$C_6H_{11}Cl$	142.6	32			1.0
1-Chloro-1,1-difluoroethane	$C_2H_3ClF_2$	-9.12		6-18%	632	351
1-Chloro-2,4-dinitrobenzene	$C_6H_3ClN_2O_4$	315	194	2.0-22%		
Chloroethane	C_2H_5Cl	12.3	-50	3.8-15.4%	519	160
2-Chloroethanol	C_2H_5ClO	126	60	4.9-15.9%	425	1.2
Chloroethene	C_2H_3Cl	-13.8	-78	3.6-33.0%	472	355
(2-Chloroethoxy)benzene	C_8H_9ClO	218.5	107			
1-Chloro-4-ethylbenzene	C_8H_9Cl	184.4	64			
2-Chloroethyl vinyl ether	C_4H_7ClO	108	27			
1-Chlorohexane	$C_6H_{13}Cl$	135.0	35			1.25
Chloromethane	CH_3Cl	-24.1		8.1-17.4%	632	574
(Chloromethyl)benzene	C_7H_7Cl	174	67	>1.1%	585	0.164
1-Chloro-3-methylbutane	$C_5H_{11}Cl$	99	<21	1.5-7.4%		
2-Chloro-2-methylbutane	$C_5H_{11}Cl$	85		1.5-7.4%	345	11.5
3-(Chloromethyl)heptane	$C_8H_{17}Cl$	171	60			
1-Chloro-2-methylpropane	C_4H_9Cl	69	-6	2.0-8.7%		19.9
2-Chloro-2-methylpropane	C_4H_9Cl	50.9	0			42.7

Name	Mol. form.	t_b/°C	FP/°C	Flam. limits	IT/°C	VP (25 °C)/ kPa
3-Chloro-2-methylpropene	C_4H_7Cl	72	-12	3.2-8.1%		16.9
1-Chloronaphthalene	$C_{10}H_7Cl$	259	121		>558	0.003
1-Chloro-4-nitrobenzene	$C_6H_4ClNO_2$	238	127			0.003
1-Chloro-1-nitroethane	$C_2H_4ClNO_2$	124.5	56			
1-Chloro-1-nitropropane	$C_3H_6ClNO_2$	142	62			
2-Chloro-2-nitropropane	$C_3H_6ClNO_2$	142	57			
1-Chloro-4-nitro-2-(trifluoromethyl)benzene	$C_7H_3ClF_3NO_2$	232	135			
1-Chlorooctane	$C_8H_{17}Cl$	183	70			0.11
1-Chloropentane	$C_5H_{11}Cl$	107.9	13	1.6-8.6%	260	4.36
2-Chlorophenol	C_6H_5ClO	173.4	64			0.308
4-Chlorophenol	C_6H_5ClO	219	121			
1-Chloropropane	C_3H_7Cl	46.2	<-18	2.6-11.1%	520	45.8
2-Chloropropane	C_3H_7Cl	35.0	-32	2.8-10.7%	593	68.9
3-Chloropropanenitrile	C_3H_4ClN	175.5	76			
2-Chloropropanoic acid	$C_3H_5ClO_2$	185	107		500	
2-Chloro-1-propanol	C_3H_7ClO	133.5	52			
1-Chloro-2-propanol	C_3H_7ClO	124.4	52			
2-Chloropropene	C_3H_5Cl	23	-37	4.5-16%		110
3-Chloropropene	C_3H_5Cl	44.8	-32	2.9-11.1%	485	48.9
Chlorotrifluoroethene	C_2ClF_3	-28.3		8.4-16.0%		1293
1-Chloro-2-(trifluoromethyl)benzene	$C_7H_4ClF_3$	153	59			
o-Cresol	C_7H_8O	191.0	81	>1.4%	599	0.041
m-Cresol	C_7H_8O	202.2	86	>1.1%	558	0.019
p-Cresol	C_7H_8O	201.9	86	>1.1%	558	0.017
Cyanamide	CH_2N_2		141			
Cyanogen	C_2N_2	-21.1		6.6-32%		
Cyclobutane	C_4H_8	12.5	<10	>1.8%		157
1,5,9-Cyclododecatriene	$C_{12}H_{18}$	240	71			
Cycloheptane	C_7H_{14}	118.8	<21	1.1-6.7%		2.90
Cyclohexane	C_6H_{12}	80.7	-20	1.3-8%	245	13.0
1,4-Cyclohexanedimethanol	$C_8H_{16}O_2$	283	167		316	
Cyclohexanethiol	$C_6H_{12}S$	158.8	43			
Cyclohexanol	$C_6H_{12}O$	160.9	68	1-9%	300	0.10
Cyclohexanone	$C_6H_{10}O$	155.4	44	1.1-9.4%	420	0.53
Cyclohexene	C_6H_{10}	82.9	-12	>1.2%	310	11.8
3-Cyclohexene-1-carboxaldehyde	$C_7H_{10}O$	164	57			
Cyclohexyl acetate	$C_8H_{14}O_2$	174	58		335	
Cyclohexylamine	$C_6H_{13}N$	133.6	31	1.9-9.4%	293	1.20
Cyclohexylbenzene	$C_{12}H_{16}$	239	99			
Cyclohexylethylamine	$C_8H_{17}N$	164	30			
Cyclohexyl formate	$C_7H_{12}O_2$	162	51			
Cyclohexylisopropylamine	$C_9H_{19}N$		34			
2-Cyclohexylphenol	$C_{12}H_{16}O$		134			
cis,cis-1,5-Cyclooctadiene	C_8H_{12}	149	35			
Cyclopentane	C_5H_{10}	49.2	-25	>1.5%	361	42.3
Cyclopentanol	$C_5H_{10}O$	140.4	51			0.294
Cyclopentanone	C_5H_8O	130.5	26			1.55
Cyclopentene	C_5H_8	44.2	-29		395	50.7
Cyclopropane	C_3H_6	-31		2.4-10.4%	498	
trans-Decahydronaphthalene	$C_{10}H_{18}$	187.3	54	0.7-5.4%	255	0.164
Decane	$C_{10}H_{22}$	174.1	51	0.8-5.4%	210	0.170
1-Decanol	$C_{10}H_{22}O$	229	82		288	0.009
1-Decene	$C_{10}H_{20}$	171	<55		235	0.210
Decylamine	$C_{10}H_{23}N$	217	99			
Decylbenzene	$C_{16}H_{26}$	298	107			
1-Decylnaphthalene	$C_{20}H_{28}$	379	177			
Decyl nitrate	$C_{10}H_{21}NO_3$		113			
Diethylene glycol dipropanoate	$C_{10}H_{18}O_5$		127			
Diacetone alcohol	$C_6H_{12}O_2$	167.9	58	1.8-6.9%	643	0.224
3,3-Diacetoxy-1-propene	$C_7H_{10}O_4$	176	82			

Name	Mol. form.	t_b/°C	FP/°C	Flam. limits	IT/°C	VP (25 °C)/ kPa
Diallyl ether	$C_6H_{10}O$	94	-7			
Diallyl phthalate	$C_{14}H_{14}O_4$		166			
4,4'-Diaminodiphenylmethane	$C_{13}H_{14}N_2$	379	220			
1,3-Diamino-2-propanol	$C_3H_{10}N_2O$		132			
Dibenzyl ether	$C_{14}H_{14}O$	298	135			
Diborane	B_2H_6	-92.49	-90	1-98%	≈40	
1,2-Dibutoxyethane	$C_{10}H_{22}O_2$	198	85			
Dibutoxymethane	$C_9H_{20}O_2$	179.7	60			
Dibutylamine	$C_8H_{19}N$	162	47	1.1-6%		0.34
Di-*sec*-butylamine	$C_8H_{19}N$	135	24			
N,N-Dibutylaniline	$C_{14}H_{23}N$	274.8	110			
2,5-Di-*tert*-butyl-1,4-benzenediol	$C_{14}H_{22}O_2$		216			
Dibutyl ether	$C_8H_{18}O$	141.6	25	1.5-7.6%	194	0.898
Dibutyl maleate	$C_{12}H_{20}O_4$	280	141			0.002
2,6-Di-*tert*-butyl-4-methylphenol	$C_{15}H_{24}O$	265	127			0.0004
Dibutyl oxalate	$C_{10}H_{18}O_4$	244	104			
Di-*tert*-butyl peroxide	$C_8H_{18}O_2$	110.0	18			3.43
Dibutyl phosphonate	$C_8H_{19}O_3P$	230	49			
Dibutyl phthalate	$C_{16}H_{22}O_4$	338	157	>0.5%	402	
Dibutyl sebacate	$C_{18}H_{34}O_4$	356	178	>0.4%	365	
Dibutyl sulfide	$C_8H_{18}S$	168	76			0.09
Dibutyl tartrate	$C_{12}H_{22}O_6$	320	91		284	
Dichloroacetyl chloride	C_2HCl_3O	108	66			
3,4-Dichloroaniline	$C_6H_5Cl_2N$	273.0	166			
o-Dichlorobenzene	$C_6H_4Cl_2$	180.2	66	2.2-9.2%	648	0.18
m-Dichlorobenzene	$C_6H_4Cl_2$	172	72			0.252
p-Dichlorobenzene	$C_6H_4Cl_2$	173.9	66			0.235
2,3-Dichloro-1,3-butadiene	$C_4H_4Cl_2$	101	10			
1,2-Dichlorobutane	$C_4H_8Cl_2$	123.9			275	2.82
1,4-Dichlorobutane	$C_4H_8Cl_2$	155	52			
2,3-Dichlorobutane, (±)-	$C_4H_8Cl_2$	119	90			
1,3-Dichloro-1-butene	$C_4H_6Cl_2$		27			
3,4-Dichloro-1-butene	$C_4H_6Cl_2$	116	45			
Dichlorodimethylsilane	$C_2H_6Cl_2Si$	70.5	<21	3.4-9.5%		18.9
Dichlorodiphenylsilane	$C_{12}H_{10}Cl_2Si$	304	142			
1,1-Dichloroethane	$C_2H_4Cl_2$	56.3	-17	5.4-11.4%	458	30.5
1,2-Dichloroethane	$C_2H_4Cl_2$	83.4	13	6.2-16%	413	10.6
1,1-Dichloroethene	$C_2H_2Cl_2$	31.6	-28	6.5-15.5%	570	80.0
cis-1,2-Dichloroethene	$C_2H_2Cl_2$	60	6	3-15%	460	26.8
trans-1,2-Dichloroethene	$C_2H_2Cl_2$	47.64	2	6-13%	460	44.2
(1,2-Dichloroethenyl)benzene	$C_8H_6Cl_2$		107			
Dichloromethane	CH_2Cl_2	39.8		13-23%	556	58.2
Dichloromethylsilane	CH_4Cl_2Si	40.9	-9	6.0-55%	316	20.1
1,1-Dichloro-1-nitroethane	$C_2H_3Cl_2NO_2$	123.5	76			
1,1-Dichloro-1-nitropropane	$C_3H_5Cl_2NO_2$	145	66			
1,5-Dichloropentane	$C_5H_{10}Cl_2$	182.9	>27			
2,4-Dichlorophenol	$C_6H_4Cl_2O$	210	114			
1,2-Dichloropropane, (±)-	$C_3H_6Cl_2$	96.4	21	3.4-14.5%	557	6.62
1,3-Dichloro-2-propanol	$C_3H_6Cl_2O$	171	74			0.125
2,3-Dichloropropene	$C_3H_4Cl_2$	93.0	15	2.6-7.8%		
Dichlorosilane	Cl_2H_2Si	8.3		4.1-99%	36	
Dicyclohexylamine	$C_{12}H_{23}N$	251	>99			0.003
Diethanolamine	$C_4H_{11}NO_2$	271.2	172	2-13%	662	<0.01
1,1-Diethoxyethane	$C_6H_{14}O_2$	102	-21	1.6-10.4%	230	3.68
1,2-Diethoxyethane	$C_6H_{14}O_2$	120.6	27		205	4.33
Diethylamine	$C_4H_{11}N$	55.4	-23	1.8-10.1%	312	30.1
2-(Diethylamino)ethyl acrylate	$C_9H_{17}NO_2$		91			
N,N-Diethylaniline	$C_{10}H_{15}N$	216	85		630	0.025
o-Diethylbenzene	$C_{10}H_{14}$	183.4	57		395	0.13
m-Diethylbenzene	$C_{10}H_{14}$	181.1	56		450	0.14

Name	Mol. form.	t_b/°C	FP/°C	Flam. limits	IT/°C	VP (25 °C)/ kPa
p-Diethylbenzene	$C_{10}H_{14}$	184	55	0.7-6.0%	430	0.13
N,N'-Diethylcarbanilide	$C_{17}H_{20}N_2O$		150			
Diethyl carbonate	$C_5H_{10}O_3$	125.9	25			1.63
N,N-Diethyldodecanamide	$C_{16}H_{33}NO$		>66			
Diethylene glycol	$C_4H_{10}O_3$	245.5	124	2-17%	224	0.001
Diethylene glycol diacetate	$C_8H_{14}O_5$	200	135			
Diethylene glycol dibenzoate	$C_{18}H_{18}O_5$		232			
Diethylene glycol dibutyl ether	$C_{12}H_{26}O_3$	255	118		310	0.005
Diethylene glycol diethyl ether	$C_8H_{18}O_3$	185	82			0.10
Diethylene glycol dimethyl ether	$C_6H_{14}O_3$	162	67			0.315
Diethylene glycol monoethyl ether	$C_6H_{14}O_3$	202	96			0.017
Diethylene glycol monoethyl ether acetate	$C_8H_{16}O_4$	218	110		425	0.029
Diethylene glycol monomethyl ether	$C_5H_{12}O_3$	194	96	1.38-22.7%	240	0.024
N,N-Diethyl-1,2-ethanediamine	$C_6H_{16}N_2$	144	46			
Diethyl ether	$C_4H_{10}O$	34.4	-45	1.9-36.0%	180	71.7
Diethyl fumarate	$C_8H_{12}O_4$	214	104			
Diethyl maleate	$C_8H_{12}O_4$	222	121		350	0.015
Diethyl malonate	$C_7H_{12}O_4$	200	93			0.048
1,3-Diethyl-5-methylbenzene	$C_{11}H_{16}$	205			455	0.074
Diethyl oxalate	$C_6H_{10}O_4$	186	76			0.030
3,3-Diethylpentane	C_9H_{20}	146.2		0.7-5.7%	290	
Diethyl phthalate	$C_{12}H_{14}O_4$	298	161	>0.7%	457	0.002
N,N-Diethyl-1,3-propanediamine	$C_7H_{18}N_2$	165	59			
Diethyl selenide	$C_4H_{10}Se$	108		>2.5%		
Diethyl succinate	$C_8H_{14}O_4$	217	90			0.15
Diethyl sulfate	$C_4H_{10}O_4S$	208	104		436	0.05
Diethyl DL-tartrate	$C_8H_{14}O_6$	281	93			
Diethyl terephthalate	$C_{12}H_{14}O_4$	303	117			
1,1-Difluoroethene	$C_2H_2F_2$	-85.5		5.5-21.3%		
Digermane	Ge_2H_6	29			≈50	
Dihexylamine	$C_{12}H_{27}N$	236	104			
Dihexyl ether	$C_{12}H_{26}O$	220	77			
3,4-Dihydro-2*H*-pyran	C_5H_8O	85.5	-18			11.4
Dihydro-2,4,6-trimethyl-4*H*-1,3,5-dithiazine, (2α,4α,6α)	$C_6H_{13}NS_2$		93			
Diisobutylamine	$C_8H_{19}N$	139.6	29			0.972
Diisobutyl phthalate	$C_{16}H_{22}O_4$	296.5	185			
Diisopropanolamine	$C_6H_{15}NO_2$	250	127		374	
Diisopropylamine	$C_6H_{15}N$	84	-1	1.1-7.1%	316	10.7
N,N-Diisopropylethanolamine	$C_8H_{19}NO$	195	79			
Diisopropyl ether	$C_6H_{14}O$	68.4	-28	1.4-7.9%	443	19.9
Diketene	$C_4H_4O_2$	127.0	34			1.41
2,5-Dimethoxyaniline	$C_8H_{11}NO_2$	270	150			
3,3'-Dimethoxybenzidine	$C_{14}H_{16}N_2O_2$		206			
1,2-Dimethoxyethane	$C_4H_{10}O_2$	85.0	-2		202	9.93
Dimethoxymethane	$C_3H_8O_2$	42.3	-32	2.2-13.8%	237	53.1
trans-1,2-Dimethoxy-4-(1-propenyl)benzene	$C_{11}H_{14}O_2$	270.5	>100			
N,N-Dimethylacetamide	C_4H_9NO	165.9	70	1.8-11.5%	490	0.075
Dimethylamine	C_2H_7N	7.3	20	2.8-14.4%	400	203
2-(Dimethylamino)ethyl methacrylate	$C_8H_{15}NO_2$		74			
3-(Dimethylamino)propanenitrile	$C_5H_{10}N_2$	177	65			
1-(Dimethylamino)-2-propanol	$C_5H_{13}NO$	124.5	35			
2,3-Dimethylaniline	$C_8H_{11}N$	223	97	>1.0%		
2,6-Dimethylaniline	$C_8H_{11}N$	211	96			0.45
N,N-Dimethylaniline	$C_8H_{11}N$	193	63		371	0.107
2,2-Dimethylbutane	C_6H_{14}	49.7	-48	1.2-7.0%	405	42.5
2,3-Dimethylbutane	C_6H_{14}	58.0	-29	1.2-7.0%	405	31.3
2,3-Dimethyl-1-butene	C_6H_{12}	55.59	<-20		360	33.6
2,3-Dimethyl-2-butene	C_6H_{12}	73.19	<-20		401	16.7
Dimethyl carbonate	$C_3H_6O_3$	90.11	19			
Dimethylcyanamide	$C_3H_6N_2$	162	71			

Safety

Name	Mol. form.	t_b/°C	FP/°C	Flam. limits	IT/°C	VP (25 °C)/ kPa
cis-1,2-Dimethylcyclohexane	C_8H_{16}	129.7	16		304	1.93
trans-1,2-Dimethylcyclohexane	C_8H_{16}	123.4	11		304	2.58
cis-1,4-Dimethylcyclohexane	C_8H_{16}	124.3	16			2.39
Dimethyl disulfide	$C_2H_6S_2$	109.72	24			3.82
Dimethyl ether	C_2H_6O	-24.8	-41	3.4-27.0%	350	
N,N-Dimethylformamide	C_3H_7NO	152.8	58	2.2-15.2%	445	0.439
2,5-Dimethylfuran	C_6H_8O	96	7			
2,6-Dimethyl-4-heptanol	$C_9H_{20}O$	193	74			
2,6-Dimethyl-4-heptanone	$C_9H_{18}O$	157	49	0.8-7.1%	396	0.23
2,3-Dimethylhexane	C_8H_{18}	115.6	7		438	3.13
2,4-Dimethylhexane	C_8H_{18}	109.4	10			4.05
1,1-Dimethylhydrazine	$C_2H_8N_2$	62.4	-15	2-95%	249	20.9
Dimethyl isophthalate	$C_{10}H_{10}O_4$	285.1	138			
Dimethyl maleate	$C_6H_8O_4$	202	113			
2,6-Dimethylmorpholine	$C_6H_{13}NO$	147	44			
trans-3,7-Dimethyl-2,6-octadien-1-ol formate	$C_{11}H_{18}O_2$	229 dec	85			
2,3-Dimethyloctane	$C_{10}H_{22}$	164	<55			
3,4-Dimethyloctane	$C_{10}H_{22}$	162	<55			
3,7-Dimethyl-6-octen-1-ol, (±)-	$C_{10}H_{20}O$	224	96			
2,3-Dimethylpentane	C_7H_{16}	89.8	-56	1.1-6.7%	335	9.18
2,4-Dimethylpentane	C_7H_{16}	80.4	-12			13.1
2,4-Dimethyl-3-pentanol	$C_7H_{16}O$	142	49			
Dimethyl phthalate	$C_{10}H_{10}O_4$	282.7	146	>0.9%	490	0.001
N,N-Dimethyl-1,3-propanediamine	$C_5H_{14}N_2$	129	38			
2,2-Dimethyl-1,3-propanediol	$C_5H_{12}O_2$	207	129		399	
2,2-Dimethyl-1-propanol	$C_5H_{12}O$	112	37			
2,5-Dimethylpyrazine	$C_6H_8N_2$	152	64			
Dimethyl sebacate	$C_{12}H_{22}O_4$	289	145			
Dimethyl sulfate	$C_2H_6O_4S$	186	83		188	0.13
Dimethyl sulfide	C_2H_6S	37.32	-37	2.2-19.7%	206	64.4
Dimethyl sulfoxide	C_2H_6OS	191.9	95	2.6-42%	215	0.084
Dimethyl terephthalate	$C_{10}H_{10}O_4$	288	153		518	
2,4-Dinitroaniline	$C_6H_5N_3O_4$		224			
Dioctyl ether	$C_{16}H_{34}O$	289	>100		205	
1,4-Dioxane	$C_4H_8O_2$	101.2	12	2.0-22%	180	4.95
1,3-Dioxolane	$C_3H_6O_2$	75.3	2			14.6
Dipentene	$C_{10}H_{16}$	176	45		237	0.259
Dipentylamine	$C_{10}H_{23}N$	204	51			
Dipentyl ether	$C_{10}H_{22}O$	187	57		170	0.13
Dipentyl maleate	$C_{14}H_{24}O_4$		132			
Dipentyl phthalate	$C_{18}H_{26}O_4$	341	118			
Dipentyl sulfide	$C_{10}H_{22}S$		85			
Diphenylamine	$C_{12}H_{11}N$	305.1	153		634	
1,1-Diphenylbutane	$C_{16}H_{18}$	287	>100			
1,3-Diphenyl-2-buten-1-one	$C_{16}H_{14}O$	342.5	177			
1,1-Diphenylethane	$C_{14}H_{14}$	285.8	>100		440	
Diphenyl ether	$C_{12}H_{10}O$	258.0	112	0.8-1.5%	618	
Diphenylmethane	$C_{13}H_{12}$	264.2	130		485	
Diphenyl phthalate	$C_{20}H_{14}O_4$		224			
1,1-Diphenylpropane	$C_{15}H_{16}$		>100			
Diphosphine	H_4P_2	63.5 dec			≈20	
Dipropylamine	$C_6H_{15}N$	107.5	17		299	3.21
Dipropyl ether	$C_6H_{14}O$	90.1	21	1.3-7.0%	188	8.35
Disilane	H_6Si_2	-14.8	-14		≈20	
Divinyl ether	C_4H_6O	28	<-30	1.7-27%	360	89.3
Dodecane	$C_{12}H_{26}$	216.3	74	>0.6%	203	0.016
1-Dodecanethiol	$C_{12}H_{26}S$	277	128			
1-Dodecanol	$C_{12}H_{26}O$	264.1	127		275	0.000016
1-Dodecene	$C_{12}H_{24}$	213.4	79			0.019
Dodecylbenzene	$C_{18}H_{30}$	329	140			

Name	Mol. form.	t_b/°C	FP/°C	Flam. limits	IT/°C	VP (25 °C)/ kPa
4-Dodecyloxy-2-hydroxybenzophenone	$C_{25}H_{34}O_3$		254			
Eicosane	$C_{20}H_{42}$	344.1	>100		232	
Epichlorohydrin	C_3H_5ClO	111.99	31	3.8-21.0%	411	2.2
1,2-Epoxybutane	C_4H_8O	63.4	-22	1.7-19%	439	31.7
2,3-Epoxypropyl acrylate	$C_6H_8O_3$		61			
Ethane	C_2H_6	-88.6		3.0-12.5%	472	
1,2-Ethanediamine	$C_2H_8N_2$	116.9	40	2.5-12.0%	385	1.62
1,2-Ethanediol	$C_2H_6O_2$	197.5	111	3.2-22%	398	0.01
1,2-Ethanediol, diacetate	$C_6H_{10}O_4$	184	88	1.6-8.4%	482	0.030
1,2-Ethanediol, diformate	$C_4H_6O_4$	174	93			
1,2-Ethanediol, monoacetate	$C_4H_8O_3$	188	102			
1,2-Ethanediyl mercaptoacetate	$C_6H_{10}O_4S_2$		202			
Ethanethiol	C_2H_6S	35.0	-17	2.8-18.0%	300	70.3
Ethanol	C_2H_6O	78.24	13	3.3-19%	363	7.87
Ethanolamine	C_2H_7NO	170.3	86	3.0-23.5%	410	0.05
Ethoxyacetylene	C_4H_6O	50	<-7			
2-Ethoxyaniline	$C_8H_{11}NO$	233	115			
4-Ethoxyaniline	$C_8H_{11}NO$	254	116			0.0007
Ethoxybenzene	$C_8H_{10}O$	169.8	63			0.204
2-Ethoxy-3,4-dihydro-2*H*-pyran	$C_7H_{12}O_2$	132	44			
2-Ethoxyethanol	$C_4H_{10}O_2$	134.7	43	3-18%	235	0.71
2-Ethoxyethyl acetate	$C_6H_{12}O_3$	156.6	56	2-8%	379	0.24
3-Ethoxypropanal	$C_5H_{10}O_2$	135.2	38			
2-Ethoxy-1-propanol	$C_5H_{12}O_2$		38			
N-Ethylacetamide	C_4H_9NO	205	110			
Ethyl acetate	$C_4H_8O_2$	77.1	-4	2.0-11.5%	426	12.6
Ethyl acetoacetate	$C_6H_{10}O_3$	180	57	1.4-9.5%	295	0.095
Ethyl acrylate	$C_5H_8O_2$	98.9	10	1.4-14%	372	5.14
Ethylamine	C_2H_7N	16.6	-16	3.5-14%	385	141
N-Ethylaniline	$C_8H_{11}N$	204	85			0.039
Ethylbenzene	C_8H_{10}	136.2	21	0.8-6.7%	432	1.28
α-Ethylbenzenemethanol	$C_9H_{12}O$	219	100			
Ethyl benzoate	$C_9H_{10}O_2$	212.5	88		490	0.04
Ethyl benzoylacetate	$C_{11}H_{12}O_3$	267 dec	141			
Ethyl bromoacetate	$C_4H_7BrO_2$	159	48			
2-Ethylbutanal	$C_6H_{12}O$		21	1.2-7.7%		
Ethyl butanoate	$C_6H_{12}O_2$	121.1	24		463	2.01
2-Ethylbutanoic acid	$C_6H_{12}O_2$	193	99		400	
2-Ethyl-1-butanol	$C_6H_{14}O$	155	57			0.206
2-Ethyl-1-butene	C_6H_{12}	64.7	<-20		315	23.4
Ethyl *trans*-2-butenoate	$C_6H_{10}O_2$	140	2			
2-Ethylbutyl acetate	$C_8H_{16}O_2$	161	54			
Ethyl *N*-butylcarbamate	$C_7H_{15}NO_2$	221	92			
Ethyl chloroacetate	$C_4H_7ClO_2$	144	64			0.640
Ethyl chloroformate	$C_3H_5ClO_2$	91	16		500	
Ethylcyanoacetate	$C_5H_7NO_2$	216	110			0.003
Ethylcyclobutane	C_6H_{12}	70	-15	1.2-7.7%	210	
Ethylcyclohexane	C_8H_{16}	131.8	35	0.9-6.6%	238	1.71
Ethylcyclopentane	C_7H_{14}	103.5	<21	1.1-6.7%	260	5.32
Ethyl decanoate	$C_{12}H_{24}O_2$	242	>100			
3-Ethyl-2,4-dimethylpentane	C_9H_{20}	123	390			1.33
Ethylene	C_2H_4	-103.8		2.7-36%	450	
Ethylene carbonate	$C_3H_4O_3$	246	143			0.003
Ethyleneimine	C_2H_5N	54	-11	3.3-54.8%	320	28.9
Ethyl formate	$C_3H_6O_2$	54.09	-20	2.8-16.0%	455	32.3
2-Ethylhexanal	$C_8H_{16}O$	161	44	0.85-7.2%	190	
2-Ethyl-1,3-hexanediol	$C_8H_{18}O_2$	243	127		360	
Ethyl hexanoate	$C_8H_{16}O_2$	165	49			0.158
2-Ethyl-1-hexanol	$C_8H_{18}O$	186.2	73	0.88-9.7%	231	0.019
2-Ethyl-2-hexenal	$C_8H_{14}O$	175	68			

Name	Mol. form.	t_b/°C	FP/°C	Flam. limits	IT/°C	VP (25 °C)/ kPa
2-Ethylhexyl acetate	$C_{10}H_{20}O_2$	200	71	0.76-8.14%	268	0.09
2-Ethylhexyl acrylate	$C_{11}H_{20}O_2$		82		252	
2-Ethylhexylamine	$C_8H_{19}N$	172	60			
N-(2-Ethylhexyl)cyclohexanamine	$C_{14}H_{29}N$		129			
2-Ethylhexyl vinyl ether	$C_{10}H_{20}O$		57			
Ethyl lactate	$C_5H_{10}O_3$	151	46	>1.5%	400	
Ethyl methacrylate	$C_6H_{10}O_2$	116	20			2.62
N-Ethyl-4-methylbenzenesulfonamide	$C_9H_{13}NO_2S$		127			
2-Ethyl-2-methyl-1,3-dioxolane	$C_6H_{12}O_2$	119	23			
Ethyl methyl ether	C_3H_8O	6	-37	2.0-10.1%	190	195
3-Ethyl-4-methylhexane	C_9H_{20}	140	24			
4-Ethyl-2-methylhexane	C_9H_{20}	134	<21	>0.7%	280	
3-Ethyl-2-methylpentane	C_8H_{18}	115.6	<21		460	3.19
Ethyl 2-methylpropanoate	$C_6H_{12}O_2$	111	13			3.25
N-Ethylmorpholine	$C_6H_{13}NO$	145	32			
1-Ethylnaphthalene	$C_{12}H_{12}$	258			480	
Ethyl nitrate	$C_2H_5NO_3$	89	10	>4%		8.56
Ethyl nitrite	$C_2H_5NO_2$	17.5	-35	4.0-50%	90	135
3-Ethyloctane	$C_{10}H_{22}$	165			230	
4-Ethyloctane	$C_{10}H_{22}$	163			229	
Ethyl octanoate	$C_{10}H_{20}O_2$	206	79			
4-Ethylphenol	$C_8H_{10}O$	217.97	104			
N-Ethyl-*N*-phenylacetamide	$C_{10}H_{13}NO$	260	52			
Ethyl phenylacetate	$C_{10}H_{12}O_2$	228	99			
5-Ethyl-2-picoline	$C_8H_{11}N$	178	68	1.1-6.6%		
Ethyl propanoate	$C_5H_{10}O_2$	98.9	12	1.9-11%	440	4.97
Ethyl propyl ether	$C_5H_{12}O$	63	<-20	1.7-9.0%		24.2
Ethyl silicate	$C_8H_{20}O_4Si$	168	52			1.17
1-Ethyl-1,2,3,4-tetrahydronaphthalene	$C_{12}H_{16}$		164			
2-Ethyltoluene	C_9H_{12}	165.1			440	
3-Ethyltoluene	C_9H_{12}	161.3			480	
4-Ethyltoluene	C_9H_{12}	162.0			475	
Ethyl *p*-toluenesulfonate	$C_9H_{12}O_3S$		158			
Ethyl vinyl ether	C_4H_8O	36	<-46	1.7-28%	202	68.8
Eucalyptol	$C_{10}H_{18}O$	176	48			0.260
Fluorobenzene	C_6H_5F	84.7	-15			10.4
Fluoroethene	C_2H_3F	-72		2.6-21.7%		
Formaldehyde	CH_2O	-19.1	85	7.0-73%	424	
Formamide	CH_3NO	217	154			0.01
Formic acid	CH_2O_2	101	50	18-57%	434	5.75
Furan	C_4H_4O	31.3	-36	2.3-14.3%		80.0
2-Furanmethanamine	C_5H_7NO	145.5	37			
2-Furanmethanol acetate	$C_7H_8O_3$	182	85			
Furfural	$C_5H_4O_2$	161.5	60	2.1-19.3%	316	0.29
Furfuryl alcohol	$C_5H_6O_2$	168	75	1.8-16.3%	491	0.097
Geraniol	$C_{10}H_{18}O$	229	>100			
Germane	GeH_4	-88.1			≈20	
Glycerol	$C_3H_8O_3$	289	199	3-19%	370	<0.01
Glycerol triacetate	$C_9H_{14}O_6$	259	138	>1.0%	433	<0.01
Glycerol tripropanoate	$C_{12}H_{20}O_6$		167			
1-Heptadecanol	$C_{17}H_{36}O$	324	154			
2-Heptadecanone	$C_{17}H_{34}O$	320	120			
Heptane	C_7H_{16}	98.38	-4	1.05-6.7%	204	6.09
2-Heptanol, (±)-	$C_7H_{16}O$	159	71			
3-Heptanol, (*S*)-	$C_7H_{16}O$	163	60			
2-Heptanone	$C_7H_{14}O$	151.0	39	1.1-7.9%	393	0.49
3-Heptanone	$C_7H_{14}O$	146	46			0.5
4-Heptanone	$C_7H_{14}O$	144	49			0.164
1-Heptene	C_7H_{14}	94	-1		260	7.52
trans-2-Heptene	C_7H_{14}	98	<0			6.56

Name	Mol. form.	t_b/°C	FP/°C	Flam. limits	IT/°C	VP (25 °C)/ kPa
Heptylamine	$C_7H_{17}N$	153	54			
Hexachloro-1,3-butadiene	C_4Cl_6	216			610	0.13
Hexadecane	$C_{16}H_{34}$	286.9	136		202	0.0002
1,4-Hexadiene (unspecified isomer)	C_6H_{10}	65	-21	2.0-6.1%		
1,5-Hexadien-3-yne	C_6H_6	85	<-20	>1.5%		10.6
Hexanal	$C_6H_{12}O$	129.6	32			1.48
Hexane	C_6H_{14}	68.72	-22	1.1-7.5%	225	20.2
Hexanedinitrile	$C_6H_8N_2$	295	93	>1.0%	550	<0.01
1,6-Hexanedioic acid	$C_6H_{10}O_4$	337.5	196		420	
2,5-Hexanediol	$C_6H_{14}O_2$	229	110			
2,5-Hexanedione	$C_6H_{10}O_2$	194	79		499	
Hexanedioyl dichloride	$C_6H_8Cl_2O_2$		72			
1,2,6-Hexanetriol	$C_6H_{14}O_3$		191			
Hexanoic acid	$C_6H_{12}O_2$	204.9	102		380	0.005
1-Hexanol	$C_6H_{14}O$	156.9	63		290	0.11
2-Hexanone	$C_6H_{12}O$	127.6	25	1-8%	423	1.54
3-Hexanone	$C_6H_{12}O$	123.5	35	1-8%		
1-Hexene	C_6H_{12}	63.4	-26	1.2-6.9%	253	24.8
cis-2-Hexene	C_6H_{12}	68.9	-21			20.0
cis-3-Hexen-1-ol	$C_6H_{12}O$	157	54			
Hexyl acetate	$C_8H_{16}O_2$	171.1	45			0.185
sec-Hexyl acetate	$C_8H_{16}O_2$	147.5	45			
Hexylamine	$C_6H_{15}N$	132	29			1.17
Hexyl methacrylate	$C_{10}H_{18}O_2$	162	82			
2-Hexyne	C_6H_{10}	84.3	-10			10.7
Hydrazine	H_4N_2	113.55	38	5-100%		1.91
Hydrogen	H_2	-252.879		4-74%		
Hydrogen cyanide	CHN	25.63	-18	6-40%	538	98.8
Hydrogen disulfide	H_2S_2	70.7	<22			
Hydrogen sulfide	H_2S	-59.55		4-44%	260	
Hydrogen telluride	H_2Te	-2			-50	
p-Hydroquinone	$C_6H_6O_2$	288	165		516	
2-Hydroxybenzoic acid	$C_7H_6O_3$		157	>1.1%	540	
2-Hydroxybiphenyl	$C_{12}H_{10}O$	281	124		530	
3-Hydroxybutanal	$C_4H_8O_2$		66		250	
7-Hydroxy-3,7-dimethyloctanal	$C_{10}H_{20}O_2$		>100			
2-Hydroxypropanenitrile	C_3H_5NO	184	77			
3-Hydroxypropanenitrile	C_3H_5NO	218	129			0.010
trans-α-Ionone, (±)-	$C_{13}H_{20}O$		>100			
trans-β-Ionone	$C_{13}H_{20}O$		>100			
Isoborneol acetate, (±)-	$C_{12}H_{20}O_2$		88			
Isobutanal	C_4H_8O	64.1	-18	1.6-10.6%	196	23.0
Isobutane	C_4H_{10}	-11.7	-87	1.8-8.4%	460	348
Isobutene	C_4H_8	-7.0		1.8-9.6%	465	300
Isobutyl acetate	$C_6H_{12}O_2$	116.9	18	1.3-10.5%	421	2.39
Isobutyl acrylate	$C_7H_{12}O_2$	137.0	30		427	
Isobutylamine	$C_4H_{11}N$	68.8	-9	2-12%	378	19.0
Isobutylbenzene	$C_{10}H_{14}$	172.7	55	0.8-6.0%	427	0.257
Isobutyl butanoate	$C_8H_{16}O_2$	157	50			0.500
Isobutylcyclohexane	$C_{10}H_{20}$	171.3			274	
Isobutyl formate	$C_5H_{10}O_2$	98.4	5	2-9%	320	5.34
Isobutyl isobutanoate	$C_8H_{16}O_2$	148	38	0.96-7.59%	432	0.552
Isobutyl phenylacetate	$C_{12}H_{16}O_2$	247	>100			
Isobutyl vinyl ether	$C_6H_{12}O$	83	-9			9.30
Isopentane	C_5H_{12}	27.83	-51	1.4-7.6%	420	91.7
Isopentyl acetate	$C_7H_{14}O_2$	141.6	25	1.0-7.5%	360	0.728
Isopentyl butanoate	$C_9H_{18}O_2$	184.8	59			0.160
Isopentyl nitrite	$C_5H_{11}NO_2$	99			210	
Isophorone	$C_9H_{14}O$	214.8	84	0.8-3.8%	460	0.06
Isopropenyl acetate	$C_5H_8O_2$	97.4	26		432	6.02

Name	Mol. form.	t_b/°C	FP/°C	Flam. limits	IT/°C	VP (25 °C)/ kPa
Isopropenylbenzene	C_9H_{10}	165.4	54	1.9-6.1%	574	0.40
2-Isopropoxyethanol	$C_5H_{12}O_2$	141.6	33			
3-Isopropoxypropanenitrile	$C_6H_{11}NO$		68			
Isopropyl acetate	$C_5H_{10}O_2$	88.6	2	1.8-8%	460	7.88
Isopropylamine	C_3H_9N	31.8	-37		402	78.0
Isopropylbenzene	C_9H_{12}	152.4	36	0.9-6.5%	424	0.61
Isopropylbenzene hydroperoxide	$C_9H_{12}O_2$	153 expl	79			0.004
Isopropyl benzoate	$C_{10}H_{12}O_2$	218	99			
2-Isopropyl-1,1'-biphenyl	$C_{15}H_{16}$		141			
Isopropylcyclohexane	C_9H_{18}	154.4			283	
Isopropyl formate	$C_4H_8O_2$	68	-6		485	18.2
Isopropyl lactate	$C_6H_{12}O_3$	167	54			
1-Isopropyl-4-methylbenzene	$C_{10}H_{14}$	177	47	0.7-5.6%	436	0.19
Isopropyl vinyl ether	$C_5H_{10}O$	56	-32			
d-Limonene	$C_{10}H_{16}$	177.6	45	0.7-6.1%	237	0.277
Linalol	$C_{10}H_{18}O$	198	71			
Maleic anhydride	$C_4H_2O_3$	202	102	1.4-7.1%	477	
2-Mercaptoethanol	C_2H_6OS	150.0	74			0.202
Mesityl oxide	$C_6H_{10}O$	129.7	31	1.4-7.2%	344	1.47
Methacrylic acid	$C_4H_6O_2$	160	77	1.6-8.8%	68	0.12
Methane	CH_4	-161.5		5.0-15.0%	537	
Methanethiol	CH_4S	6.0	-18	3.9-21.8%		202
Methanol	CH_4O	64.5	11	6.0-36%	464	16.9
2-Methoxyaniline	C_7H_9NO	221	118			0.013
2-Methoxybenzaldehyde	$C_8H_8O_2$	243	118			
3-Methoxy-1-butanol	$C_5H_{12}O_2$	159	74			
2-Methoxyethanol	$C_3H_8O_2$	124.3	39	1.8-14%	285	1.31
(2-Methoxyethoxy)ethene	$C_5H_{10}O_2$	107	18			
2-[2-(2-Methoxyethoxy)ethoxy]ethanol	$C_7H_{16}O_4$	243	118			
2-Methoxyethyl acetate	$C_5H_{10}O_3$	142	49	1.5-12.3%	392	0.67
2-Methoxyethyl acrylate	$C_6H_{10}O_3$		82			
4-Methoxyphenol	$C_7H_8O_2$	253	132		421	
3-Methoxy-1-propanamine	$C_4H_{11}NO$	113	32			
3-Methoxypropanenitrile	C_4H_7NO	165.4	65			
Methyl abietate	$C_{21}H_{32}O_2$		180			
4-Methylacetanilide	$C_9H_{11}NO$	307	168			
Methyl acetate	$C_3H_6O_2$	56.7	-10	3.1-16%	454	28.8
Methyl acetoacetate	$C_5H_8O_3$	168	77		280	0.241
4-Methylacetophenone	$C_9H_{10}O$	225	96			
Methyl acrylate	$C_4H_6O_2$	80.1	-3	2.8-25%	468	11.0
2-Methylacrylonitrile	C_4H_5N	90	1	2-6.8%		8.26
Methylamine	CH_5N	-6.4	0	4.9-20.7%	430	353
Methyl 2-aminobenzoate	$C_8H_9NO_2$	256	>100			0.002
2-Methylaniline	C_7H_9N	200.0	85		482	0.043
4-Methylaniline	C_7H_9N	201	87		482	1.74
4-Methylanisole	$C_8H_{10}O$	175	60			
2-Methyl-1,4-benzenediol	$C_7H_8O_2$	283	172			
α-Methylbenzenemethanol	$C_8H_{10}O$	205	93			
α-Methylbenzenemethanol, acetate	$C_{10}H_{12}O_2$		91			
Methyl benzoate	$C_8H_8O_2$	199	83			0.052
α-Methylbenzylamine, (±)-	$C_8H_{11}N$	193	79			
2-Methylbiphenyl	$C_{13}H_{12}$	258	137		502	
2-Methyl-1,3-butadiene	C_5H_8	34.0	-54	1.5-8.9%	395	73.4
3-Methyl-2-butanethiol	$C_5H_{12}S$	109.8	3			
Methyl butanoate	$C_5H_{10}O_2$	101.9	14			4.30
3-Methylbutanoic acid	$C_5H_{10}O_2$	176.5			416	0.067
2-Methyl-1-butanol, (±)-	$C_5H_{12}O$	129.0	50		385	0.416
3-Methyl-1-butanol	$C_5H_{12}O$	130.8	43	1.2-9.0%	350	0.315
2-Methyl-2-butanol	$C_5H_{12}O$	102.4	19	1.2-9.0%	437	2.19
3-Methyl-2-butanol, (±)-	$C_5H_{12}O$	113.7	38			1.20

Name	Mol. form.	t_b/°C	FP/°C	Flam. limits	IT/°C	VP (25 °C)/ kPa
2-Methyl-1-butene	C_5H_{10}	31.1	-20			81.4
3-Methyl-1-butene	C_5H_{10}	20.1	-7	1.5-9.1%	365	120
2-Methyl-2-butene	C_5H_{10}	38.5	-20			62.1
3-Methyl-3-buten-2-one	C_5H_8O	97		1.8-9.0%		
2-Methyl-1-buten-3-yne	C_5H_6	33	<-7			
Methyl chloroacetate	$C_3H_5ClO_2$	130	57	7.5-18.5%		1.0
Methylcyclohexane	C_7H_{14}	100.9	-4	1.2-6.7%	250	6.18
cis-2-Methylcyclohexanol	$C_7H_{14}O$	165	65		296	
trans-2-Methylcyclohexanol, (±)-	$C_7H_{14}O$	168.4	65		296	
cis-3-Methylcyclohexanol, (±)-	$C_7H_{14}O$	168	70		295	
trans-3-Methylcyclohexanol, (±)-	$C_7H_{14}O$	167	70		295	
cis-4-Methylcyclohexanol	$C_7H_{14}O$	174	70		295	
trans-4-Methylcyclohexanol	$C_7H_{14}O$	175	70		295	
4-Methylcyclohexene	C_7H_{12}	103	-1			5.43
Methylcyclopentane	C_6H_{12}	71.8	-29	1.0-8.35%	258	18.3
2-Methyldecane	$C_{11}H_{24}$	189.2			225	
1-Methyl-2,4-dinitrobenzene	$C_7H_6N_2O_4$	300 dec	207			
N-Methyl-2-ethanolamine	C_3H_9NO	159.24	74			
Methyl formate	$C_2H_4O_2$	31.6	-19	4.5-23%	449	78.1
3-Methylfuran	C_5H_6O	66	-30			
2-Methylhexane	C_7H_{16}	90.0	-1	1.0-6.0%	280	8.78
3-Methylhexane	C_7H_{16}	92	-4		280	
5-Methyl-2-hexanone	$C_7H_{14}O$	139	36	1.0-8.2%	191	0.691
Methylhydrazine	CH_6N_2	83	-8	2.5-92%	194	6.61
Methyl isocyanate	C_2H_3NO	38.3	-7	5.3-26%	534	57.7
Methyl lactate, (±)-	$C_4H_8O_3$	144.8	49	>2.2%	385	0.62
Methyl methacrylate	$C_5H_8O_2$	100.6	10	1.7-8.2%		5.10
4-Methylmorpholine	$C_5H_{11}NO$	121.4	24			20.4
1-Methylnaphthalene	$C_{11}H_{10}$	244.4			529	0.009
4-Methyl-2-nitroaniline	$C_7H_8N_2O_2$		157			
2-Methylnonane	$C_{10}H_{22}$	167			210	
Methyl octadecanoate	$C_{19}H_{38}O_2$	353	153			
Methyloxirane	C_3H_6O	35	-37	3.1-27.5%	449	71.7
cis-2-Methyl-1,3-pentadiene	C_6H_{10}	80	-12			
4-Methyl-1,3-pentadiene	C_6H_{10}	76	-34			
2-Methylpentanal	$C_6H_{12}O$	118	17		199	
2-Methylpentane	C_6H_{14}	60.21	<-29	1.0-7.0%	264	28.2
3-Methylpentane	C_6H_{14}	63.3	-7	1.2-7.0%	278	25.3
2-Methyl-2,4-pentanediol	$C_6H_{14}O_2$	197.9	102	1-9%	306	<0.01
2-Methylpentanoic acid, (±)-	$C_6H_{12}O_2$	195	107		378	
2-Methyl-1-pentanol	$C_6H_{14}O$	157	54	1.1-9.65%	310	0.236
4-Methyl-2-pentanol	$C_6H_{14}O$	132.0	41	1.0-5.5%		0.698
4-Methyl-2-pentanone	$C_6H_{12}O$	115.7	18	1.2-8.0%	448	2.64
2-Methyl-1-pentene	C_6H_{12}	62.1	-28		300	26.0
4-Methyl-1-pentene	C_6H_{12}	54	-7		300	36.1
2-Methyl-2-pentene	C_6H_{12}	67.3	<-7			21.0
4-Methyl-*cis*-2-pentene	C_6H_{12}	56.4	-32			32.5
4-Methyl-*trans*-2-pentene	C_6H_{12}	58.58	-29			29.7
Methyl 2-phenylacetate	$C_9H_{10}O_2$	215	91			
1-Methylpiperazine	$C_5H_{12}N_2$	135	42			
2-Methylpropanenitrile	C_4H_7N	102	8		482	
2-Methyl-1-propanethiol	$C_4H_{10}S$	88.5	2			9.30
2-Methyl-2-propanethiol	$C_4H_{10}S$	64.2	<-29			24.2
Methyl propanoate	$C_4H_8O_2$	78.6	-2	2.5-13%	469	11.5
2-Methylpropanoic acid	$C_4H_8O_2$	154.4	56	2.0-9.2%	481	0.17
2-Methylpropanoic anhydride	$C_8H_{14}O_3$	183	59	1.0-6.2%	329	
2-Methyl-1-propanol	$C_4H_{10}O$	107.84	28	1.7-10.6%	415	1.39
2-Methyl-2-propanol	$C_4H_{10}O$	82.3	11	2.4-8.0%	478	5.52
2-Methyl-2-propenol	C_4H_8O	114.5	33			
2-(2-Methylpropoxy)ethanol	$C_6H_{14}O_2$	155	58			

Name	Mol. form.	t_b/°C	FP/°C	Flam. limits	IT/°C	VP (25 °C)/ kPa
Methyl propyl ether	$C_4H_{10}O$	38.5	-20	2.0-14.8%		60.9
2-Methylpyrazine	$C_5H_6N_2$	129	50			
2-Methylpyridine	C_6H_7N	129.4	39		538	1.5
4-Methylpyridine	C_6H_7N	145.3	57			0.759
1-Methylpyrrole	C_5H_7N	112.7	16			
N-Methylpyrrolidine	$C_5H_{11}N$	206.1	-14			13.5
N-Methyl-2-pyrrolidinone	C_5H_9NO	204.2	96	1-10%	346	0.04
Methyl salicylate	$C_8H_8O_3$	222.6	96		454	0.015
2-Methylstyrene	C_9H_{10}	170	53	0.8-11.0%	538	0.245
3-Methylstyrene	C_9H_{10}	170	53	0.8-11.0%	538	0.236
4-Methylstyrene	C_9H_{10}	172	53	0.8-11.0%	538	0.245
2-Methyltetrahydrofuran	$C_5H_{10}O$	80	-11			12.6
3-(Methylthio)propanal	C_4H_8OS		61			
Methyl vinyl ether	C_3H_6O	6			287	180
Morpholine	C_4H_9NO	128.2	37	1.4-11.2%	290	1.34
4-Morpholineethanol	$C_6H_{13}NO_2$	220	99			
4-Morpholinepropanamine	$C_7H_{16}N_2O$	220	104			
Naphthalene	$C_{10}H_8$	218.0	79	0.9-5.9%	526	0.011
2-Naphthol	$C_{10}H_8O$	286	153			
1-Naphthylamine	$C_{10}H_9N$	300.7	157			
Neopentane	C_5H_{12}	9.50	-65	1.4-7.5%	450	171
4-Nitroaniline	$C_6H_6N_2O_2$	328	199			
Nitrobenzene	$C_6H_5NO_2$	210.7	88	>1.8%	482	0.03
Nitrocyclohexane	$C_6H_{11}NO_2$	205	88			
Nitroethane	$C_2H_5NO_2$	114.1	28	3.4-17%	414	2.79
Nitromethane	CH_3NO_2	101.19	35	>7.3%	418	4.79
1-Nitropropane	$C_3H_7NO_2$	131.2	36	>2.2%	421	1.36
2-Nitropropane	$C_3H_7NO_2$	120.2	24	2.6-11.0%	428	2.3
2-Nitrotoluene	$C_7H_7NO_2$	220.9	106			0.0014
3-Nitrotoluene	$C_7H_7NO_2$	232.1	106			0.03
4-Nitrotoluene	$C_7H_7NO_2$	238.66	106			0.0007
1-Nitro-3-(trifluoromethyl)benzene	$C_7H_4F_3NO_2$	202.8	103			
Nonadecane	$C_{19}H_{40}$	330	>100		230	
2-Nonadecanone	$C_{19}H_{38}O$		124			
Nonane	C_9H_{20}	150.8	31	0.8-2.9%	205	0.570
2-Nonanone	$C_9H_{18}O$	194	60	0.9-5.9%	360	
1-Nonene	C_9H_{18}	146.9	26			0.714
Nonyl acetate	$C_{11}H_{22}O_2$	225	68			
Nonylbenzene	$C_{15}H_{24}$	280	99			
2,5-Norbornadiene	C_7H_8	90	-21			8.97
Octadecane	$C_{18}H_{38}$	316	>100		227	
Octadecanoic acid	$C_{18}H_{36}O_2$	371	196		395	
1-Octadecanol	$C_{18}H_{38}O$	351			450	
cis-9-Octadecenoic acid	$C_{18}H_{34}O_2$	360	189		363	0.000001
Octahydroindene	C_9H_{16}	164			296	
Octanal	$C_8H_{16}O$	174	52			0.321
Octane	C_8H_{18}	125.62	13	1.0-6.5%	206	1.86
1-Octanethiol	$C_8H_{18}S$	199	69			
1-Octanol	$C_8H_{18}O$	194.7	81		270	0.01
2-Octanol	$C_8H_{18}O$	179	88		265	
2-Octanone	$C_8H_{16}O$	173	52			0.12
Octanoyl chloride	$C_8H_{15}ClO$	195	82			
1-Octene	C_8H_{16}	121.3	21		230	2.30
Octylamine	$C_8H_{19}N$	178.6	60			
4-Octylphenyl salicylate	$C_{21}H_{26}O_3$		216		416	
1,4-Oxathiane	C_4H_8OS	149	42			
2-Oxetanone	$C_3H_4O_2$	161	74	>2.9%		0.3
Oxirane	C_2H_4O	10.4	-20	3.0-100%	429	175
Paraformaldehyde	$(CH_2O)x$		70	7.0-73%	300	1.40
Paraldehyde	$C_6H_{12}O_3$	124	36	>1.3%	238	1.6

Name	Mol. form.	t_b/°C	FP/°C	Flam. limits	IT/°C	VP (25 °C)/ kPa
Pentaborane(9)	B_5H_9	60.10	30	>0.4%	35	
Pentamethylbenzene	$C_{11}H_{16}$	232	93		427	
Pentanal	$C_5H_{10}O$	103	12		222	4.58
Pentane	C_5H_{12}	36.06	-40	1.4-8.0%	260	68.3
1,5-Pentanediol	$C_5H_{12}O_2$	241	129		335	0.001
2,4-Pentanedione	$C_5H_8O_2$	140.7	34		340	1.02
1-Pentanethiol	$C_5H_{12}S$	126.6	18			1.83
Pentanoic acid	$C_5H_{10}O_2$	186.1	96		400	0.024
1-Pentanol	$C_5H_{12}O$	137.6	33	1.2-10.0%	300	0.259
2-Pentanol	$C_5H_{12}O$	119.1	34	1.2-9.0%	343	0.804
3-Pentanol	$C_5H_{12}O$	123	41	1.2-9.0%	435	1.10
2-Pentanone	$C_5H_{10}O$	102.2	7	1.5-8.2%	452	4.97
3-Pentanone	$C_5H_{10}O$	101.9	13	>1.6%	450	4.72
1-Pentene	C_5H_{10}	30.0	-18	1.5-8.7%	275	85.0
cis-2-Pentene	C_5H_{10}	36.9	<-20			66.0
trans-2-Pentene	C_5H_{10}	36.3	<-20			67.4
Pentyl acetate	$C_7H_{14}O_2$	149.4	16	1.1-7.5%	360	0.60
sec-Pentyl acetate, (*R*)-	$C_7H_{14}O_2$	142	32			
Pentylamine	$C_5H_{13}N$	104.7	-1	2.2-22%		4.00
4-*tert*-Pentylaniline	$C_{11}H_{17}N$	260.5	102			
Pentylbenzene	$C_{11}H_{16}$	203	66			
Pentyl butanoate	$C_9H_{18}O_2$	180	57			
Pentylcyclohexane	$C_{11}H_{22}$	204			239	
Pentyl formate	$C_6H_{12}O_2$	126	26			
Pentyl lactate	$C_8H_{16}O_3$		79			
1-Pentylnaphthalene	$C_{15}H_{18}$	305	124			
Pentyl propanoate	$C_8H_{16}O_2$	166	41		378	
Pentyl salicylate	$C_{12}H_{16}O_3$	270	132			
Pentyl stearate	$C_{23}H_{46}O_2$		185			
1-Pentyne	C_5H_8	39.9	<-20			58.1
Peroxyacetic acid	$C_2H_4O_3$	110	41			1.93
β-Phellandrene	$C_{10}H_{16}$	177	49			0.189
Phenanthrene	$C_{14}H_{10}$	338.4	171			
Phenol	C_6H_6O	181.8	79	1.8-8.6%	715	0.055
2-Phenoxyethanol	$C_8H_{10}O_2$	246	121			0.001
Phenyl acetate	$C_8H_8O_2$	195	80			
N-Phenyl-*N,N*-diethanolamine	$C_{10}H_{15}NO_2$		196	>0.7%	387	
N-Phenylethanolamine	$C_8H_{11}NO$	280	152			
Phenylhydrazine	$C_6H_8N_2$	244	88			0.003
4-Phenylmorpholine	$C_{10}H_{13}NO$		104			
Phenyloxirane	C_8H_8O	194.1	74		498	
1-Phenyl-1-propanone	$C_9H_{10}O$	217.4	99			
1-Phenyl-2-propylamine, (±)-	$C_9H_{13}N$	198	<100			0.06
Phorone	$C_9H_{14}O$	197	85			
Phosphine	H_3P	-87.75		>1.8%	100	
Phosphorus (white)	P	280.5			38	
Phthalic acid	$C_8H_6O_4$	dec	168			
Phthalic anhydride	$C_8H_4O_3$	285.3	152	1.7-10.5%	570	
α-Pinene	$C_{10}H_{16}$	156.3	33		255	0.64
β-Pinene	$C_{10}H_{16}$	165.8	38		275	0.61
Piperazine	$C_4H_{10}N_2$	148.63	81			
1-Piperazineethanamine	$C_6H_{15}N_3$	225	93			
1-Piperazineethanol	$C_6H_{14}N_2O$	259	124			
Piperidine	$C_5H_{11}N$	106.19	16	1-10%		4.28
Propanal	C_3H_6O	48.0	-30	2.6-17%	207	42.2
Propane	C_3H_8	-42.11	-104	2.1-9.5%	450	939
1,3-Propanediamine	$C_3H_{10}N_2$	139.2	24			
1,2-Propanediol	$C_3H_8O_2$	187.3	99	2.6-12.5%	371	0.02
1,3-Propanediol	$C_3H_8O_2$	214.7			400	0.007
Propanenitrile	C_3H_5N	97.3	2	3.1-14%	512	6.14

Name	Mol. form.	t_b/°C	FP/°C	Flam. limits	IT/°C	VP (25 °C)/ kPa
Propanoic acid	$C_3H_6O_2$	141.5	52	2.9-12.1%	465	0.553
Propanoic anhydride	$C_6H_{10}O_3$	168	63	1.3-9.5%	285	0.45
1-Propanol	C_3H_8O	97.04	23	2.2-13.7%	412	2.76
2-Propanol	C_3H_8O	82.21	12	2.0-12.7%	399	6.02
Propanoyl chloride	C_3H_5ClO	80	12			
Propargyl alcohol	C_3H_4O	113	36			
Propene	C_3H_6	-47.6		2.0-11.1%	455	
Propyl acetate	$C_5H_{10}O_2$	101.0	13	1.7-8%	450	4.49
Propylamine	C_3H_9N	47.21	-37	2.0-10.4%	318	42.1
Propylbenzene	C_9H_{12}	159.2	30	0.8-6.0%	450	0.45
2-Propyl-1,1'-biphenyl	$C_{15}H_{16}$		>100			
Propyl butanoate	$C_7H_{14}O_2$	144	37			0.618
Propylcyclohexane	C_9H_{18}	156.7			248	
Propylcyclopentane	C_8H_{16}	130.9			269	1.64
Propylene carbonate	$C_4H_6O_3$	241.6	135			0.05
Propyl formate	$C_4H_8O_2$	80.6	-3		455	10.9
Propyl nitrate	$C_3H_7NO_3$	110	20	2-100%	175	3.12
Propyl propanoate	$C_6H_{12}O_2$	122.2	79			1.88
Propyne	C_3H_4	-23.2		2.1-12.5%		581
Pyridine	C_5H_5N	115.2	20	1.8-12.4%	482	2.76
Pyrocatechol	$C_6H_6O_2$	246	127			
Pyrrole	C_4H_5N	129.74	39			1.10
Pyrrolidine	C_4H_9N	86.6	3			8.40
2-Pyrrolidone	C_4H_7NO	251.2	129			
Quinoline	C_9H_7N	237.1			480	0.011
Resorcinol	$C_6H_6O_2$	280	127	>1.4%	608	
Safrole	$C_{10}H_{10}O_2$	235	100			0.01
Salicylaldehyde	$C_7H_6O_2$	208	78			0.075
Silane	H_4Si	-111.9	-112	>1.4%	≈20	
Styrene	C_8H_8	145.3	31	0.9-6.8%	490	0.81
Succinonitrile	$C_4H_4N_2$	266	132			0.001
Sulfolane	$C_4H_8O_2S$	286	177			<0.01
L-Tartaric acid	$C_4H_6O_6$		210		425	
Terephthalic acid	$C_8H_6O_4$	300 sp	260		496	
o-Terphenyl	$C_{18}H_{14}$	337	163			
m-Terphenyl	$C_{18}H_{14}$	375	191			
α-Terpineol	$C_{10}H_{18}O$	218	90			
1,2,4,5-Tetrachlorobenzene	$C_6H_2Cl_4$	247	155			
Tetradecane	$C_{14}H_{30}$	253.5	112	>0.5%	200	0.002
1-Tetradecanol	$C_{14}H_{30}O$	295.8	141			
1-Tetradecene	$C_{14}H_{28}$	251.1	110		235	
Tetra(2-ethylbutyl) silicate	$C_{24}H_{52}O_4Si$		199			
Tetraethylene glycol	$C_8H_{18}O_5$	315	182			0.000001
Tetraethylene glycol dimethyl ether	$C_{10}H_{22}O_5$	274	141			
Tetraethylenepentamine	$C_8H_{23}N_5$	341.5	163		321	
Tetrafluoroethene	C_2F_4	-76		10.0-50.0%	200	
Tetrahydrofuran	C_4H_8O	66.0	-14	2-11.8%	321	21.6
Tetrahydrofurfuryl alcohol	$C_5H_{10}O_2$	176.3	75	1.5-9.7%	282	0.100
Tetrahydrofurfuryl oleate	$C_{23}H_{42}O_3$		199			
1,2,3,4-Tetrahydronaphthalene	$C_{10}H_{12}$	207.2	71	0.8-5.0%	385	0.05
Tetrahydropyran	$C_5H_{10}O$	88.0	-20			9.54
Tetrahydro-2*H*-pyran-2-methanol	$C_6H_{12}O_2$	189	93			
1,1,3,3-Tetramethoxypropane	$C_7H_{16}O_4$	183	77			
1,2,3,4-Tetramethylbenzene	$C_{10}H_{14}$	205	74		427	
1,2,3,5-Tetramethylbenzene	$C_{10}H_{14}$	198	71		427	
1,2,4,5-Tetramethylbenzene	$C_{10}H_{14}$	197	54			
2,2,3,3-Tetramethylpentane	C_9H_{20}	140.2	<21	0.8-4.9%	430	
2,2,3,4-Tetramethylpentane	C_9H_{20}	133.0	<21			
Tetramethylstannane	$C_4H_{12}Sn$	76.8	-12	>1.9%		14.6
Tetramethylurea	$C_5H_{12}N_2O$	177.1	77			0.138

Name	Mol. form.	t_b/°C	FP/°C	Flam. limits	IT/°C	VP (25 °C)/ kPa
Tetraphenylstannane	$C_{24}H_{20}Sn$	420	232			
Thiophene	C_4H_4S	84.1	-1			10.6
Toluene	C_7H_8	110.60	4	1.1-7.1%	480	3.79
Toluene-2,4-diisocyanate	$C_9H_6N_2O_2$	251	127	0.9-9.5%		0.003
p-Toluenesulfonic acid	$C_7H_8O_3S$		184			
Tribromosilane	Br_3HSi	109			≈20	
Tributylamine	$C_{12}H_{27}N$	207	63	1-5%		0.01
Tributyl borate	$C_{12}H_{27}BO_3$	233.8	93			
Tributyl phosphate	$C_{12}H_{27}O_4P$	289	146			
Tributyl phosphite	$C_{12}H_{27}O_3P$		120			
Tributyrin	$C_{15}H_{26}O_6$	307.5	180	>0.5%	407	
1,2,4-Trichlorobenzene	$C_6H_3Cl_3$	213.5	105	2.5-6.6%	571	0.057
1,1,1-Trichloroethane	$C_2H_3Cl_3$	74.02	-1	8-10.5%	500	16.5
1,1,2-Trichloroethane	$C_2H_3Cl_3$	113	32	6-28%	460	3.1
Trichloroethene	C_2HCl_3	86.8	32	8-10.5%	420	9.91
Trichloroethylsilane	$C_2H_5Cl_3Si$	98.7	22			6.29
(Trichloromethyl)benzene	$C_7H_5Cl_3$	221	127		211	0.35
Trichloromethylsilane	CH_3Cl_3Si	66	-9	7.6->20%	>404	22.5
Trichlorooctadecylsilane	$C_{18}H_{37}Cl_3Si$		89			
Trichloropentylsilane	$C_5H_{11}Cl_3Si$	172	63			
Trichlorophenylsilane	$C_6H_5Cl_3Si$	201	91			
1,2,3-Trichloropropane	$C_3H_5Cl_3$	158	71	3.2-12.6%		0.492
Trichloropropylsilane	$C_3H_7Cl_3Si$	123.5	37			
Trichlorosilane	Cl_3HSi	33	-50		104	
Trichlorovinylsilane	$C_2H_3Cl_3Si$	90.9	21			8.79
Tri-*o*-cresyl phosphate	$C_{21}H_{21}O_4P$	410	225		385	0.0000002
Tridecane	$C_{13}H_{28}$	235.4	79			0.005
1-Tridecanol	$C_{13}H_{28}O$	287	121			
2-Tridecanone	$C_{13}H_{26}O$	268	107			
1-Tridecene	$C_{13}H_{26}$	232.8	79			0.0047
Tri(decyl) phosphite	$C_{30}H_{63}O_3P$		235			
Tris(2-hydroxyethyl)amine	$C_6H_{15}NO_3$	350	179	1-10%		<0.01
Triethylamine	$C_6H_{15}N$	88.8	-7	1.2-8.0%	249	7.70
1,2,4-Triethylbenzene	$C_{12}H_{18}$	217	83			
Triethyl citrate	$C_{12}H_{20}O_7$	294	151			
Triethylene glycol	$C_6H_{14}O_4$	288.6	177	0.9-9.2%	371	0.0002
Triethylene glycol dimethyl ether	$C_8H_{18}O_4$	218	111			
Triethyl phosphate	$C_6H_{15}O_4P$	216	115		454	
(Trifluoromethyl)benzene	$C_7H_5F_3$	102.0	12			5.14
Triisopropanolamine	$C_9H_{21}NO_3$		160		320	
Triisopropyl borate	$C_9H_{21}BO_3$	139.6	28			
Trimethylamine	C_3H_9N	2.8	-5	2.0-11.6%	190	215
1,2,3-Trimethylbenzene	C_9H_{12}	176.0	44	0.8-6.6%	470	0.20
1,2,4-Trimethylbenzene	C_9H_{12}	169.4	44	0.9-6.4%	500	0.30
1,3,5-Trimethylbenzene	C_9H_{12}	164.7	50	1-5%	559	0.33
Trimethyl borate	$C_3H_9BO_3$	67.4	-8			17.2
2,2,3-Trimethylbutane	C_7H_{16}	80.8	<0		412	28.5
2,3,3-Trimethyl-2-butanol	$C_7H_{16}O$	128.3	<0		375	
Trimethylchlorosilane	C_3H_9ClSi	57.6	-28		395	30.7
2,5,5-Trimethylheptane	$C_{10}H_{22}$	153	<55			
2,2,5-Trimethylhexane	C_9H_{20}	124	13			2.21
3,5,5-Trimethyl-1-hexanol	$C_9H_{20}O$	193	93			
2,6,8-Trimethyl-4-nonanol	$C_{12}H_{26}O$	225.4	93			
2,6,8-Trimethyl-4-nonanone	$C_{12}H_{24}O$		91			
Trimethylolpropane	$C_6H_{14}O_3$		149			
2,4,4-Trimethyl-2-pentanamine	$C_8H_{19}N$		33			
2,2,3-Trimethylpentane	C_8H_{18}	109.8	<21		346	4.28
2,2,4-Trimethylpentane	C_8H_{18}	99.2	-12		418	6.50
2,3,3-Trimethylpentane	C_8H_{18}	114.7	<21		425	3.60
2,2,4-Trimethyl-1,3-pentanediol	$C_8H_{18}O_2$	230.1	113		346	

Name	Mol. form.	t_b/°C	FP/°C	Flam. limits	IT/°C	VP (25 °C)/ kPa
2,3,4-Trimethyl-1-pentene	C_8H_{16}		<21			
2,4,4-Trimethyl-1-pentene	C_8H_{16}	101.3	-5	0.8-4.8%	391	5.96
2,4,4-Trimethyl-2-pentene	C_8H_{16}	104.9	2		305	4.80
Trimethyl phosphate	$C_3H_9O_4P$	197.2	107			0.11
Trimethyl phosphite	$C_3H_9O_3P$	110	54			
Trinitroglycerol	$C_3H_5N_3O_9$	218 exp			270	0.00005
1,3,5-Trioxane	$C_3H_6O_3$	116	45	3.6-29%	414	
Tripentylamine	$C_{15}H_{33}N$	242.5	102			
Triphenylmethane	$C_{19}H_{16}$	359	>100			
Triphenyl phosphate	$C_{18}H_{15}O_4P$		220			
Triphenylphosphine	$C_{18}H_{15}P$		180			
Triphenyl phosphite	$C_{18}H_{15}O_3P$	360	218			
Tripropylamine	$C_9H_{21}N$	153	41			
Trisilane	H_8Si_3	52.9	<0		≈20	
Undecane	$C_{11}H_{24}$	195.9	69			0.05
2-Undecanol	$C_{11}H_{24}O$	231	113			
2-Undecanol, (±)-	$C_{11}H_{24}O$	228	113			
2-Undecanone	$C_{11}H_{22}O$	233.1	89			
Vinyl acetate	$C_4H_6O_2$	72.6	-8	2.6-13.4%	402	15.4
Vinyl butanoate	$C_6H_{10}O_2$	116.7	20	1.4-8.8%		
Vinyl *trans*-2-butenoate	$C_6H_8O_2$		26			
4-Vinylcyclohexene	C_8H_{12}	130	16		269	1.87
Vinyloxirane	C_4H_6O	68	<-50			
Vinyl propanoate	$C_5H_8O_2$	94.8	1			
1-Vinyl-2-pyrrolidinone	C_6H_9NO		98			
o-Xylene	C_8H_{10}	144.4	32	0.9-6.7%	463	0.88
m-Xylene	C_8H_{10}	139.1	27	1.1-7.0%	527	1.13
p-Xylene	C_8H_{10}	138.3	27	1.1-7.0%	528	1.19

MATERIALS COMPATIBLE WITH AND RESISTANT TO 72% PERCHLORIC ACID

Thomas J. Bruno and Paris D. N. Svoronos

Perchloric acid is used in the preparation of ion pairing agents for HPLC (see Section 8 dealing with this topic) and must be handled with great care because it can be a very powerful oxidizing agent. Cold perchloric acid at a concentration of 70% (mass/mass) or less is not considered a very strong oxidizing agent. At concentrations of 73% or higher, or at lower concentrations but at higher temperatures, perchloric acid is a powerful oxidant. The following table provides some guidance in handling this material in the laboratory (Ref. 1).

Reference

1. Bruno, T. J., and Svoronos, P. D. N., *CRC Handbook of Basic Tables for Chemical Analysis, Third Edition*, CRC Press, Boca Raton, FL, 2011 <https://doi.org/10.1201/b10385>.

Material	Comments
Elastomers	
Gum rubber	Each batch must be tested to determine compatibility
Vitons	Slight swelling only
Metals and alloys	
Tantalum	Excellent
Titanium (chemically pure grade)	Excellent
Zirconium	Excellent
Columbium (Niobium)	Excellent
Hastelloy	Slight corrosion rate
Plastics	Adequate
Polyvinyl chloride	
Teflon	
Polyethylene	
Polypropylene	
Kel-F	
Vinylidine fluoride	
Saran	
Epoxies	
Others	Adequate
Glass	
Glass-lined steel	
Alumina	
Fluorolube	
Incompatible	
Plastics	
Polyamide (nylon)	
Modacrylic ester, Dynel (35–85%) acrylonitrile	
Polyester (dacron)	
Bakelite	
Lucite	
Micarta	
Cellulose-based lacquers	
Metals	
Copper	
Copper alloys (brass, bronze, etc.) for very shock-sensitive perchlorate salts	
Aluminum (dissolves at room temperature)	
High nickel alloys (dissolves)	
Others	
Cotton	
Wood	
Glycerin-lead oxide (letharge)	

SELECTION OF LABORATORY GLOVES

Thomas J. Bruno, Paris D. N. Svoronos, and Sonja G. Ringen

The following tables provide guidance in the selection of protective gloves for laboratory use (Refs. 1–6). If protection from more than one class of chemicals is required, double gloving should be considered. Table 1 covers general hand protection from scrapes, burns, ergonomic issues, cuts, and abrasions. Table 2 identifies specific glove materials and the chemicals and categories of chemicals they are resistant to. Table 3 identifies specific chemicals and characterizes how well specific glove materials provide protection against these chemicals. The ratings are abbreviated as follows: Very Good (VG); Good (G); Fair (F); Poor (P). Gloves ranked poor are not recommended. Chemicals marked with an asterisk (*) are for limited service. In selecting the appropriate protection, one should identify the hazard and whenever possible use engineering controls (e.g., drip controls on bottles, elimination of sharps, the use of ergonomically designed tools, etc.).

References

1. Garrod, A. N., Martinez, M., and Pearson, J., *Ann. Occup. Hyg.* 43, 543, 1999 <https://doi.org/10.1016/S0003-4878(99)00043-5>.
2. Garrod, A. N., Phillips, A. M., and Pemberton, J. A., *Ann. Occup. Hyg.* 45, 55, 2001.
3. Mockelsen, R. L., and Hall, R. C., *Am. Ind. Hyg. Assoc. J.* 48, 941, 1987 <https://doi.org/10.1080/15298668791385859>.
4. OSHA, Federal Register, Vol. 59, No. 66, 29 CFR 1910, 16334-16364, 1994.
5. Bruno, T. J., and Svoronos, P. D. N., *CRC Handbook of Basic Tables for Chemical Analysis, Third Edition,* CRC Press, Boca Raton, FL, 2011 <https://doi.org/10.1201/b10385>.
6. OSHA, Personal Protective Equipment, OSHA 3151-12R, 2004.

TABLE 1. General Hand Protection Selection Criteria

Glove	Application examples
Cotton	Weighing, glass handling (avoiding contamination with skin oils); note that these gloves can be used as a first layer when also using other gloves such as for chemical hazards
Leather	Moderate hot or cold material handling; moving equipment
Gel-filled (anti-vibration)	Operation of vibrating equipment
Kevlar or fine mesh	Work with sheet metal, glass, or heavy cutting; note that these will not protect against punctures
Chemical resistant	See Table 2 below for specifics
Insulated for heat	Furnace work; handling hot glass or metal
Insulated for cold	Cryogenic work, filling Dewars, replenishing NMR magnets, etc.

TABLE 2. Hand Protection for Chemical Hazards

Glove material	Resistant to
Viton	PCBs, chlorinated solvents, aromatic solvents
Viton/Butyl	Acetone, toluene, aromatics, aliphatic hydrocarbons, chlorinated solvents, ketones, amines, and aldehydes
SilverShield and 4H (PE/EVAL)	Morpholine, vinyl chloride, acetone, ethyl ether, many toxic solvents and caustics
Barrier	Wide range of chlorinated solvents, aromatic acids
PVA	Ketones, aromatics, chlorinated solvents, xylene, MIBK, trichloroethylene; **DO NOT USE WITH WATER/ AQUEOUS SOLUTIONS**
Butyl	Aldehydes, ketones, esters, alcohols, most inorganic acids, caustics, dioxane
Neoprene	Oils, grease, petroleum-based solvents, detergents, acids, caustics, alcohols, solvents
PVC	Acids, caustics, solvents, solvents, grease, oils
Nitrile	Oils, fats, acids, caustics, alcohols
Latex	Body fluids, blood, acids, alcohols, alkalis
Vinyl	Body fluids, blood, acids, alcohols, alkalis
Rubber	Organic acids, some mineral acids, caustics, alcohols; **NOT RECOMMENDED FOR AROMATIC SOLVENTS, CHLORINATED SOLVENTS**

TABLE 3. Chemical Resistance Selection for Protective Gloves

Chemical	Neoprene	Latex/ Rubber	Butyl	Nitrile	Chemical	Neoprene	Latex/ Rubber	Butyl	Nitrile
Acetaldehyde*	VG	G	VG	G	Benzene*	P	P	P	F
Acetic acid	VG	VG	VG	VG	Butyl acetate	G	F	F	P
Acetone*	G	VG	VG	P	Butyl alcohol	VG	VG	VG	VG
Ammonium hydroxide	VG	VG	VG	VG	Carbon disulfide	F	F	F	F
Amyl acetate	F	P	F	P	Carbon tetrachloride*	F	P	P	G
Aniline	G	F	F	P	Castor oil	F	P	F	VG
Benzaldehyde*	F	F	G	G	Chlorobenzene*	F	P	F	P

Safety

Chemical	Neoprene	Latex/ Rubber	Butyl	Nitrile
Chloroform*	G	P	P	F
Chloronaphthalene	F	P	F	F
Chromic acid (50%)	F	P	F	F
Citric acid (10%)	VG	VG	VG	VG
Cyclohexanol	G	F	G	VG
Dibutyl phthalate*	G	P	G	G
Diisobutyl ketone	P	F	G	P
Dimethylformamide	F	F	G	G
Dioctyl phthalate	G	P	F	VG
Dioxane	VG	G	G	G
Epoxy resins, dry	VG	VG	VG	VG
Ethanol	VG	VG	VG	VG
Ethyl acetate	G	F	G	F
Ethyl ether*	VG	G	VG	G
Ethylene dichloride*	F	P	F	P
Ethylene glycol	VG	VG	VG	VG
Formaldehyde	VG	VG	VG	VG
Formic acid	VG	VG	VG	VG
Furfural*	G	G	G	G
Glycerin	VG	VG	VG	VG
Hexadecanoic acid	VG	VG	VG	VG
Hexane	F	P	P	G
Hydrazine (65%)	F	G	G	G
Hydrochloric acid	VG	G	G	G
Hydrofluoric acid (48%)	VG	G	G	G
Hydrogen peroxide (30%)	G	G	G	G
Hydroquinone	G	G	G	F
Isooctane	F	P	P	VG
Lactic acid (85%)	VG	VG	VG	VG
Lauric acid (36%)	VG	F	VG	VG
Lineolic acid	VG	P	F	G
Linseed oil	VG	P	F	VG
Maleic acid	VG	VG	VG	VG
Methanol	VG	VG	VG	VG
Methyl bromide	G	F	G	F
Methyl chloride*	P	P	P	P
Methyl ethyl ketone*	G	G	VG	P
Methyl isobutyl ketone*	F	F	VG	P
Methyl methacrylate	G	G	VG	F
Methylamine	F	F	G	G
Monoethanolamine	VG	G	VG	VG
Morpholine	VG	VG	VG	G
Naphthalene	G	F	F	G
Nitric acid, red and white fuming	P	P	P	P
Nitric acid*	G	F	F	F
Nitromethane*	F	P	F	F
Nitropropane	F	P	F	F
Octanol	VG	VG	VG	VG
Oleic acid	VG	F	G	VG
Oxalic acid	VG	VG	VG	VG
Perchloric acid (60%)	VG	F	G	G
Phenol	VG	F	G	F
Phosphoric acid	VG	G	VG	VG
Potassium hydroxide	VG	VG	VG	VG
Propan-2-ol	VG	VG	VG	VG
Propanol	VG	VG	VG	VG
Propyl acetate	G	F	G	F
Sodium hydroxide	VG	VG	VG	VG
Styrene	P	P	P	F
Sulfuric acid	G	G	G	G
Tetrachloroethene	F	P	P	G
Tetrohydrofuran	P	F	F	F
Toluene diisocyanate	F	G	G	F
Toluene*	F	P	P	F
1,1,2-Trichloroethene	F	F	P	G
Triethanolamine	VG	G	G	VG
Xylene*	P	P	P	P

SELECTION OF PROTECTIVE LABORATORY GARMENTS

Thomas J. Bruno and Paris D. N. Svoronos

The following table provides guidance in the selection of special protective garments that are used in the laboratory for specific tasks (Refs. 1 and 2).

References

1. Mount Sinai School of Medicine Personal Protective Equipment Guide, 2013 <www.mssm.edu/biosafety/policies>.
2. Bruno, T. J., and Svoronos, P. D. N., *CRC Handbook of Basic Tables for Chemical Analysis, Third Edition,* CRC Press, Boca Raton, FL, 2011.

Material	Type of garment	Common use
Cotton/Natural Fiber/Blends	Coveralls; lab coats; sleeve protectors; aprons	For dry dusts, particulates and aerosols
Tyvek	Coveralls; lab coats; sleeve protectors; aprons; hoods	For dry dusts and aerosols
Saranax/Tyvek SL	Coveralls; lab coats; sleeve protectors; aprons; hoods; level B suits	Aerosols; liquids; solvents
Polyethylene	Barrier gowns; aprons	Body fluids
Polypropylene	Clean room suits; coveralls; lab coats	For dry dusts; non-toxic particulates
Polyethylene/Tyvek (QC)	Coveralls; aprons; lab coats; shoe covers	Moisture; solvents
Polypropylene	Coveralls; lab coats; shoe covers; caps; clean room suits	Non-toxic particulates; dry dusts
Tychem BR; Tychem TK	Full level A and level B suits	Highly toxic particulates; dry dusts
CPF	Full level A and level B suits; splash suits	Highly toxic chemicals; gases; aerosols
PVC	Full level A suits	Highly toxic chemicals; gases; aerosols

SELECTION OF RESPIRATORS AND RESPIRATOR CARTRIDGES AND FILTERS

Thomas J. Bruno, Paris D. N. Svoronos, and Maria J. Doa

Respirators are sometimes desirable or required when performing certain tasks in the laboratory. The identity of the chemical hazard and its airborne concentrations need to be determined before choosing a respirator. This assessment should be done by experienced safety personnel or by an industrial hygienist (Ref. 1). Several types of respirators are available for use under different conditions. These include particulate respirators, chemical cartridge/gas mask respirators, powered air-purifying respirators, and self-contained breathing apparatus. The choice of a respirator depends critically upon the chemical substance, the level of that chemical substance in the air, and whether oxygen deficiency conditions require the use of a powered air-purifying respirator or a self-contained breathing apparatus. Respirator choices must be made carefully in consultation with professional expertise. The discussion below is for guidance purposes only.

Particulate Respirators

Particulate respirators are the simplest and least protective of the respirator types available. These respirators only filter out dusts, fumes, and mists. They do not protect against chemicals, gases, or vapors, and are intended only for low hazard levels.

Chemical Cartridge/Gas Mask Respirators

Chemical cartridge/gas mask respirators are also known as "air-purifying respirators" because they filter or clean chemical gases out of the air as you breathe. This respirator type includes a facepiece or mask and a cartridge or canister. Straps secure the facepiece to the head. The cartridge may also have a filter to remove particles. This type of respirator is effective only if used with the correct cartridge or filter (these terms are often used interchangeably) for a specific chemical substance.

Powered Air-Purifying Respirator (PAPR)

Powered air-purifying respirators use a fan to draw air through the filter to the user and are easier to breathe through. However, they need a fully charged battery to work properly. They use the same type of filters/cartridges as other air-purifying respirators.

Self-Contained Breathing Apparatus (SCBA)

A self-contained breathing apparatus provides clean air from a portable air tank. It is used when the air is too dangerous to breathe. It also protects against higher concentrations of dangerous chemicals. However, this type of respirator is very heavy (30 pounds or more) and requires very special training on its use and maintenance. Also, the air tanks typically last an hour or less depending upon their rating and your breathing rate.

There is a standardized color code system used by all manufacturers for the specification and selection of the cartridges and filters that are used with respirators (Ref. 1). Cartridges are available that protect against more than one hazard, but there is no "all-in-one" cartridge that protects against all substances. The cartridges are color-coded to help in the selection of the appropriate cartridge(s). Note that more than one cartridge may be required to protect against multiple hazards. It is important to know what the hazard is and how much of it is in the air in order to select the proper filters/cartridges. Table 1 provides guidance in the selection of the proper cartridge using the color code.

The use of respirators in the laboratory is subject to regulation and management. In the United States, for example, before a respirator can be used, a medical evaluation must be performed to establish that the user has sufficient lung capacity and that the use of the respirator will not cause health problems. This evaluation must be performed by a physician or nurse, and if performed by a nurse, the evaluation must be approved and signed by a physician.

The specific respirator selected must be based upon the hazard to the worker. It should be properly fitted and periodically refitted by an industrial hygienist or other trained personnel, and that fitting procedure must adhere to an established protocol with approved equipment to be effective. It must be regularly maintained and replaced in accord with the manufacturer's specifications; properly removed and disposed of to avoid contamination of self, others, or the environment; and if reusable, properly removed, cleaned, disinfected, and stored. Finally, training by an industrial hygienist or other appropriate personnel must be provided.

In addition to the cartridges specified in the table, particulate filters are available that can be used alone or in combination with those cartridges specified.

Reference

1. OSHA Bulletin: General Respiratory Protection Guidance for Employers and Workers, OSHA 3514, 2011.

Color Coding for Gas Mask Chemical Cartridges and Canisters

Contaminant	Color coding on cartridge/canister
Acid gases	White
Hydrocyanic acid gas	White with 1/2-inch green stripe completely around the canister near the bottom
Hydrocyanic acid gas and chlorpicrin vapor	Yellow with 1/2-inch blue stripe completely around the canister near the bottom
Ammonia gas	Green
Organic vapors	Black
Chlorine gas	White with 1/2-inch yellow stripe completely around the canister near the bottom
Acid gases and ammonia gas	Green with 1/2-inch white stripe completely around the canister near the bottom
Acid gases and organic vapors	Yellow
Carbon monoxide	Blue

Contaminant	Color coding on cartridge/canister
Pesticides	Organic vapor canister (black) plus a particulate filter (see color codes below)
Radioactive materials, except tritium and noble gases	Purple (magenta)
Multi-contaminants and chemical, biological, radiological, nuclear (CBRN) agents	Olive
Any particulates - P100	Purple
Any particulates - P95, P99, R95, R99, R100	Orange
Any particulates free of oil - N95, N99, N100	Teal

PROTECTIVE CLOTHING LEVELS

Thomas J. Bruno and Paris D. N. Svoronos

In the United States, OSHA defines various levels of protective clothing, and sets parameters that govern their use with chemical spills and in environments where chemical exposure is a possibility. A summary of the definitions is provided below (Refs. 1 and 2).

References

1. OSHA Technical Manual, Section VIII, Chapter 1, Chemical Protective Clothing, 2003.
2. Bruno, T. J., and Svoronos, P. D. N., *CRC Handbook of Basic Tables for Chemical Analysis, Third Edition*, CRC Press, Boca Raton, FL, 2011 <https://doi.org/10.1201/b10385>.

Level A

Vapor protective suit (meets NFPA 1991), pressure-demand, full-face SCBA, inner chemical-resistant gloves, chemical-resistant safety boots, two-way radio communications systems.

Protection Provided: Highest available level of respiratory, skin, and eye protection from solid, liquid, and gaseous chemicals.

Used When: The chemical(s) have been identified and have a high level of hazards to the respiratory system, skin, and eyes; substances are present with known or suspected skin toxicity or carcinogenicity; operations must be conducted in confined or poorly ventilated areas.

Limitations: Protective clothing must resist permeation by the chemical or mixtures present.

Level B

Liquid splash-protective suit (meets NFPA 1992), pressure-demand, full-facepiece SCBA, inner chemical-resistant gloves, chemical-resistant safety boots, two-way radio communications systems.

Protection Provided: Provides same level of respiratory protection as Level A, but somewhat less skin protection. Liquid splash protection is provided, but not protection against chemical vapors or gases.

Used When: The chemical(s) have been identified but do not require a high level of skin protection; the primary hazards associated with site entry are from liquid and not vapor contact.

Limitations: Protective clothing items must resist penetration by the chemicals or mixtures present.

Level C

Support function protective garment (meets NFPA 1993), full-facepiece, air-purifying, canister-equipped respirator, chemical-resistant gloves, safety boots, two-way radio communications systems.

Protection Provided: The same level of skin protection as Level B, but a lower level of respiratory protection; liquid splash protection but no protection to chemical vapors or gases.

Used When: Contact with site chemical(s) will not affect the skin; air contaminants have been identified and concentrations measured; a canister is available which can remove the contaminant; the site and its hazards have been completely characterized.

Limitations: Protective clothing items must resist penetration by the chemicals or mixtures present; chemical airborne concentration must be less than IDLH levels; the atmosphere must contain at least 19.5% oxygen.

Not Acceptable for Chemical Emergency Response.

Level D

Coveralls, safety boots/shoes, safety glasses, or chemical splash goggles.

Protection Provided: No respiratory protection, minimal skin protection.

Used When: The atmosphere contains no known hazard; work functions preclude splashes, immersion, potential for inhalation, or direct contact with hazard chemicals.

Limitations: The atmosphere must contain at least 19.5% oxygen.

Not Acceptable for Chemical Emergency Response.

Optional items may be added to each level of protective clothing. Options include items from higher levels of protection, as well as hard hats, hearing protection, outer gloves, a cooling system, etc.

SELECTION OF HEARING PROTECTION DEVICES

Sonja G. Ringen and Thomas J. Bruno

There are many sources of noise in the laboratory, and generally employers are required to identify employees exposed to noise in excess of 85 decibels (dB) averaged over eight working hours. It is best to apply engineering and administrative controls to mitigate these exposures. If, after applying engineering and administrative controls, sound levels in the local environment are higher than regulatory or recommended levels, hearing protection devices (HPDs) should be used to reduce noise exposure (Ref. 1).

The following table provides a range of sound levels (in decibels, dB) for common laboratory equipment.

Equipment	Sound level range (dB)
Air compressor	75–95
Shaker table	75–85
Pressure-relief valves	75–120
Pressurized vortex tubes	75–85
Sonicator (probe type)	70–110
Fume hood	30–70
Vacuum pump	50–70
Room air conditioner	30–65

In addition to the noise generated by various devices in the laboratory, additional noise can be generated by devices not directly related to the function of the lab. Noise from radios, piped-in music, HVAC systems, and telephones are also present in many laboratories.

It is not necessarily optimal to use hearing protection that simply maximizes sound attenuation; it remains important for workers to communicate with one another. Thus, devices that prevent intelligible speech can be problematic. Moreover, one must be able to recognize laboratory equipment alarms (such as low flow alarms on fume hoods, temperature alarms on ovens, and in some environments, backup alarms on forklifts, etc.). In many cases, personnel will remove hearing protection to communicate with one another or to listen for an alarm, even though removal of an HPD in a high noise environment can substantially reduce hearing protection. By choosing the right hearing protection, removal of the HPD by personnel can be minimized. In the United States and other industrialized countries, sound levels must be determined so that the correct noise reduction rating may be determined, and training must be provided by an industrial hygienist or other appropriate personnel.

Clearly, hearing protection can only be effective if it is actually used, so user preference (between earplugs and earmuffs, for example) must be considered. Because the anatomy of the ear varies with each individual, the effectiveness of earplugs might not be universal, especially if inserted incorrectly. Moreover, in some environments, personnel inserting and removing earplugs through the course of a workday can introduce dirt and bacteria and cause ear infections. When a higher noise reduction rating (NRR) is needed, sometimes using two styles of HPDs concurrently is effective, e.g., earmuffs and disposable foam earplugs, although the NRR is not simply additive. The following table provides guidance in choosing the appropriate HPD.

Note in addition to the equipment shown below, there are combination devices available. One can obtain, for example, a passive earmuff combined with a protective hard hatand various types of face shields.

Reference

1. Schulz, T., and Madison, T., *Hearing Conservation Manual, Fifth Edition*, Council for Accreditation in Occupational Hearing Conservation, Milwaukee, WI, 53202, 2014.

Type	Illustration	Notes
Passive Earmuffs		Easy to fit; good for intermittent noise exposure, multiple headband types are available to accommodate other protective equipment such as hard hats; may be uncomfortably hot or heavy; more expensive than earplugs.
Disposable Foam Earplugs	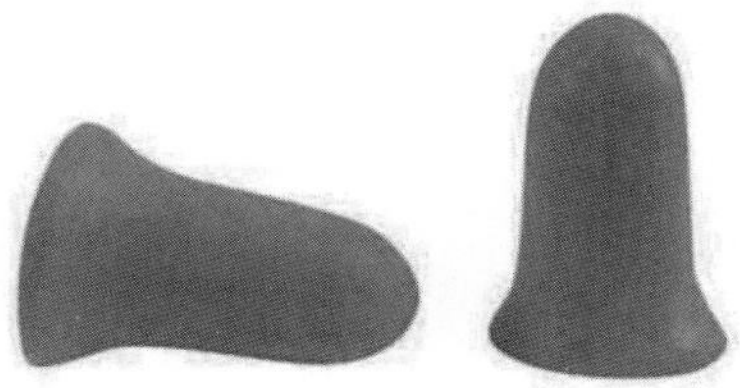	Cooler than earmuffs, and often more comfortable for extended use; can provide most attenuation; multiple sizes for different ear canals; attenuation depends on proper fit; proper fitting may be difficult to achieve or learn; hygiene issues in dirty environments; can absorb perspiration; single-use only.

Type	Illustration	Notes
Pre-Molded Reusable Earplugs	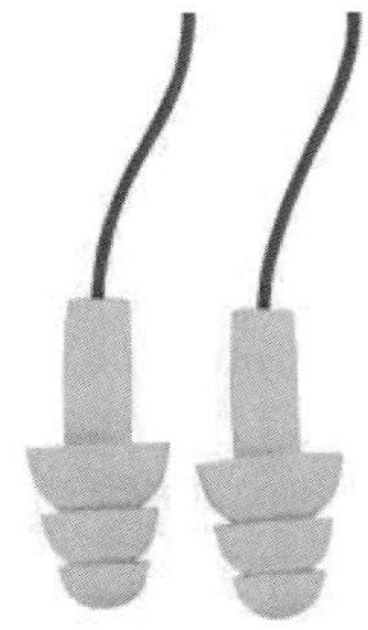	Comfortable for extended use; washable and reusable; do not absorb perspiration; multiple sizes for different ear canals; attenuation depends on proper fit; slightly more expensive per unit than the disposable earplugs above; must be cleaned between uses; may become loose with talking and chewing.
Canal Caps/Semi-Insert Earplugs	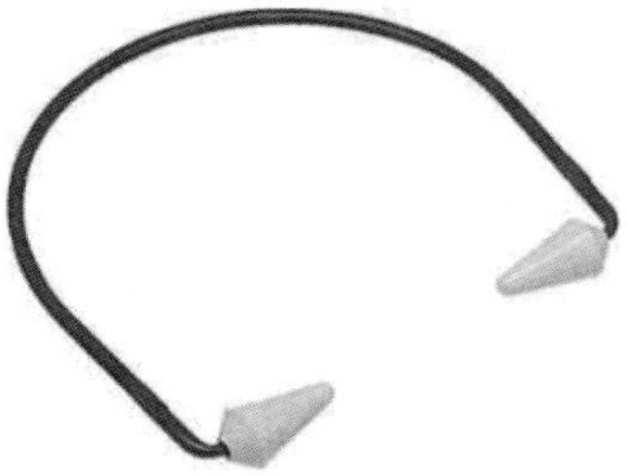	Convenient for intermittent use; various noise reduction ratings available; hangs around neck when not in use; lower attenuation than most earplugs; lower noise reduction ratings than earmuffs or earplugs; wearer's voice sounds louder (occlusion effect).
Custom Earplugs	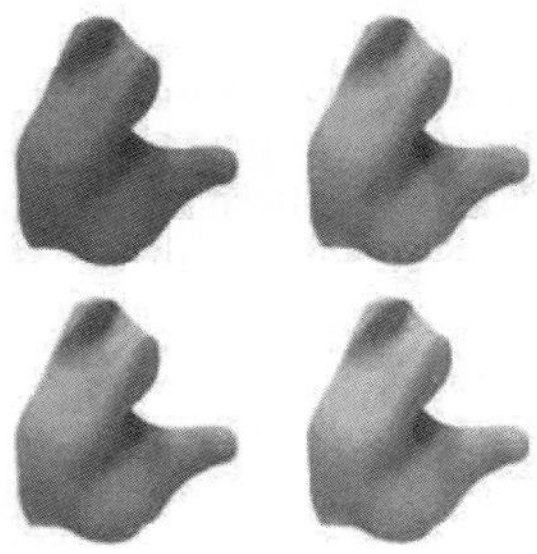	Comfortable with proper insertion; active sound management; variability in attenuation; comfort and fit variable with changes in ear canal; higher initial cost than other earplugs; quality can be variable.
Combination Earmuff		Easy to fit; face shield or face screen available; might be less adjustable than individual components; may be uncomfortably hot or heavy; more expensive than earmuffs or earplugs; often more expensive than individual components.

CHEMICAL FUME HOODS AND BIOLOGICAL SAFETY CABINETS

Thomas J. Bruno

Engineered safety equipment is preferred over the reliance on personal protective equipment, and the fume hood is one such engineered safety device that is nearly ubiquitous in laboratories. Laboratories concerned with biological specimens (microbes, spores) also commonly are equipped with biological safety cabinets. The following section provides basic information on the function and application of these devices. The purpose here is not to provide design or installation instructions, since most users will find this equipment already installed in their workspaces. Rather, this information is to allow optimal use to be made of the installation that is preexisting.

Types of Chemical Fume Hoods

Most of the chemical fume hoods considered here consist of a cabinet or enclosure set at waist level (above a table or storage cabinet) that is connected to a blower located above the hood or external to the hood through a duct system. The cabinet has an open side (or sides) to allow a user to perform work within. A movable transparent sash separates the user from the work. Most chemical fume hoods have a sill that functions as an airfoil at the work surface below the sash. The connection to the blower might be by use of a v-belt, or it may be direct drive. This allows provision of a smooth flow of air with minimal turbulence. In some installations, axially mounted blowers are used, especially if multiple hoods are ducted into a common blower. Baffles located in the rear of the cabinet provide control of the airflow patterns and can usually be adjusted to provide the best airflow around the experiment or procedure being performed. Many chemical fume hoods are equipped with airflow indicators, low-flow monitors and alarms, and differential pressure sensors to allow the user to operate safely. The major types of chemical fume hoods include the standard/conventional, walk-in, bypass, variable air volume, auxiliary air, or ductless types. Additional types include snorkels and canopies that are portable. Each type must be understood to be operated most efficiently within specifications (see discussion below on Chemical Fume Hood Operations).

Standard or Conventional

The standard chemical fume hood utilizes a constant speed motor, and for this reason the volume of air drawn into the hood will change with movement of the sash position. As the sash is lowered, the velocity of the air drawn into the hood will increase.

Bypass

The bypass chemical fume hood is very similar to the standard/conventional hood except that as the sash is lowered, a vent is opened above the sash to allow additional airflow into the hood. This prevents a large increase in velocity in the working area inside the hood.

Variable Air Volume

The variable volume chemical fume hood controls the volume of air drawn into the hood as a function of sash position, while maintaining the face velocity of the air at a constant rate, within the specifications required. These types of chemical fume hoods are more energy efficient than the standard or bypass hoods because they minimize costs incurred by laboratory heating and cooling.

Auxiliary Air

The auxiliary air chemical fume hood includes an additional blower that injects air into or at the face of the hood, providing additional flow inside the enclosed cabinet. These types of hoods are rarely installed in renovations or new construction but may be encountered in older laboratories. They are less desirable than the standard/conventional, bypass or variable volume types because they require a great deal of energy to operate (although the early designs featured the addition of an auxiliary airstream that was not air conditioned). These devices are mechanically more complex than other types, and consequently more prone to maintenance problems.

Walk-In Hood

The walk-in hood is a chemical fume hood that is mounted directly on the laboratory floor or a slightly raised chemical-resistant platform. It is used for the ventilation of larger pieces of equipment, with the advantage that these pieces of equipment can be wheeled in and out of the walk-in hood. The walk-in hood typically uses two separate sashes.

Ductless Hoods

This type of chemical fume hood does not duct the airflow to outside the laboratory, but rather the airflow is returned to the room or interstitial space after passing through a means to remove contaminants. The contaminants may be removed by HEPA filters, activated carbon cartridges, adsorbents, or catalyst beds. The means of contaminant removal must be inspected and serviced at regular intervals.

Laminar Flow Hood (Clean Bench)

Related at least in principle to ductless hoods are laminar flow hoods, sometimes called clean hoods. These are devices intended to protect the work being performed from particulates in the air, which is accomplished by bathing the work area with HEPA-filtered air either blown at low velocity over the work area or blown from the bottom of the hood as an air curtain. Only approximately 10% of the airflow is through the face of the hood. These are intended to protect the work or samples inside the hood, not primarily the user. These units should not be used in place of chemical fume hoods, rather they are used to protect the work from dust or pollen.

Snorkels and Canopies

Snorkels are flexible ducts routed from a blower duct that can be placed temporarily atop or near an experiment to provide some measure of protection. A canopy is similar, but it incorporates an additional bell-shaped collector that might be suspended above an experiment, but for the same purpose as that for the snorkel.

Chemical Fume Hood Operations

While the operation of chemical fume hoods is straightforward, safe operating practices must be observed. The face velocities should be optimized at between 80 and 120 fpm. Face velocities in excess of 125 fpm can cause turbulent flow and allow outflow of contaminants from the hood and potentially expose the user to hazards. Face velocities are checked periodically by facilities managers. Ideally, chemical fume hoods are located in low traffic areas of the laboratory, away from entry doors. Safe operation of chemical fume hoods requires observance of the following practices, divided into primary (applicable to all installations) and secondary (applicable on a case-by-case basis).

Primary Guidelines

- Before using the chemical fume hood, ensure that it is in working order. A simple airflow test can be done with a laboratory wipe, or one can observe the flow monitor if one is present. Be aware of clattering sounds that might indicate a broken belt, or screeching sounds that might indicate a failed or failing bearing.
- The motor of the chemical fume hood should be running at all times, except for maintenance. If a switch is located inside the laboratory, it should be equipped with a lockout to protect maintenance staff.
- Users should perform all work at least 6 inches inside the plane of the sash.
- The user should avoid placing his/her head inside the chemical fume hood, beyond the plane of the sash.
- Laboratory occupants should avoid traffic in front of the chemical fume hood to minimize turbulence.
- Users should avoid rapid movements in front of the chemical fume hood; this includes rapidly raising and lowering the sash.
- Keep the sash closed down as far as possible.
- The baffles of the chemical fume hood should be free of obstruction.
- If a heating device such as a hot plate is being used in the hood, the interior airflow can be dramatically changed by convection. In those instances, the lower baffles of the hood should be closed or minimized, and the middle baffles fully opened to accommodate the convection.
- Equipment located inside the cabinet that can potentially disrupt the airflow should be raised above the level of the lower baffles by a small shelf or blocks.
- Hoses and power cords that must be run into the chemical fume hood from the outside should be run through the airfoil beneath the sash.

Secondary Guidelines

- Some chemical fume hoods have a small cup sink located inside the cabinet. Water should be run into this sink periodically to maintain it free of obstruction and to keep the P-trap full and thereby prevent sewer gas backup.
- It is imperative to prevent the discharge of chemicals into the cup sink.
- Some chemical fume hoods have compressed air lines that allow the use of air-operated equipment such as vortex tubes (for heating and cooling). When using vortex tubes inside a fume hood, the sash should be fully closed because additional airflow inside the hood is present.
- In a power outage, lower the sash to within an inch of the fully closed position.
- Evaporations and digestions with perchloric acid must only be done in a specifically designed and designated perchloric acid chemical fume hood. The same considerations apply for the use of radionuclides and infectious agents.

Types of Biological Safety Cabinets (BSCs)

Biological safety cabinets, as distinct from chemical fume hoods, are enclosed ventilated cabinets intended to provide both a clean and a safe working environment for aerosols and biological hazards. All exhaust air is HEPA filtered as it exits the biosafety cabinet, removing harmful bacteria and viruses. The Centers for Disease Control (U.S.) lists three classes of biological safety cabinets, Classes I, II, and III.

Class I cabinets are open-front negative-pressure cabinets that provide personnel and environmental protection but do not provide protection to the sample or media (the product) being used in the cabinet. There is airflow into the cabinet (at a face velocity of 75 fpm) that can potentially cause sample contamination. Class I cabinets are often used to enclose specific equipment (centrifuges, harvesting equipment, or fermenters) or ongoing procedures (cultures) that potentially generate aerosols. BSCs of this class are either ducted (connected to the building exhaust system) or unducted (recirculating HEPA-filtered exhaust back into the laboratory, provided there is an interlock with the building exhaust system). Some Class I cabinets are used for animal cage changing and these typically require frequent HEPA filter changes due to odoriferous compounds saturating the filter.

A Class II biological safety cabinet provides protection for both the worker and the sample or product, making it suitable as a sterile compartment for cell culture. This type of cabinet is the most versatile and most common, with face velocities similar to those of Class I. There are four types of Class II cabinets, the main features of which are discussed below. Note that there are additional differences among the types in Class II (Types A1, A2, B1, and B2), primarily concerning the geometry of the airflows and placement of HEPA filters.

Type A1 (formerly Type A) does not have to be duct vented (although it is possible to connect to building ventilation systems by use of canopy exhaust connections), which makes it suitable for use in laboratories inaccessible to ductwork. This cabinet can be used for low to moderate hazard agents that do not include volatile toxic chemicals and volatile radionuclides. The supply air is HEPA filtered to present the sample or media being used with a particulate free airstream with a face velocity of at least 75 fps. This type of BSC cannot be used for volatile and toxic compounds and solvents because small quantities of these materials can quickly load the filter. Type A2 differs from Type A1 in that protection of the operator and the environment is only afforded if the exhaust line is canopy vented to the building exhaust. The face velocities of these units are at least 100 fps.

Type B1 cabinets must be hard vented, with 50% of the air exhausted from the cabinet while 50% can be recirculated back into the room. This cabinet may be used with etiologic agents and treated with volatile and toxic chemicals and radionuclides required as an adjunct to microbiological studies (if the work is done in the directly exhausted portion of the cabinet). The air intake velocity of the B1 type is specified to be 100 fps. Type B2 cabinets must be 100% exhausted through a dedicated duct.

The Class III cabinet is designed for highly infectious microbial agents. It is entirely gas tight, with a non-operating view window (cannot be opened). Access to the interior is through a dunk tank accessible through the floor of the cabinet or through a double-door system. Both the supply and exhaust gas streams pass through HEPA filters. Heavy-duty rubber gloves are used for manipulations in the interior of the cabinet.

GAS CYLINDER SAFETY AND STAMPED MARKINGS

Thomas J. Bruno and Paris D. N. Svoronos

The graphic below describes the permanent, stamped markings that are used on high-pressure gas cylinders commonly found in laboratories (Ref. 1). Note that individual jurisdictions and institutions have requirements for marking the cylinder contents as well. These requirements are in addition to the stamped markings, which pertain to the cylinder itself rather than to the fill contents (Ref. 2).

There are four fields of markings on cylinders that are used in the United States, labeled 1-4 on the figure.

Field 1 – Cylinder Specifications

DOT stands for the United States Department of Transportation, the agency that regulates the transport and specification of gas cylinders in the United States. The next entry, for example, 3AA, is the specification for the type and material of the cylinder. The most common cylinders are 3A, 3AA, 3AX, 3AAX, 3T, and 3AL. All but the last refer to steel cylinders, while 3AL refers to aluminum. The individual specifications differ mainly in chemical composition of the steel, and the gases that are approved for containment and transport. The 3T deals with large bundles of tube trailer cylinders.

The next entry in this field is the service pressure, in psig.

Field 2 – Serial Number

This is a unique number assigned by the manufacturer.

Field 3 – Identifying Symbol

The manufacturer identifying symbol historically can be a series of letters or a unique graphical symbol. In recent years, the DOT has standardized this identification with the "M" number, for example, M1004. This is a number issued by DOT that identifies the cylinder manufacturer.

Field 4 – Manufacturing Data

The data of manufacture is provided as a month and year. With this date is the inspector's official mark, for example, H. In recent years, this letter has been replaced with an IA number, for example IA02, pertaining to an independent agency that is approved by DOT as an inspector. If "+" is present, the cylinder qualifies for an overfill of 10% in service pressure. If "★" is present, the cylinder qualifies for a 10-year rather than a 5-year retest interval.

In some cases, the original insignia of the inspector that performed the first hydrostatic test is found below Field 4. Also stamped on the cylinder will be the retest dates. A cylinder must have a current (that is, within 5 or 10 years) test stamp. On the collar of the cylinder, the owner of the cylinder may be stamped.

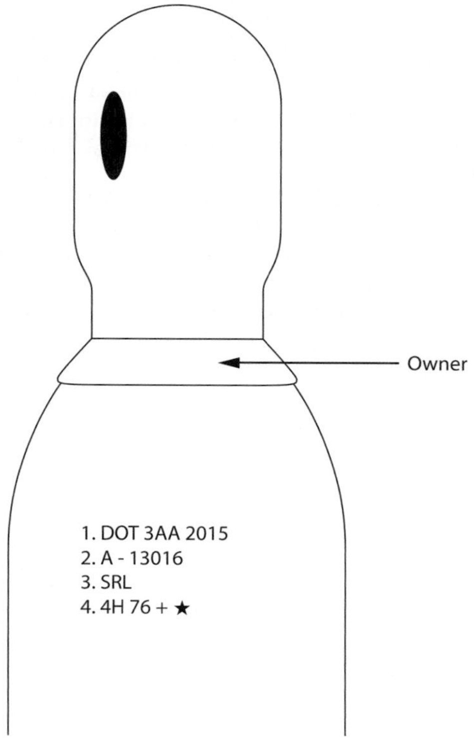

References

1. Hazardous Materials: Requirements for Maintenance, Requalification, Repair and Use of DOT Specification Cylinders, 49 CFR Parts 107, 171, 172, 173, 177, 178, 179, and 180; [Docket No. RSPA-01-10373 (HM-220D)]RIN 2137-AD58, August 8, 2002.
2. Bruno, T. J., and Svoronos, P. D. N., *CRC Handbook of Basic Tables for Chemical Analysis, Third Edition*, CRC Press, Boca Raton, FL, 2011.

COMPRESSED AIR SAFETY

Thomas J. Bruno and Paris D. N. Svoronos

Compressed air is commonly used in laboratories and is typically piped in as a utility. One commonly sees utility pods with laboratory valves or cocks for air, hot and cold water, natural gas, lab vacuum, and occasionally steam. Compressed air may be hard-plumbed into instrumentation for valve actuation, vortex-tube chilling, and as a source of air for flame ionization detectors. Specific hazards are associated with the use of compressed air sources and lines in laboratories, and safety precautions must be observed. Some important guidelines follow.

- All pipes, hoses, and fittings must have at least the rating of the maximum pressure of the compressor. Compressed air pipelines should be identified as such with labeling affixed to the pipe (high-pressure air or plant air are also acceptable labels).
- The maximum working pressure should be known and not exceeded. Typically, the pressure of laboratory compressed air lines is no more than 125 psig (862 kPa), and reduced pressures are provided by diaphragm regulators.
- Isolation valves (on/off ball-cock valves) should be provided for each instrument plumbed or connected to compressed air.
- Flexible hoses used for compressed air delivery must be appropriately rated, must be kept free from grease and corrosives, and must be secured to fixtures with hose clamps. The hoses themselves must be secured to prevent hose-whip in the event of a rupture or detachment. Care must be taken to prevent kinking of flexible air hoses.
- Compressed air must not be used to clean dirt and dust from clothing or personnel. Laboratory compressed air used for cleaning should be regulated to 15 psi (103 kPa) unless the blowguns used are equipped with diffuser nozzles to lower exit pressure and velocity.
- Static electricity can be generated through the use of pneumatic devices and tools. All equipment of these types must be grounded or bonded if it is used where fuel, flammable vapors, or explosive atmospheres are present.

All sources of laboratory or plant compressed air will contain water vapor, even if coalescence filters and particulate filters are provided inline. This water vapor can condense into liquid or ice, and can cause problems if the air contacts electrical or electronic devices. If the laboratory air supply is dried with a refrigerated conditioner, the dew point is still typically only dropped to 40 °F (4 °C).

In some older facilities, compressed air systems can be severely contaminated with oil (usually from the compressor), rust (from steel piping), or liquid water. In these older installations, it is very important to provide coalescence filters and particulate filters inline.

SAFETY IN THE USE OF CRYOGENS

Thomas J. Bruno

Cryogens (liquified gases or cryogenic liquids) are used extensively in laboratories and in analytical instruments. A cryogen is typically defined as a liquefied gas with a normal boiling temperature of no higher than –90 °C (–130 °F, different sources cite differing upper limits). Dry ice (solid carbon dioxide, which sublimes at 194.65 K, –78.5 °C) is also used to achieve low temperatures in the laboratory but is not considered to be a cryogen. The hazards associated with the use of cryogens are actually twofold. There are hazards associated with the cryogenic fluids themselves, and there are hazards associated with the containers used to store and transport the cryogenic fluids. Here, unlike other sources, we will treat these separately (Refs. 1–7).

Cryogens

The major hazards associated with the use of cryogens stem from their low temperatures, their high liquid-to-gas expansion ratios, toxicity, and air displacement. Table 1 has some important properties germane to the handling of common cryogens. The gas-to-liquid expansion ratios have a variability of up to ±5 depending on the local ambient temperature. The boiling temperatures are at atmospheric pressure (101.325 kPa). The column definitions are as follows.

Column heading	Definition
Name	Cryogen name
Mol. form.	Molecular formula of cryogen
***R*(approx.)**	Approximate gas-to-liquid expansion ratio, volume liquid:volume gas::1:R
T_b(K)	Boiling temperature, in K
T_b(C)	Boiling temperature, in °C

TABLE 1. Properties of Common Cryogens

Name	Mol. form.	*R*(approx.)	T_b(K)/K	T_b(C)/°C
Argon	Ar	847	87.302	-185.848
Helium (^{4}He)	He	757	4.2238	-268.928
Hydrogen	H_2	850	20.271	-252.879
Nitrogen	N_2	696	77.355	-195.795
Oxygen	O_2	860	90.188	-182.962

Low Temperature

The primary purpose of the use of a cryogen stems from the low boiling points. The potential of severe frostbite burn is very high if a cryogen comes into contact with skin. While a thin layer of vapor formation will protect the skin initially, if a cryogen is allowed to pool (such as in clothing), the frostbite danger is very high. Protective clothing is essential and should include cryogen gloves and safety glasses. A full face shield is strongly recommended. In addition, canvas shoes are discouraged because pooling can occur in the case of spillage. Pooling can also occur in cuffs on pant legs, which should be avoided. Pants should not be tucked into shoes. A lab coat or shirt cuffs should be tucked under glove gauntlets. If skin should come in contact with a cryogen, it should be rinsed with cold water; do not apply dry heat to the affected area. If clothing has frozen to an individual due to cryogen exposure, cold water should be used to free the clothing, and emergency personnel must be summoned.

Another hazard due to the low temperature results from the ability of these fluids to embrittle materials including hoses, floor mats, and other laboratory surfaces. Of special concern is the embrittlement of electrical insulation, which can lead to a fire hazard.

Because the boiling temperature of liquid nitrogen is below that of liquid oxygen, it is possible for oxygen to condense on all surfaces or vessels cooled by liquid nitrogen. Liquid oxygen is an oxidizer that can enhance the flammability characteristics of liquids and solids that it contacts. The liquid air that is seen dripping from lines transferring liquid nitrogen can be up to 50% oxygen. If a blue tint is observed in a vessel being used with liquid nitrogen, the presence of liquid oxygen must be assumed. Additional hazards of liquid oxygen are discussed below.

Asphyxiation

The high liquid-to-gas expansion ratios listed in Table 1 show that when a cryogen is vaporized, it has the potential to displace air in a laboratory. While most cryogens are not toxic per se, they can act as simple asphyxiants. Laboratories that use cryogens must be adequately ventilated. In labs with large containers of cryogen, it is important to have an oxygen monitor with an audible alarm. Personnel, including rescue workers, should not enter areas where the oxygen concentration is below 19.5% (vol/vol) unless provided with a self-contained breathing apparatus or air-line respirator. The safe range as indicated on an oxygen monitor is between 19.5% and 23% (vol/vol). Personnel in an area of low oxygen concentration may be unaware of the condition, thus monitoring is critical.

If a person seems to become dizzy or loses consciousness while working with cryogens, they should be moved to a well-ventilated area immediately. If breathing has stopped, apply CPR and emergency personnel must be summoned.

Overpressure

Related to the liquid-to-gas expansion ratio is the overpressure that can result if a cryogen is allowed to warm within an enclosure. Indeed, no cryogen can remain liquid within a container; some venting must be provided. If a vent becomes disabled or is not present, a warming cryogen vaporizes and produces very high pressures based on the PVT surface of the fluid.

Reactivity and Toxicity

Some cryogens pose specific hazards or handling requirements based on their chemistry.

Liquid oxygen (LOx) cannot be permitted to contact organic materials; common organic materials include solvents and vacuum pump oil. Organic materials can be readily ignited by spark or shock after exposure to LOx, including fingerprints on a surface. Clothing saturated with oxygen is readily ignitable and vigorously burns. If LOx spills on an asphalt surface, do not walk over or roll equipment over that surface for at least one hour. While not having specific toxicity issues, if LOx is exposed to high-energy electromagnetic radiation, it can produce ozone, which solidifies at LOx temperatures. Solid ozone is unstable and toxic (upon vaporization), and explodes if disturbed.

Liquid hydrogen handling requires all of the precautions used for hydrogen gas. Liquid hydrogen should not be transferred in an atmosphere of air as it readily condenses in the liquid hydrogen, resulting in a potential explosive mixture. Liquid hydrogen must be transferred by helium pressurization in properly designed vacuum-insulated transfer lines pre-purged with helium or gaseous hydrogen. Liquid hydrogen, like liquid helium, can solidify air, which can block vents and safety relief devices. Dewars and other containers made of glass should not be used for liquid hydrogen service. Breakage makes the possibility of explosion too hazardous to risk.

Cryogenic Liquid Containers

Several different types of cryogenic liquid containers are encountered in the laboratory during routine chemical analyses, and each has their own associated hazards and precautions. It is common parlance to refer to all of these containers as Dewars, but this terminology is imprecise. The small portable containers used to assemble laboratory cold baths and the small transport containers (with loose-fitting lids and carry handles) are also known as Dewars. These containers are used at ambient pressure. The larger supply containers (from which Dewars are filled) are called liquid cylinders. These containers are pressurized, with different pressure ratings available.

Liquid Cylinders

Liquid cylinders are large heavy containers with integrated casters or a dolly to facilitate movement. At least two of these casters should be equipped with a braking mechanism. Typical volumes and weights are provided below for liquid cylinders for nitrogen, oxygen, and argon.

TABLE 2. Weights of Filled Liquid Cylinders of Common Cryogens

Volume capacity (nominal)	160 L	180 L	230 L
Tare weight, kg (lb)	114 (250)	118 (260)	141 (310)
N2 filled weight, kg (lb)	233 (513)	253 (556)	303 (667)
O2 filled weight, kg (lb)	301 (662)	285 (627)	375 (825)
Ar filled weight, kg (lb)	316 (695)	342 (753)	425 (936)

Liquid helium cylinders, which often incorporate a liquid nitrogen jacket, are usually heavier than those for the common cryogens listed in Table 2. Moreover, there is a larger variety of available sizes, ranging from 50 to 500 L. The weights listed in Table 2 are typical as-filled weights. Some losses occur in transport, and the losses for helium liquid cylinders can be considerable.

The weight of these cylinders can make them challenging to handle. The personal protective equipment discussed above (cryogens) must be used when handling liquid cylinders. Cylinders should be moved by pushing, not pulling, to reduce the potential of an upset. In locations of frequent transport of liquid cylinders, bottom door sills should be removed to eliminate the potential of bouncing or rough handling. If a cylinder must be transported by elevator, a freight elevator is preferred. Personnel should not ride in the elevator car with the cylinder. The cylinder should be transported in the elevator with no personnel and be met at the receiving floor. A placard reading "CRYOGEN TRANSFER—DO NOT ENTER ELEVATOR" should be posted facing the door if the elevator is to travel more than one story.

If liquid cylinders must be transported between buildings, it is critical to use ramps and to ensure that there are no large cracks in paving that must be traversed. Note also that the casters commonly found on liquid cylinders are not rated for travel along long distances of pavement. In the event loss of control of a liquid cylinder occurs during transport, such as the cylinder begins to fall, it is usually best to simply let it go and summon qualified help as defined in the organization's standard operating procedures.

Liquid cylinders are pressurized and can contain up to 350 psi (2411 kPa), depending on the cylinder specifications. Pressure specifications on cylinders are sometimes confusing. The commonly encountered specifications are:

- psia (pounds-force per square inch absolute) gauge pressure plus local atmospheric pressure
- psid (psi difference) difference between two pressures, specified on the cylinder label
- psig pounds (force) per square inch, gauge
- psi-vg (psi-vented gauge) difference between the measuring point and the local pressure
- psi-sg (psi-sealed gauge) difference between a chamber of air sealed at atmospheric pressure and the pressure at the measuring point

Pressure relief devices are integral to all liquid cylinders and must remain unobstructed with frost. If the outlet fitting on a pressure relief valve is facing the same direction as the liquid or gas-dispensing valve, a fitting directing vented gas away from users should be added to protect personnel. Over-pressurization of liquid cylinders is a serious hazard; cylinders can rupture if a pressure relief valve becomes impaired or inoperative.

Dewar Flasks

Dewar flasks or simply, Dewars, are small cryogen containers used at atmospheric pressure, with or without a loose-fitting cover or cap. Smaller Dewars are usually made from an evacuated silvered-glass insert set into a metal jacket. Any exposed glass should be taped to prevent flying glass in the event of a catastrophic rupture.

When filling a small Dewar from a liquid cylinder, it is best to pre-cool the interior of the Dewar with a small amount of cryogen first, before completing the fill. Boiling and splashing generally occur when filling a warm container, so all personnel should stand clear and wear appropriate personal protective equipment as discussed above. The flask should be clean and dry before filling.

A Dewar flask should not be filled to beyond 80% of its capacity. Overfilling increases the risk of splashing and spillage. A beverage thermos bottle is NOT a substitute for a Dewar flask in the laboratory under any circumstances.

When carrying a small Dewar flask, it must be the only item being carried. Dewar flasks should be held as far away from the face as possible. Be aware of other personnel in the area.

Small Dewar flasks with liquid nitrogen are often used as cold traps in the laboratory. When instrument components are placed in the filled Dewar cold bath, it is important to insert components slowly to avoid splashing and excessive boiling. Any Pyrex wool insulation placed around the flask must not dip into the cryogen or become a vapor barrier. If liquid nitrogen acquires a blue tint, it has become contaminated with liquid oxygen, and the discussion of LOx hazards above applies.

When a Dewar cold trap is used in association with a vacuum pump, the trap must be carefully emptied periodically to avoid exposure to toxic chemicals and to prevent over pressurization should the trap run dry. Note also that venting liquid nitrogen near a vacuum pump v-belt can embrittle the belt and shorten its service life. Likewise, if the venting is near electrical cables, embrittlement of the electrical insulation can result in a fire hazard.

References

1. Lemmon, E.W., Properties of Cryogenic Fluids, in *CRC Handbook of Chemistry and Physics, 100th Edition*, 2019.
2. Safetygram 27, Cryogenic Liquid Containers, Air Products and Chemicals, Allentown, PA. <900-13-080-US-cryogenic-liquid-containers-safetygram-27.pdf>
3. Safetygram 8, Liquid Argon, Air Products and Chemicals, Allentown PA. <900-13-080-US-argon-safetygram-8.pdf>
4. Safetygram 16, Safe Handling of Cryogenic Liquids, Air Products and Chemicals, Allentown PA. <900-13-080-US-safe-handling-of-cryo-liquids-safetygram-16.pdf>
5. Safetygram 7, Liquid Nitrogen, Air Products and Chemicals, Allentown, PA. <900-13-080-US-liquid-nitrogen-safetygram-7.pdf>
6. Laboratory Safety Chemical Hygiene Plan (CHP), OSHA Fact Sheet, U.S. Occupational Safety and Health Administration, Washington, DC. <www.osha.gov/Publications/laboratory/OSHAfactsheet-laboratory-safety-chemical-hygiene-plan.pdf>
7. Laboratory Safety and Chemical Hygiene Plan, Queensborough Community College of City University of New York, Bayside, NY, 2012. <www.qcc.cuny.edu/ehs/docs/ChemicalHygienePlan2.pdf>

NANOMATERIAL SAFETY GUIDELINES

Thomas J. Bruno and Beverly L. Smith

The classification of nanomaterials includes nano-objects and nanoparticles; nano-objects are materials with at least one dimension (length, width, height, and/or diameter) that is between 1 nm (1×10^{-9} m) and 100 nm, and nanoparticles are materials in which all three dimensions are on this scale. [Note that the ASTM definition allows for two dimensions between 1 nm and 100 nm.] Beyond scale, nanomaterials can be classified as natural, incidental, and engineered, depending on origin. Natural nanomaterials include volcanic products, viruses, sea spray, and mineral aerosols, and are ubiquitous in nature at appreciable concentrations. Incidental nanomaterials include metal vapors produced during welding, sandblasting dust and other industrial effluents, cooking smoke, and diesel engine particulates. The environmental health and safety aspects of natural nanomaterials have received some study, and among the incidental nanomaterials, welding vapors and diesel fuel particulates have received extensive study. In recent years, however, there has been a great emphasis on engineered nanomaterials, and it is this class, which includes metal nanoparticles, nanorods, nanowires, nanotubes, buckyballs, nanocapsules, and quantum dots that are the main concern here. Study of the environmental health and safety risks of engineered nanomaterials remains an active area of research that is receiving increasing attention due to the widespread use of these materials in numerous applications ranging from medicine to energy storage. While much is still unknown regarding the fate and toxicity of this class of materials, here, we provide some general guidelines for the safe handling of nanomaterials. We begin with some simplified definitions or terms used in nanotechnology, needed for understanding of these safety guidelines as well as those provided elsewhere.

Aerodynamic diameter: An indirect measure of particle diameter defined as the diameter of a sphere with a density of 1000 kg/m^3, having the same settling velocity of a particle of interest.

Agglomerate: A group of particles (which may include nanoparticles) held together in a loose cluster by weak forces that may include van der Waals forces, surface tension, and electrostatic forces. Agglomerates are often resuspendable.

Aggregate: A heterogeneous particle held together with relatively strong forces such that the particle is not easily disassembled. Aggregates are typically not resuspendable.

Buckyballs: Spherical carbon (C60) fullerenes.

Fullerenes: Molecules composed entirely of carbon, usually in the form of a hollow sphere, ellipsoid, or tubes.

Graphene: A one atom thick sheet of carbon.

Multi-walled carbon nanotube: Multiple sheets of sheet graphene wrapped into a tube of nanoscale dimensions.

Nanoaerosol: A collection of nanomaterials suspended in a gas.

Nanocolloid: A nanomaterial suspended in a gel or other semi-solid substance.

Nanocomposite: A solid material composed of two or more nanomaterials having different physical characteristics.

Nanohydrosol: A nanomaterial suspended in a solution.

Nanotube: A seamless tube with a diameter on the order of nanometers.

Nanowire: A wire of dimensions on the order of nanometers.

Quantum dot: A nanomaterial or nanoparticle that confines the motion of conduction band electrons, valence band holes, or excitons (pairs of conduction band electrons and valence band holes) in all three spatial directions.

Single-walled carbon nanotube: A single-sheet graphene wrapped into a tube of nanoscale dimensions.

Ultrafine particle: A (usually) airborne particle with a diameter less than 100 nm.

Safety Issues and Exposure Routes

The unique safety issues posed by nanomaterials result from the potential of deep penetration into tissue, the potential of passing through the blood-brain barrier, and the possible ability to translocate between organs. Biological effects result from the size, shape, polarity/charge, adsorptive capacity, and surface composition, and the ability to bind biological proteins and receptors. Nanomaterials have a higher reactivity than the parent compounds, often having catalytic effects and often presenting greater flammability or explosion risks. For example, bulk elemental gold is considered inert, but gold nanoparticles below 5 nm are catalytic toward a number of oxidation reactions.

The most obvious exposure route of nanomaterials is respiratory; particles depositable in the air exchange region of the lungs are considered respirable. Ingestion can occur from unintentional hand-to-mouth transfer. Finally, nanoparticles can be absorbed through skin or cuts/abrasions to the skin.

Guidelines for Safe Handling of Nanomaterials

The safe handling of nanomaterials will generally follow the usual laboratory safety grid:

Elimination – A change in the experimental design to avoid the hazard

Substitution – The use of a surrogate of lower hazard

Engineering Controls – The use of enclosures, fume hoods, etc.

Administrative Controls – Adherence to standard procedures and protocols

Personal Protective Equipment (PPE) – The last line of safety, including gloves, clothing, respirators, etc.

Clearly, elimination and substitution are most useful for nanoparticles of incidental origin. Research with engineered nanoparticles must make use of engineering controls, administrative controls, and PPE.

The use of non-regenerating general ventilation systems, such as fume hoods and high-efficiency particulate air (HEPA) dust collection systems, are critical to safe handling. Where possible, installation of ultralow particulate air (ULPA) filters should be used, since they are widely viewed as being more effective for engineered nanomaterials. The lab in which nanomaterials are handled should be under negative pressure relative to the surroundings (corridor, service galley, etc.). The entry on "Chemical Fume Hoods and Biological Safety Cabinets" in this section provides additional information on fume hood selection and operation.

Where possible, manipulations should be conducted in solution (or in a liquid phase) to minimize the potential of aerosol formation.

Solution phase nanomaterials should be handled wearing gloves. Gloves should be compatible with the solvent used to disperse the nanomaterials in solution. In general, nitrile gloves are recommended, and double gloving is advisable for heavy usage or prolonged usage. Gloves with cuffs, clothing with full sleeves to protect wrists, or a laboratory coat are recommended. Some glove materials may have reactivity with certain nanomaterials, and this must be considered before selecting the glove material. Liquid or solution phase nanomaterial manipulations are best conducted in a fume hood or biological safety cabinet, especially when employing nanomaterials dispersed in open containers, in solvents with known health risks, or when higher-risk activities such as sonication, agitation, and vortex mixing are involved. Manipulations conducted in closed containers need not be performed in a fume hood or biological safety cabinet; however, a closed container should be opened inside such an enclosure because aerosols could be released upon opening. Disposable bench covers should be used where spillage is possible. Any spillage should be cleaned up immediately. Contaminated gloves should be removed and replaced immediately.

Manipulation of dry nanomaterials must be performed in a fume hood or biological safety cabinet. Transport of dry nanomaterials from place to place in the lab must be done in closed containers.

For manipulations of air-sensitive nanomaterials, a glove box or glove bag is required.

Hand washing must be done after manipulations of nanomaterials. Work areas should be cleaned after completion of tasks. Adequate consideration should be given to tasks involved with the maintenance of equipment or instrumentation used in work on nanomaterials. Such maintenance should be done with the assumption of the presence of nanomaterials.

Waste Disposal

Though the fate and toxicity of nanomaterials remains largely unknown and is still an area of active investigation, nanomaterials and any by-products from their synthesis should be treated as potentially hazardous waste. Nanomaterials should be properly disposed of based on their nanomaterial and solvent compositions. Nanomaterials containing heavy metals should be treated accordingly and separately from other waste streams. When possible, it is advisable to collect or fully dissolve nanomaterials that are present in solution to limit the volume of waste generated. Often nanomaterials can be aggregated or precipitated from solution with an anti-solvent and filtered off to be recycled or collected as solid waste. This is called "crashing out" in laboratory vernacular. Alternatively, adsorbents such as activated charcoal can often be used to remove certain types of nanomaterials from solution upon filtration and collection of the adsorbent following exposure to the nanomaterial-containing solution. Other types of nanomaterials such as metal oxides can often be dissolved completely with strong acids.

THRESHOLD LIMITS FOR AIRBORNE CONTAMINANTS

Several organizations recommend limits of exposure to airborne contaminants in the workplace. These include the United States National Institute for Occupational Safety and Health (NIOSH)-recommended exposure limits (RELs), which are intended to limit exposure to hazardous substances in workplace air to protect workers, and the non-governmental organization, American Conference of Governmental Industrial Hygienists (ACGIH). Threshold Limit Values (TLVs®) are the airborne concentrations of chemical substances under which it is believed that nearly all workers may be repeatedly exposed, day after day, over a working lifetime, without adverse effects.

In contrast to the ACGIH- and NIOSH-recommended limits of exposure, the United States Occupational Safety and Health Administration (OSHA) has established legal occupational exposure limits known as permissible exposure limits (PELs) which are based on a consideration of factors in addition to the adverse effects associated with airborne contaminants. When establishing PELs, OSHA takes into account technical feasibility and economic considerations for employers in addition to the adverse effects associated with the airborne contaminant. While these PELs are legal exposure limits, OSHA recognizes that many of its PELs are outdated and inadequate for ensuring protection of worker health. Most of OSHA's PELs were issued shortly after adoption of the Occupational Safety and Health (OSH) Act in 1970 and have not been updated since that time. OSHA has stated that alternate occupational exposure limits such as NIOSH RELs and ACGIH TLVs may serve to better protect workers.

The following table gives threshold limit values for substances that may be encountered in the atmosphere of a chemical laboratory or industrial facility. The table includes NIOSH RELs and ACGIH TLVs, but not OSHA PELs given OSHA's caution about PELs.

NIOSH RELs and the ACGIH TLVs are given in two forms:

- Time-weighted average (TWA) concentration for a normal 8-hr workday and 40-hr workweek.
- Short-term exposure limit (STEL), which should not be exceeded for more than 15 minutes.

All values refer to the concentration in air at 25 °C and normal atmospheric pressure. Data for gases are given in parts per million by volume (ppm). Values for liquids refer to mists or aerosols, and those for solids to dusts or fumes; both are stated in mass concentration units (mg/m^3). In some cases, a "Ceiling value" is provided which indicates a ceiling limit that should not be exceeded even for very brief periods because of acute toxic effects of the substance. The notation "levels as low as possible" in the Comments column indicates such a high degree of hazard that no safe limit can be recommended. Expressions such as "inhalable fraction and vapor" and "inhalable particulate matter" that appear after a TLV indicate that the TLV value has been adjusted to reflect the influence of particle size on the respiratory hazard. See Refs. 1–3 for details.

Substances are listed alphabetically by systematic name. Synonyms are also provided. The Comments provide further information on the physical form of the substance and the basis to which the limit is referred. The Formula column gives the molecular formula in the Hill convention for organic compounds and the customary line formula for inorganic compounds. The ACGIH and NIOSH TWA, STEL and Ceiling limit are provided if available for a substance.

Note that the entries in this table are for substances that have undesirable physiological effects on humans. Many other substances should be avoided in the workplace because of explosion or asphyxia hazards.

References

1. *2018 TLVs and BEIs*, American Conference of Governmental Industrial Hygienists, 1330 Kemper Meadow Drive, Cincinnati, OH 45240-1634, 2018 <www.acgih.org>.
2. *NIOSH Pocket Guide to Chemical Hazards*, U.S. Department of Health and Human Services, National Institute for Occupational Health and Safety, 2016 <www.cdc.gov/niosh/npg/>.
3. OSHA Annotated PELs, TABLE Z-1, U.S. Department of Labor, Occupational Safety and Health Administration, 2017 <www.osha.gov/dsg/annotated-pels/tablez-1.html>.

Threshold Limits for Airborne Contaminants

				ACGIH			NIOSH			
Name	Comments	Formula	CAS Reg. No.	Time-weighted average	Short-term exposure limit	ACGIH Ceiling	Time-weighted average	Short-term exposure limit	Ceiling	Basis
Abate	[Temephos]	$C_{16}H_{20}O_6P_2S_3$	3383-96-8	1 mg/m^3			10 mg/m^3 (total); 5 mg/m^3 (resp)			Cholinesterase inhibitor
Acetaldehyde	[Ethanal]	CH_3CHO	75-07-0			25 ppm				Eye and upper respiratory tract irritation
Acetamide		CH_3CONH_2	60-35-5	1 ppm (inhalable fraction and vapor)						Liver cancer and damage
Acetic acid	[Ethanoic acid]	CH_3COOH	64-19-7	10 ppm	15 ppm		10 ppm	15 ppm		Upper respiratory tract and eye irritation; pulmonary function
Acetic anhydride	[Acetyl acetate]	$C_4H_6O_3$	108-24-7	1 ppm	3 ppm				5 ppm	Eye and upper respiratory tract irritation
Acetone	[2-Propanone]	$(CH_3)_2CO$	67-64-1	250 ppm	500 ppm		250 ppm			Eye and upper respiratory tract irritation; central nervous system impairment

				ACGIH			NIOSH			
Name	**Comments**	**Formula**	**CAS Reg. No.**	**Time-weighted average**	**Short-term exposure limit**	**ACGIH Ceiling**	**Time-weighted average**	**Short-term exposure limit**	**Ceiling**	**Basis**
Acetone cyanohydrin	as CN	C_4H_7NO	75-86-5			5 mg/m³			4 mg/m³ (1 ppm) [15-minute]	Upper respiratory tract irritation; headache; hypoxia/cyanosis
Acetonitrile	[Methyl cyanide]	CH_3CN	75-05-8	20 ppm			20 ppm			Lower respiratory tract irritation
Acetophenone	[Methyl phenyl ketone]	$C_6H_5C{=}OCH_3$	98-86-2	10 ppm						Upper respiratory tract irritation; central nervous system impairment; pregnancy loss
2-(Acetyloxy)benzoic acid	[Acetylsalicylic acid (Aspirin)]	$C_9H_8O_4$	50-78-2	5 mg/m³			5 mg/m³			Skin and eye irritation
Acrolein	[2-Propenal]	$CH_2{=}CHCHO$	107-02-8			0.1 ppm	0.1 ppm		0.3 ppm	Eye and upper respiratory tract irritation; pulmonary edema; pulmonary emphysema
Acrylamide	[2-Propenamide]	C_3H_5NO	79-06-1	0.03 mg/m³ (inhalable fraction and vapor)			0.03 mg/m³			Central nervous system impairment
Acrylic acid	[2-Propenoic acid]	$C_3H_4O_2$	79-10-7	2 ppm			2 ppm			Upper respiratory tract irritation
Acrylonitrile	[Propenenitrile]	$CH_2{=}CHCN$	107-13-1	2 ppm			1 ppm		10 ppm [15-minute]	Upper respiratory tract irritation
Alachlor	[Acetamide, 2-chloro-*N*-(2,6-diethylphenyl)-*N*-(methoxymethyl)-]	$C_{14}H_{20}ClNO_2$	15972-60-8	1 mg/m³ (inhalable fraction and vapor)						Hemosiderosis (liver, spleen, kidney)
Aldrin		$C_{12}H_8Cl_6$	309-00-2	0.05 mg/m³ (inhalable fraction and vapor)			0.25 mg/m³			Central nervous system impairment; liver and kidney damage
Allyl alcohol	[2-Propen-1-ol]	C_3H_6O	107-18-6	0.5 ppm			2 ppm	4 ppm		Eye and upper respiratory tract irritation
Allyl glycidyl ether	[AGE]	$C_6H_{10}O_2$	106-92-3	1 ppm			5 ppm	10 ppm		Upper respiratory tract irritation; eye and skin irritation; dermatitis
Allyl propyl disulfide		$C_6H_{12}S_2$	2179-59-1	0.5 ppm			2 ppm	3 ppm		Upper respiratory tract and eye irritation
Aluminum	metal dust & insoluble compounds	Al	7429-90-5	1 mg/m³			10 mg/m³ (total); 5 mg/m³ (resp)			Pneumoconiosis; lower respiratory tract irritation; neurotoxicity
4-Aminobiphenyl	[*p*-Biphenylamine]; levels as low as possible	$C_{12}H_{11}N$	92-67-1							Bladder and liver cancer
4-Amino-3,5,6-trichloro-2-pyridinecarboxlic acid	[Picloram]	$C_6H_3Cl_3N_2O_2$	1918-02-1	10 mg/m³						Liver and kidney damage
Ammonia		NH_3	7664-41-7	25 ppm	35 ppm		25 ppm	35 ppm		Eye damage; upper respiratory tract irritation
Ammonium chloride	[Sal ammoniac]; fume	NH_4Cl	12125-02-9	10 mg/m³	20 mg/m³		10 mg/m³	20 mg/m³		Eye damage; upper respiratory tract irritant
Ammonium perfluorooctanoate		$C_8H_4F_{15}NO_2$	3825-26-1	0.01 mg/m³						Liver damage
Ammonium sulfamate		$NH_4NH_2SO_3$	7773-06-0	10 mg/m³			10 mg/m³ (total); 5 mg/m³ (resp)			-
Aniline	[Benzenamine]	$C_6H_5NH_2$	62-53-3	2 ppm						Methemoglobinemia
Antimony	and compounds, as Sb	Sb	7440-36-0	0.5 mg/m³			0.5 mg/m³			Skin and upper respiratory tract irritation
Arsenic	and inorganic compounds, as As	As	7440-38-2	0.01 mg/m³					0.002 mg/m³ [15-minute]	Lung cancer
Arsine	[Arsenic hydride]	AsH_3	7784-42-1	0.005 ppm					0.002 mg/m³ [15-minute]	Peripheral nervous system and vascular system impairment; kidney and liver impairment
Asbestos	all forms; limit is 0.1 fibers/mL		1332-21-4							Pneumoconiosis; lung cancer; mesothelioma
Atrazine	[6-Chloro-*N*-ethyl-*N'*-(1-methylethyl)-1,3,5-triazine-2,4-diamine]	$C_8H_{14}ClN_5$	1912-24-9	2 mg/m³			5 mg/m³			Hematologic, reproductive and developmental effects
Azinphos-methyl		$C_{10}H_{12}N_3O_3PS_2$	86-50-0	0.2 mg/m³ (inhalable fraction and vapor)			0.2 mg/m³			Cholinesterase inhibitor
Barium	and soluble compounds, as Ba	Ba	7440-39-3	0.5 mg/m³			0.5 mg/m³			Eye, skin, and GI irritation; muscular stimulation

				ACGIH			NIOSH			
Name	Comments	Formula	CAS Reg. No.	Time-weighted average	Short-term exposure limit	ACGIH Ceiling	Time-weighted average	Short-term exposure limit	Ceiling	Basis
Barium sulfate	[Barite]	$BaSO_4$	7727-43-7	5 mg/m^3			10 mg/m^3 (total); 5 mg/m^3 (resp)			Pneumoconiosis
Benomyl	No established NIOSH REL	$C_{14}H_{18}N_4O_3$	17804-35-2	1 mg/m^3 (inhalable particulate matter)						Upper respiratory tract irritation; male reproductive, testicular, and embryo/fetal damage
Benz[*a*]anthracene	[1,2-Benzanthracene]; levels as low as possible	$C_{18}H_{12}$	56-55-3							Skin cancer
Benzene		C_6H_6	71-43-2	0.5 ppm	2.5 ppm		0.1 ppm	1 ppm		Leukemia
1,2-Benzenediamine	[*o*-Phenylenediamine]	$C_6H_8N_2$	95-54-5	0.1 mg/m^3						Anemia
1,3-Benzenediamine	[*m*-Phenylenediamine]	$C_6H_8N_2$	108-45-2	0.1 mg/m^3						Liver damage; skin irritation
1,4-Benzenediamine	[*p*-Phenylenediamine]	$C_6H_8N_2$	106-50-3	0.1 mg/m^3			0.1 mg/m^3			Upper respiratory tract irritation; skin sensitization
1,3-Benzenedimethanamine	[*m*-Xylene diamine]	$C_8H_{12}N_2$	1477-55-0			0.1 mg/m^3				Eye, skin, and GI irritation
Benzenethiol	[Phenyl mercaptan]	C_6H_5SH	108-98-5	0.1 ppm					0.1 ppm [15-minute]	CNS impairment; eye and skin irritation
p-Benzidine	[[1,1'-Biphenyl]-4,4'-diamine]; levels as low as possible	$C_{12}H_{12}N_2$	92-87-5							Bladder cancer
Benzo[*b*]fluoranthene	[Benz[e]acephenanthrylene]; levels as low as possible	$C_{20}H_{12}$	205-99-2							Cancer
Benzo[*a*]pyrene	[2,3-Benzopyrene]; levels as low as possible	$C_{20}H_{12}$	50-32-8							Cancer
p-Benzoquinone	[2,5-Cyclohexadiene-1,4-dione]	$C_6H_4O_2$	106-51-4	0.1 ppm			0.4 mg/m^3 (0.1 ppm)			Eye irritation; skin damage
Benzoyl chloride	[Benzoic acid, chloride]	C_6H_5COCl	98-88-4			0.5 ppm				Upper respiratory tract and eye irritation
Benzoyl peroxide		$C_{14}H_{10}O_4$	94-36-0	5 mg/m^3			5 mg/m^3			Upper respiratory tract and skin irritation
Benzyl acetate	[(Acetoxymethyl)benzene]	$C_9H_{10}O_2$	140-11-4	10 ppm						Upper respiratory tract irritation
Beryllium	and compounds, as Be	Be	7440-41-7	0.00005 mg/m^3			0.0002 mg/m^3		0.002 mg/m^3	Beryllium sensitization; chronic beryllium disease (berylliosis)
Biphenyl	[Diphenyl]	$C_{12}H_{10}$	92-52-4	0.2 ppm			0.2 ppm			Pulmonary function
Bis(2-aminoethyl)amine	[Diethylenetriamine]	$C_4H_{13}N_3$	111-40-0	1 ppm			1 ppm			Upper respiratory tract and eye irritation
Bis(2-chloroethyl) ether	[Dichloroethyl ether]	$C_4H_8Cl_2O$	111-44-4	5 ppm	10 ppm		5 ppm	10 ppm		Upper respiratory tract and eye irritation; nausea
Bis(chloromethyl) ether	[Chloromethyl ether]	$C_2H_4Cl_2O$	542-88-1	0.001 ppm						Lung cancer
Bis(2-dimethylaminoethyl) ether	[2,2'-Oxybis[*N,N*-dimethylethanamine (DMAEE)]; NIOSH: levels as low as possible	$C_8H_{20}N_2O$	3033-62-3	0.05 ppm	0.15 ppm					Upper respiratory tract, eye, and skin irritation
Bis(2-ethylhexyl) phthalate	[Di-*sec*-octyl phthalate (DEHP)]	$C_{24}H_{38}O_4$	117-81-7	5 mg/m^3			5 mg/m^3	10 mg/m^3		Lower respiratory tract irritation
Bismuth telluride	[Tetradymite]	Bi_2Te_3	1304-82-1	10 mg/m^3 (undoped); 5 mg/m^3 (Se-doped)			10 mg/m^3 (total); 5 mg/m^3 (resp)			Lung damage
Boric acid	[Orthoboric acid]; and inorganic borate compounds	H_3BO_3	10043-35-3	2 mg/m^3	6 mg/m^3		5 mg/m^3			Upper respiratory tract irritation
Boron oxide	[Boric oxide]	B_2O_3	1303-86-2	10 mg/m^3			10 mg/m^3			Eye and upper respiratory tract irritation
Boron tribromide	[Tribromoborane]	BBr_3	10294-33-4			0.7 ppm			1 ppm (10 mg/m^3)	Respiratory tract irritation; pneumonitis
Boron trichloride	[Trichloroborane]	BCl_3	10294-34-5			0.7 ppm				Respiratory tract irritation; pneumonitis
Boron trifluoride	[Trifluoroborane]	BF_3	7637-07-2	0.1 ppm		0.7 ppm			1 ppm (3 mg/m^3)	Respiratory tract irritation; pneumonitis
Bromacil	[5-Bromo-3-*sec*-butyl-6-methyluracil]	$C_9H_{13}BrN_2O_2$	314-40-9	10 mg/m^3			1 ppm (10 mg/m^3)			Thyroid effects
Bromine		Br_2	7726-95-6	0.1 ppm	0.2 ppm		0.1 ppm	0.3 ppm		Upper respiratory tract and lower respiratory tract irritation; lung damage
Bromine pentafluoride		BrF_5	7789-30-2	0.1 ppm			0.1 ppm			Eye, skin, and upper respiratory tract irritation
Bromochloromethane	[Halon 1011]	CH_2BrCl	74-97-5	200 ppm			200 ppm			CNS impairment; liver damage
2-Bromo-2-chloro-1,1,1-trifluoroethane	[Halothane]	$C_2HBrClF_3$	151-67-7	50 ppm						Liver damage; CNS impairment; vasodilation

				ACGIH			NIOSH			
Name	Comments	Formula	CAS Reg. No.	Time-weighted average	Short-term exposure limit	ACGIH Ceiling	Time-weighted average	Short-term exposure limit	Ceiling	Basis
Bromoethane	[Ethyl bromide]; NIOSH: No established REL	C_2H_5Br	74-96-4	5 ppm						Liver damage; CNS impairment
Bromoethene	[Vinyl bromide]	$CH_2{=}CHBr$	593-60-2	0.5 ppm						Liver cancer
Bromomethane	[Methyl bromide]	CH_3Br	74-83-9	1 ppm						Upper respiratory tract and skin irritation
1-Bromopropane	[Propyl bromide]	C_3H_7Br	106-94-5	0.1 ppm						CNS impairment; peripheral neuropathy; hematological effects; developmental and reproductive toxicity (female and male)
3-Bromopropene	[Allyl bromide]	C_3H_5Br	106-95-6	0.1 ppm	0.2 ppm					Eye and upper respiratory tract irritation
Bromotrifluoromethane	[Halon-1301]	CF_3Br	75-63-8	1000 ppm			1000 ppm			Central nervous system impairment; cardiac impairment
1,3-Butadiene	[Divinyl]	C_4H_6	106-99-0	2 ppm						Cancer
Butane	both isomers	C_4H_{10}	106-97-8		1000 ppm		800 ppm			Central nervous system impairment
2,3-Butanedione	[Diacetyl]	$C_4H_6O_2$	431-03-8	0.01 ppm	0.02 ppm					Lung damage (bronchiolitis obliterans-like illness)
1-Butanethiol	[Butyl mercaptan]	$C_4H_{10}S$	109-79-5	0.5 ppm					0.5 ppm [15-minute]	Upper respiratory tract irritation
1-Butanol	[Butyl alcohol]	C_4H_9OH	71-36-3	20 ppm			50 ppm			Eye and upper respiratory tract irritation
2-Butanol	[*sec*-Butyl alcohol]	$C_4H_{10}O$	78-92-2	100 ppm			100 ppm	150 ppm		Upper respiratory tract irritation; central nervous system impairment
2-Butanone	[Methyl ethyl ketone (MEK)]	C_4H_8O	78-93-3	200 ppm	300 ppm		200 ppm	300 ppm		Upper respiratory tract irritation; central nervous system impairment; peripheral nervous system impairment
2-Butanone peroxide	[Methyl ethyl ketone peroxide]	$C_8H_{16}O_4$	1338-23-4			0.2 ppm			0.2 ppm	Eye and skin irritation; liver and kidney damage
trans-2-Butenal	[*trans*-Crotonaldehyde]	C_4H_6O	123-73-9			0.3 ppm	2 ppm			Eye and upper respiratory tract irritation
1-Butene	[1-Butylene]	C_4H_8	106-98-9	250 ppm						Body weight effects
cis-2-Butene		C_4H_8	590-18-1	250 ppm						Body weight effects
trans-2-Butene		C_4H_8	624-64-6	250 ppm						Body weight effects
3-Buten-2-one	[Methyl vinyl ketone]	C_4H_6O	78-94-4			0.2 ppm				Upper respiratory tract and eye irritation; central nervous system impairment
2-Butoxyethanol	[Ethylene glycol monobutyl ether (EGBE)]	$C_6H_{14}O_2$	111-76-2	20 ppm			5 ppm			Eye and upper respiratory tract irritation
2-Butoxyethyl acetate	[Ethylene glycol monobutyl ether acetate (EGBEA)]	$C_8H_{16}O_3$	112-07-2	20 ppm			5 ppm			Hemolysis
Butyl acetate		$C_6H_{12}O_2$	123-86-4	50 ppm	150 ppm		150 ppm	200 ppm		Eye and upper respiratory tract irritation
sec-Butyl acetate	[1-Methylpropyl acetate]	$C_6H_{12}O_2$	105-46-4	50 ppm	150 ppm		200 ppm			Eye and upper respiratory tract irritation
tert-Butyl acetate	[1,1,-Dimethylethyl acetate]	$C_6H_{12}O_2$	540-88-5	50 ppm	150 ppm		200 ppm			Eye and upper respiratory tract irritation
Butyl acrylate	[Butyl 2-propenoate]	$C_7H_{12}O_2$	141-32-2	2 ppm			10 ppm			Irritation
Butylamine	[1-Butanamine]	$C_4H_9NH_2$	109-73-9			5 ppm			5 ppm	Headache; upper respiratory tract irritation; eye irritation
tert-Butyl chromate	as CrO_3	$C_8H_{18}CrO_4$	1189-85-1			0.1 mg/m³			0.001 mg Cr(VI)/m³	Lower respiratory tract irritation; skin irritation
tert-Butyl ethyl ether	[Ethyl *tert*-butyl ether (ETBE)]	$C_6H_{14}O$	637-92-3	25 ppm						Upper respiratory tract and lower respiratory tract irritation; central nervous system impairment
Butyl glycidyl ether	[BGE]	$C_7H_{14}O_2$	2426-08-6	3 ppm					5.6 ppm [15-minute]	Reproductive toxicity; sensitization

Name	Comments	Formula	CAS Reg. No.	ACGIH Time-weighted average	ACGIH Short-term exposure limit	ACGIH Ceiling	NIOSH Time-weighted average	NIOSH Short-term exposure limit	NIOSH Ceiling	Basis
Butyl lactate		$C_7H_{14}O_3$	34451-18-8	5 ppm			5 ppm			Headache; upper respiratory tract irritation
1-*tert*-Butyl-4-methylbenzene	[4-*tert*-Butyltoluene]	$C_{11}H_{16}$	98-51-1	1 ppm			10 ppm	20 ppm		Eye and upper respiratory tract irritation; nausea
2-*sec*-Butylphenol		$C_{10}H_{14}O$	89-72-5	5 ppm			5 ppm			Upper respiratory tract, eye and skin irritation
Cadmium	metal	Cd	7440-43-9	0.01 mg/m³						Kidney damage
Cadmium	compounds, as Cd	Cd	7440-43-9	0.002 mg/m³						Kidney damage
Cadusafos		$C_{10}H_{23}O_2PS_2$	95465-99-9	0.001 mg/m³ (inhalable fraction and vapor)						Cholinesterase inhibitor
Calcium chromate	as Cr	$CaCrO_4$	13765-19-0	0.0002 mg/m³			0.0002 mg/m³ [8-hour]			Lung and sinonasal cancer; respiratory tract irritation; asthma
Calcium cyanamide	[Calcium carbimide]	$CaCN_2$	156-62-7	0.5 mg/m³			0.5 mg/m³			Eye and upper respiratory tract irritation
Calcium hydroxide	[Portlandite]	$Ca(OH)_2$	1305-62-0	5 mg/m³			5 mg/m³			Eye, skin, and upper respiratory tract irritation
Calcium metasilicate	[Parawollastonite]; synthetic, nonfibrous	$CaSiO_3$	1344-95-2	1 mg/m³			10 mg/m³ (total); 5 mg/m³ (respirable)			Pneumoconiosis; pulmonary function
Calcium oxide	[Lime]	CaO	1305-78-8	2 mg/m³			2 mg/m³			Upper respiratory tract irritation
Calcium sulfate	[Anhydrite]	$CaSO_4$	7778-18-9	10 mg/m³			10 mg/m³ (total); 5 mg/m³ (respirable)			Nasal symptoms
Camphor, (+)	[1,7,7-Trimethylbicyclo[2.2.1] heptan-2-one, (1*R*)-]	$C_{10}H_{16}O$	464-49-3	2 ppm	3 ppm		2 mg/m³			Eye and upper respiratory tract irritation; anosmia
Caprolactam	[6-Hexanelactam]	$C_6H_{11}NO$	105-60-2	5 mg/m³ (inhalable fraction and vapor)			1 mg/m³	3 mg/m³		Upper respiratory tract irritation
Captafol	[Difolatan]	$C_{10}H_9Cl_4NO_2S$	2425-06-1	0.1 mg/m³ (inhalable fraction and vapor)			0.1 mg/m³			Liver and kidney damage; dermal sensitization
Captan		$C_9H_8Cl_3NO_2S$	133-06-2	5 mg/m³ (inhalable particulate matter)			5 mg/m³			Skin irritation
Carbaryl		$C_{12}H_{11}NO_2$	63-25-2	0.5 mg/m³ (inhalable fraction and vapor)			5 mg/m³			Cholinesterase inhibitor; male reproductive toxicity; embryo damage
Carbofuran	[7-Benzofuranol, 2,3-dihydro-2,2-dimethyl-, methylcarbamate]	$C_{12}H_{15}NO_3$	1563-66-2	0.1 mg/m³ (inhalable fraction and vapor)			0.1 mg/m³			Cholinesterase inhibitor
Carbon (graphite)	[Graphite]; except fibers	C	7782-42-5	2 mg/m³ (resp)			2.5 mg/m³ (resp)			Pneumoconiosis
Carbon black	[Carbon (amorphous)] inhalable fraction	C	1333-86-4	3 mg/m³ (inhalable particulate matter)			3.5 mg/m³ Ca 0.1 mg PAHs/m³ (carbon black in presence of polycyclic aromatic hydrocarbons)			Bronchitis
Carbon dioxide	[Carbonic anhydride]	CO_2	124-38-9	5000 ppm	30,000 ppm		5000 ppm	30,000 ppm		Asphyyxia
Carbon disulfide	[Carbon bisulfide]	CS_2	75-15-0	1 ppm			1 ppm	10 ppm		Peripheral nervous system impairment
Carbon monoxide	[Carbon oxide]	CO	630-08-0	25 ppm			35 ppm		200 ppm	COHb-emia
Carbon oxysulfide	[Carbonyl sulfide]	OCS	463-58-1	5 ppm						Central nervous system impairment
Carbonyl chloride	[Phosgene]	$COCl_2$	75-44-5	0.1 ppm			0.1 ppm		0.2 ppm [15-minute]	Upper respiratory tract irritation; pulmonary edema; pulmonary emphysema
Carbonyl fluoride		COF_2	353-50-4	2 ppm	5 ppm		2 ppm	5 ppm		Lower respiratory tract irritation; bone damage
Cellulose			9004-34-6	10 mg/m³			10 mg/m³ (total); 5 mg/m³ (resp)			Upper respiratory tract irritation

				ACGIH			NIOSH			
Name	Comments	Formula	CAS Reg. No.	Time-weighted average	Short-term exposure limit	ACGIH Ceiling	Time-weighted average	Short-term exposure limit	Ceiling	Basis
Cesium hydroxide		CsOH	21351-79-1	2 mg/m³			2 mg/m³			Upper respiratory tract irritation; skin and eye irritation
Chlordane	[1,2,4,5,6,7,8,8-Octachloro-2,3,3a,4,7,7a-hexahydro-4,7-methano-1*H*-indene]	$C_{10}H_6Cl_8$	57-74-9	0.5 mg/m³			0.5 mg/m³			Liver damage
o-Chlorinated diphenyl oxide			31242-93-0	0.5 mg/m³						Chloracne; liver damage
Chlorine		Cl_2	7782-50-5	0.5 ppm	1 ppm				0.5 ppm [15-minute]	Respiratory tract irritation; airway hyper-reactivity; pulmonary edema
Chlorine dioxide		ClO_2	10049-04-4	0.1 ppm	0.3 ppm		0.1 ppm	0.3 ppm		Respiratory tract irritation; pulmonary edema
Chlorine trifluoride		ClF_3	7790-91-2			0.1 ppm	0.1 ppm			Eye and upper respiratory tract irritation; lung damage
Chloroacetaldehyde	[2-Chloro-1-ethanal]	C_2H_3ClO	107-20-0			1 ppm			1 ppm	Upper respiratory tract and eye irritation
Chloroacetic acid		$CH_2ClCOOH$	79-11-8	0.5 ppm (inhalable fraction and vapor)						Upper respiratory tract irritation
Chloroacetone		C_3H_5ClO	78-95-5			1 ppm				Eye and upper respiratory tract irritation
α-Chloroacetophenone	[2-Chloroacetophenone]	C_8H_7ClO	532-27-4	0.05 ppm			0.05 ppm			Eye, upper respiratory tract, and skin irritation
Chloroacetyl chloride		$C_2H_2Cl_2O$	79-04-9	0.05 ppm	0.15 ppm		0.05 ppm			Upper respiratory tract irritation
Chlorobenzene	[Phenyl chloride]	C_6H_5Cl	108-90-7	10 ppm						Liver damage
o-Chlorobenzylidene malononitrile		$C_{10}H_5ClN_2$	2698-41-1			0.05 ppm	0.05 ppm			Upper respiratory tract irritation; skin sensitization
2-Chloro-1,3-butadiene	[Chloroprene]	C_4H_5Cl	126-99-8	1 ppm					1 ppm (15-minute)	Lung cancer; upper respiratory tract and eye irritation
Chlorodifluoromethane	[HCFC-22]	$CHClF_2$	75-45-6	1000 ppm			1000 ppm	1250 ppm		Central nervous system impairment; asphyxia; cardiac sensitization
Chloroethane	[Ethyl chloride]	C_2H_5Cl	75-00-3	100 ppm						Liver damage
2-Chloroethanol	[Ethylene chlorohydrin]	C_2H_5ClO	107-07-3			1 ppm			1 ppm	Central nervous system impairment; liver and kidney damage
Chloroethene	[Vinyl chloride]	CH_2=CHCl	75-01-4	1 ppm						Lung cancer; liver damage
Chloromethane	[Methyl chloride]	CH_3Cl	74-87-3	50 ppm	100 ppm					Central nervous system impairment; liver, kidney and testicular damage; teratogenic effects
(Chloromethyl)benzene	[Benzyl chloride]	C_7H_7Cl	100-44-7	1 ppm					1 ppm (15-minute)	Eye, skin, and upper respiratory tract irritation
Chloromethyl methyl ether	levels as low as possible	C_2H_5ClO	107-30-2							Lung cancer
1-Chloro-4-nitrobenzene	[*p*-Chloronitrobenzene]	$C_6H_4ClNO_2$	100-00-5	0.1 ppm						Methemoglobinemia
1-Chloro-1-nitropropane		$C_3H_6ClNO_2$	600-25-9	2 ppm			2 ppm			Eye and upper respiratory tract irritation; pulmonary edema
Chloropentafluoroethane	[CFC-115]	CF_3CF_2Cl	76-15-3	1000 ppm			1000 ppm			Cardiac sensitization
2-Chloropropanoic acid	[2-Chloropropionic acid]	$C_3H_5ClO_2$	598-78-7	0.1 ppm						Male reproductive damage
2-Chloro-1-propanol	[Propylene chlorohydrin]	C_3H_7ClO	78-89-7	1 ppm						Liver damage
1-Chloro-2-propanol	[*sec*-Propylene chlorohydrin]	C_3H_7ClO	127-00-4	1 ppm						Liver damage
3-Chloropropene	[Allyl chloride]	C_3H_5Cl	107-05-1	1 ppm	2 ppm		1 ppm	2 ppm		Eye and upper respiratory tract irritation; liver and kidney damage
2-Chlorostyrene		C_8H_7Cl	2039-87-4	50 ppm	75 ppm		50 ppm	75 ppm		Central nervous system impairment; peripheral neuropathy
2-Chlorotoluene	[1-Chloro-2-methylbenzene]	C_7H_7Cl	95-49-8	50 ppm			50 ppm	75 ppm		Upper respiratory tract, eye, and skin irritation
Chlorpyrifos	[Phosphorothioic acid, *O,O*-diethyl *O*-(3,5,6-trichloro-2-pyridinyl) ester]	$C_9H_{11}Cl_3NO_3PS$	2921-88-2	0.1 mg/m³ (inhalable fraction and vapor)			0.2 mg/m³	0.6 mg/m³		Cholinesterase inhibitor

				ACGIH			NIOSH			
Name	**Comments**	**Formula**	**CAS Reg. No.**	**Time-weighted average**	**Short-term exposure limit**	**ACGIH Ceiling**	**Time-weighted average**	**Short-term exposure limit**	**Ceiling**	**Basis**
Chromium	metal	Cr	7440-47-3	0.5 mg/m^3			0.5 mg/m^3			Respiratory tract irritation
Chromium	Cr(III) compounds, as Cr (III)	Cr	7440-47-3	0.003 mg/m^3 (inhalable particulate matter)			0.5 mg/m^3			Respiratory tract irritation; asthma
Chromium	soluble Cr(VI) compounds, as Cr (VI)	Cr	7440-47-3	0.0002 mg/m^3 (inhalable particulate matter)	0.0005 mg/m^3		0.0002 mg/m^3			Lung and sinonasal cancer; respiratory tract irritation; asthma
Chromium	insoluble Cr(VI) compounds, as Cr	Cr	7440-47-3	0.01 mg/m^3						
Chromium(VI) dichloride dioxide	[Chromyl chloride] as CrVI	CrO_2Cl_2	14977-61-8	0.025 ppm (inhalable fraction and vapor)			0.001 mg/m^3			Lung and sinonasal cancer; respiratory tract irritation; asthma
Chrysene	[Benzo[a]phenanthrene]; levels as low as possible	$C_{18}H_{12}$	218-01-9							Cancer
Clopidol		$C_7H_7Cl_2NO$	2971-90-6	3 mg/m^3 (inhalable fraction and vapor)			10 mg/m^3 (total); 5 mg/m^3 (resp)	20 mg/m^3 (total)		Mutagenic effects
Coal	dust, anthracite			0.4 mg/m^3 (respirable particulate matter)			1 mg/m^3			Lung damage; pulmonary fibrosis
Coal	dust, bituminous			0.9 mg/m^3			1 mg/m^3			Lung damage; pulmonary fibrosis
Coal tar	volatiles		65996-93-2	0.2 mg/m^3 (as benzene soluble aerosol)			0.1 mg/m^3 (cyclohexane-extractable fraction)			Cancer
Cobalt	metal and inorganic compounds, as Co	Co	7440-48-4	0.02 mg/m^3			0.05 mg/m^3			Pulmonary function
Cobalt	thoracic particulate matter in hard metals containing Co and tungsten carbide, as Co	Co	7440-48-4	0.005 mg/m^3						Pneumonitis
Cobalt carbonyl	[Dicobalt octacarbonyl]; as Co	$Co_2(CO)_8$	10210-68-1	0.1 mg/m^3			0.1 mg/m^3			Pulmonary edema; spleen damage
Cobalt hydrocarbonyl	[Tetracarbonylhydrocobalt]; as Co	C_4HCoO_4	16842-03-8	0.1 mg/m^3			0.1 mg/m^3			Pulmonary edema; lung damage
Copper	fume as Cu	Cu	7440-50-8	0.2 mg/m^3			0.1 mg/m^3			Irritation; gastrointestinal; metal fume fever
Copper	dusts & mists, as Cu	Cu	7440-50-8	1 mg/m^3			1 mg/m^3			Irritation; gastrointestinal; metal fume fever
Cotton	dust			0.1 mg/m^3			0.2 mg/m^3			Byssinosis; bronchitis; pulmonary function
Coumaphos		$C_{14}H_{16}ClO_5PS$	56-72-4	0.05 mg/m^3 (inhalable fraction and vapor)						Cholinesterase inhibitor
o-Cresol	[2-Methylphenol]	C_7H_8O	95-48-7	20 mg/m^3 (inhalable fraction and vapor)			10 mg/m^3			Upper respiratory tract irritation
m-Cresol	[3-Methylphenol]	C_7H_8O	108-39-4	20 mg/m^3 (inhalable fraction and vapor)			10 mg/m^3			Upper respiratory tract irritation
p-Cresol	[4-Methylphenol]	C_7H_8O	106-44-5	20 mg/m^3 (inhalable fraction and vapor)			10 mg/m^3			Upper respiratory tract irritation
Crufomate		$C_{12}H_{19}ClNO_3P$	299-86-5	5 mg/m^3			5 mg/m^3	10 mg/m^3		Cholinesterase inhibitor
Cyanamide	[Cyanogenamide]	H_2NCN	420-04-2	2 mg/m^3			2 mg/m^3			Skin and eye irritation
Cyanide anion	cyanide salts, as CN	CN^-	57-12-5			5 mg/m^3			5 mg/m^3 [10-minute]	Upper respiratory tract irritation; headache; nausea; thyroid effects
Cyanogen		C_2N_2	460-19-5			5 ppm	10 ppm			Eye and upper respiratory tract irritation
Cyanogen bromide	[Bromine cyanide]	BrCN	506-68-3			0.3 ppm				Eye and respiratory tract irritation; pulmonary edema
Cyanogen chloride	[Chlorine cyanide]	ClCN	506-77-4			0.3 ppm			0.3 ppm	Pulmonary edema; eye, skin, and upper respiratory tract irritation
Cyclohexane	[Hexahydrobenzene]	C_6H_{12}	110-82-7	100 ppm			300 ppm			Central nervous system impairment

Name	Comments	Formula	CAS Reg. No.	ACGIH Time-weighted average	ACGIH Short-term exposure limit	ACGIH Ceiling	NIOSH Time-weighted average	NIOSH Short-term exposure limit	NIOSH Ceiling	Basis
Cyclohexanol	[Cyclohexyl alcohol]	$C_6H_{12}O$	108-93-0	50 ppm			50 ppm			Eye irritation; central nervous system impairment
Cyclohexanone	[Pimelic ketone]	$C_6H_{10}O$	108-94-1	20 ppm	50 ppm		25 ppm			Eye and upper respiratory tract irritation
Cyclohexene	[Tetrahydrobenzene]	C_6H_{10}	110-83-8	300 ppm			300 ppm			Upper respiratory tract and eye irritation
Cyclohexylamine	[Cyclohexanamine]	$C_6H_{13}N$	108-91-8	10 ppm			10 ppm			Upper respiratory tract and eye irritation
1,3-Cyclopentadiene	[Pyropentylene]	C_5H_6	542-92-7	0.5 ppm	1 ppm		75 ppm			Upper respiratory tract, lower respiratory tract, and eye irritation; central nervous system effects
Cyclopentane	[Pentamethylene]	C_5H_{10}	287-92-3	600 ppm			600 ppm			Upper respiratory tract, eye, and skin irritation; central nervous system impairment
Cyhexatin	[Tricyclohexylhydroxystannane]	$C_{18}H_{34}OSn$	13121-70-5	5 mg/m³			5 mg/m³			Upper respiratory tract irritation; body weight effects; kidney damage
Decaborane(14)		$B_{10}H_{14}$	17702-41-9	0.05 ppm	0.15 ppm		0.05 ppm	0.15 ppm		Central nervous system convulsions; cognitive decrement
Demeton	[Systox]	$C_8H_{19}O_3PS_2$	8065-48-3	0.05 mg/m³ (inhalable fraction and vapor)			0.01 mg/m³			Cholinesterase inhibitor
Demeton-*S*-methyl	[Phosphorothioic acid, *S*-[2-(ethylthio)ethyl] *O,O*-dimethyl ester]	$C_6H_{15}O_3PS_2$	919-86-8	0.05 mg/m³ (inhalable fraction and vapor)			0.5 mg/m³			Cholinesterase inhibitor
Diacetone alcohol	[4-Hydroxy-4-methyl-2-pentanone]	$C_6H_{12}O_2$	123-42-2	50 ppm			50 ppm			Upper respiratory tract and eye irritation
4,4'-Diaminodiphenylmethane	[4,4'-Methylenedianiline]	$C_{13}H_{14}N_2$	101-77-9	0.1 ppm						Liver damage
Diazinon		$C_{12}H_{21}N_2O_3PS$	333-41-5	0.01 mg/m³			0.01 mg/m³			Cholinesterase inhibitor
Diazomethane		$CH_2{=}N{\equiv}N$	334-88-3	0.2 ppm			0.2 ppm			Upper respiratory tract and eye irritation
Diborane		B_2H_6	19287-45-7	0.1 ppm			0.1 ppm			Upper respiratory tract irritation; headache
Dibromodifluoromethane		CBr_2F_2	75-61-6	100 ppm			100 ppm			Upper respiratory tract irritation; central nervous system impairment; liver damage
2-Dibutylaminoethanol		$C_{10}H_{23}NO$	102-81-8	0.5 ppm			2 ppm			Eye and upper respiratory tract irritation
2,6-Di-*tert*-butyl-4-methylphenol	[Butylated hydroxytoluene (BHT)]	$C_{15}H_{24}O$	128-37-0	2 mg/m³ (inhalable fraction and vapor)			10 mg/m³			Upper respiratory tract irritation
Dibutylphenyl phosphate		$C_{14}H_{23}O_4P$	2528-36-1	0.3 ppm						Cholinesterase inhibitor; upper respiratory irritation
Dibutyl phosphate		$C_8H_{19}O_4P$	107-66-4	5 mg/m³ (inhalable fraction and vapor)			5 mg/m³	10 mg/m³		Bladder; eye and upper respiratory irritation
Dibutyl phthalate	[Butyl phthalate]	$C_{16}H_{22}O_4$	84-74-2	5 mg/m³			5 mg/m³			Testicular damage; eye and upper respiratory tract irritation
Dichloroacetic acid		$CHCl_2COOH$	79-43-6	0.5 ppm						Upper respiratory tract and eye irritation; testicular damage
Dichloroacetylene		C_2Cl_2	7572-29-4			0.1 ppm			0.1 ppm	Nausea; peripheral nervous system impairment
o-Dichlorobenzene	[1,2-Dichlorobenzene]	$C_6H_4Cl_2$	95-50-1	25 ppm	50 ppm				50 ppm	Upper respiratory tract and eye irritation; liver damage
p-Dichlorobenzene	[1,4-Dichlorobenzene]	$C_6H_4Cl_2$	106-46-7	10 ppm						Eye irritation; kidney damage
3,3'-Dichloro-*p*-benzidine	[3,3'-Dichloro[1,1'-biphenyl]-4,4'-diamine]; levels as low as possible	$C_{12}H_{10}Cl_2N_2$	91-94-1							Bladder cancer; eye irritation
cis-1,4-Dichloro-2-butene		$C_4H_6Cl_2$	1476-11-5	0.005 ppm						Upper respiratory tract and eye irritation

Name	Comments	Formula	CAS Reg. No.	ACGIH: Time-weighted average	ACGIH: Short-term exposure limit	ACGIH Ceiling	NIOSH: Time-weighted average	NIOSH: Short-term exposure limit	Ceiling	Basis
trans-1,4-Dichloro-2-butene		$C_4H_6Cl_2$	110-57-6	0.005 ppm						Upper respiratory tract and eye irritation
Dichlorodifluoromethane	[CFC-12]	CF_2Cl_2	75-71-8	1000 ppm			1000 ppm			Liver damage
1,3-Dichloro-5,5-dimethyl hydantoin		$C_5H_6Cl_2N_2O_2$	118-52-5	0.2 mg/m³	0.4 mg/m³		0.2 mg/m³	0.4 mg/m³		Upper respiratory tract irritation
1,1-Dichloroethane	[Ethylidene dichloride]	CH_3CHCl_2	75-34-3	100 ppm			100 ppm			Upper respiratory tract and eye irritation; liver and kidney damage
1,2-Dichloroethane	[Ethylene dichloride]	CH_2ClCH_2Cl	107-06-2	10 ppm			1 ppm	2 ppm		Liver damage; nausea
1,1-Dichloroethene	[Vinylidene chloride]	$CH_2{=}CCl_2$	75-35-4	5 ppm						Liver and kidney damage
cis-1,2-Dichloroethene	[*cis*-1,2-Dichloroethylene]	$C_2H_2Cl_2$	156-59-2	200 ppm			200 ppm			Central nervous system impairment; eye irritation
trans-1,2-Dichloroethene	[*trans*-1,2-Dichloroethylene]	$C_2H_2Cl_2$	156-60-5	200 ppm			200 ppm			Central nervous system impairment; eye irritation
Dichlorofluoromethane	[Refrigerant 21]	$CHCl_2F$	75-43-4	10 ppm			10 ppm			Liver damage
Dichloromethane	[Methylene chloride]	CH_2Cl_2	75-09-2	50 ppm						COHb-emia; central nervous system impairment
1,1-Dichloro-1-nitroethane	[Ethide]	$C_2H_3Cl_2NO_2$	594-72-9	2 ppm			2 ppm			Upper respiratory tract irritation
(2,4-Dichlorophenoxy)acetic acid	[2,4-D]	$C_8H_6Cl_2O_3$	94-75-7	10 mg/m³ (inhalable particulate matter)			10 mg/m³			Thyroid effects; kidney tubular damage
1,2-Dichloropropane, (±)-	[Propylene dichloride]	$C_3H_6Cl_2$	78-87-5	10 ppm						Upper respiratory tract irritation; body weight effects
2,2-Dichloropropanoic acid	[2,2-Dichloropropionic acid]	$C_3H_4Cl_2O_2$	75-99-0	5 mg/m³			6 mg/m³			Eye and upper respiratory tract irritation
cis-1,3-Dichloropropene	[*cis*-1,3-Dichloropropylene]	$C_3H_4Cl_2$	10061-01-5	1 ppm			1 ppm			Kidney damage
trans-1,3-Dichloropropene	[*trans*-1,3-Dichloropropylene]	$C_3H_4Cl_2$	10061-02-6	1 ppm			1 ppm			Kidney damage
1,2-Dichloro-1,1,2,2-tetrafluoroethane	[CFC-114]	$C_2Cl_2F_4$	76-14-2	1000 ppm			1000 ppm			Pulmonary function
Dichlorvos	[Phosphoric acid, 2,2-dichloroethenyl dimethyl ester]	$C_4H_7Cl_2O_4P$	62-73-7	0.1 mg/m³ (inhalable fraction and vapor)			0.1 mg/m³			Cholinesterase inhibitor
Dicrotophos		$C_8H_{16}NO_5P$	141-66-2	0.05 mg/m³ (inhalable fraction and vapor)			0.25 mg/m³			Cholinesterase inhibitor
o-Dicyanobenzene	[*o*-Phthalodinitrile]	$C_8H_4N_2$	91-15-6	1 mg/m³ (inhalable fraction and vapor)						Central nervous system convulsions; body weight effects
m-Dicyanobenzene	[*m*-Phthalodinitrile]	$C_8H_4N_2$	626-17-5	5 mg/m³ (inhalable fraction and vapor)			5 mg/m³			Eye and upper respiratory tract irritation
Dicyclopentadiene		$C_{10}H_{12}$	1755-01-7	5 ppm			5 ppm			Upper respiratory tract, lower respiratory tract, and eye irritation
Dieldrin		$C_{12}H_8Cl_6O$	60-57-1	0.1 mg/m³ (inhalable fraction and vapor)			0.25 mg/m³			Liver damage;reproductive effects; central nervous system effects
Diesel fuel	as total hydrocarbons		68334-30-5	100 mg/m³ (inhalable fraction and vapor)						Dermatitis
Diethanolamine	[Bis(2-hydroxyethyl)amine]	$C_4H_{11}NO_2$	111-42-2	1 mg/m³ (inhalable fraction and vapor)			15 mg/m³			Liver and kidney damage
Diethylamine	[*N*-Ethylethanamine]	$(C_2H_5)_2NH$	109-89-7	5 ppm	15 ppm		10 ppm	25 ppm		Upper respiratory tract, eye, and skin irritation
2-Diethylaminoethanol		$C_6H_{15}NO$	100-37-8	2 ppm			10 ppm			Upper respiratory tract irritation; central nervous system convulsions
Diethylene glycol monobutyl ether	[2-(2-Butoxyethoxy)ethanol]	$C_8H_{18}O_3$	112-34-5	10 ppm (inhalable fraction and vapor)						Hematologic, liver, and kidney effects

Name	Comments	Formula	CAS Reg. No.	ACGIH Time-weighted average	ACGIH Short-term exposure limit	ACGIH Ceiling	NIOSH Time-weighted average	NIOSH Short-term exposure limit	Ceiling	Basis
Diethyl ether	[Ethyl ether]	$(C_2H_5)_2O$	60-29-7	400 ppm	500 ppm					Central nervous system impairment; upper respiratory irritation
3,3-Diethylpentane	[Tetraethylmethane]	C_9H_{20}	1067-20-5	200 ppm						Central nervous system impairment
Diethyl phthalate		$C_{12}H_{14}O_4$	84-66-2	5 mg/m^3			5 mg/m^3			Upper respiratory tract irritation
1,1-Difluoroethene	[Vinylidene fluoride]	$CH_2{=}CF_2$	75-38-7	500 ppm			1 ppm		5 ppm	Liver damage
Diglycidyl ether	[Bis(2,3-epoxypropyl) ether (DGE)]	$C_6H_{10}O_3$	2238-07-5	0.1 ppm			0.1 ppm			Eye and skin irritation; male reproductive damage
Diisopropylamine	[*N*-Isopropyl-2-propanamine]	$C_6H_{15}N$	108-18-9	5 ppm			5 ppm			Upper respiratory tract irritation; eye damage
Diisopropyl ether	[Isopropyl ether]	$C_6H_{14}O$	108-20-3	250 ppm	310 ppm		500 ppm			Eye and upper respiratory tract irritation
Dimethoxymethane	[Methylal]	$C_3H_8O_2$	109-87-5	1000 ppm			1000 ppm			Eye irritation; central nervous system impairment
N,N-Dimethylacetamide	[*N,N*-Dimethylethanamide]	C_4H_9NO	127-19-5	10 ppm			10 ppm			Liver, embryo, and fetal damage; reproductive, renal, and teratogenic effects
Dimethylamine	[*N*-Methylmethanamine]	$(CH_3)_2NH$	124-40-3	5 ppm	15 ppm		10 ppm			Upper respiratory tract and GI irritation
Dimethylaniline (unspecified isomer)	[Xylidine (unspecified isomer)]; all isomers	$C_8H_{11}N$	1300-73-8	0.5 ppm	10 ppm		2 ppm			Methemoglobinemia
2,3-Dimethylaniline	[2,3-Xylidine]	$C_8H_{11}N$	87-59-2	0.5 ppm			2 ppm			Methemoglobinemia
2,4-Dimethylaniline	[2,4-Xylidine]	$C_8H_{11}N$	95-68-1	0.5 ppm			2 ppm			Methemoglobinemia
2,5-Dimethylaniline	[2,5-Xylidine]	$C_8H_{11}N$	95-78-3	0.5 ppm			2 ppm			Methemoglobinemia
2,6-Dimethylaniline	[2,6-Xylidine]	$C_8H_{11}N$	87-62-7	0.5 ppm			2 ppm			Methemoglobinemia
3,4-Dimethylaniline	[3,4-Xylidine]	$C_8H_{11}N$	95-64-7	0.5 ppm			2 ppm			Methemoglobinemia
3,5-Dimethylaniline	[3,5-Xylidine]	$C_8H_{11}N$	108-69-0	0.5 ppm			2 ppm			Methemoglobinemia
N,N-Dimethylaniline	[*N,N*-Dimethylbenzenamine]	$C_8H_{11}N$	121-69-7	5 ppm	10 ppm		5 ppm	10 ppm		Methemoglobinemia
2,2-Dimethylbutane	[Neohexane]	C_6H_{14}	75-83-2	500 ppm	1000 ppm		100 ppm		510 ppm	Central nervous system impairment; upper respiratory and eye irritation
2,3-Dimethylbutane	[Diisopropyl]	C_6H_{14}	79-29-8	500 ppm	1000 ppm		100 ppm		510 ppm	Central nervous system impairment; upper respiratory and eye irritation
Dimethylcarbamic chloride	[Dimethylcarbamoyl chloride]	C_3H_6ClNO	79-44-7	0.005 mg/m^3						Nasal cancer; upper respiratory tract irritation
Dimethyl disulfide	[Methyl disulfide]	$C_2H_6S_2$	624-92-0	0.5 mg/m^3						Upper respiratory tract irritation; central nervous system impairment
N,N-Dimethylformamide	[DMF]	C_3H_7NO	68-12-2	5 ppm			10 ppm			Liver damage; eye and upper respiratory tract irritation
2,2-Dimethylheptane		C_9H_{20}	1071-26-7	200 ppm						Central nervous system impairment
2,3-Dimethylheptane		C_9H_{20}	3074-71-3	200 ppm						Central nervous system impairment
2,4-Dimethylheptane		C_9H_{20}	2213-23-2	200 ppm						Central nervous system impairment
2,5-Dimethylheptane		C_9H_{20}	2216-30-0	200 ppm						Central nervous system impairment
2,6-Dimethylheptane		C_9H_{20}	1072-05-5	200 ppm						Central nervous system impairment
3,3-Dimethylheptane		C_9H_{20}	4032-86-4	200 ppm						Central nervous system impairment
3,4-Dimethylheptane		C_9H_{20}	922-28-1	200 ppm						Central nervous system impairment
3,5-Dimethylheptane		C_9H_{20}	926-82-9	200 ppm						Central nervous system impairment
4,4-Dimethylheptane		C_9H_{20}	1068-19-5	200 ppm						Central nervous system impairment
2,6-Dimethyl-4-heptanone	[Diisobutyl ketone]	$C_9H_{18}O$	108-83-8	25 ppm			25 ppm			Upper respiratory tract and eye irritation
2,2-Dimethylhexane		C_8H_{18}	590-73-8	300 ppm						Upper respiratory tract irritation
2,3-Dimethylhexane		C_8H_{18}	584-94-1	300 ppm						Upper respiratory tract irritation
2,4-Dimethylhexane		C_8H_{18}	589-43-5	300 ppm						Upper respiratory tract irritation

Name	Comments	Formula	CAS Reg. No.	ACGIH Time-weighted average	ACGIH Short-term exposure limit	ACGIH Ceiling	NIOSH Time-weighted average	NIOSH Short-term exposure limit	NIOSH Ceiling	Basis
2,5-Dimethylhexane	[Biisobutyl]	C_8H_{18}	592-13-2	300 ppm						Upper respiratory tract irritation
3,3-Dimethylhexane		C_8H_{18}	563-16-6	300 ppm						Upper respiratory tract irritation
3,4-Dimethylhexane		C_8H_{18}	583-48-2	300 ppm						Upper respiratory tract irritation
1,1-Dimethylhydrazine	[UDMH]	$C_2H_8N_2$	57-14-7	0.01 ppm					0.06 ppm [2-hour]	Upper respiratory tract irritation; nasal cancer
Dimethyl mercury	[Mercury dimethyl]	$Hg(CH_3)_2$	593-74-8			0.03 mg/m³	0.01 mg/m³	0.03 mg/m³		Central nervous system and peripheral nervous system impairment; kidney damage
trans-3,7-Dimethyl-2,6-octadienal	[Citral]	$C_{10}H_{16}O$	141-27-5	5 ppm (inhalable fraction and vapor)						Body weight effects; upper respiratory tract irritation; eye damage
Dimethyl phthalate	[Methyl phthalate]	$C_{10}H_{10}O_4$	131-11-3	5 mg/m³						Eye and upper respiratory tract irritation
2,2-Dimethyl-1-propanol acetate		$C_7H_{14}O_2$	926-41-0	50 ppm	100 ppm					Upper respiratory tract irritation
Dimethyl sulfate		$C_2H_6O_4S$	77-78-1	0.1 ppm			0.1 ppm			Eye and skin irritation
Dimethyl sulfide	[2-Thiapropane]	$(CH_3)_2S$	75-18-3	10 ppm						Upper respiratory tract irritation
1,2-Dinitrobenzene	[*o*-Dinitrobenzene]	$C_6H_4N_2O_4$	528-29-0	0.15 ppm			1 mg/m³			Methemoglobinemia; eye damage
1,3-Dinitrobenzene	[*m*-Dinitrobenzene]	$C_6H_4N_2O_4$	99-65-0	0.15 ppm			1 mg/m³			Methemoglobinemia; eye damage
1,4-Dinitrobenzene	[*p*-Dinitrobenzene]	$C_6H_4N_2O_4$	100-25-4	0.15 ppm			1 mg/m³			Methemoglobinemia; eye damage
1,4-Dioxane	[1,4-Dioxacyclohexane]	$C_4H_8O_2$	123-91-1	20 ppm					1 ppm [30-minute]	Liver damage
Dioxathion		$C_{12}H_{26}O_6P_2S_4$	78-34-2	0.1 mg/m³ (inhalable fraction and vapor)			0.02 mg/m³			Cholinesterase inhibitor
1,3-Dioxolane	[1,3-Dioxacyclopentane]	$C_3H_6O_2$	646-06-0	20 ppm						Hematologic effects
Diphenylamine	[*N*-Phenylbenzenamine]	$(C_6H_5)_2NH$	122-39-4	10 mg/m³			10 mg/m³			Liver and kidney damage; hematologic effects
Diphenyl ether	[Oxybisbenzene]	$(C_6H_5)_2O$	101-84-8	1 ppm (vapor fraction)	2 ppm (vapor fraction)		1 ppm			Upper respiratory tract and eye irritation; nausea
4,4′-Diphenylmethane diisocyanate	[Methylene diphenyl diisocyanate (MDI)]	$C_{15}H_{10}N_2O_2$	101-68-8	0.005 ppm			0.005 ppm		0.020 ppm [10-minute]	Respiratory sensitization
Dipropylene glycol monomethyl ether	[1-(2-Methoxyisopropoxy)-2-propanol (DPGME)]	$C_7H_{16}O_3$	34590-94-8	100 ppm	150 ppm		100 ppm	150 ppm		Eye and upper respiratory tract irritation; central nervous system impairment
Diquat		$C_{12}H_{12}N_2$	2764-72-9	0.5 mg/m³ (inhalable particulate matter); 0.1 mg/m³ (respirable particulate matter)			0.5 mg/m³			Lower respiratory tract irritation; cataract
Disulfiram	Precautions should be taken to avoid concurrent exposure to ethylene dibromide.	$C_{10}H_{20}N_2S_4$	97-77-8	2 mg/m³			2 mg/m³			Vasodilation; nausea
Disulfoton	[Phosphorodithioic acid, *O,O*-diethyl *S*-[2-(ethylthio) ethyl] ester]	$C_8H_{19}O_2PS_3$	298-04-4	0.05 mg/m³ (inhable fraction and vapor)			0.1 mg/m³			Cholinesterase inhibitor
Diuron		$C_9H_{10}Cl_2N_2O$	330-54-1	10 mg/m³			10 mg/m³			Upper respiratory tract irritation
o-Divinylbenzene	[1,2-Divinylbenzene]	$C_{10}H_{10}$	91-14-5	10 ppm			10 ppm			Upper respiratory tract irritation
m-Divinylbenzene	[1,3-Divinylbenzene]	$C_{10}H_{10}$	108-57-6	10 ppm			10 ppm			Upper respiratory tract irritation
p-Divinylbenzene	[1,4-Divinylbenzene]	$C_{10}H_{10}$	105-06-6	10 ppm			10 ppm			Upper respiratory tract irritation
1-Dodecanethiol	[Dodecyl mercaptan]	$C_{12}H_{26}S$	112-55-0	0.1 ppm					0.5 ppm [15-minute]	Upper respiratory tract irritation
Endosulfan		$C_9H_6Cl_6O_3S$	115-29-7	0.1 mg/m³ (inhalable fraction and vapor)			0.1 mg/m³			Lower respiratory tract irritation; liver and kidney damage

Name	Comments	Formula	CAS Reg. No.	ACGIH Time-weighted average	ACGIH Short-term exposure limit	ACGIH Ceiling	NIOSH Time-weighted average	NIOSH Short-term exposure limit	Ceiling	Basis
Endrin		$C_{12}H_8Cl_6O$	72-20-8	0.1 mg/m³			0.1 mg/m³			Liver damage; central nervous system impairment; headache
Enflurane	Note: REL for exposure to waste anesthetic gas	$C_3H_2ClF_5O$	13838-16-9	75 ppm					2 ppm [60-minute]	Central nervous system impairment; cardiac impairment
Epichlorohydrin	[(Chloromethyl)oxirane]	C_3H_5ClO	13403-37-7	0.5 ppm						Upper respiratory tract irritation; male reproductive
1,2-Epoxy-4-(epoxyethyl) cyclohexane	[4-Vinyl-1-cyclohexene dioxide]	$C_8H_{12}O_2$	106-87-6	0.1 ppm			10 ppm			Female and male reproductive damage
1,2-Ethanediamine	[Ethylenediamine]	$C_2H_8N_2$	107-15-3	10 ppm			10 ppm			Upper respiratory tract irritation
1,2-Ethanediol	[Ethylene glycol]	$(CH_2OH)_2$	107-21-1	25 ppm (vapor fraction)	50 ppm (vapor fraction); 10 mg/m³ (inhalable particulate matter; aerosol only)					Upper respiratory tract irritation
1,2-Ethanediol, dinitrate	[Ethylene glycol dinitrate (EGDN)]	$C_2H_4N_2O_6$	628-96-6	0.05 ppm				0.1 mg/m³		Vasodilation; headache
Ethanethiol	[Ethyl mercaptan]	C_2H_5SH	75-08-1	0.5 ppm					0.5 ppm [15-minute]	Upper respiratory tract irritation; central nervous system impairment
Ethanol	[Ethyl alcohol]	C_2H_5OH	64-17-5		1000 ppm		1000 ppm			Upper respiratory tract irritation
Ethanolamine	[Glycinol; 2-Aminoethanol]	$CH_2OHCH_2NH_2$	141-43-5	3 ppm	6 ppm		3 ppm	6 ppm		Eye and skin irritation
Ethion	[Phosphorodithioic acid, *S,S'*-methylene *O,O,O',O'*-tetraethyl ester]	$C_9H_{22}O_4P_2S_4$	563-12-2	0.05 mg/m³ (inhable fraction and vapor)			0.4 mg/m³			Cholinesterase inhibitor
Ethoxydimethylsilane	[Dimethylethoxysilane]	$C_4H_{12}OSi$	14857-34-2	0.5 ppm	1.5 ppm					Upper respiratory tract and eye irritation; headache
2-Ethoxyethanol	[Ethylene glycol monoethyl ether (EGEE)]	$C_4H_{10}O_2$	110-80-5	5 ppm			0.5 ppm			Male reproducive and embryo/fetal damage
2-Ethoxyethyl acetate	[Ethylene glycol monoethyl ether acetate (EGEEA)]	$C_6H_{12}O_3$	111-15-9	5 ppm			0.5 ppm			Male reproductive damage
Ethyl acetate		$C_4H_8O_2$	141-78-6	400 ppm			400 ppm			Upper respiratory tract and eye irritation
Ethyl acrylate	[Ethyl propenoate]	$C_5H_8O_2$	140-88-5	5 ppm	15 ppm					Upper respiratory tract, eye, and GI irritation; central nervous system impairment; skin sensitization
Ethylamine	[Ethanamine]	$C_2H_5NH_2$	75-04-7	5 ppm	15 ppm		10 ppm			Upper respiratory tract irritation
Ethylbenzene	[Phenylethane]	C_8H_{10}	100-41-4	20 ppm			100 ppm	125 ppm		Upper respiratory tract irritation; kidney damage (nephropathy); cochlear impairment
Ethyl 2-cyanoacrylate	[Ethyl 2-cyano-2-propenoate]	$C_6H_7NO_2$	7085-85-0	0.2 ppm	1 ppm					Eye and upper respiratory tract irritation; asthma
3-Ethyl-2,2-dimethylpentane		C_9H_{20}	16747-32-3	200 ppm			200 ppm			Central nervous system impairment
3-Ethyl-2,3-dimethylpentane		C_9H_{20}	16747-33-4	200 ppm			200 ppm			Central nervous system impairment
3-Ethyl-2,4-dimethylpentane		C_9H_{20}	1068-87-7	200 ppm			200 ppm			Central nervous system impairment
Ethylene	[Ethene]	$CH_2{=}CH_2$	74-85-1	200 ppm						Asphyxia
Ethyleneimine	[Aziridine]	C_2H_5N	151-56-4	0.05 ppm	0.1 ppm					Upper respiratory tract irritation; liver and kidney damage
Ethyl formate		C_2H_5OCHO	109-94-4		100 ppm		100 ppm			Upper respiratory tract irritation
3-Ethylheptane		C_9H_{20}	15869-80-4	200 ppm			200 ppm			Central nervous system impairment
4-Ethylheptane		C_9H_{20}	2216-32-2	200 ppm			200 ppm			Central nervous system impairment
3-Ethylhexane		C_8H_{18}	619-99-8	300 ppm						Upper respiratory tract irritation
2-Ethylhexanoic acid		$C_8H_{16}O_2$	149-57-5	5 mg/m³ (inhalable fraction and vapor)						Teratogenic effects
N-Ethyl-*N*-hydroxyethanamine	[*N,N*-Diethylhydroxylamine]	$C_4H_{11}NO$	3710-84-7	2 ppm						Upper respiratory tract irritation
5-Ethylidene-2-norbornene	[5-Ethylidenebicyclo[2.2.1] hept-2-ene]	C_9H_{12}	16219-75-3	2 ppm	4 ppm				5 ppm	Upper respiratory tract and eye irritation

Safety

Name	Comments	Formula	CAS Reg. No.	ACGIH			NIOSH			Basis
				Time-weighted average	Short-term exposure limit	ACGIH Ceiling	Time-weighted average	Short-term exposure limit	Ceiling	
Ethyl isocyanate		C_3H_5NO	109-90-0	0.02 ppm	0.06 ppm					Upper respiratory tract and eye irritation
3-Ethyl-2-methylhexane		C_9H_{20}	16789-46-1	200 ppm			200 ppm			Central nervous system impairment
3-Ethyl-3-methylhexane		C_9H_{20}	3074-76-8	200 ppm			200 ppm			Central nervous system impairment
3-Ethyl-4-methylhexane	[2,3-Diethylpentane]	C_9H_{20}	3074-77-9	200 ppm			200 ppm			Central nervous system impairment
4-Ethyl-2-methylhexane		C_9H_{20}	3074-75-7	200 ppm			200 ppm			Central nervous system impairment
3-Ethyl-2-methylpentane	[2-Methyl-3-ethylpentane]	C_8H_{18}	609-26-7	300 ppm						Upper respiratory tract irritation
3-Ethyl-3-methylpentane	[3-Methyl-3-ethylpentane]	C_8H_{18}	1067-08-9	300 ppm						Upper respiratory tract irritation
N-Ethylmorpholine		$C_6H_{13}NO$	100-74-3	5 ppm			5 ppm			Upper respiratory tract irritation; eye damage
O-Ethyl *O*-*p*-nitrophenyl benzenethiophosphonate	[EPN]	$C_{14}H_{14}NO_4PS$	2104-64-5	0.1 mg/m³			0.5 mg/m³			Cholinesterase inhibitor
Ethyl silicate	[Tetraethoxysilane]	$Si(OC_2H_5)_4$	78-10-4	10 ppm			10 ppm			Upper respiratory tract and eye irritation; kidney damage
Fenamiphos		$C_{13}H_{22}NO_3PS$	22224-92-6	0.05 mg/m³ (inhalable fraction and vapor)			0.1 mg/m³			Cholinesterase inhibitor
Fensulfothion	[Phosphorothioic acid, *O,O*-diethyl *O*-[4-(methylsulfinyl)phenyl] ester]	$C_{11}H_{17}O_4PS_2$	115-90-2	0.01 mg/m³ (inhalable fraction and vapor)			0.1 mg/m³			Cholinesterase inhibitor
Fenthion	[Phosphorothioic acid, *O,O*-dimethyl *O*-[3-methyl-4-(methylthio)phenyl] ester]	$C_{10}H_{15}O_3PS_2$	55-38-9	0.05 mg/m³ (inhalable fraction and vapor)						Cholinesterase inhibitor
Ferbam	[Iron, tris(dimethylcarbamodithioato-*S,S'*)-, (OC-6-11)-]	$C_9H_{18}FeN_3S_6$	14484-64-1	5 mg/m³ (inhalable particulate matter)			10 mg/m³			Central nervous system impairment; body weight effects; spleen damage
Ferrocene	[Bis(cyclopentadienyl)iron]	$Fe(C_5H_5)_2$	102-54-5	10 mg/m³			10 mg/m³ (total); 5 mg/m³ (respirable)			Liver damage
Ferrovanadium	dust		12604-58-9	1 mg/m³	3 mg/m³		1 mg/m³	3 mg/m³		Eye, upper respiratory tract, and lower respiratory tract irritation
Flour	dust			0.5 mg/m³ (inhalable fraction)						Asthma; upper respiratory irritation; bronchitis
Fluoride ion [F⁻]	fluoride salts, as F	F^-	16984-48-8	2.5 mg/m³						Bone damage; fluorosis
Fluorine		F_2	7782-41-4	0.1 ppm		0.5 ppm	0.1 ppm			Fluorosis; eye irritation
Fluorine monoxide	[Oxygen difluoride]	F_2O	7783-41-7			0.05 ppm			0.05 ppm	Headache; pulmonary edema; upper respiratory irritation
Fluoroethene	[Vinyl fluoride]	$CH_2{=}CHF$	75-02-5	1 ppm			1 ppm		5 ppm	Liver cancer; liver damage
Folpet		$C_9H_4Cl_3NO_2S$	133-07-3	1 mg/m³ (inhalable particulate matter)						Liver damage; body weight effects
Fonofos	[Phosphonodithioic acid, ethyl-, *O*-ethyl *S*-phenyl ester]	$C_{10}H_{15}OPS_2$	944-22-9	0.01 mg/m³ (inhalable fraction and vapor)			0.1 mg/m³			Cholinesterase inhibitor
Formaldehyde	[Methanal]	HCHO	50-00-0	0.1 ppm	0.3 ppm		0.016 ppm		0.1 ppm [15-minute]	Upper respiratory tract and eye irritation; upper respiratory tract cancer
Formamide	[Methanamide]	$HCONH_2$	75-12-7	10 ppm			10 ppm			Eye and skin irritation; kidney and liver damage
Formic acid	[Methanoic acid]	HCOOH	64-18-6	5 ppm	10 ppm		5 ppm			Upper respiratory tract, eye, and skin irritation
Furfural	[2-Furaldehyde]	$C_5H_4O_2$	98-01-1	0.2 ppm						Upper respiratory tract and eye irritation
Furfuryl alcohol	[2-Furanmethanol]	$C_5H_6O_2$	98-00-0	0.2 ppm			10 ppm	15 ppm		Upper respiratory tract and eye irritation
Gallium arsenide		GaAs	1303-00-0	0.0003 mg/m³ (respirable particulate matter)						Lower respiratory tract irritation

Name	Comments	Formula	CAS Reg. No.	ACGIH Time-weighted average	ACGIH Short-term exposure limit	ACGIH Ceiling	NIOSH Time-weighted average	NIOSH Short-term exposure limit	NIOSH Ceiling	Basis
Gasoline			8006-61-9	300 ppm	500 ppm					Upper respiratory tract and eye irritation; central nervous system impairment
Germane	[Germanium tetrahydride]	GeH_4	7782-65-2	0.2 ppm			0.2 ppm			Hematologic effects
Glyoxal		CH(O)CH(O)	107-22-2	0.1 mg/m^3 (inhalable fraction and vapor)						Upper respiratory tract irritation; larynx metaplasia
Grain	dust			4 mg/m^3						Bronchitis; upper respiratory tract irritation; pulmonary function
Hafnium	metal & compounds, as Hf	Hf	7440-58-6	0.5 mg/m^3			0.5 mg/m^3			Upper respiratory tract and eye irritation; liver damage
Heptachlor		$C_{10}H_5Cl_7$	76-44-8	0.05 mg/m^3			0.5 mg/m^3			Liver damage
Heptachlor epoxide		$C_{10}H_5Cl_7O$	1024-57-3	0.05 mg/m^3						Liver damage
Heptane	all isomers	C_7H_{16}	142-82-5	400 ppm	500 ppm		85 ppm		440 ppm [15-minute]	Central nervous system impairment; upper respiratory tract irritation
2-Heptanone	[Methyl pentyl ketone]	$C_7H_{14}O$	110-43-0	50 ppm			100 ppm			Eye and skin irritation
3-Heptanone	[Ethyl butyl ketone]	$C_7H_{14}O$	106-35-4	50 ppm	75 ppm		50 ppm			Central nervous system impairment; eye and skin irritation
4-Heptanone	[Dipropyl ketone]	$C_7H_{14}O$	123-19-3	50 ppm			50 ppm			Upper respiratory tract irritation
Hexachlorobenzene	[Perchlorobenzene]	C_6Cl_6	118-74-1	0.002 mg/m^3						Porphyrin effects; skin damage; central nervous system impairment
Hexachloro-1,3-butadiene	[Perchlorobutadiene]	C_4Cl_6	87-68-3	0.02 ppm			0.02 ppm			Kidney damage
1,2,3,4,5,6-Hexachlorocyclohexane, (1α,2α,3β,4α,5α,6β)	[Lindane]	$C_6H_6Cl_6$	58-89-9	0.5 mg/m^3			0.5 mg/m^3			Liver damage; central nervous system impairment
Hexachloro-1,3-cyclopentadiene	[Perchlorocyclopentadiene]	C_5Cl_6	77-47-4	0.01 ppm			0.01 ppm			Upper respiratory tract irritation
Hexachloroethane	[Perchloroethane]	CCl_3CCl_3	67-72-1	1 ppm			1 ppm			Liver and kidney damage
Hexachloronaphthalene (unspecified isomer)	all isomers	$C_{10}H_2Cl_6$	1335-87-1	0.2 mg/m^3			0.2 mg/m^3			Liver damage; chloracne
Hexahydro-1,3-isobenzofurandione	[Hexahydrophthalic anhydride] all isomers	$C_8H_{10}O_3$	85-42-7			0.005 mg/m^3				Sensitization
Hexahydro-1,3,5-trinitro-1,3,5-triazine	[Cyclonite]	$C_3H_6N_6O_6$	121-82-4	0.5 mg/m^3			1.5 mg/m^3	3 mg/m^3		Liver damage
Hexamethylene diisocyanate		$C_8H_{12}N_2O_2$	822-06-0	0.005 ppm			0.005 ppm		0.020 ppm [10-minute]	Upper respiratory tract irritation; respiratory sensitization
Hexane		C_6H_{14}	110-54-3	50 ppm			50 ppm			Central nervous system impairment; peripheral neuropathy; eye irritation
1,6-Hexanediamine	[Hexamethylenediamine]	$C_6H_{16}N_2$	124-09-4	0.5 ppm						Upper respiratory tract and skin irritation
Hexanedinitrile	[Adiponitrile]	$C_6H_8N_2$	111-69-3	2 ppm			4 ppm			Upper respiratory tract and lower respiratory tract irritation
1,6-Hexanedioic acid	[Adipic acid]	$C_6H_{10}O_4$	124-04-9	5 mg/m^3						Upper respiratory tract irritation; central nervous system impairment
2-Hexanone	[Butyl methyl ketone]	$C_6H_{12}O$	591-78-6	5 ppm	10 ppm		1 ppm			Peripheral neuropathy; testicular damage
1-Hexene		C_6H_{12}	592-41-6	50 ppm						Central nervous system impairment
sec-Hexyl acetate	[4-Methyl-2-pentyl acetate]	$C_8H_{16}O_2$	108-84-9	50 ppm			50 ppm			Eye and upper respiratory tract irritation
Hydrazine		N_2H_4	302-01-2	0.01 ppm					0.03 ppm [2-hour]	Upper respiratory tract cancer
Hydrazoic acid	[Hydrogen azide] vapor.	HN_3	7782-79-8			0.11 ppm				
Hydrogen bromide	[Hydrobromic acid]	HBr	10035-10-6			2 ppm			3 ppm	Upper respiratory tract irritation
Hydrogen chloride	[Hydrochloric acid]	HCl	7647-01-0			2 ppm			5 ppm	Upper respiratory tract irritation
Hydrogen cyanide	[Hydrocyanic acid]	HCN	74-90-8			4.7 ppm		4.7 ppm		Upper respiratory tract irritation; headache; nausea; thyroid effects

				ACGIH			NIOSH			
Name	**Comments**	**Formula**	**CAS Reg. No.**	**Time-weighted average**	**Short-term exposure limit**	**ACGIH Ceiling**	**Time-weighted average**	**Short-term exposure limit**	**Ceiling**	**Basis**
Hydrogen fluoride	[Hydrofluoric acid]	HF	7664-39-3	0.5 ppm		2 ppm				Upper respiratory tract, lower respiratory, skin and eye irritation; fluorosis
Hydrogen peroxide		H_2O_2	7722-84-1	1 ppm			1 ppm			Eye, upper respiratory tract and skin irritation
Hydrogen selenide		H_2Se	7783-07-5	0.05 ppm			0.05 ppm			Upper respiratory tract and eye irritation; nausea
Hydrogen sulfide		H_2S	7783-06-4	1 ppm	5 ppm		1 mg/m³			Upper respiratory tract irritation; central nervous system impairment
p-Hydroquinone	[1,4-Benzenediol]	$C_6H_6O_2$	123-31-9	1 mg/m³					2 mg/m³ [15-minute]	Eye irritation; eye damage
2-Hydroxypropyl acrylate		$C_6H_{10}O_3$	999-61-1	0.5 ppm			0.5 ppm			Eye and upper respiratory tract irritation
Indene	[Indonaphthene]	C_9H_8	95-13-6	5 ppm			10 ppm			Liver damage
Indium	metal & compounds, as In	In	7440-74-6	0.1 mg/m³			0.1 mg/m³			Pulmonary edema; pneumonitis; dental erosion; malaise
Iodine	and volatile iodides.	I_2	7553-56-2	0.01 ppm (inhalable fraction and vapor)	0.1 ppm (vapor)				0.1 ppm	Hypothyroidism; upper respiratory tract irritation
Iodomethane	[Methyl iodide]	CH_3I	74-88-4	2 ppm			2 ppm			Eye damage; central nervous system impairment
Iron ion [Fe^{+2}]	soluble ferrous salts, as Fe	Fe^{+2}	15438-31-0	1 mg/m³			1 mg/m³			Upper respiratory tract and skin irritation
Iron ion [Fe^{+3}]	soluble ferric salts, as Fe	Fe^{+3}	20074-52-6	1 mg/m³			1 mg/m³			Upper respiratory tract and skin irritation
Iron(III) oxide	[Hematite]; dust & fume, as Fe	Fe_2O_3	1309-37-1	5 mg/m³						Pneumoconiosis
Iron pentacarbonyl	[Iron carbonyl [$Fe(CO)_5$]]; as Fe	$Fe(CO)_5$	13463-40-6	0.1 ppm	0.2 ppm		0.1 ppm	0.2 ppm		Pulmonary edema; central nervous system impairment
Isobutane	[2-Methylpropane]	$(CH_3)_3CH$	75-28-5		1000 ppm (explosion hazard)		800 ppm			Central nervous system impairment
Isobutene	[2-Methyl-1-propene]	$(CH_3)_2C{=}CH_2$	115-11-7	250 ppm						Upper respiratory tract irritation; body weight effects
Isobutyl acetate	[2-Methylpropyl acetate]	$C_6H_{12}O_2$	110-19-0	150 ppm			150 ppm			Eye and upper respiratory tract irritation
Isobutyl nitrite		$C_4H_9NO_2$	542-56-3			1 ppm (inhalable fraction and vapor)				Vasodilation; methemoglobinemia
Isopentane	[2-Methylbutane]	C_5H_{12}	78-78-4	1000 ppm						Narcosis; respiratory tract irritation
Isopentyl acetate	[Isoamyl acetate]	$C_7H_{14}O_2$	123-92-2	50 ppm	100 ppm		100 ppm			Upper respiratory tract irritation
Isophorone	[3,5,5-Trimethyl-2-cyclohexen-1-one]	$C_9H_{14}O$	78-59-1			5 ppm	4 ppm			Eye and upper respiratory tract irritation; central nervous system impairment; malaise; fatigue
Isophorone diisocyanate		$C_{12}H_{18}N_2O_2$	4098-71-9	0.005 ppm			0.005 ppm		0.02 ppm	Respiratory sensitization
Isopropenylbenzene	[α-Methyl styrene]	C_9H_{10}	98-83-9	10 ppm			50 ppm	100 ppm		Upper respiratory tract irritation; kidney and female reproductive damage
2-Isopropoxyethanol		$C_5H_{12}O_2$	109-59-1	25 ppm						Hematologic effects
Isopropyl acetate	[1-Methylethyl acetate]	$C_5H_{10}O_2$	108-21-4	100 ppm	200 ppm					Upper respiratory tract and eye irritation; central nervous system impairment
Isopropylamine	[2-Propanamine]	$(CH_3)_2CHNH_2$	75-31-0	5 ppm	10 ppm					Upper respiratory tract irritation; eye damage
N-Isopropylaniline		$C_9H_{13}N$	768-52-5	2 ppm			2 ppm			Methemoglobinemia
Isopropylbenzene	[Cumene]	C_9H_{12}	98-82-8	1 ppm			50 ppm			Liver damage
Isopropyl glycidyl ether	[(1-Methylethoxy)methyloxirane (IGE)]	$C_6H_{12}O_2$	4016-14-2	50 ppm	75 ppm				50 ppm [15-minute]	Upper respiratory tract and eye irritation; dermatitis
Kaolin			1332-58-7	2 mg/m³ (respirable particulate matter)			10 mg/m³ (total); 5 mg/m³ (respirable)			Pneumoconiosis

Name	Comments	Formula	CAS Reg. No.	ACGIH Time-weighted average	ACGIH Short-term exposure limit	ACGIH Ceiling	NIOSH Time-weighted average	NIOSH Short-term exposure limit	NIOSH Ceiling	Basis
Kerosene			8008-20-6	200 mg/m^3			100 mg/m^3			Skin and upper respiratory tract irritation; central nervous system impairment
Ketene		$CH_2=C=O$	463-51-4	0.5 ppm	1.5 ppm		0.5 ppm	1,5 ppm		Upper respiratory tract irritation; pulmonary edema
Lead	metal & inorganic compounds, as Pb	Pb	7439-92-1	0.05 mg/m^3			0.05 mg/m^3			Central nervous system and peripheral nervous system impairment; hematologic effects
Lead(II) arsenate	as As	$Pb_3(AsO_4)_2$	3687-31-8	0.01 mg/m^3						Lung cancer
Lead(II) chromate	[Crocoite]; as Cr (VI)	$PbCrO_4$	7758-97-6	0.0002 mg/m^3	0.00052 mg/m^3					Lung and sinonasal cancer; respiratory tract irritation; asthma
Lead(II) chromate	[Crocoite], as Pb	$PbCrO_4$	7758-97-6	0.05 mg/m^3			0.05 mg/m^3			Central nervous system and peripheral nervous system impairment; hematologic effects
Lithium hydride		LiH	7580-67-8			0.05 mg/m^3 (inhalable particulate matter)	0.025 mg/m^3			Eye and respiratory tract irritation
Magnesium oxide	[Magnesia]	MgO	1309-48-4	10 mg/m^3 (inhalable particulate matter)						Upper respiratory tract irritation; metal fume fever
Magnesium stearate	[Magnesium octadecanoate]	$Mg(C_{18}H_{35}O_2)_2$	557-04-0	10 mg/m^3 (inhalable particulate matter); 3 mg/m^3 (respirable particulate matter)						Lower respiratory tract irritation
Malathion		$C_{10}H_{19}O_6PS_2$	121-75-5	1 mg/m^3 (inhalable fraction and vapor)			10 mg/m^3			Cholinesterase inhibitor
Maleic anhydride		$C_4H_2O_3$	108-31-6	0.01 mg/m^3 (inhalable fraction and vapor)			1 mg/m^3			Respiratory sensitization
Manganese	metal & inorganic compounds, as Mn. inhalable fraction is 0.1 mg/m^3; see Ref. 1	Mn	7439-96-5	0.02 mg/m^3 (respirable particulate matter); 0.1 mg/m^3 (inhalable particulate matter)			1 mg/m^3	3 mg/m^3		Central nervous system impairment
Manganese cyclopentadienyl tricarbonyl	as Mn	$C_8H_5MnO_3$	12079-65-1	0.1 mg/m^3			0.1 mg/m^3			Skin irritation; central nervous system impairment
Manganese 2-methylcyclopentadienyl tricarbonyl	as Mn	$C_9H_7MnO_3$	12108-13-3	0.2 mg/m^3			0.2 mg/m^3			Central nervous system impairment; lung, liver, and kidney damage
Mercury	[Quicksilver]; metal & inorganic compounds, as Hg	Hg	7439-97-6	0.025 mg/m^3			0.05 mg/m^3 (vapor) (skin)		0.1 mg/m^3 (other) (skin)	Central nervous system impairment; kidney damage
Mercury	[Quicksilver]; alkyl compounds, as Hg	Hg	7439-97-6	0.01 mg/m^3	0.03 mg/m^3		0.01 mg/m^3	0.03 mg/m^3 (skin)		Central nervous system and peripheral nervous system impairment; kidney damage
Mercury	[Quicksilver]; aryl compounds, as Hg	Hg	7439-97-6	0.1 mg/m^3						Central nervous system impairment; kidney damage
Mesityl oxide	[Isobutenyl methyl ketone]	$C_6H_{10}O$	141-79-7	15 ppm	25 ppm		10 ppm			Eye and upper respiratory tract irritation; central nervous system impairment
Methacrylic acid	[2-Methylpropenoic acid]	$C_4H_6O_2$	79-41-4	20 ppm			20 ppm			Skin and eye irritation
Methanethiol	[Methyl mercaptan]	CH_3SH	74-93-1	0.5 ppm					0.5 ppm [15-minute]	Liver damage
Methanol	[Methyl alcohol]	CH_3OH	67-56-1	200 ppm	250 ppm		200 ppm	250 ppm		Headache; eye damage; dizziness; nausea

Name	Comments	Formula	CAS Reg. No.	ACGIH Time-weighted average	ACGIH Short-term exposure limit	ACGIH Ceiling	NIOSH Time-weighted average	NIOSH Short-term exposure limit	NIOSH Ceiling	Basis
Methomyl	[Acetimidic acid, *N*-[(methylcarbamoyl)oxy]thio-, methyl ester]	$C_5H_{10}N_2O_2S$	16752-77-5	0.2 mg/m³ (inhalable fraction and vapor)			2.5 mg/m³			Cholinesterase inhibitor; male reproductive damage; hematologic effects
2-Methoxyaniline	[*o*-Anisidine]	C_7H_9NO	90-04-0	0.5 mg/m³			0.5 mg/m³			Methemoglobinemia
4-Methoxyaniline	[*p*-Anisidine]	C_7H_9NO	104-94-9	0.5 mg/m³			0.5 mg/m³			Methemoglobinemia
Methoxychlor		$C_{16}H_{15}Cl_3O_2$	72-43-5	10 mg/m³						Liver damage; central nervous system impairment
2-Methoxyethanol	[Ethylene glycol monomethyl ether (EGME)]	$C_3H_8O_2$	109-86-4	0.1 ppm			0.1 ppm			Hematologic and reproductive effects
2-Methoxyethyl acetate	[Ethylene glycol monomethyl ether acetate (EGMEA)]	$C_5H_{10}O_3$	110-49-6	0.1 ppm			0.1 ppm			Hematologic and reproductive effects
2-Methoxy-2-methylbutane	[Methyl *tert*-pentyl ether (TAME)]	$C_6H_{14}O$	994-05-8	20 ppm						Central nervous system impairment; embryo/fetal damage
4-Methoxyphenol	[Mequinol]	$C_7H_8O_2$	150-76-5	5 mg/m³			5 mg/m³			Eye irritation; skin damage
1-Methoxy-2-propanol	[1,2-Propylene glycol monomethyl ether (PGME)]	$C_4H_{10}O_2$	107-98-2	50 ppm	100 ppm		100 ppm	150 ppm		Eye and upper respiratory tract irritation
Methyl acetate		$C_3H_6O_2$	79-20-9	200 ppm	250 ppm		200 ppm	250 ppm		Headache; dizziness; nausea; eye damage (degeneration of ganglion cells in the retina)
Methyl acrylate	[Methyl propenoate]	$C_4H_6O_2$	96-33-3	2 ppm			10 ppm			Eye, skin, and upper respiratory tract irritation; eye damage
2-Methylacrylonitrile	[2-Methylpropenenitrile]	C_4H_5N	126-98-7	1 ppm			1 ppm			Central nervous system impairment; eye and skin irritation
Methylamine	[Methanamine]	CH_3NH_2	74-89-5	5 ppm	15 ppm		10 ppm			Eye, skin, and upper respiratory tract irritation
2-Methylaniline	[*o*-Toluidine]	C_7H_9N	95-53-4	2 ppm						Methemoglobinemia; skin, eye, kidney, and bladder irritation
3-Methylaniline	[*m*-Toluidine]	C_7H_9N	108-44-1	2 ppm						Eye, bladder, and kidney irritation; methemoglobinemia;
4-Methylaniline	[*p*-Toluidine]	C_7H_9N	106-49-0	2 ppm						Methemoglobinemia
N-Methylaniline	[*N*-Methylbenzenamine]	C_7H_9N	100-61-8	0.5 ppm			0.5 ppm			Methemoglobinemia; central nervous system impairment
3-Methyl-1-butanol	[Isopentyl alcohol (Isoamyl alcohol)]	$C_5H_{12}O$	123-51-3	100 ppm	125 ppm		100 ppm	125 ppm		Eye and upper respiratory tract irritation
2-Methyl-1-butanol acetate		$C_7H_{14}O_2$	624-41-9	50 ppm	100 ppm		100 ppm			Upper respiratory tract irritation
3-Methyl-2-butanol acetate		$C_7H_{14}O_2$	5343-96-4	50 ppm	100 ppm		100 ppm			Upper respiratory tract irritation
3-Methyl-2-butanone	[Methyl isopropyl ketone]	$C_5H_{10}O$	563-80-4	20 ppm			200 ppm			Embryo/fetal damage; neonatal toxicity
Methyl *tert*-butyl ether	[*tert*-Butyl methyl ether (MTBE)]	$C_5H_{12}O$	1634-04-4	50 ppm						Upper respiratory tract irritation; kidney damage
Methyl 2-cyanoacrylate	[Mecrylate]	$C_5H_5NO_2$	137-05-3	0.2 ppm	1 ppm		2 ppm	4 ppm		Eye and upper respiratory tract irritation; asthma
Methylcyclohexane		C_7H_{14}	108-87-2	400 ppm			400 ppm			Upper respiratory tract irritation; central nervous system impairment; liver and kidney damage
1-Methylcyclohexanol		$C_7H_{14}O$	590-67-0	50 ppm			50 ppm			Upper respiratory tract and eye irritation
cis-2-Methylcyclohexanol		$C_7H_{14}O$	615-38-3	50 ppm			50 ppm			Upper respiratory tract and eye irritation
trans-2-Methylcyclohexanol,(±)-		$C_7H_{14}O$	615-39-4	50 ppm			50 ppm			Upper respiratory tract and eye irritation
cis-3-Methylcyclohexanol, (±)-		$C_7H_{14}O$	5454-79-5	50 ppm			50 ppm			Upper respiratory tract and eye irritation
trans-3-Methylcyclohexanol,(±)-		$C_7H_{14}O$	7443-55-2	50 ppm			50 ppm			Upper respiratory tract and eye irritation
cis-4-Methylcyclohexanol		$C_7H_{14}O$	7731-28-4	50 ppm			50 ppm			Upper respiratory tract and eye irritation
trans-4-Methylcyclohexanol		$C_7H_{14}O$	7731-29-5	50 ppm			50 ppm			Upper respiratory tract and eye irritation

Name	Comments	Formula	CAS Reg. No.	ACGIH Time-weighted average	ACGIH Short-term exposure limit	ACGIH Ceiling	NIOSH Time-weighted average	NIOSH Short-term exposure limit	NIOSH Ceiling	Basis
2-Methylcyclohexanone, (±)-		$C_7H_{12}O$	24965-84-2	50 ppm	75 ppm		50 ppm	75 ppm		Upper respiratory tract and eye irritation; central nervous system impairment
Methyl demeton		$C_6H_{15}O_3PS_2$	8022-00-2	0.05 mg/m³ (inhalable fraction and vapor)			0.5 mg/m³			Cholinesterase inhibitor
2-Methyl-3,5-dinitrobenzamide	[Dinitolmide]	$C_8H_7N_3O_5$	148-01-6	1 mg/m³			5 mg/m³			Liver damage
1-Methyl-2,3-dinitrobenzene	[2,3-Dinitrotoluene]	$C_7H_6N_2O_4$	602-01-7	0.2 mg/m³			1.5 mg/m³			Cardiac impairment; reproductive effects
1-Methyl-2,4-dinitrobenzene	[2,4-Dinitrotoluene]	$C_7H_6N_2O_4$	121-14-2	0.2 mg/m³			1.5 mg/m³			Cardiac impairment; reproductive effects
1-Methyl-3,5-dinitrobenzene	[3,5-Dinitrotoluene]	$C_7H_6N_2O_4$	618-85-9	0.2 mg/m³			1.5 mg/m³			Cardiac impairment; reproductive effects
2-Methyl-1,3-dinitrobenzene	[2,6-Dinitrotoluene]	$C_7H_6N_2O_4$	606-20-2	0.2 mg/m³			1.5 mg/m³			Cardiac impairment; reproductive effects
2-Methyl-1,4-dinitrobenzene	[2,5-Dinitrotoluene]	$C_7H_6N_2O_4$	619-15-8	0.2 mg/m³			1.5 mg/m³			Cardiac impairment; reproductive effects
4-Methyl-1,2-dinitrobenzene	[3,4-Dinitrotoluene]	$C_7H_6N_2O_4$	610-39-9	0.2 mg/m³			1.5 mg/m³			Cardiac impairment; reproductive effects
2-Methyl-4,6-dinitrophenol	[4,6-Dinitro-*o*-cresol]	$C_7H_6N_2O_5$	534-52-1	0.2 mg/m³			0.2 mg/m³			Basal metabolism
4,4′-Methylenebis[2-chloroaniline]	[3,3′-Dichloro-4,4′-diaminodiphenylmethane (MBOCA)]	$C_{13}H_{12}Cl_2N_2$	101-14-4	0.01 ppm			0.003 mg/m³ (skin)			Bladder cancer; methemoglobinemia
Methylenebis(4-cyclohexylisocyanate)		$C_{15}H_{22}N_2O_2$	5124-30-1	0.005 ppm			0.01 ppm			Respiratory sensitization; lower respiratory tract irritation
Methyl formate		CH_3OCHO	107-31-3	50 ppm	100 ppm		100 ppm	150 ppm		Central nervous system impairment; upper respiratory tract irritation; eye damage
2-Methylheptane		C_8H_{18}	592-27-8	300 ppm						Upper respiratory tract irritation
3-Methylheptane		C_8H_{18}	589-81-1	300 ppm						Upper respiratory tract irritation
4-Methylheptane		C_8H_{18}	589-53-7	300 ppm						Upper respiratory tract irritation
6-Methyl-1-heptanol	[Isooctyl alcohol]	$C_8H_{18}O$	1653-40-3	50 ppm			50 ppm (skin)			Upper respiratory tract irritation
5-Methyl-3-heptanone	[Ethyl 2-methylbutyl ketone]	$C_8H_{16}O$	541-85-5	10 ppm			25 ppm			Neurotoxicity
5-Methyl-2-hexanone	[Methyl isopentyl ketone]	$C_7H_{14}O$	110-12-3	20 ppm	50 ppm		50 ppm			Central nervous system impairment; upper respiratory tract irritation
Methylhydrazine		CH_6N_2	60-34-4	0.01 ppm					0.04 ppm [2-hour]	Upper respiratory tract and eye irritation; lung cancer; liver damage
Methyl isocyanate		$CH_3N{=}C{=}O$	624-83-9	0.02 ppm	0.06 ppm		0.02 ppm (skin)			Upper respiratory tract and eye irritation
Methyl methacrylate	[Methyl 2-methyl-2-propenoate]	$C_5H_8O_2$	80-62-6	50 ppm	100 ppm		100 ppm			Upper respiratory tract and eye irritation; body weight effects; pulmonary edema
1-Methylnaphthalene		$C_{11}H_{10}$	90-12-0	0.5 ppm						Lower respiratory tract irritation; lung damage
2-Methylnaphthalene		$C_{11}H_{10}$	91-57-6	0.5 ppm						Lower respiratory tract irritation; lung damage
2-Methyl-5-nitroaniline	[5-Nitro-*o*-toluidine]	$C_7H_8N_2O_2$	99-55-8	1 mg/m³						Liver damage
2-Methyloctane		C_9H_{20}	3221-61-2	200 ppm						Central nervous system impairment
3-Methyloctane		C_9H_{20}	2216-33-3	200 ppm						Central nervous system impairment
4-Methyloctane		C_9H_{20}	2216-34-4	200 ppm						Central nervous system impairment
Methyloxirane	[1,2-Propylene oxide]	C_3H_6O	75-56-9	2 ppm						Eye and upper respiratory tract irritation
Methyl parathion		$C_8H_{10}NO_5PS$	298-00-0	0.02 mg/m³ (inhalable fraction and vapor)			0.2 mg/m³ (skin)			Cholinesterase inhibitor
2-Methylpentane	[Isohexane]	C_6H_{14}	107-83-5	500 ppm	1000 ppm		100 ppm		510 ppm	Central nervous system impairment: upper respiratory tract and eye irritation

				ACGIH			NIOSH			
Name	Comments	Formula	CAS Reg. No.	Time-weighted average	Short-term exposure limit	ACGIH Ceiling	Time-weighted average	Short-term exposure limit	Ceiling	Basis
3-Methylpentane		C_6H_{14}	96-14-0	500 ppm	1000 ppm		100 ppm		510 ppm	Central nervous system impairment; upper respiratory tract and eye irritation
2-Methyl-2,4-pentanediol	[Hexylene glycol]	$C_6H_{14}O_2$	107-41-5	25 ppm (vapor fraction)	50 ppm (vapor fraction); 10 mg/m³ (inhalable particulate matter; aerosol only)		25 ppm			Eye and upper respiratory tract irritation
4-Methyl-2-pentanol	[Methyl isobutyl carbinol]	$C_6H_{14}O$	108-11-2	25 ppm	40 ppm		25 ppm	40 ppm (skin)		Upper respiratory tract and eye irritation; central nervous system impairment
4-Methyl-2-pentanone	[Isobutyl methyl ketone]	$C_6H_{12}O$	108-10-1	20 ppm	75 ppm		50 ppm	75 ppm		Upper respiratory tract irritation; dizziness; headache
2-Methyl-1-propanol	[Isobutyl alcohol]	$C_4H_{10}O$	78-83-1	50 ppm			50 ppm			Skin and eye irritation
2-Methyl-2-propanol	[*tert*-Butyl alcohol]	$(CH_3)_3COH$	75-65-0	100 ppm			100 ppm	150 ppm		Central nervous system impairment
Methyl silicate	[Tetramethoxysilane]	$Si(OCH_3)_4$	681-84-5	1 ppm			1 ppm			Upper respiratory tract irritation; eye damage
2-Methylstyrene	[2-Vinyl toluene]	C_9H_{10}	611-15-4	50 ppm	100 ppm		100 ppm			Upper respiratory tract and eye irritation
3-Methylstyrene	[3-Vinyl toluene]	C_9H_{10}	100-80-1	50 ppm	100 ppm		100 ppm			Upper respiratory tract and eye irritation
4-Methylstyrene	[4-Vinyl toluene]	C_9H_{10}	622-97-9	50 ppm	100 ppm		100 ppm			Upper respiratory tract and eye irritation
N-Methyl-*N*,2,4,6-tetranitroaniline	[Tetryl]	$C_7H_5N_5O_8$	479-45-8	1.5 mg/m³			1.5 mg/m³ (skin)			Upper respiratory tract irritation
Metribuzin		$C_8H_{14}N_4OS$	21087-64-9	5 mg/m³			5 mg/m³			Liver damage; hematologic effects
Mevinphos, *E*		$C_7H_{13}O_6P$	7786-34-7	0.01 ppm			0.01 ppm	0.03 ppm (skin)		Cholinesterase inhibitor
Mica		$(K,H)Al_3(SiO_4)_3$	12001-26-2	3 mg/m³			3 mg/m³			Pneumoconiosis
Mineral oil	mist			5 mg/m³	10 mg/m³					Upper respiratory tract irritation
Molybdenum	metal & insoluble compounds, as Mo	Mo	7439-98-7	10 mg/m³ (inhalable particulate matter)			15 mg/m³			Lower respiratory tract irritation
Molybdenum	soluble compounds, as Mo	Mo	7439-98-7	0.5 mg/m³ (respirable particulate matter)						Lower respiratory tract irritation
Monocrotophos	[*trans*-Dimethyl 1-methyl-3-(methylamino)-3-oxo-1-propenyl phosphate]	$C_7H_{14}NO_5P$	6923-22-4	0.05 mg/m³ (inhable fraction and vapor)			0.25 mg/m³			Cholinesterase inhibitor
Morpholine	[Tetrahydro-1,4-oxazine]	C_4H_9NO	110-91-8	20 ppm			20 ppm	30 ppm (skin)		Eye damage; upper respiratory tract irritation
Naled	[1,2-Dibromo-2,2-dichloroethylphosphoric acid, dimethyl ester]	$C_4H_7Br_2Cl_2O_4P$	300-76-5	0.1 mg/m³			3 mg/m³			Cholinesterase inhibitor
Naphtha			8030-30-6	400 ppm			100 ppm			
Naphthalene		$C_{10}H_8$	91-20-3	10 ppm			10 ppm	15 ppm		Upper respiratory tract irritation; cataracts; hemolytic anemia
1-Naphthalenylthiourea	[ANTU]	$C_{11}H_{10}N_2S$	86-88-4	0.3 mg/m³			0.3 mg/m³			Thyroid effects; nausea
2-Naphthylamine	[β-Naphthylamine]; levels as low as possible	$C_{10}H_9N$	91-59-8							Bladder cancer
Neopentane	[2,2-Dimethylpropane]	$C(CH_3)_4$	463-82-1	1000 ppm						Narcosis; respiratory tract irritation
Nickel	metal	Ni	7440-02-0	1.5 mg/m³ (inhalable particulate matter)			0.015 mg/m³			Dermatitis; pneumoconiosis
Nickel	soluble compounds, as Ni	Ni	7440-02-0	0.1 mg/m³ (inhalable particulate matter)			0.015 mg/m³			Lung damage; nasal cancer
Nickel	insoluble compounds, as Ni	Ni	7440-02-0	0.2 mg/m³ (inhalable particulate matter)			0.015 mg/m³			Lung cancer
Nickel carbonyl [$Ni(CO)_4$]	[Nickel tetracarbonyl] as Ni. Ceiling value	$Ni(CO)_4$	13463-39-3			0.05 ppm	0.001 ppm			Lung irritation
Nickel subsulfide	[Heazlewoodite]; as Ni	Ni_3S_2	12035-72-2	0.1 mg/m³ (inhalable particulate matter)			0.015 mg/m³			Lung cancer

Name	Comments	Formula	CAS Reg. No.	ACGIH			NIOSH			Basis
				Time-weighted average	Short-term exposure limit	ACGIH Ceiling	Time-weighted average	Short-term exposure limit	Ceiling	
L-Nicotine	[3-(1-Methyl-2-pyrrolidinyl) pyridine, (*S*)-]	$C_{10}H_{14}N_2$	54-11-5	0.5 mg/m³			0.5 mg/m³			GI damage; central nervous system impairment; cardiac impairment
Nitrapyrin	[Pyridine, 2-chloro-6-(trichloromethyl)-]	$C_6H_3Cl_4N$	1929-82-4	10 mg/m³	20 mg/m³		5 mg/m³ (respirable); 10 mg/m³ (total)	20 mg/m³ (total)		Liver damage
Nitric acid		HNO_3	7697-37-2	2 ppm	4 ppm		2 ppm	4 ppm		Upper respiratory tract and eye irritation; dental erosion
Nitric oxide		NO	10102-43-9	25 ppm			25 ppm			Hypoxia/cyanosis; nitrosyl-hemoglobinemia formation
4-Nitroaniline		$C_6H_6N_2O_2$	100-01-6	3 mg/m³			3 mg/m³ (skin)			Methemoglobinemia; liver damage; eye irritation
Nitrobenzene		$C_6H_5NO_2$	98-95-3	1 ppm			1 ppm			Methemoglobinemia
4-Nitrobiphenyl	levels as low as possible	$C_{12}H_9NO_2$	92-93-3							Bladder cancer
Nitroethane		$C_2H_5NO_2$	79-24-3	100 ppm			100 ppm			Upper respiratory tract irritation; central nervous system impairment; liver damage
Nitrogen dioxide		NO_2	10102-44-0	0.2 ppm			1 ppm			Lower respiratory tract irritation
Nitrogen trifluoride		NF_3	7783-54-2	10 ppm			10 ppm			Methemoglobinemia; liver and kidney damage
Nitromethane	[Nitrocarbol]	CH_3NO_2	75-52-5	20 ppm						Thyroid effects: upper respiratory tract irritation; lung damage
1-Nitropropane		$C_3H_7NO_2$	108-03-2	25 ppm			25 ppm			Upper respiratory tract and eye irritation; liver damage
2-Nitropropane	[Isonitropropane]	$C_3H_7NO_2$	79-46-9	10 ppm						Liver damage; liver cancer
N-Nitrosodimethylamine	[Dimethylnitrosamine]; levels as low as possible	$C_2H_6N_2O$	62-75-9							Liver and kidney cancer; liver damage
2-Nitrotoluene	[1-Methyl-2-nitrobenzene]	$C_7H_7NO_2$	88-72-2	2 ppm			2 ppm (skin)			Methemoglobinemia
3-Nitrotoluene	[1-Methyl-3-nitrobenzene]	$C_7H_7NO_2$	99-08-1	2 ppm			2 ppm (skin)			Methemoglobinemia
4-Nitrotoluene	[1-Methyl-4-nitrobenzene]	$C_7H_7NO_2$	99-99-0	2 ppm			2 ppm (skin)			Methemoglobinemia
Nitrous oxide	REL for exposure to waste anesthetic gas	N_2O	10024-97-2	50 ppm			25 ppm			Central nervous system impairment; hematologic effects; embryo/fetal damage
3,3,4,4,5,5,6,6,6,-Nonafluoro-1-hexene	[Perfluorobutylethene]	$C_6H_3F_9$	19430-93-4	100 ppm						Hematologic effects
Nonane	all isomers	C_9H_{20}	111-84-2	200 ppm						Central nervous system impairment
Octachloronaphthalene	[Perchloronaphthalene]	$C_{10}Cl_8$	2234-13-1	0.1 mg/m³	0.3 mg/m³		0.1 mg/m³	0.3 mg/m³		Liver damage
Octadecanoic acid	[Stearic acid]	$C_{18}H_{36}O_2$	57-11-4	10 mg/m³ (inhalable particulate matter); 3 mg/m³ (respirable particulate matter)						Lower respiratory tract irritation
Octane	all isomers	C_8H_{18}	111-65-9	300 ppm	375 ppm		75 ppm		385 ppm [15-minute]	Upper respiratory tract irritation
Osmium(VIII) oxide	[Osmic acid (Osmium tetroxide)]	OsO_4	20816-12-0	0.0002 ppm	0.0006 ppm		0.0002 ppm	0.0006 ppm		Eye, upper respiratory tract, and skin irritation
Oxalic acid	and dihydrate	$H_2C_2O_4$	144-62-7	1 mg/m³	2 mg/m³		1 mg/m³	2 mg/m³		Upper respiratory tract, eye, and skin irritation
2-Oxetanone	[β-Propiolactone]	$C_3H_4O_2$	57-57-8	0.5 ppm						
Oxirane	[Ethylene oxide]	C_2H_4O	75-21-8	1 ppm						Cancer; central nervous system impairment
Oxiranemethanol, (±)-	[Glycidol]	$C_3H_6O_2$	61915-27-3	2 ppm			25 ppm			Upper respiratory tract, eye, and skin irritation
4,4′-Oxybis(benzenesulfonyl hydrazide)		$C_{12}H_{14}N_4O_5S_2$	80-51-3	0.1 mg/m³ (inhalable particulate matter)						Teratogenic effects
Ozone	depends on workload	O_3	10028-15-6	0.1 ppm					0.1 ppm	Pulmonary function
Paraquat	as the cation	$C_{12}H_{14}N_2$	4685-14-7	0.05 mg/m³ (inhalable particular matter)			0.1 mg/m³ (respirable) (skin)			Lung damage; upper respiratory irritation

Name	Comments	Formula	CAS Reg. No.	ACGIH			NIOSH			Basis
				Time-weighted average	Short-term exposure limit	ACGIH Ceiling	Time-weighted average	Short-term exposure limit	Ceiling	
Parathion		$C_{10}H_{14}NO_5PS$	56-38-2	0.05 mg/m^3 (inhalable fraction and vapor)			0.05 mg/m^3 (skin)			Cholinesterase inhibitor
Pentaborane(9)		B_5H_9	19624-22-7	0.005 ppm	0.015 ppm		0.005 ppm	0.015 ppm		Central nervous system convulsions and impairment
Pentachloronaphthalene (unspecified isomer)	all isomers	$C_{10}H_3Cl_5$	1321-64-8	0.5 mg/m^3			0.5 mg/m^3 (skin)			Liver damage; chloracne
Pentachloronitrobenzene	[Quintozene]	$C_6Cl_5NO_2$	82-68-8	0.5 mg/m^3						Liver damage
Pentachlorophenol		C_6HCl_5O	87-86-5	0.5 mg/m^3 (inhalable fraction and vapor)	1 mg/m^3 (inhalable fraction and vapor)		0.5 mg/m^3 (skin)			Upper respiratory tract and eye irritation; central nervous system impairment; cardiac impairment
Pentaerythritol		$C_5H_{12}O_4$	115-77-5	10 mg/m^3			10 mg/m^3 (total); 5 mg/m^3 (respirable)			GI irritation
Pentanal	[Valeraldehyde]	$C_5H_{10}O$	110-62-3	50 ppm			50 ppm			Eye, skin, and upper respiratory tract irritation
Pentane	all isomers	C_5H_{12}	109-66-0	1000 ppm						Narcosis; respiratory tract irritation
Pentanedial	[Glutaraldehyde]	$C_5H_8O_2$	111-30-8			0.05 ppm			0.2 ppm	Upper respiratory tract, skin, and eye irritation
3-Pentanol acetate		$C_7H_{14}O_2$	620-11-1	50 ppm	100 ppm		100 ppm			Upper respiratory tract irritation
2-Pentanone	[Methyl propyl ketone]	$C_5H_{10}O$	107-87-9		150 ppm		150 ppm			Pulmonary function; eye irritation
3-Pentanone	[Diethyl ketone]	$C_5H_{10}O$	96-22-0	200 ppm	300 ppm		200 ppm			Upper respiratory tract irritation; central nervous system impairment
Pentyl acetate	[Amyl acetate]; all isomers	$C_7H_{14}O_2$	628-63-7	50 ppm	100 ppm		100 ppm			Upper respiratory tract irritation
sec-Pentyl acetate, (*R*)-	[*sec*-Amyl acetate, (*R*)-]	$C_7H_{14}O_2$	54638-10-7	50 ppm	100 ppm		100 ppm			Upper respiratory tract irritation
Perchloryl fluoride	[Chlorine trioxide fluoride]	ClO_3F	7616-94-6	3 ppm	6 ppm		3 ppm	6 ppm		Lower respiratory tract and upper respiratory tract irritation; methemoglobinemia; fluorosis
Perfluoroacetone	[Hexafluoroacetone]	C_3F_6O	684-16-2	0.1 ppm			0.1 ppm (skin)			Testicular and kidney damage
Perfluoroisobutene	[Perfluoroisobutylene]	C_4F_8	382-21-8							Upper respiratory tract irritation; hematologic effects
Perfluoropropene	[Hexafluoropropene]	C_3F_6	116-15-4	0.1 ppm						Kidney damage
Peroxyacetic acid	[Peracetic acid]	$C_2H_4O_3$	79-21-0		0.4 ppm (inhalable fraction and vapor)					Upper respiratory tract, eye, and skin irritation
Phenol	[Hydroxybenzene]	C_6H_5OH	108-95-2	5 ppm			5 ppm		15.6 ppm [15-minute] (skin)	Upper respiratory tract irritation; lung damage; central nervous system impairment
10*H*-Phenothiazine	[Thiodiphenylamine]	$C_{12}H_9NS$	92-84-2	5 mg/m^3			5 mg/m^3 (skin)			Eye photosensitization; skin irritation
Phenyl glycidyl ether	[PGE]	$C_9H_{10}O_2$	122-60-1	0.1 ppm					1 ppm [15-minute]	Testicular damage
Phenylhydrazine		$C_6H_8N_2$	100-63-0	0.1 ppm					0.14 ppm [2-hour] (skin)	Anemia; upper respiratory and skin irritation
Phenyl isocyanate		C_6H_5NCO	103-71-9	0.005 ppm	0.015 ppm					Upper respiratory tract irritation
N-Phenyl-2-naphthalenamine	[*N*-Phenyl-β-naphthylamine]; levels as low as possible	$C_{16}H_{13}N$	135-88-6							Cancer
Phenylphosphine	[Monophenylphosphine]	$C_6H_5PH_2$	638-21-1						0.05 ppm	Dermatitis; hematologic effects; testicular damage
Phorate	[Phosphorodithioic acid, *O,O*-diethyl *S*-[(ethylthio) methyl] ester]	$C_7H_{17}O_2PS_3$	298-02-2	0.05 mg/m^3	0.2 mg/m^3		0.05 mg/m^3	0.2 mg/m^3 (skin)		Cholinesterase inhibitor
Phosphine	[Phosphorus hydride]	PH_3	7803-51-2	0.05 ppm	0.15 ppm		0.3 ppm	1 ppm		Respiratory tract irritation; pulmonary edema
Phosphoric acid	[Orthophosphoric acid]	H_3PO_4	7664-38-2	1 mg/m^3	3 mg/m^3		1 mg/m^3	3 mg/m^3 1 mg/m^3		Upper respiratory tract, eye, and skin irritation

				ACGIH			NIOSH			
Name	Comments	Formula	CAS Reg. No.	Time-weighted average	Short-term exposure limit	ACGIH Ceiling	Time-weighted average	Short-term exposure limit	Ceiling	Basis
Phosphorus (white)	[Yellow phosphorus]	P	7723-14-0	0.1 mg/m^3			0.1 mg/m^3			Lower respiratory tract, upper respiratory tract, and GI irritation; liver damage
Phosphorus(III) chloride	[Phosphorus trichloride]	PCl_3	7719-12-2	0.2 ppm	0.5 ppm		0.2 ppm	0.5 ppm		Upper respiratory tract, eye, and skin irritation
Phosphorus(V) chloride	[Phosphorus pentachloride]	PCl_5	10026-13-8	0.1 ppm			1 mg/m^3			Upper respiratory tract and eye irritation
Phosphorus(V) sulfide	[Phosphorus pentasulfide]	P_2S_5	1314-80-3	1 mg/m^3	3 mg/m^3		1 mg/m^3	3 mg/m^3		Upper respiratory tract irritation
Phosphoryl chloride	[Phosphorus oxychloride]	$POCl_3$	10025-87-3	0.1 ppm			0.1 ppm	0.5 ppm		Upper respiratory tract irritation
Phthalic anhydride		$C_8H_4O_3$	85-44-9	0.002 mg/m^3 (inhalable fraction and vapor)	0.005 mg/m^3 (inhalable fraction and vapor)		6 mg/m^3			Respiratory sensitization; asthma
α-Pinene	[2-Pinene]	$C_{10}H_{16}$	80-56-8	20 ppm			100 ppm			Lung irritation
β-Pinene	[Nopinene]	$C_{10}H_{16}$	127-91-3	20 ppm			100 ppm			Lung irritation
Piperazine	and salts, as piperazine	$C_4H_{10}N_2$	110-85-0	0.03 ppm (inhalable fraction and vapor)			5 mg/m^3			Respiratory sensitization; asthma
2-Pivaloyl-1,3-indandione	[Pindone]	$C_{14}H_{14}O_3$	83-26-1	0.1 mg/m^3			0.1 mg/m^3			Coagulation
Platinum	metal dust	Pt	7440-06-4	1 mg/m^3			1 mg/m^3			Asthma; upper respiratory irritation
Platinum	soluble salts, as Pt	Pt	7440-06-4	0.002 mg/m^3			0.002 mg/m^3			Asthma; upper respiratory irritation
Polychlorinated biphenyls (42% chlorine)	[PCBs]		53469-21-9	1 mg/m^3			0.001 mg/m^3			Liver damage; eye irritation; chloracne
Polychlorinated biphenyls (54% chlorine)	[PCBs]		11097-69-1	0.5 mg/m^3			0.001 mg/m^3			Upper respiratory tract irritation; liver damage; chloracne
Poly(vinyl chloride)	[PVC]		9002-86-2	1 mg/m^3 (respirable particulate matter)						Pneumoconiosis: upper respiratory tract irritation: pulmonary function changes
Portland cement	no asbestos		65997-15-1	1 mg/m^3 (respirable particulate matter)			10 mg/m^3 (total); 5 mg/m^3 (respirable)			Pulmonary function; respiratory symptoms; asthma
Potassium hydroxide		KOH	1310-58-3			2 mg/m^3			2 mg/m^3	Upper respiratory tract, eye, and skin irritation
Propanal	[Propionaldehyde]	C_2H_5CHO	123-38-6	20 ppm						Upper respiratory tract irritation
1,3-Propane sultone	[1,2-Oxathiolane, 2,2-dioxide]; levels as low as possible	$C_3H_6O_3S$	1120-71-4							Cancer
Propanoic acid	[Propionic acid]	$C_3H_6O_2$	79-09-4	10 ppm			10 ppm	15 ppm		Eye, skin, and upper respiratory tract irritation
1-Propanol	[Propyl alcohol]	$CH_3CH_2CH_2OH$	71-23-8	100 ppm			200 ppm	250 ppm (skin)		Eye and upper respiratory tract irritation
2-Propanol	[Isopropyl alcohol]	$CH_3CHOHCH_3$	67-63-0	200 ppm	400 ppm		400 ppm	500 ppm		Eye and upper respiratory tract irritation; central nervous system impairment
Propargyl alcohol	[3-Hydroxy-1-propyne (2-Propyn-1-ol)]	C_3H_4O	107-19-7	1 ppm			1 ppm (skin)			Eye irritation; liver and kidney damage
Propene	[Propylene]	$CH_3CH{=}CH_2$	115-07-1	500 ppm						Asphyxia; upper respiratory tract irritation
Propoxur	[Phenol, 2-(1-methylethoxy)-, methylcarbamate]	$C_{11}H_{15}NO_3$	114-26-1	0.5 mg/m^3 (inhalable fraction and vapor)			0.5 mg/m^3			Cholinesterase inhibitor
Propyl acetate	all isomers	$C_5H_{10}O_2$	109-60-4	100 ppm	150 ppm		200 ppm	250 ppm		Upper respiratory tract and eye irritation; central nervous system impairment
1,2-Propylene glycol dinitrate		$C_3H_6N_2O_6$	6423-43-4	0.05 ppm			0.05 ppm (skin)			Headache; central nervous system impairment
Propyleneimine	[2-Methylaziridine]	C_3H_7N	75-55-8	0.2 ppm	0.4 ppm		2 ppm (skin)			Upper respiratory tract irritation; kidney damage
Propyl nitrate		$C_3H_7NO_3$	627-13-4	25 ppm	40 ppm		25 ppm	40 ppm		Nausea; headache
Propyne	[Methylacetylene]	$CH_3C{\equiv}CH$	74-99-7	1000 ppm (explosion hazard)			1000 ppm			Central nervous system impairment

Safety

Name	Comments	Formula	CAS Reg. No.	ACGIH Time-weighted average	ACGIH Short-term exposure limit	ACGIH Ceiling	NIOSH Time-weighted average	NIOSH Short-term exposure limit	NIOSH Ceiling	Basis
Pyrethrin I	[Pyrethrum]	$C_{21}H_{28}O_3$	121-21-1	5 mg/m³			5 mg/m³			Liver damage; lower respiratory tract irritation
2-Pyridinamine	[2-Aminopyridine]	$C_5H_6N_2$	504-29-0	0.5 ppm			0.5 ppm			Headache; nausea; central nervous system impairment; dizziness
Pyridine	[Azine]	C_5H_5N	110-86-1	1 ppm			5 ppm			Skin irritation; liver and kidney damage
Pyrocatechol	[1,2-Benzenediol (Catechol)]	$C_6H_6O_2$	120-80-9	5 ppm			5 ppm (skin)			Eye and upper respiratory tract irritation; dermatitis
Resorcinol	[1,3-Benzenediol]	$C_6H_6O_2$	108-46-3	10 ppm	20 ppm		10 ppm	20 ppm		Eye and skin irritation
Rhodium	metal & insoluble compounds, as Rh	Rh	7440-16-6	1 mg/m³			0.1 mg/m³			Metal = upper respiratory irritation; Insoluble compounds = lower respiratory tract irritation
Rhodium	soluble compounds, as Rh	Rh	7440-16-6	0.01 mg/m³			0.001 mg/m³			Asthma
Ronnel		$C_8H_8Cl_3O_3PS$	299-84-3	5 mg/m³ (inhalable fraction and vapor)			10 mg/m³			Cholinesterase inhibitor
Rotenone		$C_{23}H_{22}O_6$	83-79-4	5 mg/m³			5 mg/m³			Upper respiratory tract and eye irritation; central nervous system impairment
Rubber	natural latex, as inhalable proteins			0.0001 mg/m³						Respiratory sensitization
Selenium	element & compounds, as Se	Se	7782-49-2	0.2 mg/m³			0.2 mg/m³			Eye and upper respiratory tract irritation
Selenium hexafluoride		SeF_6	7783-79-1	0.05 ppm			0.05 ppm			Pulmonary edema
Sesone	[Sodium 2-(2,4-dichlorophenoxy)ethyl sulfate]	$C_8H_7Cl_2NaO_5S$	136-78-7	10 mg/m³			10 mg/m³ (total); 5 mg/m³ (respirable)			GI irritation
Silane	[Silicon tetrahydride]	SiH_4	7803-62-5	5 ppm			5 ppm			Upper respiratory tract irritation
Silicon carbide (hexagonal)	[Moissanite]; lower limits for fibrous SiC	SiC	409-21-2	10 mg/m³ (inhalable particulate matter); 3 mg/m³ (respirable particulate matter)			10 mg/m³ (total); 5 mg/m³ (respirable)			Non-fibrous forms = Upper respiratory tract irritation; Fibrous forms = Mesothelioma; cancer
Silicon dioxide (α-quartz)	[Silica]	SiO_2	14808-60-7	0.025 mg/m³ (respirable particulate matter)			0.05 mg/m³ (respirable)			Pulmonary fibrosis; lung cancer
Silicon dioxide (cristobalite)	[Cristobalite]	SiO_2	14464-46-1	0.025 mg/m³ (respirable particulate matter)			0.05 mg/m³ (respirable)			Pulmonary fibrosis; lung cancer
Silver	metal	Ag	7440-22-4	0.1 mg/m³			0.01 mg/m³			Argyria
Silver	soluble compounds, as Ag	Ag	7440-22-4	0.01 mg/m³			0.01 mg/m³			Argyria
Simazine			122-34-9	0.5 mg/m³ (inhalable particulate matter)						Hematologic effects
Sodium azide	[Smite]	NaN_3	26628-22-8		0.29 mg/m³	0.11 ppm (as hydrazoic acid vapor); 0.29 mg/m³ (as sodium azide)			0.1 ppm (as hydrazoic acid) (skin); 0.3 mg/m³ (as sodium azide) (skin)	Cardiac impairment; lung damage
Sodium fluoroacetate		$C_2H_2FNaO_2$	62-74-8	0.05 mg/m³			0.05 mg/m³	0.15 mg/m³ (skin)		Central nervous system impairment; cardiac impairment; nausea
Sodium hydrogen sulfite	[Sodium bisulfite]	$NaHSO_3$	7631-90-5	5 mg/m³			5 mg/m³			Skin, eye, and upper respiratory tract irritation
Sodium hydroxide	[Caustic soda]	NaOH	1310-73-2						2 mg/m³	Upper respiratory tract, eye, and skin irritation
Sodium metabisulfite	[Sodium pyrosulfite]	$Na_2S_2O_5$	7681-57-4	5 mg/m³			5 mg/m³			Upper respiratory tract irritation
Sodium stearate	[Sodium octadecanoate]	$NaC_{18}H_{35}O_2$	822-16-2	10 mg/m³						Lower respiratory tract irritation

Name	Comments	Formula	CAS Reg. No.	ACGIH Time-weighted average	ACGIH Short-term exposure limit	ACGIH Ceiling	NIOSH Time-weighted average	NIOSH Short-term exposure limit	NIOSH Ceiling	Basis
Sodium tetraborate decahydrate	[Borax]	$Na_2B_4O_7 \cdot 10H_2O$	1303-96-4	2 mg/m^3 (inhalable particulate matter)	6 mg/m^3 (inhalable particulate matter)		5 mg/m^3			Upper respiratory tract irritation
Starch			9005-25-8	10 mg/m^3			10 mg/m^3 (total); 5 mg/m^3 (respirable)			Dermatitis
Stibine	[Antimony(III) hydride]	SbH_3	7803-52-3	0.1 ppm			0.1 ppm			Hemolysis; kidney damage; lower respiratory tract irritation
Stoddard solvent			8052-41-3	100 ppm			350 mg/m^3		1800 mg/m^3 [15-minute]	Eye, skin, and kidney damage; nausea; central nervous system impairment
Strontium chromate	as Cr	$SrCrO_4$	7789-06-2	0.0002 mg/m^3 (inhalable particulate matter)	0.0005 mg/m^3 (inhalable particulate matter)		0.0002 mg/m^3			Lung and sinonasal cancer; respiratory tract irritation; asthma
Strychnine		$C_{21}H_{22}N_2O_2$	57-24-9	0.15 mg/m^3			0.15 mg/m^3			Central nervous system impairment
Styrene	[Vinylbenzene]	C_8H_8	100-42-5	2 ppm			50 ppm	100 ppm		Central nervous system impairment; upper respiratory tract irritation; peripheral neuropathy; ototoxicity; visual disorders
Subtilisins	as crystalline active enzyme		9014-01-1			0.00006 mg/m^3		0.00006 mg/m^3 [60-minute]		Asthma; skin, upper respiratory tract and lower respiratory tract irritation
Sucrose		$C_{12}H_{22}O_{11}$	57-50-1	10 mg/m^3			10 mg/m^3 (total); 5 mg/m^3 (respirable)			Dental erosion
Sulfometuron methyl		$C_{15}H_{16}N_4O_5S$	74222-97-2	5 mg/m^3						Hematologic effects
Sulfotep	[Tetraethyl thiodiphosphate (TEDP)]	$C_8H_{20}O_5P_2S_2$	3689-24-5	0.2 mg/m^3 (inhalable fraction and vapor)			0.2 mg/m^3 (skin)			Cholinesterase inhibitor
Sulfur chloride [ClSSCl]	[Sulfur monochloride (Disulfur dichloride)]	S_2Cl_2	10025-67-9			1 ppm			1 ppm	Eye, skin, and upper respiratory tract irritation
Sulfur decafluoride	[Disulfur decafluoride (Sulfur pentafluoride)]	S_2F_{10}	5714-22-7			0.01 ppm			0.01 ppm	Upper respiratory tract irritation; lung damage
Sulfur dioxide		SO_2	7446-09-5		0.25 ppm		2 ppm	5 ppm		Pulmonary function; lower respiratory irritation
Sulfur hexafluoride		SF_6	2551-62-4	1000 ppm			1000 ppm			Asphyxia
Sulfuric acid	[Oil of vitriol]	H_2SO_4	7664-93-9	0.2 mg/m^3 (thoracic particulate matter)			1 mg/m^3			Pulmonary function
Sulfur tetrafluoride		SF_4	7783-60-0			0.1 ppm			0.1 ppm	Eye and upper respiratory tract irritation; lung damage
Sulfuryl fluoride		SO_2F_2	2699-79-8	5 ppm	10 ppm		5 ppm	10 ppm		Central nervous system impairment
Sulprofos		$C_{12}H_{19}O_2PS_3$	35400-43-2	0.1 mg/m^3 (inhalable fraction and vapor)			1 mg/m^3			Cholinesterase inhibitor
Talc	no asbestos	$3MgO \cdot 4SiO_2 \cdot H_2O$	14807-96-6	2 mg/m^3 (respirable particulate matter)			2 mg/m^3			Pulmonary fibrosis; pulmonary function
Tantalum	dust	Ta	7440-25-7	5 mg/m^3			5 mg/m^3	10 mg/m^3		
Tantalum(V) oxide	[Tantalum pentoxide]; dust, as Ta	Ta_2O_5	1314-61-0	5 mg/m^3			5 mg/m^3	10 mg/m^3		
Tellurium	and compounds, as Te (except H_2Te)	Te	13494-80-9	0.1 mg/m^3			0.1 mg/m^3			Halitosis
Tellurium hexafluoride		TeF_6	7783-80-4	0.02 ppm			0.02 ppm			Lower respiratory tract irritation
Terbufos		$C_9H_{21}O_2PS_3$	13071-79-9	0.01 mg/m^3						Cholinesterase inhibitor
Terephthalic acid	[1,4-Benzenedicarboxylic acid]	$C_8H_6O_4$	100-21-0	10 mg/m^3						-
o-Terphenyl		$C_{18}H_{14}$	84-15-1						5 mg/m^3	Upper respiratory tract and eye irritation

				ACGIH			NIOSH			
Name	**Comments**	**Formula**	**CAS Reg. No.**	**Time-weighted average**	**Short-term exposure limit**	**ACGIH Ceiling**	**Time-weighted average**	**Short-term exposure limit**	**Ceiling**	**Basis**
m-Terphenyl		$C_{18}H_{14}$	92-06-8						5 mg/m³	Upper respiratory tract and eye irritation
p-Terphenyl		$C_{18}H_{14}$	92-94-4			5 mg/m³			5 mg/m³	Upper respiratory tract and eye irritation
1,1,2,2-Tetrabromoethane	[Acetylene tetrabromide]	$C_2H_2Br_4$	79-27-6	0.1 ppm						Eye and upper respiratory tract irritation; pulmonary edema; liver damage
Tetrabromomethane	[Carbon tetrabromide]	CBr_4	558-13-4	0.1 ppm	0.3 ppm		0.1 ppm	0.3 ppm		Liver damage; eye, upper respiratory tract, and skin irritation
1,1,1,2-Tetrachloro-2,2-difluoroethane	[Tetrachloro-1,1-difluoroethane]	$C_2Cl_4F_2$	76-11-9	100 ppm			500 ppm			Liver and kidney damage; central nervous system impairment
1,1,2,2-Tetrachloro-1,2-difluoroethane	[Tetrachloro-1,2-difluoroethane]	$C_2Cl_4F_2$	76-12-0	50 ppm			500 ppm			Liver and kidney damage; central nervous system impairment
1,1,2,2-Tetrachloroethane	[Acetylene tetrachloride]	$C_2H_2Cl_4$	79-34-5	1 ppm			1 ppm (skin)			Liver damage
Tetrachloroethene	[Perchloroethylene]	C_2Cl_4	127-18-4	25 ppm	100 ppm					Central nervous system impairment
Tetrachloromethane	[Carbon tetrachloride]	CCl_4	56-23-5	5 ppm	10 ppm			2 ppm [60-minute]		Liver damage
1,2,3,4-Tetrachloronaphthalene		$C_{10}H_4Cl_4$	20020-02-4	2 mg/m³			2 mg/m³			Liver damage
Tetraethyl lead	as Pb	$C_8H_{20}Pb$	78-00-2	0.1 mg/m³			0.075 mg/m³ (skin)			Central nervous system impairment
Tetraethyl pyrophosphate	[TEPP]	$C_8H_{20}O_7P_2$	107-49-3	0.01 mg/m³ (inhalable fraction and vapor)			0.5 mg/m³			Cholinesterase inhibitor
Tetrafluoroethene	[Tetrafluoroethylene]	$F_2C{=}CF_2$	116-14-3	2 ppm						Kidney and liver damage; liver and kidney cancer
Tetrahydrofuran	[Tetramethylene oxide (Oxolane)]	C_4H_8O	109-99-9	50 ppm	100 ppm		200 ppm	250 ppm		Upper respiratory tract irritation; central nervous system impairment; kidney damage
Tetrakis(hydroxymethyl) phosphonium chloride	also the sulfate	$C_4H_{12}ClO_4P$	124-64-1	2 mg/m³						Liver damage
2,2,3,3-Tetramethylbutane		C_8H_{18}	594-82-1	300 ppm						Upper respiratory tract irritation
Tetramethyl lead	as Pb	$C_4H_{12}Pb$	75-74-1	0.15 mg/m³			0.075 mg/m³ (skin)			Central nervous system impairment
2,2,3,3-Tetramethylpentane		C_9H_{20}	7154-79-2	200 ppm						Central nervous system impairment
2,2,3,4-Tetramethylpentane		C_9H_{20}	1186-53-4	200 ppm						Central nervous system impairment
2,2,4,4-Tetramethylpentane	[Di-*tert*-butylmethane]	C_9H_{20}	1070-87-7	200 ppm						Central nervous system impairment
2,3,3,4-Tetramethylpentane		C_9H_{20}	16747-38-9	200 ppm						Central nervous system impairment
Tetramethylsuccinonitrile	[Tetramethylbutanedinitrile]	$C_8H_{12}N_2$	3333-52-6	0.5 mg/m³ (inhalable fraction and vapor)			3 mg/m³ (skin)			Central nervous system convulsions; hypoglycemia
Tetranitromethane		$C(NO_2)_4$	509-14-8	0.005 ppm			1 ppm			Eye and upper respiratory tract irritation; upper respiratory cancer
Thallium	and compounds, as Tl	Tl	7440-28-0	0.02 mg/m³			0.1 mg/m³ (skin)			GI damage; peripheral neuropathy
4,4′-Thiobis(6-*tert*-butyl-*m*-cresol)	[Bis(5-*tert*-butyl-4-hydroxy-2-methylphenyl) sulfide]	$C_{22}H_{30}O_2S$	96-69-5	1 mg/m³ (inhalable particulate matter)			10 mg/m³ (total); 5 mg/m³ (respirable)			Upper respiratory tract irritation
Thioglycolic acid		$C_2H_4O_2S$	68-11-1	1 ppm			1 ppm (skin)			Eye and respiratory tract irritation
Thionyl chloride	[Sulfinyl dichloride]	$SOCl_2$	7719-09-7			0.2 ppm			1 ppm	Upper respiratory tract irritation
Thiram		$C_6H_{12}N_2S_4$	137-26-8	0.05 mg/m³ (inhalable fraction and vapor)						Body weight and hematologic effects
Tin	metal	Sn	7440-31-5	2 mg/m³ (inhalable particulate matter)			2 mg/m³			Pneumoconiosis
Tin	inorganic compounds, as Sn	Sn	7440-31-5	2 mg/m³			2 mg/m³			Pneumoconiosis

Name	Comments	Formula	CAS Reg. No.	ACGIH Time-weighted average	ACGIH Short-term exposure limit	ACGIH Ceiling	NIOSH Time-weighted average	NIOSH Short-term exposure limit	Ceiling	Basis
Tin	organic compounds, as Sn	Sn	7440-31-5	0.1 mg/m^3	0.2 mg/m^3		0.1 mg/m^3 (skin)			Eye and upper respiratory tract irritation; headache; nausea; central nervous system and immune effects
Titanium(IV) oxide (rutile)	[Rutile (Titanium dioxide)]	TiO_2	1317-80-2	10 mg/m^3						Lower respiratory tract irritation
Toluene	[Methylbenzene]	$C_6H_5CH_3$	108-88-3	20 ppm			100 ppm	150 ppm		Visual impairment; female reproductive effects; pregnancy loss
Toluene-2,4-diisocyanate		$C_9H_6N_2O_2$	584-84-9	0.001 ppm (inhalable fraction and vapor)	0.005 ppm (inhalable fraction and vapor)					Asthma; pulmonary function; eye irritation
Toluene-2,6-diisocyanate		$C_9H_6N_2O_2$	91-08-7	0.001 ppm (inhalable fraction and vapor)	0.005 ppm (inhalable fraction and vapor)					Asthma; pulmonary function; eye irritation
Toxaphene	[Polychlorocamphene]	$C_{10}H_{10}Cl_8$	8001-35-2	0.5 mg/m^3	1 mg/m^3					Central nervous system convulsions; liver damage
1*H*-1,2,4-Triazol-3-amine	[Amitrole]	$C_2H_4N_4$	61-82-5	0.2 mg/m^3			0.2 mg/m^3			Thyroid effects
Tribromomethane	[Bromoform]	$CHBr_3$	75-25-2	0.5 ppm			0.5 ppm			Liver damage; upper respiratory tract irritation; eye irritation
Tributyl phosphate	[Butyl phosphate]	$C_{12}H_{27}O_4P$	126-73-8	5 mg/m^3 (inhalable fraction and vapor)			2.5 mg/m^3			Bladder, eye, and upper respiratory sensitization
Trichlorfon	[2,2,2-Trichloro-1-hydroxyethylphosphonic acid, dimethyl ester]	$C_4H_8Cl_3O_4P$	52-68-6	1 mg/m^3						Cholinesterase inhibitor
Trichloroacetic acid		CCl_3COOH	76-03-9	0.5 ppm			1 ppm			Eye and upper respiratory tract irritation
1,2,4-Trichlorobenzene		$C_6H_3Cl_3$	120-82-1			5 ppm			5 ppm	Eye and upper respiratory tract irritation
1,1,1-Trichloro-2,2-bis(4-chlorophenyl)ethane	[Dichlorodiphenyltrichloroethane (DDT)]	$C_{14}H_9Cl_5$	50-29-3	1 mg/m^3			0.5 mg/m^3			Liver damage
1,1,1-Trichloroethane	[Methyl chloroform]	CH_3CCl_3	71-55-6	350 ppm	450 ppm				350 ppm [15-minute]	Central nervous system impairment; liver damage
1,1,2-Trichloroethane	[Vinyl trichloride]	$CHCl_2CH_2Cl$	79-00-5	10 ppm			10 ppm (skin)			Central nervous system impairment; liver damage
Trichloroethene	[Trichloroethylene]	C_2HCl_3	79-01-6	10 ppm	25 ppm		100 ppm		200 ppm	Central nervous system impairment; cognitive decrements; renal toxicity
Trichlorofluoromethane	[CFC-11]	CCl_3F	75-69-4			1000 ppm			1000 ppm	Cardiac sensitization
Trichloromethane	[Chloroform]	$CHCl_3$	67-66-3	10 ppm				2 ppm [60-minute]		Liver and embryo/fetal damage; central nervous system impairment
Trichloromethanesulfenyl chloride	[Perchloromethyl mercaptan]	CCl_4S	594-42-3	0.1 ppm			0.1 ppm			Eye and upper respiratory tract irritation
(Trichloromethyl)benzene	[Benzotrichloride]	$C_6H_5CCl_3$	98-07-7			0.1 ppm				Eye, skin, and upper respiratory tract irritation
Trichloronaphthalene (unspecified isomer)	all isomers	$C_{10}H_5Cl_3$	1321-65-9	5 mg/m^3			5 mg/m^3			Liver damage; chloracne
Trichloronitromethane	[Chloropicrin]	CCl_3NO_2	76-06-2	0.1 ppm			0.1 ppm			Eye irritation; pulmonary edema
2,4,5-Trichlorophenoxyacetic acid	[2,4,5-T]	$C_8H_5Cl_3O_3$	93-76-5	10 mg/m^3			10 mg/m^3			Peripheral nervous system impairment
1,2,3-Trichloropropane	[Allyl trichloride]	$C_3H_5Cl_3$	96-18-4	0.005 ppm			10 ppm (skin)			Cancer
1,1,2-Trichloro-1,2,2-trifluoroethane	[CFC-113]	$CFCl_2CF_2Cl$	76-13-1	1000 ppm	1250 ppm		1000 ppm	1250 ppm		Central nervous system impairment
Tri-*o*-cresyl phosphate	[Tri-*o*-tolyl phosphate]	$(C_7H_7)_3PO_4$	78-30-8	0.02 mg/m^3 (inhalable fraction and vapor)			0.1 mg/m^3 (skin)			Neurotoxicity; cholinesterase inhibitor
Tris(2-hydroxyethyl)amine	[Tris(2-hydroxyethyl)amine]	$C_6H_{15}NO_3$	102-71-6	5 mg/m^3						Eye and skin irritation
Triethylamine	[*N,N*-Diethylethanamine]	$(C_2H_5)_3N$	121-44-8	0.5 ppm	1 ppm					Visual impairment; upper respiratory tract irritation

				ACGIH			NIOSH			
Name	**Comments**	**Formula**	**CAS Reg. No.**	**Time-weighted average**	**Short-term exposure limit**	**ACGIH Ceiling**	**Time-weighted average**	**Short-term exposure limit**	**Ceiling**	**Basis**
1,3,5-Triglycidyl-*s*-triazinetrione	[1,3,5-Tris(oxiranemethyl)-1,3,5-triazine-2,4,6(1*H*,3*H*,5*H*)-trione]	$C_{12}H_{15}N_3O_6$	2451-62-9	0.05 mg/m³						Male reproductive damage
Triiodomethane	[Iodoform]	CHI_3	75-47-8	0.2 ppm (inhalable fraction and vapor)			0.6 ppm			Central nervous system and cardiac system impairment; liver and kidney damage
Trimellitic anhydride	[1,2,4-Benzenetricarboxylic anhydride]	$C_9H_4O_5$	552-30-7	0.0005 mg/m³ (inhalable fraction and vapor)	0.002 mg/m³ (inhalable fraction and vapor)		0.005 ppm			Respiratory sensitization
Trimethylamine	[*N,N*-Dimethylmethanamine]	$(CH_3)_3N$	75-50-3	5 ppm	15 ppm		10 ppm	15 ppm		Upper respiratory tract, eye, and skin irritation
1,2,3-Trimethylbenzene	[Hemimellitene]	C_9H_{12}	526-73-8	25 ppm			25 ppm			Central nervous system impairment; asthma; hematologic effects
1,2,4-Trimethylbenzene	[Pseudocumene]	C_9H_{12}	95-63-6	25 ppm			25 ppm			Central nervous system impairment; asthma; hematologic effects
1,3,5-Trimethylbenzene	[Mesitylene]	C_9H_{12}	108-67-8	25 ppm			25 ppm			Central nervous system impairment; asthma; hematologic effects
2,2,3-Trimethylhexane		C_9H_{20}	16747-25-4	200 ppm						Central nervous system impairment
2,2,4-Trimethylhexane		C_9H_{20}	16747-26-5	200 ppm						Central nervous system impairment
2,2,5-Trimethylhexane		C_9H_{20}	3522-94-9	200 ppm						Central nervous system impairment
2,3,3-Trimethylhexane		C_9H_{20}	16747-28-7	200 ppm						Central nervous system impairment
2,3,4-Trimethylhexane		C_9H_{20}	921-47-1	200 ppm						Central nervous system impairment
2,3,5-Trimethylhexane		C_9H_{20}	1069-53-0	200 ppm						Central nervous system impairment
2,4,4-Trimethylhexane		C_9H_{20}	16747-30-1	200 ppm						Central nervous system impairment
3,3,4-Trimethylhexane		C_9H_{20}	16747-31-2	200 ppm						Central nervous system impairment
2,2,3-Trimethylpentane	[2-*tert*-Butylbutane]	C_8H_{18}	564-02-3	300 ppm						Upper respiratory tract irritation
2,2,4-Trimethylpentane	[Isooctane]	C_8H_{18}	540-84-1	300 ppm						Upper respiratory tract irritation
2,3,3-Trimethylpentane		C_8H_{18}	560-21-4	300 ppm						Upper respiratory tract irritation
2,3,4-Trimethylpentane		C_8H_{18}	565-75-3	300 ppm						Upper respiratory tract irritation
Trimethyl phosphite		$C_3H_9O_3P$	121-45-9	2 ppm			2 ppm			Eye irritation; cholinesterase inhibitor
Trinitroglycerol	[Nitroglycerin]	$C_3H_5N_3O_9$	55-63-0	0.05 ppm				0.1 mg/m³		Vasodilation
2,4,6-Trinitrophenol	[Picric acid]	$C_6H_3N_3O_7$	88-89-1	0.1 mg/m³			0.1 mg/m³	0.3 mg/m³ (skin)		Skin sensitization; dermatitis; eye irritation
2,4,6-Trinitrotoluene	[2-Methyl-1,3,5-trinitrobenzene (TNT)]	$C_7H_5N_3O_6$	118-96-7	0.1 mg/m³			0.5 mg/m³ (skin)			Methemoglobinemia; liver damage; cataracts
Triphenyl phosphate		$C_{18}H_{15}O_4P$	115-86-6	3 mg/m³			3 mg/m³			Cholinesterase inhibitor
Tungsten	[Wolfram]; and compounds, as W	W	7440-33-7	3 mg/m³ (respirable particulate matter)			5 mg/m³	10 mg/m³		Lung damage
Turpentine	and selected monoterpenes		8006-64-2	20 ppm			100 ppm			Lung irritation
Uranium	metal & compounds, as U	U	7440-61-1	0.2 mg/m³	0.6 mg/m³		0.2 mg/m³ (insoluble compounds); 0.05 mg/m³ (soluble compounds)	0.6 mg/m³ (insoluble compounds)		Kidney damage
Vanadium(V) oxide	[Vanadium pentoxide]; as V	V_2O_5	1314-62-1	0.05 mg/m³ (inhalable particular matter)					0.05 mg/m³ [15-minute]	Upper respiratory tract and lower respiratory tract irritation
Vinyl acetate		$C_4H_6O_2$	108-05-4	10 ppm	15 ppm				4 ppm [15-minute]	Upper respiratory tract and eye irritation
4-Vinylcyclohexene		C_8H_{12}	100-40-3	0.1 ppm						Female and male reproductive damage
1-Vinyl-2-pyrrolidinone		C_6H_9NO	88-12-0	0.05 ppm						Liver damage

Name	Comments	Formula	CAS Reg. No.	ACGIH Time-weighted average	ACGIH Short-term exposure limit	ACGIH Ceiling	NIOSH Time-weighted average	NIOSH Short-term exposure limit	NIOSH Ceiling	Basis
Warfarin	[Coumadin]	$C_{19}H_{16}O_4$	81-81-2	0.01 mg/m³ (inhalable particulate matter)			0.1 mg/m³			Bleeding; teratogenic
Wood	dust			1 mg/m³ (inhalable particulate matter); (0.5 mg/m³ for western red cedar)			1 mg/m³			Pulmonary function; upper respiratory tract and lower respiratory tract irritation
o-Xylene	[1,2-Dimethylbenzene]	$C_6H_4(CH_3)_2$	95-47-6	100 ppm	150 ppm		100 ppm	150 ppm		Upper respiratory tract and eye irritation; central nervous system impairment
m-Xylene	[1,3-Dimethylbenzene]	$C_6H_4(CH_3)_2$	108-38-3	100 ppm	150 ppm		100 ppm	150 ppm		Upper respiratory tract and eye irritation; central nervous system impairment
p-Xylene	[1,4-Dimethylbenzene]	$C_6H_4(CH_3)_2$	106-42-3	100 ppm	150 ppm		100 ppm	150 ppm		Upper respiratory tract and eye irritation; central nervous system impairment
Yttrium	metal & compounds, as Y	Y	7440-65-5	1 mg/m³			1 mg/m³			Pulmonary fibrosis
Zinc chloride	fume	$ZnCl_2$	7646-85-7	1 mg/m³	2 mg/m³		1 mg/m³	2 mg/m³		Lower respiratory tract and upper respiratory tract irritation
Zinc chromate	as Cr	$ZnCrO_4$	13530-65-9	0.0002 mg/m³ (inhalable particulate matter)			0.0002 mg/m³ [8-hour]			Lung and sinonasal cancer; respiratory tract irritation; asthma
Zinc oxide	[Zincite]	ZnO	1314-13-2	2 mg/m³ (respirable particulate matter)	10 mg/m³ (respirable particulate matter)		5 mg/m³	10 mg/m³ (fume)	15 mg/m³ (dust)	Metal fume fever
Zinc stearate	[Zinc octadecanoate]	$Zn(C_{18}H_{35}O_2)_2$	557-05-1	10 mg/m³ (inhalable particulate matter)			10 mg/m³ (total); 5 mg/m³ (respirable)			Lower respiratory tract irritation
Zirconium	metal & compounds, as Zr	Zr	7440-67-7	5 mg/m³	10 mg/m³		5 mg/m³	10 mg/m³		Respiratory irritation

CHEMICAL CARCINOGENS

Maria J. Doa

The following chemical substances are listed in the *14th Report on Carcinogens*, 2016, released by the National Institute of Environmental Health Sciences (NIEHS) under the National Toxicology Program (NTP) (Ref. 1); or in the World Health Organization's International Research on Cancer (IARC) Monographs on the Identification of Carcinogenic Hazards to Humans (Volumes 1–124) (Ref. 2); or in both.

Substances are grouped in two classes as defined in the NTP report:

- **Known to be human carcinogens:** There is sufficient evidence of cancer from studies in humans showing a cause-and-effect relationship between exposure to the substance and human cancer.
- **Reasonably anticipated to be human carcinogens:** There is limited evidence of carcinogenicity from studies in humans, which indicates that causal interpretation is credible, but that alternative explanations, such as chance, bias, or confounding factors, could not adequately be excluded,

  OR

  there is sufficient evidence of carcinogenicity from studies in experimental animals, which indicates there is an increased incidence of malignant and/or a combination of malignant and benign tumors (1) in multiple species or at multiple tissue sites, or (2) by multiple routes of exposure, or (3) to an unusual degree with regard to incidence, site, or type of tumor, or age at onset,

  OR

  there is less than sufficient evidence of carcinogenicity in humans or laboratory animals; however, the agent, substance, or mixture belongs to a well-defined, structurally related class of substances whose members are listed in a previous Report on Carcinogens as either known to be a human carcinogen or reasonably anticipated to be a human carcinogen, or there is convincing relevant information that the agent acts through mechanisms indicating it would likely cause cancer in humans.

  Conclusions regarding carcinogenicity in humans or experimental animals are based on scientific judgment, with consideration given to all relevant information. Relevant information includes, but is not limited to, dose response, route of exposure, chemical structure, metabolism, pharmacokinetics, sensitive sub-populations, genetic effects, or other data relating to mechanism of action or factors that may be unique to a given substance. For example, there may be substances for which there is evidence of carcinogenicity in laboratory animals, but there are compelling data indicating that the agent acts through mechanisms which do not operate in humans and would therefore not reasonably be anticipated to cause cancer in humans.

IARC classifies substances as:

- **Group 1—Carcinogenic to humans**: This category applies whenever there is sufficient evidence of carcinogenicity humans. In addition, this category may apply when there is both strong evidence in exposed humans that the agent exhibits key characteristics of carcinogens and sufficient evidence of carcinogenicity in experimental animals.
- **Group 2A—Probably carcinogenic to humans:** This category generally applies when at least two of the following evaluations have been made, including at least one that involves either exposed humans or human cells or tissues:
  - Limited evidence of carcinogenicity in humans,
  - Sufficient evidence of carcinogenicity in experimental animals,
  - Strong evidence that the agent exhibits key characteristics of carcinogens.

If there is inadequate evidence regarding carcinogenicity in humans, there should be strong evidence in human cells or tissues that the agent exhibits key characteristics of carcinogens. If there is limited evidence of carcinogenicity in humans, then the second individual evaluation may be from experimental systems (i.e., sufficient evidence of carcinogenicity in experimental animals or strong evidence in experimental systems that the agent exhibits key characteristics of carcinogens).

- **Group 2B—Possibly carcinogenic to humans**
- **Group 3—Not classifiable as to its carcinogenicity to humans**

Only substances classified by IARC as Groups 1 and 2A are included in this list.

The NTP report and IARC monographs also list many poorly defined materials such as soots, tars, mineral oils, and coke oven emissions, as well as viruses, sunlight, ionizing radiation, etc. These substances are not included here.

Additional substances are evaluated by NTP on an ongoing basis but are not added to the list below until issued in its next formal report. See details in Ref. 1.

The table is ordered by the substance name normally used in the *CRC Handbook*. In many cases the primary name given here is different from that used in the NTP report and/or the IARC monographs; however, names used in the NTP report and/or the IARC monographs appear in the Synonym or Comments columns. Additional information and acronyms for the compounds are given in the Comments column. The substances classified by NTP as known carcinogens and those classified by IARC as Class 1 Carcinogenic to Humans are listed first.

Extensive details on each substance and of the classification approaches are given in the references.

References

1. *14th Report on Carcinogens*, <http://ntp.niehs.nih.gov/pubhealth/roc/index-1.html>, 2019.
2. IARC Monographs on the Identification of Carcinogenic Hazards to Humans, Vols. 1–124, <https://monographs.iarc.fr/agents-classified-by-the-iarc/>, 2019.

Safety

Chemicals Classified by NTP and IARC

Name	Synonym	Mol. form.	CAS Reg. No.	Comments	NTP Classification	IARC Classification
NTP Known to Be Human Carcinogens and/or IARC Group 1						
Aflatoxins			1402-68-2		Known to be human carcinogens	Group 1
4-Aminobiphenyl	*p*-Biphenylamine	$C_{12}H_{11}N$	92-67-1		Known to be human carcinogens	Group 1
Aristolochic acids					Known to be human carcinogens	Group 1
Arsenic		As	7440-38-2	and inorganic arsenic compounds	Known to be human carcinogens	Group 1
Asbestos			1332-21-4	IARC: all forms, including actinolite, amosite, anthophyllite, chrysotile, crocidolite, tremolite)	Known to be human carcinogens	Group 1
Azathioprine	6-[(1-Methyl-4-nitro-1*H*-imidazol-5-yl) thio]-1*H*-purine	$C_9H_7N_7O_2S$	446-86-6		Known to be human carcinogens	Group 1
Benzene	[6]Annulene	C_6H_6	71-43-2		Known to be human carcinogens	Group 1
p-Benzidine	[1,1'-Biphenyl]-4,4'-diamine	$C_{12}H_{12}N_2$	92-87-5	includes dyes metabolized to benzidene	Known to be human carcinogens	Group 1
Benzo[*a*]pyrene	2,3-Benzopyrene	$C_{20}H_{12}$	50-32-8		Reasonably anticipated to be human carcinogens	Group 1
Beryllium	Glucinium	Be	7440-41-7	and beryllium compounds	Known to be human carcinogens	Group 1
Bis(2-chloroethyl) sulfide	Mustard gas	$C_4H_8Cl_2S$	505-60-2	listed by NTP as mustard gas; listed by IARC as sulfur mustard	Known to be human carcinogens	Group 1
Bis(chloromethyl) ether	Chloromethyl ether	$C_2H_4Cl_2O$	542-88-1	and technical grade chloromethyl methyl ether	Known to be human carcinogens	Group 1
1,3-Butadiene	Divinyl	C_4H_6	106-99-0		Known to be human carcinogens	Group 1
1,4-Butanediol dimethylsulfonate	Busulfan	$C_6H_{14}O_6S_2$	55-98-1		Reasonably anticipated to be human carcinogens	Group 1
Cadmium		Cd	7440-43-9	and cadmium compounds	Known to be human carcinogens	Group 1
Chlorambucil		$C_{14}H_{19}Cl_2NO_2$	305-03-3		Known to be human carcinogens	Group 1
Chloroethene	Vinyl chloride	C_2H_3Cl	75-01-4		Known to be human carcinogens	Group 1
1-(2-Chloroethyl)-3-(4-methylcyclohexyl)-1-nitrosourea	Semustine	$C_{10}H_{18}ClN_3O_2$	13909-09-6	also known as MeCCNU	Known to be human carcinogens	Group 1
Chloromethyl methyl ether		C_2H_5ClO	107-30-2		Known to be human carcinogens	Group 1
Chromium		Cr	7440-47-3	hexavalent compounds only	Known to be human carcinogens	Group 1
Cyclophosphamide	Cyclophosphane	$C_7H_{15}Cl_2N_2O_2P$	50-18-0		Known to be human carcinogens	Group 1
Cyclosporin A	Cyclosporine		59865-13-3		Known to be human carcinogens	Group 1
1,2-Dichloropropane, (±)-	Propylene dichloride	$C_3H_6Cl_2$	78-87-5			Group 1
trans-Diethylstilbestrol		$C_{18}H_{20}O_2$	56-53-1		Known to be human carcinogens	Group 1
Erionite			66733-21-9		Known to be human carcinogen	Group 1
Estrogens, steroidal					Known to be human carcinogens	Group 1
N-(4-Ethoxyphenyl)acetamide	Phenacetin	$C_{10}H_{13}NO_2$	62-44-2	listed by IARC as phenacetin; listed by NTP as phenacetin and analgesic mixtures containing phenacetin	Known to be human carcinogens	Group 1
Formaldehyde	Methanal	CH_2O	50-00-0		Known to be human carcinogens	Group 1
1,2,3,4,5,6-Hexachlorocyclohexane, (1α,2α,3β,4α,5α,6β)	Lindane	$C_6H_6Cl_6$	58-89-9	also known as γ-Hexachlorocyclohexane	Reasonably anticipated to be human carcinogens	Group 1
Melphalan	L-Phenylalanine,4-[bis(2-chloroethyl)amino]-	$C_{13}H_{18}Cl_2N_2O_2$	148-82-3		Known to be human carcinogens	Group 1
Methoxsalen	9-Methoxy-7*H*-furo[3,2-*g*][1] benzopyran-7-one	$C_{12}H_8O_4$	298-81-7	with UV-A therapy (PUVA)	Known to be human carcinogens	Group 1
2-Methylaniline	*o*-Toluidine	C_7H_9N	95-53-4	also the hydrochloride	Known to be human carcinogens	Group 1

Name	Synonym	Mol. form.	CAS Reg. No.	Comments	NTP Classification	IARC Classification
4,4'-Methylenebis[2-chloroaniline]	3,3'-Dichloro-4,4'-diaminodiphenylmethane	$C_{13}H_{12}Cl_2N_2$	101-14-4	also known as MBOCA	Reasonably anticipated to be human carcinogens	Group 1
2-Naphthylamine	β-Naphthylamine	$C_{10}H_9N$	91-59-8		Known to be human carcinogens	Group 1
Nickel compounds				both inorganic and organic compounds	Known to be human carcinogens	Group 1
4-(*N*-Nitrosomethylamino)-1-(3-pyridyl)-1-butanone	Ketone, 3-pyridyl-3-(*N*-methyl-*N*-nitrosamino)propyl	$C_{10}H_{13}N_3O_2$	64091-91-4	also known as NNK	Reasonably anticipated to be human carcinogens	Group 1
N-Nitrosonornicotine	*N*'-Nitroso-3-(2-pyrrolidinyl)pyridine	$C_9H_{11}N_3O$	16543-55-8		Reasonably anticipated to be human carcinogens	Group 1
Oxirane	Ethylene oxide	C_2H_4O	75-21-8		Known to be human carcinogens	Group 1
Pentachlorophenol		C_6HCl_5O	87-86-5	and synthesis byproducts	Reasonably anticipated to be human carcinogens	Group 1
Plutonium		Pu	7440-07-5			Group 1
Polychlorinated biphenyls	PCBs		1336-36-3	IARC dioxin-like	Reasonably anticipated to be human carcinogens	Group 1
Radium		Ra	7440-14-4	Radium-224 and its decay products: Radium-226 and its decay products; Radium-226 and its decay products		Group 1
Radon		Rn	10043-92-2	source of ionizing radiation	Known to be human carcinogens	
Silicon dioxide (α-quartz)	Silica	O_2Si	14808-60-7	respirable size	Known to be human carcinogens	Group 1
Silicon dioxide (tridymite)	Tridymite	O_2Si	15468-32-3	respirable size	Known to be human carcinogens	
Silicon dioxide (cristobalite)	Cristobalite	O_2Si	14464-46-1	respirable size	Known to be human carcinogens	Group 1
Sulfuric acid	Oil of vitriol	H_2O_4S	7664-93-9	in strong acid mists	Known to be human carcinogens	Group 1
Tamoxifen		$C_{26}H_{29}NO$	10540-29-1		Known to be human carcinogens	Group 1
2,3,7,8-Tetrachlorodibenzo-*p*-dioxin	Dioxin	$C_{12}H_4Cl_4O_2$	1746-01-6	also known as TCDD	Known to be human carcingnes	Group 1
Thorium(IV) oxide	Thoria	O_2Th	1314-20-1	source of ionizing radiation	Known to be human carcinogens	
Thorium		Th	7440-29-1	Thorium-232 and its decay products		Group 1
Trichloroethene	Trichloroethylene	C_2HCl_3	79-01-6	also known as TCE	Known to be human carcinogens	Group 1
Triethylenethiophosphoramide	Thiotepa	$C_6H_{12}N_3PS$	52-24-4		Known to be human carcinogens	Group 1
NTP Reasonably Anticipated to Be Human Carcinogens and/or IARC Group 2A						
Acetaldehyde	Ethanal	C_2H_4O	75-07-0		Reasonably anticipated to be human carcinogens	Group 2B
2-(Acetylamino)fluorene		$C_{15}H_{13}NO$	53-96-3		Reasonably anticipated to be human carcinogens	
Acrylamide	2-Propenamide	C_3H_5NO	79-06-1		Reasonably anticipated to be human carcinogens	Group 2A
Acrylonitrile	Propenenitrile	C_3H_3N	107-13-1		Reasonably anticipated to be human carcinogens	Group 2B
4-Allyl-1,2-dimethoxybenzene	Methyleugenol	$C_{11}H_{14}O_2$	93-15-2		Reasonably anticipated to be human carcinogens	Group 2B
2-Amino-9,10-anthracenedione	2-Aminoanthraquinone	$C_{14}H_9NO_2$	117-79-3		Reasonably anticipated to be human carcinogens	Group 3
1-Amino-2,4-dibromo-9,10-anthracenedione	1-Amino-2,4-dibromoanthraquinone	$C_{14}H_7Br_2NO_2$	81-49-2		Reasonably anticipated to be human carcinogens	Group 2B
2-Amino-3,4-dimethylimidazo[4,5-*f*]quinoline	Me-IQ	$C_{12}H_{12}N_4$	77094-11-2		Reasonably anticipated to be human carcinogens	Group 2B

Safety

Name	Synonym	Mol. form.	CAS Reg. No.	Comments	NTP Classification	IARC Classification
2-Amino-3,8-dimethylimidazo[4,5-*f*] quinoxaline	MeIQx	$C_{11}H_{11}N_5$	77500-04-0		Reasonably anticipated to be human carcinogens	Group 2B
1-Amino-2-methyl-9,10-anthracenedione	1-Amino-2-methylanthraquinone	$C_{15}H_{11}NO_2$	82-28-0		Reasonably anticipated to be human carcinogens	Group 3
2-Amino-3-methyl-3*H*-imidazo(4,5-*f*) quinoline	IQ	$C_{11}H_{10}N_4$	76180-96-6		Reasonably anticipated to be human carcinogens	Group 2A
2-Amino-1-methyl-6-phenylimidazo[4,5-*b*]pyridine	PhIP	$C_{13}H_{12}N_4$	105650-23-5	PhIP	Reasonably anticipated to be human carcinogens	Group 2B
Azacitidine	4-Amino-1-β-D-ribofuranosyl-1,3,5-triazine-2(1*H*)-one	$C_8H_{12}N_4O_5$	320-67-2		Reasonably anticipated to be human carcinogens	Group 2A
Benz[*a*]anthracene	1,2-Benzanthracene	$C_{18}H_{12}$	56-55-3		Reasonably anticipated to be human carcinogens	Group 2B
Benzo[*b*]fluoranthene	Benz[*e*]acephenanthrylene	$C_{20}H_{12}$	205-99-2		Reasonably anticipated to be human carcinogens	Group 2B
Benzo[*j*]fluoranthene	Dibenzo[*a,jk*]fluorene	$C_{20}H_{12}$	205-82-3		Reasonably anticipated to be human carcinogens	Group 2B
Benzo[*k*]fluoranthene	2,3,1',8'-Binaphthylene	$C_{20}H_{12}$	207-08-9		Reasonably anticipated to be human carcinogens	Group 2B
2(3*H*)-Benzothiazolethione	2-Mercaptobenzothiazole	$C_7H_5NS_2$	149-30-4			Group 2A
2,2'-Bioxirane	Diepoxybutane	$C_4H_6O_2$	1464-53-5		Reasonably anticipated to be human carcinogens	
2,2-Bis(bromomethyl)-1,3-propanediol	Pentaerythritol dibromide	$C_5H_{10}Br_2O_2$	3296-90-0		Reasonably anticipated to be human carcinogens	Group 2B
2-Chloro-*N*-(2-chloroethyl)-*N*-methylethanamine	Mechlorethamine	$C_5H_{11}Cl_2N$	51-75-2	Nitrogen mustard		Group 2A
Bis(2-chloroethyl)methylamine hydrochloride	Nitrogen mustard hydrochloride	$C_5H_{12}Cl_3N$	55-86-7	also known as Mechlorethamine	Reasonably anticipated to be human carcinogens	
N,N'-Bis(2-chloroethyl)-*N*-nitrosourea	Carmustine	$C_5H_9Cl_2N_3O_2$	154-93-8	also known as BCNU	Reasonably anticipated to be human carcinogens	Group 2A
Bis[4-(dimethylamino)phenyl]methane	Michler's Base	$C_{17}H_{22}N_2$	101-61-1		Reasonably anticipated to be human carcinogens	Group 2B
1,3-Bis(2,3-epoxypropoxy)benzene	Diglycidyl resorcinol ether	$C_{12}H_{14}O_4$	101-90-6		Reasonably anticipated to be human carcinogens	Group 2B
Bis(2-ethylhexyl) phthalate	Di-*sec*-octyl phthalate	$C_{24}H_{38}O_4$	117-81-7	also known as DEHP	Reasonably anticipated to be human carcinogens	Group 2B
Bromodichloromethane		$CHBrCl_2$	75-27-4		Reasonably anticipated to be human carcinogens	Group 2B
Bromoethene	Vinyl bromide	C_2H_3Br	593-60-2		Reasonably anticipated to be human carcinogens	Group 2A
1-Bromopropane	Propyl bromide	C_3H_7Br	106-94-5		Reasonably anticipated to be human carcinogens	Group 2B
tert-Butyl-4-hydroxyanisole	Butylated hydroxyanisole	$C_{11}H_{16}O_2$	25013-16-5		Reasonably anticipated to be human carcinogens	Group 2B
Captafol	Difolatan	$C_{10}H_9Cl_4NO_2S$	2425-06-1		Reasonably anticipated to be human carcinogens	Group 2A
Chloral hydrate		$C_2H_3Cl_3O_2$	302-17-0	2,2,2-Trichloroethane-1,1-diol		Group 2A
Chloramphenicol		$C_{11}H_{12}Cl_2N_2O_5$	56-75-7		Reasonably anticipated to be human carcinogens	Group 2A
Chlorendic acid	1,4,5,6,7,7-Hexachloro-5-norbornene-2,3-dicarboxylic acid	$C_9H_4Cl_6O_4$	115-28-6		Reasonably anticipated to be human carcinogens	Group 2B

Safety

Name	Synonym	Mol. form.	CAS Reg. No.	Comments	NTP Classification	IARC Classification
Chlorinated paraffins (C_{12}, 60% Cl)			108171-26-2		Reasonably anticipated to be human carcinogens	Group 2B
4-Chloro-1,2-benzenediamine	4-Chloro-*o*-phenylenediamine	$C_6H_7ClN_2$	95-83-0		Reasonably anticipated to be human carcinogens	Group 2B
2-Chloro-1,3-butadiene	Chloroprene	C_4H_5Cl	126-99-8		Reasonably anticipated to be human carcinogens	Group 2B
1-(2-Chloroethyl)-3-cyclohexyl-1-nitrosourea	Lomustine	$C_9H_{16}ClN_3O_2$	13010-47-4	also known as CCNU and Belustine	Reasonably anticipated to be human carcinogens	Group 2A
4-Chloro-2-methylaniline	*p*-Chloro-*o*-toluidine	C_7H_8ClN	95-69-2	also the hydrochloride	Reasonably anticipated to be human carcinogens	Group 2A
1-Chloro-2-methylpropene	Dimethylvinyl chloride	C_4H_7Cl	513-37-1		Reasonably anticipated to be human carcinogens	Group 2B
3-Chloro-2-methylpropene		C_4H_7Cl	563-47-3		Reasonably anticipated to be human carcinogens	Group 2B
Chlorozotocin	2-[[[(2-Chloroethyl)nitrosoamino] carbonyl]amino]-2-deoxy-D-glucose	$C_9H_{16}ClN_3O_7$	54749-90-5		Reasonably anticipated to be human carcinogens	Group 2A
Cobalt		Co	7440-48-4	and Co compounds that release Co ions *in vivo*	Reasonably anticipated to be human carcinogens	Group 2B
Cobalt-tungsten carbide	Co/WC			powders and hard metal	Reasonably anticipated to be human carcinogens	Group 2A
Cupferron		$C_6H_9N_3O_2$	135-20-6		Reasonably anticipated to be human carcinogens	
Dacarbazine	5-(3,3-Dimethyl-1-triazenyl)-1*H*-imidazole-4-carboxamide	$C_6H_{10}N_6O$	4342-03-4		Reasonably anticipated to be human carcinogens	Group 2B
Decabromobiphenyl		$C_{12}Br_{10}$	13654-09-6		Reasonably anticipated to be human carcinogens	
2,4-Diaminoanisole sulfate	1,3-Benzenediamine, 4-methoxy, sulfate	$C_7H_{12}N_2O_5S$	39156-41-7	IARC classified the free amine	Reasonably anticipated to be human carcinogens	Group 2B
4,4'-Diaminodiphenyl ether	4,4'-Oxydianiline	$C_{12}H_{12}N_2O$	101-80-4		Reasonably anticipated to be human carcinogens	Group 2B
4,4'-Diaminodiphenylmethane	4,4'-Methylenedianiline	$C_{13}H_{14}N_2$	101-77-9	also the dihydrochloride	Reasonably anticipated to be human carcinogens	Group 2B
4,4'-Diaminodiphenylmethane dihydrochloride	4,4'-Methylenedianiline dihydrochloride	$C_{13}H_{16}Cl_2N_2$	13552-44-8	IARC classified the free amine	Reasonably anticipated to be human carcinogens	Group 2B
4,4'-Diaminodiphenyl sulfide	4,4'-Thiodianiline	$C_{12}H_{12}N_2S$	139-65-1		Reasonably anticipated to be human carcinogens	Group 2B
cis-Diamminedichloroplatinum	Cisplatin	$Cl_2H_6N_2Pt$	15663-27-1		Reasonably anticipated to be human carcinogens	Group 2A
Diazinon		$C_{12}H_{21}N_2O_3PS$	333-41-5			Group 2A
Dibenz[*a,h*]acridine		$C_{21}H_{13}N$	226-36-8		Reasonably anticipated to be human carcinogens	Group 2B
Dibenz[*a,j*]acridine	7-Azadibenz[*a,j*]anthracene	$C_{21}H_{13}N$	224-42-0		Reasonably anticipated to be human carcinogens	Group 2A
Dibenz[*a,h*]anthracene	1,2:5,6-Dibenzanthracene	$C_{22}H_{14}$	53-70-3		Reasonably anticipated to be human carcinogens	Group 2A
7*H*-Dibenzo[*c,g*]carbazole		$C_{20}H_{13}N$	194-59-2		Reasonably anticipated to be human carcinogens	Group 2B
Dibenzo[*a,e*]pyrene	Naphtho[1,2,3,4-*def*]chrysene	$C_{24}H_{14}$	192-65-4		Reasonably anticipated to be human carcinogens	Group 3
Dibenzo[*a,h*]pyrene	Dibenzo[*b,def*]chrysene	$C_{24}H_{14}$	189-64-0		Reasonably anticipated to be human carcinogens	Group 2B

Name	Synonym	Mol. form.	CAS Reg. No.	Comments	NTP Classification	IARC Classification
Dibenzo[*a,i*]pyrene	Benzo[*rst*]pentaphene	$C_{24}H_{14}$	189-55-9		Reasonably anticipated to be human carcinogens	Group 2B
Dibenzo[*a,l*]pyrene	Dibenzo[*def,p*]chrysene	$C_{24}H_{14}$	191-30-0		Reasonably anticipated to be human carcinogens	Group 2A
1,2-Dibromo-3-chloropropane		$C_3H_5Br_2Cl$	96-12-8		Reasonably anticipated to be human carcinogens	Group 2B
1,2-Dibromoethane	Ethylene dibromide	$C_2H_4Br_2$	106-93-4		Reasonably anticipated to be human carcinogens	Group 2A
2,3-Dibromo-1-propanol	DBP	$C_3H_6Br_2O$	96-13-9		Reasonably anticipated to be human carcinogens	Group 2B
2,3-Dibromo-1-propanol, phosphate (3:1)	Tris(2,3-dibromopropyl) phosphate	$C_9H_{15}Br_6O_4P$	126-72-7		Reasonably anticipated to be human carcinogens	Group 2A
p-Dichlorobenzene	1,4-Dichlorobenzene	$C_6H_4Cl_2$	106-46-7		Reasonably anticipated to be human carcinogens	Group 2B
3,3'-Dichloro-*p*-benzidine	3,3'-Dichloro[1,1'-biphenyl]-4,4'-diamine	$C_{12}H_{10}Cl_2N_2$	91-94-1		Reasonably anticipated to be human carcinogens	Group 2B
3,3'-Dichloro-*p*-benzidine dihydrochloride	3,3'-Dichloro-[1,1'-biphenyl]-4,4'-diamine dihydrochloride	$C_{12}H_{12}Cl_4N_2$	612-83-9		Reasonably anticipated to be human carcinogens	
1,2-Dichloroethane	Ethylene dichloride	$C_2H_4Cl_2$	107-06-2		Reasonably anticipated to be human carcinogens	Group 2B
Dichloromethane	Methylene chloride	CH_2Cl_2	75-09-2		Reasonably anticipated to be human carcinogens	Group 2A
1,3-Dichloropropene (unspecified isomer)		$C_3H_4Cl_2$	542-75-6	technical grade	Reasonably anticipated to be human carcinogens	Group 2B
Diethyl sulfate		$C_4H_{10}O_4S$	64-67-5		Reasonably anticipated to be human carcinogens	Group 2A
2,3-Dihydro-6-propyl-2-thioxo-4(1*H*)-pyrimidinone	Propylthiouracil	$C_7H_{10}N_2OS$	51-52-5	also known as PROP	Reasonably anticipated to be human carcinogens	Group 2B
1,8-Dihydroxy-9,10-anthracenedione	Danthron	$C_{14}H_8O_4$	117-10-2		Reasonably anticipated to be human carcinogens	Group 2B
3,3'-Dimethoxybenzidine	Dianisidine	$C_{14}H_{16}N_2O_2$	119-90-4	and dyes metabolized to 3,3'-Dimethoxybenzidine	Reasonably anticipated to be human carcinogens	Group 2B
4-(Dimethylamino)azobenzene	Methyl Yellow	$C_{14}H_{15}N_3$	60-11-7		Reasonably anticipated to be human carcinogens	Group 2B
2',3-Dimethyl-4-aminoazobenzene	4-*o*-Tolylazo-*o*-toluidine	$C_{14}H_{15}N_3$	97-56-3	also known as *o*-Aminoazotoluene	Reasonably anticipated to be human carcinogens	Group 2B
Dimethylcarbamic chloride	Dimethylcarbamoyl chloride	C_3H_6ClNO	79-44-7		Reasonably anticipated to be human carcinogens	Group 2A
1,1-Dimethylhydrazine	UDMH	$C_2H_8N_2$	57-14-7		Reasonably anticipated to be human carcinogens	Group 2A
Dimethyl sulfate		$C_2H_6O_4S$	77-78-1		Reasonably anticipated to be human carcinogens	Group 2A
1,6-Dinitropyrene		$C_{16}H_8N_2O_4$	42397-64-8		Reasonably anticipated to be human carcinogens	Group 2B
1,8-Dinitropyrene		$C_{16}H_8N_2O_4$	42397-65-9		Reasonably anticipated to be human carcinogens	Group 2B
1,4-Dioxane	1,4-Dioxacyclohexane	$C_4H_8O_2$	123-91-1		Reasonably anticipated to be human carcinogens	Group 2B
1,2-Diphenylhydrazine	Hydrazobenzene	$C_{12}H_{12}N_2$	122-66-7		Reasonably anticipated to be human carcinogens	

Name	Synonym	Mol. form.	CAS Reg. No.	Comments	NTP Classification	IARC Classification
1,3-Diphenyl-1-triazene	Diazoaminobenzene	$C_{12}H_{11}N_3$	136-35-6		Reasonably anticipated to be human carcinogens	
Disperse Blue No. 1	1,4,5,8-Tetraamino-9,10-anthracenedione	$C_{14}H_{12}N_4O_2$	2475-45-8		Reasonably anticipated to be human carcinogens	Group 2B
Doxorubicin hydrochloride	Adriamycin	$C_{27}H_{30}ClNO_{11}$	25316-40-9		Reasonably anticipated to be human carcinogens	Group 2A
Epichlorohydrin	(Chloromethyl)oxirane	C_3H_5ClO	13403-37-7		Reasonably anticipated to be human carcinogens	Group 2A
1,2-Epoxy-4-(epoxyethyl)cyclohexane	4-Vinyl-1-cyclohexene dioxide	$C_8H_{12}O_2$	106-87-6		Reasonably anticipated to be human carcinogens	
Ethyl carbamate	Urethane	$C_3H_7NO_2$	51-79-6		Reasonably anticipated to be human carcinogens	Group 2A
Ethyl methanesulfonate		$C_3H_8O_3S$	62-50-0		Reasonably anticipated to be human carcinogens	Group 2B
N-Ethyl-*N*-nitrosourea	*N*-Nitroso-*N*-ethylurea	$C_3H_7N_3O_2$	759-73-9	also known as ENU	Reasonably anticipated to be human carcinogens	Group 2A
Fluoroethene	Vinyl fluoride	C_2H_3F	75-02-5		Reasonably anticipated to be human carcinogens	Group 2A
Fuchsin	C.I. Basic Red 9, monohydrochloride	$C_{19}H_{18}ClN_3$	569-61-9		Reasonably anticipated to be human carcinogens	Group 2B
Furan	Oxacyclopentadiene	C_4H_4O	110-00-9		Reasonably anticipated to be human carcinogens	Group 2B
Glass wool fibers (inhalable)				certain types	Reasonably anticipated to be human carcinogens	Group 3
Glyphosate	*N*-(Phosphonomethyl)glycine	$C_3H_8NO_5P$	1071-83-6			Group 2A
Hexabromobiphenyl (unspecified isomer)	Firemaster FF-1	$C_{12}H_4Br_6$	67774-32-7		Reasonably anticipated to be human carcinogens	
Hexachlorobenzene	Perchlorobenzene	C_6Cl_6	118-74-1		Reasonably anticipated to be human carcinogens	Group 2B
1,2,3,4,5,6-Hexachlorocyclohexane, (1α,2α,3β,4α,5β,6β)	α-Hexachlorocyclohexane	$C_6H_6Cl_6$	319-84-6		Reasonably anticipated to be human carcinogens	Group 2B
1,2,3,4,5,6-Hexachlorocyclohexane, (1α,2β,3α,4β,5α,6β)	β-Hexachlorocyclohexane	$C_6H_6Cl_6$	319-85-7		Reasonably anticipated to be human carcinogens	Group 2B
Hexachlorocyclohexane (unspecified isomer)		$C_6H_6Cl_6$	608-73-1	all isomers	Reasonably anticipated to be human carcinogens	Group 2B
Hexachloroethane	Perchloroethane	C_2Cl_6	67-72-1		Reasonably anticipated to be human carcinogens	Group 2B
Hexamethylphosphoric triamide	Tris(dimethylamino)phosphine oxide	$C_6H_{18}N_3OP$	680-31-9	also known as HMPA	Reasonably anticipated to be human carcinogens	Group 2B
Hydrazine		H_4N_2	302-01-2		Reasonably anticipated to be human carcinogens	Group 2A
Hydrazine sulfate		$H_6N_2O_4S$	10034-93-2		Reasonably anticipated to be human carcinogens	
2-Imidazolidinethione	Ethylene thiourea	$C_3H_6N_2S$	96-45-7		Reasonably anticipated to be human carcinogens	Group 3
Indeno[1,2,3-*cd*]pyrene	1,10-(1,2-Phenylene)pyrene	$C_{22}H_{12}$	193-39-5		Reasonably anticipated to be human carcinogens	Group 2B
Indium phosphide		InP	22398-80-7			Group 2A
Isopropylbenzene	Cumene	C_9H_{12}	98-82-8		Reasonably anticipated to be human carcinogens	Group 2B

Name	Synonym	Mol. form.	CAS Reg. No.	Comments	NTP Classification	IARC Classification
Kepone	Chlordecone	$C_{10}Cl_{10}O$	143-50-0		Reasonably anticipated to be human carcinogens	Group 2B
Lead		Pb	7439-92-1	and lead compounds	Reasonably anticipated to be human carcinogens	Group 2B
Malathion		$C_{10}H_{19}O_6PS_2$	121-75-5			Group 2A
2-Methoxyaniline	*o*-Anisidine	C_7H_9NO	90-04-0	also the hydrochloride	Reasonably anticipated to be human carcinogens	Group 2B
2-Methoxy-5-methylaniline	5-Methyl-*o*-anisidine	$C_8H_{11}NO$	120-71-8	also known as *p*-Cresidine	Reasonably anticipated to be human carcinogens	Group 2B
5-Methoxypsoralen	Bergaptene	$C_{12}H_8O_4$	484-20-8			Group 2A
4-Methyl-1,3-benzenediamine	Toluene-2,4-diamine	$C_7H_{10}N_2$	95-80-7	also known as 2,4-Diaminotoluene	Reasonably anticipated to be human carcinogens	Group 2B
2-Methyl-1,3-butadiene	Isoprene	C_5H_8	78-79-5		Reasonably anticipated to be human carcinogens	Group 2B
5-Methylchrysene		$C_{19}H_{14}$	3697-24-3	also known as 5-MC	Reasonably anticipated to be human carcinogens	Group 2B
Methyl methanesulfonate		$C_2H_6O_3S$	66-27-3		Reasonably anticipated to be human carcinogens	Group 2A
N-Methyl-*N'*-nitro-*N*-nitrosoguanidine		$C_2H_5N_5O_3$	70-25-7		Reasonably anticipated to be human carcinogens	Group 2A
N-Methyl-*N*-nitrosourea	*N*-Nitroso-*N*-methylurea	$C_2H_5N_3O_2$	684-93-5		Reasonably anticipated to be human carcinogens	Group 2A
Methyloxirane	1,2-Propylene oxide	C_3H_6O	75-56-9		Reasonably anticipated to be human carcinogens	Group 2B
Metronidazole	2-Methyl-5-nitro-1*H*-imidazole-1-ethanol	$C_6H_9N_3O_3$	443-48-1		Reasonably anticipated to be human carcinogens	Group 2B
Mirex	Hexachloropentadiene dimer	$C_{10}Cl_{12}$	2385-85-5		Reasonably anticipated to be human carcinogens	Group 2B
Naphthalene		$C_{10}H_8$	91-20-3		Reasonably anticipated to be human carcinogens	Group 2B
Nickel		Ni	7440-02-0	metallic (nickel compounds are known human carcinogens)	Reasonably anticipated to be human carcinogens	Group 2B
Nitrilotriacetic acid	*N,N*-Bis(carboxymethyl)glycine	$C_6H_9NO_6$	139-13-9		Reasonably anticipated to be human carcinogens	Group 2B
2-Nitroanisole	1-Methoxy-2-nitrobenzene	$C_7H_7NO_3$	91-23-6		Reasonably anticipated to be human carcinogens	Group 2B
Nitrobenzene		$C_6H_5NO_2$	98-95-3		Reasonably anticipated to be human carcinogens	Group 2B
6-Nitrochrysene		$C_{18}H_{11}NO_2$	7496-02-8		Reasonably anticipated to be human carcinogens	Group 2A
Nitrofen	2,4-Dichloro-1-(4-nitrophenoxy)benzene	$C_{12}H_7Cl_2NO_3$	1836-75-5		Reasonably anticipated to be human carcinogens	Group 2B
Nitromethane	Nitrocarbol	CH_3NO_2	75-52-5		Reasonably anticipated to be human carcinogens	Group 2B
2-Nitropropane	Isonitropropane	$C_3H_7NO_2$	79-46-9		Reasonably anticipated to be human carcinogens	Group 2B
1-Nitropyrene		$C_{16}H_9NO_2$	5522-43-0		Reasonably anticipated to be human carcinogens	Group 2A
4-Nitropyrene		$C_{16}H_9NO_2$	57835-92-4		Reasonably anticipated to be human carcinogens	Group 2B

Name	Synonym	Mol. form.	CAS Reg. No.	Comments	NTP Classification	IARC Classification
N-Nitrosodibutylamine	Dibutylnitrosamine	$C_8H_{18}N_2O$	924-16-3		Reasonably anticipated to be human carcinogens	Group 2B
N-Nitrosodiethanolamine	2,2'-(Nitrosoimino)ethanol	$C_4H_{10}N_2O_3$	1116-54-7		Reasonably anticipated to be human carcinogens	Group 2B
N-Nitrosodiethylamine	Diethylnitrosamine	$C_4H_{10}N_2O$	55-18-5	also known as DEN	Reasonably anticipated to be human carcinogens	Group 2A
N-Nitrosodimethylamine	Dimethylnitrosamine	$C_2H_6N_2O$	62-75-9	also known as DMN	Reasonably anticipated to be human carcinogens	Group 2A
N-Nitroso-*N*-methylvinylamine	*N*-Methyl-*N*-nitrosoethenamine	$C_3H_6N_2O$	4549-40-0		Reasonably anticipated to be human carcinogens	Group 2B
4-Nitrosomorpholine	*N*-Nitrosomorpholine	$C_4H_8N_2O_2$	59-89-2		Reasonably anticipated to be human carcinogens	Group 2B
N-Nitrosopiperidine	1-Nitrosopiperidine	$C_5H_{10}N_2O$	100-75-4		Reasonably anticipated to be human carcinogens	Group 2B
N-Nitrosodipropylamine	*N*-Nitroso-*N*-propyl-1-propanamine	$C_6H_{14}N_2O$	621-64-7		Reasonably anticipated to be human carcinogens	Group 2B
N-Nitrosopyrrolidine		$C_4H_8N_2O$	930-55-2		Reasonably anticipated to be human carcinogens	Group 2B
N-Nitrososarcosine	*N*-Methyl-*N*-nitrosoglycine	$C_3H_6N_2O_3$	13256-22-9		Reasonably anticipated to be human carcinogens	Group 2B
2-Nitrotoluene	1-Methyl-2-nitrobenzene	$C_7H_7NO_2$	88-72-2		Reasonably anticipated to be human carcinogens	Group 2A
Norethisterone	19-Norpregn-4-en-20-yn-3-one, 17-hydroxy-, (17 α)-	$C_{20}H_{26}O_2$	68-22-4		Reasonably anticipated to be human carcinogens	Group 2A
Ochratoxin A		$C_{20}H_{18}ClNO_6$	303-47-9		Reasonably anticipated to be human carcinogens	Group 2B
Octabromobiphenyl (unspecified isomer)		$C_{12}H_2Br_8$	61288-13-9		Reasonably anticipated to be human carcinogens	
2-Oxetanone	β-Propiolactone	$C_3H_4O_2$	57-57-8		Reasonably anticipated to be human carcinogens	Group 2B
Oxiranemethanol, (±)-	Glycidol	$C_3H_6O_2$	61915-27-3		Reasonably anticipated to be human carcinogens	Group 2A
Oxymetholone	Androstan-3-one, 17-hydroxy-2-(hydroxymethylene)-17-methyl-	$C_{21}H_{32}O_3$	434-07-1		Reasonably anticipated to be human carcinogens	Group 2A
Phenazopyridine hydrochloride	3-(Phenylazo)-2,6-pyridinediamine, monohydrochloride	$C_{11}H_{12}ClN_5$	136-40-3		Reasonably anticipated to be human carcinogens	Group 2B
Phenolphthalein	3,3-Bis(4-hydroxyphenyl)-1(3*H*)-isobenzofuranone	$C_{20}H_{14}O_4$	77-09-8		Reasonably anticipated to be human carcinogens	Group 2B
Phenoxybenzamine hydrochloride		$C_{18}H_{23}Cl_2NO$	63-92-3		Reasonably anticipated to be human carcinogens	Group 2B
Phenyloxirane	Styrene-7,8-oxide	C_8H_8O	96-09-3		Reasonably anticipated to be human carcinogens	Group 2A
Phenytoin	5,5-Diphenyl-2,4-imidazolidinedione	$C_{15}H_{12}N_2O_2$	57-41-0	also Phenytoin sodium; also known as Dilantin	Reasonably anticipated to be human carcinogens	Group 2B
Polybrominated biphenyls	PBBs				Reasonably anticipated to be human carcinogens	Group 2A
Procarbazine		$C_{12}H_{19}N_3O$	671-16-9	also the hydrochloride	Reasonably anticipated to be human carcinogens	Group 2A
Progesterone	Pregn-4-ene-3,20-dione	$C_{21}H_{30}O_2$	57-83-0		Reasonably anticipated to be human carcinogens	Group 2A

Name	Synonym	Mol. form.	CAS Reg. No.	Comments	NTP Classification	IARC Classification
1,3-Propane sultone	1,2-Oxathiolane, 2,2-dioxide	$C_3H_6O_3S$	1120-71-4		Reasonably anticipated to be human carcinogens	Group 2A
Propyleneimine	2-Methylaziridine	C_3H_7N	75-55-8		Reasonably anticipated to be human carcinogens	Group 2B
Reserpine		$C_{33}H_{40}N_2O_9$	50-55-5		Reasonably anticipated to be human carcinogens	Group 3
Riddelline		$C_{18}H_{23}NO_6$	23246-96-0		Reasonably anticipated to be human carcinogens	Group 2B
Safrole	5-(2-Propenyl)-1,3-benzodioxole	$C_{10}H_{10}O_2$	94-59-7		Reasonably anticipated to be human carcinogens	Group 2B
Selenium monosulfide		SSe	7446-34-6		Reasonably anticipated to be human carcinogens	
Silicon carbide (cubic)		CSi	409-21-2			Group 2A
Streptozotocin	D-Glucopyranose, 2-deoxy-2-[[(methylnitrosoamino)carbonyl]amino]-	$C_8H_{15}N_3O_7$	18883-66-4		Reasonably anticipated to be human carcinogens	Group 2B
Styrene	Vinylbenzene	C_8H_8	100-42-5		Reasonably anticipated to be human carcinogens	Group 2A
Sulfallate	*N,N*-Diethyldithiocarbamic acid, 2-chloroallyl ester	$C_8H_{14}ClNS_2$	95-06-7		Reasonably anticipated to be human carcinogens	Group 2B
Teniposide		$C_{32}H_{32}O_{13}S$	29767-20-2			Group 2A
Tetrachloroethene	Perchloroethylene	C_2Cl_4	127-18-4		Reasonably anticipated to be human carcinogens	Group 2A
Tetrachloromethane	Carbon tetrachloride	CCl_4	56-23-5		Reasonably anticipated to be human carcinogens	Group 2B
Tetrafluoroethene	Tetrafluoroethylene	C_2F_4	116-14-3		Reasonably anticipated to be human carcinogens	Group 2A
N,N,N',N'-Tetramethyl-4,4'-diaminobenzophenone	Michler's ketone	$C_{17}H_{20}N_2O$	90-94-8	also known as 4,4'-Methylenebis[*N,N*-dimethylaniline]	Reasonably anticipated to be human carcinogens	Group 2B
Tetranitromethane		CN_4O_8	509-14-8		Reasonably anticipated to be human carcinogens	Group 2B
Thioacetamide	Ethanethioamide	C_2H_5NS	62-55-5		Reasonably anticipated to be human carcinogens	Group 2B
Thiourea	Thiocarbamide	CH_4N_2S	62-56-6		Reasonably anticipated to be human carcinogens	Group 3
o-Tolidine	3,3'-Dimethylbenzidine	$C_{14}H_{16}N_2$	119-93-7	and dyes metabolized to *o*-Tolidine	Reasonably anticipated to be human carcinogens	Group 2B
Toluene diisocyanate (unspecified isomer)		$C_9H_6N_2O_2$	26471-62-5	includes both 2,4- and 2,6- isomers; also known as TDI	Reasonably anticipated to be human carcinogens	Group 2B
Toxaphene	Polychlorocamphene	$C_{10}H_{10}Cl_8$	8001-35-2		Reasonably anticipated to be human carcinogens	Group 2B
1*H*-1,2,4-Triazol-3-amine	Amitrole	$C_2H_4N_4$	61-82-5		Reasonably anticipated to be human carcinogens	Group 3
1,1,1-Trichloro-2,2-bis(4-chlorophenyl) ethane	Dichlorodiphenyltrichloroethane (DDT)	$C_{14}H_9Cl_5$	50-29-3		Reasonably anticipated to be human carcinogens	Group 2A
Trichloromethane	Chloroform	$CHCl_3$	67-66-3		Reasonably anticipated to be human carcinogens	Group 2B
(Trichloromethyl)benzene	Benzotrichloride	$C_7H_5Cl_3$	98-07-7		Reasonably anticipated to be human carcinogens	Group 2A
2,4,6-Trichlorophenol		$C_6H_3Cl_3O$	88-06-2		Reasonably anticipated to be human carcinogens	Group 2B
1,2,3-Trichloropropane	Allyl trichloride	$C_3H_5Cl_3$	96-18-4		Reasonably anticipated to be human carcinogens	Group 2A

LASER HAZARDS IN THE LABORATORY

Thomas J. Bruno and Paris D. N. Svoronos

Lasers are commonly used in the laboratory, although in many instruments, most lasers are embedded in instrumentation and are therefore shielded or protected by optical barriers and interlocks that, when functioning properly, prevent accidental exposure. Care must be exercised when performing maintenance or when changing samples in such instruments. In this section we provide basic information on laser safety and hazards (Refs. 1 to 3). This is by no means exhaustive nor is it meant to substitute for an understanding of the specific safety requirements of instrumentation, or applicable law or regulations. The special case of common laser pointers has received considerable attention recently and is treated separately (Ref. 4). We note that as of 2007, the general practice in the United States is to use the IEC definitions.

References

1. American National Standard for Safe Use of Lasers, American National Standards Institute, ANSI Z136.1, 2007
2. Safety of Laser Products – Part 1: Equipment Classification and Requirements, International Electrotechnical Commission, IEC 60825-1, 2nd Ed., 2007.
3. Bruno, T. J., and Svoronos, P. D. N., *CRC Handbook of Basic Tables for Chemical Analysis*, 3rd. Ed., CRC Press, Boca Raton, FL, 2011.
4. Hadler, J., and Dowell, M., Accurate, Inexpensive Testing of Laser Pointer Power for Safe Operation, *Meas. Sci. Technol.* 24, 1, 2013.

Classes of Lasers

The following is a summary for the laser classes following the ANSI guidelines used in the United States:

Class I

Class I lasers are inherently safe with no possibility of eye damage under conditions of normal use. The safety can result from a low output power (in which case eye damage is impossible even after prolonged exposure), or due to an enclosure preventing user access to the laser beam during normal operation, such as in CD players, laser printers, surveying transits, or measurement instruments.

Class II

The blink reflex of the human eye will prevent eye damage, unless the person deliberately stares into the beam for an extended period. Thus, a Class II laser can cause some eye damage if this is done. Output power may be up to 1 mW. This class includes only lasers that emit visible light. Some laser pointers are in this category.

Class IIIa

Lasers in this class are mostly dangerous in combination with certain optical instruments that change the beam diameter or power density. Output power does not exceed 5 mW. Beam power density may not exceed 2.5 mW cm^{-2}. Many laser sights for firearms and some laser pointers are included in this category.

Class IIIb

Lasers in this class may cause damage if the beam enters the eye directly. This generally applies to lasers powered from 5 mW to 500 mW. Lasers in this category can cause permanent eye damage with exposures of 1/100th of a second or less depending on the strength of the laser. A diffuse reflection (on paper or from a matte surface) is generally not hazardous but a specular reflection from a highly reflective surface can be just as dangerous as direct exposure. Protective eyewear is recommended when direct beam viewing of Class IIIb lasers may occur. Lasers at the high power end of this class may also present a fire hazard and can lightly burn skin.

Class IV

Lasers in this class have output powers of more than 500 mW in the beam and may cause severe, permanent damage to eye or skin without being magnified by optics of eye or instrumentation. Diffuse reflections of the laser beam can be hazardous to skin or eye within the Nominal Hazard Zone. Many industrial, scientific, military, and medical lasers are in this category.

The following is a summary of the laser classes following the IEC guidelines:

Class 1

A Class 1 laser is safe under all conditions of normal use, with no known biological hazard present. This class includes high-power lasers within an enclosure that prevents exposure to the radiation and that cannot be opened without shutting down the laser. This typically requires an interlocking.

Class 1M

A Class 1M laser is safe for all conditions of normal use except when passed through magnifying optics such as microscopes, telescopes, or on optical benches. Class 1M lasers typically produce large-diameter beams, or beams that are divergent. The classification of a Class 1M laser must be changed if the emergent light is refocused.

Class 2

A Class 2 laser is safe for all conditions of normal use because the blink reflex will limit the exposure to no more than 0.25 seconds. It only applies to visible-light lasers (400 nm to 700 nm) limited to 1 mW continuous wave, or more if the emission time is less than 0.25 s or if the light is not spatially coherent. Intentional suppression of the blink reflex could lead to eye injury. Many laser pointers are Class 2.

Class 2M

A Class 2M laser is similar to a Class 2, but it is used in an instrument that may focus the beam. This laser is safe because of the blink reflex provided the beam is not viewed through optical instruments as described above for Class 1M.

Class 3R

A Class 3R laser is considered safe if handled carefully, with restricted beam viewing. These lasers can be hazardous to the human eye if the beam is viewed for extended periods of time or under fixated conditions. Continuous beam Class 3R lasers operating in the visible region are limited in power output to 5 mW. For other wavelengths and for pulsed lasers, other limits will apply.

Class 3B

A Class 3B laser is hazardous if the eye is exposed directly, but diffuse reflections such as from paper surfaces are not harmful. Continuous lasers in the wavelength range from 315 nm to the far infrared are limited in power output to 0.5 W. For pulsed lasers between 400 nm and 700 nm, the limit is 30 mJ. Other limits apply to other wavelengths and to short pulse lasers. Protective eyewear is typically required where direct viewing of a class 3B laser beam may occur. Class 3B lasers must be equipped with a key switch and a safety interlock.

Class 4

Class 4 lasers include all lasers with beam power greater than those covered in class 3B. By definition, a Class 4 laser can burn the skin, in addition to causing severe and permanent eye damage. This eye damage can result from of direct or diffuse beam viewing. These lasers may ignite combustible materials, and thus may represent a fire risk. Class 4 lasers must be equipped with a key switch and a safety interlock. Many industrial, scientific, military, and medical lasers are in this category.

Laser Pointers

Laser pointers, ubiquitous at meetings, shows, and in the classroom, deserve separate consideration because of recent work on actual observed power output. For purposes of classification into the levels discussed above, the typical laser pointer is classified as 3R. The human light aversion response can generally protect against 3R lasers; however, this response is less sensitive in the near infrared range (700 nm to 1400 nm). Thus, to prevent retinal burns, laser pointers must not emit hazardous levels of infrared, and must be Class 1 compliant in terms of accessible emission level (AEL) in that wavelength range. In a testing program undertaken at NIST, cited in Ref. 4, it was found that of the laser pointers randomly chosen and tested, all but two pointers failed to comply by more than 15% of the specified AEL, and 48% emitted more than twice the specified AEL at one or more specified wavelength. This indicates a risk of 3B exposure from these devices that are nominally classified as 3R.

GENERAL CHARACTERISTICS OF IONIZING RADIATION FOR THE PURPOSE OF PRACTICAL APPLICATION OF RADIATION PROTECTION

Thomas W. Grove and Thomas J. Bruno

The following table provides practical information to allow the design and implementation of radiation protection in laboratory and industrial environments. Additional information and details can be found in the references.

References

1. Johnson, T. E., and Birky, B. K., *Health Physics Radiological Health, Fourth Edition,* Lippincott, Williams & Wilkins, Baltimore, MD, 2012.
2. Cember, H., and Johnson, T. E., *Introduction to Health Physics, Fourth Edition,* McGraw-Hill, New York, 2011.
3. 1990 Recommendations of the International Commission on Radiological Protection. ICRP Publication 60. Ann. ICRP 21 (1-3), 1990.

Type (symbol)	Physical properties	Range	Shielding	Biological hazards	Comments
Alpha particle (α)	Very large mass (2 protons, 2 neutrons, 0 electrons) +2 charge	Very short 3 cm to 6 cm (~ 1 inch to 2 inches) in air	Few centimeters of air Sheet of paper Dead (outer) layer of skin	Internally, the source of alpha radiation is in close contact with live body tissue. It can deposit large amounts of energy in a small amount of body tissue Rarely an external hazard	Alpha particles of at least 7.5 MeV are required to penetrate the epidermis, the protective layer of skin, which is about 0.07 mm (70 μm) thick The range R of most particles of common emitters (4.5 MeV to 5.5 MeV) is 3 cm to 4 cm in air Range in air = $0.322E^{3/2}$ cm, when E is expressed in MeV
Beta particle (β)	Small mass -1 or +1 charge -1 charge particle is an electron; +1 charge particle is a positron	Short 6 cm to 600 cm (1 inch to 20 feet) in air	Low atomic number materials Plastic Glass Aluminum	Internal hazard Externally, may be hazardous to the skin and eyes	Beta particles of at least 70 keV are required to penetrate the epidermis. $R_{air} \approx 3.65$ m/MeV The range of beta particles in material in g cm^{-2} (thickness in cm multiplied by the density in g cm^{-3}) is approximately half the maximum energy in MeV. ($R \approx E_{max}/2$) Dose rate (in rad/hr) at 1 cm from a beta point source; ≈ 300 rad/hr per Curie Dose rate (in rad/hr) in a solution ≈ $2.12\bar{E}C/\rho$; where $\bar{E}$ = average energy in MeV, C = concentration in μCi cm^{-3}, and ρ = density of the solution in g cm^{-3}; the dose rate is about one-half this value at the surface An aqueous solution of 1 Curie ^{32}P in a glass vial typically produces 3 mrad/hr at 1 meter from Bremsstrahlung Shielding causes Bremsstrahlung radiation similar to x-rays and gamma rays; generally the higher the atomic number of a material, the more intense is the Bremsstrahlung radiation
Neutron (n or n^0)	Large mass No charge	Very far in air Easily travel several hundred meters High penetrating power due to lack of charge	High hydrogen content material Water Concrete Plastic	External whole body exposure hazard May be external and/or internal hazard Depends on whether source is inside or outside the body Energy dependent	Shielding can be provided by hydrogen-rich materials such as hydrocarbons, water, waxes, high water concrete Can cause neutron activation, in which radionuclides are formed Particularly damaging to soft tissue such as the cornea
X-Ray, Gamma ray (γ)	No mass No charge Electromagnetic wave X-rays and gamma rays are similar, but place of origin and energy levels may differ	Very far in air Easily travel several hundred meters Very high penetrating power since it has no mass & no charge	High atomic number materials Depleted uranium Lead Steel Concrete Water	Whole body exposure hazard May be external and/or internal hazard Can be a skin and/or eye hazard Depends on whether source is inside or outside the body Energy dependent	Shielding requires large mass and density materials; lead or depleted uranium are commonly used Doubling the distance from a point source will result in a reduction of exposure (or dose) by a factor of four Protective clothing and other PPE can effectively guard against ingestion or absorption of radioactive material but is not usually practical for protecting against x-rays or gamma rays

RADIATION SAFETY UNITS

Thomas J. Bruno and Paris D. N. Svoronos

Ionizing radiation, consisting of x-rays, gamma rays, alpha particles, beta particles, and neutrons, is measured and quantified in units of the radioactivity source and dose (Refs. 1–4). The radioactivity measures the strength of a source in terms of events of emission per second. Dose is a measure of the energy that is actually absorbed into matter.

Radioactivity

In the SI system, the bequerel (Bq) has replaced the curie (Ci) as the accepted unit of radioactivity (or simply activity). One Bq is one event of radiation emission (such as a disintegration) per second. It is related to the older unit by the following equations.

$$1\ \text{Ci} = 3.7 \times 10^{10}\ \text{Bq}$$

$$1\ \text{Ci} = 37\ \text{GBq} = 37\ 000\ \text{MBq}$$

The following chart provides a practical guide between the two units:

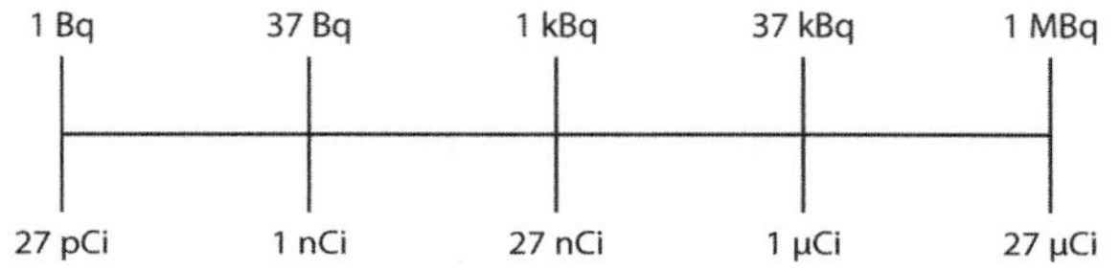

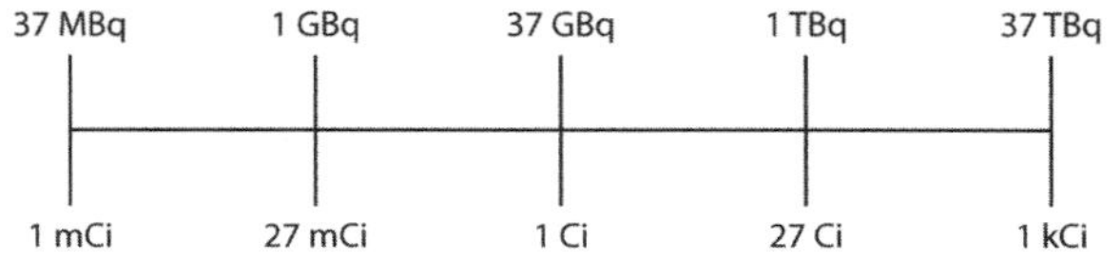

Class A radionuclides: 0.3 Bq/cm^2 = 8.1 pCi/cm^2
Class B radionuclides: 3 Bq/cm^2 = 81 pCi/cm^2
Class C radionuclides: 30 Bq/cm^2 = 810 pCi/cm^2

Energy

For ionizing radiation, the energy is measured in electron volts (eV), which is related to other energy quantities by:

$$1\ \text{eV} = 1.602\ 176\ 634 \times 10^{-19}\ \text{J (exact)}$$

Dose

The older unit of dose, which is defined as the energy that is actually absorbed, is the radiation absorbed dose (RAD). The RAD was defined as the dose that would cause 0.01 J to be absorbed in 1 kg of matter (or 100 ergs per gram). The modern SI unit is the Gray (Gy):

$$100\ \text{RAD} = 1\ \text{Gy}$$

Equivalent Dose

The committed dose, or more properly the committed dose equivalent, $H_{T,50}$, is the total dose accumulated over a 50-year period after ingestion or inhalation. The equivalent dose (also called the dose equivalent or biological dose) describes the effect of radiation on human tissue rather than the physical effects of the radiation alone. This quantity is expressed in Sieverts (Sv), and is found by multiplying the absorbed dose, in grays, by a dimensionless quality factor Q (which depends on the radiation type), and by another dimensionless factor N (the tissue weighting factor). Q is also called the Relative Biological Effectiveness (*RBE*). The factor N depends upon the part of the body irradiated, the time and volume over which the dose was spread, and the species of the subject.

The currently accepted, approximate Q factors are provided in Table 1.

The currently accepted values for the tissue-weighting factor N for human body parts are provided in Table 2.

Relative to the effect on humans, N factors for non-human animals and plants are fairly difficult to determine (Ref. 5). Table 3 contains the N factors that have been suggested for representative non-human organisms.

In terms of the older unit, rem (roentgen equivalent in man):

$$1\ \text{rem} = 0.01\ \text{Sv, assuming}\ Q = 1$$

The following chart provides a practical guide between the two units:

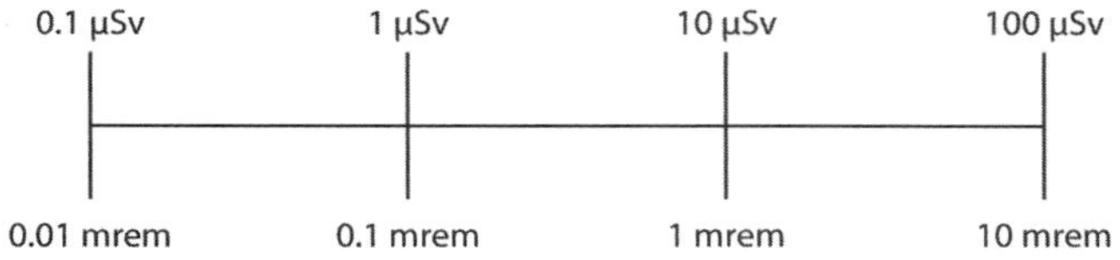

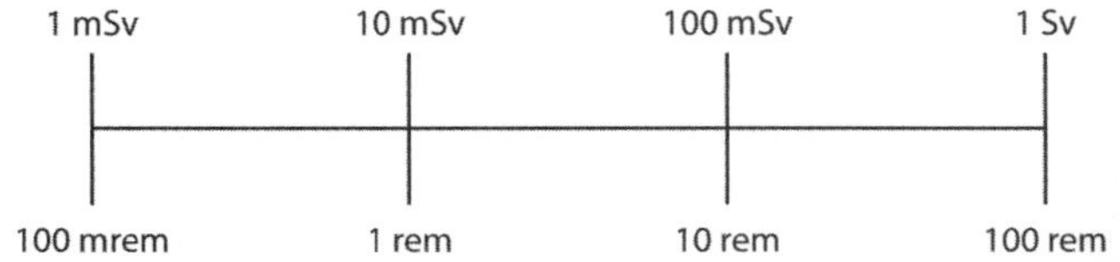

The approximate effects of full-body dosages are summarized in Table 4.

The relationship between nuclide half-lives elapsed and the remaining radioactivity is given in Table 5.

References

1. Radiation—Quantities and Units of Ionizing Radiation, Canadian Centre for Occupational Health and Safety, OSH Answer List Series, 2008.
2. Radioactivity Units, Health Physics Society, <www.hps.org/>, 2013.
3. Furr, A. K., *CRC Handbook of Laboratory Safety, Fifth Edition,* CRC Press, Boca Raton, FL, 2000.
4. Bruno, T. J., and Svoronos, P. D. N., *CRC Handbook of Basic Tables for Chemical Analysis, Third Edition,* CRC Press, Boca Raton, FL, 2011. <https://doi.org/10.1201/b10385>
5. Higley, K. A., Kocher, D.C., Real, A. G. , and Chambers, D. B., *Ann. ICRP*, 41, 233, 2012. <doi:10.1016/j.icsp.2012.06.014>.

TABLE 1. Relative Biological Effectiveness Factors (Q) for Different Radiation Types

Radiation type	Q
X-rays	1
Gamma rays	1
Beta particles	1
Thermal neutrons (<10 keV)	5
Fast neutrons (10 keV to 100 keV)	10
Fast neutrons (100 keV to 2 MeV)	20
Fast neutrons (2 MeV to 20 MeV)	10
Fast neutrons (>20 MeV)	5
Protons (>2 MeV)	5
Alpha particles	20
Other atomic nuclei	20

TABLE 2. Tissue-Weighting Factors N for Body Parts

Body part	N
Gonads	0.20
Bone marrow	0.12
Colon	0.12
Lung	0.12
Stomach	0.12
Bladder	0.05
Brain	0.05
Breast	0.05
Kidney	0.05
Liver	0.05
Muscle	0.05
Esophagus	0.05
Pancreas	0.05
Small intestine	0.05
Spleen	0.05
Thyroid	0.05
Uterus	0.05
Bone surface	0.01
Skin*	0.01
Organs not listed above, collectively	0.05
Whole body (scale definition)	≡1

* The weighting factor for skin implies a whole body exposure.

TABLE 3. Tissue-Weighting Factors N for Non-Human Organisms

Organism	N
Viruses	0.03 to 0.0003
Bacteria	0.03 to 0.0003
Single-cell organisms	0.03 to 0.0003
Insects	0.1 to 0.002
Mollusks	0.06 to 0.006
Plants	2 to 0.02
Fish	0.75 to 0.03
Amphibians	0.4 to 0.14
Reptiles	1 to 0.075
Birds	0.6 to 0.15
Humans (scale definition)	≡1

TABLE 4. Radiation Effects on Humans

Dose/Sv	Effect
1	Nausea
2–5	Hair loss, hemorrhage, death is possible
>3	Death is likely in 50% of cases within 30 days
6	Death is likely in all cases

TABLE 5. Remaining Activity for Different Elapsed Half-Lives

Half-lives elapsed	Percent remaining
0	100
1	50
2	25
3	12.5
4	6.25
5	3.125
n	$(1/2)^n(100\%)$

RELATIVE DOSE RANGES FROM IONIZING RADIATION

Thomas J. Bruno

It is important to place in perspective the relative ionizing radiation dose acquired in common laboratory settings. The most commonly encountered source is a ^{63}Ni source used in gas chromatographic electron capture detectors (ECD) and in ion mobility spectrometers (IMS) (Ref. 1). In both instruments, the source is sealed and has a radioactivity of 15 mCi. The exposures cited refer only to normal operation; it does not consider exposures if the device is dismantled or allowed to overheat.

Background

Natural background consists of the highly variable sum of all ubiquitous sources of ionizing radiation encountered on the planet (Ref. 2). Background in general can be divided into the following four major contributions:

Contribution	Average dose, mrem/year, United States
Terrestrial contribution	21
Cosmic contribution	33
Airborne radon (and daughter) contribution	228
Internal consumption contribution	29
Total Natural Background	**311**

In the United States, the average natural background ionizing radiation level is 311 mrem. This is variable due primarily to differences in altitude and primordial radionuclides and their daughters. For example, the averages in the United Kingdom and Finland are 200 mrem and 700 mrem, respectively. Higher levels are found at higher altitudes and regions with a larger radon budget. Within the United States, for example, the background in Denver, Colorado is approximately 450 mrem, while in most of Florida, it is closer to 230 mrem. The terrestrial contribution primarily arises from radionuclides of potassium, uranium, and thorium, and their daughters. The cosmic contribution arises primarily from muons, neutrons, and electrons, and varies with terrestrial magnetic field and altitude. The internal contribution results from consuming radionuclides of potassium and carbon in food and water. By far the largest contribution is from radon and radon daughters. The radon budget results from terrestrial sources of uranium (Ref. 3). Within the United States, the action level requiring indoor radon mitigation is reached when a measurement results in 4 pCi/L (150 Bq/m3) or higher. This level of radioactivity results in a dose of between 300 mrem and 700 mrem, assuming 80% indoor occupancy. The range cited results from different dose conversion coefficients and dosimetric models used by different agencies (Ref. 4).

Typical Incremental Increases above Background

Exposure to ionizing radiation in the laboratory results in a dose level in excess of the background levels discussed above (Ref. 5). The following charts place in perspective the additional dose received above background for some common exposures. Since in Figure 1 the ranges are dwarfed by tobacco use, Figure 2 is presented with this contribution removed. This is significant

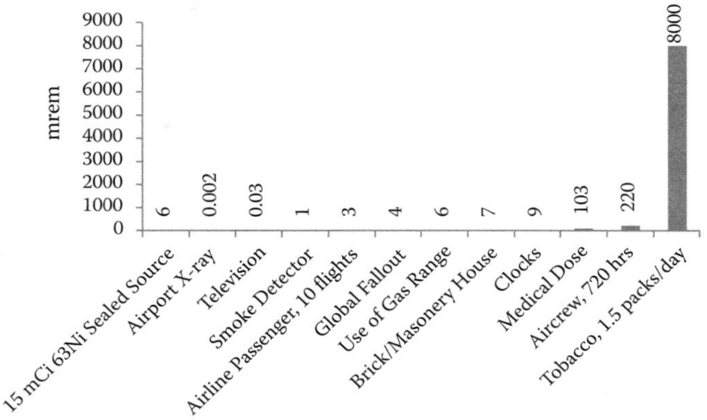

FIGURE 1. A comparison of increments to natural background levels, explicit for the 15 mCi ^{63}Ni sources used in electron capture detectors and ion mobility spectrometers.

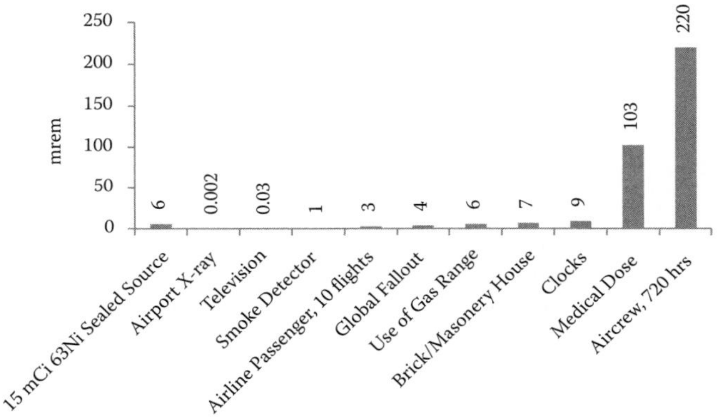

FIGURE 2. The same comparison as in Figure 1, but with tobacco use removed. Thus, the data shown above are all whole body dosages.

because the tobacco dose is specific to the lungs, and not the whole body. A more relevant comparison for Figure 1 would be obtained by multiplying the listed 8000 mrem by the tissue-weighting factor for the lungs, 0.12.

The high incremental level associated with tobacco use (1 mrem/hr while smoking) results from the accumulation of 210Po and 210Pb (radon daughters that are alpha and gamma emitters) on tobacco leaves. The incremental dose accrued by air travel is dependent on altitude, with higher levels associated with higher altitudes, and will range between 0.3 mrem/hr and 0.5 mrem/hr. The incremental dose due to medical imaging or radiation treatment can be misleading. For many individuals the dose can be close to zero, but in the case of radiation treatment it can be much higher. Indeed, patients given certain radiation treatments become incremental sources themselves, resulting in incremental dosage above background to attending medical personnel, caregivers, and the general public. The entry for clocks in both Figures 1 and 2 is for older (even antique) clocks that have dials coated with Ra/ZnS paint to provide illumination.

In instrumentation such as ECDs and IMSs, the devices are sealed sources in shielded enclosures, and are covered by general licenses in the United States. They are designed with inherent radiation safety features so that they can be used by persons with no radiation training or experience (Ref. 6).

References

1. Bruno, T. J., and Svoronos, P. D. N., *CRC Handbook of Basic Tables for Chemical Analysis, Third Edition,* CRC Press, Boca Raton, FL, 2011 <https://doi.org/10.1201/b10385>.
2. Metting, N. F., Ionizing Radiation Dose Ranges, Office of Science, U.S. Department of Energy, <www.lowdose.energy.gov>, 2010.
3. *NCRP Report 160,* Ionizing Radiation Exposure of the Population of the United States, Recommendations of the National Council on Radiation Protection and Measurements, Bethesda, MD, 2009.
4. *Background Information on "Update on Perspectives and Recommendations on Indoor Radiation,"* position statement of the Health Physics Society, 2009.
5. Johnson, T. E., and Fellman, A., *Estimated Dose and Risk from 15mCi ^{63}Ni Sealed Source Type NR-348-D-111-B,* Report Prepared for Hewlett Packard Co., CSI Radiation Safety, Gaithersburg, MD, 1999.
6. <https://www.nrc.gov/materials/miau/general-use.html>, accessed July 2018.

ANNUAL LIMITS ON INTAKES OF RADIONUCLIDES

K. F. Eckerman

The following table lists the recommended annual limits for workers on oral and inhalation intakes (ALI) for selected radionuclides based on the occupational radiation protection guidance of the International Commission on Radiological Protection (Refs. 1 and 2). An intake of one ALI corresponds to an annual whole body dose of 0.02 Sv (2 rem).

The ALI is expressed in the SI unit of activity, the becquerel (Bq), and in the conventional unit, the microcurie (μCi); 1 μCi = $3.7 \cdot 10^4$ Bq for both inhalation intakes (inh) and oral intakes (orl).

The chemical form of inhaled radionuclides is, in most instances, stated in terms of the rate of absorption to blood from the lungs and the fractional absorption from the small intestine. Types F, M, and S denote chemical forms that are absorbed from the lungs at rates characterized as fast, moderate, and slow, respectively. The time to absorb 90% of the deposited radionuclide, in the absence of radioactive decay, corresponds to about 10 minutes, 150 days, and 7000 days for Type F, M, and S compounds, respectively. Type F compounds can be considered to be more soluble than M or S, S being the most insoluble. Chemical form consideration for ingestion is specified by the fractional absorption from the small intestine, denoted as f_1. The f_1 values range from 10^{-5} to 1. Higher fractional absorption is associated with greater solubility of the compound.

The notation 1.1E+13 means $1.1 \cdot 10^{13}$. Column definitions for the table are as follows.

Column heading	Definition
Radionuclide	Symbol for radionuclide
Physical half-life	Half-life of the radionuclide; y = years; d = days; h = hours
Chemical form f_1	Chemical form consideration for ingestion as specified by the fractional absorption from the small intestine, denoted as f_1; see discussion above
ALI(inh)/Bq	Annual limit intake for inhalation, in units Bq
ALI(inh)/μCi	Annual limit intake for inhalation, in units μCi
ALI(orl)/Bq	Annual limit intake for oral ingestion, in units Bq
ALI(orl)/μCi	Annual limit intake for oral ingestion, in units μCi

References

1. *1990 Recommendations of the International Commission on Radiological Protection, ICRP Publication 60, Annals of the ICRP 21, (1–3)*, Pergamon Press, Oxford, 1991.
2. *Dose Coefficients for Intakes of Radionuclides by Workers, ICRP Publication 68, Annals of the ICRP*, 24(4), Pergamon Press, Oxford, 1995.

Annual Limits on Intakes of Radionuclides

Radionuclide	Physical half-life	Chemical form f_1 (inh)	ALI(inh)/Bq	ALI(inh)/μCi	Chemical form f_1 (orl)	ALI(orl)/Bq	ALI(orl)/μCi
^{3}H	12.3 y	HT gas	1.1E+13	3.0E+08	1.000	1.1E+13	3.0E+08
		HTO vapor	1.1E+09	3.0E+04			
^{11}C	0.340 h	CO	1.7E+10	4.5E+05	1.000	8.3E+08	2.3E+04
		CO_2	9.1E+09	2.5E+05			
		Organic compounds	6.2E+09	1.7E+05			
^{14}C	5730 y	CO	2.5E+10	6.8E+05	1.000	3.4E+07	9.3E+02
		CO_2	3.1E+09	8.3E+04			
		Organic compounds	3.4E+07	9.3E+02			
^{18}F	1.83 h	F 1.000	3.7E+08	1.0E+04	1.000	4.1E+08	1.1E+04
		M 1.000	2.2E+08	6.1E+03			
		S 1.000	2.2E+08	5.8E+03			
^{22}Na	2.60 y	F 1.000	1.0E+07	2.7E+02	1.000	6.3E+06	1.7E+02
^{24}Na	15.0 h	F 1.000	3.8E+07	1.0E+03	1.000	4.7E+07	1.3E+03
^{32}P	14.3 d	F 0.800	1.8E+07	4.9E+02	0.800	8.3E+06	2.3E+02
		M 0.800	6.9E+06	1.9E+02			
^{35}S	87.4 d	Inorganic compounds					
		F 0.800	2.5E+08	6.8E+03	0.800	1.4E+08	3.9E+03
		M 0.800	1.8E+07	4.9E+02	0.100	1.1E+08	2.8E+03
		Vapor	1.7E+08	4.5E+03			
		Organic compounds			1.000	2.6E+07	7.0E+02
^{42}K	12.4 h	F 1.000	1.0E+08	2.7E+03	1.000	4.7E+07	1.3E+03
^{43}K	22.6 h	F 1.000	7.7E+07	2.1E+03	1.000	8.0E+07	2.2E+03
^{45}Ca	163 d	M 0.300	8.7E+06	2.4E+02	0.300	2.6E+07	7.1E+02
^{47}Ca	4.53 d	M 0.300	9.5E+06	2.6E+02	0.300	1.3E+07	3.4E+02
^{51}Cr	27.7 d	F 0.100	6.7E+08	1.8E+04	0.100	5.3E+08	1.4E+04
		M 0.100	5.9E+08	1.6E+04	0.010	5.4E+08	1.5E+04
		S 0.100	5.6E+08	1.5E+04			

Radionuclide	Physical half-life	Chemical form f_1 (inh)	ALI(inh)/Bq	ALI(inh)/µCi	Chemical form f_1 (orl)	ALI(orl)/Bq	ALI(orl)/µCi
^{54}Mn	312 d	F 0.100	1.8E+07	4.9E+02	0.100	2.8E+07	7.6E+02
		M 0.100	1.7E+07	4.5E+02			
^{52}Fe	8.28 h	F 0.100	2.9E+07	7.8E+02	0.100	1.4E+07	3.9E+02
		M 0.100	2.1E+07	5.7E+02			
^{55}Fe	2.70 y	F 0.100	2.2E+07	5.9E+02	0.100	6.1E+07	1.6E+03
		M 0.100	6.1E+07	1.6E+03			
^{59}Fe	44.5 d	F 0.100	6.7E+06	1.8E+02	0.100	1.1E+07	3.0E+02
		M 0.100	6.3E+06	1.7E+02			
^{57}Co	271 d	M 0.100	5.1E+07	1.4E+03	0.100	9.5E+07	2.6E+03
		S 0.050	3.3E+07	9.0E+02	0.050	1.1E+08	2.8E+03
^{58}Co	70.8 d	M 0.100	1.4E+07	3.9E+02	0.100	2.7E+07	7.3E+02
		S 0.050	1.2E+07	3.2E+02	0.050	2.9E+07	7.7E+02
^{60}Co	5.27 y	M 0.100	2.8E+06	7.6E+01	0.100	5.9E+06	1.6E+02
		S 0.050	1.2E+06	3.2E+01	0.050	8.0E+06	2.2E+02
^{64}Cu	12.7 h	F 0.500	2.9E+08	7.9E+03	0.500	1.7E+08	4.5E+03
		M 0.500	1.3E+08	3.6E+03			
		S 0.500	1.3E+08	3.6E+03			
^{59}Ni	75000 y	F 0.050	9.1E+07	2.5E+03	0.050	3.2E+08	8.6E+03
		M 0.050	2.1E+08	5.8E+03			
		Vapor	2.4E+07	6.5E+02			
^{63}Ni	96.0 y	F 0.050	3.8E+07	1.0E+03	0.050	1.3E+08	3.6E+03
		M 0.050	6.5E+07	1.7E+03			
		Vapor	1.0E+07	2.7E+02			
^{65}Zn	244 d	S 0.500	7.1E+06	1.9E+02	0.500	5.1E+06	1.4E+02
^{67}Ga	3.26 d	F 0.001	1.8E+08	4.9E+03	0.001	1.1E+08	2.8E+03
		M 0.001	7.1E+07	1.9E+03			
^{68}Ga	1.13 h	F 0.001	4.1E+08	1.1E+04	0.001	2.0E+08	5.4E+03
		M 0.001	2.5E+08	6.7E+03			
^{68}Ge	288 d	F 1.000	2.4E+07	6.5E+02	1.000	1.5E+07	4.2E+02
		M 1.000	2.5E+06	6.8E+01			
^{75}Se	120 d	F 0.800	1.4E+07	3.9E+02	0.800	7.7E+06	2.1E+02
		M 0.800	1.2E+07	3.2E+02	0.050	4.9E+07	1.3E+03
^{79}Se	65000 y	F 0.800	1.3E+07	3.4E+02	0.800	6.9E+06	1.9E+02
		M 0.800	6.5E+06	1.7E+02	0.050	5.1E+07	1.4E+03
^{86}Rb	18.6 d	F 1.000	1.5E+07	4.2E+02	1.000	7.1E+06	1.9E+02
^{85}Sr	64.8 d	F 0.300	3.6E+07	9.7E+02	0.300	3.6E+07	9.7E+02
		S 0.010	3.1E+07	8.4E+02	0.010	6.1E+07	1.6E+03
^{87m}Sr	2.80 h	F 0.300	9.1E+08	2.5E+04	0.300	6.7E+08	1.8E+04
		S 0.010	5.7E+08	1.5E+04	0.010	6.1E+08	1.6E+04
^{89}Sr	50.5 d	F 0.300	1.4E+07	3.9E+02	0.300	7.7E+06	2.1E+02
		S 0.010	3.6E+06	9.7E+01	0.010	8.7E+06	2.4E+02
^{90}Sr	29.1 y	F 0.300	6.7E+05	1.8E+01	0.300	7.1E+05	1.9E+01
		S 0.010	2.6E+05	7.0E+00	0.010	7.4E+06	2.0E+02
^{99}Mo	2.75 d	F 0.800	5.6E+07	1.5E+03	0.800	2.7E+07	7.3E+02
		S 0.050	1.8E+07	4.9E+02	0.050	1.7E+07	4.5E+02
^{99m}Tc	6.02 h	F 0.800	1.0E+09	2.7E+04	0.800	9.1E+08	2.5E+04
		M 0.800	6.9E+08	1.9E+04			
^{99}Tc	213000 y	F 0.800	5.0E+07	1.4E+03	0.800	2.6E+07	6.9E+02
		M 0.800	6.3E+06	1.7E+02			
^{106}Ru	1.01 y	F 0.050	2.0E+06	5.5E+01	0.050	2.9E+06	7.7E+01
		M 0.050	1.2E+06	3.2E+01			
		S 0.050	5.7E+05	1.5E+01			
^{111}In	2.83 d	F 0.020	9.1E+07	2.5E+03	0.020	6.9E+07	1.9E+03
		M 0.020	6.5E+07	1.7E+03			
^{113m}In	1.66 h	F 0.020	1.1E+09	2.8E+04	0.020	7.1E+08	1.9E+04
		M 0.020	6.3E+08	1.7E+04			
^{113}Sn	115 d	F 0.020	2.5E+07	6.8E+02	0.020	2.7E+07	7.4E+02
		M 0.020	1.1E+07	2.8E+02			
^{123}I	13.2 h	F 1.000	1.8E+08	4.9E+03	1.000	9.5E+07	2.6E+03
		Vapor	9.5E+07	2.6E+03			

Radionuclide	Physical half-life	Chemical form f_1 (inh)	ALI(inh)/Bq	ALI(inh)/μCi	Chemical form f_1 (orl)	ALI(orl)/Bq	ALI(orl)/μCi
^{125}I	60.1 d	F 1.000	2.7E+06	7.4E+01	1.000	1.3E+06	3.6E+01
		Vapor	1.4E+06	3.9E+01			
^{129}I	$1.57{\cdot}10^7$ y	F 1.000	3.9E+05	1.1E+01	1.000	1.8E+05	4.9E+00
		Vapor	2.1E+05	5.6E+00			
^{131}I	8.04 d	F 1.000	1.8E+06	4.9E+01	1.000	9.1E+05	2.5E+01
		Vapor	1.0E+06	2.7E+01			
^{129}Cs	1.34 d	F 1.000	2.5E+08	6.7E+03	1.000	3.3E+08	9.0E+03
^{134}Cs	2.06 y	F 1.000	2.1E+06	5.6E+01	1.000	1.1E+06	2.8E+01
^{136}Cs	13.1 d	F 1.000	1.1E+07	2.8E+02	1.000	6.7E+06	1.8E+02
^{137}Cs	30.0 y	F 1.000	3.0E+06	8.1E+01	1.000	1.5E+06	4.2E+01
^{141}Ce	32.5 d	M 5.0E-04	7.4E+06	2.0E+02	5.0E-04	2.8E+07	7.6E+02
		S 5.0E-04	6.5E+06	1.7E+02			
^{144}Ce	284 d	M 5.0E-04	8.7E+05	2.4E+01	5.0E-04	3.8E+06	1.0E+02
		S 5.0E-04	6.9E+05	1.9E+01			
^{133}Ba	10.7 y	F 0.100	1.1E+07	3.0E+02	0.100	2.0E+07	5.4E+02
^{140}Ba	12.7 d	F 0.100	1.3E+07	3.4E+02	0.100	8.0E+06	2.2E+02
^{169}Yb	32.0 d	M 5.0E-04	9.5E+06	2.6E+02	5.0E-04	2.8E+07	7.6E+02
		S 5.0E-04	8.3E+06	2.3E+02			
^{198}Au	2.69 d	F 0.100	5.1E+07	1.4E+03	0.100	2.0E+07	5.4E+02
		M 0.100	2.0E+07	5.5E+02			
		S 0.100	1.8E+07	4.9E+02			
^{198m}Au	2.30 d	F 0.100	3.4E+07	9.2E+02	0.100	1.5E+07	4.2E+02
		M 0.100	1.0E+07	2.7E+02			
		S 0.100	1.1E+07	2.8E+02			
^{197}Hg	2.67 d	Inorganic compounds					
		F 0.400	2.4E+08	6.4E+03	1.000	2.0E+08	5.5E+03
					0.400	1.2E+08	3.2E+03
		Vapor	4.5E+06	1.2E+02			
		Organic compounds					
		F 0.020	2.0E+08	5.4E+03	0.020	8.7E+07	2.4E+03
		M 0.020	7.1E+07	1.9E+03			
^{203}Hg	46.6 d	Inorganic compounds					
		F 0.400	2.7E+07	7.2E+02	1.000	1.1E+07	2.8E+02
					0.400	1.8E+07	4.9E+02
		Vapor	2.9E+06	7.7E+01			
		Organic compounds					
		F 0.020	3.4E+07	9.2E+02	0.020	3.7E+07	1.0E+03
		M 0.020	1.1E+07	2.8E+02			
^{201}Tl	3.04 d	F 1.000	2.6E+08	7.1E+03	1.000	2.1E+08	5.7E+03
^{210}Pb	22.3 y	F 0.200	1.8E+04	4.9E-01	0.200	2.9E+04	7.9E-01
^{207}Bi	38.0 y	F 0.050	2.4E+07	6.4E+02	0.050	1.5E+07	4.2E+02
		M 0.050	6.3E+06	1.7E+02			
^{210}Po	138 d	F 0.100	2.8E+04	7.6E-01	0.100	8.3E+04	2.3E+00
		M 0.100	9.1E+03	2.5E-01			
^{224}Ra	3.66 d	M 0.200	8.3E+03	2.3E-01	0.200	3.1E+05	8.3E+00
^{226}Ra	1600 y	M 0.200	1.7E+03	4.5E-02	0.200	7.1E+04	1.9E+00
^{228}Ra	5.75 y	M 0.200	1.2E+04	3.2E-01	0.200	3.0E+04	8.1E-01
^{228}Th	1.91 y	M 5.0E-04	8.7E+02	2.4E-02	5.0E-04	2.9E+05	7.7E+00
		S 2.0E-04	6.3E+02	1.7E-02	2.0E-04	5.7E+05	1.5E+01
^{230}Th	77000 y	M 5.0E-04	7.1E+02	1.9E-02	5.0E-04	9.5E+04	2.6E+00
		S 2.0E-04	2.8E+03	7.5E-02	2.0E-04	2.3E+05	6.2E+00
^{232}Th	$1.40{\cdot}10^{10}$ y	M 5.0E-04	6.9E+02	1.9E-02	5.0E-04	9.1E+04	2.5E+00
		S 2.0E-04	1.7E+03	4.5E-02	2.0E-04	2.2E+05	5.9E+00
^{234}U	$2.44{\cdot}10^5$ y	F 0.020	3.1E+04	8.4E-01	0.020	4.1E+05	1.1E+01
		M 0.020	9.5E+03	2.6E-01	0.002	2.4E+06	6.5E+01
		S 0.002	2.9E+03	7.9E-02			
^{235}U	$7.04{\cdot}10^8$ y	F 0.020	3.3E+04	9.0E-01	0.020	4.3E+05	1.2E+01
		M 0.020	1.1E+04	3.0E-01	0.002	2.4E+06	6.5E+01
		S 0.002	3.3E+03	8.9E-02			

Radionuclide	Physical half-life	Chemical form f_1 (inh)	ALI(inh)/Bq	ALI(inh)/μCi	Chemical form f_1 (orl)	ALI(orl)/Bq	ALI(orl)/μCi
^{238}U	$4.47 \cdot 10^9$ y	F 0.020	3.4E+04	9.3E-01	0.020	4.5E+05	1.2E+01
		M 0.020	1.3E+04	3.4E-01	0.002	2.6E+06	7.1E+01
		S 0.002	3.5E+03	9.5E-02			
^{237}Np	$2.14 \cdot 10^6$ y	M 5.0E-04	1.3E+03	3.6E-02	5.0E-04	1.8E+05	4.9E+00
^{239}Np	2.36 d	M 5.0E-04	1.8E+07	4.9E+02	5.0E-04	2.5E+07	6.8E+02
^{238}Pu	87.7 y	M 5.0E-04	6.7E+02	1.8E-02	5.0E-04	8.7E+04	2.4E+00
		S 1.0E-05	1.8E+03	4.9E-02	1.0E-05	2.3E+06	6.1E+01
					1.0E-04	4.1E+05	1.1E+01
^{239}Pu	24100 y	M 5.0E-04	6.3E+02	1.7E-02	5.0E-04	8.0E+04	2.2E+00
		S 1.0E-05	2.4E+03	6.5E-02	1.0E-05	2.2E+06	6.0E+01
					1.0E-04	3.8E+05	1.0E+01
^{241}Pu	14.4 y	M 5.0E-04	3.4E+04	9.3E-01	5.0E-04	4.3E+06	1.2E+02
		S 1.0E-05	2.4E+05	6.4E+00	1.0E-05	1.8E+08	4.9E+03
					1.0E-04	2.1E+07	5.6E+02
^{241}Am	432 y	M 5.0E-04	7.4E+02	2.0E-02	5.0E-04	1.0E+05	2.7E+00
^{244}Cm	18.1 y	M 5.0E-04	1.2E+03	3.2E-02	5.0E-04	1.7E+05	4.5E+00
^{252}Cf	2.64 y	M 5.0E-04	1.5E+03	4.2E-02	5.0E-04	2.2E+05	6.0E+00

Appendix A
Sources of Physical and Chemical Data

Sources of Physical and Chemical Data A-1

SOURCES OF PHYSICAL AND CHEMICAL DATA

In addition to the primary research journals, there are many useful sources of property data of the type contained in the *CRC Handbook*. A selected list of these is presented here, with emphasis on print and electronic sources whose contents have been subject to a reasonable level of quality control.

A. Data Journals

1. ***Journal of Physical and Chemical Reference Data*** — Published jointly by the National Institute of Standards and Technology and the American Institute of Physics, this quarterly journal contains compilations of evaluated data in chemistry, physics, and materials science. It is available in print and on the Web <http://aip.scitation.org/journal/jpr>.
2. ***Journal of Chemical and Engineering Data*** — This bimonthly journal of the American Chemical Society publishes articles reporting original experimental measurements carried out under carefully controlled conditions. The main emphasis is on thermochemical and thermophysical properties. Review articles with evaluated data from the literature are also published <pubs.acs.org/journals/jceaax/index.html>.
3. ***Journal of Chemical Thermodynamics*** — This journal publishes original research papers that include highly accurate measurements of thermodynamic and thermophysical properties <www.sciencedirect.com/science/journal/00219614>.
4. ***Atomic Data and Nuclear Data Tables*** — This is a bimonthly journal containing compilations of data in atomic physics, nuclear physics, and related fields <www.sciencedirect.com/science/journal/aip/0092640X>.
5. ***Journal of Phase Equilibria and Diffusion*** — This journal presents critically evaluated phase diagrams, kinetic properties, and related data on alloy systems. It is now published by Springer and is the successor to the previous ASM periodical *Bulletin of Alloy Phase Diagrams* <www.springer.com/materials/journal/11669>.

B. Data Centers

This section lists selected organizations that perform a continuing function of compiling and critically evaluating data in specific fields of science.

1. **National Institute of Standards and Technology** — Under its Standard Reference Data program, NIST supports a number of data centers in chemistry, physics, and materials science. Topics covered include thermodynamics, fluid properties, chemical kinetics, mass spectroscopy, atomic spectroscopy, fundamental physical constants, ceramics, and crystallography. Address: Office of Standard Reference Data, National Institute of Standards and Technology, Gaithersburg, MD 20899 <www.nist.gov/srd/>.
2. **Thermodynamics Research Center** — Now located at the National Institute of Standards and Technology, TRC maintains an extensive archive of data covering thermodynamic, thermochemical, and transport properties of organic compounds and mixtures. Data are distributed in both print and electronic form. Address: Mail code 838.00, 325 Broadway, Boulder, CO 80305-3328 <www.trc.nist.gov>.
3. **Design Institute for Physical Property Data** — Under the auspices of the American Institute of Chemical Engineers, DIPPR offers evaluated data on industrially important chemical compounds. The largest project deals with physical, thermodynamic, and transport properties of pure compounds <http://www.aiche.org/dippr/>.
4. **Dortmund Data Bank** — Maintains extensive databases on thermodynamic and transport properties of pure compounds and mixtures of industrial interest. The data are distributed through DECHEMA, FIZ CHEMIE, and other outlets. Address: DDBST GmbH, Industriestr. 1, 26121 Oldenburg, Germany <www.ddbst.de>.
5. **Cambridge Crystallographic Data Centre** — Maintains the Cambridge Structural Database of over 950,000 organic compounds. The data files and manipulation software are distributed in several ways. Address: 12 Union Rd., Cambridge CB2 1EZ, U.K. <www.ccdc.cam.ac.uk>.
6. **FIZ Karlsruhe** — In addition to many bibliographic databases, FIZ Karlsruhe maintains the Inorganic Crystal Structure Database in collaboration with the National Institute of Standards and Technology. The ICSD contains the atomic coordinates and related data on over 200,000 inorganic crystals. Address: Fachinformationszentrum (FIZ) Karlsruhe, Hermann-von-Helmholtz-Platz 1, D-76344 Eggenstein-Leopoldshafen, Germany <www.fiz-karlsruhe.de/leistungen/kristallographie/kristallstrukturdepot.html>.
7. **International Centre for Diffraction Data** — Maintains and distributes the Powder Diffraction File (PDF), a file of over 890,000 x-ray powder diffraction patterns used for identification of crystalline materials. Address: 12 Campus Blvd., Newton Square, PA 19073-3273 <www.icdd.co>.
8. **Research Collaboratory for Structural Bioinformatics** — Maintains the Protein Data Bank (PDB), a file of 3-dimensional structures of proteins and other biological macromolecules. Address: Department of Chemistry and Chemical Biology, Rutgers University, 610 Taylor Road, Piscataway, NJ 08854-8087 <www.rcsb.org>.
9. **Toth Information Systems** — Maintains the Metals Crystallographic Data File (CRYSTMET) <cds.dl.ac.uk/cds/datasets/crys/mdf/llmdf.html>.
10. **Atomic Mass Data Center** — Collects and evaluates high-precision data on masses of individual isotopes and maintains a comprehensive database. Address: C.S.N.S.M (IN2P3-CNRS), Batiment 108, F-91405 Orsay Campus, France <amdc.impcas.ac.cn>.
11. **Particle Data Group** — International center for data of high-energy physics; maintains a database of properties of fundamental particles that is published in both print and electronic form. Address: MS 50-308, Lawrence Berkeley National Laboratory, Berkeley, CA 94720 <pdg.lbl.gov>.
12. **National Nuclear Data Center** — Maintains databases on nuclear structure and reactions, including neutron cross sections. The NNDC is the U.S. node in an international network of nuclear data centers. Address: Brookhaven National Laboratory, Upton, NY 11973-5000 <www.nndc.bnl.gov>.
13. **International Union of Pure and Applied Chemistry** — Address: PO Box 13757, Research Triangle Park, NC

27709-3757 <www.iupac.org>. IUPAC supports a number of long-term data projects, including these examples:

a. **Solubility Data Project** — Carries out evaluation of all types of solubility data. The results are published in the Solubility Data Series, whose current outlet is the *Journal of Physical and Chemical Reference Data* <srdata.nist.gov/solubility/>.
b. **Kinetic Data for Atmospheric Chemistry** — Maintains a comprehensive database on the kinetics of reactions important in the chemistry of the atmosphere <http://iupac.pole-ether.fr/>.
c. **Stability Constants Database** — Collection of metal-ligand stability constants and associated software <www.acadsoft.co.uk>.

C. Major Multi-Volume Handbook Series

1. ***CRC Chemical Dictionaries*** — These originally appeared in print form as the Dictionary of Organic Compounds, Dictionary of Natural Products, etc. They are now published in electronic form and are available on the Web <www.chemnetbase.com>. The consolidated version, called the Combined Chemical Dictionary, has data on more than 660,000 compounds spanning all branches of chemistry. The coverage includes physical properties, biological sources, hazard information, uses, and literature references.
2. ***Properties of Organic Compounds*** — Originally published in three editions as the *Handbook of Data on Organic Compounds*, it is now in electronic form as Properties of Organic Compounds. The database includes about 30,000 compounds; physical properties and spectral data (mass, infrared, Raman, ultraviolet, and NMR) are covered. It is offered via online access <www.chemnetbase.com>.
3. ***Beilstein Handbook of Organic Chemistry*** — The classic source of data on organic compounds, dating from the 19th century, *Beilstein* was converted to electronic form in the last decade of the 20th century. Over 8 million compounds and 10 million chemical reactions were covered, with a broad range of physical properties as well as synthetic methods and ecological data. The database is now accessed through Reaxys <www.elsevier.com/solutions/reaxys>.
4. ***Gmelin Handbook of Inorganic and Organometallic Chemistry*** — A subset of the information in the print series has been converted to electronic form and is now distributed in the same manner as *Beilstein*. In addition to the standard physical properties, the coverage includes a wide range of optical, magnetic, spectroscopic, thermal, and transport properties for about 1.4 million compounds <www.elsevier.com/solutions/reaxys>.
5. ***DECHEMA Chemical Data Series*** — DECHEMA distributes the DETHERM database, which emphasizes data used in process design in the chemical industry, including thermodynamic and transport properties of about 60,000 pure compounds and 163,000 mixtures. Access is available through in-house databases and via the Web <www.dechema.de>.
6. ***Landolt-Börnstein Numerical Data and Functional Relationships in Science and Technology*** — Landolt-Börnstein covers a very broad range of data in physics, chemistry, crystallography, materials science, biophysics, astronomy, and geophysics. Hard-copy volumes are no longer published, but most of the entire collection is available online <https://materials.springer.com/about-springer-materials-interactive>.

Appendix

D. Selected Single-Volume Handbooks

The following handbooks offer broad coverage of high-quality data in a single volume. This list is only representative.

1. ***American Institute of Physics Handbook*** — Although an old book, it contains much data that are still useful, especially in acoustics, mechanics, optics, and solid state physics. (Dwight E. Gray, ed., McGraw-Hill, New York, 1972)
2. ***Constants of Inorganic Substances*** — This book presents physical constants, thermodynamic data, solubility, reactivity, and other information on over 3000 inorganic compounds. Because it draws heavily on Russian literature, it contains a great deal of data that do not make their way into most U.S. handbooks. (R. A. Lidin, L. L. Andreeva, and V. A. Molochko, Begell House, New York, 1995)
3. ***Handbook of Chemistry and Physics*** — Now in the 101st Edition, the *CRC Handbook* covers data from most branches of chemistry and physics. The annual revisions permit regular updating of the information. Also available on the Web <hbcponline.com>. (John Rumble, ed., CRC Press, Boca Raton, FL, 2020)
4. ***Handbook of Inorganic Compounds, Second Edition*** — This book covers physical constants and other properties for about 3300 inorganic compounds. (CRC Press, Boca Raton, FL, 2011)
5. ***Handbook of Physical Properties of Liquids and Gases*** — This is a valuable source of data on all types of fluids, ranging from liquid and gaseous hydrocarbons to molten metals and ionized gases. Detailed tables of physical, thermodynamic, and transport properties are given for temperatures from the cryogenic region to 6000 K. Western and Russian literature is covered. (N. B. Vargaftik, Y. K. Vinogradov, and V. S. Yargin, Begell House, New York, 1996)
6. ***Handbook of Physical Quantities*** — The range of coverage is somewhat similar to the *CRC Handbook of Chemistry and Physics*, but with a stronger emphasis on physics than on chemistry. Solid state physics, lasers, nuclear physics, geophysics, and astronomy receive considerable attention. (Igor S. Grigoriev and Evgenii Z. Meilikhov, eds., CRC Press, Boca Raton, FL, 1997)
7. ***Kaye & Laby Tables of Physical and Chemical Constants*** — Kaye & Laby dates from 1911, and the 16th Edition was prepared in 1995 by a committee of experts. The coverage extends to almost every field of physics and chemistry; data on a limited number of representative substances or materials are given for each topic. Now available online at <http://www.kayelaby.npl.co.uk/>.
8. ***Lange's Handbook of Chemistry*** — Provides broad coverage of chemical data; last updated in 2016. (James G. Speight, ed., McGraw-Hill, New York, 2016)
9. ***Recommended Reference Materials for the Realization of Physicochemical Properties*** — This IUPAC book emphasizes highly accurate data on substances and materials that can be used as calibration standards. It covers physical, thermal, optical, and electrical properties. (K. N. Marsh, ed., Blackwell Scientific Publications, Oxford, 1987)
10. ***The Merck Index*** — Now in its 15th Edition, *The Merck Index* is a widely used source of data on over 10,000 compounds, chosen particularly for their importance in biology, medicine, and ecology. A short monograph on each compound gives information on the synthesis and uses as well as physical and toxicological properties. Now available

online at <https://www.rsc.org/merck-index>. (Maryadele J. O'Neil, ed., RSC Publishing, 2013)

E. Summary of Useful Web Sites for Physical and Chemical Properties

Most of the Web sites in the following list provide direct access to factual data on physical and chemical properties. However, the list also includes portals that link to different property databases or describe the procedure for gaining access to electronic sources of property data. There are also a few chemical directory sites that are useful for obtaining formulas, synonyms, and registry numbers for substances of interest.

Useful Web Sites

Web site	Address	Comments
ACD/Labs Spectral Data	www.acdlabs.com/products/adh/	Infrared, Raman, and NMR spectra collections from Coblentz Society and other sources
Advanced Chemistry Development	www.acdlabs.com	Chemical directory, with programs for estimating physical and spectral properties
ASM Alloy Center Database	https://www.asminternational.org/materials-resources/online-databases/-/journal_content/56/10192/15468704/DATABASE	Physical, electrical, thermal, and mechanical properties of alloys
American Mineralogist Crystal Structure Database	www.geo.arizona.edu/AMS/amcsd.php	Lattice constants of minerals
Atomic Mass Data Center	amdc.impcas.ac.cn	See B.10
Beilstein Database	www.elsevier.com/solutions/reaxys	Properties and reactions of organic compounds. See C.3
Biocatalysis/Biodegradation Database	umbbd.ethz.ch/	Biocatalytic reactions, biodegradation of chemical compounds
BioCyc	biocyc.org/	Metabolic pathways of microorganisms
Biological Macromolecule Crystallization Database	bmcd.ibbr.umd.edu	Crystal data and crystallization conditions for proteins, nucleic acids, and complexes
BRENDA	www.brenda-enzymes.info/	Enzyme nomenclature and properties
Cambridge Structural Database	www.ccdc.cam.ac.uk	See B.5
Carbon Dioxide Information Center	ess-dive.lbl.gov/	Data on atmospheric carbon dioxide; combined into the U.S. Department of Energy (DOE) Environmental Systems Science Data Infrastructure for a Virtual Ecosystem (ESS-DIVE)
ChemExper	www.chemexper.com/	Consolidated chemical catalogs from various suppliers; provides physical properties and safety data; links to molfiles and MSDS
Chemical Entities of Biological Interest (ChEBI)	www.ebi.ac.uk/chebi	Dictionary of molecules and fragments, with identifiers and structures
ChemIDplus	chem.sis.nlm.nih.gov/chemidplus/	Chemical directory
ChemIndustry	www.chemindustry.com/chemicals/	Chemical directory
CHEMnetBASE	www.chemnetbase.com	Portal to CRC Chemical Dictionaries, Handbook of Chemistry and Physics, Properties of Organic Compounds, etc.
ChemSpider	www.chemspider.com	Aggregation of chemical structures and other information from many public sources; references to synthesis; limited property data
ChemSynthesis Chemical Database	www.chemsynthesis.com	References to syntheses; limited property data
CODATA Databases	www.codata.info/resources/databases/	Thermodynamic key values and fundamental constants
Comparative Toxicogenomics Database (CTD)	ctdbase.org/	Chemical – gene/protein interactions
Cool Prop	coolprop.sourceforge.net	Thermophysical properties of 110 pure fluids and pseudo-pure fluids including humid air of interest to the organic Rankine cycle community
CRC Combined Chemical Dictionary	www.chemnetbase.com/	See C.1
Crystallography Open Database (COD)	www.crystallography.net	Crystal data on 400,000+ compounds
DECHEMA (DETHERM)	www.dechema.de/en/Databases.html	See C.5
DIPPR Pure Compound Database	www.aiche.org/dippr	See B.3
Dortmund Data Bank	www.ddbst.de	See B.4
eMolecules	www.emolecules.com	Portal to databases of chemical suppliers
eNanoMapper	http://www.enanomapper.net/	Data and modeling tools for assessing safety and properties of nanomaterials
Enzyme Nomenclature Database	www.expasy.ch/enzyme/	IUBMB nomenclature for enzymes
European Bioinformatics Institute	www.ebi.ac.uk/services	Nucleotide and protein sequences, protein structures, enzyme nomenclature and reactions
FDM Reference Spectra Databases	www.fdmspectra.com/	Infrared, Raman, and mass spectra

Web site	Address	Comments
FIZ Karlsruhe — ICSD	www.fiz-karlsruhe.de	ICSD crystal structure databases
Fundamental Physical Constants	physics.nist.gov/cuu/constants	CODATA fundamental constants
Gmelin Database	www.elsevier.com/solutions/reaxys	Properties and reactions of inorganic compounds. See C.4
Handbook of Chemistry and Physics	hbcponline.com	Online version of CRC Handbook of Chemistry and Physics
Hazardous Substances Data Bank	toxnet.nlm.nih.gov/cgi-bin/sis/htmlgen?HSDB	Physical and toxicological properties of chemicals of health or environmental importance
HITRAN Database	www.cfa.harvard.edu/hitran/	High-resolution spectroscopic data for constituents of the atmosphere; parameters for calculating atmospheric transmission
Human Metabolome Database	www.hmdb.ca	Chemical and biological data on small molecule metabolites in humans
Infotherm	www.infotherm.com	Physical and thermal properties of pure compounds and mixtures
International Centre for Diffraction Data	www.icdd.com	See B.7
Ionic Liquids Database (ILThermo)	ilthermo.boulder.nist.gov/	Thermodynamic and thermophysical properties of ionic liquids and mixtures
IUBMB	https://www.qmul.ac.uk/sbcs/iubmb/	Enzyme and nucleic acid nomenclature
IUCr Data Activities	www.iucr.org/resources/data	Portal to crystallographic databases
IUPAC Home Page	www.iupac.org/	See B.13
IUPAC Kinetics Data	iupac.pole-ether.fr/	See B.13.2
IUPAC Nomenclature Rules	iupac.org/what-we-do/nomenclature/	Useful site for organic and biochemical nomenclature
IUPAC-NIST Solubility Database	srdata.nist.gov/solubility/	See B.13.1
Knovel.com	app.knovel.com/web/index.v	Portal to Lange's Handbook, Perry's Chemical Engineers' Handbook, etc.
Kyoto Encyclopedia of Genes and Genomes (KEGG)	www.genome.ad.jp/kegg/	Includes data on drugs and other biochemical compounds
Landolt-Börnstein Online	materials.springer.com	See C.6
Lipidat	www.lipidat.tcd.ie	Structures and thermodynamic properties of lipids; crystal polymorphic transitions
MatWeb	www.matweb.com	Thermal, electrical, and mechanical properties of engineering materials
NASA Chemical Kinetics Data	jpldataeval.jpl.nasa.gov	Kinetic and photochemical data for stratospheric modeling
National Center for Biotechnology Information	www.ncbi.nlm.nih.gov	Portal to GenBank and other sequence databases
National Nuclear Data Center	www.nndc.bnl.gov	See B.12
National Toxicology Program	ntp.niehs.nih.gov	Chemical health and safety data
NIST Atomic Spectra Database	www.nist.gov/pml/atomic-spectra-database	Energy levels, wavelengths, and transition probabilities of atoms and atomic ions
NIST Ceramics Webbook	www.nist.gov/mml/mmsd/webbook	See B.1
NIST Chemistry Webbook	webbook.nist.gov	Broad range of physical, thermal, and spectral properties
NIST Data Gateway	srdata.nist.gov/gateway/	Portal to all NIST data systems; see B.1
NIST Physical Reference Data	www.nist.gov/pml/productsservices/physical-reference-data	Atomic and molecular spectra, cross sections, x-ray attenuation, and dosimetry data
NLM Gateway	www.nlm.nih.gov/databases	Portal to all National Library of Medicine databases
NMR Shift DB	nmrshiftdb.nmr.uni-koeln.de/	NMR data submitted by users
Nucleic Acid Database	ndbserver.rutgers.edu/	Crystal structures of nucleic acids
Particle Data Group	pdg.lbl.gov	See B.11
Pauling File	paulingfile.com	Phase diagrams, crystal structure, and physical property data used for materials design
Polymers — A Property Database	poly.chemnetbase.com	Properties of commercial polymers
Powder Diffraction File	www.icdd.com	See B.7
Properties of Organic Compounds	www.chemnetbase.com/	See C.2
Protein Data Bank	www.rcsb.org	See B.8
PubChem	pubchem.ncbi.nlm.nih.gov/	Chemical directory with links to biological information
Reaxys	www.elsevier.com/solutions/reaxys	Access to Beilstein and Gmelin databases
SABIO-Reaction Kinetics Database	sabio.villa-bosch.de/	Data on kinetics of biochemical reactions
Sigma-Aldrich	www.sigmaaldrich.com/	Chemical catalogs; includes some physical property data
Spectral Database for Organic Compounds	https://sdbs.db.aist.go.jp/sdbs/cgi-bin/cre_index.cgi	MS, NMR, IR, Raman, and ESR spectra; 32,000 compounds measured at AIST, Japan
Wiley Spectra Lab	wileyspectralab.com/	IR, NMR, and mass spectra
Spectra Online	www.ftirsearch.com/	FTIR and Raman spectra
SPRESI-web	www.spresi.de/	Structures, reactions, and some physical properties
SpringerMaterials	materials.springer.com	The online version of Landolt-Börnstein Tables. See C.6

Web site	Address	Comments
STN Easy	stneasy.cas.org	Chemical directory (and access to Chemical Abstracts databases)
STN Easy-Europe	stneasy.fiz-karlsruhe.de	European node of STN Easy
STN Easy-Japan	stneasy-japan.cas.org	Japanese node of STN Easy
Swissprot	enzyme.expasy.org	Enzyme nomenclature and related information
Thermodynamics of Enzyme-Catalyzed Reactions	randr.nist.gov/enzyme	Equilibrium constants of biochemical reactions
Thermodynamics Research Center	trc.nist.gov	See B.2
TOXCAST	https://www.epa.gov/chemical-research/toxicity-forecasting	EPA's Toxicity Forecaster (ToxCast) generates data and predictive models on thousands of chemicals of interest to the EPA.
TOXNET	toxnet.nlm.nih.gov	Portal to HSDB and other databases on hazardous chemicals

Index

The most efficient way to use this index is to look for the pertinent *property* (e.g., vapor pressure, entropy), *process* (e.g., shielding, calibration), or *general concept* (e.g., units, radiation). Most primary entries are subdivided into several secondary entries, e.g., under heat capacity there are 24 secondary entries such as air, metals, water, etc. Primary entries will be found for certain *classes of substances*, such as alloys, elements, organic compounds, refrigerants, semiconductors, etc. Primary entries are also given for the individual chemical elements and for a few compounds such as water and carbon dioxide. However, only the most important tables are listed under these substances. Therefore, the user will find in most cases that it is best to look first for the property of interest, then examine the table or tables that are referenced.

Entries in boldface type are the titles of tables as they appear in the Table of Contents.

The reference given for each index term is the inclusive pages of the pertinent table (e.g., 8-45 to 55). The introduction to each table describes the method of ordering the substances within that table.

The editor would be grateful for comments and suggestions on this index.

INDEX

A

AAS, definition, 12-1 to 5
Abbreviations
  amino acids, 7-3 to 4
  analytical chemistry, 8-1 to 5
  laboratory hazards, 16-3 to 4
  physical quantities, 2-1 to 12
  polymers, 13-1 to 2
  scientific terms, 2-24 to 36
  units, 1-18 to 22
Abbreviations and Symbols Used in Analytical Chemistry, 8-1 to 5
Abbreviations Used in the Assessment and Presentation of Laboratory Hazards, 16-3 to 4
Abbreviations Used in Polymer Science and Technology, 13-1 to 2
Absorption
  infrared, by earth's atmosphere, 14-34
  light, by elemental solids, 12-113 to 116
  light, by other solids, 12-117 to 121
  microwave power, by water, 6-14
  sound, by air, 14-41
Abundance, isotopic, 1-12 to 15, 8-71, 11-3 to 42
Abundance of Elements in the Earth's Crust and in the Sea, 14-21
Acceleration Due to Gravity, 14-17
  at poles and equator, 14-14 to 15
  at various latitudes, 14-17
  on the sun, moon, and planets, 14-2 to 6
  standard value, 1-1 to 9
Acid dissociation constant
  amino acids, 7-1 to 2
  biological buffers, 7-33
  inorganic acids and bases, 5-108
  organic acids and bases, 5-109 to 118
  purine and pyrimidine bases, 7-5
Acids
  activity coefficients, 5-98 to 100, 5-121 to 125
  electrical conductivity, 5-94
  enthalpy of dilution, 5-126
  fatty, 7-7 to 8
  indicators, 8-88 to 89
  inorganic, dissociation constant, 5-108
  organic, dissociation constant, 5-109 to 118
Acoustics
  sound velocity, 14-39 to 40, 14-41, 14-42
  sound velocity in air, 6-15 to 20
  sound velocity in fluids, 6-21 to 39, 6-40 to 50
  sound velocity in water and steam, 6-1 to 4
Acronyms, definitions, 2-24 to 36
Actinium: *see also* Elements
  electron configuration, 1-16
  heat capacity, 4-65
  history, occurrence, uses, 4-1 to 39
  ionization energy, 10-112 to 115
  isotopes and their properties, 11-3 to 42
  physical properties, 4-62 to 64
  thermal properties, 12-129 to 130
Activation energy for diffusion in semiconductors, 12-89 to 96
Activity coefficients of electrolytes, 5-98 to 100, 5-121 to 125
AES, definition, 12-1 to 5
AFM, definition, 12-1 to 5
Air
  absorption of sound, 14-41
  cryogenic properties, 6-161 to 162
  density, 6-15 to 20
  diffusion of gases in, 6-249 to 250
  enthalpy, 6-15 to 20
  entropy, 6-15 to 20
  heat capacity, 6-15 to 20
  index of refraction, 10-153
  mean free path, 6-54
  permittivity (dielectric constant), 6-216
  speed of sound, as function of humidity and frequency, 14-41
  speed of sound, as function of temperature, 6-15 to 20, 14-42
  thermal conductivity, 6-15 to 20, 6-240 to 242
  thermodynamic properties, 6-15 to 20
  U.S. Standard Atmosphere, 14-23 to 28
  vapor pressure, 6-15 to 20
  viscosity, 6-15 to 20, 6-233 to 234
Airborne contaminants, threshold limits, 16-48 to 75
Albedo
  planets, 14-2 to 6
  satellites of the planets, 14-6 to 8
ALI for radionuclides, 16-93 to 96
Alkali halides, secondary electron emission, 12-112
Alkali metals: *see* entries for Lithium, Sodium, etc.
Alloys
  composition, 12-97 to 104
  electrical resistivity, 12-28 to 29
  eutectic temperatures, 15-38 to 39
  magnetic properties, 12-97 to 104
  superconducting properties, 12-42 to 57
  thermal conductivity, 12-122 to 127
Alpha particles, range and shielding, 16-88
Aluminum: *see also* Elements
  critical constants, 6-82 to 84
  electrical resistivity, 12-27
  electron configuration, 1-16
  heat capacity, 4-65
  history, occurrence, uses, 4-1 to 39
  ionization energy, 10-112 to 115
  isotopes and their properties, 11-3 to 42
  magnetic susceptibility, 4-73 to 78
  molten, density, 4-71 to 72
  physical properties, 4-62 to 64
  thermal conductivity, 12-131
  thermal properties, 12-129 to 130
  thermodynamic properties, 5-1 to 2
  thermophysical properties, 12-122 to 127
  vapor pressure, 6-87 to 116
  vapor pressure, high temperature, 4-66 to 68
Ambient (definition), 1-26
Americium: *see also* Elements
  electron configuration, 1-16
  history, occurrence, uses, 4-1 to 39
  ionization energy, 10-112 to 115
  isotopes and their properties, 11-3 to 42
  vapor pressure, high temperature, 4-66 to 68
Amino acids
  abbreviations and symbols, 7-3 to 4
  in blood, 7-50 to 60
  dissociation constants, 7-1 to 2
  physical properties, 7-1 to 2
  solubility, 7-1 to 2
  structure, 7-3 to 4
Ampere
  definition, 1-18 to 22
  maintained value, 1-1 to 9
Analytical procedures
  abbreviations and symbols, 8-1 to 5
  calibration methods, 8-10 to 16
  figures of merit, 8-17
  instrumental techniques, 8-6 to 9
  organic reagents, 8-95 to 108
  preparation of reagents, 8-90 to 94
  reduction of weighings, 8-85
  solids and surfaces, 12-1 to 5
  standardization, 8-10 to 16
Analytical Standardization and Calibration, 8-10 to 16
Ångström, definition, 1-18 to 22
Annual Limits on Intakes of Radionuclides, 16-93 to 96
Antiferroelectric crystals, Curie temperature, 12-41
Antiferromagnetic elements and compounds, 12-97 to 104
Antimony: *see also* Elements
  electrical resistivity, 12-27
  electron configuration, 1-16
  heat capacity, 4-65
  history, occurrence, uses, 4-1 to 39
  ionization energy, 10-112 to 115
  isotopes and their properties, 11-3 to 42
  magnetic susceptibility, 4-73 to 78
  molten, density, 4-71 to 72
  physical properties, 4-62 to 64
  thermal properties, 12-129 to 130
  vapor pressure, 6-87 to 116
Apparent Equilibrium Constants for Enzyme-Catalyzed Reactions, 7-17 to 19
Apparent Equilibrium Thermodynamics of Protein-Ligand Binding Reactions, 7-21 to 29
Appearance potential, molecules, 10-116 to 133
APS, definition, 12-1 to 5
Aqueous Solubility and Henry's Law Constants of Organic Compounds, 5-152 to 182
Aqueous Solubility of Inorganic Compounds at Various Temperatures, 5-183 to 188
Aqueous solutions
  activity coefficients, 5-98 to 100, 5-121 to 125
  concentrative properties, 5-133 to 148
  density, 5-133 to 148
  diffusion of ions, 5-98 to 100
  diffusion of non-electrolytes, 6-252 to 253

electrical conductivity, 5-97, 5-98 to 100
enthalpy, 5-127
freezing point depression, 5-133 to 148
heat capacity, 5-65 to 66
hydrohalogen acids, conductivity, 5-96
index of refraction, 5-133 to 148
solubility product constant, 5-194 to 195
surface tension, 6-187
thermodynamic properties, 5-65 to 66
vapor pressure, 6-126
viscosity, 5-133 to 148
Argon: *see also* Elements
critical constants, 6-82 to 84
cryogenic properties, 6-161 to 162
electron configuration, 1-16
entropy, 5-1 to 2
history, occurrence, uses, 4-1 to 39
ionization energy, 10-112 to 115
isotopes and their properties, 11-3 to 42
magnetic susceptibility, 4-73 to 78
mean free path, 6-54
permittivity (dielectric constant), 6-193 to 215, 6-216
physical properties, 4-62 to 64
solubility in water, 5-149 to 150
thermal conductivity, 6-21 to 39, 6-240 to 242
thermodynamic properties at high temperature, 5-41 to 63
thermophysical properties, 6-21 to 39
vapor pressure, 6-87 to 116, 6-117 to 125
viscosity, 6-21 to 39, 6-233 to 234
Arsenic: *see also* Elements
critical constants, 6-82 to 84
electron configuration, 1-16
heat capacity, 4-65
history, occurrence, uses, 4-1 to 39
ionization energy, 10-112 to 115
isotopes and their properties, 11-3 to 42
magnetic susceptibility, 4-73 to 78
molten, density, 4-71 to 72
physical properties, 4-62 to 64
vapor pressure, 6-87 to 116
Astatine: *see also* Elements
electron configuration, 1-16
history, occurrence, uses, 4-1 to 39
ionization energy, 10-112 to 115
isotopes and their properties, 11-3 to 42
vapor pressure, 6-87 to 116
Astronomical Constants, 14-1
Astronomical unit, 14-1
Atmosphere
carbon dioxide concentration, 14-35
chemical reactions, 5-69 to 91
cosmic ray background, 11-56 to 61
infrared absorption, 14-34
mass, 14-14 to 15
planetary, composition, 14-2 to 6
standard (unit), 1-18 to 22
U.S. Standard, various properties, 14-23 to 28
Atmospheric Concentration of Carbon Dioxide, 1959–2020, 14-35
Atomic Masses and Isotopic Abundances, 1-12 to 15
Atomic masses, 1-12 to 15, 11-3 to 42
Atomic mass unit (amu), 1-1 to 9, 1-18 to 22
Atomic Radii of Elements, 9-56 to 57
Atomic radius, rare-earth elements, 4-52 to 61
Atomic spectra
analytical chemistry applications, 8-6 to 9
elements, line spectra, 10-1 to 50
ionization energies, 10-112 to 115
transition probabilities, 10-51 to 53
wavelengths, 10-1 to 50
Atomic Transition Probabilities, 10-51 to 53
Atomic weights, 1-10 to 11
Atoms
electron affinity, 10-54 to 75
electron configuration, 1-16
ionization energies, 10-112 to 115
masses, 1-12 to 15
photon cross sections, 10-134 to 139
polarizability, 10-95 to 111
radii, 9-56 to 57
spectra, 10-1 to 50
ATR, definition, 12-1 to 5
Attenuation and Speed of Sound in Air as a Function of Humidity and Frequency, 14-41
Autoignition temperature, 16-12 to 29
Avogadro constant, 1-1 to 9, 1-18 to 22
Azeotropic Data for Binary Mixtures, 6-217 to 232

B

Bands, electromagnetic (classification), 10-140 to 141
Barium: *see also* Elements
electrical resistivity, 12-27
electron configuration, 1-16
heat capacity, 4-65
history, occurrence, uses, 4-1 to 39
ionization energy, 10-112 to 115
isotopes and their properties, 11-3 to 42
magnetic susceptibility, 4-73 to 78
molten, density, 4-71 to 72
physical properties, 4-62 to 64
thermal properties, 12-129 to 130
vapor pressure, 6-87 to 116
vapor pressure, high temperature, 4-66 to 68
Barn, definition, 1-18 to 22
Barometer corrections, 15-33
Barriers to internal rotation, 9-68 to 72
Bases
activity coefficients, 5-98 to 100, 5-121 to 125
electrical conductivity, 5-94
indicators, 8-88 to 89
inorganic, dissociation constant, 5-108
organic, dissociation constant, 5-109 to 118
purine and pyrimidine, 7-5
Basic Instrumental Techniques of Analytical Chemistry, 8-6 to 9
Becquerel, definition, 1-18 to 22
Berkelium: *see also* Elements
electron configuration, 1-16
history, occurrence, uses, 4-1 to 39
isotopes and their properties, 11-3 to 42
Beryllium: *see also* Elements
electrical resistivity, 12-27
electron configuration, 1-16
heat capacity, 4-65
history, occurrence, uses, 4-1 to 39
ionization energy, 10-112 to 115
isotopes and their properties, 11-3 to 42
magnetic susceptibility, 4-73 to 78
molten, density, 4-71 to 72
physical properties, 4-62 to 64
thermal properties, 12-129 to 130
thermodynamic properties, 5-1 to 2
thermophysical properties, 12-122 to 127
vapor pressure, 6-87 to 116
vapor pressure, high temperature, 4-66 to 68
BET, definition, 12-1 to 5
Billion, milliard (definition), 8-18
Binary prefixes, 1-18 to 22
Binding energy in molecules, 9-73 to 105
Biochemical nomenclature, references, 2-15
Biochemical reactions, redox potentials, 7-14 to 16
Biological Buffers, 7-33
Biological buffers, 7-30 to 32
Biological materials and tissues
cosmic ray effects, 11-56 to 61
ionizing radiation effects, 16-88, 16-89 to 90
pH, 7-34
Biological safety cabinets, 16-37 to 38
Biosphere, mass of, 14-14 to 15
Bismuth: *see also* Elements
electrical resistivity, 12-27
electron configuration, 1-16
heat capacity, 4-65
history, occurrence, uses, 4-1 to 39
ionization energy, 10-112 to 115
isotopes and their properties, 11-3 to 42
magnetic susceptibility, 4-73 to 78
molten, density, 4-71 to 72
physical properties, 4-62 to 64
thermal properties, 12-129 to 130
vapor pressure, 6-87 to 116
Blackbody Radiation, 10-143 to 144
Blood
chemical composition, 7-50 to 60
pH, 7-34, 7-50 to 60
Bohr magneton, 1-1 to 9
Bohr radius, 1-1 to 9
Bohrium (element 107), 4-1 to 39, 11-3 to 42
Boiling point
correction to standard pressure, 15-28 to 29
cryogenic fluids, 6-161 to 162
D_2O, 6-9
elements, 4-62 to 64
elevation of, 15-30
halocarbons, 6-164 to 166
inorganic compounds, 4-40 to 51, 6-85 to 86, 6-127 to 143
metals, 12-129 to 130
organic compounds, 3-1 to 55, 6-56 to 81, 6-127 to 143
pressure dependence, 15-28 to 29
rare-earth elements, 4-52 to 61
solvents, 15-13 to 22
Boltzmann constant, 1-1 to 9, 1-18 to 22
Bond Dissociation Energies, 9-73 to 105
Bond energy, 9-73 to 105
Bond lengths
characteristic, 9-55
diatomic molecules, 9-111 to 116
gas-phase molecules, 9-19 to 54
organic crystals, 9-1 to 16
organometallic compounds, 9-17 to 18
Bond Lengths in Crystalline Organic Compounds, 9-1 to 16
Bond Lengths in Organometallic Compounds, 9-17 to 18
Bonds, chemical
disruption energy, 9-73 to 105

dissociation energy (enthalpy), 9-73 to 105
lengths and angles, 9-1 to 16, 9-17 to 18, 9-19 to 54
strength, 9-73 to 105
stretching force constants, 9-107
Boron: *see also* Elements
electron configuration, 1-16
heat capacity, 4-65
history, occurrence, uses, 4-1 to 39
ionization energy, 10-112 to 115
isotopes and their properties, 11-3 to 42
magnetic susceptibility, 4-73 to 78
molten, density, 4-71 to 72
physical properties, 4-62 to 64
thermodynamic properties, 5-1 to 2
vapor pressure, 6-87 to 116
Brass
thermal conductivity, 12-132 to 133
thermophysical properties, 12-122 to 127
Bravais lattices, 12-6 to 10
Breakdown voltage, 15-45 to 49
Bromine: *see also* Elements
critical constants, 6-82 to 84
electron configuration, 1-16
heat capacity, 4-65
history, occurrence, uses, 4-1 to 39
ionization energy, 10-112 to 115
isotopes and their properties, 11-3 to 42
physical properties, 4-62 to 64
thermodynamic properties, 5-1 to 2
thermodynamic properties at high temperature, 5-42 to 64
vapor pressure, 6-87 to 116, 6-117 to 125
Buffer Solutions Giving Round Values of pH at 25 °C, 5-132
Buffers
biological, 7-30 to 32, 7-33
for round values of pH, 5-132
standard solutions, 5-128 to 131

C

^{13}C-NMR Absorptions of Major Functional Groups, 8-66 to 67
^{13}C-NMR Chemical Shifts of Common Organic Solvents, 8-68
Cadmium: *see also* Elements
electrical resistivity, 12-27
electron configuration, 1-16
heat capacity, 4-65
history, occurrence, uses, 4-1 to 39
ionization energy, 10-112 to 115
isotopes and their properties, 11-3 to 42
magnetic susceptibility, 4-73 to 78
molten, density, 4-71 to 72
physical properties, 4-62 to 64
thermal properties, 12-129 to 130
thermodynamic properties, 5-1 to 2
vapor pressure, 6-87 to 116
vapor pressure, high temperature, 4-66 to 68
Calcium: *see also* Elements
electrical resistivity, 12-27
electron configuration, 1-16
heat capacity, 4-65
history, occurrence, uses, 4-1 to 39
ionization energy, 10-112 to 115
isotopes and their properties, 11-3 to 42
magnetic susceptibility, 4-73 to 78
molten, density, 4-71 to 72
physical properties, 4-62 to 64
thermal properties, 12-129 to 130
thermodynamic properties, 5-1 to 2
vapor pressure, 6-87 to 116
vapor pressure, high temperature, 4-66 to 68
Calibration
barometers, 15-33
boiling points, to standard pressure, 15-28 to 29
in chemical analysis, 8-10 to 16
conductivity cells, 5-95
index of refraction, 10-152
pH, 5-128 to 131
relative humidity, 15-37
temperature scale, 1-17, 15-10 to 11
thermocouples, 15-1 to 9
weighings in air, 8-85
Californium: *see also* Elements
electron configuration, 1-16
history, occurrence, uses, 4-1 to 39
isotopes and their properties, 11-3 to 42
Candela, definition, 1-18 to 22
Carbohydrate Names and Symbols, 7-12 to 13
Carbon: *see also* Elements
dielectric constant, 12-31 to 39
electron configuration, 1-16
heat capacity, 4-65
history, occurrence, uses, 4-1 to 39
ionization energy, 10-112 to 115
isotopes and their properties, 11-3 to 42
magnetic susceptibility, 4-73 to 78
physical properties, 4-62 to 64
thermodynamic properties, 5-1 to 2
vapor pressure, 6-87 to 116
Carbon dioxide
atmospheric concentration (historical), 14-35
critical constants, 6-82 to 84
global warming potential, 14-37 to 38
mean free path, 6-54
permittivity (dielectric constant), 6-193 to 215, 6-216
release by combustion of fuels, 5-68
solubility in water, 5-149 to 150
solubility in water at various pressures, 5-151
speed of sound, 14-39 to 40
standard thermodynamic properties, 5-3 to 41
sublimation pressure, 6-21 to 39, 6-85 to 86
thermal conductivity, 6-21 to 39, 6-240 to 242
thermodynamic properties at high temperature, 5-42 to 64
thermophysical properties, 6-21 to 39
vapor pressure, 6-117 to 125
viscosity, 6-21 to 39, 6-233 to 234
Carcinogenic chemicals, list, 16-76 to 85
Carrier gas properties, for chromatography, 8-20
CARS, definition, 12-1 to 5
CAS Registry Numbers
inorganic compounds, 4-40 to 51
organic compounds, 3-1 to 55
Celsius temperature, definition, 1-18 to 22
Ceramics
breakdown voltage, 15-45 to 49
permittivity (dielectric constant), 12-31 to 39
thermal conductivity, 12-136 to 137
Cerenkov light, in cosmic ray showers, 11-56 to 61
Ceres, orbital data and dimensions, 14-2 to 6
Cerium: *see also* Elements
electrical resistivity, 12-27
electron configuration, 1-16
heat capacity, 4-65
history, occurrence, uses, 4-1 to 39
ionization energy, 10-112 to 115
isotopes and their properties, 11-3 to 42
magnetic susceptibility, 4-73 to 78
molten, density, 4-71 to 72
physical properties, 4-62 to 64
thermal properties, 12-129 to 130
vapor pressure, 6-87 to 116
vapor pressure, high temperature, 4-66 to 68
Cesium: *see also* Elements
electrical resistivity, 12-27
electron configuration, 1-16
heat capacity, 4-65
history, occurrence, uses, 4-1 to 39
ionization energy, 10-112 to 115
isotopes and their properties, 11-3 to 42
magnetic susceptibility, 4-73 to 78
molten, density, 4-71 to 72
physical properties, 4-62 to 64
thermal properties, 12-129 to 130
thermodynamic properties, 5-1 to 2
vapor pressure, 6-87 to 116
vapor pressure, high temperature, 4-66 to 68
CFCs, global warming potential, 14-37 to 38
CFCs, various properties, 6-164 to 166
Characteristic Bond Lengths in Free Molecules, 9-55
Characteristics of Infrared Detectors, 10-145
Characteristics of Particles and Particle Dispersoids, 15-40
Characterization of materials, 12-1 to 5
Charge, elementary, 1-1 to 9
Chauvenet's criterion, 8-19
Chemical Abstracts Service nomenclature, 2-15
Chemical Abstracts Service Registry Numbers: *see* CAS Registry Numbers
Chemical Carcinogens, 16-76 to 85
Chemical Composition of the Human Body, 7-61
Chemical Constituents of Human Blood, 7-50 to 60
Chemical Fume Hoods and Biological Safety Cabinets, 16-39 to 40
Chemical kinetics, atmospheric reactions, 5-69 to 91
Chemical nomenclature, 2-15
Chemical Reaction Rate Constants for Atmospheric Studies, 5-69 to 91
Chemical Safety, 16-1 to 2
Chemical shifts, NMR
for contaminants in deuterated solvents, 8-65
for ^{13}C, 8-66 to 67
for ^{15}N, 8-69 to 70
for protons, 8-53 to 57, 8-58
of solvents for NMR, 8-68, 8-65
Chemical structure representation, 2-22 to 23
Chlorine: *see also* Elements
critical constants, 6-82 to 84
electron configuration, 1-16
heat capacity, 4-65

history, occurrence, uses, 4-1 to 39
ionization energy, 10-112 to 115
isotopes and their properties, 11-3 to 42
physical properties, 4-62 to 64
thermodynamic properties, 5-1 to 2
thermodynamic properties at high temperature, 5-42 to 64
vapor pressure, 6-87 to 116, 6-117 to 125
Chlorine–Bromine Combination Isotope Intensities in Mass Spectral Patterns, 8-84
Chlorofluorocarbon refrigerants, 6-164 to 166
Chromatography
carrier gas properties, 8-20
combination methods with mass spectrometry, 8-29 to 30
description and applications, 8-6 to 9
detectors for gas chromatography, 8-27 to 28
detectors for liquid chromatography, 8-37
instability of solvents, 8-36
minimum recommended injector split ratios for capillary columns, 8-35
phase ratios for capillary columns, 8-34
porous-layer open tubular (PLOT) columns, 8-28
pressure drop in open tubular gas chromatographic columns, 8-33
retention indices, 8-25 to 26, 8-31 to 32
stationary phases, 8-28, 8-25 to 26
symbols used in schematic diagrams, 8-21
Chromium: *see also* Elements
electrical resistivity, 12-27
electron configuration, 1-16
heat capacity, 4-65
history, occurrence, uses, 4-1 to 39
ionization energy, 10-112 to 115
isotopes and their properties, 11-3 to 42
magnetic susceptibility, 4-73 to 78
molten, density, 4-71 to 72
physical properties, 4-62 to 64
thermal properties, 12-129 to 130
vapor pressure, 6-87 to 116
vapor pressure, high temperature, 4-66 to 68
Classification of Electromagnetic Radiation, 10-140 to 141
Clathrate hydrates, 6-167 to 177
Clausius-Mosotti equation, 12-13 to 14
Cobalt: *see also* Elements
electrical resistivity, 12-27
electron configuration, 1-16
heat capacity, 4-65
history, occurrence, uses, 4-1 to 39
ionization energy, 10-112 to 115
isotopes and their properties, 11-3 to 42
magnetic susceptibility, 4-73 to 78
molten, density, 4-71 to 72
physical properties, 4-62 to 64
thermal properties, 12-129 to 130
vapor pressure, 6-87 to 116
vapor pressure, high temperature, 4-66 to 68
CODATA Key Values for Thermodynamics, 5-1 to 2
CODATA Recommended Values of the Fundamental Physical Constants: 2018, 1-1 to 9
Coefficient of Friction, 15-50
Coercivity, magnetic materials, 12-97 to 104
Collision diameter of gases, 6-54
Collision frequency, common gases, 6-54
Combustion, heat of
fuels, 5-68
various compounds, 5-67
Common Mass Spectral Fragmentation Patterns of Organic Compound Families, 8-72 to 73
Common Mass Spectral Fragments Lost, 8-74
Common Spurious Infrared Absorption Bands, 8-48
Common Spurious Signals Observed in Mass Spectrometers, 8-83
Common Symbols Used in Gas and Liquid Chromatographic Schematic Diagrams, 8-21
Composition
atmosphere, 14-2 to 6
earth's crust, 14-21
glasses, 12-138 to 140
human body, 7-61
magnetic materials, 12-97 to 104
planetary atmospheres, 14-2 to 6
seawater, 14-21
U.S. Standard Atmosphere, 14-23 to 28
Composition and Properties of Common Oils and Fats, 7-9 to 11
Compressed Air Safety, 16-42
Compressibility
ice, 6-11
liquids, 6-154 to 155, 6-156 to 160
semiconductors, 12-70 to 82
Compressibility and Expansion Coefficients of Liquids, 6-154 to 155
Compton wavelength (electron, proton, neutron), 1-1 to 9
Concentrative Properties of Aqueous Solutions 5-133 to 148
Conductance: *see* Conductivity, electrical
Conductivity, electrical
aqueous solutions of acids, bases, salts, 5-94
calibration standards, 5-95
electrolyte solutions, 5-97
hydrohalogen acids, 5-96
ions, at infinite dilution, 5-98 to 100
potassium chloride solutions, 5-95
seawater, 14-19 to 20
standard solutions, 5-95
water, 5-93
Conductivity, thermal: *see* Thermal conductivity
Constant Humidity Solutions, 15-36
Constantan, thermal conductivity, 12-132 to 133
Construction materials
density, 15-41
thermal conductivity, 12-136 to 137
Conversion Factors for Energy Units, 1-24
Conversion Factors for Pressure Units, 1-24
Conversion Factors for Thermal Conductivity Units, 1-26
Coolants for Cryotrapping, 8-24
Copernicium (element 112), 4-1 to 39, 11-3 to 42
Copper: *see also* Elements
electrical resistivity, 12-27
electron configuration, 1-16
heat capacity, 4-65
history, occurrence, uses, 4-1 to 39
ionization energy, 10-112 to 115
isotopes and their properties, 11-3 to 42
magnetic susceptibility, 4-73 to 78
molten, density, 4-71 to 72
physical properties, 4-62 to 64
thermal conductivity, 12-131
thermal properties, 12-129 to 130
thermodynamic properties, 5-1 to 2
vapor pressure, 6-87 to 116
vapor pressure, high temperature, 4-66 to 68
Copper-constantan thermocouple tables, 15-1 to 9
Correction of Barometer Readings to 0 °C Temperature, 15-33
Correlation charts for infrared spectra, 8-42 to 47
Correlation Table for Ultraviolet Active Functionalities, 8-39 to 41
Cosmic Radiation, 11-56 to 61
Cosmic rays, 11-56 to 61
Coulomb, definition, 1-18 to 22
Covalent radii of Elements, 9-56 to 57
Critical constants
cryogenic fluids, 6-161 to 162
elements, 4-62 to 64, 6-82 to 84
H_2O and D_2O, 6-9
halocarbons, 6-164 to 166
inorganic compounds, 6-82 to 84
organic compounds, 6-56 to 81
Critical Constants of Inorganic Compounds, 6-82 to 84
Critical Constants of Organic Compounds, 6-56 to 81
Critical micelle concentration (CMC), 16-46 to 47
Cross section, x-ray and gamma ray, 10-134 to 139
Crust
composition, 14-21
density, pressure, gravity, 14-17
Cryogenic fluids
for cryotrapping, 8-24
liquid helium properties, 6-163
safety, 16-43 to 45
thermophysical properties, 6-21 to 39
vapor pressure, 6-117 to 125
various properties, 6-161 to 162
Cryogenic materials, 12-122 to 127
Cryoscopic Constants for Calculation of Freezing Point Depression, 15-31
Crystal elastic constants, 12-22 to 26
Crystal ionic radii, 12-11 to 12
Crystal lattice energy, 12-21
Crystal optical properties
elements, 12-113 to 116
inorganic compounds, 10-146 to 149
minerals, 4-80 to 86
various materials, 12-117 to 121
Crystal structure
elements, 4-87 to 94, 12-15 to 18
elements, phase transitions, 12-19 to 20
gas clathrate hydrates, 6-167 to 177
inorganic compounds, 4-40 to 51, 4-87 to 94
magnetic materials, 12-97 to 104
minerals, 4-80 to 86, 4-87 to 94
rare-earth elements, 4-52 to 61
semiconductors, 12-70 to 82
superconductors, 12-42 to 57, 12-58 to 67

Crystal Structures and Lattice Parameters of Allotropes of Elements, 12-15 to 18
Crystal symmetry, 12-6 to 10
Crystallographic Data on Minerals, 4-87 to 94
Curie temperature
  antiferroelectric crystals, 12-41
  ferroelectric crystals, 12-40
  magnetic materials, 12-97 to 104
  rare-earth elements, 4-52 to 61
Curie Temperature of Selected Ferroelectric Crystals, 12-40
Curie, definition, 16-89 to 90
Curium: *see also* Elements
  electron configuration, 1-16
  history, occurrence, uses, 4-1 to 39
  isotopes and their properties, 11-3 to 42
  vapor pressure, high temperature, 4-66 to 68
Cylinders, gas,
  safety and stamped markings, 16-41
  standard fittings, 8-22

D

D_2O (heavy water)
  boiling point, 6-9
  critical constants, 6-9
  density, 6-10
  dissociation constant, 5-93
  fixed-point properties, 6-9
  heat capacity, 6-10
  ionization constant, 5-93
  surface tension, 6-10
  thermal conductivity, 6-10, 6-240 to 242
  triple-point constants, 6-9
  vapor pressure, 6-10
  viscosity, 6-10, 6-233 to 234
Dalton, definition, 1-18 to 22
Darmstadtium (element 110), 4-1 to 39, 11-3 to 42
Data sources, A-1 to 5
Debye equation, 6-14
Debye temperature
  rare-earth elements, 4-52 to 61
  semiconductors, 12-70 to 82
Decay mode, nuclides, 11-3 to 42
Definition of Ambient, 1-26
Definitions
  abbreviations and acronyms, 2-24 to 36
  ambient, 1-26
  chemical analysis terms, 8-10 to 16
  SI base units, 1-18 to 22
  thermodynamic functions, 2-37
Density
  air, 6-15 to 20
  aqueous solutions, 5-133 to 148
  atmosphere, as function of altitude, 14-23 to 28
  carrier gases for chromatography, 8-20
  common fluids, as function of temperature and pressure, 6-21 to 39
  common fluids, at saturation, 6-40 to 50
  construction materials, 15-41
  cryogenic fluids, 6-21 to 39, 6-161 to 162
  D_2O, 6-9
  earth, 14-14 to 15
  earth, as function of depth, 14-17
  elements, 4-71 to 72
  ethanol–water mixtures, 15-41
  ice, 6-11
  inorganic compounds, 4-40 to 51
  liquid elements and salts, 4-71 to 72
  liquids, pressure and temperature dependence, 6-154 to 155, 6-156 to 160
  metallic elements, 12-129 to 130
  minerals, 4-80 to 86
  miscellaneous materials, 15-41
  molten salts, 4-71 to 72
  oils and fats, 7-9 to 11
  organic compounds, 3-1 to 55
  planets, 14-2 to 6
  polymer melts, 13-18 to 21
  polymers, 13-3 to 4
  rare-earth elements, 4-52 to 61
  rocks, 15-41
  satellites, 14-6 to 8
  seawater, 14-19 to 20
  semiconductors, 12-70 to 82
  solvents, 15-13 to 22
  solvents, as function of temperature, 15-27
  steam, 6-1 to 4
  sulfuric acid, 15-42 to 43
  water, 6-1 to 4, 6-7 to 8
  water, supercooled, 6-11
  wood, 15-41
Density of Ethanol–Water Mixtures, 15-41
Density of Molten Elements and Representative Salts, 4-71 to 72
Density of Solvents as a Function of Temperature, 15-27
Density of Sulfuric Acid, 15-42 to 43
Density of Various Solids, 15-41
Density, Pressure, and Gravity as a Function of Depth within the Earth, 14-17
Dependence of Boiling Point of Organic Liquids on Pressure, 15-28 to 29
Depression of the freezing point, 5-133 to 148, 15-31, 15-32
Descriptive Terms for Solubility, 1-25
Detection of Outliers in Measurements, 8-19
Detectors for Gas Chromatography, 8-27 to 28
Detectors for Liquid Chromatography, 8-37
Detectors, infrared radiation, 10-145
Determination of Relative Humidity from Dew Point, 15-34
Determination of Relative Humidity from Wet and Dry Bulb Temperatures, 15-35
Deuterium
  solubility in water, 5-149 to 150
  viscosity, 6-233 to 234
Dew point and relative humidity, 15-34
Diamagnetic susceptibility
  elements, 4-73 to 78
  inorganic compounds, 4-73 to 78
  organic compounds, 3-56 to 59
Diamagnetic Susceptibility of Selected Organic Compounds, 3-56 to 59
Diamond
  dielectric constant, 12-31 to 39
  optical properties, 12-113 to 116
  thermal conductivity, 12-131
Diatomic molecules
  bond lengths, 9-111 to 116
  bond strengths, 9-73 to 105
  electron affinity, 10-54 to 75
  force constants, 9-107
  polarizability, 10-95 to 111
  spectroscopic constants, 9-111 to 116
  vibrational frequencies, 9-111 to 116
Dielectric constant
  common fluids, as function of temperature and pressure, 6-21 to 39
  common fluids, at saturation, 6-40 to 50
  cryogenic fluids, 6-21 to 39
  crystals, 12-31 to 39
  gases, 6-216
  glass, 12-41
  ice, 6-11
  liquids, 6-193 to 215
  plastics, 13-17
  quartz, 12-41
  rubbers, 13-17
  semiconductors, 12-70 to 82
  solids, 12-31 to 39
  solvents, 8-38, 15-13 to 22
  vacuum, 1-1 to 9
  water and steam, temperature and pressure dependence, 6-1 to 4
  water, frequency dependence, 6-14
Dielectric Constant of Selected Polymers, 13-17
Dielectric Constants of Glasses, 12-41
Dielectric Strength of Insulating Materials, 15-45 to 49
Diffusion
  in air, 6-249 to 250
  gases, 6-249 to 250
  gases in water, 6-251
  ions in solution, 5-98 to 100
  liquids, 6-252 to 253
  semiconductors, 12-89 to 96
Diffusion Coefficients in Liquids at Infinite Dilution, 6-252 to 253
Diffusion Data for Semiconductors, 12-89 to 96
Diffusion in Gases, 6-249 to 250
Diffusion of Gases in Water, 6-251
Dipole moment
  electric, of molecules, 9-58 to 67
  magnetic, of nuclides, 11-3 to 42
  solvents, 15-13 to 22
Dipole Moments, 9-58 to 67
Dissociation constant
  acid-base indicators, 8-88 to 89
  amino acids, 7-1 to 2
  biological buffers, 7-30 to 32, 7-33
  D_2O, 5-93
  inorganic acids and bases, 5-108
  inorganic salts in water, 5-194 to 195
  organic acids and bases, 5-109 to 118
  purine and pyrimidine bases, 7-5
  water, 5-92, 5-93
Dissociation Constants of Inorganic Acids and Bases, 5-108
Dissociation Constants of Organic Acids and Bases, 5-109 to 118
Dissociation energy of chemical bonds, 9-73 to 105
Distillation, azeotropes, 6-217 to 232
Drugs, properties, and functions, 7-35 to 56
  controlled substances, properties, 7-57 to 7-58
DSC, definition, 12-1 to 5
DTA, definition, 12-1 to 5
Dubnium (element 105), 4-1 to 39, 11-3 to 42
Dysprosium: *see also* Elements
  electrical resistivity, 12-27

electron configuration, 1-16
heat capacity, 4-65
history, occurrence, uses, 4-1 to 39
ionization energy, 10-112 to 115
isotopes and their properties, 11-3 to 42
magnetic susceptibility, 4-73 to 78
molten, density, 4-71 to 72
physical properties, 4-62 to 64
thermal properties, 12-129 to 130
vapor pressure, 6-87 to 116
vapor pressure, high temperature, 4-66 to 68

E

Earth
age, 14-14 to 15
area of land and oceans, 14-14 to 15
atmospheric composition, 14-2 to 6
composition of crust, 14-21
density as function of depth, 14-17
dimensions, 14-1, 14-14 to 15
gravity in interior, 14-17
mass and density, 14-14 to 15
orbital and rotational parameters, 14-2 to 6, 14-14 to 15
pressure in interior, 14-17
Earthquakes, historical data, 14-30 to 33
Ebullioscopic Constants for Calculation of Boiling Point Elevation, 15-30
ECR, definition, 12-1 to 5
EELS, definition, 12-1 to 5
Einsteinium: *see also* Elements
electron configuration, 1-16
history, occurrence, uses, 4-1 to 39
isotopes and their properties, 11-3 to 42
Elastic Constants of Single Crystals, 12-22 to 26
Elastic modulus, rare-earth elements, 4-52 to 61
Electrical conductance: *see* Conductivity, electrical
Electrical Conductivity of Aqueous Solutions, 5-94
Electrical Conductivity of Water, 5-93
Electrical resistivity: *see* Resistivity
Electrical Resistivity of Graphite Materials, 12-30
Electrical Resistivity of Pure Metals, 12-27
Electrical Resistivity of Selected Alloys, 12-28 to 29
Electrochemical Series, 5-101 to 107
Electrode potential general table, 5-101 to 107
Electrolytes
activity coefficients, 5-98 to 100, 5-121 to 125
diffusion of ions, 5-98 to 100
electrical conductivity, 5-97, 5-98 to 100
enthalpy of solution, 5-127
freezing point lowering, 5-133 to 148, 15-32
Electromagnetic radiation, classification of bands, 10-140 to 141
Electron
charge, mass, other properties, 1-1 to 9
in cosmic ray showers, 11-56 to 61
mean free path in solids, 12-106 to 107
range and shielding, 16-88
secondary, emission by metals, 12-112
Electron Affinities, 10-54 to 75
Electron configuration
neutral atoms, 1-16
rare-earth elements, 4-52 to 61
Electron Configuration and Ionization Energy of Neutral Atoms in the Ground State, 1-16
Electron Inelastic Mean Free Paths, 12-106 to 107
Electron Stopping Powers, 12-108 to 109
Electron volt, 1-1 to 9, 1-18 to 22
Electron Work Function of Elements, 12-110 to 111
Electronegativity, 9-106
Elementary charge, 1-1 to 9
Elements, The, 4-1 to 39
Elements
abundance of isotopes, 1-12 to 15, 11-3 to 42
atomic mass, 1-12 to 15, 11-3 to 42
atomic radii, 9-56 to 57
atomic spectrum, 10-1 to 50
atomic transition probability, 10-51 to 53
atomic weight, 1-10 to 11
boiling point, 4-62 to 64
critical temperature, 4-62 to 64
crystal ionic radii, 12-11 to 12
crystal phase transitions, 12-19 to 20
crystal structure, 4-87 to 94, 12-15 to 18
density, 4-71 to 72
in the earth's crust, 14-21
electrical resistivity, 12-27
electron affinity, 10-54 to 75
electron configuration, 1-16
electronegativity, 9-106
enthalpy of fusion, 6-144 to 153
enthalpy of vaporization, 6-127 to 143
gamma ray cross sections, 10-134 to 139
gamma ray emission, 11-3 to 42
general information, 4-1 to 39
heat capacity, 4-65
historical information, 4-1 to 39
in human blood, 7-50 to 60
in the human body, 7-61
line spectrum, 10-1 to 50
magnetic susceptibility, 4-73 to 78
melting point, 4-62 to 64
Periodic Table of Elements: *see* Inside front cover
photon attenuation coefficients, 10-134 to 139
polarizability, 10-95 to 111
radii of ions, 12-11 to 12
reference states, 5-3 to 41
in seawater, 14-21
semiconducting properties, 12-70 to 82
superconducting properties, 12-42 to 57
thermal conductivity, 6-240 to 242, 12-120 to 121, 12-129 to 130
thermal properties, 12-129 to 130
thermodynamic properties, 5-3 to 41
transition probabilities, 10-51 to 53
triple-point, 4-62 to 64
van der Waals radii, 9-56 to 57
vapor pressure, 6-87 to 116
vapor pressure at high temperature, 4-66 to 68, 4-69 to 70
work function, 12-110 to 111
x-ray cross sections, 10-134 to 139
Elevation of the boiling point, 15-30
Eluotropic Values of Solvents on Octadecylsilane and Octylsilane, 8-35
Emission, secondary electrons, 12-112
Emissivity, rare-earth metals, 4-52 to 61
ENDOR, definition, 12-1 to 5
Energy
conversion factors, 1-24
crystal lattice, 12-21
Fermi, 12-143 to 144
spectrum of cosmic rays, 11-56 to 61
Energy, activation, for diffusion in semiconductors, 12-89 to 96
Energy Content of Fuels, 5-68
Energy gap
semiconductor solid solutions, 12-83 to 84
semiconductors, 12-70 to 82, 12-85 to 88
superconductors, 12-58 to 67
Energy states of solids, 12-1 to 5
Enthalpy
air, 6-15 to 20
common fluids, as function of temperature and pressure, 6-21 to 39
common fluids, at saturation, 6-40 to 50
steam, 6-1 to 4
water, 6-1 to 4
Enthalpy of combustion, 5-67, 5-68
Enthalpy of Dilution of Acids, 5-126
Enthalpy of formation
aqueous ions, 5-65 to 66
CODATA Key Values, 5-1 to 2
free radicals, 9-73 to 105
gaseous atoms, 9-73 to 105
high temperature, 5-42 to 64
inorganic compounds, 5-3 to 41
ions, 10-116 to 133
organic compounds, 5-3 to 41
protein-ligand complexes, 7-21 to 29
semiconductors, 12-70 to 82
standard state values, 5-3 to 41
Enthalpy of fusion
cryogenic fluids, 6-161 to 162
elements, 6-144 to 153
ice, 6-11
inorganic compounds, 6-144 to 153
metals, 12-129 to 130
organic compounds, 6-144 to 153
rare-earth elements, 4-52 to 61
Enthalpy of Fusion, 6-144 to 153
Enthalpy of Solution of Electrolytes, 5-127
Enthalpy of vaporization
cryogenic fluids, 6-161 to 162
elements, 6-127 to 143
ice, 6-11
inorganic compounds, 6-127 to 143
organic compounds, 6-127 to 143
rare-earth elements, 4-52 to 61
water, 6-5 to 6
Enthalpy of Vaporization, 6-127 to 143
Entropy
air, 6-15 to 20
aqueous ions, 5-65 to 66
CODATA Key Values, 5-1 to 2
common fluids, as function of temperature and pressure, 6-21 to 39
common fluids, at saturation, 6-40 to 50
high temperature, 5-42 to 64
rare-earth elements, 4-52 to 61
standard state values, 5-3 to 41
steam, 6-1 to 4
water, 6-1 to 4
Enzyme-catalyzed reactions, equilibrium constants, 7-17 to 19
EPMA, definition, 12-1 to 5
EPR, definition, 12-1 to 5

Equation of state
Tait (for polymer melts), 13-18 to 21
virial, 6-51 to 53
Equilibrium constant of formation, 5-42 to 64
Equilibrium constant, biochemical reactions, 7-14 to 16, 7-17 to 19, 7-21 to 29
Equivalent conductance: *see* Conductivity, electrical
Erbium: *see also* Elements
electrical resistivity, 12-27
electron configuration, 1-16
heat capacity, 4-65
history, occurrence, uses, 4-1 to 39
ionization energy, 10-112 to 115
isotopes and their properties, 11-3 to 42
magnetic susceptibility, 4-73 to 78
molten, density, 4-71 to 72
physical properties, 4-62 to 64
thermal properties, 12-129 to 130
vapor pressure, 6-87 to 116
vapor pressure, high temperature, 4-66 to 68
Eris, orbital data and dimensions, 14-2 to 6
Errors in measurement, 2-13 to 14, 8-19
ESCA, definition, 12-1 to 5
ESD, definition, 12-1 to 5
Europium: *see also* Elements
electrical resistivity, 12-27
electron configuration, 1-16
heat capacity, 4-65
history, occurrence, uses, 4-1 to 39
ionization energy, 10-112 to 115
isotopes and their properties, 11-3 to 42
magnetic susceptibility, 4-73 to 78
molten, density, 4-71 to 72
physical properties, 4-62 to 64
thermal properties, 12-129 to 130
vapor pressure, 6-87 to 116
vapor pressure, high temperature, 4-66 to 68
Eutectic temperatures, low-melting alloys, 15-38 to 39
EXAFS, definition, 12-1 to 5
EXELFS, definition, 12-1 to 5
Expansion coefficient
liquid helium, 6-163
metals, 12-129 to 130
polymers, 13-3 to 4
rare-earth elements, 4-52 to 61
semiconductors, 12-70 to 82
Explosion (Shock) Hazards, 16-7
Explosion hazards of laboratory chemicals, 16-7
Explosive limits, 16-12 to 29
Exposure limits
airborne contaminants, 16-48 to 75
radionuclides, 16-93 to 96
Expression of Uncertainty of Measurements, 2-13 to 14
Extinction coefficient, in solids, 12-113 to 116

F

Farad, definition, 1-18 to 22
Faraday constant, 1-1 to 9
Fats, composition and properties, 7-9 to 11
Fatty acids and methyl esters, physical properties, 7-7 to 8
Fehling's solution, preparation, 8-90 to 94
FEM, definition, 12-1 to 5
Fermi Energy and Related Properties of Metals, 12-143 to 144
Fermium: *see also* Elements
electron configuration, 1-16
history, occurrence, uses, 4-1 to 39
isotopes and their properties, 11-3 to 42
Ferrimagnetic materials, organic, 12-105
Ferrites, magnetic properties, 12-97 to 104
Ferroelectric crystals, Curie temperature, 12-40
Ferromagnetic materials
organic, 12-105
various properties, 12-97 to 104
Ferromagnetic moment, rare-earth elements, 4-52 to 61
Figures of Merit, 8-17
Filters for respirators, 16-34 to 35
FIM, definition, 12-1 to 5
Fine structure constant, 1-1 to 9
First radiation constant, 1-1 to 9
Fixed-point properties
cryogenic fluids, 6-161 to 162
water and heavy water, 6-9
Fixed-Point Properties of H_2O and D_2O, 6-9
Flammability
chemical substances, general, 16-12 to 29
organic solvents, 15-13 to 22, 16-10 to 11
Flammability Hazards of Common Solvents, 16-10 to 11
Flammability of Chemical Substances, 16-12 to 29
Flash point: *see* Flammability
Flattening factor for the earth, 14-1
Flerovium (element 114), 4-1 to 39, 11-3 to 42
Fluorine: *see also* Elements
critical constants, 6-82 to 84
electron configuration, 1-16
heat capacity, 4-65
history, occurrence, uses, 4-1 to 39
ionization energy, 10-112 to 115
isotopes and their properties, 11-3 to 42
physical properties, 4-62 to 64
thermodynamic properties, 5-1 to 2
thermodynamic properties at high temperature, 5-42 to 64
vapor pressure, 6-87 to 116, 6-117 to 125
Fluorocarbon refrigerants, 6-164 to 166
Foods, pH, 7-34
Force Constants for Bond Stretching, 9-107
Formation, heat of: *see* Enthalpy of formation
Fossils, age of, 14-14 to 15
Francium: *see also* Elements
electron configuration, 1-16
history, occurrence, uses, 4-1 to 39
ionization energy, 10-112 to 115
isotopes and their properties, 11-3 to 42
vapor pressure, 6-87 to 116
Free energy: *see* Thermodynamic properties
Free radicals
dipole moment, 9-58 to 67
enthalpy of formation, 9-73 to 105
ionization energy, 10-116 to 133
vibrational frequencies, 9-108 to 110
Freezing point: *see also* Melting point
depression of, 5-133 to 148, 15-31
pressure dependence, 6-55
seawater, 14-19 to 20
Freezing point depression
aqueous solutions, 5-133 to 148
cryoscopic constants for various liquids, 15-31
electrolytes, 15-32
Freezing Point Lowering by Electrolytes in Aqueous Solution, 15-32
Frequency
electromagnetic radiation bands, 10-140 to 141
NMR resonances, 8-49 to 51, 8-52
Friction, coefficient of, 15-50
FTIR, definition, 12-1 to 5
Fume hoods, 16-39 to 40
Fundamental constants, 1-1 to 9
Fundamental particles, 11-1 to 2
Fundamental Physical Constants—Frequently Used Constants: *see* Inside back cover; 1-25
Fundamental Vibrational Frequencies of Small Molecules, 9-108 to 110
Fusion: *see* Enthalpy of fusion

G

g-Factor of the electron, 1-1 to 9
Gadolinium: *see also* Elements
electrical resistivity, 12-27
electron configuration, 1-16
heat capacity, 4-65
history, occurrence, uses, 4-1 to 39
ionization energy, 10-112 to 115
isotopes and their properties, 11-3 to 42
magnetic susceptibility, 4-73 to 78
molten, density, 4-71 to 72
physical properties, 4-62 to 64
thermal properties, 12-129 to 130
vapor pressure, 6-87 to 116
vapor pressure, high temperature, 4-66 to 68
Gallium: *see also* Elements
electrical resistivity, 12-27
electron configuration, 1-16
heat capacity, 4-65
history, occurrence, uses, 4-1 to 39
ionization energy, 10-112 to 115
isotopes and their properties, 11-3 to 42
magnetic susceptibility, 4-73 to 78
molten, density, 4-71 to 72
physical properties, 4-62 to 64
thermal properties, 12-129 to 130
vapor pressure, 6-87 to 116
vapor pressure, high temperature, 4-66 to 68
Gamma rays
in cosmic ray showers, 11-56 to 61
cross sections, for Elements, 10-134 to 139
energy, of nuclides, 11-3 to 42
photon attenuation coefficients, 10-134 to 139
protection against, 16-88
Garments, for protection in the laboratory, 16-33, 16-36
Gas Chromatographic Retention Indices, 8-31 to 32
Gas chromatography, carrier gas properties, 8-20
Gas clathrate hydrates, 6-167 to 177
Gas constant, 1-1 to 9, 1-25
Gas cylinders
safety and stamped markings, 16-41
standard fittings, 8-22
Gas Cylinder Safety and Stamped Markings, 16-41
Gas-phase basicity, 10-76 to 94
Gases
average velocity, 6-54
breakdown voltage, 15-45 to 49

collision diameter, 6-54
dielectric constant, 6-216
diffusion, 6-249 to 250
diffusion in water, 6-251
dipole moment, 9-58 to 67
mean free path, 6-54
permittivity, 6-216
refractive index, 10-154
solubility in water, 5-149 to 150
speed of sound in, 14-39 to 40
thermal conductivity, 6-240 to 242
threshold limits, 16-48 to 75
virial coefficients, 6-51 to 53
viscosity, 6-233 to 234
Gaussian gravitational constant, 14-1
GDMS, definition, 12-1 to 5
General Characteristics of Ionizing Radiation for the Purpose of Practical Application of Radiation Protection, 16-88
Genetic Code, The, 7-6
Geographical and Seasonal Variations in Solar Radiation, 14-29
Geological Time Scale, 14-16
Geophysical constants, 14-14 to 15
Germanium: *see also* Elements
dielectric constant, 12-31 to 39
electron configuration, 1-16
heat capacity, 4-65
history, occurrence, uses, 4-1 to 39
ionization energy, 10-112 to 115
isotopes and their properties, 11-3 to 42
magnetic susceptibility, 4-73 to 78
molten, density, 4-71 to 72
optical properties, 12-113 to 116
physical properties, 4-62 to 64
semiconducting properties, 12-70 to 82
thermal conductivity, 12-131
thermodynamic properties, 5-1 to 2
vapor pressure, 6-87 to 116
Gibbs energy of formation
aqueous ions, 5-65 to 66
biochemical species, 7-14 to 16, 7-21 to 29
high temperature, 5-42 to 64
protein-ligand complexes, 7-21 to 29
standard state values, 5-3 to 41
Glass Transition Temperature for Selected Polymers, 13-10 to 16
Glasses
compatibility with 72% perchloric acid, 16-30
composition, 12-138 to 140
density, 15-41
dielectric constant, 12-41
index of refraction, 10-150
loss factor, 12-41
resistivity, 12-41
speed of sound in, 14-39 to 40
thermal conductivity, 12-134 to 135
transmittance, 10-150
Global Temperature Trend, 1880–2020, 14-36
Global warming
atmospheric carbon dioxide concentration, 14-35
mean temperatures, global, 14-36
Global Warming Potential of Greenhouse Gases, 14-37 to 38
Gloves, resistance to chemicals, 16-31 to 32
Glucose, aqueous solution properties, 5-133 to 148
Gold: *see also* Elements
electrical resistivity, 12-27
electron configuration, 1-16
heat capacity, 4-65
history, occurrence, uses, 4-1 to 39
ionization energy, 10-112 to 115
isotopes and their properties, 11-3 to 42
magnetic susceptibility, 4-73 to 78
molten, density, 4-71 to 72
physical properties, 4-62 to 64
thermal conductivity, 12-131
thermal properties, 12-129 to 130
vapor pressure, 6-87 to 116
vapor pressure, high temperature, 4-66 to 68
Graphite
electrical resistivity, 12-30
heat capacity, 12-128
heat of combustion, 5-67
sublimation pressure, 6-87 to 116
thermal conductivity, 12-131
Gravitational constant, 1-1 to 9, 14-1
Gravitational potential, 14-2 to 6
Gravity, acceleration of
in interior of earth, 14-17
at poles and equator, 14-14 to 15
standard value, 1-1 to 9
at various latitudes, 14-17
Gray, definition, 1-18 to 22, 16-89 to 90
Greenhouse gases
carbon dioxide concentration, 14-35
carbon dioxide, from fuel combustion, 5-68
global warming potential, 14-37 to 38
infrared absorption, 14-34
radiative efficiency and lifetime, 14-37 to 38

H

Hafnium: *see also* Elements
electrical resistivity, 12-27
electron configuration, 1-16
heat capacity, 4-65
history, occurrence, uses, 4-1 to 39
ionization energy, 10-112 to 115
isotopes and their properties, 11-3 to 42
magnetic susceptibility, 4-73 to 78
molten, density, 4-71 to 72
physical properties, 4-62 to 64
thermal properties, 12-129 to 130
vapor pressure, 6-87 to 116
vapor pressure, high temperature, 4-66 to 68
Hall coefficient, rare-earth elements, 4-52 to 61
Hall density, superconductors, 12-58 to 67
Hall resistance, quantized, 1-1 to 9
Halocarbon refrigerants, 6-164 to 166
Hardness
minerals, 4-80 to 86
semiconductors, 12-70 to 82
Hartree energy, 1-1 to 9
Hassium (element 108), 4-1 to 39, 11-3 to 42
Hazards, 16-7
Hearing protection, 16-37 to 38
Heat capacity
air, 6-15 to 20
aqueous ions, 5-65 to 66
carrier gases for chromatography, 8-20
common fluids, as function of temperature and pressure, 6-21 to 39
common fluids, at saturation, 6-40 to 50
cryogenic fluids, 6-161 to 162
cryogenic materials, 12-122 to 127
elements, 4-65
high temperature, 5-42 to 64
ice, 6-11
liquid helium, 6-163
metals, 12-122 to 127, 12-128, 12-120 to 121
plastics, 12-122 to 127
rare-earth elements, 4-52 to 61
seawater, 14-19 to 20
semiconductors, 12-70 to 82
solids, 12-122 to 127
solvents, 15-13 to 22
stainless steel, 12-122 to 127
standard state values, 5-3 to 41
steam, 6-1 to 4
water, 6-1 to 4
Heat Capacity of Elements at 25 °C, 4-65
Heat Capacity of Selected Solids, 12-128
Heat conductivity: *see* Thermal conductivity
Heat of Combustion, 5-67
Heat of dilution: *see* Enthalpy of dilution
Heat of formation: *see* Enthalpy of formation
Heat of fusion: *see* Enthalpy of fusion
Heat of solution: *see* Enthalpy of solution
Heat of vaporization: *see* Enthalpy of vaporization
Heavy water (D_2O): *see* D_2O
Helium: *see also* Elements
critical constants, 6-82 to 84
cryogenic properties, 6-161 to 162, 6-163
density, 6-163
electron configuration, 1-16
enthalpy of vaporization, 6-163
entropy, 5-1 to 2
expansion coefficient, 6-163
heat capacity, 6-163
history, occurrence, uses, 4-1 to 39
ionization energy, 10-112 to 115
isotopes and their properties, 11-3 to 42
liquid properties, 6-163
magnetic susceptibility, 4-73 to 78
mean free path, 6-54
permittivity (dielectric constant), 6-163, 6-193 to 215, 6-216
physical properties, 4-62 to 64
solubility in water, 5-149 to 150
surface tension, 6-163
thermal conductivity, 6-21 to 39
thermal conductivity (gas), 6-240 to 242
thermal conductivity (liquid), 6-163
thermophysical properties, 6-21 to 39
vapor pressure, 6-87 to 116, 6-117 to 125, 6-163
viscosity, 6-21 to 39
viscosity (gas), 6-233 to 234
viscosity (liquid), 6-163
Henry, definition, 1-18 to 22
Henry's Law constant, 5-152 to 182
Hertz, definition, 1-18 to 22
High-Temperature Superconductors, 12-58 to 67
Hindered Internal Rotation, 9-68 to 72
HITRAN Molecular Spectroscopy Database, 14-34
Holmium: *see also* Elements
electrical resistivity, 12-27
electron configuration, 1-16
heat capacity, 4-65

history, occurrence, uses, 4-1 to 39
ionization energy, 10-112 to 115
isotopes and their properties, 11-3 to 42
magnetic susceptibility, 4-73 to 78
molten, density, 4-71 to 72
physical properties, 4-62 to 64
thermal properties, 12-129 to 130
vapor pressure, 6-87 to 116
vapor pressure, high temperature, 4-66 to 68
Human body
chemical composition, 7-61
pH of fluids, 7-34
sensitivity of eye to light, 10-142
Humidity, relative
from wet and dry bulb temperatures, 15-35
relation to dew point, 15-34
solutions for calibration, 15-37
solutions for constant humidity, 15-36
Hydrocarbons
heat of combustion, 5-67
solubility in seawater, 5-197
thermophysical properties, 6-21 to 39, 6-40 to 50
Hydrogen: *see also* Elements
critical constants, 6-82 to 84
cryogenic properties, 6-161 to 162
electron configuration, 1-16
enthalpy of fusion, 6-144 to 153
enthalpy of vaporization, 6-127 to 143
heat of combustion, 5-67
history, occurrence, uses, 4-1 to 39
ionization energy, 10-112 to 115
isotopes and their properties, 11-3 to 42
magnetic susceptibility, 4-73 to 78
mean free path, 6-54
permittivity (dielectric constant), 6-193 to 215, 6-216
physical properties, 4-62 to 64
solubility in water, 5-149 to 150
thermal conductivity, 6-21 to 39, 6-240 to 242
thermodynamic properties, 5-1 to 2
thermodynamic properties at high temperature, 5-42 to 64
thermophysical properties, 6-21 to 39
vapor pressure, 6-87 to 116, 6-117 to 125
viscosity, 6-21 to 39, 6-233 to 234
Hydrophilic lipophilic balance (HLB), 16-46 to 47
Hydrosphere, mass of, 14-14 to 15
Hysteresis, in magnetic materials, 12-97 to 104

I

Ice
compressibility, 6-11
crystal structure, 4-87 to 94
density, 6-11
dielectric constant, 6-11, 12-31 to 39
heat capacity, 6-11
melting point, pressure dependence, 6-13, 6-55
phase transitions, 6-11, 6-13
thermal conductivity, 6-11
thermal expansion coefficient, 6-11
vapor pressure, 6-12
ICPMS, definition, 12-1 to 5
Ignition temperature
chemical substances, general, 16-12 to 29
solvents, 15-13 to 22
InChI representation of chemical structures, 2-22 to 23
Incompatible Chemicals, 16-5 to 6
Index of refraction
air, 10-153
aqueous solutions, 5-133 to 148
gases, 10-154
glass, 10-150
inorganic crystals, 10-146 to 149
inorganic liquids, 4-79
liquid elements, 4-79
liquids, for calibration, 10-152
metals, 12-113 to 116
minerals, 4-80 to 86
oils and fats, 7-9 to 11
organic compounds, 3-1 to 55
semiconductors, 12-70 to 82, 12-113 to 116, 12-117 to 121
solids, as function of wavelength, 10-146 to 149, 12-113 to 116, 12-117 to 121
water, 10-151
Index of Refraction of Air, 10-153
Index of Refraction of Gases, 10-154
Index of Refraction of Inorganic Crystals, 10-146 to 149
Index of Refraction of Inorganic Liquids and Liquid Elements, 4-79
Index of Refraction of Liquids for Calibration Purposes, 10-152
Index of Refraction of Water, 10-151
Indicators
acid-base, 8-88 to 89
preparation, 8-90 to 94
Indicators for Acids and Bases, 8-88 to 89
Indium: *see also* Elements
electrical resistivity, 12-27
electron configuration, 1-16
heat capacity, 4-65
history, occurrence, uses, 4-1 to 39
ionization energy, 10-112 to 115
isotopes and their properties, 11-3 to 42
magnetic susceptibility, 4-73 to 78
molten, density, 4-71 to 72
physical properties, 4-62 to 64
thermal properties, 12-129 to 130
thermophysical properties, 12-122 to 127
vapor pressure, 6-87 to 116
vapor pressure, high temperature, 4-66 to 68
Influence of Pressure on Freezing Points, 6-55
Infrared Absorption by the Earth's Atmosphere, 14-34
Infrared detectors, 10-145
Infrared spectrum
analytical chemistry applications, 8-6 to 9
characteristic group frequencies, 8-42 to 47
correlation charts, 8-42 to 47
earth's atmosphere, 14-34
HITRAN Database, 14-34
spurious bands, 8-48
vibrational frequencies of molecules, 9-108 to 110
Inorganic compounds
activity coefficients, 5-98 to 100, 5-121 to 125
bond lengths and angles, 9-19 to 54
characteristic infrared frequencies, 8-42 to 47
crystal structure, 4-40 to 51, 4-87 to 94
dielectric constant, 12-31 to 39
dipole moment, 9-58 to 67
dissociation constant in water, 5-194 to 195
electrical conductivity, 5-97
enthalpy of formation, 5-3 to 41
enthalpy of fusion, 6-144 to 153
enthalpy of solution, 5-127
enthalpy of vaporization, 6-127 to 143
entropy, 5-3 to 41
Gibbs energy of formation, 5-3 to 41
heat capacity, 5-3 to 41
index of refraction, 4-79
magnetic susceptibility, 4-73 to 78
nomenclature, 2-16 to 21
permittivity, 12-31 to 39
physical properties, 4-40 to 51
polarizability, 10-95 to 111
reagents for determination of ions, 8-95 to 108
solubility as a function of temperature, 5-183 to 188
solubility product constant, 5-194 to 195
solubility, qualitative rules, 5-200 to 201
standard thermodynamic properties, 5-3 to 41
surface tension, 6-182 to 186
INS, definition, 12-1 to 5
Instability of HPLC Solvents, 8-36
Insulation, thermal conductivity of, 12-136 to 137
Insulators, breakdown voltage, 15-45 to 49
Interatomic distances
diatomic molecules, 9-111 to 116
gas-phase molecules, 9-19 to 54
organic crystals, 9-1 to 16
organometallic compounds, 9-17 to 18
Internal rotation in molecules, 9-68 to 72
International System of Units (SI), 1-18 to 22
International Temperature Scale of 1990 (ITS-90), 1-17
International Temperature Scale (ITS-90)
definition and fixed-points for ITS-90, 1-17
reference points for ITS-90, 15-10 to 11
International Union of Pure and Applied Chemistry: *see* IUPAC
Interstellar Molecules, 14-9 to 13
Iodine: *see also* Elements
critical constants, 6-82 to 84
dielectric constant, 12-31 to 39
electron configuration, 1-16
heat capacity, 4-65
history, occurrence, uses, 4-1 to 39
ionization energy, 10-112 to 115
isotopes and their properties, 11-3 to 42
magnetic susceptibility, 4-73 to 78
physical properties, 4-62 to 64
thermodynamic properties, 5-1 to 2
thermodynamic properties at high temperature, 5-42 to 64
vapor pressure, 6-87 to 116
Iodine value, oils and fats, 7-9 to 11
Ionic Conductivity and Diffusion at Infinite Dilution, 5-98 to 100
Ionic Radii in Crystals, 12-11 to 12

Ionic radii
in crystals, 12-11 to 12
rare-earth elements, 4-52 to 61
Ionization constant
biological buffers, 7-30 to 32
D_2O, 5-93
inorganic acids and bases, 5-108
inorganic compounds in water, 5-194 to 195
organic acids and bases, 5-109 to 118
water, 5-92, 5-93
Ionization Constant of Normal and Heavy Water at Saturated Vapor Pressure, 5-93
Ionization Constant of Water at Various Pressures and Temperatures, 5-92
Ionization Energies of Atoms and Atomic Ions, 10-112 to 115
Ionization Energies of Gas-Phase Molecules, 10-116 to 133
Ionization energy
atoms and ions, 10-112 to 115
molecules, 10-116 to 133
neutral atoms, 1-16
rare-earth elements, 4-52 to 61
Ionization gauges, sensitivity, 15-12
Ionization potential: *see* Ionization energy
Ionizing radiation, protection against, 16-88, 16-89 to 90
Ions
diffusion in aqueous solutions, 5-98 to 100
electrical conductivity in aqueous solutions, 5-98 to 100
enthalpy of formation, 10-116 to 133
heat capacity, aqueous solutions, 5-65 to 66
magnetic properties, 12-97 to 104
polarizability, 12-13 to 14
radii, in crystals, 12-11 to 12
thermodynamic properties, aqueous solutions, 5-65 to 66
Iridium: *see also* Elements
electrical resistivity, 12-27
electron configuration, 1-16
heat capacity, 4-65
history, occurrence, uses, 4-1 to 39
ionization energy, 10-112 to 115
isotopes and their properties, 11-3 to 42
magnetic susceptibility, 4-73 to 78
molten, density, 4-71 to 72
physical properties, 4-62 to 64
thermal properties, 12-129 to 130
vapor pressure, 6-87 to 116
vapor pressure, high temperature, 4-66 to 68
Iron: *see also* Elements
electrical resistivity, 12-27
electron configuration, 1-16
heat capacity, 4-65
history, occurrence, uses, 4-1 to 39
ionization energy, 10-112 to 115
isotopes and their properties, 11-3 to 42
magnetic susceptibility, 4-73 to 78
molten, density, 4-71 to 72
physical properties, 4-62 to 64
thermal conductivity, 12-131
thermal properties, 12-129 to 130
vapor pressure, 6-87 to 116
vapor pressure, high temperature, 4-66 to 68
Iron-constantan thermocouple tables, 15-1 to 9
Irradiance of the sun, 14-22
Isoelectric point, amino acids, 7-1 to 2
Isotopes, summary of properties, 11-3 to 42
Isotopic abundance, 1-12 to 15, 11-3 to 42
ITS-90
definition and fixed points, 1-17
reference points, 15-10 to 11
IUPAC
atomic weights, 1-10 to 11
chemical structure representation (InChI), 2-22 to 23
nomenclature for carbohydrates, 7-12 to 13
nomenclature for inorganic chemistry, 2-16 to 21
nomenclature for polymers, 13-5 to 8
pH scale, 5-128 to 131
symbols for physical quantities, 2-1 to 12

J

Josephson ratio, 1-1 to 9
Joule, definition, 1-18 to 22
Jupiter, orbital data and dimensions, 14-2 to 6

K

Katal, definition, 1-18 to 22
Kelvin, definition, 1-18 to 22
Kibibyte, kilobyte, etc., 1-18 to 22
Kilogram, definition, 1-18 to 22
Kinetics of atmospheric reactions, 5-69 to 91
Kovats retention index, 8-31 to 32
Krypton: *see also* Elements
critical constants, 6-82 to 84
cryogenic properties, 6-161 to 162
electron configuration, 1-16
entropy, 5-1 to 2
history, occurrence, uses, 4-1 to 39
ionization energy, 10-112 to 115
isotopes and their properties, 11-3 to 42
magnetic susceptibility, 4-73 to 78
mean free path, 6-54
permittivity (dielectric constant), 6-193 to 215, 6-216
physical properties, 4-62 to 64
solubility in water, 5-149 to 150
thermal conductivity, 6-240 to 242
vapor pressure, 6-87 to 116, 6-117 to 125
viscosity, 6-233 to 234

L

Laboratory garments, protective, 16-33, 16-36
Laboratory reagents
abbreviations, 16-3 to 4
preparation of, 8-90 to 94
properties, 15-13 to 22
Laboratory Solvents and Other Liquid Reagents, 15-13 to 22
Lanthanum: *see also* Elements
electrical resistivity, 12-27
electron configuration, 1-16
heat capacity, 4-65
history, occurrence, uses, 4-1 to 39
ionization energy, 10-112 to 115
isotopes and their properties, 11-3 to 42
magnetic susceptibility, 4-73 to 78
molten, density, 4-71 to 72
physical properties, 4-62 to 64
thermal properties, 12-129 to 130
vapor pressure, 6-87 to 116
vapor pressure, high temperature, 4-66 to 68
Laser Hazards in the Laboratory, 16-86 to 87
Lattice constants
elements, 4-87 to 94, 12-15 to 18
inorganic compounds, 4-87 to 94
minerals, 4-87 to 94
rare-earth elements, 4-52 to 61
semiconductors, 12-70 to 82
Lattice energy, 12-21
Lawrencium: *see also* Elements
electron configuration, 1-16
history, occurrence, uses, 4-1 to 39
isotopes and their properties, 11-3 to 42
Lead: *see also* Elements
electrical resistivity, 12-27
electron configuration, 1-16
heat capacity, 4-65
history, occurrence, uses, 4-1 to 39
ionization energy, 10-112 to 115
isotopes and their properties, 11-3 to 42
magnetic susceptibility, 4-73 to 78
molten, density, 4-71 to 72
physical properties, 4-62 to 64
thermal conductivity, 12-131
thermal properties, 12-129 to 130
thermodynamic properties, 5-1 to 2
thermophysical properties, 12-122 to 127
vapor pressure, 6-87 to 116
vapor pressure, high temperature, 4-66 to 68
LEED, definition, 12-1 to 5
Lifetime
all nuclides, 11-3 to 42
hazardous radionuclides, 16-93 to 96
Ligands
nomenclature, 2-16 to 21
reactions with proteins, 7-21 to 29
Light, speed of, 1-1 to 9, 1-18 to 22
LIMS, definition, 12-1 to 5
Line spectra of elements, 10-1 to 50
Line strengths in atomic spectra, 10-51 to 53
Liquid air, thermodynamic properties, 6-15 to 20
Liquid helium properties, 6-163
Liquid metals, density, 4-71 to 72
Liquids
breakdown voltage, 15-45 to 49
dielectric constant, 6-193 to 215
diffusion, 6-252 to 253
flammability, 15-13 to 22
index of refraction, 3-1 to 55, 4-79, 10-152
permittivity, 6-193 to 215
speed of sound in, 14-39 to 40
surface tension, 6-182 to 186
thermal conductivity, 6-243 to 248
viscosity, 6-235 to 239
Liter, definition, 1-18 to 22
Lithium: *see also* Elements
electrical resistivity, 12-27
electron configuration, 1-16
heat capacity, 4-65
history, occurrence, uses, 4-1 to 39
ionization energy, 10-112 to 115
isotopes and their properties, 11-3 to 42
magnetic susceptibility, 4-73 to 78
molten, density, 4-71 to 72
physical properties, 4-62 to 64
thermal properties, 12-129 to 130
thermodynamic properties, 5-1 to 2

vapor pressure, 6-87 to 116
vapor pressure, high temperature, 4-66 to 68
Lithosphere, mass of, 14-14 to 15
Livermorium (element 116), 4-1 to 39, 11-3 to 42
Log *P*, 5-189 to 193
Loss factor, glasses, 12-41
Loss tangent, 6-14
Low-Temperature Baths for Maintaining Constant Temperature, 15-37
Lumen, definition, 1-18 to 22
Lutetium: *see also* Elements
electrical resistivity, 12-27
electron configuration, 1-16
heat capacity, 4-65
history, occurrence, uses, 4-1 to 39
ionization energy, 10-112 to 115
isotopes and their properties, 11-3 to 42
magnetic susceptibility, 4-73 to 78
molten, density, 4-71 to 72
physical properties, 4-62 to 64
thermal properties, 12-129 to 130
vapor pressure, 6-87 to 116
vapor pressure, high temperature, 4-66 to 68
Lux, definition, 1-18 to 22

M

Madelung Constant and Crystal Lattice Energy, The, 12-21
Magnesium: *see also* Elements
electrical resistivity, 12-27
electron configuration, 1-16
heat capacity, 4-65
history, occurrence, uses, 4-1 to 39
ionization energy, 10-112 to 115
isotopes and their properties, 11-3 to 42
magnetic susceptibility, 4-73 to 78
molten, density, 4-71 to 72
thermal conductivity, 12-131
thermal properties, 12-129 to 130
thermodynamic properties, 5-1 to 2
vapor pressure, 6-87 to 116
vapor pressure, high temperature, 4-66 to 68
Magnetic materials, composition and properties, 12-97 to 104
Magnetic moment
electron, proton, other particles, 1-1 to 9
nuclei important in NMR, 8-49 to 51
all nuclides, 11-3 to 42
rare-earth elements, 4-52 to 61
Magnetic phase transitions, 12-19 to 20
Magnetic properties
alloys, 12-97 to 104
organic magnets, 12-105
rare-earth elements, 4-52 to 61
superconductors, 12-42 to 57, 12-58 to 67
Magnetic susceptibility
elements, 4-73 to 78
inorganic compounds, 4-73 to 78
organic compounds, 3-56 to 59
rare-earth elements, 4-52 to 61
semiconductors, 12-70 to 82
various materials, 12-97 to 104
Magnetic Susceptibility of Elements and Inorganic Compounds, 4-73 to 78
Magnetism, symbols and units, 1-23, 12-97 to 104
Magneton (nuclear, Bohr), 1-1 to 9
Magnetostriction, 12-97 to 104
Major Reference Masses in the Spectrum of Heptacosafluorotributylamine (Perfluorotributylamine), 8-75
Major World Earthquakes, 14-30 to 33
Makemake, orbital data and dimensions, 14-2 to 6
Manganese: *see also* Elements
electrical resistivity, 12-27
electron configuration, 1-16
heat capacity, 4-65
history, occurrence, uses, 4-1 to 39
ionization energy, 10-112 to 115
isotopes and their properties, 11-3 to 42
magnetic susceptibility, 4-73 to 78
molten, density, 4-71 to 72
physical properties, 4-62 to 64
thermal properties, 12-129 to 130
thermophysical properties, 12-122 to 127
vapor pressure, 6-87 to 116
vapor pressure, high temperature, 4-66 to 68
Mars, orbital data and dimensions, 14-2 to 6
Mass
atmosphere, oceans, and crust, 14-14 to 15
atomic mass unit, 1-1 to 9
atomic, of nuclides, 11-3 to 42
earth, moon, and sun, 14-1
electron, proton, neutron, 1-1 to 9
planets, 14-2 to 6
planets, relative to sun, 14-1
satellites, 14-6 to 8
sun, 14-2 to 6
Mass spectra
chlorine–bromine isotope intensities, 8-84
common spurious signals, 8-83
fragmentation patterns, 8-72 to 73, 8-74
organic solvents, 8-76 to 82
peaks of prominent isotopes, 8-71
reference masses in spectrum of perfluorotributylamine, 8-74
Mass Spectral Peaks of Common Organic Solvents, 8-76 to 82
Mass, Dimensions, and Other Parameters of the Earth, 14-14 to 15
Mass- and Volume-Based Concentration Units, 8-18
Materials characterization, techniques, 12-1 to 5
Materials Compatible with and Resistant to 72% Perchloric Acid, 16-30
Mean Activity Coefficients of Electrolytes as a Function of Concentration, 5-121 to 125
Mean free path
common gases, 6-54
electrons in solids, 12-106 to 107
molecules in the atmosphere, 14-23 to 28
Mean Free Path and Related Properties of Gases, 6-54
Mebibyte, megabyte, etc., 1-18 to 22
Mechanical properties
gas clathrate hydrates, 6-167 to 177
polymers, 13-3 to 4
rare-earth elements, 4-52 to 61
Meitnerium (element 109), 4-1 to 39, 11-3 to 42
Melting, Boiling, Triple, and Critical Points of Elements, 4-62 to 64
Melting point
alloys (eutectics), 15-38 to 39
amino acids, 7-1 to 2
cryogenic fluids, 6-161 to 162
D_2O, 6-9
depression of, 15-31
elements, 4-62 to 64
fatty acids, 7-7 to 8
halocarbons, 6-164 to 166
ice, as function of pressure, 6-13
inorganic compounds, 4-40 to 51, 6-144 to 153
ionic liquids, 6-178 to 181
metals, 12-129 to 130
oils and fats, 7-9 to 11
organic compounds, 3-1 to 55, 6-144 to 153
polymers, 13-3 to 4
pressure dependence, 6-55
rare-earth elements, 4-52 to 61
semiconductors, 12-70 to 82
solvents, 15-13 to 22
Melting Point of Ice as a Function of Pressure, 6-13
Mendelevium: *see also* Elements
electron configuration, 1-16
history, occurrence, uses, 4-1 to 39
isotopes and their properties, 11-3 to 42
Mercury: *see also* Elements
critical constants, 6-82 to 84
electrical resistivity, 12-27
electron configuration, 1-16
heat capacity, 4-65
history, occurrence, uses, 4-1 to 39
ionization energy, 10-112 to 115
isotopes and their properties, 11-3 to 42
magnetic susceptibility, 4-73 to 78
physical properties, 4-62 to 64
thermal conductivity, 6-243 to 248
thermal properties, 12-129 to 130
thermodynamic properties, 5-1 to 2
vapor pressure, high temperature, 4-66 to 68
Mercury (planet), orbital data and dimensions, 14-2 to 6
Metal oxides, secondary electron emission, 12-112
Metals
coefficient of friction, 15-50
compatibility with 72% perchloric acid, 16-30
crystal phase transitions, 12-19 to 20
crystal structure, 12-15 to 18
elastic constants, 12-22 to 26
electrical resistivity, 12-27
electron inelastic mean free path, 12-106 to 107
extinction coefficient, 12-113 to 116
Fermi energy, 12-143 to 144
heat capacity, 12-128
index of refraction, 12-113 to 116
optical properties, 12-113 to 116
reflection coefficient, 12-113 to 116
secondary electron emission, 12-112
speed of sound in, 14-39 to 40
sublimation pressure, 6-87 to 116
superconducting properties, 12-42 to 57
thermal conductivity, 12-131
thermal properties, 12-129 to 130
thermoelectric properties, 12-141 to 142
Metals and Alloys with Low-Melting Temperature, 15-38 to 39
Meter, definition, 1-18 to 22
Microwave bands, classification, 10-140 to 141

Middle-Range Infrared Absorption Correlation Charts, 8-42 to 47
Milliard, billion (definition), 8-18
Minerals
- chemical formulas, 4-80 to 86, 4-87 to 94
- crystal structure, 4-87 to 94
- elastic constants, 12-22 to 26
- index of refraction, 4-80 to 86
- physical constants, 4-80 to 86
- semiconducting properties, 12-70 to 82
- solubility as a function of temperature, 5-183 to 188

Minimum Recommended Injector Split Ratios for Capillary Columns, 8-35
- thermal conductivity, 12-136 to 137

Miscibility of Organic Solvents, 15-23 to 26
Mobility in semiconductors, 12-70 to 82, 12-85 to 88
Molar Electrical Conductivity of Electrolytes in Aqueous Solution, 5-97
Molar Electrical Conductivity of Aqueous HF, HCl, HBr, and HI, 5-96
Mole, definition, 1-18 to 22
Molecular weight
- amino acids, 7-1 to 2
- inorganic compounds, 4-40 to 51
- organic compounds, 3-1 to 55

Molecules
- appearance potential, 10-116 to 133
- barriers to internal rotation, 9-68 to 72
- bond lengths, 9-1 to 16, 9-17 to 18, 9-55
- bond lengths and angles, 9-19 to 54
- bond strengths, 9-73 to 105
- electron affinity, 10-54 to 75
- force constants, 9-107
- fundamental vibrational frequencies, 9-108 to 110
- ionization energy, 10-116 to 133
- polarizability, 10-95 to 111
- proton affinity, 10-76 to 94

Molybdenum: *see also* Elements
- electrical resistivity, 12-27
- electron configuration, 1-16
- heat capacity, 4-65
- history, occurrence, uses, 4-1 to 39
- ionization energy, 10-112 to 115
- isotopes and their properties, 11-3 to 42
- magnetic susceptibility, 4-73 to 78
- molten, density, 4-71 to 72
- physical properties, 4-62 to 64
- thermal properties, 12-129 to 130
- thermophysical properties, 12-122 to 127
- vapor pressure, 6-87 to 116
- vapor pressure, high temperature, 4-66 to 68

Moon
- orbital constants and other parameters, 14-2 to 5, 14-6 to 8
- ratio of mass to earth's mass, 14-1

Moscovium (element 115), 4-1 to 39, 11-3 to 42
Muon
- in cosmic ray showers, 11-56 to 61
- properties, 1-1 to 9

N

^{15}N-NMR Chemical Shifts of Major Chemical Families, 8-69 to 70
NAA, definition, 12-1 to 5
Nanomaterial Safety Guidelines, 16-46 to 47
Natural Abundance of Important Isotopes, 8-71
Néel temperature
- magnetic materials, 12-97 to 104
- rare-earth elements, 4-52 to 61

Neodymium: *see also* Elements
- electrical resistivity, 12-27
- electron configuration, 1-16
- heat capacity, 4-65
- history, occurrence, uses, 4-1 to 39
- ionization energy, 10-112 to 115
- isotopes and their properties, 11-3 to 42
- magnetic susceptibility, 4-73 to 78
- molten, density, 4-71 to 72
- physical properties, 4-62 to 64
- thermal properties, 12-129 to 130
- vapor pressure, 6-87 to 116
- vapor pressure, high temperature, 4-66 to 68

Neon: *see also* Elements
- critical constants, 6-82 to 84
- cryogenic properties, 6-161 to 162
- electron configuration, 1-16
- entropy, 5-1 to 2
- history, occurrence, uses, 4-1 to 39
- ionization energy, 10-112 to 115
- isotopes and their properties, 11-3 to 42
- magnetic susceptibility, 4-73 to 78
- mean free path, 6-54
- permittivity (dielectric constant), 6-193 to 215, 6-216
- physical properties, 4-62 to 64
- solubility in water, 5-149 to 150
- thermal conductivity, 6-240 to 242
- vapor pressure, 6-87 to 116, 6-117 to 125
- viscosity, 6-233 to 234

Neptune, orbital data and dimensions, 14-2 to 6
Neptunium: *see also* Elements
- electron configuration, 1-16
- history, occurrence, uses, 4-1 to 39
- ionization energy, 10-112 to 115
- isotopes and their properties, 11-3 to 42
- magnetic susceptibility, 4-73 to 78
- physical properties, 4-62 to 64
- vapor pressure, high temperature, 4-66 to 68

Neutron
- properties, 1-1 to 9
- range and shielding, 16-88
- scattering and absorption, 11-43 to 55

Neutron cross sections, 11-43 to 55
Neutron resonance integrals, 11-43 to 55
Neutron Scattering and Absorption Properties, 11-43 to 55
Newton, definition, 1-18 to 22
Nickel: *see also* Elements
- electrical resistivity, 12-27
- electron configuration, 1-16
- heat capacity, 4-65
- history, occurrence, uses, 4-1 to 39
- ionization energy, 10-112 to 115
- isotopes and their properties, 11-3 to 42
- magnetic susceptibility, 4-73 to 78
- molten, density, 4-71 to 72
- physical properties, 4-62 to 64
- thermal properties, 12-129 to 130
- vapor pressure, 6-87 to 116
- vapor pressure, high temperature, 4-66 to 68

Nihonium (element 113), 4-1 to 39, 11-3 to 42
Niobium: *see also* Elements
- electrical resistivity, 12-27
- electron configuration, 1-16
- heat capacity, 4-65
- history, occurrence, uses, 4-1 to 39
- ionization energy, 10-112 to 115
- isotopes and their properties, 11-3 to 42
- magnetic susceptibility, 4-73 to 78
- physical properties, 4-62 to 64
- thermal properties, 12-129 to 130
- vapor pressure, 6-87 to 116
- vapor pressure, high temperature, 4-66 to 68

Nitrogen: *see also* Elements
- critical constants, 6-82 to 84
- cryogenic properties, 6-161 to 162
- electron configuration, 1-16
- enthalpy of vaporization, 6-127 to 143
- history, occurrence, uses, 4-1 to 39
- ionization energy, 10-112 to 115
- isotopes and their properties, 11-3 to 42
- magnetic susceptibility, 4-73 to 78
- mean free path, 6-54
- permittivity (dielectric constant), 6-193 to 215, 6-216
- physical properties, 4-62 to 64
- solubility in water, 5-149 to 150
- thermal conductivity, 6-21 to 39, 6-240 to 242
- thermodynamic properties, 5-1 to 2
- thermodynamic properties at high temperature, 5-42 to 64
- thermophysical properties, 6-21 to 39
- vapor pressure, 6-87 to 116, 6-117 to 125
- viscosity, 6-21 to 39, 6-233 to 234

NMR spectrum
- analytical chemistry applications, 8-6 to 9
- characteristic ^{13}C chemical shifts, 8-66 to 67
- characteristic ^{15}N chemical shifts, 8-69 to 70
- characteristic proton chemical shifts, 8-53 to 57, 8-58, 8-65
- nuclear moments and resonance frequencies, 8-49 to 51, 8-52
- solvents, ^{13}C shifts, 8-68
- solvents, proton shifts, 8-59 to 64, 8-65

Nobel Laureates in Chemistry and Physics, 2-38 to 40
Nobelium: *see also* Elements
- electron configuration, 1-16
- history, occurrence, uses, 4-1 to 39
- isotopes and their properties, 11-3 to 42

Nomenclature
- carbohydrates, 7-12 to 13
- chemical, references, 2-15
- inorganic chemistry, 2-16 to 21
- minerals, 4-80 to 86
- physical quantities, 2-1 to 12
- polymers, 13-5 to 8

Nomenclature for Chemical Compounds, 2-15
Nomenclature for Organic Polymers, 13-5 to 8
Nomenclature of Inorganic Chemistry, 2-16 to 21
NQR, definition, 12-1 to 5
NRA, definition, 12-1 to 5
Nuclear magnetic resonance: *see* NMR
Nuclear magneton, 1-1 to 9
Nuclear spins and moments

for all nuclides, 11-3 to 42
for important nuclei in NMR, 8-49 to 51, 8-52
Nuclear Spins, Moments, and Other Data Related to NMR Spectroscopy, 8-49 to 51
Nucleic acids
genetic code, 7-6
purine and pyrimidine bases, 7-5
Nuclides, summary of properties, 11-3 to 42

O

Ocean Pressure as a Function of Depth and Latitude, 14-18
Oceans
abundance of chemical elements, 14-21
pressure as a function of depth and latitude, 14-18
Octanol–Water Partition Coefficients, 5-189 to 193
Oganesson (element 118), 4-1 to 39, 11-3 to 42
Ohm
definition, 1-18 to 22
maintained value, 1-1 to 9
Oils, composition and properties, 7-9 to 11
Optical materials
index of refraction, 10-150
glass, 10-150
human eye, 10-142
inorganic crystals, 10-146 to 149
metals, 12-113 to 116
polytetrafluoroethylene, 12-117 to 121
semiconductors, 12-113 to 116, 12-117 to 121
solids, as function of wavelength, 10-146 to 149
Optical Properties of Selected Elements, 12-113 to 116
Optical Properties of Selected Inorganic and Organic Solids, 12-117 to 121
Orbital data for planets and satellites, 14-2 to 5, 14-6 to 8
Organic Analytical Reagents for the Determination of Inorganic Ions, 8-95 to 108
Organic compounds
bond lengths (in crystals), 9-1 to 16
bond lengths and angles (in gas-phase), 9-19 to 54
bond strengths, 9-73 to 105
characteristic ^{13}C chemical shifts, 8-66 to 67
characteristic ^{15}N chemical shifts, 8-69 to 70
characteristic proton chemical shifts, 8-53 to 57, 8-58
dipole moment, 9-58 to 67
enthalpy of fusion, 6-144 to 153
enthalpy of vaporization, 6-127 to 143
heat of combustion, 5-67
infrared correlation charts, 8-42 to 47
magnetic susceptibility, 3-56 to 59
mass spectral fragmentation patterns, 8-72 to 73, 8-74
mass spectral peaks, 8-76 to 82
NMR shifts, 8-59 to 64
physical properties, 3-1 to 55
polarizability, 10-95 to 111
solubility, aqueous, 5-152 to 182
solubility, aqueous at high temperature, 5-198 to 199
sublimation pressure, 6-85 to 86
superconducting properties, 12-68 to 69
surface tension, 6-182 to 186
thermal conductivity, 6-243 to 248
thermodynamic properties, 5-3 to 41
Organic Magnets, 12-105
Organic semiconductors, 12-85 to 88
Organic Superconductors, 12-68 to 69
Organometallic compounds, bond lengths, 9-17 to 18
Oscillator strengths in atomic spectra, 10-51 to 53
Osmium: *see also* Elements
electrical resistivity, 12-27
electron configuration, 1-16
heat capacity, 4-65
history, occurrence, uses, 4-1 to 39
ionization energy, 10-112 to 115
isotopes and their properties, 11-3 to 42
magnetic susceptibility, 4-73 to 78
molten, density, 4-71 to 72
physical properties, 4-62 to 64
thermal properties, 12-129 to 130
vapor pressure, 6-87 to 116
vapor pressure, high temperature, 4-66 to 68
Outliers in measurements, 8-19
Oxidation-reduction potentials
biochemical species, 7-14 to 16
general table, 5-101 to 107
Oxygen: *see also* Elements
critical constants, 6-82 to 84
cryogenic properties, 6-161 to 162
electron configuration, 1-16
enthalpy of fusion, 6-144 to 153
enthalpy of vaporization, 6-127 to 143
history, occurrence, uses, 4-1 to 39
ionization energy, 10-112 to 115
isotopes and their properties, 11-3 to 42
magnetic susceptibility, 4-73 to 78
mean free path, 6-54
permittivity (dielectric constant), 6-193 to 215, 6-216
physical properties, 4-62 to 64
solubility in water, 5-149 to 150
thermal conductivity, 6-21 to 39, 6-240 to 242
thermodynamic properties, 5-1 to 2
thermophysical properties, 6-21 to 39
vapor pressure, 6-87 to 116, 6-117 to 125
viscosity, 6-21 to 39, 6-233 to 234

P

Palladium: *see also* Elements
electrical resistivity, 12-27
electron configuration, 1-16
heat capacity, 4-65
history, occurrence, uses, 4-1 to 39
ionization energy, 10-112 to 115
isotopes and their properties, 11-3 to 42
magnetic susceptibility, 4-73 to 78
molten, density, 4-71 to 72
physical properties, 4-62 to 64
thermal properties, 12-129 to 130
vapor pressure, 6-87 to 116
vapor pressure, high temperature, 4-66 to 68
Paramagnetic moment, rare-earth elements, 4-52 to 61
Paramagnetic susceptibility, elements and inorganic compounds, 4-73 to 78
Partial molar volume, amino acids, 7-1 to 2
Particle size, 15-40
Pascal, definition, 1-18 to 22
Pauling electronegativity scale, 9-107
Pearson symbols, 12-6 to 10, 12-15 to 18
Peltier effect, 12-141 to 142
Periodic Table of Elements: *see* Inside front cover; 1-25
Permeability, magnetic alloys, 12-97 to 104
Permeability of vacuum, 1-1 to 9
Permittivity (dielectric constant)
cryogenic fluids, temperature and pressure dependence, 6-21 to 39
crystals, 12-31 to 39
gases, 6-216
glass, 12-41
ice, 6-11
liquid helium, 6-163
liquids, 6-193 to 215
plastics, 13-17
quartz, 12-41
rubbers, 13-17
semiconductors, 12-70 to 82
solids, 12-31 to 39
solvents, 8-38
vacuum, 1-1 to 9
water and steam, temperature and pressure dependence, 6-1 to 4
water, frequency dependence, 6-14
Permittivity (Dielectric Constant) of Gases, 6-216
Permittivity (Dielectric Constant) of Inorganic Solids, 12-31 to 39
Permittivity (Dielectric Constant) of Liquids, 6-193 to 215
Permittivity (Dielectric Constant) of Water at Various Frequencies, 6-14
Peroxide hazards, tests, 16-8, 16-9
Persistent Lines of Neutral Atomic Elements, 10-1 to 50
pH
acid-base indicators, 8-88 to 89
biological buffers, 7-33
biological materials and tissues, 7-34
blood, 7-34
definition of pH scale, 5-128 to 131
foods, 7-34
solutions giving round values, 5-132
standards, 5-128 to 131
pH Scale for Aqueous Solutions, 5-128 to 131
Phase Ratios for Capillary Columns, 8-34
Phase transitions
enthalpy of fusion, 6-144 to 153
enthalpy of vaporization, 6-127 to 143
ice, 6-11, 6-13
polymers, glass to crystal, 13-10 to 16
rare-earth elements, 4-52 to 61
solid elements, 12-19 to 20
Phase Transitions in the Solid Elements at Atmospheric Pressure, 12-19 to 20
Phonon-electron coupling, rare-earth elements, 4-52 to 61
Phosphorus: *see also* Elements
critical constants, 6-82 to 84
dielectric constant, 12-31 to 39
electron configuration, 1-16
heat capacity, 4-65

history, occurrence, uses, 4-1 to 39
ionization energy, 10-112 to 115
isotopes and their properties, 11-3 to 42
magnetic susceptibility, 4-73 to 78
physical properties, 4-62 to 64
thermodynamic properties, 5-1 to 2
vapor pressure, 6-87 to 116
Photon Attenuation Coefficients, 10-134 to 139
Photopic spectral luminous efficiency function, 10-142
Physical and Optical Properties of Minerals, 4-80 to 86
Physical Constants of Inorganic Compounds, 4-40 to 51
Physical Constants of Organic Compounds, 3-1 to 55
Physical constants, fundamental, 1-1 to 9
Physical properties
amino acids, 7-1 to 2
gas clathrate hydrates, 6-167 to 177
inorganic compounds, 4-40 to 51
minerals, 4-80 to 86
oils and fats, 7-9 to 11
organic compounds, 3-1 to 55
polymers, 13-3 to 4
semiconductors, 12-70 to 82
solvents, 15-13 to 22
Physical Properties of the Rare-Earth Metals, 4-52 to 61
Physical Properties of Selected Polymers, 13-3 to 4
Physical quantities, terminology, and symbols, 2-1 to 12
PIXE, definition, 12-1 to 5
p*K*
acid-base indicators, 8-88 to 89
amino acids, 7-1 to 2
biological buffers, 7-30 to 32, 7-33
inorganic acids and bases, 5-108
organic acids and bases, 5-109 to 118
purine and pyrimidine bases, 7-5
Planck constant, 1-1 to 9, 1-18 to 22, 8-24
Planets
atmospheric composition, 14-2 to 6
general properties, 14-2 to 6
orbital parameters, 14-2 to 6
satellites, 14-6 to 8
Plastics
breakdown voltage, 15-45 to 49
compatibility with 72% perchloric acid, 16-30
density, 15-41
dielectric constant, 13-17
physical properties, 13-3 to 4
speed of sound in, 14-39 to 40
thermal conductivity, 12-122 to 127
thermophysical properties, 12-122 to 127
Platinum: *see also* Elements
electrical resistivity, 12-27
electron configuration, 1-16
heat capacity, 4-65
history, occurrence, uses, 4-1 to 39
ionization energy, 10-112 to 115
isotopes and their properties, 11-3 to 42
magnetic susceptibility, 4-73 to 78
molten, density, 4-71 to 72
physical properties, 4-62 to 64
thermal conductivity, 12-131
thermal properties, 12-129 to 130
thermophysical properties, 12-122 to 127
vapor pressure, 6-87 to 116
vapor pressure, high temperature, 4-66 to 68
Platinum-rhodium thermocouple tables, 15-1 to 9
PLOT (porous-layer open tubular) stationary phase, 8-28
Pluto, orbital data and dimensions, 14-2 to 6
Plutonium: *see also* Elements
electron configuration, 1-16
heat capacity, 4-65
history, occurrence, uses, 4-1 to 39
ionization energy, 10-112 to 115
isotopes and their properties, 11-3 to 42
magnetic susceptibility, 4-73 to 78
molten, density, 4-71 to 72
physical properties, 4-62 to 64
thermal properties, 12-129 to 130
vapor pressure, 6-87 to 116
vapor pressure, high temperature, 4-66 to 68
Point groups of small molecules, 9-108 to 110
Polarizabilities of Atoms and Molecules, 10-95 to 111
Polarizability of Atoms and Ions in Solids, 12-13 to 14
Pollutants
airborne, limits in the workplace, 16-48 to 75
Henry's Law constants, 5-152 to 182
octanol–water partition coefficients, 5-189 to 193
solubility, 5-152 to 182
Polonium: *see also* Elements
electrical resistivity, 12-27
electron configuration, 1-16
heat capacity, 4-65
history, occurrence, uses, 4-1 to 39
ionization energy, 10-112 to 115
isotopes and their properties, 11-3 to 42
physical properties, 4-62 to 64
thermal properties, 12-129 to 130
vapor pressure, 6-87 to 116
Polymers
abbreviations, 13-1 to 2
breakdown voltage, 15-45 to 49
cohesive energy density, 13-27 to 29
density, 13-3 to 4
density of melts, 13-18 to 21
dielectric constant, 13-17
electron inelastic mean free path, 12-106 to 107
glass transition temperature, 13-10 to 16
mechanical and thermal properties, 13-3 to 4
melting point, 13-3 to 4
molar volume, 13-18 to 21
nomenclature, 13-5 to 8
solubility parameters, 13-27 to 29
solvent activities, 13-22 to 26
solvents for, 13-9
thermophysical properties, 12-122 to 127
vapor pressure of solutions, 13-22 to 26
Potassium: *see also* Elements
electrical resistivity, 12-27
electron configuration, 1-16
heat capacity, 4-65
history, occurrence, uses, 4-1 to 39
ionization energy, 10-112 to 115
isotopes and their properties, 11-3 to 42
magnetic susceptibility, 4-73 to 78
molten, density, 4-71 to 72
physical properties, 4-62 to 64
thermal properties, 12-129 to 130
thermodynamic properties, 5-1 to 2
thermodynamic properties at high temperature, 5-42 to 64
vapor pressure, 6-87 to 116
vapor pressure, high temperature, 4-66 to 68
Potential oxidation-reduction, 5-101 to 107
Praseodymium: *see also* Elements
electrical resistivity, 12-27
electron configuration, 1-16
heat capacity, 4-65
history, occurrence, uses, 4-1 to 39
ionization energy, 10-112 to 115
isotopes and their properties, 11-3 to 42
magnetic susceptibility, 4-73 to 78
molten, density, 4-71 to 72
physical properties, 4-62 to 64
thermal properties, 12-129 to 130
vapor pressure, 6-87 to 116
vapor pressure, high temperature, 4-66 to 68
Preparation of Special Analytical Reagents, 8-90 to 94
Pressure
atmosphere, as function of altitude, 14-23 to 28
conversion factors, 1-24
earth, as function of depth, 14-17
effect on boiling point, 15-28 to 29
effect on freezing point, 6-55
ocean, as function of depth, 14-18
planetary atmospheres, 14-2 to 6
Pressure Drop in Open Tubular Gas Chromatographic Columns, 8-33
sensitivity of ionization gauges, 15-12
Pressure-Volume-Temperature Relationships for Polymer Melts, 13-18 to 21
Promethium: *see also* Elements
electrical resistivity, 12-27
electron configuration, 1-16
heat capacity, 4-65
history, occurrence, uses, 4-1 to 39
ionization energy, 10-112 to 115
isotopes and their properties, 11-3 to 42
physical properties, 4-62 to 64
thermal properties, 12-129 to 130
Properties and Functions of Common Drugs, 7-35 to 56
Properties of Amino Acids, 7-1 to 2
Properties of Antiferroelectric Crystals, 12-41
Properties of Carrier Gases for Gas Chromatography, 8-20
Properties of Controlled Substances, 7-57 to 7-58
Properties of Common Cross-Linked Silicone Stationary Phases, 8-25 to 26
Properties of Cryogenic Fluids, 6-161 to 162
Properties of Fatty Acids and Their Methyl Esters, 7-7 to 8
Properties of Gas Clathrate Hydrates, 6-167 to 177
Properties of Ice and Supercooled Water, 6-11
Properties of Ionic Liquids, 6-178 to 181
Properties of Important NMR Nuclei, 8-52

Properties of Liquid Helium, 6-163
Properties of Magnetic Materials, 12-97 to 104
Properties of Organic Semiconductors, 12-85 to 88
Properties of Purine and Pyrimidine Bases, 7-5
Properties of Refrigerants, 6-164 to 166
Properties of Saturated Liquid D_2O, 6-10
Properties of Seawater, 14-19 to 20
Properties of Selected Materials at Cryogenic Temperatures, 12-122 to 127
Properties of Semiconductors, 12-70 to 82
Properties of Superconductors, 12-42 to 57
Properties of the Solar System, 14-2 to 6
Protactinium: *see also* Elements
- electrical resistivity, 12-27
- electron configuration, 1-16
- heat capacity, 4-65
- history, occurrence, uses, 4-1 to 39
- ionization energy, 10-112 to 115
- isotopes and their properties, 11-3 to 42
- magnetic susceptibility, 4-73 to 78
- physical properties, 4-62 to 64
- thermal properties, 12-129 to 130
- vapor pressure, high temperature, 4-66 to 68

Protective Clothing Levels, 16-36
Protein-ligand reactions, 7-21 to 29
Proton
- in cosmic ray showers, 11-56 to 61
- mass, magnetic moment, other properties, 1-1 to 9

Proton Affinities, 10-76 to 94
Proton Chemical Shifts of Contaminants in Deuterated Solvents, 8-65
Proton NMR Absorption of Major Chemical Families, 8-53 to 57
Proton NMR Correlation Chart for Major Organic Functional Groups, 8-58
Proton NMR Shifts of Common Organic Liquids, 8-59 to 64
PSD, definition, 12-1 to 5
psia and psig, definition, 1-24
Purine bases, properties of, 7-5
Pyrimidine bases, properties of, 7-5
Pyrophoric Compounds—Compounds That Are Reactive with Air, 16-9

Q

Q-Test, 8-19
Quadrupole moments
- all nuclides, 11-3 to 42
- important nuclei for NMR, 8-49 to 51

Quartz
- crystallographic data, 4-87 to 94
- dielectric constant, 12-41
- loss factor, 12-41
- optical properties, 4-80 to 86
- thermal conductivity, 12-134 to 135

R

Rad, definition, 16-89 to 90
Radiation
- blackbody, 10-143 to 144
- electromagnetic, classification, 10-140 to 141
- microwave, classification of bands, 10-140 to 141

Radiation Safety Units, 16-89 to 90
Radiation, ionizing
- dose ranges, 16-91 to 92
- from nuclear decay, 11-3 to 42
- natural background, 16-91 to 92
- permissible intake of radionuclides, 16-93 to 96
- protection against, 16-88

Radiation, solar
- by month and latitude, 14-29
- by wavelength, 14-22
- flux, solar constant, 14-2 to 6

Radiative transition probability, 10-51 to 53
Radicals, free: *see* Free radicals
Radicals, nomenclature
Radii
- atomic, 9-56 to 57
- molecules in gases, 6-54

Radioastronomy, 14-9 to 13
Radionuclides, permissible intake, 16-93 to 96
Radium: *see also* Elements
- electron configuration, 1-16
- heat capacity, 4-65
- history, occurrence, uses, 4-1 to 39
- ionization energy, 10-112 to 115
- isotopes and their properties, 11-3 to 42
- physical properties, 4-62 to 64
- thermal properties, 12-129 to 130
- vapor pressure, 6-87 to 116

Radius of ions in crystals, 12-11 to 12
Radon: *see also* Elements
- contribution to background radiation, 16-91 to 92
- critical constants, 6-82 to 84
- electron configuration, 1-16
- history, occurrence, uses, 4-1 to 39
- ionization energy, 10-112 to 115
- isotopes and their properties, 11-3 to 42
- physical properties, 4-62 to 64
- solubility in water, 5-149 to 150
- vapor pressure, 6-87 to 116, 6-117 to 125

Rare-earth metals, general properties, 4-52 to 61
Rate constants, chemical reactions, 5-69 to 91
RBS, definition, 12-1 to 5
Reagents
- organic, for chemical analysis, 8-95 to 108
- preparation, 8-90 to 94

Reduction of Weighings in Air to Vacuo, 8-85
Reduction potentials
- biochemical species, 7-14 to 16
- general table, 5-101 to 107

Reference Points on the ITS-90 Temperature Scale, 15-10 to 11
Reference states of elements, 5-3 to 41
Reflection coefficient of solids, 12-113 to 116
Refractive index: *see* Index of refraction
Refractive Index and Transmittance of Representative Glasses, 10-150
Refractory materials, thermal conductivity, 12-136 to 137
Refrigerants
- thermal conductivity, 6-240 to 242
- various properties, 6-164 to 166

Relative biological effectiveness of radiation, 16-89 to 90
Relative Dose Ranges from Ionizing Radiation, 16-91 to 92
Relative humidity
- from wet and dry bulb temperatures, 15-35
- relation to dew point, 15-34
- solutions for calibration, 15-37
- solutions for constant humidity, 15-36

Relative Sensitivity of Bayard-Alpert Ionization Gauges to Various Gases, 15-12
Relaxation time, in water, 6-14
Rem, definition, 16-89 to 90
Remanence, magnetic materials, 12-97 to 104
Representation of Chemical Structures with the IUPAC International Chemical Identifier (InChI), 2-22 to 23
Resistivity, electrical
- alloys, 12-28 to 29
- elements, 12-27
- glasses, 12-41
- graphite and related materials, 12-30
- pure metals, 12-27
- quartz, 12-41
- rare-earth elements, 4-52 to 61
- semiconducting minerals, 12-70 to 82
- semiconductors, 12-70 to 82, 12-85 to 88
- superconductors, 12-58 to 67

Respirator cartridges and filters, 16-34 to 35
Retention indices, in gas chromatography, 8-25 to 26, 8-31 to 32
RHEED, definition, 12-1 to 5
Rhenium: *see also* Elements
- electrical resistivity, 12-27
- electron configuration, 1-16
- heat capacity, 4-65
- history, occurrence, uses, 4-1 to 39
- ionization energy, 10-112 to 115
- isotopes and their properties, 11-3 to 42
- magnetic susceptibility, 4-73 to 78
- molten, density, 4-71 to 72
- physical properties, 4-62 to 64
- thermal properties, 12-129 to 130
- vapor pressure, 6-87 to 116
- vapor pressure, high temperature, 4-66 to 68

Rhodium: *see also* Elements
- electrical resistivity, 12-27
- electron configuration, 1-16
- heat capacity, 4-65
- history, occurrence, uses, 4-1 to 39
- ionization energy, 10-112 to 115
- isotopes and their properties, 11-3 to 42
- magnetic susceptibility, 4-73 to 78
- molten, density, 4-71 to 72
- physical properties, 4-62 to 64
- thermal properties, 12-129 to 130
- vapor pressure, 6-87 to 116
- vapor pressure, high temperature, 4-66 to 68

Rochelle salts, 12-40
Rocks
- age, 14-14 to 15
- density, 15-41
- thermal conductivity, 12-136 to 137

Roentgen, definition, 16-89 to 90
Roentgenium (element 111), 4-1 to 39, 11-3 to 42
Rotational constants, diatomic molecules, 9-111 to 116
Rubbers
- breakdown voltage, 15-45 to 49
- compatibility with 72% perchloric acid, 16-30

density, 15-41
dielectric constant, 13-17
speed of sound in, 14-39 to 40
thermal conductivity, 12-136 to 137
Rubidium: *see also* Elements
electrical resistivity, 12-27
electron configuration, 1-16
heat capacity, 4-65
history, occurrence, uses, 4-1 to 39
ionization energy, 10-112 to 115
isotopes and their properties, 11-3 to 42
magnetic susceptibility, 4-73 to 78
molten, density, 4-71 to 72
physical properties, 4-62 to 64
thermal properties, 12-129 to 130
thermodynamic properties, 5-1 to 2
vapor pressure, 6-87 to 116
vapor pressure, high temperature, 4-66 to 68
Ruthenium: *see also* Elements
electrical resistivity, 12-27
electron configuration, 1-16
heat capacity, 4-65
history, occurrence, uses, 4-1 to 39
ionization energy, 10-112 to 115
isotopes and their properties, 11-3 to 42
magnetic susceptibility, 4-73 to 78
molten, density, 4-71 to 72
physical properties, 4-62 to 64
thermal properties, 12-129 to 130
vapor pressure, 6-87 to 116
vapor pressure, high temperature, 4-66 to 68
Rutherfordium (element 104)
electron configuration, 1-16
history, occurrence, uses, 4-1 to 39
isotopes and their properties, 11-3 to 42
Rydberg constant, 1-1 to 9
Sackur-Tetrode constant, 1-1 to 9

S

Safety
chemical carcinogens, 16-76 to 85
compressed air, 16-42
cryogens, 16-43 to 45
explosion hazard, 16-7
flammable chemicals, 16-10 to 11, 16-12 to 29
gas cylinders, 16-41
incompatible chemicals, 16-5 to 6
laser, 16-86 to 87
nanomaterials, 16-46 to 47
pyrophoric compounds, 16-9, 16-93 to 96
radiation, 16-88, 16-89 to 90,
water-reactive chemicals, 16-8
Safety in the Use of Cryogens, 16-43 to 45
SALI, definition, 12-1 to 5
Salinity scale for seawater, 14-19 to 20
Salts
activity coefficients, 5-98 to 100, 5-121 to 125
electrical conductivity, 5-94, 5-97
enthalpy of solution, 5-127
molten, density of, 4-71 to 72
solubility as a function of temperature, 5-183 to 188, 5-196
vapor pressure of aqueous solutions, 6-126
Samarium: *see also* Elements
electrical resistivity, 12-27
electron configuration, 1-16
heat capacity, 4-65
history, occurrence, uses, 4-1 to 39
ionization energy, 10-112 to 115
isotopes and their properties, 11-3 to 42
magnetic susceptibility, 4-73 to 78
molten, density, 4-71 to 72
physical properties, 4-62 to 64
thermal properties, 12-129 to 130
vapor pressure, 6-87 to 116
vapor pressure, high temperature, 4-66 to 68
SANS, definition, 12-1 to 5
Saponification value, oils and fats, 7-9 to 11
Satellites of the Planets, 14-6 to 8
Saturn, orbital data and dimensions, 14-2 to 6
Scandium: *see also* Elements
electrical resistivity, 12-27
electron configuration, 1-16
heat capacity, 4-65
history, occurrence, uses, 4-1 to 39
ionization energy, 10-112 to 115
isotopes and their properties, 11-3 to 42
magnetic susceptibility, 4-73 to 78
molten, density, 4-71 to 72
physical properties, 4-62 to 64
thermal properties, 12-129 to 130
vapor pressure, 6-87 to 116
vapor pressure, high temperature, 4-66 to 68
Scientific Abbreviations, Acronyms, and Symbols, 2-24 to 36
Seaborgium (element 106), 4-1 to 39, 11-3 to 42
Seawater
composition (elemental), 14-21
composition (ions), 14-19 to 20
density, 14-19 to 20
electrical conductivity, 14-19 to 20
freezing point, 14-19 to 20
pressure as a function of depth, 14-18
salinity scale, 14-19 to 20
solubility of hydrocarbons in, 5-197
specific heat, 14-19 to 20
speed of sound in, 14-39 to 40
viscosity, 14-19 to 20
Second, definition, 1-18 to 22
Second radiation constant, 1-1 to 9
Secondary Electron Emission, 12-112
Seebeck effect, 12-141 to 142
Selected Properties of Semiconductor Solid Solutions, 12-83 to 84
Selection of Hearing Protection Devices, 16-37 to 38
Selection of Laboratory Gloves, 16-31 to 32
Selection of Protective Laboratory Garments, 16-33
Selection of Respirators and Respirator Cartridges and Filters, 16-34 to 35
Selenium: *see also* Elements
critical constants, 6-82 to 84
electron configuration, 1-16
heat capacity, 4-65
history, occurrence, uses, 4-1 to 39
ionization energy, 10-112 to 115
isotopes and their properties, 11-3 to 42
magnetic susceptibility, 4-73 to 78
molten, density, 4-71 to 72
physical properties, 4-62 to 64
vapor pressure, 6-87 to 116
SEM, definition, 12-1 to 5
Semiconductors
crystal structure, 12-15 to 18, 12-70 to 82
diffusion in, 12-89 to 96
effective mass, 12-70 to 82
elastic constants, 12-22 to 26
electrical properties, 12-70 to 82, 12-85 to 88
extinction coefficient, 12-113 to 116
index of refraction, 12-113 to 116
minerals, resistivity of, 12-70 to 82
optical properties, 12-113 to 116, 12-117 to 121
organic, 12-85 to 88
physical properties, 12-70 to 82
reflection coefficient, 12-113 to 116
solid solutions, 12-83 to 84
thermal conductivity, 12-70 to 82
thermoelectric properties, 12-141 to 142
Sensitivity of the Human Eye to Light of Different Wavelengths, 10-142
Shielding, from ionizing radiation, 10-134 to 139, 16-88
Shock-sensitive chemicals, 16-7
SI units, definitions and symbols, 1-18 to 22
Siemens, definition, 1-18 to 22
Sievert, definition, 1-18 to 22, 16-89 to 90
Silicon: *see also* Elements
dielectric constant, 12-31 to 39
electron configuration, 1-16
heat capacity, 4-65
history, occurrence, uses, 4-1 to 39
ionization energy, 10-112 to 115
isotopes and their properties, 11-3 to 42
magnetic susceptibility, 4-73 to 78
molten, density, 4-71 to 72
optical properties, 12-113 to 116
physical properties, 4-62 to 64
semiconducting properties, 12-70 to 82
thermal conductivity, 12-131
thermodynamic properties, 5-1 to 2
thermodynamic properties at high temperature, 5-42 to 64
thermophysical properties, 12-122 to 127
vapor pressure, 6-87 to 116
Silver: *see also* Elements
electrical resistivity, 12-27
electron configuration, 1-16
heat capacity, 4-65
history, occurrence, uses, 4-1 to 39
ionization energy, 10-112 to 115
isotopes and their properties, 11-3 to 42
magnetic susceptibility, 4-73 to 78
molten, density, 4-71 to 72
physical properties, 4-62 to 64
thermal conductivity, 12-131
thermal properties, 12-129 to 130
thermodynamic properties, 5-1 to 2
vapor pressure, 6-87 to 116
vapor pressure, high temperature, 4-66 to 68
SIMS, definition, 12-1 to 5
SLAM, definition, 12-1 to 5
SMOW (standard mean ocean water), density, 6-7 to 8
SNMS, definition, 12-1 to 5
Sodium: *see also* Elements
electrical resistivity, 12-27
electron configuration, 1-16
heat capacity, 4-65
history, occurrence, uses, 4-1 to 39
ionization energy, 10-112 to 115
isotopes and their properties, 11-3 to 42
magnetic susceptibility, 4-73 to 78
molten, density, 4-71 to 72

physical properties, 4-62 to 64
thermal properties, 12-129 to 130
thermodynamic properties, 5-1 to 2
vapor pressure, 6-87 to 116
vapor pressure, high temperature, 4-66 to 68
Sodium chloride
activity coefficients, 5-98 to 100, 5-121 to 125
aqueous solutions, concentrative properties, 5-133 to 148
aqueous solutions, relative humidity, 15-37
enthalpy of solution, 5-127
standard thermodynamic properties, 5-3 to 41
Solar constant, 14-2 to 6
Solar Irradiance at the Earth, 14-22
Solar radiation
by month and latitude, 14-29
by wavelength, 14-22
Solar system, 14-2 to 6
Solder, thermal conductivity, 12-132 to 133
Solids, characterization and analysis, 12-1 to 5
Solubility
amino acids, 7-1 to 2
carbon dioxide in water, 5-151
drugs, 7-35 to 56
gases in water, 5-149 to 150
hydrocarbons in seawater, 5-197
inorganic compounds, 4-40 to 51
inorganic compounds, as function of temperature, 5-183 to 188, 5-196
inorganic compounds, qualitative rules, 5-200 to 201
inorganic compounds, sparingly soluble, 5-194 to 195
octanol–water partition coefficients, 5-189 to 193
organic compounds, 3-1 to 55
organic compounds in water, 5-152 to 182
organic compounds in water at high temperature, 5-198 to 199
polymers, 13-27 to 29
purine and pyrimidine bases, 7-5
salts in water, 5-196
terminology, 1-25
Solubility Chart for Inorganic Salts, 5-200 to 201
Solubility of Carbon Dioxide in Water at Various Temperatures and Pressures, 5-151
Solubility of Common Salts at Various Temperatures, 5-196
Solubility of Hydrocarbons in Seawater, 5-197
Solubility of Organic Compounds in Pressurized Hot Water, 5-198 to 199
Solubility of Selected Gases in Water, 5-149 to 150
Solubility Parameters of Selected Polymers, 13-27 to 29
Solubility Product Constants of Inorganic Salts, 5-194 to 195
Solutions
aqueous, concentrative properties, 5-133 to 148
density, 5-133 to 148
diffusion of ions, 5-98 to 100
enthalpy, for common electrolytes, 5-127
freezing point depression, 5-133 to 148
index of refraction, 5-133 to 148
ionic conductivity, 5-98 to 100
for round values of pH, 5-132
viscosity, 5-133 to 148
Solvents
azeotropic data, 6-217 to 232
density, as function of temperature, 15-27
dielectric constant, 8-38, 15-13 to 22
dipole moment, 15-13 to 22
eluotropic values, 8-35
flammability, 15-13 to 22
heat capacity, 15-13 to 22
ionic liquids, 6-178 to 181
for liquid chromatography, 8-36
mass spectral peaks, 8-76 to 82
miscibility, 15-23 to 26
NMR shifts, ^{13}C, 8-68
NMR shifts, proton, 8-59 to 64
physical properties, 15-13 to 22
for polymers, 13-9
threshold limit in air, 15-13 to 22
for ultraviolet spectrophotometry, 8-38
vapor pressure, 15-13 to 22
viscosity, 15-13 to 22
wavelength cutoff (UV), 8-38
Solvents for Common Polymers, 13-9
Solvents for Ultraviolet Spectrophotometry, 8-38
Sound velocity
air, as function of humidity and frequency, 14-41
air, as function of temperature, 14-42
air, as function of temperature and pressure, 6-15 to 20
atmosphere, as function of altitude, 14-23 to 28
fluids, 6-21 to 39, 6-40 to 50
seawater, 14-19 to 20
various solids, liquids, and gases, 14-39 to 40
water and seawater, 14-39 to 40
water and steam, 6-1 to 4
Sources of Physical and Chemical Data, A-1 to 5 A-1 to 5
Space group
elements, 12-15 to 18
notation, 12-6 to 10
Specific gravity: *see* Density
Specific heat: *see* Heat capacity
Spectroscopic Constants of Diatomic Molecules, 9-111 to 116
Spectrum, infrared
correlation charts, 8-42 to 47
fundamental vibrational frequencies, 9-108 to 110
Spectrum, line, of Elements, 10-1 to 50
Speed of light in vacuum, 1-1 to 9, 1-18 to 22
Speed of sound
air, as function of frequency, 14-41
air, as function of humidity, 14-41
air, as function of temperature, 14-42
air, as function of temperature and pressure, 6-15 to 20
atmosphere, as function of altitude, 14-23 to 28
fluids, 6-21 to 39, 6-40 to 50
various solids, liquids, and gases, 14-39 to 40
water and seawater, 14-39 to 40
water and steam, 6-1 to 4
Speed of Sound in Dry Air, 14-42
Speed of Sound in Various Media, 14-39 to 40
Spin
nuclides, 11-3 to 42
nuclides of NMR interest, 8-49 to 51, 8-52
ordering in magnetic materials, 12-105
SPM, definition, 12-1 to 5
SSMS, definition, 12-1 to 5
Standard Atmosphere (U.S.), 14-23 to 28
Standard Atomic Weights (alphabetical), 1-10 to 11
Standard Atomic Weights (by atomic number): *see* Inside back cover; 1-25
Standard Density of Water, 6-7 to 8
Standard ITS-90 Thermocouple Tables, 15-1 to 9
Standard KCl Solutions for Calibrating Electrical Conductivity Cells, 5-95
Standard Salt Solutions for Humidity Calibration, 15-37
Standard solutions, for pH measurement, 5-128 to 131
Standard Thermodynamic Properties of Chemical Substances, 5-3 to 41
Standard Transformed Gibbs Energies of Formation for Biochemical Reactants, 7-14 to 16
Standards
CODATA thermodynamic values, 5-1 to 2
for chemical analysis, 8-10 to 16
index of refraction, 10-152
laboratory weights, 8-86 to 87
temperature, 1-17
Standards for Laboratory Weights, 8-86 to 87
Stationary Phases for Porous-Layer Open Tubular Columns, 8-23
Steam, thermophysical properties, 6-1 to 4
Steel, thermophysical properties, 12-122 to 127
Stefan-Boltzmann constant, 1-1 to 9
STEM, definition, 12-1 to 5
STM, definition, 12-1 to 5
Standard Fittings for Compressed Gas Cylinders, 8-22
Stopping power, for electrons in solids, 12-108 to 109
Stratosphere, properties, 14-23 to 28
Strontium: *see also* Elements
electrical resistivity, 12-27
electron configuration, 1-16
heat capacity, 4-65
history, occurrence, uses, 4-1 to 39
ionization energy, 10-112 to 115
isotopes and their properties, 11-3 to 42
magnetic susceptibility, 4-73 to 78
molten, density, 4-71 to 72
thermal properties, 12-129 to 130
vapor pressure, 6-87 to 116
vapor pressure, high temperature, 4-66 to 68
Structure
amino acids, 7-3 to 4
bond lengths in organic crystals, 9-1 to 16
bond lengths in organometallic compounds, 9-17 to 18
characteristic ^{13}C chemical shifts, 8-66 to 67
characteristic infrared frequencies, 8-42 to 47
characteristic ^{15}N chemical shifts, 8-69 to 70

characteristic proton chemical shifts, 8-53 to 57, 8-58
crystal, of elements, 12-15 to 18
crystal, of superconductors, 12-42 to 57, 12-58 to 67
force constants, 9-107
formulas for organic compounds, 3-1 to 55
fundamental vibrational frequencies, 9-108 to 110
gas-phase molecules, 9-19 to 54
geometry of small molecules, 9-108 to 110
InChI identifier, 2-22 to 23
solids, characterization techniques, 12-1 to 5
Structure of Free Molecules in the Gas Phase, 9-19 to 54
Structures of Common Amino Acids, 7-3 to 4
Sublimation Pressure of Solids, 6-85 to 86
Sugars
aqueous solution properties, 5-133 to 148
nomenclature, 7-12 to 13
Sulfur: *see also* Elements
critical constants, 6-82 to 84
dielectric constant, 12-31 to 39
electron configuration, 1-16
heat capacity, 4-65
history, occurrence, uses, 4-1 to 39
ionization energy, 10-112 to 115
isotopes and their properties, 11-3 to 42
magnetic susceptibility, 4-73 to 78
molten, density, 4-71 to 72
physical properties, 4-62 to 64
thermodynamic properties, 5-1 to 2
thermodynamic properties at high temperature, 5-42 to 64
vapor pressure, 6-87 to 116
Sulfuric acid
activity coefficients, 5-98 to 100, 5-121 to 125
concentrative properties, 5-133 to 148
constant humidity solutions, 15-36
density, 15-42 to 43
electrical conductivity, 5-94
vapor pressure, 6-87 to 116
Summary Tables of Particle Properties, 11-1 to 2
Sun
mass, dimensions, and other properties, 14-2 to 6
radiative properties, 14-2 to 6
spectral irradiance, 14-22
Superconductors
electrical and magnetic properties, 12-42 to 57, 12-58 to 67
organic, 12-68 to 69
rare-earth elements, 4-52 to 61
transition temperature, 12-42 to 57, 12-58 to 67, 12-68 to 69
Superconductors, high temperature, 12-58 to 67
Supercooled water, 6-11
Surface Active Chemicals (Surfactants), 6-188 to 192
Surface characterization and analysis, 12-1 to 5
Surface tension
aqueous mixtures, 6-187
liquid helium, 6-163
liquid rare-earth metals, 4-52 to 61
various liquids, 6-182 to 186
water, 6-5 to 6
Surface Tension of Aqueous Mixtures, 6-187
Surface Tension of Common Liquids, 6-182 to 186
Surfactants, 6-188 to 192
Susceptibility: *see* Magnetic susceptibility
Symbols
amino acids, 7-3 to 4
carbohydrates, 7-12 to 13
magnetism, 12-97 to 104
physical quantities, 2-1 to 12, 2-24 to 36
SI units, 1-18 to 22
units, 2-24 to 36
Symbols and Terminology for Physical and Chemical Quantities, 2-1 to 12
Symmetry of Crystals, 12-6 to 10

T

Table of the Isotopes, 11-3 to 42
Tait equation, 6-156 to 160, 13-18 to 21
Tantalum: *see also* Elements
electrical resistivity, 12-27
electron configuration, 1-16
heat capacity, 4-65
history, occurrence, uses, 4-1 to 39
ionization energy, 10-112 to 115
isotopes and their properties, 11-3 to 42
magnetic susceptibility, 4-73 to 78
molten, density, 4-71 to 72
physical properties, 4-62 to 64
thermal properties, 12-129 to 130
vapor pressure, 6-87 to 116
vapor pressure, high temperature, 4-66 to 68
Technetium: *see also* Elements
electron configuration, 1-16
heat capacity, 4-65
history, occurrence, uses, 4-1 to 39
ionization energy, 10-112 to 115
isotopes and their properties, 11-3 to 42
magnetic susceptibility, 4-73 to 78
physical properties, 4-62 to 64
thermal properties, 12-129 to 130
vapor pressure, 6-87 to 116
Techniques for Materials Characterization, 12-1 to 5
Tellurium: *see also* Elements
dielectric constant, 12-31 to 39
electron configuration, 1-16
history, occurrence, uses, 4-1 to 39
ionization energy, 10-112 to 115
isotopes and their properties, 11-3 to 42
magnetic susceptibility, 4-73 to 78
molten, density, 4-71 to 72
physical properties, 4-62 to 64
vapor pressure, 6-87 to 116
Temperature
atmosphere, as function of altitude, 14-23 to 28
baths for temperature control, 8-24, 15-37, 15-38 to 39
calibration, ITS-90, 1-17, 15-10 to 11
Celsius and absolute, definitions, 1-18 to 22
glass transition, in polymers, 13-10 to 16
International Temperature Scale (ITS-90), 1-17
mean global, 14-36
planetary atmospheres, 14-2 to 6
superconducting transition, 12-42 to 57
thermocouple tables, 15-1 to 9
Temperature and Pressure Dependence of Liquid Density, 6-156 to 160
Tennessine (element 117), 4-1 to 39, 11-3 to 42
Tensile strength
polymers, 13-3 to 4
rare-earth elements, 4-52 to 61
Terbium: *see also* Elements
electrical resistivity, 12-27
electron configuration, 1-16
heat capacity, 4-65
history, occurrence, uses, 4-1 to 39
ionization energy, 10-112 to 115
isotopes and their properties, 11-3 to 42
magnetic susceptibility, 4-73 to 78
molten, density, 4-71 to 72
physical properties, 4-62 to 64
thermal properties, 12-129 to 130
vapor pressure, 6-87 to 116
vapor pressure, high temperature, 4-66 to 68
Terminology
inorganic ions and ligands, 2-16 to 21
physical quantities, 2-1 to 12
polymers, 13-5 to 8
solubility, qualitative terms, 1-25
Tesla, definition, 1-18 to 22
Testing Requirements for Peroxidizable Compounds, 16-8
Tests for the Presence of Peroxides, 16-9
TGA, definition, 12-1 to 5
Thallium: *see also* Elements
electrical resistivity, 12-27
electron configuration, 1-16
heat capacity, 4-65
history, occurrence, uses, 4-1 to 39
ionization energy, 10-112 to 115
isotopes and their properties, 11-3 to 42
magnetic susceptibility, 4-73 to 78
molten, density, 4-71 to 72
physical properties, 4-62 to 64
thermal properties, 12-129 to 130
vapor pressure, 6-87 to 116
vapor pressure, high temperature, 4-66 to 68
Thermal and Physical Properties of Pure Metals, 12-129 to 130
Thermal conductivity
air, 6-15 to 20
alloys, 12-122 to 127
argon, liquid and gas, 6-21 to 39
atmosphere, as function of altitude, 14-23 to 28
carbon dioxide, 6-21 to 39
carrier gases for chromatography, 8-20
ceramics, 12-136 to 137
commercial metals, 12-132 to 133
common fluids, as function of temperature and pressure, 6-21 to 39
common fluids, at saturation, 6-40 to 50
construction materials, 12-136 to 137
conversion factors, 1-26
cryogenic materials, 12-122 to 127
crystalline solids, 12-134 to 135
dielectric crystals, 12-134 to 135
gases, at low pressure, 6-240 to 242
glasses, 12-138 to 140
helium, liquid, 6-163
helium, liquid and gas, 6-21 to 39
ice, 6-11
insulation, 12-136 to 137
liquids, 6-243 to 248

mercury, 6-243 to 248
metals, 12-129 to 130
minerals, 12-136 to 137
nitrogen, liquid and gas, 6-21 to 39
organic compounds, 6-243 to 248
oxygen, liquid and gas, 6-21 to 39
plastics, 12-122 to 127
quartz, 12-134 to 135
rare-earth elements, 4-52 to 61
refractory materials, 12-136 to 137
rocks, 12-136 to 137
rubber, 12-136 to 137
semiconductors, 12-70 to 82
stainless steel, 12-122 to 127
superconductors, 12-58 to 67
water, 6-243 to 248
water and steam, 6-1 to 4
wood, 12-136 to 137
Thermal Conductivity of Alloys as a Function of Temperature, 12-132 to 133
Thermal Conductivity of Ceramics and Other Insulating Materials, 12-136 to 137
Thermal Conductivity of Crystalline Dielectrics, 12-134 to 135
Thermal Conductivity of Gases and Refrigerants, 6-240 to 242
Thermal Conductivity of Glasses, 12-138 to 140
Thermal Conductivity of Liquids, 6-243 to 248
Thermal Conductivity of Metals and Semiconductors as a Function of Temperature, 12-131
Thermal expansion coefficient
commercial metals and alloys, 12-122 to 127
ice, 6-11
liquids, 6-154 to 155, 6-156 to 160
metals, 12-122 to 127
rare-earth elements, 4-52 to 61
semiconductors, 12-70 to 82
Thermal neutron cross sections, 11-43 to 55
Thermocouple calibration tables, 15-1 to 9
Thermodynamic Functions and Relations, 2-37
Thermodynamic properties: *see also* Enthalpy, Heat capacity, etc.
air, 6-15 to 20
aqueous ions, 5-65 to 66
argon, 6-21 to 39
biochemical species, 7-14 to 16, 7-17 to 19, 7-30 to 32
CODATA Key Values, 5-1 to 2
common fluids, as function of temperature and pressure, 6-21 to 39
common fluids, at saturation, 6-40 to 50
gas clathrate hydrates, 6-167 to 177
helium, 6-21 to 39
high temperature, 5-42 to 64
hydrogen, 6-21 to 39
inorganic compounds, 5-3 to 41
nitrogen, 6-21 to 39
organic compounds, 5-3 to 41
oxygen, 6-21 to 39
rare-earth elements, 4-52 to 61
standard state values, 5-3 to 41
temperature dependence, 5-42 to 64
water and steam, 6-1 to 4
Thermodynamic Properties as a Function of Temperature, 5-42 to 64
Thermodynamic Properties of Aqueous Ions, 5-65 to 66
Thermodynamic Quantities for the Ionization Reactions of Buffers in Water, 7-30 to 32
Thermodynamic relations, 2-37
Thermoelectric power (thermal emf), 12-141 to 142
Thermoelectric Properties of Metals and Semiconductors, 12-141 to 142
Thermometers, wet and dry bulb, 15-35
Thermophysical Properties of Air, 6-15 to 20
Thermophysical Properties of Fluids, 6-21 to 39
Thermophysical Properties of Selected Fluids at Saturation, 6-40 to 50
Thermophysical Properties of Water and Steam, 6-1 to 4
Thomson effect, 12-141 to 142
Thorium: *see also* Elements
electrical resistivity, 12-27
electron configuration, 1-16
heat capacity, 4-65
history, occurrence, uses, 4-1 to 39
ionization energy, 10-112 to 115
isotopes and their properties, 11-3 to 42
magnetic susceptibility, 4-73 to 78
physical properties, 4-62 to 64
thermal properties, 12-129 to 130
thermodynamic properties, 5-1 to 2
vapor pressure, 6-87 to 116
vapor pressure, high temperature, 4-66 to 68
Threshold limits
airborne contaminants, 16-48 to 75
halocarbon refrigerants, 6-164 to 166
solvents, 15-13 to 22
Threshold Limits for Airborne Contaminants, 16-48 to 75
Thulium: *see also* Elements
electrical resistivity, 12-27
electron configuration, 1-16
heat capacity, 4-65
history, occurrence, uses, 4-1 to 39
ionization energy, 10-112 to 115
isotopes and their properties, 11-3 to 42
magnetic susceptibility, 4-73 to 78
molten, density, 4-71 to 72
physical properties, 4-62 to 64
thermal properties, 12-129 to 130
vapor pressure, 6-87 to 116
vapor pressure, high temperature, 4-66 to 68
Time
astronomical units, 14-1
geological scale, 14-16
Tin: *see also* Elements
electrical resistivity, 12-27
electron configuration, 1-16
heat capacity, 4-65
history, occurrence, uses, 4-1 to 39
ionization energy, 10-112 to 115
isotopes and their properties, 11-3 to 42
magnetic susceptibility, 4-73 to 78
molten, density, 4-71 to 72
physical properties, 4-62 to 64
thermal conductivity, 12-131
thermal properties, 12-129 to 130
thermodynamic properties, 5-1 to 2
vapor pressure, 6-87 to 116
vapor pressure, high temperature, 4-66 to 68
Titanium: *see also* Elements
electrical resistivity, 12-27
electron configuration, 1-16
heat capacity, 4-65
history, occurrence, uses, 4-1 to 39
ionization energy, 10-112 to 115
isotopes and their properties, 11-3 to 42
magnetic susceptibility, 4-73 to 78
molten, density, 4-71 to 72
physical properties, 4-62 to 64
thermal properties, 12-129 to 130
thermodynamic properties, 5-1 to 2
vapor pressure, 6-87 to 116
vapor pressure, high temperature, 4-66 to 68
Transition probability, atomic, 10-51 to 53
Transition temperature
glass, in polymers, 13-10 to 16
superconductors, 12-42 to 57, 12-58 to 67, 12-68 to 69
Transport properties: *see* Thermal conductivity, Viscosity, Diffusion
Triple-point constants
carbon dioxide, 6-85 to 86
cryogenic fluids, 6-161 to 162
D_2O, 6-9
elements, 4-62 to 64
various compounds, 6-85 to 86
water, 6-9
Tungsten: *see also* Elements
electrical resistivity, 12-27
electron configuration, 1-16
heat capacity, 4-65
history, occurrence, uses, 4-1 to 39
ionization energy, 10-112 to 115
isotopes and their properties, 11-3 to 42
magnetic susceptibility, 4-73 to 78
molten, density, 4-71 to 72
physical properties, 4-62 to 64
thermal conductivity, 12-131
thermal properties, 12-129 to 130
vapor pressure, 6-87 to 116
vapor pressure, high temperature, 4-66 to 68
Typical pH Values of Biological Materials and Foods, 7-34

U

U.S. Standard Atmosphere (1976), 14-23 to 28
Ultraviolet spectrophotometry
analytical chemistry applications, 8-6 to 9
chromophoric functional groups, 8-39 to 41
solvents for, 8-38
Uncertainty, means of expression, 2-13 to 14
Units
concentration of solutions, 8-18
conversion factors, 8-18
ionizing radiation, 16-89 to 90
magnetic quantities, 1-23, 12-97 to 104
pH, 5-128 to 131
SI, definitions and symbols, 1-18 to 22
Units for Magnetic Properties, 1-23
UPS, definition, 12-1 to 5
Uranium: *see also* Elements
electrical resistivity, 12-27

electron configuration, 1-16
heat capacity, 4-65
history, occurrence, uses, 4-1 to 39
ionization energy, 10-112 to 115
isotopes and their properties, 11-3 to 42
magnetic susceptibility, 4-73 to 78
molten, density, 4-71 to 72
physical properties, 4-62 to 64
thermal properties, 12-129 to 130
thermodynamic properties, 5-1 to 2
vapor pressure, 6-87 to 116
vapor pressure, high temperature, 4-66 to 68
Uranus, orbital data and dimensions, 14-2 to 6

V

Values of the Gas Constant in Different Unit Systems, 1-25
Van der Waals radii of Elements, 9-56 to 57
Vanadium: *see also* Elements
electrical resistivity, 12-27
electron configuration, 1-16
heat capacity, 4-65
history, occurrence, uses, 4-1 to 39
ionization energy, 10-112 to 115
isotopes and their properties, 11-3 to 42
magnetic susceptibility, 4-73 to 78
molten, density, 4-71 to 72
physical properties, 4-62 to 64
thermal properties, 12-129 to 130
vapor pressure, 6-87 to 116
vapor pressure, high temperature, 4-66 to 68
Vapor pressure
air, 6-15 to 20
aqueous salt solutions, 15-37
at temperatures below 300 K, 6-117 to 125
carbon dioxide, 6-85 to 86
cryogenic fluids, 6-117 to 125
elements, 6-87 to 116
elements, high temperature, 4-66 to 68, 4-69 to 70
general table, 6-87 to 116
helium, 1-17, 6-117 to 125, 6-163
ice, 6-12
inorganic compounds, 6-87 to 116
metals, at high temperatures, 4-66 to 68, 4-69 to 70
organic compounds, 6-87 to 116
polymer solutions, 13-22 to 26
rare-earth elements, 4-52 to 61
rare gases, 6-117 to 125
salt solutions, 6-126
solids, 6-85 to 86
solvents, 15-13 to 22
water, 6-1 to 4, 6-5 to 6
water, over salt solutions, 6-126, 15-36, 15-37
Vapor Pressure of Compounds and Elements, 6-87 to 116
Vapor Pressure and Other Saturation Properties of Water, 6-5 to 6
Vapor Pressure of Fluids at Temperatures below 300 K, 6-117 to 125
Vapor Pressure of Ice, 6-12
Vapor Pressure of Saturated Salt Solutions, 6-126
Vapor Pressure of the Metallic Elements—Data, 4-69 to 70
Vapor Pressure of the Metallic Elements—Equations, 4-66 to 68
Vapor Pressures (Solvent Activities) for Binary Polymer Solutions, 13-22 to 26
Vaporization: *see* Enthalpy of vaporization
Varieties of Hyphenated Gas Chromatography with Mass Spectrometry, 8-29 to 30
Velocity of light, 1-1 to 9
Velocity of sound
air, as function of frequency, 14-41
air, as function of humidity, 14-41
air, as function of temperature, 14-42
atmosphere, as function of altitude, 14-23 to 28
fluids, 6-21 to 39, 6-40 to 50
various solids, liquids, and gases, 14-39 to 40
Velocity, mean, in gases, 6-54
Venus, orbital data and dimensions, 14-2 to 6
Vibrational force constants, 9-107
Vibrational frequencies of molecules, 9-108 to 110
Virial Coefficients of Selected Gases, 6-51 to 53
Viscosity
air, 6-15 to 20
aqueous solutions, 5-133 to 148
argon, liquid and gas, 6-21 to 39
atmosphere, as function of altitude, 14-23 to 28
carrier gases for chromatography, 8-20
common fluids, as function of temperature and pressure, 6-21 to 39
common fluids, at saturation, 6-40 to 50
gases, at low pressure, 6-233 to 234
helium, liquid, 6-163
helium, liquid and gas, 6-21 to 39
ionic liquids, 6-178 to 181
liquids, 6-235 to 239
methane, liquid and gas, 6-21 to 39
nitrogen, liquid and gas, 6-21 to 39
oxygen, liquid and gas, 6-21 to 39
seawater, 14-19 to 20
solvents, 15-13 to 22
water and steam, 6-1 to 4
Viscosity of Gases, 6-233 to 234
Viscosity of Liquids, 6-235 to 239
Volt
definition, 1-18 to 22
maintained value, 1-1 to 9

W

Water
azeotropic mixtures, 6-217 to 232
compressibility, 6-154 to 155
critical constants, 6-9
density, as function of pressure and temperature, 6-1 to 4
density, 6-7 to 8
density (supercooled), 6-11
dielectric constant, as function of frequency, 6-14
dielectric constant, as function of temperature and pressure, 6-1 to 4
dielectric constant, vapor, 6-216
diffusion of gases, 6-251
dissociation constant, 5-92, 5-93
electrical conductivity, 5-93
enthalpy of fusion, 6-144 to 153
enthalpy of vaporization, 6-5 to 6
fixed-point properties, 6-9
freezing point, pressure dependence, 6-55
index of refraction, 10-151
ionization constant, 5-92, 5-93
octanol–water partition coefficients, 5-189 to 193
permittivity, as function of frequency, 6-14
permittivity, as function of temperature and pressure, 6-1 to 4
speed of sound, 6-1 to 4, 14-39 to 40
surface tension, 6-5 to 6
thermal conductivity, 6-240 to 242
thermal expansion coefficient, 6-154 to 155
thermodynamic properties at high temperature, 5-42 to 64
thermophysical properties, 6-1 to 4
triple-point constants, 6-9
vapor pressure, 6-5 to 6
vapor pressure over salt solutions, 15-36
viscosity, 6-233 to 234
Water (D_2O) (heavy water)
boiling point, 6-9
critical constants, 6-9
density, 6-10
dissociation constant, 5-93
fixed-point properties, 6-9
heat capacity, 6-10
ionization constant, 5-93
surface tension, 6-10
thermal conductivity, liquid, 6-10
thermal conductivity, vapor, 6-240 to 242
triple-point constants, 6-9
vapor pressure, 6-10
viscosity, liquid, 6-10
viscosity, vapor, 6-233 to 234
Water-Reactive Chemicals, 16-8
Watt, definition, 1-18 to 22
Wavelengths
atomic spectra, 10-1 to 50
electromagnetic radiation bands, 10-140 to 141
sensitivity of eye, 10-142
Weber, definition, 1-18 to 22
Weighings, reduction from air to vacuum, 8-85
Weights, laboratory, 8-86 to 87
Wien displacement law constant, 1-1 to 9
Wood
density, 15-41
speed of sound in, 14-39 to 40
thermal conductivity
Work function, of Elements, 12-110 to 111

X

Xenon: *see also* Elements
critical constants, 6-82 to 84
cryogenic properties, 6-161 to 162
electron configuration, 1-16
entropy, 5-1 to 2
history, occurrence, uses, 4-1 to 39
ionization energy, 10-112 to 115
isotopes and their properties, 11-3 to 42
magnetic susceptibility, 4-73 to 78
mean free path, 6-54
permittivity (dielectric constant), 6-193 to 215
physical properties, 4-62 to 64
solubility in water, 5-149 to 150
thermal conductivity, 6-240 to 242

vapor pressure, 6-87 to 116
viscosity, 6-233 to 234
X-rays
analytical chemistry applications, 8-6 to 9
attenuation coefficients, 10-134 to 139
cross sections, for elements, 10-134 to 139
shielding, 16-88

Y

Young's modulus
cryogenic materials, 6-161 to 162
rare-earth elements, 4-52 to 61
Ytterbium: *see also* Elements
electrical resistivity, 12-27
electron configuration, 1-16
heat capacity, 4-65
history, occurrence, uses, 4-1 to 39
ionization energy, 10-112 to 115
isotopes and their properties, 11-3 to 42
magnetic susceptibility, 4-73 to 78
molten, density, 4-71 to 72
physical properties, 4-62 to 64
thermal properties, 12-129 to 130
vapor pressure, 6-87 to 116
vapor pressure, high temperature, 4-66 to 68
Yttrium: *see also* Elements
electrical resistivity, 12-27
electron configuration, 1-16
heat capacity, 4-65
history, occurrence, uses, 4-1 to 39
ionization energy, 10-112 to 115
isotopes and their properties, 11-3 to 42
magnetic susceptibility, 4-73 to 78
molten, density, 4-71 to 72
physical properties, 4-62 to 64
thermal properties, 12-129 to 130
vapor pressure, 6-87 to 116
vapor pressure, high temperature, 4-66 to 68

Z

Zeotropes, 6-217 to 232
Zinc: *see also* Elements
electrical resistivity, 12-27
electron configuration, 1-16
heat capacity, 4-65
history, occurrence, uses, 4-1 to 39
ionization energy, 10-112 to 115
isotopes and their properties, 11-3 to 42
magnetic susceptibility, 4-73 to 78
molten, density, 4-71 to 72
physical properties, 4-62 to 64
thermal properties, 12-129 to 130
thermodynamic properties, 5-1 to 2
vapor pressure, 6-87 to 116
vapor pressure, high temperature, 4-66 to 68
Zirconium: *see also* Elements
electrical resistivity, 12-27
electron configuration, 1-16
heat capacity, 4-65
history, occurrence, uses, 4-1 to 39
ionization energy, 10-112 to 115
isotopes and their properties, 11-3 to 42
magnetic susceptibility, 4-73 to 78
molten, density, 4-71 to 72
physical properties, 4-62 to 64
thermal properties, 12-129 to 130
vapor pressure, 6-87 to 116
vapor pressure, high temperature, 4-66 to 68

PIERCE COLLEGE LIBRARY
PUYALLUP WA 98374
LAKEWOOD WA 98498

OMIC WEIGHTS IN ATOMIC NUMBER ORDER

Atomic number	Element	Symbol	Atomic weight
1	Hydrogen*	H	1.008 [1.00784, 1.00811]
2	Helium	He	4.002602(2)
3	Lithium*	Li	6.94 [6.938, 6.997]
4	Beryllium	Be	9.0121831(5)
5	Boron*	B	10.81 [10.806, 10.821]
6	Carbon*	C	12.011 [12.0096, 12.0116]
7	Nitrogen*	N	14.007 [14.00643, 14.00728]
8	Oxygen*	O	15.999 [15.99903, 15.99977]
9	Fluorine	F	18.998403163(6)
10	Neon	Ne	20.1797(6)
11	Sodium	Na	22.98976928(2)
12	Magnesium*	Mg	24.305 [24.304, 24.307]
13	Aluminum	Al	26.9815384(3)
14	Silicon*	Si	28.085 [28.084, 28.086]
15	Phosphorus	P	30.973761998(5)
16	Sulfur*	S	32.06 [32.059, 32.076]
17	Chlorine*	Cl	35.45 [35.446, 35.457]
18	Argon	Ar	39.948 [39.792, 39.963]
19	Potassium	K	39.0983(1)
20	Calcium	Ca	40.078(4)
21	Scandium	Sc	44.955908(5)
22	Titanium	Ti	47.867(1)
23	Vanadium	V	50.9415(1)
24	Chromium	Cr	51.9961(6)
25	Manganese	Mn	54.938043(2)
26	Iron	Fe	55.845(2)
27	Cobalt	Co	58.933194(3)
28	Nickel	Ni	58.6934(4)
29	Copper	Cu	63.546(3)
30	Zinc	Zn	65.38(2)
31	Gallium	Ga	69.723(1)
32	Germanium	Ge	72.630(8)
33	Arsenic	As	74.921595(6)
34	Selenium	Se	78.971(8)
35	Bromine*	Br	79.904 [79.901, 79.907]
36	Krypton	Kr	83.798(2)
37	Rubidium	Rb	85.4678(3)
38	Strontium	Sr	87.62(1)
39	Yttrium	Y	88.90584(1)
40	Zirconium	Zr	91.224(2)
41	Niobium	Nb	92.90637(1)
42	Molybdenum	Mo	95.95(1)
43	Technetium**	Tc	
44	Ruthenium	Ru	101.07(2)
45	Rhodium	Rh	102.90549(2)
46	Palladium	Pd	106.42(1)
47	Silver	Ag	107.8682(2)
48	Cadmium	Cd	112.414(4)
49	Indium	In	114.818(1)
50	Tin	Sn	118.710(7)
51	Antimony	Sb	121.760(1)
52	Tellurium	Te	127.60(3)
53	Iodine	I	126.90447(3)
54	Xenon	Xe	131.293(6)
55	Cesium	Cs	132.90545196(6)
56	Barium	Ba	137.327(7)
57	Lanthanum	La	138.90547(7)
58	Cerium	Ce	140.116(1)
59	Praseodymium	Pr	140.90766(1)
60	Neodymium	Nd	144.242(3)
61	Promethium**	Pm	
62	Samarium	Sm	150.36(2)
63	Europium	Eu	151.964(1)
64	Gadolinium	Gd	157.25(3)
65	Terbium	Tb	158.925354(8)
66	Dysprosium	Dy	162.500(1)
67	Holmium	Ho	164.930328(7)
68	Erbium	Er	167.259(3)
69	Thulium	Tm	168.934218(6)
70	Ytterbium	Yb	173.045(10)
71	Lutetium	Lu	174.9668(1)
72	Hafnium	Hf	178.486(6)
73	Tantalum	Ta	180.94788(2)
74	Tungsten	W	183.84(1)
75	Rhenium	Re	186.207(1)
76	Osmium	Os	190.23(3)
77	Iridium	Ir	192.217(2)
78	Platinum	Pt	195.084(9)
79	Gold	Au	196.966570(4)
80	Mercury	Hg	200.592(3)
81	Thallium*	Tl	204.38 [204.382, 204.385]
82	Lead	Pb	207.2(1)
83	Bismuth	Bi	208.98040(1)
84	Polonium**	Po	
85	Astatine**	At	
86	Radon**	Rn	
87	Francium**	Fr	
88	Radium**	Ra	
89	Actinium**	Ac	
90	Thorium	Th	232.0377(4)
91	Protactinium	Pa	231.03588(1)
92	Uranium	U	238.02891(3)
93	Neptunium**	Np	
94	Plutonium**	Pu	
95	Americium**	Am	
96	Curium**	Cm	
97	Berkelium**	Bk	
98	Californium**	Cf	
99	Einsteinium**	Es	
100	Fermium**	Fm	
101	Mendelevium**	Md	
102	Nobelium**	No	
103	Lawrencium**	Lr	
104	Rutherfordium**	Rf	
105	Dubnium**	Db	
106	Seaborgium**	Sg	
107	Bohrium**	Bh	
108	Hassium**	Hs	
109	Meitnerium**	Mt	
110	Darmstadtium**	Ds	
111	Roentgenium**	Rg	
112	Copernicium**	Cn	
113	Nihonium**	Nh	
114	Flerovium**	Fl	
115	Moscovium**	Mc	
116	Livermorium**	Lv	
117	Tennessine**	Ts	
118	Oganesson**	Og	

* The first value for this element is the conventional value to be used if there is no information on the origin of the material. This is followed by the interval in which the atomic weights in natural terrestrial materials are known to fall. See "Standard Atomic Weights" in Section 1 for more details.

** Because the element has no stable isotopes and no characteristic isotopic composition, the atomic weight is not defined.